MANUAL OF

Environmental Microbiology

THIRD EDITION

MANUAL OF Environmental Microbiology

THIRD EDITION

EDITOR IN CHIEF

Christon J. Hurst
Department of Biology, Xavier University, Cincinnati, Ohio
and
Facultad de Ingeniería, Universidad del Valle,
Ciudad Universitaria Meléndez, Santiago de Cali, Colombia

EDITORS

Ronald L. Crawford
Environmental Biotechnology Institute
University of Idaho
Moscow, Idaho

Jay L. Garland
Dynamac Corporation
Kennedy Space Center, Florida

David A. Lipson
Department of Biology
San Diego State University
San Diego, California

Aaron L. Mills
Department of Environmental Sciences
University of Virginia
Charlottesville, Virginia

Linda D. Stetzenbach
Department of Environmental and Occupational Health
School of Public Health
University of Nevada, Las Vegas
Las Vegas, Nevada

ASM PRESS
Washington, D.C.

Address editorial correspondence to ASM Press,
1752 N St. NW, Washington, DC 20036-2904, USA

Send orders to ASM Press, P.O. Box 605, Herndon, VA 20172, USA
Phone: (800) 546-2416 or (703) 661-1593
Fax: (703) 661-1501
E-mail: books@asmusa.org
Online: estore.asm.org

Library of Congress Cataloging-in-Publication Data

Manual of environmental microbiology / editor in chief, Christon J. Hurst ;
editors, Ronald L. Crawford ... [et al.].—3rd ed.
p. ; cm.
Includes bibliographical references and index.
ISBN-13: 978-1-55581-379-6 (hardcover)
ISBN-10: 1-55581-379-8 (hardcover)
1. Microbial ecology—Laboratory manuals. 2. Sanitary microbiology—Laboratory manuals. I. Hurst,
Christon J. II. Crawford, Ronald L., 1947–
[DNLM: 1. Environmental Microbiology. QW 55 M294 2007]

QR100.M36 2007
579′.17—dc22

2006034637

10 9 8 7 6 5 4 3 2 1

Printed in the United States of America

Contents

Editorial Board

Contributors

MORTEZA ABBASZADEGAN
Department of Civil and Environmental Engineering, Arizona State University, Tempe, AZ 85287

UNDINE ACHILLES-DAY
Culture Collection of Algae and Protozoa, Scottish Association for Marine Science Dunstaffnage Marine Laboratory, Dunbeg, Argyll PA37 1QA, United Kingdom

STEVEN D. ALLISON
Department of Ecology and Evolutionary Biology, University of California, Irvine, Irvine, CA 92697

CHARLES W. BACON
Toxicology and Mycotoxin Research Unit, Russell Research Center, Agricultural Research Service, U.S. Department of Agriculture, P.O. Box 5677, Athens, GA 30613

MICHELE BAHR
Bay Paul Center for Comparative Molecular Biology and Evolution, Marine Biological Laboratory, Woods Hole, MA 02543

TAMAR BARKAY
Department of Biochemistry and Microbiology, Rutgers University, New Brunswick, NJ 08901

MORTON A. BARLAZ
Department of Civil, Construction, and Environmental Engineering, North Carolina State University, Box 7908, Raleigh, NC 27695-7908

FELIX BÄRLOCHER
Department of Biology, Mount Allison University, 63B York St., Sackville, New Brunswick E4L 1G7, Canada

DENNIS A. BAZYLINSKI
School of Life Sciences, University of Nevada, Las Vegas, 4505 S. Maryland Pkwy., Las Vegas, NV 89154

J. OLE BECKER
Department of Nematology, University of California, Riverside, CA 92521

JAMIE W. BECKER
Hawaii Institute of Marine Biology, School of Ocean and Earth Science and Technology, University of Hawaii at Manoa, Kaneohe, HI 96744

ELIZABETH BENT
Department of Plant Pathology, University of California, Riverside, CA 92521

JAMES D. BEVER
Department of Biology, Indiana University, 1001 E. Third St., Bloomington, IN 47405-3700

TERRY J. BEVERIDGE
Department of Molecular and Cellular Biology, College of Biological Science, University of Guelph, Guelph, Ontario N1G 2W1, Canada

DANIEL R. BOND
Department of Microbiology and BioTechnology Institute, University of Minnesota, 140 Gortner Laboratory, 1479 Gortner Ave., St. Paul, MN 55108

JAMES BORNEMAN
Department of Plant Pathology, University of California, Riverside, CA 92521

JUAN JOSÉ BORREGO
Department of Microbiology, Faculty of Sciences, University of Malaga, 29071 Malaga, Spain

MARINA L. BRANDON
Hawaii Institute of Marine Biology, School of Ocean and Earth Science and Technology, University of Hawaii at Manoa, Kaneohe, HI 96744

SUSAN BROWN
CEH Dorset, Winfrith Technology Centre, Dorchester, Dorset DT2 8ZD, United Kingdom

ROBERT S. BURLAGE
Health Sciences Department, University of Wisconsin—Milwaukee, Milwaukee, WI 53201

MARK P. BUTTNER
Harry Reid Center for Environmental Studies, University of Nevada, Las Vegas, 4505 S. Maryland Pkwy., Las Vegas, NV 89154-4009

COLIN D. CAMPBELL
The Macaulay Institute, Craigiebuckler, Aberdeen AB15 8QH, Scotland, United Kingdom

DAVID A. CARON
Department of Biological Sciences, University of Southern California, 3616 Trousdale Pkwy., AHF 301, Los Angeles, CA 90089-0371

HECTOR CASTRO
Microbial Ecology and Physiology Group, Environmental Sciences Division, Oak Ridge National Laboratory, Bldg. 1505, MS-6038, 1 Bethel Valley Rd., P.O. Box 2008, Oak Ridge, TN 37831-6038

ASHVINI CHAUHAN
Environmental Sciences Institute, Frederick Humphries Science Research Building, Suite 305D, Florida A&M University, 1520 S. Bronough St., Tallahassee, FL 32307

GERARDO CHIN-LEO
The Evergreen State College, Olympia, WA 98505

LUDMILA CHISTOSERDOVA
Department of Chemical Engineering, University of Washington, Box 355014, Seattle, WA 98195

BRIAN G. CLEMENT
Department of Environmental and Biomolecular Systems, OGI School of Science and Engineering, Oregon Health and Science University, Beaverton, OR 97006

JOHN D. COATES
Department of Plant and Microbial Biology, University of California, Berkeley, CA 94720-3102

TOM COENYE
Laboratory of Microbiology and Laboratory of Pharmaceutical Microbiology, Ghent University, B-9000 Ghent, Belgium

FREDERICK S. COLWELL
College of Oceanic and Atmospheric Sciences, 104 COAS Administration Bldg., Oregon State University, Corvallis, OR 97331-5503

JONNA COOMBS
Department of Biology, Adelphi University, Garden City, NY 11530

ROBERT C. COOPER
BioVir Laboratories, Inc., 685 Stone Rd. #6, Benicia, CA 94510

ANA LUCÍA CÓRDOVA-KREYLOS
Department of Land, Air, and Water Resources, University of California, 1 Shields Ave., Davis, CA 95616

RONALD CORSTANJE
Environmetrics Group, Bioinformatics and Biomathematics Division, Rothamsted Research, Harpenden, Hertfordshire AL5 2JQ, United Kingdom

ELIZABETH K. COSTELLO
Department of Ecology and Evolutionary Biology, University of Colorado—Boulder, Boulder, CO 80309

MATTHEW T. COTTRELL
College of Marine Studies, University of Delaware, Lewes, DE 19958

DON COWAN
Advanced Research Centre for Applied Microbiology, Department of Biotechnology, University of the Western Cape, Bellville 7535, Cape Town, South Africa

ISABELLE M. COZZARELLI
U.S. Geological Survey, 431 National Center, Reston, VA 20192

RONALD L. CRAWFORD
Environmental Biotechnology Institute, University of Idaho, Moscow, ID 83844-1052, and Department of Microbiology, Molecular Biology, and Biochemistry, University of Idaho, Moscow, ID 83844-3052

A. J. CROSS-SMIECINSKI
Nevada System of Higher Education, University of Nevada, Las Vegas, Las Vegas, NV 89154-4009

PATRICIA CRUZ
Harry Reid Center for Environmental Studies, University of Nevada, Las Vegas, 4505 S. Maryland Pkwy., Las Vegas, NV 89154-4009

ALEXANDER I. CULLEY
Department of Botany, University of British Columbia, 1461 Biosciences, 6270 University Blvd., Vancouver, British Columbia V6T 1Z4, Canada

DAVID E. CUMMINGS
Department of Biology, Point Loma Nazarene University, 3900 Lomaland Dr., San Diego, CA 92106

RICHARD E. DANIELSON
BioVir Laboratories, Inc., 685 Stone Rd. #6, Benicia, CA 94510

JOHN G. DAY
Culture Collection of Algae and Protozoa, Scottish Association for Marine Science Dunstaffnage Marine Laboratory, Dunbeg, Argyll PA37 1QA, United Kingdom

FRANK B. DAZZO
Department of Microbiology and Molecular Genetics and Centers for Microbial Ecology, Microbial Pathogenesis, and Renewable Organic Products, Michigan State University, East Lansing, MI 48824

JODY W. DEMING
School of Oceanography, Box 357940, University of Washington, Seattle, WA 98195

RICHARD DEVEREUX
Gulf Ecology Division, U.S. Environmental Protection Agency, 1 Sabine Island Dr., Gulf Breeze, FL 32561

GREGORY J. DICK
Department of Environmental Science, Policy, and Management, University of California, Berkeley, Berkeley, CA 94720

SCOT E. DOWD
USDA-ARS Livestock Issues Research Unit,
Lubbock, TX 79403

CAROLINE DUCHAINE
Department of Biochemistry and Microbiology,
Université Laval, Québec, Québec G1K 7P4, Canada

DAVID EMERSON
American Type Culture Collection, Manassas, VA 20110

JEANETTE THURSTON ENRIQUEZ
Agricultural Research Service, U.S. Department of
Agriculture, 120 Keim Hall, East Campus,
University of Nebraska, Lincoln, NE 68583-0934

COLLEEN T. EVANS
School of Oceanography, Box 357940,
University of Washington, Seattle, WA 98195

BARRY S. FIELDS
Division of Bacterial and Mycotic Diseases,
National Center for Infectious Diseases,
Centers for Disease Control and Prevention,
Atlanta, GA 30333

JAMES K. FREDRICKSON
Pacific Northwest National Laboratory, K4-06,
P.O. Box 999, Richland, WA 99352

JED A. FUHRMAN
Department of Biological Sciences and Wrigley Institute,
University of Southern California,
Los Angeles, CA 90089-0371

JAMES R. FUXA
Department of Entomology, Louisiana Agricultural
Experiment Station, Louisiana State University Agricultural
Center, Baton Rouge, LA 70803

JAY L. GARLAND
Dynamac Corporation, Mail Code DYN-3, Kennedy Space
Center, FL 32899

TRACY B. GARTNER
Department of Biology and Environmental Science,
Carthage College, Kenosha, WI 53140

TERRY J. GENTRY
Environmental Sciences Division, Oak Ridge National
Laboratory, Oak Ridge, TN 37831-6038

DIRK GEVERS
Laboratory of Microbiology and Bioinformatics and
Evolutionary Genomics Division, Ghent University/VIB,
B-9000 Ghent, Belgium

BARRY GOODELL
Wood Science and Technology, University of Maine,
Orono, ME 04469

SERGEY A. GRINSHPUN
Center for Health-Related Aerosol Studies, Department of
Environmental Health, University of Cincinnati,
3223 Eden Ave., P.O. Box 670056,
Cincinnati, OH 45267-0056

JOHN F. GRIFFITH
Southern California Coastal Water Research Project
Authority, 3535 Harbor Blvd., Suite 110,
Costa Mesa, CA 92626

ANN E. HAJEK
Department of Entomology, Comstock Hall, Garden Ave.,
Cornell University, Ithaca, NY 14853-2601

KEVIN B. HALLBERG
School of Biological Sciences, University of Wales, Bangor,
Memorial Building, Deiniol Rd., Bangor LL57 2UW,
United Kingdom

JO HANDELSMAN
Department of Plant Pathology, University of Wisconsin—
Madison, 1630 Linden Dr., Madison, WI 53706

HAUKE HARMS
Department of Environmental Microbiology, Centre for
Environmental Research, Permoserstrasse 15, D-04105
Leipzig, Germany

STEVE H. HARRIS, JR.
U.S. Geological Survey, 3215 Marine St., Boulder, CO 80303

ANTON HARTMANN
Institute of Soil Ecology, Department of Rhizosphere
Biology, GSF-National Research Center for Environment
and Health, 85764 Neuherberg/Munich, Germany

RONALD W. HARVEY
U.S. Geological Survey, 3215 Marine St., Boulder, CO 80303

ZHILI HE
Institute for Environmental Genomics, Department of
Botany and Microbiology, University of Oklahoma,
101 David L. Boren Blvd., Norman, OK 73019

DAVID B. HEDRICK
Center for Biomarker Analysis,
University of Tennessee, 10515 Research Dr., Suite 300,
Knoxville, TN 37932-2575

THOMAS F. HESS
Department of Biological and Agricultural Engineering,
University of Idaho, Moscow, ID 83844-0904

HELENE H. HILGER
Department of Civil Engineering, University of North
Carolina—Charlotte, Charlotte, NC 28223

MARK E. HINES
Department of Biological Sciences, University of
Massachusetts Lowell, 1 University Ave.,
Lowell, MA 01854

DOROTHY M. HINTON
Toxicology and Mycotoxin Research Unit,
Russell Research Center, Agricultural Research Service,
U.S. Department of Agriculture,
Athens, GA 30613

JOHN E. HOBBIE
The Ecosystems Center, Marine Biological Laboratory, Woods
Hole, MA 02543

JON L. HOBMAN
School of Biosciences, The University of Birmingham, Edgbaston, Birmingham B15 2TT, United Kingdom

KERI HOLLAND
Department of Ecology, Evolution, and Marine Biology, University of California, Santa Barbara, Santa Barbara, CA 93106

MEREDITH A. J. HULLAR
Fred Hutchinson Cancer Research Institute, Seattle, WA 98112

CHRISTON J. HURST
Department of Biology, Xavier University, Cincinnati, OH 45207, and Facultad de Ingeniería, Universidad del Valle, Ciudad Universitaria Meléndez, Santiago de Cali, Colombia

M. KHALID IJAZ
Clorox Services Company, 7200 Johnson Dr., Pleasanton, CA 94588

HIROSHI ISHIZAKI
Department of Grazing Animal Production, National Institute of Livestock and Grassland Science, Nishinasuno, Tochigi 329-2793, Japan

WESLEY CLINTON JACKSON, JR.
Marine Biomedicine and Environmental Sciences Program, The Medical University of South Carolina, Charleston, SC 29403

LEE-ANN JAYKUS
Food Science Department, North Carolina State University, Box 7624, Raleigh, NC 27695-7624

JODY JELLISON
Department of Biological Sciences, University of Maine, Orono, ME 04469

GARY E. JENNEMAN
ConocoPhillips, 224 Geosciences Bldg., Bartlesville Technology Center, Bartlesville, OK 74004

ECKARDT JOHANNING
Occupational and Environmental Life Science, FRG, Inc., 4 Executive Park Dr., Albany, NY 12203

D. BARRIE JOHNSON
School of Biological Sciences, University of Wales, Bangor LL57 2UW, United Kingdom

RYAN T. JONES
Department of Ecology and Evolutionary Biology, University of Colorado—Boulder, Boulder, CO 80309

MARINA G. KALYUZHNAYA
Department of Microbiology, University of Washington, Box 355014, Seattle, WA 98195

WITTAYA KAONONGBUA
Department of Biology, Indiana University, 1001 E. Third St., Bloomington, IN 47405-3700

DAVID M. KARL
School of Ocean and Earth Science and Technology, University of Hawaii, Honolulu, HI 96822

SHELLY D. KELLY
Argonne National Laboratory, Argonne, IL 60439-4843

KENNETH M. KEMNER
Argonne National Laboratory, Argonne, IL 60439-4843

THOMAS L. KIEFT
Department of Biology, New Mexico Institute of Mining and Technology, Socorro, NM 87801

STAFFAN KJELLEBERG
Centre for Marine Biofouling and Bio-Innovation, School of Biotechnology and Biomolecular Sciences, University of New South Wales, Sydney, New South Wales 2052, Australia

DAVID B. KNAEBEL
Environmental Science Center, Syracuse Research Corporation, 7502 Round Pond Rd., North Syracuse, NY 13212

GUY R. KNUDSEN
Soil and Land Resources Division, University of Idaho, Moscow, ID 83844-2339

HASAN KOC
Proteomics and Mass Spectrometry Core Facility, The Pennsylvania State University, University Park, PA 16802

DARREN R. KORBER
Department of Applied Microbiology and Food Science, University of Saskatchewan, 51 Campus Dr., Saskatoon, Saskatchewan S7N 5A8, Canada

CHUNDAKKADU KRISHNA
Diversa Corporation, 4955 Directors Pl., San Diego, CA 92121

YASUHISA KUNIMI
Department of Applied Biological Sciences, Faculty of Agriculture, Tokyo University of Agriculture and Technology, Saiwai, Fuchu-shi, Tokyo 183, Japan

RICHARD T. LAMAR
EarthFax Development Corporation, 1770 N. Research Park Way, North Logan, UT 84341

LEE LANDKAMER
U.S. Geological Survey, 3215 Marine St., Boulder, CO 80303

BRIAN LANOIL
Department of Environmental Sciences, University of California, Riverside, CA 92521

JOHN R. LAWRENCE
National Water Research Institute, 11 Innovation Blvd., Saskatoon, Saskatchewan S7N 3H5, Canada

EDWARD R. LEADBETTER
Department of Marine Chemistry and Geochemistry, Woods Hole Oceanographic Institution, MS 52, Woods Hole, MA 02543

JOHAN H. J. LEVEAU
Netherlands Institute of Ecology, Heteren, The Netherlands

ESTELLE LEVETIN
Faculty of Biological Science, University of Tulsa, Tulsa, OK 74105

MARY E. LIDSTROM
Department of Chemical Engineering and Department of Microbiology, University of Washington, Box 355014, Seattle, WA 98195

STEVEN E. LINDOW
Department of Plant and Microbial Biology, University of California, Berkeley, CA 94720

DAVID A. LIPSON
Department of Biology, San Diego State University, San Diego, CA 92182-4614

FENG LIU
Department of Biochemistry and Molecular Biology, The Pennsylvania State University, University Park, PA 16802

WEN-TSO LIU
Department of Environmental Science and Engineering, National University of Singapore, Block E1A, #07-23, 9 Engineering Dr. 1, Singapore 117576

JONATHAN R. LLOYD
School of Earth, Atmospheric and Environmental Sciences and The Williamson Research Centre for Molecular Environmental Science, University of Manchester, Manchester M13 9PL, United Kingdom

JOYCE E. LOPER
Horticultural Crops Research Laboratory, Agricultural Research Service, U.S. Department of Agriculture, 3420 NW Orchard Ave., Corvallis, OR 97330

TIMOTHY S. MAGNUSON
Department of Biology, Idaho State University, Pocatello, ID 83209

DAVIN MALASARN
Division of Biology, California Institute of Technology, Pasadena, CA 91125

AARON B. MARGOLIN
Department of Microbiology, Biological Sciences Center, University of New Hampshire, Durham, NH 03824

ANDREW P. MARTIN
Department of Ecology and Evolutionary Biology, University of Colorado—Boulder, Boulder, CO 80309

DIANE McDOUGALD
Centre for Marine Biofouling and Bio-Innovation, School of Biotechnology and Biomolecular Sciences, University of New South Wales, Sydney, New South Wales 2052, Australia

GORDON A. McFETERS
Department of Microbiology, Montana State University, Bozeman, MT 59717

MICHAEL J. McINERNEY
Department of Botany and Microbiology, University of Oklahoma, 770 Van Vleet Oval, Norman, OK 73019

BRIAN McSPADDEN GARDENER
Department of Plant Pathology, Ohio Agricultural Research and Development Center, Ohio State University, Wooster, OH 44691

AARON L. MILLS
Laboratory of Microbial Ecology, Department of Environmental Sciences, University of Virginia, P.O. Box 400123, Charlottesville, VA 22904-4124

CHRISTINE L. MOE
Hubert Department of Global Health, Rollins School of Public Health, Emory University, 1518 Clifton Rd. NE, Atlanta, GA 30322

ALAN JEFF MOHR
Life Sciences Division, West Desert Test Center, U.S. Army Dugway Proving Ground, CSTE-DTC-DP-WD-LS, Building 2029, Dugway, UT 84022-5000

KATHERINE MORRIS
Institute of Geological Sciences, University of Leeds, Leeds LS2 9JT, United Kingdom

DAVID D. MYROLD
Department of Crop and Soil Science, Oregon State University, 3017 Agricultural and Life Science Bldg., Corvallis, OR 97331-7306

DIANA R. NEMERGUT
Institute of Arctic and Alpine Research and Environmental Sciences Program, University of Colorado, Boulder, CO 80309

OLIVIER NERCESSIAN
CEA/Cadarache, DSV/DEVM/LEMiR, Bâtiment 163, 13108 St-Paul lez Durance, France

T. R. NEU
Department of River Ecology, UFZ Centre for Environmental Research Leipzig-Halle, Brückstrasse 3a, 39114 Magdeburg, Germany

RACHEL T. NOBLE
Institute for Marine Sciences, University of North Carolina at Chapel Hill, 3431 Arendell St., Morehead City, NC 28557

SEÁN P. O'CONNELL
Department of Biology, 132 Natural Science Bldg., Western Carolina University, Cullowhee, North Carolina 28723

ANDREW OGRAM
Soil and Water Science Department, University of Florida, P.O. Box 110290, 2169 McCarty Hall, Gainesville, FL 32611-0290

RABIU OLATINWO
Department of Plant Pathology, University of California, Riverside, CA 92521

HANS W. PAERL
Institute of Marine Sciences, University of North Carolina at Chapel Hill, 3431 Arendell St., Morehead City, NC 28557

PIERRE PAYMENT
INRS, Institut Armand-Frappier, Université du Québec, 531 boulevard des Prairies, Laval, Québec H7V 1B7, Canada

AARON D. PEACOCK
Center for Biomarker Analysis, University of Tennessee, 10515 Research Dr., Suite 300, Knoxville, TN 37932-2575

IAN L. PEPPER
Environmental Research Laboratory, Department of Soil, Water, and Environmental Science, The University of Arizona, 2601 E. Airport Dr., Tucson, AZ 85706

TOMMY J. PHELPS
Oak Ridge National Laboratory, P.O. Box 2008, Oak Ridge, TN 37831-6036

WESLEY O. PIPES (retired)
Department of Civil Engineering, Drexel University, Philadelphia, PA 19104

DAVID A. POLYA
School of Atmospheric and Environmental Sciences and The Williamson Research Centre for Molecular Environmental Science, University of Manchester, Manchester M13 9PL, United Kingdom

KENNETH M. PORTIER
Department of Statistics and Evalution, National Home Office, American Cancer Society, Atlanta, GA 30329-4251

LAURA PRESLEY
Department of Plant Pathology, University of California, Riverside, CA 92521

YUHUI QIAN
Wood Science and Technology, University of Maine, Orono, ME 04469

DHANA RAO
Centre for Marine Biofouling and Bio-Innovation, School of Biotechnology and Biomolecular Sciences, University of New South Wales, Sydney, New South Wales 2052, Australia

MICHAEL S. RAPPÉ
Hawaii Institute of Marine Biology, School of Ocean and Earth Science and Technology, University of Hawaii at Manoa, Kaneohe, HI 96744

KELLY A. REYNOLDS
Department of Soil, Water, and Environmental Science, University of Arizona, Tucson, AZ 85706

ANNA-LOUISE REYSENBACH
Department of Biology, Portland State University, Portland, OR 97201

SCOTT RICE
Centre for Marine Biofouling and Bio-Innovation, School of Biotechnology and Biomolecular Sciences, University of New South Wales, Sydney, New South Wales 2052, Australia

CHRISTIAN S. RIESENFELD
Division of Earth and Ecosystem Sciences, Desert Research Institute, 2215 Raggio Pkwy., Reno, NV 89512

ERIC E. RODEN
Department of Geology and Geophysics, University of Wisconsin, Madison, WI 53706

CHAD W. SALTIKOV
Department of Environmental Toxicology, University of California, Santa Cruz, 1156 High St., Santa Cruz, CA 95064

SYED A. SATTAR
Centre for Research on Environmental Microbiology, Faculty of Medicine, University of Ottawa, Ottawa, Ontario K1H 8M5, Canada

MARION SAVILL
Institute of Environmental Science and Research Ltd. (ESR), Christchurch Science Centre, 27 Creyke Rd., P.O. Box 29-181, Christchurch, New Zealand

CHRISTOPHER W. SCHADT
Environmental Sciences Division, Oak Ridge National Laboratory, Oak Ridge, TN 37831-6038

FRANK W. SCHAEFER III
Microbiological and Chemical Exposure Assessment Research Division, National Exposure Research Laboratory, U.S. Environmental Protection Agency, 26 W. Martin Luther King Dr., Cincinnati, OH 45268

MICHAEL SCHMID
Institute of Soil Ecology, Department of Rhizosphere Biology, GSF-National Research Center for Environment and Health, 85764 Neuherberg/Munich, Germany

ASTRID SCHNETZER
Department of Biological Sciences, University of Southern California, Los Angeles, CA 90089

DIRK SCHÜLER
Max Planck Institute for Marine Microbiology, Celsiusstrasse 1, 28359 Bremen, Germany

KATE M. SCOW
Department of Land, Air, and Water Resources, University of California, 1 Shields Ave., Davis, CA 95616

ALEXANDRA J SCUPHAM
National Animal Disease Center, Ames, IA 50010

ROBERT L. SINSABAUGH
Department of Biology, University of New Mexico, 167A Castetter Hall, Albuquerque, NM 87131

RICHARD L. SMITH
U.S. Geological Survey, 3215 Marine St., Boulder, CO 80303

WILLIAM STAFFORD
Advanced Research Centre for Applied Microbiology, Department of Biotechnology, University of the Western Cape, Bellville 7535, Cape Town, South Africa

DAVID A. STAHL
Department of Civil and Environmental Engineering, University of Washington, 302 More Hall, Box 352700, Seattle, WA 98195-2700

LINDA D. STETZENBACH
Department of Environmental and Occupational Health, School of Public Health, University of Nevada, Las Vegas, Las Vegas, NV 89154-3063

JANICE L. STRAP
Faculty of Science, University of Ontario Institute of Technology, 2000 Simcoe St. N., Oshawa, Ontario L1H 7K4, Canada

KERRY L. SUBLETTE
Department of Chemical Engineering, University of Tulsa, 600 S. College Ave., Tulsa, OK 74101

JOSEPH M. SUFLITA
Department of Botany and Microbiology, University of Oklahoma, 770 Van Vleet Oval, Norman, OK 73019

CURTIS A. SUTTLE
Departments of Botany, Earth and Ocean Sciences, and Microbiology and Immunology, University of British Columbia, 1461 Biosciences, 6270 University Blvd., Vancouver, British Columbia V6T 1Z4, Canada

DAVID M. SYLVIA
Department of Crop and Soil Sciences, 116 ASI Bldg., The Pennsylvania State University, University Park, PA 16802-3504

RALPH S. TANNER
Department of Botany and Microbiology, University of Oklahoma, 770 Van Vleet Oval, Norman, OK 73019

MICHAEL W. TAYLOR
Abteilung für Mikrobielle Ökologie, Institut für Ökologie und Naturschutz, Universität Wien, Althanstrasse 14, A-1090 Vienna, Austria

BRADLEY M. TEBO
Department of Environmental and Biomolecular Systems, OGI School of Science and Engineering, Oregon Health and Science University, 20000 NW Walker Rd., Beaverton, OR 97006

ANDREAS TESKE
Department of Marine Sciences, University of North Carolina at Chapel Hill, 351 Chapman Hall, CB 3300, Chapel Hill, NC 27599

TORSTEN THOMAS
Centre for Marine Biofouling and Bio-Innovation, School of Biotechnology and Biomolecular Sciences, University of New South Wales, Sydney, New South Wales 2052, Australia

PETER S. THORNE
Department of Occupational and Environmental Health, College of Public Health, The University of Iowa, 100 Oakdale Campus, 176 IREH, Iowa City, IA 52246

MING TIEN
Department of Biochemistry and Molecular Biology, The Pennsylvania State University, 408 Althouse Laboratory, University Park, PA 16802

GARY A. TORANZOS
Department of Biology, University of Puerto Rico, P.O. Box 23360, San Juan, PR 00931

LEA VALINSKY
Central Laboratories, Israeli Ministry of Health, Jerusalem 94467, Israel

FREDERIK P. J. VANDECASTEELE
Environmental Biotechnology Institute, University of Idaho, Moscow, ID 83844-1052, and Department of Biological and Agricultural Engineering, University of Idaho, Moscow, ID 83844-0904

DAVID J. VAUGHAN
School of Atmospheric and Environmental Sciences and The Williamson Research Centre for Molecular Environmental Science, University of Manchester, Manchester M13 9PL, United Kingdom

PIETER T. VISSCHER
Department of Marine Science, University of Connecticut, Avery Point Campus, Groton, CT 06340

GERRIT VOORDOUW
Department of Biological Sciences, University of Calgary, 2500 University Dr. NW, Calgary, Alberta T2N 1N4, Canada

BESS B. WARD
Department of Geosciences, Princeton University, Princeton, NJ 08544

ALAN WARREN
Department of Zoology, Natural History Museum, Cromwell Rd., London SW7 5BD, United Kingdom

JEREMY S. WEBB
School of Biological Sciences, University of Southampton, Southampton SO16 7PX, United Kingdom

KARRIE A. WEBER
Department of Plant and Microbial Biology, University of California, Berkeley, CA 94720-3102

MICHAEL WEINTRAUB
Department of Earth, Ecological, and Environmental Sciences, University of Toledo, Toledo, OH 43606

STEPHEN B. WEISBERG
Southern California Coastal Water Research Project Authority, 3535 Harbor Blvd., Suite 110, Costa Mesa, CA 92626

JOHANNA V. WEISS
U.S. Geological Survey, 431 National Center, Reston, VA 20192

DAVID C. WHITE (deceased)
Center for Biomarker Analysis, University of Tennessee, 10515 Research Dr., Suite 300, Knoxville, TN 37932-2575

R. B. WHITE
EarthFax Development Corporation, 7324 S. Union Park Ave., Midvale, UT 84047

KLAUS WILLEKE
21 Donald Dr., Orinda, CA 94563-3646

LYNN L. WILLIAMSON
Department of Plant Pathology, University of Wisconsin—Madison, 1630 Linden Dr., Madison, WI 53706

GIDEON M. WOLFAARDT
Department of Chemistry and Biology, Ryerson University, 350 Victoria St., Toronto, Ontario M5B 2K3, Canada

CHIN S. YANG
P & K Microbiology Services, Inc., 1936 Olney Ave., Cherry Hill, NJ 08003

MARYLYNN V. YATES
Department of Environmental Sciences, University of California, Riverside, CA 92521

BEI YIN
Department of Plant Pathology, University of California, Riverside, CA 92521

JIZHONG ZHOU
Institute for Environmental Genomics, Department of Botany and Microbiology, University of Oklahoma, 101 David L. Boren Blvd., Norman, OK 73019

JAMES T. ZMUDA (retired)
Research and Development Laboratory, Metropolitan Water Reclamation District of Greater Chicago, 6001 Pershing Rd., Cicero, IL 60650

Preface

Environmental microbiology is a fascinating field of science which captured my imagination when I first got out my trowel 32 years ago to collect soil from beneath a suburban tree. While following the trail which ensued, I have sampled the drainage from coal mines, dug infiltration basins in the desert, dredged sediments from bays and rivers, and chopped holes through surface ice. I even have learned that sometimes the sample bottle gets filled in unusual ways, such as when the surf washed over my head or when I slipped sideways into a river. The latter event was observed by three of my students, who were courteous enough not to laugh too loudly. The *Manual of Environmental Microbiology* is an effort to combine the results and my excitement from my own discoveries with those of many colleagues. I feel honored to be a part of this collaborative effort. It is an idea which began handwritten on the back of an envelope in January 1991, and I still cherish that envelope.

The process of organizing, writing, and finally publishing this manual has been a joint volunteer effort by many hundreds of people, each of whom has contributed generously of their own experiences and expertise. Together, as a group of 136 scientists, we saw our efforts come to initial fruition with publication of the first edition of the *Manual* in the fall of 1996. By bringing to you this third edition, we now present a complete update of our field. Jay Garland, David Lipson, and Aaron Mills have been added as new volume editors for this edition, replacing Michael McInerney and Guy Knudsen, who now have become emeritus editors. Four of the section editors from the second edition, Robert Christian, Steven Newell, David Stahl, and Linda Thomashow, likewise have become emeritus editors. It is with pride that I welcome in their stead Meredith Hullar, Jonathan Lloyd, Seán O'Connell, and Ming Tien, who have joined the editorial board as new section editors for this edition. Many of the contributing authors from the first two editions have been called in other directions, and those departing contributors now are our alumni, each of whom knows that his or her efforts were appreciated by our readers. Two hundred seven scientists have contributed to this third edition of the *Manual*, including more than 100 new authors, and I welcome each of them. I offer special appreciation for the fact that nearly all of the former editors have remained among our group as authors.

The most difficult part of compiling such a volume is deciding which topics will receive center stage and which must be given a reduced representation. It also is important to recognize that the research of any given field flows and sometimes changes its path over time. The authors and editors of this manual are dedicated to helping you, our readers, keep a balanced perspective. We always will be certain that the basic foundations of environmental microbial methodology are thoroughly covered and also try to keep you up to date with the front of that flow. To keep that objective and meet our responsibility to our readers, we have updated all of those chapters that were carried over from the second edition, and many of them have been completely rewritten. You also will find that this edition contains numerous chapters on new topics.

In addition to thanking the authors and editors, without whose selfless volunteer efforts the American Society for Microbiology could not bring forth this manual, I offer particular thanks to several specific individuals. Jeff Holtmeier, the director of ASM Press, has lent us his utmost support. Ken April of ASM Press served as our production editor. Ken, most of all, has coordinated the entire process and kept us on course. Although his job is not easy, Ken somehow manages to seem calm and cheerful while surviving it all. The editors also thank the following individuals who assisted us by serving as ad hoc reviewers for this edition of the *Manual:* Anne Bernhard, Haluk Beyenal, Rima Franklin, Jim Fredrickson, Robert Genter, Tim Griffin, Jo Handelsman, Heribert Insam, Scott Kelley, George Kling, Mike Lehman, John Lindquist, Henry Mainwaring, Andrew Martin, Andrzej Paszczynski, Gary Sayler, Kathleen T. Scott, Claus Sternberg, Ben Van Mooy, and Darla Wise.

CHRISTON J. HURST

INTRODUCTION TO ENVIRONMENTAL MICROBIOLOGY

I

VOLUME EDITOR
JAY L. GARLAND

Introduction to Environmental Microbiology

CHRISTON J. HURST

1

Environmental microbiology is the study of microorganisms which exist in natural or artificial environments. The origin of scientific research in this field rests in the observations of Antony van Leeuwenhoek that were published in 1677 (6). van Leeuwenhoek used a microscope of his own creation to discover what he termed "animalcula," or the "little animals" which lived and replicated in rainwater, well water, seawater, and water from snow melt. During the intervening centuries, the expansion of our knowledge regarding environmental microorganisms has been based on increasingly detailed observations and experimentation, in which we have been aided by advancements in microscopy and the development of biochemical and mathematical tools. However, many of the individual topics which we have since come to better recognize and characterize in environmental microbiology were initially described in van Leeuwenhoek's publication.

Many of van Leeuwenhoek's observations were based on the examination of environmental samples which he variously maintained in a wine glass, porcelain dish, and glass bottle in his house. In current terminology, we would say that he had created small artificial environments, or microcosms. While we often tend to think of artificial environments as taking the form of small laboratory systems, the term also encompasses both production-scale fermentations and anthropogenic structures emplaced in the natural world. Field-emplaced anthropogenic structures are often open to the environment and thus conceptually merge with the spectrum which represents natural environments.

It is an interesting coincidence that van Leeuwenhoek's first discovery of life was in water because that is almost certainly where the earth's life began. We have come to understand that the evolution of life on this planet would have depended on water because the bipolar nature of that molecule induces the three-dimensional configuration of our biomolecules. Also, we have continued to examine both the presence and metabolic activities of microorganisms found in surface waters and in the hyporheic world which underlies it. Since the studies of van Leeuwenhoek, we have discovered that microbial life is amazingly diverse (chapter 3). Microorganisms literally cover our planet. They live in surface environments even as extreme as the fumaroles of volcanoes and in sedimentary rocks within the dry valleys of Antarctica. Microbes can be found as deep as several kilometers both in glacial ice sheets and in bedrock. In fact, it seems that microorganisms can be found living in any site where the combination of physical conditions is such that water could occur in its liquid state. At deep-ocean thermal vents, where the temperature of the water can reach far above its normal boiling point, the high barometric pressure keeps water in its liquid state and microbial life abounds. The microbes living there provide the biological basis for the surrounding ocean floor life. When the ocean's surface layer freezes during polar winters, the dissolved salts become excluded in minute vacuoles and channels within the forming ice. The water contained within these minute spaces remains in a liquid state because of the high salt concentration. There, within these "brine channel" spaces, microbial life likewise abounds and provides the basis for some elaborate sea ice endofaunal food chains. Since many types of microorganisms have developed resistance to prolonged freezing and even to desiccation, the microbes seem to have much of our planet available to them. Amazingly, albeit perhaps not surprisingly, microbes even live on and within paintings and other artworks (1), where their digestive processes show that the microbes clearly have a taste for art, although they may not appreciate these objects in the same sense as we do. While we study these many wonders, we also further our recognition that microbes comprise the basis of our biological legacy. They represent the very origin of the biofilm on Earth, and we are but a part of that biofilm. Just as the microbes are the critics that seem to have the final say about our works of art, in the end they undoubtedly will have the final word regarding life on this planet.

In his paper, van Leeuwenhoek mentioned an observation that the types and relative concentrations of organisms present in his environmental samples changed and fluctuated with respect to time. We have continued his observation that microbial colonization of environments is a process which involves a succession of organisms, and we now know that this type of succession is brought about by waves of organisms which differ in their ecological requirements. The microorganisms chemically interact with their physical environment, and their most notable effect has been the creation of an oxidizing atmosphere on this planet. By way of these chemical interactions, microbes remain crucial to the biogeochemical cycling which supports the continuation of life on our planet, turning over the elements that represent the basic ingredients of life such as carbon, hydrogen, nitrogen, oxygen, phosphorus, and sulfur.

Over the course of millennia, we have learned to use naturally occurring microbial products for our benefit, including petroleum and some of the fungal metabolites which have antimicrobial properties. While expanding the extent to which we understand microbial processes, during the last few decades we have begun learning how to harness microbial biosynthetic and degradative activities. This harnessing, including the intentional manipulation of microbial activities, constitutes the basis of microbial biotechnology, whereby we direct the activity of microorganisms within both natural and artificial environments for a variety of purposes. As one example, we utilize microorganisms to help us achieve goals such as the production of materials which are beneficial to our existence, including numerous antibiotics, vitamins, fuels such as biogas and ethanol, and perhaps eventually usable electricity (4). Microorganisms also are used as tools to help us intentionally degrade both natural and anthropogenic materials in wastewater digestors, composters, landfills, natural terrestrial environments, and natural or artificial aquatic environments. Sometimes we use microorganisms as tools to achieve agricultural goals such as protecting plants from insect damage. Furthermore, microbial processes such as using microorganisms to leach metals from ores and to enhance the recovery of petroleum from wells have been used as a means of minimizing the application of hazardous chemicals in geochemical recovery operations. Just as we sometimes use our knowledge of beneficial microbial processes to optimize their usefulness, at other times we try to prevent natural microbial activities such as those which contribute to the biofouling, corrosion, and decay of objects exposed to the environment.

As scientists, we show a typical human penchant for trying to organize and categorize things. We extend this habit even to the microbes which surround us. Presently, we use a microbial classification scheme which divides the microorganisms into three overarching taxonomic groups. Two of these are considered to be cellular, meaning that they possess cell membranes. These are the Eukaryota, or Eucarya, and the Prokaryota, or Procarya (5). The Eukaryota contains those cellular organisms with membrane-bounded intracellural structures, and its members more traditionally are known as the algae, fungi, and protozoa. The Prokaryota contains those cellular organisms which lack membrane-bounded internal structures, and its members are more traditionally known as the bacteria plus those subdivided out into the archaea. The third overarching group consists of acellular microbes, the viruses and some of their biological relatives, which previously never were allowed shelter within any biological kingdom but could fit into the proposed domain *Akamara* (3). However, from the viewpoint of environmental microbiology, the most important aspect is our increasing understanding that, within ecosystems, these groups of microorganisms naturally organize among themselves as they go about their interactions both with one another and with the macroorganisms on this planet. Perhaps the greatest question remains whether the minimal functional unit of life truly is an individual member of a species or the entire ecosystem within which that individual exists. The interactions between the members of an ecosystem (chapter 2) occur and can be studied on many levels: spatially, biochemically, and even genetically (2). We have tended to divide our descriptions of these interactions into categories based on whether the relationships are neutral, positive, or negative with regard to the organisms involved. Neutral relationships are ones in which there is neither perceived harm nor advantage for the involved organisms, as when two coexisting populations ignore the presence of one another either because their numbers are sparse or because they occupy different ecological niches. Commensal relationships occur when one organism is perceived to benefit while the other is unaffected, as when microbes which colonize the external surfaces of animals or plants derive energy by metabolizing exudates produced by their hosts. There are also mutually advantageous relationships in which both populations benefit, examples of which are synergism and mutualism. Synergism is a voluntary interaction wherein both populations of organisms could exist on their own but do better when living together. Certain intestinal bacteria which are harbored within human hosts provide an example of synergism. In return for being sheltered, these bacteria produce vitamins that the host cannot manufacture on its own and would otherwise have to acquire from its diet. Mutualism, or symbiosis, occurs when neither organism seems capable of living on its own in that particular environment. An example of symbiosis is the interrelationship between certain chemoautotrophic bacteria and their hosts, the *Riftia* marine tube worms which inhabit ocean floor thermal vents. Through a highly coevolved relationship, those chemoautotrophic bacteria are housed within and provide nutrients for the tube worms. Lastly, there are relationships which are termed competitions, and these are harmful to at least one of the involved species. In the vast majority of instances, these interactions involve independently living organisms which suffer by competing in the open for scarce resources and arise because those organisms attempt to coexist within a common ecological niche when their combined numbers are too great to be supported by that niche. The parasitism represented by infectious diseases is a specialized form of competition, in which the scarce resource is produced within the body of the victim. The interactions between organisms are not fixed in nature, and a relationship which normally is benign may become detrimental, as is the case with some opportunistic pathogens which normally exist as benign members of the skin surface microflora but can cause disease if they gain access to the interior of an immunologically weakened host. This list of interactions ends with predation, in which one organism devours and digests another as food.

Not only have we tried to elucidate the natural fate of microorganisms in the environment, but also we have often attempted to eliminate from the environment some microorganisms which are pathogenic to either humans or the plants and animals on which we depend for our sustenance. Some of these pathogens are indigenous environmental organisms, while various others are of human origin or from animals and plants and are released into the environment through natural processes. The study of their persistence in environmental media, including survival and transport in soil, water, and air, provides clues that help us anticipate and control their populations.

Readers of this manual will find that the core sections are structured with regard to the type of environmental medium being discussed. The subject of water, the hydrosphere, has been divided into two sections, one containing chapters which address the fact that water often serves as a vehicle in the transmission of pathogenic microbes (section III) and the other containing chapters on general aquatic ecology (section IV). The terrestrial environments of the lithosphere have been divided into soil and plant zone interactions (section V) and the microbiology of deeper subsurface environments and landfills (section VI). While microbes are not known to colonize the atmosphere, air serves as one of the vehicles by which both they and their often toxic by-products are transported (section VII). The subject area of microbially mediated chemical transformations bridges the hydrosphere and lithosphere due to the relatedness of the involved microbial metabolic processes, which often are performed by either the same or related genera of organisms. For this reason, the topics addressing biotransformation and biodegradation have largely been grouped into a common section (section VIII). Likewise, the basic principles of environmental microbiology (section I) and general analytical methodologies (section II) tend to be common across the range of environments that we study, and the last two subject areas have been placed in the front of the manual because of their primary importance.

I am not sure whether Antony van Leeuwenhoek could have foreseen where his discoveries have led: to the diversity of environmental microbiology subjects that we now study and the wealth of knowledge that we have accumulated. However, just as I have always enjoyed reading his account of environmental microorganisms, I feel that he would enjoy our efforts at once again summarizing all of environmental microbiology in a single document as represented by this manual. I thank the numerous microbiologists who have collaborated in creating this manual, many as contributors and some as editors. I also thank those giants in microbiology upon whose shoulders we have stood, for we could not have accomplished the creation of this manual without the advantage that those giants have afforded us.

REFERENCES

1. **Dixon, B.** 2005. The microbiology of art. *ASM News* **71:**212–213.
2. **Dorigo, U., L. Volatier, and J.-F. Humbert.** 2005. Molecular approaches to the assessment of biodiversity in aquatic microbial communities. *Water Res.* **39:**2207–2218.
3. **Hurst, C. J.** 2000. An introduction to viral taxonomy and the proposal of Akamara, a potential domain for the genomic acellular agents, p. 41–62. *In* C. J. Hurst (ed.), *Viral Ecology*. Academic Press, San Diego, Calif.
4. **Logan, B. E., C. Murano, K. Scott, N. D. Gray, and I. M. Head.** 2005. Electricity generation from cysteine in a microbial fuel cell. *Water Res.* **39:**942–952.
5. **Pennisi, E.** 1999. Is it time to uproot the tree of life? *Science* **284:**1305–1307.
6. **van Leeuwenhoek, A.** 1677. Observations, communicated to the publisher by Mr. Antony van Leewenhoeck, in a Dutch letter of the 9th of Octob. 1676. here English'd: concerning little animals by him observed in rain- well- sea- and snow-water; as also in water wherein pepper had lain infused. *Philos. Trans. R. Soc. Lond.* **11:**821–831.

Neighborhoods and Community Involvement: No Microbe Is an Island

CHRISTON J. HURST

2

> No man is an *Iland*, intire of it selfe; every man is a peece of the *Continent*, a part of the *maine*; if a *Clod* bee washed away by the *Sea*, *Europe* is the lesse, as well as if a *Promontorie* were, as well as if a *Mannor* of thy *friends* or of *thine owne* were; any mans *death* diminishes *me*, because I am involved in *Mankinde*; And therefore never send to know for whom the *bell* tolls; It tolls for *thee*.
>
> John Donne, Meditation XVII, original publication in *Devotions upon Emergent Occasions*, 1624; quoted from reference 15

The most important thing which must be understood about microorganisms in their environments is that no microbe exists by itself. Extracting a parallel out of the above quotation from John Donne, we should consider that in the environment no microbe is an island existing independently. There are only a few, extremely rare instances when a single species exists naturally in pure culture. The interactions and interdependence between members of different species are a consequence of the evolution of those species into their occupation of specific niches. The second most important thing to recognize is that, as we look around nature's neighborhoods and observe the types of interactions that exist within the communities of macrobial organisms which live there, all of those types of interactions likewise exist in the microbial world. These interactions include such things as peaceful coexistences, competitions, hunting and hiding, and the use of chemical attractants, repellents, and toxicants. In fact, these approaches to surviving in community life would have existed within the microbial realm long before the first macroorganism evolved.

Microbes form the basis of our biological heritage as humans and indeed are the basis upon which all macroorganisms, including vascular plants and animals, evolved. Microbes form the understructure which supports what we perceive as being the macrobial realm. The fact that this assemblage of microbes and macrobes has evolved together binds it together. No single species selected out of this assemblage could survive for very long on its own in the wild without the biological activity provided by those species which occupy the connecting niches. This is a guiding concept in species ecology, and occasionally we will refer back to this concept as we progress through this chapter.

We also know that the life which exists on this planet evolved by using the hydrogen bonding bipolar nature of water molecules in their liquid state to serve as an organizing force. Through this force, biomolecules are arranged into their functional configurations by virtue of their having evolved hydrophilic and hydrophobic zones. Thus, the life on this planet requires water. The pervasiveness of this biology is such that it seems that on this planet active indigenous microbial life can be found in any environmental site where the laws of physics can allow water to exist in its liquid state (see chapter 46). The additional fact that many types of organisms have evolved mechanisms which allow them to survive periods of freezing or even desiccation means that much of this planet is available for supporting life. Microbes would have formed the initial basis of the food chains for all of this planet, serving as its primary producers. Vascular plants have taken over the role of primary production in those land surface areas of the planet where the air temperature is above the freezing point of water for at least part of the year. Vascular plants also have taken over the primary production role in many of the shallow aquatic areas where sunlight reaches in abundance. However, deep inside the earth the microbes still hold sway (19), and phytoplankton in the oceans is assumed to serve as the earth's primary source of fixed carbon. Microbes also perform the primary production role in such intriguing places as the communities of sea ice endofaunal organisms, where microfauna and meiofauna (small size category of macroorganisms) live in the brine channels which exist within the Arctic and Antarctic pack ice (23, 24) and in the unconsolidated layers of platelet ice which underlie the pack ice (26). Microbes apparently live within the water of Lake Vostok which exists in Antarctica beneath about 4 km of glacial ice and due to the absence of sunlight must serve as the primary producers for that lake (30). At the other end of the production line, microbes are the key to the recycling of available nutrients, including the important aspect of cellulose recycling within the gut of termites (41) and the recycling which occurs in deep-sea sediments (52).

Microbial communities function as interacting, coevolved assemblages (11, 17, 22, 25, 53). Of course, as we first begin to look at microbial communities and the interactions among microbial populations, we immediately are faced with three significant problems. First, an enormous number of different microorganisms exist in a wide array of highly dissimilar environments (48) and communities (20). These microorganisms include archaea and bacteria (see chapter 35), fungi (see chapters 38, 51, and 52), algae and cyanobacteria (see chapters 30 and 37), protozoa (chapter 37), and viruses (chapter 36). Second is the enormously complex task of understanding and profiling the interactions among microbial populations. Such interactions not only are numerous and dissimilar but often are poorly characterized or totally undefined. Furthermore, there is difficulty in understanding the interactive role and major impact of invertebrate and vertebrate animals and vascular plant populations on microorganisms, a subject which is not within the usual scope of interest of microbiologists. For the rest of this chapter, I will try to limit our focus to just the microbes, although the more complicated reality is that microbial populations often are intimately linked with coevolved invertebrates (21, 27, 33, 54), some of which in turn consume microorganisms; vascular plants (35, 46) that in turn provide readily available organic nutrients; and vertebrate animals (3, 36) that may consume all of the above. At times it does seem that evolution works its way down strange pathways. Third, there exists a woeful lack of agreement about the various definitions that are appropriate for characterizing microbial populations, communities, environments, and interactions.

This chapter is not designed to be encompassing. The topic of microbial communities and interactions would easily fill an entire monograph. Rather, the text will serve as a prelude, presenting an introduction to some of the principal issues of environmental microbiology.

TERMINOLOGY: AS IF MERE WORDS COULD DEFINE HOME

> "When I use a word," Humpty Dumpty said, . . . "it means just what I choose it to mean—neither more nor less." "The question is," said Alice, "whether you *can* make words mean so many different things." "The question is," said Humpty Dumpty, "which is to be master—that's all."
>
> Lewis Carroll, original publication in *Through the Looking-Glass, and What Alice Found There*, 1872; quoted from reference 9

Environmental microbiology is blessed, or possibly cursed, with a multitude of terms that often are contradictory, sometimes misused, and frequently misunderstood even by the specialists. The ability to use language properly is an art and helps us to understand one another. But semantic sensitivity is frequently not the hallmark of the experimentalist, and this lack of sensitivity often is the basis for confusion and disagreement. The purpose of this section is to briefly define some of the more critical terms used in this chapter. As Lewis Carroll might like it stated, the words are defined even if the reader does not agree totally or even partially with the definitions used. Whether it is the word, the author, or the reader which is master will remain an open question (as Humpty Dumpty may have wished it to be).

Defining the ecology of a species is a complex issue, but the issue can be divided into different ecological aspects. The first aspect can be consideration of the terms "boundary" and "barrier." The descriptions provided here are brief, but the concepts of boundaries and barriers will be taken up a bit more extensively later in the chapter. The word "boundary," as used here, represents an abstract concept which describes the environmental limits beyond which the physiological requirements and capabilities of a given species do not allow its members to function (C. J. Hurst, "The Consequences of Crossing Barriers and Boundaries," presented at the 100th General Meeting of the American Society for Microbiology, 21 to 25 May 2000, Los Angeles, Calif.). This boundary encompasses the overlap of a combination of physically and chemically defined environmental characteristics and is absolute. The members of a species are obliged to live within their boundary because of the fact that their metabolism ceases to actively function beyond that boundary. While they may venture outside, any individual's existence in a live state for very long outside the boundary would require either artificial environmental support or the ability of that individual to enter temporary metabolic stasis. Often, the boundaries of different species overlap and this allows for biological interactions between the members of those species. If the boundaries of two species do not overlap, then it is unlikely that the members of those two species will biologically interact with one another. This concept of a boundary is not something marked out as if it were a physically defined territory on a map. However, we may be able to mark on a map spatial zones, territories, or physical locations which meet the requirements for being encompassed by the boundary of a species. Those spatial zones would represent potentially habitable sites for the members of that species. Most physical interaction between the members of species would occur within areas of spatial overlap. These potential habitats may be either contiguous or separated by barriers. "Barriers" are obstacles, and they can be classified by their nature as either physical, chemical, or biological (Table 1). Their nature as obstacles relates to the boundary of some particular species which is being considered. What constitutes a barrier to one species may not be a barrier to some other species. Barriers can restrict or prevent the interaction of two species even if those species have identical boundaries. This contributes to the reason why parallel evolution can produce two similar species which may occupy nearly identical niches but never encounter one another.

We now can move forward to the next ecological aspect, which is to consider that the ecology of a species has two traditionally defined components: its "niche" and its "habitat." Each of these components can be defined both in a potential perspective and in a more limited operational perspective. The major difference between these components is that, whereas a niche is biologically defined, a habitat is physically defined. The word "habitat" describes only the place where a species lives. The word "niche" describes how this species fits into that place, representing its interactions and accommodations with respect to the abiotic and biotic conditions found there. Thus, for any particular species, the description of its niche consists largely of the interactions between the members of that particular species and other species which occupy connecting niches. However, the definition of niche also includes changes in the environment effected by the species' biological activity. Among ecologists, the niche of a species represents what that species does; that is, its function in its natural habitat. The potential niche of an organism, expressed in terms of its total, genetically provided capacity for biological interactions, may be a marvelously broad thing and enables potential evolution of that species. In nature, however, competition and other interactions with and against different species narrow this potential niche to a smaller, more functionally restricted operational niche. The physically defined term "habitat" brings us to the point of identifying territories, places to which we could point and sometimes even identify

TABLE 1 Examples of barriers applying to different species[a]

Physical
- Thermal
- Acoustic (usually ultrasonic)
- Pressure (barometric, hydrostatic, osmotic)
- Radiation (electronic, neutronic, photonic, protonic)
- Impaction (includes gravitational settling)
- Adhesion (adsorption, electrostatic, van der Waals)
- Filtration (size exclusion)
- Geographic features
- Atmospheric factors (include meteorological aspects such as humidity, precipitation, and prevailing winds)

Chemical
- Ionic (includes pH and salinity)
- Surfactant
- Oxidant
- Alkylant
- Desiccant
- Denaturant

Biological
- Immunological (includes specific as well as nonspecific)
- Naturally induced (intrinsic response)
- Naturally transferred (lacteal, transovarian, transplacental, etc.)
- Artificially induced (includes cytokine injection and vaccination)
- Artificially transferred (includes injection with antiserum and tissue transfers such as transfusion and grafting)
- Biomolecular resistance (not immune related)
- Lack of receptor molecules
- Molecular attack mechanisms (includes nucleotide-based restrictions)
- Antibiotic compounds (metabolic inhibitors, either intrinsic or artificially supplied)
- Competitive (other species in ecological competition), including predation or parasitism against the subject species by members of some other species or, if the subject species is an infectious agent, competition against a vector which the subject species might need in order to achieve transmission between host organisms

[a]Information adapted from reference 29.

on a map. A habitat is a site, generally having some physical uniformity, which possesses those characteristics that appear to be of ecological importance for supporting residence by the members of a species. The potential habitat of a species can be very large but operationally may be more restricted due to deterring interactions such as competition against or predation by the members of other species. Some species end up being restricted to perhaps only a single and frequently unique habitat, whereas other species are cosmopolitan and successfully reside in a wide variety of seemingly dissimilar habitats. An individual habitat may be a specific layer in the depth profile of a characterizable soil type, the intestinal tract of members of a particular animal species, the outer surfaces of a plant root, or some zone within lake bottom sediments. Frequently, however, owing to the small size of microorganisms, the concern of the microbial ecologist is the microhabitat (49), inasmuch as individual propagules, filaments, or cells often are restricted to a site that is no larger than several millimeters or sometimes even micrometers. Thus, an organism may be found to exist only within a specific depth range in the water column of a particular lake and interact only with the other species found in that vicinity. This may be the case even though the boundary of that species might be broad enough that the organism could exist at other depths and in different bodies of water. Yet that one place then represents the operational habitat of that species, even if its potential habitat could be much larger. The interactions of this species with other organisms existing within that small vicinity are a defining part of this species' operational niche, even if its potential niche could be far larger. The reasons for restriction to that operational habitat and operational niche may be the existence of barriers, which may include a biological inability to successfully compete against, or may result from predation by, the members of other species found at other depths within that water column or found within other bodies of water.

Within a particular site, we find what we call a community. Perhaps we could consider the term "neighborhood" to define the site itself, and "community" then could define those which live therein. The community comprises all members of all species that occupy a particular site. These members of the community generally represent different populations. A "population" of a particular organism is an assemblage of individual organisms having common characteristics. In microbiology, the term "population" variously may be used to represent all individuals of a single species, subspecies, variety, race, or other subspecies designation. Within a population, there may be a large number of individual cells, or a population may be composed of a multicellular filament, such as is found among the fungi and algae. An appropriate term for an individual microbe belonging to a population is "propagule." However, the reader should be aware that many specialists use the term "community" to encompass only a particular category of organisms. For this reason, some would describe an environment as having separate communities of bacteria, fungi, algae, plants, or animals. The community of a soil in general is somewhat different from that of the subsoil or underlying aquifer, and the community in the water column is not the same as that inhabiting the underlying sediment. A term synonymous with "community" but not widely used in English is "biocoenosis." Although clear distinctions between communities and populations thus exist, the words are frequently and mistakenly used as synonyms.

"Consortium" is a term used more commonly for bacteria in associations than for assemblages of other organisms (18), but this concept does not have to be limited to just bacteria. It is a useful word when it refers to a collection of organisms that have some functional association with one another (see chapter 9) and when we talk about the properties which describe particular communities (see chapter 11). For example, one species may provide growth factors for a second species or serve to eliminate inhibitory compounds that affect a second species. To some degree, a consortium may represent a commensal or protocooperative interaction whose basis has not yet been well established. Unfortunately, however, mixtures of bacteria that have yet to be separated into pure cultures are frequently designated consortia; in these instances, the mixture may not represent a functional association but, rather, reflects the lack of isolation of individual components of the mixture. "Facilitation" is a term which describes situations when the activity of one species may facilitate a similar or different activity by another species.

UNDERSTANDING THE WAYS TO BEHAVE

The degree of species diversity within communities can vary enormously. Communities typically may have several, many, or innumerable species. Although microbiologists are prone to work with pure cultures, communities having only one species are rare in nature, except in those monospecific communities associated with disease processes or in environments so extreme that few species can survive or proliferate. There often is some benefit gained by one species from coexistence with other species. Although the benefits are unquestionable, the basis for those benefits and the mechanisms associated with the interactions in natural communities are rarely understood and rarely studied. The purpose of this section is to describe the kinds of interactions which have been observed between species.

We know that in order to become a member of a community, the propagules of a species not only must reach the environment where that community exists, but also must successfully cope with the detrimental interactions occurring there, including competition with the species which already exist in that site. Some sites seem not only readily accessible to many different types of microorganisms, but also supportive towards the survival and proliferation of those new arrivals. These receptive sites thus often end up with a high species diversity. Soils, surface waters, sediments, and other environments such as sites rich in decaying organic materials support a multitude of species of bacteria, fungi, protozoa, and sometimes algae. On the other hand, sites with high salinity, extreme pH, low nutrient levels, or high light intensity often are characterized by low species diversity, which may be reflective of a low receptivity.

An old truism states that microorganisms are potentially everywhere. That truism is little more than a platitude, and as a statement it takes on a degree of ecological inaccuracy when it is used as the basis for assuming that species or genera that are widely disseminated are also widely established. Potentially cosmopolitan species frequently are able to grow, in culture at least, very rapidly in the absence of other organisms so that a few propagules or a small biomass can multiply to reach a high cell density or large biomass in a matter of hours or days. Such largely unchecked proliferation rarely occurs in nature, however, and effective checks exist to prevent the unbridled multiplication of widely disseminated propagules. The operation of these various checks is the basis for the appropriate second half of the truism, namely, that the environment selects. Hence, the full and ecologically appropriate statement is that microorganisms are potentially everywhere, but the environment selects.

The impact of that selection is evidenced by the characteristic communities of individual habitats. Many of the bacteria, fungi, and protozoa present in surface waters are thus not characteristic of soils, and the types of heterotrophs (organisms which derive their operating energy from organic compounds found in their surrounding environment, as contrasted with autotrophic organisms, which derive their operating energy either from photosynthesis [photoautotrophic] or chemosynthesis [chemoautotrophic]) found in soils are often quite different from those that would be found on the surfaces of leaves. The forces of selection are often nonbiological, and an area that has a low pH, is exposed to high light intensity, has no available oxygen, or contains low concentrations of readily available carbon sources supports a community that is able to cope with these restraints or limitations. Such abiotic (not biological in origin) factors are often reasonably easy to demonstrate. However, more difficult to establish and perhaps more interesting from the biologist's perspective are the biotic (of biological origin) stresses that are operating in environments in which major abiotic stresses do not determine community composition.

Involvement and Homeostasis Are Evolutionary Consequences

The indigenous populations that make up the community are responsible for the biotic balance that is maintained. They regulate the population densities or biomass of the individual component species of the community, and they act to prevent the establishment of invading species. These various biotic interactions determine the stability of established microbial communities, and they reflect what is designated "homeostasis." From a biological perspective, homeostasis is the numerical and biochemical equilibrium resulting from the activities of the different species which comprise the community. From an ecological standpoint, homeostasis is the capacity of a community to maintain its stability and integrity in an environment subject to abiotic and biotic modifications. These homeostatic mechanisms are constantly operating, and although they are often modified somewhat, they generally are not altered appreciably unless an enormous perturbation occurs. But stable does not mean permanently unchanging. Both the levels of activity and the proportional numbers of the component species forming a community may fluctuate and even respond cyclically as a response to abiotic or biotic perturbations. That concept includes seasonally related changes. Characteristically, it is abiotic perturbations that may totally upset the community. Only in rare instances is a biotic influence, such as the mere introduction of propagules of a nonindigenous microorganism, sufficient to permanently alter the community to an appreciable extent. Indeed, in many instances, the new arrival is not even allowed to become established.

Ignoring the potency, as well as the limitations, of homeostatic mechanisms has led to a number of viewpoints that run counter to observations in nature. For example, individuals who want to introduce bacteria or fungi into soils, subsoils, or aquifers for bioremediation assume that the introduced species will become established and bring about the destruction of a contaminant of concern. In doing so, it is assumed that just because humans judge that the introduced microorganism has a beneficial trait (in this case, the capacity to grow on and thus destroy an organic pollutant), that organism indeed will become established and perform what is desired of it. While the humans performing the experiment may think that this important trait of the introduced microbe is desirable from their own perspective, the humans are not in charge of the situation. The introduced organism may not be accepted by the existing microbial community. Why might the microbe not be accepted? The capacity of the microbe to use a particular substrate for growth, although of paramount importance in culture, simply may not be sufficient for ecological success: the organism also must be able to cope with the various factors associated with homeostasis. An attribute that is a necessary requirement for growth is not a priori an attribute sufficient for establishment, and both necessary and sufficient traits must be present. Success requires that an introduced organism be able to compete effectively for limited resources other than the specific organic compound which laboratory studies may have proven can serve as its carbon source. The introduced organism also must be able to cope with the stresses associated with predation and parasitism, which are of great importance in many environments. In contrast with the vendors of microorganisms, who always hope for success from their introductions, there are other individuals fearfully concerned that the environmental introduction of genetically engineered microorganisms could result in successful establishment. Many of the latter individuals must optimistically presume that all introductions will fail and that modest changes in the genotype of an existing organism will not result in its establishment. Again, it helps to emphasize that natural events often prove humans to have been wrong in their assumptions. It is necessary to consider not only the strengths but also the limitations of homeostatic mechanisms. While the operation of homeostatic mechanisms does eliminate most introductions, these mechanisms are not omnipotent, and some introduced organisms do succeed in becoming established. Borrowing an analogy from the world of macroorganisms, farmers know that sowing a new variety of plant, one that has major beneficial attributes, is not sufficient to obtain high yields of that variety. The introduced crop species frequently is unable to cope with competition by weeds, parasitism by insects and plant pathogens, poor soil structure, and other stresses. Few introduced species become established, regardless of whether the introduction has the capacity to do good or is potentially injurious. Yet in some instances, an alien species does become successfully established, as evidenced by the major upsets that have occurred because of invasion by plants and animals. Attempts have been made to predict the capacity of an introduced organism to become established (32, 47), and studies have been directed towards establishing the traits that might be used to predict the outcome of an introduction. Our inability to predict the success or failure of introductions is a reflection of incomplete knowledge of the various components of homeostasis.

Being Kind and Helpful

To the general public, microorganisms are frequently considered to be solely harmful. This is evident in the use of the word "germ" (commonly referring to a microorganism thought to cause disease but perhaps also definable as a microbe unwelcome in the place where it has been found). Admittedly, it is true that when we examine the world of microorganims we find competition, parasitism, and predation to be of great importance and that disease-producing microorganisms are widespread. And yet altruism also is a widespread attribute among all major categories of microorganisms. Despite the fact that detrimental interactions are inherent to biology at all levels, we should not overlook the good that many heterotrophic and autotrophic microbial populations do for their neighbors, including the natural and often protective microfauna associated with plants and animals. In truth, it is the neutral relationships which likely are the most common in biology. "Neutrality" describes those relationships in which there is neither perceived harm nor advantage for either of two species being considered relative to one another. Interestingly, though, there sometimes is a tenuous balance between neutrality and a harmful state. That balance may hinge even on something apparently so minor as ambient temperature (4).

Using and Abusing

Some apparent types of beneficence are commonly categorized, from the ecological viewpoint at least, by the terms "commensalism," "protocooperation," and "symbiosis." Although semantically sensitive readers might expect a clear distinction among those three terms, in fact a continuum exists so that the range of each of these types of beneficial interaction merges into the next. Indeed, it is likely that there is a continuous evolution so that organisms which at one time were commensals evolve to exist in protocooperative relationships and those that are involved in protocooperation will, with time, evolve into highly dependent and mutually beneficial symbioses.

"Commensalism" defines an association in which one species benefits a second but the first gets no apparent advantage in return. The second species which is provided with something essential for its replication is deemed the commensal. The mechanisms underlying commensal associations are numerous, but only a few have been characterized. Such interactions are evident between algae excreting photosynthetically fixed carbon and heterotrophic bacteria which as commensals consume that carbon (10). Commensal relationships apparently also exist between populations jointly contributing to many biochemical transformations and to degradative processes (see Section VIII). In these, the first population or species may convert a compound which otherwise metabolically is unavailable to a second population into some product that can be used as a nutrient source by the second species. Sometimes the second population is auxotrophic, meaning that its members cannot grow in the absence of the supplied compounds, and those required compounds are termed to be the second population's growth factors. An organism which excretes a growth factor may allow for the development of this auxotroph as a commensal. Studies of water and soils indicate that a large number of indigenous species are able to excrete one or more vitamins or amino acids which may be capable of serving as growth factors. Another type of commensalism is evident in environments containing organic or inorganic inhibitors, and a species that destroys an organic inhibitor or somehow detoxifies an inorganic ion serves an altruistic role for its sensitive neighbor.

The fact that terrestrial and aquatic communities contain a high percentage of microorganisms that are auxotrophic means that commensalism must be biologically favorable in those communities. An organism that relies on its neighbors for the synthesis of a carbon source, growth factors, or detoxifying enzymes does not need to expend energy of its own to either make or detoxify those substances. Thus, although the commensal is dependent upon the helpful neighbors found in that environment, given all other factors to be equal the commensal has a competitive energetic advantage over some other organism that might reject dependence by instead choosing to expend energy to make all of its own vital compounds or to detoxify by itself all harmful substances. In turn, a chain or network of commensal reliances can evolve. And if evolution thus energetically benefits a commensal, we can understand that it would also likely favor a development in which two species in contact with one another mutually develop a greater biochemical interreliance. This is a gradient along which facilitative relationships conceptually may lead into a kind of loose confederation called "mutualism" (3). Mutually, each of the two species provides some substance needed by the other, and thus by being helpful the two species cooperate and energetically as confederates they gain in competition against otherwise similar microbial populations whose individual member species instead develop their own full armament of enzymes and physiological mechanisms. In this way, for as long as those two interreliant species decide to cooperatively coexist, they jointly would have an evolutionarily selective advantage over other partially or fully independent populations. This facilitative interaction conceptually can become the basis for a stronger interaction called "protocooperation," an association in which each of the two interactants needs and benefits the other. When functioning as a unit, a two-member protocooperative association has greater fitness than would the two species functioning independently. Still, the associate species involved in a protocooperation are not fixed in that relationship, in the sense that neither is it obligatory nor is it specific that only these two species can fulfill those roles for one another. Well-studied protocooperative relationships include those in which there is interspecies electron transfer (14), degradation of aromatic compounds (34), and the fermentation of polysaccharides (40). In a simple protocooperation, only a single benefit needs to be conferred from one population upon the second. However, if a simple protocooperation gives the interactants additional fitness, so too might greater degrees of integration. Each population may thus evolve to contribute more than one requisite to the second population. "Synergism" refers to a more closely interactive process in which two species clearly do better and appear to actively cooperate for the purpose of surviving within a particular environment. This interaction may involve jointly causing physical or biochemical changes within that environment which otherwise neither species could perform alone, or else it could be that these changes would occur much more slowly when the two species are not together. The synergants could, however, live in that environment independently of one another. The basis for the benefit-promoting synergism may be that the two interactants function better together because one provides something to the other, such as a carbon or energy source, a growth factor, O_2, or detoxifying enzymes. It often seems that synergism results from protocooperation. But, when the change to synergism occurs, the association becomes somewhat less flexible, and the identities of the two interactants become somewhat more restricted because each must provide the full complement of benefits required by the second. A further progression of synergism would seem to be the basis for the tight interactions that characterize symbiotic associations.

"Symbiosis" represents those instances when it seems essentially obligatory for different species, termed symbionts, to live together as one in a particular environment. Symbioses may involve two or more species of microorganisms, or else microbial species interacting with either plant or animal species. The most famous symbioses in microbiology are the lichens (42), each representing the association of a cyanobacterium species with a fungus species, and whose stability as symbionts has risen to the point that in the past some lichens were considered to be species. Endosymbiotic associations represent an even greater commitment to symbiosis in which one of the symbionts, termed an endosymbiont, lives within the body of the other, which is termed a host. Several endosymbioses involving plants (see chapter 52) have been intensively investigated. In some plant-microbe relationships the endosymbionts are segregated to nodule-like compartments, as is the case for the associations between legumes and members of the genera *Rhizobium* and *Bradyrhizobium* which generate fixed nitrogen, essentially creating a self-fertilizing plant. Agricultural studies of these, and the fungus-root endosymbioses termed mycorrhiza (46), have provided huge benefits to human society. Luminescent organs in deep ocean fish are another notable example of microbial endosymbionts housed within specialized tissue compartments created by the host. There also are endosymbiotic relationships wherein bacteria reside within protozoans, and it has been speculated that this type of endosymbiosis led to the evolution of intracellular organelles.

The Hunters and the Hunted

Many types of microorganisms feed upon other microorganisms. Often, the same microbes which did the feeding then serve as prey and hosts for other microorganisms and macroorganisms in elaborate food chains. There is a vast body of literature on laboratory studies of some of these general kinds of relationships, but far less is known about the ecological role played in nature by those species that parasitize or prey upon microorganisms. This shortfall in knowledge surely affects us, because it is almost certain that ultimately some of the predator-prey relationships must affect either human health or farm production. At least in part, this shortfall in terms of understanding how microbial predator-prey relationships affect natural systems probably exists because it is easier to carry out laboratory studies with only two organisms living in culture media than to establish the very significant but far more complex interactions which naturally occur in highly heterogeneous microbial communities. Sadly, however, our understanding of the importance of specific predator-prey systems also may be biased because all too often it depends on the interest of the researcher and the ease of working with some organisms as compared to others. Thus, the more visible and stepwise simpler topics and trophic chains tend to be better examined. The complex and more difficult but potentially far more important topics then get ignored. In turn, the relative abundance of literature may lead the nonecologist to conclude that a particular group of parasites or predators is indeed very important in nature and that all other groups are unimportant. The truth, then, as to which players and roles are the most significant is not yet revealed.

The viruses are subcellular microbial predators and they are obligate intracellular parasites (29). Among the cellular microbial predators are protozoa, myxobacteria, acrasiomycetes, and a number of chlorophyll-containing flagellates that are often but not always classified as algae. Of these various groups of cellular organisms, there is incontrovertible evidence for the importance of at least the protozoa as contributing factors in determining the composition, structure, and perhaps even the the function of microbial communities (5, 6, 7, 31, 44). Wastewaters, many surface waters, and the rhizosphere of actively growing plants often contain large numbers of protozoa, and a high percentage of the individual protozoa are present in a trophic stage that is associated with active feeding on bacteria. A single protozoan cell, upon division, consumes 10^3 to 10^4 bacteria. Moreover, by use of eukaryotic inhibitors, it has been shown that inhibition of protozoan growth is associated with the maintenance of large numbers of potential prey individuals and often the existence within a community of populations that otherwise would have been suppressed. Thus, eliminating protozoa can alter homeostasis.

Nevertheless, despite their abundance and activity, protozoa generally do not eliminate their prey. It may appear that a predator which is able to markedly reduce the size of a large population is somehow unable to effectively destroy a small population. Instead, however, it seems to be the case that the protozoa and most other predators are prudent. The elimination of prey by an obligate predator would result in the subsequent elimination of the predator species itself. This prudence in feeding appears to be density dependent rather than due to altruism and often has been interpreted in terms of the balance between energy needed for hunting and energy gained by the hunt. Thus, when a predator population is getting more energy by feeding on the prey, as would occur at high prey densities, than it needs to hunt for those prey, it will continue to graze. On the other hand, at low prey densities, the predator must use considerable energy to move to the few surviving prey cells or to bring those survivors to itself for consumption, and it thus will not continue to feed on that species (1). The options then available to the predatory protozoan would be to begin feeding upon another species, die of starvation, or encyst and hope for a happier future when its prey species would again be more abundant. This is not to say that the predator cannot eliminate a single prey species, which it does if the density of alternative bacterial prey is sufficient to sustain feeding (38). Thus, the effects of protozoa may be viewed in terms of reducing the size of the bacterial community and also, under circumstances when the bacterial community is large, possibly eliminating individual species. Still, soils contain higher bacterial densities, at least by total counts, than are predicted on the basis of the capacity of the indigenous protozoa to reduce the bacterial densities. For example, the total count of bacteria in soil may be in the vicinity of $10^9/cm^3$, but in solution the protozoa may reduce the bacterial density to approximately $10^6/cm^3$. Therefore, it is likely that there is a refuge in soil that possibly results from the inability of protozoa to penetrate pores large enough for bacteria but too small for the predator or from the inability of protozoa to feed on bacteria that are adsorbed to particulate surfaces. Adsorption onto particulates would effectively convert the bacteria into objects too large to be engulfed by the protozoa. It also is possible that susceptibility to grazing by protozoa is checked by other factors, perhaps including evolutionary pressure favoring potential prey organisms which can effectively conceal themselves by changing their "signature" of secreted compounds.

Less defined and still subject to extensive discussion are the roles of virus and *Bdellovibrio* species in nature. The viruses are acellular; bdellovibrios are cellular and in appearance are much like small bacteria. The bacteria, algae, fungi, protozoa, and archaea, plus all species of plants and animals, are affected by viruses, which are obligate intracellular parasites. Bdellovibrios can replicate in very complex defined media, but presumably in nature they exist as obligate intracellular parasites of bacteria. The ecology of their host, or prey, is integral in the ecology of both virus (2, 8) and *Bdellovibrio* species. Indeed, as intracellular parasites, the viruses and bdellovibrios rely on a host organism for their development and even their very existence in nature. The interactions which comprise the ecology of viruses have been examined and described by Hurst and Lindquist (29). Much less is understood about the ecology of *Bdellovibrio* (37).

WHY THEY ARE WHERE THEY ARE WHEN THEY ARE THERE

We know that in fact all of the life on earth represents a single interconnected biofilm. But the participant species involved in even similar events and interactions are not static with respect to space or time. The purpose of this section is to help the reader understand why the observed things may be happening in any given place and time.

Boundaries and Barriers

One of the concepts which is important in consideration of the ecology of the members of a species is that in nature each species has a boundary. As indicated earlier in this chapter, boundaries are operationally defined, and the physiological limitations of a species do not allow its members to actively function in the conditions found beyond that species' boundary. For example, we know that almost no species of fish can survive for very long when out of water. Thus, the interface between the atmosphere and the surface of the water represents a boundary for those species of fish. Equally, most terrestrial species cannot survive unaided for very long when submerged in water, so the water's surface likewise serves as a boundary aspect for those terrestrial species. The ecological boundaries of many amphibians and reptiles cross the water's surface, attested to by the fact that those organisms have both aquatic and terrestrial capabilities. The concept of a boundary includes a complicated mixture of factors, sometimes very highly species specific. However, we often can identify both biotic and abiotic components for some of those factors; i.e., oxygen requirements (biotic) may restrict the altitude or depth (abiotic) at which the members of a species can function. A species cannot move its boundary except through evolution. For most species, this evolution must be biological in nature. Humans, however, can move their boundary by cultural evolution. For example, although we lack effective body insulation by fur, the development of clothing and weather-resistant shelters allows us to live even in polar environments, places whose physical conditions normally would make them beyond humans' boundary. The development of the aqualung, an early portable self-contained underwater breathing apparatus (scuba), and modern rebreathing technology represents a cultural evolution which allows humans to exist underwater at least temporarily. Within the confines of its boundary, a species will find at least one habitat in which its members successfully can reside (otherwise, it would become extinct). While a particular species cannot cross and survive beyond its own boundary, its habitat may be sufficiently diverse that, while moving within that habitat, the members of that species may cross the boundaries of many other species. This overlap of boundaries allows biological interactions between the members of different species.

The parts of a species' habitat may be contiguous or may be separated by barriers (Table 1). These barriers can be physically defined and are determinable by physical measurements. They also may be tangible. For example, a species' oxygen requirements may turn geographical features into barriers. Barriers are not fixed with regard to time and space. Rather, they can appear, disappear, and move with time (e.g., as mountain ranges rise and fall, glaciers advance and retreat, continents shift toward either more polar or more equatorial zones, and competing species evolve or become extinct). Species tend to evolve survival mechanisms which allow their members to successfully move as barriers move and even evolve mechanisms which allow them to cross those barriers. Indeed, some species have evolved to a situation in which cyclical migrations between parts of their habitat are necessary for continuance of the species. For example, seasonal migrations occur among many species of bats, birds, butterflies, caribou, salmon, and whales. Even maple trees migrate within the northern hemisphere, albeit on a longer time scale, advancing and retreating with the glacial cycle (43). In the cases of monarch butterflies and maple trees, no individual member of the species completes the full migration. The capability of a species' members to migrate also facilitates colonization of new areas. Cyclical migrations could be viewed as highly evolved forms of dispersal. And, as the macroorganisms migrate, so too must there be a corresponding migration of those microbe species whose lives are either dependent upon or interdependent with the migrating macrobes.

Staying at Home versus Venturing to New Places

What does the importance of microorganisms as major pollutants have to do with dispersal? Dispersal of the host population can reduce the general incidence of disease. For example, sometimes wealthier humans can flee to other, presumably more healthful areas. In those preferred destinations, costly medical capabilities such as prophylaxis and chemotherapy have some impact upon either reducing the incidence of or ameliorating the misery resulting from microbially induced diseases. Sadly, these luxurious options do not exist in remote or poor areas of developing countries. For people in much of the world, the major means of controlling or preventing these infectious diseases is by interfering with microbial dispersal. Dispersal thus not only is important for populations of humans, animals, and plants as they try to survive the onslaught by pathogenic microbes but also is critical for viruses and other types of infectious or otherwise dependent microorganisms in their own efforts to maintain their existence by keeping up with their dispersed hosts. Dispersal also is a key factor for free-living archaea, bacteria, fungi, protozoa, and algae, which must seek new home territories in order to avoid competition against their parental populations for available space and nutrients.

Some of the scientific interest in microbial dispersal comes from basic research. There have been major breakthroughs in the use of genetic and biochemical techniques to understand chemotaxis, which is movement in response to a chemical stimulus (39). In addition, a number of very useful mathematical models have been developed in aerobiology and for use in predicting microbial dissemination through aquifers and soils (28). Interest in microbial dispersal also comes from concern with bacteria and viruses that cause diseases of importance to humans (12) and with fungi that, as a consequence of their aerial movement, are contributors to declines in food and feed production. The disposal on land of agricultural and urban wastes containing pathogens has also resulted in considerable research and monitoring of microbial movement.

Potential new habitats regularly become available for microbial colonization. For example, habitats appear as seedlings emerge from seeds and with the extension of growing root or shoot tissues, during wounding, which creates access to the interior tissues of multicellular beings (including damage to fruits), with the first appearance of a newborn infant, and at the site of bruised tissues. All of these habitats contain sterile sites which could be inhabited by a variety of microorganisms that might find the sites through successful dispersion. Similarly, possibilities occur when nutrients arrive at new locations in water bodies because of vertical and horizontal mixing of water, and new environments are created both by soil erosion and by the building of dams and ponds. The first half of the earlier statement that microorganisms are everywhere, or at least potentially everywhere, is clearly a platitude when one considers that the issue is not whether a microorganism will appear but whether members of a particular species will have reached a site in which they can grow. However, the organism also must have evolved the capability of staying at that site long enough to grow. Staying around to grow often involves adhesion to the existing surfaces (13, 51) and the formation of complex biofilms (see chapter 45). I would suggest acknowledgment that humans are just another part of the biofilm which has evolved and spread itself over and into the nooks and crannies of this planet. Finally, after adhering, the organism would have to successfully compete for nutrients and defend itself against any organisms already found in that site.

Sometimes microbial communities completely consume the supply of limiting nutritional elements or other limiting resources available in the specific sites where they reside. If there is no further input or regeneration of a limiting nutrient or resource, then the species that make up those communities will die away at those sites. Should this happen in all of the ecosystems within which some particular rare species resides, and if the members of those populations have not evolved adequate survival mechanisms, extinction will occur. Avoiding extinction thus can require that a species have evolved a means of escape either in time (via a persistence mechanism) or in space (via a dispersal mechanism). Escape in time can occur by the establishment of starvation existence, wherein an individual shuts down unneeded metabolic activity and hopes for the arrival of nutrients before it dies from starvation. Starvation modes are common among aquatic bacterial species. Escape in time also can occur by the combined development of metabolic inactivity and protection of an environmentally resistant structure. This combined approach is represented by the spores formed by fungi and by some species of bacteria, the cysts and oocysts formed by many types of protozoa, and the nucleocapsid structure formed by viruses. Escape in space dictates dispersal, leaving in search of a more favorable environment.

Traveling by Air, Sea, or Land

Microorganisms have developed many means for dispersal, and each species must have one or more methods for accomplishing its migration. A few microbial groups have evolved specialized structures to launch their cells into the air or send them swirling into the water. These organisms depend upon those vehicles to deliver their progeny to a new home. There are parasitic species which possess mechanisms which cause changes in the behavior of their animal or plant hosts that result in dispersal of the parasite (29). Still other microorganisms utilize very efficient vectors, such as mosquitoes and biting flies, which almost unerringly deliver the progeny microbes to the correct address (29). Some means of dispersal have a higher risk of failure than others, and species with more efficient dispersal mechanisms do not require as great a level of production and shedding of new propagules because each of their released propagules has a greater likelihood of success at encountering a hospitable site.

Active dispersal, in which the physiology of the microorganism controls its transport (automobility), has been the subject of considerable research, probably because it is more comforting to the microbiologist to have his or her pet control its own fate. Active dispersal, either by motility or by growth of filaments, may be somewhat random rather than resulting from a specifically directed movement (taxis) or growth in a specific direction (tropism). Still, the motility of protozoa and the growth of either bacterial chains or the filaments of fungi in soil and algae in water are sufficient to result in those species successfully encountering and colonizing new environments (28). As a consequence, considerable research has been done on the effects of chemical stimuli. In the short term, active microbial motility generally is restricted to limited distances because of the energy requirement for movement either toward or away from a chemical stimulus. In the long term, the result can be global dispersals and migrations.

In contrast with active dispersal, there also exists passive dispersal, during which the microbe is transported by a carrier. Passive dispersal is something which we normally might think of as occurring over short distances (45), but it can result in relatively quick dissemination of an organism to locations meters, kilometers, or hundreds of kilometers away from its original reservoir. Aerial dispersal is frequently of considerable importance to fungi (see chapter 73). For organisms dispersed through the atmosphere to be successful, they must produce enormous numbers of propagules to survive the random chance that one or several might alight in an environment in which that species can stay around and grow. A propagule that is transported through the air must have mechanisms to overcome three major hazards: radiation, desiccation, and extremes of temperature. It is not clear in all instances what physiological adaptations are responsible for successful resistance against such environmental hazards. However, the presence of thick walls and dark pigments in many fungi transported through the air and of carotenoid pigments in many aerially dispersed bacteria suggests that these features are important adaptations for this mode of transport. The extent of migration of some of these organisms is truly impressive. For example, in a single year *Helminthosporium maydis*, a pathogen of corn, may spread over areas of thousands of square kilometers, doing enormous financial damage to the intensively planted corn crops of North America. Similarly, spores of the fungal genus *Hemileia* apparently have spread from Angola to Brazil via trade winds which cross the Atlantic Ocean, a distance of more than 1,000 km.

Passive dispersal in water (see reference 7 and section III) and soils (see chapter 50) has also been the subject of considerable inquiry, in part because of public health problems. For example, algae associated with red tides move for some distances and then suddenly create blooms (population explosions visible to the eye because of the pigmentation of these organisms [see chapter 30]), often only to be decimated by viruses (see chapter 36). Still, the toxic products from those blooms can bioaccumulate in the food chain and may have major consequences for aquatic fauna and people who consume seafood animals from the area of the algal bloom. The interest in passive dispersal through soil is often a result of concern with the vertical migration of bacteria or viruses that cause human disease and the entry of those microorganisms into aquifers underlying soils into which the microorganisms were inadvertently or deliberately introduced (see chapter 72). However, appreciable research on the vertical movement important to microbial colonization of roots has also been conducted.

Bacteria, viruses, or fungi that are transmitted by living vectors typically have efficient dispersal mechanisms and require fewer propagules for the species to be maintained. This mode of transmission is considered outside the scope of the *Manual of Environmental Microbiology*, but the general concept of vector transmission is described elsewhere by Hurst and Lindquist (29).

Prepared for the Climate? And What about the Location?

Climate? Even geography? Yes, climate is an important consideration for microbes, and there is a microbial geography. Restricted distributions on a macroscale as well as on a microscale characterize all groups of organisms, and the microbes are no exception. The literature dealing with the geography of microbial groups is often unknown to laboratory scientists, but an investigation of microbial communities in natural environments quickly shows marked and sometimes extreme localization of microbial groups. For example, geographical distributions are evident among aquatic and terrestrial algae; free-living and pathogenic fungi; protozoa in marine water, freshwater, or soil; and bacteria in countless habitats. Many genera, and often even species, seem cosmopolitan, but their widespread distribution does not mask the restricted nature of their occurrence within particular regions, sites, or microenvironments (16, 50, 55).

A key element for any particular species of alga, archaeon, bacterium, fungus, or protozoan to successfully exist in an environment is the ability to endure all of the abiotic stresses characteristic of that environment. These stresses include essentially the same list of things which can serve as either physical or chemical barriers: factors such as unsuitable pH or temperature (what is just fine for one species may be too high or too low for another species), occasional drying or freezing in some environments, intense solar radiation, high pressures deep in the ocean, or salinity in certain terrestrial or aquatic ecosystems. The combination of these factors often produces amazing outcomes. Some algae can be found broadly distributed over the surfaces of lakes and oceans but exist in patches at each site. Some of these patches are no more than a few centimeters across, whereas other patches extend over areas of more than 300 km^2. Well-known to phycologists are the limited distributions of diatoms. Some diatom species are present only in subtropical or tropical waters, whereas others characteristically are found only in the Arctic or Antarctic and even there often are stratified by depth (24). Other algae have snowfields as their habitats. Bacteria such as *Beijerinckia* species similarly have a restricted distribution, being commonly but not solely present in soils of the tropics or subtropics. Often, the biogeography of a microorganism that is transmitted by a living vector or that is an obligate parasite is determined largely or exclusively by the biogeography of its vector or host. This is true of both the protozoan genera *Plasmodium* and *Trypanosoma*, for example, and the explanations for their distributions thus are quite simple. This also is true for vector-borne viruses, which can exist only within the same geographical areas where both their host and the vectoring species can be found (29). In the case of insect-vectored protozoans and likewise insect-vectored viruses, success requires a community interaction of the residential predatory parasitic species (remember that despite the terminology often used by protozoologists, both the protozoans and the viruses are parasitic) which lives inside the vector, a predatory but nonresidential parasitic vectoring species (the mosquito), and the free-living host species which simultaneously serves as prey for both of the predators. Not quite as simple to explain are rare distribution patterns of pathogenic microorganisms that do not have a known vector and whose distribution does not seem obligately associated with a particular known host organism. A notable example is *Coccidioides immitis*, which has a unique distribution in the Western Hemisphere (55). This organism causes disease in humans, but although that host is largely global, the fungus is restricted to certain localized geographical and climatic sites. This fungus is characteristically found in the soils of certain semiarid regions that typically are exposed to high temperatures, receive little rainfall, and frequently have high salinities. Although *C. immitis* is transmitted by wind, wind movement alone cannot account for its biogeography. Even if the fungus is cold intolerant, that possible sensitivity would not explain why it is not present in warm, humid areas to which it may be carried by the wind. Having failed to find a climatic reason for this geographical restriction of the habitat of *C. immitis*, we may in time find that the answer lies in an examination of biotic factors. It is possible that this narrow operational habitat of *C. immitis* is related to the need for a supporting community composition and interactive community structure, which then determines the sites where the organism is successful and thus is likely to be found. Conversely, then, an unfavorable community composition or structure may prohibit the organism from residing in other potential habitats.

Natural geographic and climatic restrictions are called "zonations." Zonation is evident even at a microenvironmental level. Marked horizontal and vertical differences in either the distributions or the occurrences of algae, bacteria, and fungi are evident in waters, sediments, and soils. Such microscale zonations almost certainly reflect biologically important differences in the physical and chemical characteristics of the environment. However, in only a few cases have the causes of this highly localized microenvironmental distribution been established by either physical or chemical analyses. Among the factors either known or postulated to be important in microscale biogeography are nutrient concentration and type, temperature, pH, oxygen, grazing by zooplankton, mechanical barriers, and inhibitory substances (some of which may be biotic, while others certainly are abiotic).

The fitness traits of the initial colonists in environments that are largely free of microorganisms or that have been drastically disturbed are often reasonably simple to predict. Such environments are not unknown, and they exist, for example, on the previously uncolonized surfaces of roots growing through soil, on plant materials that become bruised, and in waters that receive sudden influxes of organic or inorganic materials. The fitness traits associated with successful colonization, and with an organism's hoped establishment, frequently represent the basis for a capacity to use the organic nutrients present at the new site. For example, with root-dwelling microbes, those nutrients may be either compounds excreted by the emerging root segments or constituents of the tissues that become bruised and thus accessible for microbial utilization. Frequently, the initial colonist preempts the site so that a propagule that arrives later but has the same enzymatic capacity will be unable to multiply. In these instances of preemptive colonization, the key fitness trait is frequently associated with dispersal (getting there first).

Finding common traits among organisms can indicate a similar ancestry, albeit horizontal gene transfer can scramble phenotypic traits and complicate our efforts at understanding how phenotypic traits and physiology relate to ancestry. Still, the very fact that species can be distinguished is evidence that each has some uniqueness with respect to its combination of biochemical, physiological, and morphological traits. Some of these distinguishing traits help explain the presence of a species in one environment but its absence in another and the relative abundance and activity of the various inhabitants which exist in a given environment. This combination of a species' traits provides the basis for natural selection, accounts for the geography of a species, and explains the role (i.e., the niche) of a species in its environment. The sum of the traits of all species present within an ecosystem represents an identifying signature for that ecosystem.

How Is the Food?

An obvious need that must be satisfied for an organism to become established is the presence of all nutrients that it requires. Often, in aquatic environments, phosphorus is a limiting nutrient for photoautotrophic organisms. Many natural terrestrial ecosystems contain most and sometimes all inorganic nutrients in concentrations sufficient to maintain a reasonably large community. Heterotrophic organisms often seem more finicky than autotrophs. Thus, when considering the needs for heterotrophic organisms and any single specific site, we often find that overall the supply of energy resources is the limiting factor and that the kinds of carbon sources available affect the selection of organisms which can maintain a presence in that site. In most environments containing readily or slowly available organic molecules, a variety of dissimilar propagules able to use those carbon sources arrive, yet only a few become established. In this instance, as in so many cases in environmental microbiology when causation is being sought, the presence of a suitable energy source is a necessary requirement but in and of itself not sufficient for microbial establishment.

How can we understand what sufficiency requires? We gain this knowledge by understanding the basis for natural selection, or selection in nature. From an environmental viewpoint, the basis upon which selection operates is the set of fitness traits that underlies ecological success and sometimes the achievement of community dominance by the members of a species. These traits are the specific biochemical, physiological, or morphological characteristics that determine a species' boundary, barriers, habitat, and niche. Nature tries many different approaches to solving a particular problem. Evolution is based upon the fact that those solutions which work are retained and hopefully will again prove adequate at some future point. The issue for scientists is to identify those attributes of organisms that are necessary (the organism could not do without them) and sufficient (with those, the organism will have a chance at competitive success) to enable the organism to survive and occasionally multiply in particular environments. After we first show the importance of tolerance to the abiotic factors that are detrimental to one or another group of organisms and secondly understand the need for a supply of inorganic nutrients and a carbon or energy source, then what? The methodologies associated with enrichment or selective culture techniques are ideal for examining natural selection under artificial conditions. But we must understand the associated limitations of these methodologies and of artificial conditions. The technical limitations focus on the fact that the use of such methods to examine the community of organisms present in an environmental sample typically results in finding only the one organism that grows fastest under those artificial conditions. This would represent selection of those isolated based upon what is termed an *r* characteristic. The *r* growth strategy is one in which organisms invest in quick reproduction. These are the microbial equivalents of the plants known as dandelions. In contrast, the competitive conditions in nature are such that many inhabitants of natural communities grow slowly. For these inhabitants, the focus is upon long-term survival of the individual, and we term this pattern to be the *k* growth strategy. These are the microbial equivalents of the sequoia trees. But, don't think that the members of some species cannot shift from an *r* to a *k* strategy if it befits them! This kind of a switch happens when some intestinal bacteria try surviving in an aquatic environment. Even more confounding to our efforts is the reality that most of the organisms existing in nature do not appear to grow at all under conditions imposed by laboratory media. The reason why they do not grow in the artificial environment which we supply simply may be that we have not adequately duplicated the natural setting into which those organisms evolved. Hopefully, newer biochemical tools will help us in sorting out many of the remaining answers, allowing us to achieve success in areas where present cultural techniques are not sufficient tools.

Settling in with Relatives (In-Laws and Outlaws) and Meeting the Neighbors

Abundant literature on the biochemistry, physiology, and morphology of microorganisms has been derived from studies of pure (axenic) cultures of organisms conducted under artificial conditions, usually in liquid media, and in the absence of any other species. However, it is far from certain which of those biochemical, physiological, or morphological properties either truly are of ecological significance or facilitate the establishment of a particular microbe in a particular site. Many facile extrapolations have been made from in vitro to in vivo conditions, but rarely have these extrapolations been verified as being ecologically relevant. We can surmise that if an organism identified in culture is unable to survive and grow when exposed to artificially simulated environmental stresses, it is unlikely to be an inhabitant of an environment in which those stresses occur naturally. Some environmental stresses are easy to establish and duplicate in the laboratory, and it is simple to show by process of elimination which of these may be important in determining the absence of a specific organism in a specific habitat. Conversely, understanding the presence of a particular organism in a particular place and time can be very difficult. Many species are transported to environments in which they are able to tolerate all the abiotic stresses, but they still do not become established. What lies behind this failure of establishment? We may surmise that it stems from the homeostatic mechanisms operating in the community in which the new arrival alights and that these mechanisms are extremely effective at eliminating many of the arrivals. Yet the exact reasons for eliminating or allowing the continued presence of the new potential inhabitants are rarely understood.

We do know that with time, the initial colonizing species present at a site can become displaced as more recently arrived organisms become abundant within the community. These displacements represent natural successions. The initially dominant species often seem to become of less significance to the community as a succession proceeds, and some of the initially dominant species may be eliminated totally. Those displaced species succeed overall by having dispersed propagules to colonize other, more recently available sites before their parental population loses its foothold on the present site. Yet, despite these successional eliminations of specific species, often the overall trend of succession with time is toward an increasing species diversity. As succession proceeds, the identities of the fitness traits associated with allowing any single species to maintain a place within the community become increasingly less certain. However, we do know that among the factors that either contribute to or determine selection during succession are the availability of nutrients that are synthesized by the temporally preceding species, the alterations in concentrations of inorganic nutrients and the formation of toxic products by those earlier arrivals, successful competition for limiting resources (especially the supply of organic carbon), and the inevitable appearance of organisms that parasitize either the pioneering species or the subsequent colonists. The last factor tells us that if a species chooses to remain in the community, it must successfully dodge or repel parasitic organisms such as viruses and grazing organisms such as protozoa and invertebrates.

The displacement of organisms which occurs during the process of colonization and succession may ultimately lead to a mature or climax community. This is the assemblage of organisms most characteristic of any habitat. Often the communities of chief concern to environmental microbiologists are the climax communities, because they tend to be the most prevalent. As might be imagined, the organisms making up a climax community interact in a variety of ways and it is a dynamic structure. Not all of the interactions between the members are peaceful, and many are outright antagonistic. The climax community tends to reproduce itself, and yet it is the net functions fulfilled by the community which may be more constant than are the relative numbers of each of the component species. At this mature stage, the nature of the interactions between species can be very complex and difficult to unravel, and it may seem impossible to discern the specific fitness traits that underlie any particular organism's position in the community hierarchy at any given point in time. Despite the fact that the relative proportions of the component species may vary, as a general rule the component members of a climax community are not eliminated and many rare species can be found there.

In some instances, a community simply seems dominated by a species that has preemptively colonized the site, and its role is associated with its presence at the site before other organisms arrived. In other instances, however, a variety of organisms endowed with appropriate physiological capacities have reached the site, and they thus may end up competing for dominance. Undoubtedly, competition is one of the major interactions in climax communities. As microorganisms usually grow readily when growth is freely allowed, one or more metabolically required factors in the environment must become limiting. These limiting factors may contribute to the bases for competition. For communities dominated by heterotrophic bacteria and fungi, the limiting factor is frequently the supply of available carbon. Many environments, such as soils and sediments, contain large amounts of organic matter, but much of that organic matter is not in a readily available (biologically utilizable) form. Thus, the heterotrophic organisms that are successful are often those that are able to make use of the less readily available organic materials. In communities containing chemoautotrophs, the limiting factor is frequently the supply of the inorganic compound or ion that serves as their energy source. In surface waters, the limiting factor for the photoautotrophic algae or cyanobacteria is often the concentration of either phosphorus or nitrogen. Fluctuations in both the net activity and community composition also may cycle with time due to factors such as climatic variables and seasonality.

By way of summation, what determines the outcome of competition and obtaining a place in the community? It is tempting to suggest that the successful competitor gains its dominance based upon having an *r* strategy, meaning that the organism simply outgrows all others. However, the fact that many of the dominant organisms in natural environments adopt instead the *k* strategy and do not grow quickly suggests that it is imprudent to extrapolate to natural environments from growth rates obtained by studies performed with pure cultures under noncompetitive circumstances. Clearly, more is involved than simply growth rate per se because of other stresses which exist in nature. Likewise, we must consider the need for an organism to be transported to sites where there is an available supply of limiting nutrients and the need to avoid predation after it has arrived. Sometimes, dominance does result simply from being the first arrival, but even those first arrivals may lose dominance and become displaced during community successions. Perhaps it is simply "peacefully fitting in" which describes the recipe for success. Alternatively, it may be subterfuge. Both thievery and enslavement of other species may be involved. The success of new arrivals and their inclusion in the community also may depend in part upon the biochemical weaponry which those species bring with them, since microbes seem to have invented the concepts of laying siege and attack.

WHAT DOES IT ALL MEAN?

Perhaps the key aspect of understanding ecology is development of a comprehension that the individual species are connected, i.e., that the ecology of one species coordinates with those of other species. This coordination usually includes both macrobes and microbes. All of these species then function together as a community. An ecologist has been euphemistically defined as an individual whose feet are firmly planted in midair, possibly because of the enormous time which must be spent pondering it all. Perhaps, like Buddhist monks, we seek some "true understanding of nature" as we regard both interrelations among the activities of different species and their interactions with the physical environment which surrounds all of them. Much of ecology, including environmental microbiology, is concerned with gaining insight through the process of basic science. Indeed, basic science can help us to find the keys which then allow us to open some of the seemingly locked doors beyond which lies a better understanding of nature. However, it is also clear that environmental microbiology is partly an applied science. It has much to offer to our knowledge regarding the maintenance and restoration of environmental quality, prevention of the transmission of diseases of animals and plants, and approaches to improving human health.

The guardianship of environmental quality relies upon understanding of the role of microorganisms in preventing pollution, destroying noxious organic materials before their concentration becomes objectionable, or destroying toxic chemicals before they have an impact on humans, animals, or plants. Indeed, using information which they have gained through their studies, environmental microbiologists now are engaged in bioremediation technologies, designing ways to enhance the capacity of microbes to bring about the destruction of pollutants.

The epidemiology of communicable diseases is, to a significant degree, an extension of evaluations of microbial dispersal. The spread of viruses, bacteria, fungi, and protozoa thus has a tremendous impact on the protection of plant, animal, and human health. Plant pathologists have long recognized the significance of information on the ecology not only of the disease-producing fungi, bacteria, and viruses but also of other microorganisms, many competitive and some even protective, that reside in the same habitats. Many plant diseases are not effectively controlled by chemical agents or by sanitation procedures, and it is the activity of those other, nonpathogenic members of microbial communities that must be used to form the basis for effective control of particular diseases. Much of environmental microbiology research is interdisciplinary. For example, aerobiologists and soil microbiologists frequently interact with plant pathologists. Preventive animal husbandry and human medicine have acquired valuable information from environmental microbiology. Notably, in the field of medical treatment, the development of antibiotics and other forms of chemotherapy has provided us with some powerful tools, many of which have been based upon knowledge of upsets and restorations in microbial communities.

Thus, knowledge of microbial community structure and community function has a key role in improving our lives. Furthermore, this knowledge will aid us in understanding ways of maintaining the environment and its microbial communities. As John Donne correctly stated, "No man is an *Iland*." Carrying his understanding a step further, humans and all macroorganisms are connected to the microbial community and have evolved from it, and our future depends upon that community.

REFERENCES

1. **Alexander, M.** 1981. Why microbial predators and parasites do not eliminate their prey and hosts. *Annu. Rev. Microbiol.* **35:**113–133.
2. **Ashelford, K. E., S. J. Norris, J. C. Fry, M. J. Bailey, and M. J. Day.** 2000. Seasonal population dynamics and interactions of competing bacteriophages and their host in the rhizosphere. *Appl. Environ. Microbiol.* **66:**4193–4199.
3. **Bäckhed, F., R. E. Ley, J. L. Sonnenburg, D. A. Peterson, and J. I. Gordon.** 2005. Host-bacterial mutualism in the human intestine. *Science* **307:**1915–1920.
4. **Banin, E., T. Israely, A. Kushmaro, Y. Loya, E. Orr, and E. Rosenberg.** 2000. Penetration of the coral-bleaching bacterium *Vibrio shiloi* into *Oculina patagonica*. *Appl. Environ. Microbiol.* **66:**3031–3036.
5. **Berninger, U.-G., B. J. Finlay, and P. Kuuppo-Leinikki.** 1991. Protozoan control of bacterial abundance in freshwater. *Limnol. Oceanogr.* **36:**139–147.
6. **Bloem, J., F. M. Ellenbroek, M. J. B. Bar-Gilissen, and T. E. Cappenberg.** 1989. Protozoan grazing and bacterial production in stratified Lake Vechten estimated with fluorescently labeled bacteria and by thymidine incorporation. *Appl. Environ. Microbiol.* **55:**1787–1795.
7. **Boehm, A. B., D. P. Keymer, and G. G. Shellenbarger.** 2005. An analytical model of enterococci inactivation, grazing, and transport in the surf zone of a marine beach. *Water Res.* **39:**3565–3578.
8. **Burroughs, N. J., P. Marsh, and E. M. H. Wellington.** 2000. Mathematical analysis of growth and interaction dynamics of streptomycetes and a bacteriophage in soil. *Appl. Environ. Microbiol.* **66:**3868–3877.
9. **Carroll, L.** 1998. *Alice's Adventures in Wonderland and Through the Looking-Glass, and What Alice Found There*, Centenary ed., p. 186. Penguin Books, London, England.
10. **Cole, J. J.** 1982. Interactions between bacteria and algae in aquatic ecosystems. *Annu. Rev. Ecol. Syst.* **13:**291–294.
11. **Costello, A. M., and M. E. Lidstrom.** 1999. Molecular characterization of functional and phylogenetic genes from natural populations of methanotrophs in lake sediments. *Appl. Environ. Microbiol.* **65:**5066–5074.
12. **Cox, C. S.** 1989. Airborne bacteria and viruses. *Sci. Prog.* (Oxford) **73:**469–499.
13. **Cunliffe, D., C. A. Smart, C. Alexander, and E. N. Vulfson.** 1999. Bacterial adhesion at synthetic surfaces. *Appl. Environ. Microbiol.* **65:**4995–5002.

14. **de Bok, F. A. M., C. M. Plugge, and A. J. M. Stams.** 2004. Interspecies electron transfer in methanogenic propionate degrading consortia. *Water Res.* **38:**1368–1375.
15. **Donne, J.** 1994. Meditation XVII, p. 440–441. *In* C. M. Coffin (ed.), *The Complete Poetry and Selected Prose of John Donne*. The Modern Library, New York, N.Y.
16. **Ducklow, H. W.** 1984. Geographical ecology of marine bacteria: physical and chemical variability at the mesoscale, p. 22–31. *In* M. J. Klug and C. A. Reddy (ed.), *Current Perspectives in Microbial Ecology*. American Society for Microbiology, Washington, D.C.
17. **Fernandez, A. S., S. A. Hashsham, S. L. Dollhopf, L. Raskin, O. Glagoleva, F. B. Dazzo, R. F. Hickey, C. S. Criddle, and J. M. Tiedje.** 2000. Flexible community structure correlates with stable community function in methanogenic bioreactor communities perturbed by glucose. *Appl. Environ. Microbiol.* **66:**4058–4067.
18. **Ferry, J. G., and R. S. Wolfe.** 1976. Anaerobic degradation of benzoate to methane by a microbial consortium. *Arch. Microbiol.* **107:**33-40.
19. **Frederickson, J. K., and T. C. Onstott.** 1996. Microbes deep inside the earth. *Sci. Am.* **275(4):**68–73.
20. **Gamo, M., and T. Shoji.** 1999. A method of profiling microbial communities based on a most-probable-number assay that uses BIOLOG plates and multiple sole carbon sources. *Appl. Environ. Microbiol.* **65:**4419–4424.
21. **Goffredi, S. K., A. Warén, V. J. Orphan, C. L. Van Dover, and R. C. Vrijenhoek.** 2004. Novel forms of structural integration between microbes and a hydrothermal vent gastropod from the Indian ocean. *Appl. Environ. Microbiol.* **70:**3082–3090.
22. **González-Toril, E., E. Llobet-Brossa, E. O. Casamayor, R. Amann, and R. Amils.** 2003. Microbial ecology of an extreme acidic environment, the Tinto River. *Appl. Environ. Microbiol.* **69:**4853–4865.
23. **Gradinger, R.** 1999. Integrated abundance and biomass of sympagic meiofauna in Arctic and Antarctic pack ice. *Polar Biol.* **22:**169–177.
24. **Gradinger, R.** 1999. Vertical fine structure of the biomass and composition of algal communities in Arctic pack ice. *Mar. Biol.* **133:**745–754.
25. **Gray, N. D., R. Howarth, A. Rowan, R. W. Pickup, J. G. Jones, and I. M. Head.** 1999. Natural communities of *Achromatium oxaliferum* comprise genetically, morphologically, and ecologically distinct subpopulations. *Appl. Environ. Microbiol.* **65:**5089–5099.
26. **Günther, S., K. H. George, and M. Gleitz.** 1999. High sympagic metazoan abundance in platelet layers at Drescher Inlet, Weddell Sea, Antarctica. *Polar Biol.* **22:**82–89.
27. **Hentschel, U., J. Hopke, M. Horn, A. B. Friedrich, M. Wagner, J. Hacker, and B. S. Moore.** 2002. Molecular evidence for a uniform microbial community in sponges from different oceans. *Appl. Environ. Microbiol.* **68:**4431–4440.
28. **Hurst, C. J. (ed.).** 1991. *Modeling the Environmental Fate of Microorganisms*. American Society for Microbiology, Washington, D.C.
29. **Hurst, C. J., and H. D. A. Lindquist.** 2000. Defining the ecology of viruses, p. 3–40. *In* C. J. Hurst (ed.), *Viral Ecology*. Academic Press, Inc., San Diego, Calif.
30. **Karl, D. M., D. F. Bird, K. Björkman, T. Houlihan, R. Shackelford, and L. Tupas.** 1999. Microorganisms in the accreted ice of Lake Vostok, Antarctica. *Science* **286:**2144–2147.
31. **Kiørboe, T., K. Tang, H.-P. Grossart, and H. Ploug.** 2003. Dynamics of microbial communities on marine snow aggregates: colonization, growth, detachment, and grazing mortality of attached bacteria. *Appl. Environ. Microbiol.* **69:**3036–3047.
32. **Knox, O. G. G., K. Killham, R. R. E. Artz, C. Mullins, and M. Wilson.** 2004. Effect of nematodes on rhizosphere colonization by seed-applied bacteria. *Appl. Environ. Microbiol.* **70:**4666–4671.
33. **Koropatnick, T. A., J. T. Engle, M. A. Apicella, E. V. Stabb, W. E. Goldman, and M. J. McFall-Ngai.** 2004. Microbial factor-mediated development in a host-bacterial mutualism. *Science* **306:**1186–1188.
34. **Lima, S. A. C., M. Filomena, J. Raposo, P. M. L. Castro, and R. M. Morais.** 2004. Biodegradation of ρ-chlorophenol by a microalgae consortium. *Water Res.* **38:**97–102.
35. **Lindow, S. E., and, M. T. Brandl.** 2003. Microbiology of the phyllosphere. *Appl. Environ. Microbiol.* **69:**1875–1883.
36. **MacDonald, T. T., and G. Monteleone.** 2005. Immunity, inflammation, and allergy in the gut. *Science* **307:**1920–1925.
37. **Madigan, M. T., J. M. Martinko, and J. Parker.** 2000. Bdellovibrio, p. 487–489. *In* P. F. Corey (ed.), *Brock Biology of Microorganisms*, 9th ed. Prentice-Hall, Upper Saddle River, N.J.
38. **Mallory, L. M., C. S. Yuk, L.-N. Liang, and M. Alexander.** 1983. Alternative prey: a mechanism for elimination of bacterial species by protozoa. *Appl. Environ. Microbiol.* **46:**1073–1079.
39. **Manson, M. D.** 1992. Bacterial motility and chemotaxis. *Adv. Microb. Physiol.* **33:**277–346.
40. **Murray, W. D.** 1986. Symbiotic relationship of *Bacteroides cellulosolvens* and *Clostridium saccharolyticum* in cellulose fermentation. *Appl. Environ. Microbiol.* **51:**710–714.
41. **Ohkuma, M., S. Noda, and T. Kudo.** 1999. Phylogenetic diversity of nitrogen fixation genes in the symbiotic microbial community in the gut of diverse termites. *Appl. Environ. Microbiol.* **65:**4926–4934.
42. **Oksanen, I., J. Jokela, D. P. Fewer, M. Wahlsten, J. Rikkinen, and K. Sivonen.** 2004. Discovery of rare and highly toxic microcystins from lichen-associated cyanobacterium *Nostoc* sp. strain IO-102-I. *Appl. Environ. Microbiol.* **70:**5756–5763.
43. **Pitelka, L. F., and the Plant Migration Workshop Group.** 1997. Plant migration and climate change. *Am. Sci.* **85:**464–473.
44. **Rønn, R., A. E. McCaig, B. S. Griffiths, and J. I. Prosser.** 2002. Impact of protozoan grazing on bacterial community structure in soil microcosms. *Appl. Environ. Microbiol.* **68:**6094–6105.
45. **Sano, E., S. Carlson, L. Wegley, and F. Rohwer.** 2004. Movement of viruses between biomes. *Appl. Environ. Microbiol.* **70:**5842–5846.
46. **Scheublin, T. R., K. P. Ridgway, J. P. W. Young, and M. G. A. van der Heijden.** 2004. Nonlegumes, legumes, and root nodules harbor different arbuscular mycorrhizal fungal communities. *Appl. Environ. Microbiol.* **70:**6240–6246.
47. **Simberloff, D.** 1989. Which insect introductions succeed and which fail?, p. 61–75. *In* J. A. Drake, H. A. Mooney, F. di Castri, R. H. Groves, F. J. Kruger, M. Rejmanek, and M. Williamson (ed.), *Biological Invasions: a Global Perspective*. John Wiley & Sons, Ltd., Chichester, England.
48. **Staley, J. T.** 1999. Bacterial biodiversity: a time for place. *ASM News* **65:**681–687.
49. **Takai, K., T. Komatsu, F. Inagaki, and K. Horikoshi.** 2001. Distribution of archaea in a black smoker chimney structure. *Appl. Environ. Microbiol.* **67:**3618–3629.
50. **Tett, P.** 1987. Modelling the growth and distribution of marine microplankton, p. 387–425. *In* M. Fletcher, T. R. G. Gray, and J. G. Jones (ed.), *Ecology of Microbial Communities*. Cambridge University Press, Cambridge, United Kingdom.
51. **van Schie, P. M., and M. Fletcher.** 1999. Adhesion of biodegradative anaerobic bacteria to solid surfaces. *Appl. Environ. Microbiol.* **65:**5082–5088.

52. **Vetriani, C., H. W. Jannasch, B. J. MacGregor, D. A. Stahl, and A.-L. Reysenbach.** 1999. Population structure and phylogenetic characterizations of marine benthic archaea in deep-sea sediments. *Appl. Environ. Microbiol.* **65:**4375–4384.
53. **Voolapalli, R. K., and D. C. Stuckey.** 1999. Relative importance of trophic group concentrations during anaerobic degradation of volatile fatty acids. *Appl. Environ. Microbiol.* **65:**5009–5016.
54. **Webster, N. S., L. D. Smith, A. J. Heyward, J. E. M. Watts, R. I. Webb, L. L. Blackall, and A. P. Negri.** 2004. Metamorphosis of a scleractinian coral in response to microbial biofilms. *Appl. Environ. Microbiol.* **70:**1213–1221.
55. **Weitzman, I.** 1991. Epidemiology of blastomycosis and coccidiomycosis, p. 51–74. *In* D. K. Arora (ed.), *Handbook of Applied Mycology*, vol. 2. Marcel Dekker, New York, N.Y.

3

Prokaryotic Diversity: Form, Ecophysiology, and Habitat

FREDERICK S. COLWELL AND EDWARD R. LEADBETTER

RECOGNITION OF PROKARYOTE DIVERSITY AND THE LIMITS OF OUR COMPREHENSION

Our perceptions that microbes as a group possess extremely diverse characteristics and live and function in quite diverse habitats (many of which seem "extreme" to those acquainted only with macrobiotas) had their origins in the penetrating observations and studies of two particular individuals—the Russian S. Winogradsky and the Hollander M. Beijerinck. As van Niel noted (146), those who studied with Beijerinck and his successors in Holland and elsewhere (the "Delft school") were among the major contributors to the breadth and depth of our comprehension of prokaryote diversity, the crucial functional roles these microbes play in the biosphere, and the recognition that in the absence of these diverse microbial activities other forms of life on this planet would promptly cease to exist. The success of enrichment culture and isolation approaches (22, 47, 63, 112, 147, 153), coupled with new analytical techniques (2, 4, 51, 54, 57, 98, 119), reveals how little we understand about so many of the presumptive prokaryotes present in nearly every habitat we explore. We continue to be inspired by the extent of microbial diversity as contemporary approaches to light, electron, and other forms of microscopy have evolved and as tools of a molecular biological nature (e.g., analysis of DNA and RNA sequences, use of RNA-targeted oligonucleotide probes, stable-isotope probing, metagenomics, and community proteomics) have been brought to bear in the examination of mixed microbial populations of many habitats (4, 73, 102, 104, 107, 117, 118, 121).

Frequent reports now cite the cultivation and characterization of organisms that had been refractory to study in the laboratory or the presence of one with apparently novel features detected in a habitat that had seemingly already been well explored or the fact that the microbial content of some novel habitat is worthy of inspection. Still, we may never be able to describe completely either the enormous diversity in features of the many prokaryotes currently recognized, the wide range of habitats in which they are known to occur, or the microbial composition of the populations present in those habitats. There is no little irony in the fact that as we learn more and more about ever more members of the *Bacteria* and *Archaea*, we come to realize that we comprehend less and less, both qualitatively and quantitatively, of the diverse features of different prokaryotes and of environmental factors affecting, and reflecting, their growth, activity, metabolic capabilities, and persistence. Twenty-five years ago, some might have surmised that perhaps as much as 40% of the prokaryotic world was recognized and understood; today it is probably overly optimistic to suggest that the figure is more than 5%, if even that. And just what we consider, or how we define, prokaryotic "species" (for details, see chapter 13) has been perplexing for decades and will likely remain enigmatic.

Readers are reminded that in this assessment of prokaryotic diversity and the range of habitats in which these organisms are known to multiply (or at least survive), the phrase "as known at present" needs to be kept uppermost in mind. This limitation challenges those interested in enlarging our understanding of microbes' contribution to life on the planet. This chapter is intended to be a descriptive outline of the *Bacteria*, the *Archaea*, and some new chemical boundaries of their habitats in the more usual environmental sense (e.g., association with soils, waters, and some extreme environments) and also includes a consideration of microbes associated with macrobes.

SOME FACTORS AFFECTING LIMITATIONS IN OUR UNDERSTANDING

Many factors are responsible for our inability to fully describe the prokaryotic world. Often, major advances in the development and application of new technologies may be required to permit significant increases in our comprehension. In addition, new findings are not always integrated with an understanding of the microhabitat dimensions and features that are so essential for understanding the context of microbial habitation in a given place. If coupled to detailed measurement of the corresponding abiotic parameters in a given environment, new microbial characterization methods carry the real promise of discerning which microbial cells are the basis for maintaining the biologically tractable chemical and physical features of the habitat. Some of the hurdles that we must overcome are noted below.

Rarely does the size or shape alone of a prokaryote permit one to identify it and equate its properties with those of an organism already described. This limitation is no less serious if one encounters a novel or "new" organism, and even determining whether an object is in fact an organism or is instead some inanimate material can be difficult. Another important consideration is that for a single cell or even a few cells to be detected in a typical field of view by light microscopy, an amount on the order of 10^6 cells per ml needs to be present. Clearly, it would be easy to overlook the presence of many less abundant organisms. Also, because of the physiological diversity of microorganisms and our lack of knowledge of their physiology it is difficult to duplicate in the laboratory the conditions that facilitate the growth of all but a small subset of the *Bacteria* or *Archaea*. In addition, in many instances cells are attached to surfaces or present in biofilms the removal from which may damage the cells. New methods that rely upon extraction of DNA from the sample and subsequent molecular characterization have made advances in this area; however, for many habitats it is not readily possible either to enumerate the prokaryote population present or to isolate and then describe thoroughly the types present. Pure cultures are rarely found in habitats other than the laboratory. Even though we continue to explore the metabolic and physiologic properties of pure (as well as some simple, mixed) cultures in controlled laboratory environments and to use these insights to make reasoned suggestions about functional attributes that might occur in natural habitats (23), we obviously lack information about the capabilities of those organisms not amenable to cultivation. The advent of large-fragment DNA libraries (also called metagenomes) constructed of DNA extracted from whole communities has permitted the simultaneous determination of key functional traits and the associated 16S rRNA classification of community members without the need to culture cells and without the bias introduced by PCR (16, 121). These and other molecular methods are covered extensively in this volume in chapter 12. Of significance are the interactions of different prokaryotes with each other, or with other biota (for examples, see references 1, 17, and 32), with the result that the behaviors of members of mixed populations differ from those predicted on the basis of the study of component pure cultures. These are but a few reasons why a particular organism's presence in a habitat does not necessarily provide evidence concerning the precise in situ activities of that organism and its contributions to habitat functions (60). New investigations that target mixed cultures and complex communities collected from the environment address the lament that too many "microbial ecology" studies deal with single species and move our science progressively towards a system level understanding of microbes in their environment.

Another complication is that many bacteria exhibit an enormous degree of metabolic flexibility (e.g., nutrients consumed, metabolic end products, and mode of energy conservation); these traits are often regulated by numerous and complex environmental factors that are difficult to ascertain with accuracy for the microscale environments occupied. In contrast, not all prokaryotes possess such a broad spectrum of traits; instead, many possess a specialized metabolism or physiology. The detection of such organisms in a habitat may allow a more confident prediction of their roles in the habitat than is possible for the more metabolically flexible ones. Similarly, the dominance of metabolically specialized microbes in a given habitat may provide a gauge of the stability of the habitat.

The enrichment culture approach has resulted in many significant contributions to our understanding of the connections between particular microbes and cause-effect relationships in many habitats. Many who have carefully employed this method to determine (i) whether organisms with already known attributes were present in samples from a habitat being examined or (ii) whether a novel transformation of compound C to compound D, or some other alteration of a habitat feature, could in fact be mediated by a microbe have recognized the need to quantify the numbers of the particular organisms present as an important aspect of establishing causal relationships (147).

Molecular probe technologies are now well accepted for describing the microbial composition of the habitat of interest, yet some caveats are associated with this methodology (2, 95). In general, the probes are designed on the basis of our knowledge of molecular signatures of well-described and understood organisms, that is, organisms that have been studied in pure cultures. As long as the habitats being examined using these probe methodologies contain recognized organisms with previously described molecular signatures, then there is a reasonable likelihood that the cells will be detected. It must be remembered that because of the inherent metabolic and/or physiologic flexibility of many prokaryotes, even if probe studies reveal the presence of (for example) sulfate-reducing bacteria in a habitat, this information alone does not establish whether it is sulfate, sulfite, thiosulfate, some other sulfur-containing electron acceptor, or even nitrate that is being anaerobically respired by the population or whether it is lactate or some other electron donor that is serving as the energy source. By coupling molecular studies with water chemistry analysis of the sample, investigators may be able to understand the actual processes that are being used by the cells. Another issue that arises when one depends upon molecular probes, or any nucleic acid-dependent characterization, is the degree to which representative nucleic acid can be cleanly extracted from the samples. Still, the potential and utility of molecular probe approaches, when used in combination with those of classic enrichment culture for habitat analysis, are likely to lead to greater comprehension. The combination of culture-based and molecular strategies has thoughtfully been suggested as a wholly new means for identifying new microbial physiologies, as was demonstrated for the recently discovered anaerobic ammonia oxidizers (139).

But perhaps the most significant factors responsible for our limited understanding of bacterial diversity relate to the relatively small numbers of analysts who have devoted themselves to these tasks and to the fact that often their revelations lead to a type of "reductionism," namely, the exploration and elucidation of the properties of the organism(s) newly isolated. Because the discovery of novel properties of such isolates continues to be both stimulating and rewarding, individuals with interests in surveys and elucidation of microbial diversity have had their attentions refocused. The best explanation for the enormous numbers of newly described prokaryotes (for an example, see reference 111) possessing newly recognized biochemical, morphological, and physiological properties is probably a real increase over the past several decades in the numbers of investigators who have pursued this type of scholarship. Even so, the total number of practitioners of this art and science is not large, and their goals, significant as they are, often have not been widely appreciated or understood, nor has the research been well funded. The continued exploration of microbial diversity may best be accomplished through integrated and balanced use of culture-based approaches and new autecology investigations in which cells are studied in the context of their microniche, a method made possible through the use of sophisticated new equipment.

Many thoughtful students of prokaryotic diversity have long recognized that only a fraction of the presumed prokaryotes seen in habitat samples or appearing in enrichment cultures have ever been brought into pure culture (or into stable mixed culture) so that traits could be evaluated and the organisms could be identified (151). Seemingly insignificant modifications of the culture medium or incubation conditions often resulted in pronounced changes in the populations that developed. Even when similar modifications were made in isolation media in attempts to further cultivate such organisms, the efforts were often unsuccessful. Clearly, the organisms' physiological needs remained unmet. Possible explanations are legion. An oft-noted situation is that the populations developing in an enrichment medium with or without agar present as a solidifying agent (an example of the accumulative versus the separative enrichment culture approach) are recognizably different. Many investigators note that some organisms that do appear in accumulative (liquid) enrichments are unable to grow in an otherwise presumably identical agar-solidified medium (55, 112). Nevertheless, progress towards growing cells that were unyielding to many cultivation regimes is occurring. Attempts to physically isolate target microbes from other microbes include optical "tweezers," cell-sorting techniques, traditional dilution-to-extinction approaches, filtration, and density-gradient centrifugation (78). Subsequent to the physical isolation of the cells, creative tools have been devised to encourage cell growth (74) including diffusion chambers (63) or miniaturization techniques using cell encapsulation in gel microdroplets (153) or new cell arrays (22). High throughput of samples is a key to success and to reducing the labor required in order to test the conditions needed to grow heretofore-uncultivatable cells. Still, those practicing the art of isolation and cultivation frequently counsel patience as the key to obtaining cultures of obstinate cells (57, 74).

The historical discord between what can be seen and what can be cultivated (140) has reached a new level. Analytical approaches that are founded upon extraction of nucleic acids from populations followed by amplification and cloning of genes for rRNA allow us now to ponder the incongruency between what can be detected and what can be cultivated. In many instances, rRNA with sequences unlike those already characterized from cultivated organisms is found, providing the initial clue of the presence of a novel, undescribed organism(s) in the habitat sampled. Now, methods like stable-isotope probing (102, 115), metagenomics (16, 51, 121), and community proteomics (117) are providing striking information on community members and their relationships. Recently, such knowledge gained from molecular studies of an extremely acidic system with low microbial diversity permitted the design of cultivation conditions such that one microbe, first known only by its molecular signature from an environmental sample, was successfully grown in culture (142). Still, these powerful new tools for determining the molecular nature of microbes or communities are not ends in themselves. Their careful use requires the accompanying measurement of physical and chemical parameters of the specific environment from which the cells are derived. If these contextual data are collected, then the molecular biology data that are becoming so accessible will yield new insight into the microbial ecology and controls on microbial diversity.

MORPHOLOGIES AND SIZES OF PROKARYOTIC CELLS

We consider prokaryotic cells to have shapes described simply as spheres, rods, vibrioids, spirals, or pleomorphic, and we now recognize cells that are nearly square, rectangular, triangular, or even star shaped. Many of these different morphological types are characterized as well by the presence of constrictions, protuberances, lobes, or other geometrically irregular aspects of their surfaces. Cells may exist as single entities or as units forming chains, clumps, or filaments. Table 1 provides a selected list of the diverse morphological traits of representative organisms.

This morphological diversity contributes to the difficulty encountered in distinguishing cells from inanimate material. Of additional significance is the fact that the morphology of a given organism can undergo change depending on the extracellular environment or the stage of growth. In addition, cells subjected to drying for purposes of staining sometimes appear different from the same cells in the living, unstained state. The range of sizes for different prokaryotic cells is quite large. One unicellular organism may be barely large enough to be resolved by and seen in a light microscope, while others can be just large enough to be seen with the unaided human eye (7). The reclassification of one "protist" as, instead, yet another large prokaryote (47) generates wonder about the actual upper size limits for prokaryotes and how frequently we may revise the estimate (128). At the other end of the size spectrum, some cell types are known to respond to starvation by becoming so-called ultramicrobacteria (5). The newly cultivated *Nanoarchaeum* (56) appears to be one of the smallest microbes. While the question of minimal permissible size of a microorganism has been the subject of contention (20, 64, 71, 144), it seems that a lower size boundary must still be large enough to contain a ribosome (92). Table 2 provides a selected listing of the cell sizes currently recorded for prokaryotes.

PHYSIOLOGICAL DIVERSITY AMONG THE PROKARYOTES

In the biological world, there are no known parallels to the abilities of the *Bacteria* and *Archaea* to utilize an enormous array of energy sources to support growth and metabolism. The recognition that beneath the widely disparate nutritional and environmental needs for the growth and sustenance of different microbes there was an underlying unity in their physiological attributes was a major conceptual contribution (69) that had a marked practical influence on the development, nature, and extent of our understanding of the significance of prokaryotic diversity (70).

This encompassing view was that energy was conserved and made available for life processes as a result of cellularly mediated, coupled oxidation-reduction reactions and that as a group the *Bacteria* exploited in this way nearly every conceivable source of energy available. Then novel, this concept of an extraordinary versatility, along with the wide range of environments in which different prokaryotes were shown to grow or remain metabolically active, led to the notion that prokaryotes are unique in their ability to inhabit and thrive in environmental extremes (i.e., those not conducive to growth of macrobes) (31, 94). One outgrowth of these considerations was the recognition that every naturally occurring organic compound is subject to attack (utilization as a growth-supporting nutrient or as a cometabolic substrate) by one or another microbe and mineralized as a result of participation in the biogeochemical cycles. We have come to realize that not all compounds synthesized by animals and plants are subject to biodegradation at identical rates (the slow attack on lignocellulosic material is one example) and that the initial degradative steps of

TABLE 1 Examples of diversity in morphology of selected prokaryotes

Morphology or other cell anatomical feature	Representative genera
Coccus	
Single	*Acidianus*, *Megasphaera*
In chains	*Lactococcus*, *Streptococcus*
In other groupings	*Pediococcus*, *Sarcina*
Coccus-rod, lobed	*Sulfolobus*
Coccus ↔ rod[a]	*Arthrobacter*
Circular (or nearly so)	*Cyclobacterium*
Cyst (or microcyst) formed[a]	*Azotobacter*, *Sporocytophaga*
Endospore formed[a]	*Bacillus*, *Clostridium*, *Acetonema*, *Sporosarcina*, *Thermoactinomyces*
Exospore formed[a]	*Methylocystis*
Irregular	*Nocardia*, *Mycobacterium*, *Streptomyces*[a]
Myxospore formed[a]	*Myxococcus*, *Stigmatella*
Rectangle	*Methanopyrus*
Rod	
Long, single	*Bacillus*
Short, single	*Pseudomonas*
Often in chains	*Bacillus*, *Lactobacillus*
Spiral, small	*Bdellovibrio*, *Desulfovibrio*, *Rhodospirillum*, *Methanospirillum*
Sheathed	*Thermotoga*, *Sphaerotilus*
Square	Not yet named
Stalked[a]	*Asticcacaulis*, *Caulobacter*
Triangle	*Pyrodictium*
Vibroid	*Bdellovibrio*, *Marinomonas*, *Vibrio*
Wall-less	*Mycoplasma*, *Thermoplasma*

[a]Cell undergoes morphogenesis.

some biosynthesized materials are carried out not by *Bacteria* or *Archaea* but instead by other microbes. The initial contributions of Beijerinck and Winogradsky (134) that established the utilization of inorganic ions and molecules for energy conservation coupled to growth have since been extended in terms of both the scope of the *Bacteria* and *Archaea* involved and the range of inorganic entities able to serve in this capacity (3, 61, 90, 125, 143).

TABLE 2 Diversity in prokaryote cell size

Size (μm)[a]	Representative organism(s)
< 0.2	Some "picoplankton"
0.3–0.5	*Veillonella*
0.4	*Nanoarchaeum*
2–3	*Megasphaera*
5–25	*Thiovulum*
0.6–1.2 × 2.5–5.8	*Bacillus*
1.1–1.5 × 2–6	*Escherichia*
5–6 × 8–12	*Chromatium*
2.5–4 × 40–100	*Thiospirillum*
1–100 × 500–200	*Beggiatoa*
80 × 600	*Epulopiscium*
100–750	*Thiomargarita*

[a]Approximate diameter (× length where appropriate).

It is customary (8, 10, 35, 108) to categorize and contrast the several ways prokaryotes employ the coupled oxidation-reduction reactions listed below. Table 3 provides examples of some additional aspects of these processes.

- Aerobic respiration: molecular oxygen serves as the oxidant in a redox reaction and appears in reduced form as water, one end product of this metabolism.
- Anaerobic respiration: in environmental conditions where molecular oxygen is absent or in limited supply, an inorganic ion such as nitrate, sulfate, or carbonate serves as the terminal oxidant and becomes reduced to dinitrogen (or ammonia), sulfide, or methane, respectively. It is now recognized that a variety of other ions such as oxidized (ferric) iron or organic molecules such as fumarate, trimethylamine oxide, dimethyl sulfoxide, or the sulfonic acids can also serve as terminal oxidants for a variety of anaerobically respiring prokaryotes.
- Fermentation: an organic compound, most often a metabolic intermediate, that results from oxidation of the organic compound serving as the energy source, serves as the terminal oxidant, and a more reduced organic molecule is a metabolic end product(s).
- Phototrophy: radiant energy is absorbed by chlorophyll-, bacteriochlorophyll-, or "accessory pigment"-containing pigment complexes, resulting in an excitation of electrons present in the complex and leading to an oxidation and charge separation. When water is employed as the ultimate reductant for the sequential reactions, the

processes are characterized by evolution of molecular oxygen ("oxygenic photosynthesis"); when either inorganic or organic compounds replace water, molecular oxygen is not formed, and the processes are termed, instead, "anoxygenic photosynthesis."

We may summarize, then, one aspect of the prokaryotes' physiological diversity by noting that some can persist and multiply only in habitats that are in regular contact with the earth's atmosphere and the molecular oxygen it contains, employing aerobic respiration (i.e., are strictly or obligatorily aerobic). Other prokaryotes function only in the absence of air (58), employing either fermentation or anaerobic respiration, and thus pursue other modes of energy acquisition and conservation; we term such organisms strictly or obligatorily anaerobic (but might also describe them as strictly fermentative or as living strictly by anaerobic respiration). Another group may possess, for example, the ability either to live by aerobic or anaerobic respiration; the adjective "facultative" is added to describe such an organism's respiration. The same term is used in describing the ability of an organism to live either by aerobic respiration or by fermentation (e.g., facultatively fermentative) and in the description of a bacterium able to live either by anoxygenic phototrophy or by aerobic respiration. Evolution has resulted in a continuum of traits rather than a set of neatly packaged ones. The inadequacy of words such as "facultative" to describe the physiologies of a bacterium able to live by either fermentation, aerobic respiration, or anoxygenic phototrophy, for example, becomes readily apparent. Despite our inability to succinctly categorize microbial metabolic strategies, we recognize that life requires an accessible and energetically sufficient oxidation-reduction gradient from which to gain energy and that this requirement is a signature feature of an environment if it is to sustain life. Conceptually, this theme has been proposed as a potential marker of extraterrestrial life (39) even if microbes display a variety of strategies for accessing such energy.

In a related sense, our understanding of the roles of prokaryotes in biosphere functions may be constrained. For example, we may not even know that a nutritional lifestyle based on acetogenesis remained to be discovered in a distinct morphological prokaryotic group (spirochetes [76]), that phosphite oxidation could be coupled to dissimilatory sulfate reduction by anaerobes (123), or that N_2-dependent growth was a trait of metal-metabolizing bacteria (13). Perhaps for such reasons it has become customary to categorize microbial physiological traits in ways other than those having to do strictly with oxidation and reduction reactions.

One such way refers to the source(s) of carbon assimilated for biosynthesis: "autotrophy" describes the ability of an organism to utilize carbon dioxide as the principal source of carbon (save perhaps the need for vitamins or an amino acid[s]), while "heterotrophy" or "organotrophy" describes the use of carbon atoms of organic molecules as the principal carbon source. When categorization is focused on the source of energy to be conserved, an organism doing so at the expense of reduced inorganic ions, or molecules such as

TABLE 3 Some nutritional aspects of physiological diversity

Electron donor(s) utilized (examples)	Electron acceptor, reduced end product(s)[a]
In aerobic respiration	
Organic molecules: carbohydrates, amino acids, purines, pyrimidines, lipids, fatty acids, alcohols, hydrocarbons (both aliphatic and aromatic), sulfonic and aromatic acids	Molecular oxygen, **water**
Inorganic molecules or ions: carbon monoxide, molecular hydrogen, metallic sulfides, ammonium, nitrite, ferrous and manganous salts, elemental sulfur	
In anaerobic respiration	
Organic molecules: much as given above; possible exceptions are some sulfonic and aromatic acids	Nitrate, nitrite; **nitrite, nitrogen gas, ammonium**
Inorganic molecules or ions: much as given above	Sulfate, sulfite, elemental sulfur; **sulfite, sulfide**
	Fumarate, **succinate;** dimethyl sulfoxide, **dimethylsulfide;** ferric salts, **ferrous salts;** trimethylamine oxide, **trimethylamine**
But methanogens typically utilize primarily hydrogen gas, formate, or acetate	Carbonate, **methane**
In fermentation	
Organic molecules: carbohydrates, purines, pyrimidines	Organic molecules, protons; **alcohols, fatty acids, ketones, hydrogen gas**
In phototrophy	
Organic molecules: alcohols, fatty acids, organic acids (e.g., malate, succinate, benzoates)	Carbonate, **cellular components**
Inorganic compounds or ions: hydrogen gas, ferric, sulfide, elemental sulfur, thiosulfate	
But, for cyanobacteria, water	

[a]Reduced end products are in boldface type.

TABLE 4 Some terms used in relation to bacterial growth or metabolic activity

Acidophiles: organisms with growth or activity optima at pH values of ca. 1–5
Aerobes: organisms that use molecular oxygen in redox reactions coupled to energy conservation
  Obligate: cells that cannot grow or that remain inactive in the absence of molecular oxygen
  Facultative (or euryoxic): cells that grow or are active in the absence of molecular oxygen
Alkaliphiles: organisms with growth or activity optima at pH values of ca. >8
Anaerobes: organisms that are unable to use (i.e., consume) molecular oxygen
  Obligate: those that cannot grow or that remain inactive in the presence of molecular oxygen
  Oxyduric: those not killed by (i.e., tolerant of) molecular oxygen
  Oxylabile: those killed by the presence of molecular oxygen
  Aerotolerant: those able to grow or remain active in the presence of molecular oxygen even though they do not use it
Microaerophiles: organisms that require molecular oxygen for growth or activity but can tolerate its presence only when present at low levels (often ca. 10% of atmospheric levels)
Mixotrophs: organisms that are both autotrophic and heterotrophic, usually simultaneously, in order to conserve energy and assimilate nutrients for growth or activity
Phototrophs: organisms that use radiant energy (light) as a source of energy for growth or activity
  Obligate: those that cannot grow or that are inactive in the absence of light
  Facultative: those able to grow or remain active by gaining energy in the absence of light
Symbiosis: two or more dissimilar organisms that interact and live together
Syntrophy: the relationship between proton-reducing organisms and other organisms that consume hydrogen gas
  Obligate: the relationship when the proton-reducing organism is unable to grow or remain active in the absence of the hydrogen-consuming one

hydrogen gas, is regarded as a "lithotroph." The term "organotroph" describes an organism utilizing organic molecules; note the use of the identical word to describe both the carbon and the energy sources. Table 4 lists examples of descriptive terms often used in categorizing what are indeed different lifestyles of prokaryotes.

Prokaryote physiological diversity extends beyond relationships to molecular oxygen or how energy is derived. Optimal growth under conditions of low pH (acidophiles) or high pH (alkaliphiles) is characteristic of many different prokaryotes. Similar diversity exists in temperature optima for growth: cold-loving organisms (psychrophiles) contrast with those that are unable to grow at temperatures less than ca. 80°C, some of which (hyperthermophiles) have been shown to grow at temperatures in excess of 110°C (135). The current high temperature record holder grows at 120°C and can briefly survive at 130°C (66). Among other traits in which prokaryotes show remarkable variation from organism to organism are the abilities to tolerate (34) or need inorganic salts (as for the strictly halophilic subgroup of the *Archaea*) or to grow only in environments with low nutrient levels (oligotrophs) or high nutrient levels (copiotrophs). Yet another trait noted for certain organisms is a requirement for pressure in excess of that at the earth's surface (barophiles [110] or piezophiles [152]). Table 5 summarizes ranges of selected traits.

As noted above, the scope of nutritional diversity among the prokaryotes is not only impressive but also important, as it affects the ability of these microbes to colonize and to thrive in nearly every imaginable habitat. Some bacteria are able to oxidize compounds containing no carbon-to-carbon bonds (C-1 compounds) and to assimilate the oxidized moiety for synthesis of molecules characteristic of all cells. While many of these organisms are also able to oxidize and assimilate methanol carbon, they are generally incapable of assimilating the carbon of other simple or complex organic molecules as the sole carbon and energy source for growth or maintenance. In contrast to these "methanotrophs,"

TABLE 5 Environmental extremes in which prokaryotes are thought to multiply

Characteristic	Value
Pressure	1 to ca. 1,000 atm[a]
Temp	−1.2 to 120°C
Ionizing radiation	Up to 15,000 Gy[b] instantaneous dose
Depth	2,000 m (terrestrial subsurface) to ca. 10,000 m (ocean floor)
Salt conc.	Up to ca. 4–5 M
Acidity or alkalinity	pH of <1 to 11–12
Available water	a_w (water activity) as low as 0.6

[a] 1 atm = 101.29 kPa.
[b] 1 Gy (gray) = the absorption of 1 J of energy by 1 kg of matter.

another group, the "methylotrophs," differs in two ways. Methylotrophs cannot utilize methane for growth but can grow or remain metabolically active at the expense of methanol, formate, and methylated amines; more strikingly, however, they can use acetate, other organic acids, and amino acids. Most of the well-studied methanotrophs are specialized in terms of nutrients utilizable for growth or activity, while the methylotrophs are much less so.

Another example of a nutritional specialization is shown by at least one *Bacillus* species which can be sustained readily with urate, a purine, or compounds such as allantoin and allantoate, both of which are intermediates in the degradative pathway for urate oxidation and assimilation. However, in contrast to the scores of other *Bacillus* species, *Bacillus fastidiosus* cannot survive by using sugars, polysaccharides, amino acids, or simple proteins. In similar fashion, some *Cytophaga* species, unable to be sustained with amino acids as carbon and energy sources when first isolated from a natural habitat, are strictly dependent upon polymers such as cellulose or chitin as a source of carbon and energy for growth. Other cytophagas, by contrast, cannot use cellulose or chitin for growth, but they can grow on complex media containing peptides, amino acids, yeast extract, or simple sugars. These specificities contrast to those of some pseudomonads and their relatives which are able to use scores of organic compounds, including simple organic acids, the range of amino acids, and benzenoid and polycyclic compounds, to support cell division.

As a group, autotrophs also may vary the reductant that they use with respect to their nutritional needs. Some oxidize hydrogen gas or may use a compound such as hydrogen sulfide from which to capture energy for growth, or they may be able to grow as organotrophs by use of organic compounds. However, not all chemolithotrophs are so versatile; some "sulfur-oxidizing" chemolithotrophs use any of several reduced sulfur compounds to acquire energy, while others display much more selectivity and specificity. For another group of autotrophs (e.g., *Nitrosomonas* spp.), ammonia serves as an energy source, while nitrite does not, yet *Nitrobacter* spp. oxidize nitrite to form nitrate but are unable to gain energy from ammonia. Reduced forms of iron are suitable electron donors in energy-conserving processes for other autotrophs.

At least two other aspects of physiological diversity are significant for the discussions that follow. In one case, even though a particular organic molecule may not function as a sole source of carbon and energy for growth of an organism, the compound may be metabolized (i.e., either oxidized or reduced in whole or in part, and all or part of the molecule may be assimilated into cellular components) as long as the organism is utilizing a different molecule as a carbon and energy source. This phenomenon, first termed "cooxidation" and now most often referred to as cometabolism, has been shown to be of more general significance and distribution in prokaryotes and is essential for many bioremediation processes (83). A second trait is syntrophy (85). In anoxic environments, the degradation of reduced organic compounds (e.g., simple alcohols, fatty acids, and certain aromatic acids), resulting in the accumulation of acetate, carbon dioxide, hydrogen gas, and formate (among other end products), is energetically unfavorable. However, if the concentrations of these products are kept sufficiently low, the overall energetics for growth become more favorable. Methanogens and sulfate-reducing bacteria (sulfidogens), commonly present in anoxic habitats, consume hydrogen gas and thus lower its partial pressure in the microhabitat. Thus, one member of a coculture functions to permit the biodegradation of compounds that would otherwise be refractory to attack by pure cultures. The term "interspecies hydrogen transfer" has also been used to describe such interactions. Although not yet fully physiologically characterized, the anaerobic methane-oxidizing community that occurs in marine sediments where methane and sulfate are present as the electron donor and acceptor, respectively, appears to represent a recent example of such a microbial consortium (17, 102).

One final observation related to the physiological diversity of microbes is that many of these cells display a range of levels at which they may be metabolically active. Although cells grown in the laboratory are often pushed to grow optimally, under environmental conditions cells can enter a state of maintenance level activity or dormancy enabling their survival for periods of time far longer than the generation times or metabolic rates deduced for them in laboratory cultures. Although dormancy is a trait often attributed to cysts or endospores, it is well recognized that dormancy is a significant phenomenon in the persistence of an enormous variety of cells (65, 96). In fact, it seems that the metabolic activity of cells in numerous environmental settings can be classified according to whether the cells are actively growing, sustained at maintenance levels of activity, or clearly dormant (114).

DIVERSITY: INSIGHTS FROM MACROMOLECULAR ANALYSES

Because prokaryotes have been on this planet for perhaps four-fifths of its age, there has been ample time for repeated mutations, their accumulation and natural selection, and an introduction and accumulation of widely varied traits in these organisms. For years, the lack of a substantial traditional fossil record impeded critical considerations of evolutionary change in the microbial world. It was not until Zuckerkandl and Pauling's (155) proposal of macromolecular traits as indicators of evolutionary changes that we could ponder events of the past. Data from protein (30), rRNA (48, 130), and high-throughput DNA processing that generates complete genome sequences (e.g., the U.S. Department of Energy's Joint Genome Institute [http://www.jgi.doe.gov/]) are now used widely for this purpose and for making inferences about phylogenetic relationships. The evolution of prokaryote diversity has become a subject of intense interest, warranted because more than one-third of the phylum level prokaryotic lineages are known only by the phylotypes or environmental clone sequences (i.e., DNA sequences in a database) that reside within them and not by cultured representatives (57).

A widely accepted phylogenetic tree (Fig. 1) that classifies life in three major categories, *Archaea*, *Bacteria*, and *Eucarya*, is based on the inferences that *Archaea* and *Eucarya* diverged from ancestors of the *Bacteria*, first as a single lineage and only later diverging and becoming separately recognizable entities. The timing of this divergence, as well as the estimates for the divergence point of the eukaryotes and prokaryotes, has been considered often (30).

The potential for lateral gene transfer in natural populations remains a matter of concern in using sequence information for phylogenetic considerations and in reconciling the concept of species in microorganisms (15, 42, 45, 48, 80, 93). As the number of known microbial genomic sequences increases, there is mounting evidence that horizontal (or lateral) gene transfer has occurred frequently in

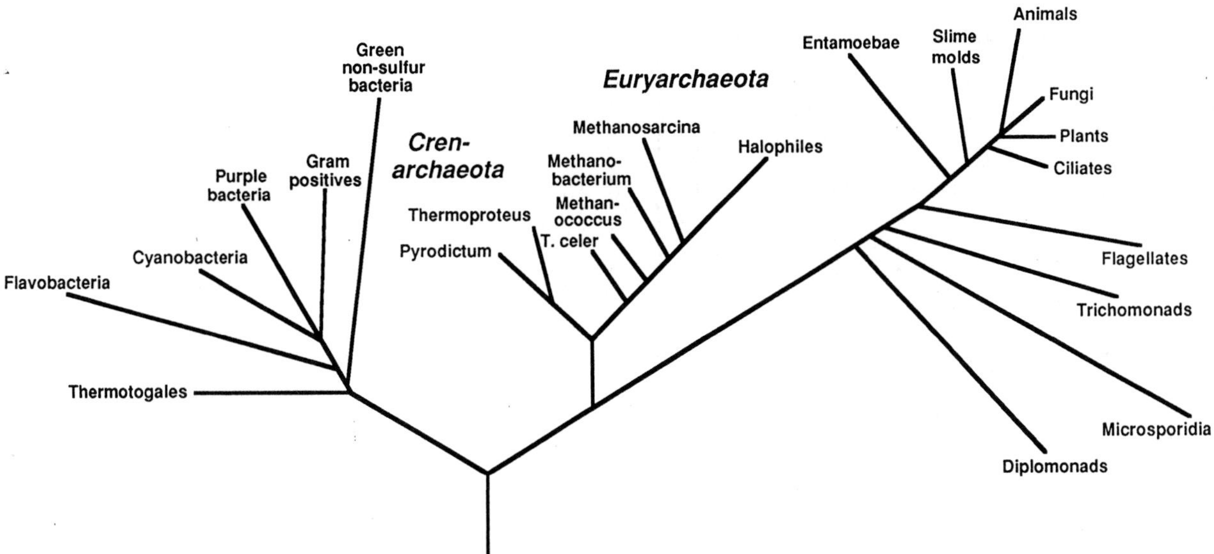

FIGURE 1 A phylogenetic tree based on evaluation of 16S rRNA sequences. The three major lineages of life (*Archaea, Bacteria,* and *Eucarya*) are shown. T. celer, *Thermococcus celer.* Reprinted from *The FASEB Journal* (99) with permission.

the course of evolution. Accordingly, the oft-accepted consensus of cellular evolution as outlined in the following paragraphs seems vulnerable to considerable revision as the details of genomic sequences become clearer. Our view of microbial evolution will be clarified by additional database surveys of gene variation and the development of models that describe how gene transfer affects evolving microbial populations (15) and ultimately, with the inclusion of experimental data that test ecologically important selection criteria (45).

In the line leading to the *Bacteria*, an initial bifurcation leads to thermophilic cells now represented by genera such as *Aquifex*, *Fervidobacterium*, and *Thermotoga* (119). It is a matter of curiosity and concern that *Aquifex*, regularly placed at the bottom of the tree, is an aerobic, respiring chemolithotroph. This represents an anomaly, as the planet presumably was devoid of molecular oxygen at the time of this evolutionary event (44, 145).

Most current *Bacteria* trees show that the metabolically versatile, green nonsulfur phototrophs were among the next groups to diverge (119). Perhaps at about the same time a group that possessed cell walls devoid of peptidoglycan emerged: the *Planctomycetales*, including genera such as *Gemmata*, *Isosphaera*, *Pirellula*, and *Planctomyces* (44, 145).

The remaining well-studied and recognized bacterial groupings designated the flavobacteria, cyanobacteria, gram-positive bacteria (of which there appear to be at least two major subgroups), and purple bacteria or proteobacteria (a group itself considered to warrant five subdivisions: alpha, beta, gamma, delta, and epsilon) appear to have emerged later as a single radiation which then underwent divergence.

Even before this refined evidence about likely events in bacterial evolution, it was suggested that thermophilic anaerobes were candidates as the most ancient of living beings. Arguments over whether autotrophs preceded heterotrophs or vice versa abounded but were based more on conjecture than on persuasive evidence. Among the candidates for the earliest of organisms were the methanogens, members of the *Archaea*, appearing later on the evolutionary scene.

Despite microscopic similarities, *Archaea* are as unlike the *Bacteria* as they are unlike the *Eucarya*. Their possession of unusual ether-linked lipids and the lack of peptidoglycan in their cell walls are among their distinguishing characteristics. The group is now divided into the subcategories of methanogens, thermo- and hyperthermophiles, and strict halophiles. The limits of our understanding of the *Archaea* are demonstrated by a series of surprising discoveries of new types of these organisms: in cold, marine environments (26, 127), as contributors to anaerobic ammonia oxidation in the oceans (36, 72), as yet uncultivated anaerobic methane oxidizers (86), as extreme acidophiles that entirely lack cell walls (46), and even as contributors to disease (77).

Although the *Eucarya* are usually cited as characteristically containing a membrane-bound nucleus, they are also usually considered to possess subcellular organelles termed chloroplasts or mitochondria. It may be significant for evolutionary considerations that in the *Eucarya*, three groups at the base of the tree—the diplomonads, the microsporidians, and the trichomonads—are devoid of mitochondria.

One of the many surprises arising from such approaches to the study of phylogeny, and of special interest for considerations of prokaryotic diversity, has been the demonstration that some biochemical and physiological traits appear in quite distantly related organisms. For example, sulfate reduction occurs both in bacteria and in archaea; the same is true for autotrophic assimilation of carbon dioxide as well as the trait of hyperthermophily.

ASPECTS OF BEHAVIORAL DIVERSITY AMONG PROKARYOTES

Direct observation of either mixed microbial populations or pure cultures reveals differences in behavior. Some organisms swim through liquids with ease and rapidity, others move more slowly; some dart about randomly as if in a frenzy, while others move for longer distances and appear to be swimming smoothly; in some, no swimming motility is ever discerned. Typically, motility reflects the presence and function of either a single flagellum or multiple flagella and the arrangement of these on the cell surface(s). Many of the *Bacteria* termed spirochetes swim much better through viscous liquids than they do in typical water samples or laboratory media; this may be a reflection of the position of their flagella (axial filaments, or endoflagella) between the cell wall and an external sheath of the cell. This location clearly does not permit the flagellum to rotate in the extracellular milieu and propel the cell in a manner similar to that of, for example, the well-studied enteric bacteria. Usually the behavior of swimming flagellated bacteria has been described as alternating "runs" and "tumbles." The observations made with the enteric bacteria may be of limited application; however, *Azospirillum* spp. (154), pseudomonads, and rhizobia are among organisms observed not only to behave differently but also to have different intracellular regulatory mechanisms involved.

A group of bacteria with rather diverse physiological properties move in a different manner, not by swimming through liquids but rather by migrating over solid surfaces (e.g., agar, cellulose, chitin, and other cells). This motility, for which locomotor organelles have not yet been visualized and for which the mechanism for motility remains unknown, has often been termed "gliding." The implication of type IV pili in one form of myxobacterial motility and in the motility termed "twitching" (characteristic of some acinetobacters and pseudomonads) is a welcome development in visualization, although the mechanistic basis (e.g., the "motor") for these functions of pili remains unclear. An earlier explanation for the gliding motility of some cyanobacteria, namely, polysaccharide extrusion through a "wall" pore complex, has recently been reassessed and supported. Cells can be seen to move (glide) on a solid surface, and the colony can spread a considerable distance (often more than 1 cm) beyond the point of cell deposition (inoculation). Such colonies usually are characterized as much thinner (very little vertical height) than those typical of enteric bacteria or pseudomonads, for example, growing on an identical nutrient-poor medium. Nutrient concentration usually exerts a significant effect on the ability of cells to glide and hence to form spreading colonies; the ability of a cell to glide is ordinarily not expressed on a nutrient-rich medium such as nutrient agar. Among the *Bacteria* best known for gliding ability are those of genera such as *Beggiatoa*, *Cytophaga*, *Flexibacter*, and *Myxococcus*; however, many others share this capability. And at least one nongliding cyanobacterium (a *Synechococcus* sp.) swims through liquids without flagella; nor have any locomotor organelles yet been discerned for this organism (109).

It has already been suggested (and previously noted) that cells alter their physiological phenotypic properties; such occurrences are myriad, and alterations in anatomical features and motility phenomena also develop. One such example is the motility of some flagellated *Bacteria* on solid surfaces referred to as "swarming." Certain *Serratia*, *Proteus*, and *Vibrio* species, among others, synthesize additional, specialized flagella as a result of contact with a solid surface, and these additional flagella enable the organism to spread. Surface features, among them the concentration of agar used in solidifying the medium, affect the cell's synthesis of these appendages for motility.

The chemotactic response of cells to concentration gradients of attractants (e.g., nutrients) or repellents by employing temporal sensing mechanisms has been extensively studied as one category of behavioral responses. Although phototrophic eukaryotic cells have long been known to respond by phototaxis to light gradients, no unambiguous instance of this phenomenon has yet been established for prokaryotic phototrophs (116, 122). Studies of the in situ behavior of prokaryotic populations have established phenomena such as diel migration (40) which undoubtedly reflect both chemo- and phototactic (and possibly other types of) behavior of members of these communities (18).

Some cells undergo either cellular or colonial morphogenesis in connection with entrance into a dormant or nongrowing ("resting") state. The formation of the remarkably resistant endospore within the cells of genera such as *Acetonema*, *Bacillus*, *Clostridium*, *Desulfotomaculum*, *Sporomusa*, *Sporosarcina*, and *Thermoactinomyces* generally is regarded as a response to nutrient-poor conditions, although this may not always be the case. Endospore production in *Epulopiscium*-like cells appears to occur at night possibly as a way of ensuring dispersal to new hosts of this intestinal symbiont of surgeonfish (33). Still, starvation seems to be an important precursor for myxospores and microcysts formed by myxobacteria and cytophagas, in which the entire cell shortens, thickens, and becomes spherical (or nearly so). This is probably the explanation for exospore formation by some methanotrophs. Other examples of changes in cell morphology resulting in specialized functions are the formation of heterocysts in some cyanobacteria and the formation of akinetes in others and stalk formation in prosthecate bacteria, as typified by *Caulobacter* spp.

Numerous investigations have verified biofilm development as a critical aspect of microbial behavior (50). Cells that are present in biofilms experience higher proximal cell densities and are subject to higher concentrations of metabolic by-products or metabolites than cells that exist in a planktonic phase (106). Cell-to-cell communication (i.e., quorum sensing) has become a key aspect in the investigation of biofilm development (24) and the term "sociomicrobiology" has been coined to represent the study of any group behavior of microbes (106). The recent discovery of so-called "nanowires" which are conductive pili used by metal-reducing microbes, possibly as an electron shunt (120), represents yet another new area of research that branches from the general field of cell attachment processes.

ENLARGING PERSPECTIVES OF THE DIVERSITY OF HABITATS EXPLOITED BY PROKARYOTES

> Everything is everywhere; but the milieu selects . . . in nature and in the laboratory.
>
> L. G. M. Baas Becking, *Geobiologie ov in eidintot de milieukunde* (9)

Why is it that prokaryotes have been known to occur in so many different habitats and that our comprehension of "new places" in which they are found is increasing?

Certainly, as may be obvious from the degree of physiological diversity outlined above, prokaryotes have the potential to exploit (81) habitats judged to be extreme in comparison with those that support the existence of animals and plants.

If we accept the dictum that prokaryotes are found virtually everywhere (although perhaps not in the totally cosmopolitan scope of the past [82, 132, 133]), then all of the habitats and the microbes therein cannot be enumerated, let alone described, in this chapter. Nonetheless, given the dynamic and exciting state of the study of microbial diversity, selected recent descriptions, reflecting either novel revelations or reconsiderations of earlier findings, are noted briefly below.

As ever-increasing numbers of unique habitats are examined by use of macromolecular sequence, stable-isotope research, and new cultivation strategies, it is ever more evident that prokaryotic diversity has been regularly underestimated by classical isolation and cultivation approaches. It is true not only in seemingly extreme habitats such as hot springs (11, 148), where this was demonstrated early, but also for habitats that are cold (25, 38, 43, 97)—as well as for those that are anoxic (e.g., the rumen [118] and the deep ocean [75]), where we encounter indications of once-unexpected population diversity and microbial impact (76, 150). Estimates derived from microbial inventories and determinations of the habitable space within the Earth's subsurface (21) suggest that among the various places where life can exist, the deep earth is the location of the greatest biomass (149). Within this subterranean realm are newly recognized habitats where cells are apparently sustained by radiolysis of water (79) or fossil hydrocarbons (52), coexist with methane hydrates (59), are able to tolerate radioactive waste (37), or are energetically driven by molecular hydrogen (19, 87, 89, 136). Although the activities of these cells are often deemed to be at or near the threshold required for cell survival (28, 68, 100), observations from some deep habitats suggest that higher levels of activity can occur at considerable sediment depths (ca. hundreds of meters) (105, 124).

The remarkable advancements in molecular characterization strategies notwithstanding, direct observational approaches to habitat analysis continue to play a large role in expanding our comprehension of the impact of prokaryotes in different habitats. The presence of purple sulfur bacteria in the form of macroscopically visible "berries" in salt marsh ponds (129) and the dense accumulation (as "plates") of another phototrophic bacterium in a meromictic saline lake (103) remind us of our limited understanding of the roles that such prokaryotes play in these and other habitats. The discovery of a green sulfur bacterium captured near a deep-sea hydrothermal vent and found to be capable of anoxygenic photosynthesis by capturing light emitted by geothermal radiation expands permissible phototrophic habitats to include those disconnected from solar input (14).

Another example of a transformation long wondered about but only recently demonstrated is that of the oxidation of ammonium under anaerobic respiratory (denitrifying) conditions (88); even more exciting is the indication that the prokaryote involved is a previously unknown member of the *Planctomycetales* (138) and the prospect that new aspects of bacterial physiology may exist in this poorly studied group. It is worth noting that continued microscopic examination played an important part in establishing cause and effect for this microbe. Although no pure cultures of these cells have been sustained, significant advancements have been made in the ability to cultivate and study them (62).

The unaided eye, too, remains a significant part of the exploration of the microbial world. Although dense mats of members of the genus *Thioploca* oxidize sulfide by reducing nitrate (35) near the coasts of Chile and Peru, these organisms were present in amounts far smaller than expected in a similar situation in Namibian shelf sediments. Instead, giant (up to 750 μm in diameter) cells of the previously unknown prokaryote given the name *Thiomargarita namibiensis* (128) were readily visible and extensively studied. Our perception of just how large a prokaryote might be was once again challenged.

A reminder that so many habitats on the planet have very limited molecular oxygen content is made evident by the emerging significance of bacterial respiration of, for example, arsenic (101) and selenium (137) and the evidence that hydrocarbons such as hexadecane can be metabolized via methanogenesis (6).

Studies of *Buchnera* spp., the prokaryotic intracellular symbionts of aphids (12), are but one reminder that not all organisms of environmental significance are free-living and that some may exist in mutualistic states. The phenomenon of bacterial luminescence, so long associated solely with the marine habitat, is associated with terrestrial biotas as well (91). Study of one squid-vibrio interaction (84) makes clear that the microbe-macrobe interaction that may result in a mutualistic (as opposed to parasitic) outcome has effects on the "normal" development of both the microbe and macrobe.

While research has determined that microhabitats are abundant and significant, there is an extreme need to reconcile the multiple impacts of microbes on large-scale Earth processes. And this last aspect of microbial diversity, how microbes from numerous distinctive habitats impact the earth through the cycling of key elements and so have the ability to scale their activities to a global dimension, has yielded some of the most startling revelations in the past few years. The long-predicted and recently verified "anammox" process may contribute 30 to 50% of the N_2 that is generated in the oceans (27). Methanogens, as they contribute to the enormous cache of methane in marine sediments, have been implicated as the governor of the globally important "gas hydrate capacitor" (29). And, of course, microbes also insinuate their way into and may exacerbate human impacts on the environment. To the now familiar explanations of such processes as microbial methylation of mercury or acid mine drainage there are new ways in which we may unknowingly work in concert with these single cells. The excess release of sulfates into the atmosphere and the subsequent deposition into some freshwater locations has caused a change in the terminal electron accepting process from carbon dioxide reduction to sulfate reduction (41). The occurrence of arsenic in groundwater in Bangladesh is now believed to be part of the biogeochemical cycling of the element (113) and an increase in the respiration of microbes in the rhizosphere of trees is thought to offset the capacity of plants to sequester carbon dioxide from the atmosphere (53). The continued systematic probing of microbial diversity and explanation of where microbes live and how they live will lead to a better understanding of the path that we are traveling together (126).

The advent of new molecular technologies heralds a time of seductively immense data sets that catalog the genomes of individuals and groups of microbes. Accordingly, microbial diversity studies that describe the communities from a range of habitats will be profuse. Yet it will be essential to acquire the corresponding environmental data along with the molecular data to enable the testing of hypotheses related to cells and their capabilities. With complete and accessible data collection, there is a good chance that even complex systems with multiple species may yield information that is comprehensible, leads to more accurate predictions, and may be used by the greater scientific community (67). In this regard, integrated teams of experimentalists and modelers working iteratively to develop a better understanding of the controls on microbial diversity should be a high priority for future research. Some excellent examples already exist (49, 131, 141).

REFERENCES

1. **Achtnich, C., A. Schuhmann, T. Wind, and R. Conrad.** 1995. Role of interspecies H_2 transfer to sulfate and ferric iron-reducing bacteria in acetate consumption in anoxic paddy soil. *FEMS Microbiol. Ecol.* **16:**61–70.
2. **Akkermans, A. D. L., M. S. Mirza, J. M. Harmsen, J. H. Biok, P. R. Herron, A. Sessitsch, and W. M. Akkermans.** 1994. Molecular ecology of microbes: a review of promises, pitfalls and true progress. *FEMS Microbiol. Rev.* **15:**185–194.
3. **Albrechtsen, H. J., G. Heron, and T. H. Christensen.** 1995. Limiting factors for microbial Fe(III)-reduction in landfill leachate polluted aquifer (Vejen, Denmark). *FEMS Microbiol. Ecol.* **16:**233–248.
4. **Amann, R. I., W. Ludwig, and K.-H. Schliefer.** 1995. Phylogenetic identification and *in situ* detection of individual microbial cells without cultivation. *Microbiol. Rev.* **59:**143–169.
5. **Amy, P. S., and R. Y. Morita.** 1983. Starvation-survival patterns of sixteen freshly isolated open-ocean bacteria. *Appl. Environ. Microbiol.* **45:**1109–1115.
6. **Anderson, R. T., and D. R. Lovley.** 2000. Hexadecane decay by methanogenesis. *Nature* **404:**722–723.
7. **Angert, E. R., K. D. Clements, and N. R. Pace.** 1993. The largest bacterium. *Nature* **362:**239–241.
8. **Atlas, R. M.** 1986. Applicability of general ecological principles to microbial ecology, p. 339–370. *In* J. S. Poindexter and E. R. Leadbetter (ed.), *Bacteria in Nature*, vol. 2. *Methods and Special Applications in Bacterial Ecology*. Plenum Press, New York, N.Y.
9. **Baas Becking, L. G. M.** 1934. *Geobiologie ov in eidintot de milieukunde*. Stockum und Zoon N. V., The Hague, The Netherlands.
10. **Balows, A., H. G. Truper, M. Dworkin, W. Harder, and K. H. Schleifer.** 1992. *A Handbook on the Biology of Bacteria: Ecophysiology, Isolation, Identification, Applications, the Prokaryotes*, 2nd ed., vol. 1–4. Springer-Verlag, New York, N.Y.
11. **Barns, S. M., R. E. Fundyga, M. W. Jeffries, and N. R. Pace.** 1994. Remarkable archaeal diversity detected in a Yellowstone National Park hot spring environment. *Proc. Natl. Acad. Sci. USA* **91:**1609–1613.
12. **Baumann, P., L. Baumann, C. Y. Lai, and D. Rouhbakhsh.** 1995. Genetics, physiology, and evolutionary relationships of the genus *Buchnera*: intracellular symbionts of aphids. *Annu. Rev. Microbiol.* **49:**55–94.
13. **Bazylinski, D. A., A. J. Dean, D. Schuler, E. J. P. Philips, and D. R. Lovley.** 2000. N_2-dependent growth and nitrogenase activity in the metal-metabolizing bacteria, *Geobacter* and *Magnetospirillum* species. *Environ. Microbiol.* **2:**266–273.
14. **Beatty, J. T., J. Overmann, M. T. Lince, A. K. Manske, A. S. Lang, R. E. Blankenship, C. L. Van Dover, T. A. Martinson, and F. G. Plumley.** 2005. An obligately photosynthetic bacterial anaerobe from a deep-sea hydrothermal vent. *Proc. Natl. Acad. Sci. USA* **102:**9306–9310.
15. **Beiko, R. G., T. J. Harlow, and M. A. Ragan.** 2005. Highways of gene sharing in prokaryotes. *Proc. Natl. Acad. Sci. USA* **102:**14332–14337.
16. **Beja, O., M. T. Suzuki, E. V. Koonin, L. Aravind, A. Hadd, L. P. Nguyen, R. Villacorta, M. Amjadi, C. Garrigues, S. B. Jovanovich, R. A. Feldman, and E. F. DeLong.** 2000. Construction and analysis of bacterial artificial chromosome libraries from a marine microbial assemblage. *Environ. Microbiol.* **2:**516–529.
17. **Boetius, A., K. Ravenschlag, C. J. Schubert, D. Rickert, F. Widdel, A. Gieseke, R. Amann, B. B. Jorgensen, U. Witte, and O. Pfannkuche.** 2000. A marine microbial consortium apparently mediating anaerobic oxidation of methane. *Nature* **407:**623–626.
18. **Caumette, P., R. Matheron, N. Raymond, and J. C. Relexans.** 1994. Microbial mats in the hypersaline ponds of Mediterranean salterns (Sahns-de-Grand, France). *FEMS Microbiol. Ecol.* **13:**273–286.
19. **Chapelle, F. H., K. O'Neill, P. M. Bradley, B. A. Methe, S. A. Ciufo, L. L. Knobel, and D. R. Lovley.** 2002. A hydrogen-based subsurface microbial community dominated by methanogens. *Nature* **415:**312–315.
20. **Cisar, J. O., D.-Q. Xu, J. Thompson, W. Swaim, L. Hu, and D. J. Kopecko.** 2000. An alternative interpretation of nanobacteria-induced biomineralization. *Proc. Natl. Acad. Sci. USA* **97:**11511–11515.
21. **Colwell, F.** 2001. Constraints on the distribution of microorganisms in subsurface environments, p. 71–95. *In* J. Fredrickson and M. Fletcher (ed.), *Subsurface Microbiology and Biogeochemistry*. John Wiley and Sons, New York, N.Y.
22. **Connon, S. A., and S. J. Giovannoni.** 2002. High-throughput methods for culturing microorganisms in very-low-nutrient media yield diverse new marine isolates. *Appl. Environ. Microbiol.* **68:**3878–3885.
23. **Conrad, R., P. Frenzel, and Y. Cohen.** 1995. Methane emission from hypersaline microbial mats: lack of aerobic methane oxidation activity. *FEMS Microbiol. Ecol.* **16:**297–306.
24. **Davies, D., M. Parsek, J. Pearson, B. Iglewski, J. W. Costerton, and E. P. Greenberg.** 1998. The involvement of cell-to-cell signals in the development of a bacterial biofilm. *Science* **280:**295–298.
25. **DeLong, E. F.** 1992. Archaea in coastal marine environments. *Proc. Natl. Acad. Sci. USA* **89:**5685–5689.
26. **DeLong, E. F., K. Y. Wu, B. B. Presellin, and R. V. M. Jovine.** 1994. High abundance of archaea in antarctic marine picoplankton. *Nature* **371:**695–697.
27. **Devol, A. H.** 2003. Nitrogen cycle: solution to a marine mystery. *Nature* **422:**575–576.
28. **D'Hondt, S., S. Rutherford, and A. J. Spivak.** 2002. Metabolic activity of subsurface life in deep-sea sediments. *Science* **295:**2067–2070.
29. **Dickens, G. R.** 2003. Rethinking the global carbon cycle with a large, dynamic and microbially mediated gas hydrate capacitor. *Earth Planet. Sci. Lett.* **213:**169–183.
30. **Doolittle, R. F., D. F. Feng, S. Tsang, G. Cho, and E. Little.** 1996. Determining divergence times of the major kingdoms of living organisms with a protein clock. *Science* **271:**470–477.
31. **Edwards, K. J., P. L. Bond, T. M. Gihring, and J. F. Banfield.** 2000. An archaeal iron-oxidizing extreme acidophile important in acid mine drainage. *Science* **287:**1796–1799.
32. **Ehrlich, H. L.** 1985. The position of bacteria and their products in food webs, p. 199–220. *In* E. R. Leadbetter and J. S. Poindexter (ed.), *Bacteria in Nature*, vol. 1. *Bacterial Activities in Perspective*. Plenum Press, New York, N.Y.
33. **Flint, J. F., D. Drzymalski, W. L. Montgomery, G. Southam, and E. R. Angert.** 2005. Nocturnal production

of endospores in natural populations of *Epulopiscium*-like surgeonfish symbionts. *J. Bacteriol.* **187:**7460–7470.

34. **Fortin, D., G. Southam, and T. J. Beveridge.** 1994. Nickel sulfide, iron-nickel sulfide and iron sulfide precipitation by a newly isolated *Desulfotomaculum* species and its relation to nickel resistance. *FEMS Microbiol. Ecol.* **14:**121–132.
35. **Fossing, H., V. A. Gallardo, B. B. Jorgensen, M. Huttel, L. P. Nielsen, H. Schulz, D. E. Canfield, S. Forster, R. N. Glud, J. K. Gundersen, J. Kuver, N. B. Ramsing, A. Teske, B. Thamdrup, and O. Ulloa.** 1995. Concentration and transport of nitrate by the mat-forming sulphur bacterium *Thioploca. Nature* **374:**713–717.
36. **Francis, C. A., K. J. Roberts, J. M. Beman, A. E. Santoro, and B. B. Oakley.** 2005. Ubiquity and diversity of ammonia-oxidizing archaea in water columns and sediments of the ocean. *Proc. Natl. Acad. Sci. USA* **102:** 14683–14688.
37. **Fredrickson, J. K., J. M. Zachara, D. L. Balkwill, D. Kennedy, S. M. W. Li, H. M. Kostandarithes, M. J. Daly, M. F. Romine, and F. J. Brockman.** 2004. Geomicrobiology of high-level nuclear waste-contaminated vadose sediments at the Hanford site, Washington State. *Appl. Environ. Microbiol.* **70:**4230–4241.
38. **Fuhrman, J. A., K. McCallum, and A. A. Davis.** 1992. Novel major archaebacterial group from marine plankton. *Nature* **356:**148–149.
39. **Gaidos, E. J., K. H. Nealson, and J. L. Kirschvink.** 1999. Life in ice-covered oceans. *Science* **284:**1631–1632.
40. **Garcia-Pichel, F., M. Mechling, and R. W. Castenholz.** 1994. Diel migrations of microorganisms within a benthic, hypersaline mat community. *Appl. Environ. Microbiol.* **60:** 1500–1511.
41. **Gauci, V., E. Matthews, N. Dise, B. Walter, D. Koch, G. Granberg, and M. Vile.** 2004. Sulfur pollution suppression of the wetland methane source in the 20th and 21st centuries. *Proc. Natl. Acad. Sci. USA* **101:**12583–12587.
42. **Gevers, D., F. M. Cohan, J. G. Lawrence, B. G. Spratt, T. Coenye, E. J. Feil, E. Stackebrandt, Y. V. de Peer, P. Vandamme, F. L. Thompson, and J. Swings.** 2005. Re-evaluating prokaryotic species. *Nat. Rev. Microbiol.* **3:**733–739.
43. **Gilichinsky, D., E. Rivkina, C. Bakermans, V. Shcherbakova, L. Petrovskaya, S. Ozerskaya, N. Ivanushkina, G. Kochkina, K. Laurinavichuis, S. Pecheritsina, R. Fattakhova, and J. M. Tiedje.** 2005. Biodiversity of cryopegs in permafrost. *FEMS Microbiol. Ecol.* **53:**117–128.
44. **Giovannoni, S. J., M. S. Rappe, D. Gordon, E. Urbach, M. Suzuki, and K. G. Field.** 1996. Ribosomal RNA and the evolution of bacterial diversity. *Symp. Soc. Gen. Microbiol.* **54:**63–85.
45. **Gogarten, J. P., and J. P. Townsend.** 2005. Horizontal gene transfer, genome innovation and evolution. *Nat. Rev. Microbiol.* **3:**679–687.
46. **Golyshina, O. V., and K. N. Timmis.** 2005. *Ferroplasma* and relatives, recently discovered cell wall-lacking archaea making a living in extremely acid, heavy metal-rich environments. *Environ. Microbiol.* **7:**1277–1288.
47. **Guerrero, R., A. Haselton, M. Sole, A. Wier, and L. Margulis.** 1999. *Titanospirillum velox*: a huge speedy, sulfur-storing spirillum from Ebro Delta microbial mats. *Proc. Natl. Acad. Sci. USA* **96:**11584–11588.
48. **Gutell, R. R., N. Larsen, and C. R. Woese.** 1994. Lessons from an evolving rRNA: 16S and 23S rRNA structures from a comparative perspective. *Microbiol. Rev.* **58:**10–26.
49. **Hallam, S. J., N. Putnam, C. M. Preston, J. C. Detter, D. Rokhsar, P. M. Richardson, and E. F. DeLong.** 2004. Reverse methanogenesis: testing the hypothesis with environmental genomics. *Science* **305:**1457–1462.
50. **Hall-Stoodley, L., J. W. Costerton, and P. Stoodley.** 2004. Bacterial biofilms: from the natural environment to infectious diseases. *Nat. Rev. Microbiol.* **2:**95–108.
51. **Handelsman, J.** 2004. Metagenomics: application of genomics to uncultured microorganisms. *Microbiol. Mol. Biol. Rev.* **68:**669–685.
52. **Head, I. M., D. M. Jones, and S. R. Larter.** 2003. Biological activity in the deep subsurface and the origin of heavy oil. *Nature* **426:**344–352.
53. **Heath, J., E. Ayres, M. Possell, R. D. Bardgett, H. I. J. Black, H. Grant, P. Ineson, and G. Kerstiens.** 2005. Rising atmospheric CO_2 reduces sequestration of root-derived soil carbon. *Science* **309:**1711–1713.
54. **Holman, H.-Y. N., D. L. Perry, and J. C. Hunter-Cevera.** 1998. Surface-enhanced infrared absorption-reflectance (SEIRA) microspectroscopy for bacteria localization on geologic material surfaces. *J. Microbiol. Methods* **34:**59–71.
55. **Holmes, A. J., N. J. P. Owens, and J. C. Murrell.** 1995. Detection of novel marine methanotrophs using phylogenetic and functional gene probes after methane enrichment. *Microbiology* **141:**1947–1955.
56. **Huber, H., M. J. Hohn, R. Rachel, T. Fuchs, V. C. Wimmer, and K. O. Stetter.** 2002. A new phylum of Archaea represented by a nanosized hyperthermophilic symbiont. *Nature* **417:**63–67.
57. **Hugenholtz, P.** 2002. Exploring prokaryotic diversity in the genomic era. *Genome Biol.* **3:**reviews0003.1–0003.8. [Epub 29 January 2002.]
58. **Hungate, R. E.** 1985. Anaerobic biotransformations of organic matter, p. 39–96. *In* E. R. Leadbetter and J. S. Poindexter (ed.), *Bacteria in Nature*, vol. 1. *Bacterial Activities in Perspective*. Plenum Press, New York, N.Y.
59. **Inagaki, F., T. Nunoura, S. Nakagawa, A. Teske, M. Lever, A. Lauer, M. Suzuki, K. Takai, M. Delwiche, F. S. Colwell, K. H. Nealson, K. Horikoshi, S. D'Hondt, and B. B. Jorgensen.** 2006. Biogeographical distribution and diversity of microbes in methane hydrate-bearing deep marine sediments on the Pacific Ocean Margin. *Proc. Natl. Acad. Sci. USA* **103:**2815–2820.
60. **Isaksen, M. F., F. Bak, and B. B. Jorgensen.** 1994. Thermophilic sulfate-reducing bacteria in cold marine sediment. *FEMS Microbiol. Ecol.* **14:**1–8.
61. **Jannasch, H. W.** 1995. Microbial interactions with hydrothermal fluids. *Geophys. Monogr.* **91:**273–296.
62. **Jetten, M., M. Schmid, K. van de Pas-Schoonen, J. S. Damste, and M. Strous.** 2005. Anammox organisms: enrichment, cultivation, and environmental analysis. *Methods Enzymol.* **397:**34–57.
63. **Kaeberlein, T., K. Lewis, and S. S. Epstein.** 2002. Isolating "uncultivable" microorganisms in pure culture in a simulated natural environment. *Science* **296:**1127–1129.
64. **Kajander, E. O., and N. Ciftcioglu.** 1998. Nanobacteria: an alternative mechanism for pathogenic intra- and extracellular calcification and stone formation. *Proc. Natl. Acad. Sci. USA* **95:**8274–8279.
65. **Kaprelyants, A. S., J. C. Gottschal, and D. B. Kell.** 1993. Dormancy in non-sporulating bacteria. *FEMS Microbiol. Rev.* **104:**271–286.
66. **Kashefi, K., and D. R. Lovley.** 2003. Extending the upper temperature limit for life. *Science* **301:**934.
67. **Kell, D. B.** 2004. Metabolomics and systems biology: making sense of the soup. *Curr. Opin. Microbiol.* **7:**296–307.
68. **Kieft, T. L., and T. J. Phelps.** 1997. Life in the slow lane: activities of microorganisms in the subsurface, p. 137–163. *In* P. S. Amy and D. L. Haldeman (ed.), *The Microbiology of the Terrestrial Deep Subsurface*. CRC Press, New York, N.Y.
69. **Kluyver, A. J.** 1931. *The Chemical Activities of Micro-Organisms*. University of London Press, Ltd., London, United Kingdom.

70. **Kluyver, A. J., and C. B. van Niel.** 1956. *The Microbe's Contribution to Biology*. Harvard University Press, Cambridge, Mass.
71. **Knoll, A. H., M. J. Osborn, J. Baross, H. C. Berg, N. R. Pace, and M. Sogin.** 1999. *Size Limits of Very Small Microorganisms*. National Research Council, Washington, D.C.
72. **Konneke, M., A. E. Bernhard, J. R. de la Torre, C. B. Walker, J. B. Waterbury, and D. A. Stahl.** 2005. Isolation of an autotrophic ammonia-oxidizing marine archaeon. *Nature* **437:**543–546.
73. **Krause, D. O., W. J. M. Smith, F. M. E. Ryan, R. I. Mackie, and C. S. McSweeney.** 1999. Use of 16S-rRNA based techniques to investigate the ecological succession of microbial populations in the immature lamb rumen: tracking of a specific strain of inoculated *Ruminococcus* and interactions with other microbial populations in vivo. *Microb. Ecol.* **38:**365–376.
74. **Leadbetter, J. R.** 2003. Cultivation of recalcitrant microbes: cells are alive, well and revealing their secrets in the 21st century laboratory. *Curr. Opin. Microbiol.* **6:**274–281.
75. **Leadbetter, J. R., and J. A. Breznak.** 1996. Physiological ecology of *Methanobacter cuticularis* sp. nov. and *Methanobacter curvatus* sp. nov., isolated from the hindgut of the termite *Reticulitermes flavipes*. *Appl. Environ. Microbiol.* **62:**3620–3631.
76. **Leadbetter, J. R., T. M. Schmidt, J. R. Graber, and J. A. Breznak.** 1999. Acetogenesis from H_2 plus CO_2 by spirochetes from termite guts. *Science* **283:**686–689.
77. **Lepp, P. W., M. M. Brinig, C. C. Ouverney, K. Palm, G. C. Armitage, and D. A. Relman.** 2004. Methanogenic Archaea and human periodontal disease. *Proc. Natl. Acad. Sci. USA* **101:**6176–6181.
78. **Liesack, W., P. H. Janssen, F. A. Rainey, N. L. Ward-Rainey, and E. Stackebrandt.** 1997. Microbial diversity in soil: the need for a combined approach using molecular and cultivation techniques, p. 375–439. *In* J. D. Elsas, J. P. Trevors, and E. M. H. Wellington (ed.), *Modern Soil Microbiology*. Marcel Dekker, New York, N.Y.
79. **Lin, L.-H., G. F. Slater, B. Sherwood Lollar, G. Lacrampe-Couloume, and T. C. Onstott.** 2005. The yield and isotopic composition of radiolytic H_2, a potential energy source for the deep subsurface biosphere. *Geochim. Cosmochim. Acta* **69:**893–903.
80. **Lorenz, M. G., and W. Wackernagel.** 1994. Bacterial gene transfer by natural genetic transformation in the environment. *Microbiol. Rev.* **58:**563–602.
81. **Lowe, S. E., M. K. Jain, and J. G. Zeikus.** 1993. Biology, ecology, and biotechnological applications of anaerobic bacteria adapted to environmental stresses in temperature, pH, salinity, or substrates. *Microbiol. Rev.* **57:**451–509.
82. **Martiny, J. B. H., B. J. M. Bohannan, J. H. Brown, R. K. Colwell, J. A. Fuhrman, J. L. Green, M. C. Horner-Devine, M. Kane, J. A. Krumins, C. R. Kuske, P. J. Morin, S. Naeem, L. Ovreas, A.-L. Reysenbach, V. H. Smith, and J. T. Staley.** 2006. Microbial biogeography: putting microorganisms on the map. *Nat. Rev. Microbiol.* **4:**102–112.
83. **McCarty, P. L.** 2000. Novel biological removal of hazardous chemicals at trace levels. *Water Sci. Technol.* **42:**49–60.
84. **McFall-Ngai, M.** 1999. Consequences of evolving with bacterial symbionts: insights from the squid-vibrio associations. *Annu. Rev. Ecol. Syst.* **30:**235–256.
85. **McInerney, M. J.** 1986. Transient and persistent associations among prokaryotes, p. 293-338. *In* J. S. Poindexter and E. R. Leadbetter (ed.), *Bacteria in Nature*, vol. 2. *Methods and Special Applications in Bacterial Ecology*. Plenum Press, New York, N.Y.
86. **Meyerdierks, A., M. Kube, T. Lombardot, K. Knittel, M. Bauer, F. O. Glockner, R. Reinhardt, and R. Amann.** 2005. Insights into the genomes of archaea mediating the anaerobic oxidation of methane. *Environ. Microbiol.* **7:**1937–1951.
87. **Morita, R. Y.** 2000. Is H_2 the universal energy source for long-term survival? *Microb. Ecol.* **38:**307–320.
88. **Mulder, A., A. A. van de Graaf, L. A. Robertson, and J. G. Kuenen.** 1995. Anaerobic ammonium oxidation discovered in a denitrifying fluidized bed reactor. *FEMS Microbiol. Ecol.* **16:**177–184.
89. **Nealson, K. H., F. Inagaki, and K. Takai.** 2005. Hydrogen-driven subsurface lithoautotrophic microbial ecosystems (SLiMEs): do they exist and why should we care? *Trends Microbiol.* **13:**405–410.
90. **Nealson, K. H., and D. Saffarini.** 1994. Iron and manganese in anaerobic respiration: environmental significance, physiology, and regulation. *Annu. Rev. Microbiol.* **48:**311–343.
91. **Nealson, K. H., T. M. Schmidt, and B. Bleakley.** 1990. Biochemistry and physiology of *Xenorhabdus*, p. 271–284. *In* R. R. Gaugler and H. K. Kaya (ed.), *Entomopathogenic Nematodes in Biological Control*. CRC Press, Boca Raton, Fla.
92. **Nealson, K. H., and D. A. Stahl.** 1997. Microorganisms and biogeochemical cycles: what can we learn from layered microbial communities?, p. 5–31. *In* J. F. Banfield and K. H. Nealson (ed.), *Reviews in Mineralogy*, vol. 35. *Geomicrobiology: Interactions between Microbes and Minerals*. The Mineralogical Society of America, Washington, D.C.
93. **Nierman, W., J. A. Eisen, and C. M. Fraser.** 2000. Microbial genome sequencing 2000: new insights into physiology, evolution and expression analysis. *Res. Microbiol.* **151:**79–84.
94. **Nordstrom, D. K., and C. N. Alpers.** 1999. Negative pH, efflorescent mineralogy, and consequences for environmental restoration at the Iron Mountain Superfund site, California. *Proc. Natl. Acad. Sci. USA* **96:**3455–3462.
95. **Ogram, A.** 1998. Isolation of nucleic acids from environmental samples, p. 273–288. *In* R. S. Burlage, R. Atlas, D. Stahl, G. Geesey, and G. Sayler (ed.), *Techniques in Microbial Ecology*. Oxford University Press, New York, N.Y.
96. **Oliver, J. D.** 2005. The viable but nonculturable state in bacteria. *J. Microbiol.* **43:**93–100.
97. **Olsen, G. J.** 1994. Archaea, archaea, everywhere. *Nature* **371:**657–658.
98. **Olsen, G. J., D. L. Lane, S. J. Giovannoni, N. R. Pace, and D. A. Stahl.** 1986. Microbial ecology and evolution: a ribosomal RNA approach. *Annu. Rev. Microbiol.* **40:**337–366.
99. **Olsen, G. J., and C. R. Woese.** 1993. Ribosomal RNA: a key to phylogeny. *FASEB J.* **7:**113–123.
100. **Onstott, T. C., T. J. Phelps, T. Kieft, F. S. Colwell, D. L. Balkwill, J. K. Fredrickson, and F. Brockman.** 1999. A global perspective on the microbial abundance and activity in the deep subsurface, p. 487–500. *In* J. Seckbach (ed.), *Enigmatic Microorganisms and Life in Extreme Environments*. Kluwer Academic Publishers, Dordrecht, The Netherlands.
101. **Oremland, R. S., T. R. Kulp, J. S. Blum, S. E. Hoeft, S. Baesman, L. G. Miller, and J. F. Stolz.** 2005. A microbial arsenic cycle in a salt-saturated, extreme environment. *Science* **308:**1305–1308.
102. **Orphan, V. J., C. H. House, K. U. Hinrichs, K. D. McKeegan, and E. F. DeLong.** 2001. Methane-consuming archaea revealed by directly coupled isotopic and phylogenetic analysis. *Science* **293:**484–487.

103. **Overmann, J., J. T. Beatty, and K. J. Hall.** 1994. Photosynthetic activity and population dynamics of *Amoebo-bacter purpureus* in a meromictic saline lake. *FEMS Microbiol. Ecol.* **15**:309–320.
104. **Parkes, R. J., B. A. Cragg, S. J. Bale, J. M. Getliff, K. Goodman, P. A. Rochelle, J. C. Fry, A. J. Weightman, and S. M. Harvey.** 1994. Deep bacterial biosphere in Pacific Ocean sediments. *Nature* **371**:410–413.
105. **Parkes, R. J., G. Webster, B. A. Cragg, A. J. Weightman, C. J. Newberry, T. G. Ferdelman, J. Kallmeyer, B. B. Jorgensen, I. W. Aiello, and J. C. Fry.** 2005. Deep sub-seafloor prokaryotes stimulated at interfaces over geological time. *Nature* **436**:390–394.
106. **Parsek, M. R., and E. P. Greenberg.** 2005. Sociomicrobiology: the connections between quorum sensing and biofilms. *Trends Microbiol.* **13**:27–33.
107. **Pernthaler, J., and R. Amann.** 2005. Fate of heterotrophic microbes in pelagic habitats: focus on populations. *Microbiol. Mol. Biol. Rev.* **69**:440–461.
108. **Pfennig, N.** 1985. Stages in the recognition of bacteria using light as a source of energy, p. 113–129. *In* E. R. Leadbetter and J. S. Poindexter (ed.), *Bacteria in Nature*, vol. 1. *Bacterial Activities in Perspective*. Plenum Press, New York, N.Y.
109. **Pitta, T. P., and H. C. Berg.** 1995. Self-electrophoresis is not the mechanism for motility in swimming cyanobacteria. *J. Bacteriol.* **177**:5701–5703.
110. **Pledger, R. J., B. C. Crump, and J. A. Baross.** 1994. A barophilic response by two hyperthermophilic, hydrothermal vent archaea: an upward shift in the optimal temperature and acceleration of growth rate at supra-optimal temperatures by elevated pressure. *FEMS Microbiol. Ecol.* **14**:233–242.
111. **Plugge, C. M., E. G. Zoetendal, and A. J. M. Stams.** 2000. *Caloramator coolhaasii* sp. nov., a glutamate-degrading, moderately thermophilic anaerobe. *Int. J. Syst. Evol. Microbiol.* **50**:1155–1162.
112. **Poindexter, J. S., and E. R. Leadbetter.** 1986. Enrichment cultures in bacterial ecology, p. 229–260. *In* J. S. Poindexter and E. R. Leadbetter (ed.), *Bacteria in Nature*, vol. 2. *Methods and Special Applications in Bacterial Ecology*. Plenum Press, New York, N.Y.
113. **Polizzotto, M. L., C. F. Harvey, S. R. Sutton, and S. Fendorf.** 2005. Processes conducive to the release and transport of arsenic into aquifers of Bangladesh. *Proc. Natl. Acad. Sci. USA* **102**:18819–18823.
114. **Price, P. B., and T. Sowers.** 2004. Temperature dependence of metabolic rates for microbial growth, maintenance, and survival. *Proc. Natl. Acad. Sci. USA* **101**: 4631–4636.
115. **Radajewski, S., I. R. McDonald, and J. C. Murrell.** 2003. Stable-isotope probing of nucleic acids: a window to the function of uncultured microorganisms. *Curr. Opin. Biotechnol.* **14**:296–302.
116. **Ragatz, L., Z. Y. Jiang, C. E. Bauer, and H. Gest.** 1995. Macroscopic phototactic behavior of the purple photosynthetic bacterium *Rhodospirillum centenum*. *Arch. Microbiol.* **163**:1–6.
117. **Ram, R. J., N. C. VerBerkmoes, M. P. Thelen, G. W. Tyson, B. J. Baker, R. C. Blake, M. Shah, R. L. Hettich, and J. F. Banfield.** 2005. Community proteomics of a natural microbial biofilm. *Science* **308**:1915–1920.
118. **Ramsak, A., M. Peterka, K. Tajima, J. C. Martin, J. Wood, M. E. A. Johnston, R. I. Aminov, J. J. Flint, and G. Avgustin.** 2000. Unravelling the genetic diversity of ruminal bacteria belonging to the CFB phylum. *FEMS Microbiol. Ecol.* **33**:69–79.
119. **Rappe, M. S., and S. J. Giovannoni.** 2003. The uncultured microbial majority. *Annu. Rev. Microbiol.* **57**:369–394.
120. **Reguera, G., K. D. McCarthy, T. Mehta, J. S. Nicoll, M. T. Tuominen, and D. R. Lovley.** 2005. Extracellular electron transfer via microbial nanowires. *Nature* **435**: 1098–1101.
121. **Rondon, M. R., P. R. August, A. D. Bettermann, S. F. Brady, T. H. Grossman, M. R. Liles, K. A. Loiacono, B. A. Lynch, I. A. MacNeil, C. Minor, C. L. Tiong, M. Gilman, M. S. Osburne, J. Clardy, J. Handelsman, and R. M. Goodman.** 2000. Cloning the soil metagenome: a strategy for accessing the genetic and functional diversity of uncultured microorganisms. *Appl. Environ. Microbiol.* **66**:2541–2547.
122. **Sackett, M. J., J. P. Armitage, E. E. Sherwood, and T. P. Pitta.** 1997. Photoresponses of the purple non-sulfur bacteria *Rhodospirillum centenum* and *Rhodobacter sphaeroides*. *J. Bacteriol.* **179**:6764–6768.
123. **Schink, B., and M. Friedrich.** 2000. Phosphite oxidation by sulphate reduction. *Nature* **406**:37.
124. **Schippers, A., L. N. Neretin, J. Kallmeyer, T. G. Ferdelman, B. A. Cragg, R. John Parkes, and B. B. Jorgensen.** 2005. Prokaryotic cells of the deep subseafloor biosphere identified as living bacteria. *Nature* **433**:861–864.
125. **Schlegel, H. G.** 1993. *General Microbiology*, 7th ed. Cambridge University Press, London, United Kingdom.
126. **Schlesinger, W. H.** 2004. Better living through biogeochemistry. *Ecology* **85**:2402–2407.
127. **Schmidt, T. M., E. F. DeLong, and N. R. Pace.** 1991. Analysis of a marine picoplankton community by 16S rRNA gene cloning and sequencing. *J. Bacteriol.* **173**: 4371–4378.
128. **Schulz, H. N., T. Brinkhoff, T. G. Ferdelman, M. H. Marine, A. Teske, and B. B. Jorgensen.** 1999. Dense population of a giant sulfur bacterium in Namibian shelf sediments. *Science* **284**:493–495.
129. **Seitz, A. P., T. H. Nielsen, and J. Overmann.** 1993. Physiology of purple sulfur bacteria forming macroscopic aggregates in Great Sippewissett Salt Marsh, Massachusetts. *FEMS Microbiol. Ecol.* **12**:225–236.
130. **Siefert, J. L., and G. W. Fox.** 1998. Phylogenetic mapping of bacterial morphology. *Microbiology* **144**:2803–2808.
131. **Spear, J. R., J. J. Walker, T. M. McCollom, and N. R. Pace.** 2005. Hydrogen and bioenergetics in the Yellowstone geothermal ecosystem. *Proc. Natl. Acad. Sci. USA* **102**:2555–2560.
132. **Staley, J. T.** 1999. Bacterial biodiversity: a time for place. *ASM News* **65**:681–687.
133. **Staley, J. T., and J. J. Gosink.** 1999. Poles apart: biodiversity and biogeography of sea ice bacteria. *Annu. Rev. Microbiol.* **53**:189–215.
134. **Stanier, R. Y.** 1951. The life-work of a founder of bacteriology. *Q. Rev. Biol.* **26**:35–37.
135. **Stetter, K. O.** 1995. Microbial life in hyperthermal environments. *ASM News* **61**:285–290.
136. **Stevens, T. O., and J. P. McKinley.** 1995. Lithoautotrophic microbial ecosystems in deep basalt aquifers. *Science* **270**:450–454.
137. **Stolz, J. F., and R. S. Oremland.** 1999. Bacterial respiration of arsenic and selenium. *FEMS Microbiol. Rev.* **23**: 615–627.
138. **Strous, M., J. A. Fuerst, E. H. M. Kramer, S. Logemann, G. Muyzer, K. T. van de Pas-Schoonen, R. Webb, J. G. Kuenen, and M. S. M. Jetten.** 1999. Missing lithotroph identified as new planctomycete. *Nature* **400**:446–449.
139. **Strous, M., J. G. Kuenen, J. A. Fuerst, M. Wagner, and M. S. M. Jetten.** 2002. The anammox case—a new experimental manifesto for microbiological eco-physiology.

Antonie Leeuwenhoek Int. J. Gen. Mol. Microbiol. **81:**693–702.

140. **Tiedje, J. M., and J. L. Stein.** 1999. Microbial biodiversity: strategies for its recovery, p. 682–692. *In* A. L. Demain, J. E. Davies, R. M. Atlas, G. Cohen, C. L. Hershberger, W.-S. Hu, D. H. Sherman, R. C. Willson, and J. H. D. Wu (ed.), *Manual of Industrial Microbiology and Biotechnology*, 2nd ed. ASM Press, Washington, D.C.
141. **Tringe, S. G., C. von Mering, A. Kobayashi, A. A. Salamov, K. Chen, H. W. Chang, M. Podar, J. M. Short, E. J. Mathur, J. C. Detter, P. Bork, P. Hugenholtz, and E. M. Rubin.** 2005. Comparative metagenomics of microbial communities. *Science* **308:**554–557.
142. **Tyson, G. W., I. Lo, B. J. Baker, E. E. Allen, P. Hugenholtz, and J. F. Banfield.** 2005. Genome-directed isolation of the key nitrogen fixer *Leptospirillum ferrodiazotrophum* sp. nov. from an acidophilic microbial community. *Appl. Environ. Microbiol.* **71:**6319–6324.
143. **Ulrich, G. A., D. Martino, K. Bureger, J. Routh, E. L. Grossman, J. W. Ammerman, and J. M. Suflita.** 1998. Sulfur cycling in the terrestrial subsurface: commensal interactions, spatial scales, and microbial heterogeneity. *Microb. Ecol.* **36:**141–151.
144. **Uwins, P. J. R., R. I. Webb, and A. P. Taylor.** 1998. Novel nano-organisms from Australian sandstones. *Am. Mineral.* **83:**1541–1545.
145. **van de Peer, Y., M. Neefs, P. de Rijk, P. de Vos, and R. de Wachter.** 1994. About the order of divergence of the major bacterial taxa during evolution. *Syst. Appl. Microbiol.* **17:**32–38.
146. **van Niel, C. B.** 1949. The "Delft school" and the rise of general microbiology. *Bacteriol. Rev.* **13:**161–174.
147. **van Niel, C. B.** 1955. Natural selection in the microbial world. *J. Gen. Microbiol.* **13:**201–217.
148. **Ward, D. M., M. J. Ferris, S. C. Nold, M. M. Bateson, E. D. Kopczynski, and A. L. Ruff-Roberts.** 1994. Species diversity in hot spring microbial mats as revealed by both molecular and enrichment culture approaches—relationship between biodiversity and community structure. *NATO ASI Ser. Ser. G* **35:**33–44.
149. **Whitman, W. B., D. C. Coleman, and W. J. Wiebe.** 1998. Prokaryotes: the unseen majority. *Proc. Natl. Acad. Sci. USA* **95:**6578–6583.
150. **Wirsen, C. O., T. Brinkhoff, J. Kuever, G. Muyzer, S. Molyneaux, and H. W. Jannasch.** 1998. Comparison of a new *Thiomicrospira* strain from the mid-Atlantic ridge with known hydrothermal vent isolates. *Appl. Environ. Microbiol.* **64:**4057–4059.
151. **Wolfe, R. S.** 1992. Foreword, p. v–vi. *In* A. Balows, H. G. Truper, M. Dworkin, W. Harder, and K. H. Schleifer (ed.), *The Prokaryotes*, 2nd ed. Springer-Verlag, New York, N.Y.
152. **Yayanos, A. A.** 1995. Microbiology to 10,500 meters in the deep sea. *Annu. Rev. Microbiol.* **49:**777–805.
153. **Zengler, K., G. Toledo, M. Rappe, J. Elkins, E. J. Mathur, J. M. Short, and M. Keller.** 2002. Cultivating the uncultured. *Proc. Natl. Acad. Sci. USA* **99:**15681–15686.
154. **Zhulin, I. G., and J. P. Armitage.** 1993. Motility, chemokinesis, and methylation-independent chemotaxis in *Azospirillum brasilense*. *J. Bacteriol.* **175:**952–958.
155. **Zuckerkandl, E., and L. Pauling.** 1965. Molecules as documents of evolutionary history. *J. Theor. Biol.* **8:**357–366.

GENERAL METHODOLOGY

II

VOLUME EDITOR
JAY L. GARLAND

SECTION EDITOR
SEÁN P. O'CONNELL

Overview: General Microbiology

JAY L. GARLAND AND SEÁN P. O'CONNELL

4

If the only tool you have is a hammer, you tend to see every problem as a nail.

Abraham Maslow

A human being should be able to change a diaper, plan an invasion, butcher a hog, conn a ship, design a building, write a sonnet, balance accounts, build a wall, set a bone, comfort the dying, take orders, give orders, cooperate, act alone, solve equations, analyze a new problem, pitch manure, program a computer, cook a tasty meal, fight efficiently, die gallantly. Specialization is for insects.

Robert A. Heinlein (quoted from *Time Enough for Love*)

The last decades of the 20th century saw both a proliferation of new tools for studying microorganisms and an increased awareness of the global importance of microbial activities. The challenge for environmental microbiology in the new millennium is to develop a predictive understanding of microbial communities in order to develop effective strategies for important global issues such as greenhouse gas mitigation, clean water generation, pollution remediation, and soil maintenance (2). Predicting the behavior of complex ecological communities, difficult even when individual elements can be clearly identified such as is the case in plant and animal systems, is further complicated by the large numbers of total individuals and different types of microbes present. A gram of soil, for example, typically contains over 10 billion individual prokaryotic cells belonging to 4,000 different "species" (3) or more, depending on the type of analysis which is used (1). A tendency toward specialization, in terms of both the types of organisms studied and the techniques employed, is natural when faced with this great diversity of organisms and the increasingly broad range of available analytical tools. However, successful solutions to our global challenges in environmental microbiology will require multiple tools and perspectives to link physiological and genetic changes at the individual level to shifts in metabolism and composition of whole communities. The goal of this general methodology section is to foster the development of generalists who will not be limited, or biased, by their knowledge of available techniques.

The following chapters are meant as intellectual portals for different methodologies, leading to more detailed information contained in both the references and subsequent sections within this manual. Rather than specific protocols and complete inventories of methods, the conceptual basis and broad classification of different approaches are emphasized. The overall section organization and brief summaries of specific chapters are presented below.

The placement of the microscopy chapter at the beginning of the section is consistent with its role in the origins of environmental microbiology (see chapter 50). Technological advances in microscope lenses, automated image capturing, computerized analyses, and novel ways to examine samples with minimal disturbance in three dimensions allow for a less biased view of microbial communities. Techniques such as confocal laser scanning microscopy (CLSM) coupled with fluorescence of discrete molecules and/or real-time gene expression allow the modern environmental microbiologist to better infer relationships between the structure of a community and its function. This gap of understanding about which species are in a community and what they are doing, termed the Heisenberg uncertainty principle as it relates to microbial ecology (2), continues to be the bane of diversity studies.

The effort to bridge the chasm of understanding between identification of species and the role that they play in the natural world has historically relied upon the attainment of a pure culture, a single species that can be described from various biochemical, physiological, and molecular approaches. Pure cultures are necessary and an ultimate goal in microbiology; however, it is estimated that 1% or fewer of Earth's microorganisms are currently culturable. This is due to difficulty in reproducing in the laboratory the complex environmental parameters that support species, including interactions between populations as well as chemical growth factors (many that science has yet to discover) produced by other species. A major challenge for environmental microbiology is in understanding the degree of "imperfection" in nature, or, in other words, how much synergism is a necessary element of function within microbial communities. This is alluded to in the two chapters describing cultivation of bacteria and fungi (chapter 6) and algae and protists (chapter 7). Additionally, the role that viruses play in mediating interactions between species needs to be better understood, and their cultivation is discussed in chapter 8.

Wolfaardt et al. (chapter 9) describe the tradition in environmental microbiology of addressing Koch's postulates for ascribing function or effect to a single species and for the use of enrichment culturing to ultimately isolate guilds performing the same function. However, they rightly suggest that the above two approaches cannot be directly applied to the interpretation of a microbial community in situ, due to the loss of interactions between populations. In the chapter on cultivation of communities, a community culture is described as one that by design keeps the structure and function of a set of interacting populations intact; this allows for the community to be resilient and adaptive, and if measured at the scale of a biofilm, for example, to be directly observed by using tools available through CLSM. One major difference between enrichment and community cultures is that the former are chemically defined at the beginning and change with growth of the populations while the latter are able to be manipulated experimentally throughout the time of culturing.

The limitation of individual-based approaches which led to community cultivation techniques was also the rationale for developing community-level methods for characterizing microbial communities. These methods all bypass the cultivation and isolation approach and instead directly assess community properties on the basis of direct extraction and analysis of unique cell constituents (i.e., cell membrane constituents, nucleic acids) or simultaneous analysis of multiple physiological properties. The multivariate profile of cell constituents or physiological properties is used as a surrogate for species abundance matrices to assess spatial or temporal changes in community state without the bias of isolation. More specific information on the function and structure of the community can be developed from these assays as well, as summarized below.

Phospholipid ester-linked fatty acids (PLFA) analysis (chapter 10) generates information on the physiological state and composition of communities based on the extraction and identification of cell membrane lipids. Overall lipid phosphate provides an estimate of total viable biomass, while distinctive PLFA "biomarkers" have been defined for a variety of different microbial groups and physiological status. Pattern analysis of PLFA markers, even if not directly linked to different bacterial types, can be used to assess dynamics in microbial community structure. Stable isotopic analysis of PLFA signatures can provide information on carbon flow within microbial communities.

Community-level physiological profiling (CLPP) involves rapid assessment of multiple sole carbon sources to characterize and classify heterotrophic microbial communities (chapter 11). The multivariate profile of carbon source use is a proven tool for assessing community dynamics in response to spatial, temporal, or experimental gradients. While this approach originally used microtiter plates from Biolog, Inc., in which respiration of 95 separate sole carbon sources is detected by the reduction of a redox-sensitive dye, studies have shown that selective enrichment in the wells provides a biased, limited view of most microbial communities. More functionally relevant and less biased CLPP approaches involving more direct measurement of respiration (i.e., CO_2 production and O_2 consumption) at lower substrate concentrations and shorter incubation times have been recently developed. These newer techniques hold promise for rapid functional assessments of microbial communities, either for generating estimates of functional richness or for examining the effects of physiochemical factors (e.g., nutrient limitation, pH effects, pollutants) on activity.

Because of the rapid expansion of molecular biology tools since the last edition of this manual, two chapters (chapters 12 and 13) are necessary to discuss the use of DNA- and RNA-based methods for assessing community structure and phylogeny. It is possible to select from a host of techniques to examine species richness, evenness, and gene diversity and expression (chapter 12) as well as more recent phylogenetic approaches for determining the relatedness of species using 16S rDNA and other gene sequences and whole-genome approaches (chapter 13). In the former chapter, approaches for obtaining high-quality nucleic acids from environmental matrices are discussed, as well as methods for community fingerprinting and techniques for determining functions of species by DNA and RNA probing. Discrimination between spatial and temporal factors affecting communities is highlighted, and deduction of ecological functioning by determining phylogeny is discussed. In the latter chapter, phylogenetic approaches to hypothesize relatedness among species are introduced. The use of genes other than 16S rDNA is expanded upon, as well as the role of secondary structure in RNA-based phylogenetic analyses. Furthermore, the use of amino acid sequences for comparison of some species groups and whole-genome approaches for others are outlined.

Advancement in molecular techniques has also enabled the manipulation of the genomes of specific organisms to provide an assayable response to environmental stimuli. The resulting bioreporters, and their fusion to miniature electronics to form biosensors, are discussed in chapter 14. An increasing number of genetically engineered organisms containing either luminescent (i.e., *lux* cassette) or fluorescent (i.e., green fluorescent protein) genes have been produced. The light response is a sensitive measure of the presence and distribution of labeled organisms, or, if the genes are fused to specific promoter regions of interest, specific microbial activities which can be of interest in themselves or as indicators of the presence of certain substrates. These approaches offer increasing promise for in situ assessment of activity, especially when linked to the advanced imaging techniques described in chapter 5 or optics-based microprobes.

The remaining three chapters of the section move away from specific methods, addressing broader issues related to the effective implementation and analysis of research tools. Chapter 15 addresses the incorporation of microbiology into the long-term ecological research at an increasing number of dedicated study sites. Long-term research, necessary to understand a range of ecosystem dynamics (e.g., natural changes, rare events, disturbances), involves manipulative and observational approaches at habitat types spanning desert and aquatic sites from polar to tropical regions. In addition to the more traditional incorporation of microbial process measurements considered important to ecosystem processes, recent application of advanced techniques for evaluating community composition provides the potential for developing predictive links between microbial community composition, microbial activities, and ecosystem processes.

Integrating modern microbial ecology into the long-term research sites has synergistic benefits beyond the application of new tools; the strong focus on research themes and ecological questions of global importance promotes the scientific rigor and relevance of microbial ecology research. While the emerging set of analytical approaches provides an improved tool set for assessing microbial community dynamics, the methods are too often applied without critical examination of the underlying theory. For example, the dogma that diversity is beneficial is largely untested, yet is a common goal of studies employing the community-level methods defined above. Linking microbial ecology to the ecological community through the long-term study sites will help strengthen the linkage between microbiologists and the leading edge of current ecological theory.

The last two chapters address issues related to the reliability of data generated through research activities, either from the perspective of the confidence in the processes by which the data are collected (i.e., quality assurance [chapter 16]) or in the conclusions drawn from the data (i.e., statistical analysis [chapter 17]). Given the expense, in both equipment and manpower, required for many of the techniques outlined in this section, researchers need to diligently protect the quality of resulting data.

Chapter 16 emphasizes the general approaches necessary for environmental microbiology laboratories to provide a product or service (i.e., data) that meets defined standards of quality. These approaches focus on both the general system by which the laboratory operates (i.e., quality assurance) and the specific measurement processes (i.e., quality control). Quality assurance elements, including management requirements, training, and assessment, are addressed in relation to both large and small laboratories. A variety of quality control methods are discussed, with emphasis on the importance of the investigator in planning and selecting appropriate approaches.

The final chapter provides an overview of the challenging task of effectively linking conceptual models of the system under study to the appropriate statistical models used for measurement and testing. Careful integration of research objectives with the appropriate study design, sampling approaches, and statistical tests is stressed. The underlying theory of descriptive, associative, and comparative and confirmative approaches is discussed with specific reference to examples related to measurement of microbial community structure, abundance, and activity. The environmental microbiologist is encouraged to view the use of statistical tools with the same rigor as the application of the other research methods discussed in this section.

Pasteur is credited with saying that the role of the infinitely small in nature (microorganisms) is infinitely large. Our ability to see microbial cells, identify them, detect their activities, and link them to communities and processes on the visible scale is increasingly limited by our creativity and not our technology. The following section should serve as an overall guide to these technological advances in environmental microbiology.

REFERENCES

1. **Gans, J., M. Wolinsky, and J. Dunbar.** 2005. Computational improvements reveal great bacterial diversity and high metal toxicity in soil. *Science* **309:**1387–1390.
2. **Madsen, E. L.** 1998. Epistemology of environmental microbiology. *Environ. Sci. Technol.* **32:**429–439.
3. **Torsvik, V., J. Goksøyr, and F. L. Daae.** 1990. High diversity in DNA of soil bacteria. *Appl. Environ. Microbiol.* **56:**782–787.

Analytical Imaging and Microscopy Techniques

J. R. LAWRENCE, D. R. KORBER, AND T. R. NEU

5

Examination of microorganisms in their natural habitat may be achieved most effectively through the application of a variety of microscope-based techniques. As indicated by Murray and Robinow (174), microscopy provides the primary approach to the study of bacteria. In ecological studies a variety of methods, including buried slides (49), soil peels (264), and capillary methods (204), have been used in conjunction with microscopy to examine the diversity of microorganisms and their associations in natural systems. These approaches have provided a clear sense of the diversity of microorganisms, the presence of uncultureable organisms, and the existence of microenvironments and have thus provided a natural basis for many fundamental questions in microbial ecology. However, without the capacity to couple microscopy, image analysis, and techniques such as those that use fluorescent reporter molecules, it has not been possible to assess, for example, the microenvironment, the state of the organism, or its taxonomic affiliation. The development of environmentally sensitive molecules, probes, and molecular techniques, in conjunction with digital imaging techniques and appropriate computer software and hardware, provides the basis for quantitative scientific visualization in environmental microbiology. Digital imaging options are now extensive, encompassing confocal laser scanning microscopy (CLSM), two-photon laser scanning microscopy (2P-LSM), 4Pi microscopy, coherent anti-Stokes Raman, synchrotron-based techniques, and other emerging technologies. On this basis, traditional microscopic observation has made the transition into the discipline of analytical imaging. Any image may be subjected to a wide range of potential analyses to quantitatively extract the information contained in the image. These approaches are increasingly used by research groups worldwide to explore critical questions in microbial ecology and applied microbiology. The reader is also referred to current and future publications by the groups of S. Molin, M. Wagner, L. Blackall, M. Hausner and S. Wuertz, and P. Stewart for a wide variety of innovative applications of LSM to questions in applied and environmental microbiology.

MICROBIOLOGICAL SPECIMEN PREPARATION AND HANDLING SYSTEMS FOR MICROSCOPIC ANALYSIS

Model Systems for Cultivation and Analysis

A major goal of the use of microscopic techniques is to achieve minimum disturbance of the system under observation. Thus, for microscopy studies a variety of devices and methods that are amenable for the culture and study of bacteria and bacterial communities have been developed. These systems all represent attempts to re-create nature in the laboratory; and all succeed and fail to various degrees for observation, enumeration, activity, or biodiversity measurements.

A significant decision regarding light microscopy is the selection of the base microscope, whether it is a standard upright microscope or a microscope in an inverted configuration; in general, this is a compromise. Inverted microscopes provide easy access to the sample for application of probes, such as microelectrodes, whereas the upright microscope offers the full range of objective lenses, including water-immersible ones, and allows viewing of the sample without a glass interface, such as a coverslip.

Details of the preparation of living cell suspensions, such as wet mounts, hanging block mounts, and drop mounts, and the use of agar or gelatin slide cultures are discussed by Murray et al. (175) and Beveridge et al. (20). Cultural methods include the use of glass slide incubation (30, 49, 100, 241), capillary techniques (204), and agar block microculture (253). These authors, as well as Marshall (160), Brock (33), and Wimpenny (274), provide a selective guide to the original literature on approaches that are fundamental to many light microscope studies of microorganisms.

A variety of flow cells and microperfusion chambers have been used to study attached growth by microorganisms (37, 74, 77, 278, 279). These systems offer the advantages of the ability to use controlled but easily changed conditions, ease of in situ observation with microscopes and microelectrodes, the ability to apply fluorescent probes, and the ability to perform in situ antibody studies. The approach also allows easy examination of spatial and temporal changes in the material under examination particularly when it is used in conjunction with computer-controlled stages (see "Analytical Approaches" and "Statistical Considerations" below). Various designs of gel-stabilized systems have been described. Early developments of this approach, as well as its advantages and limitations, have been reviewed by Wimpenny (274). Rotating annular bioreactors (RABs) with removable substrata (82, 136, 182) are also useful systems for microscopy-centered studies. A variety of different substrata can be inserted into these reactors for the cultivation of biofilms, thereby modeling a diverse range of potential habitats or systems. In addition, Lawrence et al. (144) mounted silicon nitride windows in an RAB for the development of biofilms for scanning transmission X-ray microscopy and correlative microscopy. Various substrata may also be placed in streamside flumes (30 m long by 0.3 m wide by 0.30 m deep) and removed for observation, as performed by Battin et al. (18). Another chapter in this volume (chapter 9) and the work of Wolfaardt et al. (277) provide information regarding culture techniques suitable for use with light or laser microscopy. Some of these devices, such as gel-stabilized gradient plates (77, 277) and flow cells (279) and RABs (136), are easily constructed or are commercially available.

Physical Sectioning Methods

Microbial communities (e.g., biofilms developed in flow cells or in drinking water distribution systems) may be subjected directly to CLSM examination. However, many environmental biofilms develop over a long period of time and become much thicker and denser. In fact, laser penetration becomes a limiting factor for CLSM, and consequently, samples must be physically cut into slices. Some biofilm or aggregate samples may have a gel- or agar-like consistency, which may allow direct cutting with a scalpel. Others must be embedded, and several examples are given below. In any case, care must be taken to avoid and minimize the creation of artifacts.

The technique of cryosectioning has been described by Yu et al. (286) for analysis of microbial biofilms. Huang et al. (107) applied this technique to view the distribution of alkaline phosphatase activity in bioiflms. Cryosectioning was also used for examination of pure-culture *Pseudomonas aeruginosa* biofilms (269), river biofilms (18), and thick phototrophic biofilms (288). Cryosectioning may be very useful for relatively solid samples. However, loose, fluffy samples may cause problems, as they may not be transferable to a slide or fall apart after they are thawed. Hydrophilic resins (Nanoplast) that stabilize structures and minimize artifacts may also be used in conjunction with both electron microscopy and CLSM or 2P-LSM (69, 148). Usually, for fluorescence in situ hybridization (FISH), the sample must be fixed and dehydrated, resulting in a loss of the three-dimensional (3-D) architecture. In order to overcome this problem, several groups described techniques for stabilizing the spatial relationships of bacterial communities in combination with FISH. This involves fixation without the use of ethanol in combination with embedding in agar or polyacrylamide (51, 59). Paraffin has also been used to stabilize structures in large 1- to 3-mm anaerobic granules prior to thick sectioning for LSM analyses (214). Another embedding technique used agarose for stabilization of aquatic flocs (59, 72).

Air-water interfaces are also an important microbial habitat, and Henk (99) developed a method for the collection of neuston biofilms which involved coating a glass slide with a 2% collodion monomer film. Samples are then obtained by dipping these coated slides onto the air-water interface. Adherent cells may be analyzed by light, confocal, or electron microscopy.

A special challenge is the analysis of sediment samples. They may be collected undisturbed as a sediment core, but the samples fall apart as soon as they are transferred onto the microscope stage. Furthermore, they are highly reflective, which limits the penetration of laser light if, for example, LSM is applied. For sediment samples the so-called Cryolander was developed in order to collect undisturbed sediment samples, which may later be subjected to cryosectioning (273). Similarly, soil samples are difficult to examine microscopically. The extraction and embedding of soil materials for microscopic examination have been described by Bae et al. (15) and Tobin et al. (251). Waid described a method for obtaining soil films which can be examined by light and electron microscopy (264). This technique may result in the introduction of various artifacts, such as the collapse of structures and distortions; however, this can be minimized with CLSM, in which drying or embedding of the peels is not required.

SAMPLING IN MICROSCOPY

The Limits to Light Microscopy

Although proper object illumination and precise alignment of the microscope's optical elements are important factors, the selection of the microscope objective lens, matching condenser elements, and appropriate interference filters best suited to a particular application is a critical consideration. The ideal lens matches the resolution and magnification required, the type of material to be examined and how the sample is prepared, the nature of the supporting material (glass slide; wet or dry mount; flow cell; oil, water, or glycerol immersion; water immersible; thickness of coverslips, etc.), and what type of information is desired (e.g., cell numbers, cellular structure, specific fluor binding patterns, and gross morphology).

When determination of image detail is a concern, it is important that the resolution not be dependent on lens magnification but, rather, that it be dependent on the numerical aperture (NA) of the lens, according to the following relationship:

$$d = 0.61 \times \lambda/\mathrm{NA}$$

where λ is the light source wavelength, and d represents the minimum distance between two point-like objects which can be separated and identified. Usually, a combination of high magnification (e.g., ×60 or ×100) and high NA (e.g., 1.2 or 1.4) is desirable for microscopic analysis of cellular and subcellular detail. For larger objects or when gross morphological structure and arrangement are of interest, excellent lower-magnification lenses (e.g., ×50 with 1.0 NA and <×40 lenses), including water-immersible lenses and water-, air-, and oil-immersion lenses, may be obtained from the major microscope manufacturers.

Depth of field, the distance between the closest and the farthest objects in focus within a field of view, is also a consideration for light microscopy applications. For high-NA lenses, the depth of field is approximately calculated by using the following equation:

$$\text{depth of field (in micrometers)} = \lambda \text{ (in micrometers)}/[2(\mathrm{NA})]^2$$

The working distance, the range of distances where the image formed by the lens is clearly focused, is also a critical factor in considering objective lenses. In general, as the NA increases, the working distance decreases; thus, there are always trade-offs in selecting the optimum objective for any application. New microscope technologies such as 4Pi microscopy (which are now commercially available) promise increased NA and axial resolution (see "Emerging Microscopy Methods" below).

Brock (34) concluded that direct microscopy is rarely able to reveal the presence of small microorganisms at the densities found in natural systems and that overestimation of the significance of aggregates is likely to occur from uncritical use of the light microscope. However, direct microscopic examination of various habitats by the techniques outlined below reveals considerably more detail than was previously thought possible. This fact does not, however, excuse the user from uncritical application of the technology.

Analytical Approaches

Traditionally, one would count by "eye" the numbers of objects stained with a specific probe, although this approach is best confined to relatively simple two-dimensional samples. The manual analysis of thick, complex samples such as aggregates or biofilms is tedious and is usually subject to serious errors of underestimation. LSM imaging provides a series of images that are amenable to automated and semiautomated analyses. In addition, a variety of software is available for extraction of data from LSM image stacks (see "Image Processing" below). However, a fundamental question exists: how are the numbers of objects in the microscope field or image estimated or counted? One solution is the use of techniques which do not require single-cell resolution. One example is to calculate the object area that is reactive with a particular probe, as has been described by Lawrence et al. (133) for the determination of algal and bacterial biomass or by Neu et al. (185) for the estimation of exopolymer biovolume by using LSM image stacks. Another solution is the application of stereological counting procedures. In this approach the volume fraction of a phase of interest (which is part of a larger reference volume) can be determined in a common microscope by using oculars equipped with point grids (249). As in the use of image area, objects stained by a specific fluorescent probe become the phase of interest. The approach is demonstrated in studies by Schmid et al. (223) and Daims et al. (61), who applied this method to analyze the bacterial communities of activated sludge flocs and other environmental samples, respectively. A related method, the point intercept method, may also be used to quantify the biovolume of filamentous objects (176).

Although the actual imaging of samples may be carried out effectively through manual adjustment of the microscope stage, there are situations in which a motorized computer-controlled stage is desirable. Software executed by a personal computer (PC) allows the user to return to sites of previous analysis with reasonable accuracy (within ±1 to 3 μm). This permits the microscopist to conduct temporal studies at a large number of sampling sites (e.g., down multiple transects). This also facilitates the conduct of microbiological analyses with molecular probes with different emission spectra, as the user may first observe a number of sites stained with one compound and then make subsequent observations after the second probe has been added. CLSM systems typically come equipped with *xz* computer-controlled stages. Thus, when 3-D positioning is used in conjunction with computer-controlled *xy* positioning systems, it may be performed accurately and precisely. In addition to saving previous positions of analysis, most stage-controlling software or the CLSM instrument software allows construction of analysis grids. This enables the recording of mosaic scans with large areas of interest, as well as the immediate reconstruction of large montages consisting of multiple 3-D image series. This capability facilitates the application of statistical or geostatistical techniques for the analysis of microbial communities or systems (125, 169, 182) (see below). Dandurand et al. (63) provide a useful example of the application of geostatistics to examine the distribution and variability of bacteria in the rhizoplane.

Statistical Considerations

Reproducibility within and between experiments involving microbial communities is a challenge. The question of what constitutes a representative sample (e.g., the number of fields and the total area of analysis) has been addressed previously (34). Indeed, Kuehn et al. (128) suggested that the limits of data variability were potentially influenced by many parameters, including bacterial species, growth inhibitors or stimulators, microbial behavioral factors, and experimental design. Heydorn et al. (101) noted that biofilm development is a stochastic process and, thus, can never be expected to be entirely reproducible. Thus, the challenge is to effectively sample and encompass the variability in microbial communities, particularly in microscopy studies.

Image processing and analysis enable detailed measurements of a range of parameters to be performed; however, most microscopy studies have relied on relatively small areas of analysis and have tended to ignore experimental reproducibility. The data thus obtained may be prone to errors, based on rare events resulting from insufficient or selective sampling. Computer-controlled stages have aided in the development of statistical methods for biofilm analysis. Korber et al. (124, 125) used a computer-controlled microscope stage and CLSM imaging to construct montages of contiguous regions of *Pseudomonas fluorescens* biofilms. Elements or subsets of the large biofilm montage were analyzed to define the area required to provide representative biofilm biomass data (124). It was found that the spatial variability contained within pure-culture *P. fluorescens* biofilms was large, requiring analysis areas exceeding $>1 \times 10^5$ μm^2 for statistically representative results. Similarly, Møller et al. (169) examined contiguous areas $>2.8 \times 10^5$ μm^2 during a CLSM architectural study of a degradative biofilm community. Murga et al. (173) also consider aspects of quantitative analyses of microscope samples, while Yang et al. (283) provide a valuable discussion of the application of image analyses to quantitative analyses of microbial biofilms. Following these principles, we have performed microscale analyses of microbial communities and used analysis of variance to detect significant changes in a number of parameters of interest (130, 139). Heydorn et al. (101) used the Comstat program to quantify and statistically analyze the structure of *Pseudomonas* biofilms using an analysis of variance model. In practice, determinations of representative samplings for microscopic study are not easily made.

Ethics in Digital Imaging

Manipulation and falsification of paintings and photographs have a long tradition. Particularly with digital images and the current programs available, the temptation to improve original data sets is very high and improvement of a data set is easy to perform (218). For this reason ethical considerations have been put forward in order to define a standard for images in journalism as well as in science. The following is a brief list of rules for scientific images (additional details may be found at http://swehsc.pharmacy.arizona.edu/exppath/micro/digimage_ethics.html):

- scientific digital images represent data;
- for comparison, images should be acquired under identical conditions;
- intensity measurements should be done with raw data;
- manipulation should be always done with a copy of the raw data set;

- simple adjustments of the entire data sets (i.e., contrast) are usually acceptable;
- cropping of a data set is usually acceptable;
- manipulations in one area of the data set are questionable;
- the use of software filters is usually not recommended;
- cloning of objects into data sets is very questionable;
- avoid the use of lossy compressions (JPEG images);
- be aware of magnification (microscope lens) and resolution (pixels in image); and
- be careful when changing the size (pixels).

It has been clearly pointed out that even the most sophisticated image-processing techniques cannot improve a poor-image data set. Furthermore, any digital image processing will change the data set and will not add any new information. Consequently, the first aim in scientific digital imaging is to collect an original data set of the highest possible quality. In this case the data set can be directly used for visualization or quantification. It is also very important to see the data set in different stages of processing, for example, in CLSM by looking at (i) the original stained sample in the microscope, (ii) the digital image after recording on the CLSM monitor, (iii) the results of digital image analysis (visualization and quantification), and (iv) the output for presentation or publication purposes. Some authors manipulate only the contrast and brightness if the color hues of the original image from a flat-screen monitor cannot be achieved during printing. Others simply include in their publications a statement that "no improvements" have been made (see reference 242).

CONVENTIONAL AND NOVEL LIGHT MICROSCOPY

The modern microscopist has an array of digital tools which extend the range and extent of possible analyses; however, the quality of the extracted data is still dependent on the quality of the primary image. Thus, the principles of basic light microscopy play a critical role in the collection of microscopic images for digital analysis. The various microscopic techniques, dark-field, phase-contrast, interference, fluorescence, and the others described below, provide various means of enhancing the contrast between microorganisms and their environment, thus enabling the investigator to visualize them. Surman et al. (245) compared various microscopic techniques of general application to microbial ecology. The reader is referred to Murray and Robinow (174) and Beveridge et al. (20) for reviews on microscopy fundamentals. Information specific to fluorescence microscopy, including autofluorescence, fluorescent staining, and specimen preparation and handling, may be found in the volumes by Rost (219). The first two editions of the *Manual of Environmental Microbiology* also provide useful details on the variety of methods available for conventional light microscopies.

Typically, images are obtained by transmitted, incident, or polarizing light microscopy techniques (160). However, a variety of novel techniques have been recorded in the literature. For example, Casida observed microbes in soil systems using the following: (i) color infrared photography for microscopic visualization of nonstained microorganisms (42), (ii) a modified metallurgical reflected light microscope with a floating stage (43), and (iii) a continuously variable-phase and amplitude-contrast system microscope (41) and an interval scanning microscope (44). Polarizing microscopy has been proposed for use in combination with phase-contrast microscopy or epifluorescence microscopy for examination of relationships between soil particles and microorganisms (160). Interference-reflectance microscopy may also provide information regarding attached microorganisms on solid substrata. A modification of this approach was used by Vigeant et al. (259), who used internal reflection aqueous fluorescence microscopy to measure the distances between bacteria and a test surface, providing an indication of the adhesion strength of the microorganisms. Surman et al. (245) outlined the application of Hoffman modulation-contrast microscopy, an adaptation of bright-field microscopy. Although these approaches are little used by microbial ecologists or environmental microbiologists, when they are coupled to current camera and image analysis systems, they may have considerable potential for use in studies of microorganisms in their natural habitats.

LASER SCANNING MICROSCOPY SYSTEMS

LSM fuses traditional light microscope hardware with a laser light source, a photomultiplier or charge-coupled device (CCD) detectors, and computerized digital imaging. Details about the lasers, detectors, and other options are best obtained from the specific manufacturers. General considerations have also been discussed by Wilkinson and Schut (272), Conn (55), and Pawley (201). In general, there are a few configurations: (i) the real-time instruments, which provide video rate image collection but not high-resolution optical sectioning, and (ii) point-scanning systems, which are slower but maximize lateral and axial resolution. Point-scanning microscopes are either single-photon CLSMs, which use classical pinholes or apertures to achieve an optical section, or 2P-LSMs, which use the physics of light absorption in the focal plane to obtain optical thin sections. The common lasers and combinations include argon, krypton, krypton-argon, helium, helium-neon, UV, and infrared. The *Handbook of Biological Confocal Microscopy* is also an excellent source of information on nearly all aspects of confocal microscopy (202). For a summary list of important operational points, see the work of Pawley (201).

CLSM Operation

The setup and operating procedures of CLSM systems vary with the model of the microscope. Lawrence and Neu (135) provide detailed instructions related to the operation and maintenance of LSM systems. In general, (i) the system should be turned on and allowed to warm up for about 30 min prior to imaging. This is critical with Kr-Ar systems to ensure a stable far-red (580-nm) excitation line. (ii) The working alignment must be confirmed prior to each use (alignment is particularly critical when one is working with bacteria, where a misalignment of 1 μm cannot be ignored). Test samples provided by the manufacturer should be used to check the alignment of mirrors, laser beams, beam intensity, and image quality. It is recommended that users make up slides consisting of fluorescent beads. Manufacturers provide a variety of microsphere reference standards to facilitate the adjustment and calibration of conventional fluorescence microscopes and CLSMs. These standards include beads of various intensities of fluorescence, wavelengths of emission, and sizes for the routine assessment of instrument performance. The use of Focal Check beads is particularly useful to ensure that images obtained from the same location but with different excitation-emission combinations (multiparameter imaging or colocalization studies) are in perfect register. (iii) The correct filters or settings should be selected relative to the fluorochromes used. (iv) The correct laser source and intensity (intensity may be controlled by neutral density filters or acousto-optically) should be used for excitation of the fluor (e.g., 488-nm excitation for fluorescein). (v) The user should ensure that the objective lens selected has sufficient magnification, resolution, and working distance for the material to be imaged. (vi) Photomultipliers and detectors should be adjusted so that their black level is at the point where the boundary between scans is just visible against the dark background. (vii) The gain and the black level of the detectors should be adjusted to use the entire grey scale. (LSM systems usually include an indicator which shows under- and oversaturated pixels in different colors on the display monitor, which allows easier adjustment.) (viii) The computer monitor should be checked to ensure that it has properly adjusted brightness and contrast with the displayed image centered and sized correctly. As pointed out by Zucker and Price (290), many simple everyday instruments are developed and tested according to national or international standards. However, highly advanced and expensive LSMs are sold without any standard in terms of instrument performance. As a consequence, a collection of useful tests for LSMs has been described by Zucker and coworkers (149, 290).

LSM Image Collection

LSM systems provide the opportunity to image samples in a variety of modes and in *xy* and *xz* dimensions as single images or image stacks for 3-D reconstruction. In addition, images may be obtained by using reflection, fluorescence, and transmitted light. Of greatest interest is the capacity to colocalize signals resulting from reflection, transmission, and multiple fluorescent stains. The major guidelines for all types of LSM imaging are to obtain an image using the lowest-intensity laser and the smallest-pinhole aperture to minimize photodamage, optimize image quality (i.e., signal-to-noise ratio), and minimize the thickness of the optical section. Examples of typical images collected in the *xy* or *xz* planes are shown in Fig. 1. These show *xy* images taken in biofilm communities by using negative staining (38) so that cells appear as dark objects (Fig. 1A) and positive staining (Fig. 1B). These images show both the *xy* (bottoms of Fig. 1A and B) and the *xz* (tops of Fig. 1A and B) images separated by a black line (the coverslip) of a biofilm and illustrate the information content (i.e., surface roughness and depth variations) that can be seen in images obtained through both *xy* and *xz* optical sectioning techniques. *xz* images lack the resolution of *xy* images because objective lenses have poor axial resolution. Although high-NA lenses (NA, 1.4) can significantly improve the axial resolution, it is still five times poorer than the lateral resolution. It is pos-

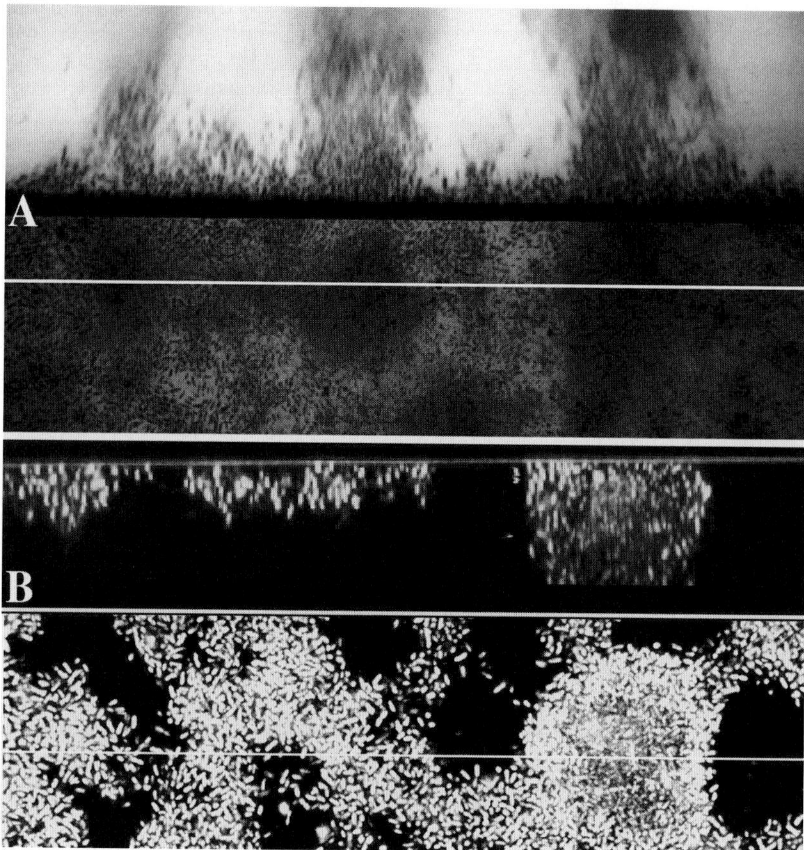

FIGURE 1 Images may be collected in the *xy* or *xz* plane, or a series of images may be collected in either the *xy* or *xz* plane. These photomicrographs show images collected in both the *xy* and *xz* planes in a biofilm community by using negative staining so that cells appear as dark objects (A) or positive staining so that cells appear as bright objects on a dark background (B). The location of the *xz* section is shown by the location of the horizontal line in each *xy* image in each panel.

sible to calculate *xz* images from *xy* serial sections, which can be advantageous since it reduces photodamage to the specimen.

CLSM systems are usually able to collect data in the fluorescence, reflection, and nonconfocal transmission modes. Typically, current systems allow the collection of data for four channels sequentially or simultaneously by using different excitation and emission wavelengths. When all the options, including reflection and transmission, are applied, data for a total of six channels, each of which reports data on a different parameter, may be obtained from a single location.

Reflection imaging may also be performed by using CLSM systems in the confocal mode. Information is obtained from regions where differences in the refractive index within the specimen result in the reflection of laser light. Confocal reflection contrast images may be obtained by CLSM and provide images superior to those provided by conventional interference microscopy. The combination of reflectance imaging with colloidal gold labeling also has been used in cell biology (231). Fig. 2 shows a reflection image of colloidal metal precipitates deposited within an exopolymer matrix. It is also possible to equip the CLSM system with a nonconfocal detector system for scanned transmitted light imaging, as illustrated by Packroff et al. (199), who used combinations of fluorescent staining and transmission imaging to observe protozoan populations. Lawrence et al. (141) provide an illustration of multichannel imaging showing the images from three separate channels of a CLSM image stack, including autofluorescence (algal chloroplasts), reflection imaging of colloidal particles, and images of exopolymer labeled with fluor-conjugated lectin (Fig. 2). One of the great benefits of dual or multichannel imaging is that the images obtained are in exact spatial register, facilitating multichannel display, fluorescence ratio imaging, stereoscopic imaging, and 3-D image reconstruction. The application of pseudocoloring can also be useful for the display of information from ratiometric images of pH, E_h, or ion concentrations or to represent the results from the application of multiple fluors.

Images may be acquired either as a direct image or through a mathematical filter. A number of options are available, including line averaging and frame averaging. However, collection through mathematical filtration reduces the information content of the image for purposes such as ratiometric analysis. Thus, the user should define the end use of the images prior to collection. Other mathematical collection filters are offered by the various manufacturers. These filters allow image enhancement through summation of signals, collection to a preset maximum signal, local contrast enhancement, median filters, linear smoothing filters, and gradient filters (202). Each of these filters attempts to reduce the noise levels in images, smooth the data set, and enhance the edges of objects within the image. Deconvolution filters may also be applied to the confocal image to further enhance their clarity (see below).

Two-Photon Laser Microscopy Option

CLSM with one-photon excitation has many advantages. Nevertheless, there are some limitations, especially with samples with the following challenges: photosensitivity, fading, scattering, laser penetration, and/or background fluorescence (187, 188). These may be overcome through the use of 2P-LSM or multiphoton LSM. 2P-LSM uses a high-peak-power infrared laser with an extremely short pulse in the femto- or picosecond range to produce a high photon density. Thereby, two or more photons are able to interact simultaneously with a fluorescent dye molecule. This effect occurs only in the focal plane, providing highly resolved optical sections without the need for a pinhole. In normal fluorescence excitation with one photon, a quantum of light is adsorbed by a molecule which is excited to a higher molecular state, followed by emission of longer-wavelength fluorescent light. In the case of two-photon microscopy, light of a longer wavelength is used during excitation such that a single photon has insufficient energy to excite the

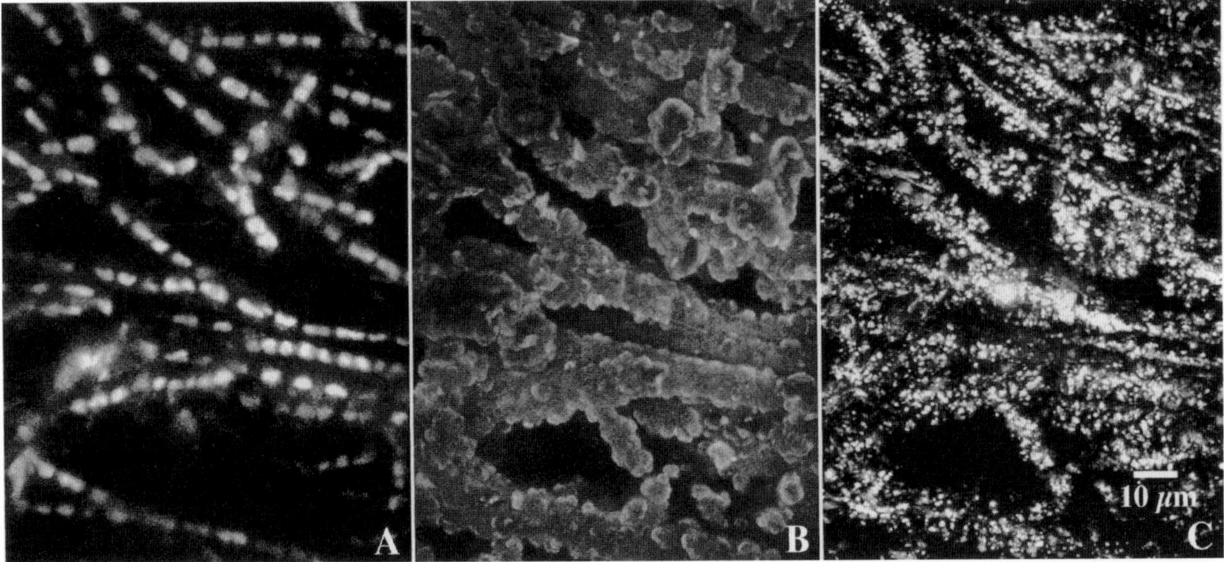

FIGURE 2 A series of confocal micrographs taken at the same location in a sample of an algal mat. These *xy* images provide an illustration of multichannel imaging showing (A) an autofluorescence signal (algal chloroplasts), (B) reflection imaging of colloidal particles, and (C) an image of exopolymer labeled with the fluor-conjugated lectin concanavalin A-FITC. Reprinted from reference 141 with permission from the publisher.

molecule to the required energy level. However, if a second photon arrives within a sufficiently short time interval, a cooperative multiphoton excitation process results. As two-photon excitation is restricted to the region of highest intensity of the incident laser light (focal plane), it minimizes photobleaching in out-of-focus areas and coincident background fluorescence. Most of the major manufacturers offer 2P-LSM as an optional setup with their conventional CLSM instruments. It should be pointed out that 2P-LSM is not a solution for every difficult sample. Certainly, the advantages of 2P-LSM can improve the imaging of microbiological samples. However, environmental samples are often very thick and dense. Indeed, if samples approach 500 μm or more, the solution is to embed and physically section the sample prior to imaging from the side (188, 269, 288).

OBJECTIVE LENSES

The objective lens selected for imaging is a major factor in successful optical sectioning. The limitation for all objective lenses, particularly when they are applied for CLSM imaging, is that the axial or *z*-dimension resolution is poor relative to the lateral resolution. The new lenses provided by most manufacturers and available for laser microscopy are corrected for spherical and chromatic aberrations from approximately 450 to 1,100 nm for typical CLSM applications and may be corrected down to 350 nm for UV applications. Chromatic and spherical aberrations have the greatest impact on confocal imaging performance, and thus, plan-apochromatic lenses (which have the least aberration) are preferred and recommended for LSM imaging. Plan-apochromats are high-NA (1.2 to 1.4) water- and oil-immersion lenses that provide the highest-resolution confocal images but have a limited working distance, although some of the water-immersion lenses designed for CLSM allow imaging through up to 220 μm of biological material. The relationship between magnification, NA, and working distance is illustrated by the following data: ×10, 0.4, and 3.1 mm, respectively (Nikon); ×20, 0.7, and 0.65 mm, respectively (Nikon); ×40, 0.85, and 0.2 mm, respectively (Nikon); ×63 water immersion, 1.2, and 0.24 mm, respectively (Zeiss); ×63 oil immersion, 1.4, and 0.18 mm, respectively (Zeiss); and ×63 water immersible, 0.95, and 2.2 mm, respectively (Zeiss). Optimal results are usually obtained only by using fixed stained materials, optically appropriate mounting media, corrections for coverslip thickness and temperature, and high-quality coverslips.

In environmental studies, it is usually necessary or desirable to apply CLSM to living fully hydrated samples; thus, the use of other lenses must be considered. It is possible to image through several hundred micrometers of biofilm with a ×60 to ×63 water-immersible lens with an NA of 0.9 to 1.0 and up to 1 mm and more when using extra-long-working-distance lenses (×20 or ×40) or ×40 water-immersible lenses. Furthermore, they can be used for direct observation without the need to use a coverslip. The ×40, 1.0-NA objective lenses have been reported to give working distances of 300 to 400 μm. When lenses with an NA of 0.2 or less are used, the thickness of the optical thin section (depth of focus, see "Sampling in Microscopy" above) is approximately 10 μm; however, this value drops rapidly to less than 1 μm when objectives with NAs of >0.6 are used. For practical purposes, images cease to be confocal when the NA of the objective lens is <0.6. Electronic enlargement (zooming) of objects up to ~100 times can be achieved by using the ×60, 1.4-NA objective lens without a decrease in resolution. To ensure Nyquist sampling during zooming operations, the pixels should be approximately two times smaller than the smallest object imaged in the specimen. Clearly, the selection of objective lenses involves trade-offs between many operational parameters, as considered by Dunn and Wang (73). The reader is referred to the work of Pawley (202) for an in-depth discussion of the choice of objective lenses for CLSM applications. Studies such as those by Neu and Lawrence (182, 183) and Lawrence et al. (130, 133, 136, 141, 144, 145) illustrate the application of water-immersible lenses to a variety of biological materials and substrata.

FLUORESCENT PROBES

Probes useful for fluorescence microscopy may be divided into three different types. Intrinsic probes are already present inside the sample (e.g., pigments such as chlorophyll or phycoerythrins and phycocyanins) (180). Extrinsic probes are those which bind directly to a target (e.g., the general nucleic acid stains, such as 4′,6′-diamidino-2-phenylindole [DAPI] or the SYTO series). Extrinsic covalently bound probes are usually high-molecular-weight molecules with a high specificity but no fluorescence (e.g., antibodies, lectins, and gene probes). A fluorochrome with a selectable excitation-emission range will be covalently bound to these probes according to the procedure outlined by the supplier of the staining kit. The potential fluorochromes may also be distinguished according to their application (e.g., binding to specific cell structures, probing of microhabitats inside or outside of cells, recording of cell or enzyme activity, and phylogenetic identification of cells). For microbiological samples, the use of a structured approach has been suggested for analysis of an unknown sample by taking advantage of sample properties and different fluorochromes (179).

Fluors have unique and characteristic spectra for excitation and emission. In most cases, there are published absorption-excitation and emission spectra that show the relative intensity of the fluorescence in relation to the wavelength of light being applied. This information is fundamental to the selection of fluors and fluor combinations for application in fluorescence and LSM studies.

In fluorescence microscopy and CLSM, there have been many potential fluorochromes applied as single stains. However, it is often desirable to use several fluorochromes for multiple staining (e.g., for three-channel imaging, with fluorescein isothiocyanate [FITC; green], tetramethyl rhodamine isocyanate [TRITC; orange], and Cy5 [far red] as one possible combination). Currently, there are numerous fluorochromes for different applications. Many of them are offered by one major supplier (93). However, many other sources offer fluorochromes, often for specific applications. Some companies offer antibodies and lectins labeled with fluorochromes. Another specific area is the synthesis and labeling of gene probes to be used for FISH (see chapter 12). Tables 1 to 3 provide lists of common fluorescent probes, their targets, and applications. There are also probes, termed quantum dots (Qdots), which have major advantages over conventional probes (e.g., they are more stable toward bleaching and have narrow emission spectra). Qdots have been applied to a variety of microbiological samples (35, 90, 116, 287). Overall, the critical factors to be considered in the selection of fluors include appropriate excitation and emission wavelengths, a detectable yield of visible light, photostability (resistance to bleaching), toxicity, penetration of the sample, and probe target specificity.

TABLE 1 Nucleic acid-specific stains and their applications

Nucleic acid stain(s)	Application(s)	Reference(s)
AO and DAPI	Various	113
TOTO-1 and TO-PRO-1	Planktonic bacteria Marine prokaryotes	
YOYO-1 and YO-PRO-1	Marine prokaryotes	
POPO-3	Bacteria in geological samples	257
PicoGreen	Marine prokaryotes Bacteria in lake and seawater	
SYBR green I and II	Marine picoplankton Marine viruses and bacteria Soil and sediment bacteria	
SYTO blue, green, and red	River biofilms Lake plankton River snow	182, 187 181, 185

IMAGE PROCESSING

In a pioneering review of digital image analysis in microbial ecology, Caldwell and coworkers summarized the early approaches for analyzing image data sets of microbial communities (39). More recently, the different approaches of image analysis in microbiology were summarized in a book (272). In microbiology digital image processing has been applied in various ways, depending on the type of sample analyzed. Basically, four types of tools have been developed: (i) analysis of single cells in liquid samples (plankton), (ii) analysis of single cells on surfaces (adhesion process, shape), (iii) analysis of filamentous structures (cyanobacteria, fungi), and (iv) analysis of microbial communities in three dimensions (aggregates, biofilms). Traditionally, planktonic samples were analyzed by microscopy in combination with a camera and image analysis. Nowadays, and more conveniently, suspended cells are analyzed by flow cytometry (for reviews, see references 64, 210, and 261). The adhesion process and the detachment of bacteria were mainly investigated in flow cells designed for direct microscopy. Early studies were focused on single cells or monolayers of bacteria (for a review, see reference 28). In the meantime, these flow cell studies have been extended to 3-D biofilms by the use of green fluorescence protein (GFP) labeling techniques with pure cultures or defined mixed cultures (see, for example, reference 51). In addition, environmental microbial communities are studied directly after they are either collected from the natural habitat or subcultivated in laboratory

TABLE 2 Exopolymer stains and their applications

Lectin	Binding specificity (major only)	Application(s)	Reference(s)
Canavalia ensiformis	Methyl-α-D-mannopyranoside D-(+)-Mannose	Marine biofilms	164
Helix pomatia	*N*-Acetylgalactosamine	Marine biofilms	164
Limulus polyphemus	*N*-Acetylneuraminic acid	Marine biofilms	164
Arachis hypogaea	D-(+)-Galactose	River biofilm	181, 182
Erythrina cristagalli	*N*-Acetyl-D-galactosamine, D-galactose	River flocs	182, 185
Ulex europaeus	L-(−)-Fucose	River flocs	182, 185
Triticum vulgaris	*N,N'*-Diacetylchitobiose, *N,N',N''*-triacetylchitotriose	Grazing of biofilms, methodology	133, 141, 187
Erythrina cristagalli	*N*-Acetylgalactosamine, D-(+)-galactose	Degradative biofilms	280
Glycine max	*N*-Acetylgalactosamine, D-(+)-galactose	Contaminants in biofilms	187, 280
Lycopersicum esculentum	*N,N'*-Diacetylchitobiose, *N,N',N''*-triacetylchitotriose	Degradative biofilms	280
Vicia faba	D-(+)-Mannose, D-(+) glucose	Degradative biofilms	280
Tetragonolobus purpureas	L-(−)-Fucose, *N*-acetyl-D-glucosamine	Biofilms	185
Abrus precatorius	D-(+)-Galactose	River snow	181
Bandeiraea simplifolia	D-(+)-Galactose, *l*-O-methyl-α-D-galactopyranoside	Biofilms	185
Lens culinaris	α-Methyl-D-mannopyranoside, D-(+)-mannose	River snow	181
Ricinus communis	D-(+)-Galactose, *N*-acetylgalactosamine	River snow	181
Nile Red	Hydrophobic polymer residues	Biofilms	144, 280
Sypro Red, Sypro Orange	Protein	River biofilms	144
Nano Orange	Flagella	Pure cultures	86

TABLE 3 Fluorescent viability stains and their applications

Fluorescent stain(s)	Application(s)	Reference(s)
Propidium iodide	Biofilms	130
Fluorescein diacetate, carboxyfluorescein diacetate	Cell viability biofilms	52
Rhodamine 123	Biofilms	162
Bis-oxonol	Mixed bacterial populations	162
$DiBAC_4$ (3)	Survival of bacteria	156
Calcofluor white	Cells	162
$DiOC_6$ (3)	Protozoa	199
LIVE/DEAD *Bac*Light kit	Biofilms	120, 182, 257
BCECF-AM	pH sensitive (fixed)	
SYTOX Green	Pure cultures	
CTC	Biofilms	216

TABLE 4 Tools developed for digital image analysis of single suspended or adherent bacteria, bacterial morphology, filamentous cells, and specific routines

Cell type or routine	Reference(s)
Suspended cells	
Planktonic bacteria	23, 65, 232, 260
Bacterial shape	24, 154
Aggregates	
Activated sludge	155
Adherent cells	
Flow chamber studies	29, 76, 168, 235, 275
Other systems or surfaces	7, 17, 85, 198, 227
Filamentous cells	
Fungi	57, 92, 170, 248
Bacteria	6, 10, 54, 176, 237
Specific routines	
Soil bacteria	25, 194
FISH	61
Microalgae	265

microcosms (for a review, see reference 188). Parallel to the different approaches of studying immobilized microbial communities, various tools for digital image analysis were developed (Tables 4 and 5).

Equipment for Image Processing and Analysis

Microscopes with a CCD camera and LSM systems represent the most common tools for obtaining a digital image of a microscopic field. Image fidelity is crucial, as the success of all image-processing and analysis steps is contingent on the quality of the primary image. The camera system must be matched with the intended end application in terms of the sensitivity, resolution, and linearity of the response. Scientific video cameras vary in terms of their horizontal resolution, light sensitivity, intrascene dynamic range, blooming characteristics, geometric distortion, and spectral response (110). The generally flexible characteristics of CCD instruments make them applicable for a wide range of microscopic imaging applications. Extremely light-sensitive, cooled (−40 to −70°C) CCD cameras with a wider range of grey-level intensities (e.g., 12- to 16-bit, 4,096 to 65,686 grey levels) may also be obtained (Photometrics Equipment, Roper Scientific, Tucson, Ariz.; Astromed Ltd., Cambridge, England; and Micro Luminetics, Inc., Los Angeles, Calif.). Cameras with photon-counting capabilities (limiting low-level illuminance, 10^{-11} lux) are suitable only for highly specialized applications (e.g., bio- or chemiluminescence and *lux* gene expression), require specialized support equipment and darkrooms, and are not generally applicable for routine lab use (200). Additional useful information on camera systems may be found in reference 110.

TABLE 5 Programs specifically developed for quantitative digital image processing of 3-D microbial biofilm data sets recorded by CLSM

Program	Requirements	Motivation and application	Reference(s)
No specific name	Quantimed	Automated image acquisition and semiautomated single-channel image analysis	128
NIH image, Scion Image, ImageJ	Freeware	Many macros and plug-ins available; programming of specific routines possible; single channel	130, 132, 141
Comstat	MatLab	3-D single-channel analysis of flow cell biofilms and microcolonies	101
ISA 2d	MatLab	2-D single-channel analysis of areal and textural parameters	283
Microstat	–	3-D analysis, improved object separation; dual-channel colocalization, automation	19
No specific name	IDL	Various tools for measurements of microbiological objects and their spatial relationships	215
GIS	Erdas Imagine, ArcView GIS	2-D analysis of marine stromatolites	206, 207
ISA 3d	MatLab	3-D single-channel analysis of textural and volumetric parameters	21
PHLIP	MatLab	3-D analysis, dual-channel colocalization and structural parameters	http://www.phlip.org
DAIME		3-D visualization and analysis of FISH-stained biofilm samples	59, http://www.microbial-ecology.net/daime

Image Analysis Software

There are two excellent freeware programs which may be downloaded over the Internet. NIH Image is a versatile freeware analysis package developed for Apple platforms (http://rsb.info.nih.gov/nih-image/), Windows-based systems (ScionImagePC at http://www.scioncorp.com), and Java applications (ImageJ; http://rsb.info.nih.gov/nih-image/). ImageTool is a PC-based freeware image analysis package available at http://ddsdx.uthscsa.edu/dig/itdesc.htlm that offers many excellent object recognition and analysis features. Other commercial packages are listed in Table 5. Despite the availability of these programs, few if any studies in environmental microbiology have used true 3-D analyses.

Basic Image-Processing Steps

Image analyses involve a series of functions linked together to form a sequence of computer operations common to most, if not all, image analysis procedures. These steps include image acquisition, image processing, image segmentation, object measurement, and data output. Readers should consult the work of Russ (220) and Gonzalez and Wintz (83) for further detail on image processing and analyses. The manner by which the primary image is digitized is dependent on the intended analysis and the software and hardware capabilities. When the specimen is stationary, image averaging improves the quality of the primary image by reducing the noise inherent to all imaging devices and by eliminating image degradation by objects which move during the image averaging process. When moving objects are of interest, single images captured at discrete intervals permit either counting (134) or motion analysis (123, 138).

As a general first step for high-quality image analysis, some form of shading correction should be used. Shading corrections typically involve the digitization and storage of a reference image, which may be subtracted from all subsequent images. Image analysis systems can detect only object pixels with grey levels different from those of neighboring pixels. Ideally, the distribution of grey levels contained in a high-contrast image spans the entire available range (e.g., 0 to 255 for an 8-bit system); however, this is seldom the case. A number of image-processing steps can facilitate the analysis of images which have poor contrast or which are "noisy." These include histogram analysis in combination with grey-level transformation and/or normalization; contrast enhancement; filtration (median, low-pass, Gaussian, Laplacian, etc.); image subtraction, addition, and multiplication; and object erosion and/or dilation (24, 25, 137, 143, 220). Digital images may also be manually edited, permitting the interactive modification of the image. These steps cannot add information to the primary image but can improve the ease of object recognition and the accuracy of object measurement. Examples of multiple-step image analysis as applied to microbiology are shown in review articles by Lawrence et al. (133, 139). For details on the use and applicability of functions available for image processing and enhancement, readers should refer to either their user manual, the work of Russ (220), or the work of Gonzalez and Wintz (83).

Images must be thresholded to create a binary image, thereby indicating which objects are to be measured by the computer during the analysis. Thresholding may be performed subjectively (with the user manually defining the grey-level cutoff value) or by using automatic thresholding functions (e.g., by use of a probabilistic model) (56, 155, 168, 233, 260). Sieracki et al. (233) evaluated nine different automatic thresholding methods, using fluorescent microspheres with known diameters as controls, to position the boundary of fluorescent objects. Although laser microscopy images are near ideally suited for thresholding, there are challenges to the accurate definition of the boundaries and sizes of fluorescent objects. These include the length of exposure to light (long exposure times result in bleaching and an apparent reduction in object size), the halo effect (which makes the fluorescent object seem larger), and user subjectivity (127). Xavier et al. (281) developed an extension of the robust automatic threshold method, known as the objective threshold selection method, for segmentation of CLSM image stacks. Yang et al. (284) examined five automated thresholding functions for the analysis of microbial biofilms imaged by CLSM and found that an iterative thresholding function performed most closely to a manual thresholding function.

Other approaches for thresholding include that used by Møller et al. (168), who used isointensity-thresholded images of fluorescently stained cells with the cell boundary positioned at a point equivalent to 20% of the smoothed, cell intensity maximum grey-level value. Blackburn et al. (24) found that applying Marr-Hildreth filtration (a combination of a Gaussian smoothing function and a Laplace filter for edge amplification) in conjunction with thresholding to a constant value provided acceptable edge detection results. Adiga and Chaudhuri (1) also deal with segmentation and counting of CLSM images of in situ fluorescence hybridization. It should be noted that for the majority of image and CLSM applications, user assessment of both image acquisition and subsequent processing is required due to factors such as low target fluorescence, background and/or debris interference, changing experimental conditions, and technological limitations.

Difference imagery is a technique which permits the selective identification of changes which occur or exist between two digital images and consequently facilitates the visualization and measurement of either microbial growth or movement. Both these approaches have previously been used to facilitate the study of microbiological growth in the presence of an inorganic matrix (36), bacterial motility (122, 138), or bacterial transport parameters (98). These applications have been more extensively reviewed by Caldwell et al. (39). This approach may be used for temporal and spatial analyses conducted by CLSM. Figures 3 and 4 illustrate the application of basic image-processing steps, including difference imagery (Fig. 4). These images are a three-channel (green, red, and far-red wavelengths) LSM image stack of a river biofilm used for the determination of algal, cyanobacterial, bacterial, and polymeric biomass. The data extracted from the image stack are also shown in Fig. 3.

Analysis and Data Output

After the objects in an image have been defined by thresholding, a number of user-defined parameters may be measured. These include the area, morphology, densiometry, number, and/or distribution and position of the objects. Any of these parameters may also be used to specifically exclude or include objects in subsequent processing steps, permitting a high degree of selectivity in the analysis. Most image analysis programs also offer flexible programming languages which may be integrated with the basic program as subroutines to provide provisions for specialized analysis and measurements. Excellent programs have been developed for automatic determination of biomass (184, 185) cell number, cell volumes (24, 128), and number of dividing

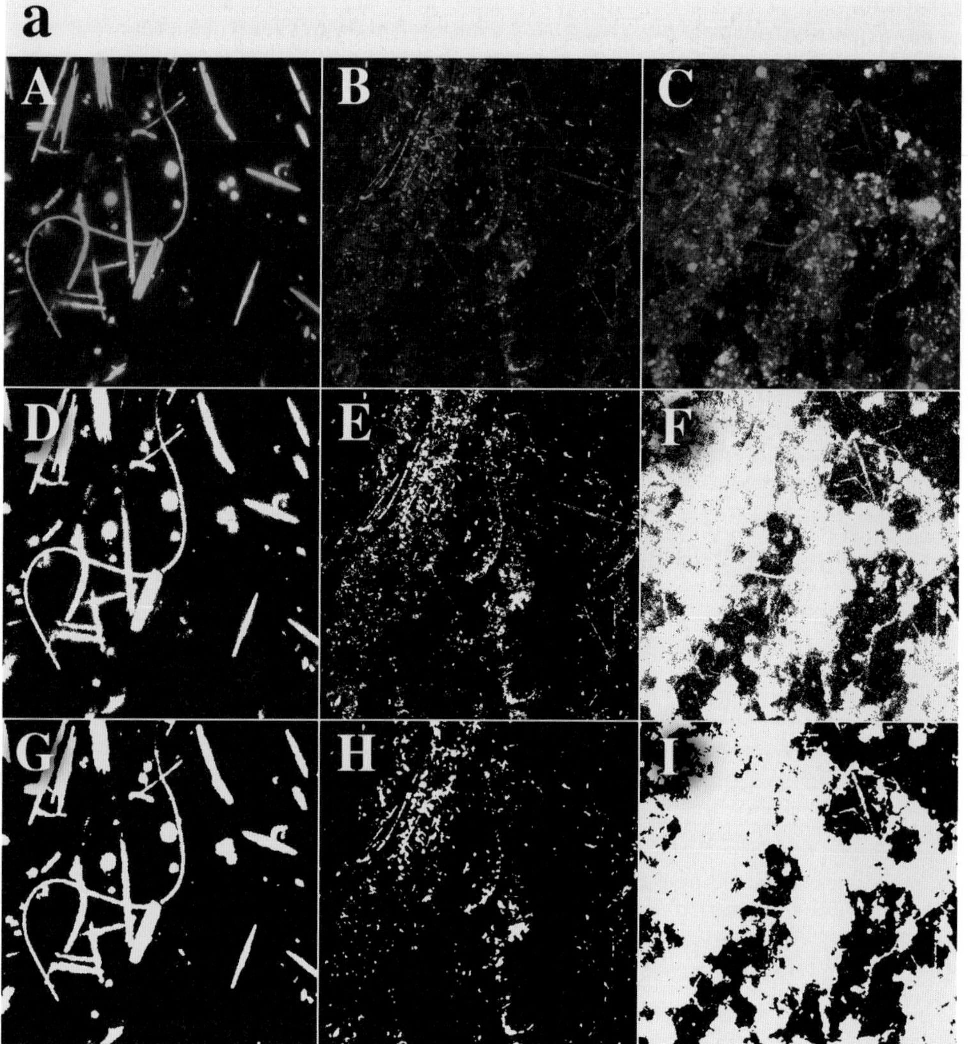

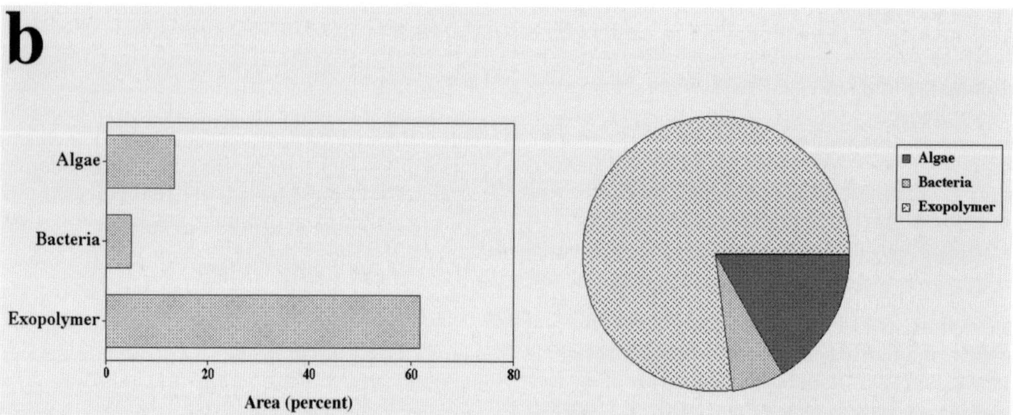

FIGURE 3 (a) Analyses of CLSM confocal image stacks illustrating the application of basic image-processing steps to the analysis of a three-channel (green, red, and far red) fluorescence emission CLSM image stack of a river biofilm. The three stacks are used for the determination of algal (A, D, and G), bacterial (B, E, and H), and polymeric (C, F, and I) biomasses. The series shows the primary image (A, B, and C), application of segmentation to identify objects (D, E, and F), and application of dilation and erosion functions to eliminate noise in the images prior to determination of the number of object pixels in each category (G, H, and I). (b) Graphs showing sample data extracted from the image stacks as a bar graph of the percent biomass in each category and a pie chart of the proportional distribution of the measured parameters.

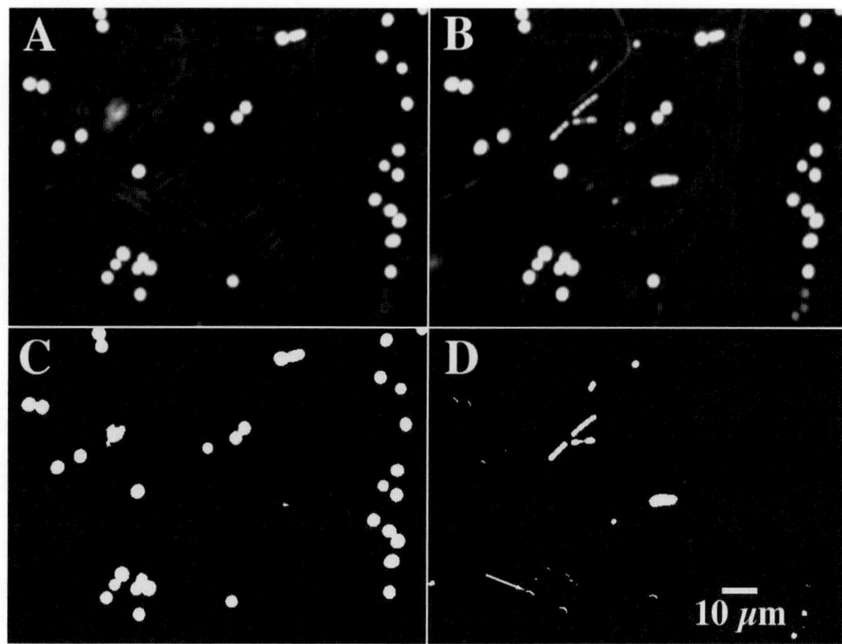

FIGURE 4 Illustration of the application of difference imagery to detect and quantify the algal and cyanobacterial signals and biomass in an LSM image series.

cells (25); quantification of growth parameters (24, 25, 37, 39, 137, 268); and determination of bacterial size and morphology (24, 25, 121, 127, 137, 232, 233). Tables 4 and 5 provide additional information on the application of various analytical software.

DATA PRESENTATION AND 3-D RECONSTRUCTION

LSM systems provide a series of perfectly registered optical thin sections that can be used to produce a variety of 3-D presentations and reconstructions. However, there is a limit to how many thin sections (some suggest 12 or fewer) may be included before the 3-D image becomes overly complex, and these problems are also specimen dependent. In addition, if there is a large space between optical sections, the computer software may simply use interpolation to insert the missing information. Since this is undesirable, sectioning must be performed at increments scaled in accordance with the required 3-D resolution. There are many options for the presentation of CLSM images: galleries showing each section, stereo pairs (140, 145, 279) (note that 10 to 20% of the population has difficulty viewing stereo pairs), red-green anaglyph projections (5, 39, 182), and three-color stereo pairs (133, 141). Figure 5 shows a three-color stereo pair depicting the distribution of algae, exopolymer, and bacteria in a river biofilm community. The image series used to construct the 3-D images is presented in Fig. 4. The use of color is also illustrated by the images in Fig. 6 and 7.

Simulated fluorescence and color coding by depth of section also provide useful means of presentation of 3-D data sets. 3-D rendering of data sets may be achieved through the

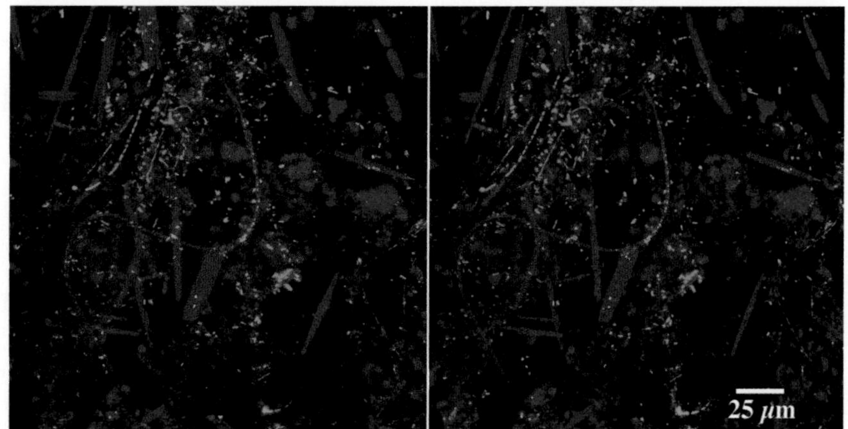

FIGURE 5 CLSM micrograph showing the three-color stereo pair created from the combination of images shown in Fig. 3. The image shows the relative positions of algal and bacterial cells and their surrounding polymer matrix.

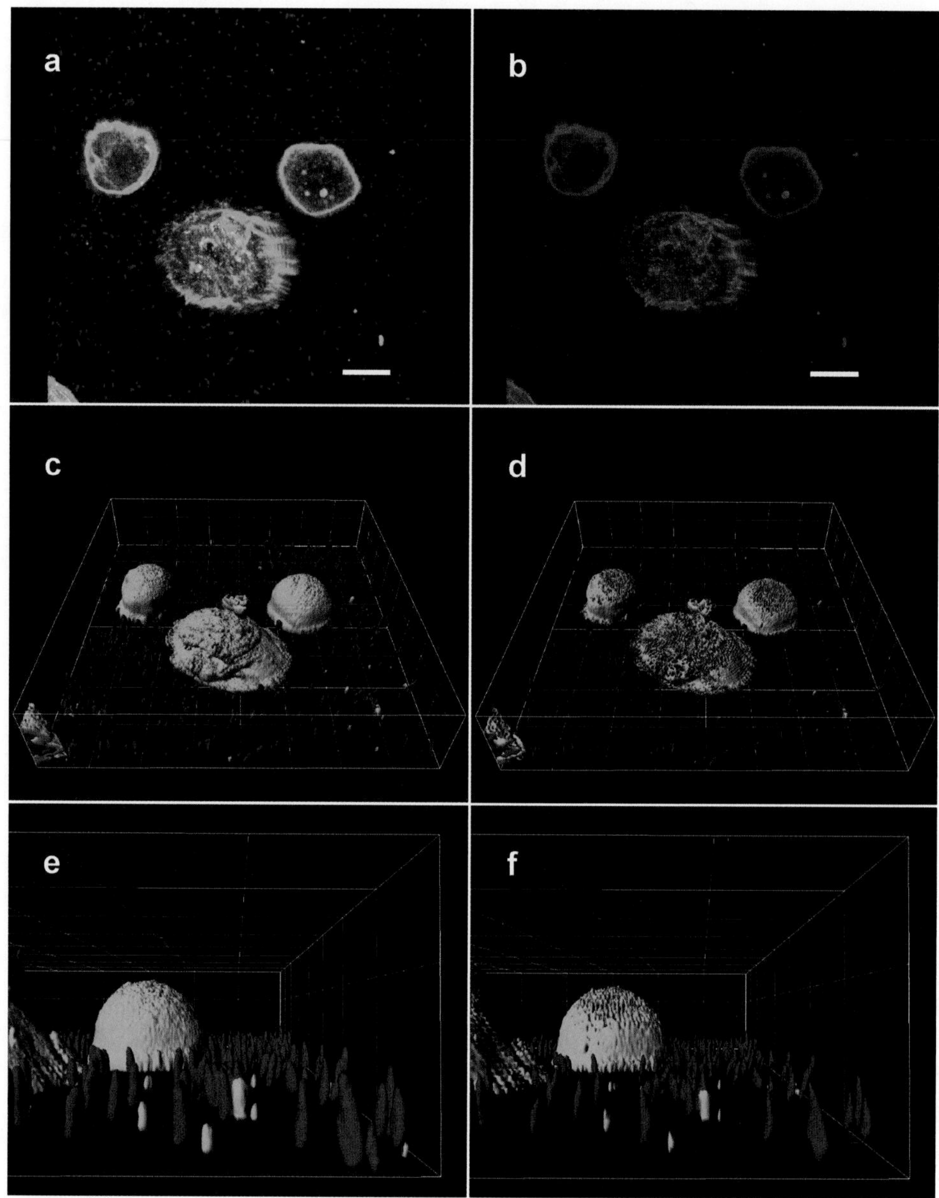

FIGURE 6 CLSM data set for protozoa grazing on a heterotrophic biofilm. The original data set is compared with the result obtained after blind deconvolution. (a) Maximum intensity projection (MIP) of original data; (b) MIP of blind deconvolution data; (c) 3-D isosurface reconstruction of original data; (d) 3-D isosurface reconstruction of deconvolved data; (e) zoomed region of interest from panel c; (f) zoomed region of interest from panel d. Color allocations are as follows: green, lectin-specific glycoconjugates stained with *Aleuria aurantia* lectin labeled with the Alexa-488 fluorochrome; red, bacterial cells stained with the nucleic acid-specific stain SYTO 60. Scale bars and grid boxes, 10 μm. Image data were recorded and deconvolution was calculated and projected by Christian Staudt.

use of ray-tracing or contour-based programs. In 3-D imaging, voxels (volume elements) rather than pixels (picture elements) are the basic imaging units. Using a series of image sections that consist of pixels, the 3-D imaging program calculates (based on the location and intensity of two-dimensional [2-D] image pixels) the corresponding space-filled model of the image slices (a 3-D array of voxels). Sequential rotations of a projection will also provide the reader with a 3-D display of the confocal data set (202).

The human eye is significantly less sensitive to changes in grey level than it is to changes in color. Color lookup tables (a user-defined function in most image analysis programs) may be applied to grey-level images, thereby improving our capacity to interpret the intensity differences contained within an image. It is also important that in CLSM imaging the output of the detectors is scaled and converted to a grey-scale image. Thus, all color images created with CLSM equipment (e.g., merged images or 3-D red-green anaglyph projections) are falsely colored so that, for example, information from the green and red channels is appropriately colored. Applications of pseudocolor have been used to show pH gradients (108), localization of rRNA

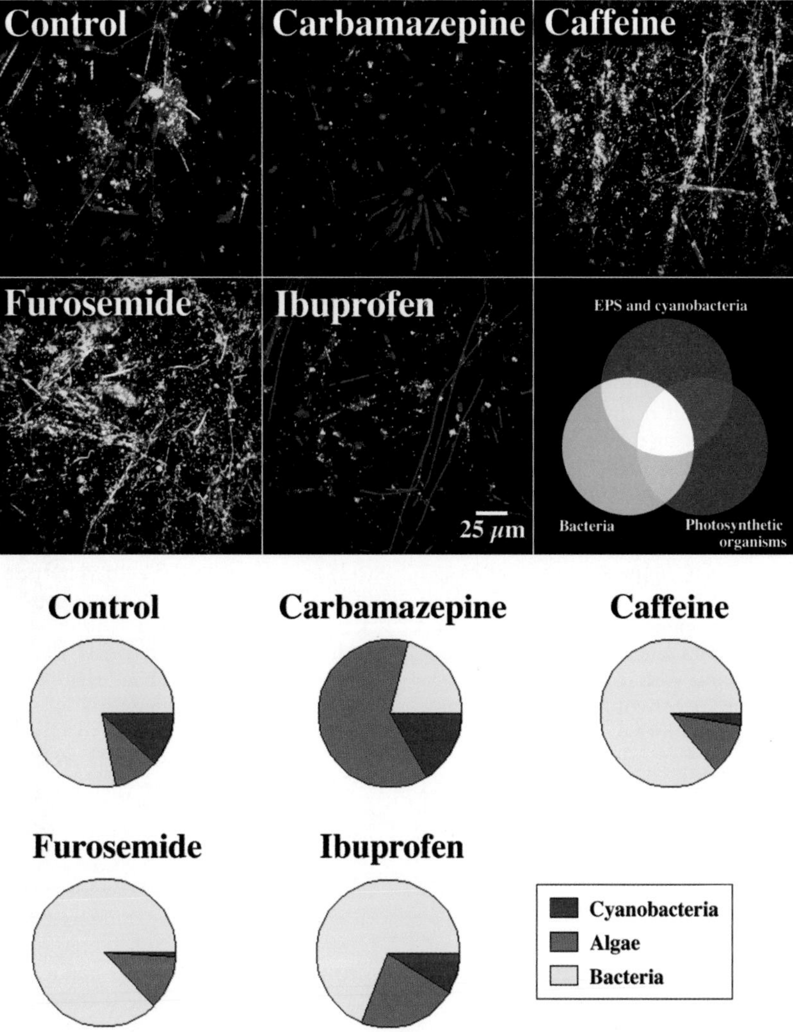

FIGURE 7 (Top) Confocal micrographs illustrating the effects of selected pharmaceuticals on biofilm composition and architecture for control, carbamazepine, caffeine, furosemide, and ibuprofen after 8 weeks of development. The color wheel indicates the assignment of colors to bacteria (green), EPS and cyanobacteria (red), and other photosynthetic organisms (blue). (Bottom) Results of image analyses of confocal laser micrographs illustrating the effects of the pharmaceuticals on the proportional distribution of biomass of algae, cyanobacteria, and bacteria in the river biofilms. Reprinted from Lawrence et al. (132) with permission from the publisher.

probe binding (2, 214), lectin positioning in complex biofilms (183, 184), and isosurface projection (59, 157) and for production of 3-D projections (5, 18, 130, 181).

Deconvolution

Deconvolution or image restoration is related to the point-spread function of all optical microscopes. It is a mathematical procedure in which an algorithm is used in order to improve the resolution of an image series collected by LSM (255). In fact, it refers to one limitation in LSM, which is the inability to equally resolve a 3-D object in the lateral direction as well as in the axial direction. Various hardware solutions for this problem have already been developed; however, most of them are available only in specialized research laboratories. Examples are 4Pi microscopy (16, 224) or the recent stimulated emission depletion microscopy (96, 115). Currently, one company offers 4Pi microscopy systems; however, it has only a limited number of such systems. Thus, deconvolution is still applied in order to mathematically improve the resolution in 2-D and 3-D images. Deconvolution may be applied for sharpening or deblurring of a single image or image series (2-D) as well as to obtain the true spatial shape of an object (3-D). The most common algorithms used for deconvolution are blind deconvolution, the nearest-neighbor method, iterative maximum-likelihood estimation, and the iterative constrained Tikhonov-Miller method. The efficiencies of the various algorithms have been compared, and the selection of a specific algorithm may depend on the content of the image data set (256, 258). Several deconvolution programs, such as Huygens (Scientific Volume Imaging, B.V., Hilversum, The Netherlands), AutoDeblur (Autoquant, Inc. Watervliet, N.Y.), and Micro-Tome and HazeBuster (VayTek, Inc., Fairfield, Iowa), are available through commercial sources. Deconvolution may also be part of the CLSM instrument software.

Care must be taken if deconvolution is to be used because if it is not properly applied it may actually create artifacts. For example, the point-spread function of the microscope must be measured by using fluorescent beads (229). This measurement must be done with each objective lens used to record images and for subsequent deconvolution. Furthermore, in thick specimens such as biofilms, the point-spread function should be measured directly in the original sample and also at different depths of the sample. A further difficulty in deconvolution is the necessity to collect images according to the Nyquist criteria (202, 266). That means that image data sets must be collected at a very high resolution (e.g., at 1,024 by 1,024 pixels [lateral] and a step size of 170 nm [axial]). This will result in extreme bleaching of the sample, it may be applicable to thin samples only, and finally, huge data sets which must be handled by the software will be created. This limitation may rule out the use of deconvolution for certain types of samples. Furthermore, deconvolution is also a procedure requiring long computing times (228). As a consequence, this method may be applied to one or only a few images; but due to time constraints, it may not necessarily be applied to numerous data sets (e.g., >100) collected over a longer experimental time period. Nevertheless, if it is properly applied, deconvolution may dramatically improve image data sets and reveal fine structures not visible in the raw data. Various examples are given at the websites of the different deconvolution software companies. An example for deconvolution in microbiology is demonstrated in Fig. 6. The data set shows a microbial biofilm stained for cells and glycoconjugates before and after deconvolution.

APPLICATIONS OF DIGITAL MICROSCOPY

Enumeration

Only a small fraction of the microorganisms in natural environments can be cultured (5). Thus, significant underestimations may occur when culturing techniques are used to enumerate microorganisms from terrestrial or aquatic samples. Direct microscopic analyses, especially epifluorescent microscopy, are therefore often used to assess the number of cells present in these environments. A review by Kepner and Pratt (113) provides a comprehensive overview of the use of epifluorescence microscopy for enumeration.

The methods described by Hobbie et al. (103) and Daley (62) and modifications thereof have been widely used to enumerate microbial populations. This method involves the placement of a known amount of an environmental sample on a known area of a slide or filter and staining of the sample with a fluorescent dye. Cells are then visualized by epi-illumination with UV or blue light. Fluorochromes traditionally applied in this procedure included acridine orange (AO), DAPI, and FITC.

A range of cell-permeant DNA stains which emit fluorescence in either the blue, green, orange, or red wavelength (e.g., the SYTO cyanine dyes; Molecular Probes, Inc., Eugene, Oreg.) are available. Various investigators have also applied the dyes TOTO, YOYO, POPO, SYBR, and PicoGreen as nucleic acid-specific stains for enumeration of bacteria in environmental samples. These stains offer improved flexibility in terms of optimizing the detection of cellular nucleic acids when they are combined with other fluors or improving the signal-to-noise ratio when background autofluorescence dictates the use of alternate excitation and emission wavelengths. The method of Hobbie et al. (103) is particularly useful for fieldwork when the required equipment is not available on site for analyses because the samples can be preserved and kept until they are returned to the laboratory. Preservatives include formaldehyde at a 3.5% final concentration (103, 211) and Lugol's iodine solution (208). Storage for up to 4 years in the dark at room temperature has been reported with Lugol's iodine solution without any significant change in bacterial numbers (208). Another advantage of this procedure is its simplicity.

Direct count procedures are now commonly used in combination with other methods to assess the ratio of the number of specific cells or populations in relation to the size of the whole community. For instance, they can be used in conjunction with autoradiography to determine the uptake of specific substrates by individuals from a community (163) or can be applied with fluorescent oligonucleotide probes to monitor the presence of defined populations in heterogeneous communities (4, 5, 61, 126, 214, 226). The direct count method has also frequently been used in tandem with other fluorescent probes, including fluorescein diacetate (52) and 5-cyano-2,3-ditotyl tetrazolium chloride (CTC) (48, 216, 230), in order to estimate the percentage of actively metabolizing cells in microbial communities. An additional approach recently developed for tracking and enumerating specific bacterial cells in natural or complex systems is the use of bacteria labeled with GFP (45, 239). This technique is suitable for the simple enumeration of cells previously transformed with plasmids or transposons containing the *gfp* gene (192) or for use with more complex gene expression reporter and monitoring systems, with which it is useful for quantifying the expression of a target gene under various conditions (see below) as well as the rates and frequencies of plasmid transfer (the acquisition of new genetic traits by recipient organisms) in microbial communities (50). Sorensen et al. (238) provide an excellent review illustrating the application of GFP and LSM to the study of horizontal gene transfer in biofilms. In any case, since GFP is produced by the cell, no staining or addition of fluorogenic substrate, other than the presence of oxygen for molecular rearrangement of the GFP into its fluorescent form, is necessary. See the work of Hansen et al. (91) for an evaluation and discussion of the limitations and the roles of oxygen and pH in the application of GFP. Appropriate genetic constructs and cloning vectors are available from commercial suppliers (living color fluorescent proteins; Becton Dickinson) or from laboratories where the GFP variants have been constructed (see the work of Christensen et al. [51] for a list of relevant GFP derivatives suitable for application to prokaryotes). Southward and Surette (239) provide an excellent overview of GFP. The application of GFP for environmental studies has also been reviewed (78).

Growth, Viability, and Metabolic Condition

The activities of bacteria vary as a function of nutrient status, pH, temperature, flow rate, etc., as well as a function of the presence of inhibitory compounds, such as cellular waste products and antimicrobial agents. In biofilms in particular, these parameters are further affected by diffusional transport processes; consequently, natural bacterial systems are typically heterogeneous with regard to cell activity. Methods based on fluorescent probes have been specifically designed either to link one or more physiological attributes of the cell to overall viability or to identify relative differences in cell activity. While it would clearly be beneficial to differentiate between fluorescent probes which report specifically on either viability or metabolic status, this remains difficult in practice, since they are frequently responsive to the same parameter. It is suggested that multiple indicators of cell condition (e.g., a combination of probes for cell membrane integrity, membrane potential, and enzymatic activity) or a ground truth of simple plate counts be considered whenever conclusions regarding cell viability are to be made.

Fluorescent viability and activity probes have been based on membrane integrity, cytoplasmic redox potential, electron transport chain activity, enzymatic activity, cell membrane potential, and gene expression levels, as well as various combinations of the above (14, 48, 216, 230). A number of studies have attempted to use the differential fluorescence of AO when it is bound to RNA versus that when it is bound to DNA to discriminate between active and inactive planktonic cells (14) as well as to delineate spatial variation in the growth rate in colonies and biofilms (268).

The relative activity of the cell's electron transport system (ETS) may also be targeted by using fluorescent chemical redox dyes. Oxidized CTC is soluble and nonfluorescent; however, when it is reduced by an active bacterial respiratory system, it accumulates intracellularly as an insoluble CTC-formazan crystal, which fluoresces in the red wavelengths. Rodriguez et al. (216) used CTC to visualize actively respiring cells in mixed bacterial populations. Yu and McFeters (285) employed CTC to estimate cell injury resulting from chlorination of biofilms in a model pipeline system. Kitaguchi et al. (114) combined CTC staining with FISH to estimate the respiratory activity of *Pseudomonas* spp. in milk. Others have also used tetrazolium salts for viability determinations using transmitted light microscopy, including 2-(*p*-iodophenyl)-3-(*p*-nitrophenyl)-5-phenyltetrazolium chloride (INT) (e.g., see references 84, 95, and 233). Like CTC, reduction of INT leads to the deposition of formazan crystals, which are identified as opaque deposits in cells counterstained with AO. Factors which need to be considered when tetrazolium salts, such as CTC, are used as indicators of ETS activity include potential toxicity (1 to 20 mM) (48, 254), temperature (48), the effects of inorganic constituents and pH (153), the growth phase (230), and assay sensitivity (191). In the last case, it was shown that CTC was able to resolve only the most active bacteria. A further concern is the chemical reduction of CTC in anaerobic environments, such as thick biofilms or flocs.

The integrity of the cell membrane has generally been accepted as a dependable indicator of cell viability. Probes sensitive to the integrity of bacterial membranes include nucleic acid compounds such as propidium iodide (PI). Because PI is relatively membrane impermeant, it does not cross functional cell membranes (93). However, PI easily penetrates cells with a ruptured membrane, and once inside the cell cytoplasm, it binds to nucleic acids, undergoes a large Stokes shift (to the red wavelengths), and results in an increase in fluorescence emission intensity and easier detection by fluorescence microscopy. A weakness of this method is that it cannot detect dead cells which have already lost their DNA, for instance, by the action of degradative enzymes.

The two-component LIVE/DEAD *Bac*Light viability kit (Molecular Probes, Inc.) is also a viability indicator based on membrane integrity. This kit contains a membrane-permeant green fluorescing DNA-labeling probe (SYTO 9) and a membrane-impermeant red fluorescing DNA-labeling probe (PI; see above). Cells with an intact cell membrane (living cells) accordingly appear green, whereas cells with compromised cell membranes fluoresce red (dead cells). A number of studies have used the LIVE/DEAD *Bac*Light system to enumerate the number of living and dead cells in experimental systems with well-defined target organisms and chemistry (12, 130). It is worth noting that the staining of natural (heterogeneous) assemblages of bacteria with the LIVE/DEAD *Bac*Light probe can give variable results due to differences in cell membrane permeability; hence, the performance of preliminary studies and the use of appropriate controls are always recommended to facilitate data interpretation. Alternative methods, such as difference imaging (36), direct determination and quantification of cell growth (119, 120), or physical viability assays (i.e., plasmolysis [119, 120] or cell elongation [95, 117]) are advised in order to verify the results obtained with the LIVE/DEAD *Bac*Light system and other chemical probes.

Fluorescein diacetate (FDA), a nonfluorescent derivative of fluorescein, is also used to detect active cells. FDA is transported across the cell membrane and hydrolyzed by a variety of esterases. The resulting intracellular release of fluorescein allows direct visualization of actively metabolizing cells by epifluorescence microscopy or LSM. This approach has been used to detect metabolic activity in bacteria (52) and fungi (236). A fluorescence detection system for monitoring alkaline phosphatase (APase) activity has been described for phosphate-limited bacterial biofilms (107, 282). Cells expressing APase activity do so only under conditions of phosphate limitation. ELF-97 (Molecular Probes, Inc.) is a water-soluble fluorogenic substrate that becomes a stable, bright yellow-green intracellular precipitate following APase cleavage. This system for monitoring metabolic activity within biofilms has potential for use with natural assemblages of bacteria. For example, Van Ommen Kloeke and Geesey (257) used ELF-97 to detect regions of APase activity in flocs from a sewage treatment system, while Nedoma et al. (177) examined APase in natural plankton. Christensen et al. (51) provide detailed protocols for the application of ELF-97 for detection of APase. Additional compounds with potential for microbiological studies include the fluorogenic enzymatic substrates (fluorescein- and carboxyfluorescein-diacetate and calcein AM), probes sensitive to DNA damage (Hoechst 33342), and probes that reflect cellular membrane potential (rhodamine 123, RH-795, JC-1, and carbocyanine derivatives).

Direct cell growth rates may be determined from rRNA, based on quantitative hybridization with ribosomal probes (168, 203, 212). This may be done very effectively by using the combination of LSM and image analysis of the fluorescence. Monitoring of the number of intracellular ribosomes as a measure of cellular activity should be used with some caution, as concerns exist with respect to ribosome stability, steady-state growth conditions, poorly characterized organisms, and growth rate versus ribosome concentration (244). Alternatively, Sternberg et al. (244) developed a novel *gfp*-based gene reporter system for monitoring cell activity. The method is based on the use of short-half-life GFPs that do not accumulate or persist intra- or extracellularly, permitting real-time analysis of gene expression and single-cell activity (8, 51, 213). Regions of activity in thick biofilms have also been determined by using short-half-life GFPs under the control of growth rate-dependent promoters (269).

Gene expression may be monitored in situ by using a number of genetically constructed expression vectors (166). GFP-based expression vector systems have been developed with stable and modifiable fluorescent characteristics, extending the potential for monitoring in situ gene expression. Møller et al. (167) used independent *gfp* gene fusions to monitor the expression of the *Pu* promoter by benzyl alcohol induction in pure and community cultures. Studies have also used GFP transcriptional fusions to examine the effect of cell-cell signaling events on amylase production in *Streptococcus gordonii-Veillonella atypica* biofilms (75).

Specific functional genes and their expression products (mRNAs) have also been visualized by using prokaryotic in situ PCR (PI-PCR) or in situ reverse transcription in combination with PI-PCR (46). The technique uses the ability of PCR to amplify either mRNA or genes (e.g., the *nahA* gene in *Pseudomonas*) of interest within bacterial cells (104). Variations of the labeling of PCR products include FLUOS-dUTP and digoxigenin-UTP (104). Note that these approaches are technically very demanding.

Microautoradiography (MAR) has been used to provide information on the metabolic activities of bacteria exposed to radiolabeled test substrates (32). Lee et al. (146) combined the single-cell resolution of FISH with MAR signals for the analysis of activated sludge flocs, which were cryosectioned and examined by confocal laser microscopy. Using this approach, the investigators visualized the uptake of organic and inorganic radiolabeled substrates in probe-defined populations from complex activated sludge communities. Nielsen et al. (191) similarly extended the application of MAR by combining it with CTC and FISH. Pernthaler and Amann (205) linked the use of FISH probes targeted against rRNA with mRNA FISH, thus providing the simultaneous identification of specific cells and expression of specific genes. Overall, an increasing suite of tools is available for in situ visualization of the structure, function, and activity of bacterial communities.

Quantification of Biodiversity

One approach for defining and quantifying the antigenic diversity contained in microbiological systems is that of using immunological fluorescent probes (fluor-conjugated poly- or monoclonal antibodies) (129). These methods were originally adapted for ecological studies (208) and have broad potential for distinguishing between morphologically similar bacteria present within complex communities (217). Details on the preparation of antibodies are provided by Lam and Mutharia (129); however, optimization for use with natural systems (e.g., biofilms, soil samples, and plant environments) will inevitably be required (202a). An excellent example of the use of antibodies in a complex environment is the examination of root colonization by Schloter et al. (222). Guggenheim et al. (87) applied species-specific fluor-labeled antibodies to study associations of bacteria in oral biofilms. Although the application of antibodies in microscopy-based research has been supplanted to an extent by oligonucleotide probes and FISH, it remains a valuable approach for examination of hydrated biofilm materials.

Gram staining techniques for fluorescence microscopy have been developed by various workers (161, 234). These approaches have been based on the use of wheat germ agglutinin, which binds to the peptidoglycan layer of gram-positive bacteria but not gram-negative bacteria, and also differential hexidium iodide and SYTO 13 staining. Commercial fluors for the identification of fungi (Fun-1), as well as the fluorescent gram staining of bacteria (*Bac*Light Gram stain; Molecular Probes, Inc.) may also be used to provide broad indications of system biodiversity. These probes have been developed and tested with pure batch cultures; therefore, the user must apply them with appropriate caution to natural communities. Karthikeyan et al. (111) used the *Bac*Light Gram stain to quantify gram-positive and gram-negative bacteria in a complex microbial community grown on various carbon sources. Similarly, Wolfaardt et al. (279) used a fluorescent Gram stain to identify the presence of a *Bacillus* sp. in a degradative microbial community.

Autofluorescence may also provide information on the compositions of microbial communities (180). The fluorescence of chlorophyll was used by Lawrence et al. (133, 141) to identify and localize algae in microbial mats and river biofilms. Similarly, Wiggli et al. (271) used autofluorescence and CLSM to examine the distribution of photosynthetic bacteria in complex mat communities. This approach has been used in a variety of microbial community studies assessing the impacts of environmental factors (47, 130, 132, 288) on photosynthetic populations.

The biodiversity of natural microbial populations may also be determined phylogenetically, allowing the identification of specific bacteria from complex systems (154a, 240, 276). FISH of oligonucleotide probes to specific bacteria has previously been used in conjunction with CLSM and epifluorescence microscopy to document microbial diversity in a range of environments, including sewage sludge, river biofilms, and the rhizosphere (3, 89, 130, 132, 158, 159, 186, 214, 225, 226). Readers are also referred to a plethora of articles by Amann and coworkers, who used FISH and related techniques to explore a variety of different habitats (2–5). Reviews dealing with the synthesis and application of oligonucleotide probes for microbiological systems have been provided by Amann et al. (5) and Perenthaler et al. (203). Zwirglmaier (292) provides an overview of the major developments and improvements in FISH as it is applied to environmental microbiology.

The application of in situ PCR (104) also represents a tool for visualization of bacteria, genes, and gene products. PCR-based methods have seen application for the in situ detection of specific bacteria within natural environments, coupling the advantages of gene amplification with direct bacterial detection (104, 247). Tani et al. (247) reported that the use of HNPP-Fast Red TR resulted in an eightfold increase in fluorescence intensity over that obtained by the use of FITC for PCR product detection. Fluor selection and optimization of the fluorescent signal are often critical to the success of these methods (154a).

Many companies custom prepare fluor-conjugated oligonucleotides if they are provided with the appropriate genetic sequence, the desired fluor conjugate, as well as the amount required. Ribosomal oligonucleotide sequences are also freely available over the Internet. Good starting points include the National Center for Biotechnological Information website (http://www.ncbi.nlm.nih.gov) and the Ribosomal Database Project II (http://rdp.cme.msu.edu/index.jsp).

Microchemical Analyses

A variety of fluorescent probes have been used in conjunction with LSM to study the forms and arrangements of bacteria, exopolysaccharides (EPSs), and voids in attached communities. Fluorescence exclusion (negative staining), in which bacteria are imaged as dark objects against a bright background, has been used to yield high-resolution optical sections of pure-culture and mixed-species biofilms (38, 169). Fluorescein solutions have been most frequently used as a negative stain, although conjugated forms of fluorescein, resazurin, FITC, TRITC, and rhodamine (i.e., conjugated to dextrans) may be used. In the majority of cases, positive staining techniques are used to study microbial communities (118, 132, 182, 186, 279). The probes used for this include, but are not limited to, AO, FITC, TRITC, SYTO dyes, Alexa dyes, and Nile Red. The quality of the fluorescence image may be influenced by a variety of factors, including pH and E_h, nonspecific fluor binding, poor probe penetration, and fluor quenching.

For a better understanding of the role of EPS in microbial processes, it is necessary to characterize the EPS in terms of spatial distribution, density, chemical composition, and charge. Commercially available fluorescent dextrans bearing a defined charge (e.g., polyanionic, cationic, and neutral) have been used (142, 169, 279) to study the charge distribution and polymer permeability within pure culture biofilms as well as in biofilm communities. Hydrophobic dyes such as Nile Red also provide information on the nature and distribution of EPS (144). Fluorescent beads with a sulfated or carboxylated surface chemistry have been used for determination of the hydrophobicity and hydrophilicity of bacterial cells and may be used for other microbial structures. Zita and Hermansson (289) describe the method by the use of beads obtained from Molecular Probes, Inc. These fluorescent beads have also been used under in situ conditions to assess the surface properties of filamentous bacteria in activated sludge flocs (190).

Proteins and other biomacromolecules may also be probed for by using fluorescent markers and microscopy. For example, Neu and Marshall (178) applied the protein specific-probe Hoechst 2495 to detect bacterial polymer footprints on surfaces. They detected the footprint material indirectly by the use of lectins and subsequent protein staining. Lawrence et al. (144) applied the protein stain SYPRO orange. They reported extensive binding to cell-associated and matrix exopolymers with colocalization of protein, lipid, and polysaccharide. Scanning transmission X-ray microscopy and X-ray absorption spectroscopy confirmed that the colocalization of these macromolecules was a valid interpretation.

Fluor-conjugated lectins, characterized by their specific binding with saccharide moieties, can be used as general stains for microbial EPS in fluorescent lectin binding assays (FLBA) (133, 141). General staining of EPS with the fluor-conjugated lectins concanavalin A-FITC and *Triticum vulgaris*-TRITC is shown in Fig. 2, 3, and 7. As noted by Neu and Lawrence (183, 184), lectins represent a useful probe for in situ techniques for 3-D examination of the distribution of glycoconjugates in fully hydrated microbial systems. The many lectins available offer a huge and diverse group of carbohydrate-specific binding molecules waiting to be used for an in situ approach (58, 243) (Table 2). Details on the use of fluor-conjugated lectins and other direct methods for the study of EPS are provided by Neu and colleagues (183, 184, 185, 243). Battin et al. (18) localized EPS in river biofilms using concanavalin A lectin, while Neu et al. (186) were able to detect clear statistically significant effects of nutrient treatments and time on the EPS compositions of river biofilms using CLSM and FLBA. Wigglesworth-Cooksey and Cooksey (270) used a combination of fluor-conjugated lectins and LSM to examine interactions in diatom-dominated biofilm communities. FLBA has been combined with FISH for the localization and identification of bacteria associated with the binding of specific lectins (27).

In addition, antibodies may be used to map the locations of specific enzymes, specific compounds such as pesticides, and a variety of other biologically relevant molecules. Lawrence et al. (131) localized the herbicide atrazine within river biofilms using antibody staining. As indicated in the publication of Wuertz et al. (280), ion- or metal-sensitive probes such as Newport Green may be used to detect the presence of metals in biofilms and within the EPS matrix. Hu et al. (106) used a zinc binding fluor and 2P-LSM to visualize the spatial distribution of Zn within biofilms.

Microenvironmental Analyses

Microelectrode studies have clearly demonstrated the presence of microenvironmental conditions and gradients, particularly within microbial aggregates and biofilms (50, 59, 165a, 196, 226). These gradients occur as a result of cellular metabolic activity, macroenvironmental conditions, and hindered diffusion.

Diffusion may be measured with nonreactive (nonbinding) fluorescent probes and imaging procedures based on CLSM (22, 67, 142, 250) and image analysis. Diffusion coefficients (D_e) may be measured by methods based on fluorescence recovery after photobleaching (FRAP) or diffusion monitoring. In FRAP, a high-intensity light source (usually a laser) is used to bleach out an area of fluorescence at a specific biofilm location. The assumption is made that molecules in the bleached area are rendered nonfluorescent. The gradual return to initial fluorescence values, which occurs only as a function of lateral fluor diffusion (commonly fluorescently labeled dextrans), is then measured temporally and is used in conjunction with the appropriate equation to calculate D_e (13, 22, 26). Lawrence et al. (142) developed an in situ fluorescence monitoring method for the estimation of one-dimensional (1-D) effective diffusion coefficients at defined locations within biofilm matrices. De Beer et al. (67) used a microinjection technique in conjunction with CLSM monitoring to determine the diffusion of a variety of fluorescent molecules in biofilms. Difficulties may arise as a consequence of specific and/or nonspecific binding of the tracer fluor to the biofilm matrix. A number of 1-D, 2-D, and 3-D equations have been used to calculate D_e values (13, 26).

The heterogeneity of diffusion inside microbial biofilms may also be determined by fluorescence correlation spectroscopy (FCS) under two-photon excitation. Compared to other techniques, FCS allows diffusion measurements within extremely small volumes (88).

Many direct measurements (e.g., by the use of microelectrodes) have indicated that microbial activity can radically alter the chemistry of an aqueous system (66, 68, 147, 225, 226). A number of fluorescent probes (e.g., fluorescein and 5+6-carboxyfluorescein) exhibit pH- and E_h-sensitive fluorescence; consequently, these probes have potential for use for in situ measurements of microbially associated chemical microenvironments. Pseudocolor images of these gradients may be seen in the review by Caldwell et al. (39). The major problems with the use of these probes for the measurement of microenvironmental conditions include (i) nonspecific interactions with biomolecules, resulting in quenching; (ii) the effects of E_h; (iii) the effects of ionic composition; and (iv) the general problem of determining the ground truth. However, Vroom et al. (262) used ratiometric and lifetime imaging to provide a more quantitative determination of pH and changes in pH with microbial biofilms, overcoming many of the concerns indicated above. Hunter and Beveridge (108) demonstrated that there was very minor interaction of the pH-sensitive probe C-SNARF-4 with the biofilm matrix when it was applied to the imaging of the pH in their biofilm system. They concluded that concerns regarding sorption effects to EPSs and cellular products were not evident.

Measurements of the intracellular pH in microorganisms have been carried out (31, 109, 197). The most common approach to the imaging of concentration gradients is ratiometric. Ratiometric methods (93) generally involve the application of a fluorescent compound that has excitation and emission lines sensitive and insensitive to its environment. Thus, the ratio of the images at these two wavelengths removes variations due to concentration, with the resultant image containing variations in fluorescence due to the parameter of interest. However, the ratiometric analyses require time-consuming, cumbersome calibration procedures for quantitative imaging. The alternative approach is that of fluorescence lifetime imaging (FLIM), which is based on the detection of differences in the decay of the fluorescence of a molecule. Its major advantage is that calibrations of the fluor response in buffer are sufficient to create a valid standard curve. A number of publications have shown that this approach can be applied to the determination of ion concentrations in bacterial cells and pH gradients in biofilms (221, 246, 262). Thus, a number of effective approaches are available for the imaging of intracellular and spatial variations in the pHs of microbial systems.

Imaging of Environmental Samples

Increasingly, CLSM is applied as a technique for the direct analysis of environmental samples. The following are selected studies which have directed CLSM imaging techniques for the examination of plant and animal surfaces, flocs and aggregates, biofilms, wastes, and soil materials. The rhizosphere has been effectively examined by using analytical imaging and molecular probes. Schloter et al. (222) studied the colonization of different plant roots by a *Rhizobium* sp. using fluorescent polyclonal antibodies. Similarly, Assmus et al. (11) used 16S rRNA probes to monitor root colonization by *Azospirillum* sp. Ramos et al. (213) used unstable GFP and CLSM to monitor activity in the rhizosphere of barley seedlings. GFP-labeled organisms, in conjunction with CLSM, have also been used to monitor the colonization, viability, and activity of inoculants associated with barley seed (252), the rhizosphere (192), and soil (89). The colonization and infection of leaf tissue have also been addressed (171, 267). Lindow and coworkers have used combinations of CLSM, GFP, and staining to examine cell distribution, colonization, and the natural microbial community on leaves (152, 189).

In animals, Nyholm et al. (195) defined the temporal and spatial characteristics of light organ colonization of the squid *Euprymna scolopes* by *Vibrio fischeri*. Those investigators used CLSM to examine the dynamics of mucus secretion during symbiont colonization. In a study related to biocontrol of the soybean cyst nematode, Nour et al. (193) examined the cyst structure and the nature of the associated microbial flora using FISH and lectin staining in combination with CLSM. Macedo et al. (157) used analytical, molecular, and CLSM techniques to study the degradation of polychlorinated biphenyl droplets by a community of bacteria isolated from a heavily contaminated soil.

Studies of aquatic habitats have also benefited from advances in CLSM imaging techniques (80). CLSM and digital image analysis were used extensively in a study examining the effects of nickel on the development of river biofilms (130). A series of 3-D projections based on three-channel image stacks is shown in Fig. 7, which illustrates the effects of selected pharmaceuticals on the development of river biofilm communities (132). The data extracted from replicate image series are shown at the bottom of Fig. 7. Similarly, Battin et al. (18) used CLSM and digital image analysis to quantify the effects of flow regimes on the architectures of river biofilms. Additional studies have investigated river biofilms (182), algal mats (141, 271), and Antarctic cyanobacterial mats (70).

One of the most active areas for the application of CLSM and a wide variety of molecular probes is in the study of industrial and naturally occurring flocs (105, 181, 263). Excellent examples of the application of CLSM in combination with novel molecular techniques have been provided (81, 223) and have been summarized by Daims (60).

Although geologic materials are not as amenable to visualization studies, these have also been examined (25, 251). Tobin et al. (251) prepared petrographic thin sections that were subsequently stained with fluorescent probes (PI and POPO-3) for the detection of RNA and DNA.

Enhanced detection and counting of *Cryptosporidium parvum* oocysts in soil, sediment, and feces has also been achieved by CLSM and immunostaining (9). A fluorogenic dye was also used to assess the viabilities of the cysts. Thus, LSM techniques not only are providing useful illustrative material but also are providing a major tool that can be used to investigate microbial communities and resolve environmental questions.

EMERGING MICROSCOPY METHODS

A number of emerging microscopy techniques, particularly in fluorescence imaging, may be applied to microbiological samples. These include FLIM, FCS, coherent anti-Stokes Raman scattering (CARS) microscopy, second harmonic imaging microscopy, 4Pi microscopy, stimulated emission depletion microscopy, and near-field scanning optical microscopy (NSOM). These techniques promise enhanced resolution and increased information content. FLIM is performed with a two-photon system. The image is based on the life of the fluorescent signal rather than on its intensity, on which conventional LSM and fluorescence imaging are based. Commercially available systems record the lifetime in the time domain mode with a pulsed two-photon laser for excitation and time-correlated single-photon counting for detection (53, 71, 79). FLIM may be used to (i) detect multiple fluorochromes, (ii) measure pH and ion (e.g., calcium) concentrations, (iii) determine the oxygen concentration, and (iv) provide information on the microenvironment from the binding status of the fluorochrome. A major advantage is that fluorochromes with similar or identical emission wavelengths can be distinguished because they have different lifetimes. Neu et al. (187) demonstrated the use of FLIM for determining the metabolic status of bacteria in microbial communities. A FLIM approach has been used to determine pH with a variety of fluorochromes, including those with sensitivities in the low-pH range (151). Lifetime imaging has considerable advantages over the more traditional ratiometric and intensity imaging of pH and the concentrations of ions, in that calibration is very straightforward. Vroom et al. (262) demonstrated pH gradients in oral biofilms using FLIM in combination with the pH-sensitive probe carboxyfluorescein. FCS can be combined with laser microscopy in order to measure microscopic fluctuations of single molecules. FCS requires a stable laser source, high-NA lenses, and either a one-photon (with a pinhole) or a two-photon (without a pinhole) instrument. Calculation of fluorescence fluctuations requires statistical analysis by means of an autocorrelation function. Further details on FCS may be found elsewhere (102, 172). Two-photon excitation may also be used for CARS microscopy, which uses vibrational contrast without fluorescent probes. CARS signals are generated in the focal region, which allows the collection only of 3-D data sets, as in 2P-LSM or one-photon LSM. CARS has seen limited use in microbiology but may be used to visualize live cells (291). Second harmonic imaging microscopy, another new technique that uses two-photon excitation, promises enhanced resolution. This approach is based on imaging events when two infrared photons are converted into a single photon of twice the energy. Further details on second harmonic imaging microscopy may be found in two recent reviews (40, 165). Stimulated emission depletion microscopy and NSOM both promise to enhance resolution. Typically, optical microscopy has a resolution in the range of half of the wavelength of the light used for imaging. In practical terms, this translates into a resolution of about 200 nm laterally and 600 nm axially. Thus, these new approaches may result in dramatic improvements in this area. By NSOM a resolution of approximately 50 nm may be achieved. NSOM has mainly been used in physics and chemistry, with few applications with fixed biological samples reported (94, 150). When 4Pi microscopy, which increases the NA and the axial resolution of the microscope by using two opposing lenses, is combined with 2P-LSM and mathematical image treatment, the resolution is nearly equal in both the xy and the xz dimensions of the sample (97). Commercial instruments are now available, but the application to microbial samples has not been carried out.

The application of synchrotron radiation, such as soft X rays in scanning transmission X-ray microscopy, can provide further increases in the information content of images through spectroscopic analyses and the use of known spectra to map the distributions of metals, other elements, and biomolecules (protein, nucleic acids, lipids, carbohydrates) in living hydrated biofilms, as demonstrated by Lawrence et al. (144). Most importantly, the method uses the intrinsic X-ray absorption properties of the sample, eliminating the need for addition of reflective, absorptive, or fluorescent probes and markers that may introduce artifacts or complicate interpretation. Hard X-ray analyses also have potential for application to biofilm-floc materials and have been used for bacterial cell-metal interaction studies (112).

COMPUTERS, MONITORS, PRINTERS, ARCHIVING, AND STORAGE

CLSM systems are usually controlled by a PC or a workstation. A common arrangement is to have linked or networked computers, with one dedicated to image collection and the other dedicated to analyses and processing. High-resolution flat-screen monitors are preferred for display, analyses, and processing. Typically, the computer systems used for image-related activities cannot have too much random-access memory (RAM), too large a hard drive, or too fast a working time for handling, processing, analysis, and storage. Archiving is an important issue, as routinely, about 1 gigabyte of data may be produced per day. Currently, DVDs represent an option with a large capacity and are a standard readable by most computers. However, for long-term storage, files may have to be transferred to a new medium type, as old computers with specific drives (e.g., zip and Jaz drives) may not be available anymore. Most systems can output to a TIFF file and read TIFF files, so unless specific reasons counter this logic, TIFF is the universally preferred storage and transfer file format. Nevertheless, the user must be aware that every CLSM system will produce a specific TIFF file that needs a specific header to access the instrument parameters. This is an issue if other software is used for subsequent image processing (visualization and quantification). In addition, the user should be cautious in the application of conversions and compression, since some formats are dead ends, meaning that the original image and information content may not be retrievable (always maintain backup copies).

The standard approach for presentation and publication has been the use of photographic reproduction techniques. These methods are discussed in full by Popkin (209) and will not be addressed in this chapter. Digital imaging opens many options in terms of image reproduction. In addition, the digital record may be extensively altered in a variety of software packages, such as Adobe Photoshop. These packages and others allow the import, manipulation, and enhancement of digital images and photographs. Digital imaging users will require image scanners (slide, page, etc.) and a capacity to import or handle a wide variety of image formats. This is required to exchange digital information and allow the development of presentation or publication graphics. Images may then be printed for publication by using a variety of high-resolution digital printers. Quality output can be expensive; however, these devices put complete production control in the hands of the researcher, which can be efficient and cost-effective.

REFERENCES

1. **Adiga, P. S. U., and B. B. Chaudhuri.** 2000. Segmentation and counting of FISH signals in confocal microscopy images. *Micron* **31:**5–15.
2. **Amann, R., R. Snaid, M. Wagner, W. Ludwig, and K. H. Schleifer.** 1996. In situ visualization of high genetic diversity in a natural microbial community. *J. Bacteriol.* **178:**3496–3500.
3. **Amann, R. I., J. Stromley, R. Devereux, R. Key, and D. A. Stahl.** 1992. Molecular and microscopic identification of sulfate-reducing bacteria in multispecies biofilms. *Appl. Environ. Microbiol.* **58:**614–623.
4. **Amann, R. I., L. Krumholz, and D. A. Stahl.** 1990. Fluorescent-oligonucleotide probing of whole cells for determinative, phylogenetic, and environmental studies in microbiology. *J. Bacteriol.* **172:**762–770.
5. **Amann, R. I., W. Ludwig, and K. H. Schleifer.** 1995. Phylogenetic identification and in situ detection of individual microbial cells without cultivation. *Microbiol. Rev.* **59:**143–169.
6. **Amaral, A. L., M. A. Pereira, M. da Motta, M.-N. Pons, M. Mota, E. C. Ferreira, and M. M. Alves.** 2004. Development of image analysis techniques as a tool to detect and quantify morphological changes in aerobic sludge. II. Application to a granule deterioration process triggered by contact with oleic acid. *Biotechnol. Bioeng.* **87:**194–199.
7. **An, Y. H., R. J. Friedman, R. A. Draughn, E. A. Smith, J. H. Nicholson, and J. F. John.** 1995. Rapid quantification of staphylococci adhered to titanium surfaces using image analyzed epifluorescence microscopy. *J. Microbiol. Methods* **24:**29–40.
8. **Andersen, J. B., C. Sternberg, L. K. Poulsen, S. P. Bjørn, M. Givskov, and S. Molin.** 1998. New unstable

variants of green fluorescent protein for studies of transient gene expression in bacteria. *Appl. Environ. Microbiol.* **64:** 2240–2246.

9. **Anguish, L. J., and W. C. Ghiorse.** 1997. Computer-assisted laser scanning and video microscopy for analysis of *Cryptosporidium parvum* oocysts in soils, sediment, and feces. *Appl. Environ. Microbiol.* **63:**724–733.
10. **Araya-Kroff, P., A. L. Amaral, L. Neves, E. C. Ferreira, M.-N. Pons, M. Mota, and M. M. Alves.** 2004. Development of image analysis techniques as a tool to detect and quantify morphological changes in anaerobic sludge. I. Application to a granulation process. *Biotechnol. Bioeng.* **87:**184–193.
11. **Assmus, B., P. Hutzler, G. Kirchhof, R. I. Amann, J. R. Lawrence, and A. Hartmann.** 1995. In situ localization of *Azospirillum brasilense* in the rhizosphere of wheat using fluorescently labeled rRNA-targeted oligonucleotide probes and scanning confocal laser microscopy. *Appl. Environ. Microbiol.* **61:**1013–1019.
12. **Auty, M. A. E., G. E. Gardiner, S. J. McBrearty, E. O. O'Sullivan, D. M. Mulvihill, J. K. Collins, G. F. Fitzgerald, C. Stanton, and R. P. Ross.** 2001. Direct in situ viability assessment of bacteria in probiotic dairy products using viability staining in conjunction with confocal scanning laser microscopy. *Appl. Environ. Microbiol.* **67:**420–425.
13. **Axelrod, A., D. E. Koppel, J. Schlessinger, E. Elsen, and W. W. Webb.** 1976. Mobility measurement by analysis of fluorescence photobleaching recovery kinetics. *Biophys. J.* **16:**1055–1069.
14. **Back, J. P., and R. G. Kroll.** 1991. The differential fluorescence of bacteria stained with acridine orange and the effects of heat. *J. Appl. Bacteriol.* **71:**51–58.
15. **Bae, H. C., E. H. Cota-Robles, and L. E. Casida, Jr.** 1972. Microflora of soil as viewed by transmission electron microscopy. *Appl. Microbiol.* **23:**637–648.
16. **Bahlmann, K., S. Jakobs, and S. W. Hell.** 2001. 4Pi-confocal microscopy of live cells. *Ultramicroscopy* **87:**155–164.
17. **Barthelson, R., C. Hopkins, and A. Mobasseri.** 1999. Quantitation of bacterial adherence by image analysis. *J. Microbiol. Methods* **38:**17–23.
18. **Battin, T. J., L. A. Kaplan, J. D. Newbold, X. Cheng, and C. Hansen.** 2003. Effects of current velocity on the nascent architecture of stream microbial biofilms. *Appl. Environ. Microbiol.* **69:**5443–5452.
19. **Bergner, S., R. Pohle, S. Al-Zubi, K. Tönnies, A. Eitner, and T. R. Neu.** 2002. Segmenting microorganisms in multi-modal volumetric datasets using a modified watershed transform, p. 429–437. *In* L. Van Gool (ed.), *Lecture Notes in Computer Science*, vol. 2449. Springer, Heidelberg, Germany.
20. **Beveridge, T. J., J. R. Lawrence, and R. G. E. Murray.** Submitted for publication.
21. **Beyenal, H., C. Donovan, Z. Lewandowski, and G. Harkin. 2004.** Three-dimensional biofilm structure quantification. *J. Microbiol. Methods* **59:**395–413.
22. **Birmingham, J. J., N. P. Hughes, and R. Treloar.** 1995. Diffusion and binding measurements within oral biofilms using fluorescence photobleaching recovery methods. *Philos. Trans. Ry. Soc. Lond. B Biol. Sci.* **350:** 325–343.
23. **Björnsen, P. K. 1986.** Automatic determination of bacterioplankton biomass by image analysis. *Appl. Environ. Microbiol.* **51:**1199–1204.
24. **Blackburn, N., A. Hagström, J. Wikner, R. Cuadros-Hansson, and P. K. Bjørnsen.** 1998. Rapid determination of bacterial abundance, biovolume, morphology, and growth by neural network-based image analysis. *Appl. Environ. Microbiol.* **64:**3246–3255.
25. **Bloem, J., M. Veninga, and J. Shepherd.** 1995. Fully automatic determination of soil bacterium numbers, cell volumes, and frequency of dividing cells by confocal laser scanning microscopy and image analysis. *Appl. Environ. Microbiol.* **61:**926–936.
26. **Blonk, J. C. G., A. Don, H. van Aalst, and J. J. Birmingham.** 1993. Fluorescence photobleaching recovery in the confocal scanning laser microscope. *J. Microsc.* **169:**363–374.
27. **Böckelmann, U., W. Manz, T. R. Neu, and U. Szewzyk.** 2002. New combined technique of fluorescent in situ hybridization and lectin-binding-analysis (FISH-LBA) for the investigation of lotic microbial aggregates. *J. Microbiol. Methods* **49:**75–87.
28. **Bos, R., H. C. van der Mei, and H. J. Busscher.** 1999. Physico-chemistry of initial microbial adhesive interactions—its mechanisms and methods for study. *FEMS Microbiol. Rev.* **23:**179–230.
29. **Bos, R., H. C. van der Mei, J. M. Meinders, and H. J. Busscher.** 1994. A quantitative method to study co-adhesion of microorganisms in a parallel plate flow chamber: basic principles of the analysis. *J. Microbiol. Methods* **20:**289–305.
30. **Bott, T. L., and T. D. Brock.** 1970. Growth and metabolism of periphytic bacteria: methodology. *Limnol. Oceanogr.* **20:**191–197.
31. **Breeuwer, P., J.-L. Drocourt, F. M. Rombouts, and T. Abee.** 1996. A novel method for continuous determination of intracellular pH in bacteria with the internally conjugated fluorescent probe 5 (and 6) carboxyfluorescein succinimidyl ester. *Appl. Environ. Microbiol.* **62:**178–183.
32. **Brock, T. D., and M. L. Brock.** 1968. Autoradiography as a tool in microbial ecology. *Nature* **209:**734–736.
33. **Brock, T. D.** 1971. Microbial growth rates in nature. *Bacteriol. Rev.* **35:**39–58.
34. **Brock, T. D.** 1984. How sensitive is the light microscope for observations on microorganisms in natural habitats? *Microb. Ecol.* **10:**297–300.
35. **Bruchez, M., Jr., M. Moronne, P. Gin, S. Weiss, and P. A. Alivisatos.** 1998. Semiconductor nanocrystals as fluorescent biological labels. *Science* **281:**2013–2016.
36. **Caldwell, D. E., and J. J. Germida.** 1985. Evaluation of difference imagery for visualizing and quantitating microbial growth. *Can. J. Microbiol.* **31:**35–44.
37. **Caldwell, D. E., and J. R. Lawrence.** 1986. Growth kinetics of *Pseudomonas fluorescens* microcolonies within the hydrodynamic boundary layers of surface microenvironments. *Microb. Ecol.* **12:**299–312.
38. **Caldwell, D. E., D. R. Korber, and J. R. Lawrence.** 1992. Imaging of bacterial cells by fluorescence exclusion using scanning confocal laser microscopy. *J. Microbiol. Methods* **15:**249–261.
39. **Caldwell, D. E., D. R. Korber, and J. R. Lawrence.** 1992. Confocal laser microscopy and digital image analysis in microbial ecology. *Adv. Microb. Ecol.* **12:**1–67.
40. **Campagnola, P. J., and L. M. Loew.** 2003. Second-harmonic imaging microscopy for visualizing biomolecular arrays in cells, tissues and organisms. *Nat. Biotechnol.* **21:** 1356–1360.
41. **Casida, L. E.** 1976. Continuously variable amplitude contrast microscopy for the detection and study of microorganisms in soil. *Appl. Environ. Microbiol.* **31:**605–608.
42. **Casida, L. E.** 1975. Infrared color photomicrography of soil microorganisms. *Can. J. Microbiol.* **21:**1892–1893.
43. **Casida, L. E.** 1969. Observation of microorganisms in soil and other natural habitats. *Appl. Microbiol.* **18:**1065–1071.
44. **Casida, L. E.** 1972. Interval scanning photomicrography of microbial cell populations. *Appl. Microbiol.* **23:**190–192.

45. **Chalfie, M., Y. Tu, G. Euskirchen, W. W. Ward, and D. C. Prasher.** 1994. Green fluorescent protein as a marker for gene expression. *Science* **263:**802–805.
46. **Chen, F., J. M. González, W. A. Dustman, M. A. Moran, and R. E. Hodson.** 1997. In situ reverse transcription, an approach to characterize genetic diversity and activities of prokaryotes. *Appl. Environ. Microbiol.* **63:**4907–4913.
47. **Chenier, M. R., D. Beaumier, R. Roy, B. T. Driscoll, J. R. Lawrence, and C. W. Greer.** 2003. Impact of seasonal variations and nutrient inputs on the cycling of nitrogen and the degradation of hexadecane by replicated river biofilms. *Appl. Environ. Microbiol.* **69:**5170–5177.
48. **Choi, J.-W., B. F. Sherr, and E. B. Sherr.** 1999. Dead or alive? A large fraction of ETS-inactive marine bacterioplankton cells, as assessed by reduction of CTC, can become ETS-active with incubation and substrate addition. *Aquat. Microb. Ecol.* **18:**105–115.
49. **Cholodny, N. 1930.** Über eine neue Method zur Untersuchung der Bodenflora. *Arch. Mikrobiol.* **1:**620–652.
50. **Christensen, B. B., C. Sternberg, J. B. Andersen, L. Eberl, S. Møller, M. Givskov, and S. Molin.** 1998. Establishment of new genetic traits in a microbial biofilm community. *Appl. Environ. Microbiol.* **64:**2247–2255.
51. **Christensen, B. B., C. Sternberg, J. B. Andersen, L. Eberl, S. Møller, M. Givskov, and S. Molin.** 1999. Molecular tools for study of biofilm physiology. *Methods Enzymol.* **310:**20–42.
52. **Chrzanowski, T. H., R. D. Crotty, J. G. Hubbard, and R. P. Welch.** 1984. Applicability of the fluorescein diacetate method of detecting active bacteria in freshwater. *Microb. Ecol.* **10:**179–185.
53. **Clegg, R. M., O. Holub, and C. Gohlke.** 2003. Fluorescence lifetime-resolved imaging: measuring lifetimes in an image. *Methods Enzymol.* **360:**509–542.
54. **Congestri, R., R. Federici, and P. Albertano.** 2000. Evaluating biomass of Baltic filamentous cyanobacteria by image analysis. *Aquat. Microb. Ecol.* **22:**283–290.
55. **Conn, M. P. (ed.).** 1999. *Methods in Enzymology*, vol. 307. *Confocal Microscopy*. Academic Press, Inc., San Diego, Calif.
56. **Corkidi, G., R. Diaz-Uribe, J. L. Folch-Mallol, and J. Nieto-Sotelo.** 1998. COVASIAM: an image analysis method that allows detection of confluent microbial colonies and colonies of various sizes for automated counting. *Appl. Environ. Microbiol.* **64:**1400–1404.
57. **Cox, P. W., G. C. Paul, and C. R. Thomas.** 1998. Image analysis of the morphology of filamentous micro-organisms. *Microbiology* **144:**817–827.
58. **Cummings, R. D.** 1994. Use of lectins in analysis of glycoconjugates. *Methods Enzymol.* **230:**66–86.
59. **Daims, H., S. Lücker, and M. Wagner.** 2005. *Daime*, a novel image analysis program for microbial ecology and biofilm research. *Environ. Microbiol.* 10.1111/j.1462-2920.2005.00880.x.
60. **Daims, H.** 2005. Microbial community structure and function of flocs, p. 317–338. *In* I. G. Droppo, G. G. Leppard, S. N. Liss, and T. M. Milligan (ed.), *Flocculation in Natural and Engineered Environmental Systems*. CRC Press, Inc., Boca Raton, Fla.
61. **Daims, H., N. B. Ramsing, K.-H. Schleifer, and M. Wagner.** 2001. Cultivation-independent, semiautomatic determination of absolute bacterial cell numbers in environmental samples by fluorescence in situ hybridization. *Appl. Environ. Microbiol.* **67:**5810–5818.
62. **Daley, R. J.** 1979. Direct epifluorescence enumeration of native aquatic bacteria: uses, limitations, and comparative accuracy, p. 29–45. *In* J. W. Costerton and R. R. Colwell (ed.), *Native Aquatic Bacteria: Enumeration, Activity, and Ecology*. STP 695. American Society for Testing and Materials, Villanova, Pa.
63. **Dandurand, L. M., D. J. Schotzko, and G. R. Knudsen.** 1997. Spatial patterns of rhizoplane populations of *Pseudomonas fluorescens*. *Appl. Environ. Microbiol.* **63:**3211–3218.
64. **Davey, H. M., and D. B. Kell.** 1996. Flow cytometry and cell sorting of heterogeneous microbial populations: the importance of single-cell analyses. *Microbiol. Rev.* **60:**641–696.
65. **David, A. W., and J. H. Paul.** 1989. Enumeration and sizing of aquatic bacteria by use of a silicon-intensified target camera linked-image analysis system. *J. Microbiol. Methods* **9:**257–266.
66. **De Beer, D., A. Glud, E. Epping, and M. Kühl.** 1997. A fast responding CO_2 microelectrode for profiling sediments, microbial mats and biofilms. *Limnol. Oceanogr.* **42:** 1590–1600.
67. **De Beer, D., P. Stoodley, and Z. Lewandowski.** 1997. Measurements of local diffusion coefficients in biofilms by microinjections and confocal microscopy. *Biotechnol. Bioeng.* **53:**151–158.
68. **De Beer, D., P. Stoodley, F. L. Roe, and Z. Lewandowski.** 1994. Effects of biofilm structures on oxygen distribution and mass transport. *Biotechnol. Bioeng.* **43:**1131–1138.
69. **Decho, A. W., and T. Kawaguchi.** 1999. Confocal imaging of in situ natural microbial communities and their extracellular polymeric secretions using Nanoplast resin. *BioTechniques* **27:**1246–1252.
70. **de los Ríos, A., C. Ascaso, J. Wierzchos, E. Fernández-Valiente, and A. Quesada.** 2004. Microstructural characterization of cyanobacterial mats from the McMurdo Ice Shelf, Antarctica. *Appl. Environ. Microbiol.* **70:**569–580.
71. **Draaijer, A., R. Sanders, and H. C. Gerritsen.** 1995. Fluorescence lifetime imaging: a new tool in confocal microscopy, p. 491–505. *In* J. B. Pawley (ed.), *Handbook of Biological Confocal Microscopy*. Plenum Press, New York, N.Y.
72. **Droppo, I. G., D. T. Flannigan, G. G. Leppard, C. Jaskot, and S. N. Liss.** 1996. Floc stabilization for multiple microscopic techniques. *Appl. Environ. Microbiol.* **62:** 3508–3515.
73. **Dunn, K. W., and E. Wang.** 2000. Optical aberrations and objective choice in multicolor confocal microscopy. *BioTechniques* **28:**542–550.
74. **Duxbury, T.** 1977. A microperfusion chamber for studying the growth of bacterial cells. *J. Appl. Bacteriol.* **42:**247–251.
75. **Egland, P. G., R. J. Palmer, Jr., and P. E. Kolenbrander.** 2004. Interspecies communication in *Streptococcus gordonii-Veillonella atypica* biofilms: signaling in flow conditions requires juxtaposition. *Proc. Natl. Acad. Sci. USA* **101:**16917–16922.
76. **Elfwing, A., Y. LeMarc, J. Baranyi, and A. Ballagi.** 2004. Observing growth and division of large numbers of individual bacteria by image analysis. *Appl. Environ. Microbiol.* **70:**675–678.
77. **Emerson, D., R. M. Worden, and J. A. Breznak.** 1994. A diffusion gradient chamber for studying microbial behavior and separating microorganisms. *Appl. Environ. Microbiol.* **60:**1269–1278.
78. **Errampali, D., K. Leung, M. B. Cassidy, M. Kostrzynska, M. Blears, H. Lee, and J. T. Trevors.** 1999. Applications of the green fluorescent protein as a molecular marker in environmental microorganisms. *J. Microbiol. Methods* **35:**187–199.
79. **Gadella, T. W. J., Jr.** 1999. Fluorescence lifetime imaging microscopy (FLIM): instrumentation and applications, p. 467–479. *In* W. T. Mason (ed.), *Fluorescent and Luminescent Probes for Biological Activity*. Academic Press, Inc., San Diego, Calif.

80. **Ghiorse, W. C., D. N. Millar, R. L. Sandoli, and P. L. Siering.** 1995. Applications of laser scanning microscopy for analysis of aquatic microhabitats. *Microsc. Res. Tech.* **33:**73–86.
81. **Ginige, M. P., P. Hugenholtz, H. Daims, M. Wagner, J. Keller, and L. L. Blackall.** 2004. Use of stable-isotope probing, full-cycle rRNA analysis, and fluorescence in situ hybridization-microautoradiography to study a methanol-fed denitrifying microbial community. *Appl. Environ. Microbiol.* **70:**588–596.
82. **Gjaltema, A., P. A. M. Arts, M. C. M. Loosdrecht, J. G. van Kuenen, and J. J. Heijnen.** 1994. Heterogeneity of biofilms in rotating annular reactors: occurrence, structure, and consequences. *Biotechnol. Bioeng.* **44:**194–204.
83. **Gonzalez, R. C., and P. Wintz.** 1977. *Digital Image Processing*. Addison-Wesley, Reading, Mass.
84. **Gribbon, L. T., and M. R. Barer. 1995.** Oxidative metabolism in nonculturable *Helicobacter pylori* and *Vibrio vulnificus* cells studied by substrate-enhanced tetrazolium reduction and digital image processing. *Appl. Environ. Microbiol.* **61:**3379–3384.
85. **Grivet, M., J.-J. Morrier, C. Souchier, and O. Barsotti.** 1999. Automatic enumeration of adherent streptococci or actinomyces on dental alloy by fluorescence image analysis. *J. Microbiol. Methods* **38:**33–42.
86. **Grossart, H.-P., G. F. Steward, J. Martinez, and F. Azam.** 2000. A simple, rapid method for demonstrating bacterial flagella. *Appl. Environ. Microbiol.* **66:**3632–3636.
87. **Guggenheim, M., S. Shapiro, R. Gmür, and B. Guggenheim.** 2001. Spatial arrangements and associative behavior of species in an in vitro oral biofilm model. *Appl. Environ. Microbiol.* **67:**1343–1350.
88. **Guiot, E., P. Georges, A. Brun, M. P. Fontaine, M.-N. Bellon-Fontaine, and R. Briandet.** 2002. Heterogeneity of diffusion inside microbial biofilms determined by fluorescence correlation spectroscopy under two photon excitation. *Photochem. Photobiol.* **75:**570–578.
89. **Hahn, D., R. I. Amann, W. Ludwig, A. D. L. Akkermans, and K.-H. Schleifer.** 1992. Detection of micro-organisms in soil after in situ hybridization with rRNA-targeted, fluorescently labelled oligonucleotides. *J. Gen. Microbiol.* **138:**879–887.
90. **Han, M., X. Gao, J. Z. Su, and S. Nie.** 2001. Quantum-dot-tagged microbeads for multiplexed optical coding of biomolecules. *Nat. Biotechnol.* **19:**631–635.
91. **Hansen, M. C., R. J. Palmer, Jr., C. Udsen, D. C. White, and S. Molin.** 2001. Assessment of GFP fluorescence in cells of *Streptococcus gordonii* under conditions of low pH and low oxygen concentration. *Microbiology* **147:** 1383–1391.
92. **Hassan, M., G. Corkidi, E. Galindo, C. Flores, and L. Serrano-Carreon.** 2002. Accurate and rapid viability assessment of *Trichoderma harzianum* using fluorescence-based digital image analysis. *Biotechnol. Bioeng.* **80:**677–684.
93. **Haugland, R. P.** 2002. *Handbook of Fluorescent Probes and Research Chemicals*. Molecular Probes, Inc., Eugene, Oreg.
94. **Haydon, P. G.** 2003. Biological near-field microscopy. *Methods Enzymol.* **360:**501–508.
95. **Heidelberg, J. F., M. Shahamat, M. Levin, I. Rahman, G. Stelma, C. Grim, and R. R. Colwell.** 1997. Effect of aerosolization on culturability and viability of gram-negative bacteria. *Appl. Environ. Microbiol.* **63:**3585–3588.
96. **Hell, S. W.** 2003. Toward fluorescence nanoscopy. *Nat. Biotechnol.* **21:**1347-1355.
97. **Hell, S. W., and M. Nagorni.** 1998. 4Pi confocal microscopy with alternate interference. *Opt. Lett.* **23:** 1567–1569.
98. **Hendry, M. J., J. R. Lawrence, and P. Maloszewski.** 1997. The role of sorption in the transport of *Klebsiella oxytoca* through saturated silica sand: effects of scale of investigation. *Ground Water* **35:**574–584.
99. **Henk, M. C.** 2004. Method for collecting air-water interface microbes suitable for subsequent microscopy and molecular analyses in both teaching and research laboratories. *Appl. Environ. Microbiol.* **70:**2486–2493.
100. **Henrici, A. T.** 1933. Studies of freshwater bacteria. I. A direct microscopic technique. *J. Bacteriol.* **25:**277–286.
101. **Heydorn, A., A. T. Nielsen, M. Hentzer, C. Sternberg, M. Givskov, B. K. Ersböll, and S. Molin.** 2000. Quantification of biofilm structures by the novel computer program COMSTAT. *Microbiology* **146:**2395–2407.
102. **Hink, M. A., J. W. Borst, and A. J. W. G. Visser.** 2003. Fluorescence correlation spectroscopy of GFP fusion proteins in living plant cells. *Methods Enzymol.* **361:**93–112.
103. **Hobbie, J. E., R. J. Daley, and S. Jasper.** 1977. Use of Nuclepore filters for counting bacteria by fluorescence microscopy. *Appl. Environ. Microbiol.* **33:**1225–1228.
104. **Hodson, R. E., W. A. Dustman, R. P. Garg, and M. A. Moran.** 1995. In situ PCR for visualization of microscale distribution of specific genes and gene products in prokaryotic communities. *Appl. Environ. Microbiol.* **61:** 4074–4082.
105. **Holloway, C. F., and J. P. Cowen.** 1997. Development of a scanning confocal laser microscopic technique to examine the structure and composition of marine snow. *Limnol. Oceanogr.* **42:**1340–1352.
106. **Hu, Z., G. Hidalgo, P. L. Houston, A. G. Hay, M. L. Shuler, H. D. Abruña, W. C. Ghiorse, and L. W. Lion.** 2005. Determination of spatial distributions of zinc and active biomass in microbial biofilms by two-photon laser scanning microscopy. *Appl. Environ. Microbiol.* **71:**4014–4021.
107. **Huang, K.-T., K. D. Xu, G. A. McFeters, and P. S. Stewart.** 1998. Spatial patterns of alkaline phosphatase expression within bacterial colonies and biofilms in response to phosphate starvation. *Appl. Environ. Microbiol.* **64:**1526–1531.
108. **Hunter, R. C., and T. J. Beveridge.** 2005. Application of a pH-sensitive fluoroprobe (C-SNARF-4) for pH microenvironment analysis in *Pseudomonas aeruginosa* biofilms. *Appl. Environ. Microbiol.* **71:**2501–2510.
109. **Imai, T., and T. Ohno.** 1995. The relationship between viability and intracellular pH in the yeast *Saccharomyces cereviseae*. *Appl. Environ. Microbiol.* **61:**3604-3608.
110. **Inoué, S.** 1997. *Video Microscopy*. Plenum Press, New York, N.Y.
111. **Karthikeyan, S., G. M. Wolfaardt, D. R. Korber, and D. E. Caldwell.** 1999. Functional and structural responses of a degradative microbial community to substrates with varying degrees of complexity in chemical structure. *Microb. Ecol.* **38:**215–224.
112. **Kelly, S. D., M. I. Boyanov, B. A. Bunker, J. B. Fein, D. A. Fowle, N. Yee, and K. M. Kemner.** 2001. XAFS determination of the bacterial cell wall functional groups responsible for complexation of Cd and U as a function of pH. *J. Synchrotron Radiat.* **8:**946–948.
113. **Kepner, R. L., and J. R. Pratt.** 1994. Use of fluorochromes for direct enumeration of total bacteria in environmental samples: past and present. *Microbiol. Rev.* **58:**603–615.
114. **Kitaguchi, A., N. Yamaguchi, and M. Nasu.** 2005. Enumeration of respiring *Pseudomonas* spp. in milk by fluorescence in situ hybridization following formazan reduction. *Appl. Environ. Microbiol.* **71:**2748–2752.
115. **Klar, T. A., S. Jakobs, M. Dyba, A. Egner, and S. W. Hell.** 2000. Fluorescence microscopy with diffraction resolution barrier broken by stimulated emission. *Proc. Natl. Acad. Sci. USA* **97:**8206–8210.

116. **Kloepfer, J. A., R. E. Mielke, W. S. Wong, K. H. Nealson, G. Stucky, and J. L. Nadeau.** 2003. Quantum dots as strain- and metabolism-specific microbiological labels. *Appl. Environ. Microbiol.* **69:**4205–4213.
117. **Kogure, K., U. Simidu, and N. Taga.** 1979. A tentative direct microscopic method for counting living marine bacteria. *Can. J. Microbiol.* **24:**415–420.
118. **Kolari, M., K. Mattila, R. Mikkola, and M. S. Salkinoja-Salonen.** 1998. Community structure of biofilms on ennobled stainless steel in Baltic Sea water. *J. Ind. Microbiol. Biotechnol.* **21:**261–275.
119. **Korber, D. R., A. Choi, G. M. Wolfaardt and D. E. Caldwell.** 1996. Bacterial plasmolysis as a physical indicator of viability. *Appl. Environ. Microbiol.* **62:**3939–3947.
120. **Korber, D. R., A. Choi, G. M. Wolfaardt, S. C. Ingham, and D. E. Caldwell.** 1997. Substratum topography influences susceptibility of *Salmonella enteritidis* biofilms to trisodium phosphate. *Appl. Environ. Microbiol.* **63:**3352–3358.
121. **Korber, D. R., G. A. James, and J. W. Costerton.** 1994. Evaluation of fleroxacin activity against established *Pseudomonas fluorescens* biofilms. *Appl. Environ. Microbiol.* **60:**1663–1669.
122. **Korber, D. R., J. R. Lawrence, B. Sutton, and D. E. Caldwell.** 1989. Effect of laminar flow velocity on the kinetics of surface recolonization by mot^+ and mot^- *Pseudomonas fluorescens*. *Microb. Ecol.* **18:**1-19.
123. **Korber, D. R., J. R. Lawrence, K. E. Cooksey, B. Cooksey, and D. E. Caldwell.** 1989. Computer image analysis of diatom chemotaxis. *Binary* **1:**155–168.
124. **Korber, D. R., J. R. Lawrence, M. J. Hendry, and D. E. Caldwell.** 1992. Programs for determining representative areas of microbial biofilms. *Binary* **4:**204–210.
125. **Korber, D. R., J. R. Lawrence, M. J. Hendry, and D. E. Caldwell.** 1993. Analysis of spatial variability within mot^+ and mot^- *Pseudomonas fluorescens* biofilms using representative elements. *Biofouling* **7:**339–358.
126. **Korber, D. R., G. M. Wolfaardt, V. Brözel, R. MacDonald, and T. Niepel.** 1999. Reporter systems for microscopic analysis of microbial biofilms. *Methods Enzymol.* **310:**3–20.
127. **Krambeck, C., H.-J. Krambeck, D. Schröder, and S. Y. Newell.** 1990. Sizing bacterioplankton: a juxtaposition of bias due to shrinkage, halos, subjectivity in image interpretation and symmetric distributions. *Binary* **2:**5–14.
128. **Kuehn, M., M. Hausner, H.-J. Bungartz, M. Wagner, P. A. Wilderer, and S. Wuertz.** 1998. Automated confocal laser scanning microscopy and semiautomated image processing for analysis of biofilms. *Appl. Environ. Microbiol.* **64:**4115–4127.
129. **Lam, J. S., and L. M. Mutharia.** 1994. Antigen-antibody reactions, p. 104–132. *In* P. Gerhardt, R. G. E. Murray, W. A. Wood, and N. R. Krieg (ed.), *Methods for General and Molecular Bacteriology*. ASM Press, Washington, D.C.
130. **Lawrence, J. R., M. Chenier, R. Roy, D. Beaumier, N. Fortin, G. D. W. Swerhone, T. R. Neu, and C. W. Greer.** 2004. Microscale and molecular assessment of the impacts of nickel, nutrients and oxygen level on river biofilm communities. *Appl. Environ. Microbiol.* **70:**4326–4339.
131. **Lawrence, J. R., G. Kopf, J. V. Headley, and T. R. Neu.** 2001. Sorption and metabolism of selected herbicides in river biofilm communities. *Can. J. Microbiol.* **47:**634–641.
132. **Lawrence, J. R., G. D. W. Swerhone, L. I. Wassenaar, and T. R. Neu.** 2005. Effects of selected pharmaceuticals on riverine biofilm communities. *Can. J. Microbiol.* **51:**655–669.
133. **Lawrence, J. R., T. R. Neu, and G. D. W. Swerhone.** 1998. Application of multiple parameter imaging for the quantification of algal, bacterial and exopolymer components of microbial biofilms. *J. Microbiol. Methods* **32:** 253–261.
134. **Lawrence, J. R., and R. A. Snyder.** 1998. Feeding behaviour and grazing impacts of a *Euplotes* sp. on attached bacteria. *Can. J. Microbiol.* **44:**623–629.
135. **Lawrence, J. R., and T. R. Neu.** Submitted for publication.
136. **Lawrence, J. R., G. D. W. Swerhone, and T. R. Neu.** 2000. Design and evaluation of a simple rotating annular reactor for replicated biofilm studies. *J. Microbiol. Methods* **42:**215–224.
137. **Lawrence, J. R., D. R. Korber, and D. E. Caldwell.** 1989. Computer-enhanced darkfield microscopy for the quantitative analysis of bacterial growth and behavior on surfaces. *J. Microbiol. Methods* **10:**123–138.
138. **Lawrence, J. R., D. R. Korber, and D. E. Caldwell.** 1992. Behavioral analysis of *Vibrio parahaemolyticus* variants in high and low viscosity microenvironments using computer-enhanced microscopy. *J. Bacteriol.* **174:**5732–5739.
139. **Lawrence, J. R., D. R. Korber, and G. M. Wolfaardt.** 2002. Image analysis of microorganisms, p. 1674–1688. *In* G. Bitton (ed.), *Encyclopedia of Environmental Microbiology*, vol. 4. John Wiley & Sons, Inc., New York, N.Y.
140. **Lawrence, J. R., D. R. Korber, B. D. Hoyle, J. W. Costerton, and D. E. Caldwell.** 1991. Optical sectioning of microbial biofilms. *J. Bacteriol.* **173:**6558–6567.
141. **Lawrence, J. R., G. D. W. Swerhone, and Y. T. J. Kwong.** 1998. Natural attenuation of aqueous metal contamination by an algal mat. *Can. J. Microbiol.* **44:**825–832.
142. **Lawrence, J. R., G. M. Wolfaardt, and D. R. Korber.** 1994. Monitoring diffusion in biofilm matrices using confocal laser microscopy. *Appl. Environ. Microbiol.* **60:** 1166–1173.
143. **Lawrence, J. R., J. A. Malone, D. R. Korber, and D. E. Caldwell.** 1989b. Computer image enhancement to increase depth of field in phase contrast microscopy. *Binary* **1:**181–185.
144. **Lawrence, J. R., G. D. W. Swerhone, G. G. Leppard, T. Araki, X. Zhang, M. M. West, and A. P. Hitchcock.** 2003. Scanning transmission X-ray, laser scanning, and transmission electron microscopy mapping of the exopolymeric matrix of microbial biofilms. *Appl. Environ. Microbiol.* **69:**5543–5554.
145. **Lawrence, J. R., Y. T. J. Kwong, and G. D. W. Swerhone.** 1997. Colonization and weathering of natural sulfide mineral assemblages by *Thiobacillus ferrooxidans*. *Can. J. Microbiol.* **43:**68–78.
146. **Lee, N., P. H. Nielsen, K. H. Adreasen, S. Juretschko, J. L. Nielsen, K.-H. Schleifer, and M. Wagner.** 1999. Combination of fluorescent in situ hybridization and microautoradiography—a new tool for structure-function analyses in microbial ecology. *Appl. Environ. Microbiol.* **65:**1289–1297.
147. **Lens, P. N. L., D. De Beer, C. C. H. Cronenberg, F. P. Houwen, S. P. P. Ottengraf, and W. H. Verstraete.** 1993. Heterogeneous distribution of microbial activity in methanogenic aggregates: pH and glucose microprofiles. *Appl. Environ. Microbiol.* **59:**3803–3815.
148. **Leppard, G. G., A. Heissenberger, and G. J. Herndl.** 1996. Ultrastructure of marine snow. I. Transmission electron microscopy methodology. *Mar. Ecol. Prog. Ser.* **135:**289–298.
149. **Lerner, J. M., and R. M. Zucker.** 2004. Calibration and validation of confocal spectral imaging systems. *Cytometry Part A* **62:**8–13.

150. **Lewis, A., H. Taha, A. Strinkovski, A. Manevitch, A. Khatchatouriants, R. Dekhter, and E. Ammann.** 2003. Near-field optics: from subwavelength illumination to nanometric shadowing. *Nat. Biotechnol.* **21:**1378–1386.
151. **Lin, H.-J., P. Herman, J. S. Kang, and J. R. Lakowicz.** 2001. Fluorescence lifetime characterization of novel low-pH probes. *Anal. Biochem.* **294:**118–125.
152. **Lindow, S. E., and M. T. Brandl.** 2003. Microbiology of the phyllosphere. *Appl. Environ. Microbiol.* **69:**1875–1883.
153. **Lisle, J. T., P. S. Stewart, and G. A. McFeters.** 1999. Fluorescent probes applied to physiological characterization of bacterial biofilms. *Methods Enzymol.* **310:**166–178.
154. **Liu, J., F. B. Dazzo, O. Glagoleva, B. Yu, and A. K. Jain.** 2001. CMEIAS: a computer-aided system for the image analysis of bacterial morphotypes in microbial communities. *Microb. Ecol.* **41:**173–194.
154a. **Liu, W.-T., and D. A. Stahl.** 2002. Molecular approaches for the measurement of density, diversity, and phylogeny, p. 114–134. *In* C. J. Hurst, R. L. Crawford, G. R. Knudsen, M. J. McInerney, and L. D. Stetzenbach (ed.), *Manual of Environmental Microbiology*, 2nd ed. ASM Press, Washington, D.C.
155. **Lopez, C., M. N. Pons, and E. Morgenroth.** 2005. Evaluation of microscopic techniques (epifluorescence microscopy, CLSM, TPE-LSM) as a basis for the quantitative image analysis of activated sludge. *Water Res.* **39:** 456–468.
156. **Lopez-Amoros, R. J. Comas, and J. Vives-Rego.** 1995. Flow cytometric assessment of *Escherichia coli* and *Salmonella typhimurium* starvation-survival in seawater using rhodamine 123, propidium iodide, and oxonol. *Appl. Environ. Microbiol.* **61:**2521–2526.
157. **Macedo, A. J., U. Kuhlicke, T. R. Neu, K. N. Timmis, and W.-R. Abraham.** 2005. Three stages of a biofilm community developing at the liquid-liquid interface between polychlorinated biphenyls and water. *Appl. Environ. Microbiol.* **71:**7301–7309.
158. **Manz, W., K. Wendt-Potthoff, T. R. Neu, U. Szewzyk, and J. R. Lawrence.** 1999. Phylogenetic composition, spatial structure, and dynamics of lotic bacterial biofilms investigated by *in situ* hybridization and confocal laser scanning microscopy. *Microb. Ecol.* **37:**225–237.
159. **Manz, W., M. Eisenbrecher, T. R. Neu, and U. Szewzyk.** 1998. Abundance and spatial organization of gram-negative sulfate-reducing bacteria in activated sludge investigated by in situ probing with specific 16S rRNA targeted oligonucleotides. *FEMS Microbiol. Ecol.* **25:**43–61.
160. **Marshall, K. C.** 1986. Microscopic methods for the study of bacterial behaviour at inert surfaces. *J. Microbiol. Methods* **4:**217–227.
161. **Mason, D. J., S. Shanmuganathan, F. C. Mortimer, and V. A. Gant.** 1998. A fluorescent gram stain for flow cytometry and epifluorescence microscopy. *Appl. Environ. Microbiol.* **64:**2681–2685.
162. **Mason, J. D., L.-A. R. Allman, J. M. Stark, and D. Lloyd.** 1995. The ability of membrane potential dyes and calcofluor white to distinguish between viable and non-viable bacteria. *J. Appl. Bacteriol.* **78:**309–315.
163. **Meyer-Reil, L.-A.** 1978. Autoradiography and epifluorescence microscopy combined for the determination of number and spectrum of actively metabolizing bacteria in natural waters. *Appl. Environ. Microbiol.* **36:**506–512.
164. **Michael, T., and C. M. Smith.** 1995. Lectins probe molecular films in biofouling: characterization of early films on non-living and living surfaces. *Mar. Ecol. Prog. Ser.* **119:**229–236.
165. **Millard, A. C., P. J. Campagnola, W. Mohler, A. Lewis, and L. M. Loew.** 2003. Second harmonic imaging microscopy. *Methods Enzymol.* **361:**47–69.
165a. **Mills, A. L., and J. L. Garland.** 2002. Application of physiological profiles to assessment of community properties, p. 135–146. *In* C. J. Hurst, R. L. Crawford, G. R. Knudsen, M. J. McInerney, and L. D. Stetzenbach (ed.), *Manual of Environmental Microbiology*, 2nd ed. ASM Press, Washington, D.C.
166. **Molin, S., and M. Givskov.** 1999. Application of molecular tools for *in situ* monitoring of bacterial growth activity. *Environ. Microbiol.* **1:**383–391.
167. **Møller, S., C. Sternberg, J. B. Andersen, B. B. Christensen, J. L. Ramos, M. Givskov, and S. Molin.** 1998. In situ gene expression in mixed-culture biofilms: evidence of metabolic interactions between community members. *Appl. Environ. Microbiol.* **64:**721–732.
168. **Møller, S., C. S. Kristensen, L. K. Poulsen, J. M. Carstensen, and S. Molin.** 1995. Bacterial growth on surfaces: automated image analysis for quantification of growth rate-related parameters. *Appl. Environ. Microbiol.* **61:**741–748.
169. **Møller, S., D. R. Korber, G. M. Wolfaardt, S. Molin, and D. E. Caldwell.** 1997. The impact of nutrient composition on a degradative biofilm community. *Appl. Environ. Microbiol.* **63:**2432–2438.
170. **Morgan, P., C. J. Cooper, N. S. Battersby, S. A. Lee, S. T. Lewis, T. M. Machin, S. C. Graham, and R. J. Watkinson.** 1991. Automated image analysis method to determine fungal biomass in soil and on solid matrices. *Soil Biol. Biochem.* **23:**609–616.
171. **Morris, C. E., J.-M. Monier, and M.-A. Jacques.** 1997. Methods for observing microbial biofilms directly on leaf surfaces and recovering them for isolation of culturable microorganisms. *Appl. Environ. Microbiol.* **63:**1570–1576.
172. **Müller, J. D., Y. Chen, and E. Gratton.** 2003. Fluorescence correlation spectroscopy. *Methods Enzymol.* **361:**69–92.
173. **Murga, R., P. S. Stewart, and D. Daly.** 1995. Quantitative analysis of biofilm thickness variability. *Biotechnol. Bioeng.* **45:**503–510.
174. **Murray, R. G. E., and C. F. Robinow.** 1994. Light microscopy, p. 8–20. *In* P. Gerhardt (ed.), *Methods for General and Molecular Bacteriology*. ASM Press, Washington, D.C.
175. **Murray, R. G. E., R. N. Doetsch, and C. F. Robinow.** 1994. Determinative and cytological light microscopy, p. 21–41. *In* P. Gerhardt (ed.), *Methods for General and Molecular Bacteriology*. ASM Press, Washington, D.C.
176. **Nedoma, J., J. Vrba, T. Hanzl, and L. Nedbalova.** 2001. Quantification of pelagic filamentous microorganisms in aquatic environments using the line-intercept method. *FEMS Microbiol. Ecol.* **38:**81–85.
177. **Nedoma, J., A. Strojsova, J. Vrba, J. Komarkova, and K. Simek.** 2003. Extracellular phosphatase activity of natural plankton studied with ELF97 phosphate: fluorescence quantification and labelling kits. *Environ. Microbiol.* **5:**462–472.
178. **Neu, T., and K. C. Marshall.** 1991. Microbial "footprints"—a new approach to adhesive polymers. *Biofouling* **3:**101–112.
179. **Neu, T. R., and J. R. Lawrence.** 2002. Laser scanning microscopy in combination with fluorescence techniques for biofilm study, p. 1772–1788. *In* G. Bitton (ed.), *The Encyclopedia of Environmental Microbiology*. John Wiley & Sons, Inc., New York, N.Y.
180. **Neu, T. R., S. Woelfl, and J. R. Lawrence.** 2004. Three-dimensional differentiation of photo-autotrophic biofilm constituents by multi-channel laser scanning microscopy (single-photon and two-photon excitation). *J. Microbiol. Methods* **56:**161–172.

181. **Neu, T. R.** 2000. *In situ* cell and glycoconjugate distribution in river snow studied by confocal laser scanning microscopy. *Aquat. Microb. Ecol.* **21:**85–95.
182. **Neu, T. R., and J. R. Lawrence.** 1997. Development and structure of microbial stream biofilms as studied by confocal laser scanning microscopy. *FEMS Microb. Ecol.* **24:**11–25.
183. **Neu, T. R., and J. R. Lawrence.** 1999. Lectin-binding analysis in biofilm systems. *Methods Enzymol.* **310:**145–152.
184. **Neu, T. R., and J. R. Lawrence.** 1999. *In situ* characterization of extracellular polymeric substances in biofilm systems, p. 21–47. *In* J. Wingender, T. R. Neu, and H.-C. Flemming (ed.), *Microbial Extracellular Polymeric Substances: Characterization, Structure and Function.* Springer International, Berlin, Germany.
185. **Neu, T. R., G. D. W. Swerhone, and J. R. Lawrence.** 2001. Assessment of lectin-binding-analysis for in situ detection of glycoconjugates in biofilm systems. *Microbiology* **147:**299–313.
186. **Neu, T. R., G. D. W. Swerhone, U. Bockelmann, and J. R. Lawrence.** 2005. Effect of CNP on composition and structure of lotic biofilms as detected with lectin-specific glycoconjugates. *Aquat. Microb. Ecol.* **38:**283–295.
187. **Neu, T. R., P. Walczysko, and J. R. Lawrence.** 2004. Two-photon imaging for studying the microbial ecology of biofilm systems. *Microbes Environ.* **19:**1–6.
188. **Neu, T. R., and J. R. Lawrence.** 2005. One-photon versus two-photon laser scanning microscopy and digital image analysis of microbial biofilms. *Methods Microbiol.* **34:**87–134.
189. **Newman, K. L., R. P. P. Almeida, A. H. Purcell, and S. E. Lindow.** 2003. Use of a green fluorescent strain for analysis of *Xylella fastidiosa* colonization of *Vitis vinifera*. *Appl. Environ. Microbiol.* **69:**7319–7327.
190. **Nielsen, J. L., L. H. Mikkelsen, and P. H Nielsen.** 2000. *In situ* detection of cell surface hydrophobicity of probe-defined bacteria in activated sludge. *Water Sci. Technol.* **43:**97–103.
191. **Nielsen, J. L., M. Aquino de Muro, and P. H. Nielsen.** 2003. Evaluation of the redox dye 5-cyano-2,3-tolyl-tetrazolium chloride for activity studies by simultaneous use of microautoradiography and fluorescence in situ hybridization. *Appl. Environ. Microbiol.* **69:**641–643.
192. **Normander, B., N. B. Hendriksen, and O. Nybroe.** 1999. Green fluorescent protein-marked *Pseudomonas fluorescens*: localization, viability, and activity in the natural barley rhizosphere. *Appl. Environ. Microbiol.* **65:** 4646–4651.
193. **Nour, S. M., J. R. Lawrence, H. Zhu, G. D. W. Swerhone, M. Welsh, T. W. Welacky, and E. Topp.** 2003. Bacteria associated with cysts of the soybean cyst nematode (*Heterodera glycines*). *Appl. Environ. Microbiol.* **69:**607–615.
194. **Nunan, N., K. Ritz, D. Crabb, K. Harris, K. Wu, J. W. Crawford, and I. M. Young.** 2001. Quantification of the *in situ* distribution of soil bacteria by large-scale imaging of thin sections of undisturbed soil. *FEMS Microbiol. Ecol.* **37:**67–77.
195. **Nyholm, S. V., B. Deplancke, H. H. Gaskins, M. A. Apicella, and M. J. McFall-Ngai.** 2002. Roles of *Vibrio fischeri* and nonsymbiotic bacteria in the dynamics of mucus secretion during symbiont colonization of the *Euprymna scolopes* light organ. *Appl. Environ. Microbiol.* **68:**5113–5122.
196. **Okabe, S., H. Naitoh, H. Satoh, and Y. Watanabe.** 2002. Biofilm composition and process—structure and function of nitrifying biofilms as determined by molecular techniques and the use of microelectrodes. *Water Sci. Technol.* **46:**233–243.
197. **Olsen, K. N., B. B. Budde, H. Siegumfeldt, K. B. Rechinger, M. Jakobsen, and H. Ingmer.** 2002. Noninvasive measurement of bacterial intracellular pH on a single-cell level with green fluorescent protein and fluorescence ratio imaging microscopy. *Appl. Environ. Microbiol.* **68:**4145–4147.
198. **O'Mahony, R., C. Basset, J. Holton, D. Vaira, and I. Roitt.** 2005. Comparison of image analysis software packages in the assessment of adhesion of microorganisms to mucosal epithelium using confocal laser scanning microscopy. *J. Microbiol. Methods* **61:**105–126.
199. **Packroff, G., J. R. Lawrence, and T. R. Neu.** 2002. *In situ* confocal laser scanning microscopy of protozoans in pure cultures and complex communities. *Acta Protozool.* **41:**245–253.
200. **Palmer, R. J., Jr., and D. C. White.** 1999. Spatially resolved, quantitative determination of luciferase activity by photon counting microscopy. *Methods Enzymol.* **310:** 152–160.
201. **Pawley, J.** 2000. The 39 steps: a cautionary tale of quantitative 3-D fluorescence microscopy. *BioTechniques* **28:** 884–888.
202. **Pawley, J. B.** 1995. *Handbook of Biological Confocal Microscopy.* Plenum Press, New York, N.Y.
202a. **Pepper, I. L., and S. E. Dowd.** 2002. PCR applications for plant and soil microbes, p. 573–582. *In* C. J. Hurst, R. L. Crawford, G. R. Knudsen, M. J. McInerney, and L. D. Stetzenbach (ed.), *Manual of Environmental Microbiology*, 2nd ed. ASM Press, Washington, D.C.
203. **Perenthaler, J., F. O. Glöckner, W. Schönhuber, and R. Amann.** 2001. Fluorescence *in situ* hybridization (FISH) with rRNA targeted oligonucleotide probes. *Methods Microbiol.* **30:**207–226.
204. **Perfil'ev, B. V., and D. R. Gabe.** 1969. *Capillary Methods of Investigating Micro-Organisms.* (Translated by J. M. Shewan.) University of Toronto Press, Toronto, Ontario, Canada.
205. **Pernthaler, A., and R. Amann.** 2004. Simultaneous fluorescence in situ hybridization of mRNA and rRNA in environmental bacteria. *Appl. Environ. Microbiol.* **70:** 5426–5433.
206. **Petrisor, A. I., T. Kawaguchi, and A.W. Decho.** 2004. Quantifying $CaCO_3$ microprecipitates within developing surface mats of marine stromatolites using GIS and digital image analysis. *Geomicrob. J.* **21:**491–496.
207. **Petrisor, A. I., and A. W. Decho.** 2004. Using geographical information techniques to quantify the specific structure of endolithic boring processes within sediment grains of marine stromatolites. *J. Microbiol. Methods* **56:**173–180.
208. **Pomroy, A. J.** 1984. Direct counting of bacteria preserved with Lugol iodine solution. *Appl. Environ. Microbiol.* **47:**1191–1192.
209. **Popkin, T. J.** 1994. Photography, p. 735–762. *In* P. Gerhardt, R. G. E. Murray, W. A. Wood, and N. R. Krieg (ed.), *Methods for General and Molecular Bacteriology.* ASM Press, Washington, D.C.
210. **Porter, J., D. Deere, M. Hardman, C. Edwards, and R. Pickup.** 1997. Go with the flow—use of flow cytometry in environmental microbiology. *FEMS Microbiol. Ecol.* **24:**93–101.
211. **Porter, K. G., and Y. S. Feig.** 1980. The use of DAPI for identifying and counting aquatic microflora. *Limnol. Oceanogr.* **25:**943–948.
212. **Poulsen, L. K., G. Ballard, and D. A. Stahl.** 1993. Use of rRNA fluorescence in situ hybridization for measuring the activity of single cells in young and established biofilms. *Appl. Environ. Microbiol.* **59:**1354–1360.
213. **Ramos, C., L. Mølbak, and S. Molin.** 2000. Bacterial activity in the rhizosphere analyzed at the single-cell

level by monitoring ribosome contents and synthesis rates. *Appl. Environ. Microbiol.* **66:**801–809.

214. **Rocheleau, S., C. W. Greer, J. R. Lawrence, C. Cantin, L. Laramee, and S. Guiot.** 1999. Differentiation of *Methanosaeta concilii* and *Methanosarcina barkeri* in anaerobic mesophilic granular sludge by fluorescent in situ hybridization and confocal scanning laser microscopy. *Appl. Environ. Microbiol.* **65:**2222–2229.
215. **Rodenacker, K., M. Hausner, and A. A. Gorbushina.** 2003. Quantification and spatial relationship of microorganisms in sub-aquatic and subaerial biofilms, p. 387–399. *In* W. E. Krumbein, D. M. Paterson, and G. A. Zavarzin (ed.), *Fossil and Recent Biofilms.* Kluwer Academic Publishers, Dordrecht, The Netherlands.
216. **Rodriguez, G. G., D. Phipps, K. Ishiguro, and H. F. Ridgway.** 1992. Use of a fluorescent redox probe for direct visualization of actively respiring bacteria. *Appl. Environ. Microbiol.* **58:**1801–1808.
217. **Rogers, J., and C. W. Keevil.** 1992. Immunogold and fluorescein immunolabelling of *Legionella pneumophila* within an aquatic biofilm visualized by episcopic differential interference contrast microscopy. *Appl. Environ. Microbiol.* **58:**2326–2330.
218. **Rossner, M., and K. M. Yamada.** 2004. What's in a picture? The temptation of image manipulation. *J. Cell Biol.* **166:**11–15.
219. **Rost, F. W. D.** 1992. *Fluorescence microscopy*, vol. I and II. Cambridge University Press, Cambridge, United Kingdom.
220. **Russ, J. C.** 2002. *The Image Processing Handbook*, 3rd ed. CRC Press, Inc. Boca Raton, Fla.
221. **Sanders, R., A. Draaijer, H. C. Gerritsen, P. M. Houpt, and Y. K. Levine.** 1995. Quantitative pH imaging in cells using confocal fluorescence lifetime imaging microscopy. *Anal. Biochem.* **227:**302–308.
222. **Schloter, M., W. Wiehe, B. Assmus, H. Steindl, H. Becke, G. Höflich, and A. Hartmann.** 1997. Root colonization of different plants by plant-growth-promoting *Rhizobium leguminosarum* bv. *trifolii* R39 studied with monospecific polyclonal antibodies. *Appl. Environ. Microbiol.* **63:**2038–2046.
223. **Schmid, M., A. Thill, U. Purkhold, M. Walcher, J. Y. Bottero, P. Ginestet, P. H. Nielsen, S. Wuertz, and M. Wagner.** 2003. Characterization of activated sludge flocs by confocal laser scanning microscopy and image analysis. *Water Res.* **37:**2043–2052.
224. **Schrader, M., K. Bahlmann, G. Giese, and S. W. Hell.** 1998. 4Pi-confocal imaging in fixed biological specimens. *Biophys. J.* **75:**1659–1668.
225. **Schramm, A., D. De Beer, J. C. van den Heuvel, S. Ottengraf, and R. Amann.** 1999. Microscale distribution of populations and activities of *Nitrosospira* and *Nitrospira* spp. along a macroscale gradient in a nitrifying bioreactor: quantification by in situ hybridization and the use of microsensors. *Appl. Environ. Microbiol.* **65:**3690–3696.
226. **Schramm, A., D. De Beer, M. Wagner, and R. Amann.** 1998. Identification and activities in situ of *Nitrosospira* and *Nitrospira* spp. as dominant populations in a nitrifying fluidized bed reactor. *Appl. Environ. Microbiol.* **64:** 3480–3485.
227. **Shakespeare, A. P., and J. Verran.** 1988. The use of automated image analysis for rapid measurement of the *in vitro* attachment of *Candida albicans* to transparent acrylic. *Lett. Appl. Microbiol.* **6:**79–83.
228. **Shaw, P. J.** 1995. Comparison of wide-field/deconvolution and confocal microscopy for 3D imaging, p. 373–387. *In* J. B. Pawley (ed.), *Handbook of Biological Confocal Microscopy*. Plenum Press, New York, N.Y.
229. **Shaw, P. J., and D. J. Rawlins.** 1991. The point-spread function of a confocal microscope: its measurement and use in deconvolution of 3-D data. *J. Microsc.* **163:**151–165.
230. **Sherr B. F., P. A. Giorgio, and E. B. Sherr.** 1999. Estimating abundance and single-cell characteristics of respiring bacteria via the redox dye CTC. *Aquat. Microb. Ecol.* **18:**117–131.
231. **Shotton, D. M.** 1989. Confocal scanning optical microscopy and its applications for biological specimens. *J. Cell Sci.* **94:**175–206.
232. **Sieracki, M. E., P. W. Johnson, and J. M. Sieburth.** 1985. Detection, enumeration and sizing of planktonic bacteria by image-analyzed epifluorescence microscopy. *Appl. Environ. Microbiol.* **49:**799–810.
233. **Sieracki, M. E., S. E. Reichenbach, and K. L. Webb.** 1989. Evaluation of automated threshold selection methods for accurately sizing microscopic fluorescent cells by image analysis. *Appl. Environ. Microbiol.* **55:**2762–2772.
234. **Sizemore, R. K., J. J. Caldwell, and A. S. Kendrick.** 1990. Alternate Gram staining technique using a fluorescent lectin. *Appl. Environ. Microbiol.* **56:**2245–2247.
235. **Sjollema, J., H. J. Busscher, and A. H. Weerkamp.** 1989. Real-time enumeration of adhering microorganisms in a parallel plate flow cell using automated image analysis. *J. Microbiol. Methods* **9:**73–78.
236. **Söderström, B. E.** 1977. Vital staining of fungi in pure cultures and in soil with fluorescein diacetate. *Soil Biol. Biochem.* **9:**59–63.
237. **Sole, A., N. Gaju, S. Mendez-Alvarez, and I. Esteve.** 2001. Confocal laser scanning microscopy as a tool to determine cyanobacteria biomass in microbial mats. *J. Microsc.* **204:**258–262.
238. **Sorensen, S. J., M. Bailey, L. H. Hansen, N. Kroer, and S. Wuertz.** 2005. Studying plasmid horizontal transfer in situ: a critical review. *Nat. Microbiol. Rev.* **3:**700–710.
239. **Southward, C. M., and M. G. Surette.** 2002. The dynamic microbe: green fluorescent protein brings bacteria to light. *Mol. Microbiol.* **45:**1191–1196.
240. **Stahl, D. A., B. Flesher, H. R. Mansfield, and L. Montgomery.** 1988. Use of phylogenetically based hybridization probes for studies of ruminal microbial ecology. *Appl. Environ. Microbiol.* **54:**1079–1084.
241. **Staley, J. T.** 1971. Growth rates of algae determined *in situ* using an immersed microscope. *J. Phycol.* **7:**13–17.
242. **Staudt, C. H. Horn, D. C. Hempel, and T. R. Neu.** 2003. Screening of lectins for staining lectin-specific glycoconjugates in the EPS of biofilms, p. 308–326. *In* P. Lens, A. P. Moran, T. Mahony, P. Stoodley, and V. O'Flaherty (ed.), *Biofilms in Medicine, Industry and Environmental Biotechnology*. IWA Publishing, London, United Kingdom.
243. **Staudt, C., H. Horn, D. C. Hempel, and T. R. Neu.** 2004. Volumetric measurements of bacterial cells and EPS glycoconjugates in biofilms. *Biotechnol. Bioeng.* **88:** 585–592.
244. **Sternberg, C., B. B. Christensen, T. Johansen, A. T. Nielsen, J. B. Andersen, M. Givskov, and S. Molin.** 1999. Distribution of bacterial growth activity in flow-chamber biofilms. *Appl. Environ. Microbiol.* **65:**4108–4117.
245. **Surman, S. B., J. T. Walker, D. T. Goddard, L. H. G. Morton, C. W. Keevil, W. Weaver, A. Skinner, K. Hanson, D. Caldwell, and J. Kurtz.** 1996. Comparison of microscope techniques for the examination of biofilms. *J. Microbiol. Methods* **25:**57-70.
246. **Szmacinski, H., and J. R. Lakowicz.** 1993. Optical measurements of pH using fluorescent lifetimes and phase modulation fluorometry. *Anal. Chem.* **65:**1668–1674.
247. **Tani, K., K. Kurokawa, and M. Nasu.** 1998. Development of a direct in situ PCR method for detection of

specific bacteria in natural environments. *Appl. Environ. Microbiol.* **64:**1536–1540.

248. **Thomas, C. R.** 1992. Image analysis: putting filamentous microorganisms in the picture. *Trends Biotechnol.* **10:** 343–348.

249. **Thompson, E.** 1930. Quantitative microscopic analysis. *J. Geol.* **38:**193.

250. **Thurnheer, T., R. Gmür, S. Shapiro, and B. Guggenheim.** 2003. Mass transport of macromolecules within an in vitro model of supragingival plaque. *Appl. Environ. Microbiol.* **69:**1702–1709.

251. **Tobin, K. J., T. C. Onstott, M. F. DeFlaun, F. S. Colwell, and J. Fredrickson.** 1999. *In situ* imaging of microorganisms in geologic material. *J. Microbiol. Methods* **37:**201–213.

252. **Tombolini, R., D. J. van der Gaag, B. Gerhardson, and J. K. Jansson.** 1999. Colonization pattern of the biocontrol strain *Pseudomonas chlororaphis* MA 342 on barley seeds visualized by using green fluorescent protein. *Appl. Environ. Microbiol.* **65:**3674–3680.

253. **Torrella, F., and R. Y. Morita.** 1981. Microcultural study of bacterial size changes and microcolony and ultramicrocolony formation by heterotrophic bacteria in seawater. *Appl. Environ. Microbiol.* **41:**518–527.

254. **Ulrich, S., B. Karrasch, H. G. Hoppe, K. Jeskulke, and M. Mehrens.** 1996. Toxic effects on bacterial metabolism of the redox dye 5-cyano-2,3-ditolyl tetrazolium chloride. *Appl. Environ. Microbiol.* **62:**4587–4593.

255. **van der Voort, H. T. M., and K. C. Strasters.** 1995. Restoration of confocal images for quantitative image analysis. *J. Microsc.* **178:**165–181.

256. **Van Kempen, G. M. P., L. J. Van Vliet, P. J. Verveer, and H. T. M. van der Voort.** 1997. A quantitative comparison of image restoration methods for confocal microscopy. *J. Microsc.* **185:**354–365.

257. **Van Ommen Kloeke, F., and G. G. Geesey.** 1999. Localization and identification of populations of phosphatase-active bacterial cells associated with activated sludge flocs. *Microb. Ecol.* **38:**201–214.

258. **Verveer, P. J., M. J. Gemkow, and T. M. Jovin.** 1999. A comparison of image restoration approaches applied to three-dimensional confocal and wide-field fluorescence microscopy. *J. Microsc.* **193:**50–61.

259. **Vigeant, M. A.-S., R. M. Ford, M. Wagner, and L. K. Tamm.** 2002. Reversible and irreversible adhesion of motile *Escherichia coli* cells analyzed by total internal reflection aqueous fluorescence microscopy. *Appl. Environ. Microbiol.* **68:**2794–2801.

260. **Viles, C., and M. E. Sieracki.** 1992. Measurement of marine picoplankton cell size by using a cooled, charge-coupled device camera with image-analyzed fluorescent microscopy. *Appl. Environ. Microbiol.* **58:**584–592.

261. **Vives-Rego, J., P. Lebaron, and G. Nebe-von Caron.** 2000. Current and future applications of flow cytometry in aquatic microbiology. *FEMS Microbiol. Rev.* **24:**429–448.

262. **Vroom, J. M., K. J. de Grauw, H. C. Gerritsen, D. J. Bradshaw, P. D. Marsh, G. K. Watson, J. J. Birmingham, and C. Allison.** 1999. Depth penetration and detection of pH gradients in biofilms by two-photon excitation microscopy. *Appl. Environ. Microbiol.* **65:** 3502–3511.

263. **Wagner, M., R. Amann, H. Lemmer, and K.-H. Schleifer.** 1993. Probing activated sludge with oligonucleotides specific for proteobacteria: inadequacy of culture-dependent methods for describing microbial community structure. *Appl. Environ. Microbiol.* **59:**1520–1525.

264. **Waid, J. S.** 1973. A method to study microorganisms on surface films from soil particles with the aid of the transmission electron microscope. *Bull. Ecol. Res. Commun. (Stockholm)* **17:**103–108.

265. **Walker, R. F., K. Ishikawa, and M. Kumagai.** 2002. Fluorescence-assisted image analysis of freshwater microalgae. *J. Microbiol. Methods* **51:**149–162.

266. **Webb, R. H., and C. K. Dorey.** 1995. The pixelated image, p. 55–67. *In* J. B. Pawley (ed.), *Handbook of Biological Confocal Microscopy*. Plenum Press, New York, N.Y.

267. **Wei, Y. D., K. N. Byer, and P. H. Goodwin.** 1997. Hemibiotrophic infection of round-leaved mallow by *Colletotrichum gloeosporioides* f. sp. *malvae* in relation to leaf senescence and reducing reagents. *Mycol. Res.* **101:** 357–364.

268. **Wentland, E. J, P. S. Stewart, C.-T. Huang, and G. A. McFeters.** 1996. Spatial variations in growth rate within *Klebsiella pneumoniae* colonies and biofilms. *Biotechnol. Prog.* **12:**316–321.

269. **Werner, E., F. Roe, A. Bugnicourt, M. J. Franklin, A. Heydorn, S. Molin, B. Pitts, and P. S. Stewart.** 2004. Stratified growth in *Pseudomonas aeruginosa* biofilms. *Appl. Environ. Microbiol.* **70:**6188–6196.

270. **Wigglesworth-Cooksey, B., and K. E. Cooksey.** 2005. Use of fluorophore-conjugated lectins to study cell-cell interactions in model marine biofilms. *Appl. Environ. Microbiol.* **71:**428–435.

271. **Wiggli, M., A. Smallcombe, and R. Bachofen.** 1999. Reflectance spectroscopy and laser confocal microscopy as tools in an ecophysiological study of microbial mats in an alpine bog pond. *J. Microbiol.* **34:**173–182.

272. **Wilkinson, M. H. F., and F. Schut.** 1998. *Digital Image Analysis of Microbes: Imaging, Morphometry, Fluorometry and Motility Techniques and Applications*. John Wiley & Sons, Inc., New York, N.Y.

273. **Wiltshire, K. H., J. Blackburn, and D. M. Paterson.** 1997. The Cryolander: a new method for fine-scale in situ sampling of intertidal surface sediments. *J. Sediment. Res.* **67:**977–981.

274. **Wimpenny, J. W. T.** 1988. *Handbook of Laboratory Model Systems for Microbial Ecosystems*, vol. 1 and 2. CRC Press, Inc., Boca Raton, Fla.

275. **Wit, P., and H. J. Busscher.** 1998. Application of an artificial neural network in the enumeration of yeasts and bacteria adhering to solid substrata. *J. Microbiol. Methods* **32:**281–290.

276. **Woese, C. R.** 1987. Bacterial evolution. *Microbiol. Rev.* **51:**221–271.

277. **Wolfaardt, G. M., J. R. Lawrence, M. J. Hendry, R. D. Robarts, and D. E. Caldwell.** 1993. Development of steady-state diffusion gradients for the cultivation of degradative microbial consortia. *Appl. Environ. Microbiol.* **59:**2388–2396.

278. **Wolfaardt, G. M., J. R. Lawrence, R. D. Robarts, and D. E. Caldwell.** 1994. Multicellular organization in a degradative biofilm community. *Appl. Environ. Microbiol.* **60:**434–446.

279. **Wolfaardt, G. M., J. R. Lawrence, R. D. Robarts, and D. E. Caldwell.** 1998. *In situ* characterization of biofilm exopolymers involved in the accumulation of chlorinated organics. *Microb. Ecol.* **35:**213–223.

280. **Wuertz, S., E. Mueller, R. Spaeth, P. Pfleiderer, and H.-C. Flemming.** 2000. Detection of heavy metals in bacterial biofilms and microbial flocs with the fluorescent complexing agent Newport Green. *J. Ind. Microbiol. Biotechnol.* **24:**116–123.

281. **Xavier, J. B., A. Schnell, S. Wuertz, R. Palmer, D. C. White, and J. S. Almeida.** 2001. Objective threshold selection procedure (OTS) for segmentation of scanning laser confocal microscope images. *J. Microbiol. Methods* **47:**169–180.

282. **Xu, K. D., P. S. Stewart, F. Xia, C.-T. Huang, and G. A. McFeters.** 1999. Spatial physiological heterogeneity in *Pseudomonas aeruginosa* biofilms is determined by oxygen availability. *Appl. Environ. Microbiol.* **64:**4035–4039.

283. **Yang, X. M., H. Beyenal, G. Harkin, and Z. Lewandowski.** 2000. Quantifying biofilm structure using image analysis. *J. Microbiol. Methods* **39:**109-119.

284. **Yang, X., H. Beyenal, G. Harkin, and Z. Lewandowski.** 2001. Evaluation of biofilm image thresholding methods. *Water Res.* **35:**1149–1158.

285. **Yu, F. P., and G. A. McFeters.** 1994. Physiological response of bacteria in biofilms to disinfection. *Appl. Environ. Microbiol.* **60:**2462–2466.

286. **Yu, F. P., G. M. Callis, P. S. Stewart, T. Griebe, and G. A. McFeters.** 1994. Cryosectioning of biofilms for microscopic examination. *Biofouling* **8:**85–91.

287. **Zhu, L., S. Ang, and W. T. Liu.** 2004. Quantum dots as a novel immunofluorescent detection system for *Cryptosporidium parvum* and *Giardia lamblia*. *Appl. Environ. Microbiol.* **70:**597–598.

288. **Zippel, B., and T. R. Neu.** 2005. Growth and structure of phototrophic biofilms under controlled light conditions. *Water Sci. Technol.* **52:**203–209.

289. **Zita, A., and M. Hermansson.** 1997. Determination of bacterial cell surface hydrophobicity of single cells in culture and in wastewater *in situ*. *FEMS Microbiol. Ecol.* **152:**299.

290. **Zucker, R. M., and O. Price.** 2001. Evaluation of confocal microscopy system performance. *Cytometry* **44:**273–295.

291. **Zumbusch, A., G. R. Holtom, and X. S. Xie.** 1999. Vibrational microscopy using coherent anti-Stokes Raman scattering. *Phys. Rev. Lett.* **82:**4142–4145.

292. **Zwirglmaier, K.** 2005. Fluorescence *in situ* hybridisation (FISH)—the next generation. *FEMS Microbiol. Lett.* **246:**151–158.

Cultivation of Bacteria and Fungi

RALPH S. TANNER

6

A brief review of the literature in applied and environmental microbiology quickly reveals a great number and variety of media used to culture bacteria and fungi in the laboratory. These range from simple formulations to commercially available media to media which require a very meticulous and skillful assembly to prepare (49, 66). Thousands of microbiological medium formulations are available from just a handful of sources (2, 5, 7, 18, 48). It may appear that cultivation of bacteria and fungi is a complicated and empirical art, requiring some skill, observation, and attention to detail (this is actually true to a significant degree). However, with knowledge of the basic concepts of medium composition and of the physical conditions which may limit microbial growth, one can enhance their ability to grow bacteria and fungi in pure culture and to enrich for, isolate, and culture many microorganisms of interest from the environment (29, 34). Many of the basic aspects of the culture of microorganisms have been outlined previously (16, 24, 28, 32). The major nutritional requirements for microbial growth, most of which were outlined by Gottschal et al. (27), that must be considered are sources of carbon; sources of energy (also called electron or hydrogen sources); electron acceptors; nitrogen sources; sources of other major nutrients, such as sulfur, phosphate, magnesium, and calcium; vitamin requirements; and trace metal requirements. Some physiochemical factors affecting growth include temperature, pH, requirement for oxygen, and salinity.

Several points should be kept in mind for use of the material in this chapter. The first is the importance and utility of considering culture conditions when working with bacteria and fungi. The research on some problems, such as the novel growth factor requirement of *Methanomicrobium mobile* (33, 57) or the biochemistry of dehalogenation in *Desulfomonile tiedjei* (17), was not really feasible until the finer points of microbial culture were determined. Consideration of this issue has also been useful for environmental studies, such as the culture of sulfate-reducing bacteria from environmental samples (55, 56, 62) or of bacteria from potable water (51). While important, however, the rationale for design of a medium or use of a specific culture condition is discussed in the literature infrequently. Secondly, many of the problems with the growth of microorganisms are due to incorrect or uncontrolled incubation temperature or culture pH, lack of a required nutrient, less than ideal provision of a carbon or energy source, and even excess of some medium component. Finally, it is not possible to thoroughly examine all of the specifics for the culture of all of the different bacteria and fungi in a single chapter. The relevant literature should be consulted for specific groups of organisms, such as thermophiles (20, 34), phototrophs (5, 22, 41), and fungi (7, 23, 34, 53, 61). However, the basics of microbial culture are reviewed in this chapter, and this information can be used in conjunction with that available from the literature.

BASIC MEDIUM PREPARATION

A recipe for a basic basal medium for cultivating microorganisms is given in Table 1. It includes a mineral solution which provides the major inorganic chemicals required by microorganisms, a vitamin solution, a trace metal solution, and a buffer for control of pH. This basal medium supports the growth of many types of microorganisms with the appropriate selection of carbon sources, energy sources, and electron acceptors (1). The selection of the various components and their roles are discussed below.

Mineral Solution

The composition of a mineral solution useful for preparing medium for many different microorganisms is given in Table 2. This mineral solution provides a source of sodium, chloride, ammonium as a nitrogen source, potassium, phosphate, magnesium, sulfate as a sulfur source, and calcium. While a mineral solution with this composition is useful for many different cultures, it may require modification for certain groups or species of microorganisms.

A requirement for sodium by most microorganisms other than marine or halophilic species has not been demonstrated. However, some phototrophs and anaerobic bacteria do require sodium for metabolism and growth (27). Many commercially prepared media contain some sodium chloride (18). The presence of a low concentration of sodium chloride in a medium does not appear to inhibit the culture of microorganisms. In contrast, potassium is required for the culture of essentially all microorganisms.

The use of ammonium as a source of available nitrogen is examined below (see "Nitrogen Sources"). It needs to be noted here that nitrate may be a preferred nitrogen source for the culture of cyanobacteria and aerobic soil microorganisms, for the enrichment and isolation of filamentous

TABLE 1 Basal microbiological medium[a]

Component	Amt/liter
Mineral solution[b]	10 ml
Vitamin solution[c]	10 ml
Trace metal solution[d]	0.5–5 ml
Buffer[e]	1–20 g
Adjust pH with N_a OH[f]	

[a]A general-purpose medium adaptable to pure culture or ecological studies with the appropriate selection of carbon sources, energy sources, and electron acceptors. A general source of growth factors, such as yeast extract (0.1 to 2.0 g/liter), must be included for many studies.

[b]Described in Table 2.

[c]Described in Table 3.

[d]Described in Table 4.

[e]The buffer used is selected for each medium: TES (pK_a 7.4) for neutral pHs, $NaHCO_3$ for CO_2-containing gas phases, MES (pK_a 6.1) for acidic pHs, phosphate for lactic acid bacteria or thiobacilli, TAPS (pK_a 8.4) or glycine (pK_a 9.8) for alkaline pHs, etc.

[f]The final pH of many media is 7.0 to 7.5. This should be adjusted as required for individual cultures or experiments.

fungi, etc. (7, 22). Ammonium nitrogen would probably suffice for many of these cultures, however, such as the fungi (43). Neither ammonium nor nitrate would be added to a medium designed to enrich for and isolate potential nitrogen-fixing bacteria.

The phosphate concentration in the basal medium described here is lower than that found in many other media but is more than enough to support the growth of most microorganisms. While cells were being labeled with radioactive phosphate for molecular systematics studies, it was observed that 20 to 50 μM phosphate supported the culture of most bacteria. Notable exceptions to this observation were the lactic acid bacteria and thiobacilli, which required millimolar levels of phosphate for culture. Phosphate at higher concentrations can function as a buffer, but there are other compounds which serve this role without the potential problem of precipitate formation or sequestering of trace metals (see below).

Magnesium is also required for the culture of bacteria and fungi. Sulfate can be used by many species as a sulfur source, but some groups or species may require sulfide or sulfur-containing amino acids as a sulfur source. Calcium may be considered more a trace metal requirement. However, it is readily incorporated in with the mineral solution. Calcium may be required at a higher concentration for the growth of some species. For example, increasing the final concentration of calcium in medium 5- to 10-fold over what is in the basal medium presented in Table 1 can greatly increase the cell yield of *Methanobacterium thermoautotrophicum* from a large-scale culture. Some marine bacteria require levels of magnesium, calcium, and iron which are higher than those provided in the basal medium described in Table 1.

TABLE 2 Mineral solution[a]

Component	Amt (g)/liter
NaCl	80
NH_4Cl	100
KCl	10
KH_2PO_4	10
$MgSO_4 \cdot 7H_2O$	20
$CaCl_2 \cdot 2H_2O$	4

[a]A solution containing the major inorganic components required for microbial growth. Add and dissolve each component in order. The mineral solution can be stored at room temperature.

Issues that should be considered for the formulation of a mineral solution or a set of stock solutions for medium preparation include precipitation, stability upon storage, and convenience. The mineral solution in Table 2 does not form a precipitate, is stable at room temperature for at least a year, and is useful for many microorganisms. The fewer the stock solutions needed to constitute a medium, the easier its preparation. The mineral solution in Table 2 was adapted from mineral solutions 1 and 2 for the culture of methanogens (3), with the idea of reducing the number of solutions and eliminating the need to weigh ammonium chloride separately for medium preparation.

Mineral stock solutions can be developed for specific groups of bacteria and fungi. For example, a 20× mineral solution for preparation of medium to recover sulfate-reducing bacteria from aquatic or terrestrial samples (56) contains the following (in grams per liter): NaCl, 100; $(NH_4)_2SO_4$, 10; $MgSO_4 \cdot 7H_2O$, 4; KH_2PO_4, 6; and $CaCl_2 \cdot 2H_2O$, 0.8. Ammonium sulfate replaced ammonium chloride so that the medium would contain more sulfate as an electron acceptor, and the amount of phosphate increased (compared to the mineral solution in Table 2) to the point that the medium would contain a precipitate after the addition of ferrous ammonium sulfate (55). A 100× mineral solution for culture of marine microorganisms could contain the following (in grams per liter): KCl, 200; NH_4Cl, 100; $MgSO_4 \cdot 7H_2O$, 40; and KH_2PO_4, 20. Sodium chloride was omitted from the stock solution since it could not be added at a concentration high enough for this stock to meet the requirements of a marine medium (20 to 28 g of sodium chloride per liter). A calcium salt was omitted from this marine mineral stock solution because it would produce a precipitate.

Omission of a calcium salt or a phosphate salt from a mineral stock solution may be required to avoid a precipitate. Use monobasic potassium phosphate (slightly acidic) rather than dibasic potassium phosphate (alkaline) to reduce problems with precipitate formation or microbial contamination in a stock solution upon storage at room temperature. The use of hydrated salts rather than anhydrous salts, such as for magnesium sulfate or calcium chloride, eases preparation of mineral stock solutions.

Vitamin Solution

A solution of water-soluble vitamins which supports or stimulates the growth of many microorganisms was adapted from the work of Wolin et al. (67) and is given in Table 3. The changes from the previous vitamin solution are the inclusion of mercaptoethanesulfonic acid, which is required by *Methanobrevibacter ruminantium* (4), and elevation of the concentration of vitamin B_{12}, which is required for the growth of *Clostridium aceticum* and *M. mobile* (57).

Other vitamins which may be required to culture microorganisms are hemin (9, 28), 1,4-naphthoquinone (17, 25), and vitamin K_1 (2-methyl-3-phytyl-1,4-naphthoquinone) (28). A 100× hemin stock solution can be prepared by dissolving 50 mg of hemin in 1 ml of 1 N sodium hydroxide and taking the dissolved hemin to a final volume of 100 ml with water (28). A 5,000× stock solution of vitamin K_1 can be prepared by dissolving 0.15 ml in 30 ml of 95% ethanol (28). It is possible that a microorganism may be stimulated by other chemical forms of a vitamin, such as niacinamide instead of nicotinic acid or pyridoxamine

TABLE 3 Vitamin solution[a]

Component	Amt (mg)/liter
Pyridoxine·HCl	10
Thiamine·HCl	5
Riboflavin	5
Calcium pantothenate	5
Thioctic acid	5
p-Aminobenzoic acid	5
Nicotinic acid	5
Vitamin B_{12}	5
MESA[b]	5
Biotin	2
Folic acid	2

[a]A solution designed to meet the water-soluble vitamin requirements of many microorganisms. Store at 4°C in the dark.

[b]MESA, mercaptoethanesulfonic acid.

rather than pyridoxine. Some strains, probably host associated, may require a preformed coenzyme such as pyridoxal phosphate or thiamine pyrophosphate for growth, but these organisms are rarely encountered by most investigators. My coworkers and I reported the discovery of a novel vitamin, 7-mercaptoheptanoylthreonine phosphate, in 1991 (33). It is likely that other novel vitamins remain to be discovered.

Trace Metal Solution

A stock solution which can supply the trace metal requirements of many microorganisms was adapted from the work of Wolin et al. (67) and is given in Table 4. Changes from the previous formulation include deletion of the sodium and magnesium salts (supplied by the mineral solution); deletion of the aluminum and borate salts (probably included since they are found in soil extracts and bacteria were once classified with plants); addition of sources of selenium, nickel, and tungsten (required for hydrogenases, formate dehydrogenases, etc.); and doubling of the concentrations of metals. In general, aerobic microorganisms grow well with only a small addition of trace metal solution (0.2 to 0.5 ml/liter of culture), while anaerobes seem to grow better with larger additions (more than 10 ml/liter for large-scale cultures). Borate (1 to 5 mg of boron per liter) is required for the culture of cyanobacteria (22).

Carbon and Energy Sources

The carbon and energy requirements of a microorganism can range from carbon dioxide and hydrogen for the chemolithotrophic methanogens and acetogens to very complex media containing amino acids, carbohydrates, nitrogen bases, and undefined components (e.g., yeast extract, meat extract, or ruminal fluid) for some bacteria. Carbon and energy requirements for microbial culture are considered together since many microorganisms that microbiologists work with are chemoorganotrophs which often use a single organic substrate to satisfy both needs. The great number of potential carbon and energy sources used for microbial culture prevents a detailed examination of them. Indeed, many of these potential substrates have probably not been examined in detail, and surprises regarding their use still await. For example, enrichments containing valerate as the carbon and energy source will probably yield just bacteria, but enrichments containing *n*-pentanol, the corresponding alcohol, may yield fungi or grazing protozoa. Some particular items with regard to the culture of chemolithotrophs and phototrophs and the use of xenobiotic or toxic substrates will be considered in this section. However, there are useful generalities in regard to carbon and energy sources which may be examined briefly.

Essentially any organic compound may be considered as a carbon and energy source for the culture of microorganisms. Some compounds, such as glucose, may be used by a great number of species, while others, such as allantoin under anoxic conditions, are used by only a few species. Lower levels of substrates are often used for the culture of aerobes (0.1 to 2 g/liter) than for the culture of anaerobes (20 g/liter for some substrates in large-scale cultures). The medium recipes given for the culture of known microorganisms are very useful as a guide for the substrates that a species and related microorganisms may use. Some ideas as to other substrates that a strain may utilize, or which may be used to improve the culture of a strain, are given below.

A good way to classify substrates is by biochemical and chemical categories: amino acids and related compounds; carboxylic acids, volatile fatty acids, and long-chain fatty acids and lipids; alcohols; nitrogen bases; carbohydrates; biological polymers; aromatic compounds and methoxylated aromatic compounds; etc. The ability of a single species to utilize any single substrate can be very specific, thus the continued role of carbohydrate and other substrate utilization patterns for phenotypic characterization of microorganisms.

Pyruvate and intermediates in the tricarboxylic acid cycle are often overlooked as substrates to support or improve the culture of microorganisms, but they have proved very useful in my own and others' hands for this purpose (17, 51). For example, the use of α-ketoglutarate as a substrate has greatly facilitated the culture of *Legionella* species (47).

Another issue with regard to carbon and energy sources is substrate metabolism without apparent growth. The monitoring of pH in substrate utilization tests in cultures that do not exhibit visible growth may indicate substrate utilization. A decrease in pH can mean the formation of acidic end products of metabolism. An increase in pH can mean the release of ammonia from a substrate or the effect of cations remaining after metabolism of an organic acid. The pH change alone may inhibit the growth of a strain, and conditions may need to be varied for the compound in question to support growth.

TABLE 4 Trace metal solution[a]

Component	Amt (g)/liter
Nitrilotriacetic acid[b]	2.0
$MnSO_4 \cdot H_2O$	1.0
$Fe(NH_4)_2(SO_4)_2 \cdot 6H_2O$	0.8
$CoCl_2 \cdot 6H_2O$	0.2
$ZnSO_4 \cdot 7H_2O$	0.2
$CuCl_2 \cdot 2H_2O$	0.02
$NiCl_2 \cdot 6H_2O$	0.02
$Na_2MoO_4 \cdot 2H_2O$	0.02
Na_2SeO_4	0.02
Na_2WO_4	0.02

[a]A solution designed to meet the trace metal requirements of many microorganisms. Store at 4°C.

[b]Adjust pH to 6 with KOH.

Some compounds may be toxic or insoluble at the levels at which certain substrates, such as formaldehyde, hydrocarbons, xenobiotics, etc. (34), are often added to media. Insoluble and volatile substrates are frequently introduced into cultures via a vapor phase (13, 34, 64). With some toxic substrates, it may be necessary to add them to a culture at a low, nontoxic level, monitor utilization, and amend the culture as the toxic substrate is consumed.

Substrates which may be inhibitory if used as pure compounds, such as oleic acid, may be added in a less available form, such as polyoxyethylenesorbitan monooleate (Tween 80). Tween 80 can be highly stimulatory for some microorganisms, both aerobes and anaerobes, for reasons not fully delineated, at a concentration of 0.2 g/liter (3, 12). Some media, including those prepared commercially, may have Tween 80 added at as high a level as 1 g/liter.

Important energy sources for chemolithotrophs include hydrogen and sulfide and other reduced sulfur compounds. Other inorganic energy sources are carbon monoxide and elemental metals (36), which may be regarded essentially as hydrogen sources, and ammonium or nitrite (21). Many chemolithotrophs can use carbon dioxide as the sole carbon source, but others may require organic carbon as a carbon source or for specific biosynthetic reactions, and organic compounds can be taken up and utilized by autotrophic microorganisms. One of the most common of these compounds is acetate, which is required by a number of methanogens and other anaerobic microorganisms (57). If it does not interfere with a study, it may be a good idea to include some acetate (1 to 5 g/liter) in media for chemolithotrophs (1). Some methanogens and ruminal bacteria require the branched-chain fatty acids isobutyrate, isovalerate, and 2-methylbutyrate for synthesis of branched-chain amino acids (9, 57). The actual branched-chain amino acids may satisfy this requirement in some microorganisms (65).

Light may be considered the energy source for the growth of phototrophic bacteria (22, 34, 41). These organisms physiologically and phylogenetically constitute a very diverse group of bacteria which includes photoautotrophs (which can use carbon dioxide as the sole carbon source) and photoheterotrophs (which use simple organic compounds such as acetate, pyruvate, malate, and ethanol as sources of reducing equivalents and cell carbon). The anoxygenic phototrophs, which include those able to use reduced sulfur compounds as sources of reducing potential, need to be cultured anaerobically (see below). I prefer to culture cyanobacteria by using the techniques required for anaerobes, since they may be gas exchanged for culture under an enriched carbon dioxide and reduced oxygen gas phase. Specific wavelengths of light, such as near infrared for the sulfur and nonsulfur purple bacteria, may be used to enrich for certain phototrophs. Some phototrophs can be grown under low light intensities, as little as 10 lx, and it is often recommended that light intensities be moderated (100 to 400 lx, for example) to enrich for, isolate, and culture many phototrophic bacteria (22, 41). However, I have successfully enriched for grown cyanobacteria and nonsulfur purple bacteria under plant growth lamps at 6,000 lx.

Nitrogen Sources

Ammonium is the inorganic nitrogen source for the basal medium described in this chapter. A lower concentration of ammonium may be used for the routine culture of aerobes, but the growth of many anaerobes is stimulated by the recommended level of ammonium (9). Nitrate may be a better source of nitrogen for the culture of many soil microorganisms and the cyanobacteria and is often recommended for the enrichment and isolation of fungi. However, ammonium should be considered a more universal nitrogen source for fungi and other microorganisms (43). If nitrate is provided as a source of nitrogen, the possibility that it can also serve as an electron acceptor should be considered. Nitrogen gas can serve as the sole nitrogen source for strains with the ability to fix nitrogen.

Urea is an organic source of nitrogen which may have use in field projects outside the laboratory. Urea should be filter sterilized for laboratory investigations to avoid possible hydrolysis to ammonia and carbon dioxide during steam sterilization. Some microorganisms require organic nitrogen compounds such as nitrogen bases or, more frequently, amino acids for their culture. Vitamin Assay Casamino Acids (Difco Laboratories, Detroit, Mich.) provides a semidefined, if incomplete, source of amino acids if this requirement needs to be examined (45). Two grams of Casamino Acids per liter of medium is a useful concentration for examining the nitrogen requirements of an isolate, but 10 to 20 g/liter should be used if the goal is to generate cell tissue. Peptone, enzymic digests of casein or soybean meal, yeast extract, etc., provide more complex nitrogen sources for microorganisms which require or are stimulated by peptides or other compounds. Microorganisms, such as the lactic acid bacteria, may require nucleic acid bases (adenine, cytosine, guanine, thymine, and/or uracil) as trace nutrients in medium. Addition of each nucleic acid base at a concentration of 1 to 10 mg/liter may fulfill this requirement.

Electron Acceptors

There are a number of electron acceptors which support microbial respiration, the more common being oxygen (reduced to water), nitrate (reduced to nitrite, ammonia, or dinitrogen), sulfate (reduced to sulfide), and carbon dioxide (reduced to methane or acetate). The importance of iron reduction (ferric to ferrous iron) to support microbial respiration in the environment, and thus in the laboratory, should be noted (37). Electron acceptors are usually added to media at a concentration of 5 to 20 mM.

There are a number of less frequently considered or more unusual electron acceptors that can be reduced by microorganisms, and many of these reductions can support metabolism and growth. These include sulfite; thiosulfate; selenate; Mn(IV); U(VI); oxyanions of tellurium, europium, and rhodium; dimethyl sulfoxide; trimethylamine *N*-oxide; and arsenate (35, 37, 38, 42, 46). Organic compounds may also be used as electron acceptors, such as fumarate (reduced to succinate). For instance, *Escherichia coli* is capable of anaerobic respiration using fumarate, malate, or aspartate as an electron acceptor (39). Unsaturated bonds in other compounds, such as ferulic acid, may be used as electron acceptors.

Undefined Media and Medium Components

Microorganisms are frequently cultured on undefined media or on media which contain undefined components, either because the exact nutritional requirements of strains are unknown or simply because of convenience (many undefined media are commercially available). Probably the most commonly used undefined medium component is yeast extract (Difco). Yeast extract is a complex mixture of acetate, amino acids, peptides, nitrogen bases, vitamins, trace metals, phosphate, etc., that supports the growth of many known microbial strains. It even contains some fermentable carbohydrate, ribose. Yeast extract in a concentration as low as 50 mg/liter can enhance the recovery of microorganisms from environmental samples in an otherwise defined medium. It is commonly used at 0.5 to 5 g/liter in a number of culture media.

All microbiologists are familiar with nutrient broth, which contains peptone and beef extract. However, nutrient broth is not a good medium for the culture of most microorganisms, primarily because although it does contain amino acids and peptides, it lacks many important potential nutrients, especially carbohydrate. Two undefined media which are more complete and useful than nutrient broth are plate count broth and tryptic soy broth (18). Both contain glucose (2 to 2.5 g/liter). Plate count broth contains yeast extract and tryptone (Trypticase) as additional components. Tryptic soy broth contains tryptone and soytone (a digest of soy meal) and is probably more commonly used than plate count broth. When preparing solid medium, note that commercial tryptic soy agar is different from solid medium prepared from tryptic soy broth in that glucose is omitted from tryptic soy agar (18).

An undefined medium commonly used for the culture of fungi is Sabouraud dextrose agar, which contains a peptone and glucose. Two other undefined media useful for the culture of fungi are potato dextrose broth, which contains glucose and potato infusion (starch and trace nutrients), and malt extract broth, which contains glucose, maltose, malt extract, and yeast extract (18, 53).

It has been noted that a higher number of cells may be recovered from environmental samples if media containing low concentrations of substrate and nutrient are used instead of rich media (40, 68). The commercially available undefined media may be used at 1/2 to 1/10 of their label-indicated concentrations for environmental studies without loss of cell recovery. A medium was formulated for potable-water studies on the basis of the principle that use of low concentrations of a wide variety of components could result in greater recovery of different heterotrophic bacteria (51).

Perhaps the ultimate in an undefined medium is one based on the use of site water, extracts of soil, sediment, and sludge; or even slurries of soil, sediment, or sludge (as the primary medium component). The use of slurries was the basis for an improved most-probable-number method to estimate sulfate-reducing bacteria in environmental samples, coupled with use of a radiotracer (62). The numbers of sulfate-reducing bacteria determined by this technique were several orders of magnitude higher than those obtained with more conventional media (62), which, however, still retain some utility (56).

Medium Preparation

Basic medium preparation consists of constituting most or all of the medium components at the proper concentrations in solution, adjusting the pH (see below), and then adding components which may interfere with pH adjustment (calcium carbonate for detection of acid production or sodium bicarbonate for a medium under a carbon dioxide-containing gas phase) or a gelling agent for solid medium. The most common gelling agent is agar, which is available in a variety of purities and forms. The extra expense of purified agar (BBL, Becton Dickinson, Cockeysville, Md.) can be justified on the basis of a greater and more consistent agar strength, higher clarity, and relative absence of interfering organic substances. Agar is commonly used in medium at a concentration of 7 to 20 g/liter. If a very pure agar is desired, Oxoid Ionagar no. 2 (Oxford, Columbia, Md.) or a high-gel-strength, electrophoretic-grade agarose may be used. Another gelling agent which is particularly useful for the culture of thermophiles is Gelrite (Scott Laboratories, Carson, Calif.), available from Sigma (St. Louis, Mo.; product G 1910). This gellan gum (5 to 12 g/liter) requires divalent cations, such as 1 g of $MgCl_2 \cdot 6H_2O$ per liter, for solidification. A mixture of $CaCl_2 \cdot 2H_2O$ and $MgCl_2 \cdot 6H_2O$ (0.8 g of each per liter) may yield a harder medium. Gelling agents should be melted in the medium prior to steam sterilization.

The major means of medium sterilization are steam sterilization (autoclaving) and filtration. For steam sterilization, it is generally considered necessary to heat the medium to 121°C at 100-kPa (15-lb/in^2) gauge pressure for 15 to 20 min. In general, the shorter the time required to sterilize a medium the better, although as the volume of medium to be sterilized increases, so does the sterilization time. Clean laboratory conditions, the use of chemically clean glassware, and the use of analytical reagent-grade or better medium components facilitate the preparation of sterile medium and materials. For some media tubed at 10 ml or less, I have been able to use routinely a 5-min steam sterilization (55, 56).

There are media or medium components which break down under the conditions of steam sterilization. These should be filter sterilized. For small volumes, the use of a prefilter-filter (pore size of 0.8 to 0.2 μm) combination, such as an Acrodisc PF syringe filter (Gelman Sciences, Ann Arbor, Mich.), is convenient. Disposable, easy-to-use filter units, such as Nalgene disposable filterware (Nalge, Rochester, N.Y.), are useful for volumes from 50 to 1,000 ml. Filter capsules which have good flow rates and filtration capacity, such as Gelman minicapsules, are available for filter sterilization of many liters of medium.

PHYSICOCHEMICAL FACTORS

Physicochemical factors which should be considered for the cultivation of bacteria and fungi include temperature, pH, relationship to oxygen, and, for some microorganisms, salinity and ionic strength. Common problems in culture of bacteria and fungi are incorrect or uncontrolled temperature or pH. If there is a problem with growing a microorganism, these factors should be checked first. Microorganisms can be particular about growth conditions. For example, strains of the same (phylogenetic) species can have 5 to 7°C differences in optimal growth temperatures.

The concept and control of incubation temperature are straightforward and will not be elaborated upon. One point that should be considered is that recovery of microorganisms from many common environmental samples is inhibited by incubation at 37°C, the temperature of many laboratory incubators. Room temperature (20 to 24°C) is better for many environmental samples. Few viable colonies are lost if these samples are incubated at 28 to 30°C (1, 51, 55, 56). Of course, different temperatures are required for samples from a hot spring or alpine stream. All of the physicochemical factors need to be considered and adjusted for recovery of bacteria and fungi from different environments.

pH

The importance of control of pH for the culture of microorganisms is often overlooked. Many bacteria, especially those associated with a host or recovered from an otherwise constant or protected environment, have fairly narrow pH optimum ranges. The pH range of *Legionella* species can be only ±0.05 pH units from optimum (47). pH can change during the culture of microorganisms. As an organic acid (added as a salt) is used as a substrate, as nitrate is reduced to ammonia or dinitrogen gas, or as carbon dioxide is consumed, the culture pH may increase. The production of organic acid from fermentation or incomplete oxidation of substrate is a common reason for a decrease in pH in culture.

Some common buffers used for microbial culture are phosphate and Tris, which have often proved unsatisfactory (26). Good's buffers are zwitterionic organic buffers with pK_as in the range useful to most microbiologists. They were developed specifically for biological research (26), and one of these, HEPES (*N*-2-hydroxyethylpiperazine-*N'*-2-ethanesulfonic acid), proved useful for tissue culture. My laboratory has found TES [*N*-tris(hydroxymethyl)methyl-2-aminoethanesulfonic acid (pK_a 7.4)] to be a better buffer for culture at neutral pHs than HEPES or PIPES [piperazine-*N*,*N'*-bis(2-ethanesulfonic acid)] (1, 55, 56). MES [2-*N*-(morpholino)ethanesulfonic acid (pK_a 6.1)] is useful for the culture of bacteria requiring slightly acidic pHs. TAPS [*N*-tris(hydroxymethyl)methyl-3-aminopropanesulfonic acid (pK_a 8.1)] can be used for microorganisms with slightly alkaline pH optima. In general, changes in pH with growth are less for aerobic cultures than for anaerobic cultures. A Good's buffer at 1 to 2 g/liter may control the pH of an aerobic culture, while 5 to 20 g/liter may be required for anaerobic cultures.

Acetate (pK_a 4.75) or citrate (pK_{a1} 3.2, pK_{a2} 5.50, pK_{a3} 6.4) is a good buffer for organisms requiring more acidic conditions. Remember that acetate and citrate may also serve as carbon and energy sources. Sodium bicarbonate is a good buffer at pH 6 to 8 for cultures incubated with carbon dioxide in the gas phase. The concentration of bicarbonate used depends on the partial pressure of carbon dioxide in the gas phase and the incubation temperature. For example, 4 to 5 g of sodium bicarbonate would be added per liter of medium for methanogens under a gas phase containing 20% carbon dioxide at 250-kPa gauge pressure (1, 3, 4, 57). This concentration of bicarbonate can be added to medium after pH adjustment and before steam sterilization, obviating the need for sterile additions of carbonate to media formulated for 100% carbon dioxide gas phases (28). Again, the use of a Good's buffer simplifies pH control under these conditions as well.

Relationship to Oxygen

Bacteria and fungi have varying requirements and tolerances for oxygen in culture and may be classified as aerobes (e.g., pseudomonads), microaerophiles (e.g., some strains of *Legionella* and *Campylobacter*), facultative anaerobes (e.g., enteric organisms such as *E. coli*), or anaerobes (e.g., clostridial species). Small-scale culture of aerobes is relatively straightforward—media are simply (aseptically) exposed to air. The culture of microaerophiles and aerotolerant anaerobes can be relatively easy with the use of a candle jar, which also provides an atmosphere enriched with carbon dioxide, or with the use of semisolid medium (32). There are several systems developed for the culture of microaerophiles and many anaerobes, particularly those of clinical interest, such as the GasPak (BBL). These systems use water for the production of the desired gas phase, which sometimes leads to problems with plate culture of environmental samples, with actively motile cells swarming over a plate's surface. Waterless gas-generating systems for the culture of microaerophiles and clinically important anaerobes have been developed (8).

The challenges of the general culture of strictly anaerobic microorganisms were alleviated in the late 1970s with the introduction of syringe transfer technique, specialized glassware, improved anaerobic glove bags and chambers, improved (handmade) plate incubators, and other methods under the umbrella classification of "Balch technique" (3, 4). The flexible chamber manufactured by Coy (Grass Lake, Mich.) can easily maintain an oxygen level of 0 to 2 ppm (compared with around 200,000 ppm in the atmosphere) and is probably the easiest system to use and maintain. In such an environment, strict anaerobes can be handled just like aerobes on an open bench. I incorporate a bed of activated charcoal just below the catalyst tray in the chamber to absorb compounds which could inactivate the oxygen-scrubbing catalyst and shallow jars of calcium chloride (anhydrous) to remove water from the chamber gas phase, which increases the interval between catalyst "activation" (drying). It seems easier to maintain an anaerobic chamber if plates and cultures are not incubated inside but are incubated in plate holders (3) outside the chamber.

The best way to prepare anaerobic medium is to boil the medium under an oxygen-free gas stream, which reduces any dissolved oxygen below a detection limit (20 ppb); seal the medium; transfer the medium into an anaerobic chamber; dispense the medium; and, finally, seal the tubes of medium in the chamber and remove them for further manipulation or sterilization. Many anaerobes, such as sulfate-reducing bacteria and most clostridia, can be cultured in "Hungate-type" anaerobic culture tubes (no. 2047; Bellco Glass, Vineland, N.J.) (1, 55, 56). These culture tubes are relatively inexpensive. The butyl rubber stopper used with the Hungate-type tubes is thin enough that skin-testing syringes can be used for transfers or sampling. These inexpensive syringes are available sterile in multipack trays (no. 305539; Becton Dickinson, Rutherford, N.J.).

Strict anaerobes and those using gases as substrates, such as methanogens and acetogens, should be cultured in aluminum seal tubes (no. 2048; Bellco Glass) (1, 3, 4). It may be necessary to use glass rather than plastic syringes for the transfer of strict anaerobes, but the syringes can be cleaned and reused. A gassing station can replace the anaerobic chamber atmosphere which was sealed in the tubes with any desired gas phase. This system (3, 4) is very useful for any experiment for which control of the culture gas phase is required.

The culture of many strict anaerobes requires poising the redox potential of the medium lower than that which mere removal of oxygen will achieve. Resazurin (50 μg/liter) can be added to medium as a redox indicator. Reducing agents for the culture of anaerobes include cysteine, sulfide, thioglycolate, dithionite, glutathione, yeast extract, and dithiothreitol (30). Ascorbic acid, elemental iron, or titanium(III) citrate may be used if sulfur compounds need to be avoided (30, 55, 56). The use of titanium(III) may result in a lower recovery of anaerobes from environmental samples (63). A good general cysteine-sulfide reducing agent which can be stored in an anaerobic chamber for at least a year is prepared as follows. Take 2 g of L-cysteine and (separately) 2 g of washed and dried crystals of $Na_2S{\cdot}9H_2O$ into an anaerobic chamber. Boil 100 ml of water under nitrogen, take it into an anaerobic chamber, dissolve the cysteine and sulfide, and dispense and seal the reducing agent. (If cysteine·HCl is used, dissolve it in the chamber in anoxic 0.1 N NaOH to neutralize the solution before the addition of sulfide.) This mixture is added at 1 to 10 ml per liter of anaerobic medium after oxygen has been removed (1, 3, 4, 17, 57).

Salinity and Ionic Strength

Many microorganisms from marine or saline environments require salt or salts for culture. For many marine bacteria, this requirement can be met simply by adding 10 to 20 g of sodium chloride per liter of medium. Seawater contains other ions, particularly magnesium. A complete seawater medium can be constituted from the recipe for Turks Island salt (10) or based on commercially available formulations such as Instant Ocean (Aquarium Systems, Mentor, Ohio). Some marine media require filter sterilization to avoid the formation of precipitates during steam sterilization. Many extreme halophiles can be cultured in medium containing elevated concentrations of sodium chloride, though some have an additional requirement for magnesium ions (60).

CULTIVATION OF FUNGI

Most microbiologists are familiar with the methods for culture of bacteria, and most of the material presented above was derived from the literature on bacteria. The fundamentals of the culture of fungi are generally the same as for the bacteria, and there are a number of useful basic references in this area (7, 11, 14, 23, 34, 43). An excellent source of further practical information for handling these microorganisms was written by K. A. Seifert (53). If fungi are routinely encountered in one's research, it would be wise to know how to identify the more common species. A useful guide is published by the American Phytopathological Society (6).

Some general trends in the media used to culture fungi include use of nitrate as a nitrogen source and use of slightly acidic pH in culture medium (7, 18, 23). Liquid enrichments should be shaken to ensure good aeration. The recovery of fungi from environmental samples can be facilitated in general by the inclusion of antibacterial agents, such as streptomycin, in the medium or by incubation at low temperatures (4 to 10°C), which permits a number of fungi to outcompete bacteria (7, 14, 23). Media with carbohydrate as the growth substrate are recommended (53), as is the use of an agent, such as rose bengal or dichloran, to slow the growth of rapidly spreading filamentous fungi (11, 53). The more common media for the culture of many fungi include Czapek's agar, malt extract agar, and potato dextrose agar (11, 53).

Many fungi can be enriched for and isolated on common bacteriological media. In fact, one problem for our work in applied and environmental microbiology is overgrowth of plates from samples, including marine specimens, by filamentous fungi. A liberal application of an inhibitory agent, such as cycloheximide (100 μg/ml), is helpful. Isolated strains of fungi can be cultured on the basal medium described at the beginning of this chapter. Once yeast cells are in pure culture, growth in liquid medium is fairly easy. Molds, on the other hand, may prefer to grow in shallow culture in tissue culture flasks, a technique that is useful for liquid culture of actinomycetes as well.

One problem with the culture of fungi is the possibility of contamination of the laboratory with fungal spores. Working properly in a biological safety hood alleviates many of these problems (53). Remember to keep the hood clean and disinfected. A useful trick for working with a mold on a plate on an open bench is to first place a lint-free towel (Kimwipes; Kimberly-Clark Corp., Roswell, Ga.) moistened with a disinfectant on the working area. Change the towel each time a different culture is manipulated.

As in the rest of biology, there are a number of exceptions to every generality given above about the cultivation of fungi. The fungi will continue to hold surprises for microbiologists as different techniques are used for their culture. Examples include the discovery of anaerobic fungi (19, 44, 59, 61) and the culture of a fungus with toluene as the sole carbon and energy source (64).

LARGER-SCALE CULTIVATION

Larger-scale (0.1- to 100-liter) culture of bacteria and fungi is, again, in principle similar to small-scale culture, with some attention given to the problems of sterilization of large volumes of medium, safe handling of large volumes of very hot liquids, use of specialized equipment such as fermentors, provision of gaseous substrate such as oxygen, etc. Oxygen is provided to larger cultures through shaking, stirring, and/or injection of sterile air in as small bubbles as possible, all to force as much oxygen as possible into solution. The supply of oxygen to aerobic cultures is a prime consideration for the scale-up of these cultures. Foaming and the use of antifoam should be considered when fermentation equipment is used. It is relatively easier to scale up the culture of many anaerobes by using Pyrex bottles (Corning Glass, Corning, N.Y.) in place of fermentors, but it will probably be necessary to have a mechanism to control the culture pH during growth. If sealed glassware is used, provision must be made to prevent a buildup of pressure due to microbial production of gases as end products of metabolism.

Large-scale culture of a microorganism may point out nutritional requirements or other factors limiting growth that were not apparent in smaller cultures. The experience with *M. thermoautotrophicum* is a good illustration of this. The medium used for tube culture of *M. thermoautotrophicum* is very simple and defined (3). When culture of this methanogen was scaled up to obtain enough cell material for biochemical studies, it was noted that production of cell mass was limited by trace metals, such as iron, cobalt, and molybdenum (52, 58). The surprising discovery that *M. thermoautotrophicum* required nickel as a result of scaling up of its culture (52) was made, and it was noted in our laboratory that cell yields could be improved if the concentration of calcium in the medium was increased 5- to 10-fold. *M. thermoautotrophicum* can take up acetate from its medium, and the cell yield was enhanced by the inclusion of 2 g of sodium acetate per liter of medium. One liter of gas (H_2:CO_2, 80:20) per min was sparged through a 100-liter fermentor of *M. thermoautotrophicum*. This gassing rate stripped some nutrients such as ammonia and sulfide, and periodic additions of sodium sulfide and ammonium hydroxide to the fermentor improved cell yield.

Like most other aspects of microbiological research, the cultivation of bacteria and fungi rests on a general foundation of basic principles but does require some mastery of a number of details. The literature is a good guide to the specifics of culture of known microorganisms. One should also rely on colleagues for advice when starting to work with unfamiliar microorganisms, such as cyanobacteria or anaerobes, or to work with new techniques, such as use of fermentation equipment. Two important things to consider for the culture of microorganisms are not to shy away from trying something new and to closely observe the culture (for such phenomena as macroscopic, microscopic, and chemical characteristics). Simple, yet novel and perhaps important, discoveries can be made on the basis of these tenets for the culture of bacteria and fungi.

CULTURING THE UNCULTURED

The importance of novel approaches to cultivation is exemplified by the routine observation that >99% of microorganisms detected by direct counting and/or other methods are not detected by culture-based methods. More importantly, entire phylogenetic lineages are not represented by a cultured organism (50). This is a challenge for a bacteriologist, but some approaches to address this issue have yielded some success. These include use of very low substrate concentrations, control of oxygen tension, and use of "in situ" incubation conditions. Some useful examples follow.

Previously uncultured representatives of *Acidobacteria* and *Verrucomicrobia* were recovered from soil and the guts of termites by a variety of culture and incubation condition manipulations such as use of microaerophilic conditions, incubation in elevated levels of carbon dioxide, low substrate concentrations, and long incubation times (54). About 4 to 7% of the total microbial community, based on direct cell counts, was recovered in this study (54). The use of low substrate concentrations and long incubation times is a common theme for improving recovery of microbes from environmental samples. Incubation under microaerophilic conditions will likely prove important in future work. This was the best single manipulation resulting in overall greater recovery of cells (54).

Up to 14% of cells present in collected seawater, including four previously uncultured lineages, were recovered by using a high-throughput method in media containing very low concentrations of substrates (e.g., about 100 μM dissolved organic carbon) (15). The cultures were initially cultured for 3 weeks, and subsequent platings were incubated for up to 8 weeks. The cultures were also initially incubated at 16°C (15). Control of temperature and pH will be important for culturing uncultured organisms, just as it is for any cultivation of bacteria. About 2,500 extinction cultures were screened over 3 years in this study (15), and this level of effort will likely be required in future efforts to culture the uncultured.

A more "in situ"-like incubation condition was used to recover microcolonies representing a number of new lineages within the *Planctomyces*, flavobacteria, and alpha subclass of *Proteobacteria* from seawater (69). Gel microdroplets containing one or a couple of cells were prepared and incubated for 5 weeks in an open system with flowing sterilized seawater or seawater amended with a low concentration of substrate (69). In another study, conditions even closer to in situ conditions were established by placing a seawater inoculum mixed with agar onto a membrane placed directly on the surface of a tidal flat sediment (31). Two novel species were recovered from this system and they grew best in the presence of each other, leading to a hypothesis that signals from different cells may be required for culture of other cells (31).

With further effort, we can ensure that progress in culturing the uncultured will continue. Remember, at various times, it was considered difficult to culture methanogens or isolates of *Legionella*, but cultivation of these microorganisms, after the initial efforts to discover appropriate techniques and media, is now relatively easy with just a modicum of experience.

REFERENCES

1. **Adkins, J. P., L. A. Cornell, and R. S. Tanner.** 1992. Microbial composition of carbonate petroleum reservoir fluids. *Geomicrobiol. J.* **10:**87–97.
2. **Atlas, R. M.** 1997. *Handbook of Microbiological Media*, 2nd ed. CRC Press, Boca Raton, Fla.
3. **Balch, W. E., L. J. Magrum, G. E. Fox, R. S. Wolfe, and C. R. Woese.** 1979. Methanogens: reevaluation of a unique biological group. *Microbiol. Rev.* **43:**260–296.
4. **Balch, W. E., and R. S. Wolfe.** 1976. New approach to the cultivation of methanogenic bacteria: 2-mercaptoethanesulfonic acid (HS-CoM)-dependent growth of *Methanobacterium ruminantium* in a pressurized atmosphere. *Appl. Environ. Microbiol.* **32:**781–791.
5. **Balows, A., H. G. Truper, M. Dworkin, W. Harder, and K.-H. Schleifer (ed.).** 1992. *The Prokaryotes*, 2nd ed., vol. I to IV. Springer-Verlag, New York, N.Y.
6. **Barnett, H. L., and B. B. Hunter.** 1998. *Illustrated Genera of Imperfect Fungi*, 4th ed. American Phytopathological Society, St. Paul, Minn.
7. **Booth, C.** 1971. Fungal culture media. *Methods Microbiol.* **4:**49–94.
8. **Brazier, J. S., and V. Hall.** 1994. A simple evaluation of the AnaeroGen system for the growth of clinically significant anaerobic bacteria. *Lett. Appl. Microbiol.* **18:**56–58.
9. **Bryant, M. P., and I. M. Robinson.** 1962. Some nutritional characteristics of predominant culturable ruminal bacteria. *J. Bacteriol.* **84:**605–614.
10. **Budavari, S. (ed.).** 1989. *Merck Index*, 11th ed. Merck & Co., Rahway, N.J.
11. **Buttner, M. P., K. Willeke, and S. A. Grinshpun.** 2002. Sampling and analysis of airborne microorganisms, p. 814–826. *In* C. J. Hurst, R. L. Crawford, G. R. Knudsen, M. J. McInerney, and L. D. Stetzenbach (ed.), *Manual of Environmental Microbiology*, 2nd ed. ASM Press, Washington, D.C.
12. **Cato, E. P.** 1983. Transfer of *Peptostreptococcus parvulus* (Weinberg, Nativelle, and Prevot 1937) Smith 1957 to the genus *Streptococcus*: *Streptococcus parvulus* (Weinberg, Nativelle, and Prevot 1937) comb. nov., rev., emended. *Int. J. Syst. Bacteriol.* **33:**82–84.
13. **Claus, D., and N. Walker.** 1964. The decomposition of toluene by soil bacteria. *J. Gen. Microbiol.* **36:**107–122.
14. **Collins, C. H., P. M. Lyne, and J. M. Grange.** 1989. *Microbiological Methods*, 6th ed. Butterworths, Boston, Mass.
15. **Connon, S. A., and S. J. Giovannoni.** 2002. High-throughput methods for culturing microorganisms in very-low-nutrient media yield diverse new marine isolates. *Appl. Environ. Microbiol.* **68:**3878–3885.
16. **Cote, R. J., and R. L. Gherna.** 1994. Nutrition and media, p. 155–178. *In* P. Gerhardt, R. G. E. Murray, W. A. Wood, and N. R. Krieg (ed.), *Methods for General and Molecular Bacteriology*. American Society for Microbiology, Washington, D.C.
17. **DeWeerd, K. A., L. Mandelco, R. S. Tanner, C. R. Woese, and J. M. Suflita.** 1990. *Desulfomonile tiedjei* gen. nov. and sp. nov., a novel anaerobic, dehalogenating, sulfate-reducing bacterium. *Arch. Microbiol.* **154:**23–30.
18. **Difco Laboratories.** 1998. *Difco Manual*, 11th ed. Becton Dickinson and Co., Sparks, Md.
19. **Dore, J., and D. A. Stahl.** 1991. Phylogeny of anaerobic rumen *Chytridiomycetes* inferred from small subunit ribosomal RNA sequence comparisons. *Can. J. Bot.* **69:**1964–1971.
20. **Edwards, C. (ed.).** 1990. *Microbiology of Extreme Environments*. McGraw-Hill Publishing Co., New York, N.Y.
21. **Focht, D. D., and W. Verstraete.** 1977. Biochemical ecology of nitrification and denitrification. *Adv. Microbiol. Ecol.* **1:**135–214.
22. **Fogg, G. E., W. D. P. Stewart, P. Fay, and A. E. Walsby.** 1973. *The Blue-Green Algae*. Academic Press, New York, N.Y.
23. **Garraway, M. O., and R. C. Evans.** 1984. *Fungal Nutrition and Physiology*. John Wiley & Sons, New York, N.Y.

24. **Gerhardt, P., and S. W. Drew.** 1994. Liquid culture, p. 224–247. *In* P. Gerhardt, R. G. E. Murray, W. A. Wood, and N. R. Krieg (ed.), *Methods for General and Molecular Bacteriology*. American Society for Microbiology, Washington, D.C.
25. **Gomez-Alarcon, R. A., C. O'Dowd, J. A. Z. Leedle, and M. P. Bryant.** 1982. 1,4–Naphthoquinone and other nutrient requirements of *Succinovibrio dextrinosolvens*. *Appl. Environ. Microbiol.* **44:**346–350.
26. **Good, N. E., G. D. Winget, W. Winter, T. N. Connolloy, S. Izawa, and R. M. M. Singh.** 1966. Hydrogen ion buffers for biological research. *Biochemistry* **5:**467–477.
27. **Gottschal, J. C., W. Harder, and R. A. Prins.** 1992. Principles of enrichment, isolation, cultivation, and preservation of bacteria, p. 149–196. *In* A. Balows, H. G. Truper, M. Dworkin, W. Harder, and K.-H. Schleifer (ed.), *The Prokaryotes*, 2nd ed., vol. I. Springer-Verlag, New York, N.Y.
28. **Holdeman, L. V., E. P. Cato, and W. E. C. Moore (ed.).** 1977. *Anaerobe Laboratory Manual*, 4th ed. Virginia Polytechnic Institute and State University, Blacksburg, Va.
29. **Holt, J. G., and N. R. Krieg.** 1994. Enrichment and isolation, p. 179–215. *In* P. Gerhardt, R. G. E. Murray, W. A. Wood, and N. R. Krieg (ed.), *Methods for General and Molecular Bacteriology*. American Society for Microbiology, Washington, D.C.
30. **Jones, G. A., and M. D. Pickard.** 1980. Effect of titanium(III) citrate as reducing agent on the growth of rumen bacteria. *Appl. Environ. Microbiol.* **39:**1144–1147.
31. **Kaeberlein, T., K. Lewis, and S. S. Epstein.** 2002. Isolating "uncultivable" microorganisms in pure culture in a simulated natural environment. *Science* **296:**1127–1129.
32. **Krieg, N. R., and P. Gerhardt.** 1994. Solid, liquid/solid, and semisolid culture, p. 216–223. *In* P. Gerhardt, R. G. E. Murray, W. A. Wood, and N. R. Krieg (ed.), *Methods for General and Molecular Bacteriology*. American Society for Microbiology, Washington, D.C.
33. **Kuhner, C. A., S. S. Smith, K. M. Noll, R. S. Tanner, and R. S. Wolfe.** 1991. 7–Mercaptoheptanoylthreonine phosphate substitutes for heat-stable factor (mobile factor) for growth of *Methanomicrobium mobile*. *Appl. Environ. Microbiol.* **57:**2891–2895.
34. **Labeda, D. P. (ed.).** 1990. *Isolation of Biotechnological Organisms from Nature*. McGraw-Hill Publishing Co., New York, N.Y.
35. **Laverman, A. M., J. S. Blum, J. K. Schaefer, E. J. P. Philips, D. R. Lovley, and R. S. Oremland.** 1995. Growth of strain SES-5 with arsenate and other diverse electron acceptors. *Appl. Environ. Microbiol.* **61:**3556–3561.
36. **Lorowitz, W. H., D. P. Nagle, Jr., and R. S. Tanner.** 1992. Anaerobic oxidation of elemental metals coupled to methanogenesis by *Methanobacterium thermoautotrophicum*. *Environ. Sci. Technol.* **26:**1606–1610.
37. **Lovley, D. R.** 1991. Dissimilatory Fe(III) and Mn(IV) reduction. *Microbiol. Rev.* **55:**259–287.
38. **Lovley, D. R., and E. J. P. Phillips.** 1992. Bioremediation of uranium contamination with enzymic uranium reduction. *Environ. Sci. Technol.* **26:**2228–2234.
39. **Macy, J., H. Kulla, and G. Gottschalk.** 1976. H_2-dependent growth of *Escherichia coli* on L-malate. *J. Bacteriol.* **125:** 423–428.
40. **Mallory, L. M., B. Austin, and R. R. Colwell.** 1977. Numerical taxonomy and ecology of oligotrophic bacteria isolated from the estuarine environment. *Can. J. Microbiol.* **23:**733–750.
41. **Mann, N. H., and N. G. Carr (ed.).** 1992. *Photosynthetic Prokaryotes*. Plenum Press, New York, N.Y.
42. **Moore, M. D., and S. Kaplan.** 1994. Members of the family *Rhodospirillaceae* reduce heavy-metal oxyanions to maintain redox poise during photosynthetic growth. *ASM News* **60:**17–23.
43. **Moore-Landecker, E.** 1982. *Fundamentals of the Fungi*. Prentice-Hall, Inc., Englewood Cliffs, N.J.
44. **Mountfort, D. O.** 1994. Anaerobic fungi: future perspectives, p. 271–279. *In* D. O. Mountfort and C. G. Orpin (ed.), *Anaerobic Fungi*. Marcel Dekker, Inc., New York, N.Y.
45. **Nolan, R. A.** 1971. Amino acids and growth factors in vitamin-free Casamino Acids. *Mycologia* **63:**1231–1234.
46. **Oremland, R., J. S. Blum, C. W. Culbertson, P. T. Visscher, L. G. Miller, P. Dowdle, and F. E. Strohmaier.** 1994. Isolation, growth, and metabolism of an obligately anaerobic, selenate-respiring bacterium, strain SES-3. *Appl. Environ. Microbiol.* **60:**3011–3019.
47. **Pasculle, A. W.** 1992. The genus *Legionella*, p. 3281–3303. *In* A. Balows, H. G. Truper, M. Dworkin, W. Harder, and K.-H. Schleifer (ed.), *The Prokaryotes*, 2nd ed., vol. IV. Springer-Verlag, New York, N.Y.
48. **Pienta, P., J. Tang, and R. Cote.** 1996. *American Type Culture Collection of Bacteria and Phages*, 19th ed. American Type Culture Collection, Rockville, Md.
49. **Rabus, R., R. Nordhaus, W. Ludwig, and F. Widdel.** 1993. Complete oxidation of toluene under strictly anoxic conditions by a new sulfate-reducing bacterium. *Appl. Environ. Microbiol.* **59:**1444–1451.
50. **Rappe, M. S., and S. J. Giovannoni.** 2003. The uncultured microbial majority. *Annu. Rev. Microbiol.* **57:**369–394.
51. **Reasoner, D. J., and E. E. Geldreich.** 1985. A new medium for the enumeration and subculture of bacteria from potable water. *Appl. Environ. Microbiol.* **49:**1–7.
52. **Schonheit, P., J. Moll, and R. K. Thauer.** 1979. Nickel, cobalt and molybdenum requirement for growth of *Methanobacterium thermoautotrophicum*. *Arch. Microbiol.* **123:**105–107.
53. **Seifert, K. A.** 1990. Isolation of filamentous fungi, p. 21–51. *In* D. P. Labeda (ed.), *Isolation of Biotechnological Organisms from Nature*. McGraw-Hill Publishing Co., New York, N.Y.
54. **Stevenson, B. S., S. A. Eichorst, J. T. Wertz, T. M. Schmidt, and J. A. Breznak.** 2004. New strategies for cultivation and detection of previously uncultured microbes. *Appl. Environ. Microbiol.* **70:**4748–4755.
55. **Tanner, R. S.** 1989. Monitoring sulfate-reducing bacteria: comparison of enumeration media. *J. Microbiol. Methods* **10:**19–27.
56. **Tanner, R. S.** 1996. Monitoring sulfide and sulfate-reducing bacteria, p. 353–362. *In* R. Bryant (ed.), *Proceedings of the Fifth International Conference on Microbial Enhanced Oil Recovery and Related Biotechnology for Solving Environmental Problems*. CONF-9509173. National Technical Information Service, Springfield, Va.
57. **Tanner, R. S., and R. S. Wolfe.** 1988. Nutritional requirements of *Methanomicrobium mobile*. *Appl. Environ. Microbiol.* **54:**625–628.
58. **Taylor, G. T., and S. J. Pirt.** 1977. Nutrition and factors limiting the growth of a methanogenic bacterium (*Methanobacterium thermoautotrophicum*). *Arch. Microbiol.* **113:** 17–22.
59. **Theodorou, M. K., D. R. Davies, and C. G. Orpin.** 1994. Nutrition and survival of anaerobic fungi, p. 107–128. *In* D. O. Mountfort and C. G. Orpin (ed.), *Anaerobic Fungi*. Marcel Dekker, Inc., New York, N.Y.
60. **Tindall, B. J.** 1992. The family *Halobacteriaceae*, p. 768–808. *In* A. Balows, H. G. Truper, M. Dworkin, W. Harder, and K.-H. Schleifer (ed.), *The Prokaryotes*, 2nd ed., vol. I. Springer-Verlag, New York, N.Y.
61. **Trinci, A. P. J., D. R. Davies, K. Gull, M. I. Lawrence, B. B. Nielsen, A. Rickers, and M. K. Theodorou.** 1994.

Anaerobic fungi in herbivorous animals. *Mycol. Res.* **98:**129–152.
62. **Vester, F., and K. Ingvorsen.** 1998. Improved most-probable-number method to detect sulfate-reducing bacteria with natural media and a radiotracer. *Appl. Environ. Microbiol.* **64:**1700–1707.
63. **Wachenheim, D. E., and R. B. Hespell.** 1984. Inhibitory effects of titanium(III) citrate on enumeration of bacteria from rumen contents. *Appl. Environ. Microbiol.* **48:**444–445.
64. **Weber, F. J., K. C. Hage, and J. A. M. de Bont.** 1995. Growth of the fungus *Cladosporium sphaerospermum* with toluene as the sole carbon and energy source. *Appl. Environ. Microbiol.* **61:**3562–3566.
65. **Whitman, W. B., E. Ankwanda, and R. S. Wolfe.** 1982. Nutrition and carbon metabolism of *Methanococcus voltae*. *J. Bacteriol.* **149:**852–863.
66. **Widdel, F., and F. Bak.** 1992. Gram-negative mesophilic sulfate-reducing bacteria, p. 3352–3378. *In* A. Balows, H. G. Truper, M. Dworkin, W. Harder, and K.-H. Schleifer (ed.), *The Prokaryotes*, 2nd ed., vol. IV. Springer-Verlag, New York, N.Y.
67. **Wolin, E. A., M. J. Wolin, and R. S. Wolfe.** 1963. Formation of methane by bacterial extracts. *J. Biol. Chem.* **238:**2882–2886.
68. **Yanagita, T., T. Ichikawa, T. Tsuji, Y. Kamata, K. Ito, and M. Sasaki.** 1978. Two trophic groups of bacteria, oligotrophs and eutrophs: their distributions in fresh and sea water areas in the central northern Japan. *J. Gen. Appl. Microbiol.* **24:**59–88.
69. **Zengler, K., G. Toledo, M. Rappe, J. Elkins, E. J. Mathur, J. M. Short, and M. Keller.** 2002. Cultivating the uncultured. *Proc. Natl. Acad. Sci. USA* **99:**15681–15686.

Cultivation of Algae and Protozoa

JOHN G. DAY, UNDINE ACHILLES-DAY, SUSAN BROWN, AND ALAN WARREN

7

Traditionally, algae and protozoa have been regarded as primitive plants and animals, respectively. Differentiation between the two groups is based primarily on the possession of plastids for photoautotrophic nutrition in the algae and a reliance on phagotrophic or osmotrophic nutritional strategies in the protozoa. However, distinctions between the algae and protozoa have never been entirely clear, and certain groups, such as the euglenids and dinoflagellates, have long been claimed both by phycologists and protozoologists. During the last 20 years, studies on the systematics and evolution of unicellular eukaryotes (algae, protozoa, and lower fungi) have been in a state of great activity. Over this period, many taxonomic boundaries, including those between the algae and protozoa, have been broken down and new relationships established. As a result, the constituent organisms are grouped together by some workers as protists, reviving the term coined by Haeckel (39), or as protoctists (60), although many systematists believe that such groups have no evolutionary or systematic validity. By contrast, other workers have proposed systems that retain the kingdom Protozoa, albeit with much-modified definitions and boundaries, while the algae are dispersed among several kingdoms, including the Plantae, Chromista, and Protozoa (14, 15). Nevertheless, the terms algae and protozoa are still useful in a functional and/or ecological sense, defining (primarily) photoautotrophic and heterotrophic protists, respectively.

In addition to the algae and protozoa, the present chapter also considers the cultivation of one group of prokaryotes, namely, the cyanobacteria. The cyanobacteria are commonly referred to as blue-green algae, a name that recalls their former inclusion within the plant kingdom as the class Cyanophyceae. "Cyanobacteria" is now the commonly accepted term for this group and is the one used here.

Algae and protozoa are essentially aquatic organisms, although many examples of both groups are found in terrestrial environments; for example, algae are present as symbionts with fungi in lichens, while both algae and protozoa are common soil inhabitants. However, even in soil they are active only when there is sufficient moisture present. The ecological importance of both groups in the aquatic environment is becoming increasingly recognized. Algae, for example, are the principal primary producers in many aquatic ecosystems, contributing up to 80% of the biomass and primary productivity in open ocean waters (reference 3 and references therein). Algae are also important bioindicators of water quality and are used to assess and monitor the health of aquatic systems. On the other hand, algae cause much concern when they reach densities sufficient to form surface scums or impart off-flavors to potable water supplies. Algal blooms, particularly those of *Microcystis*, *Planktothrix*, and other genera of potential toxin-producing cyanobacteria, are a serious problem to those charged with supplying drinking water and may significantly impair the amenity value of a water body.

Protozoa have been identified as major consumers of primary producers, both algal and bacterial, in pelagic food chains and are a vital component of the microbial loop (5). They, in turn, constitute an important food resource for metazoan plankters. Likewise, protozoa often dominate as consumers of microalgae and of bacteria in sands and sediments, organically polluted waters, and biological aerobic wastewater treatment processes (21, 22, 30). They may also be used as indicators of organic pollution and effluent quality (22, 35). In addition, marine protozoa, such as the ciliate *Mesodinium rubrum* and the dinoflagellate *Ptychodiscus brevis*, cause red tides in many parts of the world, with their attendant toxicological problems. In all these cases, the accurate identification of the organisms involved is essential, and it frequently relies on an ability to culture the organisms in the laboratory. Stable, pure cultures are the bedrock of many aspects of fundamental and applied research, as well as of the commercial exploitation of algae and protozoa. In addition, the ex situ maintenance of organisms is essential for the preservation of microbial diversity. Conservation of organisms, including of algae and protozoa, has been adopted as the basis of international agreements, including Article 9 of the Convention on Biological Diversity (Rio Convention, 1992) and Microbiological Diversity 21, as well as of a range of other international initiatives and programs (40). Therefore, successful cultivation is an important first step in many types of study of algae and protozoa.

Algae are generally assumed to be photoautotrophs, and in most cases cultures will grow in the absence of organic carbon. However, many species are capable of utilizing organic substrates (29, 41, 70), and some require trace quantities of vitamins or other organic molecules for normal growth (92). Achlorophyllous mutants or colorless organisms closely related to pigmented algae require an organic carbon source. It is generally provided in culture media by yeast extract, acetate, glucose, soil extract, or other undefined medium components (see "Protozoa," below). Media and culture conditions may be designed to suit individual strains, although in general, standard media that are satisfactory for a range of organisms are used. They are discussed in greater detail below.

Protozoa are generally assumed to be heterotrophic, although some may be autotrophic or both (mixotrophic). They utilize a wide range of sources of carbon and other nutrients. Many are obligately or facultatively parasitic, endocommensal, or epibiontic. These types fall outside the scope of this chapter, and some are dealt with elsewhere (see chapters 21 and 25). Information on their cultivation may also be found elsewhere in the literature (references 56 and 60 and references therein). Free-living protozoa may obtain carbon and other nutrients in either dissolved or particulate form. In most cases the nutrient source is external to the cell and often takes the form of other (prey) organisms. These may be eukaryotes, prokaryotes, or viruses (88). Alternatively, protozoa may feed on detritus and on other particulate or dissolved organic matter in the environment. Since most protozoa are selective feeders, cultures must seek to provide appropriate food. Some protozoa have plastids (e.g., dinoflagellates) or harbor endosymbionts (e.g., "*Zoochlorella*" species in *Paramecium bursaria*) that may provide internal sources of nutrients. In these cases, the culture methods employed for algae (see below) might be more appropriate for their maintenance in the laboratory.

Traditionally, free-living protozoa have been divided into three main groups according to their morphology: flagellates, amoebae, and ciliates. Of these, only the ciliates are a truly natural (monophyletic) group; the flagellates and amoebae are both polyphyletic and include groups that may be only distantly related. Nevertheless, from a practical viewpoint, it is still sometimes useful to refer to these groupings, since the isolation and culture methods used are often the same within each group.

In order to maintain cultures of protozoa over the long term, it is necessary to provide a medium that suits each species and a supply of appropriate food. Various publications provide comprehensive information or refer to medium preparations for protozoa (56, 60, 71, 90). However, certain isolation techniques, growth media, and culture conditions suit a wide range of organisms. Some of them are discussed below.

ISOLATION METHODS

Methods for the isolation and purification of algae and protozoa are broadly similar. The standard objective is to produce a representative uniprotistan culture, preferably clonal, derived from a single protozoan or algal unit, e.g., cell, filament, trichome, or colony. For some applications use of axenic strains is preferable, in which case all contaminating organisms must be removed and the culture must be maintained in an axenic state. However, for many protozoa a more realistic goal is to produce either a monoxenic culture, in which only one other species (usually the food organism) is present in addition to the one that has been purposefully isolated, or a polyxenic culture, in which multiple food species may be present.

The methods employed are comparable to those used for other microorganisms and can be broadly classified as enrichment methods, dilution methods, or physical and chemical methods. In all cases good microbiological practices and aseptic techniques should be employed. Furthermore, where possible all initial manipulations and transfers should be performed in media with pH and osmotic potential similar to those of the site of isolation. It is also optimal to ensure that suitable temperature, light, and oxygen tension regimens are maintained during transportation and during initial manipulations of field samples. It should be noted, however, that some protozoa fail to grow in fresh culture media even when such precautions are taken. In these cases, it may be necessary to use a conditioned culture medium, i.e., one that has been previously used for culturing other protozoa and that has been conditioned by the exudation of substances by these protozoa. Likewise, when certain protozoa are subcultured, greater success may be attained if a significant amount of "spent" medium, which may contain key nutrients and/or growth factors, is transferred to the new culture.

Enrichment

Enrichment is the inoculation of a field sample into an equal or greater volume of suitable medium and incubation under favorable conditions. By inoculation of parallel cultures in a range of media, different organisms will be selected. Transfer to mineral medium with no available nitrogen (e.g., BG 11 omitting $NaNO_3$ [see Table 2]) will select for nitrogen-fixing cyanobacteria. Organic substrates such as yeast extract (primarily for vitamins), casein (for amino acids), or urea (for nitrogen) can be added when one is isolating osmophilic protists, although amounts should be kept small to avoid bacterial overgrowth. For bacterivorous protozoa, the simplest way to enrich a sample is to add boiled grains of barley, wheat, or rice, which will promote the growth of bacteria and thereby produce a food source for the protozoa. Enrichment does not guarantee the production of a uniprotozoan culture but may be useful in increasing the numbers of cells of the desired organism for further isolation and demonstrate that the target cells can grow in the medium.

Dilution

Dilution methods are most effective for use on preponderantly uniprotistan samples. Material is sequentially diluted in appropriate medium and incubated under favorable conditions. The greatest dilution in which growth occurs is likely to be uniprotistan. The chances of success with this technique are obviously increased if the species in question is one of the most abundant, since unwanted organisms will be diluted more quickly. For very small or abundant protists, a dipping-wire method, in which the number of cells transferred at each dilution step is significantly reduced, has been developed (18). Other variations in serial dilution techniques have been described elsewhere (87). Dilution methods cannot guarantee that the cultures obtained are clonal, and it is most unlikely that an axenic strain will be produced. Once isolated by this method, it may be important to reduce the volume of liquid in which the cell is contained. The ciliate *Tetrahymena thermophila*, for example, dies abruptly if a single cell is placed in a volume of 10 μl or more of culture medium. This result appears to be a form of programmed cell death that can be prevented by reducing the volume of liquid or by increasing the cell concentration, thereby initiating a quorum-sensing mechanism (16).

Physical Methods

Physical methods involve the selection of individual protistan units and their transfer to appropriate sterilized medium and environmental conditions. This may result in an axenic and clonal culture. Micropipetting with thin capillary pipettes, working under a dissecting microscope, can be used for a wide range of organisms, particularly those that are relatively large and/or slow. This technique generally involves a series of transfers in a multiwell plate or series of sterile drops of the appropriate medium on a slide. This method, however, requires dexterity and patience and is not applicable to small protists. Other methods, such as those described below, should be tried first. The additional sequential transfers of an isolated cell through a series of washes may result in the generation of an axenic culture. An alternative instrument that may be successfully employed is a micromanipulator (36, 89), while for small flagellates, Cowling (18) describes an agar cavity-channel isolation technique that may be used in combination with such micromanipulatory methods.

Other methods by which protists may be physically isolated include selective filtration, silicone oil plating, density gradient centrifugation, and flow cytometry. The silicone oil plating technique may be used for a wide range of protists and relies on the isolation of clone-founding cells within microdroplets formed from vortex-mixed oil/culture emulsions (84). Density gradient centrifugation (8, 37, 96) has, from the authors' experience, been problematic, as excessive centrifugation may damage cells. Furthermore, motile strains move rapidly after centrifugation ends, even when density gradients are employed. Simple settling of nonswimming or dense cells may be an effective technique but will rarely result in a uniprotistan culture. This approach, in common with the use of filters or mesh, which may be used to remove large cells, filaments, and detritus, has limited potential to purify cultures. However, it is a useful first step in the purification process.

Flow cytometry (9, 20) has significant potential as an automated means of discriminatory cell sorting and strain isolation, particularly for small algal cells, on the basis of cell size, density, or other cell attributes. It has been particularly successful for the purification of bacterized cultures (81). Sorting from field samples has also been successful (82). The approach has some limitations, as the configuration of the equipment makes it unsuitable for large cells or for filamentous or colonial organisms. It is less suitable for heterotrophic protists that do not contain pigments that give a fluorescent signal. These can still be detected on the basis of side-scatter, but other organisms and inert material may produce similar signals. Keller et al. (50) used fluorescent food vacuole contents to successfully isolate protozoa. An alternative approach has been used by the authors with vital staining. Although the cells can be readily sorted, this approach has not yet successfully been employed to isolate axenic protozoa.

Many anaerobic protists are sensitive to oxygen and have to be isolated and maintained in an oxygen-free environment. Cowling (18) devised a microscope slide ring chamber that permits observation of anaerobic protists and their withdrawal by micropipetting without compromising the oxygen-free conditions within the chamber. A somewhat more complicated apparatus is described elsewhere (98).

Agar Plating

Agar plating methods rely on discerning colony growth of isolated clones on agar surfaces or within agar. These techniques have been used mostly for algae, amoebae, and some flagellates. Certain algae may be isolated by streaking a drop of algal culture onto the surface of an agar plate, using a sterile loop or capillary pipette; discrete colonies may arise from single cells or groups of cells. After an appropriate incubation interval (generally in the range of 7 to 21 days), cells should be aseptically transferred from a colony and resuspended in liquid medium, after which the process may be repeated. A variation of this technique that may be employed for motile strains involves the use of pour plates (i.e., the addition of molten agar to aliquots of a dilution series of the algal suspension). The resulting colonies are excised and transferred to fresh medium, and the process is repeated. This technique can give rise to clonal, axenic isolates of many algal species. Spray-plating techniques, which achieve separation of individual cells by directing a fine spray of cell suspension onto an agar surface, have also been used successfully (74, 97).

Agar plates can also be used to remove epiphytes from filamentous algae. Normally, by use of a metal hook, an algal filament can be dragged through semisolid agar; bacteria and other epiphytes may be removed by this procedure. The filament is then transferred to fresh sterile medium. An additional method used for some motile taxa is to excise a block of agar from a "lawn" of algae or protozoa. This block is then inverted and placed on a sterile plate of agarized medium. The motile organisms will then move across the surface of the agar, establishing a new lawn. This approach has worked for purifying strains of *Planktothrix* but was unsuccessful on application to samples containing motile bacteria or fungi.

Amoebae are commonly isolated by placing one or two drops of sample onto a nonnutrient agar plate that has been streaked with a suitable food organism and then incubating the materials (see "Amoebae" below). Sufficient room (>1 cm) should be allowed to ensure that the amoebae migrate away from their contaminants. The majority of amoeba strains, as the amoebae exhaust the local food supply, migrate over the agar surface, thereby isolating themselves from other microorganisms in the sample. Isolated amoebae may then be picked off with a sterile scalpel blade and subcultured.

In many cases the above procedures are employed in conjunction with other physical and chemical techniques, some of which are listed in Table 1. These methods are discussed in detail elsewhere; for additional information see references 7, 32, 74, 79, and 87.

Filtration

Picoplankton (<2 μm in diameter) can be successfully isolated by using filtration systems (see chapter 20). After a measured volume of sample has been passed through the filter unit, the filter is removed and placed on an agar plate of the appropriate medium. Resulting colonies may be removed and further purified by streaking on agar, or other approaches may be used if necessary.

Electromigration

Electromigration is a method for obtaining concentrated suspensions of ciliated and flagellated protozoa relatively free of bacteria and other organisms (12, 80). It works on the principle that many ciliates and flagellates orient themselves in a direct current and migrate toward the cathode. This technique can be applied to mud, sediments, and vegetation, as well as to liquid environmental samples and cultures (10). Therefore, it is applicable both to the initial isolation of species from environmental samples and to the adaptation of cultured strains to axenic cultivation (94).

TABLE 1 Additional techniques employed to isolate and purify algae

Process	Rationale or method outline
Use of antibiotics	To kill contaminating bacteria and cyanobacteria
Use of nonspecific bactericides	To kill contaminants. Methods include 2% (vol/vol) formaldehyde for 2 min or 1% (vol/vol) sodium lauryl sulfate for 12 h (for the purification of cyst/aplanospore-forming algae). Sequential 0.5, 1, 2, 5, and 10% (vol/vol) for sodium hypochlorite washes 5 min each.
Use of germanium dioxide	Incubation in medium containing 5 to 10 mg of germanium dioxide liter^{-1} kills diatoms.
UV irradiation	Low doses kill bacterial contaminants. It may also damage or kill the algal cells and may induce mutations.
Fragmentation/ultrasonication	Disruption of clumps of cells, dislodging of adhering contaminants, to produce a uniform dispersion.
Low-speed centrifugation	Centrifugation (100 to 200 $\times$ g) for 10 min should not affect gas vacuolate cyanobacteria; most contaminants are removed by centrifugation.
Phototactic response	Use of unidirectional light source. Motile algae may move on or in agar toward light away from contaminants.
High incubation temperature	Selection of thermophiles

MAINTENANCE METHODS

Algae

The choice of medium and maintenance regimen is dependent on the growth requirements for the algal isolate(s) as well as on the resources that are available locally. The trend in larger collections is to minimize the number of media and growth regimens utilized and, in addition, to maximize the subculture interval, without adversely affecting culture quality. To this end, many planktonic strains commonly isolated from aquatic environments are maintained on solidified media.

Media

Although standardization of methods and media has been achieved for some purposes, particularly ecotoxicity testing (72), in general the medium composition depends both on the requirements of the algae (e.g., diatoms require the inclusion of a silica source) and on the preferences of the researcher. Consequently a wide variety of media have been developed and are in common use; details of medium composition and their suitability for particular algae are available in specialist literature (65, 71, 79, 86, 95). Many isolates may be successfully maintained on or in a relatively broad-spectrum medium. Some of the most commonly used are detailed below. In general, isolates will be most successfully cultured and manipulated in liquid medium. However, for some applications (and in order to increase the time between subcultures), solidified media may be more appropriate. Solidified medium is produced by the addition of 15 g of bacteriological agar per liter to standard medium formulations before the medium is dispensed into tubes, universal bottles, or other appropriate vessels. After sterilization, vessels are placed at an acute angle to maximize the surface area of the agar upon gelation.

BG 11 (Table 2), a medium developed for culturing cyanobacteria (which generally prefer alkaline conditions), is also suitable for a wide range of eukaryotic microalgae.

Jaworski's medium (Table 3) is used for a wide range of freshwater and terrestrial eukaryotic algae (26, 90). It may also be used to maintain diatoms upon addition of 57 mg of $NaSiO_3 \cdot 9H_2O$ per liter. In addition, Jaworski's medium supplemented with organic material is most commonly used at the United Kingdom's Culture Collection of Algae and Protozoa (CCAP) to maintain axenic algal strains (26, 90). The additional organic material in the form of sodium acetate and complex nitrogen sources increases the potential yield of cultures that can grow mixotrophically or pho-

TABLE 2 BG 11 (blue-green algal [cyanobacterial] medium)[a]

Stock solution(s)	Amt/liter[b]
1. $NaNO_3$	15.0 g[c]
2. $K_2HPO_4 \cdot 3H_2O$	4.0 g
3. $MgSO_4 \cdot 7H_2O$	7.5 g
4. $CaCl_2 \cdot 2H_2O$	3.6 g
5. Citric acid	0.6 g
6. Ferric ammonium citrate	0.6 g
7. EDTA (disodium salt)	0.1 g
8. Na_2CO_3	2.0 g
9. Trace metal mixture:	
H_3BO_3	2.86 g
$MnCl_2 \cdot 4H_2O$	1.81 g
$ZnSO_4 \cdot 7H_2O$	0.222 g
$Na_2MoO_4 \cdot 2H_2O$	0.39 g
$CuSO_4 \cdot 5H_2O$	0.079 g
$Co(NO_3)_2 \cdot 6H_2O$	0.0494 g

[a]Based on reference 85.

[b]Add 100 ml of solution 1, 10 ml of solutions 2 through 8, and 1 ml of solution 9 to distilled or deionized water to obtain a total volume of 1 liter. pH is adjusted to 7.8 prior to sterilization.

[c]May be omitted for nitrogen-fixing cyanobacteria.

TABLE 3 Jaworski's medium[a]

Stock solution(s)	Amt/200 ml[b]
1. $Ca(NO_3)_2 \cdot 4H_2O$	4.0 g
2. KH_2PO_4	2.48 g
3. $MgSO_4 \cdot 7H_2O$	10.0 g
4. $NaHCO_3$	3.18 g
5. EDTA FeNa	0.45 g
EDTA Na_2	0.45 g
6. H_3BO_3	0.496 g
$MnCl_2 \cdot 4H_2O$	0.278 g
$(NH_4)_6Mo_7O_{24} \cdot 4H_2O$	0.2 g
7. Cyanocobalamin (vitamin B_{12})	0.008 g
Thiamine HCl (vitamin B_1)	0.008 g
Biotin	0.008 g
8. $NaNO_3$	16.0 g
9. $Na_2HPO_4 \cdot 12H_2O$	7.2 g

[a]Based on reference 90.
[b]Add 1 ml of solutions 1 through 9 to distilled or deionized water to obtain a total volume of 1 liter of medium.

toheterotrophically. Also, any bacterial contaminants in previously axenic cultures rapidly become obvious, as the medium will appear cloudy within 2 days.

Guillard's f/2 medium (Table 4) is used for a wide spectrum of marine microalgae (90). It may be used for marine diatoms on addition of 40 mg of $Na_2SiO_3 \cdot 9H_2O$ per liter. Like Jaworski's medium, it is suitable for maintaining axenic cultures. The addition of 1.0 g of yeast extract and 1.0 g of glucose or sodium acetate per liter increases the potential yield of cultures that can grow mixotrophically or photoheterotrophically.

Maintenance Conditions

Optimal temperatures depend on the environment from which the cultures were isolated. Most commonly cultured strains can be maintained at 15 to 20°C, although some organisms from extreme environments, both hot and cold, cannot survive within this temperature range. These cultures should be maintained at a temperature as close as possible to that of their habitat. Care should be taken to avoid exceeding the maximum temperature, and it is advisable to use illuminated incubators with good temperature control (e.g., ± 2°C). Light regimens, in many cases, depend on the facilities available. Usually a mixture of cool-white and warm-white fluorescent tubes is appropriate, with a light-dark regimen ranging from 12 h:12 h to 16 h:8 h. A wide range of light-dark periods may be employed with reasonable success. Alternatively, natural light from a north-facing window is suitable. In general, light levels of approximately 50 μmol photon m^{-2} s^{-1} are satisfactory, although some organisms, particularly cyanobacteria, will grow better at light levels of <25 μmol photon m^{-2} s^{-1}. In general, the objective is to extend the interval between serial transfers, so lower temperatures and light levels are employed, e.g., 10 to 15°C, at a light level of <25 μmol photon m^{-2} s^{-1}.

Once an algal culture is established, it requires periodic subculturing by the aseptic transfer of an inoculum (generally 0.5 to 10% [vol/vol] of the original culture) to sterile medium. The duration of intervals between subculturing episodes depends on the growth characteristics of the isolate as well as on the medium and culture conditions used. The duration of intervals generally ranges between 2 weeks and 6 months.

Protozoa

For flagellates and ciliates, the isolation techniques, growth media, and culture conditions are often identical or similar. Therefore, these two groups will be dealt with together, while the amoebae will be discussed separately.

Flagellates and Ciliates

The choice of culture medium will depend largely upon what the protozoan feeds. Many flagellates and ciliates feed upon bacteria, and in these cases nonselective media, designed to encourage the growth of bacterial populations, may be used. For many species, isolates may be cultured in the presence of mixed bacterial flora that coexisted with the target organism in its original habitat. Alternatively, selective cultures may be obtained by incubating the protozoa in an inorganic salt solution along with an appropriate food organism. For bacterivorous forms common, nonpathogenic laboratory cultures of bacteria may be used. In some cases, however, it may be necessary to isolate, characterize, and identify bacteria from the original sample and then use one or more of these strains as selected food organisms.

For omnivores and carnivores, an examination of the contents of the food vacuoles of the isolate may give an indication of its preferred food. Attempts should then be made to feed the isolate on something similar in shape, size, and chemical composition to its normal prey. If all else fails, it may be necessary to carry out a replicated feeding experiment by using a range of food organisms in order to determine which will support the growth of the isolate.

For some species axenic cultures may be established. In these cases, the culture medium, which may be defined or semidefined, is usually rich in organic compounds.

Culture media may be categorized into four main types: plant infusions, soil extract-based media, inorganic salt solutions, and specific (organically rich) media.

TABLE 4 Guillard's f/2 medium[a]

Stock solution(s)	Amt/liter[b]
1. $NaNO_3$	7.5 g
2. $NaH_2PO_4 \cdot 2H_2O$	5.65 g
3. Na_2EDTA	4.36 g
$FeCl_3 \cdot 6H_2O$	3.15 g
$CuSO_4 \cdot 5H_2O$	0.01 g
$ZnSO_4 \cdot 7H_2O$	0.022 g
$CoCl_2 \cdot 6H_2O$	0.01 g
$MnCl_2 \cdot 4H_2O$	0.18 g
$Na_2MoO_4 \cdot 2H_2O$	0.006 g
4. Cyanocobalamin (vitamin B_{12})	0.5 mg
Thiamine HCl (vitamin B_1)	100.0 mg
Biotin	0.5 mg
5. $Na_2SiO_3 \cdot 9H_2O$	40.18 g

[a]Based on references 61 and 90.
[b]Add 10 ml of solution 1 and 1 ml of solutions 2 to 4 to filtered natural seawater to obtain a total volume of 1 liter. Adjust pH to 8.0 prior to sterilization. If the medium is to be used for diatoms, 1 ml of solution 5 is added to the medium prior to pH adjustment and sterilization.

Plant Infusions

Plant infusions (Table 5) have long been used for the culture of bacterivorous flagellates (e.g., *Bodo* spp.) and ciliates (e.g., *Colpidium* and *Paramecium* spp.). The principle is that organic compounds leach out of, or are extracted from, plant material and that these compounds support bacterial growth. The simplest to use are polished rice, boiled wheat, or barley grains. Likewise, hay or lettuce infusions have been used successfully for many years. Dehydrated, powdered cereal leaf preparations (e.g., Wards Natural Science product 944W8602) are commercially available. These may be prepared as 0.1% to 0.25% (wt/vol) infusions either in an inorganic salt solution, mineral water, or in distilled water. Care should be taken not to add amounts of infusion larger than the recommended 0.1 to 0.25%, since they could lead to excessive growth of bacteria, which almost always inhibits protozoan growth. On the other hand, an absence of bacteria may be equally undesirable. Wheat grass powder infusions, for example, have been shown to produce hydrogen peroxide, which is lethal to the ciliate *Paramecium* in the absence of bacteria. However, the hydrogen peroxide is decomposed by bacterium-derived catalases in bacterized media and thus detoxified (66). Plant infusions may also be used for culturing marine flagellates and ciliates. Here, alternative sources of enrichment may be used such as the seaweeds *Ulva* and *Enteromorpha* spp., which should be autoclaved before use. For marine species, the infusions are made in either natural or artificial seawater (90).

Soil Extract-Based Media

As with plant infusions, the underlying principle is to extract organic compounds from the soil that will support the growth of bacterial food organisms. There are several ways of doing this, one of which is as follows: sterilize (autoclave 1 h at 15 lb/in^2) 1 part air-dried, sieved soil—preferably a sandy loam, pH 7.0, or beech leaf/lime soil compost—with 2 to 10 parts distilled water to produce a supernatant extract stock. This medium is autoclaved twice to ensure sterility and to guarantee that all fungal and bacterial spores are killed. The medium can be used with the soil remaining as a so-called "biphasic" (soil-water) culture medium. A biphasic tube culture method (79) has been used as an effective and convenient means of maintaining many flagellate and ciliate strains in long-term batch culture. Alternatively, the supernatant may be decanted and filtered for use as a liquid culture medium, either directly or diluted and with inorganic salts added. The addition of calcium carbonate or cereal grains to culture dishes or tubes of soil extract media may promote faster bacterial growth and improve culture longevity of saprophytic flagellates and ciliates. As with plant infusions, care must be taken not to allow excessive growth of bacteria, including that of actinomycetes, which might inhibit protozoa. Soil extracts are therefore normally diluted to 3 to 10% (vol/vol).

Inorganic Salt Solutions

Inorganic salt solutions provide a balanced medium, the ionic composition of which is suitable for the growth of many protozoa. Such solutions, however, contain negligible quantities of dissolved organic matter, so the addition of food organisms, and/or a carbon source for the growth of food organisms, is essential. Carnivorous ciliates (e.g., *Didinium*) generally grow well in inorganic salt solutions, provided that they have sufficient prey (e.g., *Paramecium*). Inorganic salt solutions may either be natural or artificial. Water can be regarded as a weak inorganic salt solution. For some species, filtered water from the sample site is the best medium. Most freshwater and soil species will also grow in commercially available, noncarbonated bottled mineral waters (e.g., Volvic [Perrier UK Ltd.]). Artificial inorganic salt solutions have the advantage that their chemical composition is known precisely. Perhaps the most commonly used is Prescott's and James's solution (Table 5).

Marine protozoa generally grow best in liquid media, and for many species the most convenient medium is seawater itself, filtered and autoclaved. Defined marine media such as artificial seawater medium, sold commercially as Ultramarine Synthetic Salt Solution (Waterlife Research Industries), and supplemented seawater medium may also be used for several flagellate and ciliate species. Formulations for both are given elsewhere (90).

Specific Media

For certain applications the ultimate objective is to produce an axenic culture. Many protozoa, including certain species of the flagellates *Peranema*, *Euglena*, and *Chilomonas* and of the ciliates *Tetrahymena* and *Paramecium*, have been grown axenically. To achieve this end, a defined or semidefined medium must be used. Such media invariably contain

TABLE 5 Culture media used for the maintenance of nonmarine flagellates and ciliates[a]

Medium	Component	Comments
Plant infusions	Hay	1–10 g of timothy hay or similar. Boil and filter.
	Lettuce	1–2 g of dried lettuce leaves. Boil and filter.
	Cereal grains	Boil barley or wheat grains (5 min). Add 1 grain per 10 ml of medium. Further sterilization not required.
	Cereal leaves	Boil 0.1–0.2 g of dehydrated cereal leaves (10 min). Filter. Adjust pH to 7.0–7.2.
Soil extract medium	Soil extract medium with added salts	Soil extract stock,[b] 100 ml; K_2HPO_4, 20 mg; $MgSO_4 \cdot 7H_2O$, 20 mg; KNO_3, 200 mg
Inorganic salt solutions	Prescott's and James's	$CaCl_2 \cdot 2H_2O$, 0.433 g; KCl, 0.162 g; K_2HPO_4, 0.512 g; $MgSO_4 \cdot 7H_2O$, 0.28 g. Make up 100 ml of stock solution of each in distilled water. Combine 1 ml of each stock solution and make up to 1 liter with distilled water.
Specific media	PPYE[c]	Proteose peptone (Oxoid L85), 10 g; yeast extract (Oxoid L21), 2.5 g. Boil until dissolved.
	Euglena medium	Sodium acetate (trihydrate), 1 g; Lab Lemco (beef extract) powder (Oxoid L21), 2 g; $CaCl_2$, 10 mg.

[a]Media components are given for 1 liter of final medium made with glass-distilled water unless otherwise indicated. All media sterilized by autoclaving for 1 h at 15 lb/in^2 unless otherwise indicated. Based mainly on references 18, 31, and 32.

[b]See text for method for preparing soil extract stock.

[c]Proteose peptone yeast extract.

relatively high concentrations of dissolved organic carbon so that the protozoa can feed osmotrophically. The ingredients of such media are usually derived from animal sources (powdered whole or defatted liver, proteose peptone, beef extract, etc.), although yeast extracts are often added as nutritional supplements. Commonly used formulations include proteose peptone yeast extract medium (for *Tetrahymena*) and *Euglena* medium (for acetate flagellates, e.g., euglenids and cryptomonads). Formulations for these media are given in Table 5. Gradual release of nutrients for prolonged axenic cultivation of tetrahymenids and phagotrophic chrysomonads has been achieved by using large screw-cap tubes containing powdered liver (0.4% to 0.8% [wt/vol]) overlaid with water (49). An alternative, inexpensive method to achieve long-term maintenance of *Tetrahymena* is the use of a garbanzo bean autoclaved in a test tube of water as the culture medium.

Maintenance Conditions

The main culture variables to consider are temperature, pH, oxygen tension, food type and concentration, culture medium type and concentration, culture vessel type and capacity, and inoculum age, size, and density (18). As a general rule, conditions should remain as close as possible to those from which the organism was isolated, although most species will grow more quickly at slightly elevated temperatures (18 to 25°C is optimal for many) and in the presence of a plentiful food supply. One of the greatest perils is overgrowth by bacteria, which can normally be avoided by diluting the culture medium or by adding a suspension of nongrowing bacteria, generally heat treated, to nonnutrient media. On the other hand, maintenance of strains at low or suboptimal temperatures can save labor and time. Sterile plastic tubes, flasks with screw caps, or petri dishes are all suitable for maintaining cultures of flagellates and ciliates.

Some protozoa are algivorous, while others are omnivores and feed on algae facultatively. In order to maintain an adequate food supply, cultures of these types should be kept in conditions of illumination as described in "Algae." By contrast, many other protozoan cultures may best be incubated in the dark in order to control algal contaminants and/or to prevent any deleterious effects of excessive light.

Many species of protozoa are known to produce cysts. These may be resting cysts, which are typified by their highly resistant cyst walls, or reproductive cysts, the walls of which are usually thinner and more permeable. The factors that cause encystment vary in different species, and the literature contains many descriptive accounts of this process (for reviews, see references 17 and 83). Usually these factors are those associated with adverse environmental conditions such as shortage or excess of food, extreme pH, increase in salt concentration, high temperatures, overcrowding, lack of oxygen, accumulation of metabolic by-products following rapid growth, or desiccation. Interestingly, there are far fewer accounts of excystation. Like encystation, excystation has been ascribed to a variety of factors, including osmotic phenomena, high food and oxygen concentrations, enzymatic dissolution of cystic membranes, and presence of protein degradation products (17). Often, simply transferring cysts to a fresh culture medium will stimulate excystation.

The techniques described above were principally designed for aerobic protozoa. Anaerobic protozoa require special techniques, since one of the most important considerations is the exclusion of oxygen from the culture. Cultures of anaerobic flagellates and ciliates have been routinely maintained by employing a modification of the Hungate technique. This uses pH-adjusted, degassed (N_2/CO_2 [95%:5%]) cereal infusion or soil extract medium in serum bottles. The headspace is replaced with nitrogen, and the bottles are sealed with butyl rubber bungs held in place by crimped aluminium caps (64). By use of this technique, the ciliate *Trimyema* has been maintained in culture for 12 years (31), and *Plagiopyla* and *Metopus* for >14 years, at the CCAP.

Amoebae

Amoebae may be cultured in liquid media, on agar, or in biphasic media. Information concerning the maintenance of individual species and genera is available in the literature (48, 71, 75-77, 90). In general, the larger amoebae (e.g., those belonging to the family Amoebidae) are grown in liquid culture and smaller amoebae (e.g., those belonging to the family Vahlkampfiidae) are grown on agar. Freshwater and soil amoebae may persist for some time in dilute salt solutions, mineral water, or even distilled water. Marine amoebae may persist in seawater diluted to 75% strength or less. These inorganic liquids form the basis of all culture media suitable for amoebae. Growth and cell division require additional nutrients either included in the media or

TABLE 6 Culture media for freshwater and soil amoebae

Representative families and genera	Culture media	Food organisms
Vahlkampfiidae, Thecamoebidae, Paramoebidae, Hartmannellidae, Vexilliferidae, Vannellidae, Leptomyxidae, Acanthamoebidae, Echinamoebidae, Flabellulidae, Cochliopodiidae, *Stachyamoeba*, *Deuteramoeba*, *Dactylamoeba*, *Rosculus*, *Cryptodifflugia*, *Nuclearia*, *Dictyostelium*	Agar, e.g., nonnutrient amoeba saline agar and cereal leaf medium[a]-Prescott agar (90)	Bacteria, e.g., *E. coli* and *K. aerogenes*
Arcrasidae, Cavosteliidae, Protosteliidae	Agar, e.g., corn meal glucose agar (90) and hay infusion agar (73)	Bacteria, e.g., *E. coli*, or yeasts, e.g., *R. mucilaginosa*
Amoebidae	Inorganic liquid media, e.g., Prescott's and Carrier's solution (90) and modified Pringsheim's solution (90)	Other protozoa, e.g., *Tetrahymena pyriformis* and *Colpidium striatum*
Axenic: Harmannelliidae, Vahlkampfiidae, Acanthamoebidae	Organic liquid media, e.g., proteose-peptone glucose (90) and PYNFH medium (54)	None
Euglyphidae, *Mayorella*	Biphasic, e.g., cereal leaf–Prescott liquid and cereal leaf–Prescott agar[a] (90)	Bacteria in coculture

[a]Cereal leaves available from Wards Natural Science (product code 944W8602).

TABLE 7 Culture media for marine amoebae

Representative families and genera	Culture media	Food organisms
Vahlkampfiidae, Hartmannellidae, Flabellulidae, Paramoebidae, Thecamoebidae, Leptomyxidae, *Stygamoeba*, *Nolandella*	Agar, e.g., malt and yeast extract–75% seawater agar[a]	Bacteria in coculture or (e.g.) *E. coli*
Gruberella, *Korotnevella*, *Parvamoeba*, *Mayorella*, *Trichosphaerium*	Liquid media, e.g., 75% seawater, Sigma cereal leaf–75% seawater,[a] modified Føyns Erdschreiber medium,[a] or biphasic medium[a]	Bacteria in coculture

[a]Based on reference 90.

from prey organisms, selected according to the needs of the strain. Tables 6 and 7 provide a guide to combinations of culture media and food organisms appropriate for the culture of a wide diversity of freshwater and marine amoebae, the majority of which are classified in the phylum Rhizopoda (for classification, see references 75 and 76).

Organic nutrients can be included in the medium itself either by the inclusion of nutrients such as glucose, serum, proteose peptone, and yeast extract or plant infusions (see above). However, apart from axenically cultured amoebae, for which dissolved nutrients are the sole source of food, the principal reason for increasing the organic content of a medium is to support the growth in the culture of bacteria on which amoebae can feed. Bacterial growth can also be stimulated by the addition of one or two unpolished rice grains that have been surface sterilized by rapid passage through a flame. Care must be taken not to make a medium too rich, or the bacteria may overgrow the amoebae. Amoebae may be feeding on bacteria that were isolated with them from their original habitat, but if suitable bacteria are not present in coculture with the amoebae, species such as *Escherichia coli* or *Klebsiella aerogenes* can be cultured separately and added directly to the culture vessel. Yeasts (e.g., *Rhodotorula mucilaginosa*) may be suitable for slime molds.

Although bacteria are an appropriate food source for many amoebae (particularly smaller amoebae), larger amoebae can also feed on other protists. By using an inorganic liquid medium with rice grains, it is possible to maintain cultures in which an adequate growth of bacteria supports a population of small protists (e.g., the ciliate *Colpidium*) that are in turn preyed upon by amoebae. Alternatively, the rice grains (and bacteria) can be omitted, and the protists (e.g., *Tetrahymena*) can be added directly at regular intervals (2 to 7 days). In this case, the food organism is cultured separately in a rich organic medium but is transferred, after centrifugation, into an inorganic medium before being added to the amoeba culture. It is important not to "overfeed" large amoebae. Some amoebae feed on algae, including cyanobacteria (e.g., *Nuclearia* feeding on *Planktothrix*). Again, the food organism is cultured separately and added to the amoeba culture containing an inorganic liquid medium.

Maintenance Conditions

The main culture variables are temperature, light, and moisture. Cultures of most free-living, nonpathogenic amoebae are best maintained at room temperature, about 18 to 22°C. However, if one wishes to maximize the interval between transfers, many amoebae can be maintained on agar plates and incubated at 15°C. At this incubation temperature, there is an increased likelihood of condensation occurring in the culture vessels, which can increase the likelihood of contamination. These problems may be alleviated by incubating the petri dish contents with the agar facing upward and/or by placing petri dishes with inoculated contents in a nonairtight container or in a cupboard. Although some species grow well at temperatures up to and above 30°C, temperatures above approximately 24°C may be lethal to many environmental isolates. Potential pathogens belonging to the genera *Naegleria* and *Acanthamoeba* are often maintained at 10 to15°C, although they may also be cultured at temperatures in the range of 37 to 45°C. In terms of health and safety strategies, such organisms should be regarded as category 2 pathogens and should be handled accordingly. Many small freshwater and soil amoebae form cysts, and these can be stored at 7°C. Cultures should never be placed in direct sunlight, and most amoebae are best maintained in the dark. Amoebae that require a light-dark cycle include those which feed on algae, those with algal endosymbionts, and the acrasids. To prevent a detrimental increase in the concentration of solutes during incubation of a culture and/or the drying out of agar, the junction between the lid and base of the culture vessel can be sealed with a narrow strip of Parafilm. This practice also reduces the risk of contamination.

Subculturing Methodology

The frequency with which a strain requires transfer to fresh medium depends on the species and the culture method. Subculturing is preceded by microscopical examination to ensure that cultures contain many viable cells and have not become contaminated. The risk of contamination can be minimized by subculturing under aseptic conditions, using standard microbiological techniques. Information concerning the safe handling of potential pathogens can be found in reference 1.

Agar cultures. Amoebae can be transferred to a new plate on an agar block cut from a dense parent culture. A suitable area is located microscopically, and its location is marked on the inverted base of the parent plate. A small block (about 9 mm^2) is excised with a scalpel and placed amoeba side down near the edge of the fresh agar. If bacteria or yeasts are required as food organisms, they are streaked across the agar first and the block is placed at one end of the streak. Amoebae from less dense cultures can be transferred by washing the cells from the parent plate with an inorganic liquid by using a Pasteur pipette and distributing the suspension over a fresh agar surface, previously spread with bacteria or yeast if necessary. Each time a strain is subcultured, two or three new cultures should be created. Amoeba strains vary widely in how rapidly they multiply on an agar surface; subculturing intervals range from 1 to 12 weeks. Cyst-forming amoebae are subcultured onto a slope of agar (with a bacterial streak) in a universal bottle and are incubated at 20°C for 2 weeks before being transferred to 7°C. At this temperature, they may remain viable for 6 to 48 months or longer. Excystment is induced by transferring cysts to an agar plate streaked with bacteria, either by washing (see above) or on a small block of agar. The new culture is incubated at 20°C. Some cysts may remain viable for considerably longer, although the proportion of the cysts capable of excystment and the probability of generating an active culture will decrease over time.

Testate amoebae, e.g. *Euglypha*, are often maintained on cereal leaf agar overlaid with a cereal leaf infusion. Fresh cultures are inoculated by using a disposable plastic pipette (Pastette) and gently scraping the surface of the agar to dislodge the biofilm of amoebae and bacteria. Approximately 1 ml is then transferred to a fresh petri dish overlaid with liquid medium. Cultures should be incubated at 15 to 20°C, and transfers should be performed every 6 to 8 weeks.

Liquid cultures. Suitable culture vessels include dishes with lids and tissue culture flasks made of glass or plastic. To ensure adequate oxygenation for aerobic organisms, the depth of liquid should not normally exceed 10 to 12 mm. However, axenic cultures and those tolerant of lower oxygen concentrations can be maintained in tubes containing approximately 10 ml of liquid medium. Intervals between subculturing transfers range from 2 to 12 weeks, and three or four new cultures should be generated each time. Strains in liquid culture can be subcultured and transferred either by pouring or by pipette. Amoebae readily attach to plastic and glass surfaces but can be resuspended by agitating the medium immediately prior to transfer, either by using a pipette or by swirling the culture vessel. In most cases, 25 to 100% of the total volume of parent culture is transferred to the new culture vessel containing fresh medium, and then the mixed culture is transferred back to the parent culture to equalize the volumes. To subculture axenic strains, only 5 to 10% (vol/vol) of culture is transferred aseptically to fresh medium.

Extremophiles

Organisms growing in conditions commonly regarded as extreme include anaerobes, thermophiles, psychrophiles, acidophiles, alkalophiles, halophiles, and barophiles. The principle of cultivating extremophiles is much the same as that for other organisms, i.e., to provide culture conditions that mimic those of the environment from which the organism was isolated. For a number of reasons, however, the technical challenge is usually greater. In extreme habitats, environmental conditions are often subject to steep gradients, so organisms that are spatially close to one another may be living under very different conditions. Characterizing the precise conditions within such relatively small habitats is often very difficult, and maintaining these conditions in culture may be problematic, since they are often subject to modification by the organisms themselves. Furthermore, it may be important that such conditions are also maintained during transportation to the laboratory, since even short-term exposure to "normal" conditions, e.g., ambient temperature and surface pressure, may be lethal to extremophiles.

Organisms that thrive either at high temperatures (thermophiles) or low temperatures (psychrophiles) can often be maintained in laboratory cultures by using standard growth media and incubating within the appropriate temperature range, i.e., in a culture vessel placed either in a constant temperature water bath or in a temperature-regulated incubator or refrigerator. However, attention must also be paid to other parameters, e.g., pH and pressure, which may also tend toward extreme values, especially in high-temperature environments such as deep-sea hydrothermal vents. Likewise, salinity is usually elevated in marine waters at low temperatures. In such cases, culture conditions should be adjusted accordingly. Wherever possible, manipulations should be carried out at the temperature of incubation, since temperature changes that are too great and/or prolonged may be detrimental, or even lethal, to the target organism. For example, the thermophilic ciliate *Trimyema minutum*, which has an optimal growth rate at 48°C and is able to grow at temperatures in the range of 28 to 52°C, rapidly becomes unviable at 55°C, although at 25°C it is able to survive for several weeks but unable to grow (6). However, for this organism, very short-term exposure to temperature variations is not necessarily problematic. Lee and Fenchel (55) reported that exposure to room temperature for intervals up to 1 min had no significant effect on cultures of the psychrophilic ciliate *Euplotes antarcticus* grown at −2°C.

Barophiles, i.e., organisms that thrive at high hydrostatic or atmospheric pressures, are usually associated with deep-sea research. A number of protozoa, including ciliates, flagellates and amoebae, have been isolated from pelagic deep-sea environments and from the ocean floor (78, 91). Ideally, the pressure at the sample site should be maintained throughout the collection and isolation procedure. In reality, however, very few institutions are equipped to do this, and samples are usually subject to decompression during retrieval to the surface, often with no discernible ill effect to the organisms concerned. Cultivation is then normally carried out at the same pressure and temperature prevailing at the site from which the sample was collected. Usually, this end is achieved by placing the culture vessel within a reactor pressure vessel, e.g., as described in reference 45. By using this technique, Atkins et al. (4) succeeded in establishing cultures of barophilic heterotrophic flagellates by growing them in normal enrichment media and also in filtered or autoclaved water from the same location to which heat-killed prey bacteria (10^8 to 10^9 cells/ml) had been added. By contrast, the growing interest in astrobiology has generated investigations into the cultivation and survival of various organisms, including protozoa, at low or zero atmospheric pressures, such as would be encountered in space (67).

Organisms that grow optimally at pH values that vary significantly from neutrality are termed either alkalophiles or acidophiles. Alkalophiles are those organisms whose optimum rate of growth is at least two pH units above neutrality, while for extreme acidophiles, the optimum pH for growth is at or below 3.0. A readily accessible source of acidophiles is acid mine drainage water. Johnson and Rang (47) succeeded in cultivating five protozoan species, including three heterotrophic flagellates, one ciliate, and one amoeba, from acid mine spoil effluents and leaching liquors with pHs of <3.0. The cultures were established by inoculating environmental samples containing mixed protozoan and bacterial populations into 20 mM ferrous sulfate medium (pH 2.0) and incubating at 30°C. Furthermore, four of the five protozoa were maintained for over 3 months by growing mixed cultures in liquid medium containing coarse-grained (61 to 200 μm) pyritic coal (47).

Halophiles are organisms that thrive in saline conditions; i.e., they grow best in media containing at least 0.2 M NaCl. A variety of algae and protozoa are classified as halophiles, perhaps the best studied being the phototroph *Dunaliella* (reference 11 and references therein), an alga that is commercially cultured for β-carotene.

TROUBLESHOOTING

For both algal and protozoan cultures, poor or no regeneration on transfer can usually be attributed to the inoculum, the medium, or the incubation conditions. As outlined above, the use of senescent inocula should be avoided. It may be possible to use a previous subculture or a duplicate sample that is more physiologically active than is the culture initially chosen. If there is no option, a larger inoculum can be used, or alternatively, the nutrient regimen and culture conditions can be altered (see below).

Media

The most common problems associated with the media are incorrect pH, high levels of precipitate, and incorrectly formulated media or the omission of a vital ingredient (e.g., Si for diatoms and vitamins for many marine algae). Most algae are tolerant of fairly large changes in pH, although if the inoculum's vigor is suboptimal, poor or no growth can result. In most cases, freshwater eukaryotic algae prefer acidic environments (pH 5 to 7), while cyanobacteria prefer alkaline environments (pH 7 to 9). High levels of precipitate can result in nutrient limitation and osmotically stressful microenvironments, and this situation has been observed to cause problems for both algae and protozoa. For poor inocula, improved recovery may be obtained by using a less defined medium in combination with the standard mineral medium (e.g., *Euglena* medium with Jaworski's medium or artificial seawater medium with natural seawater). Alternatively for axenic strains, supplementation of the medium with low concentrations of proteose peptone, yeast extract, vitamins, or soil extract may assist recovery. Sometimes a transfer from agarized to liquid media, or vice versa, will enhance growth. Undefined ingredients like soil extract might be replaced by a new stock preparation, as the quality may vary. Changes in soil source may have a radical influence on culture performance. Changes in pH, level of organic material, trace nutrients, soil texture, etc., can have a positive or deleterious influence on growth. Soil should never be sourced from an area that has been treated by artificial fertilizers, herbicides, etc.

Incubation Temperature and Localized Heating Effects

This situation is unlikely to be the cause of growth failure, assuming that the parent culture was originally grown under the same conditions. It is widely recognized that transferring cultures into an incubator set at a higher, physiologically suitable temperature stimulates the growth rate. This is often used to ensure a dense culture for rapid dispatch from collections (including at the CCAP) where slow-growing members of the Chlorophyta and cyanobacteria, normally incubated at 15°C, may be grown at 23°C. Lights in close proximity to the culture may result in the culture temperature's being above the temperature maximum for the strain. If illuminated incubators are not available, a common problem occurs when cultures are placed in direct sunlight. Unless sufficient care is taken to control the intensity of light reaching the culture and the heating effects of the sun, the culture is invariably killed.

Incorrect Light Levels and Unsuitable Lighting Regimen

Cyanobacterial and red algal cultures are particularly susceptible to photoinhibition, bleaching, and death attributed to excessive light levels. The same is true for some protozoan strains. Measuring the light intensity at the culture's actual location excludes overillumination. It should be noted that, in dense cultures, self-shading can have a significant effect on the intensity of light reaching an individual cell, so subcultures need to be initially incubated at relatively low light levels. Low light levels can have a limiting effect on final culture density, but assuming that light levels were sufficient to support the growth of the parent culture, they are unlikely to prevent growth of the subculture. Assuming that an appropriate light-dark regimen is used, this is unlikely to cause problems, although it is worth checking that the timing or switching mechanism is functioning effectively. Tungsten filament lights are generally not satisfactory, and only fluorescent lights, GrowLux lights, or natural light should be used.

Where extremely poor inocula are employed, it is often advantageous to incubate at a slightly higher temperature, e.g., 20°C instead of 15°C for *Tetrahymena*. If extremely low levels of viable cells or cysts are inoculated, it may be necessary to leave the inoculated tube contents alone for an extended period to allow any "healthy" cells to reproduce. It may also be disadvantageous to employ too short an interval between transfers, as this practice has been observed to result in deterioration of cultures, particularly in the case of amoebae.

Changes in culture conditions are required on transferring a stock culture for subsequent use in the classroom or in an experiment. Transfer to fresh medium is required; however, in most cases the same medium can be used as for low maintenance. The freshly transferred culture should be maintained at optimal conditions, i.e., at higher light intensities and at higher temperatures (e.g., change from 16 to 20°C and from 10 to 30 μmol m^{-2} s^{-1} to 50 to 100 μmol·photons·m^{-2}·s^{-1}) sometimes accompanied by an increase in the length of the light period (e.g., from light-dark 12 h:12 h to 14 h:10 h), or even permanent light, for 1 or 2 weeks prior to use. In order to ensure typical morphology, a change of the medium may be required. Many flagellates need to be cultured in liquid media to induce full motility, i.e., a transfer from agar slant to an Erlenmeyer flask with the liquid medium. Media with low, or no, nitrogen are required to induce optimal heterocyst formation in many cyanobacterial strains. Changes in culture conditions are required to induce reproductive stages, and one needs to experiment with a particular strain for some time to find out which conditions are most appropriate. For example, filaments of *Tribonema* grown in flasks with liquid medium at 12°C for long-term maintenance were transferred to plates with agarized medium and subsequently incubated at 16°C with an increased length of the light period. After about 3 weeks, the filaments on agar were submerged in fresh liquid culture medium, and after 1 or 2 days, about 1 to 3 hours after the beginning of the light period, extensive zoospore formation occurred (58).

QUALITY CONTROL

In most cases, assessment of quality control is based on cell morphology and culture appearance. Obvious growth is the first factor to be noted. The gross morphology of the culture and of the individual cells is observed microscopically. If no major differences between the parent culture and the subculture are observed, then quality is generally assumed to be satisfactory. The presence of any contaminating organisms in previously axenic cultures is checked by phase-contrast microscopy. In addition, a small aliquot of culture (0.1 to 0.5 ml) is spread onto organically rich media such as nutrient agar or tryptone soy agar. The plate contents are incubated at 20 to 25°C for 48 to 72 h and are then examined, by eye, for the presence of any bacterial or fungal growth.

Biochemical and molecular methods could be adopted for quality control. A standard "fingerprint" of information on chemical composition, biochemical characteristics (e.g., isoenzyme analysis), or genetic sequence data (e.g., GenBank) could be maintained. A new approach, with the potential to discriminate between protistan strains at subspecies level, has been developed by using the amplified fragment length polymorphism technique (69). Data derived by any of the above approaches could be compared with the maintained organism or new isolates. Tests such as these lend themselves to automation and are technically feasible, at a cost. However, a number of problems presently prevent the adoption of these types of system. Firstly, for many of the above-mentioned tests, strains should be axenic, a condition often difficult to achieve and requiring a great deal of time. In reality, the vast majority of algal and protozoan cultures either have one protist present or are dominated by one organism. Second, the stability of the measured characteristic should be known, and it should not be affected by the maintenance regimen; this parameter would have to be established for each strain. In addition, the cost of developing and running such a quality control system makes it unlikely to be adopted other than in specialist units or major collections.

Documentation should be maintained on each isolate. It should include name to species level and authority if appropriate; any strain number or culture collection code or number; indication whether it is a type culture; origin; isolator's name and date of isolation; medium and maintenance regimen; preservation methods; any unique or representative biochemical or other characteristics; and any regulatory conditions applying (e.g., quarantine, containment levels, and patent status). In addition, a maintenance schedule giving details on culture conditions, dates of subculture, and status of culture at each transfer should be recorded. This documentation can most easily be maintained as a computerized database, although hard copy in the form of card indices, catalogues, etc. can be used as alternatives. In some situations, such as in culture collections, it may prove beneficial to incorporate maintenance regimens into formal quality assurance systems such as ISO 9000 or the UKNCC Quality Assurance system (93). Such formal quality assurance helps minimize the problems associated with procedural limitations and can improve traceability, should problems occur.

LIMITATIONS TO PROCEDURES

Although serial subculture is the standard method for maintaining laboratory algal and protozoan strains, it is widely recognized that this technique is suboptimal. By its nature it is selective, as only those strains that can be cultured ex situ can be maintained. Some organisms obviously change when cultured on artificial media or under axenic conditions. This phenomenon is most commonly observed in cyanobacterial cultures, where loss of the ability to produce colonies, heterocysts, gas vacuoles, or akinetes can occur. In addition, changes in pigmentation and loss of toxicity have been observed. Eukaryotic microalgae may also change morphology in culture. These changes include loss of ability to produce spines in *Micractinium* species and apparently irreversible shrinkage in many diatom cultures. Cyst-forming protozoa have been observed to lose their ability to form viable cysts, while vahlkampfiid amoebae have lost their ability to form flagellates. In some of the above examples, the changes are apparently irreversible; in others, the lost attribute can be regained by changing the medium or other environmental factors. There is as yet little evidence of any major shifts in the genetic complement of cultures which have been maintained under laboratory conditions for prolonged periods. Also, in the vast majority of organisms retained in microbial collections, most commonly observed or measured characteristics appear to be stable upon serial subculture. The alternative to serially subculturing laboratory strains, which is the reisolation and purification of specific algal or protozoan strains each time that they are required for a laboratory-based study, is obviously not possible. However, the use of long-term preservation methods has the potential to prevent any major changes in phenotype or genotype of cultured specimens.

In addition to the foregoing limitations, there are the dangers of mislabelling, contamination of cultures, and other handling problems (e.g., the transfer of an inoculum to an inappropriate medium). The likelihood of these latter problems may be minimized by insuring rigid adherence to maintenance protocols and quality control procedures and by providing careful documentation.

OTHER USEFUL OR ALTERNATIVE PROCEDURES

To conserve genetic integrity and to minimize the maintenance requirements, long-term preservation methods have been developed for a wide range of biological materials (25, 52, 53). In general, the methods employed depend on the reduction, removal, or rendering unavailable of extra- and intracellular water. Protocols which have been developed involve drying, freeze-drying, and freezing. These processes remove water or make it biologically unavailable, resulting in the reduction or, in some cases, complete cessation of metabolic activity.

Drying, generally air drying, may be used successfully for a wide range of cyst-forming protozoa, and many strains are commonly transported as dried material on filter paper (2, 71). Some algal strains, particularly those which produce resistant aplanospores, have the potential for storage in dried form. Dried *Haematococcus pluvialis* may remain viable for at least 27 years (57). However, drying has not been widely applied as a method of long-term conservation of algae, primarily because of the low levels of recovery for some organisms and the short shelf life of stored material (27). More recent research using a controlled drying protocol demonstrated that the method has some potential for a number of types of green algae (59).

Freeze-drying or lyophilization is the method of choice by which many scientists preserve bacterial and fungal strains. For algae, viability levels may be extremely low in some cases, and they may decrease further upon prolonged storage (27, 42); this result may lead to difficulties in regenerating an active culture. In addition, it is theoretically possible that extremely low viability levels obtained may result in the selection of a nonrepresentative, freeze-drying-tolerant subpopulation. Although freeze-drying has not been widely adopted for the conservation of algae or protozoa, it has been used to preserve cyanobacterial cultures for up to 5 years with no reduction in viability (43). Furthermore, it is used at the American Type Culture Collection to successfully preserve a wide range of organisms (2, 71).

Freezing or cryopreservation is the optimal method of long-term storage, since high postthaw viability can be guaranteed. At low subzero temperatures (<−135°C), no further deterioration of stored material can occur and viability is effectively independent of storage duration (38, 68). A range of freezing protocols have been developed. Most utilize a two-step system with controlled or semicontrolled cooling from room temperature to an intermediate holding temperature (−40°C being commonly used), allowing cryodehydration of the cells to occur before the cells are plunged into liquid nitrogen (−196°C) (2, 23, 24, 44, 51, 62, 63). The frozen material is stored in either liquid or vapor-phase nitrogen in an appropriate liquid nitrogen storage system. Although some organisms can be successfully cryopreserved and stored at higher subzero temperatures, viability levels rapidly fall during storage (13, 42, 46). It is therefore necessary to maintain frozen cultures at extremely low temperatures, optimally in liquid nitrogen at −196°C. Where cryostore temperatures are carefully monitored and maintained, viability levels are effectively uninfluenced by the duration of the storage. This pattern has been demonstrated for a range of protists recovered after up to 22 years of storage in liquid nitrogen (28).

The methods described above all have the advantage of requiring little routine maintenance of the preserved material and, in the case of cryopreservation, can effectively guarantee genetic stability (38, 69). The obvious disadvantage is the cost of specialist equipment and personnel. In addition, the preserved cultures require a period of recovery and growth before they are usable; this period may be as little as a few days for fast-growing amoebae (13) but may exceed 4 weeks for some strains (90). The main limitation to the implementation of these techniques in large culture collections, where the equipment for preservation is available, is the recalcitrance of many organisms to conventional preservation protocols. Alternative cryopreservation strategies that show some potential include the use of exogenous antioxidants (34), vitrification, and encapsulation in calcium alginate prior to cryopreservation (33). Research to improve techniques and to expand the range of organisms that can be maintained by using long-term preservation methods is ongoing.

REFERENCES

1. **Advisory Committee on Dangerous Pathogens.** 1995. *Categorisation of Biological Agents According to Hazard and Categories of Containment.* Her Majesty's Stationery Office, London, United Kingdom. [Updates at http:www.open .gov.uk/hse/hthdir/agents.htm.]
2. **Alexander, M., P.-M. Daggett, R. Gherna, S. Jong, and F. Simione.** 1980. *American Type Culture Collection Methods. I. Laboratory Manual on Preservation Freezing and Freeze-Drying as Applied to Algae, Bacteria, Fungi and Protozoa.* American Type Culture Collection, Rockville, Md.
3. **Andersen, R. A.** 1992. Diversity of eukaryotic algae. *Biodivers. Conserv.* **1:**267–292.
4. **Atkins, M. S., O. R. Anderson, and C. O. Wirsen.** 1998. Effect of hydrostatic pressure on the growth rates and encystment of flagellated protozoa isolated from a deep-sea hydrothermal vent and a deep shelf region. *Mar. Ecol. Prog. Ser.* **171:**85–95.
5. **Azam, F., T. Fenchel, J. G. Field, J. S. Gray, L. A. Meyer-Reil, and F. Thingstad.** 1983. The ecological role of water-column microbes in the sea. *Mar. Ecol. Prog. Ser.* **10:**257–263.
6. **Baumgartner, M., K. Statter, and W. Foissner.** 2002. Morphological, small subunit rRNA and physiological characteristics of *Trimyema minutum* (Kahl, 1931), an anaerobic ciliate from submarine hydrothermal vents growing from 28°C to 52°C. *J. Eukaryot. Microbiol.* **49:** 227–238.
7. **Belcher, H., and E. Swale.** 1988. *Culturing Algae: a Guide for Schools and Colleges.* Culture Collection of Algae and Protozoa, Ambleside, United Kingdom.
8. **Berk, S. G., P. Guerry, and R. R. Colwell.** 1976. Separation of small ciliate protozoa from bacteria by sucrose gradient centrifugation. *Appl. Environ. Microbiol.* **31:**450–452.
9. **Bertz, J. W., W. Aretz, and W. Hartel.** 1984. Use of flow cytometry in industrial microbiology for strain improvement programs. *Cytometry* **5:**145–150.
10. **Borkott, H.** 1975. A method for quantitative isolation and preparation of particle-free suspensions of bacteriophagous ciliates from different substrates for electronic counting. *Arch. Protistenkd.* **117:**261–268.
11. **Borowitzka, M. A., and L. J. Borowitzka.** 1988. *Dunalliella,* p. 27–58. *In* M. A. Borowitzka and L. J. Borowitzka (ed.), *Micro-algal Biotechnology.* Cambridge University Press, Cambridge, United Kingdom.
12. **Broers, C. A. M., H. L. F. Molhuizen, C. K. Stumm, and G. D. Vogels.** 1992. An electromigration technique, applied for the concentration of anaerobic protozoa from mass cultures. *J. Microbiol. Methods* **14:**217–220.
13. **Brown, S., and J. G. Day.** 1993. An improved method for the long-term preservation of *Naegleria gruberi. Cryo-Letters* **14:**347–352.
14. **Cavalier-Smith, T.** 2003. Protist phylogeny and the high-level classification of Protozoa. *Eur. J. Protistol.* **39:**338–348.
15. **Cavalier-Smith, T.** 2004. Only six kingdoms of life. *Proc. R. Soc. Lond. B* **271:**1251–1262.
16. **Christensen, S. T., H. Sorensen, N. H. Beyer, K. Kristiansen, L. Rasmussen, and M. I. Rasmussen.** 2001. Cell death in *Tetrahymena thermophila:* new observations on culture conditions. *Cell Biol. Int.* **25:**509–519.
17. **Corliss, J. O., and S. C. Esser.** 1974. Comments on the role of the cyst in the life cycle and survival of free-living protozoa. *Trans. Am. Microsc. Soc.* **93:**578–593.
18. **Cowling, A. J.** 1991. Free-living heterotrophic flagellates: methods of isolation and maintenance, including sources of strains in culture, p. 477–491. *In* D. J. Patterson and J. Larson (ed.), *The Biology of Free-Living Heterotrophic Flagellates.* Systematics Association, special vol. no. 45. Clarendon Press, Oxford, United Kingdom.
19. **Cowling, A. J., and H. G. Smith.** 1987. Protozoa in the microbial communities of maritime Antarctic wellfields, p. 205–213. *Deuxième Colloque sur les Ecosystèmes Terrestres Subantarctiques. Comité National Français des Recherches Antarctiques,* Paimpont, France.
20. **Cunningham, A., and J. W. Leftley.** 1986. Application of flow cytometry to algal physiology and phytoplankton ecology. *FEMS Microbiol. Rev.* **32:**159–164.
21. **Curds, C. R.** 1973. The role of protozoa in the activated-sludge process. *Am. Zool.* **13:**161–169.
22. **Curds, C. R.** 1992. *Protozoa in the Water Industry.* Cambridge University Press, Cambridge, United Kingdom.
23. **Day, J. G., and J. J. Brand.** 2005. Cryopreservation methods for maintaining algae, p. 165–188. *In* R. A. Andersen (ed.), *Algal Culturing Techniques.* Elsevier Academic Press, Burlington, Mass.

24. **Day, J. G., and M. M. DeVille.** 1995. Cryopreservation of algae, p. 81–89. *In* J. G. Day and M. R. McLellan (ed.), *Cryopreservation and Freeze-Drying Protocols*. Humana Press, Inc., Totowa, N.J.
25. **Day, J. G., and M. R. McLellan (ed.).** 1995. *Cryopreservation and freeze-drying protocols*. Humana Press, Inc., Totowa, N.J.
26. **Day, J. G., and M. R. McLellan.** 1995. Conservation of algae, p. 75–98. *In* B. Grout (ed.), *Genetic Preservation of Plant Cells In Vitro*. Springer-Verlag, Heidelberg, Germany.
27. **Day, J. G., I. M. Priestley, and G. A. Codd.** 1987. Storage reconstitution and photosynthetic activities of immobilized algae, p. 257–261. *In* C. Webb and F. Mavituna (ed.), *Plant and Animal Cells, Process Possibilities*. Ellis Harwood Ltd., Chichester, United Kingdom.
28. **Day, J. G., M. M. Watanabe, G. J. Morris, R. A. Fleck, and M. R. McLellan.** 1997. Long-term viability of preserved eukaryotic algae. *J. Appl. Phycol.* **9:**121–127.
29. **Droop, M. R.** 1974. Heterotrophy of carbon, p. 530–559. *In* W. D. P. Stewart (ed.), *Algal Physiology and Biochemistry*. Blackwell, London, United Kingdom.
30. **Fenchel, T.** 1969. The ecology of marine microbenthos. IV. Structure and function of the benthic ecosystem. *Ophelia* **6:**1–182.
31. **Finlay, B. J., T. M. Embley, and T. Fenchel.** 1993. A new polymorphic methanogen, closely related to *Methanocorpusculum parvum*, living in stable symbiosis within the anaerobic ciliate *Trimyema* sp. *J. Gen. Microbiol.* **139:** 371–378.
32. **Finlay, B. J., A. Rogerson, and A. J. Cowling.** 1988. *A Beginner's Guide to the Collection, Isolation, Cultivation and Identification of Freshwater Protozoa*. Culture Collection of Algae and Protozoa and Freshwater Biological Association, Ambleside, Cumbria, United Kingdom.
33. **Fleck, R. A.** 1998. The assessment of cell damage and recovery in cryopreserved freshwater protists. Ph.D. thesis. University of Abertay, Dundee, United Kingdom.
34. **Fleck, R. A., E. E. Benson, D. H. Bremner, and J. G. Day.** 1999. Studies of free radical-mediated cryoinjury in the unicellular green alga *Euglena gracilis* using a non-destructive hyroxyl radical assay: a new approach for developing protistan cryopreservation strategies. *Free Radic. Res.* **32:**157–170.
35. **Foissner, W., and H. Berger.** 1996. A user-friendly guide to the ciliates (Protozoa, Ciliophora) commonly used by hydrobiologists as bioindicators in rivers, lakes and waste waters, with notes on their ecology. *Freshw. Biol.* **35:**375–482.
36. **Frohlich, J., and H. Konig.** 1999. Rapid isolation of single microbial cells from mixed natural and laboratory populations with the aid of a micromanipulator. *Syst. Appl. Microbiol.* **22:**249–257.
37. **Griffiths, B. S., and K. Ritz.** 1988. A technique to extract, enumerate and measure protozoa from mineral soils. *Soil Biol. Biochem.* **20:**163–173.
38. **Grout, B. W. W.** 1995. Introduction to the in vitro preservation of plant cells, tissues and organs, p. 1–20. *In* B. Grout (ed.), *Genetic Preservation of Plant Cells In Vitro*. Springer-Verlag, Heidelberg, Germany.
39. **Haeckel, E.** 1866. *Generelle Morphologie der Organismen*. G. Reimer, Berlin, Germany.
40. **Hawksworth, D. L., and B. Aguirre-Hudson.** 1994. International initiatives in microbial diversity, p. 65–72. *In* B. Kirsop and D. L. Hawksworth (ed.), *The Biodiversity of Microorganisms and the Role of Microbial Resource Centres*. World Federation for Culture Collections and United Nations Environment Programme, Egham, United Kingdom.
41. **Hellebust, J. A., and J. Lewin.** 1977. Heterotrophic nutrition. *Bot. Monogr.* **13:**169–197.
42. **Holm-Hansen, O.** 1967. Factors affecting viability of lyophilized algae. *Cryobiology* **4:**17–23.
43. **Holm-Hansen, O.** 1973. Preservation by freezing and freeze-drying, p. 195–206. *In* J. Stein (ed.), *Handbook of Phycological Methods: Culture Methods and Growth Measurements*. Cambridge University Press, Cambridge, United Kingdom.
44. **James, E. R.** 1991. Maintenance of parasitic protozoa by cryopreservation, p. 209–226. *In* B. Kirsop and A. Doyle (ed.), *Maintenance of Microorganisms and Cultured Cells*. Academic Press, London, United Kingdom.
45. **Jannasch, H. W., and C. O. Wirsen.** 1984. Variability of pressure adaptation in deep sea bacteria. *Arch. Microbiol.* **139:**281–288.
46. **John, D. T., and R. A. John.** 1996. Viability of pathogenic *Acanthamoeba* and *Naegleria* and virulence of *N. fowleri* during long-term cryopreservation. *Folia Parasitol.* **43:** 43–46.
47. **Johnson, D. B., and L. Rang.** 1993. Effects of acidophilic protozoa on populations of metal-mobilizing bacteria during the leaching of pyritic coal. *J. Gen. Microbiol.* **139:** 1417–1423.
48. **Kalinina, L. V., and F. C. Page.** 1992. Culture and preservation of naked amoebae. *Acta Protozool.* **31:**115–126.
49. **Keenan, K., E. Erlich, K. H. Donnelly, M. B. Basel, S. H. Hutner, R. Kassoff, and S. A. Crawford.** 1978. Particle-based axenic media for tetrahymenids. *J. Protozool.* **25:**385–387.
50. **Keller, M. D., L. P. Shapiro, E. M. Haugen, T. L. Cucci, E. B. Sherr, and B. F. Sherr.** 1994. Phagotrophy of fluorescently labeled bacteria by an oceanic phytoplankter. *Microb. Ecol.* **28:**39–52.
51. **Kilvington, S.** 1995. Cryopreservation of pathogenic and non-pathogenic free-living amoebae, p. 63–70. *In* J. G. Day and M. R. McLellan (ed.), *Cryopreservation and Freeze-Drying Protocols*. Humana Press, Inc., Totowa, N.J.
52. **Kirsop, B., and A. Doyle (ed.).** 1991. *Maintenance of Microorganisms and Cultured Cells*. Academic Press, London, United Kingdom.
53. **Kirsop, B. E., and J. J. S. Snell (ed.).** 1984. *Maintenance of Microorganisms*. Academic Press, London, United Kingdom.
54. **Laverde, A. V., and M. M. Brent.** 1980. Simplified soluble media for the axenic cultivation of *Naegleria*. *Protistologica* **16:**11–15.
55. **Lee, C. C., and T. Fenchel.** 1972. Studies on ciliates associated with sea ice from Antarctica. *Arch. Protistenk.* **114:** 237–244.
56. **Lee, J. J., and A. T. Soldo (ed.).** 1992. *Protocols in Protozoology*. Society of Protozoologists, Lawrence, Kans.
57. **Leeson, E. A., J. P. Cann, and G. J. Morris.** 1984. Maintenance of algae and protozoa, p. 131–160. *In* B. E. Kirsop and J. J. S. Snell (ed.), *Maintenance of Microorganisms*. Academic Press, London, United Kingdom.
58. **Lokhorst, G. M.** 2003. The genus *Tribonema* (Xanthophyceae) in the Netherlands. An integrated field and culture study. *Nova Hedwigia* **77:**19–53.
59. **Malik, K. A.** 1993. Preservation of unicellular green-algae by liquid-drying. *J. Microbiol. Methods* **18:**41–46.
60. **Margulis, L., J. O. Corliss, M. Melkonian, and D. J. Chapman (ed.).** 1989. *Handbook of Protoctista*. Jones and Bartlett, Boston, Mass.
61. **McLachlan, J.** 1973. Growth media—marine, p. 25–52. *In* J. Stein (ed.), *Handbook of Phycological Methods: Culture Methods and Growth Measurements*. Cambridge University Press, Cambridge, United Kingdom.
62. **McLellan, M. R., A. J. Cowling, M. F. Turner, and J. G. Day.** 1991. Maintenance of algae and protozoa, p. 183–208. *In* B. Kirsop and A. Doyle (ed.), *Maintenance of*

Microorganisms and Cultured Cells. Academic Press, London, United Kingdom.

63. **Menrath, P., R.-M. Brame, C. Oger, and J.-M. Delattre.** 1995. A new method of preservation of axenic amoeba of the genera *Naegleria* and *Acanthamoeba*. *Eur. J. Protistol.* **31:**73–76.
64. **Miller, T. L., and M. J. Wolin.** 1974. A serum bottle modification of the Hungate technique for cultivating obligate anaerobes. *Appl. Microbiol.* **27:**985–987.
65. **Miyachi, S., O. Nakayama, Y. Yokohama, Y. Hara, M. Ohmori, K. Komogata, H. Sugawara, and Y. Ugawa (ed.).** 1989. *World Catalogue of Algae*. Japan Scientific Societies Press, Tokyo, Japan.
66. **Mizobuchi, N., K. Yokoigawa, T. Haramuto, H. Fujisawa, and Y. Takagi.** 2003. Catalase is the bacteria-derived detoxifying substance against paramecia-killing toxin in wheat grass powder infusion. *J. Eukaryot. Microbiol.* **50:**299–303.
67. **Mogami, Y., N. Tokunaga, and S. A. Baba.** 1999. Longevity in space; experiment on the life span of *Paramecium* cell clone in space. *Adv. Space Res.* **23:**2087–2090.
68. **Morris, G. J.** 1981. *Cryobiology*. Institute of Terrestrial Ecology, Cambridge, United Kingdom.
69. **Müller J., T. Friedl, D. Hepperle, M. Lorenz, and J. G. Day.** 2005. Distinction between multiple isolates of *Chlorella vulgaris* (Chlorophyta, Trebouxiophyceae) and testing for conspecificity using amplified fragment length polymorphism and ITS RDNA sequences. *J. Phycol.* **4:** 1236–1247.
70. **Neilson, A. H., and R. A. Lewin.** 1974. The uptake and utilization of organic carbon by algae: an essay in comparative biochemistry. *Phycologia* **13:**227–264.
71. **Nerad, T. A.** 1991. *ATTC Catalogue of Protists*. American Type Culture Collection, Rockville, Md.
72. **OECD.** 1984. *Guidelines for Testing Chemicals. Section 2. Effects on Biotic Systems. No. 201. Algal Growth Inhibition Test*. Paris, France.
73. **Olive, L. S.** 1967. The Prostelida—a new order of the Mycetozoa. *Mycologia* **59:**1–29.
74. **Packer, L., and A. N. Glazer (ed.).** 1988. *Methods in Enzymology*, vol. 167. *Cyanobacteria*. Academic Press, London, United Kingdom.
75. **Page, F. C.** 1983. *Marine Gymnamoebae*. Institute of Terrestrial Ecology, Cambridge, United Kingdom.
76. **Page, F. C.** 1988. *A New Key to Freshwater and Soil Gymnamoebae*. Freshwater Biological Association, Ambleside, United Kingdom.
77. **Page, F. C.** 1991. Nackte Rhizopods, p. 3–170. *In* D. Matthes (ed.), *Protozoenfauna 2*. Gustav Fischer Verlag, Stuttgart, Germany.
78. **Patterson, D. J., K. Nygaard, G. Steinberg, and C. M. Turley.** 1993. Heterotrophic flagellates and other protists associated with oceanic detritus throughout the water column in the mid-North Atlantic. *J. Mar. Biol. Assoc. U. K.* **73:**67–95.
79. **Pringsheim, E. G.** 1946. *Pure Cultures of Algae*. Cambridge University Press, Cambridge, United Kingdom.
80. **Schmidt, H. J.** 1982. New methods for cultivating, harvesting and purifying mass cultures of the hypotrich ciliate *Euplotes aediculatus*. *J. Protozool.* **29:**132–135.
81. **Sensen, C. W., K. Heimann, and M. Melkonian.** 1994. The production of clonal and axenic cultures of microalgae using fluorescence-activated cell sorting. *Eur. J. Phycol.* **28:**93–97.
82. **Sieracki, M., N. Poulton, and N. Crosbie.** 2005. Automated isolation techniques for microalgae, p. 101–116. *In* R. A. Andersen (ed.), *Algal Culturing Techniques*. Elsevier Academic Press, Burlington, Mass.
83. **Sleigh, M. A.** 1989. *Protozoa and Other Protists*. Edward Arnold, London, United Kingdom.
84. **Soldo, A., and S. A. Brickson.** 1980. A simple method for plating and cloning ciliates and other protozoa. *J. Protozool.* **27:**328–331.
85. **Stanier, R. Y., R. Kunisawa, M. Mandel, and G. Cohen-Bazire.** 1971. Purification and properties of unicellular blue-green algae (Order Chroococales). *Bacteriol. Rev.* **35:**171–205.
86. **Starr, R. C., and J. A. Zeikus.** 1993. UTEX—the culture collection of algae at the University of Texas at Austin. *J. Phycol.* **29:**1–106.
87. **Stein, J.** (ed.). 1973. *Handbook of Phycological Methods: Culture Methods and Growth Measurements*. Cambridge University Press, Cambridge, United Kingdom.
88. **Suttle, C. A., and F. Chen.** 1992. Mechanisms and rates of decay of marine viruses in seawater. *Appl. Environ. Microbiol.* **58:**3721–3729.
89. **Throndsen, J.** 1973. Special methods—micromanipulators, p. 139–144. *In* J. Stein (ed.), *Handbook of Phycological Methods: Culture Methods and Growth Measurements*. Cambridge University Press, Cambridge, United Kingdom.
90. **Tompkins, J., M. M. DeVille, J. G. Day, and M. F. Turner.** 1995. *Culture Collection of Algae and Protozoa Catalogue of Strains*. Culture Collection of Algae and Protozoa, Ambleside, United Kingdom.
91. **Turley, C. M., and M. Carstens.** 1991. Pressure tolerance of oceanic flagellates: implications for remineralization of organic matter. *Deep-Sea Res.* **38:**403–413.
92. **Turner, M. F.** 1979. Nutrition of some marine microalgae with special reference to vitamin requirements and utilization of nitrogen and carbon sources. *J. Mar. Biol. Assoc. U. K.* **59:**535–552.
93. **United Kingdom National Culture Collection.** 2006. http://www.ukncc.co.uk.
94. **van Wagtendonk, W. J., and A. T. Soldo.** 1970. Methods used in the axenic cultivation of *Paramecium aurelia*, p. 117–130. *In* D. M. Prescott (ed.), *Methods in Cell Physiology*, vol. IV. Academic Press, New York, N.Y.
95. **Watanabe, M. M., and H. Nozaki.** 1994. *NIES—Collection List of Strains*. NIES, Tsukuba, Japan.
96. **Whitelam, G. C., T. Lanaras, and G. A. Codd.** 1983. Rapid separation of microalgae by density gradient centrifugation in percoll. *Br. Phycol. J.* **18:**23–28.
97. **Wiedeman, V. E., P. L. Walne, and F. R. Trainor.** 1964. A new technique for obtaining axenic cultures of algae. *Can. J. Bot.* **42:**958–959.
98. **Zhukov, B. F., and A. P. Mylnikov.** 1983. Cultivation of free-living colourless flagellates from waste water treatment plants. *Protozoologiya* **8:**142–152.

Cultivation and Assay of Animal Viruses

PIERRE PAYMENT

8

Viruses from infected individuals can be disseminated by fecal material, saliva, nasal secretions, skin lesions, blood, etc. They can thus be found in most environments: water, soil, and air as well as surfaces are often contaminated. Because viruses are obligate parasites, they need a suitable host to multiply and propagate; their isolation or propagation in artificial media such as those used for bacteria and fungi is not possible. Methods for their detection are therefore more limited. Viruses can be detected by various methods based on protein or nucleic acid detection, but the demonstration of their infectivity involves the use of live animals or cell cultures. Cell cultures have become the method of choice for isolating, propagating, and detecting viruses in environmental samples. Live animals require special facilities, they are often costly, and because most animal viruses are species specific, several species would be required. During the last 50 years, virologists have been able to select cell cultures that are susceptible to many commonly encountered viruses and environmental virologists have developed methods to detect very low numbers of viruses in various environments. This chapter reviews these methods and their practical use and describes basic quality control procedures to maximize their level of sensitivity.

Table 1 lists the viruses that can potentially be found in the environment. Almost every known virus can be present in air, soil, water, or fomites. Contrary to clinical virology, where a large number of viruses are often present in the sample (e.g., fecal material, sputum, secretions, etc.), in environmental samples viruses have been diluted or disseminated and they are relatively scarce. The viruses must be concentrated or extracted from thousands of liters of water and air or from several hundred grams of soil or sludge (13, 36, 46). These methods are described in the respective chapters concerning the specific environment under study. This chapter assumes that a concentrate has been obtained by these methods. The main objective of the cultivation and assay of viruses is to optimize detection methods to a level where even a single infectious unit can be detected with confidence (2, 13).

PRECAUTIONARY NOTE

All viruses must be considered pathogenic for humans or animals. Any attempt to manipulate contaminated samples or concentrate, propagate, or assay viruses should be performed under the proper biosafety conditions and by trained personnel only. I refer the reader to institutional, national, or international biosafety guidelines and other appropriate readings (48).

DETECTION METHODS

Several methods for detection of viruses in environmental samples have been described: visualization of the virus by microscopy, detection of viral antigens, detection of viral nucleic acids, and detection of viral infectivity. In most cases, the detection of viral infectivity is the main objective; it enables the researcher to estimate a health risk, a level of contamination, the efficacy of disinfection or sterilization procedures, etc. If the objective is solely to assess whether a given environment has been contaminated by viruses, the detection of viral antigens or nucleic acids could be sufficient. These two methods have the advantage of being available for a larger number of viruses than the detection of infectivity. However, they are not very precise and quantitation is difficult.

Electron Microscopy

Visualization of viral particles by electron microscopy is the most basic method. It is not very sensitive, since the observation of a single viral particle requires at least 10^6 viral particles per ml before viral particles can be observed in a single field. Electron microscopy is most often used on samples obtained after the viruses have been grown in cell culture; viral morphology can then provide precious information of the type of viruses present, and immunoelectron microscopy may even provide type-specific identification (17, 38).

Immunoassays

Antigen detection methods based on immunoassays can be used to detect viral proteins. Enzyme-linked immunosorbent assays are available for many viruses and can be used when relatively large amounts of viral antigens are present in a sample. They have been used successfully in clinical microbiology for the detection of viruses in clinical material (17) and with some success for the detection of enteric viruses in water samples (5, 11, 44). These methods are not very sensitive and are very susceptible to interference from extraneous material in environmental concentrates. This method does not provide any information on the infectivity of the viruses detected.

TABLE 1 Viruses that can potentially be found in the environment (surfaces, air, water, or soil)

Family or subfamily	Genus	Common species
Picornaviridae	*Enterovirus*	Polioviruses
		Coxsackieviruses, groups A and B
		Echoviruses
		Enteroviruses 68 through 71
	Heparnavirus	Hepatitis A virus
	Rhinovirus	Virus types infecting humans
	Aphthovirus	Foot-and-mouth-disease viruses
Caliciviridae	*Calicivirus*	Norwalk, gastroenteritis virus
Reoviridae	*Reovirus*	Reovirus
	Orbivirus	17 subgroups
	Rotavirus	Human rotaviruses
Flaviviridae	*Flavivirus*	Hepatitis C virus
Orthomyxoviridae	*Influenzavirus*	Influenza virus
Paramyxoviridae	*Paramyxovirus*	Parainfluenza virus, mumps virus
	Morbillivirus	Measles virus
	Pneumovirus	Human respiratory syncytial virus
Coronaviridae	*Coronavirus*	Human coronavirus
Bunyaviridae	*Hantavirus*	Hantaan virus
Lentivirinae	*Lentivirus*	Human immunodeficiency virus
Parvoviridae	*Parvovirus*	Human parvovirus B19
Papovaviridae	*Papillomavirus*	Human papillomaviruses (warts)
Adenoviridae	*Mastadenovirus*	Human adenoviruses
Hepadnaviridae	*Hepadnavirus*	Human hepatitis B virus
Herpesviridae	*Simplexvirus*	Human herpes simplex virus
	Varicellovirus	Varicella-zoster virus (herpesvirus 3)
	Cytomegalovirus	Human cytomegalovirus (herpesvirus 5)
	Lymphocryptovirus	Epstein-Barr virus (herpesvirus 4)
Poxviridae	*Orthopoxvirus*	Vaccinia virus, smallpox virus (variola)

Nucleic Acid Probes

Nucleic acid probes are the basis for the most specific methods for the detection and identification of viruses (39). They are being developed for a large number of viruses and are being applied with some success to the detection of viruses in environmental samples (12, 14, 22, 24, 25, 40, 47). These methods are the most sensitive but are also susceptible to various inhibitors found in the environmental concentrates; false-negative results and sample contamination during laboratory manipulations are the major issues to be resolved for these methods. When used for the direct detection of viruses in environmental samples or concentrates, these methods provide no information on the infectivity of the viruses detected. However, they can complement cell culture methods if they are used to detect the viruses after a few hours to several days in culture (29, 41). Growth of the virus in cell culture confirms infectivity and permits quantitation, while the high sensitivity of molecular methods shortens the time required to obtain a result.

Cell Culture

Preparation of Samples

The detection of infective viruses in environmental samples still relies mainly on cell culture as the method of choice. The method does not allow detection of all viruses, as many do not readily grow in cell culture. This is also true of all other methods; any method used for the detection of viruses provides an indication for the presence of only some viruses.

Detection of viruses on cell culture is not an easy process. Concentrates from environmental samples often contain organic and inorganic compounds that can be toxic to cell culture. Microbial contaminants such as bacteria and fungi, many types of planktonic microorganisms, and some invertebrates are always present in water and can still be present in the concentrate. When concentrated from samples as large as several thousand liters, they can account for a significant problem that has to be resolved before even attempting to deposit this sample on a cell culture, where bacteriological sterility is of utmost importance.

Isotonicity and pH

Samples must be isotonic and free of toxic compounds and microbial contaminants. Viruses adsorbed to surfaces, flocs, or collection filters are often eluted with alkaline proteinaceous solutions, such as beef extract or buffered solutions (13). The eluates obtained from filters or concentrated material must be neutralized (pH 7.0 to 7.5) immediately to prevent viral inactivation; this can be achieved by dropwise addition of 1 N hydrochloric acid. Concentrated (10×) balanced salt solutions or tissue culture medium can be used to adjust isotonicity and pH.

Decontamination and Detoxification

Because they contain large numbers of bacteria and fungi, as well as large amounts of extraneous material, samples must generally be decontaminated and detoxified before being applied to cell cultures (13). A preliminary test on a few cell cultures in flasks or microwells using a small amount of the sample rapidly indicates the presence of toxic substances. Toxic samples usually destroy cells in 24 h or less (cytotoxic effect). Bacterial and fungal contaminants rapidly multiply in cell culture medium and destroy all cells in a few days, giving a cloudy appearance or large mycelium. Microbial contamination of cell cultures is easily recognized by microscopy, as bacilli, cocci, yeasts, or fungi can be seen in the culture medium at ×400 magnification.

Partial decontamination of samples is most easily achieved by centrifugation at 4,000 to 10,000 × *g* for 15 to 30 min. Filtration on sterile 0.45- or 0.2-μm-pore-size membrane filters is an alternative method. Since viruses are smaller than the smallest bacteria, membranes with these pore sizes will not retain them. Precautions should be taken, however, to prevent adsorption of the viruses to the membrane: membranes can be pretreated with a small amount of fetal calf serum (2 to 3 ml of a 10% solution) or a 1% bovine serum albumin solution.

When filtration and centrifugation are not feasible or can result in significant loss of viruses, decontamination can be achieved through the addition to the sample of a mixture of antibiotics at high concentrations (penicillin, 1,000 U/ml; streptomycin, 1,000 μg/ml; gentamicin, 50 μg/ml; amphotericin B [Fungizone], 2.5 μg/ml). A combination of these three methods is often used to achieve optimal results for problematic samples. Do not exceed these antibiotic concentrations, as they can then become toxic for cell cultures and reduce the sensitivity of the cells to the viruses.

Both filtration and centrifugation can result in significant loss of viruses, and these steps should be avoided if possible. If unavoidable, laboratory experiments with seeded viruses should help establish the losses due to this processing.

Some level of detoxification can be achieved by extraction of the sample with trichlorofluoroethane. A one-half volume of the chemical is added, and the solution is mixed vigorously for several minutes and centrifuged at 1,000 × *g* to obtain two phases. The aqueous phase is collected for viral assay.

Toxicity can also be reduced after the adsorption period by removing the sample and washing the cells twice with phosphate-buffered saline (PBS) before adding fresh culture medium. While this practice can slightly reduce the efficiency of the method, it is preferable to a total loss of the sample.

Choice of Cell Cultures

There is no single cell system for the detection of viruses. Primary human cell cultures are very sensitive to virus infection and have been a substrate of choice for virologists (23, 42a). The difficulty in obtaining human cell cultures has been partly solved by the use of monkeys as the source of cells. Even this source has become more and more erratic and expensive and is now relatively unused for routine work, especially since these cultures can be contaminated by adventitious agents.

Many comparisons of the sensitivities of various cell lines to viruses have been reported (5, 8, 9, 15, 16, 20, 30, 37, 43, 47, 49). The appropriate cell line for the detection of the virus under study should be determined by using these references as guidelines. For viruses of animal species other than human, homologous cell cultures are often the best choice (35). Detailed procedures for cell culture and the required solutions are described elsewhere (2, 23, 33, 34, 42a). Environmental virologists have debated, and are still debating, which cell line is the most suitable for detecting with the greatest sensitivity the viruses of concern to environmental virology. Established kidney cell cultures from rhesus or green monkeys such as BSC-1, MA-104, Vero, and BGM cell lines are widely used, but human cell cultures such as HeLa and RD (rhabdomyosarcoma) are also used (7, 15, 30, 38, 43, 49). It is up to the researcher to select the most suitable cell culture. There are several sources for cell cultures that can be used in virology. The American Type Culture Collection (Manassas, Va.) maintains a large selection of cell culture sources and provides a quality-controlled source of cells. Each cell type has its own specific nutritional requirements that are determined not only by the experimental conditions but also by the adaptation of cell populations to the medium utilized. There are almost as many variations in the composition and the use of culture media as there are cell types in culture. Any researcher using cell cultures must maintain culture conditions at an optimal level and apply the ultimate care in their maintenance; failure to do so will lead to genetic modifications of the cells and often a decrease in their sensitivity to virus growth. This decrease would lead to a significant decrease in the efficiency of virus recovery. Such a decrease is usually detected through proper quality assurance procedures (45).

Cell cultures can occasionally be obtained through researchers and colleagues. However, cell cultures bearing the same name can in fact be entirely different cells. Cell cultures can easily be contaminated by another cell type when two are being manipulated at the same time; a good practice is to handle only one cell type at a time in the working area. Cell cultures can also be "selected" by the researcher's cell culture practices: different multiplication patterns, growing temperature, medium, serum concentration, etc. Such practices account for the fact that the same cell line at the same passage level in two different laboratories will have different sensitivities to viruses (e.g., different growth ability). Interlaboratory comparisons should therefore be performed using the same cells and according to strict protocols of quality control and quality assurance.

Cell cultures must also be tested for the presence of adventitious agents (e.g., microorganisms naturally infecting these cells). Mycoplasmas are a major problem, since they are easily introduced in cell cultures by insufficiently tested biologicals (serum or trypsin) as well as poor laboratory practices such as mouth pipetting. Detailed methods on using cell culture for the detection of viruses can be found elsewhere (2, 10, 23, 34, 42a).

Material for Cell Culture and Virus Assays

All procedures should preferably be performed by a trained virologist working with a class II type I biological safety cabinet for the protection of both the cell cultures and the analysts (48). All material used for cell culture and virus assays must be sterile and clean. All the glassware for solutions or media that will come in contact with cells must be washed, rinsed with high-quality distilled water, and sterilized in an autoclave or sterilizing oven. Any trace of detergent or residues is toxic to the cells and should be removed. Cells can be grown on almost any surface; glass and disposable plastic ware have been used in a wide variety of formats. Plastic ware has become popular, and specially treated plastics are available for cell culture work. The most commonly used sizes are 25-, 75-, and 150-cm^2 flasks sterilized by irradiation by the manufacturer. For virus enumeration, it is often convenient to use multiwell plastic plates commercially available in 4-, 12-, 24-, and 96-well configurations. Only material labeled "sterile, for cell culture" should be used.

Most cell culture media used today are synthetic and are commercially available. Chemically defined (synthetic) media are often supplemented with fetal bovine or calf serum that contains nutrients not available otherwise. Serum is added to the medium to a final concentration of 0.1 to 10%, the highest concentrations being used for cell growth and the lowest concentrations for cell culture maintenance or virus detection. When the viruses to be detected are susceptible to inhibitors or antibodies present in serum, the media can be supplemented with 1 to 10% lactalbumin hydrolysate in place of serum.

Cell cultures can easily be kept in good condition if a few simple rules are followed: store all solutions at the optimal temperature (follow the manufacturer's recommendation), give preference to freshly prepared solutions, store all solutions in the dark, check the chemical purity of the distilled water on a regular basis, and always prewarm to 37°C any medium or solution that will be in contact with cell cultures. However, do not leave medium or solutions at 37°C for several hours, as some nutrients can be inactivated.

Use of Antibiotics

Antibiotic and antimycotic solutions may be incorporated into media to reduce contamination introduced by environmental samples. Their use should serve not to mask bad laboratory practices but to maintain the cell cultures in good condition. The recommended concentrations should not be increased for cell culture maintenance. In fact, it is preferable not to use antibiotics to maintain cell culture stocks: any contamination will be detected earlier and will not jeopardize experiments or sample analysis. Their use should be restricted to decontamination of samples and prevention of contamination of experimental cell cultures. Penicillin (100 to 1,000 U/ml), streptomycin (100 to 1,000 μg/ml), gentamicin (50 μg/ml), and amphotericin B (2.5 μg/ml) are generally used. Stock solutions in water (100×) should be prepared in 5-ml aliquots and can be stored at −20°C for a year. Combinations of antibiotics at concentrations of up to 10 times the normal concentration are used to suppress or control bacterial contamination found in environmental samples. These high antibiotic concentrations can be toxic for some cell lines; prior testing is imperative.

Quality Control

A regular quality control and quality assurance control program should be in place. Both cell cultures and viruses can occasionally behave erratically. The quality of the reagent-grade water (type I) used in the laboratory must be checked frequently. Simple tests such as pH and conductivity determination rapidly identify deteriorating water quality and failing purification apparatus. A routine program could include monthly verification of the susceptibility of cells to viruses using standardized viral suspensions frozen at −70°C in aliquots (poliovirus type 1, reovirus, echovirus 11, or any other virus with the same characteristics as the viruses being sought). Titration of the viruses should be performed with maximum sensitivity (i.e., using a large number of microwells to obtain a precise virus count).

VIRUS QUANTITATION

The end point that is used for the quantitation of viruses affects the number of viruses detected. Cytopathic effect, plaque formation, or detection of viral products is the usual means used by the environmental virologist.

When a virus multiplies, it can either destroy the cell it has infected or use the cell metabolism to replicate itself without obviously affecting these cells. If the cells are destroyed, this is readily assessed by microscopic observation: the dead cells are dislodged from the flask surface, often after rounding up or showing characteristic morphological changes. These characteristic changes are called the cytopathic effect. When the cells are covered by a light agar medium to prevent them from being dislodged, the virus migrates from one cell to another, producing characteristic round holes in the cell monolayer. These holes are called plaques and come from a single initial virus particle. If the virus does not destroy the cells, indirect methods of detection must be used.

Plaque Assay

Some viruses, such as the enteroviruses, produce plaques or zones of lysis in cell monolayers overlaid with solidified nutrient medium. These plaques originate from a single infectious virus particle; thus, the titer of virus may be estimated very precisely. This method and the required solutions are described elsewhere (2, 33, 34). Plaque assay is used for a very limited number of viruses and has the lowest sensitivity of the available methods (32). Cultivable enteroviruses generally produce plaques, but other enteric viruses such as adenoviruses and reoviruses do not. The main advantages of plaque assay are that each individual virus (or aggregate) forms a single plaque and that each plaque is rarely a mixture of several virus types. To detect plaque-forming viruses, serial 10-fold dilutions of the concentrate in maintenance medium (with a pipette change at each dilution) are inoculated into 2 to 10 flasks (25 or 75 cm^2) per dilution, using not more than 1 ml of inoculum per flask. The flasks are incubated at 37°C for 1 h, with gentle rocking of the flasks every 15 min to promote virus adsorption. An agar overlay nutrient medium is then added to each flask. The temperature of the overlay must be high enough to maintain the medium as a liquid, but without "burning" the cells. Most cell cultures can tolerate up to 42°C for a very short period, but it is a good practice to first deposit the overlay on the side of the flask opposite to the cells. After solidification, incubate the culture at 37°C in an inverted position (cell monolayer up) in the dark. The plaques are counted daily for 3 to 5 days, or until the cell monolayers begin to degrade. The number of viruses in the sample or solution is the average number of plaques per flask, corrected for a specific volume (usually 1 ml) and multiplied by the dilution factor, if any. The result is expressed as PFU per unit of volume tested. The viral origin of a plaque must often be confirmed. The plaque can be extracted from the agar with a curved Pasteur pipette and transferred to a fresh confluent cell monolayer; if a cytopathic effect develops, the viral origin of the plaque is confirmed. The virus can be identified at this step, using this infected cell culture as the virus source for the procedures.

Some viruses do not form plaques except on a semisolid medium. Other substances can be used, such as 1% (wt/vol) methylcellulose, which is semisolid at 37°C and semiliquid at 4°C. This substance is sterilized by autoclaving (it then has a thick pasty appearance) and can be solubilized by being placed at 4°C or on ice for several hours. Other viruses produce a cytopathic effect but, unfortunately, one that is slow or subtle.

Neutral red is a vital stain that is picked up by the cells and permits visualization of the plaque as it is formed; the killed cells release the stain and become clear, forming the characteristic circular hole in the monolayer. Neutral red becomes toxic for cells during incubation periods of longer than 4 to 5 days. In these cases, it is preferred to use a two-step approach: a first incubation (optimized for the virus tested) without neutral red, followed by a second incubation for 24 to 48 h after addition of a second overlay containing neutral red. Alternatively, the neutral red can be totally omitted, with the agar removed after incubation and the monolayer stained with crystal violet (34).

Quantal Assays

Enteroviruses (polioviruses, coxsackievirus types B1 to B6 and A7 and A9, and echovirus types 1 to 34) induce a characteristic cytopathic effect: rounding of the cells is followed by their detachment from the cell monolayer. Adenoviruses produce rounding of the cells, an increase in refractivity, and a strong tendency to aggregate. The titration of these viruses may be performed either by placing the sample on an established cell monolayer or by mixing the virus with cells in suspension before seeding flasks. Increased sensitivity can be achieved by the addition of trypsin to the medium to promote infection (13).

The microtechnique is performed in multiwell plates and is useful when large numbers of viruses are expected (e.g., from wastewaters, heavily polluted surface waters, etc.). The macrotechnique is used for samples containing low levels of viruses (e.g., from treated waters, minimally polluted surface waters, groundwaters, etc.) when the volume of sample to be used to inoculate the cell cultures is larger. The maximum volume of sample that can be applied to a cell culture is 1 ml per 25 cm^2; larger volumes result in loss of sensitivity.

Some laboratories might have access to an electron microscope to detect directly the presence of viruses in the supernatants of infected flasks or wells. This method is expensive but has the advantages of detecting virus types on the basis of their morphology and of detecting multiple virus growths. In order to enhance the detection of slow-growing viruses, part of the infected cells or the supernatant can be placed onto fresh cell cultures. This is usually done after the original culture is frozen at −20°C to break apart the cells and then 0.1 to 0.5 ml is transferred to the fresh cell monolayers. This can be performed several times to increase the number of viruses present and demonstrate a cytopathic effect, and the virus can be identified by serum neutralization with specific antisera (42a). Most detection and identification methods are based on immunological reactions with specific or polyvalent antisera that are fluorescein or peroxidase tagged (see "Immunoassays" and "Identification of Viruses" below).

Macrotechnique (in Flasks)

A few days in advance, prepare the appropriate cell culture in growth medium (e.g., minimal essential medium [MEM]-Earle salts with 10% serum) in 25-cm^2 flasks so that the cells are confluent when they are used. Prepare serial dilutions of the virus (or sample) to be titrated in maintenance medium containing antibiotics (change pipette at each dilution to prevent virus carryover). Remove the culture medium from each flask, and inoculate not more than 1 ml of each dilution per 25-cm^2 flask; the larger the number of flasks, the more precise the titer (31). Use 10 to 20 flasks if there are few viruses in the sample. If a higher number of viruses is expected, the sample can be diluted and used to inoculate 5 flasks per dilution. Preferably, place the flasks on a rocking platform to slowly distribute the inoculum on the full monolayer for 60 min, add maintenance medium (e.g., MEM-Earle salts with 1% serum), and incubate at 37°C. Observe daily and record the number of flasks showing a cytopathic effect. Once the cytopathic effect has stopped progressing or when the controls begin to show aging, estimate the viral titer by the most probable number (MPN) method. The MPN approach has been in use for many years in bacteriology, with predefined tables for a number of experimental settings (3, 5, or 10 tubes). Their use in virology has evolved from the use of tables to simplified formulas for the calculation of the virus density in a sample (3, 4), and more recently, with the advent of microcomputers, it is now possible to use a more precise statistical approach to estimate the number of viruses in a sample. The computer programs for the calculations of the mpniu (MPN of infectious units) and their confidence intervals are available from the literature (18, 19, 21, 27, 42, 50).

The simplest form of the mpniu estimation equation for a single dilution is:

$$\text{mpniu} = -\ln (q/n)$$

where q is the number of negative cultures and n is the total number of cultures inoculated.

For more than one dilution, apply the following formula using results from dilutions with both positive and negative results (e.g., exclude dilutions that are 100% positive):

$$\text{mpniu/ml} = \frac{P}{(NQ)^{1/2}}$$

where P is the total number of cultures from all dilutions, N is the total milliliters of sample inoculated for all dilutions, and Q is the total milliliters of samples in negative cultures.

Microtechnique (in Multiwell Plates)

A few days in advance, seed the cells in multiwell plates and incubate until a confluent monolayer is obtained (note that multiwell plates are not tight fitting and must be incubated in an incubator with a 5% CO_2 atmosphere). Prepare serial dilutions of the sample to be titrated in maintenance medium (change pipette at each dilution to prevent virus carryover). Remove medium and inoculate 25 μl of each dilution into appropriate wells (with cells) of a flat-bottom 96-well plate. Use each dilution to inoculate at least four wells; the larger the number of wells, the more precise the titer estimation by the MPN approach. If the sample contains very few viruses, all wells from a plate can be inoculated with 25 μl of undiluted sample or of sample diluted 1:2 or 1:5 in maintenance medium. The use of 24-well plates is an alternative that has also been shown useful in many laboratories: their larger surface culture area increases the relative volume of sample that can be inoculated.

Once the cytopathic effect has stopped progressing or when the controls begin to show aging, estimate the virus concentration by using the MPN approach.

Immunoassays

The detection of viral products in infected cells can also be achieved by using specific or polyvalent antisera that are available commercially or can be prepared in-house (1). Both immunofluorescence and immunoperoxidase assays can be used (1, 33, 34). In the immunofluorescence assay, the antibodies fixed on viral products in the cells are detected by use of a commercial preparation of antibodies conjugated to a fluorochrome (usually fluorescein) that can be observed with a fluorescence microscope equipped with appropriate filters. In the immunoperoxidase assay, the antibodies fixed on viral products in the cells are detected by using a commercial preparation of antibodies conjugated to an enzyme, peroxidase, which can be detected with an appropriate substrate and observed with a regular microscope.

The first step in these assays is to inoculate the sample or dilutions of the sample to appropriate cell culture (BGM or MA-104, for example) in 25-cm^2 flasks (not more than 1 ml per 25 cm^2) or directly in 24- or 96-well plates if high virus concentrations are expected. Incubate for 7 to 10 days at 37°C. Observe the cells after 24 h to detect possible toxic effects; note wells where cytotoxicity is observed. After the incubation period, freeze-thaw the cell cultures and transfer 0.25 ml of the supernatant of each flask to the corresponding well of a new tissue culture plate containing the same cells and incubate again for 7 to 10 days to increase the viral titer if present. Drain the supernatant, wash twice with PBS, fix the cells by covering them for 15 min with absolute methanol containing 1% hydrogen peroxide, and finally rinse twice with PBS. Add to each well 0.3 ml of a 1:200 dilution of human immune serum globulin preparation (Cutter Biologicals, Berkeley, Calif., or equivalent from another source). Use only "immune serum globulins" and not "purified gamma globulins" that do not contain a significant amount of antibodies. Incubate for 1 to 2 hours at 37°C. Wash twice with PBS.

For detection by the immunoperoxidase assay, add 0.3 ml of an optimally diluted protein A-peroxidase conjugate per well (see the manufacturer's instructions), incubate for 1 to 2 h at 37°C, and wash four times with PBS. Prepare a fresh 0.025% solution of substrate (3,3′ diaminobenzidine tetrahydrochloride) in 100 ml of Tris-HCl (0.05 M, pH 7.6, filtered on a Whatman filter paper to remove debris). Just before use, add 30 μl of hydrogen peroxide (30% solution) per 100 ml of substrate. Substrate reactivity can be checked by adding a small drop of the diluted conjugate to 1 ml of substrate in a separate tube: a dark brown reaction should appear rapidly, indicating that the reagents are properly prepared. Add 0.5 ml of the freshly prepared substrate to each well and allow a reaction to occur for 5 to 15 min. The reaction can be observed under the microscope and stopped by washing the plate with tap water when the maximum difference between the infected and noninfected controls is observed and the background level is still low. Wash the monolayers with tap water and observe wet with an inverted microscope. Infected cells will appear dark brown, often with clearly visible nuclear or cytoplasmic inclusions. All wells in which a specific reaction or a cytopathic effect is observed are considered positive; record both the presence of a positive immunoperoxidase reaction and a cytopathic effect.

For the immunofluorescence test, replace the protein A-peroxidase conjugate with a protein A- or an anti-specific immunoglobulin-fluorescein conjugate. There is then no need for a substrate; simply carefully wash the monolayers and examine with a fluorescence microscope (1, 33, 34).

The number of viruses in the original sample can be evaluated by the MPN approach, using the volume of sample inoculated at the first passage and the number of wells at each dilution that have finally been shown to contain viruses.

IDENTIFICATION OF VIRUSES

Viruses isolated by cell culture can be propagated further in cell culture and identified by a variety of methods, including electron microscopy (33, 34), serum neutralization (1, 33, 34, 42a), molecular methods (28), and immunoassays (1, 33, 34). While in a specific outbreak situation it is useful to identify the causing agent, it is often not worth the cost of identifying these viruses in monitoring environmental samples; these viruses are pathogens and should never be present in the environment unless a major fecal contamination has occurred. This, however, remains a decision to be made by each researcher according to the objective of the ongoing research.

The viruses that are most commonly found in the environment are the polioviruses, coxsackieviruses (group B), adenoviruses, and reoviruses. By use of molecular biology methods and gene probe technology, all other viruses will eventually be detected.

REFERENCES

1. **Balows, A., W. J. Hausler, K. L. Herrmann, H. D. Isenberg, and H. J. Shadomy (ed.).** 1991. *Manual of Clinical Microbiology*, 5th ed. American Society for Microbiology, Washington, D.C.
2. **Berg, G., R. S. Safferman, D. R. Dahling, D. Berman, and C. J. Hurst.** 1983. *USEPA Manual of Methods for Virology*. EPA-600/4–84—013. Environmental Protection Agency, Cincinnati, Ohio.
3. **Chang, S. L.** 1965. Statistics of the infective units of animal viruses, p. 219–234. *In* G. Berg (ed.), *Transmission of Viruses by the Water Route*. Interscience Publishers, New York, N.Y.
4. **Chang, S. L., G. Berg, K. A. Busch, R. E. Stevenson, N. A. Clarke, and P. W. Kabler.** 1958. Application of the most probable number method for estimating concentration of animal viruses by the tissue culture technique. *Virology* **6:**27–42.
5. **Dahling, D. R., B. A. Wright, and F. P. Williams.** 1993. Detection of viruses in environmental samples—suitability of commercial rotavirus and adenovirus test kits. *J. Virol. Methods* **45:**135–147.
6. **Dahling, D. R., and B. A. Wright.** 1986. Optimization of the BGM cell line culture and viral assay for monitoring viruses in the environment. *Appl. Environ. Microbiol.* **51:** 790–812.
7. **Dahling, D. R., G. Berg, and D. Berman.** 1974. BGM, a continuous cell line more sensitive than primary rhesus and African green kidney cells for the recovery of viruses from water. *Health Lab. Sci.* **11:**275–282.
8. **Dahling, D. R., R. S. Safferman, and B. A. Wright.** 1984. Results of a survey of BGM cell culture practices. *Environ. Int.* **10:**309–311.
9. **Davis, P. M., and R. J. Phillpots.** 1974. Susceptibility of the Vero line of African green monkey kidney cells to human enteroviruses. *J. Hyg. Lond.* **72:**23–30.
10. **Freshney, R.** 1987. *Culture of Animal Cells: a Manual of Basic Techniques*, 2nd ed. Alan R. Liss Inc., New York, N.Y.
11. **Genthe, B., G. K. Idema, R. Kfir, and W. O. K. Grabow.** 1991. Detection of rotavirus in South African waters: a comparison of a cytoimmunolabeling technique with commercially available immunoassays. *Water Sci. Technol.* **24:** 241–244.
12. **Genthe, B., M. Gericke, B. Bateman, N. Mjoli, and R. Kfir.** 1995. Detection of enteric adenoviruses in South African waters using gene probes. *Water Sci. Technol.* **31:** 345–350.
13. **Gerba, C. P., and S. M. Goyal.** 1982. *Methods in Environmental Virology*. Marcel Dekker, Inc., New York, N.Y.
14. **Girones, R., M. Puig, A. Allard, F. Lucena, G. Wadell, and J. Jofre.** 1995. Detection of adenovirus and entero-

virus by PCR amplification in polluted waters. *Water Sci. Technol.* **31:**351–357.

15. **Guttman-Bass, N.** 1987. Cell cultures and other host systems for detecting and quantifying viruses in the environment, p. 195–228. *In* G. Berg (ed.), *Methods for Recovering Viruses from the Environment.* CRC Press Inc., Boca Raton, Fla.
16. **Hasler, P., and R. Wigand.** 1978. The susceptibility of Vero cell cultures for human adenoviruses. *Med. Microbiol. Immunol.* **164:**267–276.
17. **Herrmann, J. E.** 1995. Immunoassay for the diagnosis of infectious diseases, p. 110–122. *In* P. R. Murray, E. S. Baron, M. A. Pfaller, F. C. Tenover, and R. H. Yolken (eds.), *Manual of Clinical Microbiology*, 6th ed. ASM Press, Washington, D.C.
18. **Hugues, B., C. Pietri, and M. André.** 1985. Estimation of virus density in sewage effluents by 2 counting techniques. Comparison of precision as a function of inoculum volume. *Zentbl. Bakteriol. B* **181:**409–417.
19. **Hugues, C., and C. Pietri.** 1985. Influence du volume d'inoculum dans la quantification des virus selon deux techniques comparées: plages et N.P.P. utilisant un grand nombre d'inoculum par dilution. *Chemosphere* **14:**149–153.
20. **Hurst, C. J.** 1986. Evaluation of mixed cell types and 5-iodo-2′-deoxyuridine treatment upon plaque assay titers of human enteric viruses. *Appl. Environ. Microbiol.* **51:**1036–1040.
21. **Husson van Vliet, J., and P. Roussel.** 1986. Estimating viral concentrations: a reliable computation method programmed on a pocket calculator. *Comput. Methods Progr. Biomed.* **21:**167–172.
22. **Jehl-Pietri, C., B. Hugues, M. Andre, J. M. Diez, and A. Bosch.** 1993. Comparison of immunological and molecular hybridization detection methods for the detection of hepatitis A virus in sewage. *Lett. Appl. Microbiol.* **17:**162-166.
23. **Jones Brando, L. V.** 1995. Cell culture systems, p. 158–165. *In* P. R. Murray, E. J. Baron, M. A. Pfaller, F. C. Tenover, and R. H. Yolken (ed.), *Manual of Clinical Microbiology*, 6th ed. ASM Press, Washington, D.C.
24. **Leguyader, F., V. Apaire-Marchais, J. Brillet, and S. Billaudel.** 1993. Use of genomic probes to detect hepatitis A virus and enterovirus RNAs in wild shellfish and relationship of viral contamination to bacterial contamination. *Appl. Environ. Microbiol.* **59:**3963–3968.
25. **Leguyader, F., D. Menard, M. Pommepuy, and H. Kopecka.** 1995. Use of RT seminested PCR to assess viral contamination in Caribbean rivers (Martinique). *Water Sci. Technol.* **31:**391–394.
26. Reference deleted.
27. **Macdonell, M. T., E. Russek, and R. R. Colwell.** 1984. An interactive microcomputer program for the computation of most probable number. *J. Microbiol. Methods* **2:**1–7.
28. **Margolin, A. B., C. P. Gerba, K. J. Richardson, and J. E. Naranjo.** 1993. Comparison of cell culture and a poliovirus gene probe assay for the detection of enteroviruses in environmental water samples. *Water Sci. Technol.* **27:**311–314.
29. **Metcalf, T. G., J. L. Melnick, and M. K. Estes.** 1995. Environmental virology: from detection of virus in sewage and water by isolation to identification by molecular biology—a trip of over 50 years. *Annu. Rev. Microbiol.* **49:**461–487.
30. **Morris, R.** 1985. Detection of enteroviruses—an assessment of 10 cell lines. *Water Sci. Technol.* **17:**81–88.
31. **Payment, P., and M. Trudel.** 1985. Influence of inoculum size, incubation temperature and cell density on virus detection in environmental samples. *Can. J. Microbiol.* **31:**977–980.
32. **Payment, P., and M. Trudel.** 1987. Detection and quantitation of human enteric viruses in waste waters: increased sensitivity using a human immune serum globulin-immunoperoxidase assay on MA-104 cells. *Can. J. Microbiol.* **33:**568–570.
33. **Payment, P., and M. Trudel.** 1989. *Manuel de techniques virologiques*, 2nd ed. Presses de l'Université du Québec/AUPELF, Québec, Canada.
34. **Payment, P., and M. Trudel.** 1993. *Methods and Techniques in Virology*. Marcel Dekker Inc., New York, N.Y.
35. **Payment, P., F. Affoyon, and M. Trudel.** 1988. Detection of animal and human enteric viruses in water from the Assomption river and its tributaries. *Can. J. Microbiol.* **34:** 967–972.
36. **Payment, P., and E. Franco.** 1993. *Clostridium perfringens* and somatic coliphages as indicators of the efficiency of drinking water treatment for viruses and protozoan cysts. *Appl. Environ. Microbiol.* **59:**2418–2424.
37. **Pietri, C., and B. Hugues.** 1985. Influence du système cellulaire sur la quantification des virus dans les eaux usées. *Microbios Lett.* **30:**67–72.
38. **Pietri, C., B. Hugues, and D. Puel.** 1988. Immune electron microscopy in the detection of viruses other than enteroviruses on cell culture in untreated sewage. *Zentbl. Bakteriol. Mikrobiol. Hyg.* **186:**67–72.
39. **Podzorsky, R. P., and D. H. Persing.** 1995. Molecular detection and identification of microorganisms, p. 130–157. *In* P. R. Murray, E. S. Baron, M. A. Pfaller, F. C. Tenover, and R. H. Yolken (ed.), *Manual of Clinical Microbiology*, 6th ed. ASM Press, Washington, D.C.
40. **Puig, M., J. Jofre, F. Lucena, A. Allard, G. Wadell, and R. Girones.** 1994. Detection of adenoviruses and enteroviruses in polluted waters by nested PCR amplification. *Appl. Environ. Microbiol.* **60:**2963–2970.
41. **Reynolds, K. A., C. P. Gerba, and I. L. Pepper.** 1996. Detection of infectious enteroviruses by an integrated cell culture PCR procedure. *Appl. Environ. Microbiol.* **62:** 1424–1427.
42. **Russek, E., and R. R. Colwell.** 1983. Computation of most probable numbers. *Appl. Environ. Microbiol.* **45:**1646–1650.

42a. **Schmidt, N. J., and R. W. Emmons (ed.).** 1989. *Diagnostic Procedures for Viral, Rickettsial, and Chlamydial Infections*, 6th ed. American Public Health Association, Washington, D.C.

43. **Schmidt, N. J., H. H. Ho, and E. H. Lennette.** 1975. Propagation and isolation of group a coxsackieviruses in RD cells. *J. Clin. Microbiol.* **2:**183–185.
44. **Sellwood, J., and P. Wynjones.** 1995. A novel method for the detection of infectious rotavirus from water. *Water Sci. Technol.* **31:**367–370.
45. **Sewell, D. L., and R. B. Schifman.** 1995. Quality assurance: quality improvement, quality control and test validation, p. 55–66. *In* P. R. Murray, E. J. Baron, M. A. Pfaller, F. C. Tenover, and R. H. Yolken (ed.), *Manual of Clinical Microbiology*, 6th ed. ASM Press, Washington, D.C.
46. **Simard, C., M. Trudel, G. Paquette, and P. Payment.** 1983. Microbial investigation of the air in an apartment building. *J. Hyg. Camb.* **91:**277–286.
47. **Sobsey, M. D., L. R. Sangermano, and C. J. Palmer.** 1993. Simple method of concentrating enteroviruses and hepatitis A virus from sewage and ocean water for rapid detection by reverse transcriptase-polymerase chain reaction. *Appl. Environ. Microbiol.* **59:**3488–3491.
48. **Strain, B. A., and D. H. M. Gröschel.** 1995. Laboratory safety and infectious waste management, p. 75–85. *In* P. R. Murray, E. J. Baron, M. A. Pfaller, F. C. Tenover, and R. H.

Yolken (ed.), *Manual of Clinical Microbiology*, 6th ed. ASM Press, Washington, D.C.

49. **Tougianidou, D., M. Jacob, K. Herbold, T. Hahn, B. Flehmig, and K. Botzenhart.** 1989. Assessment of various cell lines (including mixed-cell cultures) for the detection of enteric viruses in different water sources. *Water Sci. Technol.* **21:**311–314.
50. **Wyshak, G., and K. Detre.** 1972. Estimating the number of organisms in quantal assays. *Appl. Environ. Microbiol.* **23:**784–790.

Cultivation of Microbial Consortia and Communities

GIDEON M. WOLFAARDT, DARREN R. KORBER, AND JOHN R. LAWRENCE

9

Microbial communities provide experimental systems that contribute to our understanding of complex processes in environmental and engineered settings, including assessment of the potential effects of environmental contaminants on microbial systems, and to making effective use of them as inoculants in industry, agriculture, and the environment (14, 19, 90). Applications include bioremediation, industrial and municipal wastewater treatment, composting, solid waste landfills, biological control agents, and biofertilizers. Microbial communities may also be useful indicators of the environmental effects of various stresses, including nutrients, organic pollutants, and metals (60).

Koch's postulates require that an organism of interest be maintained in isolation while demonstrating that it is a causative agent—a suitable approach in cases such as the study of pathogens and when industrial strains are used in pure culture fermentations. Although some extreme environments, such as hot springs, were once thought to contain pure cultures, it is now known that there are no environments in which evolution resulted in a pure culture. Indeed, aquatic and terrestrial environments typically contain communities in which evolution has resulted in interactive systems with multiple levels of organization (19). Because germ theory and a pure culture approach constrain cause-and-effect relationships to the organismal level, they may not be sufficient for analyzing environmental processes when the causative mechanism is at the community level or at multiple levels of organization.

Many free-living microorganisms are effective only when they are present in association with other groups of organisms. For example, this is the case for the degradation of pesticides, polychlorinated biphenyls, and other halogenated organic compounds. The same is also true for the organisms involved in interspecies hydrogen transfer (96) within methanogenic systems and in the case of nitrification (12), which requires a set of ammonia-oxidizing organisms and another set of nitrite-oxidizing organisms for complete oxidation of ammonia to nitrate.

Model systems may be used for cultivation of microbial communities, and they offer the advantages of relative simplicity, experimental control, and replication. However, there is a range of concerns regarding their use, including oversimplification, a contrived structure, and a lack of appropriate scale (temporal and spatial). These concerns have contributed historically to the relatively limited use of model systems, although they have increasing applications for important questions in microbial ecology.

PURE CULTURES VERSUS COMMUNITY CULTURES

A community is not an isolated cell line, nor is it a mixture of unrelated organisms. Thus, the criteria for obtaining and characterizing communities are different from those used to obtain isolated cell lines or mixed cultures. Each community is an individual network of organisms whose members are normally adapted to propagate both as individuals and as part of one or more associations. Individuals associate through specific behavioral adaptations which allow them to position themselves to interact physically, chemically, and biologically (41, 57, 61, 62). These adaptive positioning mechanisms may involve, among others, motility, chemoreceptors, chemotaxis, chemoadherence, production of exopolymers, and phagocytosis. Cultivation sometimes requires that community inputs and outputs be maintained to prevent the depletion of substrates and accumulation of products and to sustain the near-steady-state dynamics of many interactive associations. Methods used for community culture are described later in this chapter. Some of the criteria used for isolating communities, and other types of cultures, are discussed briefly below.

Pure Cultures

Pure culture studies remain the mainstay of microbiology and provide an essential element for exploring the molecular foundations of microbial life. Pure cultures are normally obtained by isolating individual cells in dilution tubes or on spread plates, using aseptic technique to maintain them in isolation while studying their properties and to obtain enough homogeneous cell material to perform the (bio)chemical (18) and molecular analyses necessary for identification and study.

There are two presumptive procedures that are normally followed to assess whether a culture free of contamination has been obtained. One is that streak plates on several different media yield a single colony type. The other is that phase microscopy and Gram staining reveal a single cell type. However, there are situations when microbial associations have been mistaken for pure cultures despite these precautions. This occurs when the organisms involved are so tightly associated that they cannot readily be cloned unless both organisms are present within each colony, as was once the case with *Methanobacterium omelianskii* (13).

Isolated cell lines may be used to generate a wide variety of useful information regarding specific bacterial strains and may be used to produce specific probes to detect and enumerate these microorganisms in environmental samples (fluor-conjugated mono- and polyclonal antibodies and 16S rRNA probes), although it is not essential to obtain a pure culture to generate RNA probes for in situ studies. They are also useful as test organisms to predict the effects of environmental contaminants on key organisms involved in nutrient cycling. Although often less effective than degradative microbial consortia, pure cultures have also been used as inoculants for bioremediation of a variety of contaminated aquatic and terrestrial environments.

Enrichment Cultures

Enrichments are frequently used as the first step in obtaining pure cultures of organisms involved in biogeochemical cycles or in the degradation of environmental contaminants. Enrichment culture methods have been reviewed by Holt and Krieg (46). They make it easier to isolate organisms in dilution tubes and on streak plates. For example, some enrichment media provide no biologically available form of nitrogen and thus enrich for nitrogen-fixing organisms. Others provide only a single halogenated organic compound as the energy source, thus enriching for specific degradative organisms.

An effective enrichment results in the predominance of a single bacterial species (9). However, enrichments have also been used to obtain degradative consortia (19), although they were not specifically designed for this purpose. The variation in environmental conditions during the time course of enrichment (depletion of substrate and accumulation of end products) generally produces corresponding variation in the species composition and spatial arrangement of organisms within consortia and communities. Cultivation of defined communities thus requires that community culture methods be used instead of enrichment culture to better define the environment in terms of the concentration and flux of both substrates and products.

The primary difference between community culture and enrichment culture is that an enrichment culture uses aseptic technique and selection procedures to enrich for a specific species and reduce biological diversity (9), while community culture attempts to maintain biological diversity. To be effective, an enrichment culture must impose a selection that allows the growth of only those organisms with desired adaptations. In the words of Beijerinck, "enrichment culture experiments can be called 'perfect' or 'imperfect'. In a perfect experiment a single species is isolated. . ." (9). In contrast, a community culture supports a diverse community of organisms that does not lead to the isolation of a single species.

Mixed Cultures

A mixed culture is produced by combining two or more pure cultures and excluding other organisms by using aseptic technique. Mixed cultures are not necessarily microbial communities. If the organisms within a mixed culture do not proliferate, survive, and adapt as a unit, then the culture may be inherently unstable, with one or more organisms quickly displacing others.

The primary value of mixed cultures is to demonstrate synergy among the isolated members of consortia, communities, and microecosystems as well as to establish the mechanisms of synergy. If the members of a community are synergistic, then various mixtures of isolated individuals should proliferate more effectively than the isolates themselves. If a specific trait of a single community member (for example, a chemoreceptor or exopolymer) is postulated to be involved in the formation of an association, then the performance of a synthetic community (a mixed culture of isolates from a native community) containing an adaptation-negative mutant can be compared to that of a synthetic community (mixed culture) containing the parental strain. The differences in performance among the individual cell lines, the adaptation-negative cell line, the native community, the synthetic community, and the synthetic community with the adaptation-negative mutant substituted for its parental strain can be used to elucidate the specific mechanisms of community-level interactions. It should be pointed out that isolation of the key community members is often problematic, a factor that will complicate delineation of the role of individual members (31). A community-level adaptation may have a detrimental effect on the proliferation of the isolated individual and a positive effect on the proliferation of its community (i.e., the adaptation-negative mutant outperforms its parent when the performances of the two are compared for isolated cell lines, but not when the strains are compared as members of an association). Experiments of this type are needed to establish a design theory for the construction of degradative and other types of microbial communities for use in agriculture, industry, and the environment.

Community Culture

Community culture refers to methods used to obtain and cultivate consortia, biofilm communities, bioaggregates, microecosystems, and other networks of interacting microorganisms growing in association with one another. Consortia consist of two or more organisms associated through physical attachment in a specific spatial arrangement (45). The term is most commonly used to describe mixtures of organisms which occur in degradative enrichments (for example, see reference 31). Communities consist of two or more cell lines associated through any of several mechanisms, including attachment, exchange of primary or secondary metabolites, and behavioral adaptations. Communities may also include one or more consortia as well as subcommunities. A microecosystem consists of a community (or set of communities and consortia) with multiple levels of biological organization exhibiting internal cycling of C, N, S, and other elements between their oxidized and reduced forms, possessing multiple trophic levels, spatial organization, and a favorable internal microenvironment(s), and existing in a nearly steady-state condition (changes in the system are relatively small compared to the rates of internal and/or external nutrient cycling).

The species composition resulting from community culture is defined through the formation of self-organizing microbial associations that are adapted to controlled environmental conditions or to perform a desired function through artificial selection. Artificial selection is the selective subculture of one community culture over another by the experimenter, based on a desired community characteristic (measured by the experimenter and used to determine which community culture will be subcultured), such as pesticide degradation, a change in pH, or the ability to promote the growth of plants (94, 95).

To confirm that a community culture has been obtained, it is thus important to demonstrate that the association of organisms produces a system which is maintained through interaction between two or more levels of organization (16). Communities often have one or more of the following characteristics that can be used to experimentally differentiate them from mixed cultures of unrelated organisms:

1. Communities are often more resistant (and resilient) to environmental stress than are cultures of isolated community members.
2. Communities often have more favorable collective growth constants (maximum specific growth rate, half-saturation constant, and/or cell yield) and more completely mineralize substrates than do cultures of isolated community members.
3. Communities often have a broader habitat range (or a different habitat range) than pure cultures of isolated community members.
4. The boundaries between community networks are often defined by ecotones occurring along spatial and temporal environmental gradients.
5. Plating efficiency is sometimes density dependent (dependent on the dilution factor) when communities are plated on agar media (50, 92).
6. Community structure (species composition) and architecture (cell positioning in biofilms) may remain constant even when challenged with pure cultures that are foreign to the community.

Communal associations where individual microorganisms acquire new characteristics that they lack as individuals have been studied and described. For instance, the *Chlorochromatium aggregatum* association between a sulfate reducer and a sulfide oxidizer allows the latter (a photosynthetic symbiont), which is the only photosynthetic bacterium with neither gas vacuoles nor flagella found in anaerobic hypolimnia, to position itself along vertical gradients of sulfide, oxygen, and light (21, 29). Numerous other heterotrophic and/or phototrophic interactions have been described (e.g., see references 47, 70, 76, and 89) and reviewed (83). Other types of consortia that have received much attention include those interactions based on hydrogen transfer (e.g., see references 13 and 32) and those involved in the degradation of refractory molecules (5, 101). Degradative networks, as opposed to isolated cell lines, may contribute to chemical communication and cometabolism between different microbial species, a larger and more flexible genetic resource pool from which metabolic pathways may evolve, the formation of favorable microenvironments, and protection within the biofilm exopolymer matrix against perturbations (62).

Community culture should be used preferentially over other culture methods when the objective is to produce a microbial association to be used as an inoculant in an in situ application or when the objective is to understand the mechanisms of community-level environmental processes and the effects of toxicants on those processes (7, 8, 25, 60, 63). Community cultures are inherently stable and durable when challenged by other microbes. They may contain predators and parasites to which pure and mixed cultures are often more vulnerable. Specific applications include degradative biofilm communities for use in bioremediation, bioaggregates and biofilms for use in restoring the activities of municipal and industrial wastewater systems following periodic bulking or exposure to toxic chemicals, inoculants for plant growth promotion, and starter cultures for silage and composting.

Following a community approach may also be important for testing the effects of pesticides, pharmaceuticals, and other chemicals on the microbial associations involved in nutrient cycling. Testing for inhibition only at the cellular level may result in false-positive results due to the relative susceptibilities of isolated cell lines compared to the durability of microbial associations. However, it may also result in false negative tests if the mechanism of inhibition blocks the interactions between organisms without necessarily inhibiting the individual members of the association when cultured in isolation. Working with river biofilm communities, Lawrence et al. (60) demonstrated specific effects on *nirS* gene-bearing populations in communities grown in the presence of nickel at 0.5 mg liter $^{-1}$, specifically a negative effect on these denitrifying populations resulting in a reduction in diversity but maintenance of the overall community.

COMMUNITY THEORY

A central question in environmental microbiology is whether microbial communities exist as units of proliferation and adaptation. If they do, then it is logical to cultivate them and study them as causative agents. If they do not exist, then, as Brefeld stated in 1881 (11), "Work with impure cultures yields nothing but nonsense." Theoretical explanations for the nature and origin of communities can be grouped into at least three categories. These include (i) competition models in which communities originate through negative interactions between individual organisms, (ii) cooperation models in which communities originate through positive synergisms between individual organisms, and (iii) a more speculative proliferation model in which communities are proposed to result from the growth of a nested series of proliferating objects. There is no consensus as to which of these explanations is most useful.

Competition

Competition is an interaction between two systems that reduces the probability of propagation for both (81). Competition theories (natural selection, group selection, and multilevel selection) generally assume that communities arise as an indirect consequence of competitive interactions between community members. From this perspective, innovation is derived exclusively from individual (30) or group (95) competition. However, competition and selection may not be efficient in explaining cooperation because each cooperative event must be derived either directly or indirectly from one or more competitive events.

Cooperation

Cooperation (mutualism and symbiosis) is an interaction between two systems that increases the probability of propagation for both (81). Cooperative theories postulate that many significant innovations in life result directly from cooperation between individual organisms or through an even larger cooperative synergy involving global chemical, physical, and biological interactions (65). The Gaia theory goes beyond the community level to postulate that the Earth is an adaptive, self-regulating system (ecosphere) that includes chemical, physical, and biological components.

Nested Proliferation

A proliferation theory has been proposed as a testable scientific thesis concerning the physical mechanism by which community-level innovation originates. This theory interprets both competition and cooperation as consequences of an innovative mechanism (15, 16). The theory is based on the idea that evolution occurs due to the growth of a nested series of proliferating objects (19). The proliferation of nucleic acids is nested within cells, the proliferation of cells is nested within organisms, the proliferation of organisms is nested within communities, the proliferation of communities is nested within ecosystems, etc. Thus, the form, function, and growth rate of each layer are calculated responses to the layers that it inhabits and that inhabit it (19), and the positioning of organisms in communities has some degree of informational functionality (68).

Implications of Community Theory

Studies have been conducted indicating that artificial selection can be used to produce ecosystems that promote the growth of plants, raise the pH of media, and degrade toxic compounds (94, 95). These studies proposed that microorganisms can thus be domesticated at the community and ecosystem levels, in much the same way that plants and animals have already been domesticated at the species level. The only difference is that whole communities and ecosystems are selected, rather than individual organisms. Such outcomes suggest that selection is also effective at higher levels of the biological hierarchy from genes to individuals to communities to ecosystems.

COMMUNITY CULTURE METHODS

Describing a community in situ does not necessarily ensure that it is understood well enough to be controlled or effectively utilized in environmental applications. If a community network has been adequately elucidated at the community level, then it should be possible to cultivate it in the laboratory under defined conditions, complete with predators and parasites, without using aseptic technique. This can be used as a criterion of success in the laboratory component of environmental microbiology, just as the isolation of a pure culture is a criterion of success in understanding and controlling disease by using the germ theory and Koch's postulates. Despite recognition of the recent advances in virtually every area of microbiological research, the need to develop a capacity that will allow prediction of the functional behaviors of communities as integrated, dynamic systems is recognized, together with the acknowledgment that the ultimate test of our understanding of genomes will require that we develop the ability to model microbial life from the cellular to the community level (for instance, see http://www.ornl.gov/sci/Genomestolife/2003meeting/GTLFacilities.pdf). Much effort, specifically focusing on the development of systems to cultivate microbial communities, is required to address this need.

For community culture, the environment defines the culture rather than aseptic technique (14, 19). Accurately controlling the environment in terms of substrate concentration and flux as well as spatial and temporal variations is thus crucial. As stated by Senior et al. (90), "It is reasonable to postulate that interacting microbial communities are common in nature; however, very few microbial associations of this type have been defined, principally because inappropriate enrichment and selection techniques have been used to isolate them from natural environments." Defining the environment requires that a steady-state system be provided which is capable of supplying the necessary inputs and outputs to sustain a dynamic microbial community (90). Thus, in addition to defining substrate concentrations, it may also be necessary to define substrate availability in terms of flux as well as spatial and temporal variability. Similarly, it is important that the end products moving from the microenvironment to the macroenvironment be continually removed as they would be in situ. Community-level studies thus involve microbiological systems which provide multidimensional environmental gradients for studies of habitat range (34, 66) as well as providing the spatial (14, 18, 98) and temporal (88) pathways necessary for communities to develop.

Batch culture systems are normally inadequate for defined studies of community cultures due to the lack of environmental control as the culture develops. Consequently, the discussion below emphasizes those culture systems which provide adequate environmental control of substrate flux and concentration throughout the time course of cultivation. Although batch systems may be used to cultivate microbial associations, they should be used and interpreted with caution.

Chemostats, nutristats, microstats, and continuous-flow slide cultures (CFSCs) are among the culture methods commonly used to cultivate consortia and communities. The chemostat provides control of substrate flux and growth rate (assuming there is no wall growth). However, the concentration of substrate cannot be controlled directly and must be determined empirically once a nearly steady-state condition has been obtained. The nutristat provides control of the concentration and temporal variation of the substrate, although the growth rate and flux must be allowed to vary so that a nearly steady state can be obtained. CFSCs provide control of the flux, concentration, and temporal variation of the substrate (but not direct control of the growth rate). The microstat provides control of substrate flux, concentration, temporal variation, and spatial variation (but not direct control of the growth rate).

Chemostats

Chemostats and other continuous cultures are among the most widely used systems for cultivating microbial communities and consortia (10, 40, 42, 73, 84, 90, 101). They provide a near-steady-state system with inputs and outputs necessary to sustain dynamic networks of interacting microorganisms (19, 90) and to study reaction kinetics (64). The volume of the culture is constant, and medium continually flows into the culture at the same rate that effluent flows out. The growth of organisms within the culture gradually reduces the concentration of the limiting nutrient until a nearly steady state is obtained in which the dilution rate (flow rate/volume) equals the specific growth rate (75). If a community is cultivated, the mean growth rate of each member within the community is equal to the dilution rate once the culture is substrate limited and has had sufficient time to reach a nearly steady-state condition.

Chemostat systems are particularly effective in providing a continuous supply of refractory compounds over the long periods necessary for degradative consortia and communities to form (26, 90, 101). They are also used to cultivate and study algal-bacterial bioaggregates (73) and to define and optimize anaerobic digestion processes (39). These applications generally require a modified outlet which ensures that flocs, granules, and bioaggregates are diluted at the same rate as suspended cells (35).

Nutristats and Gradostats

Rutgers et al. (88) devised the nutristat, a continuous culture providing defined concentrations of pentachlorophenol (PCP) and gradually increasing the PCP concentration over time. This approach, based on a temporal substrate gradient, resulted in higher growth rates and shorter acclimation times in the selection of PCP degraders than do batch or traditional chemostat systems. In most continuous culture systems, the substrate concentration is allowed to vary while the growth rate (dilution rate and flux of substrate) is held constant. The concentration of substrate is continuously monitored and controlled by use of an automated system. Gradually increasing the concentration of substrate in this system provides sufficient time for organisms and associations to gradually adapt. More recently, Muller and Babel (77) used nutristat experiments to evaluate microbial responses to various levels of dichlorophenoxyacetic acid and pH.

The gradostat (44, 66) is a multistage chemostat that is used to provide spatial gradients of concentration, as opposed to the temporal gradients of the nutristat. It consists of a series of bidirectionally linked chemostats which provide a physical pathway for community development. The gradostat normally consists of five separate continuous-culture vessels connected in series. A medium containing one substrate is pumped into the first stage, and a second substrate is pumped into the last stage. Effluent is removed at equal rates from both ends. This produces opposing gradients in one dimension, and all possible proportions of the two substrates are represented. To represent all concentrations, as well as all proportions of the two substrates (as in the case of the microstat described below), would require a five-by-five array of linked chemostats with substrates supplied from two adjacent edges of the array and with effluent being removed from the two opposing edges. Codeco and Grover (27) used gradostats to study competition for two resources (phosphorus and carbon) between bacteria and algae.

Continuous Culture for Study of Surface Growth

Although mathematical models of continuous culture often assume that no wall growth occurs, some chemostat studies have focused on biofilms of attached organisms (36, 52, 66, 69, 97, 102). A dual-dilution continuous culture was used by Caldwell and Lawrence (20) as a behavioral enrichment to select an aggressive surface colonizer from soil that was not only effective at colonizing surfaces but also highly efficient at emigrating from surface microenvironments and colonizing new surface environments. In this culture system, the aqueous phase is continually diluted (replaced) at a constant rate, while the solid phase is intermittently replaced. This system was used by Korber et al. (56) to evaluate the role of motility in colonization by comparing the performance of nonmotile mutants to that of parental strains. The system consists of 50-ml round-bottomed flasks containing 3-mm-diameter glass beads. The aqueous phase is maintained as a typical continuous culture with sufficient ventilation and with a peristaltic pump and silicon tubing for medium supply and effluent removal, respectively. The flasks are placed on a shaker operating at a speed low enough to prevent agitation of the beads and resulting shear of biofilms but sufficient to ensure adequate mixing of the aqueous phase. The flasks are routinely replaced with sterile flasks that are inoculated with colonized glass beads removed from the previous series of flasks at intervals determined by the experimenter, thus allowing independent dilution of both the aqueous and solid (particulate) phases at different rates. The glass beads are carefully rinsed to remove all unattached cells before being introduced to the sterile flasks. Because the flasks are regularly replaced, wall-associated populations do not have sufficient time to become a stable source of sloughed cells, and thus with increased aqueous-phase dilution rates, there is an increased selection pressure to obtain communities which are capable of rapidly emigrating and colonizing new surfaces as they become available. Examples where this adaptation would be of value include the colonization of growing root hairs and tips in the rhizosphere and the recovery of biofilm communities after periodic sloughing or grazing by protozoa and invertebrates.

RABs

One of the best-characterized continuous culture systems for studying surface-associated organisms is the rotating annular bioreactor (RAB) (24). This often consists of two (one fixed and one rotatable) concentric cylinders. The resistance to rotation caused by biofilm growth provides data on friction and shear. If biofilm accumulation is of interest, then rates of dilution are set to exceed the growth rate of the organisms. In this case, only biofilm-derived cells are detected in the planktonic phase. Alternatively, if both biofilm and planktonic growth is of interest, the dilution rate is set below the maximum specific growth rate, and planktonic cells are able to grow and persist in the aqueous phase. Biofilms are normally studied by removing and observing coupons (93, 97). RABs are useful for cultivating biofilm communities from industrial systems, such as rotating biological contactors, and from environmental systems (see below). They also provide sufficient cell material for direct analysis of biofilm material by use of physical, chemical, biological, and molecular tools.

RABs allow the establishment of biofilms under nearly steady-state conditions (22, 23). The system also provides a well-mixed liquid phase, turbulent flow, and constant shear fields and may be suitable for cultivation of biofilms from flowing environments. For instance, the inoculum may be provided through circulation of river water or sample water from any environment. Slides can be removed at any time for destructive sampling, or they can be analyzed or subsampled and returned to the reactor, although it should be pointed out that the latter may potentially introduce artifacts that can influence subsequent analysis. This reactor type has been used to study various aspects of biofilms, including structural features (37, 38), sorption kinetics (58), metabolism in biofilms (59), effects of biocides (91), degradation of contaminants (1, 4), and general aspects of river biofilms (25, 60, 63, 78). There are a number of commercially available rotating annular reactors. One of the limitations in the use of RABs in biofilm research can be the cost of the basic system, controllers, and other peripheral equipment. This factor may also limit important experimental criteria, such as the number of treatment effects to be examined and the number of true replicates available for analysis. Lawrence et al. described a simple RAB which may be manufactured easily at low cost and applied in studies of river microbial biofilm community development (25, 60). Figure 1 shows a bank of reactors arranged for a replicate experiment and the details of the reactor core, showing the polycarbonate slides and attached biofilm. An alternative system that shares many of the features of RABs is the rotating disk reactor (85).

Gjaltema and Griebe (38) suggested the following eight points to consider in the development of reactors for biofilm cultivation: hydrodynamics, mixing, nutrient supply and oxygen supply, development of gradients, biofilm sampling, biofilm monitoring, septic operation, and nature of the substratum. In addition, the reactor or model system must maintain the community structure in terms of species composition and the community architecture or spatial heterogeneity and must provide the capacity to create replicate systems. Various analytical approaches have revealed some

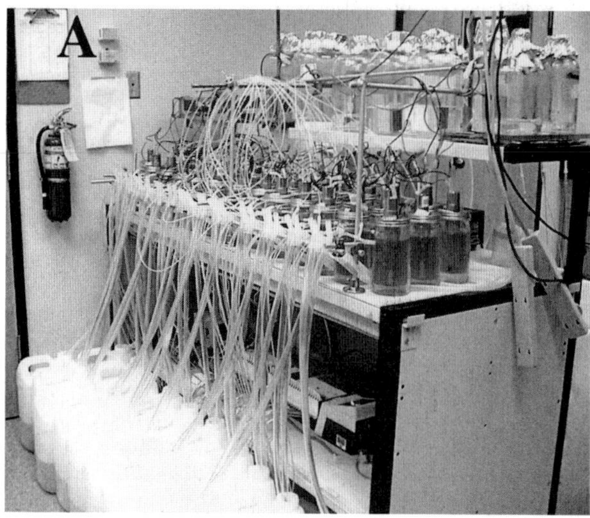

FIGURE 1 Photographs of a RAB setup, showing general arrangement of multiple reactors and plumbing (A) and details of the inner rotating cylinder with 12 removable polycarbonate strips for biofilm development and subsequent analyses (B).

common problems with culture in RABs, including heterogeneity or patchiness in the biofilm and sloughing (37). The development of unwanted gradients within RAB-cultured biofilms has also been noted. Neu and Lawrence (78) showed a typical time course of river biofilm community development in a rotating annular reactor system. Manz et al. (67) documented changes in river community structure through the application of group-level rRNA probes. Studies of the effects of environmental stress, for example, nickel and nutrients (60) or pharmaceuticals (63), have also been performed by using RAB systems.

CDFF

An alternative to the RAB fermentor is the constant depth film fermenter (CDFF) (53, 87, 98). The CDFF was originally used in studies of oral biofilms (28) but can also be used in many other situations. Biofilms are cultivated in pans that vary in depth and composition, depending on the nature of the study. Scraper blades pass over these pans, thus limiting the thickness of the biofilm to the depth of the pan. This results in the generation of a steady-state biofilm with a controlled thickness or depth.

Robbins Device

The Robbins device can be used to study biofilm communities in situ or under laboratory conditions (51). It is one of the best-known and most widely used systems for studying biofilm communities and pure culture biofilms in applications ranging from medical (79) to industrial (3, 33) and food (6) research. Since the original Robbins device was described by McCoy et al. (71), a number of modifications have been made to the design, and the modified Robbins device described by Nickel et al. (80) is probably the one used most often. In essence, the device consists of a pipe or tube with specially constructed plugs screwed through the wall and flush with the inner surface of the tube. These provide a removable surface that can be assayed for biofilm growth and composition. The design varies depending primarily on the hydraulic pressure of the fluid. The Robbins device is available through Tyler Research Corporation in Edmonton, Alberta, Canada. Diagrams and photographs are available from Tyler Research. Tyler Research manufactures a range of devices for the growth and analysis of biofilms on surfaces. These include modifications of the original Robbins device as well as an extensive line of medium- and high-pressure Tyler biofilm systems, all of which are intended primarily for in-line growth, sampling, and analysis of biofilms in industrial and laboratory environments.

CFSC

The primary advantage of CFSC over other methods of community culture is that it provides the opportunity to view the development of communities by using various forms of microscopy (chapter 5). This, coupled with the use of fluorescent molecular probes, provides an effective means of studying community-level processes nondestructively. Another advantage is that the concentration and flux of substrates are precisely and directly controlled. In the chemostat, the concentration of the substrate varies until it reaches a nearly steady-state concentration which must be empirically measured. In CFSC, substrate concentrations can also be changed almost instantaneously, permitting the use of pulses of fluorescent probes and test compounds. This culture system is most appropriate when real-time analysis of biofilms is required. A drawback of the approach is that the quantity of biomass available for conventional chemical and biochemical analyses is very small. This drawback can sometimes be overcome by using a column of glass beads to provide a sufficient attachment surface and, consequently, cell material for traditional analysis while using CFSC, connected in series or parallel, for nondestructive analysis.

CFSC has been particularly helpful in expanding the scope of laboratory studies to include microbial communities. It allows the study of microorganisms at multiple levels of organization, including the gene (43) and plasmid (49, 74) levels as well as the cell, consortium, community, and microecosystem levels (14, 18, 19). Studies with CFSCs have shown that spatial organization, including intrageneric and intergeneric coaggregation, is necessary to optimize the activities of degradative and other biofilm communities (54, 99, 101, 102). In CFSC, a flow cell is mounted on a microscope stage and irrigated with stream water, groundwater, wastewater, the effluent from a chemostat, a defined medium, etc. The community is analyzed during the time course of its development by use of fluorescent molecular probes, epifluorescence microscopy, confocal laser microscopy, and digital image analysis (18). These methods allow quantification of biochemistry, genetic sequences, genetic expression, growth, metabolism, etc., as well as the corresponding three-dimensional relationships among them. Information concerning confocal laser microscopy, digital imaging, and fluorescent molecular probes is provided in chapter 5. Commercial flow cells are available from Stoval Life Science Inc. These are gamma-irradiated, sterile flow cells for biofilm growth and microscopic examination of biofilms. The system also includes a bubble trap design to reduce peristaltic pulsation in the flow cell. Square glass capillaries similar to those used by Perfil'ev and Gabe (86) are available commercially (Friedrich and Dimmock Inc., Millville, N.J.) in sizes as fine as an internal size of 50 μm with a 25-μm wall thickness. However, conventional coverslips assembled with flowable self-leveling (RTV 110) silicone adhesive provide a larger and more uniform viewing area (20). In most cases, multichannel flow cells are required for experimental treatments, replication, and controls. These are constructed by using a polycarbonate base containing several irrigation channels, as shown in Fig. 2. Coverslips are bonded to cover the base such that the gaps between the coverslips are located between channels (55, 102). UV-transparent coverslips (zinc titanium glass; Corning Glass code 0211) and oversize coverslips are available by special order from Corning Glass Works (Corning, N.Y.).

Microstats

Microstats are used for the cultivation of biofilm communities, as opposed to planktonic communities. They do not lend themselves as well to microscopy as do CFSCs, but they do provide spatial pathways which allow the formation of communities that might otherwise not be able to develop. Spatial gradients also provide the opportunity for ecotones to form. Ecotones confirm that communities have been obtained, define the habitat range of a specific community, and can be used to determine whether the same community is responsible for the degradation of two different compounds. Although pulses of test compounds or fluorescent molecular probes can be used in the irrigation solution, the gradients cannot be turned on and off. Thus, the environment cannot be changed quickly as in the case of CFSCs.

Microstats can be used to provide one- or two-dimensional steady-state gradients of test molecules (e.g., a carbon source) within a surface microenvironment (14, 17, 19, 100). In essence, diffusion of the test compound through a porous matrix such as agarose gel leads to the development of a stable concentration gradient that is maintained by the overlying aqueous medium, which flows in the opposite direction from that of the concentration gradient (Fig. 3 left), thereby removing the test compound as it diffuses from the porous matrix. The system can be inoculated and irrigated according to the same procedure as that used for flow cells. The concentration of the test compound is maintained in the reservoir either by providing an excess supply (e.g., when supplied as crystals or powder) or by means of a peristaltic pump. This makes it possible to quantify the effect of environmental variation on the formation of communities and to quantify the distribution of various community networks with regard to habitat range. It also provides spatial pathways of community development. For example, a community degrading toxicants under unfavorable conditions may be unable to form unless a gradient of toxicant allows the community to form initially at low concentrations and gradually expand toward higher concentrations

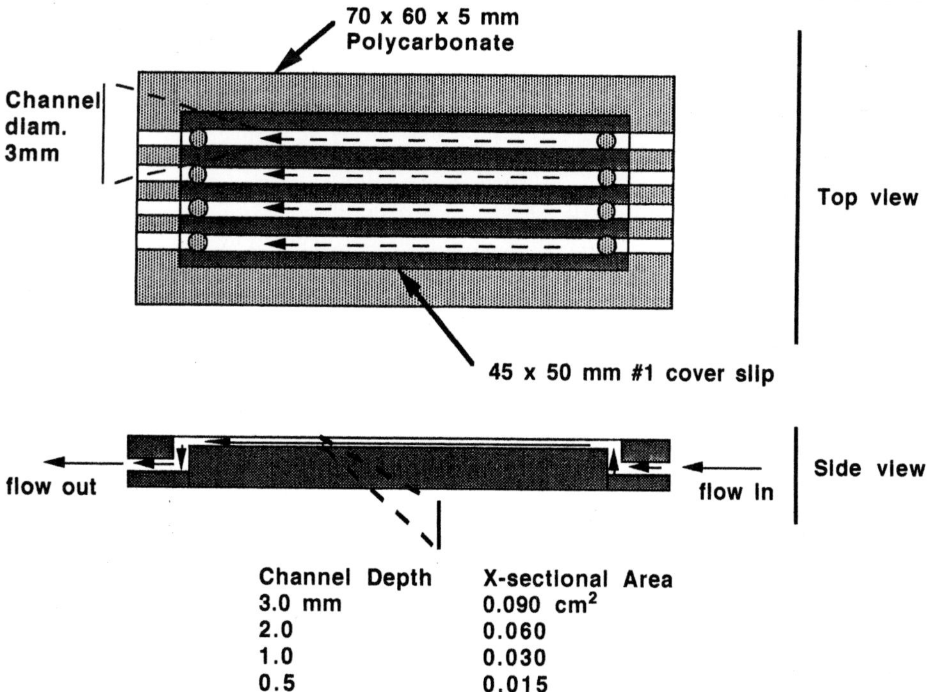

FIGURE 2 Schematic diagram of a multichannel flow cell. The use of flow cells containing multiple channels simplifies multiple comparisons, replication, and controls.

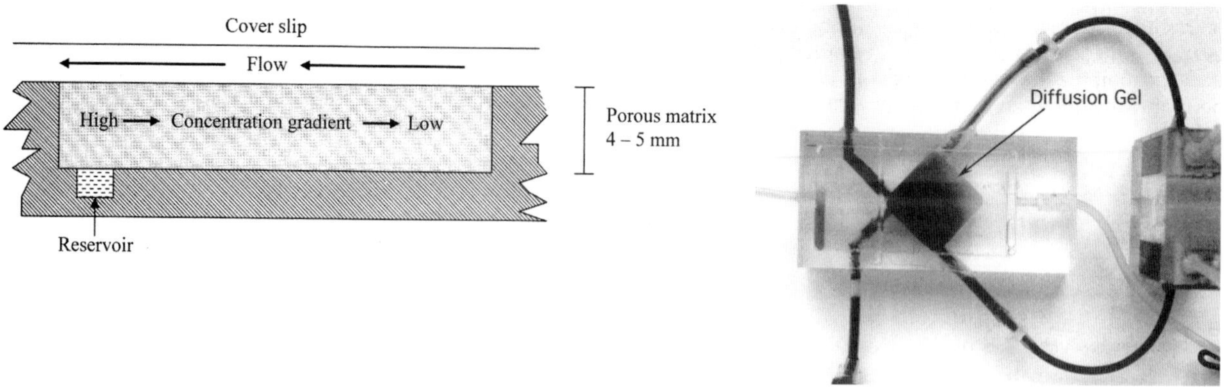

FIGURE 3 (Left) Schematic cross section of a microstat perpendicular to the concentration gradient that develops through the porous matrix. The flow over the surface against the direction of the concentration gradient removes molecules as they diffuse out of the porous medium, thereby maintaining a steady state. Adapted from reference 100. (Right) Photograph of a microstat showing two-dimensional steady-state diffusion gradients that were created by diffusion through an agarose gel. In the case of recalcitrant compounds, microbial communities are provided spatial and temporal gradients on the surface of the gel, thereby allowing organization of community composition and biofilm structure that may facilitate improved utilization of the substrate.

(once it is optimally configured in terms of community structure and biofilm architecture). In some cases, a primer gradient consisting of a labile toxicant analog may be necessary perpendicular to the primary toxicant gradient. Primer gradients provide a physical pathway along which the community may reorganize and adapt and in which labile compounds permit the cometabolism of more refractory compounds. Temporal gradients, as described by Rutgers et al. (88), are also effective in producing degradative consortia. However, using two-dimensional gradients allows the community to determine its own time frame and pathway for development. A photograph of a microstat is shown in Fig. 3 (right). In this case, two stains with different colors and an agarose gel were used as the diffusion medium to demonstrate the establishment of two-dimensional steady-state gradients. Methods for determining diffusion coefficients through the gel, as well as measurement of concentration gradients, mathematical simulation of the gradients, and quantification of the response of a degradative microbial community to the concentration gradients, were described in detail by Wolfaardt et al. (100).

In Situ Cultivation of Microbial Communities

Model substrata are also used in situ in an attempt to cultivate microbial communities and extract them for microscopic examination and other analyses. In this instance, and indeed with the laboratory reactors discussed above, the critical factor is the selection of the substratum, which influences the resultant community. Microbial communities develop at various interfaces, including solid-liquid and air-liquid boundaries such as those at stones, leaves, wood, shells, and other surfaces. If the decision is made to investigate them in situ, then the device and the substratum provided for growth and development must be selected. Important parameters to be considered include the chemical composition and surface heterogeneity of the substratum. Artificial substrata have been proposed and applied for monitoring community development and structure, particularly in the biofouling and periphyton literature (2). Surprisingly few studies have examined the effects of these parameters on colonization by microbial communities. Hunt and Perry (48) examined the effects of substratum roughness and flow rate on the in situ development of a freshwater biofilm community. They suggested significant interactive effects of surface topography and flow rate but concluded that only bacteria appeared to be notably influenced by the presence of increasing surface roughness.

The ideal substratum should conform to the following guidelines: (i) the coefficient of variation for the parameter measured should be less than or equal to that for naturally occurring substrata, (ii) the time for community development should be within the time frame of the experiment, (iii) the substratum should be easily retrievable, and (iv) community composition and architecture should be retained. A variety of substrates have been proposed, including glass, rocks, tiles and bricks, Styrofoam, polycarbonate, plastic, and nutrient-diffusing substrates. Olapade and Leff (82) examined how different bacterial taxa in biofilms responded to dissolved organic matter and inorganic nutrients by using nutrient-diffusing artificial substrates (clay flowerpots), an approach commonly used to demonstrate nutrient limitation in streams.

In addition to knowledge of the physical environment, such as the attachment substratum, the chemical environment should also be considered in the design of in situ systems. Battin et al. (7) described the impact of biofilms on the physical and chemical microhabitat in aquatic ecosystems and presented compelling evidence that links biofilm structure and function with stream ecosystem biogeochemistry. The system described by these authors (7, 8) is a good example of in situ experimentation and exemplifies the complex nature of natural systems, a fact that needs to be considered in experimental design. Their system consisted of streamside flumes (30 m long by 0.3 m wide by 0.3 m deep) that were continuously supplied at a constant flow with stream water in a once-through mode. The flumes were packed with ~21 mm-grain-size cobble and pea gravel, and the slopes were adjusted to result in different flow velocities. Unglazed ceramic coupons (1 by 1.5 by 0.3 cm) were placed in the gravel and randomly sampled for microscopic analysis of biofilms, enumeration of cell numbers, and characterization of exopolysaccharides. This information was subsequently compared with the uptake of dissolved organic carbon and hydrodynamic transient storage of suspended particles. A further important consideration during experimental design is how the analyses described elsewhere in this book can be integrated to analyze the communities that develop by use of these in situ approaches.

Storage and Preservation of Community Cultures

Community cultures are defined by their structure and architecture as well as by the genetic coding of their component cell lines. Storage of communities under conditions suitable for preserving pure cultures may be inadequate for the preservation of communities. For example, the freezing or lyophilization of a community inevitably results in the preferential survival of one cell line versus another, depending on the resistance of each to the method of preservation. Thus, while genetic structure is retained during storage procedures, community structure may not be as readily conserved (72). Since it is difficult, if not impossible, to preserve intact microbial communities in such a way that they will return to their original state following storage, a useful approach is to fingerprint the community prior to storage (by denaturing gradient gel electrophoresis, fatty acid methyl ester analysis, terminal restriction fragment length polymorphism analysis, etc.) and after resuscitation in order to ascertain whether the community remains intact.

REFERENCES

1. **Alleman, B. C., B. E. Logan, and R. L. Gilbertson.** 1995. Degradation of pentachlorophenol by fixed films of white rot fungi in rotating tube reactors. *Water Res.* **29:**61–67.
2. **Aloi, J. E.** 1990. A critical review of recent freshwater periphyton field methods. *Can. J. Fish. Aquat. Sci.* **47:** 656–670.
3. **Araujo, J. C., R. Mortara, J. R. Campos, and R. F. Vazoller.** 2004. Development and analysis of anaerobic biofilms onto hydrophobic and hydrophilic surfaces. *Environ. Technol.* **25:**809–817.
4. **Arcangeli, J.-P., and E. Arvin.** 1995. Cometabolic transformation of *o*-xylene in a biofilm system under nitrate reducing conditions. *Biodegradation* **6:**19–27.
5. **Ayala-del-Río, H. L., S. J. Callister, C. S. Criddle, and J. M. Tiedje.** 2004. Correspondence between community structure and function during succession in phenol- and phenol-plus-trichloroethene-fed sequencing batch reactors. *Appl. Environ. Microbiol.* **70:**4950–4960.
6. **Bagge, D., M. Hjelm, C. Johansen, I. Huber, and L. Grami.** 2001. *Shewanella putrefaciens* adhesion and biofilm formation on food processing surfaces. *Appl. Environ. Microbiol.* **67:**2319–2325.
7. **Battin, T. J., L. A. Kaplan, J. D. Newbold, X. Cheng, and C. Hansen.** 2003. Contributions of microbial biofilms to ecosystem processes in stream mesocosms. *Nature* **426:** 439–442.
8. **Battin, T. J., L. A. Kaplan, J. D. Newbold, X. Cheng, and C. Hansen.** 2003. Effects of current velocity on the nascent architecture of stream microbial biofilms. *Appl. Environ. Microbiol.* **69:**5443–5452.
9. **Beijerinck, M. W.** 1901. Enrichment culture studies with urea bacteria. *Centblatt. Bakteriol. II* **7:**33–61.
10. **Bradshaw, D. J., P. D. Marsh, K. M. Schilling, and D. Cummins.** 1996. A modified chemostat system to study the ecology of oral biofilms. *J. Appl. Bacteriol.* **80:**124–130.
11. **Brefeld, O.** 1881. *Botanische Untersuchungen über Schimmelpilze: Culturemethoden.* Leipzig, Germany.
12. **Bremner, J. M., and G. C. McCarthy.** 1993. Inhibition of soil nitrification by allelochemicals derived by plants and plant residues, p. 181–218. *In* Jean-Marc Bollag and G. Stotzky (ed.), *Soil Biochemistry.* vol. 8. Marcel Dekker Inc., New York, N.Y.
13. **Bryant, M. P., E. A. Wolin, M. J. Wolin, and R. S. Wolfe.** 1967. *Methanobacillus omelianskii,* a symbiotic association of two species of bacteria. *Arch. Mikrobiol.* **59:**20–31.
14. **Caldwell, D. E.** 1995. Cultivation and study of biofilm communities, p. 1–15. *In* H. Lappin-Scott and J. W. Costerton (ed.), *An Introduction to Bacterial Biofilms.* Cambridge University Press, Cambridge, United Kingdom.
15. **Caldwell, D. E.** 1999. Post-modern ecology—is the environment the organism? *Environ. Microbiol.* **1:**279–281.
16. **Caldwell, D. E.** 2000. Is DNA the only code of life, or the first to be understood? *Environ. Microbiol.* **2:**3.
17. **Caldwell, D. E., and P. Hirsch.** 1973. Growth of microorganisms in two-dimensional steady-state diffusion gradients. *Can. J. Microbiol.* **19:**53–58.
18. **Caldwell, D. E., D. R. Korber, and J. R. Lawrence.** 1992. Confocal laser microscopy and computer image analysis. *Adv. Microb. Ecol.* **12:**1–67.
19. **Caldwell, D. E., D. R. Korber, G. M. Wolfaardt, and J. R. Lawrence.** 1996. Do bacterial communities transcend Darwinism? *Adv. Microb. Ecol.* **15:**1–72.
20. **Caldwell, D. E., and J. R. Lawrence.** 1986. Growth kinetics of *Pseudomonas fluorescens* microcolonies within the hydrodynamic boundary layers of surface microenvironments. *Microb. Ecol.* **12:**299–312.
21. **Caldwell, D. E., and J. M. Tiedje.** 1975. A morphological study of anaerobic bacteria from the hypolimnia of two Michigan lakes. *Can. J. Microbiol.* **21:**362–376.
22. **Cao, Y. S., and G. J. Alaerts.** 1995. Influence of reactor type and shear stress on aerobic biofilm morphology, population and kinetics. *Water Res.* **29:**107–118.
23. **Characklis, W. G.** 1990. Laboratory biofilm reactors, p. 55–89. *In* W. G. Characklis and K. C. Marshall, (ed.), *Biofilms.* John Wiley and Sons, New York, N.Y.
24. **Characklis, W. G., G. A. McFeters, and K. C. Marshall.** 1990. Physiological ecology in biofilm systems, p. 341–393. *In* W. G. Characklis and K. C. Marshall (ed.), *Biofilms.* John Wiley and Sons, New York, N.Y.
25. **Chenier, M. R., D. Beaumier, R. Roy, B. T. Driscoll, J. R. Lawrence, and C. W. Greer.** 2003. Impact of seasonal variations and nutrient inputs on the cycling of nitrogen and the degradation of hexadecane by replicated river biofilms. *Appl. Environ. Microbiol.* **69:**5170–5177.
26. **Cinar, O.** 2004. The impact of feed composition on biodegradation of benzoate under cyclic (aerobic/anoxic) conditions. *FEMS Microbiol. Lett.* **231:**59–65.
27. **Codeco, C. T., and J. P. Grover.** 2001. Competition along a spatial gradient of resource supply: a microbial experimental model. *Am. Nat.* **157:**300–315.
28. **Coombe, R. A., A. Tatevossian, and J. W. T. Wimpenny.** 1981. Bacterial thin films as in vitro models for dental plaque, *In* R. M. Frank and S. A. Leach (ed.), *Surface and Colloid Phenomena in the Oral Cavity: Methodological Aspects.* IRL Press, London, United Kingdom.
29. **Croome, R. L., and P. A. Tyler.** 1984. The microanatomy and ecology of *Chlorochromatium aggregatum* in two meromictic lakes in Tasmania, Australia. *J. Gen. Microbiol.* **130:**2717–2724.
30. **Darwin, C.** 1859. *The Origin of Species by Means of Natural Selection or the Preservation of Favoured Races in the Struggle for Life.* New American Library, New York, N.Y.
31. **Dejonghe, W., E. Berteloot, J. Goris, N. Boon, K. Crul, S. Maertens, M. Höfte, P. De Vos, W. Verstraete, and E. M. Top.** 2003. Synergistic degradation of linuron by a

bacterial consortium and isolation of a single linuron-degrading *variovorax* strain. *Appl. Environ. Microbiol.* **69:** 1532–1541.
32. **Drzyzga, O., and J. C. Gottschal.** 2002. Tetrachloroethene dehalorespiration and growth of *Desulfitobacterium frappieri* TCE1 in strict dependence on the activity of *Desulfovibrio fructosivorans*. *Appl. Environ. Microbiol.* **68:** 642–649.
33. **Elvers, K. T., K. Leeming, and H. M. Lappin-Scott.** 2002. Binary and mixed population biofilms: time-lapse image analysis and disinfection with biocides. *J. Ind. Microbiol. Biotechnol.* **29:**331–338.
34. **Emerson, D., R. M. Worden, and J. A. Breznak.** 1994. A diffusion gradient chamber for studying microbial behavior and separating organisms. *Appl. Environ. Microbiol.* **60:** 1269–1278.
35. **Feuillade, J., and M. Feuillade.** 1979. A chemostat device adapted to planktonic *Oscillatoria* cultivation. *Limnol. Oceanog.* **24:**562–574.
36. **Food and Agricultural Organization of the United Nations.** 1963. Specifications for identity and purity of food additives, p. 136. *In Food Colors*, vol. II. Food and Agricultural Organization of the United Nations, Rome, Italy.
37. **Gjaltema, A., P. A. M. Arts, M. C. M. van Loosdrecht, J. G. Kuenen, and J. J. Heijnen.** 1994. Heterogeneity of biofilms in rotating annular reactors: occurrence, structure, and consequences. *Biotechnol. Bioeng.* **44:**194–204.
38. **Gjaltema, A., and T. Griebe.** 1995. Laboratory reactors and on-line monitoring: report of the discussion session. *Water Sci. Technol.* **32:**257–261.
39. **Goel, R., T. Tokutom, H. Yasui, and T. Noike.** 2003. Optimal process configuration for anaerobic digestion with ozonation. *Water Sci. Technol.* **48:**85–96.
40. **Gottschal, J. C., and L. Dijkhuizen.** 1988. The place of continuous culture in ecological research, p. 19–49. *In* J. W. T. Wimpenny (ed.), *CRC Handbook of Laboratory Model Systems for Microbial Ecology Research*, vol. 1. CRC Press, Boca Raton, Fla.
41. **Haefele, D. M., and S. E. Lindow.** 1987. Flagellar motility confers epiphytic fitness advantages upon *Pseudomonas syringae*. *Appl. Environ. Microbiol.* **53:**2528–2533.
42. **Harder, W., J. G. Kuenen, and A. Matin.** 1977. Microbial selection in continuous culture. *J. Appl. Bacteriol.* **43:** 1–24.
43. **Hendrickx, L., M. Hausner, and S. Wuertz.** 2003. Natural genetic transformation in monoculture *Acinetobacter* sp. strain BD413 biofilms. *Appl. Environ. Microbiol.* **69:**1721–1727.
44. **Herbert, R. A.** 1988. Bidirectional compound chemostats: applications of compound diffusion-linked chemostats in microbial ecology, p. 99–115. *In* J. W. T. Wimpenny (ed.), *CRC Handbook of Laboratory Model Systems for Microbial Ecology Research*, vol. 1. CRC Press, Boca Raton, Fla.
45. **Hirsch, P.** 1984. Microcolony formation and consortia, p. 373–393. *In* K. C. Marshall (ed.), *Microbial Adhesion and Aggregation*. Springer-Verlag, New York, N.Y.
46. **Holt, J. G., and N. R. Krieg.** 1994. Enrichment and isolation, p. 179–204. *In* P. Gerhardt, R. G. E. Murray, W. A. Wood, and N. R. Krieg (ed.), *Methods for General and Molecular Bacteriology*. American Society for Microbiology, Washington, D.C.
47. **Humenik, F. J.** 1970. Respiratory relationships of a symbiotic algal-bacterial culture for wastewater nutrient removal. *Biotechnol. Bioeng.* **12:**541–560.
48. **Hunt, A. P., and J. D. Perry.** 1998. The effect of substratum roughness and river flow rate on the development of a freshwater biofilm community. *Biofouling* **12:**287–303.
49. **Imran, M., D. Jones, and H. Smith.** 2005. Biofilms and the plasmid maintenance question. *Math. Biosci.* **193:**183–204.
50. **Karthikeyan, S., G. M. Wolfaardt, D. R. Korber, and D. E. Caldwell.** 1999. Identification of synergistic interactions among microorganisms in biofilms by digital image analysis. *Int. Microbiol.* **2:**241–250.
51. **Kharazmi, A., B. Giwercman, and N. Høiby.** 1999. Robbins device in biofilm research. *Methods Enzymol.* **310:**207–215.
52. **Kieft, T. L., and D. E. Caldwell.** 1984. Chemostat and in-situ colonization kinetics of *Thermothrix thiopara* on calcite and pyrite surfaces. *Geomicrobiol. J.* **3:**217–229.
53. **Kinniment, S. L., J. W. T. Wimpenny, D. Adams, and P. D. Marsh.** 1996. Development of a steady-state oral microbial biofilm community using the constant-depth film fermenter. *Microbiology* **142:**631–638.
54. **Kolenbrander, P. E., and J. London.** 1992. Ecological significance of coaggregation among oral bacteria. *Adv. Microb. Ecol.* **12:**183–217.
55. **Korber, D. R., G. A. James, and J. W. Costerton.** 1994. Evaluation of fleroxacin activity against established *Pseudomonas fluorescens* biofilms. *Appl. Environ. Microbiol.* **60:** 1663–1669.
56. **Korber, D. R., J. R. Lawrence, and D. E. Caldwell.** 1994. Effect of motility on surface colonization and reproductive success of *Pseudomonas fluorescens* in dual-dilution continuous culture and batch culture systems. *Appl. Environ. Microbiol.* **60:**1421–1429.
57. **Korber, D. R., J. R. Lawrence, B. Sutton, and D. E. Caldwell.** 1989. Effects of laminar flow velocity on the kinetics of surface recolonization by mot^+ and mot^- *Pseudomonas fluorescens*. *Microb. Ecol.* **18:**1–19.
58. **Kuballa, J., and T. Griebe.** 1995. Sorption kinetics of tributyltin on Elbe river biofilms. *Fresenius J. Anal. Chem.* **353:**105–106.
59. **Larsen, T. A., and P. Harremoes,** 1994. Combined reactor and microelectrode measurements in laboratory grown biofilms. *Water Res.* **28:**1435–1441.
60. **Lawrence, J. R., M. Chenier, R. Roy, D. Beaumier, N. Fortin, G. D. W. Swerhone, T. R. Neu, and C. W. Greer.** 2004. Microscale and molecular assessment of the impacts of nickel, nutrients and oxygen level on river biofilm communities. *Appl. Environ. Microbiol.* **70:**4326-4339.
61. **Lawrence, J. R., D. R. Korber, and D. E. Caldwell.** 1992. Behavioral analysis of *Vibrio parahaemolyticus* variants in high- and low-viscosity microenvironments using digital image processing. *J. Bacteriol.* **174:**5732–5739.
62. **Lawrence, J. R., D. R. Korber, and D. E. Caldwell.** 1995. Surface colonization strategies of biofilm-forming bacteria. *Adv. Microb. Ecol.* **14:**1–57.
63. **Lawrence J. R., G. D. W. Swerhone, L. I. Wassenaar, and T. R. Neu.** 2005. Effects of selected pharmaceuticals on riverine biofilm communities. *Can. J. Microbiol.* **51:** 655–669.
64. **Lovanh, N., C. S. Hunt, and P. J. J. Alvarez.** 2002. Effect of ethanol on BTEX biodegradation kinetics: aerobic continuous culture experiments. *Water Res.* **36:**3739–3746.
65. **Lovelock, J.** 1999. A way of life for agnostics. *Gaia Circ.* **2:**6–9.
66. **Lovitt, R. W., and J. W. T. Wimpenny.** 1981. Physiological behavior of *Escherichia coli* grown in opposing gradients of oxidant and reductant in the gradostat. *J. Gen. Microbiol.* **127:**269–276.
67. **Manz, W., K. Wendt-Potthoff, T. R. Neu, U. Szewzyk, and J. R. Lawrence.** 1999. Phylogenetic composition, spatial structure and dynamics of lotic bacterial biofilms investigated by fluorescent in situ hybridization and confocal scanning laser microscopy. *Microb. Ecol.* **37:**225–237.

68. **Margalef, R.** 1963. On certain unifying principles in ecology. *Am. Nat.* **117:**357–373.
69. **Marvin, W., E. Brown, and R. J. C. McLean.** 1997. An inexpensive chemostat apparatus for the study of microbial biofilms. *J. Microbiol. Methods* **30:**125–132.
70. **Massieux, B., M. E. Y. Boivin, F. P. van den Ende, J. Langenskiöld, P. Marvan, C. Barranguet, W. Admiraal, H. J. Laanbroek, and G. Zwart.** 2004. Analysis of structural and physiological profiles to assess the effects of Cu on biofilm microbial communities. *Appl. Environ. Microbiol.* **70:**4512–4521.
71. **McCoy, W. F., J. D. Bryers, J. Robbins, and J. W. Costerton.** 1981. Observations of fouling biofilm formation. *Can. J. Microbiol.* **27:**910–917.
72. **McLean, R. J. C., M. Whiteley, B. C. Hoskins, P. D. Majors, and M. M. Sharma.** 1999. Laboratory techniques for studying biofilm growth, physiology, and gene expression in flowing systems and porous media. *Methods Enzymol.* **310:**248–264.
73. **Mindl, B., B. Sonntag, J. Pernthaler, J. Vrba, R. Psenner, and T. Posch.** 2005. Effects of phosphorus loading on interactions of algae and bacteria: reinvestigation of the 'phytoplankton-bacteria paradox' in a continuous cultivation system. *Aquat. Microb. Ecol.* **38:**203–213.
74. **Molin, S., and T. Tolker-Nielsen.** 2003. Gene transfer occurs with enhanced efficiency in biofilms and induces enhanced stabilization of the biofilm structure. *Curr. Opin. Biotechnol.* **14:**255–261.
75. **Monod, J.** 1949. The growth of bacterial cultures. *Annu. Rev. Microbiol.* **3:**371–394.
76. **Montgomery, L., and T. M. Vogel.** 1992. Dechlorination of 2,3,5,6 tetrachlorobiphenyl by a phototrophic enrichment culture. *FEMS Microbiol. Lett.* **94:**247–250.
77. **Muller, R. H., and W. Babel.** 2004. *Delftia acidovorans* MC1 resists high herbicide concentrations—a study of nutristat growth on (*RS*)–2–(2,4–dichlorophenoxy)propionate and 2,4–dichlorophenoxyacetate. *Biosci. Biotechnol. Biochem.* **68:**622–630.
78. **Neu, T. R., and J. R. Lawrence.** 1997. Development and structure of microbial stream biofilms as studied by confocal laser scanning microscopy. *FEMS Microbiol. Ecol.* **24:**11–25.
79. **Neut, D., E. P. de Groot, R. S. Z. Kowalski, J. R. van Horn, H. C. van der Mei, and H. J. Busscher.** 2005. Gentamicin-loaded bone cement with clindamycin or fusidic acid added: biofilm formation and antibiotic release. *J. Biomed. Mater. Res.* 73A:165–170.
80. **Nickel, J. C., I. Ruseska, J. B. Wright, and J. W. Costerton.** 1985. Tobramycin resistance of *Pseudomonas aeruginosa* cells growing as a biofilm on urinary catheter material. *Antimicrob. Agents Chemother.* **27:**619–624.
81. **Odum, E. P.** 1971. *Fundamentals of Ecology*, 3rd ed. W. B. Saunders, Philadelphia, Pa.
82. **Olapade, O. A., and L. G. Leff.** 2005. Seasonal response of stream biofilm communities to dissolved organic matter and nutrient enrichments. *Appl. Environ. Microbiol.* **71:**2278–2287.
83. **Paerl, H. W., and J. L. Pinckney.** 1995. Microbial consortia: their roles in aquatic production and biogeochemical cycling. *Microb. Ecol.* **12:**1–20.
84. **Parkes, R. J., and E. Senior.** 1988. Multistage chemostats and other models for studying anoxic ecosystems, p. 51–71. In J. W. T. Wimpenny (ed.), *CRC Handbook of Laboratory Model Systems for Microbial Ecosystems*, vol. 1. CRC Press Inc., Boca Raton, Fla.
85. **Parsek, M. R., and E. P. Greenberg.** 1999. Quorum sensing signals in development of *Pseudomonas aeruginosa* biofilms. *Methods Enzymol.* **310:**45.
86. **Perfil'ev, B. V., and D. R. Gabe.** 1969. *Capillary Methods of Investigating Microorganisms*. University of Toronto Press, Toronto, Canada.
87. **Peters, A. C., and J. W. T. Wimpenny.** 1988. A constant depth laboratory model film fermentor. *Biotechnol. Bioeng.* **32:**263–270.
88. **Rutgers, M., J. J. Bogte, A. M. Breure, and J. G. van Andel.** 1993. Growth and enrichment of pentachlorophenol-degrading microorganisms in the nutristat, a substrate concentration-controlled continuous culture. *Appl. Environ. Microbiol.* **59:**3373–3377.
89. **Schiefer, G. E., and D. E. Caldwell.** 1982. Synergistic interaction between *Anabaena* and *Zoogloea* spp. in carbon dioxide-limited continuous cultures. *Appl. Environ. Microbiol.* **44:**84–87.
90. **Senior, E., A. T. Bull, and J. H. Slater.** 1976. Enzyme evolution in a microbial community growing on the herbicide Dalapon. *Nature* **263:**476–479.
91. **Srinivasan, R., P. S. Stewart, T. Griebe, C.-I. Chen, and X. Xu.** 1995. Biofilm parameters influencing biocide efficacy. *Biotechnol. Bioeng.* **46:**553–560.
92. **Stevens, T. O., and B. S. Holbert.** 1995. Variability and density-dependency of bacteria in terrestrial subsurface samples: implication for enumeration. *J. Microbiol. Methods* **23:**283–292.
93. **Stewart, P. S., B. M. Peyton, W. J. Drury, and R. Murga.** 1993. Quantitative observations of heterogeneities in *Pseudomonas aeruginosa* biofilms. *Appl. Environ. Microbiol.* **59:**327–329.
94. **Swenson, W., J. Arendt, and D. S. Wilson.** 2000. Artificial selection of microbial ecosystems for 3-chloroaniline degradation. *Environ. Microbiol.* **2:**564–571.
95. **Swenson, W., D. S. Wilson, and R. Elias.** 2000. Artificial ecosystem selection. *Proc. Natl. Acad. Sci. USA* **97:**9110–9114.
96. **Thiele, J. H., C. M. Thartrain, and J. G. Zeikus.** 1988. Control of interspecies electron flow during anaerobic digestion: role of floc formation in syntrophic methanogenesis. *Appl. Environ. Microbiol.* **54:**10–19.
97. **Trulear, M. G., and W. G. Characklis.** 1982. Dynamics of biofilm processes. *J. Water Pollut. Control Fed.* **54:**1288–1301.
98. **Wilson, M.** 1999. Use of constant depth film fermentor in studies of biofilms of oral bacteria. *Methods. Enzymol.* **310:**264–279.
99. **Wolfaardt, G. M., J. R. Lawrence, J. V. Headley, R. D. Robarts, and D. E. Caldwell.** 1994. Microbial exopolymers provide a mechanism for bioaccumulation of contaminants. *Microb. Ecol.* **27:**279–291.
100. **Wolfaardt, G. M., J. R. Lawrence, M. J. Hendry, R. D. Robarts, and D. E. Caldwell.** 1993. Development of steady-state diffusion gradients for the cultivation of degradative microbial consortia. *Appl. Environ. Microbiol.* **59:**2388–2396.
101. **Wolfaardt, G. M., J. R. Lawrence, R. D. Robarts, and D. E. Caldwell.** 1994. The role of interactions, sessile growth and nutrient amendment on the degradative efficiency of a bacterial consortium. *Can. J. Microbiol.* **40:**331–340.
102. **Wolfaardt, G. M., J. R. Lawrence, R. D. Robarts, S. J. Caldwell, and D. E. Caldwell.** 1994. Multicellular organization in a degradative biofilm community. *Appl. Environ. Microbiol.* **60:**434–446.

Lipid Analyses for Viable Microbial Biomass, Community Composition, Metabolic Status, and In Situ Metabolism

DAVID B. HEDRICK, AARON D. PEACOCK, AND DAVID C. WHITE*

10

The particular power of lipid biomarker analysis is that lipids can be quantitatively extracted from almost any sample matrix and analyzed by the mature techniques of chromatography and mass spectroscopy, allowing the quantification of lipids representative of an entire microbial community. Accurate quantitative data allow the application of statistics to authenticate differences across an environment or between treatments. This review covers how phospholipid fatty acids (PLFA) and other lipids can be used to estimate total viable microbial biomass, community composition, and metabolic status as well as the use of in situ ^{13}C incorporation to determine metabolic pathways in environmental samples. An earlier review (52) described the interpretation of lipid biomarkers.

TOTAL VIABLE MICROBIAL BIOMASS

Total viable microbial biomass is an important property of microbial communities. The viable microbial biomass increases with the availability of metabolizable substrates and may decrease after their exhaustion. The rates of biogeochemical transformations such as methane production, carbon sequestration, or contaminant detoxification are all proportional to the viable microbial biomass.

The viable biomass is simply the total weight of the living organisms. In macroecology (nonmicrobial ecology), the technology for the enumeration of organisms can be as simple as a pen and notepad to record the number of plants or animals observed within a defined area. In microbial ecology, however, the very small size of the organisms, their huge numbers, and the relative absence of defining morphology require more sophisticated technology. The classical approaches of enumeration of the microbial community by microscopy, culture, and measures of metabolic activity all share limitations of large systematic error from underestimation and large random error (8, 144, 145).

Phospholipid Fatty Acids

All intact cells contain polar lipids, which in microbes are primarily phospholipids. With cell death, exogenous and endogenous phospholipases rapidly transform the polar lipids in the cell membranes to neutral lipid diglycerides by removing polar phosphate-containing head groups (130). Diglycerides apparently disappear in soils less rapidly than phospholipids, since diglycerides are detectible in many natural environments (141). The viable microbial biomass can be determined by quantifying organic phosphate from the polar lipid fraction of the lipid extract by using a relatively simple colorimetric analysis (34, 39, 145). The sensitivity of the classical colorimetric analysis for lipid phosphate (LP) as initially proposed for environmental samples (145) has been improved considerably with a dye-coupled reaction to sensitivities of 1 nmol of LP, corresponding to about 10^7 bacteria (34). Gas chromatography-mass spectrometry (GC-MS) of PLFA as their methyl esters provides greater sensitivity and specificity than the LP method (102) and additionally provides detailed information on the microbial community composition and metabolic status.

Phospholipid-based biomass measurements have been applied successfully to a multitude of environments. Since the lipid recovery procedure involves extensive solvent extraction and product concentration, there are few environments for which it cannot be applied. The assay even has been used to determine the biomass of microbes in sludges of petroleum storage tanks, although extra purification steps were required to remove the neutral lipid hydrocarbon components (D. B. Ringelberg, personal communication). Samples from an enormous variety of matrices have been analyzed for lipid content, including soils (12, 13, 26, 39), soil rhizosphere (131), clinical specimens (99, 102), ice cores (105), sediments (32, 34, 35, 44, 71, 106, 125, 139), subsurface materials (8, 126, 140, 141), bioprocessing units (49, 91), rocks (5), estuarine fungi (144), and marine detritus with and without grazing (94).

Biomass-to-Cell-Number Conversions

One of the challenges with biochemical biomass measures is that the results are determined as the quantity of a component (e.g., picomoles of PLFA or micromoles of LP), while microbiologists describe biomass in terms of cell abundance. A simple conversion can be made by determining the biomarker content of monocultured cells and then counting cells to determine a value per cell. There are, however, documented accuracy problems with acridine orange direct counts (34), and there is no universally applicable conversion factor for PLFA or LP per bacterial cell or the number of cells per gram (dry weight) of bacteria (14, 33, 34, 145). Bacterial biovolumes can vary over 3 orders of magnitude (from 0.001 to 7 μm^3) (44, 101) due to differences in nutritional status, growth phase, or survival strategy. In bacterial enrichments

*Deceased.

and in isolates or mixed cultures from the sea, the LP content varied from 34 to 380 μmol of LP per g of carbon for aerobic organisms, compared to 118 to 250 μmol of LP per g of carbon for anaerobic cultures (14). *Arthrobacter crystallopoietes* showed a 30% decrease in PLFA per cell after 2 weeks of starvation (76), and *Vibrio cholerae* showed up to a 99.8% decrease in PLFA per cell after 7 days of starvation (35), with loss of culturability but not membrane integrity (55). It is important to note that the number of cells per gram (dry weight) can vary by up to 1 order of magnitude (33). Brinch-Iverson and King (14) stated that the conversion factor for bacteria of 100 μmol of LP per g of carbon (100 μmol of PLFA per g [dry weight]) (with carbon as one-half the dry weight) based on earlier work (143) was reasonably applicable to sediments containing a significant proportion of anaerobes. Smaller bacteria have a higher dry weight-to-volume ratio than larger bacteria (101). These authors developed an allometric relationship between dry weight and volume, whereby biomass equals a conversion factor times the volume raised to an exponential scaling factor. Scaling factors and conversion factors are size class dependent (101). Conversion factors for eukaryotic PLFA to biomass or biomass to cell number are even more problematic. Quantifying fungal biomass based on PLFA and sterol content presents a problem since mycelia often exist as large multinucleated cells with a huge biomass, much of which is not active (see reference 38).

Converting LP or PLFA values to cell-based carbon content or cell numbers is problematic, and results should be interpreted cautiously. Since detailed information on the distribution of cell volumes and the amount of phospholipid per cell in an environmental sample are usually unknown and typically have much greater error than the PLFA measurement, precise conversion factors are unavailable. The best solution when a conversion is needed is to choose a reasonable conversion factor and state it explicitly when publishing the results. The constant of proportionality used in this laboratory is 2.5×10^4 cells/pmol of PLFA (8).

Analysis of Intact Phospholipids

Advances in MS and liquid chromatography (LC)-MS have provided the opportunity to analyze intact phospholipids, rather than lipid phosphate or phospholipid fatty acids. Elimination of the chemical degradation step, which converts intact phospholipids to fatty acid methyl esters, eliminates time and manipulation requirements and increases the sensitivity of the analysis. The quantitation of different phospholipid classes and their plasmalogen analogues has been demonstrated (15).

A system utilizing reversed-phase liquid chromatography, electrospray ionization with piperidine added post-column to enhance negative ions, and tandem quadrupole MS demonstrated a limit of detection of 20 fmol/μl and a limit of quantitation of 60 fmol/μl in selected ion monitoring mode (85). (See "Analytical Sensitivity and Limitations of Lipid Analyses" below for a description of limits of detection and quantitation.)

MICROBIAL COMMUNITY COMPOSITION

Phospholipid Fatty Acids

PLFA are found in lipids. In glycerol-based lipids, the polar lipid head group is traditionally on the third carbon (Fig. 1). The fatty acids or ether-linked alkyl groups are on carbons 1 and 2 except for sphingolipids, which can be thought of as a glycerol with an amide on carbon 2 and both a hydroxyl and an acyl group on the next carbon in the traditional carbon 1 position, which is numbered carbon 3 in these lipids (Fig. 1).

In addition to precise estimates of microbial biomass, analysis of the PLFA provides information on microbial community composition. The PLFA profile of an environmental microbial community is dictated by the proportions of different organisms present and their metabolic state, though this information cannot usually be decoded to a list of organisms and abundances (32). There is a large body of literature, heroically compiled into a two-volume compendium edited by Ratledge and Wilkinson (114), on microbial lipid composition, which can greatly facilitate community lipid profile analysis. The standard reference on lipid structures and basic analytical methods is by Kates (65). Some specific fatty acid and other lipid biomarkers for community structure are listed in Table 1. Fatty acid desig-

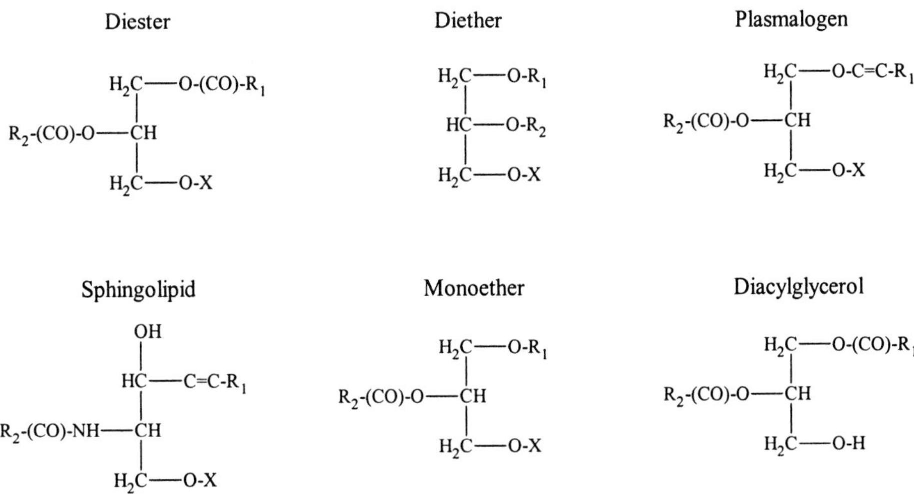

FIGURE 1 Variation in the glycerol moiety of phospholipids. The glycerol carbons are traditionally numbered with the top carbon in glycerol-based lipids as carbon 1 with the glycerol carbon attached to the polar head group of the lipid (X) as the third carbon. R_1 and R_2 are alkyl chains. For the sphingolipid, the carbon attached to the polar head group (X) is the first carbon.

TABLE 1 Examples of signature lipids and their cellular location

Genus, organism, or group	Lipid biomarker	Cellular location[a]	Reference
Archaea	Ether-linked isopranoids	PL, GL, residue	50
Actinobacteria			
Arthrobacter	a15:0	PLFA	76
Rhodococcus, *Mycobacterium*	10Me18:0	PLFA	22
Frankia	17:1ω8c	PLFA	133
Mycobacterium	Mycocerosic acids, 2OH alcohols	NL	4
Mycobacterium	2Me fatty acids	PLFA	22
Corynebacterium	C20 to C36 mycolic acids	Residue	16
Nocardia and *Rhodococcus*	C36 to C66 mycolic acids	Residue	16
Mycobacterium	C60 to C90 mycolic acids	Residue	16
Bacteroidetes/Chlorobi group			
Bacteroides and *Sphingobacterium*	*iso*-branched sphinganines	PL	148
Chloroflexi (green nonsulfur bacteria)			
Dehalococcoides	Furan fatty acids	PL	146
Cyanobacteria (blue-green algae)			
Nostoc commune	18:2ω6, 18:3ω3	PLFA	112
Firmicutes (gram-positive bacteria)			
Clostridium	Plasmalogen dimethyl acetals	PLFA	73
Acidophilic thermophilic *Bacillus*	ω-cyclohexyl fatty acids	PLFA	104
Planctomycetes			
Planctomyces	19:1ω10, 3oh20:0	PLFA and residue	69
Alphaproteobacteria			
Sphingomonas	18- and 20-carbon straight-chain sphinganines	PL	80
Betaproteobacteria			
Methanotrophs, type I	16:1ω8c, 16:1ω5c	PLFA	46
Methanotrophs, type II	18:1ω8c, 18:1ω8t	PLFA	46
Gammaproteobacteria			
Francisella tularensis	20:1ω11, 22:1ω13, 24:1ω15, 26:1ω17	PLFA	99
Legionella	3oh-i14:0, 2,3dioh-i14:0, 27oxo28:0, 27:2,3oh22:0	Residue	136
Delta/epsilon subdivision			
Desulfovibrio	i17:1ω7c	PLFA	74
Desulfobacter	10Me16:0	PLFA	74
Desulfobulbus	17:1ω6c	PLFA	74
Desulfomonile	br3oh19:0, br3oh21:0, br3oh22:0	Residue	118
Geobacter	3oh16:0, 9oh16:0, 10oh16:0	Residue	84
Thermodesulfobacteriales			
Thermodesulfotobacterium commune	*iso*- and *anteiso*-branched glycerol diethers	PL	78
Thermotogales			
Thermotoga maritima	Long-chain diacids	PL	111
Thermotoga maritima	15,16-Dimethyl-30-glyceryloxytriacontanoic acid	PL	28
Thermosipho africanus	Long-chain diacids	PL	57
Eukaryotes			
Fungi	18:2ω6, 18:3ω6, 18:3ω3	PLFA	144
Diatoms	16:1ω9c	PLFA	19
Higher plants	18:1ω11, 18:3ω3, 20:5ω3, 26:0	PLFA	133
Protozoa	20:2ω6, 20:3ω6, 20:4ω6	PLFA	138

[a]Abbreviations; NL, neutral lipids; GL, glycolipids; PL, polar lipids; residue, lipid-extracted residue. See text for fatty acid naming conventions.

nations follow the pattern "A:BωC" where A is the total number of carbon atoms in the fatty acid chain, B is the number of double bonds, and C is the position of the closest double bond to the methyl (omega [ω]) end of the molecule. The geometric configuration of the double bond is indicated by the suffix c for a *cis* or t for a *trans* double bond. For example, 16:1ω7c is a fatty acid with 16 carbons and one double bond in the *cis* configuration located 7 carbons from the ω end of the molecule. Fatty acids with methyl branches are designated by the prefix "XMe," where X represents the distance of the methyl group from the carboxylic end of the molecule (e.g., 10Me16:0), except when the methyl branch is one or two carbons from the ω end, when it is indicated by the prefix "i" for *iso* or "a" for *anteiso* branching, respectively, (e.g., i15:0). A methyl branch at an undetermined position in the molecule is indicated by the prefix "br." Cyclopropyl fatty acids are designated by the prefix "Cy" followed by the total number of carbons (e.g., Cy17:0). The position of a hydroxyl group is numbered from the carboxyl end of the fatty acid with "oh" as a prefix (e.g., 3oh16:0). A possible source of confusion is that the methyl group is included in the count of carbons for *iso*, *anteiso*, indeterminate branched, and cyclopropyl fatty acids, but not when the position of the methyl group is explicit. For example, i17:0, a17:0, br17:0, Cy17:0, and 10Me16:0 all have 17 carbons.

PLFA profiles can be treated as "fingerprints" of the microbial communities, using multivariate statistical methods such as factor analysis (52), CANOCO (129), or neural networks (3) to determine the relationships between samples. Alternatively, the PLFA composition of each sample can be summarized as the sum of fatty acids in each of several structural groups such as saturates, monounsaturates, terminally branched, midchain branched, cyclopropyl, and polyunsaturated fatty acids (see reference 52 for a discussion of how to do this).

Some PLFA are indicative of particular phylogenic groups, and considered together with the environment from which the sample was retrieved, can be helpful in interpreting results. Terminally branched saturated PLFA are common to *Firmicutes* (low-G+C gram-positive bacteria) and *Actinobacteria* (high-G+C gram-positive bacteria) but are also found in anaerobic *Proteobacteria*, such as the sulfate-reducing bacteria, and some proteobacterial facultative anaerobes (64). The anaerobic *Proteobacteria* have mostly 16-carbon and the *Actinobacteria* mostly 18-carbon terminally branched fatty acids. Branched-chain monoenoic PLFA are common in the anaerobic *Desulfovibrio*-type sulfate-reducing bacteria, both in culture and in manipulated sediments (106). Monoenoic PLFA are common in most bacteria and microeukaryotes, with the notable exception of some *Chloroflexi* (146). Specific groups of bacteria form monoenoic PLFA with the unsaturation in an atypical position, such as 18:1ω8c in the type II methane-oxidizing bacteria (46). Polyenoic PLFA are biomarkers for microeukaryotes and cyanobacteria (112, 114). Those with 18 carbons or fewer are more common in cyanobacteria (115), and those with 20 carbons or more are due to eukaryotic input (126). Recently, polyunsaturated PLFA have also been found in some Antarctic and deep-sea microorganisms (120). The fatty acid 18:2ω6 is prominent in fungi but is also found in algae, protozoa, and humans in phosphatidylcholine (PC). Polyenoic PLFA with the first unsaturation in the ω6 position are classically considered to be of animal origin, whereas organisms with the first unsaturation in the ω3 position are generally considered to be of either plant or algal origin. Normal saturated PLFA longer than 20 carbons are usually from eukaryotes. If the relative proportions of bacteria and eukaryotes are of interest, the fatty acids 16:0, 18:0, and 18:1ω9c (abundant in both bacteria and eukaryotes) may be removed from the data set prior to analysis.

The combination of lipid biomarker analysis (quantitative but not specific) and nucleic acid-based analyses (specific but not quantitative) greatly expands the specificity and scope of community compositional determinations (80).

Sterols

Sterols are very helpful in identifying and classifying microeukaryotes, especially when combined with PLFA results (125, 144). Ergosterol has been used to quantify fungi in environmental samples (96). Algae contain a diversity of sterols which have proved useful in describing taxonomic relationships (17), such as the 4-methyl sterols found in dinoflagellates (89). The identification of sterols by GC-MS has been reviewed (62).

Plasmalogen Lipids

Plasmalogen polar lipids are similar to diacyl polar lipids, except that one of the acyl groups is replaced by a vinyl ether (Fig. 1). They are major components of the polar lipids of the anaerobic *Clostridium*-like *Firmicutes* as well as minor components in other *Firmicutes* and many eukaryotes. Plasmalogen lipids have been detected in the *Clostridiales Clostridium acetobutylicum* (7), *Veillonella parvula* (61), *Megasphaera elsdenii* (61), *Selenomonas ruminantium* (63), and *Clostridium butyricum* (73); the *Lactobacillales Syntrophococcus sucromultans* (29), the *Mollicutes* member *Anaeroplasma* (79), and *Bacillales* member *Bacillus firmus* (23). Examples of bacteria in which plasmalogens were not detected include the *Alphaproteobacteria* member *Methylosinus trichosporium* (137) and the *Deltaproteobacteria* member *Desulfovibrio* (87).

A modification of the standard PLFA preparation method provides the plasmalogen side chains as dimethyl acetals, which are then determined in the same GC analysis as the fatty acid methyl esters (90). Besides its use as a community composition biomarker, the dimethyl acetal/PLFA ratio increases with increasing proportions of obligate anaerobes and anaerobic metabolism (107). Intact plasmalogen lipids have also been determined by electrospray ionization-tandem mass spectrometry (15).

Sphingolipids

Sphingolipids (Fig. 1) are found in the *α-Proteobacteria* member *Sphingomonas* (37) and in members of the *Bacteroidetes* group such as *Bacteroides* and *Sphingobacterium* (148), as well as at low levels in most eukaryotes. Detection of sphingolipids has been used to supplement PLFA and molecular data to examine the role of *Sphingomonas* in bioremediation of organic compounds in the subsurface (80).

Lipopolysaccharide Fatty Acids

Lipopolysaccharide hydroxy fatty acids are found in all bacterial phyla except the *Firmicutes* and *Actinomycetes* (33, 69) and have been used in bacterial identification, classification, and environmental studies (46, 74).

BIOMARKERS FOR MICROBIAL COMMUNITY METABOLIC STATUS

Biomarkers for the metabolic status of microbial communities include those for starvation and toxicity, unbalanced growth, and aerobic versus anaerobic growth, among others. Some practical considerations for the calculation of metabolic status biomarkers have recently been reviewed (52, 139).

Ratios of PLFA

Bacteria respond to starvation or toxic conditions by modifying their fatty acids to decrease membrane permeability. Stationary-phase growth, starvation, or exposure to solvents, alcohols, or acids causes the conversion of monoenoic PLFA to cyclopropane PLFA and the conversion of *cis* monoenoics to *trans* monoenoics (45). The *cis*→*trans* isomerization appears to be a rapid response (70), while the monoene→cyclopropyl is a more long-term adaptation. *trans*/*cis* ratios greater than 0.1 have been shown to indicate starvation in bacterial isolates (45). This value is usually 0.05 or less in healthy, nonstressed populations. The monoenoic PLFA 16:1ω7c and 18:1ω7c are increasingly converted to the cyclopropyl fatty acids Cy17:0 and Cy19:0, respectively, in *Proteobacteria* as they move from a logarithmic to a stationary phase of growth. This ratio usually falls within the range of 0.05 (log phase) to 2.5 or greater (stationary phase) (91). An increase in cyclopropyl PLFA formation has also been associated with increased anaerobic metabolism in facultative heterotrophic bacteria in monoculture studies, but Bossio and Scow (12) did not observe such a shift in agricultural fields with and without flooding.

Poly-β-Hydroxy-Alkanoates

The proportion of poly-β-hydroxyalkanoic acid (PHA) in bacteria (100) relative to PLFA provides a measure of nutritional and physiological status. Some bacteria undergo unbalanced growth and cannot divide when exposed to adequate carbon and terminal electron acceptors because of other limitations such as a lack of essential nutrients (e.g., phosphate, nitrate, and trace metals) and form PHA as a carbon storage compound. When the essential component becomes available, these bacteria catabolize PHA and form PLFA as they grow and divide. PHA/PLFA ratios can range from 0 (dividing cells) to over 40 (carbon storage). For example, the PHA/PLFA ratio in rhizosphere microbes from *Brassica napus* was <0.0001, compared with 6.6 for bacteria not associated with rootlets (131). Ratios greater than 0.2 usually indicate the beginnings of unbalanced growth in at least part of the microbial community.

Respiratory Quinones

It can be useful to determine in situ proportions of aerobic and anaerobic metabolism in a microbial community (48). In bioremediation, for example, organic contaminants are more rapidly degraded under aerobic conditions (41), while reductive dehalogenation of chlorinated ethenes (26) or immobilization of U^{6+} (20) requires anaerobic conditions. Bacteria conserve energy by three processes: photosynthesis, respiration, or fermentation. Photosynthesis can be detected by the use of unique quinones (151), the type of respiration can be detected by the amount and type of respiratory quinones, and fermentation can be detected by the lack of respiratory quinones (48). Respiratory quinones shuttle electrons between components of the electron transport chain within the cell membrane. Ubiquinones (UQ) are biomarkers for oxic respiration that mediate electron transport to oxygen and nitrate, while menaquinones (MK) can carry electrons to any electron acceptor (oxygen, nitrate, sulfate, iron, manganese, carbon dioxide, etc.). However, MK are less efficient than UQ at energy conservation when using oxygen, so bacteria utilizing UQ have an energetic advantage under oxic conditions. Bacteria fixing energy by fermentation do not make respiratory quinones. Since respiratory quinones are metabolically expensive, they are synthesized when and as needed, and their abundance reflects the dominant energy conservation process in situ (48).

Respiratory quinone profiles can be summarized by two variables, UQ/(UQ+MK) and (UQ+MK)/PLFA (Fig. 2) (48). UQ/(UQ+MK) decreases with decreasing redox level of the microbial environment but may not distinguish anaerobic respiration of sulfate or iron from samples in which methanogenesis and fermentation dominate. The variable (UQ+MK)/PLFA is used to distinguish respiring communities from those in which fermentation dominates. It does, however, often increase slightly from the oxic horizon to regions dominated by anaerobic respiration. Since there is less energy available per electron from anaerobic respiration than oxygen respiration, it requires more MK to support a given amount of biomass.

The respiratory quinone profiles also contain useful community structure information (54), and the quinone composition of many bacteria has been reviewed (24). Besides the major respiratory quinones UQ and MK, *Bacteria* and *Archaea* also contain desmethylmenaquinones, methionaquinones, plastoquinones, rhodoquinones, and caldariellaquinones, which can also vary in the length of the side chain and, in the case of MK, in the degree of hydrogenation of the side chain. The first analysis of respiratory quinones from environmental samples was done by liquid chromatography with electrochemical detection (48), which was difficult to maintain for routine analyses. Samples with sufficient biomass can be analyzed for quinones by LC with UV-visible detection (54). Recently, LC or flow injection analysis coupled to tandem MS has been applied to the analysis of respiratory quinones from environmental samples (42).

Diglycerides

When cells rupture and die, phospholipids are attacked by phosphatases, releasing diglycerides, which are not present in the membranes of living cells. A rapidly growing microbial community shows a diglyceride fatty acid/PLFA ratio of close to zero, while in less ideal environments the ratio can exceed 3 (141). The diglyceride-to-PLFA ratio in a subsurface sediment increased from <0.2 at the surface to over 2.0 at approximately 200 m of depth (141).

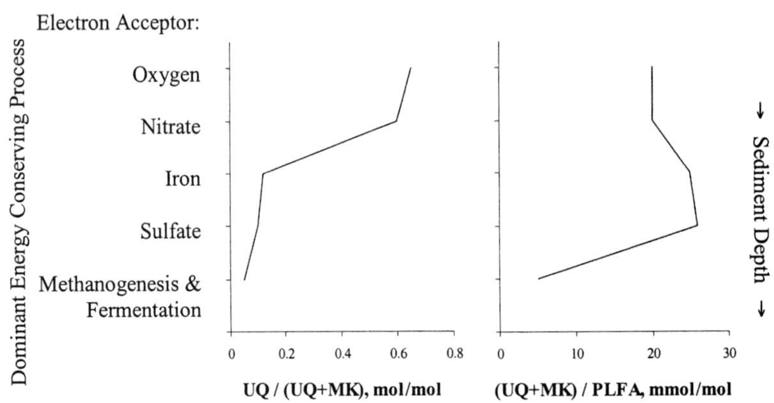

FIGURE 2 Expected responses of the biomarkers UQ/(UQ+MK) and (UQ+MK)/PLFA in an idealized sedimentary transect. Adapted from reference 48.

Other Biomarkers of Microbial Metabolic Status

A wide variety of additional biomarkers have been proposed to measure particular characteristics of microbial communities. For example, several bacteria have been shown to form acyl-ornithine lipids in lieu of phospholipids when growing under phosphate limitation including the α-proteobacteria *Paracoccus denitrificans* (150), *Brucella abortus*, and *Brucella melitensis* (93), *Hyphomicrobium vulgare* (9), and *Rhodopseudomonas viridis* (82); the β-proteobacterium *Burkholderia cepacia* (68); the γ-proteobacteria *Pseudomonas stutzeri* and *Pseudomonas aeruginosa* (68) and *Pseudomonas fluorescens* (92); and the actinobacterium *Actinomyces olivaceus* (75). Other bacteria shown to contain ornithine or other amino acid-linked lipids include the α-proteobacterium *Mesorhizobium* (21); the β-proteobacteria *Thiobacillus thiooxidans* (72) and *Bordetella pertussis* (66); the γ-proteobacterium *Pseudomonas cepacia* (25); the δ-proteobacterium *Desulfovibrio gigas* (88); the actinobacterium *Mycobacterium* sp. (77); and the bacteroidete *Flavobacterium* sp. (67). The wide phylogenic distribution of this metabolic adaptation makes it a potentially valuable biomarker for phosphate limitation. Limitation in bioavailable phosphate could be detected by the shift from phospholipids to ornithine lipids directly by LC-MS, or by a reduction in the ratio of polar lipid phosphate to polar lipid fatty acids.

In soils and sediments in certain situations, aerobic versus anaerobic metabolism can be estimated from the ratio biomarker lipids such as *iso*-branched to *anteiso*-branched saturated PLFA. Actinobacterial aerobes such as *Arthrobacter* and *Micrococcus* have i17:0/a17:0 ratios of approximately 0.2 (76), whereas proteobacterial anaerobes such as *Desulfovibrio* have i17:0/a17:0 ratios of greater than 5 (33). Other biomarkers for aerobes include 10Me18:0 for *Actinomycetes* (22) and polyunsaturates for eukaryotes (138, 144). The biomarker for methanotrophs 18:1ω8c (46) is often elevated at the aerobe-anaerobe interface. Straight-chain sphinganines are biomarkers for aerobic *Sphingomonas* (80), while branched-chain sphinganines may be due to anaerobic *Bacteroides* (148).

While many bacteria store carbon as PHA (see "Poly-β-Hydroxy-Alkanoates" above), many microeukaryotes store carbon as triglycerides (40), which can be determined as the ester-linked fatty acids in the neutral lipid fraction.

Exposure of bacteria to hypochlorite (the "chlorine" in drinking water and swimming pools) transforms monounsaturated fatty acids to oxirane fatty acids by addition of an oxygen atom across the fatty acid double bond (124). Oxirane fatty acids have been proposed as biomarkers for hypochlorite exposure.

ISOTOPICALLY LABELED LIPIDS TO TRACE IN SITU METABOLISM

Tracing In Situ Metabolism by the Fractionation of Stable Isotopes in Lipids

Stable isotopes such as ^{2}H, ^{13}C, ^{15}N, ^{18}O, and ^{34}S have long been useful research tools in metabolic studies, ecology (119), and archaeology. If a chemical bond to carbon is broken or formed in the rate-limiting step of a metabolic pathway, then the pathway will discriminate against the heavier isotope of carbon. Diffusion and evaporation can also discriminate against the heavier isotope. This allows the fate of atoms to be followed through complex systems. Recently, isotope ratio MS (IR-MS) has been developed, a method whereby organic compounds are separated by GC and catalytically combusted to CO_2 and the abundances of $^{13}CO_2$ and $^{12}CO_2$ are determined for each chromatographic peak (81).

In order to be able to use the levels of ^{13}C enrichment in lipids to identify which microbial group is represented by that lipid, many specific microbes must be examined in pure culture for their ^{13}C discriminations. In one study (1), 7 bacterial and 10 fungal strains (all aerobic heterotrophs) were fed defined or complex media, and the $^{13}C/^{12}C$ discrimination of several fatty acids in the polar lipid, glycolipid, and bound lipid fractions were determined. Differences in $\delta^{13}C$ were found for different organisms and substrates and for different fatty acids within an organism and growth substrate, but no dependence on growth phase was found. The $\delta^{13}C$ values for four species of sulfate-reducing bacteria grown under autotrophic, mixotrophic, or heterotrophic conditions have been reported (83). Anaerobic ammonium oxidizers were found to have a very large discrimination against ^{13}C of 47 per mille versus CO_2 (123). The variations in carbon isotopic signatures of methanotrophs with substrate and environmental conditions have been examined (59). The $\delta^{13}C$ values for lipids of estuarine plants (18), of marine extreme piezophiles (31), of a marine bacterium under oxic versus anoxic conditions (128), of a cyanobacterium (121), of various algae (122, 127), of basidiomycetes (53), of the coccolithophorid *Emiliania huxleyi* (116), of the thermophiles *Thermotoga maritima* and *Persephonella marina* (155), of the iron reducers *Geobacter metallireducens* and *Shewanella algae* (154), and of green and purple sulfur bacteria (134) have also been determined.

This technique has been applied to many ecosystems; a few examples follow. A suite of methods including IR-MS of fatty acids and alcohols was used to determine that the green sulfur bacterial epibionts of the phototrophic consortia "*Pelochromatium roseum*" fixed carbon photoautotrophically (43). PLFA analysis and IR-MS of PLFA documented different microbial communities and metabolisms along a transect leading downstream of the outflow from a spring with travertine deposits (152). $\delta^{13}C$ analysis of fatty acids, sterols, and hopanols was used to establish that marine cold seep mussels obtained fixed carbon from methanotrophic bacteria (58), while fatty acid depletion was used to determine that the hydrothermal vent mussel *Bathymodiolus* depends on both methane and sulfide oxidation (110). At another cold seep gas hydrate site, sulfate-reducing bacteria were shown to oxidize methane and *Beggiatoa* was shown to have grown by consumption of crude oil hydrocarbons (153). The metabolic roles of cyanobacteria and green nonsulfur bacteria in hot spring microbial mats have been examined (135). Carbon transfer from four plant species to the arbuscular mycorrhizal fungus *Glomus intraradices* was estimated (103). The dynamics of incorporation of urea carbon into microbial lipids in cattle urine patches has been examined (109).

Stable Isotope-Labeled Substrates To Trace In Situ Metabolism

IR-MS can be utilized to measure the incorporation of label into cellular components when supplied with a ^{13}C-labeled substrate (6). This provides incontrovertible evidence of biodegradation. Detection limits and carryover effects of the method have been considered (95). Incorporation of ^{13}C into nucleic acids (20) has also been reviewed (113).

^{13}C-labeled methane incorporation into microcosms has been used to determine the activities, distributions, and type of methanotrophs in landfill soil caps (27). In a soil microcosm experiment (47), it was found that ^{13}C-labeled toluene was incorporated into only 16 of the 59 PLFA detected and that the PLFA enriched in ^{13}C matched the profile of a toluene-degrading bacterium isolated from that environment. IR-MS in conjunction with several other methods was able to show that phenanthrene was degraded by *Sphingomonas* and an unidentified β-proteobacterium in two industrial soils, by an actinomycete in a roadside soil, and by an unknown bacterium in an agricultural soil (60). *Mycobacterium frederiksbergense* incorporated ^{13}C-anthracene into cellular lipids in preference to glucose, PLFA profiles responded to substrate and growth phase much more than glycolipid fatty acids, and the hydrophobicities of the mycolic acids increased when grown on anthracene (149).

In order to determine actual in situ activities, it is necessary to add the ^{13}C-labeled substrate and later recover a sample of the microbiota with minimal disturbance to the microbial community. For this purpose, the biotrap sampling device can be loaded with ^{13}C-labeled substrate, incubated in an aqueous environment such as down a subsurface monitoring well, and directly analyzed for microbial lipids (107, 147). A biotrap consists of approximately 50 Bio-Sep beads in a Teflon container with holes to allow free contact with groundwater. Bio-Sep beads are made of 25% aramid polymer (Nomex) and 75% particulate-activated carbon. They are surrounded by a layer of plastic similar to an ultrafiltration membrane with pores of 1 to 10 μm diameter and have an internal surface area of approximately 600 m^2/g. Biotraps have been used to determine that benzene and toluene are degraded by different microbial communities in the same aquifer (41).

PRACTICAL CONSIDERATIONS AND PROTOCOLS

Sample Handling

To minimize sampling artifacts, microbial activity must be stopped immediately upon sample collection. It is best to freeze environmental samples as quickly as possible, preferably at −20 to −80°C. Then, the samples should be lyophilized (freeze-dried) to further inhibit metabolism and stored under nitrogen in the dark if they are not processed immediately. Once lyophilized, dry weight can be determined. If cells have been grown in culture, the media should be centrifuged and the resulting cell pellet should be rinsed twice with 0.05 M phosphate buffer (pH 7.5) before lyophilization. In the field, samples should not be held on ice any longer than absolutely necessary, since rock samples held at 4°C showed rapid and significant changes in biomass and community composition (5). Dry ice is satisfactory for holding frozen samples. Preserving samples with buffered formaldehyde or gluteraldehyde is not as satisfactory as frozen storage because these preservatives can damage some of the less stable unsaturated lipids.

PERFORMING THE SIGNATURE LIPID BIOMARKER ANALYSIS

Meticulous technique must be practiced to ensure contaminant-free analyses. Lipids from fingers, hair, stopcock grease, oils, and paint are all potential contaminants. Plasticware cannot be used in lipid analysis. The one exception is the Teflon liner of screw caps after they have been rinsed with acetone. Glassware must be cleaned scrupulously. Once glassware is used, it is immediately fully immersed in a washtub full of hot water and detergent. The cleaning process is sufficiently effective that phosphate-containing detergent can be used. The glassware is scrubbed with a brush and rinsed five times each with cold tap water and then deionized water. Glassware is dried completely before being wrapped in aluminum foil and heated in a clean muffle furnace for a minimum of 4 h at 450°C. It is best to load the muffle furnace at the end of the day and have it turned off by an automatic timer. This allows the furnace to slowly cool down, avoiding thermal stress of the glassware, which can cause even Pyrex to crumble in your hands. New disposable glassware, such as pipettes and silicic acid columns, need not be washed before baking in the muffle furnace. If glass wool is used to trap silicic acid in a chromatography column, it is convenient to install it before firing the columns. No materials other than fired glass and acetone-rinsed Teflon may come into contact with lipid solvents or samples.

Samples can be extracted at room temperature but should be protected from light, especially fluorescent light, if photosensitive lipids (such as quinones) are being analyzed. The extractant consists of a single-phase chloroform-methanol mixture (1:2, vol/vol), after the method of Bligh and Dyer (10), which can be modified to accept a phosphate buffer (145). Investigation has shown that buffer addition can increase the recovery of PLFA from soils with high clay content (39). Recovery standards (e.g., 50 pmol of C19:0 [phosphatidylcholine nonadecanoate] or 23:0 methyl-ester per μl, which are rarely found in natural samples) can be added to samples prior to lipid extraction to estimate lipid recovery efficiency. Solvent blanks must be processed with each sample set. Samples can be extracted in glass centrifuge bottles and then centrifuged at 6,000 × *g* for 30 min. The liquid phase is then decanted into a separatory funnel for addition of buffer and chloroform to force phase separation. Generally, the phase separation is cleaner and the recovery of lipid is higher if solids are removed before phase separation. With most sediments and sufficient Bligh-Dyer extractant volume, it is usually not necessary to wash the sediment for a quantitative recovery. To isolate the lipid fraction of the extracted sample, equal volumes of chloroform and distilled water (or buffer) are added and the sample is shaken. Two phases form, with the aqueous over the organic, which is then centrifuged or allowed to separate passively overnight. The lower organic phase (containing the bacterial lipids) is collected after being filtered through a fluted Whatman 2V filter that has been preextracted with $CHCl_3$. The solvent phase is removed by rotary evaporation at 37°C.

The dried total lipid extract is dissolved in a minimum volume of chloroform and transferred to a silicic acid column and separated into neutral lipid, glycolipid, and polar lipid fractions (44) by elution with solvents of increasing polarity: chloroform, and then acetone, and then methanol, respectively. The neutral lipid fraction is analyzed for lipids such as free fatty acids, sterols, respiratory quinones, triglycerides, and diglycerides (71, 125). The glycolipid fraction can be analyzed for PHA (100). The polar lipid fraction, containing the phospholipids, can be analyzed in one of two ways depending on the purpose of the analysis. High-performance LC-electrospray ionization (ESI)-MS analysis of intact phospholipids (without further lipid processing) gives increased sensitivity over GC-MS analysis of fatty acid methyl esters. However, HPLC-ESI-MS analysis cannot provide the detailed structural information obtained by GC-MS, as described below, by HPLC-ESI-MS-MS is used. For a detailed structural analysis, the phospholipid fraction can be processed to obtain fatty acid cleaved from the glycerol phosphate head group. To do this, the polar lipid fraction is subjected to a transesterification by a mild alkaline methanolysis protocol (45) resulting in fatty acid methyl esters, which are then separated, quantified, and tentatively identified by capillary GC. Individual components can then be definitively identified by their mass spectra. Monoenoic PLFA double-bond positions are determined by GC-MS analysis of dimethyl disulfide adducts (97). More-detailed procedural information is presented in reference 142.

The lipopolysaccharide hydroxy fatty acids from the lipid A of gram-negative bacteria can be recovered from the lipid-extracted residue. This residue is hydrolyzed in acid, and the lipid components released by the hydrolysis are reextracted. After centrifugation at 6,000 × *g* for 30 min, the chloroform phase is recovered, evaporated to dryness, and methylated using "magic" methanol (methanol:chloroform:concentrated HCl; 10:1:1, vol/vol/vol) (99). The hydroxy-fatty acid methyl esters are recovered, and the solvent is removed under a stream of nitrogen. They can be purified by thin-layer chromatography (developed in hexane:diethyl ether [1:1, vol/vol]), recovered in chloroform:methanol (1:1, vol/vol), and then derivatized with *bis*(trimethylsilyl)trifluoroacetamide (BSTFA) prior to GC-MS analysis.

A modification to the standard lipid purification procedure useful for low biomass samples is described by Tunlid et al. (132). Much less solvent is used, and the samples are fractionated using very small "champagne" columns (Supelco, Bellefonte, Pa.). The samples were analyzed with positive ion chemical ionization mass spectrometry.

The percent recovery of 19:0 should be calculated when a PC standard is added, but generally the final amount of phospholipid in a sample is not adjusted to reflect that recovery. The reason for this is that PC is the most difficult phospholipid to recover and so correcting for its loss would cause other phospholipid components to be overestimated. The internal standards are best used as indicators of problems with a particular sample in a set or with a particular sample matrix.

The diversity of a community is a basic concept from macroecology (86). In order to be able to apply this concept to microbial communities, a method has been developed to calculate measures of microbial community diversity from PLFA data (51).

Once compiled, PLFA and other lipid profiles can be entered into a spreadsheet and subjected to statistical analysis. In addition to analysis of variance, two other multivariate statistical applications have grown in popularity to investigate similarities between PLFA profiles. In one approach, dendrograms from a hierarchical cluster analysis are constructed from arcsine [square root(X)]-transformed PLFA mole percent values, with similarities based on modified Euclidean distances (117). In a second approach, two-dimensional plots are generated from a principal-components analysis; they not only quantify the similarities between profiles but also identify which PLFA are responsible for the clustering (i.e., coefficients of loadings) (118, 140, 141). An adaptation of this approach is canonical correspondence analysis (129), applications of which have recently been described in references 13 and 108. Artificial neural networks have also been used to analyze signature lipid biomarker data to better effect than the linear technique of principal-components analysis (3). Artificial neural network analysis corrects for biases introduced by interdependencies among individual lipids. A neural network must be trained with a fraction of an original data set, and then the trained network can be used to analyze additional data sets, provided the input variables do not change.

Validation

The use of signature lipid biomarker analysis in determining the in situ viable microbial biomass, community composition, and nutritional-physiological status has been validated in a series of experiments (138). The induction of microbial community compositional shifts by altering the microenvironment resulted in changes that were often predictable, given past experience with microbial communities. For example, seawater biofouling communities were manipulated by incubation at altered pH and with antibiotics and specific nutrients in order to produce a community dominated by fungi, while other conditions resulted in a community dominated almost exclusively by bacteria (144). Similar experiments showed that light induced the expected changes in signature lipid biomarkers and in terminal electron acceptors (11). A second type of validation came from the isolation of a specific organism or groups of organisms with the subsequent detection of the same organisms by signature lipid analysis in consortia under conditions where their growth was induced. It was possible to induce a "crash" in methanogenesis in a bioreactor by inducing the growth of sulfate-reducing bacteria (91) or by adding traces of chloroform or oxygen (49). These crashes were accompanied by shifts in the signature lipid biomarkers that correlated with the changes in the microbial populations. Specific methane-oxidizing populations (98) or propane-oxidizing actinomycetes (117) can be induced to grow by the addition of appropriate substrates. Again, all of these community shifts were detected by measurable changes in patterns of lipid signatures. A third type of validation came from the induction of shifts in microbial community nutritional status by generating conditions of unbalanced growth in which cell growth, but not cell division, was possible. This was accomplished by chelating trace metals in the presence of tannins on epiphytic microbiota (96) and by disturbing anaerobic sediments with oxygenated seawater (35). Under these conditions, the ratio of PHA to PLFA biosynthesis increased dramatically, just as it does in monocultures of appropriate bacteria under laboratory conditions. A fourth type of validation was the detection of specific shifts in microbial communities as a result of specific grazing by predators. The sand dollar *Mellita quinquiesperforata* was shown to selectively remove nonphotosynthetic microeukaryotes from sandy sediments, by examination of the morphology of the organisms in its feeding apparatus and the signature lipid biomarker patterns of the sediment before and after grazing by the echinoderm (36). Another example involved the amphipod *Gammarus mucronatus*, which exhibited a relatively nonspecific grazing of the estuarine detrital microbiota. This organism removed the microeukaryotes that were then replaced, to a large extent, by bacteria (94). Results of the signature lipid biomarker analysis agreed with the cellular morphologies present as shown by scanning electron microscopy.

Analytical Sensitivity and Limitations of Lipid Analyses

Limits of detection and quantification depend on lipid recovery, the amount of background lipid contamination (in solvents, on glassware, etc.), and on the sensitivity of the analytical equipment. Measuring microbial biomass with a colorimetric analysis of phospholipid-derived organic phosphate is straightforward and requires little specialized equipment other than a spectrophotometer (145). However, this analysis is relatively insensitive, having limits of detection in the micromolar range or approximately 10^{10} bacteria with the stable colorimetric analysis (145) or approximately 10^7 bacteria the size of *Escherichia coli* with the dye-coupled assay (34).

Conditions under which the sensitivity of biomass detection is defined must be carefully considered, especially when the chromatographic detector is a mass spectrometer. The first consideration is that the signal-to-noise ratio must be >3 for the response to be acceptable. Sensitivity is defined in terms of limits of detection (LOD), which is the mean value of the background divided by three times the standard deviation, and limits of quantitation (LOQ), which is the mean value of the background divided by ten times the standard deviation (2). With MS, the full scan of target m/z (mass/charge) values gives less sensitivity (but a greater number of ions detected) than selected ion monitoring of one or a few selected ions. The longer the MS can have to analyze at a given m/z (dwell time), the more accurate the response becomes and the greater is the sensitivity in the signal response. Full scanning decreases the duty cycle time devoted to each ion. The time for MS analysis of a component is limited by the width of the chromatographic peak. Usually, only the peak is analyzed, and the shoulders of that peak are used to correct for the baseline drift or to detect overlapping components eluting at the same time.

Sample ionization during MS analysis also affects sensitivity. Electron impact ionization produces many fragments, which decreases the sensitivity as the biomass component is divided into many fragments. A full scan with ions from $m/z = 50$ to above the molecular ion of $m/z = 312$ has an LOQ at least 50 times greater than the selected ion monitoring. PLFA analysis using GC-chemical ionization mass spectrometry, which is a much softer ionization, produces far fewer ion fragments, thereby increasing the LOQ. In-source collisional activated dissociation leads to the formation of fatty acid fragment ions in the detector in negative ion mode. With HPLC-ESI-MS-MS, the precursor ions of each phospholipid molecule species can be selected in the first quadrupole for fragmentation in the collisional activated dissociation, and then the product ions of that specific progenitor can be scanned in the third quadrupole. With tandem quadrupole mass spectrometry, the LOQ can be 50 times lower than with the single quadrupole system.

For practical reasons, though, LODs and LOQs for environmental samples are necessarily much greater than for pure lipids used for establishing analytical equipment LODs and LOQs. First, lipid recovery from environmental samples depends on the sample matrix. Secondly, Tunlid et al. (132) demonstrated that recovering lipids from low-biomass samples is inherently difficult. This is in part due to lipids adsorbing onto glassware and the silica gel during sample processing. Thirdly, if the environmental sample has a very low bacterial content, then a large amount of matrix needs to be processed and that requires a large volume of extractant solvents, increasing the amount of contamination. Contaminants in purified solvents for GC-MS analysis procedures can be between 4 and 20 pmol of PLFA (132). Also, in PLFA analysis one of the important sources of contamination is sodium hydroxide. The contamination is almost exclusively in the PC, which is relatively rare in bacteria. One of the advantages in terms of sensitivity of HPLC-ESI-MS is that the fatty acids can be examined in specific lipids so that phosphatidylethanolamine and phosphatidylglycerol, which are much less likely to be from eukaryotic contamination of the solvents, can be measured. In addition, there is more specificity in analysis of each phospholipid than in the total patterns for all the phospholipids as the GC-MS method requires. Also, the HPLC- based methods do not require derivatization by transmethylation. HPLC-ESI-MS-MS magnifies these advantages of specificity and reduction of chemical noise for even greater sensitivity.

Currently, GC-MS LODs are approximately 10^5 bacterial cells and HPLC-MS LODs are approximately 10^4 bacterial cells (30). HPLC-MS LOQs are approximately 10^5 cells. For bacterial PLFA analysis, approximately 10^7 to 10^9 bacterial cells are sufficient to obtain results that include some rare PLFA with good signal-to-noise ratio during GC or GC-MS analysis.

We acknowledge Cory A. Lytle and Margaret Gan (Center for Biomarker Analysis, The University of Tennessee) for providing the LOQ and LOD data reported herein.

This work was supported by the Department of Energy, DEFG02-04ER63939, and the Department of Energy (Prime Sponsor) through Oregon State University, sub award G0089A-B, Office of Science, Environmental Remediation Science Program.

REFERENCES

1. **Abraham, W.-R., C. Hesse, and O. Pelz.** 1998. Ratios of carbon isotopes in microbial lipids as an indicator of substrate usage. *Appl. Environ. Microbiol.* **64:**4202–4209.
2. **ACS Committee on Environmental Improvement and Subcommittee on Environmental Analytical Chemistry.** 1980. Guidelines for data acquisition and data quality evaluation in environmental chemistry. *Anal. Chem.* **52:** 2242–2249.
3. **Almeida, J. S., M. T. Barreto Crespo, J. J. Figueiredo Marques, P. A. Noble, S. J. Macnaughton, J. R. Stephen, D. C. White, and M. J. T. Carrondo.** 1999. *Microbial Typing for Management of Remediation in Contaminated Soils. Proceedings of the African International Environmental Protection Symposium (AIEPS-99).*
4. **Alugupali, S., M. K. Sikka, L. Larsson, and D. C White.** 1998. Gas chromatography-mass spectrometry methods for the analysis of mycocerosic acids present in Mycobacterium tuberculosis. *J. Microbiol. Methods* **31:**143–150.
5. **Amy, P. A., D. L. Halderman, D. Ringelberg, and D. C. White.** 1994. Changes in bacteria recoverable from subsurface volcanic rock samples during storage at 4°C. *Appl. Environ. Microbiol.* **60:**2679–2703.
6. **Arao, T.** 1999. In situ detection of changes in soil bacterial and fungal activities by measuring ^{13}C incorporation into soil phospholipid fatty acids from ^{13}C acetate. *Soil Biol. Biochem.* **31:**1015–1020.
7. **Baer, S. H., H. P. Blaschek, and T. L. Smith.** 1987. Effect of butanol challenge and temperature on lipid composition and membrane fluidity of butanol-tolerant *Clostridium acetobutylicum*. *Appl. Environ. Microbiol.* **53:**2854–2861.
8. **Balkwill, D. L., F. R. Leach, J. T. Wilson, J. F. McNabb, and D. C. White.** 1988. Equivalence of microbial biomass measures based on membrane lipid and cell wall components, adenosine triphosphate, and direct counts in subsurface sediments. *Microb. Ecol.* **16:**73–84.
9. **Batrakov, S. G., and D. I. Nikitin.** 1996. Lipid composition of the phosphatidylcholine-producing bacterium *Hyphomicrobium vulgare* NP-160. *Biochim. Biophys. Acta* **26:**129–137.
10. **Bligh, E. G., and W. J. Dyer.** 1959. A rapid method of total lipid extraction and purification. *Can. J. Biochem. Physiol.* **31:**911–917.

11. **Bobbie, R. J., J. S. Nickels, G. A. Smith, S. D. Fazio, R. H. Findlay, W. M. Davis, and D. C. White.** 1981. Effect of light on biomass and community structure of estuarine detrital microbiota. *Appl. Environ. Microbiol.* **42:** 150–158.
12. **Bossio, D. A., and K. M. Scow.** 1998. Impacts of carbon and flooding on soil microbial communities: phospholipid fatty acid profiles and substrate utilization patterns. *Microb. Ecol.* **35:**265–278.
13. **Bossio, D. A., K. M. Scow, N. Gunapala, and K. J. Graham.** 1998. Determinations of soil microbial communities: effects of agricultural management, season, and soil type on phospholipid fatty acid profiles. *Microb. Ecol.* **36:** 1–12.
14. **Brinch-Iverson, J., and G. M. King.** 1990. Effects of substrate concentration, growth state, and oxygen availability on relationships among bacterial carbon, nitrogen and phospholipid phosphorous content. *FEMS Microbiol. Ecol.* **74:**345–356.
15. **Brügger, B., G. Erben, R. Sandhoff, F. T. Wieland, and W. D. Lehmann.** 1997. Quantitative analysis of biological membrane lipids at the low picomole level by nano-electrospray ionization tandem mass spectrometry. *Proc. Natl. Acad. Sci. USA* **94:**2339–2344.
16. **Butler, W. R., D. G. Ahearn, and J. O. Kilburn.** 1986. High-performance liquid chromatography of mycolic acids as a tool in the identification of *Corynebacterium*, *Nocardia*, *Rhodococcus*, and *Mycobacterium* species. *J. Clin. Microbiol.* **23:**182–185.
17. **Canuel, E. A., J. E. Cloen, D. B. Ringelberg, J. B. Guckert, and G. H. Rau.** 1995. Molecular and isotopic tracers used to examine sources of organic matter and its incorporation into the food webs of San Francisco Bay. *Limnol. Oceanogr.* **40:**67–81.
18. **Canuel, E. A., K. H. Freeman, and S. G. Wakeham.** 1997. Isotopic compositions of lipid biomarker compounds in estuarine plants and surface sediments. *Limnol. Oceanogr.* **42:**1570–1583.
19. **Cerón García, M. C., J. M. Fernández Sevilla, F. G. Acién Fernández, E. Molina Grima, and F. García Camacho.** 2000. Mixotrophic growth of *Phaeodactylum tricornutum* on glycerol: growth rate and fatty acid profile. *J. Appl. Phycol.* **12:**239–248.
20. **Chang, Y.-J., P. E. Long, R. Geyer, A. D. Peacock, C. T. Resch, K. Sublette, S. Pfiffner, A. Smithgall, R. T. Anderson, H. A. Vrionis, J. R. Stephen, R. Dayvault, I. Ortiz-Bernad, D. R. Lovley, and D. C. White.** 2005. Microbial incorporation of ^{13}C labeled acetate at the field scale: detection of microbes responsible for reduction of U(VI). *Environ. Sci. Technol.* **39:**9039–9048.
21. **Choma, A., and I. Komaniecka.** 2002. Analysis of phospholipids and ornithine-containing lipids from *Mesorhizobium* spp. *Syst. Appl. Microbiol.* **25:**326–331.
22. **Chou, S., P. Chedore, and S. Kasatiya.** 1998. Use of gas chromatographic fatty acid and mycolic acid cleavage product determination to differentiate among *Mycobacterium genavense*, *Mycobacterium fortuitum*, *Mycobacterium simiae*, and *Mycobacterium tuberculosis*. *J. Clin. Microbiol.* **36:**577–579.
23. **Clejan, S., A. A. Guffanti, M. A. Cohen, and T. A. Krulwich.** 1989. Mutation of *Bacillus firmus* OF4 to duramycin resistance results in substantial replacement of membrane lipid phosphatidylethanolamine by its plasmalogen form. *J. Bacteriol.* **171:**1744–1746.
24. **Collins, M. D., and D. Jones.** 1981. Distribution of isoprenoid quinone structural types in bacteria and their taxonomic implication. *Microbiol. Rev.* **45:**316–354.
25. **Cox, A. D., and S. G. Wilkinson.** 1989. Polar lipids and fatty acids of *Pseudomonas cepacia*. *Biochim. Biophys. Acta* **1001:**60–67.
26. **Cox, E. E., D. W. Major, D. W. Acton, T. J. Phelps, and D. C. White.** 1994. Evaluating trichloroethylene biodegradation by measuring the in situ status and activities of microbial populations, p. 37–49. *In* R. E. Hinchee, A. Leeson, L. Semprini, and S. K. Ong (ed.), *Bioremediation of Chlorinated Polycyclic Aromatic Compounds*. Lewis Publishers, Ann Arbor, Mich.
27. **Crossman, Z. M., F. Abraham, and R. P. Evershed.** 2004. Stable isotope pulse-chasing and compound specific stable carbon isotope analysis of phospholipid fatty acids to assess methane oxidizing bacterial populations in landfill cover soils. *Environ. Sci. Technol.* **38:**1359–1367.
28. **De Rosa, M., A. Gambacorta, R. Huber, V. Lanzotti, B. Nicolaus, K. O. Stetter, and A. Trincone.** 1988. A new 15,16-dimethyl-30-glyceryloxytriacontanoic acid from lipids of *Thermotoga maritima*. *J. Chem. Soc. Chem. Comm.* **19:**1300–1301.
29. **Doré, J., and M. P. Bryant.** 1989. Lipid growth requirement and influence of lipid supplement on fatty acid and aldehyde composition of *Syntrophococcus sucromultans*. *Appl. Environ. Microbiol.* **55:**927–933.
30. **Fang, J., and M. J. Barcelona.** 1998. Structural determination and quantitative analysis of bacterial phospholipids using liquid chromatography/electrospray ionization/mass spectrometry. *J. Microbiol. Methods* **33:**23–35.
31. **Fang, J., M. J. Barcelona, T. Abrajano, Y. Nogi, and C. Kato.** 2002. Isotopic composition of fatty acids of extremely piezophilic bacteria from the Mariana Trench at 11,000 m. *Mar. Chem.* **80:**1–9.
32. **Federle, T. W., M. A. Hullar, R. J. Livingston, D. A. Meeter, and D. C. White.** 1983. Spatial distribution of biochemical parameters indicating biomass and community composition of microbial assemblies in estuarine mud flat sediments. *Appl. Environ. Microbiol.* **45:**58–63.
33. **Findlay, R. H., and F. C. Dobbs.** 1993. Quantitative description of microbial communities using lipid analysis, p. 271–284. *In* P. F. Kemp, B. F. Sherr, E. B. Sherr, and J. J. Cole (ed.), *Handbook of Methods in Aquatic Microbial Ecology*. Lewis Publishers, Boca Raton, Fla.
34. **Findlay, R. H., G. M. King, and L. Watling.** 1989. Efficiency of phospholipid analysis in determining microbial biomass in sediments. *Appl. Environ. Microbiol.* **55:** 2888–2893.
35. **Findlay, R. H., M. B. Trexler, J. B. Guckert, and D. C. White.** 1990. Laboratory study of disturbance in marine sediments: response of a microbial community. *Mar. Ecol. Prog. Ser.* **61:**121–133.
36. **Findlay, R. H., M. B. Trexler, and D. C. White.** 1990. Response of a benthic microbial community to biotic disturbance. *Mar. Ecol. Prog. Ser.* **61:**135–148.
37. **Fredrickson, J. K., D. L. Balkwill, G. R. Drake, M. F. Romine, D. B. Ringelberg, and D. C. White.** 1995. Aromatic-degrading *Sphingomonas* isolates from the deep subsurface. *Appl. Environ. Microbiol.* **61:**1917–1922.
38. **Frostegård, Å., and E. Bååth.** 1996. The use of phospholipid fatty acid analysis to estimate bacterial and fungal biomass in soil. *Biol. Fertil. Soils* **22:**59–65.
39. **Frostegård, Å., A. Tunlid, and E. Bååth.** 1991. Microbial biomass measured as total lipid phosphate in soils of different organic content. *J. Microbiol. Methods* **14:**151–163.
40. **Gehron, M. J., and D. C. White.** 1982. Quantitative determination of the nutritional status of detrital microbiota and the grazing fauna by triglyceride glycerol analysis. *J. Exp. Mar. Biol.* **64:**145–158.
41. **Geyer, R., A. D. Peacock, A. Miltner, H.-H. Richnow, D. C. White, K. L. Sublette, and M. Kästner.** 2005. In situ assessment of biodegradation potential using biotraps amended with ^{13}C-labeled benzene or toluene. *Environ. Sci. Technol.* **39:**4983–4989.

42. **Geyer, R., A. D. Peacock, D. C. White, C. Lytle, and G. J. Van Berkel.** 2004. Atmospheric pressure chemical ionization and atmospheric pressure photoionization for simultaneous mass spectrometric analysis of microbial respiratory ubiquinones and menaquinones. *J. Mass Spectrom.* **39:**922–929.

43. **Glaeser, J., and J. Overmann.** 2003. Characterization and in situ carbon metabolism of phototrophic consortia. *Appl. Environ. Microbiol.* **69:**3739–3750.

44. **Guckert, J. B., C. P. Antworth, P. D. Nichols, and D. C. White.** 1985. Phospholipid, ester-linked fatty acid profiles as reproducible assays for changes in prokaryotic community structure of estuarine sediments. *FEMS Microbiol. Ecol.* **31:**147–158.

45. **Guckert, J. B., M. A. Hood, and D. C. White.** 1986. Phospholipid, ester-linked fatty acid profile changes during nutrient deprivation of *Vibrio cholerae*: increases in the *trans/cis* ratio and proportions of cyclopropyl fatty acids. *Appl. Environ. Microbiol.* **52:**794–801.

46. **Guckert, J. B., D. B. Ringelberg, D. C. White, R. S. Henson, and B. J. Bratina.** 1991. Membrane fatty acids as phenotypic markers in the polyphasic taxonomy of methylotrophs within the Proteobacteria. *J. Gen. Microbiol.* **137:**2631–2641.

47. **Hanson, J. R., J. L. Macalady, D. Harris, and K. M. Scow.** 1999. Linking toluene degradation with specific microbial populations in soil. *Appl. Environ. Microbiol.* **65:** 5403–5408.

48. **Hedrick, D. B., and D. C. White.** 1986. Microbial respiratory quinones in the environment: a sensitive liquid chromatographic method. *J. Microbiol. Methods* **5:**243–254.

49. **Hedrick, D. B., J. B. Guckert, and D. C. White.** 1991. The effect of oxygen and chloroform on microbial activities in a high-solids, high-productivity biomass reactor. *Biomass Bioenergy* **1:**207–212.

50. **Hedrick, D. B., J. B. Guckert, and D. C. White.** 1991. Archaebacterial ether lipid diversity analyzed by supercritical fluid chromatography: integration with a bacterial lipid protocol. *J. Lipid Res.* **32:**659–666.

51. **Hedrick, D. B., A. Peacock, J. R. Stephen, S. J. Macnaughton, J. Brüggemann, and D. C. White.** 2000. Measuring soil microbial community diversity using polar lipid fatty acid and denaturing gradient gel electrophoresis data. *J. Microbiol. Methods* **41:**235–248.

52. **Hedrick, D. B., A. Peacock, and D. C. White.** 2005. Interpretation of fatty acid profiles of soil microorganisms, p. 251–259. *In* R. Margesin and E. Schinner (ed.), *Manual for Soil Analysis—Monitoring and Assessing Soil Bioremediation*. Springer-Verlag, Berlin, Germany.

53. **Henn, M. R., G. Gleixner, and I. H. Chapela.** 2002. Growth-dependent stable carbon isotope fractionation by basidiomycete fungi: ^{13}C pattern and physiological process. *Appl. Environ. Microbiol.* **68:**4956–4964.

54. **Hiraishi, A.** 1999. Isoprenoid quinones as biomarkers of microbial populations in the environment. *J. Biosci. Bioeng.* **88:**449–450.

55. **Hood, M. A., J. B. Guckert, D. C. White, and F. Deck.** 1986. Effect of nutrient deprivation on the levels of lipid, carbohydrate, DNA, RNA, and protein levels in *Vibrio cholerae*. *Appl. Environ. Microbiol.* **52:**788–793.

56. **Huber, R., T. A. Langworthy, H. König, M. Thomm, C. R. Woese, U. B. Sleytr, and K. O. Stetter.** 1986. *Thermotoga maritima* sp. nov. represents a new genus of unique extremely thermophilic eubacteria growing up to 90°C. *Arch. Microbiol.* **144:**324–333.

57. **Huber, R., C. R. Woese, T. A. Langworthy, H. Fricke, and K. O. Stetter.** 1989. *Thermosipho africanus* gen. nov., represents a new genus of thermophilic eubacteria within the "Thermotogales." *Syst. Appl. Microbiol.* **12:** 32–37.

58. **Jahnke, L. L., R. E. Summons, L. M. Dowling, and K. D. Zahiralis.** 1995. Identification of methanotrophic lipid biomarkers in cold-seep mussel gills: chemical and isotopic analysis. *Appl. Environ. Microbiol.* **61:**576–582.

59. **Jahnke, L. L., R. E. Summons, J. M. Hope, and D. J. Des Marais.** 1999. Carbon isotopic fractionation in lipids from methanotrophic bacteria. II. The effects of physiology and environmental parameters on the biosynthesis and isotopic signatures of biomarkers. *Geochim. Cosmochim. Acta* **63:**79–93.

60. **Johnsen, A. R., A. Winding, U. Karlson, and P. Roslev.** 2002. Linking of microorganisms to phenanthrene metabolism in soil by analysis of ^{13}C-labeled cell lipids. *Appl. Environ. Microbiol.* **68:**6106–6113.

61. **Johnston, N. C., and H. Goldfine.** 1982. Effects of growth temperature on fatty acid and alk-1-enyl group compositions of *Veillonella parvula* and *Megasphaera elsdenii* phospholipids. *J. Bacteriol.* **149:**567–575.

62. **Jones, G. J., P. D. Nichols, and P. M. Shaw.** 1994. Analysis of microbial sterols and hopanoids, p. 163–195. *In* M. Goodfellow and A. G. O'Donnell (ed.), *Chemical Methods in Prokaryotic Systematics*. John Wiley & Sons, New York, N.Y.

63. **Kamio, Y., and H. Takahashi.** 1980. Isolation and characterization of outer and inner membranes of *Selenomonas ruminantium*: lipid compositions. *J. Bacteriol.* **141:**888–898.

64. **Kaneda, T.** 1991. Iso and anteiso fatty acids in bacteria: biosynthesis, function, and taxonomic significance. *Microbiol. Rev.* **55:**288–302.

65. **Kates, M.** 1986. *Techniques of Lipidology: Isolation, Analysis and Identification of Lipids*. Elsevier, Amsterdam, The Netherlands.

66. **Kawai, Y., and I. Yano.** 1983. Ornithine-containing lipid of *Bordetella pertussis*, a new type of hemagglutinin. *Eur. J. Biochem.* **136:**531–538.

60. **Kawai, Y., I. Yano, and K. Kaneda.** 1988a. Various kinds of lipoamino acids including a novel serine-containing lipid in an opportunistic pathogen *Flavobacterium*. Their structures and biological activities on erythrocytes. *Eur. J. Biochem.* **171:**73–80.

68. **Kawai, Y., I. Yano, K. Kaneda, and E. Yabuuchi.** 1988b. Ornithine-containing lipids of some *Pseudomonas* species. *Eur. J. Biochem.* **175:**633–641.

69. **Kerger, B. D., C. A. Mancuso, P. D. Nichols, D. C. White, T. Langworthy, M. Sittig, H. Schlessner, and P. Hirsch.** 1988. The budding bacteria, *Pirellula* and *Planctomyces*, with a typical 16S-rRNA and absence of peptidoglycan, show eubacterial phospholipids and unusually high proportions of long-chain beta-hydroxy fatty acids in the lipopolysaccharide lipid A. *Arch. Microbiol.* **149:**255–260.

70. **Keweloh, H., and H. J. Heipieper.** 1996. Trans unsaturated fatty acids in bacteria. *Lipids* **31:**129–137.

71. **Kieft, T. L., D. B. Ringelberg, and D. C. White.** 1994. Changes in ester-linked phospholipid fatty acid profiles of subsurface bacteria during starvation and desiccation in a porous medium. *Appl. Environ. Microbiol.* **60:**3292–3299.

72. **Knoche, H. W., and J. M. Shively.** 1972. The structure of an ornithine-containing lipid from *Thiobacillus thiooxidans*. *J. Biol. Chem.* **247:**170–178.

73. **Koga, Y., and H. Goldfine.** 1984. Biosynthesis of phospholipids in *Clostridium butyricum*: kinetics of synthesis of plasmalogens and the glycerol acetal of ethanolamine plasmalogen. *J. Bacteriol.* **159:**597–604.

74. **Kohring, L. L., D. B. Ringelberg, R. Devereux, D. Stahl, M. W. Mittelman, and D. C. White.** 1994. Comparison of phylogenetic relationships based on phospholipid fatty acid profiles and ribosomal RNA sequence similarities among dissimilatory sulfate-reducing bacteria. *FEMS Microbiol. Lett.* **119:**303–308.

75. **Konova, I. V., S. K. Kasymbekova, L. L. Mityushina, S. G. Batrakov, and M. N. Bekhtereva.** 1978. Influence of phosphorus on the composition of polar lipids of *Actinomyces olivaceus*. *Biol. Bull. Acad. Sci. USSR* **5:**217–223.
76. **Kostiw, L. L., C. W. Boylen, and B. J. Tyson.** 1972. Lipid composition of growing and starving cells of *Arthrobacter crystallopoietes*. *J. Bacteriol.* **94:**1868–1874.
77. **Laneelle, M. A., D. Prome, G. Laneelle, J. C. Prome.** 1990. Ornithine lipid of *Mycobacterium tuberculosis*: its distribution in some slow- and fast-growing mycobacteria. *J. Gen. Microbiol.* **136:**773–778.
78. **Langworthy, T. A., G. Holzer, J. G. Zeikus, and T. G. Tornabene.** 1983. Iso- and anteiso-branched glycerol diethers of the thermophilic anaerobe *Thermodesulfotobacterium commune*. *Syst. Appl. Microbiol.* **4:**1–17.
79. **Langworthy, T. A., W. R. Mayberry, P. F. Smith, and I. M. Robinson.** 1975. Plasmalogen composition of *Anaeroplasma*. *J. Bacteriol.* **122:**785–787.
80. **Leung, K. T., Y.-J. Chang, Y.-D. Gan, A. Peacock, S. J. Macnaughton, J. R. Stephen, R. S. Burkhalter, C. A. Flemming, and D. C. White.** 1999. Detection of *Sphingomonas* spp. in soil by PCR and sphingolipid biomarker analysis. *J. Ind. Microbiol.* **23:**252–260.
81. **Lichtfouse, E.** 2000. Compound-specific isotope analysis. Application to archaelogy, biomedical sciences, biosynthesis, environment, extraterrestrial chemistry, food science, forensic science, humic substances, microbiology, organic geochemistry, soil science and sport. *Rapid Comm. Mass Spectrom.* **14:**1337–1344.
82. **Linscheid, M., B. W. Diehl, M. Overmohle, I. Riedl, and E. Heinz.** 1997. Membrane lipids of *Rhodopseudomonas viridis*. *Biochim. Biophys. Acta* **1347:**151–163.
83. **Londry, K. L., L. L. Jahnke, and D. J. Des Marais.** 2004. Stable carbon isotope ratios of lipid biomarkers of sulfate-reducing bacteria. *Appl. Environ. Microbiol.* **70:**745–751.
84. **Lovley, D. R., S. J. Giovannoni, D. C. White, J. E. Champine, E. J. P. Phillips, Y. A. Gorby, and S. Goodwin.** 1992. *Geobacter metallireducens* gen. nov. sp. nov., a microorganism capable of coupling the complete oxidation of organic compounds to the reduction of iron and other metals. *Arch. Microbiol.* **159:**336–344.
85. **Lytle, C. A., Y.-D. M. Gan, and D. C. White.** 2000. Electrospray ionization/mass spectrometry compatible reversed-phase separation of phospholipids: piperidine as a post column modifier for negative ion detection. *J. Microbiol. Methods* **41:**227–234.
86. **Magurran, A. E.** 1988. *Ecological Diversity and Its Measurement.* Princeton University Press, Princeton, N.J.
87. **Makula, R. A., and W. R. Finnerty.** 1974. Phospholipid composition of *Desulfovibrio* species. *J. Bacteriol.* **120:**1279–1283.
88. **Makula, R. A., and W. R. Finnerty.** 1975. Isolation and characterization of an ornithine-containing lipid from *Desulfovibrio gigas*. *J. Bacteriol.* **123:**523–529.
89. **Mansour, M. P., J. K. Volkman, A. E. Jackson, and S. I. Blackburn.** 1999. The fatty acid and sterol composition of five marine dinoflagellates. *J. Phycol.* **35:**710.
90. **Mayberry, W. R., and J. R. Lane.** 1993. Sequential alkaline saponification/acid hydrolysis/esterification: a one tube method with enhanced recovery of both cyclopropane and hydroxylated fatty acids. *J. Microbiol. Methods* **18:**21–32.
91. **Mikell, A. T., Jr., T. J. Phelps, and D. C. White.** 1987. Phospholipids to monitor microbial ecology in anaerobic digesters, p. 413–444. *In* W. H. Smith and J. R. Frank (ed.), *Methane from Biomass, a Systems Approach.* Elsevier Pub. Co., New York, N.Y.
92. **Minnikin, D. E., and H. Abdolrahimzadeh.** 1974. The replacement of phosphatidylethanolamine and acidic phospholipids by ornithine-amide lipid and a minor phosphorus-free lipid in *Pseudomonas fluorescens* NCMB129. *FEBS Lett.* **43:**257–260.
93. **Moriyon, I., and I. Lopez-Goni.** 1998. Structure and properties of the outer membranes of *Brucella abortus* and *Brucella melitensis*. *Int. Microbiol.* **1:**19–26.
94. **Morrison, S. J., and D. C. White.** 1980. Effects of grazing by estuarine gammaridean amphipods on the microbiota of allochthonous detritus. *Appl. Environ. Microbiol.* **40:**659–671.
95. **Mottram, H. R., and R. P. Evershed.** 2003. Practical considerations in the gas chromatography/combustion/isotope ratio monitoring mass spectrometry of ^{13}C-enriched compounds: detection limits and carryover effects. *Rapid Commun. Mass Spectrom.* **17:**2669–2674.
96. **Newell, S. Y.** 1993. Membrane-containing fungal mass and fungal specific growth rate in natural samples, p. 579–586. *In* P. F. Kemp, B. F. Sherr, E. B. Sherr, and J. J. Cole (ed.), *Handbook of Methods in Aquatic Microbial Ecology.* Lewis Publishers, Boca Raton, Fla.
97. **Nichols, P. D., J. B. Guckert, and D. C. White.** 1986. Determination of monounsaturated fatty acid double-bond position and geometry for microbial monocultures and complex consortia by capillary GC-MS of their dimethyl disulphide adducts. *J. Microbiol. Methods* **5:**49–55.
98. **Nichols, P. D., C. A. Mancuso, and D. C. White.** 1987. Measurement of methanotroph and methanogen signature phospholipids for use in assessment of biomass and community structure in model systems. *Org. Geochem.* **11:**451–461.
99. **Nichols, P. D., W. R. Mayberry, C. P. Antworth, and D. C. White.** 1985. Determination of monounsaturated double bond position and geometry in the cellular fatty acids of the pathogenic bacterium *Francisella tularensis*. *J. Clin. Microbiol.* **21:**738–740.
100. **Nickels, J. S., J. D. King, and D. C. White.** 1979. Poly-beta-hydroxybutyrate accumulation as a measure of unbalanced growth of the estuarine detrital microbiota. *Appl. Environ. Microbiol.* **37:**459–465.
101. **Norland, S.** 1993. The relationship between biomass and volume of bacteria, p. 303–307. *In* P. F. Kemp, B. F. Sherr, E. B. Sherr, and J. J. Cole (ed.), *Handbook of Methods in Aquatic Microbial Ecology.* Lewis Publishers, Boca Raton, Fla.
102. **Odham, G., A. Tunlid, G. Westerdahl, L. Larsson, J. B. Guckert, and D. C. White.** 1985. Determination of microbial fatty acid profiles at femtomolar levels in human urine and the initial marine microfouling community by capillary gas chromatography-chemical ionization mass spectrometry with negative ion detection. *J. Microbiol. Methods* **3:**331–344.
103. **Olsson, P. A., I. M. van Aarle, M. E. Gavito, P. Bengtson, and G. Bengtsson.** 2005. ^{13}C incorporation into signature fatty acids as an assay for carbon allocation in arbuscular mycorrhiza. *Appl. Environ. Microbiol.* **71:**2592–2599.
104. **Oshima, M., and T. Ariga.** 1975. ω-cyclohexyl fatty acids in acidophilic thermophilic bacteria. *J. Biol. Chem.* **250:**6963.
105. **Palmisano, A. C., M. P. Lizotte, G. A. Smith, P. D. Nichols, D. C. White, and C. W. Sullivan.** 1988. Changes in photosynthetic carbon assimilation in Antarctic sea-ice diatoms during a spring bloom: variations in synthesis of lipid classes. *J. Exp. Mar. Biol. Ecol.* **116:**1–13.

106. **Parkes, R. J., N. J. E. Dowling, D. C. White, R. A. Herbert, and G. R. Gibson.** 1992. Characterization of sulfate-reducing bacterial populations within marine and estuarine sediments with different rates of sulfate reduction. *FEMS Microbiol. Ecol.* **102:**235–250.
107. **Peacock, A. D., Y. J. Chang, J. D. Istok, L. Krumholz, R. Geyer, B. Kinsall, D. Watson, K. L. Sublette, and D. C. White.** 2004. Utilization of microbial biofilms as monitors of bioremediation. *Microb. Ecol.* **47:**284–292.
108. **Pennanen, T., J. Liski, E. Baath, V. Kitunen, J. Uotila, C. J. Westman, and H. Fritze.** 1999. Structure of the microbial communities in coniferous forest soils in relation to site fertility and stand development stage. *Microb. Ecol.* **38:**168–179.
109. **Petersen, S. O., P. Roslev, and R. Bol.** 2004. Dynamics of a pasture soil microbial community after deposition of cattle urine amended with [13C]urea. *Appl. Environ. Microbiol.* **70:**6363–6369.
110. **Pond, D. W., M. V. Bell, D. R. Dixon, A. E. Fallick, M. Segonzac, and J. R. Sargent.** 1998. Stable-carbon-isotope composition of fatty acids in hydrothermal vent mussels containing methanotrophic and thiotrophic bacterial endosymbionts. *Appl. Environ. Microbiol.* **64:**370–375.
111. **Pond, J. L., T. A. Langworthy, and G. Holzer.** 1986. Long-chain diols: a new class of membrane lipids from a thermophilic bacterium. *Science* **231:**1134–1136.
112. **Potts, M., J. J. Olie, J. S. Nickels, J. Parsons, and D. C. White**. 1987. Variations in phospholipid ester-linked fatty acids and carotenoids of desiccated *Nostoc commune* (cyanobacteria) from different geographic locations. *Appl. Environ. Microbiol.* **53:**4–9.
113. **Radajewski, S., I. R. McDonald, and J. C. Murrell.** 2003. Stable-isotope probing of nucleic acids: a window to the function of uncultured microorganisms. *Curr. Opin. Biotechnol.* **14:**296–302.
114. **Ratledge, C., and S. G. Wilkinson.** 1988. *Microbial Lipids*. Academic Press Ltd., London, United Kingdom.
115. **Rězanka, T., I. Dor, A. Prell, and V. M. Dembitsky.** 2003. Fatty acid composition of six freshwater wild cyanobacterial species. *Folia Microbiol.* **48:**71–75.
116. **Riebesell, U., A. T. Revill, D. G. Holdsworth, and J. K. Volkman.** 2000. The effects of varying CO_2 concentration on lipid composition and carbon isotope fractionation in *Emiliania huxleyi*. *Geochim. Cosmochim. Acta* **64:**4179–4192.
117. **Ringelberg, D. B., J. D. Davis, G. A. Smith, S. M. Pfiffner, P. D. Nichols, J. B. Nickels, J. M. Hensen, J. T. Wilson, M. Yates, D. H. Kampbell, H. W. Reed, T. T. Stocksdale, and D. C. White.** 1988. Validation of signature polar lipid fatty acid biomarkers for alkane-utilizing bacteria in soils and subsurface aquifer materials. *FEMS Microbiol. Ecol. 62:*39–50.
118. **Ringelberg, D. B., G. T. Townsend, K. A. DeWeerd, J. M. Suflita, and D. C. White.** 1994. Detection of the anaerobic dechlorinating microorganism *Desulfomonile tiedjei* in environmental matrices by its signature lipopolysaccharide branched-long-chain hydroxy fatty acids. *FEMS Microbiol. Ecol.* **14:**9–18.
119. **Rundel, P. W., J. R. Ehleringer, and K. A. Nagy (ed.).** 1989. *Stable Isotopes in Ecological Research.* Springer Verlag, New York, N.Y.
120. **Russell, N. J., and D. S. Nichols.** 1999. Polyunsaturated fatty acids in marine bacteria—a dogma rewritten. *Microbiology* **145:**767–779.
121. **Sakata, S., J. M. Hayes, A. R. McTaggart, R. A. Evans, K. J. Leckrone, and R. K. Togasaki.** 1997. Carbon isotopic fractionation associated with lipid biosynthesis by a cyanobacterium: relevance for interpretation of biomarker records. *Geochim. Cosmochim. Acta* **61:**5379–5389.
122. **Schouten, S., W. C. M. Klein Breteler, P. Blokker, N. Schogt, W. I. C. Rijpstra, K. Grice, M. Baas, and J. S. Sinninghe Damasté.** 1998. Biosynthetic effects on the stable carbon isotopic compositions of algal lipids: implications for deciphering the carbon isotopic biomarker record. *Geochim. Cosmochim. Acta* **62:**1397–1406.
123. **Schouten, S., M. Strous, M. M. M. Kuypers, W. I. C. Rijpstra, M. Baas, C. J. Schubert, M. S. M. Jetten, and J. S. Sinninghe Damsté.** 2004. Stable carbon isotopic fractionations associated with inorganic carbon fixation by anaerobic ammonium-oxidizing bacteria. *Appl. Environ. Microbiol.* **70:**3785–3788.
124. **Smith, C. A., C. B. Phiefer, S. J. Macnaughton, A. Peacock, R. S. Burkhalter, R. Kirkegaard, and D. C. White.** 2000. Quantitative lipid biomarker detection of unculturable microbes and chlorine exposure in water distribution system biofilms. *Water Res.* **34:**2683–2688.
125. **Smith, G. A., P. D. Nichols, and D. C. White.** 1989. Triglyceride and sterol composition of sediment microorganisms from McMurdo Sound, Antarctica. *Polar Biol.* **9:**273–279.
126. **Smith, G. A., J. S. Nickels, B. D. Kerger, J. D. Davis, S. P. Collins, J. T. Wilson, J. F. McNabb, and D. C. White.** 1986. Quantitative characterization of microbial biomass and community structure in subsurface material: a prokaryotic consortium responsive to organic contamination. *Can. J. Microbiol.* **32:**104–111.
127. **Sternberg, L. D. L., M. J. DeNiro, and H. O. Ajie.** 1986. Isotopic relationships between saponifiable lipids and cellulose nitrate prepared from red, brown and green algae. *Planta* **169:**320–324.
128. **Teece, M. A., M. L. Fogel, M. E. Dollhopf, and K. H. Nealson.** 1999. Isotopic fractionation associated with biosynthesis of fatty acids by a marine bacterium under oxic and anoxic conditions. *Org. Geochem.* **30:**1571–1579.
129. **ter Braak, C. J. F.** 1987. Ordination, p. 91–169. *In* R. H. G. Jongman, C. J. F. ter Braak, and O. F. R. van Tongeren (ed.), *Data Analysis in Community and Landscape Ecology.* Centre for Agricultural Publishing and Documentation, Wageningen, The Netherlands.
130. **Tollefson, T. S., and R. B. McKercher.** 1983. The degradation of ^{14}C-labelled phosphatidyl choline in soil. *Soil Biol. Biochem.* **15:**145–148.
131. **Tunlid, A., B. H. Baird, M. B. Trexler, S. Olsson, R. H. Findlay, G. Odham, and D. C. White.** 1985. Determination of phospholipid ester-linked fatty acids and poly beta hydroxybutyrate for the estimation of bacterial biomass and activity in the rhizosphere of the rape plant *Brassica napus* (L). *Can. J. Microbiol.* **31:**1113–1119.
132. **Tunlid, A., D. Ringelberg, T. J. Phelps, C. Low, and D. C. White.** 1989. Measurement of phospholipid fatty acids at picomolar concentrations in biofilms and deep subsurface sediments using gas chromatography and chemical ionization mass spectrometry. *J. Microbiol. Methods* **10:**139–153.
133. **Tunlid, A., N. A. Schultz, D. R. Benson, D. B. Steele, and D. C. White**. 1989. Differences in the composition between vegetative cells and nitrogen-fixing vesicles of *Frankia* spp. strain Cp11. *Proc. Natl. Acad. Sci. USA* **86:**3399–3403.
134. **van der Meer, M. T. J., S. Schouten, and J. S. Sinnighe Damsté.** 1998. The effect of the reversed tricarboxylic cycle on the ^{13}C contents of bacterial lipids. *Org. Geochem.* **28:**527–533.

135. **van der Meer, M. T. J., S. Schouten, J. S. Sinninghe Damsté, J. W. de Leeuw, and D. M. Ward.** 2003. Compound-specific isotopic fractionation patterns suggest different carbon metabolisms among *Chloroflexus*-like bacteria in hot-spring microbial mats. *Appl. Environ. Microbiol.* **69:**6000–6006.
136. **Walker, J. T., A. Sonesson, C. W. Keevil, and D. C. White.** 1993. Detection of *Legionella pneumophila* in biofilms containing a complex microbial consortium by gas chromatography-mass spectrometric analysis of genus-specific hydroxy fatty acids. *FEMS Microbiol. Lett.* **113:**139–144.
137. **Weaver, T. L., M. A. Patrick, and P. R. Dugan.** 1975. Whole-cell and membrane lipids of the methylotrophic bacterium *Methylosinus trichosporium*. *J. Bacteriol.* **124:** 602–605.
138. **White, D. C.** 1988. Validation of quantitative analysis for microbial biomass, community structure, and metabolic activity. *Adv. Limnol.* **31:**1–18.
139. **White, D. C.** 1995. Chemical ecology: possible linkage between macro- and microbial ecology. *Oikos* **74:**177–184.
140. **White, D. C., and D. B. Ringelberg.** 1995. Utility of signature lipid biomarker analysis in determining in situ viable biomass, community structure, and nutritional/physiological status of the deep subsurface microbiota, p. 119–136. *In* P. S. Amy and D. L. Halderman (ed.), *The Microbiology of the Terrestrial Deep Subsurface*. CRC Lewis Press, Boca Raton, Fla.
141. **White, D. C., and D. B. Ringelberg.** 1996. Monitoring deep subsurface microbiota for assessment of safe long-term nuclear waste disposal. *Can. J. Microbiol.* **42:**375–381.
142. **White, D. C., and D. B. Ringelberg.** 1998. Signature lipid biomarker analysis, p. 255–272. *In* R. S. Burlage, R. Atlas, D. Stahl, G. Geesey, and G. Sayler (ed.), *Techniques in Microbial Ecology*. Oxford University Press, New York, N.Y.
143. **White, D. C., R. J. Bobbie, J. S. Herron, J. D. King, and S. J. Morrison.** 1979. Biochemical measurements of microbial mass and activity from environmental samples, p. 69–81. *In* J. W. Costerton and R. R. Colwell (ed.), *Native Aquatic Bacteria: Enumeration, Activity and Ecology*. ASTM STP 695. American Society for Testing and Materials, Philadelphia, Pa.
144. **White, D. C., R. J. Bobbie, J. S. Nickels, S. D. Fazio, and W. M. Davis.** 1980. Nonselective biochemical methods for the determination of fungal mass and community structure in estuarine detrital microflora. *Bot. Mar.* **23:**239–250.
145. **White, D. C., W. M. Davis, J. S. Nickels, J. D. King, and R. J. Bobbie.** 1979. Determination of the sedimentary microbial biomass by extractable lipid phosphate. *Oecologia* **40:**51–62.
146. **White, D. C., R. Geyer, A. D. Peacock, D. B. Hedrick, S. S. Koenigsberg, Y. Sung, J. He, and F. E. Löffler.** 2005. Phospholipid furan fatty acids and ubiquinone-8: lipid biomarkers that may protect *Dehalococcoides* strains from free radicals. *Appl. Environ. Microbiol.* **71:**8426–8433.
147. **White, D. C., J. S. Gouffon, A. D. Peacock, R. Geyer, A. Biernacki, G. A. Davis, M. Pryor, M. B. Tabacco, and K. L. Sublette.** 2003. Forensic analysis by comprehensive rapid detection of pathogens and contamination concentrated in biofilms in drinking water systems for water resource protection and management. *Environ. Forensics* **4:**63–74.
148. **White, D. C., A. N. Tucker, and C. C. Sweeley.** 1969. Characterization of the iso-branched sphinganines from the ceramide phospholipids of *Bacteroides melaninogenicus*. *Biochim. Biophys. Acta* **187:**527–532.
149. **Wick, L. Y., N. Pasche, S. M. Bernasconi, O. Pelz, and H. Harms.** 2003. Characterization of multiple-substrate utilization by anthracene-degrading *Mycobacterium frederiksbergense* LB501T. *Appl. Environ. Microbiol.* **69:**6133–6142.
150. **Wilkinson, B. J., M. R. Morman, and D. C. White.** 1972. Phospholipid composition and metabolism of *Micrococcus denitrificans*. *J. Bacteriol.* **112:**1288–1294.
151. **Xiong, J., and C. E. Bauer.** 2002. Complex evolution of photosynthesis. *Annu. Rev. Plant Biol.* **53:**503–521.
152. **Zhang, C. L., B. W. Fouke, G. T. Bonheyo, A. D. Peacock, D. C. White, Y. Huang, and C. S. Romanek.** 2004. Lipid biomarkers and carbon-isotopes of modern travertine deposits (Yellowstone National Park, USA): implications for biogeochemical dynamics in hot-spring systems. *Geochim. Cosmochim. Acta* **68:**3157–3169.
153. **Zhang, C. L., Y. Li, J. D. Wall, L. Larsen, R. Sassen, Y. Huang, Y. Wang, A. Peacock, D. C. White, J. Horita, and D. R. Cole.** 2002. Lipid and carbon isotopic evidence of methane-oxidizing and sulfate-reducing bacteria in association with gas hydrates from the Gulf of Mexico. *Geology* **30:**239–242.
154. **Zhang, C. L., Y. Li , Q. Ye, J. Fong, A. D. Peacock, E. Blunt, J. Fang, D. R. Lovley, and D. C. White.** 2003. Carbon isotope signatures of fatty acids in *Geobacter metallireducens* and *Shewanella algae*. *Chem. Geol.* **195:**17–28.
155. **Zhang, C. L., Q. Ye, D. Goetz, A.-L. Reysenbach, A. Peacock, D. C. White, J. Horita, D. R. Cole, J. Fang, L. Pratt, J. Fang, and Y. Huang.** 2001. Carbon isotopic fractionations associated with thermophilic bacteria *Thermotoga maritima* and *Persephonella marina*. *Environ. Microbiol.* **4:**58–64.

Physiological Profiling of Microbial Communities

JAY L. GARLAND, COLIN D. CAMPBELL, AND AARON L. MILLS

11

INTRODUCTION

Rationale for Community-Level Physiological Profiling (CLPP) in Microbial Ecology

The study of microbial ecology centers on the functional interactions that microorganisms have with the environment that surrounds them. For much of its history, the discipline progressed by examining the behavior of individual strains or small assemblages of microbes in isolation from their natural habitat. As understanding advanced, the realization that few processes occur as a result of a single microbe has led to an increased interest in microbial communities as a basic ecological unit. Because most of the measurable reactions in any environment outside a laboratory monoculture represent community-level processes, it seems reasonable to examine variability among habitats at the level of the community rather than at the level of specific populations. This suggestion recognizes that community-level activities are the sum of reactions carried out by the populations composing the community (including synergistic and antagonistic modifications of that sum), but because the individual reactions are not easily (if at all) separable in soil or water samples, it is the community-level reaction that is almost always measured. The large increase in the number of research reports that examine the community through a variety of methods, both molecular and physiological, testifies to the acceptance of the community as a practical and desirable unit of study in microbial ecology.

A complete definition of the structure of any community would require an accurate enumeration of the number of different types (strains, genotypes, and cell lines) of microorganisms present, coupled with an accurate accounting of the number of individuals representing each type. Obviously, this goal is completely unattainable at this point in time. Numbers of individual bacterial cells in environmental samples are often around 10^6 ml of surface water^{-1} to 10^9 ml of active sediment $^{-1}$ (69). Furthermore, the few data that exist suggest that a gram of soil may contain as many as 10^4 different genotypes (63). Additional sources indicate that many of these types of organisms have neither been identified nor cultured, so attempts to measure the relative abundance of the types would be fruitless.

While novel methods to discover new types and characterize microbial diversity are valuable, alternative approaches for describing microbial communities in ecologically relevant terms are needed to provide insight into the spatial and temporal patterns in both natural and man-made ecosystems. Macroecologists, even though working with plant or animal communities for which complete cataloguing of members is much less daunting, often combine structural units (i.e., species) into functional aggregates to describe relevant patterns in ecological systems. While there are different ways of defining functional groups, the application of feeding guilds is a common one that has been useful in understanding ecosystem responses for a range of trophic levels. For example, Vannote et al. (64) classified stream macroinvertebrates into functional groups (i.e., shredders, collectors, grazers, and predators) to aid their conceptualization of stream ecosystems as a continuum of species replacements to maximize energy capture. Similar approaches have been used for other macroorganisms and smaller organisms such as nematodes (76) and protozoa (53), but for the heterotrophic microorganisms that rely on direct enzymatic action to release nutrients and energy, such approaches are not possible because they depend on identifying species differences or feeding organs that allow appropriate classification. However, the role of microorganisms in the cycling of nutrients, transformation of pollutants, and gas exchange with the atmosphere depends largely on their ability to metabolize a wide range of organic C compounds. Thus, studying the ability to degrade different C sources can be considered an approach equivalent to creating functional groups based on feeding guilds and may describe the physiological status of the community.

Garland and Mills (24) introduced a methodological approach based on rapid visualization of microbial communities as a multivariate composite of their physiological traits for the dual purposes of characterizing community functional attributes and assessing community dynamics. The CLPP approach initially assessed carbon substrate use based on redox dye chemistry within Biolog microplates but has more recently been modified by a variety of researchers to assess community activities while using either CO_2 production or O_2 consumption. The procedural basis, historical context, and highlighted examples of the various approaches to CLPP are presented, followed by a general description of data analysis approaches for the resulting multivariate data. The chapter concludes with a synthesis of the various CLPP approaches, including comparison of their relative strengths and weaknesses, discussion of appropriate applications, and definition of future areas of research.

ALTERNATIVE APPROACHES FOR CLPP

CLPP Based on Redox Dye Reduction (i.e., Biolog Approach)

The original test system used for CLPP was the Biolog GN microplate. Each plate consists of 96 individual wells of about 0.17-ml volume containing nutrients, a carbon or energy source, and the redox dye tetrazolium violet (TV). Each of the wells within the GN plates is identical, except that each contains a different carbon or energy source, including a variety of simple organic molecules (i.e., sugars, amino acids, short chain fatty acids, and alcohols). One well without any carbon or energy source serves as a control for endogenous respiration (or organic matter contamination of samples in the present use).

The original Biolog approach calls for inoculation of the plate with a heavy suspension of cells (from a clonal suspension), followed by a 6-h incubation (46). In this mode, the cells respire those carbon sources for which they have enzymes and transfer the electrons from the substrate oxidation to the TV, reducing it to the colored formazan. The large number of cells in each well ensures that the reaction will occur rapidly; this approach does not require growth of the cells in the well, merely respiration of the added substrate. Garland and Mills (24) used the Biolog technology to examine communities by inoculating whole environmental samples (i.e., raw water, soil suspensions, etc.) directly into the wells without first isolating any of the organisms from the sample. After incubation of the plate contents for a few days, growth of cells in the wells of the plates yielded color formation in many of the wells (Fig. 1). The patterns of color formation were used to discriminate among the communities being evaluated. In addition to inoculation of the well contents with the mixture of organisms comprising the community as opposed to a monoculture of cells, CLPP differs from the conventional Biolog approach in that the inoculum density used for CLPP is nearly always below that needed to obtain sensible reduction of the TV dye, about 10^5 to 10^6 cells (48, 71). Typical inoculum densities for the CLPP assay are from a few hundred to a few thousand cells per well (note that each well gets about 0.15 ml of suspension); thus, growth of the cells in the wells is required for color development. The requirement for cell growth has been viewed by some as a reason to eliminate CLPP as an approach to characterizing microbial communities; however, grouping of samples based on the (quantitative) ability of cells to grow in the various wells of the Biolog plates has proven to be a sensitive means of discriminating among communities in a wide variety of environments (for examples, see reference 37). Because of the requirement for growth of the cells, the information arising from the CLPP results cannot be used to infer ecological function. At best, the results might imply a potential function; there are obviously cells present that are capable of using the substrate under the incubation conditions. There may also be cells capable of using the substrate in the wells showing no color development (i.e., no growth), but those cells can do so only under different conditions (perhaps including those of the original source of the samples). Mills and Bouma (48) suggested that using a high inoculum density ($>10^7$ cells ml^{-1}) combined with short incubation times (<24 h) might indicate the functions possessed by the community in situ, but their study involved the use of constructed communities with at most five members. It is not clear how the suggestion might be applied to a complex natural community.

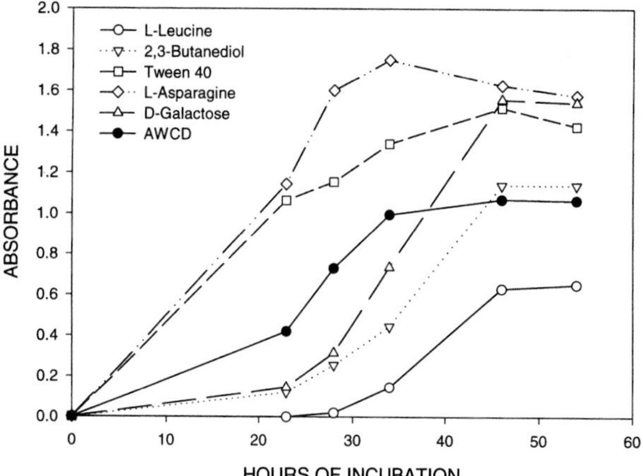

FIGURE 1 Time course of color development for several wells in a Biolog plate plotted along with the AWCD. Although only a few of the wells are shown, the AWCD at each point was calculated as the average of all 95 wells in the plate.

Several microtiter plates are available from Biolog; the GN plate was originally used, but other plates have been increasingly employed. The GP plate contains a different set of carbon sources and a recipe for growth-supporting nutrients tailored for the identification of organisms. The GP plate has been used to increase the number of carbon sources tested as part of CLPP, but no extensive comparison of GN and GP responses has been performed. SF-N plates, designed to identify fungal isolates, have been modified for community-level use with fungi (14). The MT plates contain redox dye and growth-supporting nutrients but no carbon source. Several researchers have used these plates to evaluate alternative carbon sources, including Campbell et al. (6), who found that a reduced set of "environmentally relevant" substrate compounds were effective at discriminating rhizosphere communities. This study, along with work by Insam (33), led to the manufacture by Biolog of an "Ecoplate" consisting of 31 substrates in triplicate.

Most of the work with alternative carbon sources has utilized the manufacturers' instructions for carbon loading (0.3 mg of carbon source into each well), although some researchers concerned with specific carbon sources have explored a wider range of substrate concentrations (41). Fulthorpe and Allen (16), working with isolates, employed repeated additions of lower substrate concentrations to help reduce toxicity associated with aromatic substrates. Similar problems with the toxicity of volatiles at high concentration can also be circumvented by using the inert carrier compound 2,2,4,4,6,8,8-heptamethylnonane to control substrate concentrations and ensure a constant substrate supply by gas-phase diffusion (7). This approach has been used to look at sole-C-source utilization of benzene, toluene, ethylbenzene, and xylene (BTEX) degradation in fuel-contaminated soils (38)

With the exception of the work on carbon sources, the medium recipe used for CLPP has never been optimized for community-level work. Microbial ecologists who extended the isolate assay to the community level have used the proprietary media supplied by Biolog for convenience. Addition of a variety of growth-supporting nutrients (i.e., amino acids and vitamins) could be attempted to maximize growth of different types of organisms. Use of a dye with a lower redox state might reduce the level of growth necessary to produce detectable color production. The interest in defining the medium recipe was recently accelerated when Biolog reformulated its product (i.e., GN2, GP2, and MT2 replacing GN, GP, and MT, respectively), resulting in changes that have a significant effect on the community-level response obtained from the plates (51). It is possible for researchers to make their own sole-C-source test plates with or without 2,3,5-triphenyltetrazolium chloride (TTC) (39), and these plates have been tested for CLPP, but the color development can be less intense than that found with Biolog plates. Yao et al. (75) used a customized approach to investigate the effect of pH on sole-C-source utilization tests. The multiwell plates were prepared by using a base characterization mineral medium (33) at two pH values of 4.7 and 7.0. No TTC was used, and substrate utilization and microbial growth were monitored by measuring absorbance at 405 nm. Individual researchers need to gauge the value of developing their own recipe against the convenience and standardization of using a commercially available product.

More recently, Biolog has expanded its product line to include the Phenotype Microarray, a set of 20 different plates testing substrate utilization (expanded to 190 tests) as well as sensitivity to growth-supporting factors (i.e., nitrogen, phosphorus, sulfur, and other nutrient supplements), pH, osmotic stressors, and chemical stressors (17) (e.g., antibiotics). Application of the Phenotype Microarray technology to community-level work has not been published to date.

Samples from surface water or groundwater can be inoculated directly into the wells of the Biolog plate. Some authors have had good success with marine or estuarine waters (24, 45), but there are questions concerning the salt effect on the reduction of the TV. Use of estuarine or marine samples should be preceded by a small exploratory evaluation. Dilution may be necessary for samples from waters with a high content of available organic carbon in order to reduce that concentration to avoid interference (for very slight interference, subtraction of the control well from each of the 95 substrate-containing wells will minimize any problems). However, dilution should be approached with caution, since extensive dilution can remove strains from the sample, resulting in a loss of positive tests from the pattern obtained (23). On the other hand, concentration of samples in a centrifuge may help to produce inocula of approximately equal cell abundance for samples with widely disparate cell densities, although we know of no documented examples. Approaches to standardizing inoculum are described further below.

Soil inocula should be added to the wells as a suspension. Because of the likelihood of carryover of particulate or dissolved organics from the soil, dilution is usually necessary. A preliminary examination of several dilution levels will inform the investigator as to the level of dilution necessary to eliminate contamination of the wells with soil organic matter. Again, a balance must be struck between dilution to remove organic matter and overdilution, which can cause a change in the substrate utilization pattern obtained.

CLPP has been applied to a variety of other habitats associated with attached or flocculated microorganisms, including rhizosphere (21), phyllosphere (15), drilling mud and subsurface cores (43), compost (34), and activated sludge (30, 66). In all cases, sample preparation involved extraction of cells followed by some degree of cell purification from the associated debris. The most commonly employed extraction method was homogenization via blending, although Victorio et al. (66) successfully tested a variety of approaches, including homogenization, gravel shake, and sonication. Brief centrifugation (700 to 1,000 × *g* for 5 to 10 min) followed by dilution or dilution alone successfully removed any background color issues, either due to color within the sample itself or control well color formation due to exogenous carbon addition.

The density of organisms inoculated into the plates will have a profound effect on the rate and extent of response and must be carefully considered when preparing, analyzing, and interpreting CLPP. If the researcher is interested in the rate of response per unit habitat (e.g., acute toxicity effects on community potential in bacterioplankton), then the importance of cell density may be minimal. However, when the research question involves dynamics per unit community, then failure to carefully consider inoculum levels will seriously confound the results. Standardizing inoculum density would appear to be the most straightforward approach to normalizing density effects; cell density could be diluted (or concentrated) to achieve a consistent level. However, this approach involves analytical challenges. The need to initially assess cell density detracts from the overall simplicity of the CLPP approach, limiting the ability to rapidly screen large numbers of samples. Furthermore, this approach assumes an accurate means of assessing the number of "CLPP-reactive" cells. Garland (19) reported that the rate of color development in GN microplates was more highly correlated to the number of respiring cells in the inoculum as estimated by reduction of 5-cyano-2,3-ditolyl tetrazolium chloride (59) than to the numbers of total and culturable cells, but the R^2 value was only 0.633 (versus 0.433 and 0.399, respectively). Despite these problems, standardization of inoculum density may be critical for applications requiring comparison of data across a wide range of samples (e.g., regional soil surveys). In such cases, standardizing accordingly to microbial biomass as estimated by substrate-induced respiration (SIR) has been proposed (35).

The most practical, effective approach for dealing with the effects of inoculum density involves (i) approximate equalization of cell density (i.e., within an order of magnitude), (ii) estimation of cell density (using various enumeration techniques, if feasible) for later use in interpreting CLPP, and (iii) multiple readings of color development to help capture potential differences in the overall rate of color development. If previous information on the cell density of the samples is not sufficient to allow for approximate normalization of inoculum density, a range of dilutions that encompasses the potential variance in cell density among samples is recommended.

Visual inspection of the plates is typically sufficient for quantifying functional richness (i.e., the number of responsive wells). A microplate reader is necessary if quantitative response data for the wells are desired. Manual plate readers allow for data quantification but can involve intensive monitoring to capture the time course of color development. The interval of sampling necessary to adequately capture the color development curves will depend on incubation temperature and inoculum density; a 2-h interval is sufficient for samples of a density of 10^5 to 10^7 total cells ml^{-1} in our studies. Given this intensity of measurement, the preferred instrumentation is a microplate reader linked to a plate-handling system enabling automated, repeated temporal analysis.

The CLPP assay is a sensitive, simple, rapid, and relatively inexpensive means of discriminating among microbial communities in environmental samples. The profiles do not, however, represent the functional potential of every member of the community; rather, the profiles comprise the activities of the organisms that can grow under the incubation conditions imposed on them, and, furthermore, represent those organisms that can grow relatively rapidly under those conditions. One must be careful in ascribing meaning to the utilization of individual compounds in the assay. The few studies that have compared in situ carbon source use with the rate of response to the same substrates in the Biolog plates have found no correlation (22, 28). If one understands and accepts this constraint, the patterns of substrate utilization obtained upon incubation of an environmental sample in a number of sole-carbon-source cultures contain a substantial amount of information that can be used to discriminate among communities.

CLPP Based on Headspace CO_2 Detection

Measuring the amount of carbon dioxide (CO_2) that is being respired by soil microorganisms that are decomposing organic substrates is a basic method used to assess the activity and health of soils (1). It is also possible to measure the SIR by measuring the CO_2 before and after the addition of a substrate such as glucose (3). Because the amount of substrate respired is proportional to the numbers of active cells present, this gives additional information on the size of the microbial biomass. The individual species that comprise the soil microbial community have differing capabilities to respire different substrates such that by adding different substrates it is possible to measure a CLPP (42) (Fig. 2).

Headspace analysis of respired CO_2 has been carried out for many years as a basic way of measuring the aerobic heterotrophic activity of microorganisms in soils, sediments, and waters. The corollary, measuring oxygen uptake or oxygen demand, has been much more popular in waters than in soil in part due to other considerations of measuring dissolved oxygen in relation to the health of fish and invertebrates. Community-level respiration from field samples must first separate any contribution from nonmicrobial sources such as algae and plants. This is done by removing plant parts and incubating the samples in the dark. To measure the CO_2 in headspace in closed chambers, one must also consider how much oxygen is available in the time period being measured and must also allow for any short-term perturbations caused by sampling and preparation. In the case of the latter, a settling or conditioning period of 3 to 10 days is often used to allow respiration to reequilibrate after the disturbance of sieving and mixing.

In early studies CO_2 was most often measured by using alkali traps that were then back titrated (2). Alternative measures of CO_2 measurement have been employed, including headspace analysis by gas chromatography or infrared gas analysis, bicarbonate solutions with pH indicator dyes to measure the equilibrium concentration of CO_2, and changes in conductivity within sodium hydroxide traps (50). Electrolytic respirometers to measure pressure changes due to O_2 uptake have all been used. The detection of $^{14}CO_2$ has been applied widely for the study of the degradation of ^{14}C-labeled substrates by using alkali traps to capture the labeled CO_2, which is then measured in the alkali directly after the addition of suitable scintillants or by gas chromatography–mass spectrometry after acidification to release the trapped CO_2.

The first studies of SIR (1) used infrared gas analysis and gas chromatography methods (67) to measure the CO_2 evolved after the addition of glucose to calculate the size of the microbial biomass. The first studies to use substrates other than glucose were by Sparling (58), who looked at the metabolism of different phenolic acids. These early studies were not measuring CLPP of multiple-C-source profiles per se but did lay the foundation for many of the principles about determining optimal C concentrations for maximal responses and also the physiological basis of the techniques that are now in use.

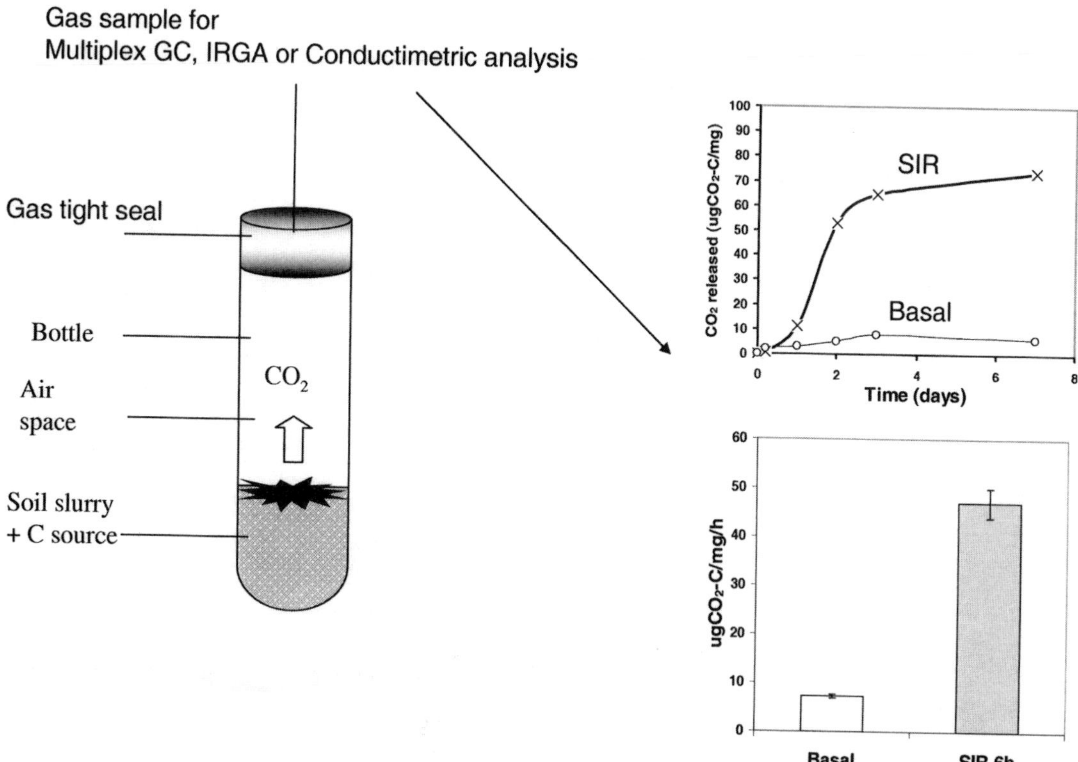

FIGURE 2 Schematic of multiple-SIR approach for performing CLPP.

Degens and Harris (11) were the first to use SIR responses of multiple substrates, essentially to carry out multiple, simultaneous SIR assays for soils with a range of substrates to measure what is termed catabolic response profiles. Measuring respiration by using these conventional methods is limited by the space and time needed to set up multiple simultaneous tests. The procedure used by Degens and Harris (11) uses headspace analysis of gastight glass jars to measure SIR of soil. Soil is weighed into 27.7-ml glass bottles, and 2 ml of a C source solution (pH adjusted to 5.8 to 6.2) is added. The resultant mixture is a slurry that ensures water is not limited and that the C source can be more easily mixed into the soil. Water is added to controls without C sources. The bottles are sealed by using septa and mixed, and their contents are then incubated at 25°C in the dark for 4 h. A gas syringe is used to withdraw headspace gas, which is measured by infrared gas analysis. The bottles are mixed once again after 2 h and also 10 min before measurement. A range of 2 carbohydrates, 6 amino acids, 2 amines, and 15 carboxylic acids has been tested with this method (9, 10). The concentrations used vary depending on the maximal SIR determined for each type of C source, i.e., 75 mM (carbohydrates), 10 mM (amines and amino acids), and 100 mM (carboxylic acids).

Clearly any of the CO_2 detection methods used for respiration studies can be used with this method, and automated online respirometry may provide additional information on the kinetic response of the SIR.

Multiple SIR (or catabolic response profiles) has been used to compare soil types (60), land use and land management effects (8, 57), waste applications to soil (13), and stress effects on soils (12).

Measuring the respiration of a large number of carbon sources, e.g., 16 to 95, can be laborious and time consuming, as most methods use between 5 and 150 g of soil in 100-ml or 2.5-liter glass jars. Consequently, the methods of measurement can be difficult to automate when processing large numbers of samples. Although methods based on the SIR of slurried soil in glass jars have been developed (11) and applied (9, 10), they are laborious compared to microtiter methods. One rapid microrespirometry method (MicroResp) that can be done within a 96-well microtiter plate format was described by Campbell et al. (5); it has the advantage of being compact, holding many samples and/or replicates, and being capable of being read on existing plate readers. Further depending on the detection system used, it can be used to test lower C concentrations than those normally used in Biolog and SIR methods and can be adapted by using glass inserts to test toxic volatile BTEX compounds found in fuel-contaminated soils (38).

The MicroResp systems consist of two microtiter plates sealed face to face (Fig. 3). The first plate is a deep-well

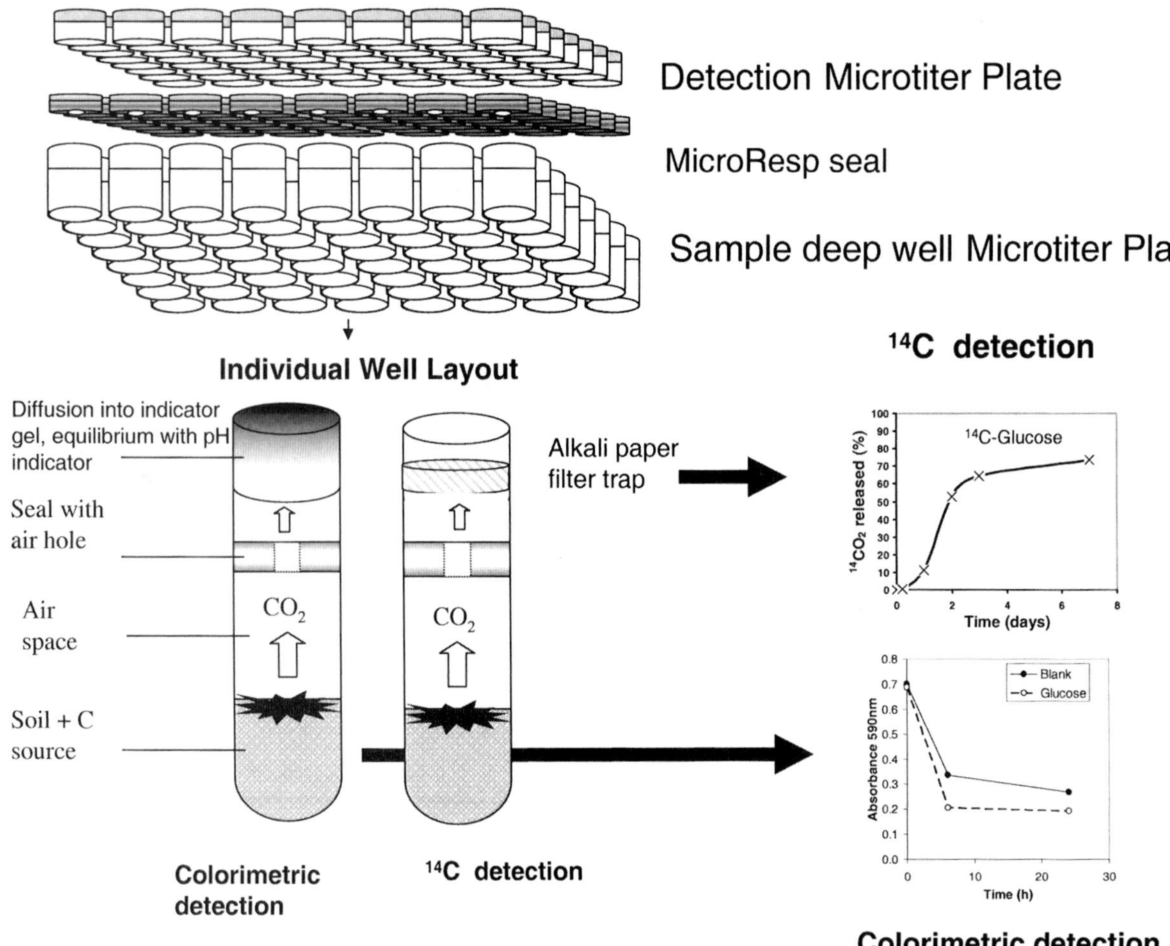

FIGURE 3 Schematic of MicroResp approach for performing CLPP.

plate, 1.2-ml capacity (Thermo LifeSciences, Basingstoke, United Kingdom), which contains the soil samples with the added carbon sources. The second plate is a normal microtiter plate that is used as the detection plate for the evolved CO_2 (Fig. 3). The two plates are sealed face to face with a silicone rubber gasket with interconnecting holes between the respective wells and can be clamped tight. Sieved soil is preadjusted to the required moisture content of 40% water-holding capacity, such that after adding the carbon source in solution, the moisture content is at 60% of the soil's water-holding capacity.

There are two detection systems for measuring evolved carbon dioxide. The first is a colorimetric method that relies on the change in pH of a solution of bicarbonate in quasiequilibrium with the well headspace. CO_2 absorption in the alkali, which is set in a gel containing a pH indicator dye, causes a color change (54). The indicator dye, cresol red (12.5 ppm [wt/wt]), potassium chloride (150 mM), and sodium bicarbonate (2.5 mM) are set in 150 μl of Noble agar (1%) in each well of the detection plate.

Each carbon source is dissolved in deionized water and prepared as a stock solution at a concentration designed to deliver 30 mg of C ml of soil water^{-1} (68) in each deep well when 25 μl was dispensed. Less soluble substrates can be prepared as stock solutions to deliver smaller amounts if required. The C source was dispensed before the soil was added to ensure that the soil contacted the C source at the same time for all wells.

The second detection system (the radioactive test) makes use of radioactive substrates and detects the radioactive carbon dioxide evolved by trapping in 40 μl of alkali (2 M NaOH). The system is assembled upside down if liquid alkali is used, or its contents can be incubated upright if alkali-soaked filter paper strips are used. After incubation, the detection-microtiter plate is removed, 200 μl of scintillant (Wallac Optiphase Supermix) is added, and the radioactivity is determined in a microplate liquid scintillation counter. Incubation is usually for 24 to 48 h, but mineralization can be detected after 5 h.

For the ^{14}C method, radiolabeled substrate was mixed with unlabeled (^{12}C) substrate to give 200 Bq well^{-1} and approximately 1 mg ml of soil water^{-1}, respectively. The actual quantities of substrate were calculated to consume approximately 100 and 70 μl of oxygen well^{-1}, assuming complete oxidation, over the 6 and 144 h of the colorimetric and ^{14}C methods, respectively. This practice ensured that the wells would not become anaerobic during extended incubation.

Soil is filled into the deep-well plate by using a filling device made from a 300-μl well microplate from which the bottom has been removed and replaced with a sliding base. Soil is filled into this microplate to a constant volume of 300 μl and is tapped gently to ensure consistent packing. The weight of soil is recorded, and the packing density is calculated. The plate is positioned over the deep-well plate, and the false bottom is removed, allowing the soil to fall into the deep wells and make contact with the C source solution. The deep-well plate is then immediately sealed with the gasket and detection plate.

For the colorimetric detection method, the detection plate is read, immediately before and after 6 h of incubation at 25°C, in a microplate reader at 590 nm. The absorbance after 6 h is normalized for any differences recorded at 0 h and is then converted to headspace CO_2 concentration by using a calibration curve. The average CO_2 evolved per sample can also be calculated and used to normalize individual C source concentrations before multivariate analysis. It is important to appreciate that the calibration curve represents the dynamic equilibration that occurs over the incubation time used and that recalibration is needed for different incubation times and bicarbonate solution concentrations.

CLPP Based on O_2 Consumption

Quantifying substrate consumption (e.g., O_2) as opposed to product appearance (e.g., CO_2) is an alternative means for performing CLPP. Detection of dissolved O_2 (DO) in the test medium may allow for improved sensitivity (i.e., testing of lower substrate concentrations), given the low solubility of O_2 in water. A fluorescence-based microplate platform, the BD Oxygen Biosensor System (BDOBS) (BD Biosciences, Bedford, Mass.), has been developed for rapidly assessing DO in a large number of simultaneous tests. The system is based on an O_2-sensitive fluorophore, 4,7-diphenyl-1,10-phenathroline ruthenium(II) chloride, adsorbed onto a silicone matrix that is permeable to oxygen (73). The fluorescence of the ruthenium dye is quenched by the presence of O_2, so the signal from the fluorophore-gel complex loaded on the bottom of the microplate wells increases in response to respiration in the overlying sample (Fig. 4). Garland et al. (28) found that CLPP could be performed with the BDOBS by using substrate concentrations as low as 0.5 to 1 mM or 10 to 100 times less than those employed in CLPP approaches using either Biolog or CO_2 monitoring. Fluorescence reaches a peak (or a plateau at higher substrate concentrations and concomitant sustained O_2 consumption) and then declines as O_2 diffusion rates exceed O_2 consumption rates (Fig. 4). The reversible nature of the dye response suggests the potential for plate reuse following cleanup and decontamination.

The configuration of the BDOBS plates confers many potential advantages to performing CLPP. Since the ruthenium dye-gel is the only component of the BDOBS technology, factors such as nutrient availability, pH, or other physicochemical properties can be readily manipulated in the assay. The location of the gel layer on the bottom of the wells allows for visibly turbid (e.g., heavy soil suspensions) or opaque (e.g., surfaces containing intact biofilms) materials to be assayed with bottom-reading plate readers. The lower substrate concentrations needed for detectable respiration than for the other detection technologies may allow for testing of naturally derived substrates; recent studies in our lab have successfully tested soil community responses to extracts from litter, root, and soil material.

DATA ANALYSIS

Binary Data (Richness)

The number of positive tests may discriminate samples based on overall richness and/or identity of the functional responses. The new approaches based on CO_2 and O_2 detection were developed to provide a more ecologically relevant measure of substrate use, making the rate of use (i.e., continuous data; see below) the preferred target response. Richness of responses within the Biolog plates has been used but can be highly confounded by density differences as discussed above. Characterization of functional richness in response to dilution of the samples may be a simple way to estimate relative structural diversity in microbial communities. Garland and Lehman (23) related functional richness (R) to inoculum density (I) using the model

$$R = R_{max} \times I/K_I + I$$

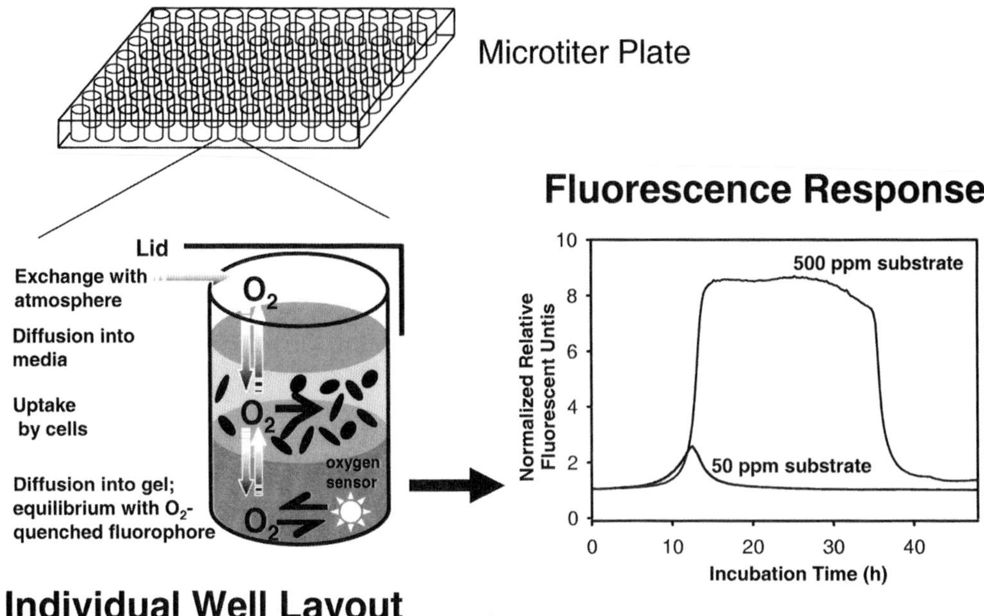

FIGURE 4 Schematic of dissolved oxygen microtiter-based systems for performing CLPP.

where R_{max} equals the maximum asymptotic level of R (which can be more than 95 but will vary among communities) and where K_I equals the value of I when R is one-half R_{max}. Relation of the model to structural diversity assumes that the rate of character loss from a mixed community upon dilution-extinction should be proportional to the diversity of biological entities in the community. This assumption is based on the idea that extinction of organisms (and their functional characteristics) will occur more rapidly as more types of organisms are distributed less evenly within a community. In other words, higher diversity will result in lower values of K_I. Very low levels of diversity caused by limited community membership or acute physiological stress may result in lower R_{max} values. The model was successfully tested with rhizosphere and bacterioplankton communities with different levels of diversity as affected by anthropogenic (metal stress) (42) or natural (community succession) processes (23). The method is very simple, involving inoculation of plates with dilution series of an inoculum, incubation of plate contents until color production is saturated (i.e., 7 days at 25°C in the study noted), and analysis of the number of positive responses. Inoculum density was estimated by both direct count (acridine orange) and plate count approaches, although any enumeration is limited in its ability to precisely detect only "CLPP-responsive" wells as noted above. A 10-fold dilution series of densities ranging from ~10^1 to 10^7 total cells ml^{-1} (~10^1 to 10^6 culturable cells ml^{-1}) was tested, although the optimal range may vary for different sample types.

Compared to techniques based on the direct measurement of the presence and relative abundance of structural elements (i.e., species or operational taxonomic units), the dilution-extinction method is a mechanistically based approach for studying diversity. A variety of theories have been produced for factors causing diversity, ranging from equilibrium concepts based on resource specialization and niche compression to regional effects related to limited or stochastic dispersal. The dilution-extinction approach is based on resource specialization and niche compression, and observed changes in the K_I, therefore, indicate that resource competition and concomitant effects on niche are active forces within the community.

The binary data also can be used as response variables to estimate the most probable number (MPN) of organisms able to utilize a particular carbon source. Gamo and Shoji (17) used six replicates of a 10-fold dilution series to generate MPN estimates from activated sludge samples and subsequently used the multivariate MPN matrix as an input for principal-component analysis (PCA) (17).

Dilution-extinction of functional richness is more readily applied to the Biolog technology than to the CO_2- and O_2-based CLPP approaches for several reasons. The ease of use of the Biolog plates (i.e., prepackaged media) allows for rapid inoculation of multiple plates. A gradient of response can be observed over a wide range of inoculum densities in the Biolog plates (i.e., as few as 10^1 culturable cells ml^{-1}), given the capacity for cell growth and eventual dye reduction in the assay. Changes in the rate or nature (e.g., shape of the fluorescence peak in the BDOBS plates) of response to a small number of targeted compounds across a relatively small range of dilutions (i.e., 10^{-1} to 10^{-2}) may provide interesting insight into the structure of given functional groups when one is using the CO_2 or O_2 approach. However, such trends have not been evaluated to date.

Continuous (Quantitative)

Quantification of the individual responses that comprise the CLPP allows for development of (i) diversity measures based on the richness and evenness of responses and (ii) profiles that can be compared by using ordination techniques. The approach for quantification is dependent on the assay type and will be discussed accordingly.

For the Biolog plates, the differences in the overall rate of color formation caused by inoculum density differences must be carefully accounted for in the acquisition and analysis of continuous data. If inoculum density was effectively standardized, analysis of plate readings at a given incubation can be performed. Since effective standardization of inoculum density is unlikely for the reasons discussed above, alternative approaches should be employed. One simple approach is based on multiple readings of the plates followed by selection of plate readings from each sample that are at a common average well color development (AWCD). AWCD is not a perfect normalization tool, since it is only a surrogate for density. However, analysis of plate readings at a common AWCD will evaluate differences in the relative pattern, rather than in overall rate, of color response. The specific value of AWCD used as a reference point does not appear to affect discrimination of samples, although lower values of AWCD (0.25 to 0.50 absorbance units) will emphasize the importance of rapidly responding wells, while higher values of AWCD (>0.50 absorbance units) will emphasize the importance of more slowly responding wells (20).

A single or relatively few plate reading times may result in samples with very different levels of AWCD, leading to subsequent analysis of the data that will be largely based on differences in the overall rate, rather than pattern, of response. Dividing the individual well values by the AWCD for the entire plate can help normalize these types of data sets (24). However, this normalization tool is limited to samples with relatively similar numbers of responsive wells, since zero values cannot be normalized.

For the multiple-SIR approaches, single-point readings of CO_2 in the headspace are used to calculate relative rates of CO_2 production. Rates can be compared directly or standardized to the size of the biomass and then used in multivariate analysis as described earlier. If kinetic data are gathered, lag phase and maximum rates may be calculated. Indices of diversity and evenness can also be calculated to summarize and compare samples. Most often the Shannon-Weiner diversity or evenness indices are used and referred to as the "functional diversity" index (77) or "catabolic diversity/evenness" (9). Wenderoth and Reber (67) calculated a "catabolic versatility" index for sole-C-source tests that used the average error instead of the standard deviation to indicate that it is multiple characteristics in a single sample that vary rather than a single characteristic measured several times. The catabolic versatility index is similar to the reciprocal of the coefficient of variation and can enhance the differences between communities more than the evenness index.

The fluorescence response in the BDOBS CLPP can be characterized via parameters describing lag (e.g., time to a minimum threshold response greater than the negative control or time to peak) or the magnitude of response (e.g., peak height or area under the curve). The lag in fluorescence response in the CLPP (as quantified by the time to peak height) was inversely correlated to the rate of in situ substrate use in rhizosphere community, suggesting that the profile may provide ecologically relevant functional information (28). In contrast, no relationship between in situ substrate use and redox dye reaction rate in the Biolog CLPP has been observed (22). The peak in fluorescence, as well as the area under the curve of the fluorescence response, should reflect the total amount of O_2 consumed. Multivariate analysis can be performed on any of the parameters defined above.

Ordination Analyses

In order to assess the similarity among communities from different environmental samples that have been analyzed for some variable, a simple correlation can be used. In the case of two variables, plotting the value of each variable on a coordinate axis set can help one view similarities among the samples upon which the measurements were made. Consider, however, how such a plot would be constructed for 95 variables. The use of ordination techniques such as PCA permits construction of simpler plots while using combinations of the variables that are associated with the largest amount of variance in the samples. These variable sets are the principal components (PC) of variance that are used to plot similarity among samples on a two (occasionally three)-dimensional axis set. Similarity among samples from different sites can be established by plotting the results of measuring two independent variables in two-dimensional coordinate space. Usually the first two, or sometimes three, components are adequate to separate samples into intelligible groupings based on the values (scores) of the individual PCs. Often an investigator wishes to obtain information from the several variables that compose each PC. In the CLPP approach, however, the plot of the samples on the axis set defined by the first two or three PCs is the visual tool used to examine the degree of similarity among each pair of samples collected. The underlying concepts of PCA and the mechanics of conducting the analysis are beyond the scope of this discussion, but PCA is a standard routine in sophisticated statistical software packages (e.g., SPSS or SAS), and instructions for applying those routines to the data are given in the documentation accompanying the software. An excellent introduction to PCA (and other multivariate techniques), including instructions on how to apply the technique while using commercial statistical software, is presented by Tabachnick and Fidell (61). The use of PCA for analysis of CLPP results has been discussed extensively in the literature (25–27, 31, 32, 36, 56, 74), and although a few alternatives have been suggested based on theoretical considerations that often involve use of the variables making up the extracted PCs for explanations of variance, simple PCA of CLPP data is the most widely used tool for interpretation of the patterns obtained from the incubations.

The PCs obtained in the analysis represent a reduced set of variables that are independent (i.e., uncorrelated) and normally distributed. As such they can be used for further statistical evaluation. In many applications, axis rotations are used to enhance interpretability of the extracted components. Orthogonal rotations (varimax, quartimax, and equamax) are available in most statistical packages that support PCA. Most CLPP practitioners do not use rotations of the component axes, as the main purpose of a rotation is to minimize the complexity of the factors to allow for interpretation of the variables within each component. Given the limitations on the significance of the individual variables (tests) in a culture-based method to the sampled environment, interpretations using the variables instead of analysis of the overall response patterns are, at best, tenuous.

One can conduct other statistical tests on the extracted components. For example, the significance of the differences among communities can be examined with multivariate analysis of variance by using the individual PC scores for a replicated sample as individuals within that sample. Use of components extracted with the covariance matrix as opposed to the correlation matrix is often preferred for biological data, since use of the correlation matrix is the equivalent of using data standardized to unit standard deviation on each variable (29). Furthermore, statistics conducted on the PCs derived from the covariance matrix are more robust, because the covariance matrix retains all of the variance in the original data. If tests are performed on the results obtained with the correlation matrix, it is essential that the amount of variance explained be specified. Most statistical packages specify the correlation matrix as the default option for PCA.

Another approach for assessing the statistical significance of differences among samples based on their CLPP is through the use of discriminant function analysis (DFA), also called canonical discriminant analysis. This approach specifically tests for differences among defined groups (e.g., experimental treatments) based on certain predictor variables (e.g., carbon source responses), an approach very different from PCA, which makes no a priori assumptions concerning relationships among samples. DFA is degraded by departures from linearity and multinormality, as well as high variable-to-sample ratios (70). Therefore, DFA is best performed on PC scores, since they represent a reduced number of linear, normally distributed variables.

Canonical correspondence analysis (CCA) has been used as a means of relating differences in CLPP to environmental variables (4). CCA is an ordination technique in which the axes are constrained to be linear combinations of environmental variables, thereby directly relating community dynamics to environmental variation (62). While this is an attractive approach that appears to be less sensitive to some of the problems associated with other multivariate approaches (e.g., skewed data and intercorrelated environmental data [52]), CCA can be easily misinterpreted and potential users should understand the underlying assumptions of the approach (49).

Kinetic Approaches

Descriptors of the kinetic profile of color development in individual Biolog wells, rather than specific absorbance values, can be used for subsequent analysis of samples. Guckert et al. (30) used the integral of color development, while Mayr et al. (47) used a modification of the integral approach by calculating a "Riemann's sum" for each well. Alternatively, the parameters of curves fit to the color development in individual wells can be used as variables to analyze microbial communities (40, 44, 65). The most commonly used model is the modified Gompertz equation (78), which yields parameters for maximum rate of color development (μ_m), lag phase (λ), and maximum extent of color development (A), although a modified logistic model yields similar parameters (44).

While the kinetic approaches are independent of incubation time, effects of inoculum density are unclear. The lag phase in color development should be inversely correlated to inoculum density, but Lindstrom et al. (44) concluded that two of the kinetic parameters (maximum rate and maximal extent of color development) were invariant with respect to inoculum density. However, their conclusions were based on dilutions and analysis of a monoculture. Dilution of mixed microbial communities will result in the extinction of rarer types of organisms that could alter functional responses.

Despite the potential utility of the kinetic approach to analysis of CLPP results, kinetic analysis does not appear to be the best approach for rapid classification of samples. Recent work by our group suggests that the use of single-point plate readings at a set point in AWCD is a better tool for classification with greater sensitivity and less dependence on inoculum than is the kinetic approach (18). Therefore, temporal monitoring of color development with subsequent selection of plate readings at a set point in AWCD appears to be the best approach when using CLPP for rapid, preliminary assessments of temporal or spatial variation in microbial communities. The potential of the kinetic approach is that a more detailed understanding of the nature of the color response may be feasible. However, this concept should be approached with caution, given the general lack of understanding of the physiological or ecological basis for differences in the kinetic parameters (λ, μ_m, and A) and the known selective bias due to growth in the plates (56). Experimental evidence suggests that the overall rate of color development in the Biolog plate is not necessarily related to in situ carbon source utilization (19, 72), so the potential for overinterpretation of changes in the nature of the color response using the kinetic parameters exists and should be assiduously avoided. Specific, hypothesis-driven experiments are needed to define and validate the information content in the kinetic parameters. Linking kinetic analysis to more sensitive methods for detecting carbon source utilization that do not rely on selective enrichment may also prove valuable.

SUMMARY AND CONCLUSIONS

The variety of techniques available for performing CLPP offers a range of approaches with relative strengths and weaknesses (Table 1). The Biolog plates are easy to use, allowing for the simultaneous evaluation of a wide range of substrates. The multivariate profile of response from Biolog plates, if carefully controlled for effects of inoculum density, is a sensitive indicator of shifts in microbial communities in response to temporal, spatial, and experimental gradients. The richness of functional response, particularly when evaluated across a range of inoculum densities (i.e., dilution-extinction), can provide unique insight into community diversity. A number of factors (high substrate concentrations, indirect evaluation of respiration via the redox chemistry, and presence of unknown proprietary ingredients) limit the ability of the Biolog CLPP to examine functionally relevant shifts in microbial communities. Alternative techniques developed to improve functional relevance involve direct measurement of respiration, either based on substrate consumption (O_2) or product appearance (CO_2), eliminating bias in the Biolog assay as a result of cells that cannot take up or are inhibited by the redox dye. The newer assays also eliminate the unknown growth factors found in the Biolog plates, allowing for manipulation of physical (e.g., pH) or chemical (e.g., nitrogen or phosphorus levels) factors as well as allowing for a complete definition of assay conditions. For toxicity studies such as measurement of the pollution-induced community tolerance (55), the more realistic incubation conditions have the major advantage over Biolog of allowing toxicant availability to be determined by in situ conditions (i.e., soil properties within the MicroResp technique). The new methods require additional manpower, since substrates and associated media are not pre-packaged and are typically more focused on a smaller set of substrates of concern.

Selection of the appropriate technique depends partially on the type of sample. While all of the methods have been applied to soil analyses, the multiple-SIR and MicroResp approaches involve the least disruption of the structure since aqueous suspensions are not required. The BDOBS system is more readily applied to soil (or other solid matrices such as compost) than is Biolog, since turbid or colored suspensions do not interfere with detection of response.

TABLE 1 Comparison of different approaches for CLPP

Approach	Assay characteristics					
	Ease of use	Assessment approach	Scope of target organisms	Ability to manipulate	Substrate concns	Required sample disruption and/or preparation
Biolog	High; prepackaged media, absorbance monitoring	Indirect; cell growth, dye reduction	Narrow; bacteria only, redox dye toxicity	Limited to substrates only, proprietary media	Nominally >100 mM	Cell suspensions needed; particulates and color cause interference
Headspace CO_2	Low; prepare own substrates, manual CO_2 measure	Direct; product formation	Heterotrophs; aerobic and potential anaerobic	Substrates and other physical and chemical factors	Intermediate, 10–100 mM	Some water addition/mixing needed to add substrate
Microplate CO_2	Intermediate; prepare own substrates, automated CO_2 measure	Direct; product formation	Heterotrophs; aerobic and potential anaerobic	Substrates and physical and chemical factors	Intermediate, 5[a]–100 mM	Some water addition/mixing needed to add substrate; soil sieving
Microplate O_2	Intermediate; prepare own substrate, automated O_2 measure	Direct; substrate consumption	Aerobic; heterotrophs and chemoautotrophs	Substrates and physical and chemical factors	Lower, 0.5–1 mM	Cell suspension used to date; sealed headspace approach feasible; no problems with particulate or color background

[a]Low substrate concentrations using ^{14}C MicroResp.

Aqueous samples (i.e., water and wastewater) can be tested by using all of the methods, although the BDOBS approach may be preferred, given the ability to use lower substrate concentrations.

A limited number of tests have compared the new CO_2- and O_2-based approaches with the Biolog CLPP, although none have directly compared the new approaches with one another. The MicroResp system was used to determine CLPP for soils treated with wastewater sludge and gave better discrimination of treatment effects than did Biolog CLPP. The utilization of the different C sources was not correlated between the two methods, but interestingly the overall pattern of treatment effects was similar when the same C sources were compared. The MicroResp CLPP gave better discrimination because the overall evenness of SIRs for the different C sources was less than in the Biolog CLPP system. Consequently, the whole-soil method may afford greater opportunity for discriminating different samples. In rhizosphere samples exposed to known concentrations of different types of surfactants, in situ rates of surfactant degradation correlated well with the rate of response to the O_2 CLPP assay but were not correlated to the response in Biolog MT plates containing surfactants (28).

The MicroResp and BDOBS systems could be adapted to measure multiple processes simultaneously, although such tests have not been developed. One can assay for gaseous products in addition to CO_2 (e.g., NH_3, volatile organic compounds, or volatile metals such as Hg or Se) in the overlying indicator gel within the MicroResp assay, provided that a suitable detection system can be found. The overlying indicator plate concept used in the MicroResp approach could be linked to the BDOBS system, allowing for simultaneous detection of DO and gaseous products.

With the newer CO_2- and O_2-based CLPP approaches, environmental microbiologists now possess methods for rapid assessment of community metabolic state that are more functionally relevant and can be readily manipulated to evaluate secondary factors such as nutrient limitation and pollutant stress. While further incorporation of new technologies to continually improve CLPP is encouraged, the existing techniques represent valuable tools for evaluating factors affecting the metabolic state of heterotrophic microbial communities.

REFERENCES

1. **Anderson, J. M.** 1987. Interactions between invertebrates and microorganisms: noise or necessity for soil processes, p. 125–146. *In* M. Fletcher, T. R. G. Gray, and J. G. Jones (ed.), *Ecology of Microbial Communities*. Cambridge University Press, Cambridge, United Kingdom.
2. **Anderson, J. M., and P. Ineson.** 1982. A soil microcosm system and its applications to measurements of respiration and nutrient leaching. *Soil Biol. Biochem.* **14:**415–416.
3. **Anderson, T. H., and K. H. Domsch.** 1978. A physiological method for the quantitative measurement of microbial biomass in soils. *Soil Biol. Biochem.* **10:**215–221.
4. **Bossio, D. A., and K. M. Scow.** 1995. Impact of carbon and flooding on the Metabolic diversity of microbial communities in soils. *Appl. Environ. Microbiol.* **61:**4043–4050.
5. **Campbell, C. D., S. J. Chapman, C. M. Cameron, M. S. Davidson, and J. M. Potts.** 2003. A rapid microtiter plate method to measure carbon dioxide evolved from carbon substrate amendments so as to determine the physiological profiles of soil microbial communities by using whole soil. *Appl. Environ. Microbiol.* **69:**3593–3599.
6. **Campbell, C. D., S. J. Grayston, and D. J. Hirst.** 1997. Use of rhizosphere carbon sources in sole carbon source tests to discriminate soil microbial communities. *J. Microbiol. Methods* **30:**33–41.
7. **Daugulis, A. J.** 1997. Partitioning bioreactors. *Curr. Opin. Biotechnol.* **8:**169-174.
8. **Degens, B. D., and M. Vojvodic-Vukovic.** 1999. A sampling strategy to assess the effects of land use on microbial functional diversity in soils. *Aust. J. Soil Res.* **37:**593–601.
9. **Degens, B. P.** 1999. Catabolic response profiles differ between microorganisms grown in soils. *Soil Biol. Biochem.* **31:**475–477.
10. **Degens, B. P.** 1998. Microbial functional diversity can be influenced by the addition of simple organic substrates to soil. *Soil Biol. Biochem.* **30:**1981–1988.
11. **Degens, B. P., and J. A. Harris.** 1997. Development of a physiological approach to measuring the catabolic diversity of soil microbial communities. *Soil Biol. Biochem.* **29:** 1309–1320.
12. **Degens, B. P., L. A. Schipper, G. P. Sparling, and L. C. Duncan.** 2001. Is the microbial community in a soil with reduced catabolic diversity less resistant to stress or disturbance? *Soil Biol. Biochem.* **33:**1143–1153.
13. **Degens, B. P., L. A. Schipper, G. P. Sparling, and M. Vojvodic-Vukovic.** 2000. Decreases in organic C reserves in soils can reduce the catabolic diversity of soil microbial communities. *Soil Biol. Biochem.* **32:**189–196.
14. **Dobranic, J. K., and J. C. Zak.** 1999. A microtiter plate procedure for evaluating fungal functional diversity. *Mycologia* **91:**756–765.
15. **Ellis, R. J., I. P. Thompson, and M. J. Bailey.** 1995. Metabolic profiling as a means of characterizing plant-associated microbial communities. *FEMS Microbiol. Ecol.* **16:**9–18.
16. **Fulthorpe, R. R., and D. G. Allen.** 1994. Evaluation of Biolog MT plates for aromatic and chloroaromatic substrate utilization tests. *Can. J. Microbiol.* **40:**1067–1071.
17. **Gamo, M., and T. Shoji.** 1999. A method of profiling microbial communities based on a most-probable-number assay that uses BIOLOG plates and multiple sole carbon sources. *Appl. Environ. Microbiol.* **65:**4419–4424.
18. **Garland, J., A. L. Mills, and J. S. Young.** 2001. Analysis of methods for numerical analysis of community-level physiological profiling of bacterial communities. *Soil Biol. Biochem.* **33:**1059–1066.
19. **Garland, J. L.** 1997. Analysis and interpretation of community-level physiological profiles in microbial ecology. *FEMS Microbiol. Ecol.* **24:**289–300.
20. **Garland, J. L.** 1996. Analytical approaches to the characterization of samples of microbial communities using patterns of potential C source utilization. *Soil Biol. Biochem.* **28:**213–221.
21. **Garland, J. L.** 1996. Patterns of potential C source utilization by rhizosphere communities. *Soil Biol. Biochem.* **28:** 223–230.
22. **Garland, J. L., K. L. Cook, and C. A. Loader.** 1996. Presented at the 96th General Meeting of the American Society for Microbiology, New Orleans, La.
23. **Garland, J. L., and R. M. Lehman.** 1999. Dilution/extinction of community phenotypic characters to estimate relative structural diversity in mixed communities. *FEMS Microbiol. Ecol.* **30:**333–343.
24. **Garland, J. L., and A. L. Mills.** 1991. Classification and characterization of heterotrophic microbial communities on the basis of patterns of community-level sole-carbon-source utilization. *Appl. Environ. Microbiol.* **57:**2351–2359.
25. **Garland, J. L., and A. L. Mills.** 1990. Presented at the 90th Annual Meeting of the American Society for Microbiology, Anaheim, Calif.

26. **Garland, J. L., and A. L. Mills.** 1999. Further comments on the interpretation of community-level physiological profiles. *Soil Biol. Biochem.* **31:**1203.
27. **Garland, J. L., and A. L. Mills.** 1999. Importance of pattern analysis in community-level physiological profiles (CLPP): a reply to the letter from P. J. A. Howard. *Soil Biol. Biochem.* **31:**1201–1202.
28. **Garland, J. L., M. S. Roberts, L. H. Levine, and A. L. Mills.** 2003. Community-level physiological profiling performed with an oxygen-sensitive fluorophore in a microtiter plate. *Appl. Environ. Microbiol.* **69:**2994–2998.
29. **Green, R. H.** 1979. *Sampling Design and Statistical Methods for Environmental Biologists.* John Wiley & Sons, Inc., New York, N.Y.
30. **Guckert, J. B., G. J. Carr, T. D. Johnson, B. G. Hamm, D. H. Davidson, and Y. Kumagai.** 1996. Community analysis by Biolog: curve integration for statistical analysis of activated sludge microbial habitats. *J. Microbiol. Methods* **27:**183-197.
31. **Howard, P. J. A.** 1997. Analysis of data from BIOLOG plates: comments on the method of Garland and Mills. *Soil Biol. Biochem.* **29:**1755–1757.
32. **Howard, P. J. A.** 1999. Analysis of inter-sample distances from BIOLOG plate data in Euclidean and simplex spaces. *Soil Biol. Biochem.* **31:**1323–1330.
33. **Insam, H.** 1997. A new set of substrates proposed for community characterization in environmental samples, p. 259–260. *In* H. Insam and A. Rangger (ed.), *Microbial Communities: Functional versus Structural Approaches.* Springer-Verlag KG, Berlin, Germany.
34. **Insam, H., K. Amor, M. Renner, and C. Crepaz.** 1996. Changes in functional abilities of the microbial community during composting of manures. *Microb. Ecol.* **31:**77–87.
35. **Insam, H., and M. Goberna.** 2004. Community level physiological profiles (Biolog substrate use tests) of environmental samples, p. 5.3.2:1–5.3.2:8. *In* A. D. L. Akkermans, J. D. van Elsas, and F. J. Debruijn (ed.), *Molecular Microbial Ecology Manual,* 2nd ed. Kluwer Academic Publishers, Dordrecht, The Netherlands.
36. **Insam, H., and W. Hitzl.** 1999. Data evaluation of community-level physiological profiles: a reply to the letter of P. J. A. Howard. *Soil Biol. Biochem.* **31:**1198–1200.
37. **Insam, H., and A. Rangger (ed.)** 1997. *Microbial Communities: Functional versus Structural Approaches.* Springer-Verlag KG, Berlin, Germany.
38. **Kaufmann, K., S. J. Chapman, C. D. Campbell, H. H. Harms, and P. Höhener.** Miniaturized test system for soil respiration induced by volatile pollutants. *Environ. Pollut.*, in press.
39. **Kennedy, A. C.** 1994. Carbon utilization and fatty acid profiles for characterization of bacteria, p. 543–556. *In* R. W. Weaver and J. S. Angle (ed.), *Methods of Soil Analysis,* part 2. *Microbiological and Biochemical Properties,* vol. 5. Soil Science Society of America, Inc., Madison, Wis.
40. **Lawley, T., and C. Bell.** 1998. Kinetic analyses of Biolog community profiles to detect changes in inoculum density and species diversity of river bacterial communities. *Can. J. Microbiol.* **44:**588–597.
41. **Lee, C., N. J. Russell, and G. F. White.** 1995. Rapid screening of bacterial phenotypes capable of degrading anionic surfactants: development of validation of a microtitre plate method. *Microbiology* **141:**2801–2810.
42. **Lehman, R. M., F. S. Colwell, and J. L. Garland.** 1997. Physiological profiling of indigenous aquatic microbial communities to determine toxic effects of metals. *Environ. Toxicol. Chem.* **16:**2232–2241.
43. **Lehman, R. M., F. S. Colwell, D. B. Ringelberg, and D. C. White.** 1995. Combined microbial community-level analyses for quality assurance of terrestrial subsurface cores. *J. Microbiol. Methods* **22:**263–281.
44. **Lindstrom, J. E., R. P. Barry, and J. F. Braddock.** 1998. Microbial community analysis: a kinetic approach to constructing potential C source utilization patterns. *Soil Biol. Biochem.* **30:**231–239.
45. **Lowit, M. B., L. K. Blum, and A. L. Mills.** 2000. Determining replication for discrimination among microbial communities in environmental samples using community-level physiological profiles. *FEMS Microbiol. Ecol.* **32:**97–102.
46. **Marello, T. A., and B. R. Bochner.** 1989. *BIOLOG® Reference Manual: Metabolic Reactions of Gram-Negative Species.* BIOLOG® and Science Tech Publishers, Hayward, Calif.
47. **Mayr, C., M. Miller, and H. Insam.** 1998. Elevated CO_2 alters community-level physiological profiles and enzyme activities in alpine grassland. *J. Microbiol. Methods* **36:**35–43.
48. **Mills, A. L., and J. E. Bouma.** 1997. Strain and function stability in gnotobiotic reactors, p. 184–194. *In* H. Insam and A. Rangger (ed.), *Microbial Communities: Functional versus Structural Approaches.* Springer-Verlag KG, Berlin, Germany.
49. **MjM Software Design.** 1999. *PC-ORD User's Guide: Multivariate Analysis of Ecological Data.* MjM Software Design, Gleneden Beach, Oreg.
50. **Nordgren, A.** 1988. Apparatus for the continuous, long-term monitoring of soil respiration rate in large numbers of samples. *Soil Biol. Biochem.* **20:**955–957.
51. **O'Connell, S. P., and J. L. Garland.** 2002. Dissimilar response of microbial communities in Biolog GN and GN2 plates. *Soil Biol. Biochem.* **34:**413–416.
52. **Palmer, M. W.** 1993. Putting things in even better order: the advantages of canonical correspondence analysis. *Ecology* **74:**2215–2230.
53. **Pratt, J. R., and J. Cairns.** 1985. Functional groups in the protozoa—roles in differing ecosystems. *J. Protozool.* **32:** 415–423.
54. **Rowell, M. J.** 1995. Colorimetric method for CO_2 measurement in soils. *Soil Biol. Biochem.* **27:**373–375.
55. **Rutgers, M., I. M. van't Verlaat, B. Wind, L. Posthuma, and A. M. Breure.** 1998. Rapid method for assessing pollution-induced community tolerance in contaminated soil. *Environ. Toxicol. Chem.* **17:**2210–2213.
56. **Smalla, K., U. Wachtendorf, H. Heuer, W.-T. Liu, and L. Forney.** 1998. Analysis of BIOLOG GN substrate utilization patterns by microbial communities. *Appl. Environ. Microbiol.* **64:**1220–1225.
57. **Sparling, G. P., L. A. Schipper, and J. M. Russell.** 2001. Changes in soil properties after application of dairy factory effluent to New Zealand volcanic ash and pumice soils. *Aust. J. Soil Res.* **39:**505–518.
58. **Sparling, G. P., and D. Vaughan.** 1981. Soil phenolic acids and microbes in relation to plant growth. *J. Sci. Food Agric.* **32:**625–626.
59. **Stahl, D.** 1997. Molecular approaches for the measurement of density, diversity, and phylogeny, p. 102–114. *In* C. J. Hurst, G. R. Knudsen, M. J. McInerney, L. D. Stetzenbach, and M. V. Walter (ed.), *Manual of Environmental Microbiology.* American Society for Microbiology, Washington, D.C.
60. **Stevenson, B. A., G. P. Sparling, L. A. Schipper, B. P. Degens, and L. C. Duncan.** 2004. Pasture and forest soil microbial communities show distinct patterns in their catabolic respiration responses at a landscape scale. *Soil Biol. Biochem.* **36:**49–55.
61. **Tabachnick, B. G., and L. S. Fidell.** 1996. *Using Multivariate Statistics,* 3rd ed. Harper-Collins College Publishers, New York, N.Y.

62. **Ter Braak, C. J. F.** 1986. Canonical correspondence analysis: a new eigenvector technique for multivariate gradient analysis. *Ecology* **67:**1167–1179.
63. **Torsvik, V., J. Goksøyr, and F. L. Daae.** 1990. High diversity in DNA of soil bacteria. *Appl. Environ. Microbiol.* **56:**782–787.
64. **Vannote, R. L., G. W. Minshall, K. W. Cummins, J. R. Sedell, and C. E. Cushing.** 1980. The river continuum concept. *Can. J. Fish. Aquat. Sci.* **37:**130–137.
65. **Verschuere, L., V. Fievez, L. Van Vooren, and W. Verstraete.** 1997. The contribution of individual populations to the Biolog pattern of model microbial communities. *FEMS Microbiol. Ecol.* **24:**353–362.
66. **Victorio, L., K. A. Gilbride, C. G. Allen, and S. N. Liss.** 1996. Phenotypic fingerprinting of microbial communities in waste-water treatment systems. *Water Res.* **30:**1077–1086.
67. **Wenderoth, D. F., and H. H. Reber.** 1999. Correlation between structural diversity and catabolic versatility of metal-affected prototrophic bacteria in soil. *Soil Biol. Biochem.* **31:**345–352.
68. **West, A. W., and G. P. Sparling.** 1986. Modifications to the substrate-induced respiration method to permit measurement of microbial biomass in soils of differing water contents. *J. Microbiol. Methods* **5:**177–189.
69. **Whitman, W. B., D. C. Coleman, and W. J. Wiebe.** 1998. Prokaryotes: the unseen majority. *Proc. Natl. Acad. Sci. USA* **95:**6578–6583.
70. **Williams, B. K., and K. Titus.** 1988. Assessment of sampling stability in ecological applications of discriminant analysis. *Ecology* **69:**1275–1285.
71. **Winding, A.** 1994. Fingerprinting bacterial soil communities using Biolog microtitre plates, p. 84–94. *In* K. Ritz, J. Dighton, and K. E. Giller (ed.), *Beyond the Biomass: Compositional and Functional Analysis of Soil Microbial Communities.* John Wiley & Sons, Inc., Chichester, United Kingdom.
72. **Winding, A., and N. B. Hendriksen.** 1997. Biolog substrate utilisation assay for metabolic fingerprints of soil bacteria: incubation effects, p. 195–205. *In* H. Insam and A. Rangger (ed.), *Microbial Communities: Functional versus Structural Approaches.* Springer-Verlag KG, Berlin, Germany.
73. **Wodnicka, M., R. D. Guardino, J. J. Hemperly, M. R. Timmins, D. T. Stitt, and J. B. Pitner.** 2000. Novel fluorescent technology platform for high throughput cytotoxicity and proliferation assays. *J. Biomolec. Screen.* **5:**141–152.
74. **Wunsche, L., and W. Babel.** 1996. The suitability of the Biolog automated microbial identification system for assessing the taxonomical composition of terrestrial bacterial communities. *Microbiol. Res.* **151:**133–143.
75. **Yao, H., Z. He, M. J. Wilson, and C. D. Campbell.** 2000. Microbial biomass and community structure in a sequence of soils with increasing fertility and changing land-use. *Microb. Ecol.* **40:**223–238.
76. **Yeates, G. W., D. A. Wardle, and R. N. Watson.** 1993. Relationships between nematodes, soil microbial biomass and weed-management strategies in maize and asparagus cropping systems. *Soil Biol. Biochem.* **25:**869–876.
77. **Zak, J. C., M. R. Willig, D. L. Moorhead, and H. G. Wildman.** 1994. Functional diversity of microbial communities: a quantitative approach. *Soil Biol. Biochem.* **26:** 1101–1108.
78. **Zweitering, M. H., I. Jongenburger, F. M. Rombouts, and K. van't Reit.** 1990. Modeling of the bacterial growth curve. *Appl. Environ. Microbiol.* **56:**1875–1881.

Molecular Approaches for the Measurement of Density, Diversity, and Phylogeny

WEN-TSO LIU AND DAVID A. STAHL

12

This chapter considers the use of molecular methods for direct measures of abundance, diversity and phylogeny of environmental populations of microorganisms. These molecular methods are mainly based on direct nucleic acid sequence recovery, genomic DNA hybridization, and nucleic acid fingerprinting. Although these methods in part cannot be separated from fundamental questions (e.g., species concept and molecular systematics) in ecology and microbial systematics, full coverage of these interrelated topics is beyond the scope of this chapter. Our primary goal is to provide a reasonably complete accounting of available technology and associated methodological biases. This emphasis also results in the exclusion of certain molecular techniques from detailed discussion, since they have not been productively applied to the description of natural systems.

NUCLEIC ACID SEQUENCE MEASURES OF COMMUNITY STRUCTURE

There are three basic formats now used to recover DNA sequence information isolated from either pure culture or environmental samples: DNA probe hybridization, restriction enzyme digestion, and chain termination sequencing of cloned (or PCR-amplified) DNA templates. The former two methods are used to identify relatively short sequence elements. For example, restriction enzymes commonly recognize four to eight nucleotide sequence elements. DNA probes, with the capacity for single-nucleotide-mismatch discrimination, are usually around 20 nucleotides in length. Longer DNA probes are commonly used to identify homologous targets but do not provide defined sequence information. However, these approaches as well as community fingerprinting techniques (see below) all have some basic limitations.

Limitations on Recovery of Nucleic Acids from Environmental Samples

The efficiency and representativeness of nucleic acid recovery (both RNA and DNA) from environmental samples are fundamental concerns in all community studies using molecular approaches. These issues have been thoroughly discussed by a number of investigators (1, 3, 47, 115, 150, 160). For DNA-based analyses, possible biases associated with DNA recovery, PCR amplification, and DNA cloning are well recognized (160, 172). Experience in our laboratories has shown that, even when one uses nucleic acids extracted from pure cultures, PCRs are sometimes inconsistent. Amplification of rRNA gene sequences by using general primers has been shown in some cases to exclude important environmental populations (6, 160, 172). The issues of PCR-generated sequence hybrids (chimeras) and the extent of sequence variation between rRNA operons of individual organisms remain to be fully evaluated but can be partially improved through a PCR conditioning step (57). Thus, the proportional recovery of specific sequences cannot be equated with abundance.

For rRNA recovery, two aspects that are distinguished by the analytical approach are reviewed here. The first is the efficiency of extraction: what fraction of total nucleic acid is recovered from the environmental matrix? For example, although the breakage technique may disrupt all microorganisms present, recovery might be reduced by degradation or adsorption of nucleic acids to matrix material (e.g., clays). The second consideration is representative rRNA recovery. Does the fractional recovery correspond to the environmental abundance of the corresponding nucleic acids present in the environment? For example, a population resistant to breakage would be fractionally underrepresented, or conversely, an exceptionally easy-to-break microorganism would be overrepresented. The use of a universal hybridization probe to evaluate relative breakage efficiency of different groups is discussed below. Even though hybridization to extracted nucleic acids should provide more direct information on abundance, the meaning of abundance as defined in molecular terms is very different from established microbiological criteria, based on CFUs or direct cell counts. Thus, comparison of different measures of total biomass should be very informative. For example, phospholipid and respiratory quinone analyses might provide a relatively independent quantification of biomass that could be related to nucleic acid-based measures (43, 46, 68).

Another important consideration is the suitability of a method for identifying and quantifying environmental nucleic acid sequences. Methods suitable for identification may not be well suited to quantification. For example, DNA restriction fragment length polymorphism (RFLP) cannot be easily interpreted when applied to environmental systems of unknown complexity and population composition. Techniques that rely upon the recovery of relatively intact DNA (e.g., for PCR amplification, cloning, or restriction digestion) generally must use less disruptive extraction techniques than do methods that analyze RNA. This is because DNA is much more sensitive to mechanical shearing than is RNA by the most disruptive of extraction methods, for example, mechanically reciprocal shaking with zirconium beads (142, 144). These are some aspects of nucleic acid recovery that must be more fully resolved before the application of any nucleic acid technology can be routinely applied to environmental studies.

Nucleic Acid Probes

Either DNA or RNA can serve as a nucleic acid probe. However, for a variety of technical reasons (e.g., ease of synthesis and stability), most studies have used DNA probes. There are two basic categories of DNA probes, functional and group specific. Functional probes targeting genes encoding specific enzymes are used to evaluate specific chemical transformations or potential activity of environmental populations (44, 67, 167). Although they will not be specifically addressed in this chapter, functional probes provide an essential connection between the different measures of environmental diversity (phylogeny) and ecology. In this regard, the use of phylogenetic groups as a measure of environmental diversity ultimately must include better understanding of unifying phenotypic characteristics of circumscribed groups. If certain traits are found to reflect membership within a group, these traits should serve to better relate community structure and function. Some examples of traits, and corresponding genes, conserved within phylogenetically defined groups include genes for nitrogen fixation (44), Ni-Fe hydrogenase (167), cellulases within some cellulolytic assemblages (67), and dissimilatory (bi)sulfite reductase (179). Although the emphasis of this chapter is molecular, it is the phenotypic reflection of molecular diversity that must remain the central consideration in environmental microbiology.

Group-specific (phylogenetic or taxonomic) probes generally target conserved biopolymers that can be used to infer phylogenetic relationships among the host organisms (5, 142). rRNA is the most widely used target molecule, and probes can be designed to target phylogenetic groups of differing evolutionary diversity, in other words, to provide explicit measures of community structure at different levels of resolution. Three types of rRNA are common to the ribosomes of prokaryotes and eukaryotes: the 16S (18S for eukaryotes), 23S, and 5S rRNAs. Since the larger rRNAs of many eukaryotes and some prokaryotes differ significantly in size, the terms small-subunit (SSU) rRNA and large-subunit (LSU) rRNA are used to refer to the 16S (18S) and 23S rRNAs, respectively, in this contribution. For a variety of technical and practical considerations (size, information content, and ease of sequencing), the SSU rRNA has become the standard measure for defining phylogenetic affiliation. Well over 160,000 SSU rRNA sequences are now available in the Ribosomal Database Project database (http://www.cme.msu.edu/RDP) (19) along with other databases for LSU and 5S rRNA sequences (24, 146, 157, 158), and the number of sequences is rapidly increasing. They have provided the most encompassing of available molecular frameworks to explore natural microbial diversity and phylogeny (50, 110, 140, 166).

There are two basic formats for using phylogenetic probes to study the environmental distribution of microorganisms. They are hybridization to total rRNA extracted from the environment (114, 116-118, 144) and hybridization to whole cells for subsequent microscopic visualization and enumeration. Nevertheless, a full presentation or review of these two methods and the rRNA-targeted probes used would require extensive referencing of recent applications to environmental, diagnostic, and determinative research. In this regard, the reader is directed to recent reviews (4, 5, 73, 142, 166) and references therein for a more complete description, and we apologize for any exceptional omissions. The following review is intended to provide an overview of key considerations on the design and optimization of rRNA probes in the quantitative analysis of natural systems.

In addition, the use of total genomic DNA probes is another category of group-specific probes for species-level identification (e.g., reverse sample genome probing [RSGP]) and is discussed below in "Genomic DNA Hybridization Measures of Community Structure."

Phylogenetic Probe Design

The essential attribute of the rRNAs with regard to probe design is the regional conservation of nucleotide sequence. Although overall conserved in sequence, these biopolymers exhibit great variation in regional sequence conservation. Some nucleotide positions and locales have remained virtually unchanged since the divergence of all existing life (universal sequences), whereas other regions vary so quickly that they can be used to differentiate among species of bacteria. In addition, the generally high copy number of rRNA per cell lends greater sensitivity to direct detection that uses hybridization techniques.

An extensive set of probes (~15 to 25-mers) has been designed to complement the rRNAs (5, 142), and among them a large fraction has been collected in two different databases (http://www.cme.msu.edu/OPD/ and http://www.probebase.net) (2, 73). Probes vary according to the region of the molecule selected as the hybridization target. The target region defines the diversity of organisms encompassed by a probe. Species-specific probes usually complement the most variable regions. More-general probes, identifying phylogenetic groups of rank greater than species, target more conserved regions of the molecule (142, 144, 166). The publications describing the development of SSU rRNA probes for clinical or environmental applications are much too numerous to address here. Thus, we will not discuss the more specialized applications but will limit discussion to the more general class of probes. Since these probes are designed to encompass larger phylogenetic groupings, they can be used to provide a phylogenetic overview to community structure. Figure 1 presents a very general overview of the character of probe design within a phylogenetic framework. It is an incomplete representation of general probes, and the reader is directed to an online resource (probeBase) for the most up-to-date and complete listing of probes (http://www.microbial-ecology.net/probebase/).

Probe Characterization: Phylogenetic Nesting of Probes

An essential aspect of probe development is the demonstration of target group specificity. To some extent, this can be demonstrated empirically by using a collection of target and nontarget group nucleic acids or fixed reference cells for studies that use whole-cell hybridization (5). Our laboratories and others have routinely used a panel of diverse rRNAs ("phylogrid") to characterize new probes (25, 66, 118). Prior to an evaluation by using the reference panel, it is essential that the temperature of dissociation (T_d) of the

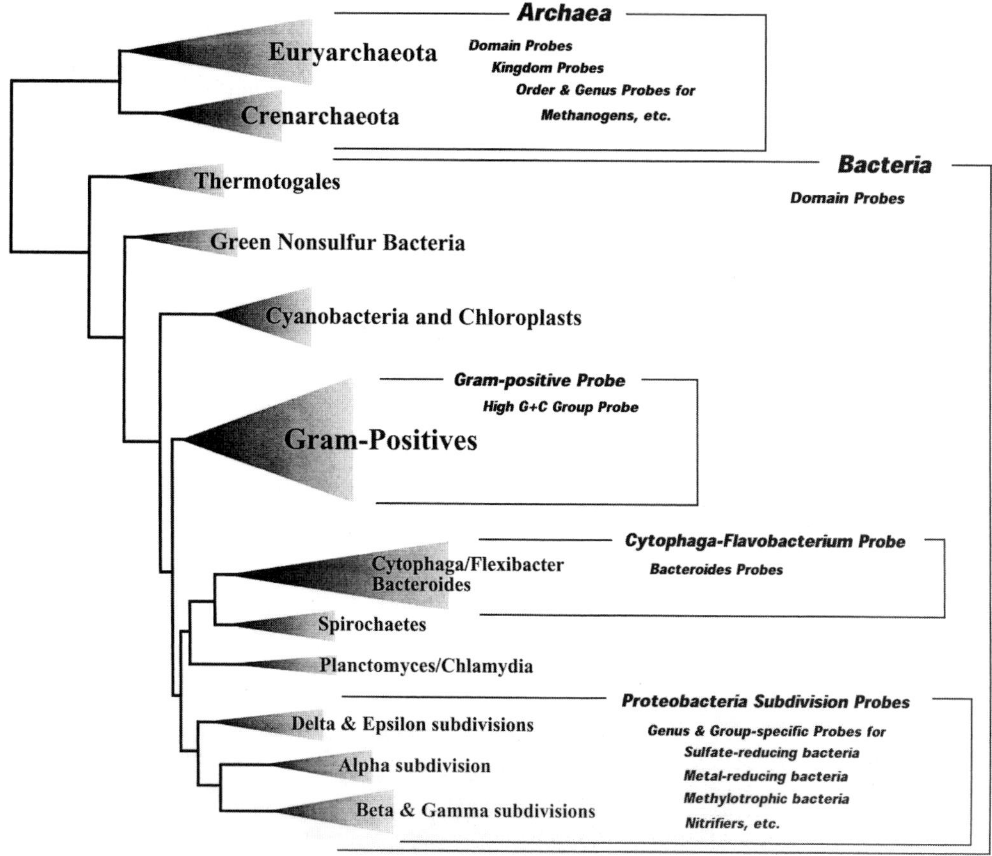

FIGURE 1 Partial listing of group-specific probes in relationship to prokaryote phylogeny. The phylogeny was adapted by Stahl (139) using the maximum likelihood analysis of Olsen, Woese, and Overbeek (107) to provide the relative branching order of the major prokaryotic lineages. The probes for the indicated target groups have been described: archaeal domain (38, 118, 143), bacterial domain (20, 38, 142), archaeal kingdoms (15), gram-positive organisms (S. Toze and D. Stahl, unpublished data), low-G+C gram-positive organisms (88), high-G+C gram-positive organisms (122), *Cytophaga-Flavobacterium* and *Bacteroides* (81), proteobacterial subdivisions (82), sulfate-reducing bacteria (25), methylotrophic bacteria (14, 154), and nitrifying bacteria (89, 165). The reader is referred to these references for more complete descriptions of probe design and characterization.

probe/target complex be experimentally determined (5, 142). It is generally insufficient to use one of the available formulas to predict T_d (142). Also, the temperature interval over which probe dissociates from the target RNA varies considerably for probes having comparable T_ds. Knowledge of the temperature range over which dissociation occurs is essential for adjusting hybridization conditions as needed to discriminate between closely related nontarget species rRNA. Thus, the initial T_d characterization should include closely related nontarget species rRNA directly extracted from pure cultures or synthesized by in vitro transcription of SSU rRNA clones (22a, 89, 113). An independent T_d evaluation must be used to characterize probes used for whole-cell hybridization, since there may be a significant difference between the T_d values and transition temperature ranges determined for the same probe by using different formats such as membrane examination, whole-cell hybridization, and DNA microarray.

There is a limitation to an empirical characterization of probe specificity. If we have only a limited appreciation of microbial diversity, it is impossible to construct a reference panel to unequivocally demonstrate specificity. However, there are a couple of additional methods to further evaluate the situation. The first is to use multiple probes, each having the same target group specificity, to quantify a single target population. For example, two probes for the domain *Archaea* (118) were used by DeLong and coworkers to independently confirm estimates of high archaeal abundance in Antarctic waters (22). Both probes hybridized to the same fraction of SSU rRNA extracted from these waters. The second approach to probe validation takes advantage of the phylogeny. As already noted, it is generally possible to construct probes for phylogenetic groups of various evolutionary depths. These probes are of a hierarchical specificity and have been described as being "nested." The use of a nested set of probes to characterize environmental diversity provides yet another consistency check. If the more specific probes fully represent the larger phylogenetic group, then the sum of the specific probe hybridization values should equal that obtained by using the more general probe. For example, the sum quantification obtained by using a complete set of species-specific probes should equal that of the corresponding genus-level probe (114, 116-118, 144). This approach was used to identify a novel lineage ("species") of cellulolytic bacteria in the equine cecum (66) and was frequently used to evaluate consistency between domain-probe summation and total population abundance determined by using a universal probe (116).

General caveats relating to the use of phylogenetic probes include the following: the probes are tools subject to refinement through experimentation. Only through general application and combination use with other methods of community analysis will they be fully evaluated or, as necessary, refined. Also, the resolution of the SSU rRNA probes is approximately at the level of species. Questions relating to the abundance and distribution of subspecies and strains will require the combination of different approaches and methods (e.g., fluorescent antibody techniques). Another concern relates to the extraction of RNA from environmental samples. Although resistant to mechanical breakage, RNA is more subject to degradation during and following extraction, generally as a consequence of endogenous nucleases or nuclease contamination. One consequence of partial degradation of sample is variable destruction of different probe target sites. For example, one of the regions used as a target site for hybridization to a "universal" probe is very sensitive to degradation (120). Thus, for methods of quantification that use extracted rRNA, it is essential that sample integrity be evaluated. This is most conveniently accomplished by using acrylamide gel electrophoresis to demonstrate recovery of high-molecular-weight species (1, 3, 128). Furthermore, the probe target position within the rRNA significantly affects the hybridization efficiency and specificity when used for hybridization to whole cells, e.g., fluorescence in situ hybridization (FISH) (36). Based on a systematic analysis that used 171 nucleotide probes (mostly 18-mers) to encompass the SSU rRNA of *Escherichia coli*, the hybridization signal (fluorescent intensity) obtained from whole-cell hybridization was shown to vary as much as 75-fold among probes. Analysis of signal intensity relative to target position suggested that hybridization intensity was highly related to higher-order structure within the rRNA and could be classified into six different "brightness" regions (36). This information is very useful for probe design with commercial packages or free software available on the Internet (e.g., ARB) (76) and for improving the whole-cell hybridization signal (35). For example, by use of unlabeled oligonucleotides (helper probes) adjacent to the target region of a fluorescently labeled probe in whole-cell hybridization, the hybridization efficiency could be improved significantly with an enhancement of signal intensities varying from 4- to 25-fold (35).

Restriction Enzymes

The most common format for using restriction enzymes to define environmental diversity is to combine digestion and fractionation of DNA extracted from an environmental sample with hybridization by using nucleic acid probes complementary to conserved gene sequences common to all or many of the organisms present in the sample. The general format was first described by Southern and is often termed "Southern blotting" (137). The rRNAs are optimal targets for hybridization (ribotypes) (41), but other conserved elements also have been used for environmental studies (e.g., nitrate reductase and formyltetrahydrofolate synthetase) (72, 135). In application, restricted DNA is separated by size on an agarose gel and transferred to a membrane support for hybridization with radiolabeled probe or a label appropriate for use with one of a variety of nonradioactive detection formats (e.g., digoxigenin) (142). The resulting population of different-sized DNA fragments hybridizing to the probe is then used to infer relationships between individual isolates or resolve different environmental populations. The separation of genes derived from different populations requires that they differ in sequence at the sites of DNA restriction or differ in length of DNA flanked by common restriction sites. For this reason, more than one restriction enzyme is generally used for restriction enzyme analysis and the resulting size distribution patterns (banding patterns) are compared.

RFLP analysis has been used to characterize extracted total DNA (164) or specific PCR-amplified DNA (92). Sequence diversity is evaluated by digesting the native or amplfied DNA with a restriction endonuclease(s) followed by size fractionation by electrophoresis on an agarose or a polyacrylamide gel. The separation of genes derived from different populations requires that they differ in sequence at the sites of DNA restriction or differ in length of DNA flanked by common restriction sites. For this reason, more than one restriction enzyme is generally used for RFLP analyses.

The primary use of RFLP analysis has been to evaluate sequence variation among rRNA gene sequences. This general method is sometimes termed "amplified rRNA restriction analysis" and is commonly used to estimate diversity among different microbial isolates (60) or rRNA clones recovered from the environment (92). It requires the use of PCR primers complementary to conserved regions of the SSU or LSU rRNA genes or the 16S-23S intergenic regions (87) and high-resolution agarose gels to effectively separate fragments with length down to approximately 50 bp. Computer-simulated RFLPs on 106 bacterial SSU rRNA sequences from representative bacterial phyla indicate that combined RFLPs of at least three separate digestions with tetrameric restriction enzymes (i.e., four-base recognition site) are required to resolve those bacterial populations to the genus level (median sequence identity $< 96.1\%$) (93). The efficacy of detecting and differentiating bacterial rRNA genes is dependent upon selection of appropriate restriction enzymes, since the RFLP patterns are not due to true restriction site polymorphism but result primarily from insertions and deletions in the SSU rRNA sequence (13). A major limitation of this approach is that it is very time consuming and cannot be used to screen a large set of isolates or clones in a cost-effective manner.

RFLP has also been used for the analysis of environmental DNA. For example, restriction digestion of rRNA gene sequences recovered by PCR amplification can produce a banding pattern (i.e., band numbers and intensity) that serves as a fingerprint of the community. The fingerprint provides a quick assessment of genotypic changes that may result from temporal or spatial changes within a habitat (86). RFLP can be combined with hybridization by using group or functional probes to better define microbial diversity in an environmental sample. However, this approach is of limited use for demonstrating the presence of a specific phylogenetic group or for estimating species richness and evenness, since each SSU rRNA gene amplicon can contribute to multiple restriction fragments (93).

To further apply restriction enzyme for quantitative analysis of microbial community structure, we highlight a new method that utilizes the endonuclease property of RNase H to cleave the RNA strand of RNA-DNA hybrid duplexes at a specified site into two fragments (156). The first step of this method is to direct the cleavage site of the RNase H on RNA molecules by hybridizing the target RNA with an olignucleotide probe or "scissor" probe. After the formation of DNA (i.e., scissor probe)-RNA hybrids,

RNase H is added to cleave the RNA strand at the probe-binding site, followed by quantification of the cleaved RNA and intact RNA by using capillary electrophoresis systems. By varying the hybridization conditions (e.g., formamide), the digestion reaction, and probe specificity (i.e., perfectly matched versus mismatched), curves for cleavage efficiency of the RNA strand of perfectly matched and mismatch probe-RNA duplexes with respect to formamide concentrations can be produced. By selecting an optimal formamide concentration, RNase H is shown to cleave the RNA strand of perfectly matched duplexes but not mismatched duplexes. By use of rRNA-based scissor probes at different levels of phylogenetic specificity (e.g., domain, group, and species), this method was demonstrated to successfully quantify bacterial and archaeal fractions in model and complex communities and to detect a specific microbial population (≥1 or 2% of total rRNA) within a microbial ecosystem. The sensitivity of detection and quantification of microbial populations is reported to be dependent on (i) the use of high-quality intact rRNA extracted from the environments studied, (ii) the mismatched position of the nontarget rRNA molecules, and (iii) the resolution of electrophoresis systems for small cleaved rRNA fragments (156).

Direct Sequence Analyses of SSU rRNA

By far, the sequence information most commonly extracted from natural systems is that for the rRNAs. There are three basic methods to recover rRNA sequence information from nucleic acid extracted from environmental samples (see reference 124 for a detailed description). They are (i) shotgun cloning, screening, and sequencing (106, 110, 129); (ii) cDNA cloning and sequencing of rRNA (168); and (iii) PCR amplification, cloning, and sequencing (6, 166). The PCR-based methods can use either DNA or RNA as template, the latter requiring the use of reverse transcriptase to generate cDNA from rRNA. These sequencing approaches all require the screening and analysis of large clone collections. The screening of a shotgun library derived from total environmental DNA is the more difficult, since only a small fraction of the clones contain part or all of the rRNA gene, occurring at about 0.125 to 0.3% of the clones (129). They can be identified by hybridization, for example by using total rRNA derived from the environmental sample as probe. Alternatively, DNA probes targeting highly conserved regions of the molecule (phylogenetic probes) may be used. This first screening step is generally not necessary for PCR-based recovery methods, since the majority of clones will contain rRNA sequences. The second phase of the analysis is the elimination of redundant clones in order to avoid expense and time associated with unnecessary sequence determinations. A variety of approaches have been used to identify redundant clones, using or combining the following strategies: complete or single-nucleotide sequencing of a small variable region, species- or group-specific phylogenetic probe screening, restriction analysis, and denaturing gradient gel electrophoresis (DGGE) (97, 106, 166). A more recently described method for rapid sequence analysis of rRNA sequence types in environmental samples employs a series of enzymatic reactions to amplify and ligate short sequences (sequence tags) from a variable region of the rRNA gene. The ligated products, each containing as many as 20 sequence tags, are then cloned and sequenced. This approach, termed serial analysis of ribosomal sequence tags, provides a much higher-throughput survey of sequence diversity but has the associated limitation of recovering much less sequence information for each rRNA variant (59, 100).

The resolution of the different screening techniques must be balanced against time and expense associated with each strategy. This determination can be made only in consideration of community diversity (number and frequency of different rRNA genes) and research objectives. The general approach as developed for community-level analyses based on rRNA sequence content could be applied to any biopolymer of appropriate conservation and community representation.

GENOMIC DNA HYBRIDIZATION MEASURES OF COMMUNITY STRUCTURE

Genomic DNA hybridization is mainly used in bacterial systematics to determine the degree of genetic similarity between genomic DNA sequences and thus provides limited information of specific sequence content. The extent to which these methods can be used to identify individual populations or to estimate genomic diversity (total number of different genomes) in a microbial community varies with method, genomic diversity, and the aforementioned questions of microbial species definition. Two methods are introduced.

RSGP

RSGP is developed by assuming that the entire genome of a microorganism can be used as a specific probe for its detection in the environment (for detailed review, see reference 40). Whole-genome probes have been used to detect *Mycobacterium*, *Mycoplasma*, *Chlamydia*, *Bacteroides*, *Pseudomonas*, *Sphingomonas*, and *Campylobacter* species (7, 28, 29, 49, 91, 134, 169). RSGP reverses the usual relationship of sample DNA and probe. The genomic DNA from different reference organisms is denatured and immobilized on a membrane support, the reference panel. DNA extracted from the environment, containing an unknown diversity of organisms, is randomly labeled and is hybridized to the reference panel. Under conditions of high stringency, whole-genome probes hybridize only to identical or closely related genotypes at the level of species or subspecies. For example, when a reference panel of DNA from different sulfate-reducing bacteria was hybridized with randomly labeled genomic DNA from any species represented on the panel, only self-hybridization or hybridization with nearly identical isolates was observed (162, 163).

The following overview addresses only key technical considerations (161–163). The method requires the antecedent isolation of reference organisms from the environment. Chromosomal DNAs are then isolated from the different strains, measured amounts are applied to membranes, and cross-hybridization among strains is evaluated by using "stringent" hybridization conditions that are empirically defined by the researcher. Strongly cross-hybridizing DNA preparations could be combined and treated as the same standard (either a single species or a set of closely related species). Different reference standards are bacterial genomes that generally show undetectable cross-hybridization signals under stringent hybridization but may have an arbitrary DNA/DNA hybridization homology below or above 70% as used for the "species" definition. The standards are then used to prepare a master filter, using bacteriophage lambda as an internal control. The amount of genomic DNA applied to the membrane varies with analytical need. For example, 20 ng is needed for analysis of DNAs obtained from pure or enrichment cultures, whereas

200 ng was appropriate for analysis of total community DNAs reported in one study. A reference concentration series of bacteriophage DNA is applied on the same membrane (e.g., 10, 20, 50, or 100 ng). Sample DNA (ca. 100 ng) and lambda DNA (ca. 200 pg) are combined, boiled, and placed on ice. A probe is prepared by random hexamer labeling by using [α-^{32}P]dCTP and Klenow polymerase. Following denaturation, the probe is hybridized to a master filter under stringent conditions. Following washing of the filter under defined conditions, bound probe is quantified (e.g., by using autoradiography), and the fraction of community DNA composed of individual component genomes (represented by the individual reference DNAs on the master membrane) is calculated from the hybridization to individual DNA standards relative to the lambda reference series (161). This practice assumes that the lambda DNA and environmentally derived DNA present in the probe mixture are labeled to the same specific activity and hybridize with comparable efficiency. One possible concern is that impurities associated with the environmental DNA may influence both relative labeling and extent of hybridization.

The technique was first applied to the analysis of microbial populations associated with oil fields (161–163). RSGP revealed a significant difference between planktonic and biofilm-associated populations in oil recovery systems. Planktonic populations were more diverse and dominated by organochemotrophs. In contrast, biofilm populations were typically dominated by one to three populations of sulfate-reducing bacteria from the *Desulfovibrionaceae*, with much lower representation by organochemotrophs. More recently, this technique was used to evaluate the impact of pollutants (e.g., benzene and toluene) on soil populations (49, 134). In these studies, a master filter was first prepared by spotting genomic DNAs of bacteria isolated from the contaminated soil. Cross-hybridization among these bacteria was first determined to normalize the environmental hybridization patterns. This study documented a clear impact of the pollutant on community structure and metabolic state of the system.

The RSGP technique is advantageous in that, once an appropriate microbial survey of the target environment has been completed, master filters can be prepared rapidly and economically in large numbers. These filters can be stored indefinitely for immediate use when new sample DNAs become available for analysis. Further, these master filters can incorporate newly isolated bacterial DNA by spotting on side strips that are hybridized with the sample DNA probe together with the master filter. Consequently, routine screening of sample DNAs against a large number of different standards is feasible when a large set of sample DNAs is simultaneously prepared.

A disadvantage of the technique is that, although the actual assay does not involve culturing, the microbial community is described only in terms of its culturable component. It generally accounts for a very small portion of the bacterial populations in a given environment. The calculated fractions can also be subject to systematic errors even though RSGP has good precision. For example, the calculated fractions are sensitive to label allocation to sample and internal standard DNA (i.e., differences in specific activity). Furthermore, since detection sensitivity of RSGP is defined by the extent of cross-hybridization of the standard DNAs, low-abundance populations that hybridize below this experimentally defined threshold may not be detected in the presence of related organisms.

DNA Reassociation

The measurement of DNA reassociation kinetics was initially used to evaluate genomic sequence complexity, revealing repetitive DNA elements in the genomes of higher eukaryotes (11). More recently it has been used to assess the diversity of natural microbial communities (152, 153). Community-level DNA sequence complexity, as inferred from the rate of DNA reassociation, is related to population complexity. This measurement of complexity is a function of the concentration of complementary strands. Under defined conditions, strand reassociation follows second-order kinetics. Thus, the rate of reassociation is proportional to the square of the nucleotide concentration of homologous DNA strands. At a given concentration of total DNA (molar concentration of nucleotides in single-stranded DNA [ssDNA]), increasing genomic complexity (larger genomes, larger numbers of genomes, and fewer repeat elements per genome) results in a reduced concentration of complementary strands and a correspondingly reduced rate of reassociation. For example, as microbial community diversity (heterogeneity) increases (e.g., greater number of unique genomes), the rate of reassociation of DNA extracted from the community decreases for the same concentration of total DNA.

Experimentally, DNA reassociation is measured over time, and the fraction of reassociated DNA (C/C_0) is expressed as a function of C_0t, where C_0 is the initial molar concentration of nucleotides in ssDNA and where t is the time in seconds (Fig. 2). The plot of this relationship is referred to as a C_0t curve. The reaction rate constant can be expressed as $1/C_0t_{,1/2}$, where $t_{1/2}$ is the time required for 50% reassociation. Under defined conditions, with temperature and monovalent ion concentration most important, $C_0t_{,1/2}$ is proportional to the complexity (e.g., number of unique genomes) of the DNA. The practical and theoretical considerations of DNA reassociation are well developed (11, 170), and the reader is referred to these readings for a more complete theoretical and practical treatment.

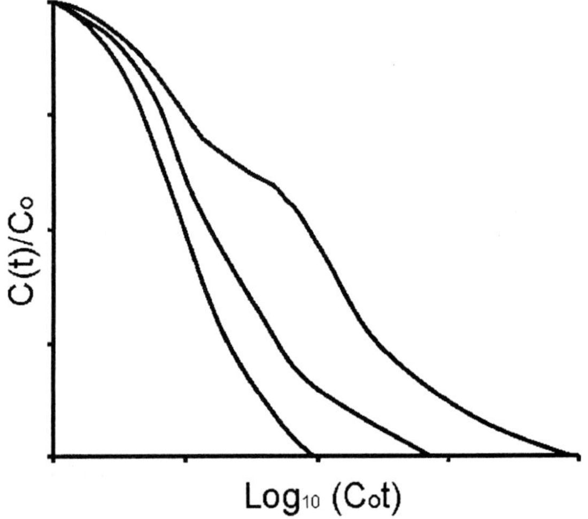

FIGURE 2 Typical C_0t curves obtained by reassociation of nucleic acids. The method can be used to estimate genome size or number by comparison to a reference genome represented by the left curve. The left and middle curves display pseudo-first-order kinetics of unique sequence DNA of different complexities, while the right curve displays pseudo-second-order kinetics.

In the application of community diversity analysis, the interpretation of DNA reassociation kinetics is made in the context of information theory, as has been developed for other diversity indices and briefly discussed above (152). It is a measure of the total amount of information in a system (richness and number of unique genomes) and the distribution of that information (evenness and abundance of individual genomes). Torsvik has expressed diversity as the number of "standard" genomes with no homology (151, 152). The results are significantly notable in that they suggest far greater diversity than anticipated. In an initial study of a soil sample taken from a beech forest (Seim, Norway), reassociation kinetics suggested the presence of approximately 4,000 genomes (grams of soil [wet wt])$^{-1}$ (152). Another study examined the effect of heavy metal contamination on the bacterial communities of soils in Braunschweig, Germany, with different levels of sludge amendments (125). Approximately 16,000, 6,400, and 2,000 bacterial genomes were estimated in the noncontaminated soil, low-metal amendments, and high-metal amendments, respectively. The number of bacterial genomes in noncontaminated soils was fourfold higher than reported previously (152). The increase in bacterial genome number in soil is mainly attributed to the use of a longer time period at a constant temperature to measure the C_0t curve. Thus, a higher $C_0t_{1/2}$ value can be obtained by taking into account not only the dominating DNA types but also the less dominating ones, as the reaction kinetics is assumed to follow pseudo-second-order reaction (Fig. 2). This study further indicated a significant reduction of genomic diversity (ca. up to eightfold) due to the metal amendments.

Although DNA reassociation provides a generally useful measure of genomic diversity, a variety of parameters must be considered, and most have yet to be systematically evaluated. One concern in the interpretation of DNA reassociation estimates is a reduction in the rate of reassociation resulting from impurities in the DNA sample. For example, Torsvik showed that the rate of reassociation increased with repeated purification of the sample DNA. This phenomenon can be evaluated in part by the addition of exogenous DNA to serve as an internal control (152). However, it is also important to more fully evaluate changes in reassociation kinetics that might result from the use of different extraction and purification techniques (150). Another consideration for any DNA-based analysis is the source of the DNA. The persistence of "inactive" DNA, either in the environment or entrained within dead or moribund cells, is essentially unanswered and is a concern in interpreting any data obtained solely from DNA. Recently, Gans and coworkers have indicated that the information theory used can significantly affect the estimation of genetic diversity using DNA reassociation by 100-fold (36a). By using computational analysis that enables direct comparison of different abundance models, this study reevaluated the results reported by Sandaa et al. (125). Approximately, 8.3×10^5, 6.4×10^4 and 7.9×10^3 genomes (g of soil [wet wt])$^{-1}$ were estimated for the noncontaminated soil, low-metal amendments, and high-metal amendments, respectively. This estimation is significantly larger than the original values, by a factor of 4 to 500. This improved computation analysis supports the use of DNA reassociation kinetics to study important questions related to the genomic diversity in different microbial ecosystems, for example, due to environmental perturbation and mapping of diversity geographically. This example further highlights the need for technical improvement.

NUCLEIC ACID FINGERPRINTS OF COMMUNITY STRUCTURE

The term "molecular fingerprint," otherwise rendered as "community fingerprint" or "phylogenetic fingerprint," is reserved for methods of analysis that generate a pattern-based characterization of community structure, most commonly represented by a banding pattern of nucleic acid fragments resolved by gel electrophoresis. Commonly used molecular fingerprint methods as summarized in Table 1 can be generalized based on the theory of pattern classification into three categories: (i) RFLP of total DNA or PCR products, (ii) fragment length heterogeneity of rRNA or PCR amplicons, and (iii) conformation or melting behavior of DNA products. Although DNA microarrays generate a type of fingerprint, the pattern of hybridization is mapped to sequence information and we treat this format separately. In principle, any gene can be explored by these fingerprint methods, but SSU and LSU rRNA genes that exhibit the trait of a molecular chronometer as aforementioned are presently the most commonly used ones. Other candidate markers include genes encoding the protein elongation factor, heat shock proteins, glutamine synthetase, ATPases, and topoisomerases, and the present databases for these genes are gradually improved.

Molecular fingerprinting methods can be used for rapid surveys using genes that provide for either phylogenetic or functional assessment of populations present in an environmental sample. Species richness and species evenness are two of those important ecological elements that can be inferred by using these methods: the former is usually estimated from the total number of different observable genetic units. The total number is determined by physical separation (e.g., electrophoretic separation of PCR amplicons by length, sequence, or conformational variation). Depending on the method and genetic marker used, each unit is often defined as a population, an operational taxonomic unit, a ribotype, a phylotype, or a genotype. However, most of these units provide little or no direct information of specific microbial population identity prior to further analysis, for example, by band purification and sequencing (68, 97). Furthermore, the extracted microbial ecology information is subject to general pitfalls associated with the use of molecular fingerprint methods (e.g., variability in cell lysis, DNA/RNA extraction, interoperon difference of SSU rRNA genes, SSU rRNA gene copy number, and PCR amplification) (see references 160 and 172). The following discussion of methods will address the theory of classification and the approaches used in the characterization of microbial communities in various environmental systems. Due to the large number of methods developed, only those commonly used ones are described in detail. Table 1 provides general advantages and disadvantages associated with individual methods.

T-RFLP

Terminal RFLP (T-RFLP) of PCR-amplified DNAs is a refined fingerprinting technique based on RFLP. The general steps include PCR amplification of a conserved target sequence (most commonly a region of the SSU rRNA gene) followed by restriction enzyme digestion and gel fractionation of resulting fragments. However, one of the two PCR primers is fluorescently labeled at the 5′ end. This labeling results in PCR amplification products that are tagged with a fluorescent dye at only one terminus. Following restriction enzyme digestion (usually with a

TABLE 1 Molecular fingerprinting methods for microbial community identification, monitoring, and tracking

Method(s)[a]	Procedure	Commonly used genetic markers	Diversity resolution	Advantages	Disadvantages	Reference(s)
T-RFLP	PCR-amplified DNA, restriction enzyme digestion, and terminal fragment size separation by a DNA autosequencer	rRNA, rRNA gene, conserved functional genes	By restriction site, length	Good gel-to-gel comparison; possible prediction of phylogenetic affiliation based on terminal fragment size and restriction enzyme(s) used	Only semiquantitative; limited resolution using unspecific primers; overestimate of diversity (>1 species from one terminal fragment)	69
PCR-ALH/ ARISA	PCR-amplification, size separation by a polyacrylamide gel or a DNA autosequencer	rRNA, rRNA gene, 16S-23S intergenic spacer	By length	Good gel-to-gel comparison; possible sequence identification of interested fragments	Only semiquantitative; limited resolution using unspecific primers; overestimate of diversity (>1 species from one fragment)	121, 145
SSCP	PCR-amplified DNA, heat denaturation, polyacrylamide gel electrophoresis, possibly by a DNA autosequencer	rRNA, rRNA gene, conserved functional genes	1-bp difference	Good gel-to-gel comparison	Only semiquantitative; limited resolution using unspecific primers; no sequence information; overestimate of diversity (multiple conformed-ssDNA bands from one dsDNA fragment)	61, 178
DGGE/ TGGE	PCR-amplified DNA, electrophoresis in a polyacrylamide gel with an increasing gradient of denaturants or temperature	rRNA, rRNA gene, conserved functional genes	1-bp difference	Inexpensive equipment setup; possible sequence identification using sequencing or hybridization	Poor reproducibility; limited resolution using unspecific primers; not quantitative; cumbersome gel preparation and staining; overestimate of diversity (formation of chimerical DNA, >1 band from one species)	96
Clone libraries	PCR-amplified DNA, clone screening, and sequencing with a DNA autosequencer	rRNA, rRNA gene, conserved functional genes	By nucleotide sequences	Fine-scale community structure analysis; identification of new genes	Expensive and labor intensive; only for dominant community members	22, 110
DNA microchip	Massively parallel hybridization among labeled targets and stationary oligonucleotide probes	RNA, rRNA gene, conserved functional genes	1-bp mismatch	High throughput; good chip-to-chip comparison; standardized procedure	Expensive for chip production and equipment setup; uncertainty with signal specificity (i.e., false-negative and false-positive signals); only for known target sequences	42, 71, 173, 174

[a] TGGE, temperature gradient gel electrophoresis.

tetrameric restriction enzyme), the restricted products are resolved by using an automated DNA sequencer equipped with a laser-induced fluorescence detector. Only the fluorescently tagged terminal fragments are detected and quantified. This is an extremely sensitive technique, since as little as 100 amol of a fluorescent tag (e.g., 6-carboxy-fluorescein) can be detected by using an ABI DNA autosequencer. Fragment lengths can be assigned by comparison to a set of size markers labeled with a different fluorescent dye and incorporated in each sample prior to electrophoresis (69). However, sizing accuracy is reported to be affected by the size calling method employed (109), the fragment length, and the type (i.e., gel-slab versus capillary) of electrophoresis systems used (99). As a whole, each T-RFLP pattern can be regarded as a fingerprint of population structure that provides information of population diversity (each terminal fragment and associated restriction site sequence define different "ribotypes") and a semiquantitative estimate of relative abundance (peak area). This information can be used for rapid comparative analysis of microbial and functional gene diversity in various microbial ecosystems.

Genes coding for the rRNA sequences are the most commonly used markers in microbial community analysis by T-RFLP. Normally, primers specific for the domains *Bacteria* (69), *Eucarya* (83), and *Archaea* (159) were used in the PCR to analyze the overall microbial diversity. Prior to the experiment, the proper combination of primer and restriction enzyme that gives a better resolution could be systematically evaluated in silico by computer simulation programs such as PatScan (http://www.unix.mcs.anl.gov/compbio/PatScan/HTML/patscan.html) (27) and TAP T-RFLP (http://35.8.164.52/html/TAP-trflp.html) (84). For example, computer-analyzed T-RFLP for 686 amplifiable sequences (8–927, *E. coli* numbering) from 1,102 complete SSU rRNA sequences of the Ribosomal Database Project indicated that those sequences could be classified into 233 different terminal restriction fragments (i.e., ribotypes) (69). Apparently, because of the high conservation of the restriction site positions with the SSU bacterial rRNA sequences, one terminal fragment may actually include one or more than one different or closely related sequences. Thus, T-RFLPs could underestimate the levels of microbial diversity by a factor of three or four, could resolve the phylogenetic population only at the level of higher-order groups, and could not correctly reflect the phylogenetic position based on the terminal restriction fragment lengths without prior knowledge of the bacterial composition (69, 84). To further increase the sensitivity of T-RFLP analysis, a hierarchical approach by employing phylum- and group-level (or below) primers is applied, for example, for the studies of sulfate-reducing bacteria (171), the *Acidobacteria* (58), the *Actinobacteria* (70), and the *Planctomycetes* (23).

A growing number of studies have demonstrated T-RFLP as a sensitive, reproducible, and robust method for the comparative analysis of microbial diversity in environmental samples. On average, 30 to 50 predominant terminal restricted fragments (i.e., different operational taxonomic units, or ribotypes) are observed within a microbial ecosystem. The least predominant account for less than 0.2% of the total amplified community rRNA (18, 54, 69). This resolution is equivalent to the screening of approximately 500 rRNA clones and was suggested to be slightly more sensitive than another popular fingerprint method, DGGE (see below), based on a comparative study of the same marine bacterioplankton community (90). Furthermore, the digitized molecular fingerprints derived from individual samples provide for a rapid and analytically based comparison of community similarity by the combined use of image analysis software (e.g., GelCompar by Applied Math, Inc.) and cluster analysis (69) or principal-component analysis (18).

T-RFLP provides an advantage over most molecular fingerprinting methods in that phylogenetic inference of predominant terminal fragment lengths within a community can be made once an appropriate database for the given sample is constructed (17, 80, 159). For example, the study by van der Maarel et al. (159) demonstrated that the observed archaeal terminal fragment lengths found in digestive tracts of marine fish could be assigned to specific marine archaeal sequences recovered from the same sample. The same approach was used in other studies (17, 80) to identify the predominant terminal fragment lengths found in archaeal or bacterial populations under changed environmental conditions. Furthermore, there are several software programs developed to infer the possible phylogenetic affiliation of predominant terminal restriction fragments within a sample. These include TAP T-RFLP (http://35.8.164.52/html/TAP-trflp.html), the T-RFLP Phylogenetic Assignment Tool (PAT) (http://trflp.limnology.wisc.edu/index.jsp), Microbial Community Analysis (MiCA) (http://mica.ibest.uidaho.edu/), and T-RFLP fragment sorter version 4.0 (http://www.oardc.ohio-state.edu/trflpfragsort/default.htm). However, the inference of phylogenetic positions based on terminal fragment length should be carefully intepretated. It is possible that the predicted fragment length can be different from observed ones due to possible variations occurred during sequencing and fragment sizing. Further, the SSU rRNA database still cannot present the total microbial diversity in microbial environments well.

In addition to community structure information, T-RFLP provides a basis for evaluating relative activity of individual populations. One of the most general measures of cellular activity is ribosome content, and the ratio of rRNA to rRNA genes generally increases with increasing growth rate (activity). Thus, comparison of T-RFLP fingerprints derived from both the rRNA gene and the rRNA of an environmental sample serves to confirm that microbial populations detected at the rRNA gene level are metabolically active and also provides some information of relative activity (80, 90). T-RFLP fingerprinting could be further used to correlate community shifts and activities with functional change by using a self-learning neural network analysis (26), community-level physiological profiles (BIOLOG) (54), and a stable-isotope probing technique (79). Since the use of an autosequencer allows for simultaneous analysis of at least 96 different samples, T-RFLP can be used as a high-throughput method to rapidly monitor community changes associated with temporal and spatial variation in the environment or occurring in response to environmental perturbations.

T-RFLP has also been extensively applied to examine variation among different functional genes in different environmental settings. The first study by Bruce (12) showed that T-RFLP could be used to rapidly characterize the genetic heterogeneity of a mercury resistance gene among different soil types (i.e., polluted and pristine) and within different areas of the same soil. Comparison of these environmentally derived T-RFLP patterns to the current sequence database suggested the presence of novel *mer* genes. Another two studies successfully used T-RFLP to reveal the diversity and expression of nitrogen fixation genes (*nif*H and *anf*) in different termite guts that exhibited different levels of nitrogen fixation or in a termite species under different feeding modes (102, 104). The other functional genes used include monooxygenase (*amoA*) (48), ammonia monooxygenase (*amoA*) gene, nitrous oxide reductase (*nosZ*) gene (123), and methyl-coenzyme reductase (*mcrA*) (77).

As for all molecular methods dependent upon DNA isolation and PCR amplification, the results of a T-RFLP characterization may not fully reflect the native population structure. It is reported that highly degenerated primers can cause a biased representation on the abundance of individual targets (78). Possible "pseudo" terminal restriction fragments can also be produced due to single-stranded amplicons in the T-RFLP but can be eliminated by digesting these pseudofragments with single-strand-specific mung bean nuclease prior to T-RFLP analysis (30). Osborn et al. (109) also demonstrated that some of the rRNA-related biases can be minimized if the key experimental parameters (e.g., sample replication and handling, PCR DNA template concentration, PCR cycle time, *Taq* polymerase types, and the minimum restriction enzyme digestion time) are systematically evaluated.

PCR-ALH and Automated Ribosomal Intergenic Spacer Analysis (ARISA)

The estimation of microbial diversity by PCR-amplicon length heterogeneity (ALH) is based on the capacity to resolve the length and length heterogeneity of the PCR amplicons from the variable regions encoding the SSU rRNA (121, 145). To do so, domain-specific PCR primers are initially used to amplify DNA fragments flanking the variable region(s). The resulting amplication products are then separated by size on polyacrylamide gels and are viewed by using a sensitive staining method, e.g., silver staining. Like T-RFLP, fluorescently tagged primers and automated analysis have been used to increase the sensitivity and reproducibility for detecting small amounts of DNA and to separate fragments differing by only 1 or 2 bp (34, 145).

The resolution of PCR-ALH is dependent on the variation within the coding regions. Among 366 marine bacterial SSU rRNA sequences examined, two variable regions corresponding to *E. coli* numbering 8–355 and 8–536 have length variations of 312 to 363 bp and 472 to 574 bp and could be used to classify microbial populations for up to 52 and 103 different categories, respectively (145). Since amplicon length generally corresponds to phylogenetic relationship, this classification has a natural order. To further infer the phylogenetic position of the organisms represented by different length amplicons, the fragments can be cloned and sequenced. Alternatively, PCR-ALH can be used to analyze a microbial community for which complete or nearly complete sequences of the target genes are available (e.g., via selective amplification, cloning, and sequencing of SSU rRNA genes). However, since most fragment-length categories contain more than one different but closely related sequences, the resolution of the different natural groupings is somewhat compromised. This approach is further limited by the precision of fragment sizing when 1-bp differentiation between two fragments is needed for discrimination or when long fragments are compared (145).

Similar to PCR-ALH, ARISA was developed for estimating microbial diversity (9, 34). It takes advantage of a range of length variation in the 16S-23S rRNA gene intergenic spacer (usually encoding tRNAs) wider than that of the coding regions for the rRNA molecules (34, 37). A recent study indicated that the measurement of microbial diversity in terms of the number of peaks, the range of peak size, and the reproducibility of ARISA fingerprints can be significantly affected by the use of different forward and reverse primers (16). This issue was likely related to the poor design of primer sets based on the limited numbers of rRNA sequences in the database previously and could be improved by using a new set of forward and reserve primers encompassing most of the 16S and 23S rRNA sequences in the database with similar anneal temperatures (16). It should be further noted that (i) 16S-23S spacer length variation does not necessarily correspond to different populations, since individual organisms generally contain multiple operons that may encode spacers of different length and (ii) intergenic DNA fragments of the same length can encompass more than one species of microorganism. Thus, these factors limit the use of ARISA for direct inference of microbial diversity. Overall, PCR-ALH and ARISA remain useful methods for comparative analysis of natural microbial populations along different temporal and spatial gradients (9, 34, 145).

SSCP

Single-strand conformation polymorphism (SSCP) is a technique that separates ssDNA fragments differing in strand length and conformation (conformers) by gel electrophoresis. SSCP has been most frequently used in conjunction with PCR to generate the target DNA fragments. SSCP-PCR was initially used to detect minor sequence variants caused by point mutations in human alleles (108) and only more recently has been used to analyze microbial community structure (61, 130, 178). Immediately before the SSCP analysis, the PCR-amplified rRNA is denatured at a high temperature (i.e., 95°C for 2 or 3 min), usually in the presence of denaturants (e.g., formamide). This process is followed by rapid cooling on ice to "trap" different structural forms. The folding of an ssDNA into different stable structures is highly related to its sequence (i.e., nucleotide sequences and composition). Thus, different sequences contribute to different structures that can be resolved on the basis of various migration rates during electrophoresis on a nondenaturing polyacrylamide gel (usually run at a constant ambient temperature). Three major structural types observed are "self-folded" ssDNA, heteroduplex double-stranded DNA (dsDNA), and homoduplex dsDNA (formed as a result of the high rate of DNA reassociation) (61). Differentiation between ssDNA and dsDNA is based upon relative migration distance (ssDNA, heteroduplex dsDNA, and homoduplex dsDNA migrating in increasing order) and by variation in silver stain color. As described above for T-RFLP analysis, the use of fluorescent-dye-labeled PCR primers in combination with analysis on an automated DNA sequencer provides for the greatest reproducibility and sensitivity (112, 178).

Domain-specific primers for the SSU rRNA gene (61, 130, 178) have been used for analysis of microbial population structure. Criteria for selecting a good primer pair include (i) adequate fragment size (~100 to ~400 bp), (ii) low homology within the target sequence region, and (iii) high fragment length heterogeneity. Computer-assisted analysis of 1,262 available SSU rRNA bacterial sequences showed that, within the variable V3 region (*E. coli* numbering of 330 to 533), there are 34 different fragment length variants distributed among 19 different bacterial phyla or groups, and 1 to 22 different variants within individual bacterial groups (61). An alternative target region flanking the V3 to V5 region (330 to 926 [*E. coli* numbering]) of bacterial SSU rRNA sequences has also been used (130). For the domain *Archaea*, primers are usually complementary to the V3 region (178). Individual bands can be retrieved and sequenced, but due to their relatively short sequence length, only approximate phylogenetic position can be inferred.

The application of SSCP as a fingerprint of microbial community structure can be limited by the uncertainty of the banding pattern obtained. For example, multiple bands can be generated from the same bacterium, e.g., via heteroduplex formation among closely related sequences. Alternatively, different ssDNAs can have a similar migration distance on the gel. One solution for eliminating multiple bands originating from a single bacterium was proposed (130). Briefly, one of the two PCR primers was labeled with a phosphate group at the 5′ terminus. Thus, the amplified DNA strands having a 5′ phosphate could be removed by digestion with λ exonuclease, leading to one conformed-ssDNA band for each microorganism (130). A similar approach is to use biotinylated primers in the PCR procedure and to separate biotinylated DNA and nonbiotinylated DNA with magnetic beads prior to SSCP analysis (131). However, both approaches require additional steps and a significant investment of time. As for the other fingerprinting methods described, SSCP has been used to characterize microbial population structure in relation to environmental conditions (61, 130, 178).

DGGE and TGGE

Methods of DGGE and temperature gradient gel electrophoresis (TGGE) are based upon the analytical separation of DNA fragments of identical or nearly identical length but of various sequence compositions. The separation is determined by the change of electrophoretic mobility of DNA fragments migrating in a gel containing a linearly increasing gradient of DNA denaturants (urea and formamide) or temperature. The change in fragment mobility is associated with partial melting of the dsDNA in discrete regions, the so-called melting domains. These methods were first developed to detect single-base changes in genes for diagnosis of human genetic diseases and in genetic linkage studies (98). More recently DGGE and TGGE have been extended to resolve environmental populations of microorganisms by separating PCR amplification products generated by using primers targeting conserved genes (97). PCR primers designed for the amplification and cloning of SSU rRNA genes were first used to demonstrate the technique as applied to environmental microbiology (96). Subsequent studies have applied the technique to characterize the diversity of genes encoding different metabolic functions (39, 95, 97, 148).

DGGE and TGGE analysis of PCR-amplified SSU rDNA fragments provides a rapid method for the characterization and monitoring of community population structure and dynamics (for a more detailed review see references 95, 97, and 127), with consideration of some of the caveats previously discussed and below. The initial study by Muyzer et al. demonstrated the presence of several distinguishable bands (between 5 and 10) in the gel separation pattern, which were most likely derived from the predominant species within those communities characterized (96). By using DGGE or TGGE, extensive studies have examined various microbial ecosystems such as microbial mats, deep-sea hydrothermal vent samples, a stratified marine water column, rhizosphere, soils, and activated sludge (95, 97, 127). And as also discussed in relation to T-RFLP analysis, DGGE can be used to evaluate relative activity of individual populations based on changing rRNA/rRNA gene ratio. The method has been used to evaluate this ratio among different natural populations by comparing the pattern and intensity of bands derived from using either rRNA genes or rRNA (using reverse transcriptase to generate cDNA) as templates (147). A similar approach was used to evaluate the expression of a Ni-Fe hydrogenase conserved among natural populations of *Desulfovibrio* spp. (167).

More-specific information of population composition can be obtained by secondary analysis on the predominant DGGE or TGGE banding pattern by using sequencing or hybridization. Sequences of individual bands (fragments) are determined following their extraction from the gel, a second round of PCR amplification, and sequencing (direct or after cloning). This sequence information can also be used to design specific oligonucleotide probes for membrane hybridization or FISH analyses. For example, this approach was used to verify the predominance of microbial populations identified by DGGE with FISH (101). Also, group- and species-specific DNA hybridization probes have been used to identify specific populations within the pattern of resolved bands following transfer of the DNA to nylon membranes (96). Group-specific PCR primers have been used to restrict population analysis to specific microbial groups (45, 56, 103). Alternatively, microbial diversity of a specific functional or phylogenetic group within a microbial ecosystem can be achieved by using PCR with group- or species-specific DGGE primers (103) or combining DGGE with nested PCR (21).

The caveats of the DGGE and TGGE methods include the following: beyond the usual concerns of representative DNA extraction, the questions of representative PCR amplification of individual populations within the target collection and formation of amplification artifacts (e.g., chimerical sequences, heteroduplexes, and polymerase error) between populations remain partly unanswered (65, 160, 172). The reader is referred to studies that have directly addressed artifacts associated with this method (51, 138). The separation of the many fragments amplified from a highly diverse bacterial community is not possible by using available technology, although resolution may be improved by using a narrower range of denaturants or two-dimensional electrophoresis (33). The phylogenetic information obtained from sequencing individual bands is limited, because only fragments up to approximately 500 bp can be well separated. Another concern associated with the technique as well as with other aforementioned molecular fingerprinting methods is the a priori assignment of individual bands to individual populations. As discussed in the introductory considerations of diversity assessments, there can be significant sequence variation among rRNA operons of an individual organism, and so individual organisms could potentially contribute to multiple bands (103).

DNA Microarrays

A DNA microarray (DNA microchip or DNA chip) is an orderly, high-density arrangement of hundreds (or thousands) of individual cDNA probes (~500 to 5,000 bp in length) or oligonucleotides (15- to 70-mers) bound directly or indirectly to a solid surface (e.g., membrane, silicon wafer, or glass) (132). Unlike the membrane hybridization format, the chip is a high-density format that allows for simultaneous hybridization of a labeled DNA/RNA target to a large set of probes, thus providing high throughputs. In recent years, the oligonucleotide-based microarray has been receiving increased attention from researchers due to its advantages over its cDNA counterpart. These include simple methodology to obtain probes, good quality control of probe manufacturing, options to select high-specificity probes to prevent nonspecific hybridization, and the potential to detect alternative spliced variants of genes. Detailed information on the types and manufacturing of DNA microchips can be found in recent articles (31, 85, 132, 141).

DNA microarray technology has emerged as a popular tool in studies of environmental microbiology. So far, different types of microarrays as classified by the type of markers used have been developed for community structure analysis, community functions, species relatedness, and community gene expression. These include phylogenetic oligonucleotide arrays (POAs) (42, 74, 173), functional genomic arrays (176), community genome array (175), whole-genome open-reading-frame array (94), and gene expression array (52). However, due to the space limitation, we will focus discussion on the use of POA for microbial community structure analysis, its present limitations and challenges, and the possible solutions to these challenges.

In general, POA uses rRNA molecules as the phylogenetic marker. Within the natural framework provided by rRNA sequence variation, a fully developed POA could include a set of probes to encompass virtually all natural microbial grouping (phylotypes) and thereby serve to simultaneously monitor the population structure at multiple levels of resolution (e.g., at the approximate taxonomic ranks of domain, kingdom, order, genus, and species) (71, 173). This situation provides a basis to adjust monitoring strategies to the resolution required and, as previously discussed, serves as a mechanism to validate hybridization signal output by requiring consistency between quantifications at different taxonomic ranks (5, 139). With this hierarchical probe design strategy, POAs have been developed and applied for the study of microbial community analysis and microbial identification, for example, for a small subset of nitrifying populations (42, 53) and for analyzing oral microbiota (136), *Bacillus anthracis* and its related species (71), all recognized lineages of sulfate-reducing prokaryotes in both natural and clinical environments (74), and the beta-proteobacterial order "Rhodocyclales" (75).

These initial studies have demonstrated the potential use of POAs for rapid and comprehensive characterization of community structure but have also identified additional areas of needed technology development (for a more detailed review see references 8, 64, and 176). For example, the high cost of special equipment and a large number of oligonucleotide probes can limit the use of DNA microchip technology in its present stage of development. Design and optimization of probes remain something of an art form because the rules for predicting duplex stability are not yet fully developed (126, 149). Further, the efficiency in achieving good discrimination between targets and nontargets, eventually at a one-mismatch resolution, is still not satisfactory. This failing is primarily related to the use of the same hybridization and washing conditions to achieve target discrimination among a large set of oligonucleotide probes that differ in sequences and thermal stability. In other words, for any two given probes, the optimal conditions required for achieving perfect-match discrimination are not always the same. Furthermore, the efficiency of mismatch discrimination is complicated by the secondary structures formed within ssDNA or single-stranded RNA targets (36, 111). As a result, the occurrence of false-positive and false-negative signals can be significant (63, 74), and consequently they reduce the detection specificity of DNA microarrays.

Likewise, quantitation by using DNA microarray remains very challenging. One reason is the inherently high variation associated with array fabrication, target labeling, hybridization, and image processing (177). For example, when targets used in hybridizations are insufficiently fragmented or are not fragmented at all, it is possible for different immobilized probes (of various taxonomic ranks or targeting different regions on the rRNA sequence) to compete for the limited number of targets, resulting in incorrect estimations that deviate significantly from the actual population numbers. Differences in the accessibility of different targets to the immobilized probes can further lead to erroneous results (36).

To mitigate some of these problems, solutions have been demonstrated. One indirect solution to confirm the hybridization specificity from DNA microarray studies is to combine this technique with other well-established molecular methods. For example, Koizumi et al. (55) applied PCR-DGGE and RNA-DNA membrane hybridization in parallel with POAs to confirm the presence of specific sulfate-reducing organisms in anaerobic aromatic-degrading communities. Loy et al. (74) validated the presence of sulfate-reducing organisms, as suggested by positive signals from multiple probes of a sulfate-reducing prokaryote-specific POA, with specific PCR amplification and cloning and sequencing of 16S rRNA genes and genes encoding dissimilatory (bi)sulfate reductase. However, these DNA-microarray-independent techniques further contribute to the time required for community analysis. Alternatively, to minimize the occurrence of false-positive signals observed in the DNA microarray, a nonequilibrium dissociation curve approach, whereby the dissociation process of all positive probe-target duplexes from low to high temperature is performed and analyzed simultaneously within a short time, is proposed (87) and demonstrated (32, 63, 71, 155). By using this approach, a false-positive signal could be identified by comparing the observed dissociation curve and the observed T_d, at which 50% of the probe-target duplex is dissociated, with those of the perfectly matched duplexes. However, further efforts are still needed to resolve issues related to the sensitivity and specificity of the DNA microarray. Eventually, the DNA microarray with its massively parallel hybridization capacity should emerge as a powerful tool for microbial ecology studies.

In summary, the development and application of molecular fingerprinting are rapidly changing our understanding of microbial communities in a wide range of microbial ecosystems. All methods provide a rapid and robust means to monitor population changes. However, as yet no single method can resolve all microbial populations in a typical habitat. We emphasize that these methods are all part of a growing molecular toolbox. Most successful applications will generally come from studies that combine them with other methods. For example, some studies have combined chemotaxonomic methods (e.g., respiratory quinone and cellular fatty acid profiling) with molecular fingerprinting methods (68, 133). Others have compared molecular fingerprints with metabolic activity measurements (e.g., Biolog, enzyme activity, microelectrode, and microautoradiography) to better resolve structure-function relationships in a community (10, 54, 62, 105). Finally, we also emphasize that this chapter's focus on molecular techniques does not reduce the need for culture-based methods to fully define the character of populations rendered to sequences and hybridization signals by molecular methods.

Research originating in our laboratories and discussed in this chapter was supported by grants from NSF (Systematics DEB 0213186,

*Ecosystems Studies DEB 9904047, Environ. Eng. BES 0332118), NIH/NIDCR (U01 DE 14955), ONR, DARPA, NASA (ROSS and NAI), DOE (GTL Contract DE-AC03-76F00098), and A*STAR-Singapore.*

REFERENCES

1. **Alm, E., and D. A. Stahl.** 2000. Critical factors influencing the recovery and integrity of rRNA extracted from environmental samples: use of an optimized protocol to measure depth-related biomass distribution in freshwater sediments. *J. Microbiol. Methods* **40:**153–162.
2. **Alm, E. W., D. B. Oerther, N. Larsen, D. A. Stahl, and L. Raskin.** 1996. The oligonucleotide probe database. *Appl. Environ. Microbiol.* **62:**3557–3559.
3. **Alm, E. W., and D. A. Stahl.** 1996. Evaluation of key parameters for extraction of native rRNA from different environmental matrices, p. 1.1.5.1–1.1.5.29. *In* A. D. L. Akkermans, J. D. van Elsas, and F. J. de Bruijn (ed.), *Molecular Microbial Ecology Manual.* Kluwer Academic Publishers, Dordrecht, The Netherlands.
4. **Amann, R., and W. Ludwig.** 2000. Ribosomal RNA-targeted nucleic acid probes for studies in microbial ecology. *FEMS Microbiol. Rev.* **24:**555–565.
5. **Amann, R. I., W. Ludwig, and K.-H. Schleifer.** 1995. Phylogenetic identification and in situ detection of individual microbial cells without cultivation. *Microbiol. Rev.* **59:**143–169.
6. **Amann, R. I., J. Stromley, R. Devereux, R. Key, and D. A. Stahl.** 1992. Molecular and microscopic identification of sulfate-reducing bacteria in multispecies biofilms. *Appl. Environ. Microbiol.* **58:**614–623.
7. **Andrew, P. W., and G. J. Boulnois.** 1990. Early days in the use of DNA probes for Mycobacterium tuberculosis and Mycobacterium avium complexes, p. 179–198. *In* A. J. L. Macario and E. Conway de Macario (ed.), *Gene Probes for Bacteria.* Academic Press, Inc., San Diego, Calif.
8. **Bodrossy, L., and A. Sessitsch.** 2004. Oligonucleotide microarrays in microbial diagnostics. *Curr. Opin. Microbiol.* **7:**245–254.
9. **Borneman, J., and E. W. Triplett.** 1997. Molecular microbial diversity in soils from eastern Amazonia: evidence for unusual microorganisms and microbial population shifts associated with deforestation. *Appl. Environ. Microbiol.* **63:**2647–2653.
10. **Brinkhoff, T., C. M. Santegoeds, K. Sahm, J. Kuever, and G. Muyzer.** 1998. A polyphasic approach to study the diversity and vertical distribution of sulfur-oxidizing *Thiomicrospira* species in coastal sediments of the German Wadden Sea. *Appl. Environ. Microbiol.* **64:**4650–4657.
11. **Britten, R. J., and D. E. Kohne.** 1968. Repeated sequences in DNA. *Science* **161:**529–540.
12. **Bruce, K. D.** 1997. Analysis of *mer* gene subclasses within bacterial communities in soils and sediments resolved by fluorescent-PCR–restriction fragment length polymorphism profiling. *Appl. Environ. Microbiol.* **63:**4914–4919.
13. **Brunk, C. F., E. Avaniss-Aghajani, and C. A. Brunk.** 1996. A computer analysis of primer and probe hybridization potential with bacterial small-subunit rRNA sequences. *Appl. Environ. Microbiol.* **62:**872–879.
14. **Brusseau, G. A., E. S. Bulygina, and R. S. Hanson.** 1994. Phylogenetic analysis and development of probes for differentiating methylotrophic bacteria. *Appl. Environ. Microbiol.* **60:**626–636.
15. **Burggraf, S., T. Mayer, R. Amann, S. Schadhauser, C. R. Woese, and K. O. Stetter.** 1994. Identifying members of the domain *Archaea* with rRNA-targeted oligonucleotide probes. *Appl. Environ. Microbiol.* **60:**3112–3119.
16. **Cardinale, M., L. Brusetti, P. Quatrini, S. Borin, A. M. Puglia, A. Rizzi, E. Zanardini, C. Sorlini, C. Corselli, and D. Daffonchio.** 2004. Comparison of different primer sets for use in automated ribosomal intergenic spacer analysis of complex bacterial communities. *Appl. Environ. Microbiol.* **70:**6147–6156.
17. **Chin, K.-J., T. Lukow, and R. Conrad.** 1999. Effect of temperature on structure and function of the methanogenic archaeal community in an anoxic rice field soil. *Appl. Environ. Microbiol.* **65:**2341–2349.
18. **Clement, B. G., L. E. Kehl, K. L. DeBord, and C. L. Kitts.** 1998. Terminal restriction fragment patterns (TRFPs), a rapid, PCR-based method for the comparison of complex bacterial communities. *J. Microbiol. Methods* **31:**135–142.
19. **Cole, J. R., B. Chai, R. J. Farris, Q. Wang, S. A. Kulam, D. M. McGarrell, G. M. Garrity, and J. M. Tiedje.** 2005. The Ribosomal Database Project (RDP-II): sequences and tools for high-throughput rRNA analysis. *Nucleic Acids Res.* **33:**D294–D296.
20. **Daims, H., A. Bruhl, R. Amann, K.-H. Schleifer, and M. Wagner.** 1999. The domain-specific probe EUB338 is insufficient for the detection of all Bacteria: development and evaluation of a more comprehensive probe set. *Syst. Appl. Microbiol.* **22:**434–444.
21. **Dar, S. A., J. G. Kuenen, and G. Muyzer.** 2005. Nested PCR-denaturing gradient gel electrophoresis approach to determine the diversity of sulfate-reducing bacteria in complex microbial communities. *Appl. Environ. Microbiol.* **71:**2325–2330.
22. **DeLong, E. F., K. Y. Wu, B. B. Prézelin, and R. V. M. Jovine.** 1994. High abundance of archaea in Antarctic marine picoplankton. *Nature* **371:**695–697.

22a. **de los Reyes, F. L., W. Ritter, and L. Raskin.** 1997. Group-specific small-subunit rRNA hybridization probes to characterize filamentous foaming in activated sludge systems. *Appl. Environ. Microbiol.* **63:**1107–1117.

23. **Derakshani, M., T. Lukow, and W. Liesack.** 2001. Novel bacterial lineages at the (sub)division level as detected by signature nucleotide-targeted recovery of 16S rRNA genes from bulk soil and rice roots of flooded rice microcosms. *Appl. Environ. Microbiol.* **67:**623–631.
24. **De Rijk, P., J. Wuyts, Y. Van de Peer, T. Winkelmans, and R. De Wachter.** 2000. The European large subunit ribosomal RNA database. *Nucleic Acids Res.* **28:**177–178.
25. **Devereux, R., S.-H. He, C. L. Doyle, S. Orkland, D. A. Stahl, J. LeGall, and W. B. Whitman.** 1990. Diversity and orgin of *Desulfovibrio* species: phylogenetic definition of a family. *J. Bacteriol.* **172:**3609–3619.
26. **Dollhopf, S. L., S. A. Hashsham, and J. M. Tiedje.** 2001. Interpreting 16S rDNA T-RFLP data: application of self-organizing maps and principal component analysis to describe community dynamics and convergence. *Microb. Ecol.* **42:**495–505.
27. **Dsouza, M., N. Larsen, and R. Overbeek.** 1997. Searching for patterns in genomic data. *Trends Genet.* **13:**497–498.
28. **Dular, R.** 1990. Gene probe detection of human and cell culture mycoplasmas, p. 417–449. *In* A. J. L. Macario and E. Conway de Macario (ed.), *Gene Probes for Bacteria.* Academic Press, Inc., San Diego, Calif.
29. **Dutilh, B., C. Bebaer, and P. A. D. Grimont.** 1990. Detection of Chlamydia trachomatis with DNA probes, p. 46–64. *In* A. J. L. Macario and E. Conway de Macario (ed.), *Gene Probes for Bacteria.* Academic Press, Inc., San Diego, Calif.
30. **Egert, M., and M. Friedrich.** 2003. Formation of pseudo-terminal restriction fragments, a PCR-related bias affecting terminal restriction fragment length polymorphism analysis of microbial community structure. *Appl. Environ. Microbiol.* **69:**2555–2562.

31. **Ekins, R., and F. W. Chu.** 1999. Microarrays: their origins and applications. *Trends Biotechnol.* **17:**217–218.
32. **El Fantroussi, S., H. Urakawa, A. E. Bernhard, J. J. Kelly, P. A. Noble, H. Smidt, G. M. Yershov, and D. A. Stahl.** 2003. Direct profiling of environmental microbial populations by thermal dissociation analysis of native rRNAs hybridized to oligonucleotide microarrays. *Appl. Environ. Microbiol.* **69:**2377–2382.
33. **Fischer, S. G., and L. S. Lerman.** 1979. Length-independent separation of DNA restriction fragments in two dimensional gel electrophoresis. *Cell* **16:**191–200.
34. **Fisher, M. M., and E. W. Triplett.** 1999. Automated approach for ribosomal intergenic spacer analysis of microbial diversity and its application to freshwater bacterial communities. *Appl. Environ. Microbiol.* **65:**4630–4636.
35. **Fuchs, B. M., F. O. Glockner, J. Wulf, and R. Amann.** 2000. Unlabeled helper oligonucleotides increase the in situ accessibility to 16S rRNA of fluorescently labeled oligonucleotide probes. *Appl. Environ. Microbiol.* **66:** 3603–3607.
36. **Fuchs, B. M., G. Wallner, W. Beisker, I. Schwippl, W. Ludwig, and R. Amann.** 1998. Flow cytometric analysis of the in situ accessibility of *Escherichia coli* 16S rRNA for fluorescently labeled oligonucleotide probes. *Appl. Environ. Microbiol.* **64:**4973–4982.
37. **García-Martinez, J., I. Bescós, J. J. Rodríguez-Sala, and F. Rodríguez-Valera.** 2001. RISSC: a novel database for ribosomal 16S-23S RNA genes spacer regions. *Nucleic Acids Res.* **29:**178–180.
38. **Giovannoni, S. J., E. F. DeLong, G. J. Olsen, and N. R. Pace.** 1988. Phylogenetic, group-specific oligonucleotide probes for identification of single microbial cells. *J. Bacteriol.* **170:**720–726.
39. **Goregues, C. M., V. D. Michotey, and P. C. Bonin.** 2005. Molecular, biochemical, and physiological approaches for understanding the ecology of denitrification. *Microb. Ecol.* **49:**198–208.
40. **Greene, E. A., and G. Voordouw.** 2003. Analysis of environmental microbial communities by reverse sample genome probing. *J. Microbiol. Methods* **53:**211–219.
41. **Grimont, F., and P. A. D. Grimont.** 1991. DNA fingerprinting, p. 249–279. *In* E. Stackebrandt and M. Goodfellow (ed.), *Sequencing and Hybridization Techniques in Bacterial Systematics.* John Wiley & Sons, Chichester, England.
42. **Guschin, D. Y., B. K. Mobarry, D. Proudnikov, D. A. Stahl, B. E. Rittmann, and A. D. Mirzabekov.** 1997. Oligonucleotide microchips as genosensors for determinative and environmental studies in microbiology. *Appl. Environ. Microbiol.* **63:**2397–2402.
43. **Hedrick, D. B., T. White, J. B. Guckert, W. J. Jewell, and D. C. White.** 1992. Microbial biomass and community structure of a phase separated methanogenic reactor determined by lipid analysis. *J. Ind. Microbiol.* **9:**193–199.
44. **Hennecke, H., K. Kaluza, B. Thony, M. Fuhrmann, W. Ludwig, and E. Stackebrandt.** 1985. Concurrent evolution of nitrogenase genes and 16S rRNA in Rhizobium species and other nitrogen fixing bacteria. *Arch. Microbiol.* **142:**342–348.
45. **Heuer, H., M. Krsek, P. Baker, K. Smalla, and E. M. Wellington.** 1997. Analysis of actinomycete communities by specific amplification of genes encoding 16S rRNA and gel-electrophoretic separation in denaturing gradients. *Appl. Environ. Microbiol.* **63:**3233–3241.
46. **Hiraishi, A., K. Masamune, and H. Kitamura.** 1989. Characterization of the bacterial population structure in an anaerobic-aerobic activated sludge system on the basis of respiratory quinone profiles. *Appl. Environ. Microbiol.* **55:**897–901.
47. **Holben, W. E.** 1994. Isolation and purification of bacterial DNA from soils, p. 727–751. *In* R. W. Weaver et al. (ed.), *Methods of Soil Analysis*, part 2. *Microbiological and Biochemical Properties.* SSSA Book Series no. 5. Soil Science Society of America, Madison, Wis.
48. **Horz, H. P., J. H. Rotthauwe, T. Lukow, and W. Liesack.** 2000. Identification of major subgroups of ammonia-oxidizing bacteria in environmental samples by T-RFLP analysis of amoA PCR products. *J. Microbiol. Methods* **39:**197–204.
49. **Hubert, C., Y. Shen, and G. Voordouw.** 1999. Composition of toluene-degrading microbial communities from soil at different concentrations of toluene. *Appl. Environ. Microbiol.* **65:**3064–3070.
50. **Hugenholtz, P., C. Pitulle, K. L. Hershberger, and N. R. Pace.** 1998. Novel division-level bacterial diversity in a Yellowstone hot spring. *J. Bacteriol.* **180:**366–376.
51. **Janse, I., J. Bok, and G. Zwart.** 2004. A simple remedy against artifactual double bands in denaturing gradient gel electrophoresis. *J. Microbiol. Methods* **57:**279–281.
52. **Johnson, M. R., C. I. Montero, S. B. Conners, K. R. Shockley, S. L. Bridger, and R. M. Kelly.** 2005. Population density-dependent regulation of exopolysaccharide formation in the hyperthermophilic bacterium Thermotoga maritima. *Mol. Microbiol.* **55:**664–674.
53. **Kelly, J. J., S. Siripong, J. McCormack, L. R. Janus, H. Urakawa, S. El Fantroussi, P. A. Noble, L. Sappelsa, B. E. Rittmann, and D. A. Stahl.** DNA microarray detection of nitrifying bacterial 16S rRNA in wastewater treatment plant samples. *Water Res.*, in press.
54. **Kerkhof, L., M. Santoro, and J. Garland.** 2000. Response of soybean rhizosphere communities to human hygiene water addition as determined by community level physiological profiling (CLPP) and terminal restriction fragment length polymorphism (TRFLP) analysis. *FEMS Microbiol. Lett.* **184:**95–101.
55. **Koizumi, Y., J. J. Kelly, T. Nakagawa, H. Urakawa, S. El-Fantroussi, S. Al-Muzaini, M. Fukui, Y. Urushigawa, and D. A. Stahl.** 2002. Parallel characterization of anaerobic toluene- and ethylbenzene-degrading microbial consortia by PCR-denaturing gradient gel electrophoresis, RNA-DNA membrane hybridization, and DNA microarray technology. *Appl. Environ. Microbiol.* **68:**3215–3225.
56. **Kowalchuk, G. A., J. R. Stephen, W. De Boer, J. I. Prosser, T. M. Embley, and J. W. Woldendorp.** 1997. Analysis of ammonia-oxidizing bacteria of the β subdivision of the class *Proteobacteria* in coastal sand dunes by denaturing gradient gel electrophoresis and sequencing of PCR-amplified 16S ribosomal DNA fragments. *Appl. Environ. Microbiol.* **63:**1489–1497.
57. **Kurata, S., T. Kanagawa, Y. Magariyama, K. Takatsu, K. Yamada, T. Yokomaku, and Y. Kamagata.** 2004. Reevaluation and reduction of a PCR bias caused by reannealing of templates. *Appl. Environ. Microbiol.* **70:**7545–7549.
58. **Kuske, C., L. Ticknor, M. Miller, J. Dunbar, J. Davis, S. Barns, and J. Belnap.** 2002. Comparison of soil bacterial communities in rhizospheres of three plant species and the interspaces in an arid grassland. *Appl. Environ. Microbiol.* **68:**1854–1863.
59. **Kysela, D. T., C. Palacios, and M. L. Sogin.** 2005. Serial analysis of V6 ribosomal sequence tags (SARST-V6): a method for efficient, high-throughput analysis of microbial community composition. *Environ. Microbiol.* **7:**356–364.
60. **Laguerre, G., M.-R. Allard, F. Reboy, and N. Amarger.** 1994. Rapid identification of rhizobia by restriction fragment length polymorphism analysis of PCR-amplified 16S rRNA genes. *Appl. Environ. Microbiol.* **60:**56–63.
61. **Lee, D.-H., Y.-G. Zo, and S.-J. Kim.** 1996. Non-radioactive method to study genetic profiles of natural bacterial

communities by PCR–single-strand conformation polymorphism. *Appl. Environ. Microbiol.* **62:**3112–3120.

62. **Lee, N., P. H. Nielsen, K. H. Andreasen, S. Juretschko, J. L. Nielsen, K. H. Schleifer, and M. Wagner.** 1999. Combination of fluorescent in situ hybridization and microautoradiography—a new tool for structure-function analyses in microbial ecology. *Appl. Environ. Microbiol.* **65:**1289–1297.
63. **Li, E. S., J. K. Ng, J. H. Wu, and W. T. Liu.** 2004. Evaluating single-base-pair discriminating capability of planar oligonucleotide microchips using a non-equilibrium dissociation approach. *Environ. Microbiol.* **6:**1197–1202.
64. **Li, E. S. Y., and W. T. Liu.** 2004. DNA microarray technology in microbial ecology studies—principle, application and current limitations. *Microb. Environ.* **18:**175–187.
65. **Liesack, W., H. Weyland, and E. Stackebrandt.** 1991. Potential risks of gene amplification by PCR as determined by 16S rDNA analysis of a mixed-culture of strict barophilic bacteria. *Microb. Ecol.* **21:**199–209.
66. **Lin, C., B. Flesher, W. C. Capman, R. I. Amann, and D. A. Stahl.** 1994. Taxon specific hybridization probes for fiber-digesting bacteria suggest novel gut-associated Fibrobacter. *Syst. Appl. Microbiol.* **17:**418–424.
67. **Lin, C., and D. A. Stahl.** 1995. Comparative analyses reveal a highly conserved endoglucanase in the cellulolytic genus *Fibrobacter. J. Bacteriol.* **177:**2543–2549.
68. **Liu, W.-T., K. D. Linning, K. Nakamura, T. Mino, T. Matsuo, and L. J. Forney.** 2000. Microbial community changes in biological phosphate-removal systems on altering sludge phosphorus content. *Microbiology* **146:**1099–1107.
69. **Liu, W.-T., T. L. Marsh, H. Cheng, and L. J. Forney.** 1997. Characterization of microbial diversity by determining terminal restriction fragment length polymorphisms of genes encoding 16S rRNA. *Appl. Environ. Microbiol.* **63:** 4516–4522.
70. **Liu, W.-T., T. L. Marsh, and L. J. Forney.** 1998. Determination of the microbial diversity of anaerobic-aerobic activated sludge by a novel molecular biological technique. *Water Sci. Technol.* **37:**417–422.
71. **Liu, W. T., A. D. Mirzabekov, and D. A. Stahl.** 2001. Optimization of an oligonucleotide microchip for microbial identification studies: a non-equilibrium dissociation approach. *Environ. Microbiol.* **3:**619–629.
72. **Lovell, C. R., and Y. Hui.** 1991. Design and testing of a functional group-specific probe for the study of natural populations of acetogenic bacteria. *Appl. Environ. Microbiol.* **57:**2602–2609.
73. **Loy, A., M. Horn, and M. Wagner.** 2003. probeBase: an online resource for rRNA-targeted oligonucleotide probes. *Nucleic Acids Res.* **31:**514–516.
74. **Loy, A., A. Lehner, N. Lee, J. Adamczyk, H. Meier, J. Ernst, K. H. Schleifer, and M. Wagner.** 2002. Oligonucleotide microarray for 16S rRNA gene-based detection of all recognized lineages of sulfate-reducing prokaryotes in the environment. *Appl. Environ. Microbiol.* **68:**5064–5081.
75. **Loy, A., C. Schulz, S. Lücker, A. Schöpfer-Wendels, K. Stoecker, C. Baranyi, A. Lehner, and M. Wagner.** 2005. 16S rRNA gene-based oligonucleotide microarray for environmental monitoring of the betaproteobacterial order "*Rhodocyclales*." *Appl. Environ. Microbiol.* **71:**1373–1386.
76. **Ludwig, W., O. Strunk, R. Westram, L. Richter, H. Meier, Yadhukumar, A. Buchner, T. Lai, S. Steppi, G. Jobb, W. Forster, I. Brettske, S. Gerber, A. W. Ginhart, O. Gross, S. Grumann, S. Hermann, R. Jost, A. Konig, T. Liss, R. Lussmann, M. May, B. Nonhoff, B. Reichel, R. Strehlow, A. Stamatakis, N. Stuckmann, A. Vilbig, M. Lenke, T. Ludwig, A. Bode, and K. H. Schleifer.** 2004. ARB: a software environment for sequence data. *Nucleic Acids Res.* **32:**1363–1371.
77. **Lueders, T., K. Chin, R. Conrad, and M. Friedrich.** 2001. Molecular analyses of methyl-coenzyme M reductase alpha-subunit (mcrA) genes in rice field soil and enrichment cultures reveal the methanogenic phenotype of a novel archaeal lineage. *Environ. Microbiol.* **3:**194–204.
78. **Lueders, T., and M. Friedrich.** 2003. Evaluation of PCR amplification bias by terminal restriction fragment length polymorphism analysis of small-subunit rRNA and *mcrA* genes by using defined template mixtures of methanogenic pure cultures and soil DNA extracts. *Appl. Environ. Microbiol.* **69:**320–326.
79. **Lueders, T., B. Pommerenke, and M. W. Friedrich.** 2004. Stable-isotope probing of microorganisms thriving at thermodynamic limits: syntrophic propionate oxidation in flooded soil. *Appl. Environ. Microbiol.* **70:**5778–5786.
80. **Lúmann, H., I. Arth, and W. Liesack.** 2000. Spatial changes in the bacterial community structure along a vertical oxygen gradient in flooded paddy soil cores. *Appl. Environ. Microbiol.* **66:**754–762.
81. **Manz, W., R. Amann, W. Ludwig, M. Vancanneyt, and K. H. Schleifer.** 1996. Application of a suite of 16S rRNA-specific oligonucleotide probes designed to investigate bacteria of the phylum cytophaga-flavobacter-bacteroides in the natural environment. *Microbiology* **142:** 1097–1106.
82. **Manz, W., R. Amann, W. Ludwig, M. Wagner, and K.-H. Schleifer.** 1992. Phylogenetic oligodeoxynucleotide probes for the major subclasses of proteobacteria: problems and solutions. *Syst. Appl. Microbiol.* **15:**593–600.
83. **Marsh, T. L., W.-T. Liu, L. J. Forney, and H. Cheng.** 1998. Beginning a molecular analysis of eukaryal community in activated sludge. *Water Sci. Technol.* **37:**455–460.
84. **Marsh, T. L., P. Saxman, J. Cole, and J. Tiedje.** 2000. Terminal restriction fragment length polymorphism analysis program, a web-based tool for microbial community analysis. *Appl. Environ. Microbiol.* **66:**3616–3620.
85. **Marshall, A., and J. Hodgson.** 1998. DNA chips: an array of possibilities. *Nat. Biotechnol.* **16:**27–31.
86. **Martínez-Murcia, A. J., S. G. Acinas, and F. Rodriguez-Valera.** 1995. Evaluation of prokaryotic diversity by restrictase digestion of 16S rDNA directly amplified from hypersaline environments. *FEMS Microbiol. Ecol.* **17:**247–256.
87. **Massol-Deya, A. A., D. A. Odelson, R. F. Hickey, and J. M. Tiedje.** 1995. Bacterial community finger printing of amplified 16S and 16-23S ribosomal DNA gene sequences and restriction endonuclease analysis (ARDRA), p. 3.3.2.1–3.3.2.8. *In* A. D. L. Akkermans, J. D. van Elsas, and F. J. de Bruijn (ed.), *Molecular Microbial Ecology Manual*. Kluwer Academic Publishers, Dordrecht, The Netherlands.
88. **Meier, H., R. Amann, W. Ludwig, and K. H. Schleifer.** 1999. Specific oligonucleotide probes for in situ detection of a major group of gram-positive bacteria with low DNA G+C content. *Syst. Appl. Microbiol.* **22:**186–196.
89. **Mobarry, B. K., M. Wagner, V. Urbain, B. E. Rittmann, and D. A. Stahl.** 1996. Phylogenetic probes for analyzing abundance and spatial organization of nitrifying bacteria. *Appl. Environ. Microbiol.* **62:**2156–2162.
90. **Moeseneder, M. M., J. M. Arrieta, G. Muyzer, C. Winter, and G. J. Herndl.** 1999. Optimization of terminal-restriction fragment length polymorphism analysis for complex marine bacterioplankton communities and comparison with denaturing gradient gel electrophoresis. *Appl. Environ. Microbiol.* **65:**3518–3525.

91. **Morotomi, M., T. Ohno, and M. Mutai.** 1988. Rapid and correct identification of intestinal *Bacteroides* spp. with chromosomal DNA probes by whole-cell dot blot hybridization. *Appl. Environ. Microbiol.* **54:**1158–1162.
92. **Moyer, C. L., F. C. Dobbs, and D. M. Karl.** 1994. Estimation of diversity and community structure through restriction fragment length polymorphism distribution analysis of bacterial 16S rRNA genes from a microbial mat at an active, hydrothermal vent system, Loihi Seamount, Hawaii. *Appl. Environ. Microbiol.* **60:**871–879.
93. **Moyer, C. L., J. M. Tiedje, F. C. Dobbs, and D. M. Karl.** 1996. A computer-simulated restriction fragment length polymorphism analysis of bacterial small-subunit rRNA genes: efficacy of selected tetrameric restriction enzymes for studies of microbial diversity in nature. *Appl. Environ. Microbiol.* **62:**2501–2507.
94. **Murray, A. E., D. Lies, G. Li, K. Nealson, J. Zhou, and J. M. Tiedje.** 2001. DNA/DNA hybridization to microarrays reveals gene-specific differences between closely related microbial genomes. *Proc. Natl. Acad. Sci. USA* **98:**9853–9858.
95. **Muyzer, G.** 1999. DGGE/TGGE: a method for identifying genes from natural ecosystems. *Curr. Opin. Microbiol.* **2:**317–322.
96. **Muyzer, G., E. C. DeWall, and A. G. Uitterlinden.** 1993. Profiling of complex microbial populations by denaturing gradient gel electrophoresis analysis of polymerase chain reaction-amplified genes coding for 16S rRNA. *Appl. Environ. Microbiol.* **59:**695–700.
97. **Muyzer, G., and K. Smalla.** 2000. Application of denaturing gradient gel electrophoresis (DGGE) and temperature gradient gel electrophoresis (TGGE) in microbial ecology. *Antonie Leeuwenhoek* **73:**127–141.
98. **Myers, R. M., T. Maniatis, and L. S. Lerman.** 1987. Detection and localization of single base changes by denaturing gradient gel electrophoresis. *Methods Enzymol.* **155:**501–527.
99. **Nachamkin, I., N. J. Panaro, M. Li, H. Ung, P. K. Yuen, L. J. Kricka, and P. Wilding.** 2001. Agilent 2100 bioanalyzer for restriction fragment length polymorphism analysis of the *Campylobacter jejuni* flagellin gene. *J. Clin. Microbiol.* **39:**754–757.
100. **Neufeld, J. D., Z. T. Yu, W. Lam, and W. W. Mohn.** 2004. Serial analysis of ribosomal sequence tags (SARST): a high-throughput method for profiling complex microbial communities. *Environ. Microbiol.* **6:**131–144.
101. **Nielsen, A. T., W.-T. Liu, C. Filipe, L. Grady, Jr., S. Molin, and D. A. Stahl.** 1999. Identification of a novel group of bacteria in sludge from a deteriorated biological phosphorus removal reactor. *Appl. Environ. Microbiol.* **65:** 1251–1258.
102. **Noda, S., M. Ohkuma, R. Usami, K. Horikoshi, and T. Kudo.** 1999. Culture-independent characterization of a gene responsible for nitrogen fixation in the symbiotic microbial community in the gut of the termite *Neotermes koshunensis*. *Appl. Environ. Microbiol.* **65:**4935–4942.
103. **Nübel, U., F. Garcia-Pichel, and G. Muyzer.** 1997. PCR primers to amplify 16S RNA genes from cyanobacteria. *Appl. Environ. Microbiol.* **63:**3327–3332.
104. **Ohkuma, M., S. Noda, and T. Kudo.** 1999. Phylogenetic diversity of nitrogen fixation genes in the symbiotic microbial community in the gut of diverse termites. *Appl. Environ. Microbiol.* **65:**4926–4934.
105. **Okabe, S., T. Itoh, H. Satoh, and Y. Watanabe.** 1999. Analyses of spatial distributions of sulfate-reducing bacteria and their activity in aerobic wastewater biofilms. *Appl. Environ. Microbiol.* **65:**5107–5116.
106. **Olsen, G. J., D. J. Lane, S. J. Giovannoni, N. R. Pace, and D. A. Stahl.** 1986. Microbial ecology and evolution: a ribosomal RNA approach. *Annu. Rev. Microbiol.* **40:** 337–365.
107. **Olsen, G. J., C. R. Woese, and R. Overbeek.** 1994. The winds of (evolutionary) change: breathing new life into microbiology. *J. Bacteriol.* **176:**1–6.
108. **Orita, M., H. Iwahana, H. Kanazawa, K. Hayashi, and T. Sekiya.** 1989. Detection of polymorphisms of human DNA by gel electrophoresis as single-strand conformation polymorphisms. *Proc. Natl. Acad. Sci. USA* **86:** 2766–2770.
109. **Osborn, A. M., E. R. B. Moore, and K. N. Timmis.** 2000. An evaluation of terminal-restriction fragment length polymorphism (T-RFLP) analysis for the study of microbial community structure and dynamics. *Environ. Microbiol.* **2:**39–50.
110. **Pace, N. R., D. A. Stahl, D. J. Lane, and G. J. Olsen.** 1986. The analysis of natural microbial populations by ribosomal RNA sequences, p. 1–55. *In* K. C. Marshall (ed.), *Advances in Microbial Ecology*. Plenum Press, New York, N.Y.
111. **Peplies, J., F. O. Glockner, and R. Amann.** 2003. Optimization strategies for DNA microarray-based detection of bacteria with 16S rRNA-targeting oligonucleotide probes. *Appl. Environ. Microbiol.* **69:**1397–1407.
112. **Pirie, F. J., D. F. York, A. A. Motala, and M. A. Omar.** 1999. Fluorescent automated single-stranded conformation (F-SSCP) analysis is able to detect a point mutation at the extreme 5′ end of a PCR product. *Clin. Biochem.* **32:**481–484.
113. **Polz, M. F., and C. M. Cavanaugh.** 1997. A simple method for quantification of uncultured microorganisms in the environment based on in vitro transcription of 16S rRNA. *Appl. Environ. Microbiol.* **63:**1028–1033.
114. **Raskin, L., W. C. Capman, M. D. Kane, B. E. Rittmann, and D. A. Stahl.** 1996. Critical evaluation of membrane supports for use in quantitative hybridizations. *Appl. Environ. Microbiol.* **62:**300–303.
115. **Raskin, L., W. C. Capman, R. Sharp, and D. A. Stahl.** 1996. Molecular ecology of gastrointestinal ecosystems, p. 243–298. *In* R. I. Mackie, B. A. White, and R. E. Isaacson (ed.), *Gastrointestinal Microbiology and Host Interactions*. Chapman & Hall, London, United Kingdom.
116. **Raskin, L., L. K. Poulsen, D. R. Noguera, B. E. Rittmann, and D. A. Stahl.** 1994. Quantification of methanogenic groups in anaerobic biological reactors using oligonucleotide probe hybridizations. *Appl. Environ. Microbiol.* **60:**1241–1248.
117. **Raskin, L., B. E. Rittmann, and D. A. Stahl.** 1996. Competition and coexistence of sulfate-reducing and methanogenic populations in anaerobic biofilms. *Appl. Environ. Microbiol.* **62:**3846–3867.
118. **Raskin, L., J. M. Stromley, B. E. Rittmann, and D. A. Stahl.** 1994. Group-specific 16S rRNA hybridization probes to describe natural communities of methanogens. *Appl. Environ. Microbiol.* **60:**1232–1240.
119. Reference deleted.
120. **Risatti, J. B., W. C. Capman, and D. A. Stahl.** 1994. Community structure of a microbial mat: the phylogenetic dimension. *Proc. Natl. Acad. Sci. USA* **91:**10173–10177.
121. **Ritchie, N. J., M. E. Schutter, R. P. Dick, and D. D. Myrold.** 2000. Use of length heterogeneity PCR and fatty acid methyl ester profiles to characterize microbial communities in soil. *Appl. Environ. Microbiol.* **66:**1668–1675.
122. **Roller, C., M. Wagner, R. Amann, W. Ludwig, and K. H. Schleifer.** 1995. In situ probing of gram-positive bacteria with high DNA G+C content using 23S rRNA-targeted oligonucleotides. *Microbiology* **141:**1267.

123. **Sakano, Y., K. Pickering, P. Strom, and L. Kerkhof.** 2002. Spatial distribution of total, ammonia-oxidizing, and denitrifying bacteria in biological wastewater treatment reactors for bioregenerative life support. *Appl. Environ. Microbiol.* **68:**2285–2293.
124. **Sambrook, J., E. F. Fritsch, and T. Maniatis.** 1989. *Molecular Cloning: A Laboratory Manual,* 2nd ed. Cold Spring Harbor Laboratory Press, Cold Spring Harbor, N.Y.
125. **Sandaa, R. A., V. Torsvik, O. Enger, F. L. Daae, T. Castberg, and D. Hahn.** 1999. Analysis of bacterial communities in heavy metal-contaminated soils at different levels of resolution. *FEMS Microbiol. Ecol.* **30:**237–251.
126. **SantaLucia, J., Jr., and D. Hicks.** 2004. The thermodynamics of DNA structural motifs. *Annu. Rev. Biophys. Biomol. Struct.* **33:**415–440.
127. **Schafer, H., and G. Muyzer.** 2001. Denaturing gradient gel electrophoresis in marine microbial ecology. *Methods Microbiol.* **30:**425–468.
128. **Schena, M., D. Shalon, R. W. Davis, and P. O. Brown.** 1995. Quantitative monitoring of gene expression patterns with a complementary DNA microarray. *Science* **270:**467–470.
129. **Schmidt, T. M., E. F. DeLong, and N. R. Pace.** 1991. Analysis of a marine picoplankton community by 16S rRNA gene cloning and sequencing. *J. Bacteriol.* **173:** 4371–4378.
130. **Schwieger, F., and C. C. Tebbe.** 1998. A new approach to utilize PCR–single-strand-conformation polymorphism for 16S rRNA gene-based microbial community analysis. *Appl. Environ. Microbiol.* **64:**4870–4876.
131. **Selvakumar, N., B. C. Ding, and S. M. Wilson.** 1997. Separation of DNA strands facilitates detection of point mutations by PCR-SSCP. *BioTechniques* **22:**604–606.
132. **Service, R. F.** 1998. Microchip arrays put DNA on the spot. *Science* **282:**396–399.
133. **Sheffield, V. C., J. S. Beck, E. M. Stone, and R. M. Myers.** 1992. A simple and efficient method for attachment of a 40-base pair, GC-rich sequence to PCR-amplified DNA. *BioTechniques* **12:**386–387.
134. **Shen, Y., L. G. Stehmeier, and G. Voordouw.** 1998. Identification of hydrocarbon-degrading bacteria in soil by reverse sample genome probing. *Appl. Environ. Microbiol.* **64:**637–645.
135. **Smith, G. B., and J. M. Tiedje.** 1992. Isolation and characterization of a nitrate reductase gene and its use as a probe for denitrifying bacteria. *Appl. Environ. Microbiol.* **58:**376–384.
136. **Smoot, L. M., J. C. Smoot, H. Smidt, P. A. Noble, M. Könneke, Z. A. McMurry, and D. A. Stahl.** 2005. DNA microarrays as salivary diagnostic tools for characterizing the oral cavity microbial community. *Adv. Dent. Res.* **18:** 6–11.
137. **Southern, E. M.** 1975. Detection of specific sequences among DNA fragments separated by gel electrophoresis. *J. Mol. Biol.* **98:**503–517.
138. **Speksnijder, A., G. A. Kowalchuk, S. De Jong, E. Kline, J. R. Stephen, and H. J. Laanbroek.** 2001. Microvariation artifacts introduced by PCR and cloning of closely related 16S rRNA gene sequences. *Appl. Environ. Microbiol.* **67:**469–472.
139. **Stahl, D. A.** 1995. Application of phylogenetically based hybridization probes to microbial ecology. *Mol. Ecol.* **4:**535–542.
140. **Stahl, D. A.** 1986. Evolution, ecology and diagnosis: unity in variety. *Bio/Technology* **4:**623– 628.
141. **Stahl, D. A.** 2004. High-throughput techniques for analyzing complex bacterial communities, *Adv. Syst. Biol.* **547:**5–17.
142. **Stahl, D. A., and R. Amann.** 1991. Development and application of nucleic acid probes in bacterial systematics, p. 205–248. *In* E. Stackebrandt and M. Goodfellow (ed.), *Sequencing and Hybridization Techniques in Bacterial Systematics*. John Wiley & Sons, Chichester, England.
143. **Stahl, D. A., R. Devereux, R. I. Amann, B. Flesher, C. Lin, and J. Stromley.** 1989. Ribosomal RNA based studies of natural microbial diversity and ecology, p. 669–673. *In* T. Hattori et al. (ed.), *Recent Advances in Microbial Ecology*. Japan Scientific Societies Press, Tokyo, Japan.
144. **Stahl, D. A., B. Flesher, H. Mansfield, and L. Montgomery.** 1988. Use of phylogenetically based hybridization probes for studies of ruminal microbial ecology. *Appl. Environ. Microbiol.* **54:**1079–1084.
145. **Suzuki, M., M. S. Rappé, and S. J. Giovannoni.** 1998. Kinetic bias in estimates of coastal picoplankton community structure obtained by measurements of small-subunit rRNA gene PCR amplicon length heterogeneity. *Appl. Environ. Microbiol.* **64:**4522–4529.
146. **Szymanski, M., M. Z. Barciszewska, J. Barciszewski, and V. A. Erdmann.** 2000. 5S ribosomal RNA database Y2K. *Nucleic Acids Res.* **28:**166–167.
147. **Teske, A., C. Wawer, G. Muyzer, and N. B. Ramsing.** 1996. Distribution of sulfate–reducing bacteria in a stratified fjord (Mariager Fjord, Denmark) as evaluated by most-probable-number counts and denaturing gradient gel electrophoresis of PCR-amplified ribosomal DNA fragments. *Appl. Environ. Microbiol.* **62:**1405–1415.
148. **Throback, I. N., K. Enwall, A. Jarvis, and S. Hallin.** 2004. Reassessing PCR primers targeting nirS, nirK and nosZ genes for community surveys of denitrifying bacteria with DGGE. *FEMS Microbiol. Ecol.* **49:**401–417.
149. **Tijssen, P.** 1993. *Hybridization with Nucleic Acid Probes*. Elsevier, Amsterdam, The Netherlands.
150. **Torsvik, V., F. L. Daae, and J. Goksoyr.** 1995. Extraction, purification, and analysis of DNA from soil bacteria, p. 29–48. *In* J. T. Trevors and J. D. van Elsas (ed.), *Nucleic Acids in the Environment: Methods and Applications*. Springer-Verlag, New York, N.Y.
151. **Torsvik, V., F. L. Daae, R. A. Sandaa, and L. Ovreas.** 1998. Novel techniques for analysing microbial diversity in natural and perturbed environments. *J. Biotechnol.* **64:** 53–62.
152. **Torsvik, V., J. Goksøyr, and F. L. Daae.** 1990. High diversity in DNA of soil bacteria. *Appl. Environ. Microbiol.* **56:**782–787.
153. **Torsvik, V., K. Salte, R. Sørheim, and J. Goksøyr.** 1990. Comparison of phenotypic diversity and DNA heterogeneity in a population of soil bacteria. *Appl. Environ. Microbiol.* **56:**776–781.
154. **Tsien, H. C., B. J. Bratina, K. Tsuji, and R. S. Hanson.** 1990. Use of oligodeoxynucleotide signature probes for identification of physiological groups of methylotrophic bacteria. *Appl. Environ. Microbiol.* **56:**2858–2865.
155. **Urakawa, H., P. A. Noble, S. El Fantroussi, J. J. Kelly, and D. A. Stahl.** 2002. Single-base-pair discrimination of terminal mismatches by using oligonucleotide microarrays and neural network analyses. *Appl. Environ. Microbiol.* **68:**235–244.
156. **Uyeno, Y., Y. Sekiguchi, A. Sunaga, H. Yoshida, and Y. Kamagata.** 2004. Sequence-specific cleavage of small-subunit (SSU) rRNA with oligonucleotides and RNase H: a rapid and simple approach to SSU rRNA-based quantitative detection of microorganisms. *Appl. Environ. Microbiol.* **70:**3650–3663.
157. **Van de Peer, Y., P. De Rijk, J. Wuyts, T. Winkelmans, and R. De Wachter.** 2000. The European small subunit ribosomal RNA database. *Nucleic Acids Res.* **28:**175–176.

158. **Van de Peer, Y., I. Van de Broeck, P. De Rijk, and R. De Wachter.** 1994. Database on the structure of small ribosomal subunit RNA. *Nucleic Acids Res.* **22:**3488–3494.
159. **van der Maarel, M. J. E. C., R. R. E. Artz, R. Haanstra, and L. J. Forney.** 1998. Association of marine archaea with the digestive tracts of two marine fish species. *Appl. Environ. Microbiol.* **64:**2894–2898.
160. **von Wintzingerode, F., U. B. Gobel, and E. Stackebrandt.** 1997. Determination of microbial diversity in environmental samples: pitfalls of PCR-based rRNA analysis. *FEMS Microbiol. Rev.* **21:**213–229.
161. **Voordouw, G., Y. Shen, C. S. Harrington, A. J. Telang, T. R. Jack, and D. W. S. Westlake.** 1993. Quantitative reverse sample genome probing of microbial communities and its application to oil field production waters. *Appl. Environ. Microbiol.* **59:**4101–4114.
162. **Voordouw, G., J. K. Voordouw, T. R. Jack, J. Foght, P. M. Fedorak, and D. W. S. Westlake.** 1992. Identification of distinct communities of sulfate-reducing bacteria in oil fields by reverse sample genome probing. *Appl. Environ. Microbiol.* **58:**3542–3552.
163. **Voordouw, G., J. K. Voordouw, R. R. Karkhoff-Schweizer, P. M. Fedorak, and D. W. S. Westlake.** 1991. Reverse sample genome probing, a new technique for identification of bacteria in environmental samples by DNA hybridization, and its application to the identification of sulfate-reducing bacteria in oil field samples. *Appl. Environ. Microbiol.* **57:**3070–3078.
164. **Vos, M., R. M. Warren, C. J. van Rensburg, J. H. Hauman, J. T. Epplen, and P. D. van Helden.** 1999. Rapid genomic diversification of Helicobacter pylori. *Electrophoresis* **20:**1476–1485.
165. **Wagner, M., G. Rath, R. Amann, H.-P. Koops, and K.-H. Schleifer.** 1995. In situ identification of ammonia-oxidizing bacteria. *Syst. Appl. Microbiol.* **18:**251–264.
166. **Ward, D. M., M. M. Bateson, R. Weller, and A. L. Ruff-Roberts.** 1992. Ribosomal RNA analysis of microorganisms as they occur in nature, p. 219–286. *In* K. C. Marshall (ed.), *Advances in Microbial Ecology*. Plenum Press, New York, N.Y.
167. **Wawer, C., and G. Muyzer.** 1995. Genetic diversity of *Desulfovibrio* spp. in environmental samples analyzed by denaturing gradient gel electrophoresis of [NiFe] hydrogenase gene fragments. *Appl. Environ. Microbiol.* **61:** 2203–2210.
168. **Weller, R., and D. M. Ward.** 1989. Selective recovery of 16S rRNA sequences from natural microbial communities in the form of cDNA. *Appl. Environ. Microbiol.* **55:**1818–1822.
169. **Wetherall, B. L., and A. M. Johnson.** 1990. Nucleic acid probes for Campylobacter species, p. 255–285. *In* A. J. L. Macario and E. Conway de Macario (ed.), *Gene Probes for Bacteria*. Academic Press, Inc., San Diego, Calif.
170. **Wetmur, J. G., and N. Davidson.** 1968. Kinetics of renaturation of DNA. *J. Mol. Biol.* **31:**349–370.
171. **Wieland, A., M. Kuhl, L. McGowan, A. Fourcans, R. Duran, P. Caumette, T. Garcia de Oteyza, J. Grimalt, A. Sole, E. Diestra, I. Esteve, and R. Herbert.** 2003. Microbial mats on the Orkney Islands revisited: microenvironment and microbial community composition. *Microb. Ecol.* **46:**371–390.
172. **Wilson, I. G.** 1997. Inhibition and facilitation of nucleic acid amplification. *Appl. Environ. Microbiol.* **63:**3741–3751.
173. **Wilson, K. H., W. J. Wilson, J. L. Radosevich, T. Z. DeSantis, V. S. Viswanathan, T. A. Kuczmarski, and G. L. Andersen.** 2002. High-density microarray of small-subunit ribosomal DNA probes. *Appl. Environ. Microbiol.* **68:**2535–2541.
174. **Wu, L. Y., D. K. Thompson, G. S. Li, R. A. Hurt, J. M. Tiedje, and J. Z. Zhou.** 2001. Development and evaluation of functional gene arrays for detection of selected genes in the environment. *Appl. Environ. Microbiol.* **67:**5780–5790.
175. **Wu, L. Y., D. K. Thompson, X. D. Liu, M. W. Fields, C. E. Bagwell, J. M. Tiedje, and J. Z. Zhou.** 2004. Development and evaluation of microarray-based whole-genome hybridization for detection of microorganisms within the context of environmental applications. *Environ. Sci. Technol.* **38:**6775–6782.
176. **Zhou, J.** 2003. Microarrays for bacterial detection and microbial community analysis. *Curr. Opin. Microbiol.* **6:**288–294.
177. **Zhou, J. Z., and D. K. Thompson.** 2002. Challenges in applying microarrays to environmental studies. *Curr. Opin. Biotechnol.* **13:**204–207.
178. **Zumstein, E., R. Moletta, and J.-J. Godon.** 2000. Examination of two years of community dynamics in an anaerobic bioreactor using fluorescence polymerase chain reaction (PCR) single-strand conformation polymorphism analysis. *Environ. Microbiol.* **2:**69–78.
179. **Zverlov, V., M. Klein, S. Lucker, M. W. Friedrich, J. Kellermann, D. A. Stahl, A. Loy, and M. Wagner.** 2005. Lateral gene transfer of dissimilatory (bi)sulfite reductase revisited. *J. Bacteriol.* **187:**2203–2208.

Phylogenetic and Genomic Analysis

DIRK GEVERS AND TOM COENYE

13

INTRODUCTION

Polyphasic Taxonomy and 16S rRNA Gene-Based Phylogeny

Polyphasic taxonomy aims at the integration of different kinds of data and information on microorganisms, and in principle all genotypic, phenotypic, and phylogenetic information may be incorporated in polyphasic taxonomy (113). Genotypic methods are directed toward DNA or RNA molecules, and these methods dominate modern taxonomy, because it is now generally accepted that a classification should reflect the natural relationships as encoded in the DNA. Genotypic methods include determination of the DNA base ratio (%G+C), DNA-DNA hybridization studies, sequence determination of rRNA genes, and a large variety of DNA-based typing methods. Determination of the number of moles of guanine + cytosine is one of the classical genotypic methods, and the moles percent G+C is considered to be part of the standard description of new bacterial taxa (113). The base composition of bacterial chromosomal DNA ranges from about 25 to 80%, and in general the moles percent G+C of a species shows a narrow range of about 1 to 3 mol% (104). The percent DNA-DNA hybridization is an indirect parameter of the sequence similarity between two entire genomes and is the "gold standard" for the delineation of species. Several methods are available, and these are reviewed by Johnson (67). Wayne et al. defined a prokaryotic species as an entity that included strains with approximately 70% or greater DNA-DNA relatedness (116). However, several examples have shown that, although data derived from DNA-DNA hybridizations are important, it is essential to be flexible about the boundaries of species demarcation to obtain an optimal classification system that facilitates identification (27, 111, 120).

There is a consensus that the phylogenetic relationships between prokaryotes can be deduced from sequence comparisons of conserved macromolecules. rRNA genes are among the best targets for these phylogenetic studies because they are universally present and functionally constant and have a mosaic structure of highly conserved and more variable domains (22, 121). The direct sequencing of the genes coding for small (16S) or large (23S) rRNA molecules by PCR technology has provided a phylogenetic framework that serves as a backbone for modern microbial taxonomy (45, 78), and this technique is now standard practice in taxonomic studies—nowadays, it is hard to imagine a new prokaryotic species being described without its 16S rRNA gene sequence being determined first! The availability of international online databases (some of them entirely devoted to rRNA gene sequences) (8, 28, 69, 126) facilitates the exchange of 16S rRNA gene sequences and has contributed significantly to their widespread use. However, there are some pitfalls associated with the use of 16S rRNA gene sequences. First of all, it should be clear that comparison of these sequences cannot replace DNA-DNA hybridizations for species delineation (43, 101). Although organisms that share less than 97% 16S rRNA gene sequence similarity will not show DNA-DNA binding values higher than 60%, isolates that share more than 97% 16S rRNA gene sequence similarity may or may not belong to the same species, and there is no threshold value for 16S rRNA sequence similarity for species recognition. Secondly, prokaryotes often contain multiple rRNA operons, and although interoperon variability of 16S rRNA gene sequences within a single genome are generally low and unlikely to affect phylogenetic analyses, exceptions exist (2, 14, 25). It should also be mentioned that overall little is known about intraspecific variation in 16S rRNA gene sequences, although several studies indicate that it may be higher than generally assumed (23, 51, 112). One of the reasons why the 16S rRNA has become the molecule of choice for phylogenetic analysis is that it was considered unlikely to be prone to horizontal gene transfer and/or recombination; however, more recently, several studies have highlighted that distinct segments along 16S rRNA gene sequences may have a reticulate evolutionary history (see, for example, reference 110). Finally, there are some characteristics of 16S rRNA genes that may lead to inaccurate trees, including (i) considerable differences in mutation rates between different lineages (123), (ii) overestimation of relatedness of species with similar nucleotide frequencies (117), and (iii) difficulties in aligning distantly related taxa.

Alternative Molecular Markers

Because of the above-mentioned limitations, alternative phylogenetic markers have been used since the early 1990s (34, 114). The choice of which additional molecule(s) to include in comparative sequence analysis is not straightforward.

Nevertheless, there seems to be agreement that these genes (i) should preferably be widely distributed, (ii) should not be transmitted horizontally frequently, (iii) should be present as a single copy, and (iv) must be long enough to contain sufficient information but short enough to allow convenient sequencing (114, 131). In addition, they should contain the "right" amount of phylogenetic information (resolution); i.e., they should neither be too conserved nor too variable. However, several studies have shown that, based on these criteria, very few (if any) genes will be found that can be applied universally, and it is most likely that the development of group-specific approaches will be necessary (24, 92, 131). Rather than presenting an exhaustive overview of studies relying on a wide range of different alternative phylogenetic markers, we would like to illustrate the use of these markers with the *recA* gene as an example.

The use of the *recA* gene in phylogenetic studies was pioneered by Lloyd and Sharp (80), Eisen (34), and Karlin et al. (70). RecA is a relatively small protein (approximately 350 amino acids) involved in homologous DNA recombination, SOS induction, and DNA-damage-induced mutagenesis, and it is capable of binding single- and double-stranded DNA, unwinding duplex DNA, and finding homologous regions (34, 70). From these early studies it was already obvious that the *recA* gene was a good candidate alternative marker: (i) it is present in most prokaryotes, (ii) some regions are conserved while others are more variable, and (iii) the extensive size and sequence conservation and the fact that *recA* is a protein-coding gene make alignments virtually unambiguous. The first large-scale study using *recA* sequences from *Bacteria* (*Proteobacteria*, *Cyanobacteria*, and gram-positive organisms) as well as from *Archaea* (34) showed that phylogenetic trees based on *recA* sequences displayed topologies and resolution similar to those of 16S-rRNA-gene-based trees. Subsequent studies have focused on various groups of bacteria.

Mycobacterial species often have very similar 16S rRNA gene sequences (e.g., *Mycobacterium kansasii* and *Mycobacterium gastri* have identical 16S rRNA gene sequences), which may hinder their accurate identification (109). Comparison of phylogenetic trees based on 16S rRNA gene and *recA* sequences revealed a general likeness in topology (11). However, unlike the 16S rRNA gene, sequence similarities of the *recA* gene are significantly lower between species (e.g., 96.2% between *M. kansasii* and *M. gastri*), allowing a more reliable species-level identification (11).

The family *Vibrionaceae* contains six genera, including the genus *Vibrio* (containing 44 species). The classification and identification of vibrios to the species level require the application of state-of-the-art genomic analyses, including amplified fragment length polymorphism fingerprinting and DNA-DNA hybridizations (108). Thompson et al. used *recA* as an alternative phylogenetic marker in the *Vibrionaceae* (106). Their data showed that there was a relatively good correlation between *recA* and 16S rRNA gene sequence data but that overall *recA* gene sequences were much more discriminatory than were 16S rRNA gene sequences. The *recA* data also showed that *Vibrio* species appear to be polyphyletic.

Species-level identification of *Burkholderia* species, and most notably of members of the *Burkholderia cepacia* complex (a group of closely related species sharing >97.7% 16S rRNA gene sequence similarity), is not straightforward (27). 16S rRNA gene sequence analysis is of limited value in the genus *Burkholderia* (especially in the *B. cepacia* complex), as its resolution is too low to allow accurate identifications. Recently a *recA*-based approach for identifying *Burkholderia* species was developed (85). With this approach, it was possible to identify all *Burkholderia* species to the species level. In particular the degree of resolution of the *recA* phylogenetic tree for members of the *B. cepacia* complex was much greater than that observed with 16S rRNA gene sequence analysis (Fig. 1). However, the presence of discrete *recA* lineages within some members of the *B. cepacia* complex adds an additional layer of complexity.

Other frequently used molecular markers include *gyrB*, *rpoB*, *rpoD*, and *hsp60* (see, for example, references 3, 19, 20, 29, 56, 75, 76, 98, and 127).

Multilocus Sequence Analysis

However, even trees based on these alternative markers have been criticized, as concern was raised that any single-gene tree cannot adequately reflect phylogenetic relationships because of the possibility of horizontal gene transfer, variable mutation rates, variable rates of recombination, and simple stochastic variation (24, 46). The ad hoc committee for the reevaluation of the species definition in bacteriology proposed that a minimum of five housekeeping genes be sequenced to achieve an adequately informative level of phylogenetic data (100). In analogy with multilocus sequence typing (a sequence-based typing method primarily used to distinguish infraspecific groups within named species), this polygenic approach has been called multilocus sequence analysis (MLSA) (46). MLSA is starting to become common practice in taxonomic studies (see, for example, references 3, 19, 20, 56, 76, 82, 83, 98, and 107), and it is anticipated that in the future it may replace DNA-DNA hybridizations (46, 100).

Classification of microorganisms by MLSA could rely on either one of two approaches. In the first approach, one may select a universal set of genes that allows for the hierarchical classification of all prokaryotes (92, 131). However, this idea may be impractical, and since we are (in general) less interested in the deeper phylogenetic relationships between genera, a more realistic approach is to look for sets of genes that can be used within all strains of a particular group (a genus or family). The reason behind this thinking is that genes that are informative within a given genus or family may not be useful or even present in other taxa. Also, genes that are conserved enough to be amplified by a common set of primers from all species may not evolve quickly enough to distinguish closely related taxa. Nevertheless, some genes may be informative in more than one group-specific set, and these more widely distributed genes could provide tools for broader comparisons. The genes used in MLSA should be ubiquitous (at least in the taxon under study) and present in a single copy, but those genes in which recombination might confer a selective advantage (e.g., virulence- and antigen-encoding genes) or closely linked genes should be avoided. Recently, a novel tool was described that automatically makes a selection of representative proteins for bacterial phylogeny (9).

The simplest approach is to concatenate the sequences of the sets of homologous genes and to use the concatenated sequences to construct a tree (see below) that can delineate clusters that can aid in the division of the genus into species. Prokaryotic identification in this scenario is a two-step process. In the first step, 16S rRNA gene sequencing is used to assign an unknown isolate to a group (genus or family). In the second step, genes and primers specific for that group are used in MLSA to assign the isolate to a species. We would like to emphasize that, at present, there are no

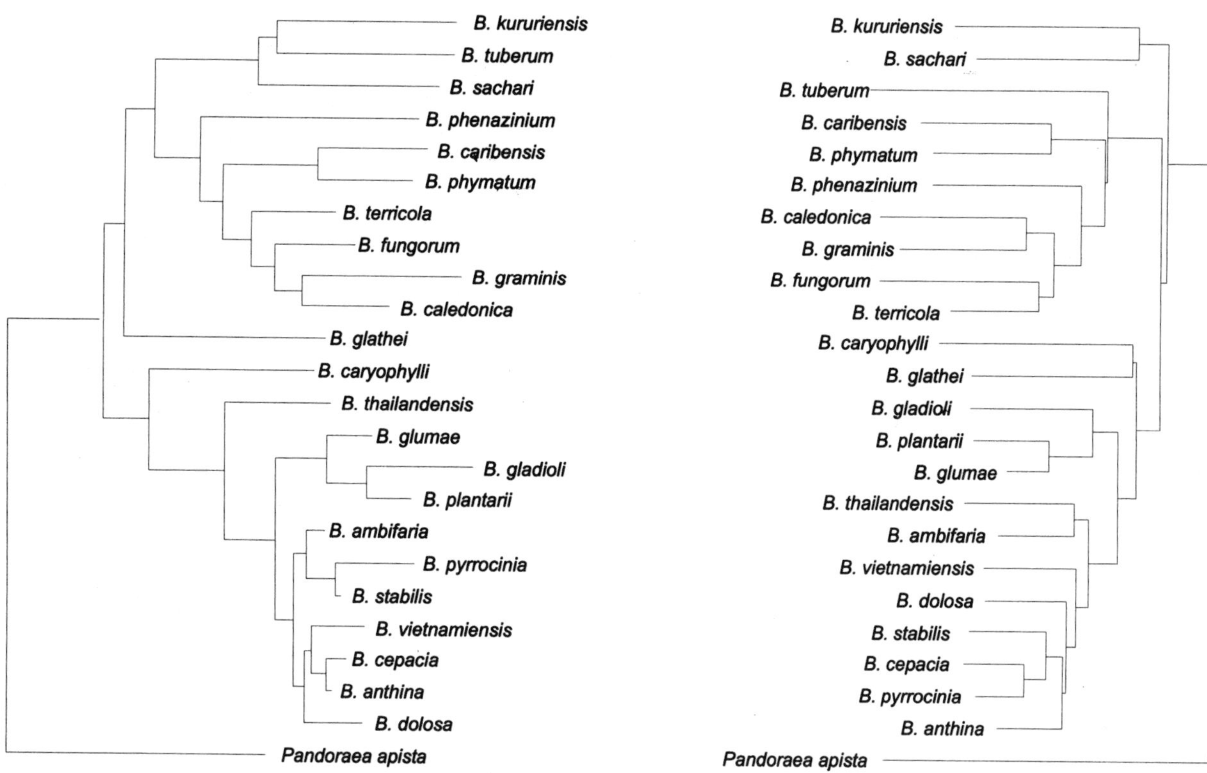

FIGURE 1 Comparison of phylogenetic trees based on 16S rRNA gene (right; scale bar indicates 1% sequence dissimilarity) and *recA* (left; scale bar indicates 5% sequence dissimilarity) sequences of a selection of *Burkholderia* species. Sequences were extracted from GenBank and were aligned by using BioNumerics 4.0 (Applied Maths, Sint-Martens-Latem, Belgium), and trees were constructed based on the NJ algorithm as implemented in BioNumerics 4.0.

guidelines to delineate species based on MLSA and that it is therefore necessary to "calibrate" novel MLSA schemes by using DNA-DNA hybridization values (76, 100). Two examples of the use of MLSA are given below.

Godoy et al. used MLSA to investigate the relationships among the closely related organisms *Burkholderia mallei*, *Burkholderia pseudomallei*, and *Burkholderia thailandensis* (47). MLSA showed the clear separation of *B. pseudomallei* isolates from those of *B. thailandensis*, a finding consistent with other data that indicate that they are different species. However, MLSA also showed that *B. mallei* is a clone within *B. pseudomallei*, raised to species status due to its ability to cause a distinctive disease (i.e., equine glanders) in horses, mules, and donkeys.

Similarly, MLSA was used to investigate the relationships among members of the *Bacillus cereus* group (including *B. cereus*, *Bacillus thuringiensis*, *Bacillus anthracis*, and *Bacillus mycoides*) (87), confirming that strains of the insect pathogen *B. thuringiensis* correspond to several distinct lineages within the phylogeny of the free-living soil bacterium *B. cereus*. *B. anthracis*, on the other hand, appears to be a specialized clone of *B. cereus* that was recognized as a single separate species because of its capacity to cause anthrax. These conclusions are confirmed by whole-genome comparisons (5, 88), in which it was shown that (i) genomes of members of this group show a high level of synteny and a high level of protein identity, (ii) very few genes are unique to one species, and (iii) much of the "specificity" in this group can be attributed to plasmid content.

PHYLOGENETIC ANALYSIS

This section aims to briefly discuss the different steps in constructing evolutionary trees, including collecting data sets of homologous sequences, generating a multiple-sequence alignment, inferring tree topology, and assessing confidence in the tree. In addition, some theoretical background is given. For further details, the reader is referred to a number of recent reviews (6, 55, 57, 119) and two recent books that highlight the topic from a more practical perspective (49, 91).

What Is a Phylogenetic Tree?

A phylogenetic tree is a tree-like diagram that depicts the evolutionary relationships between different genes or organisms. All trees have branches and nodes. External nodes are often called operational taxonomic units (OTUs) and are the terminal points where the evolutionary process has yielded the sequences under analysis. Internal nodes are the points where branches split from each other and may be called hypothetical taxonomic units to emphasize that they are the hypothetical progenitors of OTUs, also referred to as the last common ancestor of everything arising from it. Molecular phylogenetic trees are usually drawn with proportional branch lengths; the length of the branches corresponds to the degree of sequence divergence, or amount of evolution, between the two nodes that they connect.

Phylogeneticists often speak of rooted and unrooted trees. An unrooted tree only positions the individual taxa relative to each other without indicating the direction of the evolutionary process. In a rooted tree, a root represents a common ancestor of all sequences analyzed and therefore implies the order of branching in the rest of the tree. A tree can be rooted if one or more of the OTUs form an external point of reference and are known as, or are believed to be, the most distantly related of the OTUs (i.e., outgroup routing).

Phylogenetic trees illustrate the relationship among the sequences analyzed; therefore, they are always gene trees. Whether these gene trees are species trees and can be interpreted as representative of the organismal phylogeny—that is, the topology that traces the history of the replicating cell lineages that transmit genes and genomes to successive generations—depends on the gene selected. An example of a robust hypothesis for organismal phylogenies based on a multigene approach and applied to the case of γ-proteobacteria has been published by Lerat et al. (77). What should be considered the organismal phylogeny, if any, is a topic beyond the scope of this chapter (31, 125).

DNA or Protein?

Nucleotide sequences may be coding or noncoding. For protein-coding genes, the alignment can be accomplished based on the nucleotide or the amino acid sequences. Because it is the DNA that contains all the information to create functional proteins, it is often thought that the DNA sequence should also be used in molecular phylogeny. However, there are reasons why it may be more appropriate to use protein sequences for such analyses (84). (i) In protein-coding sequences, the first and second nucleotide of each codon are less prone to the incorporation of mutations because it almost always leads to a change in amino acid. When one compares sequences that have diverged for possibly hundreds of millions of years, it is likely that the third codon position has become saturated, resulting in positions with no phylogenetic signal. This pitfall is avoided by looking at amino acid sequence or by excluding the third position from each codon in the alignment. (ii) Because DNA is composed of only four different units, two randomly chosen aligned sequences will have on average 25% identical residues if gaps are not allowed. This percentage increases even up to 50% when gaps are allowed. This situation may obscure any genuine relationship that may exist between two gene sequences. By contrast, the alignment of proteins with their 20 amino acids is less cumbersome. On average, 5% of residues in two randomly chosen and aligned sequences would be identical. Even with gaps, still only 10 to 15% of residues are identical. This situation makes protein sequences easier to align and allows the signal-to-noise ratio to improve significantly.

In general, it is strongly recommended to analyze both DNA and protein data sets. For a group of closely related species or taxa, DNA-based analysis is probably a good method because problems like differences in codon bias or saturation of the third position of codons can be avoided. In case of ambiguity in the alignment of gene sequences, it is recommended to first translate the sequences to their corresponding protein sequence and then to align and determine the position gaps in the DNA sequence according to the protein alignment.

Collecting a Set of Homologous Sequences

In an evolutionary analysis, it is absolutely required to study homologous sequences, i.e., sequences that share a common origin, as this requirement is a basic premise for phylogenetic analyses. It makes no sense at all to generate a phylogenetic tree of unrelated sequences (38). Homologs can be orthologous, paralogous, or xenologous (66). Orthologs duplicate only when their host divides, i.e., along with the rest of the genome. They are strictly vertically transmitted, so their phylogeny traces that of their host lineage. Paralogs and xenologs are members of multigene families that arise by intragenomic gene duplication and by horizontal gene transfer, respectively. Keeping this distinction in mind is important, particularly when the object of a phylogenetic reconstruction is to establish evolutionary relationships between organisms. If paralogous or xenologous genes are unknowingly considered, recovery of incorrect species relationships is likely.

Homology between sequences is an all-or-none property (i.e., two sequences are either homologous or not) and is always a hypothesis. Therefore, an indirect method is necessary to assess experimentally whether two sequences are homologous. In most cases the level of sequence similarity is the criterion used (e.g., E-value or similarity score cutoff). However, sequences might have diverged to the extent that their common origin is untraceable by a direct sequence comparison. It becomes very difficult to correctly detect homology for pairs with a pairwise sequence identity between 20 and 30%, the so-called twilight zone (90). One should remember that highly divergent sequences should be excluded in the evolutionary analysis, as they prevent a reliable alignment.

Once one has decided which phylogenetic marker will be studied (see above), the first practical step of building the data set often consists of recovery of reference sequences from public databases. One search strategy for finding a set of related sequences is a keyword search in public databases (e.g., using Entrez or SRS). Although this strategy is easy and seems more intuitive, it is far from exhaustive. Difficulty arises mostly because many data entries are incorrectly or badly annotated and thus hard to find with keyword searches. To compile a comprehensive data set, sequence similarity searching (e.g., using BLAST or FASTA) is strongly recommended. A consideration to be made in the selection of sequence sampling is that the range of organisms sampled should accurately reflect the total diversity present within the range of organisms under consideration.

A well-chosen outgroup should not be too distantly related because the sequences may have become saturated with multiple mutations, by which information may have been erased. It should also not be too closely related, to make sure it is a true outgroup. In the absence of an outgroup, the best guess is to place the root in the middle of the tree (i.e., midpoint rooting), or better, not to root it at all (6).

Creating a Good Sequence Alignment

Multiple-sequence alignments are the essential prerequisite for most phylogenetic analyses. Any phylogenetic inference based on molecular data begins by comparing the homologous residues (i.e., those that descend from a common ancestral residue) among different sequences. The best way to do this is to align sequences on top of each other, so that homologous residues from different sequences line up in the same column.

As an optimal sequence alignment by a simultaneous comparison of all N sequences is essentially impossible for four or more sequences (due to the enormous computer memory and time needed), most multiple-sequence alignments are constructed by the method known as "progressive sequence alignment" (37). That is, an alignment is built up stepwise, starting with the most similar sequences and progressively adding the more dissimilar ones. CLUSTAL is without doubt the software program most widely used to align a set of more than two sequences (17) and constructs a multiple alignment in three steps. The first step involves performing all pairwise comparisons between the sequences and generating a distance matrix representing pairwise sequence similarities. Secondly, based on this matrix, a guide tree is constructed by using the neighbor-joining (NJ) method (see below). Note that this is a "quick and dirty" tree and is unsuitable for serious phylogenetic inference. Finally, the alignment is built up progressively by a series of pairwise alignments following the branching order of the guide tree.

Progressive alignment is fast but heuristic; i.e., it does not guarantee finding the most optimal solution. The major problem with progressive sequence alignment programs is the dependence of the ultimate multiple-sequence alignment on the initial pairwise sequence alignment. Match errors during early steps in the alignment protocol are accumulated and propagated, leading to further errors in later steps. This problem is also referred to as the "once a gap always a gap" problem (37). Gaps can only be added or enlarged, never moved or removed. The latter actions would make the alignment process much slower. This predicament obviously results in errors that need manual adjustment to minimize insertion and/or deletion events and to improve the quality of the alignment, a widespread and fairly well-accepted correction (6).

Once an alignment has been created and manually checked or adjusted, it is necessary to select which positions will be used for subsequent analyses. Since an alignment makes statements about the homology of amino acids or nucleotides present at each position, it is important to include only unambiguously aligned sites. If there are gaps in the alignment, it can be difficult to say confidently that all positions are correctly aligned. Furthermore, the presence of incomplete sequences and variations in length of terminal regions of genes can mean that some alignment positions are poorly sampled, with missing data. These are preferentially positions to throw away, referred to as "stripping of the alignment" or "masking of data." It is a complicated and controversial part of phylogenetic analyses, but it is clear that in most cases masking increases the accuracy of tree reconstruction (57). Programs such as Gblocks (16) objectively assess which parts of the alignment are sufficiently conserved and useful for phylogenetic analysis.

The use of structural information can improve an alignment substantially, because secondary structural elements of functional RNAs and structural features of proteins are often more conserved than are primary sequence features (53, 90). As almost all RNA molecules form secondary structures, one can outperform the classical approaches by taking structure information into account. The software tool StructMiner efficiently detects and aligns conserved structural patterns (128). If a three-dimensional structure of the protein is available, the secondary structure can be deduced by programs such as DSSP (68) or STRIDE (44). In the absence of tertiary structures for a query set of protein sequences, information about secondary structure can be gathered only by using secondary structure prediction programs. Applications that integrate structural information into the alignment process have been published (e.g., PRALINE [96], and MASS [32]). When DNA sequences from protein-coding genes are aligned, one can also obtain an improved alignment by first aligning the inferred amino acid sequences and then performing a codon-by-codon back alignment of the corresponding DNA sequences. Software packages that automatically perform this task are available (e.g., transAlign and RevTrans) (10, 118).

In conclusion, shortcomings of most methods to keep in mind are that (i) they are not guaranteed to find the optimal alignment, (ii) the optimal alignment does not necessarily correspond to the alignment of homologous positions, and (iii) an alignment is generated also for random sequences and in regions of the proteins where variability is too high to be reliable.

Inferring Tree Topology

Reconstructing the phylogeny from a sequence alignment is not straightforward (102). There are no uniquely correct methods for inferring phylogenies, many methods exist, and it is rarely possible to verify that one has arrived at the true phylogenetic tree. There are currently four primary methods for constructing phylogenies from protein and nucleic acid sequence alignments: the distance-based NJ method and the character-based methods, including maximum parsimony (MP), maximum likelihood (ML), and Bayesian inference.

Distance-Based Methods

Distance-based methods (e.g., NJ) are relatively simple and straightforward; starting from a sequence alignment, the differences between all pairwise combinations of OTUs (DNA or protein sequences) are converted into a distance matrix that represents an estimate derived from applying a so-called evolutionary or substitution model of the evolutionary distance between sequences (see further). These distances are than assembled into a tree. A disadvantage of distance methods is that they reduce the phylogenetic information to one number. The major advantage is that they are much less computer intensive, which is important when many taxa have to be compared.

Character-Based Methods

Character-based methods examine each column of the alignment separately (each position in the aligned sequences is a "character"). They look for the tree that best accommodates all of this information.

MP methods select the tree that explains the data observed in terms of the minimal number of possible substitutions. It is often the case that there are several trees, typically differing only slightly, that are consistent with the same number of events and that are therefore equally parsimonious. This method is becoming less popular mainly because MP generally uses a simple model of sequence substitution (all changes are equally probable). Moreover, it has been shown that, for more divergent sequences, MP is prone to recover incorrect trees (72).

ML methods seek to identify the single most probable tree on a statistical basis, given the chosen model of sequence evolution. It is a method that allows correcting for multiple mutational events at the same site and therefore is more suited to accurately reconstructing the relationships between sequences that have been separated for a long time or are evolving rapidly. In ML, all possible mutational pathways that are compatible with the data are considered, and the tree that has the highest probability (i.e., the likelihood of the tree) of producing the observed sequences is preferred. The main advantage of ML over MP is that ML permits the inference of phylogenetic trees by using different (more complex) evolutionary models (see below). This truth of course implies that the obtained result depends on the accuracy and assumptions included in that model. Although such models are simpler than is the true process underlying sequence evolution, they seem to be relatively robust to violation of their simplifying assumptions (60). The main obstacle of ML is the computational burden; i.e., the number of different tree topologies that have to be evaluated increases enormously as a function of the number of sequences. Even with adaptations of the ML method, such as TREE-PUZZLE (95) or the fastDNAml program (4), to the most modern computer technologies, constructing trees of 40 or more sequences becomes very impractical. Bayesian estimation of phylogeny holds promise as an alternative to ML, particularly for large molecular-sequence data sets (61). Bayesian inference of phylogeny is based upon a quantity called the posterior probability distribution of trees, which is the probability of a tree conditioned on the observations. The conditioning is accomplished by using Bayes's theorem. For more comprehensive information, it is impossible to come up with a mathematical definition; therefore, we refer elsewhere (61). The advantages of this method are numerous. Firstly, like ML methods, it is based on the likelihood function, so it inherits many of the desirable statistical properties of ML. Secondly, it allows one to incorporate prior information (if available), e.g., a systematist's prior confirmed conviction about the phylogeny of the group under study. Thirdly, it has major computational advantages and therefore allows one to study large data sets and implement complex models of sequence evolution. Finally, it not only produces a tree estimate but also measures phylogenetic uncertainty for the groups on the tree (comparable to bootstrap values on ML trees; see below). These measures are more intuitive, as they represent the posterior probability that the group is true, and are thus more easily interpreted.

Evolutionary Models

In order to estimate the genetic or evolutionary distances between pairs of sequences, a mathematical model is necessary. Simply counting the observed differences between sequences does not accurately reflect the evolutionary distances between them. It fails to take into account superimposed multiple changes at individual sites (including back mutations), different mutational rates depending on nucleotides or amino acids or among sites of alignment, and different rates of mutation for different lineages and therefore makes a correction obligatory. An evolutionary model is a set of assumptions about the process of nucleotide or amino acid substitutions. They describe the different probabilities of change from one nucleotide or amino acid to another, with the aim of correcting for unseen changes along the phylogeny. A good visualization illustrating the different models is shown by Whelan et al. (119). Various models have been developed to try to estimate the true difference between sequences based on their present states, such as amino acid substitution matrices (e.g., Dayoff, Blossom, etc.) or gamma corrections (considering among-site rate variation), etc. (for a recent overview, see reference 13). The choice of an appropriate model is paramount to accurate evolutionary reconstruction. Good fit to an accurate model should provide a robust analysis. Good fit to an inappropriate model can be seriously misleading. Whereas MP implicitly assumes a model of evolution, distance and ML methods estimate parameters according to an explicit model of evolution. However, whereas distance methods estimate only a single parameter (substitutions per site) given the model, ML can estimate all the relevant parameters of the substitution model. One should note that a model is always simplified and often makes assumptions just to turn a complex problem into a computationally tractable one. But a model becomes a powerful tool when, despite its simplified assumptions, it can fit the data and make accurate predictions about the problem at hand. Programs exist to help one in selecting the best-fit model for nucleotide substitution (MODELTEST) or protein evolution (PROTTEST); both are also available as Web tools (1, 86).

Bootstrap Analysis

How well can one trust the tree that one has constructed? A popular way of assessing the robustness of the topology of a tree is nonparametric bootstrapping (35). The bootstrap analysis tests whether the whole data set is supporting the tree or if the tree is just a marginal winner among many nearly equal alternatives. In practice, such analysis goes as follows: from the original alignment, columns in the sequence alignment are chosen at random, until a new alignment is constructed with the same size as the original one. Some characters will not be included at all in a given bootstrap replication, while others might be included more than once (i.e., the "sampling with replacement" principle). For each artificial data set, a tree is constructed and compared with the tree based on the original alignment. The number of times that a cluster, as defined in the original tree, is also found in the bootstrap tree, is recorded, and the resulting bootstrap values are superimposed on the original tree.

One difficulty with this analysis is the precise interpretation of what bootstrap values represent (99); higher is clearly better, but what is a reasonable cutoff? Some have concluded that values of 70% or higher are likely to indicate reliable groupings (54). But this cutoff is not generally accepted. More discussion of the interpretation of bootstrap values has been published elsewhere (36).

The major advantage of the bootstrap technique is that it can be applied to basically all tree construction methods, although one has to keep in mind that applying the bootstrap method multiplies the computer time needed by the number of bootstrap samples requested. This drawback is not a concern when a fast analysis (like NJ or parsimony) is employed, but it can be an obstacle when ML is used. A Bayesian analysis takes as long as an ML analysis, but as implemented in the software package MrBayes, it does not have the same drawback because bootstrapping is not necessary. Instead, it provides Bayesian posterior probabilities as indicators of branch support (55).

General Guidelines

There are three main reasons why phylogenies may be incorrect. First, a random error occurs when the informative sites (or data points) are limited, resulting in any tree that could be generated. We can cope with this problem by applying enough data and use bootstrap values as an indication of the extent of the random error. Having many data is not enough, as a second problem is caused by bias. An estimation (tree) is biased if the data set is not representative of the underlying distribution. For example, two sequences can be clustered together just because they both share an unusually high G+C content. And thirdly, there is a systematic error, i.e., tackling the problem in the wrong way by using an incorrect model (modeling error). Methods can be misleading no matter how many data one has. Therefore, no guarantee exists that one can produce the one tree with the correct topology. In order to be aware of the reliability of the topology of the resulting tree, one or all of the following should be done:

1. Apply different tree-building methods to the data set.
2. Vary the parameters used by the different programs.
3. Apply different evolutionary models for matrix construction.
4. Add or remove one or more OTUs to see the influence on tree topology.
5. Include an outgroup that may serve as a root for the tree.
6. Apply bootstrap analysis to the data set.

A tree should be considered robust and thus reliable only when widely different methods infer similar or identical tree topologies and when such topologies are supported by good bootstrap values.

Phylogenies from Multiple Genes

Until recently, phylogenetic analyses have been routinely based on homologous sequences of a single phylogenetic marker, i.e., the 16S rRNA gene among bacteria (122). Given the vast number of genome sequences now available, it is possible to compute trees from whole genomes (see "Use of Whole Genome Sequences for Phylogeny" below). However, researchers will not have access to full genome data from thousands of species in the immediate future. Given these constraints on data availability, we focus here on the use of multiple genes for constructing phylogenies, i.e., MLSA (see "Multilocus Sequence Analysis" above). Two main reasons for using multiple loci follow: (i) sequence-based approaches to organism phylogeny require loci that evolve more rapidly than do rRNA genes, in order to increase phylogenetic accuracy (89), and (ii) multiple genes provide a buffer against the distorting phylogenetic signals at a single locus, such as effects of recombination, gene conversion, and horizontal gene transfer (77). A software package (VisRD) allows a graphical inspection of the phylogenetic content of a sequence alignment to detect recombination and recombination breakpoints (42).

The main question that this section addresses is how to infer a phylogenetic tree from multiple genes. There are two fundamentally different ways. A simple approach is to concatenate multiple genes head to tail to form a super-gene alignment, assuming that all positions are independent and identically distributed, and construct a tree. Although there are definitely arguments in favor of this approach (30), one cannot ignore the specific evolutionary features for each of the genes (94). Estimation of a phylogeny always assumes a model of evolution, but because different genes likely have different evolutionary constraints and/or pressures, the parameters (including tree topology, branch lengths, rate heterogeneity among sites, and substitution probabilities) may change from gene to gene. For example, the substitution model may vary from gene to gene because of differences in G+C content. At the extreme, different genes may even support different tree topologies because of horizontal gene transfer. Therefore, it might be appropriate to describe the evolution of each gene by its own set of parameters. This method is a second approach, in which phylogenies are inferred separately for each gene and the resulting gene trees are used to generate a consensus phylogeny. This strategy gives a more conservative and a safer estimate of evolution, because it produces only high resolution in the branching pattern when there is at least a majority consensus among the different genes.

A test (the likelihood ratio test) to examine whether different genes support congruent trees and whether we can ignore gene-specific effects via concatenation is available (115). When faced with different histories, there are at least three obvious ways in which one might represent phylogeny: (i) as the collection of individual histories, (ii) as a tree representing the single dominant pattern among data sets, or (iii) as a network. A consensus network attempts to represent all phylogenetic signals present in the given set of gene trees, simultaneously, up to a given level of complexity. In the resulting network, regions of the evolutionary history that are undisputed within the set of gene trees appear tree-like, whereas regions containing conflicts are shown as a box of parallel edges (a split), the "dimensionality" of which reflects the number of conflicting signals. A comprehensive software package for analyzing and visualizing a multiple-gene data set is SplitsTree (62).

USE OF WHOLE GENOME SEQUENCES FOR PHYLOGENY

Genome Tree Approach

Since the publication of the complete genome sequence of *Haemophilus influenzae* Rd (41), over 350 completely sequenced microbial genomes have been published, and many more are under way (for an overview, see http://www.genomesonline.org/). Although there has been a general tendency to focus on organisms with particularly interesting properties (most notably human pathogens), by now the available complete genome sequences give a more or less adequate picture of the genomic diversity observed in culturable prokaryotes. A good starting point for browsing through completed genomes is the CBS Genome Atlas Database (50), which is available from http://www.cbs.dtu.dk/services/GenomeAtlas/. While the availability of an increasing number of completely sequenced genomes has significantly facilitated the search for alternative molecular markers, there are other ways in which these sequences can be used to deduce phylogenetic relationships between taxa. An overview of these different approaches is given elsewhere (24). Below, we focus on one of the most promising novel approaches to taxonomy, based on gene content.

When the gene contents of organisms are compared, the simplest approach is to consider genomes "bags of genes," and then to compare the contents of different "bags" (65). The identification of orthologous genes (orthologs are homologous sequences in different species that arose from a common ancestral gene during speciation) is pivotal in this approach and largely depends on the definition of orthology (125). Most studies use a minimal definition, terming putative orthologs as genes that have the highest level of significant pairwise identity when the genes are compared between genomes (i.e., they are identified as those homologous genes that show the largest identity of several possibilities above a certain threshold) (7, 26, 65). From the first analyses performed on a limited number of genomes, a few general trends emerged (65, 97). It was observed that (i) large genomes have many genes in common, (ii) the number of genes that two genomes have in common depends on their evolutionary distance, (iii) the fraction of shared orthologous genes decreases rapidly in evolution, faster than does the identity between the shared orthologs, and (iv) the evolution of gene content shows non-tree-like aspects, as phylogenetically closely related species do not necessarily share orthologous genes that either of them shares with a phylogenetically more distant third species. The observation that large genomes have many genes in common (irrespective of phylogenetic distance) and the observations that smaller genomes (e.g., from strictly parasitic organisms) are dominated by essential genes and share a greater fraction of their genomes with other species (7, 129) suggest that it is useful to normalize or weigh the data (i.e., to correct for differences in genome size) before further analysis. Interestingly, Snel et al. also observed that trees based on gene content do not correlate well with phenotype-based trees (97). This finding is somewhat unexpected, as it has always been assumed that differences in observed phenotypes are the result of differences in gene content (see, for example, reference 64).

Several slightly different approaches to compare gene contents are used as well. The presence and absence of families of protein-encoding genes in sequenced genomes have been used to reconstruct the relationships between a number of organisms (40, 58). In this approach, proteins are grouped together in families if their pairwise similarity is greater than a preset value, thereby eliminating the need for the identification of putative orthologs in each genome and the need for specific alignments. Tekaia et al. constructed genome trees based on whole-proteome comparisons, using hierarchical classification of genomes (taking into account genome size, levels of ancestral gene redundancy due to duplications, and net gene gain or loss) (105). Protein folds are protein families that share the same basic molecular shape but not necessarily sequence similarity. The presence or absence of these features was used by Wolf et al. and by Lin and Gerstein to build genome trees (79, 124). Protein folds are considered by some to be ideal characteristics for building phylogenetic trees, as they represent fundamental molecular units used by organisms. Yang et al. focused on the fold superfamily level (129). The advantage of using the fold superfamily level instead of the fold level is that it offers a higher level of certainty that the members of each group share common ancestry.

In the concept of "extended gene content" (introduced by Gu and Zhang [48]), the status of a gene family in a given genome is recorded as absent, present as a single copy, or present as a duplicate (instead of merely being recorded as absent or present). Also, "hybrid" measures have been defined, in which both gene content and sequence conservation are expressed. "Genome blast distance phylogeny," an approach proposed by Henz et al., starts with an all-against-all pairwise comparison of genomes (52). Subsequently a distance matrix is calculated from the resulting high-scoring pairs. Kunin et al. derived a new composite measure (called "genome conservation") from the sum of alignment scores between all proteins for every pair of organisms (74).

Methods relying on shared gene content for reconstructing phylogenies have been criticized because of the tendency of gene content convergence (due to horizontal gene transfer and gene loss). Nevertheless, trees based on gene content generally correspond well to trees based on 16S rRNA gene sequences (although it should be noted that some discrepancies observed in various studies remain at present unexplained). This result indicates that, despite horizontal gene transfer, gene duplications, and gene loss, there is a strong phylogenetic signal in gene content. However, to reduce the possible impact of these genetic processes on trees based on gene content, several methods have been developed to filter out the "noise" that is associated with them. The reader can consult references 15, 21, and 33 for further details.

There are several approaches to transform the fraction of shared genes (shared protein families and protein folds, etc.) into a genome distance matrix, subsequently construct a tree or a network from these matrices, and perform statistical tests (24, 48, 63, 73, 93, 132). While a detailed discussion of these methods is outside the scope of the text, it is worth mentioning that several studies indicate that ML and MP approaches outperform distance methods for constructing trees based on gene content data (63, 132).

Comparative Microbial Genomics with DNA Microarrays

Differences and/or similarities between microorganisms can also be studied by using DNA microarrays. These methods have the advantage of not requiring the availability of a whole-genome sequence of all organisms being studied. Hybridization of DNA to whole-genome microarrays has been used to study the genetic diversity of a wide range of bacteria (see references 39 and 130 for recent reviews). The microarray technology has also been used for a number of specialized applications in taxonomy and identification. Cho and Tiedje proposed a new approach to identify bacteria based on genomic DNA-DNA similarity by employing microarray technology (18). This method (so far evaluated only with four *Pseudomonas* species) does not require laborious cross-hybridizations, and the resulting hybridization profiles can be used in statistical procedures to identify test strains and can be stored in a database.

Microarrays for the identification of specific bacteria were also developed. An overview is presented in reference 24.

Microarrays have also been developed to study specific bacterial populations and consortia, including sulfate-reducing bacteria (81), toluene- and ethylbenzene-degrading consortia (71), and methanotrophs (12, 103).

REFERENCES

1. **Abascal, F., R. Zardoya, and D. Posada.** 2005. ProtTest: selection of best-fit models of protein evolution. *Bioinformatics* **21:**2104–2105.
2. **Acinas, S. G., L. A. Marcelino, V. Klepac-Ceraj, and M. F. Polz.** 2004. Divergence and redundancy of 16S rRNA sequences in genomes with multiple *rrn* operons. *J. Bacteriol.* **186:**2629–2635.
3. **Adekambi, T., and M. Drancourt.** 2004. Dissection of phylogenetic relationships among 19 rapidly growing *Mycobacterium* species by 16S rRNA, *hsp*65, *sod*A, *rec*A and *rpo*B gene sequencing. *Int. J. Syst. Evol. Microbiol.* **54:**2095–2105.
4. **Alsen, G. J., H. Matsuda, R. Hagstrom, and R. Overbeek.** 1994. fastDNAmL: a tool for construction of phylogenetic trees of DNA sequences using maximum likelihood. *Comput. Appl. Biosci.* **10:**41–48.
5. **Anderson, I., A. Sorokin, V. Kapatral, G. Reznik, A. Bhattacharya, N. Mikhailova, H. Burd, V. Joukov, D. Kaznadzey, T. Walunas, M. d'Souza, N. Larsen, G. Pusch, K. Liolios, Y. Grechkin, A. Lapidus, E. Goltsman, L. Chu, M. Fonstein, S. D. Ehrlich, R. Overbeek, N. Kyrpides,**

and N. Ivanova. 2005. Comparative genome analysis of Bacillus cereus group genomes with Bacillus subtilis. *FEMS Microbiol. Lett.* **250:**175–184.

6. **Baldauf, S. L.** 2003. Phylogeny for the faint of heart: a tutorial. *Trends Genet.* **19:**345–351.
7. **Bansal, A. K., and T. E. Meyer.** 2002. Evolutionary analysis by whole genome comparisons. *J. Bacteriol.* **184:** 2260–2272.
8. **Benson, D. A., I. Karsch-Mizrachi, D. J. Lipman, J. Ostell, and D. L. Wheeler.** 2005. GenBank. *Nucleic Acids Res.* **33:**D34–D38.
9. **Bern, M., and D. Goldberg.** 2005. Automatic selection of representative proteins for bacterial phylogeny. *BMC Evol. Biol.* **5:**34.
10. **Bininda-Emonds, O. R.** 2005. transAlign: using amino acids to facilitate the multiple alignment of protein-coding DNA sequences. *BMC Bioinformatics* **6:**156.
11. **Blackwood, K. S., C. He, J. Gunton, C. Y. Turenne, J. Wolfe, and A. M. Kabani.** 2000. Evaluation of *rec*A sequences for identification of *Mycobacterium* species. *J. Clin. Microbiol.* **38:**2846–2852.
12. **Bodrossy, L., N. Stralis-Pavese, J. C. Murrell, S. Radajewski, A. Weilharter, and A. Sessitsch.** 2003. Development and validation of a diagnostic microbial microarray for methanotrophs. *Environ. Microbiol.* **5:**566–582.
13. **Bos, D. H., and D. Posada.** 2005. Using models of nucleotide evolution to build phylogenetic trees. *Dev. Comp. Immunol.* **29:**211–227.
14. **Boucher, Y., C. J. Douady, A. K. Sharma, M. Kamekura, and W. F. Doolittle.** 2004. Intragenomic heterogeneity and intergenomic recombination among haloarchaeal rRNA genes. *J. Bacteriol.* **186:**3980–3990.
15. **Brown, J. R., C. J. Douady, M. J. Italia, W. E. Marshall, and M. J. Stanhope.** 2001. Universal trees based on large combined protein sequence data sets. *Nat. Genet.* **28:**281–285.
16. **Castresana, J.** 2000. Selection of conserved blocks from multiple alignments for their use in phylogenetic analysis. *Mol. Biol. Evol.* **17:**540–552.
17. **Chenna, R., H. Sugawara, T. Koike, R. Lopez, T. J. Gibson, D. G. Higgins, and J. D. Thompson.** 2003. Multiple sequence alignment with the Clustal series of programs. *Nucleic Acids Res.* **31:**3497–3500.
18. **Cho, J. C., and J. M. Tiedje.** 2001. Bacterial species determination from DNA-DNA hybridization by using genome fragments and DNA microarrays. *Appl. Environ. Microbiol.* **67:**3677–3682.
19. **Christensen, H., P. Kuhnert, J. E. Olsen, and M. Bisgaard.** 2004. Comparative phylogenies of the housekeeping genes *atp*D, *inf*B and *rpo*B and the 16S rRNA gene within the *Pasteurellaceae*. *Int. J. Syst. Evol. Microbiol.* **54:** 1601–1609.
20. **Cladera, A. M., A. Bennasar, M. Barcelo, J. Lalucat, and E. Garcia-Valdes.** 2004. Comparative genetic diversity of *Pseudomonas stutzeri* genomovars, clonal structure, and phylogeny of the species. *J. Bacteriol.* **186:**5239–5248.
21. **Clarke, G. D., R. G. Beiko, M. A. Ragan, and R. L. Charlebois.** 2002. Inferring genome trees by using a filter to eliminate phylogenetically discordant sequences and a distance matrix based on mean normalized BLASTP scores. *J. Bacteriol.* **184:**2072–2080.
22. **Clarridge, J. E., III.** 2004. Impact of 16S rRNA gene sequence analysis for identification of bacteria on clinical microbiology and infectious diseases. *Clin. Microbiol. Rev.* **17:**840–862.
23. **Clayton, R. A., G. Sutton, P. S. Hinkle, Jr., C. Bult, and C. Fields.** 1995. Intraspecific variation in small-subunit rRNA sequences in GenBank: why single sequences may not adequately represent prokaryotic taxa. *Int. J. Syst. Bacteriol.* **45:**595–599.
24. **Coenye, T., D. Gevers, Y. Van de Peer, P. Vandamme, and J. Swings.** 2005. Towards a prokaryotic genomic taxonomy. *FEMS Microbiol. Rev.* **29:**147–167.
25. **Coenye, T., and P. Vandamme.** 2003. Intragenomic heterogeneity between multiple 16S ribosomal RNA operons in sequenced bacterial genomes. *FEMS Microbiol. Lett.* **228:**45–49.
26. **Coenye, T., and P. Vandamme.** 2003. Extracting phylogenetic information from whole-genome sequencing projects: the lactic acid bacteria as a test case. *Microbiology* **149:**3507–3517.
27. **Coenye, T., P. Vandamme, J. R. Govan, and J. J. LiPuma.** 2001. Taxonomy and identification of the *Burkholderia cepacia* complex. *J. Clin. Microbiol.* **39:**3427–3436.
28. **Cole, J. R., B. Chai, R. J. Farris, Q. Wang, S. A. Kulam, D. M. McGarrell, G. M. Garrity, and J. M. Tiedje.** 2005. The Ribosomal Database Project (RDP-II): sequences and tools for high-throughput rRNA analysis. *Nucleic Acids Res.* **33:**D294–D296.
29. **Dauga, C.** 2002. Evolution of the *gyr*B gene and the molecular phylogeny of *Enterobacteriaceae*: a model molecule for molecular systematic studies. *Int. J. Syst. Evol. Microbiol.* **52:**531–547.
30. **de Queiroz, A., M. J. Donoghue, and J. Kim.** 1995. Separate versus combined analysis of phylogenetic evidence. *Ann. Rev. Ecol. Syst.* **26:**657–681.
31. **Driskell, A. C., C. Ane, J. G. Burleigh, M. M. McMahon, B. C. O'Meara, and M. J. Sanderson.** 2004. Prospects for building the tree of life from large sequence databases. *Science* **306:**1172–1174.
32. **Dror, O., H. Benyamini, R. Nussinov, and H. Wolfson.** 2003. MASS: multiple structural alignment by secondary structures. *Bioinformatics* **19** (Suppl. 1)**:**i95–i104.
33. **Dutilh, B. E., M. A. Huynen, W. J. Bruno, and B. Snel.** 2004. The consistent phylogenetic signal in genome trees revealed by reducing the impact of noise. *J. Mol. Evol.* **58:**527–539.
34. **Eisen, J. A.** 1995. The RecA protein as a model molecule for molecular systematic studies of bacteria: comparison of trees of RecAs and 16S rRNAs from the same species. *J. Mol. Evol.* **41:**1105–1123.
35. **Felsenstein, J.** 1985. Confidence-limits on phylogenies—an approach using the bootstrap. *Evolution* **39:**783–791.
36. **Felsenstein, J., and H. Kishino.** 1993. Is there something wrong with the bootstrap on phylogenies—a reply. *Syst. Biol.* **42:**193–200.
37. **Feng, D. F., and R. F. Doolittle.** 1987. Progressive sequence alignment as a prerequisite to correct phylogenetic trees. *J. Mol. Evol.* **25:**351–360.
38. **Fitch, W. M.** 2000. Homology: a personal view on some of the problems. *Trends Genet.* **16:**227–231.
39. **Fitzgerald, J. R., and J. M. Musser.** 2001. Evolutionary genomics of pathogenic bacteria. *Trends Microbiol.* **9:**547–553.
40. **Fitz-Gibbon, S. T., and C. H. House.** 1999. Whole genome-based phylogenetic analysis of free-living microorganisms. *Nucleic Acids Res.* **27:**4218–4222.
41. **Fleischmann, R. D., M. D. Adams, O. White, R. A. Clayton, E. F. Kirkness, A. R. Kerlavage, C. J. Bult, J. F. Tomb, B. A. Dougherty, J. M. Merrick, et al.** 1995. Whole-genome random sequencing and assembly of *Haemophilus influenzae* Rd. *Science* **269:**496–512.
42. **Forslund, K., D. H. Huson, and V. Moulton.** 2004. VisRD—visual recombination detection. *Bioinformatics* **20:**3654–3655.
43. **Fox, G. E., J. D. Wisotzkey, and P. Jurtshuk, Jr.** 1992. How close is close: 16S rRNA sequence identity may not

be sufficient to guarantee species identity. *Int. J. Syst. Bacteriol.* **42:**166–170.

44. **Frishman, D., and P. Argos.** 1995. Knowledge-based protein secondary structure assignment. *Proteins* **23:**566–579.
45. **Garrity, G. M., and J. G. Holt.** 2001. The road map to the Manual, p. 119–166. *In* D. R. Boone and R. W. Castenholz (ed.), *Bergey's Manual of Systematic Bacteriology*. Springer-Verlag, New York, N.Y.
46. **Gevers, D., F. M. Cohan, J. G. Lawrence, B. G. Spratt, T. Coenye, E. J. Feil, E. Stackebrandt, Y. Van de Peer, P. Vandamme, F. L. Thompson, and J. Swings.** 2005. Opinion: re-evaluating prokaryotic species. *Nat. Rev. Microbiol.* **3:**733–739.
47. **Godoy, D., G. Randle, A. J. Simpson, D. M. Aanensen, T. L. Pitt, R. Kinoshita, and B. G. Spratt.** 2003. Multilocus sequence typing and evolutionary relationships among the causative agents of melioidosis and glanders, *Burkholderia pseudomallei* and *Burkholderfia mallei*. *J. Clin. Microbiol.* **41:**2068–2079.
48. **Gu, X., and H. Zhang.** 2004. Genome phylogenetic analysis based on extended gene contents. *Mol. Biol. Evol.* **21:**1401–1408.
49. **Hall, B. G.** 2001. *Phylogenetic Trees Made Easy: a How-To Manual for Molecular Biologists*. Sinauer Associates, Inc., Sunderland, Mass.
50. **Hallin, P. F., and D. W. Ussery.** 2004. CBS Genome Atlas Database: a dynamic storage for bioinformatic results and sequence data. *Bioinformatics* **20:**3682–3686.
51. **Harrington, C. S., and S. L. W. On.** 1999. Extensive 16S rRNA gene sequence diversity in *Campylobacter hyointestinalis* strains: taxonomic and applied implications. *Int. J. Syst. Bacteriol.* **49:**1171–1175.
52. **Henz, S. R., D. H. Huson, A. F. Auch, K. Nieselt-Struwe, and S. C. Schuster.** 2004. Whole-genome prokaryotic phylogeny. *Bioinformatics* **21:**2329–2335.
53. **Heringa, J.** 1999. Two strategies for sequence comparison: profile-preprocessed and secondary structure-induced multiple alignment. *Comput. Chem.* **23:**341–364.
54. **Hillis, D. M., M. W. Allard, and M. M. Miyamoto.** 1993. Analysis of DNA sequence data: phylogenetic inference. *Methods Enzymol.* **224:**456–487.
55. **Holder, M., and P. O. Lewis.** 2003. Phylogeny estimation: traditional and Bayesian approaches. *Nat. Rev. Genet.* **4:**275–284.
56. **Holmes, D. E., K. P. Nevin, and D. R. Lovley.** 2004. Comparison of 16S rRNA, *nifD*, *recA*, *gyrB*, *rpoB* and *fusA* genes within the family *Geobacteraceae* fam. nov. *Int. J. Syst. Evol. Microbiol.* **54:**1591–1599.
57. **Horner, D. S., and G. Pesole.** 2004. Phylogenetic analyses: a brief introduction to methods and their application. *Expert Rev. Mol. Diagn.* **4:**339–350.
58. **House, C. H., and S. T. Fitz-Gibbon.** 2002. Using homolog groups to create a whole-genomic tree of free-living organisms: an update. *J. Mol. Evol.* **54:**539–547.
59. Reference deleted.
60. **Huelsenbeck, J. P.** 1995. The robustness of two phylogenetic methods: four-taxon simulations reveal a slight superiority of maximum likelihood over neighbor joining. *Mol. Biol. Evol.* **12:**843–849.
61. **Huelsenbeck, J. P., F. Ronquist, R. Nielsen, and J. P. Bollback.** 2001. Bayesian inference of phylogeny and its impact on evolutionary biology. *Science* **294:**2310–2314.
62. **Huson, D. H.** 1998. SplitsTree: analyzing and visualizing evolutionary data. *Bioinformatics* **14:**68–73.
63. **Huson, D. H., and M. Steel.** 2004. Phylogenetic trees based on gene content. *Bioinformatics* **20:**2044–2049.
64. **Huynen, M., T. Dandekar, and P. Bork.** 1998. Differential genome analysis applied to the species-specific features of Helicobacter pylori. *FEBS Lett.* **426:**1–5.
65. **Huynen, M. A., and P. Bork.** 1998. Measuring genome evolution. *Proc. Natl. Acad. Sci. USA* **95:**5849–5856.
66. **Jensen, R. A.** 2001. Orthologs and paralogs—we need to get it right. *Genome Biol.* **2:**interactions1002.1–interactions1002.3.
67. **Johnson, J. L.** 1991. DNA reassociation experiments, p. 21–40. *In* E. Stackebrandt and M. J. Goodfellow (ed.), *Nucleic Acid Techniques in Bacterial Systematics*. John Wiley & Sons Ltd., Chichester, United Kingdom.
68. **Kabsch, W., and C. Sander.** 1983. Dictionary of protein secondary structure: pattern recognition of hydrogen-bonded and geometrical features. *Biopolymers* **22:**2577–2637.
69. **Kanz, C., P. Aldebert, N. Althorpe, W. Baker, A. Baldwin, K. Bates, P. Browne, A. van den Broek, M. Castro, G. Cochrane, K. Duggan, R. Eberhardt, N. Faruque, J. Gamble, F. G. Diez, N. Harte, T. Kulikova, Q. Lin, V. Lombard, R. Lopez, R. Mancuso, M. McHale, F. Nardone, V. Silventoinen, S. Sobhany, P. Stoehr, M. A. Tuli, K. Tzouvara, R. Vaughan, D. Wu, W. Zhu, and R. Apweiler.** 2005. The EMBL Nucleotide Sequence Database. *Nucleic Acids Res.* **33:**D29–D33 .
70. **Karlin, S., G. M. Weinstock, and V. Brendel.** 1995. Bacterial classifications derived from RecA protein sequence comparisons. *J. Bacteriol.* **177:**6881–6893.
71. **Koizumi, Y., J. J. Kelly, T. Nakagawa, H. Urakawa, S. El-Fantroussi, S. Al-Muzaini, M. Fukui, Y. Urushigawa, and D. A. Stahl.** 2002. Parallel characterization of anaerobic toluene- and ethylbenzene-degrading microbial consortia by PCR-denaturing gradient gel electrophoresis, RNA-DNA membrane hybridization, and DNA microarray technology. *Appl. Environ. Microbiol.* **68:**3215–3225.
72. **Kolaczkowski, B., and J. W. Thornton.** 2004. Performance of maximum parsimony and likelihood phylogenetics when evolution is heterogeneous. *Nature* **431:**980–984.
73. **Korbel, J. O., B. Snel, M. A. Huynen, and P. Bork.** 2002. SHOT: a web server for the construction of genome phylogenies. *Trends Genet.* **18:**158–162.
74. **Kunin, V., D. Ahren, L. Goldovsky, P. Janssen, and C. A. Ouzounis.** 2005. Measuring genome conservation across taxa: divided strains and united kingdoms. *Nucleic Acids Res.* **33:**616–621.
75. **Kwok, A. Y., and A. W. Chow.** 2003. Phylogenetic study of *Staphylococcus* and *Macrococcus* species based on partial *hsp60* gene sequences. *Int. J. Syst. Evol. Microbiol.* **53:**87–92.
76. **La Scola, B., Z. Zeaiter, A. Khamis, and D. Raoult.** 2003. Gene-sequence-based criteria for species definition in bacteriology: the *Bartonella* paradigm. *Trends Microbiol.* **11:**318–321.
77. **Lerat, E., V. Daubin, and N. A. Moran.** 2003. From gene trees to organismal phylogeny in prokaryotes: the case of the γ-Proteobacteria. *PLoS Biol.* **1:**E19.
78. **Lilburn, T. G., and G. M. Garrity.** 2004. Exploring prokaryotic taxonomy. *Int. J. Syst. Evol. Microbiol.* **54:**7–13.
79. **Lin, J., and M. Gerstein.** 2000. Whole-genome trees based on the occurrence of folds and orthologs: implications for comparing genomes on different levels. *Genome Res.* **10:**808–818.
80. **Lloyd, A. T., and P. M. Sharp.** 1993. Evolution of the *recA* gene and the molecular phylogeny of bacteria. *J. Mol. Evol.* **37:**399–407.
81. **Loy, A., A. Lehner, N. Lee, J. Adamczyk, H. Meier, J. Ernst, K. H. Schleifer, and M. Wagner.** 2002. Oligonucleotide microarray for 16S rRNA gene-based detection of all recognized lineages of sulfate-reducing prokaryotes in the environment. *Appl. Environ. Microbiol.* **68:**5064–5081.
82. **Naser, S., F. L. Thompson, B. Hoste, D. Gevers, K. Vandemeulebroecke, I. Cleenwerck, C. C. Thompson,**

M. Vancanneyt, and J. Swings. 2005. Phylogeny and identification of enterococci by *atpA* gene sequence analysis. *J. Clin. Microbiol.* **43:**2224–2230.

83. **Naser, S. M., F. L. Thompson, B. Hoste, D. Gevers, P. Dawyndt, M. Vancanneyt, and J. Swings.** 2005. Application of multilocus sequence analysis (MLSA) for rapid identification of *Enterococcus* species based on *rpoA* and *pheS* genes. *Microbiology* **151:**2141–2150.
84. **Opperdoes, F. R.** 2003. Phylogenetic analysis using protein sequences, p. 207–235. *In* M. Salemi and A. M. Vandamme (ed.), *The Phylogenetic Handbook: a Practical Approach to DNA and Protein Phylogeny*. Cambridge University Press, Cambridge, United Kingdom.
85. **Payne, G. W., P. Vandamme, S. H. Morgan, J. J. Lipuma, T. Coenye, A. J. Weightman, T. H. Jones, and E. Mahenthiralingam.** 2005. Development of a *recA* gene-based identification approach for the entire *Burkholderia* genus. *Appl. Environ. Microbiol.* **71:**3917–3927.
86. **Posada, D., and K. A. Crandall.** 1998. MODELTEST: testing the model of DNA substitution. *Bioinformatics* **14:**817–818.
87. **Priest, F. G., M. Barker, L. W. Baillie, E. C. Holmes, and M. C. Maiden.** 2004. Population structure and evolution of the *Bacillus cereus* group. *J. Bacteriol.* **186:**7959–7970.
88. **Rasko, D. A., M. R. Altherr, C. S. Han, and J. Ravel.** 2005. Genomics of the *Bacillus cereus* group of organisms. *FEMS Microbiol. Rev.* **29:**303–329.
89. **Rokas, A., and S. B. Carroll.** 2005. More genes or more taxa? The relative contribution of gene number and taxon number to phylogenetic accuracy. *Mol. Biol. Evol.* **22:**1337–1344.
90. **Rost, B.** 1999. Twilight zone of protein sequence alignments. *Protein Eng.* **12:**85–94.
91. **Salemi, M., and A. M. Vandamme (ed.).** 2003. *The Phylogenetic Handbook: a Practical Approach to DNA and Protein Phylogeny*. Cambridge University Press, Cambridge, United Kingdom.
92. **Santos, S. R., and H. Ochman.** 2004. Identification and phylogenetic sorting of bacterial lineages with universally conserved genes and proteins. *Environ. Microbiol.* **6:**754–759.
93. **Sawa, G., J. Dicks, and I. N. Roberts.** 2003. Current approaches to whole genome phylogenetic analysis. *Brief Bioinform.* **4:**63–74.
94. **Schierup, M. H., and J. Hein.** 2000. Consequences of recombination on traditional phylogenetic analysis. *Genetics* **156:**879–891.
95. **Schmidt, H. A., K. Strimmer, M. Vingron, and A. von Haeseler.** 2002. TREE-PUZZLE: maximum likelihood phylogenetic analysis using quartets and parallel computing. *Bioinformatics* **18:**502–504.
96. **Simossis, V. A., and J. Heringa.** 2005. PRALINE: a multiple sequence alignment toolbox that integrates homology-extended and secondary structure information. *Nucleic Acids Res.* **33:**W289–W294.
97. **Snel, B., P. Bork, and M. A. Huynen.** 1999. Genome phylogeny based on gene content. *Nat. Genet.* **21:**108–110.
98. **Soler, L., M. A. Yanez, M. R. Chacon, M. G. Aguilera-Arreola, V. Catalan, M. J. Figueras, and A. J. Martinez-Murcia.** 2004. Phylogenetic analysis of the genus *Aeromonas* based on two housekeeping genes. *Int. J. Syst. Evol. Microbiol.* **54:**1511–1519.
99. **Soltis, P. S., and D. E. Soltis.** 2003. Applying the bootstrap in phylogeny reconstruction. *Stat. Sci.* **18:**256–267.
100. **Stackebrandt, E., W. Frederiksen, G. M. Garrity, P. A. Grimont, P. Kampfer, M. C. Maiden, X. Nesme, R. Rossello-Mora, J. Swings, H. G. Truper, L. Vauterin, A. C. Ward, and W. B. Whitman.** 2002. Report of the ad hoc committee for the re-evaluation of the species definition in bacteriology. *Int. J. Syst. Evol. Microbiol.* **52:**1043–1047.
101. **Stackebrandt, E., and B. M. Goebel.** 1994. Taxonomic note: a place for DNA-DNA reassociation and 16S rRNA sequence analysis in the present species definition in bacteriology. *Int. J. Syst. Bacteriol.* **44:**846–849.
102. **Stevens, J. R., and C. J. Schofield.** 2003. Phylogenetics and sequence analysis—some problems for the unwary. *Trends Parasitol.* **19:**582–588.
103. **Stralis-Pavese, N., A. Sessitsch, A. Weilharter, T. Reichenauer, J. Riesing, J. Csontos, J. C. Murrell, and L. Bodrossy.** 2004. Optimization of diagnostic microarray for application in analysing landfill methanotroph communities under different plant covers. *Environ. Microbiol.* **6:**347–363.
104. **Tanaoka, J.** 1994. Determination of the DNA base composition, p. 463–470. *In* M. Goodfellow and A. G. J. O'Donnell (ed.), *Chemical Methods in Prokaryotic Systematics*. John Wiley & Sons Ltd., Chichester, United Kingdom.
105. **Tekaia, F., A. Lazcano, and B. Dujon.** 1999. The genomic tree as revealed from whole proteome comparisons. *Genome Res.* **9:**550–557.
106. **Thompson, C. C., F. L. Thompson, K. Vandemeulebroecke, B. Hoste, P. Dawyndt, and J. Swings.** 2004. Use of *recA* as an alternative phylogenetic marker in the family *Vibrionaceae*. *Int. J. Syst. Evol. Microbiol.* **54:**919–924.
107. **Thompson, F. L., D. Gevers, C. C. Thompson, P. Dawyndt, S. Naser, B. Hoste, C. B. Munn, and J. Swings.** 2005. Phylogeny and molecular identification of vibrios based on multilocus sequence analysis. *Appl. Environ. Microbiol.* **71:**5107–5115.
108. **Thompson, F. L., T. Iida, and J. Swings.** 2004. Biodiversity of vibrios. *Microbiol. Mol. Biol. Rev.* **68:**403–431.
109. **Tortoli, E.** 2003. Impact of genotypic studies on mycobacterial taxonomy: the new mycobacteria of the 1990s. *Clin. Microbiol. Rev.* **16:**319–354.
110. **van Berkum, P., Z. Terefework, L. Paulin, S. Suomalainen, K. Lindström, and B. D. Eardly.** 2003. Discordant phylogenies within the *rrn* loci of rhizobia. *J. Bacteriol.* **185:**2988–2998.
111. **Vancanneyt, M., P. Vandamme, and K. Kersters.** 1995. Differentiation of *Bordetella pertussis*, *B. parapertussis*, and *B. bronchiseptica* by whole-cell protein electrophoresis and fatty-acid analysis. *Int. J. Syst. Bacteriol.* **45:**843–847.
112. **Vandamme, P., C. S. Harrington, K. Jalava, and S. L. On.** 2000. Misidentifying helicobacters: the *Helicobacter cinaedi* example. *J. Clin. Microbiol.* **38:**2261–2266.
113. **Vandamme, P., B. Pot, M. Gillis, P. Devos, K. Kersters, and J. Swings.** 1996. Polyphasic taxonomy, a consensus approach to bacterial systematics. *Microbiol. Rev.* **60:**407–438.
114. **Vandamme, P. A. R.** 2003. Taxonomy and classification of bacteria, p. 271–285. *In* P. R. Murray, E. J. Baron, J. H. Jorgensen, M. A. Pfaller, and R. H. Yolken (ed.), *Manual of Clinical Microbiology*, 8th ed. ASM Press, Washington, D.C.
115. **Waddell, P. J., H. Kishino, and R. Ota.** 2000. Rapid evaluation of the phylogenetic congruence of sequence data using likelihood ratio tests. *Mol. Biol. Evol.* **17:**1988–1992.
116. **Wayne, L. G., D. J. Brenner, R. R. Colwell, P. A. D. Grimont, O. Kandler, M. I. Krichevsky, L. H. Moore, W. E. C. Moore, R. G. E. Murray, E. Stackebrandt, M. P. Starr, and H. G. Truper.** 1987. Report of the ad hoc committee on reconciliation of approaches to bacterial systematics. *Int. J. Syst. Bacteriol.* **37:**463–464.
117. **Weisburg, W. G., S. J. Giovannoni, and C. R. Woese.** 1989. The *Deinococcus-Thermus* phylum and the effect of

rRNA composition on phylogenetic tree construction. *Syst. Appl. Microbiol.* **11:**128–134.

118. **Wernersson, R., and A. G. Pedersen.** 2003. RevTrans: multiple alignment of coding DNA from aligned amino acid sequences. *Nucleic Acids Res.* **31:**3537–3539.
119. **Whelan, S., P. Lio, and N. Goldman.** 2001. Molecular phylogenetics: state-of-the-art methods for looking into the past. *Trends Genet.* **17:**262–272.
120. **Willems, A., E. Falsen, B. Pot, E. Jantzen, B. Hoste, P. Vandamme, M. Gillis, K. Kersters, and J. De Ley.** 1990. *Acidovorax*, a new genus for *Pseudomonas facilis*, *Pseudomonas delafieldii*, E. Falsen (EF) group 13, EF group 16, and several clinical isolates, with the species *Acidovorax facilis* comb. nov., *Acidovorax delafieldii* comb. nov., and *Acidovorax temperans* sp. nov. *Int. J. Syst. Bacteriol.* **40:** 384–398.
121. **Woese, C. R.** 1987. Bacterial evolution. *Microbiol. Rev.* **51:**221–271.
122. **Woese, C. R.** 1991. The use of ribosomal RNA in reconstructing evolutionary relationships among bacteria, p. 1–24. *In* R. K. Selander, A. G. Clark, and T. S. Whittam (ed.), *Evolution at the Molecular Level*. Sinauer Associates, Inc., Sunderland, Mass.
123. **Woese, C. R., E. Stackebrandt, and W. Ludwig.** 1985. What are mycoplasmas—the relationship of tempo and mode in bacterial evolution. *J. Mol. Evol.* **21:**305–316.
124. **Wolf, Y. I., S. E. Brenner, P. A. Bash, and E. V. Koonin.** 1999. Distribution of protein folds in the three superkingdoms of life. *Genome Res.* **9:**17–26.
125. **Wolf, Y. I., I. B. Rogozin, N. V. Grishin, and E. V. Koonin.** 2002. Genome trees and the tree of life. *Trends Genet.* **18:**472–479.
126. **Wuyts, J., G. Perriere, and Y. Van De Peer.** 2004. The European ribosomal RNA database. *Nucleic Acids Res.* **32:**D101–D103.
127. **Yamamoto, S., H. Kasai, D. L. Arnold, R. W. Jackson, A. Vivian, and S. Harayama.** 2000. Phylogeny of the genus *Pseudomonas*: intrageneric structure reconstructed from the nucleotide sequences of *gyrB* and *rpoD* genes. *Microbiology* **146:**2385–2394.
128. **Yang, Q., and M. Blanchette.** 2004. StructMiner: a tool for alignment and detection of conserved secondary structure. *Genome Inform. Ser. Workshop Genome Inform.* **15:**102–111.
129. **Yang, S., R. F. Doolittle, and P. E. Bourne.** 2005. Phylogeny determined by protein domain content. *Proc. Natl. Acad. Sci. USA* **102:**373–378.
130. **Ye, R. W., T. Wang, L. Bedzyk, and K. M. Croker.** 2001. Applications of DNA microarrays in microbial systems. *J. Microbiol. Methods* **47:**257–272.
131. **Zeigler, D. R.** 2003. Gene sequences useful for predicting relatedness of whole genomes in bacteria. *Int. J. Syst. Evol. Microbiol.* **53:**1893–1900.
132. **Zhang, H., and X. Gu.** 14 November 2004, posting date. Maximum likelihood for genome phylogeny on gene content. *Stat. Appl. Genet. Mol. Biol.* **3:**no. 1, article 31. [Online.] http://www.bepress.com/sagmb/vol3/iss1/art31.

Bioreporters, Biosensors, and Microprobes

ROBERT S. BURLAGE

14

The study of microbial communities is dependent on the ability to discern individual bacterial species and microbial activities in a complex matrix. Small variations in constituents can determine whether a microorganism occupies a specific niche or whether a specific gene is active. Therefore, the tools that are used to study microbial environments must be very sensitive and the scale on which they measure might be very small indeed, perhaps the area surrounding a single microorganism. For this purpose a variety of analytical devices have been produced; they can be conveniently organized into bioreporters, microprobes, and biosensors. All of these technologies have been used in other areas of biology, notably in cell biology, but this chapter concentrates on their uses and potential uses for environmental microbiology. The danger in doing this is that a promising technology may be omitted because it is not currently used for environmental microbiology, which is certainly likely in such a fast-moving field. However, there is also an opportunity to create new tools that are relevant to specific problems in environmental microbiology.

This chapter stresses techniques that are significantly different from other molecular techniques in that they are suitable for complex environments inhabited by a diverse collection of bacteria. Other molecular assays are valuable for pure-culture work, and they certainly have great utility in environmental microbiology, but environmental microbiologists need assays that have other attributes. The assays should be nondestructive of sample material. Heterogeneity of environmental samples may be very significant, and therefore assays that examine the same sample are very valuable. The assays should be real-time procedures, since conditions can change quickly within mixed communities. In addition, the assays should be able to report conditions continuously rather than at selected assay times. Fluctuations within complex communities may be very slight yet may still be significant in understanding the ecology of the system.

Biosensors of environmental conditions and bioreporter genes of genetic activity have been developed to augment our ability to detect, identify, and quantify microorganisms in complex ecological settings and to analyze their microniches. Biosensors are devices that fuse two technologies, electronic and biological, into a unique analytical tool. Bioreporters are genes that produce a product that is easily assayed and that relates to the genetic activity of the host cell. These techniques are likely to become more generally applied as the technology matures and as commercial opportunities arise. The field has now grown so large that a comprehensive examination of the techniques is not possible here. Reference is made to relevant review articles for readers seeking a more comprehensive understanding of the technology.

BIOREPORTER TECHNOLOGY

The analysis of genetic expression is made more difficult by the lack of suitable assays for most gene products. Bioreporter genes play a surrogate role, supplying an assayable gene product when an assay for the gene product of interest is not available or very difficult to perform. Bioreporter genes have been used extensively with pure cultures to demonstrate the expression of specific genes. However, many of the most useful bioreporter genes, such as *lacZ* (encoding the β-galactosidase enzyme) and *xylE* (encoding catechol 2,3-oxygenase), are usually unsuitable for ecological studies. While any single bioreporter gene might not be present in a particular species, the presence of the gene in other members of a community is not unlikely, potentially creating a significant background problem. This can be true in a monoculture as well, such as when a *lacZ* bioreporter is used with a lac^{+} strain. To circumvent this problem, researchers have used bioreporter genes with no homologs in the host strain, such as the chitobiase gene (38). This gene, derived from a deep-sea *Vibrio harveyi* strain, catalyzes a reaction that gives a colorimetric response on agar plates and therefore can be easily detected. Background is essentially absent in *Escherichia coli*, although the use of this gene in other species has been limited so far. Another interesting example is the ice nucleation gene, *inaZ*. This gene was isolated from the plant pathogen *Pseudomonas syringae* and allows ice droplet formation at higher temperatures. It is assayed by determining the temperature at which a droplet containing the strain freezes. If it freezes at a higher temperature, then the gene product is present and gene expression has occurred. This gene has been used in several experiments, often in concert with other bioreporters (1, 5, 48, 49).

Another drawback of conventional bioreporters is that performing a biochemical assay often requires the destruction of a sample of the community. Nondestructive assays allow repeated experiments to be performed on the same sample, so that changes (development) of a microbial community can be observed. Biochemical assays for conventional bioreporters require time for sampling, processing, and signal development, and these steps introduce a significant time delay between the microbial event and the analysis of the data. There is a distinct advantage in having an assay that gives real-time results.

While there are several bioreporter genes that might be used, bioreporters that make use of light for bioreporting have significant advantages. The use of either bioluminescent or fluorescent bioreporters is now an established technology, and the uses have expanded greatly over the years. They both represent nondestructive, noninvasive means of detecting gene expression. Light can be measured with great sensitivity and precision, allowing the detection of a single cell under some conditions and with appropriate light-gathering equipment (76). Bioreporters for bioluminescent and fluorescent gene products are described below.

Bioluminescence

Bioluminescence is the production of visible light by a biochemical process. Unlike most chemical reactions, which produce heat as the main by-product, these reactions also generate enough light to be detected by conventional photodetectors. This phenomenon is easily observed in fireflies (*Photinus pyralis*), although many other species are capable of producing light. The genes for these light-producing reactions have been isolated by researchers, and they are now available on cloning vectors. When genetically fused (using either a transcriptional or protein fusion) to appropriate genes from a host bacterium, these strains produce light under defined conditions. The cloned firefly luciferase gene (*luc*) has been used to observe gene expression in animals (68), plants (42, 59), and bacteria (60). Firefly luciferase is a powerful tool for genetic analysis, although it is difficult to use in microbial ecology experiments since the reaction requires the substrate luciferin, which must be added exogenously. Often these assays utilize extracts from samples, rather than whole cells, and therefore destructive sampling is required.

Several genera and species of bioluminescent bacteria are known, although *V. harveyi* and *V. fischeri* have received the most attention. These bacteria contain *lux* genes, which are responsible for bioluminescence. The *lux* operon is a complex pathway of five genes, *luxCDABE*, and efficient expression of all of these genes in the host is required for appropriate functioning of the bioreporter. Only two genes, *luxA* and *luxB*, encoding the heterodimeric luciferase enzyme, are needed for the actual bioluminescent reaction. The *luxCDE* genes have been implicated in the recycling of the required aldehyde substrate, so that a pool of substrate is continuously available (Fig. 1). The *lux* genes, comprising about 7 kbp of DNA, have been cloned and sequenced and are available for analysis of microbial consortia. Several excellent reviews describe the genetics and physiology of bacterial bioluminescence (46, 47) and the use of these fusions (11).

Advantages and Disadvantages

The advantages of bioluminescent bioreporters lie primarily in the relative ease of light measurement. Light can be measured accurately and with great sensitivity. Since light radiates in all directions from a point source, light detection can be performed in three dimensions, giving a more sophisticated analysis of an object's position in space. It can be measured quickly (in real time) and without perturbing or destroying the sample. For instance, the light detector can be introduced into the sample and left there for an extended period, or it can detect light that passes through the glass wall of a bioreactor vessel. Bacterial interactions can thus be examined in real time, for example, to study predator-prey or symbiotic relationships. There is usually no need to add any substrates or reagents for the bioluminescence assay (although the requirement for an aldehyde substrate is discussed below). In most consortia of interest to environmental microbiologists, bioluminescence is a rare trait, and therefore a background problem is unlikely as long as ambient light can be excluded from the reaction vessel.

The *lux* genes are especially useful if a qualitative analysis is sufficient, i.e., determining whether light is being produced at a given time. For these experiments, light output can be expressed as relative light units. Relative light units are acceptable when all work is performed with the same light-measuring apparatus, under identical conditions (distance from detector, temperature, composition of the vessel holding the sample), and where no comparisons are drawn to other published results. If quantification is desired, photodetectors can be calibrated (see below).

The advantages described above must be weighed against several disadvantages that are inherent to the bioluminescent bioreporters. The bioluminescent reaction requires molecular oxygen, without which the bioreporter is inoperative. Oxygen limitation results in a lower intensity of light generated, perhaps to a level that would escape detection. The physiological state of the cell can thus have a major impact on the strength of the bioluminescent signal: cells respiring at a high rate will quickly exhaust their available oxygen and leave little for the bioluminescence reaction. This is easily seen by swirling a rapidly growing culture and watching the light intensity increase as the culture medium becomes aerated. In addition, if the host strain does not have a suitable aldehyde substrate for the bioluminescent reaction, an aldehyde must be added exogenously. Few strains make sufficient aldehyde for prolonged light production, and strains utilizing *luxAB* vectors must always have an aldehyde supplement in the medium. The substrate, usually *n*-decanal at a final concentration of 0.1 to 1.0% (vol/vol), penetrates

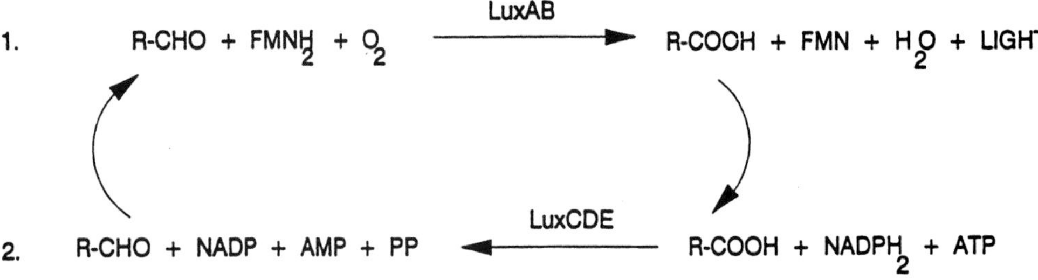

FIGURE 1 Genes and chemical intermediates involved in the bioluminescence reaction of *Vibrio fischeri*.

the cells readily, although it can be toxic at relatively low doses. Finally, the luciferase enzyme of *V. fischeri* is heat labile and is not recommended for use above 30°C. However, the luciferase from *V. harveyi* is stable at 37°C. An excellent discussion of the weaknesses of bacterial bioreporters and the means by which these techniques may be improved is available (83).

Applications

Many genetic constructions involving the *lux* genes have been prepared, most of which utilize the genes from a *Vibrio* species. Some are now available commercially. The intact *luxCDABE* cassette is available on a plasmid cloning vector, pUCD615, so that expression of bioluminescence can be placed under the control of a host promoter (70). Use of the full cassette ensures that the aldehyde substrate will be regenerated for continuous availability, although the host cell must have a suitable long-chain aldehyde (e.g., decanal) present as substrate for the luciferase reaction. The presence of such an aldehyde can be determined only empirically, and fluctuations of the aldehyde concentration are always possible. If such fluctuations are suspected, *n*-decanal can be added at the concentrations listed above to ensure an adequate supply.

The inclusion of the full cassette (i.e., the *luxA* and *luxB* luciferase and the *luxCDE* genes) obviates the need for exogenously added aldehyde, at least in strains that have a suitable aldehyde substrate to start with. The luciferase enzyme uses molecular oxygen to convert the aldehyde substrate to a carboxylic acid, with the resulting light being a by-product of the reaction. The action of the *luxCDE* genes is to recycle the carboxylic acid product of the light-producing reaction, giving a continuous supply of aldehyde. It has been observed that, at peak light emissions, the light intensity of such a strain can be boosted by the addition of aldehyde, although not greatly. This suggests that there is always an aldehyde limitation in the cell.

Alternatively, a construction utilizing only the *luxA* and *luxB* genes can be used if the aldehyde substrate is added exogenously. The *luxA* and *luxB* genes have been fused into a single open reading frame by Escher et al. (26), although the resulting *luxAB* fusion luciferase is more temperature sensitive than the native enzyme. Transposons carrying the *lux* genes are also available, such as the Tn*4431* transposon (75). This construct carries the intact *luxCDABE* cassette and a gene for antibiotic selection of transposon insertion. This construct is a valuable means of generating transposon mutants quickly. The *luxAB* genes are available on a Tn*5* derivative (9) and on a mini-Tn*5* derivative (23). There have been many applications of this technology in a variety of bacterial species. A *Salmonella* strain was used for genotoxicity assays (80) with the inducible *umu* gene promoter. A lambda phage was genetically engineered to carry the bioluminescence genes into *E. coli* cells in environmental settings, increasing the detection limits for this important pathogen (67). Hassler and Twiss (31) modified a *Synechococcus* strain to serve as a bioluminescent reporter of iron availability, another important environmental factor. The yeast *Saccharomyces cerevisiae* has been modified with *lux* genes to create a novel bioreporter of estrogenic compounds (71). Novel applications extend the trend of miniaturization of technology, which is also evident in the biosensors. Nivens et al. (56) combined bioreporter bacteria with a miniaturized optics detection device to make an autonomous detector. This is an important development since the environmental applications of bioreporters are dependent on inexpensive, independently monitored systems. There is little commercial appeal for a system that requires a visit from a human operator for every data point. Coupling this technology with a communications capability and the appropriate data storage and retrieval system would make it applicable to many field environments.

Light Measurement

Unlike many assays, in which a standard procedure is used to describe the results, light can be measured by a variety of means. Visualization of bacterial colonies may be sufficient for screening of clones during genetic construction, although the observer must be in a darkened room, and this method is not at all quantitative. Photographic film can be exposed to the light being emitted from colonies, although this technique is usually cumbersome. Several types of electronic equipment are suitable for the measurement of light. ATP photometers or luminometers, which are used for measurement of ATP concentrations by the luciferase assay, are common in laboratories. Liquid scintillation counters are also common. A liquid scintillation counter must be very sensitive in order to detect photons resulting from radioactive decay, and so these counters make good photodetectors for bioluminescence, although the coincidence channel should be disconnected prior to use. The coincidence channel eliminates background during its measurement of radiation, but it is a hindrance for bioluminescence work since light emanating from a single cell might not be detected by both photodetectors simultaneously. These methods are sensitive but are not designed specifically for bioluminescence work. Accordingly, there are problems in introducing representative samples to the photodetectors as well as in determining incubation conditions for the samples. That is, the samples would have to fit inside ordinary scintillation vials, which might not provide adequate aeration or mixing.

Commercial photomultipliers (e.g., those from Oriel, Stratford, Conn.) are recommended for remote sampling of light, including bioreactors and soil microcosms. These usually include flexible fiber-optic cables, which have a high efficiency of light transmittance, an important feature in measuring low levels of light. For extremely low levels of light, such as would be expected from single bacterial cells, charge-coupled devices (CCD) can be used (e.g., those from Hamamatsu, Hamamatsu City, Japan). The added sensitivity is reflected in the increased cost of this equipment, and few laboratories have access to one. A CCD can be used, however, to visualize signals that are seen through a microscope, and thus it has the potential to describe the physiological response of single cells, although integration of weak signals can delay output for several minutes. Accordingly, samples that move or drift during the integration time give a blurred image, if the image is detected at all.

The lack of standardization is a major shortcoming of bioluminescent reporter work, and its greatest impact is on the quantification of results. The output of a bioluminescent strain must be expressed in terms of specific activity to allow comparisons between laboratories. Units of light production would ideally be expressed as photons (quanta) of light per minute per milligram of total protein. However, each photodetector system has a different efficiency of light detection (for instance, different sensitivities for different wavelengths), as well as a different detector window geometry. A method to standardize photodetectors using a light-producing biochemical reaction has been described (58), and this method should be applied more generally. Calibration of a photodetector using a standard light source is possible, although the equipment is expensive and not generally available in laboratories.

Fluorescence

There are now several fluorescent proteins that can be used as bioreporters in bacterial cells, but the first successful one was green fluorescent protein (GFP). GFP addresses many of the disadvantages of the bioluminescent bioreporters, including the following: (i) the bioluminescence reaction requires oxygen and is unreliable under conditions of reduced oxygen tension (although GFP does require small amounts of oxygen for proper chromophore formation); (ii) bioluminescence requires functioning of the luciferase enzyme, which requires correct synthesis and folding of the protein; (iii) the luciferase enzyme requires a substrate, which might not be available in the cell; and (iv) the luciferase from *V. fischeri* is heat labile and is useless at 37°C. Increased stability at high temperatures would be an asset for all luciferases. GFP has the advantage that it is measured on the basis of its intrinsic properties and not on the basis of its biological activity in a certain milieu. Generally speaking, the newer versions of GFP and the new fluorescent proteins all have these attributes, and researchers typically choose a fluorescent protein based on its fluorescence characteristics (excitation and emission wavelengths).

The original GFP was a huge asset to the biological sciences, not only for microbiology but especially for cell biology. The ability to discern gene expression in a three-dimensional system, with great specificity, has been a boon for developmental biology (82). The success of GFP inspired a search for other fluorescent proteins, and now there are several commercially available proteins (Table 1). Typically these products include the purified protein, expression vectors for several systems (plant, animal, and bacterial), and perhaps antibodies that are specific for the fluorescent protein. The original GFP, often referred to as avGFP (for its origin in *Aequorea victoria*), must now compete with a variety of other fluorescent products that have been isolated from many sources, particularly deep-sea corals. Fluorescent colors are usually blue, green, yellow, and red, and different manufacturers tout the excitation and emission spectra of their products, particularly in regard to their ability to be used in tandem. Besides fluorescent characteristics, other attributes are desirable. The DsRed gene, from the deep-sea coral *Discosoma*, was originally thought to have great potential as a fluorescent marker comparable to GFP, but in most biological systems the protein folded so slowly that it had little practical value. Recent work with mutations of DsRed have yielded variants that are much improved, with a folding time reduced to less than 1 h (7), which is a fraction of the time for the wild-type protein.

The GFP gene has been cloned and sequenced, and the protein has been extensively characterized (61, 63). The protein that is synthesized from the GFP gene autocyclizes (16), producing a chromophore that is brightly fluorescent (Fig. 2). When the GFP gene is expressed in a cell (either prokaryotic or eukaryotic), it fluoresces a bright green after cyclization of the chromophore (13). The fluorescence makes the cell easy to detect with UV light (excitation, 395 nm) and conventional light-gathering equipment.

Advantages and Disadvantages

As with the measurement of bioluminescence, fluorescence can be measured accurately and with great sensitivity. Detection is dependent on the ability of the researcher to expose the GFP molecule to the excitation wavelength, and this can be performed with flexible fiber-optic cables introduced into a microbial ecosystem. Measurement is rapid, and there is no need to add any substrates or reagents. The problems of sample perturbation and destruction are therefore avoided.

Fluorescence of GFP is very bright, and individual bacterial cells can easily be seen by epifluorescence microscopy. GFP appears to be very slow in forming the chromophore (typically taking several hours), and the speed at which it forms seems to vary with different organisms and different growth conditions, although a comprehensive analysis of this phenomenon is lacking. The protein is extremely stable and is largely unaffected by treatment with detergents, proteases, glutaraldehyde, or organic solvents. It is also very stable over a pH range of 6 to 12 and at high (65°C) temperatures. GFP may be useful in genetic analysis of ther-

TABLE 1 Commercially available fluorescent proteins

Name	Supplier	Excitation and emission wavelengths (nm)	Color	Source organism
Aequorea victoria green fluorescent protein (GFP)	Amersham	395, 509	Green	Jellyfish
Ptilosarcus green fluorescent protein	Lux Biotechnology	485, 508	Green	Sea pen
Renilla green fluorescent protein	Lux Biotechnology	485, 506	Green	Sea pansy
Ptilosarcus green fluorescent protein	Nanolight Technology	NA[a]	Green	Sea pen
Renilla green fluorescent protein	Nanolight Technology	NA	Green	Sea pansy
Anthomedusae JRed	Evrogen	584, 610	Red	Jellyfish
Pantellina TurboGFP	Evrogen	482, 502	Green	Plankton
Heteractis HcRed	Evrogen	590, 637	Red	Sea anemone
Phialidium PhiYFP	Evrogen	525, 537	Yellow	Jellyfish
Anemonia KFP-Red	Evrogen	580, 600	Red	Sea anemone
Discosoma DsRed	Clontech	556, 586	Red	Coral
Aequorea corulescens green fluorescent protein	Clontech	475, 505	Green	Jellyfish
Anthozoa AmCyan	Clontech	458, 489	Blue	Coral
Anthozoa ZsYellow	Clontech	529, 539	Yellow	Coral
Heteractis HcRed	Clontech	588, 618	Red	Coral

[a]NA, not applicable.

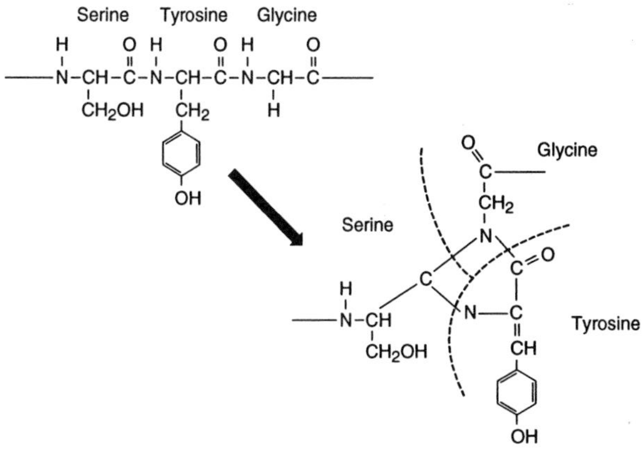

FIGURE 2 The chromophore of *Aequorea* GFP. Amino acids 65, 66, and 67 of GFP form a cyclical structure by an autocatalytic reaction. This chromophore is the source of the bright fluorescence seen with this protein. The dotted lines delineate the separate amino acids in the chromophore.

mophiles and other extremophiles. Once the protein is made, it does not degrade quickly in the cell, and therefore assays of the dynamics of gene expression, such as have been performed with the *lux* genes, are not possible with wild-type GFP. Andersen et al. (3) were able to attach a peptide tag to the carboxy terminus of GFP. This extra peptide made the mutant GFP susceptible to bacterial housekeeping proteases that recognize the carboxy ends of proteins, which resulted in a GFP with a reduced half-life, in this case ranging from 40 min to several hours, depending on the strain and the growth conditions. Because of this development, it is possible to use the bright fluorescence of GFP as an unparalleled bioreporter of real-time gene expression. Many other mutations have been introduced into GFP, producing useful variants. While wild-type GFP tends to form inclusion bodies, which limit the amount of fluorescence seen, GFP mutants have been developed that avoid this problem and which result in greatly amplified fluorescence (17).

Formation of the GFP chromophore requires molecular oxygen, although not in great amounts, and therefore is unsuitable for use under completely anaerobic conditions. Hansen et al. (30) demonstrated that GFP could be formed and detected in biofilms when dissolved oxygen was present at 0.1 ppm. Only the most oxygen-intolerant microorganisms would be unable to live under these conditions. No fluorescence is seen when cells are grown in an anaerobic environment, although once the chromophore is formed it continues to fluoresce in an anaerobic environment. The presence of GFP in bacteria does not appear to have deleterious effects on the host, although a comprehensive analysis has not been performed. Interestingly, individual molecules of GFP do not remain fluorescent at all times, but instead appear to "blink"; that is, they stop fluorescing for a few seconds and then start again (24). In any aggregate this is undetectable since the vast majority of molecules are fluorescing, but it does suggest other possible uses for GFP on a nanoscale.

Applications

Use of GFP is still a relatively new technique, and the construction of convenient cloning vectors is continuing. However, the number of applications of GFP is impressive, guaranteeing that more vectors will soon become available. These vectors typically contain the GFP gene within a polylinker region, allowing convenient manipulation of the gene for transcriptional fusions. The intact GFP gene has been inserted into a derivative of Tn*5*, and therefore random mutations with GFP are possible (12). This transposon, Tn*5GFP1*, can be introduced into a variety of gram-negative species by electroporation.

Mutations have been introduced into the GFP gene to produce fluorescent signals with altered properties. The red-shifted GFP was isolated in this manner (22). The name refers to the shift of the excitation wavelength toward the red end of the spectrum. The protein fluoresces at approximately the same wavelength (the maximum is at 505 nm instead of 510 nm) but excites at 490-nm instead of 395 nm. This shift is expected to be helpful, since the 490-nm excitation wavelength is beyond the wavelengths of excitation for cellular-protein fluorescence (due to their aromatic amino acids). A mutant GFP developed by Heim et al. (32) results in the production of a blue color instead of a green color. Multiple site mutations introduced by Anderson et al. (4) had the effect of increasing the brightness of the fluorescence. These mutants also had an altered excitation profile, allowing both mutant genes to be used in the same cell. Although they were initially tested only in eukaryotic cells, it is reasonable to expect that they will work in prokaryotes as well. Novel uses have been described for the detection of nitrates in soil systems (19) and for arsenic detection (85). In a clever adaptation of the technology, this latter group was able to detect single bioreporter cells as they responded to arsenic in a water sample, demonstrating the strength of the GFP fluorescence.

Fluorescence Measurement

Bacterial colonies expressing GFP can be easily detected on exposure to a UV light. This is easily accomplished with the UV source that is used in most molecular biology laboratories to visualize DNA in agarose gels, although an inexpensive hand-held UV light works just as well. Fluorescent bacteria can also be easily seen using epifluorescence microscopy. An appropriate filter set should be used; the filter for fluorescein detection has proved to be very useful for this purpose. A xenon or mercury lamp can be used as a source of UV excitation. For discrimination of their mutant GFPs in a fluorescence-activated cell sorter, Anderson et al. (4) used a krypton ion laser for one variant and an argon ion laser for the other. Since both proteins produced the same emission wavelength, the same filter and detector could be used.

Fluorescence spectrometry facilitates the detection of GFP fluorescence. Fluorescence spectrometers vary in sensitivity and versatility, although in general they should be able to detect GFP expression in bacteria. Quantification of bacteria in the sample is possible when a standard is examined contemporaneously. Digital imaging spectroscopy (28, 89) is an excellent means of detecting and characterizing fluorescent signals, although the expense of the system makes it unavailable to all but a few researchers.

MICROPROBES AND BIOSENSORS

Microprobes

A microprobe is a device that measures a specific physical or chemical property in a microenvironment. For instance, microprobes can be devised to test for pH, temperature, or the concentration of ionic species (72). The quality that makes these probes different from other probes is their small size, which makes them more suitable for analysis at submillimeter resolution. The development of microprobes for the examination of microbial environments has proceeded rapidly thanks to innovative construction techniques. Microprobes have been described for ammonium (20), nitrate (36, 79), oxygen (65), denitrification (by nitrous oxide production) (15), and sulfate reduction (64). Conventional technology is used to measure the analyte; the critical development is the miniaturization of the electrode. The probe tip can be in the range of 1 to 10 μm in diameter, although 10 to 50 μm is more common. The signal is usually reported as a current on an ammeter. Delicate construction of the probe is required, as is the manipulation of the probe in three dimensions with a micromanipulator and dissecting microscope. Microprobes are especially useful to study the microecology of biofilms, including formation and activity at various depths inside the biofilm. An excellent review is available (66). The main obstacle to the general use of microprobes is that there are a limited number of commercially available microprobes (with a few notable exceptions, such as Microelectrodes [Bedford, N.H.], Unisense [Aarhus, Denmark), Abtech [Richmond, Va.], and Microprobe [Gaithersburg, Md.]), and the microprobe often must be handmade by the researcher. Techniques for construction are available, although they require skill and patience. As the usefulness of microprobes becomes better appreciated, the number of manufacturers will certainly increase.

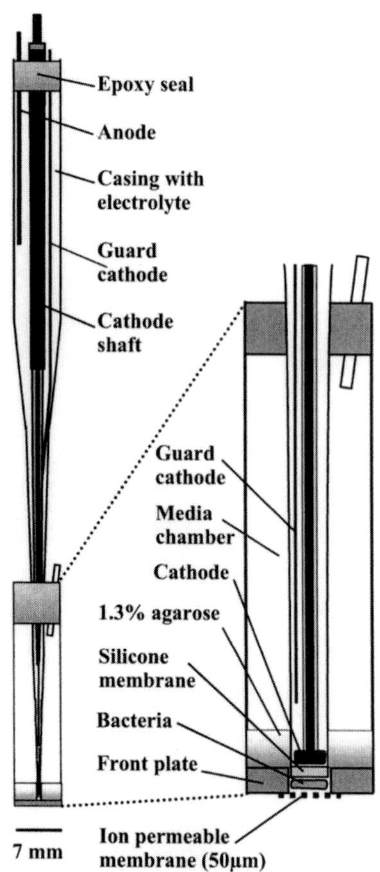

FIGURE 3 Cross-section of a typical microprobe. Reprinted from reference 55 with permission from the publisher.

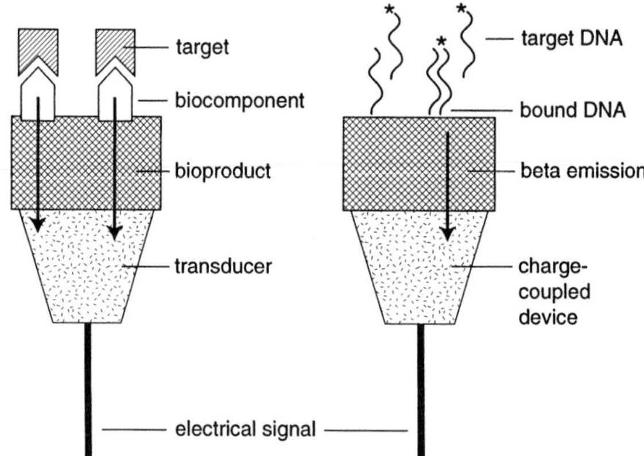

FIGURE 4 Biosensors. (Left) A generalized scheme for a biosensor. Interaction of the target analyte with the biological component results in a signal, which is transmitted to the transducer. The transducer senses the signal and converts it to an electrical signal. (Right) A DNA biosensor. The hybridization event brings the labeled DNA in contact with the transducer, a CCD camera. The CCD camera detects beta emission from ^{32}P decay and converts it to an electrical signal.

The oxygen microprobe described by Revsbech (65) provides a good example of the current microprobe technology. This microprobe has a tip that is approximately 10 μm in diameter and is sensitive to oxygen concentrations in the micromolar range. It incorporates a guard cathode that removes oxygen diffusing toward the sensor tip from the electrolyte solution, permitting stable signal acquisition from the sample. This microprobe is suitable for examination of biofilm ecology or aquatic microbiology, with the ability to discern microbial processes at the water interface. Although it is extraordinarily small for an analytical instrument, its presence is likely to disturb or influence the surrounding environment, however slightly. The prudent researcher will be attentive to possible effects from the use of these tools. Figure 3 shows a cross-section of a typical microprobe.

Biosensors

A biosensor is a type of probe in which a biological component, such as an enzyme, antibody, or nucleic acid, interacts with an analyte, which is then detected by an electronic component and translated into a measurable (electronic) signal (Fig. 4). Biosensor probes are possible because of a fusion of two technologies: microelectronics and biotechnology. Their greatest impact is in the clinical area, in which rapid test results are needed (86). However, they are also applicable to environmental analysis, and in recent years several substantial improvements have allowed them to be used more generally. A review of the application of biosensors for environmental study is available (69). There are several components to a biosensor. Different researchers may contribute to the development or optimization of a particular part or may incorporate several different technologies into one tool. For convenience, these components can be summarized as follows: the biological component (or biomolecule), the attachment method, the microfluidics, the computational component, and the electronic (sensing) component, which is also called the transducer (27).

The biological component is the molecule that interacts with the analyte of interest. A variety of substances can be used as the biological component, including nucleic acids, proteins (particularly antibodies and enzymes), lectins (plant proteins that bind sugar moieties), and complex materials (organelles, tissue slices, and microorganisms). In each case, it is the specificity of the biological component for an analyte (or group of related analytes) that makes the biomolecule attractive for use in sensing technology. For example, a single strand of DNA hybridizes only to its complementary strand under the appropriate conditions. The conditions of the assay are very important, especially where reversibility of binding is a key factor. This is apparent in antibody-based biosensors, in which the affinity of the antibody for the antigen affects the sensitivity of the biosensor. This is particularly true for the measurement of dissociation constants of antibodies, which may vary substantially.

In most cases, isolation of the biological component is necessary to ensure that only the molecule of interest is bound or immobilized on the electronic component. In some cases this is easy, such as the isolation of DNA. It is possible to have a commercial vendor make specific oligonucleotides that are pure and that are already labeled. Antibody or enzyme extraction and purification are much more complex procedures, although crude extracts can sometimes be used. The stability of the biological component is also critical, since it is being used outside of its usual biological environment. A labile protein usually makes a poor candidate for a biosensor. Koblizek et al. (40) isolated particles from *Synechococcus elongatus* that contain the photosystem II enzymes. These particles were then trapped on an oxygen electrode with a dialysis membrane. If the enzymes are working normally, oxygen is generated and is detected by the electrode. In the presence of inhibitors of the enzymes (the authors used certain herbicides), oxygen generation is diminished and the effect becomes quantifiable. Although this device is neither field-hardy nor particularly long-lived, it is a clever example of the use of a natural product to create a reliable tool.

While it is possible for these biomolecules to be free in solution in order to function, it is more common to bind them to a surface with a known location, and therefore the attachment of the biomolecule is important. This has numerous benefits. The precise location of the biomolecules will be known. This is especially important when constructing arrays of sensors, in which a different biomolecule is attached at a discrete address on a surface and can be interrogated individually by the electronic component. And, of course, having the biomolecule tethered to a site allows other reagents to be introduced by using fluid transfer technology. The objective is to bind the biomolecule in place without disrupting its biological activity. That is, the enzyme must still catalyze a reaction, the antibody must bind its antigen, and the nucleic acid must allow hybridization with its complementary strand. A number of attachment protocols have been described, particularly for the attachment to glass or silica particles. This may involve a silane cross-bridge to which biomolecules can be attached. The need to deposit organic substances on transducers in a predictable manner has been addressed by the work of Decher et al. (21). This group demonstrated the sequential construction of layers on typical transducer surfaces, such as glass, silicon wafers, and quartz. This technique should improve biosensors by increasing the uniformity of results. Amino-derivatized oligonucleotides can be attached to glass (SiO_2) surfaces such as fiber-optic cables, glass beads, or microscope slides though covalent binding with a chemical linker. Some techniques result in the nonspecific attachment of oligonucleotides to the surface, which is an impediment to the hybridization that is required for the detection by the transducer. Losses of efficiency such as this can be avoided through careful attachment of the oligonucleotide via a modified 3′ or 5′ end of the oligomer. Graham et al. (29) used a simple procedure to attach oligomers to an evanescent-wave biosensor array. Maskos and Southern (45) described the synthesis of oligonucleotides on derivatized glass bead supports. Glycol spacers of various dimensions can be added between the oligonucleotide and the glass support, which facilitates hybridization. Other methods include immobilization within carbon paste or polymers and stabilization within hydrogels or sol-gels (90). The streptavidin-biotin interaction is often used. Ultrathin applications of biological material are usually deposited on transducers by the Langmuir-Blodgett (8) or molecular self-assembly (51) techniques.

Fortunately, attachment protocols are usually uncomplicated and very reliable. Quantification of the material on a surface, or the density of the material, is another problem entirely. A number of techniques have been developed to detect mass on a surface, including total internal reflectance fluorescence, quartz crystal vibration analysis, and optical reflectometry. These techniques often take advantage of the fluorescent properties of proteins (enzymes or antibodies) for quantification. In some cases it is possible to use an intact microbial cell as the biomolecule (77), which avoids the problem of orientation but does necessitate the creation of a suitable microenvironment.

The target of the biosensor must be brought to the biomolecule by some method. Typically this is done by suspending the target in an aqueous solution and flooding the biomolecule area with it. This is easy enough to do if the volume required is in the microliter or greater range; micropipettors do an excellent job of delivering fluids. However, the trend in biosensors has been to miniaturize, both to save on space and weight and to examine very small volumes. The sample size can therefore be very small, and waste concerns are minimal. To produce a biosensor of this size, a means of transferring fluids is needed. This is the technology of microfluidics, which is very important to the design of biosensors because fluids at very low volumes are difficult to move in a uniform manner.

The computational component is the hardware and software that interprets and reports on the signal received from the biosensor; as such, it is beyond the scope of this chapter. With some systems there is merely a digital readout, while in other systems (such as DNA microarrays) the amount of information and the need to address the data points make computer control essential.

The final component is the transducer, which is also the component that is the most unfamiliar to microbiologists. Essentially, this is a source of energy which is directed at the biomolecule and which is changed by the biomolecule in some way. This change is detected electronically and reported. Since energy is supplied to the system, it is typical to consider a type of wave that is introduced (light, electromagnetic, or sound). Generally, the transducers fall into distinct categories: electrochemical, optical, piezoelectric, and calorimetric (74).

Electrochemical transducers report changes in voltage when the current is held constant (potentiometric) or report changes in current when the voltage is held constant (amperometric). These are by far the most common electrochemical transducers, although transducers based on conductance and capacitance have also been described (74). In each case, the interaction of the analyte with the biological component causes a change in potential that is detected by the sensor.

Optical biosensors use a fiber-optic probe to receive specific wavelengths of light. The versatility of fiber-optics is due to their capacity to transmit signals that report on changes in wavelength, wave propagation time, intensity, distribution of the spectrum, or polarity of the light. In general, the signal is acquired from these devices through flexible cables which can transmit light to the biological component (such as an excitation light for fluorescence) and receive light back from the sample (such as for light generation by the sample or light absorption or reflection). Light conductance can be accomplished with great efficiency (less than 1% loss over short distances), and loss of signal is usually not a problem. Acquisition of the light signal by the detector can be a considerable problem if the light source is very weak, and the difference in light signals may be small. Light propagation over longer distances (>1 m) is usually accompanied by a loss of conductance efficiency unless laser light is used.

One of the techniques which has been more fully developed, and which now has been commercialized, is the use of evanescent-wave excitation. This is dependent on an unusual principle of physics and requires some explanation to make it relevant to the biologist. When a beam of light strikes an interface between two transparent interfaces (e.g., glass and water) the light beam is reflected off the surface and refracted through the new medium. Light can be directed through a waveguide, such as a fiber-optic cable. If the angle of incidence of the light is at the critical angle, the light is refracted at 90° from the normal, essentially following the interface of the two media. If the light enters the waveguide at an angle greater than the critical angle, it undergoes total internal reflection. This creates an evanescent wave at the interface between the two media. An evanescent wave is essentially an electromagnetic wave that decays exponentially with distance from its source. Thus, they are very weak effects and are significant only within a small distance from the waveguide. This is actually a very handy attribute, since biosensor materials attached to the waveguide are affected by the evanescent wave while other materials only a short distance away are not. Typically, the affected distance is on the order of nanometers, which is optimal for most biomolecules. An excellent review is available (81).

A related technology is surface plasmon resonance (SPR). In this technique, the surface of the waveguide is coated with a thin layer of gold (43). When light hits the gold at a certain angle, there is a decrease in the reflected energy due to the creation of an evanescent wave and its interaction with surface plasmons. Plasmons are another difficult physical principle. They are quasiparticles resulting from quantization of plasma oscillations. They can be created by reflecting a photon off a thin metal film (Fig. 5). The reflected light shows an energy loss equal to integral multiples of the plasmon energy. The decrease in reflected light is measured with a CCD camera and is related to the quantity of matter that interacts on the biomolecular side of the instrument. A particularly lucid example is presented by Mullett et al. (52) and is recommended for those new to the field. Miura et al. (50) used SPR and a tethered antibody to measure benzo[*a*]pyrene. They were able to reuse the sensor multiple times. Later models incorporated a dual-chamber system to provide a reference electrode (54). This enables faster, more accurate detection of analytes. Nelson et al. (53) constructed a DNA hybridization SPR platform and demonstrated that it can be used to detect 16S rRNA specifically. The Biacore company has successfully commercialized SPR technology. A variety of biomolecules can be immobilized to the sensor surface. Introduction of target molecules leads to binding (e.g., antibody-antigen, chelator-metal, and DNA-DNA). Signal strength is determined by a CCD camera and is correlated with binding and concentration. The Biacore X is a manual model that should be suitable for many environmental experiments.

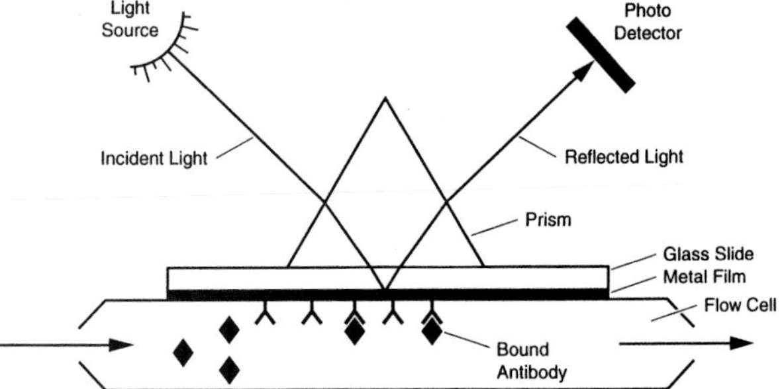

FIGURE 5 The principle of surface plasmon resonance. The sensing surface is on the opposite side of the metal film from the illuminated surface. Here an antibody-antigen-type biosensor is shown, with the sensing surface incorporated into a flow cell. The light source can be either polarized or laser light or an electron stream. The photodetector must be able to record subtle changes in light intensity.

A type of optical sensor that is related to SPR is the resonant-mirror biosensor. Instead of a metal layer as the sensing surface, a material with a high refractive index [such as titanium (IV) oxide] is used. This material is overlaid with a medium of low refractive index (such as silica), which is then connected to one side of a prism. When light enters the prism, it is totally reflected from the sensing surface. The low-index layer is typically so thin that the light may reach the high-index layer through an evanescent field, although this is dependent on the correct angle of incident light and phase matching of the resonant modes of the high-index material. When light enters the high-index medium, it propagates for a short distance before exiting through the prism. This makes the angle of the reflected light very susceptible to changes at the surface of the high-index (sensing) layer. Instead of changes in light intensity, as seen with SPR, modifications to the sensing surface are seen as phase changes and are recorded as changes in arc seconds of the reflected light. A thorough description of this technology is available (18). The resonant-mirror biosensor is quite sensitive (comparable to SPR) but is dependent on having the right materials for the sensing layer. Therefore, it is important to have a reliable attachment protocol for your molecule of interest. Just as for SPR, immunoassays have been performed with resonant-mirror biosensors (10). Complex materials can also be conjugated to the surface, and binding events can be studied in near real time (33, 34).

Piezoelectric biosensors measure changes in mass. A piezoelectric material, such as quartz crystal, oscillates at a certain frequency when a potential is applied across its surface. If the mass at the surface changes because, for example, a bound antibody complexes with a specific antigen, the frequency of oscillation will change, and this change is detectable. Lu et al. (44) used this method to detect an enzyme, glutathione S-transferase, at concentrations as low as 0.2 mg/ml. It is expected that sensitivity will continue to improve with other technological developments. The surface acoustic wave biosensor depends on the propagation of an acoustic wave through a medium. The medium is attached to a piezoelectric detector. If the energy of the wave is confined near the surface, the velocity of the wave and the effects detected by the piezoelectric detector are influenced by the mass the wave encounters. Therefore, a change of mass such as antibody-antigen binding is directly measurable.

Calorimetric transducers are comparable in function to optical transducers, except that heat generation is measured instead of light. Microcalorimeters can measure very small fluctuations in temperature and are sensitive enough to detect heats of molecular interaction (such as ligand binding), substrate use by microorganisms, and responses of immune cells to antigens. The microcantilever has great potential to sensitively and selectively detect compounds of interest. The technology is based on a mechanical stress principle. The reactive end of the device resembles a tiny diving board, a thin metal (e.g., gold) platform to which biomolecules can be attached. Typically the metal platform is less than 1 μm thick, and approximately 100 μm long. When a target molecule binds to the attached biomolecule, a mechanical stress occurs in the thin metal. A laser is focused on the free end of the metal platform, where the deformation is greatest. Deviation of the reflection is determined by reflection of the laser onto a sensitive photodetector. Alternatively, the reaction can be detected as a change in resonance frequency. Although the instruments were initially rather large devices, in recent years they have been reduced to a convenient size. The number of possible applications is very great, since anything that causes a change in the metal can be measured. In addition to measuring changes in mass, this technique is very accurate at temperature detection, and thus possibly can be used to indicate exothermic and endothermic reactions. The use of microcantilevers for temperature sensing has the potential to detect changes of as little as 10^{-6}°C. Fast response times are possible (e.g., within seconds), although care must be taken to avoid temperature fluctuations due to external conditions. Dual-material metal platforms (e.g., aluminum and silicon) have also been used. Analytes that interact differently with the two materials will deform the platform in a measurable way, allowing the interaction to be detected and quantified. A biosensor of heavy metals has been described that utilizes metal-binding proteins as the biomolecule (14). The latest models are very compact and will probably be commercialized in the near future.

Applications

The essence of the biosensor is in matching the appropriate biological and electronic components to produce a relevant signal during analysis. For example, antibodies can be attached to a piezoelectric transducer so that the binding of the antigen is recorded as a change in the attached mass. Alternatively, the antibodies can be attached to an optical fiber and the antigen binding can be recorded by evanescent-wave detection (8, 52). This technique is particularly suitable for immunoassays because the evanescent waveform is operational for only a short distance from the surface of the optical electrode, and this distance is approximately equal to the size of the immune complex (74). Immunosensors that use other detection systems have been described previously (84).

Biosensors need not be overly sophisticated: bacterial cells can be immobilized on the tip of an oxygen electrode (78) or within an online column (57). Occasionally, whole yeast cells (6) can be used. The electrode then senses activity on the basis of the change in oxygen concentration. With this type of biosensor, it is important to design a reference standard that distinguishes background activity from the activity of interest; otherwise, specificity for the analyte is lost. Both electrochemical and optical electrodes are very useful for the detection of signals from attached enzymes. The enzymatic reaction can cause a potential change that is detected by the electrochemical electrode or can cause a change in one of the components of the enzyme system that can be detected by the optical electrode. Scheper et al. (73) used the latter concept for their biosensor. They attached the glucose-fructose oxidase from *Zymomonas mobilis* to a fiber-optic cable attached to a fluorimeter. The enzyme complex contained bound $NADP^+$. When the enzyme oxidizes glucose, it reduces $NADP^+$ to NADPH, which is a fluorescent molecule. The change in fluorescence is therefore proportional to the concentration of glucose. There is also a critical weakness associated with this assay, in that it is dependent on the availability of $NADP^+$. When the supply is exhausted, the biosensor no longer functions. Resupplying the biosensor with an essential cofactor is often technically challenging but is required for long-term monitoring of the environment. A clever modification of the optical sensor was reported by Zhou et al. (91). They used conventional fluorescent antibodies to find either labeled microspheres or bacterial cells, but they then used a standing acoustic wave to concentrate the signal in one known area. This results in an increase in signal because very little is outside the range of the fluorescent sensor.

Nucleic acid biosensors depend on the ability of a single-stranded nucleic acid to hybridize with another fragment of DNA by complementary base pairing. The growing field of DNA microarray technology is testament to the strength of this type of biosensor. This same technology on a smaller scale can be used for detection of very specific nucleic acids. The biosensor described by Eggers et al. (25) integrates microelectronics, molecular biology, and computational science in an optical electrode format. Their device can detect hybridization and report on the spatial configuration of the hybridization signal on a glass surface (25) or a silicon wafer (41), to which the DNA probes are attached. Several different DNA oligomers can be attached to the optical electrode at different locations. The DNA on the biosensor is then hybridized to DNA that is free in solution. The free DNA must be labeled, usually with a fluorescent, luminescent, or radioisotope decay (^{32}P) signal. The signal is detected by a CCD camera, which is extremely sensitive. The computer identifies the location of the affected pixels and forms the signal into a recognizable array. Not only is this technology suitable for rapid DNA sequencing, but it is also applicable to the rapid detection of many different gene sequences from DNA extracted from a consortium. Because of the labeling requirement for standard microarrays, evanescent-wave technology may be more attractive. A nucleic acid biosensor that uses evanescent-wave technology has been described by Graham et al. (29). They used short fragments of nucleic acids that are small enough to reside within the field of the evanescent wave. They were able to detect fluorescein-labeled DNA hybridizing to their complementary immobilized probes in a flow cell. Fluorescence was monitored and reported as a change in the output voltage. As with many other technologies, biosensors are being miniaturized. This presents special challenges arising from the use of novel materials and assembly techniques which are beyond the scope of this chapter. However, a good review is available (37).

EMERGING TECHNOLOGIES

Other technologies are now available whose use in environmental microbiology is speculative, although the opportunities are clearly apparent. Xu et al. (88) used silver nanoparticles in a study of membrane transport. These particles, 10 to 80 nm in diameter, were used to measure the porosity of membranes under various environmental conditions, as a surrogate for antibiotic uptake. Silver appears very bright under dark-field microscopy, allowing tracking of the particles through the cells. Won et al. (87) modified the surface of a commercial magnetic nanoparticle with a small peptide and a fluorescein tagant and were able to introduce these nanoparticles into HeLa cells. Although this is a eukaryotic cell, the application is intriguing. The nanoparticles could be pulled through the cell by means of a magnet, with the progress visualized using the fluorescent tag and confocal microscopy. The application was created to determine which host molecules will interact with the attached peptide. While these applications are somewhat limited, they demonstrate that many experiments might be possible if more versatile tagants were available.

Quantum dots (also called Qdots) are nanometer-sized crystalline clusters. Typically they are made from semiconductor materials (e.g., CdTe, InP, and PbSe) at such small scales (1 to 10 nm in diameter) that they could possibly be used with bacterial cells. The construction of a Qdot is very complex, with a metal core surrounded by an inorganic "shell" material (e.g., zinc sulfide). Organic polymers can be attached to the shell, allowing complexing with various biomolecules such as antibodies or enzymes. To date the bulk of Qdot work in the biosciences has been with eukaryotic cells. Their use in environmental microbiology remains speculative, although they possess certain advantages that may make them valuable for specific tasks.

Qdots are used as fluorescent tagants, in much the same way that fluorescein is used. However, the spectral properties of Qdots are dependent on the actual size of the particle, as well as its shape and composition. With an increase in size, the color of the fluorescence shifts in a predictable manner. At first this was a disadvantage, but with refined manufacturing processes that create standard-sized particles, it becomes a great advantage. This unusual property means that Qdots might be available in an enormous range of fluorescent colors, giving workers multiple tags for environmental uses.

Qdots have several additional advantages. They have long fluorescent lifetimes (10 to 50 ns), allowing a fluorimeter to avoid the excitation wavelength and thus resulting in a more defined signal. Excitation is performed with a standard light source, eliminating the need for specific wavelengths of light. They emit exceptionally bright fluorescence, and, as already noted, they are available in a range of colors. Generally speaking, the smaller the Qdot, the more blue it is. Their emission spectra are typically very narrow and symmetric, avoiding the tailing of fluorescence seen in other tagants. They are also very stable molecules, and therefore long-term experiments can be carried out using the same tagant. In addition, they do not "photobleach" as other fluorescent compounds are prone to do.

However, Qdots also have certain disadvantages that must be acknowledged. They are nonbiological and cannot replace fluorescent bioreporter genes such as the GFP gene. Therefore they cannot be used for gene expression in vivo. They are also fabricated from heavy metals that may have toxic properties and are inherently hydrophobic. Both of these disadvantages are normally overcome by conjugating organic molecules to the surface as a means to attach reactive groups such as antibodies or biotin molecules. These modifications result in a water-soluble particle that can be more easily handled and is nontoxic. There are several protocols for attachment, many of which resemble the protocols for attachment to glass surfaces or silica particles (above). Even though they are in the nanometer range, they are still very large compared to biomolecules.

Most of the work with Qdots has been with eukaryotic cells, since they are larger and more complex internally. Using a biomolecule conjugated to a Qdot allows the tagant to either bind to a surface receptor and tag the cell for a long period or enter the cell and permit tracking of the biomolecule's progress through the cell. Excellent reviews of these applications are available (2, 35). The worker must have access to a suitable fluorescence imaging system, however, which requires skilled use (62).

Qdots are now available commercially. The first reported use of Qdots in microorganisms was by Kloepfer et al. (39). They used a CdSe Qdot that was conjugated to either a lectin or human transferrin, and they demonstrated the applicability of this technique with several different genera. As sizes and spectral properties become more standardized, the number of uses for Qdots will increase. It is suggested that they would be ideal for biofilm studies in which three-dimensional imaging of specific cells becomes important.

SUMMARY

As predicted in the previous edition of this chapter, the technologies described here have advanced and are becoming more commonplace in environmental microbiology. This is especially true for the biosensors, and the growth in commercially available products has certainly helped in this area. With the continued emphasis on multidisciplinary approaches to research topics, the use of these technologies will be invaluable. In the area of fluorescent proteins, a spectrum of excitation and emission wavelengths might soon be available that would allow the use of several bioreporter genes in one species or the use of bioreporters to distinguish individual species in a community. Multigene analysis will have a substantial impact on the understanding of genetic control. In the area of biosensors, the trend toward miniaturization and commercialization will continue. It is expected that fieldable biosensors will have a great impact on biowarfare monitoring and long-term ecological studies.

REFERENCES

1. **Afendra, A. S., C. Vargas, J. J. Nieto, and C. Drainas.** 2004. Gene transfer and expression of recombinant proteins in moderately halophilic bacteria. *Methods Mol. Biol.* **267:**209–223.
2. **Alivisatos, A. P., W. Gu, and C. Larabell.** 2005. Quantum dots as cellular probes. *Annu. Rev. Biomed. Eng.* **7:**55–76.
3. **Andersen, J. B., C. Sternberg, L. K. Poulsen, S. P. Bjorn, M. Givskov, and S. Molin.** 1998. New unstable variants of green fluorescent protein for studies of transient gene expression in bacteria. *Appl. Environ. Microbiol.* **64:** 2240–2246.

4. **Anderson, M. T., I. M. Tjioe, M. C. Lorincz, D. R. Parks, L. A. Herzenberg, G. P. Nolan, and A. Herzenberg.** 1996. Simultaneous fluorescence activated cell sorter analysis of two distinct transcriptional elements within a single cell using engineered green fluorescent protein. *Proc. Natl. Acad. Sci. USA* **93:**8508–8511.
5. **Arvanitis, N., C. Vargas, G. Tegos, A. Perysinakis, J. J. Nieto, A. Ventosa, and C. Drainas.** 1995. Development of a gene reporter system in moderately halophilic bacteria by employing the ice nucleation gene of *Pseudomonas syringae*. *Appl. Environ. Microbiol.* **61:**3821–3825.
6. **Baronian, K. H. R.** 2004. The use of yeast and moulds as sensing elements in biosensors. *Biosens. Bioelectron.* **19:** 953–962.
7. **Bevis, B. J., and B. S. Glick.** 2002. Rapidly maturing variants of the *Discosoma* red fluorescent protein (DsRed). *Nat. Biotechnol.* **20:**83–87.
8. **Blodgett, K. B., and I. Langmuir.** 1937. Build-up films of barium stearate and their optical properties. *Phys. Rev.* **51:** 964–982.
9. **Boivin, R., F. P. Chalifour, and P. Dion.** 1988. Construction of a Tn5 derivative encoding bioluminescence and its introduction in *Pseudomonas*, *Agrobacterium*, and *Rhizobium*. *Mol. Gen. Genet.* **213:**50–55.
10. **Buckle, P. E., R. J. Davies, T. Kinning, D. Yeung, P. R. Edwards, and D. Pollard-Knight.** 1993. The resonant mirror: a novel optical sensor for direct sensing of bimolecular interactions. II. Applications. *Biosens. Bioelectron.* **8:** 355–363.
11. **Burlage, R. S., and C. Kuo.** 1994. Living biosensors for the management and manipulation of microbial consortia. *Annu. Rev. Microbiol.* **48:**291–309.
12. **Burlage, R. S., Z. K. Yang, and T. Mehlhorn.** 1996. A transposon for green fluorescent protein transcriptional fusions: application for bacterial transport experiments. *Gene* **173:** 53–58.
13. **Chalfie, M., Y. Tu, G. Euskirchen, W. W. Ward, and D. C. Prasher.** 1994. Green fluorescent protein as a marker for gene expression. *Science* **263:**802–805.
14. **Cherian, S., R. K. Gupta, B. C. Mullin, and T. Thundat.** 2003. Detection of heavy metal ions using protein-functionalized microcantilever sensors. *Biosens. Bioelectron.* **19:**411–416.
15. **Christensen, P. D., L. P. Nielsen, N. P. Revsbech, and J. Sorensen.** 1989. Microzonation of denitrification activity in stream sediments as studied with a combined oxygen and nitrous oxide microsensor. *Appl. Environ. Microbiol.* **55:**1234–1241.
16. **Cody, C. W., D. C. Prasher, W. M. Westler, F. G. Prendergast, and W. W. Ward.** 1993. Chemical structure of the hexapeptide chromophore of the *Aequorea* green-fluorescent protein. *Biochemistry* **32:**1212–1218.
17. **Crameri, A., E. A. Whitehorn, E. Tate, and W. P. C. Stemmer.** 1996. Improved green fluorescent protein by molecular evolution using DNA shuffling. *Nat. Biotechnol.* **14:**315–319.
18. **Cush, R., J. M. Cronin, W. J. Stewart, C. H. Maule, J. Molloy, and N. J. Goddard.** 1993. The resonant mirror: a novel optical sensor for direct sensing of biomolecular interactions. I: Principle of operation and associated instrumentation. *Biosens. Bioelectron.* **8:**347–353.
19. **DeAngelis, K. M., P. Ji, M. K. Firestone, and S. E. Lindow.** 2005. Two novel bacterial biosensors for detection of nitrate availability in the rhizosphere. *Appl. Environ. Microbiol.* **71:**8537–8547.
20. **DeBeer, D., and J. C. Van den Heuvel.** 1988. Response of ammonium-selective microelectrodes based on the neutral carrier nonactin. *Talanta* **35:**728–730.
21. **Decher, G., B. Lehr, K. Lowack, Y. Lvov, and J. Schmitt.** 1994. New nanocomposite films for biosensors: layer-by-layer adsorbed films of polyelectrolytes, proteins or DNA. *Biosens. Bioelectron.* **9:**677–684.
22. **Delagrave, S., R. E. Hawtin, C. M. Silva, M. M. Yang, and D. C. Youvan.** 1995. Red-shifted excitation mutants of the green fluorescent protein. *Bio/Technology* **13:**151–154.
23. **DeLorenzo, V., M. Herrero, U. Jakubzik, and K. N. Timmis.** 1990. Mini-Tn5 transposon derivatives for insertion mutagenesis, promoter probing, and chromosomal insertion of cloned DNA in gram-negative bacteria. *J. Bacteriol.* **172:**6568–6572.
24. **Dickson, R. M., A. B. Cubitt, R. Y. Tsien, and W. E. Moerner.** 1997. On/off blinking and switching behavior of single molecules of green fluorescent protein. *Nature* **388:** 355–358.
25. **Eggers, M., M. Hogan, R. K. Reich, J. Lamture, D. Ehrlich, M. Hollis, B. Kosicki, T. Powdrill, K. Beattie, S. Smith, R. Varma, R. Gangadharan, A. Mallik, B. Burke, and D. Wallace.** 1994. A microchip for quantitative detection of molecules utilizing luminescent and radioisotope reporter groups. *BioTechniques* **17:**516–524.
26. **Escher, A., D. J. O'Kane, J. Lee, and A. A. Szalay.** 1989. Bacterial luciferase αβ fusion protein is fully active as a monomer and highly sensitive *in vivo* to elevated temperature. *Proc. Natl. Acad. Sci. USA* **86:**6528–6532.
27. **Gauglitz, G.** 2005. Direct optical sensors: principles and selected applications. *Anal. Bioanal. Chem.* **381:**141–155.
28. **Goldman, E. R., and D. C. Youvan.** 1992. An algorithmically optimized combinatorial library screened by digital imaging spectroscopy. *Bio/Technology* **10:**1557–1561.
29. **Graham, C. R., D. Leslie, and D. J. Squirrell.** 1992. Gene probe assays on a fibre-optic evanescent wave biosensor. *Biosens. Bioelectron.* **7:**487–493.
30. **Hansen, M. C., R. J. Palmer, C. Udsen, D. C. White, and S. Molin.** 2001. Assessment of GFP fluorescence in cells of *Streptococcus gordonii* under conditions of low pH and low oxygen concentration. *Microbiology* **147:**1383–1391.
31. **Hassler, C. S., and M. R. Twiss.** 2006. Bioavailability of iron sensed by a phytoplanktonic Fe-bioreporter. *Environ. Sci. Technol.* **40:**2544–2551.
32. **Heim, R., D. C. Prasher, and R. Y. Tsien.** 1994. Wavelength mutations and posttranslational autoxidation of green fluorescent protein. *Proc. Natl. Acad. Sci. USA* **91:**12501–12504.
33. **Hirmo, S., E. Artursson, G. Puu, T. Wadstrom, and B. Nilsson.** 1998. Characteristics of *Helicobacter pylori* interactions with sialylglycoconjugates using a resonant mirror biosensor. *Anal. Biochem.* **257:**63–66.
34. **Hirmo, S., E. Artursson, G. Puu, T. Wadstrom, and B. Nilsson.** 1999. *Helicobacter pylori* interactions with human gastric mucin studied with a resonant mirror biosensor. *J. Microbiol. Methods* **37:**177–182.
35. **Hotz, C. Z.** 2005. Applications of quantum dots in biology. *Methods Mol. Biol.* **303:**1–17.
36. **Jensen, K., N. P. Revsbech, and L. P. Nielsen.** 1993. Microscale distribution of nitrification activity in sediment determined with a shielded microsensor for nitrate. *Appl. Environ. Microbiol.* **59:**3287–3296.
37. **Jianrong, C., M. Yuqing, H. Nongyue, W. Xiaohua, and L. Sijiao.** 2004. Nanotechnology and biosensors. *Biotechnol. Adv.* **22:**505–518.
38. **Kalabat, D. Y., J. M. Froelich, T. K. Phuong, R. A. Forsyth, V. G. Newman, and J. W. Zyskind.** 1998. Chitobiase, a new reporter enzyme. *BioTechniques* **25:**1030–1035.
39. **Kloepfer, J. A., R. E. Mielke, M. S. Wong, K. H. Nealson, G. Stucky, and J. L. Nadeau.** 2003. Quantum dots as strain- and metabolism-specific microbiological labels. *Appl. Environ. Microbiol.* **69:**4205–4213.

40. **Koblizek, M., J. Masojidek, J. Komend, T. Kucera, R. Pilloton, A. K. Mattoo, and M. T. Giardi.** 1998. A sensitive photosystem II-based biosensor for detection of a class of herbicides. *Biotechnol. Bioeng.* **60:**664–669.
41. **Lamture, J. B., K. L. Beattie, B. E. Burke, M. D. Eggers, D. J. Ehrlich, R. Fowler, M. A. Hollis, B. B. Kosicki, R. K. Reich, S. R. Smith, R. S. Varma, and M. E. Hogan.** 1994. Direct detection of nucleic acid hybridization on the surface of a charge coupled device. *Nucleic Acids Res.* **22:**2121–2125.
42. **Langridge, W. H. R., K.. J. Fitzgerald, C. Koncz, J. Schell, and A. A. Szalay.** 1989. Dual promoter of *Agrobacterium tumefaciens* mannopine synthase genes is regulated by plant growth hormones. *Proc. Natl. Acad. Sci. USA* **86:**3219–3223.
43. **Liedberg, B., C. Nylander, and I. Lundstrom.** 1983. Surface plasmon resonance for gas detection and biosensing. *Sens. Activators* **4:**299–304.
44. **Lu, H., H. Chen, Y. Lin, and J. Lin.** 2000. A reusable and specific protein A-coated piezoelectric biosensor for flow injection immunoassay. *Biotechnol. Prog.* **16:**116–124.
45. **Maskos, U., and E. M. Southern.** 1992. Oligonucleotide hybridizations on glass supports: a novel linker for oligonucleotide synthesis and hybridization properties of oligonucleotides synthesized in situ. *Nucleic Acids Res.* **20:**1679–1684.
46. **Meighen, E. A.** 1991. Molecular biology of bacterial bioluminescence. *Microbiol. Rev.* **55:**123–142.
47. **Meighen, E. A.** 1994. Genetics of bacterial bioluminescence. *Annu. Rev. Genet.* **28:**117–139.
48. **Miller, W. G., M. T. Brandl, B. Quinones, and S. E. Lindow.** 2001. Biological sensor for sucrose availability: relative sensitivities of various reporter genes. *Appl. Environ. Microbiol.* **67:**1308–1317.
49. **Miller, W. G., J. H. Leveau, and S. E. Lindow.** 2000. Improved GFP and *inaZ* broad-host-range promoter-probe vectors. *Mol. Plant-Microbe Interact.* **13:**1243–1250.
50. **Miura, N., M. Sasaki, K. V. Gobi, C. Kataoka, and Y. Shoyama.** 2003. Highly sensitive and selective surface plasmon resonance sensor for detection of sub-ppb levels of benzo[a]pyrene by indirect competitive immunoreaction method. *Biosens. Bioelectron.* **18:**953–959.
51. **Moaz, R., L. Netzer, J. Gun, and J. Sagiv.** 1988. Self-assembling monolayers in the construction of planned supramolecular structures and as modifiers of surface properties. *J. Chim. Phys.* **85:**1059–1065.
52. **Mullett, W., E. P. C. Lai, and J. M. Leung.** 1998. Immunoassay of fumonisins by a surface plasmon resonance biosensor. *Anal. Biochem.* **258:**161–167.
53. **Nelson, B. P., M. R. Liles, K. B. Frederick, R. M. Corn, and R. M. Goodman.** 2002. Label-free detection of 16S ribosomal RNA hybridization on reusable DNA arrays using surface plasmon resonance imaging. *Environ. Microbiol.* **4:**735–743.
54. **Nenninger, G. G., J. B. Clendenning, C. E. Furlong, and S. S. Yee.** 1998. Reference-compensated biosensing using a dual channel surface plasmon resonance sensor system based on planar lightpipe configuration. *Sens. Actuators B* **51:**38–45.
55. **Nielsen, N., L. H. Larsen, M. S. M. Jetten, and N. P. Revsbech.** 2004. Bacterium-based NO_2^- biosensor for environmental applications. *Appl. Environ. Microbiol.* **70:**6551–6558.
56. **Nivens, D. E., T. E. McKnight, S. A. Moser, S. J. Osborn, M. L. Simpson, and G. S. Sayler.** 2004. Bioluminescent bioreporter integrated circuits: potentially small, rugged and inexpensive whole-cell biosensors for remote environmental monitoring. *J. Appl. Microbiol.* **96:**33–46.
57. **Nomura, Y., K. Ikebukaro, K. Yokoyama, T. Takeuchi, Y. Arikawa, S. Ohno, and I. Karube.** 1998. Application of a linear alkylbenzene sulfonate biosensor to river water monitoring. *Biosens. Bioelectron.* **13:**1047–1053.
58. **O'Kane, D. J., M. Ahmad, I. B. C. Matheson, and J. Lee.** 1986. Purification of bacterial luciferase by high-performance liquid chromatography. *Methods Enzymol.* **133:**109–128.
59. **Ow, D. W., K. V. Wood, M. DeLuca, J. R. deWet, D. R. Helinski, and S. H. Howell.** 1986. Transient and stable expression of the firefly luciferase gene in plant cells and transgenic plants. *Science* **234:**856–859.
60. **Palomares, A. J., M. A. DeLuca, and D. R. Helinski.** 1989. Firefly luciferase as a reporter enzyme for measuring gene expression in vegetative and symbiotic *Rhizobium meliloti* and other gram-negative bacteria. *Gene* **81:**55–64.
61. **Perozzo, M. A., K. B. Ward, R. B. Thompson, and W. W. Ward.** 1988. X-ray diffraction and time-resolved fluorescence analyses of *Aequorea* green fluorescent protein crystals. *J. Biol. Chem.* **263:**7713–7716.
62. **Pinaud, F., X. Michalet, L. A. Bentolila, J. M. Tsay, S. Doose, J. J. Li, G. Iyer, and S. Weiss.** 2006. Advances in fluorescence imaging with quantum dot bio-probes. *Biomaterials* **27:**1679–1687.
63. **Prasher, D. C., V. K. Eckenrode, W. W. Ward, F. G. Prendergast, and M. J. Cormier.** 1992. Primary structure of the *Aequorea victoria* green-fluorescent protein. *Gene* **111:**229–233.
64. **Ramsing, N. B., M. Kuhl, and B. B. Jorgensen.** 1993. Distribution of sulfate-reducing bacteria, O_2, and H_2S in photosynthetic biofilms determined by oligonucleotide probes and microelectrodes. *Appl. Environ. Microbiol.* **59:**3840–3849.
65. **Revsbech, N. P.** 1989. An oxygen microsensor with a guard cathode. *Limnol. Oceanogr.* **34:**474–478.
66. **Revsbech, N. P., and B. B. Jorgensen.** 1986. Microelectrodes: their use in microbial ecology. *Adv. Microb. Ecol.* **9:**293–352.
67. **Ripp, S., P. Jegier, M. Birmele, C. M. Johnson, K. A. Daumer, J. L. Garland, and G. S. Sayler.** 2006. Linking bacteriophage infection to quorum sensing signaling and bioluminescent bioreporter monitoring for direct detection of bacterial agents. *J. Appl. Microbiol.* **100:**488–499.
68. **Rodriguez, J. F., D. Rodriguez, J. Rodriguez, E. B. McGowan, and M. Esteban.** 1988. Expression of the firefly luciferase gene in vaccinia virus: a highly sensitive gene marker to follow virus dissemination in tissues of infected animals. *Proc. Natl. Acad. Sci. USA* **85:**1667–1671.
69. **Rogers, K. R.** 1995. Biosensors for environmental applications. *Biosens. Bioelectron.* **10:**533–541.
70. **Rogowsky, P. M., T. J. Close, J. A. Chimera, J. J. Shaw, and C. I. Kado.** 1987. Regulation of the *vir* genes of *Agrobacterium tumefaciens* plasmid pTiC58. *J. Bacteriol.* **169:**5101–5112.
71. **Sanseverino, J., R. K. Gupta, A. C. Layton, S. S. Patterson, S. A. Ripp, L. Saidak, M. L. Simpson, T. W. Schultz, and G. S. Sayler.** 2005. Use of *Saccharomyces cerevisiae* BLYES expressing bacterial bioluminescence for rapid, sensitive detection of estrogenic compounds. *Appl. Environ. Microbiol.* **71:**4455–4460.
72. **Santegoeds, C. M., A. Schramm, and D. deBeer.** 1998. Microsensors as a tool to determine chemical microgradients and bacterial activity in wastewater biofilms and flocs. *Biodegradation* **9:**159–167.
73. **Scheper, T., C. Muller, K. D. Anders, F. Eberhardt, F. Plotz, C. Schelp, O. Thordsen, and K. Schugerl.** 1994. Optical sensors for biotechnological applications. *Biosens. Bioelectron.* **9:**73–83.

74. **Sethi, R. S.** 1994. Transducer aspects of biosensors. *Biosens. Bioelectron.* **9:**243–264.
75. **Shaw, J. J., and C. I. Kado.** 1987. Direct analysis of the invasiveness of *Xanthomonas campestris* mutants generated by Tn4431, a transposon containing a promoterless luciferase cassette for monitoring gene expression, p. 57–60. *In* D. P. S. Verma and N. Brisson (ed.), *Molecular Genetics of Plant-Microbe Interactions.* Martinus Nijhoff, Dordrecht, The Netherlands.
76. **Silcock, D. J., R. N. Waterhouse, L. A. Glover, J. I. Prosser, and K. Killham.** 1992. Detection of a single genetically modified bacterial cell in soil by using charge coupled device-enhanced microbiology. *Appl. Environ. Microbiol.* **58:**2444–2448.
77. **Skladal, P., N. O. Morozova, and A. N. Reshetilov.** 2002. Amperometric biosensors for detection of phenol using chemically modified electrodes containing immobilized bacteria. *Biosens. Bioelectron.* **17:**867–873.
78. **Svitel, J., O. Curilla, and J. Tkac.** 1998. Microbial cell-based biosensor for sensing glucose, sucrose or lactose. *Biotechnol. Appl. Biochem.* **27:**153–158.
79. **Sweerts, J. R. and D. DeBeer.** 1989. Microelectrode measurements of nitrate gradients in the littoral and profundal sediments of a meso-eutrophic lake (Lake Vechten, The Netherlands). *Appl. Environ. Microbiol.* **55:**754–757.
80. **Taguchi, K., Y. Tanaka, T. Imaeda, M. Hirai, S. Mohri, M. Yamada, and Y. Inoue.** 2004. Development of a genotoxicity detection system using a biosensor. *Environ. Sci.* **11:**293–302.
81. **Taitt, C. R., G. P. Anderson, and F. S. Ligler.** 2005. Evanescent wave fluorescence biosensors. *Biosens. Bioelectron.* **20:**2470–2487.
82. **Tsien, R.** 1998. The green fluorescent protein. *Annu. Rev. Biochem.* **67:**509–544.
83. **Van der Meer, J. R., D. Tropel, and M. Jaspers.** 2004. Illuminating the detection chain of bacterial bioreporters. *Environ. Microbiol.* **6:**1005–1020.
84. **Vo-Dinh, T., M. J. Sepaniak, G. D. Griffin, and J. P. Alarie.** 1993. Immunosensors: principles and applications. *Immunomethods* **3:**85–92.
85. **Well, M., M. Gosch, R. Rigler, H. Harms, T. Lasser, and J. R. van der Meer.** 2005. Ultrasensitive reporter protein detection in genetically engineered bacteria. *Anal. Chem.* **77:**2683–2689.
86. **Wise, D. L., and L. B. Wingard (ed.).** 1991. *Biosensors with Fiberoptics.* Humana Press, Clifton, N.J.
87. **Won, J., M. Kim, Y. Yi, Y. H. Kim, N. Jung, and T. K. Kim.** 2005. A magnetic nanoprobe technology for detecting molecular interactions in live cells. *Science* **309:**121–125.
88. **Xu, X., W. J. Brownlow, S. V. Kyriacou, Q. Wan, and J. J. Viola.** 2004. Real-time probing of membrane transport in living microbial cells using single nanoparticle optics and living cell imaging. *Biochemistry* **43:**10400–10413.
89. **Youvan, D.C.** 1994. Imaging sequence space. *Nature* **369:**79–80.
90. **Yu, J., S. Liu, and H. Ju.** 2003. Mediator-free phenol sensor based on titania sol-gel encapsulation matrix for immobilization of tyrosinase by a vapor deposition method. *Biosens. Bioelectron.* **19:**509–514.
91. **Zhou, C., P. Pivarnik, A. G. Rand, and S. V. Letcher.** 1998. Acoustic standing wave enhancement of a fiber-optic *Salmonella* biosensor. *Biosens. Bioelectron.* **13:**495–500.

Ecology at Long-Term Research Sites: Integrating Microbes and Ecosystems

JOHN E. HOBBIE, MICHELE BAHR, AND ANNA-LOUISE REYSENBACH

15

Long-term study sites for ecological research exist in a variety of habitats, including forests, deserts, streams, and oceans, providing the observational framework for studying microbial ecological processes over time and space. Each site has built up a record of observational data on environmental factors, such as temperature, soil and water chemistry, and moisture, and on the biota, including species, biomass, and productivity. Microbial processes studied include decomposition, nitrification, sulfate reduction, and respiration. In addition, many sites carry out decade-long experimental manipulations such as temperature change, added nutrients, soil moisture changes, and changes in the biota. Although these experiments allow microbiologists to, for example, link changes in an environmental factor to changes in the microbial biota, the vast opportunities for microbial ecological experimentation at these long-term study sites remain relatively untapped. In this chapter, we describe:

1. The advantages for microbial ecologists of studies at long-term ecological research sites.
2. The opportunities of various types of research sites and locations available for microbial research.
3. The integration of microbial and ecosystem ecology.

LONG-TERM STUDIES IN ECOLOGY

Long-Term Sites and Ecological Questions

Many fundamental ecological issues can be understood only by a long-term approach. Examples are slow processes (soil development, forest succession, life histories of long-lived organisms, effects of acid rain, drought), rare events (rare infrequent disturbances such as fire, recruitment of organisms that reproduce episodically), disturbances that result in biotic change, and subtle changes that can be discerned only over many decades (5). Furthermore, long-term study sites allow microbial ecologists to make use of existing environmental data to study the distribution of microbes across gradients of temperature, moisture, or type of vegetation. At some long-term sites, ecological experiments with changed nutrients, water, grazing pressure, and soil temperature have been run for 30 years or more, producing a wonderful opportunity to investigate factors that influence microbial dynamics and communities.

To explore these and other ecological issues, long-term study sites were set up by various government agencies, including the U.S. Forest Service (Forest Experiment Stations), the National Park Service, the Department of Energy, the National Oceanic and Atmospheric Agency (National Estuarine Research Reserve System), and the National Science Foundation (NSF). Of these, the NSF's Long Term Ecological Research sites (LTER sites) (see http://www.LTERnet.edu for links) are the most extensive. For a description, see overview articles in a special section on the U.S. Long Term Ecological Research Network in *BioScience* (4a, 5).

NSF LTER Program

From its start in 1980 with six sites, the LTER program has grown to include 26 LTER sites in 2005. Habitats include forests, grasslands, deserts, alpine and arctic tundra, mangrove wetlands, salt marshes, estuaries, lakes, streams, a coral reef, a kelp bed, the California Coastal Current, urban areas, and an agricultural site. It is expected that research at the sites will continue for 30 to 50 years.

Each site addresses its own set of research themes or ecological questions, including climate forcing and climate change, disturbance, human-ecosystem interactions, legacies of human activity, atmosphere-land-water interactions, biodiversity, controls of food web structure, invasive species effects, succession, physical forcing factors, and human infrastructure (built environments) effects on ecosystems. Any given site may address a number of these themes. In addition, to encourage comparative cross-site research, the sites are strongly encouraged to collect so-called core data on six topics (i.e., climate, primary productivity, nutrient cycling, higher trophic levels, decomposition, and effects of disturbance).

A typical LTER or other long-term research site collects some types of data continuously (e.g., climate, stream flow, soil moisture), some types every week (e.g., stream chemistry, plant phenology, greenhouse gas flux), and some types once a year or at longer intervals (e.g., annual plant production by biomass, plant community changes, movement of invasive species). All sites make their data publicly available on the Internet. A summary report (7) gives examples from LTER sites of how observations and analysis of temporal and spatial variability of ecological parameters and processes are the only way to answer certain questions and of how these types of studies have increased the understanding of ecological phenomena.

Long-Term Experiments

The long-term data collections described above allow the results of natural experiments, such as the annual variability in the Alaskan stream, to be analyzed. Another research tool at LTER sites is a manipulation experiment in which the effect of a single factor or of a disturbance can be monitored for decades. Examples include fire, bison grazing, continuous addition of phosphate to a stream, fertilization of lakes, selective removal of plant species or plant functional groups, soil heating, soil cooling by snow removal during the winter, and the deliberate creation of hurricane-like damage to a forest.

The details of LTER experiments may be found through links on the LTERnet website (www.LTERnet.edu). Experiments give valuable information on changes in rates of ecological processes, on changes in plant growth, and on changes in microbial communities. Predictive simulation models can be developed, such as one calibrated for vegetation control plots as well as experimental plots with soil heating, shading, and nutrient addition (11). In addition, simulation models can be tested in the experimental plots for how well they work for predictions under changed conditions, as long as the experimental data were not used for developing the model.

Terrestrial Sites and Experiments

Within the LTER program, there are three temperate forest sites (Hubbard Brook, N.H.; Harvard Forest, Mass.; Coweeta, N.C.), a tropical forest (Luquilla, P.R.), an old-growth forest in the Pacific Northwest (H. J. Andrews, Oreg.), a boreal forest (Bonanza Creek, Alaska), two tundra sites (Arctic Alaska; Niwot Ridge, Colo.), three grassland sites (Cedar Creek, Minn.; Konza Prairie, Kans.; Shortgrass Steppe in Colorado), two desert sites (Sevilleta, N.Mex.; Jornada Basin, N.Mex.), an agricultural site (Kellogg Biological Station, Mich.), and two urban sites (Baltimore Ecosystem Study, Md.; Central Arizona-Phoenix, Ariz.). There are also long-term research locations associated with the Department of Energy's Free-Air Carbon Dioxide Enrichment (FACE) program, with experimental sites in an aspen forest (Wisconsin), pine forest (North Carolina), agronomic crops (Arizona), grassland (Minnesota), desert (Nevada), soybeans (Illinois), and a sweet gum forest (Tennessee). For an entrée into the home pages of the FACE sites, see http://aspenface.mtu.edu/.

Long-term experiments in these terrestrial systems include clear-cutting of an entire forested watershed, soil heating, nitrogen additions to simulate acid deposition, the simulation of hurricane-like tree blowdowns, reduction of plant diversity by selective removal of species, till versus no-till agriculture, and the effects of herbicides, pesticides, and fungicides.

Aquatic Sites and Experiments

Within the LTER program, there are two salt marsh and estuary sites (Georgia Coastal Ecosystem, Ga.; Plum Island Ecosystem, Mass.), an Everglades mangrove site (Coastal Everglades, Fla.), streams and lakes in the Arctic (Arctic Alaska), a temperate-lake site (North Temperate Lakes, Wis.), a marine barrier island site (Virginia Coast Reserve, Va.), a kelp bed (Santa Barbara Coastal Calif.), a coastal current site (California Current Ecosystem), a coral reef in French Polynesia (Moorea Coral Reef), and two Antarctic sites, one on an inland lake with a permanent surface ice cover and one in the coastal waters along the Antarctic Peninsula (McMurdo Dry Valleys, Palmer Station). There are also 29 sites in the National Oceanic and Atmospheric Agency's National Estuarine Research Reserve System (see http://nerrs.noaa.gov/) that encircle the United States and include sites in Alaska, Hawaii, the Great Lakes, and the East, West, and Gulf Coasts. Finally, there are the NSF-funded BATS (Bermuda Atlantic Time Series Studies, http://www.bbsr.edu/cintoo/bats/bats.html) and HOTS (Hawaii Ocean Time Series; biological component at http://hahana.soest.hawaii.edu/hot/hot_jgofs.html) projects in the Pacific and Atlantic Oceans, respectively. Both BATS and HOTS are part of the very extensive Ocean Research Interactive Observatory Networks (http://www.orionprogram.org/), which will continue to provide real-time and long-term data for oceanographic research. The BATS project began in 1991 and makes monthly measurements at a deepwater site of ocean physics, chemistry, and biology, with special emphasis on biotic processes controlling the cycling of inorganic and organic dissolved carbon. The HOTS project makes similar monthly measurements at a station in 4,700 m of water, with special emphasis on microbial biomass and algal pigments.

Long-term experiments in aquatic systems include the effects of clear-cuts on stream chemistry (Hubbard Brook LTER site, N.H.), of fertilization of whole lakes (Arctic LTER site, Alaska), of large woody debris in streams (H. J. Andrews LTER site, Oreg.), of changes in the predator controls of food webs (e.g., changes in fish in Arctic Alaska), of stream fertilization and recovery from fertilization (Arctic Alaska), and of the fertilization of salt marshes both by direct addition to the surface of the marsh and by enrichment of the incoming tides (Plum Island Ecosystem LTER site, Mass.). At the University of Notre Dame Environmental Research Center (northern Wisconsin), a number of whole-lake treatment studies are under way, including ecological effects of changing pH, dissolved organic carbon (DOC), fish, and nutrient concentrations (http://underc.nd.edu/).

Integrating Microbial and Ecosystem Studies

The Ecosystem Viewpoint

This chapter is written from the point of view that studies of the ecology of microbes are a necessary part of gaining a predictive knowledge of ecosystems. This is a different viewpoint from that of investigating the ecology of species or communities of microbes because it interjects the overarching question: what do ecosystem processes and ecosystem energy and nutrient budgets tell us about which microbial processes are important? Of course, in this case the importance is determined by whether or not it is a large-scale ecosystem process or control.

Ecosystem scientists have, in fact, long been able to measure microbial processes, including decomposition, the oxidation of ammonia, and bacterial growth by measuring the changes in amount of plant litter or in the rate of production of nitrate. However, linking the processes with the microbial community composition was not possible; it was as if scientists were dealing with a black box that could not be opened. In the case of a few specialized microbes, such as those that reduce sulfate to hydrogen sulfide in marine sediments, bacteria could be grown in culture and identified.

However, even here, there was no way to know if the cultured species represented 1% or 99% of the species actually carrying out the process in nature. Despite the black-box approach to microbial processes, ecosystem scientists have built up a large number of measurements of microbial rates in all ecosystems and have carried out many experiments testing the environmental controls of the bulk rates. This empirical knowledge about the controls of microbial processes includes information for many habitats on microbial response to temperature, to various organic substrates, to changes in vegetation, and to changes in soil moisture. Yet the connection between the species and ecosystem functioning was lacking.

Opening the Microbial Black Box: the Crucial Questions

The key to the black box is the use of molecular phylogenetic approaches to access microbial diversity. These approaches, both those currently in hand and those on the horizon, should finally allow the merging of microbial ecology and ecosystem ecology.

The merger must include answers to the following questions.

1. What microbes are present and in what numbers?
2. What microbes or communities of microbes are carrying out which processes?
3. What are the environmental controls on the presence and activity of the important microbes? These controls should be described with a response surface or other mathematical representation.
4. How will the microbial communities and activity respond to changed conditions?

Measurements Necessary To Integrate Microbial and Ecosystem Ecology

One of the greatest assets that the long-term study sites provide microbial ecologists is access to this vast database of process and environmental data. However, these data seldom include any measurements of the microbes themselves. The necessary measurements of the microbes range from tried and true to perhaps possible in the future. Many of the well-developed methods are described in other chapters of this manual, including microbial biomass by direct counts or by extraction of phospholipids as well as community composition studies by extraction of DNA, amplification by PCR, and denaturing gradient gel electrophoresis or terminal restriction fragment length polymorphism (TRFLP). These measures can provide incidence matrices (presence and absence of operational taxonomic units) and relative abundances and therefore can be integrated into more traditional ecological analyses (such as principal component analysis) that provide insights into factors affecting microbial community dynamics.

Examples of the Use of Long-Term Research Sites

Below, we provide some examples of where microbial ecologists have started to use the extensive long-term research framework to address the fundamental roles that microbes play in ecosystem function. There are three types of studies illustrated: correlation between microbial ecology and environmental factors, correlation studies with added data on measurements of a microbial process, and correlation studies that make use of a large-scale and long-term experimental manipulation. Each type of study has an aquatic and a terrestrial example.

EXAMPLES

Correlation Studies

Introduction

Microbial ecologists have made use of the extensive data from long-term ecological research sites to investigate the possible effects of the environment on microbial community dynamics. The results illustrate large changes in the abundance and community composition over different time scales, from weeks to years. Presumably, the species changes reflect changes in function, but correlation studies can only indicate possibilities.

Below, we provide an aquatic example from an LTER lake site and a terrestrial example from an agricultural LTER site.

Aquatic

One goal of the North Temperate Lakes LTER (http://lter.limnology.wisc.edu) is to examine the composition and dynamics of lake bacterial communities over annual time scales. One hypothesis being tested is that the abundance of nutrient N, P, or DOC controls the productivity of bacteria; this is bottom-up control. Several types of lakes lie within the site, and among them three are characterized as oligotrophic, eutrophic, or humic. Detailed chemical and physical limnological data is available for the lakes, including measurements of total nitrogen, nitrate, nitrite, ammonia, total phosphorous, and chlorophyll. Microbiological samples were taken biweekly during the ice-free season for 2 years (6, 20). The humic lake was sampled intensively during the third year, with samples taken every other day from mid-May to mid-August and daily in early June. Bacterial, nanoflagellate, phytoplankton, and zooplankton abundances were determined for the samples. Automated ribosomal intergenic spacer analysis was used as a molecular "fingerprint" of the bacterial community.

Repeated annual patterns in seasonal dynamics were present, despite interannual differences. The bacterial community was relatively stable in early spring and fall and underwent a rapid change in both species richness and abundance in the summer. Two diversity minima were observed, one in early June and the second in late summer. When the data were examined for indications of bottom-up or top-down control, no correlation between bacterial community changes and nutrient concentrations (N, P, or DOC) was found. However, the abundance of taxa of different trophic levels was correlated positively or negatively. Bacterial numbers decreased, and the percent of filamentous bacteria increased, as heterotrophic and mixotrophic flagellates increased early in the summer. Presumably, the flagellates grazed down the bacteria, and the bacteria responded to the grazing with morphological changes. Later, as *Daphnia* organisms increased in abundance, the flagellates decreased, and bacterial numbers went up again. Finally, a late-summer bloom of the dinoflagellate *Peridinium* correlated with a decrease in bacterial species richness, although the bacterial community composition varied from year to year. It was hypothesized that *Peridinium* releases a substrate that affects different bacterial species in various degrees. The authors concluded that drivers external to the bacterial community structure that community.

Terrestrial

In a study of a terrestrial habitat (2, 3), the bacterial groups most commonly found in soil microbial communities at the LTER site at Michigan State University's Kellogg Biological Station (http://lter.kbs.msu.edu) were determined. Replicate plots included cultivated fields, fields abandoned from cultivation, and fields with no history of cultivation; different plant communities were studied within each management history type. Analyses of 16S rRNA genes demonstrated a richness of ribotypes, and that sequences related to eight groups were present in a majority of sites (the alpha, beta, and gamma groups of the *Proteobacteria* and the *Actinobacteria*, *Cytophaga*, *Acidobacteria*, *Planctomycea*, and *Verrumicrobiae*).

The second study sought to determine patterns of microbial distribution by examining the relative abundance of rRNA molecules compared to total RNA. Radiolabeled oligonucleotide probes were used to track seven bacterial groups (gammaproteobacteria excluded) plus *Eukarya* in 85 samples over a 2-year period. The groups encompassed a high degree of functional diversity, but their simultaneous measurement provided a way of looking at changes in the community over time. The eight groups accounted for 59% ± 23% of the total rRNA present in the soil microbial communities. The dominant group observed was the alphaproteobacteria (25%), followed in abundance by the *Actinobacteria* (11%) and *Eukarya* (10%). Microbial community composition did not differ significantly between agricultural fields and those that had not been cultivated for 9 years. It was only in fields that had not been cultivated for >45 years that microbial communities were similar in structure to those in fields that had never been cultivated. These data point to the long-term influence of land management practices. Temporal variability was observed at all sites, with relative rRNA abundance of *Eukarya*, *Verrucomicrobiae*, and *Actinobacteria* highest in May, while all proteobacteria and planctomycetes have peaks of abundance in June and October. Specific environmental parameters that influence the abundance of microbial groups were not identified. Finally, significant differences were observed with soil depth, even in tilled soils. Verrucomicrobia relative RNA was more abundant in 0- to 5-cm surface soil, while alpha- and betaproteobacteria rRNA was more abundant in deeper (0- to 10-cm) cores.

Correlation Study with Process Rates

Introduction

In the following examples, the approach of correlation of changes in the microbial community is extended to situations where the environmental data include bulk measurements of a microbial process. Both of the examples make use of the well-studied genes of the nitrogen cycle to investigate community changes of the microbes. The processes measured include potential nitrification and net nitrification.

The aquatic examples are from the intensively studied Monterey Bay of California, while the terrestrial example is from the LTER site located at the Forest Service Experimental Station in old-growth forest of Oregon.

Aquatic

Transformation of nitrogen in the ocean is a key influence on primary production. This study (13, 19) focused on the nitrogen cycle in a California coastal environment, investigating potential links between environmental variation, biogeochemical function, and the diversity of ammonia-oxidizing bacteria (AOB). Physical variables (temperature, salinity, and nutrient concentrations) and biological parameters (ammonium assimilation rates, chlorophyll *a* concentrations, bacterial abundance and productivity, nitrification and potential nitrification) were measured at a saltwater station and several estuarine sites over a 2-year period. DNA was also extracted from the samples and amplified with primers specific for 16S rRNA and ammonia monooxygenase (*amoA*) genes and then subjected to cloning and phylogenetic analysis of AOB genes.

In Monterey Bay, bimonthly samples were taken throughout the euphotic zone (0.1% light level) at a 900-m-deep mid-bay station. Hydrographic variation in temperature and salinity was small and followed seasonal upwelling patterns. Biological variability as indicated by ammonium assimilation rates, chlorophyll *a* concentrations, and bacterial abundance was large, ranging 10- to 100-fold over the 2-year study. One biological exception was nitrification rates, which varied only 4.5-fold (18 to 79 nM day^{-1}). Rate maxima were often found near the bottom of the euphotic zone. Multiple regression analysis showed nitrification to be significantly correlated with only percent surface light intensity and depth and suggested that bacterial community composition might be a significant factor involved in nitrification.

Almost 90% of the 16S rRNA sequences were similar to gene sequences of *Nitrospira*-like bacteria and clustered with sequences of uncultured marine ribotypes collected from the Arctic Ocean and the Mediterranean Sea. A second sequence cluster found in only one sample (October 1999) was identified as similar to *Nitrosomonas europa* 16S rRNA. Phylogenetic analysis of the functional gene *amoA* displayed more complexity and fell into four major groups. The majority of sequences (58%) clustered with phylotypes from Chesapeake Bay and Kysing Fjord sediments. Three of the four groups are again most closely related to *Nitrospira*-like *amoA* sequences. The fourth group of two sequences came from the October 1999 sample and clustered with *Nitrosomonas europa* gene sequences. Overall, the Monterey Bay AOB population was dominated by *Nitrospira*-like sequences that are distinct from cultured representatives, with the October 1999 sample showing the greatest variation. However, no correlation was found between changes in AOB community composition and changes in nitrification rates.

In the Elkhorn Slough, a nearby estuary, samples were taken monthly at five sites, two of which received high nutrient runoff from agricultural fields while three received runoff from grasslands and woodlands (4). Both intact core nitrification rates and potential nitrification rates were highest at the head of the estuary in late winter and fall, but seasonal and site variations could not be correlated with any single factor. The authors suggest that salinity and oxygen and ammonium concentrations influenced by tidal flushing may interact in controlling nitrification rates. 16S rRNA and *amoA* gene sequences from the sediments displayed consistent phylogenetic relationships. The majority of sequences were related to the gene sequences of *Nitrosomonas* and were most closely related to *Nitrosomonas marina*. Sequence similarities to *Nitrosolobus*, *Nitrospina*, and *Nitrospira* were also identified. Site-specific patterns were not observed in the analysis of either gene.

Although Monterey Bay has an extensive history of research allowing investigation of the link between environmental and biological factors, the authors concluded that the dynamic nature of biological processes requires an even more intense sampling regimen, coupled with activity-based assessments such as mRNA microarrays, in order to resolve the forces driving microbial functions.

Terrestrial

The Microbial Observatory at the H. J. Andrews LTER site (http://www.fsl.orst.edu/lter/) is dedicated to the study of bacteria and fungi central to biogeochemical processes in coniferous forest ecosystems in the Central Cascade Mountains of Oregon. Because nitrogen (N) is the most nutrient limiting to tree growth in this ecosystem, researchers (12, 15) have focused on the functional diversity of microorganisms that perform N cycling processes. The influence of grassland meadow and coniferous vegetation on the functional diversity of key N cycling microorganisms and processes is being determined. A transect study was established to investigate nitrogen cycling microorganisms along natural transitions in vegetation. The objectives were to (i) determine links between vegetation type and microbial communities and processes, (ii) examine the spatial variability along meadow-to-forest transects, (iii) correlate microbial community structure and ecosystem functioning, and (iv) identify key and potentially novel nitrifying and denitrifying bacteria. The study was carried out at two high-elevation sites in the H. J. Andrews LTER site with adjacent grassland and coniferous forest vegetation (Carpenter site and Lookout site; elevation, 1,500 m). Three transects were established perpendicular to and spanning the boundary between meadow and forest at each site.

This pair of studies examined bacterial community composition and rates of nitrification and denitrification, two processes key to nitrogen cycling. The authors examined spatial gradients along the transects and the relationship of community composition to ecosystem properties such as vegetation types, soil characteristics, and process rates. Community composition was measured with TRFLP profiles of *nosZ*, a key gene in the denitrification pathway for nitrous oxide reductase, and *amoA*, a key gene in the nitrification pathway for ammonia monooxygenase subunit A. Gene fragments were cloned and sequenced for phylogenetic analysis.

Multivariate analysis (Fig. 1) indicated strong functional differences between meadow and forest soils. Levels of both potential denitrifying enzyme activity (DEA) and potential nitrification were substantially higher in the meadow soils. DEA and net nitrification values were similar in magnitude. Soil C/N ratios were highest in forest sites.

Denitrifier communities formed distinct groups according to vegetation type and site as evidenced by TRFLP data. Sequencing revealed that the majority of *nosZ* gene fragments were related to genes of the metabolically diverse alpha-proteobacterial family *Rhizobiaceae*. *amoA* genes cloned from the nitrifying bacteria at 47 of 48 sites were most similar to genes from a single cluster of *Nitrospira*.

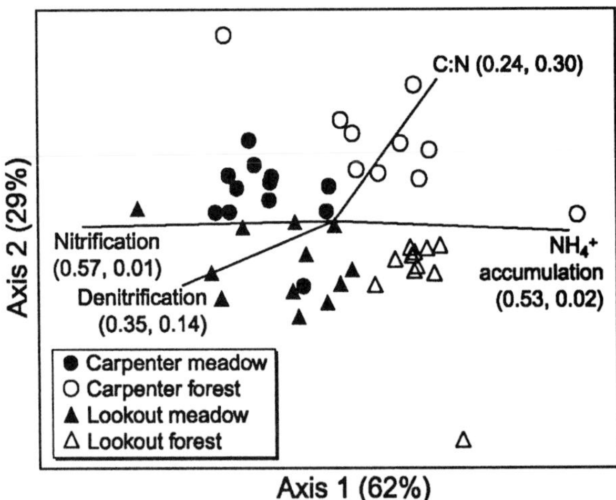

FIGURE 1 Ordination (NMS) of denitrifying community composition along a meadow-to-forest transect at the H. J. Andrews LTER site in Oregon (15). Community composition was measured with TRFLP profiles of *nosZ*, a key gene in the denitrification pathway. The percentage of variation represented by each axis is indicated (in parentheses). Vectors show the directions and relative magnitudes of correlation coefficients (r^2) between NMS axes and functional variables; r^2 values for the correlation between the functional variable and axis 1 or 2 (in that order) are shown in parentheses. Nitrification is the $\log_{10}$ of nitrification potential, denitrification is the $\log_{10}$ of denitrifying enzyme activity, C:N is of the soil, and NH_4^+ is ammonification. (Reprinted from reference 15 with permission.)

Correlation Studies in Whole-System Experiments

Introduction

In the final two examples, microbial ecologists made use of large-scale experimental manipulations to investigate the effects of environmental change on microbial communities and processes. Long-term and large-scale manipulations are a part of many LTER and similar projects. Some manipulations, such as the soil fertilization at the Arctic LTER site and the soil heating at the Harvard Forest LTER site, are now decades long.

The aquatic example is from a whole-lake fertilization experiment at the Notre Dame field station. The terrestrial example is from scientists working at the Arctic LTER site who increased the insulating snow cover with a snow fence and thus increased winter soil temperatures.

Aquatic

Bacteria in freshwaters utilize organic matter from autochthonous (within the water body) and allochthonous (outside the water body) sources. It is likely that the terrestrial humic compounds that make up much of the allochthonous organic matter are much less available to the bacteria than the alga-derived autochthonous organic matter. All lakes and streams contain organic matter from both sources, but it is difficult to characterize organic matter and to examine use of compounds within a complex mixture.

This study was based at the University of Notre Dame Environmental Research Center—East (http://www.nd.edu/~underc/east/publications/) in northern Wisconsin (8, 9). Two small adjacent lakes were used in an experiment, one as a control and one fertilized with N and P. An addition of $NaH^{13}CO_3$ to both lakes allowed the separation of the DOC sources for bacterial growth into alga-derived and terrestrially derived.

The fertilization of Peter Lake produced, relative to Tuesday Lake, an increase in algal production leading to increased rates of production of high-quality DOC. The evidence for these changes was large amounts of chlorophyll *a* and an increase in bacterial production as well as a decrease in the pCO_2 (Fig. 2). The DOC input to unfertilized Tuesday Lake was 92 to 96% terrestrial, but 22 to 24%

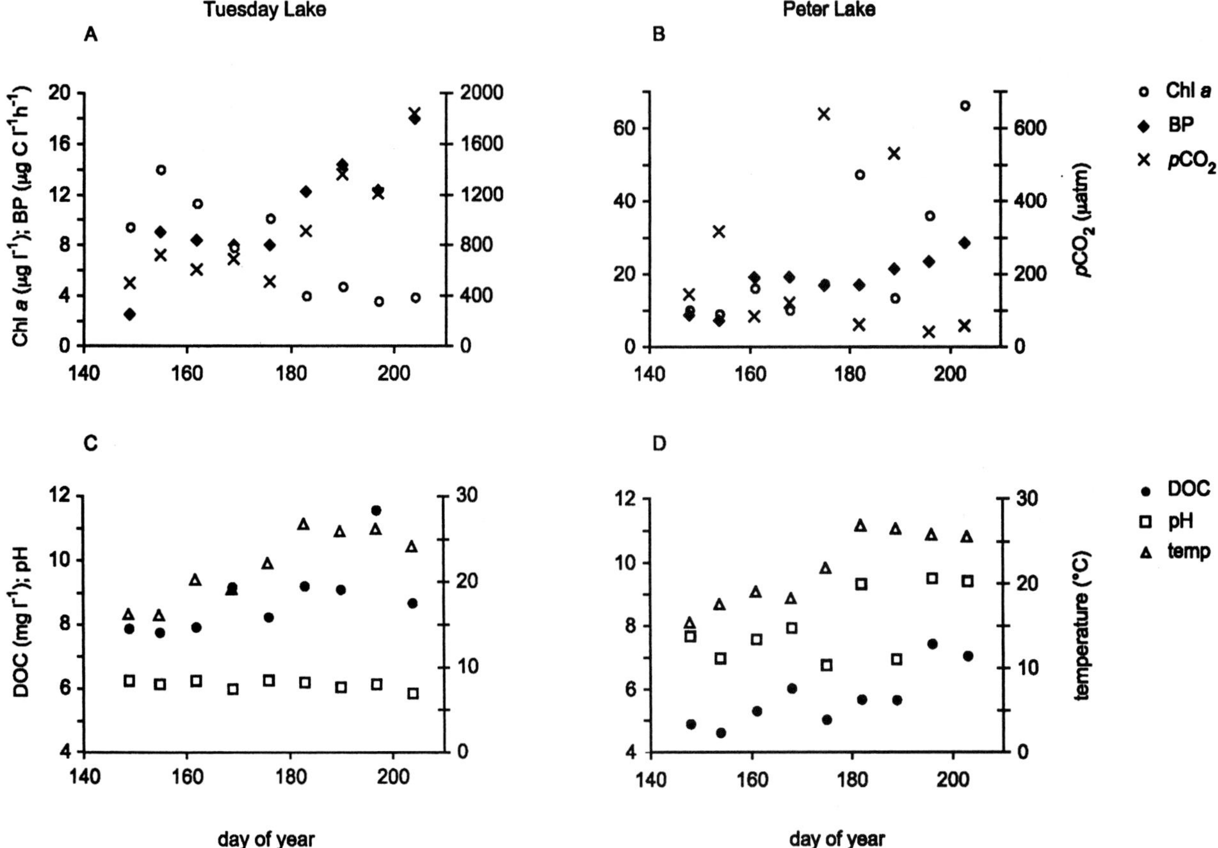

FIGURE 2 Physical (temperature), chemical (chlorophyll *a* [Chl a], DOC, pH, pCO_2), and biological (bacterial production [BP]) changes in the control Tuesday Lake and the fertilized experimental Peter Lake. (Reprinted from reference 9 with permission.)

of the biomass produced by bacteria came from algal carbon. In contrast, in fertilized Peter Lake, 55 to 70% of the DOC input was terrestrial while bacterial growth was 54 to 56% algal in origin. Clearly, there was preferential use by the bacteria of the autochthonous DOC produced by algae. It was suggested (9) that although the algal carbon production regulated the magnitude of the bacterial production, terrestrial carbon also supported bacterial productivity in both systems. The authors have no firm explanation of the preference but mention that terrestrially derived DOC typically consists of high-molecular-weight compounds with a high degree of aromaticity while alga-derived DOC is mainly low-molecular-weight compounds highly available for bacterial uptake.

The composition of bacterial communities from the two lakes was analyzed by TRFLP techniques. Nonmetric multidimensional scaling (NMS) analysis of the binary coded TRFLP data was used to assess the effects of fertilization on the composition of the microbial communities. The bacterial community composition between each lake differed and, within the lakes, the communities varied seasonally (Fig. 3).

In summary, this study documents the changes in bacterial communities resulting from one type of environmental change, fertilization. The experimental manipulation indicated that the source of the organic matter in lake ecosystems, here a terrestrial versus an aquatic source, has a significant effect on the species in the microbial community.

Terrestrial

In the Arctic, microbial activity continues throughout the winter because the snow cover insulates the soil from the extreme cold. Soil temperatures are below 0°C, but it is well known that microbial activity continues as long as there is

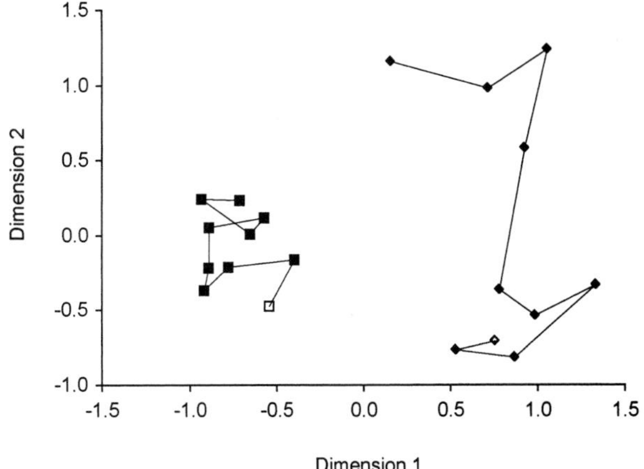

FIGURE 3 NMS of matrix calculated from binary-coded TRFLP patterns in Tuesday Lake (squares) and Peter Lake (diamonds) during the summer. Open symbols represent the results from the first sampling date. (Reprinted from reference 9 with permission.)

super-cooled liquid water (above ~−7°C). The example (16) describes the effect of snow depth as it regulates soil temperature and affects microbes. In the experiment, a snow fence was constructed at the Toolik Field Station site of the Arctic LTER site (http://ecosystems.mbl.edu/ARC/), located in the northern foothills of the Brooks Range, Alaska. The fence effectively increased winter snow depth over a 5-year period and allowed a comparison of microbial activities at ambient and increased snow levels. Intact cores were sampled through the winter and the following growing season in dry heath and moist tussock tundra. Snowpack developed earlier and deeper behind the fence, and the insulating effect of the snow resulted in significantly higher soil temperatures, which did not go below −7°C at either site. At ambient snow depth, temperatures reached −25°C (tussock) and −30°C (heath). The deeper snow also resulted in increased CO_2 fluxes, indicating higher microbial respiration rates.

Nitrogen cycling was also affected by snow depth and soil temperature. Nitrogen mineralization (Fig. 4) was stimulated to relatively constant rates throughout the winter in tussock tundra; N mineralization was lower in the heath, and net N mineralization occurred only in late winter. The amount of nitrogen was large enough in both instances, however, to substantially meet the aboveground N demand of the vegetation. In all sites, this nitrogen was subsequently immobilized at thaw, thus altering the amount and timing of plant-available N in tundra ecosystems. The nature of microbial substrate processing under frozen and thawed conditions remains uncertain.

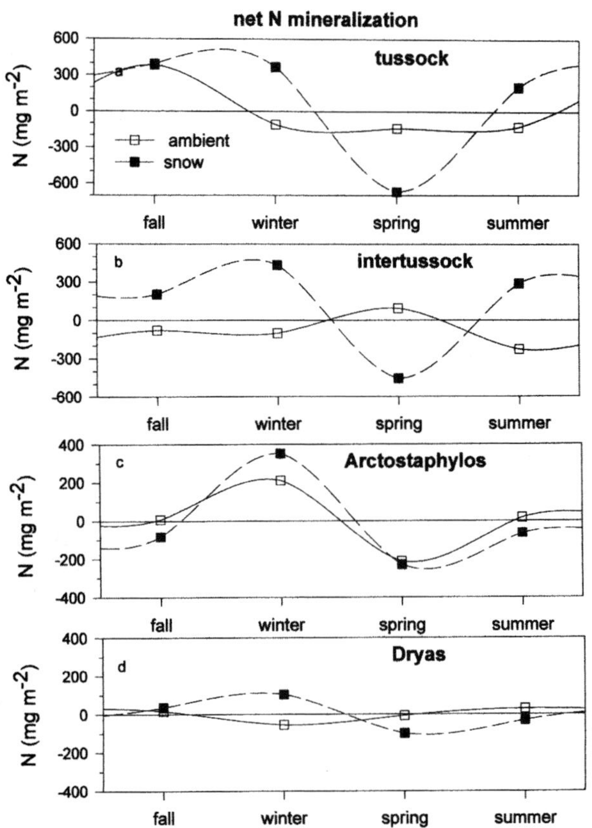

FIGURE 4 The effects of increased snow cover, caused by a snow fence, on net N mineralization at the Arctic LTER site in Alaska. The values were calculated as the changes in average soil N contents over a season; the units are milligrams of N per square meter per season. Thus, for the control (ambient), the seasons are fall (September to November), winter (November to March), spring (March to May), and summer (May to August). Results for two types of soils are illustrated, moist tundra (tussock, intertussock) and dry heath (*Arctostaphylos* and *Dryas*) soils. (Reprinted from reference 16 with permission.)

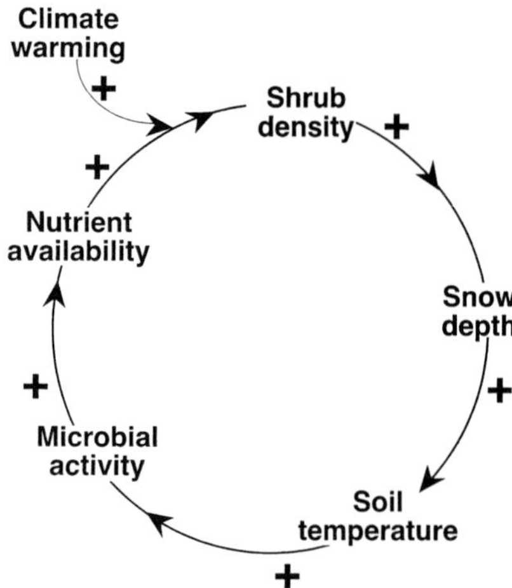

FIGURE 5 The snow-shrub-soil-microbe feedback loop. Soil temperatures in and around shrubs are higher than in shrub-poor locations, resulting in enhanced winter microbial activity throughout more of the winter. As a result, there is more net nitrogen mineralization during the winter and higher shrub leaf nitrogen content and plant growth in the summer. Larger and more abundant shrubs trap more snow and reduce winter sublimation losses, leading to deeper snow cover and still higher soil temperatures. (Reprinted from reference 17 with permission.)

These data were used in a model (17) that explores the effects of arctic climate change. The authors noted that arctic air temperatures have been warming at a rate of 0.5°C/decade over the past 30 years and that there is an increase in shrub abundance. They predict a positive feedback loop (Fig. 5) whereby global warming facilitates the shift from typical tundra to shrub-like plants. The shrubs themselves increase snow depth, promoting higher winter soil temperatures, greater microbial activity, and more plant-available nitrogen. This is a good illustration of how perturbations in an ecosystem may affect primarily microbial processes yet can have ripple effects at a global scale.

NEXT STEPS

In this chapter, we have highlighted a few of the studies that have started to use the legacy of available environmental data and experiments from LTER and other long-term study sites. There is even a special program, NSF-Microbial Observatories/Microbial Interactions and Processes, which encourages microbial research using long-term environmental monitoring data. Furthermore, there are numerous new directions and approaches being developed for monitoring diversity and processes that will continue to grow and develop to provide rapid and meaningful tools for microbial ecosystem research. With the relative ease of high-throughput sequencing, metagenomic information from different environments is becoming more readily attainable. Already, one can begin to identify habitat-specific genomic modules that provide information about potential microbial interactions and processes (18). DNA microarrays targeting 16S rRNA genes (PhyloChips) are also ambitious projects that will enable routine and rapid analysis of relative microbial diversity and abundances (10). These can be complemented with functional gene arrays (14) which can identify species expressing a gene for a particular process through probing of the RNA of a microbial community. In this way, changes in the function of the microbial community can be linked to species. Another way to link function to species is through incorporation of radiolabeled substrates in the RNA of members of complex microbial communities (1). Perhaps the next step is to combine this technology with stable isotope probing approaches for exploring microbial processes and correlate these data with more traditional process measurements.

There are no doubt many opportunities and areas for method development that provide ecosystem-level approaches for studying microbial dynamics. However, critical to all of these studies is the importance of experimental design as well as integration of diversity data, process data, and environmental measurements in meaningful ways.

REFERENCES

1. **Adamczyk, J., M. Hesselsoe, N. Iversen, M. Horn, A. Lehner, P. Halkjaer Nielsen, M. Schloter, P. Roslev, and M. Wagner.** 2003. The isotope array, a new tool that employs substrate-mediated labeling of rRNA for determination of microbial community structure and function. *Appl. Environ. Microbiol.* **69:**6875–6887.
2. **Buckley, D. H., and T. M. Schmidt.** 2001. The structure of microbial communities in soils and the lasting impact of cultivation. *Microb. Ecol.* **42:**11–21.
3. **Buckley, D. H., and T. M. Schmidt.** 2003. Diversity and dynamics of microbial communities in soils from agroecosystems. *Environ. Microbiol.* **5:**441–452.
4. **Caffrey, J. M., N. Harrington, I. Solem, and B. B. Ward.** 2003. Biogeochemical processes in a small California estuary. 2. Nitrification activity, community structure and role in nitrogen budgets. *Mar. Ecol. Prog. Ser.* **248:**27–40.

4a. **Hobbie, J. E.** 2003. Scientific accomplishments of the Long Term Ecological Research program: an introduction. *BioScience* **53:**17–20.

5. **Hobbie, J. E., S. R. Carpenter, N. B. Grimm, J. R. Gosz, and T. T. Seastedt.** 2003. The US Long Term Ecological Research program. *BioScience* **53:**21–32.
6. **Kent, A. D., S. E. Jones, A. C. Yannarell, J. M. Graham, G. H. Lauster, T. K. Kratz, and E. W. Triplett.** 2004. Annual patterns in bacterioplankton community variability in a humic lake. *Microb. Ecol.* **48:**550–560.
7. **Kratz, T. K., L. A. Deegan, M. E. Harmon, and W. K. Lauenroth.** 2003. Ecological variability in space and time: insights gained from the US LTER program. *BioScience* **53:**57–67.
8. **Kritzberg, E. S., J. J. Cole, M. L. Pace, W. Granéli, and D. R. Bade.** 2004. Autochthonous versus allochthonous carbon sources of bacteria: results from whole-lake ^{13}C addition experiments. *Limnol. Oceanogr.* **49:**588–596.
9. **Kritzberg, E. S., S. Langenheder, and E. S. Lindström.** 2006. Influence of dissolved organic matter source on lake bacterioplankton community structure and function—implications for seasonal dynamics. *FEMS Microbiol. Ecol.* **56:**406–417.
10. **Loy, A., K. Kusel, A. Lehner, H. L. Drake, and M. Wagner.** 2004. Microarray and functional gene analyses of sulfate-reducing prokaryotes in low-sulfate, acidic fens reveal cooccurrence of recognized genera and novel lineages. *Appl. Environ. Microbiol.* **70:**6998–7009.
11. **McKane, R., E. Rastetter, G. Shaver, K. Nadelhoffer, A. Giblin, J. Laundre, and F. Chapin.** 1997. Climatic effects on tundra carbon storage inferred from experimental data and a model. *Ecology* **78:**1170–1187.
12. **Mintie, A. T., R. S. Heichen, K. Cromack, Jr., D. D. Myrold, and P. T. Bottomley.** 2003. Ammonia-oxidizing bacteria along meadow-to-forest transects in the Oregon Cascade Mountains. *Appl. Environ. Microbiol.* **69:**3129–3136.
13. **O'Mullan, G. D., and B. B. Ward.** 2005. Comparison of temporal and spatial variability of ammonia-oxidizing bacteria to nitrification rates in Monterey Bay, California. *Appl. Environ. Microbiol.* **71:**697–705.
14. **Rhee, S.-K., X. Liu, L. Wu, S. C. Chong, X. Wan, and J. Zhou.** 2004. Detection of genes involved in biodegradation and biotransformation in microbial communities by using 50-mer oligonucleotide microarrays. *Appl. Environ. Microbiol.* **70:**4303–4317.
15. **Rich, J. J., R. S. Heichen, P. T. Bottomley, K. Cromack, Jr., and D. D. Myrold.** 2003. Community composition and functioning of denitrifying bacteria from adjacent meadow and forest soils. *Appl. Environ. Microbiol.* **69:** 5974–5982.
16. **Schimel, J., C. Bilbrough, and J. M. Welker.** 2004. Increased snow depth affects microbial activity and nitrogen mineralization in two Arctic tundra communities. *Soil Biol. Biochem.* **36:**217–227.
17. **Sturm, M., J. Schimel, G. Michaelson, J. M. Welker, S. F. Oberbauer, G. E. Liston, J. Fahnestock, and V. E. Romanovsky.** 2005. Winter biological processes could help convert arctic tundra to shrubland. *BioScience* **55:** 17–26.
18. **Tringe, S. G., C. von Mering, A. Kobayashi, A. A. Salamov, K. Chen, H. W. Chang, M. Podar, J. M. Short, E. J. Mathur, J. C. Detter, P. Bork, P. Hugenholtz, and E. M. Rubin.** 2005. Comparative metagenomics of microbial communities. *Science* **308:**554–557.
19. **Ward, B. B.** 2005. Temporal variability in nitrification rates and related biogeochemical factors in Monterey Bay, California, USA. *Mar. Ecol. Prog. Ser.* **292:**97–109.
20. **Yannarell, A. C., A. D. Kent, G. H. Lauster, T. K. Kratz, and E. W. Triplett.** 2003. Temporal patterns in bacterial communities in three temperate lakes of different trophic status. *Microb. Ecol.* **46:**391–405.

Quality Assurance

A. J. CROSS-SMIECINSKI

16

INTRODUCTION

The purpose of this chapter is to discuss the principles of quality assurance (QA) and to provide a discussion of techniques that can be used to train personnel to QA requirements. Starting with the definition of QA and its counterparts, quality control (QC) and quality improvement, the chapter then moves into a description of general QA program components. A brief section on ethics in research ends the chapter. A primary goal of this chapter is to describe requirements in familiar terms and in the process illustrate how an effective QA program can benefit an environmental microbiology project. This chapter is an overview. It is not necessarily intended to be implemented in its entirety.

Quality Costs

A QA program typically requires 10 to 20% of an investigator's time and generally 15% of total measurement time (5). Likewise, the industry standard budget of an independent staff of quality professionals is 10 to 15% of the technical budget. The cost depends on the extent of the specific QA requirements. Some programs, for example, those supporting enforcement, litigation, or nuclear licensing activities, require extensive documentation, procurement regulation, and record-keeping. Other programs supporting basic or applied research emphasize careful record-keeping practices, controls on field and laboratory processes, and review to ensure accuracy, completeness, and repeatability of the work. As time goes by, the trend in QA program requirements is merging and more sponsors are requiring accountability from the researchers in compliance to QA requirements and the tracking of deliverables.

Small organizations can run into the situation of receiving funding from an agency with a QA program that has extensive requirements. The sponsoring agency that has a large QA staff may find it difficult to understand why the two-person research group is overwhelmed by its QA requirements. For this reason, the ideal solution is to work under a graded approach to QA, whereby the QA requirements are tailored to the nature of the project (19). QA controls can be implemented based on the importance of the results, complexity of the study, reliability of the processes, etc., and the consequences of failure, such as risk to public health. A graded approach to QA is one way of achieving the maximum benefits of a QA program at the least cost.

Whether a research group is establishing its own QA program or implementing that of a sponsor, the added cost of the time required to implement a QA program must be covered in cost proposals. It is necessary to be fully aware of the extent of the QA required for the study before submitting a cost proposal. Although there are added costs, establishment of a QA program can attract customers and open up opportunities for work with those agencies that require QA standards of their funding recipients.

Definitions

QA

John K. Taylor defines QA as "a system of activities whose purpose is to provide to the producer or user of a product or a service the assurance that it meets defined standards of quality with a stated level of confidence" (20). In general terms QA is a philosophy used to ensure the continuous reliability of data. QA articulates a program of planning, control, and improvement efforts coordinated to meet data quality requirements.

Today most organizations sponsoring scientific research and laboratory production require the implementation of a QA program. As a taxpayer or other type of customer, each person understands the reason for this. The QA program is designed to "ensure that the money invested in the project will yield scientifically valid conclusions related to a principle hypothesis"(5).

QC

The quality of the measurement process (performance) is evaluated by comparison to standards (6). QC is the set of materials and practices used throughout the measurement process to ensure that each discrete part contributes a minimum amount of error to the results. QC requirements may be applied to any field, sampling, and measurement aspect of the project that affects the data quality. The choice of QC requirements is dependent on the study or project and usually is tied closely to the field and measurement methods. Quality control is a required part of good laboratory QA programs.

Note: the terms QA and QC are often used interchangeably. However, they have different meanings. As with QA, QC is used to ensure the reliability of data, but on a different level. Whereas QA is concerned with the quality of the information provided by the data and proceeds by controlling the system that affects the data, QC is specifically concerned with the quality of the measurement process.

Quality Improvement

Quality improvement is the evolutionary process used to enhance and prevent a stagnant quality program. Quality improvement utilizes the ingenuity of the personnel to solve problems or improve processes in the workplace, thereby promoting a climate of continuous improvement. Organizations use groups of employees, called "quality circles," that are each assigned a problem to solve. They meet, troubleshoot, and recommend changes to management that result in improvement to products, services, etc. Quality improvement works on the premise that employees have a positive interest in efficiency and want to improve their work practices, products, and environment (12).

GENERAL QA REQUIREMENTS

QA and Environmental Microbiology

Quality assurance is not new to environmental microbiology (3, 9). However, with the emphasis in environmental sciences being the detection of chemical contaminants and resulting engineered solutions (e.g., Superfund), environmental QA program requirements have focused on the chemistry and engineering fields. The environmental microbiologist therefore often finds it necessary to adapt a QA program from these fields.

Management

A goal of the QA program is to give management the opportunity to provide input and take responsibility in the planning, implementation, and assessment stages of the environmental microbiology project. The organization's policies and management commitment regarding QA must be explicit. The requirement to incorporate QA in the initial planning stages of a study or that management and subordinates be involved in certain QA decisions are examples of an organization's QA policies.

QA Program

Organizations establish a QA program in a number of ways, depending on management support, funding and organizational structure, and the likelihood for ongoing QA requirements from funding sources. A research team can implement QA requirements without having a QA staff within their organization. In such cases, oversight and compliance assessment would then be provided by the funding sponsor's QA personnel. An organization that has been directed as part of a contract to establish a QA program or has a large number of its studies needing QA might establish an independently funded staff of QA professionals to conduct these activities.

The organization funding a microbiology study usually specifies the QA requirements that must be met in terms of a QA standard or in terms of its own program. The microbiology group being funded establishes or alters its QA program to meet these requirements. However, microbiology companies, organizations, or laboratories may elect to establish a QA program that will attract the customers they want to do business with, will help ensure successful qualification as a supplier, or prevent proposal rejection for lack of a QA program. Often they choose one of the International Standard Organization (ISO) Standards, especially if they do business worldwide; but an environmental group might want to choose the Environmental Protection Agency (EPA) standard if they are interested in doing work for that agency or a nuclear standard if they want to propose work with the Department of Energy, and so on. Some examples of QA standards include ISO standards, U.S. EPA R-2 (24), and U.S. DOE orders NQA-1 (1) and 5700.6c (21). Elements that are common to most QA standards and programs are described below.

- Management reviews—periodically evaluate the effectiveness of the QA program in achieving adequate data quality
- QA plans or procedures—describe the QA requirements and how to meet them
- Technical plans and procedures—stepwise technical work instructions including those steps that satisfy QA requirements
- Audits or assessments—evaluations of performance or compliance to requirements, plans, and procedures

Organization

All components of an organization, especially the QA staff and their relationship to the management and technical staff, are important to ensure coordinated efforts and a logical flow of information. For this reason, the organizational structure and its members' respective roles and responsibilities are defined so they understand how they interact with others. Specifying the differences between informal and formal communications is important. Divisions of responsibility among the technical staff, management, supervisory staff, and contractors should be clear. The responsibility for quality lies with all personnel.

A standard requirement is for the QA staff to be independent of organizational, cost, and scheduling constraints. For this reason, QA staff are not supervised by the same individual for whom they are overseeing compliance to the QA program. Additionally, QA staff are not supported by an account managed by the same personnel being audited or assessed. Even in organizations of the highest integrity, to do otherwise would constitute the appearance of a conflict of interest. Additionally, the QA manager holds a position at a level that is the same as or higher than that of the technical work managers. In a very small organization, independence might be difficult to accomplish. A consultant might have to be contracted periodically to assess and document compliance to the QA program. Also, organizations can partner to share resources.

Quality Planning

Certain activities are included in the planning process, which can be documented in a study plan. The sponsor's technical and quality requirements need to be translated into project specifications to achieve the desired results. Costs and time constraints need to be considered during planning. Planning elements should include the following as appropriate.

- General overview, background of the project
- Scope of work, study purpose, objective or hypotheses to be tested, and a listing of primary tasks involved, including estimated resource requirements, deliverables, and their schedules
- Scientific approach or technical method used to collect, evaluate, analyze, or study results

- Standards and acceptance criteria
- Applicable procedures
- Equipment, instruments, and software
- Provisions for documenting the work performed and providing required records
- Prerequisites, special controls, specific environmental conditions, processes, skills, or staff training and certification requirements
- Data to be collected and what information will be recorded
- Field and laboratory process flow charts
- Summary tables
- How results will be presented and the resulting deliverables
- Time line of project tasks and subtasks
- Uses and limitations of the results

Personnel Qualifications and Training

Qualification and training requirements are identified in order to ensure that the work is performed safely, in compliance with the QA requirements, and efficiently. Experience and education are verified to ensure that personnel can perform assigned work. Technical, project orientation, and QA training requirements for project personnel are specified. A training program need not be an overwhelming task for a small study group. While large organizations may be able to provide employees with a training facility and staff, small organizations can accomplish training with in-house presentations, demonstrations, workshops, and assigned reading.

Typically training consists of some form of instruction followed by documentation that the instruction has been completed, occurring before personnel start conducting work. Variability is in the manner of instruction. The success of QA training can be influenced by the characteristics of the population being trained. If the trainees have prior experience in QA and they have learned to appreciate its value, or if they are particularly curious and conscientious, the training will be more successful because these individuals will strive to comply. If the trainees view QA as being an unnecessary burden, the training might not be as successful. For these reasons, it is important that instructors provide clear, concise instruction in a friendly but firm presentation. Instructors must be sympathetic and patient with individuals who are having difficulty grasping the QA program.

Typically, QA training and the presentation of the QA program are accomplished through a classroom session. Instructors employed to present the QA program can emphasize problem areas and are present to answer questions. Sometimes a written test is given. The primary disadvantages to classroom training are (i) variability among the sessions, because of differences among the instructors or the sessions; and (ii) costs associated with providing instruction personnel.

Another way to accomplish training is by reading assignment. The advantage to training by reading assignment is that the need for an instructor is eliminated, saving costs. The disadvantages of training by reading are that (i) there is no assurance that personnel have read the assigned material and (ii) there is no immediate opportunity to have questions answered.

Training to QA requirements months or weeks prior to the need to comply with them does not always ensure that individuals will remember the requirements when they are needed. For this reason, retention of the QA requirements is improved if they are reviewed at times during the work. This has been shown to be successful and can be accomplished in a number of ways, including staggered and refresher training. It can also be accomplished by conducting short surveillances starting very early in the work and scheduled periodically thereafter. In addition to reinforcing training, it builds a rapport between study personnel and the QA staff fostering a greater likelihood that study personnel will contact QA staff with questions and concerns—and work as a team.

Work Processes: Procedures and Scientific Notebooks

Work processes must be documented to ensure that the work is performed as planned and that the work can be repeated if necessary. Work processes for most environmental studies are controlled by, or recorded in, technical procedures, the scientific notebook, or a combination of the two. Generally, a process which is performed the same way each time is a work process that is appropriate for use as a written procedure. A process that employs "trial-and-error" or a high degree of professional judgment is appropriately documented in a scientific notebook. Two formats (I and II) for written procedures follow, illustrating the differences in QA program requirements.

Format I is written to a national QA standard, *Quality Assurance Requirements for Nuclear Facility Applications* (1) and includes the following.

- Technical, QA, or other requirements, including reference to the applicable method, standards, and criteria
- A description of the laboratory equipment, including manufacturer, operating manuals, model, and serial number
- Identification of data acquisition computer hardware and software, including software title and version
- Special controls, precautions, environmental conditions, and hold points
- Quality control and determinations of accuracy, precision, and bias
- A sequential description of the work to be performed including controls for altering the sequence of steps
- Quantitative or qualitative acceptance criteria
- Information to be recorded and where
- Provision for recording deviations from the procedure
- Records to be generated by use of the procedure

Format II is written for a production lab, such as a lab that tests water samples for pathogens, and includes the following.

- Parameter
- Reference
- Method
- Reagents and equipment
- Procedure steps
- QC requirements
- Data sheets, comments, etc.

Scientific notebooks are used for trial-and-error work, in combination with procedures for repetitive processes. If using scientific notebooks, establish document controls concerning identification and assignments of notebooks to be used.

Most guidelines for controlling scientific notebooks have specific requirements (13) for the manner in which entries are made and corrected and to ensure notebook security, in order that they may be admissible as legal documents. Proceduralize scientific notebook use, including organization, formatting, entry and attachment techniques, types of reviews, and archival protocols, to ensure uniform practices.

Document Control

Document control includes a system for handling documents, such as plans and procedures, and records to ensure that only the most recent version is being used and that they accurately reflect completed work (24).

- Documents that specify quality requirements or prescribe work activities should be reviewed, approved, and controlled.
- Reviews should address technical adequacy, compliance with requirements, correctness, and completeness prior to approval and issuance.
- Effective dates are placed on documents.
- The disposition of obsolete documents is ensured, to avoid their inadvertent use.
- Procedures are established and maintained to provide the current revision of a controlled document.
- Procedures are established defining the distribution of controlled documents.

Who designates the documents that are to be controlled for each study and which types of documents qualify to be controlled should be established. As an illustration, in most studies the principal investigator would probably designate which documents are to be used and would designate those documents by title and revision number. For example, QA plans and technical procedures are often controlled.

Record Keeping

Record-keeping requirements address the types of information to be documented, as well as the techniques to use. Most guidelines for controlling record keeping have specific requirements for the manner in which errors are corrected, for prohibiting the use of pencils and correction fluid (obliteration), for ensuring traceability to organism identifications, materials, equipment, procedures, and raw and reduced electronic data. Work performed should be sufficiently documented to verify that all data collected are scientifically sound and defensible as well as repeatable. Record-keeping practices affect report preparation, data evaluation, and repeatability and help to ensure that documents will be admissible in court.

Procurement of Items and Services

The organization's procurement system ensures that suppliers (vendors) provide the items and services required. One way to fulfill this goal is to use only suppliers that have a QA program consistent with the organization's own or other accepted QA program. Suppliers can be evaluated (by document reviews or by verifying implementation in a site visit) to establish an approved supplier list. The evaluation process can be conducted long before items or services are needed to avoid assessment delays when purchases are needed. Supplier programs are verified periodically thereafter, to maintain the approved supplier list. Sometimes organizations allow procurement from a supplier on another organization's (that has the same or similar supplier qualification requirements) approved-supplier list in order to save assessment costs. A thorough and effective procurement process should define the following.

- Procurement planning
- Procurement document preparation
- Selection of procurement sources
- Access for reviews
- Proposal/bid evaluation and award
- Purchaser evaluation of supplier performance
- Acceptance of the item or service
- Control of nonconformances

A streamlined procurement process could address submittal of technical and quality requirements to the supplier followed by documented acceptance of items or services received and a process for correcting items or services received that do not meet requirements.

Software and Electronic Data Control

A QA program should require establishment of procedures that are designed to prevent defects in data caused by computer software, programs, routines, macros, reduction, and other manipulations. Software qualification requirements may be extensive depending on the software used and whether it was written or purchased. Software used for graphics, presentations, or in other ways that do not affect data quality does not normally require any type of qualification. If the software is used with equipment or instrumentation and is controlled by standardization steps such as calibration, it is normally exempt from the requirement to be qualified. Such software is used and documented with practices consistent with its intended use.

Certification from the manufacturer indicates that the software has been thoroughly tested and that the manufacturer stands behind the quality of its product. Database and spreadsheet software functions usually don't need to be qualified. However, software and programs, etc., used to reduce raw data need to be verified to ensure that they accomplish their intended purpose accurately. A hand-held calculator can be used to check some software calculations. Another way to check the software is to process a known by using the software to see if it produces the correct response. The use of macros, cell formulae, or routines requires verification of the results.

Computer programs and software may need to meet extensive qualification requirements if they involve any of the following.

- The computer code was written or procured by an individual in the organization
- The program or software is being used in a sensitive study, such as national health or nuclear licensing

Qualification requirements for these types of programs may include the following.

- Software specifications, such as program needs and capabilities, data source, definition of valid input, and error checking on the data input (15)
- Establishing a staff of individual(s) responsible for software configuration management
- A description of the approach to, and documentation of, software, such as development acquisition, testing, and use

- Software life cycle processes and controls, such as classification, requirements, development, design, code development, installation testing, and other details (11, 22)

To prevent purposeful or inadvertent modification and loss of electronic data, controls need to be established. These may specify use of a limited-access system accomplished by use of a secure network drive, locked doors, or password protection. Backup of files on a regular basis and maintaining copies of unique software prevent loss caused by computer and media failures. In addition, documenting software identity, data, location, and special commands can prevent data loss caused by a change in personnel and helps coworkers continue work during personnel absences.

Audits and Assessments

Audits and assessments are defined according to the individual needs and philosophy of the QA program. Audit and assessment activities can be designed to be a review of routine data or may involve visits to the laboratory or field for an on-site evaluation. Auditors may use a general check sheet that applies to similar laboratory or field operations or possibly the study's QA plan and written procedures.

Audits or assessments emphasize to project personnel the importance of complying with the QA program, plans, and procedures. The assessment process provides the opportunity to review good laboratory practices and make other changes in the lab such as updating procedures and adding equipment or personnel. Such changes serve to benefit the work team and data quality.

Audits and assessments fall generally into two major categories, internal and external. Internal audits or self-assessments are performed within the organization and can be contracted out when independence is not possible. External audits are performed by personnel external to the defined organization such as the sponsor's QA personnel or the project leader's corporate office auditors. An environmental microbiology group may experience an audit as the subject of an audit or as an auditor, in the case of having to evaluate the QA program of a service or supplier.

Terms for some specific types of audits and assessments are described as follows.

- Audit: inspection of objective evidence (evidence that is written) for compliance to written instructions, plans, and procedures. Auditors request and inspect records and documents for evidence of compliance. A report is written detailing specific instances of noncompliance found by the auditors.
- Surveillance: observation of the work as it is being performed (23) for the purpose of process control or product acceptance (1). A report is written detailing specific instances of noncompliance found by the surveyors.
- Performance evaluation: evaluation of a system, process, or work group based on actual performance, usually against a blind performance sample. A sample is submitted to the laboratory for measurement. The laboratory's sample results are evaluated statistically in comparison to the other participants in the study or to the known value of the sample. A written report is sent to the participant detailing this comparison. If the laboratory has produced an unacceptable result, corrective action is needed and often requested by the organization that submitted the performance sample to the laboratory.
- Data quality assessment (23): evaluation of a set of data against specific criteria, contained in a check sheet. A data package is evaluated against technical, QA, and QC requirements to establish if the data is valid and if the laboratory that produced the data is in compliance with requirements.

Since many individuals have not experienced an audit or surveillance, it is useful for the project leader or in-house QA staff to prepare the study personnel. A description, with as much detail as possible, should be given as to what the audit will involve. Although good things often come out of an audit, the process usually presents a tense situation for those being audited. It is a valuable exercise to prepare personnel for an external audit or assessment by having a preaudit or by performing a self-assessment.

Nonconformance and Corrective Action

Nonconformance is defined as a deficiency in characteristic, documentation, or procedure that renders the quality of an item or activity unacceptable or indeterminate (1). A nonconformance is therefore a potentially serious deficiency, such as measuring a sample with an instrument that does not meet calibration requirements.

Discovery of a nonconformance is followed by tagging, segregating, or otherwise isolating the nonconforming item or condition so that it is not used again until it is corrected. An example of this is to indicate with a label that the out-of-calibration instrument cannot be used until it meets calibration requirements.

Corrective action is defined as the measure(s) taken to correct conditions adverse to quality and, where necessary, to prevent recurrence (1). An example of a corrective action would be to remeasure affected samples after the instrument has met calibration requirements. Corrective actions are documented with a technical justification that explains why the corrective action to be implemented is the most logical choice.

Impact evaluation is the documented analysis of the effect that a nonconformance has on items or processes. For example, in the case of the out-of-calibration instrument, it must be determined whether it has affected measurements collected since its last valid calibration.

Records

A QA record is a completed document that furnishes evidence of the quality of items or activities (1) or study results. Records are prepared, reviewed, and maintained to reflect the processes resulting in the completed work. Examples of QA records include plans and procedures used during the study to direct work, raw and reduced data, calibration reports, and chain-of-custody forms. Those specific documents and data and the time at which they become QA records are specified in procedures. Normally, documents and data become records when personnel complete a segment of work and no longer need to refer to the documents. Controls should be sufficient to ensure that records are legible, accurate, complete, and secure. Provisions should be made for storage, minimum retention times, preservation, and disposition of records. This includes provisions for proper storage of electronic data considering the longevity of the various media. Controls must be in place to ensure that records are traceable to samples, sites, data, reports, and the study of origin.

Some QA programs require records to be stored in fire-rated cabinets or in dual storage. Dual storage is defined as the storage of items sufficiently distant from one another so that one particular damaging occurrence could not destroy both sets. One way to avoid the cost that either measure would incur is to submit the records to the research sponsor or its records center as soon as possible after completion. Requirements for record retrieval and transfers should also be defined.

STUDY QUALITY

Sampling

Sampling is an integral part of the study design. The selection of sampling sites and sampling strategies is based upon the study objectives and hypotheses. Therefore, the sampling strategy should be one that ensures the achievement of the quality of data needed to reach conclusions about the experiment. Entire documents have been prepared and published (25) regarding the subject of sampling, since it is generally found to contribute the largest proportion of error to environmental sample measurements (3, 4, 8, 9, 14, 19).

Statisticians are valuable in planning the sampling activity. Study questions are most effectively supported by a predetermined number and variety of samples to be collected or measured. In turn, the proper statistical evaluation of resulting data lends confidence to the study conclusions. Statistical support, if available, or use of a good practical text (5, 20) is strongly suggested in planning sampling activities.

Sampling Site

Specify the requirements that will ensure that the study objectives are met. If sampling does not involve a geographic location, it may have other requirements, such as age, sex, or species-sample population. Consider specifying the following when planning the sampling activity.

- Sample matrices (e.g., air, water, soil, and body fluids) and target organisms
- The useful measurement levels expected or required from the site

If the site is mandated rather than selected for the study, determine the effects that the site will have on data and the project objectives.

- The reason (statistical or scientific) for choosing the sampling site and sampling frequencies
- Physical, behavioral, and other factors at the site that may affect sampling (e.g., elevation, permit requirements, or physiological obstacles) and sample quality (e.g., sources of contamination or extreme weather)
- Some or all measurements may need to be taken in the field (e.g., pH and temperature of fluid samples)
- Site modifications that may be necessary before sampling can occur (e.g., use of special transportation or construction)

Sampling Procedures

Make specific plans for collection, preservation, storage, record keeping, and transportation.

- Measurement parameters and sample volumes to be collected for each
- Sampling methods to be used (e.g., composite or grab). Use validated or (preferably) standard methods
- Preparation and cleaning of sampling equipment, containers, reagents, and supplies
- Calibration of field equipment (e.g., pH meter or vacuum pump)
- Sample processing may include preservation techniques (e.g., use of formalin or azide), transportation (e.g., overnight delivery), and storage (e.g., refrigeration) of samples before laboratory use
- Plan for container labeling, choosing materials that will not be affected by field and storage conditions
- Provision for the control of field failures (e.g., extra batteries, tubing, or sampling containers)
- Holding times, which are important in planning collection of microbiological samples (consider the maximum length of time samples can be kept before the measurement integrity is affected)
- Forms, notebooks, and procedures to be used to record sample collection, sampling conditions, and measurements
- Use of a combination of lab and field or travel blanks (travel blanks carried or shipped with samples to detect contamination during the trip), splits, duplicates, and other QC sample types
- The requirement that all sample personnel use the same procedure and be proficient

Chain-of-custody procedures employ the use of signatures to record individuals responsible (custodians) for the sample during its life. In addition, samples are kept in locked storage, in a limited access area, or in custodial presence. Use of chain-of-custody procedures ensures sample protection by eliminating the possibility of accidental or purposeful tampering and is necessary for work related to litigation. If chain-of-custody is not necessary, sample control to keep track of sample identity, purpose, holding times, scheduling, analysis types, etc., is helpful especially when processing a large number of samples, for more than a few sampling sites or sampling teams, and for multiple studies.

Measurement Methods

Measurement methods are chosen by the environmental microbiologist for the analytical parameters or target organisms in the specific matrix at the anticipated levels. Methods need to agree with the qualitative and quantitative objectives established for the study. Investigators may be limited in choices by resources and the sponsor requirements. In planning measurement methods, consider the following topics.

- Sample preparation methods
- Decontamination procedures and resulting waste
- Reagents and materials needed and their cost and availability
- Performance requirements
- Lab and technology resources
- Measurement confirmation

Measuring and Test Equipment Use and Calibration

The investigator must demonstrate that the measurement systems are operating properly. The equipment and instrumentation are calibrated and checked at a prescribed frequency commensurate with the type and history of the measurement system as well as the accuracy and precision needs of the project.

Instruments such as colony counters and analytical balances have internally automated or computerized calibrations. It is necessary, however, to verify that the internal calibration is accurate using a calibration check. For measurements that need a multipoint calibration to establish a concentration curve (calibration curve), a periodic calibration check can be used as an alternative to full recalibration to verify that the curve remains valid.

Calibration methods and frequencies are planned for instruments and equipment used to obtain measurement data (e.g., pH meters, analytical balances, pipettes, counters, PCR analyzers) that may affect the quality of the project data (23). Consider including the following when planning calibrations.

- Specific calibration procedures
- When possible and practical, certify calibrated equipment and standards including source, traceability, and purity. If a nationally recognized standard (e.g., National Institute of Standards and Technology) does not exist, designate the standards to be used. Maintain procedure and standards records traceable to the item being calibrated (23).
- Record-keeping system for achieving traceability from the calibrated item (instrument or equipment) to the standard used
- Frequency of initial calibrations (multipoint concentration curves), calibration checks (single-point checks), or recalibrations (criteria for establishing new curves)
- Calibration curve control limits and acceptance criteria (19). Examples: correlation coefficient requirement of 0.995 for initial calibration curve, or calibration check must agree within 10% of its original concentration.

Preventive Maintenance

It is critical to a project that field and laboratory instruments and equipment are operational and reliable. It can be useful to clarify exactly where each scientist's responsibility for instruments and equipment maintenance begins and ends. Establishing the division of responsibility between users and service is important in a setting in which there are multiple users and service contracts on instruments.

Preventive maintenance procedures are best addressed by reference to instrument and equipment procedures. These are often the manufacturer-supplied manuals and instructions. In addition to the following list, any maintenance issues (e.g., specific instrument problems or parts or service needs) that the investigator has noted previously should be considered in an effective preventive maintenance program.

- A schedule of preventive maintenance performed in-house and by a service representative
- An inventory of critical spare parts and supplies and who is responsible for this supply
- Maintenance contract information for critical measurement systems (e.g., current phone numbers and contract numbers) needed to facilitate service. These may be placed in procedures, manuals, in the instrument log, or even attached to the instrument.
- Archival storage and location of one-of-a-kind manuals and instructions including all versions of computer programs used to reduce data
- Inspections and acceptance testing performance and record keeping
- User training
- Continuity in case of a breakdown
- Record keeping requirements, such as what details, by whom, and when and where maintenance is recorded

QC in the Field and Laboratory

QC is used to document system performance and reliability. The use of QC determinations adds strength to a study by providing information to the data user of the conditions under which the study was conducted. These data can reveal, for example, that there is a very low level of background contamination, that the recovery rate is as high as 90%, or even that a system is extremely stable. It can also show where there are weaknesses in the study processes that must be overcome in order to advance the investigation. QC lends credibility to a study and ultimately to an investigator's reputation, by making it clear that the investigator is willing to present the reliability of study results.

The investigator must decide which parameters, measurement systems, and sets of data are critical and to what extent problems can be detected using various indicators. The investigator then determines what types of QC can be used in the study to monitor each parameter or data set, documenting data quality based on system performance. He plans and proceduralizes QC frequency, acceptance limits, corrective actions, and record keeping.

When there are sufficient historical data for the processes being used, objectives and acceptance/rejection criteria may be established. In this situation, results of the QC determinations must fall within certain limits of acceptance to support the quality of specific measurements. If QC measurements do not meet acceptable limits, they become "real-time" indicators flagging unacceptable, marginally acceptable, warning limit, etc., categories of data. Immediate corrective action can then be performed, ensuring continued collection of optimal data.

Some measurement method performance criteria may not be standardizable. Consequently, routine QC requirements may not be practical and a certain amount of trial-and-error may be necessary to pinpoint the most effective QC regime. Often a standard or published method will provide QC requirements; however, older methods often do not address QC. The principal investigator needs to determine during planning stages what critical parameters need to be controlled and establish an appropriate QC regime. The study can implement that regime until enough data are collected to establish acceptance criteria, if needed. QC results may be presented even if they were not used to grade the data, as indicators of accuracy, precision, contamination, etc. Such presentation applies toward the purpose of reporting and publication, or when the study is of short duration.

Some examples of QC used for the determination of accuracy, for qualitative identification, and for controlling production laboratory processes are given below.

- Reference materials: a substance with one or more of its properties established well enough to be used for checking the accuracy of a measurement method. An example is the use of a standard culture (e.g., American Type Culture Collection strain [5]).
- Confirmatory measurements with a second method: the use of a second measurement method to confirm a measurement that is not completely selective. An example is the use of phase-contrast microscopy to verify the identification of a *Giardia* cyst detected with an epifluorescent assay.

- Other QC methods include the use of control charts to indicate trends in instrument or system stability; proficiency testing of analysts to document individual skills; and multilaboratory performance programs, a form of proficiency testing usually associated with certification programs.

Standards versus Subjective Determinations

Unfortunately, for many aspects of environmental microbiology there are no standards or knowns to be used for QC and often measurements are subjective; for example, the assignment of color to microbial colonies and visual identification of individual organisms. In these cases, the scientist must be creative in developing QC standards. Ornithologists use a standard color chart for checking visual accuracy (5) that could be used for visual color determinations in other fields.

Industry uses reliability indices in several ways to treat subjective measurements. However, two techniques can be used to deal with almost any subjective situation. For classifications of nonquantitative (attribute) data, kappa techniques (7) are appropriate. For ratings based on a scale, such as 1 to 10, methods based on the intraclass correlation (7) are a good choice.

Data Reduction

Considerable time and effort go into planning a study and collecting data in a manner that meets a predetermined set of objectives. It is important to ensure that errors are not introduced into the data during the last phases of the study: data reduction, verification, validation, and reporting. Although it may seem labor-intensive to plan these activities, an investigator does not want to jeopardize data quality this late in the game. Data reduction is changing raw data into a more useful form (16). It includes calculations, tabulations, and transfers that result in interpretation and presentation of data, and it is during these manipulations that errors are likely to occur. A listing of personnel, procedures, statistical approaches, and products that result from data reduction should be planned to stabilize the data reduction phase. Consider the following.

- Data reduction procedures, including software or computer programs
- Statistical approach for reducing data
- Data sheets or formats
- Treatment of QC measurements
- Products of data reduction

Data verification determines if measurements have been made in accordance with prescribed, approved procedures and whether data have been quantified, recorded, and transcribed accurately. Completeness and consistency are assessed. Data verification is a routine activity performed on small sets of data as they are produced (18) and includes the following.

- Criteria used to accept, reject, or qualify data (23)
- Critical control points, points at which results are so important as to require compliance prior to further progress (5)
- Types of verifications, forms, and checklists to be used

The data validation process determines if the data are adequate for their intended use by comparison to a predetermined set of criteria (16), such as the following.

- Procedures for determining outliers and criteria for flagging data
- Acceptance criteria used to validate data (18)

ETHICS IN RESEARCH

Discussion

Unethical practices in the scientific community are generally defined as scientific misconduct. The issues associated with these practices cover a wide range, from honesty in research to basic moral questions concerning topics that can overlap into religious beliefs. The ethics issues that are associated with quality should be generally limited to those concerned with scientific conduct or misconduct that affect the quality of scientific results.

The National Academy of Sciences limits the definition of scientific misconduct. Basically, there are three types of error in research: honest errors, errors from negligence, and those involving a concerted effort to deceive (http://www.nap.edu/readingroom/books/obas). The third category includes creation of data, "dry-labbing" (fabrication), changing data to meet specific needs or desires (falsification), and using others' work without giving the credit due to the originator (plagiarism).

Of the three types of error, honest error or human error and/or conclusions based on various interpretations of data can and should be brought to light at the time of discovery, publicizing the corrections to potential users, probably through a notice in the same publication (17). In correcting the mistake, the intent of the researcher to conduct research responsibly is made clear. The second and third types of errors may be termed scientific misconduct and can have far-reaching effects, especially in the health fields (10), resulting in potential harm to humans at the worst and reflecting poorly on the research field at the least. And although sacrificing quality to the pressures of deadlines and volume of published work competition may seem justified at the time, the repercussions to the scientist and/or research organization for poor research outweigh the original motivations.

The National Science Foundation and the U.S. Public Health Service add "other serious deviations from accepted research practices" to falsification, fabrication, and plagiarism, so that other "questionable research practices" may be considered scientific misconduct (http://www.nap.edu/readingroom/books/obas). The additional definition is needed to prosecute unethical behavior that is not strictly covered under falsification, fabrication, and plagiarism. However, the flexibility of the definition could cause a scientist to be prosecuted for using new unusual methods. Other unethical practices include cover-ups, harassment of whistle blowers and other individuals, sabotage of others' work, padding resumes, misuse of funds and materials, and breaking grant regulations. The category of "other unethical practices" is not necessarily considered to fall under scientific misconduct, and individuals involved should be handled in the same way as they would be in other fields.

Case Studies

Case studies are not specifically provided here, but many exist in the literature and are available routinely via the media. Well-publicized examples of misconduct in clinical trials concern withholding research evidence of the harmfulness of a product in order to "prove" its harmlessness and provide marketability. Published case studies are available in *Science and Engineering Ethics* (5a) and in other issues of the same journal (for an example, see reference 2). Fictionalized cases of work with graduate students and postdocs are described in three volumes by the Association for Practical and Professional Ethics (B. E. Schrag, personal communication).

CONCLUSION

QA, QC, and quality improvement are integral parts of a valid scientific study. QA is not necessarily new to microbiologists, but the emphasis of QA requirements in environmental science has been on chemical and engineering studies. The chapter defines and provides examples, when possible, of typical QA requirements (subsection headings) that the environmental microbiologist might implement. Discussions of QA training and ethics in research are also presented.

REFERENCES

1. **ASME Nuclear Quality Assurance Committee.** 1994. *Quality Assurance Requirements for Nuclear Facility Applications*. ASME NQA-1-1994. The American Society of Mechanical Engineers, New York, N.Y.
2. **Bird, S. J., and D. Hoffman-Kim (ed.).** 1998. *Science and Engineering Ethics*, vol. 4, no. 1. *Whistle Blowing and the Scientific Community*. Opragen Publications, Surrey, United Kingdom.
3. **Bordner, R., and J. Winter.** 1978. *Microbiological Methods for Monitoring the Environment, Water, and Wastes*. EPA 600/8-78-017. U.S. Environmental Protec-tion Agency, Washington, D.C.
4. **Britton, L. J., and P. E Greeson (ed.).** 1989. Methods for collection and analysis of aquatic biological and microbiological samples. *In U.S. Geological Survey Techniques of Water Resources Investigations*, book 5. U.S. Geological Survey, Reston, Va.
5. **Cross-Smiecinski, A. J., and L. D. Stetzenbach.** 1994. *Quality Planning for the Life Science Researcher*. CRC Press Inc., Boca Raton, Fla.

5a. **DeMets, D.** 1999. Statistics and ethics in medical research. *Sci. Eng. Ethics* **5**:97–117.

6. **Deming, S., and S. L. Morgan.** 1993. *Experimental Design: A Chemometric Approach*, vol. 11, 2nd ed. Elsevier Science Publishing Co., New York, N.Y.
7. **Futrell, D.** 1995. *Quality Progress*, p. 81–86. American Society of Quality Control, Milwaukee, Wis.
8. **Ghiorse, W. C., and D. L Balkwill.** 1983. Enumeration and morphological characterization of bacteria indigenous to subsurface environments. *Dev. Ind. Microbiol.* **24**:213–224.
9. **Greenberg, A. E., R. R. Trussell, and L. S. Clesceri (ed.).** 1992. *Standard Methods for the Examination of Water and Wastewater*, 17th ed. American Public Health Association, Washington, D.C.
10. **Gurudevan, S. V., and W. R. Mower.** 1996. Misrepresentation of research publications among emergency medicine residency applicants. *Ann. Emergency Med.* **27**:327–330.
11. **Harris, S. D.** 1992. Software quality assurance requirements of a nuclear program written to meet current software quality assurance standards, p. 6A3.2. *In Proceedings of the Nineteenth Annual National Energy & Environmental Division Conference*. American Society for Quality Control, Milwaukee, Wis.
12. **Juran, J. M., F. M. Gryna, Jr., and R. S. Bingham, Jr.** 1979. *Quality Control Handbook*, 3rd ed. McGraw-Hill, Inc., New York, N.Y.
13. **Kanare, H. M.** 1985. *Writing the Laboratory Notebook*, p. 74. American Chemical Society, Washington, D.C.
14. **Labeda, D. P. (ed.).** 1990. *Isolation of Biotechnological Organisms from Nature*. McGraw-Hill, New York, N.Y.
15. **Morris, C. R.** 1993. Computer validation. *Network News* **7**:13.
16. **Parker, S. P.** 1989. *McGraw-Hill Dictionary of Scientific and Technical Terms*, 4th ed. McGraw-Hill Book Co., New York, N.Y.
17. **Price, A. R.** 1994. Definitions and boundaries of research misconduct. *J. Higher Education* **65**:286–297.
18. **Ramos, S. J.** 1992. A project manager's primer on data validation, p. 4C4.4–4C4.6. *In Proceedings of the Nineteenth Annual National Energy & Environmental Division Conference*. American Society for Quality Control, Milwaukee, Wis.
19. **Simes, G.** 1991. *Preparation Aids for the Development of Category I-IV Quality Assurance Project Plans*. EPA 600/8-91/003-006. Risk Reduction Engineering Laboratory, U.S. Environmental Protection Agency, Cincinnati, Ohio.
20. **Taylor, J. K.** 1987. *Quality Assurance of Chemical Measurements*. Lewis Publishers, Inc., Chelsea, Mich.
21. **U.S. Department of Energy.** 1991. *Order 57006.c*. Offices of Nuclear Energy and Environment, Safety, and Health, Washington, D.C.
22. **U.S. Department of Energy.** 2000. *Quality Assurance Requirements Document*. DOE/RW-0333P, revision 10. Office of Civilian Radioactive Waste Management, U.S. Department of Energy, Washington, D.C.
23. **U.S. Environmental Protection Agency.** 1992. *Revised Draft: EPA Requirements for Quality Assurance Project Plans for Environmental Data Operations*. EPA QA/R-5, U.S. Environmental Protection Agency, Washington, D.C.
24. **U.S. Environmental Protection Agency.** 1992. *Interim Draft: EPA Requirements for Quality Management Plans*. EPA QA/R-2. U.S. Environmental Protection Agency, Washington, D.C.
25. **U.S. Environmental Protection Agency.** 1989. *Soil Sampling Quality Assurance User's Guide*. EPA/600/8-89/046. Environmental Monitoring Systems Laboratory, U.S. Environmental Protection Agency, Las Vegas, Nev.

Issues of Study Design and Statistical Analysis for Environmental Microbiology

KENNETH M. PORTIER AND RONALD CORSTANJE

17

The practice of good science requires a concise description of study objectives and a study design that matches its objectives. A good study design requires (i) definition of the population or factors of interest in the study, (ii) identification of study units, (iii) collection of representative measurements, and (iv) a statistical analysis matched to study objectives and data characteristics. An effective data analysis requires fitting the conceptual model of the population under study to the statistical model used to measure and test those populations. None of these tasks are simple. The goal of this chapter is to help you see the statistical issues involved in study design and data analysis and guide you to good decisions. It does not provide everything you need to guarantee a perfect study design or data analysis. The recommendations made here along with your own knowledge of statistics should be sufficient to find appropriate solutions for most simple environmental microbiological studies. For more complex studies, this chapter places you in a better position to take full advantage of the advice you receive when you discuss these issues with a statistically trained colleague. When in doubt, seek expert help to avoid ending up with a completed study that cannot answer your most important questions.

The discussions of study design and statistical data analysis are separated in this chapter, but in reality the two tasks are highly interrelated. The decisions you make when identifying sampling units or replicates, how samples are selected, and what characteristics are measured will drive the data analysis. Inversely, the needs of specific statistical analysis techniques will, in turn, drive certain characteristics of the study design, such as the number of samples or replicates needed. While standard experimental designs and analysis models are available, these "off the shelf" designs will rarely fit your study exactly. You will have to modify these standard designs to fit the particular needs of the study. The changes you make should also be reflected in the associated statistical analysis. Sometimes these changes are so minor that they will not change the analysis plans. Other times, small changes will have a very large impact on the resulting data analysis. For example, theory can help determine how many samples are needed to achieve a specified confidence in the final analysis results, but to effectively use this theory you have to make assumptions. Minor changes in these assumptions can result in large changes in sample size requirements. If your assumptions are later found to be wrong, it will be too late to make changes in the study design but not too late to change analysis plans.

STUDY DESIGN ISSUES

From a statistical point of view, study objectives can be classified into three general types: descriptive (describing some aspect of microbial or environmental populations), associative (examining populations for associations among characteristics), and comparative/confirmative (formally comparing the effects of experimental treatments or testing the significance of differences in characteristics among populations). Studies in environmental microbiology often encompass more than one of these types in their objectives, but typically the study design is focused around one central objective and it is that objective's statistical design type that defines the study design. Study designs to describe some characteristic of a population (such as average biomass, species diversity, or average chemical concentration) or explore associations among a set of characteristics typically specify sampling methods and concentrate on obtaining representative samples from which the responses of interest are measured. Sampling designs are discussed in the next section. Study designs for comparison are of two types. Experiments aim to compare two or more manipulations and focus on repeating each manipulation on replicated samples from the same population in order to get a fair comparison. Comparative surveys, on the other hand, focus on getting representative samples from two or more environmental or microbial populations that are to be compared. Descriptive, associative, and comparative surveys all depend on good sampling designs. We will return to experimental study design at the end of this section.

The broad issues related to sampling protocols are discussed next. Certain aspects of sampling environmental and microbial populations, such as specific methodologies for sampling soil microbes, are discussed as they arise in other chapters of this book. For the basics of sampling, refer to any basic textbook on the topic (6, 18, 23, 34, 41).

Sampling Methods and Descriptive Statistics

Microbiological studies typically target one or more of three general aspects of microbes: (i) community structure, (ii) abundance, and (iii) activities. Related studies generally concentrate on measuring characteristics of the chemical or physical aspect of the environment in which these microbes live. The vast majority of the information about these aspects of microbial populations and associated environments is obtained through taking samples. A sample unit generated by repeating a specified process (i.e., drilling a soil core or collecting an aliquot of water) can also be referred to as a replicate. The word sample in and of itself is often confused in that it can be used two ways. In one sense it refers to a particular individual, replicate, or collection of material selected from the population for measurement, i.e., the physical sample. In this chapter we refer to this as a sample unit, observation, or replicate. In its second meaning, a sample is a collection of individual samples, observations, or replicates. One sample contains many replicates, and when we talk about multiple samples, we are referring to multiple sets of replicates or observations.

What Is a Random Variable?

Any measurable characteristic of a microbial or environmental population is referred to as a random variable or simply a variable if its value varies from sample unit to sample unit. The analytical methods typically employed to generate information on microbial populations result in an assortment of random variable types, from presence/absence measures to quantities such as respiration rate. The type and characteristics of the variables resulting from a particular measurement process have a major effect on final analyses steps.

In general terms, variables are grouped according to the scale of measurement (44). Amount data, such as respiration rates or concentrations of chemicals, are measured on a continuous scale and are referred to as continuous random variables. Some variables such as counts are quantities that are measured on discrete scales and are referred to as discrete random variables. Finally, some characteristics such as color are measured on discrete scales that cannot be easily quantified and are referred to as nominal data. Nominal scales are often used to classify conditions of the system which is being sampled or to identify experimental groups. Between the discrete quantity scale and the nominal scale are ordinal scales which are useful in describing nominal scale measurements for which a relative importance or preference can be assigned (e.g., a preference scale whereby 1 = very poor to 7 = very good). Ordinal scale data are sometimes analyzed as if they were continuous data and sometimes as discrete data.

Is the Population of Interest Clearly Defined?

In sampling, the population of interest is referred to as the target population. The population that is defined by a chosen sampling methodology or protocol is referred to as the sampling frame. For example, suppose the study objective is to determine the diversity of nitrogen-fixing bacteria in soil. The target population of this study would be all nitrogen-fixing bacterial species, of which there could easily be thousands. A culture-based technique for isolating these bacteria (the sampling methodology) might be capable of separating out only a few hundred of these bacterial species. Here the sampling frame is a small subset of the target population. The differences between the population of interest and the target population limit the extent to which subsequent statistical analysis can answer the objectives of the study. The measure of diversity obtained from a sample may not be a good estimate of the total diversity in the soil because of sampling frame limitations and not necessarily a failure of the statistical analysis. Biases in the sampling methodology limit the inferences that can be drawn. This and other issues related to bias, such as confounding, are discussed later in this chapter.

What Is the Sample Unit?

A sample is a collection of sample units (replicates or observations), and the reason we use samples is because the populations that we deal with are too large to measure entirely. It is very important that you understand what constitutes the sample unit for your study population. Any population can conceptually be considered to be made up of individuals, but more often than not in microbial studies, the number of possible individuals is so large and the individuals themselves are so small that collecting and measuring individuals is not always reasonable. For microbial communities, some potential sample units such as colonies or species are themselves composed of large numbers of individuals. In environmental studies, the concept of an individual is not a natural one since often we are dealing with populations that are defined continuously in three-dimensional space, such as the concentration of a toxic metal in the upper 5 cm of the soil surface. In these cases, the sampling units used are artificially defined composites, the form of which is defined by the collection method. A soil core or a water sample is an example of an artificially defined sample unit. Each consists of a collection of individuals that together provide the snapshot of the microbial community at a given spatial location and time. Further processing of the sample is performed to measure such characteristics as microbial community structure, individual species abundances, or various aspects of microbial activity. Sample selection methods are based on the chosen sample unit, and how the sample unit is defined has an effect on the final data type and data analysis. It is the combination of sample unit and sample selection protocol that provides the opportunity to get a good representative picture of the target population and drives all subsequent statistical analyses. A poor understanding of a study's sample unit can lead to major problems in the statistical analysis (see the discussion on pseudoreplication, for example).

What Is a Representative Sample?

A statistically representative sample has sample characteristics, such as mean, standard deviation, distributional form, diversity of species, association among characteristics, etc., that are close to what you would get if you could measure the whole population. Close is a relative term. How close is close enough depends on the objectives of the proposed study. Your objective is to create efficiencies in the sampling protocol by limiting the total sample size but still produce accurate estimates of the population characteristics. Thus, if the study objectives require an estimate of the diversity of a microbial population, you want the fewest samples necessary to provide an estimate that is close to this value. You do not want a selection protocol that will systematically over- or underestimate certain components of the microbial community, resulting in a measure of diversity that does not provide a good index of the true community structure. A selection protocol is unbiased if the chances of overestimation are the same as those for underestimation. With an unbiased estimate, as the sample size increases, the estimate will get progressively closer to the true population value. Biased sampling protocols do not have this characteristic.

How To Get an Unbiased Estimate

Unbiased estimation and hence statistical representativeness are guaranteed if random selection of sample units is used in the selection protocol. There are a large number of ways of incorporating random selection into the sampling protocol. General discussions of random sampling designs are available in the literature (23, 34, 41), and there is a specific discussion of sampling designs for microbial communities in chapter 48. For an example of estimation in environmental microbiology research, please see references 1 and 30.

In simple random sampling, every possible sample that could be selected from the target population has the same probability of being selected. The easiest way to accomplish this is to ensure that every possible sample unit has an equal probability of selection. The most efficient approach is using a randomization method to determine which units are included in the sample, in which each possible sample unit can be assigned a unique numeric identification number, such as when the study area is divided into nonoverlapping square quadrats of equal size. This is much harder if the sample unit is small, such as a soil or sediment core or a water sample. Random selection for spatially identified units typically requires placing an *x*-*y* coordinate system over the study area and then randomly choosing *x* and *y* location values that identify where the cores should be taken (Fig. 1a). A more efficient spatial sampling approach is multistage or nested random sampling in which the *x*-*y* coordinate system may be defined at different scales, from macro to micro with random selection of areas occurring in stages. For example the first stage might see the study area divided into large areas (hectares in size) with random selection of some fraction (say 20%) of the areas for sampling. In the second stage a finer coordinate system that further subdivides the selected large areas into smaller areas is defined. Again, a random sample of these finer defined areas is selected to move to the next stage. Any number of stages can be used to allow final determination of the exact location for selection of the sample.

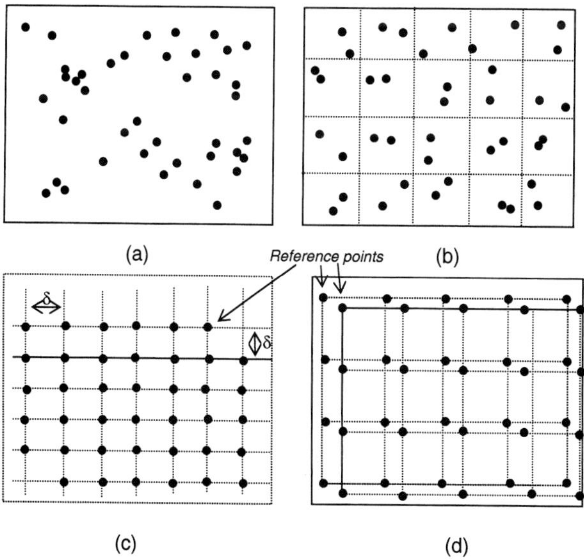

FIGURE 1 Spatial sampling plans, each containing 40 locations. (a) Simple random locations; (b) random locations within cells of a systematic grid (spatial stratification); (c) systematic locations with a random starting point and fixed between-sample distance, δ; and (d) overlapping grid samples of equal between-sample distance and two random starting points.

How Can I Use Information about the Population To Improve the Final Descriptive Statistics?

In descriptive studies there are situations in which limited knowledge of the population under study can be exploited to provide more precise and representative estimates of the statistics of interest. For example, with a spatially defined population there may be factors, such as vegetative cover or soil moisture level, that are known to affect the statistic of interest, say microbial community abundance or diversity. Sample units selected within an area of common vegetative cover would be assumed to be similar. Dividing the total study area into subareas of common factor levels produces units referred to as strata (Fig. 1b). An efficient estimation method would be to use simple random sampling to get a good estimate of the parameter of interest for each stratum and then to use these stratum estimates weighted by the fraction of the study area in each stratum to produce a more precise overall estimate. This is the fundamental thinking behind stratified sampling. Done properly, stratified sample estimates are typically much more precise than what would be obtained from a simple random sample with the same number of sample units. As an added benefit, you also have the information provided by the estimates from each stratum. A stratum can be defined as a combination of multiple factors.

Wouldn't It Be Easier To Simply Select Spatial Samples by Using a Regular Grid?

Selecting from a regular grid of locations creates a systematic sample (Fig. 1c). Systematic samples are efficient in covering the study area and typically also provide a representative picture of average values as well as variability in responses for the area (6). In addition, spatial systematic samples, typically used in conjunction with area stratification, provide the information needed to produce accurate maps. But, if the distance between adjacent samples matches the cycle of a characteristic's repeating pattern, the spatial samples will produce very biased estimates. Moreover, this approach might mask nested or scale effects since it only targets a specific distance interval. For example, assume the study area is an old agricultural field in which row crops were planted on 1-m centered rows. Further assume that the residual from the agricultural activity remains in the field and affects the density of the microbial community of interest. Overlaying a systematic sampling scheme with interlocation distances that are a multiple of 1 m will produce density responses that have less variability than is actually present in the field.

There are two things one should do when using systematic samples. The first is to use two or more overlapping grids (Fig. 1d). The second is to always use randomization in locating the reference point for each grid. Using overlapping grids allows for three or more interpoint grid spacing distances, therefore reducing the potential for biasing of the overall mean and ensuring a more representative estimate of the overall variance. The second adds randomization to the sampling to reduce the effects of designer biases.

What Is a Composite/Pooled Sample, and When Should Composites Be Used?

A composite sample is a sample unit that is created by combining or pooling other sample units. For example, if our objective is to get a good estimate of the average concentration of a toxic metal in a large area of land, individual soil cores (replicates) selected from a smaller geographic area could be physically combined to create one sample unit that is sent for analysis. The benefits of using composite sample units include (i) reduced analysis costs, (ii) better spatial representation and a potentially smaller bias in the estimate of the average, and (iii) values that are expected to more closely follow a normal distribution and hence closely meet the assumptions of standard statistical tests. On the other hand, composite samples yield much less information on the spatial variability of the characteristic and the act of physically compositing can mask important characteristics. In our toxic metal example, by spatially compositing replicate cores we lose the ability to identify locations where the concentration of toxic metal is very elevated. If the objective of the study was specifically to find sites of elevated toxin, we would not use composite sample units. Finally, note that compositing can be done either in the field or in the lab and is not limited to soil sampling. Appropriate use of composite samples can be a cost-effective way of producing precise and accurate estimates of average levels.

Controlled Experimental Protocols and Comparative Statistics

The majority of studies in environmental microbiology involve some form of comparisons since it is through comparisons that we learn where differences exist and/or what factors can influence microbial populations. Experimental designs specify the nature and extent of comparisons that are of interest in a particular study. Study factors are those environmental conditions or treatments that define the populations or the characteristics that are being compared. In comparative surveys, the study factors are typically the characteristics that uniquely identify or distinguish among the two or more populations of interest. An example would be a study to compare average microbial diversity at three sites that are defined by different vegetative communities. Vegetative community type is the experimental factor, with site being one level or replicate of the factor. Here multiple sites would be used to allow estimation of the uncertainty of the comparison of average diversity under replicated conditions. In the case of a manipulative experiment, the different manipulations are called the treatments. Treatments are defined by the levels of one or more experimental factors. For example, we may be interested in how the abundance of methane-producing microbes (methanogens) change as a function of fertilization in rice paddies. We might define six experimental treatments by combining each of three levels of nitrogen fertilizer (factor N) with two levels of phosphorus (factor P). The objective of the study may simply be to compare average abundance among the six treatments, essentially testing whether there is no change in average abundance across treatments. These treatments might be randomly applied to sample units (e.g., cores of soil with the soil all coming from the same population) that may also be called replicates or simply experimental units. For basic information on experimental designs, refer to any of the standard textbooks (18, 35, 36, 44).

What Are the Three Components of a Good Comparative Experimental Study?

All good comparative experiments have the same three components: randomization, replication, and blocking. Let us take each of these in turn.

The definition of randomization depends on the type of comparative study being performed. In a comparative survey, randomization refers to the random selection of sampling units (replicates) to be measured. Any of the random sampling designs discussed above can be used to identify individuals for measurements for each population being compared. The goal of randomization in this case is to ensure that the samples being compared are representative of the populations you have targeted for comparison. In an experimental study, randomization refers to the random assignment of treatments to experimental units (replicates). The goal of the randomization in this case is to ensure that subjective biases are kept out of the treatment comparisons as much as possible.

As you see from the above discussion, the definition of a replicate also depends on the type of comparative study. In comparative surveys, the number of replicates can be equated to the number of sample units collected for each group. The goal of replication is to ensure sufficient measurements from appropriately defined sample units to allow for precise estimation of the mean and variance from each of the populations being compared. In an experimental study, the number of replicates is defined as the number of experimental units that receive the same treatment. The goal of replication here is to ensure sufficient measurements to estimate the mean and variance for each treatment group with sufficient precision to establish a high probability of finding significant differences among the group means if the differences actually exist or of finding no significance when in fact there are no differences in the group means.

Blocking in comparative studies is a tool used by researchers to reduce some of the known sources of variability in measurements that can weaken treatment comparisons and which represent factors that are extraneous to the study objectives. By blocking replicates into groups that are homogeneous with respect to an important extraneous factor, the researcher hopes to produce blocks of replicates that would be likely to produce very similar responses if they were subjected to the same treatment. A typical study protocol would have all treatments represented in each block. At the analysis phase we remove the block "effect," thereby removing the effect of the extraneous factor leaving residual values that have reduced total variability in measurements and hence a higher probability of identifying significant differences among the treatment "effects." For a comparative survey, a priori stratification of the population is a form of blocking. In experimental studies, blocking is related to the organization of experimental units into groups of homogeneous units. We strive to create homogeneous blocks with sufficient experimental units to accommodate all treatments.

Note that typically blocking factors are conditions that are known to affect the responses of interest but are also assumed to not interact with the study treatment factors. That is, we expect the relative differences among treatment responses to be roughly the same within each block. An extraneous factor that produces relative differences among treatment responses that are very different at different levels of the extraneous factor is referred to as a confounding factor. Failure to identify and account for confounding factors is a major cause of failed studies. Confounding factors cannot be removed by blocking; they either have to be included in the study design as another treatment factor, fixed at a constant value for all replicates used in the study, or measured as a covariate or additional response. For example, suppose that in the experiment examining the change in abundance of methane-producing microbes (methanogens) as a function of fertilization in rice paddies we manipulate only the nitrogen level in the soil. A potential confounding factor for this experiment would be soil phosphorus level since it is known that nitrogen and phosphorus interact to affect microbe abundance. Either the soil P level would have to be adjusted to one common value for all experimental units or the P level should be utilized as a second treatment factor. A third option is to directly measure P level at the same time as microbe abundance is measured, with P level being used as a covariate in the subsequent statistical analysis.

What is Pseudoreplication and How Do I Avoid It?

In comparative studies, pseudoreplication (13, 37) arises typically when multiple measurements (subsampling units) made on the same replicate are used as if they were actual replicates. If the replicate of the study is a plate of agar, then each plate receives only one treatment. Suppose measurements are made on three bacterial colonies randomly selected from each plate. The replicate in this case is the plate, not the colony. If you make the mistake of identifying the colony as the replicate in the subsequent data analysis, you will be guilty of using pseudoreplicates. If at the design phase you made the mistake of wrongly identifying the measurement unit as the replicate, you might be tempted to use only one agar plate for each treatment, assuming that the multiple colony measurements from each plate are your replication. This would be disastrous since you would in essence have a study with no true replication and hence would not be able to test statistical hypotheses on these data.

Pseudoreplication is most easily identified and avoided in experimental studies in which the researcher has control over what is the experimental unit and what is the measurement unit. It is more difficult to identify pseudoreplication in comparative surveys. In these studies, the determination of a true replicate depends on the extent of the inferences to be made from the results of the study. For example, suppose you are interested in comparing microbial community diversity between pasture and pine forest habitats. If multiple soil core samples are taken from one pasture and from one adjacent pine forest, the inferences from the statistical tests on the resulting data will be limited to pastures and pine forests in the immediate area to the study (tens of hectares). On the other hand, multiple (replicate) pastures and multiple (replicate) forests must be sampled if you intend the results to apply to a broader geographic area (say hundreds or thousands of hectares) or across wider categories (say pastures in the subtropics). In the latter study, the sample unit is the pasture or forest and the soil cores are the measurement units. The soil cores would be pseudoreplicates if they were used directly to test the hypotheses of differences between pasture and forest microbial communities. This is primarily because the inferences are to be made to populations consisting of pasture and forest individuals, and hence the analysis requires estimation of variability among pastures and forest replicates.

What Other Things Should Be Considered When Designing Comparative Studies?

As the complexity of the comparisons increases, the need to get expert advice on study design increases as well. Experimental studies with multiple factors, situations in which measurements are taken over time (referred to as repeated measures data), and designs that call for nested sampling require careful design and complex analysis models. Reviewing your study design with a statistically trained colleague can help not only sort out the design but also outline an analysis approach that matches the design and resulting data characteristics.

See reference 14 for an example of a controlled experiment in environmental microbiology examining the effects of addition to soil plots of carbon, nitrogen, and phosphorus (the factors) on the soil microbial community.

STATISTICAL ANALYSIS

With a well-planned design and data collection complete, the next step in the process is to analyze the resulting data. The first step in any data analysis is to decide on the objective and possible subordinated objectives which the experiment was designed to achieve. In one sense this is not a statistical issue since it relates directly to models, conjectures, or paradigms that drive the study. On the other hand, at some point the model objectives have to be related to one or more sets of measures (statistics) that summarize the data collected and help address the questions raised by the study objectives. Effective translation of study objectives to statistics of interest is critical to the success of the study.

Statistics are simply functions of the data collected. Although the potential pool of useful statistics is large, in general statistics can be organized into three categories: (i) statistics that are used to describe the distributions of measurements/variables (descriptive statistics), (ii) statistics that describe the associations between and among variables, and finally (iii) statistics that are useful for formally testing hypotheses (test statistics). All three categories are routinely used in environmental microbiology, such as mean plate counts (distribution), correlation coefficients between respiratory activity and microbial abundance (association), and tests to determine if certain concentrations of a pollutant inhibit microbial growth (*P* values, testing hypothesis). We first discuss these three groups of statistics separately and then discuss how they are typically used in data analysis.

Before moving on to discuss statistics, we must define a few terms more carefully. Common usage and statistical use of these words will often differ and can lead to confusion. A variable is a quantity that may take on more than one value, from which follows variability, which means the potential to vary, and variate for a set of observed values of a variable. Variability is the realization of an abstract random process of which the variate is simply one expression/realization. The word variation is used to describe the observed population or the population to be sampled. A parameter is a quantity that is constant within the scope of the sample or experimental study. An observation is the actual observed value of a variable from one sample unit.

Descriptive Statistics

The goal of data collection, be it in a sampling study or an experimental design, is to obtain a set of measures from a population to gain insight into how values for a particular population characteristic vary from sample unit to sample unit. Conceptually, each characteristic is a random variable and one possible goal is to describe the associated distribution of measurements on that random variable. A distribution can be thought of as a function that tells us something about the values that populate the distribution, such as which values are most common, what is the range in values, etc. Descriptive statistics are almost always some function of the observed values that tells us something about the expected distribution of the underlying variable. Once we review some of the general types of data distributions, we will discuss how, for a given set of data, one determines the most appropriate distribution to use.

Statistics as defined previously can take the form of some graphics and tables. Graphs are pictorial descriptions or summaries of data and are therefore statistics. Plots, charts, and other diagrams are very useful tools for understanding data distributions and making the correct choice of numerical statistics. Tables of counts of individuals in specific response classes are very effective descriptors of the distribution of a discrete random variable.

Finally, it is always of interest to know something about the center and spread of a distribution. The statistic most commonly used to describe the center of a distribution is the mean. The statistic most commonly used to describe the spread of the distribution is the standard deviation. In many cases, the sample mean and standard deviation are the best estimates for the population mean and standard deviation. As we will see, for some types of data, other, better estimates can be derived once the form of the data distribution is known.

Discrete Random Variables

The simplest distribution type is the Bernoulli distribution, which is for a random variable that can take only two possible values, coded as 0 and 1, where 0 might represent a "false" or "no" response and 1 a "true" or "yes" response. By definition, the Bernoulli distribution parameter, P, is the proportion of all possible sample units that have the characteristic of interest (response equal to 1). An appropriate statistic for a Bernoulli random variable would be the simple average of the 0/1 responses, since this average represents the sample proportion of "true" or "yes" responses.

The Bernoulli distribution is just one of a number of distributions appropriate for count data for which a table is an appropriate descriptive statistic. Another distribution that is used very often is the binomial distribution. The binomial describes a random variable which can be thought of as the sum of a number of Bernoulli random variables. For example, suppose $n = 20$ small core samples are taken in a specific area. For each core you determine if a target species of microbe is present or not, coding the result for each core as a 0 (absent) or 1 (present). The total number of cores that test positive for the microbe follows a binomial distribution. The appropriate statistic for this distribution may also be the sample mean, which estimates the expected probability of observing the microbe in a randomly selected core.

The Poisson distribution describes counts of individuals or events in fixed-sized areas called quadrats that are randomly located in space. For example, determination of most probable number of microbes in a water sample might involve placing a small quantity of the solution on a microscope slide printed with a fine grid. If you counted the number of microbes in 30 randomly selected grid cells, the resulting counts would be expected to follow a Poisson distribution. The average count would be the statistic used to describe this distribution.

A geometric distribution is often used to describe the abundances of species in a resource limited environment. The abundance of each species, ranked from most abundant to least abundant, follows a geometric series (25).

Typical shapes for these distributions are given in Fig. 2. Also included is the relationship between the parameter that described the distribution and the expected mean and variance of that distribution. More information on discrete distributions can be found in reference 8.

Continuous Random Variables

Continuous random variables can describe amounts such as weights, heights, biomass, time, etc. There are many different functional forms for distributions of continuous random variables, but only a few are used extensively. The foremost continuous distribution is the normal (Gaussian), whose shape and spread are defined by the mean and variance. Related to the normal is Student's t distribution, which finds its greatest use in the two-sample statistical tests for the mean. Both the normal and Student's t are symmetric distributions; the shape of the distribution to the left of the mean is the mirror image of the shape of the distribution to the right of the mean.

Not all continuous distributions are symmetric. Some random variables are bounded; for example, concentration amount cannot be negative, and this constraint produces a build-up of probability mass close to zero, resulting in a distribution that is said to be skewed to the right. Examples of right-skewed distributions are lognormal, Weibull, and type I extreme value; they are typically used to describe random variables for concentrations, minimum heights, or minimum times. The gamma distribution is typically used to describe the time or distance to an event, is bounded by a minimum, and hence is right skewed. The F and chi-square distributions are used as distributions for test statistics; they are both bounded below by zero and hence are both right skewed. Less common are random variables that are bounded by a maximum possible value and produce left-skewed distributions, like the type II extreme value distribution, which is used for example to describe survival times. Finally, an extraordinary kind of distribution, the beta, describes data that are bounded both below by a minimum value and above by a maximum value and hence might be used to describe the distribution of a proportion value, such as percent vegetative cover.

Like discrete distributions, continuous distributions are defined by one to three parameters. For symmetric distributions, the two parameters define the center or mean and variance or spread of the distribution. For nonsymmetric distributions, the parameters are more likely to describe particular aspects of the shape of the distribution and hence are less easy to describe. Typical shapes for symmetric and nonsymmetric distributions are given in Fig. 3 along with how their parameters relate to the population mean and variance of the distribution. More information on continuous distributions can be found in reference 8.

Why Are Statistical Distributions of Concern to an Environmental Microbiologist?

As can be seen from the equations in Fig. 2 and 3, best estimates for the mean and variance depend on the distribution type most appropriate for the data of interest. If you can determine the form of the statistical distribution of a random variable in a biological or physical population, you have specified everything there is to know statistically about the object or process being investigated. Just relying on the sample mean and standard deviation as the only set of descriptive statistics limits your understanding of the random variable and can lead to suboptimal statistical hypothesis tests, i.e., you are not getting your full money's worth from the information present in your data.

How Do I Determine Which Distribution My Data Follow?

This is a difficult question to answer, as the ability to determine which form of distribution applies to your data depends on a number of factors, and a major factor is the number of observations that you have for that random variable. For some distributions, as few as 10 observations from random samples are enough to determine the distribution; for others, especially distributions that are nonsymmetric,

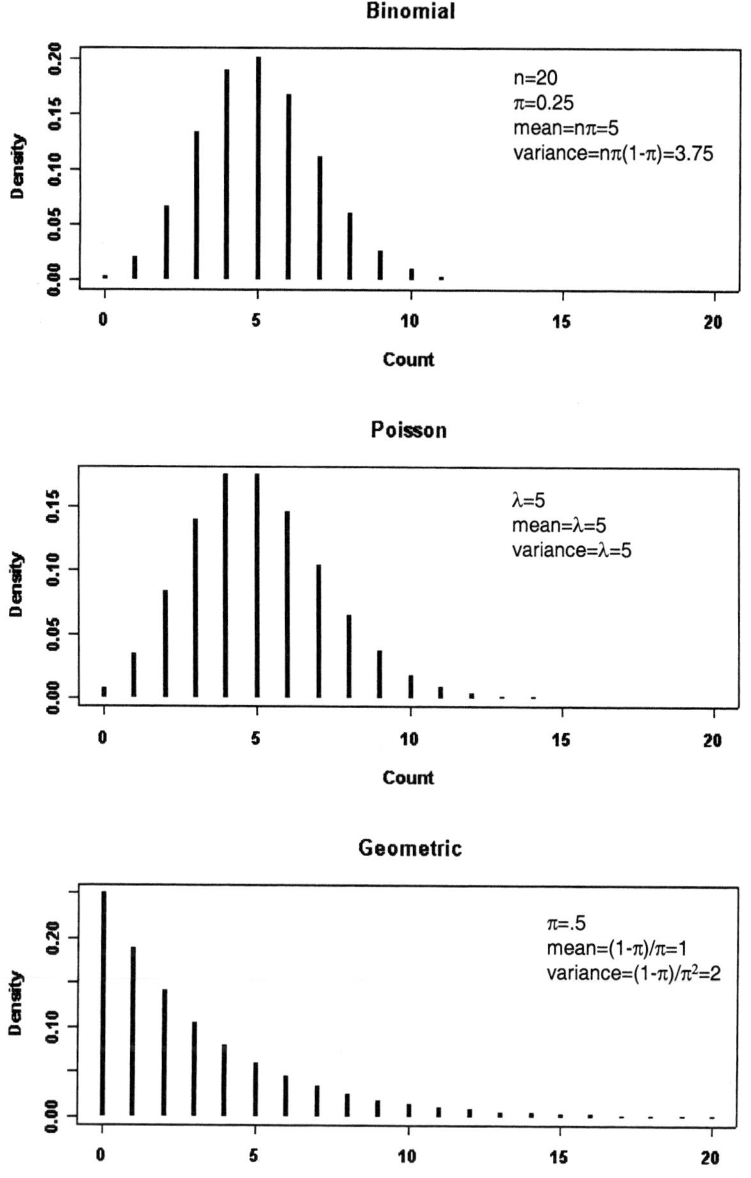

FIGURE 2 Typical shapes for common discrete statistical distributions. Distribution parameters: n, number of experiments, ranging from 2 to infinity; π, probability of success, ranging between 0 and 1; $\lambda > 0$, Poisson mean.

30 or more observations are needed. For discrete random variables, appropriate sample sizes are 100 and above.

There are two general approaches to determining the appropriate distribution forms from a given set of data. For continuous (amount) random variables a very powerful statistical analysis tool is the quantile-quantile plot. For discrete distributions and to some extent also continuous distributions, a histogram is used.

A quantile plot is simply a very specific kind of scatterplot in which the quantiles, derived from the distribution we assume the data are drawn from, are plotted on the x axis and the rank-ordered observations are plotted on the y axis. We denote the sorted list of n observations of the response by $y_{[1]} \leq y_{[2]} \leq \ldots \leq y_{[n]}$. Associated with the sorted observations are the quantities $p_i = (i - 0.375)/(n + 0.25)$ for $i = 1, 2, \ldots, n$, which are referred to as the plotting positions. These positions are used to locate n points on equal intervals over the range 0 to 1 exclusive of 0 or 1 as a plotting position. The candidate cumulative probability distribution function for the random variable x is denoted by $F(x \mid \theta)$. The symbol θ is there to remind us that the particular form of this distribution may depend on some unknown parameters. The quantiles for the candidate distribution are the values, $x_{[1]} \leq x_{[2]} \leq \ldots \leq x_{[n]}$ such that $F(x[i]) = p_i$. These values are not easy to compute, but most statistical analysis packages and even spreadsheet programs have functions that compute these values. Finally we plot the point pairs $(x_{[i]}, y_{[i]})$ in a scatter plot. If the data truly follow the candidate distribution function, the plotted points should effectively resemble a straight line. If the wrong distribution is used, a systematic deviation of the points from a straight line will be observed. The value for distribution parameters, θ, used in

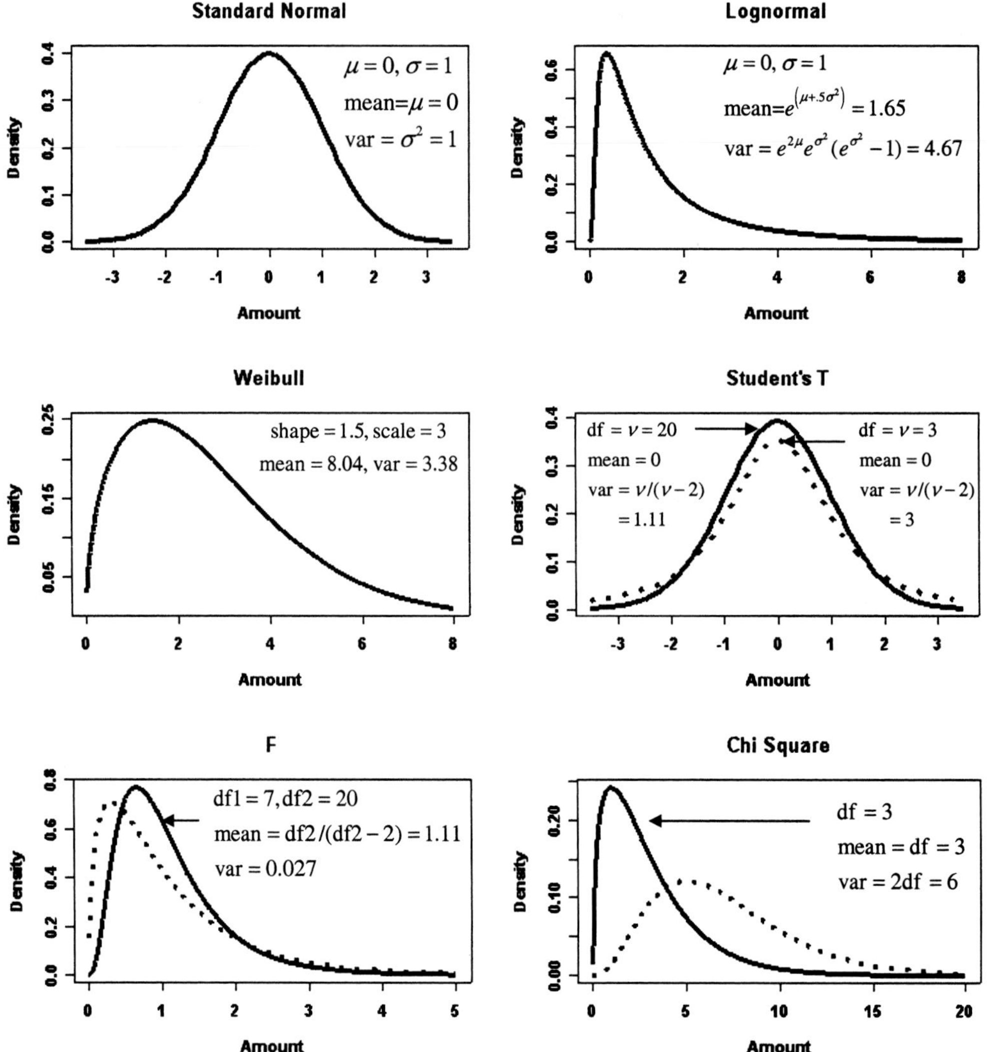

FIGURE 3 Typical shapes for common continuous statistical distributions. Distribution parameters are displayed as Greek letters. For the normal and lognormal distributions, μ and σ are the mean and standard deviation controlling the center and spread, respectively. The Weilbull distribution is controlled by a shape and a scale parameter. The Student t, F, and chi-square distributions are controlled by one or two parameters referred to as the degrees of freedom, $1 \leq$ df, df1, df2 $<$ infinity. For these distributions, changing the degrees of freedom changes both the shape and spread of the distribution as demonstrated by the dotted lines.

the computation of the quantiles, typically is the best estimate from the available data. Further discussion on quantile plots can be found in references 8 and 40.

Examples of quantile plots are given in Fig. 4. Note that when normally distributed data are plotted against normal quantiles the points follow a straight line but when lognormal data are plotted against normal quantiles a recognizable nonlinear point pattern emerges.

Histograms are essentially graphical depictions of frequency tables. Histograms can be produced with different scales for the vertical axis, with the choices being the frequency, relative frequency, or density values. Each separate bar in a histogram is related to a unique value for a discrete random variable or a bin for a continuous random variable. Thus, frequency histograms assign the number of sample individuals having a specific discrete value or a continuous value in the bin to the height of the bar. A relative frequency histogram assigns the relative frequency in a bin to the height of the bar. For a discrete random variable, the relative frequency histogram also describes the expected probability of obtaining the particular response. Density histograms can be used only with continuous random variables and have the vertical axis scaled such that the total area of all bars sums to 1.0. Density histograms are a type of nonparametric probability density function estimate. Examples of a frequency histogram and a density histogram for normally distributed data are given in Fig. 5.

Another graphic that provides distributional information but that also facilitates comparisons among groups is the box plot (40). A box plot (Fig. 6) can identify and facilitate comparisons of a number of distribution statistics, including the median, the upper and lower quartiles (75th and 25th percentiles), and the interquartile range (IQR). Variations on the box plot abound, with some providing

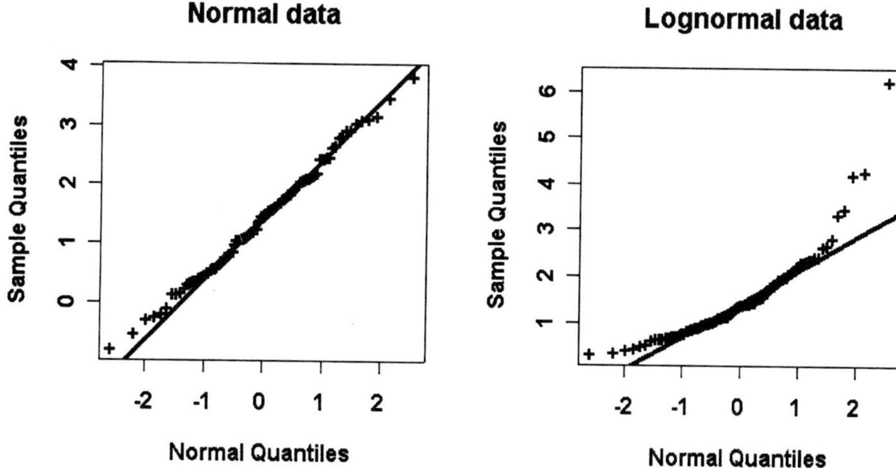

FIGURE 4 Examples of normal quantile plots. Closeness to the straight line is a measure of the normality of the sample data.

additional information on potential outliers and/or displaying the sample mean and sample standard deviation.

Exploring Associations

In natural systems microbes are found in heterogeneous environments. A large number of studies have as their goal the characterization of the complex and dynamic interactions between characteristics of the microbial community and levels of various characteristics in the surrounding environment. These biotic and abiotic characteristics interrelate and vary over spatial and temporal scales. The goal of an association analysis is to identify and quantify these interrelationships.

How Do We Analyze Associations among Continuous Variables?

The simplest type of association analysis involves the calculation of a correlation coefficient to quantify the strength of the relationship between two continuous variables. A number of correlation coefficient measures have been developed, but the two most commonly used are the Pearson product moment correlation and the Spearman rank order correlation. The Pearson correlation measures the strength of the linear relationship between two variables, obtaining a value of 1 for an exact linear relationship of positive slope, −1 for an exact linear relationship of negative slope, and 0 for no linear relationship at all. For $\{x_i, y_i\}$ pairs of data the Pearson correlation is computed as

$$r_{xy}^{p} = \frac{\left(\sum_{n}(x_i - \bar{x})(y_i - \bar{y})\right)}{\left(\sqrt{\sum_{n}(x_i - \bar{x})^2 \sum_{n}(y_i - \bar{y})^2}\right)}$$

The Spearman rank order correlation is a nonparametric measure of association that is based on the ranks of the data. Essentially, the x_i values are replaced with their rank in an

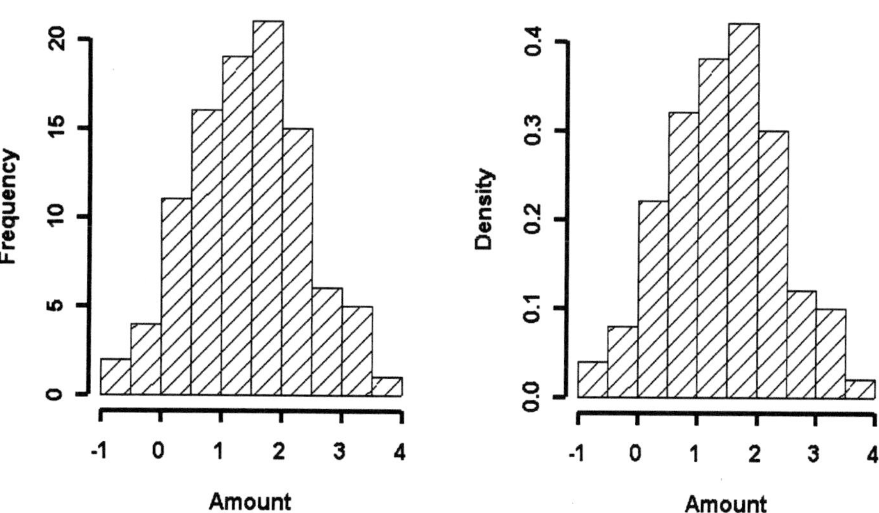

FIGURE 5 Frequency and density histograms for normally distributed data. Note that the shapes of the distribution are similar and only the vertical axis scale is changed.

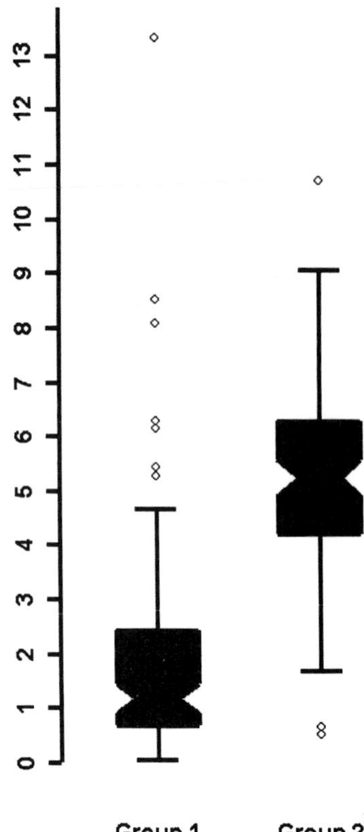

FIGURE 6 Comparative box plots indicating the median (middle of notch), upper and lower quartiles (solid box top and bottom), interquartile range (IQR = height of box), upper and lower fence values (line terminators = median ± 3 IQR), and potential outliers (circle dots).

ordered list of the *x* values and the y_i values are replaced by their rank in an ordered list of the *y* values; these rank values are then used in the Pearson correlation above to produce the Spearman correlation, r^S_{xy}. When looking at more than two variables, a correlation matrix composed of the correlation coefficients between every possible pair of variables is used. In certain multivariate methods, the correlation matrix is the fundamental starting point. These methods are often used to describe the underlying structure of the associations among characteristics represented by these variables. Multivariate techniques, such as principal components analysis and factor analysis (16, 24, 38), are aimed at explaining high correlations among variables in terms of a smaller set of unobservable factors, typically referred to as latent factors. These latent factors are actually linear combinations of observed variables that produce an index value that can be used to order the sample units on association scales. They are dimension reduction techniques, in that they identify a couple of important dimensions in the data that capture most of the association structure in the data. A plot of the data (Fig. 7) in the two or three most important dimensions (principal components or factors) should identify any inherent or hidden associations or grouping in the data.

How Do We Analyze Associations among Categorical Variables?

Analysis of the association structure among categorical (nominal scale) variables is much more difficult than with continuous variables, primarily because there is no index, such as the Pearson product-moment correlation coefficient, to capture the degree of association. To measure association among categorical variables, we must rely upon the chi-square statistic as measured for cross-tabulations of the categorical variables of interest. One method analogous to principal components is multiple correspondence analysis (CA) (22, 39). This analysis can result in a plot of principal axes that help depict the correspondence between levels of one categorical variable to levels of another categorical variable. While there are statistical packages that easily produce this analysis, expert help may be needed in properly interpreting the final graphs. When there are only a few categorical variables to study, a properly designed mosaic plot (10) should also be examined. Variations of correspondence analysis include the following.

- Detrended correspondence analysis, which is employed to correct for some of the flaws that have been found in correspondence analysis (12)
- Two-way indicator species analysis (TWINSPAN), a classification method based on dividing the CA axes (11)
- Canonical correspondence analysis, an approach to associate continuous variables (such as environmental factors) to CA scores generated for locations or categories (39). This method is useful in that it allows a certain category (e.g., sites or environments) to be classified by the categorical variables (presence/absence of species). This characterization can then be extended through regression to identify which continuous variables are important in the categorization of the classes.

How Do I Relate One Group of Continuous Variables to Another Group, Such as When Comparing Certain Microbial Population Characteristics to Environmental Characteristics?

A very common study goal is to examine the association of environmental characteristics (refer to these as the X group) to microbial community characteristics (the Y group) (31, 32). If the characteristics are continuous in both groups, a canonical correlations analysis (38) can be used. In canonical correlations analysis we seek one or two linear combinations of the X group variables that are highly correlated with corresponding linear combinations of the Y group variables. These sets of linear combinations, referred to as the canonical correlation factors, are found in such a way that the first pair of factors has the highest correlation, the second pair has the second highest correlation, etc. A plot of the first pair of canonical factors typically demonstrates the degree to which variables in X and Y covary.

Canonical correlations analysis is related to multiple regression models (2, 16) and structural equation models/path analysis (17, 19). In multiple regression analysis we attempt to explain the variation in one Y variable by using a linear combination of the X variables. A structural equation model attempts to simultaneously explain the variation between and among the X and Y variable sets with a set of common linear relationships (called latent factors). Structural equation modeling is sometimes referred to as confirmatory factor analysis because one has to explicitly specify the number and form of the linear relationships among the X and Y variables before they can be fit. While multiple regression models can easily accommodate categorical variables in the X variable set, a different model approach is needed if the Y variable is categorical. The equivalent of multiple regression for a categorical response

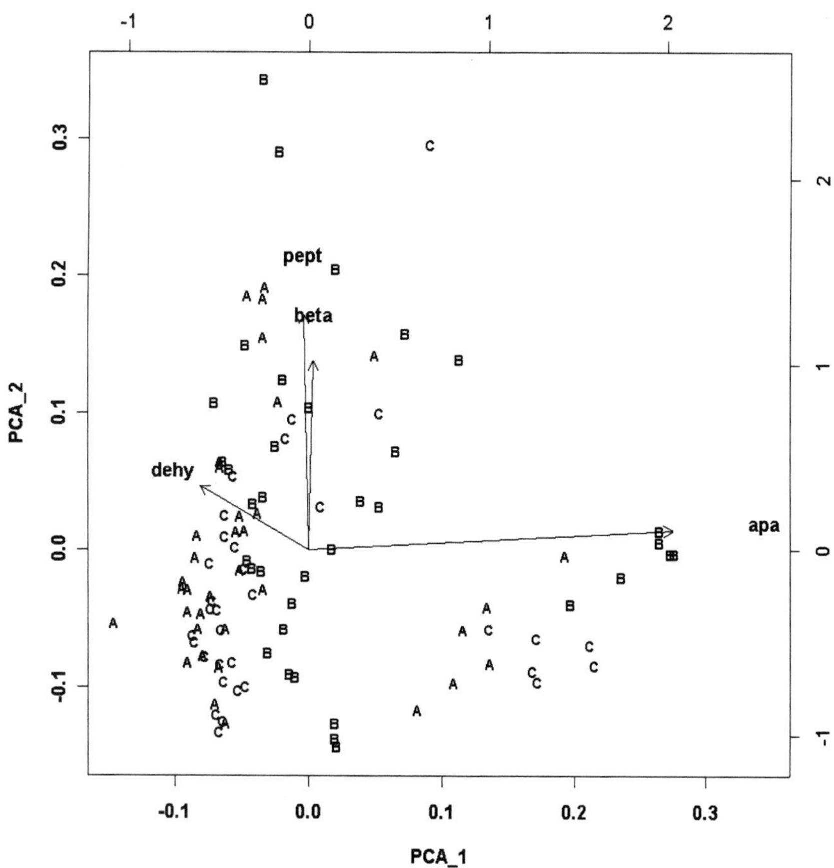

FIGURE 7 Biplot of the first two principal components of a 4-variate analysis which is known to have three groups, identified as A, B, and C.

is multiple logistic regression (16, 36). Incorporating categorical variables in the X and Y variable sets into a structural equation model analysis is not easily accomplished. An easier and more promising approach to this task is via neural network models (31). Neural network models are very flexible, allowing nonlinearity in the relationship between X and Y variable sets as well as in the definition of latent factors. The building and interpretation of such models are not straightforward tasks, and you should consult an expert early in the analysis process.

What If I Think the Association Is Expressed through Grouping of the Data?

It is very often interesting to see if the microbial or environmental data obtained via random samples from an environment naturally form groups. The focus of the association analysis changes from that of looking at correlations among variables to looking at how far the individual sample units are from each other in the multidimensional space defined by the variables measured. In statistics, the process of forming these groups comes under the heading of cluster analysis (28). In the biological sciences, cluster analysis may also be referred to as classification analysis. There are a large number of ways of clustering observations. Cluster method names have modifiers like model-based, hierarchical agglomerative, hierarchical divisive, k-means, or nonparametric density method; names that describe the approach taken in creating clusters. For data with clearly defined groups, all of these approaches produce the same groupings. For data with less well defined groups, different methods produce slightly different groupings. If you are new to cluster analysis, you should start your study with a k-means clustering simply because the approach is simple to understand and results are easy to interpret.

All clustering methods depend on how you decide to measure the distance between observations. If all the variables are continuous, a Euclidean measure or Mahalanobis measure (16, 24) is typically used. If all the variables are categorical, a number of similarity measures have been proposed which measure the degree of agreement between two observations in the individual categorical levels. When the variable set includes both continuous and categorical measures, either the categorical measures are converted to a continuous scale (say by using a correspondence analysis factor in place of the continuous scales) or two separate cluster analyses are performed and the results are compared.

Finally, there are methods, such as the previously mentioned TWINSPAN protocol (11), that allow you to simultaneously group observations and variables. This analysis allows you to see which observations are similar to each other and on which variables they have similar responses.

Do Any of These Methods Help Me Determine Which Particular Characteristics Help Define the Final Groups?

Conceptually, we typically assume that some subset of characteristics can be used to define the organization of microbial communities in the environment. The TWINSPAN method is useful in identifying variable sets that help characterize groups. The other clustering methods do not have the built-in tools for dimension reduction. As a follow-up to a cluster analysis it is sometimes very useful to perform a discriminate analysis (4, 16) of the groups generated by the cluster analysis. Discriminate analysis methods specifically look for functions of the variables that can effectively discriminate among the groups. There are stepwise variable selection approaches available for discriminate analysis that allow you to determine a more effective (parsimonious) subset of variables with which to define clusters. Sometimes discriminate analysis is also referred to as classification analysis since one use of the final model resulting from the analysis is to predict the group or class to which a new observation should belong. The confusion in the use of the "classification" name is avoided if you use the "cluster" and "discriminate analysis" terms instead. Finally, some of the more formal hypothesis tests discussed in the next section can be used to determine if particular variables have different distributions for the different groups. Any variable that does not differ in distribution across the groups is a good candidate for removal from the cluster analysis.

More-complex multivariate statistical techniques are available and are increasingly being employed to generate associations within measures of microbial community structure, abundance, or presence. For example in reference 20 a form of unsupervised neural networks (self-organizing maps) is used to establish associations between bacterial communities and Biolog (Biolog, Inc., Hayward, Calif.) microplate results (sole carbon utilization). Neural networks have also been employed (31) to generate associations between the distributions of certain genes (nitrate reductase and dissimilatory sulfite reductase) and environmental factors. However, models of this complexity require not only large data sets for parameter estimation but also a separate data set to validate the results.

Testing Hypotheses

Environmental microbiology is by definition and by designation a field of study that crosses disciplines. It often combines observational work with classical experimental research. The advent of more-powerful measurement techniques and tools, such as molecular probes, has allowed environmental microbiology to become more of an experimental science. Practitioners are increasingly capable of testing theories regarding organization, composition, abundance, and activities of microbes in their natural environments. These tests, whether carried out in controlled settings, e.g., in specially constructed micro-, meso-, or macrocosms, or in natural settings, often require complex statistical processing. Researchers often fail to examine their data for violations of the assumptions that are required to hold if the statistical tests are to be valid. In addition, the adverse effects of missing data, data lost due to sampling or biological limitations, on estimation and testing procedures often go unrecognized. The need for adequate replication can also be hidden by budget constraints, especially in large-scale environmental experiments.

How Do I Determine Which Statistical Test Should Be Used?

There is a large pool of statistical tests available in the literature. Naïve users of statistics believe that there are just a select, few tests that they are required to understand and that these tests will adequately cover all the situations they will encounter. However, the extensive pool of methods is the result of a large number of special situations in which the basic statistical tests perform very poorly. So the best way to determine which statistical test to use is to properly establish the correct conditions for the test. This requires assessing the following questions within the context of the study design and objectives.

1. Are the data measured on a continuous, discrete, or categorical scale?
2. What is being compared in the test? More specifically, what statistic is being used? Is it a mean, a median, a variance, a correlation, the predictions from a fitted model, or some other function of the data (e.g., species diversity)?
3. What are the factors, and what levels (i.e., groups) within these factors are being compared?
4. What is the null hypothesis and what is the associated alternative hypothesis?
5. What assumptions am I willing to make about the data and/or assumptions that are likely to hold? In particular are the distributions of responses in the comparison groups similar in shape? In variability? Can the central limit theorem (CLT) be expected to hold? Are there likely to be large outliers?

Often the best way of identifying the appropriate statistical test is to research the published literature for similar studies that might generate answers to the above questions. Using their statistical testing approach as a starting point, you will be able to research the statistics literature for similar tests. This approach also greatly facilitates the ability of statistically trained colleagues to help you determine an appropriate test. However, be prepared to find that other researchers might have slightly different answers to the above questions and often come to different conclusions as to which is the most appropriate test (5, 15).

What Exactly Are Null and Alternative Hypotheses?

Every statistical test has four parts (36, 44), of which the null and alternative hypotheses statements are two. Of primary interest to the researcher is the alternative hypothesis. The alternative hypothesis is a statement of what the researcher is hoping to show, for instance, that there are differences among the means of the treatment groups, that the different populations have different means, medians, variances, correlations, diversities, spatial patterns, etc. The null hypothesis is often the complement of the alternative hypothesis. If the alternative hypothesis states that there are differences, the null hypothesis states that there are no differences.

The other two parts of a statistical test consist of (i) a test statistic whose sampling distribution is known when the null hypothesis is assumed to be true and (ii) identification of which values of the test statistic will lead to the rejection of the null hypothesis. The sampling distribution of any statistic is typically defined via a theoretical argument. The derivation of most sampling distributions is usually linked in some fashion to the CLT (5, 36, 44).

What Are the Sampling Distribution of a Statistic and Its Relationship to the CLT?

Suppose we define a population of a very large number of sample units. From this population we can select a random sample of n sample units and from this sample compute the value of a statistic (descriptive, associative, or test statistic). Now suppose it were possible to record the value of the statistic for every possible sample of n units from the population. The larger the initial size of the population, the larger the number of possible samples will be. Because the sample (observed) value of the statistic varies from sample set to sample set, the statistic behaves as a random variable and therefore has a distribution. The distribution of the target statistic is called its sampling distribution. In the real world, rarely more than one set of the sample units is selected, generating typically only one realization of the target statistic. We cannot directly define the sampling distribution of this statistic from one sample.

The CLT allows us to use the information from the one set of sample units to define the sampling distribution of the target statistic, as long as the target statistic is a mean or function of the sum of observed values. Essentially, the CLT says that if you have or use a (descriptive, associative, or test) statistic that is an average or sum of observed values of a variable, with a distribution with finite mean and variance, then as the size of the sample increases, the sampling distribution of this statistic will approach that of the normal (Gaussian) distribution. A more useful form of the CLT states that if we normalize the statistic by subtracting the population mean and dividing it by the square root of the population variance divided by the sample size (i.e., the standard error), this normalized statistic will approach a standard normal distribution (e.g., one with a mean of zero and a standard deviation equal to one). For small-to-moderate sample sizes ($3 < n < 40$), the sampling distribution of the normalized statistic is closer to a Student's *t* distribution whose shape parameter, the degrees of freedom, depends on the number of observations used to estimate the statistic less the number of sample means (or sums) that have been estimated.

The CLT is crucial because with it we can develop statistical tests using only one sample. Without it, researchers would have to collect many repeated samples from the sample population in order to directly estimate the sampling distribution of the target statistic. How good the standard normal or Student's *t* distributions are for specific target statistics typically depends on whether the data come from continuous, symmetric population distributions or not. The less the data distribution looks like a normal distribution, the less the true sampling distribution looks like a normal for small sample sizes. For many statistics, the fit to a normal or Student's *t* distribution is quite good for even small sample sizes.

In recent years there has been a trend to replace the CLT-derived standard normal or Student's *t* sampling distribution with a sampling distribution that is derived via sampling with replacement from the sample data. This approach is referred to as bootstrap estimation or testing (7, 27). Statistical inference based on the bootstrap replaces the CLT-derived sampling distribution with a data-derived sampling distribution, essentially replacing theoretical derivation with computationally intensive sampling. The method works best for small-to-moderate sample sizes and has strong supporters. In practice, bootstrap estimation and tests work as well as the CLT when the data come from symmetric distributions (the best case for the CLT). The bootstrap works better than the CLT in those situations in which the distribution of the underlying data is very skewed or has other characteristics such as bimodality. Since computational complexity increases dramatically as the underlying statistical tests become more complex, bootstrap estimation and testing have been limited to the more simple sampling and experimental studies.

In the section on sampling methods we defined a composite sample unit as a unit that is created by combining or pooling other sample units or replicates. When creating composite sample units, we are essentially physically creating a mean. Thus, the values obtained from composite sample units should be more normally distributed than are the values obtained from the individual replicates (which you do not actually measure) because we are using the CLT. For this reason, composite sample data are more likely to satisfy the distributional assumptions for parametric statistical tests.

What Do Accuracy and Precision Have To Do with the Sampling Distribution of a Statistic and Associated Statistical Tests?

A statistic is a function of the data that is used to estimate a particular population parameter and/or as the foundation for a statistical test. A natural question is to ask "How accurately does this statistic measure the true (but unknown) population parameter?" Accuracy is defined as the proximity of the measured or computed value to its true value, i.e., how close the measurement is to reality. Accuracy is composed of two components, precision and bias (35). Precision is defined as the proximity of values obtained from repeatedly measuring the same quantity (e.g., how close are sample means from multiple samples taken from the same population). Precision is described in terms of the spread or variance of the sampling distribution. The smaller the standard error of an estimator is, the more precise it is. Bias, on the other hand, measures how close the statistic would be to the true population value if you could compute it using the whole population. An unbiased statistic has zero bias. Some statistics may end up too high or too low as measured against the true value. The best statistics are precise and unbiased and hence are accurate. Next best is a statistic that is precise but biased, because if we can determine the degree of bias we can derive an accurate estimate. The worst statistics are biased and have low precision and hence are considered inaccurate.

The sampling distribution of a statistic defines the precision of the estimator but does not tell us whether the statistic is biased. Clearly the best statistical tests are those based on precise and unbiased estimates. Many test statistics are composed of functions of other simpler statistics (means and variances). A good test statistic does not need every component statistic to be unbiased since for many test statistics, any biases in the component statistics are cancelled out when they are combined.

We mentioned earlier that failure to incorporate a confounding extraneous factor into a comparative study often results in a failed study. The main reason for this failure is that the absence of the confounding factor results in a set of biased comparisons with no way of cancelling out this bias. The result is that we cannot fairly test the effects of most importance to the study and hence fail in our study objective.

What Are Type I and Type II Errors, and Why Should I Worry about These?

Statistical tests are used to help researchers decide whether the differences that have been observed among different sampled populations or different treatment groups in their mean, medians, variances, or overall distributions are significantly greater than would be expected if the data were random samples from the same common population. Tests are not infallible, and there is always the chance that you will make the wrong decision. There are two ways you can make an error in your decision on any test. In a type I error, you conclude that there are significant differences among the comparison groups, when the reality is that the groups all have the same underlying distribution. You reject the null hypothesis (there are no differences among groups) when in fact the null hypothesis is a true statement of the underlying reality. The type II error occurs when you do not reject the null hypothesis when the reality is that the alternative hypothesis describes the true state. A properly designed study seeks to keep the probability of either of these errors occurring very small. If the type I error rate is too large, you have a good chance of reporting a difference which later research will discover is not really very large. As a result, you may allocate resources exploring differences or treatment effects that are not really there. If the type II error rate is too large, you have the chance of missing true differences. This represents opportunity lost since you may never go back and compare the same populations or treatments and hence may have missed something that would have been beneficial.

Type I and type II error probabilities are tied to sample size, the underlying variability of the response of interest, and what is specified as a true biologically significant effect. The relationships among these five items can be simple to define, as for a two-sample *t* test, or quite complex, but in general the following will always be true.

- If you establish the type I error probability (typically at 0.05 or 0.01) and specify most likely values for the underlying variability and the significant effect, then as you increase sample size the type II error probability will decrease.
- If you establish the type I error probability, specify the significant effect, and fix the sample size at some value, then as the variance of the underlying material decreases, the type II error probability will also decrease.
- If you establish the type I error probability, specify the underlying variability, and fix the sample size at some value, then as you increase the significant effect of interest to you, the type II error probability will subsequently decrease.

As the type II error probability decreases, we increase the power of the comparisons. Hence, to increase the power of a comparison in an experimental study you can do one or more of the following activities: increase sample size (number of replicates) in each treatment group; choose more homogeneous material by using blocking, for example; or use treatments that are expected to produce large differences. For a sampling study, it is not feasible to assume that you will be able either to influence the variability of the populations being sampled or to control the size of the significant effects; so you will often be limited to increasing sample sizes if you want to increase power. Sampling design structures, such as stratification or multistage selection, which should improve the overall precision of the final statistics of interest, improves the power of the comparisons as well.

Finally, it is often tempting to improve the power of a comparison by increasing the acceptable value for the type I error probability. This can be false economy since what you are doing is trading one type of error for another. So you decrease the probability of saying there is no difference when in fact there are differences; but at the same time you now have to worry that the differences you do find will not be sustainable in future research. If the study design is being used to screen treatments or groups for subsequent in-depth study, then there may be some benefit to a higher type I error rate. Any incorrectly specified treatments from the screening study will be discovered at a later stage of the study. But a large fraction of environmental research is not replicated; hence, the misconceptions generated by type I errors may take a long time to be discovered, resulting in the use of much more resources and time to correct the mistake than would have been needed to avoid the mistake in the first place.

What Is a *P* Value, and How Does This Relate to Statistical Testing?

In classical statistical testing, the observed value of the test statistic is typically compared to one or more threshold/critical values that identify the boundaries between the acceptance and rejection regions. The critical values are determined by the sampling distribution of the test statistic under the assumption that the null hypothesis is valid. Let T be a test statistic, for example, in the one-sample *t* test,

$$T = \frac{(\bar{x} - \mu_0)}{\left(s/\sqrt{n}\right)}$$

with sampling distribution (the CDF) under the null hypothesis defined by the function $0 \le F[t] = \Pr[T \le t] \le 1$. (In the example, $F[t]$ is a *t* distribution with $n-1$ degrees of freedom.) Let $F[t_p] = p$, then t_p is the expected value of T such that $100p\%$ of the distribution is to the left of it. The rejection region might typically be written as "Reject Ho if $|T| > t_{1-\alpha}$," where α is the prespecified probability of committing a type I error and $t_{1-\alpha}$ is the critical value of the test. Another approach is to take the observed value of the test statistic, call it $t(x)$, directly to the assumed sampling distribution, $F[t]$, to determine the probability of observing that value if the null hypothesis were true. The *P* value for the test can be computed from this probability; for our example we use P value $= 1 - F[|t(x)|]/2$. A high *P* value (typically $\ge \alpha$ or $\alpha/2$) indicates that the observed test statistic has a very good chance of coming from the reality defined by the null hypothesis. A low *P* value ($<\alpha$ or $\alpha/2$) is an indicator that the test statistic is unlikely to result from the reality defined by the null hypothesis and hence it might be best if we began considering the true reality as defined by the alternative hypothesis. Reporting a *P* value in the results from a statistical test allows external reviewers of your study the opportunity to choose for themselves a type I error probability (α) with which to examine your study results. One reader may have in mind a value of $\alpha = 0.05$ for the type I error, whereas another typically uses a value of 0.01. If your *P* value is reported as 0.03, then one reviewer will accept that there are significant differences being reported and the other reviewer will be less impressed.

What Is an ANOVA?

ANOVA is an acronym for analysis of variance. Many researchers have reported that they "did an ANOVA" or "performed an ANOVA test," but both of these phrases are inadequate to describe exactly what was done in the statistical analysis. The ANOVA table is simply a formal table of sums of squares and degrees of freedom that together provide all the information needed to compute the *F* statistic related to most CLT-based comparisons of group means. A second related ANOVA table is used to examine the significance of partial slope coefficients in a multiple regression. The ANOVA is an integral part of any general linear models analysis of sample or experimental data. The rows of an ANOVA table reflect the structure of your study design and the comparisons among study factor level effects that are important to satisfying the study objectives.

Is Every Statistical Analysis Some Form of the General Linear Model?

While it is true that many of the more widely used statistical analysis methods (linear and polynomial regression, analysis models for traditional experimental designs, logistic regression for categorical responses) derive from the general linear model, it is not true for all methods. Statisticians like using a linear model for a variety of reasons. Linear combinations of random variables have nice mathematical and statistical properties, and hence, there is extensive theory to support the effectiveness of linear models. Linear models can conceptually be thought of as first-order approximations of the nonlinear relationships that are most likely the true state in natural systems. Linear approximations that include simple polynomial terms (e.g., squares and cross-products of explanatory variables) are very effective and appropriate when a description is needed for responses that cover a small fraction of the range of a nonlinear function. Linear functions are easy to understand and interpret, making them very useful in solving practical problems.

The advent of high-speed personal computers has opened up whole new approaches to data analysis, many of which no longer depend on linear models and its associated assumptions of normally distributed and homogeneous residuals. The bootstrap methods mentioned previously are but one example. Methods that in past decades were computationally expensive, such as the replacement of one linear model with a series of interconnected polynomials (e.g., smoothing splines and loess curves) to give an adaptive nonlinear fit to the data are now readily available. Many of these new tools are used at the exploratory phase of the data analysis to inform on whether a linear model analysis will be reasonable. Their use in many fields, including environmental microbiology, is increasing. No general text on these approaches is available, but a good starting point is reference 27.

What Are Nonparametric Tests and When Are They Used?

Nonparametric statistical tests have been developed to provide the data analyst alternatives to methods that are based on a strong assumption regarding the distribution of the response variable and/or the residuals from a formal model. Using a bootstrap-derived sampling distribution for a test statistic is one example. Another is the use of ranks of the observed data in the place of the original observations, since the distribution of ranks under the null hypothesis is often easily determined and does not depend on the form of the underlying population distribution. Under some situations, the loss (if any) in power that results in using a nonparametric test versus a parametric test is replaced by increased confidence that the results of the test are likely more conservative and hence believable.

Traditionally, nonparametric test have been used in situations where sample sizes are too small to effectively use or check whether the sampling distribution suggested by the CLT is valid.

How Do I Handle a Situation in Which the Response Variable Is Categorical?

It is quite common to find study objectives that seek to predict a categorical response as a function of other categorical or continuous factors. The presence of a microbial species in samples of soils from a region having a range of chemical characteristics and the survival of a colony under different nutrient availability scenarios are two frequent examples. The statistical methods of choice for analysis in these situations include multiple logistic regression and any of a large number of discriminate analysis methods (4, 16, 26). These methods seek not to predict the value of the response but to predict the likelihood or probability of a given response under different combinations of values of the predictors. In general these models allow characterization of the conditions under which a particular categorical response is most likely to occur.

What Are Repeated Measures, and How Is the Analysis of Repeated-Measures Data Different?

In many research studies, obtaining sample units is the most expensive and difficult part of the study. Because of this, once we have a sample unit we want to maximize the amount of information we get from each. One way of doing this is to take multiple measurements, often of the same characteristic, over time. One issue that arises with repeated-measures data is that the repeated observations on the same individual are more correlated with each other than with observations from different sample units. With repeated-measures data the actual increase in usable information that results from taking repeated observations on the same individual is often less than if the same number of observations were taken independently. This interindividual correlation has to be taken into account, and an adjustment to P values must be made. At present most of the research on repeated-measures modeling and testing is occurring within the class of general linear models (42). Finding the correct model and properly fitting all parameters, including the extra variance and correlation terms added to account for intraindividual associations, are still involved processes that require the advice, if not the help, of a colleague who has training in mixed-effects linear models. Repeated-measures data are multivariate in nature. A common mistake is to treat repeated-measures data as a univariate response in a linear model with time as an explanatory factor. This approach ignores the correlation among re-sponses from the same unit and can lead to biased results and wrong conclusions.

Assessing Temporal and Spatial Patterns

As most of you recognize, microbial communities are not uniformly distributed throughout the environment. They often vary in composition, abundance, and activity. As understanding of these communities grows, the need to quantify these characteristics over time and space increases.

What Statistics Are Used To Assess Spatial Distribution Patterns?

Most researchers think that assessing spatial pattern is primarily a mapping task and that statistics is a small part of this process. Statisticians, on the other hand, look at a generated map and wonder about its accuracy and precision. The precision of the map is a function of how much sampling was done and how it was performed.

Understanding scaling factors is important when investigating biological interactions among or between microbial communities, via the interpretation of distribution patterns in the landscape (9, 32). An inappropriate sampling scale may underestimate spatial heterogeneity, thus leading to erroneous conclusions. Moreover, different ecosystem processes occur at different scales, each differentially affecting the microbial communities. At one scale, microvariations in the physicochemical environment can function as a structuring agent for the short-scale (<1-mm) spatial patterns. At a larger scale, spatial patterns of the plant communities, soil/substrate characteristics, and landscape position affect microbial community composition, abundance, and activities. Key factors influencing the soil microbial communities are different at different spatial scales, often resulting in very different conclusions about the factors that structure the communities. There are therefore nested scales at which different factors operate on the microbial community structure.

The nature of microbial community structure data, on the other hand, is often a multidimensional data array ranging from genomic information to multiple measures of abundance and activities. Understanding the relationship between microbial community structure and function often requires an appreciation of the inherent multivariate structure on both sides of the equation. This results in the need for a series of statistical methods and computational tools in order to simplify and elucidate composition, its drivers, and its relation to function. Furthermore, underlying relationships are often complex and can be nonlinear. Multivariate analysis of the microbial measures results in a form of integration that is significant both in terms of the effects of the community structure composition on abundance and activity of microbial communities and in terms of the next step, which is describing the relationships between the environment and microbial community composition (33).

Temporal-Spatial Correlations and Modeling Approaches

Most environmental microbiology field experiments have some spatial component that has its own distinctive set of difficulties. Spatial variability can often function as a considerable source of noise as the environment is heterogeneous at many scales (3, 21). Spatial measurements are often crude and done at widely divergent scales (e.g., microbial communities in landscape features). Often, natural variability overwhelms any treatment effect or relationships that you are trying to establish. Some of this variability may be due to other spatially distributed factors that need to be addressed as confounding factors in comparative studies.

If we approach the temporal/spatial environment as fundamentally being a continuous heterogeneous entity and neglect strong discontinuities or boundaries, we can make the fairly intuitive step to approach this continuous heterogeneity by proposing that in most circumstances, samples taken at times or locations close together tend to have more similar values, and therefore are more positively correlated, than those that are further apart. This concept or tendency is termed autocorrelation. Microbial communities exhibit a significant amount of autocorrelation at a variety of time and distance intervals, from microsecond/micrometer to kilosecond/kilometer, depending on the temporal/spatial extent studied (9). Autocorrelation can introduce another significant difficulty when executing or analyzing field studies, as most statistical analyses assume that the factor levels are independent. For example, a field study that compares the effects of two plant species on the microbial communities in their respective rhizospheres might ignore the fact that the presence of these plant species might be a function of some other underlying factor or even nearness to each other (the levels of the different plant species are not independent). Any conclusions are therefore limited to the circumstances in which the field study was carried out and cannot be extrapolated. There are methods that use the underlying temporal/spatial variation or correlation to estimate treatment contrasts that are more effective than the classical approach that we described previously. Some approaches use the underlying spatial variability to estimate treatment contrasts (3, 42). Another approach is a temporal/spatial visualization technique that is increasingly used in the environmental sciences (6, 43). The charge for temporal data is to extract mean time patterns from stochastic (random) patterns. The charge for spatial data is to generate the best possible map (mean spatial pattern) of the variable of interest from local stochastic patterns. When modeling across spatial scales, a mixture model is used to describe the spatial variability observed in the sampled data. The model is then used to compute spatial predictions at previously selected but unsampled locations which are then used to generate the final map (often in a geographic information systems environment). As this approach requires significant data support, it has not been used extensively in environmental microbiology (9, 29). This will certainly change in the near future as the cost of field studies decreases and "maps" of microbial community composition, abundance, or activities become an in-creasingly attractive way to depict spatial variation in these variables. These statistical techniques are fairly complex, and if you find yourself in the situation of designing a field experiment in which you suspect that there is significant spatial variability, you would be well advised to seek help from a (geo)statistically trained colleague.

CONCLUSIONS

In this chapter we have attempted to raise your awareness of the role and function of statistics at both the study design and the data analysis steps of environmental microbiological studies. Statistics are research tools and should be considered in the same way as are the other methods presented in this manual. As you read other chapters, think about the role and function of the statistical methods mentioned, about the statistical aspects of the study designs proposed, and about how the resulting data should be statistically analyzed. Most of the references provided here are readable by anyone with a basic statistics background and should be used as the first step in determining your statistical analysis plan. Ask for help from a statistician or statistically trained colleague when you feel you have exceeded your current skill level.

REFERENCES

1. **Becker, S., P. Böger, R. Oehlmann, and A. Ernst.** 2000. PCR bias in ecological analysis: a case study for quantitative Taq nuclease assays in analyses of microbial communities. *Appl. Environ. Microbiol.* **66:**4945–4953.
2. **Chatterjee, S., and B. Price.** 1977. *Regression Analysis by Example*. John Wiley & Sons, Inc., New York, N.Y.
3. **Cressie, N.** 1991. *Statistics for Spatial Data*. John Wiley & Sons, Inc., New York, N.Y.
4. **Duda, R. O., P. E. Hart, and D. G. Stork.** 2001. *Pattern Classification*, 2nd ed. John Wiley & Sons, Inc., New York, N.Y.
5. **Gardiner, W. P.** 1997. *Statistics for the Biosciences*. Prentice-Hall, Inc., London, United Kingdom.
6. **Gilbert, R. O.** 1987. *Statistical Methods for Environmental Pollution Monitoring*. Van Nostrand Reinhold, New York, N.Y.
7. **Dixon, P. M.** 2001. The bootstrap and the jackknife; describing the precision of ecological indices, p. 267–288.

In S. M. Scheiner and J. Gurevitch (ed.), *Design and Analysis of Ecological Experiments*. Oxford University Press, Oxford, United Kingdom.

8. **Evans, M., N. Hastings, and B. Peacock.** 2000. *Statistical Distributions*, 3rd ed. John Wiley & Sons, Inc., New York, N.Y.
9. **Franklin, R. B., and A. L. Mills.** 2003. Multiscale variation in spatial heterogeneity for microbial community structure in an eastern Virginia agricultural field. *FEMS Microbiol. Ecol.* **44:**335–346.
10. **Friendly, M.** 1994. Mosaic displays for multi-way contingency tables. *J. Am. Stat. Assoc.* **89:**190–200.
11. **Hill, M. O.** 1979. *TWINSPAN—a FORTRAN Program for Arranging Multivariate Data in an Ordered Two-Way Table by Classification of the Individuals and Attributes*. Ecology and Systematics, Cornell University, Ithaca, N.Y.
12. **Hill, M. O., and H. G. Gauch.** 1980. Detrended correspondence analysis: an improved ordination technique. *Vegetatio* **42:**47–58.
13. **Hurlbert, S. H.** 1984. Pseudoreplication and the design of ecological field experiments. *Ecol. Monogr.* **54:**187–211.
14. **Joergensen, R. G., and S. Scheu.** 1999. Response to soil microorganisms to the addition of carbon, nitrogen and phosphorus in a forest Rendzina. *Soil Biol. Biochem.* **31:** 859–866.
15. **Johnson, D. H.** 1999. The insignificance of statistical significance testing. *J. Wildl. Manag.* **63:**763–772.
16. **Johnson, R. A., and D. W. Wichern.** 1989. *Applied Multivariate Statistical Analysis*, 2nd ed. Prentice-Hall, Inc., Englewood Cliffs, N.J.
17. **Kingsolver, K. M., and D. G. Schemske.** 1991. Analyzing selection with path analysis. *Trends Ecol. Evol.* **6:**276–280.
18. **Krebs, C. J.** 1999. *Ecological Methodology*, 2nd ed. Benjamin/Cummings, Menlo Park, Calif.
19. **Krivtsov, V.** 2004. Investigations of indirect relationships in ecology and environmental sciences: a review and the implications for comparative theoretical ecosystem analysis. *Ecol. Model.* **174:**37–54.
20. **Laflaive, J., R. Cereghino, M. Danger, G. Lacroix, and L. Ten-Hage.** 2005. Assessment of self-organizing maps to analyze sole-carbon source utilization profiles. *J. Microbiol. Methods* **62:**89–102.
21. **Legendre, P.** 1993. Spatial autocorrelation: trouble or new paradigm? *Ecology* **74:**1659–1673.
22. **Legendre, P., and L. Legendre.** 1998. *Numerical Ecology*, 2nd ed. Elsevier, Amsterdam, The Netherlands.
23. **Levy, P. S., and S. Lemeshow.** 1999. *Sampling of Populations, Methods and Applications*, 3rd ed. John Wiley & Sons, Inc., New York, N.Y.
24. **Ludwig, J. A., and J. F. Reynolds.** 1988. *Statistical Ecology*. John Wiley & Sons, Inc. New York, N.Y.
25. **Magurran, A. E.** 1988. *Ecological Diversity and Measurement*. Princeton University Press, Princeton, N.J.
26. **Manley, B. F. J.** 1994. *Multivariate Statistical Methods: A Primer*. Chapman & Hall/CRC Press, Boca Raton, Fla.
27. **Manley, B. F. J.** 1997. *Randomization, Bootstrap and Monte Carlo Methods in Biology*, 2nd ed. Chapman & Hall/CRC Press, Boca Raton, Fla.
28. **McCune, B., J. B. Grace, and D. L. Urban.** 2002. *Analysis of Ecological Communities*. MjM Software Design, Gleneden Beach, Oreg.
29. **Morris, S. J.** 1991. Spatial distribution of fungal and bacterial biomass in southern Ohio hardwood forest soils: fine scale variability and microscale patterns. *Soil Biol. Biochem.* **31:**1375–1386.
30. **Ogram, A.** 2000. Soil molecular microbial ecology at age 20: methodological challenges for the future. *Soil Biol. Biochem.* **32:**1499–1504.
31. **Palumbo, A. V., J. C. Schryver, M. W. Fields, C. E. Bagwell, J. Z. Zhou, T. Yan, X. Liu, and C. C. Brandt.** 2004. Coupling of functional gene diversity and geochemical data from environmental samples, *Appl. Environ. Microbiol.* **70:**6525–6534.
32. **Pennanen, P., J. Liski, E. Baath, V. Kitunen, J. Uotila, C. J. Westman, and H. Fritze.** 1999. Structure of the microbial communities in coniferous forest soils in relation to site fertility and stand development stage. *Microb. Ecol.* **38:**168–179.
33. **Reynoldson, T. B., R. H. Norris, V. H. Resh, K. E. Day, and D. M. Rosenberg.** 1997. The reference condition: a comparison of multimetric and multivariate approaches to assess water quality impairment using benthic macroinvertebrates. *J. N. Am. Benthol. Soc.* **16:**833–852.
34. **Scheaffer, R. L., W. Mendenhall III, and R. L. Ott.** 1996. *Elementary Survey Sampling*, 5th ed. Duxbury Press/ Wadsworth Publishing Co., Belmont, Calif.
35. **Snedecor, G. W., and W. G. Cochran.** 1989. *Statistical Methods*. Iowa State University Press, Ames.
36. **Sokal, R. R., and F. J. Rohlf.** 1995. *Biometry*, 3rd ed. W. H. Freeman, New York, N.Y.
37. **Stewart-Oaten, A., and W. W. Murdoch.** 1986. Environmental impact assessment: "pseudoreplication" in time. *Ecology* (Tempe) **67:**929–940.
38. **Tabachnik, B. G., and L. S. Fidell.** *Using Multivariate Statistics*, 4th ed. Allyn & Bacon, Boston, Mass.
39. **Ter Braak, C. J. F.** 1986. Canonical correspondence analysis: a new eigenvector technique for multivariate direct gradient analysis. *Ecology* **67:**1167–1179.
40. **Thode, H. C., Jr.** 2002. *Testing for Normality*. Marcel Dekker, Inc., New York, N.Y.
41. **Thompson, S. K.** 2002. *Sampling*, 2nd ed. John Wiley & Sons, Inc., New York, N.Y.
42. **Verbeke, G., and G. Molenberghs.** 2000. *Linear Mixed Models for Longitudinal Data*. Springer-Verlag, New York, N.Y.
43. **Webster, R., and M. A. Oliver.** 2001. *Geostatistics for Environmental Scientists*. John Wiley & Sons, New York, N.Y.
44. **Zar, J. H.** 1996. *Biostatistical Analysis*, 3rd ed. Prentice-Hall, Inc., Upper Saddle River, N.J.

WATER MICROBIOLOGY IN PUBLIC HEALTH

VOLUME EDITOR
CHRISTON J. HURST

SECTION EDITOR
GARY A. TORANZOS

Overview of Water Microbiology as It Relates to Public Health

CHRISTON J. HURST

18

One of the most important aspects of water microbiology, from a human perspective, is the fact that we acquire numerous diseases from microorganisms found in water (11). Some of these diseases represent intoxications, which result from the ingestion of toxins produced by microorganisms. One such category of intoxications is linked to drinking water which contains toxins produced by cyanobacteria such as *Anabaena* and *Microcystis* (19). A second category of intoxications is associated with dinoflagellates such as *Gambierdiscus*, *Gonyaulax*, and *Ptychodiscus*; these are aquatic microorganisms which produce neurotoxins that can become biologically concentrated in the tissues of reef fish and shellfish (12a). Humans acquire this second category of toxins by ingesting the affected animals.

The majority of human diseases associated with water are, however, infectious in nature. The magnitude of human morbidity and mortality associated with these infectious diseases has led to the development of epidemiological surveillance studies (2, 6). The associated pathogens include numerous bacteria (2, 3, 11, 14, 15), viruses (4, 9, 10, 11), and protozoa (9, 10, 11). These water-related infectious hazards can be characterized according to various schemes. One such approach divides the hazards into four categories based upon the source of the involved pathogen and the route by which human recipients come in contact with that pathogen. Those categories could be defined as follows.

Infections that are waterborne: These are recognized as resulting from physical contact with microbially contaminated water. The contact may occur by immersion of the body as associated with bathing, recreational, or occupational aquatic activities. Alternatively, the contact may occur by ingestion of either contaminated water or ice that has been made from contaminated water. There also is a direct connection between waterborne infections and foodborne infections, when the latter are acquired from food items that have come into contact with either contaminated water or contaminated ice (2, 10, 11).

Infections caused by aquatic organisms: These result from pathogenic microorganisms or worms (by definition macroorganisms) which naturally spend at least a part of their normal life cycle either living directly in the water or else within intermediate vertebrate or invertebrate hosts that reside in aquatic environments. These diseases are acquired either by ingestion of or immersion in the water or by inadvertent ingestion of infested intermediate host animals (11, 14, 15).

Infections which have water-related insect vectors: The associated pathogens are acquired not from water but rather as a consequence of humans being bitten by invertebrate vectors (usually mosquitoes) whose life cycle depends upon access to water either on the land surface or in uncovered containers. Thus, the incidence of the resulting diseases may increase during periods of land surface flooding and during monsoon seasons (5).

Infections associated with inadequate hygiene: These are caused by pathogens which are not acquired from water but rather are acquired by either contact with or ingestion of microbially contaminated material (generally fecally contaminated). The resulting disease hazards could be ameliorated by using clean water for sanitation purposes, including bodily cleansing and the washing of plates and drinking glasses (2, 16).

The chapters in this section of the *Manual of Environmental Microbiology* address various issues related to the routes by which humans acquire waterborne infections. These routes of infection are diagrammed in Fig. 1.

The reservoirs for pathogenic microorganisms found in environmental waters can be humans, animals, or the environment itself (4, 10, 11, 18), as summarized in Table 1. However, it commonly is presumed that many of those microorganisms that infect humans and are found in our aquatic resources originate from human sources. This anthropogenic contamination can occur during either defecation in water or recreational activities conducted in water (11). In addition, domestic wastewater is of particular importance as a contributor of the pathogenic contaminants found in aquatic environments (9, 13, 17); the attendant public health concerns have resulted in the development of methods for studying and reducing the levels of pathogens in wastewater (1). Effective drinking water disinfection practices can alleviate a great deal of the human illness associated with water (8). But drinking water treatment alone does not solve the entire problem. Wastewater treatment efforts may help to reduce the incidence of problems resulting from the discharge of wastewater into

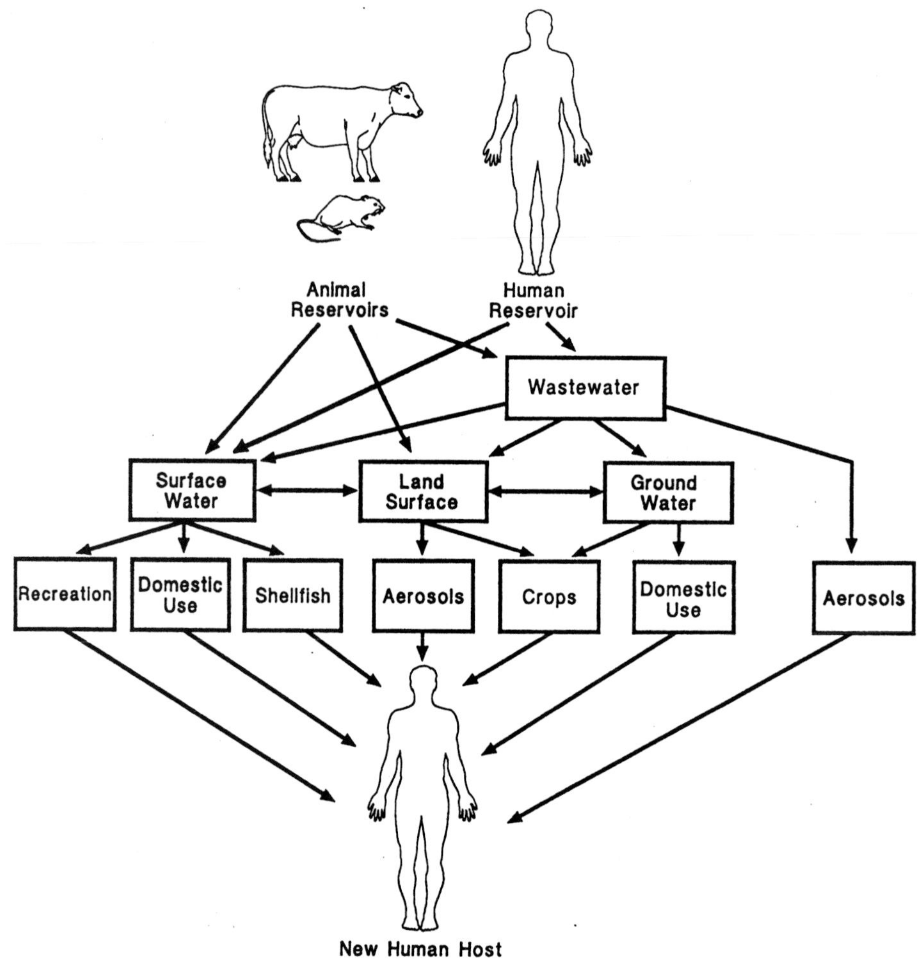

FIGURE 1 Water-related environmental routes by which infectious agents are transmitted to susceptible individuals. The animals symbolically represented are cows as a source of the microorganism that causes cryptosporidiosis and rodents as a source of those microorganisms that cause giardiasis and campylobacteriosis (11).

TABLE 1 Examples of infectious disease hazards associated with water

Source of microorganisms	Disease	Causative microbial genus
Human reservoirs	Cholera	*Vibrio*
	Dracunculiasis	*Dracunculus*
	Encephalitis	*Enterovirus*
	Entamoebiasis	*Entamoeba*
	Gastroenteritis	*Astrovirus*, *Calicivirus*, *Coronavirus*, *Norovirus*, and *Rotavirus*
	Hepatitis	*Calicivirus* and *Hepatovirus*
	Meningitis	*Enterovirus*
	Paratyphoid	*Salmonella*
	Shigellosis	*Shigella*
	Tularemia	*Francisella*
	Typhoid fever	*Salmonella*
Animal reservoirs	Campylobacteriosis	*Campylobacter*
	Cryptosporidiosis	*Cryptosporidium*
	Enteric fever	*Salmonella*
	Giardiasis	*Giardia*
	Leptospirosis	*Leptospira*
	Schistosomiasis	*Schistosoma*
	Tularemia	*Francisella*
Environmental reservoirs	Encephalitis	*Naegleria*
	Cholera	*Vibrio*
	Legionellosis	*Legionella*
	Wound infections	*Vibrio*

environmental surface water (9, 11), including illness among swimmers, contamination of drinking water, and disease that can occur when bivalve molluscan shellfish harvested from contaminated waters are subsequently consumed by humans (7, 12). Wastewater treatment will reduce aquifer contamination, which can result either indirectly following the percolation of surface-applied wastewater into the subsurface or directly during subsurface injection of wastewater (4). The treatment of wastewater also is intended to reduce the contamination of crops that may occur when wastewater is eventually discharged onto land surfaces (1).

The goal of this section of the manual is to represent and summarize current knowledge on public health aspects of water microbiology. The more generalized aspects of water microbiology are presented in section IV of this manual.

REFERENCES

1. **Acher, A. J., E. Fischer, and Y. Manor.** 1994. Sunlight disinfection of domestic effluents for agricultural use. *Water Res.* **28:**1153–1160.
2. **Acosta, C. J., C. M. Galindo, J. Kimario, K. Senkoro, H. Urassa, C. Casals, M. Corachán, N. Eseko, M. Tanner, H. Mshinda, F. Lwilla, J. Vila, and P. L. Alonso.** 2001. Cholera outbreak in Southern Tanzania: risk factors and patterns of transmission. *Emerg. Infect. Dis.* **7**(Suppl. 3)**:** 583–587.
3. **Anda, P., J. S. del Pozo, J. M. D. García, R. Escudero, F. J. G. Peña, M. C. L. Velasco, R. E. Sellek, M. R. J. Chillarón, L. P. S. Serrano, and J. F. M. Navarro.** 2001. Waterborne outbreak of tularemia associated with crayfish fishing. *Emerg. Infect. Dis.* **7**(Suppl. 3)**:**575–582.
4. **Borchardt, M. A., P. D. Bertz, S. K. Spencer, and D. A. Battigelli.** 2003. Incidence of enteric viruses in groundwater from household wells in Wisconsin. *Appl. Environ. Microbiol.* **69:**1172–1180.
5. **Cotton, P.** 1993. Health threat from mosquitoes rises as flood of the century finally recedes. *JAMA* **270:**685–686.
6. **de Wit, M. A. S., M. P. G. Koopmans, L. M. Kortbeek, N. J. van Leeuwen, A. I. M. Bartelds, and Y. T. H. P. van Duynhoven.** 2001. Gastroenteritis in sentinel general practices, the Netherlands. *Emerg. Infect. Dis.* **7:**82–91.
7. **Guyader, F. L., L. Haugarreau, L. Miossec, E. Dubois, and M. Pommepuy.** 2000. Three-year study to assess human enteric viruses in shellfish. *Appl. Environ. Microbiol.* **66:**3241–3248.
8. **Haas, C. N., and G. R. Finch.** 2001. *Methodologies for the Determination of Disinfection Effectiveness.* AWWA Research Foundation and American Water Works Association, Denver, Colo.
9. **Hurst, C. J.** 2001. Disinfection of water: drinking water, recreational water, and wastewater, p. 1023–1047. *In* S. S. Block (ed.), *Disinfection, Sterilization, and Preservation,* 5th ed. Lippincott, Williams and Wilkins, Philadelphia, Pa.
10. **Hurst, C. J.** 2001. Prevention of infection from food and water, p. 1191–1203. *In* S. S. Block (ed.), *Disinfection, Sterilization, and Preservation,* 5th ed. Lippincott, Williams and Wilkins, Philadelphia, Pa.
11. **Hurst, C. J., and P. A. Murphy.** 1996. The transmission and prevention of infectious disease, p. 3–54. *In* C. J. Hurst (ed.), *Modeling Disease Transmission and Its Prevention by Disinfection.* Cambridge University Press, Cambridge, United Kingdom.
12. **Kingsley, D. H., and G. P. Richards.** 2001. Rapid and efficient extraction method for reverse transcription-PCR detection of hepatitis A and Norwalk-like viruses in shellfish. *Appl. Environ. Microbiol.* **67:**4152–4157.

12a. **Litaker, R. W., and P. A. Tester.** 2002. Molecular methods for detecting and characterizing harmful phytoplankton, p. 342–353. In C. J. Hurst, R. L. Crawford, G. R. Knudsen, M. J. McInerney, and L. D. Stetzenbach (ed.), *Manual of Environmental Microbiology,* 2nd ed. ASM Press, Washington, D.C.

13. **Matsuura, K., M. Ishikura, H. Yoshida, T. Nakayama, S. Hasegawa, S. Ando, H. Horie, T. Miyamura, and T. Kitamura.** 2000. Assessment of poliovirus eradication in Japan: genomic analysis of polioviruses isolated from river water and sewage in Toyama prefecture. *Appl. Environ. Microbiol.* **66:**5087–5091.
14. **Mouriño-Pérez, R. R., A. Z. Worden, and F. Azam.** 2003. Growth of *Vibrio cholerae* 01 in red tide waters off California. *Appl. Environ. Microbiol.* **69:**6923–6931.
15. **Pfeffer, C. S., M. F. Hite, and J. D. Oliver.** 2003. Ecology of *Vibrio vulnificus* in estuarine waters of Eastern North Carolina. *Appl. Environ. Microbiol.* **69:**3526–3531.
16. **Rosenberg, T., O. Kendall, J. Blanchard, S. Martel, C. Wakelin, and M. Fast.** 1997. Shigellosis on Indian reserves in Manitoba, Canada: its relationship to crowded housing, lack of running water, and inadequate sewage disposal. *Am. J. Public Health* **87:**1547–1551.
17. **Sandhya, S., T. S. Uma, and K. Subbarao.** 1999. Dip slide technique for rapid qualitative estimation of fecal coliforms in water and wastewater. *Water Res.* **33:**989–994.
18. **Scott, T. M., J. B. Rose, T. M. Jenkins, S. R. Farrah, and J. Lukasik.** 2002. Microbial source tracking: current methodology and future directions. *Appl. Environ. Microbiol.* **68:**5796–5803.
19. **Westrick, J. A.** 2003. Everything a manager should know about algal toxins but was afraid to ask. *J. Am. Water Works Assoc.* **95**(9)**:**26–34.

Waterborne Transmission of Infectious Agents

CHRISTINE L. MOE

19

Waterborne transmission is a highly effective means for spreading infectious agents to a large portion of the population. Large quantities of enteric organisms can be introduced into the aquatic environment through the discharge of infected persons' (or animals') feces into sewers or unprotected waterways. In contrast to person-to-person transmission, infectious agents even from bedridden infected persons can play a role in waterborne disease transmission because pathogens in soiled bedding and clothing may be released into water during washing. When such fecal contamination mixes with unprotected and/or inadequately treated drinking water, large numbers of susceptible hosts can be exposed and become infected. Historically, this process has been documented repeatedly since the time of John Snow in the mid-19th century. Waterborne-disease outbreaks infecting tens to hundreds of thousands of people have been reported, and Ewald has observed that those pathogens that are more frequently waterborne tend to be the most virulent enteric pathogens (85).

Numerous infectious agents have been transmitted by ingestion, contact, or inhalation of water (Tables 1 and 2). Disease outcomes associated with waterborne infections include mild to life-threatening gastroenteritis, hepatitis, skin infections, wound infections, conjunctivitis, respiratory infections, and generalized infections. Most waterborne infections are acute; however, some have chronic sequelae such as hemolytic-uremic syndrome (HUS) associated with *Escherichia coli* O157:H7; myocarditis and diabetes associated with coxsackieviruses; peptic and duodenal ulcers and stomach cancer associated with *Helicobacter pylori*; reactive arthritis associated with *Yersinia*, *Shigella*, and *Salmonella*; pericarditis, orchitis, and splenic or liver abscesses due to *Salmonella enterica* serovar Typhi; and Guillain-Barré syndrome associated with *Campylobacter* (105, 129). Some waterborne microorganisms are frank pathogens, some are opportunistic pathogens, and some are toxigenic. The total number of potentially pathogenic microorganisms is unknown but may be in the thousands. New infectious agents of disease continue to be recognized.

There are five critical elements in the transmission of infectious agents through water: (i) the source of the infectious agents, (ii) specific water-related modes of transmission, (iii) attributes of the organism that allow it to survive and possibly multiply and to move into and within the aquatic environment, (iv) infectious dose and virulence factors of the organism, and (v) host susceptibility factors.

Some agents of waterborne diseases are indigenous aquatic organisms, such as *Legionella* species, *Vibrio* species, *Aeromonas hydrophila*, and *Pseudomonas aeruginosa*. Control of these infections may depend on controlling exposure to water containing such organisms or, when possible, treating the water to remove or inactivate the infectious agents. Most microbial waterborne pathogens of concern originate in the enteric tract of humans or animals and enter the aquatic environment via fecal contamination. The concentration of these pathogens in a community water supply depends in part on the number of infected persons and/or animals in the community and the opportunities for feces from these individuals to enter the water supply. Control of these diseases rests on sanitation measures and wastewater treatment to prevent the introduction of feces containing these organisms into drinking water supplies or recreational waters and on adequate water treatment to remove or inactivate these organisms in drinking water.

Several water-related modes of transmission of infectious agents are discussed in the following section. Many enteric pathogens transmitted via ingestion of feces-contaminated water can also be transmitted person to person by contact with feces-contaminated hands or fomites or by consumption of feces-contaminated food. In situations with poor sanitation and hygiene in which the pathogens are endemic, the risk attributable to water may be difficult to determine because of the risks from many other transmission routes of infection.

Different classes of organisms have specific attributes, such as size and charge, which determine their movement and survival in the aquatic environment and their susceptibility to various water and wastewater treatment processes. These are discussed in more detail in the following section by group of microorganism. Knowledge of these attributes can aid in the design of effective barriers or control strategies.

The infectious dose (ID) is the number of organisms required to cause infection and varies considerably by type of organism. In general enteric viruses and protozoa have low IDs, typically between 1 and 50 tissue culture infectious units, PFU, cysts, or oocysts. Bacterial pathogens tend to require a larger dose to cause infection. Observed median infectious dose (ID_{50} = number of organisms resulting in a 50% infection rate) data for enteric bacteria range from

TABLE 1 Illnesses acquired by ingestion of water

Agent	Source	Incubation period	Clinical syndrome	Duration
Viruses				
Astrovirus	Human feces[a]	1–4 days	Acute gastroenteritis	2–3 days; occasionally 1–14 days
Norovirus (Norwalk virus, Snow Mountain agent, and other related viruses)	Human feces[a]	1–3 days	Acute gastroenteritis with predominant nausea and vomiting	1–3 days
Enteroviruses (polioviruses, coxsackieviruses, echoviruses)	Human feces	3–14 days (usually 5–10 days)	Febrile illness, respiratory illness, meningitis, herpangina, pleurodynia, conjunctivitis, myocardiopathy, diarrhea, paralytic disease, encephalitis, ataxia, diabetes	Variable
Hepatitis A virus	Human feces	15–50 days (usually 25–30 days)	Fever, malaise, jaundice, abdominal pain, anorexia, nausea	1–2 wk to several months
Hepatitis E virus	Human feces	15–65 days (usually 35–40 days)	Fever, malaise, jaundice, abdominal pain, anorexia, nausea	1–2 wk to several months
Rotavirus A	Human feces[a]	1–3 days	Acute gastroenteritis with predominant nausea and vomiting	5–7 days
Rotavirus B	Human feces[a]	2–3 days	Acute gastroenteritis	3–7 days
Bacteria				
Aeromonas hydrophila	Fresh water	Unknown	Watery diarrhea	Avg 42 days
Campylobacter jejuni	Human and animal feces	3–5 days (1–7 days)	Acute gastroenteritis, possible bloody and mucoid feces, possible Guillain-Barré syndrome	1–4 days, occasionally >10 days
Enterohemorrhagic *E. coli* O157:H7	Human and animal feces	3–8 days	Watery, then grossly bloody diarrhea, vomiting, possible HUS	1–12 days (usually 7–10 days)
Enteroinvasive *E. coli*	Human feces	2–3 days	Possible dysentery with fever	1–2 wk
Enteropathogenic *E. coli*	Human feces	2–6 days	Watery to profuse watery diarrhea	1–3 wk
Enterotoxigenic *E. coli*	Human feces	12–72 h	Watery to profuse watery diarrhea	3–5 days
Plesiomonas shigelloides	Fresh surface water, fish, crustaceans, wild and domestic animals?	1–2 days	Bloody and mucoid diarrhea, abdominal pain, nausea, vomiting	Avg 11 days
Salmonellae	Human and animal feces	8–48 h	Loose, watery, occasionally bloody diarrhea, possible reactive arthritis	3–5 days
Salmonella enterica serovar Typhi	Human feces and urine	7–28 days (avg 14 days)	Fever, malaise, headache, cough, nausea, vomiting, abdominal pain, possible pericarditis, orchitis and splenic or liver abscesses	Weeks to months

(Continued on next page)

TABLE 1 Illnesses acquired by ingestion of water *(Continued)*

Agent	Source	Incubation period	Clinical syndrome	Duration
Shigellae	Human feces	1–7 days	Possible dysentery with fever, possible reactive arthritis	4–7 days
Vibrio cholerae O1	Human feces	9–72 h	Profuse, watery diarrhea, vomiting, rapid dehydration	3–4 days
Vibrio cholerae non-O1	Human feces	1–5 days	Watery diarrhea	3–4 days
Yersinia enterocolitica	Animal feces and urine	2–7 days	Abdominal pain, mucoid, occasionally bloody diarrhea, fever, possible reactive arthritis	1–21 days (avg, 9 days)
Parasites				
Balantidium coli	Human and animal feces	Unknown	Abdominal pain, occasional mucoid or bloody diarrhea	Unknown
Cryptosporidium parvum	Human and animal feces	1–2 wk	Profuse, watery diarrhea	4–21 days
Entamoeba histolytica	Human feces	2–4 wk	Abdominal pain, occasional mucoid or bloody diarrhea	Weeks to months
Giardia lamblia	Human and animal feces	5–25 days	Abdominal pain, bloating, flatulence, loose, pale, greasy stools	1–2 wk to months and years
Algae				
Cyanobacteria (*Anabaena* spp., *Aphanizomenon* spp., *Microcystis* spp.)	Cyanobacterial blooms in water	A few hours	Toxin poisoning (blistering of mouth, gastroenteritis, pneumonia)	Variable
Helminths				
Dracunculus medinensis (guinea worm)	Larvae discharged from worms protruding from skin of infected person	8–14 mo (usually 12 mo)	Blister, localized arthritis of joints adjacent to site of infection	Months

[a]Animal strains of these viruses are believed to be nonpathogenic for humans.

TABLE 2 Illnesses acquired by recreational contact with water[a]

Agent	Source	Incubation period	Clinical syndrome	Duration
Viruses				
Adenovirus (serotypes 3, 7, 1, 4, 14)	Humans	4–12 days	Conjunctivitis, pharyngitis, fever	7–15 days
Bacteria				
Aeromonas hydrophila	Fresh and brackish water	8–48 h	Wound infections	Weeks to months
Legionellae	Freshwater, soil	Legionnaires' disease: 2–14 days (usually 5–6 days); Pontiac fever: 5–66 h (usually 24–48 h)	Legionnaires' disease: pneumonia with anorexia, malaise, myalgia and headache, rapid fever and chills, cough, chest pain, abdominal pain and diarrhea; Pontiac fever: fever, chills, myalgia, headache	Legionnaires' disease: variable (usually weeks to months); Pontiac fever: 2–7 days
Leptospira spp.	Urine from infected domestic and wild animals	2–20 days (usually 7–12 days)	Leptospirosis (headache, chills, fever, myalgia, nausea, neck or joint pain)	A few days to 3 wk
Mycobacterium spp. (*M. marinum*, *M. balnei*, *M. platy*, *M. kansasii*, *M. szulgai*)	Marine or brackish waters, freshwater	2–4 wk	Lesions of skin or subcutaneous tissues	Months
Pseudomonas spp.	Water	Unknown	Dermatitis, ear infections, conjunctivitis	Unknown
Vibrio spp. (*V. alginolyticus*, *V. parahaemolyticus*, *V. vulnificus*, *V. mimicus*)	Marine water	*V. vulnificus*, 24 h; *V. parahaemolyticus*, 4–48 h	*V. vulnificus*: acute gastroenteritis, wound infections, septicemia *V. parahaemolyticus*: acute gastroenteritis, wound infections Ear infections	*V. vulnificus*: septicemia fatal in 2–4 days *V. parahaemolyticus*: usually 3 days
Other				
Cyanobacteria (*Anabaena*, *Aphanizomenon*, and *Microcystis* species)	Cyanobacterial blooms in marine water or freshwater	A few hours 3–7 days	Dermatitis	10 days
Naegleria fowleri	Freshwater in warm climates, soil, decaying vegetation	A few minutes to hours	Meningoencephalitis, headache, anorexia, fever, nausea and vomiting; usually fatal	
Acanthamoeba species	Water		Subcutaneous abscesses, conjunctivitis	8 days to several months
Schistosoma species	Feces and urine of infected animals and birds		Dermatitis, prickly sensation, itching	Years

[a]Agents acquired through ingestion of water are not included in this table.

10^2 to 10^8. Because such data are derived from studies of healthy, adult human volunteers under controlled conditions (vehicle of administration and gastric acidity), care must be taken when extrapolating these findings to other populations, such as malnourished children in developing countries, or human immunodeficiency virus (HIV)-infected persons. For example, studies of *Vibrio cholerae* indicate that 10^6 organisms, when ingested with water by fasting volunteers, caused no illness. However, the same inoculum ingested with food or sodium bicarbonate caused illness in 90 to 100% of the exposed volunteers because these vehicles decreased the protective gastric acidity (202). The virulence of microorganisms varies by type and strain and route of infection. Microbial virulence factors include the ability of the organism to attach to and penetrate host cells; spread and multiply within the host; cause damage to host cells via lytic infection, production of toxins, or activities which may cause initiation of an injurious host response (such as loss of blood clotting); evade or resist host defense mechanisms (ranging from stomach acidity to various components of the immune system); and resist antimicrobial therapy.

Infection and development of clinical symptoms depend on a number of specific and nonspecific host factors, such as age, immune status, gastric acidity, nutritional status, vitamin A deficiency, and possibly genetic predisposition. The majority of poliovirus and hepatitis A virus infections in young children are asymptomatic. For enteric viruses like group A rotavirus and astrovirus, symptomatic infections are common among children under 2 years of age. *E. coli* O157:H7 infections are more severe in young children and the elderly because of their weaker immune systems and are more likely to result in HUS.

CLASSIFICATION OF WATER-RELATED DISEASES

The classification of water-related diseases by Bradley (19) provides a valuable framework for understanding the relationship between infectious-disease transmission and water. This classification system facilitates the planning of effective prevention and control measures for a variety of water-related diseases, depending on the type of agent and type of transmission route involved. Bradley described four main categories of water-related infections: waterborne infections, water-washed infections, water-based infections, and infections with water-related insect vectors.

The waterborne infections are those classically recognized as waterborne disease, such as typhoid and cholera, in which an enteric microorganism enters the water source through fecal contamination and transmission occurs by ingestion of contaminated water. Transmission by this route depends on (i) the amount of fecal contamination in the water, the concentration of pathogens in the fecal contamination (determined by the number of infected persons in the community), and the survival of the pathogenic organism in water; (ii) the infectious dose of the organism; and (iii) individual ingestion (exposure) of the contaminated water. Control of these infections is generally through improvement of microbiological water quality, through water treatment and/or through source protection.

Water-washed infections are diseases due to poor personal and/or domestic hygiene. These diseases are not due to the presence of infectious agents in water but rather to the lack of readily accessible water. This limits washing of hands and utensils and thus permits the transmission of infectious agents, such as *Shigella* species, by feces-contaminated hands and utensils. Transmission is again related to the presence of feces from an infected individual, the infectious dose, the amount of fecal contamination on the hand or surface, and the survival of the organism on surfaces. Lack of water for bathing also facilitates the spread of diseases that affect the eye and skin, such as trachoma, conjunctivitis, and scabies. Control of these diseases is through provision of greater quantities of water, closer, easier access to water, and education to improve personal and domestic hygiene.

Water-based infections are worm infections in which the pathogen must spend a part of its life cycle in the aquatic environment. This category is further subdivided into diseases acquired by ingestion of water and diseases acquired by contact with water. The prototype infections in this category are dracunculiasis, due to ingestion of water contaminated with the guinea worm (*Dracunculus medinensis*), and schistosomiasis, which is transmitted by contact with water contaminated with species of the trematode genus *Schistosoma*. The original source of the guinea worm is larvae discharged from the female worm that lies in a vesicle usually on the lower leg or foot of an infected human. The larvae are discharged when the vesicle is immersed in water and are then ingested by a copepod (genus *Cyclops*), where they develop into the infective stage. Humans become infected when they ingest water containing the copepods (19). The eggs of schistosome worms enter the aquatic environment from the urine or feces of an infected human. The eggs hatch in the water to produce miracidia that infect snails, develop into the infective stage, and are shed by snails into the water over a period of months. Humans become infected when the free-swimming infective larvae penetrate the skin during water contact (19). Although schistosomiasis is typically considered a tropical disease, there are reports of schistosomal dermatitis (swimmer's itch) in the United States that appear to be associated with *Schistosoma* species (31). Control of dracunculiasis and schistosomiasis is through protection of the water source and the user by limiting skin contact with water and by eradication of intermediate hosts.

The types of water contact diseases most frequently encountered in industrialized countries are those associated with recreational water exposure to contaminated marine water, freshwater lakes, ponds, creeks, or rivers, and occasionally treated water in swimming pools, wave pools, hot tubs, and whirlpools. While many recreational water outbreaks are associated with enteric organisms and ingestion of water with fecal contamination, there are some diseases of the ear, eye, and skin that are associated with actual water contact and systemic illness associated with penetration of a pathogen through an open wound or abrasion. Reported recreational water outbreaks have involved noroviruses, *Giardia*, *Cryptosporidium*, *Shigella sonnei*, and *E. coli* O157:H7, which presumably entered the gastrointestinal tract via ingestion (31, 38, 41). Other recreational water outbreaks have involved ingestion, contact, or inhalation of indigenous aquatic organisms such as *Naegleria*, *Pseudomonas*, *Legionella* (31), several *Vibrio* species, and several *Mycobacterium* species (73). Epidemiological and microbiological studies indicate that *Staphylococcus aureus* skin and ear infections are often associated with recreational use of water, and the source of these organisms may be from other bathers or from the water (44, 45). *Vibrio vulnificus* can cause serious wound infections when a skin injury occurs in marine water or from contact of preexisting wounds with marine water (156). Cyanobacterial toxins have been associated with contact irritation after bathing in marine water or freshwaters (49). An additional cause of recreational water infections are the *Leptospira* species that are neither enteric organisms nor aquatic organisms but enter water via the urine of infected domestic and wild animals (73).

Infections with water-related insect vectors are those transmitted by insects that breed in water, such as mosquito vectors of malaria, or insects that bite near water, like the tsetse flies that transmit sleeping sickness. Control of these infections is through the application of pesticides, destruction of breeding grounds, and construction of piped water supplies.

Two additional modes of transmission that are water related are the transmission of infectious agents by inhalation of water aerosols and by the consumption of raw or undercooked shellfish or contaminated fish. The major pathogens associated with aerosol transmission are *Legionella* species, especially *L. pneumophila*, the etiologic agent of Legionnaires' disease and Pontiac fever. Legionellae are ubiquitous in water and soil and are capable of prolonged survival and reproduction in the aquatic environment. Growth within free-living amoebae appears to enhance survival and provide protection from routine disinfection (289). Outbreaks of legionellosis have been associated with aerosols from cooling towers and evaporative condensers of large buildings or with hot and cold water systems in hospitals, hotels, and other institutions. Legionellae can proliferate in hot-water tanks maintained at 30 to 54°C (276), and exposure occurs from aerosols created from showerheads. Control of these infections is through minimizing exposure to contaminated aerosols and routine cleaning and disinfection of water systems with adequate doses of chlorine or ozone (198, 288). Recent work suggests that the use of monochloramine is more effective than free chlorine as a residual disinfectant to protect against legionellae (158). Waterborne-disease outbreaks associated with *Legionella* were first included in the Centers for Disease Control and Prevention (CDC) surveillance system for waterborne disease outbreaks in 2001, and during 2001 and 2002, six outbreaks that caused illness in 80 people (41 hospitalizations) were associated with drinking water systems and additional outbreaks were associated with recreational waters and fountains (41).

The potential for aerosol transmission of *Mycobacterium avium* and other nontuberculous mycobacteria and the risk to the immunocompromised population continue to be a concern. Like *Legionella*, these organisms are frequently isolated in environmental and treated water systems and are able to colonize and propagate within water distribution systems (56, 136). Wendt et al. (282) reported the isolation of nontuberculous mycobacteria from aerosol samples near the James River in Virginia. The isolates (mostly *Mycobacterium intracellulare*) were biochemically similar to those recovered from human clinical specimens, suggesting that airborne mycobacteria derived from freshwater might be a significant source of infection.

Bivalve molluscan shellfish serve as vehicles of enteric-disease transmission because of their ability to concentrate enteric organisms from feces-contaminated water in their tissue. Numerous outbreaks have been attributed to the consumption of raw or undercooked oysters, clams, and mussels (196). Many pathogens, including hepatitis A and E viruses, noroviruses (NoV), pathogenic *E. coli*, *Salmonella enterica* serovar Typhi, and species of *Shigella*, *Vibrio*, *Plesiomonas*, and *Aeromonas*, have been implicated in shellfish-borne disease (106, 107). Shellfish and some species of fish may also serve as vehicles for algal toxins. Toxic species of *Gonyaulax* and *Gymnodinium* are concentrated by filter-feeding mollusks and can cause paralytic shellfish poisoning among shellfish consumers (26). Reef-feeding fish can concentrate toxic dinoflagellates of the genus *Gambierdiscus* that cause ciguatera seafood poisoning among consumers (26).

STUDIES OF WATERBORNE DISEASE

Epidemic Waterborne Disease

Most of the information on the risk factors and etiologic agents of waterborne disease comes from investigations of waterborne-disease outbreaks by state and local health departments and the surveillance program maintained by the CDC and the Environmental Protection Agency (EPA). It is difficult to recognize waterborne-disease outbreaks because many infections due to waterborne agents are relatively mild and self-limited, so the infected persons may not seek medical care. For those who do receive medical care, many waterborne infections may not be diagnosed or reported to public health authorities by a health care provider or laboratory (94). The outbreak surveillance system is based on voluntary reporting by state health departments and clearly represents only a fraction of the true incidence of waterborne-disease outbreaks. Data on waterborne outbreaks in the United States have been summarized by Craun (57, 58) and others (28, 29, 31–33, 38, 129). These data indicate that since the early 1980s, the overall number of waterborne-disease outbreaks associated with drinking water has been declining. This may be due to improved water treatment practices and the Surface Water Treatment Rule, which reduced the risk from waterborne protozoa. However, the proportion of drinking water outbreaks associated with problems in the distribution system has been increasing since the late 1980s and has focused new concern about the integrity of aging water distribution systems and posttreatment contamination of drinking water. Most of the drinking water outbreaks in the past 10 years have involved groundwater systems and especially the use of untreated groundwater (32, 34, 36, 38, 40, 41). The number of waterborne-disease outbreaks associated with recreational water has been steadily increasing, and the number of recreational water outbreaks reported in 2000 and 2001 is the highest since 1978, when the CDC first started collecting these data (41).

Until recently, the majority of waterborne outbreaks were classified by this surveillance system as acute gastrointestinal illness of unknown etiology (AGI). However, as outbreak investigations and diagnostic methods improve, etiologic agents are being identified in a greater proportion of outbreaks. Stool examinations by hospital laboratories typically include culture for *Salmonella*, *Shigella*, and *Campylobacter*. In addition, many laboratories test for group A rotavirus in specimens from young children and for *Giardia* and *Cryptosporidium* at the request of a physician. Protozoa were the major etiologic agents associated with drinking water outbreaks in the United States during the 1980s and early 1990s and continue to be the leading cause of recreational water outbreaks associated with treated waters because of their resistance to chlorination. Clinical symptoms suggest that many of the AGI outbreaks may be due to viral agents, such as NoV. Recent advances in the detection of NoV in clinical and environmental samples have led to increased testing and confirmation of their importance in epidemic gastroenteritis. However, testing for other viral agents is still rarely performed. In addition, there are a number of newly recognized etiologic agents for which there is some evidence of an association with waterborne disease, such as *Cyclospora cayetanensis* (292), cyanobacteria (26), *H. pylori* (155), *Mycobacteria* (136), microsporidia (54) and *Aeromonas* (239). These agents are discussed in "Emerging Waterborne Pathogens" below.

Endemic Waterborne Disease

Intervention Studies

The primary concern with infectious agents in drinking water is acute gastrointestinal illness. In industrialized countries, there have been few studies of endemic gastrointestinal disease associated with the consumption of drinking water. Two randomized intervention trials were conducted in Canada to examine the risk of gastrointestinal illness associated with the consumption of conventionally treated municipal drinking water that met current microbiological standards (224). In the first study, 606 households were recruited into the study, and 299 of these households were supplied with reverse-osmosis filters that provided additional in-home water treatment. Gastrointestinal symptoms were recorded in family health diaries. Water samples from the surface water source, treatment plant, distribution system, and study households were analyzed for several indicator bacteria and culturable viruses. Over a 15-month period, a 35% higher rate of gastrointestinal symptoms was observed in the 307 study households drinking municipal tapwater without in-home treatment than in the 299 study households supplied with reverse-osmosis filters. Symptomatology and serologic evidence suggested that much of this increased illness might have been due to low levels of enteric viruses in the municipal water supply that originated from a river contaminated by human sewage. Using a similar design, a second intervention study was conducted in the same community with 1,400 families randomly allocated to four groups of 350. One group consumed conventionally treated water that met current North American drinking water standards. The second group consumed tap water from a continuously purged tap. The third group consumed tap water that was bottled at the treatment plant, and the fourth group consumed purified bottled water (tap water treated by reverse osmosis or spring water). The health of the families was monitored for a 16-month period. The groups consuming tap water and continuously purged tap water experienced 14 and 19% more illness, respectively, than did the families consuming purified bottled water. Greater illness rates were observed in children 2 to 4 years of age. The authors concluded that 14 to 40% of the reported gastrointestinal illnesses were attributable to tap water that met current water standards and that contamination in the water distribution system was partly responsible for these illnesses (221, 225).

A similar study in Melbourne, Australia, with 600 families found no difference in self-reported gastroenteritis rates between the 300 families with a water treatment device (1-μm-pore-size absolute filter cartridge and UV treatment chamber) and the 300 families with a sham device (1, 115). Melbourne draws its drinking water from a protected forest watershed and has an unfiltered surface water supply that uses only free chlorine treatment. Free chlorine levels in the distribution system ranged from 0 to 0.94 mg/liter, with a median of 0.05 mg/liter, and 90% of samples had <0.20 mg/liter. Total coliform bacteria were detected in 18.9% of 1,167 routine 100-ml water samples, but fecal coliform bacteria were not detected. All study families had at least two children 1 to 15 years of age and completed a weekly health diary reporting gastrointestinal symptoms during the 68-week observation period. The overall rate of highly credible gastroenteritis was 0.80 cases per person-year. Children <10 years of age accounted for 40% of the total study population and for 55% of the cases of highly credible gastroenteritis. However, there was no significant difference in the illness rates between the intervention and control groups for children or for adults. Analysis of 795 fecal specimens from participants with gastroenteritis did not reveal any difference in pathogen detection rates between the two study groups.

Two recent intervention studies have been conducted in the United States to examine endemic waterborne disease. As a preliminary trial to a subsequent epidemiology study, a randomized, triple-blinded, home drinking water intervention trial of 77 households was conducted for 4 months in California (50). The drinking water was treated using a device placed under the kitchen sink that incorporated UV light and microfiltration. Although the purpose of the trial was to evaluate the "blinding" of the study (e.g., to determine if the participating households could identify the active and identical-looking placebo devices), analysis of the data showed that the incidence rate ratio of disease (adjusted for clustering) was 1.32, with a 95% confidence interval of 0.75 to 2.33. Given the small study size, the higher rate of gastrointestinal disease among the placebo group was not statistically significant; the authors concluded, however, that the relative rates of illnesses were similar to those observed by Payment et al. (224).

The full-scale intervention study was conducted in Davenport, Iowa, to determine the incidence of gastrointestinal illness associated with consumption of drinking water meeting all federal and state treatment guidelines (51, 165). The municipal water system used a single source (the Mississippi River) of water, which was treated at a single plant with conventional methods consisting of coagulation, flocculation, sedimentation, filtration (dual filters with granular activated carbon and sand), and chlorination. A total of 456 households with 1,296 participants were randomized into two groups. One group received a household water treatment device with a 1-μm-pore-size absolute ceramic filter and UV light with 35,000 to 38,000 μW-s/cm^2 output. The other group received a sham device that was identical to the active device but had an empty filter chamber and a UV light that was shielded to block the transmission of radiation but still generated the same light and heat as the active unit. Each study household had an active device for 6 months and a sham device for 6 months and was blinded to the status of their device during the study. Study participants recorded the occurrence of any symptoms in daily health diaries. Highly credible gastrointestinal illness (HCGI) was defined as in the previous studies by Payment et al., Hellard et al., and Colford et al. as (i) vomiting, (ii) watery diarrhea, (iii) soft diarrhea and abdominal cramps, or (iv) nausea and abdominal cramps. Incidence of HCGI ranged from 1.82 to 2.42 per person-years at risk. Multivariate analyses showed no effect of the household water treatment device on illness rates during the 12-month study period. As in the studies by Payment et al., the highest illness rates were in children 5 years of age and younger. The overall conclusion was that less than 11%

of the gastrointestinal illness observed in this community was due to drinking water. Unlike the studies by Payment et al. and Hellard et al., this study included households without children, and it is possible that the number of young children in the study was too small to permit detection of an effect in this more vulnerable group.

Other Epidemiologic Studies of Endemic Waterborne Disease

Microbial contamination in parts of the distribution system may also play a role in risks of endemic illness. The studies by Payment et al. (224, 225) suggest that the distribution system may have contributed to the difference between the gastrointestinal illness rates observed in study households that drank tap water and those of the study households that drank tap water with additional treatment or bottled water. A recent study conducted in Wales and northwest England from 2001 to 2002 found a very strong association ($P < 0.001$) between self-reported diarrhea and reported low water pressure at the home tap based on a postal survey of 423 subjects (130). Although there has been concern about possible health risks from pressure loss and pathogen intrusion in water distribution systems (167), this is the first study to provide solid evidence of that risk.

Untreated groundwater is a predominant cause of waterborne disease outbreaks in the United States. A longitudinal study of French alpine villages that used untreated groundwater for their drinking water supplies observed a weak relationship between rates of endemic acute gastrointestinal disease and the presence of fecal streptococcus indicator bacteria in the public water system over a 15-month study period (293). Illness data were collected through active surveillance by physicians, pharmacists, and schoolteachers. Weekly water samples were collected from frequently used taps in the distribution system of each village and were analyzed for several bacterial indicator organisms. Two studies in Ontario, Canada, have examined the relationship between *E. coli* in well water and acute gastrointestinal illness in households using the water for drinking (232, 255). In the first study of 181 households with untreated well water, water samples were collected five times during the 1-year study, and *E. coli* was detected in 20% of the household wells. From 38 to 42% of the study households had at least one child 12 years old or younger. Each study household maintained a family health diary, and an overall rate of 0.40 gastrointestinal episodes per person-year was observed. There was a significant association between wells that were *E. coli* positive and individual risk of gastrointestinal illness after adjusting for the age of the individual, the depth of the well, and the distance from the well to the septic tank. The association between the risk of gastrointestinal illness at a family level and *E. coli* detection in their well water was not statistically significant (odds ratio, 2.11, $P = 0.09$). The second study included 235 households in four rural communities and also used a prospective cohort design (255). However, the follow-up period was only 28 days. Each household well was sampled twice, and study households recorded gastroenteritis symptoms in a health diary over the 28-day study period. As in the previous study, 20% of the households had at least one water sample that exceeded the national standards for total coliforms or *E. coli*. However, there was no statistically significant association between the risk of gastrointestinal illness and the detection of indicator bacteria in the well water.

Ecological studies have attempted to find a relationship between endemic rates of hepatitis A infection reported in 75 counties and municipal source water quality and/or level of water treatment for all water supplies in the county (9). However, no statistically significant associations were observed. A time-series study (240) attempted to link turbidity in the Philadelphia water supply with hospital emergency room visits for gastrointestinal symptoms. However, a number of methodological problems with this investigation make the study results questionable. Two case-control studies have found the consumption of unfiltered surface water or shallow well water to be a risk factor for endemic giardiasis (47, 65). A cross-sectional survey of 226 HIV-infected individuals in San Francisco (with an unfiltered surface water supply) found that 53% consumed regular tap water and that the consumption of tap water was not associated with increased diarrheal illness (76).

MICROBIAL AGENTS ASSOCIATED WITH WATERBORNE DISEASE

The commonly recognized waterborne pathogens consist of several groups of enteric and aquatic bacteria, enteric viruses, and three enteric protozoa (Table 1).

Enteric and Aquatic Bacteria

In the 19th century, *Vibrio cholerae* and *Salmonella enterica* serovar Typhi were the first waterborne pathogens to be recognized, and they have been responsible for tremendous morbidity and mortality worldwide. *Vibrio* species are aquatic bacteria that are well adapted to both the estuarine environment and the intestinal tract. Waterborne enteric bacteria include both human-associated and zoonotic species. *Campylobacter* and *Salmonella* species are found in the intestinal tract of numerous domestic and wild animals. Therefore, contamination of water by animal feces also poses a human health risk. For other enteric bacteria, such as *Salmonella* serovar Typhi and *Shigella*, infections are generally limited to humans. The ID for enteric bacteria depends on several host factors, including gastric acidity, and the vehicle of transmission. Volunteer studies have shown that, when ingested with milk, the ID_{50} is approximately 10^2 CFU for *Shigella* species, approximately 10^7 CFU for *Salmonella* serovar Typhi, and approximately 10^8 CFU for enterotoxigenic *E. coli* (202). Diagnosis of these infections is typically achieved by culturing the microorganisms from clinical specimens, detecting bacterial antigens or antibodies by enzyme immunoassays, or by detecting specific genes by using molecular biology methods such as nucleic acid probe hybridization.

In the developing world, classic waterborne bacterial infections due to *V. cholerae* and *Salmonella* serovar Typhi continue to be a problem. In the United States, recent drinking water outbreaks due to bacterial agents have predominantly been due to *Shigella* species, followed by *Campylobacter* and toxigenic *E. coli*. The majority of recreational water outbreaks due to bacterial agents have been dermatitis infections caused by *Pseudomonas*, followed by enteric infections with *Shigella* and toxigenic *E. coli* infection (28, 29, 31, 32, 35, 43).

Data on the occurrence of pathogenic bacteria in water have been summarized by Emde et al. (78). However, many of the data on microbial occurrence in raw water and treated water are believed to be obsolete because they were gathered by insensitive analytical methods with selective recovery media that underestimate the bacterial levels as much as 100- to 1,000-fold (251). The isolation of bacterial pathogens from water is difficult because of low concentrations, their stressed condition, interference from competing aquatic microorganisms, and the fastidious nutrient requirements of some pathogens. Enrichment procedures are often used to promote growth of the organisms before they are transferred to selective media for identification. However, this limits the ability to quantify the pathogens in the sample (82).

The persistence of enteric bacteria in the aquatic environment depends on the species and on a variety of environmental factors (temperature, pH, sunlight, predation and competition by indigenous aquatic microorganisms, dissolved organics, attachment to particulates, association with vectors such as amoebae, protozoa, or copepods, and the presence of salts and other solutes), which have been reviewed (99, 209). The classic study by McFeters et al. (182) demonstrated that the time for a 50% reduction in the population of several enteric bacteria in well water at 20°C ranged from 2.4 h for some *Salmonella* species to 26.8 h for some *Shigella* species. More recently, it was recognized that some enteric bacteria (*Vibrio* spp., *E. coli*, *Campylobacter* spp., and *Salmonella* spp.) have the ability to enter a dormant state, referred to as "viable but nonculturable," in which they can survive for long periods of time (277). The public health significance of bacteria in this state is uncertain, although there is some evidence that they can still be infectious at high doses (203, 277). The difficulty inherent in recovering stressed bacteria from water samples may lead to false-negative results when evaluating microbial water quality. Some enteric bacteria have the ability to multiply in the aquatic environment under favorable nutrient and temperature conditions (113).

The majority of bacterial pathogens are removed or inactivated by standard water treatment practices. Removal capabilities of specific processes have been reviewed (165, 173). Most of the recent drinking water outbreaks associated with bacterial pathogens have been due to consumption of untreated groundwater. Enteric bacteria are relatively susceptible to disinfection. The CT (i.e., the product of disinfectant concentration in milligrams per liter and contact time in minutes) values for 99% inactivation by free chlorine (pH 6 to 7) vary from 0.034 to 0.05 for enteric bacteria like *E. coli* at 5°C (48) to values of 15 and 28 for aquatic bacteria such as *L. pneumophila* at 20°C and *Mycobacterium fortuitum*, respectively (250).

Enteric Viruses

The enteric viruses are more recently recognized waterborne pathogens. Human rotaviruses, NoV (Norwalk virus and other human caliciviruses), and human astroviruses were first described in the early to mid-1970s. Depending on the virus, diagnosis of these infections may be by commercially available enzyme immunoassays for antigen or antibodies (group A rotavirus, enteric adenovirus types 40 and 41, and hepatitis A virus), electron microscopy, tissue culture (enterovirus, adenovirus, astrovirus, and rotavirus), or molecular methods such as PCR and probe hybridization (NoV, astrovirus, rotavirus, adenovirus, enterovirus, and hepatitis E virus). Unlike what has been observed with enteric bacteria, concern about waterborne transmission of enteric viruses is generally limited to the strains that have humans as their natural reservoir. Although there are animal strains of many of these viruses, animal-to-human transmission is believed to be uncommon. The ID of these agents is low, typically in the range of 1 to 10 infectious units. Human volunteer studies with group A human rotavirus estimated the median ID to be between 5 and 6 focus-forming units (280).

Enteric viruses tend to be more persistent in the aquatic environment than most enteric bacteria; however, their survival depends on numerous physical, chemical, and microbial characteristics of the water as well as the virus type. The estimated time for about a 50% reduction of hepatitis A virus in groundwater at 25°C was approximately 14 days (253). Their prolonged survival and small size enable viruses to move greater distances in soil and water. Unlike aquatic or enteric bacteria, it is not possible for enteric viruses to multiply in the environment. Cultivable enteric viruses have been detected in surface waters, groundwaters, and treated drinking waters in concentrations ranging from 647 PFU per liter (241) to 1 PFU per 1,000 liters (222). Virus recovery from water samples is relatively poor, and some important enteric viruses (NoV) cannot be cultured in vitro. However, the development of new molecular amplification techniques and nucleic acid hybridization has improved the sensitivity of virus detection in water (132). A groundwater survey of 448 wells in 35 states reported that 31% of the sites were positive for at least one virus, and enterovirus RNA was detected in approximately 15%, rotavirus RNA in 14%, and hepatitis A virus RNA in 7% of the wells by reverse transcription (RT)-PCR (2). Groundwater sites in geological formations classified as alluvial or unconsolidated had higher percentages of virus-positive samples. Fout et al. (92) examined 321 samples from 29 groundwater sites by RT-PCR and reported that 72% of the sites were virus positive. Borchardt et al. (18) collected monthly samples from four municipal wells in one city in Wisconsin for a 12-month period and detected enteric viruses by RT-PCR in 50% of the samples.

Inactivation and/or removal of enteric viruses by water treatment processes has been reviewed (131, 165, 222, 250), and the results vary by virus type and treatment conditions. In general, enteric viruses are more resistant than enteric bacteria to free chlorine, and CT values for 99% inactivation range from 2 to >30 (165). Reported CT values for 99% inactivation of poliovirus 1 and rotavirus at 5°C with free chlorine (pH 6 to 7) are 1.1 to 2.5 and 0.01 to 0.05, respectively. However, the CT values for 99% inactivation of these same viruses by using chloramine (pH 8 to 9) are 768 to 3,740 (poliovirus 1) and 3,806 to 6,476 (rotavirus) (48). Experiments by Sobsey et al. (252) demonstrated that the CT value for 99.99% inactivation of hepatitis A virus (pH 6, 5°C) was 2.3 for dispersed virus and 29 for cell-associated virus. A human volunteer study that examined the infectivity of Norwalk virus in water treated with different doses of chlorine indicated that Norwalk virus remained infectious after a 3.75-mg/liter dose of chlorine and 30 min of contact time (153). However, more-recent studies (243, 244) indicate that Norwalk virus is no more-resistant to monochloramine or ozone disinfection than poliovirus 1 and bacteriophage MS2. However, these results should be interpreted with caution because they are based on RT-PCR detection of viral RNA rather than infectivity assays.

Inadequate diagnostic technology has limited the detection of many enteric viruses in both clinical and environmental samples. Consequently, enteric viruses have not been frequently identified as the etiologic agent of waterborne-disease outbreaks. From 1986 through 2002, there were 20 reported drinking water outbreaks associated with viral agents (5 with hepatitis A virus and 15 with NoV or related viruses) in the United States (28, 29, 31). These outbreaks involved over 8,300 total cases, and most were attributed to the ingestion of untreated or inadequately treated groundwater. In addition, group A rotavirus was implicated in a large waterborne-disease outbreak in Colorado due to the mechanical breakdown of a chlorinator (121). Many outbreaks currently reported as AGI are thought to be due to viral agents because of the epidemiological and clinical features of the outbreaks. Recent advances in the diagnosis and detection of enteric viruses through the application of molecular techniques are increasing the recognition of waterborne outbreaks due to viral agents. For example, 9 of the 15 NoV drinking water outbreaks reported since 1986 occurred between 1999 and 2002, and eight recreational water outbreaks associated with NoV were also reported during this period (38, 41). There have also been increasing reports of waterborne outbreaks associated with NoV in Europe (180).

Enteric Protozoa

Since 1981, enteric protozoa have been the leading cause of waterborne-disease outbreaks in which an etiologic agent was identified. In recent years, the proportion of drinking water outbreaks associated with protozoa has declined, but protozoa, especially *Cryptosporidium*, are still the major cause of recreational water outbreaks of gastroenteritis. The enteric protozoa are relatively recently recognized waterborne pathogens. Human cryptosporidiosis was first described in 1976 (184, 205) and the first reported waterborne outbreak occurred in 1984 (61). Recent evidence indicates that *Cryptosporidium* is the third most common enteric pathogen worldwide (60). Diagnosis is typically by immunofluorescent microscopy or enzyme immunoassay; however, cryptosporidiosis is probably greatly underdiagnosed. A prospective surveillance study in one state indicated that only 0.5 to 1% of all fecal samples submitted were examined for *Cryptosporidium* and that 3.7 to 8.5% of stool samples contained oocysts (15). Although both trophozoites and cysts/oocysts are shed in feces, the cysts are the infective form and, like the enteric viruses, the ID for these agents is low. Human challenge studies with three different isolates of *Cryptosporidium* indicate substantial variation in their infectivity for humans, with the median infectious dose ranging from 9 to 1,042 oocysts (259). Water contamination by both human and animal feces is an important mode of transmission for *Giardia* and *Cryptosporidium*. Humans are the only host for *Entamoeba histolytica*.

The thick-walled protozoan cysts and oocysts are environmentally resistant, but there are few data on their survival in water (236). Their size (*Giardia* cysts are 7 to 14 μm, and *Cryptosporidium* oocysts are 4 to 6 μm) tends to limit extensive migration through soil, but groundwater has become contaminated by protozoan cysts during flooding. Methods for the recovery and detection of *Giardia* and *Cryptosporidium* cysts or oocysts in water have improved in recent years and include concentration via a capsule filter followed by immunomagnetic separation and analysis by immunofluorescent microscopy (method 1623) as well as an integrated cell culture-PCR strategy to assess the viability of *Cryptosporidium* oocysts (166). Surveys of raw and treated water supplies indicate that the occurrence of *Cryptosporidium* oocysts is widespread. Two surveys of almost 300 surface water supplies in the United States revealed that 55 to 77% of surface water samples had *Cryptosporidium* oocysts (234, 237). Water analysis at 66 U.S. and Canadian surface water treatment plants demonstrated that up to 27% of treated drinking water samples had low levels of *Cryptosporidium* oocysts (164). Some detection methods may underestimate the occurrence of these organisms in water due to poor recovery efficiency or overestimate the risk to human health due to inability to assess the viability of the cysts/oocysts in a way that relates to their potential for human infectivity (236). Recent studies by LeChevallier suggest that approximately 37% of *Cryptosporidium* oocysts detected in raw water by the immunofluorescence method were viable and infectious (166). Like enteric viruses, it is not possible for enteric protozoa to multiply in the environment.

The cysts and oocysts of enteric protozoa are persistent in the environment for months in cool, moist conditions and are also relatively resistant to chlorine disinfection (87). CT values for *Giardia lamblia* cysts for 99% inactivation at pH 6 to 7 by free chlorine range from 47 to 150 (48). The CT value for 99% inactivation of *Cryptosporidium parvum* by free chlorine (pH 7) at 25°C is estimated to be 7,200 (48). Therefore, removal of these organisms by water treatment processes depends on effective coagulation/flocculation, sedimentation, and filtration. A recent case study of three surface water treatment facilities that used coagulation/flocculation, sedimentation, and filtration observed $\log_{10}$ reductions of 2.24 to 2.78 for *Giardia* and 2.3 to 2.45 for *Cryptosporidium* (163). UV light has proved to be particularly effective against *Cryptosporidium* and *Giardia* and is reported to achieve a 3-log inactivation at doses of less than 10 $mW\text{-}s/cm^2$ (165). Ozonation of drinking water can reduce *Cryptosporidium* and *Giardia* by 2 $\log_{10}$ (13). The removal and inactivation of *Cryptosporidium* and *Giardia* by drinking-water treatment processes have been comprehensively reviewed by Betancourt and Rose (13) and LeChevallier and Au (165).

As diagnostic methods for *Giardia* and *Cryptosporidium* improved, the recognition of their role in waterborne-disease outbreaks increased dramatically. From 1986 through 1996, 21% of the 172 outbreaks associated with drinking water were attributed to parasitic agents (29, 31, 32, 35). In recent years, the proportion of drinking water outbreaks attributed to protozoa has slightly declined (from 18% in 1999–2000 to 16% in 2001–2002) (38, 41). Numerous recreational water outbreaks have also been associated with *Giardia* and *Cryptosporidium*—especially outbreaks involving treated water. One striking aspect of waterborne cryptosporidiosis outbreaks is the large number of cases involved. This may be due to the fact that several outbreaks have involved larger community water systems, but it also suggests that there is widespread susceptibility to *Cryptosporidium* infection in the general population. In 1993, *Cryptosporidium* was implicated in the largest recorded waterborne-disease outbreak in United States, in which it was estimated that over 400,000 people were infected (175). The majority of drinking water outbreaks attributed to enteric protozoa have been associated with surface water supplies that were either unfiltered or whose flocculation and filtration processes were inadequate. *Cryptosporidium* outbreaks tend to have high attack rates, and fecal contamination from cattle or human sewage is usually suspected to be the source of the oocysts (236). There have been no reported waterborne outbreaks of *E. histolytica* in the United States since 1971 (57, 58).

EMERGING WATERBORNE PATHOGENS

Emerging infectious diseases have been defined as those whose incidence in humans has increased within the past 2 decades or threaten to increase in the near future (135).

There are a number of newly recognized infectious agents that have recently been associated with outbreaks of waterborne disease or appear to have the potential for waterborne transmission. Emerging waterborne pathogens are receiving increased recognition by water utilities and regulators (5). The Safe Drinking Water Act requires the EPA to identify new contaminants that may require regulation in the future, and the EPA must periodically release a list of candidate contaminants that are being considered for regulation. The most recent list of candidate contaminants was released in February 2005 and includes nine microorganisms: adenoviruses, caliciviruses, coxsackieviruses, echoviruses, *Aeromonas hydrophila*, *M. avium* and *M. intracellulare* (together referred to as the *M. avium* complex [MAC]), *Helicobacter pylori*, cyanobacteria (blue-green algae) and other toxin-producing freshwater algae, and microsporidia (*Enterocytozoon* and *Septata*) (268).

Recently Recognized Waterborne Pathogens and Suspected Waterborne Pathogens

Enteric Viruses

Human caliciviruses comprise two genera in the family *Caliciviridae* (NoV and sapoviruses) and include Norwalk virus and other viruses previously named "small round structured viruses." NoV are further classified as either genogroup I (including Norwalk and Southampton viruses and others) or genogroup II (including Lordsdale virus, Snow Mountain virus, Hawaii virus, Mexico virus, Toronto virus, and others). These viruses are the leading cause of epidemic viral gastroenteritis in the United States and acquire their names from the locations where outbreaks have occurred. Numerous NoV outbreaks linked to drinking water, recreational water, ice, shellfish, various food items, and environmental contamination have been documented (117, 149, 150, 154, 193). Until recently, insufficient diagnostic technology limited the study of the role of these viruses in both epidemic and endemic gastroenteritis. The cloning and sequencing of Norwalk virus in 1990 (138) led to rapid advances in the development of sensitive new molecular diagnostic methods (64, 139, 141, 142) and increased recognition of the importance of NoV in waterborne and foodborne disease.

Seroprevalence data suggest that epidemic and endemic NoV infections are common worldwide. Early studies in the United States reported that 50 to 70% of adults have antibodies to Norwalk and related viruses by the fifth decade of life (16, 104). More-recent studies have reported high NoV seroprevalence in England (102, 178, 279), Kuwait (69), Brazil (96, 257), Sweden (118), Japan (120), Mexico (140), Finland (170), Norway (199), Kenya (201), Chile (214), Venezuela (229), and South Africa (248, 258). These high rates of antibody prevalence indicate that much of the population is frequently exposed to NoV and may experience recurrent NoV-associated gastroenteritis because serum antibodies do not appear to confer protection from illness (192, 219).

Approximately 15,400 documented cases of waterborne disease associated with these viruses were reported in the United States between 1971 and 2002 (3, 28–32, 34, 37, 39, 42), and this is likely to be a gross underestimate. Most NoV outbreaks are probably not recognized. In 1982 Kaplan et al. (150) estimated that 42% of acute nonbacterial gastroenteritis may be due to Norwalk and related viruses on the basis of serologic evidence from 74 outbreaks investigated by the CDC from 1976 to 1980. Examining outbreak characteristics, Kaplan et al. (149) suggested that Norwalk virus and related viruses may be responsible for 23% of waterborne outbreaks of acute gastroenteritis in the United States. The low ID and occurrence of mild and asymptomatic infections facilitates waterborne transmission and high rates of subsequent secondary spread in a community (192).

With the development of new detection methods for NoV in water (broadly reactive RT-PCR primers, quantitative real-time RT-PCR, better concentration methods, and strategies to control RT-PCR inhibitors), there has been a surge in the number of reports of NoV detection in tap water (109), bottled water (14), swimming pool water (179), surface waters (172, 230, 247, 265), groundwater (18, 208), seawater (151), and sewage (172, 230, 265, 270). Reported NoV detection levels in environmental samples in The Netherlands ranged from 4 to 4.9×10^3 PCR-detectable units (PDU) per liter of river water to 9.0×10^2 to 7.5×10^3 PDU/liter in treated sewage and 5.1×10^3 to 8.5×10^5 PDU/liter in raw sewage (172). The average reported NoV levels in wastewater effluent and river water downstream of a wastewater treatment plant in Germany ranged from 1.8×10^4 to 9.7×10^5 genome equivalents per liter by Taqman quantitative real-time RT-PCR (230).

Enteric adenoviruses (serotypes 40 and 41, also known as subgenus F) are DNA viruses associated with about 5 to 12% of pediatric diarrhea and a median duration of 8.6 and 12.2 days, respectively (275). Diagnosis is by commercially available enzyme immunoassay or by PCR (275). To date, there is no evidence of transmission of enteric adenoviruses via drinking water. However, nonenteric adenoviruses (serotypes 3, 7, 1, 4, and 14) have been associated with recreational water outbreaks of pharyngoconjunctival fever and conjunctivitis. Conventional and real-time PCR methods have been developed to detect adenoviruses in environmental samples (147, 228). Adenoviruses have been detected in sewage, environmental waters, swimming pools, and drinking water (in Korea and South Africa) (93, 169, 217, 226, 272). Adenoviruses appear to be stable in water (80) and particularly resistant to UV inactivation because of their double-stranded DNA genome (207, 260, 261). Typical UV doses used for drinking water disinfection are not adequate to inactivate adenoviruses (165), and there is concern about whether adenovirus should be the target pathogen for UV disinfection guidelines.

Coxsackieviruses and echoviruses are small, single-stranded, positive-sense RNA viruses that are classified as enteroviruses. Enteroviruses are among the most common human viruses and are estimated to infect 50 million people annually in the United States (216). The vast majority of these enterovirus infections are asymptomatic or cause general febrile illnesses with or without respiratory symptoms. Coxsackieviruses and echoviruses have also been associated with aseptic meningitis, encephalitis, paralysis, herpangina, pericarditis, epidemic conjunctivitis, exanthem, hand-foot-and-mouth syndrome, and neonatal diarrhea (190). In addition to acute illnesses, enterovirus infections have also been associated with several serious chronic diseases, including myocarditis and dilated cardiomyopathy, type 1 diabetes mellitus, and neuromuscular diseases (216). The prevalence of enterovirus infections varies by serotype, season, and geographic location. Both endemic and epidemic patterns have been described (216). Enteroviruses are transmitted both through the fecal-oral route and by respiratory secretions (person-to-person and via fomites). Depending on the virus serotype, virus may be shed in respiratory secretions for about 1 month or in feces for as long as 5 months (190).

Although coxsackieviruses and echoviruses are commonly detected in water and sewage, the risk of waterborne transmission of these agents remains poorly characterized. In the past 20 years, no drinking water outbreaks associated with enteroviruses in the United States have been reported by the CDC waterborne-disease surveillance system (28-32, 34, 37, 39, 42). However, there is some recent evidence of enterovirus outbreaks possibly associated with contaminated drinking water in a summer camp in Alaska (183) and a city in Belarus (7). In a recent assessment of the risks of pediatric enterovirus infection from recreational water contact, Parkin et al. (218) concluded that the evidence against waterborne transmission of these viruses outweighs the suggestive or definitive evidence that supports waterborne transmission.

Enteroviruses can be detected in water samples by cell culture, RT-PCR, or a combination of cell culture and RT-PCR to detect infectious virus (168, 273). Surveys of viruses in various types of environmental waters and sewage indicate that the occurrence of enteroviruses in water is widespread, and enteroviruses tend to be detected more frequently than other enteric viruses. Abbaszadegan et al. (1) reported the detection of enterovirus in 15% of 448 groundwater wells in the United States. Borchardt et al. (18) detected enteroviruses by RT-PCR in 41% of monthly water samples collected before chlorination from four municipal wells in Wisconsin over a 1-year period. Others have reported the detection of enteroviruses in treated drinking water samples (74, 273).

Enteroviruses are stable in the environment, especially at neutral pH and low temperatures (218). Although generally more resistant than enteric bacteria to chlorination and UV disinfection, enteroviruses fall within the range of inactivation reported for other enteric viruses (165).

Currently, there is no evidence that other emerging enteric viral pathogens or putative pathogens, such as astrovirus, coronavirus, picobirnavirus, parvovirus, polyomavirus, circovirus, torovirus, or Aichi virus, are transmitted via drinking water or recreational water. Some of these viruses have been detected in sewage (207), which raises concern about the possibility for fecal-oral transmission via water.

Enteric and Aquatic Bacteria

Enterohemorrhagic *E. coli* O157:H7 is a pathogenic strain of *E. coli* that produces two potent toxins. This organism causes bloody diarrhea, and 2 to 7% of infections result in HUS, in which the red blood cells are destroyed and the kidneys fail (4). In many parts of the United States and Canada, *E. coli* O157:H7 is the second or third most commonly isolated enteric bacterial pathogen (103). Children and the elderly are most susceptible to HUS complications. This severe disease was first recognized in 1982 and has the highest mortality rate of all waterborne diseases in the United States. The ID is believed to be low, like that of *Shigella*, and the incubation period is 12 to 60 h. The reservoir appears to be healthy cattle, and transmission often occurs by ingestion of undercooked beef or raw milk. Person-to-person transmission is important among families and in childcare centers. Diagnosis is by culture followed by a commercially available latex agglutination test for *E. coli* O157 antigen (103).

Several outbreaks of *E. coli* O157:H7 associated with drinking water have been reported in the United States. The first recognized outbreak was associated with drinking water in a Missouri community in 1989 in which there were 243 cases. Of these patients, one-third had bloody diarrhea, 32 were hospitalized, 2 had HUS, and 4 persons died (116). The *E. coli* strain was isolated from clinical specimens but not from water samples. Unchlorinated well water and breaks in the water distribution system were considered to be contributing factors to the outbreak. In May 2000, a waterborne outbreak of *E. coli* O157:H7 in Walkerton, Ontario, Canada (population, 4,800), resulted in 2,300 cases, 65 hospitalizations, 27 cases of HUS, and seven deaths. Extensive investigation concluded that the source of the *E. coli* was cow manure that was spread near one of the water supply wells for the community and revealed that the water supply operators had failed to properly chlorinate the well and monitor for chlorine residual (123). Heavy rainfall and local flooding also contributed to the source contamination. The magnitude of this outbreak, evidence of falsification of water testing records by the plant operators, loss of public trust in local authorities and the provincial government, a boil-water advisory that lasted over 6 months, an estimated $12 million to decontaminate the water system, a total economic impact on the community of more than $64 million, and a full-scale public inquiry illustrate the serious consequences of a waterborne-disease outbreak—even in a small public system. *E. coli* O157:H7 outbreaks have also been associated with recreational water exposure (31, 32, 35). Most of these recreational water outbreaks involved lake water; however, several outbreaks have occurred in treated swimming pools (38). Prolonged survival of *E. coli* O157:H7 in water has been reported by Geldreich et al. (98), who observed only a 2-log reduction after 5 weeks at 5°C.

Aeromonas species are commonly found in water and soil. Reported densities in water range from 10^4 per ml in river water to 1 to 10 per ml in groundwater (226). There are reports of *Aeromonas* spp. in chlorinated waters, and evidence suggests that chlorination has only a temporary effect on *Aeromonas* and does not prevent regrowth in distribution systems (129). High population densities appear to be related to fecal pollution and temperature, and aeromonads proliferate in domestic and industrial wastewaters (239). There is some evidence to suggest that a high proportion of environmental isolates may produce enterotoxins (178), and several reports have suggested an association between gastroenteritis and *Aeromonas* in drinking water (23, 24, 239). However, the health significance of detecting aeromonads in drinking water is not well understood. A 2-year clinical study in Iowa concluded that three strains of *Aeromonas* were capable of causing diarrhea and that consumption of untreated water was a risk factor for *Aeromonas* infection (197). Studies in London reported a correlation between water and fecal isolates of *Aeromonas sobria* (204). A case-control study in the United States of *A. hydrophila* gastroenteritis reported that drinking untreated water was a significant risk factor for infection (119). High summer prevalence of nosocomial *A. hydrophila* infection was linked to high counts in hospital water storage tanks in France (227). A study in The Netherlands found that patients infected with cytotoxigenic strains of *Aeromonas* were more likely to have had recreational water contact during the week prior to onset of illness (162). However, other studies in The Netherlands and London found little similarity between aeromonads isolated from diarrheal feces and those found in drinking water (110, 186). Concern about the possible health effects from these organisms in The Netherlands has led to the establishment of drinking water guidelines of <20 CFU/100 ml for drinking water leaving the treatment plant and <200 CFU/100 ml for drinking water in the distribution system (271).

A wide range of atypical (nontuberculosis) mycobacteria occur in the environment. Generally, these organisms cause illness mainly in immunosuppressed populations. Although opportunistic mycobacteria can infect almost any site in the body, they are most commonly associated with pulmonary disease, cervical lymphadenopathy, and localized skin and soft tissue infections (136). Cutaneous lesions have been associated with recreational water exposure to *M. marinum* (also called *M. balnei*) (73, 136). MAC organisms have been detected in soil, wastewater, surface water, groundwater, and drinking water (4). There are numerous reports that MAC organisms are present in tap water and can persist and colonize drinking water supply systems. A U.S. survey of drinking water samples from 21 states detected nontuberculous mycobacteria in 54% of ice samples and 35% of public drinking water supplies but not in bottled water or cisterns (56). Reported concentrations of MAC in water distribution systems range from 0.08 to 45,000 CFU/100 ml (4). MAC organisms can grow over a wide temperature range (15 to 45°C) and persist in water distribution systems for years because of their ability to survive in waters with low nutrient concentrations and produce or live in biofilms (269). Several reports suggest that environmental mycobacteria have prolonged survival in water systems because of their association with free-living amoeba and other protozoa (269). Mycobacteria are resistant to routine water chlorination because of their thick cell walls with high lipid content (4, 226). The CT values for free chlorine for a 99.9% reduction of *M. avium* are 580 to 2,300 times higher than those for *E. coli*. Little is known about the susceptibility of these organisms to other disinfectants. Chlorine dioxide and chloramines appear to be more effective in reducing mycobacteria than free chlorine (269). Studies of *M. avium* survival in a model distribution system concluded that reducing biodegradable organic matter in drinking water, controlling corrosion, and maintaining an effective disinfectant residual and elevated hot water temperatures help reduce the occurrence of *M. avium* in drinking water biofilms (205).

The incidence of MAC infection, which typically causes pulmonary and disseminated disease, in HIV-infected patients has dramatically increased, causing further speculation about the possible role of water in the transmission of this agent (246). Epidemiology studies have not identified drinking water as a risk factor for MAC in healthy hosts. However, there is some evidence that suggests waterborne nosocomial transmission of MAC to immunocompromised hosts. A recent study that compared clinical mycobacterium isolates with environmental isolates from hospital hot water systems suggested that exposure to these water sources was the transmission route for several groups of AIDS patients (274). Other species (*M. kansasii* and *M. xenopi*) have been isolated from domestic and hospital water supplies; however, the health significance of this finding is uncertain (136).

H. pylori, formerly referred to as "*Campylobacter pylori*," was first described in 1982 by Marshall and Warren (17). This gram-negative, microaerophilic spiral-shaped bacterium resides in the mucus layer lining of the stomach and upper gastrointestinal tract and is associated with indigestion and abdominal pain. However, asymptomatic infections are common. Current evidence suggests that these organisms are the principal cause of duodenal ulcers and gastric ulcers (226). Chronic *H. pylori* infection is associated with gastric cancer, and infected individuals have a three- to sixfold higher risk of developing gastric cancer than uninfected individuals (226). *H. pylori* infections occur throughout the world, and the prevalence of infection increases with age. Individuals in developing countries are more likely to become infected earlier in life, and by adulthood, infection rates range from approximately 40% in developed countries to 80% in developing countries. Effective antibiotic treatment can successfully eradicate the infection in most cases, but there is no universally effective treatment regimen (226). If untreated, infection becomes chronic and probably persists for life (55).

Fecal-oral transmission of *H. pylori* infection has been suggested by several studies that implicated crowding, socioeconomic status, contact with sheep, and consumption of raw, sewage-contaminated vegetables as risk factors for infection (122, 185, 189). Studies in Peru, Colombia, and Ethiopia have identified the type of water supply as a risk factor for infection and in some cases found that the water source appeared to be a more important risk factor than socioeconomic status (11, 101, 155, 171). Other studies in Peru have argued that food prepared under unhygienic conditions is the major transmission route of *H. pylori* in developing countries (11). A seroprevalence survey of 245 healthy children in Arkansas did not find a relation between *H. pylori* seropositivity and the type of water supply (municipal or well) (89). Cross-sectional studies of U.S. and Mexican children living in the U.S.-Mexican border region also did not find that the type of water source or the frequency of drinking "purified water" was associated with *H. pylori* infection (211). It is likely that the differences between the magnitude of fecal contamination in the Peruvian and Ethiopian water sources and that of water sources in Arkansas, as well as misclassification of reported drinking water quality, may explain these conflicting reports.

Because it is a fastidious organism, *H. pylori* has been difficult to culture from environmental sources. Laboratory studies (242, 283) demonstrated that *H. pylori* survives in fresh water and sterile, distilled water (7°C) for 10 and 14 days, respectively. These studies also found evidence that *H. pylori* can survive for prolonged periods in a viable, nonculturable state. Klein et al. (155) reported difficulty detecting *H. pylori* from water samples because the culture plates were overgrown before the 3 to 4 days required to isolate *H. pylori*. Enroth and Engstrand (81) reported the development of conventional PCR methods to detect *H. pylori* in water. More recently, quantitative real-time PCR methods have been described for *H. pylori* detection at low concentrations in drinking water (181). PCR has been used to detect *Helicobacter* DNA in private wells, municipal tap water, and wastewater in Sweden (125), drinking water samples in Peru (126), and river water in Japan (95). Another study used monoclonal anti-*H. pylori* antibody and microscopy to detect actively respiring *H. pylori* in surface and shallow groundwater in Pennsylvania and Ohio and found that there was no significant correlation between the detection of either total coliforms or *E. coli* and *H. pylori* in water (114). Disinfection studies have shown that three strains of *H. pylori* were inactivated by more than 4 orders of magnitude after exposure to 0.5 mg of free chlorine per liter for 80 s (145). These findings suggest that *H. pylori* should be readily controlled by disinfection practices normally used for drinking water treatment.

Protozoa, Fungi, and Algae

Cyanobacteria (blue-green algae) occur naturally in fresh and brackish waters worldwide. Although these are not infectious agents, some species produce toxins during algal blooms that are triggered by nutrient enrichment from natural waters, agricultural fertilizer runoff, or domestic or industrial effluents (49). Approximately 46 species of cyanobacteria have been associated with toxic effects in vertebrates (10, 285). The most common freshwater toxic forms are *Microcystis* spp., *Planktothrix* (syn. *Oscillatoria*) *rubescens*, *Planktothrix* (syn. *Oscillatoria*) *agardhii*, *Anabaena* spp., *Aphanizomenon* spp., some *Oscillatoria* spp., *Cylindrospermopsis raciborskii*, *Synechococcus* spp., *Gloeotrichia* spp., *Lyngbya* spp., *Nostoc* spp., *Schizothrix* spp., and *Synechocystis* spp. (285). Toxic blooms have been reported in many parts of the United States, Canada, Europe, Southern Africa, Asia, Australia, and New Zealand (26, 49). The conditions that influence the toxicity of a bloom are not known (77). However, temperature is believed to be one key factor (26). Surveys in several countries indicate that 75% of cyanobacterial blooms contained toxins (10, 285).

Cyanobacterial toxins are of three main types: lipopolysaccharide endotoxins, hepatotoxins, and neurotoxins. Hepatotoxins appear to be more common than neurotoxins (285). Acute health effects in humans include gastroenteritis, liver damage, nervous system damage, pneumonia, sore throat, earache, and contact irritation of skin and eyes (49, 264). The potential chronic health effects of long-term exposure to cyanobacterial toxins in drinking water are unknown. It has been suggested that high rates of liver cancer in parts of the People's Republic of China may be linked to cyanobacterial hepatotoxins in drinking water (25). A 1976 outbreak of intestinal illness in Pennsylvania was associated with a cyanobacterial bloom in a municipal water supply and affected 62% of the population (26). In addition, there are several reports of adverse health effects related to contact with recreational water contact in the United States (26).

Although cyanobacterial poisoning has been long recognized, in recent years there have been numerous reports of cyanobacterial poisoning associated with surface water ingestion and contact in Australia (112, 238, 254). One outbreak of hepatoenteritis in the Palm Island Aboriginal settlement on the Australian northeast coast in 1979 affected 138 children and 10 adults, the majority of whom required hospitalization (49, 111). The outbreak was linked to a dense algal bloom in the drinking water reservoir for the island, which was treated with copper sulfate that caused lysis of the cyanobacteria and consequent release of toxins. An outbreak of gastroenteritis affecting 121 of 304 inhabitants in a Swedish village was associated with an accidental cross-connection between untreated river water and the drinking water supply. The river water contained high concentrations of *Planktothrix agardhii* that contained microcystins (285). A case-control study in South Australia reported that ingestion of chlorinated river water was a significant risk factor for gastrointestinal symptoms compared to drinking rainwater and that the weekly number of gastroenteritis cases in the study area was correlated with mean log cyanobacterial cell counts in the river water (238). An earlier epidemiologic study in New South Wales, Australia, examined levels of hepatic enzymes in routine blood specimens submitted to a local hospital and found elevated levels of gamma-glutamyltransferase (indicating damage to liver cell membranes) among city residents during a heavy algal bloom (*Microcystis aeruginosa*) in the municipal water reservoir (86). These levels were significantly higher than those measured in specimens collected from city residents 1 month before and after the bloom and compared to levels in country residents who used other water supplies.

The control of cyanobacteria is problematic because several studies indicate that the toxins can remain potent for days after the organisms have been destroyed by copper sulfate or chlorination (111, 238). Activated carbon and ozone have both been shown to be effective methods for removing these toxins. Lime treatment is also an effective control method because it can precipitate phosphorus and sediment the cyanobacterial cells (5). Based on toxicity data from mouse bioassays, the Engineering and Water Supply Department of South Australia developed interim guidelines for acceptable numbers of cyanobacteria in water supplies (238). Recently WHO suggested a provisional drinking water guideline for microcystin-LR of 1 μg/liter (286). New methods of chemical analysis, high-performance liquid chromatography, and commercially available immunoassays and enzyme assays are now available for detection of microcystins and saxitoxins (5, 285). However, further research is needed on the acute and chronic toxicity of cyanobacterial toxins, and better methods need to be developed for monitoring the types and concentrations of cyanobacterial toxins in natural and treated water supplies (77).

Cyclospora cayetanensis, formerly called "cyanobacterium-like bodies" or "big *Cryptosporidium*," was identified as a new protozoan pathogen of humans in 1993 (212). A member of the family *Eimeriidae*, it was first described by Soave and colleagues in 1986 (249) in stools from patients who had lived or traveled in developing countries and is frequently found in AIDS patients with prolonged diarrhea (212). Under light microscopy it appears as a refractile sphere, 8 to 10 μm in diameter. By fluorescent microscopy, the oocysts autofluoresce blue with a 365-nm dichromatic filter and mint green with a 450- to 490-nm dichromatic filter (176). Diagnosis is by microscopic observation of oocysts in stool samples or by conventional or real-time PCR for a segment of rRNA (176). *C. cayetanensis* infection has been associated with prolonged, self-limited watery diarrhea with an average duration of 40 days. Cyclospora infection can be effectively treated by oral trimethoprim-sulfamethoxazole. Cyclospora oocysts are shed in feces of infected individuals but are not immediately infectious until they are fully sporulated. Sporulation requires one to several weeks depending on environmental conditions (176). Ortega reported that, under laboratory conditions, *C. cayetanensis* sporulation occurred within 7 to 13 days at 22 to 32°C (213). This organism occurs in tropical climates worldwide and has been identified as a cause of endemic and epidemic diarrheal disease in North, Central and South America, Caribbean countries, Southeast Asia, and Eastern Europe (157). Unlike what is observed with *Cryptosporidium* and *Giardia*, humans appear to be the only natural host for *C. cayetanensis*, and there is no evidence of zoonotic transmission (176).

Evidence for waterborne transmission comes from a case-control study of a diarrhea outbreak among foreign residents in Nepal, in which consumption of untreated water was identified as a risk factor (245). A case-control study of *Cyclospora* infection in Guatemala also reported that the consumption of untreated water, river water, or spring water was significantly associated with risk of infection (12). Cross-sectional studies of *Cyclospora* prevalence and risk factors in Haiti suggested that the use of specific artesian wells was the only risk factor associated with *Cyclospora* infection, and *Cyclospora* was detected by both epifluorescence microscopy and PCR in one of the wells in the study area (174). The Nepal and Guatemala studies and two prospective cohort studies of Peruvian children (212) indicate a seasonal pattern of infection with peak incidence occurring in warm summer months. In the United States, most cyclosporosis outbreaks have involved foodborne transmission and were associated with imported produce from areas where the disease is endemic. The first documented waterborne outbreak of *Cyclospora* diarrhea occurred among the house staff of a Chicago hospital in the summer of 1990 (31). Contaminated open-air, rooftop water storage tanks that supplied the house staff dormitory were believed to be the source of the outbreak. However, *Cyclospora* species could not be detected in water samples from the tanks. Another waterborne outbreak of cyclosporiasis was linked to the consumption of water from water coolers on a golf course in New York in June 1995 (27). Two case reports have described *Cyclospora* infections in immunocompetent hosts. In one case the host developed prolonged diarrhea after exposure to sewage-contaminated water in his home (108). Sewage from a neighboring dairy backed up into the basement of the patient's home, and *Cyclospora* organisms were detected in sewage effluent samples taken from the pipe that served the dairy. In the other case, *Cyclospora* infection was associated with consumption of well water in New Hampshire after heavy snow melts had caused local flooding (210). *Cyclospora* oocysts have been detected by microscopy and PCR in sewage lagoons near areas of endemic cyclosporiasis in Lima, Peru (256) and in drinking water sources in rural areas in Guatemala and Haiti (72, 174). *Cyclospora* oocysts have also been detected in uncovered water tanks and one swimming pool in Egypt by using direct fluorescence monoclonal antibodies and Ziehl-Neelsen stain. As with *Cryptosporidium*, *Cyclospora* oocysts are not affected by chlorine and must be removed by filtration. Currently, there are no data on UV inactivation of *Cyclospora* oocysts.

"Microsporidia" is a general term that describes a large group of primitive, obligate, intracellular protozoa commonly found in animals. Infection and transmission of microsporidiosis in humans and animals have been recently reviewed by Didier (67, 68). The first human infection was reported in 1985 (66). Currently, species of six genera, *Nosema*, *Encephalitozoon*, *Pleistophora*, *Enterocytozoon*, *Vittaforma*, *Trachipleistophora*, and the general grouping "microsporidium" (for unclassified microsporidia) are known to cause human infections (97). With the exception of *Enterocytozoon bieneusi*, there seems to be little host specificity, as human infections with nonmammalian genera have been described, and these organisms can infect a variety of tissues and organs (intestinal tract, eyes, liver, kidney, heart, and central nervous system) (177). *E. bieneusi* and *Encephalitozoon intestinalis* (formerly referred to as *Septata intestinalis*) parasitize primarily the intestine and can cause persistent diarrhea. The vast majority of reported cases of microsporidial infections have occurred among persons infected with HIV (22). A total of 5,692 stool specimens collected from 1,454 people from 1993 to 1996 were submitted to a reference laboratory in Lyon, France, for assessment of microsporidia (54). Intestinal microsporidia were identified in 338 persons: 261 persons infected with HIV (0.1% prevalence), 16 transplant patients, and 61 others. Epidemiological and laboratory studies indicate that *E. bieneusi* is an important cause of chronic diarrhea in AIDS patients, and there is no effective treatment against this organism (62, 177, 281). Studies of seasonal variation in enteric microsporidiosis among HIV-infected persons in New Orleans, La., and southern California did not find any significant seasonal trends in the prevalence of microsporidiosis that could be evidence of waterborne transmission (53, 159).

The role of microsporidia in waterborne disease is not known. Microsporidia are excreted in the urine and feces of infected humans and animals. Transmission of microsporidia is believed to be primarily fecal-oral (22). Infection is acquired through ingestion of small (1- to 2-μm) spores with thick walls that make them environmentally resistant and allow them to remain infective for up to 4 months in the environment (278). Because of the apparent lack of host specificity, water contaminated with feces or urine from animals as well as humans could serve as a vehicle of transmission. One purported waterborne outbreak of intestinal microsporidiosis has been reported (54). The monthly incidence of intestinal microsporidiosis detected at the reference lab in Lyon was monitored from May 1993 to December 1996. Prevalence in HIV-infected patients rose from a baseline level of about 0.1% to a sharp peak of 1% in July and August 1995. Analyses of risk factors associated with microsporidiosis during the period before the peak did not indicate any relationship between microsporidiosis and the water distribution subsystem. However, during the period of peak prevalence, diagnosis of microsporidiosis was strongly associated with living in an area served by the "medium-elevation water distribution system." The water supply in this area is river water "naturally filtered by alluvia," chlorinated at two sites, and distributed to three independent subsystems with different elevations. Standard water quality measures did not indicate fecal contamination, and increased prevalence of *Cryptosporidium parvum* or *Giardia intestinalis* was not observed during the study period. Epidemiologic studies of risk factors associated with human microsporidiosis have identified exposure to swimming pools and hot tubs as potential transmission routes (63, 133).

Methods to recover and detect microsporidium spores by immunofluorescent assays and PCR are still under development (70). The first documented report of microsporidia in water in 1987 described the repeated detection of microsporidium spores by microscopy from ditch water samples collected in Florida during a 1-year period (8). Spore concentrations of up to 3,000 per ml of water were observed, and two of the five genera identified in the samples were groups that include human pathogens (*Nosema* and *Pleistophora*). More recently, Dowd et al. (71) reported PCR amplification of microsporidium DNA in 7 of 14 water concentrates and detected *Encephalitozoon intestinalis* in tertiary sewage effluent, surface water, and groundwater; *Enterocytozoon bieneusi* in surface water; and *Vittaforma corneae* in tertiary effluent. Others have reported the detection of microsporidia in swimming pools (90), river water (91), and irrigation waters (262).

Gerba et al. (100) investigated the removal of *E. intestinalis* by conventional drinking water treatment and reported a 2.47 $\log_{10}$ reduction by coagulation, sedimentation, and mixed-media filtration. The recent development of cell culture infectivity assays (144) has made it possible to examine inactivation of microsporidia by various disinfectants. Studies of the inactivation of *E. intestinalis* by chlorine and ozone indicate that the CT values for 2 $\log_{10}$ of inactivation ranged from 12.8 to 68.8 for chlorine (pH 6 and 8, respectively) and the CT values for ozone were 0.59 to 0.84 mg-min/liter (143). A reduction of 3 $\log_{10}$ or greater of *E. intestinalis* by UV light has been reported (124, 144).

Recently Recognized Waterborne Pathogens in Developing Countries

Hepatitis E Virus

Hepatitis E virus (HEV) is the only known agent of enterically transmitted non-A, non-B infectious hepatitis, although some evidence suggests that there may be others (263). HEV is a small, 30-nm virus with single-stranded, positive-sense RNA with three open reading frames. Recently classified in the genus *Hepevirus*, HEV appears to have a single serotype and at least four major genotypes (148). The recent development of serologic tests for HEV antibody based on recombinant DNA technology has enabled further study of the epidemiology of this virus, which is clinically and epidemiologically similar to hepatitis A virus (263). However, unlike hepatitis A virus, the majority of cases in epidemics occur among young adults (15 to 40 years of age), and there is a high case-fatality rate (up to 30%) among pregnant women (59). There is no specific treatment or vaccine against HEV infection. Large outbreaks, involving thousands of cases, have been reported in developing areas in Africa, Asia, and Mexico and have been linked to feces-contaminated water and inadequate chlorination. Refugee settlements in Sudan and Chad and war-torn Iraqi cities with decimated water and sewer systems have been the settings for massive waterborne outbreaks in recent years (79). In 1991, the largest documented waterborne HEV outbreak affected an estimated 79,000 persons in Kanpur, India, and was associated with contaminated surface water (200). HEV RNA was first detected in environmental samples from sewage treatment plants in India by using RT-PCR and nucleic acid probe hybridization (146). The recent development of a broadly reactive, sensitive real-time RT-PCR assay for HEV should enable better studies of the occurrence of HEV in the environment (148). Currently, there are no data on the inactivation of this virus in water by disinfection, nor are there any reports of HEV detection in water. HEV has been detected in swine manure, and it has been suggested that pigs may provide a nonhuman reservoir for HEV (291).

Group B Rotavirus

Group B rotavirus was first reported in connection with a waterborne-disease outbreak in China in 1984 (128) and differs from the group A rotavirus strains that commonly cause pediatric diarrhea. Group B rotavirus infections occur more frequently in adults than children, are associated with severe, cholera-like illness, and have been reported only in China, India, and Bangladesh (137). By electron microscopy, the virus is morphologically similar to group A rotavirus but is antigenically distinct. Diagnosis is by electron microscopy and genome electrophoresis, enzyme immunoassay, or RT-PCR. Seroprevalence surveys for group B rotavirus antibody have reported rates of up to 41% positivity in China, 5 to 18% positivity in Thailand, Burma, and Hong Kong, 1 to 12.5% positivity in the United Kingdom, United States, and Canada, and 10% positivity in Kenya (20). More than 1 million cases were reported in China in 1982 and 1983 (127), and many outbreaks involving tens of thousands of cases were attributed to feces-contaminated water (128). However, it is puzzling why so few cases of group B rotavirus have been identified since the 1980s. The detection of group B rotaviruses in water has not yet been reported.

Vibrio cholerae O139

V. cholerae O139 is a new toxigenic strain of epidemic *V. cholerae* that has been called the "Bengal" strain. This is the first non-O1 *V. cholerae* strain ever reported to be associated with epidemic cholera and is causing a new pandemic in Asia, which began in India in October 1992 (46) and is currently spreading into the Middle East. The clinical illness is indistinguishable from cholera caused by *V. cholerae* O1. However, this new epidemic is affecting persons of all ages in areas where most of the population, except for young children, have acquired immunity to *V. cholerae* O1. The Bengal strain is also significant in the history of cholera because recent evidence suggests that this strain resulted from genetic recombination and horizontal gene transfer that allowed the acquisition of unique DNA (52). As with other strains of *V. cholerae*, transmission occurs through feces-contaminated water and food.

Waterborne Disease of Unknown Etiology

Recent waterborne-disease surveillance data in the United States indicate that no etiologic agent could be determined for about one-third of the reported drinking water disease outbreaks from 1993 to 2002 (28, 29, 31). Part of this may be due to the lack of appropriate clinical and/or environmental samples or limited diagnostic techniques for many of the newly recognized pathogens. However, there are some outbreaks that, despite thorough investigation, could not be attributed to any known etiologic agent and suggest the existence of unrecognized agents of waterborne disease.

"Brainerd diarrhea" was first described in an outbreak with 122 cases that occurred in Brainerd, Minn., in 1983 (215). The illness was characterized by chronic diarrhea with acute onset, marked urgency, lack of systemic symptoms, and failure to respond to antimicrobial therapy. The distinctive feature of the illness is the average duration of 12 to 18 months. In the Brainerd outbreak, transmission was linked to raw milk that had passed through hoses rinsed with poor-quality water. Since the outbreak in Brainerd, eight other outbreaks have been classified as Brainerd diarrhea, and four of these were linked to drinking water (187). An outbreak in Illinois in 1987 was clearly associated with the consumption of untreated well water that had sporadic problems with fecal coliform counts (220). Several other possible outbreaks have been suspected in Texas in 1985 and five other states, as well as one that occurred among 58 passengers on a cruise ship, who participated in group tours to the Galapagos Islands in 1992 (188). In the Galapagos cruise ship outbreak, illness began 11 days (median) after the passengers boarded the ship and lasted 7 to 42 months. Epidemiological evidence linked the outbreak to the water supply on the ship, which was inadequately treated (188). Despite intense microbiological analyses of several outbreaks involving a total of some 300 cases, no etiologic agent for this syndrome has been identified.

PREVENTION AND CONTROL OF WATERBORNE DISEASE

Historical surveillance data (58) and epidemiologic/microbiologic studies of water and health provide an evidence base indicating that improvements in water supply and water treatment practices result in a reduction in morbidity and mortality associated with waterborne diseases. As discussed in "Classification of Water-Related Diseases" above, water-related transmission of infectious agents is associated not only with microbiological water quality but also with water availability, sanitation, and hygiene. The classic review by Esrey et al. (83) examined 144 studies of water quality, water quantity, and hygiene and concluded that improvements in microbiological water quality resulted in a 16% median reduction of diarrhea morbidity (range, 0 to 90%) and that improvements in both water quality and availability resulted in a 37% median reduction of diarrhea morbidity (range, 0 to 82%). Since the mid-1990s, a number of studies have focused on better understanding and evaluating the health impact of household water treatment practices that improve microbiological water quality in the home for households in developing countries that collect and store water because of the evidence that significant water contamination occurs during collection and storage. This recent focus on household level water quality is also relevant for industrialized countries because of the potential for contamination in the water distribution system. The recent review and meta-analysis by Fewtrell and Colford (88) of 60 studies found that in developing countries, interventions to improve household level water quality were very effective in reducing diarrheal disease, with a pooled estimate of reduction of approximately 40%. The health impact of interventions in the water supply varied (9 to 38% reduction in diarrhea morbidity), depending in part on whether the level of service was a household connection or standpipe connection and whether the improved water supply included water treatment (improvement in microbiological water quality and availability) or no treatment (improvement mainly in water availability) (88).

Reviews of water intervention studies have identified many limitations in the studies and interventions and illustrate the complexity of designing and implementing effective water intervention projects and evaluating their impact (84, 88, 191). More accurate and rapid methods to measure microbiological water quality and to identify and evaluate risk factors for waterborne disease would be valuable study tools and improve our understanding of waterborne-disease transmission and how to prevent it.

Measurements of Microbiological Water Quality

Detection of Infectious Agents in Water

Many waterborne pathogens are difficult to detect and/or quantify in water and for most of the newly recognized agents, methods to detect them in environmental samples have still to be developed. Analytical methods for microbiological water quality testing have been comprehensively reviewed by Köster at al. (160) and also a recent National Research Council (NRC) report (203). For enteric organisms, the concentrations are much lower in water than in clinical specimens; thus, their detection in water starts with a concentration process such as filtration, immunocapture, flocculation, or centrifugation. This is followed by a process to recover the pathogen from the filter, magnetic beads, floc, or sediment, and then by an enrichment or amplification process either by culture or molecular biology methods. Typically the recovery efficiencies of these procedures are low, making it difficult to estimate the original concentration of the infectious agent in the water. Also, some methods, such as immunofluorescent techniques used to detect *Giardia* cysts and *Cryptosporidium* oocysts or PCR amplification techniques to detect NoV, do not give an indication of the viability or infectivity of the organisms. Furthermore, many of these laboratory techniques are limited to specialized research or reference laboratories and are not done on a routine basis.

In most investigations of waterborne-disease outbreaks, water is identified as the vehicle of transmission by epidemiologic evidence rather than by the detection of the infectious agent in water samples. For example, of the 32 waterborne infectious disease outbreaks reported in the United States between 1991 and 1992, the etiologic agents were detected in the water in only four outbreaks and were identified as *Shigella* (1), *Giardia* (2), and *Cryptosporidium* (1, 31). This may be because the contamination of the water supply was temporary and the infectious agent died off or was flushed out of the water system before the outbreak was recognized and appropriate water samples were collected. The longer the incubation period of the infectious agent is, the longer it may be before the outbreak is recognized and water is suspected as the vehicle. The size and timing of the contamination event, the lag time until recognition of the outbreak, the survival characteristics and transport patterns of the agent, and the sensitivity and efficiency of the laboratory methods affect the likelihood that an infectious agent will be detected in water (235).

Microbial Indicator Organisms

Regulatory agencies, water utilities, and researchers have typically relied on the use of microbial indicator organisms to assess microbiological water quality and safety. This practice and the strengths and limitations of currently used, as well as proposed, indicator organisms have been recently evaluated by the NRC (203). EPA and WHO standards for microbiological water quality are expressed in terms of "total coliforms" and "fecal coliforms." These are groups of bacteria that are excreted by healthy humans and animals and serve as indicators of fecal contamination (267, 284). Laboratory tests for total and fecal coliforms in water are much easier to perform than tests to detect pathogenic microorganisms in water. Because these indicator organisms are excreted in high numbers by all individuals, the concentrations of these indicators are likely to be higher and a more constant fraction of the community fecal waste "pool." By contrast, specific pathogens are excreted only by infected individuals and their numbers in the community fecal waste pool depend on the excretion level of the particular pathogen and on the number of infected individuals in the community.

Ideally, microbial indicators should provide a measure of health risk associated with the ingestion or contact with water. Total and fecal coliforms have many limitations as predictors of risk of waterborne disease. Because of their shorter survival times in water and their greater susceptibility to water treatment processes, these indicator organisms tend to be poor models for enteric protozoa and viruses. Outbreaks of waterborne disease, especially protozoal outbreaks such as the Milwaukee epidemic (175), have been associated with water that met total and fecal coliform standards. Moreover, there are nonfecal sources for these indicator organisms and, in contrast to most enteric pathogens, total and fecal coliforms may multiply in aquatic environments with sufficient nutrients and optimal temperatures. Such characteristics may result in false-positive reports of water contamination. The performance of total and fecal coliforms as reliable indicators of tropical water quality is especially problematic and has been investigated and reviewed by Hazen (113). This is of particular concern in tropical developing countries in which there are a great number of untreated, contaminated water sources and high morbidity and mortality associated with waterborne disease (113).

Many alternative indicator organisms have been investigated, and their advantages and limitations have been reviewed (152, 203). *E. coli* and enterococci were included in the revision of the EPA recreational water standards in 1986 (266). *Clostridium perfringens* and male-specific coliphage have been proposed as potential indicators of drinking water quality that may better model the survival and disinfection resistance of enteric protozoa and viruses (223). Epidemiologic/microbiologic studies have demonstrated that *E. coli* and enterococci were better indicators of tropical drinking water quality than fecal coliforms (195). There is increasing recognition that no single organism can serve as an adequate indicator for all types of water, all routes of exposure, and all monitoring purposes. The NRC report (203) recommended a three-level, phased monitoring framework for selecting indicator organisms and evaluating risk from waterborne pathogens. Level I uses rapid, simple, low-cost screening tests to provide early warning of a change in background conditions or a possible health risk. Indicators that are sensitive and broadly applicable are most suitable for this purpose. When the screening indicators persist at high levels without a clear source of contamination and cannot be controlled, then level II monitoring is instituted. Level II involves more-detailed testing to confirm whether a real health risk exists—perhaps by measuring additional, more specific indicators that are better surrogates for the sources, persistence, and transport of the pathogens of interest or directly assaying for pathogens. Level III monitoring may involve more-detailed studies, such as tracer studies, to determine the sources of microbial contamination and how to mitigate them or studies of pathogen detection and assessment of infectivity. The overall purpose of this microbiological water quality monitoring framework is to better support risk management.

Multiple-Barrier Approach

In developed countries, waterborne-disease prevention and control are based on a multiple-barrier approach that involves source water protection, water treatment, and distribution system management and protection (3). Waterborne-disease outbreaks in the United States usually involve (i) source contamination and the breakdown of one or more of the treatment barriers (disinfection or filtration), or (ii) contamination of the distribution system, or (iii) the use of untreated water. The multiple-barrier approach emphasizes identification of all available barriers, assessing the degree of vulnerability of each barrier to the passage of pathogens, recognizing and anticipating conditions under which pathogen risk increases, and maintaining barriers at high levels of effectiveness (3). For waterborne diseases associated with aquatic microorganisms, approaches to prevention and control may depend on treatment technology or controlling exposure to untreated waters.

Risk Assessment Approaches

Risk assessment approaches have been useful to systematically identify, analyze, quantify, and characterize the risk of specific waterborne illnesses (105, 251). These models are based on field data on the occurrence of specific microorganisms in raw and treated water supplies, experimental data on removal or inactivation by various water treatment processes, and experimental dose-response data (105, 233). However, the shape of the dose-response curve, especially in the low-dose region representative of waterborne exposure, is ill defined. Furthermore, it is difficult to model variation in microbial virulence factors and host-specific characteristics, such as age and immune status, that may affect individual exposure and susceptibility to infection and disease. Finally, for infectious agents with multiple transmission routes, it may be difficult to determine the attributable risk associated with waterborne transmission compared to other routes of transmission, especially in areas where waterborne diseases are endemic (21). Newer risk assessment approaches include the use of dynamic disease transmission models of epidemic and endemic waterborne disease that incorporate multiple transmission routes and provide more useful information for decision making (75).

Recommendations and Research Needs

The provision of safe water and the control of diseases associated with contaminated water remain one of the most serious challenges facing the world today. It is currently estimated that 1.1 billion people in the world lack access to improved water supplies and 2.6 billion people lack adequate sanitation—which in turn leads to further water contamination (287). An estimated 4,000 to 6,000 children around the world die each day from diseases associated with lack of access to safe drinking water, inadequate sanitation, and poor hygiene (290). One of the United Nations Millennium Development Goals is to reduce by one-half the proportion of people without sustainable access to safe drinking water and basic sanitation by 2015. Although some parts of the world are making encouraging progress in meeting this goal, serious disparities remain. Global challenges in providing water and sanitation have recently been reviewed by Moe and Rheingans (194), who identified four major barriers to progress in the water and sanitation sector: (i) inadequate investment in water and sanitation infrastructures, (ii) lack of political will to make the provision of water and sanitation a priority, (iii) the tendency to avoid new technological or implementation approaches and not to involve the community in decisions about environmentally appropriate and sustainable water and sanitation systems, and (iv) failure to conduct evaluations of water and sanitation interventions to determine whether they are successful and sustainable.

Although the incidence of waterborne diseases has been greatly reduced in areas with effective water treatment, control of waterborne transmission of infectious agents continues to be a global problem and an important challenge for public health research. Multidisciplinary research in water and wastewater microbiology, environmental engineering, watershed management, human hygiene behavior, water economics, risk assessment, environmental and public health policy, and sustainable development can contribute to international efforts to meet the water and sanitation Millennium Development Goals. Methods to detect many waterborne pathogens in clinical and environmental samples need to be developed or improved. There is a critical need for rapid methods for those pathogens typically associated with waterborne disease as well as for other microbial pathogens that may be inadvertently or deliberately added to a water supply (134). New technologies such as DNA microarray chips (GeneChip Array Technology; bioMerieux [www.bioMerieux.com]), immunomagnetic separation and solid-phase laser scanning cytometry (231), and automated cell counting of immunofluorescent-tagged cells (Rapid Bacteria Detector RBD-2000; Advanced

Analytical Technologies [www.advanced-analytical.com]) may have potential for rapid, automated water quality analyses. There is also a need to develop simple, economical methods to detect fecal contamination in water so that they can be readily used in developing countries with limited laboratory facilities and resources. Information on the occurrence and persistence of enteric and aquatic pathogens in various types of water supplies needs to be updated, and the factors that contribute to virulence and waterborne-disease transmission need to be better characterized. For many waterborne pathogens, we need to improve our understanding of the effectiveness of various water treatment processes and disinfectants to remove and/or inactivate these microorganisms.

There is also a need for better surveillance of waterborne diseases, which would lead to earlier recognition and investigation of epidemic and endemic waterborne disease. Infection with *Giardia*, *Cryptosporidium*, and *E. coli* O157:H7 are now nationally notifiable diseases (www.cdc.gov/epo/dphsi/phs/infdis2006.htm). Routine stool examination for *Cryptosporidium*, NoV, and other pathogens would greatly further our understanding of the disease burden associated with various enteric pathogens, but this requires a shift from the current practice of stool examination solely for individual treatment purposes to the practice of stool examination for the benefit of public health. Increased awareness among health professionals of the symptoms associated with more recently described waterborne illnesses, such as those caused by *Cyclospora* or cyanobacteria, may lead to greater reporting of these conditions and provide opportunities to study their waterborne transmission. Because the health risks posed by many waterborne infectious agents are still unknown, water utilities and public health professionals need to work together to maintain vigilance for any indication of waterborne transmission of disease in their communities.

Finally, as we enter the new International Decade for Action: Water for Life 2005–2015, it is vital from a global perspective that we routinely evaluate new and existing water and sanitation systems. Such evaluations can collect the necessary information to support good policy decisions and investments and ensure that we protect our critical water resources and human health in both industrialized and developing country settings.

REFERENCES

1. **Abbaszadegan, M., M. W. LeChevallier, and C. P. Gerba.** 2003. Occurrence of viruses in U.S. groundwaters. *J. Am. Water Works Assoc.* **95:**107–120.
2. **Abbaszadegan, M., P. Stewart, M. LeChevallier, J. Rosen, and C. Gerba.** 1999. *Occurrence of Viruses in Ground Water in the United States.* American Water Works Association Research Foundation, Denver, Colo.
3. **American Water Works Association.** 1994. *Preventing Waterborne Disease: Is Your System at Risk?* Satellite Teleconference. American Water Works Association Research Foundation, Denver, Colo.
4. **American Water Works Association, Research Division Microbiological Contaminants Research Committee.** 1999. Emerging pathogens—bacteria. *J. Am. Water Works Assoc.* **91:**101–109.
5. **American Water Works Association, Research Division Microbiological Contaminants Research Committee.** 1999. Emerging pathogens—viruses, protozoa and algal toxins. *J. Am. Water Works Assoc.* **91:**110–121.
6. Reference deleted.
7. **Amvros'eva, T. V., Z. F. Bogush, O. N. Kazinets, O. V. D'iakonova, N. V. Poklonskaia, G. P. Golovneva, and R. M. Sharko.** 2004. Outbreak of enteroviral infection in Vitebsk during pollution of water supply by enteroviruses. *Vopr Virusol.* **49**(1)**:**30–34. (In Russian).
8. **Avery, S., and A. Undeen.** 1987. The isolation of microsporidia and other pathogens from concentrated ditch water. *J. Am. Mosq. Control Assoc.* **3:**54–58.
9. **Batik, O., G. F. Craun, R. W. Tuthill, and D. F. Kraemer.** 1980. An epidemiologic study of the relationship between hepatitis A and water supply characteristics and treatment. *Am. J. Public Health* **70:**167–168.
10. **Baxter, P.** 1991. Toxic marine and freshwater algae: an occupational hazard? *Br. J. Ind. Med.* **49:**505–506.
11. **Begue, R., J. Gonzales, H. Correa–Gracian, and S. Tang.** 1998. Dietary risk factors associated with the transmission of *Helicobacter pylori* in Lima, Peru. *Am. J. Trop. Med. Hyg.* **59:**637–640.
12. **Bern, C., B. Hernandez, M. B. Lopez, M. J. Arrowood, M. A. de Mejia, A. M. de Merida, A. W. Hightower, L. Venczel, B. L. Herwaldt, and R. E. Klein.** 1999. Epidemiologic studies of *Cyclospora cayetanensis* in Guatemala. *Emerg. Infect. Dis.* **5:**766–774.
13. **Betancourt, W. Q., and J. B. Rose.** 2004. Drinking water treatment processes for removal of Cryptosporidium and Giardia. *Vet. Parasitol.* **126:**219–234.
14. **Beuret, C., D. Kohler, and T. Luthi.** 2000. Norwalk-like virus sequences detected by reverse transcription-polymerase chain reaction in mineral waters imported into or bottled in Switzerland. *J. Food Prot.* **63:**1576–1582.
15. **Binford, L., M. Pentella, and B. Kwa.** 1991. A survey to determine the presence of *Cryptosporidium* infection associated with diarrhea in west central Florida, 1985–1992. *Fl. J. Public Health* **3:**18–21.
16. **Blacklow, N. R., G. Cukor, M. K. Bedigian, P. Echeverria, H. B. Greenberg, D. S. Schreiber, and J. S. Trier.** 1979. Immune response and prevalence of antibody to Norwalk enteritis virus as determined by radioimmunoassay. *J. Clin. Microbiol.* **10:**903–909.
17. **Blaser, M.** 1996. The bacteria behind ulcers. *Sci. Am.* **274:**104–107.
18. **Borchardt, M. A., N. L. Haas, and R. J. Hunt.** 2004. Vulnerability of drinking-water wells in La Crosse, Wisconsin, to enteric-virus contamination from surface water contributions. *Appl. Environ. Microbiol.* **70:**5937–5946.
19. **Bradley, D.** 1977. Health aspects of water supplies in tropical countries, p. 3–17. *In* R. Feachem, M. McGarry, and D. Mara (ed.), *Water, Wastes and Health in Hot Climates.* John Wiley & Sons, London, United Kingdom.
20. **Bridger, J.** 1994. Non-group A rotaviruses, p. 369–407. *In* A. Kapikian (ed.), *Viral Infections of the Gastrointestinal Tract.* Marcel Dekker, New York, N.Y.
21. **Briscoe, J.** 1984. Intervention studies and the definition of dominant transmission routes. *Am. J. Epidemiol.* **120:**449–455.
22. **Bryan, R., A. Cali, R. Owen, and H. Spencer.** 1991. Microsporidia: opportunistic pathogens in patients with AIDS, p. 1–26. T. Sun (ed.), *In Progress in Clinical Parasitology.* Field and Wood, Philadelphia, Pa.
23. **Burke, V., J. Robinson, M. Gracey, D. Peterson, N. Meyer, and V. Haley.** 1984. Isolation of *Aeromonas* spp. from an unchlorinated domestic water supply. *Appl. Environ. Microbiol.* **48:**367–370.
24. **Burke, V., J. Robinson, M. Gracey, D. Peterson, and K. Partridge.** 1984. Isolation of *Aeromonas hydrophila* from a metropolitan water supply: seasonal correlation with clinical isolates. *Appl. Environ. Microbiol.* **48:**361–366.
25. **Carmichael, W.** 1994. The toxins of cyanobacteria. *Sci. Am.* **270:**78–86.

26. **Carmichael, W., C. Jones, N. Mahmood, and W. Theiss.** 1985. Algal toxins and water-based diseases. *Crit. Rev. Environ. Control* **15:**175–313.
27. **Carter, R., F. Guido, G. Jacquette, and M. Rapoport.** 1996. Outbreak of cyclosporiasis associated with drinking water. *Abstr. 36th Intersci. Conf. Antimicrob. Agents Chemother. 1996*, abstr. K52.
28. **Centers for Disease Control and Prevention.** 1988. Water-related disease outbreaks, 1985. *Morb. Mortal. Wkly. Rep.* **37**(SS-2)**:**15–24.
29. **Centers for Disease Control and Prevention.** 1990. Waterborne disease outbreaks, 1986–1988. *Morb. Mortal. Wkly. Rep.* **39:**1–13.
30. **Centers for Disease Control and Prevention.** 1992. Waterborne disease in the United States: 1989–1990. *Morb. Mortal. Wkly. Rep. CDC Surveill. Summ.* **40:**1–22.
31. **Centers for Disease Control and Prevention.** 1993. Surveillance for waterborne-disease outbreaks—United States, 1991–1992. *Morb. Mortal. Wkly. Rep.* **42:**1–22.
32. **Centers for Disease Control and Prevention.** 1996. Surveillance for waterborne-disease outbreaks—United States, 1993–1994. *Morb. Mortal. Wkly. Rep.* **45:**1–33.
33. **Centers for Disease Control and Prevention.** 1998. Outbreak of acute febrile illness among athletes participating in triathlons—Wisconsin and Illinois, 1998. *Morb. Mortal. Wkly. Rep.* **47:**585–588. (Erratum, **47:**619.)
34. **Centers for Disease Control and Prevention.** 1998. Surveillance for waterborne-disease outbreaks—United States, 1995–1996. *Morb. Mortal. Wkly. Rep.* **47:**1–34.
35. **Centers for Disease Control and Prevention.** 1998. Update: leptospirosis and unexplained acute febrile illness among athletes participating in triathlons—Illinois and Wisconsin, 1998. *Morb. Mortal. Wkly. Rep.* **47:**673–676.
36. **Centers for Disease Control and Prevention.** 2000. Outbreaks of Norwalk-like viral gastroenteritis—Alaska and Wisconsin, 1999. *Morb. Mortal. Wkly. Rep.* **49:**207–211.
37. **Centers for Disease Control and Prevention.** 2000. Surveillance for waterborne-disease outbreaks in the United States, 1997–1998. *Morb. Mortal. Wkly. Rep. CDC Surveill. Summ.* **49:**1–35.
38. **Centers for Disease Control and Prevention.** 2002. Multistate outbreaks of salmonella serotype poona infections associated with eating cantaloupe from Mexico—United States and Canada, 2000–2002. *Morb. Mortal. Wkly. Rep.* **51:**1044 –1047.
39. **Centers for Disease Control and Prevention.** 2002. Surveillance for waterborne-disease outbreaks—United States, 1999–2000. *Morb. Mortal. Wkly. Rep. CDC Surveill. Summ.* **51**(8)**:**1–47.
40. **Centers for Disease Control and Prevention.** 2003. Hepatitis A outbreak associated with green onions at a restaurant—Monaca, Pennsylvania, 2003. *Morb. Mortal. Wkly. Rep.* **52:**1155–1157.
41. **Centers for Disease Control and Prevention.** 2004. Outbreak of cyclosporiasis associated with snow peas—Pennsylvania, 2004. *Morb. Mortal. Wkly. Rep.* **53:**876–878.
42. **Centers for Disease Control and Prevention.** 2004. Surveillance for waterborne-disease outbreaks associated with drinking water—United States, 2001–2002. *Morb. Mortal. Wkly. Rep. CDC Surveill. Summ.* **53:**46.
43. **Centers for Disease Control and Prevention.** 19 November 2004, posting dates. *Bioterrorism Agents/Diseases.* Centers for Disease Control and Prevention, Atlanta, Ga. [Online.] http://www.bt.cdc.gov/agent/agentlist-category .asp. Accessed 23 February 2006.
44. **Charoenca, N., and R. Fujioka.** 1993. Assessment of *Staphylococcus* bacteria in Hawaii's marine recreational waters. *Water Sci. Technol.* **27:**283–289.
45. **Charoenca, N., and R. Fujioka.** 1995. Association of staphylococcal skin infections and swimming. *Water Sci. Technol.* **31:**11–17.
46. **Cholera Working Group, International Centre for Diarrhoeal Disease Research Bangladesh.** 1993. Large epidemic of cholera-like disease in Bangladesh caused by *Vibrio cholerae* O139 synonym Bengal. *Lancet* **342:**387–390.
47. **Chute, C., R. Smith, and J. Baron.** 1987. Risk factors for endemic giardiasis. *Am. J. Public Health* **77:**585–587.
48. **Clark, R., C. Hurst, and S. Regli.** 1993. Costs and benefits of pathogen control in drinking water, p. 181–198. *In* G. Craun (ed.), *Safety of Water Disinfection: Balancing Chemical and Microbial Risks*. ILSI Press, Washington, D.C.
49. **Codd, G., S. Bell, and W. Brooks.** 1989. Cyanobacterial toxins in water. *Water Sci. Technol.* **21:**1–13.
50. **Colford, J. M., Jr., J. R. Rees, T. J. Wade, A. Khalakdina, J. F. Hilton, I. J. Ergas, S. Burns, S. Benker, C. Ma, C. Bowen, D. C. Mills, D. Vugia, D. D. Juranek, and D. Levy.** 2002. Participant blinding and gastrointestinal illness in a randomized, controlled trial of in-home drinking water intervention. *Emerg. Infect. Dis.* **8:**29–36.
51. **Colford, J. M., Jr., T. J. Wade, S. K. Sandhu, C. C. Wright, S. Lee, S. Shaw, K. Fox, S. Burns, S. Benker, M. A. Brookhart, M. van der Laan, and D. Levy.** 2005. A randomized, controlled trial of in-home drinking water intervention to reduce gastrointestinal illness. *Am. J. Epidemiol.* **161:**472–482.
52. **Colwell, R. R.** 1996. Global climate and infectious disease: the cholera paradigm. *Science* **274:**2025–2031.
53. **Conteas, C., O. Berlin, M. Lariviere, S. Pandhumas, C. Speck, R. Porschen, and T. Nakaya.** 1998. Examination of the prevalence and seasonal variation of intestinal microsporidiosis in the stools of persons with chronic diarrhea and human immunodeficiency virus infection. *Am. J. Trop. Med. Hyg.* **58:**559–561.
54. **Cotte, L., M. Rabodonirina, F. Chapuis, F. Bailly, F. Bissuel, C. Raynal, P. Gelas, F. Persat, M.-A. Piens, and C. Trepo.** 1999. Waterborne outbreak of intestinal microsporidiosis in persons with and without human immunodeficiency virus infection. *J. Infect. Dis.* **180:**2003–2008.
55. **Cover, T., and M. Blaser.** 1995. *Helicobacter pylori*: a bacterial cause of gastritis, peptic ulcer disease and gastric cancer. *ASM News* **61:**21–26.
56. **Covert, T. C., M. R. Rodgers, A. L. Reyes, and G. N. Stelma, Jr. 1999.** Occurrence of nontuberculous mycobacteria in environmental samples. *Appl. Environ. Microbiol.* **65:**2492–2496.
57. **Craun, G.** 1986. Recent statistics of waterborne disease outbreaks (1981–1983), p. 161–168. *In* G. Craun (ed.), *Waterborne Diseases in the United States*. CRC Press, Boca Raton, Fla.
58. **Craun, G.** 1986. Statistics of waterborne outbreaks in the US (1920–1980), p. 73–152. *In* G. Craun (ed.), *Waterborne Diseases in the United States*. CRC Press, Boca Raton, Fla.
59. **Cubitt, W.** 1994. Caliciviruses, p. 549–568. *In* A. Kapikian (ed.), *Viral Infections of the Gastrointestinal Tract.* Marcel Dekker, New York, N.Y.
60. **Current, W., and L. Garcia.** 1991. Cryptosporidiosis. *Clin. Microbiol. Rev.* **4:**325–358.
61. **D'Antonio, R., R. Winn, J. Taylor, T. Gustafson, W. Current, M. Rhodes, G. Gary, and R. Zajac.** 1985. A waterborne outbreak of cryptosporidiosis in normal hosts. *Ann. Intern. Med.* **103:**886–888.
62. **Dascomb, K., R. Clark, J. Aberg, J. Pulvirenti, R. Hewitt, P. Kissinger, and E. Didier.** 1999. Natural history of intestinal microsporidiosis among patients infected with human immunodeficiency virus. *J. Clin. Microbiol.* **37:** 3421–3422.

63. **Dascomb, K., T. Frazer, R. A. Clark, P. Kissinger, and E. Didier.** 2000. Microsporidiosis and HIV. *J. Acquir. Immune Defic. Syndr.* **24:**290–292.
64. **DeLeon, R., S. Matsui, R. Baric, J. Herrmann, N. Blacklow, H. Greenberg, and M. Sobsey.** 1992. Detection of Norwalk virus in stool specimens by reverse transcriptase–polymerase chain reaction and nonradioactive oligoprobes. *J. Clin. Microbiol.* **30:**3151–3157.
65. **Dennis, D., R. Smith, J. Welch, C. Chute, B. Anderson, J. Herndon, and C. vonReyn.** 1993. Endemic giardiasis in New Hampshire: a case-control study of environmental risks. *J. Infect. Dis.* **167:**1391–1395.
66. **Desportes, I., Y. Le Charpentier, A. Galian, F. Bernard, B. Cochand-Priollet, A. Lavergne, P. Ravisse, and R. Modigliani.** 1985. Occurrence of a new microsporidan: Enterocytozoon bieneusi n.g., n. sp., in the enterocytes of a human patient with AIDS. *J. Protozool.* **32:**250–254.
67. **Didier, E. S.** 2005. Microsporidiosis: an emerging and opportunistic infection in humans and animals. *Acta Trop.* **94:**61–76.
68. **Didier, E. S., M. E. Stovall, L. C. Green, P. J. Brindley, K. Sestak, and P. J. Didier.** 2004. Epidemiology of microsporidiosis: sources and modes of transmission. *Vet. Parasitol.* **126:**145–166.
69. **Dimitrov, D. H., S. A. Dashti, J. M. Ball, E. Bishbishi, K. Alsaeid, X. Jiang, and M. K. Estes.** 1997. Prevalence of antibodies to human caliciviruses (HuCVs) in Kuwait established by ELISA using baculovirus-expressed capsid antigens representing two genogroups of HuCVs. *J. Med. Virol.* **51:**115–118.
70. **Dowd, S., C. Gerba, M. Kamper, and I. Pepper.** 1999. Evaluation of methodologies including immunofluorescent assay (IFA) and the polymerase chain reaction (PCR) for detection of human pathogenic microsporidia in water. *J. Microbiol. Methods* **35:**43–52.
71. **Dowd, S., C. Gerba, and I. Pepper.** 1998. Confirmation of the human-pathogenic microsporidia *Enterocytozoon bieneusi*, *Encephalitozoon intestinalis*, and *Vittaforma corneae* in water. *Appl. Environ. Microbiol.* **64:**3332–3335.
72. **Dowd, S. E., D. John, J. Eliopolus, C. P. Gerba, J. Naranjo, R. Klein, B. Lopez, M. de Mejia, C. E. Mendoza, and I. L. Pepper.** 2003. Confirmed detection of Cyclospora cayetanesis, Encephalitozoon intestinalis and Cryptosporidium parvum in water used for drinking. *J. Water Health* **1:**117–123.
73. **Dufour, A.** 1986. Diseases caused by water contact, p. 23–41. *In* G. Craun (ed.), *Waterborne Diseases in the United States*. CRC Press, Boca Raton, Fla.
74. **Ehlers, M. M., W. O. Grabow, and D. N. Pavlov.** 2005. Detection of enteroviruses in untreated and treated drinking water supplies in South Africa. *Water Res.* **39:**2253–2258.
75. **Eisenberg, J. N., M. A. Brookhart, G. Rice, M. Brown, and J. M. Colford, Jr.** 2002. Disease transmission models for public health decision making: analysis of epidemic and endemic conditions caused by waterborne pathogens. *Environ. Health Perspect.* **110:**783–790.
76. **Eisenberg, J. N., T. J. Wade, S. Charles, M. Vu, A. Hubbard, C. C. Wright, D. Levy, P. Jensen, and J. M. Colford, Jr.** 2002. Risk factors in HIV-associated diarrhoeal disease: the role of drinking water, medication and immune status. *Epidemiol. Infect.* **128:**73–81.
77. **Elder, G., P. Hunter, and G. Codd.** 1993. Hazardous freshwater cyanobacteria (blue-green algae). *Lancet* **341:**1519–1520.
78. **Emde, K., H. Mao, and G. Finch.** 1992. Detection and occurrence of waterborne bacterial and viral pathogens. *Water Environ. Res.* **64:**641–647.
79. **Emerson, S. U., and R. H. Purcell.** 2004. Running like water—the omnipresence of hepatitis E. *N. Engl. J. Med.* **351:**2367–2368.
80. **Enriquez, C. E., C. J. Hurst, and C. P. Gerba.** 1995. Survival of the enteric adenoviruses 40 and 41 in tap, sea, and waste water. *Water Res.* **29:**2548–2553.
81. **Enroth, H., and L. Engstrand.** 1995. Immunomagnetic separation and PCR for detection of *Helicobacter pylori* in water and stool samples. *J. Clin. Microbiol.* **33:**2162–2165.
82. **Ericksen, T., and A. Dufour.** 1986. Methods to identify waterborne pathogens and indicator organisms, p. 195–214. *In* G. Craun (ed.), *Waterborne Diseases in the United States*. CRC Press, Boca Raton, Fla.
83. **Esrey, S., R. Feachem, and J. Hughes.** 1985. Interventions for the control of diarrhoeal diseases among young children: improving water supplies and excreta disposal facilities. *Bull. W. H. O.* **63:**757–772.
84. **Esrey, S. A., and J. P. Habicht.** 1986. Epidemiologic evidence for health benefits from improved water and sanitation in developing countries. *Epidemiol. Rev.* **8:**117–128.
85. **Ewald, P.** 1994. When water moves like a mosquito, p. 67–86. *In* P. Ewald (ed.), *Evolution of Infectious Disease*. Oxford University Press, New York, N.Y.
86. **Falconer, I., A. Beresford, and M. Runnegar.** 1983. Evidence of liver damage by toxin from a bloom of the blue-green alga, Microcystis aeruginosa. *Med. J. Aust.* **1:**511–514.
87. **Fayer, R.** 2004. *Cryptosporidium*: a water–borne zoonotic parasite. *Vet. Parasitol.* **126:**37–56.
88. **Fewtrell, L., and J. M. Colford, Jr.** 2004. *Water, Sanitation and Hygiene: Interventions and Diarrhoea—a Systematic Review and Meta-Analysis*. World Bank, Washington, D.C.
89. **Fiedorek, S., H. Malaty, D. Evans, C. Pumphrey, H. Casteel, D. Evans, and D. Graham.** 1991. Factors influencing the epidemiology of *Helicobacter pylori* infection in children. *Pediatrics* **88:**578–582.
90. **Fournier, S., S. Dubrou, O. Liguory, F. Gaussin, M. Santillana-Hayat, C. Sarfati, J. M. Molina, and F. Derouin.** 2002. Detection of Microsporidia, cryptosporidia and Giardia in swimming pools: a one-year prospective study. *FEMS Immunol. Med. Microbiol.* **33:**209–213.
91. **Fournier, S., O. Liguory, M. Santillana-Hayat, E. Guillot, C. Sarfati, N. Dumoutier, J. Molina, and F. Derouin.** 2000. Detection of microsporidia in surface water: a one-year follow-up study. *FEMS Immunol. Med. Microbiol.* **29:**95–100.
92. **Fout, G. S., B. C. Martinson, M. W. N. Moyer, and D. R. Dahling.** 2003. A multiplex reverse transcriptase-PCR method for detection of human enteric viruses in groundwater. *Appl. Environ. Microbiol.* **69:**3158–3164.
93. **Foy, H.** 1991. Adenoviruses, p. 77–94. *In* A. Evans (ed.), *Viral Infections of Humans: Epidemiology and Control*. Plenum Medical Book Company, New York, N.Y.
94. **Frost, F. J., G. F. Craun, and R. L. Calderon.** 1996. Waterborne disease surveillance. *J. Am. Water Works Assoc.* **88:**66–75.
95. **Fukuyama, M., M. Arimatu, K. Sakamato, et al.** 1999. PCR detection of *Helicobacter pylori* in water samples collected from rivers in Japan. *In 10th Int. CHRO Workshop*, Baltimore, Md.
96. **Gabbay, Y. B., R. I. Glass, S. S. Monroe, C. Carcamo, M. K. Estes, J. D. Mascarenhas, and A. C. Linhares.** 1994. Prevalence of antibodies to Norwalk virus among Amerindians in isolated Amazonian communities. *Am. J. Epidemiol.* **139:**728–733.
97. **Garcia, L., R. Shimizu, and D. Bruckner.** 1994. Detection of microsporidial spores in fecal specimens from patients diagnosed with cryptosporidiosis. *J. Clin. Microbiol.* **32:**1739–1741.
98. **Geldreich, E., K. Fox, J. Goodrich, E. Rice, R. Clark, and D. Swerdlow.** 1992. Searching for a water supply connection in the Cabool, Missouri, disease outbreak of *Escherichia coli* O157:H7. *Water Res.* **26:**1127–1137.

99. **Gerba, C., and G. Bitton.** 1984. Microbial pollutants: their survival and transport pattern to groundwater, p. 65–88. *In* G. Bitton and C. Gerba (ed.), *Groundwater Pollution Microbiology*. John Wiley & Sons, New York, N.Y.
100. **Gerba, C. P., K. R. Riley, N. Nwachuku, H. Ryu, and M. Abbaszadegan.** 2003. Removal of Encephalitozoon intestinalis, calicivirus, and coliphages by conventional drinking water treatment. *J. Environ. Sci. Health A Tox. Hazard. Subst. Environ. Eng.* **38:**1259–1268.
101. **Goodman, K. J., P. Correa, H. J. Tengana Aux, H. Ramirez, J. P. DeLany, O. Guerrero Pepinosa, M. Lopez Quinones, and T. Collazos Parra.** 1996. Helicobacter pylori infection in the Colombian Andes: a population-based study of transmission pathways. *Am. J. Epidemiol.* **144:**290–299.
102. **Gray, J. J., X. Jiang, P. Morgan-Capner, U. Desselberger, and M. K. Estes.** 1993. Prevalence of antibodies to Norwalk virus in England: detection by enzyme-linked immunosorbent assay using baculovirus-expressed Norwalk virus capsid antigen. *J. Clin. Microbiol.* **31:**1022–1025.
103. **Gray, L.** 1995. *Escherichia*, *Salmonella*, *Shigella*, and *Yersinia*, p. 450–452. *In* P. Murray et al. (ed.), *Manual of Clinical Microbiology*, 6th ed. ASM Press, Washington, D.C.
104. **Greenberg, H., J. Valdesuso, A. Kapikian, R. Chanock, R. Wyatt, W. Szmuness, J. Larrick, J. Kaplan, R. Gilman, and D. Sack.** 1979. Prevalence of antibody to the Norwalk virus in various countries. *Infect. Immun.* **26:**270–273.
105. **Haas, C. N., J. B. Rose, and C. P. Gerba.** 1999. *Quantitative Microbial Risk Assessment*. John Wiley, New York, N.Y.
106. **Hackney, C., and M. Potter.** 1994. Animal-associated and terrestrial bacterial pathogens, p. 172–209. *In* C. Hackney and M. Pierson (ed.), *Environmental Indicators and Shellfish Safety*. Chapman and Hall, New York, N.Y.
107. **Hackney, C., and M. Potter.** 1994. Human-associated bacterial pathogens, p. 154–171. *In* C. Hackney and M. Pierson (ed.), *Environmental Indicators and Shellfish Safety*. Chapman and Hall, New York, N.Y.
108. **Hale, D., W. Aldeen, and K. Carroll.** 1994. Diarrhea associated with cyanobacterialike bodies in an immunocompetent host. *JAMA* **271:**144–145.
109. **Haramoto, E., H. Katayama, and S. Ohgaki.** 2004. Detection of noroviruses in tap water in Japan by means of a new method for concentrating enteric viruses in large volumes of freshwater. *Appl. Environ. Microbiol.* **70:**2154–2160.
110. **Havelaar, A. H., F. M. Schets, A. van. Silfhout, W. H. Jansen, G. Wieten, and D. van der Kooij.** 1992. Typing of *Aeronomas* strains from patients with diarrhoea and from drinking water. *J. Appl. Bacteriol.* **72:**435–444.
111. **Hawkins, P., M. Runnegar, A. Jackson, and I. Falconer.** 1985. Severe hepatotoxicity caused by the tropical cyanobacterium (blue-green alga) *Cylindrospermapsis raciborskii* (Wiloszynska) Seenaya and Subba Raju isolated from a domestic water supply reservoir. *Appl. Environ. Microbiol.* **50:**1292–1295.
112. **Hayman, J.** 1992. Beyond the Barcoo—probable human tropical cyanobacterial poisoning in outback Australia. *Med. J. Aust.* **157:**794–796.
113. **Hazen, T.** 1988. Fecal coliforms as indicators in tropical waters: a review. *Toxicity Assess.* **3:**461–477.
114. **Hegarty, J., M. Dowd, and K. Baker.** 1999. Occurrence of *Helicobacter pylori* in surface water in the United States. *J. Appl. Microbiol.* **87:**697–701.
115. **Hellard, M. E., M. I. Sinclair, A. B. Forbes, and C. K. Fairley.** 2001. A randomized, blinded, controlled trial investigating the gastrointestinal health effects of drinking water quality. *Environ. Health Perspect.* **109:**773–778.
116. **Herwaldt, B. L., G. F. Craun, S. L. Stokes, and D. D. Juranek.** 1991. Waterborne-disease outbreaks, 1989–1990. *Morb. Mortal. Wkly. Rep. CDC Surveill. Summ.* **40:**1–21.
117. **Herwaldt, B. L., J. F. Lew, C. L. Moe, D. C. Lewis, C. D. Humphrey, S. S. Monroe, E. W. Pon, and R. I. Glass.** 1994. Characterization of a variant strain of Norwalk virus from a food-borne outbreak of gastroenteritis on a cruise ship in Hawaii. *J. Clin. Microbiol.* **32:** 861–866.
118. **Hinkula, J., J. M. Ball, S. Lofgren, M. K. Estes, and L. Svensson.** 1995. Antibody prevalence and immunoglobulin IgG subclass pattern to Norwalk virus in Sweden. *J. Med. Virol.* **47:**52–57.
119. **Holmberg, S., W. Schell, G. Fanning, I. Wachsmuth, F. Hickman–Brenner, P. Blake, D. Brenner, and J. Farmer.** 1986. Aeromonas intestinal infections in the United States. *Ann. Intern. Med.* **105:**683–689.
120. **Honma, S., S. Nakata, K. Numata, K. Kogawa, T. Yamashita, M. Oseto, X. Jiang, and S. Chiba.** 1998. Epidemiological study of prevalence of genogroup II human calicivirus (Mexico virus) infections in Japan and Southeast Asia as determined by enzyme-linked immunosorbent assays. *J. Clin. Microbiol.* **36:**2481–2484.
121. **Hopkins, R., G. Gaspard, T. Williams, R. Karlin, C. Cukor, and N. Blacklow.** 1984. A community waterborne gastroenteritis outbreak: evidence for rotavirus as the agent. *Am. J. Public Health* **74:**263–265.
122. **Hopkins, R., P. Vial, C. Ferreccio, J. Ovalle, P. Prado, V. Sotomayor, R. Russell, S. Wasserman, and J. Morris.** 1993. Seroprevalence of *Helicobacter pylori* in Chili: vegetables may serve as one route of transmission. *J. Infect. Dis.* **168:**222–226.
123. **Hrudey, S. E. and E. J. Hrudey.** 2004. *Safe Drinking Water—Lessons from Recent Outbreaks in Affluent Nations*. IWA Publishing, London, United Kingdom.
124. **Huffman, D. E., A. Gennaccaro, J. B. Rose, and B. W. Dussert.** 2002. Low- and medium-pressure UV inactivation of microsporidia Encephalitozoon intestinalis. *Water Res.* **36:**3161–3164.
125. **Hulten, K., H. Enroth, T. Nystrom, and L. Engstrand.** 1998. Presence of Helicobacter species DNA in Swedish water. *J. Appl. Microbiol.* **85:**282–286.
126. **Hulten, K., S. Han, H. Enroth, P. Klein, A. Opekun, R. Gilman, D. Evans, L. Engstrand, D. Graham, and F. El-Zaatari.** 1996. *Helicobacter pylori* in the drinking water in Peru. *Gastroenterology* **110:**1031–1035.
127. **Hung, T., G. Chen, C. Wang, R. Fan, R. Yong, J. Chang, R. Dan, and M. Ng.** 1987. Seroepidemiology and molecular epidemiology of the Chinese rotavirus, p. 49–62. *In* G. Bock and J. Wheelan (ed.), *Novel Diarrhoea Viruses*. John Wiley & Sons, Chichester, United Kingdom.
128. **Hung, T., G. M. Chen, C. C. Wang, and X. Chang.** 1984. Waterborne outbreak of rotavirus diarrhoea in adults caused by novel rotavirus. *Lancet* **i:**1139–1142.
129. **Hunter, P. R.** 1997. *Waterborne Disease: Epidemiology and Ecology*. John Wiley, Chichester, United Kingdom.
130. **Hunter, P. R., R. M. Chalmers, S. Hughes, and Q. Syed.** 2005. Self-reported diarrhea in a control group: a strong association with reporting of low-pressure events in tap water. *Clin. Infect. Dis.* **40:**e32–e34.
131. **Hurst, C.** 1991. Presence of enteric viruses in freshwater and their removal by the conventional drinking water treatment processes. *Bull. W. H. O.* **69:**113–119.
132. **Hurst, C., W. Benton, and R. Stetler.** 1989. Detecting viruses in water. *J. Am. Water Works Assoc.* **81:**71–80.
133. **Hutin, Y. J., M. N. Sombardier, O. Liguory, C. Sarfati, F. Derouin, J. Modai, and J. M. Molina.** 1998. Risk factors for intestinal microsporidiosis in patients with

human immunodeficiency virus infection: a case-control study. *J. Infect. Dis.* **178:**904–907.

134. **ILSI.** 1999. *Early Warning Monitoring to Detect Hazardous Events in Water Supplies.* International Life Sciences Institute (ILSI) Press, Washington, D.C.
135. **Institute of Medicine.** 1992. *Emerging Infections: Microbial Threats to Health in the United States.* National Academies Press, Washington, D.C.
136. **Jenkins, P. A.** 1991. Mycobacteria in the environment. *J. Appl. Bacteriol.* **70:**137S–141S.
137. **Jiang, B., Y. Wang, R. I. Glass, and Z. Y. Fang.** 2005. The evolution of human group B rotaviruses: correction and an update. *J. Clin. Virol.* **34:**158–159.
138. **Jiang, X., D. Graham, K. Wang, and M. Estes.** 1990. Norwalk virus genome cloning and characterization. *Science* **250:**1580–1583.
139. **Jiang, X., D. O. Matson, G. M. Ruiz-Palacios, J. Hu, J. Treanor, and L. K. Pickering.** 1995. Expression, self-assembly, and antigenicity of a snow mountain agent-like calicivirus capsid protein. *J. Clin. Microbiol.* **33:**1452–1455.
140. **Jiang, X., D. O. Matson, F. R. Velazquez, J. J. Calva, W. M. Zhong, J. Hu, G. M. Ruiz-Palacios, and L. K. Pickering.** 1995. Study of Norwalk-related viruses in Mexican children. *J. Med. Virol.* **47:**309–316.
141. **Jiang, X., J. Wang, D. Y. Graham, and M. K. Estes.** 1992. Detection of Norwalk virus in stool by polymerase chain reaction. *J. Clin. Microbiol.* **30:**2529–2534.
142. **Jiang, X., M. Wang, D. Y. Graham, and M. K. Estes.** 1992. Expression, self-assembly, and antigenicity of the Norwalk virus capsid protein. *J. Virol.* **66:**6527–6532.
143. **John, D. E., C. N. Haas, N. Nwachuku, and C. P. Gerba.** 2005. Chlorine and ozone disinfection of *Encephalitozoon intestinalis* spores. *Water Res.* 39:2369–2375.
144. **John, D. E., N. Nwachuku, I. L. Pepper, and C. P. Gerba.** 2003. Development and optimization of a quantitative cell culture infectivity assay for the micro-sporidium Encephalitozoon intestinalis and application to ultraviolet light inactivation. *J. Microbiol. Methods* **52:**183–196.
145. **Johnson, C., E. Rice, and D. Reasoner.** 1997. Inactivation of *Helicobacter pylori* by chlorination. *Appl. Environ. Microbiol.* **63:**4969–4970.
146. **Jothikumar, N., K. Aparna, S. Mamatchiammal, R. Paulmurugan, S. Saravanadevi, and P. Khanna.** 1993. Detection of hepatitis E virus in raw and treated wastewater with the polymerase chain reaction. *Appl. Environ. Microbiol.* **59:**2558–2562.
147. **Jothikumar, N., T. L. Cromeans, V. R. Hill, X. Lu, M. D. Sobsey, and D. D. Erdman.** 2005. Quantitative real-time PCR assays for detection of human adenoviruses and identification of serotypes 40 and 41. *Appl. Environ. Microbiol.* **71:**3131–3136.
148. **Jothikumar, N., T. L. Cromeans, B. H. Robertson, X. J. Meng, and V. R. Hill.** 2006. A broadly reactive one-step real-time RT-PCR assay for rapid and sensitive detection of hepatitis E virus. *J. Virol. Methods* **131:**65–71.
149. **Kaplan, J. E., R. Feldman, D. S. Campbell, C. Lookabaugh, and G. W. Gary.** 1982. The frequency of a Norwalk-like pattern of illness in outbreaks of acute gastroenteritis. *Am. J. Public Health* **72:**1329–1332.
150. **Kaplan, J. E., G. W. Gary, R. C. Baron, N. Singh, L. B. Schonberger, R. Feldman, and H. B. Greenberg.** 1982. Epidemiology of Norwalk gastroenteritis and the role of Norwalk virus in outbreaks of acute nonbacterial gastroenteritis. *Ann. Intern. Med.* **96:**756–761.
151. **Katayama, H., A. Shimasaki, and S. Ohgaki.** 2002. Development of a virus concentration method and its application to detection of enterovirus and Norwalk virus from coastal seawater. *Appl. Environ. Microbiol.* **68:** 1033–1039.
152. **Kator, H., and M. Rhodes.** 1994. Microbial and chemical indicators, p. 30–91. *In* C. Hackney and M. Pierson (ed.), *Environmental Indicators and Shellfish Safety.* Chapman and Hall, New York, N.Y.
153. **Keswick, B. H., T. K. Satterwhite, P. C. Johnson, H. L. DuPont, S. L. Secor, J. A. Bitsura, G. W. Gary, and J. C. Hoff.** 1985. Inactivation of Norwalk virus in drinking water by chlorine. *Appl. Environ. Microbiol.* **50:**261–264.
154. **Khan, A. S., C. L. Moe, R. I. Glass, S. S. Monroe, M. K. Estes, L. E. Chapman, X. Jiang, C. Humphrey, E. Pon, and J. K. Iskander.** 1994. Norwalk virus–associated gastroenteritis traced to ice consumption aboard a cruise ship in Hawaii: comparison and application of molecular method-based assays. *J. Clin. Microbiol.* **32:**318–322.
155. **Klein, P., D. Y. Graham, A. Gaillour, A. R. Opekun, E. O. Smith, et al.** 1991. Water source as risk factor for *Helicobacter pylori* infection in Peruvian children. *Lancet* **337:**1503–1506.
156. **Klontz, K., S. Lieb, M. Schreiber, H. Janowski, L. Baldy, and R. Gunn.** 1988. Syndromes of *Vibrio vulnificus* infections: clinical and epidemiologic features in Florida cases, 1981–1987. *Ann. Intern. Med.* **109:**318–323.
157. **Knight, P.** 1995. One misidentified human parasite is a cyclosporan. *ASM News* **61:**520–522.
158. **Kool, J. L., J. C. Carpenter, and B. S. Fields.** 1999. Effect of monochloramine disinfection of municipal drinking water on risk of nosocomial Legionnaires' disease. *Lancet* **353:**272–277.
159. **Kopicko, J., T. Frazer, K. Dascomb, R. Clark, E. Didier, and P. Kissinger.** 1999. Influence of seasonal variation with enteric microsporidiosis among HIV-infected individuals. *J. Acquir. Immune Defic. Syndr.* **22:**408.
160. **Köster, W., T. Egli, N. Ashbolt, K. Botzenhart, N. Burlion, T. Endo, P. Grimont, E. Guillot, C. Mabilat, L. Newport, M. Niemi, P. Payment, A. Prescott, P. Renaud, and A. Rust.** 2003. Analytical methods for microbial water quality testing, p. 237–292. *In Assessing Microbial Safety of Drinking Water: Improving Approaches and Methods.* OECD, World Health Organization, Geneva, Switzerland.
161. Reference deleted.
162. **Kuijper, E., P. Bol, M. Peeters, A. Steigerwalt, H. Zanen, and D. Brenner.** 1989. Clinical and epidemiological aspects of members of *Aeromonas* hybridization groups isolated from human feces. *J. Clin. Microbiol.* **27:**1531–1537.
163. **LeChevallier, M., and W. Norton.** 1993. Treatments to address source water concerns: protozoa, p. 145–164. *In* G. Craun (ed.), *Safety of Water Disinfection: Balancing Chemical and Microbial Risks.* ILSI Press, Washington, D.C.
164. **LeChevallier, M. W., W. D. Norton, and R. G. Lee.** 1991. *Giardia* and *Cryptosporidium* spp. in filtered drinking water supplies. *Appl. Environ. Microbiol.* **57:**2617–2621.
165. **LeChevallier, M. W., and K.-K. Au.** 2004. *Water Treatment and Pathogen Control Process Efficiency in Achieving Safe Drinking Water.* IWA Publishing, London, United Kingdom.
166. **LeChevallier, M. W., G. D. Di Giovanni, J. L. Clancy, Z. Bukhari, S. Bukhari, J. S. Rosen, J. Sobrinho, and M. M. Frey.** 2003. Comparison of method 1623 and cell culture-PCR for detection of *Cryptosporidium* spp. in source waters. *Appl. Environ. Microbiol.* **69:**971–979.
167. **LeChevallier, M. W., R. W. Gullick, M. R. Karim, M. Friedman, and J. E. Funk.** 2003. The potential for health risks from intrusion of contaminants into the distribution system from pressure transients. *J. Water Health* **1:**3–14.
168. **Lee, C., S. H. Lee, E. Han, and S. J. Kim.** 2004. Use of cell culture-PCR assay based on combination of A549

and BGMK cell lines and molecular identification as a tool to monitor infectious adenoviruses and enteroviruses in river water. *Appl. Environ. Microbiol.* **70:**6695–6705.

169. **Lee, S.-H., and S.-J. Kim.** 2004. Detection of infectious enteroviruses and adenoviruses in tap water in urban areas in Korea. *Water Res.* **36:**248–256.

170. **Lew, J. F., J. Valdesuso, T. Vesikari, A. Z. Kapikian, X. Jiang, M. K. Estes, and K. Y. Green.** 1994. Detection of Norwalk virus or Norwalk-like virus infections in Finnish infants and young children. *J. Infect. Dis.* **169:** 1364–1367.

171. **Lindkvist, P., F. Enquselassie, D. Asrat, I. Nilsson, L. Muhe, and J. Giesecke.** 1999. *Helicobacter pylori* infection in Ethiopian children: a cohort study. *Scand. J. Infect. Dis.* **31:**475–480.

172. **Lodder, W. J., and A. M. de Roda Husman.** 2005. Presence of noroviruses and other enteric viruses in sewage and surface waters in The Netherlands. *Appl. Environ. Microbiol.* **71:**1453–1461.

173. **Logsdon, G., and J. Hoff.** 1986. Barriers to the transmission of waterborne disease, p. 255–274. *In* G. Craun (ed.), *Waterborne Diseases in the United States.* CRC Press, Boca Raton, Fla.

174. **Lopez, A. S., J. M. Bendik, J. Y. Alliance, J. M. Roberts, A. J. da Silva, I. N. Moura, M. J. Arrowood, M. L. Eberhard, and B. L. Herwaldt.** 2003. Epidemiology of *Cyclospora cayetanensis* and other intestinal parasites in a community in Haiti. *J. Clin. Microbiol.* **41:** 2047–2054.

175. **MacKenzie, W., N. Hoie, M. Proctor, M. Gradus, K. Blair, D. Peterson, J. Kazmierczak, D. Addiss, K. Fox, J. Rose, and J. Davis.** 1994. A massive outbreak in Milwaukee of *Cryptosporidium* infection transmitted through the public water supply. *N. Engl. J. Med.* **331:**161–167.

176. **Mansfield, L. S., and A. A. Gajadhar.** 2004. *Cyclospora cayetanensis*, a food- and waterborne coccidian parasite. *Vet. Parasitol.* **126:**73–90.

177. **Marshall, M., D. Naumovitz, Y. Ortega, and C. Sterling.** 1997. Waterborne protozoan pathogens. *Clin. Microbiol. Rev.* **10:**67–85.

178. **Mascher, F., F. Reinthaler, D. Stunzner, and B. Lamberger.** 1988. *Aeromonas* species in a municipal water supply of a central European city: biotyping of strains and detection of toxins. *Zentbl. Bakteriol. Mikrobiol. Hyg.* **186:**333–337.

179. **Maunula, L., S., Kalso, C. H. Von Bonsdorff, and A. Ponka.** 2004. Wading pool water contaminated with both noroviruses and astroviruses as the source of a gastroenteritis outbreak. *Epidemiol. Infect.* **132:**737–743.

180. **Maunula, L., I. T. Miettinen, and C. H. von Bonsdorff.** 2005. Norovirus outbreaks from drinking water. *Emerg. Infect. Dis.* **11:**1716–1721.

181. **McDaniels, A. E., L. Wymer, C. Rankin, and R. Haugland.** 2005. Evaluation of quantitative real time PCR for the measurement of Helicobacter pylori at low concentrations in drinking water. *Water Res.* **39:**4808–4816.

182. **McFeters, G., G. Bissonnette, J. Jezeski, C. Thomson, and D. Stuart.** 1974. Comparative survival of indicator bacteria and enteric pathogens in well water. *J. Appl. Microbiol.* **27:**823–829.

183. **McLaughlin, J. B., B. D. Gessner, T. V. Lynn, E. A. Funk, and J. P. Middaugh.** 2004. Association of regulatory issues with an echovirus 18 meningitis outbreak at a children's summer camp in Alaska. *Pediatr. Infect. Dis. J.* **23:**875–877.

184. **Meisel, J., D. Perera, C. Meligro, and C. Rubin.** 1976. Overwhelming watery diarrhea associated with a *Cryptosporidium* in an immunosuppressed patient. *Gastroenterology* **70:**1156–1160.

185. **Mendall, M., P. Goggin, N. Molineaux, J. Levy, T. Toosy, D. Strachan, and T. Northfield.** 1992. Childhood living conditions and *Helicobacter pylori* in seropositivity in adult life. *Lancet* **339:**896–897.

186. **Millership, S., J. Stephenson, and S. Tabaqchali.** 1988. Epidemiology of *Aeromonas* species in a hospital. *J. Hosp. Infect.* **11:**169–175.

187. **Mintz, E.** 2003. A riddle wrapped in a mystery inside an enigma: Brainerd diarrhoea turns 20. *Lancet* **362:**2037–2038.

188. **Mintz, E., J. Weber, D. Guris, N. Puhr, J. Wells, J. Yashuk, M. Curtis, and R. Tauxe.** 1998. An outbreak of Brainerd diarrhea among travelers to the Galapagos Islands. *J. Infect. Dis.* **177:**1041–1045.

189. **Mitchell, H., Y. Li, P. Hu, Q. Liu, M. Chen, G. Du, Z. Wang, A. Lee, and S. Hazell.** 1992. Epidemiology of *Helicobacter pylori* in Southern China: identification of early childhood as the critical period for acquisition. *J. Infect. Dis.* **166:**149–153.

190. **Modlin, J. F.** 2000. Coxsackieviruses, echoviruses and newer enteroviruses, p. 1904–1919. *In* G. L. Mandell, J. E. Bennett, and R. Dolin (ed.), *Mandell, Douglas, and Bennett's Principles and Practice of Infectious Diseases.* Churchill Livingston, Philadelphia, Pa.

191. **Moe, C.** 2002. Prospective studies of endemic waterborne disease in developing countries, p. 197–206. *In* P. R. Hunter, M. Waite, and E. Ron (ed.), *Drinking Water and Infectious Diseases Establishing the Links.* CRC Press, Boca Raton, Fla.

192. **Moe, C., D. Rhodes, S. Pusek, F. Tseng, W. Heizer, C. Kapoor, B. Gilliam, M. Harb, P. Stewart, S. Miller, M. Sobsey, J. Hermann, N. Blacklow, and R. Calderon.** 1998. Determination of Norwalk virus dose-response in human volunteers, abstr. C-384. *Abstr. 98th Gen. Meet. Am. Soc. Microbiol.* American Society for Microbiology, Washington, D.C.

193. **Moe, C. L., J. Gentsch, T. Ando, G. Grohmann, S. S. Monroe, X. Jiang, J. Wang, M. K. Estes, Y. Seto, C. Humphrey, et al.** 1994. Application of PCR to detect Norwalk virus in fecal specimens from outbreaks of gastroenteritis. *J. Clin. Microbiol.* **32:**642–648.

194. **Moe, C. L., and R. D. Rheingans.** 2006. Global challenges in water, sanitation and health. *J. Water Health* **4**(Suppl. 1):41–57.

195. **Moe, C. L., M. D. Sobsey, G. P. Samsa, and V. Mesolo.** 1991. Bacterial indicators of risk of diarrhoeal disease from drinking-water in the Philippines. *Bull. W. H. O.* **69:**305–317.

196. **Morse, D., J. Guzewich, J. Hanrahan, R. Stricof, M. Shayegani, R. Deilbel, J. Grabau, N. Nowak, J. Herrmann, G. Cukor, and N. Blacklow.** 1986. Widespread outbreaks of clam- and oyster-associated gastroenteritis: role of Norwalk virus. *N. Engl. J. Med.* **314:**678–681.

197. **Moyer, N.** 1987. Clinical significance of *Aeromonas* species isolated from patients with diarrhea. *J. Clin. Microbiol.* **25:**2044–2048.

198. **Muraca, P., V. Yu, and A. Goetz.** 1990. Disinfection of water distribution systems for Legionella: a review of application procedures and methodologies. *Infect. Control. Hosp. Epidemiol.* **11:**79–88.

199. **Myrmel, M., E. Rimstad, M. Estes, E. Skjerve, and Y. Wasteson.** 1996. Prevalence of serum antibodies to Norwalk virus among Norwegian military recruits. *Int. J. Food Microbiol.* **29:**233–240.

200. **Naik, S., R. Aggarwal, P. Salunke, and N. Mehrotra.** 1992. A large waterborne viral hepatitis E epidemic in Kanpur, India. *Bull. W. H. O.* **70:**597–604.

201. **Nakata, S., S. Honma, K. Numata, K. Kogawa, S. Ukae, N. Adachi, X. Jiang, M. K. Estes, Z. Gatheru, P. M. Tukei, and S. Chiba.** 1998. Prevalence of human calicivirus infections in Kenya as determined by enzyme

immunoassays for three genogroups of the virus. *J. Clin. Microbiol.* **36:**3160–3163.

202. **Nataro, J. P., and M. M. Levine.** 1994. Bacterial diarrheas, p. 697–752. *In* A. Z. Kapikian (ed.), *Viral Infections of the Gastrointestinal Tract.* Marcel Dekker, New York, N.Y.
203. **National Research Council.** 2004. Evaluation of risk: a phased approach to monitoring microbial water quality, p. 13–16. *In Indicators for Waterborne Pathogens.* The National Academies Press, Washington, D.C.
204. **Nazer, H., E. Price, G. Huntr, U. Patel, and J. Walker-Smith.** 1990. Isolation of *Aeromonas* spp. from canal water. *Indian J. Pediatr.* **57:**115–118.
205. **Nime, F., D. Page, M. Holscher, and J. Yardley.** 1976. Acute enterocolitis in a human being infected with the protozoan *Cryptosporidium. Gastroenterology* **70:**592–598.
206. **Norton, C. D., M. W. LeChevallier, and J. O. Falkinham III.** 2004. Survival of *Mycobacterium avium* in a model distribution system. *Water Res.* **38:**1457–1466.
207. **Nwachcuku, N., and C. P. Gerba.** 2004. Emerging waterborne pathogens: can we kill them all? *Curr. Opin. Biotechnol.* **15:**175–180.
208. **Nygard, K., M. Torven, C. Ancker, S. B. Knauth, K. O. Hedlund, J. Giesecke, Y. Andersson, and L. Svensson.** 2003. Emerging genotype (GGIIb) of norovirus in drinking water, Sweden. *Emerg. Infect. Dis.* **9:**1548–1552.
209. **Olson, B.** 1993. Pathogen occurrence in source waters: factors affecting survival and growth, p. 83–97. *In* G. Craun (ed.), *Safety of Water Disinfection: Balancing Chemi-cal and Microbial Risks.* ILSI Press, Washington, D.C.
210. **Ooi, W., S. Zimmerman, and C. Needham.** 1995. *Cyclospora* species as a gastrointestinal pathogen in immunocompetent hosts. *J. Clin. Microbiol.* **33:**1267–1269.
211. **O'Rourke, K., K. J. Goodman, M. Grazioplene, T. Redlinger, and R. S. Day.** 2003. Determinants of geographic variation in Helicobacter pylori infection among children on the US-Mexico border. *Am. J. Epidemiol.* **158:**816–824.
212. **Ortega, Y., C. Sterling, R. Gilman, V. Cama, and F. Diaz.** 1993. Cyclospora species—a new protozoan pathogen of humans. *N. Engl. J. Med.* **328:**1308–1312.
213. **Ortega, Y. R., C. R. Sterling, and R. H. Gilman.** 1998. *Cyclospora cayetanensis. Adv. Parasitol.* **40:**399–418.
214. **O'Ryan, M. L., P. A. Vial, N. Mamani, X. Jiang, M. K. Estes, C. Ferrecio, H. Lakkis, and D. O. Matson.** 1998. Seroprevalence of Norwalk virus and Mexico virus in Chilean individuals: assessment of independent risk factors for antibody acquisition. *Clin. Infect. Dis.* **27:**789–795.
215. **Osterholm, M., K. Macdonald, K. White, J. Wells, J. Spika, M. Potter, J. Forfang, R. Sorenson, P. Milloy, and P. Blake.** 1986. An outbreak of a newly recognized chronic diarrhea syndrome associated with raw milk consumption. *JAMA* **256:**484–490.
216. **Palacios, G., and M. S. Oberste.** 2005. Enteroviruses as agents of emerging infectious diseases. *J. Neurovirol.* **11:** 424–433.
217. **Papapetropoulou, M., and A. C. Vantarakis.** 1995. Detection of adenovirus outbreak at a municipal swimming pool by nested PCR amplification. *J. Infect.* **36:** 101–103.
218. **Parkin, R. T., J. A. Soller, and A. W. Olivieri.** 2003. Incorporating susceptible subpopulations in microbial risk assessment: pediatric exposures to enteroviruses in river water. *J. Exp. Anal. Environ. Epidemiol.* **13:**161–168.
219. **Parks, C. G., C. L. Moe, D. Rhodes, A. Lima, L. Barrett, F. Tseng, R. Baric, A. Talal, and R. Guerrant.** 1999. Genomic diversity of "Norwalk like viruses" (NLVs): pediatric infections in a Brazilian shantytown. *J. Med. Virol.* **58:**426–434.
220. **Parsonnet, J., S. Trock, C. Bopp, C. Wood, D. Addiss, F. Alai, L. Gorelkin, N. Hargrett-Bean, R. Gunn, and R. Tauxe.** 1989. Chronic diarrhea associated with drinking untreated water. *Ann. Intern. Med.* **110:**985–991.
221. **Payment, P.** 1997. Epidemiology of endemic gastrointestinal and respiratory diseases: incidence, fraction attributable to tap water and costs to society. *Water Sci. Technol.* **35:**7–10.
222. **Payment, P., and E. Franco.** 1989. Virus removal by drinking water treatment processes. *Crit. Rev. Environ. Control* **19:**15–31.
223. **Payment, P., and E. Franco.** 1993. *Clostridium perfringens* and somatic coliphages as indicators of the efficiency of drinking water treatment for viruses and protozoan cysts. *Appl. Environ. Microbiol.* **59:**2418–2424.
224. **Payment, P., L. Richardson, J. Siemiatycki, R. Dewar, M. Edwardes, and E. Franco.** 1991. A randomized trial to evaluate the risk of gastrointestinal disease due to consumption of drinking water meeting current miocrobiological standards. *Am. J. Public Health* **81:**703–708.
225. **Payment, P., J. Siemiatycki, L. Richardson, G. Renaud, E. Franco, and M. Prevost.** 1997. A prospective epidemiological study of gastrointestinal health effects due to the consumption of drinking water. *Int. J. Environ. Health Res.* **7:**5–31.
226. **Percival, S. L., R. M. Chalmers, M. Embrey, P. R. Hunter, J. Sellwood, and P. Wyn-Jones.** 2004. The Mycobacterium avium complex, p. 155–171. *In Microbiology of Waterborne Diseases.* Elsevier Academic Press, Philadelphia, Pa.
227. **Picard, B., and P. Goullet.** 1987. Seasonal prevalence of nosocomial *Aeromonas hydrophila* infection related to aeromonas in hospital water. *J. Hosp. Infect.* **10:**152–155.
228. **Pina, S., M. Puig, F. Lucena, J. Jofre, and R. Girones.** 1998. Viral pollution in the environment and in shellfish: human adenovirus detection by PCR as an index of human viruses. *Appl. Environ. Microbiol.* **64:**3376–3382.
229. **Pujol, F. H., G. Vasquez, A. M. Rojas, M. E. Fuenmayor, C. L. Loureiro, I. Perez-Schael, M. K. Estes, and F. Liprandi.** 1998. Norwalk virus infection in Venezuela. *Ann. Trop. Med. Parasitol.* **92:**205–211.
230. **Pusch, D., et al.** 2005. Detection of enteric viruses and bacterial indicators in German environmental waters. *Arch. Virol.* **150:**929–947.
231. **Pyle, B., S. Broadaway, and G. McFeters.** 1999. Sensitive detection of *Escherichia coli* O157:H7 in food and water by immunomagnetic separation and solid-phase laser cytometry. *Appl. Environ. Microbiol.* **65:**1966–1972.
232. **Raina, P., F. L. Pollari, G. F. Teare, M. J. Goss, D. A. J. Barry, and J. B. Wilson.** 1999. The relationship between *E. coli* indicator bacteria in well-water and gastrointestinal illnesses in rural families. *Can. J. Public Health* **90:** 172–175.
233. **Regli, S., J. Rose, C. Haas, and C. Gerba.** 1991. Modeling the risk from *Giardia* and viruses in drinking water. *J. Am. Water Works Assoc.* **83:**76–84.
234. **Rose, J.** 1988. Occurrence and significance of Cryptosporidium in water. *J. Am. Water Works Assoc.* **80:**53–58.
235. **Rose, J.** 1990. Environmental sampling of waterborne pathogens: overview of methods, application limitations and data interpretation, p. 223–234. *In* G. Craun (ed.), *Methods for the Investigation and Prevention of Waterborne Disease Outbreaks.* Office of Research and Development, U.S. Environmental Protection Agency, Washington, D.C.
236. **Rose, J.** 1993. Enteric waterborne protozoa: hazard and exposure assessment, p. 115–126. *In* G. Craun (ed.), *Safety of Water Disinfection: Balancing Chemical and Microbial Risks.* ILSI Press, Washington, D.C.

237. **Rose, J., C. Gerba, and W. Jakubowski.** 1991. Survey of potable water supplies for *Cryptosporidium* and *Giardia*. *Environ. Sci. Technol.* **25:**1393–1400.
238. **Saadi, O. E., A. Esterman, A. Cameron, and D. Roder.** 1995. Murray River water, raised cyanobacterial cell counts, and gastrointestinal and dermatological symptoms. *Med. J. Aust.* **162:**122–125.
239. **Schubert, R.** 1991. Aeromonads and their significance as potential pathogens in water. *J. Appl. Bacteriol.* **70:**131S–135S.
240. **Schwartz, J., R. Levin, and K. Hodge.** 1997. Drinking water turbidity and pediatric hospital use for gastrointestinal illness in Philadelphia. *Epidemiology* **8:**607–609.
241. **Sellwood, J., and J. Dadswell.** 1991. Human viruses and water, p. 29–45. *In* P. Morgan-Capner (ed.), *Current Topics in Clinical Virology*. The Laverham Press, Salisbury, England.
242. **Shahamat, M., C. Paszko-Kolva, H. Yamamoto, U. Mia, A. Pearson, and R. Colwell.** 1989. Ecological studies of *Campylobacter pylori*. *Klin. Wochenschr.* **67:**62–63.
243. **Shin, G. A., and M. D. Sobsey.** 1998. Reduction of Norwalk virus, poliovirus 1 and coliphage MS2 by monochloramine disinfection of water. *Water Sci. Technol.* **38:**151–154.
244. **Shin, G. A., and M. D. Sobsey.** 2003. Reduction of Norwalk virus, poliovirus 1, and bacteriophage MS2 by ozone disinfection of water. *Appl. Environ. Microbiol.* **69:**3975–3978.
245. **Shlim, D., M. Cohen, M. Eaton, R. Rajah, E. Long, and B. Ungar.** 1991. An alga-like organism associated with an outbreak of prolonged diarrhea among foreigners in Nepal. *Am. J. Trop. Med. Hyg.* **45:**383–389.
246. **Singh, N., and V. Yu.** 1994. Potable water and *Mycobacterium avium* complex in HIV patients: is prevention possible? *Lancet* **343:**1110–1111.
247. **Skraber, S., B. Gassilloud, and C. Gantzer.** 2004. Comparison of coliforms and coliphages as tools for assessment of viral contamination in river water. *Appl. Environ. Microbiol.* **70:**3644–3649.
248. **Smit, T. K., P. Bos, I. Peenze, X. Jiang, M. K. Estes, and A. D. Steele.** 1999. Seroepidemiological study of genogroup I and II calicivirus infections in South and southern Africa. *J. Med. Virol.* **59:**227–231.
249. **Soave, R., J. Dubey, L. Ramos, and M. Tummings.** 1986. A new intestinal pathogen? *Clin. Res.* **34:**533A.
250. **Sobsey, M.** 1989. Inactivation of health-related microorganisms in water by disinfection processes. *Water Sci. Technol.* **21:**179–195.
251. **Sobsey, M., A. Dufour, C. Gerba, M. LeChevallier, and P. Payment.** 1993. Using a conceptual framework for assessing risks to health from microbes in drinking water. *J. Am. Water Works Assoc.* **85:**44–48.
252. **Sobsey, M., T. Fuji, and R. Hall.** 1991. Inactivation of cell-associated and dispersed hepatitis A virus in water. *J. Am. Water Works Assoc.* **83:**64–67.
253. **Sobsey, M., P. Shields, F. Hauchman, A. Davis, V. Rullman, and A. Bosch.** 1988. Survival and persistence of hepatitis A virus in environmental samples, p. 121–124. *In* A. Zuckerman (ed.), *Viral Hepatitis and Liver Disease*. Alan R. Liss, New York, N.Y.
254. **Soong, F., E. Maynard, K. Kirke, and C. Luke.** 1992. Illness associated with blue-green algae. *Med. J. Aust.* **156:**67.
255. **Strauss, B., W. King, A. Ley, and J. R. Hoey.** 2001. A prospective study of rural drinking water quality and acute gastrointestinal illness. *BMC Public Health* **1:**8.
256. **Sturbaum, G., Y. Ortega, R. Gilman, C. Sterling, L. Cabrera, and D. Klein.** 1998. Detection of *Cyclospora cayetanensis* in wastewater. *Appl. Environ. Microbiol.* **64:**2284–2286.
257. **Talal, A. H., C. L. Moe, A. A. Lima, K. A. Weigle, L. Barrett, S. I. Bangdiwala, M. K. Estes, and R. L. Guerrant.** 2000. Seroprevalence and seroincidence of Norwalk-like virus infection among Brazilian infants and children. *J. Med. Virol.* **61:**117–124.
258. **Taylor, M. B., S. Parker, W. O. Grabow, and W. D. Cubitt.** 1996. An epidemiological investigation of Norwalk virus infection in South Africa. *Epidemiol. Infect.* **116:**203–206.
259. **Teunis, P. F., C. L. Chappell, and P. C. Okhuysen.** 2002. *Cryptosporidium* dose-response studies: variation between hosts. *Risk Anal.* **22:**475–485.
260. **Thurston-Enriquez, J. A., C. N. Haas, J. Jacangelo, and C. P. Gerba.** 2003. Chlorine inactivation of adenovirus type 40 and feline calicivirus. *Appl. Environ. Microbiol.* **69:**3979–3985.
261. **Thurston-Enriquez, J. A., C. N. Haas, J. Jacangelo, K. R. Riley, and C. P. Gerba.** 2003. Inactivation of feline calicivirus and adenovirus type 40 by UV radiation. *Appl. Environ. Microbiol.* **69:**577–582.
262. **Thurston-Enriquez, J. A., P. Watt, S. E. Dowd, R. Enriquez, I. L. Pepper, and C. P. Gerba.** 2002. Detection of protozoan parasites and microsporidia in irrigation waters used for crop production. *J. Food Prot.* **65:**378–382.
263. **Ticehurst, J.** 1995. Hepatitis E virus, p. 1056–1067. *In* P. Murray, et al. (ed.), *Manual of Clinical Microbiology*, 6th ed. ASM Press, Washington, D.C.
264. **Turner, P., A. Gammie, K. Hollinrake, and G. A. Codd.** 1990. Pneumonia associated with contact with cyanobacteria. *Br. Med. J.* **300:**1440–1441.
265. **Ueki, Y., D. Sano, T. Watanabe, K. Akiyama, and T. Omura.** 2005. Norovirus pathway in water environment estimated by genetic analysis of strains from patients of gastroenteritis, sewage, treated wastewater, river water and oysters. *Water Res.* **39:**4271–4280.
266. **U.S. Environmental Protection Agency.** 1986. *Ambient Water Quality for Bacteria—1986*. EPA 44015-84-002. U.S. Environmental Protection Agency, Washington, D.C.
267. **U.S. Environmental Protection Agency.** 1994. *National Primary Drinking Water Standards*. EPA 810-F-94-001A. Office of Water, U.S. Environmental Protection Agency, Washington, D.C.
268. **U.S. Environmental Protection Agency.** 2005. *Fact Sheet: The Drinking Water Contaminant Candidate List—the Source of Priority Contaminants for the Drinking Water Program*. EPA 815-F-05-001. Office of Water, U.S. Environmental Protection Agency, Washington, D.C.
269. **Vaerewijck, M. J., G. Huys, J. C. Palomino, J. Swings, and F. Portaels.** 2005. Mycobacteria in drinking water distribution systems: ecology and significance for human health. *FEMS Microbiol. Rev.* **29:**911–934.
270. **van den Berg, H., W. Lodder, W. van der Poel, H. Vennema, and A. M. de Roda Husman.** 2005. Genetic diversity of noroviruses in raw and treated sewage water. *Res. Microbiol.* **156:**532–540.
271. **van der Kooij, D.** 1993. Importance and assessment of the biological stability of drinking water in the Netherlands, p. 165–179. *In* G. Craun (ed.), *Safety of Water Disinfection: Balancing the Chemical and Microbial Risks*. ILSI Press, Washington, D.C.
272. **Van Heerden, J., M. M. Ehlers, W. B. Van Zyl, and W. O. Grabow.** 2003. Incidence of adenoviruses in raw and treated water. *Water Res.* **37:**3704–3708.
273. **Vivier, J. C., M. M. Ehlers, and W. O. Grabow.** 2004. Detection of enteroviruses in treated drinking water. *Water Res.* **38:**2699–2705.
274. **von Reyn, C., J. Maslow, T. Barber, J. Falkinham, and R. Arbeit.** 1994. Persistent colonisation of potable water

as a source of *Mycobacterium avium* infection in AIDS. *Lancet* **343:**1137–1141.

275. **Waddell, G., A. Allard, M. Johansson, L. Svensson, and I. Uhnoo.** 1994. Enteric adenoviruses, p. 519–547. *In* A. Kapikian (ed.), *Viral Infections of the Gastrointestinal Tract.* Marcel Dekker, New York, N.Y.
276. **Wadowsky, R., R. Yee, L. Mezmar, E. Wing, and J. Dowling.** 1982. Hot water systems as sources of *Legionella pneumophila* in hospital and nonhospital plumbing fixtures. *Appl. Environ. Microbiol.* **43:**1104–1110.
277. **Walch, M., and R. Colwell.** 1994. Detection of nonculturable indicators and pathogens, p. 258–273. *In* C. Hackney and M. Pierson (ed.), *Environmental Indicators and Shellfish Safety.* Chapman and Hall, New York, N.Y.
278. **Waller, T.** 1979. Sensitivity of *Encephalitozoon cuniculi* to various temperatures, disinfectants and drugs. *Lab Anim.* **13:**227–230.
279. **Wang, J., X. Jiang, H. P. Madore, J. Gray, U. Desselberger, T. Ando, Y. Seto, I. Oishi, J. F. Lew, and K. Y. Green.** 1994. Sequence diversity of small, round-structured viruses in the Norwalk virus group. *J. Virol.* **68:**5982–5990.
280. **Ward, R., D. Bernstein, and E. Young.** 1986. Human rotavirus studies in volunteers: determination of infectious dose and serological response to infection. *J. Infect. Dis.* **154:**871–880.
281. **Weber, R., R. Bryan, R. Owen, C. Wilcox, L. Gorelkin, G. Visvesvara, et al.** 1992. Improved light-microscopical detection of microsporidia spores in stool and duodenal aspirates. *N. Engl. J. Med.* **326:**161–166.
282. **Wendt, S., K. George, B. Parker, H. Graft, and J. Falkinham.** 1980. Epidemiology of infection by nontuberculous mycobacteria. *Am. Rev. Respir. Dis.* **122:**259–263.
283. **West, A., M. Millar, and D. Tompkins.** 1990. Survival of *Helicobacter pylori* in water and saline. *J. Clin. Pathol.* **43:**609.
284. **WHO.** 1993. *Guidelines for Drinking-Water Quality,* 2nd ed., vol. 1. *Recommendations.* WHO, Geneva, Switzerland.
285. **WHO.** 1998. Freshwater algae and cyanobacteria. *In Guidelines for Safe Recreational-Water Environments.* WHO, Geneva, Switzerland.
286. **WHO.** 1998. *Guidelines for Drinking Water Quality.* WHO, Geneva, Switzerland.
287. **WHO.** 2004. *Meeting the MDG Drinking Water and Sanitation Target: a Mid-Term Assessment of Progress.* UNICEF/WHO, Geneva, Switzerland.
288. **Winn, W.** 1995. Legionella, p. 533–544. *In* P. Murray et al. (ed.), *Manual of Clinical Microbiology,* 6th ed. ASM Press, Washington, D.C.
289. **Woode, G.** 1994. The toroviruses: bovine (Breda virus) and equine (Berne virus) and the torovirus-like agents of humans and animals, p. 581–602. *In* A. Kapikian (ed.), *Viral Infections of the Gastrointestinal Tract.* Marcel Dekker, New York, N.Y.
290. **WSSCC.** 2004. *The Campaign: WASH Facts and Figures.* Water Supply and Sanitation Collaborative Council, Geneva, Switzerland.
291. **Wu, J., C. Chen, T. Chiang, I. Sheen, J. Chen, W. Tsai, Y. Huang, and S. Lee.** 2000. Clinical and epidemiological implications of swine hepatitis E virus infection. *J. Med. Virol.* **60:**166–171.
292. **Wurtz, R.** 1994. *Cyclospora*: a newly identified intestinal pathogen of humans. *Clin. Infect. Dis.* **18:**620–623.
293. **Zmirou, D.** 1987. A follow-up study of gastro-intestinal diseases related to bacteriologically substandard drinking water. *Am. J. Public Health* **77:**582–587.

Detection of Microorganisms in Environmental Freshwaters and Drinking Waters

GARY A. TORANZOS, GORDON A. McFETERS,
JUAN JOSÉ BORREGO, AND MARION SAVILL

20

The presence of some microorganisms in waters is used as an indication of possible contamination and as an index of quality deterioration. Heavy reliance has been placed on the coliform group of bacteria to determine the safety of potable water, while coliforms and other organisms are also used to signal conditions that might lead to adverse health consequences in recreational and shellfish-harvesting waters. Although this practice is far from perfect and there is considerable variety in the ways that different indicator microorganisms are applied in various geographical areas and situations, public health concerns have generally been well served for over a century. The presence of indicator organisms will likely continue to be used as a criterion of water quality that will be of value if attention is given to the development and use of optimal methods for the recovery of these microorganisms. Data also exist indicating that there is limited utility for indicator microorganisms under certain circumstances.

Gastroenteritis is the most common affliction associated with waterborne pathogens. Although, for most of the population in developed countries, minor gastroenteritis may simply mean hours of discomfort, in developing countries up to 10 million people die every year as a direct result of the consumption of contaminated water (67). However, the continuing increase in the proportion of the immunologically compromised and elderly populations in developed countries has focused more attention on the impact of waterborne disease outbreaks. The seriousness of this reality is compounded by the fact that determining the microbiological quality of waters is more complex than previously thought.

The presence of enteric pathogens in drinking and recreational waters is of great concern, and thus, it is important to determine the microbiological safety of these waters. The ideal manner for doing this would be to analyze the waters for the presence of specific enteric pathogens; however, hundreds of different microorganisms have been shown to be involved in waterborne disease outbreaks; thus, it would be impractical to look for every pathogen potentially present in water samples. Culture methods are usually used for the detection of bacteria and cell culture techniques for the detection of viruses, while microscopic methods have been used for the detection of protozoa. However, the target bacteria may not grow in culture media since they are frequently injured as a result of exposure to environmental stressors such as disinfectants used during water treatment. Additionally, several enteric viruses cannot be cultured in the laboratory, and methods for the detection of protozoan pathogens are notoriously inefficient.

As a result, analyses for the presence of waterborne pathogens become complex and do not ensure complete safety for the consumer. Although some molecular methods have been proposed, none has yet been shown to be applicable to routine sampling and analysis. Thus, some of the following groups of microorganisms are still used to determine the biological safety of the waters.

COMMONLY USED INDICATORS

Total Coliforms

By definition, coliform bacteria are facultatively anaerobic, gram-negative, non-spore-forming, rod-shaped bacteria that ferment lactose with gas and acid production in 24 to 48 h at 35°C (41). Coliform bacteria belong to the family *Enterobacteriaceae* and include *Escherichia coli* as well as various members of the genera *Enterobacter, Klebsiella,* and *Citrobacter*. That definition of the coliform bacteria has classically been translated into specific biochemical reactions or the appearance of characteristic colonies on commonly used media. However, the development of enzyme-specific media and tests has allowed the application of cytochrome oxidase (negative) and β-galactosidase (positive) as additional criteria for the coliform group (41).

These bacteria were classically used as indicators of possible fecal contamination or water pollution from sewage and thus are of sanitary significance, and although they originate from the intestinal tracts of homeothermic animals, other bacteria numerically dominate that type of microbial community (51). In addition, experience has demonstrated that members of the coliform group can originate from nonenteric environments such as wastes from the wood industry and surfaces of redwood water tanks (146), biofilms within drinking water distribution systems (52, 89), and epilithic algal mat communities in pristine streams (5, 100). The persistence of these bacteria in aquatic systems is comparable to that of some of the waterborne bacterial pathogens (110), although they are much less persistent than enteric viruses and protozoa. In the case of some aquatic environments, total coliforms present a survival rate lower than that of pathogenic microorganisms (104). Likewise, bacteria belonging to the coliform group are somewhat like many of the waterborne bacterial pathogens with respect to disinfection susceptibility but unlike the more persistent viruses and protozoa. These factors illustrate why experience and discretion are often vital in the correct interpretation of coliform data obtained from some aquatic environments.

There are serious shortcomings in the general use of these microorganisms as indicators, such as the ability of some coliforms to grow in natural waters, the lack of correlation between the number of coliforms and those of pathogenic microorganisms, and the detection of atypical strains (11, 45).

The coliform bacteria that comprise what has traditionally been termed the "total coliform" group form the primary standards for potable water in most of the world. This designation is usually interpreted as the totality of bacteria that conform to the classical, nontaxonomic coliform definition, although the inclusion of the term "total" has often led to the false assumption that all viable coliform bacteria are detected by a given method specified for the analysis of coliform bacteria.

Thermotolerant Coliforms (Fecal Coliforms)

The subset of the more comprehensive coliform, or total coliform, group that is more definitive as an indicator of homeothermic fecal contamination consists of what are termed the fecal coliforms (49). However, "thermotolerant coliform" is the scientifically accurate term for this group (32, 144). These bacteria conform to all of the criteria used to define total coliforms plus the requirement that they grow and ferment lactose with the production of gas and acid at 44.5 ± 0.2°C. Dufour (38) suggested the redefinition of the fecal coliform group with the addition of several biochemical characteristics, such as their inability to use citrate as the only carbon source, the presence of tryptophanase activity and the lack of urease activity, the negative Voges-Proskauer test, and the positive methyl red test. Bacteria in this coliform subgroup have been found to have an excellent positive correlation with fecal contamination from warm-blooded animals (111). The physiological basis of the elevated temperature phenotype in the fecal coliforms has been described as a thermotolerant adaptation of proteins to, and their stability at, the temperatures found in the enteric tracts of animals (31) that are both constant and higher than temperatures in most aquatic and terrestrial environments. However, some thermotolerant coliform bacteria that conform to this definition also belong to the genus *Klebsiella* (38, 83) and have been isolated from environmental samples in the apparent absence of fecal pollution. Such observations have been made on water receiving high levels of carbohydrate-rich industrial effluent and in contact with plant material. Similarly, in the last 20 years other members of the thermotolerant coliform group, including *E. coli*, have been detected in some pristine areas of the world (120) and associated with regrowth events in potable water distribution systems (89). Thus, caution needs to be exercised when deciding whether the presence of indicator microorganisms does indeed represent fecal contamination and thus a threat to public health.

E. coli

Among the thermotolerant coliforms, *E. coli* deserves further discussion. This bacterium has been shown to be a more specific indicator for the presence of fecal contamination than is the thermotolerant coliform group of bacteria (38). In addition, *E. coli* conforms to taxonomic as well as functional identification criteria and is enzymatically distinguished by the lack of urease and presence of β-glucuronidase, enzymes that form the basis for recently developed differential methods that will be discussed later in this chapter. One disadvantage associated with this organism as an indicator is that it has been consistently found in pristine tropical rain forest aquatic and plant systems as well as soils and so may not be a reliable signal of fecal contamination in those environments (67, 88b). Additionally, it seems to survive for short periods in aquatic environments in temperate environments (16, 104). *E. coli* has been effectively used for some time in Europe and has recently been incorporated into U.S. drinking water regulations as a specific indicator of fecal contamination. New data show that *E. coli* can survive for extended periods in lake sediments (23a).

Fecal Streptococci and Enterococci

Fecal streptococci and enterococci, which are gram-positive bacteria, have received widespread acceptance as useful indicators of microbiological water quality, since (i) they show a high and close relationship with health hazards associated with bathing in aquatic environments, mainly for gastrointestinal symptoms; (ii) they are not as ubiquitous as coliforms; (iii) they are always present in feces of warm-blooded animals; (iv) they are unable to multiply in sewage-contaminated waters; and (v) their die-off is less rapid than that of coliforms in water, and persistence patterns are similar to those of potential waterborne pathogenic bacteria (15, 16, 24, 39, 51, 118, 127, 128).

Fecal streptococci comprise species of different sanitary significance and survival characteristics; in addition, the proportions of the species of this group are not the same in animal and human feces (111). Therefore, a clearer definition of fecal streptococci is necessary to establish a specific standard methodology of enumeration. The taxonomy of this group has been subject to extensive revision (73, 91, 127). According to Borrego and Figueras (14), the following species of the *Enterococcus* and *Streptococcus* genera may be included in the fecal streptococcal group: *Enterococcus faecalis*, *E. faecium*, *E. durans*, *E. hirae*, *E. avium*, *E. gallinarum*, *E. cecorum*, *Streptococcus bovis*, *S. equinus*, *S. alactolyticus*, *S. intestinalis*, *S. hyointestinalis*, and *S. acidominimus*.

Enterococci include all the species described as members of the genus *Enterococcus*, which also fulfill Sherman's criteria: growth at 10 and 45°C, resistance to 60°C for 30 min, growth at pH 9.6 and at 6.5% NaCl, and the ability to reduce 0.1% methylene blue. These microorganisms possess a closer association with human feces than do other fecal streptococci (10, 44). This property contrasts with other members of the fecal streptococcal group such as *S. bovis* and *S. equinus* that are somewhat characteristic of fecal contamination from specific animals (49), although these species do not survive in water as well as do many of the other classical indicator bacteria and some pathogens (50, 100).

Staphylococci

Several epidemiological studies show high incidences of nondiarrheal diseases, such as eye, ear, mucous, and skin infections associated with swimming (25, 96). In this regard, many swimmers acquire staphylococcal skin infections after swimming in recreational waters (29, 96), and thus, there is a need for an indicator of water quality that addresses skin infections rather than gastrointestinal infections.

Evidence to support the use of coagulase-positive staphylococci, particularly *Staphylococcus aureus*, as an indicator of water quality includes the following: (i) this microorganism is stable and shows a great resistance to environmental conditions, especially in marine and chlorinated swimming pool waters; and (ii) its concentration in water has been shown to represent the load of microorganisms being shed by swimmers (30, 46).

Bacteriophages

The use of bacteriophages as indicators is not a new concept, since they have been previously proposed as fecal and viral indicators (84, 85, 86, 131). The term bacteriophage is a wide concept; the bacteriophage groups proposed as indicators are somatic coliphages, F(male)-specific RNA bacteriophages (FRNA phages), and phages of *Bacteroides fragilis* (62, 77, 105, 125).

Somatic coliphages are specific viruses of *E. coli* and have been the most studied group. They are commonly used as indicators of fecal and/or sewage pollution, on the basis of their direct correlation with the presence of enteric viruses in marine waters, freshwater ecosystems, and sewage effluents (17, 18, 40, 70, 137). Borrego (13) reviewed the indicator potential of somatic coliphages. He concluded that, considering the differences in origin and ecology between enteric viruses and somatic coliphages, it is doubtful that this phage group could successfully be used in all situations as enteric virus indicators; however, its members are good indicators of fecal pollution and of the microbiological water quality.

Several authors have proposed other bacteriophage groups as alternate indicators. Primrose et al. (112) proposed the use of FRNA phages as fecal pollution indicators on the basis of their inability to replicate in the water ecosystem (pili are not expressed at temperatures below 30°C). Havelaar and Hogeboom (61) developed a host bacterium specific for the detection of this bacteriophage group and proposed its members as model viruses in water hygiene because (i) the group comprises viruses similar in size, shape, and genetic makeup to human enteric viruses, which are responsible for most waterborne diseases; (ii) it represents viruses that are more stable than human enteroviruses in environmental waters and more resistant to disinfection; and (iii) the concentrations found in environmental waters correlate with sewage contamination (64).

The low incidence of the FRNA phages in feces, particularly in human feces (35, 63, 109), and their high concentration in sewage suggest that FRNA phages should multiply in the sewerage system (75). Hence, the presence of FRNA phages in water may be primarily an index of sewage pollution rather than fecal pollution (37). FRNA phages have been proposed as indicators of the environmental behavior of enteric viruses (e.g., noroviruses) in water and wastewater treatment rather than as general indicators of fecal pollution (60). FRNA phages have been placed in five different groups based on their serological characteristics, with some groups (III and IV) being more closely associated with human pollution.

B. fragilis is a strict anaerobe found in high concentrations in the human intestinal tract and dies rapidly when discharged into environmental waters. A phage of the strain HSP 40 of *B. fragilis* has been proposed as a specific index of human fecal pollution of waters (132), because (i) phages against this bacterial strain are human specific and are not isolated from the feces of other homeothermic animals; (ii) *B. fragilis* HSP 40 phages are consistently isolated from sewage, fecally polluted waters, and their sediments but not from unpolluted samples; (iii) the levels of phages are related to the degree of pollution; (iv) *B. fragilis* phages always outnumber human enteric viruses; and (v) in model experiments, no replication of these phages has been observed under simulated environmental conditions (77, 93, 133, 134). The low prevalence of these phages in waters with low and moderate levels of fecal pollution and the complex methodology for their recovery are the main drawbacks for the general use of these viruses as an indicator group (33, 34); nonetheless, efforts are under way to improve the methods of detection.

OTHER COMMONLY USED INDICATORS

Sulfite-Reducing Clostridia

The use of the sulfite-reducing members of the genus *Clostridium* (*C. perfringens* and *C. welchii*) as indicators of fecal pollution was originally proposed in the late 1800s by Houston. The presence of these microorganisms in the feces of all warm-blooded animals is the basis for this practice, although they are considered ubiquitous in aquatic sediments and the spore form explains their persistence (130). As a result of the longevity of the spores, these bacteria can be considered indicators of remote fecal pollution (143) and as possible indicators of disinfection processes of water (121).

Pseudomonas spp.

Members of the genus *Pseudomonas* are possibly the microorganisms most often isolated from bodies of water. However, contrary to the previously discussed indicators, their presence does not necessarily indicate a possible risk to public health. Some species have been linked to infections associated with exposure to recreational waters and thus have been proposed as indicators of recreational water quality (56a). *Pseudomonas aeruginosa* was found to be more resistant than acid-fast bacteria during ozonation processes (56), which demonstrates its resistance to chemical disinfection and thus its usefulness in the analysis of recreational waters such as swimming pools, which receive chemical disinfection. This microorganism has also been isolated from drinking waters. For reviews of *Pseudomonas*, its ecology, and public health, the reader is directed to references 59 and 137a.

Mycobacteria

The mycobacteria belong to a group of microorganisms considered emerging pathogens of increasing importance. Their importance as etiological agents of waterborne disease is still not completely understood. Although some members of the genus *Mycobacterium* have been known and studied for a long time as the causative agents of tuberculosis (*M. tuberculosis*, *M. africanum*, and *M. bovis*) the non-tuberculous, or atypical mycobacteria have gained notoriety as a result of an increased number of AIDS patients infected with these bacteria. About 50 species of atypical mycobacteria have been identified to date (36). There is no evidence of person-to-person transmission for any of these atypical species, and they are known to be widely distributed in the soil and water environments. The two species most commonly associated with human disease are *Mycobacterium intracellulare* and *Mycobacterium avium*, which together form the group known as MAC (*M. avium* complex). The MAC are slowly growing microorganisms, which makes it difficult to isolate them in the laboratory. In fact several decontamination techniques have been proposed in an attempt to eliminate the nontarget microorganisms in environmental samples, but these methods may also affect the MAC. The reader should read Dailloux et al. (36) for a thorough review on this topic. Neumann et al. (108) conducted a comparison of 12 employed methods for the isolation of MAC from different water samples and concluded that the method to be used depended on the type of water being analyzed.

H_2S Producers

Although most of the members of the total and fecal coliform groups are not H_2S producers, Manja et al. (95) developed a simple method which has been tested in developing countries (87) and rural areas of North America (124). Although no correlation was observed between the presence of coliforms and H_2S producers (129), other studies have shown that the presence of the H_2S producers does correlate with other criteria of microbiological water quality. Thus, this test may be a good alternative for rural areas, as well as for remote areas of the world, since it does not necessitate incubators, sophisticated equipment, or trained personnel (136). In fact, Castillo et al. (26) proposed the use of beehives as a means of keeping the temperature constant during incubation. There are unpublished results comparing the usefulness of this test to traditional indicators.

Aeromonas

The taxonomy of the genus *Aeromonas* remained unclear for many years, although the molecular approaches have established 14 species and a number of unnamed hybridization groups or genomospecies (27, 79). However, there are a number of problems associated with several described *Aeromonas* species: (i) the existence of mesophilic and psychrophilic strains; (ii) the limited number of strains that have been analyzed for taxonomic purposes; and (iii) the source of strains studied in defining a species. Another confounding problem involves strains that previously fell within the *Aeromonas hydrophila* complex (hybridization groups 1 to 3). This group has been found to be much more complex than originally thought, including organisms that share phenotypic features but do not belong to the same genomospecies (1). These organisms could in essence represent new species but from a practical and medical standpoint should probably be referred to as genomovars or nomenspecies (140).

Several species of the genus *Aeromonas* (*A. hydrophila*, *A. caviae*, *A. veronii* bt. sobria, *A. veronii* bt. veronii, *A. jandaei*, and *A. schubertii*) have been associated with different human diseases, such as gastroenteritis, and several nonenteric diseases. Most *Aeromonas*-associated infections seem to be related to water exposure of open wounds as well as to the consumption of contaminated water or food (80, 88). Underlying illnesses (diabetes, cirrhosis, and an immunosuppressive state) play a major role in the acquisition and outcome of the diseases produced by *Aeromonas*.

Mesophilic members of the genus *Aeromonas* have been detected and isolated from brackish, fresh, and estuarine waters; sewage; treated effluents; and chlorinated and nonchlorinated water supplies throughout the world (72). In recent years, several studies have been carried out to determine the occurrence and levels of mesophilic aeromonads in both polluted and unpolluted water systems, as well as to determine their human health significance (6, 7, 8, 9, 65, 71). Aeromonads were isolated from surface freshwater with temperatures of 4 to 45°C, with the highest densities at 35°C, but could not be isolated at temperatures above 45°C (72). No relationship with pH was observed and aeromonads were isolated over the pH range of 5.2 to 9.8, although laboratory studies indicated that *A. hydrophila* was unable to grow below pH 4 or above pH 10 (68). *Aeromonas* can be readily isolated from drinking water distribution systems, where they appear to survive well when associated with pipe biofilms. Within water distribution systems, aeromonads are able to proliferate at low temperatures (4°C) (72). A highly significant correlation between thermotolerant aeromonads and the trophic state of freshwater in lakes in the United States was found (119). However, other authors were unable to predict the trophic status of several lakes in relation to these microorganisms (117). In coastal or estuarine waters, halotolerant aeromonads are associated with sewage discharges (6, 8, 68).

Aeromonads are not considered to be normal inhabitants of the human gastrointestinal tract; thus, their presence in raw sewage and in treated effluents at concentrations equivalent to those of thermotolerant coliforms indicates that they are able to multiply in these waters (9, 72). Several outbreaks of *Aeromonas*-associated disease linked to either drinking water or recreational water contact have been reported associated with gastrointestinal and extraintestinal infections (23, 53, 58, 66, 82); however, the evidence is tenuous at best (107), although some rare strains isolated from drinking waters in Spain (45a) seem to contain virulence factors. The public health importance of this finding remains to be seen.

HPC

The heterotrophic plate count (HPC) method takes into consideration the enumeration of aerobic bacteria capable of growing in some commonly used media such as R_2A, HPC agar, or plate count agar. The level of bacteria determined by HPC indicates the overall microbiological status of the system and not necessarily the possibility of risk to public health, although such bacteria have been recently implicated as potential pathogens in drinking water (94, 118). Nonetheless, there are conflicting reports regarding the usefulness of HPC in waters.

Sterols and Other Chemical Indicators of Sewage Contamination

Alternative analytical approaches have been sought in lieu of classically used indicator microorganisms as indicators of fecal pollution. This search continues because of inadequacies associated with the failure of cultivation-based methods to detect indicator bacteria as well as concerns regarding the validity of their detection. The use of specific saturated sterols such as 5,β-cholestan-3,β-ol (coprostanol) has been investigated as a molecular signature of fecal contamination in water (41, 146). Although not frequently employed, this approach has been useful as a marker of fecal pollution in sediments because the compound is specific to the feces of higher animals, including humans, and is biodegradable and because there is a relationship between its concentration and the degree of fecal pollution (141, 145, 146).

Caffeine and urobilin have also been proposed as indicators of the presence of sewage, since the only source of these compounds is domestic sewage (110a). However, since these are chemical methods and are beyond the scope of this chapter, they will not be discussed further.

SOURCE IDENTIFICATION

An emphasis on source water protection and watershed management has occurred from changes to the Safe Drinking Water Act and Clean Water Act. Knowledge of land uses and source water quality is necessary to manage human activities in the watershed. Health concerns with respect to water quality and the presence of many different users of the water in the watershed have caused conflicts between the different groups with respect to the land and water use within the different catchments.

All countries would like either now or in the future to know the source of pollution in their watershed. This knowledge allows the assessment of potential health risks and therefore how to reduce potential problems from different fecal sources. Such problem reduction assists the quality of rivers, lakes, reservoirs, and groundwater used as drinking water sources. It is important for general environmental quality, food safety, and regulatory purposes.

Many microbial and chemical methods are currently in use; a "toolbox" (Table 1) approach is used, and no one target or method is presently able to answer the question.

Chemicals

A range of chemicals is also being used, such as nonprescription drugs, caffeine (see above), fecal sterols and stanols (see above), cotinine, detergents, fluorescent whitening agents, sodium triphosphates, and triclosan. It should be noted that all of these chemicals are either of fecal origin or sewage origin, since most chemicals are used in detergents.

Others

Other approaches include antibiotic resistance analysis, carbohydrate utilization profiling, and *E. coli* serotyping.

The majority of the current methods are DNA based and thus require the preexistence of a large database composed of a library of collected "patterns" to identify the unknowns. All these are susceptible to errors as a result of possible nonreproducibility (e.g., randomly amplified polymorphic DNA [RAPD]) and thus are not readily transferable between labs. Certain techniques such as multilocus sequence typing are very expensive and time-consuming.

In short, certain methods stand out as providing a certain level of discrimination at a price with a certain level of difficulty and reproducibility in different laboratories. Methods for typing *Bacteroides* (human) and *Rhodococcus coprophilus* (animal and herbivore) and for detecting fecal sterols and stanols appear to be the most promising.

The degree of resolution required by the different catchments, watersheds, and countries needs to be determined. This may well differ depending on the philosophy behind the problem, ability to assist the polluters, and where and how the issues will be addressed and solved, as well as cost involved. However, the methods need more rigorous testing, especially in regard to the reproducibility aspects.

We do need rigorous methods that can be transferred between labs and personnel and especially methods that stand up to legal challenges. The analyst needs to know the strengths and limitations of each method. A case-by-case approach may be the most useful to achieve a degree of certainty.

FC/FS Ratio

The ratio of fecal coliforms to fecal streptococci (FC/FS ratio) was proposed as an indicator of the origin of the contamination (52) because of the different concentrations at which these microorganisms are present in different animals and humans. A ratio of greater than 4 is characteristic of human fecal contamination, whereas a ratio of less than 0.7 suggests animal waste. The index has been used in the past, but its use is complicated by the differential die-off kinetics of the two bacterial groups as well as differences among individual species within those categories (15). Specifically, studies show that the fecal streptococci have a greater survival than do fecal coliforms in water and that certain enterococci such as *Streptococcus boels* and *S. equinus* die off much more rapidly than the other fecal streptococci (100). Geldreich and Kenner (52) suggested that the ratio would be valid only for less than 24 h following the discharge of feces into the water being tested. The ratio is also influenced by the methods and media used; hence, the FC/FS ratio is not generally recommended as an unambiguous means of differentiating human fecal pollution from animal

TABLE 1 Summary of microbial tracking methods[a]

Organism	Method
B. fragilis phages	Amplified fragment length polymorphism
Bacteroides-Prevotella spp.	Denaturing gradient gel electrophoresis
Bacteroidales	16S rDNA PCR
Bifidobacterium spp.	Enterotoxin biomarkers
C. perfringens	Multilocus sequence typing
Coliphages	Genotyping
Enterococcus spp.	Plasmid typing
E. coli	Colilert, BOX PCR[b]
Fecal bacterium ratios	Terminal RFLP
Fecal coliforms	
Fusobacterium	Pulsed-field gel electrophoresis
F+ DNA coliphages	
Human- and animal-specific viruses	Sequencing organism-specific genes
Peptococcus	RAPD
R. coprophilus	Repetitive element PCR (BOX- and ERIC-PCR)[b]
Ruminococcus	RFLP
S. bovis	Ribotyping

[a]Data presented at the Water Environment Research Foundation Microbial Source Tracking Workshop, San Antonio, Tex., 2005.
[b]BOX and ERIC sequences are considered palindromic sequences.

fecal pollution (5, 6a, 111), although it may have value in some instances if care is used in interpreting the results.

SIGNIFICANCE OF INDICATORS TO PUBLIC HEALTH

Analysis for the presence of indicators is a shortcut attempt to determine the microbiological quality and public health safety of waters. Very few water bodies are completely free of indicator microorganisms. High concentrations of total coliforms can be found associated with plant material; thus, their presence in surface waters is not necessarily a cause for concern. Under some circumstances, even the presence of thermotolerant coliforms (*Klebsiella* spp. and *E. coli*) can be expected in surface waters. In these instances, other indicators, such as those discussed above, should be used to determine the possible threat to public health. However, the difference between untreated and treated waters in terms of the microbiota should be emphasized. The detection of thermotolerant coliforms in treated drinking waters should be a cause for concern, since present drinking water treatment processes successfully eliminate indicator microorganisms.

Current Regulations Regarding Drinking and Surface Waters

The U.S. Environmental Protection Agency and the World Health Organization as well as the European Economic Community standards allow for a maximum number of microorganisms that can be present in a given volume of drinking water (54, 116). Table 2 gives some of the U.S. and international standards for drinking and bathing waters. These standards are believed to provide optimal public health protection. In the United States, the maximum contaminant levels are the maximum permissible concentrations of these organisms and are legally enforceable. In Canada, the maximum acceptable concentrations are nonenforceable guidelines.

GENERAL METHODS

Sample Collection, Transport, and Storage

It is just as important to use appropriate care in obtaining a representative sample as it is to analyze the sample correctly. The sample should be collected in sterile glass or polypropylene bottles or bags. Whenever it is suspected that a disinfectant is present, the sample should be amended with a solution of sodium thiosulfate ($Na_2S_2O_3$), which inactivates any residual halogen compounds present in the sample. An $Na_2S_2O_3$ concentration of 18 mg/liter neutralizes up to 5 mg of free (residual) chlorine per liter (4).

Some environmental samples may also contain high concentrations of zinc and copper. In such cases, it is recommended that the sampling containers be amended with EDTA (Na_2EDTA) at a concentration of 372 mg/liter. The presence of this chelating agent reduces metal toxicity and is especially important if more than 4 h elapse between collection and analysis of the sample (4).

Both compounds ($Na_2S_2O_3$ and Na_2EDTA) can be added together or separately to the sample bottle before sterilization by autoclaving. It is recommended that the reader consult *Standard Methods for the Examination of Water and Wastewater* (4) for more detailed information.

Additionally, the elapsed time allowed between sample collection and analysis should not exceed 24 h. The samples should be kept in a refrigerated container (or icebox) at a temperature below 10°C during transport and storage. The sample will not necessarily reach this temperature. If sample analysis within the 24-h time limit is not possible, in situ analysis with portable equipment should be considered. Care should be taken, and the sample containers should not be in contact with the ice being used to keep the sample cool.

Water samples can be analyzed for total coliform bacteria by a range of techniques, including the most probable number (MPN) using the multiple-tube fermentation test, presence absence (P-A), and membrane filtration (MF), as discussed below. Because of this variety of analytical approaches, there are a number of media that are commonly used.

MPN (Multiple-Tube Technique)

MPN analysis is a statistical method based on the random (Poisson) dispersion of microorganisms in a given sample. Classically, this assay has been performed as a multiple-tube fermentation test. Although the technique is rather time-consuming, most technicians around the world are fully trained in this technique and prefer it to other methods of water analysis. The MPN technique is highly recommended for high-turbidity waters.

TABLE 2 Bacteriological drinking water and recreational freshwater standards or guidelines

Standards established by:	No./100 ml				Enterococci (recreational)	Turbidity (NTU[b])
	Total coliforms[a]		Thermotolerant coliforms			
	Drinking	Recreational[c]	Drinking	Recreational[c]		
World Health Organization	1–10		0			<1–5
Canada	<10		0	200[d]	35[e]	<1–5
European Economic Community	0	<10,000[f]				0–4
United States	0	200[g]		<2,000[f]		1 (monthly)

[a]In systems analyzing <40 samples per month, the maximum contaminant level specifies that no more than one sample per month may be total coliform positive. In systems analyzing >40 samples per month, the maximum contaminant level specifies that no more than 5% of the monthly samples may be total coliform positive (22).
[b]NTU, nephelometric turbidity units.
[c]Recreational refers to primary contact (swimming) waters.
[d]Geometric mean of at least five samples (68a) when experience has shown that greater than 90% of the thermotolerant coliforms are *E. coli*.
[e]Geometric mean of at least five samples taken during a period not to exceed 30 days (68b).
[f]Compulsory limit. If exceeded in more than 20% of samples with at least 14 days of sampling, then bathing is prohibited (116).
[g]This is a U.S. Environmental Protection Agency criterion. Since no uniform national standards exist, it may vary from state to state.

Classically, lauryl tryptose broth is used with the MPN method, where a small inverted tube is included to facilitate the detection of gas formation and 0.01 g of bromocresol purple per liter (final concentration) is used to determine acid production. The concentration of this medium is prepared so that the addition of water sample volumes of 10, 20, or 100 ml to the liquid medium will not reduce the ingredient concentration below that of the specified medium. Typi-cally, five identical (20-ml) aliquots of drinking water are analyzed as a set. Each tube is vigorously mixed with the medium, taking care not to introduce air bubbles into the inverted tubes. These samples are then incubated at 35°C and examined for growth and for gas and acid production after 24 and 48 h. The production of gas or acidic growth after 48 h constitutes a positive presumptive reaction. Confirmation of presumptive reactions is done by inoculating the positive aliquots into brilliant green lactose bile broth tubes containing an inverted tube, incubating the samples for 24 and 48 h at 35°C, and scoring tubes with gas as representing confirmed samples that contain coliform bacteria. Bacterial density and the 95% confidence limits can be estimated with the use of MPN tables for the volumes and number of aliquots used (4).

The MPN method can also be used for the direct detection of coliforms and *E. coli* by using enzyme-specific tests as mentioned below. The characteristic color and/or fluorescent endpoints specified by the manufacturer should be used to determine the presence or level of the target bacteria. The U.S. Environmental Protection Agency has indicated that confirmation is not needed with the commercially available 4-methylumbelliferyl-β-D-glucuronide-based media; however, there are data indicating the possibility of false positives (G. A. Toranzos, unpublished data). Additionally some other bacteria have been shown to produce β-D-glucuronidase, and thus there may be a problem with false positives if these bacteria are present. For the examination of surface waters, inoculate a series of medium tubes with a 10-fold dilution series of the sample. Using five- or three-tube series, inoculate each series with one dilution. If low concentrations of indicator bacteria are present, inoculate 100 ml of water into flasks containing 100 ml of the appropriate medium at double strength or, alternatively, 10-ml volumes directly into tubes containing 10 ml of the appropriate medium at double strength. If high concentrations of bacteria are suspected, a 10-fold dilution series should be performed. A sterile solution such as phosphate buffer (pH 7.2) or peptone water (0.1% final solution, pH 6.8) should be used as the diluent.

Inoculate media as indicated in Table 3 for the presumptive test. Subsequently, all positive tubes should be subjected to the confirmation test by inoculation into a second medium as indicated in Table 4.

Membrane Filtration

MF is possibly the most widely used method in North America and Europe and because of its simplicity is being used almost universally. It is a simple test and lends itself well to the in situ analysis of samples because of the portability of the necessary equipment. The technique is based on the entrapment of the bacterial cells by a membrane filter (pore size of 0.45 μm). After the water is filtered, the membrane is placed on an appropriate medium and incubated. Discrete colonies with typical appearance are counted after 24 to 48 h. This technique is more precise than the multiple-tube (MPN) technique. The greatest limitation of the MF test is that it is useful only for low-turbidity waters and for waters that have low concentrations of nontarget (i.e., background) microorganisms. Samples of water that possess high turbidity clog the filter, and high concentrations of nontarget microorganisms mask the presence of the target colonies. Tables 5 and 6 outline the basic MF method, including confirmation procedures.

Total Coliforms

The MF analysis of water for total coliform bacteria can involve the use of LES Endo agar (Difco), m-Endo broth (Difco), or m-Coliform broth (BBL) as well as other media. If liquid medium is used, ca. 2 ml of the broth can be added to certified absorbent pads, available from commercial sources, in small sterile petri dishes, or 1.5% agar is added to the broth to make it semisolid. Details for carrying out the MF procedure are given elsewhere (56a). Characteristic colonies appear pink to dark red with a unique metallic green sheen (viewed at a magnification of ×10 to ×15 if necessary). Tergitol agar has been used to detect coliform bacteria as a means to follow the movement of the plume resulting from a marine sewage outfall (99). Only plates containing between 20 and 80 typical colonies should be counted. Confirmation can be done as described earlier by the selection of both representative typical and atypical colonies. The population density of the target bacteria, usually described as number of CFU per 100 ml, in the original sample can be calculated from the volume filtered and dilutions used, if any. The percentage of the typical colonies that are confirmed as positive can be used to determine the verified coliform density.

Thermotolerant Coliforms (Fecal Coliforms)

m-FC medium is used to quantify thermotolerant coliforms in water samples when the MF technique is used. A

TABLE 3 Analysis of drinking water or surface water (including treated wastewater) by the multiple-tube (MPN) technique

Type of analysis	Medium	Vol (ml) inoculated into 5- or 3-tube series	Positive reaction
Total coliforms (incubate at 35 ± 0.5°C)	Lactose broth or lauryl tryptose broth	10, 1, 0.1	Gas and/or acid
Thermotolerant coliforms			
Incubate at 44.5 ± 0.2°C	EC medium	10, 1, 0.1	Gas and/or acid
Incubate at 35 ± 0.5°C for 3 h and transfer to 44.5 ± 0.2°C for 21 h	A-1[a]	10, 1, 0.1	Gas
Enterococci (incubate at 35 ± 0.5°C)	Azide dextrose broth	10, 1, 0.1	Growth

[a]Although A-1 medium is recommended by the American Public Health Association only for analysis of marine waters and treated wastewater, several laboratories have tested it for analysis of surface and drinking waters with excellent results. Additionally, A-1 medium does not require a confirmation test.

TABLE 4 Confirmation procedures for the MPN technique

Type of analysis	Medium	Positive reaction
Total coliforms (incubate at 35 ± 0.5°C)	Brilliant green lactose bile broth and lauryl tryptose broth	Gas and/or acid
		Gas and/or acid
Thermotolerant coliforms (incubate at 44.5 ± 0.2°C)	EC medium	Gas and/or acid
Enterococci (incubate at 35 ± 0.5°C)	Pfizer selective *Enterococcus* medium[a]	Growth
		Brownish colonies (with halos)[b]

[a]No longer commercially available and therefore must be prepared separately.

[b]Brownish colonies confirm the presence of fecal streptococci. Colonies are transferred to brain heart infusion broth containing 6.5% NaCl. Growth in the latter medium confirms the presence of enterococci.

solution of 1% rosolic acid (dissolved in 0.02 N NaOH) can be added to the media. However, the use of rosolic acid has been questioned by some European researchers in terms of its possible toxicity to organisms present in water already stressed because of chlorine or ozone disinfection. m-FC can be used as an agar-based medium (1.5% agar added to the broth), or broth can be added to certified absorbent pads. Petri dishes containing filters are incubated at 44.5 ± 0.2°C. Incubation with this degree of accuracy can best be achieved by using incubators or water baths for this specific purpose. Petri dishes can be incubated submerged with weights after they have been placed inside two tightly sealed plastic bags. Typical thermotolerant coliform colonies will appear in various shades of blue, although atypical *E. coli* may be pale yellow, whereas nonthermotolerant coliform colonies range in color from grey to cream. Again, magnification of ×10 to ×15 can be used to observe colonies, and the desired range is 20 to 60 colonies per plate.

Typical colonies sometimes lose the characteristic appearance (green sheen and bluishness for total and thermotolerant coliforms, respectively) after 24 h; thus, reading plates at 24 h is highly recommended.

Enterococci and Fecal Streptococci

m-E medium is used for the detection of enterococci in freshwater and marine waters (the latter topic is covered in chapter 22). Although KF medium was used in the past, it has been found to be susceptible to false-positive results (3, 47; G. A. Toranzos, unpublished observations) and is no longer recommended. The m-E agar is prepared by heating the basal ingredients (10 g of peptone, 15 g of NaCl, 30 g of yeast extract, 1 g of esculin, 0.05 g of cycloheximide, 0.15 g of NaN_3, and 15 g of agar) and cooling the mixture to <46°C. Then 0.25 g of nalidixic acid is mixed in 5 ml of reagent-grade water, a few drops of 0.1 N NaOH are added to dissolve the antibiotic, and the whole volume is added to the basal medium. Finally, 0.15 g of 2,3,5-triphenyl tetrazolium chloride is added and dissolved, and the entire mixture is dispensed into petri dishes. It should be noted that the quantities specified above are for the preparation of a 1-liter volume of medium. After processing of the sample, the membrane filter is placed on the solidified medium; its contents are incubated at 41°C and observed after 24 and 48 h. After incubation, the membrane is transferred very carefully to a petri dish containing EIA medium (1 g of esculin, 0.5 g of ferric citrate, 15 g of agar, and 1 liter of water; dissolve and sterilize by autoclaving) and is incubated at 41°C for 20 min. All pink to red colonies developing a black or reddish brown precipitate on the underside of the filter should be counted.

For the detection of fecal streptococci in freshwaters and marine waters, m-Enterococcus medium (20 g of tryptone, 5 g of yeast extract, 2 g of glucose, 4 g of K_2HPO_4, 0.4 g of NaN_3, 0.1 g of 2,3,5-triphenyl tetrazolium chloride, 10 g of agar, and 1 liter of water) is used. The sample is processed, and the petri dish contents are incubated at 35°C for 24 to 48 h. Colonies should be counted after 24 and 48 h, as some of the pinpoint colonies are not readily visible after 24 h.

TABLE 5 MF procedures for analysis of drinking water or surface water (including treated wastewater)

Type of analysis	Medium	Vol (ml) filtered (depending on type of sample)[a]	Positive reaction
Total coliforms (incubate at 35 ± 0.5°C)	LES Endo agar or	100, 10, 1.0	Colonies with metallic green sheen
	M-Endo medium	100, 10, 1.0	Colonies with metallic green sheen
Thermotolerant coliforms (incubate at 44.5 ± 0.2°C)	M-FC	100, 10, 1.0	Pale- to deep-blue colonies
Fecal streptococci (incubate at 35 ± 0.15°C)	M-Enterococcus	100, 10, 1.0	Pink to deep-red colonies
Enterococci (incubate at 41°C for 24 to 48 h; then transfer membrane to EIA medium and incubate for 20 min at 41°C)	m-E	100, 10, 1.0	Pink to red colonies with a black or reddish brown precipitate on the underside after transfer of the membrane of EIA medium

[a]When treated wastewater or waters affected by sewage are being sampled, several dilutions may be analyzed, and the 100- and 10-ml volumes may be skipped altogether, with higher dilutions used instead.

TABLE 6 Confirmation procedures for the MF technique

Type of analysis	Medium	Positive reaction
Total coliforms (incubate at 35 ± 0.5°C)	Brilliant green lactose bile broth and lauryl tryptose broth	Gas and/or acid
Thermotolerant coliforms (incubate at 44.5 ± 0.2°C)	EC medium	Gas and/or acid
Fecal streptococci[a]	Bile esculin (incubate at 35 ± 0.5°C)	Growth
	BHI–6% NaCl (incubate at 35 ± 0.5°C)	Growth
	BHI (incubate at 45 ± 0.5°C)	Growth

[a]Growth in brain heart infusion broth (BHI) at 45 ± 0.5°C and bile esculin confirms the presence of fecal streptococci. Growth in bile esculin and brain heart infusion broth containing 6.5% NaCl confirms the presence of enterococci.

Recently, Audicana et al. (10) designed and tested a modification of the KEA agar, named oxolinic acid-aesculin-azide (OAA) agar, to improve the selectivity in the enumeration of fecal streptococci from water samples by MF. OAA agar showed higher specificity, selectivity, and recovery efficiencies than those obtained by using m-Enterococcus and KF agars. In addition, no confirmation of typical colonies was needed when OAA agar was used, which significantly shortens the time of sample processing (only 24 h) and increases the accuracy of the method.

Aeromonas

MF conjointly with selective and differential recovery media is used for the enumeration of aeromonads in water samples. Although an optimal and universal simple recovery medium is not available at present (48, 57, 106), mA, DNase-toluidine blue-ampicillin, mADA (O/129), mSA (O/129), ampicillin blood, ampicillin-dextrin, starch-glutamate ampicillin-penicillin, glutamate starch penicillin, pril-ampicillin-dextrin-ethanol, and starch ampicillin can be used. The use of media containing blood agar also allows for the detection of enterotoxigenic *Aeromonas* strains (76, 138). Lye and Dufour (94) designed an in situ cytotoxic activity assay of gram-negative bacteria isolated from water. The procedure consists of the transfer of the membrane filter to the surface of Y-1 mouse adrenal cells overlaid with 1% agar. After 15 min, the filter is removed, and then Y-1 cells are incubated for an additional 24 h. The release of putative cytotoxic and cytotonic products from the bacterial colonies is recognized by zones of cellular lysis and injury of Y-1 cells.

Molecular techniques have also been applied for the rapid detection and identification of aeromonads from several sources (19, 20, 58). The mere presence of aeromonads does not indicate a public health problem; additionally, the analyses for the presence of virulence factors and their expression may be needed to determine any public health importance.

Presence-Absence

The microbiological monitoring of drinking water has historically relied on MPN and MF approaches to estimate population densities of indicator organisms. Although those methods have been useful, disadvantages such as space and time requirements prompted the idea of simply testing for the presence or absence of indicator bacteria in a standardized volume (i.e., 100 ml) of water, using a liquid medium resembling many employed in the MPN approach, as described by Clark (31). This approach is easy to perform and very useful in small potable water systems where microbiological problems are more frequent, and it permits larger systems to analyze greater numbers of samples to gain a more comprehensive microbiological evaluation of their distribution network. Therefore, the presence-versus-absence concept of indicator organism occurrence has replaced the specified monthly average for compliance with U.S. regulations in the current total coliform rule (43). Although it can yield valuable information on the prevalence of microbiological problems within a system, a weakness is that this approach fails to provide data on the magnitude of such occurrences. In addition, the common connotation of "absence" can be misleading in the case of injured bacteria that are frequently present in treated drinking water systems and fail to produce a positive test on established media (98), as discussed in more detail below.

The P-A analysis of drinking water for total coliforms can entail 100 ml of sample added to 50 ml of triple-strength P-A broth (single strength is as follows: 13 g of lactose broth, 17.5 g of lauryl tryptose broth, 0.0085 g of bromocresol purple, and 1 liter of water) in 250-ml bottles. Bottles containing aliquots of the water sample to be tested have their contents subjected to incubation, and resulting endpoints are determined as described above. A distinct yellowness results from the fermentation of lactose, and gas formation can be detected as bubbles with gentle shaking. Confirmation of such presumptive results can be done as described above and in Table 4.

Enzyme-Specific Tests

New criteria have been added to the traditional definition of coliform bacteria, presented earlier, which had been based on classical microbiological characteristics and phenotypes. This new approach uses the presence of characteristic enzymatic activities to permit the differentiation of the coliform group of bacteria and *E. coli* in the determination of microbiological water quality. The definition of a coliform or a fecal coliform basically relies on the activity of a single enzyme (β-galactosidase). The new enzymatic definition of total coliform bacteria is based on the presence of β-galactosidase, and that of *E. coli* is based on the enzymatic action of β-glucuronidase. Although the common use of media that are based on these specific enzymatic activities to identify indicator bacteria in water has gained wide acceptance only in the past few years and is listed as an alternative since the 18th edition of *Standard Methods for the Examination of Water and Wastewater* (56a), enzymatic criteria have been accepted since the 1985 edition of that book (3), and microbiological media based on the same principle have been available for over two decades. In addition, a verification method using cytochrome oxidase and β-galactosidase has proved to be superior to the more traditional technique for confirming the presence of total coliforms in water (90). The endpoint characteristically incorporated into media designed for the enzymatic detection of coliforms and *E. coli* in water is the development of a specific color or fluorescence. The activity of the enzyme β-galactosidase results in the hydrolysis of substrates such as ortho-nitrophenyl-β-D-galactoside, which is colorless, to a colored product indicating the presence of coliform bacteria. *E. coli* is characterized by the production of a fluorescent end product following the hydrolysis of 4-methylumbelliferyl-β-D-glucuronide by the action of the enzyme β-glucuronidase. A wide range of other chromogenic and fluorogenic substrates acted upon by β-glucuronidase that have been incorporated into various media are also available. Exhaustive comparative testing of at least one commercially available enzyme-based medium indicated that it performed as well as accepted media in the detection of both total coliforms and *E. coli* from drinking water (42) and *E. coli* following chlorine injury, using mixed natural bacterial suspensions (97). However, chlorine-mediated injury results in the somewhat delayed development of both enzymatic endpoints in two of the commercially available media (unpublished results) because of the extended lag phase that is characteristic of stressed bacteria (98).

Currently, a number of different media based on this principle have been developed for use in the MF, MPN, and P-A techniques. All of these media can be used in a P-A or MPN manner, depending on the needs of the analyst.

Bacteriophages

The enumeration of bacteriophages is one of the simplest methods to perform. This technique utilizes a host strain of *E. coli* since viruses (phages) are obligate intracellular parasites. Several host strains can be used; each has its own limitations. One of the most widely used hosts is *E. coli* C (ATCC 13706). This host strain allows for the replication of somatic phages (those phages having their receptor sites located on the bacterial cell wall). However, *E. coli* C3000 (ATCC 15597) is a strain which has sex pili and thus allows for the replication of male-specific coliphages (those phages having the receptor sites on the sex pilus) as well as somatic phages. A third host is a genetically manipulated *Salmonella* strain that has the ability to produce sex pili but does not contain the receptor sites for most somatic coliphages on its cell wall, allowing only male-specific (F+ or F-specific) coliphages to replicate (75).

For the analysis of surface and drinking waters, a single-layer, direct plaque assay as outlined by Grabow and Coubrough (55) can be used. In addition, the method enables the analysis of 100-ml volumes, which allows for a numerical comparison with bacterial indicators. The medium consists of 14 g of meat extract, 4 g of yeast extract, 4 g of NaCl, 12 g of peptone, 1 g of sodium carbonate, 1 g of magnesium chloride, 12 g of agar, and 1 liter of water; a 13% (wt/vol) $CaCl_2$ solution is prepared separately and autoclaved. The medium is autoclaved in 100-ml volumes and kept liquid at ca. 48°C, and 1 ml of the $CaCl_2$ solution is added. Five milliliters of an overnight culture of the appropriate host is added to the medium, and finally a 100-ml volume of the sample is mixed carefully and thoroughly (taking care not to create air bubbles) and is poured into large petri dishes. The plate contents are incubated at 35°C for 6 to 24 h and read for the presence of small clear areas that represent viral PFU. Care should be taken to include negative controls (i.e., plates that contain only the medium and the bacterial host). The total number of viral plaques is counted and expressed as the number of PFU per 100 ml.

It should be noted that there are many more host bacteria being used in different laboratories around the world; however, the analyst should be aware that data obtained by using one bacterial host may not be correlated to data obtained with another. Thus, when making comparisons to databases, the host used should be taken into consideration.

Rapid Tests

The need for tests to quickly determine the possible existence of water contamination of public health significance is long-standing. This need exists because the time required to perform the current tests is greater than the mean residence time for water within many potable distribution networks. Significant factors associated with that dilemma in drinking water and other relatively high-quality aquatic environments and systems include the low densities of ambient bacteria plus the need for at least 18 h of bacterial growth before an observable endpoint is obtained with the classical techniques. The constraints associated with other systems containing higher concentrations of bacteria are clearly less demanding. However, for most applications, the rapid methods available should be viewed as research tools since at present they either lack sufficient sensitivity or are unsuitable for use in routine monitoring laboratories. However, a variety of rapid methods are in the developmental stage, and some appear very promising. Those seeking such a method need to critically evaluate which of the available methods are compatible with their specific needs. A variety of analytical approaches have been proposed for the rapid detection of bacteria in water and wastewater (56a), although most are limited by sensitivity with respect to analysis of water of good microbiological quality. The following is a list of some of the more attractive rapid methodological options that have been described in the literature and a brief statement about potential applications and noteworthy limitations of each.

1. The direct total microbial count or acridine orange direct count is often used to determine the total bacterial population in an aquatic system (56a). This analytical approach requires a fluorescence microscope with appropriate optical filters and is of limited value in determining bacterial viability unless known organisms are used under defined conditions (103). The use of 5-cyano-2,3-ditolyl tetrazolium chloride (a fluorescent compound) and a contrasting counterstain allows the determination of the total bacterial population as well as the fraction respiring in water (123) and the assessment of biofilm disinfection (74). This method provides little information of value in bacterial identification and is tedious, although image analysis technology has proven useful in making the technique less problematic and allows statistical concerns to be addressed more easily (126).
2. A rapid test that detects thermotolerant coliforms after 7 h of incubation is available (114). This method has proven of value in the examination of surface waters as well as of unchlorinated sewage and might serve as an emergency test for detection of sewage or fecal contamination in potable water. However, the presence of injured bacteria in such systems might represent an explanation for false-negative results because of the prolonged lag that is characteristic of stressed bacteria in environmental samples (98).
3. ATP detection assays have been useful in the determination of bacterial population density. This approach is rather insensitive, requiring the presence of at least 1,000 cells, and does not identify organisms, but it has been used in ecological and wastewater studies (78, 81).

4. Radiometry with labeled substrates can be both rapid and sensitive in the detection of organisms that metabolize the labeled nutrient to CO_2. This method does not discriminate among bacteria phylogenetically, although it provides information of functional significance. Despite the potential of this method, very few reports document its application in environmental analysis (115).

5. Fluorescently labeled antibodies have been used for the detection and identification of bacteria in environmental studies for some time. This method also requires a fluorescence microscope and is tedious unless image analysis is used, and it is constrained by the specificity of the antibodies used and does not indicate viability or physiological activity. However, this approach can be used in a variety of applications, including tracking a specific organism in the environment or detecting enteric bacteria (92, 143). Because of its lack of sensitivity, this method should not be used for the direct detection of pathogens in waters. Large volumes of water need to be concentrated before this technique becomes useful for pathogen detection.

6. A hybrid method that incorporates the use of fluorescent antibodies to identify bacteria along with 5-cyano-2,3-ditolyl tetrazolium chloride to determine cellular respiratory activity has recently been introduced (113). Although this method requires the use of a fluorescence microscope and can be tedious, it is unique in that it allows the rapid, simultaneous detection of specific bacteria and the discrimination of respiratory activity at the cellular level.

Recovery of Injured Bacteria

Indicator bacteria become injured in water and wastewater following sublethal exposure to a wide variety of chemical and physical environmental stressors, including disinfectants, metals, and UV irradiation (98). Such bacteria are unable to form colonies on most selective media, and between 10 and 90% of the coliform bacteria in treated drinking water may be injured (98, 102). As a consequence, injured cells are undetected in water, leading to an underestimation of a contamination event, and injured bacteria may also pass undetected into finished drinking water distribution networks (22), where they may eventually recuperate and colonize the system. A medium, m-T7, was developed to detect injured bacteria in drinking water and is commercially available (22). Other guidelines associated with the detection of injured bacteria are given in *Standard Methods for the Examination of Water and Wastewater* (56a). It should be noted that the use of media and methods designed to detect injured bacteria does not always result in the detection of greater numbers of indicator bacteria in all systems. The detection of injured bacteria by these methods in treated waters may be indicative of the potential regrowth under conditions of the distribution system, especially in the presence of high levels of nutrients, and thus may provide guidance in the diagnosis of problems within water distribution systems experiencing unexplained occurrences of excessive indicator bacteria.

Molecular Methods

The PCR provides an option for the detection of microorganisms in aquatic environments (12, 118) as well as in sewage and sludge (139). Caution is needed in the interpretation of PCR results, since this test allows for the amplification and detection of DNA sequences without regard for bacterial viability (2). In addition, materials in natural samples can interfere and inhibit the required reactions (1a), and false-negative results may be obtained.

Host-specific 16S rRNA genetic markers distinguish members of mixtures of bacterial gene sequences by detecting differences in the number of base pairs in a certain gene fragment. The process includes both length heterogeneity PCR and terminal restriction fragment length polymorphism analysis (RFLP), which restricts the amplified fragments. Repetitive DNA sequencing uses PCR to amplify palindromic DNA sequences coupled with electrophoresis. Ribotyping takes a Southern blot of restricted genomic DNA and probes it with ribosomal sequences. Pulsed-field gel electrophoresis represents genomic DNA restricted with certain enzymes and coupled with electrophoresis. This technique is very useful for fingerprinting isolates from disease outbreaks (PulseNet [27a]). Denaturing-gradient gel electrophoresis is the use of electrophoresis of PCR products based on melting properties of the amplified DNA sequences. This technique has been used to study the survival of microorganisms in the environment (88b). RAPD analysis uses a single 10-bp primer (10-mer) of arbitrary sequence to amplify DNA fragments from a genomic template. No prior knowledge of genomic sequence is required (122a). Phylogenetic identification of specific bacteria or groups of organisms can be achieved without cultivation through the use of fluorescently labeled oligonucleotide probes for rRNA sequences. This approach has gained wide application in ecological studies (2) and has been used to detect an opportunistic pathogen (*Vibrio vulnificus*) in waters after concentration on membrane filters (69).

All these methods are being used for different purposes; however, none is yet ready for routine use in water laboratories.

SURFACE FRESHWATERS

It should be realized that the microbiology of untreated waters will be drastically different from that of treated drinking waters. Thus, strong emphasis should be placed on the use of the appropriate indicators for each type of water. In the case of recreational waters (excluding swimming pools and spas, which receive fecal contamination from bathers), it should be determined if the recreational areas are being subjected to fecal pollution, perhaps using source-tracking methods. In any case the presence of bathers in the water will result in fecal deposition, and thus fecal contamination is likely. Thus, the use of the appropriate indicators to determine risk of enteric or other types of disease is of extreme importance. Risk can be determined most accurately if the area is known thoroughly. Indicators such as total coliforms may not be of much value in analyses of recreational waters, in spite of possible statistical correlations. *E. coli* and coliphages may be more reliable indicators of the presence of fecal pollution (and thus possible risk), unless the recreational waters are part of a tropical rain forest (67). In the latter case, the use of coliphages may be more advisable.

DRINKING WATERS

Traditional treatment is based on physical and chemical barriers. Thus, indicator microorganisms should not be present in finished waters. The presence of any indicator (total or thermotolerant coliforms, coliphages, or *E. coli*) in the finished waters suggests that one of the barriers is not functioning. The distribution network may be susceptible to contamination by intrusion, and therefore the presence of thermotolerant coliforms in treated drinking water should be a cause for concern. Indicators have also been shown to regrow in potable water distribution systems (89) as a result of the bacterial biofilms on pipe surfaces. The analyst should be aware of the different types of indicators, and proper use of these tools should be emphasized.

SUMMARY

We have given an overall view of the currently used methods as well as those that are now being developed for the microbiological analysis of drinking waters and surface freshwaters. The practice of using groups of bacteria or specific organisms such as *E. coli* as indicators of the possible presence of fecal contamination or as indicators of the possible presence of pathogenic microorganisms is and has been extremely useful to protect public health. However, it should be kept in mind that indicators are only tools in this endeavor. There exist numerous misunderstandings regarding the use of indicators. For example, analysis of levels of coliforms and thermotolerant coliforms is considered by many to be a simplistic method for determining and guaranteeing water quality. Such a dogmatic approach to water quality is counterproductive and may place public health in jeopardy.

There are exciting new developments in the area of rapid methods as well as the direct detection of pathogenic microorganisms. The reader is urged to become familiar with the advantages and disadvantages of these new advances. However, there is no better way of approaching water quality management and analysis than being familiar with the watershed or distribution system under investigation. This knowledge and the correct and timely application of suitable analytical methods allow for early detection of anomalous conditions within the system. The resulting data can then be used to better manage the system and thus protect the public health against not only enteric but also other types of diseases.

REFERENCES

1. **Ali, A., A. M. Carnahan, M. Atwegg, J. Luthy-Hottenstein, and S. W. Joseph.** 1996. *Aeromonas bestiarum* sp. nov. (formerly genomospecies DNA group 2 *A. hydrophila*), a new species isolated from non-human sources. *Med. Microbiol. Lett.* **5:**156–165.

1a. **Alvarez, A. J., and G. A. Toranzos.** 1997. Basic methodology for DNA and RNA amplification, p. 37–61. *In* G. A. Toranzos (ed.), *Environmental Applications of Nucleic Acid Amplification Techniques*. Technomic Publishing Co., Inc., Lancaster, Pa.

2. **Amann, R. I., W. Ludwig, and K.-H. Schleifer.** 1995. Phylogenetic identification and in situ detection of individual microbial cells without cultivation. *Microbiol. Rev.* **59:**143–169.

3. **American Public Health Association.** 1985. *Standard Methods for the Examination of Water and Wastewater*, 16th ed. American Public Health Association, Washington, D.C.

4. **American Public Health Association.** 1989. *Standard Methods for the Examination of Water and Wastewater*, 17th ed., p. 9–31, 9–32. American Public Health Association, Washington, D.C.

5. Reference deleted.

6. **Araujo, R. M., R. Pares, and F. Lucena.** 1990. The effect of terrestrial effluents on the incidence of *Aeromonas* spp. in coastal waters. *J. Appl. Bacteriol.* **69:**439–444.

7. **Araujo, R. M., R. M. Arribas, and R. Pares.** 1991. Distribution of *Aeromonas* species in waters with different levels of pollution. *J. Appl. Bacteriol.* **71:**182–186.

8. **Arcos, M. L., A. de Vicente, M. A. Moriñigo, P. Romero, and J. J. Borrego.** 1988. Evaluation of several selective media for recovery of *Aeromonas hydrophila* from polluted waters. *Appl. Environ. Microbiol.* **54:**2786–2792.

9. **Ashbolt, N. J., A. Ball, M. Dorsch, C. Turner, P. Cox, A. Chapman, and S. M. Kirov.** 1995. The identification and human health significance of environmental aeromonads. *Water Sci. Technol.* **31:**263–269.

10. **Audicana, A., I. Perales, and J. J Borrego.** 1995. Modification of kanamycin-esculin-azide agar to improve selectivity in the enumeration of fecal streptococci from water samples. *Appl. Environ. Microbiol.* **61:**4178–4183.

11. **Avila, M. J., M. A. Moriñigo, R. Cornax, P. Romero, and J. J. Borrego.** 1989. Comparative studies of coliform enumeration media from seawater samples. *J. Microbiol. Methods* **9:**175–193.

12. **Bej, A. K., S. C. McCarty, and R. M. Atlas.** 1991. Detection of coliform bacteria and *Escherichia coli* by multiplex polymerase chain reaction: comparison with defined substrate and plating methods for water quality monitoring. *Appl. Environ. Microbiol.* **57:**2429–2432.

13. **Borrego, J. J.** 1995. Coliphages as faecal pollution indicators, p. 117–159. *In* B. C. Rana (ed.), *Pollution and Biomonitoring*. McGraw-Hill, New York, N.Y.

14. **Borrego, J. J., and M. J. Figueras.** 1997. Microbiological quality of natural waters. *Microbiologia SEM* **13:**413–426.

15. **Borrego, J. J., F. Arrabal, and P. Romero.** 1982. Study of the microbiological pollution of a Malaga littoral area. II. Relationship between faecal coliforms and faecal streptococci. *J. Etud. Pollut.* **4:**561–564.

16. **Borrego, J. J., F. Arrabal, A. de Vicente, L. F. Gomez, and P. Romero.** 1983. Study of microbial inactivation in the marine environment. *J. Water Pollut. Control Fed.* **55:**297–302.

17. **Borrego, J. J., M.A. Moriñigo, A. de Vicente, R. Cornax, and P. Romero.** 1987. Coliphages as an indicator of faecal pollution in water. Its relationship with indicator and pathogenic microorganisms. *Water Res.* **21:**1473–1480.

18. **Borrego, J. J., R. Cornax, M. A. Moriñigo, E. Martinez-Manzanares, and P. Romero.** 1990. Coliphages as an indicator of faecal pollution in water. Their survival and productive infectivity in natural aquatic environments. *Water Res.* **24:**111–116.

19. **Borrell, N., S. G. Acinas, M. J. Figueras, and A. Martinez-Murcia.** 1997. Identification of *Aeromonas* clinical isolates by restriction fragment length polymorphism of PCR-amplified 16S rRNA genes. *J. Clin. Microbiol.* **35:** 1671–1674.

20. **Borrell, N., M. J. Figueras, and J. Guarro.** 1998. Phenotypic identification of *Aeromonas* genomospecies from clinical and environmental sources. *Can. J. Microbiol.* **44:**7–12.

21. **Brenner, K. P., C. C. Rankin, Y. R. Roybal, G. N. Stelma, P. V. Scarpino, and A. P. Dufour.** 1993. New medium for the simultaneous detection of total coliforms and *Escherichia coli* in water. *Appl. Environ. Microbiol.* **59:**3534–3544.

22. **Bucklin, K. E., G. A. McFeters, and A. Amirtharaja.** 1991. Penetration of coliforms through municipal drinking water filters. *Water Res.* **25:**1013–1017.

23. **Burke, V., J. Robinson, M. Gracey, D. Peterson, and K. Partridge.** 1984. Isolation of *Aeromonas hydrophila* from a metropolitan water supply: seasonal correlation with clinical isolates. *Appl. Environ. Microbiol.* **48:**361–366.

23a. **Byappanahalli, M., M. Fowler, D. Shively, and R. Whitman.** 2003. Ubiquity and persistence of *Escherichia coli* in a Midwestern coastal stream. *Appl. Environ. Microbiol.* **69:**4549–4555.

24. **Cabelli, V. J., A. P. Dufour, L. J. McCabe, and M. A. Levin.** 1982. Swimming-associated gastroenteritis and water quality. *Am. J. Epidemiol.* **115:**606–616.

25. **Calderon, R. L., E. W. Mood, and A. P. Dufour.** 1991. Health effects on swimmers and nonpoint sources of contaminated water. *Int. J. Environ. Health Res.* **1:**21–31.
26. **Castillo, G., A. Vargas, C. Etcheberrigaray, and J. A. Millañir.** 1995. Development of an innovative incubation method using beehives for determining the safety of drinking water in remote areas, p. 178. *In Abstracts of the 7th International Symposium on Microbial Ecology*.
27. **Carnahan, A. M., and M. Altwegg.** 1996. Taxonomy, p. 1–38. *In* B. Austin, M. Altwegg, P. J. Gosling and S.W. Joseph (ed.), *The Genus* Aeromonas. John Wiley & Sons, Ltd., Chichester, United Kingdom.

27a. **Centers for Disease Control and Prevention.** 1997. Standardized molecular subtyping of *Escherichia coli* O157:H7 by pulsed-field electrophoresis: a training manual. Foodborne and Diarrheal Diseases Branch, Division of Bacterial and Mycotic Diseases, National Center for Infectious Diseases, Centers for Disease Control and Prevention, Atlanta, Ga.

28. **Chang, G. W., and R. I. Lum.** 1994. mX, a simple, economical membrane filter medium for *E. coli* and total coliforms, p. 1397–1400. *In Proceedings of the AWWA/WQTC*. American Water Works Association, Denver, Colo.
29. **Chang, W. J., and F. D. Pien.** 1986. Marine-acquired infections: hazards of the ocean environment. *Postgrad. Med.* **80:**30–33.
30. **Charoenca, N., and R. S. Fujioka.** 1991. Assessment of *Staphylococcus* bacteria in Hawaii's marine recreational waters. *Water Sci. Technol.* **27:**283–289.
31. **Clark, J. A.** 1990. The presence-absence test for monitoring water quality, p. 399–411. *In* G. A. McFeters (ed.), *Drinking Water Microbiology*. Springer-Verlag, New York, N.Y.
32. **Comité Coordinador Regional de Instituciones de Agua Potable y Saneamiento de Centroamerica, Panamá y República Dominicana (CAPRE).** 1993. Normas de Calidad del Agua para Consumo Humano, 1st ed. CAPRE, San José, Costa Rica.
33. **Cornax, R., M. A. Moriñigo, I. G. Paez, M. A. Muñoz, and J. J. Borrego.** 1990. Application of direct plaque assay for detection and enumeration of bacteriophages of *Bacteroides fragilis* from contaminated-water samples. *Appl. Environ. Microbiol.* **56:**3170–3173.
34. **Cornax, R., M. A. Moriñigo, M. C. Balebona, D. Castro, and J. J. Borrego.** 1991. Significance of several bacteriophage groups as indicators of sewage pollution in marine waters. *Water Res.* **25:**673–678.
35. **Cornax, R., M. A. Moriñigo, F. Gonzalez-Jaen, M. C. Alonso, and J. J. Borrego.** 1994. Bacteriophages presence in human faeces of healthy subjects and patients with gastrointestinal disturbances. *Zentbl. Bakteriol.* **281:**214–224.
36. **Dailloux, M., C. Laurain, M. Weber, and P. H. Hartemann.** 1999. Water and nontuberculous mycobacteria. *Water Res.* **33:**2219–2228.
37. **DeBartolomeis, J., and V. J. Cabelli.** 1991. Evaluation of an *Escherichia coli* host strain for enumeration of F male-specific bacteriophages. *Appl. Environ. Microbiol.* **57:** 1301–1305.
38. **Dufour, A. P.** 1977. *Escherichia coli*: the fecal coliform, p. 48–58. *In* A. W. Hoadley and B. J. Dutka (ed.), *Bacteria Indicators/Health Hazards Associated with Water*. ASTM STP 635. American Society for Testing and Materials, Philadelphia, Pa.
39. **Dufour, A.P.** 1984. *Health Effects Criteria for Fresh Recreational Waters*. EPA-600/1-84-004. U.S. Environmental Protection Agency, Washington, D.C.
40. **Dutka, B. J., A. El-Shaarawi, M. T. Martins, and P. S. Sanchez.** 1987. North and South American studies on the potential of coliphage as a water quality indicator. *Water Res.* **21:**1127–1134.
41. **Dutka, B. J., A. S. Y. Chau, and J. Coburn.** 1974. Relationship between bacterial indicators of water pollution and fecal sterols. *Water Res.* **8:**1047–1055.
42. **Edberg, S. C., M. J. Allen, and D. B. Smith.** 1988. National field evaluation of a defined substrate method for the simultaneous detection of total coliforms and *Escherichia coli* from drinking water: comparison with the standard multiple-tube fermentation method. *Appl. Environ. Microbiol.* **54:**1559–1601.
43. **Federal Register.** 1989. Drinking water; national primary drinking water regulations: total coliforms, final rule. *Fed. Regist.* **54:**27544–27568.
44. **Figueras, M. J., I. Inza, F. L. Polo, M. T. Feliu, and J. Guarro.** 1996. A fast method for the confirmation of fecal streptococci from M-Enterococcus medium. *Appl. Environ. Microbiol.* **62:**2177–2178.
45. **Figueras, M. J., F. L. Polo, I. Inza, and J. Guarro.** 1997. Past, present and future perspective of the EU bathing water directive. *Marine Pollut. Bull.* **34:**148–156.

45a. **Figueras, M. J., A. Suarez-Franquet, M. R. Chacon, L. Soler, M. Navarro, C. Alejandre, B. Grasa, A. J. Martinez-Murcia, and J. Guarro.** 2005. First record of the rare species *Aeromonas culicicola* from a drinking water supply. *Appl. Environ. Microbiol.* **71:**538–541.

46. **Fujioka, R. S.** 1997. Indicators of marine recreational water quality, p. 176–183. *In* C. J. Hurst, G. R. Knudsen, M. J. McInerney, L. D. Stetzenbach, and M. V. Walter (ed.), *Manual of Environmental Microbiology*. ASM Press, Washington, D.C.
47. **Fujioka, R. A., A. A. Ueno, and O. T. Narikawa.** 1984. Recovery of false positive fecal streptococcus on KF agar from marine recreational waters. Technical report no. 168. Water Resources Research Center, University of Hawaii at Manoa, Honolulu.
48. **Gavriel, A., and A. J. Lamb.** 1995. Assessment of media used for selective isolation of *Aeromonas* spp. *Lett. Appl. Microbiol.* **21:**313–315.
49. **Geldreich, E. E.** 1967. Fecal coliform concepts in stream pollution. *Water Sewage Works* **114:**98–110.
50. **Geldreich, E. E.** 1970. Applying bacteriological parameters to recreational water quality. *J. Am. Water Works Assoc.* **62:**113–120.
51. **Geldreich, E. E.** 1978. Bacterial pollution and indicator concepts in feces, sewage, stormwater and solid wastes, p. 51–97. *In* G. Berg (ed.), *Indicators of Viruses in Water and Food*. Ann Arbor Science, Ann Arbor, Mich.
52. **Geldreich, E. E., and B. A. Kenner.** 1969. Concepts of fecal streptococci in stream pollution. *J. Water Pollut. Control Fed.* **41:**355–362.
53. **Ghanem, E. H., M. E. Mussa, and H. M. Eraki.** 1993. *Aeromonas*-associated gastro-enteritis in Egypt. *Zentbl. Mikrobiol.* **148:**441–447.
54. **Gleick, P. H. 1993.** Water quality and contamination, p. 225. *In* P. H. Gleick (ed.), *Water in Crisis: a Guide to the World's Fresh Water Resources*. Oxford University Press, Oxford, United Kingdom.
55. **Grabow, W. O. K., and P. Coubrough.** 1986. Practical direct plaque assay for coliphages in 100-ml samples of drinking water. *Appl. Environ. Microbiol.* **52:**430–433.
56. **Grabow, W. O. K., J. S. Burger, and E. M. Nupen.** 1980. Evaluation of acid-fast bacteria, *Candida albicans*, enteric viruses and conventional indicators for monitoring wastewater reclamation systems. *Prog. Water Technol.* **12:**803–817.

56a. **Greenberg, A. E., L. S. Clesceri, and A. D. Eaton (ed.).** 1992. *Standard Methods for the Examination of Water and Wastewater*, 18th ed. American Public Health Association, Washington, D.C.

57. **Handfield, M., P. Simard, and R. Ltarte.** 1996. Differential media for quantitative recovery of waterborne *Aeromonas hydrophila*. *Appl. Environ. Microbiol.* **62:**3544–3547.

58. **Hanninen, M.-L., and A. Siitonen.** 1995. Distribution of *Aeromonas* phenospecies and genospecies among strains isolated from water, foods, or from human clinical samples. *Epidemiol. Infect.* **115:**39–50.
59. **Hardalo, C., and S. C. Edberg.** 1997. *Pseudomonas aeruginosa*: assessment of risk from drinking water. *Crit. Rev. Microbiol.* **23:**47–75.
60. **Havelaar, A. H.** 1987. Bacteriophages as model organisms in waste treatment. *Microbiol. Sci.* **4:**362–364.
61. **Havelaar, A. H., and W. H. Hogeboom.** 1984. A method for the enumeration of male-specific bacteriophages in sewage. *J. Appl. Bacteriol.* **56:**439–447.
62. **Havelaar, A. H., and W. H. Pot-Hogeboom.** 1988. F-specific RNA-bacteriophages as model viruses in water hygiene: ecological aspects. *Water Sci. Technol.* **20:**399–407.
63. **Havelaar, A. H., K. Furuse, and W. H. Hogeboom.** 1986. Bacteriophages and indicator bacteria in human and animal faeces. *J. Appl. Bacteriol.* **60:**252–262.
64. **Havelaar, A. H., W. H. Pot-Hogeboom, W. M. Koot, and R. Pot.** 1987. F-specific bacteriophages as indicators of the disinfection efficiency of secondary effluent with ultraviolet radiation. *Ozone Sci. Eng.* **9:**353–368.
65. **Havelaar, A H., J. F. M. Versteegh, and M. During.** 1990. The presence of *Aeromonas* in drinking water supplies in the Netherlands. *Zentbl. Hyg. Umweltmed.* **190:** 236–256.
66. **Havelaar, A. H., F. M. Schets, A. Van Silfhout, W. H. Jansen, G. Wielen, and D. van der Kooij.** 1992. Typing of *Aeromonas* strains from patients with diarrhoea and from drinking water. *J. Appl. Bacteriol.* **72:**435–444.
67. **Hazen, T. C., and G. A. Toranzos.** 1990. Tropical source water, p. 32–54. *In* G. A. McFeters (ed.), *Drinking Water Microbiology*. Springer-Verlag, New York, N.Y.
68. **Hazen, T. C., C. B. Fliermans, R. P. Hirsch, and G. W. Esch.** 1978. Prevalence and distribution of *Aeromonas hydrophila* in the United States. *Appl. Environ. Microbiol.* **36:**731–738.
68a. **Health and Welfare Canada.** 1992. *Guidelines for Canadian Drinking Water Quality*. Canadian Government Publishing Centre, Ottawa, Canada.
68b. **Health and Welfare Canada.** 1993. *Guidelines for Canadian Drinking Water Quality*. Canadian Government Publishing Centre, Ottawa, Canada.
69. **Heidelberg, J. F., K. R. O'Neil, D. Jacobs, and R. R. Colwell.** 1993. Enumeration of *Vibrio vulnificus* on membrane filters with a fluorescently labeled oligonucleotide probe specific for kingdom-level 16S rRNA sequences. *Appl. Environ. Microbiol.* **59:**3474–3476.
70. **Hernández-Delgado, E. A., M. L. Sierra, and G. A. Toranzos.** 1991. Coliphages as alternate indicators of fecal contamination in tropical waters. *Environ. Toxicol. Water Qual.* **6:**131–143.
71. **Holmes, P., and L. N. Niccolls.** 1995. Aeromonads in drinking water supplies—their occurrence and significance. *J. Chartered Inst. Water Environ. Manag.* **5:**464–469.
72. **Holmes, P., N. H. Niccolls, and D. P. Sartorius.** 1996. The ecology of mesophilic *Aeromonas* in the aquatic environment, p. 127–150. *In* B. Austin, M. Altwegg, P. J. Gosling, and S. W. Joseph (ed.), *The Genus* Aeromonas. John Wiley & Sons, Ltd., Chichester, United Kingdom.
73. **Holt, J. G. (ed.).** 1994. *Bergey's Manual of Determinative Bacteriology*, 9th ed. Williams & Wilkins, Baltimore, Md.
74. **Huang, C.-T., F. P. Yu, G. A. McFeters, and P. S. Stewart.** 1995. Nonuniform spatial patterns of respiratory activity within biofilm during disinfection. *Appl. Environ. Microbiol.* **61:**2252–2256.
75. **IAWPRC Study Group on Health-Related Water Microbiology.** 1991. Bacteriophages as model viruses in water quality control. *Water Res.* **25:**529–545.
76. **Jesudason, M. V., M. John, and G. Koshi.** 1989. Comparison of three assays to detect enterotoxigenicity of *Aeromonas* isolates from patients with diarrhoea in Vellore, South India. *J. Diarrhoeal Dis. Res.* **7:**77–80.
77. **Jofre, J., A. Bosch, F. Lucena, R. Girones, and C. Tartera.** 1986. Evaluation of *Bacteroides fragilis* bacteriophages as indicators of the virological quality of water. *Water Sci. Technol.* **18:**167–177.
78. **Jorgensen, P. E., T. Eriksen, and B. K. Jensen.** 1992. Estimation of viable biomass in wastewater and activated sludge by determination of ATP, oxygen utilization rate and FDA hydrolysis. *Water Res.* **11:**1495–1501.
79. **Joseph, S. W., and A. M. Carnahan.** 1994. The isolation, identification, and systematics of the motile *Aeromonas* species. *Annu. Rev. Fish Dis.* **4:**315–343.
80. **Joseph, S. W., A. M. Carnahan, P. R. Brayton, G. R. Fanning, R. Almazan, C. Drabick, E. W. Trudo, and R. R. Colwell.** 1991. *Aeromonas jandaei* and *Aeromonas veronii* dual infection of a human wound following aquatic exposure. *J. Clin. Microbiol.* **29:**565–569.
81. **Karl, D. M.** 1980. Cellular nucleotide measurements and applications in microbial ecology. *Microbiol. Rev.* **44:**739–796.
82. **Kelly, K. A., J. M. Koehler, and L. R. Ashdown.** 1993. Spectrum of extraintestinal disease due to *Aeromonas* species in tropical Queensland, Australia. *Clin. Infect. Dis.* **16:**574–579.
83. **Knittel, M. D., R. J. Seidler, C. Eby, and L. M. Cabe.** 1977. Colonization of the botanical environment by *Klebsiella* isolates of pathogenic origin. *Appl. Environ. Microbiol.* **34:**557–563.
84. **Kott, Y.** 1966. Estimation of low numbers of *Escherichia coli* bacteriophage by use of the most probable number method. *Appl. Microbiol.* **14:**141–144.
85. **Kott, Y.** 1984. Coliphages as reliable enteric virus indicators, p. 171–174. *In* J. L. Melnick (ed.), *Viruses in Water*. Karger, Basel, Switzerland.
86. **Kott, Y., N. Roze, S. Sperber, and N. Betzer.** 1974. Bacteriophages as viral pollution indicators. *Water Res.* **8:** 165–171.
87. **Kromoredjo, P., and R. S. Fujioka.** 1991. Evaluating three simple methods to assess the microbial quality of drinking water in Indonesia. *Environ. Toxicol. Water Qual.* **6:**259–270.
88. **Krovacek, K., S. Dumontet, E. Eriksson, and S. B. Baloda.** 1995. Isolation and virulence profiles of *Aeromonas hydrophila* implicated in an outbreak of food poisoning in Sweden. *Microbiol. Immunol.* **39:**655–661.
88a. **Lasalde, C., R. Rodriguez, and G. A. Toranzos.** 2005. Statistical analyses: possible reasons for unreliable source tracking efforts. *Appl. Environ. Microbiol.* **71:** 4690–4695.
88b. **Lasalde, C., R. Rodriguez, H. H. Smith, and G. A. Toranzos.** 2005. Heterogeneity of uidA gene in environmental Escherichia coli populations. *J. Water Health* **3:** 297–304.
89. **LeChevallier, M. W.** 1990. Coliform regrowth in drinking water: a review. *J. Am. Water Works Assoc.* **82:**74–86.
90. **LeChevallier, M. W., S. C. Cameron, and G. A. McFeters.** 1983. Comparison of verification procedures for the membrane filtration total coliform technique. *Appl. Environ. Microbiol.* **45:**1126–1128.
91. **Leclerc, M., L. A. Devriese, and D. A. A. Mossel.** 1996. Taxonomical changes in intestinal (faecal) enterococci and streptococci: consequences on their use as indicators of faecal contamination in drinking water. *J. Appl. Bacteriol.* **81:**459–466.
92. **Levasseur, S., M.-O. Husson, R. Leitz, F. Merlin, F. Laurent, F. Peladan, J.-L. Drocourt, H. Leclerc, and M. Van Hoegaerden.** 1992. Rapid detection of members of

the family *Enterobacteriaceae* by a monoclonal antibody. *Appl. Environ. Microbiol.* **58:**1524–1529.

93. **Lucena, F., J. Lasobras, D. McIntosh, M. Forcadell, and J. Jofre.** 1994. Effect of distance from the polluting focus on relative concentrations of *Bacteroides fragilis* phages and coliphages in mussels. *Appl. Environ. Microbiol.* **60:** 2272–2277.
94. **Lye, D. E., and A. P. Dufour.** 1991. A membrane filter procedure for assaying cytotoxic activity in heterotrophic bacteria isolated from drinking water. *J. Appl. Bacteriol.* **70:** 89–94.
95. **Manja, E. S., M. S. Maurya, and D. M. Rao.** 1982. A simple field test for the detection of fecal pollution in drinking water. *Bull. W. H. O.* **60:**797–801.
96. **Mariño, F. J., M. A. Moriñigo, E. Martinez-Manzanares, and J. J. Borrego.** 1995. Microbiological-epidemiological study of selected marine beaches in Malaga (Spain). *Water Sci. Technol.* **31:**5–9.
97. **McCarty, S. C., J. H. Standridge, and M. C. Stasiak.** 1992. Evaluating a commercially available defined-substrate test for recovery of commercially available *Escherichia coli. J. Am. Water Works Assoc.* **84:**91–97.
98. **McFeters, G. A.** 1990. Enumeration, occurrence and significance of injured indicator bacteria in drinking water, p. 478–492. In G. A. McFeters (ed.), *Drinking Water Microbiology*. Springer-Verlag, New York, N.Y.
99. **McFeters, G. A., J. P. Barry, and J. Howington.** 1993. Distribution of enteric bacteria in Antarctic seawater surrounding a sewage outfall. *Water Res.* **27:**645–650.
100. **McFeters, G. A., G. K. Bissonnette, J. J. Jezeski, C. A. Thomson, and D. G. Stewart.** 1974. Comparative survival of indicator bacteria and enteric pathogens in well water. *Appl. Microbiol.* **27:**823–829.
101. **McFeters, G. A., S. C. Broadaway, B. H. Pyle, M. Pickett, and Y. Egozy.** 1995. Comparative performance of Colisure and accepted methods in the detection of chlorine injured total coliforms and *Escherichia coli. Water Sci. Technol.* **31:**259–261.
102. **McFeters, G. A., J. S. Kippin, and M. W. LeChevallier.** 1986. Injured coliforms in drinking water *Appl. Environ. Microbiol.* **51:**1–5.
103. **McFeters, G. A., A. Singh, S. Byun, S. Williams, and P. R. Callis.** 1991. Acridine orange staining reaction as an index of physiological activity in *Escherichia coli. J. Microbiol. Methods* **13:**87–97.
104. **Moriñigo, M. A., R. Cornax, D. Castro, E. Martinez-Manzanares, and J. J. Borrego.** 1990. Viability of *Salmonella* spp. and indicator microorganisms in seawater using membrane diffusion chambers. *Antonie Leeuwenhoek* **57:**109–117.
105. **Moriñigo, M. A., D. Wheeler, C. Berry, C. Jones, M. A. Muñoz, R. Cornax, and J. J. Borrego.** 1992. Evaluation of different bacteriophage groups as faecal indicators in contaminated natural waters in southern England. *Water Res.* **26:**267–271.
106. **Moyer, N. P.** 1996. Isolation and enumeration of aeromonads, p. 39–84. *In* B. Austin, M. Altwegg, P. J. Gosling, and S. W. Joseph (ed.), *The Genus* Aero-monas. John Wiley & Sons, Ltd., Chichester, United Kingdom.
107. **Moyer, N. P., G. Martinetti, J. Luthy-Hottenstein, and M. Altwegg.** 1992. Value of rRNA gene restriction patterns of *Aeromonas* spp. for epidemiological investigations. *Curr. Microbiol.* **24:**15–21.
108. **Neumann, M., R. Schulze-Robbecke, C. Hagenau, and K. Behringer.** 1997. Comparison of methods for isolation of mycobacteria from water. *Appl. Environ. Microbiol.* **63:** 547–552.
109. **Osawa, S., K. Furuse, and Y. Watanabe.** 1981. Distribution of ribonucleic acid coliphages in animals. *Appl. Environ. Microbiol.* **41:**164–168.
110. **Payment, P., L. Richardson, J. Siemiatycki, R. Dewar, M. Edwardes, and E. Franco.** 1991. A randomized trial to evaluate the risk of gastrointestinal disease due to consumption of drinking water meeting current microbiological standards. *Am. J. Public Health* **81:**703–708.

110a. **Piocos, E. A., and A. A. de la Cruz.** 2000. Solid phase extraction and high performance liquid chromatography with photodiode array detection of chemical indicators of human fecal contamination in water. *J. Liquid Chromatogr.* **23:**1281–1291.

111. **Pourcher, A.-M., L. A. Devriese, J. F. Hernández, and J. M. Delattre.** 1991. Enumeration by a miniaturized method of *Escherichia coli, Streptococcus bovis* and enterococci as indicators of the origin of faecal pollution of waters. *J. Appl. Bacteriol.* **70:**525–530.
112. **Primrose, S. B., N. D. Seeley, K. B. Logan, and J. W. Nicolson.** 1982. Methods for studying aquatic bacteriophage ecology. *Appl. Environ. Microbiol.* **43:**694–701.
113. **Pyle, B. H., S. C. Broadaway, and G. A. McFeters.** 1995. A rapid, direct method for enumerating respiring enterohemorrhagic *Escherichia coli* O157:H7 in water. *Appl. Environ. Microbiol.* **61:**2614–2619.
114. **Reasoner, D. J., J. C. Blannon, and E. E. Geldreich.** 1979. Rapid seven-hour fecal coliform test. *Appl. Environ. Microbiol.* **38:**229–236.
115. **Reasoner, D. J., and E. E. Geldreich.** 1989. Detection of fecal coliforms in water by using [^{14}C]mannitol. *Appl. Environ. Microbiol.* **55:**907–911.
116. **Rheinheimer, G.** 1992. *Aquatic Microbiology*, 4th ed., p. 279–282. John Wiley & Sons, New York, N.Y.
117. **Rhodes, M. W., and H. Kator.** 1994. Seasonal occurrence of mesophilic *Aeromonas* spp. as a function of biotype and water quality in temperate freshwater lakes. *Water Res.* **28:**2241–2251.
118. **Richardson, K. J., M. H. Stewart, and R. L. Wolfe.** 1991. Application of gene probe technology to the water industry. *J. Am. Water Works Assoc.* **83**(Sept.)**:**71–81.
119. **Rippey, S. R. and V. J. Cabelli.** 1989. Use of the thermotolerant *Aeromonas* group for the trophic state classification of freshwaters. *Water Res.* **23:**1107–1114.
120. **Rivera, S. C., T. C. Hazen, and G. A. Toranzos.** 1988. Isolation of fecal coliforms from pristine sites in a tropical rain forest. *Appl. Environ. Microbiol.* **54:**513–517.
121. **Sanchez, J. M., M. C. Alonso, A. Luque, and F. Egea.** 1992. Metodología para el aislamiento e identificación de especies del género *Clostridium*, p. 267–297. *In* J. J. Borrego (ed.), *Métodos Microbiológicos Rápidos para Análisis de Aguas y Alimentos*. SPICUM, Universidad de Málaga, Málaga, Spain.
122. **Sartory, D. P., and L. Howard.** 1992. A medium detecting β-glucuronidase for the simultaneous membrane filtration enumeration of *Escherichia coli* and coliforms from drinking water. *Lett. Appl. Microblol.* **15:**273–276.

122a. **Saunders, G. C., and H. C. Parkes (ed.).** 1999. *Analytical Molecular Biology*. The Royal Society of Chemistry, Cambridge, United Kingdom.

123. **Schaule, G., H.-C. Flemming, and H. F. Ridgway.** 1993. Use of 5-cyano-2, 3-ditolyl tetrazolium chloride for quantifying planktonic and sessile respiring bacteria in drinking water. *Appl. Environ. Microbiol.* **59:** 3850–3857.
124. **Seidl, P.** 1990. Microbiological investigations of drinking and recreational waters from an Indian reserve in Can-ada, p. 59–65. *In* G. Castillo, V. Campos, and L. Hertera (ed.), *Proceedings of the Second Biennial Water Quality Symposium*. Editorial Universitaria, Santiago, Chile.
125. **Simkova, A., and J. Cervenka.** 1981. Coliphage as ecological indicators of enteroviruses in various water systems. *Bull. W. H. O.* **59:**611–618.

126. **Singh, A., B. H. Pyle, and G. A. McFeters.** 1989. Rapid enumeration of bacteria by image analysis epifluorescence microscopy. *J. Microbiol. Methods* **10:**91–102.
127. **Sinton, L. W., A. M. Donnison, and C. M. Hastie.** 1993. Faecal streptococci as faecal pollution indicators: a review. Part I: taxonomy and enumeration. *N. Z. J. Mar. Freshwater Res.* **27:**101–115.
128. **Sinton, L. W., A. M. Donnison, and C. M. Hastie.** 1993. Faecal streptococci as faecal pollution indicators: a review. Part II: sanitary significance, survival, and use. *N. Z. J. Mar. Freshwater Res.* **27:**117–137.
129. **Sivaborborn, K.** 1988. *In Abstracts of the First Biennial Water Quality Symposium, Banff, Canada.*
130. **Sorensen, D. L., S. G. Eberl, and R. A. Dicksa.** 1989. *Clostridium perfringens* as a point source indicator in non-point polluted streams. *Water Res.* **23:**191–197.
131. **Stetler, R.E.** 1984. Coliphages as indicators of enteroviruses. *Appl. Environ. Microbiol.* **48:**668–670.
132. **Tartera, C., and J. Jofre.** 1987. Bacteriophages active against *Bacteroides fragilis* in sewage-polluted waters. *Appl. Environ. Microbiol.* **53:**1632–1637.
133. **Tartera, C., J. Jofre, and F. Lucena.** 1988. Relationship between numbers of enteroviruses and bacteriophages infecting *Bacteroides fragilis* in different environmental samples. *Environ. Technol. Lett.* **9:**407–410.
134. **Tartera, C., F. Lucena, and J. Jofre.** 1989. Human origin of *Bacteroides fragilis* bacteriophages present in the environment. *Appl. Environ. Microbiol.* **55:**2696–2701.
135. Reference deleted.
136. **Toranzos, G. A.** 1991. Current and possible alternate indicators of fecal contamination in tropical waters: a short review. *Environ. Toxicol. Water Qual.* **6:**121–130.
137. **Toranzos, G. A., C. P. Gerba, and H. Hanssen.** 1988. Enteric viruses and coliphages in Latin America. *Toxic Assess.* **5:**491–510.
137a. **Toranzos, G. A.** 2002. *Pseudomonas*, p. 2632–2639. *In* G. Bitton (ed.), *Encyclopedia of Environmental Microbiology.* Wiley, New York, N.Y.
138. **Tsai, G. J., and T. H. Chen.** 1996. Incidence and toxigenicity of *Aeromonas hydrophila* in seafood. *Int. J. Food Microbiol.* **31:**121–131.
139. **Tsai, Y.-L., C. J. Palmer, and L. R. Sangermano.** 1993. Detection of *Escherichia coli* in sewage and sludge by PCR. *Appl. Environ. Microbiol.* **59:**353–357.
140. **Ursing, J. B., R. A. Rossello-Mora, E. Garcia-Valdes, and J. Lalucat.** 1995. Taxonomic note: a pragmatic approach to the nomenclature of phenotypically similar genomic groups. *Int. J. Syst. Bacteriol.* **45:**604.
141. **Venkatesan, M. I., and I. R. Kaplan.** 1990. Sedimentary coprostanol as an index of sewage addition in Santa Monica Basin, southern California. *Environ. Sci. Technol.* **24:**208–214.
142. **Walter, K. S., E. J. Fricker, and C. R. Fricker.** 1994. Observations on the use of a medium detecting β-glucuronidase activity and lactose fermentation for the simultaneous detection of *Escherichia coli* and coliforms. *Lett. Appl. Microbiol.* **19:**47–49.
143. **Winkler, J., K. N. Timmis, and R. A. Snyder.** 1995. Tracking the response of *Burkholderia cepacia* G4 5223-Prl. *Appl. Environ. Microbiol.* **61:**448–455.
144. **World Health Organization.** 1989. *Guidelines for Drinking-Water Quality*, 2nd ed., p. 8–29. World Health Organization, Geneva, Switzerland.
145. **Writer, J. H., J. A. Leenheer, L. B. Barber, G. A. Amy, and S. C. Chapra.** 1995. Sewage contamination in the upper Mississippi River as measured by the fecal sterol, coprostanol. *Water Res.* **29:**1427–1436.
146. **Wun, C. K., R. W. Walker, and W. Litsky.** 1976. The use of XAD-2 resin for the analysis of coprostanol in water. *Water Res.* **10:**955–959.

Detection of Protozoan Parasites in Source and Finished Drinking Water

FRANK W. SCHAEFER III

21

Protozoa are unicellular microorganisms that, unlike bacteria and viruses, possess membrane-bound genetic material or nuclei and other assorted cellular organelles. They exhibit various forms of locomotion and reproduction that have been used to categorize them into broad groups; however, the taxonomy of the protozoa is a dynamic, evolving process about which there is little agreement.

Free-living protozoa are ubiquitous in natural waters and moist soils and fall into three broad groups: amoebae, flagellates, and ciliates. They are often abundant in both surface and groundwater supplies as part of the normal aquatic community. Free-living protozoa are of little concern to water plant operators, since they cause no treatment problems or health impacts. Identification of free-living protozoa to genus and species levels requires a great deal of skill and training. Keys, line drawings, and pictures in *An Illustrated Guide to the Protozoa* (31) aid in identifying unknown organisms at least to ordinal and familial levels. Included as representatives of free-living protozoa found in source waters are *Hartmannella* spp., *Acanthamoeba* spp., *Echinamoeba* spp., *Naegleria fowleri*, *Euglena* spp., *Tetrahymena* spp., and *Paramecium* spp.

Parasitic protozoa, which live in or on another organism like either a plant or animal, may or may not be of concern to the water treatment plant operator depending on the organism on which they live. This chapter deals primarily with human enteric protozoa and more specifically *Giardia* and *Cryptosporidium*. Within the human small and large intestine, a number of protozoa, including amoebae, flagellates, coccidians, and ciliates, can be found. Many of these organisms are not recognized as pathogenic and consequently are dismissed as commensals. *Entamoeba coli* and *Trichomonas hominis* are recognized as examples of commensal organisms that live in another organism without harming it. Parasitic protozoa, on the other hand, are known to be harmful when living in or on another organism. Included as representatives of enteric protozoa are *Entamoeba histolytica*, *Giardia lamblia*, *Cryptosporidium parvum*, *Cyclospora* spp., *Toxoplasma gondii*, *Enterocytozoon bieneusi*, *Isospora belli*, *Isospora hominis*, and *Balantidium coli*.

Protozoans, both free-living and parasitic, are classified into several taxonomic groups. Amoebae belong to the phylum Sarcomastigophora and generally possess a vesicular nucleus and pseudopods, which are retractable cytoplasmic protrusions used both for locomotion and feeding. Thecate amoebae are distinguished from athecate or naked forms, the other main subgrouping, by a closely fitting envelope or shell secreted by the trophozoite. Many of the athecate amoebae are able to withstand adverse environmental conditions by transforming from a vegetative trophozoite form to a dormant, resistant cyst form. Cysts of certain amoebae have been isolated from the air associated with dust particles (43, 62). *Hartmannella* spp., *Acanthamoeba castellanii*, *Echinamoeba* spp., and *N. fowleri* are athecate examples which can produce cysts. *Hartmannella* spp., *A. castellanii*, and *Echinamoeba* spp. are known to harbor and amplify *Legionella* spp. (18). *Acanthamoeba* spp. and *N. fowleri* are normally free-living forms but under the right conditions become opportunistic pathogens which are responsible for primary amoebomeningoencephalitis in humans. Furthermore, *Acanthamoeba* spp. also are well-documented causes of corneal keratitis in contact lens wearers.

Flagellates also belong to the phylum Sarcomastigophora and have vesicular nuclei but move by hair- or whip-like cylindrical organelles which are approximately 0.25 μm wide and are known as flagella. Flagellar length and number in this group are variable. Members of this group may also possess chloroplasts, thecal plates, basal bodies, and collars around the flagellum. Some genera exhibit multicellular colonial forms. *Euglena* spp. are examples of free-living flagellates which contain chloroplasts. *Giardia* spp. is an example of a parasitic flagellate.

Ciliates, which belong to the phylum Ciliophora, are distinguished from the other protozoan groups by their unique nuclei. They have both a large macronucleus which regulates cellular metabolism and a small micronucleus which is involved in genetics and sexual recombination. Many ciliates are phagotrophic, which means that they ingest nutrients through a mouth or cytostome. These organisms are covered with cilia which are used both in feeding and movement. Cilia are organelles similar in structure to flagella. However, they differ from flagella in being generally shorter and in being interconnected through their intracytoplasmic basal structure. As a result of the interconnection, ciliary movement can be coordinated. *Tetrahymena* spp. and *Paramecium* spp. are examples of free-living ciliates, while *B. coli* is an example of a parasitic form.

The phylum Apicomplexa is sometimes referred to as the phylum Sporozoa. This group is exclusively parasitic and is responsible for both human and veterinary disease.

Malaria and coccidian diseases are well-known examples from humans and animals, respectively. The apicomplexa are characterized by possession of an anterior apical complex. In addition, they have complicated life cycles. Unlike the phyla mentioned above, they possess no obvious locomotor organelles. Most members of this group have a vesicular nucleus, a Golgi apparatus, and mitochondria. Moreover, there is usually a sexual phase to the life cycle in this group. Examples of this group include *Cryptosporidium*, *Cyclospora*, *Isospora*, and *Toxoplasma*.

Microsporidians, a nontaxonomic term, are obligate intracellular parasites belonging to the class Microspora of the protozoan phylum Microsporidia. They are ubiquitous parasites infecting a variety of vertebrate and invertebrate hosts. Presently, there are over 100 genera encompassing around 1,000 species. From an anthropocentric point of view, insects of commercial significance like honeybees and silkworms are impacted by this group. Moreover, moving up the phylogenetic tree, snails and commercial fish such as salmon, flounder, and monkfish are prone to microsporidian diseases. Only with the advent of the AIDS epidemic was this group of pathogens recognized as a cause of human disease. A number of microsporidian genera have been recovered from humans: *Encephalitozoon*, *Nosema*, *Pleistophora*, *Trachipleistophora*, *Brachiola*, *Enterocytozoon*, *Vittaforma*, and "*Microsporidium*," a genus for all forms as yet unclassified.

BIOLOGY AND ECOLOGY

Free-living protozoa are found in air, water, and soil. For instance, large numbers of bactivorous ciliates and amoebae can be found in sewage and other organic wastes. Furthermore, they may be found on the surface, bottom, banks, and slopes of most bodies of fresh and marine water. Free-living forms may withstand temperatures ranging from freezing up to 28°C. Certain ecological stimuli may cause amoebae to encyst or to differentiate a flagellum.

Most enteric protozoans have two stages in their life cycles. The trophozoite is an actively feeding, growing, and reproducing stage. Stimuli in the host's intestinal tract induce most of the enteric protozoans to produce a resistant, dormant transmission form, which is referred to as either a cyst, oocyst, or spore. Whether commensal or parasitic, these protozoans have simple, direct life cycles and are transmitted as fecal contaminants of food and/or water. In addition, person-to-person transmission also is known to occur. Trophozoites generally do not survive outside their host, unless they are propagated in a specialized culture medium (40, 54). On the other hand, cysts, oocysts, and spores are different in that they are known to survive for long periods outside the host, especially under cool, moist conditions. Ambient temperatures around 25°C or higher and desiccation are known to rapidly reduce the time that a cyst, oocyst, or spore can survive outside the host. Unlike clinical specimens, only cysts, oocysts, and spores of parasitic forms are of concern in environmental samples. Most troubling for the water treatment industry is that many of these forms are more resistant to chemical disinfection than are enteric bacteria and viruses.

Enteric parasitic protozoa are known to produce gastrointestinal distress, including diarrhea, flatulence, cramps, anorexia, and weight loss. However, they can produce a continuum of pathologies ranging from no symptoms to extremes of illness requiring hospitalization.

Presently, *Giardia* and *Cryptosporidium* are of great concern to the water treatment industry, because they are known to be the etiologic agents responsible for a number of episodes of waterborne gastroenteritis. Their significance is also increased, as they have a number of mammalian reservoir hosts that they can use besides humans. This results in an amplification in the numbers of cysts and/or oocysts which are potential challenges to water impoundments and treatment plants.

DETECTION AND QUANTITATION

Whether one is looking for free-living or parasitic forms, no one collection method suffices for all protozoa. Free-living forms are usually in sufficient density, if present at all, so that they can be collected by a variety of grab sample techniques from either sediment or water above the sediment. Parasitic forms, on the other hand, most likely are suspended in the water and are usually very sparse. Consequently, they must be concentrated from large volumes of water.

Free-Living Protozoa

After collection, field samples of free-living forms may sometimes be observed directly under the microscope or they may be cultured. Each free-living form has optimal cultural requirements which cannot be covered in detail here. Many of the media used by the American Type Culture Collection to culture free-living forms are listed in the back of its catalog (12; http://www.atcc.org). In addition, older protozoology books by Jahn and Jahn (23) and Kudo (27) contain valuable information on methods for collecting and culturing free-living protozoans. Parasitic forms, like *Giardia* and *Cryptosporidium*, can be cultured by using very exacting aseptic procedures (40, 54) which do not lend themselves to routine use with field samples. Whatever the organism of interest, the investigator must pay close attention to pH, nutrients present, toxic substances present like hydrogen sulfide, temperature, osmotic effects, and other organisms in the sample.

Initial identification of free-living protozoans is done with taxonomic keys; however, a basic knowledge of protozoan morphology is required to use them. Fundamental keys have been developed (23) but now are somewhat out of date taxonomically. A more recent, taxonomically accepted key has been published (31) but requires considerable knowledge of protozoa on the part of the user. In many cases the only way to accurately identify some free-living protozoans to genus and species is to collaborate with experts in the field.

Detection and quantitation of protozoa usually require microscopic techniques, including epifluorescence, brightfield, phase-contrast, and differential interference contrast (DIC) microscopy. Moreover, some protozoa are identified based upon exacting staining techniques. Very large forms may require observation with a dissecting microscope. Many free-living forms are very motile and must be slowed down to be seen at all. Substances like Protoslo (Carolina Biological Supply, Burlington, N.C.), 10% methylcellulose, Detain (Ward's Natural Science Establishment, Inc.), 1% aqueous solution of a polyether, and 3% polyacrylamide hydrazide (Sigma Chemical Co.) have been used to increase the viscosity of the medium and slow protozoan movement. Whenever a microscope slide is to be observed for a protracted period, the coverslip must be sealed to the slide with nail polish, Vaspar, or petroleum jelly to prevent evaporation. Air bubbles under a sealed coverslip can provide a limited source of oxygen for a living preparation. For photomicrography of motile living material, a flash attachment for the microscope is essential.

Free-living amoebae, like *Acanthamoeba* spp., *Hartmannella* spp., *Naegleria* spp., and others, are usually grab sampled. Volumes of the samples are generally small, ranging from 1 to 4 liters of fluid or 10 to 100 ml of sediment. Fluid samples are concentrated by centrifugation at 250 × *g* for 5 to 8 min when necessary, while sediment samples are used directly. The fluid concentrates and sediments are then cultured on nonnutrient agar lawns of *Escherichia coli*, which the amoebae use as a food source. Detection, purification, and cloning of the free-living amoebae are done by isolating organisms from plaques, places where the amoebae have consumed all the bacteria (26). Identification is based upon cyst and trophozoite morphology. Collaboration with a free-living protozoan expert is required in many instances, from which antibodies and genetic primers and probes can be obtained to confirm preliminary identifications.

Parasitic Protozoa

Detection of *G. lamblia* and *C. parvum* in source and finished water has been and continues to be of great interest. *G. lamblia* is one of the most frequently reported parasitic waterborne pathogens (11). Furthermore, *C. parvum* and *Cryptosporidium hominis* also have been responsible for numerous waterborne outbreaks of gastroenteritis, with the most notable being in Carrollton, Ga., in 1987 (20), Talent, Oreg., in 1992 (59), and Milwaukee, Wis., in 1993 (33), where 13,000, 80,000, and 403,000 individuals were affected, respectively. The method used at that time to detect these protozoan pathogens is known as the ICR method (3). It is an immunofluorescence detection procedure (3) performed after concentration of protozoans from large volumes of water. Initially, concentration of *Giardia* cysts and *Cryptosporidium* oocysts from water is done by retention on a nominal porosity yarn-wound filter (38, 56). Retained particulates are eluted from the filter and concentrated by centrifugation. The pelleted *Giardia* cysts and *Cryptosporidium* oocysts are separated to some extent from other particulate debris by flotation on a Percoll-sucrose solution with a specific gravity of 1.1. A portion of the water layer–Percoll-sucrose interface is then placed on a membrane filter to form a monolayer, indirectly stained with fluorescent antibody (FA), and examined with an epifluorescence microscope. Cysts and oocysts are classified according to specific criteria (immunofluorescence, size, shape, and internal morphological characteristics), and the results are reported in terms of presumptive and confirmed cysts and oocysts per 100 liters. Presumptive cysts and oocysts are defined as objects having the right size and shape as well as fluorescence characteristics. Confirmed cysts and oocysts are those in which internal morphological characteristics like nuclei, axonemes, and median bodies can be demonstrated in *Giardia* cysts and in which sporozoites can be seen in *Cryptosporidium* oocysts by either phase-contrast or DIC microscopy (Fig. 1).

Because recoveries of *Giardia* cysts and especially *Cryptosporidium* oocysts in seeded samples were so low with the ICR method described above, a great deal of effort went into the development of an improved method that now is known as "Method 1623: *Cryptosporidium* and *Giardia* in water by Filtration/IMS/FA" (57). The first version of this method, known as "Method 1622: *Cryptosporidium* in water by Filtration/IMS/FA," detected only *Cryptosporidium* oocysts, because the reagents necessary for detecting *Giardia* cysts were not commercially available. Key changes made during the development of Methods 1622 and 1623 included use of an absolute porosity filter to concentrate the sample, reduction of the sample size from 100 to 10 liters, selective purification of the parasites from other particulates with immunomagnetic separation (IMS), and counterstaining with 4′,6′-diamidino-2-phenylindole (DAPI) to elucidate nuclei in addition to traditional staining with fluorescein isothyocyanate (FITC)-labeled antibodies, and use of rigorous quality assurance (QA). Recoveries for *Cryptosporidium* oocysts during validation studies of method 1623 (57) were around 40%, in contrast to the 11% for the ICR method.

Quality Control

Laboratories using method 1623 are expected to operate a formal QA program. The minimum requirements of this program consist of an initial demonstration of laboratory capability through performance of the initial performance and recovery (IPR) test, analysis of spiked samples to evaluate and document data quality, and analysis of standards and blanks as tests of continued performance. Periodic ongoing precision and recovery (OPR) and matrix spike (MS) samples must be spiked with flow cytometer-counted *Giardia* cysts and *Cryptosporidium* oocysts and processed within 24 h of spiking. Flow cytometer-counted cysts and oocysts can be obtained from the Wisconsin State Laboratory of Hygiene (Environmental Health Division Facility, 2601 Agriculture Dr., Madison, WI 53718) and Biotechnology Frontiers (http://www.btfbio.com/; P.O. Box 599, North Ryde BC, NSW 1670, Australia).

Data that will allow an independent reviewer to validate and trace processing and analysis steps leading to the final result are required. At a minimum the following information is required: (i) sample numbers and other identifiers; (ii) source of spiking suspensions, as well as lot number and date received; (iii) spike enumeration date; (iv) sample spiking dates and times; (v) volume filtered; (vi) filtration and concentration dates and times; (vii) initial pellet volume and resuspended pellet volume; (viii) staining completion dates and times; (ix) staining control results; (x) all required examination and confirmation information; (xi) examination and confirmation dates and times; (xii) lot numbers of elution, IMS, and staining reagents; (xiii) copies of bench sheets (a record on paper of information which accompanies the field sample), logbooks, and other recordings of raw data; and (xiv) any links to other related data outputs, which must be included in the final report.

Before using method 1623 for routine analysis, analysts in the laboratory need to demonstrate control over the analytical process and to generate acceptable precision and recovery data. To accomplish this task, each laboratory analyst needs to perform the IPR test. For the IPR, use around 100 flow cytometer-counted cysts and oocysts to spike each 10-liter Cubitainer of reagent water; four replicates are required. Then filter, elute, concentrate, purify by IMS, stain, and examine microscopically each of the four 10-liter reagent water samples. Using results of the four analyses, compute the average percent recovery (X) and the relative standard deviation of the recovery (s_r) for *Cryptosporidium*. Compare s_r and X with the corresponding limits for initial precision and recovery in Tables 1 and 2. If s_r and X meet the acceptance criteria, system performance is acceptable and analysis of blanks and field samples may begin. If s_r or X falls outside the acceptance range for recovery, system performance is unacceptable. In this event, correct the problem and repeat the test with additional samples of spiked reagent water.

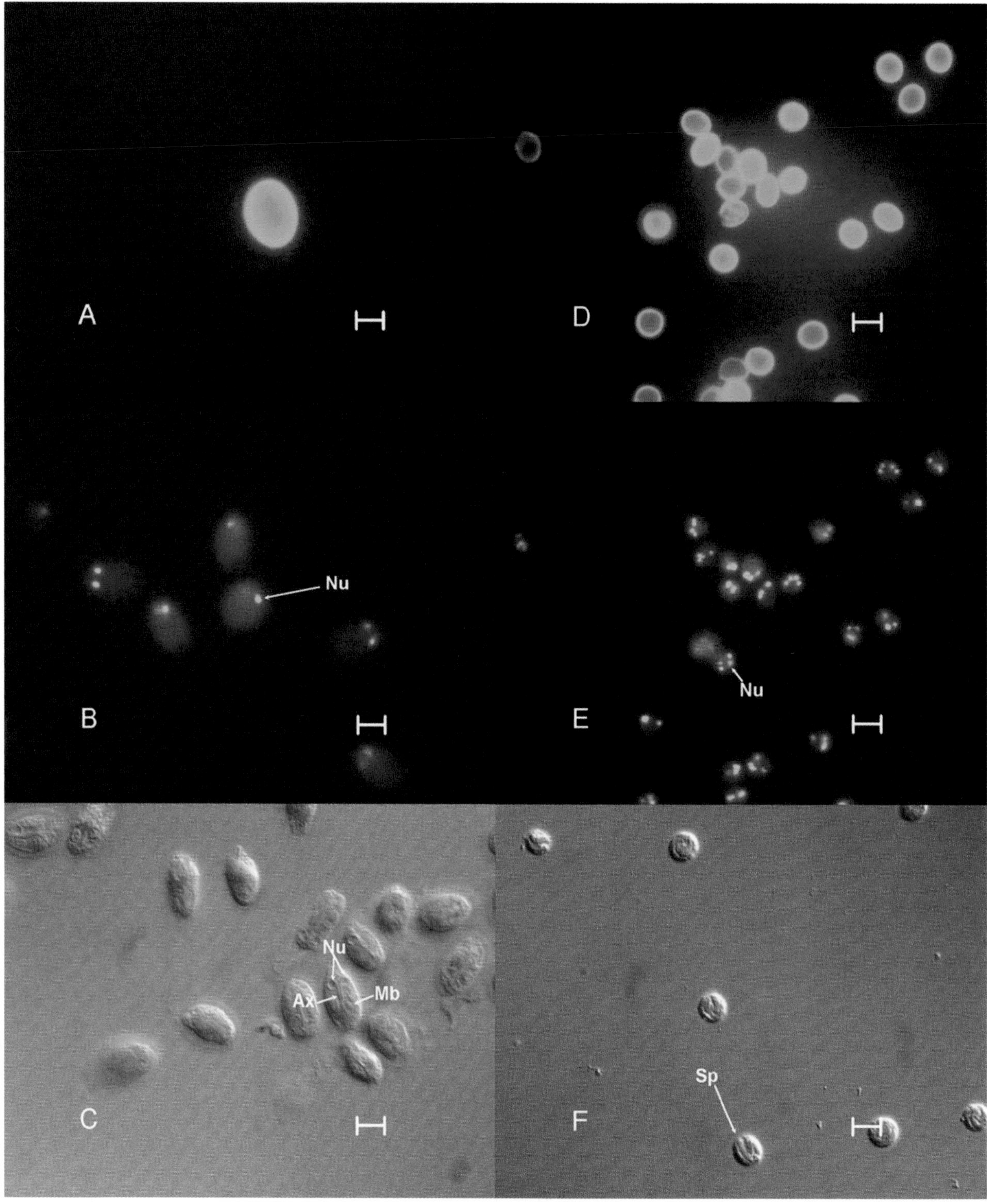

FIGURE 1. *Giardia cysts* and *Cryptosporidium* oocysts were stained by immuno-FA and DAPI and were also viewed by Nomarski DIC microscopy. (A) Immuno-FA-stained *G. lamblia* cyst. (B) DAPI-stained *G. lamblia* cysts. Nu, nucleus. (C) Nomarski DIC photomicrograph of *G. lamblia* cysts. Mb, median body; Ax, axonemes. (D) Immuno-FA-stained *C. parvum* oocysts. (E) DAPI-stained *C. parvum* oocysts. (F) Nomarski DIC photomicrograph of *C. parvum* oocyst. S, sporozoite (three showing). Bars = 5 μm. (Photomicrographs courtesy of Michael W. Ware, U.S. Environmental Protection Agency.)

The laboratory should analyze one MS sample to determine the effect of the matrix on the method's oocyst and cyst recovery, when field samples of a particular matrix type are first received from a source for which the laboratory has never before analyzed samples. In this case, the MS analysis is performed on an additional (second or duplicate) field sample. If the laboratory routinely analyzes field samples from one or more sources, then one MS analysis should be performed per

20 field samples. For example, when a laboratory receives the first field sample from a given site, the laboratory must obtain a second aliquot of this field sample to be used for the MS. When the laboratory receives the 21st field sample from this site, a separate aliquot of this 21st field sample must be collected, spiked, processed, and analyzed.

An alternative to analyzing an MS in addition to the unspiked sample is to use Biotechnology Frontiers' ColorSeed as an internal standard in the environmental sample. A vial of ColorSeed that contains 100 *Giardia* cysts and 100 *Cryptosporidium* oocysts that have been modified with a red fluorochrome is added to the environmental sample before it is processed. At the end of the protocol during microscopic observation and enumeration, the analyst differentiates between indigenous cysts and oocysts, which fluoresce only green, and the ColorSeed-spiked cysts and oocysts, which fluoresce both green and red. The ColorSeed allows the analyst to determine a percent recovery for that sample. The advantage to using ColorSeed as an internal standard is that a percent recovery can be determined for every sample in which it is used. Moreover, this determination is obtained by analyzing one rather than two separate samples. ColorSeed *Cryptosporidium* oocysts and flow cytometer-counted live *Cryptosporidium* oocysts were seeded into 29 water samples and 1 effluent sample from various locations in the United States. The method 1622 recoveries ranged from 3.0 to 59% for the ColorSeed oocysts and from 2.0 to 61% for the live oocysts. The recoveries of these two seeding procedures were highly correlated ($r = 0.802$) and were not statistically different (19).

In addition, analysis of method blanks is required to demonstrate that reagents and processing are not introducing contamination. The blank (negative control sample) is an unspiked 10-liter reagent water sample. It is processed prior to analysis of the IPR test and OPR test and prior to analysis of any of the field samples for any given week to demonstrate that reagents and processing are not introducing contamination. The laboratory should, on an ongoing basis, demonstrate through analysis of the OPR (positive control) samples that the analysis system is within acceptable control limits. At least once a week spike around 100 flow cytometer-counted cysts and oocysts into 10 liters of reagent water, and then filter, elute, concentrate, separate (purify), stain, and examine the sample to verify all performance criteria. The laboratory must analyze one OPR sample for every 20 field samples, if more than 20 field samples are analyzed in a week. Adjustment and/or recalibration of the analytical system shall be performed until all performance criteria are met. Only after all performance criteria are met may field samples be analyzed. For each organism, compute the percent recovery with the following equation:

$$P = 100 \times N/T$$

where P = percent recovery, N = the number of (oo)cysts or cysts detected, and T = the number of (oo)cysts or cysts spiked. Compare the percent recovery with the limits for OPR in Tables 1 and 2. If the recovery meets the acceptance criteria, system performance is acceptable and analysis of blanks and field samples may proceed. If, however, the recovery falls outside the range given, then system performance is unacceptable. In this event, there may be a problem with the microscope, filtration, or separation systems. To resolve this, reanalyze the OPR sample and recollect and reanalyze field samples. All samples must be associated with an OPR that passes the acceptance criteria.

The laboratory should analyze one method blank and one OPR sample each week during which samples are analyzed, if 20 or fewer field samples are analyzed during this period. The laboratory should analyze one laboratory blank and one OPR sample for every 20 field samples if more than 20 field samples are analyzed in a week. If *Cryptosporidium* oocysts, *Giardia* cysts, or any potentially interfering organism or material is found in the blank, analysis of additional field samples is halted until the source of contamination is eliminated and a blank shows no evidence of contamination. Any field sample in a batch associated with a contaminated blank that shows the presence of one or more oocysts or cysts is assumed to be contaminated and should be recollected, if possible. Any method blank in which oocysts or cysts are not detected is assumed to be uncontaminated and may be reported.

The laboratory should add results that pass the method acceptance criteria to initial and previous ongoing data and update a quality control (QC) chart to form a graphic representation of continued laboratory performance. The laboratory should develop a statement of laboratory accuracy (for reagent water, raw surface water, etc.) by calculating the average percent recovery (R) and the standard deviation of percent recovery (s_r). Express the accuracy as a recovery interval from $R - 2\ s_r$ to $R + 2\ s_r$. For example, if $R = 95\%$ and $s_r = 25\%$, the accuracy is 45 to 145%. If available, the laboratory should periodically analyze an external QC sample, such as a performance evaluation or standard reference material. The laboratory also should periodically participate in interlaboratory comparison studies with this protozoan recovery method. The specifications contained in method 1623 can be met if the analytical system is under control. The standards used for IPR and OPR should be identical, so that the most precise results will be obtained.

Internal and external calibration records of instruments and equipment must be kept on file in the laboratory's QA logbook.

TABLE 1 QC acceptance criteria for *Cryptosporidium*

Performance test	Acceptance criteria
IPR	
Mean recovery (%)	24–100
Precision (as maximum relative SD)	55
OPR (%)	11–100
MS/MS duplicate (for method modifications)	
Mean recovery (%)	13–111
Precision (as maximum relative % difference)	61

TABLE 2 QC acceptance criteria for *Giardia*

Performance test	Acceptance criteria
IPR	
Mean recovery (%)	24–100
Precision (as maximum relative SD)	49
OPR (%)	14–100
MS/MS duplicate (for method modifications)	
Mean recovery (%)	15–118
Precision (as maximum relative % difference)	30

Sampling

Field samples are collected in disposable plastic 10-liter Cubitainers that are shipped to the laboratory for processing. Sample containers should be labeled immediately with the name of the site, the date and time of sampling, and the name of the person taking the sample. If residual disinfectant is present in the water, add 100 ml of 2% (wt/vol) aqueous thiosulfate to the Cubitainers before adding the 10-liter sample. Samples should be shipped to the laboratory on or during the day when they are collected and must arrive within 24 h of sample collection. Store the field sample Cubitainers at 4 to 8°C between collection and shipment to the laboratory and upon receipt at the laboratory until ready for filtration. Do not allow the sample to freeze. Although it is desirable to ship field samples on wet or chemical ice, it is not a requirement. Recent studies reinforcing this recommendation have shown that exposure of live *C. parvum* oocysts to ambient or higher temperatures over time results in their deterioration (64).

It should be noted that the U.S. Department of Transportation regulations (49 CFR 172) prohibit interstate shipment of more than 4 liters of solution known to contain infectious materials. Similar state regulations may apply to intrastate commerce. This procedure requires a minimum sample volume of 10 liters. However, if a sample is not known to contain *Giardia*, *Cryptosporidium*, or other infectious agent, then the samples can be shipped as noninfectious. When field samples are known to contain infectious agents or are suspected of containing infectious agents, it is recommended that the field samples be filtered in the field and the filter be shipped to the laboratory to avoid violating Department of Transportation regulations.

Holding Times

Once the field sample arrives in the laboratory, various holding times apply. Periodic OPR and MS samples must be spiked and processed within 24 h of spiking with flow cytometer-counted *Giardia* cysts and *Cryptosporidium* oocysts. Laboratory filtration, elution, and concentration of a field sample must be completed within 72 h of sample collection. The concentrate must be stored at 4 to 8°C if it is not proceeding immediately to IMS. IMS and sample staining must be completed within 24 h of sample concentration. Stained slides must be stored at 4 to 8°C in the dark. Although the immunofluorescence assay, DAPI, and DIC microscopic examination and confirmation should be performed immediately after staining is complete, laboratories have up to 72 h from completion of sample staining to complete the examination and confirmation of samples. However, if fading or diffusion of DAPI staining is noticed, the laboratory must reduce this holding time and/or adjust the concentration of the DAPI staining solution, so that fading or diffusion does not occur. Never allow freezing of any of these samples or reagents.

Sample Filtration and Elution

Presently, the Pall Gelman (Ann Arbor, Mich.) Envirochek and Envirochek HV capsule filters, the IDEXX (Westbrook, Maine) Filta-Max filter, and the Whatman (Clifton, N.J.) Nuclepore CrypTest filter have been validated and found to be acceptable for concentrating 10-liter water samples with method 1623. Each of these filters has an upper limit on the volume of water that it can efficiently filter. For the sake of brevity, only the Gelman Envirochek filter is discussed here. The manufacturer's directions should be followed if either of the other filter options is used.

The steps for sample filtration and elution are as follows: first, assemble the sample concentration components, consisting of tubing, a pump, flow controller, capsule filter, and sample container as illustrated in Fig. 2. To prevent cross-contamination between samples, use the tubing between the sample container and the Gelman Envirochek filter capsule only once and throw it away. Connect the sampling system, minus the capsule, to a carboy filled with reagent water. Turn on the pump and adjust the flow rate to 2.0 liters/min., allowing 2 to 10 liters of reagent water to flush the system. Turn off the pump, when the flow rate has been adjusted, and install the capsule filter in the line, securing the inlet and outlet ends with the appropriate clamps and/or fittings.

Record the sample number, sample turbidity, and the name of the analyst filtering the sample on a bench sheet and on the capsule filter. Connect the sampling system to the field carboy of sample water, or transfer the sample water to the laboratory carboy used to regulate the flow rate above. It should be noted that, if the field sample is transferred to a laboratory carboy, the laboratory carboy must be cleaned and disinfected before it is used with another field sample. Place the drain end of the sampling system tubing into an empty graduated container with a capacity of 10 to 15 liters, calibrated at 9.0, 9.5, 10.0, 10.5, and 11.0 liters. This container is used to determine the sample volume filtered. Alternately, a totalizing meter can be placed on the drain end of this apparatus to determine the sample volume filtered.

Allow the carboy discharge tube and filter capsule to fill with sample water. Vent residual air from the filter capsule by using the bleed valve or vent port built into the filter capsule. Turn on the pump to start water flowing through the filter and verify that the flow rate is 2 liters/min. After the sample has passed through the filter, turn off the pump. Allow the pressure to decrease until flow stops. Disconnect the inlet end of the capsule filter assembly while maintaining the level of the inlet fitting above the level of the outlet fitting to prevent backwashing and the loss of oocysts and cysts from the filter. Restart the pump just long enough to allow as much water to drain from the filter capsule as possible. Based on the water level in the graduated container or the reading on the totalizing meter, record the volume filtered on a bench sheet and the capsule filter label to the nearest quarter-liter. Discard the contents of the graduated container. Loosen the outlet fitting on the capsule filter and then close the inlet and outlet fittings with the caps provided by the manufacturer.

Assemble the laboratory shaker (Lab-Line shaker, model 3589 equipped with Lab-Line side arms, model 3587-4; Lab-Line, Dubuque, Iowa) with the clamps aligned vertically so that the filters will be aligned horizontally. Prepare suffi-

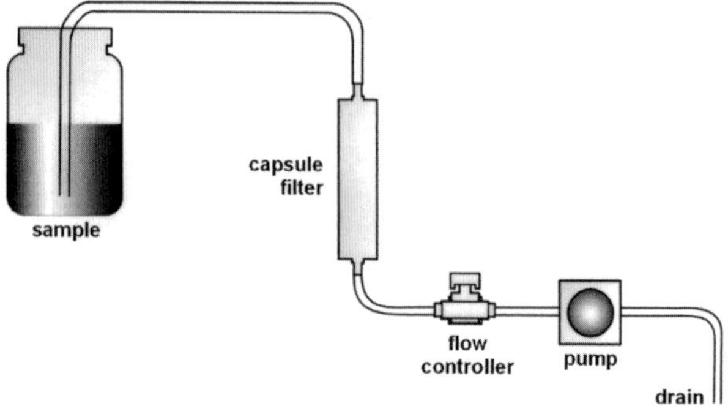

FIGURE 2 Water sampling apparatus.

cient elution buffer, which contains laureth-12, Tris solution, EDTA, and antifoam A, so that all samples to be eluted that day can be processed with the same batch of buffer. To prepare elution buffer, add 10 ml of 10% (wt/vol) aqueous laureth-12 (PPG Industries, Gurnee, Ill.) to a 1,000-ml graduated cylinder. Also add 10 ml of 1 M Tris, pH 7.4, 2.0 ml of 0.5 M disodium EDTA, pH 8.0, and 150 μl of antifoam A (Sigma-Aldrich, St. Louis, Mo.) to the graduated cylinder and dilute to 1,000 ml with reagent water. Elution may require up to 275 ml of buffer per sample. Designate at least one 250-ml conical centrifuge tube, which will be used to receive and process the filter eluate, for each sample and label it with the sample number.

Record the name of the analyst performing the elution on the bench sheet. Using a ring stand or other means, clamp each filter capsule to be eluted in a vertical position with the inlet end up. Remove the inlet cap and allow the residual liquid level to stabilize. Pour elution buffer through the inlet fitting. Enough elution buffer must be added to just cover the pleated white membrane of the filter. Replace the inlet cap and transfer the filter capsule to a clamp on the shaker. Adjust the filter capsule in the clamp on the laboratory shaker so that the bleed valve is positioned at the top of the vertical axis (in the 12 o'clock position). Turn on the shaker and set the speed to 100% of maximum (approximately 900 rpm) and agitate the capsule for 5 min. Remove the filter capsule from the shaker, remove the inlet cap, and pour the contents of the capsule into the 250-ml conical centrifuge tube. Again clamp the capsule vertically with the inlet end up, and again add enough elution buffer through the inlet fitting to just cover the pleated filter membrane. Replace the inlet cap, transfer the capsule to the shaker clamp, and adjust the capsule in the camp so this time the bleed valve is positioned on a horizontal axis at the 4 o'clock position. Turn on the shaker and agitate the capsule for five more minutes. Leaving the elution solution in the capsule, reposition it in the shaker clamp so that the bleed valve is at the 8 o'clock position. After agitating the capsule for five additional minutes, pour the contents of the capsule into the 250-ml centrifuge tube.

A variation of the filter extraction approach incorporating postfiltration sequential 5% (wt/vol) sodium hexametaphosphate and distilled water washes before the laureth-12 washes has been reported (37). Percent recoveries of *Cryptosporidium* oocysts from tap water improved from an average of 4% without the sodium hexametaphosphate wash steps to 46% with the addition of the sodium hexametaphosphate wash steps.

Centrifuge the 250-ml centrifuge tube containing the capsule filter eluate at 1,500 × *g* for 15 min by using a swinging bucket rotor. Allow the centrifuge to coast to a stop without using the brake. Record the initial pellet volume (volume of solids) and the date and time that concentration was completed on a bench sheet. What follows assumes that the entire pellet will be analyzed. Using a Pasteur pipette, carefully aspirate off the supernatant down to 5 ml above the pellet. If the sample is reagent water (e.g., an IPR or OPR sample), extra care must be taken to avoid aspirating oocysts and cysts during this step. If the packed pellet volume is less than or equal to 0.5 ml, completely resuspend the pellet by vortexing. However, if the packed pellet volume is greater than 0.5 ml, use the following formula to determine the total volume required in the centrifuge tube:

$$\text{total volume (in milliliters) required} = (\text{pellet volume}/0.5 \text{ ml}) \times 5 \text{ ml}$$

(For example, if the packed pellet volume is 0.8 ml, the total volume required is 8 ml.) Add reagent water to the centrifuge tube to bring the total volume to the level calculated above. Vortex the tube for 10 to 15 s to resuspend the pellet. Record this resuspended volume on a bench sheet. Add additional reagent water to the centrifuge tube to bring the volume to an amount evenly divisible by 10, and then vortex the tube for an additional 10 to 15 s. (For example, if the resuspended volume measured is 8 ml, add 2 ml of reagent water to bring the volume to 10 ml.) Record this final resuspended volume on a bench sheet. Process the sample above as multiple, independent 5-ml subsamples from here onward, including the preparation and examination of separate slides for each aliquot.

The IMS procedure is based on the Dynal combo IMS kit (Dynal, Inc., Lake Success, N.Y.) and should be performed on a bench top with all materials at room temperature, ranging from 15 to 25°C. Prepare a 1× dilution of SL-buffer-A from the 10× SL-buffer-A (clear, colorless solution) supplied in the kit by using reagent water. For every 1 ml of 1× SL-buffer-A required, take 100 μl of 10× SL-buffer-A and adjust the volume to 1 ml with the reagent water. A volume of 1.5 ml of 1× SL-buffer-A will be required per sample or subsample on which the Dynal IMS procedure is performed.

To a flat-sided Leighton sample tube (60 by 10 mm), add 1 ml of the 10× SL-buffer-A (as supplied—not the diluted 1× SL-buffer-A). Then add 1 ml of the 10× SL-buffer-B (supplied—magenta solution) to the Leighton sample tube containing the 10× SL-buffer-A. Quantitatively transfer 5 ml of the water sample concentrate to the Leighton sample tube containing the two SL-buffers. Rinse the centrifuge tube twice with 2.5 ml of reagent water and add the rinsate to the Leighton sample tube after each rinse. After addition of the sample and the rinses to the Leighton tube, the volume should not exceed 10 ml. Label the Leighton tube(s) with the sample number. Vortex the Dynabeads *Crypto*-Combo vial from the IMS kit for approximately 10 s to suspend the beads. Ensure that the beads are fully resuspended by inverting the tube and making sure that there is no residual pellet at the bottom. Add 100 μl of the resuspended Dynabeads *Crypto*-Combo beads to each Leighton sample tube containing the water sample concentrate and SL-buffers. Vortex the Dynabeads *Giardia*-Combo vial from the IMS kit for approximately 10 s to suspend the beads. Ensure that the beads are fully resuspended by inverting the tube and making sure that there is no residual pellet at the bottom. Add 100 μl of the resuspended Dynabeads *Giardia*-Combo beads to each Leighton sample tube containing the water sample concentrate, Dynabeads *Crypto*-Combo beads, and SL-buffers. Attach the Leighton sample tube to a Dynal rotating mixer and rotate at approximately 18 rpm for 1 h at room temperature.

After the 1-h incubation period, remove the Leighton sample tube from the rotating mixer and place the tube in the magnetic particle concentrator (MPC-1; Dynal, Inc.) with the flat side of the Leighton tube toward the magnet. Without removing the sample tube from the MPC-1, place the magnet side of the MPC-1 downward, so that the tube is horizontal and the flat side of the tube is facing down. Gently hand-rock the sample tube end to end through approximately 90°, tilting the cap end and base end of the tube up and down in turn. Continue the tilting action for 2 min with approximately one tilt per s. Ensure that the tilting action is continued throughout this period to prevent binding of low-mass, magnetic, or magnetizable material. If the sample in the MPC-1 is allowed to stand motionless for more than 10 s, repeat the tilting oscillation before continuing. Return the MPC-1 to the upright position, with the Leighton sample tube vertical and tube cap on top. Immediately remove the cap and pour off all of the liquid from the tube held in the MPC-1 into a suitable container. Do not shake the tube, and do not remove the tube from MPC-1 until all the liquid has been poured off. Now remove the Leighton sample tube from the MPC-1 and resuspend the sample by using 1 ml of the 1× SL-buffer-A (prepared above from the 10× SL-buffer-A stock). Mix very gently to resuspend all material in the Leighton tube. Do not vortex. Quantitatively transfer all the liquid from the Leighton tube to a labeled, 1.5-ml microcentrifuge tube. Ensure that all of the liquid and beads are transferred.

Place the microcentrifuge tube in the second magnetic particle concentrator (MPC-M; Dynal, Inc.) with its magnetic strip in place. Without removing the microcentrifuge tube from MPC-M, gently rock the tube through 180° by hand. Continue this action for approximately 1 min with approximately one 180° rock per s. At the end of this step, the beads should produce a distinct brown dot at the back of the tube. Immediately aspirate the liquid from the microcentrifuge tube and cap held in the MPC-M. If more than one sample is being processed, conduct three 90° rocking actions before removing the supernatant from each tube. Take care not to disturb the material attached to the wall of the tube adjacent to the magnet. Do not shake the tube. Do not remove the tube from MPC-M while conducting these steps.

To acid dissociate the parasites from the paramagnetic beads, remove the magnetic strip from the MPC-M. Add 50 μl of 0.1 N HCl to each microcentrifuge tube and then vortex vigorously for 10 to 15 s. Place the microcentrifuge tube(s) back in the MPC-M without the magnetic strip in place and allow the unit to stand in a vertical position for at least 10 min at room temperature. Then again remove each microcentrifuge tube from the MPC-M and vortex it vigorously for 5 to 10 s. Ensure that, for each tube, all of the sample is at the base of the tube(s) and again place the microcentrifuge tube back in the MPC-M. Replace the magnetic strip in MPC-M and allow the tube to stand undisturbed for approximately 10 s. Label a Dynal spot-on well slide for sample screening. Add 5 μl of 1.0 N NaOH to a slide sample well. Without removing the microcentrifuge tube from the MPC-M, transfer all of the sample from the microcentrifuge tube in the MPC-M to the sample well with the NaOH. Do not disturb the beads at the back wall of the tube. Ensure that all of the fluid is transferred. This completes the first magnetic bead dissociation.

Air dry the sample(s) onto the well slide(s). If a second magnetic bead dissociation is required (this may enhance recovery of oocysts and cysts in some cases), do not discard the beads or microcentrifuge tube after transferring the sample to a well slide. Instead, at this point, again add 50 μl of 0.1 N HCl to the microcentrifuge tube and then vortex vigorously for 10 to 15 s. Continue the procedure as described in the previous paragraph through transferring the second dissociation product to an unused well on an appropriately labeled well slide and air dry.

A more effective procedure for dissociating the paramagnetic beads from the parasite using heat dissociation has been reported (63). In this procedure the sample in the microcentifuge tube is exposed to a 10-min incubation at 80°C before being placed in the MPC-M. When heat was used instead of acid dissociation, average *Cryptosporidium* oocyst recoveries increased from 49 to 93% in seeded reagent water and from 48 to 73% in seeded river water samples.

Positive and negative staining controls must be prepared for each batch of well slides being stained. For the positive control, pipette 10 μl of positive antigen or 200 to 400 intact *Cryptosporidium* oocysts and 200 to 400 intact *Giardia* cysts to the center of a well and spread them over the well area with a pipette tip. For the negative control, pipette 50 μl of phosphate-buffered saline (PBS) (formulation below) into the center of a well and spread them over the well area with a pipette tip. For each sample also prepare a well on a slide. Air dry the sample and control slides. Apply one drop of a *Giardia* and *Cryptosporidium* combination FITC-labeled monoclonal antibody kit (e.g., Aqua-Glo) to each well. If specified in the antibody kit, apply one drop of counterstain to each well. Spread the staining fluid(s) over the entire well area with an applicator stick, if necessary. Use a different applicator stick for each well. Do not allow the stick to scratch the treated surface of the slide. Place the slides in a humid chamber in the dark, and incubate their contents at room temperature for approximately 30 min. The humid chamber consists of a tightly sealed plastic container containing damp paper towels on top of which the slides are placed. Apply one drop of 1× wash buffer to each well. Tilt each slide on a clean paper towel, long edge down. Using a clean Pasteur pipette, gently aspirate the excess detection reagent and counterstain from below each well. Avoid disturbing the samples. Do not allow the slides to dry. Apply 50 μl of working DAPI staining solution to each well. Working DAPI stain is prepared daily just before use by diluting 10 μl of a stock solution (2 mg of DAPI/ml absolute methanol; store in dark at 4° to 8°C) to 50 ml with PBS (add 1.07 g of Na_2HPO_4, 0.39 g $NaH_2PO_4 \cdot 2H_2O$, and 8.5 g of NaCl to 800 ml of reagent water; dissolve salts and adjust volume to 1 liter with reagent water; adjust pH to 7.2 with NaOH or HCl, as required). DAPI staining efficiency was improved from 49 to 93% in reagent water samples and from 48 to 73% in river water samples when heat rather than acid dissociation was used (63). Allow the DAPI stain to stand at room temperature for approximately 1 min. Then apply one drop of 1× wash buffer or PBS to each well. Tilt each slide on a clean paper towel, long edge down. Gently aspirate the excess DAPI staining solution from below each well by using a clean Pasteur pipette. Avoid disturbing the sample. Do not allow the slides to dry. Add one drop of mounting medium [dissolve 2 g of 1,4-diazabicyclo(2.2.2)octane (DABCO) in 95 ml of warm glycerol-PBS (60% glycerol, 40% PBS); adjust the volume to 100 ml with additional glycerol-PBS after DABCO dissolves] to each well and apply a coverslip. Use a Kimwipe to remove excess mounting fluid from the edges of the coverslip. Seal the coverslips to the respective slides with clear nail polish. Record the date and time that staining was completed on the bench sheet. If slides will not be read immediately, store them in a humid chamber in the dark at 4 to 8°C until ready for examination.

Microscopic Examination

This section assumes use of a microscope, capable of epifluorescence and DIC or Hoffman modulation optics, with stage and ocular micrometers and 20× (numerical aperture = 0.6) to 100× (numerical aperture = 1.3) oil immersion objectives. This section also assumes that the ocular micrometer has been calibrated. The epifluorescence portion of the microscope should be equipped with appropriate excitation and band-pass filters for examining FITC-labeled specimens (exciter filter, 450 to 490 nm; dichroic beam-splitting mirror, 510 nm; and barrier or suppression filter, 515 to 520 nm) and DAPI-labeled specimens (exciter filter, 340 to 380 nm; dichroic beam-splitting mirror, 400 nm; and barrier or suppression filter, 420 nm). If one is using ColorSeed, an additional filter set configured for Texas Red is required. The slides are read by scanning each well in a systematic fashion. Either an up-and-down or a side-to-side scanning pattern may be used (Fig. 3). Before examination of the sample wells, ensure that the positive and negative staining controls reacted appropriately. Make sure that the positive staining control contains oocysts and cysts within the numerical range expected and at the appropriate fluorescence for both FA and DAPI and that the negative staining control does not contain any oocysts or cysts. Use epifluorescence to scan the entire well for each sample at not less than ×200 total magnification for apple-green fluorescence of oocyst and cyst shapes.

When possible *Cryptosporidium* oocysts are found, brilliant apple-green fluorescing ovoid or spherical objects 4 to 6 μm in diameter are observed with brightly highlighted edges. Switch the microscope to the UV filter block for DAPI and examine these objects, and then switch to DIC optics and examine these same objects. If the UV filter block for DAPI is used, the object may exhibit one of the following characteristics: (i) up to four distinct, sky-blue nuclei, (ii) intense blue internal staining, or (iii) light blue internal staining (no distinct nuclei) and a green rim. Conditions (i) and (ii) are recorded as DAPI positive, while (iii) is recorded as DAPI negative. Using DIC, look for external or internal morphological characteristics atypical of *Cryptosporidium* oocysts (e.g., spikes, stalks, appendages, pores, one or two large nuclei filling the cell, red fluorescing chloroplasts, crystals, spores, etc.). If atypical structures are not observed, then categorize each apple-green fluorescing object as being (i) an empty *Cryptosporidium* oocyst-like object, (ii) a *Cryptosporidium* oocyst-like object with amorphous structure, or (iii) a *Cryptosporidium* oocyst with internal structure (one to four sporozoites or nuclei/oocyst). Record the shape and measurements to the nearest 0.5 μm at ×1,000 total magnification for each such object. Although not a defining characteristic, surface oocyst folds may be observed in some specimens (44).

When possible *Giardia* cysts are found, brilliant, apple-green fluorescing, round to oval objects (8 to 18 μm long by 5 to 15 μm wide) are observed with brightly highlighted edges. Switch the microscope to the UV filter block for DAPI and examine these objects, and then switch to DIC optics and examine these same objects. If the UV filter block for DAPI is used, the object may exhibit one or more of the following characteristics: (i) two to four sky-blue nuclei, (ii) intense blue internal staining, or (iii) light blue internal staining (no distinct nuclei) and a green rim. Conditions (i) and (ii) are recorded as DAPI positive, while (iii) is recorded as DAPI negative. Using DIC, look for external or internal morphological characteristics atypical of *Giardia* cysts (e.g., spikes, stalks, appendages, pores, one or two large nuclei filling the cell, red fluorescing chloroplasts, crystals, spores, etc.). If these atypical structures are not observed, then identify such apple-green fluorescing objects of the aforementioned size and shape as presumptive *Giardia* cysts. Record the shape and measurements (to the nearest 0.5 μm at ×1,000 magnification) for each such object as part of the presumptive count. If two or more internal morphological structures are observed at this point, record the object as a confirmed *Giardia* cyst as well. Record presumptive and confirmed results for *Cryptosporidium* oocysts on a *Cryptosporidium* report form and presumptive and confirmed results for *Giardia* cysts on a *Giardia* report form.

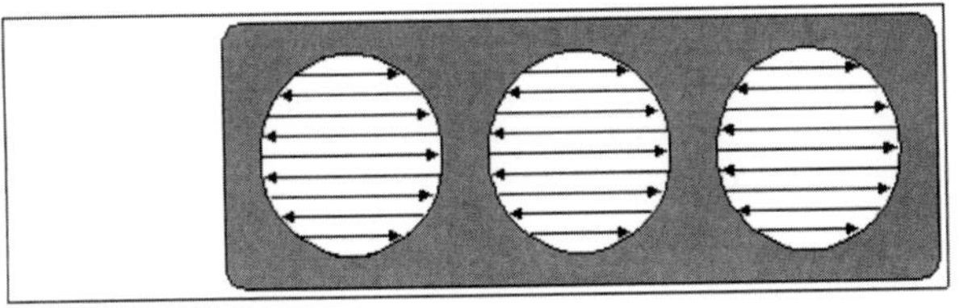

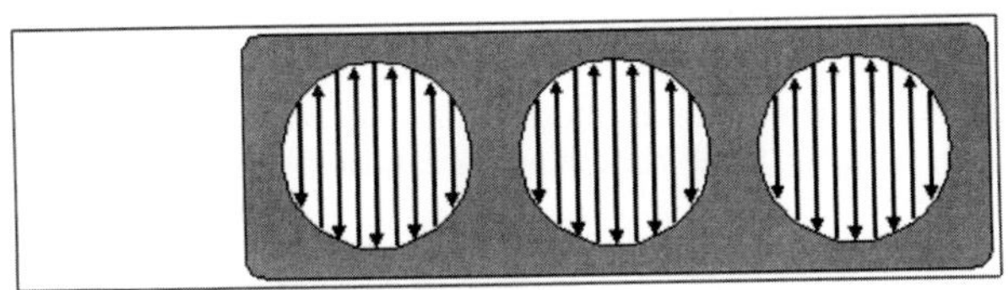

FIGURE 3 Methods for scanning a well slide.

Limitations

Overall, there are several limitations associated with this immunofluorescence detection method. (i) Organism recovery is low. The method's recovery efficiency is influenced by water quality and in particular by the water matrix being sampled (19, 49, 50). (ii) The procedure takes about a half a day to run one sample and requires specialized, expensive equipment. (iii) Nonspecificity of the monoclonal antibodies used in the test kits allows for a number of cross-reactions and potential false positives for both *Giardia* and *Cryptosporidium*. For example, algal cells of the same size and shape as *Giardia* cysts and *Cryptosporidium* oocysts react with the monoclonal antibodies used in this procedure (45). Furthermore, the *Cryptosporidium* and *Giardia* monoclonal antibodies are not species specific. (iv) The method does not determine viability, much less the infectivity of the detected organisms. (v) The procedure does not identify the host of origin. Consequently, it is not known whether the particular isolate is a significant human threat (66). (vi) Large amounts of algae and debris sometimes are recovered from water samples during this procedure, which can physically obscure the observation of the cysts and oocysts on the microscope slides. (vii) Many algae and other animate and inanimate objects have pigments that autofluoresce at the excitation wavelength for FITC and other common fluorochromes. This autofluorescence interferes with cyst and oocyst detection and makes reading of slides more time-consuming and tedious. (viii) This method is dependent on the skill, experience, and training of the microscopist. After detecting objects of the right size, shape, and fluorescence characteristic, the microscopist must demonstrate internal morphological characteristics by DIC or Hoffman modulation microscopy. Unfortunately, the microscopic portion of the analysis can be quite subjective even among experts. Moreover, this type of microscopic work is intense, demanding, and tiring. When the microscopist becomes too tired, the reliability of results becomes questionable.

While these methods have been used successfully on some occasions to confirm epidemiological evidence that a waterborne outbreak has occurred and in surveys of raw (29) and finished (28) water, they do not lend themselves to easy, routine monitoring (24). In an effort to address the limitations of the monitoring methods, the U.S. Environmental Protection Agency convened a Protozoa, Virus and Coliphage Monitoring Workshop of experts in Cincinnati, Ohio, in August 1993. Additional workshops on the protozoan method occurred in Cocoa Beach, Fla., in January 1994 and in Cincinnati in May 1995. Out of these workshops came the recognition that a negative count and low detection limit do not ensure pathogen-free water. In the report from the 1993 workshop are the following recommendations: (i) It was recommended that results obtained by using this method be interpreted with extreme caution. (ii) A training course of 3 or 4 days' duration is required before a laboratory can begin the laboratory approval process. (iii) Microscopic work must be supervised by a senior analyst. Moreover, microscopic work by a single analyst should not exceed 4 h/day or more than 5 consecutive days/week. Intermittent rest periods during the 4 h/day are encouraged.

Alternate Approaches

Recognition of the *Giardia-Cryptosporidium* immunofluorescence assay limitations has stimulated research on using other approaches to detection and identification, including the enzyme-linked immunosorbent assay (ELISA), confocal microscopy, fluorescence in situ hybridization (FISH), the fluorescence-activated cell sorter (FACS), and PCR. Presently, with the exception of ELISA, none of these approaches has been critically validated with environmental and finished water samples.

A number of clinical ELISA kits are commercially available and have been evaluated for their ability to detect spiked *Giardia* cysts and *Cryptosporidium* oocysts in concentrated environmental water samples (14). The detection limit reported for ELISA is 10 cysts or oocysts. Like the antibody-based immunofluorescence assay for these organisms, detection by ELISA suffers from cross-reactions with some algae and is adversely impacted by turbidity. However, because the ELISA technique can be done more rapidly than the immunofluorescence procedure and does not require a skilled microscopist to read the result, it has potential for screening water samples to determine which ones might require more rigorous testing by a more exacting method.

Confocal microscopy in conjunction with FISH has been used to detect *Giardia* cysts and identify them to species level (17). Oligomeric probes for *Giardia* 16S rRNA, which were conjugated to either carboxymethylindocyanine dyes (Cy3 and Cy5) or fluorescein, were used to distinguish among *G. lamblia*, *Giardia muris*, and *Giardia ardeae* cysts. When the fluorochrome FITC was used in conjunction with a detection monoclonal antibody and the rRNA probes, it was possible to detect cysts of *G. lamblia* in fecal samples as well as in environmental samples taken from a sewage lagoon. Observation of the preparations was done by using a krypton-argon laser-equipped confocal microscope, which allowed the simultaneous visualization of all the fluorochromes. The FISH assay can also be evaluated with a standard epifluorescence microscope. A FISH assay also has been developed for *Cryptosporidium* (13). Once oocysts are isolated, they are treated with 50% ethanol and are incubated at 80°C for 10 min to make them permeable. Subsequently the oocysts are incubated with the FISH probe at 48°C for 30 min. Used in conjunction with an FITC-labeled monoclonal antibody, the FISH assay was able to confirm the identity of the oocysts down to species level. The major advance in using this approach is that the occurrence of false positives is reduced and that the need for DIC optics is eliminated. The drawback, however, is the great expense and the skill needed to operate either an epifluorescence microscope or a confocal microscope. If the investigator can forgo the simultaneous observation of two or more fluorochrome labels, then all that is needed is a conventional epifluorescence microscope equipped with the requisite number of filter blocks for the various fluorochrome labels. If a procedure exploiting FISH in conjunction with fluorescent detection antibodies could be developed for use in a FACS, then the microscope time required for examining water samples would be greatly reduced if not eliminated.

A FACS is a very elaborate particle counter with a laser. The FACS, besides being able to sort particles, is able to sense the fluorescence of labeled particles or cells as well as to determine their size. This instrumentation has been used to detect *Giardia* and *Cryptosporidium* in water samples in the United Kingdom and currently is being used in Australia (60) in environmental samples. In combination with calcium carbonate flocculation (61), recoveries of *Giardia* cysts and *Cryptosporidium* oocysts are reported to range from 92 to 104% in spiked samples of wastewater and reservoir and river water. In some instances, the FACS is capable of looking at much more of the sample than the immunofluorescence assay procedure can. Because fluorescent antibodies are used in the FACS procedure and because cross-reactions can occur with other objects, it is recommended that sorted fluorescent objects of the right size and shape be confirmed on an epifluorescence microscope equipped with DIC optics. Microscopic observation is much less fatiguing after FACS processing, as most of the contaminating debris and background fluorescence are eliminated by the FACS. This approach allows most of the sample to be sorted in 5 to 15 min. Moreover, since a greater percentage of the sample is observed than in the immunofluorescence assay procedure, the probability of finding the organisms of interest is greatly increased. The major disadvantages of this technique are the skill required of the FACS operator and the microscopist, the high cost of the instrument, and the quality and reliability of the antibodies.

Solid-phase cytometry is a variation of FACS or flow cytometry. However, unlike flow cytometry, in which particles in a fluid pass by a laser, the particles in solid-phase cytometry are immobilized on either a membrane or a glass slide, which is scanned with a movable laser beam. Image analysis software used in conjunction with this instrumentation discriminates between the various fluorescent particles found during the scan of the solid support and keeps a coordinate record of all the potential positive images. If one uses the microscope attached to this instrument, all the potential positive coordinates can be examined to confirm or deny the detected object. This technology has been used to detect *Cryptosporidium* oocysts in water (32, 42, 46). Solid-phase cytometric analysis replaced traditional microscopic fluorescence analysis in a variation of method 1622 in one study. When 100 oocysts were seeded at the beginning of the solid-phase cytometric variation of method 1622 and the traditional microscopic version of method 1622, average recovery rates were estimated to be 73% for solid-phase cytometry and 49% for the traditional microscopic version of method 1622. Another study suggested that the recovery efficiency of solid-phase cytometry ranged from 82.3 to 86.3%. An advantage of this technique is that it is much faster to perform than the conventional epifluorescence microscopic assay is. Disadvantages include the expense of the instrumentation, the drying and curling of the support membranes before analysis can be completed on some samples, and the inability of the software to deal with samples containing large numbers of organisms.

The detection of *Giardia* cysts (34, 35) and *Cryptosporidium* oocysts by amplifying specific regions of DNA with PCR has been reported. A single *Giardia* cyst was detected by using DNA coding for giardin, an immunodominant surface antigen, as the target for PCR. Moreover, total mRNA increased significantly after induction of excystation in living *Giardia* cysts, making this the first successful molecular viability determination reported for this organism. Discrimination between *Giardia* species which are pathogenic and those which are not pathogenic to humans was also possible with PCR. Primers and probes have been reported for detecting *C. parvum* oocysts as well. Theory suggests that this procedure is just what is needed for detecting organisms in environmental samples that are usually low in protozoan density. However, analysis of environmental samples by this technique can be compromised by the presence of inhibitors like humic acids. Consequently, PCR works well only in highly purified samples. To overcome this problem, it has been suggested that water sample concentrates should be purified by chromatography through Sephadex G-100 and Chelex-100 spin columns (1). To date, successful application of this purification procedure has been reported only for enteric viruses. Attempts to use column purification and sample dilution in conjunction with PCR detection of *Giardia* have not been very successful, and it has been concluded that other inhibitors of PCR are present besides organic compounds like humic acids. Recent purification of water samples by using IMS has helped overcome the inhibitor problem in the detection of both *Giardia* cysts and *Cryptosporidium* oocysts (15, 16, 25, 30, 36, 50, 52, 65). Other problems with PCR are that it is not quantitative and is susceptible to false-positive reactions from organisms that are long dead and deteriorated (34). Sturbaum et al. (52) caution that PCR primers for the *C. parvum* 18S rRNA gene must be very carefully selected; otherwise, false positives from closely related *Cryptosporidium* spp. or dinoflagellates may result. In addition, Champliaud et al. (10) studied the specificities of eight different previously published *C. parvum* sequences targeting 18S rRNA. Their results showed that PCR amplification failed to differentiate *C. parvum* from *Cryptosporidium meleagridis* when the eight DNA sequences were tried. Seeding studies by Reed et al. (41) with various ratios of *C. parvum* and *C. hominis* showed that both species could not always be detected and that amplification of a single species is not a guarantee that the sample contains only that species. These studies indicate that great caution should be used in interpreting PCR results.

A great deal of emphasis has been placed on determining viability by using various fluorochromes like DAPI (9), propidium iodide (47), fluorescein diacetate (48), and SYTO-9 and SYTO-59 (39). Although a correlation between fluorescence intensity and the ability of *C. parvum* oocysts to infect neonatal CD-1 mice in the study with SYTO-9 and SYTO-59 was found, fluorescence profiles varied significantly between batches of oocysts. Another study compared *C. parvum* viability following ozone treatment of the oocysts through use of both the neonatal mouse CD-1 animal infectivity assay and in vitro fluorochrome assays like the SYTO-59 assay (8). In this investigation, which was conducted in laboratories both in the United States and the United Kingdom, in vitro fluorochrome assays of oocysts exposed to ozone showed marginal reduction of oocyst viability. In contrast, the neonatal CD-1 mouse assay demonstrated significant inactivation of the oocysts and failed to correlate with any of the in vitro fluorochrome assays. At best, fluorochrome viability testing appears to be unreliable and variable between laboratories. The search for a good oocyst viability assay has now turned to coupling IMS purification of the water sample with cell culture followed either by fluorescence assay (51) or PCR analysis (16) of developmental stages. Bukhari et al. (7) studied *C. parvum* oocysts counted by flow cytometry in a cell culture FA. Since the flow cytometer-counted oocysts were prepared in another laboratory, they were of infectivity unknown to analysts performing the *C. parvum* cell culture assay. The results indicated a high degree of correlation ($r^2 = 0.89$; $n = 26$) with the values produced by the flow cytometer. Many of the approaches mentioned above have not been multilaboratory validated as of this time for robustness and precision, and numerous questions regarding their sensitivity and specificity exist.

Microscopic particulate analysis (MPA) (55) is a method that has evolved from the immunofluorescence microscopy methods used to detect *Giardia* and *Cryptosporidium*. Initially this method was developed to determine if groundwater was under the influence of surface water. In this protocol 500 gal (1,893 liters) of water is sampled by filtering over an 8- to 24-h period with a depth filter. The particulates retained by the filter fibers then are extracted by using particle-free water containing Tween, either by hand washing or using a stomacher blender (Tekmar-Dohrmann, Cincinnati, Ohio). The recovered particulates subsequently are concentrated by centrifugation. Samples with particulate volumes of less than 20 μl/100 gal (379 liters) are observed directly by microscopic examination, while samples with particulate volumes of greater than 20 μl/100 gal are purified by flotation on Percoll-sucrose (specific gravity = 1.15) prior to microscopic examination. Unlike what is done in the parent immunofluorescence procedure, the particulates obtained by this procedure are not stained with monoclonal antibodies. Instead, staining with Lugol's iodine may or may not be done depending upon whether bright-field microscopy is being used. Bright-field microscopy definitely requires the use of Lugol's iodine staining. Observation of the particulates is done by using bright-field, phase-contrast, or DIC microscopy. All microorganisms in the 1- to 400-μm size range are counted and measured with a calibrated ocular micrometer. These microorganisms are referred to as bioindicators in this method and include diatoms, algae, *Giardia* cysts, coccidian oocysts, plant debris, pollen, rotifers, crustaceans, amoebae, nematodes, insect parts, and insect larvae. The bioindicators are identified to at least taxonomic phylum and hopefully to class. In addition, inanimate particulates are tallied. Identification of *Giardia* cysts, coccidian oocysts, and/or enteric helminths definitely is indicative of groundwater being under the influence of surface water. The occurrence of pigment bearing diatoms and algae also is indicative of groundwater being under the influence of surface water. While the list of bioindicators above is long, there was little agreement among individuals using this method as to what level of risk that they represent. To overcome this problem, a consensus MPA protocol (55) categorizes each bio-indicator into relative risk factor groups. On the basis of each bioindicator's density and risk factor category, the water is mathematically rated as to the level of risk for being under the influence of surface water. Recently the MPA method has been adapted to evaluate water filtration plant efficiency. This is done by looking at pre- and postfilter water samples. In experienced hands, this method has great value. It, however, requires much greater expertise than does the immunofluorescence assay method from which it evolved. Analysts, besides needing extensive microscopic experience, need a strong background in limnology, parasitology, phycology, and invertebrate zoology. Even though this method is very laborious, it still provides a more detailed analysis of the water than is possible from more automated approaches like particle counting.

EMERGING PROTOZOANS OF POTENTIAL HEALTH SIGNIFICANCE IN WATER

Of all the cases of waterborne gastroenteritis, only about 50% are attributable to a specific microorganism, toxin, or chemical. Until the 1970s *Giardia* was thought to be an innocuous commensal, and *Cryptosporidium* was basically an unknown. With vigilant epidemiology and the ability to monitor these protozoans, we have come to appreciate their significance. Like the emergent *Cryptosporidium* concerns in the 1980s, there are now clinical reports of gastroenteritis caused by microsporidians, *Cyclospora*, *Isospora* spp., and *Blastocystis hominis* in both immunocompetent and immunocompromised individuals. Unlike the aforementioned parasites, which remain within the gastrointestinal tract, *T. gondii* starts in humans by ingestion but becomes systemic. Since the majority of these organisms are enteric, the question is whether these emerging protozoans are of potential health significance in water? At present, other than clinical detection approaches, there are no protocols for detecting them in environmental samples. This, however, does not preclude adaptation of existent techniques for *Giardia* and *Cryptosporidium* to look for them in water.

Although rare, microsporidian infections are now being reported from immunocompetent people. Presently their classification is done on the basis of spore size (1.5 to 5 μm), nuclear configuration, the number of polar tube coils within the spore and developing forms, and the host cell-parasite relationship. The life cycle of these parasites is thought to be direct by either ingestion, inhalation, or inoculation. Of particular interest are the *Enterocytozoon* spp., which in addition to being found in humans have also been reported from wild mammals that serve as reservoir hosts (53). Two of these enteric forms, *E. bieneusi* and *Encephalitozoon intestinalis*, may be transmitted by the water route. Furthermore, *Encephalitozoon cuniculi* and *Encephalitozoon hellem* involve a more complicated life cycle, as they have been recovered from the central nervous system and eye, respectively. They can disseminate to the urinary tract from which spores can be released to the environment. Consequently, *Encephalitozoon* spp. are likely candidates for waterborne transmission as well. Since microsporidians have been detected in clinical specimens with indirect fluorescent poly- and monoclonal antibodies (2), this approach could be adapted for detecting them in water samples in a fashion similar to the immunofluorescence assay used for *Giardia* and *Cryptosporidium*. A FISH probe has been reported for detecting *E. hellem* in water samples (21). In addition, real-time PCR assays for the detection of *E. hellem*, *E. intestinalis*, and *E. cuniculi* have been published (22). Unfortunately, these advances are assays and not complete environmental detection methods. Consequently, utilization of these techniques will require further development.

Cyclospora, which is an apicomplexan, has been incriminated as a cause of prolonged diarrhea in the Third World and more recently in the developed countries. The oocysts are 8 to10 μm in diameter and have the acid-fast staining characteristic, as do *Cryptosporidium* and *Isospora*. However, the oocysts of *Cyclospora* are autofluorescent when viewed by ultraviolet microscopy. Although a real-time PCR assay for *Cyclospora cayetanensis* is now available (58), it has not been validated.

Three species of the apicomplexan *Isospora* are reported to be parasitic in humans. They are *Isospora natalensis*, *I. hominis*, and *I. belli*. *I. natalensis* is rare, occurs only in South Africa, and will not be discussed further here. Confusion exists regarding the status of *I. belli* and *I. hominis*, as some investigators believe them to be the same organism. There is even a report which suggests that *I. hominis* is really a species of *Sarcocystis*. Unlike *C. parvum* oocysts, *I. belli* oocysts are not fully sporulated at the time of passing from the host. When sporulation is complete, which requires about 48 h at room temperature, the oocyst has two sporocysts, each containing four crescent-shaped sporozoites. Overall, *I. belli* oocysts are elongated oval structures, measuring 20 to 30 μm in length by 10 to 19 μm in width, with both ends being somewhat narrow. Oocysts of *I. belli* can be detected on the basis of their shape and ability to be acid-fast stained. Like *Cryptosporidium* oocysts and *Giardia* cysts, *I. belli* oocysts can be viewed with either phase-contrast or DIC microscopy. Although the theory is unproven, dogs have been suspected of being a reservoir host for *I. belli*.

I. hominis oocysts, which are ovoid and 25 to 33 μm in length, are fully sporulated and contain two sporocysts before being passed from the host. At the time of passage, the oocyst has broken open, and ripe sporocysts containing four sporozoites are passed in the feces. The sporocysts measure around 14 μm in length. Viewing them can be aided by acid-fast staining for bright-field microscopy, or they can be viewed unstained by either phase-contrast or DIC microscopy.

T. gondii is a coccidian parasite reported to be the leading cause of morbidity and mortality in immunocompromised individuals, particularly in AIDS patients and unborn fetuses. Waterborne transmission has been reported in Panama; Brazil; and British Columbia, Canada (4, 5, 6). Oocysts of *T. gondii* are 10 or 11 μm in size and are shed from felids, the definitive host. Under UV light, they fluoresce. After sporulation, *T. gondii* oocysts contain two sporocysts, each of which contains four sporozoites. Although there are molecular and other assays for detecting this pathogen in clinical samples, methods for detecting *Toxoplasma* oocysts in water have yet to be developed.

B. hominis is a yeast-like organism of questionable taxonomic status. Depending upon the source or expert consulted, *B. hominis* either may or may not be classified as a pathogen. Generally those classifying it as a pathogen do so because drug treatment eliminates this organism as well as the mild gastroenteritis that is present, when no other etiologic agent can be detected. Often confused with amoebic cysts, *B. hominis* is found in various sizes and has a large vacuole surrounded by peripheral granules in addition to a single nucleus. Visualization can be by trichrome stain in conjunction with bright-field microscopy.

SUMMARY

Protozoans are eukaryotic organisms that maintain either a free-living or parasitic existence. Some free-living forms, under the right conditions, can become opportunistic parasites. Enteric pathogenic protozoans, like *Giardia* and *Cryptosporidium*, that are now known to be transmitted by water have been responsible for numerous waterborne outbreaks of gastroenteritis. The primary means for their environmental detection, since density levels in water are low, involves processing a large volume of water by filtration, extracting the particulates from the filter, concentrating the organisms from the particulates, and assaying for the pathogens. The most widely used method for detecting protozoans has been the indirect immunofluorescence assay. While method 1623 has improved upon the utility of the immunofluorescence assays for *Giardia* and *Cryptosporidium*, the procedure is still labor-intensive and highly dependent on the skill of the microscopist. Even with the improvements to date, this technique is known to have a number of deficiencies, including false positives, inability to determine the viability and species of the detected organisms, and low average percent recovery of cysts and oocysts.

Various attempts have been made to improve the immunofluorescence detection method. Rather than sampling by filtration and buoyant density centrifugation, efforts to purify the organisms by means of carbonate flocculation are reported to improve recoveries. PCR, cell culture, FACS use, and solid-phase cytometry are currently being evaluated as alternate test procedures. As each of these approaches is relatively new and much more research is needed, it remains to be seen whether they will be equal to or better than the current fluorescence assay procedure.

The U.S. Environmental Protection Agency, through its Office of Research and Development, funded and managed the research described here. It has been subjected to the agency's administrative review and approved for publication as an EPA document.

REFERENCES

1. **Abbaszadegan, M., M. S. Huber, C. P. Gerba, and I. L. Pepper.** 1993. Detection of enteroviruses in groundwater with the polymerase chain reaction. *Appl. Environ. Microbiol.* **59:**1318–1324.
2. **Aldras, A. M., J. M. Orenstein, D. P. Kotler, J. A. Shadduck, and E. S. Didier.** 1994. Detection of microsporidia by indirect immunofluorescence antibody test using polyclonal and monoclonal antibodies. *J. Clin. Microbiol.* **32:**608–612.
3. **American Society for Testing and Materials.** 1992. Proposed method for *Giardia* cysts and *Cryptosporidium* oocysts in low-turbidity water by a fluorescent antibody procedure, p. 925–935. *In Annual Book of ASTM Standards 11.01.* American Society for Testing and Materials, Philadelphia, Pa.
4. **Bahia-Oliveira, L. M., J. L. Jones, J. Azevedo-Silva, C. C. Alves, F. Orefice, and D. G. Addiss.** 2003. Highly endemic, waterborne toxoplasmosis in north Rio de Janeiro state, Brazil. *Emerg. Infect. Dis.* **9:**55–62.
5. **Benenson, M. W., E. T. Takafuji, S. M. Lemon, R. L. Greenup, and A. J. Sulzer.** 1982. Oocyst-transmitted toxoplasmosis associated with ingestion of contaminated water. *N. Engl. J. Med.* **307:**666–669.
6. **Bowie, W. R., A. S. King, D. H. Werker, J. L. Isaac-Renton, A. Bell, S. B. Eng, S. A. Marion, et al.** 1997. Outbreak of toxoplasmosis associated with municipal drinking water. *Lancet* **350:**173–177.
7. **Bukhari, Z., D. M. Holt, M. W. Ware, and F. W. Schaefer III.** 2005. Evaluating in vitro infectivity for measuring UV disinfection of *Cryptosporidium parvum* oocysts in finished water, p. 1–11. *In Proceedings of the 2005 Water Technology Conference, Quebec City, Quebec, Canada, November 6–10, 2005.* American Water Works Association, Denver, Colo.
8. **Bukhari, Z., M. M. Marshall, D. G. Korich, C. R. Fricker, H. V. Smith, J. Rosen, and J. L. Clancy.** 2000. Comparison of *Cryptosporidium parvum* viability and infectivity assays following ozone treatment of oocysts. *Appl. Environ. Microbiol.* **66:**2972–2980.
9. **Campbell, A. T., L. J. Robertson, and H. V. Smith.** 1992. Viability of *Cryptosporidium parvum* oocysts: correlation of in vitro excystation with inclusion or exclusion of fluorogenic vital dyes. *Appl. Environ. Microbiol.* **58:**3488–3493.
10. **Champliaud, D., P. Gobet, M. Naciri, O. Vagner, J. Lopez, J. C. Buisson, I. Varga, G. Harly, R. Mancassola, and A. Bonnin.** 1998. Failure to differentiate *Cryptosporidium parvum* from *C. meleagridis* based on amplification of eight DNA sequences. *Appl. Environ. Microbiol.* **64:**1454–1458.
11. **Craun, G. F.** 1990. Waterborne giardiasis, p. 267–293. *In* E. A. Meyer (ed.), *Giardiasis.* Elsevier Science Publishers, Amsterdam, The Netherlands.
12. **Dagget, P.-M.** 1982. Algae and protozoa, p. 31–56, 601–656. *In* H. D. Hall (ed.), *Catalog of Strains, I,* 15th ed. American Type Culture Collection, Rockville, Md.
13. **Deere, D., G. Vesey, M. Milner, K. Williams, N. Ashbolt, and D. Veal.** 1998. Rapid method for fluorescent in situ ribosomal RNA labelling of *Cryptosporidium parvum. J. Appl. Microbiol.* **85:**807–818.
14. **de la Cruz, A. A., and M. Sivaganesan.** 1995. Detection of *Giardia* and *Cryptosporidium* spp. in source water samples by commercial enzyme-immunosorbent assay, p. 543–554. *In Proceedings of the 1994 Water Technology Conference, November 6–10, 1994.* American Water Works Association, Denver, Colo.
15. **Deng, M. Q., D. O. Cliver, and W. T. Mariam.** 1997. Immunomagnetic capture PCR to detect viable *Cryptosporidium parvum* oocysts from environmental samples. *Appl. Environ. Microbiol.* **63:**3134–3138.
16. **Di Giovanni, G. D., F. H. Hashemi, N. J. Shaw, F. A. Abrams, M. W. LeChevallier, and M. Abbaszadegan.** 1999. Detection of infectious *Cryptosporidium parvum* oocysts in surface and filter backwash water samples by immunomagnetic separation and integrated cell culture-PCR. *Appl. Environ. Microbiol.* **65:**3427–3432.
17. **Erlandsen, S. L., H. van Keulen, A. Gurien, W. Jakubowski, F. W. Schaefer III, P. Wallis, D. Feely, and E. Jarroll.** 1994. Molecular approach to speciation and detection of *Giardia*: fluorochrome-rDNA probes for identification of *Giardia lamblia, Giardia muris,* and *Giardia ardeae* in laboratory and environmental samples by in situ hybridization, p. 64–66. *In* R. C. A. Thompson, J. A. Reynoldson, and A. J. Lymbery (ed.), *Giardia: from Molecules to Disease.* CAB International, Wallingford, Oxon, United Kingdom.
18. **Fields, B. S., G. N. Sanden, J. M. Barbaree, W. E. Morrill, R. M. Wadowsky, E. H. White, and J. C. Feeley.** 1989. Intracellular multiplication of *Legionella pneumophila* in amoebae isolated from hospital hot water tanks. *Curr. Microbiol.* **18:**131–137.
19. **Francy, D. S., O. D. Simmons III, M. W. Ware, E. J. Granger, M. D. Sobsey, and F. W. Schaefer III.** 2004. Effects of seeding procedures and water quality on recovery of *Cryptosporidium* oocysts from stream water by using U.S. Environmental Protection Agency method 1623. *Appl. Environ. Microbiol.* **70:**4118–4128.
20. **Hayes, E. B., T. D. Matte, T. R. O'Brien, T. W. McKinley, G. S. Logsdon, J. B. Rose, B. L. P. Ungar, D. M. Word, P. F. Pinsky, M. L. Cummings, M. A. Wilson, E. G. Long, E. S. Hurwitz, and D. D. Juranek.** 1989. Large community outbreak of cryptosporidiosis due to contamination of a filtered public water supply. *N. Engl. J. Med.* **320:**1372–1376.
21. **Hester, J. D., H. D. Lindquist, A. M. Bobst, and F. W. Schaefer III.** 2000. Fluorescent in situ detection of *Encephalitozoon hellem* spores with a 6-carboxyfluorescein-labeled ribosomal RNA-targeted oligonucleotide probe. *J. Eukaryot. Microbiol.* **47:**299–308.
22. **Hester, J. D., M. Varma, A. M. Bobst, M. W. Ware, H. D. A. Lindquist, and F. W. Schaefer III.** 2002. Species-specific detection of three human-pathogenic microsporidial species from the genus *Encephalitozoon* via fluorogenic 5′ nuclease PCR assays. *Mol. Cell. Probes* **16:**435–444.

23. **Jahn, T. L., and F. F. Jahn.** 1979. *How to Know the Protozoa.* William C. Brown Publishers, Dubuque, Iowa.
24. **Jakubowski, W.** 1984. Detection of *Giardia* cysts in drinking water, p. 263–286. *In* S. L. Erlandsen and E. A. Meyer (ed.), *Giardia and Giardiasis: Biology, Pathogenesis, and Epidemiology.* Plenum Press, New York, N.Y.
25. **Johnson, D. W., N. J. Pienaszek, D. W. Griffin, L. Misener, and J. B. Rose.** 1995. Development of a PCR protocol for sensitive detection of *Cryptosporidium* oocysts in water. *Appl. Environ. Microbiol.* **61:**3849–3855.
26. **Krogstad, D. J., G. S. Visvesvara, K. W. Walls, and J. W. Smith.** 1985. Blood and tissue protozoa, p. 612–630. *In* E. H. Lennette, A. Balows, W. J. Hausler, Jr., and H. J. Shadomy (ed.), *Manual of Clinical Microbiology,* 4th ed. American Society for Microbiology, Washington, D.C.
27. **Kudo, R. R.** 1966. *Protozoology, 5th ed.* Charles C Thomas, Springfield, Ill.
28. **LeChevallier, M. W., W. D. Norton, and R. G. Lee.** 1991. *Giardia* and *Cryptosporidium* spp. in filtered drinking water supplies. *Appl. Environ. Microbiol.* **57:**2617–2621.
29. **LeChevallier, M. W., W. D. Norton, and R. G. Lee.** 1991. Occurrence of *Giardia* and *Cryptosporidium* spp. in surface water supplies. *Appl. Environ. Microbiol.* **57:**2610–2616.
30. **LeChevallier, M. W., W. D. Norton, J. E. Siegel, and M. Abbaszadegan.** 1995. Evaluation of the immunofluorescence procedure for detection of *Giardia* cysts and *Cryptosporidium* oocysts in water. *Appl. Environ. Microbiol.* **61:** 690–697.
31. **Lee, J. J., S. H. Hutner, and E. C. Lee.** 1985. *An Illustrated Guide to the Protozoa.* Society of Protozoologists, Lawrence, Kans.
32. **Lindquist, H. D. A., M. W. Ware, R. E. Stetler, L. J. Wymer, and F. W. Schaefer III.** 2001. A comparison of four fluorescent antibody-based methods for purifying, detecting, and confirming *Cryptosporidium parvum* in surface waters. *J. Parasitol.* **87:**1124–1131.
33. **MacKenzie, W. R., N. J. Hoxie, M. E. Proctor, M. S. Gradus, K. A. Blair, D. E. Peterson, J. J. Kazmierczak, D. G. Addiss, K. R. Fox, J. B. Rose, and J. P. Davis.** 1994. A massive outbreak in Milwaukee of *Cryptosporidium* infection transmitted through the public water supply. *N. Engl. J. Med.* **331:**161–167.
34. **Mahbubani, M. H., A. K. Bej, M. Perlin, F. W. Schaefer III, W. Jakubowski, and R. M. Atlas.** 1991. Detection of *Giardia* cysts by using polymerase chain reaction and distinguishing live from dead cysts. *Appl. Environ. Microbiol.* **57:**3456–3461.
35. **Mahbubani, M. H., A. K. Bej, M. H. Perlin, F. W. Schaefer III, W. Jakubowski, and R. M. Atlas.** 1992. Differentiation of *Giardia duodenalis* from other *Giardia* spp. by using polymerase chain reaction and gene probes. *J. Clin. Microbiol.* **30:**74–78.
36. **Mahbubani, M. H., F. W. Schaefer III, D. D. Jones, and A. K. Bej.** 1998. Detection of *Giardia* in environmental waters by immuno-PCR amplification methods. *Curr. Microbiol.* **36:**107–113.
37. **McCuin, R. M., T. M. Hargey, J. E. Amburgy, and J. L. Clancy.** 2001. Improving methods for isolation of *Cryptosporidium* oocysts and *Giardia* cysts from source and filtered water, p. 1–6. *In Proceedings of AWWA's Water Quality Technology Conference, November 11–15, 2001, Nashville, Tenn.* American Water Works Association, Denver, Colo.
38. **Musial, C. E., M. J. Arrowood, C. R. Sterling, and C. P. Gerba.** 1987. Detection of *Cryptosporidium* in water by using polypropylene cartridge filters. *Appl. Environ. Microbiol.* **53:**687–692.
39. **Neumann, N. F., L. L. Gyurek, L. Gammie, G. R. Finch, and M. Belosevic.** 2000. Comparison of animal infectivity and nucleic acid staining for assessment of *Cryptosporidium parvum* viability in water. *Appl. Environ. Microbiol.* **66:**406–412.
40. **Radulescu, S., and E. A. Meyer.** 1990. In vitro cultivation of *Giardia* trophozoites, p. 99–110. *In* E. A. Meyer (ed.), *Giardiasis.* Elsevier Science Publishers, New York, N.Y.
41. **Reed, C., G. D. Sturbaum, P. J. Hoover, and C. R. Sterling.** 2002. *Cryptosporidium parvum* mixed genotypes detected by PCR-restriction fragment length polymorphism analysis. *Appl. Environ. Microbiol.* **68:**427–429.
42. **Reynolds, D. T., R. B. Slade, N. J. Sykes, A. Jonas, and C. R. Fricker.** 1999. Detection of *Cryptosporidium* oocysts in water: techniques for generating precise recovery data. *J. Appl. Microbiol.* **87:**804–813.
43. **Rivera, F., G. Roy-Ocotla, I. Rosas, E. Ramirez, P. Bonilla, and F. Lares.** 1987. Amoeba isolated from the atmosphere of Mexico City and environs. *Environ. Res.* **42:**149–154.
44. **Robertson, L. J., A. T. Campbell, and H. V. Smith.** 1993. Induction of folds or sutures on the walls of *Cryptosporidium parvum* oocysts and their importance as a diagnostic feature. *Appl. Environ. Microbiol.* **59:**2638–2641.
45. **Rodgers, M. R., C. M. Bernardino, and W. Jakubowski.** 1993. A comparison of methods for extracting amplifiable *Giardia* DNA from various environmental samples. *Water Sci. Technol.* **27:**85–88.
46. **Rushton, P., B. M. Place, and N. F. Lightfoot.** 2000. An evaluation of a laser scanning device for the detection in *Cryptosporidium parvum* in treated water samples. *Lett. Appl. Microbiol.* **30:**303–307.
47. **Sauch, J. F., D. Flanigan, M. L. Galvin, D. Berman, and W. Jakubowski.** 1991. Propidium iodide as an indicator of *Giardia* cyst viability. *Appl. Environ. Microbiol.* **57:**3243–3247.
48. **Schupp, D. G., and S. L. Erlandsen.** 1987. A new method to determine *Giardia* cyst viability: correlation of fluorescein diacetate and propidium iodide staining with animal infectivity. *Appl. Environ. Microbiol.* **53:**704–707.
49. **Simmons, O. D., III, M. D. Sobsey, F. W. Schaefer III, D. S. Francy, R. Nally, and C. D. Heaney.** 2001. Evaluation of USEPA Method 1622 for detection of *Cryptosporidium* oocysts in stream waters. *J. Am. Water Works Assoc.* **93:**78–87.
50. **Simmons, O. D., III, M. D. Sobsey, C. D. Heaney, F. W. Schaefer III, and D. S. Francy.** 2001. Concentration and detection of *Cryptosporidium* oocysts in surface water samples by method 1622 using ultrafiltration and capsule filtration. *Appl. Environ. Microbiol.* **67:**1123–1127.
51. **Slifko, T. R., D. Friedman, J. B. Rose, and W. Jakubowski.** 1997. An in vitro method for detecting infectious *Cryptosporidium* oocysts with cell culture. *Appl. Environ. Microbiol.* **63:**3669–3675.
52. **Sturbaum, G. D., P. T. Klonicki, M. M. Marshall, B. H. Jost, B. L. Clay, and C. R. Sterling.** 2002. Immunomagnetic separation (IMS)-fluorescent antibody detection and IMS-PCR detection of seeded *Cryptosporidium parvum* oocysts in natural waters and their limitations. *Appl. Environ. Microbiol.* **68:**2991–2996.
53. **Sulaiman, I. M., R. Fayer, A. A. Lal, J. M. Trout, F. W. Schaefer III, and L. Xiao.** 2003. Molecular characterization of microsporidia indicates that wild mammals harbor host-adapted *Enterocytozoon* spp. as well as human-pathogenic *Enterocytozoon bieneusi. Appl. Environ. Microbiol.* **69:**4495–4501.
54. **Upton, S. J., M. Tilley, and D. B. Brillhart.** 1994. Comparative development of *Cryptosporidium parvum* (Apicomplexa) in 11 continuous host cell lines. *FEMS Microbiol. Lett.* **118:**233–236.
55. **U.S. Environmental Protection Agency.** 1992. Consensus method for determining groundwaters under the direct influence of surface water using microscopic particulate

analysis (MPA). EPA 910/9–92–029. U.S. Environmental Protection Agency, Washington, D.C.

56. **U.S. Environmental Protection Agency.** 1996. ICR protozoan method for detecting *Giardia* cysts and *Cryptosporidium* oocysts in water by a fluorescent antibody procedure, p. VII-1–VII-44. *In ICR Microbial Laboratory Manual.* EPA/600/R–95/1/8. U.S. Environmental Protection Agency, Washington, D.C.
57. **U.S. Environmental Protection Agency.** 2001. Method 1623: *Cryptosporidium* and *Giardia* in water by filtration/IMS/FA. EPA-821-R-01-025. U.S. Environmental Protection Agency, Washington, D.C.
58. **Varma, M., J. D. Hester, F. W. Schaefer III, M. W. Ware, and H. D. A. Lindquist.** 2003. Detection of *Cyclospora cayetanensis* using a quantitative real-time PCR assay. *J. Microbiol. Methods* **53:**27–36.
59. **Vasconcelos, J. (U.S. Environmental Protection Agency, Region X).** 1992. Personal communication.
60. **Vesey, G., P. Hutton, A. Champion, N. Ashbolt, K. L. Williams, A. Warton, and D. Veal.** 1994. Application of flow cytometric methods for the routine detection of *Cryptosporidium* and *Giardia* in water. *Cytometry* **16:**1–6.
61. **Vesey, G., J. S. Slade, M. Byrne, K. Shepard, and C. R. Fricker.** 1993. A new method for the concentration of *Cryptosporidium* oocysts from water. *J. Appl. Bacteriol.* **75:**82–86.
62. **Walker, P. L., P. Prociv, W. G. Gardiner, and D. E. Moorhouse.** 1986. Isolation of free-living amoebae from air samples and an air conditioner filter in Brisbane. *Med. J. Aust.* **145:**3–4.
63. **Ware, M. W., L. Wymer, H. D. A. Lindquist, and F. W. Schaefer III.** 2003. Evaluation of an alternative IMS dissociation procedure for use with Method 1622: detection of *Cryptosporidium* in water. *J. Microbiol. Methods* **55:**575–583.
64. **Ware, M. W., and F. W. Schaefer III.** 2005. The effects of time and temperature on flow cytometry enumerated live *Cryptosporidium parvum* oocysts. *Lett. Appl. Microbiol.* **41:** 385–389.
65. **Webster, K. A., J. D. E. Pow, M. Giles, J. Catchpole, and M. J. Woodward.** 1993. Detection of *Cryptosporidium parvum* using a specific polymerase chain reaction. *Vet. Parasitol.* **50:**35–44.
66. **Zhou, L., R. Fayer, J. M. Trout, U. M. Ryan, F. W. Schaefer III, and L. Xiao.** 2004. Genotypes of *Cryptosporidium* species infecting fur-bearing mammals differ from those of species infecting humans. *Appl. Environ. Microbiol.* **70:**7574–7577.

Microbial Indicators of Marine Recreational Water Quality

STEPHEN B. WEISBERG, RACHEL T. NOBLE, AND JOHN F. GRIFFITH

22

Recreational swimming is a popular activity, with more than 600 million visitor-days to the beach annually in California alone (48). However, coastal waters are susceptible to contamination from urban runoff or other discharges and can serve as a vehicle for transmission of microbial pathogens (65). As a result, considerable resources are expended each year to measure water quality at recreational beaches (77). The need for such monitoring in the United States was recently formalized in the Beaches Environmental Assessment and Coastal Health (BEACH) Act, which requires coastal states to monitor beach water quality and warn the public when beach waters contain dangerously high levels of disease-causing microorganisms.

Most of this monitoring is based on measurement of fecal indicator bacteria. Indicator bacteria are not necessarily pathogenic but are found abundantly in human fecal material, where pathogenic organisms, such as viruses, are likely to exist. Indicator bacteria are often the main component of recreational water quality monitoring programs because they are cheaper, safer, and easier to measure than pathogens.

Historically, coliform bacteria have been the indicator of choice. This dates back to the late 1800s when Escherich first determined that coliform bacteria were numerous and always detected in feces and sewage. Klein and Houston (50) subsequently reported that they could dilute sewage more than 10,000-fold and still detect the presence of coliform bacteria, documenting a higher sensitivity and specificity as a marker of sewage than any other known physical or chemical tests. Stevenson (85) later conducted the first epidemiological study in the United States at sites on Lake Michigan, the Ohio River, and Long Island Sound. While there was an inconsistent correlation between illness and levels of coliform bacteria across study sites, data from the Lake Michigan beaches suggested increased swimming-associated health risks at geometric mean total coliform (TC) levels of 2,300 per 100 ml, which became the basis for the first bacteriological water quality criteria.

More recently, *Enterococcus* has become the preferred indicator organism for marine waters in the United States. This resulted from epidemiological studies conducted by the U.S. Environmental Protection Agency (USEPA) in the 1970s, which used more-rigorous definitions of illness than the earlier Stevenson studies and compared the efficacy of numerous candidate microbial indicators (89). These studies found that *Escherichia coli* correlated best with gastrointestinal illness in fecally contaminated fresh waters, while *Enterococcus* correlated best with illness in marine water. This led the USEPA to recommend a geometric mean criterion of 35 *Enterococcus* organisms per 100 ml in marine water, corresponding to a gastrointestinal illness rate of 19 per 1,000 swimmers (89). Prüss (72) and Wade et al. (91) subsequently conducted reviews of beachgoer epidemiology studies around the world and found that *Enterococcus* had the best relationship to health risk among presently used indicators for marine water, though less than one-half of the studies found a significant health relationship for *Enterococcus* and the dose-response curves were highly variable. This inconsistency among epidemiology study results may be due to geographic variability and differences in the sources of the contamination among studies.

The adoption of *Enterococcus* as an indicator has led to an increase in the number of beach warnings compared to when coliforms were used as the indicator. Noble et al. (67) measured TC, fecal coliform (FC) or *E. coli*, and *Enterococcus* simultaneously at California beaches and found that there was a sevenfold-higher rate of instances of water quality criteria being exceeded associated with *Enterococcus* than with the other indicators. Kinzelman et al. (49) had similar findings when comparing *Enterococcus* and *E. coli* in freshwater. This may be because *Enterococcus* survives longer in the environment than *E. coli* does, which is a positive attribute if its survival is more similar to that of pathogens.

One concern with the use of *Enterococcus* as an indicator is that the epidemiology studies on which it was based were conducted at locations where sewage was the primary source of fecal contamination. However, with improved sewage treatment systems, most failures of beach water quality standards in the United States are now associated with nonpoint source contamination (78). Since nonpoint sources generally have a higher percentage of animal fecal contributions and animals can shed bacterial indicators without some of the accompanying human pathogens, there is considerable uncertainty in extrapolating present standards to nonpoint source situations. Colford et al. (14) conducted an epidemiology study at a location where almost 20% of the *Enterococcus* measurements exceeded standards, but most of the fecal contamination was found to be from avian sources. They found no relationship between *Enterococcus* concentration and human illness. Similarly, poor correlations between bacterial indicator and virus concentrations have been found in areas where nonpoint sources are dominant (44, 66) while *Enterococcus* has been significantly correlated with viruses when a human source, such as septic tanks, has been present (57).

As a result, several states use more than one bacterial indicator, employing a toolbox approach advocated by the NRC (65). For instance, Hawaii also measures *Clostridium perfringens*, which is thought to be more specific to human sources there since *Enterococcus* has been found to grow in tropical soils (9, 83). The state of California measures three indicators, *Enterococcus*, FC and TC, based on a California-specific epidemiological study conducted at sites subject to nonpoint fecal sources (34).

PRESENT BACTERIAL MEASUREMENT METHODS

Present methods to enumerate indicator bacteria in marine coastal waters rely on growth-based assays. Three classes of growth-based assays are currently in use: multiple-tube fermentation (MTF), membrane filtration (MF), and defined substrate (DS) methods. These methods are similar in that they all depend on a metabolic endpoint reached after a period of incubation. They differ in the length of the incubation required, the materials and equipment needed to perform the test, their ease of use, their ability to measure highly turbid samples, the different subgroups of indicator bacteria they detect, and the manner in which bacterial concentrations are calculated.

Multiple-Tube Fermentation

MTF methods use probability statistics to determine the mean concentration of bacteria as the most probable number (MPN) per 100 ml. MTF is generally conducted in three sequential phases (presumptive, confirmatory, and completed), with each phase requiring 1 to 2 days of incubation. In the initial or presumptive phase, 3 volumes of samples (10, 1, and 0.1 ml) are inoculated into 3, 5, or 10 tubes containing bacteriological medium constituted to allow the target bacteria to grow. The test assumes that any single viable target bacterium in the inoculum will result in growth and a positive reaction in the medium. To recover coliform bacteria, fermentation tubes containing lactose-amended lauryl tryptose broth are inoculated; after 1 to 2 days of incubation at 35 ± 0.5°C, tubes showing turbidity, acidity, and gas (bubbles in fermentation tubes) are considered presumptively positive for coliform bacteria. These readings are considered presumptive because false-positive and false-negative reactions often occur as a result of growth and interference by nontarget bacteria. As a result, subsamples from presumptively positive tubes are then inoculated in a more selective medium to confirm the presence of the target bacteria. For TC, the confirmation test is growth of target bacteria in brilliant green lactose bile broth at 35°C within 1 to 2 days; the confirmation test for FC is the growth of target bacteria in *E. coli* media at 44.5°C within 1 to 2 days (1). The confirmation test is reliable evidence, but not proof, that the target bacteria have been detected. Therefore, subsamples of the confirmed positive reactions are inoculated onto a selective agar medium, and the target bacterium is physically recovered and Gram stained. This completed test is generally conducted on 10% of the positive confirmed samples as a quality control measure. For practical purposes, the number of positive and negative tubes in the confirmed phase of the test is typically used to determine the MPN of the target bacteria either by using a standard formula or by referring to the tables of positive and negative tube reactions as shown in *Standard Methods* (1).

Historically, the five-tube MTF test has been used most often to determine whether marine waters meet the recreational water quality standards, ignoring the inherent imprecision of the MPN method. More recently, with the advent of legislatively imposed single-sample standards, the use of MTF for making regulatory decisions has received more scrutiny. The reason for concern is that the wide confidence interval surrounding the MPN makes it likely that health warnings may be posted when indicator levels are actually below the water quality standard or, worse, that warnings may not be posted when the standard is exceeded. For example, Griffith et al. (31) found that in order to be 95% confident that a sample actually contained >104 *Enterococcus* organisms per 100 ml, the MPN produced by the MTF method would need to be at least twice the standard, while to be 95% confident that the true value was below the standard, the MPN would have to be <50 *Enterococcus* organisms per 100 ml. This variability, along with the time needed to obtain results, makes the MTF method less than optimal for many applications, including management of recreational beaches.

Using more tubes can increase the precision of the MTF method, but this must be balanced against the additional resources required. The analyst can select among the 3-tube (total of 9 tubes per test), the 5-tube (total of 15 tubes per test), or the 10-tube (total of 30 tubes per test) test format. The three-tube test requires less material, labor, and time but is the least precise. The 10-tube test is more accurate, but its requirement for material, labor, and time per test often makes its use prohibitive when a large number of samples must be processed.

Membrane Filtration

The MF test involves passing water samples through a membrane with a pore size (typically 0.45 μm) small enough to physically retain indicator bacteria on the filter surface. After filtration, the membrane is placed onto a solid growth medium (agar) selective for growth of the target bacteria. The filter membrane used in this assay is made of a material, such as cellulose nitrate, which allows nutrients from the agar to pass from the underside to the surface of the membrane. This allows viable bacteria trapped on its surface to grow and form visible colonies. Following incubation at a favorable temperature, visible colonies on the membrane's surface exhibiting the proper characteristics (color, size, and shape) are counted. The number of colonies counted is directly related to the concentration of the bacteria in the water. For example, if 100 ml of water is assayed and 50 colonies are counted, the concentration of the target bacteria will be determined as 50 CFU per 100 ml of water. CFU is used because this test assumes that each colony resulted from the growth of only one viable bacterium, though the discrete particle from which the colony formed could actually be an aggregate of several bacteria.

The reliability of the method is dependent on the selectivity of the bacteriological medium. It is recommended that 10% of the target colonies counted be streaked for isolation on another selective medium and that the isolate be Gram stained to verify that the target bacteria were recovered. Since reliable counts are obtained after one major step and after 1 day of incubation, the MF method is considered a one-step method, as opposed to the multistep MTF method.

For TC, an example method involves placing the processed filter onto m-Endo (mE) agar medium and incubating it 1 day at 35°C (1). The target colonies, characterized as red with a metallic green sheen, are counted. For FC, the processed filter is placed onto mFC agar in a tight-fitting petriplate and incubated submerged in a water bath at 44.5°C for 1 day (1). The target colonies, characterized by various shades of blue, are counted. Since coliform bacteria in marine waters are often in a stressed or injured state, they sometimes do not grow when incubated directly onto selective medium. Consequently, a resuscitation step, in which the processed membrane is initially placed onto nonselective medium (phenol red lactose agar) and incubated for 4 h at 35°C, may be necessary to allow the injured or stressed bacteria to recover. The membrane is subsequently placed onto mFC agar for overnight incubation at 44.5°C (1).

When MF is used to test for *Enterococcus*, either a two-step mE-esculin iron agar or one-step membrane-*Enterococcus* indoxyl-β-D-glucoside (mEI) method is typically used. In the two-step method, the processed filter is placed onto mE agar and incubated for 2 days at 41 ± 0.5°C. The first step involves counting all pink to red colonies on the filter, which represent fecal streptococcus bacteria, of which *Enterococcus* organisms are a subset. The entire filter is then transferred to an esculin iron agar medium and incubated for 20 min at 41°C. The formation of pink to red colonies, which develop a black or reddish brown precipitate on the underside of the filter, demonstrates that the colonies are able to hydrolyze esculin and confirms that the fecal streptococci are *Enterococcus* (1). The one-step mEI method substitutes a modified mE medium in which concentrations of triphenyltetrazolium chloride have been reduced and indoxyl β-D-glucoside has been added. As in the mE method, the processed filter is placed onto the mEI medium and incubated at 41 ± 0.5°C, but for only 1 day. The formation of blue colonies demonstrates that the colonies are able to hydrolyze indoxyl β-D-glucoside and confirms them as *Enterococcus* (62).

Defined Substrate

Like the MTF procedure, the DS method is based on an MPN statistical approach to quantification. Unlike MTF, the DS method is semiautomated, utilizing a disposable, compartmentalized pouch or microtiter plate (8). The wells in the pouch or microtiter plate are analogous to the tubes used in MTF, but they are more numerous, leading to greater method precision. Additionally, the semiautomated nature of the test reduces the DS method to a single step, thus reducing time and labor cost.

The most prevalent present DS applications are the Colilert and Colilert-18 prepackaged test kits (IDEXX, Westbrook, Maine), both of which combine a chromogenic (ONPG σ-nitrophenyl-β-D-galactopyranoside]) and fluorogenic substrate (MUG [methylumbelliferyl-β-glucuronide]) in the same medium. Sample water is diluted 10:1 with deionized water containing the dissolved medium, which contains the ONPG, MUG, and substrate and chemicals that specifically suppress the growth of competitor bacteria. The compartmentalized pouch is then sealed and incubated for 18 to 24 h at 35°C. A yellow color indicates the presence of TC in the well, while fluorescence in confirmed yellow wells, when exposed to 365-nm light, indicates the presence of *E. coli*. The result is read by counting the number of compartments that exhibit a characteristic color change. On the basis of the number of positive wells, MPN tables and a dilution factor are then used to determine the density of the target indicator bacteria per 100 ml of sample (43). The procedure for enumerating *Enterococcus* organisms is similar using the Enterolert prepackaged kit. As above, sample water is diluted and combined with powdered media. The procedure differs only in the fluorogenic substrate, which is specific for *Enterococcus* (4-methylumbelliferyl-β-D-glucuronide [MUD]), and in the incubation period, which is 24 to 28 h at 41°C.

The reliability of the MPN obtained from DS varies with the target bacteria. Griffith et al. (31) found that DS tended to overestimate TC compared to MF and MTF in California waters but was comparable to these methods for *E. coli* and *Enterococcus*. Pisciotta et al. (71) found that DS overestimated *E. coli* levels compared to MF in Florida waters. Evidence implicating nontarget bacteria as able to metabolize defined media was observed in both cases. Several studies have demonstrated that DS methods are comparable to MF for enumerating *E. coli* and *Enterococcus* organisms in cold marine waters (8, 31). However, the efficacy of these methods for regulatory purposes in warmer subtropical or tropical waters is suspect due to the presence of numerous nontarget bacteria capable of hydrolyzing ONPG, MUG, or MUD.

Comparison of the MTF, MF, and DS Methods

Despite considerable differences in terms of time, materials, cost, and protocols, all three methods generally produce comparable results (8, 31, 68). However, there are situations in which one method may be preferred over another due to the physical or biological properties of the sample. When selecting a method, one must keep in mind that each of these methods relies on a different metabolic endpoint to enumerate its target. Thus, each may enumerate specific subsets of the target bacteria in a particular environment or be susceptible to false-positive results from nontarget bacteria (49, 71). Further differences exist in the ability of the methods to process turbid samples and in the confidence interval surrounding the results.

The major advantages of the MTF method are that it can be used for both clear and turbid samples; inherently allows for the resuscitation and growth of injured bacteria; produces an endpoint that can be easily read by personnel with minimal skill; and requires minimal preparation time and effort to start the test, so that processing of samples can be initiated at any time of the day. There are also several disadvantages inherent to the MTF method, including the labor and material cost required; the substantial increases in reagents, tubes, incubation space, and cleanup required when multiple samples must be analyzed or when the sample volume must be increased to 100 ml; and the length of time (2 to 4 days) to reach a final result. Finally, MTF exhibits far greater variability than does either MF or DS, making it less suitable for monitoring in locales where assessments are based on a single-sample standard.

Major advantages of the MF method include savings in terms of time, labor, and cost compared with the MTF method; direct determination of the concentrations of bacteria with high precision; the formation of the target bacteria as colonies which can be purified for further identifi-cation and characterization; the ability to process large volume samples; and the ability to obtain a final result within 24 h. Disadvantages of MF include inapplicability of the method to turbid samples, which can clog the membrane or prevent the target bacteria from growing on the membrane; false-negative results due to the inability of viable but injured or stressed bacteria in environmental waters to grow with standard MF methods; and false-positive results when nontarget bacteria form colonies similar to the target colonies (21, 23).

DS shares some of the advantages and disadvantages of both the MTF and MF methods. Major advantages of the DS method include savings in terms of time and expense compared to both MF or MTF; the ability to enumerate a broad range of concentrations (e.g., <10 to 2,419 MPN/100 ml) with a single dilution; the ability to enumerate bacteria in turbid water and sediments; and the ability to produce final results in 18 to 24 h. Disadvantages of the DS method include lesser precision than MF and possible geographic limitations due to metabolization of the defined substrate by nontarget bacteria in subtropical environments (71).

NEW RAPID MEASUREMENT METHODS FOR MICROBIAL CONTAMINANTS

Present beach water quality monitoring programs are compromised because the standard methods for enumerating indicator bacteria are too slow to provide adequate protection to the public, i.e., results are generally available the day after the public was exposed to water of poor microbiological water quality. Furthermore, current routine measurements of indicator bacteria may not accurately predict the presence of viral and protozoan pathogens of concern because their persistence and survival characteristics in aquatic environments differ from that of the indicator bacteria. Current USEPA-approved methods (MF, MTF, and DS technologies) to measure indicator bacteria in recreational waters require an 18- to 96-h incubation period, while several studies have shown that temporal changes in indicator bacteria levels in beach water occur on much shorter time scales (6, 56). Thus, contaminated beaches remain open during the laboratory incubation period and are often clean by the time warnings are posted. The period of time taken for producing results also inhibits tracking of contamination sources, as a frequently used tracking approach is to look for differential bacterial concentrations at the confluence of upstream tributaries. However, the fecal-contamination signal can dissipate or disperse while samples are incubating, making it difficult to use this approach to successfully track fecal contamination sources.

The limiting factor for most of the currently used routine methods is that they rely on culturing techniques that either measure a metabolic endpoint or determine growth of a microorganism after an extended incubation period. New genetic and immunological methods that allow direct measurement of genes composed of RNA or DNA, or antigens or antigenic properties using antibodies, can be conducted with little or no incubation period, allowing them to be completed in only a few hours (69). These new technologies also allow expansion of the number and types of microbiological indicators that can be measured. Beach monitoring programs are presently based primarily on *E. coli* and *Enterococcus* because they are inexpensively cultured and have been shown to be useful proxies for pathogens that are more difficult and expensive to culture. The newer molecular technologies do not require culturing and can be as readily applied to pathogens, such as viruses, as they are to bacterial indicators.

Genetic Approaches

There are a wide array of technologies currently being developed that rely upon the complementarity of nucleic acid sequences, specifically the binding of oligonucleotide primers and probes to gene targets of interest. The thermodynamically favorable complementarity of DNA and RNA bases to one another allows nucleic acid priming to serve as the foundation for techniques such as PCR, microarrays, electrochemical RNA hybridization, and transcription-mediated amplification (TMA), which is similar to nucleic acid sequence-based amplification (NASBA). Primers and/or oligonucleotide probes complementary to a gene target can be designed for a wide array of microorganisms, allowing these methods to be as general or as highly specific as desired.

PCR utilizes a combination of reagents and temperature change schemes to anneal and denature nucleic acid sequences for exponential amplification of the gene of interest (76). Quantitative PCR (Q-PCR) is a primer-based molecular technique that combines the specificity of conventional PCR with the quantitative measurement of fluorescence for determining the presence of specific types of nucleic acid in environmental samples. One of the most common approaches employs the use of dual-labeled oligonucleotide probes that typically bear a 5′ fluorescent reporter dye and a "dark" quencher group in the 3′ position. The probe has a unique structure designed to hybridize to a specific target sequence. When the probe hybridizes to its target sequence, the structure is disrupted and the 5′ reporter is physically separated from the 3′ quencher, allowing fluorescent emission to be detected and measured quantitatively (39). For example, TaqMan-based Q-PCR methods have been used to estimate the number of *E. coli* organisms in environmental samples by using primer sequences complementary to specific genes (*uidA*) in *E. coli* (24). Haugland et al. (36) developed a TaqMan-based *Enterococcus* assay for recreational waters that can be done in less than 2 h, and has shown a strong relationship to the routinely used USEPA Method 1600 and significant relationships to human health effects when used during epidemiological studies (92). Further work with similar Q-PCR methods using other probe chemistries, such as Scorpions primer-probe complexes, have also demonstrated strong relationships to existing culture-based methods for both *E. coli* and *Enterococcus* (A. D. Blackwood, S. Yu, J. B. Gregory, and R. T. Noble, *Abstr. 106th Gen. Meet. Am. Soc. Microbiol.*, abstr. Q-494, p. 567, 2006).

Microarrays (or "microchips") involve the attachment of a sequence-specific probe on a slide, or array, where hybridization of the sequence of interest occurs after a series of linking and wash steps, and subsequent color change indicates positive detection. Microarrays can involve the use of cDNA probes, often greater than 200 nucleotides in length, or smaller oligonucleotides (20 to 100 nucleotides in length) that are fashioned to glass supports, nylon strips, or silica wafers (2, 94). With a single microarray, thousands of microorganisms can be analyzed at a single time. Upon hybridization of a target to the probe of interest, fluorescence is generated as data on the surface of each slide or "chip." This fluorescence can be read in a number of ways. A planar waveguide can be used for successful imaging of surface-confined fluorescence, in conjunction with the use of a cooled charge-coupled device (CCD) camera (75). Laser scanning and wide-field high-aperture microscopy (with CCD cameras) can both be used for reading the fluorescent signals. Microarrays, while having the potential to offer vast amounts of information as to the types of microbial contaminants present in a water sample (some arrays can be designed to probe for upward of 100 microorganisms in a single sample), are limited in that at best they can provide only semiquantitative information. In the future, integrated microfluidics, more advanced probe technologies, and integration with PCR approaches may provide the quantification desired for application of this technology in environmental samples (60). At this time, microarrays have a role in screening and detection of a wide array of specific viral, protozoan, and bacterial pathogens of interest, particularly emerging pathogens of concern. For example, it is possible that microarrays could be used as a screening tool at urban recreational beaches with proximity to sources of human fecal contamination and very high visitation.

TMA and NASBA are identical, but proprietary, technologies of Gen-Probe, Inc., and Biomerieux, Inc., respectively. They are isothermal based methods of RNA amplification that were developed nearly simultaneously and applied originally to clinically based human immunodeficiency virus type 1 detection (47). Both technologies are isothermal amplification methods based on transcription. Instead of utilizing a thermostable DNA polymerase, as PCR does, RNA is amplified using an enzyme mixture at a fixed temperature. NASBA has shown tremendous potential for use in environmental samples (15). For both TMA and NASBA, there are a wide variety of diagnostic and commercial applications, particularly in clinical circumstances, using these assays for detection of specific infectious agents. One advantage of NASBA over most PCR-based rapid detection methods is that a thermal cycler is not needed, improving portability. Another advantage of RNA-based approaches is that they generally measure only viable, RNA-producing cells. A reasonable level of equivalency has been demonstrated when comparing routine *Enterococcus* enumeration methods with the TMA method (29). Handheld, sensitive, rapid NASBA technologies have also been developed for detection of enteroviruses in environmental waters.

Electrochemical RNA hybridization involves the use of pulse voltammetry to track the oxidation of guanine nucleotides in the rRNA of the target organism (54). In this method, target rRNA from lysed cells is first hybridized to DNA probes bound to magnetic beads. As with TMA, the magnetic beads are then purified via magnetic separation and washing steps. Guanine bases are then released into solution from the DNA probe by treatment with sulfuric acid and detected by a graphite probe using pulse voltammetry to monitor their oxidation. While still in development, this method shows promise for use in autonomous in situ monitoring devices because it is relatively inexpensive compared to other genetic methods and involves a smaller number of steps.

Immunological Approaches

Antibody (Ab)-based approaches take advantage of the binding affinities of Abs to specific antigens and have been widely implemented in the food safety industry because of their specificity and relative simplicity. Abs are typically produced in the laboratory or can be purchased commercially and can be designed to be specific for a single strain (or serotype) of bacteria (e.g., *E. coli* O157:H7), or produced to detect a single species (*E. coli*, many strains) or groups of organisms (*Enterococcus* spp.), although the latter two approaches are generally more difficult. Once produced and tested for specificity, Abs are typically mounted onto a support system. Well-documented support systems include polystyrene waveguides, nylon supports, glass slides, and cantilevers (17, 19, 51). After antigen attachment to the primary Ab, remaining excess material is typically washed away, and fluorescently labeled secondary Abs are used for detection. Immunomagnetic capture, in which organisms are captured using an Ab-antigen-magnet complex, can also be employed (81). With use of an external magnet, the bound material can be effectively separated from the remainder of the sample. This approach can be highly effective for use in capturing target organisms from complex environmental samples. One advantage of Ab-based approaches is that captured bacteria are still viable and can be further studied, having the potential to yield more specific information about the sources of the bacteria through assessment of genotypic or phenotypic information (79). Often, however, Ab approaches are susceptible to interference from the complex sample types that are found in aquatic environments.

POTENTIAL ALTERNATIVE INDICATORS OF MARINE WATER QUALITY

Dufour (20) summarized five criteria for selecting a water quality indicator: (i) the indicator should be consistently present in feces at higher concentrations than the pathogens; (ii) it should not be present or multiply outside the human intestinal tract; (iii) it should be as resistant as, or more resistant than, the pathogens to environmental conditions and to disinfection; (iv) it must be assayed by means of a simple and reliable test; and (v) its concentrations in water should correlate with concentrations of feces-borne pathogens or with a measurable health hazard.

None of the current bacterial indicators presently used for marine recreational standards fulfills all of these criteria. Of greatest concern is that the present bacterial indicators are not unique to human fecal material and thus may overstate health risk, particularly when the fecal source is land-based runoff that contains animal input. As a result, many researchers are evaluating other microorganisms as alternative indicators of water quality that are thought to be more human specific. Some of these are bacterial indicators, while others are pathogens that cannot be cost-effectively measured using culture-based methods but have become feasible indicator candidates because the molecular methods described in the previous section have the potential to increase the speed and reduce the cost of their measurement.

Potential Alternative Bacterial Indicators

Clostridium perfringens

C. perfringens is a potentially useful water quality indicator because (i) it is consistently present in moderate concentrations in human feces as well as in sewage, (ii) it is more stable in environmental waters than most pathogens, (iii) it can be measured easily by most laboratories because it is similar to the MF assays presently in widespread use, and (iv) it has been successfully used to identify sewage contamination in several aquatic environments (26, 41, 84). One limitation to the use of *C. perfringens* is the extreme stability of this bacterial spore to environmental conditions and the possibility that its detection indicates a pollution event that occurred a long time ago. Thus, *C. perfringens* is a conservative indicator of fecal pollution in that one can be assured that the quality of that recreational water is good if there are low levels of *C. perfringens*, but high levels do not necessarily indicate that a current health risk exists.

A two-step MF method developed by Bisson and Cabelli (5) is available to assay for concentrations of *C. perfringens*. In this assay, the water samples are filtered through membranes by using the procedure used to assay for *Enterococcus* or fecal coliform. The processed membranes are then placed onto m-CP agar (5) and incubated at 44°C under anaerobic conditions. After 1 day of incubation, the membranes are immediately exposed to fumes of ammonium hydroxide. All yellow colonies that turn pink to red are acid phosphatase positive, confirming that the colonies are *C. perfringens*.

Staphylococcus aureus

Epidemiological studies have frequently demonstrated that swimmers are susceptible to nondiarrheal diseases, such as eye, ear, nose, and skin infections, which are not well correlated with presently used bacterial indicators. One alternative is *S. aureus*, which is associated with skin infections (11). Evidence to support the use of *S. aureus* as indicator of water quality includes the following: (i) the bacterium is stable in marine waters, (ii) its concentration in water has been shown to represent the load of microorganisms being shed by swimmers (12), and (iii) an effective MF method is available for marine waters (12). The limitations in the use of this indicator system are the multistep nature of the method to recover and to enumerate the concentrations of *S. aureus* and the still unresolved question as to whether recreational water serves as a vector for the transmission of this pathogen to uninfected humans (a public health problem) or whether the bacterium is present on human skin and recreational use of marine waters causes opportunities for self-infection (a personal problem).

The MF method of Charoenca and Fujioka (12) is a one-step method in which the processed membrane is placed onto Vogel-Johnson agar supplemented with 0.005% sodium azide. After 1 to 2 days of incubation at 35°C, the target black colonies representing total staphylococci are counted; 10% of the black colonies from a given site are then tested for coagulase to determine the concentrations of *S. aureus* in the water sample. The MF method utilized by Stotts et al. (88) involves filtration of seawater samples onto sterile, 47-mm-diameter, 0.45-μm-pore-size, mixed cellulose ester filters and overlay of filters onto the selective medium CHROMagar Staph aureus (CSA) (CHROMagar Microbiology, Paris, France) or onto CSA supplemented with oxacillin (4 $\mu g\ ml^{-1}$), sodium azide (0.0035%), or sodium azide plus glycine (12 g $liter^{-1}$). Alternatively, filters can be incubated in nutrient broth with oxacillin to enrich for *S. aureus* prior to plating on solid media as above. Putative *S. aureus* colonies are picked from the filters and purified by serially picking and streaking individual colonies on fresh CSA or tryptic soy agar plates.

Bacteroidales

The Order *Bacteroidales* and species such as *Bacteroides thetaiotaomicron* are good candidate indicators because they make up approximately one-third of the human fecal microflora, considerably outnumbering the FC in human feces by weight (93). While found in high numbers in human feces, they occur in low numbers in animal feces (93). The *Bacteroides* are nonspore-forming, gram-negative, obligate anaerobes, so there is little concern over regrowth in the environment. However, one study has shown that *Bacteroides distasonis* seeded in the environment persisted for 23 days, as determined by conventional PCR (53).

Development of GasPak EZ Pouch Anaerobe Systems (Becton Dickinson) has made culturing of the anaerobic *Bacteroidales* group more feasible, but the greater advances in measuring this group have been in the use of PCR-based assays (4, 18). These methods involve filtering a water sample, extracting bacterial DNA from the filter, and analyzing the extract by PCR, using published primer sets (4). Bernhard and Field succeeded in developing assays for a human-specific marker and a bovine-specific marker within *Bacteroides*, both implemented originally as conventional PCR assays and not fully quantitative. Since the publication of these assays, several groups have attempted to transform these specific assays into Q-PCR assays, but in doing so, many have created assays that are subject to cross-reactivity from other types of fecal contamination. Several sets of cross-reactivity studies have been conducted to ensure that *Bacteroides* spp. thought to stem from human feces are not coming from other sources (52). Layton et al. (55) published new Q-PCR assays specifically designed for bovines and humans. Even though these assays appear to be an improvement over other previously published Q-PCR-based organism-specific assays thought to be useful for water quality monitoring, there are still problematic instances of cross-reactivity with other types of fecal contamination (55).

Promising Viral Indicators

The survival of bacterial indicators in the ambient environment generally differs from that of viral pathogens, and thus, it may be more desirable to measure viral indicators directly. Two types of viral indicators are presently employed in some monitoring programs and epidemiology studies. The first is coliphages, which are viruses that infect coliform bacteria. The coliphages most frequently studied are F-specific coliphages and *Bacteroides fragilis* phages. The second group is enteric viruses, the measurement of which is only recently becoming cost-effective through use of PCR.

The most likely coliphage group as a candidate water quality indicator is the F (male)-specific RNA (FRNA) coliphage, because (i) it represents viruses similar in size, shape, and genetic makeup to human enteric viruses; (ii) it is more stable than human enteroviruses; and (iii) its concentrations in environmental waters have been reported to correlate with sewage contamination (37, 38). There are currently four recognized F-specific RNA coliphage types with varying levels of specificity: types II and III were reported to be associated with human and domestic sewage, whereas types I and IV predominated in animal waste (27, 42).

The assay is similar to the MF method in that the concentration of virus is based on the formation of PFU (modified EPA 1601 method). Each plaque is a visible area of infection that is similar to a bacterial CFU because it is assumed that each plaque results from infection by one virus. However, since viruses are specific, intracellular parasites, a live cell culture is required. For the assay of FRNA bacterial virus, two reliable strains of piliated bacteria which can support the growth of FRNA virus have been used. The first strain is a mutant of *Salmonella enterica* serovar Typhimurium (WG-49) developed by Havelaar and Pot-Hogeboom (38); the second, an *E. coli* mutant called HS(pFamp)R, has been developed by Debartolomeis and Cabelli (16).

A virus or phage which infects *B. fragilis* has been proposed as a good indicator of water quality, because (i) the virus multiplies only in the intestinal tracts of humans and therefore shares a major characteristic with human enteric virus; (ii) it is not found in feces of most animals, and therefore its presence is evidence of human and not animal fecal contamination; and (iii) it survives well in the environment and can be detected by using *B. fragilis* bacteria as a host. The method used to recover this phage is similar in principle to the method used to recover FRNA phage except that a specific strain (HSP40) of *B. fragilis* is required for recovery and enumeration. Use of *B. fragilis* as a host has resulted in promising data showing that this bacteriophage can be related to fecal contamination in marine waters (45, 46, 58). However, a significant problem with *B. fragilis* phage use for water quality assessments is that the phage is found in human sewage only in limited geographic areas of the world, e.g., the phage is not found in appreciable numbers in human sewage in the United States (46).

Human-Pathogenic Viruses

Human viruses are likely waterborne pathogens, making them a leading candidate for inclusion in recreational water monitoring programs. In addition, human-pathogenic viruses are more specific to humans than the presently used bacterial indicators. The limitations for widespread use of virus measurements have been the cost, laboratory equipment requirements, personnel training, and time for laboratory tests to be completed using tissue culture methods. In addition, different groups of viruses require that cells be cultured, and some viruses are not culturable. Within the last decade, however, molecular methods have been developed to detect the presence of human-pathogenic viruses, such as adenoviruses, enteroviruses, and noroviruses, in environmental samples (44, 66, 73). Many of the newest developments have been with virus assays utilizing Q-PCR, quantitative reverse transcriptase-PCR, and NASBA, with rapid assays specifically targeting the presence of human viruses in recreational waters and shellfish harvesting samples (10, 13, 25, 28, 69). The current limitation of these approaches is that they do not differentiate between infective and noninfective viruses (only infective viruses are of public health significance). Choi and Jiang (13) recently demonstrated that while adenovirus incidence was high in southern California river samples, few of the viruses found were likely to be infectious.

Tracers of Human Fecal Sources

Fecal contamination can result from numerous sources of fecal pollution, including human sewage, manure from livestock operations, indigenous wildlife, and urban runoff. All of these sources can harbor pathogens, but water with human fecal contamination is more likely to pose a public health risk. The most frequently used bacterial indicator organisms, *E. coli* and *Enterococcus*, are ubiquitous among warm-blooded animals and do not provide insight as to the source of fecal contamination. As a result, many researchers have developed laboratory techniques that allow for source identification (79, 82). These tools allow managers to prioritize areas with human fecal contamination for public health protection and to identify which fecal sources are the most appropriate targets for mitigation.

A wide range of source identification approaches have been developed, incorporating indicator bacteria, pathogens, and even chemicals (caffeine, fragrance materials, and fluorescent whitening agents) as targets. The techniques targeting bacterial indicators generally fall into two categories: genotypic (based on genetic information) or phenotypic (based upon a physiologically expressed trait of an organism). These approaches are based on the presumption that fecal bacteria in the intestines of animals coevolve to become host specific. Over time, this process produces both identifiable phenotypic traits and changes in gene sequences. In domesticated animals and humans, this animal-specific specialization may be enhanced by the introduction of different types of antibiotics and feed material among species.

The most frequently used phenotypic method is multiple antibiotic resistance profiling (33, 35, 70). This method is based on growing isolates on selective media containing a suite of antibiotics of varying types and concentrations. The resulting resistance "fingerprints" constitute a library. Resistance patterns of indicator bacteria isolated from ambient samples are then matched against the library to determine probable sources. Carbon source utilization is a similar approach, except that the fingerprints are based on differential growth in various carbon source growth media (32).

Many of the genotypic methods also rely on a library of isolates in the same way as the phenotypic methods, except that the library is based on genetic sequences. The most widely used library-dependent genotypic methods include pulsed-field gel electrophoresis, ribotyping, and repetitive-intergenic DNA sequence PCR, though other methods such as denaturing gradient gel electrophoresis are also feasible (82). All of these methods require that a large number of isolates be cultured and characterized as a reference library (61).

There are also several genetic source identification methods based on characteristics that can be identified without culturing isolates (22, 80). Most prominent among these are host-specific molecular markers focusing on terminal restriction fragment length polymorphism to members of *Bacteriodes-Prevotella* (3, 4). PCR-based methods have also been proposed for human-specific *Bifidobacterium* spp. (59), and a multiplex PCR with strain-specific primer sets detected *Bifidobacterium adolescentis* and *Bifidobacterium dentium* exclusively in human sewage samples (7).

While these tools have shown considerable promise in selected applications, there have been three recent multilaboratory evaluations that suggest that further development of source identification methods will be required before they are adopted for routine application (30, 63, 87). Griffith et al. (30) tested 12 methods employed by 22 researchers and found that no method performed consistently well across all types of samples and against all evaluation criteria. Library-independent methods generally outperformed library-based methods in their ability to identify or exclude samples with respect to human fecal contamination, but these methods are presently unable to resolve all of the sources and/or produce only a presence/absence result. Library-based methods produced quantitative results for all sources of fecal contamination but exhibited a high rate of false positives and often assigned a large percentage of contamination to sources not present in a sample. Stoeckel et al. (87) limited their evaluation to library-based methods and found that three of the eight methods that they tested provided less than a 25% correct classification for blind isolates. Like Griffith et al., they found that genetic methods outperformed those based on phenotypic approaches. Moore et al. (63) drew similar conclusions, stating that none of the source identification methods that they tested were ready yet for routine application to environmental samples.

There is also little standardization within the source identification methods at this time (86). This is particularly true for library-dependent methodologies, in which the restriction enzyme or combination of antibodies used to develop the library seems to be investigator specific. There is also no clear agreement as to which bacterial species is most host specific, though there is growing evidence that *Enterococcus* is a better target than *E. coli* (64, 87). Even the statistical methods used to interpret the results differ considerably among practitioners and have the potential to influence results (74). As such, source identification methods are more prevalent in the research community than in production laboratories at the present time.

Despite their imperfections, source identification methods provide some useful information. USEPA (90) recently produced a guidance document for water quality managers striving to understand sources of fecal contamination. Similar to Stewart et al. (86) and NRC (65), the USEPA recommends a tiered approach involving use of several methods as a means of resolving imperfections in each individual method. They also suggest that the level of effort allocated to such studies be proportional to the extent of the problem and the complexity of possible sources in the watershed. The document is intended as a reference guide for water quality evaluations and total maximum daily load implementation activities and represents a good starting point for advancement of tools for tracking fecal contamination sources.

REFERENCES

1. **American Public Health Association, American Water Works Association, and Water Pollution Control Federation.** 1998. *Standard Methods for the Examination of Water and Wastewater*, 20th ed. American Public Health Association, Washington, D.C.
2. **Bavykin, S. G., J. P. Akowski, V. M. Zakhariev, V. E. Barsky, A. N. Perov, and A. D. Mirzabekov.** 2001. Portable system for microbial sample preparation and oligonucleotide microarray analysis. *Appl. Environ. Microbiol.* **67:**922–928.
3. **Bernhard, A. E., T. Goyard, M. T. Simonich, and K. G. Field.** 2003. A rapid method for identifying fecal pollution sources in coastal waters. *Water Res.* **37:**909–913.
4. **Bernhard, A. E., and K. G. Field.** 2000. A PCR assay to discriminate human and ruminant feces based on host differences in *Bacteroides-Prevotella* 16S rRNA. *Appl. Environ. Microbiol.* **66:**4571–4574.
5. **Bisson, J. W., and V. J. Cabelli.** 1979. Membrane filtration enumeration method for *Clostridium perfringens*. *Appl. Environ. Microbiol.* **37:**55–66.
6. **Boehm, A. B., S. B. Grant, J. H. Kim, S. L. Mowbray, C. D. McGee, C. D. Clark, D. M. Foley, and D. E. Wellman.** 2002. Decadal and shorter period variability and surf zone water quality at Huntington Beach, California. *Environ. Sci. Technol.* **36:**3885–3892.
7. **Bonjoch, X., E. Balleste, and A. R. Blanch.** 2004. Multiplex PCR with 16S rRNA gene-targeted primers of *Bifidobacterium* spp. to identify sources of fecal pollution. *Appl. Environ. Microbiol.* **70:**3171–3175.
8. **Budnick, G. E., R. T. Howard, and D. R. Mayo.** 1996. Evaluation of enterolert for enumeration of enterococci in recreational water. *Appl. Environ. Microbiol.* **62:**3881–3884.
9. **Byappanahalli, B., and R. Fujioka.** 2004. Indigenous soil bacteria and low moisture may limit but allow faecal bacteria to multiply and become a minor population in tropical soils. *Water Sci. Technol.* **50:**27–32.
10. **Casper, E. T., S. S. Patterson, M. C. Smith, and J. H. Paul.** 2005. Development and evaluation of a method to detect and quantify enteroviruses using NASBA and internal control RNA (IC-NASBA). *J. Virol. Methods* **124:**149–155.
11. **Chang, W. J., and F. D. Pien.** 1986. Marine-acquired infections: hazards of the ocean environment. *Postgrad. Med.* **80:**30–33.
12. **Charoenca, N., and R. Fujioka.** 1991. Assessment of staphylococcus bacteria in Hawaii's marine recreational waters. *Water Sci. Technol.* **27:**283–289.
13. **Choi, S., and S. C. Jiang.** 2005. Real-time PCR quantification of human adenoviruses in urban rivers indicates genome prevalence but low infectivity. *Appl. Environ. Microbiol.* **71:**7426–7433.
14. **Colford, J. M., Jr., T. J. Wade, K. C. Schiff, C. C. Wright, J. F. Griffith, S. K. Sandhu, S. Burns, M. Sobsey, G. Lovelace, and S. B. Weisberg.** Water quality indicators and the risk of illness at beaches with non-point sources of fecal contamination. *Epidemiology*, in press.
15. **Collins, R. A., L.-S. Ko, K.-Y. Fung, K.-Y. Chan, J. Xing, L.-T. Lau, and A. C. H. Yu.** 2003. Rapid and sensitive detection of avian influenza virus subtype H7 using NASBA. *Biochem. Biophys. Res. Commun.* **300:**507–515.
16. **Debartolomeis, J., and V. J. Cabelli.** 1991. Evaluation of an *Escherichia coli* host strain for enumeration of F male-specific bacteriophages. *Appl. Environ. Microbiol.* **57:**1301–1305.
17. **DeMarco, D. R., and D. V. Lim.** 2002. Detection of *Escherichia coli* 157:H7 in 10- and 25-gram ground beef samples using an evanescent wave biosensor with silica and polystyrene waveguides. *J. Food Prot.* **65:**596–602.
18. **Dick, L., and K. Field.** 2004. Rapid estimation of numbers of fecal *Bacteroidetes* by use of a quantitative PCR assay for 16S rRNA genes. *Appl. Environ. Microbiol.* **70:**5695–5697.
19. **Dubitsky, A., D. DeCollibus, and G. A. Ortolano.** 2002. Sensitive fluorescent detection of protein on nylon membranes. *J. Biochem. Biophys. Methods* **51:**47–56.
20. **Dufour, A. P.** 1984. Bacterial indicators of recreational water quality. *Can. J. Public Health* **75:**49–56.
21. **Ferguson, D. M., D. F. Moore, M. A. Getrich, and M. H. Zhowandai.** 2005. Enumeration and speciation of enterococci found in marine and intertidal sediments and coastal water in southern California. *J. Appl. Microbiol.* **99:**598–608.
22. **Field, K. G., E. C. Chern, L. K. Dick, J. Fuhrman, J. Griffith, P. A. Holden, M. G. LaMontagne, B. Olson, and M. T. Simonich.** 2003. A comparative study of culture-independent, library-independent genotypic methods of fecal source tracking. *J. Water Health* **1:**181–194.
23. **Figueras, M. J., F. Polo, I. Inza, and J. Guarro.** 1994. Poor specificity of m-Endo and m-FC culture media for the enumeration of coliform bacteria in sea water. *Lett. Appl. Microbiol.* **19:**446–450.
24. **Frahm, E., and U. Obst.** 2003. Application of the fluorogenic probe technique (TaqMan PCR) to the detection of *Enterococcus* spp. and *Escherichia coli* in water samples. *J. Microbiol. Methods* **52:**123–131.
25. **Fuhrman, J. A., X. Liang, and R. T. Noble.** 2005. Rapid detection of enteroviruses from small volumes of natural waters by real time QRT-PCR. *Appl. Environ. Microbiol.* **71:**4523–4530.
26. **Fujioka, R. S., and L. K. Shizumura.** 1985. *Clostridium perfringens*, a reliable indicator of stream water quality. *J. Water Pollut. Control Fed.* **57:**986–992.
27. **Gerba, C. P.** 1987. Phages as indicators of fecal pollution, p. 197–209. *In* S. M. Goyal, C. P. Gerba, and G. Bitton (ed.), *Phage Ecology*. John Wiley and Sons, New York, N.Y.
28. **Gregory, J. B., W. A. Litaker, and R. T. Noble.** 2006. A rapid one-step quantitative reverse transcriptase PCR assay for enterovirus with a competitive internal positive control. *Appl. Environ. Microbiol.* **72:**3960–3967.
29 **Griffith, J., and S. B. Weisberg.** 2006. *Evaluation of Rapid Microbiological Methods for Measuring Recreational Water Quality*. Report no. 485. Southern California Coastal Water Research Project,. Westminster, Calif.
30. **Griffith, J. F., S. B. Weisberg, and C. D. McGee.** 2003. Evaluation of microbial source tracking methods using mixed fecal sources in aqueous test samples. *J. Water Health* **1:**141–152.
31. **Griffith, J. F., L. A. Aumand, I. M. Lee, C. D. McGee, L. L. Othman, K. J. Ritter, K. O. Walker, and S. B. Weisberg.** 2006. Comparison and verification of bacterial water quality indicator measurement methods using am-

bient coastal water samples. *Environ. Monit. Assess.* **116:** 335–344.

32. **Hagedorn, C., J. B. Crozier, K. A. Mentz, A. M. Booth, A. K. Graves, N. J. Nelson, and R. B. Reneau.** 2003. Carbon source utilization profiles as a method to identify sources of faecal pollution in water. *J. Appl. Bacteriol.* **94:** 792–799.
33. **Hagedorn, C. S., S. L. Robinson, J. R. Filtz, S. M. Grubbs, T. A. Angier, and R. B. Reneau, Jr.** 1999. Using antibiotic resistance patterns in the fecal streptococci to determine sources of fecal pollution in a rural Virginia watershed. *Appl. Environ. Microbiol.* **65:**5522–5531.
34. **Haile, R. W., J. S. Witte, M. Gold, R. Cressey, C. D. McGee, R. C. Millikan, A. Glasser, N. Harawa, C. Ervin, P. Harmon, J. Harper, J. Dermand, J. Alamillo, K. Barrett, M. Nides, and G. Wang.** 1999. The health effects of swimming in ocean water contaminated by storm drain runoff. *J. Epidemiol.* **104:**355–363.
35. **Harwood, V. J., J. Whitlock, and V. H. Withington.** 2000. Classification of the antibiotic resistance patterns of indicator bacteria by discriminant analysis: use in predicting the source of fecal contamination in subtropical Florida waters. *Appl. Environ. Microbiol.* **66:**3698–3704.
36. **Haugland, R. A., S. C. Siefring, L. J. Wymer, K. P. Brenner, and A. P. Dufour.** 2005. Comparison of *Enterococcus* measurements in freshwater at two recreational beaches by QPCR and membrane filter culture analysis. *Water Res.* **39:**559–568.
37. **Havelaar, A. H., K. Furuse, and W. H. Hogeboom.** 1986. Bacteriophages and indicator bacteria in human and animal faeces. *J. Appl. Bacteriol.* **60:**255–262.
38. **Havelaar, A. H., and W. M. Pot-Hogeboom.** 1988. F-specific RNA-bacteriophages as model viruses in water hygiene: ecological aspects. *Water Sci. Technol.* **20:**399–407.
39. **Heid, C. A., J. Stevens, K. J. Livak, and P. M. Williams.** 1996. Real time quantitative PCR. *Genome Res.* **6:**986–994.
40. **Hernandez-Lopez, J., and F. Vargas-Albores.** 1994. False-positive coliform readings using membrane filter techniques for seawater. *Lett. Appl. Microbiol.* **19:**483–485.
41. **Hill, R. T., L. T. Knight, M. S. Anikis, and R. R. Colwell.** 1993. Benthic distribution of sewage sludge indicated by *Clostridium perfringens* at a deep-ocean dump site. *Appl. Environ. Microbiol.* **59:**47–51.
42. **Hsu, F.-C., Y.-S. Shieh, J. van Duin, M. J. Beekwilder, and M. D. Sobsey.** 1995. Genotyping male-specific RNA coliphages by hybridization with oligonucleotide probes. *Appl. Environ. Microbiol.* **61:**3960–3966.
43. **Hurley, M. A., and M. E. Roscoe.** 1983. Automated statistical analysis of microbial enumeration by dilution series. *J. Appl. Bacteriol.* **55:**26–164.
44. **Jiang, S., R. T. Noble, and W. Chu.** 2001. Human adenoviruses and coliphages in urban runoff-impacted coastal waters of southern California. *Appl. Environ. Microbiol.* **67:**179–184.
45. **Jofre, J., A. Bosch, F. Lucena, R. Girones, and C. Tartera.** 1986. Evaluation of Bacteroides fragilis bacteriophages as indicators of the virological quality of water. *Water Sci. Technol.* **18:**167–177.
46. **Jofre, J., M. Blasi, A. Bosch, and F. Lucena.** 1989. Occurrence of bacteriophages infecting *Bacteroides fragilis* and other viruses in polluted marine sediments. *Water Sci. Technol.* **21:**15–19.
47. **Kievits, T., B. vanGemen, D. van Strijp, R. Schukkink, M. Dircks, H. Adriaanes, L. Malek, R. Sooknanan, and P. Lens.** 1991. NASBA™ isothermal enzymatic in vitro nucleic acid amplification optimized for the diagnosis of HIV-1 infection. *J. Virol. Methods* **35:**273–286.
48. **King, P. G.** 1999. *The Fiscal Impacts of Beaches in California.* Public Research Institute, San Francisco State University, San Francisco, Calif.
49. **Kinzelman, J., C. Ng, E. Jackson, S. Gradus, and R. Bagley.** 2003. Enterococci as indicators of Lake Michigan recreational water quality: comparison of two methodologies and their impacts on public health regulatory events. *Appl. Environ. Microbiol.* **69:**92–96.
50. **Klein, E., and A. C. Houston.** 1899. Further reports on bacteriological evidence of recent and therefore dangerous sewage pollution, p. 498–504. *In Metropolitan Water Supply 28th Annual Report of Local Government Board. Supplementary Report of Medical Officer for 1898-1899. Appendix B.* London Government Office, London, United Kingdom.
51. **Kooser, A., K. Manygoats, M. P. Eastman, and T. L. Porter.** 2003. Investigation of the antigen antibody reaction between anti-bovine serum albumin (a-BSA) and bovine serum albumin (BSA) using piezoresistive microcantilever based sensors. *Biosens. Bioelectron* **19:**503–508.
52. **Kreader, C. A.** 1995. Design and evaluation of *Bacteroides* DNA probes for the specific detection of human fecal contamination. *Appl. Environ. Microbiol.* **61:**1171–1179.
53. **Kreader, C. A.** 1998. Persistence of PCR-detectable *Bacteroides distasonis* from human feces in river water. *Appl. Environ. Microbiol.* **64:**4103–4105.
54. **LaGier, M. J., C. A. Scholin, J. W. Fell, J. Wang, and K. D. Goodwin.** 2005. An electrochemical hybridization assay for detection of the fecal indicator bacterium *Escherichia coli. Mar. Pollut. Bull.* **50:**1251–1261.
55. **Layton, A., L. McKay, et al.** 2006. Development of *Bacteroides* 16S rRNA gene TaqMan-based real-time PCR assays for estimation of total, human, and bovine fecal pollution in water. *Appl. Environ. Microbiol.* **72:**4214–4224.
56. **Leecaster, M. K., and S. B. Weisberg.** 2001. Effect of sampling frequency on shoreline microbiology assessments. *Mar. Pollut. Bull.* **42:**1150–1154.
57. **Lipp, E. K., N. Schmidt, M. E. Luther, and J. B. Rose.** 2001. Determining the effects of El Nino-Southern Oscillation events on coastal water quality. *Estuaries* **24:**491–497.
58. **Lucena, F., J. Lasobras, D. Mclntosh, M. Forcadell, and J. Jofre.** 1994. Effect of distance from the polluting focus on relative concentrations of *Bacteroides fragilis* phages and coliphages in mussels. *Appl. Environ. Microbiol.* **60:**2272–2277.
59. **Matsuki, T., K. Watanabe, R. Tanaka, and R. Oyaizu.** 1998. Rapid identification of human intestinal bifidobacteria by 16S rRNA-targeted species and group-specific primers. *Microbiol. Lett.* **167:**113–121.
60. **Maynard, C., F. Berthiaume, K. Lemarchand, J. Harel, P. Payment, P. Bayardelle, L. Masson, and R. Brousseau.** 2005. Waterborne pathogen detection by use of oligonucleotide-based microarrays. *Appl. Environ. Microbiol.* **71:** 8548–8557.
61. **McLellan, S. L.** 2004. Genetic diversity of *Escherichia coli* isolated from urban rivers and beach water. *Appl. Environ. Microbiol.* **70:**4658–4665.
62. **Messer, J., and A. P. Dufour.** 1998. A rapid and specific membrane filtration procedure for enumeration of enterococci in recreational water. *Appl. Environ. Microbiol.* **62:**3881–3884.
63. **Moore, D. F., V. J. Harwood, D. M. Ferguson, J. Lukasik, P. Hannah, M. Getrich, and M. Brownell.** 2005. Evaluation of antibiotic resistance analysis and ribotyping for identification of fecal pollution sources in an urban watershed. *J. Appl. Microbiol.* **99:**618–628.
64. **Myoda, S. P., C. A. Carson, J. J. Fuhrmann, B. Hahm, P. G. Hartel, H. Yampara-Iquise, L. Johnson, R. L. Kuntz, C. H. Nakatsu, M. J. Sadowsky, and M.**

Samadpour. 2003. Comparison of genotypic-based microbial source tracking methods requiring a host origin database. *J. Water Health* **1:**167–180.

65. **National Research Council (NRC).** 2004. *Indicators for Waterborne Pathogens*. National Academies Press, Washington, D.C.
66. **Noble, R. T., and J. A. Fuhrman.** 2001. Enteroviruses detected by reverse transcriptase polymerase chain reaction from the coastal waters of Santa Monica Bay, California: low correlation to bacterial indicator levels. *Hydrobiologia* **460:**175–184.
67. **Noble, R. T., D. F. Moore, M. Leecaster, C. D. McGee, and S. B. Weisberg.** 2003. Comparison of total coliform, fecal coliform, and enterococcus bacterial indicator response for ocean recreational water quality testing. *Water Res.* **37:**1637–1643.
68. **Noble, R. T., M. Leecaster, C. D. McGee, S. B. Weisberg, and K. Ritter.** 2004. Comparison of bacterial indicator analysis methods in stormwater-affected coastal waters. *Water Res.* **38:**1183–1188.
69. **Noble, R. T., and S. B. Weisberg.** 2005. A review of technologies being developed for rapid detection of bacteria in recreational waters. *J. Water Health* **3:**381–392.
70. **Parveen, S., R. L. Murphree, L. Edminston, C. W. Kaspar, K. M. Portier, and M. L. Tamplin.** 1997. Association of multiple-antibiotic-resistance profiles with point and nonpoint sources of *Escherichia coli* in Apalachicola Bay. *Appl. Environ. Microbiol.* **63:**2607–2612.
71. **Pisciotta, J. M., D. F. Rath, P. A. Stanek, D. M. Flanery, and V. J. Harwood.** 2002. Marine bacteria cause false-positive results in the Colilert-18 rapid identification test for *Escherichia coli* in Florida waters. *Appl. Environ. Microbiol.* **68:**539–544.
72. **Prüss, A.** 1998. Review of epidemiological studies on health effects from exposure to recreational water. *Int. J. Epidemiol.* **27:**1–9.
73. **Reynolds, K. A., C. P. Gerba, and I. L. Pepper.** 1995. Detection of enteroviruses in marine waters by direct RT-PCR and cell culture. *Water Sci. Technol.* **31:**323–328.
74. **Ritter, K. J., E. Carruthers, C. A. Carson, R. D. Ellender, V. J. Harwood, K. Kingsley, C. Nakatsu, M. Sadowsky, B. Shear, B. West, J. E. Whitlock, B. A. Wiggins, and J. D. Wilbur.** 2003. Assessment of statistical methods used in library-based approaches to microbial source tracking. *J. Water Health* **1:**209–223.
75. **Rowe-Taitt, C. A., J. P. Golden, M. J. Feldstein, J. J. Cras, K. E. Hoffman, and F. S. Ligler.** 2000. Array biosensor for detection of biohazards. *Biosens. Bioelectron.* **14:**785–794.
76. **Saiki, R. K.** 1985. Enzymatic amplification of beta globin genomic sequences and restriction site analysis for diagnosis of sickle cell anemia. *Science* **230:**1350–1354.
77. **Schiff, K. C., S. B. Weisberg, and V. E. Raco-Rands.** 2002. Inventory of ocean monitoring in the Southern California Bight. *Environ. Manag.* **29:**871–876.
78. **Schiff, K. C., J. Morton, and S. B. Weisberg.** 2003. Retrospective evaluation of shoreline water quality along Santa Monica Bay beaches. *Mar. Environ. Res.* **56:**245–254.
79. **Scott, T. M., J. B. Rose, T. M. Jenkins, S. R. Farrah, and J. Lukasik.** 2002. Microbial source tracking: current methodology and future directions. *Appl. Environ. Microbiol.* **68:**5796–5803.
80. **Scott, T. M., T. M. Jenkins, J. Lukasik, and J. B. Rose.** 2005. Potential use of a host associated molecular marker in *Enterococcus faecium* as an index of human fecal pollution. *Environ. Sci. Technol.* **39:**283–287.
81. **Shelton, D. R., J. S. van Kessel, M. R. Wachtel, K. T. Belt, and J. S. Karns.** 2003. Evaluation of parameters affecting quantitative detection of *Escherichia coli* O157 in enriched water samples using immunomagnetic electrochemiluminescence. *J. Microbiol. Methods* **55:**717–725.
82. **Simpson, J. M., J. W. SantoDomingo, and D. J. Reasoner.** 2002. Microbial source tracking: state of the science. *Environ. Sci. Technol.* **36:**5279–5288.
83. **Solo-Gabriele, H. M., M A. Wolfert, T. R. Desmarais, and C. J. Palmer.** 2000. Sources of *Escherichia coli* in a coastal subtropical environment. *Appl. Environ. Microbiol.* **66:**230–237.
84. **Sorensen, D. L., S. G. Eberi, and R. A. Dicksa.** 1989. *Clostridium perfringens* as a point source indicator in non-point polluted streams. *Water Res.* **23:**191–197.
85. **Stevenson, A. H.** 1953. Studies of bathing water quality and health. *Am. J. Public Health* **43:**529–538.
86. **Stewart, J. R., R. D. Ellender, J. A. Gooch, S. Jiang, M. Samadpour, and S. B. Weisberg.** 2003. Recommendations for microbial source tracking: lessons from a methods comparison study. *J. Water Health* **1:**225–231.
87. **Stoeckel, D. M., M. V. Mathes, K. E. Hyer, C. Hagedorn, H. Kator, J. Lukasik, T. O'Brien, T. Fenger, M. Samadpour, K. M. Strickler, and B. A. Wiggins.** 2004. Comparison of seven protocols to identify fecal contamination sources using *Escherichia coli*. *Environ. Sci. Technol.* **38:**6109–6117.
88. **Stotts, S. N., O. D. Nigro, T. L. Fowler, R. S. Fujioka, and G. F. Steward.** 2005. Virulence and antibiotic resistance gene combinations among *Staphylococcus aureus* isolates from coastal waters of Oahu, Hawaii. *J. Young Investig.* **12:**4. [Online.] http://www.jyi.org/research/re.php?id=148.
89. **U. S. Environmental Protection Agency.** 1986. *Ambient Water Quality Criteria for Bacteria—1986*. EPA 440/5-84-002. U.S. Environmental Protection Agency, Washington, D.C.
90. **U. S. Environmental Protection Agency.** 2005. *Microbial Source Tracking Guide Document*. EPA 600–R–05–064. June 2005. U.S. Environmental Protection Agency, Washington, D.C.
91. **Wade, T. J., N. Pai, J. N. S. Eisenberg, and J. M. Colford, Jr.** 2003. Do US EPA water quality guidelines for recreational waters prevent gastrointestinal illness? A systematic review and meta-analysis. *Environ. Health Perspect.* **111:**1102–1109.
92. **Wade, T. J., R. L. Calderon, E. Sams, M. Beach, K. P. Brenner, A. H. William, and A. P. Dufour.** 2006. Rapidly measured indicators of recreational water quality are predictive of swimming-associated gastrointestinal illness. *Environ. Health Perspect.* **114:**24–28.
93. **Wang, R. F., W. W. Cao, and C. E. Cerniglia.** 1996. PCR detection and quantitation of predominant anaerobic bacteria in human and animal fecal samples. *Appl. Environ. Microbiol.* **62:**1242–1247.
94. **Wu, C., J. J. Valdes, W. E. Bentley, and J. W. Sekowski.** 2003. DNA microarray for discrimination between pathogenic O157:H7 EDL933 and non-pathogenic *Escherichia coli* strains. *Biosens. Bioelectron.* **19:**1–8.

Detection of Viruses in Environmental Waters, Sewage, and Sewage Sludges

CHRISTON J. HURST AND KELLY A. REYNOLDS

23

A great many different groups of viruses are found in environmental waters. These include the many types of viruses whose hosts are natural aquatic organisms. There also are groups of viruses present in environmental waters which represent exogenous contaminants, whose hosts are non-aquatic plants and animals. Those water-associated viruses which are of greatest concern from the viewpoint of human public health are viruses which replicate in cells of the human gastrointestinal tract of infected individuals. These are referred to as human enteric viruses, and they are shed in fecal material. The principal transmission route for enteric viruses is fecal-oral, meaning that they can cause illness when a susceptible host ingests fecally contaminated water or food (9). Septic tank effluents and wastewater from human populations doubtlessly contribute many of the human enteric viruses that are found in surface waters and groundwaters (1, 7, 21, 23, 25). The presence of human enteric viruses in surface water can also result from human recreational activities performed either in or around those bodies of water.

At least some of the human enteric viruses can cross-infect, meaning that they are capable of also replicating within animals other than humans. As an example, acquisition of the disease poliomyelitis, caused by a member of the genus *Enterovirus*, has been documented in a wild population of chimpanzees that were living in relative proximity to a human community (26). Other genera of enteric viruses which should be considered to have the potential for cross-infection of related mammalian host species are *Orthoreovirus*, *Rotavirus*, and *Mastadenovirus*. Likewise, some enteric viruses of animals likely can cross-infect humans. Sources of such viruses are fecal wastes from animals, including farm wastewater which can drain into surface waters, overland runoff containing animal fecal wastes, and fecal wastes directly deposited in the water by animals. It is difficult, and in some instances it may be impossible, to determine the host species from which many of these potentially cross-infective enteric viruses originate. Thus, for the purpose of protecting public health, the presence in water of any virus capable of replicating in mammalian host animals or cultured mammalian cells is presumed to represent a pathogenic hazard to humans.

Enteric viruses are capable of surviving improperly or inadequately operated conventional drinking-water treatment processes and may therefore be found in potable water (8, 27). This knowledge has resulted in the development of methods for detecting both human enteric viruses and also fecally associated bacteriophages in water (10). These bacteriophages are not considered to be enteric in nature, because they infect the intestinal bacteria rather than the animal host, and correspondingly they themselves do not pose a public health threat. Bacteriophages are considered unsuitable in terms of meeting regulatory requirements for assessing enteric virus pollution due to problems with lack of specificity and sensitivity for many of the assay techniques and the lack of an association between detection of bacteriophage and disease occurrence in human populations (15). Instead, these bacteriophages have been studied as potential indicators of the presence of fecal material and on the premise that they may prove useful as indicators by which to gauge the fate of human enteric viruses within the natural environment. They also may prove useful as indicators of the removal or destruction of human enteric viruses during the course of water treatment processes (8, 19). Selectively choosing the bacterial hosts employed in the assay process, including the decision as to whether the bacterial host strains used would be considered indigenous or nonindigenous to the environment being sampled, is an approach which may be used to detect specific bacteriophage populations (22). Conversely, microscopic methods (28) and even automated techniques such as flow cytometry have been used to detect total bacteriophage populations in samples of water (16).

The levels of bacteriophages in environmental waters, particularly those that receive sewage effluent, are often sufficiently high that the water can be directly assayed for the presence of these bacteriophages without need for resorting to concentration techniques. This represents a desirable advantage, since viral concentration techniques have a measure of inherent inefficiency which results in some loss of virus. Also, some virus inactivation may occur during concentration processes as a result of such factors as changes in pH and the addition of chelating chemicals like EDTA. Because of these concerns, it is recommended that although the use of sodium hydroxide and sulfuric acid washes to recover viruses from filters may not be extremely deleterious to viral detection as determined by nucleic acid analysis (6, 7, 13), such washes should be considered only with due caution when viral detection is to be performed by infectivity analysis. The levels of human enteric viruses present in environmental waters often are so low as to require that they be concentrated from the water in order to perform a successful, cost-effective assay. This chapter describes techniques that can be used for concentrating human enteric viruses from environmental water, drinking water, raw wastewater, wastewater sludges, and wastewater effluents.

MECHANISMS INVOLVED IN VIRAL CONCENTRATION METHODS

A large variety of approaches have been used for concentrating viruses contained in water samples (4, 6, 7, 10, 13, 20, 27). These can be separated into five major categories based upon the mechanisms involved (10).

Passive Adsorption

Passive adsorption relies upon unaided entrapment of viruses via adsorption or absorption into pads of gauze and cotton, used in stationary or flowthrough configurations. This is followed by expression of the entrapped fluids, either with or without the supplemental use of an eluant solution, to recover the viruses.

Directed Adsorption

Directed adsorption entails adsorption of viruses to the surface of filter materials or granular solids, facilitated in many instances by pretreatment of the water. Processes which have been evaluated as pretreatments include using charged resins to remove dissolved organics from the water, adjustment of the water sample pH, and addition of salts to the water. Efforts have also been made to pretreat the adsorbent material itself in order to enhance the efficiency of viral adsorption. Examples of the latter approach include the binding of metal precipitates or charged polymers to the matrix of microporous filters. Subsequent recovery of the bound viruses from the filter materials or granular solids used as adsorbents is achieved by exposing the adsorbents to an eluant solution.

Ultrafiltration

Ultrafiltration involves retention of viruses in the original water sample during a reduction of its volume achieved by pore size exclusion. This generally is accomplished by either recirculating the water, under pressure, through hollow-fiber filters, or over the surface of flat-sheet filters in either recirculating or nonrecirculating systems. In some cases an eluant is subsequently passed through the same concentration unit to facilitate virus recovery.

Direct Physicochemical Flocculation and Phase Separation

Direct physicochemical flocculation and phase separation are representative of processes which use chemicals that cause a water sample to partition into two parts. One of those parts contains the viruses, which in effect results in concentration of the viruses. This category of viral concentration methodology relies upon (i) physicochemical flocculation (the viruses' associating with a precipitate which forms when chemicals are added to the water); (ii) polymeric phase separation (the viruses' selectively partitioning with one of two aqueous phases which are generated and gravimetrically separate when soluble polymeric substances are added to the water sample); or (iii) hydroextraction (placing the water sample into a tubing made of water-permeable membrane, with that tubing in turn packed in a hydroscopic polymeric substance which draws water out through the pores of the membrane, to result in the viruses and only a small amount of residual water being retained within the tubing).

Affinity Chromatography

Affinity chromatography is the process whereby viruses are retained during passage of the water sample through a column of polysaccharide gel particles which bear covalently linked antibodies that specifically recognize and bind a particular virus type. Viruses are then released from the column by alterations in the composition or ionic strength of the buffer fluid used for maintaining hydration of the polysaccharide gel.

THEORETICAL EXPLANATION OF DIRECTED ADSORPTION, ELUTION, AND RECONCENTRATION OF VIRUSES

Directed adsorption has generally replaced the other types of concentration techniques for use in recovering viruses from large sample volumes of water. The different types of solid adsorbents that have successfully been utilized for this can be divided into two groups: filters and granular solids. The types of filters that have been tested are composed of either yarn fiber wound around a hollow core to form a depth filtration cartridge, sheet filter materials configured either as flat layers or in the form of cartridges composed of pleated sheets of filter material, or cartridges prepared as hollow tubes of filter material. Viruses adsorb onto the filter matrix during the passage of virus-containing water samples through these filters. Recovery of the adsorbed viruses is normally achieved by subsequently passing an eluant through the filters or by dissolution of the filter if it is made of alginate. Granular solids generally are used as viral adsorbents in one of three modes: batch utilization by which the granules are mixed into the water sample and then recovered by using either filtration or magnetic attraction, the latter requiring that the granules be magnetic; as layers of granules either supported on or sandwiched between sheets of non-virus-adsorbing flat filter material through which the water sample is passed; or as columns or fluidized beds of granules exposed to the water sample by using a flowthrough configuration. Subsequent desorption of viruses from the granular solids occurs during exposure of the granules to an eluant using one of these same three modes (10).

Of the many different types and configurations of virus adsorbents, the ones that currently seem to be preferred for use in recovering viruses from large volumes of water are wound-yarn cartridge filters; pleated-sheet cartridge filters based upon either glass microfiber or nylon, including the use of sheet filter material types that are modified to have a more electropositive charge; and columns of glass powder. Filters seem to offer some advantage over the use of glass powder columns in that the filters are easily transported to the field (even by postal shipment) and readily used in the field, after which, if packed properly, they may be shipped back to the processing laboratory with very little concern about detrimentally affecting either the filter or the adsorbed viruses. This chapter presents the use of cartridge filter-based methods for concentrating viruses from environmental water, drinking water, and sewage effluent. The methods presented in this chapter for isolating viruses from raw sewage (raw wastewater) and wastewater sludges effectively utilize a process of directed adsorption and elution but differ in that they rely upon the wastewater solids to serve as an in situ adsorbent.

The adsorption of viruses onto filters, and presumably other solid adsorbents as well, is governed by both electrostatic interactions, which predominate at lower pH levels, and hydrophobic interactions, which predominate at higher pH levels. The chemical compositions of the solid adsorbent and the fluid in which the viruses are suspended influence the extent to which either attraction or repulsion occurs between the viruses and the adsorbent. Two other important factors that relate to virus adsorption from water include the relative proximity and amount of contact time allowed between the adsorbent and those viruses contained in the water sample. The latter set of factors are influenced by the rate at which the water being sampled flows either past or through the solid adsorbent and the ratio of the diameter of the virus particle to the diameter of any pores in the adsorbent material through which the water may pass. Virus adsorption can be inhibited by the presence of added proteins in the input virus suspension, possibly as a result of competition between the proteins and viruses for those adsorption sites available on the solid matrix. Other types of soluble organics such as humic and fulvic acids may also interfere with virus adsorption. Means of facilitating the adsorption of viruses onto solid matrices include first removing dissolved organic materials from the water samples by passing the water through a resin column and then adding salts, including chlorides of either sodium, magnesium, or, more effectively, aluminum. Virus adsorption to filters can also be facilitated by adjusting (usually lowering) the pH of a water sample, a process that can be performed by adding HCl either through a batch operation or using an in-line injector configured ahead of the filter (3). A pH of 3.5 to 4.0 seems to be preferred for use with negatively charged filters (2, 10), whereas near neutrality is preferred for the more positively charged filters (10). There may also be optimal pH levels for adsorption of viruses to the different types of granular solids.

The adsorption process is reversible, and the reverse process is termed elution. Elution is accomplished by exposing the adsorbent to an aqueous solution termed an eluant, following which the eluant is termed an eluate and contains at least some of the viruses which had been adsorbed to the solid matrix. Eluants can be divided into at least two categories based on their mode of action (10). The first category includes proteinaceous materials which simply compete with the proteins of the virus particles for binding sites on the adsorbent. The second category consists of compounds that alter the favorability of adsorption. Beef extract is now the most predominantly used eluant material of the first category and is recommended in this chapter, although other protein products may prove to be suitable substitutes. The second category of eluants includes solutions that contain various active substances, among which are chaotropic agents like glycine or trichloroacetic acid, detergents like Tween 80, and EDTA, which serves as a chelating agent for metal cations. Both categories of eluant may utilize elevated pH, which would help to decrease the electrostatic attraction between the viruses and adsorbents. Eluants containing viruses which have been desorbed from filters generally are reduced in volume by a secondary, or second-step, concentration technique prior to their being assayed for the presence of viruses. The secondary concentration method customarily used for beef extract-based eluants is organic flocculation as introduced by Katzenelson and coworkers (14), which is performed by lowering the pH of the eluate, during which the proteinaceous material supplied by the beef extract spontaneously precipitates. Viruses contained in the beef extract become associated with this precipitate, which can subsequently be collected by centrifugation and dissolved in a small amount of higher-pH salt-based buffer, such as sodium phosphate or glycine. A pH of approximately 3.5 appears optimal for recovering viruses from beef extract eluates via organic flocculation (10).

SAMPLING APPARATUSES AND PROCESSING EQUIPMENT

A variety of sampling apparatuses may be needed when detecting viruses in environmental waters. These include not only the usual sampling jars or bottles but often also buckets, carbuoys, pumps, cartridge filters and their holders, and hoses. If it is necessary to store a volume of sample water which is too large to fit in a standard jug or carbuoy, then even a new sterilized plastic garbage can with a tightly fitting cover can be used as a water storage container. Whenever possible, it is preferable that containers used for water samples be made of polypropylene, which can reduce unintentional adsorption of viruses to container walls (18). Alternatively, it is suggested that the apparatus be made of either stainless steel, brass, or a chemically resistant polymer. This aids in terms of ruggedness for use in field operations, cleanability, and resistance to the chemical corrosion that can be caused by salt or extremes of pH.

It is necessary to neutralize any chemical disinfectant present in the water being sampled to prevent disinfectant-related die-off of the viruses. Neutralization must be done before the water sample is processed on-site or transported for off-site (i.e., laboratory) processing. This necessity often occurs with drinking water and wastewater. The chemical disinfectants most commonly used by water treatment facilities are chlorine, chlorine dioxide, and monochloramine, any of which can be neutralized by the addition of sodium thiosulfate. If concentration of viruses from such water samples is to be performed in the field, then a compact portable chlorine test kit can be used to confirm that, by adding the directed quantity of sodium thiosulfate solution, you have successfully neutralized all free chlorine or other chlorine compounds present in the water sample. If concentration of viruses from water samples in the field is to be done using cartridge filters, then a portable pH meter may be needed, since it might be necessary to adjust the pH of the water samples prior to filtration. As a generalization, those virus-adsorbing filters that are considered electropositive in nature are effective at adsorbing viruses from natural waters whose pH is approximately 7.5 or below. Thus, when using the electropositive filters, you generally need to adjust the pH of only those alkaline waters which are above approximately 7.5. The pH of those alkaline waters should be brought down to approximate neutrality. If negatively charged filters are used, then the sample water must be adjusted to a pH of approximately 3.5 and supplemented with aluminum chloride prior to filtration. Positively charged filters thus offer a tremendous advantage in terms of their ease of use. However, the positively charged filters are more expensive and do not work well for concentrating viruses from saline waters. You should use only negatively charged filters for concentrating viruses from saline waters.

Cartridge Filtration Apparatus

The types of apparatus used for concentrating viruses from large volumes of environmental waters, drinking water, or sewage effluent by means of cartridge filtration are not standard equipment for most environmental microbiology laboratories. Thus, the equipment used with the cartridge filters is described here in some detail. Figure 1 presents the type of equipment configuration used for concentrating viruses by means of cartridge filters. A lightweight portable electric or gasoline-powered water pump fitted for use with either brass or stainless steel quick-disconnect plumbing adapters or garden hose couplings can be used to supply pressure for passing water or sewage effluent through cartridge filters. Such a pump is not needed if the water or wastewater is being supplied directly via a tap from a pressurized source. Figure 1 also shows two filter holders designed for use with standard 10-in.-long cartridge water filters, fitted for use with either quick-disconnect plumbing adapters or garden hose couplings. The first holder contains a non-virus-adsorbing prefilter that may be needed to prevent clogging of the virus-adsorbing filter, which is contained in the second filter holder. If it is necessary to modify the water pH or to add aluminum chloride to aid viral adsorption, then an in-line injector can be configured between the pump (or pressurized tap) and the first filter holder. An alternative to using an injector is to first pump the water sample into a large temporary storage tank, make any necessary chemical adjustments (supplementations) to the water while it is in that tank, and then pump the adjusted water from the tank through the filters. The filter holders should have clear base sections so that the filtration process can be visually monitored, particularly for air pockets which occasionally develop within the filter holders and tend to impede the water flow. Suitable holders should be available from the manufacturers and distributors from which the cartridge filters are obtained. Water being processed through these filters flows from the outside of the filter, through the filter material, into the hollow core. It is helpful if the top of each filter holder is fitted with a stainless steel finger-operated air pressure release valve connecting to the inflow side of the filter holder. The installation of air pressure release valves in this manner allows easy elimination of air pockets from the filter holders. Figure 1 also shows a portable water meter fitted for use with either quick-disconnect plumbing adapters or garden hose couplings attached to the outlet end of the second filter holder. Suitable meters are available from plumbing suppliers and used for metering the amount of water which is passed through the filters. Four lengths of fiber-reinforced garden hose, likewise fitted for use with either quick-disconnect plumbing adapters or garden hose couplings, are used for connecting the pump, filter holders containing their cartridge filters, and water meter to form the virus concentration apparatus shown in Fig. 1. The fourth piece of hose is used to direct outflow from the water meter. Also, one length of strong-walled water supply hose, fitted for use with either quick-disconnect plumbing adapters or garden hose couplings, is required for leading water into the pump. A standard garden hose may not be suitable for this usage, since many types of garden hose would tend to collapse inward during operation of the pump. It may be helpful to have a stainless steel or brass wire strainer fitted to the intake end of the water supply hose when concentrating environmental waters by pumping directly from their source through the filters. Use of a strainer in this instance helps to prevent the water supply hose from clogging. The suggested maximum force for driving a water sample through these filters is 30 lb/in^2. If the pump being used is particularly powerful, it may be necessary to install a flow-restricting device ahead of the first filter holder. An ice chest containing either wet ice or commercial ice packs is needed for transport (or shipment via overnight delivery service) of the used virus-adsorbing filter, inside a polypropylene container of eluant, to the laboratory where the elution and secondary concentration steps are to be performed.

Sterilization Requirements

All reagents that are to be used during any of these methods must be sterilized prior to their use. Solutions of HCl or NaOH which are at least 1 M in strength are considered to be self-sterilizing, as are solutions of at least 0.5% (wt/vol) sodium hypochlorite or calcium hypochlorite. Containers used to store water or wastewater samples should be presterilized. All parts of equipment that will come into contact with the samples should likewise be visibly clean and sterile. The only parts of a pump or fluid injector which need to be sterile are those which come into contact with the water sample. These parts can be treated using a hypochlorite solution sterilization technique. The choice of methods to be used for sterilizing different items of equipment depends on the size of the equipment and the item's resistance to

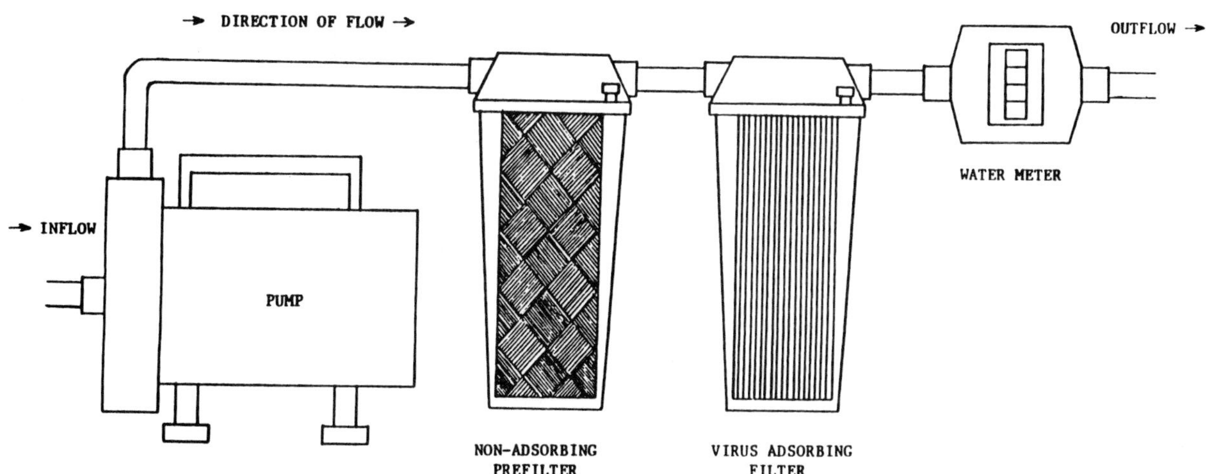

FIGURE 1 Diagram of apparatus used for concentrating viruses by using cartridge filters.

heat. Many types of apparatus can be sterilized in the laboratory before their use in the field, either by autoclaving, ethylene oxide gas, or, if accessible, gamma irradiation. All of the cartridge filters described in this chapter can be presterilized within their holders by using ethylene oxide gas treatment before they are transported to the field. Alternatively, some of the filters have sufficient resistance to heat and steam that they can be sterilized by autoclaving.

Techniques for Field Sterilization of Sampling Equipment

Field sterilization of metal or borosilicate glass objects can be performed by using a flame from a portable gas torch (such as a propane or butane torch) to heat the sample collection objects to red incandescence. Objects less resistant to high heat, such as nonborosilicate glass and some polymer materials, can be surface sterilized by dousing or immersing them in commercial 95% ethanol and then igniting the alcohol with a flame. Nearly all the objects described in this chapter, with the exception of filters and liquids, can be subjected to field sterilization techniques, which consist of chlorine-based chemical treatment using solutions such as calcium hypochlorite or sodium hypochlorite followed by the use of sodium thiosulfate solution to neutralize the residual disinfectant activity.

If the water sample is to be placed into a storage container for any reason, then that container should be visibly clean, must be sterile, and preferably should not previously have been used for any purpose other than the storage of water samples. Large (up to 50-gal and some even larger) polyethylene containers with lids are available from scientific supply companies and can be used for storage of water samples. In an emergency situation, as mentioned earlier in this chapter, a new plastic garbage container with a lid can be used for this purpose. Field sterilization of containers can be done by first completely filling them with clear water and then adding calcium hypochlorite solution (0.5%, wt/vol) to the water at a rate of 3.8 ml per gal. A 10% solution of standard household liquid bleach (sold as 5.25% sodium hypochloride by weight) can be used as a substitute for the calcium hypochlorite solution. The hypochlorite solution should be throughly mixed into the water, and this water should then be allowed to remain in the storage container for at least 15 min. The chlorinated water should then be emptied from the storage container, and the container should be rinsed very thoroughly with clear (lacking obvious turbidity) water and again completely filled with clear water. A sterile sodium thiosulfate solution (50%, wt/vol) should be added to this second filling of water at a rate of 10 ml per gal. The sodium thiosulfate solution should thoroughly be mixed into the water, and the water should then be allowed to remain in the storage container for at least 5 min. The container should then be emptied and filled with the intended water sample.

All of the equipment used in the virus concentration apparatus (Fig. 1), except for the filters, can be field sterilized using this type of chlorine-based technique. The apparatus (shown in Fig. 1) should be disassembled prior to treatment by the field sterilization technique in order to make certain that all surfaces that may later come into contact with the sample water will have received a thorough exposure to the chlorine disinfectant and that all residual hypochlorite is then neutralized before the water sample is processed through the system. Hoses and assembled filter holders (without filters), with their air pressure release valves open, should be carefully submerged in a solution of clear water containing hypochlorite prepared as described above for the treatment of storage containers. Be certain that there are no pockets of trapped air in either the hoses or filter holders, as this may preclude thorough treatment. Keep the hoses and filter holders completely submerged in this chlorinated water for a minimum of 15 min, and then remove and drain the hoses and filter holders. Next, similarly treat the hoses and filter holders by submersion for a minimum of 5 min in clear water containing sodium thiosulfate solution prepared as described above for the treatment of sample storage containers. The hoses and filter holders should be drained and can then be used immediately, or, if they are to be used at some later time, their openings should be wrapped with sterile aluminum foil or some other type of sterilized covering to prevent contamination. Field sterilization of pumps and injectors can be done by first continuously recirculating clear water containing hypochlorite solution through them for at least 15 min. Residual chlorinated water should then be drained from the pump or injector and its necessarily attached hoses. Clear water containing sodium thiosulfate solution, likewise prepared as described above, should then be recirculated through the pump or injector and attached hoses for at least 5 min. Residual water containing thiosulfate solution should then be drained from the pump or injector and attached hoses. The pump, injector, and hoses can then immediately be used to process a water sample, or, if not used immediately, their openings should receive a sterile covering as described above to prevent contamination. When properly performed, this chlorine-based field sterilization technique should be sufficient to destroy the infectiousness of viral contaminants on the treated surfaces of the equipment. This technique may not be sufficient to destroy some encysted protozoans which are resistant to chlorine.

DETECTION OF VIRUSES

Detecting viruses in water involves a three-stage process. The first stage consists of either collecting a grab sample which will be processed in the laboratory or performing a field operation in which the water being sampled is passed through cartridge filters, with those filters usually then returned to the laboratory for processing. The grab sampling approach is likely to be utilized when either the level of viruses or the content of solids is relatively high, as is the case for raw wastewater or wastewater sludges. If the level of viruses and the concentration of solids are expected to be relatively low—as is the case with most environmental surface water, groundwater, treated drinking water, and wastewater effluents—then the cartridge filtration approach is likely to be used. The second stage consists of laboratory elution processing, performed by either of two methods. The first method is used for unfiltered grab samples and consists of concentrating viruses onto the solids naturally present in a sample and then eluting the viruses from those solids. The second method, used for cartridge filters, consists of eluting adsorbed viruses from those filters. Usually either type of eluate will be further reduced in volume through a subsequent concentration process. The third stage consists of assaying viruses contained in the eluates.

Viral Concentration Techniques

The viral concentration techniques presented in this chapter are categorized according to the type of environmental sample being examined. These are environmental waters (fresh or saline), treated drinking water, wastewater effluent, raw wastewater, and wastewater sludges.

Environmental Water, Drinking Water, and Sewage Effluent

The procedure described below can be used to concentrate suspected human enteric viruses from environmental surface waters, including brackish and marine waters; groundwater; tap water; and sewage effluents. The produced concentrated sample materials can then be examined for the presence of viruses. The viruses are first adsorbed onto cartridge membrane filters. A beef extract-based eluant fluid is then used to desorb viruses from the filters. Viruses contained in the eluate are then secondarily concentrated by low-pH organic flocculation. The volume of water sample that is processed when looking for the presence of human enteric viruses depends upon the level of viruses likely to be present in the water. For sampling sewage effluents, the minimum useful sample volume may range from 15 to 30 liters. For general surface waters, a minimum useful volume would probably be 100 liters, with 150 to 200 liters preferred. For water that has very low turbidity, the apparatus described here could be used for processing volumes as great as 1,000 liters, as may be necessary when examining either groundwater or tap water that has been processed by sand filtration.

If the intended water sample consists of wastewater effluent, tap water, or some other water that may have received a chemical disinfectant, then the collected water sample should be dosed with a sodium thiosulfate solution (50%, wt/vol) at a rate of 10 ml per gal, and a sample of the dosed water should be checked to be certain that all residual disinfectant activity has been neutralized. This neutralization of residual disinfectant activity must be done before the water is either stored as a bulk sample for processing off-site or processed on-site for concentrating any viruses which the water may contain. If possible, the water sample should be processed by the virus concentration technique immediately upon its collection. If it is necessary to store a water sample before it is processed, then the storage container must first be adequately treated by a sterilization technique, and after the sample water is added, the container should be kept covered to prevent contamination. The sample should also be kept in a cool location to reduce viral inactivation until the water sample can be processed. Prolonged storage of water samples should be done in a refrigerated room (no more than approximately 4°C) and preferably for no longer than 2 days.

For each water sample that is to be processed using cartridge filtration you need a virus-adsorbing filter. You also may need a prefilter (if the water has such a high turbidity that it cannot easily be passed directly through the virus-adsorbing filter). Neither of these filter types should be reused. The prefilter should be a 3-μm-porosity (nominal porosity) wound polypropylene yarn filter of the type available as stock number M27R10S from Process Filtration Division, Parker Hannifin Company (Indianapolis, Ind.). When using cartridge filters that are considered to be positively charged for the purpose of concentrating viruses from nonsaline environmental waters, it should not be necessary to chemically modify the waters' pH prior to passing them through the filters except for lowering the pH of naturally alkaline waters to below the range of 7.5 to 8.5. This is the maximum pH range which should be used when adsorbing viruses onto the positively charged filters. The recommended positively charged virus-adsorbing filters are the pleated 0.45-μm-porosity glass and cellulose fiber membrane type, presently available as the Zeta Plus Virosorb 1MDS filter cartridge from Cuno Division of 3M (Meriden, Conn.), and the pleated 0.20-μm-porosity charged nylon membrane type, presently available as a Posidyne (formerly designated Posidyne N-66) filter cartridge from Pall Corporation (East Hills, N.Y.) (10). Both of these virus-adsorbing filters are positively charged relative to untreated cellulose ester or glass microfiber filters and are therefore suitable for concentrating viruses from water at most ambient pH levels (approximately 5 to 7.5). If the 1MDS type of charged fiber filter is used, then it is preferable to use dilute hydrochloric acid to adjust the pH of any alkaline water samples to less than 7.5 before filtration (10). The charged nylon filter type may be preferable for use with water samples of moderately alkaline pH level, as these filters can be used effectively for concentrating viruses from water at pH levels of up to approximately 8.5 (10). A possible alternative choice for the use of a positively charged filter would be the cellulose, glass, and melamine matrix filter presently available as the Cuno Zeta Plus MK (Micro Klean) filter cartridge from Cuno Div. You can also use a more electronegatively charged glass microfiber filter, of the type available as a 0.25-μm-porosity Duo-Fine E series filter cartridge from Pall Corp. (10). In fact, the use of this type of filter is preferred when concentrating viruses from either brackish or marine water, as the more positively charged filters seem to be inefficient when used on saline waters. When you use the more negatively charged filters, such as the Duo-Fine series filter cartridge, as your virus-adsorbing filters, it is necessary to both adjust the pH of the sampled water to 3.5 by addition of hydrochloric acid and add aluminum chloride before passing the sample water through the filter. The recommended final concentration of aluminum chloride varies depending on the nature of the water sample, from 0.0005 M (freshwater) to 0.0015 M (full seawater) (20).

The various reagents, namely, aluminum chloride, hydrochloric acid, and sodium thiosulfate, which may need to be added to the water sample before the sample passes through the filters (2, 10, 20), may be prepared as stock reagent solutions and injected into the flowing sample in between the pump (or pressurized source) and the prefilter (3). Alternatively, a batch adjustment technique can be used. For this technique, the water sample is collected into a sterilized temporary storage container, chemical supplementation of the water is performed while the sample is in that storage container, and the water is then pumped through the filters.

Assemble the cartridge filter virus concentration apparatus as shown in Fig. 1, making certain to properly align the pump, filter holders containing their respective filters, and water meter with respect to their indicated directions of water flow. If injectors are to be used for chemical supplementation, they should be installed ahead of the prefilter. Be certain that the tops of the filter holders are fully tightened to their bottom sections, that the filter holders contain the correct filters, and that the air pressure release valves on the filter holders are closed. Connect one end of the water supply hose to the pump inlet, and place the other end of this hose into the water being tested, whether it is being taken directly from an environmental water source, a tap water source, or a storage container prepared as described above. Note the reading on the water meter (flow meter), as these meters usually cannot be reset; in order to know how much water sample you have filtered, it is necessary to read the meter both before and after the filtration.

Turn on the pump and let the sample water pass through the filters and meter. It may be necessary to occasionally release trapped air pockets from the filter holders. The presence of air pockets that form within the filter holders can easily be seen by looking at the clear bottom parts of the holders. Such air pockets should be released, as they otherwise may reduce the rate at which water is able to flow through the filters. To release air, first make sure that the holder in question is being held upright so that the trapped air can pass through the release valve. While keeping the holder upright, slightly open the valve until the air has passed through, and then close the valve to keep the water sample from leaking out. If the holder does not contain a release valve, it sometimes is possible to eliminate the trapped air by rapidly inverting the filter holder. This maneuver is done by first placing the filter holder into an upright vertical position and then quickly inverting the filter holder, followed by rapidly restoring the holder to its normal upright position. It may be necessary to repeat this flipping maneuver. Alternatively, while making certain to hold the filter holder upright, the top of the filter holder may be loosened slightly where the top connects with the bottom by means of screw threads, to release trapped air. The top and bottom sections should then be tightened back together when water rather than air begins to seep out between the screw threads. After the water sample has been filtered, disassemble the apparatus and allow any remaining water to drain from the filter holders.

There are several procedures available for eluting viruses from the virus-adsorbing cartridge filters. This chapter describes two which allow elution to begin in the field, with eluant presterilized ahead of time in the laboratory. For each water sample to be processed, you need 1,600 ml of filter eluant solution consisting of sterile, pH 7.0, 3% (wt/vol) beef extract solution contained in a separate watertight, 1-gal-capacity, wide-mouth, screw-cap autoclavable container. A polypropylene container is preferred for this purpose because viruses do not readily adsorb to polypropylene. The eluant solution can be sterilized by autoclaving it inside these polypropylene containers. Containers of this type are available as stock number 2121-0010 from Nalge Nunc International (Rochester, N.Y.). The eluant (beef extract) solution should consist of 48 g of microbiological grade beef extract powder. The type of beef extract powder chosen should be one which produces an easily visible amount of precipitate when, dissolved in distilled water, its pH is adjusted to 3.5. It may be helpful if at least one-half of this powdered beef extract is of the type sold as Bacto Beef Extract, Desiccated Dried Form, by BD (formerly Becton Dickinson and Company, Franklin Lakes, N.J.). This helps to ensure that later on an easily visible precipitate will be produced during subsequent secondary concentration of the filter eluant. Performing either of these two elution procedures can be done more easily if a supporting device is made to maintain in an upright position the filter holders containing the filters. We have seen these types of supporting device made in many, and sometimes imaginative, ways. A simple suggestion is to take a short three-legged wooden stool and cut two round holes in the seat, with the diameter of those holes being slightly greater than the diameter of the filter holders. Placing the filter holders into these holes then maintains the filter holders upright. Because the three legs serve as a tripod, this type of support should be able to sit in a steady manner on nearly any horizontal surface.

The first field elution procedure requires that the filter holders then be opened carefully. The virus-adsorbing membrane filter should aseptically be placed into the container of beef extract solution (eluant). If a prefilter was used, both it and one-half of the eluant should be placed into an additional sterile 1-gal-capacity polypropylene container. These filters should be kept in their containers of eluant and, along with wet ice or ice packs, immediately placed into the insulated container. The insulated container of samples should then be sealed and immediately be transported to the laboratory or else shipped there, preferably by an overnight delivery service. These filters must be kept in their containers of eluant until such time as the filters are processed by the virus elution technique. The virus-adsorbing filter (and prefilter, if one was used) and filter eluant should be processed as soon as possible (preferably immediately) upon their receipt by the laboratory, using the elution and secondary concentration processes described below. If immediate elution is not possible, then the filters (and also any prefilters that have been used) should be stored under refrigeration at a temperature not greater than about 4°C until such time as the elution and secondary concentration processes can be performed. Storage of the filters inside their respective containers of eluant is done to prevent drying of the filters, since drying causes most waterborne viruses to die off very quickly. Please note that excessive bacterial growth in the filters and containers of eluant may result if the total period of transport and storage is longer than 2 days. Therefore, if lengthy storage must be done, then storage should be performed at −70°C, although with the understanding that some viruses lose viability when frozen. Frozen storage at a relatively warmer temperature, such as −20°C, would be less preferred because viruses tend to die off more rapidly in parallel with the increased sublimation which occurs at this relatively warmer temperature. In particular, members of the family *Inoviridae* seem extremely susceptible to loss of infectivity when frozen, even at temperatures of about −70°C (C. J. Hurst, unpublished observation).

To continue this first elution procedure, open the container of eluant and aseptically remove the virus-adsorbing filter from its container of eluant and place the filter into a filter holder that has been presterilized. The top of the filter holder should be fully tightened to its bottom section, and the air pressure release valve should be closed. The outlet side of the filter holder should then be connected by a hose to the outlet of a stainless steel pressure vessel (1-gal capacity, of the type sold by Millipore Corporation, Billerica, Mass.), which will hold the eluant. If a prefilter was used, then the prefilter likewise should be placed into a presterilized holder, and both holders (respectively containing the virus-adsorbing filter and prefilter) should be connected together in series so that the filters which they contain will be eluted simultaneously. Next, while continuously stirring the full 1,600-ml volume of filter eluant, adjust it to a pH of 9.5 by dropwise addition of 1 M sodium hydroxide and hydrochloric acid as necessary. The eluant then should be placed into the pressure can and from there be passed through the filter (and prefilter, if one was used) three times sequentially. The flow direction of the eluant during this elution process should be opposite to the flow direction used for the water sample during the virus adsorption. Air pressure supplied to the inlet side of the pressure can at approximately 15 lb/in^2 provides the necessary force for driving the eluant through the filters. After passing through the filter (and prefilter, if used), the eluate then should be collected into a sterile polypropylene container. This elution procedure represents a reversal of the virus adsorption technique described above.

The eluted filters should then be discarded. Any eluant remaining in the filter holder (or both holders, if a prefilter was used), pressure can, or hoses should then be added to the eluate that has passed through the filters. At this point, the pH of the eluate should immediately be neutralized by the dropwise addition of 1 M hydrochloric acid.

Alternatively, the second elution procedure begins in the field by reassembling the virus concentration apparatus as shown in Fig. 1 but without either the pump or the water meter. A hand-pumped, positive-pressure spray tank (formerly called Spray Doc; Gilmour Manufacturing, Somerset, Pa.) fitted with quick-release connectors is connected in place of the water meter in such a way as to provide air pressure into the outflow side of the holder which contains the virus-adsorbing filter. In this manner, the direction of flow of the eluant is the reverse of the flow direction used during the virus adsorption process. This type of spray vessel is available at many hardware stores that sell products for use in homes and gardens. It is important to note here that the eluant solution is not to be placed into the spray vessel, as the spray vessel is used only to provide a gentle source of air pressure to facilitate the elution process. The 1,600-ml volume of 3% beef extract eluant contained in the polypropylene container is then adjusted to pH 9.5. As much of this beef extract eluant solution as can fit (approximately 900 ml) then is poured into the filter housing which already contains the virus-adsorbing filter. Using air pressure provided by the spray vessel, this eluant is then forced from the filter holder, through the prefilter in its housing if a prefilter was used, and collected in the polypropylene container. Correct operation of this elution process is performed by gently pumping the spray vessel until the pressure increases and the beef extract solution flows evenly through the filter holder. This process should then be repeated five more times, each time pouring as much of the eluant solution as will fit into the holder which contains the virus-adsorbing filter and using air pressure provided by the spray vessel to gently force that eluant from the filter holder with the eluant collected in the polypropylene container. Then, adjust the beef extract eluant back to pH 7 and return it to the laboratory in the polypropylene container. This provides an elution which numerically is roughly equivalent to passing all 1,600 ml of eluant three times through the virus-adsorbing filter as was described above for the first elution procedure.

Back in the laboratory, a portion of the pH-neutralized filter eluate can be removed to a separate container, and this portion of the eluate can then be saved and subsequently assayed for the presence of bacteriophage. The enteric viruses contained in the filter eluate tend to have a greater resistance to brief low pH exposure than do many types of bacteriophage. Enteric viruses contained in the filter eluate (produced by following either of the two elution procedures described herein) then can be subjected to a secondary, or second-step, concentration procedure. This second concentration step, as described by Katzenelson et al. (14), represents a 20-fold concentration of the filter eluate. To begin the Katzenelson procedure, lower the pH of the filter eluate to 3.5 by dropwise addition of 1 M hydrochloric acid with continuous stirring. During this pH adjustment the eluate should become cloudy as an organic precipitate forms. The pH 3.5 eluant should be stirred for 30 min and then centrifuged for 10 min at 3,000 × *g* to collect the precipitate. The supernatant resulting from this centrifugation should be discarded, and the precipitate from all 1,600 ml of eluant should be dissolved in place (within the centrifuge bottles) using a total volume of 80 ml of sterile, 0.15 M Na_2HPO_4 prepared in distilled water. The pH of the dissolved precipitate, now referred to as final virus concentrate, should then be checked and if not already between pH 7.0 and 7.2 adjusted to within that range by adding 1 M sodium hydroxide or 1 M hydrochloric acid. Antibiotics can be added to the final virus concentrate if desired. Suggested final antibiotic concentrations are 200 U/ml for penicillin G, 200 μg/ml for streptomycin sulfate, and 2.5 μg/ml for amphotericin B. These concentrates can then be stored at −70°C prior to assay if the viruses of interest are stable when frozen in this solution. Otherwise, the concentrates should be stored under refrigeration (no greater than about 4°C). These same storage conditions should be used for any pH-neutralized filter eluate that was kept for bacteriophage analysis.

The same types of filter material used in fabricating the cartridge filters are usually available commercially as circular flat-sheet filters which can be used with other types of commercial filter holders. These flat-sheet filters and filter holders are often suitable for processing smaller volumes of sample and can be eluted by passing a correspondingly smaller volume of the same type of beef extract-based eluant through the filters (2, 10).

Raw Wastewater

The following procedure was developed for the detection of viruses in 8-liter-volume samples of raw sewage (12). Following collection of the sample, adjust it to a final concentration of 0.05 M $MgCl_2$ by addition of a 4 M $MgCl_2$ stock solution. The sample should then be stored overnight under refrigeration (preferably at 1°C) to allow the solids to settle, following which approximately the top one-half of the supernatant volume is decanted and discarded. The remainder of the sample is then centrifuged in aliquots for 20 min at 9,500 × *g* in order to pellet the solids. All of the pelleted solids from the sample are then combined and resuspended to approximately 120 ml with phosphate-buffered saline (pH 7.2, 0.01 M phosphate, prepared using sodium phosphates) containing 2% (vol/vol) fetal bovine serum. The resuspended solids are then extracted with an equal volume of chloroform (trichloromethane) and centrifuged at 1,200 × *g* for 30 min to separate the two phases. The upper (aqueous) phase from this extraction is collected and centrifuged for 20 min at 9,500 × *g* to remove residual suspended solids. Then, the supernatant from this last 9,500 × *g* centrifugation step is adjusted to a final concentration of 3.0% (wt/vol) beef extract by addition of a 30% (wt/vol) beef extract solution and is subjected to concentration by organic flocculation at pH 3.5, using the Katzenelson procedure (14) as defined earlier in this chapter, and the low-pH precipitate is dissolved in 10 ml (final volume) of 0.15 M Na_2HPO_4. The dissolved precipitate should be adjusted to pH 7.0 to 7.2 if it is not already within that range. Also, the small amount of residual suspended solids which had been present in the upper phase following chloroform extraction should be eluted by mixing them in a 5-ml volume of buffered 10% beef extract elution solution (containing, per liter, 100 g of commercial powdered beef extract, 13.4 g of $Na_2HPO_4 \cdot 7H_2O$, and 1.2 g of citric acid), followed by centrifugation for 10 min at 2,800 × *g* to repellet the solids and filtration of the resulting supernatant through a sterile 0.25-μm-pore-size (rated by absolute filtration) microbiological filter. In order to yield a final concentrated sample, the dissolved precipitate resulting from organic flocculation of the chloroform-extracted supernatant should be combined with the filtered eluate from the residual solids. The volume of the final concentrated sample ranges from approximately 11 to 14 ml (because there is some volume loss associated with processing the eluant of the residual suspended solids), representing an overall sample concentration factor of approximately 6,000-fold. These concentrated samples can then be stored at −70°C prior to assay if the viruses of interest are stable when frozen in this solution. Otherwise, the concentrated samples should be stored under refrigeration (preferably at not more than approximately 4°C). The beef extract used should be of a type that produces a visible amount of precipitate when adjusted to pH 3.5.

Wastewater Sludge

Viruses can be isolated from wastewater sludges using the following procedure (29). An appropriate sample size is approximately 500 ml of primary wastewater sludge, or 2,000 ml of secondary (mixed-liquor, activated) wastewater sludge. Each sludge sample should be adjusted to pH 3.5 with 1 M HCl and then 0.05 M $AlCl_3$ added to a final concentration of 0.0005 M. This can be done while stirring the sludge in a large beaker. The adjusted sample should then be centrifuged at 1,400 × *g* for 15 min, and the supernatant should be discarded. Resuspend the solids pellet in an amount of buffered 10% beef extract elution solution equal to five times the pellet volume. The beef extract used should be of a type that produces a visible amount of precipitate when adjusted to pH 3.5. The solids pellet should be stirred in the eluant solution for 30 min. The resuspension should then be centrifuged at 7,000 × *g* for 30 min, and the supernatant yielded from this centrifugation should then be passed in series through sterilized filters capable of effectively removing particles of diameters to less than 0.25 μm based upon nominal porosity (a suggestion being AP40, followed by either AP25 or AP20 and then AP15 series glass fiber filters from Millipore Corporation, obtainable as 142-mm-diameter disks) to remove remaining sludge solids and contaminating bacteria. The resulting filtrate can be assayed for bacteriophages. If the goal is to detect enteric viruses, then the filtrate should be diluted with a volume of distilled water equal to 2.3 times the filtrate volume (to achieve a final effective beef extract concentration of 3% by weight) and concentrated using the Katzenelson technique (14) as defined above. The supernatant resulting from the Katzenelson procedure is then discarded, and the precipitate is resuspended using a volume of 0.15 M Na_2HPO_4, equal to 1/20 the volume of the diluted filtrate. These concentrates can be stored at −70°C prior to the assay if the viruses of interest are stable when frozen in this solution. It has been noted that at least some types of viruses die off even when stored at −70°C in these processed eluates (11). If there is suspicion regarding the stability of viruses when they are frozen in any type of sample, then storage should be done under refrigeration (recommended to be approximately 4°C or less) and for the least possible amount of time. In general, the colder the storage temperature is, the greater the stability of the viruses will be (except for those few virus groups whose members are extremely susceptible to the effects of freezing). Modifications of this procedure such as processing of the beef extract eluant by fluorocarbon extraction (17) following viral elution or alternatively eluting the viruses by an enzymatic method (24) may prove helpful when the viral assay technique is to be done by either PCR or reverse transcriptase-PCR, as those modifications may reduce the possibility of the beef extract materials causing interference in the nucleic acid enzymatic assays. Alternate sludge-processing methods that were collaboratively tested have been published by Goyal et al. (5), and a good comparative testing was published by Monpoeho et al. (17).

VIRAL ASSAY TECHNIQUES

The infectivity of enteric viruses contained in environmental samples can be examined by inoculating the sample into cultures of either human or animal cells that are prepared in the laboratory as opposed to inoculating them into live animals. The simplest way to score the results from this type of test relies upon looking for virally induced changes in the inoculated cells, either in the form of a cytopathogenicity assay technique (looking for virus-induced cytopathogenic effects) or a plaque formation assay technique (looking for the development of focal areas of cell death). Methods for performing these two types of assays and also molecular approaches for detecting enteric viruses contained in environmental samples are described in chapter 48. It may prove necessary to use toxicity reduction methods when assaying concentrated environmental samples in cultured human or animal cells. The cell monolayer washing procedure described in chapter 48 can be used for reducing cytotoxic effects. Methods for using plaque formation assays to detect bacteriophages contained in environmental samples are likewise described in chapter 48. General approaches used for alternative types of viral assay techniques are described in chapter 8.

REFERENCES

1. **Abbaszadegan, M., M. LeChevallier, and C. Gerba.** 2003. Occurrence of viruses in US groundwaters. *J. Am. Water Works Assoc.* **95**(9)**:**107–120.
2. **Biziagos, E., J. Passagot, J.-M. Crance, and R. Deloince.** 1989. Hepatitis A virus concentration from experimentally contaminated distilled, tap, waste and seawater. *Water Sci. Technol.* **21:**255–258.
3. **Dahling, D. R., and B. A. Wright.** 1987. Comparison of the in-line injector and fluid proportioner used to condition water samples for virus monitoring. *J. Virol. Methods* **18:**67–71.
4. **Donaldson, K. A., D. W. Griffin, and J. H. Paul.** 2002. Detection, quantitation and identification of enteroviruses from surface waters and sponge tissue from the Florida Keys using real-time RT-PCR. *Water Res.* **36:**2505–2514.
5. **Goyal, S. M., S. A. Schaub, F. M. Wellings, D. Berman, J. S. Glass, C. J. Hurst, D. A. Brashear, C. A. Sorber, B. E. Moore, G. Bitton, P. H. Gibbs, and S. R. Farrah.** 1984. Round robin investigation of methods for recovering human enteric viruses from sludge. *Appl. Environ. Microbiol.* **48:**531–538.
6. **Haramoto, E., H. Katayama, and S. Ohgaki.** 2004. Detection of noroviruses in tap water in Japan by means of a new method for concentrating enteric viruses in large volumes of freshwater. *Appl. Environ. Microbiol.* **70:**2154–2160.
7. **Haramoto, E., H. Katayama, K. Oguma, H. Yamashita, E. Nakajima, and S. Ohgaki.** 2005. One-year monthly monitoring of Torque teno virus (TTV) in wastewater treatment plants in Japan. *Water Res.* **39:**2008–2013.
8. **Hurst, C. J.** 1991. Presence of enteric viruses in freshwater and their removal by the conventional drinking water treatment process. *Bull. W. H. O.* **69:**113–119.
9. **Hurst, C. J.** 2001. Prevention of infection from food and water, p. 1191–1203. *In* S. S. Block (ed.), *Disinfection, Sterilization and Preservation*, 5th ed. Lippincott Williams & Wilkins, Baltimore, Md.

10. **Hurst, C. J., W. H. Benton, and R. E. Stetler.** 1989. Detecting viruses in water. *J. Am. Water Works Assoc.* **81**(9)**:**71–80.
11. **Hurst, C. J., and T. Goyke.** 1986. Stability of viruses in waste water sludge eluates. *Can. J. Microbiol.* **32:**649–653.
12. **Hurst, C. J., K. A. McClellan, and W. H. Benton.** 1988. Comparison of cytopathogenicity, immunofluorescence, and *in situ* DNA hybridization as methods for the detection of adenoviruses. *Water Res.* **22:**1547–1552.
13. **Katayama, H., A. Shimasaki, and S. Ohgaki**. 2002. Development of a virus concentration method and its application to detection of enterovirus and Norwalk virus from coastal seawater. *Appl. Environ. Microbiol.* **68:**1033–1039.
14. **Katzenelson, E., B. Fattal, and T. Hostovesky.** 1976. Organic flocculation: an efficient second-step concentration method for the detection of viruses in tap water. *Appl. Environ. Microbiol.* **32:**638–639.
15. **Leclerc, H., S. Edberg, V. Pierzo, and J. M. Delattre.** 2000. Bacteriophages as indicators of enteric viruses and public health risk in groundwaters. *J. Appl. Microbiol.* **88:** 5–21.
16. **Marie, D., C. P. D. Brussaard, R. Thyrhaug, G. Bratbak, and D. Vaulot.** 1999. Enumeration of marine viruses in culture and natural samples by flow cytometry. *Appl. Environ. Microbiol.* **65:**45–52.
17. **Monpoeho, S., A. Maul, B. Mignotte-Cadiergues, L. Schwartzbrod, S. Billaudel, and V. Ferré.** 2001. Best viral elution method available for quantification of enteroviruses in sludge by both cell culture and reverse transcription-PCR. *Appl. Environ. Microbiol.* **67:**2484–2488.
18. **Moore, R. S., D. H. Taylor, L. S. Sturman, M. M. Reddy, and G. W. Fuhs.** 1981. Poliovirus adsorption by 34 minerals and soils. *Appl. Environ. Microbiol.* **42:**963–975.
19. **Nelson, K. L., B. J. Cisneros, G. Tchobanoglous, and J. L. Darby**. 2004. Sludge accumulation, characteristics, and pathogen inactivation in four primary waste stabilization ponds in central Mexico. *Water Res.* **38:**111–127.
20. **Nestor, I.** 1983. Méthodologie de détection des virus entéritiques dans l'eau de mer. *Rev. Epidemiol. Sante Publique* **31:**21–38.
21. **Paul, J. R., J. D. Trask, and S. Gard.** 1940. II. Poliomyelitic virus in urban sewage. *J. Exp. Med.* **71:**765–777.
22. **Puig, A., N. Queralt, J. Jofre, and R. Araujo.** 1999. Diversity of *Bacteroides fragilis* strains in their capacity to recover phages from human and animal wastes and from fecally polluted wastewater. *Appl. Environ. Microbiol.* **65:** 1772–1776.
23. **Rab, M. A., M. K. Bile, M. M. Mubarik, H. Asghar, Z. Sami, S. Siddiqi, A. S. Dil, M. A. Barzgar, M. A. Chaudhry, and M. I. Burney.** 1997. Water-borne hepatitis E virus epidemic in Islamabad, Pakistan: a common source outbreak traced to the malfunction of a modern water treatment plant. *Am. J. Trop. Med. Hyg.* **57:**151–157.
24. **Sano, D., K. Fukushi, Y. Yoshida, and T. Omura**. 2003. Detection of enteric viruses in municipal sewage sludge by a combination of the enzymatic virus elution method and RT-PCR. *Water Res.* **37:**3490–3498.
25. **Taylor, M. B., N. Cox, M. A. Vrey, and W. O. K. Grabow**. 2001. The occurrence of hepatitis A and astroviruses in selected river and dam waters in South Africa. *Water Res.* **35:**2653–2660.
26. **van Lawick-Goodall, J.** 1971. *In the Shadow of Man*, p. 214–224. Houghton Mifflin, Boston, Mass.
27. **Vivier, J. C., M. M. Ehlers, and W. O. K. Grabow**. 2004. Detection of enteroviruses in treated drinking water. *Water Res.* **38:**2699–2705.
28. **Wen, K., A. C. Ortmann, and C. A. Suttle**. 2004. Accurate estimation of viral abundance by epifluorescence microscopy. *Appl. Environ. Microbiol.* **70:**3862–3867.
29. **Williams, F. P., Jr., and C. J. Hurst.** 1988. Detection of environmental viruses in sludge: enhancement of enterovirus plaque assay titers with 5-iodo-2′-deoxyuridine and comparison to adenovirus and coliphage titers. *Water Res.* **7:**847–851.

Detection of Bacterial Pathogens in Wastewater and Sludge

RICHARD E. DANIELSON AND ROBERT C. COOPER

24

The number and variety of bacteria present in wastewater and associated solids are legion. Their sources are the excrement of humans and animals, other waste materials that find their way into domestic sewage, and the microbial flora in the source water. This great diversity and the associated variety of required growth conditions hamper attempts to isolate, identify, and enumerate most bacterial members of this microcosm. The presence of pathogens and the number of pathogenic bacteria that may be present in wastewater and biosolids are a function of the disease morbidity in the community from which the waste materials are derived and the degree of treatment the waste received. Relative to the total number of bacteria present, the pathogens will normally represent but a minor part. In most instances, these pathogens will play a passive role in the dynamics of the microbial ecosystem; the waste environment is hostile, and as a result the number of pathogens present tends to decrease over time. The isolation, identification, and enumeration of pathogens from this milieu are replete with all the difficulties listed above. Molecular methods (PCR) offer great promise for the rapid, direct identification of specific pathogens in wastewater and associated solids; however, at the present time problems of matrix interference (sample cleanup) remain a major hurdle.

Representative genera of bacterial pathogens that may be found in domestic wastewater and sludge include *Salmonella*, *Shigella*, *Vibrio*, *Escherichia*, *Campylobacter*, and *Yersinia*. The genus *Salmonella* includes more than 1,800 species and serovars; the genus *Shigella* is represented by four species encompassing 34 serovars; *Vibrio cholerae* O1 and non-O1 can be present; the pathogenic varieties of *Escherichia coli* include at least five serovars; *Campylobacter jejuni* is the most common *Campylobacter* species found in wastewater, followed by *C. coli*; and there are reported to be more than 50 serovars of *Yersinia enterocolitica* (19, 22, 45, 48).

Because of the difficulties in the isolation and detection of bacterial pathogens in wastewater and sludges, the use of surrogate (indicator) bacteria has been standard practice in water quality monitoring. Historically, the coliform group, the enterococci, and *Clostridium perfringens* have, in descending priority, been the bacterial indicators of choice. In all these cases, the indicator bacteria are assumed to be indigenous to feces and thus their presence in environmental samples is indicative of fecal contamination. The presence, and in some instances the absence, of these indicators is not an absolute indication of the presence of bacterial pathogens. Rather, these bacteria indicate the potential for the presence of pathogens because of the likelihood that infectious feces are present in wastewater or sludge. The bacterial indicators' greatest weakness as a public health monitoring tool for water and wastewater is that they have greater sensitivity to disinfection than do viruses and, in the case of chlorination, the cysts of protozoan parasites. The absence of indicator bacteria is not a guarantee that other more resistant microbial forms are not present. Because of these problems, there have been ongoing efforts to find better indicators of the presence of microbial pathogens in environmental samples. The ideal would be to monitor for the presence of all microbial pathogens that may be present in a liquid or solid sample, but this ideal is not likely to be realized any time in the foreseeable future.

There are a number of instances in which there is a need for direct measurements of the presence and number of bacterial pathogens in water and solids. The direct-measurement approach can be most useful in epidemiologic studies of waterborne disease, in evaluations of the efficacy of water and solids treatment processes for the reduction of pathogens, and in situations in which the sanitary significance of high indicator numbers is in question. The latter situation has occurred in association with composting sewage sludge in which coliform aftergrowth can take place, the significance of which is determined by the direct measurement of levels of *Salmonella* species (20). If a specific pathogen type were to be used as an indicator of pollution, one would still have the problem of selecting which would be the ideal surrogate. The majority of work done thus far on measuring levels of bacterial pathogens in water and solids has been directed towards salmonellae, and they have become a standard pathogen by fiat.

TRADITIONAL METHODS FOR DETECTION AND ENUMERATION OF BACTERIAL PATHOGENS

To date, methods for the detection and enumeration of bacterial pathogens from wastewater and sludge have used a cumbersome approach which includes enrichment, isolation, and identification. The numbers of bacterial pathogens found in wastewater and sludges are usually lower than the numbers of nonpathogens present. For example, the numbers of salmonellae detected in activated sludge and digested sludge range from none detected to a most probable number (MPN) of 400/g (dry weight) of solids (87). The numbers reported for raw sewage range from 7 to 8,000 per 100 ml (19). Because of these small numbers, some form of enrichment, which may be preceded by a sample concentration step, is required.

Determining the concentration of bacterial pathogens in wastewater and biosolids is a cumbrous task. The traditional method of choice is to estimate the numbers of bacteria by the MPN method. The MPN is determined by placing a series of dilutions of a wastewater sample, sample concentrate, or sludge solids extract into tubes of enrichment broth and determining whether the target bacterium has grown in any of the inoculated tubes. This requires that each tube (dilution) be screened for the presence of the pathogen by completing the isolation and identification steps.

SAMPLING REQUIREMENTS, TRANSPORTATION, AND PRESERVATION

Raw (untreated) sewage does not require the concentration of large-volume samples (typically 250 ml to 1 liter). Sterile glass or polypropylene bottles can be used to collect the samples. All samples should be refrigerated or stored on ice (not frozen) and processed in the laboratory as soon as possible, not to exceed 24 h after collection.

The detection of bacterial pathogens in treated sewage effluent (primary, secondary or better) may require the collection of larger sample sizes (1 to 10 liters) depending upon the final quality of the wastewater effluent. If the collected wastewater represents an effluent which has been treated by a tertiary or better process, large samples (10 to 20 liters or more) must be collected. For molecular investigations, depending upon the final water quality, bacteria may be concentrated directly onto filters in the field for recovery of their nucleic acid (44, 72). Furthermore, if a chlorinated effluent is discharged, the sample must be dechlorinated by adding to the sample bottle an appropriate concentration of sodium thiosulfate, usually enough to give a final concentration of 100 mg/liter prior to sterilization (25).

Sewage sludge samples should be collected in accordance with guidelines given in reference 20. The tested sample comprises subsamples taken throughout the same pile or source of sludge. All samples should be refrigerated or held on ice and processed within 24 h of collection. Results are reported as MPN per gram (dry weight) of sludge.

Wastewater samples generally are concentrated by filtration or centrifugation of known volumes of sample, whereby the bacteria present are captured on a filter or in a centrifuged pellet. Filtration using 0.45-μm-pore-size membranes is most applicable to clean (filterable) water. In the case of sewage, the utility of filtration is restricted because clogging limits the sample volume. If necessary, coarser filters such as spun glass can be employed, although they will be less efficient as bacterium collectors (16). In the case of wastewater and biosolids extracts, centrifugation can be employed but there will be some obvious limitations in the size of the sample that can be examined. A nonquantitative pseudofiltration method has been used (49) in which a gauze pad (sanitary napkins work well) is suspended in sewage flow for a period of time, after which the entire pad is placed in a suitable enrichment broth. Bacteria can be isolated from such pads up to four times more frequently than from grab samples (42). This method can also be used to isolate enteric viruses from wastewater. Other concentration methods using diatomaceous earth, fiberglass, and membrane filtration are described in *Standard Methods for the Examination of Water and Wastewater* (25). Success in using these methods of concentration prior to enrichment can be problematic in that these methods concentrate all manner of materials and microbes as well as the target pathogen; thus, while concentrating the sample one also significantly increases the amount of interference that will be encountered in subsequent steps of the isolation and identification process.

ENRICHMENT AND ISOLATION

The enrichment process employs a broth medium that allows the target bacteria to multiply to numbers large enough to facilitate their isolation and identification. In many instances, the medium to be used has been chosen directly from those used in clinical laboratories where isolation would be from the stools of infected individuals. In these samples one would expect large numbers of pathogens. The enrichments are devised to limit the growth of interfering bacteria, such as coliforms and *Proteus* species, and encourage the growth of the pathogen. Frequently the formulations used are less inhibitory to the target than to the unwanted bacteria and at the same time are not optimum for the growth of the pathogen. In stools from infected individuals, and in many cases infected food, the target bacteria are robust enough and present in sufficient numbers that they multiply despite the suboptimal growth conditions. In wastewater, sewage solids, and treated effluents, the pathogenic bacteria are in a hostile environment, are not robust, and are present in small numbers. In these instances, the selective enrichment used in the clinical setting may be too restrictive for successful enhancement. Less restrictive or nonrestrictive media can be used to allow the growth of the target pathogen and other bacteria, followed by secondary enrichment in a restrictive medium. After enrichment for the appropriate time and at the appropriate temperature, material from the broth is streaked or plated onto a solid medium for bacterial isolation. There is a large variety of such media which are selective, differential, or both. Typical colonies are selected and further identified using standard biochemical (triple sugar iron agar, lysine-iron agar, and urea agar, etc.) and immunological methods. The use of gene probes for the identification of bacterial isolates would be most applicable at this stage of the process.

SALMONELLA DETECTION

Because of the great diversity of pathogens that may be found in wastewater and solids, the diversity of media and incubation regimes required, and in most instances, the paucity of information available, the following discussion is limited to the salmonellae. Most of the available literature and reported experience in the sphere of wastewater and sludge have been directed towards the detection and enumeration of salmonellae. Much of the methodology used has been adapted from food microbiology, an area of intense activity in the detection of salmonellae.

A representative list of selective enrichment media is presented in Table 1. These media fall into three ingredient categories: (i) selenium based, (ii) tetrathionate based, and (iii) malachite green-magnesium salt based (Rappaport-Vassiliadis broth [RVB]). The conditions presented in these enrichment media do not constitute optimum growth conditions for salmonellae and are selective in that they inhibit the growth of interfering bacteria, such as *Proteus* species and coliforms, and

TABLE 1 List of selective enrichment media used for detection of *Salmonella* in wastewater and sludge

Enrichment medium	Reference
Selenite-F broth	47
Selenite brilliant green	57
Selenite brilliant green sulfa broth	57
Selenite cysteine broth[a]	55
Selenite dulcitol broth[a]	43
Tetrathionate broth[a]	53
Tetrathionate brilliant green broth	35
RVB	83

[a]The medium selected in reference 25.

hopefully provide an opportunity for the ascendancy of any salmonellae that may be present. A number of additional ingredients have been incorporated into *Salmonella* cultivation media in an attempt to enhance selectivity. These include brilliant green dye, cysteine, sulfapyridine, and novobiocin. The latter antibiotic, in concentrations of up to 80 μg/ml, has been found to be very useful in the suppression of interfering bacteria (65) and is incorporated into RVB (83). Because these enrichment media can exert a significant inhibitory effect upon stressed salmonellae, some form of preenrichment medium, such as buffered peptone broth, can be used prior to inoculation into a selective medium.

The incubation temperature can have a significant impact on the recovery efficiency of a selective medium. Harvey and Price (27), in their review of salmonellae isolation methods, indicate that the enhancement of *Salmonella* recovery by using incubation temperatures between 40 and 43°C has been recognized since the beginning of the 20th century. The use of elevated temperatures must be matched with the enrichment system used. These authors point out that results with elevated temperatures can vary among laboratories, probably because of the lack of standardization of methods. In general, elevated temperatures do not enhance the usefulness of selenite brilliant green broth or tetrathionate-based medium samples. These authors also point out that elevated-temperature methods may not detect *Salmonella enterica* serovar Typhi. The elevated-temperature regime works well with the RVB medium and is used in the secondary enrichment scheme of Hussong et al. (35) and in methods incorporating selenite dulcitol broth.

SALMONELLA ISOLATION MEDIA

There are a number of formulations of isolation agar for the detection of *Salmonella*. The major components of the most common agars are shown in Table 2. Modified semisoft Rappaport-Vassiliadis (MSRV) medium has been included because of the success in the isolation of *Salmonella* from wastewater and biosolids with the use of this formulation. These agars are selective and differential. Selectivity is brought about through the use, singly or in combination, of brilliant green dye, bile salts, or deoxycholate. Variations of this theme include the use of bismuth sulfite along with brilliant green in bismuth sulfite agar and the use of malachite green (a homologue of brilliant green), a relatively high concentration of magnesium chloride, and novobiocin in MSRV medium.

Differentiation among colony types is, for the most part, based on carbohydrate fermentation and associated pH change, as shown by indicator dyes. For example, the fermentation of lactose, present in five of the seven agars, is an attribute of coliforms but not the salmonellae; colonies of the former assume the characteristic color of the pH indicator used. A good illustration of the use of these activities to differentiate pathogens from nonpathogens is the makeup of xylose-lysine-deoxycholate (XLD) agar. As shown in Table 2, the medium contains three sugars: lactose, saccharose, and xylose, with the latter in a lower concentration than the first two. The pH indicator used is phenol red, which is yellow in acid. Coliform colonies are yellow because of fermentation of all the carbohydrates present; nonpathogenic, nonlactose fermenters, such as *Proteus* species, are detected by the fermentation of the other carbohydrates present, while salmonellae, which can ferment xylose, neutralize the acid produced through the decarboxylation of lysine, resulting in a colony of red. The presence of excess lactose and saccharose prevents lysine decarboxylase-active coliforms from neutralizing the acid produced. The production of hydrogen sulfide is indicated by reaction with ferric salts in the medium to produce red-black or yellow-black colonies, the former being common to a number of *Salmonella* species. In contrast, bismuth sulfite agar, which contains only dextrose, relies on the production of hydrogen sulfide, with the concomitant precipitation of iron sulfide producing a black metallic sheen on suspected *Salmonella* colonies. The medium relies on the inhibitory effect of brilliant green to control the growth of interfering coliforms.

TABLE 2 Major components of selected *Salmonella* isolation agars[a]

Class	Ingredient	SS	XLD	XLBG	BGA	HEA	BSA	MSRV
Carbohydrates	Dextrose	−	−	−	−	−	+	−
	Lactose	+	+	+	+	+	−	−
	Saccharose	−	+	+	+	+	−	−
	Xylose	−	+	+	−	−	−	−
Amino acid	Lysine	−	+	+	−	−	−	−
Inhibitors	Bile salts	−	−	−	−	+	−	−
	Brilliant green	+	−	+	+	−	+	−
	Deoxycholate	−	+	−	−	−		−
	Bismuth sulfate	−	−	−	−	−	+	−
	Malachite green	−	−	−	−	−	−	+
	$MgCl_2$	−	−	−	−	−	−	+
	Novobiocin	−	−	−	−	−	−	+

[a]SS, salmonella-shigella agar; XLBG, xylene-lysine brilliant green agar; BGA, brilliant green agar; HEA, Hektoen enteric agar; BSA, bismuth sulfite agar; +, present; −, absent.

MSRV agar does not contain any carbohydrate but relies on the motility of salmonellae as a differential characteristic. The use of motility as a means of separating salmonellae from background bacteria has been reported by numerous authors throughout the years (e.g., reference 27). De Smedt et al. (12) and De Smedt and Bolderdijk (13) developed a modification of the Rappaport-Vassiliadis enrichment medium in semisoft (0.8%) agar onto which drops of enrichment culture are placed; the sample is incubated at 42°C, and motility (cloudy growth) away from the spot is considered to be an indication of the presence of salmonellae. This limits the test to those salmonellae that are motile. There are very few nonmotile salmonellae, notably *S. enterica* serovar Pullorum (45).

The considerable variety of enrichment and isolation media available poses a problem for the investigator who wishes to select the most efficacious method for the isolation of salmonellae from wastewater and biosolids. Part of this decision is made based upon the investigator's past experience and sample matrix. Morinigo et al. (50–52) evaluated the use of various enrichment media and isolation agars for their effectiveness in the isolation of salmonellae from environmental samples. In one study on river water (50), they examined 10 different enrichment media which included RVB with four concentrations of novobiocin, selenite cysteine at two different incubation temperatures, and Selenite-F with and without novobiocin at two incubation temperatures. All of the samples were preenriched in buffered peptone broth and isolated onto XLD and brilliant green agars. The researchers concluded that the best recoveries of *Salmonella* were made using the RVB containing novobiocin at 10 μg/ml. In another study, the authors (51) again evaluated this enrichment medium with the inclusion of tetrathionate broth by using laboratory cultures of bacteria including salmonellae and species known to interfere with the isolation of salmonellae. They found that selenite-containing media were less effective at restraining the growth of gram-positive interference bacteria. Salmonellae grew poorly in tetrathionate broth and media containing brilliant green. Stressed (by exposure to seawater) salmonellae did well in Selenite-F, selenite cysteine, and RVB. When natural water samples were used, the RVB was deemed the best for enrichment. The authors also evaluated a great variety of isolation agars, excluding MSRV, for their ability to grow *Salmonella* as well as to inhibit interfering bacteria. Laboratory cultures of bacteria, both stressed and unstressed, were used. The results indicated that XLD agar was of selectivity equal to the best of the others and was best for the recovery of stressed *Salmonella*. From their work, the RVB with novobiocin is the enrichment medium of choice and XLD agar is an acceptable isolation agar for use with polluted water samples.

There has been increased interest in the ability to isolate *Salmonella* from sewage sludge, particularly with the adoption of a *Salmonella* standard for treated sludge that can be applied, unrestrictedly, to land (20). Yanko et al. (87) evaluated five different procedures for the isolation and enumeration of salmonellae in biosolids: two standard methods (25) which use selenite dulcitol and tetrathionate enrichments, respectively, followed by XLD and brilliant green isolation agars; Kenner and Clark's method using selenite dulcitol enrichment broth incubated at 40°C and isolation on XLD agar; the method of Hussong et al. (35) which uses a preenrichment in buffered peptone at 36°C followed by a secondary enrichment in tetrathionate brilliant green broth incubated at 43°C and isolation on xylose-lysine brilliant green agar; and a method using a selenite brilliant green enrichment broth incubated at 37°C with confirmation on XLD agar. This evaluation was conducted with digested sludge, activated sludge, and composted sludge. The method of Hussong et al. (35) and the selenite brilliant green method recovered more salmonellae than did the other three, and the first two methods were determined to be inadequate for use in compliance testing.

In none of these evaluations was MSRV agar used. In our laboratory, we have compared three methods for the enumeration (MPN) of salmonellae in biosolids and raw sewage. These methods include the selenite brilliant green enrichment and selenite dulcitol enrichment, both followed by isolation on XLD agar, and a nonselective enrichment in tryptic soy broth (15) followed by isolation on MSRV agar. All the enrichment media were incubated at 35°C for 24 to 48 h. The MSRV agar was then spotted with up to 0.2 ml (1 drop per spot) of enrichment broth and incubated for 24 h at 42°C . At the end of this period, the MSRV agar plates were examined and motility away from the spots was recorded. Material taken from the edge of a motile spot was either restreaked onto XLD agar or used directly for biochemical screening. Our experience has been that this material contains, with few exceptions, a pure culture of *Salmonella*. In our hands, this procedure is two to three times more effective than the other two methods tested (our unpublished data). In no instance have we recovered *Salmonella* by either of the first two methods and not made a recovery from the MSRV agar. The U.S. Environmental Protection Agency has recently completed a method evaluation of the application of this medium to the monitoring of biosolids for salmonellae. Round-robin testing has been completed, and the use of MSRV medium has been found to be superior to the methods now being used (82a). One reason for the success of this method may be the use of a large inoculum (0.2 ml) of enrichment culture relative to that used on a streak plate. Harvey and Price (27) point out that salmonellae growing in an enrichment culture tend not to be uniformly distributed and a streak plate may well come up negative even when salmonellae are present in the enrichment medium. Our observations seem to confirm this phenomenon since frequently not all of the six spots on a positive MSRV agar plate are motile. The use of a nonselective enrichment medium may also be of value in resuscitating stressed salmonellae.

OTHER BACTERIAL PATHOGENS

The recently published literature on the detection of bacterial pathogens other than *Salmonella* species in wastewater and sludge is relatively sparse. The culture method for the detection of *V. cholerae* is well established (25). The use of alkaline peptone water as an enrichment medium followed by isolation on thiosulfate-citrate-bile salts-sucrose agar and confirmation of typical colonies by biochemical and serological tests is an effective method. In our laboratory, we have had good success in isolating non-O1 *V. cholerae* from estuarine sediments by using this procedure.

Reports on the detection of *Shigella* have not been found. *Standard Methods* (25) suggests that methods similar to those used for the detection of *Salmonella* should be employed for the detection of *Shigella* and points out that the methodology is qualitative and low in sensitivity. Reports on the isolation and enumeration of *Yersinia* species in wastewater and sludge were not found. A membrane filter method for the detection of *Yersinia* in water is described in *Standard Methods* (25).

Methods for the isolation and enumeration of thermophilic *Campylobacter* species from the water environment are reported by Arimi et al. (1) and Jones et al. (39, 40). These authors report successful isolation and MPN enumeration using Preston campylobacter selective enrichment broth incubated at 43°C and isolation on Preston campylobacter selective agar incubated under microaerophilic conditions.

It is known that *Aeromonas* is associated with human enteric disease (32). Because of this understanding, the distribution of this autochthonous aquatic bacterium in surface water, wastewater, and sludge has been reported. Nishikawa and Kishi (54) describe a qualitative method using an alkaline peptone broth enrichment followed by isolation on bile salts brilliant green starch agar. Amylase-positive colonies on the latter medium were considered positive presumptive *Aeromonas* isolates. Further identification was based upon biochemical differentiation. Poffe and Op de Beek (63) reported good results in the isolation and enumeration of *Aeromonas hydrophila* from sewage and sludge by using direct detection methods. In the case of wastewater, they employed the membrane filter method using mA agar (67) which contains ampicillin, trehalose, and deoxycholate. In the case of sludge samples, suspensions of various dilutions were spread onto the surface of mA agar. In both instances, the suspect colonies were yellow, a result of trehalose fermentation. These colonies were further identified using biochemical differentiation. Ribas et al. (66) compared four similar media for their usefulness in the detection of *Aeromonas* spp. in polluted water. The media were of various compositions that contained ampicillin, complex carbohydrates, and in one case, glutamate; mA agar was included. The authors concluded that all of the media gave similar results but that the starch-glutamate-ampicillin medium was the simplest to use. There is a membrane filter method using ampicillin-dextrin agar containing bromthymol blue. *Aeromonas* species produce yellow colonies on this agar. The applicability of this method to wastewater is unknown. In our laboratory, we have had success with a presence-absence method in which samples are filtered through a 0.45-μm-pore-size cellulose acetate filter which is subsequently immersed in alkaline peptone water at 35°C for 24 h. This enriched material is then spread onto Rimmler-Schotts agar. *Aeromonas* isolates produce yellow colonies (70).

There has been increasing interest in the occurrence of *Helicobacter pylori* in the aquatic environment (37; J. Handwerker, G. Fox, and D. E. Schauer, *Abstr. 95th Gen. Meet. Am. Soc. Microbiol. 1995*, abstr. Q-203, p. 435, 1995). The role of drinking water in the transmission of this bacterium to humans remains problematic. At the present time, no viable method is available for the isolation or enumeration of *H. pylori* microorganisms from water and wastewater. Their presence in these media has been confirmed by molecular biological (PCR) methods (23, 24, 29, 33; L. D. Sutton, W. W. Wilke, N. A. Lynch, and R. N. Jones, *Abstr. 95th Gen. Meet. Am. Soc. Microbiol. 1995*, abstr. C-39, p. 69, 1995).

MOLECULAR BIOLOGICAL METHODS

There can be insurmountable difficulties in the use of culture methods for the isolation of certain pathogenic bacteria from wastewater and sludges. Therefore, the use of molecular techniques for the analysis of environmental samples has become common. Molecular techniques have transitioned from the research laboratory to wastewater agencies (44, 58, 78, 80, 81, 87).

The primary advantage to the application of molecular methods to the detection of pathogenic microorganisms from environmental samples is the ability to specifically and rapidly detect the organism of interest without having to actually isolate it on growth media. It may even be possible to apply molecular methods to detect those target organisms that exhibit characteristics commonly associated with viability (the presence of RNA species). Given the large background of competing bacteria, these methods hold promise for the future of monitoring for either specific or selected groups of pathogenic bacteria in wastewater and sludges.

PCR

In the section that follows, it is assumed that the reader is familiar with the basic theory behind PCR. The details of the PCR are not discussed here but can be found in various textbooks (2, 18, 36, 69). The PCR involves a few critical steps; the following are highlights of those procedures.

Primer Selection

PCR primers can be chosen as specific sequences adjacent to predetermined target sequences, or a group of random hexamers can be used as primers to create PCR-amplified molecules that contain the target sequence. If specific primer sequences are to be chosen, one can determine the preferred primer sequences either by searching the literature or nucleic acid libraries or by sequencing areas of DNA or RNA of the bacteria of interest. Primer sets can be highly specific for only one particular gene that, one hopes, is unique to a single organism, or they can be universal, amplifying related sequences that cross taxonomic family or kingdom boundaries. In addition, depending upon the amplification conditions, multiple primer sets may be simultaneously run in a nested form or as a suite of primers (5–7, 84). Target sequences of high tertiary complexity (e.g., superhelices) may be more difficult to amplify due to the structural geometry of the finished sequence (2, 82). It is important to limit the size of the final target (<3 kbp) because, as the PCR is used to replicate longer and longer pieces of DNA, there is greater chance of error (e.g., in base pair matching and the formation of primer dimers, etc.) (2, 80).

PCR Conditions

PCR is performed with a thermal cycler, an instrument that is capable of rapidly cycling temperature changes over a wide range generally between 4 to 100°C. Thermal cycler settings must be optimized for each primer set and desired product and often are determined empirically through trials. In addition, the various commercial thermal cyclers may perform slightly differently. Therefore, in order for thorough quality control, it is imperative that PCR conditions be optimized on one instrument and subsequently reproven if the investigator switches to another brand or type of cycler. The optimal annealing temperature is critical and can be determined either empirically by using a series of experiments or by consulting computer-based software which can predict melting and annealing temperatures based upon the base pair sequences (2).

Molecular Probes and Detection Techniques

There are many methods available for the detection of the PCR product. One of the easiest methods is gel electrophoresis, whereby the products of the PCR amplification are separated into bands based upon their sizes. Some drawbacks to this method include the following: (i) if the target was present in an extremely low concentration, even an amplified product may be difficult to visualize; (ii) other amplification products may interfere with the resolution of the electrophoretic bands; and (iii) if only a single band is generated for a target organism, then a single electrophoretic image may not be definitive proof of successful amplification of a specific target. The first two problems can be resolved by using specific complementary probes, which can be labeled either isotopically or nonisotopically. One may then visualize extremely low concentrations of products and avoid interfering bands which do not bind to the probes. As an aid in solving the third problem, it is advisable to amplify more than a single target site for any organism of interest, since the presence of multiple products from the same organism provides a more specific test for that organism.

Isotopic versus Nonisotopic Techniques

For many years, the use of isotopically labeled (e.g., ^{32}P-labeled) probes was the method of choice given the probes' sensitivity for detecting nucleic acid sequences. Methods using isotopically labeled probes can measure as little as femtogram quantities of products (5–7). However, because of the necessary precautions associated with the handling and cleanup of radioactive material, many investigators now prefer to use nonisotopic detection procedures whenever possible. Biotin-streptavidin, digoxigenin, or fluorescent labels can be used for obtaining colorimetric results (C. T. Yamashiro, M. Rogers, W. Jakubowski, and C. Paszko-Kolva, *Abstr. 95th Gen. Meet. Am. Soc. Microbiol. 1995*, abstr. Q-209, p. 436, 1995). Methods for nonisotopically labeled probes may not be as sensitive as those which employ isotopically labeled probes, but they are improving. The final product from the reaction can be detected by immobilizing the target sequences onto nylon filters and hybridizing them in place with a known, labeled probe, or the known probe can be immobilized and used to capture the labeled target material through hybridization as the sample is placed into contact with the filter.

The specificity of binding between the nucleic acid probe and the complementary target nucleic acid sequences versus other sequences which are similar, but not identical, is known as stringency and is dictated by the chemical conditions and temperature under which the hybridization is performed (2). The selection of hybridization conditions and the format of test presentation (Southern hybridization, dot blots, etc.) vary depending upon the makeup of the probe and the typical yield of the PCR. Consequently, the best hybridization conditions must be determined empirically for each complementary set of nucleic acid primer sequences.

Direct Probing

If the microorganisms of interest are present in sufficient quantity, PCR amplification of the target nucleic acid sequences may not be necessary. Recovery of total nucleic acid from an environmental source such as sewage, sludge, or filter concentrates may yield sufficient material for direct probing. Sommerville et al. (72) and Knight et al. (44) have demonstrated that some pathogens may be present in sufficient quantity but may not be culturable. In these instances, total nucleic acids were harvested and known nucleic acid probes specific for the bacterial pathogens were used as a detection system.

Quantification

It is possible to use PCR to quantify the number of bacteria present in the sample. Internal nucleic acid amplification controls can be run simultaneously, and the intensity of the hybridization product between the probe and the PCR amplification can be compared with that of the product of hybridization with known quantities of target nucleic acid (4, 6, 59, 60). Picard et al. (62) prepared multiple PCR tubes containing undiluted and diluted DNA prepared from soil extracts and used the resulting test reactions as a basis for constructing MPN result. Yamashiro et al. (*Abstr. 95th Gen. Meet. Am. Soc. Microbiol. 1995*), used a technique based on the method of Holland et al. (31), whereby the PCR-amplified product is fluorescently labeled as the PCR proceeds. To arrive at a quantitative result, the increasing concentration of fluorescently labeled target DNA is compared to the increasing concentration of fluorescently labeled control DNA within the same reaction tube. This procedure has become commercialized as TaqMan (PE Applied Biosystems, Inc., Foster City, Calif.).

Quality Control

Perhaps the most critical aspect of using molecular methods for environmental analysis is the application of stringent quality control methods. The U.S. Environmental Protection Agency has published a guidance document for laboratories performing PCR analyses of environmental samples (82b). Included in this document are measures which describe the physical configuration of the testing laboratory that is needed in order to control the occurrence of amplicon contamination throughout the laboratory (false-positive results). There are several quality control measures that can be employed to test for successful PCR and hybridization. If the same source of wastewater or sludge is to be repeatedly tested, spiking samples with either the organism in question or target DNA will provide information on the efficiency of sampling techniques and the effects of inhibitory compounds. In addition to using a spiked study, the investigator must ensure that every reaction tube contains a known piece and quantity of control DNA that will indicate whether PCR has been inhibited and to what extent. Specificity of the hybridization can be measured by using known labeled DNA that differs from the target by one or two base pairs. Therefore, the level of stringency first must be empirically determined for each primer set and then the established amplification regimen can be rigorously followed. Sensitivity of the test results can be determined by adding either target DNA or the test organism in the desired wastewater or sludge matrix in such a manner as to conduct dilution-to-extinction experiments.

Sample Preparation

Currently, the primary disadvantage to the application of molecular methods to environmental microbiology is the presence of naturally occurring compounds such as humic acids and minerals which may be inhibitory to the molecular procedure (8, 79; P. W. Stewart and M. Abbaszadegan, *Abstr. 95th Gen. Meet. Am. Soc. Microbiol. 1995*, abstr. Q-394, p. 469, 1995). Therefore, removing these inhibitory agents is of primary importance when designing sample preparation and concentration techniques. The following discussion will highlight a variety of methods that have been employed to isolate the DNA or RNA of the target bacteria in question.

DNA is a double-stranded nucleic acid which can form very complex superhelical structures. These structures can be quite recalcitrant to environmental effects (41, 81, 86). Treated wastewater may contain disinfectants which are indiscriminate oxidizing compounds and can damage DNA molecules. Some treated sludges can have extremely high pHs from liming processes which can denature DNA. Therefore, if molecular techniques are to be applied to limed sludges, it may be necessary to adjust (lower) the pHs of suspensions before testing.

Use of RNA as a molecular target is important because it may be helpful as a qualitative measure of viability (i.e., by measuring the presence of mRNA or rRNA). In addition, RNA is far more abundant within bacterial cells and will be easier to detect with fewer cycles of amplification in a PCR. Reverse transcriptase PCR has been used for the detection of enteric viruses in environmental waters and wastewaters (11, 81), and techniques used to preserve viral RNA should be applicable to bacterial RNA. Although RNA is more plentiful than DNA, RNA is a single-stranded nucleic acid which has been shown to be quite labile in environmental matrices (82). Therefore, it is important that steps be taken to improve the survival of RNA molecules during sample manipulation (2).

As stated above, one of the major obstacles in amplifying bacterial nucleic acids from environmental sources is the presence of inhibitory agents such as humic acid (79). Since environmental matrices can be complex and undefined, it is difficult to apply a "one cleanup step fits all" procedure. For example, approaches used for the recovery of bacterial nucleic acids from wastewater differ from those used for recovery from sludge. Palmer et al. (59, 60) used the following techniques to recover bacterial (*Legionella* sp.) nucleic acids from wastewater and reclaimed wastewater. First, the sample was concentrated onto Teflon filters (Durapore; Millipore Corp., Bedford, Mass.). Each filter was then vortexed in a lysis buffer and heated to release the bacterial DNA. The nucleic acids were further purified by centrifugation through a Spin-X filter (Costar, Cambridge, Mass.). A carrier reagent (homopolymer A RNA) was added to aid subsequent precipitation of the DNA in isopropanol. Following centrifugation, the DNA pellet was washed twice in isopropanol. Finally, the pellet was resuspended in PCR water (PE Applied Biosystems) and an aliquot was submitted to amplification by PCR. Although one manufacturer (PE Applied Biosystems) of an environmental PCR kit uses Durapore filters, other investigators (5, 58) have reported that greater yields were achieved with Fluoropore filters (both filter types are from Millipore).

Environmental water samples may also be centrifuged to concentrate bacteria into a pellet, and the pelleted cells are then lysed in a silica-guanidinium thiocyanate lysis buffer, followed by several washing steps (4, 78). These crude extracts have been shown to be sufficient to add directly to a PCR amplification.

For sludges, Tsai et al. (80) and Hussong et al. (35) reported that rapid freeze-thawing of samples followed by phenol-chloroform extraction and cleanup through a Sephadex-200 spin column provided adequate recovery of target bacterial DNA. Typically, small volumes (about 25 to 30 μl) of the extracted nucleic acids are used in 100-μl reaction mixes; therefore, rare targets may be missed. Analysis of larger-volume samples by PCR has been described by Stewart and Abbaszadegan (P. W. Stewart and M. Abbaszadegan, *Abstr. 95th Gen. Meet. Am. Soc. Microbiol. 1995*) who applied it to the detection of viruses in the environment. The basic volumes were adjusted to accommodate 100 μl of extract for a total volume of 300 μl. Because the larger volumes may introduce more inhibitory compounds, Stewart and Abbaszadegan employed cleanup through a Sephadex G-100 column (for humic acids) and through a column of Chelex-100 (for ionic inhibitors).

Recently, magnetic capture beads have been employed in a wide variety of clinical, food, and environmental applications for the recovery of target nucleic acids that are subsequently amplified by PCR (56). Hultén et al. (33) used paramagnetic beads in an attempt to isolate *H. pylori* from environmental samples to reduce the effects of background chemicals and other microorganisms on the PCR. This technique holds promise as a mechanism for capturing the target nucleic acid while incorporating wash steps which help to remove potential inhibiting agents.

Higgins et al. (30) performed an evaluation of four commercial DNA preparation products for bacteria, the Prepman (PE Applied Biosystems), Instagene (Bio-Rad, Richmond, Calif.), aXtra Amp (Ansys Diagnostics, Lake Forest, Calif.), and Isocode paper (Schleicher and Schuell, Keene, N.H.). Each of these preparation methods yielded PCR quality DNA from *E. coli* and proved to be rapid (less than 1 h) and relatively inexpensive.

FA TECHNIQUES

Fluorescently labeled antibodies specific for a target organism have been used to screen wastewater and reclaimed wastewater for target microorganisms (14, 59, 60). Since there are many microorganisms in wastewater and sludge, there may be some that share similar antigenic reaction sites (epitopes) with which the antibody stain may cross-react. Therefore, when searching through diverse bacteria, it is best to limit the occurrence of cross-reactivity by using monoclonal fluorescent antibodies (FA). The biggest advantage to using monoclonal FA stains is that they can be used to produce a rapid and relatively inexpensive test (once the monoclonal antibody has been made). The disadvantages of the use of monoclonal FA stains for environmental samples are that (i) there remains a probability that cross-reactivity may occur; (ii) the FA stain can be used to detect the presence of the target organism but does not measure viability; (iii) the array of antigens exhibited by a bacterium may differ depending on the conditions in its environment and, for this reason, depending upon the physiological state of the bacteria used to prepare the monoclonal cell line (which in turn produces the antibodies used in the detection reaction), the monoclonal antibody(ies) chosen may not recognize the key target bacterial epitopes that are expressed on bacteria in environmental waters; and (iv) the number of bacteria required for microscopic detection is substantial and, especially in environmental samples, the level of test organisms may be below this minimum number. A negative test result, therefore, should be reported as "none detected" rather than "none present."

ELISA

Enzyme-linked immunosorbent assays (ELISA) are also antibody based. Brigmon et al. (10) applied an ELISA to detect the presence of *S. enterica* serovar Enteritidis in raw sewage, sludges, and wastewater. The advantage associated with applying the ELISA technique directly to environmental samples is that it can provide a mechanistic measure (via a microtiter dish reader) with only minor manipulation of the sample. However, one of the major disadvantages to this technique is the lack of sensitivity. Brigmon et al. (10) reported that a minimum of 10^5 *S. enterica* serovar Enteritidis cells/ml were required to generate a clear response above background. Tamanai-Shacoori et al. (78) demonstrated that PCR was superior to ELISA for screening environmental samples for enterotoxigenic *E. coli*. Given the current standard of <3 *Salmonella* spp. per 4 g of sludge (20), this technology has not yet reached the sensitivity necessary for performing the testing required for regulatory compliance. Other disadvantages of this technique include those outlined for FA stains.

MST

The purpose of conducting microbial source tracking (MST) studies is to better define the sources of microbes of fecal origin. What is the "source" in MST? In a watershed, this can be wildlife, domestic animals, and/or results of human activity (e.g., leaking septic systems). Why conduct analysis for these sources? By differentiating between human and nonhuman sources, various watershed management practices can be put into place that may help control and/or limit fecal contamination of a potential drinking water source. The drinking water treatment process downstream of the watershed will have fewer potentially disease-causing organisms to remove from the source water, and thus the risk of infectious disease through the drinking water route will also be less.

MST is based upon the belief that a direct correlation exists between an organism found in the environment and the specific animal source of that specific organism. However, MST has also been used in a broader context to define a group of microorganisms that may have originated with groups or subgroups of host organisms. Methods used for MST can be broadly classified into two major strategies, library independent and library dependent.

Library-independent methods rely upon the detection of a specific target organism that has been directly linked to a specific host. For example, Bernhard and Field (8) have proposed that using bacteria of the *Bacteroides* and *Prevotella* genera can differentiate human fecal contamination from that of ruminants. Bonjoch et al. (9) targeted specific molecular markers of *Bifidobacterium* spp. thought to be most common in the human intestine. The sources of bacteriophages are also believed to be discernible dependent upon the phage type (86).

Library-dependent methods are based upon characterizing numerous isolates from a watershed by either phenotypic or genotypic classification. These isolates must be gathered from known sources (e.g., human septic systems and wild or domestic animal feces) within the watershed, and once the library has been constructed, subsequent isolates from unknown sources found in the watershed can be matched (by statistical algorithms) with those isolates previously described. Library-dependent methods can be generally subdivided into two categories, phenotypic and genotypic methods.

Typically, phenotypic methods are based on measuring the presence of certain enzymatic or metabolic activities by using certain substrates (e.g., those from BIOLOG, Hayward, Calif.) (21) or characterizing antibiotic resistance patterns (17, 26, 28, 29). The primary disadvantage of applying these methods is that the isolates must be expressing the specific metabolic trait of interest. It was reported by Smalla et al. (73) that this technique did not accurately measure the metabolic activity of the dominate species in a community.

Genotypic methods employ a variety of molecular biological techniques to generate a comparative pattern. These techniques include pulsed-field gel electrophoresis, PCR, and ribotyping. Pulsed-field gel electrophoresis can produce highly resolved "fingerprints" of specific sequences of target DNA in order to discern subtle differences between isolates (64, 74). PCR is used to amplify specific sequences directly from environmental samples, usually without isolating specific microorganisms. These sequences are then resolved to detect target organisms and/or compare the differences within the same species of organisms (38, 71). Other PCR methods such as rep-PCR, enterobacterial repetitive intergenic consensus PCR, and BOX-PCR have also been used to distinguish between isolates of the same species (34, 71, 75, 85). Ribotyping relies on the separation and analysis of specific well-conserved DNA fragments to identify a specific strain. Ribotyping requires isolated organisms from the study area for developing a library of organisms from that area (61).

Given the above, there have been many documented difficulties in applying these methods to MST studies. Method comparison studies have been carried out, and the results of the various methods rarely agreed with one another and in many cases could not be replicated with the same isolates (76, 77). Also, the results from library-based methods appear to be geographical-site specific. The data generated from a watershed study in one locale cannot be used for other locations (71). There is debate regarding the selection of the target organism to base a library. Many studies rely on *E. coli;* however, Barnes and Gordon (3) contend that there is too much variability even within clonal lines of this organism. They have suggested that enterococcus bacteria may present a more stable set of genetic markers. The selection of the method for statistical analysis of the data is critical for library-dependent methods. It has been demonstrated that simply changing the type of statistical analysis alone can reclassify an isolate from human to animal (68). It has been recommended that future MST studies utilize a combination of the methods described above (76). Regardless, caution must be used when applying and interpreting the data generated from these methods.

CONCLUSION

There is growing interest in the direct measurement of levels of specific bacterial pathogens in environmental samples. This interest is stimulated by concern about the reliability of using standard indicator bacteria in setting public health standards, the need for better epidemiologic information relative to the use of reclaimed wastewater and the disposal of biosolids, and the evaluation of pathogen reduction in waste treatment processes. Methods for the isolation and enumeration of pathogenic bacteria from wastewater and sludge are not well developed. There are a myriad of difficulties ranging from the huge diversity of microorganisms involved to the considerable variety of methodological approaches one can pursue. At this time, development of methods for the detection and enumeration of salmonellae is the most advanced, followed by increasing interest in the development of culture methods for *Campylobacter* and *Aeromonas* species, among others. The more recent developments in molecular biological methods offer great possibilities for the future.

REFERENCES

1. **Arimi, S. M., C. R. Fricker, and R. W. A. Park.** 1988. Occurance of thermophilic campylobacters in sewage and their removal by treatment processes. *Epidemiol Infect.* **101:**279–286.

2. **Ausubel, F. M., R. Brent, R. E. Kingston, D. D. Moore, J. G. Seidman, J. A. Smith, and K. Struhl (ed.).** 1989. *Current Protocols in Molecular Biology*. John Wiley and Sons, New York, N.Y.
3. **Barnes, B., and D. M. Gordon** 2004. Coliform dynamics and the implications for source tracking. *Environ. Microbiol.* **6:**501–509.
4. **Bej, A. K., J. L. DiCesare, L. Haff, and R. M. Atlas.** 1991. Detection of *Escherichia coli* and *Shigella* spp. in water by using the polymerase chain reaction and gene probes for *uid*. *Appl. Environ. Microbiol.* **57:**1013–1017.
5. **Bej, A. K., M. H. Mahbubani, R. Miller, J. L. DiCesare, L. Haff, and R. M. Atlas.** 1990. Multiplex PCR amplification and immobilized capture probes for detection of bacterial pathogens and indicators in water. *Mol. Cell. Probes* **4:**353–365.
6. **Bej, A. K., S. C. McCarty, and R. M. Atlas.** 1991. Detection of coliform bacteria and *Escherichia coli* by multiplex polymerase chain reaction: comparison with defined substrate and plating methods for water quality monitoring. *Appl. Environ. Microbiol.* **57:**2429–2432.
7. **Bej, A. K., M. H. Mahbubani, J. L. Dicesare, and R. M. Atlas.** 1991. Polymerase chain reaction-gene probe detection of microorganisms by using filter-concentrated samples. *Appl. Environ. Microbiol.* **57:**3529–3534.
8. **Bernhard, A. E., and K. G. Field.** 2000. Identification of nonpoint sources of fecal pollution in coastal waters by using host-specific 16S ribosomal DNA genetic markers from fecal anaerobes. *Appl. Environ. Microbiol.* **66:**1587–1594.
9. **Bonjoch, X., E. Balleste, and A. R. Blanch.** 2004. Multiplex PCR with 16S rRNA gene-targeted primers of *Bifidobacterium* spp. to identify sources of fecal pollution. *Appl. Environ. Microbiol.* **70:**3171–3175.
10. **Brigmon, R. L., S. G. Zam, G. Bitton, and S. R. Farrah.** 1992. Detection of *Salmonella enteritidis* in environmental samples by monoclonal antibody-based ELISA. *J. Immunol. Methods* **152:**135–142.
11. **De Leon, R., C. Shieh, R. S. Baric, and M. D. Sobsey.** 1990. Detection of enteroviruses and hepatitis A virus in environmental samples by gene probes and polymerase chain reaction, p. 833–853. *In Advances in Water Analysis and Treatment. Proceedings of the Water Quality Technology Conference*. American Water Works Association, Denver, Colo.
12. **De Smedt, J. M., R. F. Bolderdijk, H. Rappold, and D. Lautenschlaeger.** 1986. Rapid Salmonella detection in foods by motility enrichment on a modified semi-solid Rappaport-Vassiliadis medium. *J. Food Prot.* **49:**510–514.
13. **De Smedt, J. M., and R. F. Bolderdijk.** 1987. Dynamics of Salmonella isolation with modified semi-solid Rappaport-Vassiliadis medium. *J. Food Prot.* **50:**658–661.
14. **Desmonts, C., J. Minet, R. Colwell, and M. Cormier.** 1990. Fluorescent-antibody method useful for detecting viable but nonculturable *Salmonella* spp. in chlorinated water. *Appl. Environ. Microbiol.* **56:**1448–1452.
15. **Difco Laboratories.** 1984. *Difco Manual*, 10th ed. Difco Laboratories, Detroit, Mich.
16. **Dutka, J. B., and J. B. Bell.** 1973. Isolation of Salmonellae from moderately polluted waters. *J. Water Pollut. Control Fed.* **45:**316–323.
17. **Ebdon, J. E., J. L. Wallis, and H. D. Taylor.** 2004. A simplified low-cost approach to antibiotic resistance profiling for faecal source tracking. *Water Sci. Technol.* **150:**185–191.
18. **Erlich, H. A.** (ed.). 1989. *PCR Technology: Principles and Applications for DNA Amplification*. Oxford University Press, New York, N.Y.
19. **Feachem, R. G., D. J. Bradley, H. Garelick, and D. D. Mara** (ed.). 1983. *Sanitation and Disease: Health Aspects of Excreta and Wastewater Management*. John Wiley and Sons, New York, N.Y.
20. **Federal Register.** 1993. Standards for the use of disposal of sewage sludge, final rules. *Fed. Regist.* **58:**32, 9387–9404.
21. **Gamo, M., and T. Shoji.** 1999. A method of profiling microbial communities based on a most-probable-number assay that uses BIOLOG plates and multiple sole carbon sources. *Appl. Environ. Microbiol.* **65:**4419–4424.
22. **Gary, L. D.** 1995. *Escherichia, Salmonella, Shigella* and *Yersinia*, p. 450–464. *In* P. R. Murray, E. J. Baron, M. A. Pfaller, F. C. Tenover, and R. H. Yolken (ed.), *Manual of Clinical Microbiology*, 6th ed. ASM Press, Washington, D.C.
23. **Gibson, J. R., M. A. Ferrus, D. Woodward, and R. J. Owen.** 1999. Genetic diversity in *Heliocobacter pullorum* from human and poultry sources identified by an amplified fragment length polymorphism technique and pulsed-field gel electrophoresis. *J. Appl. Microbiol.* **87:**602–610.
24. **Gibson, J. R., E. Slater, J. Xerry, D. E. S. Tompkins, and R. J. Owen.** 1998. Use of an amplified-fragment length polymorphism technique to fingerprint and differentiate isolates of *Helicobacter pylori*. *J. Clin. Microbiol.* **36:**2580–2585.
25. **Greenberg, A. E., et al. (ed.).** 1992. *Standard Methods for the Examination of Water and Wastewater*, 18th ed. American Public Health Association, American Water Works Association, and Water Environment Federation, Washington, D.C.
26. **Hagedorn, C., S. L. Robinson, J. R. Filtz, S. M. Grubs, T. A. Angier, and R. B. Reneau.** 1999. Determining sources of fecal pollution in a rural Virginia watershed with antibiotic resistance patterns in fecal streptococci. *Appl. Environ. Microbiol.* **65:**5522–5531.
27. **Harvey, R. W. S., and T. H. Price.** 1979. Principals of *Salmonella* isolation. *J. Appl. Bacteriol.* **46:**27–56.
28. **Harwood, V. J., B. Wiggins, C. Hagedorn, R. D. Ellender, J. Gooch, J. Kern, M. Samadpour, A. C. H. Chapman, B. J. Robinson, and B. C. Thompson.** 2003. Phenotypic library-based microbial source tracking methods: efficacy in the California collaborative study. *J. Water Health* **01.4:** 153–166.
29. **Hegarty, J. P., M. Dows, and K. H. Baker.** 1999. Occurrence of *Helicobacter pylori* in surface water in the United States. *J. Appl. Microbiol.* **87:**697–701.
30. **Higgins, J. A., M. C. Jenkins, D. R. Shelton, R. Fayer, and J. S. Karns.** 2001. Rapid extraction of DNA from *Escherichia coli* and *Cryptosporidium parvum* for use in PCR. *Appl. Environ. Microbiol.* **67:**5321–5324.
31. **Holland, P. M., R. D. Abramson, R. Watson, and D. H. Gelfand.** 1991. Detection of specific polymerase chain reaction product by utilizing the 5′ to 3′ exonuclease activity of *Thermus aquaticus* DNA polymerase. *Proc. Natl. Acad. Sci. USA.* **88:**7276–7280.
32. **Holmberg, S. E., and J. J. Farmer.** 1984. *Aeromonas hydrophila* and *Plesiomonas shigelloides* as causes of intestinal infections. *Rev. Infect. Dis.* **6:**633–639.
33. **Hultén, K., H. Enroth, T. Nyström, and L. Engstrand.** 1998. Presence of *Helicobacter* species DNA in Swedish water. *J. Appl. Microbiol.* **85:**282–286.
34. **Hulton, C. S. J., C. F. Higgins, and P. M. Sharp.** 1991. ERIC sequences: a novel family of repetitive elements in the genome of *Escherichia coli, Salmonella typhimurium* and other enteric bacteria. *Mol. Microbiol.* **5:**825–834.
35. **Hussong, D., W. D. Burge and N. K. Enkiri.** 1985. Occurrence, growth, and suppression of *Salmonella* in composted sewage sludge. *Appl. Environ. Microbiol.* **50:**887–893.
36. **Innis, M. A., D. H. Gelfand, J. J. Sninsky, and T. J. White** (ed.). 1990. *PCR Protocols: a Guide to Methods and Applications*. Academic Press, Inc., New York, N.Y.
37. **Johnson, C. H., E. W. Rice, and D. J. Reasoner.** 1997. Inactivation of *Helicobacter pylori* by chlorination. *Appl. Environ. Microbiol.* **63:**4969–4970.

38. **Johnson, J. M., S. D. Weagant, K. C. Jinneman, and J. L. Bryant.** 1995. Use of pulsed-field gel electrophoresis for epidemiological study of *Escherichia coli* O157:H7 during a food-borne outbreak. *Appl. Environ. Microbiol.* **61:**2806–2808.
39. **Jones, K., M. Betaieb, and D. R. Telford.** 1990. Seasonal variation of thermophilic campylobacters in sewage sludge. *J. Appl. Bacteriol.* **69:**185–189.
40. **Jones, K., M. Betaieb, and D. R. Telford.** 1990. Comparison between environmental monitoring of thermophilic campylobacters in sewage effluent and the incidence of *Campylobacter* infection in the community. *J. Appl. Bacteriol.* **69:**235–240.
41. **Karl, D. M., and M. D. Bailiff.** 1989. The measurement and distribution of dissolved nucleic acids in aquatic environments. *Limnol. Oceanogr.* **34:**543–558.
42. **Kelley, S., and W. W. Sanderson.** 1960. Density of enteroviruses in sewage. *J. Water Pollut. Control Fed.* **32:**1269–1273.
43. **Kenner, B. A., and H. A. Clark.** 1974. Determination and enumeration of *Salmonella* species and *Pseudomonas aeruginosa*. *J. Water Pollut. Control Fed.* **46:**2163–2171.
44. **Knight, I. T., S. Shults, C. W. Kaspar, and R. R. Colwell.** 1990. Direct detection of *Salmonella* spp. in estuaries by using a DNA probe. *Appl. Environ. Microbiol.* **56:**1059–1066.
45. **Krieg, N. R., and J. G. Holt** (ed.). 1984. *Bergey's Manual of Systematic Bacteriology*, vol. 1. Williams and Wilkins, Baltimore, Md.
46. **Lang, A. L., Y.-L. Tsai, C. L. Mayer, K. C. Patton, and C. J. Palmer.** 1994. Multiplex PCR for detection of the heat-labile toxin gene and shiga-like toxin I and II genes in *Escherichia coli* isolated from natural waters. *Appl. Environ. Microbiol.* **60:**3145–3149.
47. **Liefson, E.** 1936. New selenite enrichment media for the isolation of typhoid and paratyphoid (salmonella) bacilli. *Am. J. Hyg.* **24:**423–432.
48. **McLaughlin, J. C.** 1995. *Vibrio*, p. 465–476. *In* P. R. Murray, E. J. Baron, M. A. Pfaller, F. C. Tenover, and R. H. Yolken (ed.), *Manual of Clinical Microbiology*, 6th ed. ASM Press, Washington, D.C.
49. **Moore, B.** 1948. The detection of paratyphoid carriers in towns by means of sewage examination. Monthly bulletin minutes. *Public Health Lab. Serv.* (Great Britain) **7:**1241.
50. **Morinigo, M. A., E. Martinez-Mananares, M. A. Munoz, and R. Cornax.** 1989. Evaluation of different plating media used in the isolation of salmonellas from environmental samples. *J. Appl. Bacteriol.* **66:**353–360.
51. **Morinigo, M. A., M. A. Munoz, R. Cornax, D. Castro, and J. J. Borrego.** 1990. Evaluation of different enrichment media for the isolation of Salmonella from polluted sea-water samples. *J. Microbiol. Methods* **11:**43–49.
52. **Morinigo, M. A., M. A. Munoz, E. Martinez-Mananares, and J. M. Sanchez.** 1993. Laboratory study of several enrichment broths for the detection of *Salmonella* spp. particularly in relation to water samples. *J. Appl. Bacteriol.* **74:** 330–335.
53. **Muller, L.** 1923. Un nouveau milieu d'enrichissement pour la recherche du bacille typhique et des paratyphiques. *C. R. Soc. Biol.* **89:**434–437.
54. **Nishikawa, Y., and T. Kishi.** 1987. A modification of bile salts brilliant green agar for isolation of motile *Aeromonas* from food and environmental specimens. *Epidemiol. Infect.* **98:**331–336.
55. **North, W. R., and M. T. Bartram.** 1953. The efficiency of selenite broth of different compositions in the isolation of *Salmonella*. *Appl. Microbiol.* **1:**130–134.
56. **Olsvik, O., T. Popovic, E. Skjerve, K. S. Cudjoe, E. Hornes, J. Ugelstad, and M. Uhlen.** 1994. Magnetic separation techniques in diagnostic microbiology. *Clin. Microbiol. Rev.* **7:**43–54.
57. **Osborne, W. W., and J. L. Stokes.** 1955. A modified selenite brilliant green medium for the isolation of Salmonella from egg products. *Appl. Microbiol.* **3:**295–299.
58. **Oyofo, B. A., and D. M. Rollins.** 1993. Efficacy of filter types for detecting *Campylobacter jejuni* and *Campylobacter coli* in environmental water samples by polymerase chain reaction. *Appl. Environ. Microbiol.* **59:**4090–4095.
59. **Palmer, C. J., G. F. Bonilla, B. Roll, C. Paszko-Kolva, L. R. Sangermano, and R. S. Fujioka.** 1995. Detection of Legionella species in reclaimed water and air with the EnviroAmp Legionella PCR kit and direct fluorescent antibody staining. *Appl. Environ. Microbiol.* **61:**407–412.
60. **Palmer, C. J., Y.-L. Tasi, C. Paszko-Kolva, C. Mayer, and L. R. Sangermano.** 1993. Detection of *Legionella* species in sewage and ocean water by polymerase chain reaction, direct fluorescent-antibody, and plate culture methods. *Appl. Environ. Microbiol.* **59**:3618–3624.
61. **Parveen, S., K. M. Portier, K. Robinson, L. Edminston, and M. Tamplin.** 1999. Discriminate analysis of ribotype profiles of *Escherichia coli* for differentiating human and nonhuman sources for fecal pollution. *Appl. Environ. Microbiol.* **65:**3132–3147.
62. **Picard, C., C. Ponsonnet, E. Paget, X. Nesme, and P. Simonet.** 1992. Detection and enumeration of bacteria in soil by direct DNA extraction and polymerase chain reaction. *Appl. Environ. Microbiol.* **58:**2717–2722.
63. **Poffe, R., and E. Op de Beek.** 1991. Enumeration of *Aeromonas hydrophila* from domestic wastewater plants and surface waters. *J. Appl. Bacteriol.* **71:**366–370.
64. **Refsum, T., E. Heir, G. Kapperud, T. Vardund, and G. Holstad.** 2002. Molecular epidemiology of *Salmonella enterica* serovar Typhimurium isolates determined by pulsed-field gel electrophoresis: comparison of isolates from avain wildlife, domestic animals, and the environment in Norway. *Appl. Environ. Microbiol.* **68:**5600–5606.
65. **Restino, L., G. S. Grauman, W. A. McCall, and W. M. Hill.** 1977. Effects of varying concentrations of novobiocin incorporated into two *Salmonella* plating media on the recovery of four *Enterobacteriaceae*. *Appl. Environ. Microbiol.* **33:**585–589.
66. **Ribas, F., J. Frias, J. M. Huguet, F. R. Ribas, and F. Lucena.** 1991. Comparison of different media for the identification and quantification of *Aeromonas* spp. in water. *Antonie Leeuwenhoek* **59:**225–228.
67. **Rippey, S. R., and V. J. Cabelli.** 1979. Membrane filter procedure for enumeration of *Aeromonas hydrophila* in fresh waters. *Appl. Environ. Microbiol.* **38:**108–111.
68. **Ritter, J. K., E. Carruthers, C. A. Carson, R. D. Ellender, V. J. Harwood, K. Kingsley, C. Nakatsu, M. Sadowsky, B. Shear, B. West, J. E. Whitlock, B. A. Wiggins, and J. D. Wilbur.** 2003. Assessment of statistical methods used in library-based approaches to microbial source tracking. *J. Water Health* **01.4:**209–233.
69. **Saiki, R. K.** 1990. Amplification of genomic DNA, p. 13–20. *In* M. A. Innis, D. H. Gelfand, J. J. Sninsky, and T. J. White (ed.), *PCR Protocols: a Guide to Methods and Applications*. Academic Press, Inc., New York, N.Y.
70. **Schott, E. B., and R. Rimmler.** 1973. Medium for the isolation of *Aeromonas hydrophila*. *Appl. Microbiol.* **25:**550–553.
71. **Seurinck, S., W. Verstraete, and S. D. Siciliano.** 2002. Use of 16S-23S rRNA intergenic spacer region PCR and repetitive extragenic palindromic PCR analyses of *Escherichia coli* isolates to identify nonpoint fecal sources. *Appl. Environ. Microbiol.* **69:**4942–4950.
72. **Sommerville, C. C., I. T. Knight, W. L. Straube, and R. R. Colwell.** 1989. Simple rapid method for the direct

islation of nucleic acids from aquatic environments. *Appl. Environ. Microbiol.* **55:**548–554.

73. **Smalla, K., U. Wachtendorf, H. Heuer, W. Liu, and L. Forney**. 1998. Analysis of BIOLOG GN substrate utilization patterns by microbial communities. *Appl. Environ. Microbiol.* **64:**1220–1225.
74. **Steele, M., B. McNab, L. Fruhner, S. Degrandis, D. Woodward, and J. A. Odumeru**. 1998. Epidemiological typing of *Campylobacter* isolates from meat processing plants by pulsed-field gel electrophoresis, fatty acid profile typing, serotyping, and biotyping. *Appl. Environ. Microbiol.* **64:**2346–2349.
75. **Stern, M. J., G. F.-L. Ames, N. H. Smith, E. C. Robinson, and C. F. Higgins**. 1984. Repetitive extragenic palindromic sequences: a major component of the bacterial genome. *Cell* **37:**1015–1026.
76. **Stewart, J. R., R. D. Ellender, J. A. Gooch, S. Jiang, S. P. Myoda, and S. Weisberg**. 2003. Recommendations for microbial source tracking: lessons learned from a methods comparison study. *J. Water Health* **01.4:**228–231.
77. **Stoeckel, D. M., M. V. Mathes, K. E. Hyer, C. Hagedorn, H. Kator, J. Lukasik, T. L. O'Brien, T. W. Fenger, M. Samadpour, K. M. Strickler, and B. A. Wiggins**. 2004. Comparison of seven protocols to identify fecal contamination sources using *Escherichia coli*. *Environ. Sci. Technol.* **38:**6109–6117.
78. **Tamanai-Shacoori, Z., A. Jolivet-Gougeon, M. Pommepuy, M. Cormier, and R. R. Colwell**. 1994. Detection of enterotoxigenic *Escherichia coli* in water by polymerase chain reaction amplification and hybridization. *Can. J. Microbiol.* **40:**243–249.
79. **Tsai, Y.-L., and B. H. Olson**. 1992. Rapid method for separation of bacterial DNA from humic substances in sediments for polymerase chain reaction. *Appl. Environ. Microbiol.* **58:**2292–2295.
80. **Tsai, Y.-L., C. J. Palmer, and L. R. Sangermano**. 1993. Detection of *Escherichia coli* in sewage and sludge by polymerase chain reaction. *Appl. Environ. Microbiol.* **59:**353–357.
81. **Tsai, Y.-L, M. D. Sobsey, L. R. Sangermano, and C. J. Palmer**. 1993. Simple method of concentrating enteroviruses and hepatitis A virus form sewage and ocean water for rapid detection by reverse transcriptase-polymerase chain reaction. *Appl. Environ. Microbiol.* **59:**3488–3491.
82. **Tsai, Y.-L., B. Tran, and C. J. Palmer**. 1995. Analysis of viral RNA persistence in seawater by reverse transcriptase-PCR. *Appl. Environ. Microbiol.* **61:**363–366.

82a. **U.S. Environmental Protection Agency**. 2003. *Method 1682: Salmonella in Biosolids by Modified Semisolid Rappaport-Vassiliadis (MSRV) Medium*. EPA-812-R-03-011. U.S. Environmental Protection Agency, Washington, D.C.

82b. **U.S. Environmental Protection Agency**. 2003. *Quality Assurance/Quality Control Guidance for Laboratories Performing PCR Analyses on Environmental Samples*. Draft for external review. U.S. Environmental Protection Agency, Washington, D.C.

83. **Vassiliadis, P., D. Trichopoulos, J. Papadakas, V. Kalapothaki, X. Zavitsanos, and C. Serie**. 1981. *Salmonella* isolation with Rappaport's enrichment medium of different compositions. *Zentbl. Bakteriol. Mikrobiol. Hyg. Abt. 1 Orig. B* **173:**382–389.
84. **Versalovic, J., T. Koeuth, and J. R. Lupski**. 1991. Distribution of repetitive DNA sequences in eubacteria and application to fingerprinting of bacterial genomes. *Nucleic Acids Res.* **19:**6823–6831.
85. **Vinjé, J., S. J. G. Oudejans, J. R. Stewart, M. D. Sobsey, and S. C. Long**. 2004. Molecular detection and genotyping of male-specific coliphages by reverse transcription-PCR and reverse line blot hybridization. *Appl. Environ. Microbiol.* **70:**5996–6004.
86. **Weinbauer, M. G., D. Fuks, and P. Peduzzi.** 1993. Distribution of viruses and dissolved DNA along a coastal trophic gradient in the northern Adriatic Sea. *Appl. Environ. Microbiol.* **59:**4074–4082.
87. **Yanko, W. A., A. S. Walker, J. L. Jackson, L. L. Liabo, and A. L. Garcia**. 1995. Enumerating *Salmonella* in biosolids for compliance with pathogen regulations. *Water Environ. Res.* **67:**364–370.

Detection of Pathogenic Bacteria, Viruses, and Parasitic Protozoa in Shellfish

LEE-ANN JAYKUS

25

INTRODUCTION

Bacterial and viral disease outbreaks have been associated with all major types of edible bivalve molluscs, and human enteric pathogens have been isolated from shellfish obtained from both opened and closed harvesting areas. This chapter contains information that pertains to those shellfish that are edible bivalve molluscs of the class Pelecypoda and include the species commonly referred to as oysters, mussels, clams, and cockles. Since most of these organisms are filter feeders, with the Gastropoda group being a notable exception, they use siphoning organelles and mucous membranes to sieve suspended food particles from the aquatic environment as a source of food. If their surrounding water is contaminated by bacteria, viruses, or parasitic protozoa, these mucous membranes may entrap the pathogens and transfer them to the digestive tract. Since these molluscan shellfish are frequently consumed whole and raw, they may act as passive carriers of human pathogens. Although the definition of shellfish can include both crustaceans and molluscs, this chapter uses the word "shellfish" in reference to only bivalve molluscs.

Illness Transmission by Bivalve Molluscs

The National Shellfish Sanitation Program was developed in 1925 to target the prevention of shellfish-associated diseases caused primarily by enteric bacteria. In the United States, this program has established bacteriological standards for shellfish and their harvesting waters based on the coliform or fecal coliform index. For approved harvesting waters, the most probable number (MPN) of coliforms must not exceed 70 per 100 ml, with no more than 10% of the samples containing coliforms at levels exceeding 230 per 100 ml. Shellfish meat may contain no more than 230 fecal coliforms per 100 g. While these standards have proven very effective in preventing outbreaks of disease due to enteric bacterial pathogens, there is little or no relationship between the levels of coliform indicator bacteria in water and shellfish and the presence human enteric viruses or naturally occurring marine *Vibrio* species. Consequently, shellfish-associated outbreaks and cases of disease due to both human enteric viruses and *Vibrio* species continue to occur in the United States. The relative importance of parasitic protozoa in shellfish-associated enteric disease is currently unknown. Microbiological standards for enteric virus, *Vibrio*, or protozoan contamination in shellfish have not been established, due in part to the lack of widely accessible detection methodology.

ETIOLOGY OF MOLLUSC-ASSOCIATED OUTBREAKS OF ILLNESS

The most recent Centers for Disease Control and Prevention (CDC) statistics (1988 to 1992) of the overall food-borne disease burden in the United States estimate that 0.7 to 2.1% of all outbreaks, approximately 1% of all cases, and up to 13.3% of food-related deaths are due to the consumption of contaminated shellfish (6). However, since few shellfish-borne diseases are reportable, this is likely to be a significant underestimate. In fact, shellfish are rarely studied as a stand-alone commodity with respect to food-borne disease. In a notable exception, Wallace et al. (98) retrospectively examined food-borne disease outbreaks in New York state and found that contaminated seafood accounted for 19% of all reported outbreaks and 10% of all illnesses, with shellfish being the most frequently implicated vehicle. Infectious diseases associated with molluscan shellfish consumption have been reviewed extensively (61, 81, 86).

Pathogenic Bacteria Transmitted by Bivalve Molluscs

Two general groups of pathogenic bacteria may be transmitted by shellfish. The first group is termed indigenous bacterial pathogens because these organisms are native to the marine environment. They consist predominantly of members of the family *Vibrionaceae*, including the genera *Vibrio*, *Aeromonas*, and *Plesiomonas*. The presence of these organisms is unrelated to fecal pollution and therefore cannot be monitored by the use of fecal coliform indices. The members of the second group, referred to as nonindigenous bacterial pathogens, are not natural marine inhabitants; and their presence in shellfish arises from either direct fecal contamination by human or animal reservoirs or poor general sanitation during harvesting, processing, or preparation of the food animals. Many of the agents implicated in shellfish-associated diseases are not reportable, and scientists debate the exact role that bacteria play in the prevalence of disease. In their review, Lipp and Rose (61) estimate that pathogens associated with fecal contamination cause only about 4% of shellfish-related food-borne disease outbreaks in the United States, while the indigenous bacterial pathogens may be responsible for as many as 20% of the cases of shellfish-associated disease. The environmental *Vibrio* species are included in the U.S. CDC Foodborne Disease Active Surveillance Network (FoodNet). A major objective of the FoodNet program is to estimate more accurately the burden of food-borne disease in the United States. Begun in 1996, the program has grown from an initial surveillance population of 14.2 million people in 5 states to 44.1 million individuals (representing 15.2% of the U.S. population) in 10 states. Interestingly, the most recent (2004) FoodNet data indicate an annual incidence of 13.2 cases of *Vibrio*-associated disease per million people, which represents a 47% increase over the 1996 to 1998 baseline estimates (12). It is clear from these and other data that both the incidence and public awareness of shellfish-associated human *Vibrio* infections have increased over the last 15 years (6, 12, 69).

Enteric Bacteria

Outbreaks of human disease associated with the presence of *Salmonella* and *Shigella* species in shellfish are well documented, even in the recent literature (6). When these agents are present, it is usually due to fecal contamination of harvesting sites, but occasionally it is due to cross-contamination by human carriers, particularly food handlers. In year 2000 summary of data collected by U.S. Food and Drug Administration (FDA) field laboratories, investigators reported that <0.5% of the domestic shellfish tested were contaminated with *Salmonella*, while imported oysters and clams had a higher prevalence of contamination (1 and 3.4% of samples tested, respectively) (42). In a 3-year shellfish market survey in Spain, Martinez-Urtaza et al. (68) reported a *Salmonella* isolation rate of 1.8%. More recently, in 2004, it was suggested that European and American environmental surveys are revealing a higher prevalence of *Salmonella* contamination in shellfish and their harvesting waters than was previously believed. For example, U.S. investigators isolated *Salmonella* from oysters harvested from all three U.S. coastlines (the East Coast, the West Coast, and the Gulf of Mexico coast) at a prevalence of 7.4%. Of these isolates, >75% were characterized as *Salmonella enterica* serovar Newport, a major human pathogen (10). In a similar Spanish environmental prevalence study (66), serovar Typhimurium was the most frequently isolated, and 3 of 23 isolates characterized were multidrug-resistant phage types DT104 or DT193 (67). Importantly, these recent prevalence studies concluded that little or no relationship existed between the presence of *Salmonella* in shellfish and the levels of fecal coliforms (10, 66). However, the public health significance of these findings has yet to be determined.

The Vibrios

Vibrios are part of the normal estuarine microbiota, have excellent survival capabilities in both marine and estuarine environments, and may be accumulated by shellfish during feeding. While the transmission of *Vibrio cholerae* by seafood is well documented throughout the world, the United States has not had a major outbreak since 1911, although sporadic cases have occurred. Most of these have been attributed to the consumption of crustacea such as crabs and shrimp; only a few have been associated with bivalve molluscs (6). However, sporadic cases in both Texas and Florida in the 1970s were linked to raw bivalve molluscs. In the 1990s, seafood imported from South American locations where cholera is endemic were associated with disease in industrialized nations.

Historically, *Vibrio parahaemolyticus* outbreaks are rare in the United States, but two highly publicized recent outbreaks have challenged this trend. In the summer of 1997, the largest outbreak of *V. parahaemolyticus* infection reported in North America occurred in the U.S. Pacific Northwest. Illness occurred in 209 people and was epidemiologically associated with the consumption of raw oysters harvested from northern Pacific waters (14). The next summer another multistate outbreak associated with the consumption of raw oysters harvested from Galveston Bay, Tex., occurred and affected as many as 296 Texas residents, with 120 additional infections occurring in 12 other states. This outbreak was caused by newly emerged *V. parahaemolyticus* serotype O3:K6 (19). Also in 1998, an outbreak associated with the consumption of bivalve molluscs harvested from Long Island Sound was associated with the O3:K6 serotype (13). In a few of these cases, mean surface water temperatures were significantly higher (1 to 5°C) than those reported in previous years. Prior to these outbreaks, which seem to have been caused by endogenous contaminants, *V. parahaemolyticus* infections in the United States were generally presumed to be due to the gross mishandling of contaminated seafood, particularly improper refrigeration, insufficient cooking, or cross-contamination.

Vibrio vulnificus has been of greater concern in the United States in recent years, since epidemiological evidence indicates that enteric exposure to this organism can result in a syndrome characterized by gastrointestinal disease followed by primary septicemia, with mortality rates approaching 50%. Those who are especially at risk include individuals with underlying liver dysfunction; those with circulatory problems, particularly circulatory problems related to diabetes; or those who are immunocompromised (45). A self-limiting gastrointestinal disease syndrome without subsequent septic disease can also be seen in healthy individuals consuming contaminated bivalve molluscs (45). Several large surveillance studies have been undertaken in the last decade. A 1996 investigation of 690 *Vibrio vulnificus* infections reported to Florida health authorities suggested that gastrointestinal disease was the predominant disease syndrome (45). However, recent U.S. CDC surveillance from 23 states indicated that the gastrointestinal syndrome was relatively uncommon (5%) compared to the rate of primary septicemia (43%) (89). Although *V. vulnificus* is most often responsible for *Vibrio*-related septic infections, other species such as *V. parahaemolyticus*, *V. cholerae* non-O1, and *Vibrio hollisae* are capable of producing morbidity similar to that caused by *V. vulnificus* (45). It is crucial to note that the majority of U.S. cases of *V. vulnificus* septic disease have occurred either in the Gulf States or as a result of the consumption of oysters harvested from those states (89). *V. vulnificus* is a leading cause of food-borne disease-related deaths, particularly in the southern states (6). Furthermore, this organism is capable of establishing the so-called viable-but-nonculturable state, which means that routine examination by conventional cultural methods may result in negative results, although viable and potentially virulent cells may be present in high numbers (77).

Enteric Viruses Transmitted by Bivalve Molluscs

Viruses are probably the most significant microbiological agents causing shellfish-associated disease; extensive reviews of shellfish-borne viral disease epidemiology have been provided by Lipp and Rose (61) and Potasman et al. (81); recent reviews of food-borne viruses have also been published (55, 79, 87). Many different enteric viruses are excreted in human feces and ultimately find their way into domestic sewage. As with the enteric bacteria, these viruses are not indigenous to the water column and can be concentrated by the molluscs during the natural filter-feeding process. The enteric viruses cause a wide variety of illnesses, including hepatitis, fever, diarrhea, gastroenteritis, meningitis, myocarditis, and others. Regardless of the large number of enteric viruses, only hepatitis A virus (HAV), hepatitis E virus (HEV), the noroviruses, the astroviruses, the rotaviruses, and other small round (featureless) viruses (SRVs) have been epidemiologically linked to various degrees to shellfish-associated viral disease. Nonetheless, sporadic cases of illness that are not detected as outbreaks may potentially be attributable to shellfish consumption.

Hepatitis A Virus

Infectious hepatitis caused by HAV is perhaps the most serious viral illness transmitted by the ingestion of contaminated shellfish. HAV is currently classified as a member of the genus *Hepatovirus* of the family *Picornaviridae* and shares many basic characteristics in common with the enterovirus group, i.e., a single-stranded RNA genome, a 28-nm diameter, and fecal-oral transmission. HAV causes a higher incidence of symptomatic infections than the human enteroviruses, with an incidence of up to 95% during outbreaks. Common symptoms of HAV infection are dark urine, nausea, vomiting, malaise, fever, chills, and jaundice. Fulminant hepatitis A infection, although rare, has a very high mortality rate (18). HAV infection is usually asymptomatic in children, but the severity can increase with age (18).

Outbreaks of shellfish-associated infectious hepatitis were not documented in the United States until the early 1960s. Since then, many cases of HAV infection associated with the ingestion of bivalve molluscs have been reported. Shellfish-associated hepatitis A is of greater concern in countries where raw bivalve molluscs have widespread consumption. For example, the largest food-borne disease outbreak ever reported was caused by HAV, where 300,000 cases attributable to bivalve mollusc consumption occurred over a 2-month period in Shangai, China, in 1988 (41). While the number of sporadic cases of HAV infection is difficult to ascertain, it is nonetheless considered to be more significant than outbreak-related cases, and some may be associated with the consumption of bivalve molluscs (18).

Human Noroviruses

Recent epidemiological estimates suggest that the human gastrointestinal viruses, predominantly the noroviruses, are a very significant cause of food-borne disease, perhaps being responsible for over 60% of all cases (69). These viruses have a characteristic morphology of 32 cup-like depressions superimposed on an icosahedral symmetry and are 28 to 34 nm in diameter (17). The Norwalk agent was the first recognized human gastroenteritis virus of clinical importance and is considered a prototype strain. This virus was discovered in 1972 in Norwalk, Ohio, and was identified by immune electron microscopy of an infectious stool filtrate (50). The first documented outbreak of shellfish-associated Norwalk virus gastroenteritis involved over 2,000 people throughout Australia (50). In the United States and Europe, bivalve mollusc-associated gastroenteritis attributed to Norwalk virus has been reported consistently since 1980; more recent outbreaks have been linked to the importation of depurated English clams, shellfish harvested from approved waters and grossly fecally contaminated waters, malfunctioning boat sewage disposal systems, and ill shellfish harvesters who routinely discharged their sewage overboard (15, 41, 85).

Early immunological studies revealed that the viral gastroenteritis agents related to Norwalk virus (the so-called small round-structured viruses) were quite antigenically diverse. More recent advances in molecular virology have increased investigators' understanding of the group; for example, the Norwalk virus genome (48) and those of related viruses (56) have been sequenced. The genome size, buoyant density, and genomic organization of these viruses are similar to those of other viruses in the family *Caliciviridae*, prompting investigators to initially classify these viruses as belonging to this virus family (50). None of these human caliciviruses have been cultivated in vitro, and there exists no practical animal model for their propagation.

In 2000, the final taxonomic status within the family *Caliciviridae*, as accepted by the International Committee on Taxonomy of Viruses, was reported (39, 95). This led to the development of genus and species definitions for four new genera within the family *Caliciviridae*. The new genera are designated *Vesivirus*, *Lagovirus*, Norwalk-like viruses (NLVs), and Sapporo-like viruses (SLVs), with the type species being vesicular exanthema of swine virus, rabbit hemorrhagic disease virus, Norwalk virus, and Sapporo virus, respectively. Since that time, the NLVs have been designated noroviruses; and at the time of this writing, the norovirus genus consists of five genogroups, genogroup I (GI; prototype, Norwalk virus), GII (prototype, Snow Mountain agent), GIII (prototype, bovine enteric calicivirus), GIV (prototype, Alphatron and Fort Lauderdale virus), and GV (prototype, Murine norovirus) (J. Vinje, personal communication). GII noroviruses are more often associated with viral gastroenteritis in humans, followed by GI (28). The SLVs, now designated sapoviruses, are genetically related to the noroviruses but are more closely related to animal caliciviruses and have occasionally caused viral gastroenteritis in humans (97).

Other Enteric Viruses

A group of SRVs has been reported to be the cause of numerous outbreaks of shellfish-associated gastroenteritis (33). These viruses do not appear to be serologically related to the noroviruses or HAV. SRVs are detected by electron microscopy, and information on the characteristics of the viruses is virtually nonexistent. Astroviruses, given their name because of the characteristic five- to six-pointed star-like form seen under an electron microscope, have also been associated with bivalve mollusc consumption (2). HEV infection is transmitted mainly by sewage-contaminated water and person-to-person contact. The etiological agent of this disease, previously believed to be a calicivirus, has now been removed from the *Caliciviridae* family and placed in an unclassified status (95). The disease caused by HEV can be more severe than that caused by HAV, with a high incidence of cholestasis and significant mortality (>15%) in pregnant women during their second or third trimester. HEV is endemic in the Middle East, Africa, and India but not in the United States. The role of shellfish in the transmission of this disease is suspected but is currently unconfirmed epidemiologically.

Enteropathogenic Protozoa

Protozoan parasites such as *Giardia* and *Cryptosporidium* species may be present in waters where shellfish grow as a result of contamination with animal farm runoff or treated and untreated sewage input. The ability of bivalve molluscs to concentrate parasitic protozoa has been demonstrated, and some have postulated a potential role for oysters in the epidemiology of *Cryptosporidium* infection (31), which was also suggested by the results of a more recent Swiss epidemiological study (5). However, confirmatory case-control or cohort studies have not yet been reported. Studies do, however, report oocyst stability for 4 to 12 weeks in artificial seawater (31). Others have detected *Cryptosporidium* oocysts in Eastern oysters harvested from commercial sites in the Chesapeake Bay (30) and along the U.S. Atlantic coast from Maine to Florida and including New Brunswick, Canada (29). *Cryptosporidium* oocysts have also been recovered from Bent mussels (*Ischadium recurvum*) (37) and from various clams, mussels, and oysters harvested in Europe (35, 64). Additional studies have demonstrated that freshwater benthic clams and zebra mussels (*Dreissena polymorpha*) are capable of capturing and concentrating *Cryptosporidium* oocysts (37). Although the latter are not used for food, these studies do indicate that even freshwater bivalve molluscs can concentrate protozoan parasites and suggest that all types of bivalve molluscs may, in fact, behave as vectors for *Giardia* and *Cryptosporidium* species. Some have even postulated the use of shellfish as bioindicators of fecal protozoan contamination (71).

METHODS FOR DETECTION OF PATHOGENS IN SHELLFISH

Many types of methodologies have been used for the detection of pathogenic bacteria, viruses, and parasitic protozoa from bivalve molluscs. This chapter presents general processing methodologies, including conventional detection approaches such as microscopic or cultural methods, as well as the molecular approaches that are increasingly being applied to the detection of these pathogens in the bivalve mollusc food matrix.

General Steps in Isolation and Detection of Pathogens from Shellfish

While the methods used for the recovery and detection of microorganisms in shellfish differ for bacteria, viruses, and parasitic protozoa, they all follow five to six general steps from sample collection to final identification (Table 1). This chapter discusses only the steps which strongly affect the assay method. Sample collection and shucking as well as meat homogenization are common to all analyses and are straightforward techniques, provided they are done aseptically. A major difference between methods designed for virus and parasitic protozoa detection and bacterial detection methods is the need to purify and concentrate the pathogens away from the food matrix prior to detection. For viruses and parasitic protozoa, the approach used for pathogen extraction and concentration not only affects the efficiency of recovery of targeted pathogens but also influences the sample-associated toxicity or inhibition of the assay. In the case of bacterial pathogens, cultural enrichment steps are used to increase target pathogen numbers. Although these methods may be considered reasonably standard, they can be complicated by high levels of background microflora or fastidious organisms, among which can be those capable of entering into a viable but nonculturable state. Molecular detection frequently involves nucleic acid extraction and concentration approaches, which affect detection sensitivity. In all cases, assay inhibitors must be removed before the application of the molecular methods. A schematic of the approaches commonly used for the detection of bacterial, viral, and parastic protozoan pathogens in shellfish is provided in Figure 1.

TABLE 1 General steps in isolation and detection of pathogens from shellfish by molecular techniques

1. Sample collection and shucking
2. Meat (or dissected tissue) homogenization
3. Cultural enrichment (*Vibrio* spp.); extraction, which includes pathogen concentration and removal of residual food matrix (human enteric viruses and parasitic protozoa)
4. Nucleic acid extraction, which provides further sample concentration and removal of inhibitors of molecular amplification
5. Detection of amplification products
6. Confirmation of molecular amplification assays

Conventional Methods for Detection of Pathogenic Bacteria in Shellfish

Examination of food products for bacterial pathogens differs from the examination of clinical samples because of the potentially lower numbers of microorganisms and the frequently poor physiological state of the organisms found in the food items. Conventional methods for the detection of pathogenic bacteria in foods rely on the ability of the organism to multiply in specific selective enrichment media, followed by defined growth, morphology, or other distinguishing features on selective plating media. Therefore, detection of pathogens is dependent on both the availability of media specific for their isolation and the ability of the organism to grow. Consequently, conventional cultural methods usually fail to detect bacteria when they are in the viable but nonculturable state.

Nonetheless, successful methods have been developed for the detection of culturable forms of *V. parahaemolyticus*, *V. vulnificus*, and *V. cholerae* in shellfish; the last of these will not be detailed in this chapter. The FDA *Bacteriological Analytical Manual* (FDA BAM) (22) cites standard procedures for the recovery of all three pathogenic *Vibrio* spp. from raw molluscan shellfish. Regardless of whether the target is *V. vulnificus* or *V. parahaemolyticus*, the method begins with the production of phosphate-buffered saline (PBS) dilutions, followed by alkaline peptone water (APW) enrichments in an MPN format. PBS enrichments can be done in parallel, if so desired. In both cases, enrichments are incubated at 35 to 37°C for 16 to 18 h. MPN tubes having turbidity are streaked onto modified cellobiose-polymyxin B-colistin (mCPC) agar for identification of *V. vulnificus* and/or thiosulfate-citrate-bile salts-sucrose (TCBS) agar for identification of *V. parahaemolyticus*. PBS enrichment broths can also be streaked onto these agars as a control, particularly in instances in which the resulting APW tubes lack turbidity. mCPC plates are incubated for 18 to 24 h at 39 to 40°C, after which they are examined for typical *V. vulnificus* colonies, which are flat and yellow with opaque centers and translucent peripheries and which are about 2 mm in diameter. TCBS agar plates are incubated for 18 to 24 h at 35 to 37°C and are examined for typical *V. parahaemolyticus* colonies, which are round, opaque, green or bluish in color, and about 2 to 3 mm in diameter. Although mCPC and TCBS may be

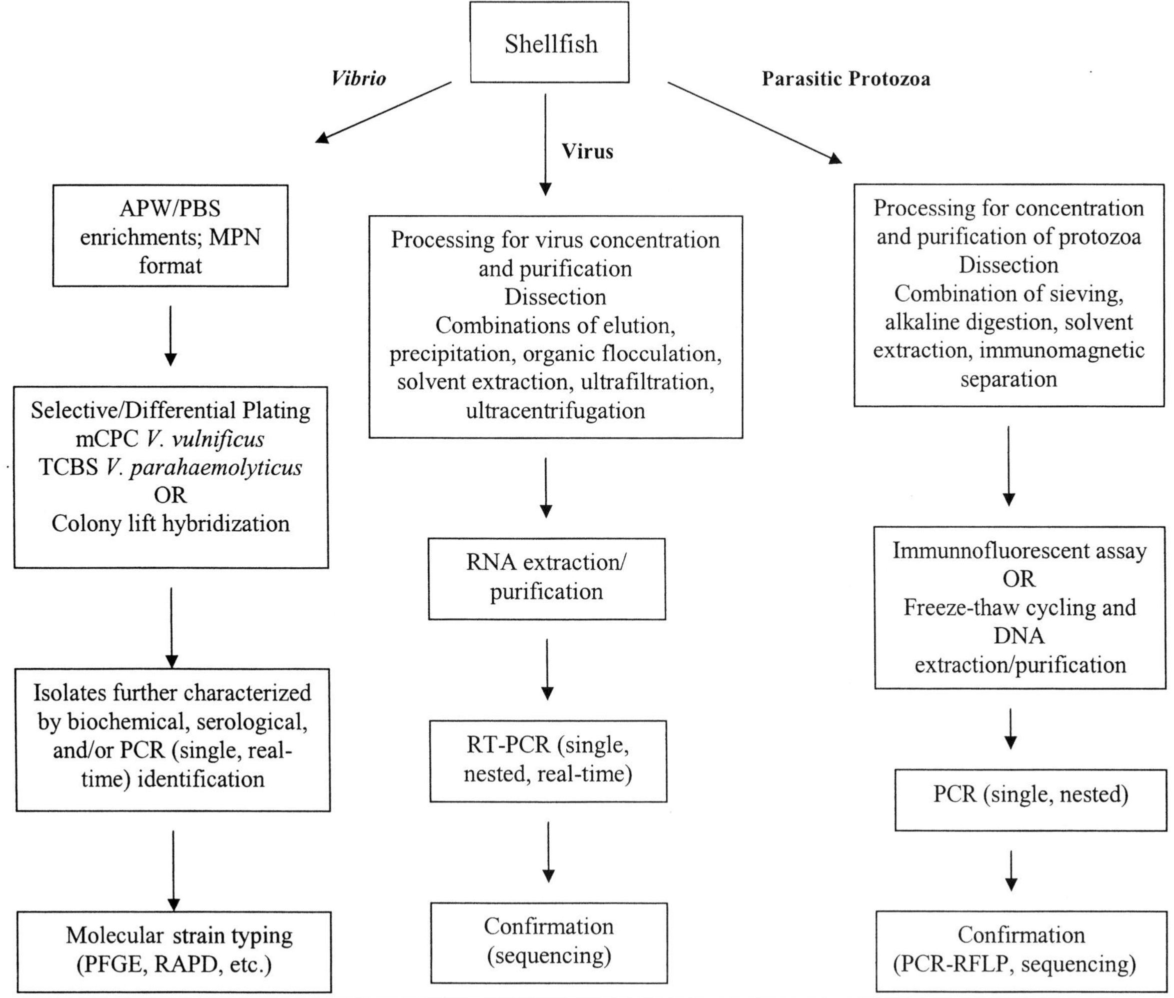

FIGURE 1 Representative scheme for sample processing and subsequent detection of pathogens (bacteria, viruses, and parasitic protozoa) in shellfish.

considered the "gold standard" bacteriological media, improvements in medium formulation are periodically reported. For biochemical identification, three or more typical colonies are selected from mCPC (*V. vulnificus*) or TCBS (*V. parahaemolyticus*) agar plates, streaked onto 1% tryptone, 1% NaCl (T_1N_1) agar, and also inoculated into gelatin agar and gelatin salt agar, followed by incubation for 18 to 24 h at 35 to 37°C. Other confirmatory biochemical tests for *V. vulnificus* and *V. parahaemolyticus* include the oxidase test, tests for motility, the arginine-glucose slant test, the triple sugar iron agar test, the O/129 Vibriostat sensitivity test, and the *o*-nitrophenyl-β-D-galactopyranoside test. In addition, FDA BAM suggests the use of species-specific alkaline phosphatase-labeled DNA probes or PCR for confirmation of *V. vulnificus* or *V. parahaemolyticus*. These will be discussed later in the chapter.

Historical Methods for Concentration and Detection of Viruses from Shellfish

Various methods, most of which were reported in the decade between 1970 and 1980, have been developed for the recovery of enteric viruses from bivalve molluscan shellfish and other seafood. These methods have recently been reviewed by Jaykus (46) and D'Souza et al. (26). Two general schemes have proven particularly successful; these are designated extraction-concentration and adsorption-elution-concentration. The goal in both cases is to separate and concentrate the viruses from shellfish meats while simultaneously providing a low-volume aqueous solution that is free of cytotoxic material and compatible with assay by mammalian cell culture. Both schemes use conditions that favor the separation of viruses from shellfish tissues, primarily through the use of filtration, centrifugation, precipitation, polyelectrolyte (organic) flocculation, and solvent extraction. However, the adsorption-elution-precipitation methods have received more use. To this end, a modified adsorption-elution-precipitation procedure (93) is described here. Briefly, 50 g of oysters is diluted seven-fold with cold distilled water and vigorously blended, and the viruses are adsorbed to the meat by reducing the conductivity to <2,000 ppm and the pH to 5.0. After centrifugation at 1,700 × *g* and 4°C for 15 min, the solids-adsorbed viruses are eluted by resuspension in 350 ml of 0.05 M glycine–0.14 M NaCl with subsequent adjustment of the pH to 7.5. Shellfish solids are removed by centrifugation at 1,700 × *g* and 4°C for 15 min, and the viruses in the supernatant are concentrated by acid precipitation at pH 4.5. The resulting floc is sedimented by centrifugation at 1,700 × *g* and 4°C for 15 min and is resuspended in 10 ml of 0.1 M disodium phosphate buffer (pH 9.3 to 9.5), followed by adjustment of the pH to 7.0 to 7.5. Cytotoxic components of the sample are precipitated by the addition of Cat-Floc T (Calgon, Ellwood City, Pa.) to a final concentration of 0.1%, and the precipitate is removed by centrifugation at 3,000 × *g* and 4°C for 20 min. The recovered supernatants are supplemented with gentamicin and kanamycin to final concentrations of 50 and 250 μg/ml, respectively, and the mixture is incubated at room temperature for 2 h. The final concentrates can be stored at −80°C until assay or further treatment. Various modifications to these protocols have included further clarification of the concentrates by polyelectrolyte flocculation, ultrafiltration, or the freeze-thaw method, followed by centrifugation.

The shellfish extracts obtained after the extraction-concentration or the adsorption-elution-concentration processes were historically assayed for viruses by using mammalian cell culture techniques. There are two basic procedures for detection and quantification of viruses in cell culture, i.e., quantal methods (determination of the 50% tissue culture infective dose or MPN) and enumerative methods (plaque assay). Quantal methods rely on the development of widespread cell monolayer deterioration (cytopathic effects), while enumerative methods detect localized cell monolayer damage under an agar overlay (a plaque). Selection of the cell culture type depends almost entirely on the viruses that one wishes to detect. Unfortunately, no single universal cell culture system detects all or even a majority of the more than 100 different human enteric viruses that can potentially contaminate shellfish. Several convenient and sensitive cell lines are commonly used for the detection of human enteroviruses, including the BGMK (buffalo green monkey kidney-derived), MA-104 (rhesus monkey kidney-derived), RD (human rhabdomyosarcoma-derived), and AGMK (primary African monkey kidney-derived) cell lines. Because of their ease of propagation in mammalian cell culture, most early studies on the prevalence of enteric viral contamination in shellfish that used cell culture propagation methods were limited to the enterovirus family. It must, however, be recognized that the epidemiologically important shellfish-borne enteric viruses do not replicate (noroviruses) or replicate poorly (HAV, astroviruses) in all currently used mammalian cell cultures. While the FRhK-4 (fetal rhesus monkey kidney-derived) cell line has proven useful for the propagation of a laboratory-adapted strain of HAV, its use for the detection of wild-type HAV is of limited value. In short, mammalian cell culture lines are not available for the detection of the noroviruses, sapoviruses, astroviruses, or SRVs. Furthermore, the lengthy assay time and the high cost seriously limit the value of this detection approach when it is available.

Historical Methods for Concentration and Detection of Human Protozoan Parasites in Shellfish

Although efforts to detect protozoan parasites in shellfish have lagged behind those for virus detection, like viruses, it is recommended that parasitic protozoa be concentrated from shellfish tissues prior to the application of detection methods. One concentration method consists of washing dissected gill tissue, followed by direct detection of parasitic protozoa (29, 64). Alternatively, whole shellfish tissue, dissected digestive glands, or gills can be homogenized (35, 64) or the hemolymph can be aspirated (71). These homogenates are sieved to remove large particulates, followed by centrifugation at low to moderate centrifugal speeds (1,000 to 2,000 × *g*) with or without diethyl ether (2:1). Sometimes these sample concentrates are processed for detection without further steps, while other investigators have further processed the resulting pellets, usually less than 0.5 g in weight, by immunomagnetic separation with commercially available beads for the capture and concentration of *Cryptosporidium* and *Giardia*. After contact for 1 h, the bead-parasite complexes are collected with a magnet and the parasites are dissociated from the beads by two sequential washes in 0.1 N HCl (for *Cryptosporidium*) or PBS (for *Giardia*). The parasite-containing supernatant is then processed for detection by one of a number of approaches.

Like viruses, *Giardia* and *Cryptosporidium* species cannot be readily propagated in the laboratory, and therefore, indirect methods need to be used for their detection. Traditional methods for the detection of parasitic agents from water and shellfish sample concentrates have been based almost exclusively on various staining approaches. A number of direct immunofluorescence assay kits are available for the detection of the parasitic protozoa; most use a fluorescein isothiocyanate-labeled monoclonal antibody and were developed for use with water samples but have been adapted to shellfish matrices (29, 35, 64, 71). Other investigators have used indirect fluorescent-antibody assays and/or various less specific staining methods (34). In all cases, microscopic examination of the smears is required.

Molecular Methods for Detection of Pathogens in Shellfish

The advent of recombinant DNA technology has made it possible to use a complementary segment of DNA or RNA for the detection of gene sequences specific to a pathogen of interest. Of particular value have been nucleic acid amplification techniques such as PCR, since this provides the most feasible alternative for achieving the sensitivity limits needed for routine assay of viral and parasitic protozoan contamination in environmental and shellfish samples. Nucleic acid amplification also offers the possibility of reducing the overall detection time for culturable bacterial pathogens by replacing lengthy cultural enrichment steps with more rapid nucleic acid enrichment. Considerable progress has been made in adapting PCR and reverse transcription-PCR (RT-PCR) methods to the detection of microbiological pathogens in shellfish.

Molecular Methods for Detection of *Vibrio* Species in Shellfish

A recent development that has facilitated the detection of pathogenic *Vibrio* spp. in shellfish has been colony lift hybridization. In fact, a direct plating colony lift hybridization method has been reported which can be used to enumerate both total and pathogenic *V. parahaemolyticus* isolates in oyster meats, with adaptations for the enumeration of *V. vulnificus* (22, 53). The *V. vulnificus* cytotoxin-hemolysin gene (*vvhA*), a putative virulence factor present in both clinical and environmental strains, can be used to detect *V. vulnificus*. For *V. parahaemolyticus*, all strains possess the thermolabile hemolysin gene (*tlh*); only pathogenic strains have the thermostable direct hemolysin gene (*tdh*), which encodes the hemolysin associated with the Kanagawa phenomenon (53). In the protocol, 1:1 oyster homogenates and 1:10 dilutions are spread plated onto the surface of dried

T_1N_3 (*V. parahaemolyticus*) or VVA (*V. vulnificus*) agar plates. After incubation at 35°C overnight, colony lifts are done by using Whatman no. 541 filters. The filters are treated with heat and ammonium acetate buffer and are then prepared for hybridization by immersion in 1× SSC (1× SSC is 0.15 M NaCL plus 0.015 M sodium citrate) supplemented with 20 μl of proteinase K at 42°C for 30 min with shaking. After the filters are rinsed three times in 1× SSC at room temperature, prehybridization is done at 54°C for 30 min with shaking, followed by hybridization in 10 ml of buffer supplemented with 5 pM of alkaline phosphatase-labeled oligonucleotide probes specific for the *tlh* and *tdh* genes (*V. parahaemolyticus*) or the *vvhA* probe (*V. vulnificus*) for 1 h at 54 or 55°C, respectively, with shaking. The filters are removed and rinsed as directed for each gene probe (53). Development and visualization are achieved by the addition of 20 ml of nitroblue tetrazolium–5-bromo-4-chloro-3-indolylphosphate solution, followed by incubation at 35°C with shaking in the dark. After the filters are dried, colony blots of bluish gray or dark brown are counted and recorded, with the results expressed as CFU per gram of oyster sample. Strains can be isolated after hybridization by aligning the hybridized filters to the initial plates and streaking each presumptive colony onto the appropriate medium (TCBS for *V. parahaemolyticus* or mCPC for *V. vulnificus*). In comparative studies, investigators have reported that the direct plating method based on colony lift hybridization performed as well as if not better than the standard MPN method for the enumeration of *V. vulnificus* and *V. parahaemolyticus* in Gulf Coast oysters (23, 36).

Members of the genus *Vibrio* can enter a viable but nonculturable state and may thus be prime candidates for detection by methods which do not rely on conventional culture. PCR has been used to identify *V. vulnificus* and *V. parahaemolyticus* in shellfish, albeit in limited studies. For example, Hill et al. (44) were able to amplify DNA sequences of *V. vulnificus* from oyster homogenates seeded at the 100-CFU/g level. Karunasagar et al. (51) reported detection limits of $>10^4$ CFU/ml when they attempted to detect *V. parahaemolyticus* by PCR amplification of a region in the *tdh* gene, as applied to lysates prepared directly from fish homogenates. Both Karunasagar et al. (51) and Dileep et al. (24) concluded that improved PCR detection sensitivity (<10 CFU/ml) could be obtained by performing amplifications after prior cultural enrichment for 6 to 8 h. Likewise, a number of multiplex PCR assays have been developed for detection of the pathogenic *Vibrio* spp. (7, 11, 99). Detection limits approach 10^0 to 10^1 CFU/g but, as with monoplex PCR, only after a prior cultural enrichment step. The requirement for a brief cultural enrichment (6 h) may not necessarily be a drawback since it can save painstaking pathogen extraction procedures and provides some indication that the organisms may be viable. The need to confirm the identities of the PCR amplification products by DNA hybridization (oligoprobing) methods cannot be overemphasized, since shellfish matrices are notorious for promoting nonspecific DNA amplification, which can result in false-positive results if the results are not confirmed (46). Unfortunately, no methods specifically targeting pathogens in the viable but nonculturable state or forgoing cultural enrichment procedures have been reported at this time. When further strain discrimination is desired, molecular typing methods such as pulsed-field gel electrophoresis (PFGE), ribotyping, and random amplified polymorphic DNA (RAPD) analysis can be used (3, 65).

Molecular Methods for Detection of Enteric Viruses in Shellfish

Shellfish meat represents a chemically complex matrix that contains many compounds which can interfere with the performance of the PCR methodology. The need to remove such inhibitory compounds is well recognized, and accordingly, some type of virion concentration step is applied before RNA extraction and subsequent nucleic acid amplification are applied for the detection of viral contamination. This can be approached in two ways, i.e., (i) nonspecific virus concentration methods, usually by use of a combination of the more traditional methods, such as filtration, precipitation, and centrifugation, and (ii) more specific capture approaches, such as immunomagnetic separation (21, 63). While the latter method might be more appealing, the difficulties associated with obtaining virus-specific antibodies that have broad reactivities, particularly for the norovirus group, have limited the applicability of this method in recent years.

Early work in this area focused on the development of methods that can be used to concentrate intact virions, with removal of enzymatic inhibitors, from oyster samples previously processed for virus isolation by adsorption-elution-precipitation methods (47). One frequently cited alternative has used polyethylene glycol (PEG) precipitation and Freon extraction to initially purify the viruses from shellfish samples as large as 50 g (57), followed by RNA extraction of 1 to 8 g weight equivalents and subsequent nucleic acid amplification by RT-PCR. In a related method, relatively small weights (1.5 to 3.0 g) of digestive diverticula are homogenized (in PBS), followed by extraction with an equal volume of chloroform-butanol (1:1 [vol/vol]) and supplementation with Cat-Floc T. After the suspension is shaken, it is centrifuged and the supernatant is collected. The viruses in this supernatant are precipitated by the addition of PEG and NaCl to reach concentrations of 10% and 0.3 M, respectively. Precipitation is facilitated by overnight storage at 4°C, followed by a second centrifugation step. The resulting pellet is resuspended in water and can be stored at −80° C for later RNA extraction and nucleic acid amplification (4, 58). While precise detection limits have not always been documented, researchers using these methods have been able to detect human enteroviruses and noroviruses in shellfish samples harvested from both pristine sites and sites that contain feces and from shellfish epidemiologically implicated in disease outbreaks (27, 57, 60, 62, 91).

More recently, a method consisting of multiple elution, PEG precipitation, and solvent extraction steps has been used in a number of laboratories (72, 90). One representative method has been applied to 25 g of shellfish meat (or digestive diverticula), which is homogenized in sterile deionized water (1:7 [wt/vol]) with subsequent pH and conductivity adjustment to 4.8 and <2,000 μS, respectively, to promote virus adsorption to shellfish tissues. The mixture is centrifuged, and the virus-containing precipitate is collected. Viruses are eluted from this pellet with 0.75 M glycine–0.15 M NaCl (pH 7.6) in a 1:7 (wt/vol) ratio. The resulting pH of the suspension is adjusted to ensure a pH ranging from 7.5 to 7.6, vortexed, and centrifuged again. In this case, the supernatant is collected and any viruses remaining in the pellet are reeluted by using 0.5 M threonine–0.15 M NaCl (pH 7.6; 1:7 [wt/vol]). Another centrifugation under identical conditions results in a second supernatant, which is pooled with the first one. The pooled eluate is adjusted to 0.3 M NaCl, and the viruses are precipitated by the addition of PEG to reach a final concentration of 8%. After incubation at 4°C for 4 h, the suspension is centrifuged, and the pellet is collected and suspended in 10 ml of PBS. This suspension is solvent extracted once with chloroform (1:1), and the resulting solvent phase is reextracted with one-half volume of 0.5 M threonine. The two supernatants resulting from the sequential solvent extractions are combined. This suspension is adjusted to 0.3 M NaCl, and the viruses are precipitated again by supplementation with 8% PEG. The pellet resulting from this final centrifugation is then extracted for RNA isolation and subsequent amplification by RT-PCR. Minor modifications of these methods have been used by others (8, 32, 73, 80).

An efficient nucleic acid extraction step is critical because the amplification efficiency is dependent on both the purity of the target template and the quantity of the target molecules obtained from the sample. The nucleic acid extraction also provides additional sample concentration. In early days, RNA extraction was done with a combination of proteinase K digestion, phenol-chloroform extraction, and CTAB (cetyltrimethylammonium bromide) precipitation (4, 58, 60, 62). Over the last decade, guanidinium thiocyanate (GuSCN) has been recognized as the most effective RNA extraction agent, as it facilitates sample lysis and also protects naked RNA from enzymatic degradation. Further RNA purification can be obtained by using secondary steps, such as binding to glass powder, silica, or oligo(dT) cellulose or by using Qiagen (Valencia, Calif.) spin columns (25, 91). These methods appear to result in the significant removal of shellfish-related inhibitors, with sample tolerances exceeding 5-g shellfish weight equivalents.

The choice of the primer used for nucleic acid amplification is another important consideration. This is particularly critical because the levels of viral contamination in shellfish are much lower than those in clinical matrices, and virus concentration and extraction methods frequently leave behind residual matrix-associated material. The result is a high degree of nonspecific amplification, which decreases both the sensitivity and the specificity of the molecular amplification. To alleviate this, primers with low degeneracies and high melting temperatures are recommended. For the detection of HAV, most primers target the viral VP1/2A junction and are broadly reactive. However, the noroviruses are genetically diverse and the design of broadly reactive primers is complicated by the need for a high degree of degeneracy. Early norovirus primers were designed to amplify conserved regions in the RNA-dependent RNA polymerase and, when used in combination, could amplify the RNA from a variety of strains (1). More recently, norovirus primers have targeted the viral capsid region (96), although most of these are degenerate. As a rule of thumb, the best primers for norovirus detection, as applied to shellfish matrices, would have reasonably high annealing temperatures, relatively low degeneracies, and broad reactivities. The interested reader is urged to consult the work of D'Souza et al. (26) for a full description of candidate HAV and norovirus primers and probes.

None of the approaches described above for both virus concentration and RNA extraction are ideal. Virion concentration provides significant sample volume reductions with recovery of infectious viruses and can be used in conjunction with a cell culture infectivity assay, when available, as long as an RNA extraction step is not yet applied. However, these concentration methods require significant sample manipulations, many of which may result in substantial virus loss, which in turn may reduce overall detection limits. In many cases, RT-PCR inhibitors are recalcitrant to both virus and RNA extraction, and the resulting sample concentrate must frequently be diluted before assay, which effectively reduces detection sensitivity.

Because of residual matrix-associated inhibitors, nonspecific products of amplification frequently occur when RT-PCR is applied to the detection of viruses in food and environmental samples. Therefore, it is critical to confirm that the nucleic acid amplification products obtained are specific to the target. Most often, the confirmation step also improves the sensitivity of the assay. Historically, the most common confirmatory tool has been Southern hybridization with specific oligonucleotide probes internal to the amplicon. DNA enzyme immunoassay methods provide an alternative to Southern hybridization (88). Investigators have also used specific nested PCRs, which use a second pair of primers internal to the first amplicon sequence, as a means to increase the amplicon copy number and provide confirmation of the results. In fact, many investigators have found it to be necessary to use nested amplifications to achieve the sensitivity needed to detect viruses in naturally contaminated products (Tables 2 and 3). Direct sequencing of the amplicon for the confirmation of RT-PCR product identity is being increasingly used and might now be considered the gold standard.

Molecular Methods for Detection of Human Protozoan Parasites in Shellfish

In general, considerably less effort has been put forth to develop molecular methods for the detection of parasitic protozoa from shellfish. Theoretically, the separation of cysts and oocysts from shellfish tissues by enzymatic digestion methods, gill washing, or a combination of homogenization, filtration, centrifugation, and solvent extraction, as described above, should provide a sample that can be extracted for nucleic acid isolation and subsequent amplification. However, the subjection of final sample concentrates to repeated cycles of washing in sterile water followed by one or more freeze-thaw cycles (29, 30, 71) is usually necessary to facilitate nucleic acid extraction, which can be done by phenol-chloroform extraction (29, 30), by GuSCN-based methods (34, 35), or with spin columns (71). Removal of excess matrix-associated inhibitors may be achieved by incorporating a silica binding or polyvinylpyrrolidone step (35). Using PCR-based methods, one study demonstrated the detection of *Cryptosporidium parvum* in commercial shellfish (30), but such detection methods required nested amplifications. A second study used PCR to detect *Cryptosporidium* in clams spiked directly or in which protozoa were allowed to bioaccumulate by exposure to artificially contaminated water; again, successful detection required nested amplifications (71). In most cases, the PCR amplification target is the 18S rRNA gene (30, 71), although the *COWP* gene has also been used (34, 35). Species and genotype-specific confirmation can be based on PCR-restriction fragment length polymorphism (RFLP) analysis or polymorphisms in the small-subunit rRNA, and oocyst infectivity may be confirmed by using a neonatal mouse bioassay (29).

Real-Time Nucleic Acid Amplification Methods for Detection of Pathogens in Shellfish

Real-time detection refers to the simultaneous detection and confirmation of amplicon identity as the amplification

TABLE 2 Representative studies reporting detection of human enteric viruses in field, market, or outbreak-implicated shellfish[a]

Agent(s)	Food	Sample type(s)	Methods	Comments	Reference(s)
NoV	Oysters (United Kingdom)	Field samples, outbreak samples, market samples (food only)	RT-PCR (single and nested), Southern hybridization, sequencing	5/31 field samples positive; 4/4 outbreak samples positive; nested RT-PCR increases detection sensitivity	57, 38
NoV	Oysters (Japan)	Outbreak samples (clinical, food)	RT-PCR (nested), cloning, sequencing	Coexistence of two different NoV genogroups in single oyster specimen	94
NoV	Oysters (United States)	Outbreak samples (food only)	RT-PCR, Southern hybridization, sequencing	2/3 recalled outbreak specimens positive (GII)	91, 92
NoV, EV, AV, RV	Oysters and mussels (France)	Field samples (food only)	RT-PCR, dot blot hybridization, sequencing	Contamination rates for oysters were 17–27%, and those for mussels were 13–52%	58
RV	Oysters (France)	Market samples (food only)	RT-PCR (nested), restriction digestion	Representative market oysters, mussels, and shrimp tested for NoVs, HAV, enterovirus, and rotavirus	40

[a]EV, human enteroviruses; NoV, noroviruses; AV, astroviruses; RV, rotavirus.

reaction is progressing. Since nucleic acid amplification and some sort of amplicon confirmation step occur simultaneously during real-time PCR, the need to perform time-consuming gel electrophoresis and hybridization methods can be circumvented. Some claim that an added advantage of real-time methods is that they can be designed to be quantitative. A number of chemistries are available for real-time amplification and detection technologies; but the three most popular ones include (i) the use of fluorophors which bind to double-stranded DNA when they are exposed to a suitable wavelength of light (e.g., SYBR green assays), (ii) the 5′ endonuclease assay (e.g., TaqMan assay),

TABLE 3 Representative studies of enteric virus prevalence in shellfish[a]

Agent(s)	Shellfish	Location, yr	Detection approach	Findings	Reference
EV, AdV, HAV	Oysters, mussels (60 samples)	Spain (Mediterranean, delta or Erbo River); different levels of fecal contamination; 1998–1999	Elution and ultracentrifugation, boom RNA extraction, RT-PCR (nested), sequencing (EV)	Positive for AdV (47%), EV (19%), and HAV (24%)	73
NoV, EV, HAV	Oysters (87 samples)	Swiss market study; shellfish originated from France, The Netherlands, and Ireland; 2001–2002	Dissection, elution, 2× PEG precipitation and solvent extraction, QIAamp extraction, RT-PCR, sequencing	Positive for NoV (9.4%), EV (4.9%), and NoV and EV (2.3%)	8
NoV	Oysters (191 samples)	Japan (approved harvesting areas of the Setouchi Sea, west of Honshu Island); 2001–2002	Dissection, solvent extraction, organic flocculation, QIAamp extraction, RT-PCR (nested and real-time), sequencing	Positive for NoV (9%); $>10^2$ genome copies estimated by real-time RT-PCR in 11/17 positive samples	76
EV	Oysters (52 samples)	France (Atlantic Ocean and English Channel sites); oysters fecally affected; 2000–2001	Dissection, solvent extraction, organic flocculation, PEG precipitation, RNA-Plus extraction, RT-PCR (nested), sequencing	Positive for EV (40.4%)	27
NoV, AdV	Mussels (681 samples)	Norwegian market survey; different levels of fecal contamination; 2000–2003	Dissection, elution, ultracentrifugation; TRIzol extraction, RT-PCR (nested), sequencing	Positive for NoV (6.8%) and AdV (18.6%)	74

[a]EV, human enteroviruses; NoV, noroviruses; AdV, human adenoviruses.

and (iii) the use of hairpin oligonucleotide probes (e.g., molecular beacons).

Over the past 2 years, a plethora of real-time PCR and RT-PCR methods have been reported for the detection of a variety of food-borne pathogens, and the shellfish-borne pathogens are no exception. Prototype real-time RT-PCR amplification technologies have been developed for the detection of HAV (16), and several are commercially available for clinical use. Likewise, prototype real-time methods have also been developed for the detection of the noroviruses in various formats, including TaqMan (70) and SYBR green (83, 84) assays. Interestingly, Beuret et al. (8) developed a multiplex real-time RT-PCR for the simultaneous detection of a panel of enteric viruses. Myrmel et al. (74) recently reported on the detection of viral contamination in shellfish using a commercial SYBR green PCR kit, while Narayanan et al. (75), Loisy et al. (62), and Jothikumar et al. (49) have all used TaqMan assays to detect noroviruses in artificially and naturally contaminated shellfish.

Real-time PCR has been applied to the detection and identification of *Vibrio* species as well. Blackstone et al. (9) were the first to report a real-time PCR method that targets the *tdh* gene and that uses the TaqMan format to detect *V. parahaemolyticus*. Recently, Davis et al. (20) developed a multiplex real-time PCR method that targets the *tlh*, *tdh*, and *trh* genes of *V. parahaemolyticus* by the use of TaqMan probes with different labels. Kaufman et al. (52) found a strong correlation between cycle threshold and log concentration when they used a TaqMan real-time PCR method that targets the *tlh* gene to detect *V. parahaemolyticus* in oyster mantle fluid. Panicker et al. (78) developed a SYBR green-based real-time PCR method which amplified a fragment of the hemolysin (*vvh*) gene of *V. vulnificus* and which was applied to shellfish and waters from the Gulf of Mexico. Prototype real-time PCR methods have been developed for the detection of the parasitic protozoan *C. parvum* (43), although these have not yet been applied to the detection of oocysts in shellfish.

CONSIDERATIONS IN CHOOSING AND DEVELOPING MOLECULAR METHODS FOR DETECTION OF PATHOGENS IN SHELLFISH

The reliability of molecular-based pathogen detection methods, as applied directly to shellfish matrices, is dependent on a number of factors, the most important of which are (i) the efficiency of the upstream sample processing method used to concentrate the agent of interest while simultaneously removing potential inhibitors, (ii) the purity and the yield of the nucleic acid used in the amplifications, and (iii) the choice of the primers used for amplification and the probes used for confirmation. As is clear from the above discussion, many upstream sample processing methods are available; the methods chosen depend on the pathogen to be detected, and many laboratories make minor modifications in methodologies from time to time. The nucleic acid extraction method can be designed to recover RNA (viruses and sometimes parasitic protozoa) or DNA (*Vibrio* spp. and parasitic protozoa), or both. The nucleic acid extraction method not only makes the target nucleic acid available for amplification and detection but also provides further sample concentration and purification, with removal of residual matrix-associated inhibitors. Fortunately, many effective nucleic acid extraction kits have become commercially available over the last decade. However, in many instances, a secondary DNA or RNA purification step, such as adsorption and elution from silica beads, is recommended. Indeed, studies have demonstrated that when viral contamination in shellfish is detected by RT-PCR, detection limits may vary by as much as 5 $\log_{10}$ units, depending on the RNA extraction method used (82). Most of the molecular approaches applied to the detection of pathogens in shellfish originated from the clinical literature and were first applied to pure cultures (bacteria) or to purified cell culture lysates or fecal suspensions (parasitic protozoa and viruses). However, for more complex sample matrices such as shellfish, the sensitivity and the specificity of molecular amplification methods are affected by the choice of primers. This is particularly significant for the norovirus group, which shows substantial genetic diversity. Interestingly, recent evidence also indicates that the choice of the specific nucleic acid amplification kit used for the detection of viruses in shellfish may also substantially affect assay detection limits (82).

Scientists have wanted to develop quantitative molecularly based assays for the detection of pathogens in shellfish for some time now. A semiquantitative MPN-RT-PCR was recently reported for estimation of the viral load in contaminated oysters implicated in an outbreak (60). Most believe that real-time nucleic acid amplification methods offer the greatest promise for the eventual development of quantitative assays. However, a word of caution is in order. While real-time methods offer advantages such as circumventing the need to perform time-consuming gel electrophoresis and hybridization, they are tricky to design and optimize. In addition, relatively little work has been done in applying these assays to the detection of pathogens at the low levels that might be anticipated in naturally contaminated shellfish. While real-time nucleic acid amplification methods are purported to be at least semiquantitative, the range at which quantification can be done is frequently limited and is often only at relatively high levels of product contamination. Clearly, much more work on these methods is necessary before they will be used consistently by multiple laboratories and for the reliable quantification of pathogen loads.

CONCLUSIONS

A number of important developments over the last decade have significantly improved the ability to detect bacterial, viral, and parasitic protozoan pathogens in shellfish. For example, dissecting shellfish and processing exclusively digestive tissue have resulted in a reduction in residual nucleic acid amplification inhibitors and have also allowed the analysis of a larger number of specimens per assay, therefore improving detection sensitivity. The use of internal amplification standards to simultaneously evaluate PCR or RT-PCR inhibition and/or to provide a semiquantitative assay has also been helpful. Most instances in which pathogens, particularly viruses and parasitic protozoa, have been detected in naturally contaminated shellfish have required the use of nested amplifications to improve assay detection limits. Because of developments such as these, the successful detection of viruses in shellfish samples epidemiologically linked to disease has been reported (54, 59, 91). In addition, a number of surveys of naturally occurring viral contamination in shellfish have recently been completed (Tables 2 and 3).

TABLE 4 Research needs to facilitate the routine use of molecular methods for detection of pathogens in shellfish

1. Simple and rapid sample preparation and concentration methods that provide high yield of the target pathogen (or associated nucleic acid) with removal of residual matrix-associated inhibitors
2. Standardized, collaboratively tested methods with appropriate technology transfer
3. Further development and optimization of quantitative (real-time) molecular amplification methods
4. Methods which are able to discriminate between infectious and noninfectious agents or between pathogenic and nonpathogenic agents

However, the current methodologies are far from perfect. Most are used exclusively by a small number of specialized laboratories, usually in the university or government research sectors. The most important reasons for the limited use of these methods include (i) their inability to discriminate between infectious and noninfectious virus; (ii) the lack of widely accepted, collaboratively tested methods; (iii) the cost and need for highly trained personnel; and (iv) the very real possibility that even after substantial upstream sample preparation, residual amplification inhibitors may persist. Research is needed in order to develop and refine the prototype protocols into collaboratively tested methods which could be used routinely and expeditiously to evaluate the microbiological safety of shellfish products (Table 4). In general, future research needs for the routine application of PCR methods to the detection of microbiological contamination in shellfish include development of the following: (i) simple, rapid, and cost-effective pathogen extraction procedures; (ii) simple and reliable methods for the removal of enzymatic inhibitors; (iii) methods which are not restricted by shellfish species, harvesting location, or season; and (iv) quantitative approaches for assessing the relative levels of contamination. Furthermore, additional experimentation is needed to establish the relationship between detection by molecular amplification techniques and the presence of infective virus particles or parasitic protozoan cysts and oocysts in shellfish and environmental samples. The practical application of real-time PCR and RT-PCR for quantification also needs to be investigated further. With consistent developmental effort, much as has been experienced over the last decade, even better methods for the rapid detection of human pathogenic microorganisms in shellfish at naturally occurring levels of contamination should be available in the future.

REFERENCES

1. **Ando, T., S. S. Monroe, J. R. Gentsch, Q. Jin, D. C. Lewis, and R. I. Glass.** 1995. Detection and differentiation of antigenically distinct small round-structured viruses (Norwalk-like viruses) by reverse transcription-PCR and Southern hybridization. *J. Clin. Microbiol.* **33:** 64–71.
2. **Appleton, H.** 1987. Small round viruses: classification and role in food-borne infections. *Ciba Found. Symp.* **128:** 108–125.
3. **Arias, C. R., M. J. Pujalte, E. Garay, and R. Aznar.** 1998. Genetic relatedness among environmental, clinical, and diseased-eel *Vibrio vulnificus* isolates from different geographic regions by ribotyping and randomly amplified polymorphic DNA PCR. *Appl. Environ. Microbiol.* **64:** 3403–3410.
4. **Atmar, R. L., H. F. Neill, J. L. Romalde, F. Le Guyader, C. M. Woodley, T. G. Metcalf, and M. K. Estes.** 1995. Detection of Norwalk virus and hepatitis A virus in shellfish tissues with the PCR. *Appl. Environ. Microbiol.* **61:** 3014–3018.
5. **Baumgartner, A., H. P. Marder, J. Munzinger, and H. H. Siegrist.** 2000. Frequency of *Cryptosporidium* spp. as cause of human gastrointestinal disease in Switzerland and possible sources of infection. *Schweiz. Med. Wochenschr.* **130:** 1252–1258.
6. **Bean, N. H., J. S. Goulding, M. T. Daniels and F. J. Angulo.** 1997. Surveillance for foodborne disease outbreaks—United States, 1988–1992. *J. Food. Prot.* **60:**1265–1286.
7. **Bej, A. K., D. P. Patterson, C. W. Brasher, M. C. L. Vickery, D. D. Jones, and C. A. Kaysner.** 1999. Detection of total and hemolysin-producing *Vibrio parahaemolyticus* in shellfish using multiplex PCR amplification of *tlh*, *tdh*, and *trh*. *J. Microbiol. Methods* **36:**215–225.
8. **Beuret, C., A. Baumgartner, and J. Schluep.** 2003. Virus-contaminated oysters: a three-month monitoring of oysters imported to Switzerland. *Appl. Environ. Microbiol.* **69:** 2292–2297.
9. **Blackstone, G. M., J. L. Nordstrom, M. C. L. Vickery, M. D. Bowen, R. F. Meyer, and A. DePaola.** 2003. Detection of pathogenic *Vibrio parahaemolyticus* in oyster enrichments by real time PCR. *J. Microbiol. Methods* **53:** 149–155.
10. **Brands, D. A., A. E. Inman, C. P. Gerba, C. J. Mare, S. J. Billington, L. A. Saif, J. F. Levine, and L. A. Joens.** 2005. Prevalence of *Salmonella* spp. in oysters in the United States. *Appl. Environ. Microbiol.* **71:**893–897.
11. **Brasher, C. W., A. DePaola, D. D. Jones, and A. K. Bej.** 1998. Detection of microbial pathogens in shellfish with multiplex PCR. *Curr. Microbiol.* **37:**101–107.
12. **Centers for Disease Control and Prevention.** 2005. Preliminary FoodNet data on the incidence of infection with pathogens transmitted commonly through food—10 sites, United States, 2004. *Morb. Mortal. Wkly. Rep.* **54:** 352–356.
13. **Centers for Disease Control and Prevention.** 1999. Outbreak of *Vibrio parahaemolyticus* infection associated with eating raw oysters and clams harvested from Long Island Sound—Connecticut, New Jersey, and New York, 1998. *Morb. Mortal. Wkly. Rep.* **48:**48–51.
14. **Centers for Disease Control and Prevention.** 1998. Outbreak of *Vibrio parahaemolyticus* infections associated with eating raw oysters—Pacific Northwest, 1997. *Morb. Mortal. Wkly. Rep.* **47:**457–462.
15. **Centers for Disease Control and Prevention.** 1997. Viral gastroenteritis associated with eating oysters—Louisiana, December 1996–January 1997. *Morb. Mortal. Wkly. Rep.* **46:**1109–1112.
16. **Costa-Mattioli, M., S. Monpoeho, E. Nicand, M.-H. Aleman, S. Billaudel, and V. Ferré.** 2002. Quantification and duration of viraemia during hepatitis A infection as determined by real-time RT-PCR. *J. Viral Hepatitis* **9:**101–106.
17. **Cubitt, W. D.** 1994. Caliciviruses, p. 549–568. *In* A. Z. Kapikian (ed.), *Viral Infections of the Gastrointestinal Tract.* Marcel Dekker, Inc., New York, N.Y.
18. **Cuthbert, J. A.** 2001. Hepatitis A: old and new. *Clin. Microbiol. Rev.* **14:**38–58.
19. **Daniels, N. A., B. Ray, A. Easton, N. Marano, E. Kahn, A. L. McShan, L. D. Rosario, T. Baldwin, M. A. Kingsley, N. D. Puhr, J. G. Wells, and F. J. Angulo.** 2000. Emergence of a new *Vibrio parahaemolyticus* serotype in raw oysters. *JAMA* **284:**1541–1546.

20. **Davis, C. R., L. C. Heller, K. K. Peak, D. L. Wingfield, C. L. Goldstein-Hart, D. W. Bodager, A. C. Cannons, P. T. Amuso, and J. Cattani.** 2004. Real-time PCR detection of the thermostable direct hemolysin and thermolabile hemolysin genes in a *Vibrio parahaemolyticus* cultured from mussels and mussel homogenate associated with a foodborne outbreak. *J. Food Prot.* **67:**1005–1008.
21. **Deng, M. Y., S. P. Day, and D. O. Cliver.** 1994. Detection of hepatitis A virus in environmental samples by antigen-capture PCR. *Appl. Environ. Microbiol.* **60:**1927–1933.
22. **De Paola, A., and C. A. Kaysner.** 2004. *Vibrio. In FDA Bacteriological Analytical Manual Online.* Association of Official Analytical Chemists, Arlington, Va. [Online.] http://www.cfsan.fda.gov/~ebam/bam-9.html. Accessed July 1, 2005.
23. **De Paola, A., M. L. Motes, D. W. Cook, J. Veazey, W. E. Garthright, and R. Blodgett.** 1997. Evaluation of an alkaline phosphatase-labeled DNA probe for enumeration of *Vibrio vulnificus* in Gulf Coast oysters. *J. Microbiol. Methods* **29:**115–120.
24. **Dileep, V., H. S. Kumar, Y. Kumar, M. Nishibuchi, I. Karunasagar, and I. Karunasagar.** 2003. Application of polymerase chain reaction for detection of *Vibrio parahaemolyticus* associated with tropical seafoods and coastal environment. *Lett. Appl. Microbiol.* **36:**423–427.
25. **DiPinto, A., V. T. Forte, G. M. Tantillo, V. Terio, and C. Buonavoglia.** 2003. Detection of hepatitis A virus in shellfish (*Mytilus galloprovincialis*) with RT-PCR. *J. Food Prot.* **66:**1681–1685.
26. **D'Souza, D. H., J. Jean, and L. Jaykus.** 2006. Methods for the detection of viral and parasitic protozoan pathogens in foods, p. 188-1–188-23. *In* Y. H. Hui, E. Castell-Perez, L. M. Cunha, I. Guerrero-Legarreta, H. H. Liang, Y. M. Lo, D. L. Marshall, W. K. Nip, F. Shahidi, F. Sherkat, R. J. Winger, and K. L. Yam (ed.), *Handbook of Food Science, Technology and Engineering,* vol. 4. CRC Press, Taylor & Francis Group, Boca Raton, Fla.
27. **Dubois, E., G. Merle, C. Roquier, A. Trompette, F. Le Guyader, C. Cruciere, and J.-J. Chomel.** 2004. Diversity of enterovirus sequences detected in oysters by RT-heminested PCR. *Int. J. Food Microbiol.* **92:**35–43.
28. **Fankhauser, R. L., J. S. Noel, S. S. Monroe, T. Ando, and R. I. Glass.** 1998. Molecular epidemiology of "Norwalk-like viruses" in outbreaks of gastroenteritis in the United States. *J. Infect. Dis.* **178:**1571–1578.
29. **Fayer, R., J. M. Trout, E. J. Lewis, M. Santin, L. Zhou, A. A. Lal, and L. Xiao.** 2003. Contamination of Atlantic coast commercial shellfish with Cryptosporidium. *Parasitol. Res.* **89:**141–145.
30. **Fayer, R., E. J. Lewis, J. M. Trout, T. K. Graczyk, M. C. Jenkins, J. Higgins, L. Xiao, and A. A. Lal.** 1999. *Cryptosporidium parvum* in oysters from commercial harvesting sites in the Chesapeake Bay. *Emerg. Infect. Dis.* **5:** 706–710.
31. **Fayer, R., T. K. Graczyk, E. J. Lewis, J. M. Trout, and C. A. Farley.** 1998. Survival in infectious *Cryptosporidium parvum* oocysts in seawater and Eastern oysters (*Crassostrea virginica*) in the Chesapeake Bay. *Appl. Environ. Microbiol.* **64:**1070–1074.
32. **Formiga-Cruz, M., A. Hundesa, P. Clemente-Casares, N. Albinana-Gimenez, A. Allard, and R. Girones.** 2005. Nested multiplex PCR assay for detection of human enteric viruses in shellfish and sewage. *J. Virol. Methods* **125:**111–118.
33. **Gill, O. N., W. D. Cubitt, O. A. McSwiffan, B. M. Watney, and C. L. R. Batlett.** 1983. Epidemic of gastroenteritis caused by oysters contaminated with small round structured viruses. *Br. Med. J.* **287:**1532–1534.
34. **Gomez-Bautista, M., L. M. Ortega-Mora, E. Tabares, V. Lopez-Rodas, and E. Costas.** 2000. Detection of infectious *Cryptosporidium parvum* oocysts in mussels (*Mytilus galloprovincialis*) and cockles (*Cerastoderma edule*). *Appl. Environ. Microbiol.* **66:**1866–1870.
35. **Gomez-Cuoso, H., F. Freire-Santos, C. F. L. Amar, K. A. Grant, K. Williamson, M. E. Ares-Mazas, and J. McLauchlin.** 2004. Detection of *Cryptosporidium* and *Giardia* in molluscan shellfish by multiplexed nested-PCR. *Int. J. Food Microbiol.* **91:**279–288.
36. **Gooch, J. A., A. DePaola, C. A. Kaysner, and D. L. Marshall.** 2001. Evaluation of two direct plating methods using nonradioactive probes for enumeration of *Vibrio parahaemolyticus* in oysters. *Appl. Environ. Microbiol.* **67:**721–724.
37. **Graczyk, T. K., R. Fayer, E. J. Lewis, J. M. Trout, and C. A. Farley.** 1999. *Cryptosporidium* oocysts in Bent mussels (*Ischadium recurvum*) in the Chesapeake Bay. *Parasitol. Res.* **85:**30–34.
38. **Green, J., K. Henshilwood, C. I. Gallimore, D. W. G. Brown, and D. N. Lees.** 1998. A nested reverse transcriptase PCR assay for detection of small round-structured viruses in environmentally contaminated molluscan shellfish. *Appl. Environ. Microbiol.* **64:**858–863.
39. **Green, K. Y., T. Ando, M. S. Balayan, T. Berke, I. N. Clarke, M. K. Estes, D. O. Matson, S. Nakata, J. D. Neill, M. J. Studdert, and H.-J. Theil.** 2000. Taxonomy of the caliciviruses. *J. Infect. Dis.* **181**(Suppl. 2)**:**S322–S330.
40. **Hafliger, D., M. Gilen, J. Luthy, and P. H. Hubner.** 1997. Semi-nested RT-PCR systems for small round structured viruses and detection of enteric viruses in seafood. *Int. J. Food Microbiol.* **37:**27–36.
41. **Halliday, M. L., L. Y. Kanf, T. K. Zhou, M. D. Hu, Q. C. Pan, T. Y. Fu, Y. S. Huang, and S. L. Hu.** 1991. An epidemic of hepatitis A attributable to the ingestion of raw clams in Shanghai, China. *J. Infect. Dis.* **164:**852–859.
42. **Heinitz, M. L., R. D. Ruble, D. E. Wagner, and S. R. Tatini.** 2000. Incidence of *Salmonella* in fish and seafood. *J. Food Prot.* **63:**579–592.
43. **Higgins, J. A., R. Fayer, J. M. Trout, L. Xiao, A. A. Lal, S. Kerby, and M. C. Jenkins.** 2001. Real-time PCR for the detection of *Cryptosporidium parvum*. *J. Microbiol. Methods* **47:**323–337.
44. **Hill, W. E., S. P. Keasler, M. W. Truckess, P. Feng, C. A. Kayser, and K. A. Lampel.** 1991. Polymerase chain reaction identification of *Vibrio vulnificus* in artificially contaminated oysters. *Appl. Environ. Microbiol.* **57:**707–711.
45. **Hlady, W. G., and K. C. Klontz.** 1996. The epidemiology of *Vibrio* infections in Florida, 1981–1993. *J. Infect. Dis.* **173:**1176–1183.
46. **Jaykus, L.** 2000. Detection of human enteric viruses in foods, p. 137–163. *In* S. Sattar (ed.), *Foodborne Diseases Handbook,* vol. 2. *Viruses, Parasites, Pathogens and HACCP.* Marcel Dekker, Inc., New York, N.Y.
47. **Jaykus, L., R. De Leon, and M. D. Sobsey.** 1996. A virion concentration method for detection of human enteric viruses in oysters by PCR and oligoprobe hybridization. *Appl. Environ. Microbiol.* **62:**2074–2080.
48. **Jiang, X. N., D. Y. Graham, K. N. Wang, and M. K. Estes.** 1990. Norwalk virus genome cloning and characterization. *Science* **250:**1580–1583.
49. **Jothikumar, N., J. A. Lowther, K. Henshilwood, D. N. Lees, V. R. Hill, and J. Vinje.** 2005. Rapid and sensitive detection of noroviruses by using TaqMan-based one-step reverse transcription-PCR assays and application to naturally contaminated shellfish samples. *Appl. Environ. Microbiol.* **71:**1870–1875.
50. **Kapikian, A. Z., M. K. Estes, and R. M. Chanock.** 1996. Norwalk group of viruses, p. 783–810. *In* B. N. Fields, D. M. Knipe, and P. M. Howley (ed.), *Fields Virology.* Lippincott-Raven, Philadelphia, Pa.

51. **Karunasagar, I., G. Sugumar, I. Karunasagar, and P. J. A. Reilly.** 1996. Rapid polymerase chain reaction method for detection of Kanagawa positive *Vibrio parahaemolyticus* in seafood. *Int. J. Food Microbiol.* **31:**317–323.
52. **Kaufman, G. E., G. M. Blackstone, M. C. L. Vickery, A. K. Bej, J. Bowers, M. D. Bowen, and A. DePaola.** 2004. Real-time PCR quantification of *Vibrio parahaemolyticus* in oysters using an alternative matrix. *J. Food Prot.* **67:**2424–2429.
53. **Kaysner, C. A., and A. DePaola.** 2001. *Vibrio*, p. 405–420. *In* F. P. Downes and K. Ito (ed.), *Compendium of Methods for the Microbiological Examinaiton of Foods*. American Public Health Association, Washington, D.C.
54. **Kingsley, D. H., G. K. Meade, and G. P. Richards.** 2002. Detection of both hepatitis A virus and Norwalk-like virus in imported clams associated with food-borne illness. *Appl. Environ. Microbiol.* **68:**3914–3918.
55. **Koopmans, M., C.-H. von Bonsdorff, J. Vinje, D. de Medici, and S. Monroe.** 2002. Foodborne viruses. *FEMS Microbiol. Rev.* **26:**187–205.
56. **Lambden, P. R., E. O. Caul, C. R. Ashley, and I. N. Clarke.** 1993. Sequence and genome organization of a human small round-structured (Norwalk-like) virus. *Science* **259:**516–519.
57. **Lees, D. N., K. Henshilwood, J. Green, C. I. Gallimore, and D. W. G. Brown.** 1995. Detection of small round structured viruses in shellfish by reverse transcription-PCR. *Appl. Environ. Microbiol.* **61:**4418–4424.
58. **LeGuyader, F., L. Haugarreau, L. Miossec, E. Dubois, and M. Pommepuy.** 2000. Three-year study to assess human enteric viruses in shellfish. *Appl. Environ. Microbiol.* **66:**3241–3248.
59. **LeGuyader, F., F. H. O'Neill, M. K. Estes, S. S. Monroe, T. Ando, and R. L. Atmar.** 1996. Detection and analysis of a small round-structured virus strain in oysters implicated in an outbreak of acute gastroenteritis. *Appl. Environ. Microbiol.* **62:**4268–4272.
60. **LeGuyader, F. S., F. H. Neill, E. Dubois, F. Bon, F. Loisy, E. Kohli, M. Pommepuy, and R. L. Atmar.** 2003. A semiquantitative approach to estimate Norwalk-like virus contamination of oysters implicated in an outbreak. *Int. J. Food Microbiol.* **87:**107–112.
61. **Lipp, E. K., and J. B. Rose.** 1997. The role of seafood in foodborne disease in the United States of America. *Rev. Sci. Tech.* **16:**620–640.
62. **Loisy, F., R. L. Atmar, P. Guillon, P. LeCann, M. Pommepuy, and F. S. LeGuyader.** 2005. Real-time RT-PCR for norovirus screening in shellfish. *J. Virol. Methods* **123:**1–7.
63. **Lopez-Sabater, E. I., M. Y. Deng, and D. O. Cliver.** 1997. Magnetic immunoseparation PCR assay (MIPA) for detection of hepatitis A virus (HAV) in American oyster (*Crassostrea virginica*). *Lett. Appl. Microbiol.* **24:**101–104.
64. **MacRae, M., C. Hamilton, N. J. C. Strachan, S. Wright, and I. D. Ogden.** 2005. The detection of *Cryptosporidium parvum* and *Escherichia coli* O157 in UK bivalve shellfish. *J. Microbiol. Methods* **60:**395–401.
65. **Marshall, S., C. G. Clark, G. Wang, M. Mulvey, M. T. Kelly, and W. M. Johnson.** 1999. Comparison of molecular methods for typing *Vibrio parahaemolyticus*. *J. Clin. Microbiol.* **37:**2473–2478.
66. **Martinez-Urtaza, J., M. Saco, J. de Novoa, P. Periz-Pineiro, J. Peiteado, A. Lozano-Leon, and O. Garcia-Martin.** 2004. Influence of environmental factors and human activity on the presence of *Salmonella* serovars in a marine environment. *Appl. Environ. Microbiol.* **70:**2089–2097.
67. **Martinez-Urtaza, J., E. Liebana, L. Garcia-Migura, P. Perez-Pineiro, and M. Saco.** 2004. Characterization of *Salmonella enterica* serovar Typhimurium from marine environments in coastal wasters of Galicia (Spain). *Appl. Environ. Microbiol.* **70:**4030–4034.
68. **Martinez-Urtaza, J., M. Saco, G. Hernandez-Cordova, A. Lozano, O. Garcia-Martin, and J. Espinosa.** 2003. Identification of *Salmonella* serovars isolated from live molluscan shellfish and their significance in the marine environment. *J. Food Prot.* **66:**226–232.
69. **Mead, P. S., L. Slutsker, V. Dietx, L. F. McCraig, J. S. Bresee, C. Shapiro, P. M. Griffin, and R. V. Tauxe.** 1999. Food-related illness and death in the United States. *Emerg. Infect. Dis.* **5:**607–625.
70. **Miller, I., R. Gunson, and W. F. Carman.** 2002. Norwalk like virus by light cycler PCR. *J. Clin. Virol.* **25:**231–232.
71. **Miller, W. A., E. R. Atwill, I. A. Gardner, M. A. Miller, H. M. Fritz, R. P. Hedrick, A. C. Melli, N. M. Barnes, and P. A. Conrad.** 2005. Clams (*Corbicula fluminea*) as bioindicators of fecal contamination with *Cryptosporidium* and *Giardia* spp. in fresh water ecosystems. *Int. J. Parasitol.* **35:**673–684.
72. **Mullendore, J. L., M. D. Sobsey, and Y. S. C. Sheih.** 2001. Improved method for the recovery of hepatitis A virus from oysters. *J. Virol. Methods* **94:**25–35.
73. **Muniain-Mujika, I., M. Calvo, F. Lucena, and R. Girones.** 2003. Comparative analysis of viral pathogens and potential indicators in shellfish. *Int. J. Food Microbiol.* **83:**75–85.
74. **Myrmel, M., E. M. M. Berg, E. Rimstad, and B. Grinde.** 2004. Detection of enteric viruses in shellfish from the Norwegian coast. *Appl. Environ. Microbiol.* **70:**2678–2684.
75. **Narayanan, J., J. Lowther, K. Henshilwood, D. N. Lees, V. R. Hill, and J. Vinje.** 2005. Rapid and sensitive detection of noroviruses by using TaqMan-based one-step reverse transcripton-PCR assays and application to naturally contaminated shellfish samples. *Appl. Environ. Microbiol.* **71:**1870–1875.
76. **Nishida, T., H. Kimura, M. Saitoh, M. Shinohara, M. Kato, S. Fukuda, T. Munemura, T. Mikami, A. Kawamoto, M. Akiyama, Y. Kato, K. Nishi, K. Kozawa, and O. Nishio.** 2003. Detection, quantitation, and phylogenetic analysis of noroviruses in Japanese oysters. *Appl. Envrion. Microbiol.* **69:**5782–5786.
77. **Oliver, J. D., and D. Wanucha.** 1989. Survival of *Vibrio vulnificus* at reduced temperatures and elevated nutrient. *J. Food Safety* **10:**79–86.
78. **Panicker, G., M. L. Meyers, and A. K. Bej.** 2004. Rapid detection of *Vibrio vulnificus* in shellfish and Gulf of Mexico water by real-time PCR. *Appl. Environ. Microbiol.* **70:**498–507.
79. **Papafragkou, E., D. H. D'Souza, and L. Jaykus.** 2006. Food-borne viruses: prevention and control, p. 289–330. *In* S. M. Goyal (ed.), *Viruses in Foods*. Springer Science, Berlin, Germany.
80. **Pina, S., M. Puig, F. Lucena, J. Jofre, and R. Girones.** 1998. Viral pollution in the environment and in shellfish: human adenovirus detection by PCR as an index of human viruses. *Appl. Environ. Microbiol.* **64:**3376–3382.
81. **Potasman, I., A. Paz, and M. Odeh.** 2002. Infectious outbreaks associated with bivalve shellfish consumption: a worldwide perspective. *Clin. Infect. Dis.* **35:**921–928.
82. **Ribao, C., I. Torrado, M. L. Vilarino, and J. L. Romalde.** 2004. Assessment of different commercial RNA-extraction and RT-PCR kits for detection of hepatitis A virus in mussel tissues. *J. Virol. Methods* **115:**177–182.
83. **Richards, G. P., M. A. Watson, R. L. Fankhauser, and S. S. Monroe.** 2004. Genogroup I and II noroviruses detected in stool samples by real-time reverse transcription-PCR using highly degenerate universal primers. *Appl. Environ. Microbiol.* **70:**7179–7184.

84. **Richards, G. P., M. A. Watson, and D. H. Kingsley.** 2004. A SYBR green, real-time RT-PCR method to detect and quantitate Norwalk virus in stools. *J. Virol. Methods* **116:**63–70.
85. **Richards, G. P.** 1985. Outbreaks of shellfish-associated enteric virus illness in the United States: requisite for development of viral guidelines. *J. Food Prot.* **48:**815–823.
86. **Rippey, S. R.** 1994. Infectious diseases associated with molluscan shellfish consumption. *Clin. Microbiol. Rev.* **7:**419–425.
87. **Sair, A. I., D. H. D'Souza, and L. Jaykus.** 2002. Human enteric viruses as causes of foodborne disease. *Comp. Rev. Food Sci. Safety* **1:**73–89.
88. **Schwab, K. J., F. H. Neill, F. Le Guyader, M. K. Estes, and R. L. Atmar.** 2001. Development of a reverse transcription-PCR-DNA enzyme immunoassay for detection of "Norwalk-like" viruses and hepatitis A virus in stool and shellfish. *Appl. Environ. Microbiol.* **67:**742–749.
89. **Shapiro, R. L., S. Altekruse, L. Hutwagner, R. Bishop, R. Hammond, S. Wilson, B. Ray, S. Thompson, R. V. Tauxe, P. M. Griffin, and the *Vibrio* Working Group.** 1998. The role of Gulf Coast oysters harvested in warmer months in *Vibrio vulnificus* infections in the United States, 1988–1996. *J. Infect. Dis.* **178:**752–759.
90. **Shieh, Y. C., R. S. Baric, J. W. Woods, and K. R. Calci.** 2003. Molecular surveillance of enterovirus and Norwalk-like virus in oysters relocated to a municipal-sewage-impacted Gulf estuary. *Appl. Environ. Microbiol.* **69:**7130–7136.
91. **Shieh, Y.-S., S. S. Monroe, R. L. Fankhauser, G. W. Langlois, W. Burkhardt III, and R. S. Baric.** 2000. Detection of Norwalk-like virus in shellfish implicated in illness. *J. Infect. Dis.* **181**(Suppl. 2)**:**S360–S366.
92. **Shieh, Y.-S. C., K. R. Calci, and R. S. Baric.** 1999. A method to detect low levels of enteric viruses in contaminated oysters. *Appl. Environ. Microbiol.* **65:**4709–4714.
93. **Sobsey, M. D.** 1987. Methods for recovering viruses from shellfish, seawater and sediments, p. 77–108. *In* G. Berg (ed.), *Methods for Recovering Viruses from the Environment.* CRC Press, Inc., Boca Raton, Fla.
94. **Sugieda, M., K. Nakajima, and S. Nakajima.** 1996. Outbreaks of Norwalk-like virus-associated gastroenteritis traced to shellfish: coexistence of two genotypes in one specimen. *Epidemiol. Infect.* **116:**339–346.
95. **van Regenmortel, M. H. V., C. M. Fauquet, D. H., L. Bishop, E. B. Carstens, M. K. Estes, S. M. Lemon, J. Maniloff, M. A. Mayo, D. J. McGeoch, C. R. Pringle, and R. B. Wickner (ed.).** 2000. *Virus Taxonomy: Classification and Nomenclature of Viruses*, p. 725–739. Academic Press, Inc., San Diego, Calif.
96. **Vinje, J., R. A. Hamidjaja, and M. D. Sobsey.** 2004. Development and application of a capsid VP1 (region D) based reverse transcription PCR assay for genotyping of genogroup I and II noroviruses. *J. Virol. Methods* **116:**109–117.
97. **Vinje, J., H. Deijl, R. van der Heide, D. Lewis, K.-O. Hedlund, L. Svensson, and M. P. G. Koopmans.** 2000. Molecular detection and epidemiology of Sapporo-like viruses. *J. Clin. Microbiol.* **38:**530–536.
98. **Wallace, B. J., J. J. Guzewich, M. Cambridge, S. Altekruse, and D. L. Morse.** 1999. Seafood-associated disease outbreaks in New York, 1980–1994. *Am. J. Prev. Med.* **17:**48–54.
99. **Wang, R. F., W. W. Cao, and C. E. Cerniglia.** 1997. A universal protocol for PCR detection of 13 species of foodborne pathogens in foods. *J. Appl. Microbiol.* **83:**727–736.

Control of Microorganisms in Source Water and Drinking Water

MORTEZA ABBASZADEGAN AND AARON B. MARGOLIN

26

There are hundreds of different enteric microorganisms that are known to infect humans. Enteric microbes are excreted in the feces of infected individuals or animals and may directly or indirectly contaminate water intended for human consumption. Such fecal contamination may originate from a variety of sources, including sewage treatment plant effluents, on-site septic waste treatment discharges, land runoff from urban, agricultural, and natural areas, and leachates from sanitary landfills. Under proper conditions, fecal microorganisms can survive environmental conditions for extended periods and have been observed to travel more than 100 m through the subsurface (36).

Since the adoption of disinfection practices by drinking water utilities in the early 1900s, the incidence of waterborne diseases has decreased dramatically (http://www.c3.org/chlorine_knowledge_center/whitepapercl.html). Still, the World Health Organization estimates that in developing countries, 2 million children die each year of diarrheal diseases traceable to contaminated water and that 1.1 billion people do not have access to safe drinking water (50). According to the American Society for Microbiology, "Water quality and quality of life are intimately intertwined. Many serious health problems could be eliminated if more countries adopted water quality practices, including the simple steps of source water protection and disinfection to ensure safe water supplies" (http://www.epa.gov/OWOW/watershed/statewide/table.htm).

All too often, the burden of controlling pathogens in drinking water has been placed on treatment plant operations. A more comprehensive approach then is required, involving the use of multiple barriers. One of the first barriers in place should be the ability to control the entrance of microbes into source waters via watershed protection. The Watershed Protection Approach (WPA) is a departure from what has been previously done and from the way federal, state, and local agencies have viewed or regulated natural resources (64). At the national level, the Safe Drinking Water Act and the Clean Water Act both directly regulate the introduction of contaminants into the nation's surface and groundwater. At the state and local levels, regulatory programs can vary but are usually designed to protect source waters by restricting and/or regulating certain activities which could degrade the quality of the source water (51).

The most cost-effective approach to potable water is a multibarrier approach. The multiple barriers should include source protection, to prevent the entry of pollutants and pathogens; proper treatment and disinfection to reduce the numbers of pathogens and remove other pollutants; and proper distribution maintenance to ensure that treated water does not become recontaminated before it reaches the consumer (45).

Traditionally, potable surface water has been prepared by a disinfection process. Chlorine and chlorine-based compounds have been around the longest and are the best known. Other methods such as the use of ozone, reverse osmosis, and distillation have been intermittently used in the United States in preparing potable water and more widely used in other countries. Recently, UV light has gained popularity for disinfection of ground and surface waters (48).

In the last decade, however, there has been a growing change in the philosophy governing the protection of our source waters. Disinfection alone is a myopic approach and is no longer considered adequate. When drinking water sources become contaminated, the cost to industry and public health is high because of the need for additional treatment processes. There is a growing recognition that effective drinking water system management includes addressing the quality and protection of water sources. On a national level, the 1996 Safe Drinking Water Act Amendments go beyond monitoring and treating contaminated water to emphasizing pollution prevention. State drinking water agencies are directed to examine the sources of drinking water as an additional layer of protection for our drinking water (66).

SOURCES OF MICROBIAL CONTAMINATION

In recent years, we have seen an increased reliance on surface waters and water recharge or reuse, which are perceived as a more vulnerable source than groundwaters in terms of the microbial quality of water. Treatment techniques can be effective at removing microbial contaminants from water. The cumbersome procedures and the prohibitive cost of monitoring human pathogens make it imperative to look for alternatives. The use of indicator microorganisms to assess the microbial quality of water has been practiced since the early years of the last century. Bacterial agents have traditionally been used as indicators of fecal contamination of waters, but no consistent relationship between fecal indicator bacteria and human enteric viruses has been shown. The possibility of finding a universal standard indicator is almost nonexistent. Therefore, environmental microbiologists today face the challenge of selecting appropriate indicators for particular groups of pathogens.

Bacteriophages have been extensively researched as water quality indicators. However, the heterogeneity of somatic and male-specific coliphages has been a major obstacle in their expanded use as indicators of microbial water quality. Male-specific and somatic groups of enteric bacteriophages have consistently been suggested as indicators of enteric viral pollution of waters because of their ease of assay, resistance to environmental stresses, and nonmultiplication in aquatic environments. Currently, the U.S. Environmental Protection Agency (EPA) is evaluating a method (EPA method 1601) to use male-specific and somatic coliphages as indicators of fecal contamination (67). Male-specific coliphages are relatively less heterogeneic than somatic coliphages, and four serotypes or groups (I through IV) of male-specific coliphages have been proposed (55). Groups IV and I are found predominantly in animal feces, whereas groups II and III are found predominantly in human feces and sewage. However, pigs are an exception to this, as their feces are reported to have all groups except III. Antisera for male-specific coliphages are not readily available; moreover, environmental isolates very often are difficult to serotype. Molecular techniques show promise for easier screening of the coliphage flora of any natural water and determination of the type of male-specific coliphages present in any sample, for a more meaningful microbial quality indicator. Genotyping of F-specific RNA coliphages has been used to track sources of fecal pollution (29), but it has been pointed out that further genetic characterization of the different phage types is needed (56). The somatic coliphages are easier to detect and have been shown to have better correlation with the occurrence of the human enteric viruses in settled water, but no such serotyping or nucleic acid typing scheme is available for them.

Antibiotic resistance analysis (ARA) of fecal coliforms and fecal streptococci has been used to determine sources of fecal pollution in natural waters (18, 25, 30, 69). In addition to ARA, a variety of phenotypic methods have been used to determine the source of fecal pollution. Scott et al. (56) provide a comprehensive review of current methods. Ribotyping has emerged as a viable method to track fecal contamination sources, although some variability has been reported (69). Parveen et al. (49) compared discriminate analysis (DA) of isolates using ARA and ribotyping. Using DA and ARA of *Escherichia coli* isolates, they classified 82 and 68% of human and nonhuman samples, respectively, with an average rate of correct classification (ARCC) of 75%. DA of ribotyping profiles correctly classified 97% of the human samples and 67% of the nonhuman samples with an ARCC of 82%.

SOURCE WATER PROTECTION

Source water protection and management can be defined as not a regulatory program, but rather a way of organizing and coordinating programs and partnerships to improve and maintain environmental quality and protect public health. Source water or a watershed can be defined as "the geographic delineation of an entire water body system and the land that drains into it" (64). The public and environmental health protection goals of these programs can be achieved by implementing proper responses and addressing increasing threats of microbiological and chemical contaminants. The benefits of such programs are to safeguard the quality of water used as a drinking water source, and the programs are predicated upon the idea that reducing significant water quality changes ultimately minimizes potabilization process treatment challenges. Any such programs should include site-specific characterization of source water and land uses and identification of potential contaminant sources. These factors can be widely variable for different water sources and may require different protection program components. A successful program for characterization of site-specific criteria should consider routine or periodic monitoring and source tracking for different contaminants. Recent advances in biotechnology and microbial source tracking methods can be applied to identification of sources of contaminants.

Water resource management in the United States began early in the history of the country. Water resource management takes into account a broad set of factors, such as the water supply for human consumption, the demand for adequate water for crops, flood control, and the maintenance of navigable waterways (15). Early attempts at watershed management followed a more single-minded, specialized approach. As uses for water were discovered or threatened, governmental authorities and private operators provided or protected water for those uses (16).

In the last decade, the EPA's WPA focused on organizing and managing states' major watersheds. In this type of approach, activities such as water quality, monitoring, planning, and permitting are coordinated on a set schedule. The WPA has four major features: targeting of priority problems, a high level of stakeholder involvement, integrated solutions that make use of the expertise and authority of multiple agencies, and measurement of success through monitoring and other data gathering (64).

Watershed management can be divided into nine subcategories: management units, management cycles, stakeholder involvement, strategic monitoring, assessment, prioritization and targeting, development of management strategies, basin or watershed plans, and implementation. Management units are areas that are delineated by the state and contain many watersheds. Management cycles pertain to monitoring of the management units. Each unit is examined cyclically so that all of a state's management units are examined in a set period, typically 5 years. Stakeholder involvement allows for more individuals and groups to become involved in the watershed management efforts. The theory behind more involvement is that the greater the number of individuals who have a concern about water quality, the more likely there will be a concerted effort to maintain that water quality (64, 66). Strategic monitoring allows for identification of a problem and determination of the extent of the problem. While increased monitoring helps to ensure watershed quality, monitoring of one watershed should not come at the price of another. Hence, monitoring should be done on a rotating basis and is contingent upon the number of watersheds and available funding. Data analysis or assessment of a watershed helps to identify problems and sources of contamination. Prioritization and targeting allow watersheds to be ranked according to resource value and degree of impairment. The next two items, development of management strategies and watershed plans, deal with the goals for the watershed and assessment of such goals. The goals for a watershed can include how the watershed will be used while the watershed quality is still maintained. It is in this part of the plan that education of the public is helpful. The last phase is implementation. Here, the plans that have been developed and brought forward are put into place (64, 66).

To augment an organizational and management plan, certain physical attributes of a watershed, including watershed characteristics, size and configuration of the reservoirs, and the flexibility of the system, must be taken into consideration. Successful watershed plans, such as that of New York City, take into consideration such factors as climate, geology, soils, physiography, hydrology, vegetation, and land use. Additionally, water quality is enhanced by the long detention times afforded by the large, cascading systems of reservoirs. In the New York City system, each reservoir can act in conjunction with other reservoirs or independently. This gives the city the ability to regulate the detention time or to bypass waters which are under environmental stress, such as with high turbidities due to heavy runoff (9).

Recreation has become a major aspect of life in the United States. To that end, there is constant pressure for multiple uses of the reservoirs, streams, and tributaries which make up any watershed. Studies have shown that the more a body of water is used for recreational activities, the greater the probability that pathogens will be introduced into the water (7). Animal waste is another threat to watersheds. Partnerships with local agriculture need to be developed to limit or restrict the introduction of farm animal waste into the watershed. This approach to watershed management addresses both the introduction of pathogens into the watershed and the individual farmers' operational and financial circumstances (9).

Microbial monitoring is a cornerstone of any watershed quality assessment. However, there are several aspects of a monitoring program which need to be considered before the program can make a substantial contribution. The monitoring program design should include frequencies for sampling analysis, protocols, and the types of analysis needed (51). Detection of pathogens alone is not an adequate monitoring tool. Pathogens occur usually in low concentrations, and the analytical techniques available for their detection are often not very reliable or sensitive. This is not to underemphasize the importance of pathogen monitoring; it is just that there still does not exist one "magic bullet," so monitoring must be a combination of evaluations of several different organisms. Data collected from any monitoring program must be interpreted carefully. While bacteria have long been used as the sole indicator for microbiological water quality, enough evidence has been established over the years to indicate that bacteria, while able to predict the presence of other bacterial contaminants or the introduction of fecal material, are not adequate predictors for the presence of viruses or protozoans. To accurately assess the presence of these organisms, they should be evaluated separately. Also, since any source water is a dynamic body and since pathogens occur usually in very low numbers, a monitoring schedule covering at least 1 year, and preferably more, should be adhered to, and then results must be evaluated at the end of that period. The monitoring program must be designed to take into account the changing properties of the source water. For instance, if the water is used for recreational activities during the summer months, the frequency of monitoring should reflect the increased usage. The same may hold true for dry versus wet weather events. Discharges or runoff may occur only during heavy rains, and the pathogens that they bring may not be detected if monitoring is done only during the dry period.

Consideration of all the parameters of watershed protection, breaking it down into many parts, ultimately trying to minimize or eliminate source water contamination completely, has many more benefits and is much more cost-effective than water treatment alone. Parts of the United States that are now practicing such an approach are seeing the increased benefits in areas such as more direct focus by stakeholders on achieving ecological goals and water quality standards; an improved basis for management decisions; enhanced program efficiency; improved coordination among regulatory agencies; enhanced public involvement; and innovative solutions to old and new problems (64). The final evidence that proactive involvement helps in eliminating pathogen introduction into a watershed can be seen by evaluating New York City's water quality. It is interesting that their source waters were found to have lower pathogen concentrations than many filtered drinking water plant effluents, as determined by both the percentage of positive samples and the maximum concentration (9).

Watershed protection does not begin and end with surface waters only. Beginning in 1984, the EPA began working with states to create groundwater protection strategies to coordinate efforts under a multitude of federal groundwater programs (64). Groundwater comprises a large percentage of available freshwater supplies. In general, groundwater supplies much of the stream flow in some localities in the United States. Groundwater and surface water are components integral and related to the hydrologic cycle, yet these resources are often managed by different programs in federal, state, and local agencies, where protection of one resource may have been achieved at the expense of the other. Additionally, in many cases, drinking water wells which are completed in shallow alluvial aquifers are influenced by surface water. Some of these wells actually draw from the hyporheic zone (the area of rivers and sediments where there is constant interchange of water) (http://www.holon.se/folke/projects/vatpark/hyporeic.shtml).

Groundwater presents different challenges from surface waters. Determination of the source of groundwater contamination can be quite difficult. Due to retention and transport times, groundwater is more susceptible to non-point source contamination. Monitoring of wells can be expensive, and sampling of underground aquifers over an extended period is labor intensive. Abbaszadegan et al. (1, 2) demonstrated that more than 33% of groundwater wells sampled were positive for fecal pollution. This rate is higher than what had originally been predicted. The EPA is planning to enact its Groundwater Protection Rule. This rule will require monitoring for fecal contamination through the detection of *E. coli*, enterococci, or coliphages in wells which draw water from sensitive aquifers. Groundwater remediation is more difficult than that of surface water, emphasizing the need for prevention. Increasing numbers of states are beginning to control contamination through coordinated efforts between their watershed protection approach and the EPA's Comprehensive State Ground Water Protection Program (65).

CHOOSING WATER TREATMENT PROCESSES

Selection of a water treatment process is a complicated task that ultimately must result in delivery of potable water which is acceptable to the consumer at a reasonable cost. Factors which must be considered include water supply, source quality, desired finished water quality, reliability of process equipment, operational requirements and personnel capabilities, flexibility in dealing with changing water quality and equipment malfunctions, available space for construction of treatment facilities, waste disposal constraints, and capital and operating costs (including chemical availability) (31).

Surface Water Sources

Most raw drinking water sources are susceptible to significant water quality changes as stresses are placed on the surrounding environment due to natural, accidental, or intentional contaminations. Source water protection and proper water treatment processes can protect consumers from potentially harmful contaminants. Treatment processes associated with surface water sources must be able to respond to rapid changes in the water quality. Runoff from rain can rapidly increase turbidity, thus decreasing efficiency in existing disinfection practices. Spills and runoff of surface contaminants such as fertilizers or insecticides can inhibit treatment plant operations. Additional problems can arise from biological contamination, such as algal blooms, insects, and animal intrusion. These issues can be addressed by placement of a plant's intake pipe at an appropriate site in the selected source water. Spring and fall turnover can turn pristine waters with low nutrient loads into turbid waters with high concentrations of organics. During the summer months, surface waters may stratify into distinct layers so that warmer water stays near the surface and cooler water is trapped below with little intermixing, resulting in anoxic conditions and solubilization of iron and manganese as well as the production of sulfur compounds from anaerobic bacterial processes (40). Treatment plants must be able to respond with additional processes, such as increased sedimentation or flocculation times and perhaps the use of ion-exchange resins to remove manganese ions, as well as additional disinfection in order to maintain potable water which is acceptable to the consumer and free of risk.

There are several advantages to using surface water as opposed to groundwater as a source water, if they can be exploited. The location and elevation of the water sources may offer the advantage of gravity flow to the treatment facility. Visualization of the water supply can yield a daily evaluation. Water levels and sources of contamination can be less costly to identify than those of groundwater (40). Finally, surface waters which are properly managed, although safe management can be an enormous task, can also be used for recreational purposes.

Groundwater

Groundwater supplies are derived primarily from wells. Water from shallow wells or "hand-dug" wells is not considered groundwater because it is usually under the influence of surface water via runoff and infiltration and hence can have many of the characteristics of surface water. "True" groundwater, in contrast, is relatively constant in its characteristics from season to season and is usually not affected by changes in the surface environment. This does not mean that variation in water quality from well to well does not exist. Depending on hydrological influences and proximity to a source of contamination, groundwater quality can vary greatly between wells. In general, groundwater is more expensive than surface water as a source, in large part because of the pumping requirements. However, groundwater usually requires less treatment than most surface water with respect to microbial pathogens, organic matter, and turbidity (51). The microbial pathogens of concern in groundwater are enteric viruses. Viruses, unlike larger microorganisms such as bacteria and protozoan cysts, are not always retained by the soil matrix and, if environmental conditions (i.e., soil type and pH, rain events, and cation concentration) allow, can migrate to underground aquifers. Additionally, viruses which are complexed with particulate matter and kept in a cold environment have been documented as migrating great distances from the source of contamination and surviving for long periods outside the human host (59).

TREATMENT AND DISINFECTION

Physical Methods

There are several types of physical methods for the treatment and/or disinfection of water, such as reverse osmosis, distillation, and use of UV light. In the last several years, UV light has been gaining popularity.

UV Light

UV radiation (types A and B) is a type of low-energy electromagnetic radiation with poor penetrating power. UV radiation (A and B) causes excitation rather than ionization of atoms by raising electrons to a state of higher energy without removing them. The UV region of the electromagnetic spectrum that is optimal for the destruction of bacteria lies between 220 and 300 nm (34). There are four types of UV technologies currently being exploited by the water industry: (i) low pressure-low intensity (LP-LI) (40 to 85 W), (ii) low pressure-medium intensity (LP-MI) (up to 170 W), (iii) medium pressure-high intensity (MP-HI) (5 to 30 kW), and (iv) pulsed UV light. LP-LI and LP-MI are monochromatic, whereas MP-HI is polychromatic and approximately 25% of the energy is in the germicidal range (35, 42).

In the 1960s, it was found that the mechanism for UV disinfection is the creation of thymine-thymine dimers, which occurs at 254 nm and breaks in the nucleic acids. More specifically, it seems that UV energy has its greatest effect on pyrimidines and is much less effective on purines. While thymine-thymine dimers are the predominant nucleotide dimers which are created, dimerization is not limited to these bases; cytosine-cytosine and cytosine-thymine dimers have also been identified in UV-irradiated bacterial cells. It has been speculated that inactivation of RNA viruses, i.e., the enteroviruses and rotavirus, occurs because of uracil dimerization (1a). Sensitivity to UV light was found to be species as well as strain dependent (24).

One problem associated with UV irradiation is that it is absorbed by proteins, and hence viruses with a tight outer capsid, such as bacteriophage MS-2, or viruses with a double-shell capsid, such as rotavirus, are usually more resistant to UV inactivation than other single-stranded RNA viruses, such as poliovirus or hepatitis A virus (70). A similar observation can be made for bacteria: the gram-negative rods are much more susceptible to inactivation than the gram-positive cocci, the latter requiring upwards of 5 to 10 times the dose of the former (19).

Since the emergence of *Cryptosporidium* spp. over the last decade, the efficacious use of UV disinfection has been called into question. Initial evaluation of *Cryptosporidium* inactivation by UV light was done by techniques such as excystation or vital stains. These techniques indicated that *Cryptosporidium* was extremely resistant to inactivation. However, it was suggested that utilization of excystation or vital stains in evaluating *Cryptosporidium* inactivation is not accurate upon comparison with mouse infectivity models (21). *Cryptosporidium* organisms which were inactivated by pulsed UV light and evaluated by mouse infectivity demonstrated a 4-log reduction at only 20 mW-s/cm^2, while evaluation by excystation or vital dyes indicated that as much as 170 mW-s/cm^2 was required for the same log reduction (21). This still needs further validation but, if true, may demonstrate that UV light does not directly kill organisms but prevents them from reproducing.

In addition to more research using *Cryptosporidium* spp., UV inactivation of viruses will have to be revisited. Originally, most of the viral inactivation studies were done with single-stranded RNA viruses and cell culture. Viruses were exposed to the UV light and then evaluated for infectivity in cell culture. Single-stranded RNA viruses may be much more susceptible to UV inactivation than double-stranded DNA viruses, such as enteric adenoviruses.

Adenoviruses are important human pathogens, and their occurrence in water as well as their role as the etiologic agents of viral gastroenteritis has probably been greatly underestimated (26). One important aspect of enteric adenoviruses which makes them unique is that they are double-stranded DNA viruses. Adenoviruses seem to be more resistant to UV irradiation than are poliovirus and coliphage MS-2, *Pseudomonas* phage PRD-1, and *Bacillus subtilis* spores (44). Other studies have demonstrated that adenoviruses 40 and 41 have a much higher degree of survival in tap water, seawater, and wastewater than poliovirus and hepatitis A virus. One reason that adenovirus survivability may exceed that of the other enteroviruses is that enteroviruses are single-stranded RNA viruses, whereas adenovirus is a double-stranded DNA virus. Since double-stranded DNA is an integral part of eukaryotic cells, there exist the mechanisms for DNA repair. Others (8, 13) have documented the host cell reactivation of radiation-damaged adenovirus in select cell lines. It would then follow that water which is disinfected by UV irradiation may still be a source of viral contamination. Obviously, much research is needed in the area of *Cryptosporidium* inactivation by UV light as well as with other protozoa and with the double-stranded DNA virus adenovirus. This research needs to examine methods for evaluating inactivation and determining the difference between viability and infectivity. Utilization of new techniques such as integrated cell culture-PCR may help to determine when an organism is no longer infectious without the use of animal models.

Measurement of the effectiveness of UV light is more difficult than that of chemical disinfectants such as chlorine and chlorine dioxide. Additionally, since UV absorption is related to water quality parameters, calculations made for disinfection efficiencies at one site may not be comparable to those made for other water sources with different water characteristics (4). Inactivation of microorganisms by UV radiation is a function of both time and intensity. Light intensity can be determined by a UV sensor, and retention time can be determined by tracer studies. Factors which can alter UV efficiency include water matrix changes, UV lamp coating, UV intensity sensor coating, or lamp age.

It is often difficult to relate the information gained by the use of sensors or actinometry to the inactivation of pathogens. Hence, an alternative method is the use of surrogate nonpathogenic organisms which have physical, chemical, and biological properties similar to those of the pathogen. This allows for the development of a calibration curve comparing the UV inactivation of a surrogate to that of a specified pathogen in a given water matrix. Inactivation of the surrogate at the pilot plant or treatment plant will then help predict the inactivation of the pathogen based upon the bench or laboratory data previously generated.

Chemical Methods

Chlorine

Chlorination has been the major line of defense against waterborne disease outbreaks in the United States since 1908 (12). Chlorination is probably the oldest and most widely used form of drinking water disinfection. It has several advantages which make it appealing and have bolstered its popularity. Chorine is relatively inexpensive and easily obtained, and it provides residual protection in distribution systems. Chlorine is a strong oxidizing agent and can be used to modify the chemical character of water. In water, chlorine gas hydrolyzes to form hypochlorous acid (HOCl). The hypochlorous acid undergoes further ionization to form hypochlorite ions (OCl^-).

Equilibrium concentrations of HOCl and OCl^- depend on the pH of the water. Alkaline pH shifts the equilibrium to the right, causing the formation of higher concentrations of HOCl. Both HOCl and OCl^- are commonly referred to as free chlorine. Upon chlorination of water, a portion of the chlorine reacts with compounds in the water, such as organics, most typically substituting an atom of chlorine in place of hydrogen. The difference between the chlorine concentration applied to the water and the free chlorine is known as the chlorine demand (i.e., the concentration of chlorine added minus the concentration of free chlorine resulting) and is one of the physical attributes used to define the needed level of water treatment.

One of chlorine's greatest attributes is ease of use. While chlorine gas requires special handling, there are two forms of chlorine which are readily available and easy to use. One form is calcium hypochlorite [$Ca(OCl)_2$], which is the predominant dry form and when dissolved in water contains approximately 70% available chlorine. Sodium hypochlorite (NaOCl) is available in liquid form at concentrations of between 5 and 15%.

When chlorine reacts with ammonia in water, the resulting reactions can form chloramines: monochloramine, dichloramine, or trichloramine. The specific reaction products that are formed depend on the pH of the water, temperature, reaction time, and the initial chlorine-to-ammonia ratio. In general, below pH 4.4, trichloramine is produced; above pH 8.5, monochloramine usually exists alone; and between pH 4.5 and 8.5, both mono- and dichloramine exist. Chloramines are thought to be less effective than hypochlorite as disinfectants (15).

The ability of hypochlorite to destroy microbial life is due predominantly to the ability of HOCl to oxidize proteins and other structures found on bacteria, viruses, and protozoa. Since HOCl is electrostatically neutral and has a relatively low molecular weight, it is able to penetrate the organism with relative ease and hence has a high germicidal activity. The biocidal activity of chlorine is greatly reduced at high pH, probably because at an alkaline pH, the predominant species of chlorine is OCl^-. The negative charge of this molecule may prevent it from penetrating or coming in close contact with surface proteins as a result of electrostatic repulsive forces that exist between the chlorine and the carboxyl ends of proteins. Organisms that are relatively large and contain a "waxy" outer coat, such as some protozoan cysts, may be more resistant to the oxidizing potential of chlorine or require much longer contact times and hence may not be adequately disinfected by chlorine CT values which have been established for the inactivation of bacteria (CT values are defined by the concentration of the disinfectant and the contact time). Leahy et al. (39) showed that 2.80 mg of chlorine per liter achieves a 99% inactivation of *Giardia muris* cysts in 16 min of contact time at pH 7.0 and 25°C. Studies by Korich et al. (38) demonstrated that 80.0 mg of chlorine per liter requires more than 90 min of contact time for a 90% inactivation of *Cryptosporidium parvum* oocysts. All of these studies were done at the bench level, in static systems. Work by Tilton (61) has demonstrated that there is very little correlation between the CT developed with a static bench-scale system and that developed with a dynamic pilot plant system, the latter usually requiring a much greater CT than is indicated by static bench-scale systems.

Advances in the microbiological examination of water have shown that hypochlorite is also less effective at inactivating viruses than at killing total and fecal coliforms, the historical indicator organisms. In its form as combined compound chloramine, chlorine's effectiveness is related to its ability to disrupt metabolic enzyme activity. The reduced effectiveness against viruses may be in part because viruses do not rely on any metabolic (enzymatic) functions for survival in the environment. Harakeh (32) reported that of six viruses tested, coxsackievirus B5 was the most resistant, with 99.99% inactivation achieved at a dose of 18.0 mg of free chlorine per liter after 5 min of contact, whereas simian rotavirus was the most sensitive, with a dose of 5.0 mg of chlorine per liter required to obtain 99.99% inactivation.

While chlorine has been widely used over the decades, it is not without its drawbacks. Since the early 1970s, it has been known that undesirable disinfection by-products (DBPs) can be produced when free chlorine interacts with selected organic precursors, such as humic substances. This interaction produces a group of halogen-substituted single-carbon compounds referred to as total trihalomethanes (THMs) (22). Because of the potential carcinogenicity of THMs, chloroform in particular, the EPA has established a maximum contaminant level of 80 μg/liter for total THMs. Approaches to control THM formation include removal of organic precursors that react with chlorine, removal of THMs from finished water, and use of alternative disinfectants (60).

Chlorine Dioxide

Chlorine dioxide (ClO_2) was first used in 1921, by the German chemist Erick Schmidt, as a bleaching agent for cellulose fibers. However, the dangers and high costs of generating chlorine dioxide delayed its widespread use until the late 1930s, when Mathieson (now Olin) Chemical Co. developed an economical ClO_2 manufacturing process.

The application of chlorine dioxide to water treatment arose from the work of Aston (10), who demonstrated that it could be used to remove taste- and odor-causing compounds such as chlorophenols. However, since the cost of chlorine dioxide for large-scale water treatment was significantly higher than the cost of chlorine, its usage was restricted to special situations until the early 1970s. In 1975, the EPA, faced with increasing public concern over THMs in drinking water, sponsored research into alternative disinfection techniques to control THMs. The earlier research of Granstrom and Lee (28) had indicated that chlorine dioxide did not form THMs, particularly chloroform, when applied at dosages similar to those used for chlorine. The implication of those studies was that ClO_2 was a viable alternative to chlorination. Preliminary studies have suggested that ClO_2 may be more effective than chlorine at inactivating protozoan cysts and enteric viruses (33, 58, 62), and it is now well documented that ClO_2 forms far smaller amounts of THMs than Cl_2 (41).

The generation and reaction of chlorine dioxide in aqueous systems are poorly understood and are a topic for significant debate. Under typical pH conditions (6.5 to 8.5) encountered in water treatment, chlorine dioxide has been shown to readily disassociate to chlorate and chlorite ions (28).

Chlorite has been detected in treated water and is of concern, since it has been shown to react with hemoglobin, causing hemolytic anemia at 50 mg/liter (3). Additionally, studies with rats indicated a decrease in sperm motility at levels above 100 mg/liter (17). The production of chlorite ions led the EPA to set a maximum contaminant level of residual chlorine dioxide together with its inorganic by-products, chlorite and chlorate ions, at 1.0 mg/liter in drinking water (68).

The oxidizing power, and therefore the germicidal effectiveness, of chlorine dioxide was once thought to be 2.5 times that of chlorine, on the basis of its full oxidation potential (E_0 = 1.95 V). However, more recent work has shown that the full oxidation potential of chlorine dioxide can be reached only under acidic conditions. At pH 6 to 8, the predominant reaction is the reduction of chlorine dioxide to chlorite ion, which has an oxidation potential of −1.16 V, which is comparable to that of chlorine (46).

Several studies comparing bactericidal (14, 43, 53) and virucidal (23) qualities of chlorine dioxide with those of chlorine have concluded that chlorine dioxide is as effective as hypochlorous acid and more effective than hypochlorite ion or chloramines in disinfection. Other significant findings show that the effectiveness of chlorine dioxide as a disinfectant is much higher than that of chlorine in waters above pH 8.

The physiological mode of inactivation of bacteria by chlorine dioxide has been shown to be the disruption of protein synthesis. The mechanism by which viruses are inactivated was partially elucidated by Noss (47), who showed that phage f2 was inactivated by chlorine dioxide reacting with tyrosine residues on the capsid protein and/or the A protein, resulting in the inhibition of viral attachment to its host bacteria.

ADVANCED OXIDATION PROCESSES

Advanced oxidation processes are processes for producing OH radicals for the oxidation of organic and inorganic impurities in water. There are a number of processes which use (i) ozone, (ii) ozone with hydrogen peroxide, or (iii) UV radiation with hydrogen peroxide (35). Ozone is the one which is used primarily for inactivation of microorganisms. The latter two, in addition to ozone, play a much greater role in the control of taste and odor as well as the reduction of DBP precursors (35).

Ozone

Ozone disinfection for drinking water is still more popular in Europe than in other areas, evidenced by the more than 1,000 European plants which utilize ozone for disinfection of drinking water. However, ozone is beginning to gain popularity in the United States. In 1990, there were approximately 40 ozone water treatment plants in operation (38a); in 1998, it was estimated that 114 plants were in operation (52). There are several factors which have limited the use of ozone compared with chlorination in the United States, among which are cost, the requirement to generate O_3 on-site (27), and the danger of exposure of plant personnel to ozone (35).

Chemically, ozone is the strongest oxidant of all the classical disinfectants. As a strong oxidant, ozone reacts with a wide variety of organics and, unlike chlorine, does not form DBPs, unless bromide is in the water; then there is the potential for the formation of bromate (http://www.chec.uiowa.edu/conf_97/conf_ses1.htm). Additionally, ozone does not impart taste and can help control organoleptic properties (odor, taste, and color). When ozone is added to water, it reacts with hydroxide ions to form hydroxyl radicals and organic radicals. Ozone attacks organic compounds by adding oxygen atoms to the unsaturated carbon-carbon bonds, yielding carboxylic acids, ketones, and aldehydes (11).

Ozone's biocidal activity is not pH dependent but decreases as the water temperature increases. This is in part because the solubility of ozone in water is a function of temperature, and as the temperature increases, the solubility decreases. Ozone treatment does not generate any residual, and hence water disinfected by ozone is susceptible to posttreatment contamination. For this reason, in the United States, ozone treatment is usually the first step, followed by the addition of chlorine.

Commercially, ozone is generated by producing a high-voltage corona discharge in a purified oxygen-containing feed gas. The ozone is then contacted with the water, and the feed gas is recycled or discharged. A typical ozone treatment plant consists of three basic subsystems: feed gas preparation, ozone generation, and ozone-water contact (20).

Ozone has been shown to be effective for the destruction of bacteria and viruses. France has adopted a standard for the use of ozone to inactivate viruses. When an ozone residual of 0.4 mg/liter can be measured 4 min after the initial ozone demand has been met, viral inactivation is satisfied (20). Kim et al. (37) demonstrated that ozone breaks the protein capsid of phage f2 into several subunits, which liberates the viral RNA and disrupts the attachment of the virion to the host pili. In the last 10 years, the overall philosophy of ozone plant design has changed. Originally, the design was for an 8- to 12-min contact time for *Giardia* inactivation. However, now plants are being designed to achieve inactivation of *Cryptosporidium* spp. This has meant a longer retention time and much higher doses of ozone (35).

Bromide ions (Br^-) in drinking water that is disinfected by ozone can lead to the formation of DBPs such as bromate (BrO_3^-). Bromide ions can enter water sources from geologic sources, saltwater intrusions, crop and soil fumigation, and salt that is applied to pavement during winter months. Unfortunately, if bromide ion is present in source waters, there are no known treatment techniques available for economically removing it (57). Additionally, complex organics in the source water which are poorly assimilated by bacteria may be broken down by ozone into more readily assimilated organics, resulting in posttreatment bacterial regrowth.

BIOLOGICAL TREATMENT

Biological treatment of water has been classically used for the conversion of organic material to biomass in wastewater. Historically, biological treatment was not looked upon favorably because of the introduction of microorganisms into the water, and drinking water is usually rendered potable by the addition of one or more disinfectants to inactivate microorganisms. As cited above, some DBPs have been associated with cancer and other abnormalities in laboratory animals; thus, treatment of drinking water by biological filters is one way to avoid the DBPs produced by more classical disinfection methods. In pilot- and full-scale biological filters, contact time is typically measured in minutes, and the water is treated in a single-pass, flowthrough mode (57). Studies by Siddiqui et al. (57) have also shown that biological treatment of water can be effectively used to produce a good-quality biostabilized water low in DBPs. Aldehyde and keto acids have also been shown to be removed by biological filters at the pilot-scale level (68).

Recently, biological treatment has been driven by the concern of increased biodegradable organic matter (BOM) in ozonated waters. Increased BOM can lead to regrowth problems. Hence, implementing biological filtration reduces the BOM and regrowth potential, so much so that stage 1 of the disinfection-DBP rule requires utilities to implement biological filtration if they ozonate their water (63).

SUMMARY

One of the greatest challenges facing societies today is provision of drinking water that is free of any pathogens and that is both palatable and affordable. Approaches for the production of potable water have traditionally centered on treatment practices. Because of economic restraints or the production of harmful DBPs, this is no longer a plausible solitary option. Development of a single alternative treatment which is economical, reliable, and safe is not likely to occur in the next decade. In fact, as we are faced with the emergence of a population whose individuals have an older average age and subpopulations that have special needs, such as the immunocompromised, as well as changing economics, we will face mandates that require our technologies in the production of potable water to become more efficient than ever before.

The EPA and many state agencies are looking to source water protection as the first step towards the production of potable water. It is much more cost-effective to prevent the introduction of pathogens into water than it is to treat the water. Mark Buehler of the Metropolitan Water District of Southern California stressed the importance of source water protection when he said that the water industry is in general agreement that "we have taken treatment as far as we can" with the available technology and that source water protection "is the future of drinking water" (5). Additionally, we must realize the differences between the problems associated with surface water and those of groundwater and understand that a system approach must be designed to address both.

REFERENCES

1. **Abbaszadegan, M., M. LeChevallier, and C. Gerba.** 2003. Occurrence of viruses in US ground waters. *J. Am. Water Works Assoc.* **95:**107–120.

1a. **Abbaszadegan, M., M. Hasan, C. P. Gerba, P. F. Roessler, B. Wilson, R. Kuennen, and E. VanDellen.** 1997. The disinfection efficacy of a point-of-use water treatment system against bacterial, viral and protozoan waterborne pathogens. *Water Res.* **31:**574–582.

2. **Abbaszadegan, M., P. Stewart, and M. LeChevallier.** 1999. A strategy for detection of viruses in groundwater by PCR. *Appl. Environ. Microbiol.* **65:**444–449.

3. **Abdul-Rahman, M. S., D. Couri, and R. J. Bull.** 1980. Kinetics of ClO_2 and effect of ClO_2 in drinking water on blood glutathione and hemolysis in rat and chicken. *J. Environ. Pathol. Toxicol.* **3:**431–449.

4. **American Public Health Association.** 1995. Chlorine, p. 36–57. *In* M. A. H. Franson (ed.), *Standard Methods for the Examination of Water and Wastewater*, 19th ed. American Public Health Association, Washington, D.C.
5. **American Water Works Association.** 1998. Source Water Protection Symposium, conference highlights. *J. Am. Water Works Assoc.* **8:**32–33.
6. Reference deleted.
7. **Anderson, M. A., M. H. Stewart, M. V. Yates, and C. P. Gerba.** 1998. Modeling the impact of body-contact recreation on pathogen concentrations in a source drinking water reservoir. *Water Res.* **11:**3293–3306.
8. **Arnold, W. R., and A. J. Rainbow.** 1996. Host cell reactivation of irradiated adenovirus in UV-sensitive Chinese hamster ovary cell mutants. *Mutagenesis* **11:**89–94.
9. **Ashendorff, A., M. A. Principe, A. Seeley, J. LaDuca, L. Beckhardt, W. Faber, Jr., and J. Mantus.** 1997. Watershed protection for New York City's supply. *J. Am. Water Works Assoc.* **3:**75–88.
10. **Aston, R. N.** 1947. Chlorine dioxide use in plants on the Niagara border. *J. Am. Water Works Assoc.* **39:**687–690.
11. **Baily, P. S.** 1975. Reactivity of ozone with various organic functional groups important to water purification, p. 263–277. *In First International Symposium on Ozone for Water and Wastewater Treatment Proceedings*.
12. **Belohlav, L. R., and E. T. McBee.** 1962. Discovery and early work, p. 137–156. *In* J. S. Sconce (ed.), *Chlorine: Its Manufacture, Properties and Uses*. ACS monograph no. 154. Van Nostrand Reinhold, New York, N.Y.
13. **Bennet, C. B., and A. J. Rainbow.** 1989. DNA damage and biological expression of adenovirus: a comparison of liquid versus frozen conditions of exposure to gamma rays. *Radiat. Res.* **120:**102–112.
14. **Bernarde, M. A., W. B. Snow, V. P. Olivicri, and B. Davidson.** 1967. Kinetics and mechanism of bacterial disinfection by chlorine dioxide. *Appl. Environ. Microbiol.* **15:**257–262.
15. **Bitton, G.** 1980. *Introduction to Environmental Virology*, p. 201–241. John Wiley & Sons, New York, N.Y.
16. **Brady, D. J.** 1996. Basic comparison of structure and functioning of legislative governmental and non-governmental bodies for water quality management in the USA and the CR: an American view. *Water Sci. Tech.* **33:**27–30.
17. **Carlton, B. D., and M. K. Smith.** 1986. Reproductive effects of alternative disinfectants and their by-products, p. 295–300. *In* R. I. Jolley, R. J. Bull, W. P. Davis, S. Katz, M. H. Roberts, Jr., and V. A. Jacolis (ed.), *Water Chlorination: Chemistry, Environmental Impact and Health Effects*. Lewis Publishers, Inc., Chelsea, Mich.
18. **Carson, C. A., L. S. Brian, M. R. Ellersieck, and A. Asfaw.** 2001. Identification of fecal *Escherichia coli* from humans and animals by ribotyping. *Appl. Environ. Microbiol.* **67:**1503–1507.
19. **Chang, J. C.** 1985. UV inactivation of pathogenic and indicator microorganisms. *Appl. Environ. Microbiol.* **49:** 1361–1365.
20. **Cheremisinoff, N. P., and P. N. Cheremisinoff.** 1993. Sterilization by radiation, p. 140–158. In WordCrafters Editorial Services Inc. (ed.), *Water Treatment and Waste Recovery Advanced Technology and Applications*. Prentice-Hall, Inc., Englewood Cliffs, N.J.
21. **Clancy, J. L.** 1998. Inactivation of Cryptosporidium parvum by ultraviolet light: dose response curves for low and medium-pressure mercury lamps and pilot scale demonstration studies. *In Proceedings of the AWWA Water Quality Technology Conference*.
22. **Craun, G. F.** 1988. Surface water supplies and health. *J. Am. Water Works Assoc.* **80:**40–52.
23. **Cronier, S., P. V. Scarpino, and M. L. Zink.** 1978. Chlorine dioxide destruction of viruses and bacteria in water, p. 651–658. *In* R. L. Jolley, H. Gorchev, and D. H. Hamiltont, Jr. (ed.), *Water Chlorination: Environmental Impacts and Health Effects*, vol. 2. Ann Arbor Science, Ann Arbor, Mich.
24. **Downes, A., and T. Blount.** 1877. Research on the effect of light upon bacteria and other organisms. *Proc. R. Soc. Lond.* **26:**488.
25. **Figueras, M. J., I. Inza, F. L. Polo, M. T. Feliu, and J. Guarro.** 1996. A fast method for the confirmation of fecal streptococci from M-Enterococcus medium. *Appl. Environ. Microbiol.* **62:**2177–2178.
26. **Girones, R., A. Allard, G. Wadell, and J. Jofre.** 1993. Application of PCR to the detection of adenoviruses in polluted waters. *Water Sci. Technol.* **27:**235–241.
27. **Glaze, W. H.** 1987. Drinking water treatment with ozone. *Environ Sci. Technol.* **21:**224.
28. **Granstrom, M. L., and G. F. Lee.** 1957. Rates and mechanisms of reaction involving oxychloro compounds. *Public Works* **80:**90–92.
29. **Griffin, D. W., C. J. Gibson, E. K. Lipp, K. Riley, J. H. Paul III, and J. B. Rose.** 1999. Detection of viral pathogens by reverse transcriptase PCR and of microbial indicators by standard methods in the canals of the Florida Keys. *Appl. Environ. Microbiol.* **65:**4118–4125.
30. **Hagedorn, C., S. L. Robinson, J. R. Filtz, S. M. Grubbs, T. A. Angier, and R. B. Beneau.** 1999. Determining sources of fecal pollution in a rural Virginia watershed with antibiotic resistance patterns in fecal streptococci. *Appl. Environ. Microbiol.* **65:**5522–5531.
31. **Hamann, C. L., Jr., J. B. McEwen, and A. G. Myers.** 1990. Guide to selection of water treatment processes, p. 157–188. *In* F. W. Pontius (ed.), *Water Quality and Treatment: a Handbook of Community Water Supplies*. McGraw-Hill, Inc., New York, N.Y.
32. **Harakeh, M. S.** 1986. Factors influencing chlorine disinfection of wastewater effluent contaminated by rotaviruses, enteroviruses, and bacteriophages, p. 681–690. *In* R. L. Jolley, R. J. Bull, V. A. Jacobs, W. P. Davis, S. Katz, and M. H. Roberts, Jr. (ed.), *Water Chlorination: Chemistry, Environmental Impact and Health Effects*, vol. 5. Lewis Publishers, Chelsea, Mich.
33. **Hoff, J. C., and E. E. Geldreich.** 1981. Comparison of the biocidal efficiency of alternative disinfectants. *J. Am. Water Works Assoc.* **73:**40–44.
34. **Huff, C. B.** 1965. Study of ultraviolet disinfection of water and factors in treatment efficiency. *Public Health Rep.* **80:**695.
35. **Issarn, N., and R. R. Trussell.** 1999. New and emerging drinking water technologies, p. 220–243. *In Identifying Future Drinking Water Contaminants*. National Academies Press, Washington, D.C.
36. **Keswick, B. H., et al.** 1982. The use of microorganisms as groundwater tracers: a review. *Groundwater* **20:**142–149.
37. **Kim, C. K., D. M. Gentile, and O. J. Sproul.** 1980. Mechanism of ozone inactivation of bacteriophage f2. *Appl. Environ. Microbiol.* **39:**210–218.
38. **Korich, D. G., J. R. Mead, M. S. Mador, N. A. Sinclair, and C. R. Sterling.** 1990. Effects of ozone, chlorine dioxide, chlorine, and monochloramine on *Cryptosporidium parvum* oocyst viability. *Appl. Environ. Microbiol.* **56:** 1423–1428.

38a. **Langlais, B., D. A. Reckhow, and D. R. Brink (ed.).** 1991. *Ozone in Water Treatment: Applications and Engineering: Cooperative Research Report*. Lewis Publishers, Chelsea, Mich.

39. **Leahy, J. G., A. J. Rubin, and O. J. Sproul.** 1987. Inactivation of *Giardia muris* cysts by free chlorine. *Appl. Environ. Microbiol.* **53:**1448–1453.

40. **Levenspiel, O.** 1972. *Chemical Reaction Engineering*. John Wiley & Sons, New York, N.Y.
41. **Malley, J. P., Jr., J. K. Edzwaid, and N. M. Ram.** 1988. Preoxidant effects on organic halide formation and removal of organic halide precursors. *Environ. Technol. Lett.* **9:**1089.
42. **Malley, J. P., Jr.** 1998. Ultraviolet disinfection, p. 213–240. *In Control of Microbes in Drinking Water*. American Society of Civil Engineers, Reston, Va.
43. **Malpas, J. F.** 1973. Disinfection of water using chlorine dioxide. *Water Treat. Exam.* **22:**209–217.
44. **Meng, Q. S., and C. P. Gerba.** 1996. Comparative inactivation of enteric adenoviruses, poliovirus, and coliphages by ultraviolet irradiation. *Water Res.* **30:**2665–2668.
45. **Morris, J.** 1994. Filtered or waivered, you need source protection. *N. Engl. Water Works Assoc.* **9:**182–191.
46. **Myhrstad, J. A., and J. E. Samdal.** 1969. Behavior and determination of chlorine dioxide. *J. Am. Water Works Assoc.* **4:**205–208.
47. **Noss, C. I.** 1986. Chlorine dioxide reactivity with proteins. *Water Res.* **20:**351–356.
48. **Parrotta, M. J., and F. Bekdash.** 1998. UV disinfection of small groundwater supplies. *J. Am. Water Works Assoc.* **90:** 71–79.
49. **Parveen, S., K. M. Portier, K. Robinson, L. Edmiston, and M. L. Tamplin.** 1999. Discriminant analysis of ribotype profiles of *Escherichia coli* for differentiating human and nonhuman sources of fecal pollution. *Appl. Environ. Microbiol.* **65:**3142–3147.
50. **Postel, S. L.** 2000. Water and world population growth. *J. Am. Water Works Assoc.* **92:**131–138.
51. **Reinert, P. E., and J. A. Hroncich.** 1990. Source water quality management, p. 189–268. *In* F. W. Pontius (ed.), *Water Quality and Treatment: a Handbook of Community Water Supplies*. McGraw Hill, Inc., New York, N.Y.
52. **Rice, R. G., and P. K. Overbech.** 1998. Impact of evolving EPA drinking water regulations on ozone use in the United States. *In Proceedings of the IOA/PAG Annual Conference*.
53. **Ridenour, G. M., and R. S. Ingols.** 1947. Bactericidal properties of chlorine dioxide. *J. Am. Water Works Assoc.* **39:**561–567.
54. Reference deleted.
55. **Schaper, M., and J. Jofre.** 2000. Comparison for detecting genotypes of F-specific RNA bacteriophages and fingerprinting the origin of fecal pollution in water samples. *J. Virol. Methods* **89:**1–10.
56. **Scott, T. M., J. B. Rose, T. M. Jenkins, S. R. Farrah, and J. Lukasik.** 2002. Microbial source tracking: current methodology and future directions. *Appl. Environ. Microbiol.* **68:**5796–5803.
57. **Siddiqui, M. S., G. L. Amy, and R. G. Rice.** 1995. Bromate ion formation: a critical review. *J. Am. Water Works Assoc.* **87:**58–70.
58. **Sobsey, M.** 1988. *Detection and Chlorine Disinfection of Hepatitis A in Water*. EPA publication CR-813-024. U.S. Environmental Protection Agency, Washington, D.C.
59. **Sobsey, M. D., and J. S. Glass.** 1981. Improved electropositive filter for concentrating viruses from large volumes of water, p. 239–245. *In* M. Goddard and M. Butler (ed.), *Viruses and Wastewater Treatment*. Pergamon Press, New York, N.Y.
60. **Stetier, R. E., R. L. Ward, and S. C. Waltrip.** 1984. Enteric virus and indicator bacteria levels in a water treatment system modified to reduce trihalomethane production. *Appl. Environ. Microbiol.* **47:**319–324.
61. **Tilton, K. S.** 1995. Comparison of chlorine and chlorine dioxide as disinfectants for surface water. Ph.D. dissertation. University of New Hampshire, Durham.
62. **U.S. Environmental Protection Agency.** 1989. Drinking water; national primary drinking water regulations; filtration, disinfection; turbidity, *Giardia lamblia*, viruses, *Legionella*, and heterotrophic bacteria. *Fed. Regist.* **54:**27486–27491.
63. **U.S. Environmental Protection Agency.** 1997. National primary drinking water regulations: disinfectants and disinfection byproducts; notice of data availability; proposed rule. *Fed. Regist.* **62:**212.
64. **U.S. Environmental Protection Agency.** 1997. *Statewide Watershed Management Facilitation*. EPA841–R-97-011. U.S. Environmental Protection Agency, Washington, D.C.
65. **U.S. Environmental Protection Agency.** 1998. National primary drinking water regulations: interim enhanced surface water treatment; final rule. *Fed. Regist.* **63:**241.
66. **U.S. Environmental Protection Agency.** 1999. *Protecting Sources of Drinking Water, Selected Case Studies in Watershed Management*. EPA 816-R-98-019. U.S. Environmental Protection Agency, Washington, D.C.
67. **U.S. Environmental Protection Agency.** 2000. *Method 1601: Male-Specific and Somatic Coliphages in Water by Two-Step Enrichment Procedure*. EPA publication 821-R-00-009. U.S. Environmental Protection Agency, Washington, D.C.
68. **Werdehoff, K. S., and P. C. Singer.** 1987. Chlorine dioxide's effect on THMFP, TOXFP, and the formation of inorganic by-products. *J. Am. Water Works Assoc.* **79:**107–113.
69. **Wiggins, B. A.** 1996. Discriminant analysis of antibiotic resistance patterns in fecal streptococci: a method to differentiate human and animal sources of fecal pollution in natural waters. *Appl. Environ. Microbiol.* **62:** 3997–4002.
70. **Wilson, B.** 1992. Coliphage MS-2 as a UV water disinfection efficacy test surrogate for bacterial and viral pathogens. Poster presentation at the American Water Works Association Water Quality Technology Conference, 8 to 11 November, Toronto, Ontario, Canada.

Assessing the Efficiency of Wastewater Treatment

WESLEY O. PIPES AND JAMES T. ZMUDA

27

INTRODUCTION

The purpose of wastewater (WW) treatment is to prevent water pollution and thereby protect public health and preserve the value of the receiving water (RW) as a resource. The physical, mechanical, biological, and chemical processes involved in WW treatment are discussed in any environmental engineering book. Reference 42 is widely recognized as the classic text for WW engineers. The quality of final effluent (FE) of a WW treatment plant (WWTP) depends upon a number of factors including the components of the untreated WW entering the WWTP, the concentrations of those components in the untreated WW, the WWTP design, and the operational condition of the WWTP.

This chapter is concerned with evaluation of the overall performance of the entire WWTP and not with the performance of individual unit processes. The methods covered are those which can be used to ascertain the strength of the untreated WW and the FE discharged to the RWs and thus the reduction in pollution achieved by treatment. Also, the methods covered are limited to microbiological methods, all of which are bioassays. Reference 1 (www.standardmethods.org) is a compilation of the generally accepted procedures for the analysis of WW and is considered the authoritative reference for WW analysts and regulatory authorities. The emphasis of this chapter is not on the methods themselves but on an overall evaluation of WWTP treatment efficiency. The objectives of WW treatment, the concepts of efficiency and effectiveness of WW treatment, and the difference between WW treatment processes and sludge treatment processes are reviewed to establish a context for this discussion.

It is practical to consider the subject of this chapter in the regulatory context (i.e., from an operator's perspective) because it helps to explain why one particular method or approach is preferred over another. As mentioned above, the design and operational conditions of a WWTP govern the quality of the FE it produces. The design of a WWTP, in turn, is governed by the intended or designated use of the FE. Designated use is a regulatory term which has its origins in the Clean Water Act (CWA). As a result of the CWA, all WWTPs must obtain National Pollutant Discharge Elimination System (NPDES) permits from the appropriate regulatory agency to discharge FE into RWs. Limits for the parameters discussed in this chapter are often conditions of NPDES permits. The limits in NPDES permits are derived by regulatory agencies in such a manner as to ensure that WWTPs produce FE suitable for discharge to RWs. As defined by the CWA, this means that FE discharges from WWTPs must not cause water quality standards (WQS) in the RWs to be exceeded.

Objectives of Treatment

The objectives of WW treatment are (i) to separate wastes from water for disposal elsewhere and (ii) to produce an FE which can be discharged to an RW body without causing pollution, i.e., violation of WQS. There are several different types of waste materials that can cause various forms of water pollution. Evaluation of the performance of WWTPs requires measurement of several parameters to provide information about the different forms of water pollution. Thus, to be a useful framework for evaluation of WW treatment, these objectives must be subdivided into specific aims related to the different forms of water pollution.

One way to subdivide objectives into specific aims is to list all of the possible biological and chemical constituents to be removed from the WW, e.g., typhoid bacteria, rotaviruses, ammonia, cyanide, and carbohydrates, and then to specify limits on the concentrations of each constituent in the FE. However, application of this approach would require analytical efforts far beyond the budget and laboratory capacity of most WWTPs. Furthermore, a complete analysis of WW is limited by the availability of analytical methods. For reasons of economy, the cost of data collection to evaluate treatment efficiency should be only a small fraction of the cost of the treatment itself.

The approach used depends not on measurement of specific waste constituents but on evaluation of parameters that are related to the various forms of pollution. Table 1 lists parameters used to evaluate the strength of pollution, along with the form of pollution associated with each of these parameters. These are aggregate parameters; that is, each measures something about the combined effect of a number of possible WW constituents which cause a particular form of pollution.

Pathogenic microorganisms which cause waterborne diseases are WW constituents, but there are so many different types of these microorganisms that detection and quantitation

TABLE 1 Parameters of water pollution

Parameter	Pollution effect	Measurement technique
Pathogenic microorganisms	Water unsuitable for human consumption or recreation	Microbiological (see index)
Indicator organisms	Indication of fecal contamination—possible presence of pathogens	Microbiological (see index)
Biodegradable organic compounds	Depletion of DO in receiving waters	Microbiological (see "BOD Testing" in this chapter) Chemical (see "COD Testing" in this chapter [not covered]) Chemical (see "TOC Testing" in this chapter [not covered])
Plant nutrients (N and P)	Stimulation of the growth of algae and aquatic plants	Microbiological (algal growth assay [see "Plant Nutrients: Algal Growth Test" in this chapter])
Toxic materials	Killing of aquatic life; interference with growth and/or reproduction	Microbiological (toxicity tests) Biological (fish and invertebrate toxicity assays [not covered]) Chemical (analyses for specific compounds [not covered])
Settleable solids	Formation of sludge banks, smothering of benthos, and anaerobic benthic conditions	Physical (not covered)
Fats, oils, and grease	Surface films—interference with water surface phenomena	Chemical extraction (not covered)

of individual pathogens are not practical means of routinely assessing WW treatment. Indicator organisms are more easily detected than pathogens, and their presence in a body of water signifies that the water may be contaminated by sewage or another source of fecal material. Evaluation of the density of one or more indicator groups in an FE is a well-established method of assessing one type of water pollution.

Any oxidizable components of WW which are not removed during the treatment process will oxidize subsequently in RW. The depletion of all dissolved oxygen (DO) in RWs is deleterious because desirable aquatic organisms are aerobic and the solubility of oxygen in water is extremely low, <12 mg/liter in the temperature range of significance to aquatic biology. Microbial utilization of only a few milligrams of DO per liter due to WW discharges can result in the elimination of the desired populations and communities in the RWs. It is not feasible to measure the concentrations of potentially numerous biodegradable organic compounds or to determine the densities of the many organisms which could be involved in depletion of DO. The biochemical oxygen demand (BOD) test is used as a measure of the potential effects of the activities of aerobic microorganisms in consuming oxygen. There are several versions of the BOD test, but the one employed for evaluation of WW treatment is the 5-day, 20°C BOD bottle test (BOD_5).

The discharge of FE with a few milligrams of nitrogen (N) and phosphorus (P) per liter in forms which are available as nutrients for algae and aquatic plants can result in excessive growth of nuisance organisms in lakes, streams, and estuaries. Chemical analysis for the total amounts of N and P in WW and FE is feasible, but the amount of growth stimulation in the RW depends on both the specific compounds present in the FE and chemical and biological factors of the RW. The algal growth assay provides an integrated assessment of this aspect of the possible pollution effects of the FE.

Potentially toxic compounds are present in many FEs, but the expression of toxicity depends on the concentration of the compound, the sensitivity of the exposed organisms, physical and chemical characteristics of the RWs, and synergistic and antagonistic interactions of the FE constituents. Since it is not feasible to measure the concentrations of all toxic compounds that might be present, toxicity bioassays are used to demonstrate the relative degree of toxicity of WW and FE. A variety of toxicity bioassay techniques using fish, invertebrates, and algae have been developed by the U.S. Environmental Protection Agency (EPA) for use in NPDES permits. The tests using fish and invertebrates are discussed in this chapter, to put everything in context. Quite a number of surrogate toxicity bioassays using microorganisms have been developed which are useful tools for detecting and investigating toxicity, although they are not used for NPDES permits.

Suspended solids, fats, oils, and grease are also measured as aggregate parameters because it would not be feasible to measure the individual chemical components of those classes of materials. The analytical procedures for those two parameters are not microbiological; for that reason, they are not covered here.

Pollutant Removal and Parameter Reduction

The two objectives of WW treatment, separating wastes from WW and preventing pollution of the RWs, are evaluated differently. Treatment efficiency depends on the extent to which specific waste materials are separated from the WW, and it can be calculated for a number of different components. Treatment effectiveness is evaluated by measuring the values of the parameters of pollution in the FE without reference to their values in the raw WW entering the WWTP. Since there are a number of different forms of water pollution, there are a number of different ways to calculate effectiveness. Specification of the WW components measured or the form of pollution evaluated is essential for describing the efficiency and effectiveness of WWTPs.

When applied to a conservative substance such as P, the term "efficiency" is used to express information about the fraction of the waste material originally present in the WW that has been separated during treatment. Most often, removal efficiency is expressed as a percentage:

$$\% \text{ Efficiency} = 100\left(\frac{Ci - Ce}{Ci}\right) \quad (1)$$

where Ci and Ce are concentrations of the waste material in the influent and effluent, respectively. Percent efficiency values are used to compare different treatment processes and to determine if a particular WWTP is accomplishing that for which it was designed.

The amount of waste material separated from the water is given by

$$\text{Waste load reduction} = (Ci - Ce)\,(Q)\,(f) \quad (2)$$

where Q is the WW volumetric flow rate and f is a unit conversion factor; e.g., when concentration is expressed in milligrams per liter and flow is expressed in cubic meters per day, an f value of 0.001 kg/g is used to give the waste load reduction in kilograms per day. Waste load reduction expresses how much is accomplished by WW treatment and also provides information about the amounts of the residual waste materials (usually called sludge) which are disposed of elsewhere.

In the WW literature, the terms "removal" and "reduction" are frequently used more or less interchangeably. Ideally, removal should be quantitatively related to both changes in the concentration of a constituent in the WW during treatment and the production of sludge, and reduction should be used in reference to the effect of treatment on possible pollution of the RW. The use of the terms removal and reduction as synonyms does not cause conceptual problems in describing the separation of a conservative material; e.g., the change in P concentration during treatment, the amount of P in the sludge, and the reduction in the potential of the FE to support algal growths in the RW are all related quantitatively. However, when applied to a parameter which is a property of one or more constituents rather than a constituent itself, the use of the term removal in place of reduction may be misleading and confuse the issues of the efficiency and effectiveness of treatment.

Toxicity is a property, not a material. It cannot be purified, weighed, or diluted to a known concentration. Toxicity is only defined by its effects on organisms. It is possible to measure the relative toxicities of WW and FE for the same treatment facilities, but the term "toxicity removal" does not make sense, whereas "toxicity reduction" does. If the toxicity of the FE is 1/100 of the toxicity of the WW, the toxicity reduction is 99%. However, the percentage of toxicity reduction does not give any information about the amount of material removed from the WW or the amount of sludge produced.

In the same manner, the exertion of oxygen demand is a property of organic compounds which can be assimilated by heterotrophic microorganisms but is not identical with the compounds themselves. In the absence of the microorganisms, oxygen demand is not exerted. If the microorganisms remain in the WW after the organic compounds are assimilated, they continue to consume oxygen. Oxygen consumption is a property of the microorganisms. In spite of this, the term "BOD_5 removal" is often used instead of "BOD_5 reduction," because it is assumed that there is a direct proportionality between the biodegradable waste components and BOD. Reduction of BOD_5 is considered a direct measure of the removal of the organic compounds which can be assimilated by microorganisms, and the production of new organisms (biological sludge) is considered to be proportional to BOD_5 reduction. Equating this property of certain waste components with the components themselves can cause confusion about the proper way to evaluate the performance of WWTPs.

WW Treatment versus Sludge Treatment

WWTPs employ two different series of treatment processes: one used for treatment of the WW and the other used for treatment of the sludge. The sludge results from the separation of the waste components and from the aggregation and settling of microorganisms which grow in biological treatment processes. Sludge treatment is not directly related to the prevention of water pollution but is intended to change the characteristics of the waste materials so that the ultimate disposal (usually to land) can be accomplished with less expense and without creating a nuisance. The biological processes used for sludge treatment are often called digestion processes and can be either anaerobic or aerobic. Sometimes, sludge digestion processes are confused with the biological process used for WW treatment.

The objectives of sludge digestion are (i) biodegradation of some of the organic matter in the sludge so that the rate of degradation of the remaining organic matter is very slow (i.e., sludge stabilization) and (ii) alteration of the properties of the sludge so that it is more easily thickened and dewatered. The focus for the design and operation of sludge digestion processes is on the changes which occur in the waste material, not on changes in water quality. The water that is separated from the sludge (usually called the supernatant) is not suitable for discharge but is mixed back into the incoming WW for further treatment.

Most of the parameters of water pollution (Table 1) are not appropriate for evaluation of the efficiency or effectiveness of sludge digestion processes. For example, in municipal WWTPs, the supernatant removed from an anaerobic sludge digester will often have a higher BOD_5 value (due to the presence of organic compounds, which are assimilated very rapidly) and a greater toxicity (due to the release of ammonia-nitrogen from proteins) than the sludge which is put into the digester. Other aggregate parameters such as volatile-suspended-solids tests and volatile-acids determinations are used for assessment of the functioning of sludge digestion processes. Those methods are not covered in this chapter.

The parameters used to measure the overall effectiveness of sludge treatment are listed and defined in Title 40 of the *Code of Federal Regulations* (40 CFR) *Standards for the Use or Disposal of Sewage Sludge* (EPA's Part 503 Rule) published in the *Federal Register* on 19 February 1993, under the authority of the CWA. Subpart D of the Part 503 Rule covers pathogens and vector attraction reduction; it was revised in 1998 and is reprinted in appendix B of reference 68. Reference 60 is a guidance document published by EPA which clarifies the Part 503 Rule. All microbiological analyses conducted for compliance with the Part 503 Rule must be done with the methods published in reference 68, although these methods are generally recognized to be inadequate and do not permit effective recovery of micro-organisms from sludge. See the appropriate chapters in this manual covering the detection of pathogens in sludge and see "Beneficial Use of Wastewater Sludge or Biosolids" below.

PATHOGENIC AND INDICATOR MICROORGANISMS

The presence of pathogenic microorganisms in FE makes the RW unsuitable for direct-contact recreational use and sometimes unsuitable for use as source water for a public supply. The absence from the FE of pathogens which can cause waterborne disease is an important measure of the effectiveness of WW treatment but is rarely determined at the present time. In this case, the efficiency of treatment is not calculated; it is not necessary to determine the total number or percentage of the pathogens removed or inactivated to evaluate treatment. Instead, WW treatment is evaluated in terms of its effectiveness in reaching the ideal condition of no pathogens in the FE.

Waterborne diseases can be caused by protozoa, bacteria, viruses, and parasitic worms. Waterborne infectious agents are discussed in other chapters in this manual. The techniques for detection of waterborne pathogens are quite varied, and some are very complex. There are few quantitatively reliable standard methods available to measure pathogens in FE. The effectiveness of WW treatment with respect to elimination of microbiological pollution is usually measured by determining the densities of indicator organisms in FE. Methods for evaluating densities of indicator bacteria are also discussed in other chapters in this manual.

For NPDES compliance, either fecal coliform (FC), *Escherichia coli*, or enterococcal densities are monitored because the permit limits are based upon WQS in the RW. As previously mentioned, permit limits for all parameters are set such that FE will not cause the WQS in the RW to be exceeded. As a result of studies conducted by Cabelli and Dufour, the EPA concluded that using FCs as an indicator of ambient microbiological water quality was inadequate and it recommended changing WQS to include limits for enterococci or *E. coli* for freshwaters and enterococci for marine waters (69). The Beaches Environmental Assessment and Coastal Health Act of 2000 (see appendix A in reference 69) also required EPA to study issues associated with pathogens and human health and, by 15 October, 2005, to publish new or revised criteria for pathogens and pathogen indicators based on these studies. EPA missed the 15 October, 2005 deadline but is expected in the near future to change all microbiological WQS including FC to include either *E. coli* or enterococci (or an alternate indicator) and to replace all FC limits in NPDES permits by either *E. coli* or enterococcal limits (or an alternate indicator), as appropriate for each WWTP.

Parenthetically, from an operator's perspective there are some weaknesses in the course EPA has taken. First, the ambient water quality criteria that EPA is expected to issue will be based largely on the results of a collaborative research study between EPA and the Centers for Disease Control and Prevention, the National Epidemiological and Environmental Assessment of Recreational Water Study (www.epa.gov/nheerl/neear/). The study was designed to find how new ways of measuring water pollution at bathing beaches can be effectively used to protect swimmers' health. It is not appropriate to apply these criteria to all waters of the United States indiscriminately. For example, it is not necessary to protect waterways designed primarily to carry barge traffic to the same extent as bathing beaches. Site-specific criteria should be developed. Second, since *E. coli* is a subset of the FC population, it is not necessary to change NPDES permits for WWTPs to require monitoring *E. coli* instead of FCs. Monitoring *E. coli* is significantly more expensive than monitoring FCs. A sensible alternative would be to give the operator an option to continue monitoring for FCs and using an appropriate site-specific correction factor (the *E. coli*-to-FC ratio) to determine new (lower) FC limits. (See reference 76 for an example.)

OXYGEN DEMAND MEASUREMENTS

Municipal WW and most industrial WW contain various types of heterotrophic microorganisms which are active and consume oxygen, if any oxygen is available. The identities of many of these organisms are not known and are not considered important (in a practical sense). For practical evaluation of WW treatment, the focus is on the potential decrease in the concentration of DO in the RW, due to the FE which is discharged and the fact that a certain degree of WW treatment can be accomplished by making use of the metabolic activities of the heterotrophic microorganisms.

A method for predicting the rate of consumption of DO by microorganisms in the RW body is essential for assessment of the potential of an FE to cause pollution. The original approach to this assessment (over 100 years ago) was to measure the DO concentration in the RW, incubate a sample of the water sealed off from contact with atmospheric oxygen for a period of time, and then measure the DO remaining to find out how much had been consumed. This approach evolved into the 20°C BOD_5 test (bottle determination), which has been in use more or less in its present form for about 65 years. In addition to measuring the pollution potential of an FE, the BOD_5 test is also used as a measure of the amount of organic material in WW, which can serve as a substrate to support the growth and metabolism of heterotrophic microorganisms active in biological WW treatment.

The BOD_5 test measures the amount of oxygen consumed by microorganisms in a 5-day period under standard conditions. Mathematical models are used to extrapolate BOD_5 results both to deoxygenation of the RW and to substrate concentration in WW. The mathematical models are applicable only when certain rather stringent conditions are met and thus do not represent either the substrate-microorganism relationships of biological WW treatment or the various phenomena of deoxygenation of RW very well. This has been widely recognized, and there have been many attempts to develop alternative analytical approaches to obtaining the needed information.

The chemical oxygen demand (COD) test was developed because the BOD_5 test requires 5 days for completion and therefore is not suitable either for operational control of treatment processes or for real-time evaluation of the efficiency of WWTP performance. The total organic carbon (TOC) test was developed as another alternative for measuring the strength of WW, based on the concept that the primary purpose of all biological and some chemical WW treatment processes is to reduce the concentration of organic material. Neither test is well suited for evaluation of the potential of the FE to cause deoxygenation of the RW, but both are used in the evaluation of the performance of WWTPs, more often as a supplement to rather than a replacement for the BOD_5 test.

BOD Testing

The BOD_5 test is the primary method for determination of the strength of WW to be treated by a biological process and the strength of FE in terms of the demand on the oxygen resources of the RWs. It is one of the most important tests for regulation of discharges via limits on the BOD_5 of the FE. BOD_5 testing of the FE is used to evaluate the efficiency of WW treatment by comparison with the BOD_5 of the influent (equation 1), and a minimum percent BOD_5 reduction efficiency is often one of the regulatory requirements. The mathematical model used for relating BOD to deoxygenation of the RW is called the DO sag equation and can be found in any technical book on water pollution or any basic environmental engineering textbook.

BOD_5 is also used to estimate the oxygen requirements of a biological treatment process and the rate of production of excess biological sludge from that process. The most widely used mathematical model of biological WW treatment is an elaboration of the chemostat model of the growth of a pure culture of a heterotrophic bacterium in a continuous culture, which is explained in most basic microbiology textbooks. Since a biological WW treatment process involves a mixed culture, with many different metabolically active bacteria, fungi, and protozoa, and because oxygen consumption is an important design parameter, the chemostat model is not adequate, and the elaboration used in WW treatment is called the kinetics of biological WW treatment.

Both the DO sag equation and the kinetics of biological WW treatment model depend on the assumption that the BOD_5 test is a measure of some fraction (e.g., two-thirds, three-fourths, or four-fifths) of the carbonaceous oxygen demand, that is, the oxygen consumed by heterotrophic microorganisms which utilize the organic matter of the waste in their metabolism. Oxygen demand exerted by autotrophic nitrifying bacteria (nitrogenous oxygen demand) is not accounted for in these old models. Since ammonia is usually present in WW, nitrification inhibitors are used to suppress the exertion of nitrogenous oxygen demand. Carbonaceous oxygen demand is called first-stage BOD, and nitrogenous oxygen demand is called second-stage BOD.

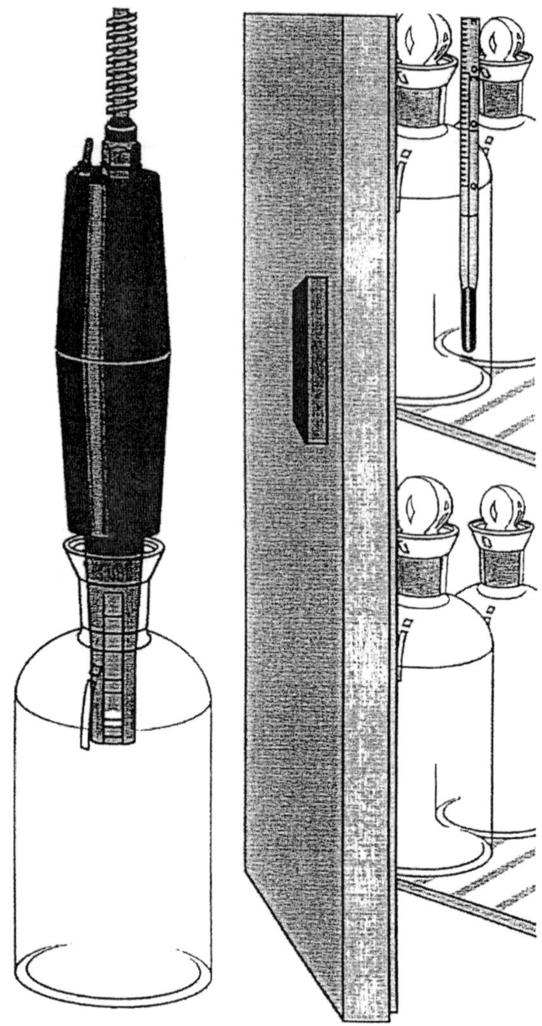

FIGURE 1 The BOD_5 test for evaluating the effectiveness of wastewater treatment is carried out by incubating diluted samples of wastewater and effluent in sealed bottles for 5 days at 20°C and measuring the oxygen depletion.

Measurement of BOD_5

Details of the determination of BOD_5 by the bottle method (Fig. 1) are described in section 5210 of reference 1. The important functions of a BOD bottle are (i) to serve as a culture vessel for the WW microorganisms and (ii) to be filled completely with a diluted sample and then sealed from contact with the atmosphere. Once the BOD bottle is sealed, the decrease in the DO over a period of 5 days provides a measure of the respiration of the microorganisms present. The conditions for the BOD bottle test were selected to give oxygen consumption measurements which can be used to compare the strength (pollution potential) of WW and FE in relation to oxygen consumption in the RW.

The BOD_5 test (Fig. 1) is a bioassay and requires the same careful attention to detail that is essential for the performance of other bioassays. The details of providing the proper conditions for exertion of oxygen demand by microbiological respiration are readily understood by microbiologists familiar with microbiological cultures and with the respirometric techniques used for so many other biochemical measurements. Extensive use of both control tests and quality assurance procedures is essential to obtain the proper results.

The rate of oxygen consumption varies with temperature. At a constant temperature, the total amount of oxygen consumed (the decrease in the DO in the BOD bottle) increases from day to day. The selections of an incubation temperature of 20°C and of an incubation period of 5 days are compromises. The temperature of 20°C is an approximation of a median summer temperature for surface waters in temperate climates. The rate of oxygen consumption decreases from day to day, and experience has shown that the oxidation of organic matter in municipal WWs can be completed, for all practical purposes, in a period of 20 days at 20°C in a BOD bottle. However, 20 days is too long to wait for results, and the conventionally selected period of 5 days allows for exertion of between 60 and 90% of the 20-day oxygen consumption of municipal WWs. In the United States, municipal WWs can be expected to have BOD_5 values in the range of 100 to 300 mg/liter, and industrial WWs are usually much stronger, with BOD_5 values at thousands of milligrams per liter and sometimes tens of thousands of milligrams per liter. Since the solubility of oxygen in water at 20°C is about 9 mg/liter, WWs have to be diluted 50- to 10,000-fold to have some DO remaining after 5 days of incubation. Effluents with BOD_5 values of >7 mg/liter also need to be diluted, usually twofold to fivefold, for the determination.

Since BOD_5 is a parameter (the daily rate of oxygen consumption averaged over 5 days) and not a material that is conserved during dilution, the effect that dilution has on the BOD_5 value is not immediately apparent. Evidence accumulated over many years demonstrates that, if the conditions in the BOD bottle are proper for the growth and metabolism of heterotrophic microorganisms, the rate of oxygen consumption will be proportional to the remaining concentrations of the pertinent waste components (organic matter and microorganisms). Thus, the decrease in the rate of exertion of oxygen demand produced by dilution is proportional to the dilution factor, because the concentrations of waste components are reduced in that proportion. When this empirical relationship holds, the BOD_5 of the undiluted WW can be calculated from the following algorithm: BOD_5 of WW = (BOD_5 of diluted sample) × (dilution factor).

The DO of the dilution water should be close to saturation before the dilutions are made because the respiration of some aerobic microorganisms is inhibited as the DO value approaches 0 during the BOD test. To provide a factor of safety, the residual DO in the BOD bottle after 5 days of incubation should be at least 2.0 mg/liter. Because the dilution water is seeded with organisms capable of assimilating the organic matter of the WW, there may be a small depletion of DO in 5 days even if no WW is added. Control bottles containing only dilution water are incubated along with the BOD_5 test bottles, and a correction is made for the DO depletion in the controls. The depletion of DO in the BOD test bottles should be at least 2 mg/liter during the incubation period to be significantly greater than the DO depletion in the control. To be sure of attaining a range of DO depletion of at least 2 mg/liter with at least 2 mg/liter remaining after the 5-day incubation period, it is necessary to prepare and incubate several dilutions of the original sample.

The composition of the solution used for dilution of samples for the BOD_5 test is an important part of proper conditions for exertion of oxygen demand. The dilution water specified previously (1) utilizes a phosphate buffer to provide nearly neutral pH during the incubation period. This dilution water also contains the required mineral nutrients in case the WW is deficient in any of them. Finally, to avoid confusing nitrogenous oxygen demand with carbonaceous oxygen demand, a compound which inhibits the nitrifying bacteria is added to the dilution water.

It is important that the dilution water not contain constituents inhibitory to the microorganisms that exert the carbonaceous oxygen demand. This must be demonstrated by running BOD_5 quality control standards prepared from pure organic compounds, usually glucose and glutamic acid. The use of controls also protects against the possibility of inhibitory material from the system being introduced during laboratory purification of water used for making up solutions.

For measurement of the BOD_5 of industrial WWs, the dilution water must be seeded (inoculated) with microorganisms which are capable of degrading the waste organic compounds. The seed may be obtained from a biological treatment process which has been acclimated to the WW or from sediments downstream from the FE discharge to the RW. In some cases, development of a laboratory culture of the seed organisms or purchase of a commercial seed may be necessary.

Standard BOD Curve and Equation

If a diluted sample of WW is incubated under the proper conditions and the oxygen consumption is recorded every day, a curve similar to that shown in Fig. 2 can be plotted to show that the rate of oxygen consumption is high initially but decreases continuously over time and approaches zero asymptotically. The fractional reduction in the amount of oxygen consumed each day is approximately the same, day by day, throughout the entire period of BOD exertion. When the rate of oxygen consumption is so small that it is no longer measurable, exertion of the first-stage BOD has been completed. In Fig. 2, which is an idealized curve, not actual data, the daily decrease in DO is no longer measurable at 20 days, and the ultimate, first-stage BOD is 5.8 mg/liter. The BOD_5 of the diluted sample is indicated as 4.4 mg/liter.

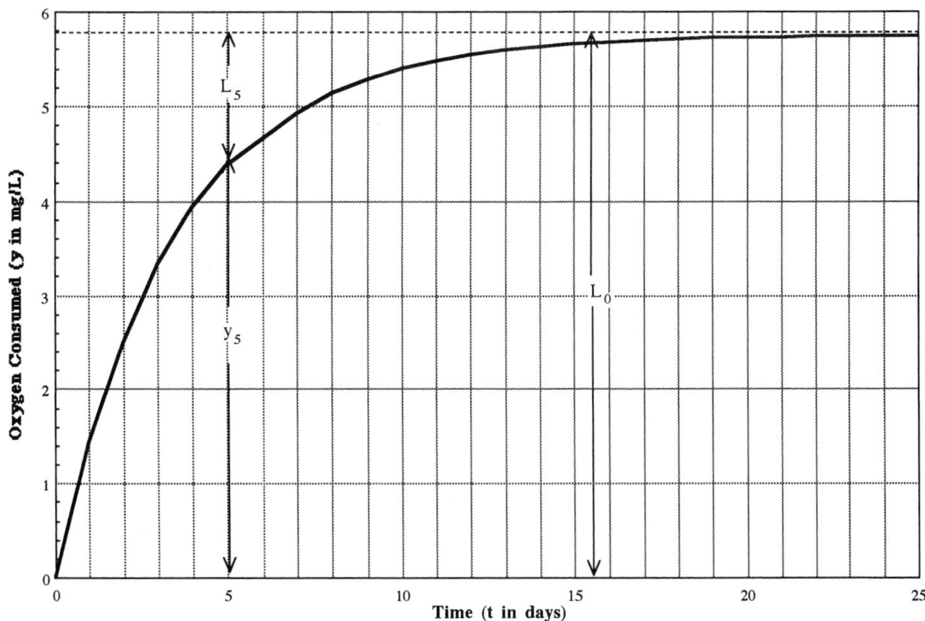

FIGURE 2 Curve representing oxygen consumption versus time during the exertion of first-stage BOD.

A mathematical equation is used as a generalization of the BOD curve. It is assumed that the rate of oxygen consumption at any time is proportional to the BOD remaining. This is expressed mathematically as $dy/dt = ky = k(L_0 - L)$, which integrates to

$$y = L_0 (1 - e^{-kt}) \tag{3}$$

where y is oxygen consumed (in milligrams), t is time (in days), and L is the BOD remaining (in milligrams per liter). The parameters are L_0, the ultimate carbonaceous BOD, and k, the BOD rate constant. This equation indicates an analogy to first-order chemical kinetics if the ultimate carbonaceous BOD remaining is regarded as a reactant. In general, L_0 and k are not constants for any WW but depend on physical and chemical conditions such as temperature, mixing, and pH.

Although there have been several attempts, a convincing theoretical rationale for the assumption that the rate of BOD exertion is proportional to the concentration of ultimate carbonaceous BOD remaining has never been developed. Thus, equation 3 is merely a mathematical description of the curve of oxygen consumption versus time. The description is reasonably accurate if the proper conditions are maintained for the BOD test. However, if the proper organisms are absent, toxic substances are present in inhibitory concentrations, or an inorganic nutrient is not available in an adequate concentration, the course of BOD exertion can be very different from that indicated by equation 3. The parameters are, in fact, empirical. In spite of this, L_0 is often interpreted as the theoretical amount of oxygen required for complete biodegradation of the organic matter present in the WW, and k is interpreted as being analogous to a rate parameter for a first-order chemical reaction. Ascribing this type of theoretical significance to L_0 and k focuses attention on the waste organic matter rather than on the microorganisms which consume the oxygen, opening the way for misinterpretations.

Microbiological Phenomena Which Result in Oxygen Consumption

A major difficulty with assuming a theoretical significance to the empirical BOD parameters L_0 and k is that there are several different microbiological phenomena which occur in a BOD bottle. These include (i) assimilation of organic matter, (ii) endogenous respiration, (iii) cryptic growth, (iv) oxygen consumption by predators, (v) oxygen consumption by nitrifying bacteria, and (vi) oxygen consumption by algae. One assumption does not describe all of these phenomena, and none of them necessarily result in a rate of oxygen consumption which is proportional to the concentration of remaining ultimate BOD; however, the overall effect is a continuous decline in the rate of oxygen consumption, as shown in Fig. 2.

Assimilation of organic compounds by a microbial consortium growing in WW results in a rate of oxygen consumption proportional to the rate at which the organic compounds are utilized. The course of oxygen consumption will be first order only if microbial assimilation is limited by the concentration of organic substrate and not by other factors such as the density of heterotrophic microorganisms, inorganic nutrients, inhibitory substances, and oxygen concentration. The BOD_5 test methodology is intended to produce the conditions under which the concentration of organic substrate is limiting.

The microorganisms which assimilate the waste organic matter are the primary heterotrophs. During growth, some substrates are used to synthesize carbohydrate or lipid compounds, which are stored as food reserves. Assimilation is the first step of BOD exertion and is followed by a period of endogenous metabolism of the primary heterotrophs, which is manifested by oxygen consumption as these microorganisms metabolize the stored food reserves. Endogenous metabolism is a mechanism by which the microbial cells meet their maintenance requirements, i.e., by replacing protein molecules and other cellular components which break down spontaneously. It does not result in an increase in the number of organisms; after the stored food reserves have been depleted, the number of viable organisms decreases because some cellular components are breaking down and are not being replaced. The rate of oxygen consumption during endogenous metabolism is proportional to the density of primary heterotrophs.

As some of the primary heterotrophs die, the cell membranes rupture, and the cell contents spill out. This organic material of the cell contents supports the growth of secondary heterotrophs. Growth when there is an overall decrease in the total number of viable cells present is called cryptic growth. Oxygen consumption due to cryptic growth cannot be differentiated from endogenous respiration of the primary heterotrophs. Secondary heterotrophs also include slow-growing bacteria and/or fungi which are able to assimilate organic compounds from the WW that are not utilized by the primary heterotrophs (so-called recalcitrant compounds). Some species of bacteria may be either primary or secondary heterotrophs, depending on the circumstances.

In addition to the bacteria and fungi which are the primary and secondary heterotrophs, WW contains protozoa and microscopic invertebrates (e.g., rotifers and microcrustacea) which feed on the bacteria and fungi. These predators also consume oxygen. Predation occurs while the heterotrophs are assimilating organic matter, but this type of oxygen consumption is quantitatively much more significant during the endogenous phase. The three simultaneous phenomena of endogenous metabolism of the primary heterotrophs, cryptic growth of the secondary heterotrophs, and predation of protozoa and invertebrates on the bacteria and fungi do not produce a rate of oxygen consumption which is proportional to the BOD.

WW contains ammonia and proteinaceous material which will release ammonia when assimilated. Ammonia can serve as an energy source for nitrifying bacteria. Nitrification exerts an oxygen demand which is independent of the organic content of the WW or FE. The oxygen demand per unit mass of ammonia is very high, theoretically, 4.57 mg of oxygen consumed per mg of ammonia oxidized; the nitrogenous oxygen demand of an FE can be a major factor in the pollution of the RW. However, nitrification in a BOD_5 test used for evaluating WW treatment is considered to be an interference with the determination of the carbonaceous oxygen demand, and nitrification inhibitors are added to the dilution water. Nitrogenous oxygen demand is not included in the model of BOD exertion illustrated in Fig. 2 and equation 3, but its exertion in the RW can have a serious pollution effect.

If biological WW treatment processes include exposure to sunlight, they often result in the growth of significant amounts of algae. This is more common for attached growth processes (e.g., trickling filters), but the effluent weirs of the settling tanks of activated sludge processes sometimes develop long stringers of filamentous algae. Any algal cells incorporated into a BOD sample will, when incubated in the dark, consume rather than produce oxygen, and this oxygen consumption is measured as part of the BOD_5. The algal contribution to the BOD_5 of an FE is not related to waste organic matter originally present in the WW and is not proportional to the ultimate carbonaceous oxygen demand of the FE.

BOD_5 and the Kinetics of Biological Treatment Processes

For kinetic models of biological WW treatment processes, BOD_5 test results are interpreted as a measure of the concentration of organic substrate which will support the growth of heterotrophs. BOD_5 is considered to be a fraction of the first-stage or ultimate BOD (L_0), which is considered to be the oxygen equivalent of the assimilable organic matter. BOD removal efficiency is calculated by equation 1, based on the rationale that organic matter can be removed, although in fact BOD_5 is a parameter which can be reduced but not removed.

For many untreated industrial WWs, the interpretation of BOD_5 as a measure of the concentration of organic substrate which will support microbial growth is a reasonable approximation. Much of the waste material is raw organic matter, and the time of flow in the sewers is so short that little microbial growth occurs before the WW reaches the WWTP. However, by the time the WW has passed through a biological treatment process, the assimilation of organic substrate is complete, and the oxygen consumption measured by the effluent BOD_5 test is the result of endogenous respiration, the activities of secondary heterotrophs, predation, and possibly algal respiration. Clearly, considering FE BOD_5 to be raw WW BOD_5 which was not removed during biological treatment is a misinterpretation.

If the BOD_5 is 70% of the ultimate carbonaceous BOD, the rate of BOD reduction is proportional to the ultimate carbonaceous BOD remaining, and the BOD_5 reduction is specified as 90%, what percentage of the BOD_5 of the raw WW has to be removed? The answer to this question is $100 \times (0.9/0.7) = 128\%$, if it is assumed that the rate constant k is the same for both the WW and FE. This exercise is trivial in terms of evaluating WW treatment; it is presented here only to emphasize the fact that substrate assimilation must be completed during biological treatment and that the BOD_5 of the FE is due to other phenomena.

Kinetic models of biological WW treatment circumvent the conceptual problem of the changing BOD phenomena by specifying that the FE samples should be filtered to remove microorganisms before BOD_5 determination is made. Thus, FE BOD_5 is considered to be a result of the same phenomenon as the WW BOD_5, i.e., assimilation of organic matter not removed during treatment. However, the whole (not filtered) FE is discharged to the RW, and evaluating WW treatment using the BOD_5 of the filtered FE uncouples the evaluation of efficiency and effectiveness of the treatment. The kinetic models used for design of biological treatment processes cannot be used to predict if the FE will be suitable for discharge without adding back the BOD_5 from respiration of microorganisms which were not removed by settling and the nitrogenous oxygen demand.

In the case of untreated municipal WW, the interpretation of BOD_5 solely as a measure of organic substrate to be assimilated during biological treatment is considerably less accurate. The time of flow in the sewers from residences to the WWTP ranges from hours to days, and a great deal of assimilation of waste organic matter can occur during that time. The WW BOD_5, as well as the FE BOD_5, may be primarily due to endogenous respiration, the activities of secondary heterotrophs, and predation. The primary accomplishment of the biological treatment process in municipal WW treatment is the aggregation of the microorganisms which grew in the sewer rather than the assimilation of organic matter.

In spite of the conceptual problems of the kinetic models of biological WW treatment, no currently developed laboratory technique is an apparent candidate to replace BOD_5 as the primary measure of both the efficiency and the effectiveness of WW treatment in relation to the possible deoxygenation of the RW. The use of BOD_5 is well established by an accumulation of decades of experience and recent regulatory usage. However, there is a need for a more widespread understanding of the BOD_5 test as a bioassay and a more sophisticated interpretation of BOD_5 results as used for evaluation of WW treatment.

COD Testing

The COD test was developed as a method for determining, in a short period of time, the oxygen equivalent of the organic content of WW. The procedures for determining COD are described in section 5220 of reference 1. Organic compounds are oxidized by dichromate ions in the presence of acid, heat, and a catalyst, which promote the breaking of aromatic ring structures. The dichromate solution is standardized in terms of equivalent molecular oxygen. Most organic compounds are oxidized completely or nearly so, but volatile compounds may be driven out of the sample before they are oxidized, and some aromatic nitrogen-containing compounds are resistant to the oxidant. Also, interference with the test may be caused by inorganic compounds which can be oxidized.

The COD test measures the oxygen equivalent not only of the waste organic matter but also of microbial cells in the WW. The oxygen demand associated with the microbial cells is only partially exerted during a BOD_5 test. Also, some organic compounds measured by the COD determination may not be metabolized by the microorganisms in either the BOD bottle or the biological treatment process. Thus, COD is not a good measure of the ultimate carbonaceous BOD. The COD value is usually higher than the BOD_5 value in both WW and FE, but in many cases, there is a consistent ratio between COD and BOD_5, although the COD/BOD_5 ratio for an FE may be different from the COD/BOD_5 ratio for given untreated WW. Thus, COD may be used as an index of BOD_5 for adjusting the operation of a biological WW treatment process as the strength of the waste varies from day to day. After 40 years of use, the COD test has not replaced the BOD_5 test for evaluation of either the efficiency or the effectiveness of WW treatment, and there is no indication that it will do so.

TOC Testing

The TOC test is a direct measure of organic content rather than the oxygen equivalent of the organic content. The procedures for determining TOC are described in section 5310 of reference 1. TOC has not replaced either BOD or COD as a WW parameter but is used for different purposes. Determination of the TOC of WW or FE gives a separate measure of the level of organic material in a sample, but it gives no indication of what fraction of this material may be biodegradable. TOC is also used for measuring concentrations of hazardous wastes in water when toxicity or resistance to biodegradation would interfere with the BOD test. The TOC test is quite sensitive and can be used to measure organic carbon levels at the level of micrograms per liter, but neither the BOD_5 test nor the COD determination can be used for that level of contamination. Concentrations in those ranges are seldom of consequence in relation to the deoxygenation of RW unless the contaminants are toxic.

PLANT NUTRIENTS: ALGAL GROWTH TEST

Compounds of N and P are the significant plant nutrients present in WW and FE. There are suitable chemical methods for determining the total amounts of N and P and also for measuring the concentrations of specific compounds containing one or the other. However, not all of the N and P in an FE is necessarily in a form which will stimulate algal growth, and there may be interactions in FE and RW which would affect the availability of the nutrients to the algae. Thus, an algal growth test has been developed to measure the relative potential of an FE to support the growth of algae and aquatic macrophytes (seed plants). Untreated WWs are seldom tested by the algal growth test; i.e., it is used as a measure of the effectiveness, rather than the efficiency, of WW treatment.

The algal growth test can be used to determine what reductions in N and P are necessary to protect the water quality downstream of the WWTP. It can also be used to determine the feasibility of nutrient criteria for the WWTP FE, e.g., a 1.0-mg/liter total P limitation. Algal growth test studies can be used to measure the biologically available N and P in the WWTP FE, as well as assess the trophic status of the RW body upstream and downstream of the point of discharge. Those waters containing >0.015 mg of biologically available P/liter and 0.165 mg of biologically available N/liter are considered to be eutrophic.

The procedure for performing the algal growth test for nutrient studies was developed by the EPA and is called the *Selenastrum capricornutum* Printz algal assay bottle test (57). *S. capricornutum* is now also called *Pseudokirchneriella subcapitata* or *Raphidocelis subcapitata*, but the EPA still uses the name *S. capricornutum*. The EPA has not updated the Printz algal assay bottle test since it was published in 1978, but the test is widely used. The Florida Department of Environmental Protection has developed extensive standard operating procedures for the test which may be viewed online (www.dep.state.fl.us/labs). Section 8111 of reference 1 describes algal assays for evaluating the nutritional status of water, and an algal growth potential test is an official standard of the American Society for Testing and Materials (ASTM) (5). These tests are based on Liebig's law of the minimum, which states that the maximum yield is proportional to the concentration of the nutrient which is present and biologically available in the lowest concentration in relation to the growth requirements of the organism.

The EPA method employs a single test alga, *S. capricornutum*, while *Standard Methods* allows the use of *S. capricornutum* for freshwater and three marine algae. The use of a single test algal species makes comparison of data obtained with different waters easier, and understanding of the growth restrictions increases as more data become available for a single test species. The benefits of using a standard test organism far outweigh the benefits of using an indigenous species, and *S. capricornutum* and indigenous algal species will produce parallel growth yield responses. It is worthwhile to point out certain considerations in conducting algal growth tests. (i) Sample preparation is dictated by the purpose of the study being conducted, but in all cases the sample must be filtered to remove indigenous algae. For nutrient studies, the sample can be autoclaved and then filtered (0.45-μm pore size filter) to determine the amount of algal biomass that can be grown from all nutrients in the water. (If a sample is autoclaved, the sample should be equilibrated after being cooled by bubbling a 1% CO_2 and air mixture to restore the CO_2 lost during autoclaving.) For a study of the effects of complex wastes, the sample is simply filtered and not autoclaved. (ii) Disodium EDTA is added as a chelating agent at a concentration of 1 mg/liter to the culture medium and test water. It was determined empirically that this concentration makes trace metals, particularly iron, biologically available while not complexing other essential nutrients, such as Ca and Mg or heavy metals which may be toxic to algae. (iii) The basic test design to determine which nutrient is limiting involves additions of P, N, or WW, depending on the specific purpose of the test. Adding materials other than the growth-limiting nutrient does not increase the growth yield. (iv) Chemical data necessary to evaluate the assay response include initial pH and total P, orthophosphate, nitrite, nitrate, ammonia, and Kjeldahl nitrogen. If the FE or RW does not support algal growth within ±20% of the limiting nutrient potential, toxic compounds may be responsible.

Biologically available N and P are determined from the dry weight of the maximum standing crop (MSC) of *S. capricornutum* after 14 days of incubation (or whenever the increase in biomass is <5% per day). The test may have to be terminated before 14 days, if the algae exhaust the nutrients in the samples and begin to die. In a typical experimental design, aliquots are taken from each test flask on days 7, 10, 11, and 14. The MSC is proportional to the initial amount of limiting nutrient available, provided that other factors are not growth limiting. The dry weight of the algal biomass can be determined gravimetrically. Alternatively, the dry weight of the algal biomass can be determined indirectly (i) with an electronic particle counter or (ii) by chlorophyll *a* measurements. If biomass is determined indirectly, the results must be converted to an equivalent dry weight using the appropriate conversion factors which have to be developed in each laboratory. Chlorophyll *a* measurement is unsatisfactory for assessing the toxic or stimulatory effects of complex wastes which may absorb and fluoresce in the same spectral region. Turbidity should not be used for algal cell measurements. Each microgram of P per liter will support 0.430 mg (±20%) (dry weight)/liter of *S. capricornutum* if other constituents are not growth limiting. Each microgram of N per liter will support 0.038 mg (±20%) (dry weight)/liter of *S. capricornutum* if other constituents are not growth limiting.

The algal growth test is not normally performed on a routine basis on FE. Results of chemical analyses provide day-to-day information on nutrient loading in the RW. The algal growth test makes it possible to differentiate the amounts of nutrients available for algal growth from the total concentrations of nutrients determined by chemical analyses. Algal growth tests can be used in studies designed to determine what reductions in the N and P content of FE would be necessary to improve water quality downstream of a WWTP. Such studies involve measuring the biologically available N and P in the FE, as well as assessing the trophic status of the RW before and after addition of the FE. A summary of the test preparations, test conditions, and test acceptability criteria for the algal growth test is shown in Table 2.

TESTING FOR EFFLUENT TOXICITY

Toxicity is a property of many WW components, which may be manifest in the RWs when a toxic waste component

TABLE 2 Summary of test preparations, test conditions, and test acceptability criteria for the green alga [*S. capricornutum (P. subcapitiata)*] Printz algal assay bottle test

Test type	Static nonrenewal
Glassware	Clean with a nonphosphate detergent, acid wash (10% HCl), neutralize with a saturated solution of Na_2CO_3, rinse with deionized water; if an electronic particle counter will be used, add a final rise of deionized water that has been filtered through a 0.22-μm-pore-size filter; dry clean glassware at 105°C in an oven
Sample	Collect in nonmetallic containers; perform autoclaving followed by filtration (0.45-μm-pore-sized membrane filter) for nutrient limitation studies; use filtration only for complex wastes such as treatment plant effluents; if pH is <6.0 or >10.0, adjust pH to 7.0 with 1.2 N HCl or 1.0 M NaOH
Sample volume required	1 to 2 liters, depending on test volume
Alga maintenance	Aseptic technique; can be maintained on medium containing 1% agar; start fresh liquid culture for each assay
Test culture flasks	Borosilicate glass, special glassware, polycarbonate, or coated glass for trace nutrient studies
Sample-to-volume ratio	20% sample-to-flask volume ratio is used for flasks shaken by hand daily; maximum permissible sample-to-volume ratios should not exceed 50% in continuously shaken flasks (100 rpm)
Temperature	24 ± 2°C
Light quality	Cool-White fluorescent lighting
Light intensity	4,304 ± 430 lux
Photoperiod	Continuous illumination
Age of test alga	7–10 days
Inoculum preparation	Inoculum suspension of alga is washed twice with deionized water and resuspended in sterile deionized water
Initial cell density in test flask	1,000 cells/ml
Sample treatments	See footnote *a*
Chemical data	Initial pH, total P, ortho-P, NO_2, NO_3, NH_3, and total Kjeldahl N
Gas exchange	Free exchange through foam plugs
Shaking rate	100 rpm or once daily by hand
pH control	pH of sample should be maintained between 6 and 10 to ensure the availability of CO_2
Test duration	14 days (or whenever the increase in biomass is <5% per day)
End point	Growth reported as the MSC in mg (dry wt)/liter which can be determined gravimetrically or indirectly using an electronic particle counter or by measuring chlorophyll *a*
Interpretation[a]	• If MSC(II) > MSC(I) and MSC(III) is not > MSC(I), then P is the primary limiting nutrient • If MSC(III) > MSC(I) and MSC(II) is not > MSC(I), then N is the primary limiting nutrient • If MSC(II) > MSC(I) and MSC(III) is not > MSC(I) and MSC(IV) > MSC(I) and MSC(VIII) > MSC(I), then N is the secondary limiting nutrient • If MSC(III) > MSC(I) and MSC(II) is not > MSC(I) and MSC(IV) > MSC(I) and MSC(VIII) > MSC(I), then P is the secondary limiting nutrient • If MSC(II) is not > MSC(I) and MSC(III) is not > MSC(I) and MSC(IV) > MSC(I) and MSC(VIII) > MSC(I), then N and P are colimited
Biologically available N	MSC(II) in milligrams per liter divided by 38 (the N yield coefficient)[b]
Biologically available P	MSC(III) in milligrams per liter divided by 430 (the P yield coefficient)[c]
Validity and accuracy	For example, for P-limited test waters MSC(I) = MSC(III) = MSC(V) = MSC(VII) (i.e., within ±20%)

[a]I = none = control (C); II = C + 0.05 mg of P/L; III = C + 1.00 mg of N/liter; IV = C + 1.00 mg of N + 0.05 mg of P/liter; V = C + 1.00 mg of Na_2EDTA/liter; VI = C + 1.00 mg of Na_2EDTA + 0.05 mg of P/liter; VII = C + 1.00 mg of Na_2EDTA + 1.00 mg of N/liter; VIII = C + 1.00 mg of Na_2EDTA + 0.05 mg of P/liter + 1.00 mg of N/liter. The averages are compared by Student's *t* test or other appropriate statistical tests.

[b]Each microgram of N per liter supports 0.038 ± 20% mg (dry weight) of *S. capricornutum*/liter (if other constituents are not growth limiting).

[c]Each microgram of P per liter supports 0.430 ± 20% mg (dry weight) of *S. capricornutum*/liter (if other constituents are not growth limiting).

is present in the appropriate concentration range. The discharge of these toxic substances in toxic amounts to the nation's waters is prohibited by the CWA. All dischargers are required to collect and report chemical data as specified in their NPDES permits to comply with limitations on the concentrations of specific chemicals in their FE. However, it is not feasible to specify limits for all the potentially toxic compounds which might be present in an FE. Furthermore, chemical data do not provide knowledge about the biological effects of various chemicals acting synergistically.

Whole-effluent toxicity (WET) is the aggregate toxic effect of an FE measured directly by an aquatic toxicity test, also referred to as a bioassay. Aquatic toxicity tests are performed by exposing aquatic organisms of a particular test species to different concentrations of an aqueous sample. In 1995, the EPA developed WET tests using algae, invertebrates, and fish for use in NPDES permits. In 2002, 10 of these methods were revised and ratified (65–67), that is, reaffirmed to be applicable for use in NPDES permits based on the results of a WET interlaboratory variability study. There are other published standard procedures for conducting bioassays on FE and RW, such as those described in references 3, 4, and 6 and in section 8000 of reference 1. However, the methods published in references 65 to 67 must be used for NPDES permit compliance monitoring.

WET tests are conducted primarily to measure toxicity to aquatic organisms at different trophic levels. These tests do not directly measure any human health effects. There are toxicity tests that can be used to study possible impacts on human health by FE components, but there are no regulatory requirements for conducting human health effect tests on FE or RW, and these tests are normally not done. In contrast to aquatic toxicity tests, living organisms are not used in human health effect tests.

The EPA is implementing an integrated approach to control the discharge of toxic materials into RWs (59). In addition to the collection of chemical-specific data to characterize FE quality, the EPA requires that WET tests be conducted when there is a reason to believe that some components of the discharge could be toxic and that the WQS of the RWs might be violated. The WET approach is a better way to protect aquatic life and human health because as already mentioned, (i) it is simply not feasible to analyze FEs for all possible industrial chemicals, (ii) WET tests can measure toxicity caused by compounds for which the EPA does not have chemical-specific numeric criteria for the protection of aquatic life, and (iii) WET testing enables prediction and avoidance of a toxic impact before the detrimental impact may occur.

EPA also developed biological criteria to be integrated into a water quality-based toxic control program (7, 61). Field surveys (or biological assessments) are performed to assess the biological integrity of the RW body and protect aquatic life. Direct measurements of aquatic life used in these surveys include aquatic vegetation, algae, fish, insects, crayfish, salamanders, frogs, worms, snails, and mussels. If biological diversity is reduced downstream of the discharge in comparison with an otherwise equivalent area upstream, the FE may contain toxic components. The results of the field surveys must be interpreted in conjunction with the results of WET testing and chemical analyses. Thus, a triad of approaches is being implemented to prevent the discharge of toxic chemicals into the nation's waterways.

WET tests may be static without renewal of the test solution, static with renewal of the test solution, or flowthrough. Flowthrough tests are generally preferable to static tests, but it is usually much more cost effective and practical to do static tests with renewals. However, static tests might not be applicable to FEs that have a high oxygen demand or contain materials that (i) are highly volatile, (ii) are rapidly biologically or chemically transformed in aqueous solutions, or (iii) are removed from test solutions in substantial quantities by the test chambers or organisms during the test. A number of factors must be considered to determine the type of test to use. These include NPDES permit requirements, the objectives of the test, FE characteristics, and fluctuations in FE toxicity. Definitive toxicity tests, as opposed to pass/fail tests, are required to assess toxicity and are conducted by exposing test organisms to a minimum of five effluent dilutions and a control. The use of pass/fail tests consisting of a single FE dilution and a control is only suitable for certain screening purposes.

Biological end points used in WET tests conducted for compliance with NPDES permit monitoring include survival, growth, and reproduction. These end points are expressed either as (i) point estimates of the concentration (fraction) of FE in a mixture of FE and dilution water or as (ii) no-observed-effect concentrations (NOECs). Point estimators can be obtained from the results of definitive testing by using a graphical method, probit analysis, or linear interpolation. NOECs are calculated by hypothesis testing.

There is an inherent variability associated with all WET tests because they involve sampling FEs and working with living organisms. Variability can produce false-positive results if the test organisms react to some factor other than toxic components of the FE. Some variability is systematic; that is, it is associated with identifiable sources such as sampling procedures, test conditions (number of test organisms, number of treatment replicates, randomization techniques, the source of test organisms, the type of food used, laboratory environmental conditions, and dilution water quality), maintenance of test organisms and laboratory equipment, analyst experience, methods of statistical analysis, and level of toxicity of the FE. Some variability is random, that is, not associated with an identifiable source. The variability associated with WET tests has not been entirely quantified, but the two categories of greatest concern are analyst experience and test organism condition and/or health (13). Three measures of variability related to WET tests are within-test (intratest) variability, within-laboratory (intralaboratory) variability, and between-laboratory (interlaboratory) variability.

In addressing variability associated with WET tests, the EPA has developed and recommended criteria for use of percent minimum significant difference (PMSD) levels for WET test methods (63). The PMSD is defined as the smallest difference between the control and another treatment that can be determined statistically in a given test. The use of PMSDs, along with other statistical tools, greatly increases the confidence that can be put in the results of a WET test when all test acceptability criteria have been met. Several commercially available software packages such as ToxCalc and CETIS (Comprehensive Environmental Toxicity Information System), both available from Tidepool Scientific Software, McKinleyville, Calif., facilitate the statistical analysis.

The EPA has maintained that accuracy of WET tests cannot be ascertained but that only the precision can be estimated. This is explained by the fact that the WET tests themselves define toxicity; that is, toxicity is a method-defined analyte, and toxicity cannot be independently measured apart from the WET tests. This means that the procedures commonly used in analytical testing to measure the performance characteristic that is termed accuracy cannot be applied to WET test methods. It does not mean that WET tests are inaccurate by their very nature. However, good quality control practices are essential to minimize variability and ensure the collection of credible data. Nevertheless, even with excellent quality control, interpretation of the results of WET tests is sometimes very difficult. The use of best professional judgment is often necessary.

Possible interferences in WET tests must be considered to avoid inaccurate results. Sources of extraneous toxic substances include contaminants in dilution water, glassware, sample hardware, and testing equipment. Improper handling of the FE sample may adversely affect test results.

Pathogens and/or planktivores in the dilution water or FE can cause effects on aquatic species similar to those of toxic FE components. High concentrations of suspended and/or dissolved solids, excessive color, and extremes of pH may mask the presence of toxic substances. In algal toxicity tests, nutrients in the FE or dilution water may confound results. The hardness of the FE and/or dilution water also may affect the results of WET tests.

Criteria that should be used in evaluating an FE for toxicity include the following. First, multiple tests should be conducted; using the results of a single test to characterize the toxicity of an FE may require unnecessarily costly corrective measures or fail to detect a toxicity problem when one exists. Second, a dose-response relationship should be demonstrated; that is, where FE toxicity exists, the rate of survival, growth, or reproduction declines as the test organisms are exposed to increasing concentrations of the toxic substance(s). Third, the effect of the FE on the existing biological conditions in the RW should be addressed.

WET tests using fish (Fig. 3) or invertebrates cannot be conducted with the same ease and frequency as routine chemical analyses or BOD_5 tests. The performance of bioassays requires the culture and maintenance of test organisms, special laboratory facilities including environmental chambers to house test organisms, and trained technicians working under supervision of a biologist or aquatic toxicologist. Each test takes from several days (acute tests) to more than a week (chronic tests) to perform. The number of bioassays that can be performed at one time in a given laboratory is very small compared to the number of chemical tests that can be performed simultaneously in the same amount of space.

Very few dischargers are able to perform conventional bioassays with fish or invertebrates in house, due to the costs of the facilities. Yet dischargers should not have to rely completely on outside testing laboratories to investigate the possible toxicity of their FEs. Cost-effective bioassays that can be performed quickly and easily using microorganisms are needed for this purpose.

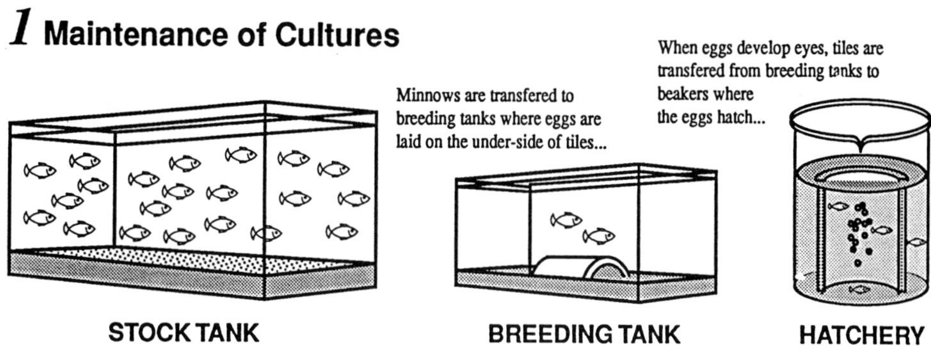

2 Conducting Bioassay 7 day Test

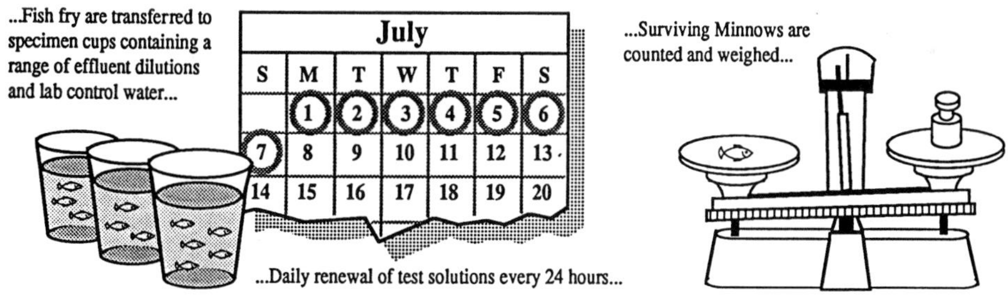

3 Calculate NOEC's for survival and growth using Dunnett statistic

...80% of controls must survive. Average weight of controls must be 0.25 mg.

FIGURE 3 Bioassays for toxicity of wastewater and effluents using fish require large amounts of laboratory space and technician time. Shown are the steps and equipment needed for a *P. promelas* larval survival and growth test.

The conditions of NPDES permits for many municipal agencies require that if toxicity is found to be associated with the effluent, the discharger may be required to conduct a toxicity identification evaluation (TIE) and a toxicity reduction evaluation (TRE). The TIE and TRE may involve a large number of bioassays to identify the cause of toxicity in an FE and approaches to reducing it (58, 62). The important information to be obtained from the bioassays for a TIE or TRE is the relative toxicities of various sources of WW discharged to the sewers. Bioassays using microorganisms are an efficient and economical way to obtain this information.

Bioassays Using Bacteria

Toxicity tests using bacteria have been developed based on growth inhibition, oxygen uptake, colony formation, motility, and bioluminescence. A modification of the standard BOD test has been used for assessing inhibitory effects of WWs (10). A reverse mutation assay using histidine-requiring mutants of *Salmonella enterica* serovar Typhimurium (the Ames test) (39) has been used to assess WW for genotoxicity (41, 51). A standard method for conducting the Ames test is published in section 8030 of reference 1. There is a genotoxicity test marketed by Strategic Diagnostics, Inc. (SDI), Mutatox, which is based on the use of a dark variant of the luminescent bacterium *Vibrio fischeri*. In the Mutatox test, mutations induce the recovery of luminescence (43).

The most widely used WET bioassays using microorganisms make use of luminescent bacteria. There are several standard methods published including (i) an official ASTM standard (2) and (ii) section 8050 of reference 1 (an abbreviated procedure for a bacterial luminescence test). There are several commercially available toxicity tests developed with the luminescent bacterium *V. fischeri*, formerly called *Photobacterium phosphoreum*, and *Photobacterium leiognathi*. Three of these commercial systems using *V. fischeri* (ToxAlert, Merck; LUMIStox, Bruno Lange; and Microtox, SDI) were evaluated by Jennings et al. (28) and found to produce very similar results when used under standardized conditions. Three of these commercially available toxicity tests (ToxScreen-II, CheckLight, Ltd.; BioTox, Hidex Oy; and Microtox) were also evaluated under the EPA Environmental Technology Verification Program (23, 25, 27). Deltatox (SDI) is a portable test system based on Microtox technology. The most thoroughly studied of these systems is the Microtox test.

The Microtox toxicity test employs a specific strain of a marine bacterium, *V. fischeri*, which emits light as an end product of its respiration. SDI cultures these bioluminescent bacteria, harvests and lyophilizes them, and markets them as Microtox reagent. The lyophilized bacteria are rehydrated with Microtox reconstitution solution to provide a ready-to-use cell suspension.

With the Microtox test, these organisms are exposed to a test sample to measure the toxic effect of the sample. The light output of the luminescent bacteria is measured before and after they have been challenged by a sample with unknown toxicity. A difference in light output is attributed to the effect of the sample on the organisms. The light of a control (reagent blank) that contains no sample is used to correct for light lost naturally (not as a result of toxicity). The end point of the Microtox bioassay is expressed in terms of an effective concentration that causes a preselected percentage of light loss at a particular time. For example, the 50% effective concentration (EC_{50}) (5 min) is the effective concentration of a sample causing a 50% decrease in the light output in 5 min.

The Microtox toxicity test system was developed to provide a standardized method for monitoring aquatic toxicity (22). The Microtox reagent (luminescent bacteria) is stored frozen until the test is performed. To perform a surrogate bioassay, the luminescent bacteria are exposed to a WW sample for 30 min. The test can be conducted by an operator after several days of training, with about 25 ft^2 of bench space. The cost of each test is a fraction of that for bioassays that utilize either fish or invertebrates. The Microtox test is an acute test. A chronic version of the test was developed, but it is no longer commercially available.

Kaiser and Palabrica (31) compared EC_{50} from Microtox acute tests with 50% lethal concentration values (LC_{50}) from fish toxicity bioassays for individual chemicals (for definitions, see Table 3, footnote *b*). They found high linear correlations of the Microtox test data with the acute toxicities of a large variety of organic chemicals for several fish, demonstrating that the Microtox test is, at least to some extent, predictive of acute fish bioassay results, making it a good test for screening. Intralaboratory coefficients of variation of 10 to 20% indicate that the Microtox basic test is reasonably reproducible (22). Intralaboratory coefficients of variation for conventional bioassays with invertebrates (*Ceriodaphnia*; Test Method 1002.0) and fish (*Pimephales promelas*, Test Method 1000) were estimated by the EPA (63) to be in the same range or slightly higher.

MetPAD is a bioassay which was developed for the determination of heavy-metal toxicity associated with industrial effluents and other samples (11). The MetPAD kit includes freeze-dried *E. coli* (bacterial reagent), dilution water (deionized and filtered through activated carbon), phosphate buffer, and three locking petri dishes, each containing a yellow filter pad (assay pad) saturated with a substrate for the enzyme β-galactosidase. To perform a surrogate assay, the bacterial reagent is rehydrated and mixed with a sample. At the end of a 90-min exposure period, a drop of the suspension is placed on the assay pad to measure enzyme activity. A purple color gives an indication of β-galactosidase activity, and the intensity of the color is inversely proportional to the sample toxicity. The spot color intensity for a given sample is compared to that of a control prepared by suspending *E. coli* in the dilution water. MetPAD is a qualitative test which employs visual detection of activity, but a semiquantitative assay can be conducted by testing bacterial responses (inhibition of β-galactosidase activity) to several dilutions of the sample.

MetPLATE is a qualitative toxicity test based on the same principle as the MetPAD (12). It is set up by exposing the *E. coli* suspension to samples in 96-well microplates. Inhibition of β-galactosidase activity by a sample is measured by reading absorbance at 575 nm with a microplate reader instead of visually scoring a response on assay pads. MetPAD and MetPLATE were demonstrated to be specific for detecting heavy-metal toxicity (11, 12, 34, 40). Organic components in aqueous samples do not inhibit β-galactosidase activity by the *E. coli* suspension. The use of these tests in conjunction with the Microtox test or other bioassays for general toxicity provides a rapid and inexpensive way to determine whether toxic components in an FE are organic materials or heavy metals.

Conventional WET tests or bioassays and bioassays using bacteria are compared in Table 3. The rationale for the use of surrogate bioassays is to allow more testing to be done, not to replace conventional bioassays. Data from short-term surrogate bioassays and chemical analyses can be analyzed to provide background information about relative

TABLE 3 Comparison of various types of WET tests

Organism (reference[s])	WET test type	Duration	Person-hours[a]	End point	Expression[b]
C. dubia (1, 65)	Acute	48 h	7	Survival	LC_{50}
C. dubia (1, 66)	Short-term chronic	7 days	28	Survival	NOEC
				Reproduction	NOEC, IC_{25}
Daphnia species (1)	Chronic	21–30 days	42–51	Survival	NOEC, IC_{25}
				Reproduction	NOEC, IC_{25}
P. promelas (1, 3, 65)	Acute	96 h	9	Survival	LC_{50}
P. promelas (1, 66)	Short-term chronic	7 days	24	Survival	NOEC
				Growth	NOEC, IC_{25}
S. capricornutum (1, 66)	Short-term chronic	96 h	12	Growth	NOEC, IC(%), S(%)
Mysidopsis bahia (67)	Short-term chronic	7 days	24	Survival	NOEC
				Growth	NOEC, IC_{25}
				Fecundity	NOEC, IC_{25}
Bacteria (ToxTrak reagent) (24)	Acute	2 h	1	Metabolic inhibition	%I
S. enterica serovar Typhimurium (39, 41, 51)	Genotoxicity	48 h	4	Induced reversions	MR
V. fischeri (2, 25, 27, 28, 30) (www.sdix.com)	Acute	5–30 min	1–2	Light loss	EC_{50}
P. leiognathi (23, 28)	Acute	90 min	2	Light loss	LD
V. fischeri (43) (Mutatox; www.sdix.com)	Genotoxicity	24 h	1–2	Light production	MR
E. coli (11)	Heavy-metal toxicity	2 h	1	Enzyme inhibition	MIC
E. coli (12)	Heavy-metal toxicity	2 h	1	Enzyme inhibition	EC_{50}
POLYTOX mixture of micro-organisms (26)	Toxicity assessment	19 and 21 min		Respiration rate (D)	%I
Nitrobacter (10)	Toxicity assessment	4 h	5	Inhibition; NO_2-to-NO_3 conversion	EC_{50}
Sewage effluent organisms; known mixed culture (synthentic sewage); *Pseudomonas* sp., *S. typhimurium*, *E. coli*; BOD inhibition test (all cited in references 10 and 18)	Toxicity screening	A few hours to 5 days	1–2	Respiration rate	EC_{50}

[a]Estimates are based on the time needed to conduct the tests in the Metropolitan Water Reclamation District of Greater Chicago. Time spent maintaining cultures is not included.

[b]LC_{50}, point estimates of the concentration required to kill 50% of the test organisms; NOEC, no-observed-effect concentration; S(%), percent stimulation; IC_{25}, point estimate of the concentration which causes a 25% reduction in the observed response of the test organism; IC(%), point estimate of the concentration which causes a given percent reduction in the observed response; MR, mutagenicity ratio; EC_{50}, point estimate of the effective concentration which causes a 50% reduction in the observed response of the test organism; %I, percent inhibition.

toxicities before WET tests, using higher organisms, are performed. The background information can be used in the design of a testing program to maximize the useful information derived from WET tests.

Comparison of the results of conventional bioassays and Microtox testing on complex FEs has validated the Microtox test as a useful means of identifying toxic FEs, even though in some instances it may be less sensitive (30) or more sensitive (44) than bioassays with higher organisms. The sensitivity of MetPAD has also been compared to that of conventional bioassays (47). EC_{50} for metals tested with the MetPLATE kit were similar to those obtained with the standard *Ceriodaphnia dubia* acute bioassay. (*Ceriodaphnia* is very sensitive to metals.) The results with the MetPLATE test compared poorly with those of bioassays conducted with higher organisms, when toxic organic compounds were tested. However, it is not necessary that surrogate tests have the same sensitivity as fish and invertebrate bioassays for them to be a useful tool in monitoring for aquatic toxicity.

The greatest amount of information about toxic effects can be gained from running a battery of bioassays with different organisms. Although some aquatic species are quite tolerant of certain toxic materials, there is no such thing as a most-sensitive species which could serve for testing the toxicity of a variety of compounds in a particular RW body. A battery of bioassays might include test organisms at three trophic levels, e.g., bacteria, *Ceriodaphnia* species, and fish. Bacterial bioassay data cannot replace data from conventional bioassays, but they are appropriate for screening and may be used in conjunction with other bioassays.

Toxicity Bioassays Using Algae

Algal toxicity bioassays involve exposing a growing culture of an alga to a range of dilutions of an FE or RW and measuring the response in terms of cell density, biomass, chlorophyll content, or optical absorbance in relation to an inoculated control of stock culture. The method developed by the EPA is entitled *Green Alga,* Selenastrum capricornutum, *Growth Test Method 1003.0* (66). In the absence of toxic materials, the algal growth response will be proportional to the initial concentration of the limiting nutrient. Macronutrients (including orthophosphate, nitrate, magnesium, calcium, and sulfate) and micronutrients (including trace minerals chelated with EDTA) are added to the FE or RW being tested so that, at a minimum, it contains the same concentrations of nutrients as the stock medium. If the growth response is significantly smaller in the cultures containing the FE or RW than in the stock culture medium, the presence of a toxic material is indicated. Conversely, a significantly greater growth response in the cultures containing the FE or RW indicates the presence of excessive amounts of nutrients.

Since the EPA developed this bioassay, there has been a debate as to whether EDTA should be included in the culture medium. The concern is that including EDTA would mask the presence of any toxic metals in the FE. In 2001, the EPA found that the use of EDTA improves test method performance by reducing the incidence of false positives and increasing test method precision (64), and it now requires the addition of EDTA to stock nutrient solutions when this bioassay is conducted for NPDES permit compliance. The EPA cautions that the addition of EDTA may cause the *S. capricornutum* growth test to underestimate the toxicity of metals. Regulatory authorities must consider this possibility when selecting test methods for monitoring FEs suspected to contain metals. When tests are being conducted for purposes other than NPDES monitoring, the tests may be conducted without EDTA.

The procedure for performing the *S. capricornutum* growth test with 125- or 250-ml flasks (Fig. 4) differs in some details from the *S. capricornutum* Printz algal assay bottle test discussed above. The *S. capricornutum* growth test is a standardized test which is incubated 96 h after initiation. The selection of the FE test concentrations should be based on the objectives of the study, but the use of a dilution factor of ≥0.5 is recommended by the EPA because it was found to improve the precision of the test (66). A more economical version of the test using 96-well microliter plates has been developed

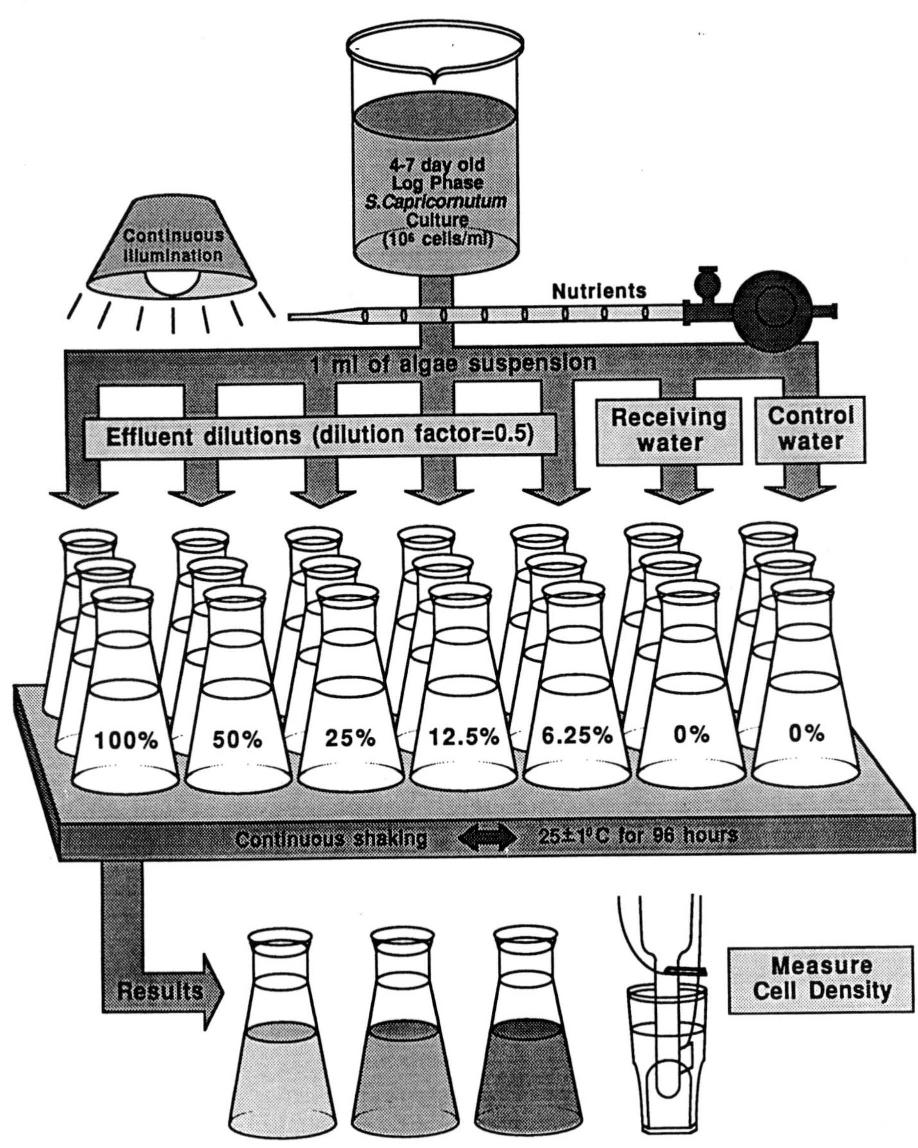

FIGURE 4 Diagram of a basic test procedure for a toxicity bioassay using an alga.

(19). Procedures for conducting toxicity tests with algae are an ASTM standard (4). For research purposes, the exposure period of an alga to the test can be varied. For example, a period shorter than 96 h may be chosen if the fate of some toxic materials (e.g., biodegradation or volatilization) in the test water is an issue. The principles of toxicity testing with algae were reviewed by Walsh (70, 71).

Data are analyzed statistically to determine the concentration of effluent with NOEC on algal growth, using a hypothesis testing approach such as Dunnett's procedure (parametric test) or Steel's many-one rank test (nonparametric test). Linear interpolation can also be used to calculate a point estimate of the effluent dilution that causes a given percentage reduction in the growth of the test organisms. The use of these statistical procedures requires a knowledge of the assumptions upon which the tests are contingent, and the assumptions must be tested. Computer programs and supporting documentation that can be used for the analysis of data from algal toxicity tests, as well as other bioassays, are available from the EPA's Environmental Monitoring and Systems Laboratory, Cincinnati, Ohio.

When biostimulation is observed, the percent stimulation [S(%)] is calculated from

$$S(\%) = 100\left(\frac{T - C}{C}\right) \quad (4)$$

where T is the mean FE or RW response and C is the mean control response. Biostimulation is often observed when algal toxicity tests are conducted on municipal WW and indicates a possible degrading enrichment effect on the RW (66). The elevated nutrient levels in municipal WW can have a confounding effect upon this toxicity test, making the results difficult to interpret and even possibly masking low-level toxicity associated with the FE. For this reason, toxicity testing using algae is seldom a condition of NPDES permits for WWTPs treating WW primarily from residences and commercial sources. Toxicity testing using algal assays is particularly useful for evaluating FEs of WWTPs having a large percentage of industrial wastes in their influents.

Some investigators have criticized the use of these standardized tests, since they may not adequately simulate the natural environment. These scientists have proposed the use of mixed natural algal populations in toxicity tests. This is mentioned in reference 10. While there may be something to this comment, there is no standardized toxicity test using mixed algal populations for WW, and it would be very difficult to develop one.

EMERGING ISSUES

The methods used to evaluate the efficiency and effectiveness of WW treatment described above are the best currently available. However, an accurate assessment of the pollutant load in WW, FE, RW, and biosolids is not currently possible for all pollutants. There is unquestionably a need to improve these methods and develop new methods to study the fate of pollutants not currently regulated under the CWA. Some of the issues being addressed today are mentioned below.

Alternate Indicators for Pathogens

The limitations associated with the use of indicator organisms are increasingly the focus of attention for water quality professionals. A major concern associated with the use of FCs as indicators for assessing the effectiveness of WW treatment is the greater susceptibility of FCs to disinfection than bacterial, protozoal, and viral pathogens. This includes *E. coli*, which is the predominant FC. Therefore, a number of alternate indicator organisms for assessing WW effectiveness have been proposed for NPDES and Part 503 regulatory compliance. These include *Clostridium perfringens*, total culturable organisms, *Bacteroides*, and somatic and male-specific coliphages. Some of these proposed indicators appear to be promising.

It is clear from the scientific literature that monitoring FCs, or any indicator, cannot predict the presence or absence of waterborne pathogens in water with certainty. There are many reported findings of pathogens in the absence of indicators. Two reports recently published by the Water Environment Research Foundation (48, 73) confirm these limitations of indicators and point out the need for methods to directly measure pathogens in WW, FE, and RW. However, it is relatively easy and cost effective to measure the density of indicator organisms, and no concentration is required, as is often necessary in measuring the density of pathogens. Therefore, despite the limitations associated with the use of indicator organisms, estimation of the densities of indicator organisms of pollution will remain a useful measure of the microbial intensity of water pollution, although the presently used indicators themselves may be replaced by alternate indicators. This practice is sound as long as the limitations of the indicators used are taken into consideration.

Enumerating Pathogens

More-sophisticated technologies including immunomagnetic separation, PCR, quantitative PCR, DNA microarray technology, proteomics (using nuclear magnetic resonance), and metabolomics are being exploited to develop methods to directly measure (i) pathogens in WW, FE, RW, and biosolids and (ii) viable but unculturable pathogens surviving in FE and RW. Reference 9 serves as an example. Detection of pathogens in WW, FE, RW, and sludge or biosolids is covered in other chapters in this manual. The technical work involved in detecting pathogens in WW and biosolid matrices includes the following hurdles: (i) determining whether the organisms detected are viable and (ii) overcoming interferences from components in the matrix. The availability of methods to measure pathogens in WW, FE, RW, and biosolids will facilitate conducting risk analyses necessary for making decisions related to WWTP design and expansion and WW reuse. Methodologies for microbial risk assessment that are applicable to WW reuse are discussed in reference 52.

Antibiotic-Resistant Bacteria

The widespread use of antibiotics in medicine and in agriculture has resulted in the contamination of environmental waters and in the emergence of antibiotic-resistant bacteria. The antibiotics and the antibiotic-resistant bacteria make their way into environmental waters by numerous routes, including sewer systems and WWTPs. Since bacteria can transfer genetic information horizontally, especially in nutrient-rich environments, there are concerns that antibiotic resistance genes are being transferred to pathogenic bacteria in the environment, including sewage. Since conventional WWTPs have not been designed to remove antibiotics and antibiotic-resistant bacteria from sewage, this is an area of active research. There are no standard methods for enumerating antibiotic-resistant bacteria in WW. However, standard methods can be modified to enumerate antibiotic-resistant bacteria in WW for research purposes. Examples are given below.

Enumeration of Antibiotic-Resistant Indicator Organisms in WW

The following approach is adapted from the work of Guardabassi and Dalsgaard (20). Standard method (SM) 9222D (1) is a membrane filter procedure for enumerating FC in water, including WW. The density of antibiotic-resistant FCs in WW is determined by incorporating antibiotics into m-FC agar when performing SM 9222D. The concentrations of antibiotics in m-FC agar are chosen in accordance with the breakpoint MICs for definition of resistance in clinical practice. The Metropolitan Water Reclamation District of Greater Chicago is studying the number of antibiotic-resistant FCs in raw WW and FE by this method.

Enumeration of Antibiotic-Resistant Pathogens in WW

An SM for the detection and enumeration of *Salmonella* and *Pseudomonas aeruginosa* in WW was published by the EPA (see appendix G of reference 68). Confirmed *Salmonella* and *P. aeruginosa* isolates obtained by this method can be investigated for antibiotic resistance by the disk diffusion method described by Bauer et al. (8). A similar approach has been used to study antibiotic-resistant *Campylobacter* spp. in WW by Koenraad et al. (32), although there are no published standard methods for enumerating *Campylobacter* spp. in WW.

Organic WW Contaminants in the Environment

Conventional WWTPs were not designed to remove pharmaceuticals, endocrine disrupters, and other organic contaminants from WW. The U.S. Geological Survey measured the concentrations of 95 WW contaminants (pharmaceuticals, hormones, and other organic WW contaminants) in 39 streams across 30 states during 1999 and 2000. These contaminants were found in 80% of the streams sampled (33). These compounds are known to have a deleterious effect on aquatic life. For example, there have been numerous reports of fish with both male and female sex tissue discovered in RWs downstream of WWTPs. How the presence of trace amounts of these compounds in the environment affects human health is not known, but questions have been raised about decreasing sperm counts in humans, the earlier onset of puberty, and subtle birth defects.

Over 100 compounds are now considered to be endocrine disrupters. These can be divided into five categories: steroid compounds (estrogens); surfactants (nonylphenol [NP] and its derivatives); pesticides, herbicides, and fungicides (dieldrin, dichlorodiphenyltrichloroethane [DDT], and 2,4-dichlorophenoxyacetic acid [2,4-D]); polyaromatic compounds (polyaromatic hydrocarbons, polychlorinated biphenols, and brominated flame retardants); and organic oxygen compounds (phthalates, dioxins, and bisphenol A). The presence of estrogen compounds and NP in the environment is causing the greatest concern, due to the potency of these compounds.

Recent studies have shown that the efficiency of antibiotic and endocrine disrupter removal by WWTPs is often very high (17, 55, 56), but it is not complete. Optimization of treatment processes (29) or redesign of WWTPs is required to achieve this goal, and reducing the introduction of these chemicals at their sources, of course, is also necessary. Because the removal of these compounds during WW treatment is accomplished at least in part by partitioning of the WW sludge, there are also concerns that these compounds represent a significant health risk . Steroid estrogens occur in WW at low concentrations (in the range of nanograms per liter) (21). Alkylphenol polyethyloxylates and their associated metabolites, including NP and octylphenol (OP), have been detected in municipal biosolids at levels ranging from 5 to 887 mg/kg and <0.5 to 12 mg/kg, respectively (35).

Bioassays are used in conjunction with chemical fractionation and analytical techniques to study compounds possessing hormone agonist or antagonist activity in WW and WW sludge for research purposes. The EPA has not approved any of these bioassays for regulatory purposes. Brief descriptions of several of these bioassays are given below. Detailed explanations of the mechanisms exploited in the use of these bioassays, the advantages and limitations associated with each type of bioassay, and a discussion of how new technologies such as microarrays will be used to derive greater benefits from the use of these types of bioassays are found in references 74 and 75. Bioassays for environmental pollutants harmful to human health will play an increasingly important role in assessing the efficiency of WW treatment.

E-Screen

The E-Screen is a cell proliferation assay (17, 54). This bioassay makes use of an estrogen-responsive breast cancer cell line (MCF-7 cells) to determine the estrogenic activity of WW sample extracts. The extracts are added to cultures of MCF-7 cells in 24-well tissue culture plates. After 5 days of incubation at 37°C, cell proliferation is measured by the sulforhodamine protein assay. The sulforhodamine assay determines total cell numbers by measuring total protein content. The plates are read at a wavelength of 515 nm on a microplate reader. Standard curves are constructed to determine the relative potency of the target chemicals: 17-β-estradiol, estrone, 17-α-ethinylestradiol, estriol, estrone, testosterone, 4-NP, 4-*tert*-op, 4-OP, and bisphenol.

Recombinant Yeast Estrogen Screen Assay

This bioassay (5, 21, 49) makes use of a recombinant yeast strain (*Saccharomyces cerevisiae*) containing the human estrogen receptor (ER) gene and an expression plasmid carrying the *lacZ* reporter gene to determine estrogenic activity in extracts of WW and biosolids. When the yeast cells are incubated for 3 to 4 days at 32°C in the presence of estrogenic compounds, the *lacZ* product, β-galactosidase, is secreted into the medium and causes the chromogenic substrate, chlorophenol red-β-D-galactopyranoside, to turn red. The color change is quantified by measuring absorbance in the 520- to 640-nm range.

CALUX Assays

See references 38 and 53. Chemical-activated luciferase gene expression (CALUX) bioassays make use of recombinant cell lines containing a nuclear receptor gene linked to the firefly luciferase reporter gene. The ER-CALUX bioassay uses T47-D human breast adenocarcinoma cells, expressing endogenous ERα and β, which are stably transfected with an estrogen-responsive luciferase reporter gene. Following exposure of the T47-D cells to an extract being tested for estrogenic activity, the cells are lysed with a lysis reagent; luciferase activity in an aliquot of the lysate is determined with luciferase reagent and a luminometer. CALUX bioassays making use of other nuclear receptors have also been designed. These include androgens, dioxin, and dioxin-like compounds.

Vitellogenin Gene Expression

See references 36 and 37. The EPA has developed a fathead minnow vitellogenin gene expression (*Vg*) assay for screening WWTP FEs for endocrine disrupters. Male and female fathead minnows (*P. promelas*) are exposed to WWTP FEs for 24 h. Then, their livers are removed, and the RNA is extracted and analyzed via reverse transcription-PCR for changes in *Vg*. Each WWTP FE assessment includes a laboratory control water and a positive control, 5 ng of the estrogen ethinylestradiol/liter. If males show *Vg* levels above controls, this indicates detection of an estrogen in the FE. If females show decreased *Vg* from the controls, this indicates detection of an androgen or antiestrogen in the FE.

Beneficial Use of Wastewater Sludge or Biosolids

The term biosolids has generally replaced the terms sewage sludge and wastewater sludge. The term biosolids is used to describe WW sludge which has been digested and stabilized. In 1996, the National Research Council (NRC) concluded that the use of biosolids in the production of crops for human consumption, when practiced in accordance with federal guidelines and regulations, presents negligible risk to the consumer, to crop production, and to the environment (45). The EPA and the WW industry view biosolids which have been properly treated and tested (referred to as excellent-quality sludge by the EPA) as a valuable resource for land application as an organic soil conditioner and partial fertilizer.

In 2002, the NRC reviewed the Part 503 Rule and issued a report addressing issues related to the land application of biosolids and public health (46). This report affirmed that there is no documented evidence that the Part 503 Rule failed to protect public health. However, the report also stated that additional scientific work was needed to reduce uncertainty about the health issues related to the land application of biosolids. In response to the NRC report, a Biosolids Research Summit was organized by the Water Environment Research Foundation. Scientists at the summit identified a number of research needs related to assessing the effectiveness of WW sludge treatment and the beneficial use of biosolids including (i) characterization of bioaerosols associated with land-applied biosolids, (ii) evaluation of the occurrence of emerging pathogens in biosolids, (iii) sampling strategies for field testing of indicators and pathogens (including bioaerosols), (iv) development of new methods to detect pathogen classes, and (v) evaluation of the possibility of using *E. coli* in place of FC determination in the Part 503 Rule (16).

Research being conducted at the University of Arizona is an example of what is being done to address issues raised by the 2002 NRC report. In response to a report that land applied biosolids may transmit *Staphylococcus aureus*, Rusin et al. (50) analyzed biosolids samples and aerosols obtained during biosolid land application for *S. aureus*. *S. aureus* was not found in any of the treated biosolids or the biosolid aerosol samples, indicating that properly treated biosolids are not a source of human infection for *S. aureus*, i.e., WW sludge treatment effectively eliminates biosolids as a source of infection with *S. aureus*.

More research along these lines is necessary but is hampered by the analytical methods currently available. These methods do not effectively recover pathogens from biosolids, and some of these methods are simply impractical to perform routinely. For example, the helminth analysis performed for Part 503 regulatory compliance (appendix I in reference 68) entails analyzing 300 g (dry weight) of biosolids. Typically, a technician must spend 12 to 15 8-h days just sitting at a microscope and examining a sample concentrate for viable *Ascaris* ova to complete this analysis. In addition to being impractical in terms of technician time, it is not a very interesting task. Thus, the pathogen load in biosolids cannot be easily determined by current methods to perform the risk analyses necessary to demonstrate that properly treated biosolids (excellent-quality sludge) are a beneficial resource and can be safely applied to land. Reference 14 describes a risk-based methodology for assessing the risk to human health from exposure to pathogens via biosolids.

SUMMARY

The efficiency of WW treatment is measured by the amount of a waste material separated from the WW and may be expressed either as a percent removal (equation 1) or as a waste load reduction (equation 2). The waste load reduction is also related to the amount of sludge produced. It is possible to calculate different treatment efficiencies and waste load reductions for different WW components.

In contrast to efficiency, the effectiveness of WW treatment is related to the possibility that the effluent discharge will cause pollution of the RW body without consideration of how much waste is separated from the original WW. Since there are several different parameters of pollution which describe different effects in the RW body, a WWTP may be effective in reducing some types of pollution but ineffective in relation to others. Both the efficiency and the effectiveness of WWTPs need to be evaluated.

Because of the large number and variety of waste materials which are carried away from our residences and businesses by water, it is not feasible to measure the concentrations of all waste materials in an FE from a WWTP, and evaluation of the effectiveness of WW treatment depends to a large extent on the use of a few parameters of pollution. These parameters are aggregate measures of specific pollution effects such as transmission of waterborne disease (indicator organism density), deoxygenation of the receiving waters (BOD_5), and toxicity to aquatic organisms (toxicity bioassay). The individual pollution effects occur independently (although two or more often occur simultaneously), and a thorough evaluation of the accomplishments of WW treatment requires the use of several parameters of pollution.

The magnitudes of the parameters of pollution are reduced during WW treatment as waste materials are removed. Parameters are properties of the WW components, not "things" which can be removed from the water. However, confusion between WW components and their properties in relation to causing pollution can result in misuse of the term removal in place of reduction. This can lead to misunderstanding of the objectives of WW treatment and misinterpretation of the data representing the accomplishments of WW treatment. The major example of this is the widespread use of the term BOD removal, which indicates confusion between waste organic compounds and their property of causing exertion of oxygen demand. Although BOD exertion is the result of microbial metabolism and at least five different metabolic phenomena are involved in BOD exertion, BOD_5 is often mistakenly treated as if it were a material which can be removed from WW.

Microbiological techniques are used almost extensively for evaluation of the effectiveness of WW treatment. The techniques for the detection of pathogenic microorganisms and indicator organisms are described in other chapters in this manual. The methods discussed in this chapter include the BOD_5 test, which is related to the potential deoxygenation of the RWs, the algal growth test, which provides an indication of possible eutrophication effects, and toxicity bioassays using algae or bacteria.

There are a number of emerging issues related to evaluating the effectiveness of WW treatment. These include (i) the need to identify alternate indicators of pathogens in WW, FE, RW, and biosolids; (ii) the need for better methods to detect pathogens themselves in WW, FE, RW, and biosolids; (iii) the challenge antibiotic-resistant bacteria represent to the WWTP industry; (iv) the need to develop bioassays to study the removal of endocrine disrupting and other organic contaminants from WW; and (v) research needs associated with the beneficial use of WW sludge or biosolids.

This chapter was revised for the third edition of this manual by James T. Zmuda. Figures 1, 3, and 4 were prepared by Stephen Brayfield of Brayfield Graphics. We thank Ali Oskouie for valuable comments and suggestions.

REFERENCES

1. **American Public Health Association.** 2005. *Standard Methods for the Examination of Water and Wastewater*, 21st ed. American Public Health Association, Washington, D.C.
2. **American Society for Testing and Materials.** 2005. Historical standard D5660-96(2004). *Standard Test Method for Assessing the Microbiological Detoxification of Chemically Contaminated Water and Soil Using a Toxicity Test with a Luminescent Marine Bacterium.* American Society for Testing and Materials, West Conshohocken, Pa.
3. **American Society for Testing and Materials.** 2005. Active standard E1192-97(2003). *Standard Guide for Conducting Acute Toxicity Tests on Aqueous Ambient Samples and Effluents with Fishes, Macroinvertebrates, and Amphibians.* American Society for Testing and Materials, West Conshohocken, Pa.
4. **American Society for Testing and Materials.** 2005. Active standard E1218-04. *Standard Guide for Conducting Static Toxicity Tests with Microalgae.* American Society for Testing and Materials, West Conshohocken, Pa.
5. **American Society for Testing and Materials.** 2005. Active standard D3978-04. *Standard Practice for Algal Growth Potential Testing with* Pseudokirchneriella subcapitata. American Society for Testing and Materials, West Conshohocken, Pa.
6. **American Society for Testing and Materials.** 2005. Active standard E1563-98(2004). *Standard Guide for Conducting Acute Toxicity Tests with Echinoid Embryos.* American Society for Testing and Materials, West Conshohocken, Pa.
7. **Barbour, M. T., J. Gerritsen, B. D. Snyder, and J. B. Stribling.** 1999. *Rapid Bioassessment Protocols for Use in Streams and Wadeable Rivers: Periphyton, Benthic Macroinvertebrates and Fish*, 2nd ed. EPA 841-B-99-002. U.S. Environmental Protection Agency, Office of Water, Washington, D.C.
8. **Bauer, A. W., W. M. M. Kirby, J. C. Sherris, and M. Turek.** 1970. Antibiotic susceptibility testing by a standardized single disk method. *Am. J. Clin. Pathol.* **53:**149–159.
9. **Berthiaume, F., R. Brousseau, K. Lemarchand, L. Masson, C. Maynard, P. Payment, P. Bayardelle, and J. Harel.** 2005. *Application of DNA Microarray Technology for Wastewater Analysis.* Report no. 01-HHE-1. Water Environment Research Foundation, Alexandria, Va.
10. **Bitton, G., and B. J. Dutka (ed.).** 1986. *Toxicity Testing Using Microorganisms*, vol. 1. CRC Press, Boca Raton, Fla.
11. **Bitton, G., B. Koopman, and O. Agami.** 1992. MetPAD: a bioassay for rapid assessment of heavy metal toxicity in wastewater. *Water Environ. Res.* **64:**834–836.
12. **Bitton, G., K. Jung, and B. Koopman.** 1994. Evaluation of a microplate assay specific for heavy metal toxicity. *Arch. Environ. Contam. Toxicol.* **27:**25–28.
13. **Burton, G. A., W. R. Arnold, L. W. Ausley, J. A. Black, G. M. DeGraeve, F. A. Fulk, J. F. Heltshe, W. H. Peltier, J. J. Pled, and J. H. Rodgers, Jr.** 1996. Effluent toxicity test variability, p. 131–156. *In* D. R. Grothe, K. L. Dickson, and D. K. Reed-Jenkins (ed.), *Whole Effluent Toxicity Testing.* SETAC Press, Pensacola, Fla.
14. **Colford, J. M., D. M. Eisenberg, J. N. S. Eisenberg, J. Scott, and J. A. Soller.** 2003. *A Dynamic Model to Assess Microbial Health Risks Associated with the Beneficial Uses of Biosolids—Phase 1.* Report no. 98-REM-1. Water Environment Research Foundation, Alexandria, Va.
15. **Desbrow, C., E. J. Routledge, G. C. Brighty, J. P. Sumpter, and M. Waldock.** 1998. Identification of estrogenic chemicals in STW effluent. 1. Chemical fractionation and in vitro biological screening. *Environ. Sci. Technol.* **32:**1549–1558.
16. **Dixon, L. G., and P. Field.** 2004. *Proceedings from the Biosolids Research Summit.* Report no. 03-HHE-1. Water Environment Research Foundation, Alexandria, Va.
17. **Drewes, J. E., J. Hemming, S. J. Ladenburger, J. Schauer, and W. Sonzogni.** 2005. An assessment of endocrine disrupting activity changes during wastewater treatment through the use of bioassays and chemical measurements. *Water Environ. Res.* **77:**12–23.
18. **Dutka, B. J., and G. Bitton (ed.).** 1986. *Toxicity Testing Using Microorganisms*, vol. 2. CRC Press, Boca Raton, Fla.
19. **Environment Canada.** 1992. *Biological Test Method: Growth Inhibition Test Using the Freshwater Alga,* Selenastrum capricornutum. Report EPS 2/RM/31. Environment Canada, Ottawa, Ontario, Canada.
20. **Guardabassi, L., and A. Dalsgaard.** 2002. *Occurrence and Fate of Antibiotic Resistant Bacteria in Sewage.* Environmental project no. 722 2002. Danish Environmental Protection Agency, Danish Ministry of the Environment, Copenhagen, Denmark.
21. **Holbrook, R. D., J. T. Novak, T. J. Grizzard, and N. G. Love.** 2002. Estrogen receptor agonist fate during wastewater and biosolids treatment processes: a mass balance analysis. *Environ. Sci. Technol.* **36:**4533–4539.
22. **Isenberg, D. L.** 1993. The Microtox™ Toxicity Test: a developer's commentary, p. 3–15. *In* M. Richardson (ed.), *Ecotoxicology Monitoring.* VCH Publishers, Weinheim, Germany.
23. **James, R., A. Dindal, Z. Willenberg, and K. Riggs.** 2003. *CheckLight, Ltd., ToxScreen-II Rapid Toxicity Testing System.* Environmental technology verification report. ETV Advanced Monitoring Systems Center, U.S. Environmental Protection Agency, Columbus, Ohio.
24. **James, R., A. Dindal, Z. Willenberg, and K. Riggs.** 2003. *Hach Company ToxTrak™ Rapid Toxicity Testing System.* Environmental technology verification report. ETV Advanced Monitoring Systems Center, U.S. Environmental Protection Agency, Columbus, Ohio.
25. **James, R., A. Dindal, Z. Willenberg, and K. Riggs.** 2003. *Hidex Oy BioTox™ Rapid Toxicity Testing System.* Environmental technology verification report. ETV Advanced Monitoring Systems Center, U.S. Environmental Protection Agency, Columbus, Ohio.

26. **James, R., A. Dindal, Z. Willenberg, and K. Riggs**. 2003. *Interlab Supply, Ltd., Polytox™ Rapid Toxicity Testing System*. Environmental technology verification report. ETV Advanced Monitoring Systems Center, U.S. Environmental Protection Agency, Columbus, Ohio.
27. **James, R., A. Dindal, Z. Willenberg, and K. Riggs**. 2003. *Strategic Diagnostics, Inc., Microtox™ Rapid Toxicity Testing System*. Environmental technology verification report. ETV Advanced Monitoring Systems Center, U.S. Environmental Protection Agency, Columbus, Ohio.
28. **Jennings, V. L. K., M. H. Rayner-Brandes, and D. J. Bird.** 2001. Assessing chemical toxicity with the bioluminescent photobacterium (*Vibrio fischeri*): a comparison of three commercial systems. *Water Res.* **35:**3448–3456.
29. **Johnson, A. C., and J. P. Sumpter.** 2001. Removal of endocrine-disrupting chemicals in activated sludge treatment works. *Environ. Sci. Technol.* **35:**4697–4703.
30. **Johnson, I., R. Buffer, R. Milne, and C. J. Redshaw.** 1993. The role of Microtox™ in the monitoring and control of effluents, p. 309–317. *In* M. Richardson (ed.), *Ecotoxicology Monitoring*. VCH Publishers, Weinheim, Germany.
31. **Kaiser, K. L. E., and V. S. Palabrica.** 1991. *Photobacterium phosphoreum* toxicity data index. *Water Pollut. Res. J. Canada* **26:**361–431.
32. **Koenraad, P. M. F. J., W. F. Jacobs-Reitsma, T. Van Der Laan, R. R. Beumer, and F. M. Rombouts.** 1995. Antibiotic suseptibility of campylobacter isolates from sewage and poultry abattoir water. *Epidemiol. Infect.* **115:** 475–483.
33. **Kolpin, D. W., E. T. Furlong, M. T. Meyer, E. M. Thurman, S. D. Zaugg, L. B. Barber, and H. T. Buxton.** 2002. Pharmaceuticals, hormones, and other organic wastewater contaminants in U.S. streams, 1999-2000: a national reconnaissance. *Environ. Sci. Technol.* **36:**1202–1211.
34. **Kong, I., G. Bitton, B. Koopman, and K. Chung.** 1995. Heavy metal toxicity testing in environmental samples. *Rev. Environ. Contam. Toxicol.* **142:**119–147.
35. **La Guardia, M. J., R. C. Hale, E. Harvey, and T. M. Mainor.** 2001. Alkylphenol ethoxylate degradation products in land-applied sewage sludge (biosolids). *Environ. Sci. Technol.* **35:**4798–4804.
36. **Lattier, D. L., T. V. Reddy, D. A. Gordon, J. M. Lazorchak, M. E. Smith, D. E. Williams, B. Weichman, R. W. Flick, A. L. Miracle, and G. P. Toth.** 2002. 17α-Ethynylestradiol-induced vitellogennin gene transcription quantified in livers of adult males, larvae, and gills of fathead minnows (*Pimephales promelas*). *Environ. Toxicol. Chem.* **21:**2385–2393.
37. **Lazorchak, J. M., and M. E. Smith.** 2005. *National Screening Survey of EDCs in Municipal Wastewater Treatment Effluents*. EPA/600/R-04/171. U.S. Environmental Protection Agency, Cincinnati, Ohio.
38. **Legler, J., C. E. van den Brink, A. Brouwer, A. J. Murk, P. T. van der Saag, A. D. Vethaak, and B. van der Burg.** 1999. Development of a stably transfected estrogen receptor-mediated luciferase reporter gene assay in the human T47D breast cancer cell line. *Toxicol. Sci.* **48:**55–66.
39. **Maron, D. M., and B. N. Ames.** 1983. Revised methods for the *Salmonella* mutagenicity test. *Mutat. Res.* **113:**173–215.
40. **Mazidji, C. N., B. Koopman, G. Bitton, and D. Neita.** 1992. Distinction between heavy metal and organic toxicity using EDTA chelation and microbial assays. *Environ. Toxicol. Water Qual.* **7:**339–353.
41. **Meier, J. R., and D. F. Bishop.** 1985. Evaluation of conventional treatment processes for removal of mutagenic activity from municipal wastewaters. *J. Water Pollut. Control Fed.* **57:**999–1005.
42. **Metcalf & Eddy, Inc.** 2003. *Wastewater Engineering: Treatment and Reuse*, 4th ed. McGraw Hill, Inc., New York, N.Y.
43. **Microbics.** 1995. *Microbics Mutatox™ Manual Genotoxicity Test System*. Azur Environmental, Hook, United Kingdom.
44. **Munkittrick, K. R., E. A. Power, and G. A. Sergy.** 1991. The relative sensitivity of Microtox, daphnid, rainbow trout and fathead minnow acute lethality tests. *Environ. Toxicol. Water Qual.* **6:**35–62.
45. **National Research Council.** 1996. *Use of Reclaimed Water and Sludge in Food Crop Production*. National Research Council, The National Academy Press, Washington, D.C.
46. **National Research Council.** 2002. *Biosolids Applied to Land: Advancing Standards and Practices*. National Research Council of the National Academies, The National Academies Press, Washington, D.C.
47. **Nelson, S. M., and R. A. Roline.** 1998. Evaluation of the sensitivity of rapid toxicity tests relative to daphnid acute lethality tests. *Bull. Environ. Contam. Toxicol.* **60:**292–299.
48. **Rose, J. B., S. R. Farrah, V. J. Harwood, A. D. Levine, J. Lukasik, P. Menendez, and T. M. Scott.** 2004. *Reduction of Pathogens, Indicator Bacteria, and Alternative Indicators by Wastewater Treatment and Reclamation Processes*. Report 00-PUM-2T. Water Environment Research Foundation, Alexandria, Va.
49. **Routledge, E. J., and J. P. Sumpter.** 1996. Estrogenic activity of surfactants and some of their degradation products assessed using a recombinant yeast screen. *Environ. Toxicol. Chem.* **15:**241–248.
50. **Rusin, P. A., S. L. Maxwell, J. P. Brooks, C. P. Gerba, and I. L. Pepper.** 2003. Evidence for the absence of *Staphylococcus aureus* in land applied biosolids. *Environ. Sci. Technol.* **37:**4027–4030.
51. **Saxena, J., and D. J. Schwartz.** 1979. Mutagens in wastewater renovated by advanced wastewater treatment. *Bull. Environ. Contam. Toxicol.* **22:**319–326.
52. **Soller, J. A., A. W. Olivieri, J. N. S. Eisenberg, R. Sakaji, and R. Danielson.** 2004. *Evaluation of Microbial Risk Assessment Techniques and Applications*. Report 00-PUM-3. Water Environment Research Foundation, Alexandria, Va.
53. **Sonneveld, E., H. J. Jansen, J. A. C. Riteco, A. Brouwer, and B. van der Burg.** 2005. Development of androgen- and estrogen-responsive bioassays, members of a panel of human cell line-based highly selective steroid-responsive bioassays. *Toxicol. Sci.* **83:**136–148.
54. **Soto, A. M., J. M. Calabro, N. V. Prechtl, A. Y. Yau, E. F. Orlando, A. Daxenberger, A. S. Kolok, L. J. Guillette, B. Le Bizec, I. G. Lange, and C. Sonnenschein.** 2004. Androgenic and estrogenic activity in water bodies receiving cattle feedlot effluent in Eastern Nebraska, USA. *Environ. Health Perspect.* **112:**346–352.
55. **Suidan, M. T., M. Esperanza, M. Zein, P. McCauley, R. C. Brenner, and A. D. Venosa.** 2005. Challenges in biodegradation of trace organic contaminants-gasoline oxygenates and sex hormones. *Water Environ. Res.* **77:**4–11.
56. **Thompson, A., P. Griffin, R. Stuetz, and E. Cartmell.** 2005. The fate and removal of triclosan during wastewater treatment. *Water Environ. Res.* **77:**63–67.
57. **U.S. Environmental Protection Agency.** 1978. *The* Selenastrum capricornutum *Printz Algal Assay Bottle Test*. EPA-600/9-78-018. U.S. Environmental Protection Agency, Corvallis, Oreg.
58. **U.S. Environmental Protection Agency.** 1991. *Methods for Aquatic Toxicity Identification Evaluations: Phase 1. Toxicity Characterization Procedures*, 2nd ed. EPA-600/6-91/003. U.S. Environmental Protection Agency, Washington, D.C.

59. **U.S. Environmental Protection Agency.** 1991. *Technical Support Document for Water Quality-Based Toxicity Control.* EPA-505/2-90/001. U.S. Environmental Protection Agency, Washington, D.C.
60. **U.S. Environmental Protection Agency.** 1994. *A Plain English Guide to the EPA Part 503 Biosolids Rule.* EPA/823/R-93/003. U.S. Environmental Protection Agency, Washington, D.C.
61. **U.S. Environmental Protection Agency.** 1996. *Biological Criteria: Technical Guidance for Streams and Small Rivers,* rev. ed. EPA-822/B-96/001. U.S. Environmental Protection Agency, Washington, D.C.
62. **U.S. Environmental Protection Agency.** 1999. *Toxicity Evaluation Guidance for Municipal Wastewater Treatment Plants.* EPA-833B-99/002. U.S. Environmental Protection Agency, Washington, D.C.
63. **U.S. Environmental Protection Agency.** 2000. *Understanding and Accounting for Method Variability in Whole Effluent Toxicity Applications under the National Pollutant Discharge Elimination System Program.* EPA 833-R-00-003. U.S. Environmental Protection Agency, Washington, D.C.
64. **U.S. Environmental Protection Agency.** 2001. *Final Report: Interlaboratory Variability Study of EPA Short-Term Chronic and Acute Whole Effluent Toxicity Test Methods,* vol. 1. EPA 821-B-01-004. U.S. Environmental Protection Agency, Washington, D.C.
65. **U.S. Environmental Protection Agency.** 2002. *Methods for Measuring Acute Toxicity of Effluents and Receiving Waters to Freshwater and Marine Organisms,* 5th ed. EPA-821-C-02-004. Environmental Monitoring Systems Laboratory, U.S. Environmental Protection Agency, Cincinnati, Ohio.
66. **U.S. Environmental Protection Agency.** 2002. *Short-Term Methods for Estimating the Chronic Toxicity of Effluents and Receiving Waters to Freshwater Organisms,* 4th ed. EPA-821-R-02-013. Office of Water, U.S. Environmental Protection Agency, Washington, D.C.
67. **U.S. Environmental Protection Agency.** 2002. *Short-Term Methods for Estimating the Chronic Toxicity of Effluents and Receiving Waters to Marine and Estuarine Organisms,* 3rd ed. EPA-821-R-02-014. Office of Water, U.S. Environmental Protection Agency, Washington, D.C.
68. **U.S. Environmental Protection Agency.** 2003. *Environmental Regulations and Technology: Control of Pathogens and Vector Attraction in Sewage Sludge.* EPA/625/R-92/013. U.S. Environmental Protection Agency, Cincinnati, Ohio.
69. **U.S. Environmental Protection Agency.** 2003. *Implementation Guidance for Ambient Water Quality Criteria for Bacteria (Draft).* EPA-823-B-03-XXX. Office of Water, U.S. Environmental Protection Agency, Washington, D.C.
70. **Walsh, G. E.** 1988. Principles of toxicity testing with marine unicellular algae. *Environ. Toxicol. Chem.* **7:**979–987.
71. **Walsh, G. E.** 1990. Algae in toxicological testing. *Appl. Phycol. Forum* **7:**5–8.
72. **Williamson, K. J., and D. G. Johnson.** 1981. A bacterial assay for assessment of wastewater toxicity. *Water Res.* **15:** 383.
73. **Yanko, W. A., R. De Leon, P. A. Rochelle, and W. Chen.** 2004. *Development of Practical Methods To Assess the Presence of Pathogens in Water.* Report no. 97-HHE-1. Water Environment Research Foundation, Alexandria, Va.
74. **Zacharewski, T.** 1997. In vitro bioassays for assessing estrogenic substances. *Environ. Sci. Technol.* **31:**613–623.
75. **Zacharewski, T. M., K. C. Fertuck, M. R. Fielden, and J. B. Mathews.** 2002. Mechanistically-based assays for the identification and characterization of endocrine disruptors, p. 559–581. *In* J. P. Vanden Heuvel, G. H. Perdew, W. Mattes, and W. F. Greenlee (ed.), *Comprehensive Toxicology,* vol. 14. Elsevier Science, Amsterdam, The Netherlands.
76. **Zmuda, J. T., R. Gore, and Z. Abedin.** 2004. *Estimation of the* Escherichia coli *to Fecal Coliform Ratio in Wastewater Effluents and Ambient Waters of the Metropolitan Water Reclamation District of Greater Chicago.* Report no. 04-10. Metropolitan Water Reclamation District of Greater Chicago, Chicago, Ill.

Modeling the Fate of Microorganisms in Water, Wastewater, and Soil

CHRISTON J. HURST

28

The natural environment is filled with microorganisms, most of which are natural residents and colonize various ecological niches. These microorganisms either live independently within the environment or live in association with various host organisms. There also are places where and times when microorganisms are released into the environment as the result of human activities. Sometimes, these microorganisms are released with the goal that they will degrade noxious or toxic wastes. At other times, microorganisms are intentionally released with the goal that they will compete against members of the resident microbiota which cause disease in plants or animals. Humans also intentionally release microorganisms such as *Bacillus thuringiensis* and polyhedrosis viruses that are intended to reduce and control populations of insects. Also released are microorganisms that are pathogenic to humans; this discharge often occurs during normal bodily functions such as sneezing, coughing, and defecation. There also is a fear that human pathogens may intentionally be released for purposes of biological warfare.

Those released microorganisms which are associated with human fecal wastes often are directly deposited on the land surface. They also may enter the terrestrial environment as infiltrates or permeates from outhouses and septic tanks, and they are present in raw municipal wastewaters, wastewater effluents, and wastewater sludges which may be discharged onto the land surface for fertilizing and irrigating crops. Wastewater effluents sometimes are applied to the land surface with the intention that eventually the water will percolate into and recharge aquifers, while at other times treated wastewater effluents are directly injected into subsurface aquifers. Just as surface-applied wastewaters percolate into groundwaters, so too can contaminated groundwater find its way back to the surface via springs and seeps as a result of subsurface flow patterns and pressure gradients. We also discharge wastewater directly into surface waters. The end result is that, regardless of where we apply the contaminants, water flow patterns can distribute them to other locations which may be very distant. Included among the microorganisms in fecal wastes are many that are pathogenic for humans; if these pathogenic organisms survive their environmental exposure long enough, then they may be encountered by new human hosts and thereby complete an environmental cycle of disease transmission: from humans through the environment and back to humans.

There is an interest in understanding the aquatic and terrestrial fate of naturally resident microorganisms. Likewise, we as humans also have an interest in monitoring the fate of those microorganisms released into the environment as a consequence of our own activities. Understanding the fate of microorganisms following either accidental or deliberate release into aquatic and terrestrial environments is of key importance when we try to understand and address the potential impact of those microorganisms upon the environment. The goal of this chapter is to explain some of the quantitative approaches which are applicable to understanding the fate of microorganisms after they are released into water and soil.

DILUTION AND ENVIRONMENTAL TRANSPORT OF CONTAMINANTS

Once microorganisms are released into the environment, their fate becomes subject to the conditions within that environment. The most immediately noticeable factor is that organisms become dispersed as they are added into water or soil. To some extent, microorganisms can be transported through the environment along with the soil, either as wind-carried particles or as material that is carried downslope in landslides. As stated above, the movement of released organisms is greatly facilitated by the flow of water, which can occur either on the surface (6, 8, 10) or in the subsurface (2, 5, 32, 45).

Because microbes have an intrinsic electrostatic charge, they adsorb to the surface of charged environmental particulates (14, 28). This is an important factor, because adsorption to particulates can enhance the ability of microorganisms to survive in the environment (20). In the subsurface, water chemistry can affect microbial travel distance (5, 7, 33, 34) and adsorption to the matrix acts to retard the transport of microorganisms (31, 35). Microbes also become impacted on the matrix while being transported by flowing subsurface waters. Additionally, microorganisms can adsorb to gas-water interfaces (40, 41) by hydrophobic mechanisms, a process which happens in unsaturated subsurface zones (42). In surface water, during periods when the water velocity is low, the tendency of microorganisms to adsorb onto suspended particulates facilitates the sedimentation of those organisms. This process carries the microorganisms from the pelagic zone (the water column) to the benthic zone (the sediment). This contributes to the fact that microbe concentrations, with viruses used here as an example, are known to often be much higher in the sediment than in the overlying water column (12). Furthermore, microbial survival may be more prolonged in the sediment than in the water column (3). Thus, the sediments then can serve as a reservoir of microbes, with subsequent scouring of stream bottoms during times of turbulent water flow resulting in resuspension of the particulate-associated microbes (44). Microbes contained in surface water flows may also be removed by impaction onto the sides or bottom of the channel.

Microorganisms released into the environment become susceptible to inactivation by a variety of physical and chemical factors. These include desiccation (21); thermal or pH-related effects upon their biomolecules (1, 20, 23, 30), which may include denaturation; radiation from sunlight (4, 24, 29, 36, 37); and effects of inorganic ions (20). Released populations of microbes may also be subjected to biochemical antagonism by microbial products (16), such as enzymes (29), and to predation (6, 15, 25, 43) by environmental microorganisms. Some released microorganisms can replicate within the natural environment, while others cannot. Under appropriate conditions, microbes may persist for several months or perhaps even years following their accidental or deliberate release into aquatic and terrestrial environments. Some excellent transport models have been developed which take these kinds of factors into consideration.

DESIGNING AND PERFORMING MICROBIAL SURVIVAL STUDIES

From a biological and ecological standpoint, it is always best to study an indigenous population which is openly exposed within the natural environment and free to move within that environment. This approach often can be used for macroorganisms. Individual plants can be identified, and reidentified as necessary, because they are held stationary by their roots; animals are identifiable either by natural or artificially applied bodily markings or by the use of tagging devices which emit electronic signals. However, despite the fact that microbes also may be tagged or stained to enhance their noticeability (13), microbes are generally too small to be monitored as individuals especially when allowed unrestricted movement within the natural environment. Thus, we often must resort to compromises when it comes to experimental design and this is particularly true for surface (6, 39) and subsurface (5, 7, 33, 35) aquatic environments.

Designing the Experiments

A series of decisions must be made when a microbial survival study is planned. The major decisions are whether to study the fate of seeded organisms or of an indigenous population and whether the organisms will be studied in place within the natural environment, in environmental material brought into the laboratory, or in an entirely artificial system. The matrix to be studied (soil, water, etc.) also affects the design of the experiment.

Physical Considerations

The most important aspect of studies designed to evaluate the fate of microorganisms is to define the study population, which implies controlling or understanding any possible movement of individuals into or out of that population. Studies of microbial persistence are most manageably performed in the laboratory, although even when this is done with natural environmental materials that have been brought into the laboratory, the observed results are not guaranteed to be identical to findings that would be obtained from observation of natural organisms freely moving in nature and interacting with other organisms within their ecosystem. The other popular alternative is to monitor natural or seeded organisms under a constrained natural exposure. This can be done by placing organisms into the environment under conditions which restrain their movement, such as membrane filter chambers (26, 27, 38) or membrane filter sandwiches (30), artificial basins (21), or even intravenous bags (9). However, it is important to understand the possible impact of any imposed barriers which would limit the metabolic and physiologic activities of the test microbial population. Barriers may affect exposure of the test population to environmental factors such as movement or loss of moisture, gas exchange, movement of chemicals, and radiation from sunlight. Conversely, one of the most important aspects of studies which are intended to mimic environmental conditions within laboratory confines is to first study the natural environment in an attempt to assess and duplicate within the laboratory the effects of natural factors.

Statistical Considerations

When the ultimate goal is to mathematically model the observed fate of microorganisms, it is best to design the experiments in such a way that the data will be amenable to statistical analyses. Also, it must be recognized that variability within the data set being modeled is a crucial factor which can cause even the best-intended analyses to produce spurious or possibly even erroneous results. Some sources of variability can be eliminated by the use of appropriate techniques in designing and performing the experiments. Techniques for eliminating unwanted error include, whenever possible, monitoring concurrently and in a single place the fate of the subject microorganisms in each of the candidate sample materials being examined and under all of the possible combinations of test conditions. This approach provides a full factorial design, which ensures the acquisition of a complete set of data upon which to base statistical models, and also helps to eliminate some of the variation which would occur if the combinations of conditions were examined at different times or in different places. If it is not possible to test all combinations of conditions simultaneously, then the order in which the conditions are tested should be chosen at random. Also, whenever possible, it is preferable to keep the experimental conditions unchanged during the course of the observation (experimentation) period. For example, if at the start of a survival study the conditions are established as a series of time point observations for monitoring the survival of organisms in soil, performed at 23°C and with 30% relative humidity, then both the relative amount of moisture in the system and the temperature should be held constant for the entire period of observation. Otherwise, even if the total amount of water is held constant, changes in temperature will vary the relative humidity. Maintaining a constancy of conditions during the observation period also greatly simplifies the subsequent task of modeling the data. Another important means of eliminating error at the design stage of experiments consists of reducing the amount of contaminating materials, such as deleterious chemicals or even potential microbial nutrients, that inadvertently may be added to the experimental system as a part of the suspending fluid in which the microorganisms are introduced.

Many sources of variability cannot be eliminated, although their impact upon the body of data may be controlled to some extent through randomization. One such source of variability is sampling error. The material being studied should be thoroughly mixed before each sample is removed. Alternatively, if the material has been aliquoted before the survival study is begun or if the quantity of material being studied is very large, then aliquots or sampling locations should be selected in a randomized manner. All samples must be treated in an identical manner, including the performance of any elutions or special handling processes. Daily variability in the sensitivity of sample processing and microbial assay techniques can contribute a substantial amount of variation to the body of data. For this reason, the order in which related samples are processed and assayed may become critical. It is preferable that all samples be processed and assayed either as a single batch or else on a completely randomized schedule. Alternatively, all samples representing the same sampling date should be processed and assayed simultaneously. For example, if three sets of experimental conditions are being studied, with each sampled at time points 0, 1, 2, 3, 4, and 5, then ideally the samples from all 18 time points should be processed and assayed simultaneously to help reduce day-to-day variation associated with the processing and assay procedures. Alternatively, the 18 samples could be processed and assayed independently of one another to help ensure that variation due to reproducibility (or sensitivity) of the processing and assaying techniques is randomly inserted within the data set. Another possibility would be to process and assay all time zero samples as a batch and likewise to batch all time 1 samples together, all time 2 samples together, etc. The last approach can result in artificially high and at times apparently erratic estimates of the changes in microbial titer that occur between sequential sampling times. However, this approach gives a truer estimation of the statistical differences between the sets of conditions. The worst alternative would be to process and assay all samples from the first combination of conditions as one batch, all samples from the second set of conditions as another batch, and all samples from the third set of conditions as yet another batch. While the last alternative yields aesthetically pleasing graphs of the observed titer changes with respect to observation time, it is faulty because it hopelessly intermixes the variability caused by differences between experimental conditions and the batch variability associated with the processing and assay techniques.

Designing the Modeling Analysis

Model equations can be of great value by helping us to understand the interactions which led to the observed results and then providing a basis for anticipating the future outcome of similar situations. The most important thing to remember in developing models is that they are only estimates and have validity within specific ranges. Other important factors which need to be considered are that models can only approximate the relationship between their dependent and independent variables and the possibility that variability which exists within the data set can interfere with attempts at successful modeling efforts. While all of the models presented in this chapter are empirical in nature, it still must be recognized that there are questionable assumptions involved in making statements about cause-and-effect relationships based upon these models. With these stipulations in mind, this section focuses on the use of regression analysis to develop statistical models that describe temporal changes in microbial titer as a function of experimental parameters and water or soil characteristics.

A microbial population can either increase, remain constant, or decrease during an observation period. These dynamics are affected by environmental factors such as temperature, desiccation, nutrients, and environmental transport. Because most microorganisms are very small, the results of microbial fate studies are commonly presented and discussed in terms of the microbial population taken as a whole. Most often, those populations of microorganisms that are released into the natural environment are fated to die there. Free-living microorganisms, including algae and most bacteria, fungi, and protozoa, that are released into the environment usually cannot successfully compete for nutrients with the populations of native organisms that already are established in that environment. Other biological entities, such as viruses and some protozoa, whose replication is dependent upon finding a suitable host organism may succeed in establishing an ongoing infection within the host population, but even then, most of the microorganisms released from an infected host are fated to lose viability before they establish suitable contact with a new host. The remainder of this chapter is devoted to mathematically examining the rate at which populations of microorganisms die off following their release into the natural environment. The simplest format for representing the die-off of populations is to chart a percent reduction with respect to time (1, 2). Mathematically, the die-off of a microbial population is analogous to the decay associated with radioisotopes, in that the frequency of individual death or decay events is assumed to occur at a statistically calculable rate. Because of this analogy, the rapidity with which the members of a microbial population die off is often expressed as an exponential decay rate described as "D" or "T_{90}" values (1, 24), either of which indicates the time required for a 90% or 1 log base 10 reduction. More progressively, we can describe this as a mathematical function calculated using an exponential decay model, and the results can be termed "microbial population decay rates" (12, 22).

The Basis of Survival Calculations

The following is a general exponential decay equation in which N represents the number, or concentration, of microorganisms, t is time, and K is a rate constant:

$$\frac{dN}{dt} = -KN \tag{1}$$

The sign associated with the rate constant is negative, indicating a net loss in titer with respect to time. According to this equation, the amount of microbial titer decrease observed within a given time period is dependent upon the number of organisms present at the start of that time period. Integration of equation 1 yields

$$\frac{N_t}{N_0} = e^{-Kt} \tag{2}$$

where N_0 is the titer at time zero (the outset of the experiment or series of observations) and N_t represents the titer at some subsequent observation time (t). Performing a logarithmic transformation of equation 2 yields equation 3:

$$\log_n \frac{N_t}{N_0} = -Kt \tag{3}$$

Equation 3 suggests that the rate of microbial die-off during the interval between times zero and t is log linear with a constant value equal to K.

Linear Regression

The relationship shown in equation 3 can be examined using simple linear regression as defined by equation 4:

$$Y = \beta_0 + \beta_1 X_1 \quad (4)$$

Whenever simple linear regression is used, Y represents the dependent, or response, variable; X represents the independent variable; β_1 is the coefficient associated with X_1, which results from regressing Y with respect to X_1; and β_0 represents the y-axis intercept value, which can also be viewed as an error term. Linear regression is a concept which is simple to understand and produces results that can be readily visualized and easily compared. It must be understood that linear regression imposes constraints on the developed model equations and does not always provide the best fit in terms of describing the relationship between the dependent and independent variables. Nevertheless, linear regression can be a very valuable tool in the study of microbial activity. Interrelations between the independent variables should be eliminated from the models, or else they should be recognized and their influence should be understood within the general framework of the developed models.

Linear regression analysis of microbial survival studies is usually performed with the dependent variable being the $\log_{10}$ reduction in titer since the initial time of observation and the independent variable being elapsed time, and the regression is performed by a least-squares technique. This application of simple linear regression to data from microbial survival studies is shown in equation 5:

$$\log_{10}\frac{N_t}{N_0} = \beta_0 + \beta_t t \quad (5)$$

This equation expresses Y in terms of titer change following a $\log_{10}$ transformation presented as $\log_{10}(N_t/N_0)$ as a function of time, t. Figure 1 shows how this equation appears when graphed. The slope of the solid line, βt, is expressed in terms of $[\log_{10}(N_t/N_0)]/t$ and is analogous to K. The dashed lines shown in Fig. 1 illustrate the effects described as shouldering, for which a lag time exists before the die-off rate becomes exponential, and tailing, for which die-off becomes less than the exponential rate at the end of the observation period. Individual plots of data from microbial die-off experiments may express shouldering, tailing, both effects, or neither effect.

Figure 2 presents the use of linear regression, as defined by equation 5, to describe the loss of titer for indigenous viruses contained in aerobically digested wastewater sludge following land surface disposal. The individual points

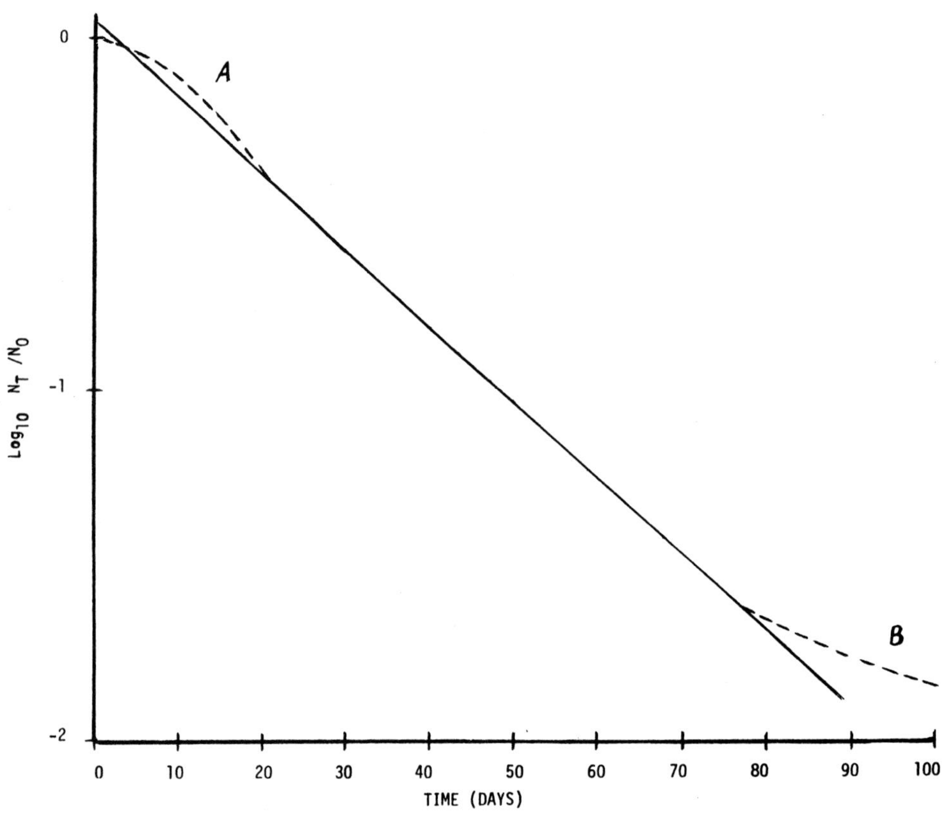

FIGURE 1 Visual presentation of equation 5. Log_{10}-transformed titer ratio values are used as the dependent variable; these values are regressed linearly with respect to time (t) as the independent variable. The solid line is the slope, β_t, which represents a rate expressed as $[\log_{10}(N_t/N_0)]/t$. β_0 is the point where the solid line intercepts the y axis. The dashed lines demonstrate deviations from log linearity due to shouldering (A) and tailing (B).

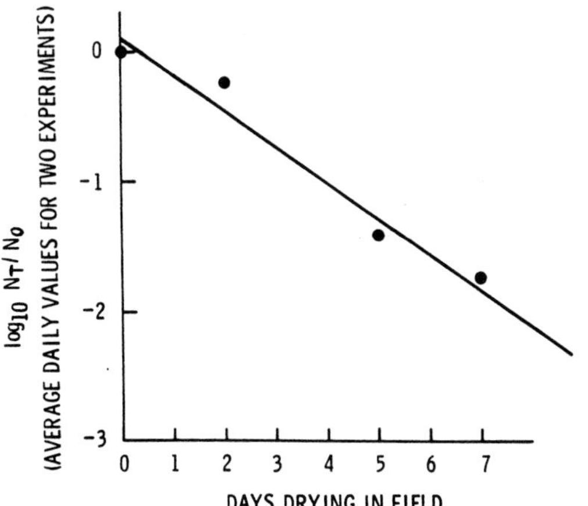

FIGURE 2 Graph of equation 5 applied to data points representing averages from two experiments that monitored the die-off of natural virus populations in aerobically digested wastewater sludge solids after land surface disposal (19).

shown on the graph represent average values from two independent trials each for days 0, 2, and 7 and a single value for day 5. Figure 2 shows that when replicate survival experiments are performed and the data are analyzed, the experimental results may be approximated reasonably well by using the log-linear approach represented by equation 5.

Figure 2 shows that when microbial survival is expressed as the $\log_{10}$ of the titer ratio N_t/N_0 (called the survival ratio), the titer at the beginning of the experiment (time zero) is plotted as an N_t/N_0 value of 0. A $\log_{10}(N_t/N_0)$ value of -1 then equals 10% of the initial titer, a value of -2 equals 1% of the initial titer, and a value of -3 equals 0.1% of the initial titer. In Fig. 2, the units of time are presented as days. However, depending upon the duration of any given experiment, the appropriate time units might range from seconds to even years.

The graphed function represented by the linear regression line shown in Fig. 2 does not pass through any of the graphed points. In fact, these graphed points possibly suggest a sigmoidal equation, the result of both shouldering and tailing. However, using the simple linear regression approach for modeling microbial die-off often suffices to describe the data and can yield a great deal of information about those environmental factors and processes which led to the observed experimental findings. A measure of the appropriateness of using this type of linear regression approach can be derived by assessing the values of r and P calculated for the regressions performed on the $\log_{10}$-transformed data from individual experiments. Very often, the calculated values for r are greater than 0.9 and can range up to greater than 0.98; the corresponding P values are often less than 0.0001 (18). In most cases, the calculated β_0 value is not 0. This probably often reflects the result of random variation in the data points and also the outcome of attempting to use a linear equation format to accommodate data that demonstrate nonlinear traits such as shouldering and tailing. Also, when the change in titer of the population is very small or the titer does not change at all with respect to time, the sign of the value β_t can sometimes be positive. This presumably results from an impact of random variation obscuring any clear trend in titer decrease.

Figure 3 presents an example of what microbial survival data can look like when the results of individual survival experiments are plotted in this fashion. Notice that the actual graphed values are connected sequentially in a dot-to-dot manner. The only time when a line that does not

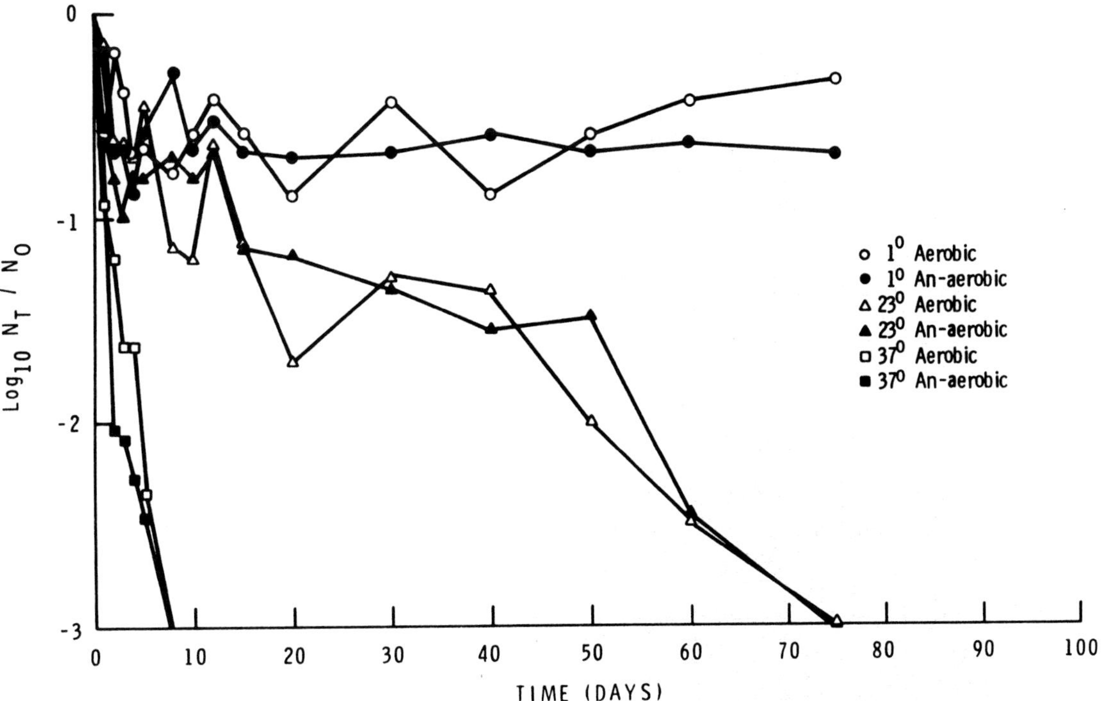

FIGURE 3 Plot of data points for the comparative survival of seeded viruses incubated in sterile soil (20).

connect the graphed points in a dot-to-dot manner should be drawn is when that line represents the plot of a calculated mathematical function. In Fig. 2, the rate of viral inactivation reflects the influence of numerous uncontrolled environmental factors such as drying, thermal inactivation, and microbial antagonism. These factors can be sorted out if the incubation conditions are experimentally controlled. Sets of slope values derived from linearly regressing the results of individual survival experiments conducted under controlled survival conditions can be compared using *t* tests and analysis of variance to determine whether differences in experimental conditions had a statistically significant effect on the rate of microbial die-off (16).

The type of linear regression shown in equation 5 can also be used to determine whether data sets derived from two different studies are comparable and thus can potentially be combined into a single data set. Table 1 shows such an example, comparing the data set represented in Fig. 2 (19) with a data set generated by Farrah and coworkers (11). Both of these studies examined the fate of indigenous viruses in aerobically digested sludge solids following land surface disposal. These two studies were performed during the same season of the year in geographical areas that have roughly similar environmental conditions. The first two regressions in Table 1 represent changes in titer for the indigenous virus populations as the dependent variable [expressed as $\log_{10}$-transformed titer ratio values, $\log_{10}(N_t/N_0)$], against days of drying in the field as the independent variable. The individual data set from the study by Farrah et al. (11) is shown in the first regression. The data set from Hurst et al. (19) is shown in the second regression. The third regression listed in Table 1 shows the result obtained by pooling the sets of viral titer data from these two different studies and then regressing the $\log_{10}$-transformed titer ratio values in that pool as the dependent variable against time of drying in the field. This table also presents the respective Pearson correlation coefficients (*r* values) and statements of probability (*P* values) that the regression values expressed are different from zero. It can be seen that the regression values for both the slope and *y* intercept generated from the pooled data set were intermediate between those corresponding values yielded when the two sets of data were regressed individually. In addition, it can be seen that regression of the combined set of data still generated a strong *r* value and produced a lower (better) *P* value than did regression of the two sets of data independently. Thus, these two sets of data appear to be compatible (17).

Thus far, the examples of data regressions presented in this chapter have utilized the titer ratio as the dependent variable versus time as the independent variable. Linear regressions, as shown in equation 4, can also be performed using the slope, β_t, from equation 5 as the dependent variable. This requires a two-step regression technique (20). The first step consists of deriving inactivation rate values in the form of $[\log_{10}(N_t/N_0)]/t$, which is accomplished by regressing $\log_{10}$-transformed titer ratios $\log_{10}(N_t/N_0)$ from various sampling dates as the dependent variable versus the length of time that the organisms were incubated in the test material as the independent variable (equation 5). The slope values (β_t) derived from this first step of the regression technique are the inactivation rate values, and these are then used as the dependent variable in a second step of analysis, in which they are linearly regressed (using equation 4) against a single independent variable, typically an environmental factor such as soil moisture level or soil pH, insolation, or a soil or water characteristic.

When an individual environmental factor, or a single characteristic of the soil or water, is regressed as the independent variable against the inactivation rate values as the dependent variable, the equation employed to model the relationship takes the form of equation 4. That individual factor or characteristic would be represented as X_1 in equation 4, with the relationship between the rate values ($[\log_{10}(N_t/N_0)]/t$) derived as β_t in equation 5 and then inserted as Y in equation 4 and that individual factor or characteristic being represented as β_1 in equation 4 (20). The resulting value for β_1 is analogous to the rate constant, *K*, as shown in equation 1.

The inactivation rate values ($[\log_{10}(N_t/N_0)]/t$) can also be regressed against a multiple number of independent variables that represent various environmental factors or characteristics of the soil or water that is being studied. One approach for doing this type of comparison is to use a multiple linear regression as presented in equation 6 (17, 20):

$$Y = \beta_0 + \beta_1 X_1 + \beta_2 X_2 + \ldots + \beta_n X_n \qquad (6)$$

In such equations, Y is the dependent, or response, variable and has the form $[\log_{10}(N_t/N_0)]/t$; β_0 represents the *y*-axis intercept; X_1 through X_n represent the different independent variables; and β_1 through β_n represent the coefficients assigned to those different independent variables. This approach thus presents the inactivation rate constant, *K*, as being the sum of several rate constants which reflect the impact of different independent variables, as described in equation 7:

$$K = k_1 + k_2 + k_3 + \ldots + k_n \qquad (7)$$

The multiple-regression equation is easiest to interpret if the different independent variables are unrelated. However, in practice, this rarely happens. For this reason, when candidate independent variables are selected for inclusion into a multiple linear regression model, it is important to first examine the set of all possible independent variables for the presence of any strong cross-correlations between them. This examination can be done quite easily by using Pearson's test (18). If two or more of the candidate variables

TABLE 1 Regression equations describing data for inactivation of indigenous viruses following land surface spreading of sludge[a]

Description	Regression equation: Slope	Regression equation: *y* intercept	*r* value for regression	*P* value for regression
Data from study by Farrah et al. (11)	0.24325	0.20603	0.984	0.002
Data from study by Hurst et al. (19)	0.28769	−0.10325	0.950	0.001
Both of the above analyzed as a single group	0.26601	0.03114	0.959	<0.0001

[a]Linear regression analysis was performed with $\log_{10}$ reduction in titer since day of spreading as the dependent variable versus number of days since sludge was placed on land as the independent variable (17).

correlate with one another in a highly significant manner, then it is probably better to allow only one of those variables to be included in a multiple-regression model. Having first examined the candidate independent variables for correlation with one another and then eliminated the potential problem of including highly cross-correlated variables in a model, one then selects which of the remaining candidate variables actually will be incorporated into the final multiple-linear-regression model equation. Two approaches can be used to make this selection. The first is to build the regression model by using a stepwise technique beginning with the candidate independent variable demonstrating the strongest correlation with the dependent variable (inactivation rate). Additional individual independent variables can then be incorporated successively to improve the model fit. When one is developing such models, it is helpful to establish criteria for deciding when to stop adding additional independent variables to the multiple regression equation. One possible criterion is to include at least one independent variable and to stop adding more variables when the next to be incorporated in sequence would increase the P value for the overall model to above 0.05, produce no improvement in the overall P value for the model, or contribute less than 0.001 to the overall r value for the model. Another approach would be to predetermine which variables to include in the equation based upon other analyses, such as the use of Pearson's correlation coefficients to assess the relationship between candidate independent variables and the inactivation rate values.

An example of using the stepwise approach for selecting multiple independent variables is shown in equation 8. This is the best stepwise regression fit for modeling virus survival based on the physical and chemical characteristics of nine soils (20).

$$\begin{aligned} Y = 0.1005 &+ (0.0025)(\text{viral adsorption to soil}) \\ &- (0.0008)(\text{extractable phosphorus}) \\ &- (0.0007)(\text{exchangeable aluminum}) \\ &- (0.0510)(\text{saturation pH}) \end{aligned} \qquad (8)$$

In this model equation, Y is represented as $[\log_{10}(N_t/N_0)]$/day, and four independent variables are included: the extent of viral adsorption to the particles of a test soil and the levels of resin-extractable phosphorus, exchangeable aluminum, and saturation pH for each of nine soils. This equation summarizes results obtained by modeling the survival of seeded viruses incubated aerobically, under nonsterile conditions, at a single temperature (23°C) in nine different characterized soils (20). Both the choice of which soil characteristics to include and the order in which those soil characteristics were inserted as variables into equation 8 were determined by stepwise multiple linear regression.

Another example of this approach, using a two-step regression technique to represent inactivation rate values as a function of multiple independent environmental factors or sample characteristics, can be shown for a study in which the survival of seeded viruses was analyzed in samples of water from five sites. Seeded viruses were incubated independently at different temperatures (−20, 1, and 22°C) in portions of water from those five different sites under aerobic, nonsterile conditions (18). Samples of the water collected from each of the different sites were chemically analyzed. The survival data from that study were first regressed using equation 5. The viral inactivation rate values, expressed in terms of $[\log_{10}(N_t/N_0)]$/day, derived from that first step of regression analysis, were then regressed as the dependent variable (equation 4) versus each of the individual water characteristics. From a list of water characteristics that were evaluated in this way, five characteristics were found to correlate with viral inactivation rates (18). These characteristics were hardness, conductivity, turbidity, suspended solids, and the capacity of indigenous nutrients present in the water samples to support the growth of seeded bacteria.

Stepwise multiple linear regression was used to determine the best fit for describing virus survival in these waters at the individual incubation temperatures of −20, 1, and 22°C. Of the five characteristics which, when tested individually, were found to correlate with the inactivation rate values, hardness and conductivity correlated so strongly with one another that only one of the two was subsequently used in any given model. Likewise, turbidity and suspended solids correlated so strongly with one another that only one of these two was subsequently allowed to occur in any given model. When a stepwise multiple linear regression was performed by using the inactivation rate values from incubation at 22°C (equation 9) as the dependent variable versus the five characteristics as potential independent variables, two of those characteristics were included in the model. They were hardness and the mean number of generations of growth which the water could support for three different bacterial species after the bacteria were independently seeded into sterilized or pasteurized samples of water from those five different sites. This second variable has been abbreviated as "mean gen PEK," since the bacterial species used were *Pseudomonas fluorescens*, *Enterobacter cloacae*, and *Klebsiella oxytoca*. All three of these bacterial species are heterotrophic organisms common to water samples.

$$\begin{aligned} Y = 2.04 \times 10^{-2} &+ (1.37 \times 10^{-3})\,(\text{hardness}) \\ &- (2.93 \times 10^{-3})(\text{mean gen PEK}) \end{aligned} \qquad (9)$$

In this model equation, Y is represented as $[\log_{10}(N_t/N_0)]$/day. This model equation suggests that the factors which significantly affected survival of the test viruses in these waters at 22°C were the level of water hardness and the level of nutrients available to support microbial growth. Stepwise multiple-linear-regression equations that were similarly derived for the other two incubation temperatures revealed that the factors which affected viral survival at 1°C were turbidity and the level of nutrients available for supporting growth of one of the three organisms, *K. oxytoca*. The only factor which appeared to have affected survival at −20°C was turbidity.

Comparing the Accuracies of Equation Formats

The linear modeling approach requires two major assumptions. The first assumption is that any differences, or distances, between the plotted data points and the graphed equation (equation 5; Fig. 2) are due to error. The second assumption is that the use of linear modeling is appropriate for the data being examined. Sources of possible error include any random variability associated with the development of ecological conditions during the course of the experiment and variability associated with sample collection, sample processing, dilutions, and the assay system. One way of testing the first assumption, and in the process comparing the basic accuracy of the equation format that is used for modeling the microbial population decay rates, is to first use the observed experimental values to generate a model and then use that model to generate a corresponding set of predicted values and to compare the sets of predicted versus observed values by means of simple linear regressions for which Y and X represent the predicted and observed values, respectively (22).

During the first part of this analysis, when the model equations are generated, Y represents the actual experimentally determined titer ratio values (N_t/N_0). During the second part of the analysis, when those model equations are used to predict the outcome of the study, the predicted titer ratio values represent estimates of the expected value of Y, $E(Y)$. If an ideal equation format existed, then the set of predicted values should be a duplicate of the set of experimentally observed values. This duplication does not occur, in part because it is impossible to identify and include all pertinent environmental variables in a mathematical model and also because it is impossible to predict the extent of those errors in each titer ratio value that would be caused by such factors as random variation, experimental variation, and variation associated with sample processing and assaying.

If an ideal equation format existed, and if it were possible to incorporate and model the effects of all sources of variability, then one could derive an ideal model. In such a case, if the set of values predicted by the model were linearly regressed against the set of experimentally observed values, then the equation yielded by that regression would have a slope of 1.0 and a y-axis intercept of zero. For this particular analysis, one could thus consider a line whose slope is 1.0 with a y-axis intercept of zero to be an ideal line. The absolute numerical difference between the slope of that ideal line and the slope of the line derived by linearly regressing the actual sets of predicted and observed values has been used to assess the basic accuracy of the equation format (22). Hurst et al. (22) published eight equation formats which were compared for their accuracy in modeling a common set of experimental data. According to that assessment, the most accurate equation format is the one which yields the lowest value for the absolute numerical difference between the calculated slope values from regression of predicted versus observed values and the ideal slope value of 1.0. An equation format termed "multiplicative error II" proved best in this regard. This equation format is shown in equation 10.

$$Y = \beta_0 X_1^{\beta 1} X_2^{\beta 2} \ldots X_n^{\beta n} t^{\beta t} \tag{10}$$

In that test, Y was the titer ratio expressed as N_t/N_0, X_1 through X_n represented experimental variables, and t represented time. β_1 through β_n represented the corresponding regression coefficients calculated for the variables X_1 through X_n, and β_t represented the coefficient calculated for time, t. Essentially, in this equation, the rate constant, K, is represented as being a product of several different contributing rates. This can be represented as equation 11:

$$K = k_1 k_2 k_3 \ldots k_n \tag{11}$$

To develop the models using the multiplicative error II equation format, equation 10 was subjected to a $\log_{10}$ transformation to yield equation 12, and the models were created by using a linear regression procedure.

$$\log_{10}\frac{N_t}{N_0} = \log_{10}\beta_0 + \beta_1 \log_{10} X_1 + \beta_2 \log_{10} X_2 + \ldots + \beta_n \log_{10} X_n + \beta_t \log_{10} t \tag{12}$$

The regression shown in equation 12 can be performed in a stepwise manner provided that the obligatory term $\beta_t \log_{10} t$ is included before $\beta_1 \log_{10} X_1$.

The accuracy of the equation format represented by equation 12 indicated that when survival data are plotted with $\log_{10}(N_t/N_0)$ as the dependent variable and time as the independent variable (equation 5; Fig. 2), the deviation (distance) between the individual graphed points and the linear regression line is not due entirely to random variability. This finding also suggests that rather than using the linear approach shown in equation 5 as the basis for modeling survival data, it may be better to use the following equation:

$$\frac{N_t}{N_0} = \beta_0 t^{\beta t} \tag{13}$$

after it has been $\log_{10}$ transformed to generate

$$\log_{10}\frac{N_t}{N_0} = \log_{10}\beta_0 + \beta_t \log_{10} t \tag{14}$$

Values of β_t derived by regressing data from individual experiments by using equation 14 could be grouped and then compared and contrasted as sets by using analytical techniques such as t tests and analysis of variance in the same manner as described above in reference to β_t values derived with equation 5. That study (22) also demonstrated that the effects of both temperature and time were best modeled by using temperature and time as independent variables, thus allowing temperature and time to have coefficients which take the form of exponents. Perhaps the most important implication from that study is that the influence of thermal effects, as measured by incubation temperature, and the relationship of titer ratio to time are both exponential rather than linear functions.

The use of ranking tests is one of the simpler approaches which can be used to determine whether data should be modeled using linear regression. If a correlation between titer ratio as the dependent variable and some independent variable such as temperature is statistically significant by rank correlation but not statistically significant by linear regression, then this strongly suggests that linear regression may be an inappropriate analytical approach for modeling that relationship; instead, the effect of the independent variable upon titer ratio should be analyzed using a nonlinear regression technique. Also, it should be noted that in all of the equations presented in this chapter, it has been assumed that all of the terms incorporated into a survival model have the same general form (i.e., as $X_1\beta_1$ or $X_1^{\beta 1}$). However, this is not a requirement. An example of a model in which the incorporated terms take different formats is that of Auer and Niehaus (4).

The technique of using a set of experimentally observed values to generate a model and then comparing values predicted by the model against that same set of experimentally observed values can be employed to give an assessment of the accuracy of the equation format which was used to construct the model. This approach has been presented both here and in the reference by Hurst et al. (22). However, this technique cannot be used to estimate the accuracy of the developed model itself, as defined by the choice of included variables and their corresponding parameter values. Performing a test of the accuracy of the developed model requires instead that the predicted values be compared to a set of experimentally observed values other than those used to generate the model. One possible approach for performing the latter type of assessment would be to divide the original data set into two subsets, or half-sets, before the model is developed. One of these half-sets of experimentally observed values could be used to generate a model, and then the predicted values from that model could be compared to the other half-set of observed values. An even better approach would be to perform two independent sets of experiments. The observed values from the first set of experiments could be used to generate a model, whose accuracy would then be assessed by using the observed values from the second set of experiments. An example of the latter approach might be to develop a model by using data from the study by Farrah et al. (11) and then assessing the model's accuracy by using it to predict what the outcome of the Hurst et al. study (19) might have been; those predicted values and the values actually observed by Hurst et al. (19) could then be compared. Unfortunately, neither of these two studies (11, 19) were considered to have yielded a large enough data set for accurately performing such a test of model accuracy.

REFERENCES

1. **Allwood, P. B., Y. S. Malik, C. W. Hedberg, and S. M. Goyal.** 2003. Survival of f-specific RNA coliphage, feline calicivirus, and *Escherichia coli* in water: a comparative study. *Appl. Environ. Microbiol.* **69:**5707–5710.
2. **Alvarez, M. E., M. Aguilar, A. Fountain, N. Gonzalez, O. Rascon, and D. Saenz.** 2000. Inactivation of MS-2 phage and poliovirus in groundwater. *Can. J. Microbiol.* **46:**159–165.
3. **Anderson, K. L., J. E. Whitlock, and V. J. Harwood.** 2005. Persistence and differential survival of fecal indicator bacteria in subtropical waters and sediments. *Appl. Environ. Microbiol.* **71:**3041–3048.
4. **Auer, M. T., and S. L. Niehaus.** 1993. Modeling fecal coliform bacteria. I. Field and laboratory determination of loss kinetics. *Water Res.* **27:**693–701.
5. **Blanford, W. J., M. L. Brusseau, T. C. J. Yeh, C. P. Gerba, and R. Harvey.** 2005. Influence of water chemistry and travel distance on bacteriophage PRD-1 transport in a sandy aquifer. *Water Res.* **39:**2345–2357.
6. **Boehm, A. B., D. P. Keymer, and G. G. Shellenbarger.** 2005. An analytical model of enterococci inactivation, grazing, and transport in the surf zone of a marine beach. *Water Res.* **39:**3565–3578.
7. **Brusseau, M. L., J. K. Oleen, J. Santamaria, L. Cheng, P. Orosz-Coghlan, A. S. Chetochine, W. J. Blanford, P. Rykwalder, and C. P. Gerba.** 2005. Transport of microsporidium *Encephalitozoon intestinales* spores in sandy porous media. *Water Res.* **39:**3636–3642.
8. **Canale, R. P., M. T. Auer, E. M. Owens, T. M. Heidtke, and S. W. Effler.** 1993. Modeling fecal coliform bacteria. II. Model development and application. *Water Res.* **27:** 703–714.
9. **Connolly, J. P., R. B. Coffin, and R. E. Landeck.** 1992. Modeling carbon utilization by bacteria in natural water systems, p. 249–276. *In* C. J. Hurst (ed.), *Modeling the Metabolic and Physiologic Activities of Microorganisms*. John Wiley & Sons, New York, N.Y.
10. **Dahling, D. R., and R. S. Safferman.** 1979. Survival of enteric viruses under natural conditions in a subarctic river. *Appl. Environ. Microbiol.* **38:**1103–1110.
11. **Farrah, S. R., G. Bitton, E. M. Hoffmann, O. Lanni, O. C. Pancorbo, M. C. Lutrick, and J. E. Bertrand.** 1981. Survival of enteroviruses and coliform bacteria in a sludge lagoon. *Appl. Environ. Microbiol.* **41:**459–465.
12. **Fischer, U. R., W. Weisz, C. Wieltschnig, A. K. T. Kirschner, and B. Velimirov.** 2004. Benthic and pelagic viral decay experiments: a model-based analysis and its applicability. *Appl. Environ. Microbiol.* **70:**6706–6713.
13. **Fuller, M. E., S. H. Streger, R. K. Rothmel, B. J. Mailloux, J. A. Hall., T. C. Onstott, J. K. Fredrickson, D. L. Balkwill, and M. F. DeFlaun.** 2000. Development of a vital fluorescent staining method for monitoring bacterial transport in subsurface environments. *Appl. Environ. Microbiol.* **66:**4486–4496.
14. **Grant, S. B., E. J. List, and M. E. Lidstrom.** 1993. Kinetic analysis of virus adsorption and inactivation in batch experiments. *Water Resour. Res.* **29:**2067–2085.
15. **Harvey, R. W.** 1991. Parameters involved in modeling movement of bacteria in groundwater, p. 89–114. *In* C. J. Hurst (ed.), *Modeling the Environmental Fate of Microorganisms*. American Society for Microbiology, Washington, D.C.
16. **Hurst, C. J.** 1988. Influence of aerobic microorganisms upon virus survival in soil. *Can. J. Microbiol.* **34:**696–699.
17. **Hurst, C. J.** 1991. Using linear and polynomial models to examine the environmental stability of viruses, p. 137–159. *In* C. J. Hurst (ed.), *Modeling the Environmental Fate of Microorganisms*. American Society for Microbiology, Washington, D.C.
18. **Hurst, C. J., W. H. Benton, and K. A. McClellan.** 1989. Thermal and water source effects upon the stability of enteroviruses in surface freshwaters. *Can. J. Microbiol.* **35:**474–480.
19. **Hurst, C. J., S. R. Farrah, C. P. Gerba, and J. L. Melnick.** 1978. Development of quantitative methods for the detection of enteroviruses in sewage sludges during activation and following land disposal. *Appl. Environ. Microbiol.* **36:**81–89.
20. **Hurst, C. J., C. P. Gerba, and I. Cech.** 1980. Effects of environmental variables and soil characteristics on virus survival in soil. *Appl. Environ. Microbiol.* **40:**1067–1079.
21. **Hurst, C. J., C. P. Gerba, J. C. Lance, and R. C. Rice.** 1980. Survival of enteroviruses in rapid-infiltration basins during the land application of wastewater. *Appl. Environ. Microbiol.* **40:**192–200.
22. **Hurst, C. J., D. K. Wild, and R. M. Clark.** 1992. Comparing the accuracy of equation formats for modeling microbial population decay rates, p. 149–175. *In* C. J. Hurst (ed.), *Modeling the Metabolic and Physiologic Activities of Microorganisms*. John Wiley & Sons, New York, N.Y.
23. **Jenkins, M. B., L. J. Anguish, D. D. Bowman, M. J. Walker, and W. C. Ghiorse.** 1997. Assessment of a dye permeability assay for determination of inactivation rates of *Cryptosporidium parvum* oocysts. *Appl. Environ. Microbiol.* **63:**3844–3850.
24. **Kay, D., C. M. Stapleton, M. D. Wyer, A. T. McDonald, J. Crowther, N. Paul, K. Jones, C. Francis, J. Watkins, J. Wilkinson, N. Humphrey, B. Lin, L. Yang, R. A. Falconer, and S. Gardner.** 2005. Decay of intestinal enterococci concentrations in high-energy estuarine and coastal waters: towards real-time T_{90} values for modelling faecal indicators in recreational waters. *Water Res.* **39:**655–667.
25. **Kinner, N. E., R. W. Harvey, K. Blakeslee, G. Novarino, and L. D. Meeker.** 1998. Size-selective predation on groundwater bacteria by nanoflagellates in an organic-contaminated aquifer. *Appl. Environ. Microbiol.* **64:**618–625.
26. **McFeters, G. A., and D. G. Stuart.** 1972. Survival of coliform bacteria in natural waters: field and laboratory studies with membrane-filter chambers. *Appl. Microbiol.* **24:** 805–811.
27. **Mezrioui, N., B. Baleux, and M. Troussellier.** 1995. A microcosm study of the survival of *Escherichia coli* and *Salmonella typhimurium* in brackish water. *Water Res.* **29:** 459–465.
28. **Moore, R. S., D. H. Taylor, M. M. M. Reddy, and L. S. Sturman.** 1982. Adsorption of reovirus by minerals and soils. *Appl. Environ. Microbiol.* **44:**852–859.
29. **Noble, R. T., and J. A. Fuhrman.** 1997. Virus decay and its causes in coastal waters. *Appl. Environ. Microbiol.* **63:** 77–83.

30. **Pesaro, F., I. Sorg, and A. Metzler.** 1995. In situ inactivation of animal viruses and a coliphage in nonaerated liquid and semiliquid animal wastes. *Appl. Environ. Microbiol.* **61:**92–97.
31. **Powelson, D. K., and C. P. Gerba.** 1994. Virus removal from sewage effluents during saturated and unsaturated flow through soil columns. *Water Res.* **28:**2175–2181.
32. **Powelson, D. K., C. P. Gerba, and M. T. Yahya.** 1993. Virus transport and removal in wastewater during aquifer recharge. *Water Res.* **27:**583–590.
33. **Quanrud, D. M., S. M. Carroll, C. P. Gerba, and R. G. Arnold.** 2003. Virus removal during simulated soil-aquifer treatment. *Water Res.* **37:**753–762.
34. **Rice, R. C., and H. Bouwer.** 1984. Soil-aquifer treatment using primary effluent. *J. Water Pollut. Control Fed.* **56**(1)**:** 84–88.
35. **Sim, Y., and C. V. Chrysikopoulos.** 1995. Analytical models for one-dimensional virus transport in saturated porous media. *Water Resources Res.* **31**(5)**:**1429–1437.
36. **Sinton, L. W., R. J. Davies-Colley, and R. G. Bell.** 1994. Inactivation of enterococci and fecal coliforms from sewage and meatworks effluents in seawater chambers. *Appl. Environ. Microbiol.* **60:**2040–2048.
37. **Sinton, L. W., R. K. Finlay, and P. A. Lynch.** 1999. Sunlight inactivation of fecal bacteriophages and bacteria in sewage-polluted seawater. *Appl. Environ. Microbiol.* **65:** 3605–3613.
38. **Smith, J. J., J. P. Howington, and G. A. McFeters.** 1994. Survival, physiological response, and recovery of enteric bacteria exposed to a polar marine environment. *Appl. Environ. Microbiol.* **60:**2977–2984.
39. **Steets, B. M., and P. A. Holden.** 2003. A mechanistic model of runoff-associated fecal coliform fate and transport through a coastal lagoon. *Water Res.* **37:**589–608.
40. **Thompson, S. S., M. Flury, M. V. Yates, and W. A. Jury.** 1998. Role of the air-water-solid interface in bacteriophage sorption experiments. *Appl. Environ. Microbiol.* **64:**304–309.
41. **Thompson, S. S., and M. V. Yates.** 1999. Bacteriophage inactivation at the air-water-solid interface in dynamic batch systems. *Appl. Environ. Microbiol.* **65:**1186–1190.
42. **Wan, J., J. L. Wilson, and T. L. Kieft.** 1994. Influence of the gas-water interface on transport of microorganisms through unsaturated porous media. *Appl. Environ. Microbiol.* **60:**509–516.
43. **Weinbauer, M. G., and M. G. Höfle.** 1998. Significance of viral lysis and flagellate grazing as factors controlling bacterioplankton production in a eutrophic lake. *Appl. Environ. Microbiol.* **64:**431–438.
44. **Wilkinson, J., A. Jenkins, M. Wyer, and D. Kay.** 1995. Modelling faecal coliform dynamics in streams and rivers. *Water Res.* **29:**847–855.
45. **Yates, M. V., and S. R. Yates.** 1991. Modeling microbial transport in the subsurface: a mathematical discussion, p. 48–76. *In* C. J. Hurst (ed.), *Modeling the Environmental Fate of Microorganisms*. American Society for Microbiology, Washington, D.C.

Estimating the Risk of Infectious Disease Associated with Pathogens in Drinking Water

CHRISTON J. HURST

29

NATURE OF THE HAZARDS ASSOCIATED WITH PATHOGENIC MICROORGANISMS IN AQUATIC ENVIRONMENTS

Most of the microorganisms present in aquatic environments seem to have no effect on human health. However, some clearly do represent a public health risk and for this reason are considered to be disease causing and are referred to as pathogens. The true definition of the term "pathogen" is "something that causes pathological changes" in the appearance of cells and tissues. Often the "pathogen" is the entire organism viewed as an infectious agent (see chapter 19). In other cases, however, the pathological change is due to a toxic substance produced by a noninfectious organism, and some toxins, such as those produced by algae and cyanobacteria, can bioaccumulate within the food chain to extremely high concentrations (see chapters 19 and 30). This chapter addresses organisms which are infectious for humans, and while some of these represent aquatic contaminants that come from either human or animal reservoirs, others represent organisms which are naturally present in the environment (see chapter 18).

ROUTES OF PATHOGEN ACQUISITION

Humans can acquire infections from water by any of three major routes: ingestion, body surface contact, and inhalation. Table 1 lists examples of the types of infectious diseases which are associated with these major routes. The route addressed in this chapter is that of ingestion. Acquisition of infection by body surface contact is addressed in chapter 19 and helpful information on recreational exposure can be found in references 14 and 15. The inhalation route is addressed in chapter 73.

DEFINING THE MAIN VARIABLES INCORPORATED IN RISK ESTIMATES FOR THE LEVEL OF INFECTIOUS DISEASE ASSOCIATED WITH INGESTION OF WATER

Five main variables must be considered when the risk of acquiring infectious disease by ingestion of water is estimated. These are the concentration of pathogenic organisms in water which are infectious by the route of ingestion, the amount of water ingested by an individual per unit of time, and the probabilities of infection, illness, and death associated with ingesting those pathogenic organisms.

Defining the Exposure Level

Two of the five main variables, the concentration of pathogenic organisms in the water which are infectious by the route of ingestion and the amount of water which an individual ingests per unit of time, act together to represent the level of an individual's exposure to the hazard. The risk estimation technique presented in this chapter could be used for any period. However, the examples presented will employ a period of 1 year, calculated as 365.25 days. The value used as an estimate for the volume of water ingested on a daily basis per individual consumer is 2 liters per day. This volume is the estimate recommended by the U.S. Environmental Protection Agency for use in calculating the risks from ingestion of water (30).

Defining the Probability of Infection

The values used for probability of infection are calculated as the inverse of the minimum or minimal infectious dose. The probability of infection thus represents the estimated likelihood that ingestion of a single measurable bacterium, protozoan, or virus will result in infection. The values for probability of infection are based on published results from studies during which human volunteers were fed different dosages of infectious microorganisms. For the purpose of this chapter, the minimum infectious dose is defined as the smallest number of microorganisms which could be found to produce infection when ingested by any of the volunteers, with smaller doses not found to produce infection. In most clinical studies, infection is defined serologically after the fact. The test for infection involves examining a pair of samples of serum collected from each individual volunteer. The first serum sample is collected at the time when that individual ingests the organism of interest, and the second serum sample is collected at some time between 4 and 6 weeks following that ingestion. If the titer (concentration) of the antibodies in the serum which were able to react with antigenic material from the strain of microorganism fed to that individual increased at least twofold during the 4- to 6-week time interval, then that individual is considered to have been infected by the ingested organism.

Some researchers who perform risk estimations would prefer to use a mathematical approximation technique based on curve fitting to modify the findings from feeding studies. However, as explained in Table 2, this type of curve

TABLE 1 Examples of infectious diseases associated with waterborne pathogens

Exposure route and type of pathogen	Type of disease	Causative microorganism(s)[a]
Ingestion[b]		
Bacterial	Enteric fever	*Salmonella* (especially *S. enterica* serovar Typhi, which causes typhoid fever)
	Enteritis	*Campylobacter*, *Shigella* (causes bacterial dysentery), *Vibrio* (especially *V. cholerae*, which causes cholera)
	Febrile syndrome	*Francisella tularensis*
	Septicemia	*Vibrio vulnificus*
Metazoan	Worm infestation	*Dracunculus medinensis*
Protozoan	Enteritis	*Cryptosporidium parvum*, *Entamoeba histolytica* (causes amebic dysentery), *Giardia lamblia*
Viral	Encephalitis	*Enterovirus*
	Gastroenteritis	*Astrovirus*, *Calicivirus*, *Coronavirus*, *Norovirus*, *Rotavirus*
	Hepatitis	*Calicivirus*, *Hepatovirus*
	Meningitis	*Enterovirus*
Body surface contact[c]		
Bacterial	Enteritis	*Vibrio cholerae*
	Nephritis	*Leptospira interrogans*
	Wound infections	*Vibrio parahaemolyticus*, *Vibrio vulnificus*
Metazoan	Worm infestation	*Schistosoma*
Protozoan	Encephalitis	*Naegleria*
	Enteritis	*Entamoeba histolytica*
Viral	Encephalitis	*Enterovirus*
	Gastroenteritis	*Astrovirus*, *Calicivirus*, *Coronavirus*, *Norovirus*, *Rotavirus*
	Meningitis	*Enterovirus*
	Pharyngoconjunctival fever	*Mastadenovirus*
Inhalation		
Bacterial	Pneumonic fever	*Legionella pneumophila*

[a]If an organism is indicated by both its genus and species names, then the disease association is with that particular species. If only a genus name is given, then the disease association is with more than one species belonging to that genus.

[b]Includes contaminated drinking water and other beverages, ice, and water-associated contamination of foods.

[c]Usually associated with either recreational or aquatic occupational activities.

TABLE 2 How the use of curve fitting can alter value estimates for the minimum infectious dose of a microorganism

Microorganism examined for infectivity	Experimentally determined minimum infectious dose[a] (reference)	Minimum infectious dose as estimated by curve fittings[c]	Underestimation of risk which would result from using curve-fitting values[d]
Echovirus 12	17 (28)	500.0	29.4
Poliovirus 1	20 (18)	110.2	5.5
Poliovirus 3	0.5[b] (13)	1.93	3.9
Rotavirus	0.9 (32)	1.62	1.8
Giardia	10.0 (27)	49.27	4.9

[a]Lowest viral dose found capable of causing infection when ingested by human volunteers. The number of viruses was determined by infectivity in cultured cells. The number of *Giardia* cells was determined by direct microscopic enumeration.

[b]Actual dose was 1 50% tissue culture infective dose ($TCID_{50}$), divided by 2 to yield an estimated $TCID_{100}$ as reported in this table, using the technique described by Hurst et al. (12).

[c]Data from reference 26. This column contains estimated minimum infectious dose values derived by performing curve-fitting operations on the actual experimental data. The curve-fitting values listed in reference 26 were presented as the estimated concentration of microorganisms per liter of drinking water which would give an annual risk of one infection per 10,000 individuals and assumed that each individual ingested 2 liters of water per day. Therefore, backtracking to obtain the estimated minimum infectious dose values which had been derived by curve fitting was done by multiplying the published values by 7.3×10^6, which was calculated as (2 liters of water/day) × (365.25 days/year) × (10,000 individuals).

[d]If the calculation of risk were done using values derived by curve fitting in place of the actual experimentally determined values, an underestimation of risk would result. The underestimation of risk (as a factor of error) which would be associated with using those curve-fitting values was calculated as follows: (the estimated minimum infectious dose as determined by Regli et al. [26] through curve fitting) ÷ (the experimentally determined minimum infectious dose initially published for that organism).

TABLE 3 Probability that a human will become infected on ingesting an enteric microorganism[a]

Type of microorganism examined (category of estimate)	Probability of infection per ingested microorganism[e]
Bacteria[b]	
Enteric pathogenic bacteria (overall estimate)	0.00001
Salmonella enterica serovar Typhi (this species in general)	0.00001
Vibrio cholerae (type O1 strains)	0.001
Vibrio cholerae (type non-O1 strains)	0.00001
Protozoa[c]	
Enteric pathogenic protozoa (overall estimate)	0.067
Cryptosporidium parvum (this species in general)	0.033
Giardia lamblia (this species in general)	0.1
Viruses[d]	
Enteric pathogenic viruses (overall estimate)	0.5
Enterovirus (this genus in general)	0.5
Enterovirus human echovirus 12 (this species in general)	0.059
Enterovirus human poliovirus 1 (this species in general)	0.5
Enterovirus human poliovirus 3 (this species in general)	2.0[f]
Rotavirus (this genus in general)	1.1[f]

[a]Probabilities were determined by volunteer feeding studies. The success of infection was determined by testing the sera of the volunteers before and after those individuals were dosed with microorganisms. The values listed in this table are medians based on data published by Hurst et al. (11). When values for the same genus or species of microorganism were available from more than a single study, an overall estimate was derived to represent that genus or species by calculating the median of the pertinent values. Likewise, overall estimates for any particular group (e.g., enteric pathogenic bacteria) of microorganisms were derived by calculating the median of the values available from studies in which members of that group had been examined.

[b]The number of bacterial organisms was determined by culture.

[c]The number of protozoa was determined as either cysts (for *Giardia*) or oocysts (for *Cryptosporidium*) by direct microscopic enumeration.

[d]The number of viruses was determined by infectivity assay in cultured cells.

[e]Probability of infection associated with each microorganism ingested. This calculation is performed as 1 ÷ minimum infectious dose.

[f]For this virus type, the number of virus particles required to cause an infection of cultured cells is greater than the number of virus particles required to cause infection of a human. Thus, the value of the probability of a human developing an infection from this virus type is higher than the titer obtained by cell culture assay of the virus.

fitting approach seems to produce "estimated" minimum infectious dose values which are higher than the actual minimum infectious dose values published by the authors of the feeding studies. The use of such curve fitting could then cause an underestimation of the true risk associated with ingesting infectious organisms. To avoid this problem of underestimating risk, the values for probability of infection used in this chapter are those listed in Table 3, and they directly represent the actual minimum infectious doses as published by the authors of the respective feeding studies.

Defining the Probability of Illness

Illness is defined as the demonstration of symptoms characteristic of disease, and not all infected individuals show symptoms. The values used in this chapter for probability of illness are listed in Table 4 and were calculated as the number of individual volunteers who developed symptoms of illness following infection by ingesting a particular microorganism divided by the total number of individual volunteers who were determined to have been infected when fed that microorganism. Thus, the probability of illness is the estimated likelihood that an incidence of infection caused by a bacterium, protozoan, or virus will progress to illness.

Defining the Probability of Death

The probability of death is the estimated likelihood that a case of illness induced by a bacterium, protozoan, or virus will progress to cause death. The values used in this chapter for probability of death are listed in Table 5 and were calculated from epidemiological data as the number of infected individuals who died from illness induced by a particular microorganism divided by the total number of individuals who were made ill by that microorganism.

CALCULATING THE RISK ESTIMATION

Format of the Equations

Equations 1 through 4 present the format used in this chapter for estimating the human health risk of infection, illness, and death on an annual basis per individual consumer.

$$\text{Number of organisms ingested annually} = \text{Number of organisms per liter of water} \times \text{Number of liters ingested per day} \times \text{365.25 days of water ingestion per year} \quad (1)$$

$$\text{Annual risk of infection} = \text{Number of organisms ingested annually} \times \text{Probability of infection per ingested organism} \quad (2)$$

$$\text{Annual risk of illness} = \text{Number of organisms ingested annually} \times \text{Probability of infection per ingested organism} \times \text{Probability of illness per incidence of infection} \quad (3)$$

$$\text{Annual risk of death} = \text{Number of organisms ingested annually} \times \text{Probability of infection per ingested organism} \times \text{Probability of illness per incidence of infection} \times \text{Probability of death per incidence of illness} \quad (4)$$

TABLE 4 Probability that infection of a human by an enteric microorganism will result in illness[a]

Type of microorganism examined (category of estimate)	Probability of illness per incidence of infection
Bacteria	
Enteric pathogenic bacteria (overall estimate)	0.4928
Salmonella enterica serovar Typhi (this species in general)	0.4928
Vibrio cholerae (type O1 strains of this species)	0.2727
Vibrio cholerae (type non-O1 strains)	0.5333
Protozoa	
Enteric pathogenic protozoa (overall estimate)	0.4386
Cryptosporidium (this genus in general)	0.5538
Cyclospora (this genus in general)	0.2200
Giardia (this genus in general)	0.4617
Giardia (species of this genus other than *G. lamblia*)	0.5011
Giardia lamblia (this species in general)	0.4617
Viruses	
Enteric pathogenic viruses (overall estimate)	0.5769
Enterovirus (this genus in general)	0.486
Enterovirus human coxsackievirus A21 (this species in general)	0.9231
Enterovirus human echovirus 12 (this species in general)	0.0508
Rotavirus (this genus in general)	0.5769

[a]Probabilities were determined by volunteer feeding studies. The success of infection was determined by testing the sera of the volunteers before and after those individuals were dosed with microorganisms. Illness was defined as the demonstration of any symptoms. The values listed in this table are medians based on data published by Hurst et al. (11). When values for the same genus or species of microorganism were available from more than a single study, an overall estimate was derived to represent that genus or species by calculating the median of the pertinent values. Likewise, overall estimates for any particular group (e.g., enteric pathogenic bacteria) of microorganisms were derived by calculating the median of the values available from studies in which members of that group had been examined.

Validation of the Risk Estimation Technique

For the purpose of this chapter, validation of equations 1 to 4 and of the values for probability of infection and illness presented in Tables 3 and 4, respectively, will be done using the findings from studies performed by Payment et al. (20, 21, 22, 23, 24). These efforts by Payment et al. include two epidemiological studies in which the occurrence of gastrointestinal illness was correlated with whether individuals ingested, within their homes, community-treated tap water as provided directly from the community distribution mains or the same water which had been further treated within their homes with reverse-osmosis water purification units. Reverse-osmosis units, when operating correctly, remove from water all microorganisms, including viruses. During their first epidemiological study, Payment et al. (22) did not include microbiological data for the water. It was possible, however, to derive appropriate microbiological data from other published studies performed by the same researchers (21, 24). The second epidemiological study (23) did include microbiological data.

Equations 1 through 4 do not include the fact that people often ingest water which has been subjected to drinking water treatment processes that would reduce its microbial concentration. If the values available for the number of organisms per liter of water represent untreated water and if the consumers are drinking untreated water, then equations 1 through 4 can be used directly without need for correction. Likewise, these equations can be used directly without need for correction if the consumers are drinking treated water and the values available for the number of organisms per liter of water represent treated water. However, if the consumers are drinking treated water and the values available for the number of organisms per liter represent untreated water, a mathematical correction will have to be made to account for the level of treatment. That correction is to multiply the estimated level of risk by a factor of [1 − (percentage of microbial reduction achieved by the water treatment process)]. These corrections are noted where appropriate in the validation exercises presented below.

For the purpose of these validation exercises, the risks of infection and illness are calculated separately for viruses capable of infecting cultured mammalian cells (and thus presumably capable of infecting humans), oocysts of *Cryptosporidium*, and cysts of *Giardia*, because the information from Payment et al. was made available in that manner. Estimations of the risks for infection and illness due to bacterial pathogens are done as back calculations and are necessarily based on an assumption that any risk not accounted for by the enumerated viruses or protozoa was due to bacteria. This assumption is required since Payment et al. reported data for the levels of indicator bacteria, which generally are not pathogenic for humans, rather than reporting the levels of pathogenic bacteria. These validation exercises do not include estimations of the number of deaths due to gastrointestinal illness which may have occurred during the epidemiological studies, because those authors did not supply information on the causes of death of the individuals who made up the different study groups.

TABLE 5 Probability that a human made ill by an enteric microorganism will die as a result of that illness[a]

Type of microorganism examined (category of estimate)	Probability of death per incidence of illness
Bacteria	
Enteric pathogenic bacteria (overall estimate)	0.0180
Campylobacter (this genus in general)	0.0003
Escherichia coli (pathogenic strains, this species in general)	0.1000
Escherichia coli (type O111 strains)	0.1111
Escherichia coli (type O157:H7 strains)	0.0582
Salmonella enterica serovar Typhi (this species in general)	0.0368
Shigella flexneri (this species in general)	0.0016
Vibrio cholerae (this species in general)	0.0195
Vibrio cholerae (type O1 strains)	0.0764
Vibrio cholerae (type non-O1 strains)	0.0137
Protozoa	
Cryptosporidium parvum (this species in general)	0.0001
Viruses	
Enteric pathogenic viruses (overall estimate)	0.0094
Calicivirus Hepatitis E virus (this species in general)	0.0280[b]
Enterovirus (this genus in general)	0.0104
Enterovirus (subset "coxsackieviruses" of this genus)	0.0075
Enterovirus (subset "echoviruses" of this genus)	0.0028
Enterovirus (subset "polioviruses" of this genus)	0.0183
Enterovirus human coxsackievirus A2 (this species in general)	0.0050
Enterovirus human coxsackievirus A3 (this species in general)	0.0741
Enterovirus human coxsackievirus A4 (this species in general)	0.0052
Enterovirus human coxsackievirus A6 (this species in general)	0.0098
Enterovirus human coxsackievirus A9 (this species in general)	0.0026
Enterovirus human coxsackievirus A10 (this species in general)	0.0171
Enterovirus human coxsackievirus A16 (this species in general)	0.0012
Enterovirus human coxsackievirus A19 (this species in general)	0.1667
Enterovirus human echovirus 6 (this species in general)	0.0029
Enterovirus human echovirus 9 (this species in general)	0.0027
Enterovirus human poliovirus 1 (this species in general)	0.0505
Enterovirus human poliovirus 2 (this species in general)	0.0182
Enterovirus human poliovirus 3 (this species in general)	0.0148
Hepatovirus Hepatitis A virus (this species in general)	0.0060
Rotavirus (this genus in general)	0.0055

[a]Probabilities were determined by epidemiological data. Illness was defined as the demonstration of any symptoms, and death was defined as cessation of self-directed metabolic activity. The values listed in this table are medians based on data published by Hurst et al. (11), with the exception of the value for *C. parvum*, which is from reference 8. When values for the same genus or species of microorganism were available from more than a single study, an overall estimate value was derived to represent that genus or species by calculating the median of the pertinent values. Likewise, overall estimates for any particular group (e.g., enteric pathogenic bacteria) of microorganisms were derived by calculating the median of the values available from studies in which members of that group had been examined.

[b]The value cited here for probability of death represents data reported for adult males. Values for females in their third trimester of pregnancy range from 0.17 to greater than 0.25.

■ Validation Exercise 1

The objective of this exercise was to use the epidemiological data published by Payment et al. (22) as a means of validating the risk assessment technique of Hurst et al. (11) (Tables 6 through 10).

Given Knowledge

Epidemiologically determined risk associated with the ingested tap water (total risk representing all infectious, pathogenic agents). The estimated risk of illness per person per year associated with microbial contaminants in the tap

TABLE 6 For validation 1: calculating the estimation of risk due to infectious, pathogenic viruses in the municipality's water

Concentration of organisms (viruses that are infectious for mammalian cells, enumerated as infectious units) per liter of raw water	3.3
Liters of water ingested per person per day	× 2.0
Days of water ingestion per year	× 365.25
Probability of infection (per ingested virus capable of infecting mammalian cells)	× 0.5
Probability of illness (per incidence of viral infection)	× 0.5769
Estimate of residual disease risk following municipal treatment of the water (this value is included because the data provided for viral levels were determined for the water prior to processing at the treatment plant), calculated as 1.000 − 0.9997	× 0.0003
Estimated annual risk of virus-induced illness per individual consumer	= 0.208606

water as supplied to consumers' individual homes by the municipal water distribution mains was 0.26. This value represents a measured overall annual incidence rate of 0.76 for gastrointestinal illness among all study participants minus the rate of 0.50 for gastrointestinal illness among "filtered-water drinkers," which represented persons who consumed municipal tap water that was treated at their home faucets by reverse-osmosis filtration units and was presented as being the control rate for all gastrointestinal illnesses acquired from other sources besides water ingested in the home (22).

Known level of infectious, pathogenic viruses in that municipality's water. A recent estimate (24) had found that the raw water entering that area's municipal drinking water treatment plants contained an average level of 3.3 viruses per liter (viruses capable of infecting cultured mammalian cells and thus presumably capable of infecting humans). The average reduction of viral load achieved by the water treatment process at those municipal plants was 99.97% (24).

Known level of infectious, pathogenic protozoa in that municipality's water. An estimate by Payment and Franco (21) found that the levels of organisms belonging to the genera *Cryptosporidium* and *Giardia* were below their limit of detection for two of that municipality's three drinking water treatment plants. However, the levels of protozoa in the finished water from the third treatment plant were 0.02 *Cryptosporidium* oocysts and 0.02 *Giardia* cysts per 100 liters. These values may have represented the limits of detection for that study.

Known level of infectious, pathogenic bacteria in that municipality's water. This level is unknown, since these researchers measured the presence of only fecal bacterial indicators rather than actually determining the level of pathogenic bacteria.

■ Validation Exercise 2

The objective of this exercise was to use the epidemiological data published by Payment et al. (23) as a means of validating the risk assessment technique of Hurst et al. (11) (Tables 11 through 15).

Given Knowledge

Epidemiologically determined risk associated with the ingested tap water (total risk representing all infectious, pathogenic agents). The estimated risk of illness per person per year associated with microbial contaminants in the tap water supplied by the municipal water distribution mains was 0.08. This value represents a measured overall annual incidence rate of 0.66 for gastrointestinal illnesses experienced by persons who drank "tap"-supplied water, minus a value of 0.58 for gastrointestinal illness among drinkers of "purified water," which represented persons who consumed either municipal tap water that was treated at their home faucets by reverse-osmosis filtration units or spring water and was presented as being the control rate for all gastrointestinal illnesses acquired from other sources besides water ingested in the home (23).

Known level of infectious, pathogenic viruses in that municipality's water. That same study (23) had found that

TABLE 7 For validation 1: calculating the estimation of risk due to *Cryptosporidium* in the municipality's water[a]

Concentration of organisms (*Cryptosporidium* oocysts, enumerated by visual inspection following immunofluorescent staining) per liter of water following municipal treatment of the water	0.0002
Liters of water ingested per person per day	× 2.0
Days of water ingestion per year	× 365.25
Probability of infection (per ingested *Cryptosporidium* oocyst)	× 0.033
Probability of illness (per incidence of cryptosporidial infection)	× 0.5538
Estimated annual risk of *Cryptosporidium*-induced illness per individual consumer	= 0.00267

[a]This calculation does not include a value for estimation of residual disease risk following municipal treatment of the water, because the data provided for *Cryptosporidium* oocyst levels were determined for the water following processing at the treatment plant.

TABLE 8 For validation 1: calculating the estimation of risk due to *Giardia* in the municipality's water[a]

Concentration of organisms (*Giardia* cysts, enumerated by visual inspection following immunofluorescent staining) per liter of water following municipal treatment of the water	0.0002
Liters of water ingested per person per day	× 2.0
Days of water ingestion per year	× 365.25
Probability of infection (per ingested *Giardia* cyst)	× 0.1
Probability of illness (per incidence of giardial infection)	× 0.4617
Estimated annual risk of *Giardia*-induced illness per individual consumer	= 0.006745

[a]This calculation does not include a value for estimation of residual disease risk following municipal treatment of the water, because the data provided for *Giardia* cyst levels were determined for the water following processing at the treatment plant.

the raw water entering the municipal drinking water treatment plants contained an average level of 413 viruses per 100 liters (4.13 viruses per liter) for viruses capable of infecting cultured mammalian cells and thus presumably capable of infecting humans. The average reduction of viral load achieved by the water treatment process at those same municipal plants was 99.9921% (4.1 $\log_{10}$ units) (23).

Known level of infectious, pathogenic protozoa in that municipality's water. That same study (23) had found that the raw water entering the municipal drinking water treatment plants contained average levels of organisms belonging to the genera *Cryptosporidium* and *Giardia* that were 14 oocysts per 100 liters (0.14 per liter) and 200 cysts per 100 liters (2.0 per liter), respectively. The average reduction of protozoan load achieved by the water treatment process at those same municipal plants was 99.9369% (3.2 $\log_{10}$ units) for *Cryptosporidium* oocysts and 99.9950% (4.3 $\log_{10}$ units) for *Giardia* cysts (23).

Known level of infectious, pathogenic bacteria in that municipality's water. This level is unknown, since those researchers measured the presence of only fecal bacterial indicators rather than actually determining the level of pathogenic bacteria.

WHAT DO ALL OF THOSE BIG RISK NUMBERS MEAN?

This section is included to help readers understand the risk values in this chapter using a simplified perspective of probability statistics. The question is this: when you look at the risk tables, the math does produce some really large numbers for those instances when a risk is calculated if water from a contaminated source were ingested without treatment. For example, in Table 11, where the raw water contained 4.13 viruses per liter, if there had been no treatment of that water such that the multiplier of 0.000079 did not apply, then the estimated risk of illness to the people who drank it would have come out to be about 870 per year. What does that mean, when we might normally think of risk probabilities as being somewhere in the range of from 0 to 1? In short, it means that if the people had drunk 2 liters of that raw source water per day, then over the course of a year each would have ingested with that water 870 times the amount of viruses required to make them ill.

The Short Answer

What it means is that if there had not been water treatment, then by drinking 2 liters a day of that water, during the course of a year each of those people would have

TABLE 9 For validation 1: overall summary of risk estimations for the known levels of infectious, pathogenic viruses and protozoans, with the remaining risk assumed to be due to infectious, pathogenic bacteria

Total epidemiologically determined risk of illness per person per year associated with microbial contaminants in the municipally supplied tap water		0.26 (100.0%)
Estimated proportion of that risk attributable to viruses in the municipally supplied tap water	0.20861 (80.2%)	
Estimated proportion of that risk attributable to *Cryptosporidium* in the municipally supplied tap water	0.00267 (1.0%)	
Estimated proportion of that risk attributable to *Giardia* in the municipally supplied tap water	0.00674 (2.6%)	
Total proportion of the epidemiologically determined risk which could be accounted for by the levels of viruses, *Cryptosporidium*, and *Giardia* present in the municipally supplied tap water	0.21802 (83.9%)	→ 0.21802 (83.9%)
Difference between the actual epidemiologically determined risk of illness and the level of risk predicted by this estimation technique using data for levels of virus and protozoa in the water		0.04198 (16.1%)

TABLE 10 For validation 1: backwards calculation to estimate the level of infectious, pathogenic bacteria in the municipality's water, assuming that bacteria accounted for the risk not attributable to viruses and protozoa[a]

Estimated annual risk of bacterially induced illness per individual consumer	0.0420
Probability of illness (per incidence of bacterial infection)	÷ 0.4928
Probability of infection (per ingested pathogenic bacterium capable of exhibiting growth on culture medium)	÷ 0.00001
Days of water ingestion per year	÷ 365.25
Liters of water ingested per person per day	÷ 2.0
Estimated concentration of organisms (pathogenic bacteria capable of exhibiting growth on culture medium) per liter of water following municipal treatment of the water	= 11.66698

[a]This back calculation does not include an estimation for the reduction of risk affected by municipal treatment of the water, because no data were provided for the extent to which pathogenic bacteria were removed by the process used at the water treatment plants.

TABLE 11 For validation 2: calculating the estimation of risk due to infectious, pathogenic viruses in the municipality's water

Concentration of organisms (viruses that are infectious for mammalian cells, enumerated as infectious units) per liter of raw water	4.13
Liters of water ingested per person per day	× 2.0
Days of water ingestion per year	× 365.25
Probability of infection (per ingested virus capable of infecting mammalian cells)	× 0.5
Probability of illness (per incidence of viral infection)	× 0.5769
Estimate of residual virus-associated disease risk following municipal treatment of the water (this value is included because the data provided for viral levels were determined for the water prior to processing at the treatment plant), calculated as 1.000000 − 0.999921	× 0.000079
Estimated annual risk of virus-induced illness per individual consumer	= 0.068749

TABLE 12 For validation 2: calculating the estimation of risk due to *Cryptosporidium* in the municipality's water

Concentration of *Cryptosporidium* oocysts (measured by visual inspection following immunofluorescent staining) per liter of raw water	0.14
Liters of water ingested per person per day	× 2.0
Days of water ingestion per year	× 365.25
Probability of infection (per ingested *Cryptosporidium* oocyst)	× 0.033
Probability of illness (per incidence of cryptosporidial infection)	× 0.5538
Estimate of residual *Cryptosporidium*-associated disease risk following municipal treatment of the water (this value is included because the data provided for *Cryptosporidium* oocyst levels were determined for the water prior to processing at the treatment plant), calculated as 1.000000 − 0.999369	× 0.000631
Estimated annual risk of *Cryptosporidium*-induced illness per individual consumer	= 0.001179

TABLE 13 For validation 2: calculating the estimation of risk due to *Giardia* in the municipality's water

Concentration of organisms (*Giardia* cysts, enumerated by visual inspection following immunofluorescent staining) per liter of raw water	2.00
Liters of water ingested per person per day	× 2.0
Days of water ingestion per year	× 365.25
Probability of infection (per ingested *Giardia* cyst)	× 0.1
Probability of illness (per incidence of infection by *Giardia*)	× 0.4617
Estimate of residual *Giardia*-associated disease risk following municipal treatment of the water (this value is included because the data provided for *Giardia* cyst levels were determined for the water prior to processing at the treatment plant), calculated as 1.000000 − 0.999950	× 0.000050
Estimated annual risk of *Giardia*-induced illness per individual consumer	= 0.003373

TABLE 14 For validation 2: overall summary of risk estimations for the known levels of infectious, pathogenic viruses and protozoans with the remaining risk assumed to be due to infectious, pathogenic bacteria

Total epidemiologically determined risk of illness per person per year associated with microbial contaminants in the municipally supplied tap water		0.08 (100.0%)
Estimated proportion of that risk attributable to viruses in the municipally supplied tap water	0.06875 (85.9%)	
Estimated proportion of that risk attributable to *Cryptosporidium* in the municipally supplied tap water	0.00118 (1.5%)	
Estimated proportion of that risk attributable to *Giardia* in the municipally supplied tap water	0.00337 (4.2%)	
Total proportion of the epidemiologically determined risk which could be accounted for by the levels of viruses, *Cryptosporidium*, and *Giardia* present in the municipally supplied tap water	0.07330 (91.6%)	→ 0.07330 (91.6%)
Difference between the actual epidemiologically determined risk of illness and the level of risk predicted by this estimation technique using data for levels of virus and protozoa in the water		0.00670 (8.4%)

ingested 870 times the amount of water required to produce a viral illness. Those people's daily risk of illness from the viruses in their water would have been 2.38 (870 divided by 365.25). In the next step, assuming that amount of risk to have derived from viruses contained in their water, as in this case, and that each person were to ingest 2 liters of the water per day, we can find our way down to a risk value of 1. In doing so, you would calculate that on average drinking 840 ml of their raw source water would be enough to get you sick, with that risk value of 1 being 2 liters divided by 2.38. All of this means that without water treatment each person is going to either die from the viruses ingested with that water or constantly be infected and sometimes be sick. Furthermore, considering the fact that the water likely contains many different types of virus, coinfection would be likely to occur, which can cause a greater than normal severity to be associated with the risk of illness for any one of the coinfecting viruses.

The risk values presented in that table actually were built up in the opposite direction, from knowing the risk per ingested infectious virus, multiplied by the number of infectious viruses per liter, number of liters ingested per day, and number of days per year.

Probability 101 (Clearly, the Less Formal Version)

A probability of 1 means that an event likely will occur within a given unit of space or time. A probability of 0 means that it likely cannot. We usually think in terms of probability values as being intermediate between 0 and 1. A more formal statistician might tell you that a probability cannot exceed 1. But formally presented statistics can be a bit daunting, so the explanation given here is on the informal side. The simple math of it all suggests that a probability of greater than 1 means that an event should occur randomly more than once in a given unit of space or time if that kind of multiple occurrence is physically possible. An example of how physical limitations can interplay with this would be that the probability of the world ending is limited to 1, since the event could not be repeated. Cracker Jack, a brand of candy-coated popcorn, advertises "a prize in every box," which means that for each box the probability of finding a prize when you look inside is at least 1. If your chance of finding a prize in a Cracker Jack box were 2 and it were physically possible that more than one prize could fit into a box, then you probably would find two prizes there. To satisfy the formal rules, you theoretically might allow one prize per half box! It just depends on what you term to be your

TABLE 15 For validation 2: backward calculation to estimate the level of infectious, pathogenic bacteria in the municipality's water, assuming that bacteria accounted for the risk not attributable to viruses and protozoa[a]

Estimated annual risk of bacterially induced illness per individual consumer	0.0067
Probability of illness (per incidence of bacterial infection)	÷ 0.4928
Probability of infection (per ingested pathogenic bacterium capable of exhibiting growth on culture medium)	÷ 0.00001
Days of water ingestion per year	÷ 365.25
Liters of water ingested per person per day	÷ 2.0
Estimated concentration of organisms (pathogenic bacteria capable of exhibiting growth on culture medium) per liter of water following municipal treatment of the water	= 1.86116

[a] This back calculation does not include an estimation for the reduction of risk affected by municipal treatment of the water, because no data were provided for the extent to which pathogenic bacteria were removed by the process used at the water treatment plants.

unit of space or time and how strictly you limit your life by formality. If you would like a little more help with understanding random probability and how it relates to risk of infection, then I would offer a suggestion that you read the analogy called "The Blindfolded Bowler" which I created for that purpose, and which is published in reference 11. If you have questions about any of these things, then please feel free to give me a phone call—the number is listed!

I could add here that although not used in this chapter, "percent probability" is an invention that is useful for those who somehow feel they like or need it. If you, as my reader, would like to think in terms of percent probability, then please know that multiplying a probability value by 100 expresses that value as a percent, and as such an event with a probability of 1 has a "percent probability of 100" or a "100% probability." Something with a 50% probability has a probability value of 0.5.

How Does This Relate at the Level of a Single Individual?

Infection, by classical definition, requires two things: microbial growth in association with the body of a host and the resulting production of a recognizable immunological response by that host. The immune response will be either an initial response, if none previously existed against that infecting organism, or an elevation to a previous response. The previous response may have been developed as the result of this host having earlier been infected either by this same species of microorganism or by an antigenically similar microorganism. A previous immune response also can have been developed by vaccination, and the potential advantage versus some potential disadvantages of this approach will be considered later in this section of the chapter from a very practical perspective. Illness means that an infection has progressed to the point that your body is demonstrating symptoms. If that illness gets bad enough, it can progress to death, which of course means that the ability to mount a response has passed. As discussed earlier in this chapter, the path that may lead to death involves three stages, of which each has its own probability. They are infection, illness, and death. They must occur in that order, but, of course and fortunately for us, not all three need to occur for every given incidence of infection. It must be presumed that reinfection and possible reestablishment of illness can occur an unlimited number of times during the life of any given person. In a practical sense, physical limitations require us to presume here that death can occur only once for each of us.

Immunity is a term often used, but is more complicated than often presented. In this chapter I address only its practical outcome, and I will begin by telling you that we must consider it as having two main parts: immunity against initial infection or reinfection, if that part can be said to exist at all, and immunity against illness. Immunity to either initial infection or reinfection, and to illness, is specific for each microbe. And there seems to be no limit to the number of times that a person can be reinfected by the same strain of a microbe. Published research has shown that reinfection can occur after less than 2 weeks. However, if you do not die from the initial infection, then your immune response normally starts up at that point and gets boostered or "reprimed" with each reinfection. The chance of your developing illness symptoms from each successive reinfection by that same organism should then progressively diminish, because this boostering increases the speed with which you could mount an effective immune response against that organism. What builds within the body from previous exposures, and is simulated by vaccination, traditionally has been called immunity. But that really is immunity to illness and not immunity to reinfection.

Immunity to reinfection would be that aspect of resistance which lasts less than 2 weeks and seemingly does not get boostered. While it may be that one's risk of infection, including reinfection by any particular microorganism, does never change, one's risk of illness from that microorganism clearly can change. This change will occur in part through that person's developing a very specific immunity to illness against that microorganism. We start life with some level of temporary immunity to illness against whatever infectious diseases were encountered by our mother during her lifetime. This maternal immunity is naturally supplied and provided by her immunoglobulins, which cross our placenta and also come to us in her breast milk. It is easy to see that this temporary immunity gives us a big survival advantage as we try to develop our own immunity to illness. Although that maternal protection is temporary, our own specific immunities to illnesses begin to develop by a natural induction process with subsequent infections or reinfections by these and other organisms. Vaccinations against organisms help us by artificially inducing immunity to illness. Immunity induced by infection often seems more persistent than does immunity induced by vaccination. However, both inductive processes require that we be immunoresponsive against each particular organism. The resulting induced immunity to illness mounted against any particular microbe seems to gradually diminish with time, and on average it lasts for perhaps 10 to 20 years before becoming ineffective. At any point in your life, the higher your level of immunity to illness against a specific microbe (and it will be high when a person's latest infection or vaccination related to that microbe was massive or relatively recent), the greater is the chance that your immunoresponsiveness as termed immunity to illness will be sufficient to protect against your experiencing any symptoms of illness at all when you next encounter that organism. With the passage of time, your immunity to illness may still exist but may have become lower to the point that your response to the next infection by that organism may not come early enough to entirely stop the infection from progression to illness, although it could at least allow a diminished level of illness. And, as your level of immunity to illness against any specific organism increases or decreases over time, so too will your corresponding risk of illness decrease or increase for that organism.

Immunocompetence is a general term which means that someone has the ability to recognize invading microbes as being foreign and to produce an immune response against them. In general, everyone is immunocompetent! Those who are not immunocompetent live only an extremely short life. But, I would like people to also consider here the more specific term immunoresponsiveness, and this term must be considered separately for each individual person and species of microorganism which they might encounter. As an example, we as humans somehow are not immunoresponsive to our normal gut microflora—unless we have something like Crohn's disease. And somebody always ends up being the loser when we discuss population risk numbers. For example, I seem not to be immunoresponsive to the poxvirus called vaccinia virus, which is used as a live virus vaccine to create a cross-reactive immunological protection against the illness smallpox, which is caused by the poxvirus called variola virus. Please note that not all of the diseases which include "pox" in their name are caused by poxviruses, and fortunately smallpox has never been shown to be waterborne! I nearly died from nephrosis due to a smallpox vaccination at the age of 2 1/2 years when the infection caused by that live virus vaccine spread throughout my internal organs. One known strain of smallpox has a risk of death of 0.01, or 1 per 100 cases of illness; the other has a risk of death that can be 0.3, or 30 per 100 cases of illness. That particular vaccination complication has an event occurrence of between only 1 in 10,000 and 1 in 20,000 smallpox vaccinations, equal to an individual probability of 0.0001 to 0.00005. For the general public, if there were an imminent threat of a smallpox outbreak, then that vaccine risk might well be worth taking! I lived only by the administration of massive amounts of antivaccinia virus serum prepared in horses and what was then the very experimental drug hydrocortisone. And I grew up to be a virologist! Whether or not I would be immunoresponsive to the actual variola virus which causes smallpox is uncertain, but clearly I seem not to have been immunoresponsive against its substitute, vaccinia virus. So my relieved mother was advised that it would be better for me to risk having smallpox than to ever get revaccinated against it. The other recommendation was that I should never be administered any antiserum made in horses because my immune reaction to that antiserum would be so strong that it could very well prove lethal. That potentially fatal immunoresponse would have been primed by having been injected with the horse immunoglobulins. I apparently have proven immunoresponsive to all other poxviruses which I have encountered in life. Thus, immunoresponsiveness is a great part of what helps us to survive in this world. And with time it can change the probability numbers for risk of illness. However, even immunoresponsiveness is not an all-or-none phenomenon and does not always lead to a protective situation. For example, each year hundreds of thousands of people die from chronic infections caused by microbes to which they may have developed high levels of immunoresponse. *Myco-bacterium tuberculosis*, which causes tuberculosis, is one of those microbes particularly adept at avoiding immunological defenses. While that mycobacterium is not waterborne, sadly other species of pathogenic mycobacteria are waterborne. The risk of infection, illness, and death can be influenced by other factors, and these are included in the following section where you will find the discussion of *caveat emptor*, or perhaps for this chapter the proper term should be *caveat potor* . . . "drinker beware"!

UNDERSTANDING THE *CAVEAT EMPTOR* PRINCIPLE AS IT SHOULD BE APPLIED TO RISK ASSESSMENT

Several important issues must be recognized and their implications must be understood when risk assessment estimations for infectious diseases are made. Six of these are listed below. The first three of these issues specifically apply to estimations for drinking water. The last three apply not only to drinking water but also to any infectious disease risks regardless of the source of the pathogenic organisms.

1. Estimations of the levels of microorganisms in samples of water are inherently underestimations. This is due to the deficiencies of the detection methodology. Among these deficiencies, it is important to note that techniques for concentrating microorganisms from environmental samples are less than 100% efficient; many of the types of microorganisms present in water samples cannot easily be cultured by the methods available; and some microorganisms which normally would be culturable, if their metabolic condition were healthy, are not readily detectable if their environmental exposure has resulted in metabolic shutdown associated with starvation.

2. To some extent, estimations of disease risk based on the levels of microbial contaminants in the municipal water distribution mains are overestimations. This is because immediately before consumption, some of the water which people ingest is treated by heating techniques that reduce the level of viable microbial contaminants in the water. This includes the heating of water to produce food items that are consumed hot, such as broths or soups, as well as the heating used in preparation of beverages like tea and coffee. However, heating water and then maintaining it warm for a long time, as occurs with water-heating tanks in homes and commercial buildings including nursing facilities, can change the risk in that additional pathogenic organisms such as bacteria of the genus *Legionella* can grow in water that is kept warm.

3. To directly estimate the level of disease risk associated with pathogenic bacteria present in water, it is best to enumerate the bacteria which actually are pathogenic. Most "indicator bacteria" are not considered pathogenic, so information regarding the level of their presence in water cannot directly be used for making risk assessment estimates.

4. If death does not occur as the outcome of an initial infection by a particular strain of a specific microorganism, then each subsequent reinfection of the same person by the same strain may carry a decreased risk that the infection would progress to illness (2, 28).

5. Immunosuppression can increase the probability that an infection will have a severe outcome. Some natural conditions, such as pregnancy, and abnormal conditions, such as malnutrition, can result in suppression of immune system function. The result of this immunosuppression is that when individuals become infected, they may have a greater probability of death as the result of that infection. A notable example of this effect is that human illness due to hepatitis E virus (11) normally has a 0.028 probability of death (which also can be stated as either 2.8%, 1 in 36, 1 death per 36 individuals, or 1/36). However, the probability of death for women in their third trimester of pregnancy at the time when they become ill from this virus ranges from approximately 0.17 (17%, or 1 in 6) to greater than 0.25 (25%, or 1 in 4). As another example, Hoxie et al. (8) observed that the overall risk of death from waterborne *Cryptosporidium*, associated with a large general population of humans who experienced an outbreak of infectious illness caused by that genus in Milwaukee, Wis., was 0.0001 (0.01%, or 1 in 10,000). However, the risk of death from *Cryptosporidium* when underlying malnutrition is present (4) has been observed to be 0.14 (14%, or 1 in 7).

6. Concurrent infections (incidences when a person is infected simultaneously with more than one pathogenic organism) can drastically increase the probability of death associated with any one of the involved pathogens. As an example, the risk of death from *Cryptosporidium* when a person concurrently has the viral infection measles (4) has been observed as 0.20 (20%, or 1 in 5), representing a 2,000-fold increase over the rate of death from *Cryptosporidium* infections in the general population (see item 5 above). By itself, measles virus is almost never fatal; however, it also seems to be the most powerfully immunosuppressive virus known to infect humans (10).

SUMMARY

Pathogenic microorganisms frequently can be found in untreated water and, as a result, can cause infections which may lead to illness and even to death among those people who ingest that water. To reduce these risks, drinking water often is treated on a community basis. However, microorganisms can survive even properly operated, community-based drinking water treatment processes, resulting in the presence in treated drinking water of pathogenic organisms belonging to three categories: bacteria (3, 20), protozoa (1, 7, 16), and viruses (9, 19, 31). These residual microbial contaminants can be sufficient to cause appreciable levels of infectious disease among consumers of the water. The potential reality of these risks is borne out both by epidemiological studies (8, 19, 22, 23, 25) and by published theoretical calculations (5, 6, 11, 17, 29).

The modeling equations presented in this chapter represent an example of how this infectious disease risk can be estimated and the extent to which the risk can be lessened by reducing the level of microorganisms in water prior to its ingestion. Two validations of these models were performed and are presented in this chapter to serve as examples of how the risk estimation technique is employed and to show how closely it approximates the actual risk which those microorganisms in drinking water impose upon the consumers of that water. These validation exercises were based on two published epidemiological studies by Payment et al. (22, 23) in which the authors examined the relationship between gastrointestinal disease in humans and the community-supplied treated drinking water which those people ingested.

The first epidemiological study by Payment et al. (22) indicated that individuals ingesting conventionally treated, community-distributed tap water had an overall annual risk of 0.76 for gastrointestinal illness. This contrasted with an annual gastrointestinal illness risk of 0.50 among a control group of consumers who drank water that originated from the same community distribution system but had gone through an in-home filtration system prior to ingestion. The difference in annual risk thereby attributed to microorganisms in the tap water was 0.26. Payment et al. had not included data regarding the load of either pathogenic bacterial, protozoan, or viral contaminants in the community-supplied treated drinking water when they published that first epidemiological study. However, based on microbial occurrence data also published by Payment et al. (20, 21, 24) for a roughly similar time frame, estimated annual ingested dosages of pathogenic viruses and protozoa per individual consumer were determined. Those dosages then were multiplied by literature-derived estimates of the probability of infection per ingested organism and for the probability of illness per incidence of infection. These values for probability of infection and probability of illness came from studies in which human volunteers knowingly had been fed microorganisms belonging to the three different categories. The validation exercises presented in this chapter did not include estimations for the risk of illness progressing to death, because mortality information was not included in the epidemiological study results that had been published by Payment et al. The validation exercise (validation 1) presented in this chapter for the first epidemiological study by Payment et al. (22) suggested that of the 0.26 annual risk of illness associated with infectious pathogenic organisms in the community-distributed tap water, viruses in the drinking water could have accounted for approximately 80.2% of the observed cases of illness, while protozoa of the genera *Cryptosporidium* and *Giardia* could have accounted for approximately 1.0 and 2.6%, respectively, of the observed annual risk of illness in that first epidemiological study by Payment and coworkers.

Several years later, after efforts to improve the efficiency of that community's drinking water treatment facilities, Payment et al. performed a second epidemiological study (23), during which they found that individuals ingesting conventionally treated, community-distributed tap water had an overall 0.66 annual risk for incidence of gastrointestinal illness. This contrasted with an annual illness risk of 0.58 among a control group of consumers who drank either bottled water or tap water that originated from the same community distribution system but had gone through an in-home filtration system prior to ingestion. The difference in annual risk thereby attributed to microorganisms in the tap water during the second epidemiological study was 0.08. The validation exercise presented in this chapter for the second epidemiological study of Payment et al. (validation 2) used virus and protozoan data directly published as a part of that second study (23). This validation suggests that of the 0.08 annual risk, viruses in the drinking water could have accounted for approximately 85.9% of the observed cases of illness, while protozoa of the genera *Cryptosporidium* and *Giardia* could have accounted for approximately 1.5 and 4.2%, respectively, of the observed illnesses. The data published by Payment et al. concerning bacterial concentrations in the tap water represent indicator bacterial groups rather than known waterborne bacterial pathogens. As such, the only way to estimate the proportion of observed infectious disease which would have been attributable to bacterial pathogens in the water during the time of the epidemiological studies is by subtraction. Knowing the total risk due to pathogens in the drinking water and then subtracting the estimates for risks attributable to viruses and protozoa results in the estimated proportion of illnesses due to bacterial pathogens in the tap water. This process reveals that for the two epidemiological studies published by Payment et al. (22, 23), bacterial pathogens could have accounted for approximately 16.1% of the observed cases of illness attributable to microorganisms in the community-distributed tap water for the first epidemiological study (22) and correspondingly could have accounted for 8.4% of the observed cases of illness for the second study (23).

The approach presented in this chapter for modeling the risks of acquiring infectious disease from contaminants in drinking water can be used to estimate the existing level of danger to a community's population. The accuracy of disease risk estimates relies on knowing the levels of pathogens in the water being ingested. Alternatively, by determining a target value for the maximum desired disease risk to a community and by knowing the levels of pathogenic microorganisms in the source water, this modeling approach could be used to estimate the level of community-based water treatment which would be needed to decrease the disease risk to that target level.

The values presented in this chapter for the risks of infection, illness, and death are general estimates assumed to represent those which would be faced by an average member of a human population. In actuality, the values for each given individual human within any population may be different from these averages, and hopefully from having read the sections "What Do All of Those Big Risk Numbers Mean?" and "Understanding the *Caveat Emptor* Principle . . . " you now can more easily understand that for any given individual the risk values may change during the course of his or her lifetime.

REFERENCES

1. **Aboytes, R., G. D. Di Giovanni, F. A. Abrams, C. Rheinecker, W. McElroy, N. Shaw, and M. W. LeChevallier.** 2004. Detection of infectious *Cryptosporidium* in filtered drinking water. *J. Am. Water Works Assoc.* **96**(9)**:**88–98.
2. **Chappell, C. L., P. C. Okhuysen, C. R. Sterling, C. Wang, W. Jakubowski, and H. L. Dupont.** 1999. Infectivity of *Cryptosporidium parvum* in healthy adults with pre-existing anti-*C. parvum* serum immunoglobulin G. *Am. J. Trop. Med. Hyg.* **60:**157–164.
3. **Covert, T. C., M. R. Rodgers, A. L. Reyes, and G. N. Stelma, Jr.** 1999. Occurrence of nontuberculous mycobacteria in environmental samples. *Appl. Environ. Microbiol.* **65:**2492–2496.
4. **Crawford, F. G., and S. H. Vermund.** 1988. Human cryptosporidiosis. *Crit. Rev. Microbiol.* **16:**113–159.
5. **Gofti-LaRoche, L., D. Demanse, J.-C. Joret, and D. Zmirou.** 2003. Health risks and parasitical quality of water. *J. Am. Water Works Assoc.* **95**(5)**:**162–172.
6. **Haas, C. N.** 1996. How to average microbial densities to characterize risk. *Water Res.* **30:**1036–1038.
7. **Hibler, C. P.** 1988. Analysis of municipal water samples for cysts of *Giardia*, p. 237–245. *In* B. R. Hammond (ed.), *Advances in Giardia Research.* University of Calgary Press, Calgary, Canada.
8. **Hoxie, N. J., J. P. Davis, J. M. Vergeront, R. D. Nashold, and K. A. Blair.** 1997. Cryptosporidiosis-associated mortality following a massive waterborne outbreak in Milwaukee, Wisconsin. *Am. J. Public Health* **87:**2032–2035.
9. **Hurst, C. J.** 1991. Presence of enteric viruses in freshwater and their removal by the conventional drinking water treatment process. *Bull. W. H. O.* **69:**113–119.
10. **Hurst, C. J., and N. J. Adcock.** 2000. Relationship between humans and their viruses, p. 519–548. *In* C. J. Hurst (ed.), *Viral Ecology*. Academic Press, Inc., San Diego, Calif.
11. **Hurst, C. J., R. M. Clark, and S. E. Regli.** 1996. Estimating the risk of acquiring infectious disease from ingestion of water, p. 99–139. *In* C. J. Hurst (ed.), *Modeling Disease Transmission and Its Prevention by Disinfection.* Cambridge University Press, Cambridge, United Kingdom.
12. **Hurst, C. J., K. A. McClellan, and W. H. Benton.** 1988. Comparison of cytopathogenicity, immunofluorescence and *in situ* DNA hybridization as methods for the detection of adenoviruses. *Water Res.* **22:**1547–1552.
13. **Katz, M., and S. A. Plotkin.** 1967. Minimal infective dose of attenuated poliovirus for man. *Am. J. Public Health* **57:** 1837–1840.
14. **Kay, D., J. Bartram, A. Prüss, N. Ashbolt, M. D. Wyer, J. M. Fleisher, L. Fewtrell, A. Rogers, and G. Rees.** 2004. Derivation of numerical values for the World Health Organization guidelines for recreational waters. *Water Res.* **38:**1296–1304.
15. **Koopman, J. S., E. A. Eckert, H. B. Greenberg, B. C. Strohm, R. E. Isaacson, and A. S. Monto.** 1982. Norwalk virus enteric illness acquired by swimming exposure. *Am. J. Epidemiol.* **115**(2)**:**173–177.
16. **LeChevallier, M. W., W. D. Norton, and R. G. Lee.** 1991. *Giardia* and *Cryptosporidium* spp. in filtered drinking water supplies. *Appl. Environ. Microbiol.* **57:**2617–2621.
17. **Mena, K. D., C. P. Gerba, C. N. Haas, and J. B. Rose.** 2003. Risk assessment of waterborne coxsackievirus. *J. Am. Water Works Assoc.* **95**(7)**:**122–131.
18. **Minor, T. E., C. I. Allen, A. A. Tsiatis, D. B. Nelson, and D. J. D'Alessio.** 1981. Human infective dose determinations for oral poliovirus type 1 vaccine in infants. *J. Clin. Microbiol.* **13:**388–389.
19. **Parshionikar, S. U., S. Willian-True, G. Shay Fout, D. E. Robbins, S. A. Seys, J. D. Cassady, and R. Harris.** 2003. Waterborne outbreak of gastroenteritis associated with a norovirus. *Appl. Environ. Microbiol.* **69:**5263–5268.
20. **Payment, P., É. Coffin, and G. Paquette.** 1994. Blood agar to detect virulence factors in tap water heterotrophic bacteria. *Appl. Environ. Microbiol.* **60:**1179–1183.
21. **Payment, P., and E. Franco.** 1993. *Clostridium perfringens* and somatic coliphages as indicators of the efficiency of drinking water treatment for viruses and protozoan cysts. *Appl. Environ. Microbiol.* **59:**2418–2424.
22. **Payment, P., L. Richardson, J. Siemiatycki, R. Dewar, M. Edwardes, and E. Franco.** 1991. A randomized trial to evaluate the risk of gastrointestinal disease due to consumption of drinking water meeting current microbiological standards. *Am. J. Public Health* **81:**703–708.
23. **Payment, P., J. Siemiatycki, L. Richardson, G. Renaud, E. Franco, and M. Prévost.** 1997. A prospective epidemiological study of gastrointestinal health effects due to the consumption of drinking water. *Int. J. Environ. Health Res.* **7:**5–31.
24. **Payment, P., M. Trudel, and R. Plante.** 1985. Elimination of viruses and indicator bacteria at each step of treatment during preparation of drinking water at seven water treatment plants. *Appl. Environ. Microbiol.* **49:**1418–1428.
25. **Rab, M. A., M. K. Bile, M. M. Mubarik, H. Asghar, Z. Sami, S. Siddiqi, A. S. Dil, M. A. Barzgar, M. A. Chaudhry, and M. I. Burney.** 1997. Water-borne hepatitis E virus epidemic in Islamabad, Pakistan: a common source outbreak traced to the malfunction of a modern water treatment plant. *Am. J. Trop. Med. Hyg.* **57:**151–157.
26. **Regli, S., J. B. Rose, C. N. Haas, and C. P. Gerba.** 1991. Modeling the risk from giardia and viruses in drinking water. *J. Am. Water Works Assoc.* **83:**76–84.
27. **Rendtorff, R. C.** 1954. The experimental transmission of human intestinal protozoan parasites. II. *Giardia lamblia* cysts given in capsules. *Am. J. Hyg.* **59:**209–220.
28. **Schiff, G. M., G. M. Stefanovic, E. C. Young, D. S. Sander, J. K. Pennekamp, and R. L. Ward.** 1984. Studies of echovirus-12 in volunteers: determination of minimal infective dose and the effect of previous infection on infectious dose. *J. Infect. Dis.* **150:**858–866.
29. **Teunis, P. F. M., and A. H. Havelaar.** 1999. *Cryptosporidium in Drinking Water: Evaluation of the ILSI/RSI Quantitative Risk Assessment Framework.* RIVM report 284-550-006. National Institute of Public Health and the Environment (RIVM), Bilthoven, The Netherlands.
30. **U.S. Environmental Protection Agency.** 1989. *Exposure Factors Handbook*, p. 2-1–2-10. EPA/600/8-89/043. Office of Health and Environmental Assessment, U.S. Environmental Protection Agency, Washington, D.C.
31. **Van Heerden, J., M. M. Ehlers, W. B. Van Zyl, and W. O. K. Grabow.** 2003. Incidence of adenoviruses in raw and treated water. *Water Res.* **37:**3704–3708.
32. **Ward, R. L., D. I. Bernstein, E. C. Young, J. R. Sherwood, D. R. Knowlton, and G. M. Schiff.** 1986. Human rotavirus studies in volunteers: determination of infectious dose and serological response to infection. *J. Infect. Dis.* **154:**871–880.

Toxic Photosynthetic Microbes

WESLEY CLINTON JACKSON, JR.

30

Algae use photosynthesis to convert the sun's energy to fuel much of the life that is sustained on Earth. It is thought that chloroplasts in higher plants were derived from cyanobacteria (algal species), and thus, in theory, most all of the energy that is used by life on Earth is attributed to algae and their plant offspring (54, 55). Besides converting the sun's energy into useful sugars through photosynthesis, algal species are also thought to have prepared Earth for higher forms of life by providing the atmosphere essential for heterotrophic respiration (4). These "grasses of the sea" are extremely important for providing the nutrients and food that we eat (85, 87). This is especially true for filter-feeding shellfish, such as oysters, mussels, scallops, and clams (32). There have also been numerous attempts to cultivate phytoplankton as a sustainable, reusable food source (12). Extracts from algae have routinely been used as vitamin supplements (12).

Of the 12,000 total algal species known, about 300 generate blooms in marine environments and cause discoloration of the surface of the water (32, 77). The name "red tide" has been given to these organisms because of their colorful bloom conditions (32), but this title is often misleading because they do not need to be red to have this distinction. Freshwater algal blooms can also cause many types of harmful effects (10). Three major types of harmful effects of algal blooms are hypoxia, resulting from their decomposition, physical interference in fish gills, and the production of toxins (Table 1). While usually posing no threat to human life, there are times when algal bloom species can cause health problems ranging from skin irritation to death (44).

Not all of the toxic effects related to algal blooms are directly related to human health. Ecosystems can be impacted negatively by high algal biomasses resulting from excessive nutrient inputs. Additions of nutrients such as phosphorous and nitrogen from farmlands, animal operations, industrial sources, lawns, golf courses, and atmospheric deposition are the main causes of eutrophication in freshwater environments (32, 49). Eutrophication encourages algal blooms by providing nutrients where, under normal conditions, these nutrients are not available to the microbes (2). The decomposition of extensive algal blooms may result in hypoxic conditions in bottom waters, leading to mortality in fish and benthic invertebrates (49, 77). Large-scale hypoxia is present in the Gulf of Mexico "dead zone," resulting from the large influx of nutrients from the Mississippi River. While not directly affecting human health, there is a significant economic impact in loss of fish and invertebrates in this region (20, 47, 67). About 300 different marine species and freshwater species of algae can produce extensive blooms leading to hypoxia (32).

The second biological process by which toxic photosynthetic microbes can be harmful is also nontoxic to humans, as indicated in Table 1. These algae are known to damage and clog the gills of fish and invertebrates in a defense against predation (32). This type of harmful algae has only been discovered due to the increase in the fish and invertebrate aquaculture industry (49). Increased pressure to limit wild fishing efforts has increased the demand for the aquaculture industry to produce seafood for international markets. Industry standards often use floating holding pens to contain fish during growth stages, and these penned fish can be exposed to blooms that wild fish would normally avoid (49). One of the largest aquaculture industry disasters occurred in 1972, when $500 million worth of yellowtail fish were killed by a bloom of *Chattonella antiqua* in the Seto Island Sea (32). Bloom-induced hypoxia and interference with gill breathing capabilities in aquatic organisms are usually not included in discussions of the harmful impacts of toxic photosynthetic microorganisms. Even though these two processes are not toxic to people, these problems are of growing concern because they continuously threaten the ecosystem fitness of the world's water.

The production of secondary metabolites is the third biological process by which algae are considered toxic (Table 1) (4, 7). Secondary metabolites produced by toxic photosynthetic algae are of the greatest concern to human health (14, 23). The effects of these toxins include nuisance skin rashes, mild to severe gastrointestinal problems, acute liver damage, and extremely potent neurotoxicity (32, 80, 90). The presence of harmful algal bloom species has been linked to increasing rates of liver cancer in some locations where drinking water is not adequately treated (23, 90). These toxins not only endanger human health but also have been found to have toxic effects on many different types of marine life (25). Generally, algae that produce these toxic secondary metabolites are divided into two different groups, the marine (primarily diatoms and dinoflagellates) and the freshwater (cyanobacteria) toxic algae. These two different groups (marine and freshwater) pose very different problems to people and the environment. Both have wide-ranging impacts, but the national focus is greater for

TABLE 1 Three general methods algae use to produce toxicity

Method of toxicity	Examples of algae that use this tactic
Hypoxia	300 species of algae cause blooms that can become hypoxic; the most notable are *Akashiwo sanguinea*, *Scripp-siella trochoidea*, and *Trichodesmium erythraeum*
Clogging/damaging gills	Many diatom, raphidophyte, dinoflagellate, and pymnesiphyte species, including *Chaetoceros convolutes*, *Karenia brevisulcata*, *Cysochromulina polyepis*, *Chattonella antiqua*, and *Heterosigma akashiwo*
Secondary metabolite toxins	Numerous eukaryotic and prokaryotic groups (see Tables 2 and 3 for details)

marine toxic algal species because they affect the international shellfish industry (32). Environmental concerns regarding marine algae are predominantly focused on organisms that may bioaccumulate toxins and impact higher trophic levels (39, 56). The marine toxins fall into six classes, most of which impact the nervous and gastrointestinal systems. Freshwater toxic algal species are mainly implicated in drinking water supplies and waters used for recreational purposes. Freshwater species produce both potent liver toxins and neurotoxins. It has been only in the last 40 years that the structure and mechanism have been understood for many of these toxins, but the adverse human and animal health effects in the presence of marine and freshwater algae have been known for thousands of years.

INTRODUCTION TO HARMFUL ALGAL BLOOMS

The first reference to a harmful algal bloom causing human and environmental consequences came from ancient Egypt, as reported in the Bible (over 1,000 years B.C.E.): ". . . all the water was changed into blood, the fish in the Nile died and the river smelled so bad that the Egyptians could not drink its water. Blood was everywhere" (Exodus 7:17–21). This water bloom was possibly a red tide, though it is not certain whether the fish kills described were due to the effects of algal toxins or hypoxic conditions (Table 1). Algae with the ability to cause human and animal health effects have been a source of concern amongst Western cultures, but other diseases and health problems with the same symptoms were overshadowing, and algae were generally overlooked as being a culprit in many ailments. It was not until the 18th century that the first clear case of death from toxic photosynthetic algae was recognized. The first recorded fatal human case occurred in 1793 among the crew of Captain George Vancouver when he landed in a small bay, later to be known as Poison Cove, in British Columbia, Canada (18). Captain Vancouver recorded that it was taboo among the native people to eat shellfish when the seawater became phosphorescent (18). However, the entire crew consumed the poisoned shellfish, resulting in the death of one crewmember and bringing about the symptoms of what is now called paralytic shellfish poisoning (PSP) in the remainder of the crew. To this day, the causative agents of PSP, saxitoxin (Fig. 1) and its derivatives, are some of the most lethal neurotoxins on the planet (13).

To date, there is no clear correlation between the concentration of harmful algal species and the magnitude of detrimental effects (32). For this reason, visual examination of water discoloration is not a useful measure of the possible toxins present. The following four areas of focus are needed to better understand the rise in toxic algal species: (i) an awareness of the ecosystem dynamics that establish blooms, (ii) an appreciation of microbial physiology and the incen-

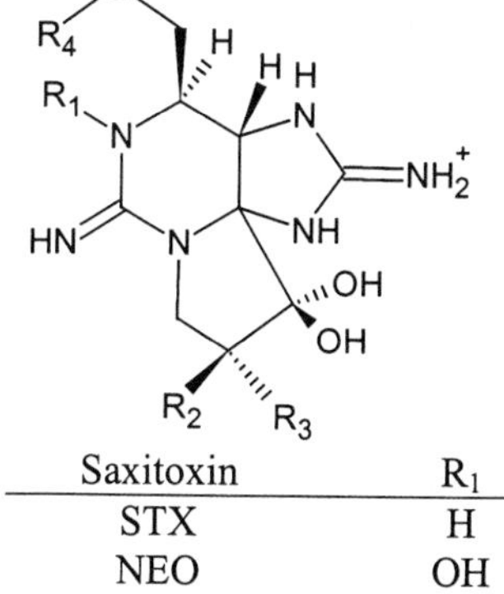

Saxitoxin	R_1	R_2	R_3	R_4
STX	H	H	H	H
NEO	OH	H	H	H
GTX1	OH	H	OSO_3-	H
GTX2	OH	H	OSO_3-	H
C2	H	OSO_3-	H	OH

FIGURE 1 Structure of saxitoxin (STX), the causative agent of PSP. Derivatives of saxitoxin include neosaxitoxin (NEO), gonyautoxins 1 and 2 (GTX1 and GTX2), and carbamoyl toxin 2 (C2).

tives gained by the production of secondary metabolites, (iii) an understanding of nutrient compositions that are conducive to algal bloom conditions, and (iv) the continuous development of high-throughput analytical methods.

MARINE TOXIC ALGAE

There are six significant types of marine poisonings which regularly occur in different parts of the world. These human poisoning events mostly result from the ingestion of fish or shellfish that have accumulated the algal toxins and are named according to the symptoms that follow, i.e., paralytic, diarrheic, amnesic, neurotoxic, and azaspiracid shellfish poisoning and ciguatera fish poisoning (PSP, DSP, ASP, NSP, AZP, and CFP, respectively) (Table 2) (31, 69). The classes of algae responsible for producing the majority of toxin poisonings in the marine environment are dinoflagellates and diatoms. These types of poisoning events and the organisms that produce them are discussed in the following sections.

Paralytic Shellfish Poisoning

Saxitoxin (and its derivatives) is one of the most potent toxins known (Fig. 1). Currently, this toxin is "banned under the chemical weapons convention and regulated under the Public Health Security and Bioterrorism Preparedness and Response Act of 2002" (69). Each year, there are over 2,000 cases worldwide of PSP, with a 5 to 15% mortality rate (32). Symptoms of PSP begin with tingling or numbness around the lips which then spreads to the face and neck. In moderate situations, this toxin can cause a prickling sensation in the extremities, headache, fever, nausea, vomiting, and diarrhea (5). In more severe cases, there can be muscular paralysis followed by respiratory difficulty, and in extreme cases, there is complete respiratory arrest (5).

The mode of action of this toxin focuses directly on the nervous system. Saxitoxins (and their derivatives) have a high affinity for the sodium channels that lie along the axons of neurons. Sodium and potassium are used to transmit the electric signal down the axon from nervous system cells in the brain to muscle and other effector cells. The blockage of potassium channels effectively stops signals from being transmitted from the brain. Saxitoxins block but do not bind sodium channels, so 24 hours after poisoning, the toxin diffuses from the body (10).

The saxitoxins and their approximately 18 known derivatives are found worldwide, with the highest concentrations of poisonings occurring in North America, Europe, and Asia (69). Millions of dollars are lost each year to the closure of shellfish beds due to the presence of this toxin. The algal species that produce these toxins include *Alexandrium catenella*, *Alexandrium minutum*, *Alexandrium tamarense*, *Gymnodinium catenatum*, and *Pyrodinium bahamense* var. *copressum* (32). This toxin is also produced by certain species of freshwater toxic photosynthetic algae, as discussed below (10).

Diarrheic Shellfish Poisoning

Northeastern portions of the United States have been plagued by periodic outbreaks of DSP. Okadaic acid (Fig. 2) and three other closely related toxins, dinophysistoxin, pectenotoxin, and yessotoxin, are all implicated in DSP intoxications (5, 69). Okadaic acid, dinophysistoxin 1 (DTX-1), and DTX-2 are potent protein phosphatase inhibitors, especially to protein phosphatase 2A, which increases the phosphorylation of many intracellular proteins (82). These toxins produce intestinal tract inflammation, diarrhea, and tumors. Pectenotoxins and yessotoxins are not normally implicated in human health effects, but they are lethal to mice in laboratory settings. Pectenotoxins have high actin-depolarizing activities, leading to the hepatotoxic effects seen after intraperitoneal (i.p.) mouse injections (38). Yessotoxins are highly potent to mice by i.p. injections, but it is unclear whether the effects of this toxin are directed to the cardiac or nervous system (26, 86). Both yessotoxins and

TABLE 2 Eukaryotic marine microalgal toxins and species that produce them

Group and syndrome	Toxin(s)	Eukaryotic microalgal species	References
Dinoflagellates			
PSP	Saxitoxin	*Alexandrium catenella*, *A. cohoriticula*, *A. fundyense*, *A. fraterculus*, *A. leei*, *minutum*, *A. tamarense*, *Gymnodinium catenatum*, *Pyrodinium bahamense* var. *copressum*	5, 10, 32, 69
NSP	Brevetoxin	*Karenia brevis*, *K. papilionacea* in editus, *K. seliformis* in editus, *K. bidigitata* in editus	1, 5, 24, 25, 32, 69
DSP	Okadaic acid, DTX-1, and DTX-2	*Dinophysis acuta*, *D. acuminate*, *D. caudate*, *D. fortii*, *D. norvetica*, *D. mitra*, *D. rotundata*, *D. sacculus*, *Prorocentrum lima*	5, 26, 32, 38, 69, 82, 86
AZP	Azaspiracid	*Protoperidinium crassipes*	7, 32, 45, 71
CFP	Ciguatera, sacritoxin, gambiertoxin, cooliatoxin, palytoxin, okadaic acid, maitotoxin	*Gambierdiscus toxicus* and other, unknown species	5, 30, 32, 37, 45, 50, 69
PEAS	Unknown agent(s)	*Pfiesteria piscicida*, *P. shumwayae*	7, 29, 48, 79, 89
Diatoms			
ASP	Domoic acid	*Pseudo-nitzschia multiseries*, *P. pungens*, *P. pseudodelicatissima*, *P. australis*, *P. seriata*, *P. delicatula*	32, 46, 49, 66, 73, 74, 81

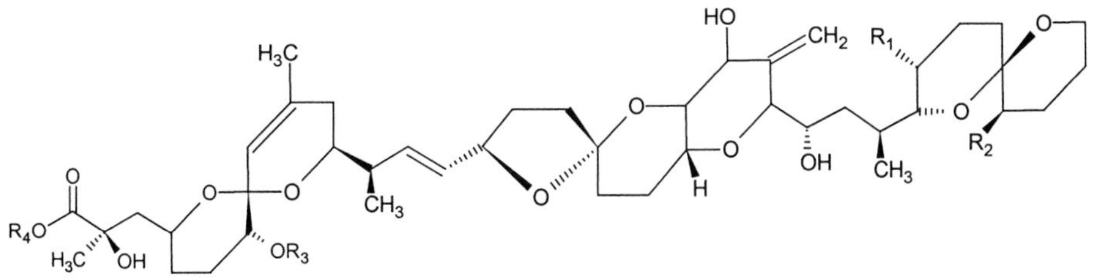

Okadaic acid	R_1	R_2	R_3	R_4
OA	CH_3	H	H	H
DTX1	CH_3	CH_3	H	H
DTX2	H	CH_3	H	H

FIGURE 2 Structure of okadaic acid (OA). Okadaic acid is the primary toxin associated with DSP events. The derivatives of okadaic acid depicted in the table are dinophysistoxins 1 and 2.

pectenotoxins are less potent in their effects to mice when administered orally (26, 86). Pectenotoxins and yessotoxins have not been implicated directly in human health effects but are often found associated with okadaic acid and dinophysistoxin in algal blooms (69).

Human effects observed during DSP intoxication usually occur within the first 12 h. Acute diarrhea, nausea, and vomiting are accompanied by intense abdominal pain (32). Patients afflicted with this poisoning recover within 3 days. Little is known about the impact of chronic exposure, but the inhibition of protein phosphatase during DSP suggests that it is a tumor promoter and may promote tumors in the digestive system (32).

The algal species that produce the toxins associated with DSP poisoning are *Dinophysis acumuinata*, *Dinophysis acuta*, *Dinophysis fortii*, *Dinophysis norvegica*, and *Prorocentrum lima*. These organisms only produce some of the toxins seen in shellfish during poisoning events. Many of these toxins are metabolized in the shellfish that ingest them, which often complicates detection efforts (82).

Neurotoxic Shellfish Poisoning

Brevetoxin is a polyether toxin responsible for NSP (Fig. 3). *Karenia brevis* (formerly known as *Gymnodinium breve* and *Ptychodiscus brevis*) is the main species responsible for producing brevetoxin. Other species that are implicated in the production of brevetoxins are *Karenia papillomacea* in editus, *Karenia seliformis* in editus, and *Karenia bidigitata* (32). Blooms of these *Karenia* species arise annually in the Gulf of Mexico, with the majority of NSP outbreaks occurring along Texas shores and the western coast of Florida. Fish kills, marine mammal deaths (bottlenose dolphins and Florida manatees), and shellfish intoxication can all occur when high concentrations of this toxin are ingested (1, 25). The effects of eating contaminated shellfish in humans occur within 1 to 3 h of ingestion and include numbness, vomiting, diarrhea, and abdominal pain. *Karenia* cells are also very fragile, breaking open easily with moderate wave action. This can cause brevetoxins to aerosolize, which can trigger acute eye irritations and respiratory complications (24, 69).

Brevetoxin binds neurotoxin site 5 on the voltage-gated sodium channels, keeping the channels open and inducing an influx of sodium inside these cells (24). This change allows the neurons to more easily depolarize, firing neural signals down the axons of neurons at a lower activation threshold. This lower threshold causes a more excitatory state for the gastrointestinal and neurological symptoms (5). The symptoms of NSP are very similar to those of CFP.

Ciguatera Fish Poisoning

CFP typically occurs because of the consumption of predatory reef fish that have concentrated toxins through biomagnification. *Gambierdiscus toxicus* is the best-known ciguatera toxin producer; other dinoflagellates thought to produce this toxin are *Ostreopsis siamensis* and *Coolia monotis* (32). These organisms live among coral reefs of Florida, the Virgin Islands, Puerto Rico, Hawaii, and many of the Pacific Islands (5, 69). *Gambierdiscus toxicus* normally lives as

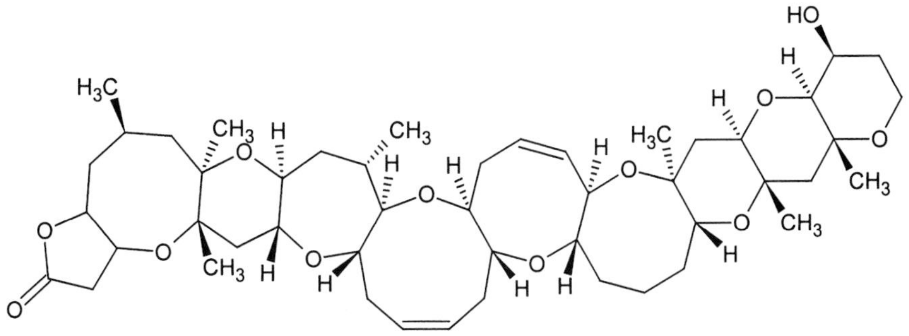

FIGURE 3 Brevetoxin structure. Derivatives of this toxin are commonly associated with NSP occurrences.

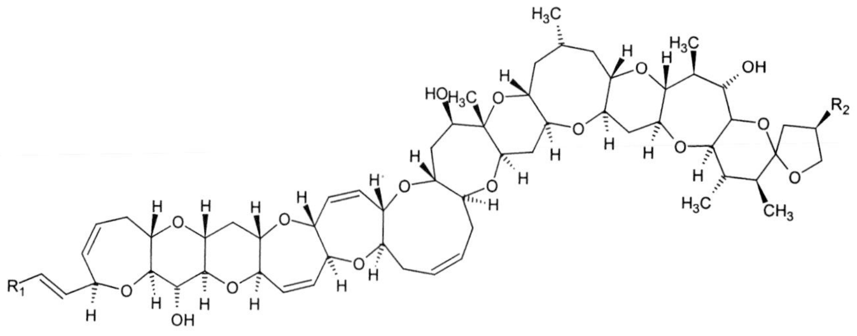

FIGURE 4 Structure of Pacific ciguatera toxin. This toxin is 10 times more potent than its Atlantic counterpart.

an epiphyte on macroscopic algae that are common to most coral reefs (45). Small reef fish feed on these macroscopic and turf algae, and these fish are then fed upon by larger fish, continuing up the food chain until the toxin is concentrated in a top predator, such as a barracuda or grouper. Over 400 different fish species have been involved in CFP (69).

The structure of ciguatera is very similar to that of brevetoxin (Fig. 4) (32), and both bind the same site on the voltage-gated sodium channels, so the symptoms they cause are very much the same (5). The symptoms of CFP are also similar to those of NSP, including diarrhea, vomiting, itching, and muscle pain; neurological effects can last from days to years (5). Toxic effects usually occur about 6 h after ingestion of toxin-containing fish. Ciguatera appears to be somewhat more toxic than brevetoxin, affecting an estimated 200,000 to 1,000,000 people annually, all in tropical and subtropical regions (5, 50).

Fish containing ciguatera often have additional toxins that are molecularly different from ciguatera and result in an increase in the toxicity of the fish (5, 50). Most of these toxins are also lipophilic, which allows them to be biomagnified in the same manner as ciguatoxin. Sacritoxins and gambiertoxins are almost identical to ciguatoxins, containing a lipophilic backbone. Other lipophilic toxins associated with CFP include cooliatoxins (neurotoxin), palytoxins (neurotoxin), and okadaic acid (DSP toxin) (30, 37, 50). Maitotoxin has been implicated in ciguatera poisonings, but this hydrophilic toxin cannot be biomagnified due to its high affinity for water (50).

Amnesic Shellfish Poisoning

The first major outbreak of ASP occurred in Canada in 1987, and since then, ASP has been a significant danger to marine wildlife and humans (49, 66, 74, 81). Domoic acid, the toxin causing ASP, is the only known toxin produced by the Bacillariophyceae, or diatoms. Diatoms are mostly unicellular organisms that can be found in almost all aquatic habitats (49). Domoic acid (Fig. 5) occurs in marine waters and is produced by *Pseudo-nitzschia multiseries*, *Pseudo-nitzschia pungens*, *Pseudo-nitzschia pseudodelicatissima*, *Pseudo-nitzschia australis*, *Pseudo-nitzschia seriata*, and *Pseudo-nitzschia delicatula* (32). Both coasts of North America have been afflicted with major ASP intoxications. Observations of low levels of domoic acid have been reported in Japan, Europe, New Zealand, and Australia (32). Human exposure to domoic acid occurs through shellfish consumption (46).

Domoic acid acts by mimicking the neuroexcitatory amino acid L-glutamic acid (49), which communicates messages across synaptic gaps to the next neuron. Domoic acid replaces L-glutamic acid in the receptor, causing the next neuron to fire. Research suggests that binding of domoic acid causes hyperexcitation targeting the hippocampus in the brain. Overstimulation by domoic acid increases the intracellular calcium ion levels in the cells, disrupting cellular homeostasis and causing cell death (46). Common symptoms of ASP include gastrointestinal distress, vomiting, cardiac fluctuations, disorientation, seizures, mild to severe memory loss (amnesia), and coma (32, 73). Besides the human implications of domoic acid poisoning, mortality in other animals, such as pelicans, cormorants, loons, grebes, sea otters, dolphins, sea lions, and humpback whales, has been attributed to domoic acid poisoning (73, 81).

Azaspiracid Poisoning

Azaspiracid is a relatively new toxin that was discovered in 1995 and named after its unusual azaspiro ring and carboxyl group (Fig. 6) (7, 32, 45, 71). The discovery of this toxin was delayed because it was misidentified as DSP toxin because of the similarity of the symptoms caused (45, 71). However, the lack of DSP toxins and the severity of symptoms led to more research on these unusual shellfish poisonings. The first confirmed case of AZP intoxication was in The Netherlands, in shellfish that were imported from the western coast of Ireland (45, 71). Since the discovery of azaspiracid on the Irish coast, the presence of this toxin has been established all along the western coast of Europe (45).

This recently emerging group of toxins is produced by the dinoflagellate *Protoperidinium crassipes* (32). Since the discovery of the original toxin (AZA1), four additional analogues have been identified (AZA2 to AZA5) (32). All of the forms of azaspiracid cause symptoms identical to those of DSP, including nausea, stomach cramping, vomiting, and severe diarrhea (32, 45, 71). One difference between DSP and AZP is that DSP toxin accumulates mainly in the hepatopancreases of shellfish, while azaspiracid toxin is found distributed throughout the shellfish.

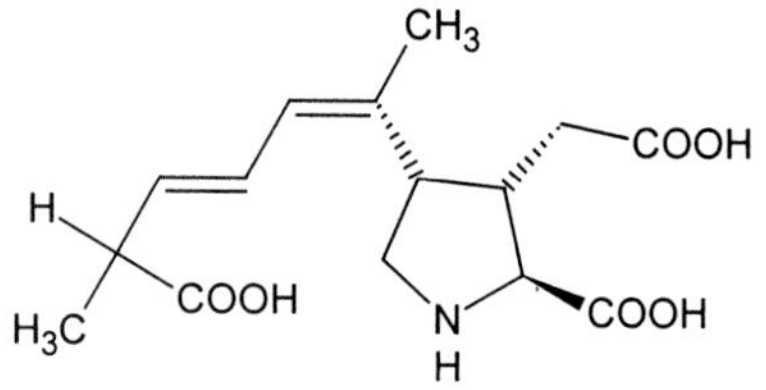

FIGURE 5 Structure of domoic acid, the toxic agent responsible for ASP.

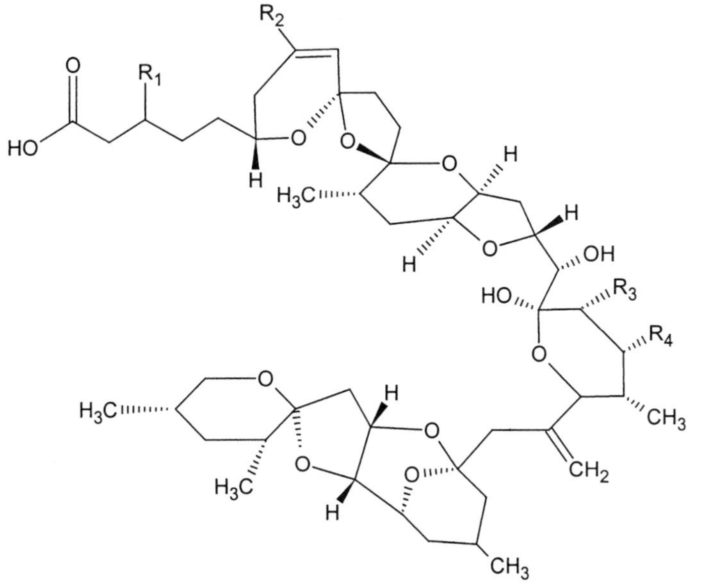

Azaspiracid	R_1	R_2	R_3	R_4
AZA	H	H	CH_3	H
AZA2	H	CH_3	CH_3	H
AZA3	H	H	H	H
AZA4	OH	H	H	H

FIGURE 6 Structure of azaspiracid. The table depicts commonly found azaspiracid derivatives.

CYANOBACTERIA (BLUE-GREEN ALGAE)

Cyanobacteria (blue-green algae) are phototrophic prokaryotes that live primarily in freshwater and terrestrial environments, while only a few species thrive in marine environments (32, 76). Many toxic (Table 3) and therapeutic agents have cyanobacterial origins (9, 32, 90). Marine cyanobacteria generally produce toxins classified as lipopolysaccharides (LPS), contact irritants, and neurotoxins. Cyanobacterial species found in brackish to freshwater typically produce hepatotoxic and neurotoxic compounds (Table 3).

Lipopolysaccharides

The surfaces of cyanobacteria and all other gram-negative bacteria have an inner phospholipid membrane, a cell wall made of peptidoglycan, and an outer membrane. LPS is part of the outer cell suprastructure anchored to this outer membrane (68). Most species of marine and freshwater cyanobacteria have LPS components as part of their outer membrane structures (32). Effects commonly seen as a result of LPS poisoning include allergenic symptoms, such as inflammation and fever, in individuals who have bathed or showered in water containing cyanobacteria (31). Compared to the LPS of other heterotrophic bacteria, the clinical effects of cyanobacterial LPS are relatively mild (31, 32, 68).

Contact Irritants

The cyanobacterial genus that is normally associated with contact irritation is *Lyngbya*. *Lyngbya* is one of the most diverse genera of toxic photosynthetic microbes. Species of *Lyngbya* are known to live in many different environments, ranging from full marine to brackish and even freshwater conditions. Over 100 novel bioactive secondary metabolites have been isolated from cultures of *Lyngbya majuscula* alone (58). The toxins specific to *Lyngbya* dominate this section. For the sake of brevity, only the most potent of the toxins produced by *Lyngbya* in marine environments are discussed here (52, 65).

Lyngbya majuscula has been found to produce over 100 unique secondary metabolites (59, 83). These toxins collectively cause what is known as swimmer's itch or seaweed dermatitis (59, 83). The first toxins that were attributed to causing contact dermatitis were aplysiatoxin and debromoaplysiatoxin (83). These toxins both cause severe erythematous (redness) and papulovesicular (erupting papules and vesicles) dermatitis. Aplysiatoxin and debromoaplysiatoxins are toxic to mice, causing bleeding through capillaries in the small intestine after i.p. injection. Aplysiatoxins are also potent tumor promoters in mice, activating protein kinase C. Other cyanobacterial species that produce aplysiatoxins are *Oscillatoria nigroviridis* and *Schizothrix calcicola* (32). Aplysiatoxin has also been implicated in food poisonings at localities where *Lyngbya* grows as an epiphyte on the leaves of the red algae *Gracilaria coronopifolia* (59, 60). Symptoms of this poisoning include abdominal pain, diarrhea, vomiting, a burning and itching sensation on the skin, mouth, and throat, loss of consciousness, convulsions, and low blood pressure (41). Several deaths have been associated with these events (41).

Another important, chemically distinct contact irritant is lyngbyatoxin A. This toxin is normally found in shallower waters than those containing aplysiatoxins (9) but causes symptoms very similar to those caused by aplysiatoxin (rash and inflammation) (42). Lyngbyatoxin A also causes severe oral and gastrointestinal inflammation (42). In mice, sublethal doses caused erosion of the stomach,

TABLE 3 Freshwater and marine prokaryotic algal toxins and cyanobacterial species known to produce them

Toxin class	Toxin(s)	Prokaryotic microalgal species	References
Toxic shock syndrome	LPS	All bacteria, some algal species	31, 32, 68
Marine			
Contact irritants	Aplysiatoxin, lyngbyatoxin A	*Lyngbya majuscula*	9, 22, 32, 41, 42, 52, 58, 59, 60, 65, 83
Neurotoxins	Antillatoxin, kalkitoxin	*L. majuscula*	3, 51, 64, 92
Brackish to freshwater			
Hepatotoxins	Microcystin, cylindrospermopsin, nodularin	*Anabaena* spp., *Microcystis* spp., *Oscillatoria (Planktothrix)* spp., *Nostoc* spp., *Cylindrospermopsis raciborskii*, *Anabaenopsis* spp., *Hapalosiphon* spp.	19, 27, 28, 34, 36, 40, 57, 61, 70, 76, 90
Freshwater			
Neurotoxins	Saxitoxin, anatoxin-A, anatoxin-A(S), *Lyngbya wollei* toxins, BMAA	*Aphanizomenon flos-aquae*, *Anabaena circinalis*, *C. raciborskii*, *L. wollei*, *Nostoc* spp.	10, 11, 53, 76, 78, 90

small intestine, cecum, and large intestine (42). Edema was also observed in the lungs, with infiltration of inflammatory cells (42). The structure of lyngbyatoxin A is identical to that of teleocidin and olivoretin metabolites, toxins produced by several strains of streptomycetes (9, 22, 32).

Marine Cyanobacterial Neurotoxins

Neurotoxins produced by the pantropical marine cyanobacterium *Lyngbya majuscula* were first discovered due to their ichthyotoxic properties (64, 92). Antillatoxin is considered one of the most potent ichthyotoxins, second only to brevetoxin (51). Antillatoxin has also been implicated in human health cases where afflicted individuals suffer from respiratory irritation, eye inflammation, and severe contact dermatitis (51). This toxin activates mammalian voltage-gated sodium channels, causing them to open and promoting sodium influx into the cell. This is similar to the action of brevetoxin (51, 64).

Kalkitoxin is another neurotoxin produced by *Lyngbya majuscula*. This toxin is structurally distinct from antillatoxin but causes many of the same effects by targeting the voltage-gated sodium channels (92). The effects observed with kalkitoxin are somewhat delayed compared to those caused by antillatoxin (3). No clinical health effects have been attributed directly to this toxin, but cytological studies suggest the potential of detrimental human health effects (3).

Hepatotoxins

Liver toxins produced by cyanobacteria are the source of many animal and human algal intoxications. Livestock and pets often suffer toxic effects by exposure to freshwater ponds where cyanobacteria are growing, most likely by ingestion of the contaminated water. The earliest report of cyanobacterial intoxication was reported in 1878 from Lake Alexandria, southern Australia (27). *Nodularia spumigena* was the cyanobacterium present and later was found to produce the seven-amino-acid cyclic hepatopeptide microcystin and the closely related five-amino-acid cyclic pentapeptide nodularin.

Microcystin is one of the most widely distributed toxins on the planet and has been found in fresh, brackish, and marine waters on all seven continents, including Antarctica (36, 76). There are over 70 congeners in the microcystin family of toxins (Fig. 7). The most potent toxin in this family is microcystin-LR. The LR designation comes from two variable portions of the amino acid structure (L stands for the amino acid leucine, and R stands for arginine) (76). All of the congeners differ with respect to the amino acids found at these two positions. Microcystin is generally found in fresh- to brackish-water environments, and if present in marine systems, is usually at very dilute concentrations. At the present time, microcystin is not recognized as a "marine" problem. However, managers of freshwater reservoirs, lakes, ponds, and estuaries are very concerned with the presence of this naturally occurring toxin.

The organisms that produce microcystins have been identified through the use of axenic cultures to ensure that the toxins are being produced by cyanobacterial strains and not by associated bacteria. Microcystin-producing cyanobacterial genera include *Anabaena*, *Microcystis*, *Oscillatoria (Planktothrix)*, *Nostoc*, and *Anabaenopsis*. The terrestrial cyanobacterial isolates from the *Hapalosiphon* genus also produce microcystins. One of the most unusual microcystin producers is symbiotic *Nostoc* organisms, which live in lichen communities and have been found to produce unusual microcystins within this symbiotic relationship (63, 76).

Nodularin (Fig. 8) is very closely related to microcystin in structure and function. There are four different structures associated with nodularin, as well as a fifth closely related toxin called motuporin that was isolated from the sponge *Theonella swinhoei* (19). Nodularin has a five-amino-acid pentapeptide structure that is produced exclusively by *N. spumigena*. Both microcystin and nodularin are potent inhibitors of class 1 and 2A protein phosphatases. The toxicity from these toxins comes from the unusual 3-amino-9-methoxy-10-phenyl-2,6,8-trimethyldeca-4,6-dienoic acid (Adda), which acts on the serine/threonine phosphatase region of protein phosphatases, rendering them useless to the cell (33, 70). Protein phosphatases are important for the

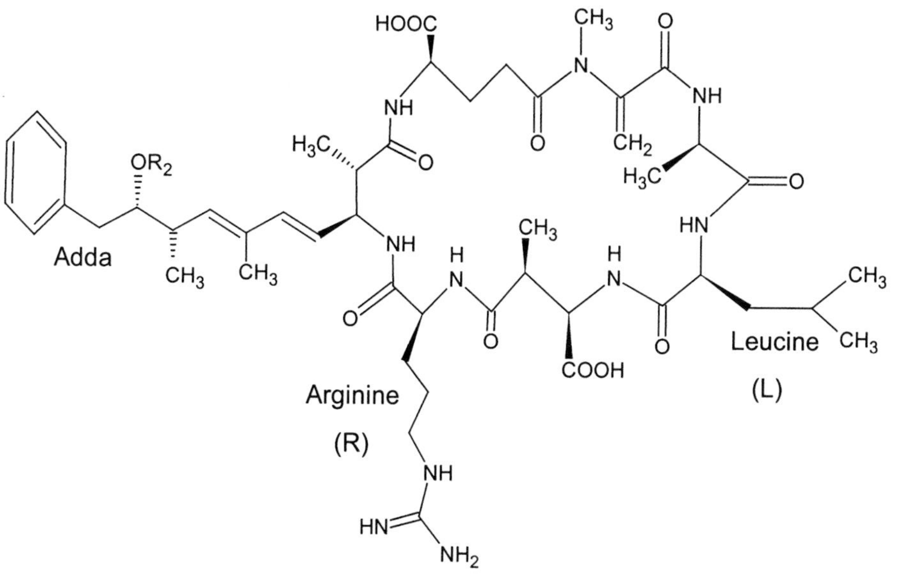

FIGURE 7 Seven-amino-acid structure of microcystin-LR. Of the 65 known microcystin derivatives, the leucine-arginine (or LR) variant is the most toxic. The toxicity of this compound originates from the Adda amino acid.

control of enzymes in eukaryotic cells, and these enzymes are either active or inactive depending on the phosphorylation state. Microcystins and nodularins preferentially target hepatocytes via the bile transport system and inhibit protein phosphatases in these cells, causing these key enzymes to be hyperphosphorylated (70). This situation leads to a perturbation in the cytoskeletal structure, resulting in a conformational change (70). When viewed under an electron microscope, hepatocytes appear to go through apoptosis (observed by blebbing of apoptotic bodies, reorganization of microfilaments, and dilation of the endoplasmic reticulum) (90). The apoptotic bodies and cellular remnants clog sinuses (microscopic arterioles, venules, and capillaries) of the liver that normally help shuttle blood cells about the organ. Clogged sinuses cause the liver to expand, causing hepatomegaly (90). In the worst cases, hemorrhagic necrosis leads to liver failure. The liver is the most obviously affected organ in the body, but microcystin and nodularin have also been found to have harmful renal and gastrointestinal effects (57, 61). Okadaic acid (described above) toxicity occurs in the same manner as microcystin and nodularin toxicity (90). Protein phosphatase-inhibiting toxins are potent tumor promoters in vivo (15, 62, 94).

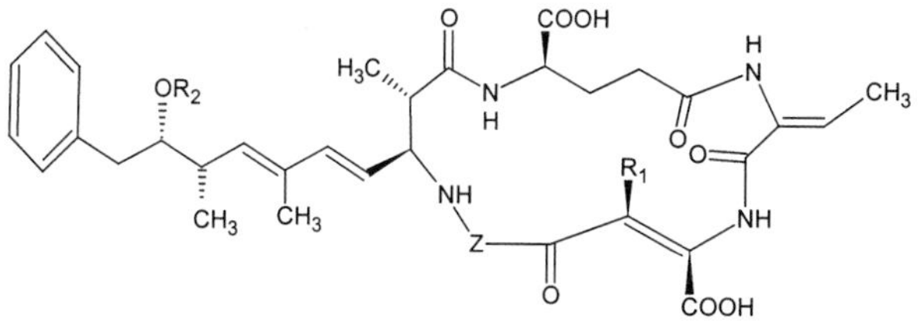

FIGURE 8 Nodularin structure. This structure is very similar to that of microcystin, except that this hepatotoxin is only made up of five amino acids. The Z represents the variable amino acid normally occupied by an arginine.

Cylindrospermopsin (Fig. 9) is another toxin of great concern which is produced by cyanobacteria in brackish and freshwater. The cyanobacterial species capable of producing cylindrospermopsin are *Cylindrospermopsis raciborskii*, *Aphanizomenon ovalisporum*, *Umezakia natans*, and *Raphidopsis curvata* (28). This toxin blocks protein synthesis of eukaryotic cells in an irreversible fashion and inhibits glutathione synthesis. Conventional blocking of protein synthesis does not produce the acute effects seen in cylindrospermopsin toxicity. Factors involving the metabolites generated by cytochrome P-450 are thought to increase toxicity, but the mechanism is not known. Presently, work is being done to elucidate these cytochrome P-450 factors involved in cylindrospermopsin toxicity (28, 40). Other target organs that are affected by cylindrospermopsin are the kidneys, spleen, intestine, heart, and thymus.

A *C. raciborskii* bloom resulted in an outbreak of hepatoenteritis in a community in Palm Island, Australia, sending a total of 149 people (139 children) to the hospital for what was then termed a "mystery disease" (28, 34, 40). Ninety-six percent of the patients treated needed intravenous therapy, while the patients with the most serious cases went into hypovolemic/acidic shock. Most patients fully recovered. The onset of symptoms happened a few days after a major water supply was treated with copper sulfate, which is a common algaecide used to control dense algal blooms (34, 40). This mystery disease was later identified as cylindrospermopsin poisoning (34).

Freshwater Cyanobacterial Neurotoxins

The first discovered neurotoxins produced by cyanobacteria were derivatives of saxitoxin, the causative agent implicated in PSP. These cyanobacterially produced saxitoxins have also been labeled aphanatoxins because they were first found in *Aphanizomenon flos-aquae* (90). The toxicity of aphanatoxins results from their affinity for sodium-gated channels existing along unmyelinated portions of axons.

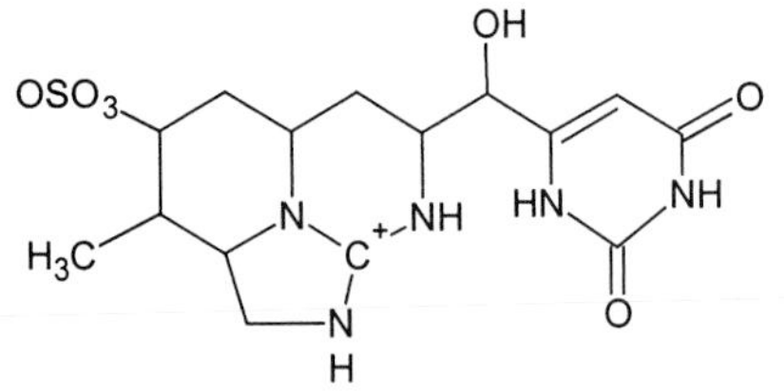

FIGURE 9 Cylindrospermopsin structure. This hepatotoxin is a potent inhibitor of protein synthesis.

Axons are covered with a fatty myelin sheath that insulates and facilitates the electrical signal as it is passed down the axon. Myelin does not cover the entire axon. There are short gaps where the axons are exposed, and these gaps are where sodium and potassium channels propagate the electrical signal. The structure, size, and charge of saxitoxins allow them to block sodium channels along the axes of neurons, which are responsible for carrying the electrical impulses from one location to another. In fatal cases, the afferent nerves to the lungs are bound with aphanatoxins, preventing the nerve impulses from reaching the diaphragm and resulting in respiratory arrest. Other cyanobacteria that produce aphanatoxins (or saxitoxin derivatives) are *Anabaena circinalis*, *Cylindrospermopsis raciborskii*, and *Lyngbya wollei*. *Lyngbya wollei*, a freshwater relative of *Lyngbya majuscula*, produces toxins that appear to be unique to this species and are called *Lyngbya wollei* toxins (90).

Anatoxins [anatoxin-A and anatoxin-A(S)] are alkaloid toxins produced by *Aphanizomenon flos-aquae*. Anatoxin-A is a potent postsynaptic cholinergic nicotinic agonist. This toxin produces a depolarizing neuromuscular blockade, inhibiting the propagation of action potentials (11). Anatoxin-A(S) also blocks signals at the neuromuscular junction, with "(S)" signifying the marked salivation that occurs when mice are exposed to this toxin. Anatoxin-A and anatoxin-A(S) disrupt similar targets to those of the aphanatoxins (PSP toxins or saxitoxins) but are different in structure and function and have different modes of action (10).

In the absence of neurotoxins, action potentials move down the neurocyte axon and approach the neuromuscular junction. The electric impulse is turned into a chemical signal by the release of acetylcholine. Acetylcholine traverses the junction and binds to the muscle cell receptor, causing the muscle to contract. Acetylcholine is then released back into the neuromuscular junction, where it is quickly broken down by acetylcholine esterase. Anatoxin-A is an acetylcholine mimic, and as it infiltrates the neuromuscular junction, it binds the muscle receptor, causing contraction (53). Anatoxin-A is not broken down by acetylcholine esterase, so it can rebind the muscle receptor and stimulate additional muscle contractions. These contractions overexert the muscle and cause fatigue and, potentially, respiratory arrest (53, 78). Anatoxin-A(S) binds covalently to the acetylcholine esterase so that the acetylcholine cannot be broken down (78). Anatoxin-A(S) is 10 times more potent than anatoxin-A in toxicological studies performed in vivo (78).

BMAA

The Chamorro people of Guam have an unusually high incidence of amyotrophic lateral sclerosis/parkinsonism-dementia. Earlier research on cycad plants revealed that these plants contain α-amino-β-methylaminopropionic acid (BMAA), a neurotoxic substance (88). The Chamorro people use the seeds of these plants to make flour for tortillas and other food items, but the flour of the cycad plant does not contain high enough concentrations of BMAA to cause the amyotrophic lateral sclerosis/parkinsonism-dementia symptoms that were observed (21). The extremely large amounts of BMAA observed in patients were proposed to come from a biomagnified source. In 2002, it was discovered that flying foxes contain very large amounts of BMAA (about 3,556 μg/g of body weight), that cycad seeds are an important part of their diet, and that the Chamorro people consider the flying fox a delicacy (16). BMAA concentrations at the roots of the fruits of the cycad plant average 0.3 μg/g (17), and BMAA is carried up through the roots to the fruits of the plant, where the seed pods contain up to 1,161 μg of BMAA/g (17). Further research found the cyanobacteria *Nostoc* spp. associated with the roots of the cycad plant and potentially responsible for the neurotoxic effects seen in humans (17). It was hypothesized that the plant uses the toxin produced by the cyanobacteria as an antiherbivore agent. Flying foxes forage on the cycad seedpods, regardless of the concentration of BMAA, and the concentration of BMAA in flying foxes is well above the range needed to cause the neurotoxic effects observed (17).

TOXIC POISONINGS UNDER INVESTIGATION

Avian Vacuolar Myelinopathy

Lakes and ponds in the southeastern United States contain a potential neurotoxin that harms avian species. Since 1994, there have been over 100 bald eagle deaths and thousands of coot deaths attributed to this presumed toxin, with the birds showing symptoms of avian vacuolar myelinopathy (AVM) (91). AVM begins with difficulty flying and a loss of coordination, and autopsies of affected birds showed widespread, bilaterally symmetrical diffuse spongy degeneration of white matter in the central nervous system (84). Feeding studies have found that the causative agent of AVM is an epiphytic cyanobacterium which grows on the leaves of the hydrilla, covering 95% of plants in areas affected by AVM (91). This epiphytic cyanobacterium has yet to be positively identified or named, but it is believed to belong to the order *Stigonematales*. Studies are under way to elucidate the structure of the toxin and determine management practices to improve the health of affected animals (91).

Possible Estuary-Associated Syndrome

The presence of the toxin associated with the possible estuary-associated syndrome (PEAS) is under intense study. In the early 1990s, fish kills along the Neuse and Pamlico estuaries of eastern North Carolina led to research which suggested that a potent toxin was responsible for the death of over a million fish, mostly menhaden (*Brevoortia tyrannus*) (7). That study found novel dinoflagellate species that were toxic to both fish and shellfish (7). The dinoflagellates were described as ambush predators with flagellated and amoeboid forms, which possess a long appendage, called a peduncle, that they use to feed on fish flesh (7). It was hypothesized that these newly discovered dinoflagellates, named *Pfiesteria piscicida* and *Pfiesteria shumwayae*, produced a potent neurotoxin that killed fish with their ambush predation tactics (6, 29, 79, 89). Controversy surrounds these two dinoflagellate strains regarding whether these organisms kill fish by producing a potent neurotoxin or solely by ambush micropredation (29, 79, 89). The human effects of exposure to this toxin in aerosolized form are thought to include mood swings, difficulty reading, and severe memory loss (48).

Trichodesmium and Tamandaré Fever

The marine cyanobacterium *Trichodesmium* is well studied outside the toxic algal arena because of its massive blooms and nitrogen fixation capabilities. This organism is thought to be responsible for one-fourth of the world's nitrogen fixation (35). This nonheterocystous cyanobacterium has also long been implicated in having toxic effects (8, 72). In 1963, large blooms of *Trichodesmium* spp. plagued the northeastern coast of Brazil, causing Tamandaré fever (72). This affliction begins with respiratory trouble followed by high fever and also includes neurotoxic effects causing muscular, articular, and postorbital pain (72). Sometimes a rash of the thorax and arms accompanies these symptoms (72). Since the 1960s, researchers have attempted to elucidate the causative toxin by using an array of biological and chemical assays (8, 32, 35, 44).

CONCLUSION

The growing human population has forced society to place higher demands on our natural resources, perhaps, as a result, pushing algae and humans into closer contact than ever before (43). Eutrophication is a direct indicator of human presence in many environments (43). Toxic photosynthetic microbes are a few significant species of algae that can cause major damage to our individual, societal, and economic health (2, 10, 23, 32, 47, 77, 90). The stimulation of algal growth through continuous disposal of nutrients into both freshwater and oceanic systems can lead to catastrophic human and economic loss if mandates and warnings are not in place to ensure the safe consumption of seafood and shellfish as well as protection of the world's drinking water supplies (2, 10, 23, 32, 47, 77, 90). Countries must work together to ensure the health of their water resources to make sure that, in turn, the water that they use will not cause them harm. For this goal, three points must be emphasized.

Firstly, reference materials must be produced for all toxins known to affect human health. Standards must be available to researchers for accurate detection and monitoring of toxic microbial species. These reference materials must be accompanied by high-throughput detection assays so that sample processing can be completed quickly enough to be preemptive and allow for early warnings. To do this, certification laboratories must be equipped to safely handle and responsibly distribute these active secondary metabolites. Also, certified detection laboratories must be equipped with accurate testing equipment. Secondly, the facts concerning toxic photosynthetic microbes must be passed to the people of the world in a morally responsible manner. Science taught from a responsible humane standpoint may be capable of educating the world to be aware of these common but periodic afflictions. As scientists, we must do our part in communicating our science and informing local, state, and national governments when there are problems. We must also be humble enough to understand and not overstate the extent to which problems may exist. We must be able to reproduce results to ensure the safety and reliability of findings. Lastly, continuing research studies must be performed to ensure that the safety of water is maintained.

One of the most interesting findings discussed here is that some toxins have very different chemical structures but have the same toxic effects. An example of this finding includes the toxins okadaic acid (which is implicated in DSP), microcystin, nodularin, and motuporin. Each of these toxins has a different chemical structure, but they have convergence with respect to their modes of action and harmful responses. Interest-ingly, in the case of microcystin, organisms do not actively pump the generated toxic compound out of the cells, as would be the case with most known toxins. Instead, this toxin is concentrated and possibly stored around the thylakoids (photosynthetic regions) of these cells (75, 93). This storage of a toxic compound may suggest that the ultimate function of microcystin and okadaic acid is to act as chemoprotectors against excessive light or to reduce reactive oxygen species within these cells (75, 93).

Appreciating and fully comparing biological information is nearly impossible at this time; that is for the future. In these times, we are learning that microbes are the biochemical toolboxes of the planet. Each group of microbes has different compounds and processes that are uniquely distinct to each species and strain. In the photosynthetic world of microbes, toxic processes are vastly unique but can be extremely dangerous to people. At the same time that we gain enlightenment about the biology of these toxic photosynthetic microbes, we improve our understanding of the effects of their toxins, and communicating this well represents the key to avoiding future human poisonings.

REFERENCES

1. **Baden, D. G., and T. J. Mende.** 1982. Toxicity of two toxins from the Florida red tide marine dinoflagellate, *Ptychodiscus brevis*. *Toxicon* **20:**457–461.
2. **Bartram, J., W. W. Carmichael, I. Chorus, G. Jones, and O. M. Skulkberg.** 1999. Introduction, p. 1–14. *In* I. Chorus and J. Bartram (ed.), *Toxic Cyanobacteria in Water*. E & FN Spon, London, United Kingdom.
3. **Berman, F. W., W. H. Gerwick, and T. F. Murray.** 1999. Antillatoxin and kalkitoxin, ichthyotoxins from the tropical cyanobacterium Lyngbya majuscula, induce distinct temporal patterns of NMDA receptor-mediated neurotoxicity. *Toxicon* **37:**1645–1648.
4. **Berman-Frank, I., P. Lundgren, and P. Falkowski.** 2003. Nitrogen fixation and photosynthetic oxygen evolution in cyanobacteria. *Res. Microbiol.* **154:**157–164.
5. **Brett, M. M.** 2003. Food poisoning associated with biotoxins in fish and shellfish. *Curr. Opin. Infect. Dis.* **16:** 461–465.
6. **Burkholder, J. M., A. S. Gordon, P. D. Moeller, J. M. Law, K. J. Coyne, A. J. Lewitus, J. S. Ramsdell, H. G. Marshall, N. J. Deamer, S. C. Cary, J. W. Kempton, S. L. Morton, and P. A. Rublee.** 2005. Demonstration of toxicity to fish and to mammalian cells by Pfisteria species: comparison of assay methods and strains. *Proc. Natl. Acad. Sci. USA* **102:**3471–3476.
7. **Burkholder, J. M., E. J. Noga, C. H. Hobbs, and H. B. Glascow.** 1992. New 'phantom' dinoflagellate is the causative agent of major estuarine fish kills. *Nature* **358:** 407–410.
8. **Capone, D. G., J. Zehr, H. Paerl, B. Bergman, and E. Carpenter.** 1997. *Trichodesmium*, a globally significant marine cyanobacterium. *Science* **276:**1221–1229.
9. **Cardellina, J. H., II, F.-J. Marner, and R. E. Moore.** 1979. Seaweed dermatitis: structure of lyngbyatoxin A. *Science* **204:**193–195.
10. **Carmichael, W. W.** 1994. The toxins of cyanobacteria. *Sci. Am.* **270:**78–86.
11. **Carmichael, W. W., D. F. Biggs, and M. A. Peterson.** 1979. Pharmacology of anatoxin-a, produced by the fresh-

water cyanophyte *Anabaena flos-aquae*. *Toxicon* **17:**229–236.

12. **Carmichael, W. W., N. R. Beetsch, J.-S. An, M. Kent, J. M. Groff, and J. A. Brock.** 1995. Microcystin and the aquaculture industry. Presented at the Seventh International Conference on Toxic Phytoplankton, Sendai, Japan.
13. **Centers for Disease Control and Prevention and U.S. Department of Agriculture.** 2006. HHS and USDA select agents and toxins. 42 CFR part 73.3B. Centers for Disease Control and Prevention, Atlanta, Ga.
14. **Codd, G. A., S. G. Bell, K. Kaya, C. J. Ward, K. A. Beattie, and J. S. Metcalf.** 1999. Cyanobacterial toxins, exposure routes and human health. *Eur. J. Phycol.* **34:**405–415.
15. **Cordier, S., C. Monfort, L. Miossec, S. Richardson, and C. Belin.** 2000. Ecological analysis of digestive cancer mortality related to contamination by diarrhetic shellfish poisoning toxins along the coasts of France. *Environ. Res.* **84:**145–150.
16. **Cox, P. A., and O. W. Sacks.** 2002. Cycad neurotoxins, consumption of flying foxes, and ALS-PDC disease in Guam. *Neurology* **58:**956–959.
17. **Cox, P. A., S. A. Banack, and S. J. Murch.** 2003. Biomagnification of cyanobacterial neurotoxins and neurodegenerative disease among the Chamorro people of Guam. *Proc. Natl. Acad. Sci. USA* **100:**13380–13383.
18. **Dale, B., and C. M. Yentsch.** 1978. Red tide and paralytic shellfish poisoning. *Oceanus* **21:**41–49.
19. **De Silva, E. D., D. E. Williams, R. J. Andersen, H. Klix, C. F. B. Holmes, and T. M. Allen.**1992. Motuporin, a potent protein phosphatase inhibitor isolated from Papua New Guinea sponge Theonella swinhoei Gray. *Tetrahedron Lett.* **33:**1561–1564.
20. **Diaz, R. J., and R. Rosenberg.** 1995. Marine benthic hypoxia: a review of its ecological effects and the behavioural responses of the benthic microfauna. *Oceanog. Mar. Biol.* **33:**245–303.
21. **Duncan, M. W., J. C. Steele, I. J. Kopin, and S. P. Markey.** 1990. 2-Amino-3-(methylamino)-propanoic acid (BMAA) in cycad flour: an unlikely cause of amyotrophic lateral sclerosis and parkinsonism-dementia of Guam. *Neurology* **40:**5767–5772.
22. **Edwards, D. J., and W. H. Gerwick.** 2004. Lyngbyatoxin biosynthesis: sequence of biosynthetic gene cluster and identification of a novel aromatic prenyltransferase. *J. Am. Chem. Soc.* **126:**11432–11433.
23. **Falconer, I. R.** 1993. *Algal Toxins in Seafood and Drinking Water*. Academic Press, New York, N.Y.
24. **Fleming, L. E., L. C. Backer, and D. G. Baden.** 2005. Overview of aerosolized Florida red tide toxins: exposures and effects. *Environ. Health Perspect.* **113:**618–620.
25. **Flewelling L. J., J. P. Naar, J. P. Abbott, D. G. Baden, N. B. Barros, G. D. Bossart, M.-Y. D. Bottein, D. G. Hammond, E. M. Haubold, C. A. Heil, M. S. Henry, H. M. Jacocks, T. A. Leighfield, R. H. Pierce, T. D. Pitchford, S. A. Rommel, P. S. Scott, K. A. Steidinger, E. W. Truby, F. M. Van Dolah, and J. H. Landsberg.** 2005. Brevetocosis: red tides and marine mammal mortalities. *Nature* **435:**755–756.
26. **Franchini, A., E. Marchesini, R. Poletti, and E. Ottaviani.** 2004. Acute toxic effect of the algal yessotoxin on Purkinje cells from cerebellum of Swiss CD1 mice. *Toxicon* **43:**347–352.
27. **Francis, G.** 1878. Poisonous Australian lake. *Nature* **18:**11–12.
28. **Froscio, S. M., A. R. Humpage, P. C. Burcham, and I. R. Falconer.** 2003. Cylindrospermopsin-induced protein synthesis inhibition and its dissociation from acute toxicity in mouse hepatocytes. *Environ. Toxicol.* **18:**243–251.
29. **Glasgow, H. B., J. M. Burkholder, S. L. Morton, and J. Springer.** 2001. A second species of ichthyotoxic Pfiesteria (Dinamoebales, Pyrrophyta). *Phycologia* **40:**234–245.
30. **Gleibs, S., and D. Mebs.** 1999. Distribution and sequestration of palytoxin in coral reef animals. *Toxicon* **37:**1521–1527.
31. **Graham, L. E., and L. W. Wilcox.** 2000. *Algae*. Prentice Hall, Upper Saddle River, N.J.
32. **Hallegraeff, G. M., D. M. Anderson, and A. D. Cembella.** 2003. *Manual on Harmful Marine Microalgae*. UNESCO Publishing, Landais, France.
33. **Hastie, C. J., E. B. Borthwick, L. F. Morrison, G. A. Codd, and P. T. W. Cohen.** 2005. Inhibition of several protein phosphatases by a non-covalent interacting microcystin and a novel cyanobacterial peptide, nostocyclin. *Biochim. Biophys. Acta* **1726:**187–193.
34. **Hawkins, P. R., M. T. C. Runnegar, A. R. B. Jackson, and I. R. Falconer.** 1985. Severe hepatotoxicity caused by the tropical cyanobacterium (blue-green alga) *Cylindrospermopsis raciborskii* (Woloszynska) Seenaya and Subba Raju isolated from a domestic water supply reservoir. *Appl. Environ. Microbiol.* **50:**1292–1295.
35. **Hawser, S. P., G. A. Codd, D. G. Capone, and E. J. Carpenter.** 1991. A neurotoxic factor associated with the bloom-forming cyanobacteria *Trichodesmium*. *Toxicon* **29:**277–278.
36. **Hitzfeld, B. C., C. S. Lampert, N. Spaeth, D. Mounfort, H. Kaspar, and D. R. Dietrich.** 2000. Toxin production in cyanobacterial mats from ponds on the McMurdo ice shelf, Antarctica. *Toxicon* **38:**1731–1748.
37. **Holmes, M. J., R. J. Lewis, A. Jones, and A. W. Hoy.** 1995. Cooliatoxin, the first toxin from Coolia monotis (Dinophyceae). *Nat. Toxins* **3:**355–362.
38. **Hori, M., Y. Matsuura, R. Yoshimoto, H. Ozaki, T. Yasumoto, and H. Karaki.** 1999. Actin depolymerizing action by marine toxin, pectenotoxin-2. *Nippon Yakurigaku Zasshi* **114:**225–229.
39. **Huges, E. O., P. R. Gorham, and A. Zhender.** 1958. Toxicity of a unialgal culture of *microcystis aeruginosa*. *Can. J. Microbiol.* **4:**225–236.
40. **Humpage, A. R., F. Fontaine, S. Froscio, P. Burcham, and I. R. Falconer.** 2005. Cylindrospermopsin genotoxicity and cytotoxicity: role of cytochrome P-450 and oxidative stress. *J. Toxicol. Environ. Health A* **68:**739–753.
41. **Ito, E., and H. Nagai.** 2000. Bleeding from the small intestine caused by aplysiatoxin, the causative agent of the red algae *Gracilaria coronopifolia* poisoning. *Toxicon* **38:**123–132.
42. **Ito, E., M. Satake, and T. Yasumoto.** 2002. Pathological effects of lyngbyatoxin A upon mice. *Toxicon* **40:**551–556.
43. **Jackson, D. F. (ed.).** 1964. *Algae and Man*. Plenum Press, New York, N.Y.
44. **Jackson, W. C.** 2002. Determination of the neurotoxic factor associated with the bloom-forming cyanobacterial genus *Trichodesmium* spp. M.S. thesis. Wright State University, Dayton, Ohio.
45. **James, K. J., A. Furey, M. Lehane, H. Ramstad, T. Aune, P. Hovgaard, S. Morris, W. Higman, M. Satake, and T. Yasumoto.** 2002. First evidence of an extensive Northern European distribution of azaspiracid poisoning (AZP) toxins in shellfish. *Toxicon* **40:**909–915.
46. **Jeffery, B., T. Barlow, K. Moizer, S. Paul, and C. Boyle.** 2004. Amnesic shellfish poison. *Food Chem. Toxicol.* **42:**545–557.
47. **Joyce, S.** 2000. The dead zone: oxygen-starved coastal waters. *Environ. Health Perspect.* **108:**A121–A125.
48. **Kaiser, J.** 2002. The science of pfiesteria: elusive, subtle, and toxic. *Science* **298:**346–349.
49. **Lee, R. E.** 1999. *Phycology*, 3rd ed. Cambridge University Press, Cambridge, United Kingdom.

50. **Lewis, R. J., and M. J. Holmes.** 1993. Origin and transfer of toxins involved in ciguatera. *Comp. Biochem. Physiol.* **C106:**615–628.
51. **Li, W. I., B. L. Marquez, T. Okino, F. Yokokawa, T. Shioiri, W. H. Gerwick, and T. F. Murray.** 2004. Characterization of the preferred stereochemistry for the neuropharmacologic actions of antillatoxin. *J. Nat. Prod.* **67:**553–568.
52. **Luesch, H., G. G. Harrigan, G. Goetz, and F. D. Horgen.** 2002. The cyanobacterial origin of potent anticancer agents originally isolated from sea hares. *Curr. Med. Chem.* **9:**1791–1806.
53. **Mahmood, N. A., and W. W. Carmichael.** 1986. Paralytic shellfish poisons produced by the freshwater cyanobacterium *Aphanizomenon flos-aquae* NH-5. *Toxicon* **24:**175–186.
54. **Martin, W., T. Rujan, E. Richly, A. Hansen, S. Cornelsen, T. Lins, D. Leister, B. Stoebe, M. Hasegawa, and D. Penny.** 2004. Evolutionary analysis of *Arabidopsis*, cyanobacterial, and chloroplast genomes reveals plastid phylogeny and thousands of cyanobacterial genes in the nucleus. *Proc. Natl. Acad. Sci. USA* **99:**12246–12251.
55. **Menand, B., L. Marechal-Drouard, W. Sakamoto, A. Dietrich, and H. Wintz.** 1998. A single gene of chloroplast origin codes for mitochondrial and chloroplastic methionyl-tRNA synthetase in *Arabidopsis thaliana*. *Proc. Natl. Acad. Sci. USA* **95:**11014–11019.
56. **Meyer, K. F., H. Sommer, and P. Schoenholz.** 1928. Mussel poisoning. *J. Prev. Med.* **2:**365–394.
57. **Milutinovic, A., M. Zivin, R. Zorc-Pleskovic, B. Sedmak, and D. Suput.** 2003. Nephrotic effects of chronic administration of microcystin-LR and -YR. *Toxicon* **42:** 281–288.
58. **Mynderse, J. S., R. E. Moore, M. Kashiwagi, and T. R. Norton.** 1977. Antileukemia activity in the Osilla-toriaceae: isolation of debromoaplysiatoxin from Lyngbya. *Science* **196:**538–540.
59. **Nagai, H., T. Yasumoto, and Y. Hokama.** 1996. Aplysiatoxin and debromoaplysiatoxin as the causative agents of a red alga *Gracilaria coronopifolia* poisoning in Hawaii. *Toxicon* **37:**753–761.
60. **Nagai, H., T. Yasumoto, and Y. Hokama.** 1997. Manauealides, some of the causative agents of a red alga *Gracilaria coronopifolia* poisoning in Hawaii. *J. Nat. Prod.* **60:**925–928.
61. **Nobre, A. C., S. M. Nunes-Monteiro, M. C. Monterio, A. M. Martins, A. Havt, P. S. Barbosa, A. A. Lima, and H.S. Monterio.** 2004. Microcystin-LR promotes intestinal secretion of water and electrolytes in rats. *Toxicon* **44:** 555–559.
62. **Ohta, T., E. Sueoka, N. Iida, A. Komori, M. Suganuma, R. Nishiwaki, M. Taematsu, S. J. Kim, W. W. Carmichael, and H. Fujiki.** 1994. Nodularin, a potent inhibitor of protein phosphatases 1 and 2A, is a new environmental carcinogen in male F344 rat liver. *Cancer Res.* **54:**6402–6406.
63. **Oksanen, I., J. Jokela, D. P. Fewer, M. Wahlsten, J. Rikkinen, and K. Sivonen.** 2004. Discovery of rare and highly toxic microcystins from lichen-associated cyanobacterium *Nostoc* sp. strain IO-102-I. *Appl. Environ. Microbiol.* **70:**5756–5763.
64. **Orjala, J., D. G. Nagle, V. L. Hsu, and W. H. Gerwick.** 1995. Antillatoxin: an exceptionally ichthyotoxic cyclic lipopeptide from the tropical cyanobacterium *Lyngbya majuscula*. *J. Am. Chem. Soc.* **117:**8281–8282.
65. **Osborne, N. J. T., P. M. Webb, and G. R. Shaw.** 2001. The toxins of *Lyngbya majuscula* and their human and ecological health effects. *Environ. Int.* **27:**381–392.
66. **Perl, T. M., L. Bedard, T. Kosatsky, J. C. Hockin, E. C. Todd, and R. S. Remis.** 1990. An outbreak of toxic encephalopathy caused by eating mussels contaminated with domoic acid. *N. Engl. J. Med.* **322:**1775–1780.
67. **Rabalais, N. N.** 2002. Hypoxia in the Gulf of Mexico. Ecosystem description. National Oceanic and Atmospheric Administration, Silver Spring, Md.
68. **Raetz, C. R. H., and C. Witfield.** 2002. Lipopolysaccharide endotoxins. *Annu. Rev. Biochem.* **71:**635–700.
69. **Ramsdell, J. S., D. M. Anderson, and P. M. Gilbert (ed.).** 2005. *HARRNESS, Harmful Algal Research and Response: a National Environmental Science Strategy 2005-2015*. Ecological Society of America, Washington, D.C.
70. **Runnegar, M. T., W. Xinhua, and S. F. Hamm-Alvarez.** 1999. Increased protein phosphorylation of cytoplasmic dynein results in impaired motor function. *Biochem. J.* **342:**1–6.
71. **Satake, M., K. Ofuji, H. Naoki, K. J. James, A. Furey, T. McMahon, J. Silke, and T. Yasumoto.** 1998. Azaspiracid, a new marine toxin having unique spiro ring assemblies, isolated from Irish mussels, *Mytilus edulis*. *J. Am. Chem. Soc.* **120:**9967–9968.
72. **Sato, S., M. N. Paranagu, and E. Eskinazi.** 1963. On the mechanism of red tide *Trichodesmium* in Recife, Northeastern Brazil, with some considerations of the relation to the human disease, "Tamandare fever." *Trans. Inst. Oceanogr. Univ. Recife* **5/6:**7–49.
73. **Scallet, A. C., L. C. Schmued, and J. N. Johannessen.** 2005. Neurohistochemical biomarkers of the marine neurotoxicant domoic acid. *Neurotoxicol. Terratol.* **27:**745–752.
74. **Scholin, C. A., F. Gulland, G. J. Doucette, S. Benson, M. Busman, F. P. Chavez, J. Cordaro, R. Delong, A. De Vogelaere, J. Harvey, M. Haulena, K. Lefebvre, T. Lipscomb, S. Loscutoff, L. J. Lowenstine, R. Martin III, P. E. Miller, W. A. McLellan, P. D. R. Moeller, C. L. Powell, T. Rowles, P. Silvagni, M. Silver, T. Spraker, V. Trainer, and F. M. Van Dolah.** 2000. Mortality of sea lions along the central California coast linked to a toxic diatom bloom. *Nature* **403:**80–83.
75. **Shi, L., W. W. Carmichael, and I. Miller.** 1995. Immunogold localization of hepatotoxins in cyanobacterial cells. *Arch. of Microbiol.* **1:**7–15.
76. **Sivonen, K., and G. Jones.** 1999. Cyanobacterial toxins, p. 41–111. *In* I. Chorus and J. Bartram (ed.), *Toxic Cyanobacteria in Water: A Guide to Their Public Health Consequences, Monitoring and Management*. E&FN Spon, London, Great Britain.
77. **Skulkberg, O. M.** 1964. Algal problems related to the eutrophication of European water supplies, and bio-assay methods to assessing fertilizing influences of pollution on inland waters. p. 262–300. *In* D. F. Jackson (ed.), *Algae and Man*. Plenum Press, New York, N.Y.
78. **Skulkberg, O. M., W. W. Carmichael, R. A. Anderson, S. Matsunaga, R. E. Moore, and R. Skulkberg.** 1992. Investigations of a neurotoxic oscillatorialean strain (Cyanophyceae) and its toxins. Isolation and characterization of homoanatoxin-a. *Environ. Toxicol. Chem.* **11:**321–329.
79. **Steidinger, K. A., J. M. Burkholder, H. B. Glasgow, Jr., C. W. Hobbs, J. K. Garrett, E. W. Truby, E. J. Noga, and S. A. Smith.** 1996. Pfiesteria piscicida gen. et sp. nov. (Pfiesteriaceae fam. nov.), a new toxic dinoflagellate with a complex life cycle and behavior. *J. Phycol.* **32:**157–164.
80. **Stommel, E. W., and M. R. Watters.** 2004. Marine neurotoxins: ingestible toxins. *Curr. Treat. Options Neurol.* **6:** 105–114.
81. **Subba Rao, D. V., M. A. Quilliam, and R. Pocklinton.** 1988. Domoic acid—a neurotoxic amino acid produced by

the marine diatom Nitzschia pungens in culture. *Can. J. Fish. Aquat. Sci.* **45:**2076–2079.

82. **Suzuki, T., T. Igarashi, K. Ichimi, M. Watai, M. Suzuki, E. Ogiso, and T. Yasumoto.** 2005. Kinetics of diarrheic shellfish poisoning toxins, okadaic acid, dinophysistoxin-1, pecentotoxin-6 and yessotoxin in scallops *Patinopecten yessoensis*. *Fish. Sci.* **71:**948–955.
83. **Thacker, R. W., and V. J. Paul.** 2004. Morphological, chemical, and genetic diversity of tropical marine cyanobacterial *Lyngbya* spp. and *Symploca* spp. (*Oscillatoriales*). *Appl. Environ. Microbiol.* **70:**3305–3312.
84. **Thomas, N. J., C. U. Meteyer, and L. Sileo.** 1998. Epizootic vacuolar myelinopathy of the central nervous system of bald eagles (Haliaeetus leucocephalus) and American coots (Fulica American) *Vet. Pathol.* **35:**479–487.
85. **Tiffany, L. H.** 1958. *Algae, the Grasses of Many Waters*. 2nd ed. Charles C Thomas, Publisher, Springfield, Ill.
86. **Tubaro, A., S. Sosa, M. Carbonatto, G. Altinier, F. Vita, M. Melato, M. Satake, and T. Yasumoto.** 2003. Oral and intraperitoneal acute toxicity studies of yessotoxin and homoyessotoxin in mice. *Toxicon* **41:**783–792.
87. **Utkilen, H., and N. Gjølme.** 1992. Toxin production by *Microcystis aeruginosa* as a function of light in continuous cultures and its ecological significance. *Appl. Environ. Microbiol.* **58:**1321–1325.
88. **Vega, A., and E. A. Bell.** 1967. α-Amino-β-methylaminopropionic acid, a new amino acid from the seeds of *Cyacas circinalis*. *Phytochemistry* **6:**759–762.
89. **Vogelbein, W. K., V. J. Lovko, J. D. Shields, K. S. Reese, P. L. Mason, L. W. Haas, and C. C. Walker.** 2002. Pfiesteria shumwayae kills fish by micropredation not exotoxin secretion. *Nature* **418:**967–970.
90. **Watanabe, M. F., K.-I. Harada, W. W. Carmichael, and H. Fujiki.** 1996. *Toxic Microcystis*. CRC Press, Boca Raton, Fla.
91. **Wilde, S. B., T. M. Murphy, C. P. Hope, S. K. Harbrun, J. Kempton, A. Birrenkott, F. Wiley, W. W. Bowerman, and A. J. Lewitus.** 2005. Avian vacuolar myelinopathy linked to exotic aquatic plants and a novel cyanobacterial species. *Environ. Toxicol.* **20:**348–353.
92. **Wu, M., T. Okino, L. M. Nogel, B. L. Marquez, R. T. Williamson, N. Sitachitta, F. W. Berman, T. F. Murray, K. McGough, R. Jacobs, K. Colsen, T. Asano, F. Yokokawa, T. Shioiri, and W. H. Gerwick.** 2000. Structure, synthesis and biological properties of kalkitoxin, a novel neurotoxin from the marine cyanobacterium *Lyngbya majuscula*. *J. Am. Chem. Soc.* **122:**12041–12042.
93. **Young, F. M., C. Thomson, J. S. Metcalf, J. M. Lucocq, and G. A. Codd.** 2005. Immunogold localization of microcystins in cryosectioned cells of Microcystis. *J. Struct. Biol.* **151:**208–214.
94. **Yu, S., N. Zhao, and X. Zi.** 2001. The relationship between cyanotoxin (microcystin, MC) in pond-ditch water and primary liver cancer in China. *Zhonghua Zhong Liu Za Zhi* **23:**96–99.

AQUATIC ENVIRONMENTS

IV

VOLUME EDITOR
CHRISTON J. HURST

SECTION EDITOR
MEREDITH A. J. HULLAR

An Overview of Methodologies in Aquatic Microbial Ecology

MEREDITH A. J. HULLAR

31

Aquatic microbial ecology is the study of the interaction of microorganisms and their aquatic environment, which encompasses both marine and freshwater systems. Microorganisms include algae, protists, fungi, bacteria, archaea, and viruses. Interactions between microorganisms and their environment include the environmental impacts on organisms and their adaptation to habitat; the organismal impacts on the environment, which include chemical, physical, and mineralogical interactions; trophic interactions; and nontrophic interactions such as competition for nutrients. These interactions involve a microbial community that is hugely varied in both diversity (species composition and distribution) and function (metabolism). Microbial diversity describes the complexity and variability at different levels of biological organization, such as genetic variability within taxa, number of taxa, relative abundance of different taxa, and functional groups, which are different taxa performing the same function. Recent advances in several areas of microbial ecology have given us insight into novel metabolisms, groups of bacteria, functional roles, and a remarkable diversity of bacteria that requires a revised view of the interaction between microorganisms and the aquatic environment.

Aquatic systems are broadly classified to include both marine and freshwater biomes. The marine biome covers about 70% of the earth's surface, while the freshwater biome, made up of rivers, lakes, and streams, is much smaller. Classification within these biomes occurs in terms of light, since a large portion of the biological productivity of aquatic systems is driven by light through photosynthesis. The water column includes the euphotic zone, or the depth of the water column to which 1% of the ambient light penetrates, and the aphotic zone, the depth of the water column which receives less than 1% of the ambient light. Zonation due to light occurs over much smaller scales in freshwater than in oceanic systems. Rivers and streams are classified in order of size with characteristics that include low light penetration in small, first-order streams due to tree canopy cover in temperate regions, to larger-order rivers with open water reaches. Estuaries are unique biomes that are characterized as zones of salt water and freshwater mixing. These highly productive systems are also influenced by light, in addition to tidal forcing and chemistry due to the mixing of salt water and freshwater.

Ecologists have pursued quantitative measures of stocks and flux of energy through ecosystems which, when integrated over many trophic levels, provide a framework to compare energy flow through different ecosystems (36). In aquatic systems, the important role of microorganisms was recognized (42) and termed the microbial loop (2). The microbial loop forms a conceptual basis in which both prokaryotic and eukaryotic microorganisms facilitate nutrient regeneration and influence productivity of aquatic ecosystems.

Microorganisms also facilitate key transformations in global geochemical cycles such as those of carbon, nitrogen, sulfur, phosphorus, and metals. The interaction of biotic and abiotic chemical transformations (biogeochemistry) occurs along chemical gradients which span a few millimeters in sediments to a few thousand meters in the world's oceans. Different functional groups of bacteria are distributed along these gradients and facilitate the transition of elements to different oxidation states. Although the overall balance for the biogeochemical cycles may be distributed over a global scale, the cycles are influenced by the metabolism occurring at the microbial level.

This section reflects the integrative nature of aquatic microbial ecology. We have structured it around areas that influence the role of the microbial community in ecosystem function such as diversity and community structure of primary and secondary producers, growth and grazing of primary and secondary producers, the role of bacteria in geochemical cycling, and specialized environments. Various habitats for bacteria in marine, estuarine, and freshwater environments are represented, including the water column (chapters 33, 34, 35, 36, 37, 39, 41, 42, and 43), biofilms (chapter 45), sediments (chapters 38, 40, 41, and 44), and the deep sea (chapter 46). The first two areas consist of six chapters associated with methods for the identification and measurement of the microbial community: chapter 32 covers new methods used to culture bacteria from the aquatic environment; chapter 33 describes measurements of the activity and structure of primary producers; the assessment of activity and linking of activity to structure of secondary producers are found in chapter 34; chapters 35 and 36 discuss analysis of bacterial, archaeal, and viral community structure; the methods for analyzing protozoa community structure are found in chapter 37; and aquatic fungal activity and community structure methodologies are found in chapter 38. The next area covers the various groups and approaches used to assess microorganisms involved in biogeochemical transformations. The carbon cycles are covered in chapters 39 (organic carbon metabolism) and 40 (C_1 metabolism). The roles of microorganisms in sulfur, nitrogen, and phosphorus cycling are covered in chapters 41 through 43. Metal cycling, which focuses on the Fe/Mn cycles, is found in chapter 44. Sampling and measuring of bacteria in unusual environments such as biofilms (chapter 45) and deep-sea environments (chapter 46) are also discussed. Chapter 15 in section II gives an overview of approaches that can be used to couple ecosystem level questions with microbial measurements and links to the databases of biogeochemical parameters that are being amassed at different long-term monitoring sites around the United States.

APPROACHES TO MICROBIAL COMMUNITY STRUCTURE ASSESSMENT: SINGLE GENE, SINGLE GENOME, TO MULTIPLE GENES, MULTIPLE GENOMES

The field of aquatic microbial ecology is being revolutionized by genetic and genomic approaches. Pioneering work in the late 1980s (37, 40), which started with community analysis based on the PCR amplification of the 16S rRNA gene from many different types of bacteria, has expanded to encompass the analysis of multiple genomes from many individuals in many populations in the community (metagenome). This overview builds and expands upon the methods presented in section IV of the second edition of this manual (8) and emphasizes the recent advances in the application of atomic radiochemical and molecular techniques, including genomics, transcriptomics, and proteomics, to aquatic microbial ecology.

The retrieval of 16S rRNA genes by PCR (21) resulted in a shift away from isolation- and cultivation-dependent approaches to the recovery of sequences representing microorganisms that could not be cultivated. Currently, it is estimated that less than 1% of microbes in aquatic systems have been cultivated (23, 46). This approach, which requires PCR amplification and cloning of 16S rRNA genes, uncovered a remarkable diversity of bacteria in nature. However, in many cases, we are unable to link the composition of the microbial community to the metabolic function of the members of the community.

New insights into the role of bacteria in aquatic systems have been discovered by coupling the distribution of 16S rRNA genes in the ocean with novel culturing techniques (see chapter 32) and sequencing of a single genome. For example, *Pelagibacter ubique* (or SAR 11) is a bacterium found in the upper reaches of the ocean that was initially identified by 16S rRNA approaches (21). However, until the organism was cultured, its physiology was unknown. Since *P. ubique* was cultured (45), the sequenced genome has unveiled other metabolic anomalies that were undetectable by the 16S rRNA approach, providing important insights into the energy flow in ocean systems. A similar approach linked a physiological basis to niche partitioning in the ocean. The distribution of different *Prochlorococcus* ecotypes within the oceanic euphotic zone correlated with different light requirements as indicated by differences in genes in the genome for photosynthesis and DNA repair mechanisms (48). Novel metabolism associated with nitrogen cycling has been attributed to Crenarchaeota, whose oceanic distribution was first described by use of 16S rRNA techniques followed by culturing (28). Chapter 40 encompasses approaches to sequencing of a single genome as well as websites, including sites for software associated with annotation genomic analysis of pure cultures.

In contrast, community genomics is the application of genomic tools and techniques to study the entire microbial community as a single unit. The entire set of genes that belongs to a given community is referred to as the metagenome (47). Analysis of the metagenome provides insight into the functions and genetic structure of the whole community without the necessity of culturing the organisms. This differs from single-genome analysis in that fragments of DNA sequences from many different organisms are analyzed. The cloned DNA fragments may contain enough sequence to include genes for species identification as well as functional genes (bacterial artificial chromosomes [BACs]) or contain smaller fragments that generally encompass fewer gene sequences (whole genome shotgun sequencing [WGS]). Microorganisms that use bacterial rhodopsins (light-dependent proton pumps) to generate energy were discovered in sea surface waters by using environmental BACs (3, 13). In a recent study using WGS and representing the largest sequencing effort of natural communities to date, increased diversity of bacterial rhodopsin genes (782 novel forms) was found in the surface waters of the Sargasso Sea (56). Recently, using metagenomic analysis, DeLong et al. (17) showed a correlation between different microbial groups distributed along light and geochemical gradients in the Pacific Ocean.

However, the analysis of fragmented DNA from multiple organisms poses unique problems since most of the analytical approaches were developed for the analysis of a single genome with multiple regions of overlapping sequence. Metagenomic data has many fragments of genome sequence from many different genomes with little overlapping sequence. Issues to consider when applying metagenomic approaches to microbial communities include how the complexity of the microbial community influences the amount of sequence required to accurately represent the community (7, 29, 59), screening of large libraries (56), and computational problems associated with genome assembly (12, 15, 16) and gene finding (14, 15, 49–51).

Metagenomic fingerprint analysis is the comparison of the patterns of DNA fragments derived from the entire set of genes in a microbial community. It promises to uncover trends between microbial groups that vary in their role in biogeochemical cycling in different ecosystems, just as the analysis of single genes in different ecosystems has identified trends in microbial community structure correlated with environmental parameters such as dissolved organic matter (11), temperature (18, 24), salinity (4), and inorganic nutrients (22). Tringe et al. (54) used environmental gene tags (EGTs), or short sequences that contain fragments of functional genes from whole genome shotgun libraries, to compare the microbial communities from three different ecosystems. Clustering by functional gene groups suggested that there was little overlap in the microbial physiology in different ecosystems, and the predicted protein complement was influenced by the characteristics of the environment. In addition, the distribution of EGTs (evenness) and number of different EGTs (richness) gave an indication of genes that were essential for microbial community function in the different environments. Despite the methodological problems, community genomic approaches hold promise for identifying differences between functional groups and the controls over biogeochemistry in different ecosystems.

APPROACHES TO ASSESS MICROBIAL ACTIVITY

In general, there are two approaches to measuring bacterial community activity in aquatic systems: (i) the direct tracer approach, which measures the conversion of a labeled substrate to a product of interest, and (ii) isotope dilution methods, in which the product pool is labeled and the dilution of the labeled product pool by new unlabeled product formation gives an estimate of the rate of production (8). These approaches use either stable or radioactive isotopes. The direct tracer approach has been applied to bacteria in natural systems to follow the flux of material through many biogeochemical cycles. Direct tracer methods have been employed for organic carbon, inorganic carbon, sulfur, nitrogen, phosphorus, and metal cycling. An overview of the isotope dilution method was discussed previously (8). Some of the problems that plague the direct tracer approach are that (i) if the protocol does not incorporate a labeling step, then a change in the product pool may go undetected since the production and consumption of substrates are often in equilibrium in natural systems; and (ii) if a labeled substrate is employed, the amount added should not dilute the unlabeled pool significantly so that the production rate is not overestimated.

Assessing the activity of the microbial community in the flux of energy and material through an aquatic system has been expanded to include phylogenetic identification of specific groups of bacteria responsible for such conversions. This allows linkage between the composition of the microorganisms (structure) to their metabolism (function) in nature without using pure culture techniques. These approaches can be divided into direct tracer or nontracer approaches. The direct tracer approach measures either radioactive or stable isotopes assimilated into biological molecules that represent a component of the cell (biomarkers) or cells. Nontracer approaches link microbial activity to phylogeny by identifying molecules associated with cell activity such as rRNA, mRNA (transcriptomics), or proteins (proteomics).

The direct tracer approach is used in conjunction with specific biomarkers, such as labeled fatty acids (5, 62), DNA (55), RNA (32, 41), or whole cells (30, 39), to identify the active members of the microbial community. The direct tracer approach coupled to biomarker analysis can be used to measure uptake and incorporation of a labeled substrate. The contribution of bacterial productivity to aquatic systems has been quantified by use of radioactive precursors required for DNA (20) or protein (27) synthesis. Recent methodological advancements have been used to dissect the distribution of the gross bacterial productivity among phylogenetically distinct populations in the microbial community (see chapter 34). In contrast, fluorescent in situ hybridization (FISH) coupled to microautoradiography measures the uptake of a compound on a per-cell basis (30, 39). This approach was used to show that the marine bacteria related to *Pelagibacter ubique* (SAR 11) preferentially use low-molecular-weight substrates and contribute to the turnover of monomers in the oceanic carbon cycle (33). They also take up labeled amino acids and dimethylsulfoniopropionate, which suggests that they are involved in the sulfur and nitrogen cycles (34). These techniques, which link productivity or uptake to specific populations in the microbial community, give us tools to understand the role of microbial communities in their habitats.

Stable isotope probing is a direct tracer approach that uses ^{13}C-labeled substrates to identify the active organisms in the microbial community (43). The ^{13}C-labeled substrate is incorporated into bacteria that are actively growing. A biomarker, usually DNA or RNA, is extracted, and the molecules enriched in ^{13}C (heavy) are separated from ^{12}C (light) molecules by density gradient centrifugation. Phylogenetic analysis of clone libraries constructed from the PCR (or reverse transcriptase PCR) amplification of the labeled (heavy) genomic DNA or rRNA can be used to identify the organisms that grew on the labeled substrate (43). This approach is outlined in chapter 40 and was used to identify freshwater bacteria that grow on C_1 compounds. A current limit of the technique is that a relatively large amount of labeled substrate must be used to achieve separation on a density centrifuge gradient (58).

Natural abundance stable isotope ratios can be altered by both physical and biological processes, resulting in an isotopic signature that can be used to identify the origin of the substrate being used by bacteria. Carbon, nitrogen, and sulfur isotopes have all been used to assess the origin of these compounds potentially available for bacterial metabolism in aquatic systems (35). While early studies used filters to concentrate and collect the bacterium-size fraction from water, this approach is plagued by contamination by debris that masks the true isotope value of bacteria. Efforts to reduce contamination led to the isotopic measurement of bacteria grown in batch cultures (9), which can select for subpopulations of the community, or measurement of microbial biomarkers such as fatty acids (5, 62), DNA (10), and rRNA (32, 41). A more direct link of function and structure combined FISH with secondary ion mass spectrometry (38). FISH was used to determine that *Methanosarcinales* anaerobically oxidize methane in a consortium of sulfate-reducing bacteria. Secondary ion mass spectrometry was used to directly measure the unique natural abundance stable isotope signature of the cells in the consortium that incorporated methane depleted in ^{13}C.

Microarray formats for both community structure analysis (for example, based on 16S rRNA genes) and transcriptomic analysis (gene expression profiling by measuring mRNA) have been developed for various groups of bacteria in aquatic systems (6, 19, 25, 53). Microarray technology is based on the ability of complementary sequences of nucleic acids to hybridize or bind to one another. This technology has the advantage of being able to detect thousands of molecules in one experiment and potentially has adequate statistical power to be used to analyze complex microbial communities from a phylogenetic or functional perspective. The isotope array, a novel approach that employs incorporation of both a radiolabeled substrate and fluorescently labeled rRNA, was used to identify diverse members of the β-proteobacteria that are able to convert nitrate to nitrite (1). However, there are still many methodological problems ranging from designing probes with the appropriate degree of resolution required in complex communities (31) to statistical treatment of the large data sets generated (57).

Metaproteomics, or the large-scale characterization of the entire protein complement of the environmental microbial community (61), has the potential to link microbial species to their physiological function. Applications of a coupled metagenomic-proteomic approach to different systems have identified microbial proteins unique to specific environments (7, 26, 44, 61). For example, analysis of a simplified microbial community associated with acid mine drainage showed that many of the proteins expressed were involved in protein refolding and response to oxidative stress, which was perhaps a survival requirement in such an acidic environment (44). Unlike substrate-specific approaches (see above), proteomic analyses can detect the response of the microbial community to changing environmental variables at a given point in time. The presence and relative abundance of proteins can provide a direct measurement of functional gene expression.

Proteomic fingerprinting (a pattern produced by hydrolysis of a protein and two-dimensional mapping of the resulting peptide fragments) was used to analyze the protein complement of water from four different environments (52). Interestingly, the phylogenetic fingerprint and dominance of different functional proteins covaried by season and biome. While the potential to elucidate the physiological role of different components of the microbial community is large, analysis of proteins from a complex environmental milieu is challenging because not only do posttranslational modifications make identification complicated, but also proteins exist in many different physical conformations. This type of variation increases analysis time because it is difficult to use one method to efficiently extract from aquatic environments all of the proteins with different physicochemical properties (26, 60, 61).

Analysis of a single genome, transcriptome, and proteome has expanded into the analysis of multiple populations of bacteria found in communities in different aquatic environments. The future challenge is going to be to integrate activity and community structure measurements with ecosystem measurements of physicochemical fluxes. This section of the *Manual of Environmental Microbiology* presents many of the tools that will be used by aquatic microbial ecologists to continue to develop an integrated view of microorganisms in aquatic systems.

REFERENCES

1. **Adamczyk, J., M. Hesselsoe, N. Iversen, M. Horn, A. Lehner, P. H. Nielsen, M. Schloter, P. Roslev, and M. Wagner.** 2003. The isotope array, a new tool that employs substrate-mediated labeling of rRNA for determination of microbial community structure and function. *Appl. Environ. Microbiol.* **69:**6875–6887.
2. **Azam, F., T. Fenchel, J. G. Field, J. S. Gray, L. A. Meyerreil, and F. Thingstad.** 1983. The ecological role of water column microbes in the sea. *Mar. Ecol. Prog. Ser.* **10:**257–263.
3. **Beja, O., E. N. Spudich, J. L. Spudich, M. Leclerc, and E. F. DeLong.** 2001. Proteorhodopsin phototrophy in the ocean. *Nature* **411:**786–789.
4. **Bernhard, A. E., T. Donn, A. E. Giblin, and D. A. Stahl.** 2005. Loss of diversity of ammonia-oxidizing bacteria correlates with increasing salinity in an estuary system. *Environ. Microbiol.* **7:**1289–1297.
5. **Boschker, H. T., W. de Graaf, M. Koster, L. Meyer-Reil, and T. E. Cappenberg.** 2001. Bacterial populations and processes involved in acetate and propionate consumption in anoxic brackish sediment. *FEMS Microbiol. Ecol.* **35:**97–103.
6. **Castiglioni, B., E. Rizzi, A. Frosini, K. Sivonen, P. Rajaniemi, A. Rantala, M. A. Mugnai, S. Ventura, A. Wilmotte, C. Boutte, S. Grubisic, P. Balthasart, C. Consolandi, R. Bordoni, A. Mezzelani, C. Battaglia, and G. De Bellis.** 2004. Development of a universal microarray based on the ligation detection reaction and 16S rRNA gene polymorphism to target diversity of cyanobacteria. *Appl. Environ. Microbiol.* **70:**7161–7172.
7. **Chen, K., and L. Pachter.** 2005. Bioinformatics for whole-genome shotgun sequencing of microbial communities. *PLoS Comput. Biol.* **1:**106–112.
8. **Christian, R. R., and D. G. Capone.** 2002. Overview of issues in aquatic microbial ecology, p. 323–328. *In* C. J. Hurst, R. L. Crawford, G. R. Knudsen, M. J. McInerney, and L. D. Stetzenbach (ed.), *Manual of Environmental Microbiology*, 2nd ed. ASM Press, Washington, D.C.
9. **Coffin, R. B., B. Fry, B. J. Peterson, and R. T. Wright.** 1989. Carbon isotopic compositions of estuarine bacteria. *Limnol. Oceanogr.* **34:**1305–1310.
10. **Coffin, R. B., D. J. Velinsky, R. Devereux, W. A. Price, and L. A. Cifuentes.** 1990. Stable carbon isotope analysis of nucleic acids to trace sources of dissolved substrates used by estuarine bacteria. *Appl. Environ. Microbiol.* **56:**2012–2020.
11. **Crump, B. C., G. W. Kling, M. Bahr, and J. E. Hobbie.** 2003. Bacterioplankton community shifts in an arctic lake correlate with seasonal changes in organic matter source. *Appl. Environ. Microbiol.* **69:**2253–2268.
12. **Dehnert, M., R. Plaumann, W. E. Helm, and M. T. Hutt.** 2005. Genome phylogeny based on short-range correlations in DNA sequences. *J. Comput. Biol.* **12:**545–553.
13. **de la Torre, J. R., L. M. Christianson, O. Beja, M. T. Suzuki, D. M. Karl, J. Heidelberg, and E. F. DeLong.** 2003. Proteorhodopsin genes are distributed among divergent marine bacterial taxa. *Proc. Natl. Acad. Sci. USA* **100:**12830–12835.
14. **Delcher, A. L., D. Harmon, S. Kasif, O. White, and S. L. Salzberg.** 1999. Improved microbial gene identification with GLIMMER. *Nucleic Acids Res.* **27:**4636–4641.
15. **Delcher, A. L., S. Kasif, R. D. Fleischmann, J. Peterson, O. White, and S. L. Salzberg.** 1999. Alignment of whole genomes. *Nucleic Acids Res.* **27:**2369–2376.
16. **Delcher, A. L., A. Phillippy, J. Carlton, and S. L. Salzberg.** 2002. Fast algorithms for large-scale genome alignment and comparison. *Nucleic Acids Res.* **30:**2478–2483.
17. **DeLong, E. F., C. M. Preston, T. Mincer, V. Rich, S. J. Hallam, N. U. Frigaard, A. Martinez, M. B. Sullivan, R. Edwards, B. R. Brito, S. W. Chisholm, and D. M. Karl.** 2006. Community genomics among stratified microbial assemblages in the ocean's interior. *Science* **311:**496–503.
18. **Donachie, S. P., S. Hou, K. S. Lee, C. W. Riley, A. Pikina, C. Belisle, S. Kempe, T. S. Gregory, A. Bossuyt, J. Boerema, J. Liu, T. A. Freitas, A. Malahoff, and M. Alam.** 2004. The Hawaiian Archipelago: a microbial diversity hotspot. *Microb. Ecol.* **48:**509–520.
19. **Dorigo, U., L. Volatier, and J. F. Humbert.** 2005. Molecular approaches to the assessment of biodiversity in aquatic microbial communities. *Water Res.* **39:**2207–2218.
20. **Fuhrman, J. A., and F. Azam.** 1982. Thymidine incorporation as a measure of heterotrophic bacterioplankton production in marine surface waters—evaluation and field results. *Mar. Biol.* **66:**109–120.
21. **Giovannoni, S. J., T. B. Britschgi, C. L. Moyer, and K. G. Field.** 1990. Genetic diversity in Sargasso Sea bacterioplankton. *Nature* **345:**60–63.
22. **Hiorns, W. D., B. A. Methe, S. A. Nierzwicki-Bauer, and J. P. Zehr.** 1997. Bacterial diversity in Adirondack mountain lakes as revealed by 16S rRNA gene sequences. *Appl. Environ. Microbiol.* **63:**2957–2960.
23. **Hugenholtz, P.** 2002. Exploring prokaryotic diversity in the genomic era. *Genome Biol.* **3:**REVIEWS0003.
24. **Hullar, M. A., L. A. Kaplan, and D. A. Stahl.** 2006. Recurring seasonal dynamics of microbial communities in stream habitats. *Appl. Environ. Microbiol.* **72:**713–722.
25. **Jenkins, B. D., G. F. Steward, S. M. Short, B. B. Ward, and J. P. Zehr.** 2004. Fingerprinting diazotroph communities in the Chesapeake Bay by using a DNA macroarray. *Appl. Environ. Microbiol.* **70:**1767–1776.
26. **Kan, J., T. E. Hanson, J. M. Ginter, K. Wang, and F. Chen.** 2005. Metaproteomic analysis of Chesapeake Bay microbial communities. *Saline Syst.* **1:**7.

27. **Kirchman, D. L., E. K'nees, and R. E. Hodson.** 1985. Leucine incorporation and its potential as a measure of protein synthesis by bacteria in natural aquatic systems. *Appl. Environ. Microbiol.* **49:**599–607.
28. **Konneke, M., A. E. Bernhard, J. R. de la Torre, C. B. Walker, J. B. Waterbury, and D. A. Stahl.** 2005. Isolation of an autotrophic ammonia-oxidizing marine archaeon. *Nature* **437:**543–546.
29. **Lander, E. S., and M. S. Waterman.** 1988. Genomic mapping by fingerprinting random clones: a mathematical analysis. *Genomics* **2:**231–239.
30. **Lee, N., P. H. Nielsen, K. H. Andreasen, S. Juretschko, J. L. Nielsen, K. H. Schleifer, and M. Wagner.** 1999. Combination of fluorescent in situ hybridization and microautoradiography—a new tool for structure-function analyses in microbial ecology. *Appl. Environ. Microbiol.* **65:**1289–1297.
31. **Loy, A., and L. Bodrossy.** 2006. Highly parallel microbial diagnostics using oligonucleotide microarrays. *Clin. Chim. Acta* **363:**106–119.
32. **MacGregor, B. J., V. Bruchert, S. Fleischer, and R. Amann.** 2002. Isolation of small-subunit rRNA for stable isotopic characterization. *Environ. Microbiol.* **4:**451–464.
33. **Malmstrom, R. R., M. T. Cottrell, H. Elifantz, and D. L. Kirchman.** 2005. Biomass production and assimilation of dissolved organic matter by SAR11 bacteria in the Northwest Atlantic Ocean. *Appl. Environ. Microbiol.* **71:** 2979–2986.
34. **Malmstrom, R. R., R. P. Kiene, M. T. Cottrell, and D. L. Kirchman.** 2004. Contribution of SAR11 bacteria to dissolved dimethylsulfoniopropionate and amino acid uptake in the North Atlantic Ocean. *Appl. Environ. Microbiol.* **70:**4129–4135.
35. **McCallister, S. L., J. E. Bauer, J. E. Cherrier, and H. W. Ducklow.** 2004. Assessing sources and ages of organic matter supporting river and estuarine bacterial production: a multiple-isotope (delta C-14, delta C-13, and delta N-15) approach. *Limnol. Oceanogr.* **49:**1687–1702.
36. **Odum, H. T.** 1988. Self-organization, transformity, and information. *Science* **242:**1132–1139.
37. **Olsen, G. J., D. J. Lane, S. J. Giovannoni, N. R. Pace, and D. A. Stahl.** 1986. Microbial ecology and evolution: a ribosomal RNA approach. *Annu. Rev. Microbiol.* **40:**337–365.
38. **Orphan, V. J., C. H. House, K. U. Hinrichs, K. D. McKeegan, and E. F. DeLong.** 2001. Methane-consuming archaea revealed by directly coupled isotopic and phylogenetic analysis. *Science* **293:**484–487.
39. **Ouverney, C. C., and J. A. Fuhrman.** 1999. Combined microautoradiography-16S rRNA probe technique for determination of radioisotope uptake by specific microbial cell types in situ. *Appl. Environ. Microbiol.* **65:**1746–1752.
40. **Pace, N., D. Stahl, D. Lane, and G. Olsen.** 1986. The analysis of natural microbial populations by rRNA sequences. *Adv. Microb. Ecol.* **9:**1–55.
41. **Pearson, A., A. L. Sessions, K. J. Edwards, and J. M. Hayes.** 2004. Phylogenetically specific separation of rRNA from prokaryotes for isotopic analysis. *Mar. Chem.* **92:**295–306.
42. **Pomeroy, L. R.** 1974. Oceans food web, a changing paradigm. *Bioscience* **24:**499–504.
43. **Radajewski, S., P. Ineson, N. R. Parekh, and J. C. Murrell.** 2000. Stable-isotope probing as a tool in microbial ecology. *Nature* **403:**646–649.
44. **Ram, R. J., N. C. Verberkmoes, M. P. Thelen, G. W. Tyson, B. J. Baker, R. C. Blake II, M. Shah, R. L. Hettich, and J. F. Banfield.** 2005. Community proteomics of a natural microbial biofilm. *Science* **308:**1915–1920.
45. **Rappe, M. S., S. A. Connon, K. L. Vergin, and S. J. Giovannoni.** 2002. Cultivation of the ubiquitous SAR11 marine bacterioplankton clade. *Nature* **418:**630–633.
46. **Rappe, M. S., and S. J. Giovannoni.** 2003. The uncultured microbial majority. *Annu. Rev. Microbiol.* **57:**369–394.
47. **Riesenfeld, C. S., P. D. Schloss, and J. Handelsman.** 2004. Metagenomics: genomic analysis of microbial communities. *Annu. Rev. Genet.* **38:**525–552.
48. **Rocap, G., F. W. Larimer, J. Lamerdin, S. Malfatti, P. Chain, N. A. Ahlgren, A. Arellano, M. Coleman, L. Hauser, W. R. Hess, Z. I. Johnson, M. Land, D. Lindell, A. F. Post, W. Regala, M. Shah, S. L. Shaw, C. Steglich, M. B. Sullivan, C. S. Ting, A. Tolonen, E. A. Webb, E. R. Zinser, and S. W. Chisholm.** 2003. Genome divergence in two Prochlorococcus ecotypes reflects oceanic niche differentiation. *Nature* **424:**1042–1047.
49. **Salzberg, S. L., A. L. Delcher, S. Kasif, and O. White.** 1998. Microbial gene identification using interpolated Markov models. *Nucleic Acids Res.* **26:**544–548.
50. **Salzberg, S. L., M. Pertea, A. L. Delcher, M. J. Gardner, and H. Tettelin.** 1999. Interpolated Markov models for eukaryotic gene finding. *Genomics* **59:**24–31.
51. **Sandberg, R., C. I. Branden, I. Ernberg, and J. Coster.** 2003. Quantifying the species-specificity in genomic signatures, synonymous codon choice, amino acid usage and G+C content. *Gene* **311:**35–42.
52. **Schulze, W. X., G. Gleixner, K. Kaiser, G. Guggenberger, M. Mann, and E. D. Schulze.** 2005. A proteomic fingerprint of dissolved organic carbon and of soil particles. *Oecologia* **142:**335–343.
53. **Taroncher-Oldenburg, G., E. M. Griner, C. A. Francis, and B. B. Ward.** 2003. Oligonucleotide microarray for the study of functional gene diversity in the nitrogen cycle in the environment. *Appl. Environ. Microbiol.* **69:**1159–1171.
54. **Tringe, S. G., C. von Mering, A. Kobayashi, A. A. Salamov, K. Chen, H. W. Chang, M. Podar, J. M. Short, E. J. Mathur, J. C. Detter, P. Bork, P. Hugenholtz, and E. M. Rubin.** 2005. Comparative metagenomics of microbial communities. *Science* **308:**554–557.
55. **van Mooy, B. A., A. H. Devol, and R. G. Keil.** 2004. Quantifying ^{3}H-thymidine incorporation rates by a phylogenetically defined group of marine planktonic bacteria (Bacteriodetes phylum). *Environ. Microbiol.* **6:**1061–1069.
56. **Venter, J. C., K. Remington, J. F. Heidelberg, A. L. Halpern, D. Rusch, J. A. Eisen, D. Wu, I. Paulsen, K. E. Nelson, W. Nelson, D. E. Fouts, S. Levy, A. H. Knap, M. W. Lomas, K. Nealson, O. White, J. Peterson, J. Hoffman, R. Parsons, H. Baden-Tillson, C. Pfannkoch, Y. H. Rogers, and H. O. Smith.** 2004. Environmental genome shotgun sequencing of the Sargasso Sea. *Science* **304:**66–74.
57. **Verducci, J. S., V. F. Melfi, S. Lin, Z. Wang, S. Roy, and C. K. Sen.** 2006. Microarray analysis of gene expression: considerations in data mining and statistical treatment. *Physiol. Genomics* **25:**355–363.
58. **Wagner, M., P. H. Nielsen, A. Loy, J. L. Nielsen, and H. Daims.** 2006. Linking microbial community structure with function: fluorescence in situ hybridization-microautoradiography and isotope arrays. *Curr. Opin. Biotechnol.* **17:**83–91.
59. **Wendl, M. C., and R. H. Waterston.** 2002. Generalized gap model for bacterial artificial chromosome clone fingerprint mapping and shotgun sequencing. *Genome Res.* **12:**1943–1949.
60. **Wilmes, P., and P. L. Bond.** 2004. The application of two-dimensional polyacrylamide gel electrophoresis and downstream analyses to a mixed community of prokaryotic microorganisms. *Environ. Microbiol.* **6:**911–920.

61. **Wilmes, P., and P. L. Bond.** 2006. Metaproteomics: studying functional gene expression in microbial ecosystems. *Trends Microbiol.* **14:**92–97.

62. **Wuchter, C., S. Schouten, H. T. Boschker, and J. S. Sinninghe Damste.** 2003. Bicarbonate uptake by marine Crenarchaeota. *FEMS Microbiol. Lett.* **219:**203–207.

Cultivating Microorganisms from Dilute Aquatic Environments: Melding Traditional Methodology with New Cultivation Techniques and Molecular Methods

JAMIE W. BECKER, MARINA L. BRANDON, AND MICHAEL S. RAPPÉ

32

Two decades ago, Carl Woese and colleagues transformed the field of microbial phylogeny when they published early records of the major eubacterial (now "bacterial") phyla and a new organization of the universal tree of life based on the comparative analysis of rRNAs (30, 31). Building on the strength of this foundation, molecular biological methods and gene sequencing have continued to reveal a great deal about the phylogenetic diversity of microorganisms. Sixteen years after the publication of Woese's seminal review of the major bacterial phyla, this early inventory of 11 or 12 major lineages had swelled to an impressive list of 52 major groups within the bacterial domain (24). The cloning and sequencing of rRNA genes directly from the environment served to propel this effort forward (11, 19, 20).

While the application of molecular biological tools to microbial communities in nature has been indispensable in providing incredible insights into microbial biodiversity and metabolic potential (for examples, see references 1, 8, and 29), these tools are not without critical limitations. Simply put, gene sequences alone do not provide conclusive information regarding the specific roles of individual microbial "species" in Earth's various biogeochemical cycles—vital information that will likely have widespread implications for biologists, chemists, and geologists alike. If scientists are to obtain clear answers to these and other questions, elusive microorganisms must be isolated, grown, and studied in a controlled laboratory environment. Indeed, questions regarding the metabolic capabilities, population genetics, and ecology of abundant microbial species are best addressed with isolated representatives in hand, yet evidence provided by the sequencing of environmental DNA supports the conclusion that less than 1% of the aquatic bacteria that appear to dominate many natural ecosystems have been cultivated in this manner (11, 12). Several factors have been identified that may help explain the relatively small number of laboratory isolates. Many microbes have slow growth rates and thus might reach low (i.e., undetectable using traditional methodologies) densities in culture. Microbes may also be inhibited by trace contaminants in the lab, or they may require interactions with other organisms to survive. Perhaps most importantly, many microbes require narrowly defined growth conditions that are not likely to be randomly recreated by microbiologists.

Traditional laboratory spread plating techniques typically utilize nutrient and organic carbon-rich media and, in addition to being time intensive and laborious, are generally considered to be biased toward the growth of a small suite of select microbes (for an example, see reference 28). In 1993, Don Button and colleagues broke new ground with their theoretical description and subsequent application of an extinction culturing method for aquatic microbial cells (4). Their approach was specifically tailored toward the isolation of oligotrophic cells by taking advantage of the unique ability those cells have to effectively compete at very low nutrient concentrations (5). They used natural seawater as the growth medium and flow cytometry to detect microbial growth with high sensitivity. Although Button's group was rewarded with cultures of novel oligotrophs (4, 26, 27), their dilution culture strategy was fairly laborious in that it required a significant amount of time and effort to yield only a small number of isolates per experiment.

As the role of uncultivated microorganisms in natural systems has become more widely appreciated, several different isolation methods have been developed and employed with varying degrees of success. Recent strategies have involved exposing cells to signaling molecules (cyclic AMP, ATP, etc.) (2, 3), providing them with an extended acclimatization period to allow for a slow transition from low to high substrate concentration (13), and encapsulating them in gel microdroplets (32, 33). Among these novel approaches is a dilution to extinction, high-throughput culturing (HTC) method, based on Button's strategy, that utilizes very low nutrient media to mimic natural substrate concentrations; in fact, growth substrates are 3 orders of magnitude less concentrated than those applied in more conventional plating techniques (7, 23). One of the fundamental differences between Button's methodology and the HTC strategy is the miniaturization and partial automation of the process. Working with small-volume cultures facilitates high rates of replication and provides greater opportunities to investigate physical and chemical variables that increase the probability of isolating microorganisms with diverse growth requirements. In order to overcome the difficulties associated with detecting small populations of cells, HTC employs an innovative procedure for arraying cell cultures on microscope slides. The procedure allows for the detection of microbe populations at densities as low as ca. 10^3 cells/ml (23). The use of the HTC method originally focused on utilizing natural sterilized water as media and incubation conditions that closely mimic the natural environment of the inoculum (for examples, see references 6, 7, 21, and 23); however, the choices of media amendments and incubation conditions are at each individual's discretion.

The value of this high-throughput technique was established when early trials of the approach yielded the first cultured representatives of many microbial groups that were previously known to exist only from environmental rRNA genes (7). Based on analyses of rRNA gene sequences, many of the cultures were more closely related to environmental clones than to any previously cultured microorganisms, the most notable isolate being the incredibly abundant and ubiquitous member of the α-*Proteobacteria Pelagibacter* (SAR11) (23). Further, in a direct comparison between traditional microbiological spread plating and the HTC extinction culturing method, the number of cells cultured in the laboratory from Oregon coastal seawater with HTC was 1.4 to 120 times greater than that which could be recovered by the conventional, nutrient-rich plating process (7). In addition to *Pelagibacter* spp., a large number of previously uncultivated cell lineages of potential significance have been cultured from both marine and freshwater environments by using the HTC method (6, 7, 21).

SAMPLE COLLECTION AND PREPARATION OF SEAWATER-BASED MEDIA

Seawater Collection

The procedure for isolating cells via extinction culturing is relatively simple, and the paramount goal of the HTC technique described below is to produce clonal aquatic microbial isolates or consortia of microorganisms for further laboratory-based experimentation (Fig. 1). Materials and containers selected for HTC are chosen to avoid toxic compounds that may hinder the growth of target microorganisms. For example, tissue culture grade (NOTE: not tissue culture treated) disposable polystyrene or polycarbonate plasticware (e.g., Nalgene Labware; Nalge Nunc International, Rochester, N.Y.) soaked overnight in 10% hydrochloric acid (HCl) and rinsed thoroughly with ultrapure water before autoclaving is ideal for both sample collection and storage of prepared media. In addition, trace metal clean techniques (9) should be approximated for the collection of water samples. A clean, Teflon-lined Niskin bottle is used for sample acquisition; the bottle should be rinsed with the seawater sample

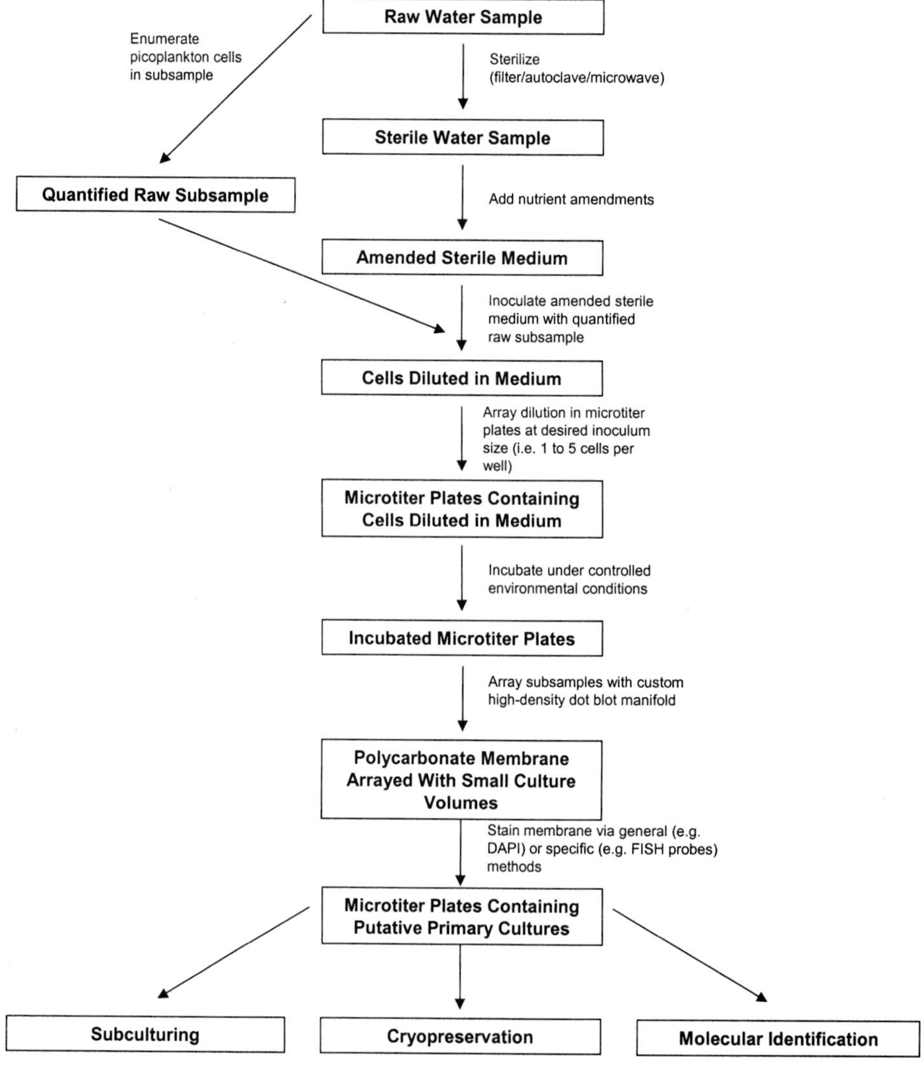

FIGURE 1 Flow diagram illustrating HTC procedures.

just prior to filling. Immediately transfer the seawater sample from the Niskin to sterile polycarbonate containers until further processing. If necessary, set aside an appropriate volume of raw sample in a sterile container to serve as the inoculant. If the inoculant is to come from a different source, the entire water sample collected may be used for medium preparation. The volume of water required for medium preparation and inoculation depends on the concentration of microbial cells in the raw sample and the design and goals of the particular cultivation experiment.

As with any technique involving the cultivation or manipulation of live microorganisms, measures taken to minimize and eliminate the risk of contamination are crucial for a successful experiment. Safeguards include frequent changing of gloves and sterile pipette tips, minimizing or eliminating the exposure of human skin during manipulations, and minimizing the time in which cultures and media are open and exposed to the researcher or the laboratory environment. All manipulations should be carried out under a laminar flow hood. Due to the limited number of times that experiments have been performed using the HTC approach, the degree to which culture contamination is an issue is not yet known. If a single cell from the inoculum's original environment found its way into a well of one of the inoculated primary microtiter plates and subsequently proliferated, it would be impossible to distinguish from a "normal" primary culture. We have, however, identified microbial isolates that we believe to be contaminants originating from the human body, including *Propionibacterium* and *Corynebacterium* spp. (unpublished data). While it is possible that these microorganisms were present in the original inoculum, they were considered putative contaminants based on the environment from which they are typically found. Outside this fairly subjective approach, no systematic method for detecting contaminants is employed.

Seawater Processing

Ultrafiltration, tangential flow filtration (TFF), treatment with UV light, microwaving, and autoclaving are just a few of the many viable methods for the sterilization of seawater. The particular method employed is at the discretion of the individual researcher and may also depend on equipment availability in the field or at sea. The sterilization of large volumes of seawater has routinely and successfully been performed using a combination of prefiltration, TFF, and autoclaving as described below.

Prefilter the raw seawater sample to be used as media through a 142-mm-diameter, 0.8-μm-pore-size Supor membrane (Pall Corp., East Hills, N.Y.) using a peristaltic pump. The prefilter membrane should be rinsed and primed prior to sample processing by passing through ca. 10 liters of ultrapure water, followed by ca. 10 liters of seawater. After priming, collect the filtrate in a sterile, acid-rinsed carboy. Carboys used for filtrate collection should be equipped with vented caps and sterile, nontoxic tubing to aid in both aseptic collection and gas sparging. Vented caps are available from Nalgene Labware (Nalge Nunc). Connect nontoxic tubing such as platinum-cured silicone Masterflex tubing (Cole-Parmer, Vernon Hills, Ill.) to the underside of the vented caps such that it hangs down inside the container ca. 1 in. from the bottom. The entire collection container is then acid and heat sterilized as described above in "Seawater Collection." Set up and prime the TFF unit and peristaltic pump as directed by the manufacturer. Pump the prefiltered seawater through the primed TFF unit and collect the permeate in a second sterile, acid-rinsed carboy equipped with a vented cap and sterile tubing. If possible, place the collection carboy inside a laminar flow hood during the filtration process to further decrease the risk of contamination. Millipore's Pellicon 2 Mini TFF unit outfitted with a 30-kDa composite regenerated cellulose filter (Millipore Corp., Billerica, Mass.) is a suitable system. Priming and cleaning of the cellulose filter is performed according to manufacturer's specifications, and the initial 2 liters of prefiltered seawater permeate that passes through the system is discarded prior to filtrate collection. Loosen all of the vented caps and autoclave the filtered seawater according to autoclave specifications for large-volume aqueous solutions. Twenty liters of liquid typically requires 4 to 4.5 h at 121°C, while 5 liters typically requires 2 to 2.5 h. Carboys should be carefully transported to a laminar flow hood for cooling after autoclaving. Vented caps should be tightened only after the water has sufficiently cooled (i.e., to room temperature) to avoid warping of the plastic containers. Using the vented caps and associated tubing apparatus described above, sparge the sterile seawater with 0.1-μm-pore-size-filtered CO_2 for 12 h, followed by filtered air for at least 24 h.

Sterile seawater is stored at 4°C to minimize potential contamination and should be inspected for continued sterility via microscopy immediately prior to each use by following either the DAPI (4′,6′-diamidino-2-phenylindole) staining protocol detailed later in this chapter or a similar procedure.

Nutrient, Organic Carbon, and Vitamin Additions

A tremendous range of medium amendments is available to target or exclude certain organisms of particular interest. For example, nitrogen, phosphorus, organic carbon compounds, and other nutrients can be added to preferentially enrich for particular bacterial species. Sterilized seawater media created by combining samples taken from several different depths or locations can also be employed to create a unique medium mixture in which novel organisms may grow. Antibiotics may also be used to prohibit the growth of certain microbes (for an example, see reference 14). Our strategy has been to enrich only slightly above ambient nutrient and carbon concentrations in order to target truly oligotrophic planktonic marine microorganisms. For example, in Kaneohe Bay on the island of Oahu, Hawaii, macronutrient concentrations typically approximate 0.1 μM for phosphate, 0.05 to 0.15 μM for nitrate, and 0.1 to 0.2 μM for ammonia (25). In general, we supplement this baseline with 1.0 μM phosphate, 1.0 μM nitrate, and 1.0 μM ammonia (unpublished data). In previous experiments, in which Oregon Coast seawater was supplemented with organic compounds (6, 7, 23), the baseline organic carbon concentration of ca. 1.0 mg/liter was supplemented with a suite of organic carbon compounds to a final concentration of 35 mg/liter. The compounds added as a supplement included D-glucose, D-ribose, succinic acid, pyruvic acid, glycerol, *N*-acetyl-D-glucosamine, and ethanol and were selected due to their previous detection in seawater, their simplicity, or their ability to be incorporated in cellular metabolic pathways.

EXTINCTION CULTURE SETUP

The desired inoculum concentration depends on the goals of the experiment, as both low and high cell numbers for inocula can be advantageous in the proper context. For instance, a larger cell-per-well inoculum might increase the number of positive primary cultures. However, repeated subculturing experiments may be necessary to obtain pure isolates. A smaller inoculum size might result in a high percentage of putatively pure primary isolates, yet many extinction wells may not exhibit any growth at all. Concentrations ranging from 1 to 10 cells per well are customarily used and often result in a relatively high number of putatively pure primary isolates. In order to determine the volume of raw seawater needed to inoculate the extinction wells, a microbial cell count of the raw sample is obtained. This ensures that the quantity of cells added to each well is known. There are many acceptable methods of obtaining a microbial cell count from seawater. The following is a routinely used protocol based on the filtration of a known volume of seawater onto a polycarbonate membrane, staining with DAPI, and enumeration via epifluorescence microscopy (22).

Add 1.25 ml of 0.2-μm-pore-size-filtered (e.g., CLS-431219 Corning syringe filter; Corning Inc., Corning, N.Y.) 10% buffered formaldehyde solution (freshly filtered) to 5 ml of raw seawater sample and invert to mix. Incubate the sample at room temperature for 15 min in the dark before vacuum filtering the mixture through a 25-mm-diameter, 0.2-μm-pore-size black polycarbonate filter (e.g., GE Osmonics Inc., Minnetonka, Minn.) with a GN-4 backing filter (Pall Corp.). Do not filter the entire sample dry; instead, leave approximately 1 ml of sample in the filter tower. Add 50 μl of a 0.2-μm-pore-size-filtered, 0.2-mg/ml DAPI stock solution to the remaining volume of sample in the filter tower and incubate the entire apparatus at room temperature in the dark. After 15 min, filter the remaining sample and allow the membrane to dry at room temperature in the dark. Once dry, mount the membrane on a microscope slide and count 15 to 20 random fields of view under UV light with an epifluorescence microscope and appropriate UV filter cubes. The average number of cells per field of view can then be used to calculate the cellular abundance of the original sample. After the cellular abundance of the inoculum has been determined, dilute the inoculum into prepared amended (if desired) media by serial dilution. If the volume of each initial culture is 1 ml, the inoculum size is calculated simply as cellular abundance per milliliter.

All aliquotting of inoculated media should be carried out under a laminar flow hood to minimize risk of contamination. Begin by transferring the desired volume of inoculated media into the wells of a microtiter culture plate. For example, distribute 1 ml into each well of a Multiwell 48-well polystyrene plate (BD Biosciences, Bedford, Mass.). Alternatively, target cells can be deposited into microtiter plate wells by fluorescence-activated cell sorting. A plate of sterile amended media is also created and maintained to serve as a negative control throughout the experiment. Prior to incubation, the outside of each microtiter plate is wrapped in Parafilm to secure the lid and to prevent evaporation of the putative primary isolates. The duration of this initial incubation is at the individual's discretion and is ideally determined experimentally via preliminary time course experiments. For temperate and subtropical seawater communities, plates are typically allowed to incubate for ca. 3 weeks. As with medium amendments, incubation parameters may be adjusted to meet individual requirements; however, this method was originally designed to utilize growth conditions (e.g., temperature, pH, light regime, etc.) that most closely resemble the natural environment from which the inocula were taken (7, 23). In this manner, the incubation conditions are directly determined by the natural conditions of the inoculum.

SCREENING ENRICHMENTS AND ISOLATES VIA EPIFLUORESCENCE MICROSCOPY

General Nucleic Acid Stains

A general screening method that scores each well as positive or negative for growth can be performed upon completion of the initial incubation period. To screen the microtiter plates for growing cells, small volumes from each well of the plates are arrayed on polycarbonate membranes (GE Osmonics Inc.) using a custom-designed dot blot manifold containing small-diameter (ca. 1 mm) wells. The number of cells in each well can be estimated by direct cell counts after staining with a DNA-binding dye such as DAPI. The purpose of screening with a nonspecific DNA-binding fluorescent dye such as DAPI is to determine which extinction wells contain growth and to ensure that all negative controls remain sterile. Occasionally, certain morphological analyses as well as information regarding the purity of individual cultures can be garnered through general screening, but cultures are typically scored simply for the presence or absence of microbial cells and for a rough estimate of cell density. In addition to the DAPI stain and fixative, it is helpful to add an appropriate dilution of 0.5-μm-diameter fluorescent microspheres (Fluoresbrite Multifluorescent; Polysciences Inc., Warrington, Pa.) to yield roughly 20 microspheres per field of view at a magnification of ×1,000. The beads fluoresce under multiple wavelengths of light and therefore aid in well location, orientation under the microscope, and size estimation of microbial cells (23).

For relatively high-throughput screening, it is useful to employ a custom-designed 48-well manifold (e.g., HyTek Plastics, Corvallis, Oreg.) (7). The manifold is used to screen one entire 48-well plate at a time, thereby circumventing the exhaustive process of making 48 individual stains in order to screen each multiwell plate. Before and after each use, the top portion of the 48-well manifold is thoroughly cleaned to avoid false positives under the microscope. To clean the manifold top, rinse it thoroughly with tap water before placing it in a heated sonicator for ca. 20 min. After sonication, rinse the manifold top with distilled water, followed by a thorough rinse with ultrapure water, ensuring that all wells are completely flushed. Shake any excess water out and orient the block on its side to dry while assembling the remainder of the manifold apparatus. Using clean forceps, load a rectangular precut piece of Whatman filter paper (Whatman Inc., Florham Park, N.J.) onto the manifold base, followed by a 48- by 60-mm 0.22-μm-pore-size black polycarbonate membrane. Check to ensure that no air is trapped beneath the membrane, as bubbles that form in the wells may obstruct filtration. Place the clean 48-well manifold top on the polycarbonate membrane and secure it into position.

After assembling the manifold, load 125 μl from each well of a multiwell plate into the corresponding well on the 48-well manifold. Add 40.0 μl of 10% buffered formaldehyde and bead solution to each well. To make 1 liter of this solution, add 270.3 ml of formalin (37% formaldehyde) to 729.7 ml of 1× phosphate-buffered saline, pH 7.4. Fresh filter (e.g., Supor-200; Pall Corp., East Hills, N.Y.) immediately prior to use and add fluorescent microspheres to a dilution appropriate to yield ca. 20 beads per field of view at a magnification of ×1,000. After adding the fixative to each well, dislodge any bubbles that form in the wells by tapping gently on the side of the apparatus or remove them using a sterile pipette tip. Incubate the manifold at room temperature for 15 min in the dark. Apply a vacuum to the manifold and filter until all wells are dry (ca. 3 to 5 min at 5 to10 in. Hg). When dismantling the apparatus, the polycarbonate membrane often remains stuck to the top portion of the manifold. Use forceps to carefully remove the membrane, mark it for orientation, and allow it to dry at room temperature for 10 to 15 min in the dark. The top portion of the manifold should be cleaned again as described above and stored in a dilute detergent solution. Affix the membrane, cell side up, to a 75- by 50-mm microscope slide (e.g., Corning Inc.) using a small piece of adhesive such as Tough-Spots (Diversified Biotech, Boston, Mass.) and transfer it to a chilled (4°C), aluminum foil-wrapped Coplin jar containing DAPI solution (5 μg of DAPI per ml) in hybridization wash (150 mM NaCl, 20 mM Tris-HCl [pH 7.4], 6 mM EDTA, 0.01% sodium dodecyl sulfate [SDS]). Incubate in the DAPI solution for 10 min at 4°C. Transfer the slide to a chilled, aluminum foil-wrapped Coplin jar containing only hybridization wash and incubate for 2 min at 4°C. Under subdued light, peel the Tough-Spots from the membrane, remove the membrane from the slide, and allow it to dry at room temperature for 10 to 15 min in the dark. Carefully mount the membrane on a clean 75- by 50-mm microscope slide using a 48- by 60-mm coverslip (e.g., Erie Scientific, Portsmouth, N.H.) and high-viscosity immersion oil (e.g., Richard-Allan Scientific, Kalamazoo, Mich.). Apply firm, even pressure to the coverslip to ensure an even viewing plane under the microscope. If the slides are not viewed immediately, they can be temporarily stored at 4°C.

FISH

Several strategies are available to determine the identity of the microorganisms contained within the initial DAPI-positive wells (hereafter referred to as "primary cultures"). If the goal of the experiment is to target one or a few discrete phylogenetic groups of microorganisms for isolation and further propagation in the laboratory, a targeted approach employing fluorescence in situ hybridization (FISH) with fluorescently labeled, rRNA-targeted oligonucleotide probes may be appropriate (for an example, see reference 24). Wells that score positive for growth via DAPI staining are consolidated (manually or robotically) into new microtiter plates, rearrayed, and hybridized to fluorescently labeled oligonucleotide probes. These probes reveal if any of the isolates are members of the phylogenetic groups targeted by the probes. By hybridizing and detecting multiple probes and fluorochromes simultaneously on high-density cell arrays, large numbers of primary isolates can be efficiently and simultaneously screened for a small number of specific target species in both a rapid and relatively inexpensive fashion.

To begin, subsample 175 μl of each primary culture into clean microtiter plates. Add 50 μl of a 0.2-μm-pore-size-filtered (e.g., CLS431219 Corning syringe filter) buffered 10% paraformaldehyde solution to each well of the microtiter plates containing the subsamples and mix thoroughly. Be careful to avoid cross-contamination by employing a fresh pipette tip per well in order to ensure that no liquid is transferred between the wells during mixing. The goal is to achieve a final concentration of 2 to 3% paraformaldehyde in each well. Incubate the plates for 6 to 10 h at 4°C, and then load the subsamples into the custom 48-well cell array manifold described above. The manifold should already be cleaned and loaded with a Whatman backing filter and 0.22-μm-pore-sized polycarbonate membrane as described previously. Add an appropriate dilution of 0.5-μm-diameter fluorescent microspheres to yield roughly 20 microspheres per field of view at a magnification of ×1,000, and filter the subsamples through the membrane. Remove the membrane from the manifold, mark it to maintain proper orientation, and place it in the dark at room temperature to dry for 10 to 15 min. Transfer the membrane to a 75- by 50-mm glass microscope slide, and affix it to the slide using a small piece of adhesive such as Tough-Spots as described previously.

All of the steps in the hybridization procedure should be performed under subdued lighting due to the light sensitivity of the fluorescently labeled oligonucleotide probes. Apply 200 μl of hybridization solution (0.9 M NaCl, 20 mM Tris-HCl [pH 7.4], 15% formamide, 0.01% SDS) containing 2 ng of fluorescently labeled oligonucleotide probe per μl to the surface of the membrane and affix a 48- by 60-mm coverslip. Incubate the membrane at 37°C in a preheated, sealed humid hybridization chamber for 4 to 12 h. Prepare four identical hybridization washes (150 mM NaCl, 20 mM Tris-HCl [pH 7.4], 6 mM EDTA, 0.01% SDS) in Coplin jars. All hybridization washes are 0.2-μm-pore-size filtered immediately prior to use. Preheat two of the washes at the final wash temperature, which is empirically determined for the selected oligonucleotide probe or probe suite (for an example, see reference 17). To one of the remaining Coplin jars containing hybridization wash, add 0.2 mg of DAPI stock per ml to a final concentration of 5 μg/ml. Wrap this and the fourth Coplin jar containing only hybridization wash with aluminum foil and place both at 4°C. After incubation, remove the coverslip and transfer the slide from the hybridization chamber to the first preheated hybridization wash and incubate for 10 min under subdued light. Transfer the slide from the first to the second preheated hybridization wash and incubate for 10 min under subdued light. Transfer the slide to the chilled Coplin jar containing DAPI solution in hybridization wash and incubate for 10 min at 4°C. Transfer the slide to the chilled Coplin jar containing only hybridization wash and incubate for 2 min at 4°C. Remove the filter from the slide and incubate in the dark at room temperature for ca. 15 min to dry. Using an antifadant such as Citifluor (Ted Pella Inc., Redding, Calif.) and a clean coverslip, mount the membrane on a clean microscope slide. Seal all edges with fingernail polish and store at −20°C if the slide is not to be viewed immediately. Score the slide by using an epifluorescence microscope containing the appropriate filter sets.

SCREENING ENRICHMENTS AND ISOLATES VIA PCR-BASED TECHNIQUES

An alternative strategy to FISH for identifying primary cultures is to characterize a phylogenetic marker such as the small-subunit (SSU) rRNA gene or adjacent internal transcribed spacer region (for examples, see references 7 and 23). In this approach, genomic DNA is extracted from small volumes (ca. 200 to 300 μl) of the primary cultures, usually in a microtiter-based format (e.g., DNeasy 96 tissue kit; Qiagen Inc., Valencia, Calif.). The PCR is then used to amplify the phylogenetic marker of choice, often in a nested or seminested fashion due to the low quantity of genomic DNA obtained from the small volumes of the primary cultures. If a large number of clonal cultures in the primary isolate pool is anticipated (e.g., by using a low initial inoculum concentration such as one cell per well), one option is to directly sequence the PCR products and identify the isolates. Alternatively, the PCR products can be grouped by restriction fragment length polymorphism (RFLP) analysis or terminal restriction fragment length polymorphism (T-RFLP) analysis (16), at which time representatives from each group can be sequenced. If a large number of mixed cultures are expected, community profiling methods such as denaturing gradient gel electrophoresis (18) or T-RFLP are useful screening tools, as they provide information on the number of different phylogenetic groups within the mixed culture, as well as information regarding the putative identity of those groups. Though affected by a wide variety of user-controlled variables, a rough estimate is that the DNA extraction-based methods described below require about 200 cells for reliable PCR amplification (7).

Genomic DNA Preparation

DNA extractions are performed using the DNeasy Tissue Kit protocol for animal tissues (Qiagen Inc.) with the following modifications. Remove 250 to 300 µl of each primary culture for DNA extraction. If not processed immediately, these aliquots may be stored at −80°C. If the samples are to be processed immediately, place them at −80°C for 10 min before continuing on with the following protocol. Remove the samples from −80°C and incubate at 70°C for 10 min. Repeat this cycle two more times. To each sample, add 32 µl of proteinase K and 290 µl of buffer AL (Qiagen's proprietary lysis buffer) and mix. To ensure optimum binding of genomic DNA to the spin column membranes, the samples must have an acidic pH. If, after testing the pH, the samples are not acidic, add 2 to 5 µl of 10% HCl to each sample and test the pH again to ensure that it has fallen below 7.0. Incubate the samples at 70°C for 10 min before adding 290 µl of 96% molecular biology grade ethanol to each sample. Mix the samples and transfer them to the DNeasy spin columns. Follow the manufacturer's instructions for the remainder of the protocol, until the final elution step. Elute the genomic DNA from the column with sterile water instead of the buffer supplied by the kit to allow for easy concentration of the genomic samples using a speed-vac, if necessary. Concentration of the genomic DNA may be required in order to supply a sufficient amount of template DNA to the subsequent PCR amplification procedure described below. Genomic DNA is quantified via fluorescence on a Spectra Max M2 plate fluorometer (Molecular Devices Corp., Sunnyvale, Calif.) using the PicoGreen dsDNA quantitation kit (Molecular Probes; Invitrogen Corp., Carlsbad, Calif.).

PCR of SSU rRNA Genes and Screening via T-RFLP

Primary cultures are frequently evaluated first via T-RFLP analysis of the SSU rRNA gene (16) in order to assess the purity of the cultures and to potentially obtain a putative identification. Due to the small volume of the primary cultures, it is often necessary to amplify the SSU rRNA gene marker in a nested or seminested fashion. The following protocol describes a typical seminested amplification.

Prepare the first reaction mixture in a total volume of 20 µl as follows: we recommend the use of slightly modified general bacterial primers such as 27F-B (5′-AGRGTTYGATYMTGGCTCAG-3′) and the reverse primer 1492R (5′-GGYTACCTTGTTACGACTT-3′) (15) in conjunction with 1 µl of genomic DNA template and a high-fidelity polymerase (e.g., PicoMaxx; Stratagene, La Jolla, Calif.). Cycle this reaction mixture for 35 rounds by using the following parameters: denaturation at 94°C for 30 s, annealing at 55°C for 1 min, and elongation at 72°C for 2 min. For the second, seminested reaction, prepare a 50-µl PCR mixture in the same manner as described above, with the following modifications: use the bacterial primers 27F-B-FAM and 519R (5′-GWATTACCGCGGCKGCTG-3′) (15). The 27F-B-FAM primer is identical to 27F-B in sequence, except that it is 5′ end-labeled with a phosphoramidite fluorochrome 5-carboxy-fluorescein (6-FAM). In addition, 2.5 µl of PCR product from the initial reaction is used as a template in the second reaction and the genomic DNA is omitted. Cycling parameters of the second reaction are identical to the first. The seminested products are purified using a commercial purification kit (e.g., QIAquick PCR Purification Kit; Qiagen Inc.).

A restriction digest of the purified PCR products is performed to create 5′ end-labeled fragments. Prepare a restriction digest mix containing 3.3 µl of molecular biology grade water, 0.2 µl of bovine serum albumin, 1 µl of 10× enzymatic reaction buffer, and 0.5 µl of HaeIII (10 units/µl) for each sample and vortex to mix. Combine 5 µl of the restriction digest mix with 5 µl of each purified PCR product and gently flick the tubes to mix. Incubate the restriction reaction mixtures at 37°C for 5 to 6 h. Store the remaining purified product at −20°C for future use. If necessary, desalt the restriction digests prior to analysis. There are multiple viable methods for desalting the digests, including ethanol precipitation or commercial kits (e.g., QIAquick Nucleotide Removal Kit; Qiagen Inc.). Finally, electrophorese the purified restriction digests on a genetic analyzer such as an ABI 3100 Genetic Analyzer from Applied Biosystems (Foster City, Calif.). If a culture appears to be pure (i.e., there is a single peak on the T-RFLP electropherogram), it can be sent for a more conclusive identification via DNA sequencing of the SSU rRNA gene. If the culture appears to be mixed (i.e., there are multiple peaks), subculturing can be performed in an attempt to separate the multiple taxa into individual cultures. It is important to remember that some microorganisms may be incapable of isolated growth in vitro and thus may require consortial cohorts. Pure cultures of interest can be sequenced on a genetic analyzer such as the ABI 3100 using the remaining unrestricted purified PCR product obtained above, while consortia may be identified by cloning and sequencing.

SUBCULTURING AND CRYOPRESERVATION

As primary isolates of interest are identified, the remaining volume of primary culture should be split between the cryopreservation of subsamples in glycerol or dimethyl sulfoxide and the subculturing of samples into 1 to 10 ml of fresh, amended, sterile medium. Ideally, this medium is identical to that used in the primary isolation step in order to maintain consistent growth conditions. The subcultures are monitored for growth as described above for the screening of primary isolates.

Cryopreservation is a common method of storing viable microbial cultures for an extended period of time (for an example, see reference 10). Primary cultures and subcultures that screen positive for growth can be stored in glycerol and thawed for later use, if necessary. It is recommended that several frozen stocks be made from each clonal isolate (if volume permits) to guard against the potential loss of a viable culture. Thorough labeling of stored isolates is an essential part of the cryopreservation process to ensure that the origin and history of each culture are known.

To cryopreserve a sample, prepare a 50% (vol/vol) glycerol solution in sterile seawater in a laminar flow hood. Autoclave the mixture and aliquot 150 µl into 1.5-ml screw-cap tubes. Nalgene System 100 screw-cap cryovials

(Nalge Nunc) are an ideal choice. To each tube, add 500 μl of viable culture in exponential growth phase and gently mix by inverting. It is preferable to freeze the stocks at a slow, controlled rate. Nalgene Cryo 1°C Freezing Containers (Nalge Nunc) are inexpensive and cool cultures at a rate of ca. −1°C/min. Prepare the container with sample tubes and isopropanol as recommended by the manufacturer and place at −80°C overnight. Remove the tubes and transfer them to a liquid-nitrogen dewar for long-term storage. If a dewar is not available, glycerol stocks may potentially be maintained for an extended period of time at −80°C, though viability may vary between isolates. While it is not possible to predict the length of time any given isolate may remain viable when cryopreserved and stored in this manner, isolates have been successfully recovered after 4 years of storage in liquid nitrogen (unpublished data).

ESTIMATING CULTURABILITY

Fundamental culturability statistics can be determined following the initial screening of primary cultures, as derived and outlined by Button and colleagues (4). The percent viability, V, is calculated using the formula $V = -\ln(1 - p)/X$, where p is the proportion of wells that scored positive for growth and X is the number of cells used for the initial inoculation. It may also be beneficial to estimate the number of pure isolates that can be expected from a particular experiment, which can be done by applying the equation $u = -n(1 - p)\ln(1 - p)$, where u is the expected number of pure isolates and n is the number of extinction wells screened. To obtain 95% confidence intervals, the exact upper and lower 95% confidence limits for p are calculated and inserted back into each equation in place of p. The result is the exact upper and lower 95% confidence limits for percent culturability and the predicted number of pure cultures.

Button and colleagues applied the viability equation in their dilution culture work in Alaska's Resurrection Bay and commonly generated culturability values ranging from 2 to 60% (4). Some of the inaugural HTC experiments, which are based on the dilution culture technique, yielded culturability values from 0.4 to 14.3%, which were, on average, higher in the summer (8.8%) than during the winter months (1.2%) (7). It should also be noted that the theoretical number of pure isolates was consistent with actual experimental results.

CONCLUDING REMARKS

The high-throughput dilution to extinction methods described here permit the investigation of a large number of chemical and physical variables simultaneously in order to identify optimal conditions that promote the growth of aquatic microbial cells with specialized growth requirements. Coupled with rapid, sensitive, and specific screening methods such as FISH, T-RFLP, or screening for the presence of specific functional genes via PCR, the HTC method allows scientists to tailor extensive cultivation efforts that target maximum overall diversity or focus their efforts on narrowly defined target microorganisms.

REFERENCES

1. **Beja, O., M. T. Suzuki, E. V. Koonin, L. Aravind, A. Hadd, L. P. Nguyen, R. Villacorta, M. Amjadi, C. Garrigues, S. B. Jovanovich, R. A. Feldman, and E. F. DeLong.** 2000. Construction and analysis of bacterial artificial chromosome libraries from a marine microbial assemblage. *Environ. Microbiol.* **2:**516–529.
2. **Bruns, A., H. Cypionka, and J. Overmann.** 2002. Cyclic AMP and acyl homoserine lactones increase the cultivation efficiency of heterotrophic bacteria from the Central Baltic Sea. *Appl. Environ. Microbiol.* **68:**3978–3987.
3. **Bruns, A., U. Nubel, H. Cypionka, and J. Overmann.** 2003. Effect of signal compounds and incubation conditions on the culturability of freshwater bacterioplankton. *Appl. Environ. Microbiol.* **69:**1980–1989.
4. **Button, D. K., F. Schut, P. Quang, R. Martin, and B. R. Robertson.** 1993. Viability and isolation of marine bacteria by dilution culture: theory, procedures and initial results. *Appl. Environ. Microbiol.* **59:**881–891.
5. **Button, D. K., B. R. Robertson, P. W. Lepp, and T. M. Schmidt.** 1998. A small, dilute-cytoplasm, high-affinity, novel bacterium isolated by extinction culture and having kinetic constants compatible with growth at ambient concentrations of dissolved nutrients in seawater. *Appl. Environ. Microbiol.* **64:**4467–4476.
6. **Cho, J. C., and S. J. Giovannoni.** 2004. Cultivation and growth characteristics of a diverse group of oligotrophic marine *Gammaproteobacteria*. *Appl. Environ. Microbiol.* **70:** 432–440.
7. **Connon, S. A., and S. J. Giovannoni.** 2002. High-throughput methods for culturing microorganisms in very-low-nutrient media yield diverse new marine isolates. *Appl. Environ. Microbiol.* **68:**3878–3885.
8. **DeLong, E. F., and D. M. Karl.** 2005. Genomic perspectives in microbial oceanography. *Nature* **439:**336–342.
9. **Fitzwater, S. E., G. A. Knauer, and J. H. Martin.** 1982. Metal contamination and its effect on primary production measurements. *Limnol. Oceanogr.* **27:**544–551.
10. **Gherna, R. L.** 1994. Culture preservation, p. 278–292. *In* P. Gerhardt, R. G. E. Murray, W. A. Wood, and N. R. Krieg (ed.), *Methods for General and Molecular Bacteriology*. American Society for Microbiology, Washington, D.C.
11. **Giovannoni, S. J., T. B. Britschgi, C. L. Moyer, and K. G. Field.** 1990. Genetic diversity in Sargasso Sea bacterioplankton. *Nature* **345:**60–63.
12. **Giovannoni, S. J., and M. Rappé.** 2000. Evolution, diversity and molecular ecology of marine prokaryotes, p. 47–84. *In* D. L. Kirchman (ed.), *Microbial Ecology of the Oceans*. John Wiley & Sons, Inc., New York, N.Y.
13. **Hahn, M. W., P. Stadler, Q. L Wu, and P. Matthias.** 2004. The filtration-acclimatization method for isolation of an important fraction of the not readily cultivable bacteria. *J. Microbiol. Methods* **57:**379–390.
14. **Konneke, M., A. E. Bernhard, J. R. de la Torre, C. B. Walker, J. B. Waterbury, and D. A. Stahl.** 2005. Isolation of an autotrophic ammonia-oxidizing marine archaeon. *Nature* **437:**543–546.
15. **Lane, D. J.** 1991. 16S/23S rRNA sequencing, p. 115–147. *In* E. Stackenbrandt and M. Goodfellow (ed.), *Nucleic Acid Techniques in Bacterial Systematics*. John Wiley & Sons, New York, N.Y.
16. **Liu, W. T., T. L. Marsh, H. Cheng, and L. J. Forney.** 1997. Characterization of microbial diversity by determining terminal restriction fragment length polymorphisms of genes encoding 16S rRNA. *Appl. Environ. Microbiol.* **63:** 4516–4522.
17. **Morris, R. M., M. S. Rappé, E. Urbach, S. A. Connon, and S. J. Giovannoni.** 2004. Prevalence of the *Chloroflexi*-related SAR202 bacterioplankton cluster throughout the mesopelagic zone and deep ocean. *Appl. Environ. Microbiol.* **70:**2836–2842.
18. **Muyzer, G., E. C. De Waal, and A. G. Uitterlinden.** 1993. Profiling of complex microbial populations by denaturing gradient gel electrophoresis analysis of polymerase chain

reaction-amplified genes encoding for 16S rRNA. *Appl. Environ. Microbiol.* **59:**695–700.

19. **Olsen, G. J., D. J. Lane, S.J. Giovannoni, N. R. Pace, and D. A. Stahl.** 1986. Microbial ecology and evolution: a ribosomal RNA approach. *Annu. Rev. Microbiol.* **40:**337–365.
20. **Pace, N. R., D. A. Stahl, D. J. Lane, and G. J. Olsen.** 1986. The analysis of natural microbial populations by ribosomal RNA sequences. *Adv. Microb. Ecol.* **9:**1–55.
21. **Page, K. A., S. A. Connon, and S. J. Giovannoni.** 2004. Representative freshwater bacterioplankton isolated from Crater Lake, Oregon. *Appl. Environ. Microbiol.* **70:**6542–6550.
22. **Porter, K. G., and Y. S. Feig.** 1980. The use of DAPI for identifying and counting aquatic microflora. *Limnol. Oceanogr.* **25:**943–948.
23. **Rappé, M. S., S. A. Connon, K. L. Vergin, and S. J. Giovannoni.** 2002. Cultivation of the ubiquitous SAR11 marine bacterioplankton clade. *Nature* **418:**630–633.
24. **Rappé, M. S., and S. J. Giovannoni.** 2003. The uncultured microbial majority. *Annu. Rev. Microbiol.* **57:**369–394.
25. **Ringuet, S.** 2003. *Biogeochemical Impacts of Storm Runoff on Water Quality in Southern Kaneohe Bay, Hawai'i.* M.S. thesis. University of Hawai'i, Manoa.
26. **Schut, F., E. J. de Vries, J. C. Gottschal, B. R. Robertson, W. Harder, R. A. Prins, and D. K. Button.** 1993. Isolation of typical marine bacteria by dilution culture: growth, maintenance, and characteristics of isolates under laboratory conditions. *Appl. Environ. Microbiol.* **59:** 2150–2160.
27. **Schut, F., J. C. Gottschal, et al.** 1997. Isolation and characterization of the marine ultramicrobacterium Sphingomonas sp. strain RB2256. *FEMS Microbiol. Rev.* **20:**363–369.
28. **Suzuki, M. T., M. S. Rappé, Z. W. Haimberger, H. Winfield, N. Adair, J. Strobel, and S. J. Giovannoni.** 1997. Bacterial diversity among small-subunit rRNA gene clones and cellular isolates from the same seawater sample. *Appl. Environ. Microbiol.* **63:**983–989.
29. **Venter, J. C., K. Remington, J. F. Heidelberg, A. L. Halpern, D. Rusch, J. A. Eisen, D. Y. Wu, I. Paulsen, K. E. Nelson, W. Nelson, D. E. Fouts, S. Levy, A. H. Knap, M. W. Lomas, K. Nealson, O. White, J. Peterson, J. Hoffman, R. Parsons, H. Baden-Tillson, C. Pfannkoch, Y. H. Rogers, and H. O. Smith.** 2004. Environmental genome shotgun sequencing of the Sargasso Sea. *Science* **304:**66–74.
30. **Woese, C. R.** 1987. Bacterial evolution. *Microbiol. Rev.* **51:**221–271.
31. **Woese, C. R., O. Kandler, and M. L. Wheelis.** 1990. Towards a natural system of organisms: proposal for the domains Archaea, Bacteria, and Eucarya. *Proc. Natl. Acad. Sci. USA* **87:**4576–4579.
32. **Zengler, K., G. Toledo, M. Rappé, J. Elkins, E. J. Mathur, J. M. Short, and M. Keller.** 2002. Cultivating the uncultured. *Proc. Natl. Acad. Sci. USA* **99:**15681–15686.
33. **Zengler, K., M. Walcher, G. Clark, I. Haller, G. Toledo, T. Holland, E. J. Mathur, G. Woodnutt, J. M. Short, and M. Keller.** 2005. High-throughput cultivation of microorganisms using microcapsules. *Methods Enzymol.* **397:**124–130.

Primary Productivity and Producers

HANS W. PAERL

33

THE PROCESS OF PRIMARY PRODUCTION AND RELEVANT MICROORGANISMS

Primary production is the biochemical conversion of inorganic to organic matter in aquatic ecosystems. The rate of primary production (primary productivity) and the subsequent accumulation of organic matter are key drivers and indicators of trophic state and water quality (7, 15). Because of the fundamental biogeochemical importance of primary production, changes in this process are used to assess ecosystem, regional, and global change in the flux of organic and inorganic carbon, caused by both natural (e.g., geochemical and climatic) and human activities, including accelerated nutrient loading and eutrophication, impacts of xenobiotic compounds and other pollutants, and hydrologic alterations.

Except for systems dominated by macroalgae and macrophytes, the bulk of production of organic matter at the base of planktonic and benthic food webs is mediated by microbial photoautotrophy or phototrophy (photosynthetically mediated CO_2 fixation), and chemolithotrophy or chemoautotrophy (chemically mediated CO_2 fixation). In these processes, light-derived or chemically derived energy is used to reduce CO_2 to organic carbon. The basic stoichiometry of CO_2 fixation is

$$nCO_2 + 2nH_2O \rightarrow n(CH_2O) + O_2 + nH_2O$$

The manner in which autotrophs derive and donate reducing power to drive this process differs between physiological groups. Phototrophs utilize light-driven photolysis of simple, readily oxidizable compounds as a source of electrons and protons to generate reducing power (NADPH) and energy (ATP) for CO_2 fixation. The two modes of phototrophy are mediated by oxygenic and anoxygenic photosynthesis. In oxygenic photosynthesis, the dominant form of primary production in oxygenated surface waters, water (H_2O) fills this role. The photolysis of H_2O results in the formation of protons, electrons, and molecular oxygen (O_2), as outlined below:

$$2H_2O \xrightarrow{h\nu} 4H^+ + 4e^- + O_2$$

This is the light reaction of photosynthesis. Combining the reducing power of the light reaction with the CO_2 fixation, or the dark reactions, of photosynthesis yields the following overall biochemical reaction:

$$6CO_2 + 12H_2O \xrightarrow{h\nu} C_6H_{12}O_6 + 6O_2 + 6H_2O$$

Oxygenic photosynthesis is conducted by all chlorophyll *a*-containing microorganisms. These include the prokaryotic blue-green algae (cyanobacteria) and eukaryotic microalgal groups (Table 1). Anoxygenic photosynthesis is confined to photosynthetic bacterial groups (Table 1). H_2S is the source of protons and electrons, with S being the product:

$$2H_2S \xrightarrow{h\nu} 4H^+ + 4e^- + 2S$$

Anoxygenic photosynthesis is restricted to O_2-devoid water columns or surface sediments. Anoxygenic photosynthesis was, in all likelihood, the prevalent form of aquatic primary production during the O_2-free Precambrian period (~2.5 billion years before the present). Oxygenic photosynthesis was ushered in with the appearance of cyanobacteria during the late Precambrian (5). Present-day oxygen levels attributed to the advent of oxygenic phototrophs have led to confinement of the anoxygenic phototrophs to stratified, illuminated bottom waters in lakes, near-surface anoxic sediments, nutrient-enriched wastewater systems (e.g., sewage treatment ponds and animal waste lagoons), laminated microbial mats and biofilms containing O_2-free microenvironments, and endosymbiotic (O_2-free) habitats.

Chemolithotrophy occurs in oxic and anoxic waters, where the oxidation of a range of inorganic and organic compounds provides reductant and energy for CO_2 fixation. Examples of aerobic chemolithotrophs commonly encountered free living in the water column and associated with oxic surface sediments include nitrifying bacteria (*Nitrosomonas* and *Nitrobacter*), which oxidize ammonium (NH_4^+) or nitrite (NO_2^-) and reduce CO_2 (Table 1). Representative facultative anaerobic and microaerophilic chemolithotrophs include the nonpigmented sulfide (S^{2-})-oxidizing bacteria (*Beggiatoa*) and the sulfur (S^0) oxidizers (*Thiobacillus*) (Table 1).

ECOLOGICAL AND BIOGEOCHEMICAL SIGNIFICANCE

Primary productivity is a measure of ecosystem fertility and trophic state (7, 15) and as such serves as a prime indicator of either natural or man-induced alteration in the productive

TABLE 1 Major functional groups of aquatic microbial autotrophs, categorized according to environmental requirements and oxic/anoxic characteristics[a]

- Oxygenic phototrophs
  - Prokaryotic
    - Cyanobacteria (P and B)
    - Photosynthetic bacteria (P and B)
      - *Erythrobacter* (O_2 tolerant, contains bacteriochlorophyll *a*)
  - Eukaryotic
    - Chlorophytes (green algae; P and B), chrysophytes (including diatoms; P and B)
    - Cryptophytes (P), dinoflagellates (colorless and pigmented; P), euglenophytes (colorless and pigmented; P and B), prasinophytes (mainly P)
- Anoxygenic phototrophs (prokaryotic: also grow photoheterotrophically)
  - Sulfide and sulfur as sole electron donor for photosynthesis
    - *Chromatium*, *Chlorobium*, *Thiocapsa*, *Thiopedia*, and *Thiospirillum* (P only in meromictic waters, B)
    - *Ectothiorhodospira* (P and B, high salinity and alkaline requirements)
    - *Prosthecochloris* (B, strictly marine)
    - *Rhodobacter* (P and B)
    - *Rhodopseudomonas*, *Helicobacter* (mainly B)
    - *Chloroflexus* (B, thermophilic)
- Aerobic chemolithotrophs (prokaryotic)
  - Colorless sulfur-oxidizing bacteria
    - *Thiobacillus* (mainly B)
    - *Thiovulum* (B, microaerophilic)
    - *Beggiatoa* (B, microaerophilic)
    - *Thiosphaera* (mainly B, microaerophilic)
  - Iron- and manganese-oxidizing bacteria
    - *Gallionella* (B, acidophilic)
    - *Planctomyces* (P and B)
    - *Leptothrix* (mainly B)
  - Nitrifying bacteria
    - *Nitrosomonas* and *Nitrosococcus* (P and B, ammonia oxidizers)
    - *Nitrobacter* and *Nitrococcus* (P and B, nitrite oxidizers)

[a]P, planktonic; B, benthic.

capacity and associated water quality of impacted waters. The rates of CO_2 fixation and O_2 production and consumption are critical facets of ecosystem metabolism, including net autotrophy versus heterotrophy and oxygen cycling. Accelerated primary production, or eutrophication, is generally accompanied by increases in phototrophic and heterotrophic biomass. Excessive production may manifest itself as unsightly, odoriferous, and toxic microalgal blooms, where production exceeds the ability of the food web to consume the vast amounts of producer biomass. The accumulation of ungrazed microalgal biomass often enhances microbial decomposition (O_2 consumption), which may yield undesirable consequences, including hypoxia (stressfully low dissolved O_2 concentrations) or anoxia (no detectable O_2), which is fatal to a broad spectrum of fauna and flora.

Short-term (hourly to weekly) changes in primary productivity reflect diel and seasonal changes in ecosystem nutrient inputs, physical alterations (e.g., changes in optical properties, vertical mixing, and water residence time), and shifts in plant community composition and activity. Long-term interannual, decadal, millennial, and geological changes may be indicative of ecological, geochemical, and climatic changes. Analyses of long-term data sets have demonstrated the utility of primary productivity measurements as indicators of either natural or anthropogenic ecosystem alterations (16, 18, 41, 64).

Productivity measurements have been coupled to other indicators of ecological and biogeochemical change (7, 8, 11, 12, 15, 41, 68). These include (i) stable C, N, S, and O isotope analyses of sources, fluxes, and fates of nutrients, (ii) biochemical indicators and markers of plant and microbe community structure and function (e.g., diagnostic pigment analyses and, as will be discussed later, cell wall carbohydrate and lipid and protein structural markers), (iii) molecular phylogenetic indicators (16s rRNA) and functional genes such as *nifH*, *nitr*, *nosZ*, and *Rbcl*, and (iv) fluorescence identification and enumeration techniques (e.g., flow cytometry). In addition, there is a suite of microscopic identification and enumeration techniques available, including various cytological stains, immuno- and autofluorescence, and microautoradiography coupled to fluorescence in situ hybridization. These techniques add to the sensitivity, specificity, relevance, and utility of productivity measurements as indicators of ecosystem structure, function, and change (41, 47, 48). The general use and application of these ancillary techniques are discussed in this chapter.

METHODS FOR MEASURING PRIMARY PRODUCTION IN AQUATIC HABITATS

Assessments of aquatic primary production are based on measuring rates of consumption of reactants or formation of products of autotrophy. The most commonly examined reactants are CO_2 (total dissolved inorganic C, or DIC, = CO_2 + HCO_3^{2-} and CO_3^{2-}) and H^+ (i.e., pH), while the products include O_2 and changes in pH. Primary production measurements are frequently supplemented with measurements of nutrient (e.g., N, P, and Fe) uptake, in order to obtain parallel estimates of biomass production and nutrient cycling.

Changes in biomass over time have also been used to approximate primary production (15, 68). This is accomplished either by microscopic determination of cell numbers or spectrophotometric or fluorometric determinations of chlorophyll *a* content. Cell counts or chlorophyll *a* content is converted to a cellular carbon (C) value by multiplying these parameters by cellular conversion factors (e.g., number of cells or amount of chlorophyll *a* multiplied by amount of carbon per cell per unit of chlorophyll *a* or biovolume). Biomass has also been estimated directly as C or N, by combusting microbial biomass and measuring C and N particulate content with a C,H,N analyzer (52, 62).

Other techniques that have been used to indirectly assess primary production include monitoring changes in stable isotopic composition (under natural abundance or enrichment conditions) of C, N, S, and O contained in microbial biomass (11, 12). One example is the use of ^{13}C for measuring CO_2 fixation and characterizing specific CO_2 fixation pathways (i.e., Calvin cycle versus C-4 and other pathways). This approach is based on the knowledge that microorganisms discriminate against heavy isotopic forms of each element during uptake, assimilatory, and growth reactions. Details of these techniques are provided in references 11 and 12 and Table 2.

CO_2 fixation can be routinely assessed by (i) measurements of CO_2 uptake by gas analysis or (ii) use of isotopic tracer techniques, specifically measuring $^{14}CO_2$ uptake.

CO_2 Uptake by Gas Analysis

CO_2 uptake can be assessed directly by gas chromatographic or infrared absorption analyses of CO_2 consumption, either in the headspace or soluble phase of gastight vessels. This approach is most often used in a laboratory or manipulated field experiments (e.g., sealed chambers or mesocosms), where access to analytical instrumentation is available.

Procedure

Populations and communities of interest are dispensed into a gastight, optically transparent vessel (normally borosilicate or Pyrex glass, polycarbonate, or Plexiglas, i.e., Lucite). The vessel aqueous and gas phases may be altered as desired. With respect to the gas phase, the vessel may be flushed with gas mixtures, either initially or continuously. Vessel contents are incubated under various illumination (including a dark treatment) regimens. At prescribed intervals, gas and/or aqueous subsamples (0.1 to several milliliters) are withdrawn through a serum stopper placed in the vessel. Vessels can also be sampled on a continuous gas-flow basis.

Analysis of $[CO_2]$ or [DIC] is either by infrared (IR) absorption (nondispersive IR detector) or gas chromatography (thermal conductivity detector). The sensitivities of these instruments are comparable. If CO_2 or DIC is the only form to be analyzed, IR analysis is most practical (especially if flowthrough procedures are used). If additional gases (e.g., O_2 and N_2) are to be analyzed, gas chromatography is preferred. CO_2 is analyzed by sampling the gas phase directly or by sparging the aqueous phase with an inert gas (He, Ar, or N) and passing the evolved CO_2 through respective instruments. To determine DIC, a 0.2-ml sample of aqueous phase is acidified in 5 to 10 ml of 30% H_3PO_4 to volatilize all inorganic C species (CO_2, HCO_3^{2-}, and CO_3^{2-}), sparged with inert gas, and analyzed by IR absorption (36). Instrument sensitivity and linearity are determined by making up a set of CO_2 or dissolved inorganic carbon (DIC) (using Na_2HCO_3 or Na_2CO_3) standards covering the range of concentrations encountered in samples.

Discussion of Technique

The gas analysis technique is most applicable to laboratory studies, where constant access to instrumentation can be ensured. If dense and active cultures (i.e., single populations) or microcosms (i.e., mixed communities) are available, this is a very effective, easily executed, and unequivocal technique. However, the ^{14}C technique is often preferred in ecosystem-level (i.e., water column or whole-lake/marine environment) studies.

While both techniques measure CO_2 uptake as an estimate of primary production, they do not necessarily account for identical processes and as a consequence may yield contrasting results. This is because the measure of CO_2 "uptake" by gas analysis includes simultaneous production (e.g., respiration or decomposition) and autotrophic consumption of CO_2. Thus, like the oxygen method of assessing primary production, this is a net measure of CO_2 flux. The ^{14}C method, on the other hand, measures only consumption of CO_2 (as $^{14}CO_2$). Small amounts of refixation of respired or excreted ^{14}C may be included in the measurement, especially during lengthy (>4-h) incubations.

The CO_2 uptake and production technique is suitable for both planktonic and benthic studies. Limitations are largely due to physical (structure and size) constraints. In particular, analyses of benthic samples can be problematic. Maintaining structural integrity of sediments, microbial biofilms, and mats is critically important. Disruption of these samples can lead to altered diffusivity and structural changes along vertical biogeochemical gradients. Such

TABLE 2 Commonly used methods for determining primary production in planktonic and benthic aquatic environments

Method	References[a]
^{14}C technique; planktonic systems	7, 9, 15, 44, 61, 68, 69
^{14}C technique; benthic systems	21, 22, 39, 54
O_2 Winkler method; planktonic systems	7, 16, 42, 68, 69
O_2 Winkler method; benthic systems	21, 22, 26, 49, 54
O_2 microelectrode method; benthic and planktonic systems	45, 53, 54
Photopigment-based methods; benthic and planktonic	23, 31, 39, 48, 71

[a]Detailed methodological and applicational references are cited.

alterations can significantly impact measurements of primary production and respiration, which in turn affect CO_2 flux (39, 54, 55). Changes in laminar flow and other forms of small-scale turbulence can profoundly affect CO_2 flux at the sediment-water interface. Therefore, samples must be stirred during measurements.

Uptake of ^{14}C-Labeled CO_2

The radioactive (β-emitting) form of carbon, ^{14}C, is a long-half-life (5,760-year) isotope commonly used to quantify CO_2 uptake and its incorporation into organic matter (15, 61). The ^{14}C method is based on several assumptions: (i) $^{14}CO_2$ biochemically behaves in a manner similar to that of the stable, dominant form of carbon, ^{12}C; however, there is a small (6%) isotopic discrimination against the heavier ^{14}C. (ii) The chemical form of ^{14}C administered (usually $NaH^{14}CO_3$) rapidly (within a few seconds after mixing) equilibrates with other forms of inorganic C (CO_2 and CO_3^{-2}) in accordance with pH and [DIC]. ^{14}C should be administered in trace quantities so that [DIC] remains unaltered by ^{14}C additions. Because of its high sensitivity, ease of use and of deployment in the field, and relatively low cost, the ^{14}C method is widely used for measuring primary production in natural waters (7, 8, 15, 61, 68, 69).

Procedure

Planktonic Samples

Water samples are collected (preferably with a nonmetallic sampler, such as a Van Dorn or Niskin bottle) and are dispensed in 100- to 250-ml incubation bottles. Transparent Pyrex, borosilicate, or polycarbonate bottles with gastight seals are preferred. If bottle contents are incubated outdoors, they should be kept shaded during filling and processing to prevent photoinhibition. For field assessments of primary productivity, at least triplicate light and single dark (opaque bottles) samples should be incubated at each sampling location. Bottles should be stored in a lighttight container until collection is complete. Parallel water samples for alkalinity (DIC or $CO_2 + HCO_3^{2-} + CO_3^{2-}$) and pH measurements are collected in well-sealed glass bottles and/or vials (completely filled). All samples should be stored in a cool, dark place until analysis.

After sample collection, $NaH^{14}CO_3$ is added to each bottle by using a preset repeating pipettor or syringe equipped with disposable tips or needles. From 0.5 to several milliliters of stock ^{14}C solution (3 to 10 μCi ml^{-1}) is added per bottle. Commercial sources of $NaH^{14}CO_3$ having specific activities of 20 to 100 μCi μmol^{-1} are available. At the above-mentioned dilutions, trace quantities of DIC (i.e., <0.1 μM) are added. Stock solutions can be made up and stored in several ways: (i) Commercially available $NaH^{14}CO_3$ solutions (usually shipped in sterile deionized water) can be diluted with sterile water (at pH 7.5 to 8.5) and stored refrigerated in a well-sealed bottle. (ii) Alternatively, the diluted stock can be dispensed in breakneck glass ampoules (i.e., "gold seal" or "blue seal" types), which are then sealed and autoclaved for long-term storage at room temperature.

Standard radioactivity protection measures, including wearing disposable gloves and lab coats and protecting the work area with disposable protective paper or plastic covers (Benchkote or equivalent material), should be used. A well-labeled set of glassware, pipettes, filtration funnels, forceps, etc. should be dedicated for repeated use with ^{14}C. Appropriate disposal and handling of solid and liquid waste (in accordance with state and federal regulations) are required.

After ^{14}C additions, bottles are sealed and agitated by repeatedly inverting (5 to 10 times) them to ensure thorough mixing of the isotope. Ensure dark bottles are truly dark by covering with layers of foil if necessary, and for in situ studies place bottles at locations and depths from which they were collected. Bottle contents are generally incubated for 2 to 4 h.

To terminate the incubation, bottles are collected and placed in a well-insulated, lighttight container for rapid transportation to the laboratory. If the time between sample collection and processing is longer than 1 h, samples should be "iced down" to restrict biological activity (i.e., respiration and decomposition of ^{14}C-labeled cell constituents). Bottle contents (partial or entire volume) are filtered under a gentle vacuum (2 to 5 lb/in^{-2} or 200 torr) on 0.45-μm-pore-size nitrocellulose (HA Millipore or equivalent) or glass fiber (Whatman GF/F or equivalent) filters. Filter funnels are rinsed with small quantities of prefiltered (0.45-μm pore size) water from which samples were obtained. Filters are then removed and placed face up under a vented hood.

Following drying, filters are placed in a well-sealed plastic container with a removable lid (0.20 by 30 by 10-cm polyethylene storage containers are suitable). A small, ~50-ml, wide-mouthed beaker half filled with concentrated (fuming) HCl is placed in the box, and the filters are exposed to HCl fumes for at least 30 min. This treatment eliminates adsorbed and abiotically precipitated inorganic ^{14}C from filtered material. Filters are then allowed to thoroughly vent and dry (usually overnight). They are placed in liquid scintillation (LS) vials to which a chemically compatible, high-efficiency LS cocktail is added. A variety of suitable cocktails is available, the most popular of which are those composed of biodegradable, nontoxic solvents (ICN Ecolume, Cytoscint, Betacount, or equivalent). Typically, ^{14}C counting efficiencies, determined from a set of quenched standards, range from 90 to 95% for ^{14}C in filtered materials (36).

Water samples stored for alkalinity measurements are analyzed (preferably on the same day) either by IR infrared absorbance or titration and pH for DIC.

The primary productivity (CO_2 fixed per unit time) is calculated as follows:

$$\text{Primary productivity (mg of C m}^{-3}\,\text{h}^{-1}) = \frac{(A_l - A_d) \times {}^{12}\text{C} \times 1.06 \times 1{,}000}{{}^{14}\text{C added} \times T}$$

where A_l is the mean counts (dpm) in light bottles, corrected for quenching and instrument background; A_d is the quench- and background-corrected counts (dpm) in a dark bottle; ^{12}C is the total DIC available, in mg of C liter^{-1}; 1.06 is a correction for the "isotope effect"; 1,000 is the conversion factor from liters to meters^{-3}; ^{14}C added is the total activity of ^{14}C added, in dpm; and T is the incubation time in hours (7, 8, 36, 61).

Primary productivity can also be calculated per square meter by utilizing integration techniques. It can additionally be "normalized" per milligram of chlorophyll *a* and per microeinstein of photosynthetically active radiation (PAR; 400 to 700 nm for oxygenic phototrophs and 400 to 850 nm for anoxygenic phototrophs) per square meter (or micromoles of photons per square meter per second). More-detailed discussion of the ^{14}C technique for freshwater and marine studies can be obtained from references by Goldman (15), Wetzel and Likens (68), and Falkowski (7, 8).

Benthic Samples

Primary productivity of benthic surface sediments, microalgal mats, and submersed biofilms can also be estimated by using the ^{14}C technique. A modification of methods described for planktonic samples is used. Sediment, mat, or biofilm subsamples are first obtained by cutting or coring small (~1- to 2-cm^2) pieces from natural assemblages. Whole cores, incubated intact, can also be used (39). Because of a high degree of heterogeneity in mats and biofilms, a large number of replicate (minimally triplicates) light samples is recommended. Sediment slurries can also be incubated in the manner described below. Replicated light and single dark subsamples are placed in cleaned 22-ml borosilicate LS vials having polypropylene-lined caps. Dark vials can be wrapped with foil. Twenty milliliters of sample water and 0.2 ml of $NaH^{14}CO_3$ (2 to 4 μCi of total activity; specific activity of 40 to 100 μCi μmol^{-1}) are added. Vials are sealed, gently mixed, and placed on their sides, with the mat surface facing up, under natural illumination and temperatures. Vials are agitated to ensure uniform distribution of isotope, nutrients, and gases. DIC is determined by IR analyses of acidified, sparged ambient water.

In the case of whole cores, $NaH^{14}CO_3$ can be directly injected into cores, which are then left to incubate under conditions described above (54). Cores are normally incubated under natural illumination and temperature conditions.

Incubations are terminated by removing sediment, mat, and core samples from vials and placing them, face up, in a fuming HCl atmosphere for at least 30 min. In the case of unconsolidated sediments, vial contents can be gently filtered onto glass fiber filters (e.g., Whatman GF/F or GF/C), which can then be processed as described for mats. Samples are air dried and processed for LS counting (LSC) by using a tissue-solubilizing, biodegradable cocktail (e.g., Cytoscint cocktail; ICN Inc.). Quenching (due to pigments or humic and other colored substances in sediments) may reduce LSC efficiencies. This is remedied by developing a quench curve, where various amounts of unlabeled sediments are amended with known quantities of calibrated [^{14}C]hexadecane or toluene (New England Nuclear, Amersham, and ICN).

The formula described for planktonic samples can be applied to benthic microalgal communities. Results are usually expressed as milligrams of C per unit surface area (square centimeter or square meter) by correcting for the surface area of mat or biofilm samples used per unit time. Data can be normalized per amount of chlorophyll *a*, particulate C, dry weight, and/or PAR flux.

Discussion of Technique

The ^{14}C technique is the technique of choice when large numbers of samples need to be rapidly analyzed and when extensive spatial-temporal (synoptic) sampling of lakes, rivers, estuaries, or oceanic regions is required. Since its introduction to marine and freshwater productivity studies in the 1950s (61, 66), this technique has been the dominant method for assessing primary production in natural waters. Despite its popularity, the ^{14}C method is not without technical and interpretational problems (13, 44, 69), which are summarized below.

Sample collection and preparation. Water samples should preferably be collected with nonmetallic samplers (e.g., Van Dorn or Niskin bottles or plastic hoses and buckets), as far away as possible from sampling platforms or vessels, while using "clean" techniques. Heavy metals, hydrocarbons (i.e., fuel and oil), and other anthropogenic contaminants (lubricants, paints, solvents, etc.) associated with sampling can alter photosynthetic rates of resident microflora (10).

The ^{14}C-$NaHCO_3$ stock solution should be made up of either distilled or deionized water or seawater free of the above-mentioned contaminants. When trace metal contamination is of concern, stock solution water is initially "cleaned up" by passing it through a Chelex column. The stock pH must be sufficiently alkaline (>pH 8.0) to guard against ^{14}C losses to the atmosphere (as $^{14}CO_2$) during preparation and dispensing.

Once samples are collected and transferred to incubation vessels, they are placed in a light-shielded environment as soon as possible. Photoinhibition, photorespiration, and (under extremely high light conditions) photooxidation of microflora may otherwise result. One should be particularly concerned about this problem in high-irradiance environments, including alpine, low-latitude tropical and subtropical regions and intertidal and shallow-water benthic regions, and when high reflectivity is encountered (i.e., on ice, snow, sand, etc.). Photoinhibition can lead to underestimates (10 to 40%) of primary productivity (17, 20).

If in situ incubations are undertaken, the investigator is faced with several choices. If samples are incubated at the depth from which they were collected, they are essentially held under constant light and turbulence conditions during the incubation period. This is the simplest and most commonly used means of incubating ^{14}C samples. In nature, phytoplankton rarely experiences nonturbulent conditions. For example, in well-mixed, illuminated near-surface waters (i.e., epilimnia in freshwater and mixed layers in estuarine and marine waters), these phototrophs experience a highly dynamic, transient light and (depending on wind) turbulence regimen during daytime.

Incubations conducted under varied light regimens yield primary productivity estimates that differ from those derived under static conditions (18, 27). This disparity holds true even when multiple-depth static incubations are integrated to yield an aerial estimate (per square meter) of primary productivity. The differences between static and dynamic incubations are particularly large in highly turbid waters, where vertical light extinction gradients are very steep in the mixed layer. To compensate, various transient light incubation schemes have been devised. These include on-deck incubators that rotate samples through variable light regimens and in situ variable light incubators, such as rotating wheels containing a range of light transmittance screens (18, 27). Sedimentation in the incubation vessels may also affect productivity measurements. To compensate, bottles should be agitated to most closely mimic natural turbulence.

Applying the ^{14}C method is particularly challenging in benthic systems. Because $^{14}CO_2$ needs to diffuse and thoroughly disperse in benthic-sediment, core, and mat samples, adequate time must be allotted (39, 54). One approach is to preincubate both light and dark samples with $^{14}CO_2$ under complete darkness prior to initiating measurements of photosynthesis. This will enable ^{14}C to diffuse into the sediment matrix. Stirring and agitation of samples during incubation also help minimize this problem. If benthic samples are highly productive, DIC limitation may occur. This creates the possibility of establishing a diffusional barrier to inward movement of DIC from the overlying water column, which will almost certainly lead to underestimates of productivity based on the ^{14}C method. This potential problem should be evaluated before using this technique on uncharacterized sediments.

Choosing appropriate incubation periods is also important. No single incubation length can be prescribed for all aquatic ecosystems. As a rule, 2 to 4 h is recommended, because this time frame should be long enough to allow for sufficient isotopic labeling of cell pools and constituents critical to the primary production process while being short enough to avoid significant ^{14}C losses due to excretion, cell death and lysis, grazing, etc., all of which ultimately affect quantification of primary productivity (20–22, 44, 69). There may also be significant seasonal differences in the above-mentioned sources of error. It is suggested that the investigator start with this time frame and adjust it according to sensitivity needs.

Time of day of productivity measurements can be critical to the determination of daily and longer-term production. It has been shown that primary productivity "tracks" irradiance. Productivity and PAR curves do not exactly overlap, however. Studies on diverse marine and freshwater systems have shown that the highest rates of primary productivity occur during mid- to late morning, prior to the period of maximum irradiance (9, 13, 20, 44). Fee (9) and others have modeled primary productivity, based on knowledge of such deviations from the diel PAR flux curve. In addition to diel variability in the relationship between PAR flux and productivity, there are day-to-day, seasonal, and spatial differences in environmental variables regulating primary productivity.

Following incubation, samples should be processed expeditiously. Unless it is necessary for specific applications (i.e., autoradiography, sample vouchers, and radioimmunological and molecular samples), one should avoid the use of preservatives, such as formaldehyde, glutaraldehyde, and Lugol's iodide. Preservation steps can lead to loss of radioactivity from samples; this is particularly true if samples are washed or treated with other solutions prior to filtration, drying, and further handling (34). If required, samples can be quick-frozen and freeze-dried but only if this is the last step prior to determining radioactivity (34).

Planktonic samples are usually concentrated by filtration prior to ^{14}C determination. One should determine appropriate filtration vacuum conditions under which successful concentrations can be achieved without loss of cellular labeled constituents. Appropriate vacuum settings depend on phytoplankton community composition and physiological state. A gentle vacuum, on the order of ~200 torr, is recommended if the above-mentioned variables are unknown. The choice of filters is critical. One should choose a filter that quantitatively retains all relevant phototrophs at any time of the year and at any location. From a practical standpoint, it is probably best to choose the smallest-pore-size filter capable of passing the volume of water incubated. Whatman GF/F (0.7-μm nominal pore size) or GF/C (1.2-μm nominal pore size) filters are most commonly used, but there are numerous equivalent filters.

As an alternative to filtration, one can collect 5 to 20 ml of incubation water, followed by direct ^{14}C analysis by LSC. Prior to counting, samples are acidified (0.1 N HCl), sparged with air or N_2 to expel nonassimilated $^{14}CO_2$, and neutralized (with dilute NaOH). This approach is most appropriate for productive waters, where large incubation volumes are not required. Relatively low LSC efficiencies (<85%) may result, because substantial amounts of water (which is a strong quenching agent) must be mixed with the LSC cocktail. An LSC fluor chemically compatible with the sample and yielding maximum ^{14}C counting efficiency is preferred. For filters, there is a wide range of high-efficiency (>95%) nonpolar organic solvent-based (toluene and xylene) and anionic-surfactant-based (Ecolume or Cytoscint; ICN) fluors available. The latter group of fluors is gaining popularity because they are nontoxic and biodegradable.

Data interpretation. Interpretation of data is a critical aspect of the ^{14}C technique. When compared to other methods of assessing productivity, specifically the O_2 method, the ^{14}C technique yields results that range from general agreement (69) to serious underestimates (59, 60). Some of the disagreement can be attributed to various methodological problems, including those for determining cellular ^{14}C losses (58). Other discrepancies may be due to the biochemistry of respective measurements. Because it accounts for the balance between CO_2 uptake and assimilation and losses via respiration, excretion, and lysis, the ^{14}C method with dark bottle correction represents net primary production. In contrast, the O_2 method accounts for total photosynthetic production of O_2 minus (only) O_2 consumption via dark-mediated respiration. It does not account for other biomass losses incurred during excretion, lysis, death, grazing, and additional light-mediated respiratory processes such as photorespiration. As such, the O_2 method is a closer estimate of gross primary production.

In practice, when these methods are compared, results are variable (13, 69). The greatest disparity between these methods occurs when primary productivity is closely linked, in time and space, to higher trophic levels. At the level of microbial interactions, autotrophy is closely linked to heterotrophy. Carbon and other nutrient (N and P) transfers can occur within minutes from host autotrophs to epiphytic heterotrophs (35). Such close metabolic coupling of primary and secondary producer components of the microbial food web is referred to as the microbial loop (2). If significant C transfer occurs within the ^{14}C incubation period, it could lead to potential overestimates of primary production, since both primary producer and secondary producer (microheterotrophs) are included in the measurement. Similarly, if micro- or mesozooplankton grazing of phytoplankton takes place during the incubation period, transfer of fixed ^{14}C can occur, leading to overestimates of net primary production. In terms of gross primary production, it can be argued, however, that ^{14}C fixed by phytoplankton and incorporated by grazers during the period of measurement should be included as primary production. To determine trophic transfer during the period of measurement, grazers can be eliminated prior to incubation to assess the quantitative significance of trophic transfer during incubations. Respira-tion, on the other hand, can lead to the loss of fixed C. If recently synthesized ^{14}C-labeled organic compounds are respired, it is possible that a portion of the ^{14}C fixed can be lost and/or recycled during incubation. This may lead to underestimates of primary production.

Oxygen-Based Methods

Oxygen evolution has long been used as a benchmark measurement of photosynthesis and, in the case of natural microbial assemblages, primary productivity (7, 13, 49). The O_2 method is often preferred over other, less direct primary production measurements. One of its main advantages is that it measures photosynthetic performance based on a key product of the light reaction of photosynthesis. Hence, measurements can be directly related to the capture of radiant energy, electron flow, and the generation of reductant. The O_2 technique is suitable for determining productivity of complex microbial assemblages, where parallel measurements of O_2 evolution and its utilization serve as useful indicators of community metabolism (12, 24). Oxygen can be easily and accurately measured without need for radioactive materials and sophisticated detection equipment. As discussed previously, diffusion limitations with the ^{14}C technique can seriously compromise the use of this method in laminated benthic systems. Such limitations are minimized with O_2 measurements (especially using microelectrodes) outlined below. In moderately to highly productive systems, where sensitivity is not limiting, the O_2 method is often preferred.

Oxygen evolution is routinely measured in two ways; (i) determination of changes in dissolved oxygen in bottles or other vessels by titration (i.e., the Winkler method) and (ii) electrochemical macro- or microelectrode determinations. The Winkler and electrochemical techniques should yield identical results, although in the latter, small amounts of O_2 are consumed during measurement. Needed supplies and measuring equipment are relatively simple and portable, making this an excellent shipboard or field technique. In addition, there are novel, experimental (i.e., nonroutine) approaches, such as the use of stable oxygen isotope uptake measurements; the reader is referred to references 8 and 12 for details of these approaches.

The Winkler Method

A divalent manganese solution is added to water samples, and then alkaline iodide is added. Under these conditions, dissolved O_2 is rapidly reduced and precipitated as manganese oxide. This step "fixes" O_2 prior to analysis. The sample is then acidified in the presence of iodide, and the oxidized manganese is converted back to its divalent state, while iodine (equivalent to the original amount of dissolved O_2 present) is liberated. The iodine is quantified either titrimetrically or spectrophotometrically.

Procedure

Duplicate 125- or 250-ml Pyrex, ground-glass-stoppered light and dark biological oxygen demand (BOD) bottles (Corning) are normally used to hold samples. Quartz bottles may be used instead if effects of UV radiation are assessed. Bottles are flushed several times with sample water and filled; investigators must make sure not to generate bubbles during filling (bubbling will affect O_2 solubility and hence concentration). This is accomplished by inserting a filling tube (Tygon or equivalent material) to the bottom of the bottle and rapidly filling the bottle, allowing it to overflow. The ground glass stopper is quickly inserted. Samples should be stored in a cool, lighttight box until all samples are collected.

After incubation, reagents may be added by pipette in the following order: 3 M manganese chloride (dissolve 480 g of $MnCl_2 \cdot 4H_2O$ in distilled water to make 1 liter) (solution 1); alkaline iodide solution, which is made by dissolving 500 g of NaOH and then 135 g of NaI in distilled water to make a volume of 1 liter (solution 2); and 10 N sulfuric acid (add 280 ml of concentrated H_2SO_4 to 500 ml of distilled water) (solution 3). Allow each solution to cool to room temperature prior to using.

The following procedure is for 125-ml bottles. Add 1 ml of solution 1 and 1 ml of solution 2. Stopper the bottles and mix well (by inverting the bottles at least 10 times). Scale the reagent additions appropriately for bottles of other sizes. Following this fixation step, bottles can be kept cool and dark for at least 1 to 2 h prior to analysis. When analyzing, allow the precipitate to settle at least halfway down the bottle.

For tritrimetric determination of $[O_2]$, the iodine liberated is titrated with a 0.01 N standardized thiosulfate solution, yielding $[O_2]$ in milligrams per liter. Solutions are as follows: for solution A, dissolve 2.9 g of analytical-grade $Na_2S_2O_3 \cdot 5H_2O$ and 0.1 g of Na_2CO_3 in 1 liter of distilled water. Add 0.2 ml of CS_2 as a preservative. For solution B, make up a 0.01 N potassium iodide standard by adding 0.3567 g of air-dried (105°C) analytical-grade KIO_3 to 250 ml of distilled-deionized water (warm if necessary to dissolve KIO_3). For solution C, make up a 1% starch solution (solubilize in dilute NaOH if necessary and neutralize with dilute HCl).

To the fixed samples in 125-ml bottles, add 1 ml of solution 3 (10 N sulfuric acid). Within 1 h of sample acidification, transfer 50 ml (using a pipette) into a 125-ml Erlenmeyer flask containing a magnetic stir bar. Titrate immediately with 0.01 N thiosulfate while stirring until the solution is slightly yellow. Then add 0.5 ml of starch solution, which should turn the sample blue. Continue titrating slowly and carefully, until the blue disappears. Titrate N_2-flushed blanks and subtract the blank to obtain a corrected titration volume (V) in milliliters. $[O_2]$ is calculated as follows:

$$\text{Milligrams of } O_2 \cdot \text{liter}^{-1} = 0.1016 \times f \times V \times 16$$

To determine the calibration factor, f, add 1 ml of concentrated sulfuric acid (solution 3) to a 125-ml BOD bottle filled with distilled water. Mix thoroughly. Add 1 ml of the $MnCl_2$ solution (solution 1) and mix again. Transfer 50 ml into the titration flask; add 50 ml of the 0.01 N KIO_3 standard (solution 2). Mix gently for 2 min, and titrate with thiosulfate. With V as the titration volume, $f = 5.00/V$. Determine the calibration factor in triplicate.

For spectrophotometric determination of $[O_2]$, allow the precipitate formed (after adding solutions 1 and 2 to sample) to settle halfway down the sample bottle. Add 1 ml of solution 3, stopper, and mix again. Wait at least 5 min for the precipitate to thoroughly dissolve. A small aliquot (5 to 10 ml) is withdrawn from the bottle and diluted to 100 ml with distilled water. Avoid violent mixing to minimize oxidation of iodine by exposure to air. Simply inverting the mixing flask will provide adequate mixing. The extinction of iodine (E_s for samples and E_b for blank) is measured at 287.5 nm in a 1-cm quartz cuvette by using a UV spectrophotometer. An N_2-flushed (to remove O_2) water sample is run in parallel as a blank. To quantify $[O_2]$ by using the spectrophotometric technique, a calibration factor must first be determined by the standard tritration technique discussed above. One can also refer to a table to determine standard O_2 saturation (percentage or milligrams of O_2 per liter) at specific temperatures (at sea level) and salinity conditions (35). This yields an O_2 solubility factor (F_s) based on the following formula:

$$F_s = [O_2]/(E_s - E_h)$$

Discussion of Technique

The Winkler technique is best suited for moderately to highly productive waters characterized by relatively high O_2 production and consumption rates. Measurements are most easily conducted on planktonic samples, where sampling volume is not restricted. The technique can also be applied to sediments, biofilms, mats, and epibiotic communities. In these systems, the method is most effectively deployed by placing domes and other enclosures directly over the surface to be examined. Specially designed domes, equipped with electric stirrers and ports for withdrawing subsamples, are routinely deployed for assays of benthic primary production, respiration, and (by subtracting respiration from production) net community metabolism (22). During measurements of benthic metabolism, corrections must be made for metabolism of the overlying water column.

Surficial O_2 production and consumption measurements suffer from methodological shortcomings, some of them similar to those incurred with ^{14}C-based methods. Perhaps most serious are diffusion problems. Oxygen, like CO_2, is a gas and is transported along diffusional gradients. In the case of very high rates of O_2 production or consumption, the rate of diffusional transport can control the rate at which O_2 is emitted from or absorbed by surficial and subsurface boundary layers. If the surface is submersed, diffusional barriers are most pronounced. Unless turbulence (as stirring or shaking) can overcome this limitation, underestimates of O_2 production and/or consumption may result. Potential underestimation due to transport of photosynthetically produced oxygen to lower depths in sediments may constitute an additional problem. Diffusional problems may be minimized by stirring or agitating samples. Care must be taken, however, to avoid disturbance and resuspension of sediments. One way to minimize diffusional problems is to employ microelectrode measurements of $[O_2]$ (see below). Microelectrodes are small enough to minimize the impact of diffusional gradients.

An interpretational limitation of the O_2 method is that, unlike the ^{14}C method, it does not account for primary productivity of anoxygenic phototrophs (e.g., photosynthetic bacteria), and chemolithotrophs (e.g., nitrifying bacteria and S-oxidizing bacteria). It does account for respiration by these and all other microorganisms (which the ^{14}C method is incapable of doing).

Electrochemical Techniques

Oxygen can be measured electrochemically by using several types of electrodes. The most widely used are cathode and Clark-style combination electrodes. In this electrode, a membrane-coated platinum cathode coated with a gold surface (to sense O_2) is embedded in a thin glass or steel cannula housing. A voltage of -0.75 V (relative to a calomel reference electrode) is applied to the cathode. The current resulting from the reduction of O_2 at the gold surface is directly proportional to $[O_2]$ in the sample. Cathode-style electrodes can be constructed with very small ($\sim$5 μm) sensing tips. Microelectrodes with small sensing tips have a high degree of spatial resolution. Furthermore, the O_2 diffusion path to the tip is fast, leading to short response times. The 90% response time (i.e., the time required to approach equilibration of external and internal $[O_2]$ conditions) is typically < 0.5 s. The construction and general application of cathode-style microelectrodes in microbial O_2 production and consumption studies are discussed in detail by Revsbech and Jørgensen (53).

In Clark-style or combination O_2 electrodes, the cathode and reference electrode are combined but physically separated in the same electrode. The cathode and electrodes are both immersed in an electrolyte (1 M KCl), which additionally serves to shield the cathode. The tip of the electrode is sealed with an O_2-permeable membrane. Clark-style electrodes vary in size and shape, depending on the scale, sensitivity, stability, response time, and durability desired. The larger BOD electrodes are typically used for environmental measurements (i.e., on a sampling line) or measurements to be made on large containerized water samples (i.e., BOD bottles, domes, and other enclosures). Response time is on the order of a few to 30 s. If high resolution, high sensitivity, and short response time are desired, smaller-diameter electrodes are needed. Minielectrodes have sensing tips of 0.5 to several millimeters in diameter, while microelectrodes have tips as small as 5 μm. The response time of Clark-style microelectrodes is similar to that for cathode microelectrodes.

The concentration of O_2 is determined by measuring changes in the current flowing through the electrode. In large Clark-style electrodes, a digital or analog ammeter serves as the detector. With all microelectrodes, the current changes are very small, and hence a very sensitive ammeter, capable of measuring down to 1 pA (10^{-12}A), is used. Portable picoammeters are commercially available (e.g., from Keithley Inc., Cleveland, Ohio) for field studies.

Mini- and microelectrodes are typically utilized for small-scale pO_2 measurements in biofilms, benthic boundary layers, sediments, and microbial mats, where strong, microscale (in micrometers) O_2 gradients are an integral component of primary production and nutrient cycling dynamics. Details of the construction and applications of mini- and micro-Clark-style electrodes are provided by Revsbech and Jørgensen (53). The procedure briefly outlined below is applicable to the use of fabricated mini- and microelectrodes for measuring pO_2 in surficial environments.

Oxygen regimens and changes therein are analyzed by mounting the micro- or minielectrode on a micromanipulator. The micromanipulator is positioned above the sample in such a way that the electrode can be rapidly inserted into the sample and $[O_2]$ measurements made at discrete intervals. Oxygen profiles are made by advancing the electrode downward into the sample at 0.25- to 0.5-mm intervals. Measurements are usually made under illuminated versus dark conditions. Once stable O_2 conditions occur, the next interval is measured. The most common way to determine photosynthetic O_2 production is to use the light-dark shift method (53–55). In this method, samples are exposed to the irradiance of interest. The microelectrode is inserted to the sampling location. $[O_2]$ is determined. Photosynthesis is then stopped by darkening the sample. This allows O_2 consumption and diffusional exchange to continue at the presumed original rate (prior to darkening). Since there is no photosynthtic O_2 production to balance consumption, $[O_2]$ decreases in proportion to the photosynthetic rate just prior to darkening. The initial slope (measured either on a recorder or data logger) of the decrease in $[O_2]$ after darkening is equal to the $[O_2]$ produced by photosynthesis.

Electrodes are standardized and calibrated by measuring current changes in O_2-depleted (N_2- or Ar-flushed) and O_2-saturated (by bubbling with air) solutions of ambient water.

Discussion of Techniques

Electrochemical measurements of $[O_2]$ have several advantages over colorimetric Winkler techniques. They are rapid and easily executed and, in the case of mini- and microelectrodes, circumvent diffusional artifacts and limitations. All electrodes require calibration against the Winkler technique. Some microelectrodes are sensitive to illumination, a problem that requires remediation by measuring current changes in response to light versus dark conditions in media free of microorganisms (e.g., distilled water). Oxygen microelectrodes are also sensitive to temperature changes, which must be assessed by using sterile or microorganism-free solutions.

Because they are very thin and constructed in large part of glass, mini- and microelectrodes are prone to breakage. Breakage is a particularly serious problem in sediments and microbial mats, where impenetrable heterogeneous substrates (sand, gravel, mucilage, slimes, etc.) can hinder their use. In addition, microelectrodes are sensitive to fouling, chemical poisoning (e.g., sulfide), and changes in sensitivity and response time, all of which necessitate extensive and repeated standardization. Unless investigators have ready access to the technology for constructing microelectrodes, replacement of broken electrodes, calibration, and standardization are expensive, time-consuming aspects of these techniques.

Possible interpretive problems include extrapolating or integrating from the short time scales required to make measurements (in minutes) to longer time scales (hour-days) required to assess productivity on community and ecosystem level. In addition, microelectrodes can potentially disrupt biofilms and mats during measurements, leading to invasion of O_2 into previously hypoxic and anoxic microzones.

Electrochemical techniques require electronic amplification (of current output) and detection and measurement accessories, the degrees of sophistication and costs varying considerably. Costs of ancillary equipment range from approximately $1,000 to $1,500 for an O_2 meter in the case of large electrodes to over $5,000 for microelectrode amplification and measurement equipment (picoammeters and electrochemical sensing equipment). In addition, a micromanipulator (~$1,500) is required to position electrodes.

Complementary Techniques and Their Applications

Standard measurements of primary productivity can be complemented by modifications, ancillary techniques, and procedures that enhance physiological, ecological, and taxonomic interpretations of productivity measurements (1, 6, 12, 18, 21, 22, 24, 44). Measurements of photopigments are highly informative in that they can serve as surrogates for rate measurements of productivity and as complementary biomass measurements. In addition, the ^{14}C method lends itself to specific techniques capable of assessing species-, group-, and community-specific rates of primary production (46, 50). Two such techniques, microautoradiography and radiolabeling of diagnostic microalgal pigments, can enhance the specificity and dimensionality of primary productivity measurements.

Pigment-Based Estimates of Primary Productivity

Changes in net phototrophic biomass over time can be used as an indirect estimate of primary productivity. This approach has been used when direct physiological rate measurements are not available or when long-term day-to-week assessments of productivity are required. The simplest and often most effective estimate of phototrophic biomass is one based on determining cellular phototopigment content. The most common pigment assessed is chlorophyll *a*, which is present in all oxygenic phototrophs. Other pigments, specific for taxonomic groups (carotenoids and phycobilins [Fig. 1]), can be used to partition production among these groups (see "Photopigment Radiolabeling") (23, 28, 30, 40, 46, 48, 50, 65, 71). All pigment-based approaches are reliant on the assumption that the cellular ratio of photopigments to dry weight or carbon content is reasonably constant. This assumption may not be valid in communities experiencing environmental change and physiological stresses. Pigment-based measurements are net estimates, since they simultaneously account for synthesis and loss of photopigments. Synthesis is largely due to growth and to production of new biomass. Losses and changes in cell pigment synthesis may be attributable to death, consumption, advective processes (sinking, sedimentation, and horizontal transport due to flushing), and changes in physiological state (i.e., photoadaptation, photobleaching, and photooxidation).

The most common measures of chlorophyll *a* are fluorometric emission and spectrophotometric absorbance of intact cells or extracted pigment (usually in 90% acetone or 90% ethanol or methanol) at either 440 or 663 nm and 750 nm (turbidity blank). Detailed methodologies for microalgal pigment extraction and spectrophotometric analyses are given by Strickland and Parsons (62), and applications to aquatic production studies are addressed by Millie et al. (29), Jeffrey et al. (23), Paerl et al. (40, 41), and Wright et al. (71). Fluorometric analyses of chlorophyll *a* (excitation maximum at 400 nm and emission maximum at 660 nm), are also commonly used in limnological or oceanographic studies. Fluorometry is more sensitive than and not as prone to optical interference as spectrophotometry (23, 30). Fluorometry can additionally be applied in situ by placing fluorometers and data logging and telemetry devices in the water column.

The above-mentioned methods are applicable to both planktonic and benthic systems. In the case of benthic measurements, accumulations of chlorophyll degradation products (e.g., pheophytin) can be significant. This can lead to interpretational problems, even if acidification corrections for pheophytin are made (23, 62). High-performance liquid chromatography (HPLC) has, however, proven effective in rapidly separating photosynthetic pigments, prior to spectrophotometric and fluorometric detection (3, 28, 65, 71).

While chlorophyll *a* is a sensitive, easy measure of microbial phototrophic biomass, a major limitation of chlorophyll *a*-based techniques is their inability to distinguish major microalgal and photosynthetic bacterial primary producer groups. To circumvent this problem, HPLC can be used to characterize and quantify phytoplankton community structure (23, 30, 40, 48). This technique is based on the use of diagnostic photopigments (Fig. 1). Distinct spectral absorbance characteristics make photopigments (including chlorophylls, carotenoids, etc.) useful and sensitive indicators of dominant phytoplankton groups, including diatoms, dinoflagellates, chlorophytes, cyanobacteria, and cryptophytes, and they can be easily detected by in-line photodiode array spectrophotometry (PDAS). HPLC can rapidly and accurately separate and quantify photopigments from mixed algal samples (23, 30, 71). HPLC-PDAS is now routinely used by researchers and water quality agencies to quantify the composition and relative biomass of phytoplankton taxonomic groups in diverse aquatic ecosystems (23, 25, 30, 40, 48, 57). Statistical procedures (ChemTax) can be applied to partition the total pool of chlorophyll *a* (total community biomass) into the major algal groups, thus allowing calculation of the relative and absolute contribution of each algal group (47, 48, 71) (Fig. 1). For the phylogenetic groups, photopigments useful as markers include chlorophyll *b* and lutein (for chlorophytes); zeaxanthin, myxoxanthophyll, and echinenone (for cyanobacteria); fucoxanthin (diatoms); peridinin (dinoflagellates); and alloxanthin (cryptomonads) (47) (Fig. 1). Protocols for this method are detailed in references 38 and 65 and described briefly in the next section ("Photopigment Radiolabeling"). Photopigment composition is usually significantly (linearly)

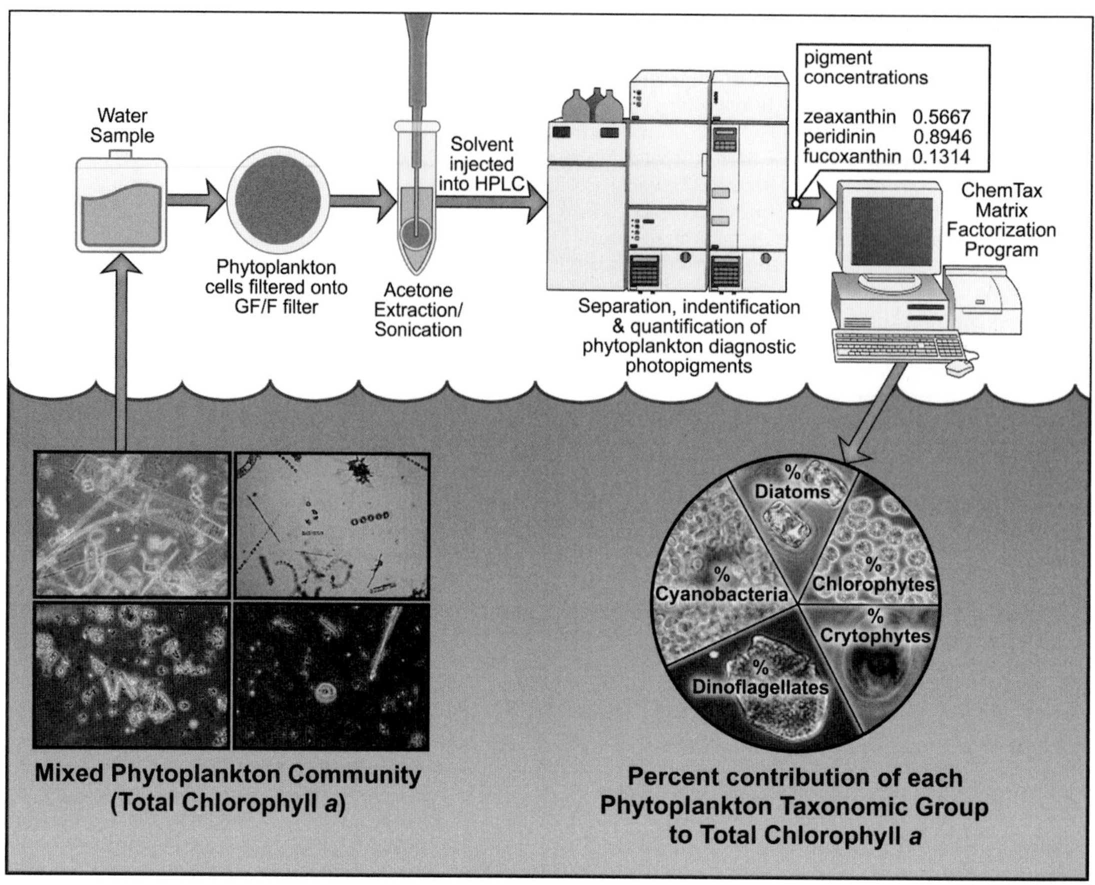

FIGURE 1 Diagram illustrating the combined use of HPLC and a matrix factorization program (ChemTax) for separating and identifying photopigments diagnostic of microalgal taxonomic groups mediating aquatic primary production.

correlated with species cell counts or biovolume estimates (6, 25, 56, 57, 64, 71).

The assessment of photopigments by using HPLC is easily executed and relatively inexpensive; it can be done in a wide variety of resource types (streams, lakes, rivers, estuaries, and oceans) and over large spatial scales. This method is able to detect significant changes in phytoplankton community composition over a broad range of time scales (<24 h to decades) and as such is well suited for monitoring programs designed to assess short- and long-term trends in water quality in response to hydrographic features (circulation or upwelling), nutrient enrichment, and climatic and hydrologic perturbations (floods or droughts). In addition top-down effects of grazing have been examined by using HPLC-based techniques. Lastly, these analyses have proven useful as a method of ground truthing remotely sensed estimates of phytoplankton bloom events (19, 31, 40, 47). This application has provided significant improvements in "scaling up," i.e., mapping the spatial distributions of phytoplankton groups over large geographical areas not amenable to routine field sampling and thus evaluating the effectiveness of nutrient management strategies. It also finds use as an early warning system for blooms of nuisance or toxic species (29, 31) and serves as a sensitive bioindicator of overall water quality conditions (40, 48).

Photopigment Radiolabeling

Photopigment radiolabeling is a potentially useful technique for measuring C-specific growth rates among diverse phototrophic taxa (15, 46, 50). This method, while neither a direct nor surrogate measure of primary productivity, provides additional dimensionality to primary productivity measurements. The conventional ^{14}C method provides a single production value for the entire community. In contrast, photopigment radiolabeling allows for simultaneous measurement of specific growth rates among individual taxonomic groups. The method is a useful extension of the ^{14}C technique coupled to HPLC separation of diagnostic photopigments. When algae are exposed to ^{14}C, the labeled C is incorporated into 3-phosphoglyceric acid and passes through the Calvin-Benson cycle and eventually into a pool of lower-molecular-weight intermediates. These labeled compounds enter the various photopigment biosynthetic pathways. After a sufficiently long incubation (2 to 24 h), the specific activity of photopigment C will equal the total phytoplankton carbon pool. The C-specific growth rate (μ, where $C_p = C_{p,o}e^{\mu t}$, C_p is C-specific biomass at time t, $C_{p,o}$ is C-specific biomass at time zero, and t is time) is a more useful parameter for describing the physiological responses of taxonomic groups of phototrophs in terms of C assimilation and allocation into biomass. A comparison of μ values obtained under different experimental conditions allows for predictions of the ultimate fate of community components over time.

By use of HPLC coupled to PDAS detection of pigments, photopigments are separated on a sequential series of C-18 HPLC columns to optimize separation of diagnostic chlorophyll and carotenoid photopigments. The sequence is as follows: a monomeric Rainin Microsorb-MV and two polymeric Vydac 201TP reverse-phase C_{18} columns with use of a binary gradient elution system (23, 65, 71). Monomeric columns provide strong retention and high efficiency, while polymeric columns select for similar compounds with minor differences in molecular structure and shape. On-line PDAS produces real-time three-dimensional chromatograms of absorbance spectra for each photopigment. Computer software allows data from individual chromatograms to be stored and processed for qualitative analyses and sample comparisons of known absorption spectra and peak retention times (23, 71). C-specific growth rates are determined by using published methods (46, 51). The ^{14}C specific activity (disintegrations per minute) of individual pigments is quantitatively determined by using an in-line radioactivity (β emission) detector. Radioactivity (disintegrations per minute) peaks are overlaid onto pigment chromatograms to identify labeled pigments. Values of community-specific growth rate (μ) and C biomass are determined by using equations given by Goericke and Welschmeyer (14) and Redalje (51). The growth rates and C biomass of taxonomic groups (cyanobacteria, diatoms, dinoflagellates, chlorophytes, etc.) are estimated from the activity of labeled diagnostic pigments.

Recent improvements of column and instrumentation technology, as well as of software applicable to data interpretation and synthesis, have streamlined this technique and have made it more broadly applicable (48).

Microautoradiography

Samples that have been incubated with $^{14}CO_2$ can be analyzed for single-cell-, taxon-, and community-specific photosynthetic capabilities and rates by using microautoradiography (51, 54). This method is based on the fact that ^{14}C assimilated by microorganisms can be detected and visualized by radiation-sensitive silver halide emulsions, which are placed over radioactive organisms and are subsequently processed by standard photographic techniques. Microautoradiography has been used to quantify primary productivity on a cell-specific basis (63, 67), and it can be combined with immunochemical and nucleic acid cellular staining/labeling techniques, such as fluorescent in situ hybridization (to phylogenetic or functional genes) to link the structural and functional aspects of microbes (41, 43).

^{14}C-labeled microorganisms are either filtered onto membrane filters or settled onto glass slides, after which radiosensitive liquid or film emulsions are applied (32–34). Following exposure and processing, both the microorganism and its radioactivity (which appears as exposed silver grains in the developed emulsion) can be examined microscopically and photographed. Radioactivity can be quantified (either as silver grains or tracks of grains) if a uniformly thick film or liquid emulsion is applied (37). For details of the technique, the reader is referred to references 32 through 34, 50, and 63. Below is a short description of the technique applied to planktonic samples.

Samples that have been incubated with ^{14}C are either fixed in 2% borate-buffered formaldehyde (pH 7 to 8) (in the case of delicate cells) or left unfixed and gently filtered (~200 torr) on 0.45-μm-pore-size Millipore HA (Millipore Corp., Bedford, Mass.) filters. Filters are then rinsed with prefiltered (0.2-μm-pore-sized filter), unlabeled sample water to remove excess radiolabel. In the case of marine samples, salts must be removed from filters, or they will crystallize upon drying and interfere with microscopic viewing. This is accomplished by rinsing filters twice with small quantities (10 to 30 ml) of 0.01 N phosphate-buffered saline, which removes salts while avoiding lysis of fragile cells. Filters are then either quick-frozen and/or air dried (for at least 4 h) and optically cleared by placing them, face up, on clean microscope slides and carefully passing them over the mouth of a 250-ml beaker containing boiling (fuming) acetone. This step will simultaneously clear filters and attach them to slides.

The following steps are conducted in the darkroom, either under a dark red ≤20-W safelight (at least 1 m away from samples) or in complete darkness, which ensures the lowest background. Slides are dipped in a melted nuclear track emulsion. Kodak NTB-2 (Kodak Chemicals, Inc., Rochester, N.Y.), a moderately sensitive, low-background, reusable emulsion, is widely recommended. The stock emulsion is kept in a lighttight container in a refrigerator at ~4°C. In darkness, it is melted at 40°C and diluted 1:1 with deionized water for use with grain density microautoradiography. If quantitative autoradiography is desired, the application of thick emulsions combined with track autoradiography or stripping film emulsions (Kodak AR-10) is recommended (37). All emulsions should be thoroughly melted at 40°C (use a water bath in darkroom or heat light-sealed containers with emulsion in an oven) before use. In both thick-layer (undiluted emulsion) track and diluted thin-layer grain density techniques, dipped and jelled slides are air dried for 30 min and exposed in lighttight slide boxes containing packets of silica gel desiccant. Exposure times typically vary from 2 to 10 days.

Exposed track and thin-layer microautoradiographs are developed in Kodak D-19 for 5 and 2 min respectively, and are transferred to a distilled-water stop bath for 2 min. They are then fixed for 5 min in Kodak rapid fixer or 15% NaS_2O_3 and are rinsed in a gentle flowing water bath for 15 min. All slides are then air dried overnight. Microautoradiographs are viewed microscopically, usually with phase-contrast oil immersion optics. For thin-layer autoradiographs, immersion oil can be directly applied to the dried emulsion. Track autoradiograph emulsions are allowed to swell (facilitating detection of tracks) by the addition of 30% glycerine followed by the application of a coverslip. Immersion oil can then be applied to the coverslip.

While microautoradiography can overcome and circumvent technical limitations of other techniques, it is time consuming, requires experience and patience, and is limited by microscopic resolution. It can also be prone to interpretational differences among investigators.

Regional and Global Studies of Primary Production: the Utility of Unattended Monitoring and Remote Sensing

Both natural perturbations (e.g., droughts, storms and hurricanes, and floods) and human activities (urban, agricultural, and industrial) can strongly impact aquatic primary production by modifying nutrient, sediment, toxin, and other xenobiotic inputs. The effects of these external stressors can be highly significant and at least in the case of human activities are increasing. The most perceptible and problematic symptoms of this trend are nutrient-enhanced eutrophication; associated increases in harmful algal blooms, hypoxia, and anoxia; and alteration of biogeochemical cycling in lakes, rivers, estuaries, and coastal waters. The scales of productivity responses to these modulators are also increasing and in the case of eutrophication range from habitat to ecosystem to larger spatial scales (regional and global). Temporally, productivity response events range from days to several weeks, depending on the timing and levels of nutrient, sediment, and toxin/xenobiotic inputs. The spatiotemporal interaction and overlay of growth- and community-altering anthropogenic inputs to aquatic ecosystems over a wide range of scales necessitate short- as well as long-term detection and characterization of productivity responses, often integrated over large areas.

These informational needs are often beyond the scope of conventional monitoring and field surveys that use the techniques discussed above on small vessels. An array of monitoring programs and platforms is available (often at little or no cost) to collect space- and time-intensive data needed to assess ecological conditions over relevant scales and to ground-truth and calibrate remote-sensing efforts aimed at scaling up. These include instrumented channel markers, bridges, piers and docks, and buoys with unattended monitoring equipment. Scientists can also deploy moorings and outfit ferries and other vessels that ply regular routes as "ships of opportunity." In particular, ferries have proven highly suitable for collecting near-real-time physical-chemical-biological data (including diagnostic photopigments and molecular markers) for assessing microbial community structural and functional responses to environmental change in large estuarine and coastal ecosystems not amenable to routine monitoring. The longest-running system (since 1993) is the Alg@line (http://www.itameriportaali.fi), operated by the Finnish Institute of Marine Research, Helsinki, Finland (50). It employs large ferries that routinely cross the Baltic Sea for collecting hydrographic, chemical (nutrient), and biological (phytoplankton and chlorophyll *a*) data. This program has successfully detected and quantified phytoplankton blooms on the Baltic Sea, and photopigment data are used to calibrate satellite imagery of such blooms. Another example is the use of the North Carolina Department of Transportation ferries for water quality monitoring. This program, FerryMon (http://www.ferrymon.org), is designed to assess and predict the relationships among human nutrient and other pollutant inputs, algal blooms, and associated water quality changes and to provide information critical to long-term water quality and fishery management. FerryMon serves as a national model for real-time assessment of coastal water quality (4).

These platforms serve as a calibration source for multispectral (to detect diagnostic phytoplankton pigments such as chlorophylls, carotenoids, and phycobilins) aircraft and satellite-based remote sensing (29). NASA's SeaWiFS satellite-based remote-sensing platform is capable of discrete spectral analyses with sufficient sensitivity and resolution. SeaWiFS includes six narrow visible bands (412, 443, 490, 510, 555, and 670 nm) and two broader near-infrared bands (765 and 865 nm). All eight bands are recorded in 10-bit resolution to take advantage of their high signal-to-noise ratios (up to 1,000). This makes for a very-high-spectral-resolution remote-sensing platform ideal for detecting and discriminating among primary production contributions from phylogenetically diverse algal groups whose diagnostic pigments can be individually characterized. In addition, aircraft-based hyperspectral systems such as NASA's AVIRIS have the potential for providing spectral resolution capable of distinguishing specific diagnostic chlorophyll, carotenoid, and phycobilin photopigments (29).

The spatial and temporal linkages of nutrient and toxic inputs with abundance of pigmented phototrophs are of critical importance in determining primary production, changes in algal biomass, bloom dynamics, and CO_2 flux in these waters. In addition, the contemporaneous impacts of autochthonous (internal; upwelling and regenerated N supplied to surface waters via mixing) and allochthonous (external; runoff, riverine, and atmospheric) nutrient and toxic mediated primary production should be assessed. Evolving remote-sensing techniques will help clarify and evaluate these nutrient-production interactions on scales appropriate for a broad range of aquatic ecosystems.

Research and logistic support was partially provided by the National Science Foundation (DEB 9815495, OCE 9905723, and OCE 0327056), U.S. Department of Agriculture NRI Project 00-35101-9981, U.S. EPA STAR Projects R82-5243-010 and R82867701, NOAA/North Carolina Sea Grant Program R/MER-43, the North Carolina Department of Natural Resources and Community Development/UNC Water Resources Research Institute (Neuse River Estuary Monitoring and Modeling Project).

I appreciate the comments, reviews, and technical assistance provided by A. Joyner, D. Millie, J. Pinckney, and B. Peierls.

REFERENCES

1. **Archer, D., and A. Devol.** 1992. Benthic oxygen fluxes on the Washington shelf and slope: a comparison of *in situ* microelectrode and chamber flux measurements. *Limnol. Oceanogr.* **37:**614–629.
2. **Azam, F., T. Fenchel, J. G. Filed, J. S. Gray, L. A. Meyer-Reil, and F. Thingstad.** 1983. The ecological role of water-column microbes in the sea. *Mar. Ecol. Prog. Ser.* **10:**257–263.
3. **Bowles, N. D., H. W. Paerl, and J. Tucker.** 1985. Effective solvents and extraction periods employed in phytoplankton carotenoid and chlorophyll determinations. *Can. J. Fish. Aquat. Sci.* **42:**1127–1131.
4. **Buzzelli, C. P., J. R. Ramus, and H. W. Paerl.** 2003. Ferry-based monitoring of surface water quality in North Carolina estuaries. *Estuaries* **26:**975–984.
5. **Cloud, P.** 1976. Beginnings of biospheric evolution and their biogeochemical consequences. *Paleobiology* **2:**351–387.
6. **Descy, J.-P., and A. Metens.** 1996. Biomass-pigment relationships in potamoplankton. *J. Plankton Res.* **18:**1557–1566.
7. **Falkowski, P. G.** 1980. *Primary Productivity in the Sea.* Plenum Press, New York, N.Y.
8. **Falkowski, P. G., and J. A. Raven.** 1997. *Aquatic Photosynthesis.* Blackwell Science, Malden, Mass.
9. **Fee, E. J.** 1975. The importance of diurnal variation of photosynthesis vs. light curves to estimate of integral primary production. *Int. Ver. Theor. Angew. Limnol. Verh.* **19:** 39–46.
10. **Fitzwater, S. E., G. A. Knauer, and J. H. Martin.** 1982. Metal contamination and its effect on primary production measurement. *Limnol. Oceanogr.* **27:**544–551.
11. **Fogel, M. L., and L. A. Cifuentes.** 1993. Isotope fractionation during primary production, p. 73–98. *In* M. H. Engel and S. A. Macko (ed.), *Organic Geochemistry.* Plenum Press, New York, N.Y.
12. **Geider, R. J., and B. J. Osborne.** 1992. *Algal Photosynthesis.* Chapman & Hall, New York, N.Y.
13. **Gieskes, W. W., and G. W. Kraay.** 1984. State-of-the-art in the measurement of primary production, p. 171–190. *In* M. J. R. Fasham (ed.), *Flow of Energy and Materials in Marine Ecosystems.* Plenum Publishing Co., New York, N.Y.
14. **Goericke, R., and N. Welschmeyer.** 1993. The chlorophyll-labeling method: measuring specific rates of chlorophyll a synthesis in cultures and in the open ocean. *Limnol. Oceanogr.* **38:**80–95.
15. **Goldman, C. R.** 1966. *Primary Productivity in Aquatic Environments.* University of California Press, Berkeley.

16. **Goldman, C. R.** 1999. Baldi Lecture: four decades of change in two subalpine lakes. *Verh. Int. Verein. Limnol.* **27:**7–26.
17. **Goldman, C. R., D. T. Mason, and J. E. Hobbie.** 1967. Two antarctic desert lakes. *Limnol. Oceanogr.* **12:**295–310.
18. **Harding, L. W., T. R. Fisher, and M. A. Tyler.** 1987. Adaptive responses of photosynthesis in phytoplankton: specificity to time-scale of change in light. *Biol. Oceanogr.* **4:**403–437.
19. **Harding, L. W., D. Degobbis, and R. Precali.** 1999. Production and fate of phytoplankton: annual cycles and interannual variability, p. 131–172. *In* T. C. Malone et al. (eds.), *Coastal and Estuarine Studies*, vol. 55. *Ecosystems at the Land-Sea Margin: Drainage Basin to Coastal Sea.* American Geophysical Union, Washinton, D.C.
20. **Harris, G. P., and B. B. Piccinin.** 1977. Photosynthesis by natural phytoplankton populations. *Arch. Hydrobiol.* **80:** 405–457.
21. **Hickman, M.** 1969. Methods for determining the primary productivity of epipelic and epipsamnic algal associations. *Limnol. Oceanogr.* **14:**936–941.
22. **Hunding, C., and B. Hargrave.** 1973. A comparison of benthic microalgal production measured by ^{14}C and oxygen methods. *J. Fish. Res. Board Can.* **30:**309–312.
23. **Jeffrey, S. W., R. C. F. Mantoura, and S. W. Wright.** 1997. *Phytoplankton Pigments in Oceanography: Guidelines to Modern Methods.* UNESCO, Paris, France.
24. **Kemp, M., P. A. Sampou, J. Garber, J. Tuttle, and W. R. Boynton.** 1992. Seasonal depletion of oxygen from bottom waters of Chesapeake Bay: roles of benthic and planktonic respiration and physical exchange processes. *Mar. Ecol. Prog. Ser.* **85:**137–152.
25. **Lewitus, A. J., D. L. White, R. G. Tymowski, M. E. Geesey, S. N. Hymel, and P. A. Noble.** 2005. Adapting the CHEMTAX method for assessing phytoplankton taxonomic composition in southeastern U.S. estuaries. *Estuaries* **28:**160–172.
26. **Lindeboom, H., A. Sandee, and H. Driessche.** 1985. A new bell jar/microelectrode method to measure changing oxygen fluxes in illuminated sediments with microalgal cover. *Limnol. Oceanogr.* **30:**693–698.
27. **Mallin, M. A., and H. W. Paerl.** 1992. Effects of variable irradiance on phytoplankton productivity in shallow estuaries. *Limnol. Oceanogr.* **37:**54–62.
28. **Mantoura, R. F. C., and C. A. Llewellyn.** 1983. The rapid determination of algal chlorophyll and carotenoid pigments and their breakdown products in natural waters by reverse-phase high-performance liquid chromatography. *Anal. Chim. Acta* **151:**297–314.
29. **Millie, D. F., M. C. Baker, C. S. Tucker, B. T. Vinyard, and C. P. Dionigi.** 1992. High-resolution, airborne remote-sensing of bloom-forming phytoplankton. *J. Phycol.* **28:** 281–290.
30. **Millie, D. F., H. W. Paerl, and J. P. Hurley.** 1993. Microalgal pigment assessments using high-performance liquid chromatography: a synopsis of organismal and ecological applications. *Can. J. Fish. Aquat. Sci.* **50:**2513–2527.
31. **Millie, D. F., O. M. Schofield, G. B. Kirkpatrick, G. Johnsen, P. A. Tester, and B. T. Vinyard.** 1997. Detection of harmful algal blooms using photopigment and absorption signatures: a case study of the Florida red-tide dinoflagellate, *Gymnodinium breve*. *Limnol. Oceanogr.* **42:**1240–1251.
32. **Paerl, H. W.** 1974. Bacterial uptake of dissolved organic matter in relation to detrital aggregation in marine and freshwater systems. *Limnol. Oceanogr.* **19:**966–972.
33. **Paerl, H. W.** 1978. Microbial organic carbon recovery in aquatic ecosystems. *Limnol. Oceanogr.* **23:**927–935.
34. **Paerl, H. W.** 1984. An evaluation of freeze-fixation as a phytoplankton preservation method for microautoradiography. *Limnol. Oceanogr.* **29:**417–426.
35. **Paerl, H. W.** 1984. Transfer of N_2 and CO_2 fixation products from *Anabaena oscillarioides* to associated bacteria during inorganic carbon sufficiency and deficiency. *J. Phycol.* **20:**600–608.
36. **Paerl, H. W.** 1987. *Dynamics of Blue-Green Algal* (Microcystis aeruginosa) *Blooms in the Lower Neuse River, North Carolina: Causative Factors and Potential Controls.* University of North Carolina Water Resources Research Institute report no. 229. University of North Carolina, Raleigh.
37. **Paerl, H. W., and E. A. Stull.** 1979. In defense of grain density autoradiography. *Limnol. Oceanogr.* **24:**1166–1169.
38. **Paerl, H. W., and D. F. Millie.** 1991. *Evaluations of Spectrophotometric, Fluorometric and High Performance Liquid Chromatographic Methods for Algal Pigment Determinations in Aquatic Ecosystems.* Workshop Report. U.S. Environmental Protection Agency, Cincinnati, Ohio.
39. **Paerl, H. W., B. M. Bebout, S. B. Joye, and D. J. Des Marais.** 1993. Microscale characterization of dissolved organic matter production and uptake in marine microbial mat communities. *Limnol. Oceanogr.* **38:**1159–1161.
40. **Paerl, H. W., L. M. Valdes, J. L. Pinckney, M. F. Piehler, J. Dyble, and P. H. Moisander.** 2003. Phytoplankton photopigments as indicators of estuarine and coastal eutrophication. *BioScience* **53:**953–964.
41. **Paerl, H. W., J. Dyble, P. H. Moisander, R. T. Noble, M. F. Piehler, J. T. Pinckney, L. Twomey, and L. M. Valdes.** 2003. Microbial indicators of aquatic ecosystem change: current applications to eutrophication studies. *FEMS Microb. Ecol.* **46:**233–246.
42. **Parsons, T. R., Y. Maita, and C. M. Lalli.** 1992. *A Manual of Chemical and Biological Methods for Seawater Analysis*. Pergamon Press, Oxford, United Kingdom.
43. **Pernthaler, A., and R. Amann.** 2004. Simultaneous fluorescence in situ hybridization of mRNA and rRNA in environmental bacteria. *Appl. Environ. Microbiol.* **70:** 5426–5433.
44. **Peterson, B. J.** 1980. Aquatic primary productivity and the ^{14}C-CO_2 method: a history of the productivity problem. *Annu. Rev. Ecol. Syst.* **11:**359–385.
45. **Pinckney, J. L., and R. G. Zingmark.** 1993. Photophysiological responses of intertidal benthic microagal communities to in situ light environments: methodological considerations. *Limnol. Oceanogr.* **38:**1373–1383.
46. **Pinckney, J. L., D. F. Millie, K. E. Howe, H. W. Paerl, and J. P. Hurley.** 1996. Flow scintillation counting of ^{14}C-labeled microalgal photosynthetic pigments. *J. Plankton Res.* **18:**1867–1880.
47. **Pinckney, J. L., H. W. Paerl, M. B. Harrington, and K. E. Howe.** 1998. Annual cycles of phytoplankton community-structure and bloom dynamics in the Neuse River Estuary, North Carolina, USA. *Mar. Biol.* **131:**371–382.
48. **Pinckney, J. L., T. L. Richardson, D. F. Millie, and H. W. Paerl.** 2001. Application of photopigment biomarkers for quantifying microalgal community composition and in situ growth rates. *Org. Geochem.* **32:**585–595.
49. **Pomeroy, L.** 1959. Algal productivity in salt marshes of Georgia. *Limnol. Oceanogr.* **4:**386–397.
50. **Rantajärvi, E., V. Gran, S. Hällfors, and R. Olsonen.** 1998. Effects of environmental factors on the planktonic community in the Gulf of Finland: unattended high frequency measurements and multivariate analyses. *Hydrobiologia* **363:**127–139.
51. **Redalje, D.** 1993. The labeled chlorophyll *a* technique for determining photoautotrophic carbon specific growth

rates and carbon biomass, p. 563-572. *In* P. Kemp, B. Sherr, E. Sherr, and J. Cole (ed.), *Handbook of Methods in Aquatic Microbial Ecology*. Lewis Publishing Co., Boca Raton, Fla.

52. **Redfield, A. C.** 1958. The biological control of chemical factors in the environment. *Am. Sci.* **46:**205–221.
53. **Revsbech, N. P., and B. B. Jørgensen.** 1986. Microelectrodes: their use in microbial ecology. *Adv. Microb. Ecol.* **9:**273–352.
54. **Revsbech, N. P., B. B. Jørgensen, and O. Brix.** 1981. Primary production of microalgae in sediments measured by oxygen microprofile, $H^{14}CO_3^-$ fixation, and oxygen evolution methods. *Limnol. Oceanogr.* **26:**717–730.
55. **Revsbech, N. P., J Nielsen, and P. K. Hansen.** 1988. Benthic primary production and oxygen profiles, p. 69–83. *In* T. H. Blackburn and J. Sørensen (ed.), *Nitrogen Cycling in Coastal Marine Environments*. John Wiley & Sons, New York, N.Y.
56. **Roy, S., J. P. Chanut, M. Gosselin, and T. Sime-Ngando.** 1996. Characterization of phytoplankton communities in the lower St. Lawrence Estuary using HPLC-detected pigments and cell microscopy. *Mar. Ecol. Prog. Ser.* **142:**55–73.
57. **Schlüter, L, F. Møhlenberg, H. Havskum, and S. Larsen.** 2000. The use of phytoplankton pigments for identifying and quantifying phytoplankton groups in coastal areas: testing the influence of light and nutrients on pigment/chlorophyll *a* ratios. *Mar. Ecol. Prog. Ser.* **192:**49–63.
58. **Sharp, J. H.** 1977. Excretion of organic matter by marine phytoplankton. Do healthy cells do it? *Limnol. Oceanogr.* **22:**381–399.
59. **Sheldon, R. W., and W. H. Sutcliffe.** 1978. Generation times of 3 h for Sargasso Sea microplankton determined by ATP analysis. *Limnol. Oceanogr.* **23:**1051–1054.
60. **Shulenberger, E., and J. L. Reid.** 1981. The Pacific shallow oxygen maximum, deep chlorophyll maximum, and primary productivity, reconsidered. *Deep Sea Res.* **28A:**29–35.
61. **Steeman Nielsen, E.** 1952. The use of radioactive carbon (^{14}C) for measuring organic production in the sea. *J. Cons. Int. Explor. Mer.* **18:**117–140.
62. **Strickland, J. D. H., and T. R. Parsons.** 1972. *A Practical Handbook of Seawater Analysis*. Fishery Research Board, Ottawa, Canada.
63. **Stull, E. A., E. De Amezaga, and C. R. Goldman.** 1973. The contribution of individual species of algae to primary productivity of Castle Lake, California. *Int. Ver. Theor. Angew. Limnol. Verh.* **18:**1776–1783.
64. **Tester, P. A., M. R. Geesey, C. Guo, H. W. Paerl, and D. F. Millie.** 1995. Evaluating phytoplankton dynamics in the Newport River Estuary (North Carolina, USA) by HPLC-derived pigment profiles. *Mar. Ecol. Prog. Ser.* **124:**237–245.
65. **Van Heukelem, L., A. Lewitus, T. Kana, and N. Craft.** 1994. Improved separations of phytoplankton pigments using temperature-controlled high performance liquid chromatography. *Mar. Ecol. Prog. Ser.* **114:**303–313.
66. **Vollenweider, R. A., and A. Nauwerck.** 1961. Some observations on the ^{14}C method for measuring primary production. *Int. Ver. Theor. Angew. Limnol. Verh.* **14:**134–139.
67. **Watt, W.** 1971. Measuring the primary production rates of individual phytoplankton species in natural mixed populations. *Deep Sea Res.* **18:**329–339.
68. **Wetzel, R. G., and G. E. Likens.** 1991. *Limnological Analyses*. Springer-Verlag, New York, N.Y.
69. **Williams, P. J. L. B., K. R. Heinemann, J. Marra, and D. A. Purdie.** 1983. Comparison of ^{14}C and O_2 measurements of phytoplankton production in oligotrophic waters. *Nature* **305:**49–50.
70. **Woitke, P., P. Schiwietz, K. Teubner, and J.-G. Kohl.** 1996. Annual profiles of photosynthetic lipophilic pigments in four freshwater lakes in relation to phytoplankton cultures as well as nutrient data. *Arch. Hydrobiol.* **137:**363–384.
71. **Wright, S. W., D. P. Thomas, H. J. Marchant, H. W. Higgins, M. D. Mackey, and D. J. Mackey.** 1996. Analysis of phytoplankton of the Australian section of the Southern Ocean: comparisons of microscopy and size frequency data with the interpretations of pigment HPLC data using the 'CHEMTAX' matrix factorization program. *Mar. Ecol. Prog. Ser.* **144:**285–298.

Bacterial Secondary Productivity

GERARDO CHIN-LEO AND COLLEEN T. EVANS

34

Planktonic heterotrophic bacteria (bacterioplankton) are now recognized to be a large and metabolically active group that contributes significantly to the biomass and to the flow of carbon in aquatic systems. Bacterial cell densities often exceed 10^9 cells liter^{-1}. Bacterial biomass in marine systems is generally greater than that of zooplankton and can exceed phytoplankton biomass in the open sea (18, 29, 30). Estimates of bacterial biomass and growth rate show that bacterial biomass turns over rapidly (in a matter of hours). Comparisons of bacterial biomass and production rates with corresponding rates of phytoplankton biomass and production rates indicate that bacteria consume a substantial fraction (20 to 40%) of the carbon fixed by phytoplankton (18, 117). In unproductive aquatic systems, bacterial respiration can exceed phytoplankton net production (25). Bacteria are now considered major secondary producers, since they convert dissolved organic matter (DOM) derived from primary producers into an abundant biomass. The consumption of bacterial biomass by specialized predators (ciliates and flagellates) may be an important pathway for the transfer of DOM to metazoan food webs (4, 88, 97).

This view of bacteria has emerged as a result of the use of modern methods for measuring the abundance and bacterial biomass production (BBP) rates in environmental samples. Bacterial abundance is measured by the enumeration of bacterial cells, using nucleic acid-specific fluorochromes and epifluorescence microscopy or flow cytometry (11, 52, 89). BBP methods involve the measurement of increases in the rate of synthesis of cellular constituents such as DNA and proteins (91). System-wide estimates of BBP determined using these independent methods agree in that bacterial production is high and consumes a substantial fraction of the algal primary production (18, 117). The importance of bacterial growth and the factors regulating this growth in ecosystems can be determined from estimates of BBP. BBP values, for example, can be used to understand the role of bacteria in trophic dynamics because BBP values reflect the amount of bacterial biomass potentially available to grazers and thus to higher trophic levels. In addition, changes in the rates of BBP can be used as an indicator of the response of bacteria to spatial and temporal fluctuations in environmental conditions. The contributions of bacteria to biogeochemical cycles can also be determined from BBP. Bacteria cycle organic matter by producing biomass or by the remineralization of organic compounds. Thus, the contribution of bacterial metabolism to organic geochemical cycles can be determined from measurements of BBP and bacterial respiration (BR). These bacterial activities can be summarized as the bacterial growth efficiency (BGE), which is the amount of bacterial biomass produced per amount of organic matter consumed. BGE can also be expressed in terms of BBP and BR as BGE = BBP/(BBP + BR). While there is an abundance of BBP data, complementary BR and BGE values are relatively sparse, due in part to the difficulties of measuring BR such as the need to separate bacteria from other organisms and the relatively long incubations needed to accurately measure bacterial oxygen utilization (26). The contribution of bacteria to carbon metabolism is often estimated from BBP and an assumed BGE or BR. While analysis of available BGE data has revealed systematic trends relating BGE to temperature and to the trophic status of a system (26, 94), BGE values vary widely, and the selection of a single value for a given system is likely to introduce uncertainties in the interpretation of the data. Furthermore, because BGE is affected by the composition of organic substrate (3, 26), BGE can vary over short time and space scales. Thus, in studies to determine the role of bacteria in carbon flow, it is important to determine BBP and BR (and thus BGE) simultaneously. This can be achieved by combining radiotracer methods of BBP with the Winkler titration (76) or with the use of oxygen microsensors (42). BR may also be accurately determined using modern coulometric techniques that measure bacterial CO_2 production (110). This technique overcomes the uncertainty associated with having to assume a given respiratory quotient value to convert oxygen utilization to CO_2 production. The coulometric method has been used effectively to concurrently measure BBP and BP (110). An analysis of available BBP and BR measured concurrently reveals that, for any given BR value, there is a large variation of BBP values (26). This variation may be due to the wide range of methods used but may also reflect the ability of bacteria to uncouple anabolism and catabolism under low-nutrient conditions in order to maximize growth and survival (27). More comparisons of BBP and BR are needed to accurately assess the contribution of bacteria to the cycling of organic carbon.

Even though there is an agreement regarding the importance of bacteria in aquatic ecosystems, the methods of assessing BBP have remained controversial and there is little agreement about the best method. This disagreement is based largely on uncertainties regarding the aspects of the taxonomy and metabolism of aquatic bacteria necessary to calibrate the methods and thus to determine the aspect of growth which is the best indicator of BBP. We know very little about the species composition and metabolic diversity of natural assemblages of bacteria. Traditional methods of laboratory microbiology, which depend on the isolation of single bacterial strains, are inadequate to study bacteria in aquatic systems. Only a small fraction of the bacteria in natural aquatic systems can be cultured (51). Therefore, current approaches for estimating BBP have focused on determining the growth of all heterotrophic bacteria rather than attempting to assess the growth rates of individual species. To use this synecological approach, assumptions must be made regarding the overall requirements and response of the entire bacterial assemblage. Furthermore, little is understood about the growth state and the growth environment of aquatic bacteria. For example, methods that measure the rate of synthesis of cellular constituents assume that bacteria have balanced, steady-state growth. Under these conditions, the rates of synthesis of all cellular constituents are equal. However, there is evidence that unbalanced growth is common in natural populations of bacteria (15, 61). During unbalanced growth, rates of synthesis of different macromolecules yield different estimates of growth. Theoretically, determinations of BBP can also be based on the utilization of DOM. However, in practice this approach can be inaccurate because bacteria in aquatic systems consume DOM, which occurs in very low concentrations (2), and the chemical composition of this DOM is largely unknown (8).

Additional challenges to developing accurate and precise methods of BBP involve the possible artifacts caused by the manipulation of natural samples. Bacteria occur in association with other organisms and nonliving particles of similar size, making it difficult to isolate them by size fractionation. Bacteria can also respond rapidly to changes in their environment; thus, containment (bottle effects) and long incubation periods may significantly modify species composition and growth (37, 98).

To properly assess the contribution of bacterial metabolism to the flow of organic matter in ecosystems, BBP needs to be converted into units of carbon or nitrogen. Because of the difficulties in isolating bacteria from surrounding debris, these values can seldom be measured directly. Carbon production is generally obtained by applying a conversion factor that translates cell production into carbon production. These conversion factors are typically derived by comparing the volume of cells to carbon content in mixed cultures of bacteria under laboratory conditions (76). Published values of the volume-to-biomass ratios vary by a factor of 14 (30), and thus, conclusions derived from BBP are very sensitive to the choice of conversion factor used. This problem is compounded by the need for other conversion factors, which translate a given measure of growth (e.g., DNA or protein synthesis) into cells produced.

Given these uncertainties and methodological limitations, it is remarkable that current BBP estimates based on a variety of independent methods show strong correlations to algal production and biomass (18, 30, 117). This suggests that when compared over large time and space scales these methods, even though they may not provide the absolute rate of BBP, do reflect actual changes in the relative rates of BBP and thus are useful to study system-wide contributions by bacteria.

However, as our knowledge of the ecology of aquatic bacteria increases, questions concerning the dynamics and controls of BBP have become more specific and demand greater accuracy and precision of the methods. Intercalibration studies have shown that differences between methods can be significant (17), making the study of the dynamics and controls of bacteria over short time and space scales problematic. Because of the various uncertainties of current methods for determining BBP, their adequacy to accurately estimate bacterial processes and rates has been questioned (57, 64). However, for many ecological questions, the current methods are still a valuable tool. For example, understanding the response of bacteria to changing conditions requires high precision, but knowledge of the absolute value of BBP is not needed (19). Additionally, there has been substantial research on the causes for differences among methods, and uncertainties can be reduced by experimental validation of the assumptions of the chosen method in the system being studied (91, 93).

Because current methods for determining BBP suffer from many uncertainties and limitations, the choice of method must follow careful thought regarding the question being addressed and the degree of accuracy that is satisfactory. For example, estimates of protein synthesis may be a better indicator of the amount of biomass available for consumption by grazers than the rate of cellular division because cells may increase in mass but not divide. The investigator must also be prepared to invest time and effort to test, in a given system, the validity of the assumptions of the chosen method.

A common approach given the difficulties in determining the accuracy of each method has been to simultaneously measure BBP by using independent methods. Agreement between these methods provides confidence that observed variations reflect real changes in rates of bacterial production. Because each method has its own assumptions and targets a different aspect of bacterial growth, differences in BBP estimates may reveal variations in environmental factors or changes in the growth state of bacteria. In this context, current methods of BBP can complement each other and provide useful information regarding the growth state and the environment of bacteria.

Of the various methods available to determine BBP, those that employ radiolabeled precursors to estimate the rate of synthesis of nucleic acids and proteins have become the most widely used and are the focus of this chapter. The rate of synthesis of these macromolecules closely reflects cell division and cellular growth. Advantages of these procedures include high specificity for bacteria, high sensitivity, small sample volumes (<100 ml), and short incubation periods (<1 h). Disadvantages include the need to test for the specificity of the label for the chosen macromolecule, the need to determine possible sources of external and internal isotope dilution, and the need for conversion factors to obtain cell or carbon production estimates from incorporation rates. Currently, the two methods most commonly used are thymidine (TdR) incorporation into DNA and leucine (Leu) incorporation into protein (6, 67).

When bacterial growth is balanced, rates of incorporation into DNA and protein give equal estimates of growth. However, uncoupling of macromolecular synthesis (unbalanced growth) occurs when bacterial growth rates shift (56). Periods of unbalanced growth may consequently reflect the response of bacteria to changes in environmental conditions (15). Simultaneous measurement of the rate of synthesis of DNA and protein are therefore not redundant and can provide information regarding the growth state of bacteria in nature (91, 92, 99).

Methods to determine BBP have been reviewed extensively (79, 91), and a manual giving step-by-step instructions of several methods is available (66). In this chapter, the rationale, advantages, and disadvantages of the most commonly used methods based on the incorporation of TdR and Leu methods are discussed. In addition, methods to determine empirically a conversion factor from thymidine or leucine incorporation to cells produced is presented as well as several procedures designed to test various assumptions of these methods. Alternative methods to determine BBP that do not rely on the uptake of radiolabeled compounds are also briefly discussed. The BBP measurement methods presented in this chapter estimate the total BBP and do not provide information on the relationship between bacterial diversity and metabolism. This chapter concludes with a review of novel methods that combine measurements of BBP with microscopy and molecular techniques to determine the proportion of total bacteria that are active, and the contribution to total BBP of specific phylogenetic groups.

THYMIDINE INCORPORATION INTO DNA

Rationale, Advantages, and Disadvantages

Of the various methods available to measure bacterial production, TdR incorporation into DNA (40) has become the most widely used in both freshwater and marine systems (18, 30, 36, 117). This method estimates the rate of bacterial DNA synthesis and consequently the rate of cell division. Nucleotides are normally synthesized de novo, but in bacteria they can also be produced via a salvage pathway (79). dTMP is synthesized de novo from dUMP by the enzyme thymidylate synthetase. In the salvage pathway, dTMP is formed through the phosphorylation of exogeneous TdR by the enzyme thymidine kinase. The TdR method assumes that TdR is added in sufficient concentration to inhibit de novo biosynthesis of nucleotides and promote the salvage mechanism. The added TdR is also assumed to be taken rapidly by bacteria, to remain stable during uptake, and to be rapidly converted into nucleotides.

The advantages of the TdR method include (i) high specificity for heterotrophic bacteria, (ii) high sensitivity and precision, and (iii) ease of use in the field. Only bacteria have the transport mechanisms for assimilating TdR and the enzyme thymidine kinase needed for nucleotide synthesis via the salvage pathway. Other organisms such as eukaryotes or cyanobacteria do not appear to incorporate TdR into DNA (9). The TdR method requires minimal manipulation of water samples. Short incubation periods (typically <1 h) are sufficient, and only nanomolar concentrations of TdR are needed. The procedures are relatively simple and do not require complex equipment, and the method can be completed quickly and is easily adapted to field use. Furthermore, the filters containing the incorporated TdR can be stored for subsequent extraction and determination of radioactivity. The basic TdR procedure has been modified for determining BBP in sediments (38, 112) and marine snow (1).

Extensive use and testing in many aquatic environments has raised concerns related to the accuracy of the method. The major concerns are as follows. (i) TdR is not always incorporated into DNA, and other macromolecules are also labeled. (ii) The conversion factors required to obtain cell or carbon production from TdR incorporation vary widely and are difficult to derive empirically. (iii) De novo synthesis may not be entirely suppressed by added TdR, leading to underestimates of DNA synthesis. (iv) Microautoradiographic studies show that not all the bacteria present incorporate TdR (28, 86); it is not clear if these unlabeled bacteria are dormant or moribund cells or are actively growing bacteria with unique metabolism or lack TdR transport systems (86, 120). (v) Bacteria in anaerobic environments such as chemolitotrophic and sulfate-reducing bacteria do not appear to incorporate TdR (41, 58), and thus, the TdR method underestimates BBP in anaerobic waters (119). The problems associated with the molecular specificity of TdR and with the conversion factor have been extensively studied. Isotope dilution caused by de novo synthesis and the fraction of the bacterial population incorporating TdR in various environments has not been investigated in as much detail.

The ^3H radioactive label of TdR often appears in macromolecules other than DNA, such as RNA and proteins indicating catabolism of the added TdR (53, 95). The fraction of the total incorporated radioactivity that appears in non-DNA macromolecules can be very variable and may lead to serious overestimates of DNA synthesis. To circumvent this problem, several purification procedures have been developed to isolate the DNA fraction (53, 62, 118). Although differences in the degree of TdR catabolism reported can be traced to differences in the extraction method used (111), examination of the degree of TdR catabolism obtained using individual methods suggests that the extent of this metabolism varies predictably with experimental and environmental conditions. For example, nitrogen starvation in seawater cultures of bacteria causes the utilization of TdR for protein synthesis (16). Differences in TdR catabolism have been observed over a salinity gradient (13) and with depth (46). In San Francisco Bay the patterns of TdR metabolism could be related to differences in the source of carbon to the system and to differences in bacterial species composition (54). Consistent patterns of TdR metabolism within a system may reflect an environmental or physiological basis. Thus, measurements of TdR catabolism can give insights into environmental conditions. Determination of the extent of TdR catabolism has become an important component of the TdR method. The procedures to isolate DNA are relatively simple and should be performed routinely. However, if the percentage of the total TdR that is incorporated into DNA is constant in a given system, it may not be necessary to perform the DNA isolation for every sample.

Another major concern of the TdR method is the need for conversion factors to translate TdR incorporation rates (picomoles per liter per hour) to cells or carbon produced (cells per liter per hour or micrograms of carbon per liter per hour). Rates of TdR incorporation are first converted into cells produced using a theoretically or empirically derived factor. The theoretical factor is calculated from (i) the percentage of thymidine in bacterial DNA, (ii) the DNA content per cell, (iii) assumptions regarding the extent of extracellular isotope dilution, and (iv) the effective inhibition of de novo synthesis (intracellular isotope dilution). The value of these factors is generally obtained from the results of laboratory studies and varies from 0.4×10^{18} to 0.5×10^{18} cells per mol of TdR incorporated. A major drawback of the theoretical conversion factor is that the information used in its computation is derived from studies using laboratory monocultures of bacteria. These values may not be representative of natural bacteria or of natural assemblages of bacteria. In studies using natural assemblages of bacteria, the amount of DNA per cell was quite variable, with smaller cells generally containing more DNA per unit of volume (16, 100). Furthermore, other assumptions related to the theoretical factor, such as the extent of isotope dilution by de novo synthesis, are very difficult to determine directly (80).

Deriving an empirical conversion factor can circumvent the difficulties of the theoretical conversion factor. The empirical conversion factor uses natural assemblages of bacteria from the study area to calibrate the TdR method and is derived by comparing, under controlled conditions, TdR incorporation with increases in bacterial numbers (32). An added advantage of these empirical conversion factors is that all possible relationships between TdR incorporation and cell production (such as TdR catabolism) are, in theory, included in the conversion factor. Published empirical conversion factors vary widely and are generally higher than the theoretical factors (30, 91). The most commonly used empirical factor is 2×10^{18} cells per mol of TdR, which is the median of 97 studies (30). Variations in the empirical conversion factor may be caused by differences in the method used to analyze the data or to differences in the experimental procedures. For example, high conversion factors relative to the commonly used factor of 2×10^{18} cells per mol of TdR may be due to underestimates of TdR incorporation caused by the lack of uptake saturation (5, 34). If uptake is not saturated, de novo synthesis leads to isotope dilution. Bell (5) reported that using higher concentrations of added TdR could reduce variations in the conversion factors. Thus, when the TdR method is used, preliminary experiments should be performed to determine the TdR concentration at which maximal uptake is achieved (saturation concentration). Conversion factors determined using the same method can vary seasonally and spatially. These variations may reflect changes in environmental parameters such as nutrient availability and temperature (23). In a study where empirical conversion factors were determined in a eutrophic lake over the course of a year, the magnitude of the conversion factor could be predicted from the regression of in situ temperature and TdR incorporation rate, suggesting that the variation in the conversion factor was constrained by environmental factors and not caused by artifacts (74).

For many questions regarding the ecological role of bacteria, BBP in units of carbon is required. Once TdR incorporation values are converted to cell production, another factor is applied to translate this information into units of carbon produced. This factor is determined by comparing cell volumes to carbon content in cultures of bacteria. A wide range of biomass conversion factors is available (76, 81). The choice of conversion factors and their combination is likely to significantly affect the calculated magnitude of bacterial carbon production based on any method using incorporation of radiolabeled compounds (30). Because of the various factors available to convert TdR incorporation rates into cell or carbon production, data are generally reported in units of TdR incorporation (typically picomoles per liter per hour).

Procedures for Estimating BBP from Incorporation Rates of TdR

The basic TdR procedure examines [methyl-^{3}H]TdR incorporation into a macromolecular fraction containing DNA, RNA, and protein, which is isolated using a cold trichloroacetic acid (TCA) extraction (6). For each water sample, replicate samples (usually 10 ml) are incubated with [methyl-^{3}H]TdR (specific activity, 40 to 80 Ci mmol^{-1}; usually a final concentration of 20 nM). Abiotic absorption of radioactivity is measured in replicate samples in which bacteria have been killed with TCA or formalin. After an incubation period (30 to 60 min), the samples are placed in an ice-cold water bath for 1 min, and then 1 ml of ice-cold 50% TCA is added to reach a final concentration of 5%. Samples are cooled for an additional 15 min. This step kills the bacteria and lyses cell membranes, allowing the removal of the TdR taken up but not incorporated into macromolecules. Following this extraction, samples are filtered through 0.45-μm-pore-size cellulose acetate membrane filters. Incubation vials are rinsed with 2 ml of ice-cold 5% TCA. Filters with the bacteria are then rinsed with the TCA used to rinse the vials and with three additional 1-ml aliquots of ice-cold 5% TCA. To remove TdR incorporated into cells but not into DNA (e.g., into lipids), the filter is rinsed further with five 1-ml aliquots of 80% ethanol. Radioactivity in the filter is measured using a scintillation counter. Quenching during radioassay, caused by the filter, is reduced by dissolving the filter with 1 ml of ethyl acetate for 30 min prior to the addition of scintillation solution.

Several procedures are available for the isolation of DNA if incorporation into other macromolecules is an issue. These include an acid/base hydrolysis (62) and a phenol-chloroform extraction (118). To purify DNA by using the phenol-chloroform procedure, lysed samples are filtered through cellulose nitrate membrane filters instead of cellulose acetate filters, since the latter are dissolved by chloroform-phenol. Following the incubation period, incorporation is stopped by addition of formaldehyde to a final concentration of 2%. RNA is extracted first by adding 5 N NaOH to reach a final concentration of 0.25 N and incubating the sample at 20 to 25°C for 60 min. The sample is then acidified with 100% TCA (1 to 2 ml/10-ml sample) to a pH of approximately 1 and stored on ice for 15 min. The sample is then filtered, and any protein remaining on the filter is extracted by rinsing with 5 ml of a 50% (wt/vol) chloroform-phenol solution (50 g of phenol in 100 ml of chloroform), leaving radiolabeled DNA on the filter. A final rinse is done with five 1-ml portions of ice-cold 80% ethanol, and the filter is prepared for radioassay as described above.

Exact sample volumes, incubation times, and final concentration of TdR may vary depending on the environment (freshwater, salt water, oligotrophic, eutrophic, etc.) and are derived from preliminary experiments. The appropriate concentration of TdR is determined from saturation experiments in which replicate water samples are incubated with various concentrations of TdR (e.g., 5, 10, 15, 20, and 25 nM). The rate of TdR incorporation is then plotted versus added TdR, and the lowest concentration at which incorporation is saturated (i.e., no increase in incorporation observed) is chosen. The length of incubation is determined by incubating replicate samples over various time intervals (e.g., 10, 20, and 30 min) and determining if the relationship between rate of incorporation and incubation time is linear. To avoid bottle effects, the shortest incubation time that yields a linear, measurable rate is chosen.

The extent of extracellular isotope dilution is determined indirectly using a standard additions method. Replicate samples are incubated with a given concentration of radioactive TdR (e.g., 5 nM) but with varying concentrations of unlabeled TdR (e.g., 0, 5, 15, 25, 35 nM). Increasing amounts of added unlabeled TdR dilute the radioactive TdR and yield decreasing incorporation rates. In the absence of external concentrations of the substrate, the plot of 1/incorporation rate versus total added substrate (labeled plus unlabeled) should pass through 0. However, if additional TdR is present in excess of what is added experimentally, the plot does not pass through 0 and the ambient concentration of TdR is determined by extrapolating the linear plot to 1/incorporation rate equals 0 ($y = 0$). The proportion of bacterial cells incorporating TdR can be determined using microautoradiography (12, 86, 105). In this technique, filters containing bacterial cells previously incubated with TdR are placed face down on a radioactivity-sensitive emulsion coating one surface of a microscope slide. The filter is then removed, and the bacteria are embedded in the emulsion. Cells that incorporated TdR "expose" the underlying emulsion, producing silver grains. The entire slide is developed by normal photographic means and additionally stained with DNA-specific fluorochromes so that silver grains and bacteria can be identified simultaneously. The preparation is then observed with bright-field illumination to count the cells with associated silver grains. By using epifluorescence the total bacterial population is determined by counting the fluorescent cells in the same sample. In addition to testing if TdR was incorporated by bacteria, microautoradiography can also be combined with immunofluorescent techniques to determine the phylogenetic identity of the bacteria incorporating TdR. This approach is discussed in a subsequent section.

LEUCINE INCORPORATION INTO PROTEIN

Rationale, Advantages, and Disadvantages

The Leu method estimates the rate of bacterial protein synthesis by measuring the incorporation rate of radioactive Leu into proteins that are isolated using a hot-TCA extraction (70, 71, 78, 100). Protein synthesis is a good indicator of BBP because proteins account for a large percentage (~60%) of bacterial biomass in natural assemblages (43) and their synthesis consumes a large percentage of cellular energy. Protein synthesis can also be estimated by using $^{35}SO_4$ as a radiolabeled precursor, but [^{3}H]Leu is the preferred substrate because algae also utilize $^{35}SO_4$, and the high concentration of sulfate in marine environments can result in significant isotope dilution. The rate of Leu incorporation is a good indicator of the rate of protein synthesis because Leu constitutes a fairly constant fraction of bacterial protein and it is not transformed into other amino acids (70). Kirchman et al. (70) demonstrated that increases in Leu incorporation agreed with increases in cell numbers and protein content. Furthermore, most of the Leu was incorporated into proteins, and little was degraded. Given the small concentrations used (usually 10 to 20 nM) and the short incubation periods (<1 h), bacteria exclusively take up the added Leu. The uptake of other amino acids such as valine has also been used to estimate BBP (59, 95), but Leu has remained the amino acid of choice.

Advantages of the Leu method include the same ease of use and application to field work as the TdR method. Furthermore, the Leu method has several advantages over the TdR method. It is more sensitive than the TdR method because more Leu (protein) is needed than TdR (DNA) in the production of new cells. The environmental and intracellular concentration of Leu, and hence the extent of isotope dilution, can be determined using high-performance liquid chromatography (100). Furthermore, Leu incorporation can be used to determine the bacterial carbon content following determination of a protein-to-carbon conversion factor (100). The Leu method has been used successfully alone and in conjunction with TdR in a wide variety of freshwater (78) and marine (92, 115) environments. Examples of other applications include measuring BBP in sediments (72, 109), riverine (42), and marine (113) aggregates, and to measure epiphytic BBP (104, 107). The procedures of the Leu method are very similar to those of the TdR method, and both methods can be performed with the same equipment and materials. Modifications have been made to reduce the cost of the procedure by using centrifugation rather than filters to concentrate the protein fraction (101). Furthermore, there is a dual-label approach that allows the simultaneous measurement of Leu and TdR in the same sample (14).

Another significant advantage of this method is that Leu incorporation can be used to determine bacterial carbon production directly because Leu composes a constant percentage of bacterial protein, and the ratio between protein and carbon over a wide range of bacterial cells is constant (100). With this information, carbon production from Leu incorporation is obtained by solving the formula

$$\text{BBP (g of C liter}^{-1}\text{ h}^{-1}) = \text{Leu incorporation (mol}^{-1}\text{ h}^{-1}) \cdot 131.2 \cdot (\%\text{Leu})^{-1} \cdot (\text{C/protein}) \cdot \text{ID},$$

where 131.2 is the formula weight of Leu, %Leu is the fraction of Leu in protein, C/protein is the ratio of cellular carbon to protein, and ID is isotope dilution. With this information derived from natural assemblages of bacteria, %Leu = 0.073, C/protein = 0.86, ID = 2 results in a conversion factor of 3.1 kg of C mol^{-1} (100).

Concerns associated with the Leu method include isotope dilution by external and internal sources. Simon and Azam (100) found a twofold internal isotope dilution, suggesting that Leu biosynthesis was not effectively inhibited by the added Leu. An advantage, nevertheless, of the Leu method is that this dilution can be measured directly. However, the procedures to determine this dilution are too complex to be performed routinely (100). Another concern related to the Leu method is that in eutrophic environments, very high concentrations of Leu (up to 200 nM) may be needed to saturate incorporation (115). Addition of very high (millimolar) concentrations of Leu may modify bacterial protein synthesis and also lead to the appearance of the radiolabel in phytoplankton or fungi (10, 39) possibly as a result of low-affinity but specific uptake systems. Furthermore, there is evidence that the cyanobacteria *Microcystis aeruginosa* and *Nodularia* spp. can take up Leu (50, 60). Thus, care must be taken to determine if blooms of cyanobacteria are present when using Leu. Finally, if bacterial internal protein turnover is rapid, the Leu method will overestimate BBP. Kirchman et al. (71) reported that bacterial protein turnover in natural assemblages of bacteria did not appear to affect the Leu method, but there have been no further published studies examining this issue.

The Leu approach has not been used as widely as the TdR approach. However, it is being increasingly applied as an alternative to the TdR method or as an independent method to complement TdR measurements. When used simultaneously, the TdR and the Leu methods are not always in agreement. For example, rates of TdR and Leu incorporation can respond differently to temperature and substrate variations (99, 108). A possible explanation for this lack of agreement is that protein and DNA synthesis respond differently to changes in growth conditions. When pure cultures of bacteria shift from one growth rate to a higher one, the rate of macromolecular syntheses increases in a specific sequence (56) with increases in protein synthesis preceding increases in DNA synthesis. This sequence is reversed when bacteria shift to a lower growth rate. The period during which the various rates of macromolecular syntheses are uncoupled is called unbalanced growth. If relative changes in the TdR and Leu incorporation indicate periods of unbalanced growth, variations in the Leu/TdR incorporation ratio over time may reflect changing growth conditions (15). These may be caused, for example, by variations in temperature or substrate (99). Therefore, simultaneous measurements of TdR and Leu can provide complementary information on the growth state of bacteria which can then be used to identify the possible limiting factors that control bacterial growth in nature (15, 72, 99).

Procedures for Estimating BBP from Incorporation Rates of Leu

The incubation and filtration procedures of the Leu method are similar to those of the TdR method (14, 67). The protein fraction is isolated following a hot TCA extraction that solubilizes nucleic acids. Replicate samples (5 to 25 ml) are incubated with [4,5-^{3}H]Leu (usually a final concentration of 10 nM, but in some cases as high as 35 nM [57], specific activity of 40 to 60 Ci mmol^{-1}). In eutrophic environments, where incorporation rates are expected to be high, a combination of radioactive and nonradioactive leucine may be added (e.g., 1 nM [4,5-^{3}H]Leu and 10 nM nonradioactive Leu) to reduce the expense of the procedure. Abiotic absorption of radioactivity is measured with formalin- or TCA-killed controls. Samples are incubated for an appropriate period of time (30 to 60 min) and then placed in ice-cold water for 1 min to end the incubation. The total macromolecular fraction is isolated by adding 1 ml of 50% TCA to reach a final concentration of 5% and then by heating samples to 80°C for 15 min, which extracts nucleic acids. Upon cooling to room temperature, samples are filtered through 0.45-μm-pore-size cellulose acetate filters to isolate the proteins. As with the TdR procedure, filters are rinsed with 5% cold TCA and with cold 80% ethanol to remove unincorporated Leu and Leu in lipids. Before radioassay, the filters are also dissolved with ethyl acetate. The exact sample volumes, incubation times, and final concentration of Leu are derived from preliminary experiments. These experiments are the same as those described in the TdR section above. Leu specificity for proteins can be tested by comparing the radioactivity on the filter with that in the hot-TCA filtrate, which contains nucleic acids. The Leu method is very specific for proteins, and nearly all the Leu taken up is assimilated into the protein fraction (70). If experimental evidence shows that this is the case in the study area, this procedure can be simplified by substituting a cold TCA extraction for the hot TCA extraction. In this case the procedures are identical to those of the TdR method. Leu incorporation can be translated directly to carbon production by using a theoretical conversion factor of 3.1 kg C mol^{-1} (100). Alternatively an empirical conversion factor can be determined using mixed cultures of bacteria from the study area.

The basic Leu procedure has been modified by using centrifugation instead of filtration to separate the bacterial proteins (101). In this modified method, the use of smaller sample volumes (1 to 2 ml) and the lack of a filtration step result in a faster, cheaper procedure with reduced radioactive waste. Briefly, samples are incubated with radiolabeled Leu in microcentrifuge tubes. After incubation and removal of unincorporated Leu, samples are centrifuged and the resulting pellets are treated with hot TCA to isolate the proteins. Scintillation fluid is then added directly to the microcentrifuge tube prior to radioassay. The Leu method using centrifugation has been further modified for use with freshwater and sediment samples. In freshwater samples, the efficiency of protein precipitation is increased by including a cold TCA extraction step, and by adding humic extracts or NaCl that act as coprecipitants for protein (73). In sediment samples, the pellets formed by centrifugation are treated to form a gel matrix that prevents the sediment from settling during the radioassay (48). The TdR method for BBP can also be applied using the centrifugation method.

While the centrifugation method has several advantages over the filtration method, care must be taken to determine the possible retention of protein in the tube because there is evidence that protein retention can vary by tube brand and manufacturer (85). Thus, the same brand of tube should be used to reduce sources of variation. Furthermore, more comparisons should be made with the filtration method as one study showed that the centrifugation method yielded lower BPP values than the filtration method (49).

Because simultaneous measurements of protein and DNA synthesis can provide complementary information regarding the growth of bacteria (92, 96), a method has been developed to concurrently measure in the same sample Leu and TdR incorporation (14). To achieve this, [^{3}H]TdR and [^{14}C]Leu are added to a single sample and the cold TCA-insoluble material is collected. [^{14}C]Leu is chosen because [2-^{14}C]TdR can undergo catabolism and still label DNA because of the retention of the label in uracil. A final concentration of 20 nM Leu (10 nM [^{14}C]Leu, specific activity of about 300 mCi mmol^{-1} and 10 nM nonradioactive Leu) is used. Termination of incubation, cold TCA extraction of macromolecules, subsequent filtration, and preparation of filters for radioassay are the same as described above for the basic Leu procedure. This dual-label method has been modified and used successfully with sediment samples (72, 109).

EXPERIMENTS TO DETERMINE EMPIRICAL CONVERSION FACTORS FOR THE TdR AND Leu METHODS

Cultures of mixed assemblages of bacteria are prepared by inoculating filter-sterilized sample water with small volumes of prefiltered sample water (1 part prefiltered to 9 parts sterilized) (31, 65). Prefiltering is accomplished using a 0.6- to 2-μm-pore-size track-etch polycarbonate filter and reduces grazers, phytoplankton, and detritus in relation to bacteria. The exact pore size used varies depending on the environment. In eutrophic systems, using a very-small-pore-size filter may exclude a significant portion of the bacteria, whereas in oligotrophic systems a large-pore-size filter may be ineffective in removing bacterivores. Sterile filtration is accomplished using 0.22-μm-pore-size Nuclepore filters. Adding the prefiltered sample to sterilized water from the same site promotes bacterial growth (2), reduces grazing by bacterivores escaping the prefiltration step, and reduces viral encounters (90). A gentle vacuum (<100 mm Hg) is used throughout to avoid release of DOM due to cell lysis. Inoculated samples (1 to 4 liters) are kept in the dark and incubated at the in situ water temperature. Subsamples (10 to 50 ml) are taken every 2 to 4 h for a total of 24 h and assayed both for bacterial abundance (52, 89) and TdR (and/or Leu) uptake by the methods described above.

There are several methods to compute a conversion factor from these experiments. The simplest and most commonly used method (integrative method) involves dividing the increase in cell numbers by the substrate incorporation integrated over time (68). The change in cellular abundance is simply the final abundance minus the initial abundance. The total substrate incorporated is computed by integrating the area under the curve described by the plot of substrate incorporation versus time. A drawback of this approach is that only two values are used to determine the increase in cell numbers. Alternatively, the data can be analyzed using an exponential growth model that considers all points (33, 68, 69). The following equation is used to compute the conversion factor: $CF = \mu\ e^B/e^b$, where μ is the growth rate determined from changes in cell abundance over time, B is the y intercept of ln(cells) versus time, and b is the y intercept of ln(substrate incorporation) versus time (68). The conversion factor yields values in units of cells. To obtain carbon production, an additional factor must be applied (see above).

OTHER METHODS

The incorporation of radiolabeled adenine into DNA and RNA has been used to measure BBP. Because adenine is incorporated into bacteria and into eukaryotic algae, the method estimates the growth of the entire microbial community. Estimates of BBP using this method therefore require a size fractionation step to isolate the bacteria. The adenine method has been applied to many water column environments (7, 65) and sediments (24) but has not been used as extensively as the TdR or Leu methods.

An advantage of this method is the ability to simultaneously measure RNA and DNA synthesis in the same incubation. Relative changes in these rates can yield information regarding the extent of unbalanced growth in mixed assemblages of bacteria. With this method it is possible to correct for isotope dilution by measuring the specific activity of the nucleic acid precursors. Adenine is incorporated into nucleic acids through various precursor molecules (ATP, dATP, ADP, and AMP). The existing pools of these compounds, in addition to inputs via de novo synthesis, dilute the specific activity of the added radioactive adenine ([2,^{3}H]adenine). In the adenine method, the immediate nucleotide triphosphate precursors are isolated and assayed for their specific activity.

Briefly, the procedure for the adenine method involves incubating replicate water samples with [^{3}H] adenine (63). Following incubation, subsamples are extracted for isolation of nucleic acids. RNA, DNA, and the precursor ATP are separated, purified, and radioassayed. From this information, rates of RNA and DNA synthesis corrected for isotope dilution are computed. If a time course experiment is performed in which replicate samples are incubated over different time intervals, the individual rates of DNA and RNA synthesis can be determined. DNA synthesis is then converted to BBP if the C/DNA ratio is known.

Several methods for the estimation of BBP that do not require radiolabeled compounds are available. The frequency of dividing cells (FDC) method, for example, estimates the growth rate of bacteria from an empirically determined relationship between growth rate, temperature, and the FDC (44, 82). By use of epifluorescence microscopy, dividing cells are identified as cells visibly showing invagination. Advantages of this method are the lack of an incubation and the ability to determine the bacterial biomass from the same sample. Disadvantages are the need to determine the relationship between temperature, growth rate, and FDC for each environment investigated, the time and tedium involved in microscopic examination of all samples, and the subjectivity involved in the identification of dividing cells.

Other methods focus on determining increases in bacterial numbers after grazers have been eliminated by size fractionation or by using eukaryotic inhibitors. The overlap of sizes of bacteria and their predators and the possible decrease of bacterial abundance by phages limit the reliability of this approach. Furthermore, metabolic inhibitors and antibiotics that are used to selectively inhibit groups of organisms are not always effective (106). Finally, the long-term incubations that are required to observe changes in bacterial numbers can introduce significant artifacts such as changes in DOM quality and concentration. Some of these bottle effects can be minimized by using dialysis bags that allow exchange of dissolved materials but maintain the bacterial population isolated from predators (47).

Another method that does not rely on radiolabeled substrates uses the TdR analog, 5-bromo-2′-deoxyuridine (BrdU) to estimate DNA synthesis (102). In this method, BrdU incorporation into bacteria is detected and measured with a chemiluminescent immunoassay. The optimum, saturating concentration of BrdU to be used in experiments can be determined using saturation experiments, as is done in ^{3}H-TdR uptake experiments. Through the development and testing of this novel procedure, BrdU incorporation rates agreed well with ^{3}H-TdR and ^{3}H-BrdU incorporation rates. The BrdU method may be a good alternative method to estimate DNA synthesis when the use of radioactive TdR is impractical. Furthermore, as discussed below, the BrdU method could be combined with molecular methods to develop a procedure that may provide information on the proportion of active cells, their phylogenetic identity, and their contribution to gross BBP.

ESTIMATING THE CONTRIBUTION OF SPECIFIC BACTERIAL GROUPS TO TOTAL SECONDARY PRODUCTIVITY

The methods discussed above represent a "black box" approach in that they estimate total BBP but do not provide any information on the phylogenetic diversity of the bacterial community or of the relative contribution of each group to total BBP. Simultaneous measurements of BBP and determination of phylogenetic diversity offer some insight into the relationship between diversity and production. However, this information is compromised because not all bacteria in the water column may be actively respiring (102). To fully understand the relationship between phylogenetic diversity and productivity, it is necessary to detect active bacteria, determine their phylogenetic identity, and measure their productivity. Currently, there is no single method that can achieve all these measurements simultaneously in situ. However, the following three methods (micro-FISH, SNAP-BAC, and BrdU uptake) represent important advances in our ability to obtain this information.

Micro-FISH is a combination of microautoradiography and fluorescence in situ hybridization (FISH) (20, 21, 22). The combination of these techniques has been used to study in situ the uptake of various substrates by specific bacterial cell types including sulfate reducers, *Thiothrix* spp., and denitrifiers (83, 84). In the micro-FISH method, microautoradiography allows visualization and enumeration of cells incorporating ^{3}H-TdR or ^{3}H-Leu, while FISH provides the ability to phylogenetically characterize both active and inactive members of the bacterial community. Thus, the micro-FISH method allows for enumeration of certain groups of bacteria and for the determination of whether or not these cells are active. This method is explained in greater detail elsewhere in this volume (chapter 39), but its implications are briefly discussed here.

Micro-FISH method has been used in a variety of environments to assess the contribution of major bacterial groups to total BBP as well as the microbial uptake of various components and size classes of DOM (20, 21, 22, 77). In one study using micro-FISH, conducted in the Delaware estuary, all major groups of marine bacteria, including alpha, beta, and gamma proteobacteria and *Cytophaga*-like bacteria, assimilated TdR and Leu (21). In addition, about 50% of the variability in the assimilation of ^{3}H-TdR and ^{3}H-Leu by bacterial groups could be explained by their abundance in the total community. This suggested that bacterial abundance and activity are partially controlled by the quantity and quality of DOM (20). Nevertheless, a large fraction of ^{3}H-TdR and ^{3}H-Leu incorporation cannot be attributable to bacterial abundance and factors affecting growth rates of the different bacterial groups such as viral lysis, protozoan grazing, nutrient limitation, etc., may explain some of this variation. Other findings resulting from the application of micro-FISH in natural waters suggest that the relative abundance of a given bacterial group does not necessarily reflect its relative contribution to total BBP. This may be due to the different capabilities and efficiencies that each bacterial group has to utilize the available substrate. If the quality of substrates or the environmental conditions that affect the bacterial utilization of these substrates vary, then the relative contribution of bacterial groups to BBP will also vary.

While micro-FISH can determine which of the active bacterial groups dominate BBP, it requires the predetermination of what groups or phyla of bacteria will be assessed. Thus, the method may not identify the dominant active bacteria if it belongs to a group not considered in the design of the experiment. Furthermore, this method does not provide information on the rates of BBP by the targeted bacteria.

A second method, selective nucleic acid polymerase biotinylation and capture (SNAP-BAC), allows quantification of ^{3}H-TdR incorporation by specific, phylogenetically defined members of a bacterial community (116). Thus, it yields BBP production for a given bacterial group. This method has the same limitations as micro-FISH in that the bacterial group to be examined is determined by the investigator and may not represent the dominant group in the study site. Nevertheless, if the experimental question focuses on determining the biomass production of a specific group, SNAP-BAC is a powerful tool. This method is carried out using an enzyme-mediated reaction. Water samples are first incubated with ^{3}H-TdR for a predetermined period (30 min in coastal waters). This step measures total TdR incorporation into bulk DNA. DNA is also extracted using enzymatic lysis of cells collected on a 0.45-pore-size-μm filter followed by a phenol-chloroform extraction and two ethanol precipitations. To determine ^{3}H-TdR incorporation by a specific group, DNA from the chosen phylum (or group, phylotype, etc.) is selectively recovered and biotinylated as follows. Aliquots of the ^{3}H-labeled DNA are combined with internal ^{33}P-labeled recovery standards. These combined samples are denatured using heat, and the resulting single-stranded DNAs are added to an enyzme-mediated reaction containing a 16S rRNA primer for a specific bacteria phylum (e.g., *Bacteriodetes* in reference 116), biotinylated dUTP, and a thermostable DNA polymerase. Cycles of annealing (binding of primer to template DNA) and extension (replication of the DNA from the primer outwards using the enzyme *Taq* polymerase) steps are used to incorporate the biotinylated dUTP into the complementary DNA strands of the targeted bacteria. The DNA is not amplified as it would be when a standard PCR is used, but becomes increasingly biotinylated during the annealing and extension cycles. After removing single-stranded overhangs by using an S1 nuclease digestion, biotinylated DNA is captured on streptavidin-coated beads. The ^{3}H and ^{33}P activity of these samples can then be quantified, and the recovery of ^{33}P is used to calculate the yield of ^{3}H-labeled DNA from the selected organism or group of organisms. This yield of ^{3}H-labeled DNA can be used to calculate the contribution of the targeted bacteria to total BBP. The use of ^{33}P internal standards is crucial and is a particular strength of the SNAP-BAC method, as it allows a quantitative measure of DNA synthesis.

The SNAP-BAC method has been used to examine the contribution of the *Bacteriodetes* phylum to total BBP in coastal Washington waters including Puget Sound (116). The *Bacteriodetes* phylum was selected because it has been shown to be a numerically important group of bacteria in estuarine and coastal waters (20). This bacterial group was found to be responsible for 56% ± 17% of community ^{3}H-TdR incorporation in the sites examined. This finding confirms that in addition to being numerically abundant in coastal marine waters, the *Bacteriodetes* are also significant in their contribution to bacterial secondary productivity. SNAP-BAC could be applied to a variety of other groups of bacteria or even single phylotypes of bacteria. It is important, however, that proper controls and stringency tests be first conducted to ensure that the primer used binds only the targeted group and that the internal standard adequately represents the DNA of the targeted bacteria.

The third method is similar to the micro-FISH procedure and uses the incorporation of the TdR analog BrdU. As described in a previous section, BrdU incorporation has been developed as a method to determine bulk BBP without the use of radioactivity. In this BBP method, a chemiluminescent immunoassay measures incorporated BrdU. To gather information on the proportion of active cells and their phylogenetic identity, the BrdU assay can be combined with FISH (87). Using this modified BrdU procedure, cells that incorporated BrdU (metabolically active) and those belonging to targeted groups can be visualized and enumerated. In a study in the North Sea using this method, 14% of bacterioplankton incorporated BrdU into their DNA after 20 h. While the relative number of cells of some species remained the same (e.g., the *Roseobacter* lineage), the proportion of BrdU-incorporating cells doubled over the course of the study. Members of the SAR86 lineages, while only representing 8% of total counted cells, constituted nearly 50% of visible BrdU-incorporating cells. The BrdU method has also been adapted to determine the growth rates of individual cells (45). In this method, cells incubated with BrdU and stained with DAPI 4′,6′-diamidino-2-phenylindole are analyzed with epifluorescence microscopy. The proportion of active cells can be determined by comparing the total DAPI counts with the counts of cells that incorporated the BrdU. Growth rates can be determined from the fluorescence signal intensity, which was found to be significantly correlated to the specific growth rate of bacterial isolates. A procedure combining BrdU and FISH thus has the potential of being able to determine the proportion of active cells, their phylogenetic identity, and their contribution to total BBP.

The methods described so far in this section determine the phylogenetic identity of active groups following a predetermination of a target bacterial group, which may or may not be important in the study site. To assess the phylogenetic structure of all active bacteria, the use of BrdU incorporation has also been employed as a means to isolate the DNA of active bacteria (BrdU-incorporating bacteria) for subsequent molecular analysis. In this application of the BrdU incorporation method, DNA is first extracted from cells incubated with BrdU. BrdU-containing DNA is then physically separated from unlabeled DNA by using antibody-coated paramagnetic beads. Unlabeled DNA is washed away, and BrdU-containing DNA remains attached to the beads. This DNA can later be separated from the beads for amplification of a selected gene by using PCR, or PCR can be done on the paramagnetic beads directly. BrdU containing DNA could also be incubated with a biotinylated antibody and separated from unlabeled DNA by using streptavidin-coated paramagnetic beads as in the SNAP-BAC method (35). The advantage of isolating BrdU-containing DNA is that this DNA originates from metabolically active bacteria. The isolated DNA can then be subjected to a variety of molecular techniques to investigate the identity and diversity of active (BrdU-incorporating) members of the bacterial community. These techniques include length heterogeneity PCR, terminal restriction fragment length polymorphism, automated rRNA intergenic spacer analysis, environmental cloning and sequencing, and quantitative PCR, among others. With this application of BrdU incorporation into the DNA of metabolically active bacteria, much can be done to determine the identity of the active members of the bacterial community and how they compare to the rest of the bacterial community. This method is particularly powerful because it can provide information on the proportion of cells in a community that are active by comparing total counts with counts of BrdU-containing cells (114), and it also provides the opportunity to examine the DNA of the active bacteria (35). This approach has advantages over the micro-FISH and SNAP-BAC methods in that the DNAs of all the active (BrdU-incorporating) bacteria are considered for phylogenetic analysis and that their total productivity can be determined in the same incubation.

In summary, micro-FISH measures the uptake of organic substrates by targeted bacteria. Using an internal (^{33}P) standard, SNAP-BAC quantitatively measures the BBP of a targeted bacterial group from the incorporation of thymidine into DNA. BrdU, like micro-FISH, measures incorporation of a substrate and thus identifies active cells. Unlike the micro-FISH method, BrdU can also provide quantitative information on BBP. Furthermore, BrdU incorporation can be coupled with FISH to examine the proportion of active cells or with molecular methods to examine the diversity of active bacteria. A limitation shared by micro-FISH, SNAP-BAC, and BrdU/FISH is that they all require prior knowledge of the bacterial community being studied. This limitation may be overcome by first isolating the DNA from active bacteria, using a modification of the BrdU method, and then analyzing it with molecular techniques such as terminal restriction fragment length polymorphism.

The three methods described in this section are relatively new and have yet to be tested in a wide range of environments and conditions. However, they exemplify how methods associated with determining total BBP (TdR, Leu, and BrdU incorporation) can be successfully combined with molecular methods to provide information on the proportion of active cells, their phylogenetic identity, and their contribution to BBP. More information of this type is essential to determining the relationship between microbial diversity and global biogeochemical cycles.

SUMMARY AND CONCLUSIONS

The most commonly used methods for the estimation of BBP are TdR incorporation into DNA and Leu incorporation into protein. These methods are easily used to determine BBP in situ, and have become routine measurements in studies of the role of bacteria in aquatic systems. Analyses of BBP measurements on a large scale have established the importance of heterotrophic bacteria as major secondary producers that consume a substantial portion of primary production. For example, a study that analyzed BBP data compiled from freshwater and marine systems found that BBP and primary production were significantly correlated. BBP ranged from 118 to 2,439 mg of C m^{-2} day^{-1}, which corresponded to an average of 30% of primary production (18). This report used carbon production values derived using the same conversion factors employed in each study. Another study, which analyzed BBP values exclusively from oceanic systems including data from long-term studies such as BATS (Bermuda Atlantic Time Series) and HOTS (Hawaiian Ocean Time Series) indicated more variation in the relationship between BBP and primary production BBP and a lower ratio of BBP to primary production than in the cross-system analysis. In this study of open ocean waters, BBP carbon values were determined from the raw data by using constant conversion factors. BBP ranged from 5.5 to 285 mg of C m^{-2} day^{-1} and was consistently between 10 and 20% of primary production (Table 5 in reference 29).

Comparisons of the relative magnitudes of BBP and primary production can also reveal environmental factors mediating the bacterial utilization of primary production and furthermore, can be used to determine if a given system is net heterotrophic or autotrophic. A comparison of BBP (using both TdR and Leu methods) and primary production along a north-south transect of the Atlantic Ocean, using high-frequency sampling (a total of 170 samples each taken every 50 to 100 nautical miles), showed that the ratio of BBP to primary production was positively correlated to temperature, reaching a maximal value of 40% near the tropics (55). This trend was interpreted to reflect increases in bacterial respiration with higher water temperatures or to indicate the effect of nutrient availability because high temperatures often coincide with oligotrophic conditions in the sea. In this study, a value of BGE of 30%, which was a median value from published values, was used to determine the amount of carbon needed to sustain the observed BBP (organic carbon demand = BPP/BGE). Areas where bacterial carbon demand exceeded primary production were considered to be net heterotrophic. Using this approach, an area in the Atlantic Ocean between 8°N and 20°S was determined to be net heterotrophic.

These large-scale studies illustrate modern applications of BBP methods and the interpretation of the data generated using these procedures. They also show how different factors can be used to convert BBP in units of substrate incorporated (TdR or Leu) to units of carbon and how BGE is often selected from a range of published values. Due to the methodological difficulties involved, conversion factors and BGE values are rarely determined for each study. Assumed values can introduce ambiguities that prevent an accurate understanding of the role of bacteria in carbon flow. It is recommended that conversion factors be determined in each study and that BR values be determined concurrently with BBP. While this may be impractical in many cases, more data on conversion factors and on BGE can provide information to further constrain these values in natural systems or to provide consistent criteria to select the best factor for a given situation.

Other concerns associated with TdR and Leu procedures include the specificity of the label for bacteria and the possible dilution of the specific activity of the added substrate by extracellular and intracellular pools. To address these concerns, the assumptions of each method should be tested in the study site. Simultaneous measurements of several independent indices of BBP are also recommended because they can help reduce the methodological limitations of individual methods. For example, because the Leu method provides carbon production estimates without the need for a carbon-per-cell conversion factor, simultaneous measurements of TdR and Leu provide estimates of both carbon and DNA synthesis. Furthermore, because balanced growth is probably not common in nature, simultaneous measurements of DNA and protein can provide important complementary information regarding the growth state and growth environment of bacteria.

The TdR and Leu procedures have been used mainly as measures of gross BBP and as such represent a "black box" approach. These methods alone do not provide any information on the diversity of microorganisms present or on the proportion of microorganisms responsible for the incorporation of the labeled substrate. New procedures that combine BBP methods with microscopy and molecular methods show that bacterial communities are phylogenetically diverse, that not all cells are active, and that the contribution of active cells to total BBP is not proportional to their relative abundance. The additional information that these methods provide allows us to examine in more detail the relationship between microbial diversity and the cycling of organic matter and to determine how each component of the bacterial community responds to changes in environmental conditions. Instead of examining the response of the entire community, the contribution of specific groups can be determined. Given the potential of these approaches to illuminating the microbial "black box," more efforts should be concentrated on refining these combined methods and on testing them in situ.

REFERENCES

1. **Alldredge, A. L.** 1993. Production of heterotrophic bacteria inhabiting marine snow, p. 531–536. *In* P. F. Kemp, B. F. Sherr, E. B. Sherr, and J. J.Cole (ed.), *Handbook of Methods in Aquatic Microbial Ecology*. Lewis Publishers, Boca Raton, Fla.
2. **Ammerman, J. W., J. A. Fuhrman, Å. Hagström, and F. Azam.** 1984. Bacterioplankton growth in seawater. I. Growth kinetics and cellular characteristics in seawater cultures. *Mar. Ecol. Prog. Ser.* **18:**31–39.
3. **Amon, R. M. W., and R. Benner.** 1996. Bacterial utilization of different size classes of dissolved organic matter. *Limnol. Oceanogr.* **4:**41–51.
4. **Azam, F., T. Fenchel, J. G. Field, J. S. Gray, L. A. Meyer-Reil, and T. F. Thingstad.** 1983. The ecological role of water-column microbes in the sea. *Mar. Ecol. Prog. Ser.* **10:**257–263.
5. **Bell, R. T.** 1990. An explanation for the variability in the conversion factor deriving bacterial cell production from incorporation of [^{3}H]-thymidine. *Limnol. Oceanogr.* **35:** 910–915.
6. **Bell, R. T.** 1993. Estimating production of heterotrophic bacterioplankton via incorporation of tritiated thymidine, p. 495–504. *In* P. F. Kemp, B. F. Sherr, E. B. Sherr, and J. J. Cole (ed.), *Handbook of Methods in Aquatic Microbial Ecology*. Lewis Publishers, Boca Raton, Fla.
7. **Bell, R. T., and B. Riemann.** 1989. Adenine incorporation into DNA as a measure of bacterial production in freshwater. *Limnol. Oceanogr.* **34:**435–444.
8. **Benner, R., J. D. Pakulski, M. McCarthy, J. I. Hedges, and P. G. Hatcher.** 1992. Bulk chemical characterization of dissolved organic matter in the ocean. *Science* **255:** 1561–1564.
9. **Bern, L.** 1985. Autoradiographic studies of (methyl-^{3}H)thymidine incorporation in a cyanobacterium (*Microcystis wesenbergii*)-bacterium association and in selected algae and bacteria. *Appl. Environ. Microbiol.* **49:**232–233.
10. **Buesing, N., and M. O. Gessner.** 2003. Incorporation of radiolabeled leucine into protein to estimate bacterial production in plant litter, sediment, epiphytic biofilms, and water samples. *Microb. Ecol.* **45:**291–301.
11. **Button, D. K., and B. R. Robertson.** 1993. Use of high-resolution flow cytometry to determine the activity and distribution of aquatic bacteria, p. 163–173. *In* P. F. Kemp, B. F. Sherr, E. B. Sherr, and J. J.Cole (ed.), *Handbook of Methods in Aquatic Microbial Ecology*. Lewis Publishers, Boca Raton, Fla.
12. **Carman, K. R.** 1993. Microautoradiographic detection of microbial activity, p. 397–404. *In* P. F. Kemp, B. F. Sherr, E. B. Sherr, and J. J. Cole (ed.), *Handbook of Methods in Aquatic Microbial Ecology*. Lewis Publishers, Boca Raton, Fla.
13. **Chin-Leo, G., and R. Benner.** 1992. Enhanced bacterioplankton production and respiration at intermediate salinities in the Mississippi River plume. *Mar. Ecol. Prog. Ser.* **87:**87–103.
14. **Chin-Leo, G., and D. L. Kirchman.** 1988. Estimating bacterial production in marine waters from the simultaneous incorporation of thymidine and leucine. *Appl. Environ. Microbiol.* **54:**1934–1939.
15. **Chin-Leo, G., and D. L. Kirchman.** 1990. Unbalanced growth in natural assemblages of marine bacterioplankton. *Mar. Ecol. Prog. Ser.* **63:**1–8.
16. **Cho, B. C., and F. Azam.** 1988. Heterotrophic bacterioplankton production measurement by the tritiated thymidine incorporation method. *Arch. Hydrobiol. Beih. Ergebn. Limnol.* **31:**153–162.
17. **Christian, R. R., R. B. Hanson, and S. Y. Newell.** 1982. Comparison of methods for measurement of bacterial growth rates in mixed batch cultures. *Appl. Environ. Microbiol.* **43:**1160–1165.
18. **Cole, J. J., S. Findlay, and M. L. Pace.** 1988. Bacterial production in fresh and saltwater ecosystems: a cross-system overview. *Mar. Ecol. Prog. Ser.* **43:**1–10.
19. **Cole, J. J., and M. L. Pace.** 1995. Why measure bacterial production? A reply to the comment by Jahnke and Craven. *Limnol. Oceanogr.* **40:**441–444.
20. **Cottrell, M. T., and D. L. Kirchman.** 2000. Natural assemblages of marine proteobacteria and members of the *Cytophaga-Flavobacter* cluster consuming low- and high-molecular-weight dissolved organic matter. *Appl. Environ. Microbiol.* **66:**1692–1697.
21. **Cottrell, M. T., and D. L. Kirchman.** 2003. Contribution of major bacterial groups to bacterial biomass production (thymidine and leucine incorporation) in the Delaware estuary. *Limnol. Oceanogr.* **48:**168–178.
22. **Cottrell, M. T., and D. L. Kirchman.** 2004. Single-cell analysis of bacterial growth, cell size, and community

structure in the Delaware Estuary. *Aquat. Microb. Ecol.* **34:**139–149.
23. **Coveney, M. F., and R. G. Wetzel.** 1988. Experimental evaluation of conversion factors for the (^{3}H)thymidine incorporation assay for bacterial secondary productivity. *Appl. Environ. Microbiol.* **54:**160–168.
24. **Craven, D. B., and D. M Karl.** 1984. Microbial RNA and DNA synthesis in marine sediments. *Mar. Biol.* **83:**129.
25. **del Giorgio, P. A., J. J. Cole, and A. Cimbleris.** 1997. Respiration rates in bacteria exceed phytoplankton production in unproductive aquatic systems. *Nature* **385:**148–151.
26. **del Giorgio, P. A., and J. J. Cole.** 1998. Bacterial growth efficiency in natural aquatic systems. *Annu. Rev. Ecol. Syst.* **29:**503–541.
27. **del Giorgio, P. A., and J. J. Cole.** 2000. Bacterial energetics and growth efficiency, p. 289–325. *In* D. Kirchman (ed.), *Microbial Ecology of the Oceans*. Wiley-Liss, Inc., New York, N.Y.
28. **Douglas, J. D., J. A. Novitsky, and R. O. Fournier.** 1987. Microautoradiography-based enumeration of bacteria with estimates of thymidine-specific growth and production rates. *Mar. Ecol. Prog. Ser.* **36:**91–99.
29. **Ducklow, H. W.** 2000. Bacterial production and biomass in the oceans, p. 85–120. *In* D. Kirchman (ed.), *Microbial Ecology of the Oceans*. Wiley-Liss, Inc., New York, N.Y.
30. **Ducklow, H. W., and C. A. Carlson.** 1992. Oceanic bacterial production. *Adv. Microb. Ecol.* **12:**113–181.
31. **Ducklow, H. W., and S. M. Hill.** 1985. Tritiated thymidine incorporation and the growth of heterotrophic bacteria in warm core rings. *Limnol. Oceanogr.* **30:**260–272.
32. **Ducklow, H. W., and D. L. Kirchman.** 1983. Bacterial dynamics and distribution during a spring diatom bloom in the Hudson River Plume. *J. Plankton Res.* **5:**333–355.
33. **Ducklow, H. W., D. L. Kirchman, and H. L. Quinby.** 1992. Determination of bacterioplankton growth rates during the North Atlantic spring phytoplankton bloom. *Microb. Ecol.* **24:**125–144.
34. **Ellenbroek, F. M., and T. E. Cappenberg.** 1991. DNA synthesis and tritiated thymidine incorporation by heterotrophic freshwater bacteria in continuous culture. *Appl. Environ. Microbiol.* **57:**1675–1682.
35. **Evans, C. T., B. Van Mooy, R. G. Keil, C. Greengrove, and G. Chin-Leo.** 2004. *Isolation of DNA from Actively Growing Heterotrophic Bacteria using 5-Bromo-2′-Deoxyuridine*. American Society of Limnology and Oceanography/The Oceanography Society, Honolulu, Hawaii.
36. **Fallon, R. D., and S. Y. Newell.** 1986. Thymidine incorporation by the microbial community of standing-dead *Spartina alterniflora*. *Appl. Environ. Microbiol.* **52:**1206–1208.
37. **Ferguson, R. L., E. N. Buckley, and A. V. Palumbo.** 1984. Response of marine bacterioplankton to differential filtration and confinement. *Appl. Environ. Microbiol.* **47:** 49–55.
38. **Findlay, S.** 1993. Thymidine incorporation into DNA as an estimate of sediment bacterial production, p. 505–508. *In* P. F. Kemp, B. F. Sherr, E. B. Sherr, and J. J. Cole (ed.), *Handbook of Methods in Aquatic Microbial Ecology*. Lewis Publishers, Boca Raton, Fla.
39. **Fischer, H., and M. Pusch.** 1999. Use of the [^{14}C]leucine incorporation technique to measure bacterial production in river sediments and the epiphyton. *Appl. Environ. Microbiol.* **65:**4411–4418.
40. **Fuhrman, J. A., and F. Azam.** 1982. Thymidine incorporation as a measure of heterotrophic bacterioplankton production in marine surface waters: evaluation and field results. *Mar. Biol.* **66:**109–120.
41. **Gilmour, C. C., M. E. Leavitt, and M. P. Shiaris.** 1990. Evidence against incorporating of exogenous thymidine by sulfate-reducing bacteria. *Limnol. Oceanogr.* **35:**1401–1409.
42. **Grossart, H., and H. Ploug.** 2000. Bacterial production and growth efficiencies: direct measurements on riverine aggregates. *Limnol. Oceanogr.* **45:**436–445.
43. **Hagström, Å.** 1984. Aquatic bacteria: measurements and significance of growth, p. 495–501. *In* M. J. Klug and C. A. Reddy (ed.), *Current Perspectives In Microbial Ecology*. American Society for Microbiology, Washington, D.C.
44. **Hagström, Å, U. Larsson, P. Hörstedt, and S. Normark.** 1979. Frequency of dividing cells, a new approach to the determination of bacterial growth rates in aquatic environments. *Appl. Environ. Microbiol.* **37:**805–812.
45. **Hamasaki, K., R. A. Long, and F. Azam.** 2004. Individual cell growth rates of marine bacteria, measured by bromodeoxyuridine incorporation. *Aquat. Microb. Ecol.* **35:**217–227.
46. **Hanson, R. B., and H. K. Lowery.** 1983. Nucleic acid synthesis in oceanic microplankton from the Drake Passage, Antarctic; evaluation of steady state growth. *Mar. Biol.* **73:**79–89.
47. **Herndl, G. J., E. Kaltenböck, and G. Müller-Niklas.** 1993. Dialysis bag incubation as a nonradiolabeling technique to estimate bacterioplankton production, p. 553–556. *In* P. F. Kemp, B. F. Sherr, E. B. Sherr, and J. J. Cole (ed.), *Handbook of Methods in Aquatic Microbial Ecology*. Lewis Publishers, Boca Raton, Fla.
48. **Hietanen, S., L. Tuominen, and J. Kuparinen.** 1999. Benthic bacterial production in the northern Baltic Sea measured using a modified [14C]leucine incorporation method. *Aquat. Microb. Ecol.* **20:**13–20.
49. **Hietanen, S., J. Kuparinen, R. J. Oja, and L. Tuominen.** 2001. Different filtration treatments and centrifugation in measuring bacterial production in brackish waters. *Boreal Environ. Res.* **6:**221–229.
50. **Hietanen, S., J. M. Lehtimaeki, L. Tuominen, K. Sivonen, and J. Kuparinen.** 2002. Nodularia spp. (Cyanobacteria) incorporate leucine but not thymidine: importance for bacterial-production measurements. *Aquat. Microb. Ecol.* **28:**99–104.
51. **Hobbie, J. E.** 1988. A comparison of the ecology of planktonic bacteria in fresh and salt water. *Limnol. Oceanogr.* **33:**750–764.
52. **Hobbie, J. E., R. J. Daley, and S. Jasper.** 1977. Use of Nuclepore filters for counting bacteria by fluorescence microscopy. *Appl. Environ. Microbiol.* **33:**1225–1228.
53. **Hollibaugh, J. T.** 1988. Limitations of the [^{3}H]thymidine method for estimating bacterial productivity due to thymidine metabolism. *Mar. Ecol. Prog. Ser.* **43:**19–30.
54. **Hollibaugh, J. T.** 1994. Relationships between thymidine metabolism, bacterioplankton community metabolic capabilities, and sources of organic matter. *Microb. Ecol.* **28:** 117–131.
55. **Hoppe, H.-G., K. Gocke, R. Koppe, and C. Begler.** 2002. Bacterial growth and primary production along a north-south transect of the Atlantic Ocean. *Nature* **416:**168–171.
56. **Ingraham, J. L., O. Maaloe, and F. C. Neidhardt.** 1983. *Growth of the Bacterial Cell*. Sinauer Associates, Sunderland, Mass.
57. **Jahnke, R. A., and D. B. Craven.** 1995. Quantifying the role of heterotrophic bacteria in the carbon cycle: a need for respiration rate measurements. *Limnol. Oceanogr.* **40:** 436–441.
58. **Johnstone, B., and R. P. Jones.** 1989. A study of the lack of (methyl-^{3}H)thymidine uptake and incorporation by chemolitotrophic bacteria. *Microb. Ecol.* **18:**73–77.
59. **Jørgensen, N. O. G.** 1992. Incorporation of [^{3}H]leucine and [^{3}H]valine into protein of freshwater bacteria: uptake

kinetics and intracellular isotope dilution. *Appl. Environ. Microbiol.* **58:**3638–3646.

60. **Kamjunke, N., and S. Jähnichen.** 2000. Leucine incorporation by *Mycrocystis aeruginosa. Limnol. Oceanogr.* **45:** 741–743.
61. **Karl, D. M.** 1981. Simultaneous rates of ribonucleic acid and deoxyribonucleic acid syntheses for estimating growth and cell division of aquatic microbial communities. *Appl. Environ. Microbiol.* **42:**802–810.
62. **Karl, D. M.** 1982. Selected nucleic acid precursors in studies of aquatic microbial ecology. *Appl. Environ. Microbiol.* **44:**891–902.
63. **Karl, D. M.** 1993. Microbial RNA and DNA synthesis derived from the assimilation of [2,^{3}H]-adenine, p. 471–482. *In* P. F. Kemp, B. F. Sherr, E. B. Sherr, and J. J. Cole (ed.), *Handbook of Methods in Aquatic Microbial Ecology*. Lewis Publishers, Boca Raton, Fla.
64. **Karl, D. M.** 1994. Accurate estimation of the microbial loop processes and rates. *Microb. Ecol.* **28:**147–150.
65. **Karl, D. M., and C. D. Winn.** 1984. Adenine metabolism and nucleic acid synthesis: applications to microbiological oceanography, p. 197–216. *In* J. Hobbie and P. Williams (ed.), *Heterotrophic Activity in the Sea*. Plenum Press, New York, N.Y.
66. **Kemp, P. F., B. F. Sherr, E. B. Sherr, and J. J. Cole (ed.).** 1993. *Handbook of Methods in Aquatic Microbial Ecology*. Lewis Publishers, Boca Raton, Fla.
67. **Kirchman, D. L.** 1993. Leucine incorporation as a measure of biomass production by heterotrophic bacteria, p. 509–512. *In* P. F. Kemp, B. F. Sherr, E. B. Sherr, and J. J. Cole (ed.), *Handbook of Methods in Aquatic Microbial Ecology*. Lewis Publishers, Boca Raton, Fla.
68. **Kirchman, D. L., and H. W. Ducklow.** 1993. Estimating conversion factors for thymidine and leucine methods for measuring bacterial production, p. 513–517. *In* P. F. Kemp, B. F. Sherr, E. B. Sherr, and J. J. Cole (ed.), *Handbook of Methods in Aquatic Microbial Ecology*. Lewis Publishers, Boca Raton, Fla.
69. **Kirchman, D. L., and M. P. Hoch.** 1988. Bacterial production in the Delaware Bay estuary estimated from thymidine and leucine incorporation rates. *Mar. Ecol. Prog. Ser.* **32:**169–178.
70. **Kirchman, D. L., E. K'nees, and R. E. Hodson.** 1985. Leucine incorporation and its potential as a measure of protein synthesis by bacteria in natural aquatic systems. *Appl. Environ. Microbiol.* **49:**599–607.
71. **Kirchman, D. L., S. Y. Newell, and R. E. Hodson.** 1989. Incorporation versus biosynthesis of leucine: implications for measuring rates of protein synthesis and biomass production by bacteria in marine systems. *Mar. Ecol. Prog. Ser.* **32:**47–49.
72. **Kirschner, A. K. T., and B. Velimirov.** 1999. Benthic bacterial secondary production via simultaneous ^{3}H-thymidine and ^{14}C-leucine incorporation, and its implication for the carbon cycles of a shallow macrophyte-dominated backwater system. *Limnol. Oceanogr.* **44:**1871–1881.
73. **Kirschner, A. K. T., and B. Velimirov.** 1999. Leucine incorporation by ^{3}H-leucine centrifugation method for determining bacterial protein synthesis in freshwater samples. *Aquat. Microb. Ecol.* **17:**201–206.
74. **Kirschner, A. K. T., P. Wihlidal, and B. Velimirov.** 2004. Variability and predictability of the empirical conversion factor for converting 3H-thymidine uptake into bacterial carbon production for a eutrophic lake. *J. Plankton Res.* **26:**1559–1566.
75. **Lee, S., and J. A. Fuhrman.** 1987. Relationships between biovolume and biomass of naturally derived marine bacterioplankton. *Appl. Environ. Microbiol.* **53:**1298–1303.
76. **Lemee, R., E. Rochelle-Newall, F. Van Wambeke, M.-D. Pizay, P. Rinaldi, and J.-P. Gattuso.** 2002. Seasonal variation of bacterial production, respiration and growth efficiency in the open NW Mediterranean Sea. *Aquat. Microb. Ecol.* **29:**227–237.
77. **Malstrom, R. T., M. T. Cottrell, H. Elifantz, and D. L. Kirchman.** 2005. Biomass production and assimilation of dissolved organic matter by Sar11 bacteria in the northwest Atlantic Ocean. *Appl. Environ. Microbiol.* **71:**2979–2986.
78. **McDonough, R. J., R. W. Sanders, K. G. Porter, and D. L. Kirchman.** 1986. Depth distribution of bacterial production in a stratified lake with an anoxic hypolimnion. *Appl. Environ. Microbiol.* **52:**992–1000.
79. **Moriarty, D. J. W.** 1986. Measurement of bacterial growth rates in aquatic systems from rates of nucleic acid synthesis. *Adv. Microb. Ecol.* **9:**243–292.
80. **Moriarty, D. J. W.** 1988. Accurate conversion factors for calculating bacterial growth rates from thymidine incorporation into DNA: elusive or illusive? *Arch. Hydrobiol. Beih. Ergebn. Limnol.* **31:**211–217.
81. **Nagata, T., and Y. Watanabe.** 1990. Carbon and nitrogen-to-volume ratios of bacterioplankton grown under different nutritional conditions. *Appl. Environ. Microbiol.* **56:**99–109.
82. **Newell, S. Y., R. D. Fallon, and P. S. Tabor.** 1986. Direct microscopy of natural assemblages, p. 1–48. *In* J. S. Poindexter and E. R. Leadbetter (ed.), *Bacteria in Nature*, vol. 2. Plenum, New York, N.Y.
83. **Nielsen, J. L., D. Christensen, M. Kloppenborg, and P. H. Nielsen.** 2003. Quantification of cell-specific substrate uptake by probe-defined bacteria under in situ conditions by microautoradiography and fluorescence in situ hybridization. *Environ. Microbiol.* **5:**202–211.
84. **Ouverney, C. C., and J. A. Fuhrman.** 1999. Combined microautoradiography-16S rRNA probe technique for determination of radioisotope uptake by specific microbial cell types in situ. *Appl. Environ. Microbiol.* **65:**1746–1752.
85. **Pace, M. L., P. Del Giorgio, D. Fischer, R. Condon, and M. Heather.** 2004. Estimates of bacterial production using the leucine incorporation method are influenced by differences in protein retention of microcentrifuge tubes. *Limnol. Oceanogr. Methods* **2:**55–61.
86. **Pedrós-Alió, C., and S. Y. Newell.** 1989. Microautoradiographic study of thymidine uptake in brackish waters around Sapelo Island, Georgia, USA. *Mar. Ecol. Prog. Ser.* **55:**83–94.
87. **Pernthaler, A., J. Pernthaler, M. Schattenhofer, and R. Amann.** 2002. Identification of DNA-synthesizing bacterial cells in coastal North Sea plankton. *Appl. Environ. Microbiol.* **68:**5728–5736.
88. **Pomeroy, L. R.** 1974. The ocean's food web: a changing paradigm. *BioScience* **9:**499–504.
89. **Porter, K. G., and Y. S. Feig.** 1980. The use of DAPI for identifying and counting aquatic microflora. *Limnol. Oceanogr.* **25:**943–948.
90. **Proctor, L. M., and J. A. Fuhrman.** 1990. Viral mortality of marine bacteria and cyanobacteria. *Nature* **343:**60–62.
91. **Riemann, B., and R. Bell.** 1990. Advances in estimating bacterial biomass and growth in aquatic systems. *Arch. Hydrobiol.* **118:**385–402.
92. **Riemann, B., R. T. Bell, and N. O. G. Jørgensen.** 1990. Incorporation of thymidine, adenine and leucine into natural bacterial assemblages. *Mar. Ecol. Prog. Ser.* **65:**87–94.
93. **Robarts, R. D., and T. Zohary.** 1993. Fact or fiction—bacterial growth rates and production rates as determined by [methyl-^{3}H]thymidine? *Adv. Microb. Ecol.* **13:**371–425.
94. **Rivkin, R. B., and L. Legendre.** 2001. Biogenic carbon cycling in the upper ocean: effects of microbial respiration. *Science* **291:**2398–2400.

95. **Servais, P.** 1995. Measurement of the incorporation rates of four amino acids into proteins for estimating bacterial production. *Microb. Ecol.* **29:**115–128.
96. **Servais, P., and G. Billen.** 1991. Bacterial production measured by ^{3}H-thymidine and ^{3}H-leucine incorporation in various aquatic ecosystems. *Arch. Hydrobiol. Beih. Ergebn. Limnol.* **37:**73–81.
97. **Sherr, E. B., and B. F. Sherr.** 1987. High rates of consumption of bacteria by pelagic ciliates. *Nature* **325:**710–711.
98. **Sherr, E. B., B. F. Sherr, and C. T. Sigmon.** 1999. Activity of marine bacteria under incubated and "in situ" conditions. *Aquat. Microb. Ecol.* **20:**213–223.
99. **Shiah, F., and H. W. Ducklow.** 1997. Bacterioplankton growth responses to temperature and chlorophyll variations in estuaries measured by thymidine:leucine incorporation ratio. *Aquat. Microb. Ecol.* **13:**151–159.
100. **Simon, M., and F. Azam.** 1989. Protein content and protein synthesis rates of planktonic marine bacteria. *Mar. Ecol. Prog. Ser.* **51:**201–213.
101. **Smith, D. C., and F. Azam.** 1992. A simple, economical method for measuring bacterial protein synthesis rates in seawater using [^{3}H]leucine. *Mar. Microb. Food Webs* **6:** 107–114.
102. **Smith, E. M., and P. A. Del Giorgio.** 2003. Low fractions of active bacteria in natural aquatic communities? *Aquat. Microbiol. Ecol.* **31:**203–208.
103. **Steward, G. F., and F. Azam.** 1999. Bromodeoxyuridine as an alternative to 3H-thymidine for measuring bacterial productivity in aquatic samples. *Aquat. Microb. Ecol.* **19:** 57–66.
104. **Suberkropp, K., and H. Weyers.** 1996. Application of fungal and bacterial production methodologies to decomposing leaves in streams. *Appl. Environ. Microbiol.* **62:** 1610–1615.
105. **Tabor, P. S., and R. A. Neihof.** 1982. Improved microautoradiographic methods to determine individual microorganisms active in substrate uptake in natural waters. *Appl. Environ. Microbiol.* **44:**945.
106. **Taylor, G. T., and M. L. Pace.** 1987. Validity of eukaryotic inhibitors for assessing production and grazing mortality of marine bacterioplankton. *Appl. Environ. Microbiol.* **53:**119–128.
107. **Thomaz, S. M., and R. G. Wetzel.** 1995. [^{3}H]Leucine incorporation methodology to estimate epiphytic bacterial biomass production. *Microb. Ecol.* **29:**63–70.
108. **Tibbles, B. J.** 1996. Effects of temperature on the incorporation of leucine and thymidine by bacterioplankton and bacterial isolates. *Aquat. Microb. Ecol.* **11:**239–250.
109. **Tibbles, B. J., C. L. Davis, J. M. Harris, and M. I. Lucas.** 1992. Estimates of bacterial productivity in marine sediments and water from a temperate saltmarsh lagoon. *Microb. Ecol.* **23:**195–209.
110. **Toolan, T.** 2001. Coulometric carbon-based respiration rates and estimates of bacterioplankton growth efficiencies in Massachusetts Bay. *Limnol. Oceanogr.* **46:**1298–1308.
111. **Torréton, J. P., and M. Bouvy.** 1991. Estimating bacterial DNA synthesis from [^{3}H]thymidine incorporation: discrepancies among macromolecular extraction procedures. *Limnol. Oceanogr.* **36:**299–306.
112. **Tuominen, L.** 1995. Comparison of leucine uptake methods and a thymidine incorporation method for measuring bacterial activity in sediment. *J. Microbiol. Methods* **24:**125–134.
113. **Turley, C. M., and E. D. Stutt.** 2000. Depth-related cell-specific bacterial leucine incorporation rates on particles and its biogeochemical significance in the Northwest Mediterranean. *Limnol. Oceanogr.* **45:**419–425.
114. **Urbach, E., K. L. Vergin, and S. J. Giovannoni.** 1999. Immunochemical detection and isolation of DNA from metabolically active bacteria. *Appl. Environ. Microbiol.* **65:**1207–1213.
115. **van Looij, A., and B. Riemann.** 1993. Measurement of bacterial production in coastal marine environments using leucine: applications of a kinetic approach to correct for isotope dilution. *Mar. Ecol. Prog. Ser.* **102:**97–104.
116. **Van Mooy, B. A. S., A. H. Devol, and R. G. Keil.** 2004. Quantifying ^{3}H-thymidine incorporation rates by a phylogenetically defined group of marine planktonic bacteria (*Bacteriodetes* phylum). *Environ. Microbiol.* **6:**1061–1069.
117. **White, P. A., J. Kalff, J. B. Rasmussen, and J. M. Gasol.** 1991. The effect of temperature and algal biomass on bacterial production and specific growth rate in freshwater and marine habitats. *Microb. Ecol.* **21:**99–108.
118. **Wicks, R. J., and R. D. Robarts.** 1987. The extraction and purification of DNA labeled with [methyl-^{3}H]thymidine in aquatic bacterial population studies. *J. Plankton Res.* **9:**1159–1166.
119. **Winding, A.** 1992. ^{3}H-thymidine incorporation to estimate growth rates of anaerobic bacterial strains. *Appl. Environ. Microbiol.* **58:**2660–2662.
120. **Zweifel, U. L., and Å. Hagström.** 1995. Total counts of marine bacteria include a large fraction of non-nucleoid-containing bacteria (ghosts). *Appl. Environ. Microbiol.* **61:**2180–2185.

Community Structure: Bacteria and Archaea

JED A. FUHRMAN

35

Community structure is generally considered to be information related to the types of organisms present in an environment and the relative proportions of those types. The various ways to determine such structure could easily fill a whole book; therefore, this brief chapter will basically provide an outline of the various approaches that are used to determine community structure and some guidelines regarding which methods may be most appropriate for specific aquatic environments and specific scientific questions. Also, it must be realized that this topic is undergoing massive revision as new molecular biological and other highly technical approaches are being brought to the fore. Therefore, I will describe some of these new methods in detail, even though they are still under development, because they are likely to become particularly important in the future.

Community structure analysis might be performed for a number of purposes, and the most appropriate techniques to use will vary with the goals of the study. Certain techniques may allow study of subsets of the microbial community in great detail but at the same time entirely miss other community members. For example, one might be interested in the portion of the community that is performing a certain function, such as photosynthesis, and it might be studied by analyzing extracted photosynthetic pigments (86) or possibly by flow cytometric analysis of pigment contents of individual cells (14) or even of genes for pigment synthesis (6, 83). Several of the techniques described below rely on the ability to identify particular preselected components of the microbial community. However, in addition to learning about known types of microorganisms, those who study the general community structure can also potentially find previously unknown microbial types, and some of them may be the most interesting. As might be expected, when one finds novel organisms, it takes considerable time to characterize such organisms.

MICROSCOPY

The oldest method for obtaining information on microbial community structure is to examine the sample with a microscope and characterize the microbes by their morphology. With some aquatic environments, one can in fact identify many types of organisms this way. An example is microbial mats: numerous types have morphologies that are readily recognizable, particularly by electron microscopy (86). Many colonial microorganisms have distinctive colony morphologies that can be recognized by standard light microscopy or even with the naked eye. However, this approach is risky because different species, due to convergent evolution, may have similar morphotypes and may thus be misidentified. More to the point, in most aquatic environments, the vast majority of bacteria have nondistinct and/or variable morphologies, so this method is inappropriate for them (84). Related to identification by morphology is the identification of certain kinds of organisms by their natural fluorescence properties, as is done with epifluorescence microscopy. For example, methanogens contain the F420 cofactor, which fluoresces green when excited with light at a 420-nm wavelength. More commonly, such identification may be based on photosynthetic pigments (often in conjunction with general size and morphology information) that can yield distinctive colors of fluorescence. For example, marine *Synechococcus* cells containing phycoerythrin have a distinctive golden color fluorescence when excited with blue light and a bright orange-red fluorescence when excited by green light. These pigment "signatures" can be more precisely determined by flow cytometry, which provides quantitative fluorescence (and other) information on many thousands of individual cells in a short time. This approach in fact was used to discover the presence in the ocean of large quantities of tiny marine prochlorophytes that had evaded detection by standard visual epifluorescence microscopy techniques (14).

CULTURING

The most common traditional method of determining community structure involves culturing the organisms from the habitat in question and identifying the cultures by standard techniques (see chapter 6). If the organisms of interest are culturable, then this approach may be suitable. It goes without saying that the culturable organisms must be viable in order to be detected, and such viable counts are often thought to avoid the problem of counting inactive organisms that may be of less interest. However, there are several important caveats. First, even for culturable organisms, it has been suggested that many individuals may be viable but still not culturable (79). A second and bigger concern is that culture conditions that have typically been used to perform such counts usually recover on the order of 1% or less of the total number of organisms in aquatic habitats (25, 47,

54), even though the majority of organisms in such habitats can be shown to be metabolically and synthetically active (27). The reasons for this inability to culture the organisms are somewhat speculative and are generally thought to be the result of culture conditions that do not adequately mimic the natural growth conditions. Possibilities include wrong substrates, trace metal sensitivity, excessive storage product formation (leading to cell damage), and viral infection, among others. Recent attempts to use dilution cultures (i.e., dilute a sample with filtered seawater such that only a few bacteria are in each tube) have greatly increased the percentage of bacteria that can be cultured (see chapter 32), but for unknown reasons many of such cultures stop growth when abundances reach about 10^5 cells per ml (11). At such abundances, the classical identification tests cannot be readily performed, but it is much better to have a pure culture than not. Linking identification to function is much easier with cultivated organisms, and recent attempts to first use molecular approaches to identify the dominant organisms in an environment and then culture them have been very valuable (15, 76, 85). In any case, many aquatic microbiologists are concerned about accepting the results of culture-based approaches for determining community structure in a comprehensive manner, and one rarely sees such studies in the literature.

IMMUNOLOGICAL APPROACHES

Immunological approaches have been used primarily to characterize and count nitrifying bacteria (100) and cyanobacteria (12). To prepare the antibodies for such a study, one must first culture the organisms and then use the culture to vaccinate an animal. The antibodies are purified from serum and then are labeled with a fluorescent tag, such as fluorescein. The antibodies are mixed with a sample, unattached antibodies are rinsed away, and the cells are then observed by epifluorescence microscopy. Of course, this method presupposes that one must have cultures of the organisms in question, and so this approach would miss organisms that are resistant to cultivation. Also, one must know something about the cross-reactivity of the antibodies to other organisms. Often, antibodies can be specific for a particular species or strain. This approach has the potential to identify nonculturable members (viable or not) of a particular serotype, which is functionally defined as the type(s) to which the antibody binds. It has also been found to be useful for organisms that might be cultured but grow slowly, such as nitrifiers.

LIPID ANALYSIS

Different groups of microorganisms have different types of lipids, and this fact has been used extensively for microbial identification. For environmental work, lipid analysis has been used as an indicator of community structure, with certain classes of lipids (or particular lipids) being used as markers for certain groups. In general, the lipids are extracted in organic solvents and are analyzed by gas chromatography. In aquatic systems, this work has been pioneered by White and colleagues (briefly reviewed in reference 102; see chapter 10). The approach provides information not only on cell type but also about nutritional status or stress. Because most such analytical methods require a substantial number of microbes of each type in each sample, this approach has been used primarily for sediments. However, with the appropriate instrumentation, it is now possible to extend such work to planktonic environments.

A particular application of lipid analysis in recent years has involved archaeal lipids. Archaea had traditionally been thought to be restricted to "extreme" environments, but it was then discovered by genetic techniques that archaea are common in marine plankton systems, especially the deep sea (30). While genetically based fluorescence in situ hybridization (FISH) techniques have been used to observe and count these archaea (22, 32), an alternative approach has been lipid analysis. It permits not only detection of the archaeal lipids themselves as indicators (20, 42), but separation of the lipids permits isotopic analysis that can be used to examine sources of carbon, from natural isotopes (42, 73) or from added tracers (7, 105), permitting differentiation of processes between different groups of organisms.

LOW-MOLECULAR-WEIGHT RNA PROFILES

The molecular size distribution pattern of low-molecular-weight RNA (including tRNAs and 5S rRNA) is thought to be unique within narrow phylogenetic groups of microorganisms. This concept has been applied to natural aquatic planktonic communities by Hofle (43, 44): the whole community RNA profile is examined on an electrophoretic gel and compared with profiles of other communities and of standard cultures. As long as the bands on the gel are well resolved, one can see patterns of similarity and dissimilarity between natural communities. It is also possible to detect patterns that suggest the presence of specific organisms or groups, and bands may be excised and sequenced for partial identification.

NON-SEQUENCE-BASED DNA APPROACHES

DNA-DNA Hybridization

There are situations in which one wishes to know the community structure to learn if two microbial communities contain the same or different organisms, yet the quantitative information on species compositions is not of particular interest. That is, one is asking simply if two communities are the same or different. In such situations, it is possible to perform a DNA-DNA hybridization assay with total DNA extracted from the two communities (52). In this approach, extracted prokaryotic DNA is labeled by nick translation, and samples are compared two at a time (reciprocal hybridizations). The results are scored by expressing the cross-hybrids as a percentage of the self-hybrids on the same filter. This percentage is expected to be the sum total of the shared common fractions between the two filters. For example, if sample 1 has 10% species A, 40% species B, and 50% species C, and sample 2 has 25% species A, 30% species C, and 45% species D, the shared common fraction is 10% + 30% = 40%. Tests with mixtures of pure culture DNAs have shown that the results are usually as expected, with 100% hybridization of identical or nearly identical samples, ranging down to about 5 to 10% hybridization between samples sharing few or no species (the 5 to 10% represents well-conserved DNA sequences that cross-hybridize between distantly related organisms). This approach has been used to see broad differences between ocean basins, between depths at stratified locations, and over seasonal scales at the same location, and it has also shown similarities between some communities over time and space (53).

Other DNA-Based Approaches

Another type of DNA-DNA hybridization is analysis of reassociation rates of single-sample DNA that is melted (strands separated). The rate is related to the complexity (or one may say diversity) of the DNA and thus can be an index of the diversity of the sample. In other words, it indicates the number of species present but not what types they are. Torsvik et al. (92) examined DNA extracted from soil microorganisms in this way, and the very slow reassociation of most of the DNA was interpreted as indicating the presence of about 4,000 completely different genomes in a 30-g sample from a deciduous forest. This was 200 times higher than the diversity from standard plate counts and graphically demonstrates the remarkably high potential overall diversity of microbial communities and its undersampling by conventional means. However, it should be pointed out that other fingerprinting methods have indicated that the large majority of some soil communities is made up of only perhaps a few dozen distinguishable taxa (69; see terminal restriction fragment length polymorphism [T-RFLP] method below).

A method to "fingerprint" a microbial community based upon the quantitative distribution of genomes with different percent G+C contents was also described (45). In this approach, community DNA is centrifuged in a density gradient that separates the DNA based upon percent G+C content, yielding a profile. Although the percent G+C is not an unambiguous identifier for particular groups, the profile of DNA along the percent G+C gradient can strongly indicate the presence of certain groups (prompting probe analysis for verification; see below). Furthermore, profiles from different samples can indicate differences between the microbial communities over time and space.

16S rRNA-BASED APPROACHES

16S rRNA Sequences from Biomass

The difficulty of culturing bacteria and the ordinary requirement that a culture be available to identify a species have presented a major dilemma to microbial ecologists. About 20 years ago, Pace et al. (72) and Olsen et al. (68) presented the elegant idea that cultures are not necessary to identify the organisms present in a natural habitat. The idea came against the backdrop of the increasing use of molecular phylogeny to help define microbial systematics (104). It was becoming clear that the nucleotide sequences in molecules like 16S rRNA are very powerful tools in determining phylogenies and consequently in microbial systematics. Large databases of such sequences have been made available; e.g., the 16S rRNA database currently contains hundreds of thousands of organisms (see http://rdp.cme.msu.edu/). The new idea was to use molecular biological techniques to obtain 16S rRNA sequences directly from the organisms freshly collected from the natural habitat without culturing them. These sequences could then be compared with those in the databases and with each other to learn how they fit into the microbial phylogenetic framework. Even if the sequence is unknown from previous work, it can be placed in relation to known organisms and other sequences from nature. A further major benefit is that the sequences can be used to make probes for quantitative composition analyses of microbial communities. This approach is proving to be very powerful and is being augmented by inclusion of 23S rRNA data as well (see chapter 11). A major aspect of the power of this approach is that the data are in the form of sequences that are universally understood and readily analyzed. Methods have focused on 16S rRNA genes because the molecule is universally distributed, has some extremely conserved segments (especially because it is not protein encoding, so there are no "silent mutations") as well as variable regions, and yields an RNA product that has many copies per cell and can be used directly as a hybridization target. The 16S rRNA sequence analyses can be done in a few ways. If one is interested in learning what types of organisms are present in a sample, it is best to use the approaches that do not restrict the results to certain groups. The initial method proposed by Pace et al. (72) called for extraction of DNA from all the microorganisms, fragmentation of that DNA with a restriction enzyme, and ligation to DNA from a bacteriophage (e.g., lambda) to make a library containing bits of DNA from all of the original organisms in the sample. Today this would be called a "metagenomic" library, because the collective genomic content of the entire assemblage is all there, but that term was not in use in 1986 (and now metagenome research is becoming a "hot" topic; see later section). In Pace et al.'s original analysis, that library is then screened for fragments coding for 16S rRNA by low-stringency hybridization to rRNA from a culture (or from cultures representing very broad groups like the domains *Bacteria, Archaea,* and *Eucarya*). The low stringency, combined with the moderately high level of conservation of this molecule, is expected to allow virtually any 16S rRNA to be detected. Screening with another rRNA, rather than with universally conserved oligonucleotide probes, is preferable because of the large number of false positives expected as a result of random matches between the oligonucleotide and the myriad genes from the natural community (in our laboratory, we found that such false positives were very common). The positive clones are then sequenced, and the sequences are aligned to sequences from a database and analyzed for phylogenetic relationships by a computer. This library approach was used successfully by Schmidt et al. (81) with marine plankton.

Ward et al. (101) took a different approach to cloning 16S rRNA sequences of microorganisms from a well-studied hot spring at Yellowstone National Park. They extracted RNA from the natural sample and performed reverse transcription with universal primers to make cDNA that was then cloned and sequenced. The length of the cloned products was usually a few hundred bases, which is suitable for identification and some general phylogenetic analyses. The results yielded numerous clone types that were not the same as the organisms that had previously been cultured from that hot spring and had been thought to be the dominant types (by microscopical observation as well as culture work). This work has been seen as good evidence that the culture-based approach finds only a subset of the natural diversity and that morphological identification of even distinctive organisms can be deceptive. More recent extension of the idea of working from rRNA directly has been comparison of clone libraries generated by amplification of DNA (by PCR; see below) to those from rRNA (by reverse transcription-PCR) extracted directly from natural samples (61, 62). These practices take note of the point that rRNA tends to reflect growth of cells (48), so the rRNA-based libraries are expected to be biased toward the more active organisms.

PCR-Based Cloning and Sequencing

One major alternative related approach to direct cloning of 16S rRNA genes has been used with great success and has in fact dominated the field: the PCR, first applied by Giovannoni et al. (34). With PCR methods, DNA is extracted from freshly collected organisms, PCR is performed, and with the primers of choice (see below), the PCR products are ligated into a phage or plasmid vector. They are cloned by standard techniques, and the clones are sequenced and analyzed phylogenetically. This approach is particularly suited to planktonic communities because there is usually very little DNA available. For example, with typical bacterial abundances of 10^6 bacteria per ml and typical bacterial DNA content of a few femtograms per cell, there are on the order of a few micrograms of bacterial DNA per liter. The geometric amplification inherent in PCR means that one can begin with 1 ng of total genomic DNA (representing roughly 10^6 bacteria) and end up with micrograms of amplified 16S rRNA genes, sufficient for a variety of subsequent assays.

The existence of regions of the 16S rRNA molecule that are essentially invariant among all known organisms means that universal primers can be used in PCR; the longest distance between such primers is about 860 bases, which is adequate (but not ideal) for most phylogenetic analyses (30, 31). The universal nature of the primers allows the broadest coverage, but it can also be a problem. For example, the high copy number of nucleus-encoded 16S rRNA genes in some eukaryotes may cause eukaryotic PCR products to swamp those of prokaryotes if they are together in the sample. One solution is to try to remove eukaryotes by filtration, as can be done in marine plankton; Fuhrman et al. (30, 31) used glass fiber filtration that removed essentially all of the eukaryotes but only about 10% of the prokaryotes. Alternatively, one can choose more specific primers to target specific groups. An example of this might be to use bacterium-specific primers to avoid amplifying eukaryotic genes when eukaryotes and prokaryotes may not be easily separated; however, one must remember that chloroplasts and mitochondria contain "bacterial" 16S rRNA sequences, so these primers do not completely avoid interference from eukaryotes (in practice, the plastids but not mitochondria tend to amplify with bacteria). On the positive side, there are nearly universal bacterial primers that allow cloning of almost the whole 16S rRNA gene (nearly 1,500 bases), maximizing the available phylogenetic information (2). However, the so-called universal primers should be used with care, in part because they are not exactly universal. Even with some ambiguous bases, such primers have mismatches with certain known groups, and one should check updated databases before embarking on this work. For example, the nearly universal bacterial primer set does not match the *Planctomyces* sequence well, and therefore one may suspect that other unknown groups may be missed. The commonly used primer at *Escherichia coli* position 1492, often treated as if it is universal, is not really so. Primer and probe matches to the current rRNA database can be examined at http://rdp.cme.msu.edu/ (Probe Match program) or with programs like ARB (58; see http://www.arb-home.de). While the PCR annealing conditions can be adjusted to allow priming in the presence of some mismatches, there still may be some rRNA genes that do not amplify well, and this possibility must be considered in interpreting the results.

Among results from PCR-based cloning and sequencing of 16S rRNA genes from marine plankton, Giovannoni et al. (34) used moderately specific PCR primers designed primarily for cyanobacteria yet still found a novel proteobacterial group (SAR11) in addition to a cyanobacterial group. Fuhrman et al. (28, 30, 31) used universal primers with microbial DNA collected from the deep sea and the euphotic zone and found novel groups of both archaea and bacteria. DeLong et al. (19) examined "marine snow" with bacterial primers and found that the clones were distinctly different from those collected from surrounding free-living bacteria. As was found by Schmidt et al. (81), who used a phage library, very few of these marine clones (with the exception of cyanobacteria) were closely related to previously known cultures. The SAR11 group was found at several marine locations and depths. Some of the clones were so distant that they are likely to be considered phyla separate from those previously studied. DeLong (17, 23) used archaeal primers and found two distinct archaeal groups in coastal temperate and polar waters; they were in the same groups as found by Fuhrman et al. (30, 31) at different locations and with universal primers. When specific primers are used, even a relatively minor component can be amplified, detected, and studied. As an example, ammonium-oxidizing bacterial 16S rRNA genes were specifically amplified from plankton by Voytek and Ward (98).

The PCR methods have been used with samples from sediments and microbial mats (5, 24, 65, 77), as well as deep-sea holothurian guts (60), often showing unexpected bacterial and archaeal diversity. Although such material is far more concentrated than plankton (more organisms per unit volume), making it easier to obtain enough material for analysis, there are many substances that can interfere with molecular analyses, so the DNA may need extensive purification. Also, such organisms may be difficult to extract. A further consideration with at least some of such samples is that they should be frozen or extracted immediately upon sampling; this is because it has been found that storage of sediments (especially if the sediments were initially anaerobic but stored in aerobic conditions) can lead to significant and rapid shifts in species compositions (78). Such rapid potential changes seem most likely in rich material with rapid potential growth.

FRAGMENT-BASED ANALYSES

Some studies have used information about the 16S rRNA clones short of partial or full sequences. These include RFLP analysis (8, 19, 64), which can be useful in grouping clones together. However, such analysis is based on only a few base positions in the sequence, and so it can lack resolution between closely related groups, particularly when universal primers are used for the initial amplification. On the other hand, it is possible to do the PCR with more specific primers, followed by restriction analysis, to indicate rather rapidly the presence of particular groups or types of interest and to compare different samples with respect to these groups. One could envision some fairly specific analyses by this approach with judicious selection of primers and restriction enzymes. Even with universal primers, the results can have resolution adequate for many types of studies.

Assemblage Fingerprinting

A related and powerful application has been the use of fingerprinting methods that give a snapshot of the entire microbial community at once, with the ability to tentatively identify different components. These methods include T-RFLP (4, 56, 69), length heterogeneity PCR (LHPCR [87]), and amplified ribosomal intergenic spacer analysis (ARISA [10, 26]). In T-RFLP, PCR is performed with one fluorescent and one nonfluorescent primer, and the products are cut with a restriction enzyme and analyzed for the size of the fluorescent end products. The presence or absence of restriction sites, as well as positions and lengths of insertions and deletions, leads to different-length products. The result looks like a chromatogram, with peaks representing different taxa. In LHPCR, the length variations of the entire PCR product (due to insertions and deletions) permit separations. In ARISA, it is variations in the spacer length between the 16S and 23S rRNA genes. In these methods, it is possible for multiple taxa to have the same length of detected product, so "identification" of a particular peak on the basis of database information is not definitive. Nevertheless, clone libraries from the environment in question can be used to find the most likely identification of peaks in the fingerprints (10, 37). In general, ARISA has the highest potential phylogenetic resolution because this spacer varies greatly and is far less conserved than the rRNA gene itself; T-RFLP has the potential for greater phylogenetic resolution than does LHPCR, as the presence or absence of restriction sites, plus the ability to choose different enzymes, adds variety. An important feature of the fingerprinting methods described above (T-RFLP, LHPCR, and ARISA) is that the results are in the form of data on the number of different PCR products of particular fragment lengths. In other words, these methods result in data that are discrete numbers such as the proportion of total fluorescence in a given fragment length. Such discrete data can be put into a database in tabular form and compared statistically, permitting exact comparisons between different gel runs and even different laboratories. This advantage allows for statistically rigorous examination of sampling strategies and the data themselves, within and between studies. It is not usually true of data from the fingerprinting methods described in the next paragraph, where the identity of an organism relates to the position of a band on a gel that can vary between different runs and particularly between different labs.

A different type of fingerprint analysis is denaturing gradient gel electrophoresis (DGGE) or thermal gradient gel electrophoresis, which are ways to separate similar-length nucleic acid molecules on the basis of small differences in the sequences. Muyzer and colleagues (66, 67) have described the use of gradient gel electrophoresis analysis of PCR products to separate different components of the PCR mixtures from natural samples. Typically, extracted DNA from an environmental sample is amplified with bacterial primers, and the products are separated by DGGE. One can get an idea of the broad diversity (number of different bacterial types) of a sample by examining the number of different bands in such an analysis. It is also possible to use probes to characterize individual bands, or the bands may be excised, cloned, and sequenced (or sometimes directly sequenced) for a detailed phylogenetic analysis. This latter ability to cut out the bands for characterization is an advantage of DGGE over the current versions of T-RFLP, LHPCR, or ARISA in that the identity of individual components can be verified. However, the T-RFLP, LHPCR, and ARISA methods yield specific size information on the products, suitable for database analysis and comparison to known sequences (for presumptive identification), whereas DGGE banding patterns are much harder to standardize. Also, in a comparison with marine communities, T-RFLP was found to detect more taxa than was DGGE (63), although this may be specific to the particular conditions of the study and the gels or fragment analyzers used. The resolution of methods that rely on examining a gel directly (like DGGE) is typically not as fine as the resolution of typical fragment analyzers used for T-RFLP or ARISA; in other words, one can typically distinguish up to about 100 different band locations on a DGGE gel, but there are several hundred distinct fragment lengths detectable by methods like ARISA. Nevertheless, when comparing fragment analysis results from different samples, gels, and labs, it is important to consider the precision of fragment size calling so that comparisons are statistically sound (39). Finally, it should be noted that fragment analysis-based fingerprinting has been reported to detect very minor components of the community, with taxa whose individuals represent about 0.1% of the community being detectable above background (40).

Choosing Cloning versus Fingerprints

Cloning and sequencing provide a great deal of detailed phylogenetic information, permitting placement of the organisms on a tree. Fingerprinting, even with matching clones to help identify individual taxa, provides a presumptive identification at best, and because there are far fewer characters that determine the identity, there is necessarily far less resolution and often some ambiguity. On the other hand, fingerprinting has the potential to show the entire bacterial assemblage in a single rapid assay, including minor components that would take hundreds or even thousands of clones to find. So it is a trade-off among resolution, cost, time for analysis, and coverage. Coverage of clone libraries can be estimated with programs like DOTUR (http://www.plantpath.wisc.edu/fac/joh/DOTUR.html) or EstimateS (http://viceroy.eeb.uconn.edu/EstimateS). Fingerprints are also more readily compared statistically, due to their relative simplicity. Therefore, in studies where compositions of many samples are to be compared, fingerprinting may have several advantages, but if one wishes an exhaustive analysis of a particular sample or a few samples with high resolution, cloning and sequencing hundreds or thousands of clones are preferred.

A non-16S rRNA method that deserves mention here is PCR from repetitive sequences in DNA to yield electrophoretic banding patterns (genomic fingerprints) that can be distinct for particular groups or strains (97). Such methods may be applicable in the future for characterizing community structure, particularly for simple communities with little complexity.

POTENTIAL BIASES

There is still a question about possible biases and errors in the molecular methods: have we replaced culturing biases with unknown biases? Potential biases could arise at a few stages. In the extraction stage, one can check microscopically to see that substantially all the cells have lysed, which has been done with some of the extraction techniques with aquatic samples (31); however, it is often not done, and this step deserves close attention. The PCR may introduce biases due to variations in primer binding or extension efficiency. There are also possible biases in the cloning step, as it is known that some sequences clone more readily than others (and some perhaps not at all). These possible problems indicate that caution is still in order when one is interpreting data on the relative amounts of different clones in libraries, and one cannot assume that clones are found in proportion to their natural abundance (but see below). There is also serious concern about possible chimeras or heteroduplexes being formed during PCR amplification (1, 3), and one needs to check clone libraries for the chimeras at least with programs like Chimera Check at rdp (http://rdp.cme.msu.edu/), bellerophon, (http://foo.maths.uq.edu.au/~huber/bellerophon.pl), mglobalCHI (http://www.hto.usc.edu/software/mglobalCHI/index.html), Pintail, or Mallard (http://www.cf.ac.uk/biosi/research/biosoft/index.html) (3, 46, 50) and also consider measures to reduce chances of heteroduplex formation (1). However, even if there are biases, they are probably quite unrelated to culturing biases, and so these approaches are yielding much new information on what organisms are present. Can we yet say if cultures are representative of "typical" aquatic prokaryotes? Direct comparisons between cultivation and cloning results have been inconclusive. Suzuki et al. (89) compared 127 cultivated organisms and 58 16S rRNA clones from the same Oregon coastal sample and reported little overlap, also noting that even the culturable marine organisms are poorly represented in sequence databases. Pinhassi et al. (75) took a different approach, using whole genome hybridization (from 48 culturable organisms) toward community DNA from the northern Baltic Sea. They reported significant genomic overlap, suggesting that the cultures are representative of a substantial part of the native community. One group readily found in 16S rRNA clone libraries and also cultures is the so-called marine alphaproteobacteria (36). On the other hand, some readily cloned sequences, such as SAR86, have not yet been cultured at the time of this writing. Therefore, it appears that cultures may represent important components of native communities but may also miss important ones as well.

There is also the question of quantitative biases. As more and more molecular analysis is done, there is a temptation to compare results from libraries or fingerprints quantitatively. Can this comparison be done at all, with the several presumed PCR biases operating in concert? There are a few studies that have examined quantitative fingerprint data, with encouraging resuts. With laboratory-created mixtures of organisms, Lueders and Friedrich (59) found that T-RFLP peak quantity was proportional to original template amounts. In an extensive field study, Brown et al. (10) found that quantitative ARISA estimates of the proportion of *Prochlorococcus* in monthly oceanic field samples (over 4 years) closely matched independent estimates of that organism by flow cytometry ($r^2 = 0.86$). While these two examples obviously cannot be applied universally to all organisms, it is interesting that these fingerprinting approaches seem to have the potential to yield accurate proportions of at least some taxa despite presumed biases. With the potential for biases, note that it is generally better to make comparisons in the relative proportion of a particular taxon between samples, rather than comparing the relative proportions of different taxa in an absolute sense.

QUANTITATIVE PROBES AND PCR

In the quantitative analysis of community structure, oligonucleotide probes are powerful tools that avoid possible biases in cloning and yield a more direct measure of the target groups of interest. One way to elicit such results is with oligonucleotide probes hybridized to bulk nucleic acid extracted from the aquatic habitat in question. Probe sequences are determined from sequence databases and can be universal or specific to certain domains, groups, or even some species (2). It is generally preferred to use RNA as the target instead of DNA because DNA is likely to yield far more false positives (unintentional hits), since mixed genomic DNA from innumerable species will have an immense variety of genes. RNA also has the benefit of being present in ribosomes and thus being much more abundant as a target and also related to the cellular growth rate (although the exact relationship in natural communities is still uncertain). Quantitative bulk hybridizations to extracted RNA from aquatic habitats have been reported by Giovannoni et al. (34) and DeLong et al. (23), showing the relative abundances of the SAR11 cluster and archaea, respectively. It should be noted that, in order to best standardize such probe binding, it is ideal to have a culture of the organism or group in question to determine the relative binding of different probes empirically.

A more recent quantitative measure of template quantity that has been applied to community structure studies is a process called real-time quantitative PCR (sometimes abbreviated QPCR or RT-QPCR), in which amplification from field samples is compared to side-by-side amplification of standards. There are a few different versions of this procedure, all of which are based on determining in which cycle during the PCR a threshold amount of specific amplification occurs. Some versions rely on fluorescent internal probes (e.g., TaqMan assay), while some simply monitor the amount of total DNA amplified (38, 49, 103). The dynamic range of the measurements often exceeds 4 orders of magnitude (sometimes reaching 10 orders of magnitude). An early application in seawater was by Suzuki and colleagues (88, 90), using 16S rRNA-based primers and probes. The approach can also be applied to a functional group, e.g., to examine the abundance in seawater of *pufM* genes that code for bacteriochlorophyll synthesis (83). It is important that, although this method avoids direct PCR bias in measurement of template amounts, interpretation of results quantitatively for the organisms themselves requires knowledge or assumptions about sampling and DNA extraction efficiency (very difficult to verify), gene copy numbers, and genome sizes. All these factors go into calculation of organism abundance from template abundance. The approach used by Schwalbach et al. (83) avoided the need to know DNA extraction efficiency. Instead, they relied on first physically separating bacteria from eukaryotes by size fractionation before DNA extraction and then using an average bacterial genome size to estimate the number of bacteria represented per nanogram of DNA in each assay extract. Had eukaryotic DNA been mixed in with the extract, the authors could not have known how much bacterial DNA was used in each assay.

FISH

Although quantitative hybridization can be readily used with natural samples, one is not usually interested in knowing the fraction of RNA coming from particular groups but instead wants to know the proportions of individuals in those groups. For such work, it is ideal to tag each cell type with specific probes that allow visual identification. The preferred mode of observation has been epifluorescence microscopy with fluorescent oligonucleotide probes to rRNA (reviewed in reference 2). Flow cytometry has also been used to automate the analysis with success (99). This area has blossomed recently. However, it has been found to work most easily with relatively rich environments, probably because slowly growing cells in relatively oligotrophic environments have few ribosomes and thus fluoresce dimly. Nevertheless, some approaches have yielded useful data with difficult samples like marine plankton. Lee et al. (55) found that multiple probes yield enough fluorescence for standard visual observation of about 75% of the bacteria in marine plankton. Fuhrman et al. (29) found that about 75% of the DAPI (4′,6′-diamidino-2-phenylindole)-countable cells from coastal marine plankton may be seen even with single fluorescent probes when video image intensification is used to boost the brightness of the images and that about 8% fluoresced with an archaeon-specific probe. Ouverney and Fuhrman (71) found that typically 90 to 95% of the total DAPI-countable cells can be tagged with a "universal" 16S rRNA probe after a marine plankton sample is treated for an hour with the antibiotic chloramphenicol. This treatment ostensibly stops protein synthesis (and prevents changes in community composition) while allowing rRNA synthesis to continue, apparently permitting particularly good binding of probes to the RNA. Bright fluorescence can also be obtained with long probes containing multiple fluorochromes, such as polyribonucleotides (22, 57, 93); however, with relatively conserved targets such as 16S rRNA, such long probes are best for tagging broad rather than narrow phylogenetic groups. Enzymatic amplification is another way to produce a bright signal with relatively few probe molecules, by using an enzyme-linked probe that reacts with a substrate to deposit brightly fluorescent product near the site of the probe, a procedure recently called CARD FISH and applicable to aquatic field samples (74, 82). There is also the possibility of using in situ PCR to produce fluorescent products within particular target cells (13, 41), although the ability to use it as a general method in natural communities is still under development. Laser confocal microscopy (2) is another promising approach with excellent sensitivity that allows for viewing of dim fluorescence and also permits examination of microbes in complex matrices (see chapter 5). This area will clearly progress rapidly.

PROBE/MICROAUTORADIOGRAPHY COMBINATION

A recent development is the ability to combine microautoradiographic characterization of the activity of single cells with 16S rRNA probes to identify those cells (16, 51, 70). With such methods, one can start to measure the activities of individual cells within natural mixed communities while at the same time learning which taxa are responsible for each measured property. This practice is a natural extension of community structure analysis. In one of the most recent and sensitive applications, Teira et al. (91) combined CARD FISH with microautoradiography to examine archaeal abundance and activity in the deep sea. This approach thus greatly reduces the need to separate and/or cultivate an organism to learn what it is doing in its natural habitat. Such an approach is a particularly powerful means to dissect the "black box" of natural microbial communities, going far beyond simply learning the taxonomic breakdown of community composition. There is little doubt that, in the future, such single-cell probe approaches will be common tools in studies of microbial community structure.

COMMUNITY METAGENOMICS

In what may be considered the ultimate expression of community structure analysis, metagenomic studies aim to characterize the genes from the entire community, ideally showing how these genes are ordered within the genomes of the constituent organisms. The original phage library concept described by Pace et al. and Olsen et al. in 1986 (68, 72) is in fact a metagenomic study, but the more recent versions aim to sequence all the DNA (or long continuous stretches at least), rather than just looking for 16S rRNA or other specific genes. This whole genome focus has been facilitated by high-throughput automated cloning and sequencing systems and most recently clone-free sequencing. Of course there are practical limits on the extent of resolution of such studies (i.e., fraction of the community that is well characterized), depending on the diversity of the community, with simple communities easier to analyze for a given amount of sequencing effort. There are two basic ways that metagenomes are investigated: shotgun approaches and large insert libraries (18). Both start with extracted community DNA. In the shotgun approach, the DNA is randomly sheared into relatively small fragments of a few thousand bases each, and the pieces are sequenced from the ends (typically around 600 to 800 bases sequenced at a time). With a sufficient number of pieces and sufficient amount of redundant sequencing, it should be possible to have enough overlap to permit computerized assembly of the fragments into continuous scaffolds and even intact genomes. Complex diverse systems with many close relatives present particular challenges for this approach, as it becomes extremely difficult and expensive to assemble long stretches of sequence when most of the fragments are "singletons" (no overlap with others), and similarities in conserved regions among close relatives may make true overlaps hard to confirm (18). In contrast, the large insert library approach does not have problems with assembling sequences within each insert, and the inserts may each represent 10% of a genome, but it is still difficult to find overlapping inserts when the sample is highly diverse. For "traditional" community structure studies, where the idea is to make lists of how many of each "kind" of organism is present in a given sample, both these types of studies have the potential to generate significant numbers of data, but it should be clear that, if this data generation is the only objective, it makes more sense to focus on particularly phylogenetically informative genes such as 16S rRNA rather than having most of the sequencing effort on other genes (a given gene typically represents 0.1% of a genome). Metagenomic studies are typically limited by low genome coverage of diverse communities due to the sheer enormity of the task. On the other hand, if one wishes to link community structure and function and does not have representative cultures to work with, the metagenomic approach is potentially very powerful (6).

There are relatively few metagenomic studies suitable for community structure analysis at the time of this writing (18, 21, 80, 94, 96). In one of the earliest of such studies (not marine), Tyson et al. (95) examined an acid mine drainage community with very low diversity and were able to assemble the genomes of a few dominant organisms and to provide evidence about a few other community members. In contrast, Venter et al. (96) examined a diverse marine plankton community from the Sargasso Sea (actually from multiple nearby samples) and reported that there were about 1,800 different kinds of bacteria present. The most abundant kinds of organisms in most samples were too diverse to have their genomes assembled (even with $>10^9$ bases sequenced), yet the data still provided extensive information on community structure based upon 16S rRNA and several other well-conserved genes (like *recA*) that provide phylogenetic information. The two genomes in one sample that were assembled by the Venter study were challenged by DeLong (18) as likely contaminants, because they appeared to be clonal on the basis of rare polymorphisms and unlike anything seen in many previous similar marine studies, while also strongly resembling terrestrial bacteria. This finding is brought up here as a reminder that contamination control in sample collection and processing is critical and not always easy. A multiple habitat (marine and terrestrial) shotgun library comparison by Tringe et al. (94) found habitat-specific patterns of many genes. Most recently, a large insert library study of a depth profile from 10 to 4,000 m in the Pacific Ocean near Hawaii was reported by DeLong et al. (21) and showed strong stratification in the distribution of genes, many of which matched sequences recently found via whole-genome sequence analysis of marine bacterial cultures. Unexpectedly, there were no reported 16S rRNA genes from the abundant SAR11 cluster in the five samples from 10- to 500-m depths, despite the presence of many other genes related to those from the one known culture of that group whose genome has been sequenced (35). This finding raises questions about unknown biases in this approach.

With regard to biases, data quality, and overall coverage of the community, genome library methods that do not use PCR avoid the potential for PCR biases (although to date, the extent of such biases has not yet been shown to be a major obstacle). But it is important that any method relying on DNA still may have sampling or extraction biases and that, with metagenomic methods that include a cloning step, certain genes may be lethal to cells harboring the vectors (an example is viral genes, but there are many other possibilities) or otherwise be excluded.

MICRODIVERSITY

A major question about community diversity relates to the extent of phylogenetic resolution that we wish to have. Recent studies indicate that many or most natural aquatic microbial communities consist of thousands or more genetically distinguishable different types of bacteria and archaea, typically consisting in large part of tight clusters of many close relatives in several broad groups (1, 9, 33). With so much sequencing power, we can distinguish very fine differences, to the point of eventually identifying each single individual if desired. The question then becomes what extent of diversity information is necessary for each question at hand? Do we want to know about bacteria or archaea with certain broad or specific functions or those that might be considered a given "species," by whatever definition one uses, or population genetics of individuals? This issue depends on each study and goes beyond the scope of this chapter, but it ultimately helps determine which approaches one wishes to use and how the results are interpreted.

I thank N. Pace, S. Giovannoni, and G. Olsen for helping me get involved in this work many years ago; S. Lee and R. Amann for helpful discussions; and R. Hicks and R. Christian for reviewing an earlier version of the manuscript.

This work was supported by NSF Microbial Observatory grant MCB0084231 and NSF grant OCE0527034.

REFERENCES

1. **Acinas, S. G., V. Klepac-Ceraj, D. E. Hunt, C. Pharino, I. Ceraj, D. L. Distel, and M. F. Polz.** 2004. Fine-scale phylogenetic architecture of a complex bacterial community. *Nature* **430:**551–554.
2. **Amann, R. I., W. Ludwig, and K. H. Schleifer.** 1995. Phylogenetic identification and in situ detection of individual microbial cells without cultivation. *Microbiol. Rev.* **59:**143–169.
3. **Ashelford, K. E., N. A. Chuzhanova, J. C. Fry, A. J. Jones, and A. J. Weightman.** 2005. At least 1 in 20 16S rRNA sequence records currently held in public repositories is estimated to contain substantial anomalies. *Appl. Environ. Microbiol.* **71:**7724–7736.
4. **Avaniss-Aghajani, E., K. Jones, D. Chapman, and C. Brunk.** 1994. A molecular technique for identification of bacteria using small subunit ribosomal RNA sequences. *BioTechniques* **17:**144–149.
5. **Barns, S. M., R. E. Fundyga, M. W. Jeffries, and N. R. Pace.** 1994. Remarkable archaeal diversity detected in a Yellowstone National Park hot spring environment. *Proc. Natl. Acad. Sci. USA* **91:**1609–1613.
6. **Beja, O., L. Aravind, E. V. Koonin, M. T. Suzuki, A. Hadd, L. P. Nguyen, S. B. Jovanovich, C. M. Gates, R. A. Feldman, J. L. Spudich, E. N. Spudich, and E. F. DeLong.** 2000. Bacterial rhodopsin: evidence for a new type of phototrophy in the sea. *Science* **289:**1902–1906.
7. **Blumenberg, M., R. Seifert, K. Nauhaus, T. Pape, and W. Michaelis.** 2005. In vitro study of lipid biosynthesis in an anaerobically methane-oxidizing microbial mat. *Appl. Environ. Microbiol.* **71:**4345–4351.
8. **Britschgi, T., and S. J. Giovannoni.** 1991. Phylogenetic analysis of a natural marine bacterioplankton population by rRNA gene cloning and sequencing. *Appl. Environ. Microbiol.* **57:**1707–1713.
9. **Brown, M. V., and J. A. Fuhrman.** 2005. Marine bacterial microdiversity as revealed by internal transcribed spacer analysis. *Aquat. Microb. Ecol.* **41:**15–23.
10. **Brown, M. V., M. S. Schwalbach, I. Hewson, and J. A. Fuhrman.** 2005. Coupling 16S-ITS rDNA clone libraries and ARISA to show marine microbial diversity: development and application to a time series. *Environ. Microbiol.* **7:**1466–1479.
11. **Button, D. K., F. Schuts, P. Quang, R. Martin, and B. R. Robertson.** 1993. Viability and isolation of marine bacteria by dilution culture: theory, procedures, and initial results. *Appl. Environ. Microbiol.* **59:**881–891.
12. **Campbell, L., E. J. Carpenter, and V. J. Iacono.** 1983. Identification and enumeration of marine chroococcoid cyanobacteria by immunofluorescence. *Appl. Environ. Microbiol.* **46:**553–559.
13. **Chen, F., B. Binder, and R. E. Hodson.** 2000. Flow cytometric detection of specific gene expression in prokaryotic cells using in situ RT-PCR. *FEMS Microbiol. Lett.* **184:**291–296.
14. **Chisholm, S. W., R. J. Olson, E. R. Zettler, J. Waterbury, R. Goericke, and N. Welschmeyer.** 1988. A novel free-living prochlorophyte abundant in the oceanic euphotic zone. *Nature* **334:**340–343.
15. **Connon, S. A., and S. J. Giovannoni.** 2002. High-throughput methods for culturing microorganisms in very-low-nutrient media yield diverse new marine isolates. *Appl. Environ. Microbiol.* **68:**3878–3885.
16. **Cottrell, M. T., and D. L. Kirchman.** 2000. Natural assemblages of marine proteobacteria and members of the *Cytophaga-Flavobacter* cluster consuming low- and high-molecular-weight dissolved organic matter. *Appl. Environ. Microbiol.* **66:**1692–1697.
17. **DeLong, E. F.** 1992. Archaea in coastal marine environments. *Proc. Natl. Acad. Sci. USA* **89:**5685–5689.
18. **DeLong, E. F.** 2005. Microbial community genomics in the ocean. *Nat. Rev. Microbiol.* **3:**459–469.
19. **DeLong, E. F., D. G. Franks, and A. A. Alldredge.** 1993. Phylogenetic diversity of aggregate-attached vs. free-living marine bacterial assemblages. *Limnol. Oceanogr.* **38:**924–934.
20. **DeLong, E. F., L. L. King, R. Massana, H. Cittone, A. Murray, C. Schleper, and S. G. Wakeham.** 1998. Dibiphytanyl ether lipids in nonthermophilic crenarchaeotes. *Appl. Environ. Microbiol.* **64:**1133–1138. (Erratum, **64:**1986.)

21. **DeLong, E. F., C. M. Preston, T. Mincer, V. Rich, S. J. Hallam, N. U. Frigaard, A. Martinez, M. B. Sullivan, R. Edwards, B. R. Brito, S. W. Chisholm, and D. M. Karl.** 2006. Community genomics among stratified microbial assemblages in the ocean's interior. *Science* **311:**496–503.
22. **DeLong, E. F., L. T. Taylor, T. L. Marsh, and C. M. Preston.** 1999. Visualization and enumeration of marine planktonic archaea and bacteria by using polyribonucleotide probes and fluorescent in situ hybridization. *Appl. Environ. Microbiol.* **65:**5554–5563.
23. **DeLong, E. F., K. Y. Wu, B. B. Prezelin, and R. V. M. Jovine.** 1994. High abundance of archaea in antarctic marine picoplankton. *Nature* **371:**695–697.
24. **Devereux, R., and G. W. Mundfrom.** 1994. A phylogenetic tree of 16S rRNA sequences from sulfate-reducing bacteria in a sandy sediment. *Appl. Environ. Microbiol.* **60:**3437–3439.
25. **Ferguson, R. L., E. N. Buckley, and A. V. Palumbo.** 1984. Response of marine bacterioplankton to differential filtration and confinement. *Appl. Environ. Microbiol.* **47:** 49–55.
26. **Fisher, M. M., and E. W. Triplett.** 1999. Automated approach for ribosomal intergenic spacer analysis of microbial diversity and its application to freshwater bacterial communities. *Appl. Environ. Microbiol.* **65:**4630–4636.
27. **Fuhrman, J. A., and F. Azam.** 1982. Thymidine incorporation as a measure of heterotrophic bacterioplankton production in marine surface waters: evaluation and field results. *Mar. Biol.* **66:**109–120.
28. **Fuhrman, J. A., and A. A. Davis.** 1997. Widespread Archaea and novel Bacteria from the deep sea as shown by 16S rRNA gene sequences. *Mar. Ecol. Prog. Ser.* **150:**275–285.
29. **Fuhrman, J. A., S. H. Lee, Y. Masuchi, A. A. Davis, and R. M. Wilcox.** 1994. Characterization of marine prokaryotic communities via DNA and RNA. *Microb. Ecol.* **28:** 133–145.
30. **Fuhrman, J. A., K. McCallum, and A. A. Davis.** 1992. Novel major archaebacterial group from marine plankton. *Nature* **356:**148–149.
31. **Fuhrman, J. A., K. McCallum, and A. A. Davis.** 1993. Phylogenetic diversity of subsurface marine microbial communities from the Atlantic and Pacific oceans. *Appl. Environ. Microbiol.* **59:**1294–1302. (Erratum, **61:**4517, 1995.)
32. **Fuhrman, J. A., and C. C. Ouverney.** 1998. Marine microbial diversity studied via 16S rRNA sequences: cloning results from coastal waters and counting of native archaea with fluorescent single cell probes. *Aquat. Ecol.* **32:**3–15.
33. **Garcia-Martinez, J., and F. Rodriguez-Valera.** 2000. Microdiversity of uncultured marine prokaryotes: the SAR11 cluster and the marine Archaea of Group I. *Mol. Ecol.* **9:**935–948.
34. **Giovannoni, S. J., T. B. Britschgi, C. L. Moyer, and K. G. Field.** 1990. Genetic diversity in Sargasso Sea bacterioplankton. *Nature* **345:**60–63.
35. **Giovannoni, S. J., H. J. Tripp, S. Givan, M. Podar, K. L. Vergin, D. Baptista, L. Bibbs, J. Eads, T. H. Richardson, M. Noordewier, M. S. Rappe, J. M. Short, J. C. Carrington, and E. J. Mathur.** 2005. Genome streamlining in a cosmopolitan oceanic bacterium. *Science* **309:** 1242–1245.
36. **González, J. M., and M. A. Moran.** 1997. Numerical dominance of a group of marine bacteria in the α-subclass of the class *Proteobacteria* in coastal seawater. *Appl. Environ. Microbiol.* **63:**4237–4242.
37. **González, J. M., R. Simo, R. Massana, J. S. Covert, E. O. Casamayor, C. Pedros-Alio, and M. A. Moran.** 2000. Bacterial community structure associated with a dimethylsulfoniopropionate-producing North Atlantic algal bloom. *Appl. Environ. Microbiol.* **66:**4237–4246.
38. **Heid, C. A., J. Stevens, K. J. Livak, and P. M. Williams.** 1996. Real time quantitative PCR. *Genome Res.* **6:**986–994.
39. **Hewson, I., and J. A. Fuhrman.** 2006. Improved strategy for comparing community fingerprints. *Microb. Ecol.* **51:** 147–153.
40. **Hewson, I., and J. A. Fuhrman.** 2004. Richness and diversity of bacterioplankton species along an estuarine gradient in Moreton Bay, Australia. *Appl. Environ. Microbiol.* **70:**3425–3433.
41. **Hodson, R. E., W. A. Dustman, R. P. Garg, and M. A. Moran.** 1995. In situ PCR for visualization of microscale distribution of specific genes and gene products in prokaryotic communities. *Appl. Environ. Microbiol.* **61:**4074–4082.
42. **Hoefs, M. J. L., S. Schouten, J. W. deLeeuw, L. L. King, S. G. Wakeham, and J. S. S. Damste.** 1997. Ether lipids of planktonic archaea in the marine water column. *Appl. Environ. Microbiol.* **63:**3090–3095.
43. **Hofle, M. G.** 1992. Bacterioplankton community structure and dynamics after large-scale release of nonindigenous bacteria as revealed by low-molecular-weight-RNA analysis. *Appl. Environ. Microbiol.* **58:**3387–3394.
44. **Hofle, M. G., and I. Brettar.** 1996. Genotyping of heterotrophic bacteria from the central Baltic Sea by use of low-molecular-weight RNA profiles. *Appl. Environ. Microbiol.* **62:**1383–1390.
45. **Holben, W. E., and D. Harris.** 1995. DNA-based monitoring of total bacterial community structure in environmental samples. *Mol. Ecol.* **4:**627–631.
46. **Huber, T., G. Faulkner, and P. Hugenholtz.** 2004. Bellerophon: a program to detect chimeric sequences in multiple sequence alignments. *Bioinformatics* **20:**2317–2319.
47. **Jannasch, H. W., and G. E. Jones.** 1959. Bacterial populations in sea water as determined by different methods of enumeration. *Limnol. Oceanogr.* **4:**128–139.
48. **Kemp, P. F., S. Lee, and J. LaRoche.** 1993. Estimating the growth rate of slowly growing marine bacteria from RNA content. *Appl. Environ. Microbiol.* **59:**2594–2601.
49. **Kim, D. W.** 2001. Real time quantitative PCR. *Exp. Mol. Med.* **33:**101–109.
50. **Komatsoulis, G. A., and M. S. Waterman.** 1997. A new computational method for detection of chimeric 16S rRNA artifacts generated by PCR amplification from mixed bacterial populations. *Appl. Environ. Microbiol.* **63:** 2338–2346.
51. **Lee, N., P. H. Nielsen, K. H. Andreasen, S. Juretschko, J. L. Nielsen, K. H. Schleifer, and M. Wagner.** 1999. Combination of fluorescent in situ hybridization and microautoradiography—a new tool for structure-function analyses in microbial ecology. *Appl. Environ. Microbiol.* **65:** 1289–1297.
52. **Lee, S., and J. A. Fuhrman.** 1990. DNA hybridization to compare species compositions of natural bacterioplankton assemblages. *Appl. Environ. Microbiol.* **56:**739–746.
53. **Lee, S., and J. A. Fuhrman.** 1991. Spatial and temporal variation of natural bacterioplankton assemblages studied by total genomic DNA cross-hybridization. *Limnol. Oceanogr.* **36:**1277–1287.
54. **Lee, S. H., and J. A. Fuhrman.** 1991. Species composition shift of confined bacterioplankton studied at the level of community DNA. *Mar. Ecol. Prog. Ser.* **79:**195–201.
55. **Lee, S. H., C. Malone, and P. F. Kemp.** 1993. Use of multiple 16S ribosomal-RNA-targeted fluorescent-probes to increase signal strength and measure cellular RNA from natural planktonic bacteria. *Mar. Ecol. Prog. Ser.* **101:**193–201.

56. **Liu, W. T., T. L. Marsh, H. Cheng, and L. J. Forney.** 1997. Characterization of microbial diversity by determining terminal restriction fragment length polymorphisms of genes encoding 16S rRNA. *Appl. Environ. Microbiol.* **63:** 4516–4522.
57. **Ludwig, W., S. Dorn, N. Springer, G. Kirchhof, and K. H. Schleifer.** 1994. PCR-based preparation of 23S rRNA-targeted group-specific polynucleotide probes. *Appl. Environ. Microbiol.* **60:**3234–3244.
58. **Ludwig, W., O. Strunk, R. Westram, L. Richter, H. Meier, Yadhukumar, A. Buchner, T. Lai, S. Steppi, G. Jobb, W. Forster, I. Brettske, S. Gerber, A. W. Ginhart, O. Gross, S. Grumann, S. Hermann, R. Jost, A. Konig, T. Liss, R. Lussmann, M. May, B. Nonhoff, B. Reichel, R. Strehlow, A. Stamatakis, N. Stuckmann, A. Vilbig, M. Lenke, T. Ludwig, A. Bode, and K. H. Schleifer.** 2004. ARB: a software environment for sequence data. *Nucleic Acids Res.* **32:**1363–1371.
59. **Lueders, T., and M. W. Friedrich.** 2003. Evaluation of PCR amplification bias by terminal restriction fragment length polymorphism analysis of small-subunit rRNA and *mcrA* genes by using defined template mixtures of methanogenic pure cultures and soil DNA extracts. *Appl. Environ. Microbiol.* **69:**320–326.
60. **McInerney, J. O., M. Wilkinson, J. W. Patching, T. M. Embley, and R. Powell.** 1995. Recovery and phylogenetic analysis of novel archaeal rRNA sequences from a deep-sea deposit feeder. *Appl. Environ. Microbiol.* **61:**1646–1648.
61. **Mills, H. J., R. J. Martinez, S. Story, and H. J. Mills.** 2005. Characterization of microbial community structure in Gulf of Mexico gas hydrates: comparative analysis of DNA- and RNA-derived clone libraries. *Appl. Environ. Microbiol.* **71:**3235–3247.
62. **Moeseneder, M. M., J. M. Arrieta, and G. J. Herndl.** 2005. A comparison of DNA- and RNA-based clone libraries from the same marine bacterioplankton community. *FEMS Microbiol. Ecol.* **51:**341–352.
63. **Moeseneder, M. M., J. M. Arrieta, G. Muyzer, C. Winter, and G. J. Herndl.** 1999. Optimization of terminal-restriction fragment length polymorphism analysis for complex marine bacterioplankton communities and comparison with denaturing gradient gel electrophoresis. *Appl. Environ. Microbiol.* **65:**3518–3525.
64. **Moyer, C. L., F. C. Dobbs, and D. M. Karl.** 1994. Estimation of diversity and community structure through restriction fragment length polymorphism distribution analysis of bacterial 16S rRNA genes from a microbial mat at an active, hydrothermal vent system, Loihi Seamount, Hawaii. *Appl. Environ. Microbiol.* **60:**871–879.
65. **Moyer, C. L., F. C. Dobbs, and D. M. Karl.** 1995. Phylogenetic diversity of the bacterial community from a microbial mat at an active, hydrothermal vent system, Loihi Seamount, Hawaii. *Appl. Environ. Microbiol.* **61:** 1555–1562.
66. **Muyzer, G., and K. Smalla.** 1998. Application of denaturing gradient gel electrophoresis (DGGE) and temperature gradient gel electrophoresis (TGGE) in microbial ecology. *Antonie Leeuwenhoek Int. J. Gen. Mol. Microbiol.* **73:**127–141.
67. **Muyzer, G., E. C. de Waal, and A. G. Uitterlinden.** 1993. Profiling of complex microbial populations by denaturing gradient gel electrophoresis analysis of polymerase chain reaction-amplified genes coding for 16S rRNA. *Appl. Environ. Microbiol.* **59:**695–700.
68. **Olsen, G. J., D. L. Lane, S. J. Giovannoni, and N. R. Pace.** 1986. Microbial ecology and evolution: a ribosomal RNA approach. *Annu. Rev. Microbiol.* **40:**337–365.
69. **Osborn, A. M., E. R. B. Moore, and K. N. Timmis.** 2000. An evaluation of terminal-restriction fragment length polymorphism (T-RFLP) analysis for the study of microbial community structure and dynamics. *Environ. Microbiol.* **2:**39–50.
70. **Ouverney, C. C., and J. A. Fuhrman.** 1999. Combined microautoradiography–16S rRNA probe technique for determination of radioisotope uptake by specific microbial cell types in situ. *Appl. Environ. Microbiol.* **65:**1746–1752.
71. **Ouverney, C. C., and J. A. Fuhrman.** 1997. Increase in fluorescence intensity of 16S rRNA in situ hybridization in natural samples treated with chloramphenicol. *Appl. Environ. Microbiol.* **63:**2735–2740.
72. **Pace, N. R., D. A. Stahl, D. L. Lane, and G. J. Olsen.** 1986. The analysis of natural microbial populations by rRNA sequences. *Adv. Microbiol. Ecol.* **9:**1–55.
73. **Pearson, A., A. P. McNichol, B. C. Benitez-Nelson, J. M. Hayes, and T. I. Eglinton.** 2001. Origins of lipid biomarkers in Santa Monica Basin surface sediment: a case study using compound-specific Delta C-14 analysis. *Geochim. Cosmochim. Acta.* **65:**3123–3137.
74. **Pernthaler, A., J. Pernthaler, and R. Amann.** 2002. Fluorescence in situ hybridization and catalyzed reporter deposition for the identification of marine bacteria. *Appl. Environ. Microbiol.* **68:**3094–3101.
75. **Pinhassi, J., U. L. Zweifel, and A. Hagstrom.** 1997. Dominant marine bacterioplankton species found among colony-forming bacteria. *Appl. Environ. Microbiol.* **63:** 3359–3366.
76. **Rappé, M. S., S. A. Connon, K. L. Vergin, and S. J. Giovannoni.** 2002. Cultivation of the ubiquitous SAR11 marine bacterioplankton clade. *Nature* **418:**630–633.
77. **Reysenbach, A.-L., G. W. Wickham, and N. R. Pace.** 1994. Phylogenetic analysis of the hyperthermophilic pink filament community in Octopus Spring, Yellowstone National Park. *Appl. Environ. Microbiol.* **60:**2113–2119.
78. **Rochelle, P. A., B. A. Cragg, J. C. Fry, R. J. Parkes, and A. J. Weightman.** 1994. Effect of sample handling on estimation of bacterial diversity in marine sediments by 16S rRNA gene sequence analysis. *FEMS Microbiol. Ecol.* **15:** 215–226.
79. **Roszak, D. B., and R. R. Colwell.** 1987. Survival strategies of bacteria in the natural environment. *Microbiol. Rev.* **51:**365–379.
80. **Schleper, C., G. Jurgens, and M. Jonuscheit.** 2005. Genomic studies of uncultivated archaea. *Nat. Rev. Microbiol.* **3:**479–488.
81. **Schmidt, T. M., E. F. DeLong, and N. R. Pace.** 1991. Analysis of a marine picoplankton community by 16S rRNA gene cloning and sequencing. *J. Bacteriol.* **173:** 4371–4378.
82. **Schonhuber, W., B. Zarda, S. Eix, R. Rippka, M. Herdman, W. Ludwig, and R. Amann.** 1999. In situ identification of cyanobacteria with horseradish peroxidase-labeled, rRNA-targeted oligonucleotide probes. *Appl. Environ. Microbiol.* **65:**1259–1267.
83. **Schwalbach, M. S., and J. A. Fuhrman.** 2005. Wide-ranging abundances of aerobic anoxygenic phototrophic bacteria in the world ocean revealed by epifluorescence microscopy and quantitative PCR. *Limnol. Oceanogr.* **50:**620–628.
84. **Sieburth, J. M.** 1979. *Sea Microbes*. Oxford University Press, New York, N.Y.
85. **Simu, K., and A. Hagstrom.** 2004. Oligotrophic bacterioplankton with a novel single-cell life strategy. *Appl. Environ. Microbiol.* **70:**2445–2451.
86. **Stolz, J. F.** 1990. Distribution of phototrophic microbes in the flat laminated microbial mat at Laguna Figueroa, Baja California, Mexico. *BioSystems* **23:**345–357.
87. **Suzuki, M., M. S. Rappe, and S. J. Giovannoni.** 1998. Kinetic bias in estimates of coastal picoplankton commu-

nity structure obtained by measurements of small-subunit rRNA gene PCR amplicon length heterogeneity. *Appl. Environ. Microbiol.* **64:**4522–4529.

88. **Suzuki, M. T., C. M. Preston, F. P. Chavez, and E. F. DeLong.** 2001. Quantitative mapping of bacterioplankton populations in seawater: field tests across an upwelling plume in Monterey Bay. *Aquat. Microb. Ecol.* **24:**117–127.
89. **Suzuki, M. T., M. S. Rappe, Z. W. Haimberger, H. Winfield, N. Adair, J. Strobel, and S. J. Giovannoni.** 1997. Bacterial diversity among small-subunit rRNA gene clones and cellular isolates from the same seawater sample. *Appl. Environ. Microbiol.* **63:**983–989.
90. **Suzuki, M. T., L. T. Taylor, and E. F. DeLong.** 2000. Quantitative analysis of small-subunit rRNA genes in mixed microbial populations via 5′-nuclease assays. *Appl. Environ. Microbiol.* **66:**4605–4614.
91. **Teira, E., T. Reinthaler, A. Pernthaler, J. Pernthaler, and G. J. Herndl.** 2004. Combining catalyzed reporter deposition-fluorescence in situ hybridization and microautoradiography to detect substrate utilization by bacteria and archaea in the deep ocean. *Appl. Environ. Microbiol.* **70:**4411–4414.
92. **Torsvik, V., J. Goksoyr, and F. L. Daae.** 1990. High diversity of DNA of soil bacteria. *Appl. Environ. Microbiol.* **56:**782–787.
93. **Trebesius, K., R. Amann, W. Ludwig, K. Muhlegger, and K. H. Schleifer.** 1994. Identification of whole fixed bacterial cells with nonradioactive 23S rRNA-targeted polynucleotide probes. *Appl. Environ. Microbiol.* **60:**3228–3235.
94. **Tringe, S. G., C. von Mering, A. Kobayashi, A. A. Salamov, K. Chen, H. W. Chang, M. Podar, J. M. Short, E. J. Mathur, J. C. Detter, P. Bork, P. Hugenholtz, and E. M. Rubin.** 2005. Comparative metagenomics of microbial communities. *Science* **308:**554–557.
95. **Tyson, G. W., J. Chapman, P. Hugenholtz, E. E. Allen, R. J. Ram, P. M. Richardson, V. V. Solovyev, E. M. Rubin, D. S. Rokhsar, and J. F. Banfield.** 2004. Community structure and metabolism through reconstruction of microbial genomes from the environment. *Nature* **428:**37–43.
96. **Venter, J. C., K. Remington, J. F. Heidelberg, A. L. Halpern, D. Rusch, J. A. Eisen, D. Y. Wu, I. Paulsen, K. E. Nelson, W. Nelson, D. E. Fouts, S. Levy, A. H. Knap, M. W. Lomas, K. Nealson, O. White, J. Peterson, J. Hoffman, R. Parsons, H. Baden-Tillson, C. Pfannkoch, Y. H. Rogers, and H. O. Smith.** 2004. Environmental genome shotgun sequencing of the Sargasso Sea. *Science* **304:**66–74.
97. **Versalovic, J., M. Schnieder, F. J. de Bruijn, and J. R. Lupski.** 1994. Genomic fingerprinting of bacteria using repetitive sequence-based polymerase chain reaction. *Methods Mol. Cell Biol.* **5:**25–40.
98. **Voytek, M. A., and B. B. Ward.** 1995. Detection of ammonium-oxidizing bacteria of the beta-subclass of the class *Proteobacteria* in aquatic samples with the PCR. *Appl. Environ. Microbiol.* **61:**1444–1450.
99. **Wallner, G., R. Erhart, and R. Amann.** 1995. Flow cytometric analysis of activated sludge with rRNA-targeted probes. *Appl. Environ. Microbiol.* **61:**1859–1866.
100. **Ward, B. B.** 1982. Oceanic distribution of ammonium-oxidizing bacteria determined by immunofluorescent assay. *J. Mar. Res.* **40:**1155–1172.
101. **Ward, D. M., R. Weller, and M. M. Bateson.** 1990. 16S rRNA sequences reveal numerous uncultured microorganisms in a natural community. *Nature* **345:**63–65.
102. **White, D. C.** 1994. Is there anything else you need to understand about the microbiota that cannot be derived from analysis of nucleic acids? *Microb. Ecol.* **28:**163–166.
103. **Wilhelm, J., and A. Pingoud.** 2003. Real-time polymerase chain reaction. *ChemBioChem* **4:**1120–1128.
104. **Woese, C. R.** 1987. Bacterial evolution. *Microbiol. Rev.* **51:**221–271.
105. **Wuchter, C., S. Schouten, H. T. S. Boschker, and J. S. S. Damste.** 2003. Bicarbonate uptake by marine Crenarchaeota. *FEMS Microbiol. Lett.* **219:**203–207.

Viral Community Structure

ALEXANDER I. CULLEY AND CURTIS A. SUTTLE

36

Viruses are abundant and ubiquitous in aquatic ecosystems and are major players in the mortality of microbial populations and in global nutrient cycles (for reviews, see references 64 and 73). Many studies are beginning to include estimates of viral abundance, while others are directly addressing the structure and diversity of viral communities and their role in the mortality of microbial ecosystems. As interest in the roles of viral communities in nature increases, so do the efforts to develop new methods. The result is a suite of new and well-established approaches that are providing a set of powerful tools with which to address questions of structure in natural virus communities.

The method used to examine viruses in nature depends on the question being addressed. Nonetheless, it is possible to identify two major approaches. The first concerns quantifying the entire viral community in terms of abundance or composition; the other approach has as its primary objective the examination of populations of viruses that infect a single host. This chapter brings together methods that can be used to count and examine the genetic diversity of communities and populations of aquatic viral communities, although many of the methods can be adapted for other environments.

COMMUNITY APPROACHES: ABUNDANCE AND DIVERSITY

Determination of Total Viral Abundance in Aquatic Samples by Nucleic Acid Staining

There are numerous instances where it is desirable to determine the abundance of viruses in aqueous samples. These include studies of factors regulating viral abundance in nature, as well as routine environmental and public health monitoring programs. The latter includes waste treatment, public water supplies, groundwater, and bathing areas. Typical concentrations of viral particles in coastal systems are in excess of 10^7 ml^{-1}. Overall, the abundance of viruses in marine water and freshwater is positively correlated with the concentration of chlorophyll and bacterial abundance and production (21, 26, 41, 44, 70). Researchers have also examined seasonal changes in the total concentration of viruses. In Tampa Bay, a productive estuary on the west coast of Florida (41) viral abundance ranged about 10-fold (from 0.2×10^7 to 2.3×10^7 ml^{-1}). The highest concentrations occurred in the summer and fall, when bacteria and chlorophyll were also most abundant. Similarly, changes in viral abundance in a large mesotrophic lake covaried with the abundance of bacteria (37). Although there are exceptions, for samples collected from a eutrophic lake in France, Bettarel et al. (4) found a weak correlation between bacterial and viral abundance and speculated that this was due to the dominance of nonbacteriophage virus-like particles (VLPs).

The first studies on viral abundance in aquatic systems used transmission electron microscopy (TEM) to count VLPs. For aquatic samples, this requires concentration of the viruses about 100- to 1,000-fold. Typically, the viruses are pelleted directly onto electron microscopy grids by ultracentrifugation (7, 21, 76), or prefiltered through 0.2-μm-pore-size filters and concentrated by ultrafiltration before being spotted and dried onto the surface of the grids (51). TEM-based methods have the advantage that viral particles are directly observed; however, no information on the infectivity of the particles is obtained, and the hosts that the viruses infect remain unknown. Moreover, problems arise in samples with high concentrations of particulate material, as nonviral particles can obscure the grid surface, making counting difficult. Consequently, estimates of viral abundance made by using TEM often have relatively large coefficients of variation. Despite these difficulties, TEM can be used to estimate the abundance of viral particles in aquatic samples. However, the cost of transmission electron microscopes and TEM-based methods, as well as the time required to prepare samples, has motivated the development of epifluorescence microscopy (EFM) for counting viruses in natural samples.

DAPI (4′,6′-diamidino-2-phenylindole) (49) is a stain which fluoresces when bound to double-stranded DNA (dsDNA) and which has been used to estimate viral abundance in aquatic samples (34, 53, 66). Estimates of viral abundance obtained by use of DAPI can be comparable to those obtained by TEM (53). However, DAPI-stained viruses are near the limit of detection by EFM and fade quickly (5). Accurate counts require excellent optical equipment. Moreover, only dsDNA viruses are visible by DAPI staining. Although marine viral communities appear to consist of primarily dsDNA viruses, this will not be true for all samples. Further details on virus counting by TEM and DAPI staining can be found elsewhere (63). More recently, virus-counting protocols that use EFM in combination with a new generation of very bright nucleic acid stains have been developed (19, 36, 48). The precision and accuracy of the counts appear to be greater than those obtained by TEM, and results suggest that viral abundance can be significantly underestimated by using TEM (5, 36).

There are several constraints to the method. Background fluorescence may prohibit use of EFM on samples with high humic content, and bacteria as well as detritus are stained. Viruses can generally be distinguished from detritus by staining characteristics, and bacteria can usually be recognized by shape, although some very small bacteria may be counted as viruses. Counting bacteria as viruses is more of a theoretical problem than a practical one. Even if all bacteria in a natural sample were counted as viruses, the error that would be introduced would typically be less than the standard deviation of the estimate of viral abundance (36). The actual error is much less, because only a very small proportion of the bacteria in a water sample are likely to be confused with viruses. However, estimates of bacterial abundance can be easily overestimated by including viruses in counts of bacteria. As is the case for TEM-based methods, no information is provided on the number of infective viruses and the viral hosts remain unknown.

Advantages of the EFM method include rapid sample preparation and the ability to process large numbers of samples in parallel. Also, EFM is a routine method in many laboratories, and the procedures are amenable for use at field stations or onboard ship. Furthermore, EFM can be adapted for enumerating viruses in sediments and soils (27, 28, 74). The use of flow cytometry (FCM) in conjunction with SYBR green I or SYBR gold has generated counts comparable to those of EFM while further increasing the speed of analysis and accuracy of counting of free aquatic viral particles (15, 44). Subsequently, FCM has been used to assay phytoplankton viability following viral infection (16). A review of FCM methodology can be found in reference 14.

The stains that are currently most widely used for EFM are YOPRO-1 (36), SYBR green I (48), and SYBR gold (19). Although the choice of stain is important, it appears that the most important step in producing accurate and reproducible results with EFM is to process samples immediately after collection. Wen et al. (78) demonstrated that viral abundance in aldehyde-fixed samples stored at 4°C decreased as much as 35% after only 4 h of storage. The authors concluded that regardless of the fluorescent stain used, slides for virus enumeration must be made immediately following sample collection. If this is impossible, samples may be fixed with glutaraldehyde and immediately flash frozen in liquid nitrogen for later processing. When samples were processed immediately after collection, there was no significant difference between estimates of viral abundance made by using YOPRO-1, SYBR green I, or SYBR gold (78).

Nevertheless, the SYBR-based and YOPRO-1-based methods both have strengths and weaknesses. YOPRO-1-fixed slides can be kept frozen for at least a year with no apparent loss in quality (78), and viruses stained with YOPRO-1 fluoresce more brightly and are more stable (5) and do not require an antifade mounting medium. However, the YOPRO-1 epifluorescence method requires a longer staining time than the SYBR method (~2 days, instead of ~30 min), although heating can be used to accelerate stain penetration and binding with YOPRO-1 (85). In a comparison of SYBR green I and SYBR gold, Chen et al. (19) concluded that SYBR gold yielded a brighter, more stable fluorescent signal. SYBR gold is also less expensive. Outlined below is the EFM method for YOPRO-1. For the SYBR green I method, see reference 47 or 48.

Outline of Method

This method is used to determine the total abundance of VLPs.

1. Dilute a stock solution of YOPRO-1 (1 mM YOPRO-1 in 1:4 dimethyl sulfoxide and water) (Molecular Probes, Inc., Eugene, Oreg.) to 50 μM YOPRO-1 in an aqueous solution of 2 mM NaCN.
2. Place a series of 80-μl drops of the solution from step 1 on the bottom of a 10-cm-diameter plastic petri dish. Place a filter paper soaked with 3 ml of an aqueous NaCl solution (0.3% [wt/vol]) in the lid to prevent evaporation of the stain.
3. Dilute unpreserved samples (100 μl) with 700 μl of sterile deionized-distilled water and place them on the surface of a 0.02-μm-pore-size Al_2O_3 Anodisc 25 membrane filter (Whatman, Florham Park, N.J.). The samples should remain within the plastic support ring of the filter by surface tension. The samples cannot be preserved, as aldehyde fixatives prevent binding of YOPRO-1, although heating can circumvent the problem somewhat for short-term storage (85). Dilution of the samples is necessary so that a reasonable volume can be filtered (800 μl) and because the high concentration of divalent cations in many natural water samples can interfere with binding of the stain.
4. Gently filter (15 kPa) each sample, using a premoistened 0.45-μm-pore-size cellulose-nitrate membrane as a backing filter.
5. Take the moist Anodisc membranes on which the samples have been filtered, and lay them (sample side up) on drops of the staining solution. Incubate in the covered petri dish for 2 days in the dark at room temperature.
6. Filter two 800-μl aliquots of deionized-distilled water through the membrane to rinse it.
7. Transfer the damp membranes to glass slides and immediately cover each with a drop of spectrophotometry-grade glycerol (Fisher, Nepean, Canada) and a coverslip and store at −20°C until processing.
8. Count >200 viruses in 20 randomly selected fields at ×1,000 using an epifluorescence microscope equipped with an acridine orange filter set (excitation, <490 nm; dichroic filter, 500 nm; barrier filter, >515 nm).
9. The number of viruses per milliliter is calculated as $P \times (A/F) \times (1{,}000/V)$ where P is the number of stained viruses in the microscope field, A is the filtration area of the filter (in square micrometers), F is the area of the microscope field (in square micrometers), and V is the volume of original sample prior to dilution or addition of stain (in microliters).

Morphological Characterization of Viral Communities by TEM

As discussed above, estimates of viral abundance based on TEM have been found to be inaccurate; however, TEM is currently the only method that permits direct determination of the morphology of individual viruses in a virus community. This method can be used to gather data on the size structure of natural viral communities (reviewed in detail in reference 73) and can even be used to determine the abundance of visibly infected cells and burst size for natural bacterial populations (reviewed in references 77 and 81).

Research investigating the size distribution of marine viruses from a variety of regions shows that viruses with capsid diameters of <60 nm dominate marine (39, 75, 82) and freshwater (37) systems. Although viruses with significantly larger capsid diameters (>110 nm) from Arctic sea ice have been reported (33), 98% of the virioplankton of Lake Superior were observed to have capsid diameters of <60 nm (68). Additionally, most viruses that have been isolated from marine systems are larger than the most abundant size classes observed in natural viral communities by TEM (6). Interestingly, although TEM images of natural communities suggest that viruses with small noncontractile tails are most abundant (3), the most frequently isolated viruses are tailed sipho- and myoviruses (50, 62). This discrepancy may be due to the method of sample preparation used in many TEM studies of natural virus communities or due to the fact that viral isolates are not representative of native viral communities.

Virus morphology has also been used to investigate the dynamics of specific components of viral communities. For example, Bratbak et al. (8) observed a substantial increase in the abundance of a large icosahedral virus coincident with the collapse of an induced phytoplankton bloom, suggesting that the virus was responsible for its demise. Other studies have shown that populations of small viruses are more dynamic than those of larger viruses (9, 35, 37) and that increases in specific viral morphotypes appear to be associated with the collapse of bacterial communities in natural systems (9, 37).

Typically, viruses are concentrated by ultracentrifugation or ultrafiltration, spotted on an electron microscopy grid, and viewed by TEM. A protocol that can be used for natural viral communities is briefly outlined below.

Outline of Method

This method is used to determine the abundance and morphology of viruses from a natural community.

1. Water samples are pelleted as soon as possible after collection (7). It is preferable that samples not be treated with fixatives or preservatives, which can cause clumping.
2. Use double-sided tape to attach electron microscopy grids coated with carbon and formvar (Ted Pella Inc., Redding, Calif.) to platforms that are flat on the surface and which fit into the bottom of an ultracentrifuge tube that can be placed in a swing-out rotor. Epoxy platforms can be made from resin that is placed in the bottom of an ultracentrifuge tube, spun in a swinging-bucket rotor to make the surface flat, and allowed to harden. For most natural water samples, 100-mm centrifuge tubes concentrate the viruses enough so that they can be viewed by TEM.
3. Place the platforms with the attached grids into the bottom of the centrifuge tubes, fill the tubes with sample, and load the tubes into the rotor. Centrifuge the samples so that particles of 80S are sedimented with 100% efficiency. For most samples, 3 h at 180,000 × *g* is suitable, but for samples of high salinity or which are cold, it may be necessary to increase the time.
4. The liquid is aspirated from the tube, the grids are removed, and the excess water is wicked from the surface of the grids with a piece of filter paper. For saltwater samples, it may also be necessary to wash the grids by placing them sample side down on a drop of distilled water. It is important not to let the surface of the grids become dry during this procedure.
5. The sample is stained by floating the grid facedown on a drop of 1% uranyl acetate for several seconds and immediately wicking the stain from the surface of the grid. This results in areas where stain has been deposited around the perimeter of the viruses, yielding a negative-stained image. This provides the best preparation for observing the morphology of viruses. If the samples are stained too long, the viruses will be positive stained, which obscures detail and causes inaccurate estimates of viral size. Negative-stained preparations are generally not adequate for estimating the abundance of viruses in aquatic samples; however, by placing several grids at the bottom of the ultracentrifuge tubes, positive-stained preparations can be made from the same water sample.
6. The grids are observed by TEM at ca. 80 kV and a magnification of ca. ×30,000. Photographic images of the preparations provide much greater resolution than the phosphorescent screen of the microscope.

Genetic Diversity of Virus Communities: Virus Community Shotgun Library Construction

It is now universally accepted that uncultured microbes far outnumber those in culture, and therefore culture-based methods do not accurately determine the composition and structure of most microbial communities. The diversity of natural communities of archaea and eubacteria has been extensively studied (52) by use of a culture-independent method based on the PCR amplification and phylogenetic analysis of rDNA genes. Unlike the case for archaea and eubacteria, there is no universally conserved, evolutionarily reflective gene encoded by all viruses, and therefore a method based on single-gene surveys cannot be used to study the diversity of the entire virus communities. Breitbart et al. (13) constructed a metagenomic library of two coastal DNA phage communities, using the linker-amplified shotgun library (LASL) method. In this study, 200 liters of seawater was prefiltered and concentrated by tangential flow filtration. The phage fraction was purified from a CsCl gradient, and the viral DNA was extracted and sheared. Linkers were added and PCR that targeted sites specific to these linkers was conducted. Amplicons were subsequently cloned and sequenced. This approach overcame the challenges of working with viruses from the environment, including the low concentration of nucleic acids per viral genome, unclonable viral genes, and modified viral DNA (29). Subsequently, this approach was used to construct libraries from marine sediment (10) and human (11) and equine (17) feces. Analysis of these libraries demonstrated that most of the sequence fragments are novel. From the results of a model based on the observed overlap of sequence fragments, it was estimated that the number of different viral genotypes ranged from approximately 1,000 in the fecal communities (11, 17) to 10^6 in the marine sediment (10). A more detailed description can be found online at http://www.sci.sdsu.edu/PHAGE/guide.html.

Outline of Method

This method is used to estimate the richness and population structure of a natural phage community.

1. Ideally, 1 μg of viral DNA is required to build a library. To acquire this amount of starting material, a large volume of water (~200 liters) is prefiltered through a 0.16-μm-pore-size filter and concentrated by tangential flow filtration. The concentrated viral community is further purified via a CsCl gradient, and the 1.35- to 1.5-g/ml size fraction is collected.
2. After DNase treatment to remove any residual free DNA, viral DNA is extracted (0.1 volume of 2 M Tris-Cl [pH

8.5]–0.2 M EDTA [TE], 1 volume of formamide, and 10 μl of glycogen) and precipitated with a 2× volume of 100% ethanol (EtOH), washed twice with 70% EtOH, resuspended in TE, and incubated for 1 h at 65°C.

3. The extracted, purified DNA is subjected to hydroshearing (HydroShear; GenMachine, San Carlos, Calif.) to disrupt genes which are potentially lethal to the cloning procedure, resulting in sequence fragments from 1 to 2 kb in size.

4. These DNA fragments are end-repaired with T4 DNA polymerase and dsNotI linkers ligated to the blunted ends.

5. Using a high-fidelity DNA polymerase (DeepVent *taq*; New England Biolabs, Ipswich, Mass.), 25 cycles of PCR are performed with primers targeting sites on the ligated dsNotI linkers.

6. The PCR products are gel fractionated. The ~2-kb size fraction is purified and cloned into a pSMART (Lucigen, Middleton, Wis.) vector.

7. Subsequently, the ligation mix is transformed into MC12 electrocompetent cells (Invitrogen, Burlington, Canada). Colonies containing inserts are sequenced.

On the basis of the Lander-Waterman equation, Breitbart et al. (13) developed a model to analyze the structure of uncultured virus communities using the sequencing data generated above. Subsequently, Angly et al. (2) have developed a user-friendly website that provides an in-depth description of the model and allows anyone with Internet access to generate estimates of evenness and richness with any shotgun library data (see http://phage.sdsu.edu/phaccs).

Genetic Diversity of Virus Communities: Fingerprints of Virus Communities with PFGE

Pulsed-field gel electrophoresis (PFGE) is another approach that can be used to examine diversity in natural viral communities. In PFGE, total viral community DNA is separated by size on an agarose gel and the number of bands on the gel is assumed to represent the number of viral species in a sample. The intensity of each band can be used to estimate the number of viral genomes per band (42). An annual study of the Chesapeake Bay (83) revealed a wide range of molecular sizes of viral DNA (ca. 12 to 314 kb). The presence of particular bands, indicating the occurrence of virioplankton with specific genome sizes, varied spatially and temporally. In a survey of a diversity of marine environments, Steward et al. (61) used PFGE to determine that on average >90% of virus-like genomes in surface seawater were in the 26- to 69-kb range, a size typical of bacteriophages. Subsequently, PFGE has been used to track changes in the virioplankton during an *Emiliania huxleyi* bloom (18), in coastal Danish waters over time (54), and over depth at an open water station in southern California (39). Although PFGE has relatively low sensitivity and resolution for detecting changes in community diversity, it has the distinct advantage of allowing the entire viral community to be examined, rather than subsets. A protocol for this method can be found in reference 60.

POPULATION APPROACHES

Abundance of Infectious Viral Populations

Community approaches are useful for studies of the total abundance of viral particles or for documentation of the dominant morphotypes in viral communities. However, these methods do not yield information about the infectivity of viruses in natural communities or about the organisms that the viruses infect. Most viruses in the ocean are infectious (80), and virus host ranges are generally strain specific (6, 46). Although free viruses are inactivated rapidly in the water column (67), some viruses can remain infectious in sediments for hundreds of years (43). Concentrations of viruses which infect specific hosts vary from less than a few viruses per milliliter to $>10^5$ ml^{-1}. Typically, investigations have found relatively low titers of viruses that infect specific isolates of heterotrophic bacteria, although $>10^4$ ml$^-$ has been reported in some instances (45). Concentrations of viruses infecting phytoplankton can be higher. For example, viruses which infect strains of the cyanobacterium *Synechococcus* can occur in excess of 250,000 infective units ml^{-1} in coastal seawater and routinely occur at $>10^3$ ml^{-1} (65). Eukaryotic phytoplankton are also subject to infection by lytic viruses, and abundances can be high; concentrations of $>10^5$ ml^{-1} have been measured in coastal seawater for viruses which infect the unicellular photosynthetic flagellate *Micromonas pusilla* (24) and surpassed 10^3 ml^{-1} during a bloom of the red-tide-forming raphidophyte *Heterosigma akashiwo* (69).

In order to infer the effect of viruses on specific taxa within the microbial community, it is necessary to measure the abundance of viruses that infect specific hosts. Typically, this requires a bioassay; therefore, the host organism(s) must be in culture. As many microorganisms are difficult to isolate and culture, this is a major limitation. For bacteria and phytoplankton that can be grown on solid media, plaque assays are relatively straightforward and give the greatest precision for estimates of the abundances of lytic viruses. Unfortunately, many isolates do not grow well on solid substrates, and a most-probable-number (MPN) assay in liquid culture must be used. The precision of MPN assays is less than that of plaque assays, but the assays are extremely sensitive and, if carefully optimized, can be used to estimate the abundance of a few infectious viruses per liter. Below are generalized descriptions of plaque and MPN assays.

The basis behind plaque assays is that viruses form visible plaques on a lawn of host cells. Typically, host cells and the sample to be assayed are mixed in molten agar which is then poured quickly and evenly over a solidified bottom layer of agar. As the bacteria or algae grow, the lawn becomes opaque, and areas where lysis of cells has occurred appear as clearings on the lawn. The number of viruses is quantified as the number of PFU in the volume of water sampled. The method has the advantages that it is relatively sensitive with a detection limit of about 5 viruses ml^{-1}, only viruses that are infectious for a specific host are enumerated, results are accurate, and the infective agent can be purified relatively easily from individual plaques. A protocol for a plaque assay can be found in reference 1.

When the host organism cannot be grown on solid medium, the number of infective viruses must be estimated by MPN assay. The principle of the method is that the sample to be titered is taken through a series of dilutions and added to liquid cultures of host cells. In those cultures to which one or more infective viruses have been added, lysis of the culture will occur. The estimate of viral titer is obtained from the number of cultures in which lysis occurs at each step in the dilution series. Typically, a series of 10-fold dilutions with 3 to 10 replicates at each dilution is added to culture tubes, and the concentration of infective viruses is estimated by comparing the number of replicates at each dilution in which lysis occurs with an MPN table. In many instances, greater accuracy and precision are obtained by conducting the assays in 96-well microtiter plates rather than culture tubes. The protocol works especially well for phytoplankton (24, 65, 69, 72). However, when greater sensitivity than ca. 50 viruses ml^{-1} is required, it is necessary to use larger cultures, so that a larger volume of sample can be screened. The microtiter procedure is outlined below; a more detailed discussion of MPN assays is provided elsewhere (79).

Outline of Method

This method is used to determine the abundance of infective viruses of a specific host.

1. Prepare the medium that will be used to construct the dilution series of the natural water sample that will be titered. The medium should allow vigorous growth of the host organism, be free of any viruses that might cause lysis, and be free of particulate material to which viruses might adsorb. Depending on the host-virus system, suitable media include natural water that has been ultrafiltered or filter sterilized and autoclaved or artificial media.
2. Transfer the assay organism into fresh medium so that it will be near the beginning of exponential growth when it is dispensed into the microtiter plates.
3. Collect the sample to be titered in a clean container, and prepare a series of 10-fold dilutions. It may be necessary to prefilter the sample through 1.0- or 0.2-μm-pore-size filters to remove other organisms that might overgrow the assay organism.
4. Transfer 200 to 250 μl of the exponentially growing culture into each well of a microtiter plate, and for each dilution add ca. 50 μl to each of 8 to 16 wells of the microtiter plate. Include 8 to 16 wells to which only medium is added to ensure that it supports vigorous growth.
5. After sufficient time has passed to complete about 10 lytic cycles, score the plates for the number of wells that have cleared at each dilution series. The plates should be monitored on a regular basis during this interval since resistant cells can grow in wells in which lysis has occurred, thereby causing incorrect estimates of MPN. Ideally, an aliquot from each well should be transferred into fresh, exponentially growing cultures to ensure that the lytic agent can be propagated, or in case a virus has not been amplified enough to cause lysis. Calculate the MPN by using tables or preferably a computer program that also provides statistical parameters for the estimates (38).

Genetic Diversity of Viral Populations: Fingerprinting of Isolates

Viruses are an abundant, dynamic, and ecologically important component of aquatic ecosystems, and there is now strong evidence that they are the most genetically diverse biological entities on the planet (29). Virus morphology provides little information about viral lifestyle, and culture-based approaches are limited by the relatively restricted number of hosts in culture; therefore, molecular tools targeting viral nucleic acids directly are the evident choice for studying the population diversity and dynamics of related groups of viruses. At present, molecular methods have been developed to describe the diversity and track temporal changes in subsets of both DNA and RNA viruses and to genetically fingerprint newly isolated virus clones (see below), and as research progresses, more precise and informative approaches are being developed at a rapid pace. A comprehensive review of molecular techniques is beyond the scope of this chapter; however, we have presented outlines for current molecular methods in marine virology that have been particularly successful and informative.

Restriction digestion of viral genomic DNA (23, 39, 41) has been used to characterize genetic variability among marine virus isolates. In this method, purified genomic DNAs from virus isolates are simply subjected to digestion by the same restriction endonucleases. Differences in restriction patterns are due to sequence differences in the virus genomes being compared. Cottrell and Suttle (23) found that although viruses isolated from within and among locations were morphologically indistinguishable by electron microscopy, all were genotypically different. A comparison of the digestion patterns of bacteriophages isolated from different depths at a Pacific Ocean station demonstrated the significantly greater diversity of phages in comparison to their hosts (39). Although this approach is relatively straightforward, it requires large amounts of purified viral DNA from an individual virus. Recently, Comeau et al. (22) described a method for the characterization of isolated viruses that requires relatively little starting DNA template (~30 ng) and takes little time to execute. Degenerate primed random amplification of polymorphic DNA (DP-RAPD) uses degenerate primers in conjunction with PCR to generate fingerprints of viral genomes. This method was shown to produce reproducible fingerprints for bacteriophage and algal viruses encompassing an array of genome sizes and virus families. Moreover, the method continued to reproduce viral fingerprints even in the presence of contaminating host DNA. A description of the method is included below.

Outline of Method

This method is used to generate fingerprints of individual viral genomes.

1. DNA from prefiltered viral lysate of an individual virus strain can be extracted by use of either phenol-chloroform or a hot-cold treatment in a thermocycler (3 cycles of 95°C for 2 min and 4°C for 2 min). Hot-cold-treated samples must be diluted 10-fold with buffer to reduce salts in the subsequent PCR mixture.
2. The DP-RAPD PCR mixture consists of 1.5 U of Platinum *Taq* DNA polymerase (Invitrogen), 1× *Taq* DNA polymerase buffer, 4.5 mM $MgCl_2$, 0.6 mM each deoxyribonucleoside triphosphate (dNTP), and 2 μM R10D primer (5′-GTCASSWSSW-3′, where S and W represent G/C and A/T, respectively) in a 50-μl volume.
3. PCR is conducted under the following thermocycler conditions: an initial denaturation step at 95°C for 1 min 30 s followed by 40 cycles of denaturation at 95°C for 45 s, annealing at 40°C for 3 min, an extension for 1 min at 72°C and then a final extension at 72°C for 10 min.
4. A 10-μl volume of each PCR product is loaded onto a 1.5% agarose gel in 0.5× TBE buffer (45 mM Tris-borate, 1 mM EDTA [pH 8.0]). The gel is run at 90 V for 3 h, stained with ethidium bromide, and visualized under UV radiation.
5. Comeau et al. (22) used GelCompar II (Applied Maths, Austin, Tex.) to quantitatively compare fingerprints. Refer to the work of Ferrari and Hollibaugh (30) for more detail concerning this program.

Genetic Diversity of Viral Populations: Degenerate PCR and Reverse Transcription-PCR Targeting Single Genes

Sensitive molecular approaches have been developed to study the diversity of specific virus groups from natural samples. Although no gene is shared universally in viruses, researchers have found that subsets of viruses share genes that produce phylogenies that are congruent with established viral taxonomy based on virus structure, host, and epidemiology. Researchers have used PCR with primers targeting evolutionarily informative genes to explore the diversity of a variety of important virus groups in the ocean. In this approach, a fragment of the target molecule is amplified from a community of extracted viral nucleic acids by PCR. The diversity of amplicons in this reaction can be assessed by a number of methods, including cloning of amplicons and sequencing (12, 25), cloning followed by a comparison of insert restriction enzyme digestion patterns (20), separation of products on a denaturing gel and comparison of community fingerprints (31, 58, 86), and endonuclease digestion of fluorescently end-labeled PCR products followed by the generation of community profiles on an automated sequence analyzer (71).

This approach was first used to investigate the diversity of viruses in the family *Phycodnaviridae*, a group of large dsDNA viruses that infect algae (20). Targeting the viral DNA polymerase, this research revealed a vast amount of genetic variation that was not represented in cultures and showed that very similar sequences were distributed on a global scale (58). A subsequent temporal study in British Columbia, Canada, showed that the algal virus community is remarkably stable, even while the host community is undergoing dramatic shifts in composition (59). On the basis of the amplification of a sequence from a capsid gene conserved among *Emiliania huxleyi* viruses (EhV) (55), it was demonstrated that although a highly diverse population of EhV was present at the beginning, only a small number of dominant virus strains participated in the final eradication of a bloom of the globally important coccolithophorid *Emiliania huxleyi*.

Fuller et al. (32) developed a PCR-based approach targeting g20, a capsid gene conserved in a subset of myoviruses. Investigations based on this method revealed the incredible diversity present in myovirus communities (31, 71, 86). Yet, despite this great diversity, nearly identical sequences were recovered from virus communities ranging from Arctic waters to freshwater catfish ponds (56). Similarly, genetically indistinguishable podovirus sequences have been found in a wide range of environments (12). Finally, degenerate reverse transcription-PCR has been used to assess the diversity of RNA virus communities in the ocean. Targeting a conserved fragment of the RNA-dependent RNA polymerase (RdRp) gene found among members of the picorna-like virus superfamily, Culley et al. (25) screened RNA virus communities from samples collected in British Columbia coastal waters. Not only did amplification occur in different years at the same station, but it also occurred in samples from a variety of environments ranging from estuarine to marine and urban to pristine. Phylogenies based on RdRp sequences demonstrated that the environmental sequences clustered outside of all established RNA virus groups and may represent members of four novel RNA virus families. While a method for the amplification and analysis of RdRp genes from aquatic RNA viral communities is outlined below, descriptions of the use of degenerate PCR and denaturing gradient gel electrophoresis to study algal virus diversity (57) and myovirus diversity (32) can be found elsewhere.

Outline of Method

This method is used to determine the diversity of RNA picorna-like viruses in a seawater sample.

1. Concentrated viral community nucleic acids are extracted with a column-based kit (QIAmp Minelute virus spin kit; QIAGEN, Mississauga, Canada) according to the protocol provided by the manufacturer and eluted in 20 μl of elution buffer.
2. Samples are treated with DNase to remove undesirable viral DNA. A reaction mixture of extracted sample, 1 U of DNase I, and 1× DNase buffer (Invitrogen) is incubated for 15 min at room temperature. After digestion, 1 μl of 25 mM EDTA is added and the reaction mixture is heated to 65°C for 15 min to terminate the reaction.
3. A first-strand synthesis reaction mixture composed of the DNase-treated sample, 0.1 μM (final concentration) primer RdRp 2 (5′-MAC CCA ACK MCK CTT SAR RAA-3′), and 0.5 mM dNTP is heated to 65°C for 5 min. The reaction mixture is placed on ice, and 7 μl of 1× first-strand buffer, 5 mM dithiothreitol, 1× RNaseOUT, and 1 U of SS III is added to make a final volume of 20 μl. In a thermocycler, this reaction mixture is incubated at 50°C for 50 min and then at 70°C for 15 min and subsequently placed on ice.
4. One unit of RNase H is added, and the reaction mixture incubated at 37°C for 20 min. This step removes the RNA strand of the RNA-DNA hybrid and may increase template availability during the PCR.
5. A PCR mixture with 5 μl of cDNA template, 1× PCR buffer, 3 mM $MgCl_2$, 0.6 mM dNTP mix, 1 μM RdRp 1 (5′-GGR GAY TAC ASC IRW TTT GAT-3′) and 1 uM RdRp 2 (5′-MAC CCA ACK MCK CTT SAR RAA-3′), and 1 U of Platinum *Taq* DNA polymerase (Invitrogen) is heated in a thermocycler under the following conditions: an initial denaturation step at 94°C for 1 min 15 s followed by 40 cycles of denaturation at 95°C for 45 s annealing at 50°C for 45 s, an extension for 1 min at 72°C, and then a final extension at 72°C for 5 min.
6. 15-μl volume of each PCR product is loaded onto a 1.5% agarose gel in 0.5× TBE buffer. The gel is run at 100 V for 45 min, stained with ethidium bromide, and visualized under UV radiation. The target product is approximately 500 bp in size.
7. The community of amplicons in this reaction can be analyzed in a variety of ways dependent on the research question. Further steps include the use of denaturing gradient gel electrophoresis (57) or T-RFLP (71) to generate community fingerprints of picorna-like viruses based on RdRp genes, and/or cloning and sequencing to produce sequence phylogenies (10, 25).

Genetic Diversity of Viral Populations: DNA Hybridization

DNA hybridization can be used to quantify the genetic differences among viruses. By labeling entire genomes from several *Micromonas purilla* virus (MpV) isolates and using them as probes against the DNA from other isolates, it was shown that there was as much diversity among MpV isolates from a single water sample as there was between isolates from different oceans (23). Hybridization was also used to examine the genetic diversity among vibriophage isolates from Florida and Hawaii. With a DNA fragment cloned from a marine vibriophage used as a probe, it was shown that virus isolates from Hawaii were less diverse than those from Florida (41). Wommack et al. (84) used DNA hybridization in conjunction with PFGE to demonstrate dramatic, episodic shifts in the virioplankton of the Chesapeake Bay. The authors suggested that these data support the hypothesis that viruses greatly influence clonal diversity in their hosts. Recently, DNA hybridization was used to investigate phage distribution at an open water station in the subtropical Pacific (39) and in a hypersaline lake (40). Interestingly, phages from the lake were present only in samples taken from the same time and place, while phages from the Pacific station were persistent over 1 year. Hybridization analysis is useful only for making comparisons among closely related isolates. Moreover, if entire genomic DNA is used as a probe, the task of doing all pairwise comparisons among the viral isolates can be daunting. Also, reciprocal probing may not yield the same result (e.g., probing DNA A with DNA B may not yield the same result as probing DNA B with DNA A). A protocol for this method can be found in reference 84.

REFERENCES

1. **Adams, M. H.** 1959. *Bacteriophages*. John Wiley & Sons, New York, N.Y.
2. **Angly, F., B. Rodriguez-Brito, D. Bangor, P. Mcnairnie, M. Breitbart, P. Salamon, B. Felts, J. Nulton, J. Mahaffy, and F. Rohwer.** 2005. PHACCS, an online tool for estimating the structure and diversity of uncultured viral communities using metagenomic information. *BMC Bioinformatics* **6:**41.
3. **Bergh, O., K. Y. Børsheim, G. Bratbak, and M. Heldal.** 1989. High abundance of viruses found in aquatic environments. *Nature* **340:**467–468.
4. **Bettarel, Y., C. Amblard, T. Sime-Ngando, J. F. Carrias, D. Sargos, F. Garabetian, and P. Lavandier.** 2003. Viral lysis, flagellate grazing potential, and bacterial production in Lake Pavin. *Microb. Ecol.* **45:**119–127.
5. **Bettarel, Y., T. Sime-Ngando, C. Amblard, and H. Laveran.** 2000. A comparison of methods for counting viruses in aquatic systems. *Appl. Environ. Microbiol.* **66:**2283–2289.
6. **Børsheim, K. Y.** 1993. Native marine bacteriophages. *FEMS Microbiol. Lett.* **102:**141–159.
7. **Børsheim, K. Y., G. Bratbak, and M. Heldal.** 1990. Enumeration and biomass estimation of planktonic bacteria and viruses by transmission electron microscopy. *Appl. Environ. Microbiol.* **56:**352–356.
8. **Bratbak, G., J. K. Egge, and M. Heldal.** 1993. Viral mortality of the marine alga *Emiliania huxleyi* (Haptophyceae) and termination of algal blooms. *Mar. Ecol. Prog. Ser.* **93:**39–48.
9. **Bratbak, G., M. Heldal, S. Norland, and T. Thingstad.** 1990. Viruses as partners in spring bloom microbial trophodynamics. *Appl. Environ. Microbiol.* **56:**1400–1405.
10. **Breitbart, M., B. Felts, S. Kelley, J. M. Mahaffy, J. Nulton, P. Salamon, and F. Rohwer.** 2004. Diversity and population structure of a near-shore marine-sediment viral community. *Proc. R. Soc. London B* **271:**565–574.
11. **Breitbart, M., I. Hewson, B. Felts, J. M. Mahaffy, J. Nulton, P. Salamon, and F. Rohwer.** 2003. Metagenomic analyses of an uncultured viral community from human feces. *J. Bacteriol.* **185:**6220–6223.
12. **Breitbart, M., J. H. Miyake, and F. Rohwer.** 2004. Global distribution of nearly identical phage-encoded DNA sequences. *FEMS Microbiol. Lett.* **236:**249–256.
13. **Breitbart, M., P. Salamon, B. Andresen, J. M. Mahaffy, A. M. Segall, D. Mead, F. Azam, and F. Rohwer.** 2002. Genomic analysis of uncultured marine viral communities. *Proc. Natl. Acad. Sci. USA* **99:**14250–14255.
14. **Brussaard, C. P. D.** 2004. Optimization of procedures for counting viruses by flow cytometry. *Appl. Environ. Microbiol.* **70:**1506–1513.
15. **Brussaard, C. P. D., D. Marie, and G. Bratbak.** 2000. Flow cytometric detection of viruses. *J. Virol. Methods* **85:**175–182.
16. **Brussaard, C. P. D., D. Marie, R. Thyrhaug, and G. Bratbak.** 2001. Flow cytometric analysis of phytoplankton viability following viral infection. *Aquat. Microb. Ecol.* **26:**157–166.
17. **Cann, A. J., S. E. Fandrich, and S. Heaphy.** 2005. Analysis of the virus population present in equine faeces indicates the presence of hundreds of uncharacterized virus genomes. *Virus Genes* **30:**151–156.
18. **Castberg, T., A. Larsen, R. A. Sandaa, C. P. D. Brussaard, J. K. Egge, M. Heldal, R. Thyrhaug, E. J. Van Hannen, and G. Bratbak.** 2001. Microbial population dynamics and diversity during a bloom of the marine coccolithophorid *Emiliania huxleyi* (Haptophyta). *Mar. Ecol. Prog. Ser.* **221:**39–46.
19. **Chen, F., J.-R. Lu, B. J. Binder, Y.-C. Liu, and R. E. Hodson.** 2001. Application of digital image analysis and flow cytometry to enumerate marine viruses stained with SYBR Gold. *Appl. Environ. Microbiol.* **67:**539–545.
20. **Chen, F., C. A. Suttle, and S. M. Short.** 1996. Genetic diversity in marine algal virus communities as revealed by sequence analysis of DNA polymerase genes. *Appl. Environ. Microbiol.* **62:**2869–2874.
21. **Cochlan, W. P., J. Wikner, G. F. Steward, D. C. Smith, and F. Azam.** 1993. Spatial distribution of viruses, bacteria and chlorophyll a in neritic, oceanic and estuarine environments. *Mar. Ecol. Prog. Ser.* **92:**77–87.
22. **Comeau, A. M., S. Short, and C. A. Suttle.** 2004. The use of degenerate-primed random amplification of polymorphic DNA (Dp-Rapd) for strain-typing and inferring the genetic similarity among closely related viruses. *J. Virol. Methods* **118:**95–100.
23. **Cottrell, M. T., and C. A. Suttle.** 1991. Wide-spread occurrence and clonal variation in viruses which cause lysis of a cosmopolitan, eukaryotic marine phytoplankter, *Micromonas pusilla*. *Mar. Ecol. Prog. Ser.* **78:**1–9.
24. **Cottrell, M. T., and C. A. Suttle.** 1995. Dynamics of a lytic virus infecting the photosynthetic marine picoflagellate, *Micromonas pusilla*. *Limnol. Oceanogr.* **40:**730–739.
25. **Culley, A. I., A. S. Lang, and C. A. Suttle.** 2003. High diversity of unknown picorna-like viruses in the sea. *Nature* **424:**1054–1057.
26. **Culley, A. I., and N. A. Welschmeyer.** 2002. The abundance, distribution, and correlation of viruses, phytoplankton, and prokaryotes along a Pacific Ocean transect. *Limnol. Oceanogr.* **47:**1508–1513.
27. **Danovaro, R., and M. Serresi.** 2000. Viral density and virus-to-bacterium ratio in deep-sea sediments of the eastern Mediterranean. *Appl. Environ. Microbiol.* **66:**1857–1861.
28. **Drake, L. A., K. H. Choi, A. G. E. Haskell, and F. C. Dobbs.** 1998. Vertical profiles of virus-like particles and bacteria in the water column and sediments of Chesapeake Bay, USA. *Aquat. Microb. Ecol.* **16:**17–25.
29. **Edwards, R. A., and F. Rohwer.** 2005. Viral metagenomics. *Nat. Rev. Microbiol.* **3:**504–510.
30. **Ferrari, V. C., and J. T. Hollibaugh.** 1999. Distribution of microbial assemblages in the central Arctic Ocean Basin studied by PCR/DGGE: analysis of a large data set. *Hydrobiologia* **401:**55–68.
31. **Frederickson, C. M., S. M. Short, and C. A. Suttle.** 2003. The physical environment affects cyanophage communities in British Columbia inlets. *Microb. Ecol.* **46:**348–357.
32. **Fuller, N. J., W. H. Wilson, I. R. Joint, and N. H. Mann.** 1998. Occurrence of a sequence in marine cyanophages similar to that of T4 g20 and its application to PCR-based detection and quantification techniques. *Appl. Environ. Microbiol.* **64:**2051–2060.
33. **Gowing, M. M., B. E. Riggs, D. L. Garrison, A. H. Gibson, and M. O. Jeffries.** 2002. Large viruses in Ross

Sea late autumn pack ice habitats. *Mar. Ecol. Prog. Ser.* **241:**1–11.
34. **Hara, S., K. Terauchi, and I. Koike.** 1991. Abundance of viruses in marine waters: assessment by epifluorescence and transmission electron microscopy. *Appl. Environ. Microbiol.* **57:**2731–2734.
35. **Heldal, M., and G. Bratbak.** 1991. Production and decay of viruses in aquatic environments. *Mar. Ecol. Prog. Ser.* **72:**205–212.
36. **Hennes, K. P., and C. A. Suttle.** 1995. Direct counts of viruses in natural waters and laboratory cultures by epifluorescence microscopy. *Limnol. Oceanogr.* **40:**1050–1055.
37. **Hennes, K., and M. Simon.** 1995. Significance of bacteriophages for controlling bacterioplankton growth in a mesotrophic lake. *Appl. Environ. Microbiol.* **61:**333–340.
38. **Hurley, M. A., and M. E. Roscoe.** 1983. Automated statistical analysis of microbial enumeration by dilution series. *J. Appl. Bacteriol.* **55:**159–164.
39. **Jiang, S., W. Fu, W. Chu, and J. A. Fuhrman.** 2003. The vertical distribution and diversity of marine bacteriophage at a station off Southern California. *Microb. Ecol.* **45:**399–410.
40. **Jiang, S., G. Steward, R. Jellison, W. Chu, and S. Choi.** 2004. Abundance, distribution, and diversity of viruses in alkaline, hypersaline Mono Lake, California. *Microb. Ecol.* **47:**9–17.
41. **Kellogg, C. A., J. B. Rose, S. C. Jiang, J. Thurmond, and J. H. Paul.** 1995. Genetic diversity of related vibriophages isolated from marine environments around Florida and Hawaii, USA. *Mar. Ecol. Prog. Ser.* **120:**89–98.
42. **Klieve, A. V., and R. A. Swain.** 1993. Estimation of ruminal bacteriophage numbers by pulsed-field gel electrophoresis and laser densitometry. *Appl. Environ. Microbiol.* **59:**2299–2303.
43. **Lawrence, J. E., A. M. Chan, and C. A. Suttle.** 2002. Viruses causing lysis of the toxic bloom-forming alga, *Heterosigma akashiwo* (Raphidophyceae), are widespread in coastal sediments of British Columbia, Canada. *Limnol. Oceanogr.* **47:**545–550.
44. **Marie, D., C. P. D. Brussaard, R. Thyrhaug, G. Bratbak, and D. Vaulot.** 1999. Enumeration of marine viruses in culture and natural samples by flow cytometry. *Appl. Environ. Microbiol.* **65:**45–52.
45. **Moebus, K., and H. Nattkemper.** 1983. Taxonomic investigations of bacteriophage sensitive bacteria isolated from marine waters. *Helgol. Meeresunters.* **36:**357–373.
46. **Nagasaki, K., Y. Tomaru, K. Nakanishi, N. Hata, N. Katanozaka, and M. Yamaguchi.** 2004. Dynamics of *Heterocapsa circularisquama* (Dinophyceae) and its viruses in Ago Bay, Japan. *Aquat. Microb. Ecol.* **34:**219–226.
47. **Noble, R. T.** 2001. Enumeration of viruses. *Methods Microbiol.* **30:**43–50.
48. **Noble, R. T., and J. A. Fuhrman.** 1998. Use of SYBR Green I for rapid epifluorescence counts of marine viruses and bacteria. *Aquat. Microb. Ecol.* **14:**113–118.
49. **Porter, K. G., and Y. S. Feig.** 1980. The use of DAPI for identifying and counting aquatic microflora. *Limnol. Oceanogr.* **25:**943–948.
50. **Proctor, L. M.** 1997. Advances in the study of marine viruses. *Microsc. Res. Technol.* **37:**136–161.
51. **Proctor, L. M., and J. A. Fuhrman.** 1990. Viral mortality of marine bacteria and cyanobacteria. *Nature* **343:**60–62.
52. **Rappe, M. S., and S. J. Giovannoni.** 2003. The uncultured microbial majority. *Annu. Rev. Microbiol.* **57:**369–394.
53. **Ricciardi-Rigault, M., D. F. Bird, and Y. T. Prairie.** 2000. Changes in sediment viral and bacterial abundances with hypolimnetic oxygen depletion in a shallow eutrophic Lac Brome (Quebec, Canada). *Can. J. Fish. Aquat. Sci.* **57:** 1284–1290.
54. **Riemann, L., and M. Middelboe.** 2002. Stability of bacterial and viral community compositions in Danish coastal waters as depicted by DNA fingerprinting techniques. *Aquat. Microb. Ecol.* **27:**219–232.
55. **Schroeder, D. C., J. Oke, M. Hall, G. Malin, and W. H. Wilson.** 2003. Virus succession observed during an *Emiliania huxleyi* bloom. *Appl. Environ. Microbiol.* **69:**2484–2490.
56. **Short, C. M., and C. A. Suttle.** 2005. Nearly identical bacteriophage structural gene sequences are widely distributed in both marine and freshwater environments. *Appl. Environ. Microbiol.* **71:**480–486.
57. **Short, S. M., and C. A. Suttle.** 2000. Denaturing gradient gel electrophoresis resolves virus sequences amplified with degenerate primers. *BioTechniques* **28:**20–26.
58. **Short, S. M., and C. A. Suttle.** 2002. Sequence analysis of marine virus communities reveals that groups of related algal viruses are widely distributed in nature. *Appl. Environ. Microbiol.* **68:**1290–1296.
59. **Short, S. M., and C. A. Suttle.** 2003. Temporal dynamics of natural communities of marine algal viruses and eukaryotes. *Aquat. Microb. Ecol.* **32:**107–119.
60. **Steward, G. F.** 2001. Fingerprinting viral assemblages by pulsed field gel electrophoresis (PFGE). *Methods Microbiol.* **30:**85–103.
61. **Steward, G. F., J. L. Montiel, and F. Azam.** 2000. Genome size distributions indicate variability and similarities among marine viral assemblages from diverse environments. *Limnol. Oceanogr.* **45:**1697–1706.
62. **Sullivan, M. B., J. B. Waterbury, and S. W. Chisholm.** 2003. Cyanophages infecting the oceanic cyanobacterium *Prochlorococcus*. *Nature* **424:**1047–1051.
63. **Suttle, C. A.** 1993. Enumeration and isolation of viruses, p. 121–134. *In* P. F. Kemp, B. F. Sherr, E. B. Sherr, and J. J. Cole (ed.), *Handbook of Methods in Aquatic Microbial Ecology*. Lewis Publishers, Boca Raton, Fla.
64. **Suttle, C. A.** 2005. Viruses in the sea. *Nature* **437:**356–361.
65. **Suttle, C. A., and A. M. Chan.** 1994. Dynamics and distribution of cyanophages and their effect on marine *Synechococcus* spp. *Appl. Environ. Microbiol.* **60:**3167–3174.
66. **Suttle, C. A., A. M. Chan, and M. T. Cottrell.** 1990. Infection of phytoplankton by viruses and reduction of primary productivity. *Nature* **347:**467–469.
67. **Suttle, C. A., and F. Chen.** 1992. Mechanisms and rates of decay of marine viruses in seawater. *Appl. Environ. Microbiol.* **58:**3721–3729.
68. **Tapper, M. A., and R. E. Hicks.** 1998. Temperate viruses and lysogeny in Lake Superior bacterioplankton. *Limnol. Oceanogr.* **43:**95–103.
69. **Tomaru, Y., K. Tarutani, M. Yamaguchi, and K. Nagasaki.** 2004. Quantitative and qualitative impacts of viral infection on a *Heterosigma akashiwo* (Raphidophyceae) bloom in Hiroshima Bay, Japan. *Aquat. Microb. Ecol.* **34:**227–238.
70. **Vanucci, S., V. Bruni, and G. Pulicano.** 2005. Spatial and temporal distribution of virioplankton and bacterioplankton in a brackish environment (Lake of Ganzirri, Italy). *Hydrobiologia* **539:**83–92.
71. **Wang, K., and F. Chen.** 2004. Genetic diversity and population dynamics of cyanophage communities in the Chesapeake Bay. *Aquat. Microb. Ecol.* **34:**105–116.
72. **Waterbury, J. B., and F. W. Valois.** 1993. Resistance to co-occurring phages enables marine *Synechococcus* communities to coexist with cyanophages abundant in seawater. *Appl. Environ. Microbiol.* **59:**3393–3399.
73. **Weinbauer, M. G.** 2004. Ecology of prokaryotic viruses. *FEMS Microbiol. Rev.* **28:**127–181.

74. **Weinbauer, M. G., C. Beckmann, and M. G. Hofle.** 1998. Utility of green fluorescent nucleic acid dyes and aluminum oxide membrane filters for rapid epifluorescence enumeration of soil and sediment bacteria. *Appl. Environ. Microbiol.* **64:**5000–5003.
75. **Weinbauer, M. G., and P. Peduzzi.** 1994. Frequency, size and distribution of bacteriophages in different marine bacterial morphotypes. *Mar. Ecol. Prog. Ser.* **108:**11–20.
76. **Weinbauer, M. G., and C. A. Suttle.** 1997. Comparison of epifluorescence and transmission electron microscopy for counting viruses in natural marine waters. *Aquat. Microb. Ecol.* **13:**225–232.
77. **Weinbauer, M. G., C. Winter, and M. G. Hofle.** 2002. Reconsidering transmission electron microscopy based estimates of viral infection of bacterio-plankton using conversion factors derived from natural communities. *Aquat. Microb. Ecol.* **27:**103–110.
78. **Wen, K., A. C. Ortmann, and C. A. Suttle.** 2004. Accurate estimation of viral abundance by epifluorescence microscopy. *Appl. Environ. Microbiol.* **70:**3862–3867.
79. **Wilhelm, S. W., and L. Poorvin.** 2001. Quantification of algal viruses in marine samples. *Methods Microbiol.* **30:**53–65.
80. **Wilhelm, S. W., M. G. Weinbauer, C. A. Suttle, R. J. Pledger, and D. L. Mitchell.** 1998. Measurements of DNA damage and photoreactivation imply that most viruses in marine surface waters are infective. *Aquat. Microb. Ecol.* **14:**215–222.
81. **Wommack, K. E., and R. R. Colwell.** 2000. Virioplankton: viruses in aquatic ecosystems. *Microbiol. Mol. Biol. Rev.* **64:**69–114.
82. **Wommack, K. E., R. T. Hill, M. Kessel, C. E. Russek, and R. R. Colwell.** 1992. Distribution of viruses in the Chesapeake Bay. *Appl. Environ. Microbiol.* **58:**2965–2970.
83. **Wommack, K. E., J. Ravel, R. T. Hill, J. Chun, and R. R. Colwell.** 1999. Population dynamics of Chesapeake Bay virioplankton: total community analysis using pulsed-field gel electrophoresis. *Appl. Environ. Microbiol.* **65:**231–240.
84. **Wommack, K. E., J. Ravel, R. T. Hill, and R. R. Colwell.** 1999. Hybridization analysis of Chesapeake Bay virioplankton. *Appl. Environ. Microbiol.* **65:**241–250.
85. **Xenopoulos, M. A., and D. F. Bird.** 1997. Virus a la sauce Yo-Pro: microwave-enhanced staining for counting viruses by epifluorescence microscopy. *Limnol. Oceanogr.* **42:** 1648–1650.
86. **Zhong, Y., F. Chen, S. W. Wilhelm, L. Poorvin, and R. E. Hodson.** 2002. Phylogenetic diversity of marine cyanophage isolates and natural virus communities as revealed by sequences of viral capsid assembly protein gene G20. *Appl. Environ. Microbiol.* **68:**1576–1584.

Protistan Community Structure

DAVID A. CARON AND ASTRID SCHNETZER

37

Protistan assemblages of aquatic ecosystems are the focus of extensive research in aquatic ecology. One stimulus for this work has been the long-standing recognition that phototrophic protists (the unicellular algae) constitute a major fraction of the primary productivity within aquatic ecosystems. Another incentive has been the realization that protozoa (heterotrophic protists) play a pivotal role in the flow of energy and elements in these communities (109, 118). Studies of the abundance, biomass, and trophic activities of protists have now been conducted in a wide range of aquatic ecosystems. In addition, many laboratory experiments have examined the general biology and physiology of various protistan species grown under carefully controlled conditions. The synthesis of this information into models of how protistan assemblages are structured and how they function in nature has advanced considerably during the last few decades.

Protozoa traditionally have been distinguished from unicellular eukaryotic algae as heterotrophic organisms that exist by absorbing dissolved organic substances (osmotrophy) or, more commonly, by engulfing prey (phagocytosis). It is important to recognize that the term "Protozoa" now has more historical significance than phylogenetic or ecological meaning. Modern phylogenetic schemes have merged many protozoan taxa with algal taxa and vice versa. For example, numerous heterotrophic species of chrysophytes and dinoflagellates exist within these traditionally algal taxa. Heterotrophic species within each group are closely related to chloroplast-bearing species based on ultrastructural features and DNA sequence data but obviously have quite different modes of nutrition from those of their photosynthetic counterparts (4, 80, 99, 115). In addition to the existence of "apochlorotic algae," chloroplast-bearing genera that are capable of phagotrophy in addition to photosynthesis exist within the chrysophyte, dinoflagellate, prymnesiophyte, cryptophyte, and euglenophyte algae (112, 129, 130). This mixotrophic behavior obscures the distinction between traditional definitions of algae and protozoa and has led ecologists to conceptualize protistan trophic activity as a continuum of nutritional modes (76).

There are also difficulties in the classification of some ciliated protozoa as phototrophs or heterotrophs. Some ciliate species ingest and digest algal prey but are able to retain the chloroplasts of their prey in a functional state, thereby providing those ciliates with limited photosynthetic ability (112, 127). Photosynthesis in green ciliates contributes significantly to the overall nutrition of these protozoa and may form a notable fraction of the primary productivity of some planktonic communities (33).

The close phylogenetic affinities of some flagellated protozoa with algal taxa, as well as the mixed nutrition of many protists, indicate the artificiality of the historical distinction between the algal and protozoan taxa (80). Therefore, use of the term "protist" in reference to both chloroplast-bearing forms (i.e., algae) and heterotrophic forms (protozoa) has gained popularity in recent years.

ASSESSING PROTISTAN COMMUNITY STRUCTURE

Protistan Species Diversity

Complete taxonomic characterization of the protistan assemblage has been achieved for few, if any, aquatic communities. There are at least three major, interrelated reasons for this situation, namely, (i) the large number of protistan species present in most aquatic environments, (ii) the tremendous ranges in size and abundance of these species, and (iii) the disparate methodologies that are required for sampling and identifying them. Overcoming the difficulties posed by these issues is central to improving the present state of our knowledge concerning the structure of natural assemblages of aquatic protists.

Literally thousands of species of protists have been described, and many, perhaps all, of these species have sympatric distributions (52). In addition, free-living protists range in size from approximately 0.8-μm prasinophytes up to some species of radiolaria that can form cylindrical gelatinous colonies of ≈1 cm in diameter and >1 m in length. Similarly, relative abundances of protistan species within a habitat may vary by several orders of magnitude. These immense ranges of size and abundance make it necessary to apply a number of sampling techniques in order to adequately sample all of the protistan species in an environment. The unique physical and chemical characteristics of different aquatic environments (e.g., planktonic versus benthic environments) also contribute to the varied protocols that are necessary for sampling of protistan assemblages (120).

Even within an environment, sampling protocols must be adapted for particular groups of protists. For example, enumerating species of benthic protozoa among the sedimentary particles in which they exist has been a long-standing problem in assessing protistan species diversity in those environments. Various methods for extracting and concentrating protozoa from sediments have relied on the mobility of the community in response to changing salinity, extraction by centrifugation, or enrichment culture (1, 48, 111, 126, 144). Such approaches have resulted in reasonable estimations of the protozoan diversity of some sediment environments, but the success of these methods is usually group specific (22). The extraction of protozoa by the sea-ice method or by centrifugation and staining may work well for large or highly mobile species, such as benthic ciliates, but this method may be less useful for slow-moving forms, such as small amoebae. For the latter forms, enrichment cultivation appears to be the most appropriate method (97).

Adjustments to sampling protocols are also necessary for adequate sampling of different protistan groups within plankton communities. Sample volumes of 200 to 500 ml are usually sufficient for oceanic waters for sampling of flagellated protists (typical abundances are hundreds to thousands per milliliter), and volumes of 0.5 to 2 liters are usually sufficient for sampling of ciliated protists (typical abundances are tens to thousands per liter), but actinopods and foraminifera usually must be concentrated by the use of plankton nets or filters. These techniques, however, are damaging to delicate species of planktonic ciliates (62), and some actinopods that lack mineralized skeletal structures may be significantly undersampled. At present, visualization and enumeration in situ may be the only tractable means of obtaining accurate counts of these forms (36). Common methods for sampling planktonic protists were recently reviewed (63).

Preservation, fixation, and other manipulations are necessary prerequisites for the identification of most protistan species once appropriate samples have been collected. Notable exceptions to this generalization are the "naked" amoebae (primarily the Gymnamoebae), as some of the characteristics that are essential for identification are present only in living specimens (97). For the remaining protistan groups, appropriate preservation is dependent on the taxon under consideration (63). For flagellated (and often ciliated) protists, aldehyde fixatives (formaldehyde and glutaraldehyde) are commonly employed, often followed by osmium tetroxide when electron microscopy is planned (79, 119). A variety of fixatives have been developed for ciliated protozoa, most of which are usually employed in combination with postfixation staining methods that are used to visualize cytological features of the cells (82, 94, 131, 144).

The preservation of some protistan taxa requires special consideration. A preservative that does not promote the dissolution of skeletal structures must be employed for species which possess such structures (e.g., some actinopods and foraminifera) because they are important diagnostic parameters. Careful adjustment of the pH of the preservative is necessary to prevent dissolution of foraminiferan tests (11), while the addition of strontium is necessary to prevent dissolution of acantharian skeletons (91). When these requirements conflict, subsamples must be preserved separately for the different groups. For example, planktonic foraminifera (which require an alkaline pH) typically would be preserved differently than planktonic ciliates, which are often preserved in acid Lugol's solution (131).

The identification of protistan species in mixed natural assemblages depends on criteria that are often as different as the methodologies used to sample and preserve these assemblages. Therefore, taxonomic expertise among protistologists is often limited to one of the numerous protistan groups (e.g., diatoms, dinoflagellates, amoebae, ciliates, and foraminifera) or some portion of one of these major categories.

Ciliates typically possess morphological features that provide sufficient taxonomic criteria for identifying species by light microscopy. Cell shape, cell size, the location and characteristics of the oral area, the presence of a lorica, and particularly the arrangement of the somatic ciliature are useful features for species identification (80). Ciliates are often easier to identify than many of the flagellated and amoeboid protists because of the presence of these features, and extensive species lists exist for various environments (55, 104).

Flagellated protists typically possess fewer morphological features that can serve as useful taxonomic criteria when they are observed by light microscopy. Cell size and shape, chloroplast arrangement, and flagellation are important criteria for identification by light microscopy. Some diagnostic features, however, are only visible by electron microscopy (e.g., flagellar hairs and body scales). Electron microscopy is often necessary for distinguishing the numerous genera and species of small heterotrophic flagellates (<10 μm). The need to establish these features by electron microscopy makes it difficult to process large numbers of samples. Moreover, many of the latter taxa have not been described adequately. There is considerable uncertainty about the validity of numerous genera (99, 102) and thus about the true species diversity of small heterotrophic flagellates in many environments.

The amoeboid protists are a polyphyletic collection of species, and the methods of identification applied to these species are heterogeneous. The naked amoebae are identified based on features of the living organisms, including cell size and shape during locomotion, arrangement and type of pseudopodia, morphology of the floating form, etc. The requirement for live material for species identification has made the determination of the species diversity of natural assemblages of amoebae a difficult topic, but new methodological approaches are slowly yielding information on the taxonomy, distribution, and general ecology of these species (6, 19, 97, 111, 123, 124). Identification of the many types of testate amoeboid protists (testacea, foraminifera, radiolaria, and others) is based largely on the skeletal structures that are present in many of these species and on features of their cellular organization. The presence of a rigid skeletal structure in many of these species makes it possible to use plankton nets or screens for collecting and concentrating these specimens from the plankton or sediment. Nevertheless, there is mounting evidence from the application of molecular biological approaches that traditional methods may be missing much of the diverse populations of these groups (70, 73, 103).

Difficulties associated with sampling and identifying the entire spectrum of protists (as described above) in natural communities hamper the documentation of the true protistan species diversity of any natural ecosystem. Exceptions to this generality might be found for environments where protistan diversity is greatly reduced due to severe environmental factors, such as anaerobic conditions (54), extreme pH (2), and some enteric environments (3). However, it is safe to generalize that the vast majority of studies of natural communities have underestimated total protistan species diversity.

Analyses of species diversity for particular taxa of protists (i.e., the ciliates, flagellates, or amoebae), however, have been more complete. The most complete information exists for plankton communities, where extensive lists of ciliated protozoa, chloroplast-bearing flagellates, and skeleton-bearing sarcodines (foraminifera and actinopods) have been obtained (10, 43, 88, 98, 100, 105).

Protistan Abundance and Biomass

Identification of the protistan species present in an aquatic environment provides useful but limited information on their potential contributions to the structure and function of the total biological community because of the tremendous size range and varied trophic activities of protistan species. A much broader understanding of their importance can be obtained by combining species lists with estimates of abundance and biomass. Most modern methods for collection and identification of protists have been designed with this goal in mind. Generally accepted methods are now emerging which minimize problems associated with the loss of cells during collection, enrichment, preservation, and sample processing for specific groups of protists, thus allowing accurate estimates of protistan abundance.

As with species identification, estimating population abundances of amoebae presents the most formidable problem. The amorphous shapes of these microorganisms in preserved samples make them difficult to recognize and enumerate in natural samples, and their association with detritus and other particles can also obscure them from view. The few abundance estimates that are available for these species have been obtained by using a most-probable-number culture technique that relies on the growth of amoebae in serial dilutions of water or sediment samples (34, 111).

Protistan abundance measurements can be used to calculate total protistan biomass (typically expressed in units of carbon), using measurements of abundance and cell volume and empirically derived carbon-volume conversion factors. Cell volume measurements obtained from microscopic studies are combined with abundance estimates to calculate the volumes of particular protistan taxa, and carbon-volume conversion factors are then applied to calculate the carbon content. Carbon-volume conversion factors must take shrinkage due to fixation into account, as well as the variable vacuolar space of protists. Shrinkage due to fixation can be both taxon and size specific. Typical values for converting the amount of volume to carbon range between 160 and 360 fg of C μm^{-3} for flagellated protists (90, 139) and 190 fg of C μm^{-3} for ciliated protists (108). Carbon-volume conversion factors for larger sarcodines (acantharia, radiolaria, and foraminifera) are based on aspects of the cells that are resistant to net collection (92). A method has been proposed for estimating the cell volume of naked amoebae that directly relates the diameter of the nucleus to total cell volume (110). Commonly used conversion factors were summarized and reviewed recently (63).

Describing Protistan Community Structure by Classical Approaches

The term "community structure" implies that organized relationships exist between protists and other microorganisms within natural ecosystems. Indeed, the "niche" concept has been applied to protistan assemblages, with the implication that the number of protistan species in an environment is indicative of the number of unique ecological roles for protists in these assemblages. Unfortunately, it is unrealistic to consider all protistan species in a community as separate entities at this time because of the great species diversity of these assemblages, but this is a problem common to many ecological studies. The limited ecological information available on the realized niches of many protistan species and the extreme difficulty in obtaining species identifications and abundance/biomass information for all protistan species in an assemblage have resulted in the use of various simplified groupings as a way of reducing the complexity of protistan assemblages into manageable (and measurable) quantities.

Various means of simplifying protistan community structure have been used. The most popular approaches have grouped protists by trophic mode (e.g., phototrophic versus heterotrophic), size, and prey type (for heterotrophs), in keeping with the trophic-level concept of Lindeman (84). Trophic categories must be somewhat more flexible than simply "phototrophy versus heterotrophy" because of the common behavior of mixed nutrition among protists (24). Nevertheless, aggregation of species into "trophospecies" (136) is still a useful and necessary procedure for partitioning the assemblage in order to allow investigations of energy and elemental flow through aquatic communities in models of manageable size (30).

For heterotrophic protists, it is a common procedure to group species according to the type of prey that they consume. Bacterivorous flagellates and ciliates in plankton communities or bacterivorous flagellates and amoebae on suspended particles may be grouped together to represent a major sink for bacterial biomass in the plankton. Similarly, ciliate species may be grouped into bacterivorous, herbivorous, and predacious species (39). Such "feeding guilds" ignore some of the details of protistan feeding behavior (such as omnivory), but they are useful for reducing the complexity of the assemblage. Feeding guilds are often treated as single species in biological or biogeochemical models of ecosystem function.

The organization of protists by size is a logical one for two reasons. The allometric dependence of growth and metabolism can be used to constrain the potential contribution of a particular size range of protists to biogeochemical cycles (9, 27, 51, 72). In addition, predator-prey relationships are typically size dependent, with larger predators consuming smaller prey. For example, many heterotrophic flagellates of 2 to 20 μm in length consume bacteria and cyanobacteria that are <2 μm long, and many ciliate species of 20 to 100 μm in length consume algae and protozoa of <20 μm long.

While this generalization is realistic, it is not absolute. Many species of heterotrophic dinoflagellates consume diatom prey that are considerably larger than themselves by employing a pseudopodial "feeding veil" (75). Similarly, some planktonic sarcodines (acantharia, radiolaria, and foraminifera) are capable of consuming metazoan prey considerably larger than themselves due to the production of a sticky pseudopodial network that entangles and immobilizes prey items (132).

Notwithstanding these exceptions, size-dependent grazing models are the most common manner of organizing protistan populations into manageable units for inclusion into models of elemental flow in aquatic ecosystems (8, 42, 74, 93). The aggregation of species into groups within models probably reduces the predictive capabilities of these models, but their outcomes thus far appear to be in reasonable agreement with field data. It remains to be seen how the reduction of species diversity in these models will affect predictions of the responses of the community to internal and external perturbations, but the gradual disaggregation of

TABLE 1 Species list showing the range of protistan diversity of an oligotrophic oceanic environment[a]

Protistan taxon	Avg size (μm)	Presence of chloroplasts	Occurrence of phagotrophy	Probable prey[b]	Representative abundance ($liter^{-1}$)
Flagellated/nonmotile protists					
Dinoflagellates					
Protoperidinium sp.	55	No	No	Dia	1×10^2
Gymnodinium sp.	50	Yes	Yes	C, Din, Dia	2×10^2
Prorocentrum micans	25 × 40	Yes	No		1×10^3
Ornithocercus magnificus	40	No[c]	Yes	Dia, Sf	1×10^2
Chrysophytes/chrysomonads					
Paraphysomonas imperforata	7.0	No	No	B, Cc, Sf	10^5
Picophagus flagellatus	1.8	No	Yes	B	10^5
Ochromonas sp.	6.0	Yes	Yes	B, Cc	10^5
Prymnesiophytes					
Chrysochromulina ericina	6.0	Yes	Yes	B, Cc	10^4
Chlorophytes					
Nannochloris atomus	4.0	Yes	No		10^5
Ostreococcus taurii	0.8	Yes	No		10^5
Bacillariophytes (diatoms)					
Minutocellus polymorphus	3.0	Yes	No		10^5
Coscinodiscus concinnus	75 × 200	Yes	No		10^{-2}
Ditylum brightwellii	20 × 100	Yes	No		10^2
Rhizosolenia clevei	200 × 500	Yes	No		10^{-3}
Ethmodiscus rex	1,000	Yes	No		10^{-4}
Choanoflagellates					
Diaphanoeca grandis	2.5	No	Yes	B	10^5
Amoeboid protists					
Gymnamoebae (naked amoebae)					
Platyamoeba weinsteini	3 × 12	No	Yes	B	1×10^0
Flabellula citata	4 × 30	No	Yes	B	5×10^{-1}
Foraminifera					
Globigerina bulloides[h]	700[f]	No	Yes	Omi, Mz	10^{-3}
Globigerinoides sacculifer[h]	700[f]	No[d]	Yes	Omi, Mz	10^{-3}
Acantharea					
Amphilonche elongata[h]	50 × 400	No[d]	Yes	Omi	1×10^{-1}
Spumellarian radiolaria					
Thalassicolla nucleata[h]	1,000[f]	No[d]	Yes	Omi, Mz	10^{-4}
Collozoum caudatum[h]	200[f,g]	No[d]	Yes	Omi, Mz	10^{-5}
Ciliated protists					
Tintinnids					
Tintinnopsis parva	20 × 40	No	Yes	B, Cc, Sf	10^2
Oligotrichs					
Strombidium sulcatum	25 × 50	Yes[e]	Yes	B, Cc, Sf	10^2
Loboea strobila	50 × 150	Yes[e]	Yes	B, Cc, Sf	10^2
Hypotrichs					
Euplotes woodruffi	65 × 120	No	Yes	C	5×10^{-1}
Hymenostomatids					
Uronema marinum	10 × 20	No	Yes	B, Cc	10^3

[a]This is only a partial list of representative species to exemplify the breadth of trophic modes in a real assemblage. Species are organized according to major taxa. Pertinent ecological information and realistic abundances are provided based on literature values.

[b]Dia, diatoms; C, ciliates; Din, dinoflagellates; Sf, small flagellated protists; B, bacteria; Cc, chroococcoid cyanobacteria; Mz, metazoan zooplankton; Omi, omnivorous on prokaryotic and eukaryotic unicells.

[c]Species harbors extracellular, symbiotic cyanobacteria that contribute to the photosynthetic nutrition of the host.

[d]Species harbors intracellular, symbiotic dinoflagellates that contribute to the photosynthetic nutrition of the host.

[e]Species that retain functional chloroplasts from ingested prey.

[f]Size does not take into account an extensive pseudopodial network.

[g]Colonial species with colonies up to 1 cm in width and 1 m in length.

[h]Considers adult specimens only.

these models into more ecologically relevant compartments should provide insight into this issue.

Example from the Plankton

A hypothetical example indicates the analytical approaches for examining protistan community structure and the limitations of these approaches. A species list of protists that is representative of an oceanic plankton community is given in Table 1. This assemblage is not meant to be realistic or complete, but rather indicative of the breadth of protistan sizes and nutritional modes that exist in this type of ecosystem. Pertinent information on cell size, photosynthetic and/or phagotrophic ability, prey type(s), and typical abundances is also provided (abundances do not take bloom conditions into account, which are common for phytoplankton). The species in this assemblage have been arranged according to major taxonomic (i.e., phylogenetic) affinities (80).

As shown in Table 1, taxonomic groupings of protists correspond poorly to the nutritional modes and ecological roles of the species. Reorganization of the same species into groups based on the nutritional modes of the species provides a very different classification of this assemblage (Table 2). This reorganization indicates the classical dichotomy between phototrophs and heterotrophs, but it also indicates the more recent realization that many protistan species possess the ability to use mixed nutrition. Note that this ability results in some of the species appearing in more than one category.

The collection of species in this assemblage also demonstrates the enormous breadth of cell sizes that can be displayed by protistan assemblages (Fig. 1). The size range is not necessarily restricted for any particular trophic mode. In this assemblage, heterotrophs range in size from ≈1.8 μm to >1 mm, phototrophs range from ≈0.8 μm to >1 mm, and mixotrophs range from ≈6.0 μm to >1 mm if one includes symbiont-bearing sarcodines. Commonly employed plankton size class designations are also shown in Fig. 1. These designations correspond to organisms of 0.2 to 2.0 μm (picoplankton), 2 to 20 μm (nanoplankton), 20 to 200 μm (microplankton), 0.2 to 20 mm (mesoplankton), and 20 to 200 mm (macroplankton) in their longest dimension (121). Protists occur in all of these size classes, as indicated in Fig. 1, although the majority of these species occur in the nano and micro classes.

One generality that is clear from Table 1 is that small planktonic protists typically occur in greater abundance than do large species. This relationship is shown clearly by 1- to 2-order-of-magnitude differences in abundance when individuals of <20, 20 to 200, and >200 μm are tallied according to plankton size class (Fig. 2A). Phototrophic, heterotrophic, and mixotrophic protists also exhibit large differences in

TABLE 2 Planktonic protistan species list from Table 1, arranged according to trophic category[a]

Phototrophs	Heterotrophs
Gymnodinium sp.	Bacterivores
Prorocentrum micans	*Paraphysomonas imperforata*
Ochromonas sp.	*Ochromonas* sp.
Chrysochromulina ericina	*Chrysochromulina ericina*
Nannochloris atomus	*Diaphanoeca grandis*
Ostreococcus tauri	*Platyamoeba weinsteini*
Minutocellus polymorphus	*Flabellula citata*
Coscinodiscus concinnus	*Amphilonche elongata*
Ditylum brightwellii	*Thalassicolla nucleata*
Rhizosolenia clevei	*Collozoum caudatum*
Ethmodiscus rex	*Tintinnopsis parva*
	Strombidium sulcatum
Mixotrophs	*Loboea strobila*
Phagotrophic algal species	*Uronema marinum*
Gymnodinium sp.	*Picophagus flagellatus*
Ochromonas sp.	Herbivores
Chrysochromulina ericina	*Protoperidinium* sp.
Chloroplast-retaining species	*Gymnodinium* sp.
Strombidium sulcatum	*Ornithocercus magnificus*
Loboea strobila	*Paraphysomonas imperforata*
Symbiont-bearing species	*Amphilonche elongata*
Ornithocercus magnificus	*Thalassicolla nucleata*
Hastigerina pelagica	*Collozoum caudatum*
Amphilonche elongata	*Tintinnopsis parva*
Thalassicolla nucleata	*Strombidium sulcatum*
Collozoum caudatum	*Loboea strobila*
	Globigerina bulloides
	Carnivores
	Paraphysomonas imperforata
	Globigerina bulloides
	Amphilonche elongata
	Thalassicolla nucleata
	Collozoum caudatum
	Euplotes woodruffi

[a]Note that some species occur in more than one category.

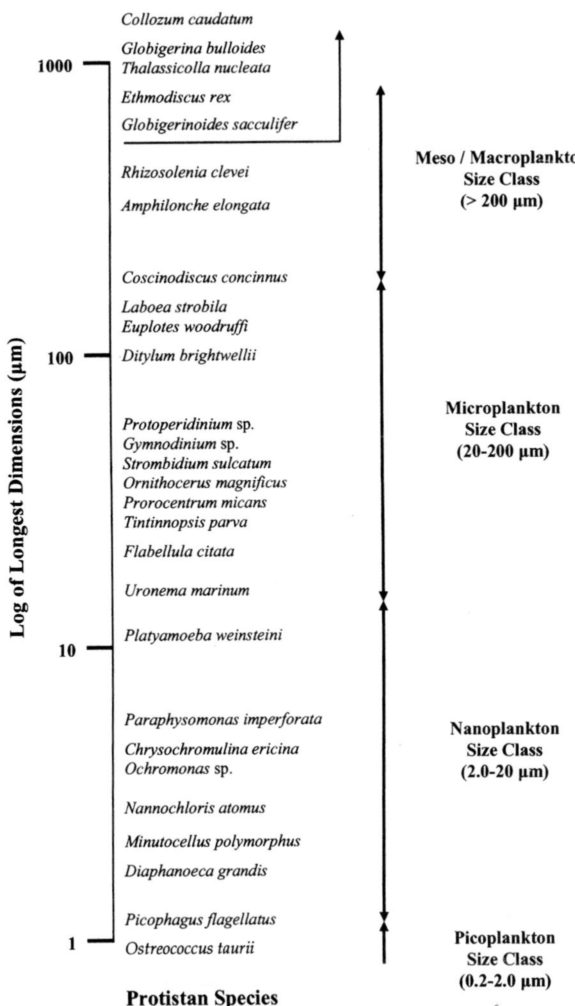

FIGURE 1 Approximate sizes (longest dimension) of the planktonic protistan species listed in Table 1. Commonly employed size class designations are shown on the right. Note that the sizes of these protists span >3 orders of magnitude. The arrow running under *Globigerinoides sacculifer* indicates that the group of five species enclosed by the arrow can be larger than 1,000 µm.

abundance between size classes, although the abundances of these groups within a size class are often similar.

The large disparity that is apparent among the abundances of protists in different size classes (Fig. 2A) is greatly reduced when the total volumes of living protists are compared (Fig. 2B). Small cell size among nanoplankton is generally balanced by high abundances of these species, while larger cellular volumes of large protists compensate for their low abundances. These general relationships of protistan abundance and biovolume are consistent with data from natural assemblages of nanoplanktonic and microplanktonic protists (18, 25, 37, 60).

The information summarized in Tables 1 and 2 and Fig. 1 and 2 can be used to construct a typical box model depicting the flow of materials from producers to consumers for this hypothetical protistan community (Fig. 3). The species have been grouped according to nutritional mode, size, and approximate predator-prey relationships. Arrows in the model indicate the presumed directions of energy and material flow (i.e., from producers to consumers and from small organisms to large consumers). These types of models are now common in the literature. They indicate the fundamental roles of bacteria, cyanobacteria, and phototrophic and heterotrophic protists as producers and consumers of organic matter in aquatic ecosystems. These depictions also recognize the important role of bacteria and their consumers as a means of converting detrital material back into living biomass, thus making it available for reentry into food webs, a fundamental role of "the microbial loop" (8).

Major goals for work on modeling microbial loop processes have been the development of working models that accurately describe energy and elemental flow within these communities and the incorporation of microbial processes into classical models of aquatic community structure and function. Models such as the one shown in Fig. 3 are appropriate for these purposes because they attempt to reduce a complex assemblage of microorganisms to a manageable number of trophic compartments and trophic interactions. These models are therefore strongly influenced by methodologies available for identifying protistan species (or trophospecies) and for investigating their trophic interactions.

The model in Fig. 3 might adequately describe energy or elemental flow in this hypothetical protistan assemblage if the biomass and flow parameters of the model could be determined. However, this type of depiction of community structure still has some inherent flaws. As mentioned earlier, predator-prey relationships that are not size dependent are difficult to represent and measure. Energy is depicted as moving from smaller to larger size classes in this model, but

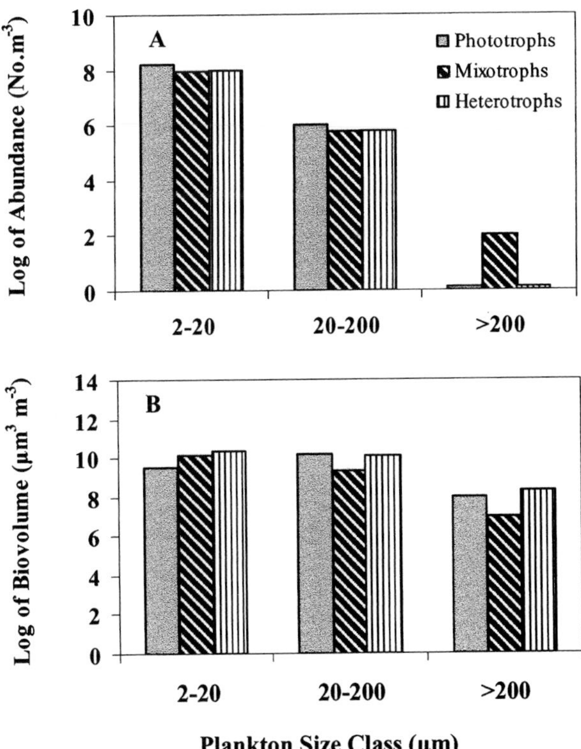

FIGURE 2 Abundance (A) and biovolume (B) relationships of the protistan assemblage listed in Table 1. The species have been grouped according to size class and trophic mode (phototrophic, mixotrophic, or heterotrophic).

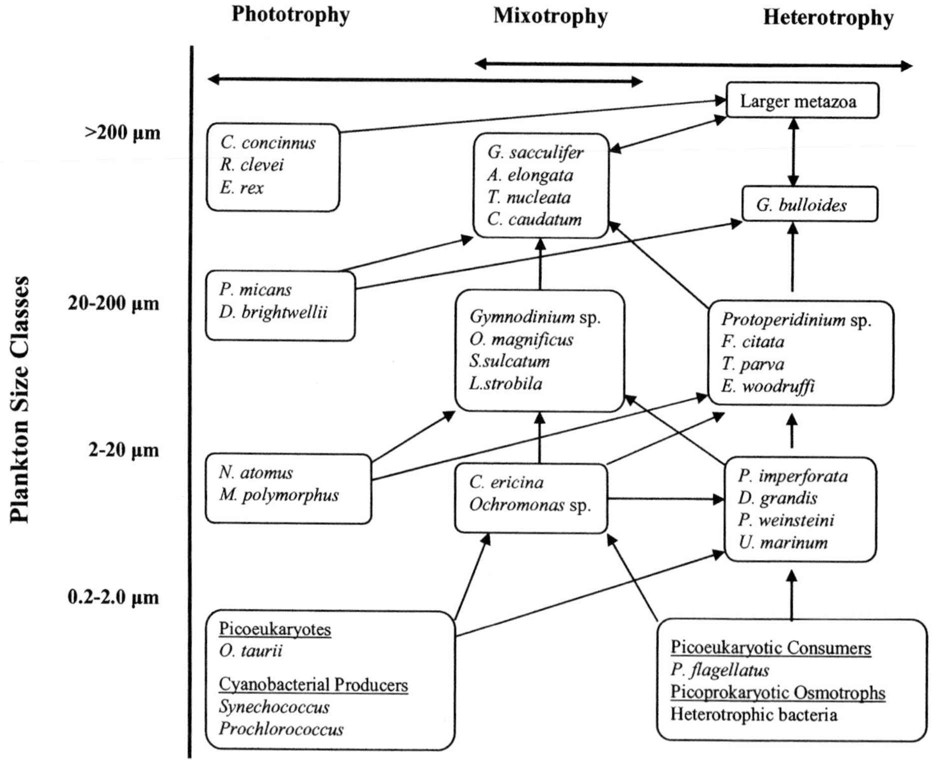

FIGURE 3 Box model approximating the major trophic interactions within the protistan assemblage listed in Table 1. The species have been grouped according to known or presumed size-dependent and trophism-dependent relationships. Arrows indicate the directions of energy flow in predator-prey interactions.

this representation is incorrect for species such as *Protoperidinium* sp., which can graze on diatoms larger than itself, and for the sarcodines *Globigerinoides sacculifer*, *Thalassicolla nucleata*, *Collozoum caudatum*, and *Globigerina bulloides*, which can consume metazoan prey. The double-headed arrows connecting these last compartments indicate the potential for the flow of energy in either direction. In practice, these measurements are difficult to make.

Similarly, selective grazing and omnivory are difficult to incorporate into this type of model. For example, *Euplotes woodruffi* is a predacious ciliate that feeds primarily on other ciliates (in this assemblage it might feed on *Uronema marinum*). On the other hand, *Tintinnopsis parva* may accept a variety of small protists and other microorganisms as prey. The distinction between these two rather different nutritional modes has been forfeited by placing them into the same trophic compartment. Clearly, if the goal of this modeling exercise were to understand the factors affecting the success or failure of either of these two species in plankton communities, then this model would be unsatisfactory. It is for reasons such as this that the appropriate conceptualization and representation of protistan community structure must take into account the goal of the investigator in constructing such models.

Box models that approximate trophic relations within planktonic communities can often be tested. For example, if we hypothesize that ciliates in the microplankton size class are the dominant consumers in the plankton, we would expect that their removal (e.g., by size fractionation) would result in an increase in the abundance of species in the nanoplankton size class. This relationship constitutes the basis of a "trophic cascade," which assumes that changes in the activities of organisms at one trophic level will result in alternating positive and negative effects on the assemblages in lower trophic levels. For example, the removal of microplankton and larger zooplankton from the community depicted in Fig. 3 would be expected to result in increases in nanoplanktonic species (2.0 to 20 µm), which would in turn result in decreases in the abundance of picoplanktonic species (0.2 to 2.0 µm). The strength with which a trophic cascade propagates through a food web has been shown to depend on the complexity of trophic relationships within the assemblage (71, 96, 106).

Describing Protistan Community Structure by Molecular Approaches

We have learned a great deal about the taxonomic composition and trophic structure of aquatic protistan communities through the application of traditional approaches of morphological analysis and culture. Nevertheless, the tremendous diversity of protistan assemblages and the varied methods required for identifying these species and their abundances, biomass, and trophic activity continue to hamper in-depth understanding of the structure and function of these communities. As a result, new conceptual and methodological approaches are coming into use that help to deal with these recalcitrant problems.

Increasingly valuable approaches are culture-independent, molecular methods that allow the characterization of community structure without the need for culture and/or extensive morphological analyses. Several approaches to accomplish this task have been pursued. Two common ones are various methods of DNA fragment analysis, which provides a quick "snapshot" of the relative diversity of the community as a whole, and the cloning and sequencing of specific genes in order to obtain detailed information on the species (phylotypes) present in an assemblage (Fig. 4). Both approaches employ the extraction of nucleic acids directly from samples, avoiding biases inherent in traditional taxonomic characterizations due to fixation, preparation, or culture. Nonetheless, most extant molecular approaches possess potential biases due to the efficiency of nucleic acid extraction and (if used) PCR amplification.

Fragment analysis approaches rely on distinguishing microbial taxa by unique characteristics (e.g., migration rate through a gel) of their genes or gene fragments when amplified from a natural assemblage. Several approaches have been employed for eukaryotic assemblages, including the common methods of denaturing gradient gel electrophoresis and terminal restriction fragment length polymorphism (T-RFLP) analysis (31, 38, 61, 138). By far, these approaches have focused primarily on rRNA genes, particularly small-subunit rRNA genes. The intergenic spacer regions have also been employed for fragment analyses, but these regions may actually resolve differences at the strain level rather than the species level. The number of DNA fragments in such analyses is indicative of the number of operational taxonomic units present in the sample. Ideally, operational taxonomic units can be equated with species, although this approach assumes that each species produces a single, unique fragment. An example of a T-RFLP pattern is shown in Fig. 5, where several cultured protists were combined into a single sample prior to extraction and analysis. Fragment sizes for the species were determined using monocultures of the protists, and thus the identity of each fragment can be assigned (as shown in the figure).

The utility of DNA fragment analysis is the relative speed with which it can be accomplished. For this reason, this approach is useful in ecological studies that generate large numbers of samples and for comparative studies where the goal is to obtain a rapid comparison between environmental samples or experimental treatments. However, DNA fragment analysis methods typically assess only the numerically dominant taxa within an assemblage (rare species may produce fragments that cannot be detected), and they usually cannot provide taxonomic information without additional work (such as sequencing fragments produced in the analyses).

The cloning and sequencing of genes from environmental samples have recently yielded a large amount of information on the structure of microbial communities, especially prokaryotic assemblages (57). This approach also constitutes a powerful method for analyzing microbial eukaryotic assemblages. Cloning and sequencing are more time- and cost-intensive than fragment analysis but provide improved phylogenetic/taxonomic resolution. Comparison of sequences with publicly available databases provides molecular-level taxonomic identities that describe the protistan diversity within a sample (Fig. 6).

Cloning and sequencing of microbial species have largely involved work on rRNA genes (usually the small subunit, but also the large subunit), in contrast to work on

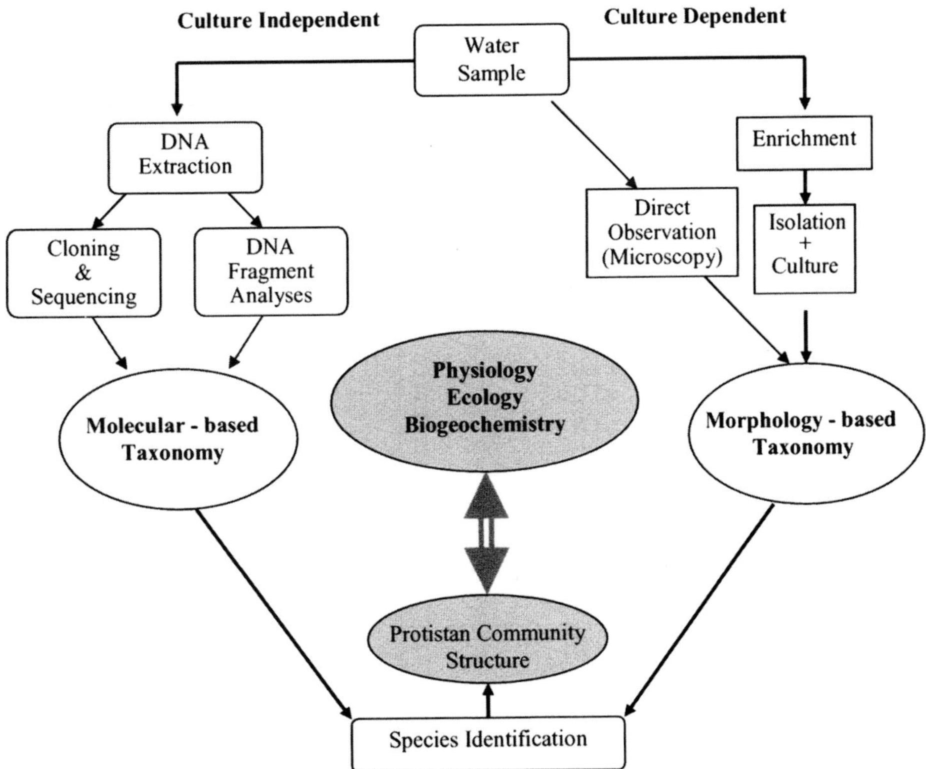

FIGURE 4 Conceptual framework for using culture-independent and -dependent methods to study protistan community structure. Arrows indicate the order in which this goal can be accomplished (see the text for details).

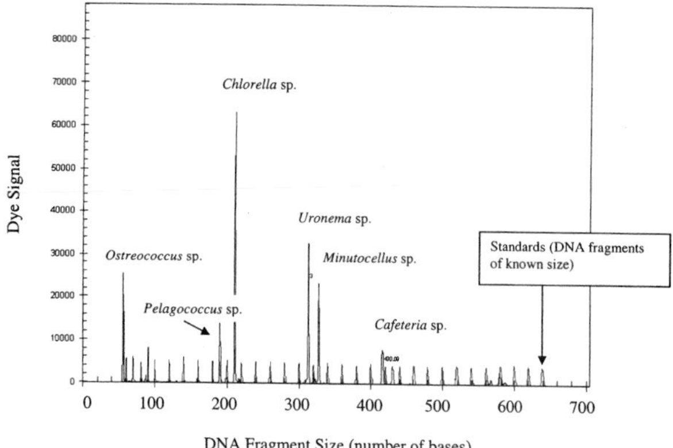

FIGURE 5 T-RFLP pattern for an artificial protistan community. The 18S rRNA genes of six known species were PCR amplified and digested with a single restriction enzyme. The unique T-RFLP fragment size for each of the species could be recognized in the mixed sample (as indicated).

higher organisms that has focused on the cytochrome oxidase I gene, such as in the Barcode of Life project. Some crossover in approaches is warranted, but few studies have been performed to date using cytochrome oxidase I genes to assess microorganismal diversity. Studies of natural assemblages have yielded exciting new information on the diversity and structure of protistan assemblages. These insights have included the detection of previously unknown diverse populations and many novel phylotypes in oceanic plankton (31, 69, 95), freshwater plankton (47), and shallow and deep-sea benthic communities (44, 70, 87). This approach has also revealed the unexpected existence of "marine" taxa in freshwater ecosystems (73) and possibly undescribed, kingdom-level diversity in anoxic ecosystems (35).

A logical extension of gene sequencing has been the use of sequence information in methods targeting specific taxa of microorganisms. These techniques do not examine protistan community structure per se, but rather attempt to quantify a particular taxon (or taxa) in a natural assemblage. Methods such as fluorescence in situ hybridization and quantitative real-time PCR allow estimates of species abundance that are highly sensitive, accurate, and often much more rapid than traditional microscopy (17, 32, 83, 107, 116). The development of DNA arrays for assaying the presence of species in natural samples is, in some ways, an extension of these approaches. Arrays greatly expand the number of taxa that can be assayed at one time. Thus far, the application of phylochips has been restricted largely to prokaryotic assemblages (45, 85, 137), but the development of protistan arrays is virtually assured as sequence information amasses for eukaryotes.

In addition to gene-based approaches, immunological methods for determining the presence and abundance of specific taxa have contributed to our knowledge of protistan species. These approaches have been applied successfully to phototrophic protists (21, 26, 117). Biochemical markers to indicate the presence, abundance, and activity of protists also may provide new methods for examining natural assemblages. Detailed pigment analyses have provided useful insights into the contributions of specific taxa to total algal biomass (146), and lipid biomarkers may also be useful for obtaining information on the biomass and nutritional status of microbial assemblages (143).

The success of using molecular (genetic/immunological) signatures for assessing the community structure of natural protistan assemblages will ultimately depend on linking these signatures to classical (morphological) species descriptions and to the physiological abilities of protistan phylotypes. Ultimately, molecular approaches, in combination with classical methods, will provide new tools for studying the emergent physiological, ecological, and biogeochemical processes that are created and/or affected by protistan community structure (Fig. 4).

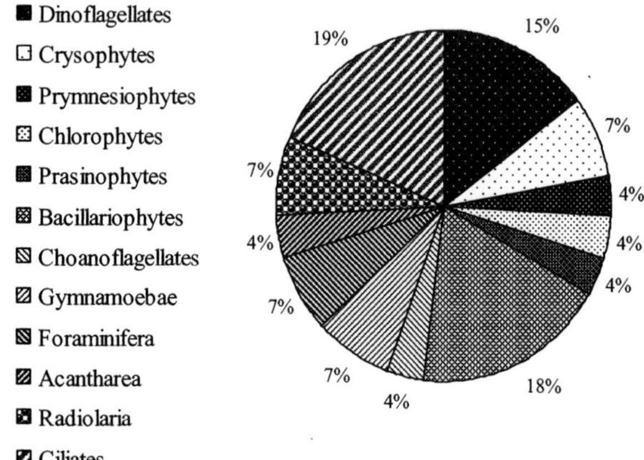

FIGURE 6 Relative contributions (percentages of total number of clones) of various protistan taxa to a theoretical clone library ($n = 27$) created for the protistan assemblage presented in Table 1.

TEMPORAL AND SPATIAL CHANGES IN COMMUNITY STRUCTURE

The most significant differences in the species composition and trophic relationships of protistan communities exist between different aquatic environments. However, there is also a rapidly increasing database on changes in community structure over seasonal and shorter timescales. These changes appear to be most significant in temperate and polar climates.

Freshwater versus Marine Ecosystems

Probably the most distinct difference between freshwater and marine protistan communities is the restriction of the larger sarcodines (acantharia, radiolaria, and foraminifera) to brackish and marine ecosystems. Adult sarcodines are often the most conspicuous macroscopic organisms of tropical and subtropical oceanic plankton communities in surface waters, while swarmer cells and juvenile specimens of these species contribute to the entire size spectrum of protozooplankton (28). In marine sediments, benthic foraminifera can constitute an important component of faunal assemblages (65, 70, 81). In addition, recent molecular analyses have indicated that many previously undetected foraminiferal taxa may exist in some freshwater ecosystems (73).

In contrast to large differences in the assemblages of sarcodine species in fresh- and salt water, there appears to be a fair degree of similarity with respect to types of ciliates and flagellates in these environments. The ecological roles of small protozoa in freshwater and marine plankton communities appear to be similar and related to bacterial production over very broad scales of examination (113). Most ciliates in the freshwater and marine plankton also appear to play grossly analogous ecological roles as consumers of small prokaryotes and eukaryotes (12, 104). Mixotrophic (phagotrophic) algae exist in both fresh and marine waters (5, 14, 16), as do species of chloroplast-retaining ciliates (40, 53, 127, 128). These generalizations do not necessarily mean that the same species of flagellates or ciliates occur in both ecosystems, but rather that similar ecological niches have been filled by protistan species in these different environments.

Benthic versus Pelagic Ecosystems

Although there are species of protists that are commonly found in both benthic and pelagic environments, there are clearly numerous species within these assemblages that are uniquely suited for one environment or the other (50). Morphological adaptations of ciliates to life between sediment particles in the benthos has resulted in the evolution of cell forms that allow movement through this medium. Common adaptations include cylindrical or flattened shapes, flexible cell walls, and patterns of ciliature that facilitate movement on surfaces and grazing on prey loosely associated with particulate material. Numerous species permanently attach to surfaces. In contrast, ciliates in pelagic environments tend to have more rounded shapes and patterns of ciliature that afford rapid swimming behavior and feeding on suspended particles.

There is great protistan biodiversity among and within benthic environments as a consequence of differences in various environmental gradients, such as sediment grain size, organic loading, and oxygen gradients. In addition, the number of microenvironments at one locale may be considerable. Spatial heterogeneity, the remoteness of many benthic environments, and difficulties associated with sampling, concentrating, and observing benthic protists continue to limit our knowledge of the protistan fauna of many benthic ecosystems. There are extremely few observations of the protistan fauna of the deep ocean benthos (7, 44, 64, 65, 86, 87, 125, 135).

The amoebae are particularly well suited for existence in benthic environments. Locomotion and feeding of these species take place on particles. Therefore, benthic environments tend to support significant assemblages of amoebae (114). Amoebae occurring in the plankton are generally assumed to be associated with suspended particulate material or with the air-water interface (19, 34, 111).

Among the larger sarcodines, there are dichotomies between pelagic and benthic assemblages. Foraminifera occur in both environments, but the species occurring in these two environments are different. The planktonic species are restricted to pelagic, oceanic ecosystems, while benthic species are common from salt marshes to abyssal depths. Most radiolaria (polycystines and phaeodaria) and acantharia are restricted to pelagic, oceanic ecosystems. There are relatively few exceptions to these generalities, making the larger sarcodine faunas of benthic and pelagic ecosystems quite distinct.

The contribution of phototrophic protists to flagellate assemblages in surface waters of pelagic ecosystems clearly differentiates them from flagellate assemblages of benthic environments, in which phototrophs may be reduced or absent. However, the heterotrophic flagellate assemblages of these two habitats also differ in composition. Many flagellated protozoa occur in both environments, but species that are capable of particle attachment or movement along surfaces (e.g., bodonid flagellates) tend to predominate in benthic environments, while forms that feed on suspended bacteria (e.g., chrysomonad flagellates and choanoflagellates) tend to predominate in pelagic ecosystems (23, 78, 101, 134, 140–142).

Pelagic environments generally are considered more homogeneous ecosystems than benthic ecosystems, but there are sources of heterogeneity in the plankton. Epibiotic (and possibly enteric) protistan assemblages have not been adequately studied, but they contribute to protistan species diversity in the water column (133). Suspended particles also create unique microhabitats in pelagic ecosystems for some protozoan species that are more characteristic of the benthos. Macroscopic detrital aggregates in marine planktonic ecosystems (so-called marine snow) may create a false benthos for benthic species by creating microenvironments with elevated abundances of bacteria and other prey (23, 122). Similar "oases" for unique protozoan assemblages in the plankton may be established by using artificial foam substrates (20). It has been demonstrated that the colonization and species succession of protozoa on these natural and artificial substrates may follow a pattern similar to that for the colonization of oceanic islands by higher organisms (147).

Depth and Seasonal Distributions

The seasonality of algal species composition and abundance in pelagic environments is well known. Distributions of the protistan algae with depth have also received considerable attention. In contrast, changes in total protozoan abundance or biomass with season or depth have been documented, but there is relatively little information on changes in species composition or community structure and function with depths. This paucity of information is not surprising given the logistical problems associated with the collection of long-term data sets or multiple samples from considerable depths.

Most studies to date have been restricted to a particular group of heterotrophic protists because of either the methodological approach or taxonomic expertise. Often, these investigations have reported only changes in broad taxonomic or ecologically relevant categories (heterotrophic flagellates, mixotrophic flagellates, ciliates, etc.). For example, depth and seasonal changes in abundance of flagellated and/or ciliated protozoa have been reported for a variety of marine and freshwater environments (13, 29, 41, 77). More detailed data on the spatiotemporal distributions are available only for taxa for which identification is more straightforward (10, 66).

It is difficult to generalize concerning changes in the community structure of protistan assemblages as a function of season from these scattered reports. For temperate communities, seasonal changes in species composition and winter reductions in the intensity of grazing activity are likely, but the extents of these changes remain largely undetermined for most environments. Temperature is a strong controlling influence on processes within the microbial loop of temperate ecosystems (145), but diverse heterotrophic and phototrophic protistan assemblages abound even in extremely low-temperature environments, such as marine habitats of Antarctica (58, 59, 61). In the latter environment, rather unique assemblages of phototrophic and heterotrophic protists have adapted to existence in association with sea ice.

The vertical distributions of protozoa typically demonstrate greater overall abundances in surface waters than at depth. These distributions of abundance are clearly related to the production of organic material in surface waters.

Fine-scale vertical distributions, however, can be complex. Elevated abundances of protozoa have been observed at the air-water interface (34), at oxic-anoxic boundaries within water columns (148), and at subsurface biological features, such as deep chlorophyll maxima (46). Vertical distributions of protists in sediments typically are related to physical and chemical gradients within the benthos (15, 49). The exploitation of these chemical/physical features within the benthos and water column can increase the diversity of protistan assemblages of an environment by providing unique microhabitats for the growth of species able to exist there.

FUTURE DIRECTIONS AND CHALLENGES

We have learned a great deal about the taxonomic composition and trophic structure of aquatic protistan communities since the discovery of these microbes by Antonie van Leeuwenhoek more than 3 centuries ago. Much of this work was descriptive in nature prior to 1970, but significant progress has been made within the last 3 decades towards an understanding of the trophic activities and ecological roles of numerous protistan species.

Despite these advances, large gaps in our knowledge still exist with respect to the overall diversity and distribution of aquatic protistan assemblages. For example, an accurate account of the overall diversity of protists has yet to be established. Modern molecular biological approaches have revealed unexpected, and as yet largely uncharacterized, protistan diversity in a wide variety of ecosystems (35, 73). The (morphological) identities of these taxa, their physiological abilities, and the overall diversity of microbial eukaryotes in nature constitute active areas of research at this time. Most importantly, linking molecular and/or morphological signatures to physiological/ecological functions will help to establish the biogeochemical roles played by protists. In this way, it will be possible to relate changes in protistan community structure to changes in nutrient and elemental cycles in nature. One recent study has indicated that the dominance structure of planktonic protistan assemblages can be highly responsive to containment and changes in environmental conditions (31). These results imply the existence of a large reservoir of rare species in natural plankton assemblages and the potential for rapid shifts in dominance (and possibly ecosystem function) as a consequence of changes in environmental conditions.

Minute protists (<10 μm) are likely the least understood of all eukaryotes with respect to diversity in natural aquatic ecosystems and will continue to receive attention in the near future (47, 69, 87, 89, 95). These species have been greatly underestimated by traditional approaches of identification (microscopy), largely because they possess few distinctive morphological characteristics. The use of molecular approaches coupled with novel culturing approaches is rapidly expanding our knowledge of the diversity and ecological roles of these taxa (67, 68).

Finally, significant uncertainty also remains regarding the distribution of protists on our planet. Finlay and Fenchel (52, 56) have speculated that most (if not all) protistan species have global distributions because their small size facilitates dispersal. The debate over whether protists have cosmopolitan or endemic distributions is, to some degree, a matter of the species concept applied to protists and the amount of intraspecific variability (morphological, DNA sequence, or physiological) that one is willing to accept. Nevertheless, the present emphasis in microbial ecology on studies of prokaryotic and protistan diversity in a wide range of ecosystems around the planet will provide us with a much better understanding of protistan biodiversity and the distribution of these species.

Support for preparation of the manuscript was provided in part by National Science Foundation grants MCB-0084231 and OPP-0125437, by The Center for Embedded Networked Sensing (CENS) under NSF cooperative agreement CCR-0120778, by National Oceanic and Atmospheric Administration grant NA160P2790, and by Environmental Protection Agency grant RD-83170501.

We are grateful to Patrick Vigil and Peter Countway for providing the T-RFLP pattern used in Fig. 5.

REFERENCES

1. **Alongi, D. M.** 1986. Quantitative estimates of benthic protozoa in tropical marine systems using silica gel: a comparison of methods. *Estuar. Coast. Shelf Sci.* **23:**443–450.
2. **Amaral Zettler, L. A., F. Gómez, E. Zettler, B. G. Keenan, R. Amils, and M. L. Sogin.** 2002. Eukaryotic diversity in Spain's River of Fire. *Nature* **417:**137.
3. **Amaral Zettler, L. A., A. D. Laatsch, E. Zettler, T. A. Nerad, J. Cole, F. C. Diaz, J. Diaz, D. H. Janzen, A. Sittenfeld, O. Masoni, and A. L. Reysenbach.** 2005. A microbial observatory of caterpillars: isolation and molecular characterization of protists associated with the saturniid moth caterpillar *Rothschildia lebeau*. *J. Eukaryot. Microbiol.* **52:**107–115.
4. **Andersen, R. A., Y. Van de Peer, D. Potter, J. P. Sexton, M. Kawachi, and T. LaJeunesse.** 1999. Phylogenetic analysis of the SSU rRNA from members of the Chrysophyceae. *Protist* **150:**71–84.
5. **Arenovski, A. L., E. L. Lim, and D. A. Caron.** 1995. Mixotrophic nanoplankton in oligotrophic surface waters of the Sargasso Sea may employ phagotrophy to obtain major nutrients. *J. Plankton Res.* **17:**801–820.
6. **Arndt, H.** 1993. A critical review of the importance of rhizopods (naked and testate amoebae) and actinopods (heliozoa) in lake plankton. *Mar. Microb. Food Webs* **7:**3–29.
7. **Arndt, H., K. Hausmann, and M. Wolf.** 2003. Deep-sea heterotrophic nanoflagellates of the Eastern Mediterranean Sea: qualitative and quantitative aspects of their pelagic and benthic occurrence. *Mar. Ecol. Prog. Ser.* **256:**45–56.
8. **Azam, F., T. Fenchel, J. G. Field, J. S. Gray, L. A. Meyer-Reil, and F. Thingstad.** 1983. The ecological role of water-column microbes in the sea. *Mar. Ecol. Prog. Ser.* **10:**257–263.
9. **Banse, K.** 1982. Cell volumes, maximal growth rates of unicellular algae and ciliates, and the role of ciliates in the marine pelagial. *Limnol. Oceanogr.* **27:**1059–1071.
10. **Bé, A. W. H.** 1977. An ecological, zoogeographic and taxonomic review of recent planktonic foraminifera, p. 1–100. *In* A. T. S. Ramsay (ed.), *Oceanic Micropaleontology*. Academic Press, London, United Kingdom.
11. **Bé, A. W. H., and O. R. Anderson.** 1976. Preservation of planktonic foraminifera and other calcareous plankton, p. 250–258. *In* H. F. Steedman (ed.), *Zooplankton Fixation and Preservation*. UNESCO Press, Paris, France.
12. **Beaver, J. R., and T. L. Crisman.** 1989. The role of ciliated protozoa in pelagic freshwater ecosystems. *Microb. Ecol.* **17:**111–136.
13. **Bernard, C., and F. Rassoulzadegan.** 1994. Seasonal variations of mixotrophic ciliates in the northwest Mediterranean Sea. *Mar. Ecol. Prog. Ser.* **108:**295–301.
14. **Berninger, U.-G., D. A. Caron, and R. W. Sanders.** 1992. Mixotrophic algae in three ice-covered lakes of the Pocono Mountains, USA. *Freshw. Biol.* **28:**263–272.

15. **Berninger, U.-G., and S. S. Epstein.** 1995. Vertical distribution of benthic ciliates in response to the oxygen concentration in an intertidal North Sea sediment. *Aquat. Microb. Ecol.* **9:**229–236.
16. **Bird, D. F., and J. Kalff.** 1986. Bacterial grazing by planktonic lake algae. *Science* **231:**493–495.
17. **Bowers, H. A., T. Tengs, H. B. Glasgow, Jr., J. M. Burkholder, P. A. Rublee, and D. W. Oldach.** 2000. Development of real-time PCR assays for rapid detection of *Pfiesteria piscicida* and related dinoflagellates. *Appl. Environ. Microbiol.* **66:**4641–4648.
18. **Buck, K. R., F. P. Chavez, and L. Campbell.** 1996. Basin-wide distributions of living carbon components and the inverted trophic pyramid of the central gyre of the North Atlantic Ocean, summer 1993. *Aquat. Microb. Ecol.* **10:**283–298.
19. **Butler, H., and A. Rogerson.** 1995. Temporal and spatial abundance of naked amoebae (Gymnamoebae) in marine benthic sediments. *J. Eukaryot. Microbiol.* **42:**724–730.
20. **Cairns, J., Jr., D. L. Kuhn, and J. L. Plafkin.** 1979. Protozoan colonization of artificial substrates, p. 34–57. *In* R. L. Wetzel (ed.), *Methods and Measurements of Periphyton Communities: a Review.* Special technical publication 690. American Society for Testing and Materials, Philadelphia, Pa.
21. **Campbell, L., L. P. Shapiro, and E. Haugen.** 1994. Immunochemical characterization of eukaryotic ultraplankton from the Atlantic and Pacific oceans. *J. Plankton Res.* **16:**35–51.
22. **Carey, P. G.** 1992. *Marine Interstitial Ciliates.* Chapman & Hall, London, United Kingdom.
23. **Caron, D. A.** 1991. Heterotrophic flagellates associated with sedimenting detritus, p. 77–92. *In* D. J. Patterson and J. Larsen (ed.), *The Biology of Free-Living Heterotrophic Flagellates.* Clarendon Press, Oxford, United Kingdom.
24. **Caron, D. A.** 2000. Symbiosis and mixotrophy among pelagic microorganisms, p. 495–523. *In* D. L. Kirchman (ed.), *Microbial Ecology of the Oceans.* Wiley-Liss, Inc., New York, N.Y.
25. **Caron, D. A., H. G. Dam, P. Kremer, E. J. Lessard, L. P. Madin, T. C. Malone, J. M. Napp, E. R. Peele, M. R. Roman, and M. J. Youngbluth.** 1995. The contribution of microorganisms to particulate carbon and nitrogen in surface waters of the Sargasso Sea near Bermuda. *Deep-Sea Res.* **42:**943–972.
26. **Caron, D. A., M. R. Dennett, D. M. Moran, R. A. Schaffner, D. J. Lonsdale, C. J. Gobler, R. Nuzzi, and T. I. McLean.** 2003. Development and application of a monoclonal antibody technique for counting *Aureococcus anophagefferens*, an alga causing recurrent brown tides in the mid-Atlantic United States. *Appl. Environ. Microbiol.* **69:**5492–5502.
27. **Caron, D. A., and J. C. Goldman.** 1990. Protozoan nutrient regeneration, p. 283–306. *In* G. M. Capriulo (ed.), *Ecology of Marine Protozoa.* Oxford University Press, New York, N.Y.
28. **Caron, D. A., and N. R. Swanberg.** 1990. The ecology of planktonic sarcodines. *Rev. Aquat. Sci.* **3:**147–180.
29. **Carrick, H. J., and G. L. Fahnenstiel.** 1990. Planktonic protozoa in Lakes Huron and Michigan: seasonal abundance and composition of ciliates and dinoflagellates. *J. Gt. Lakes Res.* **16:**319–329.
30. **Christian, R. R.** 1994. Aggregation and disaggregation of microbial food webs. *Microb. Ecol.* **28:**327–329.
31. **Countway, P. D., R. J. Gast, P. Savai, and D. A. Caron.** 2005. Protistan diversity estimates based on 18S rDNA from seawater incubations in the western N. Atlantic. *J. Eukaryot. Microbiol.* **52:**95–106.
32. **Coyne, K. J., D. A. Hutchins, C. E. Hare, and S. C. Cary.** 2001. Assessing temporal and spatial variability in *Pfiesteria piscicida* distributions using molecular probing techniques. *Aquat. Microb. Ecol.* **24:**275–285.
33. **Crawford, D. W.** 1989. *Mesodinium rubrum*: the phytoplankter that wasn't. *Mar. Ecol. Prog. Ser.* **58:**161–174.
34. **Davis, P. G., D. A. Caron, and J. M. Sieburth.** 1978. Oceanic amoebae from the North Atlantic: culture, distribution, and taxonomy. *Trans. Am. Microsc. Soc.* **96:**73–88.
35. **Dawson, S. C., and N. R. Pace.** 2002. Novel kingdom-level eukaryotic diversity in anoxic environments. *Proc. Natl. Acad. Sci. USA* **99:**8324–8329.
36. **Dennett, M. R., D. A. Caron, A. F. Michaels, S. M. Gallager, and C. S. Davis.** 2002. Video plankton recorder reveals high abundances of colonial radiolaria in surface waters of the central North Pacific. *J. Plankton Res.* **24:**797–805.
37. **Dennett, M. R., S. Mathot, D. A. Caron, W. O. Smith, Jr., and D. J. Lonsdale.** 2001. Abundance and distribution of phototrophic and heterotrophic nano- and microplankton in the southern Ross Sea. *Deep-Sea Res. II* **48:**4019–4037.
38. **Diez, B., C. Pedrós-Alió, T. L. Marsh, and R. Massana.** 2001. Application of denaturing gradient gel electrophoresis (DGGE) to study the diversity of marine picoeukaryotic assemblages and comparison of DGGE with other molecular techniques. *Appl. Environ. Microbiol.* **67:**2942–2951.
39. **Dolan, J. R.** 1991. Guilds of ciliate microzooplankton in the Chesapeake Bay. *Estuar. Coast. Shelf Sci.* **33:**137–152.
40. **Dolan, J. R.** 1992. Mixotrophy in ciliates: a review of *Chlorella* symbiosis and chloroplast retention. *Mar. Micro. Food Webs* **6:**115–132.
41. **Dolan, J. R., and D. W. Coats.** 1990. Seasonal abundances of planktonic ciliates and microflagellates in mesohaline Chesapeake Bay waters. *Estuar. Coast. Shelf Sci.* **31:**157–175.
42. **Ducklow, H. W.** 1994. Modeling the microbial food web. *Microb. Ecol.* **28:**303–319.
43. **Dworetzky, B. A., and J. J. Morley.** 1987. Vertical distribution of radiolaria in the eastern equatorial Atlantic: analysis of a multiple series of closely-spaced plankton tows. *Mar. Micropaleontol.* **12:**1–19.
44. **Edgcomb, V. P., D. T. Kysela, A. Teske, A. de Vera Gomez, and M. L. Sogin.** 2002. Benthic eukaryotic diversity in the Guaymas Basin hydrothermal vent environment. *Proc. Natl. Acad. Sci. USA* **99:**7658–7662.
45. **El Fantroussi, S., H. Urakawa, A. E. Bernhard, J. J. Kelly, P. A. Noble, H. Smidt, G. M. Yershov, and D. A. Stahl.** 2003. Direct profiling of environmental microbial populations by thermal dissociation analysis of native rRNAs hybridized to oligonucleotide microarrays. *Appl. Environ. Microbiol.* **69:**2377–2382.
46. **Fairbanks, R. G., and P. H. Wiebe.** 1980. Foraminifera and chlorophyll maximum: vertical distribution, seasonal succession, and paleoceanographic significance. *Science* **209:**1524–1526.
47. **Fawley, M. J., K. P. Fawley, and M. A. Buchheim.** 2004. Molecular diversity among communities of freshwater microchlorophytes. *Microb. Ecol.* **48:**489–499.
48. **Fenchel, T.** 1967. The ecology of marine microbenthos. I. The quantitative importance of ciliates as compared with metazoans in various types of sediments. *Ophelia* **4:**121–137.
49. **Fenchel, T.** 1969. The ecology of marine microbenthos. IV. Structure and function of the benthic ecosystem, its chemical and physical factors and the microfauna communities with special reference to the ciliated protozoa. *Ophelia* **6:**1–182.
50. **Fenchel, T.** 1987. *Ecology of Protozoa.* Science Tech/Springer-Verlag, Madison, Wis.

51. **Fenchel, T., and B. J. Finlay.** 1983. Respiration rates in heterotrophic, free-living protozoa. *Microb. Ecol.* **9:**99–122.
52. **Finlay, B. J.** 2002. Global dispersal of free-living microbial eukaryote species. *Science* **296:**1061–1063.
53. **Finlay, B. J., U.-G. Berninger, L. J. Stewart, R. M. Hindle, and W. Davison.** 1987. Some factors controlling the distribution of two pond-dwelling ciliates with algal symbionts (*Frontonia vernalis* and *Euplotes daidaleos*). *J. Protozool.* **34:**349–356.
54. **Finlay, B. J., K. E. Clarke, E. Vicente, and M. R. Miracle.** 1991. Anaerobic ciliates from a sulfide-rich solution lake in Spain. *Eur. J. Protistol.* **27:**148–159.
55. **Finlay, B. J., G. F. Esteban, and T. Fenchel.** 1998. Protozoan diversity: converging estimates of the global number of free-living ciliate species. *Protist* **149:**29–37.
56. **Finlay, B. J., and T. Fenchel.** 2004. Cosmopolitan metapopulations of free-living microbial eukaryotes. *Protist* **155:**237–244.
57. **Fuhrman, J. A.** 2002. Community structure and function in prokaryotic marine plankton. *Antonie Leeuwenhoek* **81:**521–527.
58. **Garrison, D. L.** 1991. An overview of the abundance and role of protozooplankton in Antarctic waters. *J. Mar. Syst.* **2:**317–331.
59. **Garrison, D. L., and M. M. Gowing.** 1993. Protozooplankton, p. 123–166. *In* E. I. Friedmann (ed.), *Antarctic Microbiology*. Wiley-Liss, Inc., New York, N.Y.
60. **Garrison, D. L., M. M. Gowing, M. P. Hughes, L. Campbell, D. A. Caron, M. R. Dennett, A. Shalapyonok, R. J. Olson, M. R. Landry, S. L. Brown, H.-B. Liu, F. Azam, G. F. Steward, H. W. Ducklow, and D. C. Smith.** 2000. Microbial food web structure in the Arabian Sea: a US JOGOFS study. *Deep-Sea Res. II* **47:**1387–1422.
61. **Gast, R. J., M. R. Dennett, and D. A. Caron.** 2004. Characterization of protistan assemblages in the Ross Sea, Antarctica, by denaturing gradient gel electrophoresis. *Appl. Environ. Microbiol.* **70:**2028–2037.
62. **Gifford, D. J.** 1985. Laboratory culture of marine planktonic oligotrichs (Ciliophora, Oligotrichida). *Mar. Ecol. Prog. Ser.* **23:**257–267.
63. **Gifford, D. J., and D. A. Caron.** 1999. Sampling, preservation, enumeration and biomass of marine protozooplankton, p. 193–221. *In* R. P. Harris, P. H. Wiebe, J. Lenz, H. R. Skjoldal, and M. Huntley (ed.), *ICES Zooplankton Methodology Manual*. Academic Press, London, United Kingdom.
64. **Gooday, A. J., and P. J. D. Lambshead.** 1989. Influence of seasonally deposited phytodetritus on benthic foraminiferal populations in the bathyal Northeast Atlantic: the species response. *Mar. Ecol. Prog. Ser.* **58:**53–67.
65. **Gooday, A. J., L. A. Levin, P. Linke, and T. Heeger.** 1992. The role of benthic foraminifera in deep-sea food webs and carbon cycling, p. 63–91. *In* G. T. Rowe (ed.), *Deep-Sea Food Chains and the Global Carbon Cycle*. Kluwer Academic, Dordrecht, The Netherlands.
66. **Gowing, M. M.** 1993. Seasonal radiolarian flux at the VERTEX North Pacific time-series site. *Deep-Sea Res.* **40:** 517–545.
67. **Guillou, L., M.-J. Chrétiennot-Dinet, S. Boulben, S. Y. Moon-van der Staay, and D. Vaulot.** 1999. *Symbiomonas scintillans* gen. et sp. nov. and *Picophagus flagellatus* gen. et sp. nov. (Heterokonta): two new heterotrophic flagellates of picoplanktonic size. *Protist* **150:**383–398.
68. **Guillou, L., M.-J. Chrétiennot-Dinet, L. K. Medlin, H. Claustre, L.-D. Goër, and D. Vaulot.** 1999. *Bolidomonas*: a new genus with two species belonging to a new algal class, the Bolidophyceae (Heterokonta). *J. Phycol.* **35:**368–381.
69. **Guillou, L., W. Eikrem, M.-J. Chrétiennot-Dinet, F. Le Gall, R. Massana, K. Romari, C. Pedrós-Alió, and D. Vaulot.** 2004. Diversity of picoplanktonic prasinophytes assessed by direct nuclear SSU rDNA sequencing of environmental samples and novel isolates retrieved from oceanic and coastal marine ecosystems. *Protist* **155:**193–214.
70. **Habura, A., J. Pawlowski, S. D. Hanes, and S. S. Bowser.** 2004. Unexpected foraminiferal diversity revealed by small-subunit rDNA analysis of Antarctic sediment. *J. Eukaryot. Microbiol.* **51:**173–179.
71. **Hairston, N. G., Jr., and N. G. Hairston, Sr.** 1997. Does food web complexity eliminate trophic-level dynamics? *Am. Nat.* **149:**1001–1007.
72. **Hansen, P. J., P. K. Bjørnsen, and B. W. Hansen.** 1997. Zooplankton grazing and growth: scaling within the 2–2,000-μm body size range. *Limnol. Oceanogr.* **42:**687–704.
73. **Holzmann, M., A. Habura, H. Giles, S. S. Bowser, and J. Pawlowski.** 2003. Freshwater foraminiferans revealed by analysis of environmental DNA samples. *J. Eukaryot. Microbiol.* **50:**135–139.
74. **Hopkinson, C. S., Jr., and J. J. Vallino.** 1994. Toward the development of generally applicable models of the microbial loop in aquatic ecosystems. *Microb. Ecol.* **28:**321–326.
75. **Jacobson, D. M., and D. M. Anderson.** 1986. Thecate heterotrophic dinoflagellates: feeding behavior and mechanisms. *J. Phycol.* **22:**249–258.
76. **Jones, R. I.** 1994. Mixotrophy in planktonic protists as a spectrum of nutritional strategies. *Mar. Microb. Food Webs* **8:**87–96.
77. **Jürgens, K., and G. Stolpe.** 1995. Seasonal dynamics of crustacean zooplankton, heterotrophic nanoflagellates and bacteria in a shallow, eutrophic lake. *Freshw. Biol.* **33:**27–38.
78. **Larsen, J., and D. J. Patterson.** 1990. Some flagellates (Protista) from tropical marine sediments. *J. Nat. Hist.* **24:**801–937.
79. **Leadbeater, B. S. C.** 1993. Preparation of pelagic protists for electron microscopy, p. 241–251. *In* P. F. Kemp, B. F. Sherr, E. B. Sherr, and J. J. Cole (ed.), *Handbook of Methods in Aquatic Microbial Ecology*. Lewis Publishers, Boca Raton, Fla.
80. **Lee, J. J., G. F. Leedale, and P. Bradbury.** 2000. *An Illustrated Guide to the Protozoa*, 2nd ed. Allen Press, Inc., Lawrence, Kans.
81. **Lee, J. J., K. Sang, B. ter Kuile, E. Strauss, P. J. Lee, and W. W. Faber, Jr.** 1991. Nutritional and related experiments on laboratory maintenance of three species of symbiont-bearing, large foraminifera. *Mar. Biol.* **109:**417–425.
82. **Lee, J. J., and A. T. Soldo.** 1992. *Protocols in Protozoology*. Allen Press, Inc., Lawrence, Kans.
83. **Lim, E. L., D. A. Caron, and M. R. Dennett.** 1999. The ecology of *Paraphysomonas imperforata* based on studies employing oligonucleotide probe identification in coastal water samples and enrichment culture. *Limnol. Oceanogr.* **44:**37–51.
84. **Lindeman, R. L.** 1942. The trophic-dynamic aspect of ecology. *Ecology* **23:**399–418.
85. **Liu, W.-T., A. Mirzabekov, and D. A. Stahl.** 2001. Optimization of an oligonucleotide microchip for microbial indentification studies: a non-equilibrium dissociation approach. *Environ. Microbiol.* **3:**619–629.
86. **López-Garcia, P., H. Philippe, F. Gail, and D. Moreira.** 2003. Autochthonous eukaryotic diversity in hydrothermal sediment and experimental microcolonizers at the Mid-Atlantic Ridge. *Proc. Natl. Acad. Sci. USA* **100:**697–702.
87. **López-Garcia, P., F. Rodriguez-Valera, C. Pedrós-Alió, and D. Moreira.** 2001. Unexpected diversity of small eukaryotes in deep-sea Antarctic plankton. *Nature* **409:**603–607.

88. **Maeda, M.** 1986. An illustrated guide to the species of the families Halteriidae and Strobilidiidae (Oligotrochida, Ciliophora), free swimming protozoa common in the marine environment. *Bull. Ocean Res. Inst. Univ. Tokyo* **21:**1–67.
89. **Massana, R., J. Castresana, V. Balague, L. Guillou, K. Romari, A. Goisillier, K. Valentin, and C. Pedros-Alio.** 2004. Phylogenetic and ecological analysis of novel marine stramenopiles. *Appl. Environ. Microbiol.* **70:**3528–3534.
90. **Menden-Deuer, S., and E. J. Lessard.** 2000. Carbon to volume relationships for dinoflagellates, diatoms, and other protist plankton. *Limnol. Oceanogr.* **45:**569–579.
91. **Michaels, A. F.** 1991. Acantharian abundance and symbiont productivity at the VERTEX seasonal station. *J. Plankton Res.* **13:**399–418.
92. **Michaels, A. F., D. A. Caron, N. R. Swanberg, F. A. Howse, and C. M. Michaels.** 1995. Planktonic sarcodines (Acantharia, Radiolaria, Foraminifera) in surface waters near Bermuda: abundance, biomass and vertical flux. *J. Plankton Res.* **17:**131–163.
93. **Moloney, C. L., and J. G. Field.** 1991. The size-based dynamics of plankton food webs. 1. A simulation model of carbon and nitrogen flux. *J. Plankton Res.* **13:**1003–1038.
94. **Montagnes, D. J. S., and D. H. Lynn.** 1993. A quantitative protargol stain (QPS) for ciliates and other protists, p. 229–240. *In* P. F. Kemp, B. F. Sherr, E. B. Sherr, and J. J. Cole (ed.), *Handbook of Methods in Aquatic Microbial Ecology*. Lewis Publishers, Boca Raton, Fla.
95. **Moon-van der Staay, S. Y., R. De Wachter, and D. Vaulot.** 2001. Oceanic 18S rDNA sequences from picoplankton reveal unsuspected eukaryotic diversity. *Nature* **409:**607–610.
96. **Pace, M. L., J. J. Cole, and E. J. Carpenter.** 1998. Trophic cascades and compensation: differential responses of microzooplankton in whole-lake experiments. *Ecology* **79:**138–152.
97. **Page, F. C.** 1983. *Marine Gymnamoebae*. Institute of Terrestrial Ecology, Cambridge, Mass.
98. **Parke, M., and P. S. Dixon.** 1976. Checklist of British marine algae, third revision. *J. Mar. Biol. Assoc. U. K.* **56:**527–594.
99. **Patterson, D. J.** 1999. The diversity of eukaryotes. *Am. Nat.* **154**(Suppl.)**:**S96–S124.
100. **Patterson, D. J., and J. Larsen (ed.).** 1991. *The Biology of Free-Living Heterotrophic Flagellates*. Clarendon Press, Oxford, United Kingdom.
101. **Patterson, D. J., and A. G. B. Simpson.** 1996. Heterotrophic flagellates from coastal marine and hypersaline sediments in Western Australia. *Eur. J. Protistol.* **32:**423–448.
102. **Patterson, D. J., and M. Zöulffel.** 1991. Heterotrophic flagellates of uncertain taxonomic position, p. 427–475. *In* D. J. Patterson and J. Larsen (ed.), *The Biology of Free-Living Heterotrophic Flagellates*. Clarendon Press, Oxford, United Kingdom.
103. **Pawlowski, J., J. J. Lee, and A. Gooday.** 1993. Microforaminifera—perspective on a neglected group of Foraminifera. *Arch. Protistenk.* **143:**271–284.
104. **Pierce, R. W., and J. T. Turner.** 1992. Ecology of planktonic ciliates in marine food webs. *Rev. Aquat. Sci.* **6:** 139–181.
105. **Pierce, R. W., and J. T. Turner.** 1993. Global biogeography of marine tintinnids. *Mar. Ecol. Prog. Ser.* **94:**11–26.
106. **Polis, G. A., A. L. W. Sears, G. R. Huxel, D. R. Strong, and J. Maron.** 2000. When is a trophic cascade a trophic cascade? *Trends Ecol. Evol.* **15:**473–475.
107. **Popels, L. C., S. C. Cary, D. A. Hutchins, R. Forbes, F. Pustizzi, C. J. Gobler, and K. J. Coyne.** 2003. The use of quantitative polymerase chain reaction for the detection and enumeration of the harmful alga *Aureococcus anophagefferens* in environmental samples along the United States East Coast. *Limnol. Oceanogr.* **1:**92–102.
108. **Putt, M., and D. K. Stoecker.** 1989. An experimentally determined carbon:volume ratio for marine "oligotrichous" ciliates from estuarine and coastal waters. *Limnol. Oceanogr.* **34:**1097–1103.
109. **Reid, P. C., C. M. Turley, and P. H. Burkill.** 1991. *Protozoa and Their Role in Marine Processes*. Springer-Verlag, Berlin, Germany.
110. **Rogerson, A., H. G. Butler, and J. C. Thomason.** 1994. Estimation of amoeba cell volume from nuclear diameter and its application to studies in protozoan ecology. *Hydrobiologia* **284:**229–234.
111. **Rogerson, A., and J. Laybourn-Parry.** 1992. The abundance of marine naked amoebae in the water column of the Clyde estuary. *Estuar. Coast. Shelf Sci.* **34:**187–196.
112. **Sanders, R. W.** 1991. Mixotrophic protists in marine and freshwater ecosystems. *J. Protozool.* **38:**76–81.
113. **Sanders, R. W., D. A. Caron, and U.-G. Berninger.** 1992. Relationships between bacteria and heterotrophic nanoplankton in marine and fresh water: an inter-ecosystem comparison. *Mar. Ecol. Prog. Ser.* **86:**1–14.
114. **Sawyer, T. K.** 1980. Marine amebae from clean and stressed bottom sediments of the Atlantic Ocean and Gulf of Mexico. *J. Protozool.* **27:**13–32.
115. **Schlegel, M.** 1994. Molecular phylogeny of eukaryotes. *Trends Ecol. Evol.* **9:**330–335.
116. **Scholin, C. A., K. R. Buck, T. Britschgi, G. Cangelosi, and F. P. Chavez.** 1996. Identification of *Pseudo-nitzschia australis* (Bacillariophyceae) using rRNA-targeted probes in whole cell and sandwich hybridization formats. *Phycologia* **35:**190–197.
117. **Shapiro, L. P., L. Campbell, and E. M. Haugen.** 1989. Immunochemical recognition of phytoplankton species. *Mar. Ecol. Prog. Ser.* **57:**219–224.
118. **Sherr, E. B., and B. Sherr.** 2002. Significance of predation by protists in aquatic microbial food webs. *Antonie Leeuwenhoek* **81:**293–308.
119. **Sherr, E. B., and B. F. Sherr.** 1993. Preservation and storage of samples for enumeration of heterotrophic protists, p. 207–212. *In* P. F. Kemp, B. F. Sherr, E. B. Sherr, and J. J. Cole (ed.), *Handbook of Methods in Aquatic Microbial Ecology*. Lewis Publishers, Boca Raton, Fla.
120. **Sieburth, J. M.** 1979. *Sea Microbes*. Oxford University Press, New York, N.Y.
121. **Sieburth, J. M., V. Smetacek, and J. Lenz.** 1978. Pelagic ecosystem structure: heterotrophic compartments of the plankton and their relationship to plankton size fractions. *Limnol. Oceanogr.* **23:**1256–1263.
122. **Silver, M. W., M. M. Gowing, D. C. Brownlee, and J. O. Corliss.** 1984. Ciliated protozoa associated with oceanic sinking detritus. *Nature* **309:**246–248.
123. **Sims, G., A. Rogerson, and R. Aitken.** 1994. Identification and taxonomy of marine amoebae using the small-subunit 18S ribosomal RNA gene. *J. Eukaryot. Microbiol.* **42**(Suppl.)**:**1A.
124. **Sims, G. P., R. Aitken, and A. Rogerson.** 2002. Identification and phylogenetic analysis of morphologically similar naked amoebae using small subunit ribosomal RNA. *J. Eukaryot. Microbiol.* **49:**478–484.
125. **Small, E. B., and M. E. Gross.** 1985. Preliminary observations of protistan organisms, especially ciliates, from the 21°N hydrothermal vent site. *Biol. Soc. Wash. Bull.* **6:**401–410.
126. **Starink, M., M.-J. Bär-Gilissen, R. P. M. Bak, and T. E. Cappenberg.** 1994. Quantitative centrifugation to extract benthic protozoa from freshwater sediments. *Appl. Environ. Microbiol.* **60:**167–173.

127. **Stoecker, D., A. E. Michaels, and L. H. Davis.** 1987. Large proportion of marine planktonic ciliates found to contain functional chloroplasts. *Nature* **326:**790–792.
128. **Stoecker, D., A. Taniguchi, and A. E. Michaels.** 1989. Abundance of autotrophic, mixotrophic and heterotrophic planktonic ciliates in shelf and slope waters. *Mar. Ecol. Prog. Ser.* **50:**241–254.
129. **Stoecker, D. K.** 1998. Conceptual models of mixotrophy in planktonic protists and some ecological and evolutionary implications. *Eur. J. Protistol.* **34:**281–290.
130. **Stoecker, D. K.** 1999. Mixotrophy among dinoflagellates. *J. Eukaryot. Microbiol.* **46:**397–401.
131. **Stoecker, D. K., D. J. Gifford, and M. Putt.** 1994. Preservation of marine planktonic ciliates: losses and cell shrinkage during fixation. *Mar. Ecol. Prog. Ser.* **110:**293–299.
132. **Swanberg, N. R., and D. A. Caron.** 1991. Patterns of sarcodine feeding in epipelagic oceanic plankton. *J. Plankton Res.* **13:**287–312.
133. **Taylor, F. J. R.** 1982. Symbioses in marine microplankton. *Ann. Inst. Océanogr.* (Paris) **58**(Suppl.)**:**61–90.
134. **Throndsen, J.** 1969. Flagellates of Norwegian coastal waters. *Nytt Mag. Bot.* **16:**161–216.
135. **Turley, C. M., and K. Lochte.** 1990. Microbial response to the input of fresh detritus to the deep-sea bed. *Palaeogeogr. Palaeoclimatol. Palaeoecol.* **89:**3–23.
136. **Turner, J. T., and J. C. Roff.** 1993. Trophic levels and trophospecies in marine plankton: lessons from the microbial food web. *Mar. Microb. Food Webs* **7:**225–248.
137. **Urakawa, H., S. El Fantroussi, H. Smidt, J. C. Smoot, E. H. Tribou, J. J. Kelly, P. A. Noble, and D. A. Stahl.** 2003. Optimization of single-base-pair mismatch discrimination in oligonucleotide microarrays. *Appl. Environ. Microbiol.* **69:**2848–2856.
138. **Van Hannen, E. J., M. P. Van Agterveld, H. J. Gons, and H. J. Laanbroek.** 1998. Revealing genetic diversity of eukaryotic microorganisms in aquatic ecosystems by denaturing gradient gel electrophoresis. *J. Phycol.* **34:** 206–213.
139. **Verity, P. G., C. Y. Robertson, C. R. Tronzo, M. G. Andrews, J. R. Nelson, and M. E. Sieracki.** 1992. Relationships between cell volume and the carbon and nitrogen content of marine photosynthetic nanoplankton. *Limnol. Oceanogr.* **37:**1434–1446.
140. **Vørs, N.** 1993. Heterotrophic amoebae, flagellates and heliozoa from Arctic marine waters (North West Territories, Canada and West Greenland). *Polar Biol.* **13:**113–126.
141. **Vørs, N.** 1992. Heterotrophic amoebae, flagellates and heliozoa from the Tvärminne area, Gulf of Finland, in 1988–1990. *Ophelia* **36:**1–109.
142. **Vørs, N.** 1993. Marine heterotrophic amoebae, flagellates and heliozoa from Belize (Central America) and Tenerife (Canary Islands), with descriptions of new species, *Luffisphaera bulbochaete* n. sp., *L. longihastis* n. sp., *L. turriformis* n. sp. and *Paulinella intermedia* n. sp. *J. Eukaryot. Microbiol.* **40:**272–287.
143. **White, D. C.** 1994. Is there anything else you need to understand about the microbiota that cannot be derived from analysis of nucleic acids? *Microb. Ecol.* **28:**163–166.
144. **Wickham, S., A. Gieseke, and U.-G. Berninger.** 2000. Benthic ciliate identification and enumeration: an improved methodology and its application. *Aquat. Microb. Ecol.* **22:**79–91.
145. **Wiebe, W. J., W. M. Sheldon, Jr., and L. R. Pomeroy.** 1992. Bacterial growth in the cold: evidence for an enhanced substrate requirement. *Appl. Environ. Microbiol.* **58:**359–364.
146. **Wright, S. W., S. W. Jeffrey, F. C. Mantoura, C. A. Llewellyn, T. Bjørnland, D. Repeta, and N. Welschmeyer.** 1991. Improved HPLC method for the analysis of chlorophylls and carotenoids from marine phytoplankton. *Mar. Ecol. Prog. Ser.* **77:**183–196.
147. **Yongue, W. H., Jr., and J. Cairns, Jr.** 1978. The role of flagellates in pioneer protozoan colonization of artificial substrates. *Pol. Arch. Hydrobiol.* **25:**787–801.
148. **Zubkov, M. V., A. F. Sazhin, and M. V. Flint.** 1992. The microplankton organisms at the oxic-anoxic interface in the pelagial of the Black Sea. *FEMS Microbiol. Ecol.* **101:**245–250.

Decomposition and Fungal Community Structure in Aquatic Environments

FELIX BÄRLOCHER

38

Traditionally, fungi have been defined as eukaryotic, spore-producing, nonphotosynthetic organisms having absorptive nutrition mechanisms and existing as a network of branched tubes (hyphae) with cell walls (56). In the last two decades, it has become increasingly clear that this describes not a single phylogenetic lineage but rather a way of life shared by organisms of different evolutionary backgrounds. This finding has led to the common practice of using the term "fungi" in a very broad sense and to reserve the term "Fungi" for the true fungi that appear to share a common ancestor (26). Typically, we recognize chromistan fungi (or pseudofungi) with cellulosic hyphal walls (phyla Oomycota or Peronosporomycota and Hyphochitriomycota) and eumycotan fungi (true fungi, or simply Fungi) with chitinous walls (phyla Chytridiomycota, Zygomycota, Glomeromycota, and Dikaryomycota, the last subdivided into Basidiomycotina and Ascomycotina) (56). Fungal phylogeny remains in a state of flux, and other classifications have been proposed (72).

Fungi tend to dominate terrestrial decomposition of plant detritus, but the realization that they may be equally important in aquatic environments is relatively recent (6, 42, 44). It is based primarily on research done in two marine (*Spartina* salt marshes and mangrove swamps [e.g., references 44, 54, 57, and 76]) and two freshwater (streams and marshes [6, 42, 104]) habitats. The emphasis of this chapter will be on leaf decomposition in streams, which is dominated by aquatic hyphomycetes. It will largely ignore marine fungi (Eumycota and Oomycota) (56), aeroaquatic fungi (39, 69), freshwater ascomycetes (on living and dead macrophytes around edges of ponds, lakes, and wetlands and on wood debris partially submerged in streams) (95, 105), and various zoosporic fungi (17, 88, 95).

Aquatic hyphomycetes (also known as Ingoldian fungi, freshwater hyphomycetes, or amphibious fungi [6]) are anamorphic (mitosporic) fungi. Their classification is based on the morphology and development of conidia (asexual spores), which are predominantly multiradiate (often tetraradiate) or sigmoid (50). Over 300 species have been described, and roughly 10% of these have been connected to teleomorphs (sexual state) (96, 110). The majority belong to the Ascomycotina, and a few belong to the Basidiomycotina. Aquatic hyphomycetes are most common on deciduous leaves decaying in fast-flowing, well-aerated streams. The history of their discovery and early research on their biology have been described elsewhere (6, 8). The group remained relatively obscure until limnologists realized that leaves and needles of riparian vegetation are often the dominant suppliers of energy to stream food webs and that aquatic hyphomycetes are important in making these substrates acceptable to stream invertebrates through a process called conditioning. Not surprisingly, considerable effort has been expended on estimating fungal biomasses and growth rates, based primarily on measuring concentrations of fungus-specific ergosterol and the rate of its synthesis from radioactive acetate. Several recent reviews have thoroughly discussed this topic (43–45), and it will not be considered here. Instead, the emphasis will be on methods used to analyze the decomposition of various substrates and on methods to characterize fungal community structure.

DECOMPOSITION

Most investigations rely on a technique introduced by terrestrial ecologists some 70 years ago: plant material is collected and dried, and preweighed portions are exposed in containers such as boxes, open-ended tubes, or most commonly, litter bags with variable mesh sizes (7, 24). Periodically, some containers are removed, the remaining detrital mass is determined, and chemical analyses are performed. This approach deviates from the natural process and can introduce artifacts concerning the course and rate of decomposition and the microbial community involved in it (7).

Sample Collection, Preparation, and Exposure

In temperate regions, most deciduous leaves are shed within a 4- to 6-week period in autumn, and most studies have been initiated in late fall or early winter (111). However, leaves naturally fall into streams throughout the year (e.g., through storm events and outbreaks of disease), and the decomposition process varies seasonally because of differences in substrate qualities and fungal communities (80, 99).

The majority of the leaves fall directly from trees into the stream (7). Leaf shedding is preceded by senescence, during which the integrity of cell membranes is generally maintained. Although drying the leaves before exposing them in streams will lower random variability of mass loss measurements, it also affects the leaching (abiotic removal) of organic and inorganic compounds (9), which in turn influences fungal colonization. To avoid these artifacts, leaves should be captured as soon as they hit the ground, kept under humid conditions, and introduced into a stream with as little delay as possible (9). Generally, the effects of predrying leaves (rapid leaching and delayed fungal colonization) are most noticeable during the early stages of decomposition.

Leaves can be exposed in a stream individually and without any constraints, loosely tied together as packs, or contained in litter bags with variable mesh sizes (7, 10, 24). Measuring decay in unattached leaves most closely approximates the natural process but is difficult and labor-intensive. It generally results in higher mass loss rates than studies with leaf packs or bags, presumably because it maximizes mechanical fragmentation. Packs held together with nylon filaments and attached within the stream allow free access of invertebrates and are considered a reasonable compromise between realism and reproducibility. However, decay rates depend on leaf pack size, and in streams, leaf accumulations continously form and break up. As a rule, mass loss in litter bags resembles that of loose, naturally entrained leaves in depositional zones, while mass loss in leaf packs is close to that of loose leaves in riffle areas (32). The use of litter bags with different mesh sizes allows size-selective exclusion of macroinvertebrates but may lower oxygen and nutrient supplies, especially if leaves are packed too tightly.

The procedures used to collect and expose leaves in streams will affect several interdependent aspects of decomposition: physical fragmentation, the availability of inorganic nutrients and oxygen, and colonization by microorganisms. In turn, the combined action of these factors can profoundly influence the course and speed of decay. As always, it is essential to define the objectives of a study before deciding on a technique. If the goal is to isolate factors that shape the fungal community (e.g., nutrients, current speed, and the presence of invertebrates), some type of restraining device is usually essential. The benefits of lower variability have to be weighed against the loss of realism.

The use of dried material in litter bags has also been the method of choice to study decay in submerged or emergent macrophytes. The latter, however, do not usually abscise leaves or stems (75–78). Their decomposition is therefore best studied by marking individual plants or leaves with electrical cable ties and measuring decay and microbial colonization in situ. Application of this technique has profoundly altered our view of the fungal involvement in macrophyte decomposition in marshes (12, 58, 59).

Statistical Analysis of Mass Loss

The standard approach for statistical analysis of mass loss is to fit remaining mass data (generally expressed as ash-free, dry mass) to the exponential decay model (10, 111). This method provides a single number (the decay coefficient, k) that characterizes the rate of mass loss. Based on daily decay coefficients, leaf decay has been classified as fast ($k > 0.01$), medium ($k = 0.005$-0.001), or slow ($k < 0.005$) (86). More elaborate models divide organic matter into chemical classes, each of which decays at a specific rate (70). The simplest case, the double exponential model, assumes a labile and a refractory class of leaf components (113). When the refractory material is completely inert, or when the decay rate gradually declines, an asymptotic model may be appropriate (100).

Traditionally, the exponential decay equation has been converted into a linear form before regression analysis by using natural log (ln) values of remaining mass. This transformation enhances errors associated with small values and is therefore no longer recommended (71). The alternative is nonlinear regression analysis, or curve fitting, where parameters are determined iteratively rather than analytically. During the fitting procedure, initial estimates are modified to increasingly improve the fit of the curve to the data. This process is continued until additional improvements fall below a predetermined threshold (71).

Often, the question is whether there are significant differences between two values of k (e.g., decay rates of maple and oak leaves). The recommended approach is to repeat the experiment several times and compare the estimates of the parameter k with a t test. This is labor-intensive and statistically conservative. If, as is usual, the experiment has only been done once, the best-fit values for two groups can be compared with a t test by using the standard error reported by the curve-fitting program. This method assumes normal error distribution, which is approximately true for the exponential decay equation (71).

Alternatively, the two data sets can be analyzed separately as well as simultaneously. The question then is whether the separate fits are significantly better than the pooled fit. This method, known as analysis of covariance, is again strictly valid only for a linear relationship between x and y (71). Another approach is based on Akaike's information criterion, which answers the following questions: Which model is more likely to have generated the data? How much more likely? The theory behind Akaike's information criterion is quite difficult; it combines maximum likelihood, information theory, and the concept of information entropy. Computations and interpretation of the results are straightforward (http://www.graphpad.com/curvefit).

Dynamics of Individual Compounds

Overall mass loss is a relatively crude indicator of fungal activity. It is often desirable to measure the disappearance or synthesis of individual compounds or compound classes and correlate their dynamics with the composition of the fungal community. Commonly measured fractions include elements such as nitrogen and phosphorus, organic compounds such as proteins and amino acids, total and soluble carbohydrates, lignin, cellulose and hemicelluloses, phenolics, and lipids (48). Lipophilic fractions of fungal mycelia have been shown to affect substrate palatability to leaf-shredding invertebrates (91), which may influence the structure and development of the fungal community.

Enzymatic Activities

Leaf decomposition by aquatic hyphomycetes depends crucially on an array of exoenzymes attacking structural polymers (48). If biodiversity studies are to yield more than lists of species or of DNA sequences, it is essential to include functional aspects. In pure culture, different species have been shown to have different arrays of hydrolytic enzymes (28, 102). The diversity of laccase gene sequences in salt marshes has been described elsewhere (66). The first step in that study was aligning published amino acid sequences of laccases and designing fungal primers specific for laccase genes. The same approach can be applied to other fungal exoenzymes; it promises to shed light on potential connections between structural (number of different strains or species) and functional diversity.

Community-wide metabolic potential can be monitored by turbidimetry in microtitration plates containing 96 wells with different substrates (e.g., SFN2, SFP2, and BIOLOG) (22). The wells are inoculated with an environmental sample (e.g., leaf particles). It is likely that this method favors rapidly growing species able to grow on a single substrate, which includes many yeasts (22).

Global antibodies allow estimation of the potential capacity for certain key processes of a heterogeneous communtiy. This approach has been applied to photosynthetic enzymes (87) but could be extended to various fungal hydrolytic enzymes. It requires knowledge of amino acid sequences of one or more key proteins from a large number of organisms.

COMMUNITY STRUCTURE

The preferred substrates of aquatic hyphomycetes, decaying leaves, are discrete, spatially separate units. Fungal dispersal is therefore likely to be based on propagules released by vegetative mycelia. Indeed, aquatic hyphomycetes invest up to 50% of their growth process producing conidia (8). Many of these conidia can be identified to the species level by their morphologies and development patterns. This ability has removed some of the pressure to look for alternative approaches when assessing diversity and community structures. Studies of both have been and continue to be based primarily on conidium identification, and molecular techniques have been introduced recently and used sparingly. Nevertheless, as with other fungi, the reliance on reproductive structures (conidia or other spores) is an indirect approach and ignores metabolically active stages (vegetative mycelia), which generally are more relevant for ecological functions. Current molecular techniques have much to offer to freshwater mycologists, and their adoption may require little more than minor adjustments and optimization. At the same time, well-established traditional techniques can provide valuable corroboration or modification of molecular data, which is often impossible with other microbial taxa.

To characterize fungal communities in streams, we can concentrate on their vegetative phase (mycelia growing on leaves) or on types and numbers of released conidia.

Counting and Identifying Conidia

Persistent cakes of foam appear in many soft-water streams below waterfalls and rapids or after heavy rainfalls and tend to accumulate the multiradiate and sigmoid conidia of aquatic hyphomycetes. Collecting this foam and examining it under a microscope is a quick way to characterize the fungal community (34). Generally, a few milliliters is scooped into a container, allowed to collapse, and mixed with an equal volume of FAA (50% ethanol, 90 ml; glacial acetic acid, 5 ml; formalin, 5 ml). Without FAA, conidia quickly germinate, which makes their identification more difficult (but may facilitate DNA extraction and amplification). This approach does not discriminate between spores produced in the stream and those with a terrestrial origin. In addition, the likelihood of being trapped in foam is species specific; these counts therefore give a biased representation of the fungal community. Finally, identification from a single conidium may be ambiguous and require confirmation by observation of spore development or by DNA analysis.

A more complete representation of the fungal community can be obtained by sucking a standard volume of stream water through a membrane filter with a pore size of 5 or 8 μm (55). Conidia are trapped on the filter, where they can be stained (0.1% cotton blue in lactic acid), counted, and identified. Again, there is no distinction between imported and autochthonous conidia (though most are likely to be autochthonous), and some cannot be identified with certainty. Major concerns are sample size and frequencies. One study reported that between 1,100 and 1,350 ml of water have to be filtered to capture 90% of all morphologically distinct conidia (47). In temperate streams, there are clear seasonal fluctuations in total spore numbers, with a maximum reached shortly after leaf fall, but the relative proportions of the various species do not remain constant from year to year (8).

For these broad-scale surveys with large sample numbers, routine use of current molecular techniques is still more labor-intensive and expensive than that of traditional microscopic surveys. Molecular techniques may be useful for resolution of doubtful identifications (due to variable spore morphology), for documentation of the presence of DNA from sources other than conidia of aquatic hyphomycetes (other spores or mycelia) (89), and for broad, phylogenetic classification of the fungal community.

Characterizing Fungi Present on Leaves

Generally, the main interest in characterizing fungi present on leaves is in documenting and quantifying metabolically active fungal stages, i.e., mycelia. These cannot be classified except into very broad categories with conventional microscopic techniques (e.g., dikaryotic Basidiomycotina if clamp connections are present). A reproductive phase has to be induced. This can be done by incubating freshly collected leaves under suitable conditions. For aquatic hyphomycetes, agitated or aerated nutrient-poor (e.g., sterilized distilled water or stream water) solutions are generally preferred. Intact leaves may be used, or leaves may be cut into small squares (e.g., 36 mm^2) (94), which are labeled, incubated, and examined separately. Provided that the leaf is reasonably thin, it can be immersed in cotton blue solution and the distribution of conidiophores or conidia can be mapped with the help of a light microscope (84, 94). If information on the spatial distribution of individual species is not important, leaves or leaf disks in nutrient-poor solution are aerated or agitated for 24 to 48 h. The released spores are then trapped on a membrane filter, stained, counted, and identified (11). This method favors the detection of aquatic hyphomycete mycelia and, among them, those that respond quickly to turbulence by initiating spore production. It will miss dormant or nonreproducing fungal structures or those that release smaller or less striking propagules. With substrates other than leaves (wood and roots), the conventional aeration period of 24 to 48 h may be much too short (98); in addition, ecological fungal groups other than aquatic hyphomycetes may be more active. Molecular methods show great promise to circumvent some of these shortcomings.

With thicker, more opaque leaves or other substrates, direct observation is often difficult or impossible. Under these circumstances, a particle-plating technique can be used: the substrate is cut into small squares (e.g., 4 mm^2) (13) which are placed on an agar medium. Fungi grow on the medium and may be isolated as pure cultures. Plugs of these cultures are submerged in distilled water; if the organisms are aquatic hyphomycetes, they will generally produce conidia within a few days and can be identified. The origin of these cultures is unknown; it may have been a spore or a metabolically active piece of mycelium. Aquatic hyphomycetes are more commonly isolated on nutrient-poor media incubated at low temperatures (5 to 15°C) (13).

The plating technique includes an intermediate step between analysis of what is present on the substrate and identification: the fungus must be able to invade an artificial medium and then reproduce when growing on it. Molecular methods allow a more direct approach.

Molecular Methods

Many of the considerations for choosing and collecting samples apply regardless of subsequent analysis. The great advantage of molecular methods is that they are based on a much greater number of unique traits than traditional approaches (primarily nucleotide sequences rather than morphological data), resulting in much higher resolution. In addition, molecular traits are essentially independent of life stage; there is no need to induce a reproductive stage to allow unequivocal identification and phylogenetic assessment. On the other hand, molecular methods do not generally discriminate between dormant and metabolically active fungi.

Depending on the objectives, the fungal community can be analyzed in situ (within the substrate) to preserve spatial information or its "blueprint" (nucleic acid) can be extracted, amplified, and analyzed externally, allowing analysis of very small samples.

In Situ Analysis

Fluorescence in situ hybridization includes fixation of target cells, hybridization with 15- to 25-bp fluorescently labeled oligonucleotide probes (typically targeting small-subunit [SSU] rRNA), removal of nonhybridized probes, and visualization and quantification by epifluorescence or confocal laser scanning microscopy (22). Since rRNAs have regions of variable sequence preservation, it is possible to design probes that are kingdom, genus, or species specific (phylogenetic stains) (33, 46). However, it has been difficult to develop consistently useful probes for fungi because of their cell walls, which often impede probe entry, and because of autofluorescence of hyphae and/or substrates. An online database for fluorescence in situ hybridization (http://www.microbial-ecology.net/probebase) (65) lists only 10 fungal probes, and one of the very few published studies describes spatial microheterogeneity of *Aureobasidium pullulans* on terrestrial leaves (3). Some success with aquatic hyphomycetes has been reported previously (15, 16, 68), although limited permeability and strong autofluorescence of fungal cells remain major problems. Both may be alleviated to some extent by pretreatment with β-glucanases (15, 101). An alternative modification uses in situ PCR (fixed substrates or cells) followed by hybridization. This approach has not been attempted with aquatic fungi, but its feasibility has been shown with pure cultures (101) and mycorrhizas (5).

Monoclonal antibodies potentially allow both quantification of mycelia through enzyme-linked immunosorbent assays and localization of mycelia through immunofluorescence. They are frequently used for plant pathogenic fungi (35) and have successfully been raised against four species of aquatic hyphomycetes (19–21). Advantages and disadvantages of this approach are discussed in references 35, 43, and 74.

Extraction and Analysis of Nucleic Acids

The most common approach for extraction and analysis of nucleic acids has been as follows: nucleic acids are extracted from pure cultures or environmental samples, amplified by using PCR, and analyzed. Variations in gene sequences are used as proxy indicators of microbial diversity. The type of gene chosen depends on the required resolution or the objectives of the study. Some genes are universal and can be used to address questions of total community diversity, and others are limited to specific groups and provide information on the diversity of functional groups. In the first category, structural RNAs (rRNA and tRNA), internal transcribed spacers (ITS), and housekeeping genes have most commonly been chosen. Molecular diversity studies have concentrated on 18S RNA, leading to the creation of large and growing databases such as GenBank (http://www.ncbi.nlm.nih.gov). This makes it possible to target phylogenetically delineated groups to the exclusion of others (e.g., Basidiomycotina versus other Eumycota) (79). The second category includes genes of secondary metabolism corresponding to specific metabolic functions (49). Molecular biology of aquatic hyphomycetes is in its infancy, but work on other mycological subareas, such as mycorrhizas (37, 52), is well advanced. Regardless of ecological group or habitat, these methods commonly reveal previously unknown fungal lineages (93, 106), indicating the presence of uncultured fungi. We can expect the same in streams and lakes, especially if we move into less thoroughly studied habitats such as the hyphorheic zone and groundwater or aquatic roots (14, 98).

Extraction

For diversity studies, the emphasis continues to be on DNA, while RNA analyses are essential for determining gene expressions. Isolation of full-length RNA is more problematic than that of DNA; the primary source of failure is contamination by very stable ribonucleases. General isolation protocols are found in references 1 and 4.

Traditional extraction procedures involve several steps (homogenization, heating with detergents, phenol-chloroform extractions, and several cleaning steps). With fungi, sodium dodecyl sulfate and CTAB (cetyltrimethylammonium bromide) have been the preferred detergents (e.g., sodium dodecyl sulfate [63, 67] and CTAB [30, 41]). In our work with DNA from pure cultures or from environmental samples, my colleagues and I have found the use of commercial kits to be more time- and cost-effective. We generally use soil DNA extraction kits from MoBio laboratories (Solana Beach, Calif.; http://www.mobio.com), which were also applied in a study of salt marsh ascomycetes (27). They provide readily amplifiable DNA. Other kits are available to remove PCR inhibitors common in environmental samples, for example, Geneclean (http://www.qbiogene.com).

Generally, up to 30 mg of pure fungal culture or up to 50 mg of environmental samples (leaves) has been used to extract DNA (79–84). Routine extraction and amplification are possible with much smaller amounts, e.g., single plugs lifted from pure cultures with a tip of a Pasteur pipette (H. McKinnon, unpublished observation). In combination with PCR primers for fungal species, species density within a leaf (how many species occur in 1 mm^2 of leaf area) may be determined and may provide information on whether species occurrences are mutually independent.

Another potential application is extraction, amplification, and sequencing of DNA from a single spore, allowing molecular identification of morphologically ambiguous conidia. These steps have been possible on a routine basis from as few as 20 spores collected on a membrane filter (the section with the spores of interest is cut out and placed in an extraction tube) (89). When applied to a single conidium, we have had a success rate of roughly 30% (H. McKinnon, G. Nikolcheva, and F. Bärlocher, unpublished observation). DNA extraction from single spores of lichens (109) and arbuscular mycorrhizal fungi (40) has been reported. More consistent results are obtained when the spore is allowed to germinate and the resulting microcolony is harvested and extracted. This approach is particularly useful in dealing with morphologically simple or rare conidia and provides a valuable check on the validity of visual identifications.

PCR

The basic concepts of PCR, as well as its potential pitfalls, are well established (22, 23, 61, 90, 108). The first step is the choice of appropriate primers; they allow enrichment of specified sequences from complex mixtures (multiple copies of an entire genome from a pure culture or multiple copies of several distinct genomes from environmental samples). The choice of primers depends on the objectives of the investigation: for phylogenetic purposes, the gene for SSU rRNA has most often been used. In eukaryotes, it occurs in the nuclear (18S rRNA) and mitochondrial (16S rRNA) genomes. More recently, other genes have been examined, e.g., β-tubulin (103), EF1-alpha, RNA polymerase, and chitinase subunit genes (4, 25). The ITS (ITS1 and ITS2, separated by 5.8S), situated between the genes for the SSU and large-subunit rRNA, 18S and 28S (112), are less conserved and commonly have nucleotide insertions and deletions. This often prevents unambiguous alignment of distantly related taxa (25, 26). A second problem is often a lack of variation among very closely related species: identical ITS sequences are not necessarily evidence of conspecificity. Nevertheless, "no other highly variable region is as easy to amplify from minute or complex samples" (25), and due to their multicopy nature and highly conserved priming site, the ITS may be the best target for the identification of unknown environmental samples. Their usefulness, of course, depends on the number of sequences available in databases like GenBank. There are fewer than 20,000 ITS sequences available from some 73,000 described fungal species, out of an estimated 1.5 million existing species (15). Of the approximately 300 species of aquatic hyphomycetes, partial sequences from 30 are known. Not surprisingly, sequencing ITS genes or gene fragments amplified from environmental samples only rarely allows identification, but the number of distinct sequences (phylotypes) gives an initial estimate of fungal diversity.

Table 1 lists some recently described fungal primers, with the regions they amplify and the assumed degrees of specificity, that have been used with environmental samples. The ITS primers generally amplify all fungal groups, including Oomycota (112). In all cases, the specificity of the primer is subject to revision as more sequences become available. Maps of frequently used genes and primers for their amplification are found, for example, in reference 112 and at http://plantbio.berkeley.edu/~bruns and http://www.lutzonilab.net/pages/primer.shtml.

Only one primer has to be specific for the group of interest; the second primer can be a more general eukaryotic primer. For aquatic hyphomycetes, the following forward/reverse primer pairs have been used: NS1/GCFung (680 bp of the SSU for terminal restriction fragment length polymorphism [T-RFLP] analysis of pure cultures) (83), F1300/D (450 bp of the SSU for T-RFLP analysis of environmental samples) (83), NS1/GCFung (for denaturing gradient gel electrophoresis [DGGE]) (83), ITS5/ITS4 (470

TABLE 1 Fungal primers for amplifying DNA from environmental samples, the adjacent genes, and presumed degrees of specificity

Primer	Gene	Specificity	Reference(s)
AU2	SSU	Fungi	106
AU4	SSU	Fungi	106
GCFung	SSU	Fungi	67
F1300	SSU	Fungi	83
ITS3	ITS2	Fungi (and some plants)	79, 112
ITS4Asco		Ascomycotina	79
ITS4Basidio		Basidiomycotina	
ITS4Chytrid		Chytridiomycota	
ITS4Zygo		Zygomycota	
ITS4Oo		Oomycota	
ITS4A	ITS2	Ascomycotina (including Oomycotina)	61
ITS4B	ITS2	Basidiomycotina (including some plants)	41
ITS1F	ITS1	Fungi	41
EF4f	SSU	Fungi	97
EF3r			
Fung5r			
FR1	SSU	Fungi	107
nLSU1221R	LSU[a]	Fungi	93
NSI	SSU	Eukarya	67
Nu-SSU-0817-5′	SSU	Fungi	23
Nu-SSU-1196-3′			
Nu-SSU-1536-5′			

[a]LSU, large-subunit rRNA.

bp, ITS1 to ITS2, for T-RFLP analysis) (79), ITS3/ITS4 (370 bp of ITS2 for DGGE) (79, 80), and the three overlapping pairs C/NS4, NS3/NS6, and NS5/D or NS5/ITS2 (the entire SSU for phylogenetic analyses) (18).

To cast a wider net and include fungi other than aquatic hyphomycetes, an attempt has been made to develop fungus-specific primers (ITS4Asco, ITS4Basidio, ITS4Chytrid, ITS4Oo, and ITS4Zygo, reverse primers used with ITS3 [79]). The products were analyzed by DGGE. An alternative approach is to use primers that capture the PCR products of a broad range of fungal phyla and follow up by analyzing cloned sequences. This has been done with the primer pairs AU2 and AU4 (approximately 1,330 bp of the SSU) (106) and ITS9 and nLSU1221R (1.8 to 2.2 kbp) (93); we are currently attempting to apply this approach to aquatic fungi.

Hebert et al. (51) suggested using the mitochondrial gene for cytochrome *c* oxidase I as a universal "bar code." One of its postulated advantages is the lack of introns, and it has shown remarkable powers of resolution for various animal groups. Unfortunately, fungi often have several introns (up to 16 in *Podospora anserina* [31]).

When intraspecific genetic variability is of interest, primers with an arbitrary sequence of 15 to 25 bp have been used for random amplification of polymorphic DNA. The products are analyzed with electrophoresis, and the resulting patterns are used as genetic markers (29, 85). Due to concerns about reproducibility, random amplification of polymorphic DNA is no longer recommended (22). A more powerful and robust alternative is amplified fragment length polymorphism analysis, which has been applied successfully to the aquatic hyphomycete *Tetrachaetum elegans* (60).

Two main sources of PCR-based errors are recognized: (i) PCR biases and artifacts and (ii) data interpretation (22, 108). Biases occur during both extraction and amplification of DNA from diverse communities. Artifacts include heteroduplexes, chimeras, and mutagenesis. They can lead to errors in sequence analyses and may result in spurious peaks or bands, and therefore overestimates of diversity, in rapid profiling techniques such as DGGE and T-RFLP analysis. The frequency of occurrence of heteroduplexes (two annealed sequences from different sources) generally increases with the diversity of similar templates, low primer concentrations, and high numbers of PCR cycles. Steps can be taken to minimize the last two factors. Another simple precaution is combining the products of several amplification reactions to minimize the effect of stochastic events during early cycles. To estimate the degree of heteroduplex formation, PCR with a mixture of DNA from known members of the community may be useful.

A recently developed alternative to traditional PCR is whole-genome amplification (53). It is a robust and accurate method of amplifying nanogram DNA quantities and allows complete representation of the entire genome with no detectable allele bias. Whole-genome amplification includes both PCR-based techniques and non-PCR linear amplification protocols. A variant, strand displacement amplification has successfully amplified DNA from single spores of *Glomus* and *Gigaspora* (40) and clearly holds promise for aquatic mycology.

Analysis

The greatest amount of information from the PCR product can be gained by sequencing. This is generally straightforward when DNA was extracted and amplified from a pure culture (see, however, reference 92). Environmental extracts typically contain DNA from several taxa; in these cases, PCR fragments are inserted into *Escherichia coli* plasmids, propagated, and sequenced (cloning). Each of these steps may distort the end result; potential sources of bias are discussed in references 22 and 36. In bacterial communities, the estimated species richness is generally reproducible with a standardized number of clones, but species composition is not and would require much larger samples than those commonly used (36). Similar considerations presumably apply to fungal communities. No such studies have been conducted with aquatic hyphomycetes, but recent studies on salt marsh fungi (26) and on terrestrial fungi (93, 106) provide useful templates.

Community Fingerprinting Approaches

If an appropriate gene is chosen, extracting and amplifying DNA from a community produce a number of sequences that differ slightly from each other. These differences may affect susceptibility to restriction enzymes or ease of denaturation by chemical gradients or temperature. The number of different sequences can be estimated by RFLP or T-RFLP analysis, DGGE, or temperature gradient gel electrophoresis. T-RFLP analyses have been applied to fungi on *Spartina alterniflora* (27), and both T-RFLP and DGGE (38, 73, 104) approaches have been used to characterize communities of aquatic hyphomycetes (79–84). With aquatic hyphomycetes, processing of samples with T-RFLP analysis was more rapid after a relatively lengthy optimization process. A total of 21 species (30 strains) were used to calibrate the two methods. When an SSU fragment was analyzed, DGGE gave 11 distinct bands; with an ITS region, it yielded 17 bands. For T-RFLP analysis, 10 and 19 distinct terminal fragments were produced with the SSU and ITS, respectively. In this particular example, traditional methods have potentially higher resolving power—provided that identifiable spores are being released. This is particularly true for scanning of filters with conidia of many different species; a skilled mycologist can detect a species represented by a single spore among thousands, but for DGGE, a species has to contribute ≥3% to the total biomass to be registered reliably (83, 89).

Single-stranded conformational polymorphism analysis is based on the fact that single-stranded DNA adopts a sequence-specific tertiary structure. It does not require gradient gels and is a relatively simple and effective method to detect small sequence variabilities. It has more commonly been used for profiling bacterial communities (64).

Quantitative Methods

The various profiling techniques indicate the presence or absence of certain taxa; they do not as a rule allow conclusions as to the abundance or biomasses of the organisms. By measuring intensities of stained bands in DGGE, relative population sizes of amplified phylotypes can be estimated (73, 79, 83). One must be aware, however, that DNA/biomass ratios vary among fungal species and that extraction from environmental samples and amplification can introduce additional biases.

DNA amounts can also be estimated by quantitative PCR techniques (22). Competitive PCR measures simultaneous amplification of a target template and an added standard. The amount of target DNA is estimated by combining it with a dilution series of the standard; the ratio of the two PCR products at the end point is used to calculate the initial target concentration.

Automated real-time PCR uses a fluorogenic probe. Detected fluorescence is proportional to the amount of accumulated PCR product and is measured in each cycle of amplification. Quantification is based on the critical threshold cycle value, which is the cycle number where the amplification product becomes detectable above background noise. The threshold value is inversely proportional to initial DNA template concentration. Absolute quantitation requires prior addition or parallel amplification of a known standard.

Real-time PCR works well when there is one well-defined target, such as a single species of an arbuscular mycorrhizal fungus in plant host tissue (2). Multiplex real-time PCR allows simultaneous analysis of up to 12 targets (22). Validating and optimizing this approach to keep track of 12 fungal species in ecosystems would be a valuable addition to currently available methods.

Microarrays are potentially useful tools in molecular ecology (22). Oligonucleotide probes are immobilized on a matrix, which is then used to capture labeled target molecules by hybridization. Their primary use has been in sequencing and genetic analyses, but combinations of taxon-specific probes on PhyloChips may allow one to rapidly take stock of a fungal community. To complement this information, metabolic activities may be characterized by functional gene arrays. These approaches remain to be tested in fungal ecology (22).

CONCLUSIONS

There is a widespread conviction among scientists that we are in the middle of a sixth mass extinction (62). In addition to esthetic and ethical considerations, there are concerns that this may jeopardize essential ecological functions and services. These concerns have provided new impetus to biodiversity studies: we need to take stock of currently existing species. Molecular biology has opened exciting new avenues of research, which bypass some of the traditional obstacles in microbial taxonomy and ecology (e.g., small size and limited morphological variation of target organisms). Currently, neither molecular nor traditional methods lead to comprehensive species lists; we therefore need statistical evaluations to extrapolate from our samples to ecosystems (8). In addition, molecular methods have their own limitations and pitfalls, and species lists are not particularly useful unless combined with information on how species contribute to ecological functions. Again, molecular methods provide powerful new approaches. The question remains whether knowledge of species and their enzymatic capabilities suffices for understanding how ecosystems function and for predicting the effects of continued species losses due to anthropogenic disturbance. Or do these require information on higher-level properties (such as the ability to reproduce and disperse and the avoidance of predators and pathogens)? To answer these questions, traditional techniques remain indispensable.

REFERENCES

1. **Akkermans, A. D. L., J. D. van Elsas, and F. J. de Bruijn (ed.).** 1995. *Molecular Microbial Ecology Manual.* Kluwer, Dordrecht, The Netherlands.
2. **Alkan, N., V. Gadkar, J. Coburn, O. Yarden, and Y. Kapulnik.** 2004. Quantification of the arbuscular mycorrhizal fungus *Glomus intraradices* in host tissue using real-time polymerase chain reaction. *New Phytol.* **161:**877–885.
3. **Andrews, J. H., R. N. Spear, and E. V. Nordheim.** 2002. Population biology of *Aureobasidium pullulans* on apple leaf surfaces. *Can. J. Microbiol.* **48:**500–513.
4. **Ausubel, F. M., R. Brent, R. E. Kingston, R. E. Moore, J. G. Seidman, J. A. Smith, and K. Struhl (ed.).** 1999. *Current Protocols in Molecular Biology*, 4th ed. Wiley Interscience, New York, N.Y.
5. **Bago, B., Y. Piche, and L. Simon.** 1998. Fluorescently-primed *in situ* PCR in arbuscular mycorrhizas. *Mycol. Res.* **102:**1540–1544.
6. **Bärlocher, F. (ed.).** 1992. *The Ecology of Aquatic Hyphomycetes*. Springer, Berlin, Germany.
7. **Bärlocher, F.** 1997. Pitfalls of traditional techniques when studying decomposition of vascular plant remains in aquatic habitats. *Limnetica* **13:**1–11.
8. **Bärlocher, F.** 2005. Freshwater fungal communities, p. 39–59. *In* J. Deighton, J. F. White, Jr., and P. Oudemans (ed.). *The Fungal Community: Its Organization and Role in the Ecosystem*. Taylor & Francis, CRC Press, Boca Raton, Fla.
9. **Bärlocher, F.** 2005. Leaching, p. 33–36. *In* M. A. S. Graça, F. Bärlocher, and M. O. Gessner (ed.), *Methods To Study Litter Decomposition*. Springer, Dordrecht, The Netherlands.
10. **Bärlocher, F.** 2005. Leaf mass loss estimated by litter bag technique, p. 37–42. *In* M. A. S. Graça, F. Bärlocher, and M. O. Gessner (ed.), *Methods To Study Litter Decomposition.* Springer, Dordrecht, The Netherlands.
11. **Bärlocher, F.** 2005. Sporulation of aquatic hyphomycetes, p. 185–188. *In* M. A. S. Graça, F. Bärlocher, and M. O. Gessner (ed.), *Methods To Study Litter Decomposition.* Springer, Dordrecht, The Netherlands.
12. **Bärlocher, F., and N. R. Biddiscombe.** 1996. Geratology and decomposition of *Typha latifolia* and *Lythrum salicaria* in a freshwater marsh. *Arch. Hydrobiol.* **136:**309–325.
13. **Bärlocher, F., and B. Kendrick.** 1974. Dynamics of the fungal population on leaves in a stream. *J. Ecol.* **62:**761–791.
14. **Bärlocher, F., L. G. Nikolcheva, K. P. Wilson, and D. D. Williams.** Fungi in the hyporheic zone of a springbrook. *Microb. Ecol.*, in press.
15. **Baschien, C.** 2003. Development and evaluation of rRNA targeted *in situ* probes and phylogenetic relationships of freshwater fungi. Ph.D. thesis. Technical University of Berlin, Berlin, Germany.
16. **Baschien, C., W. Manz, T. R. Neu, and U. Szewzyk.** 2001. Fluorescence *in situ* hybridization of freshwater fungi. *Int. Rev. Hydrobiol.* **86:**371–381.
17. **Beakes, G. W.** 2003. Lower fungi: a review of microscopical techniques for the taxonomic and ecological study of zoosporic freshwater fungi, p. 51–79. *In* C. K. M. Tsui and K. D. Hyde (ed.), *Freshwater Mycology*. The Fungal Diversity Press, Hong Kong.
18. **Belliveau, M. J.-R., and F. Bärlocher.** 2005. Molecular evidence confirms multiple origins of aquatic hyphomycetes. *Mycol. Res.* **109:**1407–1417.
19. **Bermingham, S., F. M. Dewey, and L. Maltby.** 1995. Development of a monoclonal antibody-based immunoassay for the detection and quantification of *Anguillospora longissima* colonizing leaf material. *Appl. Environ. Microbiol.* **61**:2606–2613.
20. **Bermingham, S. L., Maltby, and F. M. Dewey.** 1997. Use of immunoassays for the study of natural assemblages of aquatic hyphomycetes. *Microb. Ecol.* **33:**223–229.
21. **Bermingham, S., F. M. Dewey, P. J. Fisher, and L. Maltby.** 2001. Use of a monoclonal antibody-based immunoassay for the detection and quantification of *Heliscus lugdunensis* colonizing alder leaves and roots. *Microb. Ecol.* **42:**506–512.
22. **Bidartando, M. I., and M. Gardes.** 2005. Fungal diversity in molecular terms: profiling, identification, and quantification

in the environment, p. 215–239. *In* J. Deighton, J. F. White, Jr., and P. Oudemans (ed.), *The Fungal Community: Its Organization and Role in the Ecosystem*. Taylor & Francis, CRC Press, Boca Raton, Fla.

23. **Borneman, J., and R. J. Hartin.** 2000. PCR primers that amplify fungal rRNA genes from environmental samples. *Appl. Environ. Microbiol.* **66:**4356–4360.
24. **Boulton, A. J., and P. I. Boon.** 1991. A review of the methodology used to measure leaf litter decomposition in lotic environments: time to turn over an old leaf? *Aust. J. Mar. Freshw. Res.* **42:**1–43.
25. **Bruns, T. D.** 2001. ITS reality. *Inoculum* **52:**2–3.
26. **Bruns, T. D., T. J. White, and J. W. Taylor.** 1991. Fungal molecular systematics. *Annu. Rev. Ecol. Syst.* **22:**525–564.
27. **Buchan, A., S. Y. Newell, J. I. L. Moreta, and M. A. Moran.** 2002. Analysis of internal transcribed spacer (ITS) regions of rRNA genes in fungal communities in a southeastern U.S. salt marsh. *Microb. Ecol.* **43:**329–340.
28. **Chamier, A.-C.** 1985. Cell-wall-degrading enzymes of aquatic hyphomycetes: a review. *Bot. J. Linn. Soc.* **91:**67–81.
29. **Charcosset, J.-Y., and M. Gardes.** 1999. Infraspecific genetic diversity and substrate preference in the aquatic hyphomycete *Tetrachaetum elegans*. *Mycol. Res.* **103:**736–742.
30. **Conway, D. R., J. C. Frankland, V. A. Saunders, and D. R. Wilson.** 2000. Effects of elevated atmospheric CO_2 on fungal competition and decomposition of *Fraxinus excelsior* litter in laboratory microcosms. *Mycol. Res.* **104:** 187–197.
31. **Cummings, R. J., F. Michel, and K. L. McNally.** 1989. DNA sequence analysis of the 24.5 kilobase pair cytochrome oxidase subunit I mitochondrial gene from *Podospora anserina*: a gene with sixteen introns. *Curr. Genet.* **16:**381–406.
32. **Cummins, K. W., G. L. Spengler, G. M. Ward, R. M. Speaker, R. W. Ovink, D. C. Mahan, and R. L. Mattingly.** 1980. Processing of confined and naturally entrained leaf litter in a woodland stream ecosystem. *Limnol. Oceangr.* **25:**952–957.
33. **DeLong, E. F., G. S. Wickham, and N. R. Pace.** 1989. Phylogenetic stains: ribosomal RNA-based probes for the identification of single cells. *Science* **243:**1360–1363.
34. **Descals, E.** 2005. Techniques for handling Ingoldian fungi, p. 129–141. *In* M. A. S. Graça, F. Bärlocher, and M. O. Gessner (ed.), *Methods To Study Litter Decomposition*. Springer, Dordrecht, The Netherlands.
35. **Dewey, F. M., C. R. Thornton, and C. A. Gilligan.** 1997. Use of monoclonal antibodies to detect, quantify and visualize fungi in soils. *Adv. Bot. Res.* **24:**275–308.
36. **Dunbar, J., S. M. Barns, L. O. Ticknor, and C. R. Kuske.** 2002. Empirical and theoretical bacterial diversity in four Arizona soils. *Appl. Environ. Microbiol.* **68:**3035–3045.
37. **Egger, K. N.** 1994. Molecular analysis of ectomycorrhizal fungal communities. *Can. J. Bot.* **73**(Suppl. 1)**:**S1415–S1422.
38. **Fischer, S. G., and L. S. Lerman.** 1983. DNA fragments differing by single base-pair substitutions are separated in denaturing gradient gels: correspondence with melting theory. *Proc. Natl. Acad. Sci. USA* **80:**1579–1583.
39. **Fisher, P. J.** 1977. New methods of detecting and studying saprophytic behaviour of aero-aquatic hyphomycetes from stagnant water. *Trans. Br. Mycol. Soc.* **68:**407–411.
40. **Gadkar, V., and M. C. Rillig.** 2005. Application of Phi29 DNA polymerase mediated whole genome amplification on single spores of arbuscular mycorrhizal (AM) fungi. *FEMS Microbiol. Lett.* **242:**65–71.
41. **Gardes, M., and T. D. Bruns.** 1993. ITS primers with enhanced specificity for basidiomycetes—application to the identification of mycorrhizae and rusts. *Mol. Ecol.* **2:**113–118.
42. **Gessner, M. O., and G. Van Ryckegem.** 2002. Water fungi as decomposers in freshwater ecosystems, p. 3353–3364. *In* G. Bitton (ed.), *Encyclopedia of Environmental Microbiology*. Wiley & Sons, New York, N.Y.
43. **Gessner, M. O., and S. Y. Newell.** 2002. Biomass, growth rate, and production of filamentous fungi in plant litter, p. 390–408. *In* C. J. Hurst, R. L. Crawford, G. R. Knudsen, M. J. McInerney, and L. D. Stetzenbach (ed.), *Manual of Environmental Microbiology*, 2nd ed. ASM Press, Washington, D.C.
44. **Gessner, M. O., K. Suberkropp, and E. Chauvet.** 1997. Decomposition of plant litter by fungi in marine and freshwater ecosystems, p. 303–322. *In* D. T. Wicklow and B. Söderström (ed.), *The Mycota*, vol. IV. *Environmental and Microbial Relationships*. Springer, Berlin, Germany.
45. **Gessner, M. O., F. Bärlocher, and E. Chauvet.** 2003. Qualitative and quantitative analyses of aquatic hyphomycetes in streams, p. 127–157. *In* C. K. M. Tsui and K. D. Hyde (ed.), *Freshwater Mycology*. The Fungal Diversity Press, Hong Kong.
46. **Giovannoni, S. J., E. F. DeLong, G. J. Olsen, and N. R. Pace.** 1988. Phylogenetic group-specific oligodeoxynucleotide probes for identification of single microbial cells. *J. Bacteriol.* **170:**720–726.
47. **Gönczöl, J., A. Révay, and P. Csontos.** 2001. Effect of sample size on the detection of species and conidial numbers of aquatic hyphomycetes collected by membrane filtration. *Arch. Hydrobiol.* **150:**677–691.
48. **Graça, M. A. S., F. Bärlocher, and M. O. Gessner** (ed.). 2005. *Methods To Study Litter Decomposition*. Springer, Dordrecht, The Netherlands.
49. **Güimil, S., H.-S. Song, T. Zhu, A. Sesma, A. Osbourn, C. Roux, V. Ionnidis, E. J. Oakley, M. Docquier, P. Descombes, S. P. Briggs, and U. Paszkowski.** 2005. Comparative transcriptomics of rice reveals an ancient pattern of response to microbial colonization. *Proc. Natl. Acad. Sci. USA* **102:**8066–8070.
50. **Gulis, V., L. Marvanová, and E. Descals.** 2005. An illustrated key to the common temperate species of aquatic hyphomycetes, p. 153–167. *In* M. A. S. Graça, F. Bärlocher, and M. O. Gessner (ed.), *Methods To Study Litter Decomposition*. Springer, Dordrecht, The Netherlands.
51. **Hebert, P. D. N., A. Cywinska, S. L. Ball, and J. R. de Waard.** 2003. Biological identifications through DNA barcodes. *Proc. R. Soc. Lond. B* **270:**313–322.
52. **Horton, T. R., and T. D. Bruns.** 2001. The molecular revolution in ectomycorrhizal ecology: peeking into the black-box. *Mol. Ecol.* **10:**1855–1871.
53. **Hughes, S., N. Arneson, S. Done, and J. Squire.** 2005. The use of whole genome amplification in the study of human disease. *Prog. Biophys. Mol. Biol.* **88:**173–189.
54. **Hyde, K. D., and S. B. Pointing (ed.).** 2000. *Marine Mycology—a Practical Approach*. The Fungal Diversity Press, Hong Kong.
55. **Iqbal, S. H., and J. Webster.** 1973. Aquatic hyphomycete spora of the River Exe and its tributaries. *Trans. Br. Mycol. Soc.* **61:**331–346.
56. **Kendrick, B.** 2001. *The Fifth Kingdom*. Focus Publishing, Newburyport, Mass.
57. **Kohlmeyer, J., B. Volkmann-Kohlmeyer, and S. Y. Newell.** 2004. Marine and estuarine mycelial Eumycota and Oomycota, p. 533–545. *In* G. M. Mueller, G. F. Bills, and M. S. Foster (ed.), *Biodiversity of Fungi. Inventory and Monitoring Methods*. Elsevier Academic Press, Amsterdam, The Netherlands.
58. **Kuehn, K. A., and K. Suberkropp.** 1998. Decomposition of standing dead litter of the emergent macrophyte *Juncus effuses*. *Freshw. Biol.* **40:**717–727.

59. **Kuehn, K.A., M. J. Lemke, K. Suberkropp, and R. G. Wetzel.** 2000. Microbial biomass and production associated with decaying leaf litter of the emergent macrophyte *Juncus effusu*. *Limnol. Oceanogr.* **45:**862–870.
60. **Laitung, B., E. Chauvet, N. Feau, K. Fève, L. Chikhi, and M. Gardes.** 2004. Genetic diversity in *Tetrachaetum elegans*, a mitosporic aquatic fungus. *Mol. Ecol.* **13:**1679–1692.
61. **Larena, I., O. Salazar, V. González, M. C. Julián, and V. Rubio.** 1999. Design of a primer for ribosomal DNA internal transcribed spacer with enhanced specificity for ascomycetes. *J. Biotechnol.* **75:**187–194.
62. **Leakey, R. A., and R. Lewin.** 1995. *The Sixth Extinction: Patterns of Life and the Future of Humankind.* Doubleday, New York, N.Y.
63. **Lee, S. B., M. G. Milgroom, and J. W. Taylor.** 1988. A rapid, high yield mini-prep method for isolation of total genomic DNA from fungi. *Fungal Genet. Newsl.* **35:**23–24.
64. **Lee, D. H., Y. G. Zo, and S. J. Kim.** 1996. Nonradioactive method to study genetic profiles of natural bacterial communities by PCR-single-strand-conformation polymorphism. *Appl. Environ. Microbiol.* **69:**327–333.
65. **Loy, A., M. Horn, and M. Wagner.** 2003. ProbeBase—an online resource for rRNA-targeted oligonucleotide probes. *Nucleic Acids Res.* **31:**514–516.
66. **Lyons, J. L., S. Y. Newell, A. Buchan, and M. A. Morgan.** 2003. Diversity of Ascomycete laccase gene sequences in a southeastern US salt marsh. *Microb. Ecol.* **45:**270–281.
67. **May, L. A., B. Smiley, and M. G. Schmidt.** 2001. Comparative denaturing gradient gel electrophoresis analysis of fungal communities associated with whole plant corn silage. *Can. J. Microbiol.* **47:**829–841.
68. **McArthur, F. A., M. O. Baerlocher, N. A. B. MacLean, M. D. Hiltz, and F. Bärlocher.** 2001. Asking probing questions: can fluorescent in situ hybridization identify and localise aquatic hyphomycetes on leaf litter? *Int. Rev. Hydrobiol.* **86:**429–438.
69. **Michaelides, J., and B. Kendrick.** 1982. The bubble-trap propagules of *Beverwykella*, *Helicoon* and other aero-aquatic fungi. *Mycotaxon* **14:**247–260.
70. **Minderman, G.** 1968. Addition, decomposition and accumulation of organic matter in forests. *J. Ecol.* **56:**355–362.
71. **Motulsky, H. J., and L. A. Ransnsas.** 1987. Fitting curves to data using nonlinear regression: a practical and nonmathematical review. *FASEB J.* **1:**365–374.
72. **Mueller, G. M., G. F. Bills, and M. S. Foster (ed.).** 2004. *Biodiversity of Fungi: Inventory and Monitoring Methods.* Elsevier Academic Press, Amsterdam, The Netherlands.
73. **Muyzer, G., E. C. de Waal, and A. G. Uitterlinden.** 1993. Profiling of complex microbial populations by denaturing gradient gel electrophoresis analysis of polymerase chain reaction-amplified genes coding for 16S rRNA. *Appl. Environ. Microbiol.* **59:**695–700.
74. **Newell, S. Y.** 1992. Estimating fungal biomass and productivity in decomposing litter, p. 521–561. *In* G. C. Carroll and D. T. Wicklow (ed.), *The Fungal Community: Its Organization and Role in the Ecosystem*, 2nd ed. Marcel Dekker, New York, N.Y.
75. **Newell, S. Y.** 1993. Decomposition of shoots of a salt-marsh grass. Methodology and dynamics of microbial assemblages. *Adv. Microb. Ecol.* **13:**301–326.
76. **Newell, S. Y.** 1996. Established and potential impacts of eukaryotic mycelial decomposers in marine/terrestrial ecotones. *J. Exp. Mar. Biol. Ecol.* **200:**187–206.
77. **Newell, S. Y., and R. D. Fallon.** 1989. Litterbags, leaf tags, and decay of nonabscised intertidal leaves. *Can. J. Bot.* **67:**2324–2327.
78. **Newell, S. Y., J. D. Miller, and R. D. Fallon.** 1989. Decomposition and microbial dynamics for standing, naturally positioned leaves of the salt-marsh grass *Spartina alterniflora*. *Mar. Biol.* **101:**471–481.
79. **Nikolcheva, L. G., and F. Bärlocher.** 2004. Taxon-specific primers reveal unexpectedly high diversity during leaf decomposition in a stream. *Mycol. Prog.* **3:**41–50.
80. **Nikolcheva, L. G., and F. Bärlocher.** 2005. Seasonal and substrate preferences of fungi colonizing leaves in streams: traditional vs. molecular evidence. *Environ. Microbiol.* **7:** 270–280.
81. **Nikolcheva, L. G., and F. Bärlocher.** 2005. Molecular approaches to estimate fungal diversity. I. Terminal restriction fragment length polymorphism (T-RFLP), p. 169–176. *In* M. A. S. Graça, F. Bärlocher, and M. O. Gessner (ed.), *Methods To Study Litter Decomposition.* Springer, Dordrecht, The Netherlands.
82. **Nikolcheva, L. G., and F. Bärlocher.** 2005. Molecular approaches to estimate fungal diversity. II. Denaturing gradient gel electrophoresis (DGGE), p. 177–184. *In* M. A. S. Graça, F. Bärlocher, and M. O. Gessner (ed.), *Methods To Study Litter Decomposition.* Springer, Dordrecht, The Netherlands.
83. **Nikolcheva, L. G., A. M. Cockshutt, and F. Bärlocher.** 2003. Diversity of freshwater fungi on decaying leaves—comparing traditional and molecular approaches. *Appl. Environ. Microbiol.* **69:**2548–2554.
84. **Nikolcheva, L. G., T. Bourque, and F. Bärlocher.** 2005. Fungal diversity during initial stages of leaf decomposition in a stream. *Mycol. Res.* **109:**246–253.
85. **Peláez, F., G. Platas, J. Collado, and M. T. Diéz.** 1996. Infraspecific variation in two species of aquatic hyphomycetes assessed by RAPD analysis. *Mycol. Res.* **100:**831–837.
86. **Petersen, R. C., and K. W. Cummins.** 1974. Leaf processing in a woodland stream. *Freshw. Biol.* **4:**343–368.
87. **Pichard, S. L., L. Campbell, K. Carder, J. B. Kang, J. Patch, F. R. Tabita, and J. H. Paul.** 1997. Analysis of ribulose biphosphate carboxylase gene expression in natural phytoplankton communities by group-specific gene probing. *Mar. Ecol. Prog. Ser.* **149:**239–253.
88. **Powell, M. J.** 1993. Looking at mycology with a Janus face: a glimpse at Chytridiomycetes active in the environment. *Mycologia* **85:**1–20.
89. **Raviraja, N. S., L. G. Nikolcheva, and F. Bärlocher.** 2005. Diversity of aquatic hyphomycete conidia assessed by microscopy and by DGGE. *Microb. Ecol.* **49:**1–7.
90. **Rodriguez, R. J., D. Cullen, C. P. Kurtzman, G. G. Khachatourians, and D. D. Hegedus.** 2004. Molecular methods for discriminating taxa, monitoring species, and assessing fungal diversity, p. 77–102. *In* G. M. Mueller, G. F. Bills, and M. S. Foster (ed.), *Biodiversity of Fungi. Inventory and Monitoring Methods.* Elsevier Academic Press, Amsterdam, The Netherlands.
91. **Rong, Q., K. R. Sridhar, and F. Bärlocher.** 1995. Food selection of three leaf-shredding invertebrates. *Hydrobiologia* **316:**173–181.
92. **Sanders, I. R., A. Koch, and G. Kuhn.** 2003. Arbuscular mycorrhizal fungi: genetics of multigenomic, clonal networks and its ecological consequences. *Biol. J. Linn. Soc.* **79:**59–60.
93. **Schadt, D. W., A. P. Martin, D. A. Lipson, and S. K. Schmidt.** 2003. Seasonal dynamics of previously unknown fungal lineages in tundra soils. *Science* **301:**1359–1361.
94. **Shearer, C. A., and L. C. Lane.** 1983. Comparison of three techniques for the study of aquatic hyphomycete communities. *Mycologia* **75:**498–508.
95. **Shearer, C. A., D. M. Langsam, and J. E. Longcore.** 2004. Fungi in freshwater habitats, p. 513–531. *In* G. M.

Mueller, G. F. Bills, and M. S. Foster (ed.), *Biodiversity of Fungi. Inventory and Monitoring Methods*. Elsevier Academic Press, Amsterdam, The Netherlands.

96. **Sivichai, S., and E. B. G. Jones.** 2003. Teleomorphic-anamorphic connections of freshwater fungi, p. 259–274. *In* C. K. M. Tsui and K. D. Hyde (ed.), *Freshwater Mycology*. The Fungal Diversity Press, Hong Kong.
97. **Smit, E., P. Leeflang, B. Glandorf, J. D. van Elsas, and K. Wernars.** 1999. Analysis of fungal diversity in the wheat rhizosphere by sequencing of cloned PCR-amplified genes encoding 18S rRNA and temperature gradient gel electrophoresis. *Appl. Environ. Microbiol.* **65:**2614–2621.
98. **Sridhar, K. R., and F. Bärlocher.** 1992. Endophytic aquatic hyphomycetes of roots of spruce, birch and maple. *Mycol. Res.* **96:**305–308.
99. **Sridhar, K. R., and F. Bärlocher.** 1993. Seasonal changes in microbial colonization of fresh and dried leaves. *Arch. Hydrobiol.* **128:**1–12.
100. **Sridhar, K. R., G. Krauss, F. Bärlocher, N. S. Raviraja, R. Wennrich, and R. Baumbach.** 2002. Decomposition of alder leaves in two heavy metal polluted streams in Central Germany. *Aquat. Microb. Ecol.* **26:**73–80.
101. **Sterflinger, K., W. E. Krumbein, and A. Schwiertz.** 1998. A protocol for PCR in situ hybridization of hyphomycetes. *Int. Microbiol.* **1:**217–220.
102. **Suberkropp, K., and M. J. Klug.** 1980. The maceration of deciduous leaf litter by aquatic hyphomycetes. *Can. J. Bot.* **58:**1025–1031.
103. **Thell, A., S. Stenroos, T. Feuerer, I. Kärnefelt, L. Myllys, and J. Hyvönen.** 2002. Phylogeny of cetrarioid lichens (Parmeliaceae) inferred from ITS and ß-tubulin sequences, morphology, anatomy and secondary chemistry. *Mycol. Prog.* **1:**335–354.
104. **Tsui, C. K. M., and K. D. Hyde (ed.).** 2003. *Freshwater Mycology*. The Fungal Diversity Press, Hong Kong.
105. **Tsui, C. K. M., K. D. Hyde, and I. J. Hodkiss.** 2003. Methods for investigating the biodiversity and distribution of freshwater ascomycetes and anamorphic fungi on submerged wood, p. 195–209. *In* C. K. M. Tsui and K. D. Hyde (ed.), *Freshwater Mycology*. The Fungal Diversity Press, Hong Kong.
106. **Vandenkoornhuyse, P., S. L. Baldauf, C. Leyval, J. Straczek, and J. P. W. Young.** 2002. Extensive fungal diversity in plant roots. *Science* **295:**2051.
107. **Vainio, E. J., and J. Hantula.** 2000. Direct analysis of wood-inhabiting fungi using denaturing gradient gel electrophoresis of amplified ribosomal DNA. *Mycol. Res.* **104:**927–936.
108. **von Wintzingerode, F., U. B. Göbel, and E. Stackebrandt.** 1997. Determination of microbial diversity in environmental samples: pitfalls of PCR-based rRNA analysis. *FEMS Microbiol. Rev.* **21:**213–229.
109. **Walser, J.-C., S. Zoller, U. Büchler, and C. Scheidegger.** 2001. Species-specific detection of *Lobaria pulmonaria* (lichenized ascomycete) diaspores in litter samples trapped in snow cover. *Mol. Ecol.* **10:**2129–2138.
110. **Webster, J.** 1992. Anamorph-teleomorph relationships, p. 99–117. *In* F. Bärlocher (ed.), *The Ecology of Aquatic Hyphomycetes*. Springer, Berlin, Germany.
111. **Webster, J. R., and E. F. Benfield.** 1986. Vascular plant breakdown in freshwater ecosystems. *Annu. Rev. Ecol. Syst.* **17:**567–594.
112. **White, T. J., T. Bruns, S. Lee, and J. W. Taylor.** 1990. Amplification and direct sequencing of fungal ribosomal RNA genes for phylogenetics, p. 315–322. *In* M. A. Innis, D. H. Gelfand, J. J. Sninsky, and T. J. White (ed.), *PCR Protocols: a Guide to Methods and Applications*. Academic Press, New York, N.Y.
113. **Wieder, R. K., and G. E. Lang.** 1982. A critique of the analytical methods used in examining decomposition data obtained from litter bags. *Ecology* **63:**1636–1642.

Bacterial Organic Carbon Cycling in Aquatic Environments

MATTHEW T. COTTRELL

39

Bacteria and dissolved organic matter (DOM) are important components of marine food webs and carbon cycling in the ocean. Although DOM comprises compounds with complex elemental compositions, this review is concerned primarily with dissolved organic carbon (DOC), which supports the growth and respiration of bacteria. The DOC pool is equal in size to the amount of carbon in the atmosphere (ca. 700×10^{15} tons) (20). The flux of carbon through the DOC pool is also substantial and supports the metabolism of heterotrophic bacteria and in turn grazers that compose the microbial loop (1). Grazing and algal excretion are believed to be the main sources of DOC available to bacteria (46). Although our understanding of the details of DOC sources is incomplete, the consensus is that about half of the carbon fixed by primary production is consumed by bacteria and processed by the microbial loop.

The goal of this chapter is to introduce some of the important aspects of bacterial organic carbon cycling and to provide an update on the molecular biological approaches currently in use for examining carbon cycling by bacteria without cultivation. The two types of analyses considered here are single-cell analysis and cloning of environmental DNA in large-insert vectors. There are obvious differences in the types of information that these approaches provide, and of course the types of questions that they can address are different as well. What they have in common is the ability to open the microbial "black box" and describe what kinds of bacteria are involved in aquatic carbon cycling and how they do it. Such investigations by microbial ecologists are motivated by the desire to understand how the composition of microbial communities dictates the way that food webs and carbon cycling function in the oceans and other aquatic environments.

The DOM pool is chemically complex and consists of many compounds that differ greatly in chemical structure, molecular weight, C:N ratio, and availability to microbes. We know the most about the chemical makeup of high-molecular-weight compounds, including biopolymers, and less about monomeric components of the DOM pool. In contrast, there are more data on the use by bacteria of low-molecular-weight DOM compounds than of high-molecular-weight compounds. Low-molecular-weight compounds, such as free amino acids and sugars, are used rapidly by bacteria, which keeps concentrations low in the nanomolar range. The assimilation of dissolved free amino acids generally supports over 20% of bacterial production (26, 42, 63). Sugars are another important component of DOM and like amino acids are maintained at low concentrations due to rapid use by bacteria. Although the sugars in seawater are probably not all used equally by bacteria, glucose alone can support 20 to 30% of bacterial production (33, 55, 61).

However, there are still considerable gaps in our understanding of the use of even low-molecular-weight compounds. Amino acids and sugars have been the focus of much work because they are the main components of plankton biomass and therefore represent the main sources of the DOM pool. However, compounds such as amino sugars, which are the building blocks of the biopolymers comprising cell walls of plankton organisms, probably deserve more attention. Amino sugar concentrations in seawater are low (25), and consumption by bacteria has received little study. One low-molecular-weight compound, glycolate, probably deserves more study because it is released by phytoplankton during photorespiration of diatoms (51). In fact, concentrations of glycolate in the tropical Atlantic range from less than 0.002 to 1 μM (35), much higher than the concentrations of sugars and amino acids. Glycolate should be an interesting compound for studying the coupling between phytoplankton DOC and bacterial consumption. The relationship between glycolate and bacteria has received some attention in microbial mat environments (23).

Other low-molecular-weight compounds of interest that are produced by phytoplankton include the sulfur-containing compounds dimethylsulfoniopropionate (DMSP) and methanethiol. It has been suggested that heterotrophic bacteria can satisfy nearly all of their sulfur requirement from DMSP, although it probably meets less than half of their carbon demand (29). One interesting aspect of DMSP degradation by bacteria is the generation of dimethylsulfide (DMS) (19), which is an important source of sulfur to the atmosphere and can increase cloud formation.

High-molecular-weight DOM compounds such as proteins and polysaccharides are important components of the DOM flux and can support a large fraction of bacterial growth (24, 27, 58, 61). However, it may be misleading to think of high-molecular-weight DOM flux as being completely separate from the flux of monomers if there is coupling between polymer hydrolysis and monomer uptake. The hydrolysis of biopolymers may contribute to the production of monomers, including free amino acids and monosaccharides. Hydrolysis is an essential step in the consumption of high-molecular-weight DOM because bacteria are restricted to transporting only compounds of less than about 500 Da (66). Since hydrolysis takes place outside the cell, it is possible that some monomers may escape immediate uptake and join the pool of free monomers to be taken up only later.

An uncoupling between hydrolysis and uptake may lead to cross feeding between bacteria when the released monomers are taken up by other cells. The ability to produce hydrolytic enzymes varies among bacteria (41), and the expression of these enzymes is typically not constitutive. Under the cross-feeding scenario, the uptake of monomers by bacteria not able to degrade certain biopolymers or not producing hydrolytic enzymes themselves may not be properly considered separate from the high-molecular-weight DOM flux.

The evidence concerning the release of monomers from high-molecular-weight DOM is not conclusive. One example is the degradation of polysaccharides. In deep waters, the addition of high-molecular-weight DOM greatly reduces glucose uptake (61). However, the data are not definitive because there is no effect of the addition on glucose uptake in surface waters. Similarly, data on the molecular compositions of high- and low-molecular-weight DOM point to different conclusions. The makeup of the free sugar pool is not consistent with high-molecular-weight DOM's being a source of free glucose. If the low-molecular-weight pool is simply produced from hydrolysis of high-molecular-weight DOM, then the compositions of low- and high-molecular-weight DOM should be the same. High-molecular-weight DOM and "combined" sugars analyzed by acid hydrolysis comprise several neutral sugars that occur in approximately the same proportions. In contrast, the free neutral monosaccharide pool is made up of only glucose at low concentrations (61). The picture is not entirely clear, though, because when total concentrations of neutral sugars are high (>50 nM), free neutral sugars other than glucose are detectable. Additional information on the biological controls of bacterial hydrolysis of biopolymers and the potential release of monomers may be useful in determining if high-molecular-weight DOM is a source of monomers.

Understanding bacterial organic carbon cycling undoubtedly requires insights into the biology and ecology of the microorganisms involved. Although a wealth of information on carbon cycling by bacteria has been obtained without considering the microbes themselves, treating the microbes as a black box implicitly assumes that DOM consumption levels are equal for all bacteria. However, we now know that bacterial communities are phylogenetically diverse and therefore probably include a variety of metabolic types. Phylogenetic diversity of uncultured microbes is likely reflected in metabolic diversity that is hypothesized to influence the linkage between DOM fluxes and bacterial growth and respiration. Metabolic tests for substrate utilization by cultivated bacteria were an important underpinning of bacterial taxonomy before the advent of molecular phylogenetic approaches. With the current focus on molecular phylogeny using 16S rRNA genes, which are not directly linked to metabolism, there is an obvious need for information on the actual metabolism of bacteria and their roles in DOM cycling.

The answers to some basic questions concerning bacterial diversity would greatly expand our current view of DOM consumption by bacteria. What kinds of bacteria are most important for carbon cycling, or are they all equally important? Do abundant bacteria contribute more to carbon cycling than less-abundant bacteria? What adaptations provide bacteria with the means to succeed in aquatic environments where growth is typically limited by organic carbon substrates? Approaches to addressing these questions concerning the relationship between the structure and function of aquatic microbial communities are outlined below.

SINGLE-CELL ANALYSIS OF DOM CONSUMPTION

Epifluorescence microscopy using nucleic acid stains is the most widely used approach to determining the abundance of prokaryotes in aquatic systems. Direct counting of microbes by fluorescence microscopy was originally developed in the 1970s using acridine orange to stain preserved samples collected on optically flat polycarbonate filters (21). Over the years, changes to the direct count procedure largely involved the introduction of new stains such as DAPI (4′,6′-diamidino-2-phenylindole) for greater accuracy of counting in particle-rich waters (53) and SYBR green for counting of virus-like particles (47). However, the introduction of fluorescently labeled oligonucleotide probes targeting rRNA sequences for enumerating different bacterial phylotypes marked a major advance (12). The introduction of these "phylogenetic stains" made it possible to enumerate different types of bacteria, not just to determine the total abundance.

Another advance in enumerating microbes in aquatic systems was the introduction of microautoradiography for counting of metabolically active cells. The assessment of microbial activity by microautoradiography is achieved by incubating cells with a radioactive compound, collecting the cells on a polycarbonate filter for microscopy, and then assaying uptake by coating the cells with a photographic film emulsion. Active cells are identified as those surrounded by silver grains in the emulsion. In the earliest applications of this approach, the uptake of thymidine demonstrated active DNA synthesis and growth of microbes in the ocean (6). One technical problem with the original method was that silver grains sometimes obstructed the view of the cells. This limitation was solved by transferring the cells from the filter to an emulsion-coated slide (65) instead of covering the filter with the emulsion. Data on thymidine and leucine uptake assayed at the single-cell level indicate that the activity and growth of most bacteria in aquatic systems can be readily assessed by microautoradiography (16, 31, 60).

The next logical step in the development of methods for examining single-cell activity was combining the fluorescent in situ hybridization (FISH) approach for examining bacterial community structure with microautoradiography. Three approaches have been taken to combine FISH and autoradiography to identify bacteria actively taking up radiolabeled compounds. Two of the approaches involve performing the FISH step first and then coating the sample with an autoradiographic emulsion. This approach was taken by Lee et al. (36) and Cottrell and Kirchman (9) by performing the FISH step with bacteria on glass coverslips and then coating the cells with autoradiographic emulsion. Cottrell and Kirchman (8) presented a simplified version of their procedure in which FISH probes are first applied to bacteria attached to polycarbonate filters instead of coverslips and then the bacteria are transferred into a photographic emulsion (65). In contrast, in the method given by

Ouverney and Fuhrman (49), the bacteria are first transferred into the autoradiographic emulsion and then the FISH probes are applied. The methods of Cottrell and Kirchman can be performed with plain glass slides, whereas those presented by Lee et al. and Ouverney and Fuhrman require special slides.

The Major Phylogenetic Groups and DOM Uptake

Microbial ecologists exploring links between bacterial community structure and function are often necessarily vague about what defines community structure. The exact meaning of community structure probably varies among microbial ecologists, but in essence it means the relative abundance of different kinds of bacteria. The trouble is pinning down what is meant by different bacteria because microbiologists have difficulty defining the fundamental unit of biological diversity, the species (7). Nevertheless, there is great interest in understanding how the phylogenetic diversity of *Bacteria* and *Archaea* (22) impacts the way organic carbon cycles in aquatic systems.

Examining the major bacterial divisions and subclasses that typically make up marine communities (18) avoids difficulties in the definition of bacterial species. The goal of one study by Cottrell and Kirchman (9) was to determine whether the relative contributions of various types of bacteria to DOM consumption depend solely on the relative abundance of these types of bacteria in the community. This study used microautoradiography and FISH to test the hypothesis that low-molecular-weight compounds are used by all bacteria and high-molecular-weight compounds are used by a smaller, less diverse group of bacteria. Single-cell analysis of bacterial usage of even simple, low-molecular-weight compounds such as amino acids and *N*-acetylglucosamine showed that usage levels differed across the major bacterial groups and did not correlate with the abundance of these groups. Biopolymers, including chitin and protein, were also not used equally by the major phylogenetic groups of bacteria. *Cytophaga*-like bacteria were overrepresented among the bacteria consuming chitin and protein. In contrast, the amino acid-consuming part of the community was dominated by alpha-proteobacteria, and the contribution of alpha-proteobacteria to protein consumption was about that expected based on their abundance in the entire community.

It was somewhat surprising that the levels of consumption of DOM compounds differed among the proteobacterial subclasses and the *Cytophaga*-like bacteria because these are such broad taxonomic groups. Given their phylogenetic diversity, these groups may potentially contain enough physiological diversity to include bacteria capable of using a broad range of DOM compounds. The differences seen in levels of DOM consumption may indicate that these groups are only modestly diverse. Data on alpha-proteobacteria support this hypothesis. The SAR11 clade accounts for 77% of the alpha-proteobacteria in the Delaware estuary where these experiments were done (32). In addition, physiological diversity may be lower than suggested by the high phylogenetic diversity of some bacterial groups. For example, even though *Cytophaga*-like bacteria are more diverse than alpha-proteobacteria in the Delaware River (32), they still are overrepresented in the portion of bacterial populations using chitin, *N*-acetylglucosamine, and protein but are underrepresented in the population consuming amino acids. Despite the high diversity of *Cytophaga*-like bacteria, this group of bacteria is not capable of using all DOM compounds equally.

Consumption of an Alga-Derived Component of DOM (DMSP)

Although it may seem obvious that bulk organic carbon cycling is the sum of the activities of different bacteria, it is unclear whether consumption of certain compounds can be ascribed to special bacteria. DMSP is an alga-derived compound that is the precursor of the climatically active gas DMS. Metabolism of DMSP generates DMS and provides carbon and sulfur to the bacteria (30). Based on work with cultivated bacteria, it was thought that bacteria of the *Roseobacter* clade, a subgroup of alpha-proteobacteria, might be the key participants in DMSP metabolism (19, 43). To test this hypothesis, Malmstrom et al. (39) identified the bacteria assimilating [^{35}S]-DMSP in a number of oceanic marine environments where about half of the bacterial community assimilated DMSP. Although alpha-proteobacteria dominated DMSP assimilation, accounting for 35 to 40% of the bacteria assimilating DMSP, other groups of bacteria were involved in the metabolism of this compound. *Cytophaga*-like bacteria and gamma-proteobacteria each accounted for 15 to 30% of DMSP-assimilating bacteria. The level of DMSP assimilation by the *Roseobacter* clade was higher on a per-cell basis than that by any other bacteria, which was consistent with the hypothesis based on culture data, but *Roseobacter* bacteria did not account for most of the DMSP assimilation. The importance of other bacteria in the assimilation of DMSP was clear because DMSP turnover was high even when *Roseobacter* bacteria were not present.

The Ubiquitous SAR11 Bacteria

The SAR11 clade is a group of alpha-proteobacteria that has been recognized as being one of the most abundant types of bacteria in the surface-mixed layer of the ocean, often accounting for 35% of the total prokaryotes (44). Despite its high abundance, the contribution of SAR11 bacteria to the turnover of organic carbon was unclear because SAR11 bacteria were reported to be very small and to have few ribosomes (44, 54), two characteristics that suggest low metabolic activity.

However, an examination of the contribution of SAR11 bacteria to dissolved DMSP and amino acid assimilation in the North Atlantic Ocean suggests that SAR11 bacteria are highly active and play a significant role in carbon cycling in the ocean (38). SAR11 bacteria were found to constitute 25% of the total prokaryotes in the North Atlantic Ocean and were typically as large as, if not larger than, other prokaryotes. Furthermore, more than half of SAR11 bacteria assimilated dissolved amino acids and DMSP, whereas about 40% of the other prokaryotes assimilated these compounds. It was estimated from the abundance and activity of SAR11 bacteria that this group of bacteria is responsible for about 50% of amino acid assimilation and 30% of DMSP assimilation. SAR11 bacteria appeared to contribute more to amino acid assimilation than would be expected based on their abundance, suggesting that SAR11 bacteria have a competitive advantage over other bacteria for these labile compounds.

Estimates of biomass production and the assimilation of DOM by SAR11 bacteria provide further insights into carbon cycling by these bacteria. Malmstrom et al. (37) used single-cell analysis to examine the assimilation of leucine, which is a measure of bacterial biomass production. In the northwest Atlantic Ocean, when SAR11 bacteria are abundant (>25% of prokaryotes), they account for about half of leucine incorporation, which suggests that SAR11 bacteria are major contributors to bacterial biomass production. Specific growth rates of SAR11 bacteria typically exceed that of the entire community.

Archaea

The discovery of abundant *Archaea* in the oceans is arguably one of the most intriguing findings in marine microbial ecology in recent years. *Archaea* are usually regarded as extremophiles thriving in high-temperature, high-salt, or extremely anaerobic environments and unable to compete with other microbes under nonextreme conditions. However, it is now known that *Archaea* can constitute a substantial fraction of the prokaryotes in marine communities (17). The abundance of *Archaea* is just a few percent in the surface-mixed layer and up to 30% below a depth of a few hundred meters (11, 45). If the metabolism of such abundant prokaryotes does not include consumption of organic matter, the estimates of the fluxes through the prokaryotes based on bulk measurements have to be reassessed.

Using a combination of microautoradiography and fluorescence microscopy with oligonucleotide probes targeting 16S rRNA, Ouverney and Fuhrman (50) demonstrated that marine planktonic *Archaea* take up amino acids. About 60% of the *Archaea* were active in the uptake of nanomolar concentrations of amino acids in the Mediterranean and Pacific, where *Archaea* comprised about 43 and 14% of the total prokaryotes, respectively. Thus, at least some *Archaea* appear to compete successfully with *Bacteria* for dissolved amino acids and are about as active as *Bacteria* based on a qualitative assessment of radiolabeling. However, several questions remain concerning the contribution of *Archaea* to organic matter cycling. It is not clear whether the activity of *Archaea* without demonstrable amino acid uptake was simply below the detection level of autoradiography or whether they do not take up amino acids or were not growing at all. Evidence of bicarbonate assimilation into the lipids of *Archaea* in the dark suggests that some may be autotrophic (67), however. The consumption of high-molecular-weight compounds by *Archaea* remains unexplored.

Major Phylogenetic Groups and Biomass Production

Data on metabolism at the single-cell level contribute to our understanding of the relationship between bacterial community structure and bacterial biomass production. A study of thymidine and leucine incorporation in the Delaware estuary indicates that bacterial abundance generally reflects the importance of the major bacterial groups to bacterial biomass production (8). Alpha-proteobacteria were the dominant substrate-active bacteria in the saline part of the estuary. In contrast, beta-proteobacteria were more important in freshwaters. The second most important group at all salinities was the *Cytophaga*-like bacteria, and gamma-proteobacteria were least important regardless of salinity. Bacterial abundance explained about half of the variation in thymidine and leucine assimilation by the major bacterial groups. Similarly, rare bacterial groups did not appear to contribute disproportionately to bacterial production. Gamma-proteobacteria have the potential for rapid growth (14), but the growth rates of gamma-proteobacteria inferred from the sizes of silver grains in thymidine and leucine autoradiography were not different from those of other bacterial groups. Furthermore, the gamma-proteobacteria group appeared to have no more highly active cells than any other group (8). However, there was substantial variation in the sizes of silver grains associated with active bacteria in all the major groups, suggesting a wide range in growth rates of the bacteria composing these phylogenetic groups.

Outline of FISH and Autoradiography Method

Fluorescence microscopy with DAPI stain and fluorescently labeled oligonucleotide probes is used to count total cells and different phylogenetic groups, respectively. Light microscopy is used to count active cells, which are identified by the silver grains they generate in autoradiographic emulsions.

Incubation with Radiolabeled Compounds

1. Fill incubation bottles with seawater (30 to 60 ml is typical). The sample volume should be sufficient to allow samples to be withdrawn for replicate filters or possibly a time course. Typical volumes filtered range from 10 to 60 ml, depending on the bacterial abundance (see below).
2. Poison the killed control with 2% paraformaldehyde. Wait 15 min before adding the radiolabeled compound to be sure that all metabolic activity has ceased.
3. Add the radiolabeled compound. It is often desirable to add the compound at a concentration low enough that it does not increase the natural abundance of the compound, e.g., subnanomolar concentrations in experiments investigating the uptake of amino acids or sugars. In other cases, saturating concentrations are added, e.g., 20 nM [^{3}H] thymidine or [^{3}H] leucine in experiments monitoring cells actively synthesizing DNA or protein. See Kemp (28) for additional guidance on measuring microbial activity by using radiolabeled compounds.
4. Incubate at an in situ temperature (typically 20 min to 4 h) in the dark. The shortest incubation times are applicable in environments with high bacterial activity, such as estuaries. Oligotrophic environments where activity is lower require longer incubations. It is best to keep the incubation as short as possible to minimize the effects of containment.
5. Transfer the sample to another bottle for fixation. Add paraformaldehyde to 2% and fix overnight at 4°C.

Sample Filtration

1. Load the filter manifold with a 0.45-μm pore-size nitrocellulose filter supporting a 0.22-μm pore-size polycarbonate filter (25 mm in diameter). The supporting filter aids in obtaining a uniform distribution of cells.
2. Filter 10 to 60 ml of the fixed sample onto a 0.22-μm pore-size filter. The volume to be filtered depends on the bacterial abundance. Ten milliliters is typical for estuaries, and as much as 60 ml is necessary in highly oligotrophic waters. Filter enough to get plenty of cells per microscopic field of view, but there should be sufficient space between cells to allow room for silver grains.
3. Rinse the filter three times with deionized water. It is important to remove the fixative because overfixation can lead to lower detection by FISH. Rinsing also removes radiolabel that was not taken up that would contribute to a background of silver grains not associated with cells.
4. Remove the polycarbonate filter still under a vacuum so that it comes off with a minimum of moisture.
5. Store the filters in 7-ml scintillation vials in the freezer.

FISH

1. Remove the polycarbonate filter from the freezer and cut a wedge-shaped slice. Pieces ranging in size from 1/8 to 1/12 of a 25-mm-diameter filter are convenient and allow several probes to be used with a single filter. Return the remaining filter to the freezer.

2. Cover a glass slide with parafilm by using tape on the back of the slide to hold the parafilm in place. Several samples can go on one slide if they are to be assayed with the same probe.

3. Prepare 30 μl of probe solution for each sample (approximately one-eighth of a 25-mm-diameter filter). Use appropriate hybridization buffers, formamide concentrations, and temperatures for the FISH probes (52).

4. Place the filter piece face down on a 30-μl drop of probe solution on the parafilm.

5. Place the slide in a 50-ml plastic centrifuge tube with a cap. The sealed tube minimizes probe evaporation during the hybridization step.

6. Close the tube and incubate from 2 to 18 h at 42°C in a temperature-controlled oven. Probe-positive cells are typically visible with a 2-h incubation, but some samples require longer hybridization to maximize the number of probe-positive cells.

7. Pour into microtiter dishes 1 ml of wash buffer for each filter piece. Warm to 48°C in the oven.

8. Move filter pieces into the warm wash solution.

9. Incubate at 48°C for 15 min. The washing step removes unbound FISH probe and reduces the fluorescent background.

10. Rinse the filter pieces by dipping them in deionized water.

11. Dry the filter pieces in the dark on filter paper.

12. Arrange the filter pieces in a labeled 96-well titer plate for transport to the darkroom. The plate is useful for keeping samples organized and readily accessible in the darkroom under a dim safelight.

Autoradiography

1. Set an aluminum block in ice. The cold surface will be used to gel the emulsion after the slides are dipped in the emulsion.

2. Set the stock emulsion in a 43°C water bath and incubate until fully melted (~30 min). The emulsion forms a gel under the recommended storage at 4°C.

3. Dilute the melted emulsion with water in a black film canister. For ^{3}H, dilute NTB-2 (Kodak, Rochester, N.Y.) with 2 parts water and for ^{14}C or ^{35}S, dilute LM1 (Amersham, Piscataway, N.J.) with 1 part water. Amersham also makes EM1 emulsion for ^{3}H. Diluted emulsion forms a thinner coating of emulsion, which can help keep the entire sample in a single plane of focus.

4. Set the diluted emulsion in a 43°C water bath.

5. Place a safelight appropriate for an autoradiographic emulsion (Kodak; catalog no. 152 1525) on the bench close to the 96-well plate with the samples. It is easiest to see the samples when they are close to the safelight, which emits a very dim red glow.

6. Dip a labeled glass slide in the melted emulsion aliquot in a 35-mm film canister. The film canister is a convenient size for slide dipping, and it is light tight when the cap is in place.

7. Working briefly under the safelight, place a filter piece (approximately one-eighth of a filter) with cells down on the emulsion. Best results are obtained when the filter is laid down in one motion. Avoid sliding the filter across the emulsion as this can lead to a high background of silver grains.

8. Immediately place the slide onto the aluminum block.

9. Allow the emulsion to gel for 15 min.

10. Transfer slides to light-tight boxes.

11. Place dark boxes in the refrigerator for the exposure time. Exposure times range from a few hours to days depending on the isotope, compound, and microbial activity. A time course of exposures using samples and killed controls is useful for identifying an exposure time that is long enough to reveal active cells but not so long that it becomes difficult to distinguish background silver grains from those generated by cells.

Development

1. Prepare Kodak Dektol developer (catalog no. 146 4726) and fixer (catalog no. 197 1749; not rapid fix) by following the instructions provided by the manufacturer.

2. Fill three containers with developer, stopper (water), and fixer in a 14°C water bath.

3. Develop slides for 2 min, stop development by immersing them for 10 s in water, fix them for 6 min, and wash them in water for 6 min.

4. Dry the slides overnight in a vacuum desiccator.

5. Trace the outline of the filter on the back of the slide. This makes it easier to find the cells when the slide is on the microscope.

6. Using forceps, peel the filter away from the emulsion and discard. Even though the cells remain in the emulsion, the filter should be treated as radioactive.

7. Mount the slide with a coverslip by using an antifade mountant (4:1 mixture of Citifluor, Ted Pella, and VectaShield; Vector Labs, Burlingame, Calif.) containing 0.5 μg of DAPI/ml.

Microscopic Analysis

Total cells stained with DAPI and cells hybridized with fluorescently labeled probes are identified under epifluorescence. Next, the illumination is changed to transmitted light to identify the cells surrounded by silver grains in the same field of view (9). The data can be acquired by manual or semiautomated microscopy and image analysis (8).

1. Use an epifluorescence microscope fitted with DAPI and Cy3 filter sets.

2. Score cells in the following categories:

Probe positive: Cells that are labeled with the fluorescently labeled FISH probe.

Active: Cells that took up the radiolabeled compound and show silver grains under transmitted light microscopy. An estimate of activity can be obtained by counting the silver grains or measuring their size (see, e.g., reference 37).

Probe positive and active: Cells that generated silver grains and are labeled with the FISH probe.

ANALYSIS OF ENVIRONMENTAL DNA ENCODING ENZYMES HYDROLYZING BIOPOLYMERS

Our understanding of the degradation of organic matter will benefit from a greater knowledge of the genes encoding enzymes involved in the hydrolysis of biopolymers. Molecular methods are needed to retrieve the genes of uncultured bacteria that degrade biopolymers in the environment because only a small fraction (<1%) of the bacteria in seawater can be cultured (13, 15, 34). Approaches using nucleic acid probes and PCR primers cannot be applied solely by using gene sequences of cultured bacteria because the genes of cultured bacteria may differ from those of bacteria yet to be cultured. The sequences of genes encoding hydrolases such as chitinases of even known cultured bacteria are very different (64), suggesting that it will be impossible to design universal PCR primers to retrieve genes from uncultured bacteria. One alternative approach that does not rely on sequence conservation uses libraries of environmental DNA that are screened for a metabolic activity of interest to retrieve genes from natural bacterial communities without cultivation (2, 56, 57, 62).

One study of a marine system focused on identifying chitinase genes of uncultured bacteria. Cottrell et al. (10) cloned genomic DNA isolated from coastal and estuarine waters into a lambda phage vector in order to isolate representative and abundant chitinase genes. The high cloning efficiency of the lambda phage vector enabled the construction of a library containing more than 2×10^5 clones. Another advantage of the phage cloning approach is that phage clones forming plaques on lawns of *Escherichia coli* can be rapidly screened by using fluorogenic or colorimetric substrate analogues. Libraries were screened for genes encoding enzymes that hydrolyze a fluorogenic analogue of chitin, MUF-diNAG (4-methylumbelliferyl β-D-*N*,*N*′-diacetylchitobioside). Such expression screening is based on the expectation that clones carrying a complete gene with a promoter or those cloned in the reading frame of the promoter on the vector will be expressed.

Plaque assay screening revealed nine MUF-diNAG-positive clones of 75,000 screened from the estuarine sample and two positive clones of 75,000 from the coastal sample. The number of clones detected with the plaque assay was consistent with estimates of the portion of culturable bacteria that degrade chitin. The somewhat unexpected conclusion of this study was that culture-dependent methods do not greatly underestimate the portion of marine bacterial communities capable of chitin degradation. Based on work with cultures, the percentage of bacteria capable of chitin degradation ranges from 0.4 to 19% of total cultured bacteria (5, 48, 59), and estimates from the cloning of environmental DNA were within this range. However, there still remains a large pool of uncultured chitin-degrading bacteria in aquatic environments. Describing their chitinases will produce a better understanding of chitin degradation in the ocean.

In another study, Brennan et al. (4) examined microbial xylanases in insect guts. Although the study did not include aquatic bacteria, the approach used in this study should be applicable to aquatic microbes with appropriate modification by using a DNA extraction method for aquatic microbial DNA (see below). This study focused on the degradation of xylan because it is the main constituent of hemicellulose, which is the second most abundant plant polymer after cellulose. DNA extracted from the gut microbes of termites and moths was cloned into a lambda phage vector, and plaques were screened for xylanase activity by using an azo dye-linked xylan. Disappearance of the colored substrate surrounding plaques was used to identify xylanase-positive clones. Several novel xylanase enzymes with unusual primary sequences and novel domains of unknown function were discovered. Biochemical analysis confirmed that the new enzymes have xylanase activity and hydrolyze a variety of substituted β-1,4-linked xylose-containing substrates, producing unique hydrolysis products. The great phylogenetic distance between the novel xylanases and the known xylanases suggests that microbes in the insect gut have evolved in isolation from other microbes that have been successfully cultivated.

Outline of Environmental DNA Cloning Method

Clone libraries of environmental DNA have a number of advantages. The inserts can be large enough to bear entire genes that potentially can be identified by activity screening. Such activity screening solves the problems of screening using primers or probes based on the sequences of previously identified genes. In addition, phylogenetic analysis of adjacent genes on large inserts can be useful in identifying the source bacterium.

1. Prefilter 5 to10 liters of seawater through a 142-mm-diameter, 0.8 μm-pore-size filter to remove larger organisms. Removing the larger organisms helps exclude eukaryotes but has the disadvantage of removing particle-attached prokaryotes as well.
2. Collect the bacteria from the filtrate by using 47-mm-diameter, 0.2-μm-pore-size Durapore (type GVWP; Millipore, Billerica, Mass.) or Supor (Gelman, East Hills, N.Y.) filters. In order to filter 5 to 10 liters you will need to use six filters. When the six filters clog, you probably have enough material (bacteria) on them. It is also possible to use 0.45-μm-pore-size filters, which filter faster and probably retain nearly all prokaryotes.
3. Place the filters in a closable plastic bag with 5 ml of sucrose buffer and store frozen (−80°C is best). Sucrose buffer is 20 mM EDTA, 400 mM NaCl, 0.75 M sucrose, and 50 mM Tris-HCl, pH 9.0.
4. Thaw the sample when you are ready to purify the nucleic acids.
5. Extract the DNA by using standard protocols for cell lysis with sodium dodecyl sulfate and proteinase K and for nucleic acid purification with phenol-chloroform extraction (3). Wearing gloves and using disposable plasticware help minimize contamination with nucleases. Wide-bore pipette tips minimize shearing of the DNA.
6. Recover the DNA from the final aqueous phase by ethanol precipitation. Add 0.1 volume of 2.5 mM sodium acetate and 2.5 volumes of 100% ethanol, mix, and then incubate on ice for 1 h. Collect the DNA precipitate by centrifugation in an SS-34 Sorvall (Du Pont, Wilmington, Del.) or equivalent rotor at 30,000 × *g* for 1 h.
7. Pour off the supernatant, fill the centrifuge tube with 70% ethanol, and centrifuge again for 10 min. Pour off the supernatant. This ethanol wash step removes traces of salt that can inhibit enzymes used later on.
8. Dry the sample, and then resuspend it in deionized water. Be careful not to dry the DNA too much because it may become difficult to dissolve. However, it is important to remove all traces of ethanol, which can inhibit enzymes used in later steps.
9. Remove the RNA by RNase A digestion. Follow the protocol supplied by the manufacturer.
10. Prepare the DNA for ligation by partial restriction digestion with the enzyme Tsp508I (New England Biolabs, Beverly, Mass.) by using 3.3 U of enzyme per μg of DNA at 65°C for 15 min.
11. Size select the restriction fragments from 2 to 10 kb by ethanol precipitation of the 30% portion of a sucrose (10 to 40%) step gradient (40,000 rpm in a Beckman TLS-55

rotor for 12 h). See Maniatis et al. (40) for details on preparing a sucrose density gradient.

12. Ligate 400 ng of restriction fragments into Lambda Zap II vector predigested with EcoRI/calf intestinal alkaline phosphatase-treated vector (Stratagene, La Jolla, Calif.) by using T4 DNA ligase (Boehringer Mannheim, Ridgefield, Conn.) according to the manufacturer's instructions.

13. Package phage using Gigapack III packaging extract (Stratagene), and determine the titer of the fraction of phage containing inserts by using blue-white color selection.

14. Amplify the library by following the procedure provided with the cloning reagents to generate clones that can be screened in more than one screening experiment.

15. Screen the library for enzyme activity by using fluorometric or colorimetric substrate analogs on 150-mm-diameter petri plates containing 4×10^4 plaques. Fluorometric analogs are applied after plaques form by spraying with 50 μM substrate in a phosphate buffer. Colorimetric substrates are combined with the top agar.

16. Inspect plaques illuminated with UV light if sprayed with fluorometric substrates. Clearing zones form around plaques assayed with colorimetric substrates, indicating degradation of the substrate.

17. Elute phages from active plaques by using a suitable phage suspension buffer and purify by two rounds of plaque isolation by following procedures outlined by the manufacturer of the cloning reagents.

REFERENCES

1. **Azam, F., T. Fenchel, J. G. Field, J. S. Gray, L. A. Meyerreil, and F. Thingstad.** 1983. The ecological role of water-column microbes in the sea. *Mar. Ecol. Prog. Ser.* **10:**257–263.
2. **Beja, O., M. T. Suzuki, E. V. Koonin, L. Aravind, A. Hadd, L. P. Nguyen, R. Villacorta, M. Amjadi, C. Garrigues, S. B. Jovanovich, R. A. Feldman, and E. F. DeLong.** 2000. Construction and analysis of bacterial artificial chromosome libraries from a marine microbial assemblage. *Environ. Microbiol.* **2:**516–529.
3. **Bostrom, K. H., K. Simu, A. Hagstrom, and L. Riemann.** 2004. Optimization of DNA extraction for quantitative marine bacterioplankton community analysis. *Limnol. Oceanogr. Methods* **2:**365–373.
4. **Brennan, Y., W. N. Callen, L. Christoffersen, P. Dupree, F. Goubet, S. Healey, M. Hernandez, M. Keller, K. Li, N. Palackal, A. Sittenfeld, G. Tamayo, S. Wells, G. P. Hazlewood, E. J. Mathur, J. M. Short, D. E. Robertson, and B. A. Steer.** 2004. Unusual microbial xylanases from insect guts. *Appl. Environ. Microbiol.* **70:**3609–3617.
5. **Brisou, J., C. Tysset, A. de Rautlin, R. de la Roy, R. Curcier, and R. Moreau.** 1964. Étude sur la chitinolyse a milieu marin. *Ann. Inst. Pasteur* **106:**469–478.
6. **Brock, T. D.** 1967. Bacterial growth rate in the sea—direct analysis by thymidine autoradiography. *Science* **155:** 81–83.
7. **Cohan, F. M.** 2002. What are bacterial species? *Annu. Rev. Microbiol.* **56:**457–487.
8. **Cottrell, M. T., and D. L. Kirchman.** 2003. Contribution of major bacterial groups to bacterial biomass production (thymidine and leucine incorporation) in the Delaware estuary. *Limnol. Oceanogr.* **48:**168–178.
9. **Cottrell, M. T., and D. L. Kirchman.** 2000. Natural assemblages of marine proteobacteria and members of the *Cytophaga-Flavobacter* cluster consuming low- and high-molecular-weight dissolved organic matter. *Appl. Environ. Microbiol.* **66:**1692–1697.
10. **Cottrell, M. T., J. A. Moore, and D. L. Kirchman.** 1999. Chitinases from uncultured marine microorganisms. *Appl. Environ. Microbiol.* **65:**2553–2557.
11. **DeLong, E. F., L. T. Taylor, T. L. Marsh, and C. M. Preston.** 1999. Visualization and enumeration of marine planktonic archaea and bacteria by using polyribonucleotide probes and fluorescent in situ hybridization. *Appl. Environ. Microbiol.* **65:**5554–5563.
12. **DeLong, E. F., G. S. Wickham, and N. R. Pace.** 1989. Phylogenetic stains—ribosomal RNA-based probes for the identification of single cells. *Science* **243:**1360–1363.
13. **Eguchi, M., and Y. Ishida.** 1990. Oligotrophic properties of heterotrophic bacteria and in situ heterotrophic activity in pelagic seawaters. *FEMS Microbiol. Ecol.* **73:**23–30.
14. **Eilers, H., J. Pernthaler, and R. Amann.** 2000. Succession of pelagic marine bacteria during enrichment: a close look at cultivation-induced shifts. *Appl. Environ. Microbiol.* **66:**4634–4640.
15. **Ferguson, R. L., E. N. Buckley, and A. V. Palumbo.** 1984. Response of marine bacterioplankton to differential filtration and confinement. *Appl. Environ. Microbiol.* **47:** 49–55.
16. **Fuhrman, J. A., and F. Azam.** 1982. Thymidine incorporation as a measure of heterotrophic bacterioplankton production in marine surface waters—evaluation and field results. *Mar. Biol.* **66:**109–120.
17. **Fuhrman, J. A., and C. C. Ouverney.** 1998. Marine microbial diversity studied via 16S rRNA sequences: cloning results from coastal waters and counting of native archaea with fluorescent single cell probes. *Aquat. Ecol.* **32:**3–15.
18. **Giovannoni, S. J., and M. S. Rappé.** 2000. Evolution, diversity and molecular ecology of marine prokaryotes, p. 47–84. *In* D. L. Kirchman (ed.), *Microbial Ecology of the Oceans*. Wiley-Liss, New York, N.Y.
19. **Gonzalez, J. M., R. P. Kiene, and M. A. Moran.** 1999. Transformation of sulfur compounds by an abundant lineage of marine bacteria in the alpha-subclass of the class *Proteobacteria*. *Appl. Environ. Microbiol.* **65:**3810–3819.
20. **Hedges, J. I., and J. M. Oades.** 1997. Comparative organic geochemistries of soils and marine sediments. *Org. Geochem.* **27:**319–361.
21. **Hobbie, J. E., R. J. Daley, and S. Jasper.** 1977. Use of Nuclepore filters for counting bacteria by fluorescence microscopy. *Appl. Environ. Microbiol.* **33:**1225–1228.
22. **Hugenholtz, P., B. M. Goebel, and N. R. Pace.** 1998. Impact of culture-independent studies on the emerging phylogenetic view of bacterial diversity. *J. Bacteriol.* **180:**4765–4774.
23. **Jonkers, H. M., and R. M. M. Abed.** 2003. Identification of aerobic heterotrophic bacteria from the photic zone of a hypersaline microbial mat. *Aquat. Microb. Ecol.* **30:**127–133.
24. **Jorgensen, N. O. G., N. Kroer, R. B. Coffin, X. H. Yang, and C. Lee.** 1993. Dissolved free amino-acids, combined amino-acids, and DNA as sources of carbon and nitrogen to marine bacteria. *Mar. Ecol. Prog. Ser.* **98:**135–148.
25. **Kaiser, K., and R. Benner.** 2000. Determination of amino sugars in environmental samples with high salt content by high performance anion exchange chromatography and pulsed amperometric detection. *Anal. Chem.* **72:**2566–2572.
26. **Keil, R. G., and D. L. Kirchman.** 1991. Contribution of dissolved free amino acids and ammonium to the nitrogen requirements of heterotrophic bacterioplankton. *Mar. Ecol. Prog. Ser.* **73:**1–10.
27. **Keil, R. G., and D. L. Kirchman.** 1999. Utilization of dissolved protein and amino acids in the northern Sargasso Sea. *Aquat. Microb. Ecol.* **18:**293–300.

28. **Kemp, P. F.** 1993. *Handbook of Methods in Aquatic Microbial Ecology*. Lewis Publishers, Boca Raton, Fla.
29. **Kiene, R. P., and L. J. Linn.** 2000. Distribution and turnover of dissolved DMSP and its relationship with bacterial production and dimethylsulfide in the Gulf of Mexico. *Limnol. Oceanogr.* **45:**849–861.
30. **Kiene, R. P., and L. J. Linn.** 2000. The fate of dissolved dimethylsulfoniopropionate (DMSP) in seawater: tracer studies using S-35-DMSP. *Geochim. Cosmochim. Acta* **64:**2797–2810.
31. **Kirchman, D., E. Knees, and R. Hodson.** 1985. Leucine incorporation and its potential as a measure of protein synthesis by bacteria in natural aquatic systems. *Appl. Environ. Microbiol.* **49:**599–607.
32. **Kirchman, D. L., A. I. Dittel, R. R. Malmstrom, and M. T. Cottrell.** 2005. Biogeography of major bacterial groups in the Delaware Estuary. *Limnol. Oceanogr.* **50:** 1697–1706.
33. **Kirchman, D. L., B. Meon, H. W. Ducklow, C. A. Carlson, D. A. Hansell, and G. F. Steward.** 2001. Glucose fluxes and concentrations of dissolved combined neutral sugars (polysaccharides) in the Ross Sea and Polar Front Zone, Antarctica. *Deep-Sea Res. Pt. II* **48:**4179–4197.
34. **Kogure, K., U. Simidu, and N. Taga.** 1980. Distribution of viable marine bacteria in neritic seawater around Japan. *Can. J. Microbiol.* **26:**318–323.
35. **Leboulanger, C., L. Oriol, H. Jupin, and C. Descolas-Gros.** 1997. Diel variability of glycolate in the eastern tropical Atlantic Ocean. *Deep-Sea Res. Pt. I* **44:**2131–2139.
36. **Lee, N., P. H. Nielsen, K. H. Andreasen, S. Juretschko, J. L. Nielsen, K. H. Schleifer, and M. Wagner.** 1999. Combination of fluorescent in situ hybridization and microautoradiography—a new tool for structure-function analyses in microbial ecology. *Appl. Environ. Microbiol.* **65:**1289–1297.
37. **Malmstrom, R. R., M. T. Cottrell, H. Elifantz, and D. L. Kirchman.** 2005. Biomass production and assimilation of dissolved organic matter by SAR11 bacteria in the Northwest Atlantic Ocean. *Appl. Environ. Microbiol.* **71:**2979–2986.
38. **Malmstrom, R. R., R. P. Kiene, M. T. Cottrell, and D. L. Kirchman.** 2004. Contribution of SAR11 bacteria to dissolved dimethylsulfoniopropionate and amino acid uptake in the North Atlantic Ocean. *Appl. Environ. Microbiol.* **70:**4129–4135.
39. **Malmstrom, R. R., R. P. Kiene, and D. L. Kirchman.** 2004. Identification and enumeration of bacteria assimilating dimethylsulfoniopropionate (DMSP) in the North Atlantic and Gulf of Mexico. *Limnol. Oceanogr.* **49:**597–606.
40. **Maniatis, T., E. F. Fritsch, and J. Sambrook.** 1982. *Molecular Cloning : a Laboratory Manual.* Cold Spring Harbor Laboratory, Cold Spring Harbor, N.Y.
41. **Martinez, J., D. C. Smith, G. F. Steward, and F. Azam.** 1996. Variability in ectohydrolytic enzyme activities of pelagic marine bacteria and its significance for substrate processing in the sea. *Aquat. Microb. Ecol.* **10:**223–230.
42. **Middelboe, M., N. H. Borch, and D. L. Kirchman.** 1995. Bacterial utilization of dissolved free amino acids, dissolved combined amino acids and ammonium in the Delaware Bay estuary: effects of carbon and nitrogen limitation. *Mar. Ecol. Prog. Ser.* **128:**109–120.
43. **Moran, M. A., J. M. Gonzalez, and R. P. Kiene.** 2003. Linking a bacterial taxon to sulfur cycling in the sea: studies of the marine Roseobacter group. *Geomicrobiol. J.* **20:** 375–388.
44. **Morris, R. M., M. S. Rappé, S. A. Connon, K. L. Vergin, W. A. Siebold, C. A. Carlson, and S. J. Giovannoni.** 2002. SAR11 clade dominates ocean surface bacterioplankton communities. *Nature* **420:**806–810.
45. **Murray, A. E., C. M. Preston, R. Massana, L. T. Taylor, A. Blakis, K. Wu, and E. F. DeLong.** 1998. Seasonal and spatial variability of bacterial and archaeal assemblages in the coastal waters near Anvers Island, Antarctica. *Appl. Environ. Microbiol.* **64:**2585–2595.
46. **Nagata, T.** 2000. Production mechanisms of dissolved organic matter, p. 121–152. *In* D. L. Kirchman (ed.), *Microbial Ecology of the Oceans*. Wiley-Liss, New York, N.Y.
47. **Noble, R. T., and J. A. Fuhrman.** 1998. Use of SYBR Green I for rapid epifluorescence counts of marine viruses and bacteria. *Aquat. Microb. Ecol.* **14:**113–118.
48. **Okutani, K. (ed.).** 1975. *Microorganisms Related to Mineralization of Chitin in Aquatic Environments*. University of Tokyo Press, Tokyo, Japan.
49. **Ouverney, C. C., and J. A. Fuhrman.** 1999. Combined microautoradiography-16S rRNA probe technique for determination of radioisotope uptake by specific microbial cell types in situ. *Appl. Environ. Microbiol.* **65:**1746–1752.
50. **Ouverney, C. C., and J. A. Fuhrman.** 2000. Marine planktonic archaea take up amino acids. *Appl. Environ. Microbiol.* **66:**4829–4832.
51. **Parker, M. S., E. V. Armbrust, J. Piovia-Scott, and R. G. Keil.** 2004. Induction of photorespiration by light in the centric diatom Thalassiosira weissflogii (Bacillariophyceae): molecular characterization and physiological consequences. *J. Phycol.* **40:**557–567.
52. **Pernthaler, J., F. O. Glockner, W. Schonhuber, and R. Amann.** 2001. Fluorescence in situ hybridization (FISH) with rRNA-targeted oligonucleotide probes. *Methods Microbiol.* **30:**207–226.
53. **Porter, K. G., and Y. S. Feig.** 1980. The use of DAPI for identifying and counting aquatic microflora. *Limnol. Oceanogr.* **25:**943–948.
54. **Rappé, M. S., S. A. Connon, K. L. Vergin, and S. J. Giovannoni.** 2002. Cultivation of the ubiquitous SAR11 marine bacterioplankton clade. *Nature* **418:**630–633.
55. **Rich, J. H., H. W. Ducklow, and D. L. Kirchman.** 1996. Concentrations and uptake of neutral monosaccharides along 140 degrees W in the equatorial Pacific: contribution of glucose to heterotrophic bacterial activity and the DOM flux. *Limnol. Oceanogr.* **41:**595–604.
56. **Riesenfeld, C. S., R. M. Goodman, and J. Handelsman.** 2004. Uncultured soil bacteria are a reservoir of new antibiotic resistance genes. *Environ. Microbiol.* **6:**981–989.
57. **Rondon, M. R., P. R. August, A. D. Bettermann, S. F. Brady, T. H. Grossman, M. R. Liles, K. A. Loiacono, B. A. Lynch, I. A. MacNeil, C. Minor, C. L. Tiong, M. Gilman, M. S. Osburne, J. Clardy, J. Handelsman, and R. M. Goodman.** 2000. Cloning the soil metagenome: a strategy for accessing the genetic and functional diversity of uncultured microorganisms. *Appl. Environ. Microbiol.* **66:**2541–2547.
58. **Rosenstock, B., and M. Simon.** 2001. Sources and sinks of dissolved free amino acids and protein in a large and deep mesotrophic lake. *Limnol. Oceanogr.* **46:**644–654.
59. **Seki, H.** 1966. Seasonal fluctuation of heterotrophic bacteria in the sea of Aburatsubo Inlet. *J. Oceanogr. Soc. Jpn.* **22:**15–26.
60. **Simon, M., and F. Azam.** 1989. Protein content and protein synthesis rates of planktonic marine macteria. *Mar. Ecol. Prog. Ser.* **51:**201–213.
61. **Skoog, A., B. Biddanda, and R. Benner.** 1999. Bacterial utilization of dissolved glucose in the upper water column of the Gulf of Mexico. *Limnol. Oceanogr.* **44:**1625–1633.
62. **Stein, J. L., T. L. Marsh, K. Y. Wu, H. Shizuya, and E. F. DeLong.** 1996. Characterization of uncultivated prokaryotes: isolation and analysis of a 40-kilobase-pair genome

fragment from a planktonic marine archaeon. *J. Bacteriol.* **178:**591–599.

63. **Suttle, C. A., A. M. Chan, and J. A. Fuhrman.** 1991. Dissolved free amino acids in the Sargasso Sea—uptake and respiration rates, turnover times, and concentrations. *Mar. Ecol. Prog. Ser.* **70:**189–199.
64. **Svitil, A. L., and D. L. Kirchman.** 1998. A chitin-binding domain in a marine bacterial chitinase and other microbial chitinases: implications for the ecology and evolution of 1,4-beta-glycanases. *Microbiology* (Reading) **144:**1299–1308.
65. **Tabor, P. S., and R. A. Neihof.** 1982. Improved microautoradiographic method to determine individual microorganisms active in substrate uptake in natural waters. *Appl. Environ. Microbiol.* **44:**945–953.
66. **Weiss, M. S., U. Abele, J. Weckesser, W. Welte, E. Schiltz, and G. E. Schulz.** 1991. Molecular architecture and electrostatic properties of a bacterial porin. *Science* **254:**1627–1630.
67. **Wuchter, C., S. Schouten, H. T. S. Boschker, and J. S. S. Damste.** 2003. Bicarbonate uptake by marine *Crenarchaeota*. *FEMS Microbiol. Lett.* **219:**203–207.

Environmental Genomics of C_1 Metabolism

MARINA G. KALYUZHNAYA, OLIVIER NERCESSIAN, MARY E. LIDSTROM, AND LUDMILA CHISTOSERDOVA

40

C_1 compounds (compounds containing no carbon-carbon bonds) are important components of the global carbon cycling in the environment. One of the major C_1 compounds is methane, originating from natural emissions and, increasingly, as a result of human activities (25). Other environmentally important C_1 compounds are methanol, methylated amines, methylated sulfur compounds, and halogenated methanes (1, 29, 34, 46, 67). The balance of C_1 compounds in the environment is achieved via specific activities of C_1-oxidizing microbes, carried out aerobically and anaerobically (32, 74). This chapter is devoted to genome-based approaches for analyzing microbial communities involved in C_1 metabolism in aerobic environments.

METHYLOTROPHY AND METHYLOTROPHS

Bacteria with a propensity to use C_1 compounds as single sources of carbon and energy are known as methylotrophs. Facultative methylotrophs are also capable of growth on multicarbon compounds, while obligate methylotrophs are not. Methylotrophs are successful in utilizing C_1 compounds due to the presence of a number of specialized enzymes and pathways (Fig. 1). Methanotrophs oxidize methane using either copper-containing membrane-bound methane monooxygenase (pMMO) or iron-containing soluble methane monooxygenase (sMMO). The two enzymes are evolutionarily unrelated and possess different substrate specificities and affinities: pMMO has a higher affinity toward methane, while sMMO is capable of using a wider range of substrates (54). Methanol dehydrogenase (MDH) is utilized for methanol oxidation. Gram-negative methylotrophs possess an MDH using pyrroloquinoline quinone as a cofactor (2), while gram-positive methylotrophs employ a nonhomologous, NAD-linked MDH (35). Methylamine is oxidized either by methylamine dehydrogenase, found mainly in terrestrial methylotrophs, or via the *N*-methylglutamate pathway typical of aquatic methylotrophs (17, 37). Specific enzyme systems for oxidation of halogenated methanes (corrinoid-linked methyltransferase, dichloromethane dehalogenase) and methanesulfonic acid (methanesulfonic acid monooxygenase [MSM]) also exist (46, 69).

Oxidation of many C_1 substrates produces formaldehyde, an extremely toxic compound. Thus, efficient formaldehyde metabolism is essential to the methylotrophic lifestyle. Three of four major formaldehyde oxidation pathways are employed by methylotrophs. Glutathione-dependent formaldehyde oxidation so far has been characterized only in some autotrophic methylotrophs, which oxidize C_1 compounds to CO_2 and then assimilate CO_2 via the Calvin-Benson-Bassham cycle (33). Tetrahydrofolate-dependent formaldehyde oxidation so far has been connected with only chloromethane (and possibly bromomethane) utilization, while the tetrahydromethanopterin (H_4MPT)-dependent pathway is widespread among various phylogenetic groups capable of methylotrophy (69). Genes for a glutathione-independent, NAD-linked formaldehyde dehydrogenase are present in some methylotroph genomes, but the role of this enzyme in methylotrophy has not been documented.

A bona fide methylotroph relies on the presence of specific C_1-assimilatory pathways to build biomass. With the exceptions of autotrophic methylotrophs and chloromethane utilizers, formaldehyde is also the compound entering either the ribulose monophosphate cycle or the serine cycle for assimilation. Part of the ribulose monophosphate cycle can also be used for formaldehyde oxidation by some methylotrophs (13).

Methylotrophs are ubiquitous in the environment and can be readily isolated in pure cultures. Common methanotrophic isolates include *Methylocystis*, *Methylosinus*, *Methylomonas*, and *Methylobacter* spp. from terrestrial and freshwater environments (3), *Methylococcus* spp. from hot springs (7), and *Methylomicrobium* spp. from saline environments (38). Common methylotrophic isolates include *Methylobacterium* spp. from plants, *Methylophilus* spp. and *Methylopila* spp. from soils and aquatic environments (20, 26), and *Methylophaga* spp. from marine environments and saline lakes (21, 67). All of these methylotrophs grow well on synthetic media and tolerate the high concentrations of C_1 substrates often used for laboratory cultivation. However, some methylotrophs have proven less amenable to cultivation and demand a range of specific adjustments to be grown in the laboratory (18, 19). Furthermore, on the basis of the culture-independent experiments described below, most methylotrophs still remain uncultivated. It is also possible that some nonmethylotrophic microorganisms are involved in cometabolism or in detoxification of C_1 compounds. Approaches other than cultivation are obviously needed to understand the composition of C_1-utilizing populations and the mechanisms that regulate and coordinate the specific activities in these communities.

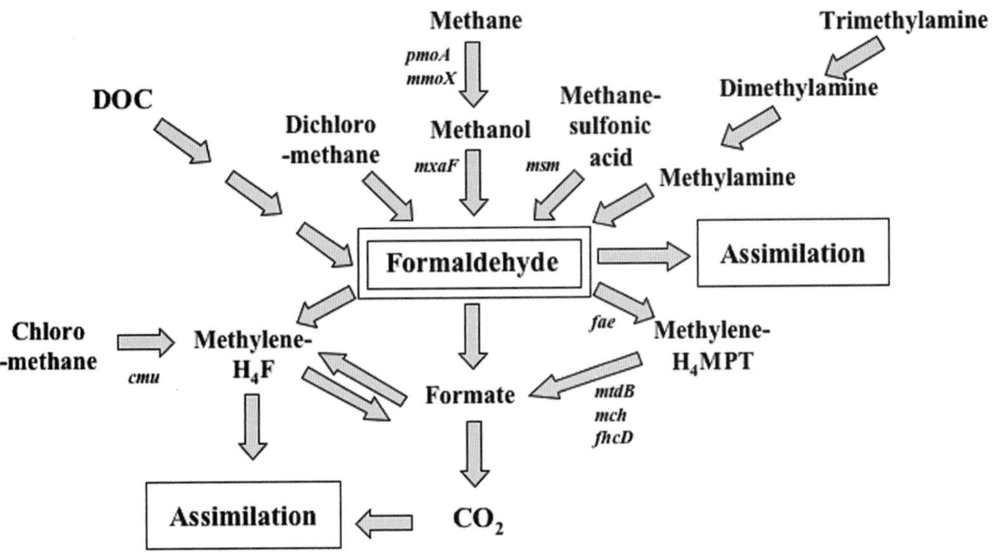

FIGURE 1 The central role of formaldehyde in C_1 metabolism. Genes routinely detected in environmental samples are shown (italic).

GENOMICS OF METHYLOTROPHIC BACTERIA

Knowledge of the content and structure of the genomes of methylotrophic bacteria is instrumental for detecting C_1 metabolism genes in the environment. So far, the genomes of five methylotrophs have been completely or almost completely sequenced, including two alphaproteobacterial facultative methylotrophs, *Methylobacterium extorquens* (12) and *Paracoccus denitrificans* (http://genome.jgi-psf.org/draft_microbes/parde/parde.home.html); a betaproteobacterial obligate methylotroph, *Methylobacillus flagellatus* (http://genome.jgi-psf.org/draft_microbes/metfl/metfl.home.html); a gammaproteobacterial methanotroph, *Methylococcus capsulatus* (78); and a betaproteobacterial facultative methylotroph, *Methylibium petroleophilum* (http://genome.jgi-psf.org/inished_microbesmetpe/metpe.home.html). In the first four cases, the genomics have been combined with extensive functional analysis (12, 13, 14, 33), thus providing information on key genes involved in C_1 metabolism in these bacteria. For M. *extorquens* and M. *capsulatus*, genome-based proteomic analyses have also been conducted (45, 48). Comparative analysis of methylotroph genomes, as well as the proteomic analyses, provides knowledge on the complement of the genes essential for C_1 metabolism in the environment. Methylotrophic ability has been previously analyzed from a modular point of view (12, 14), in which there are a number of discrete functional metabolic modules that are necessary for methylotrophy. Enzyme systems such as pMMO, sMMO, and MDH and pathways such as the assimilatory serine or ribulose monophosphate cycles represent such functional modules. At least one module needs to be present for primary oxidation, formaldehyde oxidation, and assimilation, but multiple modules may be present for each function. For example, M. *capsulatus* possesses all three known modules for C_1 assimilation, and all of them seem to be expressed (14, 45, 78). In the known genomes, genes involved in the same functional module are usually physically connected (12, 40, 78). For the methylotroph genomes analyzed so far, on the order of 100 key methylotrophy genes have been identified, which corresponds to only 1.5 to 2.5% of the total number of genes in the genomes. The current knowledge of the methylotrophy modules, however, remains incomplete. For example, for the recently identified methanol utilizer M. *petroleophilum* (55), the identity of the enzyme (or the genes) responsible for methanol oxidation remains unknown, as no typical *mox* gene cluster (12) is present.

ENVIRONMENTAL DETECTION OF C_1 OXIDIZERS

PCR-Based Surveys

Detection of C_1 utilizers based on PCR amplification from total environmental DNA is routine (Fig. 2). Three types of oligonucleotide primers for PCR amplification are currently in use. The first type consists of primers targeting 16S rRNA genes of the well-known methanotrophic groups (23). The second type of primers targets functional genes encoding the highly conserved enzymes for primary oxidation of methylotrophic substrates, including pMMO (*pmoA* [3, 11]), sMMO (*mmoX* [3, 49]), MDH (*mxaF* [52]), MSM (*msmA* [5]), and the corrinoid-linked methyltransferase (*cmuA* [9, 51, 64]). No tools exist at this time for detection of methylamine oxidation capacity. The third type of primers for environmental detection targets functional genes for formaldehyde oxidation (*fae*, *mtdB*, *mch*, and *fhcD*), specifically for the H_4MPT-linked pathway widespread in C_1 utilizers, thus allowing for broader detection of C_1 oxidation capacity (39, 41). This last type of primers differs from the former two also with respect to the conservation of the genes targeted. The genes for the H_4MPT-linked oxidative pathway are known to be much less conserved than the primary oxidation genes, consistent with the long evolutionary history of these genes. Accordingly, these primers have been designed to target genes significantly diverging in sequence (39, 41), increasing the potential for detection of novel genes belonging to uncultivated bacteria.

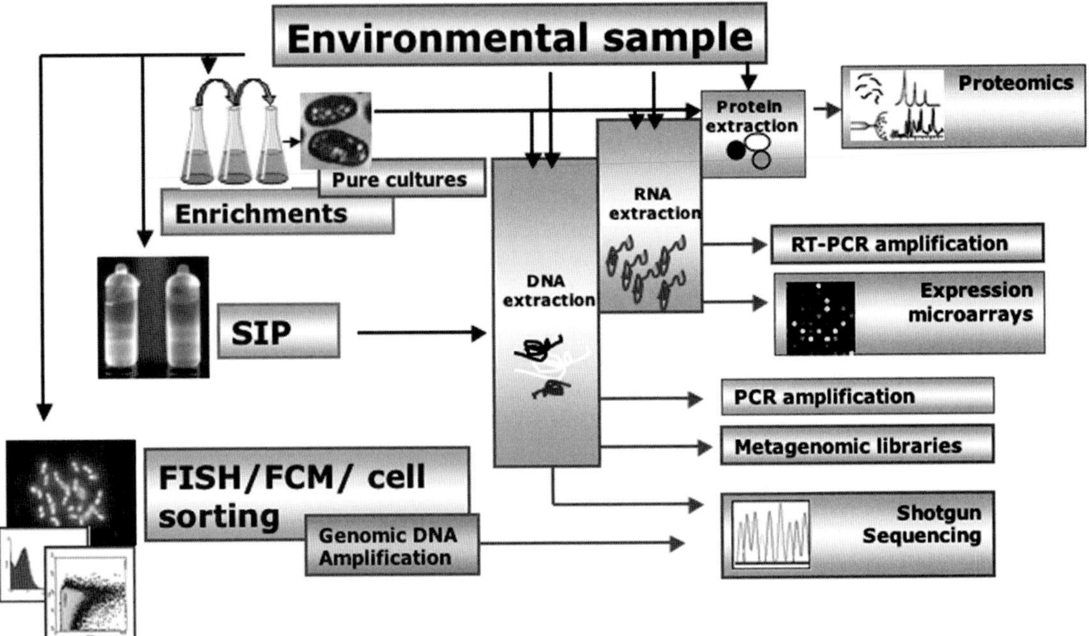

FIGURE 2 Strategies for detecting C_1-metabolizing populations in the environment.

The use of these broad detection primers has uncovered the existence in environments of a variety of sequences diverging from the sequences of known methylotrophs (39, 41). Many of these sequences fell within the *Proteobacteria* and planctomycetes. Interestingly, some of the sequences diverged very deeply from all known sequences, and these have been proposed to belong to a new bacterial phylum (39, 41, 43). These deeply diverging sequences for the H_4MPT-linked C_1 transfer pathway seem to be the dominant type in Lake Washington sediment, suggesting a potential function in C_1 metabolism.

Fluorescent In Situ Hybridization (FISH) and FISH-Based Flow Cytometric (FCM) Separation

16S rRNA-targeted FISH has been successfully employed for methylotroph detection and enumeration. A number of group-specific probes targeting methanotrophic bacteria of type I (gammaproteobacteria) and type II (alphaproteobacteria) that allow the comparison of the relative abundance of the two types in environmental samples have been developed (10, 19, 23). Recently, CARD (catalyzed fluorescent reporter deposition)-FISH, a more sensitive protocol that allows simultaneous detection of 16S rRNA and a functional gene (*pmoA*) mRNA in environmental samples, has been implemented (57, 66). While originally developed for detection of pMMO, this protocol may be adapted to detect other genes encoding key enzymes of C_1 metabolism. This approach opens new opportunities for not only detecting key players in C_1 metabolic processes by detecting major C_1 utilization genes, but also obtaining clues about their phylogenetic identities. In addition, this method provides a means for detection and quantification of gene expression directly in environmental samples.

FCM and FCM-based cell sorting are another powerful means for culture-independent characterization of environmental microbes, as they allow direct extraction of specific subpopulations from complex communities. The method has been widely used for detection and accurate enumeration of specific bacterial populations that are parts of complex microbial communities, including in bacterioplankton or soil communities (58, 66, 70). A typical protocol for FCM separation includes (i) cell extraction, (ii) fixation and/or staining, and (iii) FCM analysis. Detection of specific groups is based on light scattering and/or autofluorescence. In the absence of natural autofluorescence, microbial cells can be differentiated by various types of fluorescence staining, using either immunofluorescent labeling or FISH. The efficiency of cell extraction and separation from particles prior to such analyses remains a central issue in cases of soil or sediment samples. However, several protocols that allow successful cell recovery from crude samples are now available (27, 58). FISH-based FCM separation using group-specific 16S ribosomal DNA-based probes targeting methanotrophic bacteria of type I (MG84, gammaproteobacteria) and type II (MA450, alphaproteobacteria) has been successfully used for extraction of specific methanotrophic populations from a lake sediment (44). The approach resulted in enrichment of both the more abundant type I methanotroph population (12 to 15% of total microbial cells) and the less abundant type II methanotroph population (2 to 5% of total microbial cells). These populations were further characterized for the types of *pmoA* expressed, via reverse transcriptase PCR (RT-PCR) amplification, demonstrating that both type I and type II populations not only expressed a *pmoA* gene typical of the respective type, but also expressed divergent *pmoA* gene types, some of which were not previously identified in this habitat (15, 16, 44).

Not only rRNA, but also mRNA molecules transcribed from highly expressed genes (such as *mxaF*, *pmoA*, *mmoX*, and *fae*) present prospective targets for FISH detection combined with FCM and cell sorting. The standard protocol for this type of detection is presented in Table 1. We applied this approach to extract bacterial populations possessing

TABLE 1 Protocol for mRNA-targeting FISH-based cell sorting

Procedure	Steps	Comments and options
Probe Synthesis	Synthesize and label probe by PCR amplification with Alexa Fluor 488-5-UTP. Collect probe by ethanol precipitation, wash in 70% ethanol two times, and resuspend in 20 μl of H_2O.	Alternatively, the probe can be amplified by PCR and labeled by random priming or nick translation. Alexa Fluor 488-5-UTP can be replaced by a large selection of fluorophore-labeled nucleotides.
Restriction	To obtain optimal size of hybridization probe (20–60 bp), digest the PCR product with an appropriate restriction enzyme. Inactivate restriction enzymes by adding EDTA and heating at 80°C for 20 min. The probe can be used directly in FISH experiments.	The choice of a restriction enzyme is sequence dependent. This step may require optimization.
Cell extraction	Extract bacterial cells by one of the following: (1) shaking for 20 min at 200 rpm at room temperature (2) homogenizing (PRO200 Homogenizer 115V, PROScientific) for 5 min (with 1-min intervals) on ice at high speed (3) vortexing at high speed in MBB-8 (Biospec product) for 2 min with 0.1-mm-diameter Zirconia Silia Beads (Biospec product) at 5°C. Follow by filtration (20-μm filter) and centrifugation at 6,500 × *g* for 3 min.	If FCM analysis is not intended, cells may be fixed without extraction. For FCM applications, separation of cells from particles is essential to avoid high background levels and/or cell aggregation.
FISH Fixation	4% formaldehyde in PBS[a] buffer, 3 h on ice. Wash twice with PBS buffer. Collect cells by centrifugation at 6,500 × *g* for 3 min.	All FISH and FCM steps must be performed with RNase-free plasticware and DEPC-treated solutions. Fixation by 70% ethanol may be used instead.
Hybridization	Resuspend cells in 100 μl of hybridization buffer, incubate at T_{hyb} [b] for 1 h, add probe (50 ng/μl), and incubate for an additional 3 h or overnight.	Hybridization buffer: 0.9 M NaCl, 20 mM Tris (pH 8.0), 0.1% sodium dodecyl sulfate, 20–40% formamide
Washes	Collect cells by centrifugation at 6,500 × *g* for 3 min, resuspend in 100 μl of wash buffer, and incubate for 15 min at T_{hyb}. Collect cells by centrifugation at 6,500 × *g* for 3 min and resuspend in 1 ml of cold PBS.	Wash buffer: 0.5 M NaCl, 20 mM Tris (pH 8.0), 5 mM EDTA, 0.1% sodium dodecyl sulfate
FCM/cell sorting	Actual settings can vary with the application and should be optimized by using BD Liquid Counting Beads, control cell solution (with and without FISH staining), and an unstained sample. Acquire events at a rate of 300–3,000/s. Gate cells based on combined parameters for FSC, SSC, and fluorescence. For data analysis and graphics, the WinMDI 2.1 software package may be used (http://facs.scripps.edu/software.html).	Recommended references: 34a, 44, 56a, 66a Recommended websites: http://www.cyto.purdue.edu/flowcyt/research/micrflow/, http://www.bdbiosciences.com/immunocytometry_systems/

[a]PBS, phosphate-buffered saline.
[b]T_{hyb}, hybridization temperature.

divergent *fae* genes from the Lake Washington sediment community. These divergent genes have been previously detected via PCR-based surveys. Two representative divergent *fae* genes were fluorochrome (fluorescein or Alexa Fluor) labeled and hybridized with cells extracted directly from the sediment and then subjected to FCM-based cell sorting. Cells displaying a high fluorescence signal were collected and used for RT-PCR amplification of *fae*. A number of *fae* sequences amplified from the sorted cells were determined and shown to be highly similar to those used for probe design. These data demonstrated the applicability of the mRNA-FISH-FCM-cell sorting approach for extracting specific subpopulations from environmental samples which express specific functional genes and thus are potentially important in environmental processes, but which resist cultivation in the laboratory. The approach described, however, is tedious and time-consuming. For each specific target gene, optimization of every step in the protocol may be required, e.g., labeling efficiency, optimum probe concentration, number and type of restriction enzymes resulting in optimal probe length, etc.

Environmental Arrays

So far, environmental arrays have targeted only methane oxidizers, via detection of the *pmoA* genes. Bodrossy and colleagues have developed an array consisting of 68 oligonucleotide probes ranging in length between 17 and 27 nucleotides, representing a range of the known *pmoA* genes (8, 68). Hybridization targets have been generated by amplification of *pmoA* genes from environmental DNA followed by in vitro transcription. The RNA polymerase reaction mixture included Cy3- or Cy5-labeled dUTP; thus, the resulting RNA was fluorescently labeled. These arrays allowed qualitative detection of a variety of *pmoA* genes in the environment but did not allow quantitative detection, as the method was PCR based. The array described by Tiquia and colleagues (71) consisted of 50-mer oligonucleotides targeting a number of environmentally important functional genes and included only three *pmoA*-specific oligonucleotides. This array was hybridized to labeled DNA (71). The detection limit for both types of arrays has been estimated at around 5% of the total microbial population, thus targeting only dominant methanotroph species (8, 68, 71). Another major limitation of this type of array is that only sequences belonging to cultured bacteria are targeted, so the true complexity of methanotrophic communities in environments may not be reflected. So far, no arrays have been employed for detecting expression of C_1 genes in the environment.

ASSESSMENT OF THE ACTIVE C_1-METABOLIZING POPULATIONS

RNA Detection

By use of the tools for PCR-based detection described above, both environmental DNA and environmental RNA may be analyzed. The advantage of analyzing RNA versus DNA is in specifically targeting populations in a habitat or under experimental conditions. Improvement in techniques for isolation of nucleic acids from environments (28) and the availability of convenient kits for rapid RNA purification make these experiments feasible. RNA is first reverse transcribed with an RT, and the resulting DNA is then used for PCR amplification. Detection of 16S rRNA genes by using universal primers (47) allows estimation of the proportion of known methylotrophs with respect to the total active population. Primers targeting 16S rRNA genes of specific groups of methylotrophs may be used to detect these groups directly. Quantitative PCR using the group-specific primers may be applied to compare the relative abundance of the groups of interest. Detection of functional genes via RT-PCR provides the most valuable information on expression of genes for specific C_1 metabolic functions (44).

SIP

Stable isotope probing (SIP) is a relatively new approach to assessing microbial populations actively involved in metabolizing a compound of interest. The technique pioneered by J. C. Murrell's group is based on feeding natural populations (microcosms) a heavy (^{13}C) isotope of a C_1 compound of interest (methane, methanol, etc.), and then separating a labeled (heavy) fraction of DNA by isopycnic centrifugation (59). The heavy DNA fraction is enriched in sequences representing the organisms that have consumed the labeled compound and turned it into cell constituents, including DNA. The nature of these organisms is revealed via PCR amplification of genes of interest, using universal or group-specific phylogenetic PCR primers, or using primers targeting functional genes. So far, the SIP technique has been employed to target populations actively metabolizing methane, methanol, methylamine, formaldehyde, formate, and chloromethane (9, 36, 49, 50, 56, 59). It is understood that SIP experiments should be conducted under conditions resembling the conditions in situ as closely as possible, to obtain a true picture of the activities and processes native to a given environment. However, the success of SIP is inherently dependent on altering the in situ conditions to a degree, since a higher than in situ concentration of the labeled substrate in question is required for successful labeling (22, 59). Furthermore, prolonged incubations are often required (22, 59), which may result in enrichment of fast-growing microbes, and sometimes incubations at higher than in situ temperatures produce optimal results. While SIP does have the potential of uncovering unexpected and unsuspected participants in certain environmental processes (50, 56, 59), SIP results need critical interpretation (22, 56). On the other hand, some of the seeming disadvantages of the SIP technique, such as cross-feeding of the labeled compounds (50, 62), suggest a new dimension to the use of this technique, for exactly the purpose of following trophic links for C_1 (or other) compound utilization by a microbial community.

THE METAGENOMIC APPROACH

Large DNA Insert Libraries

The metagenomic approach, or isolation, cloning, and analysis of large (up to 100-kb) DNA fragments directly from the environment, is becoming an approach of choice for understanding the biology of as yet uncultured organisms (31). Metagenomics were first applied to search for novel activities and functions possessed by uncultured microbes (63) and later served to link a function of interest to the phylogeny of uncultured microbes (6). We have screened a large-insert metagenomic library constructed from DNA isolated from the top layer of Lake Washington sediment for the presence of genes involved in H_4MPT-linked reactions by PCR amplification (43). We estimated that at least one *fhcD* gene and at least one *fae* gene is present per every 39 and 34 Mbp of the metagenome, respectively. Comparing these numbers to the number of 16S rRNA genes detected in the same metagenome, we calculated that between 20 and 25% of the total population inhabiting Lake Washington sediment must contain genes for H_4MPT-linked reactions (43). Sequencing of *fae* and *fhcD* genes detected in the metagenome revealed that few of them belonged to known organisms; instead, most belonged to uncultivated *Proteobacteria*, planctomycetes, and novel bacterial divisions.

Metagenomics are especially powerful in combination with other detection methods, such as those described earlier in this chapter. By analyzing large genomic fragments, "interesting" (divergent, novel) sequences discovered via PCR-based analyses may be placed into a metabolic or phylogenetic context (6, 65) or may lead to the discovery of other divergent and novel genes (43, 65). Identification of the functional genes of interest in specific environments may be simplified and accelerated by constructing metagenomic libraries from DNA enriched in specific functional genes, for example, ^{13}C-labeled DNA generated in a SIP experiment (22). With all the advantages large-insert metagenomic libraries provide, they may still be prone to cloning biases, even when constructed in low-copy-number vectors; thus, some sequences may be artificially underrepresented while others are overrepresented in these libraries (43). Despite these limitations, the metagenomic approach provides an extra layer of information and supplements the data from other culture-dependent and culture-independent approaches.

WGS Community Sequencing

The recently pioneered approach of whole-genome shotgun (WGS) community sequencing is a small DNA insert-based metagenomic (environmental genomics) approach in which DNA is isolated from an environment (72, 73, 76), sometimes after appropriate sample fractionation (30), and is shotgun sequenced just like DNA from an individual organism (4). This approach has the potential of becoming the ultimate one to describe a given community gene pool and structure. Currently, at least two serious problems that prevent broad implementing of this approach exist. The first is the prohibitive cost, and the second is the computational difficulty of assembling the large databases of random DNA sequences (72, 76). For example, the success of the assembly of the acid mine drainage metagenome heavily relied on the available data for completely sequenced and assembled genomes of the microbes composing the community (73). Similarly, in the Sargasso Sea metagenome, only genomes aligned with related sequenced genomes were assembled into large scaffolds (76). To enable the analysis of environmental genomes versus separate, nonaffiliated gene sequences, sufficient sequence coverage is required. Venter and colleagues have calculated that to cover every sequence in the Sargasso Sea sample they used, a 10 times larger sequencing effort would be required (amounting to 20 million sequencing reactions) (76). Of course, in environments experiencing seasonal or spatial dynamic changes, this effort will need to be further extended. Nevertheless, databases such as the one generated by Venter and colleagues provide rich resources for detecting genes of interest. For example, the abundance of genes potentially involved in C_1 metabolism in the Sargasso Sea metagenomic database has been examined, via BLAST analyses, using known methylotrophy genes as queries (42). While no genes bearing a high degree of similarity to the genes found in known methylotrophs were detected in the database, consistent with low abundance of methanotrophs and methylotrophs belonging to previously characterized groups in the sample, genes for H_4MPT-linked formaldehyde oxidation were present with high abundance and revealed clustering patterns typical of these in known methylotrophs. These genes, however, diverged significantly from the genes characterized from cultured representatives, and the phylogenetic affiliations of the species possessing these genes could not be determined.

Targeted WGS Sequencing (Functional Metagenomics)

One approach that should allow for better coverage in environmental WGS sequencing would be to limit the sequencing effort to a population of interest, or a functional group (functional metagenomics). Some communities are naturally enriched in certain functional groups (30), while others may be enriched in the laboratory. In collaboration with the Joint Genome Institute, our group has recently initiated a project that involves such group-targeted environmental sequencing of the enriched C_1-utilizing communities from the Lake Washington sediment. Five separate microcosms have been set up, each exposed to a ^{13}C-labeled C_1 substrate (methane, methanol, methylamine, formaldehyde, or formate). ^{13}C-labeled DNA from the microcosms has been separated as described above and used for shotgun library construction and sequencing. On the basis of our preliminary analyses, the community diversity reflected by the ^{13}C-labeled DNA was significantly simplified compared to the diversity found in the ^{12}C-DNA fraction. It is expected that sequencing these microcosm community DNAs will result in sequence coverage high enough to allow for assembly into contigs and scaffolds of significant length. The available methylotroph genomes will be useful at this stage for ordering the contigs and scaffolds along the chromosomes, potentially resulting in (almost) complete genomic sequences.

FUTURE DIRECTIONS

Many of the novel approaches for environmental detection, manipulation, and analysis of native populations of C_1 utilizers are just emerging and are subject to further development and improvement. The expected progress in applying these novel techniques will inevitably bring about new dimensions to which these techniques may be expanded. We have already mentioned a potential of the SIP technique for tracing synergistic trophic activities in the environment. As with organismal genomics, the progress in environmental (functional) genomics will result in a platform for environmental (functional) transcriptomics. We envision a novel type of environmental expression array that specifically targets a local community (such as a Lake Washington sediment community) and is based on a specific sequence database (functional metagenome). The transcriptomics efforts will then be followed by and combined with environmental metaproteomics, based on the same specific metagenome. Environmental proteomics have already been implemented in an artificially simple natural community (61) but not yet in a community of significant complexity. The progress in environmental genomics will bring about new clues as to the possible means of cultivating microbes that have so far resisted cultivation. Alternatively, genomics of specific organisms may become feasible without the cultivation step. Cells of interest may be labeled with a specific probe (functional or phylogenetic) and enriched from the community via FCM separation, as discussed above, followed by whole-genome amplification (24, 60) and shotgun sequencing. The field of environmental genomics is still in its infancy, and it is easy to predict that the future will be full of surprises.

REFERENCES

1. **Anbar, A. D., Y. L. Yung, and F. P. Chavez.** 1996. Methyl bromide: ocean sources, ocean sinks, and climate sensitivity. *Global Biogeochem. Cyc.* **10:**175–190.
2. **Anthony, C.** 2004. The quinoprotein dehydrogenases for methanol and glucose. *Arch. Biochem. Biophys.* **428:**2–9.
3. **Auman, A. J., S. Stolyar, A. M. Costello, and M. E. Lidstrom.** 2000. Molecular characterization of methanotrophic isolates from freshwater lake sediment. *Appl. Environ. Microbiol.* **66:**5259–5266.
4. **Bankier, A. T.** 2001. Shotgun DNA sequencing. *Methods Mol. Biol.* **167:**89–100.
5. **Baxter, N. J., J. Scanlan, P. De Marco, A. P. Wood, and J. C. Murrell.** 2002. Duplicate copies of genes encoding methanesulfonate monooxygenase in *Marinosulfonomonas methylotropha* strain TR3 and detection of methanesulfonate utilizers in the environment. *Appl. Environ. Microbiol.* **68:**289–296.
6. **Beja, O., L. Aravind, E. V. Koonin, M. T. Suzuki, A. Hadd, L. P. Nguyen, S. B. Jovanovich, C. M. Gates, R. A. Feldman, J. L. Spudich, E. N. Spudich, and E. F. DeLong.** 2000. Bacterial rhodopsin: evidence for a new type of phototrophy in the sea. *Science* **289:**1902–1906.

7. **Bodrossy, L., J. C. Murrell, H. Dalton, M. Kalman, L. G. Puskas, and K. L. Kovacs.** 1995. Heat-tolerant methanotrophic bacteria from the hot water effluent of a natural gas field. *Appl. Environ. Microbiol.* **61:**3549–3555.
8. **Bodrossy, L., N. Stralis-Pavese, J. C. Murrell, S. Radajewski, A. Weilharter, and A. Sessitsch.** 2003. Development and validation of a diagnostic microbial microarray for methanotrophs. *Environ. Microbiol.* **5:**566–582.
9. **Borodina, E., M. J. Cox, I. R. McDonald, and J. C. Murrell.** 2005. Use of DNA-stable isotope probing and functional gene probes to investigate the diversity of methyl chloride-utilizing bacteria in soil. *Environ. Microbiol.* **7:**1318–1328.
10. **Bourne, D. G., A. J. Holmes, N. Iversen, and J. C. Murrell.** 2000. Fluorescent oligonucleotide rDNA probes for specific detection of methane oxidising bacteria. *FEMS Microbiol. Ecol.* **31:**29–38.
11. **Bourne, D. G., I. R. McDonald, and J. C. Murrell.** 2001. Comparison of *pmoA* PCR primer sets as tools for investigating methanotroph diversity in three Danish soils. *Appl. Environ. Microbiol.* **67:**3802–3809.
12. **Chistoserdova, L., S. W. Chen, A. Lapidus, and M. E. Lidstrom.** 2003. Methylotrophy in *Methylobacterium extorquens* AM1 from a genomic point of view. *J. Bacteriol.* **185:**2980–2987.
13. **Chistoserdova, L., L. Gomelsky, J. A. Vorholt, M. Gomelsky, Y. D. Tsygankov, and M. E. Lidstrom.** 2000. Analysis of two formaldehyde oxidation pathways in *Methylobacillus flagellatus* KT, a ribulose monophosphate cycle methylotroph. *Microbiology* **146:**233–238.
14. **Chistoserdova, L., J. A. Vorholt, and M. E. Lidstrom.** 2005. A genomic view of methane oxidation by aerobic bacteria and anaerobic archaea. *Genome Biol.* **6:**208.
15. **Costello, A. M., A. J. Auman, J. L. Macalady, K. M. Scow, and M. E. Lidstrom.** 2002. Estimation of methanotroph abundance in a freshwater lake sediment. *Environ. Microbiol.* **4:**443–450.
16. **Costello, A. M., and M. E. Lidstrom.** 1999. Molecular characterization of functional and phylogenetic genes from natural populations of methanotrophs in lake sediments. *Appl. Environ. Microbiol.* **65:**5066–5074.
17. **Davidson, V. L.** 2000. Methylamine dehydrogenase. Structure and function of electron transfer complexes. *Subcell. Biochem.* **35:**119–143.
18. **Dedysh, S. N., N. S. Panikov, W. Liesack, R. Grosskopf, J. Zhou, and J. M. Tiedje.** 1998. Isolation of acidophilic methane-oxidizing bacteria from northern peat wetlands. *Science* **282:**281–284.
19. **Dedysh, S. N., M. Derakshani, and W. Liesack.** 2001. Detection and enumeration of methanotrophs in acidic sphagnum peat by 16S rRNA fluorescence in situ hybridization, including the use of newly developed oligonucleotide probes for *Methylocella palustris*. *Appl. Environ. Microbiol.* **67:**4850–4857.
20. **Doronina, N. V., Y. A. Trotsenko, B. B. Kuznetsov, T. P. Tourova, and M. S. Salkinoja-Salonen.** 2002. *Methylobacterium suomiense* sp. nov. and *Methylobacterium lusitanum* sp. nov., aerobic, pink-pigmented, facultatively methylotrophic bacteria. *Int. J. Syst. Evol. Microbiol.* **52:**773–776.
21. **Doronina, N. V., T. D. Darmaeva, and Y. A. Trotsenko.** 2003. *Methylophaga alcalica* sp. nov., a novel alkaliphilic and moderately halophilic, obligately methylotrophic bacterium from an East Mongolian saline soda lake. *Int. J. Syst. Evol. Microbiol.* **53:**223–229.
22. **Dumont, M. G., and J. C. Murrell.** 2005. Stable isotope probing—linking microbial identity to function. *Nat. Rev. Microbiol.* **3:**499–504.
23. **Eller, G., S. Stubner, and P. Frenzel.** 2001. Group-specific 16S rRNA targeted probes for detection of type I and type II methanotrophs by fluorescent in situ hybridization. *FEMS Microbiol. Lett.* **198:**91–97.
24. **Erwin, D. P., I. K. Erickson, M. E. Delwiche, F. S. Colwell, J. L. Strap, and R. L. Crawford.** 2005. Diversity of oxygenase genes from methane- and ammonia-oxidizing bacteria in the Eastern Snake River Plain aquifer. *Appl. Environ. Microbiol.* **71:**2016–2025.
25. **Etiope, G., and R. W. Klusman.** 2002. Geologic emissions of methane to the atmosphere. *Chemosphere* **49:**777–789.
26. **Gallego, V., M. T. García, and A. Ventosa.** 2005. *Methylobacterium hispanicum* sp. nov. and *Methylobacterium aquaticum* sp. nov., isolated from drinking water. *Int. J. Syst. Evol. Microbiol.* **55:**281–287.
27. **Gough, H. L., and D. A. Stahl.** 2003. Optimization of direct cell counting in sediment. *J. Microbiol. Methods* **52:**39–46.
28. **Griffiths, R. I., A. S. Whiteley, A. G. O'Donnell, and M. J. Bailey.** 2000. Rapid method for coextraction of DNA and RNA from natural environments for analysis of ribosomal DNA- and rRNA-based microbial community composition. *Appl. Environ. Microbiol.* **66:**5488–5491.
29. **Guenther, A.** 2002. The contribution of reactive carbon emissions from vegetation to the carbon balance of terrestrial ecosystems. *Chemosphere* **49:**837–844.
30. **Hallam, S. J., N. Putnam, C. M. Preston, J. C. Detter, D. Rokhsar, P. M. Richardson, and E. F. DeLong.** 2004. Reverse methanogenesis: testing the hypothesis with environmental genomics. *Science* **305:**1457–1462.
31. **Handelsman, J.** 2004. Metagenomics: application of genomics to uncultured microorganisms. *Microbiol. Mol. Biol.* **68:**669–685.
32. **Hanson, R. S., and T. E. Hanson.** 1996. Methanotrophic bacteria. *Microbiol. Rev.* **60:**439–471.
33. **Harms, N., W. N. M. Reijnders, S. Koning, and R. J. M. van Spanning.** 2001. Two-component system that regulates methanol and formaldehyde oxidation in *Paracoccus denitrificans*. *J. Bacteriol.* **183:**664–670.
34. **Harper, D. B., J. T. Hamilton, V. Ducrocq, J. T. Kennedy, A. Downey, and R. M. Kalin.** 2003. The distinctive isotopic signature of plant-derived chloromethane: possible application in constraining the atmospheric chloromethane budget. *Chemosphere* **52:**433–436.

34a. **Hawley, T. S., and R. G. Hawley (ed.).** 2004. *Methods in Molecular Biology*, vol. 263. *Flow Cytometry Protocols*. Humana Press Inc., Totowa, N.J.

35. **Hektor, H. J., H. Kloosterman, and L. Dijkhuizen.** 2002. Identification of a magnesium-dependent NAD(P)(H)-binding domain in the nicotinoprotein methanol dehydrogenase from *Bacillus methanolicus*. *J. Biol. Chem.* **277:**46966–46973.
36. **Hutchens, E., S. Radajewski, M. G. Dumont, I. R. McDonald, and J. C. Murrell.** 2004. Analysis of methanotrophic bacteria in Movile Cave by stable isotope probing. *Environ. Microbiol.* **6:**111–120.
37. **Jones, J. G., and E. Bellion.** 1991. In vivo ^{13}C and ^{15}N NMR studies of methylamine metabolism in *Pseudomonas* species MA. *J. Biol. Chem.* **266:**11705–11713.
38. **Kalyuzhnaya, M., V. Khmelenina, B. Eshinimaev, N. Suzina, D. Nikitin, A. Solonin, J. L. Lin, I. McDonald, C. Murrell, and Y. Trotsenko.** 2001. Taxonomic characterization of new alkaliphilic and alkalitolerant methanotrophs from soda lakes of the Southeastern Transbaikal region and description of *Methylomicrobium buryatense* sp. nov. *Syst. Appl. Microbiol.* **24:**166–176.
39. **Kalyuzhnaya, M. G., M. E. Lidstrom, and L. Chistoserdova.** 2004. Utility of environmental primers targeting ancient enzymes: methylotroph detection in Lake Washington. *Microb. Ecol.* **48:**463–472.

40. **Kalyuzhnaya, M. G., N. Korotkova, G. Crowther, C. J. Marx, M. E. Lidstrom, and L. Chistoserdova.** 2005. Analysis of gene islands involved in methanopterin-linked C_1 transfer reactions reveals new functions and provides evolutionary insights. *J. Bacteriol.* **187:**4607–4614.
41. **Kalyuzhnaya, M. G., O. Nercessian, M. E. Lidstrom, and L. Chistoserdova.** 2005. Development and application of PCR primers based on *fhcD* for environmental detection of methanopterin-linked C_1 metabolism in bacteria. *Environ. Microbiol.* **7:**1269–1274.
42. **Kalyuzhnaya, M.G., O. Nercessian, A. Lapidus, and L. Chistoserdova.** 2005. Fishing for biodiversity: novel methanopterin-linked C_1 genes deduced from the Sargasso Sea metagenome. *Environ. Microbiol.* **7:**1909–1916.
43. **Kalyuzhnaya, M. G., S. Bowerman, O. Nercessian, M. E. Lidstrom, and L. Chistoserdova.** 2005. Highly divergent genes for methanopterin-linked C_1 transfer reactions in Lake Washington, assessed via metagenomic analysis and mRNA detection. *Appl. Environ. Microbiol.* **71:**8846–8854.
44. **Kalyuzhnaya, M. G., R. Zabinsky, S. Bowerman, D. R. Baker, M. E. Lidstrom, and L. Chistoserdova.** 2006. Fluorescence in situ hybridization-flow cytometry-cell sorting-based method for separation and enrichment of type I and type II methanotroph populations. *Appl. Environ. Microbiol.* **72:**4293–4301.
45. **Kao, W. C., Y. R. Chen, E. C. Yi, H. Lee, Q. Tian, K. M. Wu, S. F. Tsai, S. S. Yu, Y. J. Chen, R. Aebersold, and S. I. Chan.** 2004. Quantitative proteomic analysis of metabolic regulation by copper ions in *Methylococcus capsulatus* (Bath). *J. Biol. Chem.* **279:**51554–51560.
46. **Kelly, D. P., and J. C. Murrell.** 1999. Microbial metabolism of methanesulfonic acid. *Arch. Microbiol.* **172:**341–348.
47. **Lane, D. J.** 1991. 16S/23S rRNA sequencing, p. 115–175. *In* E. Stackebrant and M. Goodfellow (ed.), *Nucleic Acid Techniques in Bacterial Systematics*. Wiley, New York, N.Y.
48. **Laukel, M., M. Rossignol, G. Borderies, U. Volker, and J. A. Vorholt.** 2004. Comparison of the proteome of *Methylobacterium extorquens* AM1 grown under methylotrophic and nonmethylotrophic conditions. *Proteomics* **4:**1247–1264.
49. **Lin, J. L., S. Radajewski, B. T. Eshinimaev, Y. A. Trotsenko, I. R. McDonald, and J. C. Murrell.** 2004. Molecular diversity of methanotrophs in Transbaikal soda lake sediments and identification of potentially active populations by stable isotope probing. *Environ. Microbiol.* **6:**1049–1060.
50. **Lueders, T., B. Wagner, P. Claus, and M. W. Friedrich.** 2004. Stable isotope probing of rRNA and DNA reveals a dynamic methylotroph community and trophic interactions with fungi and protozoa in oxic rice field soil. *Environ. Microbiol.* **6:**60–72.
51. **McAnulla, C., I. R. McDonald, and J. C. Murrell.** 2001. Methyl chloride utilising bacteria are ubiquitous in the natural environment. *FEMS Microbiol. Lett.* **201:**151–155.
52. **McDonald, I. R., and J. C. Murrell.** 1997. The methanol dehydrogenase structural gene *mxaF* and its use as a functional gene probe for methanotrophs and methylotrophs. *Appl. Environ. Microbiol.* **63:**3218–3224.
53. **McDonald, I. R., K. L. Warner, C. McAnulla, C. A. Woodall, R. S. Oremland, and J. C. Murrell.** 2002. A review of bacterial methyl halide degradation: biochemistry, genetics and molecular ecology. *Environ. Microbiol.* **4:**193–203.
54. **Murrell, J. C., B. Gilbert, and I. R. McDonald.** 2000. Molecular biology and regulation of methane monooxygenase. *Arch. Microbiol.* **173:**325–332.
55. **Nakatsu, C. H., K. Hristova, S. Hanada, X.-Y. Meng, J. R. Hanson, K. M. Scow, and Y. Kamagata.** 2006. *Methylibium petroleiphilum* gen. nov., sp. nov., a new methyl *tert*-butyl ether-degrading methylotroph of the *Betaproteobacteria*. *Int. J. Syst. Evol. Microbiol.* **56:**983–989.
56. **Nercessian, O., E. Noyes, M. G. Kalyuzhnaya, M. E. Lidstrom, and L. Chistoserdova.** 2005. Bacterial population active in metabolism of C_1 compounds in the sediment of Lake Washington, a freshwater lake. *Appl. Environ. Microbiol.* **71:**6885–6899.

56a. **Ormerod, M. G. (ed.).** 2000. *Flow Cytometry: A Practical Approach*, 3rd ed. Oxford University Press, New York, N.Y.

57. **Pernthaler, A., and R. Amann.** 2004. Simultaneous fluorescence in situ hybridization of mRNA and rRNA in environmental bacteria. *Appl. Environ. Microbiol.* **70:**5426–5433.
58. **Porter, J., R. Pickup, and C. Edwards.** 1997. Evaluation of flow cytometric methods for the detection and viability assessment of bacteria from soil. *Soil Biol. Biochem.* **29:**91–100.
59. **Radajewski, S., P. Ineson, N. R. Parekh, and J. C. Murrell.** 2000. Stable-isotope probing as a tool in microbial ecology. *Nature* **403:**646–649.
60. **Raghunathan, A., H. R. Ferguson, Jr., C. J. Bornarth, W. Song, M. Driscoll, and R. S. Lasken.** 2005. Genomic DNA amplification from a single bacterium. *Appl. Environ. Microbiol.* **71:**3342–3347.
61. **Ram, R. J., N. C. Verberkmoes, M. P. Thelen, G. W. Tyson, B. J. Baker, I. R. C. Blake, M. Shah, R. L. Hettich, and J. F. Banfield.** 2005. Community proteomics of a natural microbial biofilm. *Science* **308:**1915–1920.
62. **Rangel-Castro, J. I., K. Killham, N. Ostle, G. W. Nicol, I. C. Anderson, C. M. Scrimgeour, P. Ineson, A. Meharg, and J. L. Prosser.** 2005. Stable isotope probing analysis of the influence of liming on root exudate utilization by soil microorganisms. *Environ. Microbiol.* **7:**828–838.
63. **Rondon, M. R., P. R. August, A. D. Bettermann, S. F. Brady, T. H. Grossman, M. R. Liles, K. A. Loiacono, B. A. Lynch, I. A. MacNeil, C. Minor, C. L. Tiong, M. Gilman, M. S. Osburne, J. Clardy, J. Handelsman, and R. M. Goodman.** 2000. Cloning the soil metagenome: a strategy for accessing the genetic and functional diversity of uncultured microorganisms. *Appl. Environ. Microbiol.* **66:**2541–2547.
64. **Schafer, H., I. R. McDonald, P. D. Nightingale, and J. C. Murrell.** 2005. Evidence for the presence of a CmuA methyltransferase pathway in novel marine methyl halide-oxidizing bacteria. *Environ. Microbiol.* **7:**839–852.
65. **Schleper, C., G. Jurgens, and M. Jonuscheit.** 2005. Genomic studies of uncultivated archaea. *Nat. Rev. Microbiol.* **3:**479–488.
66. **Sekar, R., B. M. Fuchs, R. Amann, and J. Pernthaler.** 2004. Flow sorting of marine bacterioplankton after fluorescence in situ hybridization. *Appl. Environ. Microbiol.* **70:**6210–6219.

66a. **Shapiro, H. M.** 2003. *Practical Flow Cytometry*, 4th ed. Wiley and Sons, Inc., New York, N.Y.

67. **Sieburth, J. M., P. W. Johnson, V. M. Church, and D. C. Laux.** 1993. C_1 bacteria in the water column of Chesapeake Bay, USA. III. Immunologic relationships of the type species of marine monomethylamine- and methane-oxidizing bacteria to wild estuarine and oceanic cultures. *Mar. Ecol. Prog. Ser.* **95:**91–102.
68. **Stralis-Pavese, N., A. Sessitsch, A. Weilharter, T. Reichenauer, J. Riesing, J. Csontos, J. C. Murrell, and L. Bodrossy.** 2004. Optimization of diagnostic microarray for application in analysing landfill methanotroph communities under different plant covers. *Environ. Microbiol.* **6:**347–363.

69. **Studer, A., C. McAnulla, R. Buchele, T. Leisinger, and S. Vuilleumier.** 2002. Chloromethane-induced genes define a third C_1 utilization pathway in *Methylobacterium chloromethanicum* CM4. *J. Bacteriol.* **184:**3476–3484.
70. **Thomas, J., M. Desrosiers, Y. St.-Pierre, P. Lirette, J. Bisaillon, R. Beaudet, and R. Villemur.** 1997. Quantitative flow cytometry detection of specific microorganisms in soil samples using rRNA targeted fluorescent probes and ethidium bromide. *Cytometry* **27:**224–232.
71. **Tiquia, S. M., L. Wu, S. C. Chong, S. Passovets, D. Xu, Y. Xu, and J. Zhou.** 2004. Evaluation of 50-mer oligonucleotide arrays for detecting microbial populations in environmental samples. *BioTechniques* **36:**664–675.
72. **Tringe, S. G., C. von Mering, A. Kobayashi, A. A. Salamov, K. Chen, H. W. Chang, M. Podar, J. M. Short, E. J. Mathur, J. C. Detter, P. Bork, P. Hugenholtz, and E. M. Rubin.** 2005. Comparative metagenomics of microbial communities. *Science* **308:**554–557.
73. **Tyson, G. W., J. Chapman, P. Hugenholtz, E. E. Allen, R. J. Ram, P. M. Richardson, V. V. Solovyev, E. M. Rubin, D. S. Rokhsar, and J. F. Banfield.** 2004. Community structure and metabolism through reconstruction of microbial genomes from the environment. *Nature* **428:**37–43.
74. **Valentine, D. L.** 2002. Biogeochemistry and microbial ecology of methane oxidation in anoxic environments: a review. *Antonie Leeuwenhoek* **81:**271–282.
75. **Van Aken, B., C. M. Peres, S. L. Doty, J. M. Yoon, and J. L. Schnoor.** 2004. *Methylobacterium populi* sp. nov., a novel aerobic, pink-pigmented, facultatively methylotrophic, methane-utilizing bacterium isolated from poplar trees (*Populus deltoides* × *nigra* DN34). *Int. J. Syst. Evol. Microbiol.* **54:**1191–1196.
76. **Venter, J. C., K. Remington, J. F. Heidelberg, A. L. Halpern, D. Rusch, J. A. Eisen, D. Wu, I. Paulsen, K. E. Nelson, W. C. Nelson, D. E. Fouts, S. Levy, A. H. Knap, M. L. W. Lomas, K. Nealson, O. White, J. Peterson, J. Hoffman, R. Parsons, H. Baden-Tillson, C. Pfannkoch, Y.-H. Rogers, and H. O. Smith.** 2004. Environmental genome shotgun sequencing of the Sargasso Sea. *Science* **304:**66–74.
77. **Volsch, A., N. F. Nader, H. K. Geiss, G. Nebe, and C. Birr.** 1990. Detection and analysis of two serotypes of ammonia-oxidizing bacteria in sewage plants by flow cytometry. *Appl. Environ. Microbiol.* **56:**2430–2435.
78. **Ward, N., Ø. Larsen, J. Sakwa, L. Bruseth, H. Khouri, A. S. Durkin, G. Dimitrov, L. Jiang, D. Scanlan, K. H. Kang, M. Lewis, K. E. Nelson, B. Methe, M. Wu, J. F. Heidelberg, I. T. Paulsen, D. Fouts, J. Ravel, H. Tettelin, Q. Ren, T. Read, R. T. DeBoy, R. Seshadri, S. L. Salzberg, H. B. Jensen, N. K. Birkeland, W. C. Nelson, R. J. Dodson, S. H. Grindhaug, I. Holt, I. Eidhammer, I. Jonasen, S. Vanaken, T. Utterback, T. V. Feldblyum, C. M. Fraser, J. R. Lillehaug, and J. A. Eisen.** 2004. Genomic insights into methanotrophy: the complete genome sequence of *Methylococcus capsulatus* (Bath). *PLoS Biol.* **2:**e303.

Sulfur Cycling

MARK E. HINES, PIETER T. VISSCHER, ANDREAS TESKE, AND RICHARD DEVEREUX

41

Sulfur (S) is an important element biochemically and geochemically. It constitutes ~1% of the dry mass of organisms, where it serves many structural and enzymatic functions. S also acts as a significant electron donor and acceptor during many bacterial metabolisms (51, 52). S can be found in a range of valence states from the highly reduced sulfide (−2) to the most oxidized form in sulfate (SO_4^{2-}) (+6). There are several intermediate valence forms of S that can act as both electron donors and acceptors, depending on environmental conditions, the most notable being elemental sulfur (S^0) and thiosulfate ($S_2O_3^{2-}$) (88). Microbial S transformations are closely linked with the carbon cycle, in which S reduction coupled with organic matter utilization is a major mineralization pathway in anoxic habitats, while S oxidations, some of which are autotrophic and/or phototrophic (52), can occur aerobically and anaerobically. Many S compounds are highly reactive, and microorganisms often must compete with abiotic reactions, including the oxidation of reduced S by O_2. This competition makes the study of certain physiological types of S bacteria difficult. In addition, the S cycle is complicated further by the reactivity of sulfide with metals and the oxidation of metal sulfides by bacteria. In general, the high chemical and biological reactivity of S compounds results in a tight coupling of the oxidative and reductive portions of the S cycle in aquatic habitats, particularly at the redoxcline, where S cycling can be extremely rapid (52).

Microorganisms of the S cycle are extremely diverse. The anaerobic sulfate-reducing bacteria (SRB), which are unique physiologically and genetically, are represented by several genera, most of which were discovered in the last 20 years (25, 79, 137). Once thought to be restricted primarily to sulfate respiration or fermentation, SRB have been shown to be capable of utilizing iron and manganese as electron acceptors (73). O_2 reduction has been demonstrated, but O_2-dependent growth has not been confirmed (124). SRB are rather nutritionally diverse, with the known suite of electron donors increasing each year (36). Chemolithotrophic S oxidation is mediated aerobically by colorless S bacteria, some purple S bacteria, and SRB. Anaerobically, nitrate-respiring chemolithotrophs oxidize sulfide, and both oxygenic and anoxygenic phototrophic bacteria use sulfide as an electron donor for photosynthesis (130, 133). Intermediates produced during sulfide oxidation can be oxidized, reduced, or fermented (disproportionated) (30, 54, 55). Although S reduction and oxidation are often studied separately for convenience, it is becoming clear that in many instances both processes must be considered simultaneously.

SAMPLE COLLECTION AND HANDLING

For enumeration, isolation, and rate measurements, it is important to process samples quickly after they are obtained and to maintain samples near ambient temperatures (50). For anaerobes, collect and handle samples anoxically. Typical water samplers include Niskin bottles and bags. To prevent contamination by O_2 that diffuses through or from plastic, water samples should be transferred to glass bottles by allowing them to overflow before sealing. If rubber stoppers are used, they should be the type that retard O_2 diffusion. Butyl rubber is best, and thick butyl stoppers (e.g., Bellco 2048-11800) can be penetrated many times with needles without risk of O_2 contamination. Since many S compounds are reactive with rubber stoppers and since most stoppers release some S compounds, Teflon-lined butyl septa (e.g., Wheaton 224168) are a useful alternative. However, they usually cannot withstand repeated needle penetration.

Sediments are obtained by using grabs or corers. Ideally, for rate measurements, use small-diameter (3-cm) whole-core (~15-cm-long) incubations in which amendments of dissolved constituents are introduced axially via microliter syringes through predrilled holes filled with a sealant. Whole cores should have the overlying water removed prior to incubation to prevent depletion of O_2 in surficial layers due to stagnant water. In aquatic sediments where O_2 penetration is less than 10 to 15 mm, most of the sulfide-oxidizing activity takes place in the upper 20 to 30 mm, so cores for S oxidation measurements should sample the upper 3 to 5 cm. Sediments removed from cores should be handled in N_2-filled glove bags or boxes. However, most relatively active sediments can withstand short exposures to laboratory air, i.e., the few minutes required to transfer samples from cores to incubation vessels (15). Cores should be used within 24 h after collection and should not be exposed to temperatures above ambient, since even short exposures to elevated temperatures will increase rates.

Sedimentary pore water samples for chemical measurements can be obtained in several ways: (i) centrifugation of sediments under N_2 or by centrifugation (2 min at 8,000 × g) of small volumes in a microcentrifuge filter unit (pore size, 0.45 μm; Gelman Sciences, Ann Arbor, Mich.); (ii) squeezing of samples under inert gas flow (N_2) by using prefabricated squeezers (100); (iii) in situ dialysis samplers referred to as "peepers" (41), which must equilibrate with pore water for ~2 weeks (these cannot be used when sediments are subjected to tides; peepers should also be preincubated in an O_2-free atmosphere for 2 weeks prior to deployment to remove O_2, which dissolves in plastic and will diffuse into anoxic sediment [14]); and (iv) in situ "sippers," which are lysimeters with porous Teflon collars that allow for the removal of pore water in a few minutes under a slight vacuum (45). In all cases, samples must not come in contact with O_2, and except for peeper samples, which do not have to be filtered, pore waters must be filtered anoxically. When sampling sites such as marshes or seagrass beds that contain macrophytes, avoid sampling techniques that disturb the sediments, since damaged roots leak dissolved material (49). Sippers are well suited for vegetated sediments (45).

SULFATE REDUCTION

SRB are a specialized group of anaerobic bacteria that are responsible for the dissimilatory reduction of SO_4^{2-} to sulfide that is linked to organic matter oxidation. They are important in the anaerobic degradation of organic matter in most aquatic habitats, where they are situated at the bottom of the anaerobic food chain. SRB are also a major source of sulfide, which is highly reactive and geochemically important and is used by S-oxidizing bacteria as described below. Because of the abundance of SO_4^{2-} in seawater, dissimilatory SO_4^{2-} reduction is considered a major process in marine sediments, yet it is also important in the decomposition of organic material in anoxic freshwater habitats. Although the reduction of SO_4^{2-} is considered to be the classic role of SRB in the environment, these bacteria are capable of diverse metabolisms, including metal and O_2 reduction, metal methylation and demethylation, organic fermentations, use of xenobiotics and petroleum by-products, S disproportionation, and the utilization of various intermediate redox states of S, including sulfonates (71). It is generally accepted that SRB oxidize products of fermentative bacteria, such as fatty acids, alcohols, some aromatic acids, a few amino acids, and hydrogen. The suite of substrates used varies among genera. For detailed information on SRB, various reviews are available (36, 88, 93, 112, 137).

Rate Measurements

Rates of SO_4^{2-} reduction have become routine measurements in studies of the biogeochemistry of anoxic aquatic environments. Rates can be determined by monitoring the loss of SO_4^{2-} in samples incubated anaerobically over time (jar experiments) or by determining the rate of production of reduced-^{35}S end products generated during the reduction of $^{35}SO_4^{2-}$ in incubated samples. The ^{35}S technique is preferable, since it uses a short incubation period (1 day or less) and undisturbed sediment. However, if one is unable to use the radiotracer technique, a jar approach provides an estimate of SO_4^{2-} reduction. Minimize manipulation in all cases, especially the mixing of oxic and anoxic layers in sediments. See reference 50 for a discussion of effects of mixing and dilution on measuring rates of sedimentary SO_4^{2-} reduction.

Jar Technique

Homogenize a sediment sample and either distribute subsamples among several vessels that are sacrificed over time or maintain one vessel in which subsamples are removed over time. For the former, use centrifuge tubes (15 to 50 ml) so pore water can be extracted easily. When using plastic, preincubate the contents of tubes in an inert atmosphere for ~2 weeks, and incubate the contents of sample-filled tubes similarly, since many plastics allow O_2 to pass and since they bleed O_2 (14). Diffusion-retardant plastic bags can be used without additional precaution (13). At selected time intervals, extract and filter pore waters and measure SO_4^{2-} content (analytical techniques are described below). Determine the wet and dry weights of known volumes of sediment so that rates can be expressed on a sediment volume basis. Incubation times should be kept to a minimum, and a linear decrease in SO_4^{2-} content is desirable, since nonlinear losses are difficult to quantify. For nearshore marine sediments, incubation times range from a few days to weeks. Since freshwaters contain much less SO_4^{2-}, incubation times are short (a few days to <2.0 h).

Consumption of SO_4^{2-} is calculated from the slope of the concentration over time ($\delta[SO_4^{2-}]/\delta t$). The consumption rate is either expressed in amount (micromoles or millimoles) per volume (e.g., cubic centimeters of sediment or milliliters of pore water) or per mass (grams of sediment) per unit of time (minutes, hours, or days). Rates are usually expressed as nanomoles per cubic centimeter of sediment per day or integrated over depth as micromoles per square meter.

^{35}S Technique

The radiotracer technique is the most widely used method for measuring rates of sulfate reduction. $^{35}SO_4^{2-}$ is introduced in tracer amounts in small volumes, and ^{35}S end products are isolated. Unlike the jar technique, which measures SO_4^{2-} loss, the accumulation of labeled end products is very sensitive, so incubation times are short and undisturbed samples can be used. Early studies measured the accumulation of acid-volatile sulfides only. However, since the discovery that significant quantities of reduced ^{35}S are recovered as non-acid-soluble phases (i.e., S^0 and FeS_2) (48), these are included by using a chromium reduction procedure (140). The rapid isotopic exchange of ^{35}S among reduced pools (33) precludes the use of this technique as a measure of the rate of production of specific reduced-S species. Therefore, rates are determined from the sum of all the inorganic reduced-S compounds. The following is a one-step distillation procedure for measuring rates of SO_4^{2-} reduction. See reference 32 for additional details.

Radiotracer (2.0 μl of $^{35}SO_4^{2-}$ in weak acid or sterile water; ~1.0 μCi) is introduced, via a microliter syringe through longitudinal holes filled with a sealant, into small-diameter (~3-cm) short (15-cm) cores while withdrawing the syringe. It is also possible to take horizontal subcores in syringes that are sealed and injected (43). For samples that are difficult to subcore, like salt marsh peat samples that contain thick rhizomes, large-diameter cores are sliced into small sections that are introduced into 5-ml syringes without homogenizing. To determine the SO_4^{2-} content of the sediments, additional cores are sectioned horizontally and pore water is extracted by centrifugation. After incubation in the dark from 2 to 24 h at in situ temperature, cores are cut into segments that are mixed with 20% zinc acetate and are then frozen (−20°C) to fix sulfides and terminate activity. Incubation time depends on the SO_4^{2-} content of the sediment, with shorter incubation periods used when SO_4^{2-} levels are low. Try to avoid consuming more than 10% of the ambient SO_4^{2-} during incubation. Thawed samples are centrifuged, and the activity of the remaining $^{35}SO_4^{2-}$ is determined by scintillation counting. When individual subcores are used, the ^{35}S activity of this supernatant is not required as long as the activity of the injected ^{35}S is known. The pellet is washed twice with seawater to remove excess $^{35}SO_4^{2-}$, transferred to a reaction vessel, and mixed with 10 ml of ethanol. The vessel is connected to a condenser and contains ports for introducing reagents and a constant stream of N_2 gas (12). The top of the condenser is connected to two traps in series, each containing 10 ml of a solution of 6% zinc acetate and 10% ammonium hydroxide. Two drops of Antifoam B (Baker Chemicals) are added to the traps to prevent foaming, which can be excessive for samples with relatively high-reduced S levels. The reaction vessel is flushed with N_2 for 20 min, and then 40 ml of $CrCl_2$ solution and 20 ml of concentrated HCl are added. The contents are gently boiled for 45 to 60 min with a constant stream of N_2. The total reduced S is converted to H_2S, which is swept into the traps and collected as ZnS. Over 98% of the ^{35}S is collected in the first trap. Half of the trap material (~5.0 ml) is mixed with a gel-forming scintillation cocktail and counted. If desired, the sulfide content of the remaining ZnS can be determined colorimetrically (16) for a measure of the pool size of total reduced inorganic S.

The reduced chromium solution is prepared by percolating 1.0 M $CrCl_3 \cdot 6H_2O$ (pH 1.0) through a column filled with amalgamated zinc granules (grain size, 0.5 to 1.5 mm). The zinc is amalgamated by soaking briefly in a solution of $HgCl_2$ (~0.25 M), transferred to a glass column (inside diameter, ~2 cm), and washed with 3 volumes of 0.1 M HCl. The $CrCl_3$ solution, which is dark green, becomes a bright blue when reduced ($CrCl_2$) and is collected within syringes or in serum bottles previously flushed with N_2. It is important that only bright blue reduced chromium be used in reaction vessels. If the solution elutes from the column as even slightly greenish blue, it should be discarded. Only freshly prepared $CrCl_3 \cdot 6H_2O$ solutions should be run through zinc columns, since solution a few days old will not reduce adequately. Also use only $CrCl_3 \cdot 6H_2O$, which is a fine green powder. $CrCl_3 \cdot 6H_2O$ in large chunks that exhibit a slight bluish hue will never completely reduce, even if recently manufactured.

An alternative to the active distillation procedure was developed by Ulrich et al. (122), in which the chromium reduction reaction is employed in a passive mode. Briefly, samples (<2 ml) incubated with ^{35}S and fixed as described above are introduced anoxically into 120-ml serum vials and sealed with thick butyl rubber stoppers. Each serum vial contains a test tube with 2.5 ml of anoxic 10% zinc acetate as a sulfide trap. Reduced chromium solution (8 ml) followed by anoxic concentrated HCl (4 ml) is injected into each vial, which is then placed on a rotary shaker. After ~30 h, the zinc acetate-filled test tubes are removed and homogenized by both vortexing and sonicating, and the radioactivity of the ZnS precipitate is then determined by scintillation counting. All solutions must be prepared and stored anoxically. Refer to reference 122 for details. This passive technique does not quantitatively recover S in S^0, so care should be taken when analyzing samples in which S^0 is a significant portion of the S generated during the incubation. In addition, calcium carbonate-containing sediments rapidly neutralize the acidic chromium solution, preventing the reduction of S species. This neutralization may be overcome by using small sediment samples. However, sediments rich in calcium carbonate cannot be analyzed with this technique unless the quantities of reagents are increased and excess pressures in vials are released.

More recently, a cold chromium distillation procedure was developed that enables the measurement of SO_4^{2-} reduction rates that are extremely low (<1 pmol cm^{-3} day^{-1}) (56). Unlike the technique of Ulrich et al. (122) described above, which uses a passive distillation, this technique actively strips reduced S into a Zn acetate trap by using a stream of N_2, but the chromous acid solution is not heated. Radiosulfate preparations contain traces of volatile ^{35}S that are carried to trap solutions, yielding a small blank that increases the detection limit of the method. This background radioactivity, the composition of which remains unknown, is essentially eliminated by using a cold distillation. Since the reduction of S^0 is incomplete without heating, *N,N*-dimethylformamide (DMF) is added to the reaction flask to facilitate S^0 reduction. The distillation setup and other details are well described in reference 56. Although this cold distillation method decreases the detection limit and is thus superior, the authors state that the method is more time-consuming and laborious and uses DMF, which is quite toxic, so they recommend its use only for samples exhibiting very low rates of SO_4^{2-} reduction.

Microscale distribution of sulfate reduction activity can be assessed by using Ag foil coated with $^{35}SO_4^{2-}$ (Y. Cohen and Y. Helman, *Abstr. 97th Gen. Meet. Am. Soc. Microbiol.*, p. 393, 1997). Ag foil (Sigma) is washed with acetone followed by application of ^{35}S (use carrier-free $^{35}SO_4^{2-}$) and air drying overnight. The coated foil is then placed on a fresh sediment sample and incubated. After rinsing off residual sulfate, activity is determined by measuring $Ag^{35}S$ formed during SO_4^{2-} reduction by using autoradiography or gel scanners (e.g., Bio-Rad Molecular Imager System GS-525) that are able to detect ^{35}S on the foil surface.

The rate of sulfate reduction is calculated as

$$\text{rate} = \frac{[SO_4^{2-}] \cdot (^{35}S\text{-TRIS}) \cdot 1.06}{(^{35}SO_4^{2-} + {}^{35}S\text{-TRIS}) \cdot t}$$

with the unit being nanomoles of SO_4^{2-} per cubic centimeter (whole sediment) per day, where $[SO_4^{2-}]$ is the SO_4^{2-} concentration in nanomoles per cubic centimeter (calculated from pore water values and water content of sediment determined from wet and dry weights), ^{35}S-TRIS is the total radioactivity of ZnS, $^{35}SO_4^{2-}$ is the total radioactivity of the supernatant (the $^{35}SO_4^{2-}$ left after incubation), t is the incubation time in days, and 1.06 corrects for isotope fractionation (50).

Cultivation, Enrichment, Enumeration, and Isolation

SRB are readily isolated from many environments, including freshwater and salt water, soils, oil-bearing shales and strata, intestinal contents, sewage, and paper mill effluents (93). In addition to the common mesophilic forms, thermophilic and hyperthermophilic forms have also been isolated. Although sulfate reduction has been shown to occur in the oxic layer of microbial mats and although SRB can survive brief exposure to air, they have so far been cultivated only under anoxic, reducing conditions. The physical removal of O_2 from a medium, by boiling or sparging with N_2, is often sufficient to permit growth of robust strains since sulfide may be carried over in the inoculum. However, the use of a prereduced, "poised" medium is usually necessary, particularly for fastidious strains or when SRB are present in low numbers as would be obtained from a dilution series. The greatest variety of SRB has been isolated by using a bicarbonate-buffered medium that contains sulfate, an electron donor, and sulfide as the reductant.

Most-probable-number (MPN) enumeration of SRB has been reportedly improved through the use of natural sediment or sludge media combined with the inclusion of $^{35}SO_4^{2-}$ as a tracer (126). Background on culturing anaerobic bacteria is provided in chapter 6. Many practical details of cultivating SRB have been described elsewhere (93, 137, 138).

Composition and Preparation of Media

The general-purpose media described below can be used to cultivate most types of mesophilic or moderately thermophilic SRB, although no one particular formulation is suited to all types. As with all anaerobes, cultures are incubated in the absence of air by filling the containers as completely as possible or by using a headspace filled with anoxic gas. Use septa or stoppers made from butyl rubber. Incubation temperatures are typically 30°C for mesophilic strains and 55°C for thermophilic strains.

Postgate's Media

Postgate's media (93) are useful for routine cultivation of *Desulfovibrio* and *Desulfotomaculum* spp. and are relatively simple to prepare. They may be solidified with 10 g of agar per liter. N_2 may be used as the headspace gas.

Postgate's medium B

KH_2PO_4	0.5 g
NH_4Cl	1.0 g
$CaSO_4$	1.0 g
$MgSO_4 \cdot 7H_2O$	2.0 g
Sodium lactate	3.5 g
Yeast extract	1.0 g
Ascorbic acid	0.1 g
Thioglycolic acid	0.1 g
$FeSO_4$	0.5 g
Tap water	1.0 liter

Adjust pH between 7.0 and 7.5 and autoclave under N_2. Add NaCl to 2.5% for marine strains or use seawater. This medium forms a precipitate that is useful for long-term maintenance of cultures, since it is thought to provide a microhabitat. The ingredients may be prepared as stock solutions. However, the reductants thioglycolate and ascorbate must be freshly added, as they readily deteriorate in air. The completed medium should be used as soon as possible; deterioration of the reductants is indicated by the transient appearance of purple. The medium diagnostically blackens upon growth of SRB.

Postgate's medium C

KH_2PO_4	0.5 g
NH_4Cl	1.0 g
$NaSO_4$	4.5 g
$CaCl_2 \cdot 6H_2O$	0.06 g
$MgSO_4 \cdot 7H_2O$	0.06 g
Sodium lactate	6.0 g
Yeast extract	1.0 g
$FeSO_4$	0.004 g
Sodium citrate · $2H_2O$	0.3 g
Distilled water	1.0 liter

Adjust pH to ~7.2 and autoclave under N_2. For marine strains, prepare with NaCl at 2.5%. The medium may be cloudy after autoclaving. However, this is generally a clear medium, the iron being chelated by the citrate, and is useful for large-scale cultures of *Desulfovibrio* or *Desulfotomaculum*.

Widdel and Pfennig Medium

Widdel and Pfennig medium, a defined medium described in detail by Widdel and Hansen (137), was used to isolate most of the newly identified genera of SRB. All SRB cultivated to date can be grown on a variation of this medium. The basal salt solution is adjusted with respect to NaCl, $MgCl_2 \cdot 6H_2O$, or $CaCl_2$ for preparation of a freshwater, brackish, or marine medium, respectively. Electron donors are selected according to the physiology of the strain to be grown or the type to be enumerated. The medium is assembled under 80% N_2–20% CO_2 from separately prepared solutions added to the basal salt solution in the order given below. To make a solid medium, it is recommended that the agar be washed three times in distilled water to remove inhibitors and growth substrates (137), added to the basal salts (10-g liter^{-1} [final concentration]), and then autoclaved. The sterile solution must be kept at ~50°C to keep the agar molten while the other additions are made.

Solution A: basal salts

	Freshwater	*Brackish*	*Marine*
Na_2SO_4	4.0 g	4.0 g	4.0 g
H_2PO_4	0.2 g	0.2 g	0.2 g
NH_4Cl	0.25 g	0.25 g	0.25 g
NaCl	1.0 g	7.0 g	20.0 g
$MgCl_2 \cdot 6H_2O$	0.4 g	1.2 g	3.0 g
KCl	0.5 g	0.5 g	0.5 g
$CaCl_2 \cdot 2H_2O$	0.1 g	0.1 g	0.15 g
Resazurin (0.1%)	1.0 ml	1.0 ml	1.0 ml
Distilled water to:	950 ml	950 ml	950 ml

The volume of water used is adjusted for subsequent additions; the amounts of salts given are for 1 liter of the completed medium. The salts are added to stirred water to prevent the formation of precipitates. The solution is prepared in a stoppered vessel for making the subsequent additions and dispensing the completed medium. After autoclaving, the solution is cooled under N_2-CO_2 and the additions are made under the anaerobic gas with mixing on a magnetic stir plate. The final pH of the medium should be between 7.0 and 7.3.

Solution B: trace elements (137)

HCl (66%, 8 M)	12.5 ml
$FeSO_4 \cdot 7H_2O$	2,100 mg
H_3BO_3	30 mg
$MnCl_2 \cdot 4H_2O$	100 mg
$CoCl_2 \cdot 6H_2O$	190 mg
$NiCl_2 \cdot 6H_2O$	24 mg
$CuCl_2 \cdot 2H_2O$	2 mg
$ZnO_4 \cdot 7H_2O$	144 mg
$Na_2MoO_4 \cdot 2H_2O$	36 mg
Distilled water	987 ml

The solution is autoclaved under N_2. Add 1.0 ml liter of medium^{-1}.

In addition, some strains require additional trace elements (prepared in 100 ml of distilled water): NaOH, 40 mg; $Na_2SeO_3 \cdot 5H_2O$, 0.6 mg; and $Na_2WO_4 \cdot 2H_2O$, 0.8 mg. The solution is autoclaved under nitrogen. Add 1.0 ml liter of medium^{-1}.

Solution C: bicarbonate (1.0 M)

$NaHCO_3$	8.4 g
Distilled water	100 ml

The solution is saturated with CO_2 by sparging or by shaking in a tightly closed bottle with one-third of the capacity left as headspace that is replenished several times with CO_2. The solution may be filter sterilized or autoclaved in bottles with secured seals under CO_2. Add 30 ml liter of medium^{-1}.

Solution D: organic substrates

These are prepared separately as neutralized solutions (usually 1.0 to 2.0 M) and autoclaved under N_2. One is selected for use depending upon the physiology of the strain to be grown or the type to be enriched. The range of substrate concentrations used is given below.

Acetate	10–20 mM
Benzoate	2–5 mM
Butyrate	5–12 mM
Caproate	2–5 mM
Ethanol	10–20 mM
Formate	10–20 mM
Lactate	10–40 mM
Propionate	10–20 mM
Succinate	10–20 mM
Pyruvate (filter sterilized)	10–40 mM

Add 5 to 15 ml of the concentrated stock solution liter of medium^{-1}.

Solution E: vitamins

These are prepared separately for stability, filter sterilized, and kept in the dark.

Mixed vitamins

4-aminobenzoic acid	4 mg
D-(+)-biotin	1 mg
Nicotinic acid	10 mg
Calcium D-(+)-pantothenate	5 mg
Pyridoxine dihydrochloride	15 mg
Sodium phosphate buffer (10 mM, pH 7.1)	100 ml

Add 1.0 ml liter of medium^{-1}.

Thiamine

Thiamine chloride	10 mg
Sodium phosphate buffer (25 mM, pH 3.4)	100 ml

Add 1.0 ml liter of medium^{-1}.

Vitamin B_{12}

Cyanocobalamin	5 mg
Distilled H_2O	100 ml

Add 1.0 ml liter of medium^{-1}.

Solution F: sulfide

$Na_2S \cdot 9H_2O$	4.8 g
Distilled water	100 ml

Sodium sulfide crystals deteriorate in air. Clean crystals are selected or rinsed with distilled water if needed. The water is sparged with and kept under N_2 while the crystals are being dissolved. The solution is autoclaved under N_2 and added at 7.5 ml liter of medium^{-1}.

The more fastidious species of SRB are stimulated by the addition of sodium dithionite, which further reduces the medium. This reductant is added after the medium has already been reduced with sulfide. Aqueous sodium dithionite rapidly breaks down with exposure to O_2 and must be prepared with O_2-free water under N_2. A small tube with a side arm to introduce a sterile stream of N_2 is recommended for preparation and storage of dithionite, which remains stable for about 5 days if not exposed to air. Since dithionite crystals are usually sterile due to their toxicity, an estimated solid amount may be added directly from the reagent bottle to the medium by using a sterile spatula fashioned from a platinum wire.

Dithionite solution

O_2-free distilled water (autoclaved under nitrogen)	10 ml
$Na_2S_2O_4$ (weighed aseptically)	0.2 g

Add 1.0 ml liter of medium^{-1}.

Inoculation, Transfer, and Storage

Use a 1 to 10% (vol/vol) inoculum (the higher concentration is used for slower-growing strains). Inoculations can be made with a syringe needle through septum seals or with a pipette for stoppered cultures. When one is using a pipette, the tip is placed beneath the surface of the medium to limit exposure of the inoculum to air and the headspace is flushed with the anaerobic gas as the stopper is replaced. Transfers are performed on a 6- to 12-week basis; a stringiness becomes apparent due to lysis of cells in aged cultures. For storage, cultures are removed from the incubator prior to attaining stationary phase and stored at 4 to 6°C in the dark. Care should be taken as some strains may expire if refrigerated. Postgate's medium B is suited to the storage of *Desulfovibrio* and *Desulfotomaculum* species. For long-term maintenance, cultures containing 5 to 10% dimethyl sulfoxide (DMSO) may be kept under liquid nitrogen.

Enumeration and Isolation, MPN, and Solid Media

Dilution of samples into liquid media can be used to enumerate SRB by the MPN technique (67). Containers should be refrigerated until the samples are processed, and samples are diluted in poised media and shaken to dislodge and separate cells, since many SRB grow in clumps. Growth of SRB is confirmed by the blackening of media that contain an iron indicator, such as Postgate's medium B, or by growth in Widdel and Pfennig medium. With the latter, growth is indicated by the appearance of turbidity, which can be measured by increase in absorbance at 600 nm against an uninoculated blank and should be confirmed by chemically testing for SO_4^{2-} reduction or sulfide production by one of the assays given below. Transfer of an inoculum from a culture obtained with the highest dilution of the series to a medium solidified with agar can be used to obtain an isolate in pure culture.

Enumerations can also be performed by counting the formation of colonies in solid media. The sample is serially diluted into tubes held at 40°C that contain the liquefied medium. A pipette is used to make a transfer and briefly distribute the inoculum, the tip being kept beneath the surface of the medium. The agar can be solidified as a layer over the inside of the tube (roll tubes) or as a plug. Tube contents are incubated in an inverted position so that condensation collects away from the forming colonies and can easily be removed. As described above, colonies of SRB blacken in the presence of the iron indicator. In media without an indicator, the colonies may contain a slight yellowish, reddish, or grayish pigment.

Pure cultures can be obtained from well-separated colonies picked from the agar with a drawn Pasteur pipette. The colony mass is suspended in a few milliliters of anoxic saline or medium and is used to inoculate a small volume (~10 ml) of medium, some being reserved for microscopic inspection.

Tests for Purity

Microscopically observe the cell suspension for uniform morphology. It may be necessary to repeat selection of a colony. Postgate (93) offers the following additional tests. Contamination of a culture by an aerobe can readily be determined by plating onto nutrient agar containing glucose and peptone; no colonies should appear after the plate contents are incubated in air at 30°C. To test for contaminant anaerobes, prepare 25 ml of molten peptone-glucose agar (pH 7.0 to 7.6) and add sterile $Fe(NH_4)_2(SO_4)_2$ to 0.05%. Cool the medium to 40°C, dispense into tubes, and make a dilution series from the test culture. Allow the agar to solidify as a plug and incubate under N_2. Black colonies should appear; colonies that are not black are contaminants.

Molecular Techniques for Density and Diversity Measurement

It is now well understood that only a very small percentage of the bacteria observed in an environmental sample can be cultivated. In recent years, molecular methods based on PCR, gene cloning, and hybridization probes have been developed to examine natural populations of bacteria directly without cultivation. SRB have historically been difficult to isolate and identify with culture techniques. Most genera have been described only fairly recently. Yet it has been determined that culturable numbers of SRB, based on MPN methods that use the defined medium of Widdel and Pfennig, could account for measured rates of sulfate reduction (55, 128). Nonetheless, strains continue to be isolated that are morphologically, physiologically, and phylogenetically unique (35, 40, 55, 106, 107, 113). Similarly, 16S rRNA genes are frequently cloned from environmental samples, indicating that many lineages of SRB have not yet been cultured (24, 105). In addition to rRNA sequences, use of the bisulfite reductase gene (18, 57, 78, 135) and of hydrogenase gene sequences (136) has provided phylogenetic information on natural communities of SRB.

rRNA-Based Methods

rRNA-based methods are based largely on oligonucleotide probes designed to hybridize to 16S rRNA sequences and take advantage of the agreement between the phylogeny and physiology of the gram-negative mesophilic SRB (22, 23, 25). Details on the utilization of rRNA probes are given elsewhere in this manual (see chapter 12). Oligonucleotide probes may be used to measure the rRNA abundance of a targeted group by hybridization to nucleic acids extracted from a natural microbial community and immobilized on a membrane support. The probes may also be used to directly observe and enumerate targeted cells in a sample by whole-cell hybridization (2, 3) and serve as PCR primers for the amplification of 16S rRNA of a targeted group from an environmental sample (4, 24, 81, 82, 105, 119). Detection of SRB in environmental samples often requires nested PCR assays of hierarchical phylogenetic specificity (19, 20).

rRNA-targeted oligonucleotide hybridization probes have been developed and described for many of the gram-negative SRB (2, 24, 35, 94). These probes have been used to examine communities and in situ distribution of SRB in various habitats, including sediments (24, 107), activated sludge (76, 108, 110), biofilms (89, 96, 97, 109), microbial mats (35, 77, 104, 118), the water column (95), and the rhizosphere of aquatic plants (44–46, 70, 105). Hybridization with oligonucleotide probes is an important tool in locating SRB populations in biogeochemical microscale gradients; understanding the distribution of SRB in microbial mat and biofilm studies in particular has been greatly aided through the use of microelectrodes to measure gradients of O_2 and H_2S (96, 108).

The current repertoire of rRNA-targeted oligonucleotides should be regarded as work in progress. Recently cultured, phylogenetically distinct SRB genera and species and novel environmental phylotypes require the design of new, specific probes for their detection and quantification. In this way, the repertoire of rRNA probes for SRB is constantly updated (35, 72, 98, 99, 105, 107).

The large 16S rRNA gene sequence database for sulfate-reducing prokaryotes has allowed the development of oligonucleotide microarrays for sulfate-reducing prokaryotes. In this technique, 16S rRNA-targeted oligonucleotide probes (18-mers) with hierarchical and parallel phylogenetic specificity are linked onto a small-scale grid surface (SRP-PhyloChip); simultaneous hybridization of the probes on the SRP-PhyloChip with environmental nucleic acid extracts allows the simultaneous detection of widely diverse sulfate-reducing prokaryotic lineages in environmental samples (74, 75).

DNA-DNA Hybridizations

A technique termed reverse sample genome probing (RSGP) has been developed and used to study SRB in oil field production waters, soil, and mining wastes (115, 134). In RSGP, genomic DNAs from pure cultures are applied to membranes and serve as standards against which radiolabeled DNA obtained from a microbial community, either by direct extraction or selective enrichment, is hybridized. The pure cultures may be obtained from collections or isolated from the environment under investigation. The genomes used as standards are chosen on the basis of having no or limited cross-hybridization. RSGP was shown to generate reproducible fingerprints of microbial communities of SRB. Quantitation of hybridization signals indicated enrichment of SRB in biofilms on metal surfaces that had contact with production water. A DNA-DNA hybridization method using rRNA-targeted oligonucleotide probes to estimate SRB cell numbers has also been described (29).

FISH

Fluorescence in situ hybridization (FISH) is shorthand for 16S rRNA-targeted hybridization of whole cells with fluorescently labeled oligonucleotides that are complementary to phylogenetically specific sequence motifs of the 16S rRNA molecules within a cell and for subsequent viewing of hybridized cells with epifluorescence microscopy (91). FISH is widely used for detection, identification, and quantification of microorganisms in the environment; the analysis of sulfur-oxidizing bacteria gives particularly interesting examples for FISH applications. Many morphologically conspicuous genera of sulfur-oxidizing bacteria have so far resisted cultivation (*Thioploca*, *Thiomargarita*, *Thiovulum*, and *Achromatium*) or include only relatively few cultured species (*Beggiatoa*); consequently, cells and samples for physiological and molecular studies (including basic 16S rRNA gene sequencing) have to be collected from environmental populations. FISH with genus- or strain-specific 16S rRNA-targeted probes is essential for ground-truthing 16S rRNA gene sequences from environmental material and for ruling out contamination; by the same token, FISH is essential for assessing the natural diversity of 16S rRNA phylotypes within genera of uncultured sulfur oxidizers. For example, FISH has defined the phylogenetic position and diversity of the genera *Thioploca*, *Beggiatoa*, and *Thiomargarita* (1, 68, 81, 111, 116, 117), of the recently identified veil-forming candidate genus and species "*Thioturbo danicus*" (84), and of the genus *Achromatium* (37). The genus *Achromatium* provides perhaps the best present-day example for using FISH techniques to investigate the ecophysiology of an uncultured sulfur bacterium. FISH hybridization combined with microautoradiography allowed researchers to identify carbon assimilation patterns of specific *Achromatium* phylotypes (38); further, FISH hybridization tracked specific *Achromatium* phylotypes under different redox regimes and enabled identification of their ecophysiological preferences (39).

The cooccurrence and close association of symbiotic sulfur-oxidizing bacteria and SRB within eukaryotic hosts (oligochaete worms) are one of the most unusual results of FISH analysis. In brief, the sulfate-reducing endosymbionts supply sulfide that is being oxidized by autotrophic, sulfur-oxidizing endosymbionts; in turn, the biomass of the sulfur oxidizers is the source of nutrients and nourishment for the gutless host animal (10, 28). The cooccurrence of sulfur-oxidizing bacteria and SRB in natural microbial associations, as detected with FISH techniques, is probably more common than expected (35).

Functional Genes of Dissimilatory Sulfate Reduction

Diversity analyses of SRB are increasingly based on phylogenetically conserved, functional key genes of dissimilatory sulfate reduction, mostly the *dsrAB* genes coding for the alpha and beta subunits of dissimilatory sulfite reductase (66, 135, 141) and the *apsA* gene coding for the alpha subunit of adenosine-5′-phosphosulfate reductase (34, 47). The enzyme adenosine-5′-phosphosulfate reductase catalyzes the reduction of sulfate (covalently bound as activated adenosine-5′-phosphosulfate) to sulfite and is specific for SRB; dissimilatory sulfite reductase catalyzes the next step in the pathway, the six-electron transfer from sulfite to sulfide. Therefore, *dsrAB* should be regarded as a marker gene for dissimilatory sulfate and sulfite reduction. Dissimilatory and assimilatory sulfite reductases are mutually related members of a diversified gene family of sulfite reductases (27) but are too divergent to be targeted by a single inclusive PCR assay. Phylogenies of SRB based on 16S rRNA genes and *dsrAB* genes show at least five deeply branching lineages of cultured sulfate-reducing prokaryotes, the mesophilic delta-proteobacterial sulfate reducers, the nonthermophilic gram-positive SRB, the thermophilic bacterial genera *Thermodesulfovibrio* and *Thermodesulfobacterium*, the hyperthermophilic archaeon *Archaeoglobus* (66, 135, 141), and the thermophilic bacterium *Thermodesulfobium* (79). Interestingly, the sulfate-reducing archaeon *Archaeoglobus* obtained its *dsrAB* gene from a deeply branching bacterial lineage and is unlikely to represent an ancestral archaeal *dsrAB* prototype or a valid outgroup for bacterial *dsrAB* genes (66). Discordant 16S rRNA and *dsrAB* gene phylogenies are found for the thermophilic bacterial genus *Thermodesulfobacterium* and several thermophilic gram-positive sulfate-reducing bacterial species. These discordances either reflect differential loss of ancestral paralogs or lateral transfer of delta-proteobacterial *dsrAB* genes to nonproteobacterial lineages (66, 141).

Cloning and sequencing surveys of PCR-amplified *dsrAB* genes have uncovered a great diversity of presently uncultured phylogenetic lineages of SRB. Novel lineages are not limited to extreme environments such as hydrothermal vents and hot springs (26, 31, 75, 85) but occur also in mesophilic marine and freshwater habitats and soil (6, 75). As a basis for these studies, primer combinations that amplify large, 1.9-kb fragments of the *dsrA* and *dsrB* genes should be preferred (66, 135). Shorter PCR fragments based on internal *dsrAB* primers produce short sequences that complicate systematic comparisons of different *dsrAB* phylotypes and require multiple phylogenies to accommodate *dsrAB* fragments of different lengths (6).

While *dsrAB* gene studies commonly analyze the environmental diversity of sulfate-reducing prokaryotes in clone library surveys, other molecular approaches can be based on *dsrAB* genes as well. For example, *dsrAB* gene expression can now be monitored by using a quantitative PCR assay based on the TaqMan approach (87).

SULFUR OXIDATION

Sulfide is produced from degradation of sulfur-containing organic matter and by dissimilatory sulfate reduction. Microbial metabolism of sulfide competes with chemical oxidation, either by O_2 or Fe^{2+}. During biological and chemical conversion of sulfide (oxidation state, −2) to SO_4^{2-} (+6), a variety of intermediates (−1 to +5) are formed. Some of these intermediates, such as polysulfides, react chemically with organic matter; during those reactions organic S compounds can be formed. Metabolic pathways discussed here include those that are of particular environmental importance. Under oxic conditions, chemolithotrophic sulfide oxidation by colorless S bacteria (thiobacilli and filamentous sulfide oxidizers) and some purple S bacteria and SRB takes place. In the absence of O_2, nitrate-respiring chemolithotrophs engage in sulfide oxidation, and both oxygenic and anoxygenic phototrophic bacteria use sulfide as an electron donor for photosynthesis. Intermediates that are produced during sulfide oxidation can either be oxidized or sometimes reduced by SRB or, alternatively, fermented (disproportionated). Various reviews provide detailed information on oxidative processes in the S cycle (51, 52, 58, 121).

During chemolithotrophic metabolism, sulfide is respired with O_2 or nitrate. Since chemical oxidation competes with microbial oxidation, thiosulfate ($S_2O_3^{2-}$) is often used as a sulfide analog because the former does not chemically oxidize spontaneously (52). Chemolithotrophic sulfide oxidation plays an important role in hydrothermal vent environments, in certain symbiotic relationships, and, most important, in (marine) sediments and water columns of stratified lakes.

Rate Measurements

Rate measurements can be determined by O_2, nitrate, and/or sulfide (or $S_2O_3^{2-}$) consumption. Problems introduced by chemical reactions of sulfide and O_2 or metals such as iron require one to exercise caution. Killed controls or the use of inhibitors is necessary to account for abiotic loss factors. When adding sulfide or $S_2O_3^{2-}$ it is important to employ concentrations as close to those in situ as possible. Sulfide can be especially toxic to bacteria at elevated concentrations, so its use in excess may cause consumption rates that deviate from those in situ.

For water samples, 5 to 50 ml is placed in a chamber or reaction vessel. The endogenous O_2 consumption is determined while the sample is stirred. A small volume (10 to 100 μl) of a concentrated sulfide or $S_2O_3^{2-}$ stock solution is added to the reaction vessel, and the disappearance of O_2 is monitored over time. Alternatively, the disappearance of sulfide or $S_2O_3^{2-}$ is monitored by measuring the concentration in subsamples. Autoclaved or filter-sterilized controls are treated identically to determine abiotic loss factors. Preparation of stock solutions of sulfide and $S_2O_3^{2-}$ (50 to 500 mM) is described below. The analytical techniques for measuring these compounds and nitrate and O_2 are described below.

Intact core measurements are the preferred method for measuring S transformations (49, 50). A major problem with sediments is abiotic reactions, such as iron and manganese sulfide mineral formation and oxidation of reduced S by O_2. The substrate (preferably $S_2O_3^{2-}$ to minimize abiotic consumption) is introduced with a glass syringe by axial microinjection. Concentrated stock solutions are added in 5- to 25-μl amounts to avoid changing pore water content. Multiple cores are incubated simultaneously, and individual ones are sacrificed at each time point. Upon sampling, the cores are sectioned and the pore water concentration of the substrate is measured as described below. To prevent use of $S_2O_3^{2-}$ by SRB, molybdate (5 to 10 mM [final concentration of Na_2MO_4]) can be injected simultaneously as a "specific" inhibitor (see below). Since this method is elaborate and makes the use of replicates impractical, it may be necessary to repeat the experiment twice or three times.

Measurements can also be made by using sediments that are slurried (e.g., 1 part sediment plus 1 to 3 parts sterilized water or artificial medium of similar ionic strength) and placed in a vessel (100 to 500 ml) or serum bottle (20 to 100 ml), where the consumption of sulfide, $S_2O_3^{2-}$, O_2, or nitrate is monitored in aliquots collected over time. The slurry should contain a gas phase (N_2 for anoxic conditions) and be stirred or shaken, and temperature should be controlled. In slurry experiments, sulfide and O_2 can be measured directly with electrodes placed in the reaction vessel.

Consumption of sulfide or $S_2O_3^{2-}$ can be calculated from the initial slope of the concentration over time (δ[sulfide]/δt or $\delta[S_2O_3^{2-}]/\delta t$). For water samples this calculation can be normalized easily to volume; for sediments the consumption rate is either expressed in amount (micromoles or millimoles) per volume (e.g., cubic centimeter of sediment or milliliter of pore water) or per mass (gram of sediment) per unit of time (minute, hour, or day). Another approach is to calculate rates of sulfide or $S_2O_3^{2-}$ consumption by assuming a steady state, in which the SO_4^{2-} reduction (i.e., sulfide production) rate equals the sulfide consumption rate or in which the $S_2O_3^{2-}$ oxidation rate equals $1/2 \times SO_4^{2-}$ reduction rate $\times$ 0.6 (which assumes that 60% of the sulfide proceeds through $S_2O_3^{2-}$ [120]).

Radiotracer Studies and Isotope Exchange

The use of [^{35}S]sulfide for measuring S oxidation creates difficulties because, in addition to chemical reaction with O_2 and Fe^{2+}, rapid isotope exchange occurs among reduced-S species (i.e., $H^{35}S^-$, $^{35}S^0$, and $Fe^{35}S$), presumably via polysulfide ($^{35}Sn^{2-}$) (33). However, [^{35}S]sulfide synthesized from $^{35}S^0$ has been applied successfully for short-term sediment incubations (55). In general, the use of $^{35}S_2O_3^{2-}$ is preferable (53). Incubation times must be short (<1 h) to prevent metabolism of the S compounds formed. Radionuclides are applied as described above for whole-core incubations, and reactions are terminated by slicing and then homogenizing sections of cores in ice-cold 1% zinc acetate. Pore water is retrieved by microcentrifugation (2 min at 8,000 $\times$ g at 4°C) and filtration through a 0.2-μm-pore-size nylon filter (Alltech, Deerfield, Ill.), and after dilution in an $S_2O_3^{2-}$ carrier solution (0.1 mM), $S_2O_3^{2-}$, polythionates, and SO_4^{2-} are separated by ion chromatography (129). The eluted fractions containing these S compounds are collected in scintillation vials, and activity is determined by using scintillation cocktails designed for aqueous samples (e.g., Ecolume or Ecolite; ICN Biomedicals, Inc.).

Cultivation, Enrichment, and Isolation

Cultivation and enumeration of sulfur oxidizers are most successful in CO_3^{2-}-buffered media (128). *Thiomicrospira* spp. from marine environments are especially sensitive to elevated phosphate concentrations. The success of isolation procedures can be enhanced by mixing mineral media with filtered (pore size, 0.45 μm) and autoclaved water from which the inoculum is taken (1:1 [vol/vol]). Gradient cultures (two-layered agar systems, in which the bottom compartment contains sulfide and the upper compartment consists of mineral agar in which O_2 can freely diffuse) are useful for the isolation and cultivation of filamentous S bacteria (*Beggiatoa* [86]). Cultivation in chemostats is also a powerful tool to determine sulfide oxidation rates (123, 127).

Medium for S Oxidizers

Mineral medium base

NH_4Cl	0.2 g
$CaCl_2 \cdot 2H_2O$	0.225 g
KCl	0.2 g
$MgCl_2 \cdot 6H_2O$	0.2 g
KH_2PO_4	0.02 g

Add 1.0 ml of Widdel and Pfennig's trace element solution (listed above for SRB). For *Thiobacillus* and *Thiomicrospira* spp., add $FeSO_4 \cdot 7H_2O$ (0.1 to 1 mg), and for marine isolates add NaCl (25 g). Dissolve salts for 1.0 liter in 800 ml and autoclave. Prepare and autoclave each of the following separately (grams liter^{-1}): Na_2CO_3, 20; $Na_2S_2O_3$, 248; $Na_2S \cdot 7\text{-}9H_2O$, dissolve 24 g liter^{-1} in boiling 2% Na_2CO_3 solution, close cap, and autoclave; CH_3COONa, 8.1; 1 M HCl; and 1 M NaOH; if desired, add yeast extract to final medium (10). Add Na_2CO_3 stock to medium base (1:9, vol/vol) and either $Na_2S_2O_3$ (1:9, vol/vol) or Na_2S (1:99, vol/vol); if required, add CH_3COONa (1:9, vol/vol) and yeast extract (1:99, vol/vol). Adjust final volume to 1.0 liter. Adjust pH to 7.5 to 8.0 with sterile HCl or NaOH. For aerobes, transfer complete medium into an Erlenmeyer flask with a cotton plug (do not shake, because this will cause the pH to change since the medium is CO_3^{2-} buffered). For anaerobes, use bottles or tubes sealed with butyl rubber stoppers or septa. Culturing of phototrophs requires incandescent light (a 15- or 40-W light bulb placed 10 to 20 cm from the culture).

Inocula and Maintenance

Inocula that have a high success rate can be obtained from a variety of environments, including marine sediments, estuaries, stratified lakes, and sulfur-containing (thermal) springs. Storage time of the inoculum should be kept to a minimum; when sediments are used, allow air exchanges for aerobes, restrict exposure to O_2 for microaerophiles (denitrifying *Thiomicrospira* spp.), and keep O_2 free for anaerobes (purple and green sulfur bacteria). Isolation can be direct (sample is transferred to a bottle or Erlenmeyer flask) or after serial 10-fold dilutions. Alternatively, a reaction vessel through which a continuous low flow (dilution rate of 0.005 to 0.02 h^{-1}) of medium is fed (simple chemostat) yields good results for slow-growing species such as certain (denitrifying) thiobacilli and *Thiomicrospira* spp. Phototrophs are isolated after dilution in liquefied agar (see above), and single colonies are picked and transferred to fresh (liquid) medium (92).

Enumeration

S oxidizers are enumerated by using MPN techniques (67). A water sample of known size is diluted by 10-fold several times to obtain a series ranging from 10^{-3} to 10^{-9} that of the original concentration. Sediment samples are diluted 100 times and sonicated for approximately 30 to 60 s (40 W) prior to the same treatment. Each dilution step is used as an inoculum (1:9, vol/vol) in fresh medium to which a pH indicator is added (e.g., bromocresol purple; 0.4 g $liter^{-1}$). Acetate, or any desired organic substrate, can be added in addition to CO_3^{2-} to include heterotrophic $S_2O_3^{2-}$ utilizers. Since a wide variety of heterotrophs use $S_2O_3^{2-}$ but produce polythionates, it is important to include a pH indicator. Three or five tube series can be used. Positive tubes are scored and transferred to fresh medium to confirm growth. Positives from the greatest dilutions are used to isolate organisms that are likely to be important in the environment. Epifluorescence microscopy (125) can be used to estimate cell counts in the inoculum. Molecular approaches described above for SRB can similarly be used to study S oxidizers. For example, oligonucleotides specific for *Thiomicrospira* spp. have been used as PCR primers and probes for investigating the distribution and diversity of this group in coastal sediments (11); denaturing gradient gel electrophoresis analysis (82) has detected *Thiomicrospira* spp. at hydrothermal vents (83).

Phototrophic Processes

Sulfide and $S_2O_3^{2-}$ are electron donors for purple S and green S bacteria (51, 133). Certain cyanobacteria can also use these electron donors under special environmental regimens and then function as sulfur oxidizers (53). In addition to sulfide oxidation experiments discussed above, chlorophyll (Chl), bacteriochlorophylls (BChl), and other pigments can be used as biomass indicators, and light of specific wavelengths can be used to selectively stimulate or inhibit certain species (101, 114). Cyanobacteria have additional pigments, phycobiliproteins, which have peak absorbance at 550 to 650 nm. Accessory pigments (carotenoids) of anoxygenic phototrophs typically absorb at wavelengths between 400 and 550 nm. Purple non-S bacteria contain spirilloxanthin, okenone, or lycopene, while green non-S bacteria contain carotene, chlorobactene, or isonieratene. By using certain light sources, either group of bacteria (oxygenic versus anoxygenic) can be selected in enrichment cultures. Photopig-ments can be measured spectrophotometrically after extraction in methanol with or without hexane (114). In methanol, absorption maxima (nanometers) and coefficients ($gram^{-1}$ · liters · centimeters) of Chl*a*, BChl*a*, BChl*b*, BChlc, BChl*d*, BChl*e*, and BChlg are 665 and 74.5; 771 and 84.1; 794 (unstable in air); 668 and 86.0; 654 and 82.3; 646 and 82.3; and 765 (coefficient in methanol unknown), respectively. Different BChls can be further differentiated upon acidification, yielding bacteriopheophytin with different absorption maxima (114). High-performance chromatography with fluorescence detection enables photopigment quantification as well (139). Photopigment concentrations can be used to calculate protein, which may be useful for specific rate measurements in situ. Although the specific photopigment concentration typically increases with de-creasing light intensity, a value of 22 to 30 mg of protein μg of $BChl^{-1}$ is typically found (128).

Organic Sulfur Oxidation

Dimethyl sulfide (DMS), DMSO, and methane thiol (CH_3SH) are consumed by a variety of thiobacilli, anoxygenic phototrophs, saltmarsh ascomycetes, and SRB (5, 65, 129–132). Metabolism can be assessed as described above for other rate measurements. Stock solutions of oxidizable substrates should be prepared fresh. DMS is prepared gravimetrically. CH_3SH is prepared from dimethyl disulfide by cleaving with tributylphosphine (60). The concentration of these organosulfur compounds can be determined by gas chromatography (GC) with flame ionization detection (FID) or flame photometric detection (FPD) (61). FPD is S specific and therefore more sensitive than FID, and measurements of volatile S compounds obtained by using GC-FPD are facilitated by doping the FPD fuel line with low levels of an S gas (i.e., carbonyl sulfide, sulfur dioxide, or carbon disulfide [80]). Purge-and-trap systems can further lower detection limits (64). Care should be taken with CH_3SH, which rapidly binds to sediments, glass, and rubber (63), and appropriate controls are necessary to account for abiotic losses. DMS can be oxidized photochemically and biologically to DMSO, which can be measured as DMS after chemical reduction with a stabilized titanium solution (62). When performing experiments with organosulfur compounds, use Teflon-lined stoppers and heat glassware overnight (300 to 450°C) prior to use.

INHIBITORS

Molybdate (MoO_4^{2-}) has been used as a "specific" competitive inhibitor of SO_4^{2-} reduction, which is useful for determining the direct role of SRB in biogeochemical transformations (90). In marine samples, MoO_4^{2-} concentrations of 20 to 28 mM are applied, although a much lower (5 mM) concentration has almost the same effect (90). For freshwater samples, less MoO_4^{2-} is required (<2 mM). In addition to SRB, MoO_4^{2-} inhibits S metabolism by thiobacilli and phototrophs by 15 and 30%, respectively (128). All of these organisms contain ADP and/or ATP sulfurylases, which may be affected by MoO_4^{2-}. Other group VI oxyanions, such as WO_4^{2-} and SeO_4^{2-}, have also been used to inhibit SO_4^{2-} reduction but are not quite as thorough. However, these latter compounds may prove useful for some applications since they interact less with free thiols than does MoO_4^{2-} (60). Anthraquinone derivatives like 1,8-dihydroxyanthraquinone (DHAQ) can also be used to selectively inhibit SO_4^{2-} reduction (17).

ANALYSES

Dissolved sulfide is measured colorimetrically (16) or with ion-specific electrodes (103). Both techniques have a detection limit of 1 to 10 μM. Alternatively, samples can be acidified in closed serum bottles, and the headspace can be assayed for H_2S by GC with FID or FPD. GC techniques have very low (nanomolar) detection limits and are not necessary for most applications. Microelectrodes can be used for measuring sulfide microprofiles (69). $S_2O_3^{2-}$ is determined colorimetrically after cyanolysis (59), which can also be used to measure polysulfides (127). The detection limit with this technique for both compounds is 10 μM. It is important to use pore water as a reagent blank to account for possible background color. Alternatively, anion-exchange chromatography and UV detection can be used to measure $S_2O_3^{2-}$ by using samples that are filtered through 0.2-μm-pore-size nylon filters (Alltech) (limit of detection, 1 to 5 μM [7]). High-performance liquid chromatography with fluorometric detection can also be used to measure oxidized and reduced S species (102). Nitrate is measured colorimetrically after reaction with sulfanilamide (1 to 5 μM, detection limit [9]) or in freshwater samples with an ion-specific electrode (30 μM = limit of detection [21]). Nitrate can also be measured by ion chromatography. O_2 is measured with an O_2 monitor consisting of a polarographic electrode (Clark type; Yellow Springs International, Yellow Springs, Ohio) attached to a strip chart recorder (8).

REFERENCES

1. **Ahmad, A, J. P. Barry, and D. C. Nelson.** 1999. Phylogenetic affinity of a wide, vacuolate, nitrate-accumulating *Beggiatoa* sp. from Monterey Canyon, California, with *Thioploca* spp. *Appl. Environ. Microbiol.* **65:**270–277.
2. **Amann, R. I., B. Binder, S. W. Chisholm, R. Olsen, R. Devereux, and D. A. Stahl.** 1990. Combination of 16S rRNA-targeted oligonucleotide probes with flow cytometry for analyzing mixed microbial populations. *Appl. Environ. Microbiol.* **56:**1919–1925.
3. **Amann, R. I., W. Ludwig, and K.-H. Schleifer.** 1995. Phylogenetic identification and in situ detection of individual cells without cultivation. *Microbiol. Rev.* **59:**143–169.
4. **Amann, R. I., J. Stromley, R. Devereux, R. Key, and D. A. Stahl.** 1992. Molecular and microscopic identification of sulfate-reducing bacteria in multispecies biofilms. *Appl. Environ. Microbiol.* **58:**614–623.
5. **Bacic, M. K., S. Y. Newell, and D. C. Yoch.** 1998. Release of dimethylsulfide from dimethylsulfoniopropionate by plant-associated salt marsh fungi. *Appl. Environ. Microbiol.* **64:**1484–1489.
6. **Bahr, M., B. C. Crump, V. Klepac-Ceraj, A. Teske, M. L. Sogin, and J. E. Hobbie.** 2005. Molecular characterization of sulfate-reducing bacteria in a New England salt marsh. *Environ. Microbiol.* **7:**1175–1185.
7. **Bak, F., A. Schuhmann, and K. H. Jansen.** 1993. Determination of tetrathionate and thiosulfate in natural samples and microbial cultures by a new, fast and sensitive ion chromatographic technique. *FEMS Microbiol. Ecol.* **12:**257–264.
8. **Beechey, R. B., and D. W. Ribbons.** 1972. Oxygen electrode measurements. *Methods Microbiol.* **6B:**25–53.
9. **Bendschneider, K., and R. J. Robinson.** 1957. A new spectrophotometric method for the determination of nitrite in sea water. *J. Mar. Res.* **11:**87–96.
10. **Blazejak, A., C. Erséus, R. Amann, and N. Dubilier.** 2005. Coexistence of bacterial sulfide oxidizers, sulfate reducers, and spirochetes in a gutless worm (Oligochaeta) from the Peru Margin. *Appl. Environ. Microbiol.* **71:**1553–1561.
11. **Brinkhoff, T., C. M. Santegoeds, K. Sahm, J. Küver, and G. Muyzer.** 1998. A polyphasic approach to study the diversity and vertical distribution of sulfur-oxidizing *Thiomicrospira* species in coastal sediments of the German Wadden Sea. *Appl. Environ. Microbiol.* **64:**4650–4657.
12. **Canfield, D. E., R. Raiswell, J. T. Westrich, C. M. Reaves, and R. A. Berner.** 1986. The use of chromium reduction in the analysis of reduced inorganic sulfur in sediments and shales. *Chem. Geol.* **54:**149–155.
13. **Canfield, D. E., B. Thamdrup, and J. W. Hansen.** 1993. The anaerobic degradation of organic matter in Danish coastal sediments: iron reduction, manganese reduction, and sulfate reduction. *Geochim. Cosmochim. Acta* **57:**3867–3883.
14. **Carignan, R., S. St. Pierre, and R. Gachter.** 1994. Use of diffusion samplers in oligotrophic lake sediments—effects of free oxygen in sampler material. *Limnol. Oceanogr.* **39:**468–474.
15. **Chambers, R. M., J. T. Hollibaugh, and S. M. Vink.** 1994. Sulfate reduction and sediment metabolism in Tomales Bay, California. *Biogeochemistry* **25:**1–18.
16. **Cline, J. D.** 1969. Spectrophotometric determination of hydrogen sulfide in natural waters. *Limnol. Oceanogr.* **14:**454–458.
17. **Cooling, F. B., III, C. L. Maloney, E. Nagel, J. Tabinowski, and J. M. Odom.** 1996. Inhibition of sulfate respiration by 1,8-dihydroxyanthraquinone and other anthraquinone derivatives. *Appl. Environ. Microbiol.* **62:**2999–3004.
18. **Cottrell, M. T., and S. C. Cary.** 1999. Diversity of dissimilatory bisulfite reductase genes of bacteria associated with the deep-sea hydrothermal vent polychaete annelid *Alvinella pompejana*. *Appl. Environ. Microbiol.* **65:**1127–1132.
19. **Daly, K., R. J. Sharp, and A. J. McCarthy.** 2000. Development of oligonucleotide probes and PCR primers for detecting phylogenetic subgroups of sulfate-reducing bacteria. *Microbiology* **146:**1693–1705.
20. **Dar, S. A., J. G. Kuenen, and G. Muyzer.** 2005. Nested PCR-denaturing gradient gel electrophoresis approach to determine the diversity of sulfate-reducing bacteria in complex microbial communities. *Appl. Environ. Microbiol.* **71:**2325–2330.
21. **de Beer, D., and J. P. R. A. Sweerts.** 1989. Measurement of nitrate gradients with an ion selective microelectrode. *Anal. Chim. Acta* **219:**351–356.
22. **Devereux, R., M. Delaney, F. Widdel, and D. A. Stahl.** 1989. Natural relationships among sulfate-reducing eubacteria. *J. Bacteriol.* **171:**6689–6695.
23. **Devereux, R., S.-H. He, C. L. Doyle, S. Orklnad, D. A. Stahl, J. LeGall, and W. B. Whitman.** 1990. Diversity and origin of *Desulfovibrio* species: phylogenetic definition of a family. *J. Bacteriol.* **172:**3609–3619.
24. **Devereux, R., and G. W. Mundfrom.** 1994. A phylogenetic tree of 16S rRNA sequences from sulfate-reducing bacteria in a sandy marine sediment. *Appl. Environ. Microbiol.* **60:**3437–3439.
25. **Devereux, R., and D. A. Stahl.** 1993. Phylogeny of sulfate-reducing bacteria and a perspective for analyzing their natural communities, p. 131–160. *In* J. M. Odom and R. Singleton, Jr. (ed.), *Sulfate-Reducing Bacteria: Contemporary Perspectives*. Springer-Verlag, New York, N.Y.
26. **Dhillon, A., A. Teske, J. Dillon, D. A. Stahl, and M. L. Sogin.** 2003. Molecular characterization of sulfate-reducing bacteria in the Guaymas Basin. *Appl. Environ. Microbiol.* **69:**2765–2772.
27. **Dhillon, A., S. Goswami, M. Riley, A. Teske, and M. L. Sogin.** 2005. Domain evolution and functional diversification of sulfite reductases. *Astrobiology* **5:**18–29.
28. **Dubilier, N., C. Mülders, T. Ferdelman, D. de Beer, A. Pernthaler, M. Klein, M. Wagner, C. Erséus, F. Thiermann, J. Krieger, O. Giere, and R. Amann.** 2001. Endosymbiotic sulphate-reducing and sulphide-oxidizing bacteria in an oligochaete worm. *Nature* **411:**298–302.
29. **Edgcomb, V. P., J. H. McDonald, R. Devereux, and D. W. Smith.** 1999. Estimation of bacterial cell numbers in humic acid-rich salt marsh sediments with probes directed to 16S ribosomal DNA. *Appl. Environ. Microbiol.* **65:**1516–1523.
30. **Elsgaard, L., and B. B. Jørgensen.** 1992. Anoxic transformations of radiolabeled hydrogen sulfide in marine and freshwater sediments. *Geochim. Cosmochim. Acta* **56:**2425–2435.

31. **Fishbain, S., J. G. Dillon, H. L. Gough, and D. A. Stahl.** 2003. Linkage of high rates of sulfate reduction in Yellowstone hot springs to unique sequence types in the dissimilatory sulfate respiration pathway. *Appl. Environ. Microbiol.* **69:**3663–3667.
32. **Fossing, H., and B. B. Jørgensen.** 1989. Measurement of bacterial sulfate reduction in sediments: evaluation of a single step chromium reduction. *Biogeochemistry* **8:**205–222.
33. **Fossing, H., S. Thode-Andersen, and B. B. Jørgensen.** 1992. Sulfur isotope exchange between S-35-labeled inorganic sulfur compounds in anoxic marine sediments. *Mar. Chem.* **38:**117–132.
34. **Friedrich, M.** 2002. Phylogenetic analysis reveals multiple lateral transfers of adenosine-5′-phosphosulfate reductase genes among sulfate-reducing microorganisms. *J. Bacteriol.* **184:**278–289.
35. **Fukui, M., A. Teske, B. Assmuss, G. Muyzer, and F. Widdel.** 1999. Physiology, phylogenetic relationships, and ecology of filamentous sulfate-reducing bacteria (Genus *Desulfonema*). *Arch. Microbiol.* **172:**193–203.
36. **Gibson, G. R.** 1990. Physiology and ecology of the sulphate-reducing bacteria. *J. Appl. Bacteriol.* **69:**769–797.
37. **Glöckner, F. O., H.-D. Babenzien, J. Wulf, and R. Amann.** 1999. Phylogeny and diversity of *Achromatium oxaliferum*. *Syst. Appl. Microbiol.* **22:**28–38.
38. **Gray, N. D., R. Howarth, R. W. Pickup, J. G. Jones, and I. M. Head.** 2000. Use of microautoradiography and fluorescence in situ hybridization to determine carbon metabolism in mixed natural communities of uncultured bacteria from the genus *Achromatium*. *Appl. Environ. Microbiol.* **66:**4518–4522.
39. **Gray, N. D., D. Comaskey, I. P. Miskin, R. W. Pickup, K. Suzuki, and I. M. Head.** 2004. Adaptation of sympatric *Achromatium* spp. to different redox conditions as a mechanism for coexistence of functionally similar sulphur bacteria. *Environ. Microbiol.* **6:**669–677.
40. **Harms, G., K. Zengler, R. Rabus, F. Aeckersberg, D. Minz, R. Rossello-Mora, and F. Widdel.** 1999. Anaerobic oxidation of *o*-xylene, *m*-xylene, and homologous alkylbenzenes by new types of sulfate-reducing bacteria. *Appl. Environ. Microbiol.* **65:**999–1004.
41. **Hesslein, R. H.** 1976. An *in situ* sampler for close interval pore water studies. *Limnol. Oceanogr.* **21:**912–914.
42. **Hines, M. E., G. T. Banta, A. E. Giblin, J. E. Hobbie, and J. T. Tugel.** 1994. Acetate concentrations and oxidation in salt marsh sediments. *Limnol. Oceanogr.* **39:**140–148.
43. **Hines, M. E., R. S. Evans, B. R. S. Genthner, S. G. Willis, S. Friedman, J. N. Rooney-Varga, and R. Devereux.** 1999. Molecular phylogenetic and biogeochemical studies of sulfate-reducing bacteria in the rhizosphere of *Spartina alterniflora*. *Appl. Environ. Microbiol.* **65:**2209–2216.
44. **Hines, M. E., and G. E. Jones.** 1985. Microbial biogeochemistry in the sediments of Great Bay, New Hampshire. *Estuar. Coast. Shelf Sci.* **20:**729–742.
45. **Hines, M. E., S. L. Knollmeyer, and J. B. Tugel.** 1989. Sulfate reduction and other sedimentary biogeochemistry in a northern New England salt marsh. *Limnol. Oceanogr.* **34:**578–590.
46. **Hines, M. E., R. S. Evans, B. R. S. Genthner, S. G. Willis, S. Friedman, J. N. Rooney-Varga, and R. Devereux.** 1999. Molecular phylogenetic and biogeochemical studies of sulfate-reducing bacteria in the rhizosphere of *Spartina alterniflora*. *Appl. Environ. Microbiol.* **65:** 2209–2216.
47. **Hipp, W. M., A. S. Pott, N. T. Schmitz, I. Faath, C. Dahl, and H. G. Trüper.** 1997. Towards the phylogeny of APS reductase and sirohaem sulfite reductases in sulfate-reducing and sulfur-oxidizing prokaryotes. *Microbiology* **143:**2891–2902.
48. **Howarth, R. W.** 1979. Pyrite: its rapid formation in a salt marsh and its importance in ecosystem metabolism. *Science* **203:**49–51.
49. **Howes, B. L., J. W. H. Dacey, and S. G. Wakeham.** 1985. Effects of sampling technique on measurements of porewater constituents in salt marsh sediments. *Limnol. Oceanogr.* **30:**221–227.
50. **Jørgensen, B. B.** 1978. A comparison of methods for the quantification of bacterial sulfate reduction in coastal marine sediments. I. Measurements with radiotracer techniques. *Geomicrobiol. J.* **1:**11–27.
51. **Jørgensen, B. B.** 1982. Ecology of the bacteria of the sulphur cycle with special reference to anoxic-oxic interface environments. *Philos. Trans. R. Soc. Lond.* **298:**543–561.
52. **Jørgensen, B. B.** 1988. Ecology of the sulphur cycle: oxidative pathways in sediments, p. 31–63. *In* J. A. Cole and S. J. Ferguson (ed.), *The Nitrogen and Sulphur Cycles*. Cambridge University Press, Cambridge, United Kingdom.
53. **Jørgensen, B. B.** 1994. Sulfate reduction and thiosulfate transformations in a cyanobacterial mat during a diel oxygen cycle. *FEMS Microbiol. Ecol.* **13:**303–312.
54. **Jørgensen, B. B.** 1990. A thiosulfate shunt in the sulfur cycle of marine sediments. *Science* **249:**152–154.
55. **Jørgensen, B. B., and F. Bak.** 1991. Pathways and microbiology of thiosulfate transformations and sulfate reduction in a marine sediment (Kattegat, Denmark). *Appl. Environ. Microbiol.* **57:**847–856.
56. **Kallmeyer, J., T. G. Ferdelman, A. Weber, H. Fossing, and B. B. Jørgensen.** 2004. A cold chromium distillation procedure for radiolabeled sulfide applied to sulfate reduction measurements. *Limnol. Oceanogr. Methods* **2:**171–180.
57. **Karkoff-Schweizer, R. R., D. P. Huber, and G. Voordouw.** 1995. Conservation of the genes for dissimilatory sulfite reductase from *Desulfovibrio vulgaris* and *Archaeoglobus fulgidis* allows their detection by PCR. *Appl. Environ. Microbiol.* **61:**290–296.
58. **Kelly, D. P.** 1989. Physiology and biochemistry of unicellular bacteria, p. 193–218. *In* H. G. Schlegel and B. Bowien (ed.), *Autotrophic Bacteria*. Springer-Verlag, Berlin, Germany.
58. **Kelly, D. P., L. A. Chambers, and P. A. Trudinger.** 1969. Cyanolysis and spectrometric estimation of trithionate in a mixture with thiosulfate and tetrathionate. *Anal. Chem.* **41:**898–901.
60. **Kiene, R. P.** 1991. Evidence for the biological turnover of thiols in anoxic marine sediments. *Biogeochemistry* **13:** 117–135.
61. **Kiene, R. P.** 1993. Measurement of dimethylsulfide (DMS) and dimethylsulfoniopropionate (DMSP) in seawater and estimation of DMS turnover rates, p. 601–610. *In* P. F. Kemp, B. F. Sherr, E. B. Sherr, and J. J. Cole (ed.), *Handbook of Methods in Aquatic Microbial Ecology*. Lewis Publishers, Boca Raton, Fla.
62. **Kiene, R. P., and G. Gerard.** 1994. Determination of trace levels of dimethylsulfoxide (DMSO) in seawater and rain water. *Mar. Chem.* **47:**1–12.
63. **Kiene, R. P., and M. E. Hines.** 1995. Microbial formation of dimethyl sulfide in anoxic *Sphagnum* peat. *Appl. Environ. Microbiol.* **61:**2720–2726.
64. **Kiene, R. P., and S. K. Service.** 1991. Decomposition of dissolved DMSP and DMS in estuarine waters: dependence on temperature and substrate concentration. *Mar. Ecol. Prog. Ser.* **76:**1–11.
65. **Kiene, R. P., and P. T. Visscher.** 1987. Production and fate of methylated sulfur compounds from methionine and

dimethylsulfoniopropionate in anoxic salt marsh sediments. *Appl. Environ. Microbiol.* **53:**2426–2434.

66. **Klein, M., M. Friedrich, A. J. Rogers, P. Hugenholtz, S. Fishbain, H. Abicht, L. L. Blackall, D. A. Stahl, and M. Wagner.** 2001. Multiple lateral transfers of dissimilatory sulfite reductase genes between major lineages of sulfate-reducing prokaryotes. *J. Bacteriol.* **183:**6028–6035.
67. **Koch, A. L.** 1994. Growth measurement, p. 248–277. *In* P. Gerhardt, R. G. E. Murray, W. A. Wood, and N. R. Krieg (ed.), *Methods for General and Molecular Bacteriology*. American Society for Microbiology, Washington, D.C.
68. **Kojima, H., A. Teske, and M. Fukui.** 2003. Morphological and phylogenetic characterizations of freshwater *Thioploca* species from Lake Biwa, Japan, and Lake Constance, Germany. *Appl. Environ. Microbiol.* **69:**390–398.
69. **Kühl, M., C. Steuckart, G. Eickert, and P. Jeroschewski.** 1998. A H_2S microsensor for profiling biofilms and sediments: application in an acidic lake sediment. *Aquat. Microb. Ecol.* **15:**201–209.
70. **Küsel, K., H. C. Pinkart, H. L. Drake, and R. Devereux.** 1999. Acetogenic and sulfate-reducing bacteria inhabiting the rhizoplane and deep cortex cells of the sea grass *Halodule wrightii*. *Appl. Environ. Microbiol.* **65:**5117–5123.
71. **Lie, T. J., J. R. Leadbetter, and E. R. Leadbetter.** 1998. Metabolism of sulfonic acids and other organosulfur compounds by sulfate-reducing bacteria. *Geomicrobiol. J.* **15:** 135–149.
72. **Llobet-Brossa, R. Rabus, M. E. Böttcher, M. Könneke, N. Finke, A. Schramm, R. L. Meyer, S. Grötzschel, R. Rossello-Mora, and R. Amann.** 2002. Community structure and activity of sulfate-reducing bacteria in an intertidal surface sediment: a multi-method approach. *Aquat. Microb. Ecol.* **29:**211–226.
73. **Lovley, D. R., and E. J. P. Phillips.** 1994. Novel processes for anaerobic sulfate production from elemental sulfur by sulfate-reducing bacteria. *Appl. Environ. Microbiol.* **60:** 2394–2399.
74. **Loy, A., A. Lehner, N. Lee, J. Adamczyk, H. Meier, J. Ernst, K.-H. Schleifer, and M. Wagner.** 2002. Oligonucleotide microarray for 16S rRNA gene-based detection of all recognized lineages of sulfate-reducing prokaryotes in the environment. *Appl. Environ. Microbiol.* **68:** 5064–5081.
75. **Loy, A., K. Kusel, A. Lehner, H. L. Drake, and M. Wagner.** 2004. Microarray and functional gene analyses of sulfate-reducing prokaryotes in low-sulfate, acidic fens reveal cooccurrence of recognized genera and novel lineages. *Appl. Environ. Microbiol.* **70:**6998–7009.
76. **Manz, W., M. Eisenbrecher, T. R. Neu, and U. Szewzyk.** 1998. Abundance and spatial organization of gram-negative sulfate-reducing bacteria in activated sludge investigated by in situ probing with specific 16S rRNA targeted oligonucleotides. *FEMS Microbiol. Ecol.* **25:**43–61.
77. **Minz, D., S. Fishbain, S. J. Green, G. Muyzer, Y. Cohen, B. E. Rittmann, and D. A. Stahl.** 1999. Unexpected population distribution in a microbial mat community: sulfate-reducing bacteria localized to the highly oxic chemocline in contrast to a eukaryotic preference for anoxia. *Appl. Environ. Microbiol.* **65:**4659–4665.
78. **Minz, D., J. L. Flax, S. J. Green, G. Muyzer, Y. Cohen, M. Wagner, B. E. Rittmann, and D. A. Stahl.** 1999. Diversity of sulfate-reducing bacteria in oxic and anoxic regions of a microbial mat characterized by comparative analysis of dissimilatory sulfite reductase genes. *Appl. Environ. Microbiol.* **65:**4666–4671.
79. **Mori, K., H. Kim, T. Kakegawa, and S. Hanada.** 2003. A novel lineage of sulfate-reducing microorganisms: Thermodesulfobiaceae fam. nov., *Thermodesulfobium narugense*, gen. nov., sp. nov., a new thermophilic isolate from a hot spring. *Extremophiles* **7:**283–290.
80. **Morrison, M. C., and M. E. Hines.** 1990. The variability of biogenic sulfur flux from a temperate salt marsh on short time and space scales. *Atmos. Environ.* **24:**1771–1779.
81. **Mussmann, M., H. N. Schulz, B. Strothmann, T. Kjaer, L. P. Nielsen, R. A. Rossello-Mora, R. I. Amann, and B. B. Jørgensen.** 2003. Phylogeny and distribution of nitrate-storing *Beggiatoa* spp. in coastal sediments. *Environ. Microbiol.* **5:**523–533.
82. **Muyzer, G., E. C. De Waal, and A. G. Uitterlinden.** 1993. Profiling of complex microbial populations by denaturing gradient gel electrophoresis analysis of polymerase chain reaction-amplified genes coding for 16S rRNA. *Appl. Environ. Microbiol.* **59:**695–700.
83. **Muyzer, G., A. Teske, C. O. Wirsen, and H. W. Jannasch.** 1995. Phylogenetic relationships of *Thiomicrospira* species and their identification in deep-sea hydrothermal vent samples by denaturing gel electrophoresis of 16S rDNA fragments. *Arch. Microbiol.* **164:**165–172.
84. **Muyzer, G., E. Yildirim, U. van Dongen, M. Kühl, and R. Thar.** 2005. Identification of "*Candidatus* Thioturbo danicus," a microaerophilic bacterium that builds conspicuous veils of sulfidic sediments. *Appl. Environ. Microbiol.* **71:**8929–8933.
85. **Nakagawa, T., J.-I. Ishibashi, A. Maruyama, T. Yamanaka, Y. Morimoto, H. Kimura, T. Urabe, and M. Fukui.** 2004. Analysis of dissimilatory sulfite reductase and 16S rRNA gene fragments from deep-sea hydrothermal sites of the Suijo Seamount, Izu-Bonin Arc, Western Pacific. *Appl. Environ. Microbiol.* **70:**393–403.
86. **Nelson, D. C., B. B. Jørgensen, and N. P. Revsbech.** 1986. Growth pattern and yield of a chemoautotrophic *Beggiatoa* sp. in oxygen-sulfide microgradients. *Appl. Environ. Microbiol.* **52:**225–233.
87. **Neretin, L. N., A. Schippers, A. Pernthaler, K. Hamann, R. Amann, and B. B. Jørgensen.** 2003. Quantification of dissimilatory (bi)sulphite reductase gene expression in *Desulfobacterium autotrophicum* using real-time RT-PCR. *Environ. Microbiol.* **5:**660–671.
88. **Odom, J. M., and R. Singleton, Jr.** 1993. *The Sulfate-Reducing Bacteria: Contemporary Perspectives*. Springer-Verlag, New York, N.Y.
89. **Okabe, S., T. Itoh, H. Satoh, and Y. Watanabe.** 1999. Analyses of spatial distributions of sulfate-reducing bacteria and their activity in aerobic wastewater biofilms. *Appl. Environ. Microbiol.* **65:**5107–5116.
90. **Oremland, R. S., and D. G. Capone.** 1988. Use of "specific" inhibitors in biogeochemistry and microbial ecology. *Adv. Microb. Ecol.* **10:**285–383.
91. **Pernthaler, J., F. O. Glöckner, W. Schönhuber, and R. Amann.** 2001. Fluorescence in situ hybridization with rRNA-targeted oligonucleotide probes. *Methods Microbiol.* **30:**207–226.
92. **Pfennig, N.** 1978. *Rhodocyclus purpureus* gen. nov. and sp. nov., a ring-shaped, vitamin B_{12}-requiring member of the *Rhodspirillaceae*. *Int. J. Syst. Bacteriol.* **28:**283–288.
93. **Postgate, J. R.** 1984. *The Sulphate-Reducing Bacteria*, 2nd ed. Cambridge University Press, Cambridge, United Kingdom.
94. **Rabus, R., M. Fukui, H. Wilkes, and F. Widdel.** 1996. Degradative capacities and 16S rRNA-targeted whole-cell hybridization of sulfate-reducing bacteria in an anaerobic enrichment culture utilizing alkylbenzenes from crude oil. *Appl. Environ. Microbiol.* **62:**3605–3613.
95. **Ramsing, N. B., H. Fossing, T. G. Ferdelman, F. Andersen, and B. Thamdrup.** 1996. Distribution of bacterial populations in a stratified fjord (Mariager Fjord, Denmark)

quantified by in situ hybridization and related to chemical gradients in the water column. *Appl. Environ. Microbiol.* **62:**1391–1404.

96. **Ramsing, N. B., M. Kühl, and B. B. Jørgensen.** 1993. Distribution of sulfate-reducing bacteria, O_2, and H_2S in photosynthetic biofilms determined by oligonucleotide probes and microelectrodes. *Appl. Environ. Microbiol.* **59:** 3840–3849.
97. **Raskin, L., B. E. Rittmann, and D. A. Stahl.** 1996. Competition and coexistence of sulfate-reducing and methanogenic populations in anaerobic biofilms. *Appl. Environ. Microbiol.* **62:**3847–3857.
98. **Ravenschlag, K., K. Sahm, C. Knoblauch, B. B. Jørgensen, and R. Amann.** 2000. Community structure, cellular rRNA content, and activity of sulfate-reducing bacteria in marine Arctic sediments. *Appl. Environ. Microbiol.* **66:**3592–3602.
99. **Ravenschlag, K., K. Sahm, and R. Amann.** 2001. Quantitative molecular analysis of the microbial community in marine Arctic sediments. *Appl. Environ. Microbiol.* **67:**387–395.
100. **Reeburgh, W. S.** 1967. An improved interstitial water sampler. *Limnol. Oceanogr.* **12:**163–165.
101. **Repeta, D. J., D. J. Simpson, B. B. Jørgensen, and H. W. Jannasch.** 1989. Evidence for anoxygenic photosynthesis from the distribution of bacteriochlorophylls in the Black Sea. *Nature* **342:**69–72.
102. **Rethmeier, J., A. Rabenstein, M. Langer, and U. Fischer.** 1997. Detection of traces of oxidized and reduced sulfur compounds in small sample quantities by combination of different HPLC methods. *J. Chromatogr. A* **760:**295–302.
103. **Revsbech, N. P., and B. B. Jørgensen.** 1986. Microelectrodes: their use in microbial ecology. *Adv. Microb. Ecol.* **9:**293–352.
104. **Risatti, J. B., W. C. Chapman, and D. A. Stahl.** 1994. Community structure of a microbial mat: the phylogenetic dimension. *Proc. Natl. Acad. Sci. USA* **91:**10173–10177.
105. **Rooney-Varga, J. N., R. Devereux, R. S. Evans, and M. E. Hines.** 1997. Seasonal changes in the relative abundance of uncultivated sulfate-reducing bacteria in a salt marsh sediment and rhizosphere of *Spartina alterniflora*. *Appl. Environ. Microbiol.* **63:**3895–3901.
106. **Rooney-Varga, J. N., B. R. S. Genthner, R. Devereux, S. G. Willis, S. D. Friedman, and M. E. Hines.** 1998. Phylogenetic and physiologic diversity of sulfate-reducing bacteria isolated from a salt marsh sediment. *Syst. Appl. Microbiol.* **21:**557–568.
107. **Sahm, K., C. Knoblauch, and R. Amann.** 1999. Phylogenetic affiliation and quantification of psychrophilic sulfate-reducing isolates in marine Arctic sediments. *Appl. Environ. Microbiol.* **65:**3976–3981.
108. **Santegoeds, C. M., L. R. Damgaard, C. Hesselink, J. Zopfi, P. Lens, G. Muyzer, and D. de Beer.** 1999. Distribution of sulfate-reducing and methanogenic bacteria in anaerobic aggregates determined by microsensor and molecular analyses. *Appl. Environ. Microbiol.* **65:** 4618–4629.
109. **Santegoeds, C. M., T. G. Ferdelman, G. Muyzer, and D. de Beer.** 1998. Structural and functional dynamics of sulfate-reducing populations in bacterial biofilms. *Appl. Environ. Microbiol.* **64:**3731–3739.
110. **Schramm, A., C. M. Santegoeds, H. K. Nielsen, H. Ploug, M. Wagner, M. Pribyl, J. Wanner, R. Amann, and D. de Beer.** 1999. On the occurrence of anoxic microniches, denitrification, and sulfate reduction in aerated activated sludge. *Appl. Environ. Microbiol.* **65:**4189–4196.
111. **Schulz, H. N., T. Brinkhoff, T. G. Ferdelman, M. Hernández Mariné, A. Teske, and B. B. Jørgensen.** 1999. Giant sulphur bacteria discovered in Namibian shelf sediments. *Science* **284:**493–495.
112. **Skyring, G. W.** 1987. Sulfate reduction in coastal ecosystems. *Geomicrobiol. J.* **5:**295–374.
113. **So, C. M., and L. Y. Young.** 1999. Isolation and characterization of a sulfate-reducing bacterium that anaerobically degrades alkanes. *Appl. Environ. Microbiol.* **65:** 2969–2976.
114. **Stal, L. J., H. van Gemerden, and W. E. Krumbein.** 1984. The simultaneous assay of chlorophyll and bacteriochlorophyll in natural microbial communities. *J. Microbiol. Methods* **2:**295–306.
115. **Telang, A. J., G. Voordouw, S. Ebert, N. Sifeldeen, J. M. Foght, P. M. Fedorak, and D. W. S. Westlake.** 1994. Characterization of the diversity of sulfate-reducing bacteria in soil and mining waste water environments by nucleic acid hybridization techniques. *Can. J. Microbiol.* **40:**955–964.
116. **Teske, A., N. B. Ramsing, J. Küver, and H. Fossing.** 1995. Phylogeny of *Thioploca* and related filamentous sulfide-oxidizing bacteria. *Syst. Appl. Microbiol.* **18:**517–526.
117. **Teske, A., M. L. Sogin, L. P. Nielsen, and H. W. Jannasch.** 1999. Phylogenetic position of a large marine *Beggiatoa*. *Syst. Appl. Microbiol.* **22:**39–44.
118. **Teske, A., N. B. Ramsing, K. Habicht, M. Fükui, J. Kuver, B. B. Jørgensen, and Y. Cohen.** 1998. Sulfate-reducing bacteria and their activities in cyanobacterial mats of Solar Lake (Sinai, Egypt). *Appl. Environ. Microbiol.* **64:**2943–2951.
119. **Teske, A., C. Wawer, G. Muyzer, and N. B. Ramsing.** 1996. Distribution of sulfate-reducing bacteria in a stratified fjord (Mariager Fjord, Denmark) as evaluated by most-probable-number counts and denaturing gradient gel electrophoresis of PCR-amplified ribosomal DNA fragments. *Appl. Environ. Microbiol.* **62:**1405–1415.
120. **Thamdrup, B., K. Finster, H. Fossing, J. W. Hansen, and B. B. Jørgensen.** 1994. Thiosulfate and sulfite distributions in porewater of marine sediments related to manganese, iron, and sulfur geochemistry. *Geochim. Cosmochim. Acta* **58:**67–73.
121. **Trüper, H. G.** 1989. Chemosynthetically sustained ecosystems in the deep sea, p. 147–166. *In* H. G. Schlegel and B. Bowien (ed.), *Autotrophic Bacteria*. Springer-Verlag, Berlin, Germany.
122. **Ulrich, G. A., L. R. Krumholz, and J. M. Suflita.** 1997. A rapid and simple method for estimating sulfate reduction activity and quantifying inorganic sulfides. *Appl. Environ. Microbiol.* **63:**1627–1630.
123. **van Gemerden, H., and H. H. Beeftink.** 1978. Specific rates of substrate oxidation and product formation in autotrophically growing *Chromatium vinosum* cultures. *Arch. Microbiol.* **119:**135–143.
124. **van Niel, E. W. J., and J. C. Gottschal.** 1998. Oxygen consumption by *Desulfovibrio* strains with and without polyglucose. *Appl. Environ. Microbiol.* **64:**1034–1039.
125. **Velji, M. I., and L. J. Albright.** 1993. Improved sample preparation for enumeration of aggregated aquatic substrate bacteria, p. 139–142. *In* P. F. Kemp, B. F. Sherr, E. B. Sherr, and J. J. Cole (ed.), *Handbook of Methods in Aquatic Microbial Ecology*. Lewis Publishers, Boca Raton, Fla.
126. **Vester, F., and K. Ingvorsen.** 1998. Improved most-probable-number method to detect sulfate-reducing bacteria with natural media and a radiotracer. *Appl. Environ. Microbiol.* **64:**1700–1707.
127. **Visscher, P. T., J. W. Nijburg, and H. van Gemerden.** 1990. Polysulfide utilization by *Thiocapsa roseopersicina*. *Arch. Microbiol.* **155:**75–81.

128. **Visscher, P. T., R. A. Prins, and H. van Gemerden.** 1992. Rates of sulfate reduction and thiosulfate consumption in a marine microbial mat. *FEMS Microbiol. Ecol.* **86:**283–294.

129. **Visscher, P. T., P. Quist, and H. van Gemerden.** 1991. Methylated sulfur compounds in microbial mats: in situ concentrations and metabolism by a colorless sulfur bacterium. *Appl. Environ. Microbiol.* **57:**1758–1763.

130. **Visscher, P. T., and B. F. Taylor.** 1993. Aerobic and anaerobic degradation of a range of alkyl sulfides by a denitrifying marine bacterium. *Appl. Environ. Microbiol.* **59:**4083–4089.

131. **Visscher, P. T., and B. F. Taylor.** 1993. Organic thiols as organolithotrophic substrates for growth of phototrophic bacteria. *Appl. Environ. Microbiol.* **59:**93–96.

132. **Visscher, P. T., and H. van Gemerden.** 1991. Photoautotrophic growth of *Thiocapsa roseopersicina* on dimethyl sulfide. *FEMS Microbiol. Lett.* **81:**247–250.

133. **Visscher, P. T., F. P. Vandenende, B. E. M. Schaub, and H. van Gemerden.** 1992. Competition between anoxygenic phototrophic bacteria and colorless sulfur bacteria in a microbial mat. *FEMS Microbiol. Ecol.* **101:**51–58.

134. **Voordouw, G., Y. Shen, C. S. Harrington, A. J. Telang, T. R. Jack, and D. W. S. Westlake.** 1993. Quantitative reverse sample genome probing of microbial communities and its application to oil field production waters. *Appl. Environ. Microbiol.* **59:**4101–4114.

135. **Wagner, M., A. J. Roger, J. L. Flax, G. A. Brusseau, and D. A. Stahl.** 1998. Phylogeny of dissimilatory sulfite reductases supports an early origin of sulfate respiration. *J. Bacteriol.* **180:**2975–2982.

136. **Wawer, C., and G. Muyzer.** 1995. Genetic diversity of *Desulfovibrio* spp. in environmental samples analyzed by denaturing gradient gel electrophoresis of [NiFe] hydrogenase gene fragments. *Appl. Environ. Microbiol.* **61:**2203–2210.

137. **Widdel, F., and T. A. Hansen.** 1991. The dissimilatory sulfate- and sulfur-reducing bacteria, p. 583–634. *In* A. Balows, H. G. Trüper, M. Dworkin, W. Harder, and K.-H. Schleifer (ed.), *The Prokaryotes*, 2nd ed. Springer-Verlag, New York, N.Y.

138. **Widdel, F., G.-W. Kohring, and F. Mayer.** 1983. Studies on dissimilatory sulfate-reducing bacteria that decompose fatty acids. III. Characterization of the filamentous gliding *Desulfonema limicola* gen. nov., sp. nov., and *Desulfonema magnum* sp. nov. *Arch. Microbiol.* **134:**286–294.

139. **Yacobi, Y. Z., W. Eckert, H. G. Trüper, and T. Berman.** 1990. High-performance liquid chromatography detection of phototrophic bacterial pigments in aquatic environments. *Microb. Ecol.* **19:**127–136.

140. **Zhabina, N. N., and I. I. Volkov.** 1978. A method of determination of various sulfur compounds in sea, sediments and rocks, p. 735–745. *In* W. E. Krumbein (ed.), *Environmental Biogeochemistry and Geomicrobiology*, vol. 3. Ann Arbor Science Publishers, Ann Arbor, Mich.

141. **Zverlov, V., M. Klein, S. Lücker, M. W. Friedrich, J. Kellermann, D. A. Stahl, A. Loy, and M. Wagner.** 2005. Lateral gene transfer of dissimilatory (bi)sulfite reductase revisited. *J. Bacteriol.* **187:**2203–2208.

Nitrogen Cycling in Aquatic Environments

BESS B. WARD

42

Nitrogen is a key element in biological processes, and its availability is often a controlling factor in the productivity of aquatic systems (16, 44). The marine nitrogen cycle includes several components to which eukaryotes contribute directly (e.g., assimilation of inorganic and organic N and decomposition of organic N), but the core transformations are predominantly or solely the domain of prokaryotes. Of these, denitrification and nitrogen fixation are the two processes that control the fixed N inventory. In the budget discussions here, the term denitrification includes both conventional denitrification (respiratory reduction of oxidized N to N_2) and anaerobic ammonia oxidation (anammox), unless otherwise stated.

The state of knowledge for methods in the evaluation of the important microbial components of the marine nitrogen cycle was reviewed by Capone in the second edition of this manual (18). The reader is referred to that chapter for an excellent treatment of the methods, including an excellent table comparing the alternative methods for each process. The present chapter will therefore focus on developments in the field since 2000, after a brief review of the basic questions and approaches in each section. The most significant advances have been in the area of assessing the diversity, abundance, and activity of microorganisms involved in N cycling, while little has changed in the realm of direct rate measurements. The content of this chapter reflects that emphasis, and the reader is referred to the earlier editions for more complete information on rate measurements.

Perhaps the most significant and pervasive advance in N cycle research since 2000 is the application of molecular methods in many formats and for many processes. These advances have had the most direct effects on our ability to evaluate and characterize the populations of microbes involved in the various N cycle processes. Molecular biology has not had a major effect on evaluation of rates of processes, but molecular methods have the capability to increase our understanding of factors that are responsible for population distributions and regulation of rates in response to environmental factors.

A second major advance in N cycle research since the time of MEM2 is the improvement in stable isotope methodology, in terms of the sensitivity and versatility of the instruments, their cost and accessibility to most researchers, and the development of new techniques and refinements of existing methods. Specific techniques will be described in sections on particular components of the N cycle. Beyond specific techniques, the integrative nature of isotopic data means that isotopic approaches are amenable to global modeling and regional assessments, without detailed measurements of specific processes. Insights from these approaches are mentioned after treatment of individual processes from the microbiological perspective.

This chapter follows the approximate format of the earlier chapter by Capone (18) and focuses on the three core N cycle processes that are performed wholly or predominantly by prokaryotes: nitrogen fixation, nitrification, and denitrification.

NITROGEN FIXATION

Biological N_2 fixation is the enzyme-mediated reduction of N_2 to ammonium. A phylogenetically and physiologically diverse array of bacteria and archaea carry out the process, using a highly conserved and complex suite of enzymes, in a variety of aquatic environments. N_2 fixation is the major source of fixed N to the biosphere and is particularly important in N-limited environments. Recent reviews on aquatic N_2 fixation and the molecular ecology of N-fixing microorganisms are available (17, 19, 69, 111).

The major advance that has occurred since the time of the last MEM chapter on marine N fixation is the shift to use of molecularly based methods. This approach has been used for population and diversity studies and to link activity by particular groups to measured rates in the ocean. Major findings include the identification of a broad diversity of N fixation genes in natural environments, which presumably imply the presence and activity of many N-fixing microbes in addition to the major players, such as *Trichodesmium*, that have long been recognized. In some cases, particular gene sequences have been linked to specific organisms whose contributions to N fixation have been previously unknown or underestimated. Varying patterns of gene expression imply differential N fixation rates by different members of the N-fixing assemblage, as well as variable responses to environmental factors such as light.

While molecular methods have allowed important advances at the scale of individual cells, equally important advances also have occurred at ecosystem scales. Satellite measurements of sea surface color have been combined with ocean circulation models to evaluate N fixation rates on the scale of ocean basins (27). Biogeochemical models have been used to constrain N cycling rates and global and regional N budgets (32).

Populations of Nitrogen-Fixing Organisms

The most important N-fixing organisms in the ocean are various strains of the nonheterocystous filamentous cyanobacterium *Trichodesmium* (19). Methods for assessing its population range from satellite identification (73, 89) to microscopic counts (20, 23) to specific molecular analyses (47). Major advances in the study of N-fixing microbial populations in recent years have also occurred in the direction of assessing the potential importance of N fixers other than *Trichodesmium*. N fixation is a trait that is found in several of the major bacterial and archaeal phyla and therefore is directly identified by the presence of genes encoding enzymes in the N fixation pathway, rather than by 16S rRNA genes. Evidence for significant horizontal gene transfer of the nitrogen fixation (*nif*) genes is, however, lacking (111), so it is possible to use 16S rRNA genes to study particular N-fixing organisms in some cases (35, 61). The most common method now in use for detecting and identifying N-fixing microbes is the molecular detection of the *nifH* gene, which encodes the structural protein dinitrogenase reductase. This method was first developed by Zehr and coworkers and has been very widely applied in the last decade in a wide range of environments. In the marine environment, its application has been instrumental in demonstrating the presence of a diverse assemblage of microbes that apparently possess the genetic capability for N fixation (113) and in identifying unicellular cyanobacteria that contribute significantly to N fixation in the open ocean (115).

The methods used to detect and describe diverse *nifH* genes from environmental samples follow the standard protocols described in chapter 12. The key to the method is the use of nested PCR primers (110, 112), introduced by Zehr and coworkers and modified by others for particular applications (e.g., reference 10). Current details on the application of PCR-based methods to study N fixation can be found in a recent methodological review (114). The most challenging aspect of the molecular analyses is often the retrieval of high-quality DNA from the environmental sample. Aquatic samples are usually the least problematic, due to the low concentration of substances that might inhibit PCR, but it is usually necessary to concentrate the particulate material by filtration (onto either flat membrane filters or capsule filters), particularly in the pelagic realm of oceans and lakes. DNA can be extracted by using any number of kits or protocols, all derived from the basic lysis and precipitation protocols. The DNeasy kit (Qiagen) and the PureGene kit (Gentra) are usually successful for extraction of DNA from filter formats.

In sediment samples, the concentration of cells is often very high, but this advantage is countered by the high concentration of organic matter that must be removed from the extract in order to achieve efficient amplification. The FastDNA kit (Bio101) for DNA extraction from soils is widely used and can be easily modified to include additional cleanup steps. Even after cleanup, sediment DNA extracts often must be diluted by 100- to 1,000-fold in order to obtain PCR amplification.

It is often desirable to characterize *nifH* gene expression instead of simply the presence of the genes in the DNA of the sample. In this case, RNA is extracted from the sample (RNeasy kit; Qiagen) and is reverse transcribed to obtain cDNA (110). The standard precautions necessary for handling RNA in natural samples are required (72).

Diversity of N-Fixing Populations

By far the widest application of the PCR approach has been for investigations of *nifH* diversity, but activity rates have also been estimated (see below). Quantitative DNA recovery is not usually possible, but in order to minimize bias from small sample size, multiple DNA extracts from a single sample and multiple PCR products from each are often pooled before cloning. A clone library is made from the amplified *nifH* products (379 bp) and screened by completely sequencing the inserts, most often by cloning into a commercial vector such as the Topo-TA kit (Invitrogen) that allows blue/white screening of transformants. The fragment obtained with the usual *nifH* PCR primers is short, and thus, it is common to sequence the fragments completely rather than use a fragment analysis approach (see "Denitrification" below). The number of clones sequenced is often arbitrary and depends more on financial considerations than on an analysis of the extent of gene diversity in the sample. A true estimate of the diversity of the sample is very difficult to obtain, because PCR is necessarily biased and can amplify only sequences with a relatively high degree of similarity. It is clear from the few analyses that have been published that *nifH* gene diversity is very broad, including at least several thousand different phylotypes. *nifH* sequences within major bacterial phyla vary by 10 to 15% at the amino acid level (22 to 26% at the nucleotide level) and within the archaea by over 40% at the amino acid level (60% at the nucleotide level) (111). A recent assessment of methods used to estimate species richness from clone libraries and metagenomic data accepted the 3% identity cutoff for 16S rRNA (87) and suggested that a 20% cutoff for functional genes yields the same resolution (77). It is difficult to apply a single cutoff to all functional genes, because the meaningful delineation is likely to vary among genes depending on physiological constraints and evolutionary history. The 20% cutoff appears to be too large for most functional genes in the N cycle (see below), including *nifH*, where it would not resolve the diversity within major phyla.

Abundance of N-Fixing Populations

Conventional approaches to determining the abundance of N-fixing microorganisms relied upon cultivation, using most-probable-number (MPN) techniques and selective media that did not contain fixed N. The variety of fundamental metabolisms (e.g., aerobic and anaerobic photoautotrophy, aerobic and anaerobic heterotrophy, etc.) requires the use of multiple media, and the oft-proved maxim that only a trivial fraction of all the microbes living in the environment can be cultivated makes it obvious that cultivation methods must yield severe underestimates. Intense cultivation efforts are invaluable for bringing previously unknown types into culture, where they can be studied and used for the development of molecular and biochemical assays, but cultivation is not the method of choice for abundance estimates.

The PCR methods used to retrieve *nifH* genes and to characterize their diversity are also the basis of new approaches to quantification of N-fixing organisms. However, even these very sensitive methods are probably not capable at present of providing an overall abundance estimate of the entire functional group. Probes based on the *nifH* gene sequence are adapted for use in quantitative (also known as real-time) PCR assays, either with species-specific probes (the TaqMan approach [25, 81]) or with nonspecific fluorescence markers (the SYBR green approach). The PCR product obtained by standard PCR protocols can also be used to quantify gene abundance by using hybridization methods with radioactive or chemiluminescent probes (10). These PCR-based methods all depend on the specificity of the primers and probes and work best when the specificity is narrowly defined. Thus, they are excellent for determination of the (relative) abundance of specific phylotypes but cannot be extrapolated to the total N-fixing assemblage.

As for all PCR-based methods, obtaining a high-quality DNA extract is often a challenge for these quantitative methods. In addition, either the extraction must be 100% efficient or there must be some way to quantify the efficiency so that the final gene copy number can be corrected for incomplete recovery. Most workers implicitly assume 100% DNA extraction efficiency or normalize to DNA yield per volume filtered (81) or to the abundance of some other gene (16S rRNA) quantified in the same way (90). The best compromise appears to be use of replicate extractions and amplification reactions with the assumption that, if DNA extraction is not 100% efficient, it is at least reproducible between samples in a given study.

Perhaps the most desirable quantification is not the total number of N-fixing microbes or even the number of a certain type of N fixers but rather the abundance or relative abundance of microbes that are actively involved in fixing N at particular times and places. This has been approached by quantifying the amount of nitrogenase enzyme present by using nitrogenase-specific antibodies (30), but the method has not been widely applied. The PCR assay approaches described above for DNA can be applied to mRNA or cDNA in order to describe the diversity of organisms expressing the gene and to quantify the number of gene copies expressed. As for DNA, both quantitative PCR (115) and hybridization (10) approaches are possible. Permutations of this approach have been used to assess in situ activity of specific *nifH* phylotypes (115) and to identify which phylotypes are actively expressing the *nifH* genes by using reverse transcriptase PCR (34, 110).

Rates of N Fixation

The $^{15}N_2$ Direct Tracer Method

The best available method for determination of N fixation rates is the ^{15}N direct tracer method. Despite advances in mass spectrometer methodology and sensitivity, however, it remains a much more difficult measurement than acetylene reduction (see next section) and thus is much less widely used. For a total rate measurement, samples to be incubated (up to 4 liters in open ocean settings) are enclosed in gastight vessels and spiked with trace concentrations of highly enriched $^{15}N_2$. At the end of the incubation, the contents of the vessel are filtered, and the ^{15}N content of the particulate material is determined by mass spectrometry or emission spectrometry (63, 64). By manually picking individual colonies, Orcutt et al. (68) were able to measure the N fixation rates specifically due to *Trichodesmium* in Sargasso Sea waters. Size fractionation has been used to exclude *Trichodesmium* in order to assess the contribution of other, unicellular N fixers (63).

Reverse transcriptase PCR has been used to detect gene expression by specific phylotypes of *nifH* in the open ocean (10, 115). This is a powerful approach for demonstrating that at least some of the measured rate of N fixation could be due to particular groups of organisms. In combination with rate measurements made by ^{15}N tracer incubations and measures of cell abundance by molecular or microscopic means, the overall measured rate may be partitioned among various members of the assemblage, but gene expression alone cannot be translated into transformation rates.

The Acetylene Reduction Assay

The most commonly used method for evaluation of N fixation rates in aquatic samples continues to be the acetylene reduction assay (88), preferred for its ease of use in many settings with many sample formats. Acetylene is an alternative substrate for nitrogenase, and the rate of ethylene production in a sample incubated with acetylene can be converted to the N_2 reduction rate with appropriate controls and conversion factors (detailed protocol in reference 15). Because of its ease of use, and the ease and sensitivity of measuring ethylene production on a gas chromatograph, acetylene reduction is the method of choice for most field applications. Calculation of actual N_2 fixation rates assumes a standard ratio of C_2H_2 to N_2 reduction. Although good agreement has often been reported between rates measured by the $^{15}N_2$ and C_2H_2 methods, the ratio has been reported to vary in some systems. Thus a calibration or method comparison is wise when quantitative N flux data are required.

NITRIFICATION

Nitrification is the two-step process by which ammonium is oxidized to nitrite and then to nitrate by a few small, phylogenetically constrained groups of bacteria. Although some partial oxidations of nitrogen have been reported in heterotrophs, nitrifying activity in the marine environment is attributed to chemolithoautotrophic bacteria. The ammonia-oxidizing bacteria (AOB) are considered to be strict autotrophs, while the nitrite-oxidizing bacteria (NOB) have a limited heterotrophic potential; growth of some strains is enhanced by metabolism of small organic acids.

The most significant advance in the study of nitrification over the last decade is the wide application of molecular ecological methods for the assessment of nitrifying populations in many natural and constructed environments. These methods have led to the identification of major uncultivated groups of conventional nitrifiers that appear to dominate the marine assemblages. Recently, genes homologous to the signature functional gene in conventional nitrifiers have been identified in association with archaeal genomes (76, 93). Cultivation of archaeal autotrophic nitrifiers was recently reported (53), and they have already been widely detected in the environment (37). Recent reviews on nitrification in the marine environment include references 98, 99, and 101.

Populations of Nitrifying Microorganisms

Molecular biological methods to identify and characterize nitrifying bacteria have provided insights into the distribution, diversity, and activity of nitrifying populations. Unlike N fixation and denitrification, however, nitrification is associated with a small and phylogenetically constrained group of organisms, so both 16S rRNA and functional genes associated with the nitrification pathways can be used to identify and characterize nitrifier populations. For ammonia oxidizers, the key functional gene is *amoA*, which encodes a subunit of the ammonia monooxygenase enzyme, the first step in ammonia oxidation. While the AOB have been a model group for molecular ecological studies (55), much less attention has been paid to the NOB. Both 16S and functional gene approaches should, however, be applicable to this group as well, and this is an area of developing research.

Diversity of Nitrifying Populations

Both 16S rRNA and functional gene-based investigations have shown that the range of microbes involved in nitrification in the ocean is much larger than suspected on the basis of the culture collection. For both AOB and NOB, it appears that the dominant organisms in the ocean, as detected in clone libraries, have not been cultivated. Earlier methods of population assessment relied on cultivation and the application of, e.g., immunological probes derived from cultivated strains (102). The immunofluorescence (IF) method was among the first to be able to differentiate among different kinds of AOB and NOB and to quantify their distribution in the water column, but its reliance on cultivation limits its power. Even though the first molecular probes for AOB genes were also derived from cultivated strains, they have succeeded in identifying uncultivated types. Whereas cultivation and isolation efforts had produced mainly *Nitrosomonas* and *Nitrosococcus* strains, molecular analyses of both 16S rRNA and *amoA* genes identify an uncultivated clade most similar to *Nitrosospira* as the most widely detected and likely most abundant AOB in marine water column and sediment environments (45, 67).

Thus, a modern assessment of nitrifier diversity now relies on molecular approaches, concomitant with the challenges of DNA extraction and purification mentioned in the preceding section on N fixation genes. For detection of 16S rRNA genes, direct PCR amplification with AOB-specific primers is usually successful (2, 45). Some authors have found that a nested approach, with initial amplification by using universal bacterial primers followed by amplification with an inner primer set that is AOB specific, increases sensitivity (95). Similarly for *amoA*, a nested approach is sometimes useful. Several different sets of primers for both genes have been published, and an excellent compendium describing their known specificities can be found in the review by Koops et al. (54). Explicit instructions for their application can be found in a recent method review by Ward and O'Mullan (106).

At the time of this writing, evidence is accumulating that an entirely novel kind of ammonia-oxidizing microbe may be abundant in the ocean. Although reports of its isolation are not yet public, an archaeon with ammonia-oxidizing capability has been reported from coastal seawater. An *amoA*-like gene se-quence identified on scaffolds containing archaeal rRNA genes was detected in the Sargasso Sea metagenome (93) and in metagenomic libraries from soils (76). By using primers designed on the basis of these sequences, Francis et al. (37) have detected this *amoA*-like gene in a wide variety of marine and sediment environments. Thus, it appears that two very important groups of ammonia oxidizers have not been previously cultivated and that this broader diversity of ammonia oxidizers necessitates a further investigation into their physiology and environmental regulation of their activities.

The diversity of NOB populations has been investigated in wastewater systems by using specific 16S rRNA-based PCR methods (78a), but these methods have not yet been applied in the marine environment.

Abundance of Nitrifying Organisms

MPN methods have long been used to enumerate and cultivate nitrifying bacteria. Because growth of both AOB and NOB can be detected with a simple assay for the appearance or disappearance, respectively, of nitrite, the approach has certain attractions and does yield isolates (which may not, however, be "representative"). MPN has been found to underestimate AOB abundances by 2,000- to 100,000-fold (41), and the overwhelming evidence that standard media do not enrich for the most important environmental phylotypes means that the MPN approach should not be relied on for quantification. It should be useful with specific medium modification for retrieval of novel types, however, and may be useful in cultivating the organisms whose genes dominate nitrifier clone libraries.

IF has been mentioned above as a method to quantify the distribution of specific strains of AOB (97, 109) and NOB (36, 102), but its cultivation-based limitations restrict its applications. IF detection of the ammonia oxygenase (AMO) and nitrite oxidoreductase (NOR) proteins themselves (3) shows promise but has not been widely applied and may be limited by narrow specificity. Fluorescence in situ hybridization (FISH) involves a protocol similar to IF, but its specificity, although still based on probes ultimately derived from cultures, is much broader and extends to organisms that have not yet been cultivated. Recent modifications with much increased sensitivity (catalyzed reporter deposition [CARD]-FISH) (70) are particularly useful for nitrifiers, whose slow growth is associated with low rRNA content. FISH assays are usually based on rRNA or rRNA genes because ribosomes can occur in very large numbers in individual cells, and the genes themselves are usually present in multiple copies in the genome. FISH has been very successfully applied in wastewater systems (e.g., references 7 and 38), where high cell number and activity favor detection. It has also been successfully applied, however, in much more challenging environments such as the permanently ice-covered lakes of Antarctica (94, 96) and seawater (80). The method is usually membrane filter based, so that the sample is somewhat concentrated ($\geq$20 ml can be filtered). The filter is affixed to a slide, and the hybridization (with oligonucleotide probes representing 16S rRNA gene sequences) and wash steps are performed in sequence. The method has also been adapted for flow cytometry, which has advantages of throughput and automation, but is probably pushing the limits of detection for nitrifiers (80). The oligonucleotide-based probe methods are much less selective than IF but are most powerful when used with highly specific probes rather than with "universal" nitrifier probes.

Quantitative PCR based on *amoA* genes has also been used to enumerate nitrifiers in aquatic sediment microcosms (72) and agricultural soils (66a), as well as in wastewater systems (41a), but to date there are no published reports about using this approach in marine systems.

Quantitative PCR based on 16S rRNA sequences has been used to quantify NOB in wastewater systems (32b, 41a, 72a). PCR in combination with MPN and IF has been used to assess the abundance of *Nitrobacter hamburgensis* in soils (36). To date, there are no published reports of these applications to NOB in the marine environment.

Nitrification Rates

^{15}N Tracer Methods

The best method for quantification of nitrification rates remains the direct ^{15}N tracer approach, and increased availability of mass and emission spectrometers should make its application more practical as well. It has been most widely applied in water column settings (60, 84, 100, 105, 107). It is also an incubation approach and has suffered in the past from the necessity to enrich the substrate pool by the addition of large concentrations of ^{15}N-ammonia or -nitrite, thus artificially enhancing the observed rate. Improvements in assay techniques and mass spectrometer sensitivity with small N masses have minimized this problem. The major challenge of the direct ^{15}N approach is the necessity to extract the labeled products and substrates from the seawater after incubation. Both nitrite and nitrate can be converted to diazo dye complexes and removed with organic solvents (59) or reduced to ammonia by using Dvarda's alloy or $TiCl_3$ (9) and captured by microdiffusion (9, 82). Use of denitrifying bacterial cultures to convert NO_3^- and NO_2^- to N_2O followed by mass spectrometric analysis of the labeled N_2O makes possible a major improvement in sensitivity to the approach (83).

The same analytical approaches can be applied in a tracer or isotope dilution format. In the former, the label is added to the substrate pool and detected as it accumulates over time in the product pool. Knowledge of the size of the initial substrate pool is essential in order to prevent perturbation by excess substrate addition. It is also important to measure the isotopic enrichment of the substrate pool over the course of the incubation because its dilution (e.g., of the ammonium pool by ammonification of in situ organic matter) influences the observed enrichment of the product pool. Failure to account for isotope dilution of the substrate pool can lead to a serious underestimation of the oxidation rate if the substrate pool is initially very small. The problem is minimized by using short incubations and minimal substrate additions.

In the isotope dilution format, the label is added to the product pool, and its dilution by addition of new product with natural abundance isotopic signature during the incubation provides an estimate of production rate. Dilution may offer an advantage over direct tracer experiments when the product pool (e.g., NO_3^-) is large and the substrate pool (e.g., NH_4^+ or NO_2^-) is small, because the isotope addition constitutes a much smaller perturbation.

The advantages of the direct ^{15}N approaches, compared to inhibitor methods, include shorter, thus less artifactual, incubations; minimal perturbations to in situ conditions (ambient light and nutrient conditions can be used); and much greater sensitivity. In a comparison of the ^{14}C and ^{15}N isotopic methods for measurement of nitrification rates, Andersson et al. (1a) reported that the ratio of the two rates varied as a function of temperature and oxygen concentration and that the necessity of excluding phytoplankton $^{14}CO_2$ uptake by performing incubations in the dark was seen as unacceptably artifactual. In surface water samples in which phytoplankton would otherwise be active, removal of their contributions to N cycling has large effects on all other aspects of N transformations.

Inhibitor-Based Nitrification Assays

Inhibitor methods depend on the ability of many compounds to interact specifically with the active site of the AMO or NOR enzyme. A large number of potential inhibitors have been used for AMO (4); allylthiourea and N-Serve continue to be popular for field studies (5, 6, 33). Chlorate is still the only specific inhibitor reported for nitrite oxidation. Replicate incubations are carried out in the dark with additions of either AMO or NOR inhibitors, and the changes in inorganic N concentrations are used to infer nitrification rates. A more sensitive permutation of this approach involves measurement of $^{14}CO_2$ fixation in the presence and absence of inhibitor, where the decrease in rate of $^{14}CO_2$ assimilation in the presence of nitrifier inhibitor is attributed to nitrification (8, 84). A conversion between CO_2 fixation and N oxidation rates is then used to estimate nitrification.

Castro-Gonzalez and Farias (22) used an inhibitor approach to study N_2O cycling by both nitrifying and denitrifying bacteria in the oxygen minimum zone off Chile. In this environment, exclusion of light was not an issue and the quantification of N_2O by gas chromatography increased the sensitivity of the approach. Gaseous inhibitors such as methylfluoride (62) and acetylene (14, 52) have the advantage of rapid diffusion over dissolved ions, which can be especially important in sediments. Methylfluoride has much greater specificity (acetylene inhibits denitrification as well) but is expensive and not easily available. Changes in concentrations of inorganic nitrogen components are a traditional way to assess nitrification potentials of sediments, either in the presence (43) or absence (13) of inhibitors.

One of the main attractions of the inhibitor approaches is their ease of use and analysis. Scintillation counters are much more common and easier to use than are instruments required for stable isotope analysis (see below), and inorganic N determinations involve analytical methods that are already standard to most laboratories. Although the inhibitor approaches are usually not appropriate for absolute rate measurements (because of uncertainty in conversion factors and perturbations during incubations), they can be very useful for spatial or temporal comparisons within studies.

DENITRIFICATION

Denitrification is the respiratory process in which bacteria reduce nitrate to N_2. The ability for conventional denitrification (reduction of nitrate or nitrite to nitrous oxide or N_2) is very widespread amongst the bacteria and has even been reported in archaea and eukaryotes (fungi) (116). Most denitrifiers are heterotrophs that utilize denitrification as an alternative respiratory mode and can grow equally well or better on oxygen. In the absence of oxygen, denitrification allows the continued decomposition of organic matter and results in loss of fixed N from the system. Dissimilatory nitrate reduction to ammonium, also referred to as nitrate fermentation, can equal or exceed respiratory denitrification rates in marine sediments and does not lead to the loss of fixed N.

The biggest change in our understanding of denitrification in recent years is the recognition of an entirely novel process that contributes to fixed N loss in both sediment and water column environments. Anammox is defined as the oxidation of ammonia to dinitrogen by nitrite under anoxic conditions (92) and is performed by a group of bacteria identified as within *Planctomycetales*. It was first identified in wastewater systems in Europe but has since been detected in marine and estuarine sediments and water columns.

Populations of Denitrifying Organisms

Unlike nitrification, denitrification is not considered a defining characteristic of a bacterial strain because it is usually a facultative trait and not associated with any particular carbon metabolism. Thus, the only way to evaluate denitrifier populations is to focus on the transformation process itself. Traditional MPN and culture methods depend on the detection of gas production that is dependent on the presence of oxidized N and occurs under anoxic conditions. As with nitrification and N fixation, many different media would be required to select all possible denitrifiers, and the result is a serious underestimate in terms of abundance and a strong selection in terms of diversity. Nevertheless, selective enrichment has led to the identification and characterization of many novel metabolisms associated with denitrification and has proven the potential for alternative (to oxygen) modes of respiration to be involved in, e.g., hydrocarbon degradation (85).

Molecular methods for characterization of denitrifying populations depend on the genes encoding enzymes in the denitrification pathway and have focused on nitrite, nitric oxide, and nitrous oxide reductases. The capacity for nitrate reduction to nitrite is more common than the ability for complete denitrification to N_2O or N_2 and is shared by organisms that perform dissimilatory denitrification to ammonium. The genes involved in the anammox reaction are not yet well known, so assessment of anammox bacterial populations relies on 16S rRNA genes.

Diversity of Denitrifying Organisms

Conventional Denitrifiers

The most widely used approach for assessment of denitrifier diversity and community structure is based on the genes that encode nitrite reductase. Two different forms of this enzyme, one a heme enzyme (encoded by *nirS*) and one containing copper (*nirK*), are widely distributed among bacteria. The most widely used PCR primers for *nirS* are those of Braker et al. (12), which amplify a great variety of *nirS* genes from *Proteobacteria*. The *nirK* primers developed by the same authors are not as successful, and newly developed *nirK* primers have proven more robust in applications in the marine environment (37a). The *nosZ* primers developed by Scala and Kerkhof (74) have also been widely applied in both aquatic and terrestrial environments.

The degree of diversity encompassed by clone libraries of *nirS*, *nirK*, and *nosZ* gene fragments indicates that a very large number of different kinds of denitrifiers is present in most environments. It is, however, common to assess the degree to which the diversity of the clone library has been assessed by using rarefaction analysis, i.e., applying operational taxonomic units that are defined by a minimal sequence distance or by fragment patterns (42). It is nearly impossible to saturate a clone library in the case of denitrification genes (as indicated by approaching an asymptote in a rarefaction curve), and it is likely that a large amount of denitrifier gene diversity remains to be discovered. This great diversity and the relatively large size of the PCR fragment obtained with the commonly used primers have encouraged the use of (terminal) restriction fragment length polymorphism ([T]-RFLP) analysis for investigation of denitrifier diversity and community composition. Although applicable in theory to any gene, fragment analysis has been applied more commonly in denitrification studies than in any other part of the N cycle. RFLP is often used to screen clone libraries prior to sequencing or to provide an alternative analysis of gene diversity for comparison to gene sequences (48). T-RFLP is a rapid, high-throughput approach for identifying major components of denitrifying assemblages and for investigating relationships between community composition and environmental conditions (11, 75). It is vastly more rapid and economical than sequencing the large number of fragments needed to characterize a single site.

In conjunction with a T-RFLP study, it is essential to sequence some of the cloned genes in order to verify that the observed fragments do indeed correspond to the target gene (i.e., to show that the desired target gene, rather than a nonspecific target of similar size, has been amplified). It is often found that a single T-RFLP operational taxonomic unit corresponds to several different sequences, which are not particularly closely related, that happen to share the same restriction site (21). T-RFLP analysis thus risks masking much of the diversity that would be revealed in a complete sequence analysis.

Anammox Organisms

The capacity for anaerobic ammonia oxidation thus far appears to be restricted to a group of bacteria in the *Planctomyctetales*, which are sufficiently closely related at the 16S rRNA level that specific 16S rRNA probes suffice to detect them and to define the functional group of anammox bacteria. On the basis of 16S rRNA sequences, there are two major groups of anammox bacteria; one group consists solely of sequences derived from wastewater and wastewater isolates, and the other contains both wastewater- and seawater-derived sequences. A nested PCR approach is usually employed, in which the first stage is performed by using *Planctomycetales*-specific 16S rRNA primers and the inner stage with anammox-specific primers. Several subsets of primers can be used to detect different types selectively (78).

Abundance of Denitrifying Organisms

MPN approaches are not recommended for quantitative purposes because of the variety of basic metabolic types that can exhibit denitrification capability. IF (polyclonal antibodies for cell surface antigens) has been applied for enumeration of individual species (103), but its specificity precludes its usefulness as a more general approach to enumerate all denitrifiers. Antibodies specific for enzymes involved in denitrification offer more general probes (29), but even they tend to be too specific and are not capable of detecting even very similar proteins from different species (104).

PCR-based methods have also been used for quantification of denitrifiers but have not been widely applied. Species-specific quantitative PCR based on the *nirS* sequence of *Pseudomonas stutzeri* illustrated the potential of the approach (40), but broader gene-specific quantitative PCR that might detect the entire *nirS*- or *nirK*-bearing denitrifier populations has not yet been reported. FISH for denitrifiers would have to be based on the denitrification genes, and because they are generally present in only one copy per genome, the assay is not sensitive enough. In situ PCR, in which the amplification occurs inside individual fixed cells, has been reported (24), but again only on a very specific basis, and has not been widely applied. Anammox bacteria have been detected and enumerated in water samples by using FISH and CARD-FISH based on 16S rRNA gene probes (56).

Rates of Denitrification

Direct N_2 Flux

Increases in N_2 during incubations are a direct indication of net denitrification, and this approach has a long history of use in sediment environments (50). Because of its inherent low sensitivity due to the large initial pool of N_2 in the overlying water, removal of N_2 from the overlying water initially, long-term incubations and stringent precautions to prevent atmospheric leaks are usually required, and even then the incubations require a week or more for a single rate measurement (66, 79). Using very-high-sensitivity mass spectrometry, Devol (32a) was able to detect denitrification in deep-sea sediments from changes in N_2 in the overlying water of an in situ benthic chamber.

The N_2 Ar method that uses membrane inlet mass spectrometry has recently been applied to measure both instantaneous and integrated denitrification rates (49, 58) in estuarine and marine sediments. This is a promising method that requires minimal perturbation (no addition of substrate) and offers increased sensitivity over measurement of N_2 flux alone, such that incubations of a day or less are sufficient.

$^{15}N_2$ Tracer Methods

Denitrification is very difficult to detect by direct ^{15}N tracer methods in the open ocean because of the dilute nature of the population and the huge background concentration of the product, N_2 gas. In recent years, the isotope-pairing approach (65) has been widely applied in sediments. For application to sediments, the nitrate in the overlying water can be very highly enriched with ^{15}N, either by replacing the ambient water with low-nitrate synthetic seawater or allowing the ambient nitrate to be depleted during preincubation prior to tracer addition. To further increase sensitivity, the headspace and overlying water can be purged of N_2 gas to enhance detection of labeled N_2 that is produced during the incubation. The isotope-pairing approach assumes that N_2 produced during denitrification results from random pairing of all NO_3^- molecules in the incubation. By knowing the isotopic enrichment of the overlying water NO_3^- pool to which the tracer is added and assuming that all NO_3^- produced by nitrification carries the natural abundance signature, the contributions of coupled nitrification/denitrification and denitrification of ambient NO_3^- can be determined independently.

The isotope-pairing approach assumes that the only source of N_2 is denitrification of NO_3^- or NO_2^-, and the discovery of anammox, in which N_2 is derived from ammonium plus NO_2^-, dealt the method a nearly fatal blow. It has been adapted, however, for the explicit purpose of measuring anammox itself by incorporating the addition of ^{15}N-labeled NH_4^+ separately or in addition to ^{15}N-labeled NO_3^- (91). Both conventional denitrification and anammox can be measured in ^{15}N incubations when three replicate incubations receive different isotope additions. The production of $^{30}N_2$ when $^{15}NO_3^-$ is added indicates conventional denitrification. The production of $^{29}N_2$ when either $^{15}NO_3^-$ or $^{15}NH_4^+$ is added indicates anammox. Initial application of the isotope-pairing approach to measure anammox included preincubation to remove ambient NO_3^- and NO_2^- and purging with He or Ar to remove ambient N_2. Subsequent applications have used fewer perturbations and more sophisticated calculations to detect the two processes in the presence of ambient N concentrations. Incubations are usually carried out in small-volume exetainers, always with large additions of labeled substrate, however, so true ambient rates have probably not been reported. Anammox has been detected in stratified water columns (31, 57) and in the turbid near-bottom waters on the Namibian African coast (56), but conventional denitrification has not yet been detected in the water column by using this method.

A few direct tracer incubations in which $^{15}N_2$ was detected after incubation with near-in situ levels of ^{15}N-enriched NO_3^- have been reported (A. H. Devol, unpublished data), and they indicate that in situ rates of denitrification in the oxygen minimum zones of the ocean are low and variable. Modification of the direct tracer approach to detect N_2O in the presence of acetylene should offer improved sensitivity.

The Acetylene Blockage Method

The most commonly used method to estimate denitrification rates is the acetylene blockage method. Acetylene specifically inhibits the nitrous oxide reductase enzyme and thus leads to the accumulation of nitrous oxide in an incubation to which acetylene has been added (86, 108). The accumulated nitrous oxide is sensitively detected with electron capture gas chromatography; the method is therefore analogous to the detection of N fixation by use of acetylene as an alternative substrate for nitrogenase and is carried out in a very similar manner. Still, the method lacks the sensitivity needed to detect in situ rates of denitrification and often requires the addition of nitrate. Thus, it should be considered a measure of potential denitrification and is very useful for comparing communities and populations under different environmental conditions but not for evaluating in situ rates.

For the water column of the open ocean, we are lacking adequate methods for quantitative evaluation of both in situ denitrification rates and denitrifying populations. Refinements in isotopic approaches are likely to resolve the first challenge in the near term, but the diversity of organisms with denitrification capability continues to complicate comprehensive population assessments. Improvements in rate measurements will also resolve the relative importance and absolute rates of both denitrification and anammox.

GLOBAL AND REGIONAL OCEANOGRAPHIC APPROACHES TO THE N CYCLE

Recent compilations of N transformations for the global ocean usually conclude that the budget is out of balance and that denitrification significantly exceeds N fixation (16, 26). Absolute balance between the inputs and outputs of fixed N need not be expected, and the relative balance likely varies over geological time. The long-range and large-scale implications of this lack of balance are beyond the scope of this method chapter, but they raise the important issue of how nitrogen transformation rates are regulated. Millennial-scale variations in the fixed N inventory must ultimately reflect the physiological constraints of the microorganisms responsible for the biochemical transformations at work in the microscale of microbial environments.

Nitrogen is considered the macroelement most likely to limit primary production, and the distribution of its various chemical forms is thus diagnostic of the dominant processes involved in its transformations. Redfield and coworkers long ago recognized that the relative concentrations of dissolved inorganic pools of N and P in the ocean tended to reflect the composition of the particulate material, whose mass was dominated by phytoplankton (71). The classic Redfield C:N:P ratio of 106:16:1 implies that biomass is produced and inorganic nutrients are regenerated according to this average ratio on a global basis. Therefore, excess inorganic N ($N:P > 16$) implies degradation of biomass derived from N_2 fixation, and excess P ($N:P < 16$) implies removal of inorganic N by denitrification. Gruber and Sarmiento (39) and Deutsch et al. (32) expanded and refined this approach with the concept of $N^* = N - 16P + 2.9$ (units = micromoles kilogram^{-1}). The absolute value of N^* is arbitrary, but deviations from the global mean are easily recognized; negative N^* values imply denitrification and positive values imply nitrogen fixation. The N^* definition relied on Anderson's (1) stoichiometry for conventional denitrification. The potential effect of anammox on the overall N stoichiometry is probably small but is still debated, partly because the biochemistry of the organisms involved is still not clear.

Deutsch et al. (32) applied the N* idea to the high-quality data set of inorganic nutrient concentrations from the World Ocean Circulation Experiment to evaluate the distributions and magnitudes of nitrogen fixation and denitrification in the Pacific Ocean. Average vertical N* profiles representing four major regions of the Pacific clearly illustrated the major features of nitrogen transformations on this large scale. The highest N* values were found in the surface of all four regions, and N* generally displayed a middepth minimum. Significant positive anomalies are characteristic of the surface waters of the northern and southern subtropical gyres, where N_2 fixation contributes to the N inventory, and significant negative anomalies characterized the denitrification zones in the eastern tropical North and South Pacific. On a basin-wide scale, Deutsch et al. (32) estimated a water column denitrification rate of 48 ± 5 Tg of N year^{-1} and a sediment rate of approximately one-third of the water column flux. After accounting for atmospheric deposition and net ocean circulation, these authors computed that a fixation rate of 59 Tg of N year^{-1} (39 mmol m^{-2} year^{-1}) would be required to balance the Pacific N budget. This value agreed well with the annual average of the time series data from the Central Pacific Station Aloha (31 to 51 mmol m^{-2} year^{-1} [51]).

Montoya et al. (63) presented the most extensive data set yet for direct measurements of N_2 fixation rates by using $^{15}N_2$ incorporation into biomass in the Pacific (transects in the eastern tropical South Pacific and in the Southwestern Pacific). Their rates ranged from very similar (46 mmol m^{-2} year^{-1}) to the previous average for the Central Pacific at Station Aloha to much higher rates in other regions of the Pacific (190 mmol m^{-2} year^{-1} in the eastern North Pacific gyre) and very much higher rates (1,458 mmol m^{-2} year^{-1}) in the pigment maximum in the Arafura Sea, a shallow coastal region north of Australia. This huge range of rates illustrates the difficulty of comparing integrative biogeochemical estimates with direct measurements. Montoya et al. (63) compiled N fixation rate estimates from various regions based on both acetylene reduction and ^{15}N assays and documented a range of daily rates greater than 1,000-fold. Highest rates overall are reported from the Atlantic, where much progress has been made in integrating remote sensing, ocean circulation modeling, and incubation-based measurements to obtain a detailed picture of nitrogen fixation on the basin scale (27, 28, 46).

CURRENT AND FUTURE DEVELOPMENTS

Direct measurements of in situ denitrification rates in the open ocean are much rarer than those of nitrogen fixation. In addition, debate is currently quite active over the relative importance of conventional denitrification and anammox. Most previous estimates of denitrification rates based on biogeochemical fluxes included both processes. Recent advances in the study of the anammox process have highlighted how poorly conventional denitrification is understood, especially in the open ocean, and have focused attention on the need to improve methods for the direct in situ measurement of denitrification rates. The next five years of N cycle research will probably resolve the issue of their relative contributions, their distributions in the marine environment, and the best way to assess and distinguish the two processes.

New developments in molecular ecology are likely to continue, and their impact may spread beyond the area of population assessment where they have so far had the biggest influence. Genomic sequences from cultivated organisms will provide the basis for assays for detection of many genes involved in different pathways of the N transformations. This will improve our ability to study regulation of activity in response to environmental variables.

Metagenomic approaches will allow us to link function, regulation, and population assessments by allowing us to identify multiple genes and operons and to link them to the identity of uncultivated organisms. While not directly related to the rates of biogeochemical processes in the environment, these kinds of genetic information do relate to regulation of rates. The net observed transformation rates depend on the environmental responses of different components of the microbial assemblage.

For N_2 fixation, *Trichodesmium* is clearly the dominant organism in oceanic systems, and much can be learned about the system by understanding *Trichodesmium*, even though there is additional vast diversity among N_2-fixing assemblages. The extent of dominance and diversity in denitrifying and nitrifying assemblages and the distribution of particular phylotypes in the ocean are not well known. Obtaining cultures of organisms that represent clades that are important in clone libraries would be a first step toward understanding how the diverse components of denitrifying and nitrifying assemblages contribute to ecosystem function.

REFERENCES

1. **Anderson, L. A.** 1995. On the hydrogen and oxygen content of marine phytoplankton. *Deep Sea Res.* A **42:**1675–1680.

1a. **Andersson, M. G. I., N. Brion, and J. J. Middelburg.** 2006. Comparison of nitrifier activity versus growth in the Scheldt estuary—a turbid, tidal estuary in northern Europe, *Aquat. Microb. Ecol.* **42:**149–158.

2. **Bano, N., and J. T. Hollibaugh.** 2000. Diversity and distribution of DNA sequences with affinity to ammonia-oxidizing bacteria of the β subdivision of the class *Proteobacteria* in the Arctic Ocean. *Appl. Environ. Microbiol.* **66:**1960–1969.

3. **Bartosch, S., I. Wolgst, E. Speick, and E. Bock.** 1999. Identification of nitrite-oxidizing bacteria with monoclonal antibodies recognizing the nitrite oxidoreductase. *Appl. Environ. Microbiol.* **65:**4126–4133.

4. **Bedard, C., and R. Knowles** 1989. Physiology, biochemistry, and specific inhibitors of CH_4, NH_4^+ and CO oxidation by methanotrophs and nitrifiers. *Microbiology* (Reading) **53:**68–84.

5. **Bianchi, M., F. Feliatra, P. Tréguer, M.-A. Vincendeau, and J. Morvan.** 1997. Nitrification rates, ammonium and nitrate distribution in upper layers of the water column and in sediments of the Indian sector of the Southern Ocean. *Deep Sea Res. Part II* **44:**1017–1032.

6. **Bianchi, M., C. Fosset, and P. Conan.** 1999. Nitrification rates in the NW Mediterranean Sea. *Aquat. Microb. Ecol.* **17:**267–278.

7. **Biesterfeld, S., P. Russell, and L. Figueroa.** 2003. Linking nitrifying biofilm structure and function through fluorescent in situ hybridization and evaluation of nitrification capacity. *Water Environ. Res.* **75:**205–215.

8. **Billen, G.** 1976. Evaluation of nitrifying activity in sediments by dark ^{14}C-bicarbonate incorporation. *Water Res.* **10:**51–57.

9. **Binnerup, S. J., K. Jensen, N. P. Revsbech, M. H. Jensen, and J. Sørensen.** 1992. Denitrification, dissimilatory reduction of nitrate to ammonium, and nitrification in a bioturbated estuarine sediment as measured with ^{15}N and microsensor techniques. *Appl. Environ. Microbiol.* **58:** 303–313.
10. **Bird, C., J. M. Martinez, A. G. O'Donnell, and M. Wyman.** 2005. Spatial distribution and transcriptional activity of an uncultured clade of planktonic diazotrophic gamma-proteobacteria in the Arabian Sea. *Appl. Environ. Microbiol.* **71:**2079–2085.
11. **Braker, G., H. L. Ayala-del-Rio, A. H. Devol, A. Fesefeldt, and J. M. Tiedje.** 2001. Community structure of denitrifiers, *Bacteria*, and *Archaea* along redox gradients in Pacific Northwest marine sediments by terminal restriction fragment length polymorphism analysis of amplified nitrite reductase (*nirS*) and 16S rRNA genes. *Appl. Environ. Microbiol.* **67:**1893–1901.
12. **Braker, G., A. Fesefeldt, and K.-P. Witzel.** 1998. Development of PCR primer systems for amplification of nitrite reductase genes (*nirK* and *nirS*) to detect denitrifying bacteria in environmental samples. *Appl. Environ. Microbiol.* **64:**3769–3775.
13. **Caffrey, J. M., N. E. Harrington, I. P. Solem, and B. B. Ward.** 2003. Biogeochemical processes in a small California estuary, Elkhorn Slough, CA: 2. Nitrification activity, community structure and role in nitrogen budgets. *Mar. Ecol. Prog. Ser.* **248:**27–40.
14. **Caffrey, J. M., and L. G. Miller.** 1995. A comparison of two nitrification inhibitors used to measure nitrification rates in estuarine sediments. *FEMS Microbiol. Ecol.* **17:** 213–219.
15. **Capone, D. G.** 1993. Determination of nitrogenase activity in aquatic samples using the acetylene reduction procedure, p. 621-631. *In* P. F. Kemp, B. F. Sherr, E. B. Sherr, and J. J. Cole (ed.), *Handbook of Methods in Aquatic Microbial Ecology*. Lewis Publishers, Boca Raton, Fla.
16. **Capone, D. G.** 2000. The marine nitrogen cycle, p. 455-493. *In* D. L. Kirchman (ed.), *Microbial Ecology of the Ocean*. Wiley-Liss, New York, N.Y.
17. **Capone, D. G.** 2001. Marine nitrogen fixation: what's the fuss? *Cur. Opin. Microbiol.* **4:**341–348.
18. **Capone, D. G.** 2002. Microbial nitrogen cycling, p. 439–449. *In* C. J. Hurst, R. L. Crawford, G. R. Knudsen, M. J. McInerney, and L. D. Stetzenbach (ed.), *Manual of Environmental Microbiology*, 2nd ed. ASM Press, Washington, D.C.
19. **Capone, D. G., J. A. Burns, J. P. Montoya, A. Subramaniam, C. Mahaffey, T. Gunderson, A. F. Michaels, and E. J. Carpenter.** 2005. Nitrogen fixation by *Trichodesmium* spp.: an important source of new nitrogen to the tropical and subtropical North Atlantic Ocean. *Global Biogeochem. Cycles* **19**. doi:10.1029/2004GB002331.
20. **Carpenter, E. J., A. Subramaniam, and D. G. Capone.** 2004. Biomass and primary productivity of the cyanobacterium *Trichodesmium* spp. in the tropical N Atlantic ocean. *Deep Sea Res.* **51:**173–203.
21. **Castro-Gonzalez, M., G. Braker, L. Farias, and O. Ulloa.** 2005. Communities of nirS-type denitrifiers in the water column of the oxygen minimum zone in the eastern South Pacific. *Environ. Microbiol.* **7:**1298–1306. [Online.] doi: 10.1111/j.1462–2920.
22. **Castro-Gonzalez, M., and L. Farias.** 2004. N_2O cycling at the core of the oxygen minimum zone off northern Chile. *Mar. Ecol. Prog. Ser.* **280:**1–11.
23. **Chang, J.** 2000. Precision of different methods used for estimating the abundance of the nitrogen-fixing marine cyanobacterium, Trichodesmium Ehrenberg. *J. Exp. Mar. Biol. Ecol.* **245:**215–224.
24. **Chen, F., and R. E. Hodson.** 2001. In situ PCR/RT-PCR coupled with in situ hybridization for detection of functional gene and gene expression in prokaryotic cells. *Methods Microbiol.* **30:**409–424.
25. **Church, M. J., B. D. Jenkins, D. M. Karl, and J. P. Zehr.** 2005. Vertical distributions of nitrogen-fixing phylotypes at Stn ALOHA in the oligotrophic North Pacific Ocean. *Aquat. Microb. Ecol.* **38:**3–14.
26. **Codispoti, L. A., J. A. Brandes, J. P. Christensen, A. H. Devol, S. W. A. Naqvi, H. W. Paerl, and T. Yoshinari.** 2001. The oceanic fixed nitrogen and nitrous oxide budgets: moving targets as we enter the anthropocene? *Sci. Mar.* **65:**85–102.
27. **Coles, V. J., R. R. Hood, M. Pascual, and D. G. Capone.** 2004. Modeling the impact of *Trichodesmium* and nitrogen fixation in the Atlantic Ocean. *J. Geophys. Res. Oceans* **109**. doi:10.1029/2002JC001754.
28. **Coles, V. J., C. Wilson, and R. R. Hood.** 2004. Remote sensing of new production fuelled by nitrogen fixation. *Geophys. Res. Lett.* **31**. doi:10.1029/2003GLO19018.
29. **Coyne, M., A. Arundkumari, B. Averill, and J. M. Tiedje.** 1989. Immunological identification and distribution of dissimilatory heme cd1 and nonheme copper nitrite reductases in denitrifying bacteria. *Appl. Environ. Microbiol.* **55:**2924–2931.
30. **Currin, C. A., H. W. Paerl, G. Suba, and R. S. Alberte.** 1990. Immunofluorescent detection and characterization of N_2-fixing microorganisms from aquatic environments. *Limnol. Oceanogr.* **35:**59–71.
31. **Dalsgaard, T., D. E. Canfield, J. Petersen, B. Thamdrup, and J. Acuna-Gonzalez.** 2003. N_2 production by the anammox reaction in the anoxic water column of Golfo Dulce, Costa Rica. *Nature* **422:**606–608.
32. **Deutsch, C., N. Gruber, R. M. Key, J. L. Sarmiento, and A. Ganachaud.** 2001. Denitrification and N-2 fixation in the Pacific Ocean. *Global Biogeochem. Cycles* **15:**483–506.
32a. **Devol, A. H.** 1991. Direct measurement of nitrogen gas fluxes from continental-shelf sediments. *Nature* **349:**319–321.
32b. **Bionisi, H. M., A. C. Layton, G. Harms, I. R. Gregory, K. G. Robinson, and G. S. Saylor.** 2002. Quantification of *Nitrosomonas oligotropha*-like ammonia-oxidizing bacteria and *Nitrospira* spp. from full-scale wastewater treatment plants by competitive PCR. *Appl. Environ. Microbiol.* **68:**245–253.
33. **Dore, J. E., and D. M. Karl.** 1996. Nitrification in the euphotic zone as a source for nitrite, nitrate, and nitrous oxide at station ALOHA. *Limnol. Oceanogr.* **41:**1619–1628.
34. **Falcon, L. I., E. J. Carpenter, F. Cipriano, B. Bergman, and D. G. Capone.** 2004. $N\text{-}_2$ fixation by unicellular bacterioplankton from the Atlantic and Pacific oceans: phylogeny and in situ rates. *Appl. Environ. Microbiol.* **70:**765–770.
35. **Falcon, L. I., F. Cipriano, A. Y. Chistoserdov, and E. J. Carpenter.** 2002. Diversity of diazotrophic unicellular cyanobacteria in the tropical North Atlantic Ocean. *Appl. Environ. Microbiol.* **68:**5760–5764.
36. **Feray, C., B. Volat, V. Degrange, A. Clays-Josserand, and B. Montuelle.** 1999. Assessment of three methods for detection and quantification of nitrite-oxidizing bacteria and *Nitrobacter* in freshwater sediments (MPN-PCR, MPN-Griess, immunofluorescence). *Microb. Ecol.* **37:** 208–217.
37. **Francis, C. A., K. J. Roberts, M. J. Beman, A. E. Santoro, and B. B. Oakley.** 2005. Ubiquity and diversity of ammonia-oxidizing archaea in water columns and sediments of the ocean. *Proc. Natl. Acad. Sci. USA.* **102:** 14683–14688.

38. **Gieseke, A., L. Purkhold, M. Wagner, R. Amann, and A. Schramm.** 2001. Community structure and activity dynamics of nitrifying bacteria in a phosphate-removing biofilm. *Appl. Environ. Microbiol.* **67:**1351–1362.
39. **Gruber, N., and J. L. Sarmiento.** 1997. Global patterns of marine nitrogen fixation and denitrification. *Global Biogeochem. Cycles* **11:**235–266.
40. **Gruntzig, V., S. C. Nold, J. Zhou, and J. M. Tiedje.** 2001. *Pseudomonas stutzeri* nitrite reductase gene abundance in environmental samples measured by real-time PCR. *Appl. Environ. Microbiol.* **67:**760–768.
41. **Hall, G. H.** 1986. Nitrification in lakes, p. 127–156. *In* J. I. Prosser (ed.), *Nitrification.* IRL Press, Oxford, United Kingdom.
41a. **Harms, G., A. C. Layton, H. M. Dionisi, I. R. Gregory, V. M. Garrett, S. A. Hawkins, K. G. Robinson, and G. S. Sayler.** 2003. Real-time PCR quantification of nitrifying bacteria in a municipal wastewater treatment plant. *Environ. Sci. Technol.* **37:**343–351.
42. **Heck, K. L., G. van Belle, and D. Simberloff.** 1975. Explicit calculation of the rarefaction diversity measurement and the determination of sufficient sample size. *Ecology* **56:**1459–1461.
43. **Henricksen, K., J. I. Hansen, and T. H. Blackburn.** 1981. Rates of nitrification, distribution of nitrifying bacteria, and nitrate fluxes in different types of sediment from Danish waters. *Mar. Biol.* **61:**299–304.
44. **Herbert, R.** 1999. Nitrogen cycling in coastal marine ecosystems. *FEMS Microbiol. Rev.* **23:**563–590.
45. **Hollibaugh, J. T., N. Bano, and H. W. Ducklow.** 2002. Widespread distribution in polar oceans of a 16S rRNA gene sequence with affinity to *Nitrosospira*-like ammonia-oxidizing bacteria. *Appl. Environ. Microbiol.* **68:**1478–1484.
46. **Hood, R. R., V. J. Coles, and D. G. Capone.** 2004. Modeling the distribution of Trichodesmium and nitrogen fixation in the Atlantic Ocean. *J. Geophys. Res. Oceans* **109.** doi:10.1029/2002JC001753.
47. **Janson, S., B. Bergman, E. J. Carpenter, S. J. Giovannoni, and K. Vergin.** 1999. Genetic analysis of natural populations of the marine diazotrophic cyanobacterium *Trichodesmium*. *FEMS Microbiol. Ecol.* **30:**57–65.
48. **Jayakumar, D. A., C. A. Francis, S. W. A. Naqvi, and B. B. Ward.** 2004. Diversity of nitrite reductase genes (*nirS*) in the denitrifying water column of the coastal Arabian Sea. *Aquat. Microb. Ecol.* **34:**69–78.
49. **Kana, T. M., M. B. Sullivan, J. C. Cornwell, and K. M. Groszkowski.** 1998. Denitrification in estuarine sediments determined by membrane inlet mass spectrometry. *Limnol. Oceanogr.* **43:**334–339.
50. **Kaplan, W., I. Valiella, and J. M. Teal.** 1979. Denitrification in a saltmarsh ecosystem. *Limnol. Oceanogr.* **24:** 726–734.
51. **Karl, D. M., R. Letelier, L. Tupas, J. E. Dore, J. Christian, and D. Hebel.** 1997. The role of nitrogen fixation in the biogeochemical cycling in the subtropical North Pacific Ocean. *Nature* **388:**533–538.
52. **Kester, R. A., L. de Boer, and H. J. Laanbroek.** 1996. Short exposure to acetylene to distinguish between nitrifier and denitrifier nitrous oxide production in soil and sediment samples. *FEMS Microbiol. Ecol.* **20:**111–120.
53. **Konnecke, M., A. E. Berhnard, J. R. de la Torre, C. B. Walker, J. B. Waterbury, and D. A. Stahl.** 2005. Isolation of an autotrophic ammonia-oxidizing marine archaeon. *Nature* **437:**543–546.
54. **Koops, H.-P., U. Purkhold, A. Pommerening-Roser, G. Timmermann, and M. Wagner.** 2003. The lithoautotrophic ammonia-oxidizing bacteria. *In* M. Dworkin (ed.), *The Prokaryotes: An Evolving Electronic Resource for the Microbiological Community.* Springer-Verlag, New York, N.Y. [Online.] http://141.150.157.117.8080/prokPUB/index.htm.
55. **Kowalchuk, G. A., and J. R. Stephen.** 2001. Ammonia-oxidizing bacteria: a model for molecular microbial ecology. *Annu. Rev. Microbiol.* **55:**485–529.
56. **Kuypers, M. M., G. Lavik, D. Woebken, M. Schmid, B. M. Fuchs, R. Amann, B. B. Jorgensen, and M. S. M. Jetten.** 2005. Massive nitrogen loss from the Benguela upwelling system through anaerobic ammonium oxidation. *Proc. Natl. Acad. Sci. USA* **102:**6478–6483. [Online.] doi/10.11073/pnas.0502088102.
57. **Kuypers, M. M. M., A. O. Sliekers, G. Lavik, M. Schmid, B. B. Jorgensen, J. G. Kuenen, J. S. S. Damste, M. Strous, and M. S. M. Jetten.** 2003. Anaerobic ammonium oxidation by anammox bacteria in the Black Sea. *Nature* **422:**608–611.
58. **Laursen, A. E., and S. P. Seitzinger.** 2002. The role of denitrification in nitrogen removal and carbon mineralization in Mid-Atlantic Bight sediments. *Cont. Shelf Res.* **22:** 1397–1416.
59. **Lipschultz, F., S. C. Wofsy, and L. E. Fox.** 1986. Nitrogen-metabolism of the eutrophic Delaware River ecosystem. *Limnol. Oceanogr.* **31:**701–716.
60. **Lipschultz, F., S. C. Wofsy, B. B. Ward, L. A. Codispoti, G. J. W. Friedrich, and J. W. Elkins.** 1990. Bacterial transformations of inorganic nitrogen in the oxygen-deficient waters of the Eastern Tropical South Pacific Ocean. *Deep Sea Res.* **37:**1513–1541.
61. **Mazard, S. L., N. J. Fuller, K. M. Orcutt, O. Bridle, and D. J. Scanlan.** 2004. PCR analysis of the distribution of unicellular cyanobacterial diazotrophs in the Arabian Sea. *Appl. Environ. Microbiol.* **70:**7355–7364.
62. **Miller, L. G., M. D. Coutlakis, R. S. Oremland, and B. B. Ward.** 1993. Selective inhibition of nitrification (ammonium oxidation) by methylfluoride and dimethyl ether. *Appl. Environ. Microbiol.* **59:**2457–2464.
63. **Montoya, J. P., C. M. Holl, J. P. Zehr, A. Hansen, T. A. Villareal, and D. G. Capone.** 2004. High rates of N-2 fixation by unicellular diazotrophs in the oligotrophic Pacific Ocean. *Nature* **430:**1027–1031.
64. **Montoya, J. P., M. Voss, P. Kaehler, and D. G. Capone.** 1996. A simple, high-precision, high-sensitivity tracer assay for N_2 fixation. *Appl. Environ. Microbiol.* **62:**986–993.
65. **Nielsen, L.** 1992. Denitrification in sediment determined from nitrogen isotope pairing. *FEMS Microbiol. Ecol.* **86:** 357–362.
66. **Nowicki, B. L.** 1994. The effect of temperature, oxygen, salinity, and nutrient enrichment on estuarine denitrification rates measured with a modified nitrogen gas flux technique. *Estuar. Coast. Shelf Sci.* **35:**137–156.
67. **O'Mullan, G. D., and B. B. Ward.** 2005. Relationship of temporal and spatial variabilities of ammonia-oxidizing bacteria to nitrification rates in Monterey Bay, CA. *Appl. Environ. Microbiol.* **71:**697–705.
68. **Orcutt, K. M., F. Lipschultz, K. Gundersen, R. Arimoto, A. F. Michaels, A. H. Knap, and J. R. Gallon.** 2001. A seasonal study of the significance of N-2 fixation by Trichodesmium spp. at the Bermuda Atlantic Time-series Study (BATS) site. *Deep Sea Res. Part II* **48:**1583–1608.
69. **Paerl, H., and J. P. Zehr.** 2000. Marine nitrogen fixation, p. 387-426. *In* D. L. Kirchman (ed.), *Microbial Ecology of the Oceans*, Wiley-Liss, New York, N.Y.
70. **Pernthaler, J., A. Pernthaler, and R. Amann.** 2002. Fluorescence in situ hybridization and catalyzed reporter deposition for the identification of marine bacteria. *Appl. Environ. Microbiol.* **68:**3094–3101.

71. **Redfield, A. C., B. H. Ketchum, and F. A. Richards.** 1963. The influence of organisms on the composition of seawater, p. 26-77. *In* M. N. Hill (ed.), *The Sea*, vol. 2. Wiley-Interscience, Hoboken, N.J.
72. **Risgaard-Petersen, N., M. H. Nicolaisen, N. P. Revsbech, and B. A. Lomstein.** 2004. Competition between ammonia-oxidizing bacteria and benthic microalgae. *Appl. Environ. Microbiol.* **70:**5528–5537.
72a. **Robinson, K. G., H. M. Dionisi, G. Harms, A. C. Layton, I. R. Gregory, and G. S. Sayler.** 2003. Molecular assessment of ammonia- and nitrite-oxidizing bacteria in full-scale activiated sludge wastewater treatment plants. *Water Sci. Technol.* **48:**119–126.
73. **Sarangi, R. K., P. Chauhan, and S. R. Nayak.** 2004. Detection and monitoring of Trichodesmium blooms in the coastal waters off Saurashtra coast, India, using IRS-P4 OCM data. *Curr. Sci.* **86:**1636–1641.
74. **Scala, D. J., and L. J. Kerkhof.** 1998. Nitrous oxide reductase (*nosZ*) gene-specific PCR primers for detection of denitrifiers and three *nosZ* genes from marine sediments. *FEMS Microbiol. Lett.* **162:**61–68.
75. **Scala, D. J., and L. J. Kerkhof.** 2001. Horizontal heterogeneity of denitrifying bacterial communities in marine sediments by terminal restriction fragment length polymorphism analysis. *Appl. Environ. Microbiol.* **66:**1980–1986.
76. **Schleper, C., G. Jurgens, and M. Jonuscheit.** 2005. Genomic studies of uncultivated Archaea. *Nat. Rev. Microbiol.* **3:**479–488.
77. **Schloss, P. D., and J. Handelsman.** 2005. Introducing DOTUR, a computer program for defining operational taxonomic units and estimating species richness. *Appl. Environ. Microbiol.* **71:**1501–1506.
78. **Schmid, M., S. Schmitz-Esser, M. S. M. Jetten, and M. Wagner.** 2001. 16A-23A rDNA intergenic spacer and 23S rDNA of anaerobic ammonium-oxidizing bacteria: implications for phylogeny and in situ detection. *Environ. Microbiol.* **3:**450–459.
78a: **Schramm, A., D. de Beer, M. Wagner, and R. Amann.** 1998. Identification and activities in situ of *Nitrosospira* and *Nitrospira* spp. as dominant populations in a nitrifying fluidized bed reactor. *Appl. Environ. Microbiol.* **64:**3480–3485.
79. **Seitzinger, S. P.** 1988. Denitrification in freshwater and coastal marine ecosystems: ecological and geochemical significance. *Limnol. Oceanogr.* **33:**702–724.
80. **Sekar, R., B. M. Fuchs, R. Amann, and J. Pernthaler.** 2004. Flow sorting of marine bacterioplankton after fluorescence in situ hybridization. *Appl. Environ. Microbiol.* **70:**6210–6219.
81. **Short, S. M., B. D. Jenkins, and J. P. Zehr.** 2004. Spatial and temporal distribution of two diazotrophic bacteria in the Chesapeake Bay. *Appl. Environ. Microbiol.* **70:**2186–2192.
82. **Sigman, D. M., M. A. Altabet, R. H. Michener, D. C. McCorkle, B. Fry, and R. M. Holmes.** 1997. Natural abundance-level measurement of the nitrogen isotopic composition of oceanic nitrate: an adaptation of the ammonia diffusion method. *Mar. Chem.* **57:**227–242.
83. **Sigman, D. M., K. L. Casciotti, M. Andreani, C. Barford, M. Galanter, and J. K. Bohlke.** 2001. A bacterial method for the nitrogen isotopic analysis of nitrate in seawater and freshwater. *Anal. Chem.* **73:**4145–4153.
84. **Somville, M.** 1978. A method for the measurement of nitrification rates in water. *Water Res.* **12:**843–848.
85. **Song, B., M. M. Haggblom, J. Z. Zhou, J. M. Tiedje, and N. J. Palleroni.** 1999. Taxonomic characterization of denitrifying bacteria that degrade aromatic compounds and description of *Azoarcus toluvorans* sp. nov. and *Azoarcus toluclasticus* sp. nov. *Int. J. Syst. Bacteriol.* **49:**1129–1140.
86. **Sorensen, J.** 1978. Denitrification rates in a marine sediment as measured by the acetylene inhibition technique. *Appl. Environ. Microbiol.* **36:**139–143.
87. **Stackebrandt, E., and B. M. Goebel.** 1994. Taxonomic note: a place for DNA-RNA reassociation and 16S rRNA sequence analysis in the present species definition of bacteriology. *Int. J. Syst. Bacteriol.* **44:**846–849.
88. **Stewart, W. D. P., G. P. Fitzgerald, and R. H. Burris.** 1967. In situ studies on N_2 fixation, using the acetylene reduction technique. *Proc. Natl. Acad. Sci. USA* **58:**2071–2078.
89. **Subramaniam, A., E. Carpenter, D. Karentz, and P. Falkowski.** 1999. Bio-optical properties of the marine diazotrophic cyanobacteria *Trichodesmium* spp. II. A reflectance model for remote sensing. *Limnol. Oceanogr.* **44:**618–627.
90. **Suzuki, M. T., L. T. Taylor, and E. F. DeLong.** 2000. Quantitative analysis of small-subunit rRNA genes in mixed microbial populations via 5′-nuclease assays. *Appl. Environ. Microbiol.* **66:**4605–4614.
91. **Thamdrup, B., and T. Dalsgaard.** 2002. Production of N_2 through anaerobic ammonium oxidation coupled to nitrate reduction in marine sediments. *Appl. Environ. Microbiol.* **68:**1312–1318.
92. **Vandegraaf, A. A., A. Mulder, P. Debruijn, M. S. M. Jetten, L. A. Robertson, and J. G. Kuenen.** 1995. Anaerobic oxidation of ammonium is a biologically mediated process. *Appl. Environ. Microbiol.* **61:**1246–1251.
93. **Venter, C. J., K. Remington, J. G. Heidelberg, A. L. Halpern, D. Rusch, J. A. Eisen, D. Wu, I. Paulsen, K. E. Nelson, W. Nelson, D. E. Fouts, S. Levy, A. H. Knap, M. W. Lomas, K. Nealson, O. White, J. Peterson, J. Hoffman, R. Parsons, H. Baden-Tillson, C. Pfannkoch, J.-H. Rogers, and H. O. Smith.** 2004. Environmental genome shotgun sequencing of the Sargasso Sea. *Science.* **304:**66–74.
94. **Voytek, M. A., J. C. Priscu, and B. B. Ward.** 1999. The distribution and relative abundance of ammonia-oxidizing bacteria in lakes of the McMurdo Dry Valley, Antarctica. *Hydrobiologia* **401:**113–130.
95. **Voytek, M. A., and B. B. Ward.** 1995. Detection of ammonium-oxidizing bacteria of the beta-subclass of the class *Proteobacteria* in aquatic samples with the PCR. *Appl. Environ. Microbiol.* **61:**1444–1450.
96. **Voytek, M. A., B. B. Ward, and J. C. Priscu.** 1998. The abundance of ammonium-oxidizing bacteria in Lake Bonney, Antarctica determined by immunofluorescence, PCR and in situ hybridization, p. 217–228. *In* J. C. Priscu (ed.), *Ecosystem Dynamics in a Polar Desert: the McMurdo Dry Valleys, Antarctica.* American Geophysical Union, Washington, D.C.
97. **Ward, B. B.** 1982. Oceanic distribution of ammonium-oxidizing bacteria determined by immunofluorescent assay. *J. Mar. Res.* **40:**1155–1172.
98. **Ward, B. B.** 2000. Nitrification and the marine nitrogen cycle, p. 427–453. *In* D. L. Kirchman (ed.), *Microbial Ecology of the Oceans.* John Wiley & Sons, Inc., New York, N.Y.
99. **Ward, B. B.** 2001. Nitrification in aquatic systems, p. 2144–2167. *In* D. G. Capone (ed.), *The Encyclopedia of Environmental Microbiology.* John Wiley & Sons, Inc., New York, N.Y.
100. **Ward, B. B.** 2005. Temporal variability in nitrification rates and related biogeochemical factors in Monterey Bay, California, USA. *Mar. Ecol. Prog. Ser.* **292:**97–109.
101. **Ward, B. B.** Nitrification in marine systems. *In* D. G. Capone, D. A. Bronk, M. R. Mulholland, and E. Carpenter

(ed.), *Nitrogen in the Marine Environment*, in press. Academic Press, San Diego, Calif.

102. **Ward, B. B., and A. F. Carlucci.** 1985. Marine ammonium- and nitrite-oxidizing bacteria: serological diversity determined by immunofluorescence in culture and in the environment. *Appl. Environ. Microbiol.* **50:**194–201.
103. **Ward, B. B., and A. R. Cockroft.** 1993. Immunofluorescence detection of the denitrifying bacterium, *Pseudomonas perfectomarina*, in seawater and intertidal sediment environments. *Microb. Ecol.* **25:**233–246.
104. **Ward, B. B., A. R. Cockroft, and K. A. Kilpatrick.** 1993. Antibody and DNA probes for detection of nitrite reductase in seawater. *J. Gen. Microbiol.* **139:**2285–2293.
105. **Ward, B. B., K. A. Kilpatrick, E. Renger, and R. W. Eppley.** 1989. Biological nitrogen cycling in the nitracline. *Limnol. Oceanogr.* **34:**493–513.
106. **Ward, B. B., and G. D. O'Mullan.** 2005. Community level analysis: genetic and biogeochemical approaches to investigate community composition and function in aerobic ammonia oxidation, p. 395–413. *In* J. R. Leadbetter (ed.), *Environmental Microbiology*. Academic Press, San Diego, Calif.
107. **Ward, B. B., and O. C. Zafiriou.** 1988. Nitrification and nitric-oxide in the oxygen minimum of the eastern tropical North Pacific. *Deep Sea Res.* A **35:**1127–1142.
108. **Yoshinari, T., R. Hynes, and R. Knowles.** 1977. Acetylene inhibition of nitrous oxide reduction and measurement of denitrification and nitrogen fixation in soil. *Soil Biol. Biochem.* **9:**177–183.
109. **Zaccone, R., G. Caruso, and M. Azzaro.** 1996. Detection of *Nitrosococcus oceanus* in a Mediterranean lagoon by immunofluorescence. *J. Appl. Bacteriol.* **80:**611–616.
110. **Zani, S., M. T. Mellon, J. L. Collier, and J. P. Zehr.** 2000. Expression of *nifH* genes in natural microbial assemblages in Lake George, New York, detected by reverse transcriptase PCR. *Appl. Environ. Microbiol.* **66:**3119–3124.
111. **Zehr, J. P., B. D. Jenkins, S. M. Short, and G. F. Steward.** 2003. Nitrogenase gene diversity and microbial community structure: a cross-system comparison. *Environ. Microbiol.* **5:**539–554.
112. **Zehr, J. P., and L. A. McReynolds.** 1989. Use of degenerate oligonucleotides for amplification of the *nifH* gene from the marine cyanobacterium *Trichodesmium theiebautii*. *Appl. Environ. Microbiol.* **55:**2522–2526.
113. **Zehr, J. P., M. T. Mellon, and S. Zani.** 1998. New nitrogen-fixing microorganisms detected in oligotrophic oceans by amplification of nitrogenase (*nifH*) genes. *Appl. Environ. Microbiol.* **6:**3444–3450.
114. **Zehr, J. P., and S. M. Short.** 2005. Quantitative analysis of *nifH* genes and transcripts from aquatic environments. *Methods Enzymol.* **397:**380–394.
115. **Zehr, J. P., J. B. Waterbury, P. J. Turner, J. P. Montoya, E. Omoregie, G. F. Steward, A. Hansen, and D. M. Karl.** 2001. Unicellular cyanobacteria fix N_2 in the subtropical North Pacific Ocean. *Nature* **412:**635–638.
116. **Zumft, W. G.** 1997. Cell biology and molecular basis of denitrification. *Microbiol. Mol. Biol. Rev.* **61:**533–616.

The Marine Phosphorus Cycle

DAVID M. KARL

43

Phosphorus (P) is an element required for the growth of all marine organisms. The delivery of P to the global ocean is controlled primarily by the combined influences of continental weathering and fluvial discharge (34, 69). These processes are balanced over geological time scales by P burial in deep-sea sediments. During the past few centuries, the marine P cycle has been perturbed by anthropogenic forcing of both inputs (riverine, sewage loading, and atmospheric deposition) and outputs (organic P accumulations in coastal zones and enhanced P export to the open ocean) to the system, with dire projections for the future (66).

Within the pelagic environment, P exists in a variety of inorganic and organic forms that are interconverted between dissolved and particulate matter, creating what is known as the marine P cycle (Fig. 1). With few exceptions, P in the marine environment is present in the pentavalent state (+5), as PO_4^{3-}, whether as free orthophosphate (P_i) or as P incorporated into organic compounds. The presence of P as either phosphine gas (PH_3), phosphide (−3), or phosphite (+3) has been reported for environmental samples (71), but these forms are probably rare in open ocean ecosystems.

Most of the P transformations and fluxes (the arrows between the boxes in Fig. 1) in the marine P cycle result from the metabolic activities of microorganisms, especially prokaryotes (bacteria and archaea). These dynamic P cycle processes are fueled ultimately by solar energy captured via photosynthesis. In some marine ecosystems, P may actually limit solar energy capture and, therefore, the total metabolism of the biome in question. Furthermore, because the marine P cycle is inextricably linked to the cycles of carbon (C), oxygen (O), nitrogen (N), sulfur (S), and selected trace elements (Fe, Zn, Co, and Cu), as well as to energy flow, it needs to be studied as an integral component of the ecosystem, ranging from genomes to biomes (28; also see http://cmore.soest.hawaii.edu).

Within cells and tissues, P is vital for both structure (e.g., membranes) and function (e.g., information storage and energy transduction); living organisms are truly built around P (24). So fundamental are the roles of phospholipids, nucleotides, and nucleic acids that it is surprising that the marine P cycle is not studied or understood more comprehensively. This current situation derives, in large part, from a historical focus on the role of N as the production rate-limiting nutrient in the sea (83). This contrasts sharply with investigations of freshwater ecosystems, where P is well established as a key environmental variable (39, 45). However, the role of P in the dynamics of marine ecosystems is beginning to change (2, 48, 91). A brief status report is that we have made significant progress, especially during the past 2 decades, in our general understanding of the marine P cycle but that equally significant challenges remain.

MARINE P CYCLE METHODOLOGIES

Field investigation of the marine P cycle requires the use of a variety of methods to measure the ambient concentrations of total dissolved and particulate (both organic and inorganic) matter to assess local inventories of P and to estimate P fluxes. The latter include the delivery to and losses from the ecosystem in question and the rates of microbial P uptake and microbial decomposition of P-containing organic matter. Some analytical procedures are nearly a century old and are still used today (5, 29); others are just being field tested (20).

The analysis of P compounds in seawater is neither simple nor straightforward (52, 53). Strickland and Parsons (86), in their classic treatise on methods of seawater analysis, defined eight different operational classes of P compounds based on reactivity to the "standard" acidic molybdate colorimetric reagents, ease of hydrolysis, and molecular weight/particle size. These pools range from "inorganic, soluble and reactive," presumably P_i, through "enzyme hydrolyzable phosphate" (P_i released following treatment with the enzyme alkaline phosphomonoesterase) to "inorganic, particulate and unreactive" (presumably the P-containing mineral fraction). Some of these classes have no accurate method of direct analysis, so concentrations must be inferred by measuring the residual P following removal of readily defined pools. These procedures can help to constrain certain pools but carry with them a large analytical uncertainty.

In seawater, the most commonly measured P pools are (i) soluble reactive P (SRP), (ii) total dissolved P (TDP), and (iii) particulate P (PP). A schematic of the stepwise analysis of these pools is presented in Fig. 2.

Soluble Reactive P

The SRP pool includes P_i plus any dissolved organic phosphorus (DOP) compounds that are susceptible to hydrolysis during the time required to form the colored 12-molybdophosphoric acid (12-MPA) complex in an acidic medium. Slight, but

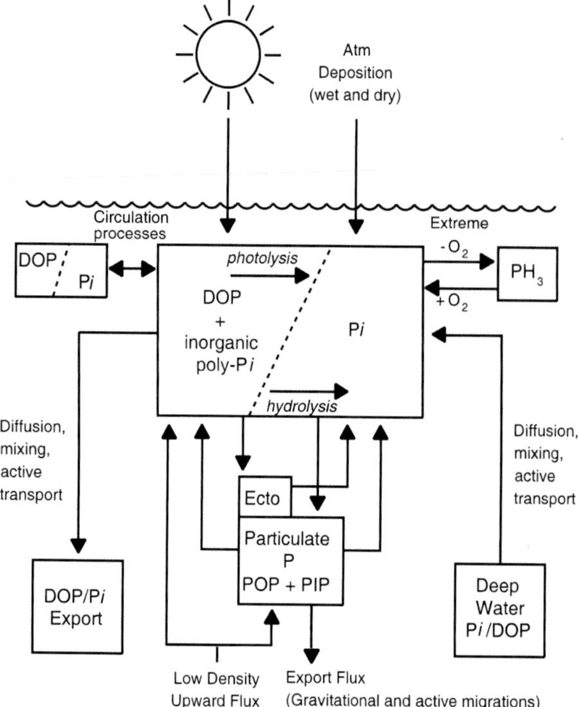

FIGURE 1 Schematic representation of the open ocean P cycle showing the major pools (boxes) and transformations and fluxes (arrows). The large box in the center, TDP, is comprised of phosphate (P_i), inorganic polyphosphate (poly-P_i), and DOP. The last group is usually the largest proportion of TDP in near-surface waters. Although the large DOP pool is poorly characterized, at least a portion of it is available as a source of P for microbial growth. Particulate P includes both living and nonliving organic P (POP) and PIP; POP is the dominant component. Ectoenzymatic activity (Ecto) may be important for the microbial assimilation of DOP into POP. Whereas euphotic-zone processes are dominated by the net conversion of P_i to organic P, subeuphotic-zone (>200 m) P cycle processes are dominated by the net remineralization of organic P to P_i. A number of physical processes, as shown, are responsible for redistributing P throughout the global oceans. In the marine environment, P is almost exclusively in the pentavalent state (+5) as PO_4^{3-}; only under highly reduced conditions is P transformed to phosphine gas (PH_3).

probably significant, differences have been observed between SRP and independent, more specific estimates of P_i for nearly every aquatic ecosystem investigated to date (60, 81), indicating the presence of labile DOP. While alternative methods of P_i detection do exist, few are as reliable or sensitive as the 12-MPA method for SRP (33, 73).

Total Dissolved P

The TDP pool includes SRP plus all P-containing but otherwise "nonreactive" (not reactive to 12-MPA color development in acidic medium, as described for SRP above) molecules or compound classes that can be hydrolyzed by initial treatment with one or more oxidizing agents (e.g., high-intensity ultraviolet photo-oxidation, persulfate chemical oxidation, or nitrate chemical oxidation) that renders them reactive to 12-MPA SRP color development. Different oxidizing reagents and reaction conditions result in different TDP yields, so it is critical to evaluate several methods for a particular habitat before deciding on the method of choice. For example, polyphosphates are not rendered reactive via the standard high-intensity UV (1,200 W, 60 to 80°C, 1- to 2-h exposure) photo-oxidation technique (3), but they are recovered by most chemical oxidation techniques. Karl and Yanagi (62) developed a method based on less rigorous photo-oxidation conditions (330 W, 40°C, 20 min) to separate total DOP into two separate fractions based on their lability in the photo-oxidation procedure, namely, ultraviolet photo-oxidation-labile compounds, consisting primarily of monophosphate esters, and ultraviolet photo-oxidation-stable compounds, consisting primarily of nucleotides and nucleic acids. Some investigators have recommended multiple, sequential oxidation treatments (e.g., ultraviolet photo-oxidation followed by persulfate oxidation [80]) in order to maximize TDP yields. The difference between paired measurements of SRP and TDP is often assumed to be equivalent to DOP (e.g., DOP = TDP − SRP) (Fig. 2); however, other constituents (e.g., inorganic polyphosphates if chemical oxidation is employed), the release of adsorbed, reactive P, or dissolution/hydrolysis of P-containing minerals may also contribute to what is more correctly termed the ultraviolet photo-oxidation (or persulfate/nitrate oxidation)-reactive pool, depending upon the method employed. However, for most low-nutrient open ocean ecosystems, the majority of P in this operationally defined pool is likely to be organic, so the term "DOP" is justified.

Particulate P

PP includes all of the P contained in particles, living and nonliving, that is above a threshold size, usually 0.5 to 1 μm in diameter. Often PP and TDP measurements are made from paired treatments of the same seawater sample, so the TDP pool might also include some submicrometer particulate matter (even small living cells and viruses) and colloids

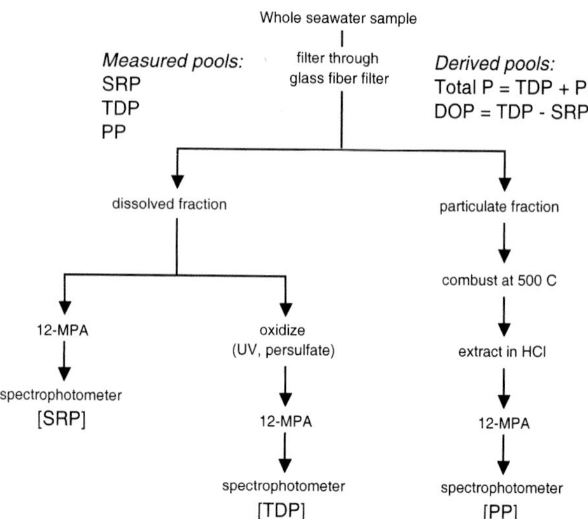

FIGURE 2 Schematic representation of the stepwise analysis of P in seawater. From three direct measurements, SRP, TDP, and PP, one can derive total P (TDP + PP) and DOP (TDP − SRP). Additional analytical procedures exist for the measurement of inorganic polyphosphates, for the separation of PP into organic and inorganic components, and for partial chemical and/or molecular characterization of the TDP and PP pools. However, SRP, TDP, and PP are the most commonly measured P cycle constituents.

that are below the size cutoff employed for the operational definitions of dissolved and particulate. If present, small particles and colloids would tend to overestimate TDP and DOP and would tend to underestimate PP, using the above-referenced "standard" procedures. For most low-nutrient open ocean ecosystems, this is likely to be a negligible source of error because the PP pool is usually <10% of the total P inventory, but this may not be true for coastal habitats. Adsorption of selected dissolved P compounds onto the filters used for PP analysis is a more likely source of error than inclusion of particles in the dissolved P inventories. Because the impact of adsorption is expected to be a nonlinear function of the volume filtered, whereas ideal collection of particles scales linearly with the volume processed (i.e., adsorption exhibits a "loading" phenomenon), the presence or impact of adsorption on PP estimation can be evaluated systematically by processing replicate PP samples derived from different volumes of seawater.

The PP pool is sometimes further divided into particulate inorganic P (PIP) and particulate organic P (POP). The distinction between PIP and POP is, at present, uncertain. There are at least three possible sources of PIP, including (i) P-containing minerals (e.g., apatite), (ii) dissolved P (inorganic or organic) that is adsorbed to the surfaces of particles, including living cells, and (iii) intracellular, metabolic pools of P_i or inorganic polyphosphate. While the last two pools are technically inorganic, they are more fundamentally part of the biogenic pool because of their association with and relevance to microbial metabolism. While it is almost certain that mineral P (e.g., apatite), if present, would contribute to the PP inventory, it is not expected to be abundant in open ocean ecosystems. Surface-bound P has been reported to range from 14 to 57% of the total cellular P for a variety of laboratory-reared phytoplankton species (35). If this observation is found to be a general phenomenon of marine microorganisms, estimated P_i uptake rates and P cell quotas (and derived C-to-N-to-P ratios) may have been systematically biased.

The total P fraction in seawater can also be divided further into broad classes of biomolecules, such as lipid-P, carbohydrate-P, and nucleic acid-P, using standard biochemical fractionation procedures or compound-specific analysis of, for example, ATP or cyanocobalamin (vitamin B_{12}). Most of these analyses require a preconcentration step to provide enough material for reliable analysis. In the case of PP, this can be achieved by sample filtration. While there are technical limits to the amount of seawater that can be processed based on the porosity and loading characteristics of the filter matrix, it is usually possible to effect a 10^3 to 10^5 concentration factor by using a standard 25-mm-diameter Whatman glass fiber filter (0.7-μm nominal porosity) for most open ocean water samples. Characterization of the DOP pool requires preconcentration by ultrafiltration, which usually targets the high-molecular-mass (usually defined as >30,000 Da) fraction (7). For open ocean environments, the high-molecular-mass fraction ranges from 19 to 38% of the total DOP pool (19), so a majority of the TDP pool, namely, the low-molecular-mass fraction (<30,000 Da), cannot be concentrated reliably by this or most other available analytical procedures. At present, <10% of the total DOP pool in any given marine environment can be identified by its molecular structure.

Isotopic Tracers

Studies of the uptake, assimilation, and transformation of selected inorganic and organic P-containing compounds are facilitated by the use of isotopic tracers. There are at least three general applications, including (i) use of naturally occurring, cosmogenically produced radioactive isotopes of P (63, 64), (ii) use of naturally occurring stable isotopes of oxygen (mainly ^{18}O) of phosphate (12), and (iii) use of exogenously supplied, radioisotopically labeled P-containing molecules and compounds. For selected applications of DOP turnover, ^3H-, ^{14}C- or ^{35}S-labeled P-containing compounds have been used, but only P-radiolabeled precursors can unambiguously trace the fate of P. There are two radioisotopes of P, namely, ^{32}P (E_{max} = 1.71 MeV; half-life = 14.3 days) and ^{33}P (E_{max} = 0.25 MeV; half-life = 25.3 days). These have been used extensively for investigations ranging from cell metabolism and energy transduction to whole-ecosystem studies (53).

Bioreporter Molecules and Macromolecules

In the past few decades, novel enzyme- and nucleic acid-based methods have been applied to the marine P cycle. Examples include (i) partial molecular characterization of the DOP by the addition of specific enzymes to whole, filtered, or partially purified seawater (87); (ii) measurement of the expression of alkaline phosphatase activity in single cells or mixed microbial assemblages as an ecophysiological indicator of P stress (31); (iii) measurement of cyclic AMP production as an indicator of the synthesis of a number of inducible catabolic proteins (1); and (iv) application of molecular markers to investigate P limitation or DOP utilization (32, 85).

SELECTED P POOL INVENTORIES AND FLUXES

In the open sea, P delivery from the deep water (>200 m) sustains photosynthesis in the surface water, which ultimately supplies reduced organic matter and energy for the growth of all heterotrophic organisms, from bacteria to whales (Fig. 1). A small portion of the P that is incorporated into organic matter in the near-surface waters "leaks" back into the deep sea through the combined processes of gravitational settling of particulate matter, active net downward movement of organic matter via active migrations of zooplankton and fish, and the downward movement of dissolved organic matter contained within sinking water parcels. Over time, this organically bound P is decomposed, mostly via the activities of deep-sea microorganisms. The organic P is released as P_i, which eventually accumulates and diffuses upward to start the photosynthetic production cycle over again. This process of the net removal of near-surface organic matter and subsequent decomposition and remineralization at depth is termed "the biological pump" because of its role in pumping organic matter (and energy) to regions where solar energy does not penetrate.

The structure and function of the marine P cycle are both time and space (vertical and horizontal) variable; a comprehensive understanding will require a four-dimensional resolution of the key P inventories and fluxes. At the present time, the ocean remains grossly undersampled, even with respect to the most basic features of the P cycle, in part because there are no convenient remote sensing techniques to provide high-frequency, broad spatial coverage observations. Nevertheless, the historical data set held at the U.S. NOAA-NESDIS National Oceanographic Data Center (NODC) includes an archive of approximately 170,000 vertical profiles of SRP that were collected on research expeditions worldwide between 1925 and 1993, with most of the data (>105,000 profiles) obtained between 1960 and 1980 (21). These data are publicly available via the online oceanographic profile database (http://www.nodc.noaa.gov). During this approximately 75-year period of observation, the general analytical method for SRP quantification, namely, the 12-MPA colorimetric assay, has remained fairly constant, although automated analyses using segmented-flow instrumentation began to replace the manual method following its development in the late 1960s (4). According to Strickland and Parsons (86), both the manual and the automated methods for SRP have a detection limit of approximately 0.03 μM P; the precision of SRP analysis at this concentration probably approaches ±100%. Consequently, near-surface waters of the central regions (gyres) of the major ocean basins are often reported as "below detection limits," and for this reason, several high-sensitivity SRP detection methods have been developed and employed (36, 59, 77). Few, if any, of these high-sensitivity SRP data sets probably exist in the NODC collection.

The NODC data set was also searched for all paired measurements of SRP and TDP so that a global ocean data set of DOP concentrations could be derived (53). The extracted and edited Global Open Ocean DOP (GOOD) data set included 139,747 measurement pairs from all major ocean basins, but with several notable gaps, especially for the eastern North Pacific Ocean, the South Pacific Ocean, and the Southern Ocean (53; the GOOD data set is available at ftp://ftp.soest.hawaii.edu/dkarl/misc/GOOD).

In comparison to these fairly extensive data sets on SRP and DOP, information about the distribution of PP or any of the major P cycle fluxes is relatively sparse. For these parameters, we have relied mostly on data collected at Station ALOHA (short for "a long-term oligotrophic habitat assessment") (22°45′N, 158°W) in the North Pacific subtropical gyre (NPSG) during the period 1988-2004 as part of the Hawaii Ocean Time-series (HOT) program (58; the HOT data set is available at http://hahana.soest.hawaii.edu).

Global Patterns of SRP, DOP, and PP

On the global ocean scale, near-surface SRP reveals a fairly predictable distribution in space and time (seasonal) that results from a combination of the physical processes of nutrient delivery, seasonal mixed-layer dynamics, and the availability of light for net community organic matter (effectively, PP) production (21) (Fig. 3). A comprehensive, global analysis of the seasonal cycle of SRP in open ocean surface waters revealed the following general trends (21): (i) surface SRP concentrations are highest in high-latitude regions in winter and lowest in subtropical regions in summer; (ii) mean SRP concentrations in the North Pacific are approximately two to three times greater than those in the North Atlantic at comparable latitudes both in summer and in winter (0.76 ± 0.29 μM versus 0.23 ± 0.10 μM and 1.44 ± 0.47 μM versus 0.56 ± 0.24 μM, respectively); (iii) for mid- and high-latitude regions (e.g., north and south of approximately 30° latitude), the surface SRP concentration is generally lowest in summer following the vernal bloom of phytoplankton (88); (iv) the highest surface ocean SRP values (>2.0 μM) are found in the Southern Ocean, where they remain high throughout the year because of a combination of deep mixing, upwelling, and a low rate of PP production; (v) midocean Pacific equatorial regions, especially

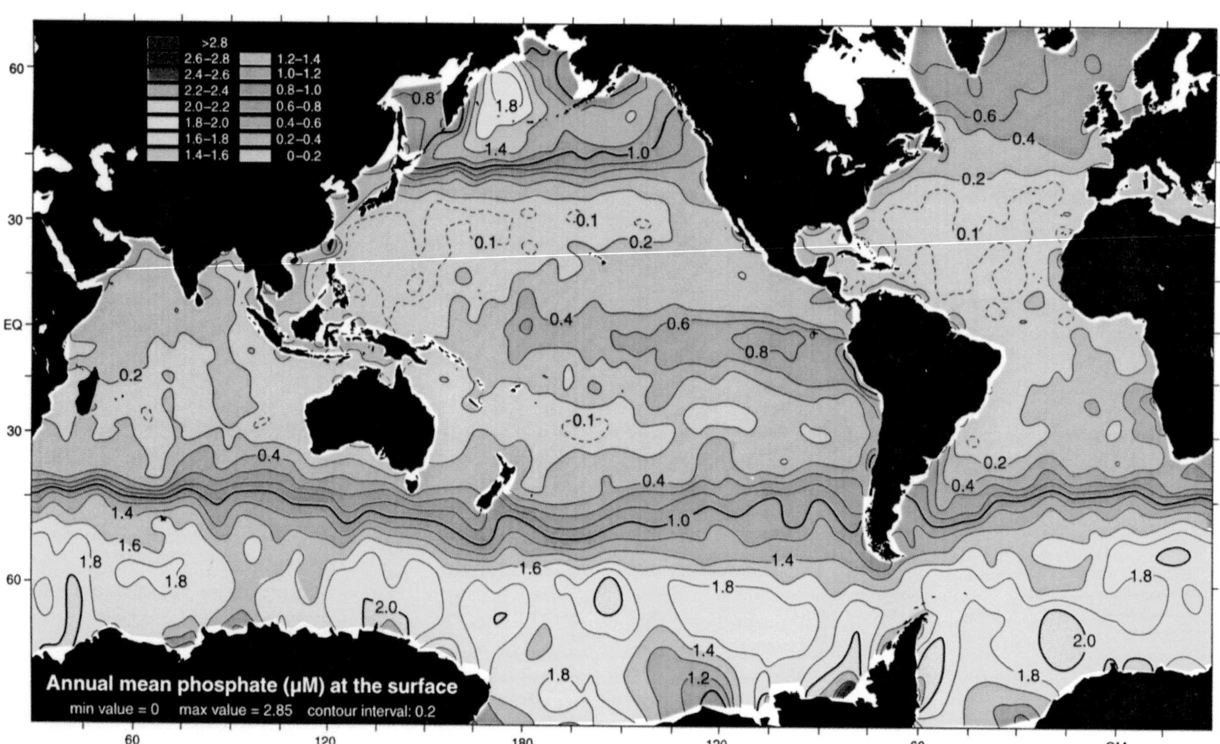

FIGURE 3 Annual mean phosphate (SRP) concentrations at the ocean's surface, based on the worldwide data set held at NODC. Surface SRP concentrations range from <0.1 μM in the oligotrophic, central gyres of the major ocean basins to >1 μM at high latitudes, especially in the Southern Ocean and North Pacific Ocean. (From the *World Ocean Atlas 2001*. Data are publicly available at http://www.nodc.noaa.gov.)

the eastern tropical Pacific, have elevated SRP concentrations (0.4 to 1.0 μM) throughout the year as a consequence of upwelling of SRP-enriched deep water; and (vi) SRP in the subtropical gyres is depleted to ≤0.2 μM (and probably much lower) throughout the year. The accumulation, either seasonally or permanently, of SRP in seawaters exposed to sunlight implies that some other nutrient may be ultimately responsible for determining the rates of net organic matter production in these regions. Other controls, including but not limited to the availability of iron, may contribute to the accumulation of SRP in certain areas of the world ocean. Regardless of the mechanism, transient (or seasonal) accumulations of SRP are indicative of ecosystems that receive more P than they consume and export. If organic matter production and export were P limited, then the near-surface ocean should have SRP concentrations of 1 nM or less. This condition has been reported for only a few selected habitats (16, 72, 95), perhaps as much a result of analytical limitations of the standard methods as a result of the absence of SRP.

Temporal variability in SRP has also been observed at Station ALOHA (59). During the first 6 years of the program, the upper-water-column (0 to 100 m) SRP concentration varied from <10 nM to >150 nM and generally conformed to one of two characteristic profiles that were distinguished by the presence or absence of distinct near-surface (0 to 30 m) SRP concentration maxima. Near-surface SRP enrichments are not unique to Station ALOHA (38). Though still unexplained, these near-surface enrichments in SRP may result from atmospheric deposition of P_i or via a more complex series of ecosystem processes involving the upward movement of buoyant P-containing particulate organic matter followed by hydrolysis and/or photolysis at the sea surface (60). A resolution of these potential pathways of P importation to the euphotic zone is critical to our understanding of the marine P cycle.

The vertical distribution of SRP, as well as that of most other required growth elements (e.g., nitrate, silicic acid, dissolved inorganic carbon, and trace elements, such as cobalt, iron, copper, and cadmium), exhibits a profile that is characterized by relative depletion near the surface and relative enrichment at greater depth (Figures 4 and 5). This profile type is produced and maintained by a balance between SRP input by upward diffusive-advective processes

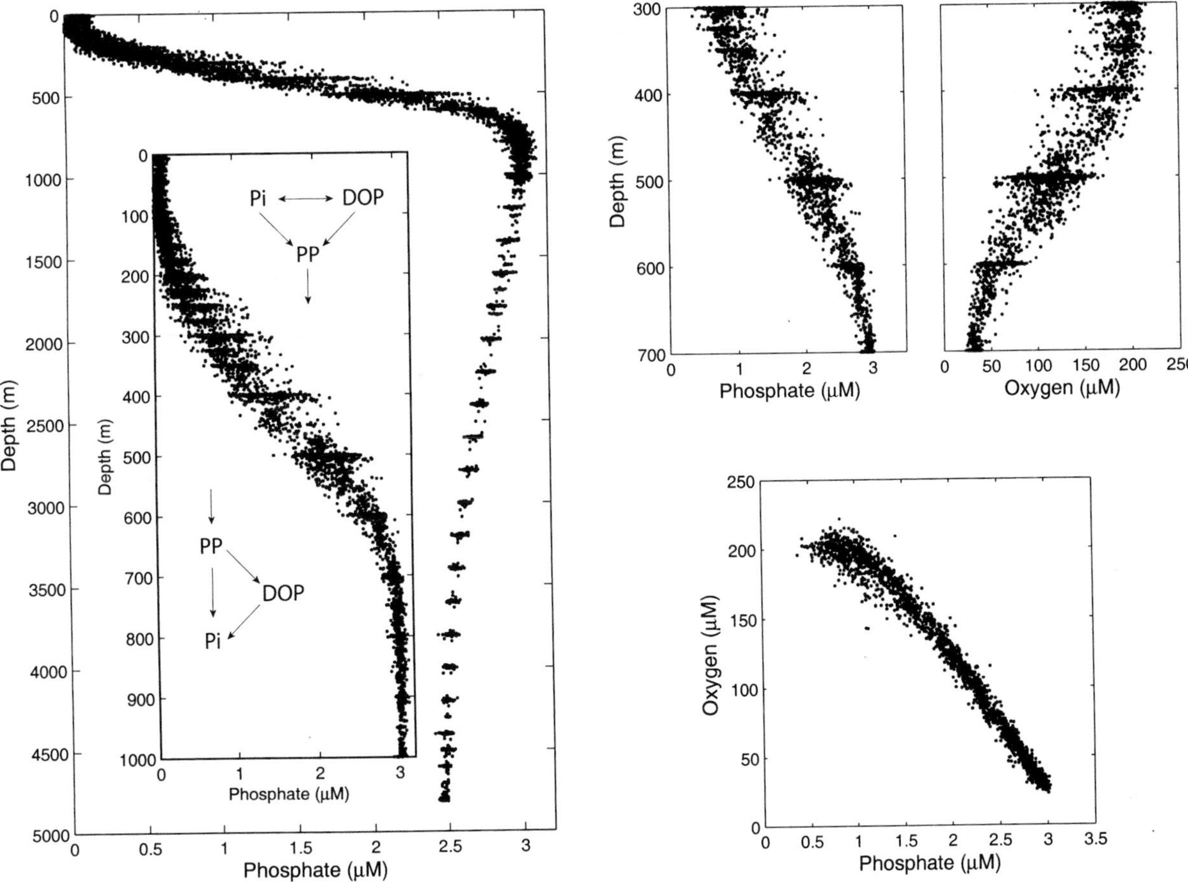

FIGURE 4 Relationship between vertical distributions of SRP and dissolved oxygen (O_2) at Station ALOHA in the North Pacific gyre. (Left) Graph of SRP (μM) versus depth (m) showing the characteristic "nutrient-like" distribution of SRP, with regions of net P_i uptake and export near the surface and net P_i remineralization at greater depths. The inset shows these main P cycle processes, which are most intense in the upper 1,000 m of the water column. (Right, top) SRP and O_2 concentration-versus-depth profile of the 300- to 700-m region of the water column at Station ALOHA showing the effects of net remineralization of organic matter. (Right, bottom) Model 2 linear regression analysis of SRP versus O_2 suggesting an average consumption of 80 μmol liter^{-1} O_2 for each 1 μmol of P_i regenerated.

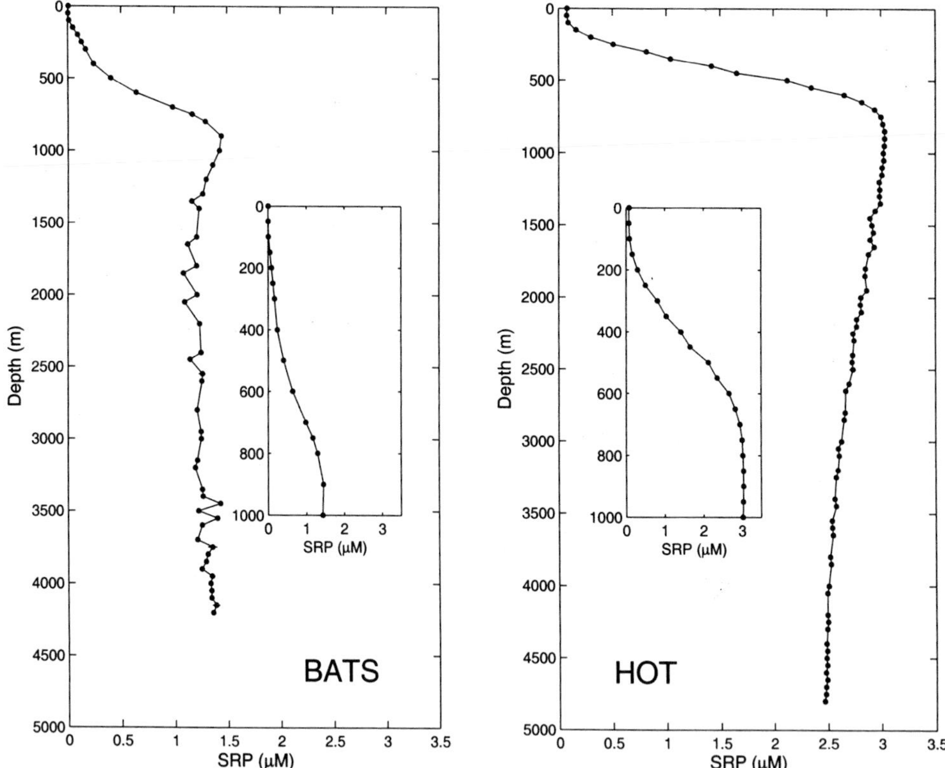

FIGURE 5 SRP-versus-depth profiles for the North Atlantic Ocean (Bermuda Atlantic Time-Series Study [BATS]) and the North Pacific Ocean (HOT) showing significant interocean differences, including a steeper phosphocline gradient (i.e., a larger change in SRP concentration per meter in the upper mesopelagic zone region) and higher deep-water (>4,000 m) SRP concentrations for HOT. These differences in SRP inventories and gradients have significant implications for SRP fluxes into the euphotic zone.

and net removal by organic matter production and export out of the euphotic zone. The large-scale, depth-dependent "reflux" of P, namely, the conversion of dissolved inorganic P to sinking particulate organic P followed by deep water remineralization back to dissolved inorganic P, defines the primary process of the open ocean P cycle. The maximum mid-water-depth SRP concentrations are always colocated with the maximum for nitrate and minimum for oxygen as a result of the aerobic decomposition of organic matter (Fig. 4). With few exceptions, and regardless of geographical location or season, the vertical distribution of SRP shows this behavior. It is important to emphasize that the decomposition of organic matter in the deep sea and the release of SRP from DOP and PP are almost exclusively results of the metabolic activities of microorganisms, specifically the heterotrophic bacteria and archaea that dominate these regions.

The SRP concentration-versus-depth gradient (also referred to as the slope of the phosphocline, reported in micromoles of P meter^{-4} over some specified depth interval) and the absolute SRP concentration of the deep-water end member vary considerably depending upon geographical location (Fig. 5). For example, both the maximum deep-water SRP concentration and the steepness of the phosphocline are greater in the North Pacific Ocean than in the North Atlantic Ocean (Fig. 5). This has important implications for the rate of nutrient delivery by mixing and other physical processes. The global trend of deep-water SRP (and all other nutrients) is a systematic increase "down" the Atlantic from north to south and "up" the Pacific from south to north (14) (Fig. 6). These systematic geographic distributions result from the fact that the deep water in the Pacific basin is much older than the deep water in the Atlantic, and the higher SRP concentrations reflect the longer time for remineralization to occur and are in accordance with the greater age of the Pacific Ocean water masses. Although the global data set for paired measurements of SRP and TDP for the estimation of DOP concentrations is not nearly as large as that for SRP only, several general trends are also evident in the extant observations (53). First, global ocean DOP concentrations have a smaller concentration range than SRP concentrations (≤0.5 μM for DOP compared to ≤3.0 μM for SRP). Second, for any selected location, SRP and DOP concentrations are negatively correlated. At Station ALOHA in the NPSG, the DOP percentage of TDP ranges from >80% in the relatively SRP-depleted surface waters to <1% at the SRP maximum (Fig. 7). These distributions are sustained by the net production of organic matter via photosynthesis (both DOM and particulate organic matter), organic matter export, and net oxidation and remineralization of P-containing dissolved and particulate matter with a net release of SRP at depth (Fig. 4). In most marine ecosystems, P "turns over" (i.e., is taken up as SRP, released as PP or DOP, and remineralized back to SRP) many times locally before it is eventually removed by a number of physical (advection/diffusion of DOP and gravitational settling of

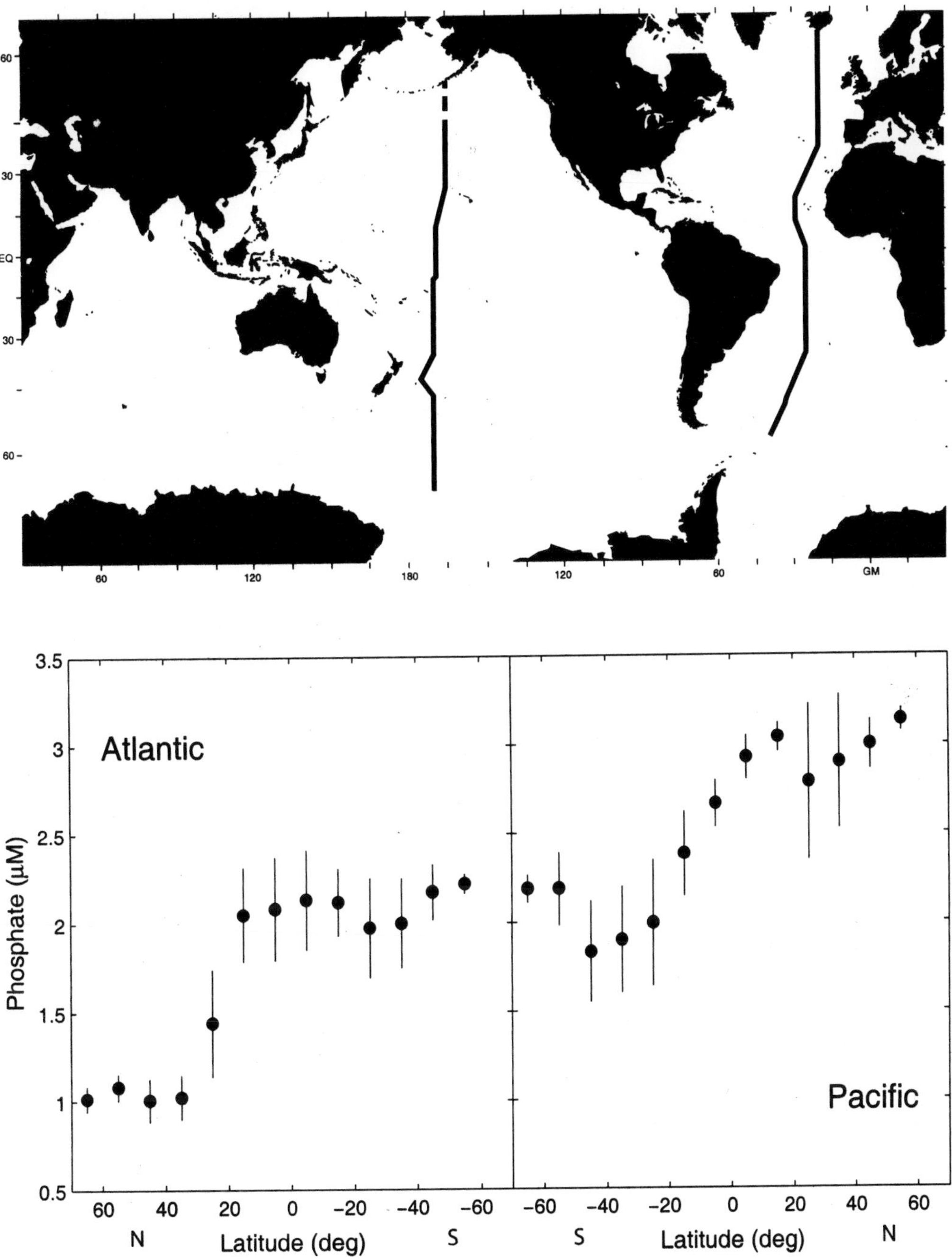

FIGURE 6 Map showing the locations of World Ocean Circulation Experiment program transects A-16 (Atlantic) and P-15 (Pacific). Data from these cruises were obtained from http://woce.nodc.noaa.gov, averaged over the depth range of 500 to 1,500 m, and then combined into 10° latitude bins and plotted as mean SRP concentrations (± 1 standard deviation). The resultant plot shows a systematic increase in SRP concentrations at midwater depths "down" the Atlantic and "up" the Pacific. This result is in accordance with the known pathways of deep water circulation and the increasing age of the deep water mass. The accumulation of SRP is a result of the time-integrated decomposition of organic matter. Deg, degrees.

PP) and biological (e.g., vertical or horizontal migrations) processes. In the subtropical gyres, there may be 20 to 50 turnover cycles before a given P atom is exported as PP from the near-surface environment.

The vertical distribution of PP shows a characteristic enrichment in surface waters and a decrease with depth to relatively low concentrations beneath the euphotic zone (Fig. 8). This pattern is sustained by the availability of light and nutrients as the major controls on the photosynthetic production of organic matter, and therefore the absolute PP concentration is expected to vary seasonally and geographically in response to changes in light and nutrient

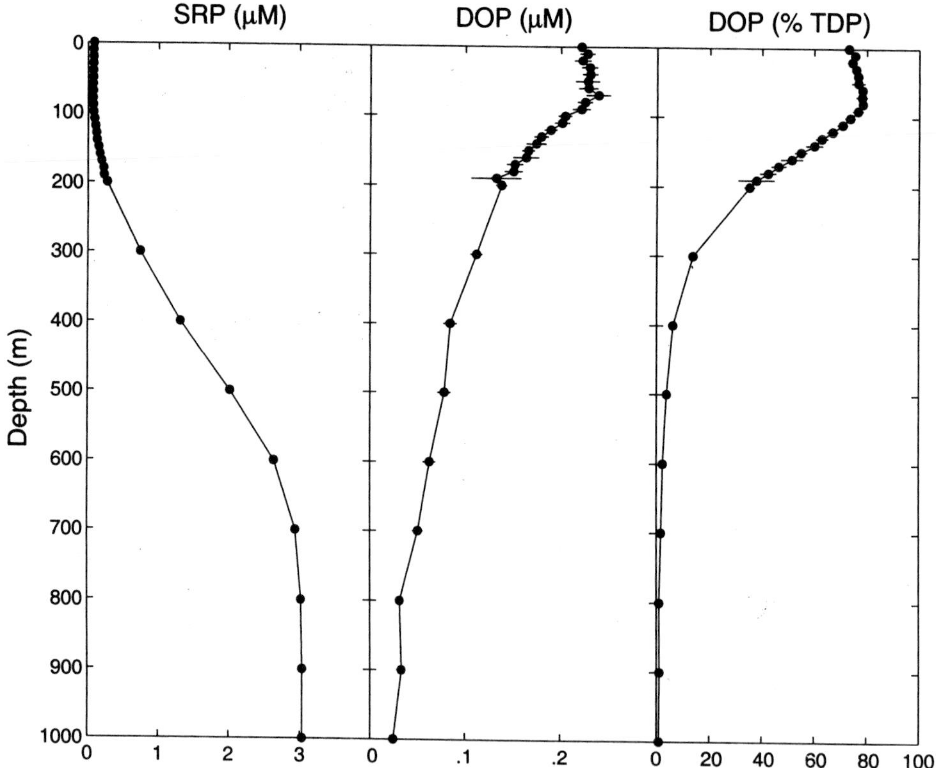

FIGURE 7 Concentration-versus-depth profiles for SRP (μM) and DOP (μM) at Station ALOHA. Also shown is DOP as a percentage of TDP.

availability. However, the current database is insufficient to provide a comprehensive global view of PP distributions. At Station ALOHA, PP in the upper 100 m of the water column is approximately 15 nmol of P liter^{-1}, compared to 300 to 500 nmol of TDP liter^{-1}, or ≤0.5% (Fig. 8). Based on simultaneous measurements of particulate ATP (P-ATP), a biomarker for living cells, and assuming a ratio of 6.1 g of P to 1 g of ATP for microbial biomass, an estimate of the amount of living PP can be obtained. At Station ALOHA, approximately 40 to 50% of the PP appears to be associated with living cells in the euphotic zone, decreasing to <20% at greater depths. The remainder of the PP is presumably associated with nonliving organic matter or PIP (most likely the former).

P_i Uptake, PP Production, and P Export

Radioisotopically labeled P_i ($^{32}P_i$ or $^{33}P_i$) has been used to track P from the dissolved inorganic form into living organisms (37). This is generally accomplished by conducting a time course incubation experiment to assess the rates of P_i uptake and incorporation into newly synthesized biomolecules. Because the ambient P_i pool can be measured with fairly high precision and because the addition of the exogenous tracer is under the control of the experimenter, the specific radioactivity (radioactivity per unit mass of P) of the P_i pool can be determined. This information can be used to convert the measured radioactivity incorporated per unit volume per unit time into a more ecologically meaningful P mass flux, assuming that the specific radioactivity does not change during the incubation procedure. If the P_i uptake rate is high, relative to the size of the ambient P_i pool (i.e., there is a rapid P_i turnover time) or if there is a net remineralization of organic P to P_i during the experiment, then this assumption of constant specific radioactivity may be invalid. Under the latter conditions, the specific radioactivity must be measured to correct for isotope dilution. Quantitative assessments of isotope dilution can also be used to estimate P_i regeneration in the same experiment as that estimating P_i uptake (37).

During the uptake and incorporation of $^{32}P_i$ and/or $^{33}P_i$ into newly synthesized PP, there is also a simultaneous production of labeled DOP in most experiments (74). Since the specific radioactivities of these PP and DOP pools are not known or easily determined, they may turn over faster or slower than the bulk P_i pool. This complicates any straightforward interpretation beyond estimating initial (effectively instantaneous) P_i uptake and incorporation rates. For these and other reasons, it is good practice to perform relatively short incubation experiments and to establish complete radiochemical mass balances for all likely P-labeled pools (i.e., P_i, DOP, and PP).

P_i uptake has been measured in the subtropical North Pacific Ocean along with other relevant microbial and biogeochemical parameters (11). P_i uptake rates in the well-lit upper water column were 3.0 to 8.2 nmol liter^{-1} day^{-1} (median, 3.5 nmol liter^{-1} day^{-1}), and P_i pool turnover times ranged from 2 to 40 days (median, 9 days). This P_i uptake is the result of the metabolic activities of both autotrophic and heterotrophic microorganisms. For the habitats investigated, the Michaelis-Menten half-saturation constant (K_m) for P_i uptake was typically higher than that of the ambient P_i pool, indicating an even higher "potential" for P_i assimilation. This

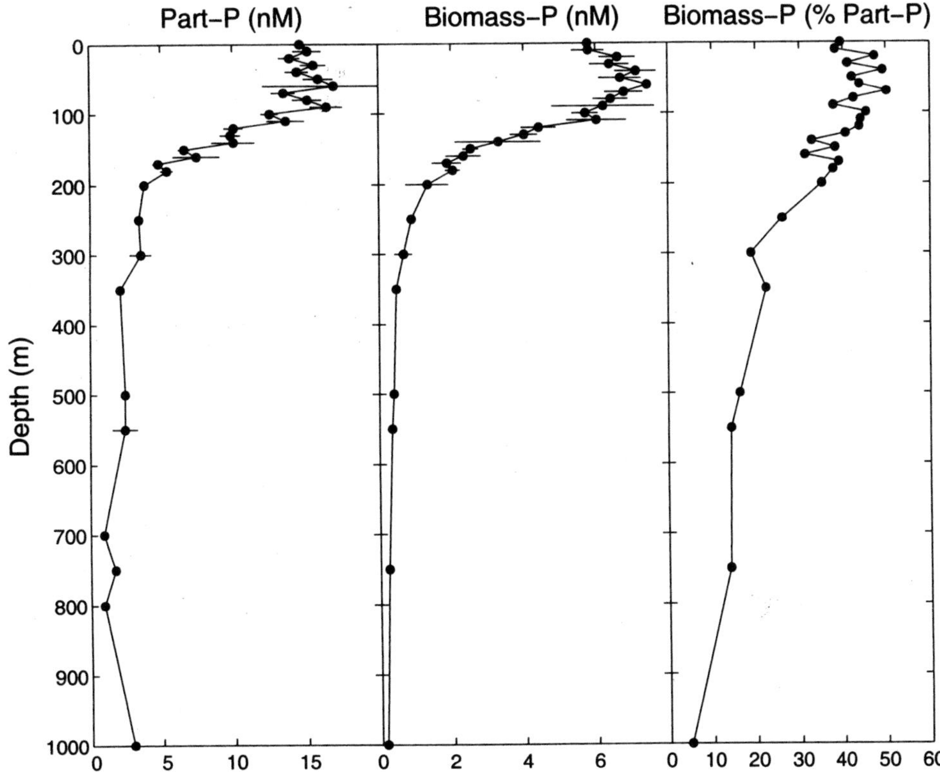

FIGURE 8 Concentration-versus-depth profiles of particulate P (nM) and biomass P (nM) at Station ALOHA. Biomass P was derived from P-ATP measurements, assuming a biomass P-to-ATP ratio of 6.1 (see the text for details). Also shown is biomass P as a percentage of PP.

was confirmed by demonstrating enhanced P_i uptake after P_i additions in the range of 25 to 250 nM (11). Furthermore, a significant percentage (10 to 40%) of the P_i taken up was found in the DOP pool, suggesting that gross uptake of P_i greatly exceeded assimilation estimated from PP production alone. The fate of the produced [^{32}P] DOP was not investigated explicitly in this study, but the authors did establish that P_i regeneration from exogenously added nucleotides (ATP and GDP) was up to 50 times more rapid than the net uptake rates. These results documented the presence of a remineralization-intensive microbial assemblage for this oligotrophic environment.

Another major process in the marine P cycle is the downward P flux, usually by gravitational settling, of recently produced particulate organic matter. This removal of P from the near-surface waters is sometimes termed "export production." In many marine environments, P export is balanced by P import, so there exists a more or less constant inventory of total P in the near-surface ocean over long temporal scales (e.g., years to decades). However, for any given marine habitat, seasonal changes in P inventories can occur (21), and for some habitats, even longer-term climate-driven changes in the P inventories have been observed (see "Selected Ecological Studies" below).

Data obtained during the decade-long Vertical Transport and Exchange (VERTEX) research program firmly established a quantitative relationship between primary production and export in open ocean ecosystems (67); export production was reported to range from 13 to 25% of contemporaneous primary production when expressed in terms of carbon (75). The amount of sinking organic matter remaining at any depth can be estimated by a normalized power function, $\text{Flux}_z = \text{Flux}_{100\text{ m}} (z/100)^b$, where z is the selected reference depth and b is the empirically determined coefficient. For open ocean ecosystems, Martin et al. (67) reported that C flux was $1.53(z/100)^{-0.858}$ mol m^{-2} year^{-1}.

This quantitative relationship has been used to estimate deep-water nutrient regeneration rates and rates of oxygen consumption. In their analysis, Martin et al. (67) also estimated P fluxes and net SRP regeneration at depth by assuming a fixed 106 C:1 P stoichiometry for sinking particles. More recently, direct measurements of PP flux revealed a higher-than-expected C-to-P ratio, with significant time and depth variations (56). The remineralization length scale for P (248 m) is much shorter than that for C (368 m), implying that PP is remineralized more rapidly, and perhaps more efficiently, than particulate C (17). These fluxes translate into turnover times of 2 to 5 days for particles sinking at a rate of approximately 100 m day^{-1}. The deep penetration of C relative to that of P has important implications for the net sequestration of atmospheric carbon dioxide via the ocean's biological pump.

Assessments of BAP

Regardless of the rigor and precision with which P-containing compound pools are measured, the ecological significance of these analytical determinations will be incomplete until reliable estimates of the biologically available P (BAP) pool are routinely available (53). In addition to P_i, which is generally the preferred substrate for microorganisms, the P contained in a variety of polymeric inorganic compounds, in monomeric and polymeric organic compounds, and in selected P-containing minerals may be available to some or all microorganisms; indeed, some microorganisms may prefer ester-linked P sources to free orthophosphate (22, 89). However, the bioavailability of most organic P pools depends on ambient P_i pool concentrations and on the expression of specific enzymes that control transport, salvage, and specific substrate hydrolysis. Because the synthesis of many of these organic-P-scavenging enzymes is induced only at low P_i concentrations, bioavailability may be a variable, time- and habitat condition-dependent parameter rather than an easily predicted or measured metric. For example, bacterial growth on a mixed P_i-plus-methylphosphonate medium showed a systematic phased response, with initial P_i utilization followed by induction and expression of the phosphonate degradation pathway (92). Therefore, environmental assessments of the BAP may depend on the time scale of consideration; for example, substrates that appear to be recalcitrant on short time scales (e.g., <1 day) may fuel longer-term (annual to decadal) microbial metabolism, especially with changes in the availability of P_i (see "Selected Ecological Studies" below).

Karl and Bossard (55) estimated BAP by using a γ-P labeling technique that they developed for ATP pool turnover. Because the intracellular ATP pool turns over rapidly, the specific radioactivity in the γ-P position of the P-ATP pool will reach equilibrium, not only with the extracellular P_i pool but also with the potentially larger pool of extracellular P that is available to the microbial assemblage at that time, i.e., the BAP pool. Bossard and Karl (13) successfully applied this ^{32}P labeling method to a variety of marine and freshwater habitats and reported that BAP was greater than or nearly equal to measured SRP. More recently, Björkman et al. (11) and Björkman and Karl (9) used the γ-P ATP labeling method to estimate the BAP pool in the oligotrophic North Pacific Ocean. In their studies, BAP was generally greater than P_i or SRP but, as reported by Bossard and Karl (13), less than the TDP pool. The BAP-to-TDP molar ratio varied in both space (vertically in the water column) and time (over the year at a given site) and generally exceeded the P_i-to-TDP and SRP-to-TDP ratios, sometimes by factors of >2. These results suggest that there exists some fraction of the ambient DOP that has a bioavailability factor equal to or greater than that of P_i in these open ocean habitats.

ATP as a "Model" DOP Compound

ATP is a biologically labile but chemically stable molecule with well-known and vital functions in cellular energetics, metabolism, and biosynthesis. It is also one of the few P-containing organic molecules that can be detected easily and reliably in both particulate and dissolved forms in seawater by use of the well-characterized firefly bioluminescence detection system (46).

Measurements of P-ATP in marine ecosystems date back to the classic work of Holm-Hansen and Booth, who developed the theoretical basis for its use in total microbial biomass estimation (41, 42). ATP's unique characteristics include the following: (i) ATP is ubiquitous in all living cells, (ii) ATP has a relatively short half-life following cell death and autolysis, and (iii) ATP is present at a fairly constant intracellular concentration, regardless of nutritional mode or growth rate. Furthermore, ATP can be extracted readily from cells and stabilized in solution by using boiling Tris or phosphate buffers and can be quantified by using the sensitive (picomolar detection limit) and precise firefly bioluminescence detection assay.

The inherent chemical stability of ATP in "sterile" seawater (approximately 8-year half-life at 21°C [44]) suggested that dissolved ATP (D-ATP) might exist in the environment as a transient after cell death and autolysis or from other production pathways. Azam and Hodson (6) were the first to report the presence of D-ATP in seawater, and this has now been confirmed for a variety of aquatic environments (see reference 53). D-ATP concentrations are typically highest in surface coastal waters, decreasing with distance from shore and with depth in the water column (6, 10, 68). These D-ATP distribution patterns suggest a close coupling between microbial biomass and ecosystem productivity. While the exact mechanisms of D-ATP production remain unclear, grazing, viral lysis, autolysis, and exudation or leakage from cells are all possible pathways (e.g., see reference 8).

Marine microorganisms assimilate D-ATP for P and for purine salvage. Using [^{14}C-]ATP, Azam and Hodson (6) estimated that <2% of the assimilated D-ATP was respired. Measured D-ATP pool turnover times are short (hours to days) for most marine ecosystems, especially compared to the predicted abiotic, chemical turnover time. As a result, ATP flux through marine microbial assemblages may be a significant step in the marine P cycle. Because ATP has a fundamental role in cellular bioenergetics and because it is one of the few (perhaps the only) P-containing organic compounds that can readily be detected in both particulate and dissolved pools and whose uptake and turnover can be assayed with convenient ^{14}C-, ^{3}H-, or $^{32}P/^{33}P$-labeled radiotracers, it can be used as a model compound in studies of the marine P cycle (6, 53).

SELECTED ECOLOGICAL STUDIES

The Ecological Enigma of "Nonliving" Nucleic Acids

Pioneering research efforts to measure DNA in the sea began in the late 1960s with the quantitative laboratory and field studies of Holm-Hansen and colleagues (41, 43). During these initial investigations, it was established that a large proportion of particulate DNA was associated with nonliving organic matter (41). Subsequent studies confirmed the presence of a large pool of detrital DNA, both in particulate (61) and in dissolved (26) fractions. Dissolved RNA (D-RNA) has also been detected in seawater (50, 84). Given the vital role of nucleic acids in cell metabolism, the existence of nonliving pools of DNA and RNA, often in excess of the total "living" fractions, appears enigmatic.

The distribution of dissolved DNA (D-DNA) follows the general pattern of DOM in the marine environment, namely, the highest concentrations are in coastal waters, with the concentration decreasing with distance from shore and with depth in the water column (27). For samples collected at an oligotrophic North Pacific station, Karl and Bailiff (50) reported higher concentrations of D-DNA than particulate DNA and D-RNA-to-D-DNA ratios ranging from 3 to 10, similar to particulate RNA-to-DNA ratios found in growing microorganisms. For a series of stations in the English Channel, Hicks and Riley (40) reported that P contained in the dissolved nucleic acid fraction (RNA plus DNA) accounted for 27 to 49% of the total DOP. A similar estimation for samples collected in Tokyo Bay and Sagami Bay yielded a mean dissolved nucleic acid P proportion of 12.9% of the total DOP (84).

The production of extracellular nucleic acids may be a manifestation of microzooplankton grazing processes or virus-induced cell lysis. The concentrations of D-DNA and nanoflagellates were found to covary at a station in the Adriatic Sea (90); quantitative estimates indicated that most of the ingested DNA was subsequently released into the environment. Brum (15) recently investigated the production of D-DNA at a station in the oligotrophic North Pacific Ocean. Her analyses indicated that free viruses constituted 49 to 63% of the total D-DNA pool; the remainder appear to be "free" DNA. In a series of experiments, the free DNA production rate was estimated to be 0.1 to 0.4 μg liter^{-1} h^{-1}, and the turnover rate was 1 to 6 h within the upper photic zone. This rapid recycling of macromolecular DOP has important implications for microbial nutrition. For example, Lorenz and Wackernagel (65) reported that the turnover of D-DNA was greater in P-limited than in N-limited habitats, suggesting that DNA, a P-enriched macromolecule, may be an important source of P for microbial assemblages.

Another important ecological implication is the possible role of free DNA as a vehicle for genetic transformation under ecologically permissive conditions. Natural transformation by extracellular DNA is a potential mechanism for lateral gene exchange in natural aquatic habitats. Bacteria and archaea are able to actively take up DNA and recombine it into their genomes in a process called natural transformation (e.g., see reference 79). The presence of gene-sized DNA fragments in seawater prompted DeFlaun and Paul (25) to look for natural transformation among marine microbial assemblages, and later, Paul et al. (76) were the first to document transformation in seawater samples under ambient environmental conditions. The D-DNA pool may therefore represent the "community genome" (25). The ecological implications of horizontal gene transfer are profound; this is an important contemporary area of research, poised for rapid progress in the next decade.

NPSG Station ALOHA

The central portions of all major ocean basins are characterized as "marine deserts" because of the small standing stocks of photosynthetic organisms and small nutrient inventories (Fig. 3). These oceanic "gyres" are the dominant biome on planet Earth, and until recently, very little was known about their biogeochemical characteristics and P cycle dynamics, in large part because they are remote and grossly undersampled. This all began to change in 1988, when two open ocean time-series stations, one in the North Atlantic gyre (also known as the Sargasso Sea) near Bermuda (Bermuda Atlantic Time-series Study) and the other in the NPSG near Hawaii (HOT), were established to observe and interpret variations in key physical and biogeochemical processes, including the marine P cycle. The NPSG benchmark location, dubbed Station ALOHA, is located at 22°45′N, 158°W, approximately 100 km north of Oahu, Hawaii, in deep water (4,800 m), outside any biogeochemical influence of the Hawaiian Ridge yet still close enough to facilitate approximately monthly sampling from shore-based facilities (58). Now, nearly 20 years into these intensive sampling programs, the NPSG has become one of the most well-studied open ocean ecosystems, providing a global reference point for tracking the health of the ocean, including the rate of carbon dioxide sequestration and related changes in acidification, and experimental platforms to study seasonal and interannual ecosystem dynamics. Results obtained, in part, during the nearly 2-decade-long HOT research program have shown the effects of climate variation on ecosystem structure and function in the North Pacific gyre. Climate-induced changes in the stratification of the surface ocean have affected nutrient and trace element budgets and have selected for N_2-fixing bacteria and *Prochlorococcus*, resulting in a domain shift from predominantly *Eukarya* to predominantly *Bacteria* (47). Numerous biological consequences, many involving the marine P cycle, have been observed, and others are expected, as detailed below.

One of the unexpected and potentially significant ecological trends observed at Station ALOHA is the long-term "disappearance" of SRP and PP and its relationship to the coupled cycles of carbon, nitrogen, and possibly iron (51, 54). At the start of the HOT program (October 1988), the near-surface (0 to 60 m) inventory of SRP was approximately 6 to 7 mmol P m^{-2}, which was considered a very low concentration at that time. Sixteen years later, the 0- to 60-m SRP inventory has decreased 80%, to approximately 1 to 2 mmol P m^{-2} (Fig. 9). Clearly, the long-term balance (years to decades) between P delivery and P export that maintains most marine ecosystems in a near-balanced steady state has been perturbed at Station ALOHA.

During the same 16-year observation period, the upper-water-column PP inventory at Station ALOHA decreased significantly, from average values of 1.2 to 1.6 mmol m^{-2} during the 1989-1992 period to approximately 0.8 mmol m^{-2} in 2004 (Fig. 10). Moreover, P appears to be progressively decoupled from the C and N cycles over time, as shown by the increased PC-to-PP and PN-to-PP ratios of the suspended particulate matter pool. At the start of the HOT observation program, the molar PC-to-PN-to-PP ratio (131 C:18 N:1 P) was close to the canonical Redfield ratio (78) (106 C:16 N:1 P), but by 2004 the stoichiometry of the particulate matter pool was significantly depleted in P relative to C and N (186 C:28 N:1 P) (Fig. 10). Although the data set on DOM is not as complete, it has been reported that both dissolved organic carbon and dissolved organic nitrogen have accumulated over recent time, but DOP has remained relatively constant (18, 57). As a result, the C-to-P and N-to-P ratios of the DOM pool have also increased to values that are well in excess of the predicted Redfield ratio (18). Finally, the export of PP—but not PC or PN—from the euphotic zone, as measured using free-drifting sediment traps (56), also systematically decreased over the observation period (Fig. 11). Again, the C:N:P elemental stoichiometry of the sinking particles is depleted in P relative to either C or N compared to that in the Redfield ratio. These temporal decreases in SRP and PP inventories and in PP export are even more surprising considering that the rate of primary production (as measured using the in situ ^{14}C technique) has increased nearly 50%, from 1989 values of approximately 400 mg C m^{-2} day^{-1} to contemporary rates of approximately 600 mg C m^{-2} day^{-1} (HOT data are available at http://hahana.soest.hawaii.edu). These observations are all consistent with the hypothesized selection for microbes (mostly bacteria) that are able to use dissolved nitrogen gas (N_2) as a source of N for cell growth (the so-called N_2-fixing microorganisms) and with a switch from an N-controlled biome to one controlled by P, or perhaps P and iron (23, 47, 49, 54, 70).

If N_2 fixation is an important pathway for the supply of new bioavailable nitrogen for the ecosystem, this process will force an already P-stressed biome into one that is even more P limited. Ecological consequences might be further alteration of the C-to-P ratio of new biomass or selection of microorganisms that do not require as much phosphorus for

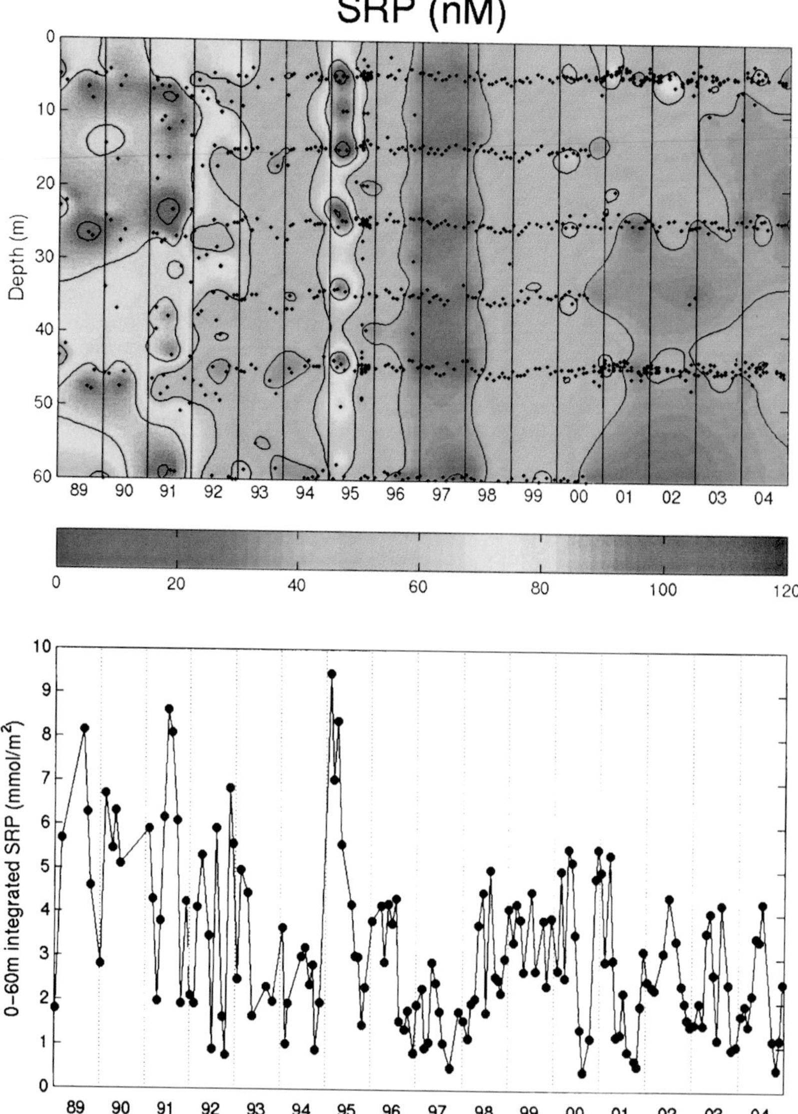

FIGURE 9 Changes in SRP inventories for samples collected at Station ALOHA during the period from 1989 to 2004. (Top) Contour plot of SRP concentrations in the upper 0 to 60 m of the water column. (Bottom) Depth-integrated (0 to 60 m) inventories of SRP showing a systematic loss of SRP from this ecosystem over the past 15 years.

growth (e.g., see reference 93). At Station ALOHA, there still appears to be an "excess" of P_i (Fig. 9), even though the inventory has been depleted over the past 2 decades. Further reduction of P_i, most likely to subnanomolar concentrations, can be expected along with selection for alternative "P capture" mechanisms, including novel enzymes that can mobilize the older, semilabile organic phosphorus residues, such as C-P-bonded phosphonates, and a further shift in the cell size and activity spectra to those of smaller, slower-growing microorganisms. This has a potential impact on the trophic structure, as a shift in the cell size of microorganisms may result in a shift in the grazer populations, which would impact the top-down grazing control of microbial populations, and therefore on the nutrient cycling rates, further exacerbating P limitation. Even the effectiveness of viral infection and lysis may be impacted by P limitation (94), and this could, in turn, impact microbial selection pressures. Without an adequate resupply of P and other nutrients, these P-stressed open ocean ecosystems could lose biomass, biodiversity, and the ability to respond to habitat variations and climate change.

A key negative feedback to enhanced N_2 fixation, decoupling of N and P cycles, and the export of organic matter with a high N-to-P ratio is the eventual buildup of a mesopelagic nutrient reservoir that has an enriched N-to-P ratio relative to cellular needs. As these regenerated nutrients slowly feed back into the euphotic zone, they will select against N_2-fixing microorganisms because there will be other sources of nitrogen available for the growth of microorganisms that do not need to convert atmospheric nitrogen to biomass. This

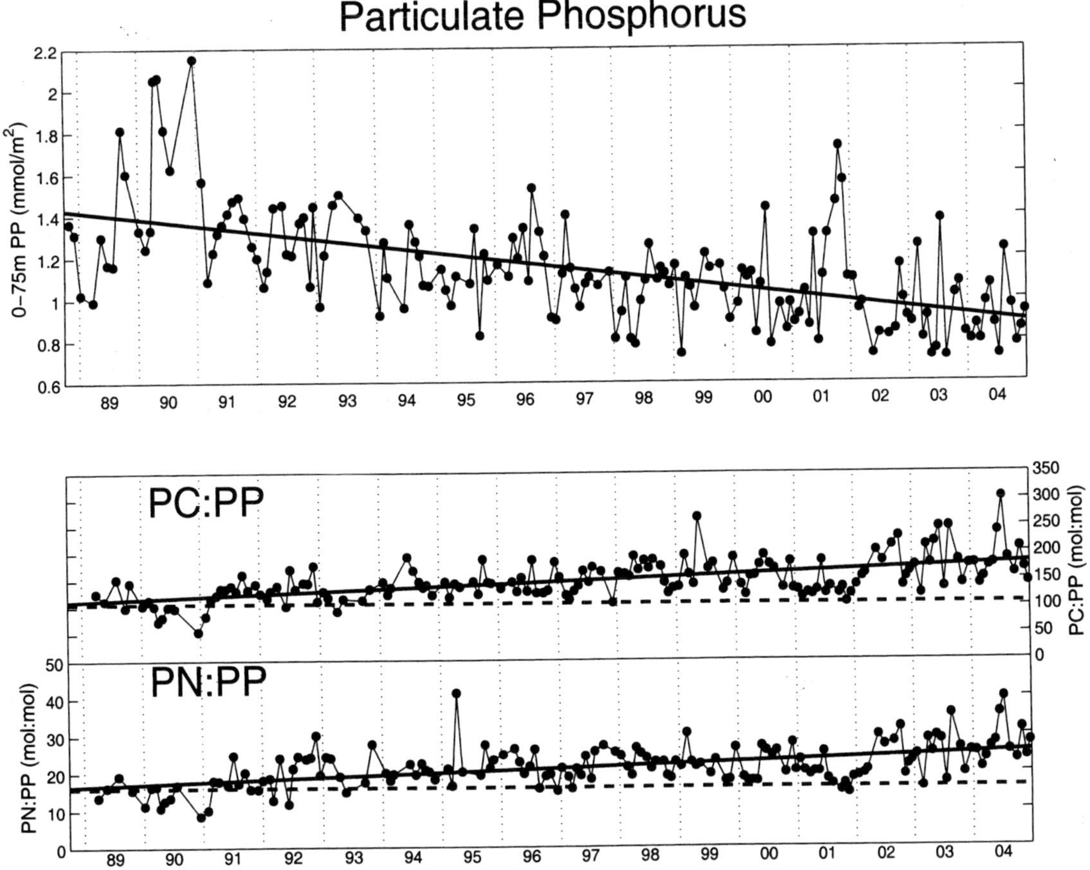

FIGURE 10 Changes in PP inventories and C:N:P stoichiometry of particulate matter collected at Station ALOHA during the period from 1989 to 2004. The solid trend lines are the best-fit linear regression analyses for each data set. The dashed lines in the PC:PP and PN:PP plots show the "ideal" elemental stoichiometry of 106 C:16 N:1 P described by Redfield et al. (78).

will lead to another shift in community structure, ecological stoichiometry, grazing control, and organic matter export under the newly established nutrient regimen. This alternation between N limitation and P (or P and Fe) limitation in the NPSG would be predicted to occur on an approximately 20-to-50-year cycle based on the estimated residence time of nutrients in the upper mesopelagic zone reservoir.

This conceptual model of temporal alternation between N and P control of production and export processes provides a reconciliation of the paradigms that were developed in the 1960s to explain the fixed Redfield ratio and N-controlled ecosystem processes of that era with the new paradigms that are necessary to accommodate contemporary observations in a sea of change (47). The extent to which greenhouse gas-induced warming and other changes to the surface ocean will have effects on the ability to switch between these two ecosystem states is currently unknown, but it seems almost certain that the dimensions of the subtropical gyres will expand and that, with that expansion, the oceans as a whole will become more stratified and more nutrient depleted.

FUTURE PROSPECTS

The "internal" marine P cycle is a complex and dynamic set of mostly microbiological processes that are responsible for the major transformations of P between and among the inorganic and organic P pools in the sea. The P cycle is solar powered and inextricably tied to the complementary cycles of C, N, and other biophilic elements. The marine P cycle is also space and time variable and is ultimately influenced by large-scale climate and physical forces that collectively act to control heat and momentum balances in the marine environment. Based on pioneering research efforts during the past decade, we now recognize that microbial community structure and microbial diversity are important aspects of the marine P cycle in their control of the ocean genotype, through natural selection pressures, and the ocean phenotype, through environmental regulation of genotype expression (30). The use of a "fixed" C-to-N-to-P ratio in modeling studies of marine biogeochemistry may be an inappropriate assumption and therefore has poor predictive value. It should be abandoned in favor of biogeochemical models that accommodate flexible ecological stoichiometry. Future studies of the marine P cycle will need to be comprehensive in their approach and draw on diverse subdisciplines, from the application of novel molecular techniques in P metabolism to individual and community-level ecophysiology and basin-scale ecosystem analysis. This cross-disciplinary approach will demand novel computational methods, from bioinformatics to

Carbon Export

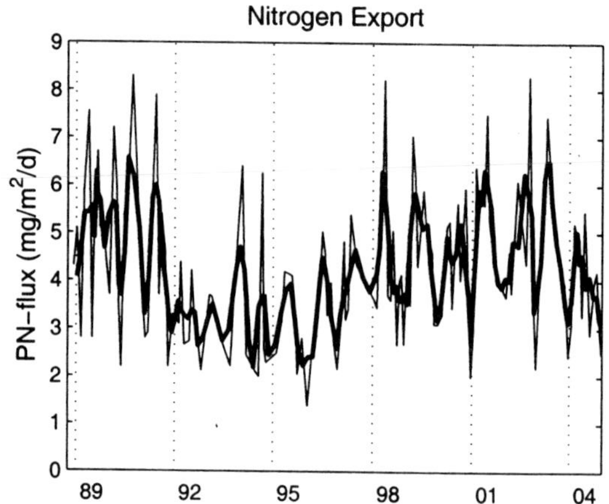

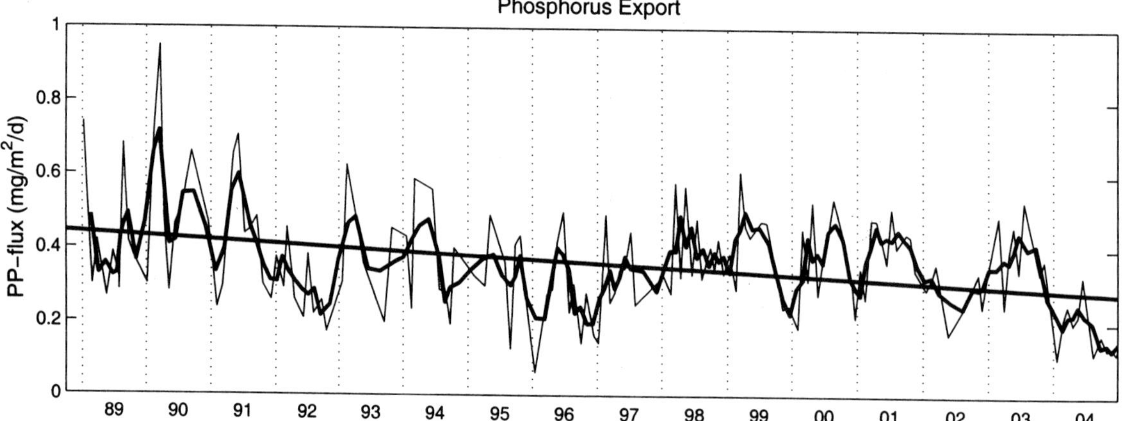

FIGURE 11 Temporal changes in the export of particulate matter at Station ALOHA during the period from 1989 to 2004, as measured by drifting sediment traps deployed at a reference depth of 150 m. The heavy lines represent the three-point running means of approximately monthly observations. Presented in the upper two plots are the PC and PN export fluxes showing coherent temporal trends. Presented in the lower plot is the PP export flux, showing a fundamentally different pattern, specifically a systematic decrease in PP export over time. d, day.

numerical simulations with a broad range of intrinsic scales (82). The next decade should be very exciting.

I thank Meredith Hullar for the cordial invitation to contribute this chapter and for her constructive editorial comments. Karin Björkman, a long-time P cycle collaborator, provided helpful comments that improved the manuscript, and Lisa Lum and Lance Fujieki provided invaluable assistance.

The original research reported in this chapter was supported by grants from the National Science Foundation and the Gordon and Betty Moore Foundation awarded to D.M.K.

REFERENCES

1. **Ammerman, J. W., and F. Azam.** 1981. Dissolved cyclic adenosine monophosphate (cAMP) in the sea and uptake of cAMP by marine bacteria. *Mar. Ecol. Prog. Ser.* **5:**85–89.
2. **Ammerman, J. W., R. R. Hood, D. A. Case, and J. B. Cotner.** 2003. Phosphorus deficiency in the Atlantic: an emerging paradigm in oceanography. *Eos* **84:**165, 170.
3. **Armstrong, F. A., P. M. Williams, and J. D. H. Strickland.** 1966. Photo-oxidation of organic matter in seawater by ultraviolet radiation, analytical and other applications. *Nature* **211:**481–483.
4. **Armstrong, F. A. J., C. R. Sterns, and J. D. H. Strickland.** 1967. The measurement of upwelling and subsequent biological processes by means of the Technicon autoanalyzer and associated equipment. *Deep-Sea Res.* **14:** 381–389.
5. **Atkins, W. R. G., and E. G. Wilson.** 1926. The colorimetric estimation of minute amounts of compounds of silicon, of phosphorus, and of arsenic. *Biochem. J.* **20:**1223–1228.
6. **Azam, F., and R. E. Hodson.** 1977. Dissolved ATP in the sea and its utilisation by marine bacteria. *Nature* **267:**696–698.
7. **Benner, R., J. D. Pakulski, M. McCarthy, J. I. Hedges, and P. G. Hatcher.** 1992. Bulk chemical characteristics of dissolved organic matter in the ocean. *Science* **255:**1561–1564.
8. **Berman, T., M. Nawrocki, G. T. Taylor, and D. M. Karl.** 1987. Nutrient flux between bacteria, bacterivorous nano-

planktonic protists and algae. *Mar. Microb. Food Webs* **2:** 69–82.

9. **Björkman, K. M., and D. M. Karl.** 2003. Bioavailability of dissolved organic phosphorus in the euphotic zone at Station ALOHA, North Pacific Subtropical Gyre. *Limnol. Oceanogr.* **48:**1049–1057.
10. **Björkman, K. M., and D. M. Karl.** 2005. Presence of dissolved nucleotides in the North Pacific subtropical gyre and their role in cycling of dissolved organic phosphorus. *Aquat. Microb. Ecol.* **39:**193–203.
11. **Björkman, K., A. L. Thomson-Bulldis, and D. M. Karl.** 2000. Phosphorus dynamics in the North Pacific subtropical gyre. *Aquat. Microb. Ecol.* **22:**185–198.
12. **Blake, R. E., J. R. O'Neil, and A. V. Surkov.** 2005. Biogeochemical cycling of phosphorus: insights from oxygen isotope effects of phosphoenzymes. *Am. J. Sci.* **305:** 596–620.
13. **Bossard, P., and D. M. Karl.** 1986. The direct measurement of ATP and adenine nucleotide pool turnover in microorganisms: a new method for environmental assessment of metabolism, energy flux and phosphorus dynamics. *J. Plankton Res.* **8:**1–13.
14. **Broecker, W. S., and T.-H. Peng.** 1982. *Tracers in the Sea.* Lamont-Doherty Geological Observatory, Columbia University, Palisades, N.Y.
15. **Brum, J. R.** 2005. Concentration, production and turnover of viruses and dissolved DNA pools at Stn ALOHA, North Pacific Subtropical Gyre. *Aquat. Microb. Ecol.* **41:**103–113.
16. **Cavender-Bares, K. K., D. M. Karl, and S. W. Chisholm.** 2001. Nutrient gradients in the western North Atlantic Ocean: relationship to microbial community structure, and comparison to patterns in the Pacific Ocean. *Deep-Sea Res.* **48:**2373–2395.
17. **Christian, J. R., M. R. Lewis, and D. M. Karl.** 1997. Vertical fluxes of carbon, nitrogen and phosphorus in the North Pacific subtropical gyre near Hawaii. *J. Geophys. Res.* **102:**15,667–15,677.
18. **Church, M. J., H. W. Ducklow, and D. M. Karl.** 2002. Multiyear increases in dissolved organic matter inventories at Station ALOHA in the North Pacific Subtropical Gyre. *Limnol. Oceanogr.* **47:**1–10.
19. **Clark, L. L., E. D. Ingall, and R. Benner.** 1998. Marine phosphorus is selectively remineralized. *Nature* **393:**426.
20. **Colman, A. S., R. E. Blake, D. M. Karl, M. L. Fogel, and K. K. Turekian.** 2005. Marine phosphate oxygen isotopes and organic matter remineralization in the oceans. *Proc. Natl. Acad. Sci. USA* **102:**13023–13028.
21. **Conkright, M. E., W. W. Gregg, and S. Levitus.** 2000. Seasonal cycle of phosphate in the open ocean. *Deep-Sea Res. I* **47:**159–175.
22. **Cotner, J. B., and R. G. Wetzel.** 1992. Uptake of dissolved inorganic and organic phosphorus compounds by phytoplankton and bacterioplankton. *Limnol. Oceanogr.* **37:**232–243.
23. **Cullen, J. J., P. J. S. Franks, D. M. Karl, and A. Longhurst.** 2002. Physical influences on marine ecosystem dynamics, p. 297–336. *In* A. R. Robinson, J. J. McCarthy, and B. J. Rothschild (ed.), *The Sea*, vol. 12. John Wiley & Sons, Inc., New York, N.Y.
24. **deDuve, C.** 1991. *Blueprint for a Cell: the Nature and Origin of Life.* Neil Patterson Publishers, Burlington, N.C.
25. **DeFlaun, M. F., and J. H. Paul.** 1989. Detection of exogenous gene sequences in dissolved DNA from aquatic environments. *Microb. Ecol.* **18:**21–28.
26. **DeFlaun, M. F., J. H. Paul, and D. Davis.** 1986. Simplified method for dissolved DNA determination in aquatic environments. *Appl. Environ. Microbiol.* **52:**654–659.
27. **DeFlaun, M. F., J. H. Paul, and W. H. Jeffrey.** 1987. Distribution and molecular weight of dissolved DNA in subtropical estuarine and oceanic environments. *Mar. Ecol. Prog. Ser.* **38:**65–73.
28. **DeLong, E. F., and D. M. Karl.** 2005. Genomic perspectives in microbial oceanography. *Nature* **437:**336–342.
29. **Denigès, G.** 1921. Détermination quantitative des plus faibles quantités de phosphates dans les produits biologiques par la méthode céruléomolybdique. *C. R. Soc. Biol. Paris* **84:**875–877.
30. **Doney, S. C., M. R. Abbott, J. J. Cullen, D. M. Karl, and L. Rothstein.** 2004. From genes to ecosystems: the ocean's new frontier. *Frontiers Ecol. Environ.* **2:**457–466.
31. **Dyhrman, S. T., and B. Palenik.** 1999. Phosphate stress in cultures and field populations of the dinoflagellate *Prorocentrum minimum* detected by a single-cell alkaline phosphatase assay. *Appl. Environ. Microbiol.* **65:**3205–3212.
32. **Dyhrman, S. T., P. D. Chappell, S. T. Haley, J. W. Moffett, E. D. Orchard, J. B. Waterbury, and E. A. Webb.** 2006. Phosphonate utilization by the globally important marine diazotroph *Trichodesmium. Nature* **439:** 68–71.
33. **Fiske, C. H., and Y. Subbarow.** 1925. The colorimetric determination of phosphorus. *J. Biol. Chem.* **66:**375–400.
34. **Froelich, P. N., M. L. Bender, N. A. Luedtke, G. R. Heath, and T. DeVries.** 1982. The marine phosphorus cycle. *Am. J. Sci.* **282:**474–511.
35. **Fu, F.-X., Y. Zhang, K. Leblanc, S. A. Sañudo-Wilhelmy, and D. A. Hutchins.** 2005. The biological and biogeochemical consequences of phosphate scavenging onto phytoplankton cell surfaces. *Limnol. Oceanogr.* **50:**1459–1472.
36. **Fuwa, K., W. Lei, and K. Fujiwara.** 1984. Colorimetry with a total-reflection long-capillary cell. *Anal. Chem.* **56:** 1640–1644.
37. **Harrison, W. G.** 1983. Uptake and recycling of soluble reactive phosphorus by marine microplankton. *Mar. Ecol. Prog. Ser.* **10:**127–135.
38. **Haury, L. R., C. L. Fey, and E. Shulenberger.** 1994. Surface enrichment of inorganic nutrients in the North Pacific Ocean. *Deep-Sea Res.* **41:**1191–1205.
39. **Hecky, R. E., and P. Kilham.** 1988. Nutrient limitation of phytoplankton in freshwater and marine environments: a review of recent evidence on the effects of enrichment. *Limnol. Oceanogr.* **33:**796–822.
40. **Hicks, E., and J. P. Riley.** 1980. The determination of dissolved total nucleic acids in natural waters including sea water. *Anal. Chim. Acta* **116:**137–144.
41. **Holm-Hansen, O.** 1969. Determination of microbial biomass in ocean profiles. *Limnol. Oceanogr.* **14:**740–747.
42. **Holm-Hansen, O., and C. R. Booth.** 1966. The measurement of adenosine triphosphate in the ocean and its ecological significance. *Limnol. Oceanogr.* **11:**510–519.
43. **Holm-Hansen, O., W. H. Sutcliffe, Jr., and J. Sharp.** 1968. Measurement of deoxyribonucleic acid in the ocean and its ecological significance. *Limnol. Oceanogr.* **13:**507–514.
44. **Hulett, H. R.** 1970. Non-enzymatic hydrolysis of adenosine phosphates. *Nature* **225:**1248–1249.
45. **Hutchinson, G. E.** 1971. *A Treatise on Limnology.* John Wiley & Sons, New York, N.Y.
46. **Karl, D. M.** 1980. Cellular nucleotide measurements and applications in microbial ecology. *Microbiol. Rev.* **44:**739–796.
47. **Karl, D. M.** 1999. A sea of change: biogeochemical variability in the North Pacific subtropical gyre. *Ecosystems* **2:** 181–214.
48. **Karl, D. M.** 2000. Phosphorus, the staff of life. *Nature* **406:**31–32.

49. **Karl, D. M.** 2002. Nutrient dynamics in the deep blue sea. *Trends Microbiol.* **10:**410–418.
50. **Karl, D. M., and M. D. Bailiff.** 1989. The measurement and distribution of dissolved nucleic acids in aquatic environments. *Limnol. Oceanogr.* **34:**543–558.
51. **Karl, D. M., N. R. Bates, S. Emerson, P. J. Harrison, C. Jeandel, O. Llinas, K.-K. Liu, J.-C. Marty, A. F. Michaels, J. C. Miquel, S. Neuer, Y. Nojiri, and C. S. Wong.** 2003. Temporal studies of biogeochemical processes determined from ocean time-series observations during the JGOFS era, p. 239–267. *In* M. J. R. Fasham (ed.), *Ocean Biogeochemistry: the Role of the Ocean Carbon Cycle in Global Change*. Springer, New York, N.Y.
52. **Karl, D. M., and K. M. Björkman.** 2001. Phosphorus cycle in seawater: dissolved and particulate pool inventories and selected phosphorus fluxes. *Methods Microbiol.* **30:** 239–270.
53. **Karl, D. M., and K. M. Björkman.** 2002. Dynamics of DOP, p. 246–366. *In* D. Hansell and C. Carlson (ed.), *Biogeochemistry of Marine Dissolved Organic Matter*. Elsevier Science, Amsterdam, The Netherlands.
54. **Karl, D. M., K. M. Björkman, J. E. Dore, L. Fujieki, D. V. Hebel, T. Houlihan, R. M. Letelier, and L. M. Tupas.** 2001. Ecological nitrogen-to-phosphorus stoichiometry at Station ALOHA. *Deep-Sea Res. II* **48:**1529–1566.
55. **Karl, D. M., and P. Bossard.** 1985. Measurement and significance of ATP and adenine nucleotide pool turnover in microbial cells and environmental samples. *J. Microbiol. Methods* **3:**125–139.
56. **Karl, D. M., J. R. Christian, J. E. Dore, D. V. Hebel, R. M. Letelier, L. M. Tupas, and C. D. Winn.** 1996. Seasonal and interannual variability in primary production and particle flux at Station ALOHA. *Deep-Sea Res. II* **43:** 539–568.
57. **Karl, D., R. Letelier, L. Tupas, J. Dore, J. Christian, and D. Hebel.** 1997. The role of nitrogen fixation in biogeochemical cycling in the subtropical North Pacific Ocean. *Nature* **388:**533–538.
58. **Karl, D. M., and R. Lukas.** 1996. The Hawaii Ocean Time-Series (HOT) Program: background, rationale and field implementation. *Deep-Sea Res. II* **43:**129–156.
59. **Karl, D. M., and G. Tien.** 1992. MAGIC: a sensitive and precise method for measuring dissolved phosphorus in aquatic environments. *Limnol. Oceanogr.* **37:**105–116.
60. **Karl, D. M., and G. Tien.** 1997. Temporal variability in dissolved phosphorus concentrations in the subtropical North Pacific Ocean. *Mar. Chem.* **56:**77–96.
61. **Karl, D. M., and C. D. Winn.** 1984. Adenine metabolism and nucleic acid synthesis: applications to microbiological oceanography, p. 197–215. *In* J. E. Hobbie and P. J. L. Williams (ed.), *Heterotrophic Activity in the Sea.* Plenum Publishing Corp., New York, N.Y.
62. **Karl, D. M., and K. Yanagi.** 1997. Partial characterization of the dissolved organic phosphorus pool in the oligotrophic North Pacific Ocean. *Limnol. Oceanogr.* **42:** 1398–1405.
63. **Lal, D., and T. Lee.** 1988. Cosmogenic ^{32}P and ^{33}P used as tracers to study phosphorus recycling in the ocean. *Nature* **333:**752–754.
64. **Lal, D., Y. Chung, T. Platt, and T. Lee.** 1988. Twin cosmogenic radiotracer studies of phosphorus cycling and chemical fluxes in the upper ocean. *Limnol. Oceanogr.* **33:**1559–1567.
65. **Lorenz, M. G., and W. Wackernagel.** 1994. Bacterial gene transfer by natural genetic transformation in the environment. *Microbiol. Rev.* **58:**563–602.
66. **Mackenzie, F. T., L. M. Ver, and A. Lerman.** 2002. Century-scale nitrogen and phosphorus controls of the carbon cycle. *Chem. Geol.* **190:**13–32.
67. **Martin, J. H., G. A. Knauer, D. M. Karl, and W. W. Broenkow.** 1987. VERTEX: carbon cycling in the Northeast Pacific. *Deep-Sea Res.* **34:**267–285.
68. **McGrath, S. M., and C. W. Sullivan.** 1981. Community metabolism of adenylates by microheterotrophs from the Los Angeles and Southern California coastal waters. *Mar. Biol.* **62:**217–226.
69. **Meybeck, M.** 1982. Carbon, nitrogen and phosphorus transport by world rivers. *Am. J. Sci.* **282:**401–450.
70. **Mills, M. M., C. Ridame, M. Davey, J. La Roche, and R. J. Geider.** 2004. Iron and phosphorus co-limit nitrogen fixation in the eastern tropical North Atlantic. *Nature* **429:**292–294.
71. **Morton, S. C., and M. Edwards.** 2005. Reduced phosphorus compounds in the environment. *Crit. Rev. Environ. Sci. Technol.* **35:**333–364.
72. **Moutin, T., T. F. Thingstad, F. Van Wambeke, D. Marie, G. Slawyk, P. Raimbault, and H. Claustre.** 2002. Does competition for nanomolar phosphate supply explain the predominance of the cyanobacterium *Synechococcus? Limnol. Oceanogr.* **47:**1562–1567.
73. **Murphy, J., and J. P. Riley.** 1962. A modified single solution method for the determination of phosphate in natural waters. *Anal. Chim. Acta* **27:**31–36.
74. **Orrett, K., and D. M. Karl.** 1987. Dissolved organic phosphorus production and turnover in surface waters. *Limnol. Oceanogr.* **32:**383–395.
75. **Pace, M. L., G. A. Knauer, D. M. Karl, and J. H. Martin.** 1987. Primary production, new production and vertical flux in the eastern Pacific Ocean. *Nature* **325:** 803–804.
76. **Paul, J. H., M. E. Frischer, and J. M. Thurmond.** 1991. Gene transfer in marine water column and sediment microcosms by natural plasmid transformation. *Appl. Environ. Microbiol.* **57:**1509–1515.
77. **Proctor, C. M., and D. W. Hood.** 1954. Determination of inorganic phosphate in sea water by an iso-butanol extraction procedure. *J. Mar. Res.* **13:**122–132.
78. **Redfield, A.C., B. H. Ketchum, and F. A. Richards.** 1963. The influence of organisms on the composition of seawater p. 26–77. *In* M. N. Hill (ed.), *The Sea, Ideas and Observations on Progress in the Study of the Seas*, vol. 2. Interscience, New York, N.Y.
79. **Redfield, R. J., M. R. Schrag, and A. M. Dean.** 1997. The evolution of bacterial transformation: sex with poor relations. *Genetics* **146:**27–38.
80. **Ridal, J. J., and R. M. Moore.** 1990. A re-examination of the measurement of dissolved organic phosphorus in seawater. *Mar. Chem.* **29:**19–31.
81. **Rigler, F. H.** 1968. Further observations inconsistent with the hypothesis that the molybdenum blue method measures orthophosphate in lake water. *Limnol. Oceanogr.* **13:** 7–13.
82. **Rothstein, L. M., J. J. Cullen, M. Abbott, E. Chassignet, K. Denman, S. Doney, H. Ducklow, K. Fennel, M. Follows, D. Haidvogel, E. Hofmann, D. Karl, J. Kindle, I. Lima, M. Maltrud, C. McClain, D. McGillicuddy, J. Olascoaga, Y. Spitz, J. Wiggert, and J. Yoder.** 2006. Modeling ocean ecosystems: the PARADIGM program. *Oceanography* **19:**17–45.
83. **Ryther, J. H., and W. M. Dunstan.** 1971. Nitrogen, phosphorus and eutrophication in the coastal marine environment. *Science* **171:**1008–1013.
84. **Sakano, S., and A. Kamatani.** 1992. Determination of dissolved nucleic acids in seawater by the fluorescence dye, ethidium bromide. *Mar. Chem.* **37:**239–255.
85. **Scanlan, D. J., and W. H. Wilson.** 1999. Application of molecular techniques to addressing the role of P as a key effector in marine ecosystems. *Hydrobiologia* **401:**149–175.

86. **Strickland, J. D. H., and T. R. Parsons.** 1972. *A Practical Handbook of Seawater Analysis*. Fisheries Research Board of Canada, Ottawa, Canada.
87. **Suzumura, M., K. Ishikawa, and H. Ogawa.** 1998. Characterization of dissolved organic phosphorus in coastal seawater using ultrafiltration and phosphohydrolytic enzymes. *Limnol. Oceanogr.* **43:**1553–1564.
88. **Sverdrup, H. U.** 1953. On conditions for the vernal blooming of phytoplankton. *J. Cons. Int. Explor. Mer* **18:** 287–295.
89. **Tarapchak, S. J., and R. A. Moll.** 1990. Phosphorus sources for phytoplankton and bacteria in Lake Michigan. *J. Plankton Res.* **12:**743–758.
90. **Turk, V., A.-S. Rehnstam, E. Lundberg, and A. Hagström.** 1992. Release of bacterial DNA by marine nanoflagellates, an intermediate step in phosphorus regeneration. *Appl. Environ. Microbiol.* **58:**3744–3750.
91. **Tyrrell, T.** 1999. The relative influences of nitrogen and phosphorus on oceanic primary production. *Nature* **400:** 525–531.
92. **Wackett, L. P., B. L. Wanner, C. P. Venditti, and C. T. Walsh.** 1987. Involvement of the phosphate regulon and the *psiD* locus in carbon-phosphorus lyase activity of *Escherichia coli* K-12. *J. Bacteriol.* **169:**1753–1756.
93. **White, A. E., Y. H. Spitz, D. M. Karl, and R. M. Letelier.** 2006. Flexible elemental stoichiometry in *Trichodesmium* spp. and its ecological implications. *Limnol. Oceanogr.* **51:**1777–1790.
94. **Wilson, W. H., N. G. Carr, and N. H. Mann.** 1996. The effect of phosphate status on the kinetics of cyanophage infection in the oceanic cyanobacterium *Synechococcus* sp. WH7803. *J. Phycol.* **32:**506–516.
95. **Wu, J., W. Sunda, E. A. Boyle, and D. M. Karl.** 2000. Phosphate depletion in the western North Atlantic Ocean. *Science* **289:**759–762.

Microbial Metal Cycling in Aquatic Environments*

ERIC E. RODEN AND DAVID EMERSON

44

The biotic and abiotic cycling of metallic elements (metals and metalloids) is an important component of the biogeochemistry of aquatic ecosystems (17). The chemical nature of metals has a fundamental influence on the way they are cycled in the environment. Table 1 provides a brief overview of the chemistries of different metal groups and their biogeochemical behaviors in aquatic systems. There are three major active biological processes that influence the cycling of metals in aquatic environments (Fig. 1), as follows: (i) uptake and incorporation into living photosynthetic and heterotrophic organisms and release during biomass decay establish a fundamental link between organic matter processing and metal cycling; (ii) enzymatically catalyzed metal redox transformations coupled to either anaerobic respiration or lithoautotrophy exert fundamental control over metal solubility and transport and create the potential for syntrophic microbial interactions based on metal redox cycling; and (iii) transfer of one or more methyl groups to a metal through biological metabolism strongly influences the toxicity of the metal to living organisms throughout aquatic food webs. In addition to these active processes, complexation of metals by microbial cell walls, extracellular polymeric substances (e.g., polysaccharides), and metal-chelating compounds (siderophores) can also influence metal transport in aquatic systems (72). This chapter focuses on microbial metal redox metabolism, with an emphasis on iron (Fe) and manganese (Mn) cycling in the water column and surface sediments. The reviews by Sunda (226) and Gadd (75) (and references cited therein) provide an introduction to the literature on metal uptake and biomethylation in aquatic systems, respectively. We deal here exclusively with metal cycling in circumneutral pH environments. Fe and Mn are by far the most abundant redox-active metals in aquatic environments, and their cycling has a direct or indirect impact on the behaviors of a wide variety of other metal elements (225). The brief consideration of cycling of other metal elements at the end of the chapter is therefore couched in terms of the connection to Fe and Mn cycling. Because of the similar biogeochemical behaviors of Fe and Mn, their cycling processes are discussed together, with important differences being emphasized whenever appropriate. Although the redox cycling of Fe and Mn in hydromorphic soils (159, 248, 249) is not explicitly addressed, the processes described here for aquatic sediments are active in such environments. Likewise, the cycling of Fe and Mn in groundwater environments is conceptually analogous to that described for aquatic systems, although the nature of the spatial gradients and time scales involved differ significantly (33).

Fe AND Mn INPUT TO AQUATIC SYSTEMS

Fe and Mn constitute approximately 4.3 and 0.1 weight %, respectively, of the Earth's continental crust (268). The majority of the Fe and Mn that enters aquatic systems comes in the form of insoluble oxides (Fe and Mn) and silicate phases (Fe only; Mn-rich silicates are uncommon), which are produced during weathering of rock-forming silicate minerals (e.g., olivines, pyroxenes, and amphiboles) in the terrestrial environment and transported to coastal marine environments and lakes by rivers and streams (175, 206). Glacial sediment and wind-borne dust transport represent minor sources of Fe and Mn inputs to the marine environment (175), although the latter is an important source of Fe for primary production in the open ocean (53). In addition to these particulate inputs, small but significant quantities of soluble reduced Fe and Mn are introduced to ocean basins through hydrothermal vent fluids (175), where they may serve as energy sources for lithotrophic microbial metabolism (64, 147). Atmospheric deposition and various kinds of anthropogenic inputs (e.g., sewage and industrial and mining wastes) are important sources of Fe, Mn, and other metals in freshwater ecosystems (124, 206).

MICROBIAL Fe AND Mn REDOX CYCLING

Overview

Under the well-oxygenated and circumneutral pH (ca. 6 to 8) conditions typical of terrestrial environments, Fe(II) and Mn(II) released to solution during mineral weathering are

*Dedicated to the memory of Robert Wetzel, a leading aquatic scientist (and friend and colleague of E.E.R.) who passed away in April 2005. Though not a specialist in metal cycling, Wetzel continuously emphasized the enormous importance of microorganisms in the functioning of aquatic ecosystems and, through his prolific research and writings (including the textbook *Limnology*), laid the foundation for a wide range of studies of microbially mediated elemental cycling in freshwater systems.

TABLE 1 Overview of metal biochemistry in aquatic environments[a]

Metal group	Representative metals[b]	Chemical behavior	Role in biological processes
"Hard" metal ions (alkali-metal and alkaline-earth cations)	Na^+, K^+, Mg^{2+}, Ca^{2+}	Participate mainly in electrostatic interactions with anions and carboxylic sites on organic compounds; generally not strongly complexed in natural waters	Na^+ and K^+ are electrolytes and osmotic pressure regulators; Ca^{2+} and Mg^{2+} stabilize biological structures (macromolecules to exoskeletons)
"Soft" metal ions	Zn^{2+}, Cd^{2+}, Hg^{2+}/Hg^0, Pb^{2+}	Form covalent bonds with N- and S-containing sites on organic compounds	Zn^{2+} is an enzyme cofactor (for RNA and DNA polymerases and DNA-binding proteins); Cd^{2+}, Hg^{2+}, Pb^{2+}, and Cu^{2+} are toxic metals that disrupt enzyme activity or modify macromolecule structure; Hg^{2+} is subject to biological methylation and to regeneration via demethylation; Hg^0 is the end product of enzymatic Hg^{2+} reduction
Transition metal ions	Fe^{3+}/Fe^{2+}, Mn^{4+}/Mn^{2+}, Co^{3+}/Co^{2+}, Cr^{6+}/Cr^{3+}, V^{5+}/V^{4+}, Tc^{7+}/Tc^{4+}, Mo^{6+}, Ni^{2+}, Cu^{2+}	Participate in both electrostatic and covalent interactions with complexants; are strongly complexed by OH groups on mineral surfaces; are generally susceptible to redox transformation	Fe^{2+}, Mn^{2+}, Co^{2+}, Cr^{3+} and Mo^{6+} are enzyme cofactors and coenzymes; Fe^{2+}, Mn^{2+}, Co^{2+}, Cr^{3+}, V^{4+}, and Tc^{4+} are end products of biotic and abiotic Fe^{3+}, Mn^{4+}, Co^{3+}, Cr^{6+}, V^{5+}, and Tc^{7+} reduction; Fe^{2+}, Mn^{2+}, and Co^{2+} are electron donors for chemolithotrophy; Ni^{2+} and Cu^{2+} are enzyme cofactors and coenzymes (toxic at high levels)
Hydrolyzed metal ions	Fe^{3+}, Mn^{4+}, Al^{3+}, Cr^{3+}	Strongly hydrolyzed at circumneutral pH; generally present as insoluble oxides with reactive OH surface groups	Fe^{3+} and Mn^{4+} are electron acceptors for anaerobic respiration; Al^{3+} is toxic when present in soluble uncomplexed form
Metalloids	As^{5+}/As^{3+}, $Se^{6+}/Se^{4+}/Se^0/Se^{2-}$	Not metals proper but share complexation and redox properties in common with transition metals	As^{3+}, Se^{4+}, Se^0, and Se^{2-} are end products of biotic and abiotic As^{5+}, Se^{6+}, Se^{4+}, and S^0 reduction; As^{5+}, As^{3+}, Se^{6+}, and Se^{4+} and are subject to biological methylation; As^{3+} and Se^0 are electron donors for chemolithotrophy
Actinides	U^{6+}/U^{4+}	Share complexation and redox properties in common with transition metals	U^{4+} is the end product of biotic and abiotic U^{6+} reduction and is an electron donor for chemolithotrophy

[a]Based on information from references 19, 145, 167, and 168.

[b]Superscripts denote charges on metal ions only; the major form of a given metal ion in natural waters is not always the free ion. Important examples include the following [the major form at neutral pH is shown in brackets; the subscript "(s)" indicates an insoluble precipitate]: Fe^{3+} [$Fe(OH)_{3(s)}$], Mn^{4+} [$MnO_{2(s)}$], Cr^{6+} [$HCrO_4^-$], Tc^{7+} [TcO_4^-], Tc^{4+} [$TcO_{2(s)}$], As^{5+} [$H_2AsO_4^-$], As^{3+} [$H_3AsO_3^0$], Se^{6+} [SeO_4^{2-}], V^{5+} [Vo_4^{3-}], V^{4+} [$VO_{2(s)}$], U^{6+} [$UO_2(CO_3)_2^{2-}$], and U^{4+} [$UO_{2(s)}$].

rapidly oxidized and precipitated as Fe(III) and Mn(IV) oxides, respectively. The oxide minerals are generally much more biologically reactive than silicate phases, although microbial redox transformation of structural Fe in clay minerals is known to occur (e.g., see the work of Kostka et al. [114] and Shelobolina et al. [201, 202]). Once solid-phase Fe(III) and Mn(IV) oxides enter an aquatic system, they are subject to downward transport through particle settling, ultimately leading to deposition in sediments. Reductive transformation of these oxide phases en route to and within aquatic sediments leads to the release of dissolved Fe(II) and Mn(II), which in turn are reoxidized through reaction with

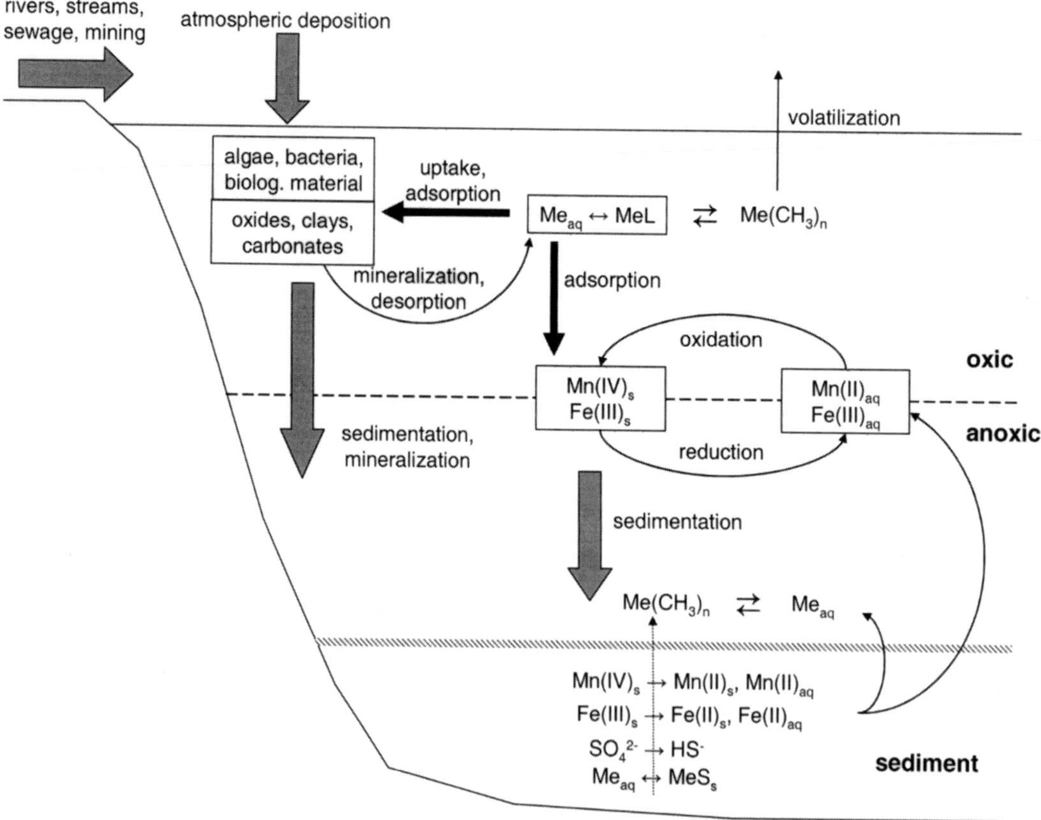

FIGURE 1 Schematic representation of the cycling of metals in aquatic ecosystems. Note that when the water column is well-oxygenated, metal redox cycling is localized at the sediment-water interface. Modified from Fig. 1 in reference 206 with permission of Routledge/Taylor & Francis Group LLC.

O_2 (and possibly other oxidants [see below]) at the aerobic-anaerobic interface within the water column or in surface sediments. These coupled processes constitute the basic pattern of metal redox cycling illustrated in Fig. 1. Although Fe and Mn are generally thought of as trace elements in open water environments, they can accumulate to quite high concentrations (e.g., tens of millimoles per liter [see Fig. 4]) in sediments and must therefore be viewed as major elements in sediment biogeochemical cycling (148, 251, 252). It is relevant in this context that Fe redox cycling is likely to have played a major role in the global biogeochemistry of Earth during the Achaean and early Proterozoic eons (109, 110, 258), when massive deposition of Fe-rich sediments (banded iron formations) and Mn deposits took place in association with the slow conversion of the atmosphere and oceans from anoxic to oxic conditions (92).

Many shallow-water aquatic systems (e.g., lakes and estuaries) undergo seasonal temperature or salinity-driven density stratification, which restricts eddy diffusive mixing and associated solute transport between surface and bottom waters. Such stratification typically leads to depletion of oxygen from bottom water and the establishment of an aerobic-anaerobic boundary across which metal redox cycling takes place. Similar conditions arise in marine basins with restricted vertical mixing, e.g., the Black Sea (155) and Orca Basin in the Gulf of Mexico (250). There is also strong zonation of redox conditions in the sediments of shallow-water aquatic systems, with aerobic processes generally taking place only within the upper few millimeters of sediment (240). Cycling of Fe and Mn between oxidized and reduced forms takes place in sediments in a manner analogous to that in stratified water bodies, although the concentrations of metals involved are much higher and the (vertical) spatial scale over which cycling takes place (generally a few to a few tens of centimeters) is highly compressed.

Microbial catalysis plays a significant role in both oxidation and reduction of Fe and Mn in natural environments. However, a variety of abiotic redox reactions can also take place, and these processes compete with microbial processes. Thus, the relative importance of biotic and abiotic processes and interactions between them are key issues in metal cycling in aquatic systems (28). Figure 2 provides a summary of the major oxidants, reductants, and pathways for Fe and Mn redox transformations in circumneutral-pH aquatic systems and is used as a framework for the discussion of metal redox cycling in this chapter.

Fe(III) and Mn(IV) Oxide Speciation

Fe and Mn oxides occur in nature in a spectrum of phases ranging from poorly crystalline materials, such as ferrihydrite [$Fe(III)_5HO_8 \cdot 4H_2O$] and vernadite [δ-Mn(IV)O_2], to well-crystallized minerals, such as goethite [α-Fe(III)OOH], hematite [α-$Fe(III)_2O_3$], and pyrolusite [β-Mn(IV)O_2] (39, 173). For our purposes, generic Fe(III) and Mn(IV) oxide surfaces that react with various organic and inorganic reductants through biotic and/or abiotic pathways are considered Fe and Mn oxides.

Although insoluble oxide phases are by far the most abundant forms of Fe(III) and Mn(IV) in neutral-pH

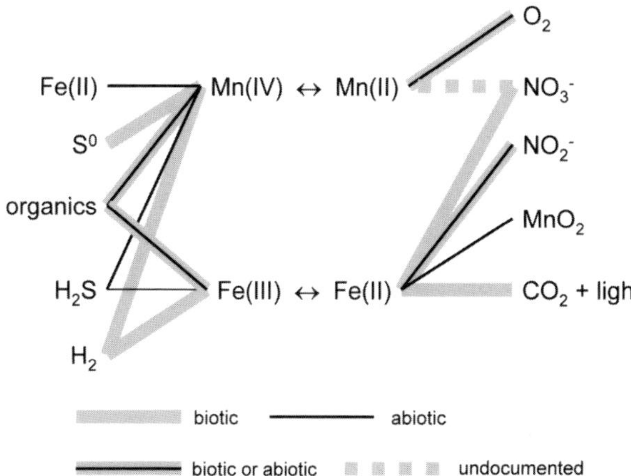

FIGURE 2 Pathways of Fe and Mn oxidation and reduction in circumneutral-pH aquatic environments. Modified from Fig. 8.4 in reference 28 with permission of the publisher.

aquatic environments, several recent studies suggested that small but significant quantities of soluble Fe(III) (undoubtedly in complex with organic ligands) exist in circumneutral sediment pore fluids (29, 144, 161, 179). Such compounds are likely to turn over very rapidly, as rates of biotic and abiotic soluble Fe(III) reduction are generally at least an order of magnitude higher than rates of solid-phase oxide reduction (130, 225). Likewise, soluble Mn(III)-containing complexes can serve as efficient abiotic oxidants as well as electron acceptors for microbial metabolism (116) and may play a role in Mn redox cycling in redox transition environments (143). Moreover, recent studies suggested that Mn(III) produced during microbial Mn(II) oxidation (see below) is likely to serve as a competitor for siderophore-bound Fe(III) in nutrient-limited environments (169, 264).

Fe(III) and Mn(IV) Oxide Reduction

Fe(III) and Mn(IV) oxides are subject to reduction by both biotic and abiotic pathways (90). For reasons explained below, major reductive transformations (both biotic and abiotic) of Fe and Mn are generally limited to anaerobic environments. However, it should be noted that light can catalyze the reductive dissolution of Fe(III) and Mn(IV) oxides under aerobic conditions, particularly in the presence of dissolved organic matter (257). Mn(IV) oxides react readily with humic substances under aerobic conditions in neutral pH waters (e.g., seawater), resulting in the production of low-molecular-weight organic compounds such as pyruvate, acetaldehyde, and formate (227). In addition, photoreduction of Fe(III)-siderophore complexes increases the availability of siderophore-bound Fe for uptake by planktonic assemblages (10), a process which could have a major influence on the role of Fe as a limiting nutrient for primary production in large areas of the oceans. Thus, abiotic photoreduction has the potential to exert a significant influence on carbon cycling in the aerobic photic zones of aquatic systems.

Reduction Coupled to Oxidation of Organics and H_2

It is well recognized that under anaerobic conditions, dissimilatory metal-reducing bacteria (DMRB) can couple the oxidation of organic compounds and H_2 to enzymatic transfer of electrons to Fe(III), Mn(IV), and a variety of other oxidized metal species (130, 131, 156). Representative stoichiometric reactions for oxidation of acetate and H_2 are given in Table 2 (reactions 1 to 4). An ever-widening range of bacteria and archaea are able to catalyze these and other complete or partial organic carbon oxidation reactions (134). Dissimilatory metal reduction differs fundamentally from both assimilatory reduction coupled to the uptake of metals for biosynthesis (e.g., that of metal-containing enzymes and electron transport chain components [145]) and dissimilatory reduction as a mechanism for metal (e.g., Hg^{2+} and As^{5+}) detoxification and resistance in bacteria (207). Dissimilatory microbial metal reduction is generally restricted to anaerobic conditions because (i) facultative organisms such as *Shewanella* do not express metal reduction activity in the presence of O_2 (8) and (ii) other DMRB such as *Geobacter*, are strongly inhibited by O_2 (83, 133) (note, however, that *Geobacter sulfurreducens* possesses genes for aerobic respiration and can grow by this process at low O_2 tensions [128]). Dissimilatory Fe(III) and Mn(IV) oxide reduction is also suppressed in the presence of NO_3^- (47, 213, 265), presumably due to preferential utilization of the soluble and more thermodynamically favorable electron acceptor (note, however, that the details of the regulation of dissimilatory metal reduction by NO_3^- have not yet been elucidated). In addition, Fe(II) can serve as an electron donor for microbially catalyzed NO_3^- reduction (see below), which could also contribute to suppression of net Fe(III) reduction in the presence of nitrate. To our knowledge, the influence of nitrate on Mn(IV) reduction by DMRB has not been evaluated. As a result of these limitations, DMRB activity is restricted to depths (in the water column and/or sediment) below which O_2 and NO_3^- are not saturating.

Although a variety of anaerobic organisms can transfer small quantities of reducing equivalents to Fe(III) during fermentative growth, the vast majority of dissimilatory metal-reducing activity in nature is linked to oxidation of organic carbon or H_2 coupled to respiration with Fe(III) or Mn(IV) (see references 130 and 132 for review). Likewise, although certain organics (e.g., reducing acids such as salicylate and ascorbate; thiols such as cysteine; and various reducing quinone-containing compounds, including quinone groups on natural humic and fulvic acids) can abiotically reduce Fe(III) and Mn(IV) oxides (90, 217), abiotic reduction by such compounds is likely to be only a minor process (in terms of total electrons transferred) compared to DMRB respiration (136). Moreover, a wide variety of organic carbon compounds that do not react spontaneously with Fe(III) oxides can be oxidized readily to CO_2 by DMRB (136). Similarly, the mere presence of a low system redox potential (e.g., as a result of microbial consumption of oxidants, such as oxygen and nitrate, or production of fermentation intermediates) does not lead to spontaneous oxide reduction (136, 153). Thus, classical geochemical models that treat metal oxide reduction as a freely reversible process dependent on the system redox potential (e.g., see the work of Ponnamperuma (172)) do not accurately represent reality (130). In fact, low redox potentials are typically the result of DMRB activity rather than the cause of Fe(III) and Mn(IV) oxide reduction, since the redox potentials of nonsulfidogenic environments are typically controlled by the Fe(II)-Fe(III) redox couple (224).

Reduction Coupled to Oxidation of Inorganic Compounds

In contrast to the minor role of organics in abiotic Fe(III) and Mn(IV) oxide reduction, reaction with hydrogen sulfide produced by dissimilatory sulfate-reducing bacteria can be a significant pathway for Fe(III) and Mn(IV) reduction

TABLE 2 Summary of reactions associated with Fe and Mn redox cycling in natural water and sediments[a]

Dissimilatory microbial Fe(III) and Mn(IV) reduction coupled to oxidation of acetate	
$CH_3COO^- + 4\,MnO_2 + 7H^+ \rightarrow 2HCO_3^- + 4Mn^{2+} + 4H_2O$	(1)
$CH_3COO^- + 8Fe(OH)_3 + 15H^+ \rightarrow 2HCO_3^- + 8Fe^{2+} + 20H_2O$	(2)
Dissimilatory microbial Fe(III) and Mn(IV) reduction coupled to oxidation of H_2	
$H_2 + MnO_2 + 2H^+ \rightarrow Mn^{2+} + 2H_2O$	(3)
$H_2 + 2Fe(OH)_3 + 4H^+ \rightarrow 2Fe^{2+} + 6H_2O$	(4)
Abiotic Fe(III) and Mn(IV) reduction coupled to oxidation of HS^- and Fe^{2+}	
$HS^- + 2Fe(OH)_3 + 5H^+ \rightarrow S^0 + 2Fe^{2+} + 6H_2O$	(5)
$HS^- + MnO_2 + 3H^+ \rightarrow S^0 + Mn^{2+} + 2H_2O$ (high $[H_2S]$)	(6)
$HS^- + 4MnO_2 + 7H^+ \rightarrow SO_4^{2-}$ + 4Mn2$^+$ + $4H_2O$ (low $[H_2S]$)	(7)
$2Fe^{2+} + MnO_2 + 4H_2O \rightarrow 2Fe(OH)_3 + Mn^{2+} + 2H^+$	(8)
$2{\equiv}O\text{-}Fe^+ + MnO_2 + 4H_2O \rightarrow {\equiv}OH + 2Fe(OH)_3 + Mn^{2+}$	(9)
Dissimilatory microbial Mn(IV) reduction coupled to oxidation of S^0	
$S^0 + 3MnO_2 + 4H^+ \rightarrow SO_4^{2-} + 3Mn^{2+} + 2H_2O$	(10)
Abiotic Fe(III) and Mn(IV) reduction coupled to microbial disproportionation of S^0	
$3S^0 + 2Fe(OH)_3 \rightarrow SO_4^{2-} + 2FeS + 2H_2O + 2H^+$	(11)
$S^0 + 3MnO_2 + 4H^+ \rightarrow SO_4^{2-} + 3Mn^{2-} + 2H_2O$	(12)
Abiotic and/or biotic Fe(II) oxidation by O_2	
$4Fe^{2+} + O_2 + 10H_2O \rightarrow 4Fe(OH)_3 + 8H^+$	(13)
$4{\equiv}O\text{-}Fe^+ + O_2 + 10H_2O \rightarrow 4{\equiv}OH + 4Fe(OH)_3 + 4H^+$	(14)
Abiotic Fe(II) oxidation by NO_2^-	
$2Fe^{2+} + NO_2^- + 4.5H_2O \rightarrow 2Fe(OH)_3 + 0.5N_2O + 3H^+$	(15)
$2{\equiv}O\text{-}Fe^+ + NO_2^- + 4H_2O \rightarrow 2{\equiv}OH + 2Fe(OH)_3 + 0.5N_2O + H^+$	(16)
Biotic Fe(II) oxidation by NO_2^-	
$3Fe^{2+} + NO_2^- + 7H_2O \rightarrow Fe(OH)_3 + 0.5N_2 + 5H^+$	(17)
$3{\equiv}O\text{-}Fe^+ + NO_2^- + 7H_2O \rightarrow 3{\equiv}OH + 3Fe(OH)_3 + N_2 + 2H^+$	(18)
$6Fe^{2+} + NO_2^- + 16H_2O \rightarrow 6Fe(OH)_3 + NH_4^+ + 10H^+$	(19)
$6{\equiv}O\text{-}Fe^+ + NO_2^- + 16H_2O \rightarrow 6{\equiv}OH + 6Fe(OH)_3 + N_2 + 4H^+$	(20)
Biotic Fe(II) oxidation by NO_3^-	
$2Fe^{2+} + NO_3^- + 5H_2O \rightarrow 2Fe(OH)_3 + NO_2^- + 4H^+$	(21)
$2{\equiv}O\text{-}Fe^+ + NO_3^- + 5H_2O \rightarrow 2{\equiv}OH + 2Fe(OH)_3 + NO_2^- + 2H^+$	(22)
$5Fe^{2+} + NO_3^- + 12H_2O \rightarrow 5Fe(OH)_3 + 0.5N_2 + 9H^+$	(23)
$5{\equiv}O\text{-}Fe^+ + NO_3^- + 12H_2O \rightarrow 5{\equiv}OH + 5Fe(OH)_3 + 0.5N_2 + 4H^+$	(24)
$8Fe^{2+} + NO_3^- + 21H_2O \rightarrow 8Fe(OH)_3 + NH_4 + 14H^+$	(25)
$8{\equiv}O\text{-}Fe^+ + NO_3^- + 21H_2O \rightarrow 8{\equiv}OH + 8Fe(OH)_3 + NH_4^+ + 6H^+$	(26)
Biotic Fe(II) oxidation by light and CO_2	
$4Fe^{2+} + CO_2 + 11H_2O \rightarrow 4Fe(OH)_3 + CH_2O + 8H^+$	(27)
Abiotic and/or biotic Mn(II) oxidation by O_2	
$2Mn^{2+} + O_2 + 2H_2O \rightarrow 2MnO_2 + 4H^+$	(28)
$2{\equiv}O\text{-}Mn^+ + O_2 + 2H_2O \rightarrow 2{\equiv}OH + 2MnO_2 + 2H^+$	(29)

[a]See the text for details and references.

in organic- and sulfur-rich coastal marine sediments (24, 25, 118, 119, 239, 243). Dissolved sulfide reacts rapidly with both Fe(III) and Mn(IV) at neutral pH (Table 2, reactions 5 to 7), resulting principally in the production of elemental sulfur (S^0). Direct microbial catalysis may also contribute to sulfide-linked Mn(VI) reduction (195), and certain groups of sulfate-reducing bacteria (e.g., *Desulfobulbus*) can couple the oxidation of S^0 to SO_4^{2-} to the reduction of Mn(IV) oxides (138) (Table 2, reaction 10). Another important microbially catalyzed process that links sulfur oxidation to Fe(III) and Mn(IV) oxide reduction is the disproportionation of S^0 and other partially oxidized sulfur compounds (26, 84, 242). Sulfur disproportionation results in the production of SO_4^{2-} and HS^- and is thermodynamically feasible only in the presence of a sink for the produced HS^-, such as Fe(III) or Mn(IV) oxides (see reactions 11 and 12 in Table 2). In addition, Fe(II) is an effective reductant for Mn(IV) reduction (137, 174) (Table 2, reactions 8 and 9), and this reaction is likely to play a major role in Mn(IV) reduction in most sediments, where Fe is generally 10 to

100-fold more abundant than Mn (28). As a result of the rapid abiotic reaction of Fe(II) with Mn(IV), it is generally difficult, if not impossible, to separate Fe(III) and Mn(IV) reduction activities in such sediments (27, 243, 246), especially since direct methods for tracing enzymatic Fe(III) and Mn(IV) reduction are not available (see below). Nevertheless, in sediments where Mn is abundant relative to Fe and where rates of organic input are modest, Mn(IV) reduction can dominate over Fe(III) reduction (1, 2, 27, 245). Chapters 5 and 8 in the new text by Canfield et al. (28) provide an up-to-date summary of the quantitative role of DMRB activity versus sulfide-linked reduction in Fe(III) and Mn(IV) oxide reduction as well as the overall role of DMRB in aquatic sediment carbon metabolism.

Fe(II) and Mn(II) Oxidation

As in the case of Fe(III) and Mn(IV) reduction, oxidation of Fe(II) and Mn(II) takes place via both abiotic and enzymatic pathways (Fig. 2). Chemical reactions for aqueous and surface-associated Fe(II) and Mn(II) oxidation are shown in Table 2 [note that surface-associated Fe(II) and Mn(II) are described by the terms $\equiv$O-Fe^{+} and $\equiv$O-Mn^{+} to indicate their likely coordination by O atoms on mineral surfaces]. Fe(II) may be oxidized by O_2, NO_2^-, NO_3^-, Mn(IV) oxide, and light-CO_2. At this time, the only confirmed pathway for Mn(II) oxidation is reaction with O_2 (Table 2, reactions 28 and 29). A growing list of prokaryotes are able to couple growth to Fe(II) oxidation at circumneutral pH either under O_2-limiting conditions (e.g., microaerophilic organisms) or anaerobically by coupling Fe oxidation to photosynthesis or nitrate reduction (see below). An even more diverse group of microbes, including fungi, can oxidize Mn(II), although the physiological benefit to the organism has yet to be elucidated (236).

Aerobic Fe(II) Oxidation

Information is scarce on the rates and pathways (biotic versus abiotic) of microbial Fe(II) oxidation under circumneutral pH conditions. One important reason for this lack is the very rapid abiotic reaction of Fe(II) with O_2 at circumneutral pH, which makes studies of biological catalysis problematic. Indeed, the difficulty of distinguishing between biotic and abiotic processes has often led to the a priori assumption that Fe(II) oxidation by O_2 is largely a chemical process, and thus the role of microbes is ignored (e.g., see the work of Davison and Seed [45] and Millero et al. [149]). Nevertheless, bacteria have been associated with circumneutral Fe(II) oxidation and Fe(III) oxyhydroxide deposition for over a century (57, 76). A review by Emerson (61) provides an overview of the history of research on circumneutral bacterial Fe(II) oxidation as well as the physiology and systematics of Fe(II)-oxidizing bacteria (FeOB).

Recent enrichment and isolation studies with Fe(II)-O_2 opposing-gradient systems (102, 120) (see below) have expanded the range of organisms known to be involved in aerobic circumneutral Fe(II) oxidation beyond the traditional stalk-forming *Gallionella* spp. and sheathed bacteria of the *Sphaerotilus-Leptothrix* group to include unicellular organisms from the α-, β-, and γ-*Proteobacteria* (54, 61, 63, 210). The involvement of unicellular bacteria in circumneutral Fe(II) oxidation is consistent with the observation of large numbers of unicellular organisms closely associated with Fe(III) oxides in microbial mats present at a groundwater Fe seep in Denmark (65) and of Fe(III) oxide-rich plaques on aquatic macrophyte roots (68). Studies by Emerson and colleagues (63, 66, 158) have clearly demonstrated the potential for chemolithotrophic growth of such organisms coupled to Fe(II) oxidation at circumneutral pH.

FeOB are likely to play a significant role in circumneutral Fe(II) oxidation at redox interfacial environments where diffusion-limited O_2 transport leads to low dissolved-O_2 partial pressures (microaerobic conditions) within the zone of Fe(II)-O_2 overlap, i.e., environments characterized by opposing diffusion gradients of O_2 and Fe(II). Examples of such environments in aquatic systems include stream sediments and groundwater Fe seeps, wetland surface and rhizosphere sediments, irrigation ditches, municipal and industrial water distribution systems, and hydrothermal vents (61). The presence of substantial numbers of cultivatable FeOB in a variety of aquatic environments (Table 3) attests to their presence and putative activity in nature. Insight into how bacterial Fe(II) oxidation can compete effectively with chemical oxidation by O_2 under microaerobic conditions can be obtained by considering the kinetics versus the thermodynamics of Fe(II) oxidation by O_2 as a function of the Fe(II) and O_2 concentrations at circumneutral pH (187). Assuming a fixed dissolved Fe(II) concentration of 100 μM at pH 7.0, thermodynamic calculations indicate that the free energy associated with Fe(II) oxidation (equation 13 in Table 2) decreases only slightly (<5%) as the dissolved O_2 concentration decreases from 300 to 1 μM; in contrast, the half-life of Fe(II) (estimated by using the rate expression for abiotic Fe^{2+} oxidation by O_2 given by Singer and Stumm (208)) increases by a factor of 300. Thus, assuming that FeOB can compete with abiotic Fe(II) oxidation under microaerobic conditions, considerable amounts of energy should still be available for chemolithotrophic metabolism. Continuous-flow reactor experiments have, in fact, demonstrated that FeOB can compete effectively with abiotic Fe(II) oxidation under microaerobic conditions (66, 158). Moreover, experiments with Fe(II)-O_2 opposing-gradient culture systems containing diffusion probes for the collection of aqueous Fe and precipitated Fe(III) oxides have shown that FeOB can account for virtually all (ca. 90%) Fe(II) oxidation at diffusion transport-limited aerobic-anaerobic interfaces (212). FeOB produce a narrow band of cells and Fe(III) oxides in the vicinity of the aerobic-anaerobic interface in which the O_2 concentrations are only a few percentage points of air saturation (12, 63, 212).

Anaerobic Fe(II) Oxidation

The first definitive evidence for O_2-independent microbial Fe(II) oxidation came from studies of anoxygenic phototrophic bacteria that were enriched and isolated from anoxic muds from either marine coastal habitats or freshwater ponds (56, 272). Additional organisms from marine (221) and freshwater (41, 88, 89, 98) environments were subsequently obtained. None of these organisms are obligate Fe(II) oxidizers: in addition to Fe(II), they can all utilize multiple electron donors for photosynthetic growth, and in some cases, their growth on Fe(II) is significantly stimulated by an additional electron donor, such as acetate or glucose (56). A recently described phototrophic Fe(II) oxidizer, *Rhodopseudomonas palustris* TIE-1, illustrates the metabolic diversity of these organisms. TIE-1 can grow photoautotrophically on Fe(II) or thiosulfate, photoheterotrophically with organic substrates, or in the dark as a heterotroph (98). To date, most anaerobic photosynthetic Fe(II) oxidizers belong to the α- or γ-*Proteobacteria*. To our knowledge, there have been no studies of the abundance of Fe(II)-oxidizing anoxygenic phototrophs in natural systems or of their contribution to Fe(II) oxidation. Shallow, Fe(II)-rich, stratified water bodies, such

TABLE 3 Studies which have enumerated the abundance of culturable Fe- and Mn-reducing and -oxidizing microorganisms in aquatic environments[a]

Organism(s)	Environment	Method	Electron donor	Electron acceptor	Reference(s)
FeRB	Unvegetated freshwater wetland sediments	Liquid culture MPN	Ac	Ferr	188, 211
	Vegetated freshwater wetland sediments	Liquid culture MPN	Ac	Ferr	269, 270
	Saltmarsh sediments	Liquid culture MPN	Ac/Lac	Ferr	118
	Metal-polluted freshwater lake sediments	Liquid culture MPN	Ac	Ferr	42
	Estuarine sediments	Liquid culture MPN	NA	FeNTA/FeCit	22
	Petroleum-contaminated marine harbor sediments	Liquid culture MPN	Ac	Ferr	36
	Saltmarsh sediments	Plate counts	Ac	FeCit	142
FeRB and MnRB	Black Sea sediments	Liquid culture MPN	Ac	Ferr/Vern	245
	Saltmarsh sediments	Plate counts	Lac	FeCit/Vern	111, 112
	Orca Basin water column	Plate counts	NA	FeCit/Vern	250
FeOB	Unvegetated freshwater wetland sediments	Opposing-gradient MPN	$FeCl_2$	O_2	211
	Vegetated freshwater wetland sediments	Opposing-gradient MPN	$FeCl_2$	O_2	269, 270
	Freshwater pond, ditch, and stream sediments	Liquid culture MPN	$FeSO_4$	NO_3^-	220
	Freshwater wetland sediments	Liquid culture MPN	$FeCL_2$	NO_3^-	265
	Freshwater lake sediments	Liquid culture MPN	$FeCL_2$	NO_3^-	267
FeOB and MnOB	Orca Basin water column	Plate counts	$FeSO_4$	O_2	250

[a]Abbreviations (in alphabetical order): Ac, acetate; FeCit, ferric citrate; FeNTA, ferric nitrilotriacetic acid; FeRB, Fe(III)-reducing bacteria; Ferr, ferrihydrite (synthetic); Lac, lactate; MnRB, Mn(IV)-reducing bacteria; NA, nutrient agar; Vern, vernadite (synthetic).

as the meromictic lake studied by Rodrigo et al. (190), would represent logical targets for such studies.

Shortly after the discovery of anaerobic Fe(II) oxidation coupled to photosynthesis, it was recognized that some bacteria can couple Fe(II) oxidation to the reduction of NO_3^- (219). The existence of such organisms was long suspected (85), based on gradients of NO_3^- and dissolved Fe(II) in marine sediments (74, 108). Because Fe(II) does not react spontaneously with NO_3^- under temperature and aqueous geochemical conditions typical of the Earth's surface (266), experimental studies of biological NO_3^--dependent Fe(II) processes are relatively straightforward compared to those of aerobic Fe(II) oxidation. A diverse group of organisms from a variety of environments are now known to perform this reaction (222). Some organisms derive energy from the process (12, 219), while others apparently do not (121). Recently, the first pure culture of a lithoautotrophic nitrate-reducing FeOB was reported (267). Given the abundance of Fe in subsurface (dark) environments (130), this process would appear to have the potential to contribute significantly to Fe redox cycling in situ. However, there have been only a few estimates of the abundance of NO_3^--reducing FeOB in nature to date (Table 3), and no quantitative studies of this process are yet available. Recent experimental studies suggested that the role of NO_3^--dependent Fe(II) oxidation is likely to be particularly significant at the interface between NO_3^- and Fe(III) reduction zones in sediments influenced by periodic fluctuations in the inputs of organic carbon and oxidants (267).

Mn(II) Oxidation

In contrast to our limited knowledge of microbial Fe(II) oxidation in circumneutral-pH environments, considerable information is available on the rates and pathways of microbial Mn(II) oxidation in natural waters. Compared to the case for Fe(II) oxidation, it has been easier to establish a direct role for microorganisms in Mn(II) oxidation because (i) Mn(II) oxidation is a strictly aerobic process (196) and (ii) the abiotic reaction of Mn(II) with O_2 is very slow at pH values lower than or equal to that of seawater (49, 154). The key role of microbial catalysis in Mn(II) oxidation has been demonstrated for (i) stratified freshwater lakes (34, 101, 150, 151), (ii) stratified marine basins (69, 196, 233, 237), (iii) hydrothermal plumes in the Pacific Ocean (147), and (iv) the Mississippi River (203). In addition, a variety of lines of evidence suggest a key role of Mn(II)-oxidizing bacteria (MnOB) in the genesis of both marine and freshwater Mn nodules (58). Studies of the depth distributions of Mn and other chemical constituents in a stratified lake ecosystem in Michigan (127, 230) and in several other marine basins (157, 231, 277) also provided a strong indication of the role of microbial catalysis in Mn(II) oxidation. Likewise, direct (e.g., see references 21 and 106) and indirect (244) studies of Mn(II) oxidation in aquatic surface sediments (methods discussed below) have demonstrated the significance of bacterial catalysis. Ironically, the specific benefits of Mn(II) oxidation to microbes have been much more difficult to ascertain, i.e., no known microbes are capable of coupling growth to Mn(II) oxidation. Nonetheless, given the ubiquity of enzymatically coupled Mn(II) oxidation in nature, it must have some biological significance. Recent reviews by Tebo and colleagues (234, 236) provide extensive coverage of the organisms involved and the importance of Mn(II) oxidation as a biogeochemical process. It is intriguing that one of the best studied examples of biological Mn(II) oxidation, involving *Bacillus* sp. spores, suggests that even dormant microbes have the potential to influence metal cycling in aquatic systems.

Coupled Fe-Mn Oxidation and Reduction: Potential for Microbial Syntrophy

The inputs of Fe(III) and Mn(IV) oxides to shallow-water aquatic sediments are typically small in relation to the deposition of organic matter. Hence, efficient recycling of Fe(II) and Mn(II) back to their oxidized forms must occur in order for Fe(III) and Mn(IV) reduction to contribute significantly to overall biogeochemical cycling (28). Molecular diffusion of porewater solutes, together with bioirrigation and bioturbation by sediment macrofauna, is the physical process responsible for promoting Fe and Mn redox cycling in sediments (85). In most situations, the Fe and Mn entering an aquatic sediment are cycled tens to hundreds of times before being permanently buried. In light of the foregoing analysis of the role of microbes in the oxidative side of the Fe and Mn cycles in aquatic environments, it is clear that Fe and Mn redox cycling is an example of microbial syntrophy analogous to the well-known syntrophic relationships among N- and S-oxidizing and -reducing microorganisms in natural systems (103). This is especially true for Mn cycling, since abiotic Mn(II) oxidation is strongly dependent on enzymatic activity at neutral pH.

In addition to providing the basic mechanistic linkage between metal oxidation and reduction (Fig. 1), recent studies of Fe(II) oxidation at circumneutral pH suggest that FeOB activity alters the microscale locus of Fe(III) oxide deposition [relative to Fe(II) oxidation] so as to delay or retard FeOB cell encrustation with Fe(III) oxides (212). These and other experimental results (187, 211) suggest that FeOB have the potential to enhance the coupling of Fe(II) oxidation and Fe(III) reduction at redox interfaces, thereby promoting rapid microscale cycling of Fe. The presence of abundant culturable Fe(II)-oxidizing and Fe(III)-reducing bacteria in redox interfacial environments, such as wetland surface sediments (211), wetland plant rhizosphere sediments (269), and neutral-pH groundwater Fe seeps (180), strongly suggests the operation of microscale Fe redox cycling in nature. It remains to be determined whether analogous interactions take place between Mn(II)-oxidizing and Mn(IV)-reducing organisms in Mn-rich environments. The presence of both types of organisms in marine (59) and freshwater (216) ferromanganese nodules suggests that rapid internal Mn cycling may take place across redox microgradients within metal-rich nodules. The establishment of experimental artificial consortia of Mn-cycling organisms would be a fruitful area of research in this context; such a consortium was recently reported for As-respiring microbes (182).

METHODS FOR ANALYZING MICROBIAL Fe AND Mn CYCLING

Chemical Profiling

Measurements of the depth distributions of Fe and Mn concentrations and redox species in both the water column (e.g., in stratified systems) and sediments provide fundamental insight into the spatial (and temporal) dynamics of metal cycling. Such measurements are critical for constraining the depth intervals at which reduction and oxidation take place in relation to other biotic and abiotic processes. Reliable analysis of the abundance and speciation of Fe and Mn (and other redox-active trace metals) across redox gradients in the water column and sediments requires specialized wet-chemical (e.g., colorimetric or chromatographic), spectroscopic (atomic absorption or inductively coupled plasma spectroscopy), and/or electrochemical (voltammetric or potentiometric) methods. The reader is referred to the recent compendium by Buffle and Horvai (18) as an entry into the vast literature on trace metal analysis in aquatic systems.

Figure 3 shows profiles of O_2, dissolved and particulate Fe, Mn, and Co species, and dissolved sulfide in a stratified mesotrophic lake (Paul Lake, Mich.) (127, 230). The measured profiles are clearly indicative of active metal redox cycling at and below the aerobic-anaerobic interface, and a key role for microbial activity in coupled Mn-Co redox cycling (see Table 5) is apparent. Figure 4 shows profiles of aqueous and solid-phase Fe and Mn (and other constituents) in the sediments of a freshwater wetland

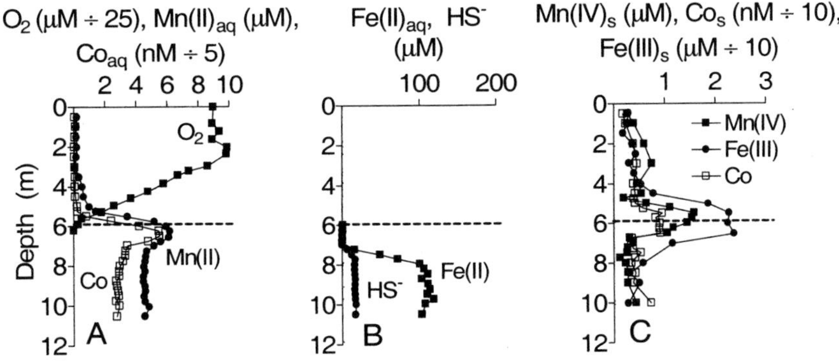

FIGURE 3 Depth profiles of dissolved Fe(II) and Mn(II), particulate Fe(III) and Mn(IV), and other constituents in a stratified lake (Paul Lake, Mich.). Particulate Fe(III) and Mn(IV) may include small amounts of sorbed Fe(II) and/or Mn(II). Dashed lines indicate approximate boundaries across which metal redox cycling takes place. Redrawn from Fig. 1 in reference 127 and Fig. 1 and 7 in reference 230 with permission of the publishers.

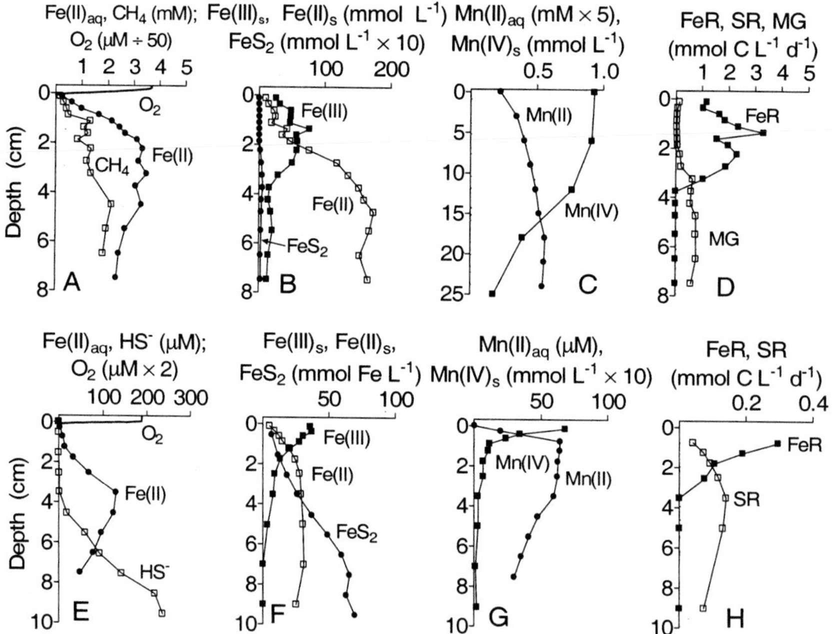

FIGURE 4 Depth profiles of aqueous and solid-phase Fe, Mn, and other constituents and rates of Fe(III) reduction (FeR), sulfate reduction (SR), and methanogenesis (MG) in the sediments of a freshwater wetland (Talladega National Forest) (top panels) and a shallow coastal embayment (Aarhus Bay, Denmark) (bottom panels). The Mn(IV) values shown in panels C and G may include small amounts of sorbed or precipitated Mn(II). Data are redrawn from Fig. 1 in reference 211, Fig. 2 in reference 189, Fig. 3 in reference 50, Fig. 2, 4, and 6 in reference 243, Fig. 2 in reference 244, and Fig. 3 in reference 239 with permission of the publishers.

(Talladega National Forest [50, 189]) and a shallow coastal embayment (Aarhus Bay, Denmark [243]). Both the water column and sediment chemical profiles illustrate several key aspects of the Fe-Mn redox cycling systems in aquatic environments. First, a significant accumulation of reduced Fe and Mn occurs only below the aerobic zone, which is several meters deep in the stratified lake but only a few millimeters deep in the two sediments. The sharp opposing gradients of O_2 and reduced metals clearly demonstrate the potential for localized Fe(II) and Mn(II) oxidation at the aerobic-anaerobic interface, which results in the regeneration of Fe(III) and Mn(IV) oxides that can serve once again as electron acceptors for biotic or abiotic processes; this is the basic metal redox recycling system discussed above. Another key aspect of the lake and marine sediment profiles is that they clearly illustrate the tendency for Mn(IV) reduction to take place prior to Fe(III) reduction in the classical thermodynamic sequence of electron acceptor utilization in diagenetic systems (74, 85). The profiles also illustrate the potential for competition between abiotic and biotic Fe(III) and Mn(IV) reduction within redox transition zones: dissolved sulfide, transported upward by eddy diffusive transport in the water column and by molecular diffusion and bioturbation in the sediment, enters the zone where metal oxides are abundant, where it can serve as a potent abiotic reductant (see above). Note that dissolved HS^- was absent in the S-poor, Fe-rich freshwater wetland sediment and that an accumulation of CH_4 rather than HS^- took place below the zone of Fe(III) reduction in these highly sulfate-limited sediments (note also the vastly lower concentration of FeS_2 [pyrite] in the freshwater versus marine sediment). Finally, the profiles also show that large quantities of dissolved and/or solid-phase Fe(II) produced by biotic or abiotic Fe(III) reduction may be transported into the Mn(IV) oxide-enriched surface layer and contribute to abiotic Mn(IV) reduction.

The Fe and Mn profiles shown in Fig. 3 and 4 were obtained by wet-chemical extraction and colorimetric or spectroscopic analysis techniques, whereas the dissolved O_2 profiles were obtained by use of a polarographic electrode in situ for the lake water column and a Clark-style polarographic microelectrode (181) for laboratory-incubated cores. These techniques are "industry standards" for analysis of Fe and Mn redox cycling in aquatic systems, and the reader is referred to the source papers and the references cited therein for detailed methodological descriptions. It should be noted that a variety of new, more advanced techniques have been developed in recent years for quantifying the fluxes as well as distributions of reactants and products of metal cycling in aquatic systems. Such techniques include automated in situ benthic flux chambers equipped with dissolved O_2 microsensors (80, 81); electrochemical microsensors for making simultaneous dissolved O_2, Mn(II), Fe(II), and HS^- measurements at a submillimeter spatial resolution (16), which can be used either in situ or in the laboratory; and gel probe techniques that allow determinations of both the concentrations and production rates of dissolved metals (and sulfide) in sediments at a submillimeter resolution (44, 46, 232). The interested reader is directed toward a recent review by Viollier et al. (255) and to the text by Buffle and Horvai (18), which provides numerous references to the primary literature on these and other new analytical methodologies for use in aquatic systems.

Process Measurements

Fe(III) and Mn(IV) Reduction

There have been numerous qualitative studies of Fe(III) and Mn(IV) oxide reduction in aquatic sediments. Such studies typically involve incubation of homogenized surface sediment or sediment slurries under anaerobic conditions in sealed bottles, with periodic analysis of reduced and oxidized solid- and aqueous-phase constituents. Such studies can be used to evaluate the potential for oxide reduction as a function of various parameters and conditions, e.g., the presence or absence of biological activity, the influence of exogenous electron donor or acceptor addition, or the abundance of oxide substrate as a function of depth in the sediment. The approaches and results of these studies have been reviewed elsewhere (20, 189, 239, 271). We focus here on methods that can be used to obtain quantitative (or at least semiquantitative) estimates of metal oxide reduction activity in sediments.

Due to isotopic exchange between reduced and oxidized metal phases, radiolabeled Fe (and probably also Mn) cannot be used to measure rates of biotic or abiotic metal reduction (1, 186). In the absence of radiotracer methods, two different approaches have been used to estimate rates of dissimilitary metal reduction in aquatic sediments. Both approaches involve incubation of homogenized sediment from various depth intervals in sealed glass bottles (e.g., 10- to 100-ml serum bottles fitted with thick rubber stoppers) or gastight plastic bags (86). The first method involves direct determination of changes in Fe(III) and/or Fe(II) pool sizes by dilute HCl extraction and colorimetric Fe analysis; an example of a rate-versus-depth profile determined by this approach is shown in Fig. 4D. This method is most useful for freshwater systems where the influence of sulfide-mediated Fe(III) reduction is minimal (135, 189), although it has also been employed successfully with Fe(III) oxide-rich marine sediments (91) as well as bioturbated saltmarsh sediments (113) and brackish tidal marsh sediments (160). In the last two studies, incubations were conducted in the presence and absence of molybdate, a specific inhibitor of microbial sulfate reduction (164), in order to constrain the role of sulfide-linked abiotic reduction. To our knowledge, this technique has not been used to estimate rates of Mn(IV) reduction, probably because any acid-soluble Fe(II) in the sediment [which is generally present in relatively large quantities relative to Mn(IV)] would be expected to react very rapidly with Mn(IV) liberated during acid extraction. Nevertheless, the accumulation of dissolved Mn(II) in sediment pore fluids has been used as a qualitative indicator of microbial Mn(IV) reduction in Mn-rich Danish coastal (27) and Black Sea (245) sediments.

The second commonly used approach for estimating rates of Fe(III) and Mn(IV) oxide reduction in sediments is an indirect one. Independently measured rates of carbon oxidation coupled to other respiratory processes (e.g., sulfate reduction and/or methanogenesis, determined by measuring porewater sulfate depletion or methane gas production or by using ^{35}S and ^{14}C trace techniques) are subtracted from total rates of organic carbon mineralization (e.g., as estimated from measurements of total dissolved inorganic carbon or ammonium production during anaerobic incubation), and the difference is attributed to other respiratory pathways, such as Fe(III) and Mn(IV) reduction (27, 239, 241). Parallel measurements of the depth distributions of O_2 and NO_3^- are used to constrain the depth intervals over which aerobic respiration and denitrification can contribute to in situ carbon oxidation, which are typically limited to the upper few millimeters and centimeters, respectively, in shallow, organic-rich aquatic sediments. Similarly, the depth distribution of Mn abundance is used as an indicator of the zone in which microbial Mn(IV) reduction can contribute to organic carbon mineralization. An example of a depth distribution of Fe(III) reduction rates determined by this approach (based on measurements of total dissolved inorganic carbon production minus rates of sulfate reduction rates measured by using $^{35}SO_4^{2-}$) is provided in Fig. 4H. As discussed above, because of the rapid reaction of Fe(II) with Mn(IV), with this approach it is generally not possible to attribute carbon oxidation to Mn(IV) reduction, except in situations where Mn is relatively abundant compared to Fe. Site S9, at a 695-m depth in the Skagerrak (located off the northeast coast of Jutland, Denmark), provides an excellent example of such conditions: Mn(IV) reduction is the only important anaerobic carbon oxidation process in the upper 10 cm of the sediments at this site (27).

Fe(II) and Fe(III) Extraction

A central issue in the direct analysis of Fe(III) oxide reduction is the methodology used to estimate reduced and oxidized Fe pool sizes. Wet-chemical extraction is commonly employed to estimate the abundances of different operationally defined fractions of solid-phase Fe and Mn. Extraction with 0.5 or 1 M HCl for ca. 1 h, followed by analysis of Fe(II) and total Fe in the extract with the Ferrozine reagent (218) or by electrochemical detection, is typically used to estimate the abundance of amorphous, microbially reducible Fe(III) oxides (135, 139), determined from the difference between total Fe and Fe(II) in the extract (115, 186, 259). Another common extraction procedure for determination of microbially reducible Fe(III) oxide and solid-phase Fe(II) is the use of 0.2 M ammonium oxalate or oxalic acid (pH 3) (115, 170, 197, 243). The extraction must be performed under anaerobic conditions to prevent oxidation of Fe(II) (which is slow but not trivial at pH 3) and in the dark to prevent light-catalyzed reduction of Fe(III) in the extract (170). As in the case of the dilute HCl extraction procedure, Fe(II) and Fe(III) in the extract are analyzed with Ferrozine in the presence and absence of NH_2OH-HCl. The ammonium oxalate extraction method is not recommended for routine use in studies of microbial Fe(III) reduction, as crystalline Fe(III) oxides are dissolved by oxalate in the presence of even trace quantities of Fe(II) (229), leading to a gross overestimation of the abundance of microbially reducible Fe(III) (171).

Fe(II) Oxidation

At present, no direct methods are available for separating biotic and abiotic Fe(II) oxidation under circumneutral-pH conditions. There are no specific inhibitors of biological Fe(II) oxidation, although general respiratory inhibitors such as sodium azide do inhibit respiration-linked Fe(II) oxidation (see below). The use of stable Fe isotopes has gained a good deal of attention recently (99), but unequivocal association of an isotopic signature with biological Fe(II) oxidation [or Fe(III) reduction] remains elusive (100). The lack of good geochemical markers or biochemical indicators of microbial Fe(II) oxidation means that the microbes themselves must be relied upon to reveal their role in the process. Fortunately, some aerobic FeOB produce morphologically unique Fe(III) oxide-bearing structures that act as biomarkers. The characteristic morphologies of Fe(III) oxide-encrusted stalks associated with *Gallionella* spp. and hollow tubular sheaths associated with *Leptothrix* spp. have been recognized as composing the fabric of Fe(III) oxide deposits in freshwater springs or wetlands since the early days of light microscopy (55, 87). Simple phase-contrast or bright-field microscopic examination of field samples at magnifications at or above ×400 is a rapid way to determine if these biomarkers are present; examples of some of these structures are shown in Fig. 5. These morphological biomarkers are confined primarily to the oxic-anoxic interface in aquatic environments (e.g., in springs, wetlands, and oceanic vents) because, to our knowledge, no anaerobic FeOB form these kinds of structures, nor have they been observed in static subsurface environments such as the rhizosphere, where O_2 enters the system by diffusion from plant roots.

Determining the specific contribution of active microbial metabolism to Fe(II) oxidation rates either in nature or in the laboratory is challenging. As yet, there are no direct methods for measuring Fe(II) oxidation rates in situ. Because of the problems of Fe isotope exchange discussed above, radiolabeled Fe(II) cannot be used to trace metal oxidation rates in materials containing abundant pools of both oxidized and reduced Fe. Rather, samples must be manipulated or removed for bulk rate measurements. In this case, it is important to first know the ambient pH and the concentrations of Fe(II) and O_2 in order to replicate these experimentally. Dissolved O_2 concentrations are typically around 1% of the air-saturated $[O_2]$ at aerobic-anaerobic interfaces. Under these conditions, purely chemical oxidation of dissolved Fe(II) is relatively slow. However, autooxidation of Fe(II) catalyzed by preformed Fe(III) oxide surfaces is much more rapid, and thus the problem becomes one of distinguishing between autooxidation and biotic oxidation. The most effective means of doing this is to inhibit all microbial respiration. Sodium azide is a highly effective respiratory inhibitor that does not seem to influence Fe redox chemistry or mineralogy (66). When applied to a pure culture in a bioreactor or to simple field microcosms, measuring Fe(II) oxidation before and after treatment with azide gives a good approximation of the overall Fe(II) oxidation rate. Recently, scanning voltammetric microelectrodes were employed to simultaneously measure O_2 and Fe(II) and to determine Fe(II) oxidation rates in both pure FeOB cultures and freshwater microbial mats (180). The electrode measurements confirmed that the $[O_2]$ was stable and in the right range, and continuous monitoring of Fe(II) concentrations allowed environmentally realistic oxidation rate estimates to be obtained.

Mn(II) Oxidation

The stability of Mn(II) in solution at neutral pH makes process-level studies of Mn(II) oxidation relatively easy compared to those of Fe(II) oxidation. As in the case of Fe(II) oxidation, sodium azide effectively inhibits microbial activity without altering other Mn(II) partitioning pathways [i.e., abiotic oxidation and sorption of Mn(II) to Mn(IV) oxide and other particle surfaces] (69, 194). Mandernack and Tebo (147) used azide together with radiotracer Mn(II) to separate rates of biotic oxidation from those of abiotic oxidation and Mn(II) sorption in hydrothermal vent fluids and plumes. In most cases, the amount (e.g., micromolar concentrations) of ^{54}Mn(II) added was small relative to the total pool of Mn(II) so that the ^{54}Mn(II) functioned as a true tracer. This appears to be a robust approach for determining Mn(II) oxidation rates in water samples. To our knowledge, ^{54}Mn(II) has not been applied to studies of Mn(II) oxidation in sediments. Based on results obtained with ^{55}Fe(III) and ^{54}Mn(IV) (see above), it seems likely that rapid partitioning of ^{54}Mn(II) independent of microbial activity would render the technique useless in studies of sediments, although this speculation needs to be verified. In any case, monitoring of bulk Mn(II) loss over time in Mn(II)-spiked sediment slurries with and without sodium azide can be used to obtain a qualitative indication of Mn(II) oxidation potentials in sediments (e.g., see the work of Kepkay [107]). As with studies of water column samples (147), extraction with excess Cu^{2+} ions (added as $CuSO_4$) is required to determine the amount of Mn(II) lost from solution via sorption as opposed to oxidation.

Two other techniques that have been used to estimate rates of Mn(II) oxidation in aquatic sediments are (i) in situ dialysis cell experiments (21, 106) and (ii) a combination of sediment-water Mn(II) efflux and porewater Mn(II) concentration gradient measurements and modeling (244). In the first approach, a dialysis cell containing Mn(II)-spiked

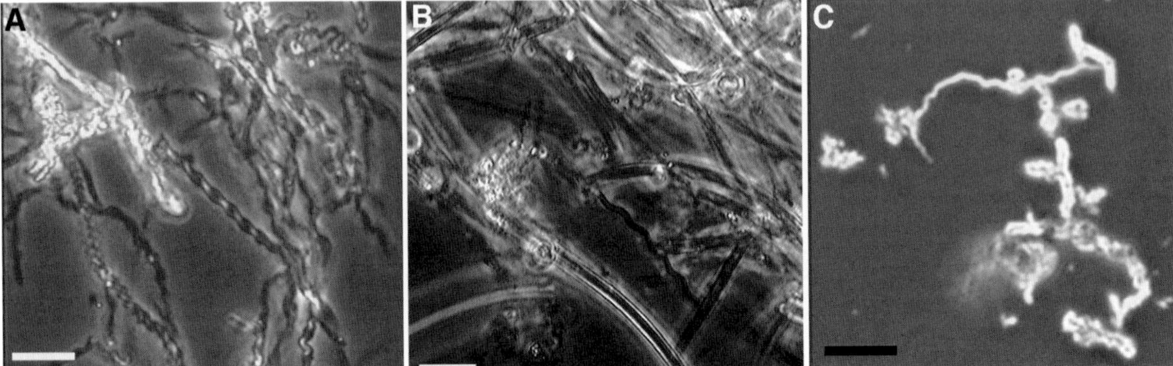

FIGURE 5 Fe(III) oxide morphologies produced by O_2-dependent FeOB activity. (A) Stalk-like structures characteristic of *Gallionella* spp. from a freshwater spring rich in Fe(II); (B) sheath structures of Fe(III) oxides, also from a freshwater spring; (C) filamentous Fe(III) oxides from a marine hydrothermal vent. In all cases, the structures are composed of amorphous ferrihydrite-like minerals that dissolve readily in ammonium oxalate. Although these structures were all formed by FeOB, the bacteria are not visible unless stained with a fluorescent dye. Bar = 10 μm.

bottom water, with or without azide, is placed into a sediment core contained within a specially designed core box. After incubation for a few days, the core box is dismantled, the sediment adjacent to the dialysis cell is sampled in thin (e.g., 0.25-cm) slices going away from the cell, and the Mn(II) content of the porewater is determined. One-dimensional diffusion-reaction modeling of the porewater Mn(II) concentration as a function of the distance from the dialysis cell surface in poisoned versus unpoisoned systems is then used to separate biological Mn(II) oxidation from abiotic oxidation and Mn(II) sorption. The modeling approach and assumptions are explained in detail by Burdige and Kepkay (21). Using this technique, Kepkay (107) estimated that microbial oxidation increased the rate of Mn(II) binding by ca. fivefold in metal-rich lake sediments.

The extent to which results from dialysis cell experiments can be translated into in situ Mn(II) oxidation at the sediment-water interface is not clear, since the initially aerobic dialysis cell is placed in contact mainly with anaerobic sediment within the core, as opposed to material at the sediment-water interface. In contrast, the combination of sediment Mn(II) efflux and porewater Mn(II) gradient measurements and modeling presented by Thamdrup et al. (244) provides a way to realistically constrain in situ rates of biotic versus abiotic Mn(II) oxidation in surface sediments. In this approach, the total rate of Mn(II) oxidation is estimated from the difference between measured sediment Mn(II) efflux rates (determined by the use of either in situ benthic chambers or laboratory-incubated sediment cores) and estimated rates of upward diffusive Mn(II) transport calculated based on measured porewater Mn(II) profiles (e.g., profiles analogous to that shown in Fig. 4F). The biotic contribution to Mn(II) oxidation is estimated in turn by diffusion-reaction modeling of the porewater Mn(II) profiles (via the same basic approach as that described in reference 21), using a second-order Mn(II) oxidation rate law that accounts for known abiotic Mn(II) oxidation kinetics (43) and which can be adjusted to account for biotic activity as required to obtain a fit to measured porewater Mn(II) concentrations in the vicinity of the sediment-water interface. The results of Thamdrup et al.'s analysis suggest that microbial activity accelerates Mn(II) oxidation rates by a factor of ca. 1,000, implying that such activity is crucial for maintaining high rates of Mn redox cycling in the surface sediment.

Cultivation-Based Methods

Fe(III) and Mn(IV) Reduction

Enrichment and isolation studies have generated a great deal of information about the diversity, phylogeny, and physiology of dissimilatory Fe(III)- and Mn(IV)-reducing prokaryotes in surface and subsurface sedimentary environments (134). However, there have been relatively few quantitative cultivation-based studies of DMRB abundance in natural systems compared to that of other groups of anaerobic respiratory prokaryotes (e.g., nitrate-reducing, sulfate-reducing, and methanogenic organisms). Table 3 provides a summary of these studies, including a brief description of the cultivation methodology employed for DMRB enumeration. To our knowledge, no systematic evaluation of the relative efficiencies of different enumeration techniques (e.g., plate count versus most-probable-number [MPN] methods) for this group of organisms has been conducted. However, comparisons of measured (or estimated) rates of Fe(III) reduction and numbers of culturable DMRB suggest that such techniques are likely to underestimate true densities by 1 to 2 orders of magnitude (118, 188). Roden and Wetzel (188) employed an MPN enumeration procedure with synthetic growth medium to estimate changes in wetland sediment DMRB population size in response to the addition of different amounts of labile organic matter. Cell-normalized rates of Fe(III) reduction computed from measured rates of Fe(III) reduction and MPN estimates obtained after Fe(III) reduction was complete (516 ± 237 fmol cell^{-1} day^{-1}) were ca. 15-fold higher than those obtained in experiments where similar sediments were inoculated with various concentrations (1.7×10^7 to 1.3×10^8 cells ml^{-1}; five total levels) of *Geobacter metallireducens* (37 ± 3 fmol cell^{-1} day^{-1}) (184). These results are not surprising since it is well known that MPN procedures employing synthetic growth media typically underestimate the abundance of anaerobic respiratory microbes, such as sulfate-reducing bacteria, by one or more orders of magnitude in both marine (79, 104, 177, 253) and freshwater (9) environments. Hence, at present, quantitative analysis of DMRB abundance by using cultivation-based approaches remains problematic. The refinement of MPN procedures (e.g., through use of natural media) could help to close the gap between true and cultivatable DMRB abundances (253).

Detailed descriptions of procedures for the enrichment, enumeration, and isolation of DMRB [including preparation of Fe(III) and Mn(IV) oxide phases] are available (35, 62, 77, 117). A new method for simultaneous enumeration and isolation of DMRB (105) based on the classical roll-tube methodology of Hungate (96) has been employed with synthetic ferrihydrite to isolate hyperthermophilic Fe(III)-reducing organisms from Obsidian Pool in Yellowstone National Park. Another potential source of solid-phase Fe(III) for use with this method is the clay size fraction of sediment (E. S. Shelobolina, H. A. Vrionis, K. H. Williams, and D. R. Lovley, unpublished data), which is concentrated by standard soil particle size separation techniques (200). The clay size fraction is enriched in Fe(III) oxides (39) and, in addition, contains adsorbed natural organic matter and other trace nutrients that can increase microbial recovery in laboratory media. However, due to the complex composition of the clay size fraction, it must be used with caution for enumeration of Fe(III)-reducing bacteria in natural samples, since organisms not capable of gaining energy for growth from Fe(III) reduction (e.g., fermentative bacteria) may also form colonies in the roll-tube medium.

Fe(II) and Mn(II) Oxidation

Ghiorse (77) provided a systematic description of methods for the cultivation and enumeration of FeOB and MnOB. Emerson and Floyd (62) recently provided an up-to-date summary of methods for cultivation of both aerobic and anaerobic neutrophilic FeOB, which have been adapted for use in MPN-based enumerations in both freshwater and marine environments. Table 3 provides a summary of studies to date which have enumerated FeOB and MnOB in aquatic systems. The recent upsurge of interest in the biogeochemical redox cycling of Fe and Mn (60, 254) is likely to bring with it increased applications of these techniques to aquatic as well as other redox-gradient-containing environments. Evaluation of the efficiencies of MPN-based FeOB and MnOB enumeration procedures is a near-term priority for quantitative studies of Fe and Mn redox cycling.

Molecular Methods

Despite the rapid expansion in our knowledge of the phylogenetic and physiological diversity of DMRB, FeOB, and

TABLE 4 Molecular biological studies of metal-reducing and metal-oxidizing microorganisms in aquatic environments[a]

Organisms	Environment	Technique	Reference(s)
FeRB	Hydrocarbon-contaminated aquifer sediments	Enrichment culture and 16S rRNA gene DGGE (universal primers), MPN-PCR (*Geobacteraceae* primers)	6, 193
FeRB	Organic carbon-amended aquifer sediments	16S rRNA gene clone libraries and DGGE (universal primers); MPN-PCR (*Geobacteraceae* primers)	209
FeRB	Landfill leachate-contaminated groundwater	16S rRNA gene DGGE (universal primers)	192
FeRB (*Geobacter*)	Organic carbon-amended aquifer sediments	16S rRNA gene RT-PCR and limiting dilution (*Geobacteraceae* primers)	223
FeRB (*Geobacteraceae*)	Metal-polluted freshwater lake sediments	16S rRNA gene clone libraries, DGGE, MPN-PCR, RT-PCR (*Geobacteraceae* primers)	42
FeRB	Acetate-amended, U(VI)-contaminated aquifer sediments	16S rRNA gene clone libraries (universal primers); MPN-PCR and RT-PCR (*Geobacteraceae* primers)	93
FeRB (*Geobacteraceae*)	Hydrocarbon-contaminated aquifer sediments	*nifD* mRNA RT-PCR (*Geobacteraceae* primers)	94
FeRB (*Geobacteraceae*)	Acetate-amended, U(VI)-contaminated aquifer sediments; hydrocarbon- and acetate-contaminated aquifer sediments; freshwater sediment microbial fuel cells	16S rRNA (cDNA) clone libraries (universal primers) and RT-PCR (*Geobacteraceae* primers); *gltA* mRNA RT-PCR (*Geobacteraceae* primers)	95
FeRB	Estuarine sediments	16S rRNA gene clone libraries (universal and *Geobacteraceae* primers)	22
FeOB and MnOB	Freshwater pond	16S rRNA FISH probes	204
MnOB	Freshwater pond	PCR detection of putative Mn(II) oxidation gene (*mofA*)	205
MnOB and FeOB	Cave Fe(III) and Mn(IV) oxides	16S rRNA gene clone libraries (universal primers)	163
MnOB	Cave Fe(III) and Mn(IV) oxides	16S rRNA gene clone libraries (universal primers)	214
MnOB	Lake Michigan Fe(III) and Mn(IV) nodules	16S rRNA gene clone libraries (universal primers)	216
MnOB	Loihi Seamount	Enrichment culture and 16S rRNA gene clone libraries (universal primers)	238
Fe- and Mn-associated bacteria	Freshwater reservoir	16S rRNA gene clone libraries (universal primers)	215
FeOB	Deep ocean basalt	16S rRNA gene clone libraries (universal primers)	247
FeOB	Loihi Seamount	TRFLP analysis of 16S rRNA genes (universal primers)	64
FeOB	Deep ocean basalt	RFLP and 16S rRNA gene clone libraries (universal primers)	191
FeOB and FeRB	Wetland sediment	16S rRNA gene clone libraries and DGGE (universal primers)	265
FeOB and FeRB	Wetland sediment	16S rRNA gene clone libraries and DGGE (universal primers)	265

[a]Abbreviations (in alphabetical order; also see Table 3, footnote *a*): DGGE, denaturing gradient gel electrophoresis; FISH, fluorescence in situ hybridization; MPN-PCR, most probable number PCR; RFLP, restriction fragment length polymorphism; RT-PCR, real-time PCR; TRFLP, terminal restriction fragment length polymorphism.

MnOB (67, 134, 236), progress in the development of molecular tools for detection and quantification of these organisms in nature has been slow compared to that for other biogeochemically significant groups of microorganisms (e.g., sulfate reducers, methanogens, nitrifiers, and methanotrophs). This can be attributed, at least in part, to the wide phylogenetic diversity of these groups of organisms, both within and outside the *Proteobacteria*, which makes it difficult or impossible to design reliable broad-spectrum 16S rRNA gene-based approaches analogous to those available for detection of aerobic nitrifiers, sulfate reducers, and methanogens (4, 5, 176, 178, 262). Our knowledge of functional genes involved in Fe and Mn redox transformations is not yet sufficient to permit the development of robust functional gene approaches like those available for analysis of the abundance and diversity of nitrate-reducing (*napA*, *narG*, *nirS*, and *nirK*) (15, 162), sulfate-reducing (*dsrAB* and *apsA*) (256), and methanogenic (*mcrA*) (73) organisms. Despite these limitations, steady progress in the development and application of molecular tools for analysis of metal-reducing and -oxidizing organisms in natural environments is taking place. Table 4 provides a summary of studies performed to date that have employed such techniques (primarily through use of the small-subunit rRNA gene) in aqueous environments.

Several new developments relevant to future molecular techniques for analysis of metal-reducing and -oxidizing organisms have taken place in recent years, including (i) the firm identification of genes for outer membrane cytochromes involved in metal reduction in DMRB (e.g., *omcB* in *Geobacter* and *mtrC* and *omcA* in *Shewanella*) (see the work of DiChristina et al. (48) for a review); (ii) the identification of a novel citrate synthase gene similar to eukaryotic genes in the *Geobacteraceae* (13), which can be used to assess the abundance and activity of this group of environmentally important DMRB (95); (iii) the isolation and characterization of the first neutrophilic Fe(II)-oxidizing organism (a β-proteobacterium designated strain TW2) capable of both lithoautotrophic and organotrophic growth (on acetate) (210), which may represent a target for genetic studies of microaerophilic Fe(II) oxidation that will be required for the development of functional gene markers of this group of organisms; (iv) the isolation and characterization of several different nitrate-reducing, Fe(II)-oxidizing organisms (including a new β proteobacterium, designated strain 2002, that is capable of fully autotrophic growth), all of which (like strain TW2) are capable of organotrophic growth and thus could be candidates for genetic studies of nitrate-dependent Fe(II) oxidation (222, 267); (v) the isolation and characterization of a genetically tractable phototrophic Fe(II)-oxidizing bacterium (*R. palustris* strain TIE-1) which shows promise for gaining insight into the genes involved in phototrophic Fe(II) oxidation (98); and (vi) ongoing advances in our knowledge of the biochemical

TABLE 5 Overview of connections between Fe and Mn redox transformations and the behaviors of other metals or metalloid elements in aquatic systems

Metal(s)	Connection to Fe and Mn redox	Reference(s)
Divalent cations (e.g., Cd^{2+}, Cu^{2+}, Ni^{2+}, Pb^{2+}, Sr^{2+}, and Zn^{2+})	Released from Fe(III) and Mn(IV) oxide surfaces during biotic or abiotic reduction in sediments	123, 130, 275
	May be incorporated into mixed Fe(III)-Fe(II) hydroxide or metal-carbonate phases produced during dissimilatory Fe(III) oxide reduction in sediments	37, 38, 185
	Adsorbed and/or coprecipitated with Fe(III) and Mn(IV) oxides produced during biotic or abiotic oxidation	235
Arsenic	As(V) and As(III) released from Fe(III) oxide surfaces during biotic and abiotic reduction in sediments	97, 166, 276
	As(V) reduced enzymatically by some DMRB, liberating more mobile As(III)	51, 166
	As(III) oxidized to As(V) and coprecipitated with Fe(III) oxides during NO_3^--dependent Fe(II) oxidation	78, 199
	As(III) sorption onto and incorporation into (immobilized) Fe-sulfide phases (e.g., FeS and FeS_2) produced beneath Fe(III) and Mn(IV) reduction zone in sediment	14, 183
Chromium	Cr(VI) reduced and immobilized to insoluble Cr(III) by some DMRB	261
	Cr(VI) reduced rapidly by biogenic Fe(II) above pH 5	273
Cobalt	Co(III) reduced to more mobile Co(II) by some DMRB	23, 82
	Co(II) released from Fe(III) and Mn(IV) oxide surfaces during biotic or abiotic reduction	125, 126, 152
	Co(II) oxidized and/or coprecipitated with Mn(IV) oxides during microbial Mn(II) oxidation	127, 228, 230
Mercury	Hg(II) reduced by some DMRB	129, 133
	Hg(II) methylated by some DMRB	71, 263
	Methyl-Hg demethylated by DMRB	263
Selenium	Se(VI) and Se(IV) released from Fe(III) and Mn(IV) oxide surfaces during biotic or abiotic reduction	11
	Se(VI) reduced to Se(IV) and (mainly) Se(0) (insoluble) by some DMRB, leading to removal of Se from solution and accumulation in sediments	165
	Se(0) oxidized to Se(IV) and Se(VI) by FeOB and MnOB	52
Uranium	U(VI) reduced to U(IV) (insoluble) by many DMRB, leading to removal of U from solution and accumulation in sediments; enzymatic U(VI) reduction to U(IV) by DMRB is the primary sink for U in marine sediments and provides a strategy for remediation of U-contaminated groundwater aquifers	7, 140, 141
	U(IV) oxidized to U(VI) and coprecipitated with Fe(III) oxides during nitrate-dependent Fe(II) oxidation	70, 122, 198

mechanisms and genes involved in enzymatic Mn(II) oxidation in model Mn(II)-oxidizing organisms (236, 264) which should eventually lead to the development of functional gene markers of this activity that can be applied to natural systems. A related noteworthy development is the identification of a robust functional gene marker (*arrA*) for dissimilatory arsenate reduction activity (146), which is useful for studies of the role of microbial As(V) reduction in the mobilization of As in groundwater. Likewise, genes involved in As(III) oxidation in both organotrophic and lithotrophic organisms are known (*aoxA* and *-B* and *aroA* and *-B*, respectively) (see reference 40 for a review) and will provide the basis for the development of tools to detect this process in aquatic systems.

Two new high-throughput molecular techniques that have been applied to the analysis of DMRB and other prokaryotic populations in natural systems include high-density microarrays of 16S rRNA gene probes (phylochips) for analysis of PCR-amplified 16S rRNA genes (274) and bead-based suspension microarrays for direct detection and quantification of 16S rRNA (30). These technologies have been used to analyze DMRB and other anaerobic respiratory microbial populations in subsurface sediments (32, 260) and show great promise for use in unraveling, among other questions in environmental microbiology (31), the spatial and temporal dynamics and community structures of metal-cycling microbial populations in aquatic environments.

CONNECTIONS TO OTHER METAL CYCLES

The redox cycling of Fe and Mn exerts direct and/or indirect effects on the behavior of a wide variety of other metallic elements in natural systems. Depending on the chemical nature of the metal, the potential exists for both mobilization and immobilization coupled to Fe and Mn redox transformations. Table 5 provides a summary of the behaviors of several metals in relation to such transformations in aquatic ecosystems, along with representative literature references.

This contribution was supported in part by grants from the U.S. National Science Foundation, the U.S. Department of Energy, and the National Aeronautics and Space Administration.

REFERENCES

1. **Aguilar, C., and K. H. Nealson.** 1994. Manganese reduction in Oneida Lake, New York—estimates of spatial and temporal manganese flux. *Can. J. Fish. Aquat. Sci.* **51:**185–196.
2. **Aller, R. C.** 1990. Bioturbation and manganese cycling in hemipelagic sediments. *Phil. Trans. R. Soc. London A* **331:**51–68.
3. **Aller, R. C., and P. D. Rude.** 1988. Complete oxidation of solid phase sulfides by manganese and bacteria in anoxic marine sediments. *Geochim. Cosmochim. Acta* **52:**751–765.
4. **Amann, R. I., L. Krumholz, and D. A. Stahl.** 1990. Fluorescent-oligonucleotide probing of whole cells for determinative, phylogenetic, and environmental studies in microbiology. *J. Bacteriol.* **172:**762–770.
5. **Amann, R. I., J. Stromley, R. Devereux, R. Key, and D. A. Stahl.** 1992. Molecular and microscopic identification of sulfate-reducing bacteria in multispecies biofilms. *Appl. Environ. Microbiol.* **58:**614–623.
6. **Anderson, R. T., J. N. Rooney-Varga, C. V. Gaw, and D. R. Lovley.** 1998. Anaerobic benzene oxidation in the Fe(III) reduction zone of petroleum-contaminated aquifers. *Environ. Sci. Technol.* **32:**1222–1229.
7. **Anderson, R. T., H. A. Vrionis, I. Ortiz-Bernad, C. T. Resch, P. E. Long, R. Dayvault, K. Karp, S. Marutzky, D. R. Metzler, A. Peacock, D. C. White, M. Lowe, and D. R. Lovley.** 2003. Stimulating in situ activity of *Geobacter* species to remove uranium from the groundwater of a uranium-contaminated aquifer. *Appl. Environ. Microbiol.* **69:**5884–5891.
8. **Arnold, R. G., M. R. Hoffmann, T. J. DiChristina, and F. W. Picardal.** 1990. Regulation of dissimilatory Fe(III) reduction activity in *Shewanella putrefaciens*. *Appl. Environ. Microbiol.* **56:**2811–2817.
9. **Bak, F., and N. Pfenning.** 1991. Sulfate-reducing bacteria in littoral sediment of Lake Konstanz. *FEMS Microbiol. Ecol.* **85:**43–52.
10. **Barbeau, K., E. L. Rue, K. W. Bruland, and A. Butler.** 2001. Photochemical cycling of iron in the surface ocean mediated by microbial iron(III)-binding ligands. *Nature* **413:**409–413.
11. **Belzile, N., Y. W. Chen, and R. R. Xu.** 2000. Early diagenetic behaviour of selenium in freshwater sediments. *Appl. Geochem.* **15:**1439–1454.
12. **Benz, M., A. Brune, and B. Schink.** 1998. Anaerobic and aerobic oxidation of ferrous iron at neutral pH by chemoheterotrophic nitrate-reducing bacteria. *Arch. Microbiol.* **169:**159–165.
13. **Bond, D. R., T. Mester, C. L. Nesbo, A. V. Izquierdo-Lopez, F. L. Collart, and D. R. Lovley.** 2005. Characterization of citrate synthase from *Geobacter sulfurreducens* and evidence for a family of citrate synthases similar to those of eukaryotes throughout the *Geobacteraceae*. *Appl. Environ. Microbiol.* **71:**3858–3865.
14. **Bostick, B. C., S. Fendorf, and B. A. Manning.** 2003. Arsenite adsorption on galena (PbS) and sphalerite (ZnS). *Geochim. Cosmochim. Acta* **67:**895–907.
15. **Braker, G., J. Z. Zhou, L. Y. Wu, A. H. Devol, and J. M. Tiedje.** 2000. Nitrite reductase genes (*nirK* and *nirS*) as functional markers to investigate diversity of denitrifying bacteria in Pacific Northwest marine sediment communities. *Appl. Environ. Microbiol.* **66:**2096–2104.
16. **Brendel, P. J., and G. W. Luther.** 1995. Development of a gold amalgam voltammetric microelectrode for the determination of dissolved Fe, Mn, O_2, and S(II) in porewaters of marine and freshwater sediments. *Environ. Sci. Technol.* **29:**751–761.
17. **Buffle, J., and R. R. DeVitre (ed.).** 1994. *Chemical and Biological Regulation of Aquatic Systems*. Lewis, Boca Raton, Fla.
18. **Buffle, J., and G. Horvai.** 2000. *In Situ Monitoring of Aquatic Systems*. John Wiley & Sons, New York, N.Y.
19. **Buffle, J., and W. Stumm.** 1994. General chemistry of aquatic systems, p. 1–43. *In* J. Buffle and R. R. DeVitre (ed.), *Chemical and Biological Regulation of Aquatic Systems*. Lewis, Boca Raton, Fla.
20. **Burdige, D. J.** 1993. The biogeochemistry of manganese and iron reduction in marine sediments. *Earth Sci. Rev.* **35:**249–284.
21. **Burdige, D. J., and P. E. Kepkay.** 1983. Determination of bacterial manganese oxidation rates in sediments using an in situ dialysis technique. 1. Laboratory studies. *Geochim. Cosmochim. Acta* **47:**1907–1916.
22. **Burke, I. T., C. Boothman, J. R. Lloyd, R. J. G. Mortimer, F. R. Livens, and K. Morris.** 2005. Effects of progressive anoxia on the solubility of technetium in sediments. *Environ. Sci. Technol.* **39:**4109–4116.
23. **Caccavo, F., D. J. Lonergan, D. R. Lovley, M. Davis, J. F. Stolz, and M. J. McInerney.** 1994. *Geobacter sulfurreducens* sp. nov., a hydrogen- and acetate-oxidizing dissimilatory metal-reducing microorganism. *Appl. Environ. Microbiol.* **60:**3752–3759.

24. **Canfield, D. E., and D. J. DesMarais.** 1993. Biogeochemical cycles of carbon, sulfur, and free oxygen in a microbial mat. *Geochim. Cosmochim. Acta* **57:**3971–3984.
25. **Canfield, D. E., B. B. Jorgensen, H. Fossing, R. Glud, J. Gundersen, N. B. Ramsing, B. Thamdrup, J. W. Hansen, L. P. Neilsen, and P. O. J. Hall.** 1993. Pathways of organic carbon oxidation in three continental margin sediments. *Mar. Geol.* **113:**27–40.
26. **Canfield, D. E., and B. Thamdrup.** 1994. The production of ^{34}S-depleted sulfide during bacterial disproportionation of elemental sulfur. *Science* **266:**1973–1975.
27. **Canfield, D. E., B. Thamdrup, and J. W. Hansen.** 1993. The anaerobic degradation of organic matter in Danish coastal sediments: iron reduction, manganese reduction, and sulfate reduction. *Geochim. Cosmochim. Acta* **57:** 3867–3883.
28. **Canfield, D. E., B. Thamdrup, and E. Kristensen.** 2005. *Aquatic Geomicrobiology*. Elsevier, San Diego, Calif.
29. **Carey, E., and M. Taillefert.** 2005. The role of soluble Fe(III) in the cycling of iron and sulfur in coastal marine sediments. *Limnol. Oceanogr.* **50:**1129–1141.
30. **Chandler, D. P., and A. E. Jarrell.** 2004. Automated purification and suspension array detection of 16S rRNA from soil and sediment extracts by using tunable surface microparticles. *Appl. Environ. Microbiol.* **70:**2621–2631.
31. **Chandler, D. P., and A. E. Jarrell.** 2005. Taking arrays from the lab to the field: trying to make sense of the unknown. *BioTechniques* **38:**591–600.
32. **Chandler, D. P., A. E. Jarrell, E. E. Roden, J. Golova, B. Chernov, M. J. Schipma, A. D. Peacock, and P. E. Long.** 2006. Suspension array analysis of 16S rRNA from Fe- and SO_4^{2-}-reducing bacteria in uranium-contaminated sediments undergoing bioremediation. *Appl. Environ. Microbiol.* **72:**4672–4687.
33. **Chapelle, F. H.** 2001. *Ground-Water Microbiology and Geochemistry*, 2nd ed. John Wiley & Sons, Inc., New York, N.Y.
34. **Chapnick, S. D., W. S. Moore, and K. H. Nealson.** 1982. Microbially mediated manganese oxidation in a freshwater lake. *Limnol. Oceanogr.* **27:**1004–1014.
35. **Coates, J. D., and L. A. Achenbach.** 2001. The biogeochemistry of aquifer systems, p. 719–727. *In* C. J. Hurst, R. L. Crawford, G. R. Knudsen, M. J. McInerney, and L. D. Stetzenbach (ed.), *Manual of Environmental Microbiology*, 2nd ed. ASM Press, Washington, D.C.
36. **Coates, J. D., R. T. Anderson, J. C. Woodward, E. J. P. Phillips, and D. R. Lovley.** 1996. Anaerobic hydrocarbon degradation in petroleum-contaminated harbor sediments under sulfate-reducing and artificially imposed iron-reducing conditions. *Environ. Sci. Technol.* **30:**2784–2789.
37. **Cooper, D. C., A. L. Neal, R. K. Kukkadapu, D. Brewe, A. Coby, and F. W. Picardal.** 2005. Effects of sediment iron mineral composition on microbially mediated changes in divalent metal speciation: importance of ferrihydrite. *Geochim. Cosmochim. Acta* **69:**1739–1754.
38. **Cooper, D. C., F. Picardal, J. Rivera, and C. Talbot.** 2000. Zinc immobilization and magnetite formation via ferric oxide reduction by *Shewanella putrefaciens* 200. *Environ. Sci. Technol.* **34:**100–106.
39. **Cornell, R. M., and U. Schwertmann.** 1996. *The Iron Oxides*. VCH Verlagsgesellschaft mbH, Weinheim, Germany.
40. **Croal, L. R., J. A. Gralnick, D. Malasarn, and D. K. Newman.** 2004. The genetics of geochemistry. *Annu. Rev. Genet.* **38:**175–202.
41. **Croal, L. R., C. M. Johnson, B. L. Beard, and D. K. Newman.** 2003. Iron isotope fractionation by anoxygenic Fe(II)-phototrophic bacteria. *Geochim. Cosmochim. Acta* **68:**1227–1242.
42. **Cummings, D. E., O. L. Snoeyenbos-West, D. T. Newby, A. M. Niggemyer, D. R. Lovley, L. A. Achenbach, and R. F. Rosenzweig.** 2003. Diversity of *Geobacteraceae* species inhabiting metal-polluted freshwater lake sediments ascertained by 16S rDNA analyses. *Microb. Ecol.* **46:**257–269.
43. **Davies, S. H. R., and J. J. Morgan.** 1989. Manganese(II) oxidation kinetics on metal oxide surfaces. *J. Colloid Interface Sci.* **129:**63–77.
44. **Davison, W., G. W. Grime, J. A. W. Morgan, and K. Clarke.** 1991. Distribution of dissolved iron in sediment pore waters at submillimetre resolution. *Nature* **353:**323–325.
45. **Davison, W., and G. Seed.** 1983. The kinetics of the oxidation of ferrous iron in synthetic and natural waters. *Geochim. Cosmochim. Acta* **47:**67–79.
46. **Davison, W., and H. Zhang.** 1994. In situ speciation measurements of trace components in natural waters using thin-film gels. *Nature* **367:**546–548.
47. **DiChristina, T. J.** 1992. Effects of nitrate on dissimilatory iron reduction by *Shewanella putrefaciens* 200. *J. Bacteriol.* **174:**1891–1896.
48. **DiChristina, T. J., J. K. Fredrickson, and J. M. Zachara.** 2005. Enzymology of electron transport: energy generation with geochemical consequences, p. 27–52. *In* J. F. Banfield, J. Cervini-Silva, and K. H. Nealson (ed.), *Molecular Geomicrobiology*, vol. 59. Mineralogical Society of America, Washington, D.C.
49. **Diem, D., and W. Stumm.** 1984. Is dissolved Mn^{2+} being oxidized by O_2 in absence of Mn-bacteria or surface catalysts? *Geochim. Cosmochim. Acta* **48:**1571–1573.
50. **Donahoe, R. J., and C. X. Liu.** 1998. Pore water geochemistry near the sediment-water interface of a zoned, freshwater wetland in the southeastern United States. *Environ. Geol.* **33:**143–153.
51. **Dowdle, P. R., A. M. Laverman, and R. S. Oremland.** 1996. Bacterial dissimilatory reduction of arsenic(V) to arsenic(III) in anoxic sediments. *Appl. Environ. Microbiol.* **62:**1664–1669.
52. **Dowdle, P. R., and R. S. Oremland.** 1998. Microbial oxidation of elemental selenium in soil slurries and bacterial cultures. *Environ. Sci. Technol.* **32:**3749–3755.
53. **Duce, R. A., and N. W. Tindale.** 1991. Atmospheric transport of iron and its deposition in the ocean. *Limnol. Oceanogr.* **36:**1715–1726.
54. **Edwards, K. J., D. R. Rogers, C. O. Wirsen, and T. M. McCollom.** 2003. Isolation and characterization of novel psychrophilic, neutrophilic, Fe-oxidizing chemolithoautotrophic α- and γ-*Proteobacteria* from the deep sea. *Appl. Environ. Microbiol.* **69:**2906–2913.
55. **Ehrenberg, C. G.** 1836. Vorlage mettheilungen ueber das wirklige vorkommen fossiler infusorien und ihre grosse verbreitung. *Poggendorfs Ann. Phys. Chem.* **38:**213–227.
56. **Ehrenreich, A., and F. Widdel.** 1994. Anaerobic oxidation of ferrous iron by purple bacteria, a new type of phototrophic metabolism. *Appl. Environ. Microbiol.* **60:**4517–4526.
57. **Ehrlich, H. L.** 1995. *Geomicrobiology*. Marcel Dekker, Inc., New York, N.Y.
58. **Ehrlich, H. L.** 2002. *Geomicrobiology*, 4th ed. Marcel Dekker, New York, N,Y.
59. **Ehrlich, H. L.** 2000. Ocean manganese nodules: biogenesis and bioleaching possibilities. *Miner. Metallurg. Proc.* **17:**121–128.
60. **Emerson, D.** 2004. Introduction to special issue on microbial Fe(II) oxidation at neutral pH. *Geomicrobiol. J.* **21:** 369.
61. **Emerson, D.** 2000. Microbial oxidation of Fe(II) and Mn(II) at circumneutral pH, p. 31–52. *In* D. R. Lovley

(ed.), *Environmental Microbe-Metal Interactions.* ASM Press, Washington, D.C.

62. **Emerson, D., and M. M. Floyd.** 2005. Enrichment and isolation of iron-oxidizing bacteria at neutral pH. *Methods Enzymol.* **397:**112–130.
63. **Emerson, D., and C. Moyer.** 1997. Isolation and characterization of novel lithotrophic iron-oxidizing bacteria that grow at circumneutral pH. *Appl. Environ. Microbiol.* **63:**4784–4792.
64. **Emerson, D., and C. Moyer.** 2002. Neutrophilic Fe-oxidizing bacteria are abundant at the Loihi Seamount hydrothermal vents and play a major role in Fe oxide deposition. *Appl. Environ. Microbiol.* **68:**3085–3093.
65. **Emerson, D., and N. P. Revsbech.** 1994. Investigation of an iron-oxidizing microbial mat community located near Aarhus, Denmark: field studies. *Appl. Environ. Microbiol.* **60:**4022–4031.
66. **Emerson, D., and N. P. Revsbech.** 1994. Investigation of an iron-oxidizing microbial mat community located near Aarhus, Denmark: laboratory studies. *Appl. Environ. Microbiol.* **60:**4032–4038.
67. **Emerson, D., and J. V. Weiss.** 2004. Bacterial iron oxidation in circumneutral freshwater habitats: findings from the field and the laboratory. *Geomicrobiol. J.* **21:**405–414.
68. **Emerson, D., J. V. Weiss, and J. P. Megonigal.** 1999. Iron-oxidizing bacteria are associated with ferric hydroxide precipitates (Fe plaque) on the roots of wetland plants. *Appl. Environ. Microbiol.* **65:**2758–2761.
69. **Emerson, S., S. Kalhorn, L. Jacobs, B. M. Tebo, K. H. Nealson, and R. A. Rosson.** 1982. Environmental oxidation rate of manganese(II): bacterial catalysis. *Geochim. Cosmochim. Acta* **46:**1073–1079.
70. **Finneran, K. T., M. E. Housewright, and D. R. Lovley.** 2002. Multiple influences of nitrate on uranium solubility during bioremediation of uranium-contaminated subsurface sediments. *Environ. Microbiol.* **4:**510–516.
71. **Fleming, E. J., E. E. Mack, P. G. Green, and D. C. Nelson.** 2006. Mercury methylation from unexpected sources: molybdate-inhibited freshwater sediments and an iron-reducing bacterium. *Appl. Environ. Microbiol.* **72:**457–464.
72. **Ford, T., and R. Mitchell.** 1992. Microbial transport of toxic metals, p. 83–101. *In* R. Mitchell (ed.), *Environmental Microbiology.* Wiley-Liss, New York, N.Y.
73. **Friedrich, M. W.** 2005. Methyl-coenzyme M reductase genes: unique functional markers for methanogenic and anaerobic methane-oxidizing Archaea. *Methods Enzymol.* **397:**428–442.
74. **Froelich, P. N., G. P. Klinkhammer, M. L. Bender, N. A. Luedtke, G. R. Heath, D. Cullen, P. Dauphin, D. Hammond, B. Hartman, and V. Maynard.** 1979. Early oxidation of organic matter in pelagic sediments of the eastern equatorial Atlantic: suboxic diagenesis. *Geochim. Cosmochim. Acta* **43:**1075–1090.
75. **Gadd, G. M.** 1993. Microbial formation and transformation of organometallic and organometalloid compounds. *FEMS Microbiol. Rev.* **11:**297–316.
76. **Ghiorse, W. C.** 1984. Biology of iron- and manganese-depositing bacteria. *Annu. Rev. Microbiol.* **38:**515–550.
77. **Ghiorse, W. C.** 1994. Iron and manganese oxidation and reduction, p. 1079–1096. *In* J. M. Bigham (ed.), *Methods of Soil Analysis,* part 2. *Microbiological and Biochemical Properties.* Soil Science Society of America, Madison, Wis.
78. **Gibney, B., and K. Neusslin.** Arsenic sequestration by nitrate respiring microbes in urban lake sediments. Submitted for publication.
79. **Gibson, G. R., R. J. Parkes, and R. A. Herbert.** 1987. Evaluation of viable counting procedures for the enumeration of sulfate-reducing bacteria in estuarine sediments. *J. Microbiol. Methods* **7:**201–210.
80. **Glud, R. N., J. K. Gundersen, B. B. Jorgensen, N. P. Revsbech, and H. D. Schulz.** 1994. Diffusive and total oxygen-uptake of deep-sea sediments in the Eastern South-Atlantic Ocean: in situ and laboratory measurements. *Deep-Sea Res. I* **41:**1767–1788.
81. **Glud, R. N., J. K. Gundersen, N. P. Revsbech, and B. B. Jorgensen.** 1994. Effects on the benthic diffusive boundary layer imposed by microelectrodes. *Limnol. Oceanogr.* **39:**462–467.
82. **Gorby, Y. A., F. Caccavo, and H. Bolton.** 1998. Microbial reduction of cobalt^{III}EDTA$^-$ in the presence and absence of manganese(IV) oxide. *Environ. Sci. Technol.* **32:**244–250.
83. **Gorby, Y. A., and D. R. Lovley.** 1991. Electron transport in the dissimilatory iron reducer, GS-15. *Appl. Environ. Microbiol.* **57:**867–870.
84. **Habicht, K. S., and D. E. Canfield.** 2001. Isotope fractionation by sulfate-reducing natural populations and the isotopic composition of sulfide in marine sediments. *Geology* **29:**555–558.
85. **Haese, R. R.** 2000. The reactivity of iron, p. 233–261. *In* H. D. Schulz and M. Zabel (ed.), *Marine Geochemistry.* Springer, New York, N.Y.
86. **Hansen, J. W., B. Thamdrup, and B. B. Jorgensen.** 2000. Anoxic incubation of sediment in gas-tight plastic bags: a method for biogeochemical process studies. *Mar. Ecol. Prog. Ser.* **208:**273–282.
87. **Harder, E. C.** 1919. Iron-depositing bacteria and thier geologic relations. *U. S. Geol. Surv. Prof. Pap.* **113:**7–89.
88. **Heising, S., L. Richter, W. Ludwig, and B. Schink.** 1999. *Chlorobium ferrooxidans* sp. nov., a phototrophic green sulfur bacterium that oxidizes ferrous iron in coculture with a "Geospirillum" sp. strain. *Arch. Microbiol.* **172:**116–124.
89. **Heising, S., and B. Schink.** 1998. Phototrophic oxidation of ferrous iron by a Rhodomicrobium vannielii strain. *Microbiology* (Reading) **144:**2263–2269.
90. **Hering, J. G., and W. Stumm.** 1990. Oxidative and reductive dissolution of minerals, p. 427–464. *In* M. F. Hochella and A. F. White (ed.), *Mineral-Water Interface Geochemistry,* vol. 23. Mineralogical Society of America, Washington, D.C.
91. **Hines, M. E., J. Faganeli, and R. Planinc.** 1997. Sedimentary anaerobic microbial biogeochemistry in the Gulf of Trieste, northern Adriatic Sea: influences of bottom water oxygen depletion. *Biogeochemistry* **39:**65–86.
92. **Holland, H. D., and J. F. Kasting.** 1992. The environment of the archaean Earth, p. 21–24. *In* J. W. Schopf and C. Klein (ed.), *The Proterozoic Biosphere: an Interdisciplinary Study.* Cambridge University Press, Cambridge, United Kingdom.
93. **Holmes, D. E., K. T. Finneran, R. A. O'Neil, and D. R. Lovley.** 2002. Enrichment of members of the family *Geobacteraceae* associated with the stimulation of dissimilatory metal reduction in uranium-contaminated aquifer sediments. *Appl. Environ. Microbiol.* **68:**2300–2306.
94. **Holmes, D. E., K. P. Nevin, and D. R. Lovley.** 2004. In situ expression of *nifD* in *Geobacteraceae* in subsurface sediments. *Appl. Environ. Microbiol.* **70:**7251–7259.
95. **Holmes, D. E., K. P. Nevin, R. A. O'Neil, J. E. Ward, L. A. Adams, T. L. Woodard, H. A. Vrionis, and D. R. Lovley.** 2005. Potential for quantifying expression of the *Geobacteraceae* citrate synthase gene to assess the activity of *Geobacteraceae* in the subsurface and on current-harvesting electrodes. *Appl. Environ. Microbiol.* **71:**6870–6877.
96. **Hungate, R. E.** 1969. A roll tube method for cultivation of strict anaerobes. *Methods Microbiol.* **3B:**117–132.
97. **Islam, F. S., A. G. Gault, C. Boothman, D. A. Polya, J. M. Charnock, D. Chatterjee, and J. R. Lloyd.** 2004. Role of metal-reducing bacteria in arsenic release from Bengal delta sediments. *Nature* **430:**68–71.

98. **Jiao, Y. Y. Q., A. Kappler, L. R. Croal, and D. K. Newman.** 2005. Isolation and characterization of a genetically tractable photo autotrophic Fe(II)-oxidizing bacterium, *Rhodopseudomonas palustris* strain TIE-1. *Appl. Environ. Microbiol.* **71:**4487–4496.
99. **Johnson, C. M., and B. L. Beard.** 2005. Biogeochemical cycling of iron isotopes. *Science* **309:**1025–1027.
100. **Johnson, C. M., B. L. Beard, E. E. Roden, D. K. Newman, and K. H. Nealson.** 2004. Isotopic constraints on biogeochemical cycling of Fe, p. 359–408. *In* C. M. Johnson, B. L. Beard, and F. Albarède (ed.), *Reviews in Mineralogy and Geochemistry*, vol. 55. *Geochemistry of Non-Traditional Stable Isotopes*. Mineralogical Society of America, Washington, D.C.
101. **Johnson, D., B. Chiswell, and K. Ohalloran.** 1995. Microorganisms and manganese cycling in a seasonally stratified freshwater dam. *Water Res.* **29:**2739–2745.
102. **Jones, J. G.** 1983. A note on the isolation and enumeration of bacteria which deposit and reduce ferric iron. *J. Appl. Bacteriol.* **54:**305–310.
103. **Jorgensen, B. B.** 1989. Biogeochemistry of chemoautotrophic bacteria, p. 117–146. *In* H. G. Schlegel and B. Bowien (ed.), *Biochemistry of Autotrophic Bacteria*. Science Tech Publishers, Madison, Wis.
104. **Jorgensen, B. B.** 1978. A comparison of methods for the quantification of bacterial sulfate reduction in coastal marine sediments. I. Measurement with radiotracer techniques. *Geomicrobiol. J.* **1:**11–28.
105. **Kashefi, K., D. E. Holmes, A. L. Reysenbach, and D. R. Lovley.** 2002. Use of Fe(III) as an electron acceptor to recover previously uncultured hyperthermophiles: isolation and characterization of *Geothermobacterium ferrireducens* gen. nov., sp. nov. *Appl. Environ. Microbiol.* **68:** 1735–1742.
106. **Kepkay, P. E.** 1985. Kinetics of microbial manganese oxidation and trace-metal binding in sediments—results from an in situ dialysis technique. *Limnol. Oceanogr.* **30:** 713–726.
107. **Kepkay, P. E.** 1985. Microbial manganese oxidation and nitrification in relation to the occurrence of macrophyte roots in a lacustrine sediment. *Hydrobiologia* **128:**135–142.
108. **Klinkhammer, G. P.** 1980. Early diagenesis in sediments from the eastern equatorial Pacific. II. Pore water metal results. *Earth Planet. Sci. Lett.* **49:**81–101.
109. **Konhauser, K. O., T. Hamade, R. Raiswell, R. C. Morris, F. G. Ferris, G. Southam, and D. E. Canfield.** 2002. Could bacteria have formed the Precambrian banded iron formations? *Geology* **30:**1079–1082.
110. **Konhauser, K. O., D. K. Newman, and A. Kappler.** 2005. The potential significance of microbial Fe(III) reduction during deposition of Precambrian banded iron formations. *Geobiology* **3:**167–177.
111. **Koretsky, C. M., C. M. Moore, K. L. Lowe, C. Meile, T. J. Dichristina, and P. Van Cappellen.** 2003. Seasonal oscillation of microbial iron and sulfate reduction in saltmarsh sediments (Sapelo Island, GA, USA). *Biogeochemistry* **64:**179–203.
112. **Koretsky, C. M., P. Van Cappellen, T. J. DiChristina, J. E. Kostka, K. L. Lowe, C. M. Moore, A. N. Roychoudhury, and E. Viollier.** 2005. Salt marsh pore water geochemistry does not correlate with microbial community structure. *Estuar. Coast. Shelf Sci.* **62:**233–251.
113. **Kostka, J. E., B. Gribsholt, E. Petrie, D. Dalton, H. Skelton, and E. Kristensen.** 2002. The rates and pathways of carbon oxidation in bioturbated saltmarsh sediments. *Limnol. Oceanogr.* **2002:**230–240.
114. **Kostka, J. E., E. Haefele, R. Vieweger, and J. W. Stucki.** 1999. Respiration and dissolution of iron(III)-containing clay minerals by bacteria. *Environ. Sci. Technol.* **33:**3127–3133.
115. **Kostka, J. E., and G. W. Luther III.** 1994. Partitioning and speciation of solid phase iron in saltmarsh sediments. *Geochim. Cosmochim. Acta* **58:**1701–1710.
116. **Kostka, J. E., G. W. Luther, and K. H. Nealson.** 1995. Chemical and biological reduction of Mn(III)-pyrophosphate complexes: potential importance of dissolved Mn(III) as an environmental oxidant. *Geochim. Cosmochim. Acta* **59:**885–894.
117. **Kostka, J. E., and K. N. Nealson.** 1998. Isolation, cultivation, and characterization of iron- and manganese-reducing bacteria, p. 58–78. *In* R. S. Burlage, R. Atlas, D. Stahl, G. Geesey, and G. Sayler (ed.), *Techniques in Microbial Ecology*. Oxford University Press, Oxford, United Kingdom.
118. **Kostka, J. E., A. Roychoudhury, and P. VanCappellen.** 2002. Rates and controls of anaerobic microbial respiration across spatial and temporal gradients in saltmarsh sediments. *Biogeochemistry* **60:**49–76.
119. **Kostka, J. E., B. Thamdrup, R. N. Glud, and D. E. Canfield.** 1999. Rates and pathways of carbon oxidation in permanently cold arctic sediments. *Mar. Ecol. Prog. Ser.* **180:**7–21.
120. **Kucera, S., and R. S. Wolfe.** 1957. A selective enrichment method for *Gallionella ferruginea*. *J. Bacteriol.* **74:** 344–349.
121. **Lack, J. G., S. K. Chaudhuri, R. Chakraborty, L. A. Achenbach, and J. D. Coates.** 2002. Anaerobic biooxidation of Fe(II) by Dechlorosoma suillum. *Microb. Ecol.* **43:**424–431.
122. **Lack, J. G., S. K. Chaudhuri, S. D. Kelly, K. M. Kemner, S. M. O'Connor, and J. D. Coates.** 2002. Immobilization of radionuclides and heavy metals through anaerobic bio-oxidation of Fe(II). *Appl. Environ. Microbiol.* **68:**2704–2710.
123. **Landa, E. R., E. J. P. Phillips, and D. R. Lovley.** 1991. Release of 226Ra from uranium mill tailings by microbial Fe(III) reduction. *Appl. Geochem.* **6:**647–652.
124. **Larocque, A. C. L., and P. E. Rasmussen.** 1998. An overview of trace metals in the environment, from mobilization to remediation. *Environ. Geol.* **33:**85–91.
125. **Lee, B. G., and N. S. Fisher.** 1993. Microbially mediated cobalt oxidation in seawater revealed by radiotracer experiments. *Limnol. Oceanogr.* **38:**1593–1602.
126. **Lee, Y., and B. M. Tebo.** 1994. Cobalt(II) oxidation by the marine manganese(II)-oxidizing *Bacillus* sp. strain SG-1. *Appl. Environ. Microbiol.* **60:**2949–2957.
127. **Lienemann, C. P., M. Taillefert, D. Perret, and J. F. Gaillard.** 1997. Association of cobalt and manganese in aquatic systems: chemical and microscopic evidence. *Geochim. Cosmochim. Acta* **61:**1437–1446.
128. **Lin, W. C., M. V. Coppi, and D. R. Lovley.** 2004. *Geobacter sulfurreducens* can grow with oxygen as a terminal electron acceptor. *Appl. Environ. Microbiol.* **70:**2525–2528.
129. **Lloyd, J. R.** 2003. Microbial reduction of metals and radionuclides. *FEMS Microbiol. Rev.* **27:**411–425.
130. **Lovley, D. R.** 1991. Dissimilatory Fe(III) and Mn(IV) reduction. *Microbiol. Rev.* **55:**259–287.
131. **Lovley, D. R.** 1993. Dissimilatory metal reduction. *Annu. Rev. Microbiol.* **47:**263–290.
132. **Lovley, D. R.** 1987. Organic matter mineralization with the reduction of ferric iron: a review. *Geomicrobiol. J.* **5:**375–399.
133. **Lovley, D. R., S. J. Giovannoni, D. C. White, J. E. Champine, E. J. P. Phillips, Y. A. Gorby, and S. Goodwin.** 1993. *Geobacter metallireducens* gen. nov. sp. nov., a microorganism capable of coupling the complete

oxidation of organic compounds to the reduction of iron and other metals. *Arch. Microbiol.* **159:**336–344.

134. **Lovley, D. R., D. E. Holmes, and K. P. Nevin.** 2004. Dissimilatory Fe(III) and Mn(IV) reduction. *Adv. Microb. Physiol.* **49:**219–286.
135. **Lovley, D. R., and E. J. P. Phillips.** 1986. Availability of ferric iron for microbial reduction in bottom sediments of the freshwater tidal Potomac River. *Appl. Environ. Microbiol.* **52:**751–757.
136. **Lovley, D. R., and E. J. P. Phillips.** 1991. Enzymatic versus nonenzymatic mechanisms for Fe(III) reduction in aquatic sediments. *Environ. Sci. Technol.* **25:**1062–1067.
137. **Lovley, D. R., and E. J. P. Phillips.** 1988. Manganese inhibition of microbial iron reduction in anaerobic sediments. *Geomicrobiol. J.* **6:**145–155.
138. **Lovley, D. R., and E. J. P. Phillips.** 1994. Novel processes for anaerobic sulfate production from elemental sulfur by sulfate-reducing bacteria. *Appl. Environ. Microbiol.* **60:**2394–2399.
139. **Lovley, D. R., and E. J. P. Phillips.** 1987. Rapid assay for microbially reducible ferric iron in aquatic sediments. *Appl. Environ. Microbiol.* **53:**1536–1540.
140. **Lovley, D. R., E. J. P. Phillips, Y. A. Gorby, and E. R. Landa.** 1991. Microbial reduction of uranium. *Nature* **350:**413–416.
141. **Lovley, D. R., E. E. Roden, E. J. P. Phillips, and J. C. Woodward.** 1993. Enzymatic iron and uranium reduction by sulfate-reducing bacteria. *Mar. Geol.* **113:**41–53.
142. **Lowe, K. L., T. J. Dichristina, A. N. Roychoudhury, and P. V. Cappellen.** 2000. Microbiological and geochemical characterization of microbial Fe(III) reduction in salt marsh sediments. *Geomicrobiol. J.* **17:**163–176.
143. **Luther, G. W., D. B. Nuzzio, and J. F. Wu.** 1994. Speciation of manganese in Chesapeake Bay waters by voltammetric methods. *Anal. Chim. Acta* **284:**473–480.
144. **Luther, G. W., P. A. Shellenbarger, and P. J. Brendel.** 1996. Dissolved organic Fe(III) and Fe(II) complexes in salt marsh porewaters. *Geochim. Cosmochim. Acta* **60:**951–960.
145. **Madigan, M. T., and J. M. Martinko.** 2006. *Brock Biology of Microorganisms*, 11th ed. Prentice Hall, Upper Saddle River, N.J.
146. **Malasarn, D., C. W. Saltikov, K. M. Campbell, J. M. Santini, J. G. Hering, and D. K. Newman.** 2004. *arrA* is a reliable marker for As(V) respiration. *Science* **306:**455.
147. **Mandernack, K. W., and B. M. Tebo.** 1993. Manganese scavenging and oxidation at hydrothermal vents and in vent plumes. *Geochim. Cosmochim. Acta* **57:**3907–3923.
148. **Megonigal, J. P., M. E. Hines, and P. T. Visscher.** 2004. Anaerobic metabolism: linkages to trace gases and aerobic processes, p. 317–424. *In* W. H. Schlesinger (ed.), *Biogeochemistry*. Elsevier-Pergamon, Oxford, United Kingdom.
149. **Millero, F. J., S. Sotolongo, and M. Izaguirre.** 1987. The oxidation kinetics of Fe(II) in seawater. *Geochim. Cosmochim. Acta* **51:**793–801.
150. **Miyajima, T.** 1992. Biological manganese oxidation in a lake. 1. Occurrence and distribution of Metallogenium sp and its kinetic properties. *Arch. Hydrobiol.* **124:**317–335.
151. **Miyajima, T.** 1992. Biological manganese oxidation in a lake. 2. A thermodynamic consideration of the habitat utilization of Metallogenium sp. *Arch. Hydrobiol.* **124:**411–426.
152. **Moffett, J. W., and J. Ho.** 1996. Oxidation of cobalt and manganese in seawater via a common microbially catalyzed pathway. *Geochim. Cosmochim. Acta* **60:**3415–3424.
153. **Munch, J. C., and J. C. G. Ottow.** 1983. Reductive transformation mechanism of ferric oxides in hydromorphic soils. *Ecol. Bull.* **35:**383–394.
154. **Murray, J. W., and P. G. Brewer.** 1977. Mechanism of removal of manganese, iron, and other trace metals from seawater, p. 291–325. *In* G. P. Glasby (ed.), *Marine Manganese Deposits*. Elsevier, Amsterdam, The Netherlands.
155. **Murray, J. W., L. A. Codispoti, and G. E. Friedrich.** 1995. Oxidation-reduction environments—the suboxic zone in the Black Sea. *Adv. Chem. Ser.* **244:**157–176.
156. **Nealson, K. H., and D. Saffarini.** 1994. Iron and manganese in anaerobic respiration: environmental significance, physiology, and regulation. *Annu. Rev. Microbiol.* **48:**311–343.
157. **Neretin, L. N., C. Pohl, G. Jost, T. Leipe, and F. Pollehne.** 2003. Manganese cycling in the Gotland Deep, Baltic Sea. *Mar. Chem.* **82:**125–143.
158. **Neubauer, S. C., D. Emerson, and J. P. Megonigal.** 2002. Life at the energetic edge: kinetics of circumneutral iron oxidation by lithotrophic iron-oxidizing bacteria isolated from the wetland-plant rhizosphere. *Appl. Environ. Microbiol.* **68:**3988–3995.
159. **Neubauer, S. C., D. Emerson, and J. P. Megonigal.** Microbial oxidation and reduction of iron in the root zone and mobility of heavy metals. *In* A. Violante, P. M. Huang, and G. Stotzky (ed.), *Biophysico-Chemical Processes of Heavy Metals and Metalloids in Soil Environments*, in press. John Wiley & Sons, Hoboken, N.J.
160. **Neubauer, S. C., K. Givler, S. K. Valentine, and J. P. Megonigal.** 2005. Seasonal patterns and plant-mediated controls of subsurface wetland biogeochemistry. *Ecology* **86:**3334–3344.
161. **Nevin, K. P., and D. R. Lovley.** 2002. Mechanisms of Fe(III) oxide reduction in sedimentary environments. *Geomicrobiol. J.* **19:**141–159.
162. **Nogales, B., K. N. Timmis, D. B. Nedwell, and A. M. Osborn.** 2002. Detection and diversity of expressed denitrification genes in estuarine sediments after reverse transcription-PCR amplification from mRNA. *Appl. Environ. Microbiol.* **68:**5017–5025.
163. **Northup, D. E., S. M. Barns, L. E. Yu, M. N. Spilde, R. T. Schelble, K. E. Dano, L. J. Crossey, C. A. Connolly, P. J. Boston, D. O. Natvig, and C. N. Dahm.** 2003. Diverse microbial communities inhabiting ferromanganese deposits in Lechuguilla and Spider caves. *Environ. Microbiol.* **5:**1071–1086.
164. **Oremland, R. S., and D. G. Capone.** 1988. Use of specific inhibitors in biogeochemistry and microbial ecology. *Adv. Microb. Ecol.* **10:**285–383.
165. **Oremland, R. S., and J. Stolz.** 2000. Dissimilatory reduction of selenate and arsenate in nature, p. 199–224. *In* D. R. Lovley (ed.), *Environmental Microbe-Metal Interactions*. ASM Press, Washington, D.C.
166. **Oremland, R. S., and J. F. Stolz.** 2005. Arsenic, microbes and contaminated aquifers. *Trends Microbiol.* **13:**45–49.
167. **Oremland, R. S., and J. F. Stolz.** 2003. The ecology of arsenic. *Science* **300:**939–943.
168. **Ortiz-Bernad, I., R. T. Anderson, H. A. Vrionis, and D. R. Lovley.** 2004. Vanadium respiration by *Geobacter metallireducens*: novel strategy for in situ removal of vanadium from groundwater. *Appl. Environ. Microbiol.* **70:**3091–3095.
169. **Parker, D. L., G. Sposito, and B. M. Tebo.** 2004. Manganese(III) binding to a pyoverdine siderophore produced by a manganese(II)-oxidizing bacterium. *Geochim. Cosmochim. Acta* **68:**4809–4820.
170. **Phillips, E. J. P., and D. R. Lovley.** 1987. Determination of Fe(III) and Fe(II) in oxalate extracts of sediments. *Soil Sci. Soc. Am. J.* **51:**938–941.

171. **Phillips, E. J. P., D. R. Lovley, and E. E. Roden.** 1993. Composition of non-microbially reducible Fe(III) in aquatic sediments. *Appl. Environ. Microbiol.* **59:**2727–2729.
172. **Ponnamperuma, F. N.** 1972. The chemistry of submerged soils. *Adv. Agron.* **24:**29–96.
173. **Post, J. E.** 1999. Manganese oxide minerals: crystal structures and economic and environmental significance. *Proc. Natl. Acad. Sci. USA* **96:**3447–3454.
174. **Postma, D.** 1985. Concentration of Mn and separation from Fe in sediments. I. Kinetics and stoichiometry of the reaction between birnessite and dissolved Fe(II) at 10°C. *Geochim. Cosmochim. Acta* **49:**1023–1033.
175. **Poulton, S. W., and R. Raiswell.** 2002. The low-temperature geochemical cycle of iron: from continental fluxes to marine sediment deposition. *Am. J. Sci.* **302:**774–805.
176. **Purdy, K. J., M. A. Munson, T. Cresswell-Maynard, D. B. Nedwell, and T. M. Embley.** 2003. Use of 16S rRNA-targeted oligonucleotide probes to investigate function and phylogeny of sulphate-reducing bacteria and methanogenic archaea in a UK estuary. *FEMS Microbiol. Ecol.* **44:**361–371.
177. **Ramsing, N. R., H. Fossing, T. G. Ferdelman, F. Andersen, and B. Thamdrup.** 1996. Distribution of bacterial populations in a stratified fjord (Mariager Fjord, Denmark) quantified by in situ hybridization and related to chemical gradients in the water column. *Appl. Environ. Microbiol.* **62:**1391–1404.
178. **Raskin, L., J. M. Stromley, B. Rittmann, and D. A. Stahil.** 1994. Group-specific 16S rRNA hybidization probes to describe natural communities of methanogens. *Appl. Environ. Microbiol.* **60:**1232–1240.
179. **Ratering, S., and S. Schnell.** 2000. Localization of iron-reducing activity in paddy soil by profile studies. *Biogeochemistry* **48:**341–365.
180. **Rentz, J. A., C. Kraiya, G. W. Luther, and D. Emerson.** 2005. Iron cycle in a neutrophilic iron seep. *Astrobiology* **5:**292.
181. **Revsbech, N. P.** 1989. An oxygen microelectrode with guard cathode. *Limnol. Oceanogr.* **34:**472–476.
182. **Rhine, E. D., E. Garcia-Dominguez, C. D. Phelps, and L. Y. Young.** 2005. Environmental microbes can speciate and cycle arsenic. *Environ. Sci. Technol.* **39:**9569–9573.
183. **Rittle, K. A., J. I. Drever, and P. J. S. Colberg.** 1995. Precipitation of arsenic during bacterial sulfate reduction. *Geomicrobiol. J.* **13:**1–11.
184. **Roden, E. E.** 2005. Unpublished data.
185. **Roden, E. E., M. R. Leonardo, and F. G. Ferris.** 2002. Immobilization of strontium during iron biomineralization coupled to dissimilatory hydrous ferric oxide reduction. *Geochim. Cosmochim. Acta* **66:**2823–2839.
186. **Roden, E. E., and D. R. Lovley.** 1993. Evaluation of ^{55}Fe as a tracer of Fe(III) reduction in aquatic sediments. *Geomicrobiol. J.* **11:**49–56.
187. **Roden, E. E., D. Sobolev, B. Glazer, and G. W. Luther.** 2004. Potential for microscale bacterial Fe redox cycling at the aerobic-anaerobic interface. *Geomicrobiol. J.* **21:**379–391.
188. **Roden, E. E., and R. G. Wetzel.** 2002. Kinetics of microbial Fe(III) oxide reduction in freshwater wetland sediments. *Limnol. Oceanogr.* **47:**198–211.
189. **Roden, E. E., and R. G. Wetzel.** 1996. Organic carbon oxidation and suppression of methane production by microbial Fe(III) oxide reduction in vegetated and unvegetated freshwater wetland sediments. *Limnol. Oceanogr.* **41:**1733–1748.
190. **Rodrigo, M. A., E. Vicente, and M. R. Miracle.** 2000. The role of light and concentration gradients in the vertical stratification and seasonal development of phototrophic bacteria in a meromictic lake. *Arch. Hydrobiol.* **148:**533–548.
191. **Rogers, D. R., C. M. Santelli, and K. J. Edwards.** 2003. Geomicrobiology of deep-sea deposits; estimating community diversity from low-temperature seafloor rocks and minerals. *Geobiology* **1:**109–118.
192. **Röling, W. F. M., B. M. van Breukelen, M. Braster, B. Lin, and H. W. van Verseveld.** 2001. Relationships between microbial community structure and hydrochemistry in a landfill leachate-polluted aquifer. *Appl. Environ. Microbiol.* **67:**4619–4629.
193. **Rooney-Varga, J. N., R. T. Anderson, J. L. Fraga, D. Ringelberg, and D. R. Lovley.** 1999. Microbial communities associated with anaerobic benzene degradation in a petroleum-contaminated aquifer. *Appl. Environ. Microbiol.* **65:**3056–3063.
194. **Rosson, R. A., B. M. Tebo, and K. H. Nealson.** 1984. Use of poisons in determination of microbial manganese binding rates in seawater. *Appl. Environ. Microbiol.* **47:**740–745.
195. **Schippers, A., and B. B. Jorgensen.** 2001. Biogeochemistry of pyrite and iron sulfide oxidation in marine sediments. *Geochim. Cosmochim. Acta* **66:**85–92.
196. **Schippers, A., L. N. Neretin, G. Lavik, T. Leipe, and F. Pollehne.** 2005. Manganese(II) oxidation driven by lateral oxygen intrusions in the western Black Sea. *Geochim. Cosmochim. Acta* **69:**2241–2252.
197. **Schwertmann, U.** 1964. Differenzierung der Eisenoxide des Bodens durch Extraktion mit Ammoniumoxalatlosung. *Z. Pflanzenernaehr. Dueng. Bodenkd.* **105:**194–202.
198. **Senko, J. M., Y. Mohamed, T. A. Dewers, and L. R. Krumholz.** 2005. Role for Fe(III) minerals in nitrate-dependent microbial U(IV) oxidation. *Environ. Sci. Technol.* **39:**2529–2536.
199. **Senn, D. B., and H. F. Hemond.** 2002. Nitrate controls on iron and arsenic in an urban lake. *Science* **296:**2373–2376.
200. **Shelobolina, E. S., R. R. Anderson, Y. N. Vodyanitskii, A. M. Sivtsov, R. Yuretich, and D. R. Lovley.** 2004. Importance of clay size minerals for Fe(III) respiration in a petroleum-contaminated aquifer. *Geobiology* **2:**67–76.
201. **Shelobolina, E. S., C. Gaw-VanPraagh, and D. R. Lovley.** 2003. Use of ferric and ferrous iron containing minerals for respiration by *Desulfitobacterium frappieri*. *Geomicrobiol. J.* **20:**143–156.
202. **Shelobolina, E. S., S. M. Pickering, and D. R. Lovley.** 2005. Fe-cycle bacteria from industrial clays mined in Georgia, USA. *Clays Clay Min.* **53:**580–586.
203. **Shiller, A. M., and T. H. Stephens.** 2005. Microbial manganese oxidation in the lower Mississippi River: methods and evidence. *Geomicrobiol. J.* **22:**117–125.
204. **Siering, P. L., and W. C. Ghiorse.** 1997. Development and application of 16S rRNA-targeted probes for detection of iron- and manganese-oxidizing sheathed bacteria in environmental samples. *Appl. Environ. Microbiol.* **63:**644–651.
205. **Siering, P. L., and W. C. Ghiorse.** 1997. PCR detection of a putative manganese oxidation gene (mofA) in environmental samples and assessment of mofA gene homology among diverse manganese-oxidizing bacteria. *Geomicrobiol. J.* **14:**109–125.
206. **Sigg, L.** 1994. Regulation of trace elements in lakes: the role of sedimentation, p. 177–197. *In* J. Buffle and R. R. DeVitre (ed.), *Chemical and Biological Regulation of Aquatic Systems*. Lewis, Boca Raton, Fla.
207. **Silver, S.** 1997. The bacterial view of the periodic table: specific functions for all elements, p. 345–360. *In* J. F. Banfield and K. H. Nealson (ed.), *Geomicrobiology:*

Interactions between Microbes and Minerals, vol. 35. Mineralogical Society of America, Washington, D.C.

208. **Singer, P. C., and W. Stumm.** 1972. Acid mine drainage—the rate limiting step. *Science* **167:**1121–1123.
209. **Snoeyenbos-West, O. L., K. P. Nevin, R. T. Anderson, and D. R. Lovley.** 2000. Enrichment of *Geobacter* species in response to stimulation of Fe(III) reduction in sandy aquifer sediments. *Microb. Ecol.* **39:**153–167.
210. **Sobolev, D., and E. Roden.** 2004. Characterization of a neutrophilic, chemolithoautotrophic Fe(II)-oxidizing β-Proteobacterium from freshwater wetland sediments. *Geomicrobiol. J.* **21:**1–10.
211. **Sobolev, D., and E. E. Roden.** 2002. Evidence for rapid microscale bacterial redox cycling of iron in circumneutral environments. *Antonie Leeuwenhoek* **181:**587–597.
212. **Sobolev, D., and E. E. Roden.** 2001. Suboxic deposition of ferric iron by bacteria in opposing gradients of Fe(II) and oxygen at circumneutral pH. *Appl. Environ. Microbiol.* **67:**1328–1334.
213. **Sorensen, J.** 1982. Reduction of ferric iron in anaerobic, marine sediment and interaction with reduction of nitrate and sulfate. *Appl. Environ. Microbiol.* **43:**319–324.
214. **Spilde, M. N., D. E. Northup, P. J. Boston, R. T. Schelble, K. E. Dano, L. J. Crossey, and C. N. Dahm.** 2005. Geomicrobiology of cave ferromanganese deposits: a field and laboratory investigation. *Geomicrobiol. J.* **22:** 99–116.
215. **Stein, L. Y., G. Jones, B. Alexander, K. Elmund, C. Wright-Jones, and K. H. Nealson.** 2002. Intriguing microbial diversity associated with metal-rich particles from a freshwater reservoir. *FEMS Microbiol. Ecol.* **42:** 431–440.
216. **Stein, L. Y., M. T. LaDuc, T. J. Grundl, and K. H. Nealson.** 2001. Bacterial and archael populations associated with freshwater ferromanganous micronodules and sediments. *Environ. Microbiol.* **3:**10–18.
217. **Stone, A. T., and J. J. Morgan.** 1987. Reductive dissolution of metal oxides, p. 221–254. *In* W. Stumm (ed.), *Aquatic Surface Chemistry*. John Wiley & Sons, New York, N.Y.
218. **Stookey, L. L.** 1970. Ferrozine—a new spectrophotometric reagent for iron. *Anal. Chem.* **42:**779–781.
219. **Straub, K. L., M. Benz, B. Schink, and F. Widdel.** 1996. Anaerobic, nitrate-dependent microbial oxidation of ferrous iron. *Appl. Environ. Microbiol.* **62:**1458–1460.
220. **Straub, K. L., and B. E. E. Buchholz-Cleven.** 1998. Enumeration and detection of anaerobic ferrous-iron-oxidizing, nitrate-reducing bacteria from diverse European sediments. *Appl. Environ. Microbiol.* **64:**4846–4856.
221. **Straub, K. L., F. A. Rainey, and F. Widdel.** 1999. *Rhodovulum iodosum* sp. nov, and *Rhodovulum robiginosum* sp. nov., two new marine phototrophic ferrous-iron-oxidizing purple bacteria. *Int. J. Syst. Evol. Microbiol.* **49:** 729–735.
222. **Straub, K. L., W. A. Schonhuber, B. E. E. Buchholz-Cleven, and B. Schink.** 2004. Diversity of ferrous iron-oxidizing, nitrate-reducing bacteria and their involvement in oxygen-independent iron cycling. *Geomicrobiol. J.* **21:**371–378.
223. **Stults, J. R., O. Snoeyenbos-West, B. Methe, D. R. Lovley, and D. P. Chandler.** 2001. Application of the 5′ fluorogenic exonuclease assay (TaqMan) for quantitative ribosomal DNA and rRNA analysis in sediments. *Appl. Environ. Microbiol.* **67:**2781–2789.
224. **Stumm, W., and J. J. Morgan.** 1996. *Aquatic Chemistry*, 2nd ed. John Wiley & Sons, Inc., New York, N.Y.
225. **Stumm, W., and B. Sulzberger.** 1992. The cycling of iron in natural environments: considerations based on laboratory studies of heterogeneous redox processes. *Geochim. Cosmochim. Acta* **56:**3233–3257.
226. **Sunda, W. G.** 2000. Trace metal-phytoplankton interactions in aquatic systems, p. 79–107. *In* D. R. Lovley (ed.), *Environmental Micorbe-Metal Interactions*. ASM Press, Washington, D.C.
227. **Sunda, W. G., and D. J. Kieber.** 1994. Oxidation of humic substances by manganese oxides yields low-molecular-weight organic substrates. *Nature* **367:**62–64.
228. **Sundby, B., L. G. Anderson, P. O. J. Hall, A. Iverfeldt, M. M. Rutgersvanderloeff, and S. F. G. Westerlund.** 1986. The effect of oxygen on release and uptake of cobalt, manganese, iron and phosphate at the sediment-water interface. *Geochim. Cosmochim. Acta* **50:**1281–1288.
229. **Suter, D., C. Siffert, B. Sulzberger, and W. Stumm.** 1988. Catalytic dissolution of iron(III) (hydr)oxides by oxalic acid in the presence of Fe(II). *Naturwissenschaften* **75:**571–573.
230. **Taillefert, M., B. J. MacGregor, J. F. Gaillard, C. P. Lienemann, D. Perret, and D. A. Stahl.** 2002. Evidence for a dynamic cycle between Mn and Co in the water column of a stratified lake. *Environ. Sci. Technol.* **36:**468–476.
231. **Taylor, G. T., M. Iabichella, T. Y. Ho, M. I. Scranton, R. C. Thunell, F. Muller-Karger, and R. Varela.** 2001. Chemoautotrophy in the redox transition zone of the Cariaco Basin: a significant midwater source of organic carbon production. *Limnol. Oceanogr.* **2001:**148–163.
232. **Teasdale, P. R., S. Hayward, and W. Davison.** 1999. In situ, high-resolution measurement of dissolved sulfide using diffusive gradients in thin films with computer-imaging densitometry. *Anal. Chem.* **71:**2186–2191.
233. **Tebo, B. M.** 1991. Manganese(II) oxidation in the suboxic zone of the Black Sea. *Deep-Sea Res.* **38:**S883–S905.
234. **Tebo, B. M., J. R. Bargar, B. G. Clement, G. J. Dick, K. J. Murray, D. Parker, R. Verity, and S. M. Webb.** 2004. Biogenic manganese oxides: properties and mechanisms of formation. *Annu. Rev. Earth Planet. Sci.* **32:**287–328.
235. **Tebo, B. M., and L. M. He.** 1999. Microbially mediated oxidation precipitation reactions, p. 393–414. *In* D. L. Sparks and T. J. Grundl (ed.), *Mineral-Water Interfacial Reactions*. American Chemical Society, Washington, D.C.
236. **Tebo, B. M., H. A. Johnson, J. K. McCarthy, and A. S. Templeton.** 2005. Geomicrobiology of manganese(II) oxidation. *Trends Microbiol.* **13:**421–428.
237. **Tebo, B. M., K. H. Nealson, S. Emerson, and L. Jacobs.** 1984. Microbial mediation of Mn(II) and Co(II) precipitation at the O_2/H_2S interfaces in two anoxic fjords. *Limnol. Oceanogr.* **29:**1247–1258.
238. **Templeton, A. S., H. Staudigel, and B. M. Tebo.** 2005. Diverse Mn(II)-oxidizing bacteria isolated from submarine basalts at Loihi Seamount. *Geomicrobiol. J.* **22:**127–139.
239. **Thamdrup, B.** 2000. Bacterial manganese and iron reduction in aquatic sediments. *Adv. Microb. Ecol.* **16:** 41–84.
240. **Thamdrup, B., and D. E. Canfield.** 2000. Benthic respiration in aquatic sediments, p. 86–103. *In* O. E. Sala, R. B. Jackson, H. A. Mooney, and R. W. Howarth (ed.), *Methods in Ecosystem Science*. Springer, New York, N.Y.
241. **Thamdrup, B., and D. E. Canfield.** 1996. Pathways of carbon oxidation in continental margin sediments off central Chile. *Limnol. Oceanogr.* **41:**1629–1650.
242. **Thamdrup, B., K. Finster, J. W. Hansen, and F. Bak.** 1993. Bacterial disproportionation of elemental sulfur

coupled to chemical reduction of iron or manganese. *Appl. Environ. Microbiol.* **59:**101–108.

243. **Thamdrup, B., H. Fossing, and B. B. Jorgensen.** 1994. Manganese, iron, and sulfur cycling in a coastal marine sediment, Aarhus Bay, Denmark. *Geochim. Cosmochim. Acta* **58:**5115 –5129.
244. **Thamdrup, B., R. N. Glud, and J. W. Hansen.** 1994. Manganese oxidation and in situ manganese fluxes from a coastal sediment. *Geochim. Cosmochim. Acta* **58:**2563–2570.
245. **Thamdrup, B., R. Rossello-Mora, and R. Amann.** 2000. Microbial manganese and sulfate reduction in Black Sea shelf sediments. *Appl. Environ. Microbiol.* **66:** 2888–2897.
246. **Thomsen, U., B. Thamdrup, D. A. Stahl, and D. E. Canfield.** 2004. Pathways of organic carbon oxidation in a deep lacustrine sediment, Lake Michigan. *Limnol. Oceanogr.* **49:**2046–2057.
247. **Thorseth, I. H., T. Torsvik, V. Torsvik, F. L. Daae, R. Pedersen, and K.-S. Party.** 2001. Diversity of life in ocean floor basalt. *Earth Planet. Sci. Lett.* **194:**31–37.
248. **vanBreemen, N.** 1988. Effects of seasonal redox processes involving iron on the chemistry of periodically reduced soils, p. 797–809. *In* J. W. Stucki, B. A. Goodman, and U. Schwertmann (ed.), *Iron in Soils and Clay Minerals*. D. Reidel Publishing Co., Boston, Mass.
249. **vanBreemen, N.** 1988. Long-term chemical, mineralogical and morphological effects of iron-redox processes in periodically flooded soils, p. 811–823. *In* J. W. Stucki, B. A. Goodman, and U. Schwertmann (ed.), *Iron in Soils and Clay Minerals*. D. Reidel Publishing Co., Boston, Mass.
250. **VanCappellen, P., E. Viollier, and A. Roychoudhury.** 1998. Biogeochemical cycles of manganese and iron at the oxic-anoxic transition of a stratified marine basin (Orca Basin, Gulf of Mexico). *Environ. Sci. Technol.* **32:** 2931–2939.
251. **VanCappellen, P., and Y. Wang.** 1996. Cycling of iron and manganese in surface sediments: a general theory for the coupled transport and reaction of carbon, oxygen, nitrogen, sulfur, iron, and manganese. *Am. J. Sci.* **296:** 197–243.
252. **VanCappellen, P., and Y. Wang.** 1995. Metal cycling in surface sediments: modeling the interplay of transport and reaction, p. 21–64. *In* H. E. Allen (ed.), *Metal Contaminated Aquatic Sediments*. Ann Arbor Press, Chelsea, Mich.
253. **Vester, F., and K. Invorsen.** 1998. Improved most-probable-number method to detect sulfate-reducing bacteria with natural media and a radiotracer. *Appl. Environ. Microbiol.* **64:**1700–1707.
254. **Villalobos, M., and B. M. Tebo.** 2005. Introduction: advances in the geomicrobiology and biogeochemistry of manganese and iron oxidation. *Geomicrobiol. J.* **22:**77–78.
255. **Viollier, E., C. Rabouille, S. E. Apitz, E. Breuer, G. Chaillou, K. Dedieu, Y. Furukawa, C. Grenz, P. Hall, F. Janssen, J. L. Morford, J. C. Poggiale, S. Roberts, T. Shimmield, M. Taillefert, A. Tengberg, F. Wenzhofer, and U. Witte.** 2003. Benthic biogeochemistry: state of the art technologies and guidelines for the future of in situ survey. *J. Exp. Mar. Biol. Ecol.* **285:**5–31.
256. **Wagner, M., A. Loy, M. Klein, N. Lee, N. B. Ramsing, D. A. Stahl, and M. W. Friedrich.** 2005. Functional marker genes for identification of sulfate-reducing prokaryotes. *Methods Enzymol.* **397:**469–489.
257. **Waite, T. D.** 1988. Photochemical effects on the mobility and fate of heavy metals in the aquatic environment. *Environ. Technol. Lett.* **9:**977–982.
258. **Walker, J. C. G.** 1984. Suboxic diagenesis in banded iron formations. *Nature* **309:**340–342.
259. **Wallmann, K., K. Hennies, I. Konig, W. Petersen, and H. D. Knauth.** 1993. New procedure for determining reactive Fe(III) and Fe(II) minerals in sediments. *Limnol. Oceanogr.* **38:**1803–1812.
260. **Wan, J., T. K. Tokunaga, E. Brodie, Z. Wang, Z. Zheng, D. Herman, T. C. Hazen, M. K. Firestone, and S. R. Sutton.** 2005. Reoxidation of bioreduced uranium under reducing conditions. *Environ. Sci. Technol.* **39:** 6162–6169.
261. **Wang, Y. T.** 2000. Microbial reduction of chromate, p. 225–235. *In* D. R. Lovley (ed.), *Environmental Microbe-Metal Interactions*. ASM Press, Washington, D.C.
262. **Ward, B. B., and G. D. O'Mullan.** 2005. Community level analysis: genetic and biogeochemical approaches to investigate community composition and function in aerobic ammonia oxidation. *Methods Enzymol.* **397:**395–413.
263. **Warner, K. A., E. E. Roden, and J. C. Bonzongo.** 2003. Microbial mercury transformations in anoxic freshwater sediments under iron-reducing and other electron-accepting conditions. *Environ. Sci. Technol.* **37:**2159–2165.
264. **Webb, S. M., G. J. Dick, J. R. Bargar, and B. M. Tebo.** 2005. Evidence for the presence of Mn(III) intermediates in the bacterial oxidation of Mn(II). *Proc. Natl. Acad. Sci. USA* **102:**5558–5563.
265. **Weber, K. A., P. F. Churchill, M. M. Urrutia, R. K. Kukkadapu, and E. E. Roden.** 2006. Anaerobic redox cycling of iron by freshwater sediment microorganisms. *Environ. Microbiol.* **8:**100–113.
266. **Weber, K. A., F. W. Picardal, and E. E. Roden.** 2001. Microbially-catalyzed nitrate-dependent oxidation of biogenic solid-phase Fe(II) compounds. *Environ. Sci. Technol.* **35:**1644–1650.
267. **Weber, K. A., J. Pollock, K. A. Cole, S. M. O'Connor, L. A. Achenbach, and J. D. Coates.** 2006. Anaerobic nitrate-dependent iron(II) bio-oxidation by a novel lithoautotrophic betaproteobacterium, strain 2002. *Appl. Environ. Microbiol.* **72:**686–694.
268. **Wedepohl, K. H.** 1995. The composition of the continental-crust. *Geochim. Cosmochim. Acta* **59:**1217–1232.
269. **Weiss, J. V., D. Emerson, S. M. Backer, and J. P. Megonigal.** 2003. Enumeration of Fe(II)-oxidizing and Fe(III)-reducing bacteria in the root zone of wetland plants: implications for a rhizosphere iron cycle. *Biogeochemistry* **64:**77–96.
270. **Weiss, J. V., D. Emerson, and J. P. Megonigal.** 2005. Rhizosphere iron(III) deposition and reduction in a *Juncus effusus* L.-dominated wetland. *Soil. Sci. Soc. Am. J.* **69:**1861–1870.
271. **Weiss, J. V., D. Emerson, and P. Megonigal.** 2004. The role of geochemical composition on the microbial reduction potential of Fe(III) pools in wetlands: comparison of the rhizosphere and bulk soil. *FEMS Microbiol. Ecol.* **48:**89–100.
272. **Widdel, F., S. Schnell, S. Heising, A. Ehrenreich, B. Assmus, and B. Schink.** 1993. Ferrous iron oxidation by anoxygenic phototrophic bacteria. *Nature* **362:**834–835.
273. **Wielinga, B., M. M. Mizuba, C. M. Hansel, and S. Fendorf.** 2001. Iron promoted reduction of chromate by dissimilatory iron-reducing bacteria. *Environ. Sci. Technol.* **35:**522–527.

274. **Wilson, K. H., W. J. Wilson, J. L. Radosevich, T. Z. DeSantis, V. S. Viswanathan, T. A. Kuczmarski, and G. L. Andersen.** 2002. High-density microarray of small-subunit ribosomal DNA probes. *Appl. Environ. Microbiol.* **68:**2535–2541.

275. **Zachara, J. M., J. K. Fredrickson, S. C. Smith, and P. L. Gassman.** 2001. Solubilization of Fe(III) oxide-bound trace metals by a dissimilatory Fe(III) reducing bacterium. *Geochim. Cosmochim. Acta* **65:**75–93.

276. **Zobrist, J., P. R. Dowdle, J. A. Davis, and R. S. Oremland.** 2000. Mobilization of arsenite by dissimilatory reduction of adsorbed arsenate. *Environ. Sci. Technol.* **34:**4747–4753.

277. **Zopfi, J., T. G. Ferdelman, B. B. Jorgensen, A. Teske, and B. Thamdrup.** 2001. Influence of water column dynamics on sulfide oxidation and other major biogeochemical processes in the chemocline of Mariager Fjord (Denmark). *Mar. Chem.* **74:**29–51.

Biofilms on Living Surfaces

JEREMY S. WEBB, MICHAEL W. TAYLOR, SCOTT RICE,
TORSTEN THOMAS, DHANA RAO, DIANE McDOUGALD,
AND STAFFAN KJELLEBERG

45

Biofilms are increasingly recognized as the predominant mode of bacterial growth in the environment. The last 10 years has witnessed a massive expansion of research in this field, with rapid advances in molecular technologies and microscopy enabling detailed studies of the biofilm mode of life across a range of systems and organisms. Biofilm research now features as a focal point in many areas of microbiology, including medical, industrial, and environmental contexts. For example, it is now realized that more than 60% of bacterially induced infections are due to biofilm formation, that biofilms are a major cause of poor process performance and material degradation in industrial settings, and that effective bioremediation in soils requires a biofilm community. Biofilms also constitute a significant portion of the bacterial biomass and activity in natural habitats. In marine systems, an increasing number of studies have revealed that bacteria form complex communities on diverse substrata. This includes a range of inanimate surfaces such as particulate matter, marine snow, and sediments. Living surfaces, such as those of marine algae and invertebrates, also harbor complex biofilm communities. In fact, on a global scale, this constitutes an immense surface area for interactions between bacteria and eukaryotic host substrata. Thus, it is proposed that microbial communities on the surfaces of marine living sessile organisms offer a powerful system for studies of prokaryote-eukaryote interactions and biofilms on living surfaces. The epiphytic and epibiotic bacterial communities interact both positively and negatively with their hosts. For example, an established bacterial community can protect the host against colonization by pathogens and fouling organisms. However, the establishment of extensive biofilms may also lead to reduced nutrient exchanges and can cause physical impairment of the host.

Recent progress in studies of the microbiology of the surfaces of marine algae and sponges has provided new insights into a series of ecological processes. This chapter presents contemporary approaches to the study of several such processes, including the establishment of community composition and diversity, gene function in complex microbial consortia, chemically mediated interactions between prokaryotes and eukaryotes, key events in bacterial interspecies competition and colonization of surfaces, and factors facilitating persistence and survival of pathogens in the environment. While several aspects of modern technologies and methods are discussed, the intent of this chapter is to familiarize the reader with such approaches in the context of the biological processes displayed by marine microbial epiphytic and epibiotic communities on living surfaces.

MICROBIAL COMMUNITY ANALYSIS OF BIOFILMS ON LIVING SURFACES

Special Considerations for Studying Biofilm Communities on Living Surfaces

The study of microbial communities associated with the surfaces of other organisms poses some interesting challenges. Foremost is the matter of whether detected microbes are stable associates (symbionts) or merely transients which happened to be present during sampling. This is a particular problem in aquatic environments, since the native flora of the water column is in perpetual contact with the surface(s) under investigation. Pretreatment of samples to remove loosely associated bacteria prior to analysis may help, although there is a risk that true symbionts are also removed by any mechanical procedure. Furthermore, for filter-feeding hosts such as marine sponges, which are permeated by intricate aquiferous systems, it is likely impossible to effectively remove transient (or food source) organisms from within. One means of addressing this conundrum lies with the sampling approach. If it can be proven that the same microorganism(s) is consistently found on a given host (i.e., with repeated sampling in space and time), then this provides solid evidence of a stable association. The possibility that the same microorganism was also present in the surrounding environment at all sampling times (and therefore acted as a contaminant in the studied host) can be excluded only if these environments are also examined. Alternatively, evidence for an intimate association could be obtained by, for example, microscopic detection of putative symbionts within host cells (e.g., as shown for bacteria in marine sponges [33]).

With the exception of direct, microscopy-linked techniques such as fluorescence in situ hybridization (FISH), nucleic acid-based methods for characterizing microbial communities first require the extraction of DNA or RNA from the system of interest. For microbial communities associated with living surfaces, this generally means that DNA from the host organism is coextracted. PCR amplification with, say, *Bacteria*-specific primers is the simplest way to circumvent any potential problems arising from this: for example, many gene libraries derived from complex environments are based on the 16S rRNA gene which occurs in prokaryotes (eukaryotes possess the homologous 18S rRNA gene). However, in some circumstances even this approach may prove problematic; for example, sequences from algal plastids commonly appear in 16S rRNA gene libraries taken from plant-dominated habitats (e.g., the seagrass *Halophila stipulacea* [107]), and caution must be taken when examining these data. The presence of microalgae in aquatic biofilms can create similar complications: microalgal plastids contain as many as 650 copies of small circular genomes (70), suggesting enormous potential for PCR bias in favor of these organisms.

Sampling Design and Analysis

The importance of sampling design, replication, and statistical testing is generally underemphasized in microbial ecology (reviewed in reference 67), yet these are critical issues in investigations of any ecological process or system. Ecological systems are intrinsically variable, and only via rigorous sampling approaches can one hope to account for the myriad sources of this variability. Questions of host association require particularly well constructed sampling schemes, since only repeated sampling will provide an insight into whether detected associations are truly stable (see also "Special Considerations for Studying Biofilms on Living Surfaces" above). Moreover, the complex, highly differentiated sample types represented by many hosts may contribute to variability in microbial distributions. Morphological differentiation in the host (e.g., different tissue types) could create additional variability in symbiont distributions, while other factors such as chemical composition may also play a role. An excellent example of the latter is the marine macroalga *Delisea pulchra*, in which bacterial community structure differs on the scale of centimeters (63; H. J. Baillie et al., unpublished data), thought to be due to a concentration gradient of bioactive halogenated furanones along the algal surface (88).

Hierarchical, nested sampling designs allow variability at many levels (both temporal and spatial) to be accounted for. Figure 1 shows a sampling scheme used in a denaturing gradient gel electrophoresis (DGGE)-based study of bacteria in three Australian marine sponges (96). The design accounts for variability at all levels of interest: (i) temporal (in this case seasonal) variation, (ii) variation among different host species, (iii) variation among individuals of the same host species, and (iv) variation within a single individual.

The observant reader may have already noticed that the presented scheme requires the collection and analysis of a total of 150 samples. Such intensive sampling precludes the use of laborious community analysis methods, such as gene library construction or FISH analysis with multiple probes. Established high-throughput fingerprinting techniques such as DGGE, terminal restriction fragment length polymorphism (T-RFLP), and automated rRNA intergenic spacer analysis (ARISA), or emerging techniques such as DNA microarrays (57, 58), thus offer the most realistic option in this respect. A useful compromise is therefore to apply rapid fingerprinting techniques on all available samples in order to get an impression of variability in the studied community. If variability appears to be minor (or is at least minor within sample types, compared with among sample types), it should then be valid to select a much smaller subset of representative samples and perform more detailed phylogenetic and quantitative analyses with them.

Formal statistical tests can be conducted subsequent to data collection, whereby the total variability in the data can be partitioned into its various component causes. The analysis of choice depends on the data type. For binary data based on the presence (=1) or absence (=0) of DGGE bands in a complex pattern, multivariate procedures such as analysis of similarity (11), clustering, and nonmetric multidimensional scaling (61) are available. If quantitative information is obtained for various phylogenetic groups (e.g., by FISH or quantitative PCR), then an analysis of variance is a straightforward way of testing for significant differences in a data set. Subsequent (post hoc) tests can then reveal where these differences actually lie (i.e., exactly which samples differ).

Molecular Methods for Analysis of Microbial Communities on Living Surfaces

High-Throughput Fingerprinting Techniques

Numerous methods exist for the rapid profiling (or fingerprinting) of microbial communities, including DGGE (68), T-RFLP (71), and ARISA (31). These methods are detailed elsewhere in this volume, and here we limit our discussion to their pros and cons for analysis of communities on living

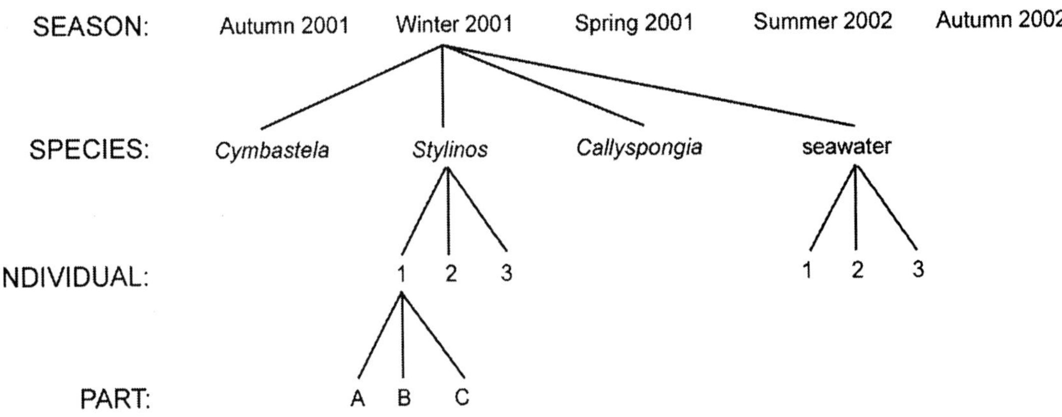

FIGURE 1 Sampling scheme for comparing microbial community compositions in three marine sponges and surrounding seawater. Reproduced with permission from reference 96.

surfaces. The major benefit of such approaches is that they offer relatively cheap, high-throughput analyses of large numbers of samples (allowing for the high levels of replication which are essential when questions of host association are being addressed). A further advantage of DGGE is that phylogenetic information can be obtained via the excision and subsequent sequencing of separated bands (Fig. 2). The presence or absence of DGGE bands (or T-RFLP peaks) can also be used to construct cluster diagrams for effective visual comparisons of the data.

Aside from the biases inherent to essentially any PCR-based method (73, 91), perhaps the most significant limitation of community fingerprinting techniques is one that has received little attention, that is, the difficulty in gauging the actual extent of phylogenetic relatedness among members of different complex communities. For example, two samples could exhibit virtually no overlap in their respective DGGE banding patterns (or T-RFLP peaks)—implying very different communities—and yet still contain phylogenetically highly similar organisms. This potential problem can be at least partially addressed by sequencing excised DGGE bands, but this is a labor-intensive process and limits the numbers of samples which can be analyzed.

Gene Libraries

Arguably, the most informative measures of microbial species richness are achieved via construction of 16S rRNA gene libraries and the subsequent analysis of clones by sequencing (see elsewhere in this volume). The limiting factors for gene library analyses are typically time and effort, particularly for highly diverse samples (e.g., as found in marine sponges [38]), for which large amounts of work are required to satisfactorily document the diversity in the library. If the goal is to compare the overall compositions of microbial communities on different host organisms, then it may be sufficient to focus sequencing efforts on documenting, for example, the genus or species level diversity in the library, at the expense of microdiversity (i.e., minor differences in 16S rRNA gene sequence). Gene libraries based on the 16S rRNA gene have been successfully generated for many living surfaces (e.g., marine sponges [39, 105; M. W. Taylor et al., unpublished results], [32], and seagrasses [106,

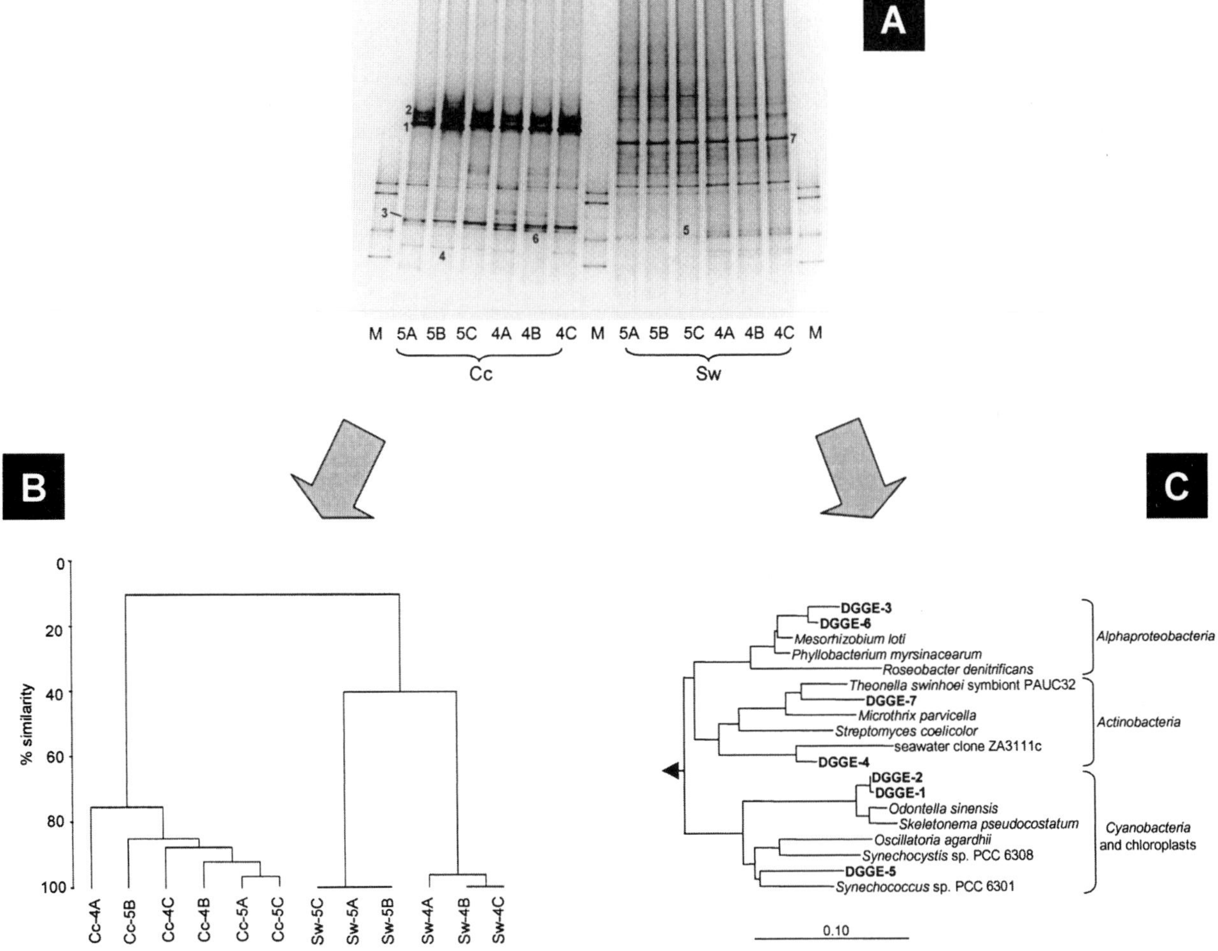

FIGURE 2 DGGE-based analysis of bacterial community composition in the marine sponge *Cymbastela concentrica* (Cc) and seawater (Sw) from southeastern Australia. (A) 16S rRNA gene DGGE gel. Sample numbers and letters refer to collection sites and replicate sponges, respectively. The bands which were excised and sequenced are indicated 1 to 7. (B) Cluster diagram representing the presence or absence of bands in panel A. (C) Phylogenetic analysis of excised DGGE bands from panel A.

107]). Continuing advances in high-throughput sequencing technologies (with associated reductions in costs) will ensure that this remains a valuable tool for microbial community ecologists.

Quantitative Methods

While various PCR biases (73, 91) prevent methods such as DGGE, T-RFLP, and gene library analysis from being truly quantitative, there do exist methods from which information on microbial abundance can be derived.

FISH offers the potential for quantifying microbial populations in a sample, at least on a relative or proportional basis. By using catalyzed reporter deposition-FISH (CARD-FISH), Tujula and coworkers showed that alphaproteobacteria comprised 70% of the epiphytic microbial community on the marine alga *Ulva lactuca*, with 17% of the organisms belonging to the cosmopolitan marine genus *Roseobacter* (N. Tujula, G. Crocetti, I. Dahllof, and S. Kjelleberg, unpublished data). The combination of FISH with spiking experiments (18) and sophisticated image analysis software (17) also creates the possibility of obtaining absolute numbers of organisms. One common problem when FISH is applied to living surfaces is autofluorescence from host cells. This may be ameliorated by quenching the sample with fluorescent light prior to FISH probe addition (105), by separating microbial and host cells via mechanical means (30), or via the amplification of probe signals above background levels by CARD-FISH (83). A major drawback with FISH is the time taken to analyze high numbers of samples, especially if multiple probes are to be used (essential when different phylogenetic groups are examined).

Many of the biases of standard multitemplate PCR are overcome by quantitative, real-time PCR. This technique, based on the continuous, "real-time" monitoring of fluorescence during the exponential stages of the PCR, has already been successfully used to quantify the abundance of the gammaproteobacterial genus *Pseudoalteromonas* on marine algal and ascidian surfaces (85).

Recommended Approach for Characterizing Microbial Communities

For a comprehensive view of microbial community structure on living surfaces, the use of gene libraries and (for quantitative information) FISH is recommended. However, these methods are impractical when large numbers of samples are involved. One approach is therefore to screen many samples by using rapid fingerprinting techniques (e.g., DGGE) to gain an impression of variability within and among sample types. Subsequent, detailed analyses of community composition and organism abundance can then be obtained for selected samples by gene libraries and FISH, respectively.

METAGENOMIC ANALYSIS OF BIOFILM COMMUNITIES ASSOCIATED WITH LIVING SURFACES

Biofilms on living surfaces represent a huge reservoir of biodiversity, and biofilms can also provide a rich source of novel (bio)chemical interactions. For example, biofilms on living surfaces often harbor high numbers of bacteria that produce extracellular inhibitory compounds against other bacteria and/or eukaryotic colonizing organisms (see "Internal Dynamics of Biofilms" below), which may provide novel bioactives for environmental and medical applications. Studies of biofilms on algal surfaces have also provided new insights into chemical signal molecules, their antagonists, and their role in marine chemical ecology (see "Chemical Interactions between Prokaryotes and Eukaryotes" below). With microbiological approaches and modern molecular tools such as fingerprinting techniques and metagenomics, we are well equipped to gain a deeper insight into the biology of these systems and open up new biotechnological opportunities.

A Brief Introduction to Metagenomics

Molecular surveys based on ribosomal or conserved functional genes have shown that the microbial diversity in the marine environment is much greater than previously expected (e.g., reference 21). Furthermore, recent studies of living surfaces clearly highlighted that these habitats harbor microbial communities distinct from those of the surrounding waters, which has significant implications for the total diversity estimates for the world's oceans (38, 96) (see "Microbial Community Analysis of Biofilms on Living Surfaces" above). While molecular fingerprinting techniques provide a means for a quick and easily replicated survey of diversity of a single gene (and maybe a few conserved genes such as *amoA* and *recA*), they do not allow for a comprehensive description of functional gene diversity. In essence, environmental fingerprinting techniques provide the phylogenetic stamp of an organism but leave much of its functional diversity unexplored.

The direct cloning of total, environmental DNA into suitable vector-host systems provides a means of capturing and accessing the total functional and genetic diversity of the environment. This approach gained popularity in the mid-1990s and has now been baptized with the term metagenomics or environmental genomics (37, 79). Recombinant libraries derived from environmental DNA (metagenomic libraries) can essentially contain (or cover) the genomes of all organisms present in the particular environment and therefore can represent both its total phylogenetic and functional gene complement. In general, the field of metagenomics employs two main strategies to analyze and explore a particular sample. In the first approach (the screening strategy), the metagenomic library is produced or transferred into a suitable expression strain (such as *Escherichia coli* BL21 or *Streptomyces lividans*) and a clone array of suitable magnitude (more than several thousand clones) is screened for the heterologous expression of certain phenotypes. Typical screens have so far targeted the production of enzymes (lipases, nucleases, and chitinases), pigments, or antibiotic activities, with the result that several new biocatalysts with no or little homology to previously described proteins have been identified (16). The second approach (the sequencing strategy) involves the sequencing of the metagenomic library followed by a bioinformatics-supported analysis of the sequence database to assemble, identify, and/or characterize new genes and genomes. In a recent study, Tyson and colleagues demonstrated the usefulness of this approach by applying high-throughput environmental random shotgun sequencing to a microbial biofilm in an acid mine drainage system (99). Due to the relatively low complexity of the biofilm community (approximately 5 to 10 phylotypes), sufficient sequence information could be practically generated to fully or partially assemble the genomes of the most abundant bacterial and archaeal representatives. Analysis of genetic microheterogeneity (i.e., the genomic variation between groups or individuals of a population) as well as predicted biochemical features gave an unprecedented insight into this natural, microbial assemblage. In another equally impressive study, more than 10^9 bp of sequence information was generated from the microbial, planktonic community of a Sargasso Sea surface water (100). Surprisingly, in this nutrient-limited habitat, over 1.2×10^6 new genes and 148 new prokaryotic phylotypes were discovered in a sample containing approximately 1,800 genomic species. These two examples highlight the sheer power and utility of metagenomics to describe the diversity of both known and unknown environmental microorganisms on a phylogenetic and functional level. This should also encourage other researchers to consider addressing fundamental issues of environmental microbiology by applying metagenomics, especially with the technologies becoming more widely available through rapid cost reduction of DNA sequencing and the development of new bioinformatics tools.

Specific Considerations for Metagenomics of Living Surfaces

The close association of microbial communities with living surfaces demands some specific practical considerations for metagenomic applications. The reader is also referred to "Special Considerations for Studying Biofilm Communities on Living Surfaces" (above); however, some additional points of particular relevance to metagenomics will be made here. First, the associated community has to be efficiently and completely recovered from the host surface to ensure appropriate representation of all microbial groups in an environmental DNA library. Second, host cells have to be removed from the sample as the generally larger genome size of eukaryotes would result in a library with a significant number of clones containing "unwanted" DNA. To illustrate this point, a clone library derived from one cell of *Ciona intestinalis* (sea squirt; genome size of approximately 116×10^6 bp) and 30 prokaryotic cells (average genome size assumed to be approximately 4×10^6 bp) would theoretically result in a library containing 50% eukaryotic DNA. Third, the DNA extraction and purification method has to provide efficient lysis for all cell types present and DNA of sufficient purity and quality for library construction. The latter point is of particular importance when the aim is to construct large-insert libraries with DNA fragments of >35 kb (such as fosmids or bacterial artificial chromosomes).

Several studies have described the successful purification of microbial cells and their nucleic acids from sponges (4, 81, 82). The methods employed in general involve the initial homogenization of the sponge tissue (by mechanical shearing and/or sponge disassembly in magnesium- and calcium-free artificial seawater) followed by a size-based separation of the microbial cells from the large eukaryotic cells and tissue. Efficient separation can be achieved by filtration with different size cutoff membranes and/or differential centrifugation. Microscopy of samples taken from the different processing stages should be routinely performed to ensure that eukaryotic cells are being removed and that no shape- or size-based selection occurs during the isolation of the microbial community (e.g., filamentous cells versus cocci). Despite the diversity of prokaryotic cells associated with sponges, standard cell lysis protocols (e.g., involving lysozyme, proteinase K, and sodium dodecyl sulfate treatment) followed by organic extraction and alcohol precipitations were sufficient to obtain high-quality DNA suitable even for the construction of large-insert libraries (81).

Isolation of microbial communities from other aquatic living surfaces such as seaweeds or marine invertebrates has been less thoroughly investigated. However, it can safely be assumed that most microbial communities are embedded on the surfaces in a polymer matrix (90). Mechanical stress such as vortexing or shaking might be sufficient to remove some biofilm material; however, one should check the efficiency of such procedures by microscopy of the surface. Often, such attempts remove only the most superficially and loosely attached cells and leave the core of the biofilm intact. Again, mechanical homogenization of the whole sample followed by filtration or differential centrifugation can be considered, but microbial biofilm clumps might be difficult to separate from eukaryotic cells or tissue. Another approach would be to degrade the biofilm matrix in order to release the microbial cells from the host surface. The composition of naturally occurring biofilm matrices is poorly understood but often consists of a complex mixture of polysaccharides, proteins, and/or nucleic acids (90). These components can be degraded by enzymes (such as polysaccharide lyases, proteases, or nucleases [55, 108]) or by mild-chemical treatments (e.g., metaperiodate for the oxidation of sugars [103]). Owing to the complexity of natural biofilms, development and fine-tuning of a specific enzyme cocktail for each biofilm type are to be expected (55). Cell integrity and survival should also be monitored by microscopy as the enzymes mentioned above can affect other cellular components of the host and the microorganisms.

Other than these specific issues associated with obtaining "good" environmental DNA, existing protocols and approaches for library production, screening, sequencing, etc., can be followed, and sufficient reference is given in recent reviews of the field (37, 79). Finally, every metagenomic project (especially when it involves whole-environment shotgun sequencing) generates large amounts of complex data that demand powerful and sophisticated information management systems and bioinformatics. A greater need for universal and effective approaches has been recognized, and initiatives to support the scientific community with this issue are well under way (9, 29, 102).

CHEMICAL INTERACTIONS BETWEEN PROKARYOTES AND EUKARYOTES

Biofilms occur as high-density, matrix-encased populations attached to surfaces. Biofilms are especially resistant to environmental stresses (e.g., nutritional and oxidative stress) and to host-mediated responses (e.g., complement proteins and phagocytes) (27, 41). Due to their increased resistance, biofilms can be highly problematic in industrial and medical settings. Therefore, it is increasingly important to understand how biofilms form and detach, including the internal and external factors that control these processes. One key feature involved in biofilm formation, and other bacterial phenotypes for a number of pathogenic bacteria, is signal-based regulatory systems (34, 86, 109). In general, these systems function via extracellular signals which bacteria use to assess their local population density. When these signals reach sufficient concentrations (e.g., at high cell densities), a feedback regulatory loop is induced, resulting in a very rapid expression of phenotypes in the population of cells. The best studied of these regulatory loops are the acylated homoserine lactone (AHL) systems in gram-negative bacteria (34, 86, 109). AHL-mediated density-dependent signalling (or quorum sensing [QS]) occurs when a threshold concentration of the AHLs (produced by the product of the synthase I gene) accumulates at a sufficient level to interact with receptor proteins (produced from the products of the R gene), which then act as transcriptional regulators by binding to the promoter of target genes (34, 86, 109). QS has been shown to control a wide range of phenotypes, including virulence factor expression, bioluminescence, and biofilm formation where QS mutants of *Pseudomonas aeruginosa*, *Serratia marcescens*, *Burkholderia cepacia*, and *Aeromonas hydrophila* were defective in biofilm formation (19, 46, 56, 59). QS mutants of *P. aeruginosa* have been shown to have reduced virulence in both in vitro and in vivo models of infection (15, 72). Since QS has been shown to regulate these important phenotypes, it has been suggested that the specific interference with QS would be a suitable strategy for the control of bacteria (e.g., biofilm formation [40, 78, 87]).

One approach for the isolation of such QS inhibitors would be to identify natural living surfaces that resist biofilm formation through the production of compounds that inhibit the QS response. Biofilms occur on practically all surfaces, and most bacteria can make biofilms. Indeed, as described above, living surfaces are not exempt from this phenomenon (see "Microbial Community Analysis of Biofilms on Living Surfaces" above). The formation of biofilms on living surfaces has consequences for the individual (plant or animal), and hence, a number of defenses have evolved to either prevent such colonization or recruit specific bacteria to their surface that thwart subsequent colonization by other bacteria. Some solutions used by marine plants to cope with biofilm formation and subsequent colonization by epiphytes include the shedding of tissue, colonization by bacteria that prevent settlement of other bacteria (see "Internal Dynamics of Biofilms" below), and the production of their own chemical defense mechanisms. This section addresses mechanisms for the identification and testing of compounds isolated from living surfaces for activity as inhibitors of biofilm formation via the specific interference with regulatory systems important for biofilm formation.

Selection of Target Organisms for the Isolation of QS Inhibitors

Marine eukaryote-prokaryote assemblages are ideal systems for the discovery of QS antagonists, both because of the diversity of host organisms and associated metabolites and because the surfaces of marine organisms are constantly challenged by the more than 10^6 microorganisms/ml that are typically found in seawater. Thus, the selection pressure in such systems to develop sophisticated strategies for the defense of an organism's surface is intense.

In our initial discovery of QS antagonists, our objective was to understand the chemically mediated inhibition of biofilm formation by marine plants and invertebrates (22, 24, 63). Thus, the selection of surfaces for analysis was driven first by an ecological understanding of the organisms that were biofilm resistant and then by the desire to subsequently identify and determine the mechanism of action for those compounds. In hindsight, the identification of QS inducers or antagonists that are produced on the surfaces of marine plants is not so startling, and this theme has now been repeated in the context of terrestrial plants (3, 97, 98). Therefore, the targeted screening of other plants may yield similar findings, and the same methods can be applied.

Isolation of Bioactive Compounds

There are a number of different bioassays that can be used to screen for QS inhibitors. Instead of recapitulating the protocols for the application of a variety of bioassays which have been extensively published and reviewed (see references 1, 5, 64, 76, and 77), some general considerations and approaches for the identification of QS signals or antagonists are presented here. The first and perhaps most important consideration is to understand the question being asked in the testing. If the objective is to relate the chemistry of a particular surface to its ability to resist fouling or inhibit QS, then the localization of the specific chemistry is essential. For example, if the organism's chemicals are internal and never reach the surface of a healthy plant, then that chemistry may not be involved in the inhibition of surface-colonizing organisms. As an example, we have shown that the QS-inhibiting compounds, furanones, produced by *Delisea pulchra* are secreted onto the surface of the plant via gland cells located near the surface where the compounds can interact with the surface bacteria (24). These experiments involved testing a variety of extraction protocols designed to remove the surface-associated molecules without causing significant lysis of the plant's epithelial cells, as determined by microscopy. The extracts were shown to contain the active compounds. Because furanones are fluorescent under UV exposure, it was also possible to examine the surface of the plant by fluorescence microscopy to identify gland cells in transverse and flat sections of the plant (24). On the other hand, if the experiments are based simply on the identification of inhibitory compounds without regard for their ecological roles, then the localization of compounds is not relevant and total extracts of plants can be used. Also note that, since almost all marine surfaces are covered by bacterial biofilms, when extractions are performed it is important to be able to attribute the metabolites to the correct source; do they come from the plant, or the surface-associated bacteria? This is essential for resampling of materials. If the bacteria contribute the specific chemistry, then during resampling it becomes important to ensure that the host tissue has the correct microbial community.

Chemical extraction of surfaces is generally performed by using an organic solvent or series of solvents that differ in their polarities. The identification of active compounds, inducers or antagonists, in a complex mixture can be accomplished by bioassay-guided fractionation of the extracts, using high-pressure liquid chromatography with an appropriate mobile phase, in which individual fractions are tested for activity in QS reporter bioassays. The active compounds are then subjected to structural elucidation via a combination of mass spectroscopy and one- and two-dimensional ^{1}H and ^{13}C nuclear magnetic resonance spectroscopy (e.g., COSY, NOESY, HSQC, HMBC, and INADEQUATE). When activity of either fractions or whole material is tested for, there are a number of options for bioassay systems that detect the presence of QS inducers or inhibitors (1, 5, 64, 76, 77). One rapid method for assessment has been to use whole material directly in an agar plate-based bioassay, such as *Chromobacterium violaceum*, and look for zones of induction or inhibition (77, 99). While not quantitative, the method is rapid to perform and can also separate growth inhibition of the reporter strain from QS inhibition effects. When whole extract, live material, or organic fractions are tested, it is essential that specific effects on gene regulation can be separated from growth inhibition of the monitor strains. As indicated above, there are a broad range of QS bioassays, strictly qualitative or semiquantitative, and chemical means for determination of signal and antagonist concentrations that are beyond the scope here but which have been reviewed extensively. It is important to choose the assay that fits the objectives of the study, and thus, a thorough understanding of the strengths and weaknesses of the assays is essential.

Conclusion

This pressure of bacterial QS-mediated colonization is not limited to marine plants and animals, but rather is also experienced by freshwater, soil, and terrestrial organisms, and hence it is not surprising to find that QS inhibitors can also be isolated from these sources (3, 97, 98). For example, compounds that stimulate as well as act as inhibitors have also been isolated from pea, crown vetch, and *Chlamydomonas reinhardtii*. The ability of plant and animal surfaces to generate such compounds suggests that there has been a strong evolutionary drive to allow these living substrata for bacterial colonization to play an active role in the recruitment of specific flora or the inhibition of settlement by antagonistic or pathogenic bacteria and therefore represent a potentially rich source of novel antagonists for the control of bacterial QS and thus bacterial colonization of surfaces.

INTERNAL DYNAMICS OF BIOFILMS: COLONIZATION AND COMPETITION

Colonization of Marine Algae

Bacteria are the primary colonizing organisms on surfaces and serve as a focus for the attachment and growth of other organisms within the marine environment, including complex fouling communities of invertebrates, sessile plants, and animals (20). Recent research has shown that the species composition of bacterial biofilms can play an important role in determining the nature of the fouling community that develops on marine surfaces. For example, it is thought that biofilm bacteria on living surfaces can benefit their host by the production of extracellular inhibitory compounds against fouling organisms (6, 44, 75). This is particularly important for certain marine algae, such as *Ulva lactuca*, which apparently do not possess physical or chemical defense mechanisms against fouling organisms. Thus, studies of the colonization and composition of bacterial biofilms on living algal surfaces are contributing exciting new insights into natural defense mechanisms against fouling organisms. Approaches and methods to study the colonization ecology of biofilms on the surface of the marine alga *Ulva lactuca* (sea lettuce) will be discussed here.

Experimental System to Study Algal Colonization

Marine bacterial strains of interest can be genetically marked with color tags by using expression plasmids that encode genes for fluorescent proteins. Fluorescent proteins allow colonization to be monitored in real time, without the use of destructive stains, by epifluorescence or confocal laser microscopy. We have successfully labeled a range of marine bacteria, using pCJS10 (75) with the green fluorescent protein gene *gfp*mut3 (14), derived from the broad-host-range vector pHRP304 (2). Other plamid-based expression vectors have also been described for the labeling of marine bacterial strains (89). For mixed-species biofilm experiments, multiple color tags can be used. For example, a red fluorescent protein color tag can be introduced into strains by using the pCJS10-derived plasmid pCJS10R. This plasmid contains the red fluorescent protein gene *dsred* (Clontech) in place of *gfp*mut3 on pCJS10 (75).

Methods for the handling and propagation of marine algae within the laboratory are provided in detail elsewhere (80). For studies of colonization and biofilm formation on plant surfaces, it is often necessary to remove the natural bacterial community from the plant surface. The method we have used to make *U. lactuca* axenic involves the excision of disks (1-cm diameter) from the algal thallus, followed by treatment of the disks with dilute NaOCl and a cocktail of antibiotics dissolved in autoclaved seawater (80). Such treatments can be harmful to the algal tissue; therefore, an assessment of the effect of each treatment regimen on plant tissue viability should be carried out. This can be done under light microscopy by staining treated *U. lactuca* disks with Evans blue (92). The blue dye penetrates into dead and damaged plant cells, while intact cells are able to exclude the dye. After the removal of the natural biofilm community, all manipulations and colonization experiments using algal disks are carried out. Labeled bacterial strains can be applied to the algal surface by immersing the disks in a bacterial suspension for a defined adhesion period. The disks can then be washed to remove unattached cells and transferred to fresh 24-well plates. Attachment and biofilm formation can then be monitored in situ by use of an inverted epifluorescence or confocal laser microscope equipped with the appropriate filter sets for visualization of fluorescent-protein tags.

Colonization Ecology of *Ulva lactuca*: Role of Antibacterial Compounds and Microcolonies

One important group of biofilm-forming marine bacteria that is often found in association with *U. lactuca* is *Pseudoalteromonas* spp. This genus produces a diverse range of biologically active compounds that specifically target marine fouling organisms (35, 42, 52, 53). For example, *Pseudoalteromonas tunicata* D2 produces at least five novel inhibitory compounds against fouling organisms, including the potent antibacterial protein AlpP, which is widely effective against gram-negative and gram-positive bacteria (25, 26, 43–45, 54). By use of fluorescent-protein color tags, studies of cocolonization and competition of *P. tunicata* and other marine bacteria on *U. lactuca* have demonstrated that inhibitory compounds produced by marine bacteria can provide a competitive advantage during growth within biofilms on the *U. lactuca* plant surface. Wild-type *P. tunicata* aggressively colonizes and dominates strains that are sensitive to the AlpP protein. In contrast, a *P. tunicata* AlpP mutant was defective in its ability to colonize and overtake biofilms under the same conditions. Moreover, marine strains that were tolerant to the AlpP protein were found to be recalcitrant to invasion and displacement by *P. tunicata* during biofilm development (75; D. Rao, J. S. Webb, and S. Kjelleberg, unpublished results). Competitive interactions between bacteria such as these may play an important role in determining the composition of such communities on algal surfaces.

Similar to other model biofilms, *P. tunicata* biofilms develop a complex architecture consisting of matrix-enclosed structures called microcolonies separated by a network of open water channels (60). Micro-colonies are often described as protective structures that enhance survival and competition in stressful environments (62, 104). Microcolonies within a biofilm can be composed of single-species populations or mixed populations with various degrees of interaction, depending on the environmental conditions under which they were formed. Using the experimental system described above, our studies and those of other groups have shown that competitive interactions between bacteria that produce antibacterial compounds often lead to biofilms with each strain forming its own microcolonies (75, 93). We have also shown that microcolony formation allows for enhanced persistence of bacterial strains during competition within biofilms. Thus, the extent and nature of microcolony formation during surface colonization may play an important role in survival and competition during the colonization of marine algal surfaces.

BIOFILM FORMATION BY *VIBRIO* SPP. ON LIVING SURFACES

Colonization of living surfaces by *Vibrio* spp. provides an excellent example of the role of biofilms in enhancing the survival and spread of bacteria within the environment. Studies of such colonization processes are important for the determination of factors affecting environmental persistence of bacteria, in particular pathogens responsible for human disease. The fact that many *Vibrio* spp. form viable but nonculturable cells in the environment also limits the use of culture techniques for the study of natural biofilms. This section will describe what is known about the attachment to and biofilm formation on living surfaces in the marine and estuarine environment by *Vibrio* spp.

Traditionally, epidemiologists considered the human intestine the only reservoir of *Vibrio cholerae* and believed that survival of *V. cholerae* in the environment was brief. Now it is known that this bacterium is present as a normal resident rather than as an occasional visitor (13) and that it is found mostly in biofilms formed on zooplankton. Laboratory studies have demonstrated that *V. cholerae* forms biofilms on copepod carapaces (23, 49, 50), and *V. cholerae*, *Vibrio parahaemolyticus*, and *Vibrio alginolyticus* have been found attached to copepods isolated from the environment (8, 48). Studies have shown that other plankton species are also colonized by *V. cholerae* (94), including cyanobacteria, diatoms, phaeophytes, and freshwater vascular aquatic plants (12, 28, 51, 84, 94). *V. cholerae* has also been cultured from adult winged chironomids (nonbiting midges) caught in the air 3 km from a water source, indicating the potential of these midges to act as a vehicle for transmission (7, 36). Techniques used to study natural biofilms on plankton isolated from the environment include culturing (23, 47, 48), fluorescent monoclonal antibody staining (48), and scanning electron microscopy (23, 49). For example, scanning electron microscopy indicated that *V. cholerae* preferentially attaches to swimming appendages of copepods (23). Studies using fluorescent monoclonal antibody staining detected *V. cholerae* in 64% of the plankton samples, while only 0.34% of the samples were positive by conventional culture methods, highlighting the fact that cells in environmental samples may be viable but nonculturable (48).

To better understand the role of biofilm formation in the environment, some studies have focused on the mechanism of attachment to living surfaces. Lectins with specificity to *N*-acetylglucosamine (GlcNAc), the sugar component of chitin, occur in *V. cholerae*, *Vibrio harveyi*, and *Vibrio furnissii* (66, 74, 110). Treatment of *V. cholerae* with pronase E reduced attachment to chitin by 70% (95), indicating that chitin binding proteins are present on the cell surface and are important for attachment. The mannose-sensitive hemagglutinin (MSHA) pilus of *V. cholerae* may play a specific role in attachment and biofilm formation in the environment, as a mutant strain of *V. cholerae* O1 El Tor lacking the MSHA pilus was significantly defective in adherence to the planktonic crustacean *Daphnia*, while a similar *V. cholerae* O139 mutant lost all ability to adhere (10). In contrast, adherence of an MSHA mutant of a *V. cholerae* O1 classical strain was unaffected, indicating that classical strains adhere by another mechanism. For both wild-type and MSHA mutant strains, the density of attached cells was approximately a log higher in nutrient-limited medium than in Luria-Bertani medium, indicating that nutrient availability may be one factor affecting biofilm formation. We have seen similar results in laboratory studies of attachment of *V. cholerae* and *Vibrio vulnificus* to inanimate surfaces using the microplate assay (D. McDougald, unpublished observations), indicating that low nutrient concentrations may be a general inducer of biofilm formation.

Biofilm formation on the surface of plankton may allow for protection of bacteria against numerous stresses encountered in the environment. Laboratory-based experiments show that chitin protects *V. cholerae* O1 from acid stress of the level encountered in the stomach (69). Alum, which is used to purify drinking water in many developing countries, induces loss of culturability in free-living *V. cholerae* cells, but not when the bacteria are attached to the surface of copepods, indicating that this treatment may not be sufficient for treatment of drinking water (69). In the environment, consumption by protozoa is considered to be a major source of bacterial mortality and as such would be expected to be a driving force in the evolution of protective phenotypes for bacteria. Matz et al. have recently demonstrated that the presence of protozoan grazers does indeed induce biofilm formation by *V. cholerae* (62). Furthermore, it was shown that biofilm cells produce an extracellular product with antipredator activity.

Probably the best-studied colonization interaction with marine *Vibrio* spp. is the symbiosis between *Vibrio fischeri* and the Hawaiian squid, *Eurprymna scolopes*. In this association, *V. fischeri* colonizes the light organ of the squid within hours of hatching (101). Within the first hours to days of symbiosis, *V. fischeri* cells become intimately associated with the crypt epithelial cells, and eventually nearly every bacterium is in contact with a host cell. This interaction may involve adhesin-glycan interactions, as indicated by the following observations: (i) mannose is the most common glycan along membranes of crypt brush border (65), (ii) addition of mannose to seawater completely inhibited initiation of symbiotic colonization, and (iii) *V. fischeri* agglutinates guinea pig red blood cells in a mannose-sensitive reaction. This type of interaction is similar to the colonization of eukaryotic host cells by pathogens.

There are many other associations of vibrios with other organisms in the environment which are likely to involve biofilm formation. For example, *V. harveyi* is known to colonize prawns and *V. anguillarum* colonizes salmonid fish, and both are responsible for high mortality rates in farm-reared stock. In addition, *V. vulnificus* is known to colonize oysters and is not easily removed by depuration, indicating that the association is not simply via ingestion of the bacteria by the oyster. Thus, biofilm formation by *Vibrio* spp. is of general interest from a medical point of view, and for these reasons it is important to understand reservoirs for their survival, mechanisms of dissemination, and resistance to environmental stresses such as grazing pressure. The mechanisms that underpin the process of biofilm formation are of interest as well as the full distribution of surfaces colonized by this important group of bacteria. Lastly, the colonization phenotypes that are important for attachment to surfaces in the environment have probably led to the ability of these bacteria to colonize the host and thus contribute to pathogenicity or to symbiosis, as is the case with *V. cholerae* and *V. fischeri*, respectively.

Clearly, the implication of the association between *V. cholerae* and plankton for the epidemiology of cholera outbreaks is worthy of more investigation. The studies highlighted here provide indications that biofilm formation on the surface of plankton may lead to increased survival of bacteria in the environment and may provide protection against mortality factors such as protozoan grazing. Thus, evolution of adaptive mechanisms, such as biofilm formation, has led to phenotypes such as pathogenesis and QS.

REFERENCES

1. **Andersen, J. B., A. Heydorn, M. Hentzer, L. Eberl, O. Geisenberger, B. Christensen, S. Molin, and M. Givskov.** 2001. *gfp*-based *N*-acyl homoserine lactone sensor systems for detection of bacterial communication. *Appl. Environ. Microbiol.* **67:**575–585.
2. **Bagdasarian, M., R. Lurz, B. Ruckert, F. C. Franklin, M. M. Bagdasarian, J. Frey, and K. N. Timmis.** 1981. Specific-purpose plasmid cloning vectors. II. Broad host range, high copy number, RSF1010-derived vectors, and a host-vector system for gene cloning in *Pseudomonas*. *Gene* **16:**237–247.
3. **Bauer, W. D., and M. Teplitski.** 2001. Can plants manipulate bacterial quorum sensing? *Aust. J. Plant Physiol.* **28:**913–921.
4. **Bewley, C. A., N. D. Holland, and D. J. Faulkner.** 1996. Two classes of metabolites from *Theonella swinhoei* are localized in distinct populations of bacterial symbionts. *Experientia* **52:**716–722.
5. **Blosser, R. S., and K. M. Gray.** 2000. Extraction of violacein from *Chromobacterium violaceum* provides a new quantitative bioassay for *N*-acyl homoserine lactone autoinducers. *J. Microbiol. Methods* **40:**47–55.
6. **Boyd, K. G., D. R. Adams, and J. G. Burgess.** 1999. Antibacterial and repellent activities of marine bacteria associated with algal surfaces. *Biofouling* **14:**227–236.
7. **Broza, M., H. Gancz, M. Halpern, and Y. Kashi.** 2005. Adult non-biting midges: possible windborne carriers of *Vibrio cholerae* non-O1 non-O139. *Environ. Microbiol.* **7:** 576–585.
8. **Carli, A., L. Pane, S. Bertone, and C. Pruzzo.** 1993. Occurrence of *Vibrio alginolyticus* in Ligurian Coast rock pools (Tyrrhenian Sea, Italy) and its association with the copepod *Tigriopus fulvus* (Fisher 1860). *Appl. Environ. Microbiol.* **59:**1960–1962.
9. **Chen, K., and L. Pachter.** 2005. Bioinformatics for whole-genome shotgun sequencing of microbial communities. *PLoS Comput. Biol.* **1:**e2.
10. **Chiavelli, D. A., J. W. Marsh, and R. K. Taylor.** 2001. The mannose-sensitive hemagglutinin of *Vibrio cholerae* promotes adherence to zooplankton. *Appl. Environ. Microbiol.* **67:**3220–3225.
11. **Clarke, K. R., and R. M. Warwick.** 1994. *Change in Marine Communities: an Approach to Statistical Analysis and Interpretation.* Plymouth Marine Laboratory, Plymouth, United Kingdom.
12. **Cockburn, T. A., and J. G. Cassanos.** 1960. Epidemiology of endemic cholera. *Public Health Rep.* **75:**791–803.
13. **Colwell, R. R., R. J. Seidler, J. Kaper, S. W. Joseph, S. Garges, H. Lockman, D. Maneval, H. Bradford, N. Roberts, E. Remmers, I. Huq, and A. Huq.** 1981. Occurrence of *Vibrio cholerae* serotype O1 in Maryland and Louisiana estuaries. *Appl. Environ. Microbiol.* **41:** 555–558.
14. **Cormack, B. P., R. H. Valdivia, and S. Falkow.** 1996. FACS-optimized mutants of the green fluorescent protein (GFP). *Gene* **173:**33–38.
15. **Cosson, P., L. Zulianello, O. Join-Lambert, F. Faurisson, L. Gebbie, M. Benghezal, C. van Delden, L. K. Curty, and T. Kohler.** 2002. *Pseudomonas aeruginosa* virulence analyzed in a *Dictyostelium discoideum* host system. *J. Bacteriol.* **184:**3027–3033.
16. **Cowan, D., A. Arslanoglu, S. Burton, G. Baker, R. Cameron, J. Smith, and Q. Meyer.** 2004. Metagenomics, gene discovery and the ideal biocatalyst. *Biochem. Soc. Trans.* **32:**298–302.
17. **Daims, H., S. Lucker, and M. Wagner.** 2006. daime, a novel image analysis program for microbial ecology and biofilm research. *Environ. Microbiol.* **8:**200–213.
18. **Daims, H., N. B. Ramsing, K. H. Schleifer, and M. Wagner.** 2001. Cultivation-independent, semiautomatic determination of absolute bacterial cell numbers in environmental samples by fluorescence in situ hybridization. *Appl. Environ. Microbiol.* **67:**5810–5818.
19. **Davies, D. G., M. R. Parsek, J. P. Pearson, B. H. Iglewski, J. W. Costerton, and E. P. Greenberg.** 1998. The involvement of cell-cell signals in the development of a bacterial biofilm. *Science* **280:**295–298.
20. **Davis, A. R., N. M. Targett, O. J. McConnel, and C. M. Young.** 1989. Epibiosis of marine algae and benthic invertebrates: natural products chemistry and other mechanisms inhibiting settlement and overgrowth. *Bioorg. Mar. Chem.* **3:**85–114.
21. **DeLong, E. F.** 1997. Marine microbial diversity: the tip of the iceberg. *Trends Biotechnol.* **15:**203–207.
22. **deNys, R., A. D. Wright, G. M. Konig, and O. Sticher.** 1993. New halogenated furanones from the marine alga *Delisea pulchra* (cf fimbriata). *Tetrahedron* **49:**11213–11220.
23. **Dumontet, S., K. Krovacek, S. B. Baloda, R. Grottoli, V. Pasquale, and S. Vanucci.** 1996. Ecological relationship between *Aeromonas* and *Vibrio* spp. and planktonic copepods in the coastal marine environment in Southern Italy. *Comp. Immunol. Microbiol. Infect. Dis.* **19:**245–254.
24. **Dworjanyn, S. A., R. de Nys, and P. D. Steinberg.** 1999. Localisation and surface quantification of secondary metabolites in the red alga *Delisea pulchra*. *Mar. Biol.* **133:** 727–736.
25. **Egan, S., S. James, C. Holmstrom, and S. Kjelleberg.** 2002. Correlation between pigmentation and antifouling compounds produced by *Pseudoalteromonas tunicata*. *Environ. Microbiol.* **4:**433–442.
26. **Egan, S., S. James, C. Holmstrom, and S. Kjelleberg.** 2001. Inhibition of algal spore germination by the marine bacterium *Pseudoalteromonas tunicata*. *FEMS Microbiol. Ecol.* **35:**67–73.
27. **Elkins, J. G., D. J. Hassett, P. S. Stewart, H. P. Schweizer, and T. R. McDermott.** 1999. Protective role of catalase in *Pseudomonas aeruginosa* biofilm resistance to hydrogen peroxide. *Appl. Environ. Microbiol.* **65:**4594–4600.
28. **Epstein, P. R.** 1993. Algal blooms in the spread and persistence of cholera. *BioSystems* **31:**209–211.
29. **Field, D., B. Tiwari, and J. Snape.** 2005. Bioinformatics and data management support for environmental genomics. *PLoS Biol.* **3:**e297.
30. **Fieseler, L., M. Horn, M. Wagner, and U. Hentschel.** 2004. Discovery of the novel candidate phylum "Poribacteria" in marine sponges. *Appl. Environ. Microbiol.* **70:**3724–3732.
31. **Fisher, M. M., and E. W. Triplett.** 1999. Automated approach for ribosomal intergenic spacer analysis of microbial diversity and its application to freshwater bacterial communities. *Appl. Environ. Microbiol.* **65:**4630–4636.
32. **Frias-Lopez, J., A. L. Zerkle, G. T. Bonheyo, and B. W. Fouke.** 2002. Partitioning of bacterial communities between seawater and healthy, black band diseased, and dead coral surfaces. *Appl. Environ. Microbiol.* **68:**2214–2228.
33. **Friedrich, A. B., H. Merkert, T. Fendert, J. Hacker, P. Proksch, and U. Hentschel.** 1999. Microbial diversity in

the marine sponge *Aplysina cavernicola* (formerly *Verongia cavernicola*) analyzed by fluorescence in situ hybridization (FISH). *Mar. Biol.* **134:**461–470.
34. **Fuqua, C., and E. P. Greenberg.** 2002. Listening in on bacteria: acyl-homoserine lactone signalling. *Nat. Rev. Mol. Cell. Biol.* **3:**685–695.
35. **Gauthier, M. J., and G. N. Flatau.** 1976. Antibacterial activity of marine violet pigmented *Alteromonas* with special reference to the production of brominated compounds. *Can. J. Microbiol.* **22:**349–354.
36. **Halpern, M., Y. B. Broza, S. Mittler, E. Arakawa, and M. Broza.** 2004. Chironomid egg masses as a natural reservoir of *Vibrio cholerae* non-O1 and non-O139 in freshwater habitats. *Microb. Ecol.* **47:**341–349.
37. **Handelsman, J.** 2004. Metagenomics: application of genomics to uncultured microorganisms. *Microbiol. Mol. Biol. Rev.* **68:**669–685.
38. **Hentschel, U., L. Fieseler, M. Wehrl, C. Gernert, M. Steinert, J. Hacker, and M. Horn.** 2003. Microbial diversity of marine sponges. *Prog. Mol. Subcell. Biol.* **37:**59–88.
39. **Hentschel, U., J. Hopke, M. Horn, A. B. Friedrich, M. Wagner, J. Hacker, and B. S. Moore.** 2002. Molecular evidence for a uniform microbial community in sponges from different oceans. *Appl. Environ. Microbiol.* **68:**4431–4440.
40. **Hentzer, M., L. Eberl, J. Nielsen, and M. Givskov.** 2003. Quorum sensing: a novel target for the treatment of biofilm infections. *Drug Dev.* **17:**241–250.
41. **Hoiby, N., H. K. Johansen, C. Moser, Z. J. Song, O. Ciofu, and A. Kharazmi.** 2001. *Pseudomonas aeruginosa* and the *in vitro* and *in vivo* biofilm mode of growth. *Microb. Infect.* **3:**23–35.
42. **Holmstrom, C., S. Egan, A. Franks, S. McCloy, and S. Kjelleberg.** 2002. Antifouling activities expressed by marine surface associated *Pseudoalteromonas* species. *FEMS Microbiol. Ecol.* **41:**47–58.
43. **Holmstrom, C., S. James, B. A. Neilan, D. C. White, and S. Kjelleberg.** 1998. *Pseudoalteromonas tunicata* sp. nov., a bacterium that produces antifouling agents. *Int. J. Syst. Evol. Microbiol.* **48:**1205–1212.
44. **Holmstrom, C., and S. Kjelleberg.** 1994. The effect of external biological factors on settlement of marine invertebrates and new antifouling technology. *Biofouling* **8:**147–160.
45. **Holmstrom, C., D. Rittschof, and S. Kjelleberg.** 1992. Inhibition of settlement by larvae of *Balanus amphitrite* and *Ciona intestinalis* by a surface-colonizing marine bacterium. *Appl. Environ. Microbiol.* **58:**2111–2115.
46. **Huber, B., K. Riedel, M. Hentzer, A. Heydorn, A. Gotschlich, M. Givskov, S. Molin, and L. Eberl.** 2001. The *cep* quorum-sensing system of *Burkholderia cepacia* H111 controls biofilm formation and swarming motility. *Microbiology* **147:**2517–2528.
47. **Huq, A., R. R. Colwell, M. A. Chowdhury, B. Xu, S. M. Moniruzzaman, M. S. Islam, M. Yunus, and M. J. Albert.** 1995. Coexistence of *Vibrio cholerae* O1 and O139 Bengal in plankton in Bangladesh. *Lancet* **345:**1249.
48. **Huq, A., R. R. Colwell, R. Rahman, A. Ali, M. A. R. Chowdhury, S. Parveen, D. A. Sack, and E. Russek-Cohen.** 1990. Detection of *Vibrio cholerae* O1 in the aquatic environment by fluorescent-monoclonal antibody and culture methods. *Appl. Environ. Microbiol.* **56:**2370–2373.
49. **Huq, A., E. B. Small, P. A. West, and R. R. Colwell.** 1984. The role of planktonic copepods in the survival and multiplication of *Vibrio cholerae* in the aquatic environment, p. 521–534. *In* R. R. Colwell (ed.), *Vibrios in the Environment.* John Wiley & Sons, Inc., New York, N.Y.
50. **Huq, A., E. B. Small, P. A. West, M. I. Huq, R. Rahman, and R. R. Colwell.** 1983. Ecological relationships between *Vibrio cholerae* and planktonic crustacean copepods. *Appl. Environ. Microbiol.* **45:**275–283.
51. **Islam, M. S., Z. Rahim, M. J. Alam, S. Begum, S. M. Moniruzzaman, A. Umeda, K. Amako, M. J. Albert, R. B. Sack, A. Huq, and R. R. Colwell.** 1999. Association of *Vibrio cholerae* O1 with the cyanobacterium, *Anabaena* sp., elucidated by polymerase chain reaction and transmission electron microscopy. *Trans. R. Soc. Trop. Med. Hyg.* **93:**36.
52. **Isnansetyo, A., and Y. Kamei.** 2003. MC21-A, a bactericidal antibiotic produced by a new marine bacterium, *Pseudoalteromonas phenolica* sp. nov. O-BC30(T), against methicillin-resistant *Staphylococcus aureus. Antimicrob. Agents Chemother.* **47:**480–488.
53. **Ivanova, E. P., T. Sawabe, Y. V. Alexeeva, A. M. Lysenko, N. M. Gorshkova, K. Hayashi, N. V. Zukova, R. Christen, and V. V. Mikhailov.** 2002. *Pseudoalteromonas issachenkonii* sp. nov., a bacterium that degrades the thallus of the brown alga *Fucus evanescens. Int. J. Syst. Evol. Microbiol.* **52:**229–234.
54. **James, S., C. Holmstrom, and S. Kjelleberg.** 1996. Purification and characterization of a novel antibacterial protein from the marine bacterium D2. *Appl. Environ. Microbiol.* **62:**2783–2788.
55. **Johansen, C., P. Falholt, and L. Gram.** 1997. Enzymatic removal and disinfection of bacterial biofilms. *Appl. Environ. Microbiol.* **63:**3724–3728.
56. **Labbate, M., S. Y. Queck, K. S. Koh, S. A. Rice, M. Givskov, and S. Kjelleberg.** 2004. Quorum sensing-controlled biofilm development in *Serratia liquefaciens* MG1. *J. Bacteriol.* **186:**692–698.
57. **Loy, A., A. Lehner, N. Lee, J. Adamczyk, H. Meier, J. Ernst, K. H. Schleifer, and M. Wagner.** 2002. Oligonucleotide microarray for 16S rRNA gene-based detection of all recognized lineages of sulphate-reducing prokaryotes in the environment. *Appl. Environ. Microbiol.* **68:**5064–5081.
58. **Loy, A., C. Schulz, S. Lucker, A. Schopfer-Wendels, K. Stoecker, C. Baranyi, A. Lehner, and M. Wagner.** 2005. 16S rRNA gene-based oligonucleotide microarray for environmental monitoring of the betaproteobacterial order "Rhodocyclales." *Appl. Environ. Microbiol.* **71:**1373–1386.
59. **Lynch, M. J., S. Swift, D. F. Kirke, C. W. Keevil, C. E. Dodd, and P. Williams.** 2002. The regulation of biofilm development by quorum sensing in *Aeromonas hydrophila. Environ. Microbiol.* **4:**18–28.
60. **Mai-Prochnow, A., F. Evans, D. Dalisay-Saludes, S. Stelzer, S. Egan, S. James, J. S. Webb, and S. Kjelleberg.** 2004. Biofilm development and cell death in the marine bacterium *Pseudoalteromonas tunicata. Appl. Environ. Microbiol.* **70:**3232–3238.
61. **Manly, B. F. J.** 2004. *Multivariate Statistical Methods: a Primer*, 3rd ed. Chapman & Hall/CRC, London, United Kingdom.
62. **Matz, C., D. McDougald, A. M. Moreno, P. Y. Yung, F. H. Yildiz, and S. Kjelleberg.** 2005. Biofilm formation and phenotypic variation enhance predation-driven persistence of *Vibrio cholerae. Proc. Natl. Acad. Sci. USA* **102:**16819–16824.
63. **Maximilien, R., R. de Nys, C. Holmstrom, L. Gram, M. Givskov, K. Crass, S. Kjelleberg, and P. D. Steinberg.** 1998. Chemical mediation of bacterial surface colonisation by secondary metabolites from the red alga *Delisea pulchra. Aquat. Microb. Ecol.* **15:**233–246.
64. **McClean, K. H., M. K. Winson, L. Fish, A. Taylor, S. R. Chhabra, M. Camara, M. Daykin, J. H. Lamb, S. Swift, B. W. Bycroft, G. Stewart, and P. Williams.** 1997. Quorum sensing and *Chromobacterium violaceum:* exploitation of

violacein production and inhibition for the detection of *N*-acylhomoserine lactones. *Microbiology* **143:**3703–3711.
65. **McFall-Ngai, M. J., C. Brennan, V. Weis, and L. Lamarcq.** 1998. Mannose adhesion-glucan interactions in the *Euprymna scolopes-Vibrio fischeri* symbiosis, p. 273–276. *In* Y. Le Gal and H. O. Halvorson (ed.), *New Developments in Marine Biotechnology*. Plenum Press, New York, N.Y.
66. **Montgomery, M. T., and D. L. Kirchman.** 1993. Role of chitin-binding proteins in the specific attachment of the marine bacterium *Vibrio harveyi* to chitin. *Appl. Environ. Microbiol.* **59:**373–379.
67. **Morris, C. E., M. Bardin, O. Berge, P. Frey-Klett, N. Fromin, H. Girardin, M. H. Guinebretiere, P. Lebaron, J. M. Thiery, and M. Troussellier.** 2002. Microbial biodiversity: approaches to experimental design and hypothesis testing in primary scientific literature from 1975 to 1999. *Microbiol. Mol. Biol. Rev.* **66:**592–616.
68. **Muyzer, G., and K. Smalla.** 1998. Application of denaturing gradient gel electrophoresis (DGGE) and temperature gradient gel electrophoresis (TGGE) in microbial ecology. *Antonie Leeuwenhock* **73:**127–141.
69. **Nalin, D. R., V. Daya, A. Reid, M. M. Levine, and L. Cisneros.** 1979. Adsorption and growth of *Vibrio cholerae* on chitin. *Infect. Immun.* **25:**768–770.
70. **Nubel, U., F. Garcia-Pichel, E. Clavero, and G. Muyzer.** 2000. Matching molecular diversity and ecophysiology of benthic cyanobacteria and diatoms in communities along a salinity gradient. *Environ. Microbiol.* **2:**217–226.
71. **Osborn, A. M., E. R. Moore, and K. N. Timmis.** 2000. An evaluation of terminal-restriction fragment length polymorphism (T-RFLP) analysis for the study of microbial community structure and dynamics. *Environ. Microbiol.* **2:** 39–50.
72. **Pearson, J. P., M. Feldman, B. H. Iglewski, and A. Prince.** 2000. *Pseudomonas aeruginosa* cell-to-cell signaling is required for virulence in a model of acute pulmonary infection. *Infect. Immun.* **68:**4331–4334.
73. **Polz, M. F., and C. M. Cavanaugh.** 1998. Bias in template-to-product ratios in multitemplate PCR. *Appl. Environ. Microbiol.* **64:**3724–3730.
74. **Pruzzo, C., R. Tarsi, M. del Mar Lleo, C. Signoretto, M. Zampini, L. Pane, R. R. Colwell, and P. Canepari.** 2003. Persistence of adhesive properties in *Vibrio cholerae* after long-term exposure to sea water. *Environ. Microbiol.* **5:** 850–858.
75. **Rao, D., J. S. Webb, and S. Kjelleberg.** 2005. Competitive interactions in mixed-species biofilms containing the marine bacterium *Pseudoalteromonas tunicata*. *Appl. Environ. Microbiol.* **71:**1729–1736.
76. **Ravn, L., A. B. Christensen, S. Molin, M. Givskov, and L. Gram.** 2001. Methods for detecting acylated homoserine lactones produced by Gram-negative bacteria and their application in studies of AHL-production kinetics. *J. Microbiol. Methods* **44:**239-251.
77. **Rice, S. A., S. Kjelleberg, M. Givskov, W. De Boer, and L. Chernin.** 2004. *In situ* detection of bacterial quorum sensing signal molecules, p. 1629–1650. *In* G. A. Kowalchuck (ed.), *Molecular Microbial Ecology Manual*. Kluwer Academic Publishers, Dordrecht, The Netherlands.
78. **Rice, S. A., D. McDougald, N. Kumar, and S. Kjelleberg.** 2005. The use of quorum sensing blockers as therapeutic agents for the control of biofilm associated infections. *Curr. Opin. Investig. Drugs* **6:**178–184.
79. **Riesenfeld, C., P. Schloss, and J. Handelsman.** 2004. Metagenomics: genomic analysis of microbial communities. *Annu. Rev. Genet.* **38:**525–552.
80. **Scheffel, A.** 2003. Protocol for generating axenic *Ulva lactuca* and studies of type IV pili mediated attachment of *Pseudoalteromonas tunicata* to the algal surface. Diploma thesis. University of New South Wales, Sydney, Australia.
81. **Schirmer, A., R. Gadkari, C. Reeves, F. Ibrahim, E. Delong, and C. Hutchinson.** 2005. Metagenomic analysis reveals diverse polyketide synthase gene clusters in microorganisms associated with the marine sponge *Discodermia dissoluta*. *Appl. Environ. Microbiol.* **71:**4840–4849.
82. **Schleper, C., E. DeLong, C. Preston, R. Feldman, K. Wu, and R. Swanson.** 1998. Genomic analysis reveals chromosomal variation in natural populations of the uncultured psychrophilic archaeon *Cenarchaeum symbiosum*. *J. Bacteriol.* **180:**5003–5009.
83. **Schonhuber, W., B. Zarda, S. Eix, R. Rippka, M. Herdman, W. Ludwig, and R. Amann.** 1999. In situ identification of cyanobacteria with horseradish peroxidase-labeled, rRNA-targeted oligonucleotide probes. *Appl. Environ. Microbiol.* **65:**1259–1267.
84. **Shukla, B. N., D. V. Singh, and S. C. Sanyal.** 1995. Attachment of non-culturable toxigenic *Vibrio cholerae* O1 and non-O1 and *Aeromonas* spp. to the aquatic arthropod *Gerris spinolae* and plants in the River Ganga, Varanasi. *FEMS Immunol. Med. Microbiol.* **12:**113.
85. **Skovhus, T. L., N. B. Ramsing, C. Holmstrom, S. Kjelleberg, and I. Dahllof.** 2004. Real-time quantitative PCR for assessment of abundance of *Pseudoalteromonas* species in marine samples. *Appl. Environ. Microbiol.* **70:**2373–2382.
86. **Smith, J. L., P. M. Fratamico, and J. S. Novak.** 2004. Quorum sensing: a primer for food microbiologists. *J. Food Prot.* **67:**1053–1070.
87. **Smith, R. S., and B. H. Iglewski.** 2003. *Pseudomonas aeruginosa* quorum sensing as a potential antimicrobial target. *J. Clin. Investig.* **112:**1460–1465.
88. **Steinberg, P. D., R. De Nys, and S. Kjelleberg.** 2002. Chemical cues for surface colonization. *J. Chem. Ecol.* **28:**1935–1951.
89. **Stretton, S., S. Techkarnjanaruk, A. M. McLennan, and A. E. Goodman.** 1998. Use of green fluorescent protein to tag and investigate gene expression in marine bacteria. *Appl. Environ. Microbiol.* **64:**2554–2559.
90. **Sutherland, I.** 2001. The biofilm matrix—an immobilized but dynamic microbial environment. *Trends Microbiol.* **9:**222–227.
91. **Suzuki, M. T., and S. J. Giovannoni.** 1996. Bias caused by template annealing in the amplification of mixtures of 16S rRNA genes by PCR. *Appl. Environ. Microbiol.* **62:**625–630.
92. **Swain, D., and D. N. De.** 1994. Vital staining, a technique for rapid screening of plant protoplast viability. *Ind. J. Exp. Biol.* **32:**501–506.
93. **Tait, K., and I. W. Sutherland.** 2002. Antagonistic interactions amongst bacteriocin-producing enteric bacteria in dual species biofilms. *J. Appl. Microbiol.* **93:**345–352.
94. **Tamplin, M. L., A. L. Gauzens, A. Huq, D. A. Sack, and R. R. Colwell.** 1990. Attachment of *Vibrio cholerae* serogroup O1 to zooplankton and phytoplankton of Bangladesh waters. *Appl. Environ. Microbiol.* **56:**1977–1980.
95. **Tarsi, R., and C. Pruzzo.** 1999. Role of surface proteins in *Vibrio cholerae* attachment to chitin. *Appl. Environ. Microbiol.* **65:**1348–1351.
96. **Taylor, M. W., P. J. Schupp, I. Dahllof, S. Kjelleberg, and P. D. Steinberg.** 2004. Host specificity in marine sponge-associated bacteria, and potential implications for marine microbial diversity. *Environ. Microbiol.* **6:**121–130.
97. **Teplitski, M., H. Chen, S. Rajamani, M. Gao, M. Merighi, R. T. Sayre, J. B. Robinson, B. G. Rolfe, and W. D. Bauer.** 2004. *Chlamydomonas reinhardtii* secretes compounds that mimic bacterial signals and interfere with

quorum sensing regulation in bacteria. *Plant Physiol.* **134:**137–146.

98. **Teplitski, M., J. B. Robinson, and W. D. Bauer.** 2000. Plants secrete substances that mimic bacterial *N*-acyl homoserine lactone signal activities and affect population density-dependent behaviors in associated bacteria. *Mol. Plant-Microbe Interact.* **13:**637–648.
99. **Tyson, G. W., J. Chapman, P. Hugenholtz, E. E. Allen, R. J. Ram, P. M. Richardson, V. V. Solovyev, E. M. Rubin, D. S. Rokhsar, and J. F. Banfield.** 2004. Community structure and metabolism through reconstruction of microbial genomes from the environment. *Nature* **428:**37–43.
100. **Venter, J., K. Remington, J. F. Heidelberg, A. L. Halpern, D. Rusch, J. A. Eisen, D. Wu, I. Paulsen, K. E. Nelson, W. Nelson, D. E. Fouts, S. Levy, A. H. Knap, M. W. Lomas, K. Nealson, O. White, J. Peterson, J. Hoffman, R. Parsons, H. Baden-Tillson, C. Pfannkoch, Y. H. Rogers, and H. O. Smith.** 2004. Environmental genome shotgun sequencing of the Sargasso Sea. *Science* **304:**66–74.
101. **Visick, K. L., and M. McFall-Ngai.** 2000. An exclusive contract: specificity in the *Vibrio fischeri-Euprymna scolopes* partnership. *J. Bacteriol.* **182:**1779–1787.
102. **Wackett, L. P.** 2005. Web alert metagenomics: an annotated selection of World Wide Web sites relevant to topics in Environmental Microbiology. *Environ. Microbiol.* **7:**2034.
103. **Wang, X., J. F. Preston III, and T. Romeo.** 2004. The *pgaABCD* locus of *Escherichia coli* promotes the synthesis of a polysaccharide adhesin required for biofilm formation. *J. Bacteriol.* **186:**2724–2734.
104. **Webb, J. S., M. Givskov, and S. Kjelleberg.** 2003. Bacterial biofilms: prokaryotic adventures in multicellularity. *Curr. Opin. Microbiol.* **6:**578–585.
105. **Webster, N. S., K. J. Wilson, L. L. Blackall, and R. T. Hill.** 2001. Phylogenetic diversity of bacteria associated with the marine sponge *Rhopaloeides odorabile*. *Appl. Environ. Microbiol.* **67:**434–444.
106. **Weidner, S., W. Arnold, and A. Puhler.** 1996. Diversity of uncultured microorganisms associated with the seagrass *Halophila stipulacea* estimated by restriction fragment length polymorphism analysis of PCR-amplified 16S rRNA genes. *Appl. Environ. Microbiol.* **62:**766–771.
107. **Weidner, S., W. Arnold, E. Stackebrandt, and A. Puhler.** 2000. Phylogenetic analysis of bacterial communities associated with leaves of the seagrass *Halophila stipulacea* by a culture-independent small-subunit rRNA gene approach. *Microb. Ecol.* **39:**22–31.
108. **Whitchurch, C. B., T. Tolker-Nielsen, P. C. Ragas, and J. S. Mattick.** 2002. Extracellular DNA required for bacterial biofilm formation. *Science* **295:**1487.
109. **Winans, S. C., and B. L. Bassler.** 2002. Mob psychology. *J. Bacteriol.* **184:**873–883.
110. **Yu, C., A. M. Lee, and S. Roseman.** 1987. The sugar-specific adhesion/deadhesion apparatus of the marine bacterium *Vibrio furnissii* is a sensorium that continuously monitors nutrient levels in the environment. *Biochem. Biophys. Res. Commun.* **149:**86–92.

Extreme High-Pressure Marine Environments

JODY W. DEMING

46

ENVIRONMENTS CONSIDERED

Aquatic microbiology has a long history permeated by the study of microorganisms from familiar local habitats under conditions considered "mild"—room temperature, atmospheric pressure, neutral pH, moderate salinity (from freshwater to seawater), and often, nonlimiting nutrients and electron acceptors. In contrast, microorganisms inhabiting extreme and often remote aquatic environments, from anoxic pockets within an exotic invertebrate to submarine hydrothermal vents and freezing polar waters, have more often than not been considered esoterica. The first book to focus attention on the subject, by assembling available information from selected extreme environments into one place, was published 26 years ago (86), with the chapters written prior to the discovery of hydrothermal vents or confirmation that cold-adapted barophilic (pressure-loving) bacteria inhabit the deep sea (168). A sea change in appreciation of all manner of "extremophiles" has occurred since then, as evidenced by the emergence of journals dedicated to these organisms and their wide-ranging implications to fields as diverse as biotechnology, climate change, and astrobiology. The observation that the vast majority of Earth's microorganisms reside in extreme aquatic environments, given the immense volume of the deep ocean, is a simple fact (158). Some combination of extreme temperature, pressure, food supply, acidity, redox conditions, or even water activity describes the norm, not the rarity, for aquatic microbial habitats.

This chapter provides some history, current trends, and specific examples of strategies and experimental protocols for studying microorganisms (*Bacteria* and *Archaea*) that inhabit the largest volume of extreme (or any inhabited) environment on the planet, the pressurized deep ocean and its subseafloor realm. The pressure class of microbe addressed most frequently is the barophilic group, composed of organisms that require elevated hydrostatic pressures to achieve optimal growth rates (see the tabulations in references 10, 38, and 56). Barophiles can also grow (more slowly) at atmospheric pressure, but obligate barophiles expire at atmospheric pressure. Barotolerant organisms (*Escherichia coli* is an example) achieve optimal growth at atmospheric pressure but, nevertheless, can grow under higher pressures. The term barophile (and its variants) is increasingly supplanted by "piezophile" (also meaning pressure-loving), based on linguistic arguments made by Yayanos (166). Continued use of the term barophile nods to history and ready understanding ("bar" is a common unit for pressure) while honoring Claude ZoBell and Richard Morita, who first reported the existence of barophiles in the cold deep sea nearly 50 years ago (173).

Because temperature determines the pressure range for microbial growth and other activities, temperature issues appear throughout this chapter. Approaches for studying both psychrophilic (optimal growth at ≤15°C and maximal growth at ≤20°C [102]) and thermophilic, particularly hyperthermophilic (optimal growth at ≥80°C and maximal growth at ≥90°C [7]), microorganisms from the deep sea, with an emphasis on natural communities, are considered. Sampling and experimental information is provided according to the following two broadly defined extreme temperature and pressure regimens in the deep sea: (i) the cold abyssal and hadal zones of the ocean, where cold means temperatures of <4°C, abyssal means depths of >3.7 km (the average depth of the global ocean), and hadal means depths of >6 km (as in island-arc trenches); and (ii) the hot habitats associated with deep-sea hydrothermal vent systems, where hot means temperatures of >90°C and deep sea means depths of >2 km. Some advances made recently at shallower depths (1 to 2 km) are also addressed when they bear on the more extreme habitats. The cold and hot deep-sea regimens have in common elevated hydrostatic pressure, which increases approximately 10^7 Pa for every kilometer increase in water depth. It increases at a greater rate below the seafloor as a result of lithostatic pressure in the subsurface biosphere (7), a relatively inaccessible extreme environment that is proving highly significant as a biosphere, both volumetrically and scientifically (9, 24, 113).

The overall goal of this chapter is to enlighten researchers new to the study of the deep ocean about the availability of equipment and approaches for evaluating its living microbial inhabitants in situ or under simulated in situ conditions. The application of gene-based molecular methods, bypassing the need to incubate a natural sample, has begun in earnest to assess microbial diversity, finally tapping into the vast unculturable populations of the hot and cold deep sea (e.g., see references 29, 30, 38, 39, 95, 122, and 139). Determining microbial activities in situ under extreme conditions, however, remains the larger challenge to meet.

BACKGROUND RATIONALE FOR SAMPLING PROCEDURES

Although few scientists have ready access to submersible operations in the deep sea or even to standard sampling expeditions by surface ships, the new investigator can find established researchers forthcoming with expertise, field samples, or cultured strains. Many research goals can be met through collaboration without ever going to sea (52, 79, 107, 156). The conditions under which the original sample or culture was obtained, however, frequently figure into the conclusions that can be drawn from subsequent research. The views of a researcher working with an established deep-sea microorganism will be sharpened by an understanding of the ecological rationale (39) and protocols applied to obtain the test strain.

Temperature Considerations

For studies of the cold deep sea, temperature is an essential parameter to consider and control. Barophiles were not readily obtained and maintained in pure culture until insulated sampling gear was used and a silica gel culturing medium that could be both inoculated and solidified at cold temperatures was developed (45). All barophiles from the cold deep sea have since proved sensitive to warming at atmospheric pressure in laboratory culture (37), although notably less so under elevated pressures (43). The fact that obligately barophilic bacteria have been isolated from samples warmed (and decompressed) during gear recovery (44) suggests that strict attention to in situ temperature during shipboard enrichments may be more critical than maintenance of in situ temperature (or pressure) during short-term sample retrieval procedures.

If the goal is not to isolate in culture but rather to study metabolic activities of the barophilic component of microbial assemblages, then strict adherence to in situ temperature during all operations appears to be essential (42). Common sense and existing data (167) dictate that sample warming will eliminate some of the psychrophilic barophiles, allowing shallow-water "intruders," already present but typically dormant in the deep sea (38), to assume dominance in incubation experiments. A systematic use of different recovery and incubation temperatures to assess indigenous versus shallow-water microbes has not been attempted, although a trend in the literature toward that goal is apparent (116, 117, 146, 150). Niche-specific mixes of thermal and pressure classes of microorganisms in the cold deep sea have been predicted (38).

Isolating thermophilic microorganisms from the hot deep sea presents an opposing situation to that for isolating psychrophilic barophiles. Samples from the cold deep sea are brought from <4°C to warmer surface waters (except at high latitudes, where the entire water column approaches 0°C) over the course of several hours, while samples from hot vents cool quickly, within minutes, from superheated temperatures (>100°C to at least 420°C) to bottom-seawater temperature (<4°C) as small samples are pulled from hot sources into the massive cold-water bath of the deep sea. In the first case, some psychrophilic barophiles may succumb to thermal stress during transit through even a mild thermal gradient. In the second case, the potential loss of hot-vent inhabitants due to thermal shifts is not known, but rapid cooling has not prevented the establishment of extensive collections of hyperthermophilic microorganisms (typically anaerobic heterotrophic or methanogenic members of the domain *Archaea*) native to deep-sea smoker vents (7, 38, 56, 139). In an unusual case, the size of the recovered hot material, an intact sulfide chimney, was sufficiently large that it was still fuming, with temperatures in excess of 100°C, upon recovery shipboard (129). The fact that the organism currently holding the upper temperature limit for growth, 121°C (79), was cultured from the walls of this chimney suggests that keeping vent samples hot during retrieval may be more important than previously realized.

In laboratory tests, anaerobic hyperthermophiles have shown tolerance of cold temperatures, even under oxygenated conditions, for periods of months to years. These observations have compelling implications for novel stress responses and long-term survival in the cold deep sea after dispersal from a vent (38). The stability of hyperthermophiles at low temperatures also means that sample chilling (immersing a hot container into an ice bath) can be a simple way to end an experimental growth or reaction period in the laboratory.

Pressure Considerations

Microorganisms respond differently to hyperbaric (gas) pressures than they do to the hydrostatic pressures of the ocean. To generalize, they are more sensitive to gas pressure, regardless of the thermal class of organisms under study (107, 144). Pressurized gas is sometimes used to generate hydrostatic pressures in incubation vessels for deep-sea studies, but the targeted organisms are exposed only to liquid under pressure. In the hottest regions of hydrothermal vents, exposure to gas under pressure during phase separation is conceivable, making hyperbaric pressure a consideration in the design of equipment for deployment on smokers (115), but this is a highly specialized case. Pressure information provided in this chapter pertains to hydrostatic pressure.

With rare exceptions (e.g., gas-vacuolate bacteria), marine bacteria and archaea exist in pressure equilibrium with their fluid surroundings, i.e., they do not implode upon decompression. Decompression-sensitive points in enzymatic pathways or membrane processes, however, may lead to time-dependent (21) impaired metabolic activities, morphological aberrations, or cell lysis (62). For microscopic assessments of deep-sea microbial biomass, samples need only be fixed as soon as possible after retrieval shipboard (and protected against temperature increases, as discussed above), not protected against decompression during gear ascent (21). When the temperature is controlled, brief decompression periods during exponential growth phase in laboratory culture do not appear to cause obvious or irreversible damage (44, 169). Bacteria isolated in the complete absence of decompression (66) have proved no different in their (lack of) sensitivity to brief periods of decompression from those obtained via traditional sampling and culturing means.

For metabolic assessments of microbial community activity in the cold deep sea that rely on samples recompressed aboard ship, the effects of decompression during retrieval remain poorly known. Methods for discriminating decompression effects from other factors have proved elusive. The problem lies with the distribution of microorganisms in the ocean: microorganisms with widely diverse origins and requirements for activity exist throughout the deep sea, even though they remain inactive under deep-sea conditions (38). Thus, higher activity rates in decompressed samples can signal the release of shallow-water bacteria from the dormancy-inducing effects of deep-sea conditions, while lower rates can mean a substrate or temperature limitation on the same organisms (12, 66). If the goal is to determine in situ rates of activity, the surest (albeit more costly) route is to protect against both temperature and pressure changes, either by using decompression-free samplers or by conducting incubation experiments on the seafloor (14, 19, 63, 160). Certainly, rates derived from incubation experiments conducted at atmospheric pressure are not reliable measures of in situ deep-sea activity; quantitative conclusions for the deep sea, drawn from observations of atmospheric incubations in a recent spate of such studies (11, 26, 54, 127, 131, 145), should be approached with skepticism. Much has and still can be learned, however, from studies of various treatment effects on cold deep-sea samples recompressed shipboard (reviewed in references 37 and 151).

For the hot deep sea, the development of approaches to study hyperthermophilic activity in situ has progressed, but much work remains. Whereas sophisticated samplers are available for studies of undecompressed samples from the cold deep sea (see below), means to maintain both in situ temperature and pressure on hot-vent samples during retrieval procedures do not exist. Instead, the emphasis has been on developing ways to measure hyperthermophilic microbial activity directly on the seafloor (see below).

Substrate and Redox Considerations

In early studies of temperature or pressure effects on bacteria from the cold deep sea, neither substrate concentration nor oxygen availability was a limiting experimental factor (66, 173). With the development of decompression-free sampling devices and the use of various levels of substrate by different groups of investigators, substrate concentration emerged as a critical variable (32). Given the oligotrophic and nonculturable state of most marine bacteria, especially those in the deep ocean (30, 38, 78, 111), the substrate level in the design of microbial experiments remains important.

Few researchers studying psychrophilic barophiles have concerned themselves with oxygen limitation; for the most part, oxygen saturates the cold deep sea, from which all such barophiles have been cultured. Many of the well-known psychrophilic barophiles, however, are facultatively anaerobic (80). A relatively unexplored area of deep-sea microbiology is the realm of obligately anaerobic psychrophilic barophiles, pertaining to deeply buried sediments and abyssal animal guts (41, 72, 95, 113) and to deep anoxic waters (92). Means to preserve redox conditions during cold-sea sampling and experimentation are available (see below).

In contrast, attention to reducing conditions lies at the forefront of microbial studies of the hot deep sea. Superheated geothermal fluids contain no free oxygen, being heavily laden instead with reduced forms of gases, sulfur, and other mineral species (7). In keeping with such highly reduced source environments, virtually all cultured vent hyperthermophiles have proven to be anaerobic (1, 7, 56; see reference 105 for an exception). Their recovery and isolation, however, have not required anaerobic conditions during sampling; many have been and continue to be derived from submersible-recovered samples exposed to oxygenated seawater either during retrieval (e.g., see references 58, 69, 90, and 140) or naturally (prior to sampling), after eruptive release from hot subseafloor habitats (58, 134). This success can be attributed to the fact that anaerobic hyperthermophiles survive oxygenated conditions if the temperature is suboptimal for growth, which is always the case for hyperthermophiles recovered through a cold deep-sea water column.

Low Water Activity

Low water activity, expressed as desiccation, is rarely a consideration in microbial studies of the deep ocean, although terrestrial spore-forming bacteria, evolved to survive periods of dessication, are known to persist in deep-sea sediments (37) and although recent work has uncovered thermophilic spore formers at hydrothermal vents (153). Low water activity as a result of extreme saltiness, however, is relevant to the microbiology of deep brine formations, such as those that occur in the Gulf of Mexico (149) and the Red Sea (47) or in subsurface regions of hydrothermal vents, where liquid-gas-phase separation leads to high salt concentrations in the fluids (28, 36). Theoretical arguments have been put forward to link high salinity with microbial stability at high temperatures and hydrostatic pressures; some experimental data from the Juan de Fuca vent field were explained by invoking microbial stability during brine formation by phase separation (7, 36). On the other hand, no hyperthermophilic halophiles (salt-loving microorganisms) have been obtained in culture, either from the hot deep sea or from other hot or salty environments. Whether a concerted effort to detect such organisms has been made is not clear from the literature (112), but recent work has revealed a plethora of mesophilic halophiles in the vicinity of hydrothermal vents (81), with some showing barophilic tendencies at selected combinations of salt concentration and temperature (82). Hyperthermophilic halophiles, including barophilic forms, may be a new class of microorganisms awaiting discovery (7). Sampling approaches and culturing methods for known (mesophilic) halophiles are well established and are reviewed elsewhere (112).

Physical Disassociation of Organisms

The physical orientations of microorganisms to each other and to surfaces in the environment can be critical determinants of in situ activity and of success in detecting activity or culturing targeted organisms. This generality applies to both the cold and the hot deep sea. In the cold ocean, attachment to particles is known to affect microbial activity in different ways, depending on the pressure requirements of the test organism, its starvation state, and the incubation temperature (123). Barophilic bacteria are predicted to favor associations with surfaces, especially organic detritus, while barotolerant or barosensitive bacteria may tend to exist in suspension (38). The selection of deep-sea inocula replete with particles (e.g., sediments, sinking aggregates, and animal guts) for culturing purposes continues to yield novel barophilic species (109, 147, 162), just as it yielded the first isolated barophiles decades ago (43, 44, 168, 169). A recent metagenomic analysis of surface to deep waters in the subtropical Pacific revealed a prevalence of genes associated with surface adhesion in the deeper waters examined (30), reinforcing the importance of physical associations of microorganisms with surfaces (or each other) in the cold deep sea. The same importance of surface association appears to hold in the hot deep sea, where hydrothermal sediments, flanges, and sulfides, as well as guts of animals feeding on hot sources of microorganisms, have all yielded hyperthermophilic members of the domain *Archaea* that are well adapted to deep-sea pressures (37). The thermal and chemical gradients established in porous matrices of sulfidic deposits at hydrothermal vents continue to guide the cultivation of members of the domain *Bacteria* versus hyperthermophilic *Archaea* (7, 39, 52, 128, 129, 140, 141).

In assessing the metabolic activities of deep-sea microbial communities, the effects of physical disassociations during sample treatment need to be considered carefully. The most common experimental treatment of deep-sea sediments is to slurry the sediment, typically after diluting it with ambient prefiltered (0.2-μm-pore-size filter) seawater (34, 46, 152). The benefit of slurrying (with or without dilution) is homogeneity in labeling and distribution of subsamples for time-course experiments. The cost is the disruption of physical associations that may determine in situ activity. An activity rate measured in slurried sediment can be expected to differ from that which occurs in situ; for example, for aerobic substrate utilization, rates are higher in oxygenated slurries than in undisturbed cores (121). For hydrolytic enzyme activity, some evidence suggests that slurried rates will be lower with greater slurry dilution factors (118). The same trend would be expected if the target activity involves anaerobic consortia. To circumvent some of these problems, methods for injecting the compound of interest directly into undisturbed sediment cores have been developed (70, 99) and adapted for pressurized incubations (2, 16, 121). The disadvantages of the whole-core injection approach include diffusional constraints on the distribution of substrate within the injected horizon of sediment and reliance on separate subcores rather than homogeneously mixed subsamples for replication and killed controls. Nevertheless, the whole-core injection approach with pressurized incubation has continued to yield insights into undisturbed abyssobenthic activities, especially when seafloor experiments have not been feasible (Table 1) (2, 16, 40, 121).

Early examples of the importance of considering physical associations of bacteria with surfaces in the deep sea include the discovery of high densities of barophilic bacteria on the hindgut linings of abyssal sediment-ingesting invertebrates (41), believed to influence carbon and nitrogen cycling in the deep sea (72), and of surface-colonizing, biofilm-forming prokaryotes of unusual morphologies in both the cold (37) and the hot (65) deep sea. The more recent use of molecular techniques to assess microbial inhabitants in close physical association with each other or with various surfaces in the deep sea has yielded exciting surprises. The long-standing mystery of anaerobic methane oxidation appears to be resolved by the discovery in ice-cold gas-hydrate-rich sediments of a physically structured archaeal-bacterial symbiosis, where the archaea, living in dense aggregates surrounded by

TABLE 1 Examples[a] of studies of microbial community activity in the cold deep sea conducted under in situ conditions or at simulated in situ temperatures and pressures[b]

Process	Parameter measured	Sample container[c]	Investigator(s) (reference[s])
Biomass production	Increase in cell number by epifluorescence microscopy	A	Deming (31), Deming and Baross (37)
		E	Lochte and Turley (91)
	Incorporation of [^{3}H]thymidine	A	Alongi (2), Deming and Carpenter (40)
		E	Turley (150), Poremba (117), Patching and Eardly (114), Boetius et al. (16), Turley and Dixon (152), Dixon and Turley (46)
Protein production	Incorporation of [^{3}H]leucine	E	Turley (150), Poremba (117), Patching and Eardly (114)
		Pressure-retaining sampler	Tamburini et al. (142, 143)
Dissolved organic carbon consumption	Incorporation and/or respiration of ^{14}C-labeled amino acids, glucose, acetate, etc.	A	Deming and Colwell (42), Deming and Carpenter (40)
		A and D	Relexans et al. (121)
		Corer on lander	Wirsen and Jannasch (160)
		Corer on submersible	Cahet and Sibuet (19)
		Pressure-retaining sampler (and A[d])	Bianchi and Garcin (12, 13), Tholosan et al. (146), Tamburini et al. (143)
Organic carbon hydrolysis	Hydrolysis of fluorogenic substrate analogs	A	Meyer-Reil and Koster (100)
		D	Deming and Baross (38)
		E	Boetius and Lochte (15)
		E	Poremba (118)
		Pressure-retaining sampler	Tamburini et al. (142, 143)
Particulate organic carbon consumption	Incorporation and/or respiration of ^{14}C-labeled phytodetritus	Corer on submersible	Cahet and Sibuet (19)
		E	Boetius et al. (16)
	Incorporation of ^{13}C-labeled phytodetritus	Corer on lander	Witte et al. (161)
	Disappearance of unlabeled phytodetritus	E	Lochte and Turley (91)
		E	Poremba (116)
Sulfate reduction	Reduction of [^{35}S]sulfate	D	Boetius et al. (16)

[a]Since 1985. See the text for discussion and citations of earlier work as well as recent incubation work that fails to consider pressure.
[b]See the text for examples of activity deduced from phylogenetic or metagenomic analyses of samples without the need to incubate.
[c]Letters refer to the containers in Fig. 2, repressurized shipboard as shown in Fig. 1, unless otherwise indicated.
[d]Stainless steel (rather than plastic) syringes were used.

sulfate-reducing bacteria, reverse the process of methanogenesis (17, 50). The classical understanding of biofilm formation is under challenge because of the detection of novel biofilms formed by a single group of methane-metabolizing archaea on carbonate rock surfaces bathed in the unusual high-temperature, high-pH hydrothermal fluids of Lost City (130), a newly discovered submarine hydrothermal system driven not by magmatic heat sources but by exothermic rock reactions (84). The upper temperature limit for life is being examined anew by the study of microorganisms sorbed to interior surfaces of sulfide formations at hydrothermal vents (51), where the natural thermal gradients within such structures can be evaluated simultaneously (128, 129). Early colonists of hot smokers appear to include the *Ignicoccus-Nanoarchaeum* symbiotic assemblage (94), whose novel, obligate physical association has also provided insight into the early evolution of the smallest forms of life on Earth (60).

Contamination Problems

Because of an inability to completely sterilize most sampling gear and the nearly universal distribution of a wide variety of microorganisms in seawater, contamination takes on special meaning in deep-sea research. Contaminants can include not only the typical airborne and skin-transferred microbes but also marine bacteria that are naturally present yet not the target of study. Novel types of microorganisms can be enriched selectively from a deep-sea sample regardless of the presence of contaminants, but aseptic technique should be used as much as possible throughout all procedures as a matter of good practice. For example, for sampling of uncompromised gut contents, deep-sea animals should be kept chilled, handled minimally, and dissected with sterile tools as soon as possible after retrieval. Sediment cores should be subsampled from the interior of the recovered mass by use of a sterile device. Fluid sampling devices, from Niskin bottles to sophisticated pressure-retaining seawater samplers or thermally tolerant titanium vent samplers, have been designed to wash liberally with the desired sample prior to capturing it. Flushing with hot geothermal fluids should be particularly effective against contaminants in situ, since only very unusual indigenous microorganisms are expected to survive exposure to superheated temperatures. To guard against contaminants, sample analyses should be as selective as possible for the molecular or metabolic types under study.

SAMPLING EQUIPMENT

Cold Deep Environments

Pressure-retaining samplers for retrieving undecompressed water samples have been developed independently by at least three laboratories. The multichambered stainless steel device developed in the 1970s by Jannasch and colleagues (68) collected seawater, recovered it shipboard, and allowed for injection of a radiolabeled substrate and later removal of subsamples, all without a pressure change. Protection against pressure changes depended on gas pressure in a chamber separated from the sample by a piston floating in a buffer chamber of sterile water. This design was later refined to allow sample concentration during collection (64) and subculturing on an agar plate in a hyperbaric chamber (67).

A second pressure-retaining water sampler was developed by Colwell and colleagues in the same time frame

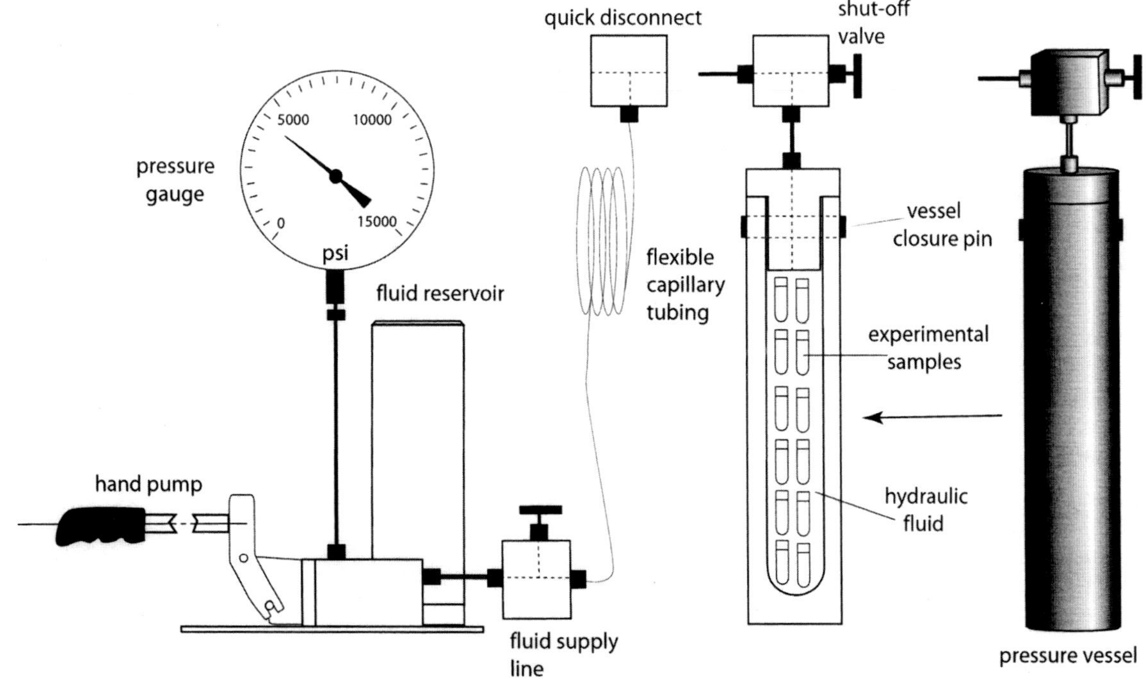

FIGURE 1 Examples of typical equipment for conducting microbial studies at elevated hydrostatic pressures, including a stainless steel hand pump, gauge, and fluid reservoir (Enerpac Division, Applied Power, Inc.), high-pressure valves and flexible capillary tubing (HIP, Erie, Pa.), and a quickly disconnecting, threadless pressure vessel and threadless cap (custom-built by Tem-Pres Division, Leco Corp., modeled on the work of Yayanos [165] and Yayanos and Van Boxtel [170]). See the text for additional details.

(138). It operates entirely on the basis of hydrostatic pressure generated by a hand pump (such as that shown in Fig. 1), thus obviating the extra precautions required to work with potentially explosive gas pressure. The use of hydrostatic pressure also demands attention to safety issues, including the potential mechanical failure of a pressure fitting or user release of pressure too quickly, which can cause a spray of fluid from the pressurized chamber. The former has not occurred in research laboratories to my knowledge; pressure equipment is built to withstand pressures twice those anticipated by the user. The latter causes little damage to the operator (you get wet), though protective eye gear is ad-vised.

A third pressure-retaining water sampler was developed more recently by Bianchi and colleagues (12, 14). It borrows from the designs of its predecessors but is simpler in operation than the first samplers, though still dependent on gas pressure. It has not yet been deployed in the cold deep sea, but at warm (13°C) Mediterranean depths it has enabled detection of pressure-adapted microbial and enzymatic activity (Table 1) (13, 142, 143, 146). For microbial studies of cold deep-sea sediments without a pressure change, researchers are dependent on free-vehicle benthic landers (63, 160, 161) or submersible operations (19). As a result, few data are available, and some results remain ambiguous; even interpretations from the latest in situ work (161) rely in part on results from recovered samples incubated at atmospheric pressure shipboard. To study microbial aspects of undecompressed deep-sea animals, a pressure-retaining trap for abyssal crustacea was developed and used successfully in the 1970s to recover live specimens under pressure (164), which, upon death and decay of the specimens under pressure, yielded the first barophilic bacterium in pure culture (168). Others developed remote means to study the responses of invertebrate intestinal floras in situ (159). Methodological difficulties in discriminating microbial from invertebrate activity, however, appear to have limited research in this direction.

For research on decompressed deep-sea samples, a wide variety of sampling devices, described in the oceanographic literature (126), have been used by microbiologists. These include Niskin samplers for seawater (13), moored sediment traps for sinking particulates (42), and large and small box corers (18) and multicorers (4) for sediments (16, 31, 34, 40, 46, 91, 152). A variety of trawls, dredges, and tethered and free-vehicle traps have been used to recover abyssal and hadal animals (42, 106).

Hot Deep Environments

Until recently, the small-scale sampling of hot deep environments at submarine hydrothermal vents for microbial purposes had always involved manned submersible operations, limiting the number of microbial sampling expeditions since the discovery of hot vents in 1979 (see Table 3 in reference 36). The recent investment of Japanese funding in deep-sea microbiology, coupled with continuing support for French and U.S. submersible work at hydrothermal vents, has increased the pace of exploration and included the use of remotely operated vehicles for small-scale sampling to study hyperthermophiles (25, 105). Sampling from a surface ship remains important for some purposes, particularly for rapid-response sampling at recent eruption sites (57, 134), detecting hydrothermal plumes (137), recovering deep waters near vents (96, 97), and tapping into the hot subsurface via deep-sea drilling cores (125, 136). The future will bring deep-sea experimentation conducted via seafloor observatories powered by fiber-optic cables (83).

Thermally tolerant sampling devices for the collection of superheated geothermal fluids under deep-sea pressure were first developed by geochemists. The basic sampling instrument, a titanium syringe, draws in hot geothermal fluid in situ and returns it shipboard cooled and decompressed. Diagrams of the original titanium syringe and an expanded sampling system incorporating multiple syringes, a flowthrough system, and rigorous temperature checks on the collected samples are provided and reviewed elsewhere (7). The latter system has proved useful for the recovery of hot fluids minimally diluted with seawater (36, 132). To collect particles that may be present in superheated geothermal fluids, microbiologists have developed instruments for emplacement directly in the flow of a smoker, including the following: (i) the smoker poker, a titanium tube with a spring-loaded closure mechanism and glass mesh filling that can be inserted deep (decimeters) into the throat of a smoker (77); and (ii) the vent cap, a conical flowthrough device that is positioned atop a smoker and offers colonizing surfaces to microorganisms in the smoker plume (77, 108). Recent variations on the smoker poker and vent cap include the vent catheter (105), the biocolumn, designed specifically to cap deep-sea drilling holes venting warm fluids (25), stainless steel probes driven directly into the ocean crust to release warm subsurface fluids (61), and titanium mesh mounted on titanium plates or within sampling chambers deployed directly in venting fluids after first removing the chimney structure (94).

Solid sulfidic materials from smoker walls and flanges (horizontal expansions of a smoker wall) are collected directly as grab samples by a submersible or remotely operated vehicle manipulator. Whole chimneys have been recovered, requiring significant engineering expertise, for precise microbial sampling shipboard while the structure is still hot (129). Coring devices for actively venting smoker walls are available (22) and currently in use for in situ experimentation by microbiologists (ongoing work). Even without precise sampling, numerous hyperthermophilic isolates and evidence for whole communities of hyperthermophilic members of the domain *Archaea* have been obtained from grab samples of smoker structures (7, 51, 56, 122, 128, 129, 139). Standard submersible push cores in the hot sediments of Guaymas Basin vents resulted in the discovery of the first known hyperthermophilic, and later barophilic (73), sulfate-reducing microorganisms (71, 155).

A few sophisticated titanium samplers that tolerate smoker extremes in temperature, pressure, and chemistry and also double as in situ incubators have been developed for microbial experimentation on actively venting smokers on the seafloor. The first was designed to capture superheated fluids, release a tracer compound into the sampling chamber, and incubate the filled chamber in the flow of the smoker. Results of deployments at Juan de Fuca smokers (7, 132) yielded indications of microbial stability (particulate DNA) under the conditions of the experiment. The sampler was not reliable against leaks, however, returning only 4 of 39 deployments with uncompromised samples. A new titanium sampler-incubator with an improved closure mechanism has since been designed, built, and successfully tested on the seafloor (115).

SIMULATING EXTREME ENVIRONMENTS IN THE LABORATORY

Cold Temperatures and Elevated Pressures

Pressure-retaining water samplers also double as laboratory chambers for simulating in situ temperature and pressure conditions in the cold deep sea. With a constant pressure maintained mechanically, the challenge frequently is to keep the sampler temperature constant during instrument retrieval shipboard and transfer from one setting to the next, e.g., from the deck to a temperature-controlled laboratory. Temperature control during transport is typically accomplished in an unsophisticated fashion, using ice-filled jackets around the sampler. The possible negative effects of temperature shifts are largely avoided simply by limiting the time the sampler spends outside a temperature-controlled chamber. An ideal location for the deployment and use of pressure-retaining water samplers without incurring a biologically relevant change in temperature would be the deep Arctic (or Southern) Ocean, where the temperature differentials between deep-sea and surface waters, and the deck atmosphere, are minor. Such polar sampling and experimentation have not been reported.

For the majority of researchers, who do not have access to pressure-retaining deep-sea samplers, decompressed samples and cultured organisms can be studied under in situ conditions simulated in simple stainless steel pressure vessels (Fig. 1) immersed in temperature-controlled water baths or incubators. The earliest pressure vessels used by microbiologists involved thick threaded caps sealed to the main vessel via multiple large bolts (20, 173). Pressure was applied through tubing that also required threaded fittings to the cap. This configuration greatly limited the rate at which a vessel could be sealed or opened; most early experiments under simulated deep-sea conditions were thus restricted to end-point incubations.

Important advances in pressure vessel design that streamlined culture work and enabled time-course experiments were made by Yayanos and colleagues (165, 170). The threaded-cap-and-vessel design was replaced by a smooth-sided cap (and vessel) held in place by a smooth-sided crossbar penetrating the walls of the vessel (Fig. 1). After pressurization of the closed vessel via introduction of a small volume of hydraulic fluid through flexible capillary tubing (Fig. 1), the pin and cap become immobilized by the pressure differential. With no pressure in the vessel, the pin slides out easily, followed by the cap. Pressurization or decompression is further facilitated by use of a quick-disconnect interface between the pump and the pressure vessel (Fig. 1). To facilitate my pressure research, I worked with engineers at the Tem-Pres Division of the Leco Corporation (Blanchard, Pa.) to produce a similar quick-disconnect device and streamlined pressure vessels; these can be ordered directly from the company.

The simplest, most transportable (for shipboard work) means of pressurizing a vessel involves a hand pump equipped with a high-pressure gauge for monitoring the rise and fall of pressure during operations (Fig. 1). Electric pumps, however, have also been used to create deep-sea hydrostatic pressures in the laboratory (93, 171). Although early pressure studies involved the use of various hydraulic oils and risked sample contamination, distilled (and sterilized, depending on the chance of sample contact) water has since replaced oil as a simple, clean, and effective hydraulic fluid for this type of operation.

Advancing beyond the approach of incubating one or more pressure vessels in a temperature-controlled bath or incubator, Yayanos et al. (172) described a pressure-temperature gradient apparatus. This instrument accommodates eight pressure vessels housed in parallel in an insulated aluminum block. A temperature gradient is established in the block via a heater at one end and a cooler at the other end and is controlled via thermostats at each end. Large multifactored data sets can be generated relatively quickly with this approach. To date, it has been used to score bacterial growth in small tubes of liquid media placed along the temperature gradient and colony formation in long glass tubes filled with solid media (43, 148, 167, 172). The potential for additional types of assays has not yet been realized.

When a microbial sample is pressurized, it must be isolated from the surrounding hydraulic fluid in a container that is pressure responsive to avoid rupture. Either the entire container or some part of it, such as a mobile piston or pliant cap, must be flexible in response to pressure. In all cases, air bubbles must be expelled from the container prior to sealing. Their presence can cause hyperbaric gas toxicity to the test organisms and leaks in containers as the gas expands and contracts during sample pressurization and decompression (unlike pressurizing gas, pressurizing a liquid reduces its volume by only a small degree). Examples of containers used by a wide cross section of researchers working on microbial issues in the cold deep sea include (i) a modified plastic syringe with a mobile, rubber-capped syringe plunger (Fig. 2A), (ii) a plastic test tube sealed with a rubber stopper that is free to move into the tube (Fig. 2B), (iii) an open-ended glass tube sealed at both ends with a rubber stopper free to move into the tube (Fig. 2C), (iv) an open-ended plastic tube (sealed at the ends as shown in Fig. 2C) with predrilled injection ports sealed individually with a silicon-based glue or collectively by heat shrinking a plastic sleeve around the injection core (Fig. 2D), (v) a polyethylene bag that is heat sealed after sample loading (Fig. 2E), and (vi) an agar-filled plastic pouch that is heat sealed after inoculation (Fig. 2F).

Modified syringes were the early choice and have been used by many researchers (Table 1), especially for work at sea, as they are calibrated and can double as the subsampling device. Note that if a Becton Dickinson syringe is used, the air space in the rubber cap of the syringe plunger should be displaced with sterile water (by injecting from the back end of the syringe) to guard against cap collapse and leakage under pressure. Glass or plastic test tubes, which do not double as subsampling devices at sea, have proved a simpler alternative for laboratory work with pure cultures in liquid media (43, 123, 172). A sterile glass bead added to a modified syringe or test tube enables sample mixing without decompression: the pressurized vessel can be rocked automatically (165) or inverted manually to effect mixing. Test tubes are the container of choice when a solid silica gel medium is used (32, 45). Open-ended glass tubes have also been used effectively with solid media and pure cultures in the pressure-temperature gradient apparatus (172) mentioned above. Open-ended plastic tubes with predrilled, sealed injection ports and a beveled edge at one end, much like those developed for coastal work (99), have been used at sea as both subsampling devices for abyssal sediments and subsequent experimental containers (38, 40, 121). Radiolabeled or fluorogenic substrates can be injected into sediment horizons, and the whole subcore can be incubated under simulated deep-sea conditions. Prior to pressurization, the full length of the injected subcore is wrapped with Parafilm to protect against leakage through an injection port under pressure.

Polyethylene bags, where the entire container flexes under pressure, are also used effectively in both field (Table 1) and pure culture (53) work. An elaborate adaptation of the bag approach, used to isolate barophiles from deep-sea fish (106), involves a planoconvex agar lens. The planar surface of the lens is inoculated, overlaid with a second agar lens, and heat sealed in a plastic pouch for subsequent

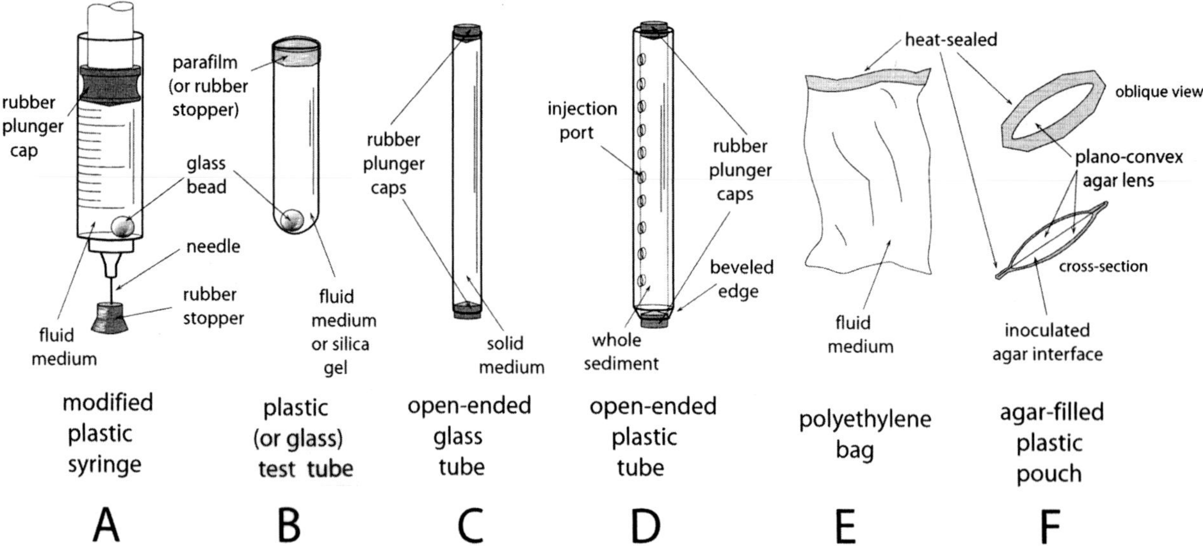

FIGURE 2 Examples of sample containers used at elevated hydrostatic pressures, i.e., with pressure-responsive (moveable) parts (containers A to D) or a completely flexible design (containers E and F). Containers A and D double as sample collectors in the field for fluid and sediment, respectively. Containers B and C (when filled with solid media [45, 172]) and container F (106) allow for colony formation under pressure. Container C, which can be of any length, is used in the pressure-temperature gradient instrument of Yayanos et al. (172). See Table 1 for examples of container use in selected studies; see the text for additional details.

pressurization. Colonies form at the interface of the two lenses.

Other clever sample containers (not featured in Fig. 2) have also been developed. For example, for some morphological or rapid-reaction studies in which the sample may need to be fixed or poisoned under pressure, the Landau-Thibodeau method originally devised for the study of pressure-sensitive nonmarine microorganisms (89) is available. A two-chambered metal device allows the process of interest to occur in one chamber separated from a fixative or poison in the other chamber by a glass coverslip. The pressurized vessel is inverted sharply, causing a small steel ball in the second chamber to break the glass partition and mix the fixative or poison with the culture or reaction solution. Samples are then decompressed for processing. The use of this approach to examine ultrastructural effects of decompression on an obligately barophilic bacterium revealed that morphological aberrations due to decompression are time dependent (21), i.e., there are no immediate effects. Other innovative adaptations of existing devices to pressure work include the use of microcupules that come on commercially available test strips for enzyme assays (133) and the use of sterile polyethylene transfer pipettes as calibrated and heat-sealable incubation containers (82).

High Temperatures and Pressures

The field of high-temperature, high-pressure laboratory research has expanded rapidly in the last decade in parallel with keen community-wide interests in the phylogeny, biochemistry, molecular genetics, and biotechnological applications of hyperthermophiles (1, 7, 36, 56, 87, 93, 101, 103, 122, 139). Several laboratories are now equipped to study microbial processes at temperatures and pressures that simulate the hot deep sea. The material complications at high temperatures, however, are greatly magnified compared with work at cold temperatures. Pressure vessels and sampling containers must be heat resistant and, frequently, resistant to acidic or reducing conditions. Pressure relief valves become essential safety features, since hydrostatic pressure increases rapidly in a closed system as the temperature increases. Liquids also vaporize under pressure, depending on the temperature and chemistry of the solution; possible phase transitions must be anticipated and either avoided or monitored. All existing high-temperature, high-pressure systems are custom-built; researchers work with experienced engineers to build effective systems and establish appropriate safety protocols.

The following two types of laboratory systems have been constructed and used for studies of microorganisms from the hot deep sea: (i) a main pressure vessel incubated with one combination of temperature and pressure conditions and subsampled over time without altering the incubation conditions of the sample and (ii) multiple pressure vessels incubated at different temperatures, in one or more ovens, for end-point experiments. The first of the time-course systems was developed by Yayanos and colleagues (171) and used by others (6, 8) to culture hyperthermophiles at superheated temperatures. An engineering colleague and I developed a second time-course system (85) that has been used to test the thermal stability of selected organic molecules (85, 119) and construction materials for experiments at superheated temperatures on the seafloor (115), as well as the role of pressure acclimation in the survival of hyperthermophiles at superheated temperatures (122). A third system, developed by Baross and colleagues, has been used with cultures of deep-sea hyperthermophiles and their enzymes (7, 39), as has a fourth system developed by Horikoshi and colleagues

(103, 140). Most of these systems are designed to accommodate temperatures up to 350°C and pressures up to about 6 × 10^7 Pa. The second and third systems were built commercially by the Tem-Pres Division of the Leco Corporation; design specifications and models are available directly from Tem-Pres. A fifth time-course system, developed by Clark and colleagues, has been used very effectively for both hydrostatic and hyperbaric pressure studies of methanogenesis at temperatures of <120°C (101, 107).

Rapid progress in the study of pressure effects on cultures of deep-sea heterotrophic hyperthermophiles has come from the use of individual pressure vessels and ovens for end-point experiments (55, 93, 135; see Table 2 in reference 36). The same vessels used for cold deep-sea research (manufactured by Tem-Pres) can be used for temperatures up to about 200°C. Any standard oven large enough to accommodate the vessels can provide the desired temperature. A setup similar to that shown in Fig. 1, but with the pump and gauge mounted outside the oven, was used to document colony formation at 120°C under deep-sea pressure (see reference 35 for further operational details). An important improvement of this system is the "wine rack," designed by Baross and colleagues and described by Baross and Deming (7). The device accommodates eight pressure vessels in separately insulated and temperature-regulated chambers, such that a matrix of temperatures and pressures can be tested in a single experiment. Pressure is created and monitored as shown in Fig. 1, but the temperature is computer controlled by independent thermoregulators inserted into each vessel. This system has been used to document barophily in a deep-sea hyperthermophile, pressure-induced upward shifts in its optimal temperature and growth rate, novel stress responses, and the thermostability of DNA polymerases (7, 36, 55, 93). The key to success in these studies was the use of a wider range of test pressures than usual, including pressures expected in the subsurface biosphere, well below vent features on the seafloor (36). A similar system of multiple pressure vessels and ovens has also been established in the Prieur laboratory, where its use has revealed pressure-induced upward shifts in the survival temperatures of hyperthermophiles (93).

Yet another variant on the theme of end-point experiments at high temperatures and pressures is the recent development in the Jørgensen laboratory of a high-pressure thermal gradient block (74), generally patterned after the system described earlier for low-temperature, high-pressure research (172). In this high-temperature system, an aluminum block houses multiple interconnected stainless steel high-performance liquid chromatography columns, each loaded with duplicate samples. The columns are pressurized and a temperature gradient is established across the block, enabling an experiment at a single pressure but multiple temperatures. Initial work with this new gradient block has revealed the presence of hyperthermophilic, barophilic sulfate-reducing bacteria in Guaymas Basin sediments, even after cold storage in the laboratory (73).

Pressure-responsive sample containers for use at high temperatures resemble the designs shown in Fig. 2A and E but are constructed of heat- and acid-tolerant materials. If temperatures of <120°C are tested, then modified, autoclavable plastic syringes (Fig. 2A) are still usable, though not repeatedly. Glass syringes, with silicon grease to facilitate movement of the frosted glass plunger under pressure, can be used up to about 200°C. Custom-made glass Hungate tubes with specialized septa have also proved useful up to about 200°C in the high-pressure thermal gradient block described above (74). Higher temperatures require custom-built titanium syringes with thermally tolerant O-rings on the plunger. The flexible-bag approach (Fig. 2E) can be used at the highest temperatures achievable in existing laboratory systems if gold is substituted for plastic (85).

SAMPLE ANALYSIS

Isolation and Culturing

A variety of culturing media and techniques have been developed specifically for high-pressure deep-sea microbial research. For the cold deep sea, the repertoire has been somewhat limited to enrichment media selective for aerobic (or facultatively anaerobic) heterotrophs that grow under high pressure. Other physiological types of psychrophilic barophiles have not yet been verified in culture, but the potential existence of manganese oxidizers, iron oxidizers, ammonia oxidizers, sulfate reducers, methane cyclers, and even nitrogen fixers that are both psychrophilic and barophilic has been suggested by culture-independent analyses (23, 27, 38, 88, 96, 97) and by psychrophilic isolates obtained but not yet examined for pressure traits (48). The development of a solid silica gel medium for aerobic heterotrophs (45) marked a turning point for cold deep-sea research, since it enabled colony formation under pressure and isolation of the first barophile (168). The advantage of the medium is that it can be inoculated as a liquid (salt-free silica sol) and solidified later by adding sea salts. The whole process can be conducted at a cold temperature so that the inoculum is never exposed to warming, as would be the case with an agar-based medium. This medium was used to characterize cardinal growth temperatures of many new aerobic heterotrophic barophiles in the years following its development (38, 42, 167, 172). Instead of adapting this silica gel medium to other types of nutritional menus, however, subsequent researchers have either developed a different solid-medium approach for colony formation under pressure (106) or continued to study existing cultures in standard marine broth media (10, 38, 59). A typical marine broth is based on peptone and yeast extract (Difco); the silica gel medium can be used with a variety of organic substrates but was originally optimized with tryptone, glucose, and yeast extract (45).

A wider menu of culturing media and protocols has been developed for hyperthermophiles from the hot deep sea, in keeping with intense community-wide interest in the diversity of known hot reducing habitats. Several reviews and publications provide complete details on medium preparation, inoculation, and general anaerobic technique at high temperatures, as well as culture maintenance protocols for the range of known deep-sea hyperthermophiles (5, 7, 56, 124). Each nutritional group and many individual species have very specific trace nutrient requirements, including metals, sulfur compounds, and organic compounds. For example, the new researcher should be aware of unusual requirements for tungsten at concentrations much higher than would be provided in a standard trace metal mixture (5, 7). The range of hyperthermophiles isolated from the hot deep sea includes sulfur-dependent heterotrophs, non-sulfur-dependent heterotrophs, iron reducers (that form magnetite), and methanogens. All are marine in origin, as indicated by their strict salt requirement; to date, none is an obligate barophile (none has been enriched initially under high pressure, which is the key to success in obtaining the first obligately barophilic psychrophiles), but each isolate that has been studied under pressure shows some form of positive response to elevated pressure (36, 55, 56, 93, 140).

Initially, because of thermostability problems, few studies of deep-sea hyperthermophiles relied on solid media; liquid media are still the rule for studies of growth and physiology, most recently at temperatures up to 130°C (79). For growth under elevated pressures, a solid medium based on the use of thermally stable Gelrite (Kelco Division, Merck & Co., Inc., San Diego, Calif.) was developed and used at temperatures up to 120°C (35). As the availability and variety of deep-sea hyperthermophiles have expanded, the desire to explore their genetics, which requires approaches based on solid media, has led to the development of stable solid media now in wide use in that arena (110).

Community Measurements

The direct analysis of the abundance or biomass of whole microbial communities in samples from the cold deep sea, without sample incubation, has involved most of the obvious techniques long available to microbial ecologists. For example, epifluorescence microscopy (38, 120, 163), the firefly luciferase ATP assay (75, 121), and lipid extraction (3) have been used to assess microbial biomass in the deepest waters and sediments of the ocean, including deep subseafloor sediments (113, 135). The recent application of a rapid and direct viable count method to deep-sea sediments holds promise for evaluating the active state of deep-sea microorganisms without the need to incubate samples (120), as does the application of molecular phylogenetic methods. For example, deep cold subseafloor sediments appear to contain large numbers of living bacteria when RNA content, evaluated microscopically by catalyzed reporter deposition-fluorescence in situ hybridization, is the measure of viability (127). Fluorescence in situ hybridization work in general has revealed that the deeper portions of the ocean, whether in the subtropical Pacific (78), anoxic Caribbean Basin (92), northern Atlantic (145), or high Arctic (157), harbor greater fractions of *Archaea* than do shallow waters, where *Bacteria* tend to dominate. Based on the use of radiolabeled substrates, many of these archaea may be chemolithotrophic nitrifiers (49, 54), but whether or not they are active under the pressures of the deep sea has not been addressed. Similarly, both microbial growth and sulfate reduction have been measured in samples of deep subseafloor sediments, again by using radiolabeled tracers, but only by incubating the samples at atmospheric pressure (113, 127). The recent apparent trend of interpreting results obtained at atmospheric pressure as representative of activities at in situ pressures needs to be reversed.

Measurements of microbial activity that require an experimental incubation period and that have been made under in situ conditions, whether directly on the seafloor or simulated in the laboratory, have also involved most of the standard techniques regularly applied to milder environments. Recent examples of processes measured, analytical approaches taken, and types of sample containers used during pressurized incubation are summarized in Table 1, along with citations that provide further methodological details. The use of various gene-based approaches that also involve sample incubation does not appear in this table because researchers newly venturing into the deep sea with these techniques have not yet conducted their experiments at in situ pressures.

Attempts to measure microbial abundance, biomass, or activity in hot deep-sea samples and, by implication, in a deep hot subsurface biosphere were initially plagued by technical difficulties and controversy over interpretation of contamination (36, 77, 132). They were also stymied by the prevailing wisdom that microorganisms were not capable of withstanding, much less responding favorably to, the superheated temperatures of submarine geothermal fluids. The issues surrounding this earlier debate have been addressed elsewhere (7). Since then, new players and approaches, especially molecular gene-based techniques, have been brought to bear on the issue (Table 2). The growing number of experiments that have been attempted with hot deep-sea samples, either in situ or under simulated in situ conditions, is represented in Table 2 according to the analytical approaches taken, types of sample containers used during pressurized incubations, and relevant citations for additional details. Unlike in the cold deep sea, where survival at low temperatures and high pressures is common to typical marine microorganisms as well as marine hyperthermophiles and even human-derived bacteria (reviewed in reference 38), surviving conditions at the upper end of the temperature scale in the hot deep sea represents a critical process that is likely unique to hyperthermophiles, perhaps especially barophilic hyperthermophiles. Survival under hot deep-sea conditions, readily measured by gene-based techniques that do not require an in situ incubation approach for environmental relevance, thus appears in Table 2 as an "activity" being measured in the hot deep sea with increasing frequency.

LIMITATIONS YET TO BE OVERCOME

The outstanding challenge to all microbial ecologists is to find ways to measure rates of in situ activity noninvasively. The problem for extreme deep-sea environments, as elsewhere, lies with the available approaches to measuring activity, not with limitations to deep-sea sampling or seafloor experimentation. The latter are limited only by resources (and perhaps motivation), since a wide variety of sampling gear and instrumentation, including pressure-retaining devices, are available for in situ study of deep-sea extremophiles. The missing link is the means to measure rates of in situ microbial activity without having to disturb (amend) or incubate the sample. The use of molecular gene-based techniques has advanced our understanding of phylogenetic diversity and the presence of activity-specific genes in deep-sea microbial communities, but acquiring rates of activity true to the in situ environment remains a challenge. Targeted gene expression can be evaluated noninvasively and without incubation (an approach not yet attempted in the deep sea), but it does not yield a rate of activity. The most promising direction for both the cold and hot deep sea will couple gene-based analyses of samples fixed in situ (no chance of negative or selective effects of shifts in temperature or pressure during recovery) with in situ incubation directly in deep waters and on (or below) the seafloor. Time-course tracking of stable isotope-labeled substrates into, for example, archaeal lipids during in situ incubation has the potential to yield meaningful and group-specific rates of activity.

Along with in situ work, expanding the repertoire of laboratory experiments and analyses of deep-sea extremophiles still holds great promise for discovery. Emerging at a rapid pace are novel microorganisms with novel properties from both the cold and hot deep sea, including evidence for nitrogen-fixing deep-sea *Archaea* (97), barophilic hyperthermophilic sulfate reducers (73), ionizing radiation-resistant hyperthermophilic *Archaea* (69), and nearly monospecific biofilms of thermophilic methane-cycling *Archaea* (130). After decades of exploration, entirely novel

TABLE 2 Examples[a] of microbial community measurements on samples recovered from the hot (>90°C) deep sea, made directly or after incubation in situ or at laboratory-simulated in situ temperatures and pressures

Process	Approach	Sample device or incubator	Investigator(s) (reference)
Cell survival	ATP assay on particles from smoker fluids	Smoker poker, deployed in situ	Karl et al. (76)
	Microscopic analysis of artificial or natural surfaces	Vent cap with colonizable surfaces, incubated in situ[b]	Karl et al. (77)
		Grab sample of sulfide chimney	Schrenk et al. (128)
	Lipid analysis of natural material	Grab sample of sulfide flange	Hedrick et al. (52), Baross and Deming (7)
	DNA extraction from particles in smoker fluids	Titanium (Ti) syringe sampler, deployed at smokers	Straube et al. (132), Takai and Horikoshi (139)
		Ti sampler-incubator, incubated in situ in smokers	Baross and Deming (7)
	DNA hybridization and microscopic analysis of cells in natural material	Grab sample of sulfide material	Harmsen et al. (51), Takai et al. (141)
		Sulfide chimney recovered whole, still fuming	Schrenk et al. (129)
		Grab sample of carbonate "beehive" material	Kelley et al. (84)
	DNA amplification from natural samples of sulfidic material, smoker fluids, or subsurface fluids	Grab sample of sulfide material or chimney recovered whole, still fuming (16S rRNA genes)	Takai et al. (141), Schrenk et al. (129)
		Vent catheter deployed in situ (16S rRNA and dissimilatory sulfite reductase genes)	Nakagawa et al. (105)
		Titanium mesh colonization plates deployed in situ (16S rRNA genes)	McCliment et al. (94)
		Ti syringe sampler, deployed at diffuse flow vents (16S rRNA genes)	Holden et al. (57), Summit and Baross (134)
Biomass production	Increase in hydrolyzable protein in mixed culture from smoker fluids	Ti syringe in laboratory-heated pressure vessel	Baross and Deming (6), Baross et al. (8)
	Incorporation of [^{3}H]thymidine into particles in smoker fluids	Glass syringes in shipboard-heated pressure vessels	Deming (33)
	Increase in extractable DNAs from particles in smoker fluids	Ti sampler-incubator, incubated in situ in smokers	Baross and Deming (7)
Sulfate reduction	Reduction of [^{35}S]sulfate in hot sediments (after cold storage)	Glass Hungate tubes in laboratory high-pressure thermal gradient block	Kallmeyer et al. (74), Kallmeyer and Boetius (73)

[a]See reference 7 for a comparative review of experimental details and results from prior to 1995; see reference 56 for the latest review.
[b]Vent caps have been deployed more recently, but only on cooler (<90°C) vents (108).

environments in the deep sea are still being discovered (84), while vast regions remain poorly sampled, including the abyssal depths in the subzero Arctic Ocean (157) and the hot subsurface biosphere (9, 24). The first whole-genome sequences for pressure-adapted psychrophiles, however, are now available to explore fully (98, 154).

REFERENCES

1. **Adams, M. W. W.** 1999. The biochemical diversity of life near and above 100°C in marine environments. *J. Appl. Microbiol.* **85**(Suppl. S)**:**108S–117S.
2. **Alongi, D. M.** 1990. Bacterial growth rates, production and estimates of detrital carbon utilization in deep-sea sediments of the Solomon and Coral Seas. *Deep-Sea Res.* **37:**731–746.
3. **Baird, B. H., D. E. Nivens, J. H. Parker, and D. C. White.** 1985. The biomass, community structure, and spatial distribution of the sedimentary microbiota from a high-energy area of the deep sea. *Deep-Sea Res.* **32:**1089–1099.
4. **Barnett, P. R. O., J. Watson, and D. Connelly.** 1984. A multiple corer for taking virtually undisturbed samples from shelf, bathyal and abyssal sites. *Oceanol. Acta* **7:**399–408.
5. **Baross, J. A.** 1993. Isolation and cultivation of hyperthermophilic bacteria from marine and freshwater habitats, p. 21–30. *In* P. F. Kemp, B. F. Sherr, E. B. Sherr, and J. J. Cole (ed.), *Handbook of Methods in Aquatic Microbial Ecology.* Lewis Publishers, Boca Raton, Fla.
6. **Baross, J. A., and J. W. Deming.** 1983. Growth of "black smoker" bacteria at temperatures of at least 250°C. *Nature* (London) **303:**423–426.
7. **Baross, J. A., and J. W. Deming.** 1995. Growth at high temperatures: isolation and taxonomy, physiology, and ecology, p. 169–217. *In* D. M. Karl (ed.), *The Microbiology of Deep-Sea Hydrothermal Vents.* CRC Press, New York, N.Y.

8. **Baross, J. A., J. W. Deming, and R. R. Becker.** 1984. Evidence for microbial growth in high pressure, high temperature environments, p. 186–195. *In* M. J. Klug and C. A. Reddy (ed.), *Current Perspectives in Microbial Ecology: Third International Symposium on Microbial Ecology.* American Society for Microbiology, Washington, D.C.
9. **Baross, J. A., W. S. D. Wilcock, D. S. Kelley, E. F. DeLong, and S. C. Cary.** 2004. The subsurface biosphere at mid-ocean ridges: issues and challenges. *Geophys. Monogr. Ser.* **144:**1–11.
10. **Bartlett, D. H.** 2002. Pressure effects on in vivo microbial processes. *Biochim. Biophys. Acta* **1595:**367–381.
11. **Bendtsen, J., C. Lundsgaard, M. Middelboe, and D. Archer.** 2002. Influence of bacterial uptake on deep-ocean dissolved organic carbon. *Global Biogeochem. Cycles* **16:** 1127. [Online.] doi:10.1029/2002GB001947.
12. **Bianchi, A., and J. Garcin.** 1993. In stratified waters the metabolic rate of deep-sea bacteria decreases with decompression. *Deep-Sea Res.* **40:**1703–1710.
13. **Bianchi, A., and J. Garcin.** 1994. Bacterial response to hydrostatic pressure in seawater samples collected in mixed-water and stratified-water conditions. *Mar. Ecol. Prog. Ser.* **111:**137–141.
14. **Bianchi, A., J. Garcin, and O. Tholosan.** 1999. A high pressure serial sampler to measure microbial activity in the deep sea. *Deep-Sea Res.* **46:**2129–2142.
15. **Boetius, A., and K. Lochte.** 1994. Regulation of microbial enzymatic degradation of organic matter in deep-sea sediments. *Mar. Ecol. Prog. Ser.* **104:**299–307.
16. **Boetius, A., T. Ferdelman, and K. Lochte.** 2000. Bacterial activity in sediments of the deep Arabian Sea in relation to vertical flux. *Deep-Sea Res. II* **47:**2835–2875.
17. **Boetius, A., K. Ravenschlag, C. J. Schubert, D. Rickert, F. Widdel, A. Gieseke, R. Amann, B. B. Jorgensen, U. Witte, and O. Pfannkuche.** 2000. A marine microbial consortium apparently mediating anaerobic oxidation of methane. *Nature* (London) **407:**623–626.
18. **Boland, G. S., and G. T. Rowe.** 1991. Deep-sea benthic sampling with the GOMEX box corer. *Limnol. Oceanogr.* **36:**1015–1020.
19. **Cahet, G., and M. Sibuet.** 1986. Activité biologique en domaine profond: transformations biochimiques in situ de composés organiques marqués au carbone-14 à l'interface eau-sédiment par 2000 m de profondeur dans le golfe de Gascogne. *Mar. Biol.* **90:**307–315.
20. **Certes, A.** 1884. Sur la culture, a l'abri des germes atmospheriques, des eaux et des sediments rapportes par les expeditions du Traveilleur et du Talisman, 1882-1883. *C. R. Acad. Sci.* **98:**690–693.
21. **Chastain, R. A., and A. A. Yayanos.** 1991. Ultrastructural changes in an obligately barophilic marine bacterium after decompression. *Appl. Environ. Microbiol.* **57:**1489–1497.
22. **Cook, T. L., and D. S. Stakes.** 1995. Biogeological mineralization in deep-sea hydrothermal deposits. *Science* **267:** 1975–1979.
23. **Cowen, J. P.** 1989. Positive pressure effect on manganese binding by bacteria in deep-sea hydrothermal plumes. *Appl. Environ. Microbiol.* **55:**764–766.
24. **Cowen, J. P.** 2004. The microbial biosphere of sediment-buried oceanic basement. *Res. Microbiol.* **155:**497–506.
25. **Cowen, J. P., S. J. Giovannoni, F. Kenig, H. P. Johnson, D. Butterfield, M. S. Rappé, M. Mutnak, and P. Lam.** 2003. Fluids from aging ocean crust that support microbial life. *Science* **299:**120–123.
26. **Danavaro, R., N. Della Croce, A. Dell'Anno, and A. Pusceddu.** 2003. A depocenter of organic matter at 7800 m depth in the SE Pacific Ocean. *Deep-Sea Res. I* **50:**1411–1420.
27. **De Angelis, M. A., J. A. Baross, and M. D. Lilley.** 1991. Enhanced microbial methane oxidation in water from a deep-sea hydrothermal vent field at simulated in situ hydrostatic pressures. *Limnol. Oceanogr.* **36:**565–570.
28. **Delaney, J. R., D. S. Kelley, M. D. Lilley, D. A. Butterfield, J. A. Baross, W. S. D. Wilcock, R. W. Embley, and M. Summit.** 1998. The quantum event of oceanic crustal accretion: impacts of diking at mid-ocean ridges. *Science* **281:**222–230.
29. **DeLong, E. F., and D. M. Karl.** 2005. Genomic perspectives in microbial oceanography. *Nature* (London) **437:** 336–342.
30. **DeLong, E. F., C. M. Preston, T. Mincer, V. Rich, S. J. Hallam, N.-U. Frigaard, A. Martinez, M. B. Sullivan, R. Edwards, B. Rodriguez Brito, S. W. Chishom, and D. M. Karl.** 2006. Community genomics among stratified microbial assemblages in the Ocean's interior. *Science* **311:**496–503.
31. **Deming, J. W.** 1985. Bacterial growth in deep-sea sediment trap and boxcore samples. *Mar. Ecol. Prog. Ser.* **25:**305–312.
32. **Deming, J. W.** 1986. Ecological strategies of barophilic bacteria in the deep ocean. *Microbiol. Sci.* **3:**205–211.
33. **Deming, J. W.** 1987. Thermophilic bacteria associated with black smokers along the East Pacific Rise, p. 325–332. *In Deuxieme Colloque International de Bacteriologie Marine, IFREMER, Actes de Colloques 3.* Centre National de la Recherche Scientifique, Brest, France.
34. **Deming, J. W.** 1993. ^{14}C tracer method for measuring microbial activity in deep-sea sediments, p. 405–414. *In* P. F. Kemp, B. F. Sherr, E. B. Sherr, and J. J. Cole (ed.), *Handbook of Methods in Aquatic Microbial Ecology.* Lewis Publishers, Boca Raton, Fla.
35. **Deming, J. W., and J. A. Baross.** 1986. Solid medium for culturing black smoker bacteria at temperatures to 120°C. *Appl. Environ. Microbiol.* **51:**238–243.
36. **Deming, J. W., and J. A. Baross.** 1993. Deep-sea smokers: windows to a subsurface biosphere? *Geochim. Cosmochim. Acta* **57:**3219–3230.
37. **Deming, J. W., and J. A. Baross.** 1993. The early diagenesis of organic matter: bacterial activity, p. 119–144. *In* M. H. Engel and S. A. Macko (ed.), *Organic Geochemistry.* Plenum Press, New York, N.Y.
38. **Deming, J. W., and J. A. Baross.** 2000. Survival, dormancy and nonculturable cells in extreme deep-sea environments, p. 147–197. *In* R. R. Colwell and D. J. Grimes (ed.), *Nonculturable Microorganisms in the Environment.* ASM Press, Washington, D.C.
39. **Deming, J. W., and J. A. Baross.** 2002. Search and discovery of microbial enzymes from thermally extreme environments in the ocean, p. 327–362. *In* R. P. Dick and R. G. Burns (ed.), *Enzymes in the Environment, Activity, Ecology, and Applications.* Marcel Dekker, Inc., New York, N.Y.
40. **Deming, J. W., and S. D. Carpenter.** Factors influencing benthic bacterial abundance and activity on the northern continental slope and abyssal plain of the Gulf of Mexico. Submitted for publication.
41. **Deming, J. W., and R. R. Colwell.** 1981. Barophilic bacteria associated with deep-sea animals. *BioScience* **31:**507–511.
42. **Deming, J. W., and R. R. Colwell.** 1985. Observations of barophilic microbial activity in samples of sediment and intercepted particulates from the Demerara abyssal plain. *Appl. Environ. Microbiol.* **50:**1002–1006.
43. **Deming, J. W., H. Hada, R. R. Colwell, K. R. Luehrsen, and G. E. Fox.** 1984. The ribonucleotide sequence of 5S rRNA from two strains of deep-sea barophilic bacteria. *J. Gen. Microbiol.* **130:**1911–1920.

44. **Deming, J. W., L. K. Somers, W. L. Straube, D. G. Swartz, and M. T. MacDonell.** 1988. Isolation of an obligately barophilic bacterium and description of a new genus, *Colwellia* gen. nov. *Syst. Appl. Microbiol.* **10:**152–160.
45. **Dietz, A. S., and A. A. Yayanos.** 1978. Silica gel media for isolating and studying bacteria under hydrostatic pressure. *Appl. Environ. Microbiol.* **36:**966–968.
46. **Dixon, J. L., and C. M. Turley.** 2000. The effect of water depth on bacterial numbers, thymidine incorporation rates and C:N ratios in northeast Atlantic surficial sediments. *Hydrobiologia* **440:**217–225.
47. **Eder, W., L. L. Jahnke, M. Schmidt, and R. Huber.** 2001. Microbial diversity of the brine-seawater interface of the Kebrit Deep, Red Sea, studied via 16S rRNA gene sequences and cultivation methods. *Appl. Environ. Microbiol.* **67:**3077–3085.
48. **Edwards, K. J., D. R. Rogers, C. O. Wirsen, and T. M. McCollom.** 2003. Isolation and characterization of novel psychrophilic, neutrophilic, Fe-oxidizing, chemolithotrophic α- and γ-*Proteobacteria* from the deep sea. *Appl. Environ. Microbiol.* **69:**2906–2913.
49. **Francis, C. A., K. J. Roberts, J. M. Beman, A. E. Santoro, and B. B. Oakley.** 2005. Ubiquity and diversity of ammonia-oxidizing archaea in water columns and sediments of the ocean. *Proc. Natl. Acad. Sci. USA* **102:**14683–14688.
50. **Hallum, S. J., N. Putnam, C. M. Preston, J. C. Detter, D. Rokhsar, P. M. Richardson, and E. F. DeLong.** 2004. Reverse methanogenesis: testing the hypothesis with environmental genomics. *Science* **305:**1457–1462.
51. **Harmsen, H. J. M., D. Prieur, and C. Jeanthon.** 1997. Distribution of microorganisms in deep-sea hydrothermal chimneys investigated by whole-cell hybridization and enrichment culture of thermophilic subpopulations. *Appl. Environ. Microbiol.* **63:**2876–2883.
52. **Hedrick, D. B., R. J. Pledger, D. C. White, and J. A. Baross.** 1992. In situ microbial ecology of hydrothermal vent sediments. *FEMS Microbiol. Ecol.* **101:**1–10.
53. **Helmke, E., and H. Weyland.** 1986. Effect of hydrostatic pressure and temperature on the activity and synthesis of chitinases of Antarctic Ocean bacteria. *Mar. Biol.* **91:**1–7.
54. **Herndl, G. J., T. Reinthaler, E. Teira, H. van Aken, C. Veth, A. Pernthaler, and J. Pernthaler.** 2005. Contribution of *Archaea* to total prokaryotic production in the deep Atlantic Ocean. *Appl. Environ. Microbiol.* **71:**2303–2309.
55. **Holden, J. F., and J. A. Baross.** 1995. Enhanced thermotolerance by hydrostatic pressure in the deep-sea hyperthermophile *Pyrococcus* strain ES4. *FEMS Microbiol. Ecol.* **18:**27–34.
56. **Holden, J. F., and R. M. Daniel.** 2004. The upper temperature limit for life based on hyperthermophile culture experiments and field observations in the subseafloor biosphere at mid-ocean ridges. *Geophys. Monogr. Ser.* **144:**13–24.
57. **Holden, J. F., M. Summit, and J. A. Baross.** 1998. Thermophilic and hyperthermophilic microorganisms in 3–30°C hydrothermal fluids following a deep-sea volcanic eruption. *FEMS Microbiol. Ecol.* **25:**33–41.
58. **Holden, J. F., K. Takai, M. Summit, S. Bolton, J. Zyskowski, and J. A. Baross.** 2001. Diversity among three novel groups of hyperthermophilic deep-sea *Thermococcus* species from three sites in the northeastern Pacific Ocean. *FEMS Microbiol. Ecol.* **36:**51–60.
59. **Horikoshi, K.** 1998. Barophiles: deep-sea microorganisms adapted to an extreme environment. *Curr. Opin. Microbiol.* **1:**291–295.
60. **Huber, H., M. J. Hohn, R. Rachel, T. Fuchs, V. C. Wimmer, and K. O. Stetter.** 2002. A new phylum of Archaea represented by a nanosized hyperthermophilic symbiont. *Nature* (London) **417:**63–67.
61. **Huber, J. A., H. P. Johnson, D. A. Butterfield, and J. A. Baross.** 2006. Microbial life in ridge flank crustal fluids. *Environ. Microbiol.* **8:**88–99.
62. **Jaenicke, R.** 1987. Cellular components under extremes of pressure and temperature: structure-function relationship of enzymes under pressure, p. 257–272. *In* H. W. Jannasch, R. E. Marquis, and A. M. Zimmerman (ed.), *Current Perspectives in High Pressure Biology*. Academic Press, London, United Kingdom.
63. **Jahnke, R. A., and M. B. Christiansen.** 1989. A free-vehicle benthic chamber instrument for sea floor studies. *Deep-Sea Res.* **36:**625–637.
64. **Jannasch, H. W., and C. O. Wirsen.** 1977. Retrieval of concentrated and undecompressed microbial populations from the deep sea. *Appl. Environ. Microbiol.* **33:**642–646.
65. **Jannasch, H. W., and C. O. Wirsen.** 1981. Morphological survey of microbial mats near deep-sea thermal vents. *Appl. Environ. Microbiol.* **41:**528–538.
66. **Jannasch, H. W., and C. O. Wirsen.** 1982. Microbial activities in undecompressed and decompressed deep-seawater samples. *Appl. Environ. Microbiol.* **43:**1116–1124.
67. **Jannasch, H. W., C. O. Wirsen, and C. D. Taylor.** 1982. Deep-sea bacteria: isolation in the absence of decompression. *Science* **216:**1315–1317.
68. **Jannasch, H. W., C. O. Wirsen, and C. L. Winget.** 1973. A bacteriological, pressure-retaining, deep-sea sampler and culture vessel. *Deep-Sea Res.* **20:**661–664.
69. **Jolivet, E., S. L'Haridon, E. Corre, P. Forterre, and D. Prieur.** 2003. *Thermococcus gammatolerans* sp. nov., a hyperthermophilic archaeon from a deep-sea hydrothermal vent that resists ionizing radiation. *Int. J. Syst. Evol. Microbiol.* **53:**847–851.
70. **Jorgensen, B. B.** 1978. A comparison of methods for the quantification of bacterial sulfate reduction in coastal marine sediments. I. Measurements with radiotracer techniques. *Geomicrobiol. J.* **1:**11–27.
71. **Jorgensen, B. B., M. F. Isaksen, and H. W. Jannasch.** 1992. Bacterial sulfate reduction above 100°C in deep-sea hydrothermal vent sediments. *Science* **258:**1756–1757.
72. **Jumars, P. A., L. M. Mayer, J. W. Deming, J. A. Baross, and R. A. Wheatcroft.** 1990. Deep-sea deposit-feeding strategies suggested by environmental and feeding constraints. *Philos. Trans. R. Soc. Lond.* A **331:**85–101.
73. **Kallmeyer, J., and A. Boetius.** 2004. Effects of temperature and pressure on sulfate reduction and anaerobic oxidation of methane in hydrothermal sediments of Guaymas Basin. *Appl. Environ. Microbiol.* **70:**1231–1233.
74. **Kallmeyer, J., T. G. Ferdelman, K.-H. Jansen, and B. B. Jørgensen.** 2003. A high-pressure thermal gradient block for investigating microbial activity in multiple deep-sea samples. *J. Microbiol. Methods* **55:**165-172.
75. **Karl, D. M.** 1978. Distribution, abundance, and metabolic states of microorganisms in the water column and sediments of the Black Sea. *Limnol. Oceanogr.* **23:**936–949.
76. **Karl, D. M., D. J. Burns, K. Orrett, and H. W. Jannasch.** 1984. Thermophilic microbial activity in samples from deep-sea hydrothermal vents. *Mar. Biol. Lett.* **5:**227–231.
77. **Karl, D. M., G. T. Taylor, J. A. Novitsky, H. W. Jannasch, C. O. Wirsen, N. R. Pace, D. J. Lane, G. J. Olsen, and S. J. Giovannoni.** 1988. A microbiological study of Guaymas Basin high temperature hydrothermal vents. *Deep-Sea Res.* **35:**777–791.
78. **Karner, M. B., E. F. DeLong, and D. M. Karl.** 2001. Archaeal dominance in the mesopelagic zone of the Pacific Ocean. *Nature* (London) **409:**507–510.

79. **Kashefi, K., and D. Lovley.** 2003. Extending the upper temperature limit for life. *Science* **301:**934.
80. **Kato, C., and Y. Nogi.** 2001. Correlation between phylogenetic structure and function: examples from deep-sea *Shewanella*. *FEMS Microbiol. Ecol.* **35:**223–230.
81. **Kaye, J. Z., and J. A. Baross.** 2000. High incidence of halotolerant bacteria in Pacific hydrothermal vent and pelagic environments. *FEMS Microbiol. Ecol.* **32:**249–260.
82. **Kaye, J. Z., and J. A. Baross.** 2004. Synchronous effects of temperature, hydrostatic pressure, and salinity on growth, phospholipid profiles, and protein patterns of four *Halomonas* species isolated from deep-sea hydrothermal vent and sea surface environments. *Appl. Environ. Microbiol.* **70:**6220–6229.
83. **Kelley, D. S., J. A. Baross, and J. R. Delaney.** 2002. Volcanoes, fluids, and life at mid-ocean ridge spreading centers. *Annu. Rev. Earth Planet. Sci.* **30:**385–491.
84. **Kelley, D. S., J. A. Karson, G. L. Früh-Green, D. R. Yoerger, T. M. Shank, D. A. Butterfield, J. M. Hayes, M. O. Schrenk, E. J. Olson, G. Proskurowski, M. Jakuba, A. Bradley, B. Larson, K. Ludwig, D. Glickson, K. Buckman, A. S. Bradley, W. J. Brazelton, K. Roe, M. J. Elend, A. Delacour, S. M. Bernasconi, M. D. Lilley, J. A. Baross, R. E. Summons, and S. P. Sylva.** 2005. A serpentinite-hosted ecosystem: the Lost City hydrothermal field. *Science* **307:**1428-1434.
85. **Kelly, R. M., and J. W. Deming.** 1988. Extremely thermophilic archaebacteria: biological and engineering considerations. *Biotechnol. Prog.* **4:**47–62.
86. **Kushner, D. J.** (ed.). 1978. *Microbial Life in Extreme Environments*. Academic Press, New York, N.Y.
87. **Laksanalamai, P., and F. T. Robb.** 2004. Small heat shock proteins from extremophiles: a review. *Extremophiles* **8:**1–11.
88. **Lam, P., J. P. Cowen, and R. D. Jones.** 2004. Autotrophic ammonia oxidation in a deep-sea hydrothermal plume. *FEMS Microbiol. Ecol.* **47:**191–206.
89. **Landau, J. V., and L. Thibodeau.** 1962. The micromorphology of *Amoeba proteus* during pressure-induced changes in the sol-gel cycle. *Exp. Cell Res.* **27:**591–594.
90. **LePage, E., E. Marguet, C. Geslin, O. Matte-Tailliez, W. Zillig, P. Forterre, and P. Tailliez.** 2004. Molecular diversity of new *Thermococcales* isolates from a single area of hydrothermal deep-sea vents as revealed by randomly amplified polymorphic DNA fingerprinting and 16S rRNA gene sequence analysis. *Appl. Environ. Microbiol.* **70:**1277–1286.
91. **Lochte, K., and C. M. Turley.** 1988. Bacteria and cyanobacteria associated with phytodetritus in the deep sea. *Nature* (London) **333:**67–69.
92. **Madrid, V. M., G. T. Taylor, M. I. Scranton, and A. Y. Chistoserdov.** 2001. Phylogenetic diversity of bacterial and archaeal communities in the anoxic zone of the Cariaco Basin. *Appl. Environ. Microbiol.* **67:**1663–1674.
93. **Marteinsson, V. T., P. Moulin, J.-L. Birrien, A. Gambacorta, M. Vernet, and D. Prieur.** 1997. Physiological responses to stress conditions and barophilic behavior of the hyperthermophilic vent archeon *Pyrococcus abyssi*. *Appl. Environ. Microbiol.* **63:**1230–1236.
94. **McCliment, E. A., K. M. Voglesonger, P. A. O'Day, E. E. Dunn, J. R. Holloway, and S. C. Cary.** 2006. Colonization of nascent, deep-sea hydrothermal vents by a novel archaeal and nanoarchaeal assemblage. *Environ. Microbiol.* **8:**114–125.
95. **McInerney, J. O., M. Wilkinson, J. W. Patching, T. M. Embley, and R. Powell.** 1995. Recovery and phylogenetic analysis of novel archaeal rRNA sequences from a deep-sea deposit feeder. *Appl. Environ. Microbiol.***61:**1646–1648.
96. **Mehta, M. P., D. A. Butterfield, and J. A. Baross.** 2003. Phylogenetic diversity of nitrogenase (*nifH*) genes in deep-sea and hydrothermal vent environments. *Appl. Environ. Microbiol.* **69:**960–970.
97. **Mehta, M. P., J. A. Huber, and J. A. Baross.** 2005. Incidence of novel and potentially archaeal nitrogenase genes in the deep Northeast Pacific Ocean. *Environ. Microbiol.* **7:**1525–1534.
98. **Methé, B. A., K. E. Nelson, J. W. Deming, B. Momen, E. Melamud, X. Zhang, J. Moult, R. Madupa, W. C. Nelson, R. J. Dodson, L. M. Brinkac, S. C. Daugherty, A. S. Durkin, R. T. DeBoy, J. F. Kolonay, S. A. Sullivan, L. Zhou, T. M. Davidsen, M. Wu, A. L. Huston, M. Lewis, B. Weaver, J. F. Weidman, H. Khouri, T. R. Utterback, T. V. Feldblyum, and C. M. Fraser.** 2005. The psychrophilic lifestyle as revealed by the genome sequence of *Colwellia psychrerythraea* 34H through genomic and proteomic analyses. *Proc. Natl. Acad. Sci. USA* **102:**10913–10918.
99. **Meyer-Reil, L.-A.** 1986. Measurement of hydrolytic activity and incorporation of dissolved organic substrates by microorganisms in marine sediments. *Mar. Ecol. Prog. Ser.* **31:**143–149.
100. **Meyer-Reil, L.-A., and M. Koster.** 1992. Microbial life in pelagic sediments: the impact of environmental parameters on enzymatic degradation of organic material. *Mar. Ecol. Prog. Ser.* **81:**65–72.
101. **Michels, P. C., D. Hei, and D. S. Clark.** 1996. Pressure effects on enzyme activity and stability at high temperatures. *Adv. Protein Chem.* **48:**341–376.
102. **Morita, R. Y.** 1975. Psychrophilic bacteria. *Bacteriol. Rev.* **39:**144–167.
103. **Moriya, K., T. Inada, M. Kyo, and K. Horikoshi.** 1995. Large-scale fermentation under high hydrostatic pressure using a newly developed deep-sea baro/thermophilic collection and cultivation system. *J. Mar. Biotechnol.* **2:**175–177.
104. **Nakagawa, S., K. Takai, K. Horikoshi, and Y. Sako.** 2004. *Aeropyrum camini* sp. nov., a strictly aerobic, hyperthermophilic archaeon from a deep-sea hydrothermal vent chimney. *Int. J. Syst. Evol. Microbiol.* **54:**329–335.
105. **Nakagawa, T., J.-I. Ishibashi, A. Maruyama, T. Yamanaka, Y. Morimoto, H. Kimura, T. Urabe, and M. Fukui.** 2004. Analysis of dissimilatory sulfite reductase and 16S rRNA gene fragments from deep-sea hydrothermal sites of the Suiyo Seamount, Izu-Bonin Arc, Western Pacific. *Appl. Environ. Microbiol.* **70:**393–403.
106. **Nakayama, A., Y. Yano, and K. Yoshida.** 1994. New method for isolating barophiles from intestinal contents of deep-sea fishes retrieved from the abyssal zone. *Appl. Environ. Microbiol.* **60:**4210–4212.
107. **Nelson, C. M., M. R. Schuppenhauer, and D. S. Clark.** 1992. High-pressure, high-temperature bioreactor for comparing effects of hyperbaric and hydrostatic pressure on bacterial growth. *Appl. Environ. Microbiol.* **58:**1789–1793.
108. **Nercessian, O., A.-L. Reysenbach, D. Prieur, and C. Jeanthon.** 2003. Archaeal diversity associated with *in situ* samplers deployed on hydrothermal vents on the East Pacific Rise (13°N). *Environ. Microbiol.* **5:**492–502.
109. **Nogi, Y., S. Hosoya, C. Kato, and K. Horikoshi.** 2004. *Colwellia piezophila* sp. nov., a novel piezophilic species from deep-sea sediments in the Japan Trench. *Int. J. Syst. Evol. Microbiol.* **52:**1527–1532.
110. **Noll, K. M., and M. Vargas.** 1997. Recent advances in genetic analyses of hyperthermophilic Archaea and Bacteria. *Arch. Microbiol.* **168:**73–80.
111. **Oliver, J. D.** 1993. Formation of viable but nonculturable cells, p. 239–272. *In* S. Kjelleberg (ed.), *Starvation in Bacteria*. Plenum Press, New York, N.Y.
112. **Oren, A.** 2002. *Halophilic Microorganisms and Their Environments*. Kluwer Academic Press, Dordrecht, The Netherlands.

113. **Parkes, R. J., G. Webster, B. A. Cragg, A. J. Weightman, C. J. Newberry, T. G. Ferdelman, J. Kallmeyer, B. B. Jørgensen, I. W. Aiello, and J. C. Fry.** 2005. Deep sub-seafloor prokaryotes stimulated at interfaces over geological time. *Nature* (London) **436:**390–394.
114. **Patching, J. W., and D. Eardly.** 1997. Bacterial biomass and activity in the deep waters of the eastern Atlantic—evidence of a barophilic community. *Deep-Sea Res.* **44:** 1655-1670.
115. **Phillips, H., L. E. Wells, R. V. Johnson II, S. Elliott, and J. W. Deming.** 2003. LAREDO: a new instrument for sampling and in situ incubation of deep-sea hydrothermal vent fluids. *Deep-Sea Res. I* **50:**1375–1387.
116. **Poremba, K.** 1994. Simulated degradation of phytodetritus in deep-sea sediments of the NE Atlantic (47°N, 19°W). *Mar. Ecol. Prog. Ser.* **105:**291–299.
117. **Poremba, K.** 1994. Impact of pressure on bacterial activity in water columns situated at the European continental margin. *Netherlands J. Sea Res.* **33:**29–35.
118. **Poremba, K.** 1995. Hydrolytic enzymatic activity in deep-sea sediments. *FEMS Microbiol. Ecol.* **16:**213–222.
119. **Qian, Y., M. H. Engel, S. A. Macko, S. Carpenter, and J. W. Deming.** 1993. Kinetics of peptide hydrolysis and amino acid decomposition at high temperature. *Geochim. Cosmochim. Acta* **57:**1271–1274.
120. **Quéric, N.-V., T. Soltwedel, and W. E. Arntz.** 2004. Application of a rapid direct viable count method to deep-sea sediment bacteria. *J. Microbiol. Methods* **57:**351–367.
121. **Relexans, J.-C., J. W. Deming, A. Dinet, J.-F. Gaillard, and M. Sibuet.** 1996. Sedimentary organic matter and micro-meiobenthos with relation to trophic conditions in the northeast tropical Atlantic. *Deep-Sea Res. I* **43:** 1343–1368.
122. **Reysenbach, A.-L., and E. Shock.** 2002. Merging genomes with geochemistry in hydrothermal ecosystems. *Science* **296:**1077–1082.
123. **Rice, S. A., and J. D. Oliver.** 1992. Starvation response of the marine barophile CNPT-3. *Appl. Environ. Microbiol.* **58:**2432–2437.
124. **Robb, F. T., K. R. Sowers, S. DasSarma, A. R. Place, H. J. Schreier, and E. M. Fleischmann (ed.).** 1995. *Archaea, a Laboratory Manual.* Cold Spring Harbor Laboratory Press, Plainview, N.Y.
125. **Rochelle, P. A., J. C. Fry, R. J. Parkes, and A. J. Weightman.** 1992. DNA extraction for 16S rRNA gene analysis to determine genetic diversity in deep sediment communities. *FEMS Microbiol. Lett.* **100:**59–66.
126. **Rowe, G. T., and M. Sibuet.** 1983. Recent advances in instrumentation in deep-sea biological research, p. 81–95. *In* G. T. Rowe (ed.), *The Sea*, vol. 8. *Deep-Sea Biology.* John Wiley & Sons, New York, N.Y.
127. **Schippers, A., L. N. Neretin, J. Kallmeyer, T. G. Ferdelman, B. A. Cragg, R. J. Parkes, and B. B. Jorgensen.** 2005. Prokaryotic cells of the deep subseafloor biosphere identified as living bacteria. *Nature* (London) **433:**861–864.
128. **Schrenk, M. O., D. S. Kelley, and J. A. Baross.** 1999. Attachment of hyperthermophilic microorganisms to mineral substrata: in situ observations and subseafloor analogs. *Proc. Geol. Soc. Am.* **31:**7.
129. **Schrenk, M. O., D. S. Kelley, J. R. Delaney, and J. A. Baross.** 2003. Incidence and diversity of microorganisms within the walls of an active deep-sea sulfide chimney. *Appl. Environ. Microbiol.* **69:**3580–3592.
130. **Schrenk, M. O., D. S. Kelley, S. A. Bolton, and J. A. Baross.** 2005. Low archaeal diversity linked to subseafloor geochemical processes at the Lost City hydrothermal field, Mid-Atlantic Ridge. *Environ. Microbiol.* **6:**1086–1095.
131. **Soltwedel, T., C. Hasemann, N.-V. Quéric, and K. von Juterzenka.** 2005. Gradients in activity and biomass of the small benthic biota along a channel system in the deep Western Greenland Sea. *Deep-Sea Res. I* **52:**815–835.
132. **Straube, W. L., J. W. Deming, C. C. Somerville, R. R. Colwell, and J. A. Baross.** 1990. Particulate DNA in smoker fluids: evidence for existence of microbial populations in hot hydrothermal systems. *Appl. Environ. Microbiol.* **56:**1440–1447.
133. **Straube, W. L., M. O'Brien, K. Davis, and R. R. Colwell.** 1990. Enzymatic profiles of 11 barophilic bacteria under in situ conditions: evidence for pressure modulation of phenotype. *Appl. Environ. Microbiol.* **56:**812–814.
134. **Summit, M., and J. A. Baross.** 2001. A novel microbial habitat in the mid-ocean ridge subseafloor. *Proc. Natl. Acad. Sci. USA* **98:**2158–2163.
135. **Summit, M., B. Scott, K. Nielsen, E. Mathur, and J. Baross.** 1998. Pressure enhances thermal stability of DNA polymerase from three thermophilic organisms. *Extremophiles* **2:**339–345.
136. **Summit, M., A. Peacock, D. Ringelberg, D. C. White, and J. A. Baross.** 2000. Phospholipid fatty acid-derived microbial biomass and community dynamics in hot, hydrothermally influenced sediments from Middle Valley, Juan de Fuca Ridge. *Proc. ODP Sci. Results* **169:** 1–19.
137. **Sunamura, M., Y. Higashi, C. Miyako, J. Ishibashi, and A. Murayama.** 2004. Two *Bacteria* phylotypes are predominant in the Suiyo Seamount hydrothermal plume. *Appl. Environ. Microbiol.* **70:**1190–1198.
138. **Tabor, P. S., J. W. Deming, K. Ohwada, H. Davis, M. Waxman, and R. R. Colwell.** 1981. A pressure-retaining deep ocean sampler for measurement of microbial activity in the deep sea. *Microb. Ecol.* **7:**51–65.
139. **Takai, K., and K. Horikoshi.** 1999. Genetic diversity of Archaea in deep-sea hydrothermal vent environments. *Genetics* **152:**1285-1297.
140. **Takai, K., A. Sugai, T. Itoh, and K. Horikoshi.** 2000. *Paleococcus ferrophilus* gen. nov., sp. nov., a barophilic, hyperthermophilic archaeon from a deep-sea hydrothermal vent chimney. *Int. J. Syst. Evol. Microbiol.* **50:**489–500.
141. **Takai, K., T. Komatsu, F. Inagaki, and K. Horikoshi.** 2001. Distribution of archaea in a black smoker chimney structure. *Appl. Environ. Microbiol.* **67:**3618–3629.
142. **Tamburini, C., J. Garcin, M. Ragot, and A. Bianchi.** 2002. Biopolymer hydrolysis and bacterial production under ambient hydrostatic pressure through a 2000 m water column in the NW Mediterranean Sea. *Deep-Sea Res. II* **49:**2109–2123.
143. **Tamburini, C., J. Garcin, and A. Bianchi.** 2003. Role of deep-sea bacteria in organic matter mineralization and adaptation to hydrostatic pressure conditions in the NW Mediterranean Sea. *Aquat. Microb. Ecol.* **32:**209–218.
144. **Taylor, C. D.** 1979. Growth of a bacterium under a high-pressure oxyhelium atmosphere. *Appl. Environ. Microbiol.* **37:**42–49.
145. **Teira, E., T. Reinthaler, A. Pernthaler, J. Pernthaler, and G. J. Herndl.** 2004. Combining catalyzed reporter deposition-fluorescence in situ hybridization and microautoradiography to detect substrate utilization by *Bacteria* and *Archaea* in the deep ocean. *Appl. Environ. Microbiol.* **70:**4411–4414.
146. **Tholosan, O., J. Garcin, and A. Bianchi.** 1999. Effects of hydrostatic pressure on microbial activity through a 2000-m deep water column in the NW Mediterranean Sea. *Mar. Ecol. Prog. Ser.* **183:**49–57.

147. **Toffin, L., A. Bidault, P. Pignet, B. J. Tindall, A. Slobodkin, C. Kato, and D. Prieur.** 2004. *Shewanella profunda* sp. nov., isolated from deep marine sediments of the Nankai Trough. *Int. J. Syst. Evol. Microbiol.* **54:** 1943–1949.
148. **Trent, J. D., and A. A. Yayanos.** 1985. Pressure effects on the temperature range for growth and survival of the marine bacterium *Vibrio harveyi*: implications for bacteria attached to sinking particles. *Mar. Biol.* **89:**165–172.
149. **Tuovila, B. J., F. C. Dobbs, P. S. LaRock, and B. Z. Siegel.** 1987. Preservation of ATP in hypersaline environments. *Appl. Environ. Microbiol.* **53:**2749–2753.
150. **Turley, C. M.** 1993. The effect of pressure on leucine and thymidine incorporation by free-living bacteria and by bacteria attached to sinking oceanic particles. *Deep-Sea Res.* **40:**2193–2206.
151. **Turley, C. M.** 2000. Bacteria in the cold deep-sea benthic boundary layer and sediment-water interface of the NE Atlantic. *FEMS Microbiol. Ecol.* **33:**89–99.
152. **Turley, C. M., and J. L. Dixon.** 2002. Bacterial numbers and growth in surficial deep-sea sediments and phytodetritus in the NE Atlantic: relationships with particulate organic carbon and total nitrogen. *Deep-Sea Res. I* **49:**815–826.
153. **Urios, L., V. Cueff, P. Pignet, and G. Barbier.** 2004. *Tepidibacter formicigenes* sp. nov., a novel spore-forming bacterium isolated from a Mid-Atlantic hydrothermal vent. *Int. J. Syst. Evol. Microbiol.* **54:**439–443.
154. **Vezzi, A., S. Campanaro, M. D'Angelo, F. Simonato, N. Vitulo, F. M. Lauro, A. Cestaro, G. Malacrida, B. Simionati, N. Cannata, C. Romualdi, D. H. Bartlett, and G. Valle.** 2005. Life at depth: *Photobacterium profundum* genome sequence and expression analysis. *Science* **307:**1459–1461.
155. **Weber, A., and B. B. Jorgensen.** 2002. Bacterial sulfate reduction in hydrothermal sediments of the Guaymas Basin, Gulf of California, Mexico. *Deep-Sea Res. I* **49:** 827–841.
156. **Welch, T. J., and D. H. Bartlett.** 1997. Cloning, sequencing and overexpression of the gene encoding malate dehydrogenase from the deep-sea bacterium *Photobacterium* species strains SS9. *Biochim. Biophys. Acta* **1350:**41–46.
157. **Wells, L. E., M. Cordray, S. Bowerman, L. Miller, W. F. Vincent, and J. W. Deming.** 2006. Archaea in particle-rich waters of the Beaufort Shelf and Franklin Bay, Canadian Arctic: clues to an allochthonous origin? *Limnol. Oceanogr.* **51:**47–59.
158. **Whitman, W. B., D. C. Coleman, and W. J. Wiebe.** 1998. Prokaryotes: the unseen majority. *Proc. Natl. Acad. Sci. USA* **95:**6578–6583.
159. **Wirsen, C. O., and H. W. Jannasch.** 1983. In situ studies on deep-sea amphipods and their intestinal microflora. *Mar. Biol.* **78:**69–73.
160. **Wirsen, C. O., and H. W. Jannasch.** 1986. Microbial transformations in deep-sea sediments: free-vehicle studies. *Mar. Biol.* **91:**277–284.
161. **Witte, U., F. Wenzhöfer, S. Sommer, A. Boetius, P. Heinz, N. Aberle, M. Sand, A. Cremer, W.-R. Abraham, B. B. Jørgensen, and O. Pfannkuche.** 2003. In situ experimental evidence of the fate of a phytodetritus pulse at the abyssal sea floor. *Nature* (London) **424:**763–766.
162. **Xu, Y., Y. Nogi, C. Kato, Z. Liang, H.-J. Rüger, D. De Kegel, and N. Glansdorff.** 2003. *Moritella profunda* sp. nov. and *Moritella abyssi* sp. nov., two psychropiezophilic organisms isolated from deep Atlantic sediments. *Int. J. Syst. Evol. Microbiol.* **53:**533–538.
163. **Yamaguchi, A., Y. Watanabe, H. Ishida, T. Harimoto, K. Furusawa, S. Suzuki, J. Ishizaka, T. Ikeda, and M. M. Takahashi.** 2002. Structure and size distribution of plankton communities down to the greater depths in the western North Pacific Ocean. *Deep-Sea Res. II* **49:** 5513–5529.
164. **Yayanos, A. A.** 1982. Recovery and maintenance of live amphipods at a pressure of 580 bars from an ocean depth of 5700 meters. *Science* **200:**1056–1059.
165. **Yayanos, A. A.** 1982. Deep-sea biophysics, p. 409-416. *In Subseabed Disposal Program Annual Report January to September 1981*, vol. II. *Appendices (Principal Investigator Progress Reports)*. Sandia National Laboratories, Albuquerque, N.Mex.
166. **Yayanos, A. A.** 1995. Microbiology to 10,500 meters in the deep sea. *Annu. Rev. Microbiol.* **49:**777–805.
167. **Yayanos, A. A., and E. F. DeLong.** 1987. Deep-sea bacterial fitness to environmental temperatures and pressures, p. 17–32. *In* H. W. Jannasch, R. E. Marquis, and A. M. Zimmerman (ed.), *Current Perspectives in High Pressure Biology*. Academic Press, London, United Kingdom.
168. **Yayanos, A. A., A. S. Dietz, and R. Van Boxtel.** 1979. Isolation of a deep-sea barophilic bacterium and some of its growth characteristics. *Science* **205:**808–810.
169. **Yayanos, A. A., A. S. Dietz, and R. Van Boxtel.** 1981. Obligately barophilic bacterium from the Mariana Trench. *Proc. Natl. Acad. Sci. USA* **78:**5212–5215.
170. **Yayanos, A. A., and R. Van Boxtel.** 1982. Coupling device for quick high-pressure connections to 100 MPa. *Rev. Sci. Instrum.* **53:**704–705.
171. **Yayanos, A. A., R. Van Boxtel, and A. S. Dietz.** 1983. Reproduction of *Bacillus stearothermophilus* as a function of temperature and pressure. *Appl. Environ. Microbiol.* **46:** 1357–1363.
172. **Yayanos, A. A., R. Van Boxtel, and A. S. Dietz.** 1984. High-pressure-temperature gradient instrument: use for determining the temperature and pressure limits of bacterial growth. *Appl. Environ. Microbiol.* **48:**771–776.
173. **ZoBell, C. E., and R. Y. Morita.** 1957. Barophilic bacteria in some deep-sea sediments. *J. Bacteriol.* **73:**563–568.

SOIL, RHIZOSPHERE, AND PHYLLOSPHERE

VOLUME EDITOR
DAVID A. LIPSON

Introduction and Overview: Soil, Rhizosphere, and Phyllosphere

DAVID A. LIPSON

47

To the uninitiated, it might seem almost paradoxical that such an outwardly humble substance as soil should bear such advanced technical treatments as those presented here: microarrays, reporter genes, molecular probes, fluorescence microscopy, metagenomics, proteomics, and others. However, soil microbial communities are arguably among the most complex biological entities, dwelling in extremely heterogeneous and complex physical environments. The bacterial diversity in soils apparently dwarfs that of aquatic and marine ecosystems (7, 19), with estimates ranging from around 6,000 up to 38,000 species per g (5, 7, 23). In soils, a myriad of habitats exist within a few millimeters of each other, ranging broadly in physical characteristics. For example, microbes may exist in anaerobic microsites within aggregates of otherwise well-aerated soils. Similarly, the diffusion of nutrients to microbial populations is affected by whether they are planktonic (floating freely in pore water) or whether they associate with particles or form biofilms on surfaces such as minerals, organic matter, and roots (3). Furthermore, soils offer a staggering array of chemical compounds (2, 22). Because of the great physical and biological complexity of soils, understanding of the ecology of microorganisms in soils has lagged behind that of larger organisms, and only recently have the most general of patterns begun to emerge. For example geographic patterns in microbial distribution, such as species-area relationships, have been detected (10, 11), though a recent study showed that variation in soil bacterial community structure and diversity over large spatial scales was best explained by pH and other soil factors, rather than by geographical distance, latitude, or climate (8). Another great challenge in environmental microbiology is connecting community structure and function (23). Progress is being made in this area by using several approaches, including those that correlate variations in community structure and process rates (14) and approaches that ascribe functions to uncultured microbes by using stable-isotope probing (15, 18) and by measuring gene expression in soils (20). There has also been some recent success in obtaining pure cultures of soil bacteria from phyla, such as the *Acidobacteria* and *Verrucomicrobia*, that are highly abundant but rarely cultured (6, 12).

The study of plant-associated microbial communities presents challenges similar to those encountered in soils. Plant surfaces provide a rich variety of habitats for microbes, offering a wide range of oxygen and nutrient content and exposure to environmental stress (1). Microbes can dwell on leaf surfaces high in the canopy and around roots tens of meters below the surface (4, 21). The microhabitats created by biofilms on these surfaces create an additional level of complexity (13, 17), and to further complicate matters, horizontal gene transfer is rampant in biofilms (16). Of all plant-associated environments, the rhizosphere is particularly elusive and challenging to study. In fact, it is difficult even to define precisely (9).

The chapters in this section are dedicated to overcoming the many challenges outlined above. The 17 chapters other than this introduction fall roughly into three categories. The first describes procedures for sampling, detecting, and isolating microorganisms from soils (chapter 48), arthropods (chapter 63), and plant surfaces (chapter 52). This group of chapters also includes chapters on sampling of viruses (chapter 50) and isolation of mycorrhizal fungi (chapter 51). Chapter 59 describes methods for identifying organisms by fluorescent in situ hybridization, and practical aspects of using PCR for studying soil microbes are covered in chapter 55. Another set of chapters deals with studying metabolic activity of microbes (chapter 57) and microbial processes such as enzymatic activity (chapter 58) and nitrogen cycling (chapter 56), as well as an approach for determining the function of specific microbial populations within complex communities (chapter 61). Chapter 60 describes the use of reporter genes for studying gene expression in microbial populations. The final set of chapters pertains to analysis of microbial communities (chapter 53) and powerful statistical and phylogenetic approaches to analyze such data (chapters 49 and 64). Chapter 62 reviews horizontal gene transfer in microbial communities, and chapter 54 presents the use of microarrays in environmental applications.

Environmental microbiology is a dynamic field that has produced many innovations in recent decades. This section presents both classical and cutting-edge techniques in an attempt to summarize the current state of the field. However, given the breadth of the field and space constraints of this volume, the techniques presented here should be supplemented by consulting the references cited within these chapters.

REFERENCES

1. **Andrews, J. H., and R. F. Harris.** 2000. The ecology and biogeography of microorganisms of plant surfaces. *Annu. Rev. Phytopathol.* **38:**145–180.

2. **Bode, H. B., B. Bethe, R. Hofs, and A. Zeeck.** 2002. Big effects from small changes: possible ways to explore nature's chemical diversity. *Chembiochem* **3:**619–627.
3. **Burns, R. G., and J. E. M. Stach.** 2002. Microbial ecology of soil biofilms: substrate bioavailability, bioremediation and complexity. *Dev. Soil Sci.* **28B:**17–42.
4. **Canadell, J., R. B. Jackson, J. R. Ehleringer, H. A. Mooney, O. E. Sala, and E. D. Schulze.** 1996. Maximum rooting depth of vegetation types at the global scale. *Oecologia* **108:**583–595.
5. **Chatzinotas, A., R. A. Sandaa, W. Shoenhuber, R. Amann, F. L. Daae, V. Torsvik, J. Zeyer, and D. Hahn.** 1998. Analysis of broad-scale differences in microbial community composition of two pristine forest soils. *Syst. Appl. Microbiol.* **21:**579–587.
6. **Chin, K.-J., D. Hahn, U. Hengstmann, W. Liesack, and P. H. Janssen.** 1999. Characterization and identification of numerically abundant culturable bacteria from the anoxic bulk soil of rice paddy microcosms. *Appl. Environ. Microbiol.* **65:**5042–5049.
7. **Curtis, T. P., W. T. Sloan, and J. W. Scannall.** 2002. Estimating prokaryotic diversity and its limits. *Proc. Natl. Acad. Sci. USA* **99:**10494–10499.
8. **Fierer, N., and R. B. Jackson.** 2006. The diversity and biogeography of soil bacterial communities. *Proc. Natl. Acad. Sci. USA* **103:**626–631.
9. **Hinsinger, P., G. R. Gobran, P. J. Gregory, and W. W. Wenzel.** 2005. Rhizosphere geometry and heterogeneity arising from root-mediated physical and chemical processes. *New Phytol.* **168:**293–303.
10. **Horner-Devine, M. C., M. Lage, J. B. Hughes, and B. J. M. Bohannan.** 2004. A taxa-area relationship for bacteria. *Nature* **432:**750–753.
11. **Hughes Martiny, J. B., B. J. M. Bohannan, J. H. Brown, R. K. Colwell, J. A. Fuhrman, J. L. Green, M. C. Horner-Devine, M. Kane, J. Adams-Krumins, C. R. Kuske, P. J. Morin, S. Naeem, L. Øvreås, A.-L. Reysenbach, V. H. Smith, and J. T. Staley.** 2006. Microbial biogeography: putting microorganisms on the map. *Nat. Rev. Microbiol.* **4:**102–112.
12. **Joseph, S. J., P. Hugenholtz, P. Sangwan, C. A. Osborne, and P. H. Janssen.** 2003. Laboratory cultivation of widespread and previously uncultured soil bacteria. *Appl. Environ. Microbiol.* **69:**7210–7215.
13. **Lindow, S. E., and M. T. Brandl.** 2003. Microbiology of the phyllosphere. *Appl. Environ. Microbiol.* **69:**1875–1883.
14. **Lipson, D. A., M. Blair, K. Grieve, G. Barron-Gafford, and R. Murthy.** 2006. Relationships between microbial community structure and soil processes under elevated atmospheric carbon dioxide. *Microb. Ecol.* **51:**302–314.
15. **Manefield, M. W., A. S. Whiteley, R. I. Griffiths, and M. J. Bailey.** 2002. RNA stable isotope probing, a novel means of linking microbial community function to phylogeny. *Appl. Environ. Microbiol.* **68:**5367–5373.
16. **Molin, S., and T. Tolker-Nielsen.** 2003. Gene transfer occurs with enhanced efficiency in biofilms and induces enhanced stabilisation of the biofilm structure. *Curr. Opin. Biotechnol.* **14:**255–261.
17. **Morris, C. E., and J. M. Monier.** 2003. The ecological significance of biofilm formation by plant-associated bacteria. *Annu. Rev. Phytopathol.* **41:**429–453.
18. **Radajewski, S., P. Ineson, N. R. Parekh, and J. C. Murrell.** 2000. Stable-isotope probing as a tool in microbial ecology. *Nature* **403:**646–649.
19. **Ritz, K., B. S. Griffiths, V. L. Torsvik, and N. B. Hendriksen.** 1997. Analysis of soil and bacterioplankton community DNA by melting profiles and reassociation kinetics. *FEMS Microbiol. Lett.* **149:**151–156.
20. **Saleh-Lakha, S., M. Miller, R. G. Campbell, K. Schneider, P. Elahimanesh, M. M. Hart, and J. T. Trevors.** 2005. Microbial gene expression in soil: methods, applications and challenges. *J. Microbiol. Methods* **63:**1–19.
21. **Schenk, H. J., and R. B. Jackson.** 2005. Mapping the global distribution of deep roots in relation to climate and soil characteristics. *Geoderma* **126:**129–140.
22. **Schulten, H.-R., and M. Schnitzer.** 1998. The chemistry of soil organic nitrogen: a review. *Biol. Fertil. Soils* **26:**1–15.
23. **Torsvik, V., and L. Øvreås.** 2002. Microbial diversity and function in soil: from genes to ecosystems. *Curr. Opin. Microbiol.* **5:**240–245.

Surface Soil Microbial Sampling Methods

DAVID B. KNAEBEL

48

We are all fairly acquainted with soils, but we are not often aware of their hidden complexity. We walk on soil; our roads traverse it; we grow our vegetables in it; we play on it or watch sports played on it. We see it, in one form or another, on a daily basis. From these vantage points, soil is a relatively mundane part of our everyday lives. These perspectives, however, belie its microscopic complexity, composition, and activity: soil possesses an incredible mineral and biological diversity and, as an entity, performs significant fractions of global mineral and gas exchange processes. The complex composition of soils—of freshly exposed and/or weathered minerals, with an assortment of organic complexes loosely or closely associated with those minerals—provides a matrix for a rich subterranean biological community, one that is comprised of a vast assortment of microorganisms. Furthermore, in addition to this microbial community of interest to the microbial ecologist, soils harbor the underground tissues of terrestrial plants as well as entire communities of macroscopic animals, invertebrates primarily, that interact with each other, with the mineral and organic matrix, and with the microbial community. While soil microbial ecologists are often interested in the dynamic effects of terrestrial plants on microbial processes, the effects of soil fauna should not be underestimated. In humid temperate soils, it is estimated that all the topsoil passes through the digestive systems of the resident earthworms every 4 years (8). These rich mineral and biological components of soils, along with the presence of a rich assortment of nutrient materials, provide a setting for an incredibly diverse microbial community. Common measures and estimates of the microbial diversity of soils propose that there are billions of individuals per gram of a typical soil, with these billions being comprised of several thousands of different types (6, 7, 23, 25, 31).

In addition to the many physical and biological interactions between the mineral, organic, macrobiological, and microbiological components of soil, one other factor also adds to its complexity, and that is the temporal component of the interactions. Different fractions of the soil interact with each other on substantially different time scales, such that some parts of dead plant materials are degraded and assimilated within a season whereas other components remain intact for decades to centuries (especially in bog soils). Some mineral weathering is enhanced by microbial activities over even longer periods of time. When summed, the interactions, and the pace of these interactions, culminate in a highly complex, dynamic environment and ecosystem that we tread upon daily, without much consideration of such complexity.

This chapter provides a first step into the analysis of this complex world of surface soil and a selection of conceptual approaches and resources for sampling surface soils so that the components of interest can be properly acquired and handled for subsequent microbial ecological studies. The focus on surface soils is one that distinguishes this chapter from others dealing with subsurface or aquifer sampling. (Operationally, the term "surface soil" refers to those soils and those soil horizons that can be collected using hand-operated tools and includes any bulk soil [barren soil] or rhizosphere [plant root-containing soil].) This chapter focuses on approaches that enable the scientist to acquire surface soil materials that are of appropriate quality and quantity for subsequent studies and provides information and resources for ensuring that soils are appropriately transported, stored, and distributed prior to use in experiments. The methods and approaches discussed in this chapter are general ones that are provided to enable scientists to address their questions. They apply to soil collection efforts, whether the subsequent experiments address any of the following.

- Understanding the fate and effects of one microbial species within a complex soil community
- Teasing apart the functional roles of microbial groups and assemblages in a complex soil community
- Understanding the roles of microbial groups within temporal and/or physical contexts
- Describing and measuring the impacts of different manipulations carried out to test hypotheses regarding soil responses to different treatments (e.g., fertilization studies, chemical fate and degradation studies, or physical manipulations)
- Other study questions of interest to the surface soil researcher

That is, the methods and approaches here are relatively irrespective of the pretreatment or subsequent use of the sample; they are based on the premise that the appropriate design of a soil collection effort and the subsequent collection of the soil samples will provide samples that can be used for the particular study design and subsequent study. As expected, there are some considerations one needs to make for collecting adequate samples for different experiments, and therefore, this chapter also provides relevant caveats and exceptions that should be considered. One primary consideration is this: the design and breadth of the experiments that are to use the soil samples are the "design drivers" for the soil collection approach and strategy. Careful consideration must take place prior to sampling, should include discussions with all team members needing access to the samples, and should touch on the experiments that they need to conduct and the data needed for subsequent analyses.

The reader is encouraged to consult other resources prior to designing experiments and collecting soil samples. There are a number of texts that provide descriptions of soil microbiology, ecology, and the soil matrix, such as *Environmental Microbiology* (16), *Principles and Applications of Soil Microbiology* (24), *Microbial Ecology: Fundamentals and Applications*, 4th ed. (1), and *Soil Microbiology and Biochemistry* (20), along with resources that will enable the collections to take place and standard methods that are available for consideration (see Table 2). One other source, Morris et al. (18), provides a rich review of the types of research that have been conducted in microbial diversity studies over the last quarter-century, and the references cited therein provide a good set of specific approaches, designs, and methods for the experimenter to consider. Finally, this chapter is not the first one written on the subject of soil sampling, and the reader is also encouraged to consult other resources and use them all to formulate an appropriate sampling plan. These resources include, among others, parallel chapters in the second edition of this manual (26) and similar sections of *Environmental Microbiology* (16).

SOIL SAMPLING STRATEGY

Surface soil sampling is a generally straightforward process that requires few sophisticated pieces of equipment but has components that need to be considered to ensure that the samples collected provide the necessary power for determining the answers to the experimental questions. The different components are described in Table 1 and expanded in the text below. The text provides considerations that serve as examples of ones that should be evaluated when selecting soil collection sites and activities to occur at those sites. The goal is to design the collection effort such that samples provide adequate discriminatory power necessary for the experimental design questions to be answered during execution. In

TABLE 1 Factors to consider in the design of soil sampling activities

Sampling issue	Considerations
Experimental design/soil collection approach	Clearly define the objectives of the study. Identify all experiments to be conducted and the number of replicates required for each experiment. Determine the number of replicates needed per site and per sample area and the volumes of soil needed for each experimental treatment and the entire experimental design.
	Define all comparisons to be made that may influence the soil collection approach. Different considerations must be given to those experiments that compare disturbed soil to intact soil (e.g., cores); bulk soil to rhizosphere; natural soil to contaminated soil; dry soil to water-logged soil; etc.
Site access considerations	Determine any access issues related to site access: are the sites public lands, private, institutional, part of a scientific study site (e.g., long-term ecological research sites, free air carbon dioxide enrichment sites, etc.), or are they regulated due to contamination or other safety issues related to the site? Contact landowner or site manager for permissions and historical data regarding the site.
Site and schedule needs	Will the experimental design call for a single sampling effort, or will the study involve repeated sample collection efforts? What is the timing of those collection efforts—is the sampling to occur weekly, monthly, seasonally, annually, etc.? If the site is to be accessed on a repeated basis, ensure that site access permissions have been granted.
	Determine the number of, and estimated schedule for, collection of all samples at each site.
	Consider methods to ensure successful sampling if there are seasonal effects, such as frozen or sun-baked soils. Consider similar approaches to ensure successful sampling during inclement weather.
Statistical needs	The sampling design and site selection design should be based on appropriate statistical needs for subsequent experiments, with enough treatment and replication to provide for detection of real differences between sites or treatments. Consult appropriate statistical references (4).
Logistic needs	Determine logistics of preparation for sampling, sampling, and processing the samples, including chain of custody and record keeping.
	Special considerations need to be developed for bulk soil versus root zone soil sampling, or for intact (e.g., core) soil versus disturbed soil sampling.
Additional considerations	What can be done to collect information about the site, either by landowner contacts, by public or institutional history research, or by map/GIS databases on the site)?
	Can each site be visited prior to the sampling effort?
	What needs to be done to prepare all the sampling equipment and support materials?
	What will happen on the day of sample collection?
	How will the procedures of the sample collection effort be recorded?
	What standard or special sample handling and transport procedures are needed?
	How will the sample be stored following collection and prior to analysis?

this case, a hypothetical scenario (see below) provides a setting to show how different factors may affect a soil sampling approach. The details of this scenario are not critical, but the considerations made for each element are worthy of consideration. In general, this scenario describes a well-planned soil sampling approach but includes one situation where an expansion of the experimental design was considered, which allowed for evaluation of a site complexity discovered late in the site visitation process. The benefits and problems associated with such expansions are discussed.

It should be noted that a list of resources (Table 2) provides sources for sampling equipment and for different standard sampling protocols that are available to the reader. The list of sampling equipment sources is not an endorsement of such equipment but provides an idea of what materials are available for collection, storage, and transport of soils. The other resources provide standard methods that have been sanctioned by government or standards agencies (e.g., the U.S. Environmental Protection Agency [EPA] and the American Society for Testing and Materials [ASTM]).

SOIL COLLECTION HYPOTHETICAL EXAMPLE

Experimental Design Objectives

The following hypothetical example provides a scenario for consideration for design of soil sample collection efforts. It serves as an example of the types of sampling that might occur but also highlights some of the logistics that need to be worked out prior to and during a sampling effort. That is, the act of sampling itself is often straightforward, but it is the experiment-specific planning and unknown situations encountered in the field and at each site that can greatly impact the success or failure of a soil sampling effort. This scenario provides a glimpse into some of these events.

One consideration that should be provided now, however, is one that simplifies the process and increases the likelihood for acceptance of the sampling method. This is the suggestion that the experimenter review and then adopt a standard, accepted method for soil sampling or use such a method as the primary basis for the collection effort. Original sampling approaches, while meeting the exact needs of the researcher, may generate samples that are exposed to inadvertent effects that could render the sample useless or of questionable status for subsequent use. For example, a novice sample collector may consider that more samples could be collected more quickly if the sampler's collection surfaces that come in contact with soils are lubricated with mineral oil. The subsequent amendment of mineral oil to the soil matrix in the sample, however, may modify the nutrient status of the soils in unknown ways, change nutrient extractability, and/or interrupt certain microbial processes. Avoiding such mistakes by the review and adoption of standard methods appropriate for the experimenter's study question increases the likelihood for successful experimentation on the samples and subsequent professional communication of the results.

Hypothetical Scenario

The objective of this particular hypothetical study is to examine the differences in soil microbial complexity as potentially affected by different periodic fertilizer treatments on commercial, agricultural turfgrass fields. The particular definition of "soil microbial complexity" is left to the reader. The definition could be stated as "the overall microbial complexity of the soil" (a dauntingly difficult determination) or could focus on the complexity of a specific subset of the microbial community. For example, the focus of the research could be on specific pesticide-degrading microorganisms, such as triazine degraders, or on subsets of the nitrogen-related community, the ammonium oxidizers.

In this scenario, there are two treatments of concern to the experimenters: (i) standard chemical fertilizer enrichments and (ii) sewage sludge amendment enrichments. Both enrichments are to be compared to turfgrass fields receiving no such amendments. Of concern to the experimenters are not only the effects of materials in the sewage sludge (e.g., metals and other contaminants) on soil microbial diversity and processes but also the existence of any distinguishable changes in the microbial complexity over time in response to these two treatments and in comparison to control sites. To address the time element, the experimenters wish to sample the soils four times over three consecutive years; this will allow them to evaluate the microbial community during year −1 (the year prior to treatment), during year 0 (both just prior to treatment and just after treatment), and during year +1 (the year following treatment).

Experiments To Be Conducted and Volumes of Soil Needed

For each site, the experimenters meet to discuss the subsequent needs and arrive at approximate masses of soils needed, as well as the statistical comparisons that they would like to conduct for each site. They determine that for each soil site to be evaluated, the experimenters will examine nucleic acid profiles (requiring 20 g of soil), determine the microbial biomass of the soil by using lipid biomarker analysis (requiring 100 g), and measure the microbial activity of the soil community by measuring uptake of ^{14}C-radiolabeled substrates into microbial biomass and their respiration to $^{14}CO_2$ (requiring 400 g). In addition, soil chemical and physical properties will be determined by appropriate soil-testing laboratories (requiring 300 g). Based on the expected volumes needed for each experimental test set, the total expected volume of soil needed per site and per collection effort is approximately 1 kg.

At each site, the experimenters will assess within-site variability and between-site variability. To do this, at each site, they will collect bulk samples from the soil immediately below each corner of a square that has 5-m-long sides and that is placed arbitrarily in a central region of each field. (The locations of the squares are photographed, and latitude and longitude coordinates are recorded.) The experimenters are interested in knowing the processes that occur both within the root zone (e.g., the rhizosphere zone) of the turfgrass and below the root zone of the turfgrass. Each sampling site will therefore consist of the removal of 1 kg of soil from the root zone of the turf immediately below each corner of the sampling square, as well as removal of 1 kg of soil approximately 10 cm below the lowest extent of the turf root zone, for a total of eight 1-kg soil samples per site. Soil samples are to be collected in large (4.6-liter) sterile, polypropylene containers.

Soil Collection Approach

The experimenters have selected commercial turfgrass fields as study sites and have located candidate sites by using U.S. Department of Agriculture (USDA) Natural Resources Conservation Service (NRCS) maps (Table 2) and by consulting with local USDA extension agents. The

TABLE 2 Soil sampling resources

Resource type	Organization	Resource and contact information[a]
Sampling equipment (e.g., augers, corers, sieves, sample containers, etc.)	AMS Inc.	105 Harrison St. American Falls, ID 83211 Phone: (800) 635-7330, (208) 226-2017 Fax: (208) 226-7280 E-mail: ams@ams-samplers.com Website: http://www.ams-samplers.com/main.shtm
	ASC Scientific	2075 Corte del Nogal, Suite T Carlsbad, CA 92011 Phone: (800) 272-4327, (760) 431-2655 Fax: (760) 431-0904 E-mail: sales@ascscientific.com Website: http://www.ascscientific.com/default.htm
	Geotechnical Services, Inc.	2852 Walnut Ave., Suite E-1 Tustin, CA 92780 Phone: (800) 524-9111, (714) 832-5610 Fax: (714) 832-8766 E-mail: info@geotechnical.net Website: http://www.geotechnical.net/rock_soil.shtml
	NASCO-Modesto	4825 Stoddard Rd. P.O. Box 3837 Modesto, CA 95352-3837 Phone: (800) 558-9595 Fax: (209) 545-1669 E-mail: custserv@enasco.com Website: http://www.enasco.com/whirlpak/
	Soil Moisture Equipment Corp.	801 S. Kellogg Ave. Goleta, CA 93117 Phone: (805) 964-3525 Fax: (805) 683-2189 E-mail: sales@soilmoisture.com Website: http://www.soilmoisture.com
	Van Walt Limited	Prestwick Lane Grayswood Haslemere, Surrey GU27 2DU United Kingdom Phone: 44 (0)1428 661 660 Fax: 44 (0)1428 656 808 E-mail: sales@vanwalt.com Website: http://www.vanwalt.co.uk/hand_augers.htm
Soil sampling methods and approaches from government sites	EPA: National Exposure Research Laboratory	*Characterization and Monitoring—Soil and Sediment Sampling Methods* http://www.epa.gov/esd/cmb/tasks/soil.htm
	EPA: Superfund	*Soil Screening Guidance* http://www.epa.gov/superfund/resources/soil/index.htm
	EPA: Superfund	*Technical Protocol for Evaluating Natural Attenuation of Chlorinated Solvents in Ground Water*, Appendix A http://www.epa.gov/superfund/resources/gwdocs/protocol.htm
	EPA: Environmental Monitoring Systems Laboratory	*Soil Sampling Quality Assurance User's Guide*, 2nd ed. (EPA 600/8-89/046) NTIS/PB89-189864 or http://clu-in.org/download/char/soilsamp.pdf
	EPA: Environmental Monitoring Systems Laboratory	*Preparation of Soil Sampling Protocols: Sampling Techniques and Strategies* (EPA 600/SR-92/128) http://nepis.epa.gov/pubtitleORD.htm (see "600SR92128 *Preparation of Soil Sampling Protocols: Sampling Techniques and Strategies. Project Summary*")

(Continued on next page)

TABLE 2 (*Continued*)

Resource type	Organization	Resource and contact information[a]
Soil sampling methods and approaches from government sites (*continued*)	ASTM	*Standard D7015-04: Standard Practices for Obtaining Undisturbed Block (Cubical and Cylindrical) Samples of Soils*
	ASTM	*Standard D5911-96 e1: Standard Practice for Minimum Set of Data Elements to Identify a Soil Sampling Site*
	ASTM	*Standard D6913-04e1: Standard Test Methods for Particle-Size Distribution (Gradation) of SoilsUsing Sieve Analysis*
	ASTM	*Standard D4700-91 e1: Standard Guide for Soil Sampling from the Vadose Zone*
	ASTM	*Standard D6640-01: Standard Practice for Collection and Handling of Soils Obtained in Core Barrel Samplers for Environmental Investigations*
	ASTM	*Standard D4220-95: Standard Practices for Preserving and Transporting Soil Samples*
	USDA Natural Resources Conservation Services (NRCS)	Parent site: http://www.nrcs.usda.gov/ Soil maps: http://websoilsurvey.nrcs.usda.gov/app/

[a]The commercial sites provided here do not constitute any kind of stated or implied endorsement. The physical, e-mail, and website addresses and other contact information were current as of the date of submission of this chapter. Some contact information may change over time.

experimenters have identified 12 potential turfgrass farms that could serve as sites to examine the two treatments along with adjacent, no-treatment control sites. Their experimental design, in consideration of treatment replication, will work if there are nine sites, with three of each type of treatment. During the landowner contact phase, it is learned that only nine landowners have fields that are at the appropriate stage of treatment (approximately 1 year before fertilizer application). These nine sites are selected, and written permissions to collect samples are acquired. Six of the nine landowners have requested to be kept apprised of the experimental results, and the experimenters record this (some landowners also desire to be kept apprised of experimental progress). The landowners have been asked to provide any records of land treatment, and all have provided reports, with varying levels of detail, regarding the applications of the different fertilizer treatments to their fields, along with the schedules for applications for the current year and the next 2 years.

Site and Scheduling Needs

In conjunction with the field treatment schedules for each farm, the site sampling activities are scheduled for the mid- to late spring of each year, such that the first sampling occurs during a year of no treatment, the second sampling occurs approximately 2 weeks prior to treatment, the third sampling occurs approximately 2 weeks following treatment, and the fourth sampling occurs 1 year following treatment. The experimenters use the number of sites (nine), the number of turf soil replicates per site (four), the number of bulk soil replicates per site (four), and the amount of soil to be collected for each replicate (1 kg) to organize the sample collection, transport, and storage needs for each site at each of the four collection times. The sampling effort will result in the collection of 36 1-kg samples at each sampling effort date and a total of 144 total samples over the four sampling dates. They realize that an enclosed van will be necessary for each date, for transport of equipment and sample collection and storage containers to the field sites and back to the laboratory. The van, in addition to providing room for the samples, can provide better temperature control than other vehicles (e.g., automobile trunks) during the sampling day. Consideration is also given to sample archiving and long-term storage, including gaining permissions to appropriate space needed in cold rooms, etc. (21).

Based on the location of the sites and the number of sites to be evaluated, it is determined that two or three sites can be sampled each day, with samples returned to the laboratory each evening; it will take 3 to 4 days to conduct all the sampling. Logistical planning for site visitation is critical to the success of their sampling, since an interruption of a few days between sampling different sites may expose the sites to different weather conditions (e.g., heavy rains) that may affect the uniformity of the soil conditions upon collection. Since sampling is to occur in the spring, the experimenters expect that the soils should be at moderate temperatures. However, immediately prior to each sampling series, weather forecasts are consulted to ensure—as much as possible—that weather conditions will be similar on the three or four consecutive collection days and that the "window" of collection days provides a consistent set of climate conditions.

Statistical Needs

In the hypothetical example, the sampling site and the experimental design call for one site to be assessed per treatment field (Fig. 1A, open circle). The experimenters have decided against compositing samples from throughout the field, as they wish to evaluate the levels of variability within each site (based on the samples collected at the different depths from each of the four corners of their 25-m^2 sampling square). That is, samples will be collected and then kept separate, not mixed together, for generating a master sample that represents the entire diversity of the particular site and depth. Rather, the experimenters are interested in assessing and understanding the within-site variability that can be afforded by such collection efforts. Each corner sample will then be subjected to replicated measurements for the assays

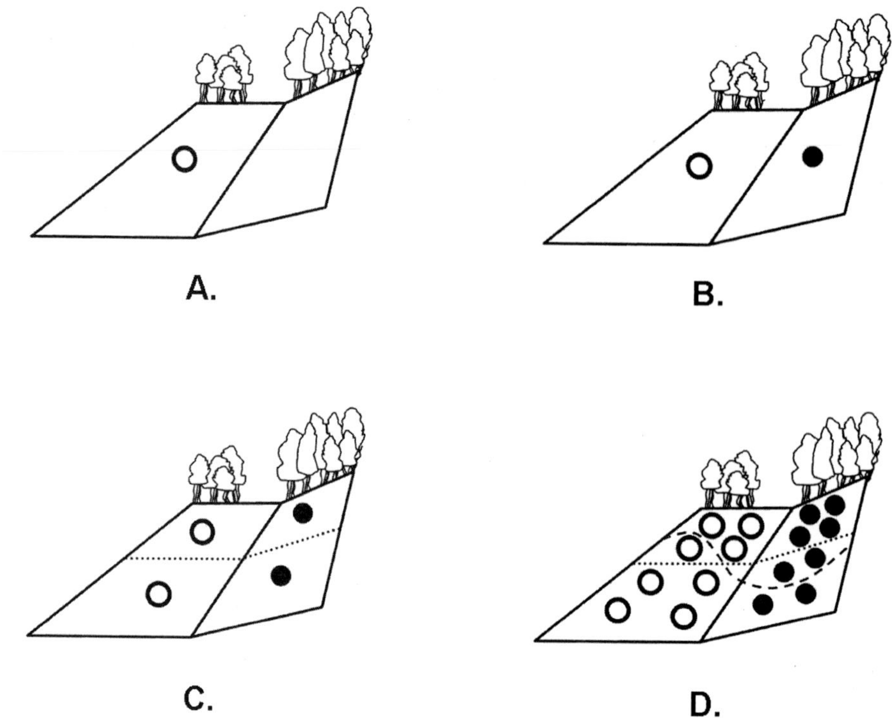

FIGURE 1 Soil sampling options for a hypothetical turfgrass microbial ecology study site (see the text).

of concern to ascertain variability within each of the four corner samples, and then these data will be applied in the comparisons between depths (rhizosphere versus nonrhizosphere soil) and sites, as well as between dates.

The experimenters know prior to arrival at their first site, based on the NRCS map analysis, that there is an area adjacent to the agricultural field on a sloping hill (based on the soil map slope designations) which does not appear to be used for farming. Upon visiting the site, however, the experimenters are informed by the field manager (not the landowner) of two additional factors: first, that the field that they were planning on sampling had been split and treated differently than expected: up until 2 years prior, the two sides had been treated with two different herbicides; and second, that up until 5 years ago, the adjacent hillside had been cropped with the same plants.

The experimenters consider the option of collecting samples to be able to evaluate the historical effects of pesticide treatments and the crop plants (on the hillside field) on the microbial ecology of that soil (now gone fallow) and consider selecting an additional site up-gradient and centered in the fallow field (Fig. 1B). This change in plan would result in two sampling locations, indicated by the open and closed circles. Eight additional sample containers would be required as well as additional time needed for sample collection.

Another consideration is raised to evaluate the effects of the herbicide treatments, and so the team, in conjunction with the landowner and the site manager, considers selecting sampling sites that are near the center of the two zones of treatment (Fig. 1C); if conducted, these sites would be matched by corresponding sample sites on the hillside. This raises the number of sampling sites to four and the total number of sample containers to 32 for this one site and again increases the time needed for this site sampling.

Finally, increasing the number of sampling sites within any one study site may increase the power of the group comparisons but also increases the time needed for sampling, along with sampling equipment and containers and experimental supplies to conduct the assays in the laboratory. That said, the expanded study site analysis may reveal underlying conditions that may significantly impact the information gained from microbial complexity determinations. In this hypothetical case, an underlying geological formation (depicted by the dashed line in Fig. 1D) that affects drainage and soil moisture is present and (unknowingly to the experimenters) leads to considerable site-specific effects on the microbial community characteristics of interest. Only when the site is exhaustively sampled is the underlying effect detected.

If all these factors were considered and then implemented, conducting such a number of assessments would expand the experimental effort by a factor of 16. This ad hoc expansion of an experimental design is not recommended, as it would be frustrating to the experimenters and likely lead to an ineffective study design. The phenomenon of such factors becoming apparent as study sites are visited and experiments are conducted, however, is common. When assessed, such information may provide the foundation for additional subsequent formal experimental designs. In this case, the experimenters decide to adhere to the original experimental design but make note of the new site information for designing subsequent site-specific experiments. The historical land use information is recorded and kept in mind for subsequent considerations during experimental analysis of the soils from this site.

Logistic Needs

For each sampling effort and prior to the collection effort being conducted, detailed sampling materials lists, plans, and standard procedures will be developed. The sampling material list will provide a means for organization of the sample containers needed (along with several spares for field errors) for efficient and adequate soil sample collection. Likewise, it will provide details of the particular sampling equipment (e.g., soil coring devices, trowels, and shovels) needed, as well as other support equipment (e.g., cameras, thermometers, field notebooks, global positioning systems [GPS], field tape measures, portable weather stations, sample storage container coolers and boxes, sterilizing solutions and waste receptacles, safety equipment, etc.) needed in the field. It will also provide information necessary for conducting the collections and for organizing materials necessary for the cleaning and sterilization of the sampling materials in the field. Development of such a materials and scheduling checklist is highly recommended.

The Site Sampling Effort

Upon arrival at each site, the experimenters will record as much information about the site as possible prior to and during sampling. Written observations will be recorded in field notebooks, and if possible, photographs will be taken of the site and the surrounding area and of the collection efforts as they proceed. The experimenters will also measure the air temperature and soil temperature and record ambient climate data. The experimenters may have access to handheld GPS units and will collect geographical coordinates for the site if available. Otherwise, maps, field tape measures, and available aerial photography data can be used to determine the coordinates of the sampling site(s).

The experimenters will also need to consider their logistical needs for aseptic collection and handling of the samples. They will need to bring an appropriate number of sterile sampling devices (along with spares) to collect enough samples at each site, or they will need to bring appropriate materials for field sterilization of the sampling materials. In the first case, samples may be collected using sterilized equipment that have been transported aseptically to the field—and once opened used quickly in the field—with each sterile sample collection device being used a single time. Alternatively, a sterile sample collection system may be used and then field sterilized between samplings. This will involve full cleaning of the sampling surfaces of the equipment by application of appropriate cleansers, followed by rinsing, followed by sterilization with field-suitable agents. Note that the rinseates from these cleaning and rinsing treatments should be collected in tubs or other suitable, sealable containers such that the cleanser and rinse residues are returned to the laboratory for appropriate disposal. The field-suitable sterilization approach could include repeated scrubbing with 80% ethanol followed by flaming of the equipment, assuming that the equipment composition (e.g., stainless steel) is unaffected by such treatments. Caution, of course, must be exercised in such efforts, and the sterilization efforts must be designed with safety of the users in mind. Consideration of wind direction, placement of the flammable materials away from the flaming process, etc., should be thoroughly thought out upon arrival at each site. For example, consideration should be given to what will happen if a site sampling effort is conducted on a very windy day, which may make a flame sterilization of surfaces dangerous or untenable. Alternative sterilization solutions and procedures, such as soaking the equipment in hydrogen peroxide or sodium hypochlorite (17), followed by several rinses with sterile water and drying with sterile wipes, may be also considered.

Soil Sampling

During sample collection at this site, sterile trowels will be used to excavate turfgrass samples, which will be removed; the soil will be gently removed from the root tissues by using clean and field-sterilized stainless steel trowels or spatulas; and the loosened soil will be carefully distributed to sterile polypropylene containers (or other suitable, inert containers). Sample containers and container material should be selected to match the volume of sample needed, as well as for maintaining sample integrity during transport, long-term storage, and resampling (if necessary). All samples will be labeled, and label records will be written in the field notebook. Soil sampling should take place efficiently and with a focus on uniform sample handling, careful distribution of soil to the sample containers, and sealing of the container when the correct volume or mass is collected.

Soil immediately beneath (10 cm) the turfgrass root zones (exposed upon removal of the turfgrass) will likewise be collected into separate sterile polypropylene containers. During the entire process, collection observations should be recorded.

Other Considerations Relevant to the Scenario

In soil collection situations where experimenters need to collect from other types of surface soils, such as forest or dense prairie sites, removal of forest or prairie floor litter (e.g., dead leaves, twigs, etc.) covering the surface of the soil will be at the discretion of the experimenter but will always be dependent on the experimental design constraints. In situations where soils are collected from sites known to be, or suspected of being, contaminated with hazardous or other chemicals of concern, alternative collection equipment, along with alternative sample containers (i.e., Teflon or glass containers, with Teflon-lined lids) should be considered for use, depending on the nature of the contaminant. Proper authorization for access to such sites and proper training for safe access to, and activity at, the sites are also needed. Finally, if the site under consideration has been treated with a particular microorganism or suite of organisms for examining their effects upon environmental processes, care should be taken to ensure that aseptic technique is used throughout the collection process and that steps are taken to minimize any introduction of other organisms or inadvertent disturbances to the sampling site. These final recommendations are good ones to follow regardless of the presence of any added microbial species and ensure that samples collected from any site are microbiologically representative of the site.

Travel Logistics

As noted above, the experimenters need to evaluate the microbial complexity of the soils collected from each site but have to collect samples from nine sites. They have calculated that based on travel time between sites (an average of 30 min) and time spent at the site (approximately 2 h each), that two or three sites can be sampled per day. Therefore, the experimenters decide to store collected samples in a cooler in the transport vehicle at reduced temperature (approximately 4°C) to lower microbial activity (21). The samples should not be frozen because the experimenters wish to conduct measurements that are based on an active microbial community. If the experimental team is large enough, however, and if each sample can be processed immediately (the day of, or the day following, collection), collected samples can be kept at ambient temperatures as long as they are protected from temperature swings, and sample analysis can begin upon return to the laboratory each day.

Generally, however, following the sampling effort, the samples are returned to the laboratory and there undergo additional treatments or manipulations prior to experimentation. Two common (but not mandatory) treatments are to partially air dry moist soils and to sieve the soils to a uniform particle size. Regardless of the treatment used, the uniform handling of samples is critical to ensure that the samples are treated in a parallel fashion prior to subsequent experiments. Moist or wet soils are commonly partially air dried under a stream of air; ideally, this is done using a sterile stream of air, such that laboratory or ambient airborne microbes are not introduced into the soils. (An exception to this practice is when the soil under consideration is a historically water-logged or saturated soil, where air drying would represent an irrelevant and confounding treatment upon the microbial community.) Sieving is commonly done to ensure that gross plant tissue, invertebrates, rock materials, or detritus are removed prior to experimentation; sieving should be performed using clean, sterile sieves to prevent cross-contamination between soil samples. The resulting treated soils should be stored in containers that are uniform in composition and size, and they should be stored identically prior to the beginning of experimentation.

Once ready for experimentation, soil subsamples should be aseptically and carefully distributed to experimental microcosms, extraction vessels, etc. The aseptic distribution of samples to the experimental chambers ensures that other ambient organisms or laboratory strains are not inadvertently distributed to soil experimental chambers and that signatures from these strains do not confound subsequent experimental analysis.

SOIL COLLECTION HYPOTHETICAL EXAMPLE SUMMARY

The objective of this illustration was to communicate that considerable care needs to be given to soil sample site selection, study design, and the use of the soils for subsequent discriminatory tests. Furthermore, site history data should assist in this study design effort, and as much information as possible about the site(s) should be collected before site visitation. Reevaluation of the study design following site visitation can provide important information that can be used either to interpret the influence of discovered factors (that the experimenters were unaware of prior to the execution of the sampling) on subsequent experimental results or to guide the design of additional sampling approaches and experiments that will provide the capability to experimentally address those factors.

ADDITIONAL SOIL SAMPLE COLLECTION CONSIDERATIONS

Bulk Soil versus Intact Sampling Considerations

The above scenario provided an illustration of collecting soil from a site by using a bulk soil collection effort; no attempt was made to maintain the physical, three-dimensional structure of the soil matrix. Experimenters interested in surface soil dynamics, however, should consider two basic approaches: sampling whereby no attempt is made to maintain the physical integrity of the soil matrix and sampling whereby an attempt is made to maintain the physical integrity of the soil structure. The reasons for this are generally that upsetting the physical integrity of the soil during collection (digging, root removal, etc.) will more than likely change the activity of the microbial community (5), affect the relationships between the microbial community and the macroorganisms present as well as the soil physical structure, and affect the soil atmosphere. Depending on the severity of the disruption, a change in the soil community composition or activity can occur. If slight shifts in these components are of concern to the experimenters, then they should consider sampling in a manner that maintains the physical structure of the soils.

Intact Soil Sample Collection

The choice of which type of sampling to use, disturbed or undisturbed, depends on the experimental design and, to some degree, on the logistics of sample handling, storage, and use. In most cases, soil integrity is desired because of subsequent experimental needs to address microbial processes or characteristics that change with depth. If this is the case, a simple soil corer (available commercially from several vendors [Table 2]) will permit the collection of soil cores that maintain such integrity. Samples should be collected using clean and sterile equipment, and if coring liners are available, the samples should be capped and stored. If liners are not available, cores can be carefully extruded into sample bags (e.g., Whirl-Pak bags) and stored in a manner to maintain the integrity of the sample until it can be analyzed. Laboratory analysis of intact samples should take place as quickly as possible to minimize the possible unknown effects—such as those arising from aeration of the soil core during the collection and extrusion process, or the compaction of the soil at the extremities of the core—upon microbial processes of interest. In all cases, these sample collection efforts require use of soil-coring devices or other devices that can remove a plug of soil without disturbing the majority of soil, except at the edges of the sample. Care must be taken during soil sample collection, storage, and transportation to minimize changes to the integrity of the soil sample.

There are several commercially available coring systems that can be used to collect samples of a range of different dimensions (e.g., length), which can be selected based on the experimental design and subsequent use of the samples (Table 2). Surface-soil-coring devices can also be custom fabricated from large-diameter pipe materials (e.g., polyvinyl chloride plastic piping and stove pipe [22]). Regardless of the coring system used, maintaining the physical integrity of the soil following collection requires planning, care, and additional effort.

Bulk Soil Sample Collection

For samples to be collected when there is not a need for physical integrity of the soil, common soil tools such as trowels, shovels, spatulas, or even spoons can be used. Collection equipment should be cleaned between each sample collection effort and sterilized prior to collection of the next sample. Since the equipment cannot generally be autoclaved in the field, it should be cleaned, chemically and/or heat sterilized, and rinsed with sterile water and aseptically dried prior to sampling.

Bulk Soils: Treatment Considerations

Soil collection efforts that do not maintain this integrity are not constrained by the same sample handling needs as those that do, but the introduction of physical changes in the soil structure may have unknown effects upon subsequent experimental results. These types of disturbance treatments have been studied for their subsequent effects on microbial community structure (21). Based on changes in phospholipid markers, sieving (a disturbance in some ways more extreme than bulk soil collection) had little overall effect on the community structure. If the use of such bulk soil collection methods is employed, it is suggested that all discrete soil samples be subjected to treatments that are uniform in nature, in order to subject the soil samples to similar amounts of disturbance and minimize sample-to-sample treatment heterogeneity. For example, identical volumes of soils should be sieved for the same lengths of time, and at the same temperature, using the same amount of hand or machine agitation.

Sample Containers for Soil Cores or Bulk Soils

Sample containers that receive the sample from the coring equipment (or any sampling equipment used) should be clean, sterile, and rugged enough for transport needs. If containers are reused, they should be washed, rinsed with deionized and/or distilled water, and sterilized. Soil samples should be transferred into the containers by using a sterile technique if possible. Consideration should be given to selection of the best sample container material and size that meet needs for transport, handling, and packing, especially if sampling is being conducted in remote field locations. Glass containers are often considered optimal due to the lack of reactivity, but their weight and general fragility can be hindrances during field collection efforts. Lids for glass containers should be Teflon or inert-plastic lined. Plastic sample containers (including sterile Whirl-Pak bags) can also be used for field soil collection efforts.

Sample Transport

Following collection and introduction of the sample into the sample container, the sample must be transported to the laboratory or to the field testing site in a manner that both is efficient and maintains the desired properties of the soil. There are several considerations for the temperature at which the samples are to be transported. Assuming that the samples were collected from a temperate soil, if the samples are to be immediately analyzed for microbial growth and/or biological activity, the samples are best kept at ambient temperature (or the soil in situ temperature), such that the microbial community undergoes little or no temperature stress prior to analysis. If, however, the samples are to undergo a significant storage period prior to analysis, then two approaches can be considered: one is to lower the temperature (to 4°C) to reduce the predominant community microbial activity; the other is to maintain the samples at the ambient temperature such that no temperature-driven shifts in the microbial community composition occur during the storage period. Finally, if the samples are to be used for analysis of nucleic acids (i.e., DNA and RNA) and the experimental design allows for it, immediate deep-freezing of the sample will assist in stabilizing the nucleic acids present in the sample for subsequent extractions and purifications and prevent any growth or changes in the microbial community that might occur during the time between field sampling and laboratory analysis.

RHIZOSPHERE SOIL SAMPLING CONSIDERATIONS

Rhizosphere versus Bulk Soil Sampling Considerations

While the two basic types of soil samples mentioned above (intact versus nonintact) are important to consider when designing the soil sample collection approach, another characteristic of soils also influences the collection design and subsequent experimentation. This is the characteristic of the presence or absence of root tissues in the soil sample. The interest in the area of close association of soil and roots, i.e., the rhizosphere, resides in the large impact that plants and plant exudates have on soils and soil microbial communities, both from a macroscopic and microscopic perspective. The full description of such impacts is beyond the scope of this chapter, but in brief, plant roots exert a generally positive influence (e.g., stimulation of growth and increase of numbers) on the soil microbial community members in close proximity to the root tissue (27). This effect is often referred to as the rhizosphere effect, and the rhizosphere is operationally described as the zone of soil, from 0 mm to a few millimeters away from plant root tissue, where the microbial community is influenced. The nature of this effect is tied directly and intimately to the presence of, and exudates from, the living plant roots in the soil and to the microorganisms flourishing as a result of such influence (9).

Because of the dynamic and biological nature of the rhizosphere effect, it is of interest to better understand the fate of chemicals that are distributed to soils containing root tissues (14, 15, 19) and is, philosophically, best studied in experiments that maintain the physical integrity and biological activity of the roots' association with the microbial community. Studies that use soils where the plant root-soil-microorganism integrity is removed, however, also remove the dynamic nature of the root zone effect, and different processes ensue. At the point of removal, any microscopic root tissues present in the soil sample die, their components are subject to degradation, labile carbon exudates from any living plant tissues cease, and the living, dynamic nature of the rhizosphere effect is lost (2, 3). Experiments that maintain such associations (14, 15) are preferable to those that do not, but there are experimental situations in which such root associations cannot be maintained. The intact systems are generally more difficult to implement and maintain than those soil microbiological studies for which no such maintenance is required. Even in cases where the rhizosphere relationship is maintained, sampling from such chambers is problematic due to the immediate disturbance and destruction of the rhizosphere zone that takes place upon sample removal. The best approach to any rhizosphere study is to use methods that analyze the microbial community either in the presence of the rhizosphere effect (e.g., in the presence of living plants) or immediately following removal of a sample from the rhizosphere zone.

ADDITIONAL LABORATORY TREATMENTS

Air Drying

As mentioned in the scenario above, bulk soils are commonly partially air dried, sieved, and then stored at ambient or cooler temperatures (e.g., 4°C). Air drying is preferably done under a stream of sterile air flow to decrease the likelihood of introduction of laboratory microbial strains or other airborne propagules into the soil samples. This requires the sample to be placed in a container wherein a sterile stream of air can be introduced, and exhaust venting of the drying air from the soil chambers is directed towards laboratory exhaust hoods. It is recommended that such air be filtered prior to exhaust from the chambers, to lessen microbe-laden aerosol/dust deposition in building or lab exhaust systems.

While air drying is common practice, it is known to change both the microbial activity and the physiological profiles of soils (10, 13, 29). A few reports, however, suggest that while such physiological changes are demonstrable, the underlying microbial composition of the soil may remain intact (10, 29). Therefore, while the metabolic potential may not be lost from air drying soils, there is a strong likelihood for a change in the expression of this metabolic potential.

Sieving

Sieving the soils to uniform particle sizes (2- or 4-mm grating size is common [10–12, 21, 28]) should be done with sieves exhaustively cleaned between samplings to reduce the likelihood of microbial cross-contamination between soil samples. The purpose of sieving is both to remove rocks and larger plant debris and invertebrates and to make the soils easier to distribute to experimental chambers (12, 30).

SUMMARY

During my career as a soil microbial ecologist, we often joked that soil sampling could blithely be described as "grabbing a shovel and a bucket and digging." This was a light-hearted oversimplification and crude representation and was not meant to place the questions of the soil microbiologist in a dim light. It is a fact that common surface soils, like most microbial ecosystems, are complex, poorly understood, and very difficult to study. The approaches and mind-set provided in this chapter are designed to assist the experimenter in developing well-planned sampling approaches as well as in considering what factors might arise during the execution of those sampling efforts. These are intended to assist the soil microbiologist to ask scientifically valid questions of the collected samples and increase the scientific community's understanding of this highly complex environment.

REFERENCES

1. **Atlas, R. M., and R. Bartha.** 1998. *Microbial Ecology: Fundamentals and Applications*, 4th ed. Benjamin Cummings Science Publishing, Menlo Park, Calif.
2. **Barber, D. A., and J. M. Lynch.** 1977. Microbial growth in the rhizosphere. *Soil Biol. Biochem.* **9:**305–308.
3. **Barber, D. A., and J. K. Martin.** 1976. The release of organic substance by cereal roots in the soil. *New Phytol.* **76:**69–80.
4. **Box, G. E., W. G. Hunter, and J. S. Hunter.** 1978. *Statistics for Experimenters: an Introduction to Design, Data Analysis, and Model Building*. John Wiley and Sons, New York, N.Y.
5. **Calderon, F. J., and L. E. Jackson.** 2002. Rototillage, disking, and subsequent irrigation: effects on soil nitrogen dynamics, microbial biomass, and carbon dioxide efflux. *J. Environ. Qual.* **31:**752–758.
6. **Curtis, T. P., W. T. Sloan, and J. W. Scannell.** 2002. Estimating prokaryotic diversity and its limits. *Proc. Natl. Acad. Sci. USA* **99:**10494–10499.
7. **Dunbar, J., S. M. Barns, L. O. Ticknor, and C. R. Kuske.** 2002. Empirical and theoretical bacterial diversity in four Arizona soils. *Appl. Environ. Microbiol.* **68:**3035–3045.
8. **Gardner, W. R.** 1991. Soil science as a basic science. *Soil Sci.* **151:**2–6.
9. **Gomes, N. C. M., O. Fagbola, R. Costa, N. G. Rumjanek, A. Buchner, L. Mendona-Hagler, and K. Smalla.** 2003. Dynamics of fungal communities in bulk and maize rhizosphere soil in the tropics. *Appl. Environ. Microbiol.* **69:**3758–3766.
10. **Griffiths, R. I., A. S. Whiteley, A. G. O'Donnell, and M. J. Bailey.** 2003. Physiological and community responses of established grassland bacterial populations to water stress. *Appl. Environ. Microbiol.* **69:**6961–6968.
11. **Ibekwe, A. M., S. K. Papiernik, J. Gan, S. R. Yates, C. H. Yang, and D. E. Crowley.** 2001. Impact of fumigants on soil microbial communities. *Appl. Environ. Microbiol.* **67:**3245–3257.
12. **Knaebel, D. B., T. W. Federle, D. C. McAvoy, and J. R. Vestal.** 1994. Effect of mineral and organic soil constituents on microbial mineralization of organic compounds in a natural soil. *Appl. Environ. Microbiol.* **60:**4500–4508.
13. **Knaebel, D. B., T. W. Federle, and J. R. Vestal.** 1990. Mineralization of linear alkylbenzene sulfonate (LAS) and linear alcohol ethoxylate (LAE) in 11 contrasting soils. *Environ. Toxicol. Chem.* **9:**981–988.
14. **Knaebel, D. B., and J. R. Vestal.** 1992. Effects of intact rhizosphere microbial communities on the mineralization of surfactants in surface soils. *Can. J. Microbiol.* **38:**643–653.
15. **Knaebel, D. B., and J. R. Vestal.** 1994. Intact rhizosphere microbial communities used to study microbial biodegradation in agricultural and natural soils: influence of soil organic matter on mineralization kinetics, p. 56–69. *In* T. A. Anderson and J. R. Coats (ed.), *ACS Symposium Series*, vol. 563. *Bioremediation through Rhizosphere Technology*. American Chemical Society, Washington, D.C.
16. **Maier, R. M., I. A. Pepper, and C. P. Gerba.** 2000. *Environmental Microbiology*. Academic Press, San Diego, Calif.
17. **Mazzola, P. G., T. C. V. Penna, and A. M. da S. Martins.** 2003. Determination of decimal reduction times (D value) of chemical agents used in hospitals for disinfection purposes. *BMC Infect. Dis.* **3:**24–34.
18. **Morris, C. E., M. Bardin, O. Berge, P. Frey-Klett, N. Fromin, H. Girardin, M.-H. Guinebretière, P. Lebaron, J. M. Thiéry, and M. Troussellier.** 2002. Microbial biodiversity: approaches to experimental design and hypothesis testing in primary scientific literature from 1975 to 1999. *Microbiol. Mol. Biol. Rev.* **66:**592–616.
19. **Mortensen, G. K., H. Egsgaard, P. Ambus, E. S. Jensen, and C. Gron.** 2001. Influence of plant growth on degradation of linear alkylbenzene sulfonate in sludge-amended soil. *J. Environ. Qual.* **30:**1266–1270.
20. **Paul, E. A., and F. E. Clark.** 1996. *Soil Microbiology and Biochemistry*, 2nd ed. Academic Press, San Diego, Calif.
21. **Peterson, S. O., and M. J. Klug.** 1994. Effects of sieving, storage, and incubation temperature on the phospholipid fatty acid profile of a soil microbial community. *Appl. Environ. Microbiol.* **60:**2421–2430.
22. **Reinhardt, C. H., C. A. Cole, and L. R. Stover.** 2000. A method for coring inland, freshwater wetland soils. *Wetlands* **20:**421–425.
23. **Rösch, C., A. Mergel, and H. Bothe.** 2002. Biodiversity of denitrifying and dinitrogen-fixing bacteria in an acid forest soil. *Appl. Environ. Microbiol.* **68:**3818–3829.
24. **Silvia, D. M., J. J. Fuhrmann, P. G. Hartel, and D. A. Zuberer.** 1998. *Principles and Applications of Soil Microbiology*. Prentice Hall, Upper Saddle River, N.J.
25. **Torsvik, V., J. Goksøyr, and F. L. Daae.** 1990. High diversity in DNA of soil bacteria. *Appl. Environ. Microbiol.* **56:**782–787.
26. **van Elsas, J. D., K. Smalla, A. K. Lilley, and M. J. Bailey.** 2002. Methods for sampling soil microbes, p. 505–515. *In* C. J. Hurst, R. L. Crawford, G. R. Knudsen, M. J. McInerney, and L. D. Stetzenbach (ed.), *Manual of Envi-ronmental Microbiology*, 2nd ed. ASM Press, Washington, D.C.
27. **Whipps, J. M., and J. M. Lynch.** 1986. The influence of the rhizosphere on crop productivity. *Adv. Microb. Ecol.* **9:**187–244.

28. **Williamson, K. E., K. E. Wommack, and M. Radosevich.** 2003. Sampling natural viral communities from soil for culture-independent analyses. *Appl. Environ. Microbiol.* **69:**6628–6633.
29. **Winding, A., S. J. Binnerup, and J. Sørensen.** 1994. Viability of indigenous soil bacteria assayed by respiratory activity and growth. *Appl. Environ. Microbiol.* **60:**2869–2875.
30. **Winding, A., R. Rønn, and N. B. Hendriksen.** 1997. Bacteria and protozoa in soil microhabitats as affected by earthworms. *Biol. Fertil. Soils* **24:**133–140.
31. **Zhou, J., B. Xia, D. S. Treves, L.-Y. Wu, T. L. Marsh, R. V. O'Neill, A. V. Palumbo, and J. M. Tiedje.** 2002. Spatial and resource factors influencing high microbial diversity in soil. *Appl. Environ. Microbiol.* **68:**326–334.

Application of Phylogenetic Techniques in Studies of Soil Microbial Communities

RYAN T. JONES, ELIZABETH K. COSTELLO, AND ANDREW P. MARTIN

49

The investigation of microbial communities increasingly relies on DNA sequence data for either descriptive or comparative purposes. Small subunit (SSU) rRNA has received most of this sequencing effort, but studies increasingly sequence functional genes from the environment. While many other molecular approaches exist as a heuristic means to describe or compare communities (e.g., denaturing gradient gel electrophoresis, [DGGE], restriction fragment length polymorphism [RFLP], automated ribosomal intergenic spacer analysis [ARISA], etc.), using DNA sequences has the distinct advantage of incorporating evolutionary history into such studies. Our goal in writing this chapter is twofold: (i) we want to persuade the researcher that investigating microbial communities by using DNA sequence data is the most appropriate method for making community comparisons or for inferring ecological processes based on community membership, and (ii) we want to present the best available analytic methods, the steps needed to use those methods, and how to interpret the results.

This chapter has three main sections: "Background," "Methods," and "Worked Examples." We give a brief background on molecular approaches applied to microbial communities and on the motivation for incorporating phylogenetics into microbial studies. The methods section contains a synopsis of how to generate phylogenies and then discusses phylogenetic methods used to answer two main types of questions: (i) are two (or more) microbial communities significantly different from one another, and (ii) what processes govern the composition and assembly of microbial communities? Finally, using a variety of the described methods, we work through two previously published data sets describing soil microbial communities and interpret the results.

BACKGROUND

Investigation of Microbial Communities in the Molecular Age

The advent of molecular techniques and cloning allowed microbiology to escape the petri dish and radically changed our understanding of microbial diversity (26). Analyses of SSU rRNA gene sequences continue to reveal unknown divisions of bacteria; less than 20 years ago, only 12 divisions were known (42, 43), and the latest published work proposes over 50 bacterial divisions (of those, only 26 have been cultured) (31). These advances point to the wealth of information gained, from a community ecology perspective, through the use of molecular methods rather than cultivation techniques. Indeed, more than 99% of microbial lineages go undetected with a cultivation approach (37). A more complete understanding of community membership gives us the grounding needed to begin investigating the ties between community membership and ecosystem function, the roles geography and niche effects play in community assembly and differentiation, and the community-wide rates and effects of gene flow (among others). These are complicated objectives, and an effective approach will necessarily include creativity, a synthesis of various disciplines within microbiology and evolutionary biology, and sound methodology.

Analyses of Molecular Data by Use of Traditional Ecological Approaches

A number of molecular approaches are used as a proxy to identify taxonomic groups present in microbial communities. These methods use differences in intergenic spacer size (IGS) (e.g., ARISA), differences in migration properties of genes through gels (e.g., denaturing gradient gel electrophoresis), or differences in size fragments of genes subjected to restriction digest enzymes (e.g., RFLP or terminal RFLP) to characterize community diversity. Community similarity is calculated from the presence or absence (and potentially relative abundance) of each taxonomic group and percent overlap (shared groups) between communities. The ARISA approach, for example, has been implemented as a way to differentiate between bacterial and fungal communities of different soil types (30). The advantage of this type of approach is that it is comparatively faster and cheaper than constructing and sequencing clone libraries. The downside is that the evolutionary signal of the lineages present in these communities is nearly or completely lost. For example, the ARISA method uses IGS profiles to differentiate between communities, but the IGSs for the Ascomycota, Basidiomycota, Zygomycota, and the Plasmodiophoromycota lineages all overlap (30). Furthermore, no large-scale database exists to compare one community to another using these methods. Despite the drawback that these methods lose the evolutionary history between community members, in many respects, these types of markers are ideal for an initial heuristic attempt at describing microbial community similarity.

While using traditional ecological indices (e.g., Sorenson's similarity index, Jaccard's index, Shannon's diversity index, etc.) may be best for many molecular techniques, using them to make biological inferences with DNA sequence data is less justified. In an era of widespread community gene sequencing, the use of operational taxonomic units (OTUs) plays a paradoxical role in microbial community analyses. An OTU is constructed by binning community DNA sequences into taxonomic units (similar to a species or family) based on sequence similarity (e.g., >97% similarity, but the level of similarity varies arbitrarily by study), and it is a convenient and easy way to characterize and compare microbial communities, using standard ecological indices based on shared and unique species. However, sequence data are ideal for generating phylogenies describing the evolutionary relationships between the community members, and binning lineages into OTUs for use in nonphylogenetic analyses ignores a wealth of information contained in the sequences. Nevertheless, a number of approaches use OTUs to estimate the diversity of microbial communities. These methods include parametric approaches that estimate diversity by fitting relative OTU abundances to a certain model (e.g., lognormal) (6, 8) and nonparametric approaches such as the Chao1 estimator, which estimates the total community diversity by comparing the number of singleton OTUs to the number of doubletons (5). The effect of land use on bacterial diversity in soils has been investigated using such a nonparametric approach (14). These approaches are reviewed in more detail elsewhere (2).

Why Phylogenetics?

Many ecological studies, microbial community ecology included, investigate community dynamics using species as the unit of measurement (e.g., species richness, shared species, etc.). However, unless the evolutionary history of the collection of species can arguably be represented as a polytomy (Fig. 1), using species as the unit is unjustified because the species are not independent (9). This is because any two species have a shared evolutionary history up to the point that they diverged from their last common ancestor. The more evolutionary history two species share, the more similar they are. Harvey and Pagel, in an excellent volume dedicated to the subject, give three reasons why phylogenetically related species are similar to each other: (i) phylogenetic niche conservatism—past and present lineages are likely to occupy similar niches, (ii) phylogenetic time lags—some amount of time must pass before a trait of a common ancestor is lost in one of the daughter lineages, and (iii) different adaptive responses—organisms with similar phenotypes are likely to respond similarly to the same selective pressures (10). Microbes, like all organisms, cannot escape their phylogenetic constraints. However, it must be remembered that microbial phylogenies are usually reconstructed from a specific gene (e.g., 16S), and a gene phylogeny does not necessarily reflect the actual phylogeny of a group of organisms (in light of horizontal gene flow, this may be especially true for microbes).

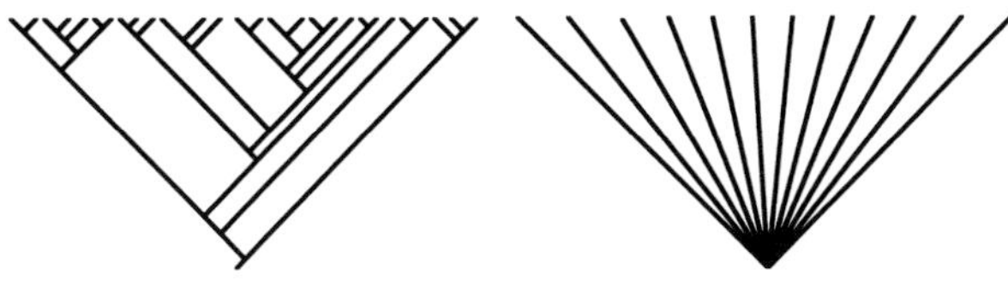

FIGURE 1 On the left is a typical phylogenetic tree. Branching events closer to the tips of the tree represent more-recent speciation events than branching events deeper in the tree. Closely related species share more life history traits than less related species. On the right is a polytomy—a type of phylogenetic tree in which all species are equally related to one another. If the evolutionary history of a collection of species can accurately be depicted as a polytomy, then species can be used as the unit of comparison.

METHODS

In this section we present a brief summary of how to construct phylogenies followed by descriptions of techniques to compare microbial communities and to infer ecological processes. A schematic demonstrating the various paths available once DNA sequences are in hand may be useful (Fig. 2).

Constructing Phylogenies

Phylogenies represent attempts to reconstruct the ancestral relationships between individuals in a collection of organisms. However, because only contemporary (in general) organisms are available to the researcher, the past must be inferred by comparing the individuals of a collection to each other based on an evolutionary model and some optimality criterion. While macroecologists have the luxury to debate the relative importance of phenotype versus genotype in making these reconstructions, microbial community ecologists generally must infer evolutionary relationships based on genotype exclusively (although genomic studies of cultured microbes can certainly help bridge the gap between microbial genotype and phenotype). Fortunately, a variety of tree-building methods exist to construct phylogenies from DNA sequence data. When using phylogenies in any type of study, it is important to think about which phylogenetic method is appropriate on a case-by-case basis. Often, the neighbor-joining algorithm, conveniently included in many sequence alignment programs, is used without any justification.

Before a tree-building method can be implemented, the community DNA sequences must be aligned. An alignment is used to infer positional homology for a group of homologous genes. Numerous automated sequence alignment programs are available including Clustal X (15, 38), ARB (20), and the NAST aligner (7). NAST and ARB are recommended for SSU rRNA genes because they allow users to align sequences according to established secondary structure models, and within a phylogenetic context. To do this, they unite searchable databases of aligned sequences and associated sequence information with tools for data import and export, alignment, masking, filtering, and phylogenetic analysis under a common graphical interface. NAST is the quickest and easiest to use, as sequences are uploaded to a website (http://greengenes.lbl.gov) and the alignment is sent back to the submitter. ARB is available at http://www.arb-home.de.

After the sequences of interest are aligned, the next step is to generate a phylogeny of the organisms. The researcher must choose between using all of the information contained within the individual nucleotide sites of the sequences and using a summary of the information contained in each site by creating a distance matrix based on the alignment.

Using a distance approach is the most computationally efficient method; a matrix describing the proportion of sequence similarity between all taxa in the collection is subjected to some clustering algorithm (e.g., UPGMA or neighbor joining [NJ]) to generate a phylogeny. These distance

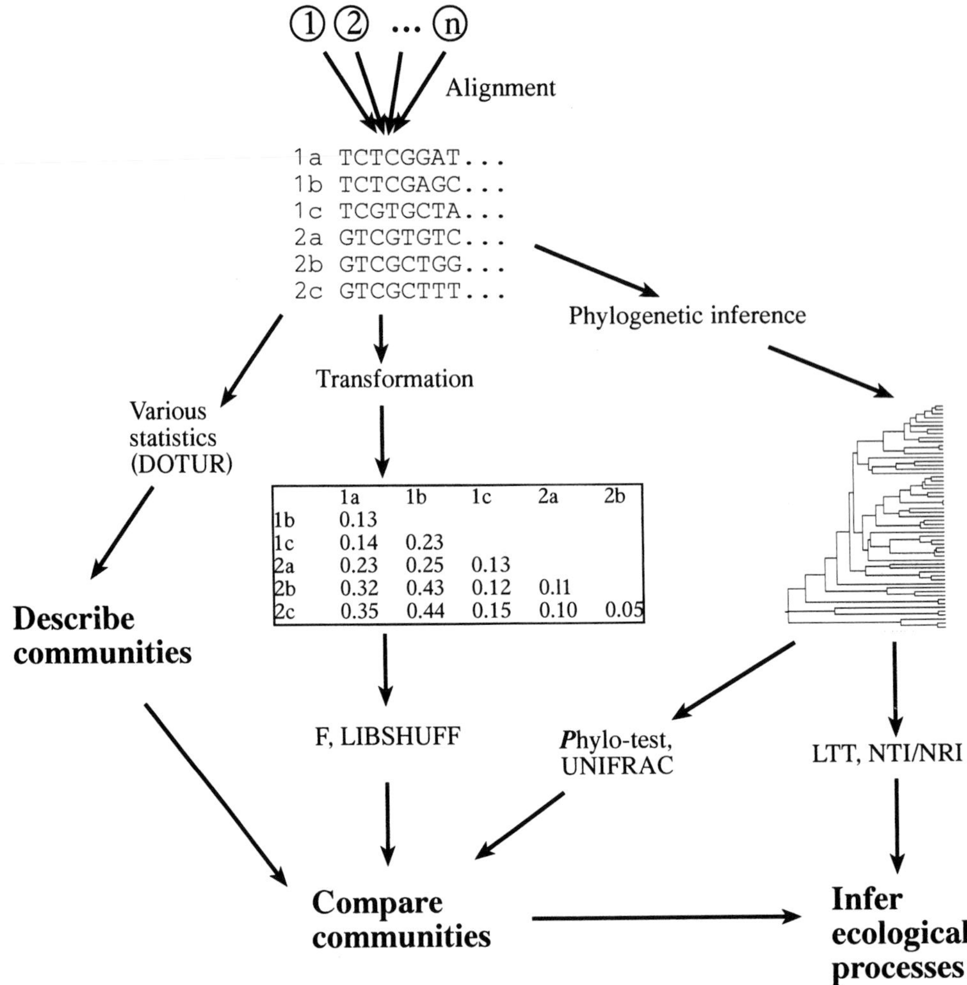

FIGURE 2 Flow chart beginning with a collection of DNA sequences and ending with making biological inferences. After DNA sequences are aligned, the alignment can be sent to descriptive statistics or transformed into a distance matrix, or phylogenetic relationships can be inferred. After the second step, outputs can be analyzed in a variety of programs to make biological inferences.

matrices can be constructed without any correction (purely based on sequence overlap) or can be corrected based on some model of evolution that incorporates differential rates of mutation from one nucleotide type to another. A disadvantage of the distance approach is the loss of information that results from the initial conversion of sequence data into a distance matrix—the information contained in the individual nucleotides at each site is lost. Furthermore, some clustering algorithms (e.g., NJ) yield one phylogenetic tree—while this may be appealing to the researcher, it creates a false sense that the true or correct tree is in hand. Detrimentally, the uncertainty of phylogenetic estimation is ignored.

If, rather than converting the aligned sequences into a distance matrix, the individual characters of the sequences remain the focus of analysis, an optimality criterion such as maximum parsimony or maximum likelihood can be used to find the best phylogenetic tree. Maximum parsimony, another computationally efficient method, is an approach that constructs phylogenies based on minimizing the number of evolutionary events (with sequence data, these events are mutations from one nucleotide to another) between individuals. Advocates of parsimony believe that the simplest explanation of the data is most likely to describe the correct evolutionary tree. This is certainly debatable; while advocates of parsimony believe in assuming little about a model of evolution, critics argue that because some evolutionary events are more likely than others (e.g., transitions versus transversions), incorporating models of evolution are vital to phylogenetic inference. Essentially, the argument against parsimony is that as time goes on after a speciation event some base pair positions will have multiple mutational events (e.g., A to C to G) and parsimony will misrepresent this evolutionary history as only one mutation event. A likelihood approach, on the other hand, takes a random evolutionary relationship and produces a probability of observing that relationship considering the model and sequence alignment [likelihood = probability (data | hypothesis)]. Maximum likelihood cycles through many, many iterations of random trees and keeps the tree with the highest probability. Zhang and Nei found that distance, parsimony, and likelihood approaches all give reliable phylogenetic reconstructions with low sequence divergence; however, when sequence divergence is high, likelihood and distance methods outperform parsimony (with accuracy less affected by correct model choice) (44).

Models of evolution are an important consideration when using distance or likelihood approaches. Evolutionary models range from the simple Jukes-Cantor model (assumes equal base frequencies and all substitutions have equal probability) to the parameter-rich general reversible model (empirical base frequencies can be used, and each of the six possible nucleotide substitutions has its own rate) (17, 32). Two additional parameters can be added to these models: a gamma distribution of substitution rates among sites (some base pair positions are more likely to experience a mutation than others) and a distribution of invariant sites (some base pair positions are not allowed to mutate, usually because of some structural demand). Adding parameters to models of evolution is potentially problematic: more-parameter-rich models generate phylogenies with higher likelihood scores whether the phylogenies are necessarily more likely or not. A particular model should be chosen because it is significantly better at explaining the data despite its parameters. MODELTEST, a program which takes an aligned sequence file as its input, tests 56 models of evolution and outputs the best model to use based on the log likelihood ratio test or on Akaike information criterion (27).

Since it is impossible to infer the exact evolutionary history of a group of organisms, it is important to incorporate the uncertainty surrounding phylogenetic inference into any tests depending on evolutionary relationships (13). Nonparametric bootstrapping and Bayesian analysis are two common techniques used to ascertain phylogenetic uncertainty. Bootstrapping is a method that constructs new sequence files by randomly sampling base pair positions (i.e., a column from a DNA sequence file) with replacement from the aligned sequence file (9). A tree is found for each of these new files, and each node is given a bootstrap support value (percentage of trees with that node). Since bootstrapping generates a collection of phylogenies through subsampling the sequence data, the output gauges the sensitivity of the inferred phylogeny to the noise in the signal sent by the data. In other words, bootstrapping estimates the uncertainty of the estimated phylogeny due to the choice of sequence data. Bayesian analysis, on the other hand, generates a collection of trees that represents the uncertainty conditional on the model of evolution and the data. Bayesian analysis works by creating a long chain of randomly selected phylogenetic trees and comparing each new tree's likelihood score to the likelihood score of the tree held over from the previous iteration; if the current tree is more likely, or at least likely enough, it will replace the held tree. The held trees are compared to one another, similarly to bootstrapping, and nodes are given posterior probability values. The posterior probability values represent the probability that a tree from the collection contains the group subtended by the branch. Posterior probabilities generated by Bayesian analysis estimate the uncertainty of phylogeny estimation due to the assumed model of evolution and the corresponding parameter values.

Methods for Determining Whether Two or More Communities Differ

Several different metrics have been developed for assessing whether two or more a priori defined groups differ. One class of methods depends on transforming the data from characters (i.e., a string of nucleotides) to pairwise differences between the sampled sequences and using the distance data without estimating a phylogenetic tree. Another class of methods requires estimation of phylogeny.

Distance-Based Metrics

Perhaps the simplest approach for assessing whether two communities differ is hierarchical analysis of variance, the F test. In this approach the average diversity within a community, usually estimated as the average sequence difference between pairs of sampled sequences, is compared to the diversity of all sampled lineages, namely $F = (D_T - D_W)/D_T$, where D_T is the diversity of all sampled sequences and D_W is the average diversity within a particular community. Significance is assessed by randomly assigning sampled sequences to a particular community from the pool of all sampled sequences without replacement, keeping sample size the same, and calculating F. Typically, a thousand replicate randomized communities provide a distribution of F under the null expectation that the communities have assembled randomly from the sampled pool of sequences.

Another approach, called ∫-LIBSHUFF, has been developed (35, 36). ∫-LIBSHUFF implements the integral form of the Cramér-von Mises statistic to test whether two gene libraries are drawn from the same, underlying distribution and whether one library is a subset of another (35). Using distance matrices, the test integrates differences between homologous coverage (C_X) and heterologous coverage (C_{XY}) across the observed genetic distances (ΔC_{XY}). The test bases homologous coverage on the percentage of nonsingleton sequences in library X as a function of distance and heterologous coverage on the percentage of sequences in library X having a similar sequence in library Y as a function of distance. The difference between homologous and heterologous coverage represents the percentage of nonsingleton, nonshared sequences. The reverse comparison (C_{YX}) switches the perspective, such that homologous coverage is calculated from library Y, and heterologous coverage is the percentage of sequences in library Y with similar sequences in library X. *P* values are found by randomly assigning sequences to libraries, maintaining library size, and recalculating the statistic; this process is repeated many times. The proportion of the distribution that falls above the observed value is the probability that the two gene libraries are sampled from the same community. The *P* value for the reverse comparison is simultaneously calculated by ∫-LIBSHUFF. Small P values for both comparisons indicate that the libraries were drawn from compositionally distinct microbial communities. A small *P* value coupled with a large *P* value for the reverse comparison suggests that one library is a subset of the other. Finally, large *P* values for both comparisons imply that the libraries were likely drawn from the same underlying distribution of microbes. ∫-LIBSHUFF is available at http://www.plantpath.wisc.edu/fac/joh/S-LIBSHUFF.html.

The distance metrics are sensitive to model choice for transformation of the sequence data into pairwise distances. Models with few parameters (e.g., Jukes-Cantor or the Kimura two-parameter model) may underestimate the genetic distances, and the error increases with increasingly divergent sequences. One consequence of an inappropriate model is a loss of power when the sampled sequences are highly divergent. Such error can be minimized by adopting an appropriate model of sequence evolution following the recommendations of Posada and Crandall and implemented in the program ModelTest (27). A more important limitation of distance-based methods is that it is not possible to incorporate uncertainty associated with estimation of evolutionary parameters. Namely, the parameter of interest is the evolutionary divergence between sampled sequences. Additionally, transformation of the data results in a loss of information because the pattern of shared DNA sequence is lost.

Phylogenetic Metrics

Phylogenetic methods provide a robust means for characterizing microbial communities based on molecular sequence data (22). One method, originally developed by Maddison and Slatkin (21), assesses the degree of covariation between phylogeny and microbial community. In this method, phylotypes are assigned character states (i.e., community type) and character state transitions are optimized across the phylogenetic tree by using parsimony; namely, the number of character state changes needed to reconcile the distribution of states across the tree measures the covariation between phylogeny and character states (in this case, community type). The null expectation is approximated by randomizing the relationships of the sequences many times and comparing the distribution of randomized values to the observed number of character state changes (21).

Martin used the method developed by Maddison and Slatkin and applied it to microbial communities; this method is named the phylogenetic test (sometimes referred to as the P-test or Phylo-test). In Martin's paper, the analysis was based on a single estimate of phylogeny; phylogenetic uncertainty was ignored (22). Schadt et al. incorporated phylogenetic uncertainty by using Bayesian phylogenetic analysis to generate a large number of estimates of phylogeny and using parsimony to estimate a distribution for the number of transitions between one community type and another (34). The estimated distribution for the observed data was compared to the null expectation to assess whether the composition of communities differed. In this case, Schadt et al. discovered that the fungal communities sampled in the winter and spring were similar but the communities sampled in the spring and summer differed significantly (34). The results suggested that successional changes associated with marked shifts from the wet, cool conditions of winter and spring to the drier and warmer conditions of summer involve the turnover of highly divergent groups rather than shifts in the relative abundance of closely related species.

Although the method adopted by Schadt et al. incorporated phylogenetic uncertainty by using Bayesian methods, mapping character state changes using parsimony assumes that the maximum number of character state changes along any lineage is one. However, character states may change multiple times along a lineage, and more importantly, the number of character state changes is likely to be positively correlated with the length of a branch—in other words, longer branches have a higher chance of a character change than shorter branches (12). With the availability of SIMMAP (http://www.simmap.com), a program that implements a Bayesian method of character evolution estimation, it is possible to incorporate the uncertainty of character state mapping when testing for differentiation of microbial communities (16). Accounting for the uncertainty of the inferred phylogenetic relationships among sampled sequences and the uncertainty of character state mapping provides a more realistic and conservative assessment of whether different microbial communities harbor phylogenetically distinct assemblages of species.

The effect of incorporating phylogenetic and character mapping uncertainty is evident in Fig. 3. When *nirK* sequences from two soil communities were analyzed, the results based on a single estimate of phylogeny indicated that the two communities differed significantly (note that the NJ-Parsimony line falls to the far left of the null distribution in Fig. 3A). However, significance vanished when

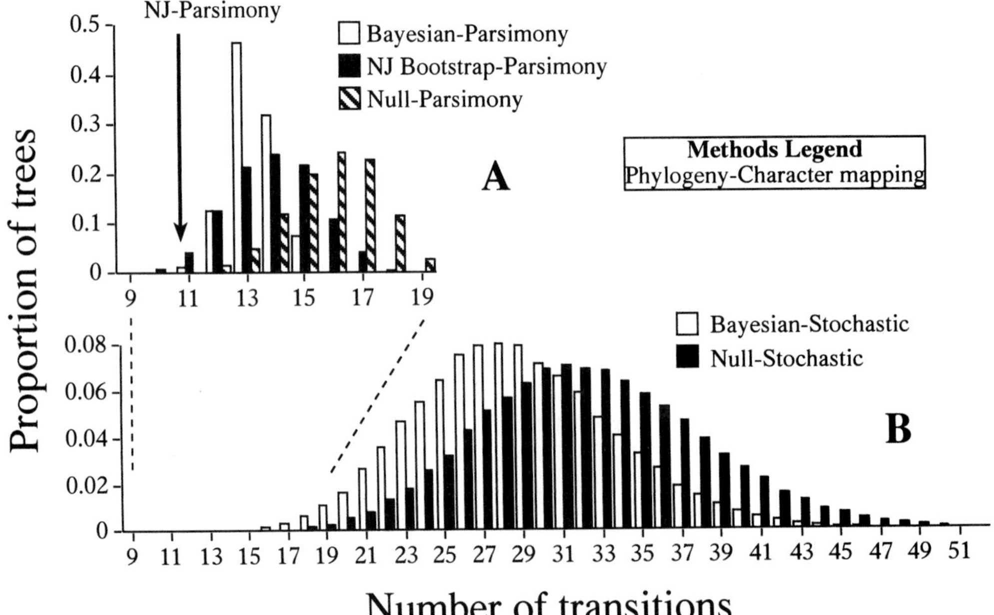

FIGURE 3 The effect of incorporating phylogenetic and character state mapping uncertainty into the Phylo-test when comparing *nirK* DNA sequences (from reference 28). This demonstrates the perils of using a single evolutionary relationship when performing the Phylo-test. (A) When a single NJ tree is used (NJ-Parsimony) the two communities are significantly different. However, incorporating phylogenetic uncertainty through a bootstrap analysis (NJ Bootstrap-Parsimony) or using Bayesian trees (Bayesian-Parsimony) reverses the inference. (B) Incorporating character state mapping uncertainty causes a nearly complete overlap of the observed (Bayesian-Stochastic) and null (Null-Stochastic) distributions. Reprinted from *Microbial Ecology* (16) with permission from the publisher.

phylogenetic uncertainty was incorporated through a bootstrap analysis or using Bayesian trees (note the overlap of the NJ Bootstrap-Parsimony and the Bayesian-Parsimony distributions with the null distribution in Fig. 3A). When both phylogenetic and character mapping uncertainties were included in the analysis, the observed data showed a nearly complete overlap with expectations, assuming that the two communities were random samples from the pool of sampled sequences (Fig. 3B).

In cases where there are three or more communities, ordination or cluster analysis can be used to summarize pairwise comparisons of communities. These ordination and clustering methods essentially create "trees of trees" that portray entire communities at the tips rather than individual organisms. A recently developed method, UNIFRAC, determines the amount of unique branch lengths separating communities, and the pairwise results can be used in ordination or cluster analysis (19). Similarly, pairwise F_{ST} comparisons could also be used in cluster analyses of communities. Finally, Nemergut et al. introduced a method for using maximum likelihood scores for trees (trees of individual communities and trees for combined communities) to calculate the distance between two communities (25). Using this method, Nemergut et al. demonstrated that young soils from a successional series in Peru were more similar to each other than to older soils and that intermediate-aged soils were as different from each other as they were from younger and older soils. As the number of communities being compared increases, ordination or cluster analysis provides a convenient means of summarizing the data.

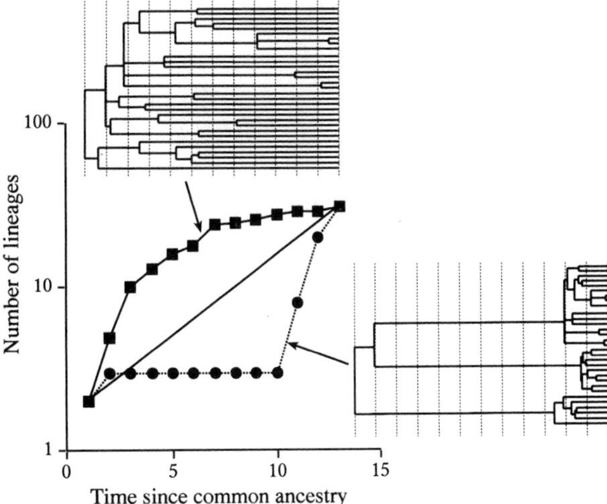

FIGURE 4 Lineage-per-time plots demonstrating an excess of highly divergent lineages (top) or an excess of closely related lineages (bottom right). Constant birth and extinction rates would yield an exponential lineage-per-time plot (indicated by the straight line). Note the logarithmic scale of the y axis. Reprinted from *Applied and Environmental Microbiology* (22) with permission from the publisher.

Methods for Inferring Processes Governing the Composition and Assembly of Communities

Several different approaches have been developed for extracting information about biological processes from the pattern of relationships evident in phylogenetic trees. Lineage-per-time plots were introduced by Nee et al. and others as a means of inferring rates of speciation and extinction from phylogenies (24). Webb developed methods for testing whether sampled phylogenetic lineages comprising ecological communities were more closely related or more distantly related to each other than expected by chance based on the available species pool (39). Both methods have been applied to the analysis of microbial communities with intriguing results. Here we describe both methods as a means of illustrating the inferences that can be gained by applying these approaches.

Martin introduced lineage-per-time plots as a means of assessing whether some form of selection can be invoked to explain the diversity of microbial communities (22). A lineage-per-time plot summarizes the distribution of divergence times for all nodes of the estimated phylogeny (Fig. 4). In the absence of selection (i.e., equal rates of speciation and extinction), the distribution of divergence times should be exponentially distributed: the time between successive divergence events progressively decreases with increasing distance from the root of the tree. Comparison of observed lineage-per-time plots with best-fit exponential regression equations provides a means of assessing whether communities harbor an excess of highly divergent or closely related species. An excess of divergent lineages suggests that selection may maintain high diversity in the community, whereas an excess of closely related lineages might be a fingerprint of a recent selective sweep of one or a few microbial types (1, 18, 22). For microbial ecologists, lineage-per-time plots may be most useful for comparing two different communities or different genes (e.g., a functional gene and a phylogenetic marker gene) from the same community.

Visualization of lineage-per-time plots requires ultrametric trees—trees in which all terminal nodes are equally distant from the common ancestor of all sampled sequences. Ultrametric trees can be constructed using maximum-likelihood or nonparametric methods. One drawback of using ultrametric trees is that they may not be an accurate representation of the evolutionary history of a set of widely divergent organisms. It may be more appropriate to use ultrametric trees for more narrowly defined phylogenetic groups (e.g., alphaproteobacteria) rather than entire communities.

The program GENIE conveniently summarizes the distribution of branch lengths that can be visualized graphically as a lineage-per-time plot. In addition, GENIE permits the calculation of a summary statistic, Y (29). Y is zero when the data match the null expectation of an exponential distribution; increasingly negative and positive values indicate an excess of divergent and closely related lineages, respectively. Martin et al. relied on Y in their study of microbial evolution and cladogenesis (23).

Inference of process depends on comparing the observed distribution of branch lengths (the lineage-per-time plot) with null expectations. One possible null expectation, as mentioned, is that the distribution of branching times is exponential. Alternatively, the null expectation for a particular sample of sequences can be generated from a larger sample of sequences that represent the local, regional, or global pool of sequences from which the sample is theoretically drawn. In this case, the approach is to randomly sample the same number of sequences from the species pool as is present in the community of interest. Repetitive sampling (1,000 replicates) provides the ability to generate a null distribution of lineage-per-time plots. Significantly more or fewer lineages in one community versus another (at a range of evolutionary time points) can be assessed by comparing the observed lineage-per-time plot with the null distribution. This approach can be implemented using the LTTR function in PHYLOCOM v3.22 (http://www.phylodiversity.net) (40).

Lineage-per-time plots are most instructive when comparing two (or more) communities or a number of different genes from a single community. Interestingly, most samples of 16S rRNA gene sequences yield exponential distributions of branching times (23). By contrast, samples of functional genes yield lineage-per-time plots that are markedly concave or convex. Such a comparison suggests that selection may play a stronger role on metabolic genes than on genes central to information processing.

Horner-Devine and Bohannan used Webb's NRI/NTI metrics for investigating the assembly and composition of microbial communities (11). The indices test whether a community of interacting organisms is significantly clustered (more related to each other than by chance) or significantly overdispersed (less related to each other than by chance) relative to the overall species pool (39, 41). NRI is the net relatedness index and compares the mean pairwise phylogenetic distance of taxa from one community to the mean pairwise phylogenetic distance of the overall species pool. NTI is the nearest taxon index and quantifies terminal clustering by analyzing only the closest phylogenetically related member. Both NRI and NTI increase with conservatism of community type with phylogeny; if NTI increases relative to NRI, there is more conservatism of traits at the tips of the phylogeny. Significance is tested by randomly assigning community identity to terminal taxa, obtaining index values from randomized data, and comparing observed values to randomized values. If the observed value is greater than 97.5% of randomized values, then the observed is significantly clustered; if the observed value is lower than 97.5% of randomized values, then the observed is significantly overdispersed. Because we used 1,000 trees in each of our tests, significance is determined based on the average of all 1,000 significance values. The null model essentially represents random community assembly with the assumption that organisms are readily able to disperse within the study system. This metric is easily implemented in PHYLOCOM v3.34, available at http://www.phylodiversity.net (40).

WORKED EXAMPLES

As a guide for illustrating the inferences that can be gained from analysis of environmental DNA sequences, we have worked through two published data sets: 16S sequences from ammonia-oxidizing bacteria (AOB) isolated from three soil types in Costa Rica, i.e., a forest, a pasture, and a tree plantation (4); and a set of large ribosomal subunit sequences of alpine soil fungi isolated during three different seasons, i.e., winter, spring, and summer (34). For all three data sets, we performed the phylogenetic test, the F test, ∫-LIBSHUFF, LTTR, and NRI/NTI.

For the phylogenetic-based tests (Phylo-test, NRI/NTI, and LTTR), we estimated the phylogeny of the sampled sequences using Bayesian methods (implemented using MrBayes, v.3.0 [33]). For all data sets, we obtained 1,000 trees that constituted the posterior probability distribution for phylogeny estimation. In other words, Bayesian analysis yielded 1,000 estimates of phylogeny. To account for uncertainty of character mapping, where character is the environment type of the sampled sequence (e.g., pasture or forest), we simulated 50 character state mappings for each tree (using SIMMAP [3]), yielding 50,000 estimates of the number of switches between two or more community types. For NRI/NTI and LTTR analyses, each of the observed 1,000 trees was compared to 1,000 randomizations (for a total of 1,000,000 randomizations).

Costa Rican AOB

Using the total number of transitions for all three communities of Costa Rican soil AOB, the Phylo-test demonstrated significant covariation between community type and phylogeny ($P = 0.006697$). Pairwise comparisons, however, revealed that only the comparison of pasture and plantation differed significantly. The F-test revealed strong differentiation between all pairwise comparisons (pasture and plantation: $F_{ST} = 0.12420$, $P < 0.00001$; pasture and forest: $F_{ST} = 0.04865$, $P < 0.00001$; and forest and plantation: $F_{ST} = 0.02980$, $P < 0.00001$). Using ∫-LIBSHUFF we found that the pasture soil community differed significantly from the tree plantation and forest soil AOB communities and that tree plantation and forest soil harbored phylogenetically similar AOB communities (Table 1).

The AOB lineages-through-time plots suggest that divergent processes may have operated between the forest community and the pasture and plantation communities (Fig. 5). The forest community has significantly more lineages at each time point than the pasture or plantation communities. This demonstrates that a wider range of phylogenetic diversity occurs at each time point, indicating that selection is maintaining diversity at all depths of the tree. This diversity could be under selection because key processes arose deep in the evolutionary past and are necessary for ammonia-oxidizing processes. Alternatively, the forest could have more deep-level diversity because the pasture and plantation communities have recently had major land use changes which caused a selective sweep favoring lineages able to colonize the newly formed habitats. The second inference seems more likely because the shape of the forest lineage per time plot is close to one with constant birth and death rates. So, the forest community does not appear to have an excess of highly divergent lineages on its own, but in comparison to the other communities it does.

NTI/NRI analysis revealed that the bacteria in the pasture are significantly more related to each other (both throughout the tree and at the tips) than to forest or plantation bacteria. In contrast, the forest bacteria are significantly overdispersed at both index levels. The tree plantation exhibits significant clustering for the NTI index but

TABLE 1 ∫-LIBSHUFF comparisons of Costa Rican soil bacterial 16S rRNA gene libraries[a]

Homologous library (X)	*P* value of ΔCxy heterologous library (Y) for Costa Rican soil vs:		
	TP	FOR	PAS
TP		0.8930	0.0000*
FOR	0.6643		0.0046*
PAS	0.0001*	0.0018*	

[a]Libraries were from tree plantation (TP), forest (FOR), and pasture (PAS) soils. The default number of randomizations (10,000) was used. To ensure a probability of less than 5% of a Type I error, minimum *P* values were corrected for multiple comparisons. *, for the six-way comparison, values were significant at *P* values of <0.0085. Data from reference 6.

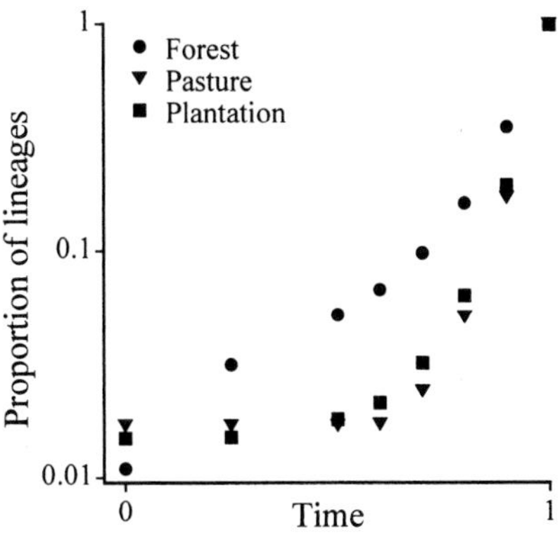

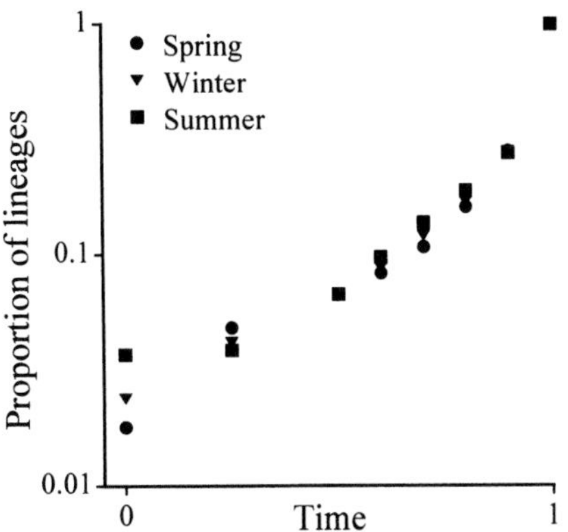

FIGURE 5 Lineage-per-time plots of Costa Rican AOB communities (16S) and alpine tundra fungal communities (18S). In the AOB comparison, the forest community has significantly more lineages at each time point. The convex lineage-per-time plots of the pasture and plantation communities demonstrate an excess of closely related community members. In the fungal community comparison, the spring community has significantly more lineages at time 0.25 and the summer community has significantly fewer lineages at time 0.9.

insignificance for the NRI index. These results indicate that the pasture has the highest "net ecological similarity" of the three communities. This could mean that niche effects or habitat filtering operate most strongly in the pasture, severely limiting the diversity of lineages able to survive. The significantly overdispersed forest community could indicate that competition across time has limited the number of closely related species that can coexist. Our results differ from a similar analysis of the same data set (11). Potential reasons for the discrepancies include a difference in the number of base pairs analyzed (Horner-Devine and Bohannan used 290 characters from the 16S DNA sequences, whereas we used 407 characters); differences in method of phylogenetic inference (they used a bootstrapped NJ tree, and we used Bayesian trees); and the design of the comparative analysis (they considered each library separately [i.e., three forest libraries, three pasture libraries, etc.], while we pooled all samples from each community type). The differences in our results demonstrate the importance that decisions made about alignment techniques, tree-building methods, or data inclusion can have on making biological inferences.

In terms of biology, these results underscore the conclusions of Carney et al. (4) that the pasture AOB community was compositionally distinct from forest and plantation AOB communities by using OTUs based on 97% sequence similarity and nonphylogenetic tests of diversity and community composition. Indeed, the F-test and LIBSHUFF revealed strong differentiation. However, when the uncertainty of phylogenetic estimation and environment assignment were incorporated into the analysis (i.e., the Phylo-test), only the pasture and plantation differed from each other—both of these communities were a subset of the forest sample. This result, coupled with evidence from the LTT and NTI/NRI for an overdispersed forest community (i.e., a highly diverse community) and coupled with phylogenetic clustering of the pasture and plantation community, suggests that each modified environment—pasture and plantation—represents a subsample of the original forest community in which only a fraction of bacteria survived and the survivors diversified, yielding clusters of closely related lineages. This process is similar to founder effects.

Alpine Tundra Fungal Communities in Winter, Spring, and Summer

Using the total number of transitions for all three communities, the Phylo-test demonstrated significant covariation between phylogeny and community type ($P = 0.044226$). However, pairwise comparisons between seasons show no significance: winter and spring, $P = 0.54746$; winter and summer, $P = 0.27016$; and spring and summer, $P = 0.37203$. These results differ from those of the original study (39). While the initial study found no covariation between winter and spring communities and phylogeny, significance between spring and summer communities and phylogeny was found. We attribute our different findings to our inclusion of character mapping uncertainty in the analysis. The initial study used parsimony to map character state transitions across the phylogeny, while we used a Bayesian procedure that takes branch length into account when mapping character state changes. While pairwise comparisons yielded insignificant results for the Phylo-test, each pairwise comparison was significant using F_{ST}: winter and spring, $F_{ST} = 0.02751$, $P = 0.02703$; winter and summer, $F_{ST} = 0.24352$, $P < 0.00001$; and spring and summer, $F_{ST} = 0.24099$, $P < 0.00001$. These results were mirrored by the LIBSHUFF analysis. The latter results support the finding by Schadt et al. (34) that the fungal community experiences a nearly complete turnover between spring and summer. This is an interesting and informative study because the distance-based comparative methods yielded significance whereas the phylogenetic methods did not, underscoring the effect of incorporating uncertainty in the analysis of community differentiation.

The three seasonal fungal communities show similar lineage per time plots throughout the different time slices (Fig. 5). However, the spring community has significantly more lineages at time 0.25, and the summer community has significantly fewer lineages at time 0.9. This indicates that the spring community has significantly more deep diversity relative to winter or summer; this could be a result of the spring community being a transitional community from winter to summer and thus containing deeply divergent lineages in decline from the winter and early colonizing summer lineages. Alternatively, the conditions of the spring soil environment could be selecting for divergent processes characterized by the divergent lineages. The significantly fewer lineages found at time 0.9 for the summer community is due to less fine-scale diversification relative to the winter or spring communities. This could be a result of competition among closely related lineages in the summer community or a higher rate of speciation in the winter and spring communities.

The NRI/NTI analysis revealed that all three seasonal communities exhibited phylogenetic clustering (winter: NRI average = 2.45857, NTI average = 1.35207; spring: NRI average = 1.68569, NTI average = 0.23711; summer: NRI average = 1.16873, NTI average = 1.49273), but only the winter NRI value was significant. This indicates that the winter community has the highest "net ecological similarity," which could be a result of stronger niche effects governing the winter community composition relative to the spring or summer communities. However, since the null hypothesis cannot be rejected for the spring or summer communities, we must infer that they assemble without regard to phylogeny.

Caveats and Warnings

We strongly recommend incorporating phylogenetic uncertainty into community analyses. However, a downside of accounting for uncertainty is that implementation of the methods demands familiarity with the methods and their implementation in various stand-alone software packages. Furthermore, the best methods, in our assessment, require understanding models of DNA sequence evolution, phylogenetic inference, and character evolution, challenging subjects and areas of active research. In the end, practitioners of these methods become, by necessity, experts in phylogenetic analysis. This should not be viewed as discouraging; rather, it should be viewed as an instigation of future collaborations between evolutionary biologists and microbial ecologists.

REFERENCES

1. **Acinas, S. G., V. Klepac-Ceraj, D. E. Hunt, C. Pharino, I. Ceraj, D. L. Distel, and M. F. Polz.** 2004. Fine-scale phylogenetic architecture of a complex bacterial community. *Nature* **430:**551–554.
2. **Bohannan, B. J. M., and J. Hughes.** 2003. New approaches to analyzing microbial biodiversity data. *Curr. Opin. Microbiol.* **6:**282–287.
3. **Bollback, J. P.** 2005. SIMMAP: stochastic character mapping of discrete traits on phylogenies, 1.0 Beta 2.0 ed.
4. **Carney, K. M., P. A. Matson, and B. J. M. Bohannan.** 2004. Diversity and composition of tropical soil nitrifiers across a plant diversity gradient and among land-use types. *Ecol. Lett.* **7:**684–694.
5. **Chao, A.** 1984. Non-parametric estimation of the number of classes in a population. *Scand. J. Stat.* **11:**265–270.
6. **Curtis, T. P., W. T. Sloan, and J. W. Scannell.** 2002. Estimating prokaryotic diversity and its limits. *Proc. Natl. Acad. Sci. USA* **99:**10494-10499.
7. **DeSantis, T. Z., I. Dubosarskiy, S. R. Murray, and G. L. Andersen.** 2003. Comprehensive aligned sequence construction for automated design of effective probes (CASCADE-P) using 16S rDNA. *Bioinformatics* **19:**1461–1468.
8. **Dunbar, J., S. M. Barns, L. O. Ticknor, and C. R. Kuske.** 2002. Empirical and theoretical bacterial diversity in four Arizona soils. *Appl. Environ. Microbiol.* **68:**3035–3045.
9. **Felsenstein, J.** 1985. Phylogenies and the comparative method. *Am. Nat.* **125:**1–15.
10. **Harvey, P. H., and M. D. Pagel.** 1991. *The Comparative Method in Evolutionary Biology*. Oxford University Press, Oxford, United Kingdom.
11. **Horner-Devine, M. C., and B. J. M. Bohannan.** Phylogenetic clustering and overdispersion in bacterial communities. *Ecology*, in press.
12. **Huelsenbeck, J. P., R. Nielsen, and J. P. Bollback.** 2003. Stochastic mapping of morphological characters. *Syst. Biol.* **52:**131–158.
13. **Huelsenbeck, J. P., B. Rannala, and J. P. Masly.** 2000. Accommodating phylogenetic uncertainty in evolutionary studies. *Science* **288:**2349–2350.
14. **Hughes, J. B., J. J. Hellmann, T. H. Ricketts, and B. J. M. Bohannan.** 2001. Counting the uncountable: statistical approaches to estimating microbial diversity. *Appl. Environ. Microbiol.* **67:**4399–4406.
15. **Jeanmougin, F., J. D. Thompson, M. Gouy, D. G. Higgins, and T. J. Gibson.** 1998. Multiple sequence alignment with Clustal X. *Trends Biochem. Sci.* **23:**403–405.
16. **Jones, R. T., and A. P. Martin.** Testing for differentiation of microbial communities using phylogenetic methods: accounting for uncertainty of phylogenetic inference and character state mapping. *Microb. Ecol.*, in press.
17. **Jukes, T. H., and C. R. Cantor.** 1969. Evolution of protein molecules, p. 21–132. *In* H. N. Munro (ed.), *Mammalian Protein Metabolism*. Academic Press, New York, N.Y.
18. **Klepac-Ceraj, V., M. Bahr, B. C. Crump, A. P. Teske, J. E. Hobbie, and M. F. Polz.** 2004. High overall diversity and dominance of microdiverse relationships in salt marsh sulphate-reducing bacteria. *Environ. Microbiol.* **6:** 686–698.
19. **Lozupone, C., and R. Knight.** 2005. UniFrac: a new phylogenetic method for comparing microbial communities. *Appl. Environ. Microbiol.* **71:**8228–8235.
20. **Ludwig, W., O. Strunk, R. Westram, L. Richter, H. Meier, Yadhukumar, A. Buchner, T. Lai, S. Steppi, G. Jobb, W. Forster, I. Brettske, S. Gerber, A. W. Ginhart, O. Gross, S. Grumann, S. Hermann, R. Jost, A. Konig, T. Liss, R. Lussmann, M. May, B. Nonhoff, B. Reichel, R. Strehlow, A. Stamatakis, N. Stuckmann, A. Vilbig, M. Lenke, T. Ludwig, A. Bode, and K. H. Schleifer.** 2004. ARB: a software environment for sequence data. *Nucleic Acids Res.* **32:**1363–1371.
21. **Maddison, W. P., and M. Slatkin.** 1991. Null models for the number of evolutionary steps in a character on a phylogenetic tree. *Evolution* **45:**1184–1197.
22. **Martin, A. P.** 2002. Phylogenetic approaches for describing and comparing the diversity of microbial communities. *Appl. Environ. Microbiol.* **68:**3673–3682.
23. **Martin, A. P., E. K. Costello, A. F. Meyer, D. R. Nemergut, and S. K. Schmidt.** 2004. The rate and pattern of cladogenesis in microbes. *Evolution* **58:**946–955.
24. **Nee, S., A. O. Mooers, and P. H. Harvey.** 1992. Tempo and mode of evolution revealed from molecular phylogenies. *Proc. Natl. Acad. Sci. USA* **89:**8322–8326.
25. **Nemergut, D. R., S. P. Anderson, C. C. Cleveland, A. P. Martin, A. E. Miller, A. Seimon, and S. K. Schmidt.** 2006. Microbial community succession in an unvegetated, recently deglaciated soil. *Microb. Ecol.*, in press.
26. **Pace, N. R.** 1997. A molecular view of microbial diversity and the biosphere. *Science* **276:**734–740.
27. **Posada, D., and K. A. Crandall.** 1998. MODELTEST: testing the model of DNA substitution. *Bioinformatics* **14:** 817–818.
28. **Prieme, A., G. Braker, and J. M. Tiedje.** 2002. Diversity of nitrite reductase (*nirK* and *nirS*) gene fragments in forested upland and wetland soils. *Appl. Environ. Microbiol.* **68:**1893–1900.
29. **Pybus, O. G., and P. H. Harvey.** 2000. Testing macro-evolutionary models using incomplete molecular phylogenies. *Proc. R. Soc. Lond. B Biol. Sci.* **267:**2267–2272.

30. **Ranjard, L., F. Poly, J. C. Lata, C. Mougel, J. Thioulouse, and S. Nazaret.** 2001. Characterization of bacterial and fungal soil communities by automated ribosomal intergenic spacer analysis fingerprints: biological and methodological variability. *Appl. Environ. Microbiol.* **67:**4479–4487.
31. **Rappe, M. S., and S. J. Giovannoni.** 2003. The uncultured microbial majority. *Annu. Rev. Microbiol.* **57:**369–394.
32. **Rodriguez, F., J. L. Oliver, A. Marin, and J. R. Medina.** 1990. The general stochastic-model of nucleotide substitution. *J. Theor. Biol.* **142:**485–501.
33. **Ronquist, F., and J. P. Huelsenbeck.** 2003. MrBayes 3: Bayesian phylogenetic inference under mixed models. *Bioinformatics* **19:**1572–1574.
34. **Schadt, C. W., A. P. Martin, D. A. Lipson, and S. K. Schmidt.** 2003. Seasonal dynamics of previously unknown fungal lineages in tundra soils. *Science* **301:**1359–1361.
35. **Schloss, P. D., B. R. Larget, and J. Handelsman.** 2004. Integration of microbial ecology and statistics: a test to compare gene libraries. *Appl. Environ. Microbiol.* **70:**5485–5492.
36. **Singleton, D. R., M. A. Furlong, S. L. Rathbun, and W. B. Whitman.** 2001. Quantitative comparisons of 16S rRNA gene sequence libraries from environmental samples. *Appl. Environ. Microbiol.* **67:**4374–4376.
37. **Staley, J. T., and A. Konopka.** 1985. Measurement of in situ activities of nonphotosynthetic microorganisms in aquatic and terrestrial habitats. *Annu. Rev. Microbiol.* **39:** 321–346.
38. **Thompson, J. D., D. G. Higgins, and T. J. Gibson.** 1994. Clustal-W—improving the sensitivity of progressive multiple sequence alignment through sequence weighting, position-specific gap penalties and weight matrix choice. *Nucleic Acids Res.* **22:**4673–4680.
39. **Webb, C. O.** 2000. Exploring the phylogenetic structure of ecological communities: an example for rain forest trees. *Am. Nat.* **156:**145–155.
40. **Webb, C. O., D. D. Ackerly, and S. W. Kembel.** 2005. Phylocom: software for the analysis of community phylogenetic structure and character evolution. Versions 3.22 and 3.34b.
41. **Webb, C. O., D. D. Ackerly, M. A. McPeek, and M. J. Donoghue.** 2002. Phylogenies and community ecology. *Annu. Rev. Ecol. Syst.* **33:**475–505.
42. **Woese, C. R.** 1987. Bacterial evolution. *Microbiol. Rev.* **51:**221–271.
43. **Woese, C. R., and G. E. Fox.** 1977. Phylogenetic structure of prokaryotic domain: the primary kingdoms. *Proc. Natl. Acad. Sci. USA* **74:**5088–5090.
44. **Zhang, J. Z., and M. Nei.** 1997. Accuracies of ancestral amino acid sequences inferred by the parsimony, likelihood, and distance methods. *J. Mol. Evol.* **44:**S139–S146.

Sampling Viruses from Soil

CHRISTON J. HURST AND KELLY A. REYNOLDS

50

Human enteric viruses are common microbial contaminants of domestic drainage and sewerage systems. As such, they are present in permeates from outhouses, as well as effluents from septic tanks and related waste disposal facilities, which generally are discharged directly into soil. These viruses are also present in municipal sewage, which is often discharged onto land after various levels of treatment. One of the major public health concerns associated with human enteric viruses in domestic drainage and sewage is that these viruses may survive after introduction into the soil environment (23) and subsequently cause illness in susceptible persons who consume virus-contaminated groundwater or foods made from contaminated crops (27). Research has been done on several topics that are pertinent to this concern, including the extent of viral adsorption to soil, the ability of viruses to persist in soil, migration of viruses through soil in association with the movement of water or wastewater, viral persistence in groundwater, and the survival of viruses on vegetables that might be contaminated by "night soil" or wastewater irrigation. In addition to human enteric viruses in soil, it must be understood that other types of viruses also are present in the terrestrial environment. These viruses include those which naturally infect other categories of microorganisms (2, 6), including bacteria, cyanobacteria, algae, fungi, and protozoa. Such viruses, all of which often are termed "phages," can interact with their microbial hosts even within the pores of consolidated rock (7). Soil also contains viruses of vascular plants (8, 10, 29) and viruses whose natural hosts are vertebrate or invertebrate animals that live in terrestrial environments (47).

This chapter focuses on the methodology used to detect animal viruses in samples of soil. This methodology generally relies on elution and subsequent concentration of viruses from the soil, after which either cytopathogenicity or plaque formation assays are used to detect the viruses. These assays are based on the use of cultured animal cells as hosts for viral replication. This chapter also describes the use of plaque formation methodology to detect bacteriophages, viruses which infect bacteria. Other types of assay procedures, such as those based on the PCR (48), have also been developed to detect viruses in soil samples. Two good general reviews on viruses in soil are those by Duboise et al. (12) and Williams et al. (51).

BACKGROUND

Viruses have a strong tendency to bind to soil and mineral particles (23, 32, 37, 49); therefore, it can be expected that most of the viruses in environmental soil samples will be adsorbed to the surface of soil particles. The adsorption of viruses to soil particles occurs by charged surface interactions (36). This adsorption is reversible, and the favorability of conditions for adsorption depends on the energy of activation for the system that is being studied (40). The electrostatic interactions which are important to viral adsorption are pH dependent, being stronger at lower pH levels (50). For this reason, the pHs of solutions used to elute (desorb) viruses from soil normally range from neutral to alkaline. Proteinaceous materials such as beef extract are often used as the basis for such eluant solutions in the hope that protein compounds in the eluant will help to displace the adsorbed viruses (5, 28, 48). The success of eluting viruses from soil depends not only on the physical and chemical characteristics of soil (22, 39) but also on the physiological and morphological characteristics of the viruses (11, 39).

SOIL-SAMPLING AND -PROCESSING EQUIPMENT

Many different types of apparatus can be used to collect soil samples. These range from spoons and spatulas to shovels and powered augers. Occasionally, even powered earth-moving equipment may be used for clearing away adjacent or overlying material before the soil that is to be sampled can be reached. It is important that those surfaces on the sampling apparatus which will come into direct contact with the final soil sample to be analyzed for viruses should not be contaminated with the types of viruses for which the soil is being examined, and it is certainly preferable that these surfaces be sterile. Collection of the final soil sample should be done in an aseptic manner, which can be facilitated by using such devices as presterilized liners for core samplers and by wearing presterilized disposable gloves. Soil samples should be kept in sterile containers until either assayed for viruses by direct inoculation or subjected to an elution process to separate the viruses from the soil particles. The available options for suitable sample containers include wide-mouthed screw-cap plastic jars and zipper closure plastic bags. Soil samples should be kept chilled to reduce thermal inactivation of the viruses. This chilled storage can be done by packing the sample containers in wet ice (H_2O) or by refrigeration. A significant loss in viral titer may occur upon freezing of soil samples (20). Thus, soil samples should neither be stored in a freezer nor packed in dry ice (CO_2). The solutions used to elute viruses from the soil are termed eluates. The eluates are frequently subjected to concentration techniques to reduce the volume of material which subsequently must be assayed. These eluates can be stored by refrigeration (1 to 4°C) before they are assayed or even stored frozen at or below −70°C if the viruses of interest are stable when frozen in the eluate. It should be noted that viral infectivity is not necessarily stable in frozen beef extract-based eluates (25).

Many types of equipment used for soil sampling can be sterilized in the laboratory before use in the field. Depending on the material composition of the equipment, this sterilization may be done by either autoclaving, ethylene oxide gas immersion, or gamma irradiation. Metal or borosilicate glass objects can be field sterilized by using flame from a portable propane or butane gas torch to heat the objects to red incandescence. Objects less resistant to high heat, such as nonborosilicate glass and some polymer materials, can be surface sterilized by dousing or immersing them in 95% ethanol and then igniting the alcohol with a flame. Objects made from many types of polymers can be field sterilized by using chlorine solutions, followed by a sodium thiosulfate solution to neutralize the residual disinfectant activity, as described in chapter 23.

REPORTING RESULTS

The results presented below, from different viral assay techniques, are reported as viral infectious units and expressed in terms of either PFU, most probable number (method 9510-G [9]), or 50% tissue culture infectious doses ($TCID_{50}$) as a 50% end point (43), depending on the nature of the viral titration assay being performed. These assay results should be reported per unit dry weight of the analyzed soil sample, expressed either in grams or kilograms of soil. Expressing the reported analytical results in terms of dry weight of the soil will avoid the problem of artifactual differences which may result when comparing viral titer reports that are based on the wet weight of soil samples.

DETECTION OF ANIMAL VIRUSES

It is possible to directly assay soil-associated enteric viruses by inoculating dilute soil suspensions onto cultures of animal host cells. Likewise, it is possible to assay soil-associated bacteriophages by directly inoculating dilute soil suspensions into cultures of bacterial host cells. One advantage of this direct-inoculation approach is that it eliminates the additional time and materials usually required for processing soil by an elution technique. However, three potential problems must be considered when directly inoculating host cell cultures with suspensions of soil particles. First, soil particles can obscure visualization of assay results. Second, soil particles may contain associated chemicals which can prove toxic to host cells. Third, other microorganisms associated with soil particles may result in contamination of the assay, which kills or overgrows the host cells. In order to minimize the adverse effects associated with these three problems, the concentration of viruses contained in a soil sample must be sufficiently high to allow for diluting a suspension of the soil prior to direct inoculation. If it is not, then it will be necessary to elute the viruses from the soil. It also may be necessary to concentrate the resulting eluates prior to their being assayed.

Titers resulting from the assay of either soil eluates or concentrates prepared from eluates cannot be considered equivalent to viral titers obtained from assaying soil suspensions by direct inoculation, because processes used for viral elution and for subsequent concentration of the eluates are not completely efficient. Since elution methods always involve some loss of virus, direct inoculation may be considered a more accurate assay of the level of viruses present in a soil sample.

TECHNIQUES FOR ELUTION AND SUBSEQUENT CONCENTRATION OF VIRUSES FROM SOIL SAMPLES

The following methods are basic and were developed for detecting human enteric viruses; with appropriate modification, they might also be useful for detecting bacteriophages and other types of viruses. Additional purification techniques such as cation-exchange chromatography could prove helpful for specific viral families, such as baculoviruses (3).

Elution of Viruses from Soil

For processing soil samples by virus elution from soil, 50 g of soil (wet weight) should be suspended in 50 ml of a buffered 10% beef extract solution (containing, per liter, 100 g of commercial powdered beef extract, 13.4 g of Na_2HPO_4, and 1.2 g of citric acid). The pH of this formulation is approximately 7.0. The beef extract used should be of a type that produces a visible amount of flocculant precipitate when the pH of solutions of the beef extract is lowered to 3.5. One suitable type is beef extract, desiccated, from BD (BD, Franklin Lakes, N.J., formerly Becton Dickinson and Co. [Sparks, Md.]). The suspensions of soil and buffer should be agitated for 30 min with a magnetic stirrer, during which time the pH of the suspensions should be maintained at 7.0 ± 0.1 by adding 1 M HCl or 1 M NaOH dropwise as needed. The suspensions should then be centrifuged for 30 min at 2,500 × g, after which the supernatant material is collected, and the soil pellet is discarded. The supernatant, now termed an eluate, should then be passed through a series of sterilized filters (which may be stacked like a sandwich) that are effective in removing particles with diameters of greater than 0.25 μm based on nominal porosity (a suggested glass fiber filter is model AP40, followed by either AP25 or AP20 and then AP15 series glass fiber filters from Millipore Corp., Billerica, Mass.) to remove remaining soil particles and reduce the level of contaminating soil bacteria. After this processing technique, the filtered supernatant (filtered eluate) may then be either directly assayed for the presence of viruses or further concentrated prior to assay by the concentration method described below. If the viruses of interest are stable when frozen, then storage of the filtered eluates prepared from soil samples should be done either at or below −70°C. Otherwise, the filtered eluates should be stored at 1 to 4°C. This soil elution procedure comes from a collaborative testing study (28).

Concentration of Viruses from Soil Eluates

If the filtered eluate needs to be concentrated, then it should first be diluted to 3% (wt/vol) beef extract by the addition of 2.3 equivalent volumes of sterile distilled water. While the diluted eluate is agitated with a magnetic stirrer, the pH of the diluted eluate should be adjusted to 3.5 ± 0.1 by dropwise addition of 1 M HCl. Stirring should be continued until either a visible flocculant precipitate has formed or 30 min has passed. The pH of the solution should be periodically checked during the course of stirring and readjusted as necessary to 3.5 ± 0.1 by adding either 1 M HCl or 1 M NaOH. The solution should then be centrifuged at 1,000 × *g* for 5 min at 4°C, after which the resulting supernatant fluid should carefully be removed by pipetting and then discarded. A magnetic stirring bar should then be added to the centrifuge bottle containing the precipitate, and the precipitate should be resuspended in 8 to 10 ml (total volume) of 0.05 M glycine (pH 11.0). The pH which results when the precipitate and glycine buffer mixture are combined at this ratio should be at least 9.5. That resulting pH should periodically be checked during this stirring and, if necessary, increased to 9.5 by addition of 1 M NaOH. After dissolution of the precipitate, the solution is centrifuged at 1,000 × *g* for 10 min at 4°C, and the resulting supernatant is collected. This supernatant is then considered a concentrated sample, and it should be adjusted to pH 7.5 by adding 1 M HCl (with a caution not to lower the pH below 7.4 to avoid formation of a troublesome precipitate), supplemented with antibiotics if desired (suggested final concentrations: 200 U of penicillin G/ml, 200 μg of streptomycin sulfate/ml, and 2.5 μg of amphotericin B/ml), and made isotonic by adding 0.5 ml of 3.0 M NaCl per 10 ml of concentrated sample. If the viruses to be studied are stable when frozen in this concentrated sample material, then the concentrated sample should be stored frozen either at or below −70°C until assayed for the presence of viruses. If the target viruses are not stable when frozen in this concentrated sample material, then the concentrated sample should be stored under refrigeration at 1 to 4°C. Not all viruses are stable when exposed to the pH 3.5 and 9.5 conditions that are used in this concentration procedure. Thus, caution is necessary when using this method for concentrating viruses whose pH stability is unknown. This procedure comes from a collaborative testing study (28).

TECHNIQUES FOR ASSAYING ANIMAL VIRUSES

The assay techniques described in this chapter were developed for human enteric viruses, and they use explanted cells that have been propagated in the laboratory to serve as hosts for virus replication. With appropriate modification, these methods might be useful for assaying other types of animal or plant viruses. Methods are presented for performing a cytopathogenicity assay, a plaque formation assay, and also a combined or integrated cell culture-PCR (CC/PCR) assay. Cytopathogenicity assays are based on the appearance of characteristic visible changes in the morphology of living host cells that result from the process of viral infection. Such changes are termed cytopathogenic effects (CPE) and can be assessed by microscopic examination of the infected cells. Plaque assays detect the production of focal areas of cell death (termed plaques) that may appear within a virally infected culture of host cells. Detection of these plaques is often facilitated by exposing the cultures of cells to a vital stain such as neutral red. Neutral red causes live cells to appear pink while dead cells appear colorless. At modest concentrations, this dye does not seem to interfere with cellular viability. The plaques normally can be detected by visual examination of the infected cell cultures without the aid of a microscope. Further background information on cytopathogenicity and plaque assay procedures, including specific information on their use with environmental samples and comparisons of their relative effectiveness for this purpose, can be found in chapter 8 of this manual and in reference 26. These viral assays should always be performed with monolayer cultures of cells that have been grown to confluency. Preparation of stock virus suspensions for use as positive controls in animal virus assays can be done as described in chapter 8.

Methods for Viral CPE Assays

BGM, a continuous cell line originating from African green monkey kidneys, can be used for detecting a large number of different human enteric virus species by CPE assays. HEp-2, a continuous human epidermoid carcinoma cell line, is useful for detecting human adenoviruses. PLC/PRF/5, a cell line derived from a human primary hepatocellular carcinoma (15), may be helpful in the specific isolation of human adenovirus types 40 and 41. The BGM cell line can be obtained from Daniel R. Dahling, U.S. Environmental Protection Agency, Cincinnati, Ohio, or purchased as Buffalo green monkey kidney (BGMK) cells from Bio-Whittaker, Walkersville, Md. The HEp-2 cell line can be purchased from the American Type Culture Collection, Manassas, Va. The PLC/PRF/5 cell line can be obtained from Kelly A. Reynolds, University of Arizona, Tucson. Production of individual tube cultures of either BGM or HEp-2 cells for use in CPE assays can be done in screw-cap, 16- by 125-mm borosilicate glass or polystyrene cell culture tubes. Individual cultures of cells for use in CPE assays can also be prepared in multiple-well polystyrene cell culture plates. Plates that contain 24 wells work well for this type of assay. Both of these cell lines should be incubated at 37°C. Culture tubes should be tightly capped during incubations. Multiple-well plates should be sealed with transparent sealing tape during incubations.

The growth medium used for routine cultivation of the BGM cell line consists of equal parts of Eagle's minimum essential medium (MEM) (13) and Leibovitz medium L-15 (33), supplemented with 10% (vol/vol) fetal bovine serum (FBS) and 5 ml of a 7.5% (wt/vol) stock sodium bicarbonate solution per liter. This same medium can also be used for preparing individual tube or plate cultures of BGM cells for use in virus CPE assays. The growth medium used for both routine cultivation of the HEp-2 cell line and preparation of tube or plate cultures of the same line should consist of MEM supplemented with 10% (vol/vol) FBS and 5 ml of sodium bicarbonate solution per liter. Either Earle's balanced salt solution or MEM, supplemented with FBS at a final level of 5% (vol/vol) and 15 ml of a 7.5% (wt/vol) stock sodium bicarbonate solution per liter, can be used with both cell lines as a maintenance medium for maintaining the viability of confluent cell culture monolayers prior to the CPE assay procedure. The type of serum used may influence the sensitivity of the viral assay techniques (19). Both the growth and the maintenance media can be supplemented with HEPES buffer at a level of 20 mM, although addition of HEPES to the medium may require a corresponding change in the amount of sodium bicarbonate that is added to the medium. Antibiotics can be added to growth and maintenance media to suppress microbial contamination; a combination of 100 U of penicillin G per ml, 30 μg of amikacin sulfate per ml, and 100 μg of streptomycin sulfate per ml is suggested.

One hour prior to inoculation of the cell cultures with sample material, the medium in the tubes or plates should be replaced with an inoculation medium, consisting of

FBS-free Hanks balanced salt solution or MEM containing 5 ml of sodium bicarbonate solution per liter and antibiotics (hereafter referred to as inoculation medium). Guanidine hydrochloride should be present in the inoculation medium at a final concentration of 100 μg/ml when cultures of HEp-2 cells are being infected for detection of adenoviruses. The guanidine is used to selectively suppress the replication of enteroviruses while allowing unrestricted replication of adenoviruses (26). Guanidine should not be present during routine passage of the cell lines or during the production of cell monolayers in the tubes or multiple-well plates, and it should not be used when one attempts to isolate enteroviruses. Following addition of the inoculation medium, the cultures should be incubated for 1 h at 37°C, and then 0.05 to 0.1 ml of environmental sample material should be added to the inoculation medium contained in each tube or well. This inoculum may first be diluted as necessary in a diluent such as inoculation medium before introducing it into the cell cultures. At 4 h postinoculation, the medium and inoculum contained in the inoculated tubes or wells should be removed and replaced with fresh inoculation medium that does not contain any sample material. The inoculated tubes or plates should then be incubated at 37°C, and the medium in the tubes or plates should additionally be changed on days 2 and 5 postinoculation (plus day 8 for adenovirus assays), using fresh inoculation medium containing FBS at a final concentration of 2% (vol/vol). Following inoculation with sample material, culture tubes can be incubated on a roller apparatus during the full postinoculation period in an effort to speed development of CPE. The cell cultures should be examined on days 2 through 7 postinoculation for the presence of characteristic enterovirus CPE or on days 4 through 10 postinoculation for the presence of characteristic adenovirus CPE. The viral titers from this type of assay are normally expressed in terms of $TCID_{50}$ (50% end points) as determined by the technique of Reed and Muench (43) or as most probable numbers (100% end points determined by using method 9510-G [9]) per gram (dry weight) of the soil sample that was examined. An estimation technique which can be used to convert 50%-end-point assay titers to 100% values has been published by Hurst et al. (26).

Method for Performance of Integrated CC/PCR Assay

The standard culture method for detection of human pathogenic viruses is expensive and time-consuming, and it may be complicated by the presence of compounds that are toxic to the cultured cells used to support viral replication. Furthermore, some enteric viruses do not readily produce either CPE or plaques in cultured cells. Direct PCR assay of environmental concentrates is rapid and specific but frequently complicated by inhibitory compounds naturally present in the environmental samples. Another problem with direct examination of environmental concentrates by PCR is that only small volumes of sample can be examined due both to the expense of PCR reagents and to the limitations of the thermal cycling equipment currently available (manually performing each temperature change in order to handle larger volumes would be so tedious as to be impractical and extremely costly). Direct examination of the environmental concentrates by PCR has been done (16) but can lead to false-negative results (45). Furthermore, one cannot distinguish between PCR amplification of infectious versus noninfectious viral sequences, which can lead to misinterpretation of the implications associated with positive test results obtained by direct PCR examination of the environmental concentrates. The resolution of many of these problems, particularly the latter, may lie in dividing each concentrated environmental sample into two aliquots. One of the aliquots is inoculated for a brief period into cultured host cells to allow for viral infection of the cells, during which time the viruses replicate their nucleic acid genomes. Both the uninoculated environmental sample and material produced by the infected culture of cells are then examined by PCR. Since only viable viruses contained in the environmental sample can reproduce their nucleic acid genomes by infecting the cultured cells, a positive PCR result for cell culture material clearly implies that at least some of the viral genomic nucleic acid contained in the environmental sample did indeed represent infectious viruses. This combined approach of processing an enviromental sample by inoculating it into cultured cells and then performing PCR analysis of the material generated by that intermediate exposure to cell cultures is termed an integrated CC/PCR technique. The CC/PCR approach has proven very useful to achieve a relatively rapid detection of culturable viruses (30, 31, 44).

By first growing infectious virus populations in cell culture and then performing PCR analysis, the integrated CC/PCR method allows for the detection of viable populations of noncytopathogenic and cytopathogenic viruses in environmental samples without the inherent problems encountered with either cell culture or PCR alone. Although highly effective, the primary limitation to the integrated CC/PCR is the cost of maintaining and applying two detection methodologies. The protocol for integrated CC/PCR is similar to that of conventional cell culture; however, the sample is incubated on cell culture monolayers just long enough to produce progeny virus and terminated prior to the development of visible CPE. The incubation times vary for different target viruses, because some virus types are known to grow more rapidly in cell culture than others do. The concentration of progeny virus increases directly with the initial virus inoculum and thus leads to more rapid detection and shorter incubation times. After minimal incubation (5 to 72 h) in the cultured cells, the infection is stopped by freezing the cells. The cells then are thawed, and during the thaw cycle the tubes or flasks containing the cells are gently shaken to allow the ice crystals to scrape the cell monolayer from the inside walls of the tube or the inside bottom of the flask. The cell lysate which results from this freezing and thawing is then centrifuged to pellet larger cell debris, and the produced supernatant is collected. A 10-μl aliquot of that supernatant serves as the material to be examined by PCR analysis.

PCR provides a rapid, sensitive, and specific method for amplifying the genomic material either of DNA viruses by PCR in its original form or of RNA viruses by reverse transcription of the RNA followed by PCR, also termed reverse transcription-PCR (RT-PCR). When PCR is used in combination with a culture assay, the problems of small reaction volumes, inhibitory compounds, and detection of noninfectious sequences, which are normally associated with either PCR or RT-PCR, are eliminated. The enzymatic amplification of viral nucleic acid sequences during the PCR process is facilitated by a series of temperature cycles and reaction buffers designed to amplify target viral sequences between two distinct primer locations, one upstream and the other one downstream from the target genome (17). Designing the sequence of the oligonucleotide primers is vital to the effectiveness and specificity of the PCR. The use of primer sequences that target a highly conserved region of the genome may allow the detection of multiple virus groups. Conversely, if desired, determining the presence of only a specific virus type or strain may be done by targeting a more specific or unique genomic sequence (38). Computer software packages are available to aid in primer design (e.g., the Wisconsin Package, version 10.0; Genetics Computer Group, Madison, Wis.). Finally, if desired, the amplified PCR product can be genomically sequenced for very effective further identification and characterization of the detected virus.

For amplifying viral RNA, we recommend that a master mix of 1.5 μl of 10× PCR buffer, 3.5 μl of 25 mM $MgCl_2$, and 4 μl of 10 mM deoxynucleoside triphosphate (dNTP) mixture be added to each 5-μl volume of supernatant (produced by centrifuging the cell lysate) to be tested by RT-PCR. This mixture should be placed into an appropriately sized polypropylene micro-test tube (these tubes are actually manufactured as microcentrifuge tubes), and the surface of the mixture should be layered with 3 drops of mineral oil. The mineral oil serves as a barrier to reduce evaporation. An aliquot (5 μl) of the supernatant containing potential target viral RNA is added to the tube containing the master mixture, and this combination is then heat extracted for 5 min at 95°C. RNA in the sample is then reverse transcribed by the addition of 0.5 μl each of random hexamers, avian mye-loblastosis virus reverse transcriptase, and RNase inhibitor. The sample mixture is then subjected to one cycle consisting of 24°C for 10 min, 44°C for 60 min, 99°C for 5 min, and 5°C for 5 min. During this cycle, cDNA is produced by transcription of the RNA, after which this cDNA is amplified by single PCR. Performance of the single PCR is done by adding to the micro-test tube 3.5 μl of 10× buffer II, 1.5 μl of 25 mM $MgCl_2$, 0.25 μl each of upstream and downstream primers (50 pmol of each), 28.75 μl of distilled water, and 0.25 μl of AmpliTaq DNA polymerase enzyme (1.25 U/reaction) and subjecting it to 30 cycles of template denaturing, primer annealing, and extension. The temperature cycles for this single PCR begin with an initial cycle which allows hybridization of the primers to the cDNA and consists of heating at 94°C for 1 min, followed by 50°C for 45 s and then 72°C for 1 min. This initial cycle is followed by 30 repetitions of the following amplification cycle: 94°C for 1 min, 55°C for 45 s, and 72°C for 1 min. Finally, the mixture is heated again at 72°C for 10 min. Samples that are negative for viral genomic sequences following this single PCR of 30 amplification cycles are subsequently analyzed by a seminested PCR. The seminested-PCR technique consists of placing a 5-μl volume of the product material generated from the single PCR into a new micro-test tube to which is added 44.5 μl of fresh reagents (all of the reagents listed above except that AmpliTaq Gold is substituted as the amplification enzyme in place of regular AmpliTaq) with a substitution of different upstream and downstream primers selected such that their hybridization positions are internal to the sequence amplified in the single PCR. The temperatures and times used to provide the amplification conditions for this seminested PCR are identical to those used in the single PCR except that the amplification cycle is performed for only 20 to 25 repetitions. All of the above protocols require about 7 h to complete with a thermal cycling device such as a Perkin-Elmer model 480 Gene Amp DNA thermal cycler (Roche Molecular Systems, Inc., Branchburg, N.J.). All of the PCR and RT-PCR reagents described here are available from Perkin-Elmer Cetus, Norwalk, Conn.

The extreme sensitivity of PCR allows for amplification of even a few molecules of target cDNA or DNA. This is the benefit of PCR technology, but at the same time it means that great care must be taken to ensure quality control of the methods used for the PCR and RT-PCR, because contaminating DNA or RNA also may be amplified. There should be separate work stations assigned for handling samples pre- and post-PCR amplification, and these two work stations should be in separate rooms. Each of the two work stations should have its own dedicated set of reagents, supplies, and pipetting devices. In addition, all buffer solutions should be prepared with high-pressure liquid chromatography (HPLC)-grade water and then autoclaved (please note that primers, dNTPs, reverse transcriptase, and DNA polymerase cannot be autoclaved). Aliquots from master mixes and other reagents are made in advance to minimize the number of repeated samplings from a single stock reagent container. Disposable gloves should be used in the manipulations and should be changed frequently. A microcentrifuge should always be used to move the liquid contents down to the bottom of the test tubes before the tubes are opened (called "spinning down the tubes"). Only aerosol-resistant pipette tips should be used when materials are added, removed, or transferred. The addition of cell culture supernatant or any other source of either RNA or DNA to a test tube should be done only after all other reagents have been added to that tube. Each tube must be tightly capped before the next is opened. Only heat-activated enzymes should be used, and only the minimum necessary number of amplification cycles should be performed. The possibility of contamination should be checked by the use of negative controls run not only at the beginning but also at the end of each day's processing of tubes by PCR techniques. These negative controls must contain all reagents and enzymes used in the PCR, but with autoclaved HPLC-grade water substituted for either the 5 μl of cell lysate supernatant or the 5 μl of product from the single PCR. Chapter 55 of this manual may be helpful in understanding how this viral cultivation-PCR technique compares with other approaches used for PCR detection of microbes in soil.

Method for Performance of Viral Plaque Formation Assays

The BGM cell line can be used for detecting a large number of different *Enterovirus* species and some other enteric viruses by plaque formation assay. Viral plaque formation assays can be performed by a soft agar overlay technique, and the assay titers should be reported in terms of PFU (a 100% end point) per gram (dry weight) of soil examined. The growth and maintenance media used for this cell line are described above for the CPE assay. The plaque formation assays can be performed by using confluent monolayer cultures of cells grown in either clear-glass prescription bottles, glass cell culture flasks, or polystyrene cell culture flasks, using the same type of procedure described in the CPE assay section. The cell culture bottles or flasks should be tightly capped during all incubations. The inoculation medium should be removed from the bottles or flasks immediately prior to introducing the inoculum onto the cell monolayers. The inoculum consists of an environmental sample that may have been diluted as necessary with a diluent such as inoculation medium. The volume of inoculum used per bottle or flask will vary, depending on the surface area of monolayer contained in the individual bottles or flasks. A typical ratio would be 0.1 to 0.2 ml of inoculum in a 25-cm^2 (cell surface area) culture bottle or flask. The inoculum should be evenly distributed over the surface of the cell monolayers by gently rocking (tilting) the bottles or flasks. After inoculation, the cultures should be incubated for 1 h at 37°C to allow viruses from the inoculum to adsorb (attach) to the cultured cells. The cultures may be kept on a gently tilting, rotating table during this adsorption period to aid even distribution of the inoculum and to prevent drying of the monolayer surface. Agar overlay medium should then be added to the cultures. A suggested overlay medium formulation is MEM lacking phenol red, containing 7.5 mg of neutral red dye per liter, and supplemented to final levels of 1% (vol/vol) FBS, 15 μl of 7.5% (wt/vol) stock sodium bicarbonate solution per liter, 7.5 g of agar per liter, and 50 mg of $MgCl_2$ per liter. The type of serum used may influence the sensitivity of the viral assay techniques (19). It is advisable to have antibiotics in the overlay medium to help suppress the growth of any possible contaminating bacteria or fungi.

It is desirable to confirm the positive status of at least 10 plaques per sample or of all plaques for aliquoted samples that yield fewer than 10 plaques. Confirmation can be accomplished by individually collecting material from each of these plaques and then inoculating the collected material from each plaque into separate, fresh cell cultures, as described above for the CPE assay. These fresh cultures should then be examined for the development of viral characteristic CPE as described for the CPE assay. Development of these CPE in the inoculated cultures is considered a confirmation that the plaque was formed by viruses capable of replicating in that cell line. The percentage of confirmable plaques should be utilized to establish a confirmed-virus level for each sample, and these confirmed values should be used for statistical analysis. The results of this type of assay are reported either as PFU or as confirmed PFU per gram (dry weight) of the soil sample that was examined.

Modifications that show promise for enhancing the sensitivity of detecting enteric viruses by plaque formation assays include pretreating the cell cultures with a medium that contains iododeoxyuridine and performing sequential inoculation (4). Sequential inoculation is done by incubating the inoculum on a culture of an initial cell line for 2 h and then transferring that inoculum to a second culture of either the same or a different cell line. For example, the best combination of cell lines for detecting species of *Enterovirus* might be BGM as the initial cell line followed by either BGM or RD, a human rhabdomyosarcoma cell line. The plaque assay titers from the two cultures are then summed to yield a total plaque assay titer (4). A dual-cell-line approach may also be used to first isolate a broad range of viruses and then select for specific wild-type strains of those viruses (34). Alternatively, two cell lines may be used together in a plaque assay with one grown as an attached monolayer and the other growing suspended in the agar overlay medium (35). Concentrated environmental samples are sometimes toxic to the cell culture lines. This can be caused by the presence of organic compounds that are eluted from the soil and concentrated along with the viruses. Several approaches have been developed to reduce these cytotoxic effects (24). The simplest approach, and one which does not result in an overall loss of viral titer, is to dilute the filtered eluate or concentrated sample material prior to introducing it onto the cultured cells. Another simple alternative is to wash the inoculum off the cell monolayers before adding the overlay medium. This is done after the sample is first allowed to incubate on the monolayers for 1 h at either room temperature or 37°C, so that virus adsorption to the cells can occur. After this exposure period, the inoculum is poured out of the cell culture bottles or flasks and replaced with 10 ml (per 75-cm^2 cell culture area) of washing solution (0.85% [wt/vol] sodium chloride containing 2% [vol/vol] FBS). The culture bottles or flasks are then gently rocked twice from side to side, which allows the washing solution to resuspend any remaining inoculum. The washing solution is then poured out, and soft agar overlay medium is dispensed into the cell culture bottles or flasks. Some loss of virus titer can be expected to result from this washing procedure (24).

DETECTION OF BACTERIAL VIRUSES (BACTERIOPHAGES)

Bacteriophages can be detected by direct assay of soil suspensions using a plaque formation technique. If the presence of soil particles in the assay causes a problem, either because the resulting turbidity obscures assessment of the results or the number of contaminating soil bacteria and fungi carried along with the soil particles complicates plaque enumeration, then the bacteriophages can be eluted from the soil particles and the eluate can be assayed. The elution method described above for use with human enteric viruses can also be used for elution of bacteriophages from soil, with assay of the filtered supernatant. Some groups of bacteriophages have a very limited resistance to pH 3.5 exposure. For this reason, it is preferable to avoid performing the low-pH-concentration step if the processed eluate is to be analyzed for the presence of bacteriophages.

Cultivation-Dependent Analyses

Two approaches used for detecting bacteriophages in environmental samples by cultivation are direct-assay and enrichment techniques. The direct-assay technique relies on the environmental sample as the inoculum. Enrichment techniques are used when the expected concentration of bacteriophages in the environmental sample is low. Enrichment consists of adding the sample, whether it be material directly collected from the environment or a processed and purified environmental material, to broth medium of the type used to grow the desired host bacteria. A culture of the host bacteria is then added to the broth which contains the sample. The resultant "enrichment" mixture is then incubated at a temperature which would support growth of the desired bacteriophage. If present, the desired bacteriophages hopefully will be able to replicate within the provided host bacteria. When successful, this technique results in an increase in the concentration of the desired bacteriophage within that enrichment mixture to a level which allows the phage to be enumerated by the plaque formation assays described below.

When one performs a bacteriophage assay, the key issue to keep in mind is that candidate bacterial host strains differ in the range of bacteriophages to which they are susceptible (21, 41, 42). There are two general approaches, both termed plaque assays, for enumerating bacteriophages by means of plaque formation. One uses a single agar layer, and the other uses a double agar layer.

The first of these assays, the double agar layer technique (18), is the more traditional approach. It uses a bottom layer of agar medium with a high concentration of agar (usually 10 to 15 g/liter) to help a top layer of agar medium with a lower agar concentration (usually 5 to 8 g/liter) adhere to the inside bottom of the petri dish when the dishes are inverted during incubation. In this double agar layer technique, the bottom layer of agar medium is added to the petri plates and allowed to congeal. The bacteriophage plaque assay is then performed in the subsequently added top layer of agar medium. The host bacteria grow in the top agar layer to form a turbid suspension, termed a lawn. The bacteriophages produce areas of cell lysis in the host lawn. These areas of lysis appear circular and are termed plaques. This traditional method can be used to assay approximately 1 ml of sample per 10-cm-diameter petri dish. The bottom agar layer is presumed to be inert, although dissolved compounds can diffuse from the bottom agar layer into the top agar layer. An example of intentional use of this diffusion action is the incorporation of crystal violet dye into the bottom agar layer (1), which is done to inhibit growth of contaminating gram-positive bacteria in the top layer. A typical formulation for the bottom agar layer would be approximately 15 ml of either tryptic soy broth or a tryptone-yeast extract medium such as modified LB broth (21), containing 10 g of agar per liter. The top layer for each 10-cm petri dish would consist of 3 ml of the same type of broth medium supplemented with salts of either calcium or magnesium or both to aid phage replication and containing 8 g of agar per liter, 1 ml of the sample or diluted sample being assayed, and 0.2 ml of a fresh overnight culture of the host bacterial strain grown in the same broth medium (21). After the top agar layer also congeals, the plates are inverted and subjected to overnight incubation to allow formation of the lawn and development of visible plaques. The final agar concentration in the top agar layer based on this formulation would be 5.7 g/liter. The composition of modified LB broth medium (per liter) is 10 g of Bacto tryptone (BD), 5 g of Bacto yeast extract (BD), 9 g of sodium chloride, 1 g of magnesium chloride hexahydrate, and 1 ml of 1 M calcium chloride. The pH of modified LB broth medium should be adjusted to 7.0 by addition of NaOH before the medium is sterilized by autoclaving prior to use.

The second bacteriophage assay approach is the single agar layer technique (14) (method 9211-D in reference 9), which relies on only one agar layer with a higher final agar concentration (approximately 8 g/liter) to eliminate the need for an underlying agar layer. This allows the assay of as much as 5 ml of water sample or diluted sample per 10-cm-diameter petri dish. A typical medium formulation for this single agar layer would be 100 ml of a double-strength broth medium, containing 16 g of agar per liter, combined with 100 ml of either sample or diluted sample plus 10 ml of a fresh overnight culture of host bacteria grown in the same type of broth medium. This mixture would be divided into 21 petri dishes, at 10 ml per dish. The final agar concentration in this single agar layer would be approximately 7.6 g/liter.

The final concentration of agar used in both the single agar layer method and the top agar layer of the traditional double agar layer method can affect the dispersion of phages through the plaque assay medium. For this reason, stock suspensions of some bacteriophage types seem to produce lower plaque assay titers in the single agar layer assay, with its higher agar concentration, than in the traditional double agar layer assay. An example of this, from personal observation, is the replication of phage 23356-B1 (ATCC) when grown in its host, *Klebsiella pneumoniae* (ATCC 23356).

Addition of tetrazolium dyes to the agar medium can aid visualization of phage plaques. Dye concentrations of up to 150 μg/ml seem to serve this purpose without causing detrimental effects on the assay titers (21). With these dyes, the live bacterial cells in the lawn appear colored because they reduce the dye to its colored formazan, while the plaque areas contain oxidized dye and thus appear relatively colorless. This dye supplementation technique can be used with both the traditional double agar layer assay technique and the single agar layer technique. In the double-layer assay, dye supplementation is accomplished by adding 0.75 ml of a 2% (wt/vol) filter-sterilized aqueous solution of either tetrazolium red dye (T-8877; Sigma-Aldrich Co., St. Louis, Mo.) or tetrazolium violet dye (T-0138; Sigma-Aldrich) per 100 ml of top agar medium. The amount of dye solution added per 100 ml of the double-strength medium used in the single agar layer technique is 1.5 ml. A recent technique for reducing the level of contaminating bacteria in samples to be assayed for the presence of bacteriophages is to treat the samples with hydrogen peroxide and then add catalase to destroy any residual peroxide (1) before assaying the samples.

Stock bacteriophage suspensions for use in comparative assays and as positive controls for bacteriophage assays on material from environmental samples can be prepared by the following technique, which is similar to that described by Hershey et al. (18). First, a double agar layer assay is performed, and petri plates showing confluent lysis of the lawn by bacteriophages are each flooded with 5 ml of broth medium. After 2 h of incubation at room temperature, the petri plates are swirled gently. The broth medium, now containing suspended phage, is then aspirated from the top agar layer with a pipette and filtered through either a 0.25- or a 0.45-μm-pore-size (rated by absolute filtration) filter to remove contaminating bacteria. If the strain of bacteriophage being examined is stable when frozen, then the bacteriophage stock suspension produced (filtered aspirate) can be stored frozen either at or below −70°C. If the strain of bacteriophage being examined is not stable when frozen, then the bacteriophage stock suspension should be stored under refrigeration at 1 to 4°C. The medium used when preparing stock bacteriophage suspensions should not contain any tetrazolium dye. The incubation temperature used for bacteriophage plaque formation assays varies, depending on the host bacterial strain being used and the group of bacteriophages whose detection is being sought.

Cultivation-Independent Analyses

There are cultivation-independent analytical techniques for phage enumeration and classification which are of increasing interest. These depend upon isolating the viruses from soil particles by an elution technique similar to that described above, followed by either light microscopy or transmission electron microscopy (52). Proteomic analyses (46) and population ecology approaches such as those presented in chapter 61 provide possible clues as to the direction in which genomics or metagenomics may lead the future of this field.

DETECTION OF PLANT VIRUSES

Many viruses that infect terrestrial plants can be found in soil. The sources of these viruses include decaying roots and fallen material such as seeds and debris. Some of these viruses are soilborne, meaning that their transmission is associated with the soil. Depending on the virus type, transmission of a soilborne viral infection may be either abiotic or biotic in nature (10). In abiotic transmission, the infection is assumed to be acquired by root contact with the soil.

Biotic transmission occurs in association with soil-dwelling organisms, including nematodes and fungi, that serve as biological vectors for the viruses. General information on the ecology of plant viruses can be found in references 8 and 29. Suggestions for methodology for detecting plant viruses contained in soil, including those viruses associated with nematodes and fungi, appear as protocols 13, 14, and 15 (10).

REFERENCES

1. **Asghari, A., S. R. Farrah, and G. Bitton.** 1992. Use of hydrogen peroxide treatment and crystal violet agar plates for selective recovery of bacteriophages from natural environments. *Appl. Environ. Microbiol.* **58:**1159–1163.
2. **Ashelford, K. E., S. J. Norris, J. C. Fry, M. J. Bailey, and M. J. Day.** 2000. Seasonal population dynamics and interactions of competing bacteriophages and their host in the rhizosphere. *Appl. Environ. Microbiol.* **66:**4193–4199.
3. **Barsoum, J.** 1999. Concentration of recombinant baculovirus by cation exchange chromatography. *BioTechniques* **26:**834–840.
4. **Benton, W. H., and C. J. Hurst.** 1990. Sequential inoculation as an adjunct in enteric virus plaque enumeration. *Water Res.* **10:**905–909.
5. **Bitton, G., M. J. Charles, and S. R. Farrah.** 1979. Virus detection in soils: a comparison of four recovery methods. *Can. J. Microbiol.* **25:**874–880.
6. **Burroughs, N. J., P. Marsh, and E. M. H. Wellington.** 2000. Mathematical analysis of growth and interaction dynamics of streptomycetes and a bacteriophage in soil. *Appl. Environ. Microbiol.* **66:**3868–3877.
7. **Chang, P. L., and T. F. Yen.** 1984. Interaction of *Escherichia coli* B and B/4 and bacteriophage T4D with Berea sandstone rock in relation to enhanced oil recovery. *Appl. Environ. Microbiol.* **410:**544–550.
8. **Chatterji, A., and C. M. Fauquet.** 2000. Ecology of plant viruses, with special reference to whitefly-transmitted geminiviruses (WTG's), p. 321–351. *In* C. J. Hurst (ed.), *Viral Ecology*. Academic Press, Inc., San Diego, Calif.
9. **Clesceri, L. S., A. E. Greenberg, and A. D. Eaton (ed.).** 1998. *Standard Methods for the Examination of Water and Wastewater*, 20th ed. American Public Health Association, Washington, D.C.
10. **Dijkstra, J., and C. P. de Jager.** 1998. *Practical Plant Virology*. Springer-Verlag KG, Berlin, Germany.
11. **Dowd, S. E., S. D. Pillai, S. Wang, and M. Y. Corapcioglu.** 1998. Delineating the specific influence of virus isoelectric point and size on virus adsorption and transport through sandy soils. *Appl. Environ. Microbiol.* **64:**405–410.
12. **Duboise, S. M., B. E. Moore, C. A. Sorber, and B. P. Sagik.** 1979. Viruses in soil systems. *Crit. Rev. Microbiol.* **7:** 245–285.
13. **Eagle, H.** 1959. Amino acid metabolism in mammalian cell cultures. *Science* **130:**432–437.
14. **Grabow, W. O. K., and P. Coubrough.** 1986. Practical direct plaque assay for coliphages in 100-ml samples of drinking water. *Appl. Environ. Microbiol.* **52:**430–433.
15. **Grabow, W. O. K., D. L. Puttergill, and A. Bosch.** 1992. Propagation of adenovirus types 40 and 41 in the PLC/PRF/5 primary liver carcinoma cell line. *J. Virol. Methods* **37:**201–208.
16. **Gratacap-Cavallier, B., O. Genoulaz, K. Brengel-Pesce, H. Soule, P. Innocenti-Francillard, M. Bost, L. Gofti, D. Zmirou, and J. M. Seigneurin.** 2000. Detection of human and animal rotavirus sequences in drinking water. *Appl. Environ. Microbiol.* **66:**2690–2692.
17. **Griffin, D. W., C. J. Gibson III, E. K. Lipp, K. Riley, J. H. Paul III, and J. B. Rose.** 1999. Detection of viral pathogens by reverse transcriptase PCR and of microbial indicators by standard methods in the canals of the Florida Keys. *Appl. Environ. Microbiol.* **65:**4118–4125.
18. **Hershey, A. D., G. Kalmanson, and J. Bronfenbrenner.** 1943. Quantitative methods in the study of the phage-antiphage reaction. *J. Immunol.* **46:**267–279.
19. **Hoyt, J. L., and A. B. Margolin.** 2000. Fortified sera and their use in environmental virology. *Appl. Environ. Microbiol.* **66:**2259–2262.
20. **Hurst, C. J.** 1979. Viral detection and persistence during the land treatment of sludge and wastewater. Ph.D. dissertation. Baylor College of Medicine, Houston, Tex.
21. **Hurst, C. J., J. C. Blannon, R. L. Hardaway, and W. C. Jackson.** 1994. Differential effect of tetrazolium dyes upon bacteriophage plaque assay titers. *Appl. Environ. Microbiol.* **60:**3462–3465.
22. **Hurst, C. J., and C. P. Gerba.** 1979. Development of a quantitative method for the detection of enteroviruses in soil. *Appl. Environ. Microbiol.* **37:**626–632.
23. **Hurst, C. J., C. P. Gerba, and I. Cech.** 1980. Effects of environmental variables and soil characteristics on virus survival in soil. *Appl. Environ. Microbiol.* **40:**1067–1079.
24. **Hurst, C. J., and T. Goyke.** 1983. Reduction of interfering cytotoxicity associated with wastewater sludge concentrates assayed for indigenous enteric viruses. *Appl. Environ. Microbiol.* **46:**133–139.
25. **Hurst, C. J., and T. Goyke.** 1986. Stability of viruses in waste water sludge eluates. *Can. J. Microbiol.* **32:**649–653.
26. **Hurst, C. J., K. A. McClellan, and W. H. Benton.** 1988. Comparison of cytopathogenicity, immunofluorescence and *in situ* DNA hybridization as methods for the detection of adenoviruses. *Water Res.* **22:**1547–1552.
27. **Hurst, C. J., and P. A. Murphy.** 1996. The transmission and prevention of infectious disease, p. 3–54. *In* C. J. Hurst (ed.), *Modeling Disease Transmission and Its Prevention by Disinfection*. Cambridge University Press, Cambridge, United Kingdom.
28. **Hurst, C. J., S. A. Schaub, M. D. Sobsey, S. R. Farrah, C. P. Gerba, J. B. Rose, S. M. Goyal, E. P. Larkin, R. Sullivan, J. T. Tierney, R. T. O'Brien, R. S. Safferman, M. E. Morris, F. M. Wellings, A. L. Lewis, G. Berg, P. W. Britton, and J. A. Winter.** 1991. Multilaboratory evaluation of methods for detecting enteric viruses in soils. *Appl. Environ. Microbiol.* **57:**395–401.
29. **Jones, R. A. C., and D. A. Nicholas.** 1998. Impact of an insidious virus disease in the legume component on the species balance within self-regenerating annual pasture. *J. Agric. Sci.* **131:**155–170.
30. **Lee, C., S.-H. Lee, E. Han, and S.-J. Kim.** 2004. Use of cell culture-PCR assay based on combination of A549 and BGMK cell lines and molecular identification as a tool to monitor infectious adenoviruses and enteroviruses in river water. *Appl. Environ. Microbiol.* **70:**6695–6705.
31. **Lee, H. K., and Y. S. Jeong.** 2004. Comparison of total culturable virus assay and multiplex integrated cell culture-PCR for reliability of waterborne virus detection. *Appl. Environ. Microbiol.* **70:**3632–3636.
32. **Lefler, E., and Y. Kott.** 1974. Enteric virus behavior in sand dunes. *Isr. J. Technol.* **12:**298–304.
33. **Leibovitz, A.** 1963. The growth and maintenance of tissue-cell cultures in free gas exchange with the atmosphere. *Am. J. Hyg.* **78:**173–180.
34. **Manor, Y., R. Handsher, T. Halmut, M. Neuman, B. Abramovitz, A. Mates, and E. Mendelson.** 1999. A double-selective tissue culture system for isolation of wild-type poliovirus from sewage applied in a long-term environmental surveillance. *Appl. Environ. Microbiol.* **65:**1794–1797.

35. **Mocé-Llivina, L., F. Lucena, and J. Jofre.** 2004. Double-layer plaque assay for quantification of enteroviruses. *Appl. Environ. Microbiol.* **70:**2801–2805.
36. **Moore, R. S., D. H. Taylor, M. M. M. Reddy, and L. S. Sturman.** 1982. Adsorption of reovirus by minerals and soils. *Appl. Environ. Microbiol.* **44:**852–859.
37. **Moore, R. S., D. H. Taylor, L. S. Sturman, M. M. Reddy, and G. W. Fuhs.** 1981. Poliovirus adsorption by 34 minerals and soils. *Appl. Environ. Microbiol.* **42:**963–975.
38. **Muscillo, M., G. La Rosa, A. Carducci, L. Cantiani, and C. Marianelli.** 1999. Molecular and biological characterization of poliovirus 3 strains isolated in Adriatic seawater samples. *Water Res.* **33:**3204–3212.
39. **Ostle, A. G., and J. G. Holt.** 1979. Elution and inactivation of bacteriophages on soil and cation-exchange resin. *Appl. Environ. Microbiol.* **38:**59–65.
40. **Preston, D. R., and S. R. Farrah.** 1988. Activation thermodynamics of virus adsorption to solids. *Appl. Environ. Microbiol.* **54:**2650–2654.
41. **Puig, A., N. Queralt, J. Jofre, and R. Araujo.** 1999. Diversity of *Bacteroides fragilis* strains in their capacity to recover phages from human and animal wastes and from fecally polluted wastewater. *Appl. Environ. Microbiol.* **65:**1772–1776.
42. **Rajala-Mustonen, R. L., and H. Heinonen-Tanski.** 1994. Sensitivity of host strains and host range of coliphages isolated from Finnish and Nicaraguan wastewater. *Water Res.* **28:**1811–1815.
43. **Reed, L. J., and H. Muench.** 1938. A simple method of estimating fifty percent endpoints. *Am. J. Hyg.* **27:**493–497.
44. **Reynolds, K. A., C. P. Gerba, and I. L. Pepper.** 1996. Detection of infectious enteroviruses by an integrated cell culture-PCR procedure. *Appl. Environ. Microbiol.* **62:**1424–1427.
45. **Reynolds, K. A., C. P. Gerba, and I. L. Pepper.** 1997. Rapid PCR-based monitoring of infectious enteroviruses in drinking water. *Water Sci. Technol.* **35:**423–427.
46. **Rohwer, F., and R. Edwards.** 2002. The phage proteomic tree: a genome-based taxonomy for phage. *J. Bacteriol.* **184:**4529–4535.
47. **Rothman, L. E., and J. H. Myers.** 2000. Ecology of insect viruses, p. 385–412. *In* C. J. Hurst (ed.), *Viral Ecology.* Academic Press, Inc., San Diego, Calif.
48. **Straub, T. M., I. L. Pepper, M. Abbaszadegan, and C. P. Gerba.** 1994. A method to detect enteroviruses in sewage sludge-amended soil using the PCR. *Appl. Environ. Microbiol.* **60:**1014–1017.
49. **Taylor, D. H., A. R. Bellamy, and A. T. Wilson.** 1980. Interaction of bacteriophage R17 and reovirus type III with the clay mineral allophane. *Water Res.* **14:**339–346.
50. **Taylor, D. H., R. S. Moore, and L. S. Sturman.** 1981. Influence of pH and electrolyte composition on adsorption of poliovirus by soils and minerals. *Appl. Environ. Microbiol.* **42:**976–984.
51. **Williams, S. T., A. M. Mortimer, and L. Manchester.** 1987. Ecology of soil bacteriophages, p. 157–179. *In* S. M. Goyal, C. P. Gerba, and G. Bitton (ed.), *Phage Ecology.* John Wiley & Sons, Inc., New York, N.Y.
52. **Williamson, K. E., K. E. Wommack, and M. Radosevich.** 2003. Sampling natural viral communities from soil for culture-independent analyses. *Appl. Environ. Microbiol.* **69:**6628–6633.

Isolation, Culture, and Detection of Arbuscular Mycorrhizal Fungi

DAVID M. SYLVIA, WITTAYA KAONONGBUA, AND JAMES D. BEVER

51

Mycorrhizae are mutualistic associations between beneficial soil fungi and plant roots and are common in natural soils. They have an important role in increasing plant uptake of P and other poorly mobile nutrients (74). The arbuscular mycorrhizal (AM) fungi colonize members of more than 90% of all vascular plant families. The term vesicular-arbuscular was originally applied to this symbiotic association, but since a major suborder lacks the ability to form vesicles in roots, AM is now the preferred acronym. Because of the obligate nature of these organisms, their manipulation is different from that of most other soilborne fungi, and much is still unknown about their biology in natural and managed ecosystems. Disturbance can change AM fungus distribution, abundance, and species composition within the ecosystem (63). Although isolation of their spores from the environment is relatively simple, multiplication of spores generally requires several months' growth under high-light conditions. Contemporary immunological, physiological, and molecular techniques to detect and identify soilborne organisms are now being applied to AM fungi and are leading to a better understanding of their distribution and function in the environment.

The purpose of this chapter is to summarize methods to (i) isolate and estimate numbers of soilborne propagules of AM fungi, (ii) propagate AM fungi by traditional and innovative methods, and (iii) detect and assess properties of these fungi by using recent biochemical and molecular technology. For further detail and explanation, the reader may consult various reviews of these methods (36, 47, 49, 73, 91) and a more extensive, step-by-step treatment of several of the procedures presented here (96). Methods for manipulating ectomycorrhizal fungi have been detailed elsewhere (72).

ISOLATION AND ESTIMATION OF NUMBERS OF SOILBORNE PROPAGULES

AM fungi are among the most common fungi in soil, but they are often overlooked because they do not grow on standard dilution-plating media. Soilborne propagules of AM fungi may include chlamydospores or azygospores, colonized roots, and hyphae. Isolating spores and quantifying root colonization are the most basic procedures for working with these fungi. Spores are required for pure, single-species cultures, and detection of colonization of roots is necessary to verify a functional association by visualization of arbuscules. The spores of AM fungi are larger than those of most other fungi, ranging from 10 to 1,000 μm in diameter. Most spores are between 100 and 200 μm in diameter and can easily be observed with a dissecting microscope. Hayman (36) and Schenck and Perez (91) have reviewed several methods for extracting spores of AM fungi from soil. The wet sieving and decanting density gradient centrifugation method is the most widely used method. The major variable in its application is in the use of single or multiple densities of sucrose. Multiple layers of different densities can provide cleaner spores and are useful for separating different species. Another method of separating spores from debris utilizes a series of sieves of various pore sizes. As with most of the techniques, it works best for sandy soils and less well for clay and organic soils. Soil samples with significant clay contents can be soaked in 6.3 mmol of sodium hexametaphosphate to disperse the clay fraction (64).

Method for Spore Isolation

Place a soil sample (50 to 100 g is usually sufficient) into a 2-liter container and add 1.5 liters of water. Vigorously mix the suspension to free spores from soil and roots. For fungal species that form spores in roots (e.g., *Glomus intraradices* and *G. clarum*), blend the soil-root sample for 1 min in 300 ml of water to free spores from roots. Next, let the suspension settle for 15 to 45 s (times vary depending on soil texture) and decant the supernatant through standard sieves. Sieves should be selected so as to capture the spores of interest. The spore sizes of most species can be found on the International Collection of (Vesicular) Arbuscular Mycorrhizal Fungi (INVAM) website at http://invam.caf.wvu.edu. Use a 425-μm-pore-size sieve over a 38-μm-pore-size sieve for unknown field samples. Examine the contents of the top sieve for sporocarps that may be up to 1 mm in diameter. For clay soils, it is advisable to repeat the decanting and sieving procedure with the settled soil. Roots may be collected from the larger-mesh sieve for evaluation of internal colonization. Transfer sieved material to 50-ml centrifuge tubes with a fine stream of water from a wash bottle, and balance opposing tubes. Centrifuge at 1,200 to 1,300 × *g* in a swinging-bucket rotor for 3 min, allowing the centrifuge to stop without braking. Remove the supernatant carefully to avoid disturbing the pellet, and then with a finger remove the organic debris that adheres to the side of the tube. Suspend soil particles in chilled 1.17 M sucrose, mix the contents with a spatula, and centrifuge the samples immediately at 1,200 to 1,300 × g for 1.5 min, applying the brake to stop the centrifuge. Pour the supernatant through the small-mesh sieve, carefully rinse the spores held on the sieve with tap water, and wash the spores into a plastic petri dish inscribed with parallel lines spaced 0.5 cm apart. Spores may be counted by scanning the dish under a dissecting microscope.

The identification of AM fungi can be difficult because the taxonomy of this group was based almost entirely on a limited number of morphological characteristics of the spores. Unless taxonomy is the major objective, we recommend that the spores be identified initially only to the genus level. Isolates of special interest should be given a unique isolate code and then classified to the species level at a later date. For authoritative detail on the identification of spores, we direct the reader elsewhere (67–69). Maintenance of good germplasm should be an essential part of any AM fungal research program; therefore, careful notes need to be taken on the sizes, colors, surface characteristics, and wall morphologies of the spore types recovered. Initiate pot cultures with each spore type, keeping detailed records on the origin and subsequent pot culture history of the isolate.

INVAM maintains a collection of AM isolates. Samples of AM fungi may be submitted to INVAM for verification of classification and possible inclusion in their collection. For additional information, refer to the website at http://invam.caf.wvu.edu. Another culture collection, the International Bank for the Glomeromycota, exists in Europe, and its Web address is http://www.kent.ac.uk/bio/beg/.

Root Colonization

The AM fungi do not cause obvious morphological changes in roots; however, they produce arbuscules and, in many cases, vesicles in roots. To observe AM structures within the root, it is necessary to clear cortical cells of cytoplasm and phenolic compounds and then to differentially stain the fungal tissue. Phillips and Hayman (83) published the oft-cited method to visualize AM fungi in roots by using 0.05% trypan blue in lactophenol, but the use of phenol is now discouraged (54). The clearing agent for nonpigmented roots is generally 10% KOH, but treatment with H_2O_2 (83) or NaOCl (31) may be necessary for pigmented roots. Decolorizing with H_2O_2 is slower than decolorizing with hypochlorite, but there is less danger of complete destruction of fungal and cortical tissue. However, NaOCl is a very fast and effective bleaching agent, and the procedure requires no heating. Alternatives to trypan blue for staining are chlorazol black E (12) and acid fuchsin (53). For nonpigmented roots, it is also possible to observe colonization nondestructively by inducing autofluorescence (3).

Method for Clearing and Staining of Roots

Place root samples (approximately 0.5 g) in perforated plastic holders (e.g., OmniSette tissue cassettes; Fisher Scientific, Pittsburgh, Pa.) and store them in cold water until they are processed. Place enough 1.8 M KOH into a beaker (without samples) to allow samples to be covered, and heat the solution to 80°C in a fume hood. Place samples in the heated KOH for the desired time, 15 min for tender roots such as onion and 30 min for other roots. If samples are still pigmented after the initial treatment, rinse them with at least three changes of water and then place them in a beaker with either 30% (wt/wt) H_2O_2 at 50°C or 3% (wt/vol) NaOCl, acidified with several drops of 5 M HCl. Times may vary from several seconds to several minutes. Check roots frequently to avoid destruction of the cortex and fungal tissue. Rinse roots with copious amounts of water as soon as samples are bleached white or become transparent, and then rinse them with five changes of tap water. Cover the samples with tap water, add 5 ml of concentrated HCl for each 200 ml of water, stir, and drain. Repeat once. Dispense enough trypan blue stain into a beaker (without samples) to cover the samples, and heat the solution to 80°C. To prepare the stain, add the following in order to a flask while stirring: 800 ml of glycerin, 800 ml of lactic acid, 800 ml of distilled water, and finally 1.2 g of trypan blue. Place samples in the stain for at least 30 min, cool overnight, and drain the stain into the large flask for reuse. Rinse samples with two changes of tap water to destain. Additional destaining in water may be necessary for some roots.

Methods for Root Colonization Assessment

Various methods have been used to estimate root colonization by AM fungi (36, 53). The grid line intersect method has the advantage of providing an estimate of both the proportion of colonized root and the total root length (30). This is important because some treatments affect root and fungal growth differently. For example, when P is applied, the total root length may increase more rapidly than the colonized root length, and thus the proportion of colonized root will decrease even though the actual length of the colonized root is increasing.

McGonigle et al. (62) argued that the grid line intersect method is somewhat subjective because arbuscules may be difficult to distinguish with a dissecting microscope. They proposed the use of a magnified-intersect method whereby roots are observed at a magnification of ×200 and arbuscules are quantified separately from vesicles and hyphae. Another limitation of the grid line intersect method is that the intensity of colonization at each location is not estimated. To obtain an estimate of intensity, one can use a morphometric technique (100) whereby a grid of dots is placed over an image of squashed roots and colonized cortical cells are counted.

To quantify root colonization, spread a cleared and stained root sample evenly in an inscribed, 10-cm-diameter plastic petri dish. A grid of squares should be inscribed on the underside of the dish as specified by Giovannetti and Mosse (30) so that the total number of root intersections is equal to the root length in centimeters. Using a dissecting microscope, scan only the grid lines and record the total number of root intersections with the grid as well as the number of intersections with colonized roots. Verify any questionable colonization with a compound microscope. To do this, cut out a small portion of the root with a scalpel, place it in water on a microscope slide, and look for AM structures at a magnification of ×100 to ×400. Remember that the stains are not specific for AM fungi; other fungi colonizing the root will also stain, and so it is important to verify the presence of arbuscules or vesicles in the root with a compound microscope.

Propagule Assays

Spore counts often underestimate the inoculum density of AM fungi, since colonized roots and hyphae can also serve as propagules. The most commonly used methods to obtain an estimate of the total number of propagules are the most probable number (MPN) and infectivity assays. The MPN assay provides estimates of propagule numbers, but confidence limits are usually very high. The infectivity assay is less complex and time-consuming than the MPN assay, but the actual propagule numbers are not estimated. Rather, the infectivity assay provides a relative comparison of propagule densities among various soils or treatments.

MPN Assay

The MPN assay was developed to estimate the density of organisms in a liquid culture (16). Porter (84) first used it to estimate the propagule density of AM fungi in soil. The general procedure for the MPN assay is to dilute natural soil with disinfested soil. Place equal portions of the dilution series in small containers (5 to 10 replications of each dilution), plant a susceptible host plant in each container, and grow plants long enough to obtain good root colonization. Plants are then washed free of soil, and roots are assessed for the presence or absence of colonization. Values for an MPN assay can be obtained from published tables (25); however, these tables restrict experimental design, thereby reducing the accuracy that can be obtained. A better approach is to program the equations into a computer and directly solve for the MPN value based on the optimal experimental design; increased replication and decreased dilution factor improve accuracy and reduce confidence limits. Numerous factors affect the outcome of an MPN assay (4, 19, 66, 103); therefore, caution should be exercised when values from different experiments are compared. Nonetheless, this assay has been a useful tool for estimating propagule numbers in field soil, pot cultures, and various forms of inocula.

Important considerations for evaluating AM fungi with the MPN assay are as follows.

1. Dilution factor. Preliminary studies should be conducted so that the lowest possible dilutions are used to bracket actual numbers found in the soil.
2. Sample processing. Samples should be kept cool and processed as soon as possible after collection. The sample soil needs to be relatively dry, and root pieces of >2 mm in diameter should be removed from the sample to allow thorough mixing with the diluent soil. These treatments will affect propagule numbers and viability, and all samples must be treated similarly.
3. Diluent soil. The soil preferably should be the same as the original sample and should be pasteurized rather than sterilized. Controls with no sample added should be set up with the pasteurized soil to ensure that all AM propagules have been eliminated.
4. Host plant. The host must be highly susceptible to colonization by AM fungi, produce a rapidly growing, fibrous root system, and be readily cleared for observation of colonization. *Zea mays* L. is a good choice.
5. Length of assay. Plants need to be grown long enough so that roots fully exploit the soil in each container. It is better to err on the conservative side and grow plants until they are pot bound. Roots with well-developed mycorrhizae are also more easily evaluated. A typical assay may run for 6 to 8 weeks.
6. Confirmation of negative results. The entire root system must be examined to confirm a negative reading.

Infectivity Assays

Plants are grown under standard conditions, and root colonization is estimated after 3 to 6 weeks (65). The amount of colonization is assumed to be proportional to the total number of AM propagules in the soil. The length of the assay is critical, and preliminary studies are needed to select the proper harvest time for a given plant-soil combination. If plants grow for too short a time, the full potential for colonization is not realized; however, plants grown for too long a time may become uniformly colonized despite differences in AM fungal populations.

An infection unit method may also be used to quantify mycorrhizal propagules (27). The principle is that a count of infection units is a more reliable measure of the number of viable propagules than are other methods. However, this method is applicable only in short-term experiments because infection units are discernible only during the initial stages (1 to 3 weeks) of colonization.

Quantification of Soilborne Hyphae

Even though the hyphae that grow into the soil matrix from the root are the functional organs for nutrient uptake and translocation, few researchers have obtained quantitative data on their growth and distribution. This is largely because of the technical difficulties in obtaining reliable data—there is no completely satisfactory method to quantify external hyphae of AM fungi in soil. Three major problems have yet to be overcome: (i) there is no reliable method to distinguish AM fungal hyphae from the myriad of other fungal hyphae in the soil, (ii) assessment of the viability and activity of hyphae is problematic, and (iii) meaningful quantification is very time-consuming. Nonetheless, clarification of the growth dynamics of external hyphae is essential to further understanding of their function in soil. Existing methods for quantifying soilborne hyphae have been reviewed elsewhere (95). Furthermore, the application of the molecular techniques described later in this chapter should lead to rapid advances in our knowledge of these structures.

CULTURE METHODS

Traditional Culture Methods

The culture of AM fungi on plants in disinfested soil, using spores, roots, or infested soil as inocula, has been the most frequently used technique for increasing propagule numbers (91). Many host plants have been used under a variety of conditions (94, 98). Examples of plants that have been used successfully are alfalfa, maize, onion, Sudan grass, and wheat. Generally, the host selected should become well colonized (>50% of the root length), produce root mass quickly, and be able to tolerate the high-light conditions required for the fungus to reproduce rapidly. Hosts that can be propagated from seed are preferable to cuttings since they are more easily disinfested. Most seeds may be disinfested with 10% household bleach (0.525% NaOCl) for 5 to 15 min followed with five washes of water.

Disinfesting the fungal propagules is a critical step because other fungi, bacteria, actinomycetes, and nematodes may be propagated with or instead of the AM fungi. Hepper (39) reviewed procedures for disinfesting and germinating spores, and Williams (102) detailed a method for reducing contamination of colonized root pieces. The most effective methods use chlorine compounds, surfactants, and combinations of antibacterial agents. One effective method is to incubate pot culture-produced spores in a solution of 2% chloramine T, 200 ppm of streptomycin sulfate, and a trace of Tween 20 for 20 min followed by at least five changes of sterile water. Spores from the field usually have higher levels of contamination, and a thorough wash with water containing a surfactant should be used prior to the disinfesting treatment.

All components of the culture system should be disinfested prior to initiation of a pot culture. The method of soil disinfestation is especially important. The objective is to kill existing AM fungi, pathogenic organisms, and weed seeds while preserving a portion of the nonpathogenic microbial community (47, 91). Several methods, including fumigation with biocides and exposure to ionizing radiation, have been used to eliminate AM fungi from soil. However, the safety and convenience of heat pasteurization makes it the preferred method. Large batches (50 to 100 kg) may be treated by heating to 85°C for two 8-h periods with 48 h between treatments in a commercial soil pasteurizer. Alternatively, smaller batches (4 kg each) may be heated to the same temperature by using three 2-min exposures to 700-W, 2,475-MHz microwave radiation. Prior to either treatment method, the soil must be passed through a 2-mm-pore-size sieve and wetted to at least 10% (wt/wt) moisture.

Culturing of AM fungi in soilless media avoids the detrimental organisms in nonsterile soil and allows control over many of the physical and chemical characteristics of the growth medium. Soilless media are more uniform in composition, weigh less, and provide aeration better than soil media. Jarstfer and Sylvia (47) have reviewed the various conducive substrates. Most soilless media do not buffer P concentration, so care must be taken to avoid high levels of P in the solution in the root zone. As discussed below for soil, several strategies may be used to regulate fertility and provide conditions conducive to the culture of AM fungi in soilless media.

Conducive environmental conditions for cultures of AM fungi are a balance of high light intensity, adequate moisture, and moderate soil temperature without detrimental additions of fertilizers or pesticides (47). Good light quality (λ, 400 to 700 nm) and high photosynthetic photon flux density are necessary for colonization and sporulation. Where natural light conditions are poor (photosynthetic photon flux density, <500 μmol m^{-2} s^{-1}), supplemental high-intensity lamps should be used. Soil moisture affects AM fungal development directly and indirectly. Directly, excessive moisture may encourage the growth of hyperparasites on spores in the culture. Indirectly, any moisture condition that inhibits primary root growth will reduce the spread of colonization. The best strategy is to apply water regularly to well-drained substrate. Likewise, soil temperature is also important directly for the fungus and indirectly for its effect on the chosen hosts. Sporulation is positively correlated with soil temperatures from 15 to nearly 30°C for many AM fungi; however, at higher temperatures sporulation may decrease as the host is stressed. Generally, soil temperatures and moisture conditions that are optimal for the host should also prove best for the fungus.

Chemical amendments can have both beneficial and detrimental effects on the development of colonized root systems and on sporulation. Responses to P and N fertilization may be isolate dependent (22) and are affected by the relative amounts of N and P supplied. Three approaches may be used to supply the plants with nutrients: (i) apply balanced nutrients except for P, which is applied at a rate 10-fold more dilute than recommended, (ii) apply dilute but balanced nutrients frequently, or (iii) mix a time release fertilizer into the substrate. Application of pesticides can also affect AM fungal colonization and sporulation. Prior to testing any pesticide with cultures of AM fungi, we recommend that previous reviews (47, 91) and the manufacturer be consulted.

In the greenhouse, pot cultures should be isolated from contaminated soil, splashing water, and crawling insects in order to prevent contamination. In addition, specific isolates of AM fungi should be kept well separated from one another. To initiate pot cultures, place a layer of inoculum 1 to 2 cm below the seed or cutting. The inoculum may consist of spores, colonized roots, or infested soil. Infested soil is often used to obtain initial isolates from the field; however, these mixed-species cultures should rapidly progress to single-species cultures initiated from 20 to 100 healthy, disinfested, and uniform spores. For critical taxonomic studies, single-spore cultures should be produced (91). For step-by-step methods of pot culturing and more extensive discussion of these methods, consult reference 47 or 96. Cultures should be grown for 3 to 6 months under the conducive conditions stated above. At that time, soil cores of approximately 5% of the container volume should be assessed for colonization and spore production. Cultures with more than 10 spores per g may be stored at 4°C. Many isolates tolerate air drying and storage at room temperature for long periods (>5 years); however, as the shelf lives of most of these fungi are uncertain, we recommend that both methods be tested to ensure survival of isolated germplasm. Spores may be conditioned to survive at −70°C by first air drying with the host (21). New isolates should be deposited with either INVAM or the International Bank for the Glomeromycota.

Aeroponic and Hydroponic Cultures

A benefit of aeroponic and some hydroponic systems is that colonized roots and spores are produced free of any substrate, permitting more efficient production and distribution of inocula. At least seven species of AM fungi have been grown in various nutrient-solution systems on hosts representing at least 21 genera (47). Usually, plants are inoculated with AM fungi and grown in sand or vermiculite before they are transferred into a culture system. The plants are grown for a period of 4 to 6 weeks under conditions conducive for colonization, after which they are washed and nondestructively checked for colonization (3); however, it is also possible to inoculate plants directly into the culture system (42). The P concentrations that have been reported to support AM fungal growth in solution cultures range from <1 to 24 μmol.

A system that applies a fine nutrient mist to roots of intact plants (aeroponic culture) produces an excellent AM fungal inoculum and concentrations of spores greater than those produced in soil-based pot cultures of the same age. Because the colonized-root inoculum produced in this system is free of any substrate, it can be sheared with the sharp blade of a food processor, resulting in high propagule densities. A detailed description of these methods and applications has been published previously (48).

Monoxenic Cultures

The growth of AM fungi in pure culture in the absence of a host has not been achieved. However, selected AM fungi colonize roots of intact plants or root-organ cultures to achieve monoxenic cultures that are useful for basic research on symbiosis (26). Root-inducing transferred DNA (Ri T-DNA)-transformed root cultures offer the most efficient method to grow axenic roots because no plant growth regulators are required for sustained growth. Fortin et al. (26) provide procedures for initiating and maintaining transformed root cultures colonized by AM fungi.

BIOCHEMICAL DETECTION AND QUANTIFICATION METHODS

Biochemical methods have been used to improve the means of detection and quantification of AM fungi in the environment. Additionally, some biochemical methods have contributed to a better understanding of phylogenetic relationships among AM fungi. Protein analyses are providing a better understanding of the symbiosis and may lead to the discovery of unique molecules with which to tag and quantify AM fungi. Other methods allow nonspecific assessment of biomass of AM fungi in roots and hyphae extracted from the soil, but proper controls are necessary for interpretation of results. None of these techniques are yet widely used or standardized, but they are developed to the point where they can be applied to environmental analyses with reasonable success. With all of them, the purity of samples and controls is an essential consideration.

Biochemical Approaches for AM Fungus Detection

Immunoassays

Serological techniques have the potential for specific detection of AM hyphae and spores in soil as well as in planta; however, few highly specific antibodies have been proven thus far. AM fungi, like many others, have poor antigenic properties (33). It appears that many immunogenic components of both spores and hyphae are internal, cell membrane-associated components of the cell walls. In addition, associated bacteria and actinomycetes provide interfering and stronger antigenic sites on the external surfaces. These are especially difficult to remove from spores even with antibiotic agents, oxidants, or sonication because they are deeply embedded in the spore walls (11, 33). When monoclonal products are screened, isolate and microbiological purity of the AM fungi should be the foremost concern; however, very clean AM fungal material is hard to come by due to the obligate biotrophic nature of these fungi. Monoxenic AM fungal cultures may be the most suitable material for generating AM-fungus-specific antibodies. As suggested by Hahn et al. (33), immunogold labeling of thin sections will show where the antibody is attaching and thus confirm its specificity. Purified cytoplasmic proteins make good antigens (105) but may be of limited value for labeling intact spores or hyphae from soil. See reference 82 for a good introduction and specifics on the methods for use of monoclonal antibodies with mycorrhizae. Polyclonal antibodies specific for surface antigens of AM fungi generated by using soluble proteins from *G. intraradices* have also been reported to give different patterns of soluble protein antigens for different AM fungi (56). Thingstrup et al. (97) have also reported similar findings on the potential use of polyclonal antisera to differentiate between *Scutellospora* species. However, polyclonal antibodies are generally less specific and can show broad cross-reactions with other non-AM fungi. Thus, one should be very cautious when using polyclonal antibodies to identify AM fungi. Despite their possible utility for detection and species-level identification of AM fungi, serological techniques appear to have been largely neglected by AM fungus researchers in recent years.

Protein and Isozyme Analyses

Qualities of proteins, in either spores or the colonized root, also may be used to detect and identify AM fungi. Protein assays can be based on total or enzyme-related protein molecules. For colonized root, identification with isozymes depends on the resolution of differences between host enzyme and diagnostic fungal enzyme banding patterns after electrophoresis (40, 99). The interactions between the particular hosts and fungi are extremely important, and some combinations yield protein bands that are indistinguishable. Careful preparation of root and hyphal materials by using protocols that prevent phenolic oxidation is necessary. With these methods, as with the other biochemical methods, isolate purity and positive controls are of utmost importance. It should also be noted that specific detection from the mycorrhizae requires an identical noncolonized root system as the control. Rosendahl and Sen (86) provide the specific methods for sample preparation and analysis, as well as a review of previous applications. Researchers have also exploited spore protein patterns of AM fungi by using either denaturing or native polyacrylamide gel electrophoreses for reasonably accurate discrimination and identification of AM fungal species and isolates (5, 89). Their results consistently indicate that neither the host species used for spore production nor different generation and storage times affect these spore protein profiles; however, the physiological states (germinating versus dormant) of spores may affect the results (5). Nevertheless, some isozyme analyses appear to be promising as complementary approaches to PCR-based methods for AM fungal species identification (6).

Fatty Acid Analyses

Fatty acid methyl ester (FAME) profiles of AM fungi have proven to be reliable measures of similarity below the family level (32). These profiles are stable across hosts and with storage and subculturing (8). Graham et al. (32) found that the fatty acid 16:1ω5 content of roots provides a reproducible index of colonization by *Glomaceae*, *Acaulosporaceae*, and *Scutellospora* and the fatty acid 16:1ω5 is not a good indicator of infection by *Gigaspora* species, *Paraglomaceae*, or *Archaeosporaceae*. *Gigaspora* can be detected by using 20:1ω9 (88). The FAME techniques for detecting and identifying AM fungi may be used with as little as 10 to 20 mg of spores (130 to 500 spores) or 15 to 30 mg of oven-dried root material (8, 32). After collecting and washing spores or root material and air or oven drying, the methods detailed by Sasser (90) are used to extract and quantify the FAME contents of AM fungi. Individual AM fungal species do not produce unique fatty acids, but some AM fungal taxa can be distinguished by distinct fatty acid profiles (18). At present, no comprehensive database of FAME profiles for AM fungi exists, and thus researchers wishing to investigate and identify AM fungal species from environmental samples based on fatty acid analysis must generate their own reference FAME profiles for each AM fungal species. Additionally, AM fungal species from a mixed sample have not been differentiated. Furthermore, it has been shown that rates of production of these lipids not only can vary between AM fungal species but also can be influenced by the physiological status of mycorrhizal symbioses (76). Although this analysis clearly can be used to detect the presence of AM fungi both in colonized roots and in soil, its suitability for AM fungal species identification may be limited only to well-controlled experiments.

Biochemical Approaches for AM Fungal Quantification

Determination of Chitin Content

Chitin content has been used to estimate fungal biomasses in roots under controlled experimental conditions (9, 38).

However, the utility of this method for natural soils is limited because chitin is ubiquitous in nature. It is found in the cell walls of most fungi and the exoskeletons of arthropods, and certain soils exhibit physical and chemical properties which interfere with the analysis (46). However, the content of chitin in roots may be used effectively to determine the fungal biomass when comparable noncolonized control plants are available. Chitin content determination has also been used to estimate the hyphal biomass in soil; however, the limitations discussed above for this method also apply here. For this method, dried soil samples are mixed with concentrated KOH, and the mixture is autoclaved for 1 h to degrade chitin into chitosan. Subsamples of the soil-KOH suspension are transferred into centrifuge tubes and assayed for chitin content as described above for roots. By subtracting the chitin content of control soil without AM fungi from that of soil containing AM fungi, the biomass of AM fungi can be estimated.

Ergosterol Detection

Ergosterol, a fungus-specific sterol, has been widely used for biomass estimation for other fungi, including ectomycorrhizal fungi. Although it has been reported to occur in AM fungi (28) and has been used to estimate the biomass of AM fungi (34), its suitability as a biomass indicator for AM fungi is questionable due to its low content in AM fungi compared to that in other fungi in the environment (75, 79), as well as variation in production of ergosterol among AM fungal species (35).

Fatty Acid Analyses

In addition to their potential use in AM fungal detection and identification, fatty acid analyses also have great potential for AM fungal biomass estimation both for colonized roots and for soil (76, 78). The signature fatty acid 16:1ω5 is rather unique because it is not usually produced by other fungi but is found in relatively large proportion in total fatty acids of many AM fungal species (32). Although some bacteria are capable of producing this particular fatty acid, as long as proper controls and additional measures are included, it should be possible to use this analysis for relatively accurate biomass estimation for AM fungi (7, 75, 77). However, since some AM fungi do not produce 16:1ω5, total biomass estimation for mixed communities of AM fungi may be problematic.

Glomalin Detection

A nonspecific soil glycoprotein called glomalin is produced by hyphae of AM fungi and has been implicated in soil aggregate development. It can be detected in situ by immunofluorescence techniques or by various specific or general protein assay methods (104, 106). Many questions remain regarding its role in the ecophysiology of AM fungi and soil genesis, yet the techniques for extraction and quantification are now available. Briefly, a soil sample of 1 g is autoclaved for 30 to 90 min with 20 or 50 mM sodium citrate. The extraction time and sodium citrate concentration determine whether recent or total glomalin is removed from the soil sample. This extract may then be further analyzed by such techniques as enzyme-linked immunosorbent assay, Bradford total protein assay, dot blot analysis, and sodium dodecyl sulfate-polyacrylamide gel electrophoresis. Because glomalin is highly recalcitrant, its utility for field application for AM fungal quantification may be limited and even for a greenhouse experiment, a proper control sample is needed. The amount of glomalin in the field may not be a good indicator of AM fungal biomass, but rates of glomalin production may reflect total AM fungal activity and possibly total biomass (61, 107).

DNA- AND PCR-BASED DETECTION AND QUANTIFICATION METHODS

Molecular Approaches for AM Fungal Detection

Molecular genetic techniques have revolutionized AM fungal systematics (69, 92) and allow identification of the species of fungi present in roots or soil that may not be detected from examination of spores. Although the application of molecular genetic tools for the detection of AM fungi has become commonplace, the approaches have not been standardized and are not trouble free. A persistent problem with all molecular approaches is the possibility of contaminant sequences. Other species of fungi as well as bacteria often inhabit spores of AM fungi (11, 41), and there are countless organisms that may be associated with field-sampled roots. As a result, molecular approaches rely on phylogenetic analysis of a targeted gene sequence to identify the PCR amplicons as belonging to AM fungi. Contamination can be reduced through the use of primers that selectively amplify sequences from AM fungi. Although AM fungi appear to be monophyletic (with the exception of *Geosiphon*), families within the group are highly divergent (93), and at present, there is not a single set of primers that selectively amplifies sequences from all AM fungal species. The widely used primer AM1 selectively amplifies a region of the small subunit of the rRNA genes of *Gigasporaceae*, *Acaulosporaceae*, and *Glomeraceae*; however, it does not amplify species in *Archaeosporaceae* and *Paraglomeraceae* (13), and even this primer also amplifies non-AM-fungal contaminants (13, 58).

A second constraint in the application of molecular markers for AM fungal species identification is the quality of publicly available databases. While the number of AM fungal sequences in GenBank is large (http://www.ncbi.nlm.nih.gov/), many AM fungal species are not represented even by the best-studied genes. Moreover, annotation of GenBank data is difficult, so that non-AM-fungal contaminant sequences remain labeled as AM fungal species and there is poor correspondence of sequences to AM fungal species for some taxa, perhaps due to inconsistent identification of spores in different studies.

The best gene to target depends on the taxonomic level being studied (i.e., resolution at the family or genus level). Most studies thus far have focused on the nuclear ribosomal sequences of AM fungi, which occur in multiple repeats within the genome and therefore require less initial DNA. To date, the small subunit of the rRNA gene has been the most popular targeting sequence among AM fungus researchers and has the most complete publicly available database and several published selective primers, including the AM1 primer mentioned above. The small subunit of the rRNA gene appears to provide relatively good resolution for differentiating between AM fungal morphospecies. The large subunit of the rRNA gene also appears to be promising for species identification. The intergenic spacer region of the rRNA gene is highly variable—even within individual AM fungal spores (81, 85)—and, therefore, is suitable for distinguishing among isolates of the same AM fungal species.

The approaches to DNA extraction and PCR amplification are now routine, with several commercial products being used successfully. Once amplicons have been obtained, they can be prescreened based on product sizes, restriction sites, or strand conformations (see below). If the targeted sequence is highly conserved or the template's purity is high, amplicons may be directly sequenced. On the other hand, if the targeted sequences are highly polymorphic or the template is of questionable purity (i.e., from a mixed community as expected from environmental samples), it may be necessary to clone these PCR products into bacteria prior to sequencing. Once amplicons have been cloned, several molecular techniques can be employed for screening of these clones (see, for example, chapters 49 and 55 in this manual).

Methods of Screening and Separation of Variants

Clones may be separated by size or variation at restriction enzyme sites (restriction fragment length polymorphism [RFLP]) (43, 80). Single-strand conformational polymorphism analysis has also been used to detect genetic variation among and within AM fungi (14, 45, 51, 52). This analysis is based on the principle that differences in sequence will result in different single-strand DNA conformations that can be detected by electrophoresis. Denaturing gradient gel electrophoresis (DGGE) and temperature gradient gel electrophoresis (TGGE) techniques have also been used routinely in microbial ecology studies (71). The main principle for both methods is that differences in sequence composition affect the melting behavior, leading to different motilities. DGGE has been used to describe the diversity in AM fungal communities (20, 55). Similarly, TGGE has also been used to characterize AM fungi (17). Although we have introduced these methods in the context of screening cloned amplicons, DGGE and TGGE could be applied directly to PCR products or to high-quality genomic DNA to assess the level of variation. Bands could then be excised from the gel, purified, cloned, and then sequenced to confirm their glomeromycotan origins.

Community Fingerprinting with T-RFLP

Terminal RFLP (T-RFLP) analysis is similar to other techniques in that it is based on PCR amplification of targeted glomeromycotan sequences. However, this technique has a different goal in that it provides a "shotgun view of community" without the need for cloning and sequencing (50, 70). Briefly, targeted sequences are amplified and end labeled with fluorescence dyes previously attached to the primer. The end-labeled PCR products are purified and then digested with a restriction enzyme, and end-labeled fragments are analyzed (15, 59). The number of peaks (different fragment sizes) in a given profile approximates the number of species in the sample. Resolution can be improved by using more than one digestion enzyme or by labeling both primers so that both ends of the PCR products can be analyzed (50). The success of this technique is dependent on the primer choice and consistent stringency of PCR to enable data from different studies to be directly compared. Given the problem of amplification of non-AM-fungal contaminants, the glomeromycotan origin of the fragments needs to be confirmed. This could be done by generating a reference T-RFLP profile database by using a high-quality DNA template derived from known AM fungal species. This approach is not definitive for environmental samples as unknown isolates may share restriction sites in similar locations. A more direct approach would be to isolate and sequence the actual fragments from the T-RFLP profile, but to our knowledge this has not been done with AM fungi.

Genome-Wide Approaches

A few investigations of AM fungi have employed a genome-wide approach that uses either arbitrary primers (1, 87, 108) or primers based on repeated motifs (29, 57, 60, 101, 109) to generate unique banding patterns for different AM fungal species or isolates. These approaches are relatively easy to employ and are economical; however, there is not a way to confirm the glomeromycotan origin of the amplified DNA, which limits utility in working with environmental samples. One of these approaches, random amplification of polymorphic DNA analysis (1, 29, 57, 60, 101, 108, 109), has been used as an initial step in the generation of specific primers for individual AM fungal species (1, 57). The AM fungal origin of these species tags would then need to be confirmed.

Molecular Approaches for AM Fungus Quantification

Quantitative studies of AM-fungal molecular ecology have commonly characterized and screened multiple cloned amplicons from individual samples. With this approach, high confidence in the AM fungal origin of the amplicons comes with great effort invested in each individual sample, thereby limiting the sample size and the statistical power of the studies. A few molecular approaches have potential for direct estimates of relative AM fungal densities that may allow more efficient measurement of large numbers of samples. Attempts have been made to use conventional PCR methods in a quantitative fashion (23); however, with this method there may be nonlinear relationships between the abundance of the PCR product and the initial sequence abundance that may prevent reliable scoring of relative density. There have also been a few attempts to quantify AM fungal root colonization by using real-time (RT) PCR (2, 24, 44). The RT-PCR has an advantage over competitive PCR in that it does not require a development of external competitor standards for quantification and it is possible to determine the amount of genomic DNA or the starting copy number of the targeted sequence (37). Isayenkov et al. (44) found that RT-PCR provided a good estimate of the density of arbuscules within roots in a pot experiment with a single species of AM fungi, but the utility of RT-PCR for field samples remains to be demonstrated.

T-RFLP

In addition to its application for molecular investigation of AM fungal communities, T-RFLP has potential to provide a quantitative analysis of the relative composition of these fungi. Instead of simply counting how many peaks there are for each T-RFLP profile, it may be possible to analyze the peak height (level of fluorescence intensity) of each end-labeled fragment to estimate the relative abundance of AM fungal species in the samples (70). However, there are potential problems with nonlinear relationships between the amplicon abundance and the initial concentrations of DNA in the sample generated by the competitive PCR amplifications. These problems may be reduced by standardization of the amount of templates used for PCR, but the potential for T-RFLP peak heights to estimate AM fungal relative abundance remains to be demonstrated.

Current Limitations and Future Directions for Molecular Approaches

Although current molecular methods improve our ability to detect AM fungi in the field, monitoring of the abundance and distribution of individual fungal species remains laborious and expensive. Moreover, the molecular approaches are currently limited by reliance on rRNA gene and molecular data sets that include only a small subset of the total AM fungal diversity (10). These limitations can be alleviated only through more extensive collection and taxonomic description of AM fungal species, along with cloning and sequencing of rRNA and single-copy genes. Over time, the molecular data sets will continue to improve, creating the potential for additional applications. In particular, more complete molecular data sets would allow the creation of microarrays for one-step characterization of AM fungal composition. This DNA bar-coding approach would improve the detection of AM fungal species and may ultimately facilitate the quantification of AM fungal species, perhaps by coupling this approach with fatty acid methods.

REFERENCES

1. **Abbas, J. D., B. A. D. Hetrick, and J. E. Jurgenson.** 1996. Isolate specific detection of mycorrhizal fungi using genome specific primer pairs. *Mycologia* **88:**939–946.
2. **Alkan, N., V. Gadkar, J. Coburn, O. Yarden, and Y. Kapulnik.** 2004. Quantification of the arbuscular mycorrhizal fungus *Glomus intraradices* in host tissue using real-time polymerase chain reaction. *New Phytol.* **161:**877–885.
3. **Ames, R. N., E. R. Ingham, and C. P. P. Reid.** 1982. Ultraviolet-induced autofluorescence of arbuscular mycorrhizal root infections: an alternative to clearing and staining methods for assessing infections. *Can. J. Microbiol.* **28:** 351–355.
4. **An, Z.-Q., J. W. Hendrix, D. E. Hershman, and G. T. Henson.** 1990. Evaluation of the "most probable number" (MPN) and wet-sieving methods for determining soil-borne populations of endogonaceous mycorrhizal fungi. *Mycologia* **82:**576–581.
5. **Avio, L., and M. Giovannetti.** 1998. The protein pattern of spores of arbuscular mycorrhizal fungi: comparison of species, isolates and physiological stages. *Mycol. Res.* **102:** 985-990.
6. **Bago, B., S. P. Bentivenga, V. Brenac, J. C. Dodd, Y. Piche, and L. Simon.** 1998. Molecular analysis of *Gigaspora* (Glomales, Gigasporaceae). *New Phytol.* **139:** 581–588.
7. **Balser, T. C., K. K. Treseder, and M. Ekenler.** 2005. Using lipid analysis and hyphal length to quantify AM and saprotrophic fungal abundance along a soil chronosequence. *Soil Biol. Biochem.* **37:**601–604.
8. **Bentivenga, S. P., and J. B. Morton.** 1994. Stability and heritability of fatty acid methyl ester profiles of glomalean endomycorrhizal fungi. *Mycol. Res.* **98:**1419–1426.
9. **Bethlenfalvay, G. J., R. S. Pacovsky, and M. S. Brown.** 1981. Measurement of mycorrhizal infection in soybeans. *Soil Sci. Soc. Am. J.* **45:**871–874.
10. **Bever, J. D., P. A. Schultz, A. Pringle, and J. B. Morton.** 2001. Arbuscular mycorrhizal fungi: more diverse than meets the eye, and the ecological tale of why. *Bioscience* **51:**923–931.
11. **Bianciotto, V., E. Lumini, P. Bonfante, and P. Vandamme.** 2003. '*Candidatus Glomeribacter gigasporarum*' gen. nov., sp nov., an endosymbiont of arbuscular mycorrhizal fungi. *Int. J. Syst. Evol. Microbiol.* **53:**121–124.
12. **Brundrett, M. C., Y. Piche, and R. L. Peterson.** 1984. A new method for observing the morphology of vesicular-arbuscular mycorrhizae. *Can. J. Bot.* **62:**2128–2134.
13. **Clapp, J. P., T. Helgason, T. J. Daniell, and J. P. W. Young.** 2002. Genetic studies of the structure and diversity of arbuscular mycorrhizal fungal communities, p. 201–224. *In* M. G. A. van der Heijden and I. R. Sanders (ed.), *Mycorrhizal Ecology*. Springer-Verlag, Berlin, Germany.
14. **Clapp, J. P., A. Rodriguez, and J. C. Dodd.** 2001. Inter- and intra-isolate rRNA large subunit variation in *Glomus coronatum* spores. *New Phytol.* **149:**539–554.
15. **Clement, B. G., L. E. Kehl, K. L. Debord, and C. L. Kitts.** 1998. Terminal restriction fragment patterns (TRFPs), a rapid, PCR-based method for the comparison of complex bacterial communities. *J. Microbiol. Methods* **31:**135–142.
16. **Cochran, W. G.** 1950. Estimation of bacterial densities by means of the "most probable number." *Biometrics* **6:**105–116.
17. **Cornejo, P., C. Azcon-Aguilar, J. M. Barea, and N. Ferrol.** 2004. Temporal temperature gradient gel electrophoresis (TTGE) as a tool for the characterization of arbuscular mycorrhizal fungi. *FEMS Microbiol. Lett.* **241:** 265–270.
18. **Declerck, S., S. Cranenbrouck, Y. Dalpe, S. Seguin, A. Grandmougin-Ferjani, J. Fontaine, and M. Sancholle.** 2000. *Glomus proliferum* sp nov.: a description based on morphological, biochemical, molecular and monoxenic cultivation data. *Mycologia* **92:**1178–1187.
19. **de Man, J. C.** 1975. The probability of most probable numbers. *Eur. J. Appl. Microbiol.* **1:**67–78.
20. **de Souza, F. A., G. A. Kowalchuk, P. Leeflang, J. A. van Veen, and E. Smit.** 2004. PCR-denaturing gradient gel electrophoresis profiling of inter- and intraspecies 18S rRNA gene sequence heterogeneity is an accurate and sensitive method to assess species diversity of arbuscular mycorrhizal fungi of the genus *Gigaspora*. *Appl. Environ. Microbiol.* **70:**1413–1424.
21. **Douds, D. D., and N. C. Schenck.** 1990. Cryopreservation of spores of vesicular-arbuscular mycorrhizal fungi. *New Phytol.* **115:**667–674.
22. **Douds, D. D., Jr., and N. C. Schenck.** 1990. Increased sporulation of vesicular-arbuscular mycorrhizal fungi by manipulation of nutrient regimes. *Appl. Environ. Microbiol.* **56:**413–418.
23. **Edwards, S. G., A. H. Fitter, and J. P. W. Young.** 1997. Quantification of an arbuscular mycorrhizal fungus, *Glomus mosseae*, within plant roots by competitive polymerase chain reaction. *Mycol. Res.* **101:**1440–1444.
24. **Filion, M., M. St. Arnaud, and S. H. Jabaji-Hare.** 2003. Direct quantification of fungal DNA from soil substrate using real-time PCR. *J. Microbiol. Methods* **53:**67–76.
25. **Fisher, R. A., and F. Yates.** 1963. *Statistical Tables for Biological, Agricultural and Medical Research*. Oliver and Boyd, Edinburgh, Scotland.
26. **Fortin, J. A., G. Becard, S. Declerck, Y. Dalpe, M. St. Arnaud, A. P. Coughlan, and Y. Piche.** 2002. Arbuscular mycorrhiza on root-organ cultures. *Can. J. Bot.* **80:**1–20.
27. **Franson, R. L., and G. J. Bethlenfalvay.** 1989. Infection unit method of vesicular-arbuscular mycorrhizal propagule determination. *Soil Sci. Soc. Am. J.* **53:**754–756.
28. **Frey, B., A. Vilariño, H. Schüepp, and J. Arines.** 1994. Chitin and ergosterol content of extraradical and intraradical mycelium of the vesicular-arbuscular mycorrhizal fungus *Glomus intraradices*. *Soil Biol. Biochem.* **26:**711–717.
29. **Gadkar, V., A. Adholeya, and T. Satyanarayana.** 1997. Randomly amplified polymorphic DNA using the M13 core sequence of the vesicular-arbuscular mycorrhizal fungi *Gigaspora margarita* and *Gigaspora gigantea*. *Can. J. Microbiol.* **43:**795–798.
30. **Giovannetti, M., and B. Mosse.** 1980. An evaluation of techniques for measuring vesicular arbuscular mycorrhizal infection in roots. *New Phytol.* **84:**489–500.
31. **Graham, J. H., D. M. Eissenstat, and D. L. Drouillard.** 1991. On the relationship between a plant's mycorrhizal

dependency and rate of vesicular-arbuscular mycorrhizal colonization. *Funct. Ecol.* **5**:773–779.

32. **Graham, J. H., N. C. Hodge, and J. B. Morton.** 1995. Fatty acid methyl ester profiles for characterization of glomalean fungi and their endomycorrhizae. *Appl. Environ. Microbiol.* **61**:58–64.
33. **Hahn, A., K. Horn, and B. Hock.** 1995. Serological properties of mycorrhizas, p. 181–201. *In* A. Varma and B. Hock (ed.), *Mycorrhiza: Structure, Function, Molecular Biology and Biotechnology*. Springer-Verlag, Berlin, Germany.
34. **Hart, M. M., and R. J. Reader.** 2002. Does percent root length colonization and soil hyphal length reflect the extent of colonization for all AMF? *Mycorrhiza* **12**:297–301.
35. **Hart, M. M., and R. J. Reader.** 2002. Taxonomic basis for variation in the colonization strategy of arbuscular mycorrhizal fungi. *New Phytol.* **153**:335–344.
36. **Hayman, D. S.** 1984. Methods for evaluating and manipulating vesicular-arbuscular mycorrhiza, p. 95–117. *In* J. M. Lynch and J. M. Grainger (ed.), *Microbiological Methods for Environomental Microbiology*. Academic Press, London, England.
37. **Heid, C. A., J. Stevens, K. J. Livak, and P. M. Williams.** 1996. Real time quantitative PCR. *Genome Res.* **6**:986–994.
38. **Hepper, C. M.** 1977. A colorimetric method for estimating vesicular-arbuscular mycorrhizal infection in roots. *Soil Biol. Biochem.* **9**:15–18.
39. **Hepper, C. M.** 1984. Isolation and culture of VA mycorrhizal (VAM) fungi, p. 95–112. *In* C. L. Powell and D. J. Bagyaraj (ed.), *VA Mycorrhiza*. CRC Press, Inc., Boca Raton, Fla.
40. **Hepper, C. M., R. Sen, and C. S. Maskall.** 1986. Identification of vesicular-arbuscular mycorrhizal fungi in roots of leek (*Allium porrum*) and maize (*Zea mays*) on the basis of enzyme mobility during polyacrylamide gel electrophoresis. *New Phytol.* **102**:529–539.
41. **Hijri, M., D. Redecker, J. Petetot, K. Voigt, J. Wostemeyer, and I. R. Sanders.** 2002. Identification and isolation of two ascomycete fungi from spores of the arbuscular mycorrhizal fungus *Scutellospora castanea*. *Appl. Environ. Microbiol.* **68**:4567–4573.
42. **Hung, L. L., D. M. O'Keefe, and D. M. Sylvia.** 1991. Use of hydrogel as a sticking agent and carrier for vesicular-arbuscular mycorrhizal fungi. *Mycol. Res.* **95**:427–429.
43. **Husband, R., E. A. Herre, S. L. Turner, R. Gallery, and J. P. W. Young.** 2002. Molecular diversity of arbuscular mycorrhizal fungi and patterns of host association over time and space in a tropical forest. *Mol. Ecol.* **11**:2669–2678.
44. **Isayenkov, S., T. Fester, and B. Hause.** 2004. Rapid determination of fungal colonization and arbuscule formation in roots of *Medicago truncatula* using real-time (RT) PCR. *J. Plant Physiol.* **161**:1379–1383.
45. **Jansa, J., A. Mozafar, S. Banke, B. A. McDonald, and E. Frossard.** 2002. Intra- and intersporal diversity of ITS rDNA sequences in *Glomus intraradices* assessed by cloning and sequencing, and by SSCP analysis. *Mycol. Res.* **106**:670–681.
46. **Jarstfer, A. G., and R. M. Miller.** 1985. Progress in the development of a chitin assay technique for measuring extraradical soilborne mycelium of V-A mycorrhizal fungi, p. 410. *In* R. Molina (ed.), *Proceedings of the 6th North American Conference on Mycorrhizae*. Forest Research Laboratory, Oregon State University, Corvallis.
47. **Jarstfer, A. G., and D. M. Sylvia.** 1992. Inoculum production and inoculation strategies for vesicular-arbuscular mycorrhizal fungi, p. 349–377. *In* B. Metting (ed.), *Soil Microbial Ecology: Applications in Agriculture and Environmental Management*. Marcel Dekker, Inc., New York, N.Y.
48. **Jarstfer, A. G., and D. M. Sylvia.** 1999. Aeroponic culture of VAM fungi, p. 427–441. *In* A. K. Varma and B. Hock (ed.), *Mycorrhiza: Structure, Function, Molecular Biology and Biotechnology*, 2nd ed. Springer-Verlag, Berlin, Germany.
49. **Jeffries, P.** 1987. Use of mycorrhizae in agriculture. *Crit. Rev. Biotechnol.* **5**:319–357.
50. **Johnson, D., P. J. Vandenkoornhuyse, J. R. Leake, L. Gilbert, R. E. Booth, J. P. Grime, J. P. W. Young, and D. J. Read.** 2004. Plant communities affect arbuscular mycorrhizal fungal diversity and community composition in grassland microcosms. *New Phytol.* **161**:503–515.
51. **Kjøller, R., and S. Rosendahl.** 2000. Detection of arbuscular mycorrhizal fungi (Glomales) in roots by nested PCR and SSCP (single stranded conformation polymorphisms). *Plant Soil* **226**:189–196.
52. **Kjøller, R., and S. Rosendahl.** 2001. Molecular diversity of glomalean (arbuscular mycorrhizal) fungi determined as distinct *Glomus* specific DNA sequences from roots of field grown peas. *Mycol. Res.* **105**:1027–1032.
53. **Kormanik, P. P., and A.-C. McGraw.** 1982. Quantification of vesicular-arbuscular mycorrhizae in plant roots, p. 37–45. *In* N. C. Schenck (ed.), *Methods and Principles of Mycorrhizal Research*. American Phytopathological Society, St. Paul, Minn.
54. **Koske, R. E. and J. N. Gemma.** 1989. A modified procedure for staining roots to detect VA mycorrhizas. *Mycol. Res.* **92**:486–505.
55. **Kowalchuk, G. A., F. A. De Souza, and J. A. Van Veen.** 2002. Community analysis of arbuscular mycorrhizal fungi associated with *Ammophila arenaria* in Dutch coastal sand dunes. *Mol. Ecol.* **11**:571–581.
56. **Kozlova, N. V., O. K. Strunnikova, N. M. Labutova, and G. S. Muromtsev.** 2001. Production and specificity of polyclonal antibodies against soluble proteins from the arbuscular mycorrhizal fungus *Glomus intraradices*. *Mycorrhiza* **10**:301–305.
57. **Lanfranco, L., P. Wyss, C. Marzachi, and P. Bonfante.** 1995. Generation of RAPD-PCR primers for the identification of isolates of *Glomus mosseae*, an arbuscular mycorrhizal fungus. *Mol. Ecol.* **4**:61–68.
58. **Lee, J. K., M. S. Tae, A. H. Eom, and S. S. Lee.** 2003. Restriction analyses of PCR amplified partial SSU ribosomal DNA to distinguish arbuscular mycorrhizal fungi from other fungi colonizing plant roots. *Mycobiology* **31**:68–73.
59. **Liu, W. T., T. L. Marsh, H. Cheng, and L. J. Forney.** 1997. Characterization of microbial diversity by determining terminal restriction fragment length polymorphisms of genes encoding 16S rRNA. *Appl. Environ. Microbiol.* **63**:4516–4522.
60. **Longato, S., and P. Bonfante.** 1997. Molecular identification of mycorrhizal fungi by direct amplification of microsatellite regions. *Mycol. Res.* **101**:425–432.
61. **Lovelock, C. E., S. F. Wright, and K. A. Nichols.** 2004. Using glomalin as an indicator for arbuscular mycorrhizal hyphal growth: an example from a tropical rain forest soil. *Soil Biol. Biochem.* **36**:1009–1012.
62. **McGonigle, T. P., M. H. Miller, D. G. Evans, G. S. Fairchild, and J. A. Swan.** 1990. A new method which gives an objective measure of colonization of roots by vesicular-arbuscular mycorrhizal fungi. *New Phytol.* **115**:495–501.
63. **Miller, R. M.** 1987. The ecology of vesicular-arbuscular mycorrhizae in grass- and shrublands, p. 135–170. *In* G. Safir (ed.), *Ecophysiology of VA Mycorrhizal Plants*. CRC Press, Inc., Boca Raton, Fla.

64. **Miller, R. M., D. R. Reinhardt, and J. D. Jastrow.** 1995. External hyphal production of vesicular-arbuscular mycorrhizal fungi in pasture and tallgrass prairie communities. *Oecologia* **103:**17–23.
65. **Moorman, T., and F. B. Reeves.** 1979. The role of endomycorrhizae in revegetation practices in the semi-arid West. II. A bioassay to determine the effect of land disturbance on endomycorrhizal populations. *Am. J. Bot.* **66:** 14–18.
66. **Morton, J. B.** 1985. Underestimation of most probable numbers of vesicular-arbuscular mycorrhizae endophytes because of non-staining mycorrhizae. *Soil Biol. Biochem.* **17:**383–384.
67. **Morton, J. B.** 1988. Taxonomy of VA mycorrhizal fungi: classification, nomenclature, and identification. *Mycotaxon* **32:**267–324.
68. **Morton, J. B., and G. L. Benny.** 1990. Revised classification of arbuscular mycorrhizal fungi (Zygomycetes): a new order, Glomales, two new suborders, Glomineae and Gigasporineae, and two new families, Acaulosporaceae and Gigasporaceae, with an emendation of Glomaceae. *Mycotaxon* **37:**471–491.
69. **Morton, J. B., and D. Redecker.** 2001. Two new families of Glomales, Archaeosporaceae and Paraglomaceae, with two new genera, *Archaeospora* and *Paraglomus*, based on concordant molecular and morphological characters. *Mycologia* **93:**181–195.
70. **Mummey, D. L., M. C. Rillig, and W. E. Holben.** 2005. Neighboring plant influences on arbuscular mycorrhizal fungal community composition as assessed by T-RFLP analysis. *Plant Soil* **271:**83–90.
71. **Muyzer, G., and K. Smalla.** 1998. Application of denaturing gradient gel electrophoresis (DGGE) and temperature gradient gel electrophoresis (TGGE) in microbial ecology. *Antonie Leeuwenhoek* **73:**127–141.
72. **Norris, J. R., D. J. Read, and A. K. Varma (ed.).** 1991. *Methods in Microbiology*, vol. 23. *Techniques for the Study of Mycorrhiza*. Academic Press Ltd., London, England.
73. **Norris, J. R., D. J. Read, and A. K. Varma (ed.).** 1992. *Methods in Microbiology*, vol. 24. *Techniques for the Study of Mycorrhiza*. Academic Press Ltd., London, England.
74. **O'Keefe, D. M., and D. M. Sylvia.** 1991. Mechanisms of the vesicular-arbuscular mycorrhizal plant-growth response, p. 35–53. *In* D. K. Arora, B. Rai, K. G. Mukerji, and G. R. Knudsen (ed.), *Handbook of Applied Mycology*. Marcel Dekker, Inc., New York, N.Y.
75. **Olsson, P. A.** 1999. Signature fatty acids provide tools for determination of the distribution and interactions of mycorrhizal fungi in soil. *FEMS Microbiol. Ecol.* **29:**303–310.
76. **Olsson, P. A., E. Bååth, and I. Jakobsen.** 1997. Phosphorus effects on the mycelium and storage structures of an arbuscular mycorrhizal fungus as studied in the soil and roots by analysis of fatty acid signatures. *Appl. Environ. Microbiol.* **63:**3531–3538.
77. **Olsson, P. A., E. Baath, I. Jakobsen, and B. Söderström.** 1995. The use of phospholipid and neutral lipid fatty acids to estimate biomass of arbuscular mycorrhizal fungi in soil. *Mycol. Res.* **99:**623–629.
78. **Olsson, P. A., R. Francis, D. J. Read, and B. Söderström.** 1998. Growth of arbuscular mycorrhizal mycelium in calcareous dune sand and its interaction with other soil microorganisms as estimated by measurement of specific fatty acids. *Plant Soil* **201:**9–16.
79. **Olsson, P. A., L. Larsson, B. Bago, H. Wallander, and I. M. van Aarle.** 2003. Ergosterol and fatty acids for biomass estimation of mycorrhizal fungi. *New Phytol.* **159:**7–10.
80. **Opik, M., M. Moora, J. Liira, U. Koljalg, M. Zobel, and R. Sen.** 2003. Divergent arbuscular mycorrhizal fungal communities colonize roots of *Pulsatilla spp.* in boreal Scots pine forest and grassland soils. *New Phytol.* **160:**581–593.
81. **Pawlowska, T. E., and J. W. Taylor.** 2004. Organization of genetic variation in individuals of arbuscular mycorrhizal fungi. *Nature* **427:**733–737.
82. **Perotto, S., F. Malavasi, and G. W. Butcher.** 1992. Use of monoclonal antibodies to study mycorrhiza: present applications and perspectives. *Methods Microbiol.* **24:**221–248.
83. **Phillips, J. M., and D. S. Hayman.** 1970. Improved procedures for clearing roots and staining parasitic and vesicular-arbuscular mycorrhizal fungi for rapid assessment of infection. *Trans. Br. Mycol. Soc.* **55:**158–161.
84. **Porter, W. M.** 1979. The 'most probable number' method for enumerating infective propagules of vesicular arbuscular mycorrhizal fungi in soil. *Aust. J. Soil Res.* **17:**515–519.
85. **Pringle, A., J. M. Moncalvo, and R. Vilgalys.** 2000. High levels of variation in ribosomal DNA sequences within and among spores of a natural population of the arbuscular mycorrhizal fungus *Acaulospora colossica*. *Mycologia* **92:** 259–268.
86. **Rosendahl, S., and R. Sen.** 1992. Isozyme analysis of mycorrhizal fungi and their mycorrhiza. *Methods Microbiol.* **24:** 169–194.
87. **Rosendahl, S., and J. W. Taylor.** 1997. Development of multiple genetic markers for studies of genetic variation in arbuscular mycorrhizal fungi using AFLP. *Mol. Ecol.* **6:** 821–829.
88. **Sakamoto, K., T. Iijima, and R. Higuchi.** 2004. Use of specific phospholipid fatty acids for identifying and quantifying the external hyphae of the arbuscular mycorrhizal fungus *Gigaspora rosea*. *Soil Biol. Biochem.* **36:**1827–1834.
89. **Samra, A., E. Dumasgaudot, V. Gianinazzi-Pearson, and S. Gianinazzi.** 1996. Soluble proteins and polypeptide profiles of spores of arbuscular mycorrhizal fungi. Interspecific variability and effects of host (myc+) and non-host (myc−) *Pisum sativum* root exudates. *Agronomie* **16:**709–719.
90. **Sasser, M. J.** 1991. Identification of bacteria through fatty acid analysis, p. 199–204. *In* F. Klement, K. Rudolf, and D. C. Sands (ed.), *Methods of Phytobacteriology*. Akademiai Kiado, Budapest, Hungary.
91. **Schenck, N. C., and Y. Perez.** 1990. Isolation and culture of VA mycorrhizal fungi, p. 237–258. *In* D. P. Labeda (ed.), *Isolation of Biotechnological Organisms from Nature*. McGraw-Hill, New York, N.Y.
92. **Schussler, A., D. Schwarzott, and C. Walker.** 2001. A new fungal phylum, the Glomeromycota: phylogeny and evolution. *Mycol. Res.* **105:**1413–1421.
93. **Schwarzott, D., C. Walker, and A. Schussler.** 2001. *Glomus*, the largest genus of the arbuscular mycorrhizal fungi (Glomales), is nonmonophyletic. *Mol. Phylogenet. Evol.* **21:**190–197.
94. **Sreenivasa, M. N., and D. J. Bagyaraj.** 1988. Selection of a suitable host for mass multiplication of *Glomus fasciculatum*. *Plant Soil* **106:**289–290.
95. **Sylvia, D. M.** 1992. Quantification of external hyphae of vesicular-arbuscular mycorrhizal fungi. *Methods Microbiol.* **24:**53–66.
96. **Sylvia, D. M.** 1994. Vesicular-arbuscular mycorrhizal (VAM) fungi, p. 351–378. *In* R. W. Weaver, J. S. Angle, P. J. Bottomley, D. Bezdicek, S. Smith, A. Tabatabai, and A. G. Wollum (ed.), *Methods of Soil Analysis*, part 2. *Microbiological and Biochemical Properties*. Soil Science Society of America, Madison, Wis.
97. **Thingstrup, I., M. Rozycka, P. Jeffries, S. Rosendahl, and J. C. Dodd.** 1995. Detection of the arbuscular mycorrhizal fungus *Scutellospora heterogama* within roots using polyclonal antisera. *Mycol. Res.* **99:**1225–1232.

98. **Thompson, J. P.** 1986. Soilless cultures of vesicular-arbuscular mycorrhizae of cereals: effects of nutrient concentration and nitrogen source. *Can. J. Bot.* **64:**2282–2294.
99. **Tisserant, B., V. Brenac, N. Requena, P. Jeffries, and J. C. Dodd.** 1998. The detection of *Glomus* spp. (arbuscular mycorrhizal fungi) forming mycorrhizas in three plants, at different stages of seedling development, using mycorrhiza-specific isozymes. *New Phytol.* **138:**225–239.
100. **Toth, R., and D. Toth.** 1982. Quantifying vesicular-arbuscular mycorrhizae using a morphometric technique. *Mycologia* **74:**182–187.
101. **Vandenkoornhuyse, P., and C. Leyval.** 1998. SSU rDNA sequencing and PCR-fingerprinting reveal genetic variation within Glomus mosseae. *Mycologia* **90:**791–797.
102. **Williams, P. G.** 1990. Disinfecting vesicular-arbuscular mycorrhizas. *Mycol. Res.* **94:**995–997.
103. **Wilson, J. M., and M. J. Trinick.** 1982. Factors affecting the estimation of numbers of infective propagules of vesicular arbuscular mycorrhizal fungi by the most probable number method. *Aust. J. Soil Res.* **21:**73–81.
104. **Wright, S. F., M. Franke-Snyder, J. B. Morton, and A. Upadhyaya.** 1996. Time-course study and partial characterization of a protein on hyphae of arbuscular mycorrhizal fungi during active colonization of roots. *Plant Soil* **181:**193–203.
105. **Wright, S. F., J. B. Morton, and J. E. Sworobuk.** 1987. Identification of a vesicular-arbuscular mycorrhizal fungus by using monoclonal antibodies in an enzyme-linked immunosorbent assay. *Appl. Environ. Microbiol.* **53:**2222–2225.
106. **Wright, S. F., and A. Upadhyaya.** 1998. A survey of soils for aggregate stability and glomalin, a glycoprotein produced by hyphae of arbuscular mycorrhizal fungi. *Plant Soil* **198:**97–107.
107. **Wright, S. F., and A. Upadhyaya.** 1999. Quantification of arbuscular mycorrhizal fungi activity by the glomalin concentration on hyphal traps. *Mycorrhiza* **8:**283–285.
108. **Wyss, P., and P. Bonfante.** 1993. Amplification of genomic DNA of arbuscular mycorrhizal (AM) fungi by PCR using short arbitrary primers. *Mycol. Res.* **97:**1351–1357.
109. **Zézé, A., E. Sulistyowati, K. Ophelkeller, S. Barker, and S. Smith.** 1997. Intersporal genetic variation of *Gigaspora margarita*, a vesicular arbuscular mycorrhizal fungus, revealed by m13 minisatellite-primed PCR. *Appl. Environ. Microbiol.* **63:**676–678.

Isolation, In Planta Detection, and Uses of Endophytic Bacteria for Plant Protection

CHARLES W. BACON AND DOROTHY M. HINTON

52

Endophytic bacteria are defined as those that dwell intercellularly in association with plants for most, if not all, of their life cycles. These organisms are therefore symbiotic and are further distinguished in that the bacterium lives within the plant or portions of it as a nonpathogen, although slight to moderate degrees of pathogenicity may be expressed. This distinction serves to remove from the following discussion the many latent and obligate pathogenic bacteria that form similar, but detrimental, associations with plants. Also not included in this definition are those bacteria whose interactions with plants are restricted, i.e., epiphytic and rhizoplane or phyllosphere colonizers and dormant or latent stages of endophytic colonizers, as well as diazotrophic bacteria that form root and stem nodules, such as the legume-colonizing species of *Rhizobium*, *Bradyrhizobium*, etc. However, the diazotrophic endophytic species that are included are those that do not form nodules on hosts, such as *Azospirillum* and some species of *Rhizobium*.

The classification of a bacterium or fungus as an endophyte is based on its recovery from surface-sterilized plant material and seeds. Endophytes are pragmatically defined, but there is one error inherent in this definition. Bacteria are often isolated from plant materials that have not been adequately surface sterilized, and the isolated bacterium is not demonstrated to reside within the association by microscopic techniques. Thus, successful detection of bacteria in plants is dependent upon a comprehensive procedure. Adequate surface sterilization is important in delimiting the endophytic habit, and particular attention must be directed to achieving this. The literature is replete with microorganisms described as being endophytic, but on scrutiny, the plant material was not properly surfaced sterilized, nor were any attempts made to show the endophytic habit microscopically. This is particularly relevant when the recovered organism is a known saprophyte. Finally, the plant must be reinfected with the endophytic strain, as well as other strains of the species, and comparisons must be made with the original endophyte-infected materials. This information will establish either the uniqueness of a strain as an endophyte or the uniqueness of endophytism within the species.

Bacterial endophytes live within the intercellular spaces of plants (Fig. 1), where they live off apoplastic nutrients (31, 50). The intercellular spaces consist of a series of connecting nonliving components of the plant, forming the apoplast, which is distinct from the intracellular space, or symplasm. The intercellular spaces are located in the cortical tissue of roots and in the parenchyma tissue of leaves. The intercellular spaces are formed by the juxtaposition of three or four cells, followed by the dissolution of the middle lamellae (Fig. 1B), producing what is referred to as schizogenous intercellular spaces. The intercellular spaces in the leaves and stems are significantly different from those in the roots, which often include both schizogenous intercellular spaces and lysigenous intercellular spaces, where spaces are formed from the absorption or disappearance of cells and their contents. The intercellular spaces of all plants have not been sufficiently fully chemically characterized to generalize about their contents, total volume, and relatedness to the symplasm. However, for the sake of discussion, we do have some information that describes this location as a significant area for the physiology of the plant, which is based on sugarcane and a few other plants. The volume occupied by schizogenous intercellular spaces consists of a significant portion of the plant axis, and in leaves, as much as 6% of the leaf tissue consists of intercellular spaces, or apoplasm (138). The intercellular spaces of roots in most plants are typically schizogenous, although in grasses they may be larger, consisting of lysigenous lacunae formed from the breakdown of surrounding cells (Fig. 1). The compositions of intercellular spaces vary, and some consist of air or dry, flaky deposits (Fig. 1A), but the majority of spaces consist of fluid, which in maize has been shown to contain the inorganic ions potassium, calcium, sulfur, phosphorus, and chlorine (30). Most bacterial endophytes occupy fluid-filled schizogenous intercellular spaces, and thus, our discussion is limited primarily to them. The size of the intercellular lacunae may be greatly reduced in various tissue types, especially in the stele, resulting in lacunae literally filled to capacity with bacterial cells. The intercellular spaces of the green portion of plants also consist of gases and fluids, supplying endophytic inhabitants with the required gases. The intercellular spaces in leaves are predominantly located in the extrafascicular plant parenchymatous tissues, which are lacking in nutrients, and consist only of oxygen and air-saturated water, suggesting an explanation for why this area is sporadically colonized by some endophytic species and entirely avoided by others. Evidence that endophytic bacteria may either enlarge or alter the nature of the intercellular space is provided by apparent remains of middle-lamella-type materials of opposing cells (Fig. 1B).

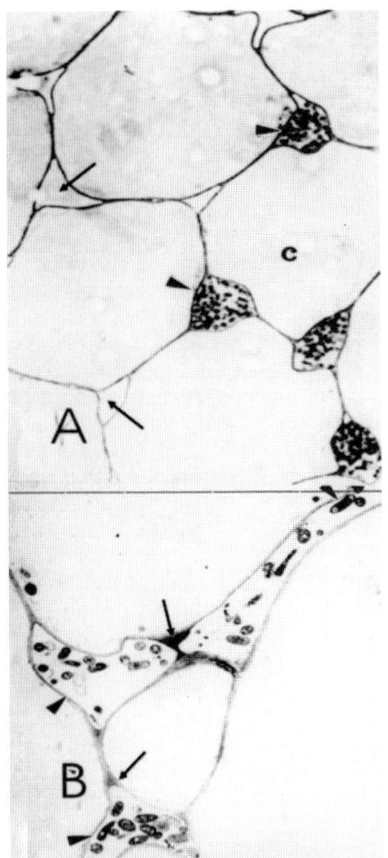

FIGURE 1 Light (A) and transmission electron (B) micrographs of *Bacillus mojavensis*, an endophytic bacterium associated with corn seedlings. Note the two types of intercellular spaces: empty spaces (arrows) which might have been filled with flaky deposits of unknown substances and are probably lysigenous intercellular spaces (A) and bacterial cells (arrowheads) located between schizogenous intercellular spaces formed from the partial breakdown of the middle lamellae of cells shown in the electron micrograph (A) (arrows) within cells of the root cortex (c) (B).

The nutrient concentrations in both the apoplasm and symplasm interact with the phloem, dispelling the earlier notion that the apoplasm is relatively free of nutrients. Indeed, current research indicates that nutrient transport within plant tissues occurs through an apoplastic route via the cell wall continuum to the symplastic route via the plasmodesmata (50, 93, 138). The concentrations of sugars available within the apoplasm are dependent on the nature of phloem loading from either the apoplasm or symplasm (61), but in plant species and even in individuals within a population, phloem loading can vary. Thus, the intercellular space is rich in substances necessary to support the growth of organisms, as demonstrated by the large number of bacterial cells observed (Fig. 1). Further, the concentrations of apoplastic sugars reported to occur in endophyte-infected apoplasm (Table 1) are sufficient to support microbial growth in squash and other plants (31, 50, 90).

The relationships between endophytic bacteria and plants may be mutualistic, resulting in benefits for specific agronomic plant characteristics. Therefore, the use of endophytic bacteria opens up new areas for biotechnological exploitation, which drives the necessity to isolate and culture these organisms, as well as to modify plants for increased volume of the endophytic niche. Endophytes are used for biological control of various plant diseases; for enhanced agronomic plant characteristics, such as increased drought tolerance and nitrogen efficiency; as delivery systems for herbicides and pharmacological agents; and for environmental decontamination. The techniques for examining and culturing bacterial endophytes are relatively old-fashioned and vary from simple to complex, but they are fundamental for in planta studies of host-parasite relationships. These microorganisms are usually obligate biotrophs, their isolation may be difficult and slow, and the techniques needed to study host-bacterial relations usually require considerable modification. Techniques useful for the necessary dissections of the bacterial components from plant hosts to determine the physiological, biochemical, and ecological contributions of bacteria to the intact association are complex. We briefly review endophytic bacteria and outline procedures useful in establishing their presence, culture, and physiological and metabolic interactions with the host and their biotechnological uses. Similar techniques used for endophytic fungi may also be helpful, and we refer you to earlier editions of this manual dealing with specific aspects of fungal endophytes.

ENDOPHYTIC SPECIES OF BACTERIA

The known bacterial endophytes include only a limited number of genera, in which several species exist. In the past, the members of these associations were described as endorhizosphere dwellers, but currently the term is used to include infections in other plant organs (Table 2). However, the distribution and behavior of bacteria within plants have not been completely documented. Endophytic bacteria consist of both gram-negative and gram-positive species, although a higher number of gram-negative species have been reported as agents of biological control (79). While highly controversial, they are considered more evolutionarily diverse and higher than the primitive *Firmicutes* (63). However, the number of species reported as endophytic is low compared to other species found on plants as epiphytes

TABLE 1 Metabolites found in the intercellular spaces of plants[a]

Class	Metabolites
Amino acids	Serine, alanine, asparagine, aspartic acid, threonine, valine, lysine, tyrosine, leucine, glycine, isoleucine, methionine, aminobutyric acid, cystine, arginine
Organic acids	Malic acid, tartaric acid, oxalic acid, fumaric acid, citric acid
Sugars	Sucrose, glucose, fructose, galactinol, inositol, galactose, raffinose, stachyose, verbacose, arabinose, xylose, cellulose, fructans
Protein	Polypeptides (15- to 32-kDa size range)
Inorganic compounds	Potassium, calcium, sulfur, phosphorus, chloride

[a]Data are from references 31, 48, 54, 62, 92, 93, 95, and 111.

or epibiotic or rhizospheric colonizers; undoubtedly this number will increase as the search for more efficient endophytic biocontrol agents continues and as more species are delimited by molecular technology. The number of bacterial endophytes is expected to increase as procedures improve for their detection. Most of the endophytic species belong to the families *Acetobacteraceae*, *Enterobacteriaceae*, *Spirillaceae*, and *Bacillaceae* (83, 97, 121, 145) (Table 2). Thus, species of the genera *Acetobacter*, *Achromobacter*, *Burkholderia*, *Campylobacter*, *Corynebacterium*, *Cytophaga*, *Bacillus*, *Brevibacterium*, *Flavobacterium*, *Leuconostoc*, *Klebsiella*, *Micrococcus*, *Enterobacter*, *Azospirillum*, and *Erwinia* have been associated with ovules, seeds, pods, roots, leaves, and stems of several plants of economic importance, and these bacteria have been shown either to have a positive role or to have no apparent effects on the survival of several plant species (35, 50, 59, 67, 68, 72, 77, 83, 86, 90, 93, 97, 98, 100, 109, 121, 125, 129). Endophytic bacteria include anaerobic, aerobic, and microaerobic species.

BACTERIAL ISOLATION

General Procedures

It is very important always to select vigorous and disease-free plant material for isolation work, as this prevents the isolation of localized pathogenic endophytic organisms, and it is usually desirable to isolate several strains of the same bacterial endophyte from the same host but from different locations. During endophyte isolation, there is a high probability that there will be several species within different tissue types of the same host. Therefore, the method used must allow the recovery and separation of mixed species of bacteria. The isolation of several taxa of fungi may present some problems, since fast-growing species prevent the growth of slower-growing species. However, the incubation temperature, medium selection, and some of the newer molecular techniques described below can be used to circumvent this problem. A high degree of success depends on the initial sample sterilizations, and there are several protocols for this (23, 36, 66, 77, 101, 119), which must be carefully followed, since isolation of endophytes depends on tissue disinfection. The general methods of surface sterilizing plant material for isolation utilize one or two sterilizers in sequence, such as commercial bleach or 1% chloramine-T (*N*-monochloro-*p*-toluenesulfonamide), followed by ethanol or 50% sulfuric acid. Either one or a combination of sterilizers accomplishes the important task of removing surface debris and microorganisms without interfering with endophytic organisms.

Prewash

A vigorous prewashing of plant material is necessary to reduce the number of surface microflora organisms and to remove debris, which increases the effectiveness of the surface sterilizer. The prewash solution may be either sterile tap water, distilled water, or one of several dilute buffers (pH 7.2). Sterile distilled water may cause a decrease in some bacterial endophytes, although most endophytes should not be affected by water quality. The plant material should consist of the proximal portions of fresh roots, stems, and leaves, which must be removed from the plant immediately before the prewash phase of this procedure. The excised roots, stems, or leaves are cut into 5- to 8-cm segments. It is not usually necessary to use detergents in any solution during the isolation of endophytes from green material.

The isolation of endophytes from seeds requires some special considerations, since seeds are dormant dehydrated structures and most systemic seed-borne bacteria are similarly dormant. Improper rehydration of seeds affects the viability of bacteria (129, 137). Most dry seeds must first be briefly wetted in sterile tap water or allowed to imbibe for at least 2 to 4 h in sterile tap water or buffer on sterilized damp filter paper before any surface sterilizers are added. The temperatures during wetting or imbibition should be maintained between 4 and 10°C, depending on the plant species, with the cooler temperatures for cereals and grasses. Alternatively, all seeds may be incubated at low temperatures, which prevents the multiplication of nonendophytic saprophytic flora on seeds. Permitting dry seeds to imbibe in sterilizers results in poor viability of systemic seed-borne microorganisms. The seeds that have imbibed can then be surface sterilized for endophyte isolation. If the seeds have been treated with various pesticides, the prewash procedure should be preceded by water washes containing a detergent, followed by several washes in 75 to 90% ethanol and/or aqueous rinses with charcoal to remove the pesticides (11). Consideration of a detergent must include the final staining procedure, since some identification is based on immunofluorescent stains, which would be affected by detergents that are also fluorescent.

Surface Sterilization

The container used to sterilize plants and seeds should be the sterile disposable type with a screw cap. Specimen cups (4 to 6 oz) are ideal for all washes and sterilizers. The prewashed material is placed in a sterile buffer and sonicated for 1 to 2 min in a sonicator bath. Most plant material can be surface sterilized by vigorous shaking on a rotary or reciprocal shaker in either a 1% chloramine-T solution for 30 min or full-strength bleach (5.25% sodium hypochlorite) solution containing 0.01% Tween 20 or 0.05% Triton X-100 for 3 to 5 min for fresh plant material or for as long as 15 min for seeds and woody stems. Ovules, ovule-bearing pods, and tender plant parts may be surface sterilized in reduced concentrations of a sterilizer but for a longer period. Thus, this type of material requires preliminary tests, and the procedure of Mundt and Hinkle (100) should be followed to determine the optimum conditions. The sterilizing solution is removed and the plant material is rinsed three times in sterile distilled water for 3 min each time. Surface sterilization of the plant material is checked by quickly sliding both surfaces of the sterilized tissue over the surface of nutrient agar and incubating the plates, with the tissue removed, at 30°C for a week or longer. Modified procedures have been published for surface sterilization and other aspects of isolating bacterial endophytes (81, 119, 136).

Surface Sterilization of Specific Tissues

Seeds and Stems

Isolation of bacterial endophytes from seeds on agar media works best if the seeds have not germinated. Germination of the seeds is prevented by the addition of 200 mg of filter-sterilized (0.23-μm pore size) cycloheximide/liter (2 ml of a 100-mg/ml stock solution in 75% ethanol) into cool medium. Some seeds may require larger amounts of cycloheximide to retard germination, which should be determined experimentally. Other fungicides that may be incorporated into media selective for bacterial-endophyte isolation include benylate, chlorothalonil, pentacloronitrobenzene, and nystatin. Woody stems are washed thoroughly in the sonication treatment described above and rinsed in 90 to 100% ethanol for 3 to 10 min, the stem is ignited, and tissue is aseptically removed from inner portions of the stem. Alternatively, full-strength bleach may be used as described above for endophytic fungal isolation. Isolation techniques may be modified to include techniques for isolating xylem-inhabiting endophytic bacteria from woody stems (141, 156), although such bacteria make poor endophytes for biotechnological uses.

Root and Leaf Tissue

Healthy roots are removed and immediately washed free of soil with sterile distilled water as described above for the sonication treatment and soaked in sterile phosphate-buffered saline, pH 7.2, for 10 min to equilibrate the osmotic pressure and prevent passive diffusion of sterilizing agents into the roots (67). A 5- to 10-mm section of root is surface sterilized in a 1% aqueous solution of chloramine-T (77, 109) for up to 30 min; a period of 1 h may be tried for woody and storage roots. The roots are then immersed for 30 min in sterile 0.05 M phosphate buffer (pH 7.2) and then rinsed several times in sterile distilled water. The roots are plated on one or more media for bacterial isolations. It may be necessary to cut off and discard the end of each root when it is removed from the chloramine-T solution. This is to prevent the bacteria from having to pass through any disinfecting solution that might have penetrated tissue and vascular bundles during their egress. Leaves are detached and placed in the 1% chloramine-T solution for 30 min or in full-strength bleach for 5 min, washed for 3 min in sterile distilled water, and plated.

Plant Tissue Incubation and Isolation

The incubation temperature during the isolation of bacterial endophytes should include a range of temperatures (25 to 45°C). If several species of bacterial endophytes are anticipated, it may be necessary to incubate the tissue at several different temperatures. The incubation time varies and is medium and endophyte species dependent, but a 2-week period is common for most bacterial endophytes. It is essential that the surface disinfection be thorough, and this is not an easy process. If the process is too long or the sterilizing solution is too strong, the disinfectant will penetrate plant tissue, killing the bacteria or lowering their number. The disinfection process can be monitored for recovery of surface contaminants by sterility checks of the solution over time (27, 133).

Two general methods are used to isolate bacterial endophytes from surface-disinfected plant material: plant tissue direct and physical extraction isolation techniques. The plant tissue direct technique involves plating tissue directly onto an agar medium, and the bacteria are recovered from the cut ends of the tissue. This technique is identical to that described above for root and leaf tissues. The plant tissue direct technique can be developed into a quantitative procedure with the use of the vital stain acridine orange (69), and a direct count of bacteria can be made to measure the number of endophytic bacteria in stem and root tissues.

The physical isolation techniques include vacuum (4, 20, 22, 60), pressure extraction (65), sonication (94), and centrifugation (49). During the physical isolation techniques, a weighed amount, usually 10 g, of surface-sterilized seeds or plant parts is ground in a sterile mortar and pestle, a sonicator, or a blender specifically designed for sterile grinding in either a sterile buffer solution, half-strength nutrient broth, or sterile water. The extracting solution should contain a fungal inhibitor, usually 0.20% cycloheximide, and 0.1 to 0.5 ml of this solution is spread over the surface of the desired medium, such as nutrient agar or King's B medium (Table 3). Alternatively, the isolation broth may be spread over each of the three differential media as described in the classic work by McInroy and Kloepper (97). The three media used were R2A medium (Difco Laboratories, Detroit, Mich.), used for oligotrophic bacteria; TSA medium (Difco Laboratories), used for culturable heterotrophic bacteria; and SC medium for fastidious organisms (Table 3) (45). With these three media, 36 genera of endophytic bacteria were identified from sweet corn and cotton, for a total of 81 species, including 46 unidentified species and 8 species not reported as endophytes (97). If quantitative results are not required and tissue is being examined for the first time for bacterial endophytes or slow-growing and fastidious species are being sought, the surface-disinfected tissue can be incubated for 3 or 4 days prior to being plated. The plates are observed daily for bacterial growth, and the total number of bacteria is determined by a typical spread plate method, usually after 7 days.

In general, the physical extraction techniques are used following surface disinfection, and the total number of bacteria isolated by this procedure may be highly specific to plant tissue types, resulting in marked quantitative differences from those obtained by the tissue direct methods (65, 97). However, there were no significant differences between the total endophytes isolated from grapevines with the homogenized-tissue and the vacuum extraction techniques (22), suggesting that plant tissue and organ types might be factors.

Culture Media

Most endophytic bacteria can be cultured in the laboratory on routine laboratory media. The contents of the endophytic niche are the source of nutrients for all endophytes, and it is for this that there is a lack of information. The organic and inorganic nutrients found within the apoplasm consist of several simple and complex sugars and related carbohydrates derived from photosynthesis, although several nitrogenous compounds, such as amino acids and amine, also occur, as well as key inorganic nutrients (Table 1). While the compounds listed in Table 1 are based on a limited number of plants (31, 51, 54, 62, 138), these apoplasmic nutrients may serve as the basis for the formulation of culture media for most bacterial endophytes.

The direct isolation of most bacterial endophytes from seeds or plant tissue can be accomplished on nutrient agar, potato dextrose agar, cornmeal agar, glucose-yeast extract, tryptic soy agar, the nitrogen-poor medium LGT, King's B medium, and MacConkey agar (Table 3) (15, 16, 50, 68, 103, 109, 148). If it is desired to isolate aerobic heterotrophic N_2-fixing bacteria, the N-poor medium (Table 3) or one of the N-free media recommended by Watanabe et al. (148) can be used. Specific carbon sources and the nitrogen status of media encourage the growth of different groups of bacteria during isolation and initial culture. Such semiselective media are available for the isolation of *Azospirillum*, *Acetobacter*, and *Klebsiella planticola* (Table 3) (15, 16, 50, 69, 103, 137). McInroy and Kloepper (98) recommend the use of three media as a routine screening procedure during the isolation of endophytic bacteria (Table 3): medium R2A (Difco) for oligotrophic bacteria, TSA (Difco) for culturable heterotrophic bacteria, and medium SC (46) for the growth of fastidious organisms. We (9) find that J agar is excellent for the isolation of *Bacillus* species (Table 3).

TABLE 2 Species of bacterial endophytes used in plant protection of different crop species and in environmental and human health

Crop	Endophytic taxa	Reference(s)
Cereals		
Maize (*Zea mays* L.)	*Arthrobacter* sp., *Aureobacterium* sp., *Bacillus mojavensis*, *B. thuringiensis*, *Bacillus* spp., *Burkholderia cepacia*, *Burkholderia* sp., *Corynebacterium* sp., *Enterobacter* sp., *Klebsiella terrigena*, *K. pneumoniae*, *Pseudomonas* sp., *Staphylococcus* sp.	9, 17, 43, 53, 55, 85, 97, 107
Wild and cultivated rice (*Oryza officinalis*, *O. barthiii*, *O. rufipogon*, *O. glandiglumis*, *O. breviligulata*, *O. sativa*)	*Agrobacterium* sp., *Azorhizobium* sp., *Azospirillum lipoferum*, *Azospirillum brasilense*,[a] *Bacillus* sp., *Bradyrhizobium* sp., *Burkholderia graminis*, *Herbaspirillum rubrisubalbicans*, *H. seropedicae*, *Ideonella dechloratans*, *Enterobacter cancerogenus*, *Pseudomonas* sp., *Rhizobium leguminosarum*	51, 117, 130, 157
Sorghum bicolor	*Herbaspirillum* sp.	105
Wheat (*Triticum aestivum*)	*Bacillus* sp., *B. polymyxa*, *Burkholderia cepacia*	16, 17, 96, 121
Forage grasses		
Kaller grass (*Leptochloa fusca*), root	*Azoarcus* sp.	118
Perennial ryegrass (*Lolium perenne*)	*Bacillus* sp.	40
Cameroon grass (*Pennisetum purpureum*)	*Acetobacter* sp.	119
Crested wheatgrass (*Agropyron cristatum*)	*B. polymyxa*	70
Fruit, nut, and forest trees		
Tangerine (*Citrus reticulata*) and sweet orange (*C. sinensis*), stem	*Curtobacterium flaccumfaciens*, *Enterobacter cloacae*, *Methylobacterium* sp., *M. zatmanii*, *Nocardia* sp., *Pantoea agglomerans*, *Xanthomonas campestris*, *Bacillus pumilus*	1
Rough lemon (*Citrus jambhiri*), root and fruit	*Achromobacter* sp., *Alcaligenes* sp., *Moraxella acinetobacter*, *Actinomyces* sp., *Arthrobacter* sp., *Bacillus* sp., *Citrobacter* sp., *Coryebacterium* sp., *Enterobacter sakazakii*, *E. agglomerans*, *Flavobacterium* sp., *Klebsiella* sp., *Providencia* sp., *Pseudomonas putida*, *Serratia liquefaciens*, *Shigella* sp., *Vibrio* sp., *Yersinia* sp., *Rickettsia*-like sp.	1, 59
Loblolly pine (*Pinus taeda*)	*Bacillus subtilis*, *B. pumilus*, *B. sphaericus*, *Paenibacillus macerans*, *Serratia marcescens*	52
Slash pine (*Pinus elliottii*)	*Bacillus subtilis*, *B. pumilus*, *B. sphaericus*, *Paenibacillus macerans*, *Serratia marcescens*	52
Black spruce (*Picea mariana*)	*Pseudomonas fluorescens*, *Arthrobacter*, *A. citreu*	21
White spruce (*Picea glauca*)	*Pseudomonas*, *Arthrobacter*	21
White spruce hybrids (*P. glauca* × *engelmannii*)	*Bacillus polymyxa*, *Pseudomonas putida*, *Hydrogenophaga pseudoflava*, *Staphyloccus hominis*	24, 41, 42, 106
Pinus sylvestris	*Arthrobacter*, *Agrobacterium*	114
Pinus banksiana	*Pseudomonas putida*	21, 38
Lodgepole pine (*Pinus contorta*)	*Bacillus polymyxa*	38, 127
Pseudotsuga menziesii	*Arthobacter oxydans*, *Pseudomonas syringae*, *Microbacterium imperiale*, *Pseudomonas aureofaciens*, *Bacillus polymyxa*	24, 39
Tsuga heterophylla	*Bacillus polymyxa*	23
Quercus serrata	*Azobacter chroococcum*, *Bacillus megaterium*	108
Malus pumila	*Pseudomonas*	28
Eucalyptus camaldulensis	*Azotobacter chroococcum*, *Bacillus megaterium*	99
Fagus sylvatica	*Agrobacterium radiobacter*	91
Casuarina cunninghamiana	*Azospirillum brasilense*	123
Valonia oak (*Quercus ithaburensis*)	*Azospirillum brasilense*	158
Chestnut (*Castanea sativa*)	*Bacillus subtilis*	149
Fruits, vegetables, and berries		
Cucumber (*Cucumis sativus* L.), root and fruit	*Agrobacterium* sp., *Bacillus* sp., *Burkholderia* sp., *Citrobacter* sp., *Clavibacter* sp., *Erwinia carotovora*, *Proteus mirabilis*, *Serratia* sp., *Xanthomonas* sp.	97

Grape (*Vitis* spp.)	*Bacillus fastidiosus*, *B. insolitus*, *Clavibacter* sp., *Comamonas* sp., *Curtobacterium* sp., *Enterobacter* sp., *Klebsiella ozaenae*, *K. pneumoniae*, *K. terrigena*, *Moraxella bovis*, *Pantoea* sp., *Pseudomonas cichorii*, *Rahnella acquatilis*, *Rhodococcus luteus*, *Staphylococcus* sp., *Xanthomonas* sp.	22
Tomato (*Lycopersicon esculentum*), stem and fruit	*Pseudomonas syringae*, *Escherichia coli*	113, 125
Peanut	*Bacillus* spp.	82
Brassica spp. (cauliflower, etc.)	*B. polymyxa*, *Achromobacter* spp.	25, 113
Beans, peas (*Phaseolus vulgaris*)	*B. polymyxa*, *Bacillus* spp.	48, 110, 146
Potato (*Solanum tuberosum*), tuber	*Acidovorax* sp., *Acinetobacter* sp., *Actinomyces*, *Agrobacterium* sp., *Alcaligenes* sp., *Arthrobacter ureafaciens*, *Bacillus alclophialus*, *B. pasteurii*, *B. sphaericus*, *Capnocytophaga* sp., *Comamonas* sp., *Corynebacterium* sp., *Curtobacterium citrenum*, *C. leteum*, *Deleya* sp., *Enterobacter* sp., *Erwinia* sp., *Flavobacterium* sp., *Kingella kingae*, *Klebsiella* sp., *Leuconostoc* sp., *Methylobacterium* sp., *Micrococcus* sp., *Pantoea* sp., *Pasteurella* sp., *Photobacterium* sp., *Pseudomonas tolaasii*, *Psychrobacter* sp., *Serratia* sp., *Serratia liquefaciens*, *Serratia plymuthica*, *Serratia proteamaculans*, *Shewanella* sp., *Sphinogomonas* sp., *Vibrio* sp., *Xanthomonas* sp.	47, 71, 135
Carrot (*Dacus carota*)	*Bacillus* spp.	136
Legumes		
Red clover (*Trifolium pretense*), leaves, stem, and root	*Acidovorax* sp., *Aerobacter cloaceae*, *Agrobacterium rhizogenes*, *A. tumefaciens*, *Arthobacter ilicis*, *Bacillus brevis*, *B. megaterium*, *Bordetella* sp., *Cellulomonaas*, *Comamonas* sp., *Curtobacterium citreum*, *C. luteum*, *Deleya* sp., *Enterobacter* sp., *Escherichia* sp., *Kingella denitrificans*, *K. kingae*, *Klebsiella* sp., *Methylobacterium* sp., *Micrococcus varians*, *Pantoea agglomerans*, *Pasteurella* sp., *Phyllobacterium* sp., *Pseudomonas cichorii*, *P. corrugata*, *P. fulva*, *P. syringae*, *P. tolaasii*, *Psychrobacter immobilis*, *Rhizobium* sp., *Serratia* sp., *Sphingomonas* sp., *Variovorax* sp., *Xanthomonas compestris*, *X. oryzae*	40, 134
Alfalfa (*Medicago sativa* L.), roots	*Erwinia*-like sp., *Pseudomonas* sp.	57
Livestock feed and miscellaneous food and fiber crops		
Coffee (*Coffea arabica* L.), root and stem	*Acetobacter* sp.	76
Lupine (*Lupinus* sp.), root	*Burkholderia cepacia*	17
Cotton (*Gossypium hirsutum* L.), root and stem	*Acinetobacter baumannii*, *Agrobacterium radiobacter*, *Bacillus* sp., *Bacillus endophyticus*, *Burkholderia cepacia*, *B. gladioli*, *B. pickettii*, *Cellulomonas* sp., *Chryseobacterium* sp., *Comamonas testosteroni*, *Curtobacterium* sp., *Enterobacter cloacae*, *Enterobacter* sp., *Escherichia coli*, *Hydrogenophaga* sp., *Klebsiella* sp., *Kluyvera* sp., *Methylobacterium* sp., *Pseudomonas* sp., *Stenotrophomonas* sp., *Ochrobacterium anthropi*, *Pantoea* sp., *Phyllobacterium* sp., *Pseudomonas saccharophila*, *P. stutzeri*, *P. chloroaphis*, *Ralstonia japonicum*, *Rhizobium japonicum*, *Rhizobium* spp., *Serratia* sp., *Sphingomonas paucimobilis*, *Staphylococcus* sp., *Xanthomonas campestris*, *Yersinia frederiksenii*	59, 98, 100, 122
Sorghum bicolor, shoot	*Herbaspirillum* sp.	75, 105
Sugar beet (*Beta vulgaris*), root	*Bacillus* sp., *Corynebacterium* sp., *Erwinia* sp., *Lactobacillus* sp., *Pseudomonas* sp., *Xanthomonas* sp.	73
Sugar cane (*Saccharum officinarum*), root and stem	*Acetobacter* sp., *Herbaspirillum* sp., *Gluconacetobacter diazotrophicus*	29, 74, 105, 121
Teosinte (*Zea luxurians*), stem and root	*Klebsiella* sp.	107
Soybean (*Glycine max*), stem, leaves, root	*Pseudomonas citronellolis*, *P. oryzihabitans*, *P. staminea*, *Erwinia* sp., *Klebsiella pneumoniae*, *K. oxytoca*, *Agrobacterium*, *Caulobacter* sp., *Enterobacter agglomerans*, *Pantoea* sp.	81
Endophytic phytoremediation	*Bacillus mojavensis*, *Bacillus novalisa*, *Bacillus megaterium*, *Paenibacillus* sp., *Burkholderia cepacia*	18, 56, 128, 147
Human health antibiotics		
Perennial ryegrass (*Lolium perenne*), snakevine (*Kennedia nigriscans*)	*Streptomyces* sp.	128, 147

[a]*Azospirillum lipoferum* is the most common isolate from surface-sterile plant parts of C-4-type grasses, while *A. brasilense* is the most common species isolated from C-3 grasses and sugarcane (29). Data are based on reported isolations of bacteria from surface-sterilized plant parts and not necessarily microscopic determinations and inoculations back into the original plant material.

TABLE 3 Isolation and culture media for endophytic bacteria

Medium (reference) and components	Amt
Nutrient peptone agar[a] (109)	
Glucose	10.0 g
Malt extract	5.0 g
Yeast extract	1.0 g
Nutrient agar	15.0 g
Bacto peptone	1.0 g
Distilled water	1,000 ml
MacConkey agar—*Klebsiella* spp.	
MacConkey agar (Difco)	40.0 g
myo-Inositol	10.0 g
Carbenicillin	0.05 g
Distilled water	1,000 ml
J agar for *Bacillus* spp. (9)	
Tryptone	5 g
Yeast extract	15 g
K_2HPO_4	3 g
Glucose	2 g
Agar	20 g
Distilled water	1,000 ml
N-poor medium (LGT) (83)	
dl-Malic acid	5.0 g
NaOH	3.0 g
$MgSO_4 \cdot H_2O$	0.2 g
NaCl	0.1 g
$CaCl_2$	0.02 g
Yeast extract	0.1 g
$FeCl_3$	10.0 mg
$NaMoO_4 \cdot 2H_2O$	2.0 mg
$MnSO_2$	2.1 mg
H_3BO_3	2.8 mg
$Cu(NO_3) \cdot 3H_2O$	0.04 mg
$ZnSO_4 \cdot H_2O$	0.24 mg
NH_4Cl	1.0 g
Phosphate buffer[b]	100 ml
Distilled water	900 ml
Yeast mannitol agar (34)	
Mannitol	0.5 g
Yeast extract	1.0 g
$MgSO_4 \cdot 7H_2O$	0.2 g
NaCl	0.1 g
K_2HPO_4	0.5 g
Agar	15.0 g
Distilled water	1,000 ml
King's B medium	
Proteose peptone no. 3	20.0 g
Glycerol	0.01 g
$MgSO_4$	1.5 g
K_2HPO_4	1.5
Distilled water	1,000 ml
SC medium for fastidious bacteria (45)	
Papain digest of soy meal	8.0 g
Cornmeal agar	17.0 g
K_2HPO_4	1.0 g
KH_2PO_4	1.0 g
$MgSO_4 \cdot H_2O$	0.2 g
Bovine hemin · Cl[c]	15.0 ml
Bovine serum[c]	10.0 ml
Glucose[c]	1.0 ml
Cysteine	1.0 g
Distilled water	974 ml

[a]All media, except where indicated, were adjusted to between pH 6.8 and 7.0.
[b]Phosphate buffer consists of (per liter of water) 6.0 g of K_2HPO_4 and 4.0 g of KH_2PO_4; the pH of the medium was adjusted to 6.8.
[c]Bovine hemin chloride was prepared as a 0.1% solution in 0.05 N NaOH, bovine serum albumin fraction 5 was prepared as a 20% aqueous solution, and glucose was prepared as a 50% aqueous solution. All three solutions were filter sterilized and added to cool autoclaved medium. The final pH of the medium was adjusted to 6.6 with NaOH or HCl (45).

Bacterial Inoculations and Plant Culture

It is important to emphasize that bacterial endophytes, e.g. *Bacillus mojavensis*, usually readily colonize plants following the topical application of the inoculum to seeds. It is not necessary to force, inject, or reapply the bacterial inoculum of this species to plants once the seeds are inoculated with the bacterium (9). Entry into the cells of the plant occurs through the roots, probably through abrasions or broken sections of the roots. This may be the exception rather than the rule for colonization of plants by other bacterial endophytes. We have no information on the endophytic colonization of plants following the application of inocula to the leaves and stems of mature plants, although it may be possible. The maize seedling assay (11) is useful as a test for endophytic colonization by the strains. Briefly, seeds are surface and internally sterilized before use, rinsed in sterile distilled water, soaked in water for 4 h, heat treated in a water bath at 60°C for 5 min, and rinsed in cool sterile water (7). The bacterial inoculum (0.01 ml; 2 × 10^6 CFU/ml) is placed on the sterilized seeds, and the seeds are incubated on moist filter paper at 25°C for 48 h. For endospore-forming species, inoculated kernels may be placed on sterile filter paper and dried, sealed, stored, and used when desired. The advantage of this procedure is synchronized germination due to the prior imbibing process. The kernels are planted in a sterile synthetic soil mix (Redi-Earth; pH 5.5 to 6.0) in 6-in plastic pots (thinned to 12 plants per pot). The plants are watered daily, and after 2 weeks, they are fertilized at weekly intervals with a liquid 20-20-20 (N-P-K) fertilizer containing micronutrients. Uninoculated seeds, used as controls, are planted as described for the treatment groups. All plants are tested for bacterial infection by one of the procedures listed above under tissue and plant isolation techniques using specific agar, depending on the growth requirements of the bacteria.

IN PLANTA VISUALIZATION, RECOGNITION, AND LOCALIZATION

Microscopy

Surface-sterilized plant materials or seeds may be examined with a light microscope for the occurrence of bacterial cells. Both gram-negative and gram-positive bacterial endophytes are stained with an aniline blue-lactic acid stain. The stain consists of lactic acid (85%), 50.0 ml; aniline blue stain, 0.1 g; and water, 100 ml. This stain is frequently used to quickly examine bacteria in tissues from fresh and dried plant materials (13, 33). Tissues used for this technique must be free from pigments, and therefore, roots, pith tissues from flower stems, and the lower portions of the first internodes from most plant materials are appropriate. As thin a section as possible is made with a razor blade, placed on a slide, and covered with the aniline blue stain (three or more drops), and the slide is gently heated in a very low flame for a few seconds to aid penetration of the stain. Excess stain is then removed by blotting with tissue, water is added, and the plant tissues are examined at ×400 or greater magnification for the presence of blue-stained bacteria (Fig. 1A). If dried plant materials are examined, 0.1% aqueous aniline blue without lactic acid may be used, and heating of the slides is omitted. The slides may be stored for 2 to 4 weeks by sealing the cover slip with clear fingernail polish before the stain decolorizes.

Endophytic bacteria in plants also may be visualized by the procedure of Patriquin and Dobereiner (109) with tetrazolium stain. This stain (PBMT) consists of 2,3,5-triphenyltetrazolium chloride (TPTZ) and is prepared as a sterile solution that consists of 0.05 M potassium phosphate buffer (pH 7.2) containing 0.625 g of malic acid and 1.5 g of TPTZ per liter. The buffer-malate mixture is autoclaved, and the TPTZ is added after the autoclaving. Chloramine-T surface-sterilized plant tissue (roots, shoots, and seeds) is incubated in tubes overnight in the PBMT stain and is cut longitudinally or in cross section by hand or with a microtome. Sections of tissue are placed as wet mounts on slides and examined at ×100. The bacterial cells are stained dark red to purple, and with practice, endophytic bacteria can be easily observed at magnifications lower than ×100.

Bacterial endophytes can be enhanced by counterstaining the PBMT-stained tissue with aniline blue for 1 min, removing the aniline blue, and then making a wet mount. This combined technique was used to establish the endophytic nature of *Bacillus mojavensis* in corn plants (9). Acridine orange is another stain that can be used for the direct counting of endophytes in situ (69), and when combined with epifluorescence microscopy (33), it becomes a vital stain. Plant particulate matter may interfere with this stain, which can be easily overcome by observing the preparation within 4 h as opposed to the 24-h period that is required for staining plant material (19). Additional vital stains useful for bacterial endophytes include calcofluor white M2R (109) and its modification celufluor, sold as Fungi-Fluor (Polysciences, Inc., Warrington, Pa.). Electron and scanning microscopy should be used as the ultimate tools for determining infection, host-microbe interaction, and localization of endophytes (7–9, 66, 100, 115).

Immunological and Nucleic Acid Hybridization Localization

Specific techniques are available that may be used for in situ identification, localization, and metabolism studies for bacterial endophytes. These techniques are not necessarily suited for isolation of unknown bacteria, but rather, they are useful for studying specific microbial endophytes. Thus, immunological techniques can be used alone (116) or combined with fluorescent-staining methods (64, 144) to identify and quantify microbial populations in tissues. These staining methods and their modifications (see, for example, references cited in reference 65) make immunological techniques powerful and very specific tools for studying microbial endophytes in planta, their association with specific tissues, and their production of specific products (26, 34, 44, 72, 112, 126).

Nucleic acid hybridization in situ (44, 64) provides a powerful and very specific tool for detecting and identifying bacteria, primarily because the nucleic acids are highly conserved for a species, and it has been used to detect a variety of bacteria (46, 64, 83, 104, 121). Detecting specific microbial DNA and RNA sequences in plant tissue is relatively new, and in combination with other methods, it promises to be helpful in detecting single cells (64) and other unexplored facets of microbial endophytes. Several procedures are being used that are based on analysis of terminal restriction fragment length polymorphism and sequencing of the 16S rRNA clones. These procedures are used to determine both species identity and interspecific interactions at the population level (see below).

Reporter Genes and Measurement of Metabolic Activity

Ecological and physiological host-bacterial studies often require that the target endophyte be tagged. Reporter gene constructs are a useful application of genome technology for tagging cells, and since they measure transcriptional activity, they can also be used as measures of metabolic activity. The commonly used reporter genes encode proteins whose expression is easily detected by rapid, sensitive, highly specific, quantitative, and reproducible measurements. These include chloramphenicol acetyltransferase, β-galactosidase, luciferase, β-gluconidase (GUS), and the green fluorescent protein or novel fluorescent proteins. Additional uses for reporter systems in microbial ecology are discussed elsewhere in this book.

The most widely exploited gene reporter systems are the green fluorescent protein and GUS assays. The gene *gusA* (formerly *uidA*) encodes a β-glucuronidase, which was originally isolated from *Escherichia coli* (151, 153). The activity of this enzyme, which cleaves a conjugated compound to release the aglycone and glucuronic acid, is utilized for visualization and monitoring of the presence and activity of transgenic organisms containing the introduced GUS gene. For histochemical localization, tissues are incubated with its specific stain (5-bromo-4-chloro-3-indolyl-β-D-glucuronic acid) and monitored for blue staining (131), while a more quantitative GUS assay involves cellular disruption, extraction of the enzyme, and fluorometric measurements of activity using the substrate 4-methylumbelliferyl-β-D-glucuronide (58, 131). For general studies of localization, biomass, and metabolic activity, the GUS gene can be transformed into organisms and expressed continuously behind a constitutive promoter. However, if specific studies involving expression of a single gene of interest are of concern, the promoter of that cloned native gene can be place in front of the GUS gene, and upon transformation of the organism with this construct, the expression pattern of the native gene can be assessed temporally and spatially by monitoring GUS activity.

Beyond the in vitro and in vivo assessments of tissue colonization and cell viability allowed by culturing or staining of endophytes, measurements of actual in planta metabolic activities of endophytes can also be performed. The metabolic activities of bacterial endophytes have been assessed extensively by directly measuring the quantity of secondary metabolites produced either in planta or in vitro (58, 151, 153). However, the production of bacterial secondary metabolites is temporally regulated by nutrient availability and stages of growth and development. Therefore, assessing their quantity in planta has limited value as an indicator of metabolic activity throughout the developmental cycle of the endophytic association. While many bacterial endophytes have been transformed with GUS to measure various ecological tracking and host relationships, a unique use of this reporter gene was the assessment of in planta metabolic activity of endophytes (150–153). Since transcription should not be a limiting factor, GUS expression would be more influenced by the protein-synthesis machinery. Thus, in a paper studying the interaction of a fungal endophyte with grasses that should have direct application to bacterial endophytes, the assumption of Herd et al. (67) was that protein synthesis (i.e., GUS activity) would be influenced by the overall metabolic state of the endophyte, so measurement of GUS activity was assessed as an indicator of overall metabolic activity. Based upon this assumption, they quantified the in planta distribution of GUS-transformed metabolic activity by removing sections of infected plant tissue, extracting the β-glucuronidase, and incubating the extract with the MUG substrate prior to fluorometric measurement. The activity decreased as progressive samples were taken toward the apical half of leaf blades, which had the least activity. Establishment of this basal-apical gradient was evident early in leaf development. A lateral gradient was also evident, in which younger plant tissues exhibited greater fungal metabolic activity than older plant tissues. This distribution was supported by their microscopic examinations, as well.

Catabolic gene probes have been used for analysis of catabolic genotypes in both cultured and noncultured endophytic populations (128). The technique is particularly attractive for those endophytes that are not culturable or that may be overgrown by species with higher growth rates. The use of specific catabolic gene probes (*alkB*, *ndoB*, *xylE*, *ntdA*, and others) can determine the roles of bacterial endophytes in phytoremediation studies as described below. In this procedure, bacteria are extracted from plant materials, serially diluted in appropriate buffers, placed on media such as tryptone agar modified with starch (128), and incubated overnight, and the colonies are counted. The colonies are lifted onto nylon membranes and lysed, and the DNA is denatured, fixed, cross-linked to membranes, and hybridized to one of several metabolic gene probes (128). This technique is applicable to nonculturable endophytes, since sufficient cells are isolated from tissues during the extraction procedures, the cells are concentrated, and the probes are sensitive enough to detect their in planta activity.

Exploitation of Bacterial Endophytes

The endophytic niche offers a unique opportunity for the control of pathogens. The endophyte is protected within the plant and will multiply within the intercellular spaces as the plant grows, thereby potentially colonizing the entire plant axis. The association is long-term, as an endophyte effect was still detected after 2 or more years (35–37), suggesting an application for perennials, such as trees (35) and grasses (118) (Table 1).

The major use of endophytic bacteria has been for the control of plant diseases, and the manner in which this is accomplished is the subject of current research. The high diversity of endophytic bacteria might explain the large number of proposed mechanisms invoked as the basis of disease control. These include the production of antimicrobial compounds (9, 86, 89, 126, 140, 143); macronutrient competition (78, 154, 155); competition for micronutrients, such as iron, by siderophore production (24, 78, 139); hyperparasitism (80); and, more recently, systemic acquired resistance (159, 160).

The use of *B. mojavensis* to control the production of the fumonisin mycotoxin by *Fusarium verticillioides* on maize was demonstrated with experiments conducted under greenhouse and plant growth room conditions (14). However, the field use of *B. mojavensis* resulted in less than favorable control of *F. verticillioides* and the production of the fumonisin mycotoxins by the fungus. Further, this species and other *Fusarium* species produced fusaric acid, particularly when *Fusarium*-infected plants were grown under abiotic soil stresses (12, 32). Fusaric acid is a potent antibiotic for *B. mojavensis*, as well as several other biocontrol species (10, 87, 102). Fusaric acid-tolerant mutant strains of biocontrol endophytes have been developed that now offer promise for the field use of these species (2, 5, 6, 120, 142).

Plant growth responses to bacteria are well established for several crop plants (88, 136). What is not known is which of the many endophytic species present are responsible for the observed changes in plant growth and what mechanisms underlie these effects. A likely explanation for some growth responses is the production of phytohormones that affect plant growth and development. Since there are so many species that can produce this effect, the causes are expected to be varied unless the mechanism is general. The most obvious reasons for growth responses include the production of phytohormones, such as ethylene, auxins, and cytokinins (3, 81), all of which have a positive effect on plant growth and development. Stimulation of plant growth and development by endophytic bacteria has also been attributed to enhanced nutrient availability due to activities such as increased mineral uptake, the solubilizing of bound soil iron and phosphorus, and providing the plant with nitrogen (42, 127, 136). Nitrogen fixation is performed by several endophytic species of bacteria, including *Azospirillum*, *Enterobacter cloacae*, *Alcaligenes*, *Acetobacter diazotrophicus*, *Herbaspirillum seropedicae*, *Ideonella dechlorantans*, and *Azoarcus* sp. Recent evidence has indicated the in planta fixation of nitrogen in rice (51) and sugarcane (74), which in the past was controversial. Ladha and Reddy (84) estimated that as much as 200 kg of N ha^{-1} $year^{-1}$ can be produced by the use of endophytes in plants. Additional growth-stimulation acting in synergy with biocontrol from bacterial endophytes includes an interaction with the natural soil bacterial population, which is based on the prior crop history and the buildup of specific soil bacteria, and the resulting benefits derived from soil chemistry, i.e., allelopathy (135, 136).

Endophytic bacteria have potential as vectors for the indirect transformation of plants for useful products that can be expressed in planta, augmenting the natural nutritional qualities of foods or producing value-added components of food crops, as well as serving as living pesticide vectors. Such surrogate transformations can be used for the production of specific medicinal compounds for human medicine and other pharmaceutical applications (132). Most of the uses of surrogate transformation have been for the delivery of pesticides, such as Bt (*Bacillus thuringiensis*), as the studies cited above indicate. However, surrogate transformation has potential for increasing plant performance and nitrogen efficiency, enhancing biotic and abiotic stress resistance, accelerating seedling emergence and subsequent plant development, increasing or improving plant nutritional qualities, and increasing herbage yield.

Finally, bacterial endophytes have uses as agents of environmental decontamination. It is well known that plants have the potential to remediate polluted water and soils (56, 124). However, only recently has it been shown that bacterial endophytes also have the potential to remediate polluted soils independently (147) or in concert with plants that detoxify polluted soils (18, 128, 147). The uses for phytoremediation with endophytic bacteria range from reducing petroleum hydrocarbon contamination in soils to reducing soil heavy-metal concentrations, including water-soluble and volatile organic xenobiotics, such as benzene, toluene, ethylbenzene, high-ammonia waste and animal manure, chloroform, dichloromethane, xylene, and other hydrophobic pollutants.

REFERENCES

1. **Araujo, W. L., A. Marcos, W. Maccheroni, J. W. L. Van Vuurde, and L. Azevedo.** 2002. Diversity of endophytic bacteria populations and their interaction with *Xyletta fastidiosa* in citrus plants. *Appl. Environ. Microbiol.* **68:**4906–4914.
2. **Arias, J. A.** 1985. Secretory organelle and mitochondrial alterations induced by fusaric acid in root cells of *Zea mays*. *Physiol. Plant Pathol.* **27:**149–158.
3. **Arshad, M., and W. T. Frankenberger.** 1991. Microbial production of plant hormones, p. 327–334. *In* B. D. Keister and P. B. Cregan (ed.), *The Rhizosphere and Plant Growth*. Kluwer Academic Publishers, Dordrecht, The Netherlands.
4. **Azevedo, R. A., C. Damerval, P. J. Lea, J. Landry, C. M. Bellato, L. W. Meinhardt, M. Le Guilloux, S. Delhaye, A. A. Toro, S. A. Gaziola, V. A. Varisi, and P. L. Gratao.** 2004. Genetic control of lysine metabolism in maize endosperm mutants. *Funct. Plant Biol.* **31:**339–348.
5. **Backmann, E.** 1956. Der Einfluss von Fusarinsaure auf die Wasserpermeabilitat von pflanzlichen Protoplasten. *Phytopathol. Z.* **27:**255–288.
6. **Bacon, C. W., D. M. Hinton, and A. Hinton.** 2006. Growth-inhibiting effects of concentrations of fusaric acid on the growth of *Bacillus mojavensis* and other biocontrol *Bacillus* species. *J. Appl. Microbiol.* **100:**185–194.
7. **Bacon, C. W., and D. M. Hinton.** 1988. Ascosporic iterative germination in *Epichloe typhina*. *Trans. Br. Mycol. Soc.* **90:**563–569.
8. **Bacon, C. W., and D. M. Hinton.** November 1999. Use of *Bacillus subtilis* as an endophyte for the control of diseases caused by fungi. U.S. patent 5,994,117.
9. **Bacon, C. W., and D. M. Hinton.** 2002. Endophytic and biological control potential of *Bacillus mojavensis* and related species. *Biol. Control* **23:**274–284.
10. **Bacon, C. W., D. M. Hinton, J. K. Porter, A. E. Glenn, and G. Kuldau.** 2004. Fusaric acid, a *Fusarium verticillioides* metabolite, antagonistic to the endophytic biocontrol bacterium *Bacillus mojavensis*. *Can. J. Bot.* **82:**878–885.
11. **Bacon, C. W., D. M. Hinton, and M. D. Richardson.** 1994. A corn seedling test for resistance to *Fusarium moniliforme*. *Plant Dis.* **78:**302–305.
12. **Bacon, C. W., J. K. Porter, W. P. Norred, and J. F. Leslie.** 1996. Production of fusaric acid by *Fusarium* species. *Appl. Environ. Microbiol.* **62:**4039–4043.
13. **Bacon, C. W., and J. F. White, Jr.** 1994. Stains, media, and procedures for analyzing endophytes, p. 47–56. *In* C. W. Bacon and J. F. White, Jr. (ed.), *Biotechnology of Endophytic Fungi*. CRC, Boca Raton, Fla.
14. **Bacon, C. W., I. E. Yates, D. M. Hinton, and F. Meredith.** 2001. Biological control of *Fusarium moniliforme* in maize. *Environ. Health Perspect.* **109:**325–332.
15. **Bagley, S. T., and R. J. Seidler.** 1978. Primary *Klebsiella* identification with MacConkey-inositol carbenicillin agar. *Appl. Environ. Microbiol.* **36:**536–538.
16. **Balandreau, J.** 1983. Microbiology of the association. *Can. J. Bot.* **29:**851–859.
17. **Balandreau, J., V. Villard, A. Provoost, L. Aeyen, and J. V. Colpaert.** 2001. *Burkholderia cepacia* genomovar III is a common plant-associated bacterium. *Appl. Environ. Microbiol.* **67:**982–985.
18. **Barac, T., S. Taghavi, B. Borremans, A. Provoost, L. Oeyen, J. V. Colpaert, J. Vangronsveld, and D. van der Lelie.** 2004. Engineered endophytic bacteria improve phytoremediation of water-soluble, volatile, organic pollutants. *Nat. Biotechnol.* **22:**583–588.
19. **Bashan, Y., and G. Holguin.** 1994. Root-to-root travel of the beneficial bacterium *Azospirillum brasilense*. *Appl. Environ. Microbiol.* **60:**2120–2131.
20. **Bazzi, C., C. Piazza, and T. J. Burr.** 1987. Detection of *Agrobacterium tumefaciens* in grapevine cuttings. *Bull. OEPP* **17:**105–112.
21. **Beall, F., and B. Tipping.** 1989. Plant growth-promoting rhizobacteria in forestry, abstr. 117. *In Forest Resource Marketing Proceedings*. Ontario Forest Resource Commis-sion, Toronto, Canada.
22. **Bell, C. R., G. A. Dickie, W. L. G. Harvey, and J. W. Y. F. Chan.** 1995. Endophytic bacteria in grapevine. *Can. J. Microbiol.* **41:**46–53.
23. **Bent, E., and C. P. Chanway.** 1998. The growth-promoting effects of a bacterial endophyte on lodgepole pine are partially inhibited by the presence of other rhizobacteria. *Can. J. Microbiol.* **44:**980–988.
24. **Bergeron, R. J., and W. R. Weimar.** 1990. Kinetics of iron acquisition from ferric siderophores by *Paracoccus denitrificans*. *J. Bacteriol.* **172:**2650–2657.
25. **Bertrand, H., C. Plassard, X. Pinochet, B. Touraine, P. Normand, and J. C. Cleyet-Marel.** 2000. Stimulation of the ionic transport system in *Brassica napus* by a plant growth promoting rhizobacterium (*Achromobacter* sp.). *Can. J. Microbiol.* **46:**229–236.
26. **Buyer, J. S., M. G. Kratzke, and L. J. Sikora.** 1993. A method for detection of pseudobactin, the siderophore produced by a plant-growth-promoting *Pseudomonas* strain, in the barley rhizosphere. *Appl. Environ. Microbiol.* **59:**677–681.

27. **Cactano-Anolles, G., G. Favelukes, and W. D. Bauer.** 1993. Optimization of surface sterilization for legume seed. *Crop Sci.* **87:**561–568.
28. **Caesar, A. J., and T. J. Burr.** 1987. Growth promotion of apple seedling and rootstocks by specific strains of bacteria. *Phytopathology* **77:**1583–1588.
29. **Calvalcante, C. P., and J. Bobereiner.** 1988. A new acid-tolerant nitrogen-fixing bacterium associated with sugarcane. *Plant Soil Sci.* **108:**23–31.
30. **Canny, M. J., and C. X. Huang.** 1993. What is in the intercellular spaces of roots? Evidence from the cryoanalytical-scanning electron microscope. *Physiol. Plant.* **87:**561–568.
31. **Canny, M. J., and M. E. McCully.** 1988. The xylem sap of maize roots: its collection, composition, and formation. *Aust. J. Plant Physiol.* **15:**557–566.
32. **Capasso, R., A. Evidente, A. Cutignano, M. Vurro, M. C. Zonno, and A. Bottalico.** 1996. Fusaric and 9,10-dehydrofusaric acids and their methyl esters from *Fusarium nygamai*. *Phytochemistry* **41:**1035–1039.
33. **Cassell, A. E.** 1965. Rapid graphical method for estimating the precision of direct microscopic counting data. *Appl. Microbiol.* **13:**293–296.
34. **Chakraborty, R. N., H. N. Patel, and S. B. Desai.** 1990. Isolation and partial characterization of catechol-type siderophore from *Pseudomonas stutzeri* RC 7. *Curr. Microbiol.* **20:**283–286.
35. **Chanway, C. P.** 1996. Endophytes: they're not just fungi. *Can. J. Bot.* **74:**321–322.
36. **Chanway, C. P.** 1996. Inoculating of tree roots with plant growth promoting soil bacteria: an emerging technology for reforestation. *Forest Sci.* **43:**99–112.
37. **Chanway, C. P.** 1998. Bacterial endophytes: ecological and practical implications. *Sydowia* **50:**149–170.
38. **Chanway, C. P., and F. B. Holl.** 1991. Biomass increase and associative nitrogen fixation of mycorrhizal *Pinus contorta* Dougl. seedlings inoculated with a plant growth promoting *Bacillus* strain. *Can. J. Bot.* **70:**1025–1031.
39. **Chanway, C. P., and F. B. Holl.** 1992. Influence of soil biota on Douglas-fir (*Pseudotsuga menziesii (6). J Franco)* seedling growth: the role of rhizosphere bacteria. *Can. J. Bot.* **70:**1025–1031.
40. **Chanway, C. P., F. B. Holl, and R. Turkington.** 1990. Specificity of association between *Bacillus* isolates and genotypes of *Lolium perenne* and *Trifolium repens* from a grass-legume pasture. *Can. J. Bot.* **68:**1126–1130.
41. **Chanway, C. P., M. Shishido, and F. B. Holl.** 1989. Bacterial inoculation of lodgepole pine, white spruce, and Douglas-fir grown in containers. *In* T. D. Landis (ed.), *Intermountain Forest Nursery Association, Proceedings.* General Technical Report RM-184:93–77. USDA Forest Service, Rocky Mountain Forest and Range Experiment Station, Bismarck, N.Dak.
42. **Chanway, C. P., M. Shishido, and F. B. Holl.** 1994. Root-endophytic and rhizosphere plant growth promoting *Bacillus* strains causes increased seedling emergence and biomass, p. 72–74. *In* M. H. Ryder (ed.), *Improving Plant Productivity with Rhizosphere Bacteria. Proceedings of the 3rd International Workshop PGPR.* CSIRO Division of Soils, Adelaide, Australia.
43. **Chelius, M. K., and E. W. Triplett.** 2000. *Dyadobacter fermentans* gen. nov., sp. nov., a novel Gram-negative bacterium isolated from surface-sterilized *Zea mays* stems. *Int. J. Syst. Bacteriol.* **50:**751–758.
44. **Davis, D. A., H. C. Lynch, and J. Varley.** 1999. The production of surfactin in batch culture by *Bacillus subtilis* ATCC 21332 is strongly influenced by the conditions of nitrogen metabolism. *Enzyme Microb. Technol.* **25:**322–329.
45. **Davis, M. J., A. G. Gillaspie, Jr., R. W. Harris, and R. H. Lawson.** 1980. Ratoon stunting disease of sugarcane: isolation of the causal bacterium. *Science* **210:**1365–1367.
46. **De Boer, S. H., and L. J. Ward.** 1995. PCR detection of *Erwinia carotovora* subsp. *atroseptica* associated with potato tissue. *Phytopathology* **85:**854–858.
47. **De Freitas, J. R., and J. J. Germida.** 1990. Plant growth promoting rhizobacteria for winter wheat. *Can. J. Microbiol.* **36:**265–272.
48. **Delrot, S., and J. Bonnemain.** 1981. Involvement of protons as a substrate for the sucrose carrier during phloem loading in *Vicia fava* leaves. *Plant Physiol.* **67:**560–564.
49. **De Wit, P. J. G. M., and G. Spikman.** 1982. Evidence for the occurrence of race and cultivar specific elicitors of necrosis in intercellular fluids of compatible interactions of *Cladosporium fulvum* and tomato. *Physiol. Plant Pathol.* **21:**1–11.
50. **Dong, Z., M. J. Canny, M. E. McCully, M. R. Roboredo, C. F. Cabadilla, E. Ortega, and R. Rodés.** 1994. A nitrogen-fixing endophyte of sugarcane stems. A new role for the apoplast. *Plant Physiol.* **105:**1139–1147.
51. **Elbeltagy, A., K. Ishoka, S. Tadashi, B. Ye, T. T. Hamada, I. Suzuki, and K. Minamisawa.** 2001. Endophytic colonization and in planta nitrogen fixation by a *Herbaspirillum* sp. isolated from wild rice species. *Appl. Environ. Microbiol.* **67:**5285–5293.
52. **Enebak, S. A., G. Wei, and J. W. Kloepper.** 1997. Effects of plant growth-promoting rhizobacteria on loblolly and slash pine seedlings. *Forest Sci.* **44:**139–144.
53. **Estrada, P., P. Mavingui, B. Cournoyer, F. Fontaine, J. Balandreau, and J. Caballero-Mellado.** 2002. A N_2-fixing endophytic *Burkholderia* sp. associated with maize plants cultivated in Mexico. *Can. J. Microbiol.* **48:**285–294.
54. **Fiala, V., C. Glad, M. Martin, R. Jolivet, and S. Derridj.** 1990. Occurrence of soluble carbohydrates on the phylloplane of maize (*Zea mays* L.): variation in relation to leaf heterogeneity and position on the plant. *New Phytol.* **115:**609–615.
55. **Fisher, P. J., O. Petrini, and H. M. L. Scott.** 1992. The distribution of some fungal and bacterial endophytes in maize (*Zea mays* L.). *New Phytol.* **122:**299–305.
56. **Folmsbee, M. J., M. J. McInerney, and D. P. Nagle.** 2004. Anaerobic growth of *Bacillus mojavensis* and *Bacillus subtilis* requires deoxyribonucleosides or DNA. *Appl. Environ. Microbiol.* **70:**5252–5257.
57. **Gagne, S., H. Richard, H. Rousseau, and H. Antoun.** 1987. Xylem-residing bacteria in alfalfa roots. *Can. J. Microbiol.* **33:**996–1000.
58. **Gallagher, S. R. (ed.).** 1992. *GUS Protocols: Using the Gus Gene as a Reporter of Gene Expression.* Academic Press, San Diego, Calif.
59. **Gardner, J. M., A. W. Feldman, and R. M. Zablotowicz.** 1982. Identity and behavior of xylem residing bacteria in rough lemon roots of Florida citrus trees. *Appl. Environ. Microbiol.* **43:**1335–1342.
60. **Gasson, M. J.** 1980. Indicator technique for antimetabolite toxin production by phytopathogenic species of *Pseudomonas*. *Appl. Environ. Microbiol.* **39:**25–29.
61. **Giaquinta, R.** 1976. Evidence for phloem loading from the apoplast: chemical modification of membrane sulfhydryl groups. *Plant Physiol.* **57:**872–875.
62. **Godfrey, B. E. S.** 1976. Leacheates from aerial parts of plants and their relation to plant surface microbial populations, p. 433–439. *In* C. H. Dickinson and T. F. Preece (ed.), *Microbiology of Aerial Plant Surfaces.* Academic Press Ltd., London, United Kingdom.
63. **Gupta, R. S.** 2002. Phylogeny of *Bacteria*: are we close to understanding it? *ASM News* **68:**284–291.
64. **Hahn, D., R. I. Amann, W. Ludwig, A. D. L. Akkermans, and K.-H. Schleifer.** 1992. Detection of microorganisms in soil after in situ hybridization with rRNA-targeted, fluorescently labeled oligonucleotides. *J. Gen. Microbiol.* **138:**878–879.

65. **Hallmann, J., J. W. Kloepper, and R. Rodriguez-Kabana.** 1997. Application of the Scholander pressure bomb to studies on endophytic bacteria of plants. *Can. J. Microbiol.* **43:**411–416.
66. **Hallmann, J., A. Quadt-Hallmann, W. F. Mahaffee, and J. W. Kloepper.** 1997. Bacterial endophytes in agricultural crops. *Can. J. Microbiol.* **43:**895–914.
67. **Herd, S., M. J. Christensen, K. Saunders, D. B. Scott, and J. Schmid.** 1997. Quantitative assessment of in planta distribution of metabolic activity and gene expression of an endophytic fungus. *Microbiology* **143:**267–275.
68. **Hinton, D. M., and C. W. Bacon.** 1995. *Enterobacter cloacae* is an endophytic symbiont of corn. *Mycopathologia* **129:**117–125.
69. **Hobbie, J. E., R. J. Daley, and S. Jasper.** 1977. Use of nuclepore filters for counting bacteria by fluorescence microscopy. *Appl. Environ. Microbiol.* **33:**1225–1228.
70. **Holl, F. B., C. P. Chanway, R. Turkinton, and R. A. Radley.** 1988. Growth response of crested wheatgrass (*Agropyron cristatum* L.), white clover (*Trifolium repens*) and perennial ryegrass (*Lolium perenne*) to inoculation with *Bacillus polymyxa*. *Soil Biol. Biochem.* **20:**19–24.
71. **Hollis, J. P.** 1951. Bacteria in healthy tissue. *Phytopathology* **41:**350–367.
72. **Hwang, J., W. S. Chilton, and D. M. Benson.** 2002. Pyrrolnitrin production by *Burkholderia cepacia* and biocontrol of Rhizoctonia stem rot of poinsettia. *Biol. Control* **25:**56–63.
73. **Jacobs, M. J., W. M. Bugbee, and D. A. Gabrielson.** 1895. Enumeration, location, and characterization of endophytic bacteria within sugar beet roots. *Can. J. Bot.* **63:** 1262–1265.
74. **James, E. K., and F. L. Olivares.** 1998. Infection and colonization of sugar cane and other graminaceous plants by endophytic diazotrophs. *Crit. Rev. Plant Sci.* **17:**77–119.
75. **James, E. K., F. L. Olivares, J. I. Baldani, and J. Döbereiner.** 1997. Herbaspirillum, an endophytic diazotroph colonizing vascular tissue in leaves of *Sorghum bicolor* L Moench. *J. Exp. Bot.* **48:**785–797.
76. **Jimenez-Salgado, T., L. E. Fuentes-Ramirez, A. Tapia-Hernandez, A. Mascarua-Esparza, E. Martinez-Romero, and J. Caballero-Mellado.** 1997. *Coffea arabica* L., a new host plant for *Acetobacter diazotrophicus*, and isolation of other nitrogen-fixing acetobacteria. *Appl. Environ. Microbiol.* **63:**3676–3683.
77. **Kloepper, J. W., and C. J. Beauchamp.** 1992. A review of issues related to measuring colonization of plant roots by bacteria. *Can. J. Microbiol.* **38:**1219–1232.
78. **Kloepper, J. W., J. Leong, M. Tientize, and M. N. Schroth.** 1980. Enhanced plant growth by siderophores produced by plant growth promoting rhizobacteria. *Nature* **286:**885–886.
79. **Kobayashi, D. Y., and J. D. Palumbo.** 2000. Bacterial endophytes and their effects on plants and uses in agriculture, p. 199–233. *In* C. W. Bacon and J. F. White, Jr. (ed.), *Microbial Endophytes*. Marcel Dekker, Inc., New York, N.Y.
80. **Kranz, J.** 1981. Hyperparasitism of biotrophic fungi, p. 327–352. *In* J. P. Blakeman (ed.), *Microbial Ecology of the Phylloplane*. Academic Press, Ltd., London, United Kingdom.
81. **Kuklinsky-Sobral, K., W. L. Araujo, C. Mendonça, L. C. Geran, A. Pískala, and J. L. Azevedo.** 2004. Isolation and characterization of soy bean-associated bacteria and their potential for plant growth promotion. *Environ. Microbiol.* **6:**1244–1251.
82. **Kumar, B. S. D.** 1996. Crop improvement and disease suppression by *Bacillus* spp. SR2 from peanut rhizosphere. *Indian J. Exp. Biol.* **34:**794–798.
83. **Ladha, J. K., W. L. Barraquio, and I. Watanabe.** 1983. Isolation and identification of nitrogen-fixing *Enterobacter cloacae* and *Klebsiella planticola* associated with rice plants. *Can. J. Microbiol.* **29:**1301–1308.
84. **Ladha, J. K., and P. M. Reddy.** 1995. Introduction: assessing opportunities for nitrogen fixation in rice—a frontier project. *Plant Soil* **194:**1–10.
85. **Lalande, R. N., N. Bissonnette, D. Coutlee, and H. Antoun.** 1989. Identification of rhizobacteria from maize and determination of their plant-growth promoting potential. *Plant Soil Sci.* **115:**7–11.
86. **Lambert, B., E. Leyns, L. Van Rooyen, F. Gossele, Y. Popon, and J. Swings.** 1987. Rhizobacteria of maize and their fungal activities. *Appl. Environ. Microbiol.* **53:** 1866–1871.
87. **Landa, B. B., J. M. Cachinero-Díaz, P. Lemanceau, R. M. Jiménez-Díaz, and C. Alabouvette.** 2002. Effect of fusaric acid and phytoanticipins on growth of rhizobacteria and *Fusarium oxysporum*. *Can. J. Microbiol.* **48:**971–985.
88. **Lazarovits, G., and J. Nowak.** 1997. Rhizobacteria for improvement of plant growth and establishment. *HortScience* **32:**188–192.
89. **Leifert, C., H. Li, S. Chidburee, S. Hampson, S. Workman, D. Sigee, H. A. S. Epton, and A. Harbour.** 1995. Antibiotic production and biocontrol activity by *Bacillus subtilis* CL27 and *Bacillus pumilus* CL45. *J. Appl. Bacteriol.* **78:**97–108.
90. **Lewis, D. H.** 1967. Sugar alcohols (polyols) in fungi and green plants. *New Phytol.* **66:**143–184.
91. **Leyval, C., and J. Berthelin.** 1989. Influence of acid-producing *Agrobacterium* and *Laccaria laccata* on pine and beech growth, nutrient uptake and exudation. *Agric. Ecosys. Environ.* **28:**313–319.
92. **Livingston, D. P., and C. A. Henson.** 1998. Apoplastic sugars, fructans, fructan exohydrolase, and invertase for winter oat: responses to second-phase cold hardening. *Plant Physiol.* **116:**403–408.
93. **Madore, M., and J. A. Webb.** 1981. Lead free space analysis and veins loading in *Cucurbita pepo*. *Can. J. Bot.* **59:**2550–2557.
94. **Mahaffee, W. F., and J. W. Kloepper.** 1997. Temporal changes in the bacterial communities of soil, rhizosphere, and endorhiza associated with field-grown cucumber (*Cucumis sativus*). *Microb. Ecol.* **34:**210–223.
95. **Marentes, E., M. Griffith, A. Mlynarz, and R. A. Brush.** 1993. Proteins accumulate in the apoplast of winter rye leaves during cold-acclimation. *Physiol. Plant.* **87:**499–507.
96. **Mavingui, P., G. Laguerre, O. Berge, and T. Heulin.** 1992. Genotypic and phenotypic diversity of *Bacillus polymyxa* in the wheat rhizosphere. *Appl. Environ. Microbiol.* **58:**1894–1903.
97. **McInroy, J. A., and J. W. Kloepper.** 1995. Survey of indigenous bacterial endophytes from cotton and sweet corn. *Plant Soil* **173:**337–342.
98. **Misaghi, I. J., and C. R. Donndelinger.** 1990. Endophytic bacteria in symptom-free cotton plants. *Phytopathology* **80:**808–811.
99. **Mohammad, G., and R. Prasad.** 1988. Influence of microbial fertilizers on biomass accumulation in polypotted *Eucalyptus camaldulensis* Dehn. seedlings. *J. Trop. For.* **4:**74–77.
100. **Mundt, J. O., and N. F. Hinkle.** 1976. Bacteria within ovules and seeds. *Appl. Environ. Microbiol.* **32:**694–698.
101. **Nairn, J. D., and C. P. Chanway.** 1999. Recovery of a rhizosphere-colonizing GEM from inside wheat roots. *Can. J. Microbiol.* **45:**612–615.
102. **Notz, R., M. Maurhofer, H. Dubach, D. Haas, and G. Défago.** 2002. Fusaric acid-producing strains of *Fusarium oxysporum* alter 2,4-diacetylphloroglucinol biosynthetic gene expression in *Pseudomonas fluorescens* CHA0 in

vitro and in the rhizosphere of wheat. *Appl. Environ. Microbiol.* **68:**2229–2235.

103. **Okon, Y., S. L. Albrecht, and R. H. Burris.** 1977. Methods for growing *Spirillum lipoferum* and for counting it in pure culture and in association with plants. *Appl. Environ. Microbiol.* **33:**85–88.

104. **Okuda, T., A. Yanagisawa, F. Fujimori, Y. Nishizuka, Y. Takehana, and M. Sugiyama.** 1995. New isolation methods and polymerase chain reaction strain discrimination techniques for natural products screening programs. *Can. J. Bot.* **73**(Suppl. 1E-H)**:**S946–S954.

105. **Olivares, F. L., E. K. James, I. Baldani, and J. Dobereiner.** 1992. Infection of mottled stripe disease-susceptible and resistant sugar cane varieties by the endophytic diazotroph *Herbaspirillum*. *New Phytol.* **135:**723–737.

106. **O'Neill, G. A., C. P. Chanway, P. E. Axelrod, R. A. Radley, and F. B. Holl.** 1992. Growth response specificity of spruce inoculated with coexistent rhizosphere bacteria. *Can. J. Bot.* **70:**2347–2353.

107. **Palus, J. A., J. Borneman, P. W. Ludden, and E. W. Triplett.** 1996. A diazotrophic bacterial endophyte isolated from stems of *Zea mays* L and *Zea luxurians* Iltis and Doebley. *Plant Soil* **186:**135–142.

108. **Pandey, R. K., R. K. Bahl, and P. R. T. Rao.** 1986. Growth stimulating effects of nitrogen fixing bacteria (biofertiliser) on oak seedlings. *Ind. For.* **112:**75–79.

109. **Patriquin, D. G., and J. Dobereiner.** 1978. Light microscopy observations of tetrazolium-reducing bacteria in the endorhizosphere of maize and other grasses in Brazil. *Can. J. Microbiol.* **24:**734–742.

110. **Petersen, D. J., M. S. Srinivasan, and C. P. Chanway.** 1996. *Bacillus polymyxa* stimulates increased *Rhizobium etli* populations and nodulation when co-resident in the rhizosphere of *Phaseolus vulgaris*. *FEMS Microbiol. Lett.* **142:**271–276.

111. **Petersen, W., and M. Böttger.** 1991. Contribution of organic acids to the acidification of the rhizosphere of maize seedlings. *Plant Soil* **132:**159–163.

112. **Phister, T. G., D. J. O'Sullivan, and L. L. McKay.** 2004. Identification of bacilysin, chlorotetaine, and iturin A produced by *Bacillus* sp. strain CS93 isolated from pozol, a Mexican fermented maize dough. *Appl. Environ. Microbiol.* **70:**631–634.

113. **Pillay, V. K., and J. Nowak.** 1997. Inoculum density, temperature, and genotype effects on in vitro growth promotion and epiphytic and endophytic colonization of tomato (*Lycopersicon esculentum* L.) seedlings inoculated with a pseudomonad bacterium. *Can. J. Microbiol.* **43:**354–361.

114. **Pokojska-Burdziej, A.** 1982. The effect of microorganisms, microbial metabolites and plant growth regulators (IAA and GA) on the growth of pine seedlings (*Pinus sylvestris* L.). *Pol. J. Soil Sci.* **15:**137–143.

115. **Quadt-Hallmann, A., J. Hallmann, and J. W. Kloepper.** 1997. Bacterial endophytes in cotton: location and interaction with other plant associated bacteria. *Can. J. Microbiol.* **43:**254–259.

116. **Quadt-Hallmann, A., and J. W. Kloepper.** 1996. Immunological detection and localization of the cotton endophyte *Enterobacter asburiae* JM22 in different plant species. *Can. J. Microbiol.* **42:**1144–1154.

117. **Reddy, P. M., K. Ladha, R. So, R. Hernandez, F. B. Dazzo, O. R. Angeles, M. C. Ramos, and F. J. de Bruijn.** 1997. Rhizobial communication with rice roots: induction of phenotypic changes, mode of invasion and extent of colonization. *Plant Soil Sci.* **194:**81–98.

118. **Reinhold-Hurek, B., and T. Hurek.** 1998. Life in grasses: diazotrophic endophytes. *Trends Microbiol.* **6:**139–144.

119. **Reis, V. M., F. L. Olivares, and J. Bobereiner.** 1994. Improved methodology for isolation of *Acetobacter diazotrophicus* and confirmation of its endophytic habitat. *World J. Microbiol. Biotechnol.* **10:**401–405.

120. **Remotti, P. C., H. J. M. Löffler, and L. Van Vloten-Doting.** 1997. Selection of cell-lines and regeneration of plants resistant to fusaric acid from *Gladiolus* × *grandiflorus* cv. 'Peter Pears'. *Euphytica* **96:**237–245.

121. **Rennie, R. J., J. R. De Freitas, A. P. Ruschel, and P. B. Vose.** 1982. Isolation and identification of N_2-fixing bacteria associated with sugar cane (*Saccharum* sp.). *Can. J. Microbiol.* **28:**462–467.

122. **Reva, O. N. S. V. V., H. Pettersson, and F. G. Priest.** 2002. *Bacillus endophyticicus* sp. nov., isolated from the inner tissues of cotton plants (*Gossypium* sp.). *Int. J. Syst. Bacteriol.* **52:**101–107.

123. **Rodríguez-Barrueco, C., E. Cervantes, N. S. Subbarao, and E. Rodríguez-Caceres.** 1991. Growth promoting effect of *Azospirillum brasilense* on *Casuarina cunninghamiana* Miq. seedlings. *Plant Soil* **135:**121–124.

124. **Salt, D. E., R. D. Smith, and I. Raskin.** 1999. Phytoremediation. *Annu. Rev. Plant Physiol. Plant Mol. Biol.* **49:**643–668.

125. **Samish, Z., R. Etinger-Tulezynska, and M. Bick.** 1961. Microflora within healthy tomatoes. *Appl. Microbiol.* **9:**20–25.

126. **Schnider-Keel, U., A. Seematter, M. Maurhofer, C. Blumer, B. K. Duffy, C. Gigot-Bonnefoy, C. Reimmann, R. Notz, G. Defago, D. Hass, and C. Keel.** 2000. Autoinduction of 2,4-diacetylphoroglucinol biosynthesis in the biocontrol agent *Pseudomonas fluorescens* CHA0 and repression by the bacterial metabolites salicylate and pyoluteorin. *J. Bacteriol.* **182:**1215–1225.

127. **Shishido, M., C. Breuil, and C. P. Chanway.** 1999. Endophytic colonization of spruce by plant growth-promoting rhizobacteria. *FEMS Microbiol. Ecol.* **29:**191–196.

128. **Siciliano, S. D., N. Fortin, A. Mihoc, G. Wisse, S. Labelle, D. Beaumier, D. Ouellette, R. Roy, L. G. Whyte, M. K. Banks, P. Schwab, K. Lee, and C. W. Greer.** 2001. Selection of specific endophytic bacterial genotypes by plants in response to soil contamination. *Appl. Environ. Microbiol.* **67:**2469–2475.

129. **Sleesman, J. P., and C. Leben.** 1976. Bacterial desiccation: effects of temperature, relative humidity and culture age on survival. *Phytopathology* **66:**1334–1338.

130. **Stolzfus, J. R., R. M. P. P. So, J. K. Ladha, and F. J. de Bruijn.** 1997. Isolation of endophytic bacteria from rice and assessment of their potential for supplying rice with biologically fixed nitrogen. *Plant Soil* **194:**25–36.

131. **Stomp, A.-M.** 1992. Histochemical localization of β-glucuronidase, p. 103–113. *In* S. R. Gallagher (ed.), *GUS Protocols: Using the GUS Gene as a Reporter of Gene Expression*. Academic Press, San Diego, Calif.

132. **Strobel, G., and B. Daisy.** 2003. Bioprospecting for microbial endophytes and their natural products. *Microbiol. Mol. Biol. Rev.* **67:**491–502.

133. **Sturz, A. V.** 1959. The role of endophytic bacteria during seed piece decay and potato tuberization. *Plant Soil* **175:**257–263.

134. **Sturz, A. V., B. R. Christie, and B. G. Matheson.** 1998. Association of bacterial endophyte populations from red clover and potato crops with potential for beneficial allelopathy. *Can. J. Microbiol.* **44:**162–167.

135. **Sturz, A. V., B. R. Christie, and J. Nowak.** 2000. Bacterial endophytes: potential role in developing sustainable systems of crop production. *Crit. Rev. Plant Sci.* **19:**1–30.

136. **Surette, M. A., A. V. Sturz, R. R. Lada, and J. Nowak.** 2003. Bacterial endophytes in processing carrots (*Daucus

carota L. var. sativus): their localization, population density, biodiversity and their effects on plant growth. *Plant Soil* **253:**381–390.

137. **Talbot, H. W., and R. J. Seidler.** 1979. Cyclitol utilization associated with the presence of klebsielleae in botanical environments. *Appl. Environ. Microbiol.* **37:** 909–915.
138. **Tetlow, I. W., and J. J. Farrar.** 1993. Apoplastic sugar concentration and pH in barley leaves infected with brown rust. *J. Exp. Biol.* **44:**929–936.
139. **Thomashow, L. S., and D. M. Weller.** 1990. Role of antibiotics and siderophores in biocontrol of take-all disease of wheat. *Plant Soil* **129:**93–99.
140. **Thomashow, L. S., and D. M. Weller.** 1996. Current concepts in the use of introduced bacteria for biological disease control: mechanisms and antifungal metabolites, p. 187–235. *In* G. Stacy and N. T. Keen (ed.), *Plant-Microbe Interactions.* Chapman and Hall, New York, N.Y.
141. **Timmer, L. W., R. H. Brlansky, R. F. Lee, and B. C. Raju.** 1983. A fastidious xylem-limited bacterium infecting ragweed. *Phytopathology* **73:**957–979.
142. **Toyoda, H., H. Hashimoto, R. Utsumi, H. Kobayashi, and S. Ouchi.** 1988. Detoxification of fusaric acid by a fusaric acid-resistant mutant of *Pseudomonas solanacearum* and its application to biological control of *Fusarium* wilt of tomato. *Phytopathology* **78:**1307–1311.
143. **Validov, S., O. Mavrodi, L. De La Fuente, A. Boronin, D. Weller, L. Thomashow, and D. Mavrodi.** 2005. Antagonistic activity among 2,4-diacetylphloroglucinol-producing fluorescent *Pseudomonas* spp. *FEMS Microbiol. Lett.* **242:**249–256.
144. **Van Vuurde, J. W. L., and N. J. M. Roozen.** 1990. Comparison of immunofluorescence colony staining in media, selective isolation on pectate medium, ELISA and immunofluorescence cell staining for detection of *Erwinia carotovora* subsp. *atroseptica* and *E. chrysanthemi* in cattle manure slurry. *Neth. J. Plant Pathol.* **96:**75–89.
145. **Vincent, J. M.** 1970. *A Manual for the Practical Study of the Root-Nodule Bacteria.* Blackwell Scientific Publications, Oxford, United Kingdom.
146. **Walker, R., A. A. Powell, and B. Seddon.** 1998. *Bacillus* isolates from the spermosphere of peas and dwarf French beans with antifungal activity against *Botrytis cinerea* and *Phythium* species. *J. Appl. Bacteriol.* **84:**791–801.
147. **Walton, B. T., and T. A. Anderson.** 1990. Microbial degradation of trichloroethylene in the rhizosphere: potential application to biological remediation of waste sites. *Appl. Environ. Microbiol.* **56:**1012–1016.
148. **Watanabe, I., W. L. Barraquio, M. R. De Guzman, and D. A. Cabrera.** 1979. Nitrogen-fixing (acetylene reduction) activity and population of aerobic heterotrophic nitrogen-fixing bacteria associated with wetland rice. *Appl. Environ. Microbiol.* **37:**813–819.
149. **Wilhelm, E., W. Arthofer, R. Schafleitner, and B. Krebs.** 1998. *Bacillus subtilis* an endophyte of chestnut (*Castanea sativa*) as antagonist against chestnut blight (*Cryphonectria parasitica*). *Plant Cell Tissue Org. Culture* **52:**105–108.
150. **Wilson, K. J.** 2005. Molecular techniques for the study of rhizobial ecology in the field. *Soil Biol. Biochem.* **27:**501–514.
151. **Wilson, K. J., S. G. Hughes, and R. A. Jefferson.** 1992. The *Escherichia coli gus* operon: induction and expression of the *gus* operon in *E. coli* and the occurrence of GUS in other bacteria, p. 7–22. *In* S. R. Gallagher (ed.), *GUS Protocols: Using the GUS Gene as a Reporter of Gene Expression.* Academic Press, San Diego, Calif.
152. **Wilson, K. J., A. Parra, and L. Botero.** 1999. Application of the GUS marker gene technique to high-throughput screening of rhizobial competition. *Can. J. Microbiol.* **45:**678–685.
153. **Wilson, K. J., A. Sessitsch, J. C. Corbo, K. E. Giller, A. D. L. Akkermans, and R. A. Jefferson.** 1995. β-glucuronidase (GUS) transposons for ecological and genetic studies of rhizobia and other Gram-negative bacteria. *Microbiology* **141:**1691–1705.
154. **Wilson, M., and S. E. Lindow.** 1994. Ecological similarity and coexistence of epiphytic ice-nucleating (Ice^+) *Pseudomonas syringae* strains and non-nucleating (Ice^-) biological control agent. *Appl. Environ. Microbiol.* **60:** 3128–3137.
155. **Wilson, M., and S. E. Lindow.** 1999. Location and survival of leaf-associated bacteria in relation to pathogenicity and potential for growth within the leaf. *Appl. Environ. Microbiol.* **65:**1435–1443.
156. **Yance, C. E., and C. J. Chang.** 1987. Detection of xylem-limited bacteria from sharpshooter leafhoppers, and their feeding hosts in peach environ monitored by culture isolations and ELISA techniques. *Environ. Entomol.* **16:**68–71.
157. **Yanni, Y. G., R. Y. Rizk, V. Corich, A. Squartini, K. Ninke, S. Philip-Hollingsworth, G. Orgambide, F. De Bruijn, J. Stoltzfus, D. Buckley, T. M. Schmidt, P. F. Mateos, J. K. Ladha, and F. B. Dazzo.** 1997. Natural endophytic association between *Rhizobium leguminosarum* cv. *trifolii* and rice roots and assessment of its potential to promote rice growth. *Plant Soil* **194:**99–114.
158. **Zaady, E. A., A. Perevolotsky, and Y. Okon.** 1993. Promotion of plant growth by inoculum with aggregated and single cell suspensions of *Azospirillum brasilense. Soil Biol. Biochem.* **25:**819–823.
159. **Zhang, S., M. S. Reddy, and J. W. Kloepper.** 2002. Development of assays for assessing induced systemic resistance by plant growth-promoting rhizobacteria against blue mold of tobacco. *Biol. Control* **23:**79–86.
160. **Zhang, S. A., A. L. Moyne, M. S. Reddy, and J. W. Kloepper.** 2002. The role of salicylic acid in induced systemic resistance elicited by plant growth-promoting rhizobacteria against blue mold of tobacco. *Biol. Control* **25:**288–296.

Methods of Soil Microbial Community Analysis

ANDREW OGRAM, HECTOR CASTRO, AND ASHVINI CHAUHAN

53

Knowledge of community composition and structure is central to many ecological and environmental studies, including studies in related disciplines such as chemical and environmental engineering, soil microbiology, biodegradation and bioremediation, and marine microbiology. A variety of techniques are currently available for analyzing community structure, and many of these can be applied with minor modifications to communities for which they were not originally developed. Even though analysis of microbial community composition and structure is one of the most important aspects of environmental microbiology, it also is one of the most technically challenging.

The majority of microorganisms present in many environments may not be readily cultivated by current technologies and therefore may not be included in most analyses. Current estimates indicate that fewer than 1% of the microorganisms present in many environments are readily culturable, indicating that many techniques based on laboratory cultivation may be significantly biased (77, 78). Soil microbial communities probably are the most complex of natural communities, and one study estimated that there may be as many as 1,000,000 "distinct genomes" per g of soil (21). Even if all species present in a community were cultivable, it is frequently not feasible to define and enumerate all species within most communities due to time and cost considerations. Many approaches either study small, well-defined groups of microorganisms, e.g., those involved in nitrification or metabolism of a particular xenobiotic compound, or use a broader-spectrum approach and define the relative numbers of individuals capable of utilizing an array of carbon sources (35) or possessing characteristic fatty acids (82).

To overcome difficulties and limitations associated with laboratory cultivation, techniques have been developed that bypass cultivation and are based on the direct extraction of indicator molecules such as nucleic acids (69) or fatty acids (82) from environmental samples. The composition of extracted nucleic acids or fatty acids may be analyzed, and depending on the nature of the analysis, information regarding the structure, activity, and in some cases, the nutritional state of the community may be obtained. No techniques are without limitations or biases, however, and the user should be familiar with the effects that a given bias will have on the particular application.

Since most environmental microbiological studies are concerned primarily with bacteria, only techniques concerned with characterization of bacteria will be covered here. This chapter is divided into two sections, the first covering approaches that require laboratory cultivation or incubation and the second addressing those based on analysis of indicator molecules that have been extracted directly from soil communities. Reduction and analysis of data by various means such as principal component analysis, hierarchical cluster analysis, and semivariance analysis (25) are useful in studying specific changes in communities (82) but are beyond the scope of this chapter. (See chapter 64 in this volume.)

SAMPLING CONSIDERATIONS

The first practical consideration encountered in analyzing the composition of soil microbial communities is the size and number of samples required. Sampling strategy is discussed elsewhere in this manual (see chapter 17), but a basic understanding of the nature of soil microbial communities may aid in the design of an appropriate sampling strategy for a given application. An ecological community is loosely defined as "a unified assemblage of populations occurring and interacting at a given location" (3). Soil microbial communities may exist not as discrete units but rather as loosely bound units along a continuum of environmental variables that can extend laterally across a soil as well as downward through a soil profile. The minimum physical boundaries for a soil microbial community in terms of mass or volume are not definite, and communities in bulk soil are likely to consist of broad zones along environmental gradients that are specific to the individual study site. Soil community structures may, however, change dramatically on relatively small scales, such as may be observed in the top few centimeters of soil profiles, within soil aggregates, and in the rhizosphere (59, 89). These microhabitats are ignored when, for example, several hundred grams of "bulk surface soil" is collected from depths of between 0 and 15 cm and mixed together. This approach may result in combining relatively distinct communities present in these microzones with more homogeneous communities in bulk soil.

Microbial communities in bulk soil can be highly variable over space and time (56), and the sampling strategy should be designed to account for this variability. Additional variability is found in soils at different depths. Soil profiles are composed of a series of horizons that may vary with depth, and for most applications, sampling of subsoils should be conducted with attention to both depth and horizonation rather than to depth alone.

METHODS BASED ON LABORATORY CULTIVATION

Many commonly used methods of community analysis are based on the isolation of individual colonies on solid medium, followed by characterization by one or more methods that will be discussed below. This general strategy is appropriate for studies requiring identification or characterization of individual isolates or for population genetics studies of a particular species. In addition, this may be the most sensitive approach available for enumeration and characterization of specific groups, if a suitable selective medium is available, of isolates possessing particular growth characteristics (such as degradation of a particular xenobiotic compound) or resistances (such as those to antibiotics or heavy metals). It also may be used in lieu of the more technically challenging and time-consuming approaches based on the characterization of nucleic acids or fatty acids isolated directly from the environmental sample. The greatest limitation of this general approach is that relatively few species in an environmental sample are readily cultivated and many isolates of interest may not be included in the analysis. This limitation is changing, however, and significant advances have been made recently in our ability to cultivate previously unculturable groups.

Traditional Strain Isolation

Prior to characterization, isolates must first be separated from the soil matrix and grown on an appropriate laboratory medium. Separation of bacteria from soil particulates may result in the loss of a significant proportion of the community due to the close association of cells with the organic matrix of soil particles. Many soil microorganisms produce extracellular polysaccharides that promote irreversible adhesion to soil particulates. Inefficient disruption of the bonds between the extracellular polysaccharide and the soil matrix decreases the efficiency of separation and may also create a bias toward recovery of those isolates that are loosely adhered to soil particles. Separation usually is accomplished by vortex mixing of 1 part soil in 9 parts of an aqueous solution buffered to a pH of between 6.5 and 7.5 and containing small amounts of NaCl. Suspension of the sample in this buffer serves two functions: it disperses the soil particles, thereby breaking up small aggregates that may decrease extraction efficiency; and it at least partially disrupts the association between the cells and the soil particles.

Following suspension, the sample may be diluted in sterile, osmotically balanced buffer such as phosphate-buffered saline and centrifuged through a density gradient established by a compound such as Nycodenz. Soil particle sediment and cells form a band in such a gradient (4, 48). Following washing and resuspension in buffer, the cells may be plated out onto an appropriate medium.

Because cultivation of environmental isolates on laboratory media may be the single largest source of bias in community structure analysis, the method of cultivation becomes of primary importance. Standard plating media and incubation conditions are not appropriate for isolation and characterization of strict anaerobes or those organisms that have unusual or unknown growth requirements. The researcher should be aware of the limitations of most plating media and consider using a variety of different media and incubation conditions to optimize the diversity of isolates.

There is no single medium of choice for cultivation of heterotrophic aerobic bacteria isolated from soils, but many include various concentrations of yeast extract (70) or soil extracts (71) supplemented with inorganic nutrients (6). Soil bacteria have a range of preferences for concentrations and types of nutrients but can be divided into two operationally defined classes: copiotrophs and oligotrophs. Copiotrophs have higher growth rates on rich complex media than on dilute media, but the growth of oligotrophs may be inhibited at rich nutrient concentrations. Oligotrophs grow optimally on similar media that have been diluted to between 1 and 10% of the amounts used by copiotrophs. To ensure that the widest diversity of aerobic heterotrophs is isolated, it is best to spread cell suspensions on two sets of agar plates containing both rich and dilute nutrient concentrations.

Plating media can be used to intentionally select against the majority of species in order to isolate a specific group (6). Cycloheximide commonly is used as a selective agent to inhibit the growth of fungi. Fungi are very common in soils and can quickly overgrow agar plates, making the isolation of bacterial colonies very difficult.

The time allowed for incubation of soil isolates is of great importance due to the variability in growth rates among soil bacteria. Some colonies may appear on solid medium after only 1or 2 days of incubation, and others may require incubation for months (32, 33, 67). This variability requires that plates be examined daily for new colonies. A limitation of this approach is that slowly growing bacteria may be overgrown by more rapidly growing bacteria, thereby resulting in the inability to recover some isolate strains.

Once bacterial isolates have been cultivated, two general strategies may be used in selecting colonies for further characterization. If the researcher is interested primarily in applications involving species richness (the number of different species present in a community), colonies differing in morphology should be chosen. If, however, the researcher is interested primarily in species diversity (species richness and evenness), colonies should be chosen randomly. There are few rules for determining the number of colonies that should be selected for further analysis, and it usually is determined by practical considerations based on plans for further analysis and the researcher's limitations and resources. One approach that may be used to determine the extent to which the number of colonies randomly selected describes the community diversity is rarefaction analysis (30). After the appropriate strategy has been decided upon, isolates should be purified by streaking each colony at least twice for isolation.

When grown on synthetic media, many environmental bacteria, particularly pseudomonads, secrete copious amounts of extracellular polysaccharides that may make separation of isolates difficult. In addition, slowly growing isolates may be overgrown and copurified with isolates of more rapidly growing strains and, hence, larger colonies.

The limit of detection of individual species by this approach may be high, particularly if no system is used to select for the species of interest. The limit of detection can be calculated from the starting mass of soil, the final dilution, and the volume spread onto the agar plates.

Comparison of Colony Morphologies

The simplest means of differentiating between isolates for the purpose of diversity measurements is by comparison of colony morphologies. Similar colony types can be grouped into morphotypes by careful observation of the shapes, sizes, colors, and levels of mucoidy at various stages of colony development. This rather crude system does not allow identification of species, but it is suitable for accurate groupings of strains or isolates (27) and can be used for diversity analyses or for reduction of the number of isolates for further characterization by more sophisticated means.

Because of the great variety of soil microorganisms, it is not possible to give detailed descriptions of all colony morphologies that one may encounter. There are, however, qualitative guidelines that aid in identifying characteristics that differ among colony types. A note of caution should be interjected: colony morphologies are highly dependent on the growth substrate and the colony characteristics observed for an isolate on one medium may be quite different from those observed on another medium. This is particularly true in comparing colony morphologies on rich versus dilute media. Differences in morphology are often more easily spotted on rich media than on dilute media. Groupings of isolates into similar morphotypes should be conducted with colonies of similar ages from the same batch of medium. There may also be some variation in the morphologies of colonies of the same isolate, and categorization of isolates based on colony characteristics should be somewhat broad and based on several aspects of morphology rather than only one or two characteristics.

The most obvious colony characteristics are color and size. Many colonies are off-white in color, but others are colored, and close attention to these shades, in combination with other characteristics, can aid in discriminating among colony morphotypes. Colony size is a useful indicator of relative growth rate and can be used in a general comparison between isolates. Precise sizing of colonies is not of great value because of possible small differences in growth rates even between colonies of the same isolate, but colonies observed at a particular time may be classified as small, medium, or large (71).

The shape of colonies is likely to be the most important discriminant. Careful attention should be paid to the shape (round, irregular, or filamentous), the edges (smooth or small versus large lobes), the elevation (flat, embedded in the agar, made up of concentric circles, convex, or showing a raised area in the center), and the level of mucoidy (dry, powdery, wet, or very wet) (71).

Approaches to Cultivation of Diverse Strains

Since fewer than 1% of the microorganism species present in soils have been cultivated to date, several strategies have been developed to increase the number of bacteria cultivable under laboratory conditions. These strategies employ variations of the following conditions: low nutrient concentrations, the presence of signal compounds involved in cell-to-cell communication, long incubation times, and the use of simulated natural environments. Cultivation of broad groups of microorganisms is a time-consuming effort such that most of these methods utilize high-throughput approaches to reduce the time required to screen large numbers of samples.

One strategy for high-throughput cultivation is to dilute samples to extinction (1 to 5 cells/ml) and culture aliquots of the dilutions in microtiter plates incubated under the desired incubation conditions. After incubation, the cultures are screened for growth by using fluorescence microscopy. This technique has been used in marine microbiology, has been reported to detect 10^3 cells/ml, and has obtained a level of culturability 1.4 to 120 times higher than traditional plating in rich agar (13). The use of robotics and automatization of this technique may improve the number of samples analyzed, allowing for more sample coverage.

Another approach, used with both marine and soil samples, is based on encapsulation of cells in gel microdroplets (91) that are incubated in a chromatography column and sorted with flow cytometry. The gel microdroplets allow communication between cells and exchange of nutrients, the cultivation is done in an open system, and the technique can be scaled up for high-throughput applications.

A particularly notable approach is the plate wash PCR method. This method was developed to target uncultivated members of the divisions *Acidobacteria* and *Verrumicrobia* from soils and termite guts (73). The method is based on variations in the compositions of atmosphere and nutrients and the presence of humic substances and catalase or acyl homoserine lactone. In this approach, the inoculum is incubated under a range of conditions, the surfaces of the plates are washed, and DNA is extracted and screened by PCR with primers specific for the groups of interest. The process is repeated until individual colonies of interest are identified.

Janssen and coworkers (32, 33, 67) have had success in cultivating previously uncultured organisms by employing long incubation times (up to 3 months), replacing agar with gellan as the solidifying agent (33), and employing polymers as carbon sources (67). The use of polymers such as xylan is thought to decrease the shock experienced by soil bacteria in the presence of relatively high nutrient concentrations.

Another approach to increase the probability of isolating previously uncultivated strains in pure culture is the use of a diffusion growth chamber in simulating the natural environment (34). The chamber is formed of a stainless steel washer sandwiched between two polycarbonate membranes, and the microorganisms are incubated in the inner space in a semisolid agar. Isolates that grow only in the presence of other microorganisms (perhaps indicating the role of compound signal molecules) may be isolated by using this approach.

METHODS FOR STRAIN IDENTIFICATION

Characterization Based on Growth Substrates

A commonly used approach for identification of environmental isolates is based on growth on specific substrates and fermentative abilities (35). This type of analysis typically is preceded by initial characterization on the basis of the Gram reaction, cell shape and size, and oxidase and catalase reactions. A number of systems are available from commercial sources, with Biolog (Hayward, Calif.) and API (Plainview, N.Y.) being among the most commonly used.

Lipid Analysis

Bacterial strains also may be identified by the presence of specific cellular lipids. There are several major classes and subclasses of lipids, including free acids, hydrocarbons, fatty alcohols, and other forms of bound fatty acids such as phospholipids, peptidolipids, and glycolipids (35). There are many different kinds of fatty acids in the lipids of microorganisms, and different organisms have different combinations of these fatty acids to form distinct characteristic, stable patterns within taxa (15). These fatty acid patterns can be analyzed quantitatively to provide taxonomic information at the species level (87). Fatty acids can easily be extracted and esterified with methanol to form fatty acid methyl esters (FAMEs) (5). Phospholipids can be purified from mixtures of lipids by solid-phase extraction (90) and form phospholipid fatty acid methyl esters (PLFAs). These FAMEs and PLFAs can be analyzed quantitatively and qualitatively by high-resolution fused-silica capillary gas chromatography (50). This technique has been used widely and can be automated.

By comparison of the fatty acid patterns of unknown isolates with those of reference strains in a computer database or library, one can identify the unknown isolates. This approach has been used widely in identifying medically important bacteria and plant pathogens (15). A few precautions must be considered when using FAME analysis for identification of bacteria. First, the purity of the isolate is very important. Second, the selection of reference strains is critical to the success of the identification. Third, standard environmental growth conditions for reference strains and isolates are needed for the determination of taxonomic relationships, because fatty acid patterns are influenced by a variety of factors including growth medium, incubation temperature and moisture, culture phase, and technique of analysis (15, 82). Finally, the database of environmental strains is still limited, and species identification may not be possible for many unknown environmental isolates.

rRNA Characterization

A more precise method of determining the phylogenetic affiliation of an isolate is by characterization of its rRNA. Characterization of rRNA genes is a very powerful and frequently used means of accurately characterizing environmental isolates to phylogenetic levels ranging from domain (*Bacteria*, *Archaea*, and *Eucarya*) (88) to more finely tuned phylogenetic placement.

The technique is based on the concept that rRNA molecules, particularly the 16S and 23S rRNA molecules of prokaryotes, are highly conserved throughout evolution and are therefore useful as specific indicators of the phylogenetic affiliations of environmental isolates (37). The nucleotide sequences of certain portions of the 16S and 23S rRNA molecules are subject to change more quickly over evolutionary time than are other portions of the same molecules. Those regions that change very slowly are characteristic of broad phylogenetic groupings, such as domains, and those regions that change somewhat more quickly are characteristic of narrower phylogenetic groupings, such as the *Proteobacteria* and high-G+C-content gram-positive bacteria, etc. The regions that change most rapidly over evolutionary time (so-called hypervariable regions) are characteristic of even narrower phylogenetic groupings (37).

Phylogenetic placement of isolates requires sequencing of individual 16S rRNA genes. In general, genomic DNA is isolated from the selected colonies and primers specific for 16S rRNA genes are used to amplify these genes by PCR. The sizes of the amplification products depend on the primers selected but in general are between 1 and 1.5 kb. The amplification products from purified colonies usually are typically purified in spin columns and sequenced.

Automatic sequencing is now relatively convenient, with automated sequencers available at many universities and commercial sites. Sequencing of 16S rRNA genes may be complicated, however, by the presence of strong secondary structures making denaturation of the template difficult under standard conditions. Sequencing of rRNA genes from some species may require alternative strategies for overcoming these problems, and the reader is referred to the review by Lane (37) on this subject.

A detailed description of phylogenetic analysis is presented elsewhere in this book, and only the general principles are presented here. As a first approximation, sequences may be compared with those in the National Center for Biotechnology Information database by BLAST (1), an online program that reports the similarity of the unknown sequences to previously reported sequences (http://www.ncbi.nlm.nih.gov/BLAST/). A number of computer programs have been developed for more rigorous phylogenetic assessment of sequences. These programs, many of which are part of the Genetics Computer Group and are available at most universities, are useful for comparisons of sequences with those in databases such as GenBank and for the construction of phylogenies. First approximations of phylogenic affiliation can easily be obtained with programs accessible from the ribosomal database project home page maintained by the Center for Microbial Ecology at Michigan State University (http://rdp.cme.msu.edu/index.jsp). The ribosomal database project home page also provides access to programs for phylogenetic placement of rRNA gene sequences, and the interested reader is encouraged to browse this site. Other commonly used phylogenetic packages include TREECON for Windows (80), PAUP for Macintosh (75), and ARB for Unix (43).

DNA Fingerprinting: Restriction Fragment Length Polymorphism (RFLP) Analysis, Arbitrarily Primed PCR (AP-PCR), and Repetitive Extragenic Palindromic PCR (REP-PCR)

A variety of DNA fingerprinting techniques can be used to rapidly differentiate closely related environmental strains. Regardless of the specific technique, the data obtained are DNA fragments that, when separated on agarose or polyacrylamide gels, yield a banding pattern specific to the genome under investigation. These banding patterns are often likened to bar codes used in product merchandising. Fingerprinting techniques typically are not used to identify environmental isolates to the level of species but rather to demonstrate small genetic differences or similarities for population genetics studies or for diversity studies of closely related species or strains. This type of approach is most appropriate for "fine-structure" analysis of specific components of microbial communities rather than a broad-spectrum approach directed toward identifying specific taxonomic groups or describing the structures of entire communities.

The oldest of these approaches, RFLP analysis, uses restriction endonucleases to digest purified DNA from individual strains in order to identify polymorphisms within individual genes (83). Polymorphisms, or differences within a specific gene, may result in different numbers of sites that are recognized by the restriction endonuclease used in each digestion. Such differences can be caused by single base pair changes or by larger changes such as those from insertions, deletions, and rearrangements. Fragments within the digested DNA are separated by agarose gel electrophoresis, transferred to a DNA binding membrane by using standard procedures, and detected by hybridization with a labeled probe specific for the gene of interest. Differences or similarities in the numbers and sizes of fragments from each isolate can then be identified. The similarity among numbers of restriction sites within individual genes also has been used to calculate evolutionary relatedness among genes from different organisms (55).

One of the primary requirements for RFLP analysis is that the target DNA be very pure; contaminants may partially inhibit restriction enzyme digestion, thereby yielding partial digestion products that confound the analysis. Chromosomal DNA should be subjected to multiple rounds of phenol-chloroform extractions, followed by at least two rounds of chloroform extractions. Depending on the strain, other steps may be required to remove contaminating substances (45). Perhaps the easiest approach to obtaining DNA suitable for RFLP analysis is amplification of specific genes by PCR. Such DNA is amplified away from any contaminants, the amplification products are digested with the enzymes of choice, and the digestion products are visualized by gel electrophoresis. This approach has the added benefit that hybridization is not required; the digestion products result only from the gene of interest and are in sufficient concentrations to be visualized by staining with ethidium bromide.

The enzymes used in RFLP analyses typically are 6-bp-recognition restriction endonucleases such as EcoRI, HindIII, and BamHI. For relatively small (≤500 bp) segments of DNA or for highly conserved genes such as 16S rRNA genes, enzymes that recognize 4 bp (and generate a greater number of fragments) such as *TaqI*, HaeIII, and Sau3A are used. For characterization of 16S rRNA genes, the 4-bp-recognition enzymes HhaI, RsaI, and BstU1 may provide the greatest estimation of diversity (51).

AP-PCR (a generic term that includes both randomly amplified polymorphic DNA analysis and DNA amplification fingerprinting) is similar to RFLP analysis in that fingerprints specific to the target genome are generated and can be used to differentiate between closely related isolates (26). Unlike RFLP analysis, which is designed to identify differences within specific genes, AP-PCR techniques screen for differences between entire genomes. The greater the number of primers used in individual amplification reactions, the greater the amount of the target genome scanned and the greater the likelihood of identifying differences among isolates. For this reason, several AP-PCR, each using a different primer, usually are performed. AP-PCR is much faster than RFLP analysis because it is PCR-based and no digestion with restriction endonucleases or hybridization is required. AP-PCR uses single PCR primers (rather than the two different primers used in most PCR applications) of random sequences between 8 and 10 nucleotides in length. A number of companies synthesize such primers. The primers amplify segments of the target DNA, typically 200 to 2,000 bp in length, which usually are separated on 2% agarose gels and visualized by staining with ethidium bromide (68).

The generation of amplification products from genomic DNA is dependent on variables, including primer sequence, primer concentration, template concentration, template quality, cycling parameters, and the thermocycler. Because of the sensitivity of AP-PCR and REP-PCR banding patterns to so many variables, a lack of reproducibility is frequently a problem. For this reason, three replicate amplifications always should be conducted. Reproducibility may be less problematic with AP-PCR patterns generated by one operator taking care to be consistent among reactions, but comparisons of patterns generated between laboratories or even between operators within the same lab may not be possible (61). Variability between fingerprints should be addressed; an example of such a treatment was suggested by McSpadden Gardener et al. (49).

REP-PCR and BOX analysis (41) resemble AP-PCR in that fingerprints characteristic of a particular isolate are generated, but there may be fewer problems with variability than in AP-PCR. REP-PCR is based on the amplification of repeated sequences that are highly conserved and widely distributed among genomes and is therefore not random (14, 41).

ANALYSIS BASED ON DIRECT EXTRACTION OF INDICATOR MOLECULES

In order to bypass the current limitations and biases associated with laboratory cultivation, alternative approaches to community structure analysis are available that are based on indicator molecules extracted directly from environmental samples. The most commonly used indicators are DNA, RNA, and phospholipids. Each of these molecules yields different information regarding the structures and activities of microbial communities, and the user should choose the molecule and the method that best suit the specific application. Methods for isolation of DNA, RNA, and phospholipids from environmental samples are presented elsewhere in this manual.

A trade-off in information and biases exists between methods based in cultivation and those based in direct extraction of indicator molecules. Analysis of nucleic acids and lipids directly extracted from environmental samples does not incur biases associated with cultivation of strains; however, these methods are limited and biased in other ways (57). Most notable among the limitations is that little regarding the physiology of the target organism may be readily deduced by most analyses currently available. Isolation of organisms allows a more complete understanding of the range of physiological functions characteristic of an organism than would be afforded by sequence analysis of one or two genes or identification of a particular lipid.

Lipid Analysis

Total fatty acids extracted from environmental samples have been used to study microbial community structures and metabolic states and to compare similarities and differences among soil microbial communities (8, 20). Certain components of microbial populations are enriched in particular fatty acids. For example, odd-numbered and branched-chain fatty acids are associated with gram-positive bacteria. Gram-negative bacteria contain higher proportions of even-numbered monounsaturated straight-chain and cyclopropane fatty acids, eubacteria generally do not contain polyunsaturated fatty acids, and plasmalogen phospholipids are enriched in anaerobic prokaryotic bacteria (82, 87).

The relative concentrations of different kinds of polar lipids in each membrane are characteristic of the type of membrane, the cell type, and the species; however, the fatty acid components of the individual membrane lipids are not fixed and may vary with nutritional state and environmental conditions in order to maintain membrane fluidity (36). By studying the changes in fatty acid profiles, especially the presence of certain phospholipid fatty acids as biomarkers for certain microbial types, an indication of the metabolic activity of the microbial community can be obtained (20, 24). The primary limitation of this method is that it is difficult to correlate the changes in PLFA patterns with the dynamics of specific groups of organisms, and more data are needed (20). Analysis of community structure by lipid analysis is discussed in greater detail in chapter 10.

Gene Probes

The type of information obtained by quantitative hybridization of gene probes to nucleic acids extracted directly from environmental samples is largely dependent on the target nucleic acid. Hybridization to DNA is useful for rapidly identifying specific aspects of community structure, such as the presence and relative concentrations of genes encoding a particular function. Hybridization to RNA is an indicator of the relative activities of the target groups. The activities of certain non-rRNA genes can be estimated by using mRNA as a target, although many mRNA molecules are too labile to be detected in this way. Regardless of the bound nucleic acid, the principles involved are essentially the same.

In hybridization of extracted nucleic acids, probe specificity and probe length become important considerations. Precise determination of concentrations of specific sequences is difficult because of the current lack of knowledge of the genetics of most soil microorganisms. If regions of the probe hybridize with nontarget genes in the sample, target gene abundance is overestimated. Alternatively, if the probe hybridizes poorly with target genes because the sequences have diverged during evolution, concentrations of target sequences may be underestimated. For these reasons, probes should ideally be relatively short (≤500-bp), conserved segments of well-characterized genes, and the washing conditions should be relatively stringent (i.e., low salt concentration and high temperature).

In general, a dilution series of known concentrations of the target gene (typically the probe DNA) is applied to a DNA binding membrane adjacent to the environmental samples. The use of triplicate samples is advised to help compensate for variability in the procedure including variability associated with some membranes. Raskin et al. (65) conducted an extensive survey of variabililties in hybridization of rRNA bound to membranes from various manufacturers, and the reader is encouraged to consult this article prior to purchasing nylon membranes for quantitative hybridizations. Inclusion of negative controls is very important.

Solutions containing target nucleic acids are applied to the membrane as a grid through a vacuum manifold. Depending on the type of manifold, the target may be applied either as circles (dot blots) or as narrow ellipses (slot blots). Slot blots are preferred for quantitative analyses because they are more easily analyzed by scanning densitometry after hybridization.

Standard procedures for labeling the probe (either radioactively or nonradioactively), prehybridization, hybridization, and washing are followed (58, 68). Nonradioactive labeling systems are becoming more widely used, and a variety of strategies are available for labeling and detection with these systems. The stringency of washing generally is kept high to avoid possible nonspecific hybridization with related, but nontarget, genes in the environmental sample (58).

The extent of hybridization usually is measured by exposing the membrane to X-ray film or by phosphorimaging. When X-ray film is used, the film should be preflashed with a photographic strobe (68) in order to increase the sensitivity and the linear range of the autoradiogram. The membrane should be wrapped in plastic wrap such as saran wrap, and the film and membrane should be placed in an autoradiography cassette. When ^{32}P-labeled probes are used, the exposure of the film is greatly enhanced by placing the film and membrane between intensifying screens that can be purchased with the cassettes. Exposure times can vary from a few minutes to a week or longer, depending on the intensity of the signal emitted from the bound probe, and optimal exposure times must be determined empirically for each application. For some labels and some kinds of intensifying screens, the sensitivity of detection is improved by carrying out the exposure at −80°C. The signal from nonradioactive probes can be amplified by using streptavidin-alkaline phosphatase conjugates (9). After an appropriate time, the film is developed and the density of the exposed regions corresponding to the placement of the samples is measured by scanning densitometry. The amount of the target sequence in each sample can then be estimated by comparison to a standard curve derived from the signal intensities from the dilution series of target genes. Alternatively, fingerprints of communities may be generated by reverse sample genome probing, in which a set of known genes of interest is applied to the membrane and the extracted sample DNA is labeled and used as a probe (84).

Phosphorimagers are instruments that detect the level of radiation associated with spots on membranes and report this information as disintegrations per unit of time. Standard curves may be generated as described above for autoradiography, and quantitative information may be obtained. Phosphorimagers have the advantage over standard X-ray film in that the results obtained can be more quantitative, they have lower limits of detection than X-ray film, and exposure times are shorter. In addition, the dynamic range of phosphorimagers is much greater than that of X-ray film such that phosphorimages are rarely overexposed.

Analysis of rRNA and rRNA Genes

Analysis of the diversity of rRNA molecules or their genes in an environmental sample can be of great value in characterizing community structures and the relative activities of phylogenetic groups of interest. As with any approach, caution should be exercised with interpretation of results of rRNA analysis. 16S rRNA gene sequence analysis is a powerful tool for discovering and evaluating the diversity of soil bacteria, but it also may mask a significant amount of diversity. Species may not be defined solely by 16S rRNA gene sequences, nor do the sequences necessarily reflect the range of metabolic functions present within a given species. In addition, any approach requiring PCR is subject to the limitations currently inherent in PCR, including generation of artifacts and deviation from quantitative amplification (discussed below).

Strategies for rRNA analysis include sequencing and phylogenetic analysis (23, 38), hybridization with group-specific probes (2), and classification by RFLP patterns for placement into different operational taxonomic units (51). Designation of RFLP patterns as operational taxonomic units is a convenient means of assessing species richness, although little about individual phylotypes is known. Total community composition analyses are conducted with rRNA genes rather than rRNA molecules because DNA includes all members of the community, regardless of activity. If, however, one is interested in the metabolically active component of the community, rRNA rather than rRNA genes should be analyzed.

The simplest approach to analysis of rRNA is hybridization with phylogenetic probes by using procedures similar to those described in "Gene Probes" above. A variety of oligonucleotide probes that have been designed to hybridize exclusively to specific phylogenetic groups are listed in reference 2. This approach is suitable for characterizing the active groups of a community but is less useful for identifying specific members because the probes are directed toward broad phylogenetic groups. For the latter purpose, cloning and sequence analysis are required. If the primary goal is assessment of phylogenetic richness rather than identification, characterization of the diversity of 16S rRNA genes within a community by denaturing gradient gel electrophoresis (DGGE) or terminal RFLP (T-RFLP) may be appropriate (see below). Regardless of the specific approach, the initial steps of these procedures require direct extraction of total RNA or DNA, reverse transcription of RNA into cDNA (if RNA is of interest), and PCR amplification by primers specific for the groups of interest (37). Because the amplification products will have been derived from mixed communities, they are assumed to be mixtures of molecules of similar sizes originating from the various organisms present in the sample.

Prior to sequence analysis, individual molecules must be segregated by cloning (38), preferably in vectors specifically designed for this purpose. The proportion of clones present in the PCR-generated library should not be assumed to be quantitatively representative of the proportions of the individual rRNA genes in the environmental sample because of primer bias and possible biases (74) in the relative rates of PCR amplification of rRNA genes resulting from differences in copy number and proximity of the operons to one another in the template DNA (17).

Once the clone library has been created, the number of clones to be sequenced may be reduced by comparison of RFLP patterns. Since rRNA genes are only between 1.5 and 2 kb in size and are highly conserved, restriction endonucleases that recognize 4 bp, rather than 6, are commonly used (51). Moyer et al. (51) found that the enzymes RsaI, HhaI and BstUI yield patterns that discriminate among the greatest numbers of species in silico. Reduction of the numbers of clones for sequence analysis by preliminary grouping based on RFLP patterns may result in an overestimation of community diversity because of the presence of multiple, divergent rRNA (*rrn*) operons within a single environmental strain. Most bacterial species harbor between 1 and 10 *rrn* operons, and small differences (microheterogeneities) that result in different RFLP patterns have been observed in some strains (64). Conversely, underestimation of diversity can result if different strains exhibit the same fingerprint. In some cases, digestion with more than one enzyme may provide greater resolution among different phylotypes than that afforded by one enzyme.

Amplification of mixtures of molecules is subject to PCR artifacts, including the formation of products composed of the 3′ end of one molecule and the 5′ end of another. These so-called chimeras are thought to be present in fewer than 10% of the clones obtained by this method but can be very difficult to identify. Sequencing the amplification products and checking for chimeras with the Chimera_Check (http://35.8.164.52/cgis/chimera.cgi?su=SSU) or Bellerophon (http://foo.maths.uq.edu.au/~huber/bellerophon.pl) program is advised, but these programs are not foolproof (66).

The most detailed information concerning the phylogenetic affiliations of the individual members of a community is obtained by sequencing all of the individual clones, but this can be an expensive and time-consuming procedure. Hence, it usually is conducted only with communities with low species diversities, such as those found in hot springs and bioreactors. It should be noted that widespread use of automated sequencers allows characterization of many more clones than was once possible, but sequencing of complete rRNA genes from all clones is still rare. Sequencing of several hundred bases from one end of the cloned sequence frequently provides sufficient information for phylogenetic characterization. Primers for sequencing may also be directed toward specific regions of a gene that are known to be less well conserved than sequences that would be included in sequencing from one end of the insert. Targeting more divergent regions for sequencing may allow greater diversity within the clone library to be identified. Care should be taken in interpreting rRNA gene sequences and inferring phylogenetic affiliations based on these sequences. rRNA gene sequence data alone cannot be used to classify sequences to the level of species (72). This subject is covered in more detail elsewhere in this manual.

A more general approach to analysis of complex communities, such as those in soils, is to determine the presence of phylogenetic groups of interest by hybridization of the individual clones with group-specific oligonucleotide probes (2, 38).

Rapid Screening Methods: DGGE, TGGE, T-RFLP Analysis, and ARISA

For estimation of phylogenetic richness, DGGE, temperature gradient gel electrophoresis (TGGE), T-RFLP analysis, and automated ribosomal intergenic spacer analysis (ARISA) are possible strategies. These PCR-based methods produce electrophoretic bands with lengths representative of the different forms of a given gene in an amplification product derived from a mixed template. In all four approaches, the number of electrophoretic bands provides an indication of the richness of different target genes in the sample. If the target gene is a 16S rRNA gene, the analysis will provide an indication of the phylogenetic diversity in the sample. These methods yield the same general type of information, although the techniques are different (54). DGGE and TGGE are based on separation of individual molecules within a mixed-template PCR amplification product by taking advantage of unique "melting domains" present within segments of the amplified DNA. Melting domains are discrete regions in which the double-stranded DNA heteroduplex melts in response to a specific temperature or a specific concentration of the denaturant, determined by the base composition within the domain. When mixtures of molecules (e.g., amplification products of 16S rRNA genes from environmental samples) are electrophoresed through a gradient of increasing concentrations of a denaturing agent such as urea, individual melting domains within the molecules begin to melt at a specific point in the gradient, thereby affecting their relative mobilities. Rather than using a gradient of the denaturant, TGGE uses a uniform concentration of denaturant in the gel, and temperature is increased uniformly with time throughout the electrophoresis. This may result in more easily reproducible separations than are common with DGGE. Amplification products typically are small (approximately 250 to 500 bp), and PCR primers typically are clamped at the 5′ end by long (between 30 and 50 bp) GC-rich sequences. The GC-rich clamp acts as a high-temperature melting domain, preventing complete denaturation of the product (53). The individual molecules within the mixture of amplification products are separated as a function of temperature or the concentration of the denaturant, yielding an indication of the richness of the target gene. DGGE and TGGE may be semiquantitative if the appropriate controls are included (18), and individual bands may be excised and sequenced for estimation of the phylogenetic placement of the species. Alternatively, the gel may be blotted and hybridized with group-specific probes (52, 76).

Optimization of DGGE is dependent on the gradient, the sizes and diversities of the amplification products, and the base compositions of the individual molecules. Pouring the gradient can be problematic (54), but commercially produced systems are available for this purpose. Selection of primers and target groups is critical; targeting of a diverse group (e.g., all bacteria in a soil sample) may yield an unwieldy number of bands such that analysis is difficult. DGGE and TGGE may be most appropriate for analyzing narrower groups such that fewer than 100 bands can be analyzed.

T-RFLP analysis has an advantage over DGGE in that it can be automated, although it may not be as sensitive as DGGE in detecting individual genotypes. T-RFLP analysis is based on the assumption that the restriction fragment profiles of individual genes (such as 16S rRNA genes) are characteristic of a given phylotype or genotype and that the lengths of the 5′-most or 3′-most fragments (terminal fragments) of digested amplification products are representative of their respective genotypes. In reality, the lengths of the terminal fragments may not be unique for a given genotype, and in some cases, they may not even be representative of broader phylogenetic groupings. The success of T-RFLP analysis must be judged on a case-by-case basis, given the target gene and the sample.

The choice of target gene obviously is dependent on the research application. In general, T-RFLPs developed for functional genes for which strong phylogenies have been developed (10), such as *dsrAB* (11) and *mcrA* (12), provide greater resolution than T-RFLPs developed for 16S rRNA genes. Problems arising with T-RFLPs based on rRNA genes include lack of resolution between phylotypes (16) and interpretation of complex patterns. Focusing on specific functional groups of microorganisms may lead to more straightforward interpretation of T-RFLPs, although a local database of target sequences should first be constructed by analysis of clone libraries of selected samples.

In practice, DNA from an environmental sample is amplified with appropriate primers, one of which is labeled at the 5′ end with a fluorescent tag such as the phosphoramidite fluorochrome 5-carboxyfluorescein. PCR conditions with fluorescence-labeled primers typically require longer amplification times and lower annealing temperatures than are optimum for nonlabeled primers. The amplification product is digested with an appropriate enzyme (such as HhaI or RsaI; see above), and the mixture is desalted, typically by a commercially available desalting column. The lengths of the fragments are analyzed by an automated sequencer or by capillary electrophoresis such that only the labeled terminal fragments are detected. The terminal fragment lengths can be compared with those found in other communities, and qualitative relationships among the communities can be inferred. Practical considerations for T-RFLPs are based mainly on accurate measurement of the fragments and interpretation of the results. An appropriate size standard should be included in every lane to maximize the accuracy of estimation of the lengths of the terminal fragments; accurate sizing of fragments decreases for fragments of greater than 600 bp (40). An extensive evaluation of T-RFLP optimization and variability is presented by Osborn et al. (60), and a program and database for designing and interpreting T-RFLPs are available at the microbial community analysis (http://mica.ibest.uidaho.edu/) and T-RFLP phylogenetic assignment tool (http://trflp.limnology.wisc.edu/index.jsp) websites.

T-RFLPs can be limited, particularly when the target gene is highly conserved, such as the 16S rRNA gene. Divergent groups may have similar terminal restriction fragments, such that those groups will not be resolved by this method. ARISA overcomes some of the limitations of T-RFLP analysis. In ARISA, DNA is PCR amplified by using primers targeting the 16S to 23S intergenic spacer region of the rRNA operon, including a forward fluorescence-labeled primer. Intergenic spacer regions are more variable than either the 16S or the 23S rRNA gene, and restriction fragments generated from within this region are more likely to differ among closely related phylotypes than fragments generated from within the genes. The PCR product sizes range from 400 to 1,200 bp. The 16S to 23S intergenic amplicons are separated and quantified by using an automated electrophoresis system. The results are compared with 16S to 23S intergenic spacer fragments from database information (19).

Stable Isotope Probing

Identification and characterization of microorganisms based on the environmental processes they perform are frequently a major focus in microbial ecology. An approach to linking metabolic networks of microorganisms in soils and characterizing organisms that utilize a specific substrate is stable isotope probing (SIP) (7, 62, 63, 85, 86). SIP takes advantage of the difference in densities between indicator molecules (such as lipids and nucleic acids) of microorganisms that have metabolized a ^{13}C-labeled compound and those that are either inactive or have consumed ^{12}C-labeled substrates. A ^{13}C-labeled substrate of interest, such as methane (31, 39), methanol (22), fatty acids (44), H_2-$^{13}CO_2$ (42), or a xenobiotic compound (85), is incubated with the soil sample for a sufficient time for the substrate to be consumed. The target molecule is extracted from the samples, and the heavy biopolymer is separated from the lighter, ^{12}C-labeled polymer. For nucleic acids, ^{13}C-labeled nucleic acid is separated from ^{12}C-labeled nucleic acid by ultracentrifugation through either a cesium chloride (CsCl) density gradient for DNA or a cesium trifluoroacetate (CsTFA) density gradient for RNA (46, 47). Ethidium bromide is incorporated into the CsCl or CsTFA prior to mixing with nucleic acids to stain the nucleic acids so that they may be visualized under ultraviolet light following centrifugation. The heavier (^{13}C-labeled DNA or RNA) fraction is composed of nucleic acids from those microorganisms that metabolized either the parent compound or metabolites of the parent compound. By using standard methods (68), the lower band is withdrawn from the centrifuge tube and the ethidium bromide and CsCl or CsTFA is removed from the ^{13}C-labeled nucleic acid fraction. The nucleic acid is then concentrated, and genes of interest (e.g., 16S rRNA genes) may be amplified by PCR, cloned, and sequenced.

A major limitation of DNA-based SIP is that the substrate must be incorporated into the DNA of the microorganisms that are involved in the process under investigation, which generally requires long incubation periods ranging from days (31, 46) to weeks (44, 62, 63). Furthermore, longer incubations may also lead to formation of labeled intermediates of the primary substrate metabolism, leading to their assimilation by cross feeders (62, 63, 86). To circumvent these issues, RNA-based SIP has been proposed (46, 47). In theory, RNA may be a more reliable marker than DNA because RNA synthesis occurs at a much higher rate and labeling can occur without DNA synthesis or replication of the organism. Reverse transcriptase PCR amplification of ^{13}C-labeled 16S rRNA can then be utilized to directly link substrate utilization with the consuming organisms (46, 47).

Metagenomic Analysis

An alternative approach to soil microbial community analysis is cloning of the soil "metagenome," defined as the genomes of the total microbiota found in soils (29). Several steps are required to perform metagenomic analysis: isolation of high-quality DNA from soil, cloning into an appropriate vector, and screening of clones. Clones are screened by using function-driven or sequence-driven analysis (28). In the sequence-driven analysis, functional or phylogenetic markers of interest can be targeted by standard molecular biology techniques. A more inclusive approach can be taken, such as obtaining the complete sequences of all inserts (81). This approach is expensive due to the cost of sequencing large numbers of clones and corresponding data analysis, although advances in high-throughput sequence analysis and data handling would clearly benefit this approach.

In the function-driven approach, clones are screened for specific activities such as antibiotic resistance or production or desired enzymatic activities. The main limitation of this approach is the requirement for heterologous gene expression in a laboratory host such as *Escherichia coli*.

An inventive method of screening clone libraries for catabolic genes from uncultured microorganisms involves the use of "substrate-induced gene expression screening" (79). This approach is based on catabolic gene expression that is induced by substrates of the metabolic pathway. A reporter system (the *gfp* gene encoding green fluorescent protein) combined with fluorescence-activated cell sorting is employed, making it a high-throughput method. This is a potentially powerful approach since various combinations and types of substrates could be tested.

We acknowledge support from the National Science Foundation (DEB 00778368) and the Department of the Interior during the writing of this chapter.

REFERENCES

1. **Altschul, S. F., W. Gish, W. Miller, E. W. Myers, and D. J. Lipman.** 1990. Basic local alignment search tool. *J. Mol. Biol.* **215:**403–410.
2. **Amann, R. I., W. Ludwig, and K.-H. Schleifer.** 1995. Phylogenetic identification and in situ detection of individual microbial cells without cultivation. *Microbiol. Rev.* **59:**143–169.
3. **Atlas, R. M., and R. Bartha.** 1993. *Microbial Ecology: Fundamentals and Applications*, 3rd ed. Benjamin/Cummings Co., New York, N.Y.
4. **Bakken, L. R., and V. Lindahl.** 1995. Recovery of bacterial cells from soil, p. 9–27. *In* J. D. van Elsas and J. T. Trevors (ed.), *Nucleic Acids in the Environment: Methods and Applications*. Springer-Verlag, Heidelberg, Germany.
5. **Bligh, E. G., and W. J. Dyer.** 1959. A rapid method of total lipid extraction and purification. *Can. J. Biochem. Physiol.* **37:**911–917.
6. **Bobbie, R. J., and D. C. White.** 1980. Characterization of benthic microbial community structure by high-resolution gas chromatography of fatty acid methyl esters. *Appl. Environ. Microbiol.* **39:**1212–1222.
7. **Boschker, H. T. S., S. C. Nold, P. Wellsbury, D. Bos, W. de Graaf, R. Pel, R. J. Parkes, and T. E. Cappenberg.** 1998. Direct linking of microbial populations to specific biogeochemical processes by ^{13}C-labelling of biomarkers. *Nature* **392:**801–804.
8. **Buyer, J. S.** 1995. A soil and rhizosphere microorganism isolation and enumeration medium that inhibits *Bacillus mycoides*. *Appl. Environ. Microbiol.* **61:**1839–1842.
9. **Cano, R. J., M. J. Torres, R. E. Klem, J. C. Palomares, and J. Casadesus.** 1992. Detection of salmonellas by DNA hybridization with a fluorescent alkaline phosphatase substrate. *J. Appl. Bacteriol.* **72:**393–399.
10. **Castro, H., S. Newman, K. R. Reddy, and A. V. Ogram.** 2005. Distribution and stability of sulfate reducing prokaryotic and hydrogenotrophic methanogenic assemblages in nutrient-impacted regions of the Florida Everglades. *Appl. Environ. Microbiol.* **71:**2695–2704.
11. **Castro, H., K. R. Reddy, and A. Ogram.** 2002. Composition and function of sulfate-reducing prokaryotes in eutrophic and pristine areas of the Florida Everglades. *Appl. Environ. Microbiol.* **68:**6129–6137.
12. **Castro, H. F., A. Ogram, and K. R. Reddy.** 2004. Phylogenetic characterization of methanogenic assemblages in eutrophic and oligotrophic areas of the Florida Everglades. *Appl. Environ. Microbiol.* **70:**6559–6568.
13. **Connon, S. A., and S. J. Giovannoni.** 2002. High-throughput methods for culturing microorganisms in very-low-nutrient media yield diverse new marine isolates. *Appl. Environ. Microbiol.* **68:**3878–3885.
14. **de Bruijn, F. J.** 1992. Use of repetitive (repetitive extragenic palindromic and enterobacterial repetitive intergenic consensus) sequences and the polymerase chain reaction to fingerprint the genomes of *Rhizobium meliloti* isolates and other soil bacteria. *Appl. Environ. Microbiol.* **58:**2180–2187.
15. **Dembitsky, V. M., E. E. Shubina, and A. G. Kashin.** 1992. Phospholipid and fatty acid composition of some basidiomycetes. *Phytochemistry* **31:**845–849.
16. **Dunbar, J., L. O. Ticknor, and C. R. Kuske.** 2000. Assessment of microbial diversity in four southwestern United States soils by 16S rRNA gene terminal restriction fragment analysis. *Appl. Environ. Microbiol.* **66:**2943–2950.
17. **Farrelly, V., F. A. Rainey, and E. Stackebrandt.** 1995. Effect of genome size and *rrn* gene copy number on PCR amplification of 16S rRNA genes from a mixture of bacterial species. *Appl. Environ. Microbiol.* **61:**2798–2801.
18. **Felske, A., A. D. L. Akkermans, and W. M. De Vos.** 1998. Quantification of 16S rRNAs in complex bacterial communities by multiple competitive reverse transcription PCR in temperature gradient gel electrophoresis fingerprints. *Appl. Environ. Microbiol.* **64:**4581–4587.
19. **Fisher, M. M., and E. W. Triplett.** 1999. Automated approach for ribosomal intergenic spacer analysis of microbial diversity and its application to freshwater bacterial communities. *Appl. Environ. Microbiol.* **65:**4630–4636.
20. **Frostegard, A., E. Baath, and A. Tunlid.** 1993. Shift in the structure of soil microbial communities in limed forests as revealed by phospholipid fatty acid analysis. *Soil Biol. Biochem.* **25:**723–730.
21. **Gans, J., M. Wolinsky, and J. Dunbar.** 2005. Computational improvements reveal great bacterial diversity and high metal toxicity in soil. *Science* **309:**1387–1390.
22. **Ginige, M. P., P. Hugenholtz, H. Daims, M. Wagner, J. Keller, and L. L. Blackall.** 2004. Use of stable-isotope probing, full-cycle rRNA analysis, and fluorescence in situ hybridization-microautoradiography to study a methanol-fed denitrifying microbial community. *Appl. Environ. Microbiol.* **70:**588–596.
23. **Giovannoni, S. J., T. B. Britschgi, C. L. Moyer, and K. G. Field.** 1990. Genetic diversity in Sargasso Sea bacterioplankton. *Nature* (London) **345:**60–63.
24. **Guckert, J. B., M. A. Hood, and D. C. White.** 1986. Phospholipid ester-linked fatty acid profile changes during nutrient deprivation of *Vibrio cholerae*: increases in the

trans/cis ratio and proportions of cyclopropyl fatty acids. *Appl. Environ. Microbiol.* **52:**794–801.

25. **Haack, S. K., H. Garchow, D. A. Odelson, L. J. Forney, and M. J. Klug.** 1994. Accuracy, reproducibility, and interpretation of fatty acid methyl ester profiles of model bacterial communities. *Appl. Environ. Microbiol.* **60:**2483–2493.
26. **Hadrys, H., M. Balick, and B. Schierwater.** 1992. Applications of random amplified polymorphic DNA (RAPD) in molecular ecology. *Mol. Ecol.* **1:**55–63.
27. **Haldeman, D. L., and P. S. Amy.** 1993. Diversity within a colony morphotype: implications for ecological research. *Appl. Environ. Microbiol.* **59:**933–935.
28. **Handelsman, J.** 2004. Metagenomics: application of genomics to uncultured microorganisms. *Microbiol. Mol. Biol. Rev.* **68:**669–685.
29. **Handelsman, J., M. R. Rondon, S. F. Brady, J. Clardy, and R. M. Goodman.** 1998. Molecular biological access to the chemistry of unknown soil microbes: a new frontier for natural products. *Chem. Biol.* **5:**R245–R249.
30. **Heck, K. L., G. van Belle, and D. Simberloff.** 1975. Explicit calculation of the rarefaction diversity measurement and the determination of sufficient sample size. *Ecology* **56:**1459–1461.
31. **Hutchens, E., S. Radajewski, M. G. Dumont, I. R. McDonald, and J. C. Murrell.** 2004. Analysis of methanotrophic bacteria in Movile Cave by stable isotope probing. *Environ. Microbiol.* **6:**111–120.
32. **Janssen, P. H., P. Yates, B. Grinton, P. Taylor, and M. Sait.** 2003. Improved culturability of soil bacteria and isolation in pure culture of novel members of the divisions *Acidobacteria, Actinobacteria, Proteobacteria,* and *Verrucomicrobia. Appl. Environ. Microbiol.* **68:**2391–2396.
33. **Joseph, S. J., P. Hugenhotlz, P. Sangwan, C. Osborne, and P. Janssen.** 2003. Laboratory cultivation of widespread and previously uncultured soil bacteria. *Appl. Environ. Microbiol.* **69:**7210–7215.
34. **Kaeberlein, T., K. Lewis, and S. S. Epstein.** 2002. Isolating "uncultivable" microorganisms in pure culture in a simulated natural environment. *Science* **296:**1127–1129.
35. **Kennedy, A. C.** 1994. Carbon utilization and fatty acid profiles for characterization of bacteria, p. 543–553. *In* R. W. Weaver, S. Angle, P. Bottomley, D. Bezdicek, S. Smith, A. Tabatabai, and A. Wollum (ed.), *Methods of Soil Analysis,* part 2. *Microbiological and Biochemical Properties.* Soil Science Society of America, Madison, Wis.
36. **Kieft, T. L., D. B. Ringelberg, and D. C. White.** 1994. Changes in ester-linked phospholipid fatty acid profiles of subsurface bacteria during starvation and desiccation in a porous medium. *Appl. Environ. Microbiol.* **60:**3292–3299.
37. **Lane, D. J.** 1991. 16S/23SrRNA sequencing, p. 115–175. *In* E. Stackebrandt and M. Goodfellow (ed.), *Nucleic Acid Techniques in Bacterial Systematics.* John Wiley & Sons, New York, N.Y.
38. **Liesack, W., and E. Stackebrandt.** 1992. Occurrence of novel groups of the domain *Bacteria* as revealed by analysis of genetic material isolated from an Australian terrestrial environment. *J. Bacteriol.* **174:**5072–5078.
39. **Lin, J. L., S. Radajewski, B. T. Eshinimaev, Y. A. Trotsenko, I. R. McDonald, and J. C. Murrell.** 2004. Molecular diversity of methanotrophs in Transbaikal soda lake sediments and identification of potentially active populations by stable isotope probing. *Environ. Microbiol.* **6:**1049–1060.
40. **Liu, W. T., T. L. Marsh, H. Cheng, and L. J. Forney.** 1997. Characterization of microbial diversity by determining terminal restriction fragment length polymorphisms of genes encoding 16S rRNA. *Appl. Environ. Microbiol.* **63:** 4516–4522.
41. **Louws, F. J., D. W. Fulbright, C. T. Stephens, and F. J. de Bruijn.** 1994. Specific genomic fingerprints of phytopathogenic *Xanthomonas* and *Pseudomonas* pathovars and strains generated with repetitive sequences and PCR. *Appl. Environ. Microbiol.* **60:**2286–2295.
42. **Lu, Y. H., T. Lueders, M. W. Friedrich, and R. Conrad.** 2005. Detecting active methanogenic populations on rice roots using stable isotope probing. *Environ. Microbiol.* **7:** 326–336.
43. **Ludwig, W., O. Strunk, R. Westram, L. Richter, H. Meier, Yadhukumar, A. Buchner, T. Lai, S. Steppi, G. Jobb, W. Forster, I. Brettske, S. Gerber, A. W. Ginhart, O. Gross, S. Grumann, S. Hermann, R. Jost, A. Konig, T. Liss, R. Lussmann, M. May, B. Nonhoff, B. Reichel, R. Strehlow, A. Stamatakis, N. Stuckmann, A. Vilbig, M. Lenke, T. Ludwig, A. Bode, and K. H. Schleifer.** 2004. ARB: a software environment for sequence data. *Nucleic Acids Res.* **32:**1363–1371.
44. **Lueders, T., B. Pommerenke, and M. W. Friedrich.** 2004. Stable-isotope probing of microorganisms thriving at thermodynamic limits: syntrophic propionate oxidation in flooded soil. *Appl. Environ. Microbiol.* **70:**5778–5786.
45. **Malik, M., J. Kain, C. Pettigrew, and A. Ogram.** 1994. Purification and molecular analysis of microbial DNA from compost. *J. Microbiol. Methods* **20:**183–196.
46. **Manefield, M., A. S. Whiteley, R. I. Griffiths, and M. J. Bailey.** 2002. RNA stable isotope probing, a novel means of linking microbial community function to phylogeny. *Appl. Environ. Microbiol.* **68:**5367–5373.
47. **Manefield, M., A. S. Whiteley, N. Ostle, P. Ineson, and M. J. Bailey.** 2002. Technical considerations for RNA-based stable isotope probing: an approach to associating microbial diversity with microbial community function. *Rapid Commun. Mass Spectrom.* **16:**2179–2183.
48. **Mayr, C., A. Winding, and N. B. Hendriksen.** 1999. Community level physiological profile of soil bacteria unaffected by extraction method. *J. Microbiol. Methods* **36:**29–33.
49. **McSpadden Gardener, B. B., K. L. Schroeder, S. E. Kalloger, J. M. Raaijmakers, L. S. Thomashow, and D. M. Weller.** 2000. Genotypic and phenotypic diversity of *phlD*-containing *Pseudomonas* strains isolated from the rhizosphere of wheat. *Appl. Environ. Microbiol.* **66:**1939–1946.
50. **Moss, C. W.** 1981. Gas-liquid chromatography as an analytical tool in microbiology. *J. Chromatogr.* **203:**337–347.
51. **Moyer, C. L., J. M. Tiedje, F. C. Dobbs, and D. M. Karl.** 1996. A computer-simulated restriction fragment length polymorphism analysis of bacterial small-subunit rRNA genes: efficacy of selected tetrameric restriction enzymes for studies of microbial diversity in nature. *Appl. Environ. Microbiol.* **62:**2501–2507.
52. **Muyzer, G., E. C. de Waal, and A. G. Uitterlinden.** 1993. Profiling of complex microbial populations by denaturing gradient gel electrophoresis analysis of polymerase chain reaction-amplified genes coding for 16S rRNA. *Appl. Environ. Microbiol.* **59:**695–700.
53. **Muyzer, G., and K. Smalla.** 1998. Application of denaturing gradient gel electrophoresis (DGGE) and temperature gradient gel electrophoresis (TGGE) in microbial ecology. *Antonie Leeuwenhoek* **73:**127–141.
54. **Myers, R. M., T. Maniatis, and L. S. Lerman.** 1987. Detection and localization of single base changes by denaturing gradient gel electrophoresis. *Methods Enzymol.* **155:**501–527.
55. **Nei, M., and W. H. Li.** 1979. Mathematical model for studying genetic variation in terms of restriction endonucleases. *Proc. Natl. Acad. Sci. USA* **76:**5269–5273.
56. **Nunan, N., K. Wu, I. M. Young, J. W. Crawford, and K. Ritz.** 2003. Spatial distribution of bacterial communities and their relationships with the micro-architecture of soil. *FEMS Microbiol. Ecol.* **44:**203–215.

57. **Ogram, A.** 2000. Soil molecular microbial ecology at age 20: methodological challenges for the future. *Soil Biol. Biochem.* **32:**1499–1504.
58. **Ogram, A. V., and D. F. Bezdicek.** 1994. Nucleic acid probes, p. 543–553. *In* R. W. Weaver, S. Angle, P. Bottomley, D. Bezdicek, S. Smith, A. Tabatabai, and A. Wollum (ed.), *Methods of Soil Analysis*, part 2. *Microbiological and Biochemical Properties*. Soil Science Society of America, Madison, Wis.
59. **Oline, D. K., and M. C. Grant.** 2002. Scaling patterns of biomass and soil properties: an empirical analysis. *Landscape Ecol.* **17:**13–26.
60. **Osborn, A. M., E. R. B. Moore, and K. N. Timmis.** 2000. An evaluation of terminal-restriction fragment length polymorphism (T-RFLP) analysis for the study of microbial community structure and dynamics. *Environ. Microbiol.* **2:**39–50.
61. **Penner, G. A., A. Bush, R. Wise, W. Kim, L. Domier, K. Kasha, A. Laroche, G. Scoles, S. J. Molnar, and G. Fedak.** 1993. Reproducibility of random amplified polymorphic DNA (RAPD) analysis among laboratories. *PCR Methods Appl.* **2:**341–345.
62. **Radajewski, S., P. Ineson, N. R. Parekh, and J. C. Murrell.** 2000. Stable-isotope probing as a tool in microbial ecology. *Nature* **403:**646–649.
63. **Radajewski, S., I. R. McDonald, and J. C. Murrell.** 2003. Stable-isotope probing of nucleic acids: a window to the function of uncultured microorganisms. *Curr. Opin. Biotechnol.* **14:**296–302.
64. **Rainey, F. A., N. L. Ward-Rainey, P. H. Janssen, H. Hippe, and E. Stackebrandt.** 1996. *Clostridium paradoxum* DSM 7308T contains multiple 16S rRNA genes with heterogeneous intervening sequences. *Microbiology* **142:**2087–2095.
65. **Raskin, L., W. C. Capman, M. D. Kane, B. E. Rittmann, and D. A. Stahl.** 1996. Critical evaluation of membrane supports for use in quantitative hybridizations. *Appl. Environ. Microbiol.* **62:**300–303.
66. **Robison-Cox, J. F., M. M. Bateson, and D. M. Ward.** 1995. Evaluation of nearest-neighbor methods for detection of chimeric small-subunit rRNA sequences. *Appl. Environ. Microbiol.* **61:**1240–1245.
67. **Sait, M., P. Hugenholtz, and P. Janssen.** 2002. Cultivation of globally distributed soil bacteria from phylogenetic lineages previously only detected in cultivation-independent surveys. *Environ. Microbiol.* **4:**654–666.
68. **Sambrook, J., E. Fritsch, and T. Maniatis.** 1989. *Molecular Cloning: a Laboratory Manual*, 2nd ed. Cold Spring Harbor Laboratory Press, Cold Spring Harbor, N.Y.
69. **Sayler, G. S., and A. C. Layton.** 1990. Environmental application of nucleic acid hybridization. *Annu. Rev. Microbiol.* **44:**625–648.
70. **Sayler, G. S., M. S. Shields, E. Tedford, A. Breen, S. Hooper, K. Sirotkin, and J. Davis.** 1985. Application of DNA-DNA colony hybridization to the detection of catabolic genotypes in environmental samples. *Appl. Environ. Microbiol.* **49:**1295–1303.
71. **Seeley, H. W., and P. VanDemark.** 1981. *Microbes in Action: a Laboratory Manual of Microbiology*, 3rd ed. W. H. Freeman and Co., New York, N.Y.
72. **Stackebrandt, E., and B. M. Goebel.** 1994. Taxonomic note: a place for DNA-DNA reassociation and 16S rRNA sequence analysis in the present species definition in bacteriology. *Int. J. Syst. Bacteriol.* **44:**846–849.
73. **Stevenson, B. S., S. A. Eichorst, J. T. Wertz, T. M. Schmidt, and J. A. Breznak.** 2004. New strategies for cultivation and detection of previously uncultured microbes. *Appl. Environ. Microbiol.* **70:**4748–4755.
74. **Suzuki, M. T., and S. J. Giovannoni.** 1996. Bias caused by template annealing in the amplification of mixtures of 16S rRNA genes by PCR. *Appl. Environ. Microbiol.* **62:**625–630.
75. **Swofford, D. L.** 1999. PAUP: Phylogenetic Analysis Using Parsimony, version 4.0. Sinauer Associates, Sunderland, Mass.
76. **Teske, A., C. Wawer, G. Muyzer, and N. B. Ramsing.** 1996. Distribution of sulfate-reducing bacteria in a stratified fjord (Mariager Fjord, Denmark) as evaluated by most-probable-number counts and denaturing gradient gel electrophoresis of PCR-amplified ribosomal DNA fragments. *Appl. Environ. Microbiol.* **62:**1405–1415.
77. **Torsvik, V., J. Goksøyr, and F. L. Daae.** 1990. High diversity in DNA of soil bacteria. *Appl. Environ. Microbiol.* **56:**782–787.
78. **Torsvik, V., K. Salte, R. Sørheim, and J. Goksøyr.** 1990. Comparison of phenotypic diversity and DNA heterogeneity in a population of soil bacteria. *Appl. Environ. Microbiol.* **56:**776–781.
79. **Uchiyama, T., T. Abe, T. Ikemura, and K. Watanabe.** 2005. Substrate-induced gene-expression screening of environmental metagenome libraries for isolation of catabolic genes. *Nat. Biotechnol.* **23:**88–93.
80. **Van de Peer, Y., and R. Wachter.** 1997. Construction of evolutionary distance trees with TREECON for Windows: accounting for variation in nucleotide substitution rate among sites. *Comput. Appl. Biosci.* **13:**227–230.
81. **Venter, J. C., K. Remington, J. F. Heidelberg, A. L. Halpern, D. Rusch, J. A. Eisen, D. Y. Wu, I. Paulsen, K. E. Nelson, W. Nelson, D. E. Fouts, S. Levy, A. H. Knap, M. W. Lomas, K. Nealson, O. White, J. Peterson, J. Hoffman, R. Parsons, H. Baden-Tillson, C. Pfannkoch, Y. H. Rogers, and H. O. Smith.** 2004. Environmental genome shotgun sequencing of the Sargasso Sea. *Science* **304:**66–74.
82. **Vestal, J. R., and D. C. White.** 1998. Lipid analysis in microbial ecology. *Bioscience* **39:**535–541.
83. **Vilgalys, R., and M. Hester.** 1990. Rapid genetic identification and mapping of enzymatically amplified ribosomal DNA from several *Cryptococcus* species. *J. Bacteriol.* **172:**4238–4246.
84. **Voordouw, G., Y. Shen, C. S. Harrington, A. J. Telang, T. R. Jack, and D. W. S. Westlake.** 1993. Quantitative reverse sample genome probing of microbial communities and its application to oil field production waters. *Appl. Environ. Microbiol.* **59:**4101–4114.
85. **Wackett, L. P.** 2004. Stable isotope probing in biodegradation research. *Trends Biotechnol.* **22:**153–154.
86. **Wellington, E. M. H., A. Berry, and M. Krsek.** 2003. Resolving functional diversity in relation to microbial community structure in soil: exploiting genomics and stable isotope probing. *Curr. Opin. Biotechnol.* **6:**295–301.
87. **White, D. C., R. J. Bobbie, J. D. King, J. Nickels, and P. Amoe.** 1979. Lipid analysis of sediments for microbial biomass and community structure, p. 87–103. *In* C. D. Litchfield and P. L. Seyfried (ed.), *Methodology for Biomass Determinations and Microbial Activities in Sediments*. ASTM STP 673. American Society for Testing and Materials, Philadelphia, Pa.
88. **Woese, C. R., O. Kandler, and M. L. Wheelis.** 1990. Towards a natural system of organisms: proposal for the domains *Archaea*, *Bacteria*, and *Eucarya*. *Proc. Natl. Acad. Sci. USA* **87:**4576–4579.
89. **Yang, J., R. D. Hammer, and R. W. Blanchar.** 1995. Microscale pH spatial distribution in the Ap horizon of Mexico silt loam. *Soil Sci.* **160:**371–375.
90. **Zelles, L., and Q. Y. Bai.** 1993. Fractionation of fatty acids derived from soil lipids by solid phase extraction and their quantitative analysis by GC-MS. *Soil Biol. Biochem.* **25:**495–507.
91. **Zengler, K., G. Toledo, M. Rappe, J. Elkins, E. J. Mathur, J. M. Short, and M. Keller.** 2002. Cultivating the uncultured. *Proc. Natl. Acad. Sci. USA* **99:**15681–15686.

Microarrays: Design and Use for Agricultural and Environmental Applications

SCOT E. DOWD, HIROSHI ISHIZAKI, AND JEANETTE THURSTON ENRIQUEZ

54

Environmental microorganisms are diverse and typically exist in dynamic community structures that are continually changing based upon many environmental factors, including nutrient levels, temperature, water content, pH, and oxygen status. The emerging microarray technology has already been used in the field of environmental microbiology to characterize microbial communities, uncover metabolic activities of specific communities, understand interactions between microorganisms and their environment, and identify specific microorganisms (6, 12, 29, 60). This chapter provides an introduction to the technology, with focus on the field of environmental microbiology.

WHAT ARE MICROARRAYS?

One of the first reports on the use of microarrays was published in *Science* in 1995 (52) and detailed a high-capacity system that could be used to monitor many genes in parallel. Microarrays are often considered the mid-1990s revolution in molecular biology that followed the very important mid-1980s development of PCR (37). In essence, and in its most common form, a microarray is a silicon- or glass-based solid surface onto which has been deposited, in defined and isolated locations, individual DNA-based elements (usually in the form of spots), each of which typically corresponds to an individual gene. Microarrays are high-density, high-throughput methods for analysis of the presence and expression of specific genes. Spots on arrays are typically 100 to 300 μm in size and are spaced about the same distance apart. Each spot is essentially used as a gene probe. By the use of this technique, arrays consisting of 30,000 different gene probes can fit onto the surface of a conventional microscope slide.

The primary use of microarrays is as a method for comparative quantification of gene expression levels between a control sample and a treatment sample. Historically, there are many other methodologies that have been used for the quantification of gene expression levels. These include Northern blots, S1 nuclease protection assays, differential display, sequencing of cDNA libraries, and serial analysis of gene expression. The basis for microarray technology likely has its origins in the techniques of Edward Southern, who described hybridization between nucleic acids when one of them is immobilized on a solid support (traditionally filter paper). Thus, Southern hybridization is a direct ancestor of microarray technology. There are two primary types of microarrays, cDNA-based and oligonucleotide-based microarrays. In this chapter, we focus primarily on oligonucleotide-based technology because of the specialized problems encountered in agricultural and environmental microbiology.

We can illustrate the concept of microarrays schematically. Figure 1 provides a schematic representation of a microarray experiment using the popular two-dye cross-hybridization technique. The goal is to derive an expression profile for a set of genes relative to that of a control sample. Thus, if a certain gene in the sample of interest is up-regulated or present in greater abundance because of the experimental treatment or specific environmental conditions, we will see a higher fluorescent signal of this gene than in the control sample. As an easy-to-understand example, if we have 45 separate genes (e.g., those noted to confer tetracycline resistance in bacteria) and we are seeking to evaluate the expression or presence of any of these genes in a soil sample after amendment with tetracycline, we might make a 9-element by 5-element array containing probes targeting each of these genes. Each element on the array would correspond to one of the 45 known tetracycline resistance genes. Total RNA can then be extracted from representative soil samples before and after amendment with tetracycline. With the use of two fluorescent dyes, Cy3 and Cy5, which are the most commonly utilized dyes in microarray experiments, mRNAs from control samples are converted to Cy3 (green)-labeled cDNAs, while the treatment cDNAs are labeled with Cy5 (red). These two independently labeled nucleic acid samples are then mixed together in equal concentrations and hybridized to the same array. The probes on the array are present at concentrations that are severalfold in excess of the labeled cDNAs, which results in noncompetitive binding of the targets to the arrayed probes. After hybridization and washing, the slide is scanned, using two different lasers adjusted to two different wavelengths. These wavelengths correspond to the two dyes used. If we consider that Cy3 is typically considered green and Cy5 is considered red, and if the treatment sample has a gene, *x*, that is up-regulated and is labeled with Cy3, then the "gene *x*" spot (element) on the array will appear more green. Thus, in our example, if the treatment sample has specific genes in higher abundance, these targets have a better chance through competitive exclusion to bind to more hybridization

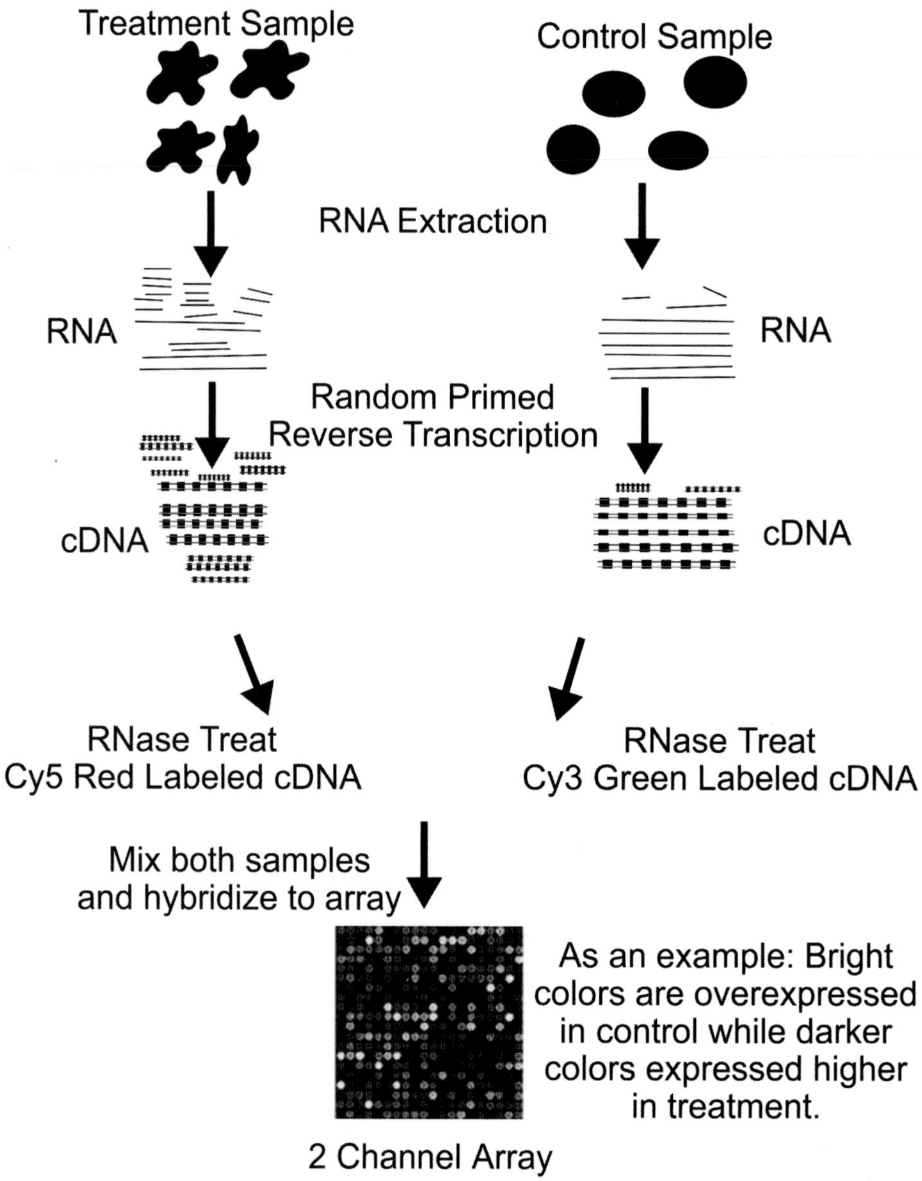

FIGURE 1 Schematic representation of a microarray experiment using the two-dye cross-hybridization technique. In this experiment, the expression profiles of a control cell and a cell of the same genetic lineage that has been exposed to a treatment are compared. The RNAs from the control are labeled during cDNA formation with Cy5, which is artificially colored red (note that the figure is printed in black and white and meant only to provide a schematic concept of the experimental procedure). The RNAs from the treated cells are extracted and labeled with Cy3 (green) during cDNA synthesis. After being labeled, the two samples are cohybridized to the microarray. Those genes that are expressed in more abundance in the control sample (or down-regulated during treatment) will show up as green spots. Conversely, those genes that are expressed in more abundance in the treatment sample will be visualized as red spots. Those genes which are expressed equally in the treatment and control samples will appear as yellow spots. In this way, one can see how the expression of an entire transcriptome responds to experimental treatments.

sites on the array. This results in a measurement of the ratio of transcript levels for each gene represented on the array.

Apart from this simplified overview of a microarray experiment, this chapter provides basic information intended to be used as a primer on the development and use of microarrays in the field of environmental microbiology. Although not exhaustive in its descriptions and discussions, this chapter touches on the major issues and pitfalls of this technology. This chapter is intended to provide enough introductory methodologies and discussions to act as a framework for the development of microarray technology for environmental applications. As noted above, most of our discussions will revolve around oligonucleotide-based array techniques. The oligonucleotide-based format will likely be of most use for applications in the field of environmental microbiology. This chapter also provides overviews of laboratory protocols, including RNA extraction, reverse transcription of mRNA into labeled cDNA, hybridization of this labeled cDNA to the microarray, scanning of the hybridized slide to generate images, image processing, and bioinformatic analyses of the generated data. Various kits, commercial hardware, and software packages are references that are useful in these processes. Additionally, the design of microarrays and examples of how to set up microarray studies are discussed.

HOW ARE MICROARRAYS DESIGNED?

The fabrication of a microarray must take into account budgetary concerns, probe design, probe size, the type of slide, and the type of spotting solution. Budgetary concerns are obviously the most important consideration in most cases. In general, the software-based design of the probe is relatively inexpensive or, in most cases, free. However, the actual synthesis of oligonucleotide probes can cost hundreds or tens of thousands of dollars, depending on the scope of the project, the experimental design, and the type of data sought. For instance, the synthesis of each 45-mer oligonucleotide can range from $6 to $12, depending on the source and the size of the order. Generally, most commercial companies will negotiate a better price if a large number of oligonucleotides are being ordered. As an example of the potential cost to synthesize an array against the genome of *Pseudomonas putida*, which has roughly 5,350 genes, it would cost $30,000 to $60,000, depending on the length of the probes. This investment should, however, provide enough oligonucleotides that, using modern technologies, a researcher could fabricate up to 10,000 slides. Thus, multiple researchers interested in the same organism can generate enough capital to buy the original microarray, with each researcher having access to as many slides as needed.

The Microarray Probe

The size of the oligonucleotide probes is important as well. The general consensus is that the best results are obtained with oligonucleotides that are approximately 70 nucleotides long. Good success utilizing smaller oligonucleotides of approximately 40 bp has also been achieved with modern surface and spotting chemistries and by the commercial company Affymetrix, which has great success utilizing much smaller oligonucleotides in the range of 20 bp. The size obviously influences the cost of each oligonucleotide, but more importantly, only oligonucleotides of the highest quality should be used for microarraying. Ramdas et al. (45) reviewed and analyzed the effects of oligonucleotide length and hybridization. Their review noted that longer oligonucleotides (70-mers) had more efficient hybridization (22). However, another study evaluated the lengths of PCR products used as probes and found that hybridization efficiency reaches a plateau at 712 bases for PCR products and decreases if the length of the PCR product increases (59).

Design of Probes

In general, oligonucleotide microarrays are designed by using some of the basic principles of primer and probe design for PCR and Southern hybridization. Probes can range from about 20 to 80 bp and are required to be very specific for the target DNA, without notable regions of potential cross-hybridization. Other parameters, such as secondary structure, melting temperature, and specificity, are also important. There are software applications that are useful in the design of microarray oligonucleotides (49, 50, 54, 63, 70). In general, the utility of software applications in the design of probes is limited by the quality of data entered into the software system.

Once a set of probes has been designed for the project (e.g., a targeted array containing all of the genes involved in biodegradation of hexavalent chromium in soil by *P. putida*), they can be synthesized by various commercial companies, such as Integrated DNA Technologies or Operon. As important as the length of the probe is, the quality of the oligonucleotide is also important. Current oligonucleotide synthesis techniques have limitations in the synthesis of long oligonucleotides. In general, the efficiency of generating full-length oligonucleotides decreases as the length increases (34). If a 100-bp oligonucleotide is synthesized, the concentration of full-length oligonucleotides produced can be as low as 37%. The oligonucleotides need not be modified but should be column purified. If probes are ordered in 96- or 384-well plates, the concentration of oligonucleotides should be normalized across all the wells to save a lot of time and labor. Both Operon and IDT-DNA will normalize concentrations as part of their service. Once oligonucleotides have been synthesized, they usually arrive lyophilized. The proper choice of reagent in which to resuspend the oligonucleotides is important and dependent on the size of the oligonucleotide, as noted below.

Probe Spotting or Printing

More effective use of small oligonucleotides has been facilitated by the recent development of new chemistries, such as epoxide coating of slides and the use of optimized spotting solutions. The lower cost of small oligonucleotides enhances their appeal by allowing more of a transcriptome to be covered at the same cost. To address this, companies such as Corning Inc. have released epoxide-coated slides along with a specific spotting solution optimized for smaller oligonucleotides. It should be emphasized that the spotting solution must be optimized along with the slide chemistry and that not all spotting solutions will work with all slide coatings. Using these newer technologies allows for efficient use of small oligonucleotide arrays (43- to 50-mers). Table 1 provides a breakdown of what types of slide-coating chemistries and spotting solutions are useful.

Always follow the microarray slide manufacturer's recommendations and directions for choice of compatible spotting solutions and spotting protocols. As noted by Corning Inc., the chemical and physical properties of the spotting solution greatly influence DNA retention, spot morphology, and hybridization efficiency. A 50% dimethyl sulfoxide (DMSO) solution tends to be stable but to result in a larger spot size when printing, thereby limiting the achievable feature density. For arrays that do not require 20,000 to 30,000 elements, this is an excellent choice. Solutions containing DMSO may eventually result in DNA aggregation, especially if the concentration of DMSO exceeds 75%. This is especially problematic with repeated use of the same source plates. Aqueous sodium citrate (SSC)- and $NaPO_4$-containing media will salt out during evaporation, which may cause quill pins to clog. This situation can lead to variability in DNA concentration among source wells. An SSC-betaine formulation provides good stability and spot size; however, betaine limits the shelf life of the microarray (Corning Inc. technical website and personal communication).

Once the slides and spotting solution have been chosen based upon the oligonucleotide length, the next step is to resuspend the oligonucleotides. The goal is to obtain the

TABLE 1 Common microarray slide coatings and associated spotting solutions

Slide surface	Recommended printed content	Binding interaction	Recommended printing solution
Epoxide-coated slides	Useful for short oligonucleotides (30-mer to 50-mer), with or without amino acid modifications	Covalent	Epoxide-specific printing solution, e.g., Pronto! epoxide spotting solution
Amino-silane-based slides (e.g., GAPS, UltraGAPS, and GAPSII)	Useful for cDNA, PCR products, and long oligonucleotides (>50-mer)	Ionic	Pronto! Universal spotting solution, 15 to 50% DMSO, 3 × SSC, 3 × SSC supplemented with 1.5 M betaine, or 150 mM $NaPO_4$ (pH 8.5)

proper concentration of oligonucleotides to be used for spotting. In general, following the manufacturer's recommendations for the spotting solution and slide chemistry is recommended. If too little oligonucleotide is applied in each spot, the sensitivity of the array is reduced dramatically (45), while too much will result in smearing and tailing of the spot during spotting and hybridization. Typically, once the oligonucleotides have been resuspended, it is common practice to run a test printing of a small number of slides prior to a full printing run of 50 or 100 slides. Once the test printing of a few dozen spots on a couple of slides is complete, the oligonucleotides can be visualized by staining with a dye, such as SYBR 555, which binds to single-stranded DNA. Invitrogen offers the Paragon DNA microarray quality control stain kit, which allows microarrays to be checked for DNA spotting consistency. An example of two slides, one of which did not have good spots, is provided in Fig. 2. The following are general concerns. Avoid direct contact with the surface of the slide to be printed. Only the print pins and processing solutions should touch the print area to avoid contamination and abrasion of the coating. Also, use the slides in a clean environment, since particles falling onto the slide surface may cause defects in the printed array and possible nuclease contamination. As described below, self-contained printing environments are highly recommended to prevent such contamination.

Basic Microarray Facility Equipment

In addition to the high cost of building a basic microarray laboratory, a good deal of equipment needs to be purchased. Many universities and research organizations have core facilities that handle microarrays, which can help small academic laboratories. There are also commercial entities that provide various specialized microarray services, such as printing and scanning, and even those that perform complete microarray studies. The following are some of the full-service companies that can assist laboratories with limited resources or limited access to equipment: TeleChem International Inc., Operon, Nimblegen Systems Inc., CombiMatrix, and Clontech.

The following provides a short list of laboratory equipment utilized in a basic microarray laboratory. This list provides the primary considerations for setting up a new microarray laboratory.

- Microarray laser scanner (large cost)
- Microarray printer (large cost, but some laboratories have constructed their own)
- Centrifuges with attachments for microscope slides
- Spectrophotometer for measurement of RNA and DNA
- Real-time thermocyclers (quantitative instruments are typically required for microarray result validation)
- Software for data acquisition (usually supplied with the scanner)
- Software for data processing (many free options are available)

One of the primary costs of a microarray facility is the device, or robot, which performs the printing of the array. Manufacturers of microarray printers include Genomic Solutions and Genetix. In general, a microarray printer is a device that has several integral features. It has a deck that holds microarray slides, a robotic arm controlled by a computer that picks up oligonucleotides and deposits them onto the slide, a deck to hold 96- or 384-well plates containing the probes, a wash station to clean the print heads between samples, a set of print heads, and, usually, environmental control features that modulate humidity and temperature and, most importantly, promote a dust-free environment. Modern microarray printers typically have 48 print heads that pick up a very small volume of DNA from each well of the 96- or 384-well plate and then deposit an even smaller volume onto defined and exact locations on the complete set of microarray slides. In this way, each slide is an exact duplicate, which promotes easier data analyses. The key to printing the microarray is that each spot must be very small, usually no more than 300 μm. Furthermore, the spots are placed onto the slide with great precision, usually no more than 300 μm from each other. Thus, the need for precise placement at this resolution necessitates the use of robotics.

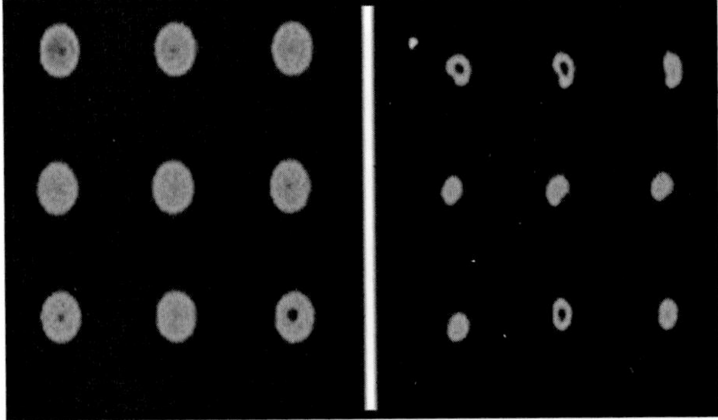

FIGURE 2 Two different slides printed and evaluated by use of a Paragon DNA microarray quality control stain kit. The slide on the left has consistent well-defined spots, while the slide on the right appears to have either too little oligonucleotide applied or an improper chemistry mixture that does not allow the spots to set and fix uniformly. A proper spotting solution must be used with the proper slide chemistry to allow for uniform spotting quality.

A typical microarray slide (1 in. by 3 in.) can usually hold 20,000 to 30,000 spots or more, depending on the printer. Modern printing also utilizes precision print heads (pins) that pick up only a few microliters from the 96- or 384-well plate and distribute this very small volume across hundreds of slides in only a few hours. This prevents needless wasting of expensive oligonucleotides and allows for consistent, reliable, and reproducible spotting across many slides.

MICROARRAY EXPERIMENTS

The general steps of a microarray experiment are as follows. (i) Prepare a DNA chip, using the chosen target DNAs (oligonucleotides, PCR products, etc.). (ii) Generate a hybridization solution containing a mixture of fluorescently labeled cDNAs. (iii) Incubate the hybridization mixture containing fluorescently labeled cDNAs with the DNA chip. (iv) Detect bound cDNAs by using laser technology and store the data in a computer. (v) Analyze the data, using computational methods. (vi) Confirm the results of a representative or statistically significant population of up- and down-regulated genes obtained from the microarray experiments. Confirmation is typically performed by using independent methods, such as quantitative PCR (QPCR) or Northern blotting. QPCR and Northern blotting confirmation can also reduce the number of arrays needed in an analysis because the cost of microarray hybridization is typically much higher. As a general rule, each sample in a microarray experiment is run at least in triplicate, with dye swaps included. QPCR is then used to validate a representative population of the regulated genes.

Careful consideration must be made to obtain enough RNA from a given sample type to allow for duplication or replication of results as well as for dye-swap controls. The minimum amount of total RNA required for standard microarray hybridization is in the range of 5 to 10 μg. Thus, with replicates and validation by QPCR, at least 25 μg of total RNA is needed for each sample. If multiple treatments will be compared to a control, then that much more RNA from the control sample is required to hybridize against each treatment. In general, a microarray experiment is no different from any other scientific investigation. Appropriate positive and negative controls are the foundation of adequate quality control and are needed to provide confidence in the results. Replicates of microarray hybridization runs should be planned to decrease experimental error and to provide statistical power.

EXTRACTING DNA OR RNA FROM SOIL OR MICROBES

DNA Extraction

DNA extraction methods are based on the total nucleic acid content of the microbial community, or community DNA. In the soil environment, one way to obtain community DNA is to separate all cells from the soil and subsequently lyse the cells and extract the nucleic acids. Intact cells may be lysed by heat, chemical, or enzymatic treatment. There are a large number of published protocols detailing various optimized DNA extraction methods (19, 27). The diversity of methods probably reflects the chemical diversity of soil types, and it is unlikely that a single method will ever be suitable for all types. After lysis, the sample undergoes one or more purification processes for subsequent analyses. Cell extraction followed by DNA analysis is still likely to result in a more representative estimate of the total bacterial community than a simple plate dilution on growth medium. A second approach for extraction of community DNA from soil is to lyse the bacterial cells in situ and subsequently extract and purify the DNA. There are several protocols for the extraction of community DNA which have been described previously (3, 20), with most relying on the use of lysozyme or sodium dodecyl sulfate as a lysing agent. Once extracted, the DNA is purified by cesium chloride density centrifugation or the use of commercial purification kits, such as the Soil Master kit (Epicentre Biotechnologies) or the FastDNA spin kit for soil from Qbiogene.

RNA Extraction

RNA typically has a relatively short half-life outside the cell. Compared with methods for DNA extraction, there have been relatively few published protocols for the extraction of RNA from soil (17, 23). Although these techniques have particular advantages in terms of product purity, specialized equipment is required for high-speed centrifugation, and the procedures are often time-consuming. Recently, a rapid method was developed for the direct extraction of DNA and/or RNA from freshwater sediments (35), with the RNA being of sufficient quality for reverse transcriptase PCR without the need for extensive purification. This type of protocol could be adapted for use as an RNA extraction protocol for microarray analysis. The commercial company Qbiogene also makes direct and indirect kits for RNA extraction from soil samples.

Labeling

RNA Labeling

Expression analysis protocols are based on the reverse transcription of mRNAs from either highly purified poly(A) mRNA (eukaryote) or total RNA (eukaryote and prokaryote) extracts. Comprehensive purification of RNA is essential to avoid all contaminating protein, polysaccharide, and other organic material, especially RNases. Most RNA extraction protocols have been developed for the extraction of high-quality RNA by the use of in-house or commercial kits and reagents. Initial protocols for target labeling depend on direct labeling. For direct labeling, reverse transcription is primed with an oligo-(dT) primer, which anneals to the poly(A) tail located at the 3′ end of the vast majority of mammalian mRNAs, and/or a random primer, which is used for cDNA transcription of total RNA from bacterial samples. cDNA transcription takes place in the presence of fluorescently labeled nucleotides (typically the carbocyanine dye Cy3 or Cy5-conjugated dCTP or dUTP). Note that these types of labeled nucleotides are bulky, making their incorporation by the use of standard enzymes very inefficient. Additionally, the rate of incorporation can differ between dyes, potentially resulting in dye biases (74). However, most commercial kits provide adequately uniform labeling if directions and optimization suggestions are carefully observed.

As an alternative to direct labeling, amino allyl labeling, an indirect labeling method, may be used. Indirect labeling circumvents the need to incorporate bulky fluorescent dyes during reverse transcription. In this method, an amino-allyl-modified dUTP is utilized instead of a prelabeled nucleotide. After reverse transcription, the free amine group on the amino allyl-dUTP can be coupled to a reactive *N*-hydroxysuccinimidyl ester fluorescent dye. This technique requires an extra step, but its benefits—including better sensitivity, an absence of dye biases, and decreased cost—seem to be worth the extra effort. There are now several *N*-hydroxysuccinimidyl dyes available, including the standard Cy3 and Cy5 dyes. Another alternative, a commercially available method called 3DNA dendrimer technology (Genisphere Inc.), has been developed. This method is directed toward increasing a predictable and consistent signal strength over that with basic direct dye incorporation without using an RNA amplification step. A range of alternative dyes, such as Alexa fluor (Molecular Probes, Invitrogen Corp.) and Oyster dye (Denovo Biolabels), exhibit enhanced fluorescence and improved photostability. Typical microarray labeling procedures require at least 2 to 4 μg of poly(A)$^+$ RNA or 10 to 50 μg of total RNA, which is equivalent to at least 10^7 cells or many milligrams of tissue per cDNA microarray (76). To obtain the required amounts of RNA from small limited samples, techniques based on isothermal linear RNA polymerase amplification have been developed by Van Gelder, Eberwine, and others (15, 31, 41, 67). In this method, poly(A)$^+$ RNA is converted into unlabeled double-stranded cDNA by reverse transcription with an oligo(dT) primer carrying a bacteriophage T7 promoter. Since prokaryotic mRNA has no stable poly(A) tail, it must be polyadenylated to be a suitable substrate for amplification (MessageAmp II bacteria kit; Ambion) before reverse transcription. The cDNA is then used as a template for in vitro transcription catalyzed by the T7-based RNA polymerase. The in vitro transcripton reaction can be configured to synthesize either labeled or unlabeled antisense RNA that can subsequently be labeled by reverse transcription with fluorescently labeled deoxynucleoside triphosphates. The latter is preferred since RNA polymerase activity is less prone to be influenced by the template sequence or template concentration. Hence, RNA amplification is less subject to bias than the PCR amplification technique (44).

Purification of Labeled Target

After target labeling, unincorporated dyes must be removed to reduce the background signal. The most useful and reliable tools are column-based purification systems, which are available in both single-spin-column-and multiwell formats from several suppliers (Amersham, Corning, QIAGEN, and TeleChem International). Ethanol precipitation of labeled targets is a money-saving alternative for purifying amine-modified cDNA from unincorporated nucleotides and hydrolyzed RNA in the indirect labeling procedure, but labeled DNA prepared in this way can fail to hybridize (58).

DNA Labeling

Labeled DNA targets made from genomic DNA can be used for statistical normalization or expression study microarray results and can also be used for comparative genomic studies. Genomic DNA can be labeled by nick translation or by random priming with the Klenow fragment of DNA polymerase (48, 51), as follows. (i) The genomic DNA samples are first sheared by mechanical means, such as sonication, in the random priming method. (ii) DNA fragments (1 to 3 kb) are purified via column-based or electrophoretic methods. (iii) The one- or two-step labeling procedures described above can then be performed. A labeling reaction with 0.5 to 2 μg of genomic DNA can typically yield enough targets for a single microarray hybridization (32).

Hybridization

Hybridization is the process of incubating the labeled target DNA with the probe DNA tethered to the microarray substrate. Fluorescent target DNA hybridizes to complementary probe DNA on the slide, and the amount of immobilized fluorescence can be determined.

Before hybridization, most slides need to be treated to block or inactivate nonspecific binding sites. The procedure, called prehybridization, depends on the slide type and spotting chemistry. For aminosilane-coated slides, a prehybridization solution containing 1% bovine serum albumin, 5× SSC, and 0.1% sodium dodecyl sulfate has been found to be effective (21). Denhardt's reagents, salmon sperm DNA, tRNA, Cot1 DNA, and poly(dA) are familiar blocking agents which are used to minimize the background signal (10) and to eliminate nonspecific hybridization due to repetitive sequences (38). Consideration of the melting temperature equation for polynucleotides dictates that reactions in 4 to 5× SSC be performed at 60 to 65°C. Hybridization reactions employing 50% formamide are typically performed at 42°C as the duplex melting temperature; however, the kinetics of hybridization in formamide is slower than that in aqueous solution (7). The use of formamide ensures adequate stringency and reduces the temperature-induced detachment of noncovalently attached target molecules. Usually, the hybridization solution containing the labeled target DNA is placed onto the array and covered with a coverslip. The slide is then placed in a humidified chamber for passive hybridization or for mixing and recirculation of the hybridization reaction mixture to ensure a homogeneous probe concentration. The temperature of hybridization and the washing conditions depend on the G+C content of the organism, with the temperature ranging between room temperature and 55°C.

Following hybridization, the array is put through a series of washing steps to remove all unbound labeled target cDNAs. Washing is one of the most critical steps in obtaining consistently low background levels. Incomplete washing causes deposition of salts or fluorescent materials that may still remain on the array. The wash buffers need to be of the highest quality available and must be free of contaminants, dust, and particles (7). The whole array must be immersed in the washing solutions, and the array must not be allowed to dry out between washes in order to avoid high background fluorescence. After the final washing step, the array is dried immediately and completely by blowing the slide with clean compressed N_2 gas or by centrifugation at a low speed for ~5 min.

Probe distribution is often problematic, resulting in variations in gene expression that are dependent on the spatial position. Many commercial instruments now have the potential to allow automated, highly reproducible hybridization. The instruments available vary from those based on simple approaches, such as a vibrating temperature-controlled platform, to those based on complex systems of proven application, with built-in liquid handling, washing, mixing, and hybridization. Although the advantages and disadvantages of the systems differ, the volume of probe required for hybridization is a consideration that is applicable to all. Frequently, the instruments require substantial dilution of the probe, with a consequent loss of signal. Other factors include the reliability of sealing of the hybridization chamber and the possibility that the sealing mechanisms may not be compatible with all array layouts.

Scanning Arrays

The binding of targets to probes is detected by scanning the array, typically by using a scanning confocal laser, and the signals corresponding to the expression state of the transcripts are quantified by image analysis software. Fluorescence intensities correspond to the levels of hybridization of the two targets to the probes spotted on the slide. Light emitted at the wavelengths of the fluorescent labels (532 nm for Cy3 and 635 nm for Cy5) is converted to an electrical signal with a photomultiplier tube. These signals are then displayed as a 16-bit tagged image file format (TIFF) image (with Cy3 in green and Cy5 in red) and given numerical values. Generally, the photomultiplier tube of the scanner needs to be adjusted to maximize its dynamic range (0 to 65,536 pixels), to prevent signal saturation, and to balance the intensities of the two excitation wavelengths (18). With large numbers of experiments, it is prudent to scan all arrays with, at the very least, the same model of scanner, as scanners can vary in sensitivity and reliability.

Results

Data Extraction from Microarray

Most commercial microarray scanners are supplied with data extraction software. The arrays can produce well-defined, clearly delineated signals resulting in images obedient to data extraction by highly developed digital image processing techniques. The image processing steps in the analysis of a microarray are as follows. First, addressing or gridding of the exact locations of the individual spots or transcript signals is performed. Next, the microarray pixels are classified as either foreground (corresponding to a spot or transcript signal) or background. Although improvement in printing processes has simplified grid measurements and the detection of spot boundaries, there is no consensus on the best approach to background subtraction, although the use of statistical methods such as the lowess method is common. Finally, the foreground data are extracted, contributions due to background intensity are subtracted, and calculations of red and green fluorescence intensities or transcript intensity are made for each spot. After image acquisition and conversion of the image to spot intensity and/or intensity ratios, this large body of data must be stored in spreadsheet form for further analysis. Graphical displays are useful for determining the overall success of a microarray experiment (57). The red-green image produced during scanning can detect any problems with color balance, hybridization, spatial effects, spot quality, or artifacts such as scratches and dust. The original image can also be used to check for differential expression of a particular gene. Before plot analysis, the raw intensity data are log transformed ($\log_2$) to spread the values more evenly across the scale, from 0 to 65,535 pixels. If any negative values for red or green foreground intensity have arisen due to a high spot background, these will be removed from the analysis on a log scale. An informative visualization tool that uses the log-transformed values is the MA plot (14). This scatter plot has M values [$M = \log_2(R/G)$] on the vertical axis and A values [$A = (1/2)\log_2(R \times G)$] on the horizontal axis (R is the red fluorescence intensity, and G is the green fluorescence intensity). The majority of genes should not be differentially expressed. An MA plot of good-quality microarray data should therefore have an elongated comet shape centered on an M value of 0. Differentially expressed genes will appear away from this line, with those having the greatest intensities appearing to the right of the plot. The storage of primary scanned data (in a TIFF image file) will potentially be important when investigators take advantage of future developments in image extraction and/or analysis software.

Normalization of Microarray Data

Normalization is essential in microarray experiments to adjust the data for systematic nonbiological effects arising from technical variation and measurement errors. Such normalization steps must be carried out before further data analysis is attempted in order to minimize false-positive results and to identify genes that are truly and differentially expressed between samples. The aim of normalization is to correct for other differences in fluorescence intensities between samples on the same slide, or between slides, which do not represent true biological variation between samples. Red intensities (Cy5) are often lower than green intensities due to the relatively less incorporation of Cy5-labeled nucleotides, and the imbalance in the red and green intensities is usually not constant across spots within and between arrays. Additionally, these dye biases can arise from a variety of factors, including physical properties of the dyes, experimental variability in the hybridization and processing procedures, differences in the amount of input RNA in each sample, and scanner-specific excitation and collection processes. There are some normalization methods to address the issue of irregular spatial distribution across the surface of an array, but no universally accepted technique for microarray data normalization exists. Overall, the literature supports the use of intensity-dependent normalization methods, such as the print-tip lowess method (40, 57, 75). This method is capable of removing biases without altering the structure of the data. Essentially, this method corrects the M values [$\log_2(R/G)$ ratios] for nonbiological spatial and intensity effects.

Analysis of Microarray Data

The next task is to determine the boundary between significantly differentially expressed and unchanged genes, given the natural and experimental variation in the experiment. Major sources of variation in the data include normal physiological gene expression variation (43) and noise that may be introduced by a lack of slide uniformity, printing irregularities, spotting fluctuations, and poor sample hybridization (69). Numerous investigations have applied an arbitrary threshold to define significant differential expression, usually a twofold change. This is essentially a ranking method which, while intuitive, suffers from the limitation of not detecting truly differentially expressed genes that have changed by less than the arbitrary cutoff.

Clustering

Clustering algorithms are often employed to group genes with similar expression patterns. The most popular clustering method is hierarchical clustering (16). This is a methodology that arranges the gene or sample profiles into a tree so that similar profiles appear close together and dissimilar profiles are further apart. The most common method for nonhierarchical cluster analysis is determining k means. However, a method termed self-organizing maps has recently been applied to expression data generated from DNA chip arrays (64). Clustering the genes enhances their similarities and makes it easy to correlate those that might have a relationship with the specific event that the investigator is analyzing (28). Cluster analysis methods also provide the ability to reveal structure within a group of array data, grouping known and unknown genes together in ways that may suggest coordinated control mechanisms and/or functions in common pathways (55). These sorts of methods have led to the classification of cell and tissue types and the postulation of functional categories.

Commercial Microarray Resources

There are many commercial companies that offer various resources specific to microarray technology. Table 2 serves as a primary resource that can be utilized to begin learning about the options and technologies available.

CURRENT USE OF MICROARRAYS IN SOIL MICROBIOLOGY DISCIPLINES

There are a multitude of uses for microarrays in environmental, agricultural, and soil microbiology disciplines. This section details a few of these uses. In the coming years, the use of microarrays will likely increase dramatically, as will their potential and actual uses in the agricultural and environmental sciences.

Microbial Community Structure

Since <1% of soil microorganisms can be cultured readily by standard techniques (2), studies involving cultural methods cannot provide a realistic view of microbial diversity and therefore lack the ability to identify all members of microbial communities. Thus, truly understanding microbial diversity and changes in community structure in complex soil systems is not possible at this time with standard cultural techniques. DNA microarrays have been applied to the collection of information on soil microbial diversity that cannot be obtained by the aforementioned cultural methods. Furthermore, investigators can monitor phylogenetically diverse microbial communities spatially and temporally and obtain valuable information pertaining to microbial community changes under various environmental conditions.

Various functional and 16S rRNA genes have been used as targets for microarray analyses to assess the overall microbial diversity and changes in community structure in soil at various sites. A microarray constructed with universal and species-specific 16S rRNA oligonucleotide probes was used by Small et al. (56) to detect species of *Geobacter* and *Desulfovibrio* in soil extracts. Loy et al. (30) designed a comprehensive 16S rRNA microarray for the specific detection of sulfate-reducing bacteria. Using this technology, they later observed stable sulfate-reducing bacterial populations within the top 30 cm of soils collected from two low-sulfate, acidic fens in Germany (29).

Unlike 16S rRNA microarrays, community genome arrays are constructed by using whole genomic DNAs from bacterial isolates. Wu et al. (73) used whole genomic DNAs from 67 closely or distantly related bacterial strains to construct their microarray. The ability of this system to detect specific bacterial species and to evaluate the overall microbial community composition was demonstrated through analyses of marine sediment, river sediment, and soil samples. Since the microarray was constructed with cultured environmental isolates or identified bacterial strains, uncultured or previously unidentified microorganisms could not be detected. A useful application of microarrays developed by Sebat et al. (53), called metagenomic profiling, enables researchers to rapidly characterize clones from genomic libraries derived from environmental DNAs and leads to the identification of uncultured species within these libraries.

Using functional gene sequences for microarray construction allows the detection of uncultivated or unrecognized microorganisms and "narrows down the analysis to a functionally defined group of microbes" (5). Wu et al. (72) designed a microarray to uncover functional gene diversity (nitrogen cycling) and distribution within marine sediment and soil samples. Bodrossy and colleagues examined landfill-site soils for methanotrophs and functionally related bacteria, using a microarray constructed with sequences covering the functional methane monooxygenase gene. Using this method, the investigators were able to identify not only the most abundant strains but also the less dominant bacteria within their collected samples. More recently, microarrays targeting the particulate methane monooxygenase genes of methanotrophs and functionally related bacteria were used by Stralis-Pavese et al. (60) to identify methanotroph community diversity in lysimeters simulating landfill sites with different plant vegetation. This group was able to identify trends and variations in methanotroph community structures characterized by depth, methane supply, and plant cover. Interestingly, the application of this microarray system to the detection or diagnosis of environmental changes was demonstrated when an unexpected community composition was observed. This very unusual methanotroph diversity pattern was then compared to the gas profile of the site and revealed that a gas leakage had occurred in the lysimeter (60). Nitrogen cycle genes have also been monitored to estimate composition differences among nitrifier and denitrifier communities between two different sites. Rhee et al. (47) assessed a microarray for screening of biodegrading bacterial populations in soil. Their microarray was constructed with genes involved in biodegradation and metal resistance. The most abundant microbial groups were identified in soils from microcosms containing or lacking polyaromatic hydrocarbon, benzene, toluene, ethylbenzene, and xylene. Changes in the microbial community in microcosm soils were also observed under different incubation conditions (47).

Functional Genes in Soils

Little is known about functional gene expression in natural environmental communities. For assessing functional gene distribution, Wu and coworkers (72) constructed, for the first time, a prototype PCR product-based microarray which involved genes in nitrogen cycling (nitrite reductase, ammonia monooxygenase, and methane monooxygenase genes) from pure cultures and those cloned from marine sediments. This functional gene array revealed differences in the apparent distributions of these gene families in sediment and soil samples. Following these findings, the same laboratory developed another type of functional gene array containing 50-bp probes constructed from public databases and their own sequence collection (62). The hybridization specificity of the new system was extremely improved, but the sensitivity was 10 times lower than that of the PCR product-based array due to the use of shorter probes. However, by using increased amounts (2 μg) of bulk community DNAs from marine sediments, they acquired very strong hybridizations on the new array. They proposed that the amount of target DNA sample should not be a major limiting factor for environmental studies, as the average DNA yields from many surface soil and sediment samples can be quite high. This new array has potential for use as a specific, sensitive, and potentially quantitative parallel tool for characterizing the composition, structure, activity, and dynamics of microbial communities in natural environments.

Microarrays targeting mRNA enable researchers to relate community structure to community function. The design of these types of microarrays allows investigators to evaluate gene expression and therefore to expose important metabolic activities of specific microbial communities.

TABLE 2 Microarray equipment and reagent supply companies

Company	Products and services
Ambion, Inc.	RNA amplification products, labeling reagents, RNA extraction kits
Amersham Biosciences	Reagents, labeling kits, software
Corning Inc.	Slides, labeling kits, hybridization equipment
Genisphere Inc.	Microarray printing, scanning, and hybridization equipment
Invitrogen Corp.	Reagents, labeling kits
QIAGEN Inc.	Reagents, labeling kits
TeleChem International Inc.	Custom arrays, various custom services
Operon	Custom arrays, oligonucleotide services, printed arrays, long oligonucleotide arrays
Nimblegen Systems Inc.	Custom array service, complete analysis service
CombiMatrix	Custom array service, complete analysis service
Clontech	RNA extraction, labeling, custom arrays, various services

While there are no current examples in the scientific literature of functional gene expression in soil observed by using microarrays, there are numerous examples of other applications. Microarrays have frequently been used to identify gene expression in one microorganism with environmental implications. A great deal of information can be obtained from this application, such as the identification of genes important for certain metabolic activities, understanding the expression of functional genes under different conditions, and understanding expression among diverse communities and within symbiotic and other microbial associations. Microarrays have been applied to monitor the expression of phase-related genes of *Histoplasma capsulatum* (24), and to determine gene regulation in mutant root nodule bacteria (9), and several other studies have targeted the functional genes of one microorganism. In particular, Hwang et al. (24) were able to identify genes involved in the mycelial and yeast phases of *H. capsulatum* and observed differential expression of these genes during the two growth phases. Chao et al. (9) were able to identify operons in the root nodule bacterium *Sinorhizobium meliloti* that are important in different microbial functions, such as metal uptake.

In addition to applying microarray technologies to monitoring of gene expression in a given microorganism, research groups have also utilized microarrays for monitoring gene expression in microbial communities (13) and between members of symbiotic relationships (13). Dennis et al. (13) developed a microarray that had many functional genes found in a diverse range of bacteria. This research group applied their multifunctional gene microarray to RNAs extracted from activated sludge. Barnett et al. (4) developed a microarray to monitor gene expression of both members of a symbiotic relationship, the soil bacterium *S. meliloti* and the plant host *Medicago truncatula*. They were able to show that *S. meliloti* adapts differently to abiotic and symbiotic environments (4). Very recently, Methe et al. (33) assessed the gene expression of *Geobacter sulfurreducens*, using a whole-genome microarray, under different environmental conditions. Using their microarray approach, these investigators were able to detect physiological changes in *G. sulfurreducens* associated with changes in the environment.

Pathogen Detection

Similar to what was discussed above (see "Microbial Community Structure" and "Functional Genes in Soils"), pathogen detection can be accomplished through assessment of specific conserved sequences, analysis of functional genes, and analysis of functional gene expression (8). Genetic composition and whole-genome expression profiles revealed by microarray analysis of important pathogens may result in a wealth of information, including the identification of specific pathogens and their phenotypes. Phenotypic information gained with microarrays can identify important virulence genes and host preferences. As with other microarray applications, microarrays targeting pathogens can reveal relatedness among pathogenic strains, pathogen response to changing environmental conditions, and host-pathogen relationships.

Pathogens can be detected with microarrays that target the 16S rRNA gene. Warsen et al. (68) were able to detect, with 100% specificity, 18 microorganisms, with 15 of these being important fish pathogens. Other genes have also been used to identify closely related pathogenic species. Panicker et al. (39) developed a microarray that can discriminate between pathogenic and nonpathogenic *Vibrio* species. Neither of these studies, however, assessed the ability of the microarray systems to detect the target pathogens in soil samples. For both of these studies, microarray detection was assessed with cultures grown in the laboratory. Panicker et al. (39) also assessed the ability of their microarray to detect and discriminate between *Vibrio* species in oyster tissue homogenates.

Another approach to pathogen detection is to target virulence genes of specific pathogens or pathogen groups. In one example, a microarray was constructed with virulence factor probes for closely related enteric bacteria (11). It was demonstrated that specific biochemical or pathogenic markers of the pathogenic bacterial components of a sample could be elucidated, leading to the identification of specific pathogens within a sample. Furthermore, microarrays targeting virulence factors can provide valuable information pertaining to phylogenetic relatedness and classification (11). Wang et al. (66) designed a microarray for the detection of 140 different viruses. This 70-mer oligonucleotide microarray provided a unique hybridization pattern for every virus, therefore allowing discrimination between even closely related viruses.

Understanding pathogen responses to environmental conditions can provide important information for gaining insight into pathogen transmission and developing strategies for their control in water, soil, air, and food supplies. Microarrays enable researchers to target and monitor specific genes to identify those important for survival. In one study, global gene expression patterns of *Campylobacter jejuni* were evaluated (36). Gene expression was monitored under stressful conditions (different oxygen and temperature conditions) in order to determine survival mechanisms of the organism. These researchers were able to identify genes that were either up- or down-regulated under specific stressful conditions, such as low and high temperatures and under anaerobic and aerobic oxygen conditions. Ren and Prescott (46) designed a microarray to evaluate the induction of virulence plasmid genes of the soil bacterium *Rhodococcus equi*, a pathogen of immunocompromised humans and equine foals. They were able to observe the expression of specific virulence genes of *R. equi* under different environmental conditions and while the organism was inside macrophages.

BENEFITS OF USING MICROARRAYS FOR APPLICATIONS IN ENVIRONMENTAL MICROBIOLOGY

Microarray technology is a recently developed and powerful genomic tool that is widely used to investigate gene expression and regulation on a genome scale and for detection of genetic polymorphisms in environmental samples (77). Microbial communities account for 50% of Earth's biomass, and 1 g of soil may harbor up to 10 billion microorganisms, potentially composed of thousands of different species. Since <1% of the microorganisms observed can be cultivated and characterized, soil ecosystems are, to a large extent, uncategorized. To date, microarray technology has been successfully applied to the study of environmental samples by the use of innovative approaches to gene expression, comparative genomics, and mixed community analysis. These studies revealed that microarray-based genomic technologies have great potential as specific, sensitive, quantitative, parallel high-throughput tools for microbial detection, identification, and characterization in environments. Novel methods and approaches related to microarray-based genomic technologies enable us to explore the vast diversity of environmental microorganisms. New tools in bioinformatics and statistical analysis enable us to handle the huge amount of data obtained through multidimensional studies that combine growth-independent molecular analyses with analyses of microbial growth, activity, and physiology and integrate measures of environmental parameters. Further investigations of microbial sequences, comparative genomics, and microarray technology will improve our understanding of structure-function relationships, evolution, and effects of abiotic and biotic factors on soil microbial communities.

Many of the studies discussed above were not carried out with environmental samples, such as soil and water, but may be useful for these types of sample matrices. Basic detection of microorganisms by using 16S rRNA gene or 16S rRNA probes, probes designed from a genomic library for a specific microorganism, or probes designed to identify functional genes is a useful tool for microarray detection of specific microorganisms in a wide range of sample types. As shown in previous studies, microarrays can be used to detect specific microorganisms and to assess community composition alongside community function in complex microbial communities. As a further example to elaborate the benefits of microarrays over other methods, Adamczyk et al. (1) designed a microarray to monitor the community composition of ammonia-oxidizing bacteria and the carbon dioxide fixation activity of these bacteria. Oligonucleotide probes targeting the 16S rRNAs of ammonia-oxidizing bacteria were designed and applied to sludge samples. Their approach overcame one of the major challenges in microbial ecology, namely, the simultaneous collection of information relating to community structure and function. These types of studies enable microbial ecologists to directly associate community members with their specific activities and functions within an ecosystem. In a study conducted by Kim et al. (25), microarrays targeting bacteria that produce foaming and bulking and those that produce flocs were applied to sludge samples. This microarray approach was suggested as a useful tool for identification of less favorable bacterial populations in activated sludge, such as *Gordonia amarae*, which is capable of producing foaming and bulking. Understanding spatial and temporal relationships between members of a microbial community and uncovering spatial and temporal differences in functional gene expression are other ways in which environmental scientists may apply microarrays. This type of microarray application would be useful to agricultural scientists interested in spatial and temporal differences in nutrient cycling and their effects on crop development and growth. In another application, spatial patterns in gene expression were monitored in the extrametrical mycelium and mycorrhizal root tips produced by the ectomycorrhizal fungus *Paxillus involutus* (71). The investigators observed different expression profiles for rhizomorphs and nutrient patch mycelium compared to that for mycorrhizal tips. They concluded that differential gene expression signifies functional specialization of tissues forming ectomycorrhizal associations. Lastly, microarrays can also uncover interactions between members of symbiotic relationships through the simultaneous evaluation of functional gene expression of both members.

Disadvantages of Using Microarrays To Study Environmental Samples

The following example illustrates the primary disadvantage of using microarrays to study soil samples. Cho and Tiedje (12) extracted DNAs from sediment and soil by using a soil DNA isolation kit (MoBio, Solana Beach, Calif.) and labeled the DNAs by using a FluoroLink Cy3-dCTP kit (Amersham Pharmacia, Piscataway, N.J.) for random priming (High Prime; Roche, Indianapolis, Ind.). They found that the microarray did not detect *nirS* sequences in either the sediment or soil samples, even when 10^5 cells containing this gene were spiked into the soil. The high observed detection limit agrees with what is to be expected. Their calculation indicates that only when genomes containing target sequences contribute more than 1/20 of the environmental DNA extracted can they be detected. Since total soil DNA (or environmental DNA) contains DNAs originating from a huge diversity of bacteria plus fungi, archaea, protozoa, plant debris, and microinvertebrates, the quantitative detection of particular genes by a microarray using current technology is problematic. This illustrates a primary disadvantage of using microarray technology for the study of environmental samples.

Cho and Tiedje suggested the following three possible solutions to improve the detection limit of microarray hybridization: (i) increasing the amount of DNA (probe) immobilized on the microarray substratum, (ii) developing a microarray signal detection system with a higher sensitivity, and (iii) developing procedures to selectively enrich bacterial genomes or genes of interest from the environmental samples. Once the issue of the high detection limit is resolved, the microarray-based method should be a powerful tool for massively parallelled analyses of important genes and microbes in the environment.

There are several issues related to the use of microarrays that limit their use in environmental studies. At the front end of this potentially powerful tool is the need for sensitive nucleic acid extraction from complicated sample matrices. Direct extraction of nucleic acids from environmental samples may coextract humic acids or other organic materials that may affect nucleic acid hybridization with microarrays. Coextraction of these substances may also inhibit nucleic acid amplification by PCR. The sensitivities of microarrays are also quite low; researchers have reported sensitivities of 7.5×10^6 cell equivalents of RNA for soil extracts, 1.3×10^7 cells for the presence of background RNA, and $>10^4$ to 10^6 cells for sludge. Thus, the efficiency of nucleic acid extraction and amplification is vital to this technology.

Because low concentrations of target nucleic acid are present in most environments and high concentrations of nucleic acid are required for microarray detection, accurate and sensitive target nucleic acid amplification may be needed. PCR amplification has been used by several groups to amplify nucleic acids prior to microarray analysis. PCR amplification has several drawbacks, including inhibition by compounds coextracted with nucleic acids, difficulties in design and unbiased amplification of multiple organisms in one reaction, and the need for sequence information a priori. In a recent study, Vora et al. (65) attempted to evaluate alternative approaches for nucleic acid amplification prior to microarray analysis. They applied PCR to amplify DNA of *Escherichia coli* O157:H7 from laboratory and environmental samples. They reported the best sensitivity of detection for random amplification PCR approaches (random-primed, isothermal Klenow fragment-based, and ϕ29 DNA polymerase-based methods) over an established multiplex PCR strategy that targets virulence genes.

Probe specificity is a well-known problem for microarray applications. The amount and quality of sequence information limit the design of specific oligonucleotide probes. For environmental samples, not all target sequences may be known. Sequence divergence in the environment may affect microarray hybridization and thus cause the microarray to miss or underestimate some members within an ecosystem. Koizumi et al. (26) stated that this is a drawback of designing specific probes for the detection of specific microorganisms. They successfully detected bacterial nucleic acids from oil-contaminated sediments by using universal probes, but no signals were observed for probes targeting specific groups of sulfate-reducing bacteria. In situations where uncharacterized organisms are present, exhaustive gene or clone libraries for the target gene need to be gathered in order to design group- or clade-specific probes for detection and distinction among all possible targets within the studied ecosystem (61).

Another issue is the reproducibility of results obtained by microarray analysis. Piper et al. (42) reported that the reproducibility of microarray results differs between laboratories and even within the same laboratory. However, in future studies, as methods improve, good intralaboratory reproducibility and performance will likely be observed.

REFERENCES

1. **Adamczyk, J., M. Hesselsoe, N. Iversen, M. Horn, A. Lehner, P. H. Nielsen, M. Schloter, P. Roslev, and M. Wagner.** 2003. The isotope array, a new tool that employs substrate-mediated labeling of rRNA for determination of microbial community structure and function. *Appl. Environ. Microbiol.* **69:**6875–6887.
2. **Amann, R. I., W. Ludwig, and K. H. Schleifer.** 1995. Phylogenetic identification and in situ detection of individual microbial cells without cultivation. *Microbiol. Rev.* **59:**143–169.
3. **Bahnweg, G., S. Schulze, E. M. Moller, H. Rosenbrock, C. Langebartels, and H. Sandermann, Jr.** 1998. DNA isolation from recalcitrant materials such as tree roots, bark, and forest soil for the detection of fungal pathogens by polymerase chain reaction. *Anal. Biochem.* **262:**79–82.
4. **Barnett, M. J., C. J. Toman, R. F. Fisher, and S. R. Long.** 2004. A dual-genome symbiosis chip for coordinate study of signal exchange and development in a prokaryote-host interaction. *Proc. Natl. Acad. Sci. USA* **101:**16636–16641.
5. **Bodrossy, L., N. Stralis-Pavese, J. C. Murrell, S. Radajewski, A. Weilharter, and A. Sessitsch.** 2003. Development and validation of a diagnostic microbial microarray for methanotrophs. *Environ. Microbiol.* **5:**566–582.
6. **Bovet, L., U. Feller, and E. Martinoia.** 2005. Possible involvement of plant ABC transporters in cadmium detoxification: a cDNA sub-microarray approach. *Environ. Int.* **31:**263–267.
7. **Bowtell, D. D.** 1999. Options available—from start to finish—for obtaining expression data by microarray. *Nat. Genet.* **21:**25–32.
8. **Bryant, P. A., D. Venter, R. Robins-Browne, and N. Curtis.** 2004. Chips with everything: DNA microarrays in infectious diseases. *Lancet Infect. Dis.* **4:**100–111.
9. **Chao, T. C., A. Becker, J. Buhrmester, A. Puhler, and S. Weidner.** 2004. The *Sinorhizobium meliloti fur* gene regulates, with dependence on Mn(II), transcription of the *sitABCD* operon, encoding a metal-type transporter. *J. Bacteriol.* **186:**3609–3620.
10. **Cheung, V. G., M. Morley, F. Aguilar, A. Massimi, R. Kucherlapati, and G. Childs.** 1999. Making and reading microarrays. *Nat. Genet.* **21:**15–19.
11. **Chizhikov, V., A. Rasooly, K. Chumakov, and D. D. Levy.** 2001. Microarray analysis of microbial virulence factors. *Appl. Environ. Microbiol.* **67:**3258–3263.
12. **Cho, J. C., and J. M. Tiedje.** 2002. Quantitative detection of microbial genes by using DNA microarrays. *Appl. Environ. Microbiol.* **68:**1425–1430.
13. **Dennis, P., E. A. Edwards, S. N. Liss, and R. Fulthorpe.** 2003. Monitoring gene expression in mixed microbial communities by using DNA microarrays. *Appl. Environ. Microbiol.* **69:**769–778.
14. **Dudoit, S., and J. Fridlyand.** 2002. A prediction-based resampling method for estimating the number of clusters in a dataset. *Genome Biol.* **3:**RESEARCH0036.
15. **Eberwine, J., H. Yeh, K. Miyashiro, Y. Cao, S. Nair, R. Finnell, M. Zettel, and P. Coleman.** 1992. Analysis of gene expression in single live neurons. *Proc. Natl. Acad. Sci. USA* **89:**3010–3014.
16. **Eisen, M. B., P. T. Spellman, P. O. Brown, and D. Botstein.** 1998. Cluster analysis and display of genome-wide expression patterns. *Proc. Natl. Acad. Sci. USA* **95:**14863–14868.

17. **Felske, A., B. Engelen, U. Nubel, and H. Backhaus.** 1996. Direct ribosome isolation from soil to extract bacterial rRNA for community analysis. *Appl. Environ. Microbiol.* **62:**4162–4167.
18. **Forster, T., D. Roy, and P. Ghazal.** 2003. Experiments using microarray technology: limitations and standard operating procedures. *J. Endocrinol.* **178:**195–204.
19. **Frostegard, A., S. Courtois, V. Ramisse, S. Clerc, D. Bernillon, G. F. Le, P. Jeannin, X. Nesme, and P. Simonet.** 1999. Quantification of bias related to the extraction of DNA directly from soils. *Appl. Environ. Microbiol.* **65:**5409–5420.
20. **Griffiths, R. I., A. S. Whiteley, A. G. O'Donnell, and M. J. Bailey.** 2000. Rapid method for coextraction of DNA and RNA from natural environments for analysis of ribosomal DNA- and rRNA-based microbial community composition. *Appl. Environ. Microbiol.* **66:**5488–5491.
21. **Hegde, P., R. Qi, K. Abernathy, C. Gay, S. Dharap, R. Gaspard, J. E. Hughes, E. Snesrud, N. Lee, and J. Quackenbush.** 2000. A concise guide to cDNA microarray analysis. *Biotechniques* **29:**548–556.
22. **Hughes, T. R., M. Mao, A. R. Jones, J. Burchard, M. J. Marton, K. W. Shannon, S. M. Lefkowitz, M. Ziman, J. M. Schelter, M. R. Meyer, S. Kobayashi, C. Davis, H. Dai, Y. D. He, S. B. Stephaniants, G. Cavet, W. L. Walker, A. West, E. Coffey, D. D. Shoemaker, R. Stoughton, A. P. Blanchard, S. H. Friend, and P. S. Linsley.** 2001. Expression profiling using microarrays fabricated by an ink-jet oligonucleotide synthesizer. *Nat. Biotechnol.* **19:**342–347.
23. **Hurt, R. A., X. Qiu, L. Wu, Y. Roh, A. V. Palumbo, J. M. Tiedje, and J. Zhou.** 2001. Simultaneous recovery of RNA and DNA from soils and sediments. *Appl. Environ. Microbiol.* **67:**4495–4503.
24. **Hwang, L., D. Hocking-Murray, A. K. Bahrami, M. Andersson, J. Rine, and A. Sil.** 2003. Identifying phase-specific genes in the fungal pathogen Histoplasma capsulatum using a genomic shotgun microarray. *Mol. Biol. Cell* **14:**2314–2326.
25. **Kim, B. C., J. H. Park, and M. B. Gu.** 2004. Development of a DNA microarray chip for the identification of sludge bacteria using an unsequenced random genomic DNA hybridization method. *Environ. Sci. Technol.* **38:** 6767–6774.
26. **Koizumi, Y., J. J. Kelly, T. Nakagawa, H. Urakawa, S. El-Fantroussi, S. Al-Muzaini, M. Fukui, Y. Urushigawa, and D. A. Stahl.** 2002. Parallel characterization of anaerobic toluene- and ethylbenzene-degrading microbial consortia by PCR-denaturing gradient gel electrophoresis, RNA-DNA membrane hybridization, and DNA microarray technology. *Appl. Environ. Microbiol.* **68:**3215–3225.
27. **Krsek, M., and E. M. Wellington.** 1999. Comparison of different methods for the isolation and purification of total community DNA from soil. *J. Microbiol. Methods* **39:**1–16.
28. **Lee, P. S., and K. H. Lee.** 2000. Genomic analysis. *Curr. Opin. Biotechnol.* **11:**171–175.
29. **Loy, A., K. Kusel, A. Lehner, H. L. Drake, and M. Wagner.** 2004. Microarray and functional gene analyses of sulfate-reducing prokaryotes in low-sulfate, acidic fens reveal cooccurrence of recognized genera and novel lineages. *Appl. Environ. Microbiol.* **70:**6998–7009.
30. **Loy, A., A. Lehner, N. Lee, J. Adamczyk, H. Meier, J. Ernst, K. H. Schleifer, and M. Wagner.** 2002. Oligonucleotide microarray for 16S rRNA gene-based detection of all recognized lineages of sulfate-reducing prokaryotes in the environment. *Appl. Environ. Microbiol.* **68:**5064–5081.
31. **Luo, L., R. C. Salunga, H. Guo, A. Bittner, K. C. Joy, J. E. Galindo, H. Xiao, K. E. Rogers, J. S. Wan, M. R. Jackson, and M. G. Erlander.** 1999. Gene expression profiles of laser-captured adjacent neuronal subtypes. *Nat. Med.* **5:**117–122.
32. **Majtan, T., G. Bukovska, and J. Timko.** 2004. DNA microarrays—techniques and applications in microbial systems. *Folia Microbiol.* (Prague) **49:**635–664.
33. **Methe, B. A., J. Webster, K. Nevin, J. Butler, and D. R. Lovley.** 2005. DNA microarray analysis of nitrogen fixation and Fe(III) reduction in *Geobacter sulfurreducens*. *Appl. Environ. Microbiol.* **71:**2530–2538.
34. **Mir, K. U., and E. M. Southern.** 1999. Determining the influence of structure on hybridization using oligonucleotide arrays. *Nat. Biotechnol.* **17:**788–792.
35. **Miskin, I. P., P. Farrimond, and I. M. Head.** 1999. Identification of novel bacterial lineages as active members of microbial populations in a freshwater sediment using a rapid RNA extraction procedure and RT-PCR. *Microbiology* **145:**1977–1987.
36. **Moen, B., A. Oust, O. Langsrud, N. Dorrell, G. L. Marsden, J. Hinds, A. Kohler, B. W. Wren, and K. Rudi.** 2005. Explorative multifactor approach for investigating global survival mechanisms of *Campylobacter jejuni* under environmental conditions. *Appl. Environ. Microbiol.* **71:** 2086–2094.
37. **Mullis, K., F. Faloona, S. Scharf, R. Saiki, G. Horn, and H. Erlich.** 1986. Specific enzymatic amplification of DNA in vitro: the polymerase chain reaction. *Cold Spring Harbor Symp. Quant. Biol.* **51:**263–273.
38. **Nguyen, C., D. Rocha, S. Granjeaud, M. Baldit, K. Bernard, P. Naquet, and B. R. Jordan.** 1995. Differential gene expression in the murine thymus assayed by quantitative hybridization of arrayed cDNA clones. *Genomics* **29:** 207–216.
39. **Panicker, G., D. R. Call, M. J. Krug, and A. K. Bej.** 2004. Detection of pathogenic *Vibrio* spp. in shellfish by using multiplex PCR and DNA microarrays. *Appl. Environ. Microbiol.* **70:**7436–7444.
40. **Park, T., S. G. Yi, S. H. Kang, S. Lee, Y. S. Lee, and R. Simon.** 2003. Evaluation of normalization methods for microarray data. *BMC Bioinformatics.* **4:**33.
41. **Phillips, J., and J. H. Eberwine.** 1996. Antisense RNA amplification: a linear amplification method for analyzing the mRNA population from single living cells. *Methods* **10:**283–288.
42. **Piper, M. D., P. Ran-Lapujade, C. Bro, B. Regenberg, S. Knudsen, J. Nielsen, and J. T. Pronk.** 2002. Reproducibility of oligonucleotide microarray transcriptome analyses. An interlaboratory comparison using chemostat cultures of Saccharomyces cerevisiae. *J. Biol. Chem.* **277:** 37001–37008.
43. **Pritchard, C. C., L. Hsu, J. Delrow, and P. S. Nelson.** 2001. Project normal: defining normal variance in mouse gene expression. *Proc. Natl. Acad. Sci. USA* **98:**13266–13271.
44. **Puskas, L. G., A. Zvara, L. Hackler, Jr., T. Micsik, and H. P. Van.** 2002. Production of bulk amounts of universal RNA for DNA microarrays. *Biotechniques* **33:**898–900, 902, 904.
45. **Ramdas, L., D. E. Cogdell, J. Y. Jia, E. E. Taylor, V. R. Dunmire, L. Hu, S. R. Hamilton, and W. Zhang.** 2004. Improving signal intensities for genes with low-expression on oligonucleotide microarrays. *BMC Genomics* **5:**35.
46. **Ren, J., and J. F. Prescott.** 2003. Analysis of virulence plasmid gene expression of intra-macrophage and in vitro grown Rhodococcus equi ATCC 33701. *Vet. Microbiol.* **94:**167–182.
47. **Rhee, S. K., X. Liu, L. Wu, S. C. Chong, X. Wan, and J. Zhou.** 2004. Detection of genes involved in biodegradation and biotransformation in microbial communities by

using 50-mer oligonucleotide microarrays. *Appl. Environ. Microbiol.* **70:**4303–4317.

48. **Richmond, C. S., J. D. Glasner, R. Mau, H. Jin, and F. R. Blattner.** 1999. Genome-wide expression profiling in Escherichia coli K-12. *Nucleic Acids Res.* **27:**3821–3835.
49. **Rimour, S., D. Hill, C. Militon, and P. Peyret.** 2005. GoArrays: highly dynamic and efficient microarray probe design. *Bioinformatics* **21:**1094–1103.
50. **Rouillard, J. M., C. J. Herbert, and M. Zuker.** 2002. OligoArray: genome-scale oligonucleotide design for microarrays. *Bioinformatics* **18:**486–487.
51. **Salama, N., K. Guillemin, T. K. McDaniel, G. Sherlock, L. Tompkins, and S. Falkow.** 2000. A whole-genome microarray reveals genetic diversity among Helicobacter pylori strains. *Proc. Natl. Acad. Sci. USA* **97:**14668–14673.
52. **Schena, M., D. Shalon, R. W. Davis, and P. O. Brown.** 1995. Quantitative monitoring of gene expression patterns with a complementary DNA microarray. *Science* **270:**467–470.
53. **Sebat, J. L., F. S. Colwell, and R. L. Crawford.** 2003. Metagenomic profiling: microarray analysis of an environmental genomic library. *Appl. Environ. Microbiol.* **69:**4927–4934.
54. **Sharma, A., G. P. Srivastava, V. K. Sharma, and S. Ramachandran.** 2004. ArrayD: a general purpose software for microarray design. *BMC Bioinformatics* **5:**142.
55. **Sherlock, G.** 2000. Analysis of large-scale gene expression data. *Curr. Opin. Immunol.* **12:**201–205.
56. **Small, J., D. R. Call, F. J. Brockman, T. M. Straub, and D. P. Chandler.** 2001. Direct detection of 16S rRNA in soil extracts by using oligonucleotide microarrays. *Appl. Environ. Microbiol.* **67:**4708–4716.
57. **Smyth, G. K., and T. Speed.** 2003. Normalization of cDNA microarray data. *Methods* **31:**265–273.
58. **Springer, A. L., L. R. Booth, M. D. Braid, C. M. Houde, K. A. Hughes, R. J. Kaiser, C. Pedrak, D. A. Spicer, and S. Stolyar.** 2003. A rapid method for manual or automated purification of fluorescently labeled nucleic acids for sequencing, genotyping, and microarrays. *J. Biomol. Tech.* **14:**17–32.
59. **Stillman, B. A., and J. L. Tonkinson.** 2001. Expression microarray hybridization kinetics depend on length of the immobilized DNA but are independent of immobilization substrate. *Anal. Biochem.* **295:**149–157.
60. **Stralis-Pavese, N., A. Sessitsch, A. Weilharter, T. Reichenauer, J. Riesing, J. Csontos, J. C. Murrell, and L. Bodrossy.** 2004. Optimization of diagnostic microarray for application in analysing landfill methanotroph communities under different plant covers. *Environ. Microbiol.* **6:**347–363.
61. **Taroncher-Oldenburg, G., E. M. Griner, C. A. Francis, and B. B. Ward.** 2003. Oligonucleotide microarray for the study of functional gene diversity in the nitrogen cycle in the environment. *Appl. Environ. Microbiol.* **69:**1159–1171.
62. **Tiquia, S. M., L. Wu, S. C. Chong, S. Passovets, D. Xu, Y. Xu, and J. Zhou.** 2004. Evaluation of 50-mer oligonucleotide arrays for detecting microbial populations in environmental samples. *Biotechniques* **36:**664–665.
63. **Tolstrup, N., P. S. Nielsen, J. G. Kolberg, A. M. Frankel, H. Vissing, and S. Kauppinen.** 2003. OligoDesign: optimal design of LNA (locked nucleic acid) oligonucleotide capture probes for gene expression profiling. *Nucleic Acids Res.* **31:**3758–3762.
64. **Toronen, P., M. Kolehmainen, G. Wong, and E. Castren.** 1999. Analysis of gene expression data using self-organizing maps. *FEBS Lett.* **451:**142–146.
65. **Vora, G. J., C. E. Meador, D. A. Stenger, and J. D. Andreadis.** 2004. Nucleic acid amplification strategies for DNA microarray-based pathogen detection. *Appl. Environ. Microbiol.* **70:**3047–3054.
66. **Wang, D., L. Coscoy, M. Zylberberg, P. C. Avila, H. A. Boushey, D. Ganem, and J. L. DeRisi.** 2002. Microarray-based detection and genotyping of viral pathogens. *Proc. Natl. Acad. Sci. USA* **99:**15687–15692.
67. **Wang, E., L. D. Miller, G. A. Ohnmacht, E. T. Liu, and F. M. Marincola.** 2000. High-fidelity mRNA amplification for gene profiling. *Nat. Biotechnol.* **18:**457–459.
68. **Warsen, A. E., M. J. Krug, S. LaFrentz, D. R. Stanek, F. J. Loge, and D. R. Call.** 2004. Simultaneous discrimination between 15 fish pathogens by using 16S ribosomal DNA PCR and DNA microarrays. *Appl. Environ. Microbiol.* **70:**4216–4221.
69. **Watson, S. J., F. Meng, R. C. Thompson, and H. Akil.** 2000. The "chip" as a specific genetic tool. *Biol. Psychiatry* **48:**1147–1156.
70. **Wernersson, R., and H. B. Nielsen.** 2005. OligoWiz 2.0—integrating sequence feature annotation into the design of microarray probes. *Nucleic Acids Res.* **33:**W611–W615.
71. **Wright, D. P., T. Johansson, Q. A. Le, B. Soderstrom, and A. Tunlid.** 2005. Spatial patterns of gene expression in the extramatrical mycelium and mycorrhizal root tips formed by the ectomycorrhizal fungus Paxillus involutus in association with birch (Betula pendula) seedlings in soil microcosms. *New Phytol.* **167:**579–596.
72. **Wu, L., D. K. Thompson, G. Li, R. A. Hurt, J. M. Tiedje, and J. Zhou.** 2001. Development and evaluation of functional gene arrays for detection of selected genes in the environment. *Appl. Environ. Microbiol.* **67:**5780–5790.
73. **Wu, L., D. K. Thompson, X. Liu, M. W. Fields, C. E. Bagwell, J. M. Tiedje, and J. Zhou.** 2004. Development and evaluation of microarray-based whole-genome hybridization for detection of microorganisms within the context of environmental applications. *Environ. Sci. Technol.* **38:**6775–6782.
74. **Yang, I. V., E. Chen, J. P. Hasseman, W. Liang, B. C. Frank, S. Wang, V. Sharov, A. I. Saeed, J. White, J. Li, N. H. Lee, T. J. Yeatman, and J. Quackenbush.** 2002. Within the fold: assessing differential expression measures and reproducibility in microarray assays. *Genome Biol.* **3:**RESEARCH0062.
75. **Yang, Y. H., S. Dudoit, P. Luu, D. M. Lin, V. Peng, J. Ngai, and T. P. Speed.** 2002. Normalization for cDNA microarray data: a robust composite method addressing single and multiple slide systematic variation. *Nucleic Acids Res.* **30:**e15.
76. **Zhao, H., T. Hastie, M. L. Whitfield, A. L. Borresen-Dale, and S. S. Jeffrey.** 2002. Optimization and evaluation of T7 based RNA linear amplification protocols for cDNA microarray analysis. *BMC Genomics* **3:**31.
77. **Zhou, J.** 2003. Microarrays for bacterial detection and microbial community analysis. *Curr. Opin. Microbiol.* **6:**288–294.

PCR: Agricultural and Environmental Applications for Soil Microbes

SCOT E. DOWD AND IAN L. PEPPER

55

THE PCR

A Brief History of PCR

PCR was invented in 1983 by Kary Mullis while he was working for the Cetus Corporation (28). PCR quickly was shown to be of incomparable usefulness in simplifying and accelerating molecular biology research (10, 26). Between 1984 and 1992, the Cetus Corporation was issued patents on the PCR process (the patents are now owned by Hoffmann-La Roche), the key enzyme (*Taq* polymerase), instrumentation, and various important methods. As testament to the impact of PCR on scientific research, Mullis received a Nobel Prize in chemistry for this development. Anyone involved in molecular biology research now realizes that it is hard to imagine performing such research without PCR.

PCR is an enzymatic reaction that allows amplification of DNA through a repetitive process. During each temperature cycle of PCR, any DNA that is present in the reaction is copied. Thus, during the course of the PCR process the amount of DNA doubles during each cycle. In practice, 25 to 30 cycles of PCR results in an approximately 10^6-fold increase in the amount of DNA present. This amplification results in a large amount of a specific DNA sequence being produced in quantities sufficient for common laboratory analyses, including visualization after gel electrophoresis and staining.

Targeted amplification of DNA results in the increased sensitivity of detection of sequences present even in trace amounts and even those contained in mixed cell populations. Although it was originally used in genetic and clinical applications, PCR is now widely used in environmental microbiology as well. Applications include the detection of microbial pathogens in environmental samples, estimates of microbial diversity (5, 23, 46, 55, 56), and even indirect measures of enzymatic activity (17, 24, 52).

This chapter introduces the primary concepts behind PCR, describes different flavors or specialized derivatives of PCR that are used to overcome various problems, gives an overview of ways in which the use of PCR can enhance soil microbiology research, and provides information on equipment and other products designed for PCR.

PCR Basic Methodology Overview

PCR is a method used to specifically amplify very small amounts of target nucleic acid in vitro to levels sufficient for common laboratory manipulations. The reaction mixture contains three essential ingredients: the first are the building blocks for DNA, which are the four nucleotides (C, A, T, and G), the second is the enzyme which is a specialized DNA polymerase which builds these nucleotides into a copy of the target nucleic acid, and the third is a pair of primers designed specifically for the type of target. An extensive discussion on the design of primers is provided later in this chapter. The PCR itself is a temperature cycling reaction which passes the components described above through three temperatures or phases. These temperature phases are denaturation, annealing, and extension. Ultimately, the PCR copies and amplifies the target sequence from the nucleic acid template, doubling the number of copies during each cycle of temperature. We further elaborate on the PCR in the following sections as appropriate, but the reader is encouraged to read several review articles that discuss the basic chemistry and principles of the PCR (44, 49, 53)

The Principles of Temperature Cycling

A typical PCR amplification reaction consists of the addition of the basic reagents and template (target DNA) to a tube and the subjection of these reagents to repeated cycling of temperatures to achieve (i) denaturation or separation of template double-stranded DNA, (ii) annealing of primers to the denatured single-stranded template DNA, and (iii) extension from the 3′ ends of primers to achieve the synthesis of a copy of each of the single strands. Table 1 provides a standard 100-μl PCR mixture that can routinely be used. In a typical protocol, a starting point for the temperature cycling during denaturation is 94°C for 1.5 min, that during primer annealing is between 50 and 60°C for 1 min, and that during the primer extension step is 72°C for 1 min. This is subsequently repeated for 25 cycles, with the annealing temperature at 55 or 60°C. At the end of 25 cycles, in order to account for all unextended target sequences, a final extension period of 7 to 10 min at 72°C is typically added. At the end of the amplification cycles and final extension, the enzyme reaction is stopped by incubation at 4°C. In the early days of PCR, the use of multiple water baths set to the appropriate temperatures and manual transfer of the reaction mixture from one bath to the next was common. These days many commercial companies have specialized pieces of equipment called thermocyclers that eliminate the need for water baths. Table 2 provides

TABLE 1 Standard PCR mixture

Reaction component	Vol (μl)
H_2O	61.5–66.5
10× reaction buffer[a]	10
dNTPs (1.25 mM)	16
Primer 1 (0.1 μg/μl)	1.0
Primer 2 (0.1 μg/μl)	1.0
Template	5–10
Enzyme	0.5
Total vol	100

[a]The buffer consists of 100 mM Tris-HCl (pH 8.3; 25°C), 500 mM KCl, 15 mM $MgCl_2$, and 0.01% (wt/vol) gelatin and is autoclaved. This buffer provides the preferred pH and ionic strength for PCR amplification.

the details on a few commercial companies that sell thermocyclers for use in most basic PCR methodologies.

Basic PCR Protocol

The various components of the reaction mixture (Table 1) are added to a 0.5- or 0.2-ml polypropylene tube. The quality of the tubes used in the reaction is important, since uniform and efficient heat transfer through the tube is critical for the cycling conditions. The components are added to the tube sequentially by using either a positive-displacement pipette or pipettes dedicated for PCR. To avoid nonspecific amplification products due to mispriming, the reaction is set up by using ice baths. When a large number of reaction tubes are being set up, it is advisable to use so-called master mixes; e.g., if 10 amplifications are being set up, 10-fold amounts of all reagents (except the template) are mixed together in a single tube in bulk and then individually aliquoted into separate PCR tubes. The template is finally added to the tubes. The use of master mixes reduces errors when extremely small amounts of the reagents, e.g., *Taq* polymerase, are dispensed into the individual tubes. To cut down on expenses, 50-μl or even 25-μl reaction mixtures can be run by proportionately reducing the volumes. Many companies now supply what is known as master mixes (Table 3) and have detailed instructions on the methods involved in PCR. These master mixes are premixed and optimized solutions containing all of the required ingredients for the PCR in buffers that enhance the efficiency of the enzymatic reactions. By adding the user's target DNA (template) and the specific primers designed to target this template, the user can typically achieve successful PCRs with little or no optimization.

DIFFERENT PCR METHODOLOGIES

Nested and Seminested PCRs

Nested PCR is a valuable two-step PCR method that can be used to amplify regions or targets. In essence, the use of nested PCR improves sensitivity without impairing specificity. In nested PCR, the initial basic PCR is followed by a second complete PCR in which a new set of oligonucleotide primers designed to anneal within the product of the first PCR is used. The product of the initial PCR can also be purified by using spin columns (e.g., PCR purification kit; QIAGEN Inc., La Jolla, Calif.), and the purified product can be used in the second reaction. Finally, an additional step is to use a small aliquot of the first reaction mixture and directly add it to a second reaction mixture containing the nested primers. These purification or dilution reactions are often performed to avoid the carryover of reaction products, such as pyrophosphates and environmental inhibitors, which can inhibit the second reaction. Another method for performing nested PCR is a one-reaction method in which you design the primers such that the outer primer pair is a longer primer with a higher melting temperature, while the inner primers are shorter with a shorter product and much lower annealing temperatures. By adding lower (limiting) concentrations of the outer primers and using higher annealing temperatures in the first 20 cycles (e.g., 60°C), the outer primers should be exhausted. Following these initial cycles, the remaining cycles are conducted at lower annealing temperatures (e.g., 45°C), which allows annealing of the smaller inner primers and subsequent amplification of the inner smaller product. Seminested PCR is similar to nested PCR, but instead of having two complete primer sets, seminested PCR involves the use of only one nested primer in the second PCR and one primer identical to that used in the first PCR.

Touchdown PCR

Touchdown PCR is a technique which can be used to increase the specificity and the sensitivity of PCR. Touchdown PCR involves the steady increase of the annealing temperature during the cycles of the PCR. This procedure eventually selects for only the most specific products and toward the end eliminates amplification of all products altogether. Alternatively, the annealing temperature can gradually be decreased with increasing numbers of cycles of the PCR to give a competitive advantage to the correct target in initial reactions over closely related nonspecific products. In later cycles the specific product has been increased such that when the annealing temperature is lowered they are present at a higher concentration than nonspecific targets and so are preferentially amplified due to sheer availability.

Hot-Start PCR

The use of a hot start during PCR is an effective method used to decrease the formation of nonspecific products, enhance amplification, and reduce the formation of primer dimers and cold fusion products (27). There are a number of ways to perform hot-start PCR. Basically, all the PCR reagents except one essential for the reaction to commence, usually the nucleotides, the DNA polymerase, or the Mg^{2+},

TABLE 2 Companies that manufacture and sell thermocyclers

Company	Product	Web address
MidSci	TC-412 Techne	www.midsci.com
Applied Biosystems	96-Well GeneAmp 9700	www.appliedbiosystems.com
Bio-Rad	DNA engine TC	www.bio-rad.com

TABLE 3 Examples of PCR master mixes

Company	Mix	Web address
Brinkmann	Eppendorf MasterMix	www.emdbiosciences.com
Invitrogen	PCR SuperMix	www.invitrogen.com
Promega	PCR Master Mix	www.promega.com

are mixed together. The reaction mixture is thoroughly heat denatured by heating at 95°C for 5 to 10 min and is then brought to 80°C. The missing ingredient is added and cycling is begun. Preferred methods of performing hot-start PCR include the use of wax beads that contain either Mg^{2+} or *Taq* polymerase, which is not released into the reaction mixture until the wax melts, anti-*Taq* antibodies, which inhibit the action of *Taq* but which denature irreversibly above 80°C; and finally, modified forms of *Taq* which do not become active until they are heated above 80°C (e.g., *AmpliTaq Gold* [Perkin-Elmer Corporation] or HotStar Taq [QIAGEN]).

Booster PCR

The PCR-induced formation of primer dimers and cold fusion products tends to be more prevalent in PCRs that begin with less than 1,000 target templates. This can be alleviated by using booster PCR. In early cycles the concentration of the primers is reduced, thus reducing primer-primer interactions. After 15 cycles the additional primers are added up to a final concentration of 0.1 μM and another 30 to 40 cycles are performed.

Two-Step PCR

In normal PCR, three-temperature cycling is performed. In most cases it is preferable to design primers with higher annealing temperatures. If appropriate PCR primers can be designed to have annealing temperatures of about 65°C or higher, then the annealing and the extension phases of the PCR can be combined, resulting in a two-temperature-cycle PCR. Not only does two-step PCR provide maximal specificity to the amplification, but the time to complete the reaction is also reduced.

Multiplex PCR

Multiplex PCR is a method for the simultaneous amplification of many targets of interest in one reaction through the optimization and use of more than one pair of primers. Multiplex PCR can be used to detect mutations and polymorphisms, for genotyping, for detection of pathogens, etc. The benefit can be seen in the specific detection of more than one type of organism in an environmental sample; however, multiplex assays can be tedious and time-consuming to design, as they require extensive optimization and validation procedures. There is currently great interest in the detection of pathogens by PCR, since the technique does not involve culturing of the organisms prior to analysis. Important here is the fact that many pathogens in environmental samples may be viable but nonculturable. Although such pathogens are not detectable by conventional culture methodologies, these pathogens are capable of infecting humans and causing disease. Multiplex PCR involves the use of multiple sets of primers, which result in multiple products of amplification. Such a multiplex system has been used to distinguish *Salmonella* spp. from *Escherichia coli* (30). The multiplex system is particularly useful when all of the products of amplification are diagnostic for a particular species, since there is then greater assurance that the pathogen of interest will be detected. Multiplex PCR has also been used with multiple sets of primers specific to different genomes. In this case, more than one organism can be detected in one PCR.

RT-PCR

Reverse transcriptase (RT) PCR differs from normal PCR in that it starts with RNA as the template and involves the use of a second enzyme known as RT. During the first reaction cycle this enzyme is used to make a cDNA copy of a single-stranded RNA sequence. Following the RT-catalyzed formation of cDNA from the RNA, the reaction can then be subjected to normal PCR with *Taq* polymerase. The cDNA is a complementary copy of the bases in the RNA, except that thymine replaces uracil. The most commonly used RT enzyme is that of avian myeloblastosis virus (AMV), although Moloney murine leukemia virus RT is also used. AMV RT is often preferred for PCR because it retains significant activity at temperatures higher than 37°C (at least 42°C), which enables stringency to be applied at the reverse transcription stage of the RT-PCR. Thermostable enzymes with RT activity and DNA polymerase activity (e.g., r*Tth* polymerase) are now available. These enzymes offer the advantage that they are suitable for doing RT-PCR in one tube without the need to transfer the reagents to a new buffer. They may also be more efficient than two-enzyme reactions when they are optimized. RT-PCR forms the basis for quantitative PCR methods that are used to estimate enzymatic potential or gene expression. Both of these procedures are discussed below.

UTILITY OF PCR FOR DETECTION OF PATHOGENS AND OTHER ENVIRONMENTAL MICROBES OF INTEREST

Estimates of Enzymatic Potential

Detection of pathogens requires the use of highly specific primers that allow detection of nucleic acid sequences at the species level. However, there are other ways of using PCR with environmental samples with primers of lower specificity that amplify the DNA from a wider array of organisms. Primers derived from conserved sequences or plasmids that code for enzymatic activity can be used to estimate the potential for that enzymatic activity in an environmental sample, even when the activity is due to a diverse number of organisms. For example, primers derived from plasmid pJP4 have been used to estimate the potential degradation of 2,4-dichlorophenoxyacetic acid (29). This plasmid has been associated with a variety of different bacteria. mRNA can also be amplified by RT-PCR. Because mRNA has a half-life of only a few minutes and is produced immediately prior to protein synthesis, detection of mRNA gives an estimate of the metabolic activity that is actually expressed. Devers et al. (8)

used mRNA-based RT-PCR to show the degradative potential of microorganisms by atrazine in soil. They found that *atz* genes are basally expressed and up-regulated in response to atrazine treatment. Such mRNA studies can be used in ecological studies to estimate the specific rates of microbial transformation or the rates of biodegradation.

PCR-Based Analysis of Microbial Communities

The microbial community composition in environmental samples can be described by PCR analysis of conserved genes, such as 16S rRNA. These analyses are critical to many recent phylogenetic studies. PCR amplification of 16S rRNA genes has been used to characterize bacteria from numerous environments (4, 9, 13, 30, 33, 41). By this approach PCR-generated 16S rRNA genes can be produced by using universal primers from conserved regions of DNA. The 16S rRNA gene PCR products can then be subcloned and sequenced. Sequence analysis of fragments allows the diversity of uncultured organisms to be determined.

In addition to describing community structure, PCR-based techniques have been used to estimate levels of diversity in microbial communities. Our views on microbial diversity in recent years have changed dramatically through the advantages that molecular methods, especially PCR, have provided (13). Many excellent reviews on the estimation of microbial diversity in environmental samples exist in the literature. Bohannan and Hughes (3) have reviewed modern molecular techniques and statistical approaches for estimation and comparisons of microbial diversity in environmental samples. Also, Kirk et al. (19) have performed in-depth reviews of molecular methods for the study of diversity.

DNA Fingerprinting by PCR

DNA fingerprinting can be used as a method not only to study individual isolates but also to study communities. In recent years there has been increased interest in the use of DNA fingerprinting methodologies to identify bacteria at the isolate level and to subtype pathogenic bacteria (16, 22). PCR can be used in several different ways to fingerprint bacteria. One method uses a single arbitrary primer and is known as arbitrarily primed PCR (AP-PCR) (38, 47). A random or arbitrary primer (random octamers or random nonamers) can prime a bacterial genome at multiple sites and thus gives rise to multiple amplification products of various sizes. When the products are separated by electrophoresis, they result in a banding pattern or fingerprint that is unique to the organism under investigation. However, when AP-PCR is used, it is critical that all PCR mixture components and conditions be controlled rigorously, since variable conditions result in variable banding patterns.

More recently, repetitive DNA sequences within bacteria have been used to generate DNA fingerprints. Primers designed from these repetitive sequences anneal to the repeat sequences and again result in multiple amplification products. Enterobacterial repetitive intergenic consensus sequence (ERIC) PCR (34) and repetitive intergenic palindromic sequence (REP) PCR (36) are excellent examples of such protocols. ERIC-PCR and REP-PCR differ from AP-PCR in that only two 20-bp primers are used in the assay. Overall, DNA fingerprinting methodologies have great potential to identify isolates of bacteria with similar DNA sequences. It is noteworthy, however, that the degree of discrimination obtained by some of these methodologies is so great that isolates can often be placed into different groups on the basis of their fingerprints, even when no functional differences are observed between the organisms.

Methods used to obtain fingerprints for entire communities of bacteria, in soils, for example, include denaturing gradient gel electrophoresis (DGGE) and temperature gradient gel electrophoresis (TGGE) (7, 12). In DGGE, by the use of universal primers for the 16S rRNA gene for eubacteria and community DNA extracted from a soil sample, the PCR nonspecifically amplifies any 16S rRNA gene template present. This, in effect, results in the formation of a wide variety of products from different bacterial species. Taken together, these products in theory represent the diversity of the microbial populations in the soil sample. Even though these products are all the same relative size and would not be separated or distinguished on a typical agarose gel, they do have different sequences and thus different temperatures of melting primarily on the basis of their relative G+C contents. This fact is used in DGGE and TGGE. By applying a gradient of temperature or chemical denaturant, those strands which have lower temperatures of melting (higher percentages of A and T) are separated first, resulting in decreased electrophoretic mobility, while those sequences with higher melting temperatures (higher G+C contents) migrate faster until a concentration of denaturant or a temperature sufficient to melt the two strands is reached. In this way, the products amplified from community DNA are separated on the basis of their melting temperature domains. Upon separation, individual bands can also be excised and sequenced, providing some information on diversity.

A more recent method, terminal-restriction fragment length polymorphism (T-RFLP) analysis, has also been used to study microbial community structures. T-RFLP analysis is based on the restriction endonuclease digestion of PCR products that have been fluorescently end labeled through the use of labeled primers. The digested product is usually mixed with a DNA size standard which is labeled with a different fluorescent dye, after which the fragments are separated by electrophoresis by using an automated laser detection capillary system like that manufactured by Applied Biosystems. Upon analysis, only the terminal, end-labeled restriction fragments are detected. The reproducibility and robustness of the method are based on a fingerprinting method that produces bands of precise sizes that can be phylogenetically assigned with the sequences in an appropriate database (32).

Q-PCR

One of the more significant developments based upon PCR is probably quantitative PCR (Q-PCR). In certain kinds of investigations, a result indicating only the presence or absence of a particular gene sequence is not sufficient and quantification of the number of transcripts of that gene may be required. For example, an evaluation of the incidence or the relative concentration of a particular biodegradative organism within a soil sample would certainly benefit from quantification of the amplified product (18, 39, 48, 54). Q-PCR allows quantification of the starting amounts of DNA, cDNA, or RNA templates. Q-PCR is based on the detection of a fluorescent reporter molecule whose signal increases as the PCR product accumulates with each cycle of amplification. This signal is typically measured in real time with a specialized thermocycler. A list of quantitative thermocyclers is provided in Table 4. The fluorescent reporter molecules include dyes that bind to double-stranded DNA, i.e., SYBR green I, or sequence-specific probes, such as Molecular Beacons from idtDNA or TaqMan probes from Applied Biosystems. Q-PCR exceeds

TABLE 4 Examples of Q-PCR machines

Company	System	Detection methodology	Web address
Bio-Rad	iCycler iQ real-time PCR detection system	CCD[a] with proprietary intensifying screen	www.bio-rad.com
Applied Biosystems	ABI Prism 7000 sequence detection system	Four-position filter wheel and CCD camera	www.appliedbiosystems.com
Stratagene	Mx3000 real-time PCR system	One scanning photomultiplier tube	www.stratagene.com

[a]CCD, charge-coupled device.

the limitations of traditional end-point PCR methods by allowing either the absolute or relative quantification of the PCR product at the end of each cycle. This ability has greatly enhanced several areas of research, including gene expression analysis and genotyping assays.

Types of Q-PCR

There are two primary forms of Q-PCR. The most popular form of Q-PCR is known as probe-based Q-PCR. The second is a dye-based method. Perkin-Elmer Corporation has developed a very accurate Q-PCR system called the TaqMan system. The TaqMan system uses two specific primers and a third oligonucleotide which is actually an internal probe labeled with two separate fluorescent molecules. The first molecule on the probe is termed the signal dye, and the second is known as the quencher dye. When the probe is not bound to DNA, its secondary structure holds the two labels in close proximity to each other, causing the quencher dye to absorb the fluorescence emitted by the signal dye. During amplification of the correct fragment the probe will bind internally to the short PCR product. The probe is modified at its 3′ end to prevent extension from it by *Taq* polymerase. When the *Taq* polymerase encounters the probe as it copies the template, it digests it base by base from its 5′ end, releasing individual nucleotides into solution. When the quencher and signal dyes are separated by this means, the signal dye's emissions become detectable. During the PCR, a laser scanner monitors the increase in signal strength. The increase in signal strength can be compared to the signals on a standard curve and has been found to be linear during the initial cycles of the PCR. The second type of Q-PCR methodology is based on SYBR green. QIAGEN has a SYBR green-based Q-PCR kit which is different from the kits used for probe-based methods because it essentially detects all double-stranded DNA molecules. SYBR green I, a commonly used fluorescent DNA binding dye, binds to all double-stranded DNA; and detection is monitored by measuring the in-crease in fluorescence throughout the cycle. SYBR green I has excitation and emission maxima of 494 and 521 nm, respectively. The specificity of Sigma's SYBR green-based Q-PCR detection is greatly enhanced by the incorporation of a hot-start-mediated *Taq* polymerase. Many commercial companies now sell PCR kits, similar to the master mixes noted above, that facilitate Q-PCR analysis. Table 5 lists some of these manufacturers and quantitative PCR kits.

ADVANCED PCR METHODOLOGIES AND CONSIDERATIONS

Design of Primers

It is probably already apparent that the choice of the primer sequences is critical for the successful amplification of a specific DNA sequence. The PCR primer sequences can be deduced from known DNA sequences as well as by the use of computer programs. The overall choice of the primers must be guided by the aim of the researcher. If detection of a target DNA that is specific to a given species or genus of a bacterium is required, then sequences unique to that bacterium must be used for the design of the primers. For example, the *lamB* gene codes for the production of an outer membrane protein in *E. coli*, and primers designed from this gene sequence detect *E. coli*. However, in some cases it may be more prudent to design primers from conserved sequences. Thus, for instance, primers designed from conserved portions of *nod* genes could be used to detect all species of *Rhizobium*. Strategy is therefore an important aspect governing the choice of the sequence from which the primers will be designed. In addition, other criteria determine the ultimate choice of primer sequences. In general, most primers are 17 to 30 bp and are separated by a few as 100 bp or as many as several thousand base pairs. Of course, the distance between the primers determines the size of the amplification product. If primers contain complementary sequences, they can hybridize to each other, producing a primer dimer. DNA sequences specific for almost any well-characterized organism are accessible in computer databases, such as GenBank.

The selection of oligonucleotides which perform well as PCR primers is a matter of identifying some simple sequence-related characteristics. DNA polymerase focuses on 3′ ends of double-stranded nucleic acid sequence with unpaired 5′ extensions. Therefore, most of the calculations

TABLE 5 Q-PCR reagent kits

Company	Mix	Web address
Stratagene	FullVelocity QRT-PCR master mix	www.stratagene.com
Applied Biosystems	TaqMan one-step RT-PCR master mix reagents kit	www.appliedbiosystems.com
Sigma-Aldrich	Quantitative RT-PCR ReadyMix	www.sigma-aldrich.com
ABgene	ABsolute QRT-PCR mix	www.abgene.com

for primer selection are based on the location of 3′ ends that are unique, pair with relatively low stability (to avoid priming of other sites), and have little internal complementarities and secondary structure (to avoid priming of primers). Beyond these simple features, the design of efficient primers requires evaluation of the free energy and melting temperature of the target sequence in comparison to those of potential primer pairs. These factors are related to the number of hydrogen bonds formed or the G+C content of the template, with GC-rich sequences having greater stability than AT-rich sequences. In addition, sequence ordering is also a component of stability. For example, the duplex TT/AA is more stable than the duplex AC/GT due to a more favorable interaction between the neighboring bases on the same strand. The stability of the duplex is based on the free energy of adjacent dinucleotides and is calculated from dinucleotide entropy and enthalpy data. The calculation of free energy, ΔG, is

$$\Delta G = \Delta H - T\Delta S \quad (1)$$

where ΔH is enthalpy, ΔS is entropy, and T is temperature. The value of T would usually default to 25°C, but this factor can be manipulated as well.

For a given sequence, $S(abcde)$, where a, b, c, d, and e are nucleotides in the sequence, ΔH is the sum of ΔH values of the dinucleotides:

$$\Delta H(abcde) = \Delta H(ab) + \Delta H(bc) + \Delta H(cd) + \Delta H(de) + \Delta H_{ini} \quad (2)$$

where ΔH_{ini} is a helix formation constant.

ΔS is the sum of the ΔS values for the dinucleotides:

$$\Delta S(abcde) = \Delta S(ab) + \Delta S(bc) + \Delta S(cd) + \Delta S(de) + \Delta S_{ini} \quad (3)$$

where ΔS_{ini} is a helix formation constant.

Good primers will have the following characteristics: (i) ΔG values of the entire primer that are near the mean ΔG value for the entire template; (ii) pentamer ΔG values of the primer at the 3′ end that are lower than the mean ΔG value for the template, and (iii) the stability profile of pentamer ΔG values for the primer that are above the mean near the 5′ end and descending in value toward the 3′ end.

In addition to free energy comparisons between the primers and the template, melting temperature is another critical factor in primer design. Efficient primers should have very similar melting temperatures in order to promote the amplification of both strands in equal amounts. The melting temperature (T_m) of primers is calculated with the following formula:

$$T_m = 81.5 + 16.6\log[\text{salt}] + \frac{41(\%\ \text{G+C content})}{100} - \frac{65(\%\ \text{formamide content})}{100} - \frac{675}{\text{length}} \quad (4)$$

where [salt] is the concentration of Na^+ in moles. As mentioned previously, the high-efficiency amplification of products requires an annealing temperature that is optimal for the specific combination of primers and the template sequence. A carefully chosen annealing temperature will significantly increase the product yield and reduce the amplification of partially polymerized nonspecific products. Annealing temperature (T_a) calculations are as follows:

$$T_a = 0.3\ T_m\ \text{primer} + 0.7\ T_m\ \text{product} - 14.9 \quad (5)$$

where T_m primer is the minimum melting temperature of the two primers and T_m product is the melting temperature of the product. Many commercially available programs that will do these calculations are available, and others are beyond the scope of this discussion. One of the more popular and efficient is PrimerSelect (DNAstar Inc, Madison, Wis.). Table 6 provides a list of primer design software that can be found on the Internet or purchased from commercial sources.

Specificity of Amplification

The degree of specificity of primers used in PCR can intentionally be varied. If unique primers are required to amplify specific DNA, then unique DNA sequences must be chosen as the target for amplification, e.g., a primer pair that allows only amplification of *Salmonella*-specific DNA and not *E. coli*-specific DNA. Since these two groups of organisms are closely related, the design of these primers is critical to distinguish between the species of the two genera. However, if the detection of all or most bacterial DNA in an environmental sample is required (e.g., in a bioassay), then primers termed universal primers are designed that allow amplification of a conserved DNA sequence present in all bacteria. Such conserved sequences exist within 16S rRNA genes.

The degree of specificity can be varied by primer design and also by changing the annealing temperature. In general, as the annealing temperature is increased, the number of base pair mismatches allowable for hybridization decreases, increasing the specificity of amplification. Increasing the annealing temperature from 50 to 55°C often decreases nonspecific amplification. However, along with increased specificity, there is an associated decrease in sensitivity. The maximum allowable annealing temperature is generally 10°C less than the melting temperature, which depends on the percent G+C content. Primers can sometimes result in several different amplification products, in addition to an amplification product specific to the target DNA. The amplification product from the correct target sequence can be identified by the use of appropriate size markers of standard DNA, but the identity must be confirmed.

Sensitivity of Detection

The sensitivity of amplification is important when a given DNA sequence is in an environmental sample but is present at a low concentration or a low copy number. In this

TABLE 6 Examples of primer design software

Company	Item	Platform	Web address
Integrated DNA Technologies	PrimerQuest	Web	scitools.idtdna.com/Primerquest
Whitehead Institute for Biomedical Research	Primer 3	Web	fokker.wi.mit.edu/cgi-bin/primer3/primer3_www.cgi
DNAstar	Primer Select	PC or Mac	www.dnastar.com

instance, the amplification protocol must be sensitive enough to detect the target DNA sequence and not result in a false negative. The level of sensitivity must be defined so that a negative result can be quantified. For example, current Environmental Protection Agency drinking water standards necessitate that less than one *Salmonella* cell per 100 ml be detectable; therefore, the use of PCR to detect *Salmonella* must somehow be capable of meeting these standards. Sensitivity can be increased by (i) optimization of amplification, (ii) increasing the number of cycles, or (iii) concentration of the amount of target DNA in an environmental sample, for example, by filtration of a large volume of water followed by elution of bacterial cells into a smaller final volume.

Sensitivity can be evaluated in terms of whole-cell lysates or pure genomic DNA preparations. In terms of whole cells, target bacteria are grown to late log phase in broth medium, serially diluted, and plated on agar to determine the numbers of CFU. Concurrently, 500 μl of each dilution is added to a PCR tube and centrifuged at 14,000 rpm for 10 min. All but 10 μl of the supernatant is discarded, and the resulting pellet is resuspended in 89.5 μl of the reaction mixture lacking enzyme. The cells are boiled for 10 min, and 0.5 μl of *Taq* polymerase is added after the tubes are cooled to room temperature. PCR amplification is then conducted. Sensitivity can also be determined in terms of pure genomic DNA by using 10-fold dilutions of stock DNA preparations as a template for PCR. The DNA concentrations in the stock preparations are quantified by using A_{260} values.

Sensitivity in terms of cell counts based on the numbers of CFU can be misleading since the total number of copies of a target sequence is always greater than the numbers of CFU. Broth cultures inevitably contain dead or lysed cells with target sequences, and each viable cell likely contains multiple copies of each genome. Therefore, equating sensitivity with the numbers of CFU tends to overestimate the actual sensitivity of the method. Sensitivity also depends on the method used to detect the amplified product. By using ethidium bromide staining of DNA, PCR can often detect 10^3 to 10^4 cells, whereas the use of ^{32}P-labeled gene probes usually increases the sensitivity by 2 orders of magnitude. The use of other DNA stains such as SYBR green or SYBR gold (Molecular Probes Inc. Eugene, Oreg.) has also resulted in increased sensitivity over that achieved by ethidium bromide staining.

In terms of pure DNA, low numbers of target cells can typically be amplified (20, 51). Consider a total genome of 4×10^6 bp, which is equivalent to approximately 9 fg of DNA, and a target sequence of 179 bp. One copy of the target DNA is therefore equivalent to $(179/4 \times 10^6) \times 9 \times 10^{-15}$ g, or 0.4 ag. Therefore, one copy of the target DNA represents 0.4 ag and the sensitivity of detection based on different studies can be determined. Overall, the issue of sensitivity is complex and must be evaluated for each individual set of primers. Such an evaluation is critical if PCR amplifications are to be used for diagnostic purposes. In addition, it must be noted that sensitivity in pure culture is always greater than that in environmental samples due to the presence of substances inhibitory to the PCR.

PCR Optimization

Adjuncts or additives such as dimethyl sulfoxide (1.1 to 5%), formamide (1.25 to 10%), and glycerol (15 to 20%) are often added to PCR mixtures to aid preferential primer-target annealing. Glycerol increases and formamide decreases the thermostability of *Taq* polymerase. The use of adjuncts can either enhance or be deleterious to PCR. Some reports indicate up to twofold yield increases; others indicate a decrease in yield, based on the use of certain adjuncts. Nonionic detergents such as Nonidet P-40, Tween 20, and Triton X-100 (0.01 to 0.1%) can also aid in the preferential annealing of primers to template over reannealing of double-stranded full-length fragments and stabilize *Taq* at high temperatures. Bovine serum albumin (BSA) was initially included in most PCR buffers (10 to 100 $\mu g\ ml^{-1}$) in order to stabilize the polymerase. In the presence of high concentrations of PCR buffer salts, BSA tends to precipitate, sequestering with it other reagents. Gelatin at approximately 0.01% is another alternative small protein used to stabilize *Taq*. Ammonium sulfate (10 to 25 mM) may also be added to attempt to enhance annealing. The single-stranded DNA binding protein has also been used to increase the PCR product yield. Many adjuncts may improve the yield of the target product by chemically decreasing the annealing temperature. Examples of optimization kits containing a range of adjuvants that can be used to optimize and increase the sensitivity and the specificity of PCRs are provided in Table 7.

Confirmation of Product DNA Sequences

The primers used in PCR delineate the amplified product size, based on the positions of primer annealing within the microbial genome. The theoretical product size can be estimated from DNA size markers or ladders and gives an indication that the correct nucleic acid sequence has been amplified. However, it is useful to definitively prove that the actual product sequence is the desired target. Gene probe techniques such as Southern hybridizations have traditionally been used to identify target sequences. The correct amplification product can also be identified by the use of end-labeled gene probes specific to an internal region of the amplified product.

An alternative approach that does not use probes is seminested PCR. In seminested PCR, the upstream external primer used in the original amplification and an internal downstream primer are used. The internal primer is designed from sequences contained within the theoretical original product sequence. Therefore, in the second seminested PCR protocol, amplification occurs only if the correct internal target sequence was initially present. Seminested PCR uses an additional 25 to 30 cycles, with the template being an aliquot of the original PCR mixture. In essence, then, this is a

TABLE 7 Examples of PCR optimization kits containing adjuncts and optimization buffers

Company	Kit	Web address
Roche Applied Science	PCR optimization kit	www.roche-applied-science.com
Stratagene	Opti-Prime PCR optimization kit	www.stratagene.com
Invitrogen	PCR Optimizer kit	www.invitrogen.com
EPICENTRE	FailSafe PCR PreMix selection kit	www.epibio.com

"double" PCR and can increase the sensitivity of detection by 1 to 2 orders of magnitude.

With the advent of modern low-cost and high-throughput sequencing technology, the use of probes and other methods for confirmation of the product sequence is now becoming a thing of the past. PCR product sequencing followed by computer database homology comparison is now the best method for the evaluation of PCR products. The process is fairly simple and uses commonly available kits. The primary drawback to this method is the high front-end cost required to obtain sequencing machines. However, most universities now have core sequencing facilities, and large commercial companies offer low-cost sequencing services. Thus, essentially for the same cost and time that it would take to confirm the identity of a product by Southern hybridization, you can obtain and confirm the entire PCR product sequence. Although it is unlikely in a well-designed PCR, it is possible that an amplicon that is of the correct size but that does not originate from the target sequence could be amplified. If no further confirmation of the identity of the product were done, a false-positive result could be recorded. Other PCR amplicon verifications commonly take the form of a Southern or dot-blot hybridization with an internal probe, use of an internal probe in a system such as TaqMan, sequencing of the product, or digestion of the product with specific restriction endonucleases. Alternatively, nested PCR can provide increased certainty that the correct amplicon has been obtained. For high-throughput laboratories, the use of systems such as TaqMan is suitable. For smaller laboratories, nested PCR or restriction digestion of the amplicon is the least expensive and most rapid means of product verification.

Limitations of PCR

The advantages of PCR are well known and include speed, sensitivity, and relatively low costs; however, there are some disadvantages. One of the major issues is that PCR detects nonviable organisms. Josephson et al. (15) demonstrated that PCR would detect nonviable bacterial pathogens, and therefore, the relevance of a PCR-positive detection was brought into question. However, that study also showed that the nucleic acids within nonviable bacteria degrade relatively quickly in the environment and, therefore, that a positive PCR result implied that the organisms either were viable or were recently viable. Analogous data for viruses were obtained by Straub et al. (40). In that study, cell culture analyses of environmental samples for enteroviruses were negative, whereas PCR analyses gave positive identifications. That study showed that noninfectious viruses could remain in the environment for long periods of time, with intact nucleic acids protected by the viral protein capsids. Therefore, for viruses, the public health hazards of PCR-positive results need to be evaluated. Note, however, that the integrated cell culture-PCR methodology (37) is much more likely to detect only infectious virus.

There are three other major limitations to the PCR detection of gene sequences in environmental samples. One is small sample size; in other words, the PCR template is limited to only a few microliters, which can limit the amount of the environmental sample that can be processed. This problem can be overcome by the use of innovative methods of sample processing. The second limitation is the presence of substances inhibitory to the PCR; and the third limitation is laboratory contamination, which results in false-positive results. However, even with these three primary limitations, we cannot ignore the fact that there are many potential sources of bias when we attempt to use PCR for the evaluation of environmental samples. Examples include the preferential annealing of primers to some targets, which results in the amplification of some sequences but the complete lack of amplication of others (43). Therefore, it is important to optimize those factors that we can control to reduce the variability and bias that are inherent with many PCR-based analyses. The factors believed to be important include sample handling, the use of replicates, pooling of samples, dilution effects, testing and choice of polymerase enzymes, primers, and reaction annealing temperatures.

Preventing Contamination

False positives arise from contamination but can be reduced by the following:

- Use pre- and post-PCR stations. To reduce the carryover of amplified DNA, samples should be prepared at a pre-PCR station in a separate room or far away from where the amplified products are analyzed. This pre-PCR station is required to be an amplified product-free zone. A different post-PCR station is used to analyze products by electrophoresis or by the use of gene probes. The single most important step required to avoid contamination is the complete separation of the PCR preparation location from anything to do with the amplified PCR product, to conduct the PCR setup in laminar-flow hoods, and to use UV light to control any template carryover before and after PCR.
- For pre-PCR use dedicated sets of pipette tips, reagents, tubes, and pipetting devices that are not used for any other purpose. Template DNA should not be handled with these dedicated pieces of equipment.
- Autoclave buffer solutions and use molecular-grade water. Note that primers, deoxynucleoside triphosphates (dNTPs), and *Taq* DNA polymerase cannot be autoclaved.
- Aliquot reagents to minimize the number of repeated samplings from given reagents. Label aliquots so that if contamination occurs, it can be traced.
- Use disposable gloves and change them frequently, particularly between pre- and post-PCR stations.
- Spin down tubes prior to opening to remove sample from tube caps and walls.
- Use positive-displacement pipettes to avoid contamination via aerosols.
- Premix reagents before dividing them into aliquots; i.e., use master mixes. Pipette the reagents for a no-DNA negative control last.
- Add all reagents before the addition of sample DNA; i.e., add DNA last, and cap each tube before proceeding to the next sample.
- Never analyze or purify PCR products on the same day before the preparation of a PCR. If necessary, always set up the PCR before analysis of the products from another PCR.
- Use the minimum number of PCR cycles possible for a given sample.
- Always use at least two negative controls with every assay. The negative control contains no added DNA as a template. One is made and capped before the template DNA is added to any of the other tubes and serves as a reagent control. The second is made last and serves as a control for cross contamination during the PCR setup.

PROCESSING OF SOIL SAMPLES FOR PCR AMPLIFICATIONS

PCR and Soils Do Not Mix

Although the technology of PCR is relatively easy to conduct with samples from a pure culture, it is often much more difficult to carry out with nucleic acid sequences derived from environmental samples, particularly soils. Researchers familiar with the manipulation of a variety of sample types believe that DNA extraction and purification from environmental samples, especially soils and sediments, are some of the most difficult tasks in molecular biology (1, 14, 25). Thus, as detailed in this section, extensive processing is usually required to remove substances that would inhibit the PCR. PCR can be inhibited either physically or chemically. Physical inhibition can result from the presence of soil colloids associated with DNA extracted from a soil or sediment sample. Some of these chemical inhibitors that are found in soils have been characterized, such as fulvic acids, humic acids, and heavy metals, while other inhibitors remain to be identified. Even trace amounts carried over from samples into the PCR can cause enzyme inhibition (2, 21, 42, 45). Compounds with the potential for inhibition can be present in any environmental sample. Examples of environmental samples include sewage sludges, soils, sediments, marine waters, groundwater, ultrapure water, and food. Because of the diverse nature of these samples, different PCR protocols and DNA extraction methods have been developed for the specific, sensitive detection of a variety of nucleic acid sequences.

Extraction of Cells from Soil

There have been many approaches that allow the PCR detection of DNA sequences derived from soil and sediments (6, 11). The first involves the extraction of intact cells from soil, followed by cell lysis and subsequent PCR analysis. This can be done by adding surfactants to aid with the removal of cells that are absorbed to colloidal or humic substances and then extracting the soil with some kind of solvent. Many extraction solutions have been tried, including soluble sodium solutions, phosphate buffers, and calcium solutions. In practice, the percent recovery of cells from soil depends on the nature of the soil, as well as the nature of the organisms. Typically, higher recoveries are obtained from coarse-textured soils than fine-textured soils or soils high in organics. Autochthonous cells that have been present in soil for long periods of time are normally sorbed more strongly than young zymogenous cells growing rapidly due to a readily metabolizable substrate. Some cells are intrinsically attached more tightly to soil particles than others due to the secretion of polysaccharides. All of these factors can result in the selective extraction of some bacterial populations over others. In addition, it is difficult to separate the cells from the soil particles themselves. Differential centrifugation has been used to overcome this obstacle (35). However, despite these problems, cell extraction still likely results in a more representative estimate of the total bacterial community than simple dilution and plating. Once the soil has been extracted, the total cell biomass can be lysed and then subjected to PCR analysis. Lysis is normally done directly in a thermocycler at a temperature of 98°C. Prior to lysis, the cells must be purified because humic acids can inhibit the PCR (6).

Direct Lysis of Bacteria in Soil

Several protocols for the extraction of community DNA rely on the use of lysozyme or sodium dodecyl sulfate as a bacterial cell lysing agent. Once it is extracted, the DNA can be purified by cesium chloride density centrifugation or by the use of commercial kits (see above). An alternative and more frequently used addition to these protocols is phenol extraction followed by ethanol precipitation as a means of preliminary purification. Once the DNA has been purified it can be subjected to PCR analysis. Although cell lysis in situ is a very attractive approach, it is not without limitations. Soils high in colloids can adsorb DNA so that the efficiency of extraction is not identical in different soils (50). Also, the efficiency of lysis is not always perfect. Finally, note that DNA extractions may include eukaryotic DNA as well as bacterial DNA. Overall, the community-based extraction methods (cells or DNA) represent a higher proportion of the bacterial population than dilution and plating alone and often result in broader estimations of diversity. They also are likely to represent a different segment of the population of bacteria than that obtained by the cultural methodology. They also have the advantage of being well suited to subsequent PCR analysis.

Commercially Available Kits for Extraction of DNA from Soil

Currently there are commercially available DNA extraction and purification kits designed for use with soils. Two soil DNA extraction kits are detailed in Table 8. These are completely self-contained, standardized kits used for the extraction of community DNA from soils. These kits are typically easy to use and the results are reproducible, but as with other methods, they may not work consistently for different types of soil.

CONCLUSIONS

PCR is a powerful molecular tool that can be used to detect gene sequences associated with plant and soil microbes. PCR applications include detection of specific microbes and estimates of microbial diversity and enzymatic activity. The key to the successful use of PCR is in the design of the

TABLE 8 Soil DNA extraction kits

Company	Kit	Quantity	Maximum yield	Method	Starting amt	Web address
EPICENTRE	SoilMaster DNA extraction kit	50 extractions	?	Spin column	100 mg of soil/column	www.epibio.com
Mo Bio Laboratories, Inc.	PowerSoil DNA isolation kit	50 preparations	20 μg	Bead beating, spin filter method	250 mg	www.mobio.com

primers, which should be directed by the research objectives of the scientist. Although PCR with isolates from pure cultures is relatively straightforward, PCR detection of gene sequences in environmental samples is not so simple. Such samples often contain substances inhibitory to the PCR, and these substances must be removed prior to PCR amplification. Different protocols are available for microbes in either plant or soil environments. Overall, PCR adds a useful new technology to aid in the study of plant and soil microbes.

REFERENCES

1. **Anderson, I. C., C. D. Campbell, and J. I. Prosser.** 2003. Diversity of fungi in organic soils under a moorland-Scots pine (Pinus sylvestris L.) gradient. *Environ. Microbiol.* **5:** 1121–1132.
2. **Bahnweg, G., S. Schulze, E. M. Moller, H. Rosenbrock, C. Langebartels, and H. Sandermann, Jr.** 1998. DNA isolation from recalcitrant materials such as tree roots, bark, and forest soil for the detection of fungal pathogens by polymerase chain reaction. *Anal. Biochem.* **262:**79–82.
3. **Bohannan, B. J., and J. Hughes.** 2003. New approaches to analyzing microbial biodiversity data. *Curr. Opin. Microbiol.* **6:**282–287.
4. **Bruce, K. D., W. D. Hiorns, J. L. Hobman, A. M. Osborn, P. Strike, and D. A. Ritchie.** 1992. Amplification of DNA from native populations of soil bacteria by using the polymerase chain reaction. *Appl. Environ. Microbiol.* **58:**3413–3416.
5. **Chelius, M. K., and E. W. Triplett.** 2001. The diversity of Archaea and Bacteria in association with the roots of Zea mays L. *Microb. Ecol.* **41:**252–263.
6. **Courtois, S., A. Frostegard, P. Goransson, G. Depret, P. Jeannin, and P. Simonet.** 2001. Quantification of bacterial subgroups in soil: comparison of DNA extracted directly from soil or from cells previously released by density gradient centrifugation. *Environ. Microbiol.* **3:**431–439.
7. **de Oliveira, V. M., H. L. Coutinho, B. W. Sobral, C. T. Guimaraes, J. D. van Elsas, and G. P. Manfio.** 1999. Discrimination of Rhizobium tropici and R. leguminosarum strains by PCR-specific amplification of 16S-23S rDNA spacer region fragments and denaturing gradient gel electrophoresis (DGGE). *Lett. Appl. Microbiol.* **28:**137–141.
8. **Devers, M., G. Soulas, and F. Martin-Laurent.** 2004. Real-time reverse transcription PCR analysis of expression of atrazine catabolism genes in two bacterial strains isolated from soil. *J. Microbiol. Methods* **56:**3–15.
9. **Diallo, M. D., M. Martens, N. Vloemans, S. Cousin, T. T. Vandekerckhove, M. Neyra, P. de Lajudie, A. Willems, M. Gillis, W. Vyverman, and K. Van der Gucht.** 2004. Phylogenetic analysis of partial bacterial 16S rDNA sequences of tropical grass pasture soil under Acacia tortilis subsp. raddiana in Senegal. *Syst. Appl. Microbiol.* **27:**238–252.
10. **Farr, C. J., R. K. Saiki, H. A. Erlich, F. McCormick, and C. J. Marshall.** 1988. Analysis of RAS gene mutations in acute myeloid leukemia by polymerase chain reaction and oligonucleotide probes. *Proc. Natl. Acad. Sci. USA* **85:** 1629–1633.
11. **Felske, A., B. Engelen, U. Nubel, and H. Backhaus.** 1996. Direct ribosome isolation from soil to extract bacterial rRNA for community analysis. *Appl. Environ. Microbiol.* **62:**4162–4167.
12. **Garbeva, P., J. A. van Veen, and J. D. van Elsas.** 2003. Predominant Bacillus spp. in agricultural soil under different management regimes detected via PCR-DGGE. *Microb. Ecol.* **45:**302–316.
13. **Hugenholtz, P., B. M. Goebel, and N. R. Pace.** 1998. Impact of culture-independent studies on the emerging phylogenetic view of bacterial diversity. *J. Bacteriol.* **180:** 4765–4774.
14. **Hurt, R. A., X. Qiu, L. Wu, Y. Roh, A. V. Palumbo, J. M. Tiedje, and J. Zhou.** 2001. Simultaneous recovery of RNA and DNA from soils and sediments. *Appl. Environ. Microbiol.* **67:**4495–4503.
15. **Josephson, K. L., C. P. Gerba, and I. L. Pepper.** 1993. Polymerase chain reaction detection of nonviable bacterial pathogens. *Appl. Environ. Microbiol.* **59:**3513–3515.
16. **Joshi, B., and S. Walia.** 1995. Characterization by arbitrary primer polymerase chain reaction of polychlorinated biphenyl (PCB)-degrading strains of Comamonas testosteroni isolated from PCB-contaminated soil. *Can. J. Microbiol.* **41:**612–619.
17. **Junca, H., and D. H. Pieper.** 2003. Amplified functional DNA restriction analysis to determine catechol 2,3-dioxygenase gene diversity in soil bacteria. *J. Microbiol. Methods* **55:**697–708.
18. **Kabir, S., N. Rajendran, T. Amemiya, and K. Itoh.** 2003. Real-time quantitative PCR assay on bacterial DNA: in a model soil system and environmental samples. *J. Gen. Appl. Microbiol.* **49:**101–109.
19. **Kirk, J. L., L. A. Beaudette, M. Hart, P. Moutoglis, J. N. Klironomos, H. Lee, and J. T. Trevors.** 2004. Methods of studying soil microbial diversity. *J. Microbiol. Methods* **58:** 169–188.
20. **Knaebel, D. B., and R. L. Crawford.** 1995. Extraction and purification of microbial DNA from petroleum-contaminated soils and detection of low numbers of toluene, octane and pesticide degraders by multiplex polymerase chain reaction and Southern analysis. *Mol. Ecol.* **4:**579–591.
21. **LaMontagne, M. G., F. C. Michel, Jr., P. A. Holden, and C. A. Reddy.** 2002. Evaluation of extraction and purification methods for obtaining PCR-amplifiable DNA from compost for microbial community analysis. *J. Microbiol. Methods* **49:**255–264.
22. **Luo, H. F., H. Y. Qi, and H. X. Zhang.** 2004. Diversity surveys of soil bacterial community by cultivation-based methods and molecular fingerprinting techniques. *J. Environ. Sci.* (China) **16:**581–584.
23. **Malkawi, H. I., I. Saadoun, F. A. Moumani, and M. M. Meqdam.** 1999. Use of RAPD-PCR fingerprinting to detect genetic diversity of soil Streptomyces isolates. *New Microbiol.* **22:**53–58.
24. **Marchesi, J. R., and A. J. Weightman.** 2003. Diversity of alpha-halocarboxylic acid dehalogenases in bacteria isolated from a pristine soil after enrichment and selection on the herbicide 2,2-dichloropropionic acid (Dalapon). *Environ. Microbiol.* **5:**48–54.
25. **Martin-Laurent, F., L. Philippot, S. Hallet, R. Chaussod, J. C. Germon, G. Soulas, and G. Catroux.** 2001. DNA extraction from soils: old bias for new microbial diversity analysis methods. *Appl. Environ. Microbiol.* **67:**2354–2359.
26. **Mullis, K., F. Faloona, S. Scharf, R. Saiki, G. Horn, and H. Erlich.** 1986. Specific enzymatic amplification of DNA in vitro: the polymerase chain reaction. *Cold Spring Harbor Symp. Quant. Biol.* **51** (Pt 1):263–273.
27. **Mullis, K. B.** 1991. The polymerase chain reaction in an anemic mode: how to avoid cold oligodeoxyribonuclear fusion. *PCR Methods Appl.* **1:**1–4.
28. **Mullis, K. B.** 1990. The unusual origin of the polymerase chain reaction. *Sci. Am.* **262:**56–61, 64–65.
29. **Neilson, J. W., K. L. Josephson, S. D. Pillai, and I. L. Pepper.** 1992. Polymerase chain reaction and gene probe detection of the 2,4-dichlorophenoxyacetic acid degradation plasmid, pJP4. *Appl. Environ. Microbiol.* **58:**1271–1275.

30. **Nogales, B., E. R. Moore, E. Llobet-Brossa, R. Rossello-Mora, R. Amann, and K. N. Timmis.** 2001. Combined use of 16S ribosomal DNA and 16S rRNA to study the bacterial community of polychlorinated biphenyl-polluted soil. *Appl. Environ. Microbiol.* **67:**1874–1884.
31. **Ogram, A., W. Sun, F. J. Brockman, and J. K. Fredrickson.** 1995. Isolation and characterization of RNA from low-biomass deep-subsurface sediments. *Appl. Environ. Microbiol.* **61:**763–768.
32. **Osborn, A. M., E. R. Moore, and K. N. Timmis.** 2000. An evaluation of terminal-restriction fragment length polymorphism (T-RFLP) analysis for the study of microbial community structure and dynamics. *Environ. Microbiol.* **2:** 39–50.
33. **Peixoto, R. S., H. L. da Costa Coutinho, N. G. Rumjanek, A. Macrae, and A. S. Rosado.** 2002. Use of rpoB and 16S rRNA genes to analyse bacterial diversity of a tropical soil using PCR and DGGE. *Lett. Appl. Microbiol.* **35:**316–320.
34. **Peters, S. M., J. Bryan, and M. F. Cole.** 2000. Enterobacterial repetitive intergenic consensus polymerase chain reaction typing of isolates of Enterobacter cloacae from an outbreak of infection in a neonatal intensive care unit. *Am. J. Infect. Control* **28:**123–129.
35. **Pillai, S. D., K. L. Josephson, R. L. Bailey, C. P. Gerba, and I. L. Pepper.** 1991. Rapid method for processing soil samples for polymerase chain reaction amplification of specific gene sequences. *Appl. Environ. Microbiol.* **57:** 2283–2286.
36. **Rahmati, A., M. Gal, G. Northey, and J. S. Brazier.** 2005. Subtyping of Clostridium difficile polymerase chain reaction (PCR) ribotype 001 by repetitive extragenic palindromic PCR genomic fingerprinting. *J. Hosp. Infect.* **60:**56–60.
37. **Reynolds, K. A., C. P. Gerba, and I. L. Pepper.** 1996. Detection of infectious enteroviruses by an integrated cell culture-PCR procedure. *Appl. Environ. Microbiol.* **62:** 1424–1427.
38. **Sato, M., S. Yoshida, K. Nagao, S. Nishizono, M. Kusaba, M. C. Hung, I. Ikeda, and K. Imaizumi.** 2000. Application of the random arbitrary primed polymerase chain reaction differential display method to isolate genes of cholesterol metabolism-related proteins from rat liver. *Biosci. Biotechnol. Biochem.* **64:**1058–1060.
39. **Schwartz, E., V. Trinh, and M. Scow.** 2000. Measuring growth of a phenanthrene-degrading bacterial inoculum in soil with a quantitative competitive polymerase chain reaction method. *FEMS Microbiol. Ecol.* **34:**1–7.
40. **Straub, T. M., I. L. Pepper, and C. P. Gerba.** 1995. Comparison of PCR and cell culture for detection of enteroviruses in sludge-amended field soils and determination of their transport. *Appl. Environ. Microbiol.* **61:**2066–2068.
41. **Sun, H. Y., S. P. Deng, and W. R. Raun.** 2004. Bacterial community structure and diversity in a century-old manure-treated agroecosystem. *Appl. Environ. Microbiol.* **70:**5868–5874.
42. **Sutlovic, D., G. M. Definis, S. Andelinovic, D. Gugic, and D. Primorac.** 2005. Taq polymerase reverses inhibition of quantitative real time polymerase chain reaction by humic acid. *Croat. Med. J.* **46:**556–562.
43. **Suzuki, M. T., and S. J. Giovannoni.** 1996. Bias caused by template annealing in the amplification of mixtures of 16S rRNA genes by PCR. *Appl. Environ. Microbiol.* **62:**625–630.
44. **Tokumoto, Y., and K. Saigo.** 1990. A short review of the polymerase chain reaction. *Rinsho Byori* **85**(Suppl.)**:**32–42. (In Japanese.)
45. **Tsai, Y. L., and B. H. Olson.** 1992. Rapid method for separation of bacterial DNA from humic substances in sediments for polymerase chain reaction. *Appl. Environ. Microbiol.* **58:**2292–2295.
46. **Utsumi, M., S. E. Belova, G. M. King, and H. Uchiyama.** 2003. Phylogenetic comparison of methanogen diversity in different wetland soils. *J. Gen. Appl. Microbiol.* **49:**75–83.
47. **van Belkum, A., W. van Leeuwen, J. Kluytmans, and H. Verbrugh.** 1995. Molecular nosocomial epidemiology: high speed typing of microbial pathogens by arbitrary primed polymerase chain reaction assays. *Infect. Control Hosp. Epidemiol.* **16:**658–666.
48. **van Elsas, J. D., A. S. Rosado, A. C. Wolters, E. Moore, and U. Karlson.** 1998. Quantitative detection of Sphingomonas chlorophenolica in soil via competitive polymerase chain reaction. *J. Appl. Microbiol.* **85:**463–471.
49. **Vas, A.** 1992. Polymerase chain reaction and other gene techniques in pharmacogenetics: an introduction and review. *Acta Physiol. Hung.* **79:**253–260.
50. **Watson, R. J., and B. Blackwell.** 2000. Purification and characterization of a common soil component which inhibits the polymerase chain reaction. *Can. J. Microbiol.* **46:**633–642.
51. **Way, J. S., K. L. Josephson, S. D. Pillai, M. Abbaszadegan, C. P. Gerba, and I. L. Pepper.** 1993. Specific detection of *Salmonella* spp. by multiplex polymerase chain reaction. *Appl. Environ. Microbiol.* **59:**1473–1479.
52. **Williamson, N., P. Brian, and E. M. Wellington.** 2000. Molecular detection of bacterial and streptomycete chitinases in the environment. *Antonie Leeuwenhoek* **78:**315–321.
53. **Wright, P. A., and D. Wynford-Thomas.** 1990. The polymerase chain reaction: miracle or mirage? A critical review of its uses and limitations in diagnosis and research. *J. Pathol.* **162:**99–117.
54. **Yang, S. S., R. A. Bellogin, A. Buendia, M. Camacho, M. Chen, T. Cubo, A. Daza, C. L. Diaz, M. R. Espuny, R. Gutierrez, M. Harteveld, X. H. Li, M. C. Lyra, N. Madinabeitia, C. Medina, L. Miao, F. J. Ollero, M. M. Olsthoorn, D. N. Rodriguez, C. Santamaria, H. R. Schlaman, H. P. Spaink, F. Temprano, J. E. Thomas-Oates, A. A. Van Brussel, J. M. Vinardell, F. Xie, J. Yang, H. Y. Zhang, J. Zhen, J. Zhou, and J. E. Ruiz-Sainz.** 2001. Effect of pH and soybean cultivars on the quantitative analyses of soybean rhizobia populations. *J. Biotechnol.* **91:**243–255.
55. **Yang, Y. H., J. Yao, and M. C. Wang.** 2004. RAPD marker and substrate utilization pattern applied to study microbial community diversity in the soil affected by agricultural chemicals. *J. Environ. Sci. Health B* **39:**125–138.
56. **Zhou, J., M. E. Davey, J. B. Figueras, E. Rivkina, D. Gilichinsky, and J. M. Tiedje.** 1997. Phylogenetic diversity of a bacterial community determined from Siberian tundra soil DNA. *Microbiology* **143**(Pt. 12)**:**3913–3919.

Quantification of Nitrogen Transformations

DAVID D. MYROLD

56

Nitrogen cycling in soils has been studied intensively for many years because of its importance to plant productivity. The complexity of the N cycle, with its many pools and processes (Fig. 1), has also caught the interest of soil scientists and microbiologists. In addition to the continued importance of N to agricultural production and the vigor of natural ecosystems, more recent concerns about soil acidification, caused partly by atmospheric inputs of N, NO_3^- movement in surface and groundwater, and contributions of N gases, such as N_2O and NO, to ozone destruction and the greenhouse effect, have required N transformations to be quantified.

This chapter describes some of the most useful methods for measuring the most important soil N transformations (N_2 fixation, N mineralization, nitrification, and denitrification) and includes briefer descriptions of less important soil N transformations (NO_3^- immobilization, heterotrophic nitrification, dissimilatory NO_3^- reduction to NH_4^+, and anammox) and newer functional gene detection methods. It is selective, not comprehensive, as entire books have been written about the techniques used to measure just one or two of the N cycle processes. Several of the chapters found in references 25, 42, and 59 are especially recommended as additional sources of information about current methods for measuring N transformation rates in soil.

DINITROGEN FIXATION

The process of N_2 fixation involves the biological reduction of N_2 to the level of NH_3, which is subsequently incorporated into amino acids. It is a reaction that requires a large amount of energy to break the triple bond of N_2. The nitrogenase enzyme that catalyzes this reaction is oxygen sensitive and is known to be present only in prokaryotes, mainly bacteria.

Some bacteria fix N_2 in a free-living state (nonsymbiotic N_2 fixation), others are closely associated with plant roots (associative N_2 fixation), and still others form a mutualistic symbiosis with plants (symbiotic N_2 fixation). Generally, the greatest rates of N_2 fixation occur in symbioses because the plants can directly supply the high energy demands of this process.

The first estimates of N_2 fixation came from measuring the increase in total N of the system under study. This method is not particularly sensitive but can work reasonably well for studying pure culture syntheses of N_2-fixing symbioses. It is more difficult to do for soil systems but has the advantage of being integrative over time and can be used to obtain field estimates (58, 60). Its sensitivity is probably limited to rates of $> \sim 20$ kg of N ha^{-1}, and doing it well requires a significant amount of sampling and labor. Thus, this method is not described in this chapter.

Greater sensitivity in measuring N_2 fixation can be obtained by using ^{15}N, a stable isotope of N that is present naturally as 0.3663% of the N_2 in the atmosphere. The definitive test for the presence of N_2-fixing activity is probably the incorporation of ^{15}N-labeled N_2. An alternative method is to label the soil with ^{15}N and estimate the amount of N_2 fixed by using the principle of isotope dilution. Each of these applications of ^{15}N techniques is described below.

The most recently developed method for estimating N_2 fixation is the use of acetylene as a surrogate substrate for the nitrogenase enzyme. This method is very sensitive and easy to do, which accounts for its widespread use. There are numerous pitfalls to its use, however; thus, the acetylene reduction assay must be used with care.

$^{15}N_2$ Gas as a Tracer

Nitrogen in the atmosphere, in soils, and in plants is largely made up of the ^{14}N isotope, with just a small proportion (about 0.37%) of the heavier ^{15}N isotope. Thus, when soil or a plant-soil system is exposed to an atmosphere that has been enriched with $^{15}N_2$, any N_2 fixed by the system will be enriched with ^{15}N, with the amount of $^{15}N_2$ fixed being proportional to the rate of N_2 fixation. This method is the definitive test for N_2 fixation and is particularly useful for calibrating the acetylene reduction method. In practice, $^{15}N_2$ fixation can only be done in the laboratory, using closed systems. An example protocol for measuring N_2 fixation with $^{15}N_2$ is given below; additional details can be found in references 33, 57, and 60.

Materials

A closed system of some sort is needed. For a soil system, this can be simply a sealed bottle, but more sophisticated designs are required for studying symbiotic N_2 fixation. Small plants can be enclosed entirely, whereas larger plants can have just their root systems sealed off from the atmosphere. For all systems, it is important to strike a balance between minimizing the gas-phase volume to reduce the amount of $^{15}N_2$ needed and still having a sufficient volume

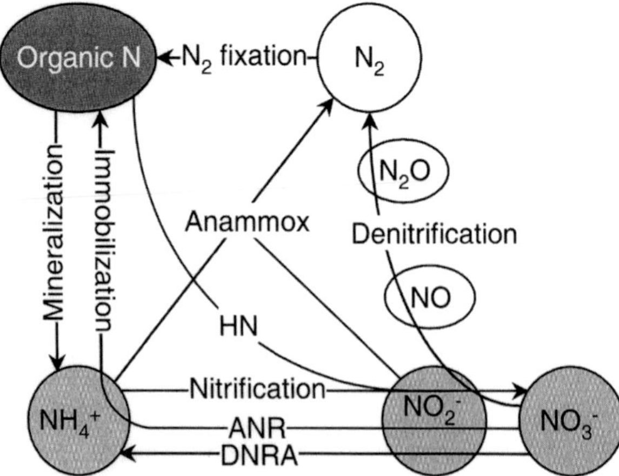

FIGURE 1 N cycle, showing pools of N and N transformations that potentially occur in soil. Dark gray shapes represent N in insoluble forms; light gray shapes represent dissolved, ionic forms of N; and white shapes represent N gases. ANR, assimilatory NO_3^- reduction, also known as NO_3^- immobilization; HN, heterotrophic nitrification; DNRA, dissimilatory NO_3^- reduction to NH_4^+, also known as NO_3^- ammonification.

for adequate aeration and dilution of any gaseous products that might affect the measurement. A ratio of 0.3 liter of gas-phase volume per g of plant tissue has been proposed as a rule of thumb (60).

$^{15}N_2$ is available from commercial sources, usually as 99 atom% $^{15}N_2$. This is the most convenient form; however, $^{15}N_2$ can also be made by oxidizing $^{15}NH_4^+$ with hypobromite (57).

Analysis of ^{15}N is done by mass spectrometry (31).

Procedure

The system to be measured is sealed, and $^{15}N_2$ is added. An unlabeled control should also be used. Often, 10 atom% $^{15}N_2$ is a good working concentration; however, one can make more exact estimates for any given system based on assumptions about N_2 fixation rates, system volumes, incubation times, and mass spectrometer sensitivities (57, 60). After a brief period to allow complete mixing of the added $^{15}N_2$, a sample of the gas phase should be taken to determine the actual ^{15}N enrichment. The length of the incubation period is dependent on the atom% ^{15}N enrichment and the amont of N_2-fixing activity. The greater sensitivity attained with a long incubation period must be balanced against the need to maintain concentrations of gases in the system that are close to atmospheric, e.g., 21% oxygen. Another gas sample should be taken at the end of the incubation period, and its atom% ^{15}N enrichment should be averaged with that of the initial sample. The soil or plant sample is then analyzed for atom% ^{15}N abundance. Grinding the entire plant for analysis is simplest if just an overall rate of N_2 fixation is desired; however, plant parts can be sampled separately, and a weighted average may be used. The atom% ^{15}N excess of the N_2-fixing system is the difference in atom% ^{15}N between the labeled and control systems. Dividing this atom% ^{15}N excess by the atom% ^{15}N excess of the atmosphere gives the fraction of N that came from N_2 fixation. Multiplying this fraction by the total N content of the plant (or soil) and dividing the result by the incubation time will yield a rate of N_2 fixation.

^{15}N Isotope Dilution

^{15}N can also be used to label the soil or nutrient solution rather than the atmosphere. In practice, this is easier to do and has the advantage that it can be applied in the field or the laboratory. It also offers an integrative measure of N_2 fixation, which is very difficult to obtain with either $^{15}N_2$ tracer or acetylene reduction methods. With the increasing availability and reduced cost of ^{15}N isotope analysis, ^{15}N isotope dilution has become the method of choice for most N_2 fixation studies. It is not without its drawbacks and difficulties, however, and must be used carefully (56, 57, 60). The general principles for field studies of N_2 fixation by ^{15}N isotope dilution are given below.

Materials

^{15}N-labeled fertilizer, most commonly urea, ammonium salts, or nitrate salts, is used. An atom% ^{15}N enrichment of 5 to 10% is usually adequate when fertilizer is added at a rate of 5 to 10 kg of N ha^{-1}.

Procedure

Well-replicated plots of sufficient size to encompass the variations in the growing plants should be used. Reasonable plot sizes range from 1 to 2 m^2 for pasture legumes to 2 to 5 m^2 for grain legumes. Substantially larger plots are needed for tree-sized plants, although paired-tree plots are often used to reduce the cost of ^{15}N fertilizer material.

A critical aspect to all ^{15}N isotope dilution measurements of N_2 fixation is the selection of a non-N_2-fixing reference plant. For some legume species, this may be a nonnodulating variety of the same species; unfortunately, nonnodulating varieties are unavailable for most plants. A suitable reference plant should not fix N_2 but should have a similar rooting pattern to that of the test plants to access the same pools of soil N, take up soil N in the same temporal pattern, and respond similarly to any treatments or environmental stresses. Sudan grass has been proposed as a reasonable choice for grain legumes, and perennial ryegrass may be used for pasture legumes (57). There is no standard for shrubs or trees.

Soil is usually labeled by dissolving the ^{15}N-labeled fertilizer in water and sprinkling the solution uniformly over the treated plots. It is sometimes advantageous to split the fertilizer into two or more applications, particularly with perennial plants, in order to have the soil more uniformly labeled over the growing season. With annual N_2-fixing plants, it may be possible to label the soil organic N pool one year and then use this relatively uniformly labeled plot to measure N_2 fixation in subsequent years. Unamended control plots should also be used as a means of assessing the background atom% ^{15}N of the N_2-fixing and reference plant species.

Plants may be harvested periodically or just at the end of a growing season. Analysis is simplified if the entire plant can be ground; however, it is possible to sample different plant parts to obtain a weighted average. The fraction of N coming from N_2 fixation is equal to 1 minus the quotient of the atom% ^{15}N excess of the fixing plant over the atom% ^{15}N excess of the nonfixing plant. These atom% ^{15}N excess values are relative to those of plants growing in control (unlabeled) soils.

Acetylene Reduction

Measuring the ethylene produced from the reduction of acetylene is easy, inexpensive, and more sensitive than ^{15}N

techniques (3, 24, 60). Although it has been used to estimate field rates over time in the field, it is not well suited for such work and is much better as a point-in-time measurement. An additional pitfall is that the ratio of reduced acetylene to fixed N_2 is not always the 3:1 or 4:1 ratio given by the ideal stoichiometry of the reaction, but instead is quite variable. This can be overcome, however, by using $^{15}N_2$ fixation to determine the ratio for the system of interest. Other problems include the disturbance related to enclosing root nodules for the assay and the sensitivity of at least some root nodules to the so-called acetylene-induced decline (27). Excised root nodules should not be used. A key use of the acetylene reduction method is to screen for N_2-fixing activity in soil and root nodule samples.

Materials

A source of acetylene is needed. Acetylene is most easily generated by adding water to calcium carbide. Commercial acetylene can also be used but should be purified by passing the gas through concentrated sulfuric acid and water traps to remove contaminants. Some type of gastight system is needed to hold the sample. This can be as simple as a flask for soil samples or a complicated cuvette system for evaluating the activity of root nodules. Ideally, a flow through system rather than a static incubation vessel is used with nodulated plants (49, 60). To assay the acetylene and ethylene, a gas chromatograph equipped with a flame ionization detector is required.

Procedure

The sample to be assayed is placed in a vessel large enough to allow an adequate supply of oxygen but small enough to enhance the sensitivity. Acetylene is added to a final concentration of 10 kPa (10% [vol/vol]) and well mixed throughout the gas phase. A control without acetylene should be used to measure background ethylene production. During the assay, gas samples are taken periodically for acetylene and ethylene analysis. Gas samples may be stored for a short period in syringes with the needles stuck into a rubber stopper; evacuated vials can be used for longer storage. Gas samples, typically 1-ml samples, are analyzed by gas chromatography, and concentrations of ethylene are calculated from a standard curve. Acetylene reduction rates are obtained by regression analysis of the ethylene concentration versus time. The amount of N_2 fixed can be estimated by dividing the acetylene reduction rate by the measured (using $^{15}N_2$) or theoretical ratio of moles of acetylene reduced per mole of N fixed.

NITROGEN MINERALIZATION AND IMMOBILIZATION

The term N mineralization is most often used to describe the net accumulation of NH_4^+ plus NO_3^- in soil and is thought to represent the amount of N available for plant uptake. It is more accurate, however, to call this net N mineralization. Net N mineralization is the result of gross N mineralization (also called ammonification) plus gross nitrification minus any assimilation (often called immobilization) of NH_4^+ or NO_3^- into organic N. The concurrent processes of NH_4^+ production and consumption are done by all heterotrophic soil microorganisms.

Numerous laboratory and field methods have been devised to measure N mineralization (4, 13, 43). A subset of these techniques for determining laboratory and field rates of net and gross N mineralization is described below.

Net N Mineralization

The basic principle in determining net N mineralization rates is to measure the change in the sum of NH_4^+ plus NO_3^- concentrations during an incubation period. Soil conditions can be adjusted to optimize N mineralization or left unchanged to simulate natural conditions.

Laboratory Measurement

Laboratory measurement will determine net N mineralization under optimal conditions; however, it can be modified easily if other environmental conditions are desired.

Soil sampled from the field is sieved (a 2-mm pore size is the typical standard, although screens with larger pores can be used for wet or clayey soils) and mixed to homogeneity, and soil water is adjusted to field capacity. The moist soil (about 25 g) is placed in a beaker or cup with a capacity of about 100 cm^3 and covered with polyethylene film that has a small hole to facilitate aeration. A minimum of three replicates should be used. A sample (equivalent to about 10 g [dry weight]) is taken at the beginning of the experiment for NH_4^+ and NO_3^- analysis. The soil is incubated at room temperature for 28 days. During incubation, the soil water content is kept within 5% of field capacity by the addition of water as uniformly as possible. At the end of the incubation period, a sample of soil is taken for NH_4^+ and NO_3^- analysis, and the remainder is used to determine the gravimetric water content.

Extraction of NH_4^+ and NO_3^- is done by shaking soil with 2 M KCl (10:1 ratio of extraction solution to soil) for 1 h. The extract can be filtered (prerinse filters to remove any traces of inorganic N) or centrifuged. The clear solution is used to measure the concentrations of NH_4^+ and NO_3^- either colorimetrically or by distillation and titration (20).

Net N mineralization is calculated from the difference between the amounts of NH_4^+ and NO_3^-–N at the end and beginning of the incubation. (The difference in amounts of NO_3^-–N yields net nitrification.)

A common modification of this method is to sample soil at several intervals over the course of incubation. From the time course of NH_4^+ and NO_3^- production, it is possible to calculate the kinetics of net N mineralization (50).

Field Measurement

Hart et al. (13) presented an excellent discussion of the advantages and disadvantages of several methods to measure net N mineralization in the field, e.g., buried bags, resin cores, open and closed cores, etc. All methods are disruptive to some degree. All sever roots, which may enhance N immobilization as the dead roots decompose and may alter soil water content by eliminating transpiration. Closed-top cores and buried bags eliminate leaching, whereas leaching can occur in open-top cores and resin cores but is measured only in resin cores. Because it is one of the less complicated methods, the following method using closed-top, solid cylinders is perhaps as good as any.

Intact soil columns are collected in thin-walled steel or polyvinyl chloride cylinders which are sharpened at one end. Cylinders of about 5 cm in diameter work well and can be cut to the desired length. A minimum of 20 cylinders per site should be used. Ten cylinders are used to measure initial NH_4^+ and NO_3^- concentrations; the other 10 are left in place and covered with loosely fitting tops. The covers prevent additions or removals of inorganic N by water. A 1-month incubation period is typically used.

Soil in the cylinders is removed and sieved (4-mm pore size). The coarse and fine fractions are weighed and sampled for gravimetric water content. A sample of the fine fraction is extracted with 2 M KCl (10:1 extractant-to-soil ratio) and analyzed for NH_4^+ and NO_3^- (the amounts of NH_4^+ and NO_3^- in the coarse fraction are assumed to be negligible). The difference in the amounts of NH_4^+ and NO_3^-–N per core between the 1-month and initial samples gives the net N mineralization (the change in amounts of NO_3^-–N gives the net nitrification), which is usually expressed on an areal basis.

The variation in field net N mineralization is always high. Some statistical efficiency can be gained by increasing the number of cylinders used and pooling soil from several cylinders to keep the number of laboratory analyses reasonable.

Gross N Mineralization

Measuring the gross rate of N mineralization (ammonification or NH_4^+ production) requires the use of $^{15}NH_4^+$ (13, 40). It also allows the gross rate of NH_4^+ consumption to be determined. The principle is the same for field or laboratory measurements; the major differences are the high variability in the field and the greater difficulty in adding the ^{15}N label uniformly to field samples without disturbance. By analogy, gross nitrification and NO_3^- immobilization (also known as assimilatory NO_3^- reduction) can be determined by using $^{15}NO_3^-$. Thus, only the method for gross N mineralization will be described.

Procedure

Add a sufficient amount of highly enriched (99 atom% ^{15}N) $^{15}NH_4^+$ to increase the native NH_4^+ pool size about 1 mg of N kg of soil^{-1}, using a small volume of solution so as not to greatly disrupt the soil water status. This may be done conveniently by making several injections with a syringe and long needle. At least six samples should be treated in this way, half of which will be extracted immediately for initial measurements of NH_4^+ concentration and isotopic composition and sampled for gravimetric water content. The other half of the samples are incubated for 24 to 48 h prior to analysis. The concentration of NH_4^+ can be measured either colorimetrically or by distillation and titration (20).

A convenient method of preparing KCl extracts for ^{15}N analysis is to diffuse the NH_4^+ in the KCl extract onto an acidified paper filter disk (7), although distillation methods (14) also work. The atom% ^{15}N of the extracted NH_4^+ is measured by mass spectrometry (31).

Calculations

There are numerous ways to calculate gross N mineralization and NH_4^+ consumption rates from the concentration and atom% ^{15}N data, depending on the assumptions that one makes about reaction order and whether remineralization of immobilized ^{15}N is allowed (13, 40). Table 1 shows the calculation procedure, assuming zero-order kinetics and no remineralization, which is usually reasonable for 24- to 48-h incubations.

Comments

Excellent analytical technique must be used in ^{15}N experiments because small amounts of contamination can greatly affect the final results. Hauck et al. (16) give sound advice on this topic.

NITRIFICATION

Nitrification is most commonly used to describe the autotrophic oxidation of NH_4^+ to NO_2^- and further to NO_3^-. The first oxidation step from NH_4^+ to NO_2^- is done by the ammonia-oxidizing bacteria. It is usually the rate-limiting step. Ammonia oxidation can be inhibited by acetylene (10 Pa) and several other compounds, e.g., nitrapyrin (N-Serve). Nitrite oxidizers convert NO_2^- to NO_3^-. This process can be selectively inhibited by chlorate (ClO_3^-).

In most soils, autotrophic nitrification is the dominant NO_3^--producing transformation; however, NO_3^- production directly from organic N by heterotrophic microorganisms can be important in some soils. The following methods are used for determining autotrophic nitrification, but modifications for heterotrophic nitrification are noted.

Potential Nitrification Activity

The maximum or potential nitrification activity of a soil can be measured in a short-term laboratory assay by providing optimal conditions for nitrification (47). Because the

TABLE 1 Calculation of gross N mineralization and NH_4^+ consumption rates, using an equation initially developed by Kirkham and Bartholomew (21)[a]

Time (day)	NH_4^+ concn (mg of N kg of soil^{-1})	NH_4^+ isotopic composition (atom% ^{15}N excess)
0	10.0	50.0
1	5.0	12.5

[a]This method is valid for zero-order kinetics, no mineralization of immobilized ^{15}N, and cases when gross N mineralization and NH_4^+ consumption rates are not equal.

Definitions:

p_A = NH_4^+ production (gross N mineralization or ammonification) rate (mg of N kg of soil^{-1} day^{-1})

c_A = NH_4^+ consumption rate (mg of N kg of soil^{-1} day^{-1})

t = time (day)

$A_{XS,0}$ = atom% ^{15}N excess of the NH_4^+ pool at time zero (%)

$A_{XS,t}$ = atom% ^{15}N excess of the NH_4^+ pool at time t (%)

atom% ^{15}N excess = atom% ^{15}N of an ^{15}N-enriched pool minus its atom% ^{15}N prior to ^{15}N addition

A_0 = NH_4^+ concentration at time zero (mg of N kg of soil^{-1})

A_t = NH_4^+ concentration at time t (mg of N kg of soil^{-1})

Calculations:

$$p_A = [(A_0 - A_t)/t] \cdot [\ln(A_{XS,0}/A_{XS,t})/\ln(A_0/A_t)]$$
$$= [(10 - 5)/1] \cdot [\ln(50/12.5)/\ln(10/5)]$$
$$= 5 \cdot (1.386/0.693)$$
$$= 10 \text{ mg of N kg of soil}^{-1} \text{ day}^{-1}$$

$$c_A = p_A - [(A_t - A_0)/t]$$
$$= 10 - [(5 - 10)/1]$$
$$= 15 \text{ mg of N kg of soil}^{-1} \text{ day}^{-1}$$

length of this assay is too short for significant growth of autotrophic nitrifiers, the potential nitrification activity is also an index of the size of the active ammonia-oxidizing population.

Procedure

An aerobic soil slurry (20 g of moist field soil in 90 ml of 0.5 mM potassium phosphate buffer [pH 7.2]) is amended with NH_4^+ (0.2 ml of a 0.25 M ammonium sulfate solution) and 1 ml of 1 M $KClO_3$. The soil slurry is shaken at room temperature, and four or five aliquots (5 ml each) are removed at 1- to 2-h intervals. The aliquots are assayed for NO_2^-, which accumulates because ClO_3^- blocks its oxidation to NO_3^-. The concentration of NO_2^- can be measured colorimetrically (20). The rate of NO_2^- increase is calculated by linear regression and divided by the soil dry weight to give the potential nitrification activity.

Comments

Acetylene at concentrations of >10 Pa can selectively inhibit autotrophic nitrification without inhibiting heterotrophic nitrification. Thus, this soil slurry method can be used with acetylene in place of chlorate as an inhibitor. In this case, the production of NO_3^- with and without acetylene is measured, and the difference is the amount of heterotrophic nitrification.

Laboratory and Field Nitrification Rates

Net nitrification rates can be measured in laboratory incubations, using the same methods as those used for net N mineralization. The difference is that only the accumulation of NO_3^- is measured. Similarly, gross rates of nitrification can be measured by utilizing the isotope dilution method, adding ^{15}N as NO_3^- and measuring the dilution of the atom% ^{15}N of the NO_3^- pool.

DENITRIFICATION

Under anaerobic conditions, some bacteria can reduce NO_3^- to the gases NO, N_2O, and N_2 by the process known as denitrification. Under most conditions, N_2 is the primary product.

Denitrification typically has been the most poorly quantified N transformation. It has often represented the N that was not accounted for in an N budget. The N budget approach is probably still the definitive standard; however, other methods have supplanted it as the method of choice for most studies. Measurements of losses of fertilizer N from denitrification are conveniently done by using ^{15}N-labeled fertilizer. In this method, the amount of ^{15}N in all ecosystem components is tallied and subtracted from the amount of ^{15}N that was added in the fertilizer. Dividing this missing amount of ^{15}N by the atom% ^{15}N of the fertilizer yields the total amount of fertilizer N lost by denitrification. One advantage of this approach is that it gives an integrative measure of denitrification. A disadvantage is that leaching of ^{15}N is assumed to be negligible; however, this may be reasonable if little ^{15}N is found in deeper soil samples.

The two most commonly used methods to measure denitrification are not integrative (15). The simplest and least expensive is the so-called acetylene block method (23), which takes advantage of the fact that acetylene inhibits the last step in the denitrification process, the reduction of N_2O to N_2. Thus, the accumulation of N_2O reflects the denitrification rate. Excellent accounts of the advantages and disadvantages of using the acetylene block method have been written (10, 19, 44). Two applications of this method are presented here, namely, an assay for denitrifier enzyme activity and field measurements using soil cores.

Denitrification has also been measured by adding highly enriched $^{15}NO_3^-$ to soils and measuring the ^{15}N-labeled gaseous products (29, 32). The use of this technique in the field is described below.

Denitrifier Enzyme Activity

Measuring the denitrifier enzyme activity of a soil is a relatively quick and easy procedure. It is useful as an index of the denitrifying activity of a soil. Denitrifier enzyme activity has even been correlated with annual denitrification losses in some soils, although it is unlikely that there is a single, universal relationship. The procedure described below is based on that given by Tiedje (53).

Procedure

Soil (10 to 25 g) is placed in a container that can be made gastight. Serum bottles or Erlenmeyer flasks work well. A solution containing readily available C (usually 1 mM glucose) and NO_3^- (1 mM KNO_3) is added in sufficient amounts to make a soil slurry. Chloramphenicol, which inhibits protein synthesis, can be added at a working concentration of 0.1 to 1 g liter^{-1}. The container is capped and flushed with an O_2-free gas. Acetylene is added to a final concentration of 10 kPa (10% [vol/vol]). The soil slurry is shaken and incubated for 1 to 2 h. During this period, at least four gas samples are taken and analyzed for N_2O concentration. A gas chromatograph equipped with an electron capture detector works well for N_2O analysis. The slope of the increase in the total amount of N_2O (gaseous and dissolved phases) (Table 2) divided by the dry weight of the soil gives the value for denitrifier enzyme activity.

Field Measurements Using the Acetylene Block Method

Acetylene inhibition of N_2O reduction has been used to measure denitrification in static cores, cores with a recirculated gas phase, and cores with soil covers (22, 28). Each of these methods has advantages and disadvantages (11, 55); however, only the static core method is detailed here because it is simpler and allows for more replication than the others. Adequate replication is one of the most important considerations in measuring denitrification rates in soils because the coefficient of variation is often very large (>100%). The acetylene block method is also the most sensitive means of measuring denitrification, with rates of a few milligrams of N per hectare per day being detectable.

TABLE 2 Use of Bunsen absorption coefficient to calculate total amount of N_2O in a system [a]

Temp (°C)	Bunsen coefficient (α)
5	1.06
10	0.88
15	0.74
20	0.63
25	0.54
30	0.47
35	0.41

[a] The total mass (M_t) of N_2O is equal to the gaseous concentration (C_g) of N_2O times the sum of the gas volume (V_g) plus the liquid volume (V_l) times the Bunsen coefficient (α), or $M_t = C_g [(V_g + V_l) \times \alpha]$.

Procedure

Soil cores should be taken with a minimum of disturbance. This is often facilitated by using an impact coring device that houses a recessed sleeve, which serves to hold the soil core upon extraction. For soils with good structural stability, the sleeves can be removed when the cores are transferred to an incubation vessel. Otherwise, it is best to leave the soil in the sleeves. The sleeves may, however, be perforated to improve gas transport, which can be a limitation of the static core method. Soil cores of 2.5 to 7.5 cm in diameter are quite typical. Larger-diameter cores tend to compact the soil less, and the greater soil volume helps to decrease sampling variability.

Soil cores are sealed gastight. It is important that the total volume of the gas phase be sufficiently large that aeration of the soil sample is not greatly altered. Purified acetylene (see the foregoing discussions of N_2 fixation methods) is added to a final concentration of 10 kPa (10% [vol/vol]) and mixed by pumping with a large syringe. The sealed system is incubated for up to 24 h at in situ temperatures. Gas samples are taken after mixing of the gas phase with a large syringe. Often only two measurements are taken, with the first taken shortly after the addition of acetylene and the other taken at the end of incubation. Additional samples can be taken at intermediate times to ensure that the production of N_2O is linear, which it should be unless a substrate (either NO_3^- or C) becomes limiting during incubation. Gas samples can be stored in evacuated tubes; however, one should always include controls for leakage and background contamination (28).

Nitrous oxide is analyzed by gas chromatography with an electron capture detector. The denitrification rate is calculated from the difference in the total amounts of N_2O sampled at two time points divided by the difference in time and the weight (or surface area) of the soil. It is important to include the amount of dissolved N_2O, which is a fairly soluble gas. This can be done by using Henry's law or, as shown in Table 2, the Bunsen absorption coefficient.

Comments

The potential shortcomings of the acetylene inhibition and static core techniques are related to gas diffusion and side effects from acetylene. For acetylene to inhibit N_2O reduction, it must reach the site of denitrification, and for N_2O to be measured, it must diffuse into the headspace. Gas transport is facilitated by exposing the sides and tops of the soil core to the atmosphere and by mixing the headspace. These practices are probably adequate for soils with low clay levels that are not saturated with water. They may not be adequate for wet, highly structured clay soils. Unfortunately, there is no good way to circumvent the gas transport problems of such soils.

Acetylene can affect other processes in soil. Contaminants in the acetylene may inhibit microbial activity or serve as a substrate and enhance microbial activity; thus, one needs to use pure acetylene. Some soil microbes can use acetylene as a C source, and long exposure times can enrich these populations. This is minimized by short (<1 day) incubations. Perhaps the biggest problem is that acetylene inhibits ammonia oxidation, thereby inhibiting nitrification and the production of NO_3^-. For many soils, there is adequate NO_3^- so that denitrification does not become limiting in 1 day. Some soils, however, may become limited much sooner. This can be detected by measuring the time course of denitrification and looking for decreased activity over time. If a decline in denitrification activity is observed, a shorter incubation time can be used. Even so, in some soils nitrification and denitrification may be tightly coupled in space, and acetylene inhibition may underestimate denitrification rates.

^{15}N Gas Emission

Dinitrogen can be composed of two ^{14}N atoms, two ^{15}N atoms, or one ^{14}N atom and one ^{15}N atom. These molecular species are distributed according to a binomial distribution. Thus, the native atmosphere, which is about 0.37% ^{15}N, has very little (0.0013%) $^{30}N_2$. By adding NO_3^- with high atom% ^{15}N enrichment to the soil, one can measure the increase in the amount of $^{30}N_2$ with a high sensitivity. This was realized in the 1950s; however, it has only been applied with any regularity since about 1980, largely because of the greater availability of sensitive mass spectrometers.

Although ^{15}N gas emission can, in principle, be used with soil cores and for laboratory incubations, its most useful application is to measure rates of denitrification in relatively undisturbed field soils. The only major disturbance is the addition of highly labeled $^{15}NO_3^-$ (>20 atom% ^{15}N). This is probably not a major issue for soils that are normally fertilized with N and can be minimized in unmanaged soils. When the NO_3^- pool is labeled with 50 atom% ^{15}N or more, rates of 5 g of N ha^{-1} day^{-1} can be measured.

Procedure

Soil microplots of a size consistent with the gas collection chamber are established. A chamber with a diameter of 10 to 20 cm is fairly typical. Although the soil chamber can be pressed into the surface few centimeters of soil, it is also possible to use a permanent base that is implanted at a deeper depth. The soil chamber is then attached to this base when measurements are being made. The chamber itself is usually constructed of polyvinyl chloride, with a rubber septum for sampling gases. Normally, the headspace of the chamber is kept small to maximize the sensitivity.

The microplot is labeled with NO_3^- enriched to 50 atom% ^{15}N or greater. Because of the high level of enrichment, only a small amount of label must be added, perhaps double or triple the native NO_3^- pool size. It is important, however, to add the label as evenly as possible. Thus, NO_3^- is usually dissolved in water and injected according to a grid pattern.

The soil chamber is set over the labeled soil, and gas samples are taken periodically during the incubation period, which might be a few hours. Samples from a cover over a nonlabeled plot are also taken to determine the background labeling of N_2 in the atmosphere. Gas samples are most conveniently stored in evacuated tubes for later mass spectroscopic analysis. A longer incubation time can be used to increase the sensitivity; however, care must be taken that a long emplacement of the soil cover does not disturb the system by decreasing aeration, increasing temperatures, or affecting plant processes.

The ratios of $^{29}N_2$ and $^{30}N_2$ to $^{28}N_2$ are measured with a mass spectrometer. These data are then used in rather complex calculations (Table 3) that give both the atom% ^{15}N of the NO_3^- pool that underwent denitrification and the fraction of the headspace gas that came from denitrification. By multiplying the fraction from denitrification by the total amount of N_2 gas in the chamber and dividing this value by the surface area and time, the denitrification flux is calculated. This flux is a function of both production (the denitrification rate) and transport, and further calculations are thus needed to calculate the denitrification rate (18).

TABLE 3 Example calculations for denitrification, using the principle of isotope distribution (1)[a]

Time (h)	Ion current ($^{29}N_2^+/^{28}N_2^+$ [r''])	Ion current ($^{30}N_2^+/^{28}N_2^+$ [r'''])	Molecular fraction of $^{29}N_2$ (^{29}x)	Molecular fraction of $^{30}N_2$ (^{30}x)	Atom fraction of ^{15}N (a)
0	0.007353	0.00001352	0.007299	0.00001342 (x_a)	0.003663 (a_a)
24	0.00742050	0.00002927	0.007365	0.00002905 (x_m)	0.003712 (a_m)

[a]Definitions:
$^{29}x = r''/(1 + r'' + r''')$
$^{30}x = r'''/(1 + r'' + r''')$
$a = (^{29}x + 2^{30}x)/2$
d = fraction of N_2 from denitrification
a_p = atom fraction of $^{15}NO_3^-$ pool
Calculations:
$d = (a_m - a_a)^2/(x_m + a_a^2 - 2a_a a_m)$
$= (0.003712 - 0.003663)^2/(0.00002905 + 0.003663^2 - 2 \cdot 0.003663 \cdot 0.003712)$
$= 0.0001545$
$a_p = a_a + (x_m + a_a^2 - 2a_a a_m)/(a_m - a_a)$
$= (0.003663 + (0.00002905 + 0.003663^2 - 2 \cdot 0.003663 \cdot 0.003712)/ (0.003712 - 0.003663)$
$= 0.3182$

Comments

The ^{15}N gas emission method suffers from gas diffusion problems similar to those of the acetylene block method. Consequently, it is not surprising that the two methods have agreed quite well when comparisons have been made (32). In theory, it can be shown that denitrification may be under- or overestimated if the NO_3^- pool is not uniformly labeled; however, empirical evidence has not shown this to be a major problem.

MINOR SOIL PROCESSES

We have already touched upon NO_3^- immobilization (see "Gross N Mineralization" above) and heterotrophic nitrification (see "Nitrification" above). Here we briefly introduce dissimilatory NO_3^- reduction to NH_4^+ (DNRA), also known as NO_3^- immobilization, which may be more prevalent in some soils than recognized previously, and the recently discovered anaerobic NH_4^+ oxidation, or anammox, process.

DNRA

Although DNRA has been studied in amended soils for many years (8), this anaerobic process was not thought likely to occur in most soils (54). Several recent reports, however, suggest that DNRA may be relatively important in some soils (2, 26, 48, 62). Measuring DNRA requires the use of ^{15}N-labeled NO_3^- and measurement of the label in the NH_4^+ pool, using the methods described for gross N mineralization.

Anammox

Although hypothesized to be thermodynamically possible several decades ago (6), anammox was discovered in a denitrifying pilot plant in 1995 (30). It was subsequently studied in other wastewater treatment facilities and in sediments (46, 51). To date, there is one report of an attempt to measure anammox in soil, and the process was found to be of minor importance (26). Measuring anammox requires adding ^{15}N-labeled NH_4^+ and measuring its appearance in N_2. $^{15}NH_4^+$ combines with unlabeled NO_2^- to form $^{29}N_2$, which is measured by mass spectrometry.

FUNCTIONAL GENE ANALYSIS

The enzymes and associated genes for the dominant soil N processes are well known and have recently been used as targets for molecular analyses by taking advantage of PCR. Details for these molecular methods are given elsewhere in this manual, so this section highlights how they have been used and provides specific information relevant to the N cycle functional genes.

Most commonly, the composition or diversity of a given functional gene has been assayed either by using fingerprinting techniques, such as denaturing gradient gel electrophoresis (DGGE) or terminal restriction fragment length polymorphism analysis, or by cloning and sequencing. More recently, functional gene copy numbers have been quantified by the use of approaches such as competitive or quantitative PCR, which should serve as indexes of the population size of the organisms harboring specific genes.

N_2 Fixation Genes

Because of sequence availability, the structural gene for dinitrogenase, *nifH*, was the first N cycle functional gene analyzed by PCR methods (63). As noted previously (64), several primer sets have been developed for *nifH*. Table 4 provides references to some of the *nifH* primer sets that have been used for soils. Each primer set undoubtedly has its advantages and disadvantages, such as its specificity (too broad or too narrow), and a recent study suggested that better coverage is obtained by using several primer pairs (9). One rather unique aspect of studies that have focused on

TABLE 4 Functional genes associated with key N cycle processes

Process and enzyme	Functional gene(s)	Reference(s)[a]
N_2 fixation		
Dinitrogenase	*nifH*	39, 61
Nitrification		
Ammonia monooxygenase	*amoA*	45
Denitrification		
Nitrate reductase	*narG*	38
Nitrite reductase	*nirS*, *nirK*	52
Nitric oxide reductase	*norB*	5
Nitrous oxide reductase	*nosZ*	52

[a]References should be consulted for primer sequences, PCR recipes, and thermocycling parameters.

nifH is that most often a nested PCR protocol has been used, which can enhance both sensitivity and specificity.

Nitrification Genes

Although autotrophic nitrification involves two groups of organisms, the NH_3 oxidizers and the NO_2^- oxidizers, only the functional genes involved in NH_3 oxidation have been used for molecular studies. The *amoA* gene, which encodes one of the subunits of NH_3 monooxygenase, has been used most extensively and is probably the most studied N cycle functional gene. Despite the large number of papers published, most have used the same *amoA* primer set (45), sometimes with minor modifications, for example, when used with DGGE (34) or quantitative PCR (35).

Denitrification Genes

Unlike the case for N_2 fixation and nitrification, multiple functional genes have been used to study the composition of denitrifying bacteria (Table 4). Recent reviews of these genes (36) and their applications to microbial ecology (37) are available. There are two dissimilatory nitrate reductase genes, but *narG* is the only one to be used in ecological studies, and its use has been limited, probably because nitrate reductase is present in nondenitrifying bacteria. Similarly, there are two nitric oxide reductase genes, of which only *norB* has been used in a few studies. The bulk of the research done with functional genes for denitrification has utilized the nitrite reductase gene, the "defining" gene in the denitrification process, and the gene for nitrous oxide reductase, which catalyzes the terminal step. A recent paper assessed many of the *nirS*, *nirK*, and *nosZ* primers that have been published in the literature (52) for use with DGGE, but remember that for other analysis techniques, other primers may be preferable; for example, different *nirS* and *nirK* primers have been designed for use in quantitative PCR (12, 17, 41).

SUMMARY

Measuring rates of N cycle transformations is not a simple task. Most measurements are disruptive in some way, and each has its advantages and disadvantages. As with all scientific analyses, the method should be selected in accordance with the experimental objectives and perhaps evaluated prior to use.

REFERENCES

1. **Arah, J. R. M.** 1992. New formulae for mass spectrometric analysis of nitrous oxide and dinitrogen emissions. *Soil Sci. Soc. Am. J.* **56:**795–800.
2. **Bengtsson, G., and C. Bergwall.** 2000. Fate of ^{15}N labeled nitrate and ammonium in a fertilized forest soil. *Soil Biol. Biochem.* **32:**545–557.
3. **Bergersen, F. J.** 1980. Methods for evaluating biological nitrogen fixation. John Wiley & Sons, New York, N.Y.
4. **Binkley, D., and S. C. Hart.** 1989. The components of nitrogen availability assessments in forest soils. *Adv. Soil Sci.* **10:**57–112.
5. **Braker, G., and J. M. Tiedje.** 2003. Nitric oxide reductase (*norB*) genes from pure cultures and environmental samples. *Appl. Environ. Microbiol.* **69:**3476–3483.
6. **Broda, E.** 1977. Two kinds of lithotrophs missing in nature. *Z. Allg. Mikrobiol.* **17:**491–493.
7. **Brooks, P. D., J. M. Stark, B. B. McInteer, and T. Preston.** 1989. Diffusion method to prepare soil extracts for automated nitrogen-15 analysis. *Soil Sci. Soc. Am. J.* **53:**1707–1711.
8. **Buresh, R. J., and W. H. Patrick.** 1978. Nitrate reduction to ammonium in anaerobic soil. *Soil Sci. Soc. Am. J.* **42:** 913–918.
9. **Bürgmann, H., F. Widmer, W. Von Sigler, and J. Zeyer.** 2004. New molecular screening tools for analysis of free-living diazotrophs in soil. *Appl. Environ. Microbiol.* **70:** 240–247.
10. **Duxbury, J. M.** 1986. Advantages of the acetylene method of measuring denitrification, p. 73–92. *In* R. D. Hauck and R. W. Weaver (ed.), *Field Measurement of Dinitrogen Fixation and Denitrification*. Soil Science Society of America, Madison, Wis.
11. **Groffman, P. M., E. A. Holland, D. D. Myrold, G. P. Robertson, and X. Zou.** 1999. Denitrification, p. 272–288. *In* G. P. Robertson, D. C. Coleman, C. S. Bledsoe, and P. Sollins (ed.), *Standard Soil Methods for Long-Term Ecological Research*. Oxford University Press, New York, N.Y.
12. **Grüntzig, V., S. C. Nold, J. Zhou, and J. M. Tiedje.** 2001. *Pseudomonas stutzeri* nitrite reductase gene abundance in environmental samples measured by real-time PCR. *Appl. Environ. Microbiol.* **67:**760–768.
13. **Hart, S. C., J. M. Stark, E. A. Davidson, and M. K. Firestone.** 1994. Nitrogen mineralization, immobilization, and nitrification, p. 985–1018. *In* R. W. Weaver, J. S. Angle, and P. J. Bottomley (ed.), *Methods of Soil Analysis*, part 2. *Microbiological and Biochemical Properties*. Soil Science Society of America, Madison, Wis.
14. **Hauck, R. D.** 1982. Nitrogen–isotope-ratio analysis, p. 735–779. *In* A. L. Page, R. H. Miller, and D. R. Keeney (ed.), *Methods of Soil Analysis*, part 2. *Chemical and Microbiological Properties*, 2nd ed. American Society of Agronomy, Madison, Wis.
15. **Hauck, R. D.** 1986. Field measurement of denitrification—an overview, p. 59–72. *In* R. D. Hauck and R. W. Weaver (ed.), *Field Measurement of Dinitrogen Fixation and Denitrification*. Soil Science Society of America, Madison, Wis.
16. **Hauck, R. D., J. J. Meisinger, and R. L. Mulvaney.** 1994. Practical considerations in the use of nitrogen tracers in agricultural and environmental research, p. 907–950. *In* R. W. Weaver, J. S. Angle, and P. J. Bottomley (ed.), *Methods of Soil Analysis*, part 2. *Microbiological and Biochemical Properties*. Soil Science Society of America, Madison, Wis.
17. **Henry, S., E. Baudoin, J. C. López-Gutiérrez, F. Martin-Laurent, A. Brauman, and L. Philippot.** 2004. Quantification of denitrifying bacteria in soils by *nirK* gene targeted real-time PCR. *J. Microbiol. Methods* **59:**327–335. (Erratum, **61:**289–290, 2005.)
18. **Hutchinson, G. L., and A. R. Mosier.** 1981. Improved soil cover method for field measurement of nitrous oxide flux. *Soil Sci. Soc. Am. J.* **45:**311–316.
19. **Keeney, D. R.** 1986. Critique of the acetylene blockage technique for field measurement of denitrification, p. 103–115. *In* R. D. Hauck and R. W. Weaver (ed.), *Field Measurement of Dinitrogen Fixation and Denitrification*. Soil Science Society of America, Madison, Wis.
20. **Keeney, D. R., and D. W. Nelson.** 1982. Nitrogen—inorganic forms, p. 643–698. *In* A. L. Page, R. H. Miller, and D. R. Keeney (ed.), *Methods of Soil Analysis*, part 2. *Chemical and Microbiological Properties*, 2nd ed. American Society of Agronomy, Madison, Wis.
21. **Kirkham, D., and W. V. Bartholomew.** 1954. Equations for following nutrient transformations in soil utilizing tracer data. *Soil Sci. Soc. Am. Proc.* **18:**33–34.
22. **Klemedtsson, L., G. Hansson, and A. Mosier.** 1990. The use of acetylene for the quantification of N_2 and N_2O

production from biological processes in soil, p. 167–180. *In* N. P. Revsbech and J. Sørensen (ed.), *Denitrification in Soil and Sediments*. Plenum Press, New York, N.Y.

23. **Knowles, R.** 1990. Acetylene inhibition technique: development, advantages, and potential problems, p. 151–166. *In* N. P. Revsbech and J. Sørensen (ed.), *Denitrification in Soil and Sediments*. Plenum Press, New York, N.Y.
24. **Knowles, R., and W. L. Barraquio.** 1994. Free-living dinitrogen-fixing bacteria, p. 179–197. *In* R. W. Weaver, J. S. Angle, and P. J. Bottomley (ed.), *Methods of Soil Analysis*, part 2. *Microbiological and Biochemical Properties*. Soil Science Society of America, Madison, Wis.
25. **Knowles, R., and T. H. Blackburn.** 1993. *Nitrogen Isotope Techniques*. Academic Press, Inc., San Diego, Calif.
26. **Matheson, F. E., M. L. Nguyen, A. B. Cooper, and T. P. Burt.** 2003. Short-term nitrogen transformation rates in riparian wetland soil determined with nitrogen-15. *Biol. Fertil. Soils* **38:**129–136.
27. **Minchin, F. R., and J. F. Witty.** 1989. Limitations and errors in gas exchange measurements with legume nodules, p. 79–96. *In* J. G. Torrey and L. J. Winship (ed.), *Applications of Continuous and Steady-State Methods to Root Biology*. Kluwer Academic Publishers, Dordrecht, The Netherlands.
28. **Mosier, A. R., and L. Klemedtsson.** 1994. Measuring denitrification in the field, p. 1047–1065. *In* R. W. Weaver, J. S. Angle, and P. J. Bottomley (ed.), *Methods of Soil Analysis*, part 2. *Microbiological and Biochemical Properties*. Soil Science Society of America, Madison, Wis.
29. **Mosier, A. R., and D. S. Schimel.** 1993. Nitrification and denitrification, p. 181–208. *In* R. Knowles and T. H. Blackburn (ed.), *Nitrogen Isotope Techniques*. Academic Press, Inc., San Diego, Calif.
30. **Mulder, A., A. A. van de Graff, L. A. Robertson, and J. G. Kuenen.** 1995. Anaerobic ammonium oxidation discovered in a denitrifying fluidized bed reactor. *FEMS Microbiol. Ecol.* **16:**177–184.
31. **Mulvaney, R. L.** 1993. Mass spectrometry, p. 11–57. *In* R. Knowles and T. H. Blackburn (ed.), *Nitrogen Isotope Techniques*. Academic Press, Inc., San Diego, Calif.
32. **Myrold, D. D.** 1990. Measuring denitrification in soils using ^{15}N techniques, p. 181–198. *In* N. P. Revsbech and J. Sørensen (ed.), *Denitrification in Soil and Sediments*. Plenum Press, New York, N.Y.
33. **Myrold, D. D., R. W. Ruess, and M. J. Klug.** 1999. Dinitrogen fixation, p. 241–257. *In* G. P. Robertson, D. C. Coleman, C. S. Bledsoe, and P. Sollins (ed.), *Standard Soil Methods for Long-Term Ecological Research*. Oxford University Press, New York, N.Y.
34. **Nicolaisen, M. H., and N. B. Ramsing.** 2002. Denaturing gradient gel electrophoresis (DGGE) approaches to study the diversity of ammonia-oxidizing bacteria. *J. Microbiol. Methods* **50:**189–203.
35. **Okano, Y., K. R. Hristova, C. M. Leutenegger, L. E. Jackson, R. F. Denison, B. Gebreyesus, D. Lebauer, and K. M. Scow.** 2004. Application of real-time PCR to study effects of ammonium on population size of ammonia-oxidizing bacteria in soil. *Appl. Environ. Microbiol.* **70:** 1008–1016.
36. **Philippot, L.** 2002. Denitrifying genes in bacterial and archaeal genomes. *Biochim. Biophys. Acta* **1577:**355–376.
37. **Philippot, L., and S. Hallin.** 2005. Finding the missing link between diversity and activity using denitrifying bacteria as a model functional community. *Curr. Opin. Microbiol.* **8:**234–239.
38. **Philippot, L., S. Piutti, F. Martin-Laurent, S. Hallet, and J. C. Germon.** 2002. Molecular analysis of the nitrite-reducing community from unplanted and maize-planted soils. *Appl. Environ. Microbiol.* **68:**6121–6128.
39. **Poly, R., L. Jocteur Monrozier, and R. Bally.** 2001. Improvement in RFLP procedure to study the community of nitrogen fixers in soil through the diversity of *nifH* gene. *Res. Microbiol.* **152:**95–103.
40. **Powlson, D. S., and D. Barraclough.** 1993. Mass spectrometry, p. 209–242. *In* R. Knowles and T. H. Blackburn (ed.), *Nitrogen Isotope Techniques*. Academic Press, Inc., San Diego, Calif.
41. **Qiu, X.-Y., R. A. Hurt, L.-Y. Wu, C.-H. Chen, J. M. Tiedje, and J.-Z. Zhou.** 2004. Detection and quantification of copper-denitrifying bacteria by quantitative competitive PCR. *J. Microbiol. Methods* **59:**199–210.
42. **Robertson, G. P., D. C. Coleman, C. S. Bledsoe, and P. Sollins.** 1999. *Standard Soil Methods for Long-Term Ecological Research*. Oxford University Press, New York, N.Y.
43. **Robertson, G. P., D. Wedin, P. M. Groffman, J. M. Blair, E. A. Holland, K. J. Nadelhoffer, and D. Harris.** 1999. Soil carbon and nitrogen availability, p. 258–271. *In* G. P. Robertson, D. C. Coleman, C. S. Bledsoe, and P. Sollins (ed.), *Standard Soil Methods for Long-Term Ecological Research*. Oxford University Press, New York, N.Y.
44. **Rolston, D. E.** 1986. Limitations of the acetylene blockage technique for field measurement of denitrification, p. 93–102. *In* R. D. Hauck and R. W. Weaver (ed.), *Field Measurement of Dinitrogen Fixation and Denitrification*. Soil Science Society of America, Madison, Wis.
45. **Rotthauwe, J.-H., K.-P. Witzel, and W. Liesack.** 1997. The ammonia monooxygenase structural gene *amoA* as a functional marker: molecular fine-scale analysis of natural ammonia-oxidizing populations. *Appl. Environ. Microbiol.* **63:**4704–4712.
46. **Schmid, M. C., B. Maas, A. Dapena, K. van de Pas-Schoonen, J. van de Vossenberg, B. Kartal, L. van Niftrik, I. Schmidt, I. Cirpus, J. G. Kuenen, M. Wagner, J. S. Sinninghe Damsté, M. Kuypers, N. P. Revsbech, R. Mendez, M. S. M. Jetten, and M. Stous.** 2005. Biomarkers for in situ detection of anaerobic ammonium-oxidizing (anammox) bacteria. *Appl. Environ. Microbiol.* **71:**1677–1684.
47. **Schmidt, E. L., and L. W. Belser.** 1994. Autotrophic nitrifying bacteria, p. 159–177. *In* R. W. Weaver, J. S. Angle, and P. J. Bottomley (ed.), *Methods of Soil Analysis*, part 2. *Microbiological and Biochemical Properties*. Soil Science Society of America, Madison, Wis.
48. **Silver, W. L., D. J. Herman, and M. K. Firestone.** 2001. Dissimilatory nitrate reduction to ammonium in upland tropical forest soils. *Ecology* **82:**2410–2416.
49. **Silvester, W. B., R. Parsons, F. R. Minchin, and J. F. Witty.** 1989. Simple apparatus for growth of nodulated plants and for continuous nitrogenase assay under defined gas phase, p. 55–66. *In* J. G. Torrey and L. J. Winship (ed.), *Applications of Continuous and Steady-State Methods to Root Biology*. Kluwer Academic Publishers, Dordrecht, The Netherlands.
50. **Stanford, G., and S. J. Smith.** 1972. Nitrogen mineralization potentials of soils. *Soil Sci. Soc. Am. Proc.* **38:**99–102.
51. **Stous, M., and M. S. M. Jetten.** 2004. Anaerobic oxidation of methane and ammonium. *Annu. Rev. Microbiol.* **58:**99–117.
32. **Throbäck, I. N., K. Enwall, Å. Jarvis, and S. Hallin.** 2004. Reassessing PCR primers targeting *nirS*, *nirK*, and *nosZ* genes for community surveys of denitrifying bacteria with DGGE. *FEMS Microbiol. Ecol.* **49:**407–417.
53. **Tiedje, J. M.** 1994. Denitrifiers, p. 245–267. *In* R. W. Weaver, J. S. Angle, and P. J. Bottomley (ed.), *Methods of Soil Analysis*, part 2. *Microbiological and Biochemical Properties*. Soil Science Society of America, Madison, Wis.
54. **Tiedje, J. M., A. J. Sexstone, D. D. Myrold, and J. A. Robinson.** 1982. Denitrification: ecological niches, competition and survival. *Antonie Leeuwenhoek* **48:**569–583.

55. **Tiedje, J. M., S. Simkins, and P. M. Groffman.** 1989. Perspectives on measurement of denitrification in the field including recommended protocols for acetylene based methods, p. 217–240. *In* M. Clarholm and L. Bergstrom (ed.), *Ecology of Arable Land*. Kluwer Academic Publishers, Dordrecht, The Netherlands.

56. **Vose, P. B., and R. L. Victoria.** 1986. Re-examination of the limitations of nitrogen-15 isotope dilution technique for the field measurement of dinitrogen fixation, p. 23–58. *In* R. D. Hauck and R. W. Weaver (ed.), *Field Measurement of Dinitrogen Fixation and Denitrification*. Soil Science Society of America, Madison, Wis.

57. **Warembourg, F. R.** 1993. Nitrogen fixation in soil and plant systems, p. 127–156. *In* R. Knowles and T. H. Blackburn (ed.), *Nitrogen Isotope Techniques*. Academic Press, Inc., San Diego, Calif.

58. **Weaver, R. W.** 1986. Measurement of biological dinitrogen fixation in the field, p. 1–10. *In* R. D. Hauck and R. W. Weaver (ed.), *Field Measurement of Dinitrogen Fixation and Denitrification*. Soil Science Society of America, Madison, Wis.

32. **Weaver, R. W., J. S. Angle, and P. J. Bottomley.** 1994. *Methods of Soil Analysis*, part 2. *Microbiological and Biochemical Properties*. Soil Science Society of America, Madison, Wis.

60. **Weaver, R. W., and S. K. A. Danso.** 1994. Dinitrogen fixation, p. 1019–1045. *In* R. W. Weaver, J. S. Angle, and P. J. Bottomley (ed.), *Methods of Soil Analysis*, part 2. *Microbiological and Biochemical Properties*. Soil Science Society of America, Madison, Wis.

61. **Widmer, F., B. T. Shaffer, L. A. Porteous, and R. J. Seidler.** 1999. Analysis of *nifH* gene pool complexity in soil and litter at a Douglas fir forest site in the Oregon Cascade mountain range. *Appl. Environ. Microbiol.* **65:** 374–380.

62. **Yin, S. X., D. Chen, L. M. Chen, and R. Edis.** 2002. Dissimilatory nitrate reduction to ammonium and responsible microorganisms in two Chinese and Australian paddy soils. *Soil Biol. Biochem.* **34:**1131–1137.

63. **Zehr, J. P., and L. A. McReynolds.** 1989. Use of degenerate oligonucleotides for amplification of the *nifH* gene from the marine cyanobacterium *Trichodesmium* spp. *Appl. Environ. Microbiol.* **55:**2522–2526.

64. **Zehr, J. P., and P. J. Turner.** 2001. Nitrogen fixation: nitrogenase genes and gene expression. *Methods Microbiol.* **30:**271–285.

Quantifying the Metabolic Activity of Soil- and Plant-Associated Microbes

GUY R. KNUDSEN

57

Virtually every facet of life on earth is profoundly affected by the metabolic activities of microorganisms. It has only been relatively recently, however, that we have seen an explosion in the diversity of methodologies for the study of microbial metabolic activity in the environment. Determining the metabolic status of microbial cells in natural habitats is significant, because metabolic status largely determines the effects that the microbes have on their environment (33). In contrast, determining the identity and distribution of microorganisms can provide information only on their potential environmental activity. Historically, a great deal of research in soil and plant microbiology has focused on biochemical processes, including microbial roles in the processes of biogeochemical cycles (51), but necessarily, many studies were performed in vitro or by using artificially constructed systems. Frequently, the potential metabolic capabilities of microbial populations in natural systems have had to be extrapolated from capabilities observed in the laboratory setting.

Traditional methods, with respirometry being the most notable example, have provided a wealth of information on total microbial metabolic activity in soils but have not always provided a deeper understanding of soil microbial community structure and function. It has long been known that the majority of soil microbes are not amenable to recovery by standard cultural techniques, and often, the environmental roles of these organisms, if they were detected at all, could only be guessed at. However, recent advances in molecular biology, fluorescent-stain technology, and microscopy have provided new and/or improved tools to detect and characterize soil microbes in situ, including their abundance, genetic makeup, and environmental activities. Particularly exciting advances have been made in our ability to observe, under natural conditions and with minimum perturbation, many aspects of microbial physiology that previously could be demonstrated only with pure cultures under highly controlled conditions. Also, by helping to escape the limitations of pure culture methods, these new techniques have provided insight into the roles that individual microbial populations play as contributing members of microbial consortia.

Microbial metabolism can be defined as the sum total of all the chemical reactions that occur in microbial cells. The seemingly infinite variety of metabolic processes and products suggests potential for a correspondingly vast array of detection methodologies, and the scope of available technology for monitoring microbial metabolic activity is increasing so rapidly that it is difficult to catalog the available methods. However, methods can be broadly categorized as those that detect end products or intermediate products of catabolic or anabolic activity (e.g., detection of carbon dioxide produced by aerobic respiration via respirometry), those that attempt to correlate fluxes in specific structural components of the microbial biomass (e.g., phospholipid ester-linked fatty acids) with metabolic activity levels, and those that are designed to detect the biochemical indicators of metabolism (e.g., RNA and specific enzymes) or the specific products of gene expression.

Soil, the rhizosphere, and the phyllosphere are populated by complex communities of organisms, including microbes (bacteria, archaea, fungi, and protists) but also typically including plant and animal components. The rhizosphere and phyllosphere are defined by the presence of a plant, so it is inherently impossible to characterize microbial activity in these habitats through the study of genetically uniform microbial strains in isolation. Determining microbial contributions to these communities may require a variety of techniques with different areas of focus. For example, it has long been known that microbes may be metabolically active even though they are not culturable (70). Furthermore, quantification of the viable microbial biomass may help elucidate the potential for specific microbial metabolic activities in a particular habitat, but the nonviable biomass also may be important as a component of the food web, despite its lack of a direct metabolic contribution. Although it is possible and common to make inferences about microbial biomass based on microbial community metabolic activity, it is noteworthy that these may not be strongly correlated; i.e., a high level of metabolic activity does not necessarily mean an actively growing, dividing microbial community (66). Also, it is important to remember that microbial physiological activities often are strongly affected by the physical arrangement of the members of a population, especially aggregations such as those that occur in biofilms (38), on root surfaces (11), and in soil microsites. Physiological and spatial heterogeneities have important implications for the ecology of microbial populations and communities (32), and the ability of recently developed molecular and microscopic methods to address heterogeneity in environmental samples is especially exciting. Below, several of the most popular methodologies for quantifying microbial metabolic activity are briefly discussed. In each case, references to original reports and more-extensive, step-by-step protocols are provided. The related subject area of bioreporter gene technology, which allows the analysis of in situ expression by supplying assayable gene products when the gene product of interest is difficult or impossible to assay, is treated in detail elsewhere in this volume. Similarly, biosensor technology is very promising for environmental microbiology; recent reviews on applications of biosensors for environmental microbiology are available (7, 56).

RESPIROMETRY METHODS

The use of various methods to measure O_2 consumption and/or CO_2 evolution from the aerobic respiration of microbial populations or communities is one of the most venerable techniques in soil microbiology. Respirometry has the advantages of being a relatively fast and easy technology. Respirometry is based on the relationship between substrate utilization, terminal electron acceptor utilization, and growth. A variety of methodologies and equipment may be used to continuously monitor oxygen consumption and/or reduced gas production by respiring microbes. The most fundamental method for measuring the respiration of the soil microbiota involves placing a chamber of known volume on the soil surface and monitoring the subsequent rate of CO_2 increase within the chamber, typically by withdrawing air samples at specified time intervals and analyzing them. One variation of this method involves continuous sampling as the air is cycled in a closed-circuit system. The methodology has been reviewed extensively elsewhere (1, 61, 62, 68). Stotzky (62) pointed out several limitations of classical respirometry methods. First, respirometry methods based on oxygen uptake measure the activity of aerobic and facultative organisms only. Commonly, experiments are performed in a laboratory setting where the necessary apparatus is available by using soil and microbial populations or communities that have been disturbed by removal from the natural environment and possibly by sieving or other treatment. Stotzky (62) noted that such physical perturbation can result in artifactual peaks of CO_2 evolution. Other ways to quantify CO_2 consumption and/or CO_2 evolution are based on gravimetric, conductimetric, manometric, titrimetric, potentiometric, calorimetric, gas chromatographic, or optical (infrared) principles and have been reviewed elsewhere (1, 41, 42, 50, 61).

Respirometry is increasingly being used for determination of biodegradation kinetics, and microcosm screening studies often are performed under controlled conditions to evaluate biodegradability potential and options for bioremediation. The use of in situ respirometry, in which microbial activity is assessed by measuring changes in O_2 and CO_2 levels in subsurface air, has proved valuable for determination of potential biodegradability and toxicity under field conditions. The carbon dioxide level typically is measured by withdrawing a small sample (e.g., 10 to 100 µl) of headspace gas with a syringe; the sample is then analyzed by gas chromatography-mass spectrometry. Oxygen levels in liquids can easily be measured by using a dissolved oxygen electrode. In situ respirometry has been used extensively to evaluate processes and potential field sites for petroleum hydrocarbon remediation (19, 44), as well as to estimate the time required to remediate these sites via bioventing. These tests typically involve aeration of a soil for approximately 24 h and then monitoring of oxygen and carbon dioxide levels in the soil gas after the air injection is stopped. Rates are compared to background respiration rates in an uncontaminated area, and then the rate of oxygen utilization is used to estimate the hydrocarbon degradation rate in the contaminated soil.

With respirometry, it is possible to measure the gross metabolic activity of the soil microbiota by adding a nonspecific substrate such as glucose. It also is possible to measure the metabolic activity of specialized populations by adding certain substrates, such as specific carbohydrates, lipids, proteins, or xenobiotic compounds, that can be mineralized by the population(s) of interest (60, 62). Rates of gross metabolic activity, obtained by introducing glucose or other nonspecific substrates, can be compared to rates of specific metabolic activity (e.g., with selective substrates such as aldehydes) to form a ratio. The ratio thus formed can be used as an indicator of whether chemical or biological alterations of the soil affect certain components of the microbiota differentially (42, 49, 62). Stotzky (62) pointed out that when aldehydes are used as specific substrates in such studies, the data should be correlated with data from nitrification studies, since both nitrification (an autotrophic process) and aldehyde mineralization (a heterotrophic process) are restricted to specific microbial species, although the two processes show similar kinetics.

Isotopic Respirometry Methods

One approach to the monitoring of metabolic activity via changes in the levels of a relevant substrate or product pool is to use substrates that are enriched either in radioactive (e.g., ^3H and ^{14}C) or stable (such as ^{13}C and ^{15}N) isotopes and to then follow the conversion of the labeled substrate to the product (10, 31). One significant advantage of using a ^{14}C-based respirometry method is that it allows the sensitive determination of minor functional groups, for example certain xenobiotic degraders. However, it has been noted that the effects of substrate pool level increases on uptake kinetics need to be considered (20, 45). A related approach involves isotope dilution methods, which are discussed in the context of aquatic microbial ecology in chapter 42. Briefly, by this method the investigator labels a product pool and then monitors the decrease in the isotope ratio within that pool over time; the rate of isotope dilution provides a measure of the rate of production of the substance. This method has the advantage of only slightly perturbing natural concentrations of the product, minimally affecting the kinetics of the system (18).

There are other variations on isotopic methods, including microautoradiography, which has proved especially useful as a tool to study the metabolic activity of aquatic microorganisms from natural samples (9). Microautoradiography typically is performed with microbes that have been concentrated on membrane filters and can effectively quantify the proportion of microorganisms in a sample that is metabolizing a particular radiolabeled compound (9). Methods for microautoradiography have been described in detail elsewhere (55, 63). One newer method, substrate-tracking autoradiographic fluorescent in situ hybridization (STARFISH) (47), is discussed below.

The INT-Formazan Method

The compound iodonitrotetrazolium formazan (INT-formazan) is deposited as discrete, dark-red intracellular granules by actively respiring aerobic bacteria. When used in conjunction with acridine orange (AO) staining and counting (acridine orange direct count [AODC]) (21), the method may be used to quantify the actively respiring proportion of a bacterial population or community. The INT-formazan method has found application in aquatic microbial ecology (48) and soil and rhizosphere microbiology (for an example, see reference 43). However, the INT-formazan method has several disadvantages. It is labor-intensive, and the results are qualitative, in that cells are scored as either metabolically active (positive) or negative (no evidence of metabolic activity) (29, 59). The method requires an incubation period with a specific substrate, and the necessary sample handling may generate artifacts related to bacterial growth rates and metabolic activity (29). Tetrazolium salts also have proved useful as vital stains to measure the relative metabolic activity of fungal spores and hyphae, including mycorrhizal fungi.

Reduction of a tetrazolium dye is the principle behind the commercially available Biolog MicroPlates, used to monitor cell respiration. When cells are able to utilize a specific nutrient present in one of the 96 wells of the plate, the respiratory production of NADH acts to reduce a tetrazolium dye, so that characteristic patterns are formed within the wells of the plate. These patterns then may be evaluated using proprietary Biolog software. Although the BIOLOG microtiter plate system originally was developed for taxonomic classification of bacterial isolates, the method has also been adapted for characterization of the functional potential of microbial communities (17, 58).

LIPID ANALYSIS METHODS

Phospholipid analysis provides a way to quantify microbiological biomass, to help quantify the abundance of specific groups of microbes, and to provide information about their physiological status. Cells that retain membrane integrity all contain polar lipids, primarily as phospholipids. Quantifying lipid phosphate or phospholipid ester-linked fatty acids gives a measurement of the microbial biomass that contains intact cellular membranes (65), since cells without intact membranes are not viable and the polar lipids in cell membranes are rapidly converted to nonpolar neutral lipid diglycerides upon death of the cell (67). As pointed out by White et al. (65), one difficulty with interpretation of biochemical biomass data is that the units of measure typically are micromoles of a component, rather than the more traditional units of numbers of cells per unit of sample. However, lipid analysis techniques may be especially useful in conjunction with other methods to measure metabolic activity and can help overcome problems of disturbance artifacts that often are associated with other methods such as respirometry (12, 14, 15, 40). Because ^{14}C can easily be incorporated into phospholipid fatty acids, tracer studies with phospholipids have great promise for microbial ecology (16).

METHODS BASED ON CELLULAR NUCLEIC ACID POOLS AND SYNTHESIS RATES

Several methods for estimating the metabolic activity levels of microbial populations involve quantification of cellular pools and rates of synthesis of specific biochemical components including RNA, DNA, ATP, and total adenine nucleotide. Because the specific growth rate of bacteria is dependent on cellular amounts and rates of rRNA synthesis, there is a strong correlation between cellular RNA content, or RNA:DNA ratio, and growth rate (29). ATP and total adenine nucleotide pool turnover rates can provide estimates of energy flux and the specific growth rate in natural microbial populations. Kemp et al. (29) described procedures that can be used to measure cell-specific quantities of rRNA and DNA in order to quantify the frequency distribution of activity among cells. By these procedures, fluorescently labeled oligonucleotide probes are hybridized to complementary 16S rRNA sequences in intact cells. Cell fluorescence, which is proportional to cellular rRNA content, is then measured and compared to measurements of DNA content obtained by fluorescence of 4′,6′-diamidino-2-phenylindole (DAPI)-stained cells, generating an rRNA:DNA ratio for individual cells. Procedures for measuring rates of stable RNA synthesis have been applied to bacterial cultures as well as aquatic bacteria and are based on the uptake and incorporation of radiolabeled adenine into cellular RNA (24). Methods for simultaneous quantification of RNA and DNA synthesis rates also have been developed and used to estimate rates of growth, cell division, and biomass in microbial communities (25–28).

Recently, Lee et al. (34) used a combination of fluorescent in situ hybridization (FISH) performed with rRNA-targeted oligonucleotide probes and microautoradiography to quantify specific substrate uptake profiles of individual bacterial cells within microbial communities. Ouverney and Fuhrman (47) proposed a new method, STARFISH, for studying the function of specific microbial groups in heterogeneous microbial communities. The STARFISH technique involves the use of group-specific fluorescent 16S rRNA probes, along with simultaneous determination of ^{3}H-substrate uptake by each of the subgroups present using microautoradiography. Total direct counts, FISH, and microautoradiography can be combined on a single slide to determine proportions of different subgroups within a microbial community, proportions of total cells in the community that take up a radioactively labeled substance, and the distribution of uptake within each microbial subgroup. Thus, the STARFISH method has the potential to simultaneously measure, in situ, the composition of a microbial community as well as the capacity of the member groups to take up specific organic nutrients (47).

METHODS INVOLVING MICROSCOPY

A number of methods for the quantification of metabolic activity and/or biomass of individuals, populations, or microbial communities involve direct microscopic observation of cells. These include simple measurement of cell size and size distributions, formation of microscopically observable metabolic products, fluorochromes used as physiological stains, and genetically based marker and reporter systems. Several of these methods are reviewed in detail elsewhere in this volume (see chapters 5, 14, 59, and 60).

Microscopic Observation of Cell Morphology

Many biological processes are linked to the biomass and size distributions of bacterial and fungal cells. Microscopic observation of bacterial morphology may in some instances provide information about metabolic status. For example, James et al. (24) examined the surface growth of an *Acinetobacter* sp. growing under different nutrient regimens. The tools that they used included continuous-flow slide culture, phase-contrast microscopy, scanning confocal laser microscopy, and computer image analysis. They observed that growth in a high-nutrient medium resulted in a transition of the bacterial cells from a coccoid to a bacillar morphology, whereas bacilliform cells reverted to cocci by reduction-division when irrigated with a starvation medium. Thus, they hypothesized that the different morphotypes represented specialized physiological adaptations for attachment and colonization in low-nutrient systems (coccoid cells) or dispersal under high-nutrient conditions (bacilliform cells) (24).

In aquatic microbiology, the use of flow cytometry for particle size determination has allowed rapid assessment of variations in microbial abundance (for an example, see reference 8). The primary advantage of flow cytometry is that very large numbers of individual cells can be measured and a straightforward estimate of microbial biomass can be obtained. However, the obvious difficulties of separating soil bacteria from soil particles and of separating plant-associated bacteria from their substrata have limited the use of this technology in soil and plant microbiology. A number of separation techniques, such as vortexing in an appropriate buffer, or centrifugation through a density gradient generated by a compound such as Nycodenz, have proved useful in this regard (see chapter 53).

Fluorescent Stains and Gene Products

As stated earlier in this chapter, knowing the metabolic status of microbial cells can provide insight into their potential effects on their environment. For this reason, there has been considerable research into methodologies for the identification of viable cells within heterogeneous microbial communities by using so-called physiological stains (33). Much of this work appears in the biofilm literature, because staining of biofilms with fluorogenic stains, followed by direct observation by epifluorescence microscopy, often can provide a quantitative picture of their physiological architecture with a minimum of physical disruption (37, 38). This approach is especially useful because natural biofilm communities typically are composed of a wide range of different organisms (64) and because bacteria from environmental samples frequently exhibit reduced culturability. Fluorochromes have long been used for direct enumeration of total microbes from environmental samples, and a number of standard methodologies have been published (30). The DNA-specific fluorescent stain DAPI has been used extensively to monitor bacterial cells in a variety of habitats (22, 23, 52). AO probably has been the fluorochrome most widely used by environmental microbiologists. AO commonly is applied to the AODC method, which is especially useful for the enumeration of planktonic bacteria. Although total direct counts per se provide no information on the viability or activity of cells, AO staining also has been interpreted with putative physiological implication, with the color of AO-stained bacterial cells used to distinguish between active and inactive cells (38). The concept is based on differential fluorescence of single-stranded versus double-stranded nucleic acids when they are complexed with AO. However, as pointed out by McFeters et al. (38), relatively little specific published information is available to positively correlate established bacterial physiological indices with results of the AO staining reaction.

A number of other fluorochromes have been used in an effort to link one or more physiological attributes to the viability of individual bacterial cells in environmental samples. Fluorescent viability indicators have been based on attributes including cytoplasmic redox potential, cell membrane potential and membrane integrity, electron transport chain activity, and enzymatic activity (4, 5, 6, 54, 63). The fluorogenic stain 5-cyano-2,3-ditolyl tetrazolium chloride (CTC) is indicative of cellular electron transport activity; cellular respiration results in the reduction of CTC and the accumulation of intracellular fluorescent crystals that can be detected microscopically (36, 54, 59). When using CTC, one must pay careful attention to the composition of the CTC-containing medium used to enumerate respiring bacteria. Pyle et al. (53) demonstrated that the level of CTC reduction by several organisms (*Klebsiella pneumoniae*, *Salmonella enterica* serovar Typhimurium, and *Escherichia coli*) decreased with increasing pH levels of the CTC medium and that increasing phosphate concentrations contributed to the inhibitory effect. The fluorogenic stain rhodamine 123 responds specifically and proportionally to cellular proton motive force (38). Yu and McFeters (71) provided a comparison of CTC and rhodamine 123 as indices of bacterial activity in biofilms of *K. pneumoniae* and compared the results with data obtained by the AODC method. A number of proprietary fluorochromes, for example, Molecular Probes' Live/Dead BacLight, have been developed specifically to target one or more indicators of cellular viability. The BacLight system comprises two nucleic acid-binding stains: SYTO 9 and propidium iodide. The SYTO 9 stain penetrates all bacterial membranes and stains the cells green, whereas propidium iodide penetrates only cells with damaged membranes, with the combination of the two stains producing red fluorescing cells. Thus, the BacLight system allows total (red or green) and viable (green only) cells to be counted simultaneously.

A major advantage of the use of specific fluorogenic stains in conjunction with epifluorescence microscopy to probe the physiological status of microbial community members is that together these two methodologies can provide valuable information on the spatial arrangement of physiological activity (69). This approach has been used to advantage by researchers studying the effects of antimicrobial treatments of biofilm communities (for examples, see references 38 and 71). Huang et al. (23) used fluorogenic stains (CTC and DAPI) in conjunction with cryoembedding and image analysis to demonstrate spatial heterogeneity in respiratory activity within bacterial biofilms following application of the disinfectant monochloramine. Epifluorescence micrographs of biofilm sections revealed gradients of respiratory activity within biofilms in response to the disinfectant treatment, and these gradients could be quantified by calculating the ratio of CTC and DAPI intensities as measured by image analysis (23).

A number of studies have combined fluorescence staining techniques with soil thin-section technology, in order to observe and quantify the distribution and/or activity of microorganisms in situ (35). There are several factors that determine the success of this approach, including soil texture and the specific procedures that are used for resin embedding of the soil (35). With any fluorochrome, attention must be paid to the possibility of nonspecific adsorption to soil particles or plant surfaces, as well as the possible shadowing effects of soil particles. Also, the autofluorescence of some soil components can be a significant limitation to the use of fluorescence microscopy for soil ecological studies (35).

The use of both marker and reporter genes to monitor microbial activity in situ has expanded greatly in recent years. For example, expression of the green fluorescent protein (GFP) gene has been utilized in bacteria and fungi as a reporter system to monitor gene expression (72) and microbe-plant interactions. The infection processes of several GFP-tagged pathogenic fungi have been monitored within host plant tissue (13, 39, 57). The biological control fungus *Clonostachys rosea* was transformed to express GFP and monitored in soil and on plant tissue by using epifluorescence and confocal scanning laser microscopy (36). Orr and Knudsen (46) used a strain of the fungus *Trichoderma harzianum* that was engineered to express GFP, along with epifluorescence microscopy and computer image analysis, to quantify the biomass of *T. harzianum* in a complex soil microbial community consisting of indigenous fungal and bacterial populations. This method also proved useful for determining the effects of the indigenous microbiota on growth and proliferation of *T. harzianum* in nonsterile soil (2, 3).

For many years we have realized that bacteria and fungi dominate the biogeochemical processes on this planet, yet an understanding of their metabolic activities in natural habitats continues to present enormous challenges to researchers. Traditional methods for assessment of microbial numbers and metabolic activity, such as respirometry, have provided much valuable information, but their lack of specificity has been a major drawback. Today, however, the availability of innovative tools for assessment of microbial metabolic activity is increasing at an extremely rapid rate. The specificity of many of these new tools allows us to peer ever more closely into a microbial world that could only be imagined a few years ago. This chapter has attempted to provide an overview of a number of these new methodologies, but it only presents a snapshot in time, as the number of new techniques and the opportunities they present for environmental microbiology continue to rapidly expand.

REFERENCES

1. **Anderson, J. P. E.** 1982. Soil respiration, p. 831–871. *In* A. L. Page, R. H. Miller, and D. R. Keeney (ed.), *Methods of Soil Analysis*, part 2. American Society of Agronomy, Madison, Wis.
2. **Back, J. P., and R. G. Kroll.** 1991. The differential fluorescence of bacteria stained with acridine orange and the effects of heat. *J. Appl. Bacteriol.* **71:**51–58.
3. **Bae, Y. S., and G. R. Knudsen.** 2000. Cotransformation of *Trichoderma harzianum* with β-glucuronidase and green fluorescent protein genes provides a useful tool for monitoring fungal growth and activity in natural soils. *Appl. Environ. Microbiol.* **66:**810–815.
4. **Bae, Y. S., and G. R. Knudsen.** 2005. Soil microbial biomass influence on growth and biocontrol efficacy of *Trichoderma harzianum*. *Biol. Control* **32:**236–242.
5. **Betts, R. P., P. Bankes, and J. G. Banks.** 1989. Rapid enumerations of viable micro-organisms by staining and direct microscopy. *Lett. Appl. Microbiol.* **9:**199–202.
6. **Bottomley, P. J., and S. P. Maggard.** 1990. Determination of viability within serotypes of a soil population of *Rhizobium leguminosarum* bv. Trifolii. *Appl. Environ. Microbiol.* **56:**533–540.
7. **Burlage, R. S.** 1997. Emerging technologies: bioreporters, biosensors, and microprobes, p. 115–123. *In* C. J. Hurst, G. R. Knudsen, M. J. McInerney, L. D. Stetzenbach, and M. V. Walter (ed.), *Manual of Environmental Microbiology*. ASM Press, Washington, D.C.
8. **Button, D. K., and B. R. Robertson.** 1989. Kinetics of bacterial processes in natural aquatic systems based on biomass as determined by high-resolution flow cytometry. *Cytometry* **10:**558–563.
9. **Carman, K. R.** 1993. Microautoradiographic detection of microbial activity, p. 397–404. *In* P. F. Kemp, B. F. Sherr, E. B. Sherr, and J. J. Cole (ed.), *Handbook of Methods in Aquatic Microbial Ecology*. Lewis Publishers, Boca Raton, Fla.
10. **Coleman, D. C., and B. Fry.** 1991. *Carbon Isotope Techniques*. Academic Press, Inc., New York, N.Y.
11. **Dandurand, L. M., D. J. Schotzko, and G. R. Knudsen.** 1997. Spatial patterns of rhizoplane populations of *Pseudomonas fluorescens*. *Appl. Environ. Microbiol.* **63:**3211–3217.
12. **Dobbs, F. C., and R. H. Findlay.** 1993. Analysis of microbial lipids to determine biomass and detect the response of sedimentary microorganisms to disturbance, p. 347–358. *In* P. F. Kemp, B. F. Sherr, E. B. Sherr, and J. J. Cole (ed.), *Handbook of Methods in Aquatic Microbial Ecology*. Lewis Publishers, Boca Raton, Fla.
13. **Dumas, B., S. Centis, N. Sarrazin, and M.-T. Esquerré-Tugayé.** 1999. Use of green fluorescent protein to detect expression of an endopolygalacturonase gene of *Colletotrichum lindemuthianum* during bean infection. *Appl. Environ. Microbiol.* **65:**1769–1771.
14. **Findlay, R. H., P. C. Pollard, D. J. W. Moriarty, and D. C. White.** 1985. Quantitative determination of microbial activity and community nutritional status in estuarine sediments: evidence for a disturbance artifact. *Can. J. Microbiol.* **31:**493–498.
15. **Findlay, R. H., M. B. Trexler, J. B. Guckert, and D. C. White.** 1990. Laboratory study of disturbance in marine sediments: response of a microbial community. *Mar. Ecol. Prog. Ser.* **61:**121–133.
16. **Findlay, R. H., and D. White.** 1987. A simplified method for bacterial nutritional status based on the simultaneous determination of phospholipid and endogenous storage lipid poly beta-hydroxy alkanoate. *J. Microbiol. Methods* **6:** 113–120.
17. **Garland, J. L., and A. L. Mills.** 1991. Classification and characterization of heterotrophic microbial communities on the basis of patterns of community-level sole-carbon-source utilization. *Appl. Environ. Microbiol.* **57:**2351–2359.
18. **Gilbert, P. M., and D. G. Capone.** 1993. Mineralization and assimilation in aquatic, sediment and wetland systems, p. 243–272. *In* R. Knowles and T. H. Blackburn (ed.), *Nitrogen Isotope Techniques*. Academic Press, Inc., New York, N.Y.
19. **Hinchee, R. E., and M. Arthur.** 1991. Bench scale studies of the soil aeration process for bioremediation of petroleum hydrocarbons. *J. Appl. Biochem. Biotechnol.* **28:**901–906.
20. **Hobbie, J. E.** 1990. Measuring heterotrophic activity in plankton. *Methods Microbiol.* **22:**235–250.
21. **Hobbie, J. E., R. J. Daley, and S. Jasper.** 1977. Use of Nuclepore filters for counting bacteria by fluorescence microscopy. *Appl. Environ. Microbiol.* **33:**1225–1228.
22. **Huang, C.-T., G. A. McFeters, and P. S. Stewart.** 1996. Evaluation of physiological staining, cryoembedding and autofluorescence quenching techniques on fouling biofilms. *Biofouling* **9:**269–277.
23. **Huang, C.-T., F. P. Yu, G. A. McFeters, and P. S. Stewart.** 1995. Nonuniform spatial patterns of respiratory activity within biofilms during disinfection. *Appl. Environ. Microbiol.* **61:**2252–2256.
24. **James, G. A., D. R. Korber, D. E. Caldwell, and W. J. Costerton.** 1995. Digital image analysis of growth and starvation responses of a surface-colonizing *Acinetobacter* sp. *J. Bacteriol.* **177:**907–915.
25. **Karl, D. M.** 1979. Measurement of microbial activity and growth in the ocean by rates of stable ribonucleic acid synthesis. *Appl. Environ. Microbiol.* **38:**850–860.
26. **Karl, D. M.** 1981. Simultaneous rates of ribonucleic acid and deoxyribonucleic acid syntheses for estimating growth and cell division of aquatic microbial communities. *Appl. Environ. Microbiol.* **42:**802–810.
27. **Karl, D. M.** 1993. Adenosine triphosphate (ATP) and total adenine nucleotide (TAN) pool turnover rates as measures of energy flux and specific growth rate in natural populations of microorganisms, p. 483–494. *In* P. F. Kemp, B. F. Sherr, E. B. Sherr, and J. J. Cole (ed.), *Handbook of*

Methods in Aquatic Microbial Ecology. Lewis Publishers, Boca Raton, Fla.

28. **Karl, D. M.** 1983. Total microbial biomass estimation derived from the measurement of particulate adenosine-5′-triphosphate, p. 359–368. *In* P. F. Kemp, B. F. Sherr, E. B. Sherr, and J. J. Cole (ed.), *Handbook of Methods in Aquatic Microbial Ecology*. Lewis Publishers, Boca Raton, Fla.
29. **Kemp, P. F., S. Lee, and J. LaRoche.** 1993. Evaluating bacterial activity from cell-specific ribosomal RNA content measured with oligonucleotide probes, p. 415–422. *In* P. F. Kemp, B. F. Sherr, E. B. Sherr, and J. J. Cole (ed.), *Handbook of Methods in Aquatic Microbial Ecology*. Lewis Publishers, Boca Raton, Fla.
30. **Kepner, R. L., Jr., and J. R. Pratt.** 1994. Use of fluorochromes for direct enumeration of total bacteria in environmental samples. *Microbiol. Rev.* **58:**603–615.
31. **Knowles, R., and T. H. Blackburn.** 1991. *Nitrogen Isotope Techniques*. Academic Press, Inc., New York, N.Y.
32. **Knudsen, G. R.** 2005. Bacteriology of soil and plants, p. 195–210. *In* P. Borriello, P. R. Murray, and G. Funke (ed.), *Topley & Wilson's Microbiology and Microbial Infections*, 10th ed., *Bacteriology*. Hodder Arnold, London, England.
33. **Lawrence, J. R., D. R. Korber, G. M. Wolfaardt, and D. E. Caldwell.** 1997. Analytical imaging and microscopy techniques, p. 29–51. *In* C. J. Hurst, G. R. Knudsen, M. J. McInerney, L. D. Stetzenbach, and M. V. Walter (ed.), *Manual of Environmental Microbiology*. ASM Press, Washington, D.C.
34. **Lee, N., P. H. Nielsen, K. H. Andreasen, S. Juretschko, J. L. Nielsen, K.-H. Schleifer, and M. Wagner.** 1999. Combination of fluorescent in situ hybridization and microautoradiography—a new tool for structure-function analyses in microbial ecology. *Appl. Environ. Microbiol.* **65:**1289–1297.
35. **Li, Y., W. A. Dick, and O. H. Tuovinen.** 2004. Fluorescence microscopy for visualization of soil microorganisms—a review. *Biol. Fertil. Soils* **39:**301–311.
36. **Lübeck, M., I. M. B. Knudsen, B. Jensen, U. Thrane, C. Janvier, and D. F. Jensen.** 2002. GUS and GFP transformation of the biocontrol strain *Clonostachys rosea* IK726 and the use of these marker genes in ecological studies. *Mycol. Res.* **106:**815–826.
37. **McFeters, G. A., A. Singh, S. Byun, P. R. Callis, and S. Williams.** 1991. Acridine orange staining reaction as an index of physiological activity in *E. coli*. *J. Microbiol. Methods* **13:**87–97.
38. **McFeters, G. A., F. P. Yu, B. H. Pyle, and P. S. Stewart.** 1995. Physiological methods to study biofilm disinfection. *J. Ind. Microbiol.* **15:**333–338.
39. **Mikkelsen, L., N. Roulund, M. Lübeck, and D. F. Jensen.** 2001. The perennial ryegrass endophyte *Neotyphodium lolii* genetically transformed with the green fluorescent protein gene (*gfp*) and visualization in the host plant. *Mycol. Res.* **105:**644–650.
40. **Moriarty, D. J. W., D. C. White, and T. J. Wassenberg.** 1985. A convenient method for measuring rate of phospholipid synthesis in seawater and sediments: its relevance to the determination of bacterial productivity and the disturbance artifacts introduced by measurements. *J. Microbiol. Methods* **3:**321–330.
41. **Nordgren, A.** 1988. Apparatus for the continuous long term monitoring of soil respiration rate in large numbers of samples. *Soil Biol. Biochem.* **20:**955–958.
42. **Nordgren, A., E. Baath, and B. Soderstrom.** 1988. Evaluation of soil respiration characteristics to assess heavy metal effects on soil microorganisms using glutamic acid as a substrate. *Soil Biol. Biochem.* **20:**949–954.
43. **Norton, J. M., and M. K. Firestone.** 1991. Metabolic status of bacteria and fungi in the rhizosphere of ponderosa pine seedlings. *Appl. Environ. Microbiol.* **57:**1161–1167.
44. **Ong, S. K., R. E. Hinchee, R. Hoeppel, and R. Schultz.** 1991. In situ respirometry for determining aerobic degradation rates, p. 541–545. *In* R. E. Hinchee and R. F. O'Fenbuttel (ed.), *In Situ Bioremediation*. Butterworth-Heinemann, Boston, Mass.
45. **Oren, A., and T. H. Blackburn.** 1979. Estimation of sediment denitrification rates at in situ nitrate concentrations. *Appl. Environ. Microbiol.* **37:**174–176.
46. **Orr, K. A., and G. R. Knudsen.** 2004. Use of GFP and image analysis to quantify proliferation of Trichoderma harzianum in nonsterile soil. *Phytopathology* **94:**1383–1389.
47. **Ouverney, C. C., and J. A. Fuhrman.** 1999. Combined microautoradiography-16S rRNA probe technique for determination of radioisotope uptake by specific microbial cell types in situ. *Appl. Environ. Microbiol.* **65:**1746–1752.
48. **Padgett, D. E.** 1993. Distinguishing bacterial from nonbacterial decomposition of *Spartina alterniflora* by respirometry, p. 465–469. *In* P. F. Kemp, B. F. Sherr, E. B. Sherr, and J. J. Cole (ed.), *Handbook of Methods in Aquatic Microbial Ecology*. Lewis Publishers, Boca Raton, Fla.
49. **Palmborg, C., and A. Nordgren.** 1993. Modelling microbial activity and biomass in forest soil with substrate quality measured using near infrared reflectance spectroscopy. *Soil Biol. Biochem.* **25:**1713–1718.
50. **Parkinson, K. J.** 1981. An improved method for measuring soil respiration in the field. *J. Appl. Ecol.* **18:**221–228.
51. **Paul, E. A., and F. E. Clark.** 1989. *Soil Microbiology and Biochemistry*. Academic Press, Inc., New York, N.Y.
52. **Pyle, B. H., S. C. Broadaway, and G. A. McFeters.** 1995. A rapid, direct method for enumerating respiring enterohemorrhagic *Escherichia coli* O157:H7 in water. *Appl. Environ. Microbiol.* **61:**2614–2619.
53. **Pyle, B. H., S. C. Broadaway, and G. A. McFeters.** 1995. Factors affecting the determination of respiratory activity on the basis of cyanoditolyl tetrazolium chloride reduction with membrane filtration. *Appl. Environ. Microbiol.* **61:**4304–4309.
54. **Rodriguez, G. G., D. Phipps, K. Ishiguro, and H. F. Ridgway.** 1992. Use of a fluorescent redox probe for direct visualization of actively respiring bacteria. *Appl. Environ. Microbiol.* **58:**1801–1808.
55. **Rogers, A. W.** 1977. *Techniques of Autoradiography*. Elsevier/North-Holland Biomedical, New York, N.Y.
56. **Rogers, K. R.** 1995. Biosensors for environmental applications. *Biosens. Bioelectron.* **10:**533–541.
57. **Rohel, E. A., A. C. Payne, B. A. Fraaije, and D. W. Hollomon.** 2001. Exploring infection of wheat and carbohydrate metabolism in *Mycosphaerella graminicola* transformants with differentially regulated green fluorescent protein expression. *Mol. Plant-Microbe Interact.* **14:**156–163.
58. **Smalla, K. U. Wachtendorf, H. Heuer, W. Liu, and L. Forney.** 1998. Analysis of BIOLOG GN substrate utilization patterns by microbial communities. *Appl. Environ. Microbiol.* 64:1220–1225.
59. **Smith, J. J., and G. A. McFeters.** 1997. Mechanisms of INT (2-(4-iodophenyl)-3-(4-nitrophenyl)-5-phenyl tetrazolium chloride), and CTC (5-cyano-2,3-ditolyl tetrazolium chloride) reduction in *Escherichia coli* K-12. *J. Microbiol. Methods*. **29:**161–175.
60. **Stenström, J., B. Stenberg, and M. Johansson.** 1998. Kinetics of substrate-induced respiration (SIR): theory. *Ambio* **27:**35–39.
61. **Stotzky, G.** 1965. Microbial respiration, p. 1550–1570. *In* C. A. Black et al. (ed.), *Methods of Soil Analysis*. American Society of Agronomy, Madison, Wis.
62. **Stotzky, G.** 1997. Quantifying the metabolic activity of microbes in soil, p. 453–458. *In* C. J. Hurst, G. R. Knudsen, M. J. McInerney, L. D. Stetzenbach, and M. V. Walter (ed.), *Manual of Environmental Microbiology*. ASM Press, Washington, D.C.

63. **Tabor, P. S., and R. A. Neihof.** 1982. Improved microautoradiographic method to determine individual microorganisms active in substrate uptake in natural waters. *Appl. Environ. Microbiol.* **44:**945–950.
64. **Ward, D. M., M. M. Bateson, R. Weller, and A. L. Ruff-Roberts.** 1992. Ribosomal RNA analysis of microorganisms as they occur in nature. *Adv. Microb. Ecol.* **12:**219–286.
65. **White, D. C., H. C. Pinkart, and D. B. Ringelberg.** 1997. Biomass measurements: biochemical approaches, p. 91–101. *In* C. J. Hurst, G. R. Knudsen, M. J. McInerney, L. D. Stetzenbach, and M. V. Walter (ed.), *Manual of Environmental Microbiology.* ASM Press, Washington, D.C.
66. **White, D. C., G. A. Smith, and G. R. Stanton.** 1984. Biomass, community structure, and metabolic activity of the microbiota in benthic marine sediments and sponge spicule mats. *Antarct. J. US* **29:**125–126.
67. **White, D. C., and A. T. Tucker.** 1969. Phospholipid metabolism during bacterial growth. *J. Lipid Res.* **10:**220–223.
68. **Winding, A., S. J. Binnerup, and J. Sorenson.** 1994. Viability of indigenous soil bacteria assayed by respiratory activity and growth. *Appl. Environ. Microbiol.* **60:**2869–2875.
69. **Xu, H.-S., N. Roberts, F. L. Singleton, R. W. Atwell, D. J. Grimes, and R. R. Colwell.** 1982. Survival and viability of nonculturable *Escherichia coli* and *Vibrio cholerae* in the estuarine and marine environment. *Microb. Ecol.* **8:**313–323.
70. **Xu, K. D., P. S. Stewart, F. Xia, C.-T. Huang, and G. A. McFeters.** 1998. Spatial physiological heterogeneity in *Pseudomonas aeruginosa* biofilm is determined by oxygen availability. *Appl. Environ. Microbiol.* **64:**4035–4039.
71. **Yu, F. P., and G. A. McFeters.** 1994. Rapid in situ assessment of physiological activities in bacterial biofilms using fluorescent probes. *J. Microbiol. Methods* **20:**1–10.
72. **Zeilinger, S., C. Galhaup, K. Payer, S. L. Woo, R. L. Mach, C. Fekete, M. Lorito, and C. P. Kubicek.** 1999. Chitinase gene expression during mycoparasitic interaction of *Trichoderma harzianum* with its host. *Fungal Genet. Biol.* **26:**131–140.

Soil Enzymes: Linking Proteomics and Ecological Processes

STEVEN D. ALLISON, TRACY B. GARTNER, KERI HOLLAND, MICHAEL WEINTRAUB, AND ROBERT L. SINSABAUGH

58

INTRODUCTION

Overview

> Enzymes are things invented by biologists that explain things which otherwise require harder thinking.
>
> Jerome Ysroael Lettvin

> Ecosystems . . . are functionally a system of stored, immobilized enzymes.
>
> Robert G. Wetzel

Extracellular enzymes are the proximate agents of organic matter transformation in soils. From a physiological perspective, these enzymes catalyze the initial steps in catabolism: the conversion of polymeric compounds into dissolved substrates that microbes can consume for growth. From an ecosystem perspective, extracellular enzymes are the mediators of decomposition, dissolved organic carbon (C) production, and nitrogen (N) and phosphorus (P) mineralization.

The production of extracellular enzymes is controlled at the organismal level by pairing low-level constitutive synthesis with induction-repression pathways tied to environmental substrate availability (21). Enzyme production is also coordinated at the population and community levels by quorum-sensing systems that have not yet been extensively studied in soil systems (6). The existence of these higher-order regulatory systems underscores the role of extracellular enzymes in the integration of microbial community function.

Once expressed, the function and fate of extracellular enzymes are environmentally controlled (Fig. 1). Most commonly, this control is considered in terms of bulk environmental conditions such as moisture, temperature, and pH. However, enzyme kinetics and enzyme turnover are also affected by soil structure, which controls enzyme immobilization and substrate accessibility (80). As extracellular enzymes interact with humic molecules, reactive oxidants, proteases, and soil particles, their function is progressively compromised.

Traditionally, activities of extracellular enzymes have been difficult to interpret due to technological difficulties and knowledge gaps (16, 38). However, these challenges are being overcome, and measurements of extracellular enzyme activity (EEA) are increasingly used in ecosystem ecology as indices of nutrient limitation, organic matter characteristics, microbial activity, and ecosystem response to environmental perturbations (4, 61, 64, 83). High-throughput protocols based on microplate technology can largely automate measurements of many enzyme activities and allow kinetic parameters to be estimated across multiple soil fractions. Data on microbial distributions and diversity are accumulating rapidly, and advances in molecular biology are providing new tools that are applicable to extracellular enzyme studies. In particular, proteomic approaches can be used to identify the extracellular enzymes that link genomic information with ecological processes.

Classification and Assay of Extracellular Enzyme Activity

Enzymes are formally classified by the chemical reaction that they catalyze (39). In soils, the extracellular enzymes of interest are those involved in the degradation of plant and microbial cell walls, the generation of humus, and the mineralization of organic N and P. These enzymes can be broadly divided into those that catalyze hydrolytic reactions, typically the cleavage of the C-O and C-N bonds that link monomers (e.g., glycosidases, peptidases, and esterases), and those that catalyze oxidative reactions, typically C-C and C-O-C bonds. The hydrolases can be further resolved into those that release monomers from the ends of polymers (exohydrolases) and those that cleave interior linkages (endohydrolases). The oxidative enzymes can be roughly divided into oxygenases and peroxidases that use molecular oxygen and hydrogen peroxide, respectively, as electron acceptors.

The vast majority of EEA studies focus on exohydrolase activities, principally because they are the easiest to assay. Soluble substrates, linked to colorigenic or fluorigenic moieties, are readily available. Because exohydrolases typically generate monomeric or dimeric products that can be directly consumed by microorganisms, these activities can often be related to other measures of microbial activity (83). Endohydrolase activities are more problematic to study. Their substrates are often insoluble, and it can be difficult to quantify the number of bonds broken (37). Theoretically, these activities are further removed from microbial metabolism and are strongly influenced by organic matter composition and soil structure.

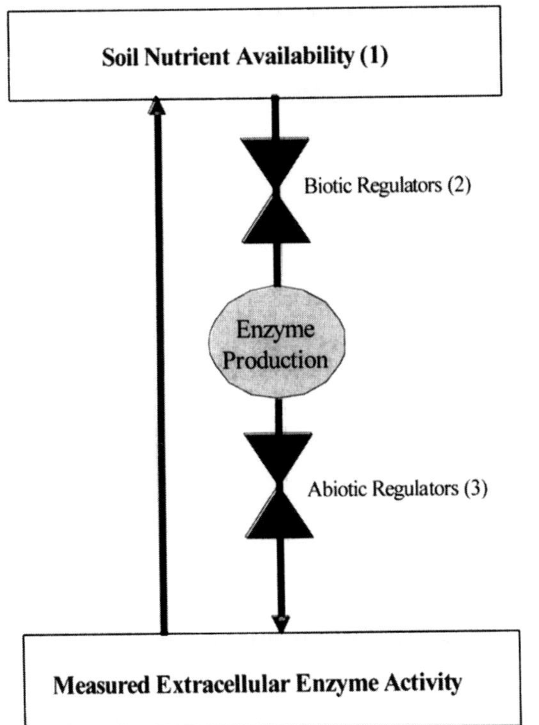

FIGURE 1 Extracellular enzyme activity catalyzes organic matter decomposition and nutrient mineralization in soil. Extracellular enzyme production is induced by signal pathways linked to substrate availability (1) and microbial community organization (2). Upon release, the function and fate of extracellular enzymes are determined by environmental conditions and soil structure (3). These abiotic variables determine the economics of microbial enzyme production in relation to microbial growth.

Studies of oxidative EEA are on the increase. Oxidative enzymes are associated with the degradation of lignin, tannin, and other secondary aromatic compounds and the formation of humic substances. These enzymes have low substrate specificity, utilize free-radical mechanisms, and catalyze both degradative and polymerization reactions (22). On ecosystem scales, these activities can be linked to dissolved organic C export and soil C sequestration (33, 34, 91).

In the sections that follow, we present brief overviews of recent advances in extracellular enzyme research.

MICROBIAL COMMUNITY CONTROLS ON ENZYME PRODUCTION

Economic Framework for Extracellular Enzyme Production

The synthesis of enzymes requires energy and consumes nutrients that are lost to the producer if the enzyme is released into the environment. Moreover, the producer has little control over the function and fate of the enzymes released. In a competitive and resource-limited environment, these factors dictate that microorganisms regulate extracellular enzyme production to limit costs and maximize resource returns. This economic view is supported by the ubiquity of signal pathways that link enzyme expression to environmental substrate availability or population density (20, 84). An economic framework is therefore a good starting point for predicting patterns of enzyme activity at the community level in relation to environmental gradients in substrate composition and nutrient availability.

There is substantial evidence that microbial communities allocate resources to enzyme production in relation to substrate availability and growth requirements (3, 20, 71). In fertilization experiments, there is an inverse relationship between phosphatase activity and high P availability (20, 61). For N, the situation is more complex because N mineralization is closely coupled to C mineralization (55) and because organic N is found in a diversity of compounds (e.g., amino sugars, polypeptides, and humus) that are degraded by different enzymes. However, there is evidence that increased inorganic N availability suppresses peptidase activities (81) and conversely that addition of organic N can induce aminohydrolase and peptidase activities (74, 95). In some systems, there is also an inverse relationship between inorganic N availability and oxidative enzyme activity (75).

Distribution of Extracellular Enzyme Activity across Groups

If the controls on enzyme expression are similar across prominent taxa, the distribution of EEA should reflect nutrient demand and optimal resource allocation at the microbial community level (57). This assumption is the basis for comparisons of EEA among systems (71, 73, 76). However, the identities of microbes that produce particular enzymes within the soil matrix are largely unknown. Most molecular analyses of community composition focus on rRNA sequences or other taxonomic tags that permit little inference about the degradative capabilities of uncultured organisms. Comparative studies of functional gene diversity within soil communities are increasing (47, 49), but many sequences cannot be matched to specific taxa, and genes that are present may not be expressed. Schulze et al. (70) addressed the latter problem by isolating and sequencing proteins extracted from soil particles, providing direct evidence for the contribution of particular taxa to extracellular function.

Consequently, most information about the taxonomic distribution of extracellular enzyme production comes from culture studies. The literature on cellulases, ligninases, and other enzymes involved in the degradation of cell walls is the subject of frequent reviews (7, 14, 42). In general, the ability to produce hydrolases that degrade the polysaccharide components of cell walls is widely distributed across bacteria and fungi (48). The expression of oxygenases and peroxidases appears to be more restricted. Many basidiomycete fungi produce peroxidative enzymes capable of degrading lignin and other recalcitrant aromatic compounds (14, 42), but other fungi and many bacteria also produce oxidative enzymes to degrade aromatic molecules (22).

At the ecosystem level, soil EEA should reflect enzyme production by the active members of the microbial community. Like macroorganisms, microorganisms show r- or K-selected traits based on the range of substrates they can utilize, their rates of growth, and their ability to prevent interference from competitors (28, 30). Based on the ability of microbes to mineralize C, Fierer et al. (30) suggest that many alphaproteobacteria and bacteroidetes may be more r selected, whereas many betaproteobacteria and gram-positive bacteria may be more K selected. K-selected microbes are more persistent and degrade recalcitrant substrates that require a large investment in extracellular enzyme production with a lower probable return (28). Furthermore, studies of enzyme evolution suggest that specialist function may be advantageous in specific environments, while generalist functions may confer a wider advantage in more heterogeneous environments (29). Therefore, successional patterns in the active microbial community driven by changes in substrate availability can determine EEA distributions (72).

PLANT CONTROLS ON EXTRACELLULAR ENZYME ACTIVITY

Enzyme Production by Roots

Plant roots produce a variety of extracellular enzymes, including phosphatase, invertase, amylase, and protease (reviewed in reference 24). Of these, phosphatase is most prevalent. In response to low ambient inorganic phosphate concentrations, phosphatase production in the root zone increases by as many as 2 orders of magnitude (79, 94).

Enzyme Production by Mycorrhizal Fungi

Root activity may be supplemented by enzyme production from mycorrhizal fungi (reviewed in reference 66). Arbuscular mycorrhizal fungi are primarily involved in P capture via production of phosphatases, accounting for 48 to 59% of total P uptake when P is supplied in organic form but only 22 to 33% when supplied in inorganic form (82). In contrast, the contribution of arbuscular mycorrhizal fungi to N mineralization is much more limited (36, 50).

The ability of ectomycorrhizal fungi (ECM) to obtain nutrients from complex organic sources remains uncertain, given that only a fraction of currently known ECM species have been examined for EEA (23). However, genes encoding lignin and manganese peroxidases are present in a number of basidiomycetes (19), and ECM show proteolytic activity in systems where N mineralization is limited (18, 78). Cellulase, phosphatase, and polyphenol oxidase activities have also been observed in ECM mats (8, 35).

Though unable to attack lignin directly with peroxidases (9–11), ericoid mycorrhizal fungi can degrade soil organic matter by producing hydrogen peroxide and hydroxyl radicals, which degrade cellulose (40) and participate in side chain oxidation and demethylation of lignin (15). Other enzymes such as laccase and catechol oxidase target hydrolyzable polyphenols (10, 11); polyphenols often link with N- and P-containing substrates in humus, so degrading polyphenols contributes to N and P mineralization.

Effects of Plant Inputs on Enzyme Production

Soil microbes are commonly thought to be C limited (86); thus, the timing of plant C inputs to the soil is one of the principal controls on microbial EEA production (85). Inputs of plant C from root exudates are typically greatest at the height of the plant-growing season, when photosynthetic rates and nutrient demand are highest. These exudates contain a complex mixture of organic acid anions, sugars, amino acids, purines, nucleosides, inorganic ions (e.g., HCO_3^-, OH^-, and H^+), and enzymes (24) and may be the principal form of C inputs to soil (13).

Invasive Species

There is good evidence that invasive plant species alter nutrient cycling processes and that these changes are sometimes mediated by EEA (27, 43). On young lava flows on the island of Hawaii, invasion by the N-fixing tree *Falcataria moluccana* dramatically increased acid phosphatase activity, a change that could alleviate secondary P limitation and facilitate further invasion (2). In a greenhouse experiment with understory plants from temperate deciduous forest, most soil EEAs were greater in pots containing an invasive stiltgrass (*Microstegium vimineum*) than in pots containing native blueberry (*Vaccinium* spp.) (43). In contrast, both P- and N-degrading activities decreased beneath pasture grass (*Cenchrus ciliaris*) compared to results with native caatinga canopy (*Ziziphus joazeiro* and *Spondias tuberosa*) in Brazil (96). Given that both positive and negative changes occur, more studies are needed before one can make general statements about enzyme responses to plant invasions and how these changes may feed back to invasibility.

ABIOTIC CONTROLS ON EXTRACELLULAR ENZYME ACTIVITY

Interpreting potential EEA measurements as indices of recent plant and microbial enzyme production on an ecosystem scale can be problematic. Soils contain a large stabilized enzyme pool with turnover times of weeks to months and possibly much longer (16, 25). Soil texture and the size distribution of soil particles largely determine the size of this abiotic pool, which can vary for different enzymes (52). Enzymes are stabilized in soils a number of ways: microencapsulation, adsorption, physical entrapment, cross-linking, copolymer formation, or ion exchange (93). The type of substrate (carbohydrate, lipid, or clay) stabilizing the enzyme determines the processes involved and the impact of stabilization on enzyme kinetic parameters (half-saturation constants, activation energy, and reaction rate) (41). For example, many researchers have observed a shift in the pH optima of clay-bound enzymes; this is thought to result from changes in enzyme conformation because of electrostatic interactions with mineral surfaces (45).

Generally, the abiotic enzyme pool is thought to be functionally passive because the enzymes or their substrates are physically occluded or kinetically compromised, although this may not always be the case (65, 67). As a result, some researchers find poor correlations between potential EEA and measures of microbial biomass or activity (e.g., references 89 and 92). Also, significant changes in active EEA as a consequence of disturbance or succession within the microbial community tend to be masked when measured in association with a larger passive pool. However, stabilization effects are not necessarily disadvantageous for enzyme producers. Stabilization may extend the functional ranges and turnover times of enzymes, resulting in a greater return on resource investment in spatially and temporally heterogeneous environments.

The kinetics of soil enzymes are also directly affected by water availability, pH, substrate availability, and temperature (54, 59, 65). In some cases, activity optima may be adapted to local conditions. For example, enzymes isolated from extreme thermophiles are stable at high temperatures, while enzymes produced by psychrophiles show high stability at low temperatures (87). Similarly, enzymes derived from halophiles are active at high salinities.

Abiotic variables establish the landscape in which enzyme producers compete. For example, diffusion rates are affected by moisture availability, soil structure, and the chemical properties of enzyme and substrates. High diffusion rates lower the return on investment to the enzyme producer and could result in the down-regulation of enzyme production (1, 28). However, if water availability is too low, the diffusion of substrates, products, and enzymes diminishes rapidly (56), and enzyme function may decline due to increasing solute concentrations and shifts in pH.

For comparative purposes, EEA potentials are generally measured under reference conditions that may be very different from those in situ. Depending on the research question, this abstraction can either complicate or facilitate the relationship of EEA to in situ process rates. The distribution of enzymes throughout the soil matrix reflects the dynamic interaction between soil structure, environmental conditions, and microbial community characteristics (Fig. 1). As a result, soil EEA profiles can integrate microbial activity over time scales that exceed biomass turnover. Thus, bulk soil EEA measurements are most useful for research questions about long-term (i.e., seasonal to annual) responses of microbes to the soil environment. To assess short-term changes in active enzyme pools, i.e., those connected most directly to the extant microbial community, it is usually necessary to isolate selected soil fractions, such as the rhizosphere.

ENZYMES AND SOIL CARBON DYNAMICS

Decomposition and Microbial Succession

The decomposition of a cohort of plant material can be described as a successional loop that links substrate, microorganisms, and extracellular enzymes: substrate composition selects for microbial populations that produce extracellular enzymes that modify the substrate, leading to changes in microbial community composition (26). Early in decomposition, litter is colonized by fast-growing, opportunistic microorganisms that have a high affinity for soluble substrates. As decomposition proceeds, this group is succeeded by slower-growing decomposers specializing in holocellulose degradation. Eventually, all nonlignified polysaccharides in the litter are consumed, and the decomposition of the remaining material is tied to the oxidative breakdown of lignin and humic condensates by slow-growing decomposers. EEA potentials, linked to both substrate composition and microbial activity through the successional loop, are versatile metrics for monitoring or comparing decomposition processes.

Enzymes and Carbon Sequestration

Litter material that does not decompose is sequestered in the soil organic matter pool as humus. The quantity of this material is a function of initial litter composition, nutrient availability, biotic interactions, and environmental conditions. In general, more C tends to be sequestered from litter with high concentrations of lignin, tannin, and other secondary compounds, largely because the oxidative degradation of these aromatics creates reactive intermediates that condense with other molecules to produce humics (12).

Enzymatic responses to increasing N availability have important consequences for soil C pools. In Michigan, N deposition stimulated C-degrading enzymes and soil C losses in forests dominated by labile litter inputs while suppressing oxidative enzymes and soil C losses in forests receiving more recalcitrant litter inputs (17, 90). Similar results have been identified for soil organic matter: decomposition of light soil fractions typically increases with N addition, but the heavier, more recalcitrant soil fractions become further stabilized, presumably as a result of lower oxidative enzyme activity (60). Such shifts in enzyme activity have been used to explain many negative or small responses of decomposition to N addition, especially in recalcitrant substrates (32, 51).

Some micronutrients, particularly Mn and Ca, can also affect decomposition rates. Mn is essential for Mn peroxidase, a lignin-degrading enzyme produced by many fungi, and is also involved in the regulation of other lignolytic enzymes, including laccase and lignin peroxidase (5, 63). Ca availability can limit the growth of lignin-degrading white rot fungi (46).

Mineral-Carbon Interactions

The storage of organic C (OC) in soil is strongly influenced by soil texture. Because of their surface area and layered structure, clay minerals provide most of the sorptive capacity of soils. Clays stabilize organic matter by shielding molecules from enzymatic attack within small pores or by altering substrate conformations (53, 97). These immobilized organic molecules may not be exposed to decomposition until protective clay minerals are degraded by weathering processes.

Another common mechanism by which clays protect OC from enzymatic attack is through multivalent-cation bridging between negatively charged organic groups and negatively charged clay particles (reviewed in reference 44). Multivalent cations cause clays to remain flocculated, reducing the exposure of OC adsorbed onto clay particles to attack by microbial enzymes. Second, they can also reduce the efficiency of enzymatic degradation of OC by altering the three-dimensional orientation of organic molecules containing carboxyl groups (44).

METHODOLOGICAL CONSIDERATIONS

Measurement Principles

The Michaelis-Menten equation describes how the rate of a simple enzymatic reaction, V, changes with substrate concentration, S:

$$V = \frac{V_{max}[S]}{K_m + [S]}$$

V_{max} is the maximum velocity of the reaction when all enzymes are functioning at capacity. K_m is a half-saturation constant, i.e., the concentration of substrate at which V is half of V_{max}, and a measure of the binding affinity between the substrate and the enzyme.

In most ecological studies, only V_{max} is measured and values are generally interpreted as comparative indicators of effective enzyme concentrations. For soils, V_{max} is typically measured in a homogenized slurry containing a buffer with added substrate. Slurries are necessary to reduce diffusion barriers and ensure that enzymes have adequate access to substrate. Enzyme activity is expressed as the rate of substrate depletion or product formation. Often, V_{max} is measured at optimal pH and temperature to maximize assay sensitivity, even if these conditions differ substantially from those in situ. However, it is also common to conduct enzyme assays at in situ pH and temperature to obtain more realistic comparisons of potential activity across systems.

Enzyme assays are sometimes performed on soil samples that have previously been air dried or frozen. These steps are likely to alter enzyme activity relative to fresh samples and should be performed only when logistical constraints preclude assaying fresh samples.

The incubation time of the enzyme assay is another consideration. Longer assays (>12 h) increase the likelihood of substrate depletion, product uptake, and enzyme turnover or synthesis, resulting in reaction rates that are not stable over time. For comparing potential activities among soils, incubations should be held to the minimum time required to measure accumulation of the reaction product to provide the best measure of initial rate of reaction. In some cases, the linearity of the reaction over time can be verified by repeated measurements of product concentration.

Recent Advances in Microplate Enzyme Assays

Recently, high-throughput, high-sensitivity assays based on microplate technology have been developed to determine the activities of a wide variety of different soil enzymes (68, 77). Many of these assays use substrates that yield fluorescent reaction products, such as methyl coumarin, which can be detected with high sensitivity (68). These methods have greatly increased the ease of measuring soil enzyme activity.

Microplate technology also enables well-known colorimetric assays to be scaled down for high-throughput analyses. Two main challenges must be overcome to reduce variability to acceptable levels in these assays. First, there must be adequate homogenization of the environmental sample to ensure that the slurry in the microplate wells is representative of the initial material. This issue is particularly relevant for fresh litter or plant samples that do not easily break into small pieces. For these types of samples, a more rigorous homogenization is required, and/or wide-orifice pipette tips can be used to transfer the sample homogenate into microplate wells.

The second challenge is that particles from the sample homogenate scatter the light beam of the microplate reader and make absorbance readings highly variable. This issue can be overcome by running a large number of replicate samples and controls (i.e., ~16). Alternatively, if the particles settle during the incubation period, the assay solution (containing the reaction product) can be carefully pipetted onto new plates before determination of the product concentration, leaving the particles behind.

The above principles can be applied to microplate procedures by using a variety of different substrates and enzymes. For example, many hydrolase assays are based on the cleavage of a *p*-nitrophenol (or *p*-nitroaniline) group from compounds analogous to naturally occurring substrates. Oxidase assays, for enzymes such as polyphenol oxidase and peroxidase, are also readily adaptable to microplate readers. For these enzymes, no separate color development step is required, since the enzymatic reaction results directly in the formation of a colored product. Other enzymes, such as urease, require more complicated chemical procedures to assay the concentration of product formed, but these procedures are also easily adapted to the microplate scale.

MODELS

Enzyme-Driven Decomposition Models

Over the past few decades, there have been successive efforts to generate heuristic models that link soil enzyme activity with decomposition processes. In most models, enzyme activities are tied to substrate degradation through Michaelis-Menten kinetics, in contrast to larger-scale models in which decomposition rates are represented as first-order functions of organic matter concentration, which in turn may be modulated by environmental parameters such as temperature, moisture, or nutrient availability (62). In enzyme-based models, enzymes act as proximal controls on C and nutrient mineralization, and changes in model parameters affect these processes by altering enzyme production or activity.

Sinsabaugh and Moorhead (76) presented a model in which enzyme activity depended on the availability of nutrients and decomposition rates depended on the activity of lignocellulose-degrading enzymes. As N and P availability increased, the microbial community was assumed to allocate more resources toward the production of enzymes that generated assimilable C, i.e., lignocellulases, thereby increasing decomposition rates. Allison (1) adopted a similar conceptual framework for modeling enzyme production but extended it to include the N cost of enzyme synthesis.

Schimel and Weintraub (69) developed a simulation that included dual enzyme-substrate controls on decomposer activity. They recognized that the enzymatic breakdown of organic matter could be limited by the availability of effective enzyme binding sites as well as by enzyme production and kinetics. This model provided a biochemical mechanism that linked two empirical observations: microbial communities can be C limited even when soil organic matter is abundant, and manipulations that increase enzyme activity do not necessarily increase substrate degradation rates.

Moorhead and Sinsabaugh (58) combined an enzyme-driven decomposition model with a multiguild representation of microbial community structure. The model included compartments for three guilds of soil microbes (opportunists, decomposers, and miners) distinguished by their substrate affinities and growth characteristics. By simulating the successional loop, the model captures the differential effects of N on decomposition and projects the humus input to soil organic matter.

Soil Structure Models

The heterogeneous spatial structure of soil imposes constraints on enzyme dynamics and the economics of microbial nutrition. Vetter et al. (88) developed a model to predict the flux of soluble compounds to bacteria foraging within particulate organic matter. Based on model results, they concluded that foraging with extracellular enzymes would be a viable strategy only under conditions where substrate was abundant within the foraging radius of the bacterium. Therefore, extracellular enzymes should be produced in a spatially heterogeneous pattern corresponding to substrate availability within soils and particulate organic matter. Allison (1) showed that rates of enzyme diffusion controlled growth rates of enzyme-producing microbes and mediated competitive relationships with cheaters (opportunists) that did not produce enzymes. If rates of enzyme diffusion were high, spatial structure in the model system declined and cheaters drove enzyme producers to extinction, resulting in low enzyme concentrations and system collapse. A model by Ekschmitt et al. (28) also predicts that enzyme-producing bacteria will survive only in environments with restricted diffusion. These models provide a mechanistic explanation for observations that most microorganisms, particularly bacteria, tend to retain extracellular enzymes within periplasmic spaces or attached to outer envelopes.

DIRECTIONS

Innovative approaches based on molecular biology are resolving long-standing questions about the mechanisms of biogeochemical processes and the controls on microbial diversity (e.g., reference 31). Much of this work focuses on genomic analyses of microbial communities to elucidate the distribution of taxa or genes associated with particular soil processes. Because few organisms have been cultured and fewer still have had their genomes sequenced, there remains a vast gap between taxonomic information and functional information on microbial communities. A similar gap isolates ecological genomics from ecological processes. Ecological proteomics—the study of ecologically relevant proteins—bridges these gaps.

Recent technological advances make it possible to pursue ecological proteomics at high levels of spatial and taxonomic resolution. Schulze et al. (70) demonstrated the utility of this approach by extracting proteins from a variety of natural habitats, including surface water, soil solution, and soil particles. For soils from a beech forest ecosystem, they found that nearly all of the 75 proteins (representing 16 taxa, ~50% bacterial) identified from soil solution were intracellular proteins. However, proteins desorbed directly from soil particles were predominantly extracellular enzymes, mostly cellulases and laccases, including both fungal and bacterial enzymes. In effect, soil particles acted as collectors and stabilizers of lignocellulose-degrading enzymes.

These types of studies demonstrate that technological impediments to ecological proteomics can be overcome to yield ecologically relevant information on microbial processes. As a result, increasing effort can be devoted to linking ecological function with proteomic and genomic information. The cost of partial protein sequencing is now comparable to that of other ecological analyses, and for many enzymes, particularly exohydrolases, kinetic analyses can be largely automated by using robotic dispensers and microplate spectrofluorometers. By combination of proteomic analyses with kinetic assays, the molecular basis of decomposition processes can be compared across systems and directly linked to microbial community composition and soil structure. Combined with nucleic acid sequencing, the potential exists to resolve the contribution of individual populations of microbes to emergent ecological processes.

REFERENCES

1. **Allison, S. D.** 2005. Cheaters, diffusion and nutrients constrain decomposition by microbial enzymes in spatially structured environments. *Ecol. Lett.* **8:**626–635.
2. **Allison, S. D., C. Nielsen, and R. F. Hughes.** 2006. Elevated enzyme activities in soils under the invasive nitrogen-fixing tree *Falcataria moluccana*. *Soil Biol. Biochem.* **38:**1537–1544.
3. **Allison, S. D., and P. M. Vitousek.** 2005. Responses of extracellular enzymes to simple and complex nutrient inputs. *Soil Biol. Biochem.* **37:**937–944.
4. **Aon, M. A., M. N. Cabello, D. E. Sarena, A. C. Colaneri, M. G. Franco, J. L. Burgos, and S. Cortassa.** 2001. I. Spatio-temporal patterns of soil microbial and enzymatic activities in an agricultural soil. *Appl. Soil Ecol.* **18:**239–254.
5. **Archibald, F., and B. Roy.** 1992. Production of manganic chelates by laccase from the lignin-degrading fungus *Trametes* (*Coriolus*) *versicolor*. *Appl. Environ. Microbiol.* **58:** 1496–1499.
6. **Barras, F., F. Vangijsegem, and A. K. Chatterjee.** 1994. Extracellular enzymes and pathogenesis of soft-rot *Erwinia*. *Annu. Rev. Phytopathol.* **32:**201–234.
7. **Beguin, P.** 1990. Molecular biology of cellulose degradation. *Annu. Rev. Microbiol.* **44:**219–248.
8. **Bending, G. D., and D. J. Read.** 1995. The structure and function of the vegetative mycelium of ectomycorrhizal plants. V. The foraging behaviour of ectomycorrhizal mycelium and the translocation of nutrients from exploited organic matter. *New Phytol.* **130:**401–409.
9. **Bending, G. D., and D. J. Read.** 1996. Effects of the soluble polyphenol tannic acid on the activities of ericoid and ectomycorrhizal fungi. *Soil Biol. Biochem.* **28:**1595–1602.
10. **Bending, G. D., and D. J. Read.** 1996. Nitrogen mobilization from protein-polyphenol complex by ericoid and ectomycorrhizal fungi. *Soil Biol. Biochem.* **28:**1603–1612.
11. **Bending, G. D., and D. J. Read.** 1997. Lignin and soluble phenolic degradation by ectomycorrhizal and ericoid mycorrhizal fungi. *Mycol. Res.* **101:**1348–1354.
12. **Berg, B.** 2000. Litter decomposition and organic matter turnover in northern forest soils. *Forest Ecol. Manag.* **133:** 13–22.
13. **Bertin, C., X. Yang, and L. A. Weston.** 2003. The role of root exudates and allelochemicals in the rhizosphere. *Plant Soil* **256:**67.
14. **Blanchette, R. A.** 1991. Delignification by wood-decay fungi. *Annu. Rev. Phytopathol.* **29:**381–398.
15. **Burke, R. M., and J. W. G. Cairney.** 2002. Laccases and other polyphenol oxidases in ecto- and ericoid mycorrhizal fungi. *Mycorrhiza* **12:**105–116.
16. **Burns, R. G.** 1982. Enzyme activity in soil: location and a possible role in microbial ecology. *Soil Biol. Biochem.* **14:** 423–427.
17. **Carreiro, M. M., R. L. Sinsabaugh, D. A. Repert, and D. F. Parkhurst.** 2000. Microbial enzyme shifts explain litter decay responses to simulated nitrogen deposition. *Ecology* **81:**2359–2365.
18. **Chalot, M., and A. Brun.** 1998. Physiology of organic nitrogen acquisition by ectomycorrhizal fungi and ectomycorrhizas. *FEMS Microbiol. Rev.* **22:**21–44.
19. **Chen, D. M., A. F. S. Taylor, R. M. Burke, and J. W. G. Cairney.** 2001. Identification of genes for lignin peroxidases and manganese peroxidases in ectomycorrhizal fungi. *New Phytol.* **152:**151–158.
20. **Chróst, R. J.** 1991. Environmental control of the synthesis and activity of aquatic microbial ectoenzymes, p. 29–59. *In* R. J. Chróst (ed.), *Microbial Enzymes in Aquatic Environments*. Springer-Verlag, New York, N.Y.
21. **Chróst, R. J., and W. Siuda.** 2002. Ecology of microbial enzymes in lake ecosystems, p. 35–72. *In* R. G. Burns and R. P. Dick (ed.), *Enzymes in the Environment*. Marcel Dekker, New York, N.Y.
22. **Claus, H.** 2004. Laccases: structure, reactions, distribution. *Micron* **35:**93–96.
23. **Courty, P.-E., K. Pritsch, M. Schloter, A. Hartmann, and J. Garbaye.** 2005. Activity profiling of ectomycorrhizal communities in two forest soils using multiple enzymatic tests. *New Phytol.* **167:**309–319.
24. **Dakora, F. D., and D. A. Phillips.** 2002. Root exudates as mediators of mineral acquisition in low-nutrient environments. *Plant Soil* **245:**35–47.
25. **De Cesare, F., A. M. V. Garzillo, V. Buonocore, and L. Badalucco.** 2000. Use of sonication for measuring acid phosphatase activity in soil. *Soil Biol. Biochem.* **32:**825–832.
26. **Dighton, J.** 2003. *Fungi in Ecosystem Processes*. Marcel Dekker, New York, N.Y.
27. **Ehrenfeld, J. G.** 2003. Effects of exotic plant invasions on soil nutrient cycling processes. *Ecosystems* **6:**503–523.
28. **Ekschmitt, K., M. Q. Liu, S. Vetter, O. Fox, and V. Wolters.** 2005. Strategies used by soil biota to overcome soil organic matter stability—why is dead organic matter left over in the soil? *Geoderma* **128:**167–176.
29. **Elena, S. F., and R. E. Lenski.** 2003. Evolution experiments with microorganisms: the dynamics and genetic bases of adaptation. *Nat. Rev. Genet.* **4:**457–469.

30. **Fierer, N., M. A. Bradford, and R. B. Jackson.** Towards an ecological classification of soil bacteria. *Ecology*, in press.
31. **Fierer, N., and R. B. Jackson.** 2006. The diversity and biogeography of soil bacterial communities. *Proc. Natl. Acad. Sci. USA* **103:**626–631.
32. **Fog, K.** 1988. The effect of added nitrogen on the rate of decomposition of organic matter. *Biol. Rev.* **63:**433–462.
33. **Freeman, C., N. Ostle, and H. Kang.** 2001. An enzymic "latch" on a global carbon store. *Nature* **409:**149.
34. **Gallo, M. E., C. L. Lauber, S. E. Cabaniss, M. P. Waldrop, R. L. Sinsabaugh, and D. R. Zak.** 2005. Soil organic matter and litter chemistry response to experimental N deposition in northern temperate deciduous forest ecosystems. *Global Change Biol.* **11:**1514–1521.
35. **Griffiths, R. P., M. A. Castellano, and B. A. Caldwell.** 1991. Ectomycorrhizal mats formed by *Gautieria monticola* and *Hysterangium setchelli* and their association with Douglas-fir seedlings: a case study. *Plant Soil* **134:**255–259.
36. **Hodge, A., C. D. Campbell, and A. H. Fitter.** 2001. An arbuscular mycorrhizal fungus accelerates decomposition and acquires nitrogen directly from organic material. *Nature* **413:**297–299.
37. **Hoppe, H.-G., C. Arnosti, and G. F. Herndl.** 2002. Ecological significance of bacterial enzymes in the marine environment. *In* R. G. Burns and R. P. Dick (ed.), *Enzymes in the Environment*. Marcel Dekker, New York, N.Y.
38. **Insam, H.** 2001. Developments in soil microbiology since the mid-1960s. *Geoderma* **100:**389–402.
39. **International Union of Biochemistry and Molecular Biology.** 1992. *Enzyme Nomenclature 1992*. Academic Press, San Diego, Calif.
40. **Joseleau, J.-P., S. Gharibian, J. Comtat, A. Lefebvre, and K. Ruel.** 1994. Indirect involvement of ligninolytic enzyme systems in cell wall degradation. *FEMS Microbiol. Lett.* **13:**255–264.
41. **Kelleher, B. P., A. J. Simpson, K. O. Willeford, M. J. Simpson, R. Stout, A. Rafferty, and W. L. Kingery.** 2004. Acid phosphatase interactions with organo-mineral complexes: influence on catalytic activity. *Biogeochemistry* **71:**285–297.
42. **Kirk, T. K., and R. L. Farrell.** 1987. Enzymatic "combustion": the microbial degradation of lignin. *Annu. Rev. Microbiol.* **41:**465–505.
43. **Kourtev, P. S., J. G. Ehrenfeld, and M. Häggblom.** 2003. Experimental analysis of the effect of exotic and native plant species on the structure and function of soil microbial communities. *Soil Biol. Biochem.* **35:**895–905.
44. **Krull, E., J. Baldock, and J. Skjemstad.** 2001. Soil texture effects on decomposition and soil carbon storage, p. 103–110. *In Net Ecosystem Exchange CRC Workshop Proceedings*.
45. **Leprince, F., and H. Quiquampoix.** 1996. Extracellular enzyme activity in soil: effect of pH and ionic strength on the interaction with montmorillonite of two acid phosphatases secreted by the ectomycorrhizal fungus *Hebeloma cylindrosporum*. *Eur. J. Soil Sci.* **47:**511–522.
46. **Lindeberg, G.** 1944. Über die Physiologie ligninabbauender Bodenhymenomyzeten. *Symb. Bot. Ups.* **8:**1–183.
47. **Luis, P., G. Walther, H. Kellner, F. Martin, and F. Buscot.** 2004. Diversity of laccase genes from basidiomycetes in a forest soil. *Soil Biol. Biochem.* **36:**1025–1036.
48. **Lynd, L. R., P. J. Weimer, W. H. van Zyl, and I. S. Pretorius.** 2002. Microbial cellulose utilization: fundamentals and biotechnology. *Microbiol. Mol. Biol. Rev.* **66:** 506–577.
49. **Lyons, J. I., S. Y. Newell, A. Buchan, and M. A. Moran.** 2003. Diversity of ascomycete laccase gene sequences in a southeastern US salt marsh. *Microb. Ecol.* **45:**270–281.
50. **Mader, P., H. Vierhailig, R. Streitwolf-Engel, T. Boller, B. Frey, P. Christie, and A. Wiemken.** 2000. Transport of 15N from soil compartments separated by a polytetrafluoroethylene membrane to plant roots via the hyphae of arbuscular mycorrhizal fungi. *New Phytol.* **146:**155–161.
51. **Magill, A. H., and J. D. Aber.** 1998. Long-term effects of experimental nitrogen additions on foliar litter decay and humus formation in forest ecosystems. *Plant Soil* **203:** 301–311.
52. **Marx, M.-C., E. Kandeler, M. Wood, N. Wermbter, and S. C. Jarvis.** 2005. Exploring the enzymatic landscape: distribution and kinetics of hydrolytic enzymes in soil particle-size fractions. *Soil Biol. Biochem.* **37:**35–48.
53. **Mayer, L. M., L. L. Schick, K. R. Hardy, R. Wagai, and J. McCarthy.** 2004. Organic matter in small mesopores in sediments and soils. *Geochim. Cosmochim. Acta* **68:**3863–3872.
54. **McClaugherty, C. A., and A. E. Linkins.** 1990. Temperature responses of enzymes in 2 forest soils. *Soil Biol. Biochem.* **22:**29–33.
55. **McGill, W. B., and C. V. Cole.** 1981. Comparative aspects of cycling of organic C, N, S and P through soil organic matter. *Geoderma* **26:**267–286.
56. **Moldrup, P., T. Olesen, T. Komatsu, P. Schjonning, and D. E. Rolston.** 2001. Tortuosity, diffusivity, and permeability in the soil liquid and gaseous phases. *Soil Sci. Soc. Am. J.* **65:**613–623.
57. **Moorhead, D. L., and R. L. Sinsabaugh.** 2000. Simulated patterns of litter decay predict patterns of extracellular enzyme activities. *Appl. Soil Ecol.* **14:**71–79.
58. **Moorhead, D. L., and R. L. Sinsabaugh.** A theoretical model of litter decay and microbial interaction. *Ecol. Monogr.*, in press.
59. **Nannipieri, P., L. Muccini, and C. Ciardi.** 1983. Microbial biomass and enzyme activities—production and persistence. *Soil Biol. Biochem.* **15:**679–685.
60. **Neff, J. C., A. R. Townsend, G. Gleixner, S. J. Lehman, J. Turnbull, and W. D. Bowman.** 2002. Variable effects of nitrogen additions on the stability and turnover of soil carbon. *Nature* **419:**915–917.
61. **Olander, L. P., and P. M. Vitousek.** 2000. Regulation of soil phosphatase and chitinase activity by N and P availability. *Biogeochemistry* **49:**175–190.
62. **Parton, W. J., D. S. Schimel, C. V. Cole, and D. S. Ojima.** 1987. Analysis of factors controlling soil organic matter levels in Great Plains grasslands. *Soil Sci. Soc. Am. J.* **51:**1173–1179.
63. **Perez, J., and T. W. Jeffries.** 1992. Roles of manganese and organic acid chelators in regulating lignin degradation and biosynthesis of peroxidases by *Phanerochaete chrysosporium*. *Appl. Environ. Microbiol.* **58:**2402–2409.
64. **Perucci, P.** 1992. Enzyme activity and microbial biomass in a field soil amended with municipal refuse. *Biol. Fertil. Soils* **14:**54–60.
65. **Quiquampoix, H., S. Servagent-Noinville, and M. Baron.** 2002. Enzyme adsorption on soil mineral surfaces and consequences for the catalytic activity, p. 285–306. *In* R. G. Burns and R. P. Dick (ed.), *Enzymes in the Environment*. Marcel Dekker, New York, N.Y.
66. **Read, D. J., and J. Perez-Moreno.** 2003. Mycorrhizas and nutrient cycling in ecosystems: a journey towards relevance? *New Phytol.* **157:**475–492.
67. **Renella, G., L. Landi, and P. Nannipieri.** 2002. Hydrolase activities during and after the chloroform fumigation of soil as affected by protease activity. *Soil Biol. Biochem.* **34:**51–60.
68. **Saiya-Cork, K. R., R. L. Sinsabaugh, and D. R. Zak.** 2002. The effects of long-term nitrogen deposition on

extracellular enzyme activity in an *Acer saccharum* forest soil. *Soil Biol. Biochem.* **34:**1309–1315.

69. **Schimel, J. P., and M. N. Weintraub.** 2003. The implications of exoenzyme activity on microbial carbon and nitrogen limitation in soil: a theoretical model. *Soil Biol. Biochem.* **35:**549–563.
70. **Schulze, W. X., G. Gleixner, K. Kaiser, G. Guggenberger, M. Mann, and E.-D. Schulze.** 2005. A proteomic fingerprint of dissolved organic carbon and of soil particles. *Oecologia* **142:**335–343.
71. **Sinsabaugh, R. L., R. K. Antibus, A. E. Linkins, C. A. McClaugherty, L. Rayburn, D. Repert, and T. Weiland.** 1993. Wood decomposition: nitrogen and phosphorus dynamics in relation to extracellular enzyme activity. *Ecology* **74:**1586–1593.
72. **Sinsabaugh, R. L., M. M. Carreiro, and S. Alvarez.** 2002. Enzyme and microbial dynamics during litter decomposition, p. 249–266. *In* R. G. Burns and R. P. Dick (ed.), *Enzymes in the Environment.* Marcel Dekker, New York, N.Y.
73. **Sinsabaugh, R. L., M. M. Carreiro, and D. A. Repert.** 2002. Allocation of extracellular enzymatic activity in relation to litter composition, N deposition, and mass loss. *Biogeochemistry* **60:**1–24.
74. **Sinsabaugh, R. L., S. Findlay, P. Franchini, and D. Fischer.** 1997. Enzymatic analysis of riverine bacterioplankton production. *Limnol. Oceanogr.* **42:**29–38.
75. **Sinsabaugh, R. L., M. E. Gallo, C. Lauber, M. P. Waldrop, and D. R. Zak.** 2005. Extracellular enzyme activities and soil organic matter dynamics for northern hardwood forests receiving simulated nitrogen deposition. *Biogeochemistry* **75:**201–215.
76. **Sinsabaugh, R. L., and D. L. Moorhead.** 1994. Resource allocation to extracellular enzyme production: a model for nitrogen and phosphorus control of litter decomposition. *Soil Biol. Biochem.* **26:**1305–1311.
77. **Sinsabaugh, R. L., H. Reynolds, and T. M. Long.** 2000. Rapid assay for amidohydrolase (urease) activity in environmental samples. *Soil Biol. Biochem.* **32:**2095–2097.
78. **Smith, S. E., and D. J. Read.** 1997. *Mycorrhizal Symbiosis,* 2nd ed. Academic Press, San Diego, Calif.
79. **Smith, V. H.** 2002. Effects of resource supplies on the structure and function of microbial communities. *Antonie Leeuwenhoek* **81:**99–106.
80. **Sollins, P., P. Homann, and B. A. Caldwell.** 1996. Stabilization and destabilization of soil organic matter: mechanisms and controls. *Geoderma* **74:**65–105.
81. **Stursova, M., C. L. Crenshaw, and R. L. Sinsabaugh.** 2006. Microbial responses to long-term N deposition in a semi-arid grassland. *Microb. Ecol.* **51:**90–98.
82. **Tarafdar, J. C., and H. Marschner.** 1994. Phosphatase activity in the rhizosphere and hydrosphere of VA mycorrhizal wheat supplied with inorganic and organic phosphorus. *Soil Biol. Biochem.* **26:**387–395.
83. **Trasar-Cepeda, C., C. Leirós, F. Gil-Sotres, and S. Seoane.** 1998. Towards a biochemical quality index for soils: an expression relating several biological and biochemical properties. *Biol. Fertil. Soils* **26:**100–106.
84. **Travisano, M., and G. J. Velicer.** 2004. Strategies of microbial cheater control. *Trends Microbiol.* **12:**72–78.
85. **Tscherko, D., U. Hammesfahr, M.-C. Marx, and E. Kandeler.** 2004. Shifts in rhizosphere microbial communities and enzyme activity of *Poa alpina* across an alpine chronosequence. *Soil Biol. Biochem.* **36:**1685–1698.
86. **Vance, E. D., and F. S. Chapin III.** 2001. Substrate limitations to microbial activity in taiga forest floors. *Soil Biol. Biochem.* **33:**173–188.
87. **van den Burg, B.** 2003. Extremophiles as a source for novel enzymes. *Curr. Opin. Microbiol.* **6:**213–218.
88. **Vetter, Y. A., J. W. Denning, P. A. Jumars, and B. B. Krieger-Brockett.** 1998. A predictive model of bacterial foraging by means of freely released extracellular enzymes. *Microb. Ecol.* **36:**75–92.
89. **Waldrop, M. P., T. C. Balser, and M. K. Firestone.** 2000. Linking microbial community composition to function in a tropical soil. *Soil Biol. Biochem.* **32:**1837–1846.
90. **Waldrop, M. P., D. R. Zak, and R. L. Sinsabaugh.** 2004. Microbial community response to nitrogen deposition in northern forest ecosystems. *Soil Biol. Biochem.* **36:**1443–1451.
91. **Waldrop, M. P., D. R. Zak, R. L. Sinsabaugh, M. Gallo, and C. Lauber.** 2004. Nitrogen deposition modifies soil carbon storage through changes in microbial enzyme activity. *Ecol. Appl.* **14:**1172–1177.
92. **Watanabe, K., and K. Hayano.** 1995. Seasonal variation of soil protease activities and their relation to proteolytic bacteria and *Bacillus* spp. in paddy field soil. *Soil Biol. Biochem.* **27:**197–203.
93. **Weetall, H.** 1975. Immobilized enzymes and their application in the food and beverage industry. *Proc. Biochem.* **10:**3–24.
94. **Weintraub, M. N., and J. P. Schimel.** 2005. The seasonal dynamics of amino acids and other nutrients in Alaskan Arctic tundra soils. *Biogeochemistry* **73:**359–380.
95. **Weintraub, M. N., and J. P. Schimel.** 2005. Seasonal protein dynamics in Alaskan Arctic tundra soils. *Soil Biol. Biochem.* **37:**1469–1475.
96. **Wick, B., H. Tiessen, and R. S. C. Menezes.** 2000. Land quality changes following the conversion of the natural vegetation into silvo-pastoral systems in semi-arid NE Brazil. *Plant Soil* **222:**59–70.
97. **Zimmerman, A. R., J. Chorover, K. W. Goyne, and S. L. Brantley.** 2004. Protection of mesopore-adsorbed organic matter from enzymatic degradation. *Environ. Sci. Technol.* **38:**4542–4548.

Immunofluorescence Microscopy and Fluorescence In Situ Hybridization Combined with CMEIAS and Other Image Analysis Tools for Soil- and Plant-Associated Microbial Autecology*

FRANK B. DAZZO, MICHAEL SCHMID, AND ANTON HARTMANN

59

Pioneering microscopes designed by van Leeuwenhoek in the 1680s first showed the existence of microbes, and microscopy performed by Pasteur, Koch, and Winogradsky in the late 1800s provided the very first scientifically based insights into the relevance of microbes in health, agricultural, and biotechnological sciences. Now we are in the postgenome area, when more information about the detailed genetic design of microbes is beginning to be known (www.ncbi.nlm.nih.gov/genomes/lproks.cgi). In addition, very powerful nucleic acid-based tools are available to obtain information about some of the phylogenetic diversity and metabolic potential of microbial communities. However, this knowledge does not indicate the abundance, distribution, or ecophysiology of the organisms at spatial scales relevant to their interaction with their habitat or other organisms in situ. The microscopy approach is therefore even more valid than ever before, because it can provide very relevant ecological information that fills major gaps in molecular microbial ecology. Furthermore, since ecological processes change with the spatial scale at which they are observed and measured, knowing how scale influences our observations and measurements is critical to defining the dimensions of the ecological niche occupied by microorganisms in situ. One meter to a microbe is approximately equivalent in length to 1,000 miles to an adult human! Microscopy can be substantially enriched to study microbes in situ by applying fluorescent labeled immunological and molecular genetic tools, referred to as immunofluorescence microscopy (IFM) and fluorescence in situ hybridization (FISH), respectively. Also, by using specific antibodies for enzymes or oligonucleotide probes against gene transcripts, selected in situ activities of microbes can be visualized. Thus, microscopy can now open the gate for real-world ecological studies of microbes in their microhabitats at the spatial scales relevant to their ecological niches, like those commonly studied for a long time in plant and animal ecology.

In this chapter, we describe the application of fluorescent molecular probes used with IFM and FISH techniques for studies of microbial autecology, with an emphasis on soil and root-associated microbes. Recent developments in computer-assisted microscopy used in conjunction with confocal laser scanning microscopy (CLSM) provided powerful tools to augment and optimize this microscopy approach in both two and three dimensions. This technology is supported by advanced digital image analysis software applications like CMEIAS (Center for Microbial Ecology Image Analysis System), DAIME (Digital Image Analysis in Microbial Ecology), and others that have been custom designed by microbial ecologists to extract, analyze, and report on ecologically relevant data from digital images acquired by these quantitative techniques, thereby strengthening microscopy-based methods for understanding microbial ecology. Also, the quantitative image analysis tools available in these software applications can be applied to various other microscopy techniques used in microbial ecology research, so long as the foreground microbes of interest can be found and segmented from background in the digital images.

This chapter does not cover the increasing application of other powerful cellular fluorescence techniques like tagging with autofluorescent proteins such as green fluorescence protein (GFP) or red fluorescence protein. The reader is referred to chapters 14 and 60 of this manual and a recent review (64) on the use of GFP to engineer bioreporter strains for microbial autecological studies. However, a word of caution. Since gene tagging of the organism of interest involves genetic and metabolic modifications, these added biosynthetic burdens may bias the organism's in situ ecophysiology and behavior and therefore limit its use for experimentation in certain habitats. In contrast, immunofluorescence and FISH can be performed with genetically unaltered wild types and can also give insights into the autecology of hitherto uncultured organisms.

ESSENTIALS AND ADVANCEMENTS IN FLUORESCENCE MICROSCOPY

The unique essential features of a fluorescence microscope are (i) a high-intensity gas-discharge light source (e.g., an HBO mercury lamp) to supply sufficient radiation of a wavelength required for fluorescence excitation of the fluorochrome; (ii) an exciter filter, which transmits only the high-energy radiation required for fluorescence excitation; and (iii) a barrier filter to remove excessive and extraneous

*This chapter is dedicated to the memory of Georges L. Truchet, our friend and colleague who advanced many pioneering contributions of microscopy in order to visualize beneficial plant-microbe interactions accurately.

excitation light not absorbed by the specimen, with selective passage of light in the fluorescence emission spectrum. Modern fluorescence microscopes are also equipped with a dichroic beam splitter to take advantage of the simpler operation, brighter fluorescence, and better image quality achieved by an incident illumination path which directs the excitation light down through the objective lens to the specimen and the emitted fluorescence light back up again through the same lens (hence, epifluorescence), rather than by dark-field transmitted illumination from beneath the specimen.

Proper filter sets must be used to match the specific optical requirements for excitation and emission of fluorescent light by different fluorochromes. Fluorescein isothiocyanate (FITC), and to a lesser extent tetramethyl rhodamine isothiocyanate (TRITC), are common fluorochromes for single labeling experiments of IFM. For FITC, the excitation and emission maxima are 490 nm and 510 to 550 nm, respectively, and for TRITC they are 560 nm and 600 to 650 nm, respectively. The optimum filter set for FITC is a band pass combination BP450-BP490 exciter filter, a dichroic chromatic beam splitter FT-510 filter, and a long wave pass LP-520 barrier filter. FITC and TRITC are not ideally suited for double labeling experiments, however, since spontaneous internal fluorescence results from the overlap in their excitation and emission curves. In this case, a better alternative for use with FITC is Texas Red, which has greater spectral separation.

Because fluorescence microscopy involves low light levels, the available light must be utilized effectively. Factors that help optimize the use of available light include locating the microscope in a dimmed darkened room to dilate the investigator's eye pupils, ensuring proper focus and light source alignment, using fluorite objectives of high numerical aperture and incident illumination, and using high-speed film for photomicrography or digital cameras with very high sensitivity and low electronic noise.

The development of the CLSM was a most significant advancement for the use of IFM and FISH to study in situ microbial autecology. Indeed, CLSM combined with specific molecular probe fluorescence staining or tagging is currently the "gold standard" to visualize specific organisms in microhabitats (4, 38, 109, 110). Its ability to eliminate stray and out-of-focus light that interferes with the formation of the object's image eliminates a major limitation of the conventional fluorescence microscope, especially when examining plant tissue, soil particles, or organic debris that emits a significant amount of objectionable background fluorescence. Also, by digitizing a sequential series of two-dimensional (2-D) images while focusing through the specimen in the third (*z*) dimension (called a "Z series"), an all-inclusive image of the flattened stack with extended focus or a three-dimensional (3-D) reconstructed image can be produced, rotated, and quantitatively analyzed. Digital phi *z*- or *xz*-cross-sections through the sample and perpendicular to the optical axis can also be constructed in silico. By spiking samples with different known amounts of bacterial cells serving as internal standard (30), or more directly by performing voxel abundance image analysis on spatially calibrated confocal Z-series stacks (24), the absolute local density of bacterial cells in a user-defined, complex sample volume can be determined in situ. Some CLSM systems can scan the entire fluorescence emission spectrum for each acquired image pixel to produce a spectral signature useful to identify and confirm the presence and location of the fluorescent dye in the sample. Dual photon excitation is useful for examining thicker specimens with reduced photobleaching (71). A computer-controlled motorized stage and software applications (e.g., Confocal Assistant) are also available to create loss-less montages of georeferenced image reconstructions for spatial distribution analyses of microorganisms (3, 24, 70). For more information on the theory, variation in design, and operation of confocal microscopy, consult chapter 5 of this manual.

PRODUCTION OF FLUORESCENT ANTIBODIES AND THEIR USE IN IFM

Antibodies are immunoglobulin glycoproteins made by higher animals in response to foreign antigens. The antibodies that are most useful for studies of microbial autecology are those which combine specifically with antigens on the surface of microorganisms. The objective of IFM is to detect antigens at single-cell resolution by methods which combine the specificity of antibodies as molecular probes that are covalently conjugated to a fluorochrome (e.g., FITC) and the awesome resolving power of microscopy using appropriate fluorescence optics to selectively detect the immobilized fluorescent molecular probe at a theoretical resolution of 0.2 μm. IFM using fluorescent antibodies (FA) is an extremely sensitive (one cell or less) method of evaluating these specific antibody-antigen interactions, providing useful autecological information on the detection, morphology, local abundance, and spatial distribution of the specific microorganism of interest, with differentiation at the strain level. An important feature of FITC as an ideal fluorochrome for FA is that the green color of its fluorescence emission spectrum lies within the human eye's most sensitive region of detection.

The technique of IFM dates from the work of Coons et al. (26), who first prepared a fluorescent conjugated antibody, and later from the work of Cherry et al. (23), who applied its use in the detection of pathogenic microorganisms for the clinical diagnosis of infectious diseases. Its use in studies of microbial autecology was pioneered by Schmidt and Bankole (113), who tested contact slides with fluorochrome-labeled antibodies to demonstrate a specific detection system for *Aspergillus flavus* in soil, and soon thereafter by Hill and Gray (62) to study the autecology of *Bacillus* in soil. IFM now has widespread application in microbiology and biotechnology, based largely on the availability of high-grade commercial fluorochrome reagents, FA conjugates, and good fluorescence microscopes. IFM is still the most sensitive and highest resolving method of choice for in situ studies of microbial autecology at the strain level. When optimized using high-titer specific antibodies and superlative resolving optics, IFM can provide very useful information on the distribution of physiologically important bacterial surface antigens at subcellular resolution (15, 39, 40).

There are two general approaches in immunofluorescence staining: direct and indirect. Both approaches involve the production, in an immunologically competent animal (e.g., rabbit), of a primary specific antibody against the antigen of interest. For the direct technique, fluorescent conjugates are prepared for each primary specific antibody under investigation. Although this involves isolating, conjugating, and purifying the antibody-fluorochrome conjugate, results from direct IFM are achieved quicker (~2 h) once the FA reagent is prepared. In the indirect immunofluorescence (so-called "sandwich") technique, the primary specific antibody is not conjugated. Instead, a purified fluorescent conjugate of a secondary antibody is used (e.g., affinity-purified goat anti-rabbit immunoglobulin G-FITC conjugate) to detect and specifically bind to the immunoglobulin glycoprotein of the animal used for primary antibody production, which itself has reacted specifically with, and therefore immobilized on, the antigen that evoked its production. The major advantages of the indirect technique are (i) higher immunospecificity in many cases; (ii) elimination of the need to conjugate the primary antibody by using high-quality, commercially available, purified conjugated secondary antibodies (universal secondary antibody for all primary antibody preparations derived from the serum of the same animal species); (iii) suitability when only small quantities of primary antibody are available; and (iv) higher luminance (fluorescence brightness per target cell), hence higher sensitivity, achieved by binding of several fluorescent secondary antibodies to each primary antibody that had already reacted immunospecifically with the antigen(s) on the microbial cell surface.

In the next section, we describe procedures to prepare rabbit polyclonal antibodies to surface antigens of microbes, their conjugation to a fluorochrome, preparation of soil and root samples for examination, and use of purified antibody-fluorochrome conjugates to detect target microorganisms by IFM. This is followed by selected examples to illustrate the use of IFM in autecological studies on *Rhizobium* and its association with higher plant roots, including its combined use with quantitative CMEIAS image analysis to measure this microbe's in situ spatial abundance and patterns of distribution on root surfaces. Finally, we make brief mention of the application of IFM to detect specific metabolic potentials of microbes in situ, using FA directed against the corresponding enzymes that catalyze those reactions. Several ancillary reviews on the use of the IFM technique in microbial ecology are available, including protocols to prepare monoclonal antibodies, which is beyond the scope of this chapter (17, 34, 42, 56, 83, 112, 130).

DEVELOPMENT OF ANTISERA

Antigen Preparation

Many components of a microorganism are antigenic. Microbial proteins and polysaccharides contain numerous epitopes that elicit an immune reaction, but some peptides as small as a few amino acids can be antigenic as well. The method of preparing conventional antisera suitable for IFM varies with the antigen used. In most studies of plant- or soil-associated bacteria, the goal is to produce antibodies suitable for strain-specific in situ detection of the organism of interest and assessment of its abundance in the environmental sample. This is best accomplished by preparing cells of the organism in ways that favor the preservation and immunogenicity of their cell surface somatic antigens (O-antigens of the lipopolysaccharide in the outer membrane of Gram negatives and lipoteichoic acids of Gram positives). These antigens generally are highly immunodominant, heat stable, and expressed on cell surfaces even when the organisms are in a quiescent, nongrowing state, as is the normal case in unamended bulk soil.

To avoid extraneous fluorescence when using the finished fluorochrome-labeled antibody preparation, care must be taken to exclude unwanted immunogenic materials that may contaminate the antigen preparation used for immunization and lead to nonspecific reactions. A typical protocol for preparing bacteria for immunization is to grow the organism in a simple defined medium whenever possible (avoid complex media containing yeast extract or animal or plant protein hydrolysates that may evoke immune responses themselves) and then wash the cells three times by centrifugation in a phosphate-buffered saline (PBS) (145 mM NaCl and 10 mM K_2HPO_4-KH_2PO_4, pH 7.4) solution that has been autoclaved and filtered (0.22-μm pore size). Resuspend the cells in autoclaved-filtered PBS to a concentration of ~10^9 cells per ml in a sterile stoppered test tube and vent with a sterile syringe needle. Place the tube into a boiling water bath for 1 h. After cooling, wash the heat-treated cells thoroughly by centrifugation with PBS, resuspend to a turbidity of approximately 10^9 cells/ml in PBS, and store frozen until used.

Protocols for preparing fungal immunogens are somewhat different (49, 68, 114, 126). Spores and hyphae from liquid cultures are washed, homogenized with a Waring blender in sterile-filtered PBS, collected by centrifugation, and resuspended to 0.5 to 0.75 mg (wet weight) per ml of sterile-filtered PBS. Preparations are used undiluted and injected intravenously. For preparing vesicular-arbuscular mycorrhizal fungi that are difficult to grow in pure culture, harvest their hyphae and spores from root-associated sand used to cultivate the host plant with the vesicular-arbuscular mycorrhizal fungi under gnotobiotic conditions. Wash the harvested fungal cells repeatedly by centrifugation in sterile water, resuspend in a minimal volume of PBS (approximately 1,000 spores per ml), disrupt by sonication on ice to avoid overheating, and then use as antigen after diluting 1:3 with PBS. Alternatively, germinate surface-sterilized spores between sterile filter disks (0.45-μm pore) and then separate the hyphae and vesicles from the spores to use as antigen.

Immunization

For the reader who is inexperienced in antibody production, Harlow and Lane (60) have prepared a comprehensive laboratory manual for antibody production. A suitable preparation of the antigen is presented to the animal's immune system such that antibodies specific to that antigen are produced. Conventional antibody production requires preparation of the injectable antigen, methods for introducing the immunogen into the animal, periodic withdrawal of small volumes of blood for testing the antibody level by using some quantifiable antigen-antibody reaction, and finally, collection and storage of the desired immune serum. Specificity rather than absolute titer is the main objective. In general, antibodies produced earlier in the immunization schedule are more specific than antibodies made after extended immunization.

The most difficult aspect of antibody production is acquiring the skill and patience necessary for working with the animals. The aid of someone experienced in handling animals, immunization, and collection of blood is invaluable in this process. Guidelines for the humane housing and care of animals should be followed.

Virgin female New Zealand White rabbits typically are used for polyclonal antibody production. It is advisable to immunize two or more animals with a given antigen preparation, since not all individuals produce the desired antibodies. Prior to injection, a preimmune draw of ~3 ml of blood should be obtained from the marginal ear vein and its serum should be tested by IFM to determine if the animal's serum already contains antibodies reactive with surface antigens on the organism of interest. Animals with preimmune serum that contains antibodies against the target organism should not be used for serum production.

Various immunization schedules have been used successfully for bacteria (34, 42, 83, 112) and fungi (49, 113, 114). For primary immunization, the antigen suspension is thoroughly emulsified 1:1 with Freund's incomplete adjuvant by drawing the mixture into a syringe and then expelling it repeatedly until the mixture is of a smooth consistency. Injection doses of antigen plus incomplete adjuvant are always administered subcutaneously or intramuscularly. Beginning a few weeks later, booster injections of the antigen suspension alone are made intravenously to increase the antibody titer by secondary immune responses. These injections should be continued until the serum is obtained with desired specificity and titer.

The mineral oil in Freund's incomplete adjuvant functions as a slow release mechanism, maintaining the antigen in the animal for extended periods to continue eliciting an immune response. This reduces the number of injections required to elicit a satisfactory immune response. Otherwise, the foreign matter of the antigen may be quickly degraded and voided by the animal. Freund's complete adjuvant contains killed cells of *Mycobacterium tuberculosis* suspended in mineral oil to aid in the stimulation of the immune system. Complete adjuvant can be used if the finished FA is intended for examinations of pure cultures or in vitro work but should be avoided when one expects to use the FA to probe environmental samples. The presence of *M. tuberculosis* in the antigen solution used for booster injections can result in the finished serum containing significant titers against other organisms, which may introduce unwanted FA reactions.

Blood Harvest and Preparation of the Collected Antiserum

Animal care facilities normally have personnel trained to collect the blood from the immunized animal at the end of the immunization schedule. Either ear bleed or cardiac puncture is used. The total volume of blood obtained per rabbit is approximately 100 to 120 ml (roughly 50 to 60 ml of crude serum per animal). Alternatively, several 30-ml harvests can be taken every 10 days. If the latter method is used, one should verify that the titer remains at an adequate level for each bleed (booster shots may be required) and the animal should also receive sterile saline to replace body fluids lost to each bleed.

Allow the blood to clot for approximately 2 h at room temperature and then refrigerate to shrink the clot. Slice the clot with a narrow spatula; incubate for 30 min at 37°C followed by overnight incubation at 4°C. Decant the serum and centrifuge at slow speed (5,000 × *g*) for 10 min to pellet unclotted blood cells. Carefully remove the supernatant fluid of cell-free serum, distribute in small aliquots, store frozen at −20°C or lyophilize, or add a bacterial inhibitor (0.1% sodium azide or 0.01% thimerosal) and store at 4°C. Avoid repeated freezing and thawing.

When feasible, commercial production of custom antibodies is a good alternative time-saver. An internet search for "custom polyclonal antibody production" conducted on 12 April 2006 found 396 commercial websites that provide that service for a fee. A two-rabbit, custom polyclonal antibody production service typically costs in the range of US$450 to $650 at the time of writing.

Purification of Antibodies and Conjugation to a Fluorochrome

Transfer 20 ml of cleared serum to a 50-ml beaker containing a small stir bar. Chill on ice or in a cold room facility. With slow stirring, add an equal volume of cold saturated $(NH_4)_2SO_4$ (preadjusted to pH 7.0 with NH_4OH). Allow the serum proteins to precipitate overnight at 4°C and then pellet by centrifuging at 10,000 × *g* for 15 min. Discard the supernatant fluid and gently redissolve the pellet to the original volume with PBS. Add an equal volume of saturated $(NH_4)_2SO_4$ again, stir for 1 h at 4°C, and pellet the resulting precipitate by centrifugation. This precipitation-centrifugation step can be repeated several times if the resuspended protein solution is still pink, indicating free hemoglobin. The final precipitate of partially purified globulins should be redissolved in one-half the original volume and then extensively dialyzed (14,000 to 16,000 molecular weight cutoff) with multiple changes of PBS to remove the $(NH_4)_2SO_4$ since it can interfere with the subsequent fluorochrome-labeling step. Centrifuge at 10,000 × *g* to clarify if necessary.

Prepare a solution of isolated rabbit immunoglobulins (10 to 20 mg of protein/ml) in less than 8 ml and dialyze overnight against 0.25 M sodium carbonate buffer (pH 9.0) containing 0.1 M NaCl and 1:10,000 thimerosal (no azide). Adjust the protein concentration to 10 mg/ml using carbonate buffer and add a high-grade, purified fluorochrome (e.g., FITC) dissolved in carbonate buffer or dimethylsulfoxide (DMSO) at the ratio of 0.05 mg of dye per mg of protein. This ratio is crucial to produce the proper stoichiometry of fluorochrome-antibody conjugation without protein denaturation. Incubate at room temperature without stirring for 3 h. Under these alkaline conditions, the isothiocyanate moiety of the fluorochrome dye covalently links to the γ-NH_2 nitrogen of lysine residues in the antibody protein.

The product of the last step is a mixture containing both conjugated and unconjugated FITC. The latter must be completely removed from the FA conjugate before it can be used, either by dialysis against PBS or by gel filtration. Gel filtration is the preferred method because it is faster and more efficient, yielding a product that is totally free of unconjugated fluorochrome. A gel filtration column of Sephadex G-25 or its equivalent is normally used for these separations, with bed dimensions of approximately 2.5 cm (diameter) by 30 cm (length). Load the FA mixture onto the bed of Sephadex G-25 equilibrated with PBS at 4°C. Gently wash the sample into the column and elute with PBS. Two yellow band fractions separate; the first, faster-running band elutes in the void volume and contains the fluorochrome dye-antibody protein conjugate, and the second, slower-migrating band is the unreacted dye. Collect the first band, refrigerate overnight, and clarify it by centrifugation at 10,000 × *g* for 15 min if necessary.

The fluorescein/protein ratio should be adjusted with PBS to between 2 and 4 but not higher to avoid protein inactivation. Determine this ratio by measuring the absorbance of the conjugate at wavelengths of 280 and 495 nm and solving the following formula:

$$\text{Molar ratio (fluorescein:protein)} = (2.87 \times A_{495}) / A_{280} - (0.35 \times A_{495}) \quad (1)$$

Sterilize the FA conjugate by passage through a 0.45-μm-pore-size filter, distribute aliquots in tubes, and store at −20°C or lower.

Testing the Immunofluorescent Titer and Specificity of the Antimicrobial Antibody

At this stage, it is important to measure the strength ("titer") of the antimicrobial antibody used for immunofluorescence (i.e., to determine how much the primary antibody solution can be diluted and still produce a positive immunofluorescence result, mathematically expressed as the inverse of the positive dilution endpoint) and, most importantly, its specificity in detecting the microbial strain of interest. This is typically done by testing dilutions of the antibody on heat-fixed cells of the bacteria used for immunization plus similar slides with other bacteria (both similar and unrelated) likely to be present in the same environmental samples. Apply a small volume of the diluted antibody to the heat-fixed smear and incubate at room temperature in a moist chamber (wetted filter paper inside a glass petri dish covered with aluminum foil) for 1 h. Then gently rinse off excess FA with PBS followed by distilled water and let air dry. Add a drop of mounting fluid formulated to prevent rapid fluorescent photobleaching (32), (Vectashield; Vector Laboratories, Inc., Burlingame, Calif.; or Citifluor; Agar Scientific Ltd., Cambridge, United Kingdom). Carefully cover the smear with a clean coverslip without introducing bubbles (practice makes perfect). Store slides horizontally at 4°C protected from light (a black microscope slide box or dish completely covered with aluminum foil works well) until the time of examination.

When scoring the IFM titer by using the pure culture smears, grade the fluorescing cells by using a subjective scale between 0 and 4 (more detail is provided below) of increasingly bright fluorescence. An ideal endpoint titer would be ≥1,000. If necessary, concentrate the sample by dialysis against polyethylene glycol to strengthen the fluorescent intensity of the FA. Transfer the conjugate to dialysis tubing (molecular weight cutoff, 12,000 to 14,000) and then place on a bed of dry polyethylene glycol 20,000. Remove the dialysis bag when the volume of the nondialyzable antibody solution has been reduced appropriately.

If present, unwanted cross-reactions need to be eliminated. Here is where use of multiple animals to produce high-titer, specific antibody successfully can be important, since prevention of cross-reacting antibody production can be somewhat of a "black box" that varies among individual animals immunized with the same antigen. Methods that can eliminate known immunological cross-reactions (listed in order of increasing difficulty) are (i) diluting the antibody prep to a level that no longer reacts with the heterologous antigen(s) while still yielding strong, positive immunofluorescence intensity with the microbe of interest (try this first, since sometimes it is all that is needed); (ii) removing the cross-reacting antibody by adsorption using steamed cells of the corresponding heterologous microbe; or (iii) affinity purifying the homologous antibody by using an immobilized antigen column (110). These methods may work well with antibody from one animal but not another. In a recent study in the author's lab, one rabbit's antiserum had a very high antibody titer of 4,000 to somatic antigens of an endophytic strain of *Rhizobium leguminosarum* bv. trifolii isolated from rice roots (131), but its cross-reactivity to another genomically distinct strain of the same bacterial species (present in the same soil habitat) could not be eliminated by either the dilution or adsorption methods indicated above (36). In contrast, all undesirable cross-reactions to other native strains of rhizobia in the same habitat, using antiserum from a second rabbit immunized with the same bacterial strain of interest, were completely eliminated by simple dilution of the immune serum to 1:1,000, while it still produced a very strong positive immunofluorescence reaction to the homologous bacterium at that dilution (36).

Use of Gelatin-RITC Conjugate Counterstain

Bohlool and Schmidt (16) developed a gelatin-rhodamine isothiocyanate (RITC) conjugate that effectively controls nonspecific staining, especially for examining microbes in soil since FAs notoriously bind nonspecifically to soil colloids (clay separates and particulate organic matter). When alkaline-hydrolyzed gelatin is applied as a conjugate with RITC, it can provide an effective counterstain that gives good color contrast to the green fluorescence of the FITC-labeled antibody. If a contrasting colored counterstain is not needed, the unconjugated alkaline preparation of autoclaved gelatin works equally well in blocking nonspecific staining of antibody to soil particles (F. Dazzo and D. Zvyagintsev, unpublished observation). To prepare the alkaline hydrolysate of gelatin, mix 2 g of gelatin with 100 ml of water, adjust the pH to 10 to 11 with 1 N NaOH, and then autoclave for 10 min at 121°C. After cooling, readjust to the original pH. To prepare the rhodamine-gelatin conjugate, dissolve 16 mg of RITC in a minimum volume of acetone, filter through a 0.45-μm-pore-size filter to remove undissolved materials, and then add to the 2 g of alkaline-treated gelatin solution. Conjugate overnight with slow stirring and remove the nonconjugated RITC by dialyzing against PBS, pH 7.2. Add merthiolate (1:10,000) and distribute into 20-ml aliquots. Working stocks can be refrigerated, and excess volumes can be stored frozen (−20°C).

CELL EXTRACTION AND RECOVERY FROM SOIL AND ROOTS

Counting efficiency by IFM is dependent upon quantitative extraction and recovery of the organism from the environmental sample. For soil and rhizosphere samples, flocculation and centrifugation techniques have been developed for maximizing cell recovery in a fashion that still allows for detection and counting by IFM. Unfortunately, given the heterogeneous nature of soils, optimization of the protocol is an empirical exercise and may not work for some soils because of their texture class and organic matter content. The flocculation technique tends to work well for soils containing a relatively high clay content, whereas sandy or silty soils may be clarified with centrifugation. While not always the case, some soils containing high organic matter content are not suitable for FA-based autecology studies.

Counting efficiency can be determined by inoculating autoclaved soil with a known number of target cells and then comparing IFM counts following different extraction protocols to the "ground truth data " of viable plate counts. Extraction techniques can also be optimized in nonsterilized soils in which IFM counts are compared with viable counts obtained with an antibiotic-resistant mutant of the same strain used for antigen development. Three protocols developed to maximize extraction of cells from soil slurries and plant roots are described below. All three methods can be modified as required for most soil types. Additionally, methods for direct detection of microbes in soil and on roots are provided.

Flocculation Method

The flocculation method has been applied to both soil and rhizosphere samples (46, 67, 87). An extractant used with good success is a solution containing hydrolyzed gelatin and $(NH_4)_2HPO_4$. A 1% gelatin (wt/vol in water) stock solution is adjusted to pH 10.5 and then partially hydrolyzed by autoclaving for 10 min. The gelatin stock solution is diluted 1:10 in 0.1 M $(NH_4)_2HPO_4$, with the final pH adjusted to 8.2 and used as the extractant.

Demezas and Bottomley (46) modified the extraction protocol of Kingsley and Bohlool (67). The cells are released from the soil particles by diluting 10 g of soil in 50 ml of the ammonium-phosphate:gelatin extractant, adding 5 g of glass beads, and shaking for 15 min using a wrist-action shaker. Another 45 ml of extractant is added, and the sample is shaken for an additional 15 min. An aqueous solution containing divalent cation is then added to flocculate the dispersed soil. Kingsley and Bohlool (67) achieved variable success by adding 0.2 g of $Ca(OH)_2$ and 0.5 g of $MgCO_3$ to the above soil-buffer mix. Demezas and Bottomley (46) compared various flocculants and found that a mixture of 0.8 g of $CaCl_2 \cdot 2H_2O$ and 0.5 g of $4MgCO_3 \cdot Mg(OH) \cdot 4H_2O$ provided optimum flocculation, sample clarity, and cell recovery for the soil used in their work. After the flocculent is added and thoroughly mixed (5 min on the wrist-action shaker), the suspension is allowed to settle for 1 h. Portions of the supernatant are then filtered (with suction) through Sudan Black-treated polycarbonate filters (see "Filter Preparation" below). A filtering manifold that accommodates multiple filters is preferred to optimize sample handling.

Centrifugation Technique

The density centrifugation technique (127) works well with a variety of soils, ranging from oxisol clays to sandy loam ultisols. This technique is recommended for soils typically dominated by high silt and sand fractions that respond poorly to the flocculation treatments described above. A 10-g soil sample is mixed with 95 ml of a 1% $CaCl_2$ solution, 1 drop of Tween 80 and 3 drops of antifoam agent are added, and the solution is vigorously mixed for 5 min using a blender. A 10-ml aliquot is then diluted with 20 ml of a 1.33-g/cm^3 sucrose solution and layered on top of a 30-ml 1.33-g/cm^3 sucrose bed. The sample is centrifuged in a swinging bucket rotor for 15 min. The centrifugal force used varies depending on the soil. Centrifugation speeds generating forces of 252 to 715 $\times$ g yielded samples with sufficient clarity and without suffering extensive reduction in cell recovery. After centrifugation, 10 ml is removed from the sucrose layer and diluted with 90 ml of distilled H_2O. This suspension is then filtered through Sudan Black-stained filters and processed for examination (described below).

Gradients for buoyant density centrifugation can be made using Nycodenz instead of sucrose to reduce adverse osmotic effects when isolating microbial communities from soil (75).

Cell Recovery from Roots

A centrifugation technique to recover *Bradyrhizobium japonicum* cells from soybean root tips also has been used (84). The soybean roots were excavated from the soil, the adhering soil was stripped away by pulling the root through the tips of sterile forceps, and the distal 2 cm of 10 to 15 roots were removed, weighed, and homogenized in 1 ml of PBS buffer per root tip. The homogenate was centrifuged for 5 min at 250 $\times$ g to sediment root debris and soil materials into a loose pellet. Portions (1.0 ml) of the supernatant were passed through filters, which were then treated with FA and examined.

FA STAINING OF EXTRACTED MICROBES

Filter Preparation

The filter should be stained black to obtain optimum background contrast for viewing the fluorescing cells. Polycarbonate filters (25-mm diameter) are preferred over cellulose filters, as their surface is more uniform and the cells tend not to become embedded and obscured from view. A pore size of 0.45 μm is commonly used because 0.22-μm pores can clog quickly. This depends on the sample, and prefiltration may be necessary to remove interfering organic materials and larger mineral particles, with only small reductions in cell recovery (18). Prestained black filters are available commercially. White filters can be stained with either 0.3% (wt/vol) Sudan Black in 70% (vol/vol) ethanol, or 0.2% Irgalan Black in 2% acetic acid. Filters are placed in staining solution (5 ml per filter), soaked for 48 h, and then washed with filtered distilled H_2O until free dye is no longer visible. Washed filters are air dried between foil spacers and stored in sealed containers.

Staining Microbes Collected on Filters

After the sample has been extracted by one of the methods discussed above, a measured volume of the sample is collected on the black filter, which is transferred to a glass slide. A 0.3-ml volume of the alkaline hydrolyzed gelatin (with or without RITC conjugate) is applied to the sample and incubated at room temperature until nearly dry. The appropriately diluted FA is applied (2 or 3 drops per filter), and the slide is placed in a moist chamber (e.g., wetted filter paper within a covered glass petri dish) and incubated at room temperature for 30 min in the dark. The stained filter is returned to the filtering manifold and washed (using vacuum) with at least 100 ml of 0.22-μm-pore-size-filtered 0.85% saline and then returned to the microscope slide.

The indirect IFM technique is similar, except that two antibodies are used. The unlabeled primary antibody preparation is applied to the filter and incubated as discussed above. A commercially available, fluorochrome-labeled secondary antibody that recognizes the primary antibody (e.g., goat anti-rabbit antibodies diluted 1:20 in PBS) is then added. The same fluorochrome-labeled secondary antibody can be used with any primary antibody preparation derived from serum of the same animal species and can be very convenient if large numbers of serologically different organisms are sought in the same environmental sample.

For both immunofluorescence procedures, add a drop of the photobleach retardant mounting fluid to the filter, allow a few minutes for the mounting fluid to spread evenly across the filter, cover with the proper thickness coverslip without introducing bubbles, and examine immediately or store at 4°C in the dark until the time of examination.

DIRECT FA STAINING OF MICROBES IN PROCESSED SOIL

Some soils can be processed and then applied directly to slides for IFM without filtration steps. In this case, combine 0.5 g of soil to 10 ml of filtered distilled water and sonicate continuously for 5 s. Allow coarse particles to settle by gravity for 42 min (optimized by Stokes' law to sediment soil particles larger than 2 μm), transfer the top 5-ml sample to a new tube, vortex, and dilute with water (the ratio varies between 1:20 and 1:200, depending on the microbial density). Afterwards, spread 4-μl samples evenly within the 6-mm-diameter wells on Teflon-coated slides. Air dry and apply hot air (with a hair dryer) to fix the cells to the slide. Apply 4 μl of the gelatin solution to each well and dry at room temperature. Then, incubate with the antibodies as described above and examine.

Immunofluorescence Reaction Score

When using IFM to study the autecology of an organism, a semiquantitative scoring of the cells' immunofluorescence reaction can be useful, particularly when other species and/or serologically related strains are also present in the sample. The following scoring system is being used in an autecological biogeography study of a candidate biofertilizer rice-endophyte strain of *R. leguminosarum* bv. trifolii sampled from root nodules on uninoculated plants of berseem clover (*Trifolium alexandrinum*) cultivated throughout the Egyptian Nile delta (36, 43, 131):

0 = no immunofluorescent cells above background

1+ = ≤1 positive immunofluorescent cell per microscope field, sufficient brightness for photomicrography

2+ = >1 to <100% positive immunofluorescent cells per microscope field

3+ = 100% of cells show ≤3+ positive immunofluorescence intensity

4+ = 100% of cells show 4+ positive immunofluorescence intensity

Unlike the 3+ reaction, the luminosity for the 4+ category is of almost blinding brightness, with saturated yellow-green fluorescent light covering the entire cell (no central dark areas) and a strong fluorescent halo surrounding them (Fig. 1, left panel). The ascending sort of mean grayscale luminosity for each cell (Fig. 1, right panel) measured by CMEIAS image analysis software (described below) shows the higher, less variable range for the 4+ scale, due to the more intense, uniform level of immunofluorescent brightness for each cell. This 4+ brightness should match the immunofluorescence intensity of the cells used for immunization when they are reacted with the homologous antibody.

DIRECT MICROSCOPICAL COUNTING OF BACTERIA IN SOIL BY USING IFM

Enumerate the brightly immunofluorescent cells in 20 randomly chosen microscopic fields. Then, solve the following formula:

$$\text{Number of specific bacteria/g of soil} = (N \times A \times D)/a \times V \quad [2]$$

where N is the average number of immunofluorescent cells per microscopic field; A is the effective area of the slide sample well or the effective filtering area of the filter; D is the dilution factor; a is the area of the microscopic field; and V is the volume of the entire diluted sample passed through the filter or placed directly into the well.

Although most microbes in natural habitats undergo rapid turnover when dead, the issue of viability always persists when conducting IFM, especially in nutrient-impoverished environments. Microautoradiography can be used in conjunction with IFM to detect substrate uptake by immunofluorescent microbes (see reference 17 and references therein). Bottomley (18) described a direct viable-count method for assessing the viability of IFM-detectable bacteria. The extracted sample containing the microbes of interest is incubated with yeast extract and nalidixic acid for 16 to 24 h and then processed for IFM counting. Viable cells elongate without cell division as a growth response to yeast extract in the presence of nalidixic acid. A comparison of results obtained by this method and the conventional immunofluorescence staining method without this treatment yields a measurement of the percentage of viable cells of the target strain reactive with the specific antibody.

Direct Examination of Microbes on Roots by Using IFM

The IFM technique can be used to study colonization of the rhizoplane environment by specific microorganisms (33, 34, 39-43, 87, 108, 109, 112). Roots of seedlings (e.g., white clover) that do not autofluoresce or whose autofluorescent light would be absorbed by the fluorescence emission filter can be processed and examined directly by IFM (42). If the root emits autofluorescent light in the range that also includes the FA fluorochrome, first apply a counterstain (e.g., Gram's crystal violet or erichrome black) to reduce plant autofluorescence to levels at which the target organism can be clearly distinguished. Prepare the erichrome black solution by dissolving 9 mg in 1.0 ml of DMSO, adding 5.0 ml of a solution containing 50 ml of DMSO, 20 ml of distilled water, 10 ml of 0.1 M aluminum chloride, and 10 ml of 1.0 M acetic acid (pH adjusted to 5.2 with NaOH), and then adding distilled water to a final volume of 100 ml. Soak the roots in the erichrome solution until they appear red, and then destain by soaking them in 0.85% saline until the color no longer changes. Place the root on a glass slide,

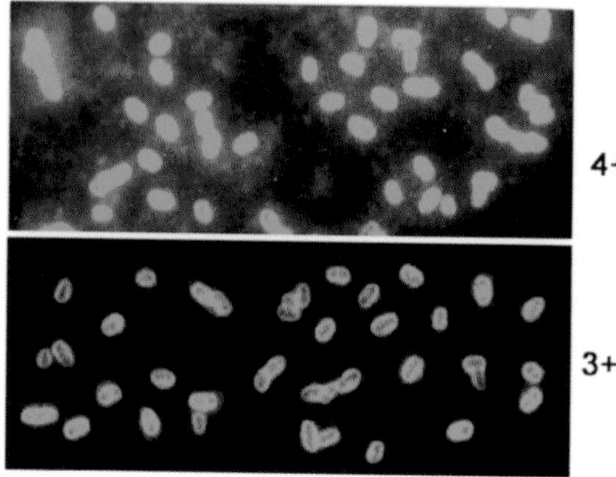

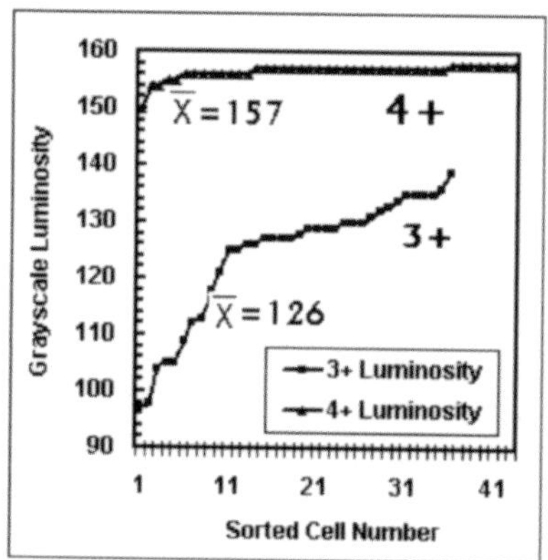

FIGURE 1 Differences in mean cell luminosity between typical 4+ and 3+ immunofluorescence reactions. Cells of *R. leguminosarum* bv. trifolii E11 are reacted with different dilutions of a strain-specific, polyvalent rabbit antiserum. (Left) Typical 4+ and 3+ reactions; (right) ascending sort of mean cell luminosity for 4+ and 3+ immunofluorescent reactions measured by CMEIAS image analysis. Note the higher, more uniform mean luminosity for the population of cells producing the 4+ reaction.

apply the (rhodamine-conjugated) alkaline-treated gelatin counterstain, and air dry. Apply the FA, incubate at room temperature, and then wash the root in three changes of filtered isotonic plant growth medium (10 min each). Add the mounting fluid containing a mixture of photobleach retardant (32) and isotonic plant growth medium, apply the coverslip, and examine microscopically.

For inoculated roots grown in hydroponic slide or plate cultures (41), gently rinse them with isotonic plant growth medium, cut them to a size that can fit under a coverslip, and (if necessary) seal the cut edge with molten paraffin to prevent infiltration of antibody into the root's central vascular system. Transfer the excised root segments to microscope slides, apply the diluted antibody solution (the volume depends on the thickness of the root), apply a coverslip, and examine.

When grown in soil, roots must be processed to remove adhering soil that obscures the rhizoplane while still retaining the adherent microflora. A washing protocol optimized for white clover (*Trifolium repens*) roots grown in lightly textured soil (35) is as follows. Gently excavate seedlings, transfer to 10-cm-diameter dishes containing isotonic plant growth medium, and soak for 30 min with occasional shaking. Then, transfer the seedlings to new dishes containing isotonic plant growth medium, seal with Parafilm, and shake on a gyrorotary platform for 2 h at 100 rpm. Avoid shear forces of strong vortexing to minimize loss of attached microbes. Remove seedlings, stain with the FA, and examine. Approximately 80% of the rhizoplane surface and its adherent microbial communities (assessed by scanning electron microscopy) can be visualized by using this procedure to dislodge soil without damaging the root hairs.

The first CLSM examination of the general rhizoplane microflora in situ was done with acridine orange-stained roots of white clover seedlings grown in soil (38). This approach eliminated the major background fluorescence due to absorption of dye by the underlying root, which makes conventional fluorescence microscopy impossible. Simultaneously, Schloter et al. (109) demonstrated the usefulness of CLSM for IFM examination of *Azospirillum* on wheat roots. They used a dual laser system to produce the green autofluorescence of the root background upon which the distinctive red immunofluorescence of *Azospirillum* (probed with TRITC-labeled monoclonal antibodies) could be easily seen. They also utilized the noninvasive optical sectioning ability of the confocal microscope to locate the *Azospirillum* cells within the root mucigel layer. More recently, optical sectioning by CLSM has been used to document the entry of rhizobia into rice roots, especially in crevices at lateral root emergence, their colonization within cortical cells and intercellular spaces of intact rice roots (100, 102), and their ascending migration up to aerial plant tissues (24). Because confocal laser scanning microscopy solves many of the problems inherent in conventional fluorescence microscopy, it is currently considered to be the best type of microscopy available to examine microbial colonization of root surfaces. To illustrate its wide application, a Google search on 19 June 2006 found 18,300 hits for confocal+root+microbe.

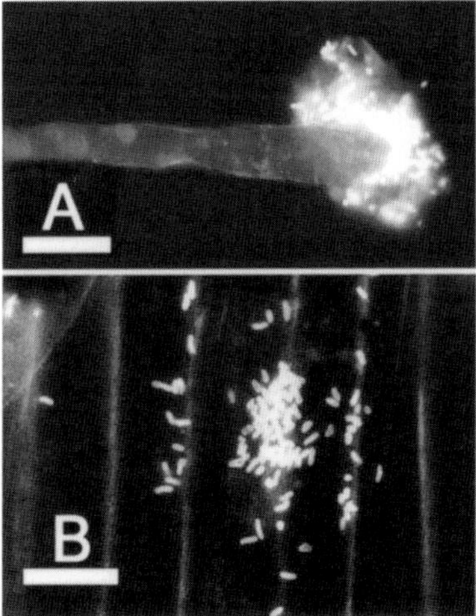

FIGURE 2 In situ detection of *R. leguminosarum* bv. trifolii on the root surface of its host, white clover (*Trifolium repens*), using indirect IFM with a monoclonal antibody specific for its lipopolysaccharide O-antigen. (A) Conventional epifluorescence. Immunofluorescent cells have aggregated at the tip of a root hair, where the host lectin, trifoliin A, accumulates. (B) Confocal laser scanning epifluorescence. Immunofluorescent cells have developed a microcolony on the rhizoplane.

Figure 2A and B illustrate the differences in image quality one obtains when examining the attachment and colonization of *R. leguminosarum* bv. trifolii to epidermal cells of white clover roots by using FA with conventional epifluorescence microscopy and CLSM, respectively. The reduction in out-of-focus fluorescence when using confocal optics is readily apparent, resulting in improved image quality with higher contrast and greater resolution of individual bacterial cells. These epifluorescence micrographs illustrate the host lectin-mediated aggregation of the bacterial symbiont on root hair tips and the formation of a microcolony in situ on the root epidermis (including the junctions between host cells).

Figure 3 illustrates the usefulness of IFM and noninvasive optical sectioning by CLSM to resolve fine structural detail during infection of legume root hairs by rhizobia. The confocal micrograph represents a single thin optisection through the intact tip region of a markedly curled root hair and provides direct evidence for the entrapment of immunofluorescently labeled rhizobial cells in the overlap of the "shepherd's crook" structure. It has been predicted that this confining cellular structure serves to concentrate in a localized region the metabolic events of microsymbiont penetration while preventing lysis of the root hair during primary host infection. This immunofluorescence micrograph was acquired with relative ease by using confocal optics but would be very difficult to obtain using conventional epifluorescence microscopy.

APPLICATION OF CMEIAS IMAGE ANALYSIS WITH IFM

Computer-assisted digital image analysis can increase the value of IFM for examining microbial colonization of roots even further, by allowing one to extract, store, retrieve, and electronically transmit numerical information regarding selected image features. Quantitative studies of microbial biogeography are greatly facilitated using computer-assisted microscopy, especially for defining the appropriate in situ spatial scale of quorum sensing, niche competition, and biofilm colonization in natural microbial habitats.

Recently, one of us (F.B.D.) has led a team of microbiologists, mathematicians, and computer scientists to develop and release CMEIAS advanced image analysis software applications designed to strengthen microscopy-based methods for

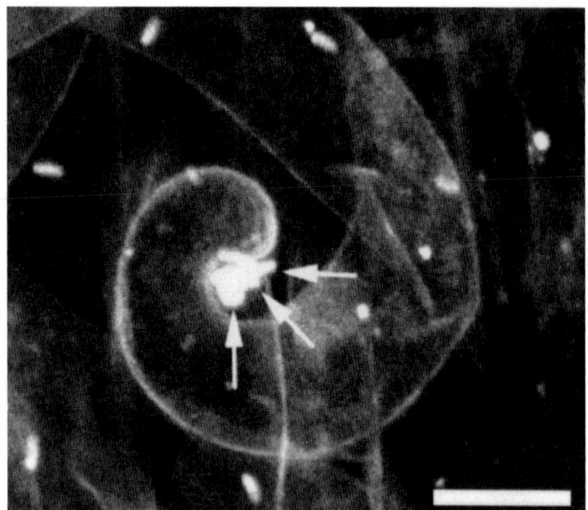

FIGURE 3 Use of optical sectioning with the laser scanning confocal microscope to detect *R. leguminosarum* bv. trifolii cells within the overlap of a markedly curled root hair of white clover. This step precedes primary host infection in the *Rhizobium*-legume symbiosis. Bacteria are stained by immunofluorescence using a monoclonal antibody against their somatic O-antigen. Bar, 10 μm.

understanding microbial ecology at spatial scales directly relevant to their ecological niche. The first-release version of CMEIAS reports on the in situ morphological diversity and abundance of microbial assemblages in digital micrographs (33, 76). Program files, a user manual, a training tutorial, and practice images for CMEIAS v.1.27 (operating in UTHSCSA ImageTool Ver. 1.27 using Windows) are available for noncommercial research and educational purposes as a free internet download at http://cme.msu.edu/cmeias.

An advanced suite of CMEIAS v. 3.0 software applications is under construction and will be released from the same website (33, 37). CMEIAS plug-in upgrades operating in UTHSCSA ImageTool version 3.0 have over 75 new features, including several computing tools to facilitate segmentation of foreground touching objects in images, classify a very high order of morphological diversity, and measure microbial abundance and spatial patterns of distribution by plotless and quadrat-based spatial point patterns and geostatistical analyses. We will soon release a CMEIAS Color Segmentation software application to segment foreground cells in complex color images of immunofluorescence, FISH, vital stain, and in situ metabolic activity stains (103, 104; C. Gross et al., unpublished data). The CMEIAS Color Segmentation program analyzes the 3-D color space and nearest neighbor spatial location of each sampled foreground pixel in the differentially colored community by using region-growing algorithms and then creates new color-segmented "object detected" images containing the population of sampled foreground objects of interest in a noise-free background. From these data plus image outputs, CMEIAS then reconstructs a new grayscale image of the selected population, with retention of each cell's representative morphology, abundance, position, size, and luminosity (102, 103; Gross et al., unpublished). A CMEIAS Quadrat Maker software application is being developed to optimize grid dimensions and produce quadrat images for plot-based and geostatistical analysis (37). A variety of stand-alone applications and Excel Add-ins are being developed to compute ecological statistics on image analysis data acquired using community-optimized CMEIAS classifiers (37). A listing of the literature on applications of CMEIAS in microbial ecology research is conveniently linked for online viewing at the following website page: http://cme.msu.edu/cmeias/publications.html. The exciting innovations in image analysis technology featured in CMEIAS v. 3.0 software will undoubtedly enhance the ecological analysis of in situ bacterial colonization using immunofluorescence and other discriminating microscopy techniques operating at single cell resolution.

To illustrate CMEIAS applied to autecological biogeography using IFM, digital image analysis was performed on the FA-labeled cells of *R. leguminosarum* bv. trifolii colonized on the white clover root epidermis (Fig. 4). In approximately 2 s of computing time, CMEIAS extracted several measurement parameters from each individual, immunofluorescent bacterial cell in Fig. 4 and reported the data, shown in Table 1, that define their spatial abundance while colonized on the root surface.

Of particular autecological biogeography interest are the image analysis data extracted by CMEIAS needed to perform distance-based and quadrat-based spatial point pattern analyses, and geostatistical analyses on spatial interactions among the bacteria. These attributes define spatial patterns of distribution, indices of dispersion, and mathematical modeling that predict bacterial colonization behavior in situ. The spatial randomness index, computed from the image area and average and standard deviations of the first and second nearest neighbor distances (Table 1), provides quantitative data indicating that the bacteria in Fig. 4 exhibit a clustered pattern of spatial distribution (values are <1.0). This indication of clumped spacing of organisms on the root surface is confirmed quantitatively by five other plot-based spatial distribution indices (variance/mean ratio, Morista's dispersion index, negative binomial K, Lloyd's mean crowding index, and Lloyd's patchiness index [Table 1]) computed from the CMEIAS data. This clumped or aggregated spatial pattern is consistent with the behavior of bacteria that remain in situ while actively growing into microcolonies on

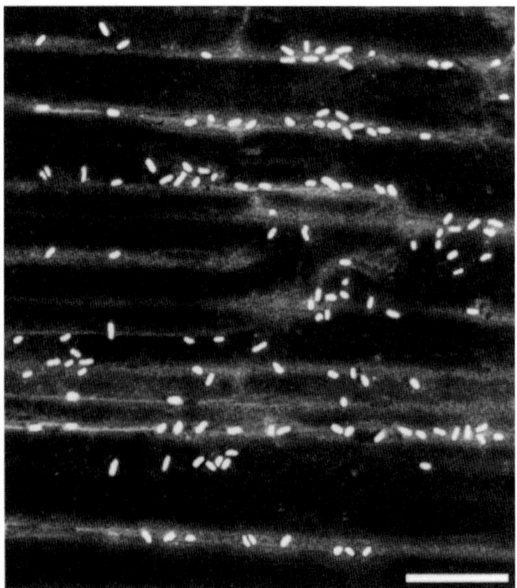

FIGURE 4 Epifluorescence micrograph of immunofluorescent cells of *R. leguminosarum* bv. trifolii 0403 colonized on the root surface (especially at junctions between epidermal cells) below the root hair region of a white clover seedling. Bar scale, 20.0 μm.

TABLE 1 CMEIAS image analysis of the in situ abundance and spatial distribution of the immunofluorescent bacteria shown in Fig. 5

Parameter measured	Value reported
Cell count	135
Microbial area coverage (μm²)	284.57
Total image area (μm²)	10,738.86
% Substratum coverage	2.7%
% Unoccupied substratum area	97.3%
Mean cell biovolume (μm³)	1.860 ± 0.650
Total microbial biovolume (μm³)	250.762
Mean cell biomass C (fg)	372 ± 130
Total microbial biomass C (fg)	50,152
Mean cell biosurface area (μm²)	8.37 ± 2.22
Total microbial biosurface area (μm²)	1,130.20
Spatial density (cells/mm²)	12,571
Mean 1st nearest neighbor distance (μm)	3.9 ± 2.5
Mean 2nd nearest neighbor distance (μm)	5.8 ± 2.9
Spatial randomness index	0.88 (clustered)
Variance/mean ratio	1.796 (clustered)
Morista's dispersion index	1.996 (clustered)
Negative binomial K index	0.950 (clustered)
Lloyd's mean crowding index	1.596 (clustered)
Lloyd's patchiness index	1.995 (clustered)

the root surface. This occurs frequently in troughs at junctions between adjacent epidermal cells just below and within the root hair region (31, 33).

This growth strategy is further indicated by the multimodal (rather than normal) distribution of the first and second nearest neighbor distances (Fig. 5, left panel), and by the abundance of clusters near the origin in 2-D scatterplots of the first versus second nearest neighbor distances (Fig. 5, right panel).

CMEIAS can also extract the information required to perform a geostatistical analysis of the spatial distribution of bacterial populations colonized on surfaces. Geostatistics is a portion of spatial statistics concerned with spatial continuity (continuous spatial variation). Its origin was designed for statistical estimation of ore reserves in the mining industry, where sampling is expensive and therefore limited (69). This method employs a user-defined, continuously variable parameter (e.g., localized density of colonized microbes) measured at various sampling points within the spatial domain to test for spatial autocorrelation, i.e., does this pattern of distribution exhibit spatial structure, and do events among cells at one location influence cells at another location? When found, the autocorrelation result produces a mathematical model that accurately connects various spatially dependent relationships derived from regionalized variable theory and makes optimal, statistically rigorous interpolation (kriging) maps of the parameter measured within that spatial domain, based on analysis of its weighted average from sampled locations. Geostatistical analysis and krig mapping require that the relevant parameter being analyzed (the so-called "Z variate") is a quantitative (nonbinary) value that is continuously distributed over the spatial domain and that the sampling sites are georeferenced, i.e., located at known X, Y coordinates relative to a landmark position in the same spatial domain.

Several powerful features of geostatistics apply to the autecological biogeography of microorganisms when using IFM. At the core of geostatistical analysis is the variogram plot, which defines the extent to which the measured Z variate exhibits spatial dependence between sample locations. The mathematical model that statistically best fits the autocorrelation data defines the range of linear radial distance that individual microbes can be separated from each other and yet still influence their neighbor's location or measured activity. The autocorrelation variogram also includes directional information to test for anisotropy in the spatial pattern, i.e., determines whether the range of spatial autocorrelation varies when examined at different compass directions, and if so, in which direction the measured Z variate is most spatially dependent. A second feature is that the kriging map output provides a statistically defendable, quantitative, continuous interpolation of the spatial variability of the parameter measured (like a weather map), even for points not physically sampled. When applied to microbes colonizing roots, this provides a powerful simulator and estimator of expected behavior of microbial cell-to-cell interactions over the entire rhizoplane surface, including regions that cannot be examined directly because they are obscured in microscopy images by overlying root hairs (33). In essence, this analysis puts in situ microbial autecology into the realm of the real-world scale relevant to the microbes themselves.

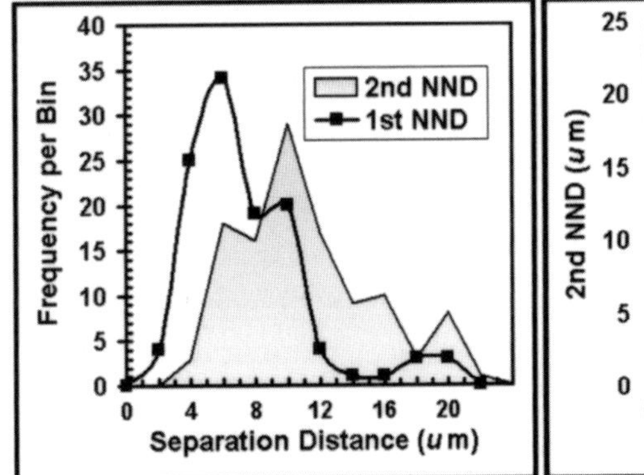

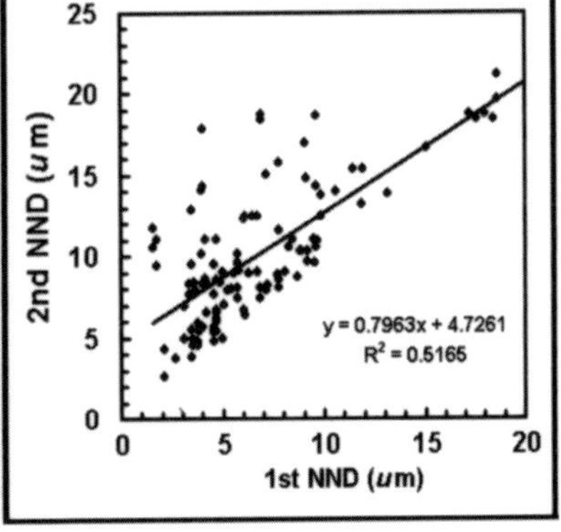

FIGURE 5 Frequency (left) and 2-D scatterplots (right) of the CMEIAS plotless spatial distribution analyses of the first and second nearest neighbor distances (NND) separating each individual bacterium shown in Fig. 4.

CMEIAS quadrat-based and geostatistical analyses were performed on the distribution of the immunofluorescent cells on the rhizoplane substratum shown in Fig. 4 and on the local density of bacteria within small, contiguous quadrats covering the root surface as the Z variate. The CMEIAS Quadrat Maker software application was used to evaluate the grid dimensions and hence optimize the quadrat sampling size in relation to the resultant sampling range of local density per quadrat to resolve small cell aggregates in this plot-based spatial analysis. The resultant, optimized grid contained 11 rows and 16 columns, producing 176 sampling quadrats, each being 8.8 μm wide and 6.4 μm in height, representing 9 and 6% of the image, respectively, and having a sampling range of local density of between 0 and 6 cells per quadrat. For geostatistical analysis, the coordinate posting of each quadrat sample was weighted by the density and position of the enclosed bacteria, and its georeferenced X, Y coordinates were measured by the CMEIAS Centroid measurement attribute using the lower left corner of the image as the 0, 0 landmark position. Geostatistical analyses of the resultant CMEIAS georeferenced data were performed using GS+ software to test for spatial autocorrelation and interpolation.

The results indicated spatial autocorrelation in the pattern of local density for the bacteria on the root surface. The best-fit semivariogram of the local spatial density is a spherical model presented in Fig. 6, left panel. In general, spherical mathematical models often best fit spatial distributions with patches of larger and smaller values of approximately equal size, and their maximum separation distance of autocorrelation is typically the diameter of the clustered patches. The specific example of Fig. 6, left panel, is characterized by a very small nugget Y intersect of 0.052, representing the very low extent of random variation within the data, a sill asymptote of 1.473, representing the structural variance present, and an effective range of separation distances of up to ~12 μm where these colonization densities are autocorrelated. The 2-D kriging map (Fig. 6, right panel) derived from this best-fit spherical autocorrelation model depicts the local density of bacteria interpolated over the entire spatial domain of the image, reminiscent of predicted radial gradients of diffusing phytohormone growth regulators and N-acylhomoserine lactone signal molecules elaborated by as few as two individual bacteria conducting cell-to-cell communication while colonizing plant roots (24, 53). This kriging interpolation analysis provides a statistically defendable prediction that the colonization pattern is a discontinuous biofilm containing spatially discrete patches of bacteria that influence colonization densities of neighboring cells over a range of separated distances of up to ~12 μm.

Inspection of these data immediately indicates that good-quality immunofluorescence micrographs of microbes in situ contain massive amounts of information on colonization behavior and that CMEIAS computing tools can efficiently extract and report on those quantitative data, with unparalleled resolution at spatial scales that are ecologically relevant to the microbes themselves. When combined with IFM, CMEIAS image analysis is a powerful method to study the autecological biogeography of microbes of interest, as is currently being done with unique rice endophytes of *R. leguminosarum* bv. trifolii that are strong candidates of effective agricultural biofertilizer inoculant strains capable of promoting crop yield in rice-berseem clover rotations in the Nile delta of Egypt (36, 43).

IN SITU IMMUNOLABELING OF CELL-ASSOCIATED ENZYMES INDICATIVE OF METABOLIC ACTIVITY

The in situ detection and localization of specific enzyme proteins in microbial cells by IFM has major usefulness in the autecological understanding of microbial metabolic functions. The approach requires highly specific antibodies directed against the enzyme(s) of interest. Fluorescently labeled antibodies may be used directly, or the primary antibody may be used together with a secondary antibody in the indirect sandwich technique. Specific antibodies are available for the *nifHDK* gene products of the nitrogenase enzyme (22, 105), the AmoB-protein of ammonia oxidizers (98), and the *dnirK*-gene product of dissimilatory copper nitrite-reductase (14, 86). Single bacterial cells expressing the enzyme of interest in situ on or inside roots could be visualized by using CLSM (85). Specific antibody labeling and 16S rRNA targeted probing could be combined (10) to obtain information on metabolic activities, autecology, and phylogenetic identity. Detailed experimental conditions have to be established empirically depending on the stability of the antigen-antibody binding.

The following protocol was used by Triplett and colleagues (22) to examine the endophytic location of *Klebsiella pneumoniae* cells expressing the enzyme nitrogenase reductase

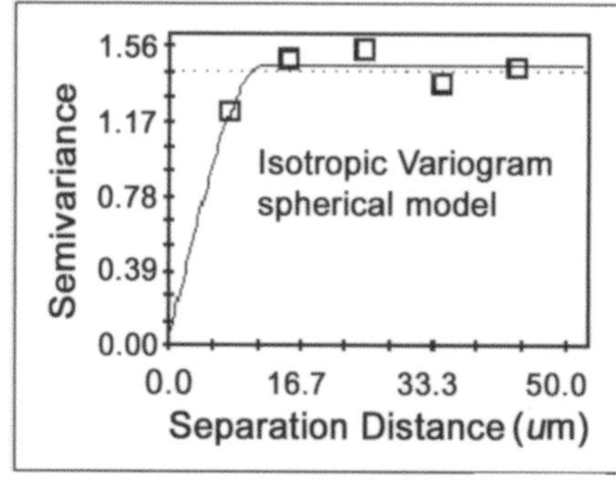

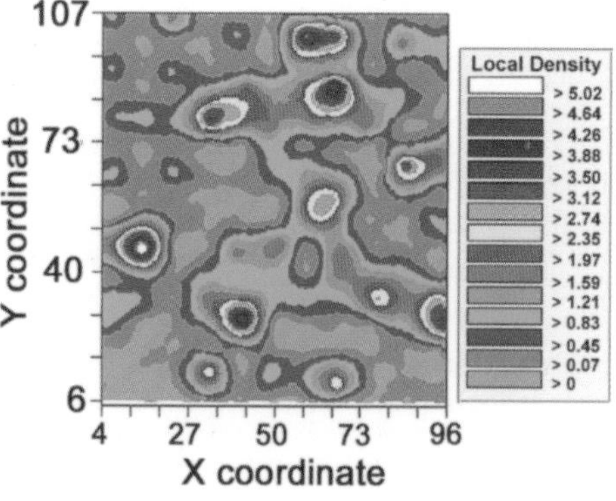

FIGURE 6 Geostatistical analysis of the bacterial colonization of the root surface shown in Fig. 4. (Left) Semivariogram autocorrelation plot and best-fit spherical model of the local spatial density; (right) 2-D kriging map derived from the autocorrelation model depicting the local density of bacteria interpolated over the entire spatial domain.

(NifH) in planta within corn and wheat roots. Inoculated roots were washed and fixed for 30 min with 4% paraformaldehyde, followed by 20% normal goat serum to block nonspecific binding of protein and then primary antibody staining (1:100 dilution) overnight at room temperature. After being washed repeatedly with PBS, roots were stained for 2 h at room temperature with a commercial fluorescent secondary antibody (rhodamine red-X conjugate) used at a 1:250 dilution and then washed again, freehand sectioned, mounted with a photobleach retardant solution, and examined by epifluorescence CLSM. In this case, the immunolabeled bacteria fluoresce yellow due to the combined emitted fluorescence of the red rhodamine fluorochrome conjugated antibody bound to the NifH enzyme and the yellow-green fluorescence of GFP, both produced intracellularly by the same bacterial test strain.

THE FISH TECHNIQUE

Introduction

Originating with the proposal of using biological macromolecules as "evolutionary clocks" for taxonomic studies (135), the RNA of the 16S ribosomal subunit has become the gold standard to analyze prokaryotes phylogenetically because it is ubiquitously distributed, has consistent function, has sequence motifs ranging from absolutely conserved to highly variable, and is supported by ever-growing searchable databases of nucleotide sequences (e.g., the National Center for Biotechnology Information [www.ncbi.nlm.nih.gov] and Ribosomal Database Project II [www.rdp.cme.msu.edu]) that are readily accessible on the internet (13, 25, 79, 81, 91, 92, 128). In addition, the 16S rRNA sequence has been used as the target molecule for preparing fluorescent "phylogenetic stains," consisting of complementary oligonucleotide probes (usually 15 to 25 nucleotides in length, fluorochrome covalently linked to the 5′ end). These fluorescent molecular probes are used in the FISH technique to specifically identify both culturable and nonculturable bacteria at the species or phylogenetically higher level in situ, revealing important additional information on their size, shape, morphology, arrangement, and spatial location, based on fluorescence microscopy (especially CLSM) (3, 5, 7, 9, 19, 30, 45, 55, 65, 78, 118, 123, 124).

Development of Oligonucleotide Probes

During the past decade, freely available software packages have been developed to assist in 16S rRNA-based studies, including specific probe design (1, 8, 25, 80, 99). A large online probeBase database (77) provides an overview of more than 700 published oligonucleotide probes and their characteristics for prokaryotic rRNAs suitable for FISH. This database also automates specificity reanalysis through the RDP-II ProbeMatch tool (25; rdp.cme.msu.edu). To avoid confusion, newly designed or reevaluated probes should be named so as to provide information on their target gene(s), taxonomic level, binding position, etc. (1).

The design of specific rRNA-targeted oligonucleotide probes is dynamic and very dependent on the continuously growing databases. This perpetual updating of 16S rRNA sequence databases carries the caveat that "what was valid in the past may not be valid today." Thus, the specificity, accuracy, and usability of the probes for FISH experiments should be periodically retested. Unfortunately, some bacterial phylogenetic probes designed in silico do not work well, due, inter alia, to variable accessibility of the probe target sites in the 3-D structure of the 16S and 23S rRNA molecules and their tight association with ribosomal proteins (11, 12, 50-52). A useful consensus map that compares the in situ accessibility of rRNA for *E. coli*, the planctomycete *Pirellula* sp. strain 1, the crenarchaeote *Metallosphaera sedula*, and the eukaryote *Saccharomyces cerevisiae* has been developed (11, 12). Since the accessibility may vary considerably for bacteria of broadly divergent lineages, it is advisable to check if a selected probe (position) may have been used successfully for related organisms in similar habitats using, for example, the tool available at the probeBase website (www.microbial-ecology.net/probebase) (77). Optimizing the incubation period during hybridization influences the accessibility of the probe during the FISH protocol (133). Also, thermodynamic effects like changes in the Gibbs free energy are important during calculation of the affinity between probe and target site (132). Sometimes, accessibility is improved by using unlabeled oligonucleotide "helper probes" that bind near the probe target sequence (50). To avoid the dissociation of the helper probe under stringent hybridization conditions, the specificity of these "helper probes" must be identical or broader than the specificity of the labeled probes and they must have a differentiation melting temperature (T_d) at least as high as the T_d of the fluorescent probe.

FISH Probes for Culturable Microorganisms in Pure Culture

The development of new FISH probes for cultured bacteria proceeds as follows. First, isolate the DNA using an adequate DNA extraction protocol. Next, PCR-amplify the 16S or 23S rRNA coding genes by using specific primer pairs that amplify almost the entire rRNA coding gene, because then the information content is the best possible for probe design (65). Accurately align the sequences, and then use computing tools like PROBE-design or PROBE_match included in the ARB software package (80) to design the probe. Finally, test the designed probes by FISH using homologous target cells and closely related nontarget cells under different hybridization temperatures and concentrations of deionized formamide in the hybridization buffer in order to test the specificity and optimize the stringency conditions.

Community "Environmental" Samples

Abundant, not-yet-cultured bacteria in community samples can be visualized and identified in situ by FISH after completing the full "cyclic rRNA approach" (7, 78, 91, 117). Most of their sequences are unknown, necessitating that their 16S rRNA be cloned from DNA or RNA extracted from environmental samples by using optimized protocols (59) or commercial DNA extraction kits. However, since many microbes adapt by differentiating to highly resistant cells in stressed habitats (e.g., unamended soil), they do not lyse quantitative in proportion to their abundance in situ. Therefore, different DNA isolation protocols should be evaluated quantitatively from environmental samples in order to obtain DNA representing as much of the diversity of microbial communities as possible (7). Unfortunately, claims that "community DNA" represents the entire diversity of sampled microbial biomass are commonplace but are often untested, hence equivocal and unproven.

The Clone-FISH approach was developed to optimize newly designed oligonucleotide probes for bacteria that currently cannot be cultured (116). Here, the cloned 16S rRNA is transcribed in vivo using plasmids with, for example, a T7 RNA polymerase promoter in *E. coli* host cells.

The heterologous rRNA transcripts are tested for their target-site accessibility and optimal hybridization conditions as probes targeting bacteria. Thus, Clone-FISH offers the fascinating possibility of rapidly screening environmental rRNA gene libraries for clones that carry the genes of interest (116). When combined with flow cytometry (see below), this technique can enrich for populations of cells that carry the cloned rRNA genes with the respective probe target site, even if low in abundance. This cloning approach could also extend to transcripts of other genes, opening the opportunity to raise the concentration of heterogeneous mRNA of genes encoding selected metabolic functions (e.g., *nifH*, *amoA*, *dsrAB*, or *cbbL*) cloned into *E. coli* at levels detectable by standard FISH procedures. This approach would allow rapid screening and sorting of functional libraries for genes of probe-defined target groups.

The General FISH Protocol

The protocol basically consists of a fixation step, a hybridization step under stringent conditions, and a stringent washing step to remove unspecific bound probes. The FISH analysis can be performed either on glass slides, in suspension, or with cells immobilized on filters.

The fixation step is commonly done by a brief exposure to heat, paraformaldehyde solution (7) and/or absolute ethanol. For the latter, combine 3 volumes of the fixative (4% paraformaldehyde dissolved in PBS buffer, pH 7.3) with 1 volume of the bacterial or sample suspension, mix by vortexing, incubate at 4°C for at least 1.5 to 12 h, remove the fixative by centrifugation, wash the cells repeatedly in PBS buffer, and resuspend in an adequate volume of a 1:1 mixture of PBS and absolute ethanol. Pretreatment of gram-positive bacteria with lysozyme or mutanolysine may be helpful or necessary for optimal probe penetration of their thick cell walls (125). However, this treatment may obliterate the cell morphology and not work well when using fine-textured soil. An alternate method is to substitute the PFA fixative with a 1:1 mixture of absolute ethanol and PBS, otherwise using the protocol described above.

Aliquots of fixed cells or samples are spotted on the slide, dried at 46°C, and dehydrated in a graded ethanol series (3 min each, 50, 80, and 100% ethanol). Hybridizations are performed for at least 1.5 h in a humid chamber at 46°C using 10 μl of hybridization buffer containing 0.9 M sodium chloride (pH 8.0), 0.01% (wt/vol) sodium dodecyl sulfate, 10 mM Tris-HCl (pH 8.0), various amounts of deionized formamide (according to the designated stringency), and either 30 ng each of Cy3- and Cy5-labeled or 50 ng each of FLUOS labeled probes. After completing the hybridization cycle, slides are rinsed with prewarmed washing buffer and incubated for 20 min at 48°C in a tempered water bath. The washing buffer usually contains 20 mM Tris-HCl (pH 8.0), 0.01% (wt/vol) sodium dodecyl sulfate, and sodium chloride ranging in concentration from 56 mM (high stringency) to 0.9 M (low stringency) depending on the stringency chosen for the hybridization buffer. Next, the slides are rinsed with deionized water to remove salts. After air drying, the slides are embedded in Citifluor AF-1 or Vectashield antifading reagent (32), sealed with a coverslip of proper thickness (specified on the objective lens), and examined by epifluorescence microscopy.

Optimization of the Hybridization Conditions

The hybridization conditions for FISH need to be optimized empirically to allow the specific detection of target organisms while minimizing false-negative labeling of nontarget organisms. Assuming sufficient accessibility within the target cells, the stringency of fluorescently labeled probes is optimized by finding the formamide concentration in the hybridization buffer that still yields bright fluorescent signals in target organisms while allowing no or only marginal detectable signals of nontarget organisms. Modern computer-assisted microscopy software can facilitate the construction of probe dissociation plots and optimal hybridization conditions in situ (28, 29) (see "Computer-Assisted Image Segmentation and Analysis of FISH Images" below).

Obstacles and Limitations of the Standard Technique

The standard FISH approach has limitations. One is that no "standard" protocol works well universally for all prokaryotes (7). Another is the occasional low sensitivity obtained when the molecular probe contains only one moiety of the fluorescence dye. For immunofluorescence, this limitation is largely overcome by using the more sensitive indirect sandwich technique, yielding far superior fluorescent brightness than does the direct technique. A third limitation is inherent to the physiological state of the microbes: nongrowing cells in nutrient-impoverished environments (e.g., unamended bulk soil) contain few ribosomes. This limitation restricts the application of standard FISH to the identification of prokaryotes with relatively high cellular ribosome content, which is favored only in nutrient-enriched environmental habitats supporting rapid, active growth, e.g., the rhizosphere or rhizoplane. By contrast, the luminosity of immunofluorescent cells often can be even brighter under nutrient-limited conditions due to increased expression and polymerization of their somatic antigens reacting with the homologous antibody. Fourth, the specific hybridization and washing conditions for unknown uncultured prokaryotes cannot be optimized accurately in complex habitats, and therefore the luminosity of fluorescence brightness per cell does not always reflect its ribosome content. Fifth, the protocols required to achieve sufficient sensitivity and specificity of FISH may disrupt the biological substrata (e.g., the rhizoplane), causing redistribution of the colonized microflora and limitation in measurements of their spatial patterns of colonization over large spatial domains. In contrast, the protocol for IFM using gentle incubation conditions works well to retain the spatial ecology of the target microorganism colonizing its substratum (Fig. 2 to 4).

A further consideration influencing the application of FISH in microbial ecology is the relatively slow mutation rate of rRNA, which does not typically create sufficient differences to distinguish strains of the same species. That task is much more effectively accomplished for in situ autecology studies using IFM with antibodies directed against strain-specific surface antigens (34, 36, 42).

RECENT DEVELOPMENTS OF IMPROVED FISH TECHNIQUES

Signal-to-Noise Ratio

Several innovative approaches to improve the signal-to-noise ratio for rRNA probe design are available (124). Fluorescent-labeled peptide nucleic acid (PNA) probes were reported to yield bright signals, targeting a site of the 16S rRNA with usually low accessibility for oligonucleotide probes (97). Since the PNA pseudopeptides have an uncharged polyamine structure, low salt and high temperatures can be applied in the hybridization. The PNA probes may also provide the advantage for improved staining of gram-positive bacteria and cyanobacteria (74, 129). Another possibility of improving the specific signal intensity of the probe conferring fluorescence staining could use photoinduced electron transfer between the fluorescent dyes used for probe labeling and the nucleotides of the probe or target molecule (82). Although detailed experimental data on quenching effects in the FISH analysis are not yet available, this effect may cause problems in the design of probes and the application in physicochemically complex environments like soil. In pure cultures, self-ligating quenched probe pairs were used to suppress unspecific binding of labeled probes to invalid sample material in FISH analysis (2, 107).

The amount of effort required to edit digital colored FISH (and immunofluorescence) images prior to image analysis can range widely in relation to the signal-to-noise ratio of the image's foreground:background overlap in color space. When that ratio is very high, splitting the FISH image into its red, green, and blue color channels may sometimes be sufficient. However, most real-color images acquired in environmental samples are complex, with too low of a signal-to-noise ratio for that image processing method to suffice alone, and therefore a significant amount of additional image processing is required to achieve full segmentation required for extraction of accurate quantitative information from microbes in color images (otherwise, garbage in, garbage out). Thus, most of the quantitative data from FISH analyses of complex samples have been extracted manually, which is very tedious. As a consequence, the high labor and time-consuming requirements of the FISH protocol have realistically restricted FISH-based studies to just a few samples. The CMEIAS Color Segmentation program (103, 104; C. Gross et al., unpublished) aims to help fill that gap by using computer vision algorithms to evaluate the physical spatial relationships and 3-D color space of pixels needed to fully segment foreground objects in complex RGB digital images.

Amplification of the Signal Intensity

Cells with low rRNA content in oligotrophic environments (e.g., soil) produce low or missing signals with FISH (20). In this case, multilabeled poly(ribo)nucleotide probes synthesized by incorporation of labeled nucleotides or haptens like digoxigenin during in vitro transcription or PCR (44, 66, 120, 125, 134, 137) permit visualization of a significantly higher percentage of prokaryotes in oligotrophic habitats than does FISH with monolabeled oligonucleotide probes (96). However, polynucleotide probes are expensive, labor-intensive to produce, and sensitive to enzymatic degradation and cannot be designed to discriminate closely related groups of prokaryotes due to their nucleotide length (200 to 1,000 bp).

The enzymatic signal amplification cascade using polyribonucleotide probes and the catalyzed reporter deposition (CARD-FISH) approach uses horseradish peroxidase-labeled oligonucleotide probes and tyramide signal amplification to enhance the signal intensities of hybridized cells (94, 115, 116). The improved sensitivity provided by CARD-FISH detected 94% of the total cell counts of plankton in the North Sea, whereas only 48% of the cells were fluorescently stained using the standard FISH protocol with the same probe carrying only one fluorescent label (95). Very recently, CARD-FISH was applied successfully to analyze microcolony-forming soil bacteria without enrichment (48). To enhance the successful diffusion and penetration of the relatively large horseradish peroxidase-labeled oligonucleotides into the target cell, a rigorous pretreatment for cell wall permeabilization is necessary (125). This undesirable consequence can be diminished by using an improved CARD-FISH protocol in which the cells are embedded (hence somewhat stabilized) in low-temperature gelling agarose before cell wall permeabilization (94). However, it remains to be determined whether thick cell wall structures of gram-positive bacteria, planctomycetes, or archaea are permeabilized efficiently enough by this technique to produce bright labeling. Also, lysis of cells of the target, nontarget, and root substrata could occur as a consequence of this harsh treatment, thus altering the detectable composition and spatial distribution of the microbial community under study.

FISH Targets Other than 16S rRNA

Sequences are more varied in the larger 23S rRNA than in the 16S rRNA, with potentially improved resolution in strain differentiation. However, the 23S rRNA sequence database is much smaller and in silico predictions are less certain. Furthermore, phylogenetic trees based on 16S versus 23S rRNA sequences are not necessarily congruent with each other (121), raising question about their extrapolation to phylogenetic relationships. A further possible alternative being explored is the transcribed intergenic spacer regions of the rRNA operon. If the target bacteria are physiologically highly active and thus many spacer transcripts are available per cell, the detection of bacteria via the internal transcribed spacer is possible (90, 111). If the transcription rate of the internal transcribed spacer is low, the in situ detection requires the use of CARD-FISH or other signal amplification methods. If the goal is to track a specific strain, then the alternative choice of IFM using strain-specific fluorescent antibodies still offers the highest probability of success, especially since it carries no requirement for disruption in integrity of the bacterial cell envelope for the molecular probe to produce a usable signal above noise in environmental samples, is relatively simple and inexpensive to perform, and most immunodominant surface antigens persist (indeed some even increase) on quiescent cells.

The tmRNA or 10Sa RNA, encoded by the *ssrA* gene in *E. coli* and involved in the degradation of truncated proteins, is also being considered as a useful target for FISH analysis (21). This rather stable RNA molecule accumulates at about 1,000 copies per cell in a wide range of bacteria. For some bacterial groups, tmRNA offers more suitable probe target sites of differentiation than does 16S rRNA. Since the cellular concentration of tmRNA is too low for detection using fluorescence-labeled oligonucleotide probes, tmRNA could be detected using CARD-FISH (94, 115). In the future, other genes or even gene fragments on plasmids may also be identified in situ by FISH, applying polynucleotide probes by using a procedure called RING-FISH (recognition of individual single genes) (136).

The FISH-MAR Technique

Microautoradiography can be combined with IFM to detect substrate uptake activity of strain-specific microbes (see reference 17 and references therein). Microautoradiography can also be combined with FISH (FISH-MAR) to provide information on the phylogeny of microbes that take up specific substrates (58, 63, 72, 73, 93, 106). Using suitable isotopically labeled substrates (especially alpha- or weak beta-emitters, such as ^{14}C or ^{3}H), these techniques are ideally suited to visualize in situ substrate-uptake patterns in probe-defined culturable and nonculturable bacteria directly in the environment. The alternative approach of combining FISH with CTC (5-cyano-2,3-tolyl-tetrazolium chloride) staining fails to detect many active cells in some environments (88). For example, the abundance of physiologically active iron-reducing bacteria could be quantified in situ and assigned to phylogenetic groups by incubation with labeled acetate under appropriate conditions (89). FISH-MAR also assessed the contribution of certain bacterial groups to the gross productivity of an estuary by measuring the uptake and assimilation of ^{3}H-thymidine and ^{3}H-leucine (27, 47). Furthermore, several novel insights into the complex phenomena in microbial ecology have also been revealed by this technique. For example, Lee and coworkers (73) found differences in substrate uptake patterns not only between different probe-defined bacterial groups but also within one type of a probe-defined bacterial group, suggesting that either differences in activity or phylogeny of the probe-defined bacterial group (meaning that the gene probe is not specific enough) were prevalent. Rossello-Morá and colleagues (106) found using the FISH-MAR technique that bacteria behaved differently in the laboratory than in their natural environment, stressing the need to rely more on in situ activity measurements of bacteria in their natural environment than on laboratory culture studies (a lesson repeated often in microbial ecology). The potential ability of FISH-MAR techniques to target the ecological niche for physiological groups of microorganisms in environmental samples may help to close the gap to the general enzymatic measurements, which are also very much increased in sensitivity (101). Recently, a quantitative approach was developed to determine the in situ substrate affinity constant (K_s) of acetate uptake by probe-defined cells of filamentous bacteria in a complex system using the FISH-MAR approach (88). Using the CARD-FISH protocol together with microautoradiography, the detection of active bacteria was further improved and archaeal cells were found to be the major physiologically active fraction of the total prokaryotic microflora that actively sequestered ^{14}C-labeled L-aspartic acid in the deep ocean (119). Nevertheless, the FISH-MAR technique also has its limitations, such as limited supply and variety of suitable radiotracers for research, and problems with interpretation of the uptake pattern, e.g., whether the uptake of the radiotracer is based on (nonbiological) pure physicochemical adsorption or on bona fide physiological nutrient uptake, and whether the detected signal is from the substrate itself or radiolabeled metabolic products derived from it. Thus, further improvements of this technique or its combination with other methods such as stable-isotope-probing, full-cycle rRNA analysis, or IFM will have to be developed for a more thorough characterization of in situ nutrient transport activity (54, 57, 122).

FISH Analysis and Cytometry

Another obstacle of microscopical analysis, even if supported by high-throughput automated quantification and image analysis, is the limited sensitivity for detection of less abundant species. Cytometric analysis in combination with FISH (6) with many thousands of cells analyzed could solve the problem, but the bacterial cells have to be desorbed, separated from the nonbiological solid-phase matrices (e.g., soil separates), and disaggregated from biofilm or microcolony assemblages. Approaches to reach this goal were reported by Hartmann et al. (61) for desorption of bacteria from root surfaces and by Dazzo (35) for desorption of rhizosphere soil to visualize the underlying adherent rhizoplane microbial community in situ by CLSM and SEM without damaging the root epidermis itself. However, modifications and improvements are obligatory to achieve best results in desorption and disaggregation of bacteria for each type of matrix. Although the protocols described earlier in this chapter (see "Cell Extraction and Recovery from Soil") should work satisfactorily for FISH analysis, extending their use to produce samples for flow cytometry would still be hampered by excessive noise in light scattering of invalid objects due to residual clay and organic particles in the same size range as the microbes.

Computer-Assisted Image Segmentation and Analysis of FISH Images

DAIME software was recently released to the scientific community (www.microbial-ecology.net/daime) and is available for both Linux and Windows 2000/XP platforms (29). Like CMEIAS (described above), DAIME software allows either manual or automated analysis of single images as well as 2-D-image series in stacks. The unique feature of DAIME of significant value for FISH studies is its ability to extract luminosity data from 3-D-image stacks for plotting melting curves in order to evaluate the optimal hybridization conditions for newly designed, fluorescence-labeled

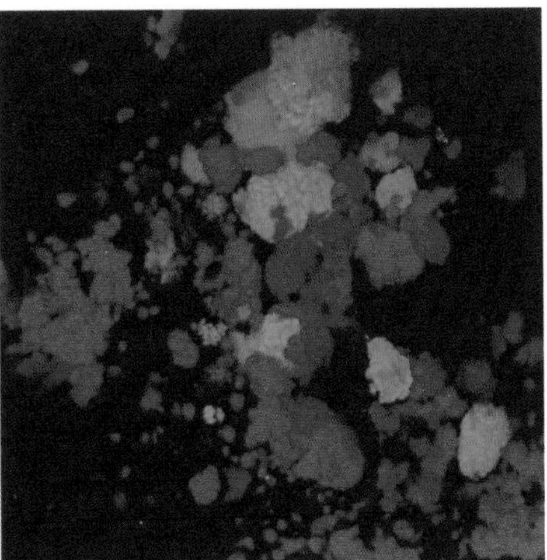

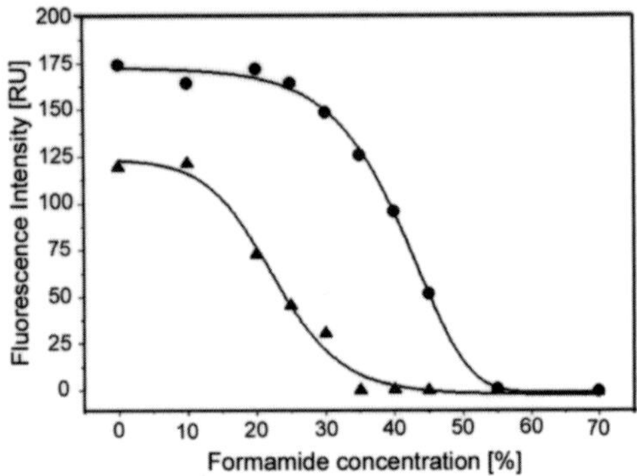

FIGURE 7 (Top) 2-D projection of a 3-D confocal image stack showing microcolonies of two different ribotypes of nitrite-oxidizing bacteria, genus *Nitrospira*, in a biofilm from a wastewater treatment plant. The image was rendered by the program DAIME and provided courtesy of Frank Maixner and Holger Daims (University of Vienna, Vienna, Austria). (Bottom) Formamide concentration series for the determination of optimal hybridization conditions for a newly designed 16S rRNA-targeted oligonucleotide probe using the program DAIME. Data courtesy of Sebastian Lücker and Holger Daims (University of Vienna, Vienna, Austria).

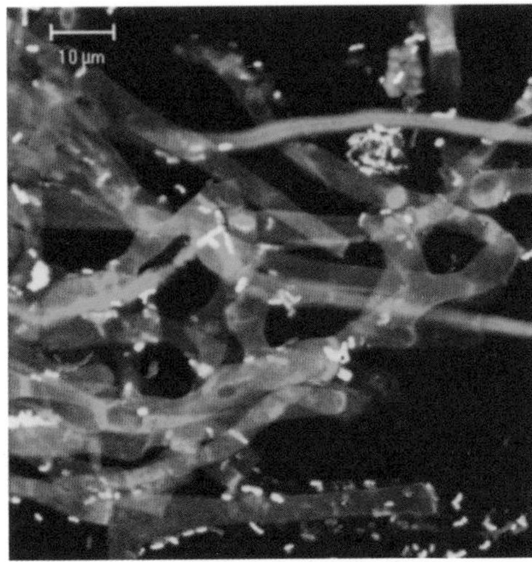

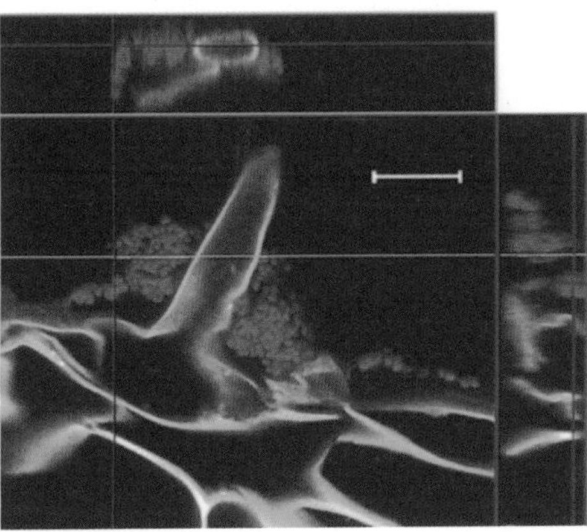

FIGURE 8 (Left) Bacteria with ectomycorrhizal hyphae. The ectomycorrhizal fungus *Fagirhiza pallida* was retrieved from roots of beech trees in the Kranzberg forest near Freising, Germany. For FISH analysis, the following probes were applied: EUB338Mix Cy3, Bet42a Fluos, and Gam42a Cy5. Due to concomitant binding of probes, beta-proteobacteria are stained yellow (Beta42a Fluos plus EUB338Mix Cy3) and other *Bacteria* (non-beta- or -gamma proteobacteria) are labeled red (EUB338Mix Cy3). Another unidentified organism that hybridized with none of the three fluorescent probes is included in the image because of its blue autofluorescence. (Right) *A. brasilense* on wheat roots. Wheat seedlings were inoculated with *A. brasilense* Sp245, and their colonization was visualized by FISH analysis. The following oligonucleotide probes were applied: EUB-338Mix Cy3 and Abras-1420-Cy5. Due to concomitant binding of both probes, the bacterial cells display magenta fluorescence. The computer-rendered orthogonal view displays the third dimension in the 3-D-*z*-stack of *xy*-optisections.

rRNA oligonucleotide probes. Figure 7, top panel, presents an example of a 2-D projection of a 3-D confocal FISH image stack rendered by the DAIME program. It shows microcolonies of nitrite-oxidizing bacteria of the genus *Nitrospira* in a biofilm from a wastewater treatment plant. FISH detected two different *Nitrospira* populations that coexisted in close proximity within the biofilm despite being potential competitors for their common substrate nitrite. To evaluate the optimal hybridization conditions for newly designed oligonucleotide probes, formamide concentration series were performed to obtain probe dissociation curves. The newly designed probe was hybridized with increasing concentrations of formamide in the hybridization buffer to a target (circles) and a nontarget (triangles) of bacteria. In this case, the 16S rRNA of the nontarget organism has one base mismatch to the probe sequence at its probe-binding site. Digital images of the labeled cells were acquired and analyzed by the DAIME program to compute the mean cell luminosity at each formamide concentration tested. The resultant data are plotted in Fig. 7, bottom panel, indicating that 35 to 40% formamide is the optimal concentration producing the highest signal:noise ratio for use of this fixative with this probe.

CMEIAS can extract further information of ecological relevance from this same FISH image. The CMEIAS Color Segmentation program was applied to reconstruct two new images segmented by the variable color ranges sampled from 30 pixels selected from each of the two differentially colored populations of bacterial ribotype targets. Assuming axial symmetry of the foreground objects, CMEIAS analysis of the biovolume computed from these projected images indicates a relative distribution of abundance among the two different *Nitrospira* ribotypes as follows: the red fluorescent bacterial aggregates represent 65.5% of the total microbial biovolume and 70.5% of the total microbial biosurface area, whereas the green fluorescent bacterial aggregates represent 34.5% of the total microbial biovolume and 29.5% of the total microbial biosurface area. These results indicate that the two different *Nitrospira* ribotypes have an approximately 2:1 ratio in relative abundance in this community. This illustrates how CMEIAS and DAIME can work together to analyze FISH images, providing rapid, direct, and precise in silico measurements of population diversity and abundance in communities in situ, without the cost and uncertainties of biased sampling and 16S rRNA gene copy number associated with quantitative real-time PCR-RNA analysis.

Figure 8 (flattened *z*-stack and *xy* Phi-*z* sections) illustrates the successful use of a multiple-probe FISH protocol for in situ identification and localization of several specific bacterial ribotypes in the same natural mycorrhizosphere community. In this case, the same hybridization protocol was applied using two oligonucleotide probes differing in

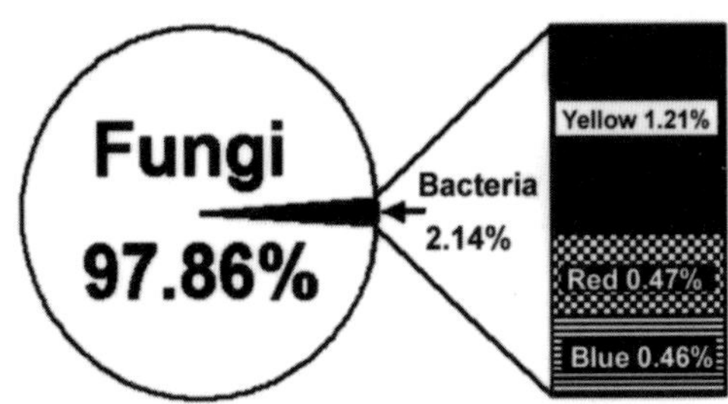

FIGURE 9 Distribution of microbial abundance in the community represented by the FISH image of Fig. 8, left panel. CMEIAS was used to segment the microbes in the image according to their foreground color ranges and then measure the cumulative cell biovolume for each microbial ribotype.

TABLE 2 CMEIAS image analysis of the spatial abundance and distribution of individual ribotype populations distinguished by multiprobe FISH in the community image shown in Fig. 8, left panel

Community attribute	Ribotype population		
	Yellow β-proteobacteria	Red γ-proteobacteria	Blue autofluorescent bacteria
No. of individuals	161 (63.6%)	73 (28.9%)	19 (35.8%)
Cumulative biovolume (μm³)	85.35 (56.3%)	33.41 (22.0%)	32.89 (21.7%)
% Fungal substratum coverage	6.09	2.47	1.34
Spatial density (cells/mm² of fungal surface)	70,153	31,808	8,279
Mean distance to nearest same ribotype (μm)	2.4 ± 2.1	2.7 ± 2.5	2.0 ± 1.6

fluorochrome label and phylogenetic specificity—from the kingdom to the group and species levels. Mixed colors display when both fluorescent probes bind to the rRNA in the same target cell.

CMEIAS can extract an abundance of quantitative information on microbial community structure from the multiprobe FISH image (Fig. 8, left panel). First, the CMEIAS Color Segmentation program is applied to analyze the color foreground objects and reconstructs a new grayscale image of the selected population, with retention of each cell's representative morphology, abundance, position, size, and luminosity. Then, CMEIAS object analysis and cumulative object analysis plug-in modules are applied to extract the data from each isolated population, reporting on attributes for each individual cell and for the corresponding cumulative population. The computed results are summarized in Fig. 9 and Table 2 and 3. These data indicate the overwhelming dominance of fungal biomass in the community, the difference in colonization of fungal hyphae by each bacterial population, and the influence of biovolume measurement as the metric of microbial abundance rather than cell counts to compute the community diversity and evenness indices. The results show that the increased size of microbes compensates for the smaller number of individuals, with the consequence being that the computed indices of diversity and evenness are higher when abundance weighted by cell biovolume is considered rather than ignored, and CMEIAS-acquired data on both size and number of individual cells work well together as an accurate measure of microbial abundance when computing various indices of microbial community structure. The same conclusions are reached from similar data trends based on a morphological analysis of community stability and ecological succession in anaerobic microbial bioreactors when perturbed by high nutrient loading (see reference 76 and references therein). It should become immediately obvious that the in situ analysis of microbial community structure can be significantly enhanced when 16S rRNA-based multiprobe FISH is combined with CMEIAS quantitative image analysis.

Concluding Remark

Fluorescent molecular probes used with IFM and FISH in conjunction with CMEIAS and other software applications of digital image analysis represent powerful research tools of exceptional specificity and sensitivity available for quantitative studies of the in situ ecology of microbial populations and communities in soil and in association with roots.

Portions of this work in the laboratory of F.B.D. were supported by MSU Research Excellence Funds for the Center for Microbial Ecology, the Center for Renewable Organic Materials, and the Center for Microbial Pathogenesis, the MSU-KBS Long-Term Ecological Research program, the Michigan Agricultural Experiment Station, and projects BIO5-001-015 and BIO10-001-011 of the US-Egypt Science and Technology Joint Program. Portions of this work in the laboratory of A.H. were supported by the Bundesministerium für Bildung und Forschung, Berlin, Germany (grant no. 0311954).

We thank Holger Daims, Department of Microbial Ecology, University of Vienna, Vienna, Austria, for providing Fig. 7. We thank Stephan Gantner, James Cole, and David Lipson for reviewing the manuscript.

TABLE 3 CMEIAS image analysis of the ribotype diversity distinguished by multiprobe FISH and weighted by biovolume abundance in the community image shown in Fig. 8, left panel

Community diversity and similarity indices	Calculation based on:		
	No. of individuals	Biovolume	No. of individuals/ biovolume ratio
Simpson's inverse dominance	2.025	2.425[a]	0.835
Shannon-Wiener's diversity	1.213	1.426[a]	0.831
Smith and Wilson evenness	0.579	0.875[a]	0.662
Simpson's (1/D) evenness	0.675	0.808[a]	0.767
Proportional similarity		85.5%	
Average Euclidean distance		102.6	

[a]Notice how the evenness (and hence the diversity) component of community structure is increased when abundance is based on cell number weighted by cell biovolume measured by CMEIAS.

REFERENCES

1. **Alm, E. W., D. B. Oerther, N. Larsen, D. A. Stahl, and L. Raskin.** 1996. The oligonucleotide probe database. *Appl. Environ. Microbiol.* **62:**3557–3559.
2. **Alm, E. W., D. Zheng, and L. Raskin.** 2000. The presence of humic substances and DNA in RNA extracts affects hybridization results. *Appl. Environ. Microbiol.* **66:** 4547–4554.
3. **Amann, R., B. M. Fuchs, and S. Behrens.** 2001. The identification of microorganisms by fluorescence in situ hybridization. *Curr. Opin. Biotechnol.* **12:**231–236.
4. **Amann, R., and W. Ludwig.** 2000. Ribosomal RNA-targeted nucleic acid probes for studies in microbial ecology. *FEMS Microbiol. Rev.* **24:**555–565.
5. **Amann, R., and K.-H. Schleifer.** 2001. Nucleic acid probes and their application in environmental microbiology, p. 67–82. *In* D. R. Boone, R. W. Castenholz, and G. M. Garrity (ed.), *Bergey's Manual of Systematic Bacteriology*, vol. 1. Springer-Verlag, New York, N.Y.
6. **Amann, R. I., B. J. Binder, R. J. Olson, S. W. Chisholm, R. Devereux, and D. A. Stahl.** 1990. Combination of 16S rRNA-targeted oligonucleotide probes with flow cytometry for analyzing mixed microbial populations. *Appl. Environ. Microbiol.* **56:**1919–1925.
7. **Amann, R. I., W. Ludwig, and K.-H. Schleifer.** 1995. Phylogenetic identification and *in situ* detection of individual microbial cells without cultivation. *Microbiol. Rev.* **59:**143–169.
8. **Ashelford, K. E., A. J. Weightman, and J. C. Fry.** 2002. PRIMROSE: a computer program for generating and estimating the phylogenetic range of 16S rRNA oligonucleotide probes and primers in conjunction with the RDP-II database. *Nucleic Acids Res.* **30:**3481–3489.
9. **Assmus, B., P. Hutzler, G. Kirchhof, R. Amann, J. R. Lawrence, and A. Hartmann.** 1995. In situ localization of *Azospirillum brasilense* in the rhizosphere of wheat with fluorescently labeled, rRNA-targeted oligonucleotide probes and scanning confocal laser microscopy. *Appl. Environ. Microbiol.* **61:**1013–1019.
10. **Assmus, B., M. Schloter, G. Kirchhof, P. Hutzler, and A. Hartmann.** 1997. Improved *in situ* tracking of rhizosphere bacteria using dual staining with fluorescence-labeled antibodies and rRNA-targeted oligonucleotides. *Microb. Ecol.* **33:**32–40.
11. **Behrens, S., B. M. Fuchs, F. Mueller, and R. Amann.** 2003. Is the in situ accessibility of the 16S rRNA of *Escherichia coli* for Cy3-labeled oligonucleotide probes predicted by a three-dimensional structure model of the 30S ribosomal subunit? *Appl. Environ. Microbiol.* **69:**4935–4941.
12. **Behrens, S., C. Ruhland, J. Inacio, H. Huber, A. Fonseca, I. Spencer-Martins, B. M. Fuchs, and R. Amann.** 2003. In situ accessibility of small-subunit rRNA of members of the domains *Bacteria*, *Archaea*, and *Eucarya* to Cy3-labeled oligonucleotide probes. *Appl. Environ. Microbiol.* **69:**1748–1758.
13. **Benson, D. A., M. S. Boguski, D. J. Lipman, J. Ostell, and B. F. Ouellette.** 1998. GenBank. *Nucleic Acids Res.* **26:**1–7.
14. **Bothe, H., G. Jost, M. Schloter, B. B. Ward, and K.-P. Witzel.** 2000. Molecular analysis of ammonia oxidation and denitrification in natural environments. *FEMS Microbiol. Rev.* **24:**673–690.
15. **Bishop, P. E, F. B. Dazzo, E. R. Appelbaum, R. J. Maier, and W. J. Brill.** 1977. Intergeneric transfer of genes involved in the *Rhizobium*-legume symbiosis. *Science* **198:** 938–940.
16. **Bohlool, B. B., and E. L. Schmidt.** 1968. Nonspecific staining: its control in immunofluorescence examination of soil. *Science* **162:**1012–1014.
17. **Bohlool, B. B., and E. L. Schmidt.** 1980. The immunofluorescence approach in microbial ecology. *Adv. Microb. Ecol.* **4:**203–241.
18. **Bottomley, P. J.** 1994. Light microscopic methods for studying soil microorganisms, p. 81–105. *In* R. Weaver, J. S. Angle, and P. J. Bottomley (ed.), *Methods of Soil Analysis*. Soil Science Society of America, Madison, Wis.
19. **Bouvier, T. C., and P. A. del Giorgio.** 2003. Factors influencing the detection of bacterial cells using fluorescence in situ hybridization (FISH): a qualitative review of published reports. *FEMS Microbiol. Ecol.* **44:**3–15.
20. **Chatzinotas, A., R. A. Sandaa, W. Schonhuber, R. Amann, F. L. Daae, V. Torsvik, J. Zeyer, and D. Hahn.** 1998. Analysis of broad-scale differences in microbial community composition of two pristine forest soils. *Syst. Appl. Microbiol.* **21:**579–587.
21. **Chauhan, A. K., and D. Apirion.** 1989. The gene for a small stable RNA (10Sa RNA) of *Escherichia coli*. *Mol. Microbiol.* **3:**1481–1485.
22. **Chelius, M., and E. W. Triplett.** 2000. Immunolocalization of dinitrogenase reductase produced by *Klebsiella pneumoniae* in association with *Zea mays*. *Appl. Environ. Microbiol.* **66:**783–787.
23. **Cherry, W. B., M. Goldman, and T. R. Carski.** 1960. *Fluorescent Antibody Techniques*. U.S. Department of Health, Education, and Welfare, Atlanta, Ga.
24. **Chi, F., Shi-Hua Shen, Hai-Ping Cheng, Yu-Xiang Jing, Y. G. Yanni, and F. B. Dazzo.** 2005. Ascending migration of endophytic rhizobia from roots to leaves inside rice plants and assessment of their benefits to the growth physiology of rice. *Appl. Environ. Microbiol.* **71:**7271–7278.
25. **Cole, J. R., B. Chai, R. J. Farris, Q. Wang, S. A. Kulam, D. M. McGarrell, G. M. Garrity, and J. M. Tiedje.** 2005. The Ribosomal Database Project (RDP-II): sequences and tools for high-throughput rRNA analysis. *Nucleic Acids Res.* **33:**294–296.
26. **Coons, A. H., H. J. Creech, R. N. Jones, and E. Berliner.** 1942. The demonstration of pneumococcal antigen in tissues by the use of fluorescent antibody. *J. Immunol.* **45:**159–164.
27. **Cottrell, M. T., and D. L. Kirchmann.** 2003. Contribution of major bacterial groups to bacterial biomass production (thymidine and leucine incorporation) in the Delaware estuary. *Limnol. Oceanogr.* **48:**168–178.
28. **Daims, H., A. Bruhl, R. Amann, K. H. Schleifer, and M. Wagner.** 1999. The domain-specific probe EUB338 is insufficient for the detection of all Bacteria: development and evaluation of a more comprehensive probe set. *Syst. Appl. Microbiol.* **22:**434–444.
29. **Daims, H., S. Lucker, and M. Wagner.** 2006. DAIME, a novel image analysis program for microbial ecology and biofilm research. *Environ. Microbiol.* **8:**200–213.
30. **Daims, H., N. B. Ramsing, K. H. Schleifer, and M. Wagner.** 2001. Cultivation-independent, semiautomatic determination of absolute bacterial cell numbers in environmental samples by fluorescence in situ hybridization. *Appl. Environ. Microbiol.* **67:**5810–5818.
31. **Dandurand, L. M., D. Schotzko, and G. Knudsen.** 1997. Spatial patterns of rhizoplane populations of *Pseudomonas fluorescens*. *Appl. Environ. Microbiol.* **63:**3211–3217.
32. **Davidson, R. S., and D. Goodwin.** 1983. An effective way of retarding the fading of fluorescence of labeled material during microscopical examination. *Proc. R. Med. Soc.* **18:**151–153.
33. **Dazzo, F. B.** 2004. Applications of quantitative microscopy in studies of plant surface microbiology, p. 503–550. *In* A. Varma, L. Abbott, D. Werner, and R. Hampp (ed.), *Plant Surface Microbiology*. Springer-Verlag, Berlin, Germany,
34. **Dazzo, F. B.** 2004. Production of anti-microbial antibodies and their utilization in studies of microbial autecology by

IFM and *in situ* CMEIAS image analysis, p. 911–932. *In* G. Kowalchuk, F. deBruijn, I. Head, A. Akkermans, J. Elsas (ed.), *Molecular Microbial Ecology Manual*, 2nd ed. Kluwer Publishers, Dordrecht, The Netherlands.
35. **Dazzo, F. B.** Visualization of rhizoplane microflora by computer-assisted microscopy and spatial analysis by CMEIAS image analysis. *In* J. Luster and R. Fidlay (ed.), *Handbook of Methods Used in Rhizosphere Research*, in press. Swiss Federal Institute for Forest, Snow, and Landscape Research, Birmensdorf, Switzerland. www.rhizo.at/handbook.
36. **Dazzo, F. B., A. R. Joseph, A. B. Gomaa, P. Robertson, and Y. G. Yanni.** 2003. Quantitative indices for the autecological biogeography of a *Rhizobium* endophyte of rice at two spatial scales. *Symbiosis* **35:**147–158.
37. **Dazzo, F. B., G. Tang, G. Zhu, C. Gross, D. Nasr, C. Passmore, E. Polone, A. Squartini, A. Prabhu, C. Reddy, R. Peretz, L. Gao, R. Bollempalli, D. Trione, E. Marshall, J. Wang, M. Li, D. McGarrell, S. Gantner, J. Liu, and Y. Yanni.** 2006. CMEIAS v3.0: advanced image analysis software to strengthen microscopy-based approaches for understanding microbial ecology. Presented at the 2006 All Investigator Meeting, Kellogg Biological Station—Long Term Ecological Research Program, Michigan State University, May 9, 2006. [Online.] lter.kbs.msu.edu/Meetings/2006ASM/ Abstracts/Dazzo.htm.
38. **Dazzo, F., P. Mateos, G. Orgambide, S. Philip-Hollingsworth, S. Squartini, N. Subba-Rao, H. S. Pankratz, D. Baker, R. Hollingsworth, and J. Whallon.** 1993. The infection process in the *Rhizobium*-legume symbiosis and visualization of rhizoplane microorganisms by laser scanning confocal microscopy, p. 259–262. *In* R. Guerrero and C. Pedros-Alio (ed.), *Trends in Microbial Ecology*, Spanish Society for Microbiology, Barcelona, Spain.
39. **Dazzo, F. B., G. L. Truchet, J. E. Sherwood, E. M. Hrabak, and A. E. Gardiol.** 1982. Alteration of the trifoliin A-binding capsule of *Rhizobium trifolii* 0403 by enzymes released from clover roots. *Appl. Environ. Microbiol.* **44:**478–490.
40. **Dazzo, F. B., G. L. Truchet, J. E. Sherwood, E. M. Hrabak, M. Abe, and H. S. Pankratz.** 1984. Specific phases of root hair attachment in the *Rhizobium trifolii*-clover symbiosis. *Appl. Environ. Microbiol.* **48:**1140–1150.
41. **Dazzo, F. B., and J. Wopereis.** 2000. Unraveling the infection process in the *Rhizobium*-legume symbiosis by microscopy, p. 295–347. *In* E. Triplett (ed.), *Prokaryotic Nitrogen Fixation: a Model System for the Analysis of a Biological Process.* Horizon Scientific Press, Norwich, United Kingdom.
42. **Dazzo, F. B., and S. Wright.** 1996. Production of antimicrobial antibodies and their use in IFM, p. 1–27. *In* A. Akkermans, J. van Elsas, and F. de Bruijn (ed.), *Molecular Microbial Ecology Manual*, vol. 4:12. Kluwer Academic Publishers, Dordrecht, The Netherlands.
43. **Dazzo, F. B., and Y. G. Yanni.** 2006. The natural *Rhizobium*-cereal crop association as an example of plant-bacterial interaction, p. 109–127. *In* N. Uphoff, A. Ball, E. Fernandes, H. Herren, O. Husson, M. Laing, C. Palm, J. Pretty, N. Sanginga, and J. Thies (ed.), *Biological Approaches to Sustainable Soil Systems*. CRC Press, Boca Raton, Fla.
44. **DeLong, E. F., L. T. Taylor, T. L. Marsh, and C. M. Preston.** 1999. Visualization and enumeration of marine planktonic archaea and bacteria by using polyribonucleotide probes and fluorescent in situ hybridization. *Appl. Environ. Microbiol.* **65:**5554–5563.
45. **DeLong, E. F., G. S. Wickham, and N. R. Pace.** 1989. Phylogenetic stains: ribosomal RNA-based probes for the identification of single cells. *Science* **243:**1360–1363.
46. **Demezas, D. H., and P. J. Bottomley.** 1986. Autecology in rhizospheres and nodulating behavior of indigenous *Rhizobium trifolii*. *Appl. Environ. Microbiol.* **52:**1014–1019.
47. **Elifantz, H., R. R. Malmstrom, M. T. Cottrell, and D. L. Kirchman.** 2005. Assimilation of polysaccharides and glucose by major bacterial groups in the Delaware estuary. *Appl. Environ. Microbiol.* **71:**7799–7805.
48. **Ferrari, B. C., N. Tujula, K. Stoner, and S. Kjelleberg.** 2006. Catalyzed reporter deposition-FISH allows for enrichment-independent detection of microcolony-forming soil bacteria. *Appl. Environ. Microbiol.* **72:**918–922.
49. **Friese, C. F., and M. F. Allen.** 1991. Tracking the fates of exotic and local VA mycorrhizal fungi: methods and patterns. *Agric. Ecosyst. Environ.* **34:**87–96.
50. **Fuchs, B. M., F. O. Glockner, J. Wulf, and R. Amann.** 2000. Unlabeled helper oligonucleotides increase the in situ accessibility to 16S rRNA of fluorescently labeled oligonucleotide probes. *Appl. Environ. Microbiol.* **66:**3603–3607.
51. **Fuchs, B. M., K. Syutsubo, W. Ludwig, and R. Amann.** 2001. In situ accessibility of *Escherichia coli* 23S rRNA to fluorescently labeled oligonucleotide probes. *Appl. Environ. Microbiol.* **67:**961–968.
52. **Fuchs, B. M., G. Wallner, W. Beisker, I. Schwippl, W. Ludwig, and R. Amann.** 1998. Flow cytometric analysis of the in situ accessibility of *Escherichia coli* 16S rRNA for fluorescently labeled oligonucleotide probes. *Appl. Environ. Microbiol.* **64:**4973–4982.
53. **Gantner, S., M. Schmid, C. Dürr, R. Schuhegger, A. Steidle, P. Hutzler, C. Langebartels, L. Eberl, A. Hartmann, and F. B. Dazzo.** 2006. In situ spatial scale of calling distances and population density-independent *N*-acylhomoserine lactone mediated communication by rhizobacteria colonized on plant roots. *FEMS Microbiol. Ecol.* **56:**188–194.
54. **Ginige, M. P., P. Hugenholtz, H. Daims, M. Wagner, J. Keller, and L. L. Blackall.** 2004. Use of stable-isotope probing, full-cycle rRNA analysis, and FISH-microautoradiography to study a methanol-fed denitrifying microbial community. *Appl. Environ. Microbiol.* **70:**588–596.
55. **Giovannoni, S. J., E. F. DeLong, G. J. Olsen, and N. R. Pace.** 1988. Phylogenetic group-specific oligodeoxynucleotide probes for identification of single microbial cells. *J. Bacteriol.* **170:**720–726.
56. **Gray, T. R. G.** 1990. Methods for studying the microbial ecology of soil. *Methods Microbiol.* **22:**309–342.
57. **Gray, N. D., and I. M. Head.** 2001. Linking genetic identity and function in communities of uncultured bacteria. *Environ. Microbiol.* **3:**481–492.
58. **Gray, N. D., R. Howarth, R. W. Pickup, J. G. Jones, and I. M. Head.** 2000. Use of combined microautoradiography and FISH to determine carbon metabolism in mixed natural communities of uncultured bacteria from the genus *Achromatium*. *Appl. Environ. Microbiol.* **66:**4518–4522.
59. **Griffiths, R. I., A. S. Whiteley, A. G. O'Donnell, and M. J. Bailey.** 2000. Rapid method for coextraction of DNA and RNA from natural environments for analysis of ribosomal DNA- and rRNA-based microbial community composition. *Appl. Environ. Microbiol.* **66:**5488–5491.
60. **Harlow, E., and D. Lane.** 1988. Antibodies: a laboratory manual. Cold Spring Harbor, New York, N.Y.
61. **Hartmann, A., R. Pukall, M. Rothballer, S. Gantner, S. Metz, M. Schloter, and B. Mooge.** 2004. Microbial community analysis in the rhizosphere by *in situ* and *ex situ* analysis molecular probing, biomarker and cultivation techniques, p. 449–469. *In* A. Varma, L. Abbott, D. Werner, and R. Hampp (ed.), *Plant Surface Microbiology*. Springer-Verlag, Berlin, Germany.
62. **Hill, I. R., and T. R. G. Gray.** 1967. Application of the fluorescent antibody technique to an ecological study of bacteria in soil. *J. Bacteriol.* **93:**1888–1897.
63. **Ito, T., J. L. Nielsen, S. Okabe, Y. Watanabe, and P. H. Nielsen.** 2002. Phylogenetic identification and substrate

uptake patterns of sulfate-reducing bacteria inhabiting an oxic-anoxic sewer biofilm determined by combining microautoradiography and fluorescent in situ hybridization. *Appl. Environ. Microbiol.* **68:**356–364.
64. **Jansson, J. K.** 2003. Marker and reporter genes: illuminating tools for environmental microbiologists. *Curr. Opin. Microbiol.* **6:**310–316.
65. **Juretschko, S., G. Timmermann, M. Schmid, K. H. Schleifer, A. Pommerening-Roser, H. P. Koops, and M. Wagner.** 1998. Combined molecular and conventional analyses of nitrifying bacterium diversity in activated sludge: *Nitrosococcus mobilis* and *Nitrospira*-like bacteria as dominant populations. *Appl. Environ. Microbiol.* **64:**3042–3051.
66. **Karner, M. B., E. F. DeLong, and D. M. Karl.** 2001. Archaeal dominance in the mesopelagic zone of the Pacific Ocean. *Nature* **409:**507–510.
67. **Kingsley, M. T., and B. B. Bohlool.** 1981. Release of *Rhizobium* spp. from tropical soils and recovery for immunofluorescence enumeration. *Appl. Environ. Microbiol.* **42:** 241–248.
68. **Kough, J. E., N. Malajczuk, and R. G. Linderman.** 1983. Use of the indirect immunofluorescent technique to study the vesicular-arbuscular fungus *Glomus epigaeum* and other *Glomus* species. *New Phytol.* **94:**57–62.
69. **Krige, D. G.** 1951. A statistical approach to some basic mine valuation problems on the Witwatersrand. *J. Chem. Metal and Mining Soc. S. Afr.* **52:**119–139.
70. **Kuehn, M., M. Hausner, H. J. Bungartz, M. Wagner, P. A. Wilderer, and S. Wuertz.** 1998. Automated confocal laser scanning microscopy and semiautomated image processing for analysis of biofilms. *Appl. Environ. Microbiol.* **64:**4115–4127.
71. Reference deleted.
72. **Lee, N., P. H. Nielsen, K. H. Andreasen, S. Juretschko, J. L. Nielsen, K. H. Schleifer, and M. Wagner.** 1999. Combination of fluorescent in situ hybridization and microautoradiography—a new tool for structure-function analyses in microbial ecology. *Appl. Environ. Microbiol.* **65:**1289–1297.
73. **Lee, N., P. H. Nielsen, H. Aspegren, M. Henze, K. H. Schleifer, and J. la Cour Jansen.** 2003. Long-term population dynamics and *in situ* physiology in activated sludge systems with enhanced biological phosphorus removal operated with and without nitrogen removal. *Syst. Appl. Microbiol.* **26:**211–227.
74. **Lehtola, M. J., E. Torvinen, I. T. Miettinen, and C. W. Keevil.** 2006. FISH using peptide nucleic acid probes for rapid detection of *Mycobacterium avium* subsp. *avium* and *Mycobacterium avium* subsp. *paratuberculosis* in potable-water biofilms. *Appl. Environ. Microbiol.* **72:**848–853.
75. **Lindahl, V., and L. R. Bakken.** 1995. Evaluation of methods for extraction of bacteria from soil. *FEMS Microbiol. Ecol.* **16:**135–142.
76. **Liu, J., F. B. Dazzo, O. Glagoleva, B. Yu, and A. Jain.** 2001. CMEIAS: a computer-aided system for the image analysis of bacterial morphotypes in microbial communities. *Microb. Ecol.* **41:**173–194. (Erratum, **42:**215.)
77. **Loy, A., M. Horn, and M. Wagner.** 2003. probeBase: an online resource for rRNA-targeted oligonucleotide probes. *Nucleic Acids Res.* **31:**514–516.
78. **Ludwig, W., S. H. Bauer, M. Bauer, I. Held, G. Kirchhof, R. Schulze, I. Huber, S. Spring, A. Hartmann, and K. H. Schleifer.** 1997. Detection and *in situ* identification of representatives of a widely distributed new bacterial phylum. *FEMS Microbiol. Lett.* **153:**181–190.
79. **Ludwig, W., O. Strunk, S. Klugbauer, N. Klugbauer, M. Weizenegger, J. Neumaier, M. Bachleitner, and K. H. Schleifer.** 1998. Bacterial phylogeny based on comparative sequence analysis. *Electrophoresis* **19:**554–568.
80. **Ludwig, W., O. Strunk, R. Westram, L. Richter, H. Meier, Yadhukumar, A. Buchner, T. Lai, S. Steppi, G. Jobb, W. Forster, I. Brettske, S. Gerber, A. W. Ginhart, O. Gross, S. Grumann, S. Hermann, R. Jost, A. Konig, T. Liss, R. Lussmann, M. May, B. Nonhoff, B. Reichel, R. Strehlow, A. Stamatakis, N. Stuckmann, A. Vilbig, M. Lenke, T. Ludwig, A. Bode, and K. H. Schleifer.** 2004. ARB: a software environment for sequence data. *Nucleic Acids Res.* **32:**1363–1371.
81. **Maidak, B. L., N. Larsen, M. J. McCaughey, R. Overbeek, G. J. Olsen, K. Fogel, J. Blandy, and C. R. Woese.** 1994. The Ribosomal Database Project. *Nucleic Acids Res.* **22:**3485–3487.
82. **Marras, S. A., F. R. Kramer, and S. Tyagi.** 2002. Efficiencies of fluorescence resonance energy transfer and contact-mediated quenching in oligonucleotide probes. *Nucleic Acids Res.* **30:**122.
83. **McDermott, T. R., and F. B. Dazzo.** 2002. Use of fluorescent antibodies for studying the ecology of soil- and plant-associated microbes, p. 615–626. *In* C. J. Hurst, R. L. Crawford, G. R. Knudsen, M. J. McInerney, and L. D. Stetzenbach (ed.), *Manual of Environmental Microbiology*, 2nd ed. ASM Press, Washington, D.C.
84. **McDermott, T. R., and P. H. Graham.** 1989. *Bradyrhizobium japonicum* inoculant mobility, nodule occupancy, and acetylene reduction in the soybean root system. *Appl. Environ. Microbiol.* **55:**2493–2498.
85. **Metz, S., W. Beisker, A. Hartmann, and M. Schloter.** 2003. Detection methods for the expression of the dissimilatory copper nitrite-reductase (*dnirK*) in environmental samples. *J. Microbiol. Methods* **55:**41–50.
86. **Metz, S., A. Hartmann, and M. Schloter.** 2002. Development and characterization of murine monoclonal antibodies specific for dissimilatory copper nitrite reductase. *Hybrid. Hybridomics* **21:**351–357.
87. **Moawad, H. A., W. R. Ellis, and E. L. Schmidt.** 1984. Rhizosphere response as a factor in competition among three serogroups of indigenous *Rhizobium japonicum* for nodulation among field-grown soybeans. *Appl. Environ. Microbiol.* **47:**607–612.
88. **Nielsen, J. L., M. Aquino de Muro, and P. H. Nielsen.** 2003. Evaluation of the redox dye 5-cyano-2,3-tolyl-tetrazolium chloride for activity studies by simultaneous use of microautoradiography and FISH. *Appl. Environ. Microbiol.* **69:**641–643.
89. **Nielsen, J. L., and P. H. Nielsen.** 2002. Enumeration of acetate-consuming bacteria by microautoradiography under oxygen and nitrate respiring conditions in activated sludge. *Water Res.* **36:**421–428.
90. **Oerther, D. B., J. Pernthaler, A. Schramm, R. Amann, and L. Raskin.** 2000. Monitoring precursor 16S rRNAs of *Acinetobacter* spp. in activated sludge wastewater treatment systems. *Appl. Environ. Microbiol.* **66:**2154-2165.
91. **Olsen, G. J., D. J. Lane, S. J. Giovannoni, N. R. Pace, and D. A. Stahl.** 1986. Microbial ecology and evolution: a ribosomal RNA approach. *Annu. Rev. Microbiol.* **40:**337–365.
92. **Ostell, J. M., and J. A. Kans.** 1998. The NCBI data model. *Methods Biochem. Anal.* **39:**121–144.
93. **Ouverney, C. C., and J. A. Fuhrman.** 1999. Combined microautoradiography-16S rRNA probe technique for determination of radioisotope uptake by specific microbial cell types in situ. *Appl. Environ. Microbiol.* **65:**1746–1752.
94. **Pernthaler, A., J. Pernthaler, and R. Amann.** 2002. FISH and catalyzed reporter deposition for the identification of marine bacteria. *Appl. Environ. Microbiol.* **68:**3094–3101.

95. **Pernthaler, A., J. Pernthaler, M. Schattenhofer, and R. Amann.** 2002. Identification of DNA-synthesizing bacterial cells in coastal North Sea plankton. *Appl. Environ. Microbiol.* **68:**5728–5736.
96. **Pernthaler, A., C. M. Preston, J. Pernthaler, E. F. DeLong, and R. Amann.** 2002. Comparison of fluorescently labeled oligonucleotide and polynucleotide probes for the detection of pelagic marine bacteria and archaea. *Appl. Environ. Microbiol.* **68:**661–667.
97. **Perry-O'Keefe, H., S. Rigby, K. Oliveira, D. Sorensen, H. Stender, J. Coull, and J. J. Hyldig-Nielsen.** 2001. Identification of indicator microorganisms using a standardized PNA FISH method. *J. Microbiol. Methods* **47:** 281–292.
98. **Pink, C., C. Coeur, P. Potier, and E. Bock.** 2001. Polyclonal antibodies recognizing the AmoB protein of ammonia oxidizers of the beta-subclass of the class *Proteobacteria*. *Appl. Environ. Microbiol.* **67:**118–124.
99. **Pozhitkov, A. E., and D. Tautz.** 2002. An algorithm and program for finding sequence specific oligonucleotide probes for species identification. *BMC Bioinformatics* **3:**9.
100. **Prayitno, J., J. Stefaniak, J. McIver, J. J. Weinman, F. B. Dazzo, J. K. Ladha, W. Baraquio, Y. G. Yanni, and B. G. Rolfe.** 1999. Interactions of rice seedlings with nitrogen-fixing bacteria isolated from rice roots. *Austr. J. Plant Physiol.* **26:**521–535.
101. **Pritsch, K., S. Raidl, E. Marksteiner, H. Blaschke, R. Agerer, M. Schloter, and A. Hartmann.** 2004. A rapid and highly sensitive method for measuring enzyme activities in single mycorrhizal tips using 4-methylumbelliferone-labelled fluorogenic substrates in a microplate system. *J. Microbiol. Methods* **58:**233–241.
102. **Reddy, P. M., J. K. Ladha, R. So, R. Hernandez, F. B. Dazzo, O. Angeles, M. Ramos, and F. de Bruijn.** 1997. Rhizobial communication with rice: induction of phenotypic changes, mode of invasion and extent of colonization. *Plant Soil* **194:**81–98.
103. **Reddy, C. K., and F. B. Dazzo.** 2004. Computer-assisted segmentation of bacteria in color micrographs. *Microsc. Anal.* **18:**5–7.
104. **Reddy, C. K., F.-I. Liu, and F. B. Dazzo.** 2003. Semiautomated segmentation of microbes in color images, p. 548–559. *In* R. Eschbach and G. Marcu (ed.), *Proceedings of the International Society for Optical Engineering (SPIE) and the Society for Imaging Science and Technology*, vol. 5008. *Color Imaging VIII: Processing, Hardcopy, and Applications.*
105. **Reinhold-Hurek, B., and T. Hurek.** 2000. Life in grasses: diazotrophic endophytes. *Trends Microbiol.* **6:**139–144.
106. **Rossello-Mora, R., N. Lee, J. Anton, and M. Wagner.** 2003. Substrate uptake in extremely halophilic microbial communities revealed by microautoradiography and FISH. *Extremophiles* **7:**409–413.
107. **Sando, S., and E. T. Kool.** 2002. Imaging of RNA in bacteria with self-ligating quenched probes. *J. Am. Chem. Soc.* **124:**9686–9687.
108. **Schank, S. C., R. L. Smith, G. C. Weiser, D. A. Zuberer, J. H. Bouton, K. H. Quesenberry, M. E. Tyler, J. R. Milam, and R. C. Littell.** 1979. Fluorescent antibody technique to identify *Azospirillum brasilense* associated with roots of grasses. *Soil Biol. Biochem.* **11:**287–295.
109. **Schloter, M., R. Borlinghaus, W. Bode, and A. Hartmann.** 1993. Direct identification and localization of *Azospirillum* in the rhizosphere of wheat using fluorescence-labeled monoclonal antibodies and confocal scanning laser microscopy. *J. Microsc.* **171:**173–177.
110. **Schmid, M., M. Rothballer, B. Aßmus, P. Hutzler, M. Schloter, and A. Hartmann.** 2004. Detection of microbes by confocal laser scanning microscopy, p. 875–910. *In* G. A. Kowalchuk, F. de Bruijn, I. M. Head, A. D. L. Akkermans, and J. D. van Elsas (ed.), *Molecular Microbial Ecology Manual II*. Kluwer Academic Publishers., Dordrecht, The Netherlands.
111. **Schmid, M., S. Schmitz-Esser, M. Jetten, and M. Wagner.** 2001. 16S-23S rDNA intergenic spacer and 23S rDNA of anaerobic ammonium-oxidizing bacteria: implications for phylogeny and *in situ* detection. *Environ. Microbiol.* **3:**450–459.
112. **Schmidt, E. L.** 1974. Quantitative autecological study of microorganisms in soil by immunofluorescence. *Soil Sci.* **118:**141–149.
113. **Schmidt, E. L., and R. O. Bankole.** 1962. Detection of *Aspergillus flavus* in soil by immunofluorescent staining. *Science* **136:**776–777.
114. **Schmidt, E. L., J. A. Biesbrock, B. B. Bohlool, and D. H. Marx.** 1974. Study of mycorrhizae by means of fluorescent antibody. *Can. J. Microbiol.* **20:**137–139.
115. **Schönhuber, W., B. Fuchs, S. Juretschko, and R. Amann.** 1997. Improved sensitivity of whole-cell hybridization by the combination of horseradish peroxidase-labeled oligonucleotides and tyramide signal amplification. *Appl. Environ. Microbiol.* **63:**3268–3273.
116. **Schramm, A., B. M. Fuchs, J. L. Nielsen, M. Tonolla, and D. A. Stahl.** 2002. FISH of 16S rRNA gene clones (Clone-FISH) for probe validation and screening of clone libraries. *Environ. Microbiol.* **4:**713–720.
117. **Snaidr, J., R. Amann, I. Huber, W. Ludwig, and K. H. Schleifer.** 1997. Phylogenetic analysis and in situ identification of bacteria in activated sludge. *Appl. Environ. Microbiol.* **63:**2884–2896.
118. **Stahl, D. A., B. Flesher, H. R. Mansfield, and L. Montgomery.** 1988. Use of phylogenetically based hybridization probes for studies of ruminal microbial ecology. *Appl. Environ. Microbiol.* **54:**1079–1084.
119. **Teira, E., T. Reinthaler, A. Pernthaler, J. Pernthaler, and G. J. Herndl.** 2004. Combining catalyzed reporter deposition-FISH and microautoradiography to detect substrate utilization by bacteria and *Archaea* in the deep ocean. *Appl. Environ. Microbiol.* **70:**4411–4414.
120. **Trebesius, K.-H., R. Amann, W. Ludwig, K. Mühlegger, and K.-H. Schleifer.** 1994. Identification of whole fixed bacterial cells with nonradioactive 23S-rRNA-targeted polynucleotide probes. *Appl. Environ. Microbiol.* **60:**3228–3235.
121. **Van Berkum, P., Z. Terefework, L. Paulin, S. Suomalainen, K. Lindström, and B. D. Eardly.** 2003. Discordant phylogenies within the *rrn* loci of rhizobia. *J. Bacteriol.* **185:** 2988–2998.
122. **Wagner, M.** 2004. Deciphering the function of uncultured microorganisms. *ASM News* **70:**63–70.
123. **Wagner, M., R. Amann, H. Lemmer, and K. H. Schleifer.** 1993. Probing activated sludge with oligonucleotides specific for proteobacteria: inadequacy of culture-dependent methods for describing microbial community structure. *Appl. Environ. Microbiol.* **59:**1520–1525.
124. **Wagner, M., M. Horn, and H. Daims.** 2003. FISH for the identification and characterization of prokaryotes. *Curr. Opin. Microbiol.* **6:**302–309.
125. **Wagner, M., M. Schmid, S. Juretschko, K. H. Trebesius, A. Bubert, W. Goebel, and K. H. Schleifer.** 1998. *In situ* detection of a virulence factor mRNA and 16S rRNA in *Listeria monocytogenes*. *FEMS Microbiol. Lett.* **160:**159–168.
126. **Wilson, J. M., M. J. Trinick, and C. A. Parker.** 1983. The identification of vesicular-arbuscular mycorrhizal fungi using immunofluorescence. *Soil Biol. Biochem.* **15:**439–445.

127. **Wollum, A. G., II, and R. H. Miller.** 1980. Density centrifugation method for recovering *Rhizobium* spp. from soil for fluorescent-antibody studies. *Appl. Environ. Microbiol.* **39:**466–469.
128. **Woese, C. R.** 1987. Bacterial evolution. *Microbiol. Rev.* **51:**221–271.
129. **Worden, A. Z., S. W. Chisholm, and B. J. Binder.** 2000. In situ hybridization of *Prochlorococcus* and *Synechococcus* (marine cyanobacteria) spp. with rRNA-targeted peptide nucleic acid probes. *Appl. Environ. Microbiol.* **66:**284–289.
130. **Wright, S. F.** 1992. Immunological techniques for detection, identification, and enumeration of microorganisms in the environment, p. 45–63. *In* M. A. Levin, R. R. Seidler, and M. Rogul (ed.), *Microbial Ecology—Principles, Methods, and Applications.* McGraw-Hill, New York, N.Y.
131. **Yanni, Y. G., R. Y. Rizk, V. Corich, A. Squartini, K. Ninke, S. Philip-Hollingsworth, G. Orgambide, F. DeBruijn, J. Stoltzfus, D. Buckley, T. M. Schmidt, P. F. Mateos, J. K. Ladha, and F. B. Dazzo.** 1997. Natural endophytic association between *Rhizobium leguminosarum* bv. *trifolii* and rice roots and assessment of its potential to promote rice growth. *Plant Soil* **194:**99–114.
132. **Yilmaz, L. S., and D. R. Noguera.** 2004. Mechanistic approach to the problem of hybridization efficiency in fluorescent in situ hybridization. *Appl. Environ. Microbiol.* **70:**7126–7139.
133. **Yilmaz, L. S., H. E. Okten, and D. R. Noguera.** 2006. Making all parts of the 16S rRNA of *Escherichia coli* accessible in situ to single DNA oligonucleotides. *Appl. Environ. Microbiol.* **72:**733–744.
134. **Zimmermann, J., W. Ludwig, and K.-H. Schleifer.** 2001. DNA polynucleotide probes generated from representatives of the genus *Acinetobacter* and their application in FISH of environmental samples. *Syst. Appl. Microbiol.* **24:**238–244.
135. **Zuckerkandl, E., and L. Pauling.** 1965. Molecules as documents of evolutionary history. *J. Theor. Biol.* **8:**357–366.
136. **Zwirglmaier, K., W. Ludwig, and K.-H. Schleifer.** 2004. Recognition of individual genes in a single bacterial cell by fluorescence in situ hybridization (RING-FISH). *Mol. Microbiol.* **51:**89–96.
137. **Zwirglmaier, K., W. Ludwig, and K.-H. Schleifer.** 2004. Improved method for polynucleotide probe-based cell sorting, using DNA-coated microplates. *Appl. Environ. Microbiol.* **70:**494–497.

Reporter Gene Systems Useful in Evaluating In Situ Gene Expression by Soil- and Plant-Associated Bacteria*

JOHAN H. J. LEVEAU, JOYCE E. LOPER, AND STEVEN E. LINDOW

60

A reporter gene system is composed of a gene that lacks its endogenous promoter and therefore is transcribed only if placed downstream from an exogenous promoter. Consequently, the product of a reporter gene indirectly "reports" the transcriptional activity of a promoter to which it is fused. A gene is selected to serve as a reporter based on several criteria, including the ease and sensitivity with which its product can be detected. Both reporter genes and marker genes encode gene products that can be detected conveniently, but a marker gene is transcribed and translated from an endogenous, constitutive promoter, whereas a reporter gene is transcribed from an exogenous promoter. All reporter genes lack their endogenous promoters, but some also lack a ribosome binding site and a translational start site. These reporter genes generate translational fusions resulting in chimeric proteins. For studies of microbial ecology, reporter gene systems that generate transcriptional fusions have been used more commonly than those that generate translational fusions. This chapter focuses on those reporter gene systems that are useful in assessing the in situ transcriptional activities of promoters in bacterial cells in soil or associated with plant tissues or surfaces.

Most genes in microorganisms do not encode products that are conveniently detected or quantified, especially in natural habitats. The concept of using a gene that confers a readily assayable product to report the activity of another gene was a major breakthrough in molecular biology. Reporter genes have been indispensable tools used by molecular biologists to develop our present understanding of gene regulation in both prokaryotes and eukaryotes. More recently, reporter gene systems that are useful in assessing in situ gene expression by bacteria in natural habitats have been described (43, 61).

CHARACTERISTICS OF REPORTER GENES THAT ARE USEFUL IN MICROBIAL ECOLOGY

To serve as useful tools for the assessment of gene expression by bacteria inhabiting natural substrates, reporter genes must meet a number of criteria that are not required of those systems used exclusively to monitor gene expression by bacteria grown in culture. The most important criteria are described below.

Detectability

The product of a reporter gene must be uncommon in the environments under evaluation. Background levels of the gene product contributed by organisms indigenous to the environment decrease the sensitivity of the reporter gene system and interfere with detection of even moderate levels of gene expression.

Biological or physical components of the environment can obscure the detection of certain reporter gene products. For example, soil particles can conceal products detected visually, and naturally occurring pigments can quench light reactions that otherwise could be detected and quantified easily.

Sensitivity

Bacteria in natural habitats are often found in microcolonies or even as individual cells rather than in the large cell numbers typical of bacterial cultures. Therefore, the sensitivity of detection is frequently an overriding consideration in selecting a reporter gene system for studies assessing in situ gene expression by bacteria inhabiting natural substrates. Sensitivity reflects the minimum concentration of a reporter gene product that can be detected in an environmental sample. The most sensitive reporter gene systems can detect gene expression in a single bacterial cell if the target promoter is actively transcribed. Sensitive reporter genes provide an opportunity to evaluate temporal and spatial heterogeneity in gene expression.

Reliable Quantification

Bacteria in natural habitats may be subjected to dramatic changes in nutrient availability and may periodically be exposed to stresses, such as those imposed by desiccation, anoxic environments, changes in osmolarity, or a variety of other factors. The optimal reporter gene system would be expressed by bacteria inhabiting variable and dynamic environments. In reality, all gene products are affected by at least some environmental or physiological factors. Therefore, the major factors influencing the stabilities and activities of reporter gene products must be defined and considered in selecting a reporter gene, designing the experiment, and interpreting experimental data.

*Publication 3657 of the Netherlands Institute of Ecology (NIOO-KNAW).

Specific Activity

Quantitative estimates of reporter gene product concentrations are most useful if they can be normalized to the number of cells containing a reporter gene fusion. Differences in reporter gene product concentrations can then be partitioned into effects contributed by changes in bacterial population size and those contributed by changes in gene expression. Certain genetic markers, such as antibiotic resistance, facilitate the quantification of culturable cells of the bacterial strain containing the reporter gene fusion. Some bacteria, such as epiphytic strains of *Pseudomonas syringae* on leaf surfaces (107), largely exist in a culturable state in their natural habitats, but the nonculturability of organisms in many other habitats complicates estimates of cell numbers that contribute to the amount of reporter gene product. For these situations, a method for quantifying bacterial cells in situ is needed to determine specific activity. Examples include the use of fluorescent in situ hybridization (FISH) to estimate the total numbers of cells in samples containing reporter strains (6, 46; also see chapter 59).

Responsiveness to Changes in Transcriptional Activity

Ideally, a reporter gene product should serve as a quantitative estimate of the immediate rate of transcription of the gene to which the reporter is fused. A stable reporter gene product accumulates over time, however, and its concentration reflects the cumulative transcriptional activity of the target promoter rather than the current transcriptional activity. For example, if transcription of a target promoter increases with time, then the reporter gene product encoded just prior to detection will constitute the majority of the product present in the cell and will provide a good estimate of current transcriptional activity. If the target promoter is undergoing repression, however, a large proportion of a stable reporter gene product will have been produced earlier, and its concentration will not reflect the current transcriptional activity of the target promoter. Responsiveness of a reporter comes at the cost of sensitivity: a labile reporter gene product may not accumulate to detectable concentrations. Therefore, a reporter gene should be selected to provide the best compromise between sensitivity and stability.

REPORTER GENE SYSTEMS THAT ARE USEFUL IN ENVIRONMENTAL MICROBIOLOGY

lacZ

The *lacZ* gene encodes the enzyme β-galactosidase, which cleaves the disaccharide lactose into glucose and galactose. β-Galactosidase can be quantified by a convenient colorimetric assay and can be detected by the development of a blue color in colonies grown on agar media supplemented with the substrate 5-bromo-4-chloro-3-indolyl-β-D-galactopyranoside (X-Gal). β-Galactosidase can also be quantified in extremely sensitive assays generating fluorescent and chemiluminescent products (8, 66, 113). Due to the ease and sensitivity of its detection and the large number of plasmid vectors and transposons available for making transcriptional and translational fusions, *lacZ* is probably the most common reporter gene used in studies of gene regulation by bacteria in culture (92). Background levels of β-galactosidase activity expressed by many indigenous bacteria (4, 11), plants, and animals limit the use of the *lacZ* reporter for assessing in situ gene expression by bacteria inhabiting natural substrates. Fortunately, bacterial and eukaryotic β-galactosidases are differentially sensitive to high pH (8) and heat (112); therefore, eukaryotic enzymes can be selectively inactivated, and the activities of microbial *lacZ* gene fusions can be assessed directly on plants (2). Pigmented and fluorescent compounds present in environmental samples (such as plant tissues or soil) may interfere with sensitive colorimetric or fluorometric assays of β-galactosidase activity (8, 9). Due to the stability of β-galactosidase, the *lacZ* reporter cannot accurately measure rapid decreases in gene expression. In spite of these difficulties, in situ gene induction in many plant-associated bacteria has been assessed successfully with *lacZ* fusions (13, 42). The relatively large numbers of bacterial cells containing the *lacZ* fusion in plant tissues or on plant surfaces apparently allowed measurements of bacterial β-galactosidase, despite background levels of enzymatic activity.

Measurements of gene expression in cells harboring *lacZ* gene fusions are normally made by estimating the β-galactosidase activity of a large population of cells ($>10^8$ cells/sample), and thus they provide an estimate of the average level of gene expression. The abundance of β-galactosidase in individual cells can be quantified from the immunofluorescence of cells probed with a fluorescently labeled antibody specific to β-galactosidase (48). While more cumbersome than the use of reporter genes encoding fluorescent proteins (see below), this approach can provide valuable information about cell-to-cell variation in gene expression.

gusA

The *gusA* gene (also called *uidA*) of *Escherichia coli* encodes β-glucuronidase, which catalyzes the hydrolysis of a wide range of glucuronides (27, 44, 45). The most common substrate for this enzyme is 5-bromo-4-chloro-3-indolyl-β-D-glucuronide (X-Gluc), which produces a blue chromophore upon cleavage (45). Other compounds that produce a fluorescent pigment upon cleavage are also available (8, 9, 27, 66). *gusA* fusions are used extensively in studies of gene expression by viral plant pathogens, plants, *Saccharomyces cerevisiae*, and filamentous fungi, all of which lack indigenous β-glucuronidase activity. *gusA* has also been useful in assessing gene expression by bacterial pathogens and symbionts of plants, especially in resolving spatial patterns of in situ transcriptional activity of bacterial cells. Its use in natural habitats containing mixed bacterial populations may be limited, however, by the presence of β-glucuronidase in certain indigenous bacteria (27).

xylE

The *xylE* gene encodes catechol-2,3-dioxygenase, which converts catechol to 2-hydroxymuconic semialdehyde, a bright yellow pigment that can be visualized or measured spectrophotometrically (114). The *xylE* gene, originally found on the TOL plasmid in *Pseudomonas putida*, is uncommon in other organisms. The stability of catechol-2,3-dioxygenase varies with oxygen availability (79) and with the physiological status of the bacterial cell (88). In stationary phase cells of *P. putida*, for example, the specific activity of the enzyme decreased 39% in 20 min, whereas the activity was relatively stable in exponentially growing cells. Therefore, quantification of *xylE* activity is subject to error unless the factors influencing the stability of catechol-2,3-dioxygenase are taken into account. Nevertheless, the *xylE* reporter was used successfully to study in situ gene expression by *P. putida* inhabiting the rhizosphere (10).

Genes Conferring Bioluminescence

Bacterial luciferases catalyze a light-emitting reaction that requires O_2, a long-chain aliphatic aldehyde substrate, and a source of reducing equivalents, usually reduced flavin mononucleotide ($FMNH_2$) (72). The active form of luciferase is a dimer encoded by the *luxAB* genes (72). In the marine bacteria *Vibrio fischeri* and *Vibrio harveyi*, substrate biosynthesis is conferred by the products of *luxCDE*. Both O_2 and $FMNH_2$ are available in most bacterial cells; therefore, virtually any bacterium can produce light in an aerobic environment if it contains the entire *luxCDABE* operon. In natural habitats, nutritional resources required for substrate production by bacteria may be limiting, so many researchers use *luxAB* as a reporter and add the aldehyde substrate exogenously. *N*-Decanal, the most commonly used substrate, can be added to cells in the vapor phase (94) or as an emulsion to cells suspended in aqueous media (17). The gene (*luc*) conferring luminescence in the North American firefly *Photinus pyralis* has also been used as a reporter (50). Several excellent reviews of the biochemistry and genetics of bioluminescence have been published (72, 108).

Genes conferring bacterial bioluminescence are useful reporters for studies of in situ gene expression for many reasons. (i) There is little background luciferase activity in most environmental samples (87). (ii) Luciferase can be quantified conveniently with scintillation counters, dedicated luminometers, or other photon-counting devices (13, 38). (iii) Luciferase can be detected very sensitively with charge-coupled device video cameras (38). In fact, the luminescence of an individual cell containing a highly expressed *lux* operon and growing under conditions where metabolic activity is high and O_2 is abundant can be detected by this method (91). (iv) Because light emission by individual cells or, more commonly, cell aggregates in environmental substrates can be detected directly and nondestructively, spatial and temporal patterns of the in situ transcriptional activity of cells can be resolved (13, 87).

Despite its virtues, the *lux* reporter system is not without limitations. Because bioluminescence requires abundant O_2 and $FMNH_2$ (72), it will not be expressed by cells growing in anaerobic environments or by metabolically inactive cells. More importantly, light emission will not be proportional to the rate of *lux* gene transcription unless cells produce $FMNH_2$ in nonlimiting amounts (i.e., unless cells are metabolically active). In fact, this phenomenon has been exploited to provide indirect estimates of the metabolic status of cells from measures of bioluminescence of constitutively transcribed *lux* genes (98, 99). Light emission from bioluminescent cells suspended in translucent aqueous environments or present on the surfaces of different substrates can be measured, but the intensity of light emitted can be quenched by particulate material (i.e., soil particles) or soluble pigments. The extent of quenching must be determined if bioluminescence of bacteria in environmental samples is to be quantified accurately. Luciferase is a stable enzyme; therefore, the *lux* reporter will not respond rapidly to changes in transcriptional activity of the target promoter. The responsiveness of the *lux* reporter can be enhanced by a heat shock treatment, which denatures the luciferase protein (94).

Fluorescent Proteins

Over the past decade, genes encoding fluorescent proteins have become increasingly important to microbial ecology as both molecular markers (25, 43) and reporters for bacteria occupying natural habitats. The green fluorescent protein (GFP) from the jellyfish *Aequorea victoria*, which absorbs violet light with a maximum absorbance at 395 nm and emits green light with a maximum at 509 nm (14), was the first of several fluorescent proteins to be used as a reporter. The abundance of GFP in an individual cell can be quantified by fluorescence spectroscopy or confocal laser microscopy. Thus, the *gfp* reporter, like *lux*, can be used to assess gene transcription in individual cells, which represents an advantage over other reporter genes (such as *lacZ*, *xylE*, and *inaZ*) that measure the average transcription in a population of cells. Cellular metabolism is not required for the fluorescence of GFP, so the *gfp* reporter is useful in studies evaluating in situ gene expression by cells that are not actively growing, as may be the case in many natural habitats. GFP is stable in the presence of many denaturants and proteases, persists at high temperatures (<65°C) and over a wide pH range, and tolerates treatment with common fixatives, such as paraformaldehyde or glutaraldehyde (14). Therefore, GFP can be detected in samples that have received harsh treatment, such as fixation for microscopic analyses (24, 77, 78, 104).

A series of different GFP mutants with altered excitation and emission wavelengths has been developed. Color variants of GFP have great potential when used with the native GFP, with each other as dual reporter genes monitoring the expression of two or more genes in the same cell, or as marker genes used in conjunction with a *gfp* reporter gene. A blue variant of GFP produces blue fluorescence, with excitation and emission maxima at around 385 and 445 nm, respectively (35). Because of the distinctive shorter wavelength at which this color variant is excited and emits light, it is attractive for use in conjunction with GFP and all other color variants. Unfortunately, the blue variant has a very low fluorescence efficiency, making it difficult to detect compared to all other color variants. A yellow variant with excitation and emission wavelengths of 513 and 527 nm, respectively, is quite fluorescent (82). Likewise, a cyan variant with excitation and emission maxima of 453 and 501 nm, respectively, has shown great utility for some studies (37). Neither the cyan nor yellow variant can be used in conjunction with the native GFP when green fluorescence is to be quantified because their spectral peaks overlap, but they offer much promise when used together. Red-shifted variant GFPs, which differ from the wild-type GFP in their optimal excitation wavelength (490 versus 395 nm) and their enhanced emission intensity, are superior to the native GFP for most applications since they are more compatible with the spectral emissions of laser light sources (such as argon laser light sources, where λ is 488 nm) that are used for confocal laser microscopy (21, 35, 36). The red-shifted GFPs, particularly the most commonly used variant (S35T), are 20 to 35 times more fluorescent than the wild-type GFP, presumably because they are more soluble and more uniformly distributed throughout the cytoplasm of cells (18). The red-shifted GFP molecules are more resistant to photobleaching than wild-type GFP at the longer blue wavelengths used for excitation, and because they are intrinsically more fluorescent, much shorter excitation times are sufficient for detection (19). A red fluorescent protein (RFP) with a structure similar to that of GFP has been isolated from corals (71), and is commercially available. The long emission wavelength of RFP allows it to be quantified independently of GFP in cells. Unfortunately, the native RFP appears to take much longer than GFP or its color variants to take on an active conformation after its production in a cell; while

GFP becomes active within minutes after production, RFP appears to take many hours (>24) to become fluorescent (J. Leveau and S. E. Lindow, unpublished data). Not only does this tend to reduce the fluorescence emission of cells harboring RFP, but it also makes this protein a poor candidate for use as a reporter gene unless more rapidly maturing variants are found. More recently, more fluorescent and fast-maturing variants of RFP have been described (34, 49, 89, 93).

Wild-type GFP is an extremely stable protein; the fluorescence of cells containing this protein does not decrease for at least 24 h after *gfp* transcription has ceased. The half life of GFP extends over bacterial generations because GFP is depleted primarily by dilution upon cell division (54). This stability places a major limitation on the use of wild-type *gfp* as a reporter gene because at a given time, the amount of GFP fluorescence emitted by a given cell reflects transcription that led to the production of GFP at some point in the history of that cell or its predecessors. Other reporter genes, such as *lacZ* and *inaZ*, also share this problem of excessive stability, which complicates studies in which gene expression decreases over time. Recently, unstable GFP variants have been described (1). These variants harbor a C-terminal peptide extension, such as AANDENYALAA, which makes them targets of proteolysis (1). By altering the final three amino acids of the C-terminal extension, Andersen et al. (1) constructed alleles such as GFP(LVA), GFP(AAV), and GFP(ASV), with half lives in *E. coli* of 40, 60, and 100 min, respectively. When introduced into other bacterial species, the stabilities of these GFP variants can differ from those in *E. coli* (54). The GFP variant with the shortest half-life would be expected to have the most utility when used to assess short-term changes in gene expression, but it also suffers from a relatively low overall fluorescence, even when actively transcribed. Given a constant level of *gfp* expression, the amount of GFP fluorescence exhibited by a cell is inversely proportional to the half-life of the GFP variant. Thus, unless the gene whose transcription is being studied is highly active, a GFP variant of intermediate stability probably offers the best compromise between sensitivity and responsiveness to changes in transcription.

GFP may also suffer from susceptibility to extreme environmental conditions, thereby interfering with the quantitative response of GFP-based biosensors to environmental signals. While *gfp* fusions to promoters such as that of *proU* have provided useful information on the levels of water availability experienced by bacteria such as *P. syringae* in natural habitats such as leaves (5), recent reports suggest that exposure of such whole-cell biosensors to solar radiation and rapid desiccation can cause a reduction in GFP fluorescence (16). Since *gfp*-based biosensors are attractive tools for the study of stressful microbial habitats, it will be important to determine if GFP sensitivity to environmental extremes commonly occurs.

Virtually all bacterial strains tolerate low levels of GFP, but some strains are intolerant of high cellular levels. The reasons for this intolerance are not known, and even strains of the same species can differ greatly in their tolerance of high GFP levels (Leveau and Lindow, unpublished data). Intolerance of GFP is often manifested as genetic instability of constructs harboring *gfp* gene fusions, with deletions and mutations in *gfp* or in promoter-containing regions resulting in decreased GFP production. The current strategy for dealing with GFP intolerance is to select another bacterial strain for investigation, but the use of reduced-half-life variants of GFP might alleviate the problem, as they are less likely to accumulate to toxic levels in the bacterial cell.

Quantification of GFP fluorescence is unimportant when *gfp* is used as a marker gene to detect a cell or the cellular location of a protein to which GFP is fused, but quantification is important when *gfp* is used as a reporter gene to estimate the rate of transcription of target genes. Methods to quantify GFP fluorescence differ in ease, sensitivity, and precision. With cultured cells, GFP fluorescence can be estimated readily in a fluorimeter. A minimum of 10^8 to 10^9 cells in a sample is typically required (depending on the level of expression of the *gfp* reporter gene) to quantify GFP fluorescence with a fluorimeter (73).

Fluorescence-based cell flow cytometers are used to detect cells exhibiting GFP fluorescence. Because of the small size of bacterial cells, particularly those from environmental samples, the resolution and sensitivity of flow cytometers for measuring GFP fluorescence of individual bacterial cells are low compared to those for larger cells. Flow cytometers can easily resolve differences in fluorescence between cells harboring large and small amounts of GFP but may not adequately distinguish smaller differences in GFP content (96). The advantage of flow cytometry in GFP quantification is that it can be combined with fluorescence-activated cell sorting to physically separate cells exhibiting high levels of GFP fluorescence from those that are dim. Such approaches have been useful in so-called "promoter trapping" experiments, such as those in which habitat-specific virulence gene expression has been noted (15, 67, 68, 100, 101, 111). Cells must be dispersed in an aqueous medium to be examined with a flow cytometer. Clearly, this dispersal destroys information on the spatial variation in gene expression in natural habitats. Likewise, since cells must be placed in very small droplets of fluid for analysis in a flow cytometer, the device is easily disrupted or disabled by particulate materials that might accompany bacterial cells extracted from environmental samples without extensive prior cleanup. Because the amount of environmental debris can often greatly exceed that of the biosensor cells to be assessed by flow cytometry, it is usually desirable to use a second, constitutively expressed fluorescent marker in such biosensors so that they might be more readily distinguished from the debris. Such a marker will enable the separate evaluation of biosensor cells from environmental debris based on a contingency decision triggered by the presence or absence of the second fluorescent marker. Since the goal is to measure GFP fluorescence, it is important that the second fluorophore not interfere with measures of green fluorescence. Various red fluorescent proteins are thus attractive markers for this purpose. Unfortunately, *rfp* and the improved DsRed have often been found to provide inadequate levels of red fluorescence for this purpose (93). Recently, new variants of *rfp* providing much higher levels of red fluorescence have been described (49, 89, 93), and they might be generally applicable as marker genes to be used in conjunction with *gfp* reporter genes.

Quantification of GFP fluorescence from digital images by fluorescence microscopy, while tedious, offers the highest sensitivity and resolution of GFP fluorescence and retains information on spatial patterns of gene expression in natural habitats. GFP can be quantified over a large range of activity because the intensity of fluorescence can be controlled by the intensity of the incident irradiation. While the details of fluorescence microscopy are not addressed here, cells can be recovered from natural habitats, such as leaves or roots, and placed on microscope slides with minimal cleanup of environmental constituents (6, 53). Alternatively, cells on plant surfaces and soil can be examined directly by epifluorescence or confocal laser scanning microscopy (5, 6, 7, 41, 46, 76). To evaluate large numbers of single cells with relative speed, digital images acquired during fluorescence microscopy can be analyzed by specialized software, such as IP Lab (Scanalytics, Rockville, Md.), Scion Image (Scion Corporation, Frederick, Md.), or the freely available NIH Image (http://rsb.info.nih.gov/nih-image/about.html) or Image/J (http://rsb.info.nih.gov/ij). Excitation and emission GFP filter sets have been designed, enabling the GFP fluorescence of cells on a variety of backgrounds to be assessed readily without interference from autofluorescent compounds.

In situ gene expression by a bacterial population can be assessed by fluorescence microscopy only if one can account for all cells harboring the *gfp* reporter gene among the assemblage of microorganisms present in natural habitats. It is not sufficient to simply examine a microbial community for the presence of cells exhibiting GFP fluorescence because this will detect only those cells whose gene expression is above the threshold activity; cells with low or no gene expression will not be detected. Staining techniques involving a chromophore with an emission spectrum different from that of GFP can be used to localize all cells containing the *gfp* reporter, irrespective of GFP fluorescence. For example, cells of a given strain can be detected by using fluorescently labeled antibodies specific to that strain or by FISH with a fluorescently labeled oligonucleotide specific to a variable region of the 16S rRNA gene (6, 46). Unfortunately, staining techniques such as FISH dislodge cells from their original positions in the microbial community, thereby negating spatial information that could be associated with cell activity. In contrast, color variants of GFP used as marker genes in combination with a *gfp* reporter can be employed to study spatial patterns of gene expression in a bacterial population on plant surfaces. For example, by using fluorescence microscopy, one might first localize cells containing both an *rfp* marker gene and a *gfp* reporter gene by producing a digital image of cells exhibiting red fluorescence. The proportion of those cells that also exhibit green fluorescence when evaluated with a GFP filter set will correspond to the proportion of cells in the *rfp*-marked population that express the *gfp* reporter gene. This approach can provide valuable information about the factors that influence the gene expression of a bacterium in a natural habitat.

Because of the great interest in the use of *gfp* as a reporter gene, a variety of genetic constructs have been made available for generating transcriptional fusions to *gfp*. A Tn*10*- and several Tn*5*-based transposon delivery vectors have been developed, principally for *gfp* marker genes, although some can produce promoterless *gfp*-target gene fusions (59, 70, 90, 95-97). Vectors designed specifically to produce transcriptional fusions to a promoterless *gfp* gene with enhanced translational efficiency have also been described (74, 75).

inaZ

Contrary to the common misconception that water freezes at the same temperature at which it melts (0°C), pure water can remain in the liquid state if cooled to −40°C (102). Supercooling to temperatures lower than −15°C rarely occurs in nature, however, due to the presence of heterologous ice nuclei. Almost all organic and inorganic substances can catalyze ice formation (i.e., serve as ice nuclei) at temperatures between −15 and −40°C, and certain compounds can serve as nuclei at temperatures as high as −6°C (102). In the absence of biological ice nuclei, however, water present in environmental samples can supercool to −6°C or lower. A number of bacterial and fungal species can serve as biological ice nuclei at warm subzero temperatures (>−5°C) (39, 58, 86). The best characterized of these is *P. syringae*, a widely distributed bacterial epiphyte of plants (58). The ice nucleation activity (INA) of *P. syringae* is conferred by a single gene (*inaZ*) that encodes an outer membrane protein (InaZ) (105). Individual ice nucleation proteins do not serve as ice nuclei, but they form large, homogeneous aggregates that collectively orient water molecules into a conformation mimicking the crystalline structure of ice, thereby catalyzing ice formation (26, 47, 109). Oriented water molecules freeze at temperatures slightly below zero (i.e., −2 to −10°C) instead of supercooling.

The transcriptional activity of a promoter fused to a promoterless *inaZ* gene can be assessed by the INA. The characteristics of the *inaZ* reporter gene system are described below.

Detection of Ice Nuclei in Environmental Samples

Ice nuclei can be detected by a tube-freezing technique (40) and a droplet freezing assay (described below), both of which detect ice formation in a supercooled aqueous solution in which an environmental sample is placed. Both methods detect ice nuclei directly from environmental samples and therefore can assess in situ gene expression by bacteria containing a transcriptional fusion to *inaZ*.

Background Levels of Ice Nuclei

A number of bacterial (40) and fungal (86) species can serve as ice nuclei, but these species generally are present in very small numbers in most environmental samples. At −5°C, there is very little background INA in the soil, rhizosphere, or water samples that we have tested. In contrast, epiphytic populations of ice nucleation-active (Ice$^+$) bacteria are common on aerial surfaces of field-grown plants collected from favorable environments (i.e., those with cool temperatures and high relative humidities or free moisture levels) (39). Certainly, large populations of Ice$^+$ bacteria on leaf surfaces of field-grown plants present background levels of ice nuclei that will interfere with the detection of ice nuclei contributed by introduced bacteria containing the ice nucleation reporter gene. Although the ice nucleation reporter gene will not be useful in this situation, background levels of ice nuclei in most environmental samples rarely limit the usefulness of this reporter (M. D. Henkels and J. E. Loper, unpublished data; S. E. Lindow, unpublished data).

Quantification of INA

INA, expressed as the number of ice nuclei per cell, can be quantified conveniently by a droplet freezing assay, which is described in detail below. Soil particles, plant material, and other constituents of environmental samples do not interfere with measurements of INA (40, 62, 63, 110). INA of bacteria containing the *inaZ* reporter gene can be quantified accurately even from soil suspensions (51, 62, 63, 64, 84).

INA is related quantitatively to the InaZ concentration in the cell. In *E. coli*, INA increases with the square of the InaZ concentration until INA reaches 10^{-1} nuclei per cell

(105, 106). Above that level, INA increases with the third power (cube) of the InaZ concentration until it reaches the maximum detectable level of one nucleus per cell. A similar second- to third-power relationship of INA to InaZ protein concentration was also established for *P. syringae* (57). This relationship is the basis of a hypothesis that a rate-limiting step of dimerization or trimerization of InaZ precedes the formation of the large protein aggregates in outer membranes of gram-negative bacteria that serve as ice nuclei (105). The second- to third-power relationship indicates that *inaZ* can serve as an "amplifiable" reporter of the transcriptional activities of promoters to which it is fused.

Sensitivity and Range of Detection

Two important and related advantages of the ice nucleation reporter system are the sensitivity with which INA can be detected and the tremendous range of INAs that can be quantified. Theoretically, a single ice nucleus can be detected simply by observing ice formation in a bacterial suspension cooled to −5°C. Because virtually every cell is ice nucleation active if the *inaZ* gene is transcribed from a very active promoter (57, 64, 105), it is conceivable that even a few Ice$^+$ cells (i.e., <10) could be detected in an environmental sample. This property, coupled with the low levels of background INA present in most environmental samples, imparts the exceptional sensitivity with which InaZ can be detected in environmental samples (73).

The lower limit of detection of INA is the inverse of the bacterial population size. If one ice nucleus is detected per milliliter of a bacterial culture containing 10^8 CFU/ml, then the INA is 10^{-8} ice nuclei per cell. It is therefore possible to estimate a range of INAs varying 10^8-fold, which corresponds to a 2×10^4-fold change in InaZ concentration. Few reporter systems can accurately estimate transcriptional activity over a 20,000-fold range.

Lack of Dependence of INA on High Metabolic Activity

Researchers have reported that INA is expressed by bacterial cells during stationary phase in culture (20, 33, 80) and, with one exception (85), also during exponential growth phase, indicating that the *inaZ* reporter is useful for assessing the gene expression of cells that are not actively growing, as may be common in natural habitats. The INA of *P. syringae* containing *inaZ* transcribed from its native promoter is commonly 10-fold to 100-fold lower during exponential growth phase than during stationary phase (21, 33, 80). A sequence similar to a "gear box" promoter, which is typically more active in stationary-phase cells than in exponentially growing cells, is present 5′ of *inaZ* in *P. syringae* (80). Therefore, the effect of physiological status on INA may be at least partially at the level of *inaZ* transcription and may vary among *inaZ* transcriptional fusions.

The ice nucleation protein is relatively rare in an Ice$^+$ bacterial cell. A cell with an actively transcribed *inaZ* gene has approximately 500 to 1,000 copies of InaZ, which is far fewer than the 10^5 copies of major outer membrane proteins typically present in gram-negative bacteria (105). Therefore, expression of INA does not place a large metabolic demand on the bacterial cell. Differences in the relative growth and survival of nearly isogenic Ice$^+$ and Ice$^-$ strains of *P. syringae* in culture or on leaf surfaces have not been detected (59), indicating that INA does not alter the bacterium's fitness, at least in certain habitats.

Responsiveness to Changes in Transcriptional Activity

If inoculum of *P. syringae* or *Pseudomonas fluorescens* containing a transcriptional fusion to *inaZ* is grown under cultural conditions where the target promoter is inactive, then gene induction can be detected within hours of inoculation of a natural substrate on which bacterial cells express *inaZ* (51, 62-64, 88). In contrast, INA (at −5°C) expressed by *P. fluorescens* containing an *inaZ* fusion does not fully reflect the repressed level of gene expression until 12 to 24 h after transfer from conducive conditions to those that are not conducive to *inaZ* transcription (62, 64). Exponentially growing and stationary-phase cells of *P. fluorescens* are similar in the rate at which INA changes in response to a shift between conducive and nonconducive conditions (62). Therefore, INA of cells that are not actively growing, which may be a common physiological state of bacteria in many natural habitats, responds to changes in the transcription of *inaZ*.

Usefulness of *inaZ* as a Reporter Gene

inaZ has many of the characteristics needed for assessing in situ gene expression by bacteria inhabiting natural habitats. Nevertheless, the *inaZ* reporter gene is not without limitations. Below, we provide background information on biological ice nucleation, elaborate on the characteristics of InaZ that are most relevant to its use as a reporter gene product, and give practical information on the methods and applications of the *inaZ* reporter gene system.

PROPERTIES OF ICE NUCLEATION RELEVANT TO ITS USE AS A REPORTER

inaZ, the InaZ Protein, and the INA Phenotype

Genes encoding the ice nucleation protein have been cloned from a number of bacterial species (105). The largest portion of each gene is composed of a hierarchy of repeated motifs of 24, 48, and 144 nucleotides. According to a structural model of InaZ aggregates (47), the repeated amino acid sequences produce a periodic array of hydrogen bond donor and acceptor groups on the protein surface that interact with neighboring InaZ proteins, membrane lipids, and water outside the cell. Individual InaZ proteins have no fixed conformation; instead, they are elongated, flexible molecules that do not function as ice nuclei. A fixed conformation is acquired when the repeated sequences of neighboring InaZ proteins interdigitate to form large, nearly planar aggregates. The number of InaZ molecules in the planar aggregate determines the temperature at which an ice nucleus is active: an aggregate of only two or three monomers is required to form an ice nucleus that is active at −10°C, whereas an aggregate of at least 50 InaZ proteins is required to nucleate ice at −2°C. An ice nucleus that is active at −5°C is formed when 20 to 40 monomers of the InaZ protein assemble cooperatively on the surface of the bacterial outer membrane (12, 31).

The relatively low abundance of InaZ in a bacterial cell (105), coupled with the requirement for aggregation of 20 to 40 InaZ proteins on the surface of the outer membrane, may explain why each cell has a low probability of nucleating ice at temperatures of −5°C and above. For example, only a fraction of cells (typically 10^{-1} or less) in a native population of Ice$^+$ *P. syringae* inhabiting leaf surfaces of plants in the field express INA at −5°C at any given time (39, 58).

Factors Influencing INA

Membrane Fluidity

Planar aggregates of InaZ proteins are stabilized through their interactions with lipids in the outer membranes of gram-negative bacteria (30). Factors that increase membrane fluidity (i.e., a shift to a higher temperature [33] or treatment with phenethyl alcohol [60]) decrease INA, presumably by destabilizing InaZ protein aggregates.

Temperature

INA is affected by the temperature at which bacterial cells are grown as well as the temperature of the cells immediately before a freezing event. INA of gram-negative bacteria is maximal at growth temperatures of 18 to 24°C (33). INA of *P. syringae* at −5°C is 1,000-fold lower in cultures grown at 30°C than in cultures grown at 20°C. Because the concentration of InaZ does not differ among cells grown within this temperature range, the influence of temperature appears to be at the posttranslational level, most likely on InaZ aggregation (33). Similarly, INA of cells that are not actively growing responds to temperature changes. For example, INA of *P. syringae* at −5°C increased from 10^{-8} to 1 nucleus/cell in response to a decrease in growth temperature from 30 to 16°C (80). These dramatic changes in INA, which occur very rapidly and do not require protein synthesis, are probably due to changes in membrane fluidity that occur when the temperature is shifted (60).

RECOMMENDATIONS FOR USE OF THE *inaZ* REPORTER GENE SYSTEM

Plasmids and Transposons for Constructing Transcriptional Fusions to *inaZ*

To date, all *inaZ* reporter gene constructs have been based on the *inaZ* gene cloned from *P. syringae* (32), which is 3,603 nucleotides long and encodes a protein of 153 kDa. Most constructs make use of a 3,678-bp fragment extending from the AhaIII site at nucleotide 775 through the EcoRI site at nucleotide 4453 of a clone that contains *inaZ* (GenBank accession number X03035). This yields transcriptional fusions to the *inaZ* gene, which possesses a ribosome binding site and a start site for translation of an InaZ protein of predictable size and activity. We refer the reader to the work of Panopoulos (83), who describes methods for constructing *inaZ* transcriptional and protein fusions in some detail.

The transposons Tn*3*-Spice (57) and Tn*3*-nice (51) are useful for mutagenesis of DNA cloned into a *polA*-independent plasmid vector. These transposons differ principally in the types of antibiotic resistance that they confer (i.e., resistance to streptomycin sulfate and spectinomycin is conferred by Tn*3*-Spice, and resistance to neomycin is conferred by Tn*3*-nice). In both transposons, the segment preceding the *inaZ* gene has translational stop codons in all three reading frames, thereby ensuring that only transcriptional fusions will be generated.

The transposon mini-Tn*5*-*inaZ* can insert randomly into the genomes of many soil- and plant-associated bacteria, generating transcriptional fusions with the genes into which it inserts. Mini-Tn*5*-*inaZ* was constructed (S. Carnegie, V. O. Stockwell, and J. E. Loper, unpublished data) by cloning the 3,678-bp fragment containing *inaZ* into mini-Tn*5*-Km in pUT/Km (22).

A series of stable, broad-host-range plasmids is available for generating transcriptional fusions of the *inaZ* reporter gene with cloned promoters (74).

Experimental Design

Evaluation of INA in Culture

INA has been expressed successfully in all species of gram-negative bacteria into which genes conferring ice nucleation have been introduced, including many strains of *E. coli*, *P. syringae*, *P. fluorescens*, *P. putida*, *Pseudomonas aureofaciens*, *Ralstonia solanacearum*, *Xanthomonas campestris*, *Agrobacterium tumefaciens*, *Rhizobium meliloti*, *Erwinia amylovora*, *Erwinia carotovora*, and a number of halophilic bacteria (3, 23, 28, 51, 57, 62-64). Nevertheless, the level of INA conferred by *inaZ* varies among bacteria, and the *inaZ* reporter will not be useful in bacteria that express INA inefficiently. Even strains within a given bacterial species vary in the efficiency with which they express INA. For example, many, but not all, strains of *E. carotovora* express INA efficiently from an introduced plasmid containing *inaZ* transcribed from a *lac* promoter (11).

Evaluation of Sensitivity

The principal advantages of the *inaZ* reporter gene are its sensitivity (73) and its range of detection. Nevertheless, certain promoters transcribe at levels that are too high for quantitative assessment by the *inaZ* reporter. For example, *P. syringae* containing a fusion of *inaZ* to an iron-regulated promoter expresses INA that is inversely related to the iron availability in a culture medium (64). Virtually every cell expresses INA in medium supplemented with 10^{-6} M $FeCl_3$, and higher levels of transcriptional activity expressed by cells growing in a medium with less iron cannot be accurately quantified with the *inaZ* reporter system. Therefore, it is important to select a reporter gene system with a range of detection that matches the transcriptional activity of the promoter to be evaluated.

Need for $InaZ^+$ Control

An *inaZ* gene transcribed from its native promoter or from any characterized promoter that is active in the test strain should be included as a control in experiments evaluating in situ gene expression by bacteria inhabiting natural substrates. This control strain allows one to separate effects attributable to transcription from those attributable to posttranscriptional effects on InaZ (i.e., translation, stability, or aggregation).

Quantification of INA by Droplet Freezing Assay

In the droplet freezing assay, environmental samples (without prior preparation) are suspended in an aqueous solution; small volumes (i.e., droplets) of dilutions of the solution are cooled to −5°C, and the fraction of those that freeze is recorded. The number of ice nuclei in the sample is calculated from the fraction of droplets that freeze. INA of a bacterial population is calculated by dividing the number of ice nuclei by the number of culturable bacterial cells in the sample. A detailed description of the droplet freezing assay follows.

Preparation of Bacterial Suspensions from Environmental Samples

Place environmental samples, such as plant material or soil, in a dilute buffer such as 10 mM potassium phosphate buffer (pH 7.0). Bacterial INA is relatively insensitive to pH values of 3 to 9 but decreases outside of this range. The aqueous solution should not contain high concentrations of salt or organic substances that could affect the freezing point of the solution. Bacterial cells can be dislodged from plant material by shaking or sonication in an ultrasonic cleaning bath, as described previously (65, 81). In our experience, neither of these procedures affects bacterial INA significantly. It is essential that samples and dilution buffer tubes are not exposed to high temperatures during the assay procedure.

Preparation of Foil

An ice nucleus-free surface on which samples of a bacterial suspension can be cooled is required for this assay. The smooth surface of aluminum foil coated with a thin film of paraffin is ideal for this purpose. Medium-weight foil (0.025 mm thick) should be cut into rectangular sheets sized to fit conveniently in the refrigerated bath that will be used for the assay. To prepare the foil surface, spray a nearly saturated solution of paraffin in xylene (approximately 5% [wt/vol]) on one side of the foil and place the foil sheets in an oven held at 65°C for 5 min. The heat will volatilize the solvent and melt the paraffin onto the sheet. After cooling to room temperature, the aluminum foil sheets will appear slightly greasy. They can be stacked and stored indefinitely at room temperature.

A suitable surface can also be prepared by applying paint sealant renewer (TRUCO Inc., Cleveland, Ohio) to foil by rubbing a saturated cotton ball evenly over the surface until the applied coating is dry and clear. Unfortunately, the chemical composition of the coating applied to the foil surface by this method is poorly characterized. Nevertheless, the method works well (M. Brodhagen and J. E. Loper, unpublished data) and provides an alternative for those who lack the equipment to safely apply a xylene-paraffin aerosol to foil surfaces.

Placement of Foil

The aluminum foil sheets are used as a solid surface to support droplets for the freezing assay. A refrigerated circulating bath is the heat sink for the assay. The temperature of the coolant in the bath should be high enough to reduce background levels of ice nuclei found in environmental samples and low enough to optimize detection of INA of the bacterial strain containing the reporter gene fusion. In our experience, −5°C is the best compromise between these two factors. We prefer ethanol as a coolant because it has a low viscosity (even at −10°C) and does not form a thick boundary layer underneath the aluminum foil. Polyethylene glycol also works well and does not pose a fire hazard, but it is messier and more viscous and has a greater tendency to foam than ethanol. Circulation of coolant in the bath must be maintained throughout the assay to ensure adequate and uniform heat transfer from the droplets to the heat sink.

To make a "boat" that can float on the surface of the coolant, fold up approximately 1 cm of each edge of the rectangular sheet of aluminum foil, with the paraffin-treated side up. Avoid trapping air bubbles underneath the foil sheets as you place them on the coolant surface. Droplets placed over air bubbles will not cool quickly to the bath temperature, and differences in droplet temperature are important sources of error in the droplet freezing assay. Bubbles can be minimized by laying one end of the boat on the surface of the bath and slowly lowering the remainder of the foil sheet, in the same manner one would use to place a coverslip on a microscope slide. To detect trapped air bubbles, direct a moist stream of air onto the floating foil; condensation will be visible on the foil surface but will not form directly over trapped air bubbles. Discard aluminum foil boats with trapped air bubbles.

Placement of Droplets on Foil

Begin placing droplets of bacterial suspensions on the surface of the foil immediately after the foil is placed on the bath. Over time, a foamy, insulating layer of aerated coolant can collect under the foil. This insulating layer reduces heat transfer, and drops may not be in temperature equilibrium with the coolant. Ideally, one should finish placing droplets on the foil within 5 min of the foil being placed on the coolant. Each droplet is a small volume (typically 10 μl), which enhances the rate of heat transfer between the droplet and the coolant. Plastic-tipped micropipettes are convenient for delivering the droplets and are unlikely to scratch the surface of the aluminum foil sheet. The number of droplets that freeze is a probability function; therefore, estimates of ice nuclei in a sample become more accurate as the number of droplets evaluated increases. We typically evaluate 40 droplets per dilution, which represents a compromise between accuracy and expediency of the assay.

Observation of Frozen Droplets

Droplets from a bacterial suspension on a boat will freeze quickly at −5°C or −10°C. If the freezing temperature of the sample is considerably higher than that of the coolant, then droplets will freeze within 10 to 20 s. If the nucleation temperature of the sample is similar to that of the coolant, then ice nucleation events may occur even 1 to 2 min after placement of the droplets on the boat. For this reason, observe the droplets for ice formation 1 to 2 min after you place droplets on the boat. If the room temperature is high or air movement over the bath is substantial, cover the bath with a piece of Plexiglas and observe ice formation through the Plexiglas cover. This will stabilize the temperature of the droplets and increase the reproducibility of the assay. Frozen droplets, which appear opaque and peaked, can easily be differentiated from clear, hemispherical unfrozen droplets. The distinction is less obvious if the unfrozen droplets are also opaque, as occurs if a soil or turbid bacterial suspension is being tested. In that case, touch droplets with a sterile toothpick or pipette tip to verify whether they are solid or liquid. Sometimes, a droplet that first appears liquid will suddenly turn solid after it is touched. Such a drop must be considered unfrozen because it is likely that the probe triggered nucleation. Keep in mind that an insulating layer of aerated coolant can form underneath the foil, in which case frozen droplets will eventually begin to melt.

From a series of dilutions of bacterial suspensions, select those dilutions for which some but not all of the droplets freeze. The number of ice nuclei will be estimated most accurately from dilutions in which >3 but fewer than <35 of the 40 droplets freeze. As the proportion of frozen droplets increases, so does the probability that a given droplet contains more than one ice nucleus. This phenomenon is addressed by calculating the number of ice nuclei from the proportion of frozen droplets.

Calculation of INA

The number of ice nuclei in the sample is calculated as proposed by Vali (103), who recognized that freezing events are approximated by a Poisson distribution (because frozen droplets may have more than one ice catalyst present, although only one catalyst is detected), as follows:

$$N = \ln[1/(1-f)]/(V^{dr} D^{s})$$

where N is the concentration of active ice nuclei in the initial solution (i.e., nuclei per milliliter), f is the fraction of frozen droplets (number of droplets that froze/total number of droplets) (note that N should be calculated from dilutions for which the fraction of droplets that froze is >0 and <1), V^{dr} is the volume of each droplet (in milliliters), and D^{s} is the factor describing the dilution of the solution from which droplets were tested.

The concentration of ice nuclei can be normalized to the density of culturable cells that contain the *inaZ* fusion, determined by the spread plate technique (29) on a selective agar medium. INA is calculated by dividing the concentration of ice nuclei/per milliliter (N) by the density of culturable bacteria (CFU/per milliliter) present in the environmental sample.

Interpretation of Results

The droplet freezing assay estimates the number of ice nuclei present in an aqueous solution. By combining this assay with methods estimating numbers of culturable bacteria, one can calculate INA of a population of bacterial cells, expressed as the number of nuclei per bacterial cell. Theoretically, the maximum INA that can be detected is one nucleus per cell. (A bacterial cell cannot be divided between droplets, so even if the cell has more than one functional ice nucleus present in the outer membrane, the second nucleus will not be detected.) In certain situations, however, more than one nucleus per cell is observed. For example, cells can retain INA for a limited time after they lose viability (58), and some bacterial species (such as *Erwinia herbicola*, also called *Pantoea agglomerans*) produce extracellular membrane vesicles that contain ice nuclei (26). Droplets containing nuclei contributed by nonviable cells and membrane vesicles will freeze even though these structures are not culturable cells.

INAs ranging from 1 to 10^{-8} nucleus per cell are typically expressed by cells containing transcriptional fusions to *inaZ*. Although one may incorrectly assume that INAs of <1 indicate heterogeneous gene expression among bacterial cells, this cannot be concluded from data obtained by the droplet freezing assay. Ice nucleation is assumed to be a stochastic process whereby the likelihood of assembling an ice nucleation site from ice nucleation proteins increases exponentially with increasing concentrations of ice nucleation proteins in the bacterial cell (and hence with increasing activities of the promoter linked to an *inaZ* reporter gene). Thus, cells with similar levels of ice nucleation proteins have similar probabilities of serving as a nucleant, even if nucleation is exhibited by only a subset of the cells in the population. *inaZ*, like most other reporter gene systems, reports the average level of transcriptional activity in a population of cells.

INTERPRETATION OF REPORTER GENE DATA

To close this chapter on the use of bioreporters in microbial ecology, we discuss issues to consider in the interpretation of bioreporter data. A more in-depth discussion of some of these issues has been presented elsewhere (55).

The most common misinterpretation of reporter data has to do with the fact that what one measures (i.e., the abundance or activity of a reporter gene product) is not what one would like to know (i.e., promoter activity). The terms reporter gene activity and promoter activity are often used synonymously, but they are not the same thing. The fundamental difference between promoter activity and reporter gene activity has been exposed quite decisively for GFP (54). GFP reporter gene activity is expressed as the number of fluorescent GFP molecules per cell, measured as fluorescence of the cell. Factors contributing to GFP reporter gene activity include (i) the activity of the promoter from which *gfp* is transcribed and translated to nonfluorescent GFP (promoter activity), (ii) the rate at which nonfluorescent GFP matures to become fluorescent (maturation), (iii) the rate at which both nonfluorescent and fluorescent GFP molecules are degraded (degradation), and (iv) the doubling rate (i.e., growth rate) of the bacterium that serves as the bioreporter. A mathematical model incorporating these four factors (54) clarifies the point that a change in promoter activity may cause a change in reporter activity, whereas the reverse assumption is not necessarily true. A change in reporter activity may be due to a change in promoter activity but may just as well be explained by changes in maturation, degradation, and/or dilution.

An analogy can be made between fluorescent GFP accumulation in bioreporter cells and the traffic pattern depicted in Fig. 1. Cars come in from the left at a rate that is

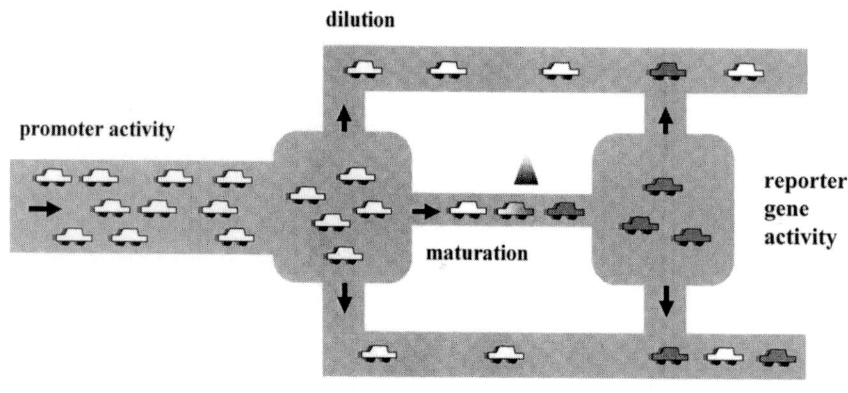

FIGURE 1 Accumulation of fluorescent GFP in bioreporter cells can be compared to the traffic situation depicted here. See the text for an explanation.

expressed as the number of cars passing per unit of time (i.e., promoter activity). At the end of this road, the cars arrive in a parking lot that allows them to go in one of three different directions (the roads to dilution, maturation, and degradation). Each of these roads has its own capacity, i.e., number of cars that it can handle per hour (comparable to the rate of dilution through doubling, maturation, or degradation). Cars that end up on the road of maturation arrive at a second parking lot with two possible exits, dilution and degradation, again with each having its own car-handling capacity. The main argument of the mathematical model (54) is that the rate at which cars enter this traffic grid (i.e., promoter activity) can be estimated accurately from the number of cars that are present in the second parking lot (i.e., reporter gene activity measured as cell fluorescence) only if one also knows the number of cars on the roads to dilution and degradation. For example, an increase in the number of cars in the second parking lot (higher reporter gene activity) could be explained as (i) more cars entering the traffic grid (higher promoter activity), (ii) a lower doubling rate of the bioreporter (less dilution), (iii) the road to degradation accepting fewer cars (less degradation of GFP), (iv) the road to maturation accommodating more cars (faster maturation of GFP to a fluorescent state), or (v) any combination of the four previous situations. Of these possibilities, only the first is usually considered in the interpretation of GFP bioreporter data.

It is important to consider the influences of dilution, degradation, and maturation when using reporter gene data to estimate promoter activity. Most of the commonly used variants of GFP have a constant rate of maturation and are very stable (i.e., have a degradation rate close to zero), so their fluorescence in bioreporters is a function only of promoter activity and the doubling rate. Yet the effect of the doubling rate is rarely taken into account and might even be misinterpreted as a change in promoter activity. The doubling rate is important because given a constant promoter activity, bioreporters that divide half as fast are, in theory, twice as bright. This is especially worrisome if one considers how data from bioreporters in natural settings are generally compared to data from calibration experiments in culture. A typical experiment involves growing the reporter cells in a so-called noninducing medium that serves the following two purposes: (i) to generate enough cells to release into the environment of study (e.g., soil, rhizosphere, or phyllosphere) and (ii) to serve as a calibration reference for comparison to the bioreporters in the field, i.e., to establish a basal level of expression. It is safe to say that in most environments the average growth rate of introduced bacteria is lower than that in a typical noninducing culture medium. We can then imagine a situation in which bioreporters are introduced into an environment that lacks the stimulus to which the bioreporter is supposed to respond. Yet because the bioreporter cells grow slower than in the noninducing medium, the cells will appear more fluorescent upon their introduction, which could easily lead to the false conclusion that promoter activity is induced and the bioreporter is conveying a positive report. In addition, the activities of many promoters are functions of the growth rate (56), and for most, if not all, of the promoters that have been used for bioreporting purposes so far, the influence of growth rate is unknown. Furthermore, the growth rate of the bioreporter may also affect the stability of the reporter gene product. For example, it has been shown that the protease activity which is apparently responsible for the instability of GFP variants GFP(AAV), GFP(ASV), and GFP(ALV) (1) varies with the growth rate (52). The influences of dilution, maturation, and degradation on interpretation of bioreporter data, exemplified for GFP reporters in this discussion, apply to all reporter genes, including *inaZ*, *lacZ*, and *lux*.

Accurate estimates of promoter activity from reporter activity depend on the use of controls that provide accurate estimates of dilution, maturation, and degradation of the reporter gene. The dilution factor can be assessed by estimating the growth rate of the bioreporter strain in the natural environment. For example, the growth rates of GFP bioreporters of fructose availability in the phyllosphere (54) were estimated through direct plating over time and by determining the ribosome contents of individual bioreporter cells, which are proportional to the growth rate. Both assessments aided in the interpretation of the GFP data as being a true response of the reporter bacteria to the availability of the sugar fructose on leaf surfaces (54). Combining a reporter gene fusion and a constitutive gene fusion (to a second reporter gene) in every individual bioreporter cell provides a particularly powerful estimate of the dilution factor. This use of internal reference reporter fusions has been successfully employed on a limited basis to date (69) but it is likely to become a common practice for the use of bioreporters in microbial ecology. A control strain that constitutively expresses the reporter gene is essential to interpretation of reporter gene experiments in microbial ecology, as these controls can provide a combined estimate of the dilution, maturation, and degradation factors. With appropriate consideration for the biology of the bioreporter and the use of proper controls, reporter gene systems represent a powerful approach for assessing the physical, chemical, and biological characteristics of natural microbial habitats.

CONCLUDING REMARKS

Reporter gene systems provide an excellent opportunity for microbiologists to gain new perspectives on the activities of bacteria inhabiting natural substrates, including their expression of specific genes, their recognition of environmental signals, their metabolic activities, and the chemical nature of the habitats that they occupy. A number of innovative and complementary reporter gene systems that are useful in environmental microbiology are now available. Our discussion focused on the ice nucleation and GFP reporter genes, which have several unique attributes that make them extremely useful in studies evaluating in situ gene expression by bacteria inhabiting natural environments. We expect that these reporter genes will continue to be important tools in future inquiries into the ecology of microorganisms.

REFERENCES

1. **Andersen, J. B., C. Steinberg, L. K. Poulsen, S. P. Bjorn, M. Givskov, and S. Molin.** 1998. New unstable variants of green fluorescent protein for studies of transient gene expression in bacteria. *Appl. Environ. Microbiol.* **64:** 2240–2246.
2. **Arsène, F., S. Katupitiya, I. R. Kennedy, and C. Elmerich.** 1994. Use of *lacZ* fusions to study the expression of *nif* genes of *Azospirillum brasiliense* in association with plants. *Mol. Plant-Microbe Interact.* **7:**748–757.
3. **Arvanitis, N., C. Vargas, G. Tegos, A. Perysinakis, J. J. Nieto, A. Ventosa, and C. Drainas.** 1995. Development of a gene reporter system in moderately halophilic bacteria by employing the ice nucleation gene of *Pseudomonas syringae*. *Appl. Environ. Microbiol.* **61:**3821–3825.

4. **Atlas, R. M., G. Sayler, R. S. Burlage, and A. K. Bej.** 1992. Molecular approaches for environmental monitoring of microorganisms. *BioTechniques* **12:**706–717.
5. **Axtell, C. A., and G. A. Beattie.** 2002. Construction and characterization of a *proU-gfp* transcriptional fusion that measures water availability in a microbial habitat. *Appl. Environ. Microbiol.* **68:**4604–4612.
6. **Brandl, M. T., B. Quiñones, and S. E. Lindow.** 2001. Heterogeneous transcription of an indoleacetic acid biosynthetic gene in *Erwinia herbicola* on plant surfaces. *Proc. Natl. Acad. Sci. USA* **98:**3454–3459.
7. **Bringhurst, R. M., Z. G. Cardon, and D. J. Gage.** 2001. Galactosides in the rhizosphere: utilization by *Sinorhizobium meliloti* and development of a biosensor. *Proc. Natl. Acad. Sci.USA* **98:**4540–4545.
8. **Bronstein, I., J. J. Fortin, P. E. Stanley, G. S. A. B. Stewart, and L. J. Kricka.** 1994. Chemiluminescent and bioluminescent reporter gene assays. *Anal. Biochem.* **219:** 169–181.
9. **Bronstein, I., J. J. Fortin, J. C. Voyta, R.-R. Juo, B. Edwards, C. E. M. Olesen, N. Lijam, and L. J. Kricka.** 1994. Chemiluminescent reporter gene assays: sensitive detection of the GUS and SEAP gene products. *BioTechniques* **17:**172–177.
10. **Buell, C. R., and A. J. Anderson.** 1993. Expression of the *aggA* locus of *Pseudomonas putida* in vitro and in planta as detected by the reporter gene, *xylE*. *Mol. Plant-Microbe Interact.* **6:**331–340.
11. **Bull, C. T., S. R. Carnegie, and J. E. Loper.** 1996. Pathogenicity of mutants of *Erwinia carotovora* subsp. *carotovora* deficient in aerobactin and catecholate siderophore production. *Phytopathology* **86:**260–266.
12. **Burke, M. J., and S. E. Lindow.** 1990. Surface properties and size of the ice nucleation site in ice nucleation active bacteria: theoretical considerations. *Cryobiology* **27:**80–84.
13. **Burlage, R. S., and C.-T. Kuo.** 1994. Living biosensors for the management and manipulation of microbial consortia. *Annu. Rev. Microbiol.* **48:**291–309.
14. **Chalfie, M., Y. Tu, G. Euskirchen, W. W. Ward, and D. C. Prasher.** 1994. Green fluorescent protein as a marker for gene expression. *Science* **263:**802–805.
15. **Chang, J. H., J. M. Urbach, T. F. Law, L. W. Arnold, A. Hu, S. Gombar, S. R. Gran, F. M. Ausubel, and J. L. Dangl.** 2005. A high-throughput, near saturating screen for type III effector genes from *Pseudomonas syringae*. *Proc. Natl. Acad. Sci. USA* **102:**2549–2554.
16. **Chen, C., and G. A. Beattie.** 2004. Critical evaluation of green fluorescent protein-based bioreporters deployed in stressful environments. *Phytopathology* **94:**S17.
17. **Cirvilleri, G., and S. E. Lindow.** 1994. Differential expression of genes of *Pseudomonas syringae* on leaves and in culture evaluated with random genomic *lux* fusions. *Mol. Ecol.* **3:**249–257.
18. **Cormack, B. P., R. H. Valdivia, and S. Falkow.** 1996. FACS-optimized mutants of the green fluorescent protein (GFP). *Gene* **173:**33–38.
19. **Cubitt, A. B., R. Heim, S. R. Adams, A. E. Boyd, L. A. Gross, and R. Y. Tsein.** 1995. Understanding, improving and using green fluorescent proteins. *Trends Biochem. Sci.* **20:**448–455.
20. **Deininger, C. A., G. M. Mueller, and P. K. Wolber.** 1988. Immunological characterization of ice nucleation proteins from *Pseudomonas syringae, Pseudomonas fluorescens,* and *Erwinia herbicola*. *J. Bacteriol.* **170:**669–675.
21. **Delagrave, S., R. E. Hawtin, C. M. Silva, M. M. Yang, and D. C. Youvan.** 1995. Red-shifted excitation mutants of the green fluorescent protein. *Bio/Technology* **13:**151–154.
22. **de Lorenzo, V., and K. N. Timmis.** 1994. Analysis and construction of stable phenotypes in gram-negative bacteria with Tn5 and Tn10-derived minitransposons. *Methods Enzymol.* **235:**386–405.
23. **Duijff, B. J., G. Recorbet, P. A. H. M. Bakker, J. E. Loper, and P. Leamanceau.** 1999. Microbial antagonism at the root level is involved in the suppression of *Fusarium* wilt by the combination of nonpathogenic *Fusarium oxysporum* Fo47 and *Pseudomonas putida* WCS358. *Phytopathology* **89:**1073–1079.
24. **Eberl, L., R. Schulze, A. Ammendola, O. Geisenberger, R. Earhart, C. Steinberg, S. Molin, and R. Amann.** 1997. Use of green fluorescent protein as a marker for ecological studies of activated sludge communities. *FEMS Microbiol. Lett.* **149:**77–83.
25. **Errampalli, D., K. Leung, M. B. Cassidy, M. Kostrzynska, M. Blears, H. Lee, and J. T. Trevors.** 1999. Applications of the green fluorescent protein as a molecular marker in environmental microorganisms. *J. Microbiol. Methods* **35:**187–199.
26. **Fall, R., and P. K. Wolber.** 1995. Biochemistry of bacterial ice nuclei, p. 63–83. *In* R. E. Lee, Jr., G. J. Warren, and L. V. Gusta (ed.), *Biological Ice Nucleation and Its Applications*. APS Press, St. Paul, Minn.
27. **Gallagher, S. R.** 1992. *GUS Protocols: Using the GUS Gene as a Reporter of Gene Expression*. Academic Press, New York, N.Y.
28. **Georgakopoulos, D. G., M. Hendson, N. J. Panopoulos, and M. N. Schroth.** 1994. Analysis of expression of a phenazine biosynthesis locus of *Pseudomonas aureofaciens* PGS12 on seeds with a mutant carrying a phenazine biosynthesis locus-ice nucleation reporter gene fusion. *Appl. Environ. Microbiol.* **60:**4573–4579.
29. **Gerhardt, P., R. G. E. Murray, W. A. Wood, and N. R. Krieg.** 1994. *Methods for General and Molecular Bacteriology*, p. 255–257. American Society for Microbiology, Washington, D.C.
30. **Govindarajan, A. G., and S. E. Lindow.** 1988. Phospholipid requirement for expression of ice nuclei in *Pseudomonas syringae* and in vitro. *J. Biol. Chem.* **263:**9333–9338.
31. **Govindarajan, A. G., and S. E. Lindow.** 1988. Size of bacterial ice-nucleation sites measured in situ by radiation inactivation analysis. *Proc. Natl. Acad. Sci. USA* **85:** 1334–1338.
32. **Green, R. L., and G. J. Warren.** 1985. Physical and functional repetition in a bacterial ice nucleation gene. *Nature* **317:**645–648.
33. **Gurian-Sherman, D., and S. E. Lindow.** 1995. Differential effects of growth temperature on ice nuclei active at different temperatures that are produced by cells of *Pseudomonas syringae*. *Cryobiology* **32:**129–138.
34. **Hakkila, K., M. Maksimow, A. Rosengren, M. Karp, and M. Virta.** 2003. Monitoring promoter activity in a single bacterial cell by using green and red fluorescent proteins. *J. Microbiol. Methods* **54:**75–79.
35. **Heim, R., A. B. Cubitt, and R. Y. Tsien.** 1995. Improved green fluorescence. *Nature* **373:**663–664.
36. **Heim, R., D. C. Prasher, and R. Y. Tsien.** 1994. Wavelength mutations and posttranslational autooxidation of green fluorescent protein. *Proc. Natl. Acad. Sci. USA* **91:** 12501–12504.
37. **Heim, R., and R. Y. Tsein.** 1996. Engineering green fluorescent protein for improved brightness, longer wavelengths and fluorescence resonance energy transfer. *Curr. Biol.* **6:**178–182.
38. **Hill, P. J., C. E. D. Rees, M. K. Winson, and G. S. A. B. Stewart.** 1993. The application of *lux* genes. *Biotechnol. Appl. Biochem.* **17:**3–14.
39. **Hirano, S. S., and C. D. Upper.** 1995. Ecology of ice nucleation-active bacteria, p. 41–61. *In* R. E. Lee, Jr., G. J.

Warren, and L. V. Gusta (ed.), *Biological Ice Nucleation and Its Applications*. APS Press, St. Paul, Minn.

40. **Hirano, S. S., L. S. Baker, and C. D. Upper.** 1985. Ice nucleation temperature of individual leaves in relation to population sizes of ice nucleation active bacteria and frost injury. *Plant Physiol.* **77:**259–265.
41. **Holden, P. A., M. G. LaMontagne, A. K. Bruce, W. G. Miller, and S. E. Lindow.** 2002. Assessing the role of *Pseudomonas aeruginosa* surface-active gene expression in hexadecane biodegradation in sand. *Appl. Environ. Microbiol.* **68:**2509–2518.
42. **Howie, W. J., and T. V. Suslow.** 1991. Role of antibiotic biosynthesis in the inhibition of *Pythium ultimum* in the cotton spermosphere and rhizosphere by *Pseudomonas fluorescens*. *Mol. Plant-Microbe Interact.* **4:**393–399.
43. **Jansson, J. K.** 2003. Marker and reporter genes: illuminating tools for environmental microbiologists. *Curr. Opin. Microbiol.* **6:**310–316.
44. **Jefferson, R. A., S. M. Burgess, and D. Hirsh.** 1986. β-Glucuronidase from *Escherichia coli* as a gene-fusion marker. *Proc. Natl. Acad. Sci. USA* **83:**8447–8451.
45. **Jefferson, R. A.** 1989. The *gus* reporter gene system. *Nature* **342:**837–838.
46. **Joyner, D., and S. E. Lindow.** 2000. Heterogeneity of iron bioavailability on plants assessed with a whole-cell GFP-based bacterial biosensor. *Microbiology* **146:**2435–2445.
47. **Kajava, A. V., and S. E. Lindow.** 1993. A molecular model of the three-dimensional structure of bacterial ice nucleation proteins. *J. Mol. Biol.* **232:**709–717.
48. **Kang, Y., E. Saile, M. A. Schell, and T. P. Denny.** 1999. Quantitative immunofluorescence of regulated *eps* gene expression in single cells of *Ralstonia solanacearum*. *Appl. Environ. Microbiol.* **65:**2356–2362.
49. **Knop, M., F. Barr, C. G. Riedel, T. Heckel, and C. Reichel.** 2002. Improved version of the red fluorescent protein (drFP583/DsRed/RFP). *BioTechniques* **33:**596–598.
50. **Koncz, C., W. H. R. Langridge, O. Olsson, J. Schell, and A. A. Szalay.** 1990. Bacterial and firefly luciferase genes in transgenic plants: advantages and disadvantages of a reporter gene. *Dev. Genet.* **11:**224–232.
51. **Kraus, J., and J. E. Loper.** 1995. Characterization of a genomic region required for production of the antibiotic pyoluteorin by the biological control agent *Pseudomonas fluorescens* Pf-5. *Appl. Environ. Microbiol.* **61:**849–854.
52. **Levchenko, I., M. Seidel, R. Sauer, and T. Baker.** 2000. A specificity-enhancing factor for the ClpXP degradation machine. *Science* **289:**2354–2356.
53. **Leveau, J., and S. E. Lindow.** 2001. Appetite of an epiphyte: quantitative monitoring of bacterial sugar consumption in the phyllosphere. *Proc. Natl. Acad. Sci. USA* **98:**3446–3453.
54. **Leveau, J. H. J., and S. E. Lindow.** 2001. Predictive and interpretive simulation of green fluorescent protein expression in reporter bacteria. *J. Bacteriol.* **183:**6752–6762.
55. **Leveau, J. H. J., and S. E. Lindow.** 2002. Bioreporters in microbial ecology. *Curr. Opin. Microbiol.* **5:**259–265.
56. **Liang, S. T., M. Bipatnath, Y. C. Xu, S. L. Chen, P. Dennis, M. Ehrenberg, and H. Bremer.** 1999. Activities of constitutive promoters in *Escherichia coli*. *J. Mol. Biol.* **292:**19–37.
57. **Lindgren, P. B., R. Frederick, A. G. Govindarajan, N. J. Panopoulos, B. J. Staskawicz, and S. E. Lindow.** 1989. An ice nucleation reporter system: identification of inducible pathogenicity genes in *Pseudomonas syringae* pv. *phaseolicola*. *EMBO J.* **8:**1291–1301.
58. **Lindow, S. E.** 1983. The role of bacterial ice nucleation in frost injury to plants. *Annu. Rev. Phytopathol.* **21:**363–384.
59. **Lindow, S. E.** 1985. Ecology of *Pseudomonas syringae* relevant to the field use of Ice deletion constructed in vitro for plant frost control, p. 23–35. *In* H. O. Halvorson, D. Pramer, and M. Rogul (ed.), *Engineered Organisms in the Environment: Scientific Issues*. American Society for Microbiology, Washington, D.C.
60. **Lindow, S. E.** 1995. Membrane fluidity as a factor in production and stability of bacterial ice nuclei active at high subfreezing temperatures. *Cryobiology* **32:**247–258.
61. **Lindow, S. E.** 1995. The use of reporter genes in the study of microbial ecology. *Mol. Ecol.* **4:**555–566.
62. **Loper, J. E., and M. D. Henkels.** 1999. Utilization of heterologous siderophores enhances levels of iron available to *Pseudomonas putida* in the rhizosphere. *Appl. Environ. Microbiol.* **65:**5357–5363.
63. **Loper, J. E., and M. D. Henkels.** 1997. Availability of iron to *Pseudomonas fluorescens* in rhizosphere and bulk soil evaluated with an ice nucleation reporter gene. *Appl. Environ. Microbiol.* **63:**99–105.
64. **Loper, J. E., and S. E. Lindow.** 1994. A biological sensor for iron available to bacteria in their habitats on plant surfaces. *Appl. Environ. Microbiol.* **60:**1934–1941.
65. **Loper, J. E., T. V. Suslow, and M. N. Schroth.** 1984. Lognormal distribution of bacterial populations in the rhizosphere. *Phytopathology* **74:**1454–1460.
66. **Manafi, M., W. Kneifel, and S. Bascomb.** 1991. Fluorogenic and chromogenic substrates used in bacterial diagnostics. *Microbiol. Rev.* **55:**335–348.
67. **Marco, M. L., J. Legac, and S. E. Lindow.** 2003. Conditional survival as a selection strategy to identify plant-inducible genes of *Pseudomonas syringae*. *Appl. Environ. Microbiol.* **69:**5793–5801.
68. **Marco, M. L., J. Legac, and S. E. Lindow.** 2005. *Pseudomonas syringae* genes induced during colonization of leaf surfaces. *Environ. Microbiol.* **7:**1379–1391.
69. **Marsoli, M., J. Feliciano, E. Michelini, S. Daunert, and A. Roda.** 2002. Internal response correction for fluorescent whole-cell biosensors. *Anal. Chem.* **74:**5948–5953.
70. **Matthysse, A. G., S. Stretton, C. Dandie, N. C. McClure, and A. E. Goodman.** 1996. Construction of GFP vectors for use in gram-negative bacteria other than *Escherichia coli*. *FEMS Microbiol. Lett.* **145:**87–94.
71. **Matz, M. V., A. F. Fradkov, Y. A. Labas, A. P. Savitsky, A. G. Zaraisky, M. L. Markelov, and S. A. Lukyanov.** 1999. Fluorescent proteins from nonbioluminescent anthozoa species. *Nat. Biotechnol.* **17:**969–973.
72. **Meighen, E. A.** 1993. Bacterial bioluminescence: organization, regulation, and application of the *lux* genes. *FASEB J.* **7:**1016–1022.
73. **Miller, W. G., M. Y. Brandl, B. Quiñones, and S. E. Lindow.** 2001. Biological sensor for sucrose availability: relative sensitivities of various reporter genes. *Appl. Environ. Microbiol.* **67:**1308–1317.
74. **Miller, W. G., J. H. J. Leveau, and S. E. Lindow.** 2000. Improved GFP and *inaZ* broad host range promoter-probe vectors. *Mol. Plant-Microbe Interact.* **13:**1243–1250.
75. **Miller, W. G., and S. E. Lindow.** 1997. An improved GFP cloning cassette designed for prokaryotic transcriptional fusions. *Gene* **191:**149–153.
76. **Møller, S., C. Steinberg, J. B. Andersen, B. B. Christensen, J. L. Ramos, M. Givskov, and S. Molin.** 1998. In situ gene expression in mixed-culture biofilms: evidence of metabolic interaction between community members. *Appl. Environ. Microbiol.* **64:**721–732.
77. **Monier, J.-M., and S. E. Lindow.** 2003. Exploring *Pseudomonas syringae* ecology via direct microscopic observations of the leaf surface, p. 29–40. *In* N. S. Iacobellis, A. Volmer, S. W. Hutcheson, J. W. Mansfield, C. E. Morris, J. Murrilo, N. W. Schaad, D. E. Stead, G. Surico, and M. S. Ullrich

(ed.), Pseudomonas syringae *and Related Pathogens: Biology and Genetic*. Kluwer Academic Publishers, Boston, Mass.

78. **Monier, J.-M., and S. E. Lindow.** 2005. Aggregates of resident bacteria facilitate survival of immigrant bacteria on leaf surfaces. *Microb. Ecol.* **49:**343–352.
79. **Morgan, J. A. W., C. Winstanley, R. W. Pickup, J. G. Jones, and J. R. Saunders.** 1989. Direct phenotypic and genotypic detection of a recombinant pseudomonad population released into lake water. *Appl. Environ. Microbiol.* **55:**2537–2544.
80. **Nemecek-Marshall, M., R. LaDuca, and R. Fall.** 1993. High-level expression of ice nuclei in a *Pseudomonas syringae* strain is induced by nutrient limitation and low temperature. *J. Bacteriol.* **175:**4062–4070.
81. **O'Brien, R. D., and S. E. Lindow.** 1989. Effect of plant species and environmental conditions on epiphytic population sizes of *Pseudomonas syringae* and other bacteria. *Phytopathology* **79:**619–627.
82. **Ormo, M., A. B. Cubitt, K. Kallio, L. A. Gross, R. Y. Tsien, and S. J. Remington.** 1996. Crystal structure of the *Aequorea victoria* green fluorescent protein. *Science* **273:** 1392–1395.
83. **Panopoulos, N. J.** 1995. Ice nucleation genes as reporters, p. 271–281. *In* R. E. Lee, Jr., G. J. Warren, and L. V. Gusta (ed.), *Biological Ice Nucleation and Its Applications*. APS Press, St. Paul, Minn.
84. **Pierson, E. A., D. W. Wood, J. A. Cannon, F. M. Blachere, and L. S. Pierson III.** 1998. Interpopulation signaling via *N*-acyl-homoserine lactones among bacteria in the wheat rhizosphere. *Mol. Plant-Microbe Interact.* **11:** 1078–1084.
85. **Pooley, L., and T. A. Brown.** 1991. Effects of culture conditions on expression of the ice nucleation phenotype of *Pseudomonas syringae*. *FEMS Microbiol. Lett.* **77:**229–232.
86. **Pouleur, S., C. Richard. J.-G. Martin, and H. Antoun.** 1992. Ice nucleation activity in *Fusarium acuminatum* and *Fusarium avenaceum*. *Appl. Environ. Microbiol.* **58:**2960–2964.
87. **Prosser, J. I., K. Killham, L. A. Glever, and E. A. S. Rattray.** 1996. Luminescence-based systems for detection of bacteria in the environment. *Crit. Rev. Biotechnol.* **16:** 157–183.
88. **Rahme, L. G., M. N. Mindrinos, and N. J. Panopoulos.** 1992. Plant and environmental sensory signals control the expression of *hrp* genes in *Pseudomonas syringae* pv. *phaseolicola*. *J. Bacteriol.* **174:**3499–3507.
89. **Shaner, N. C., R. E. Campbell, P. A. Steinbach, B. N. G. Giepmans, A. E. Palmer, and R. Y. Tsien.** 2004. Improved monomeric red, orange and yellow fluorescent proteins derived from *Discosoma* sp. red fluorescent protein. *Nat. Biotechnol.* **22:**1567–1572.
90. **Shimada, T., H. Makinoshima, Y. Ogawa, T. Miki, M. Maeda, and A. Ishihama.** 2004. Classification and strength measurements of stationary-phase promoters by use of a newly developed promoter cloning vector. *J. Bacteriol.* **186:**7112–7122.
91. **Silcock, D. J., R. N. Waterhouse, L. A. Glover, J. I. Prosser, and K. Killham.** 1992. Detection of a single genetically modified bacterial cell in soil by using charge-coupled device-enhanced microscopy. *Appl. Environ. Microbiol.* **58:**2444–2448.
92. **Slauch, J. M., and T. J. Silhavy.** 1991. Genetic fusions as experimental tools. *Methods Enzymol.* **204:**213–248.
93. **Sorensen, M., C. Lippuner, T. Kaiser, A. Misslitz, T. Aebischer, and D. Bumann.** 2003. Rapidly maturing red fluorescent protein variants with strongly enhanced brightness in bacteria. *FEBS Lett.* **552:**110–114.
94. **Stewart, G. S. A. B., and P. Williams.** 1992. *lux* genes and the applications of bacterial bioluminescence. *J. Gen. Microbiol.* **138:**1289–1300.
95. **Stretton, S., S. Techkarnjanaurk, A. M. McLennan, and A. E. Goodman.** 1998. Use of green fluorescent protein to tag and investigate gene expression in marine bacteria. *Appl. Environ. Microbiol.* **64:**2554–2559.
96. **Tombolini, R., A. Unge, M. E. Davy, F. J. deBruijn, and J. Jansson.** 1997. Flow cytometric and microscopic analysis of GFP-tagged *Pseudomonas fluorescens* bacteria. *FEMS Microbiol. Ecol.* **22:**17–28.
97. **Tomlin, K. L., R. D. Clark, and H. Ceri.** 2004. Green and red fluorescent protein vectors for use in biofilm studies of the intrinsically resistant *Burkholderia cepacia* complex. *J. Microbiol. Methods* **57:**95–106.
98. **Unge, A., and J. Jansson.** 2001. Monitoring population size, activity, and distribution of *gfp-luxAB*-tagged *Pseudomonas fluorescens* SBW25 during colonization of wheat. *Microb. Ecol.* **41:**290–300.
99. **Unge, A., R. Tombolini, L. Molbak, and J. K. Jansson.** 1999. Simultaneous monitoring of cell number and metabolic activity of specific bacterial populations with a dual *gfp-luxAB* marker system. *Appl. Environ. Microbiol.* **65:** 813–821.
100. **Valdivia, R. H., and S. Falkow.** 1997. Fluorescence-based isolation of bacterial genes expressed within host cells. *Science* **277:**2007–2011.
101. **Valdivia, R. H., and S. Falkow.** 1996. Bacterial genetics by flow cytometry: rapid isolation of *Salmonella typhimurium* acid-inducible promoters by differential fluorescence induction. *Mol. Microbiol.* **22:**367–378.
102. **Vali, G.** 1995. Principles of ice nucleation, p. 1–39. *In* R. E. Lee, Jr., G. J. Warren, and L. V. Gusta (ed.), *Biological Ice Nucleation and Its Applications*. APS Press, St. Paul, Minn.
103. **Vali, G.** 1971. Quantitative evaluation of experimental results on the heterogeneous freezing nucleation of supercooled liquids. *J. Atmos. Sci.* **28:**402–409.
104. **Ward, W. W., C. W. Cody, R. C. Hart, and M. J. Cormier.** 1980. Spectophotometric identity of the energy transfer chromophores in Renilla and Aequorea green-fluorescent proteins. *Photochem. Photobiol. Rev.* **31:**611–615.
105. **Warren, G. J.** 1995. Identification and analysis of *ina* genes and proteins, p. 85–99. *In* R. E. Lee, Jr., G. J. Warren, and L. V. Gusta (ed.), *Biological Ice Nucleation and Its Applications*. APS Press, St. Paul, Minn.
106. **Watanabe, N. M., M. W. Southworth, G. J. Warren, and P. K. Wolber.** 1990. Rates of assembly and degradation of bacterial ice nuclei. *Mol. Microbiol.* **4:**1871–1879.
107. **Wilson, M., and S. E. Lindow.** 1992. Relationship of total and culturable cells in epiphytic populations of *Pseudomonas syringae*. *Appl. Environ. Microbiol.* **58:**3908–3913.
108. **Wilson, T., and J. W. Hastings.** 1998. Bioluminescence. *Annu. Rev. Cell. Dev. Biol.* **14:**197–230.
109. **Wolber, P. K.** 1993. Bacterial ice nucleation. *Adv. Microb. Physiol.* **34:**203–237.
110. **Wright, C. A., and G. A. Beattie.** 2004. *Pseudomonas syringae* pv. *tomato* cells encounter inhibitory levels of water stress during the hypersensitive response of *Arabidopsis thaliana*. *Proc. Natl. Acad. Sci. USA* **101:** 3269–3274.
111. **Yang, S., N. T. Perna, D. A. Cooksey, Y. Okinaka, S. E. Lindow, A. M. Ibekwe, N. T. Keen, and C.-H. Yang.** 2004. Genome-wide identification of plant-upregulated genes of *Erwinia chrysanthemi* 3937 using a GFP-based IVET leaf array. *Mol. Plant-Microbe Interact.* **17:**999–1008.

112. **Young, D. C., S. D. Kingsley, K. A. Ryan, and F. J. Dutko.** 1993. Selective inactivation of eukaryotic β-galactosidase in assays for inhibitors of HIV-1 TAT using bacterial β-galactosidase as a reporter enzyme. *Anal. Biochem.* **215:**24–30.

113. **Zhang, Y.-Z., J. J. Naleway, K. D. Larison, Z. Huang, and R. P. Haugland.** 1991. Detecting *lacZ* gene expression in living cells with new lipophilic, fluorogenic β-galactosidase substrates. *FASEB J.* **5:**3108–3113.

114. **Zukowski, M. M., D. F. Gaffney, D. Speck, M. Kauffmann, A. Findeli, A. Wisecup, and J.-P. Lecocq.** 1983. Chromogenic identification of genetic regulatory signals in *Bacillus subtilis* based on expression of a cloned *Pseudomonas* gene. *Proc. Natl. Acad. Sci. USA* **80:**1101–1105.

Identifying Microorganisms Involved in Specific In Situ Functions: Experimental Design Considerations for rRNA Gene-Based Population Studies and Sequence-Selective PCR Assays

JAMES BORNEMAN, J. OLE BECKER, ELIZABETH BENT, BRIAN LANOIL, BRIAN McSPADDEN GARDENER, RABIU OLATINWO, LAURA PRESLEY, ALEXANDRA J SCUPHAM, LEA VALINSKY, AND BEI YIN

61

This chapter examines experimental design considerations for a population-based approach for identifying microorganisms involved in specific in situ functions. In this chapter, the term "function" is used in its broadest sense and may refer to any number of defined biochemical, physiological, or ecological phenomena. Although this chapter focuses on a particular population-based approach, many of the experimental design considerations discussed here apply to a wide range of rRNA gene-based population studies and sequence-selective PCR assays.

BACKGROUND

Attempts to identify microorganisms or consortia of microorganisms involved in specific in situ functions are hindered by numerous factors, including the difficulties of culturing most taxa, interactions among different species in biofilms, variations in chemical conditions across microenvironments, and differences in physical scale between the investigators and subjects. Several experimental approaches have been developed to address some of these difficulties. To identify organisms involved in the metabolism of specific substrates, approaches utilizing isotopes (1, 16, 60, 77, 79, 84) and bromodeoxyuridine (12) can be employed. To identify metabolically active organisms, rRNA (30, 41, 59, 96) and bromodeoxyuridine (104) analyses can be used. To obtain associations among functional genes and microorganisms involved in defined processes, several approaches can be utilized, including examining the diversity and abundance of specific genes (6, 43, 69), analyzing large DNA fragments obtained from environmental samples (8), and genome shotgun sequencing of entire microbial communities (108).

Another approach is to correlate the abundance of microbial populations with the level of a specific functional parameter. Such experiments can be performed by identifying or creating a series of environmental samples with various levels of the functional parameter and then examining the microbial community composition in each of the samples. Abundant organisms or phylogenetically distinct markers, such as rRNA genes, that positively correlate with the functional parameter are considered candidate organisms to be examined in subsequent validation experiments. Functional gradients can be created by manipulating the microbial composition with methods such as differential heat treatments, targeted antimicrobial agents, and nutritional amendments. Components of the microbial community can be identified by rRNA gene analysis, and the relative abundance of each taxon can then be associated with the level of the functional parameter. Taxa exhibiting the strongest correlations represent organisms putatively involved in the defined in situ function and will be further investigated in subsequent validation experiments. This approach has been used to examine, for example, plant disease (66, 70) and suppressiveness of soilborne pests (112, 119, 120).

This chapter examines an experimental approach that uses the population-based strategy described above. The approach has the following three phases: (i) identifying populations of rRNA genes whose abundances correlate with the functional parameter, (ii) validating the rRNA gene correlates identified in phase I by using an independent quantitative assay, and (iii) isolating the microorganisms identified by the rRNA gene correlates and reintroducing them into the environment to assess their functions in situ.

This approach was recently used to identify microorganisms that suppress the population development of the plant-parasitic nematode *Heterodera schachtii* in a southern California soil (113, 115). One important feature of this suppressive soil was the ability of *H. schachtii* cysts isolated from this soil to transfer suppressiveness to nonsuppressive soils (114). This meant that the organisms involved in the suppressiveness inhabited the cysts, allowing these studies to focus on the microorganisms within the cysts rather than the more complex communities found in soil. In phase I, oligonucleotide fingerprinting of rRNA genes (OFRG) (13, 105-107) was used to identify bacterial and fungal rRNA genes found in *H. schachtii* cysts isolated from soils possessing various levels of suppressiveness (119, 120). Five soil treatments with various levels of suppressiveness, generated by mixing different amounts of suppressive and fumigation-induced nonsuppressive soil, were planted with *Brassica juncea* and infested with *H. schachtii* juveniles. After two nematode generations, rRNA gene sequences that were more abundant in cysts isolated from the highly suppressive soils than in cysts from the less suppressive soils were considered candidate sequences. In phase II, sequence-selective quantitative PCR was used to determine the amounts of the candidate sequences in cysts from soils possessing various levels of suppressiveness, which were produced by mixing various quantities of suppressive and fumigation-induced nonsuppressive soil and by biocidal treatments. Phases I and II of this project led to the identification of genes with high sequence identity to α-proteobacterial and *Dactylella oviparasitica* rRNA genes which consistently and positively correlated with high levels of *H. schachtii* suppressiveness (119, 120). These studies also identified a gene with high sequence identity to a *Fusarium oxysporum* rRNA gene which correlated with minimal to moderate levels of suppressiveness. In phase III, strains of *D. oviparasitica* and *F. oxysporum* were isolated from *H. schachtii* cysts collected from the suppressive soil. These strains were introduced into fumigation-induced nonsuppressive soil, planted with *B. juncea*, and infested with *H. schachtii* juveniles. After two nematode generations, *F. oxysporum* consistently reduced *H. schachtii* populations, but not in a statistically significant manner, while *D. oviparasitica* lowered *H. schachtii* populations to the levels observed in the naturally suppressive soil (76). These results support the underlying principle of the experimental approach, i.e., microorganisms involved in a specific function can be identified by correlating the abundances of their rRNA genes with the levels of the specific functional parameter.

EXPERIMENTAL DESIGN CONSIDERATIONS

In the material below, we examine experimental design considerations for a three-phase population approach for identifying microorganisms involved in specific in situ functions. Many of the experimental design considerations described below are also applicable to a wide range of rRNA gene-based population studies and sequence-selective PCR assays.

Phase I: Correlation Studies

In phase I, rRNA genes that positively correlate with the functional parameter of interest are identified. This is accomplished by identifying or creating a series of environmental samples with various levels (or gradients) of the functional parameter and then examining the microbial community composition in each of these samples. Functional gradients are created by manipulating the microbial community with methods such as differential heat treatments, targeted antimicrobial agents, and nutritional amendments. Microbial community composition is examined by rRNA gene analysis. Experimental design considerations for phase I include (i) the use of multiple functional gradients, (ii) the ability to obtain thorough descriptions of microbial composition, (iii) the attempt to examine all types of microorganisms via broad-range primer selection and design, and (iv) examining activity levels and functional genes.

Multiple Functional Gradients

In this chapter, the phrase "multiple functional gradients" refers to the development of more than one set of environmental samples, each created by a different method of manipulating the microbial community (heat, antimicrobial agents, nutrient amendments, etc.) and each possessing a series of samples with various levels of the functional parameter. Note that such gradients can also be identified from natural environmental samples instead of being created by manipulation. Multiple functional gradients can be used in several ways.

First, they can be used to reduce the number of candidate organisms and limit the number of false-positive results produced in phase I correlation studies. These tasks are accomplished by focusing the subsequent phase II and III studies on those rRNA genes (and associated organisms) that consistently correlate with the parameter of interest for all functional gradients. The rationale here is that although any single method may lead to the identification of rRNA gene correlates, some of the correlates may be artifacts of that particular method and would not lead to the identification of organisms involved in the function in the environment.

Second, multiple functional gradients can be used to increase the likelihood of identifying the target organism(s). This may be important because the PCR-based methods used in phase I likely produce depictions of microbial community composition that do not accurately reflect the number and variety of rRNA genes that exist in the environment (11, 20, 25, 33, 37, 47, 62, 64, 78, 80, 81, 87, 95, 98, 99, 101, 102, 110, 117). Since the depictions of rRNA genes produced by such methods are associated with factors such as the relative amplification efficiencies of the target sequence and the other rRNA genes in the sample, examining a series of samples which possess different microbial communities should increase the likelihood that the true relative amount of the target sequence will be revealed. In addition, as described in the aforementioned references, results from such amplification reactions will more likely reflect the number and variety of rRNA genes that exist in the environment if certain recommendations are followed, such as using the fewest PCR cycles possible and using high-molecular-weight DNA as the template.

Creating Gradients

Methods for creating functional gradients include (i) using selective antimicrobial agents (2, 26, 52, 54, 97, 119, 120); (ii) applying broad-spectrum soil fumigants, such as methyl bromide (89), methyl iodide (113), and metam sodium (113); (iii) applying gamma-irradiation (2, 52); (iv) using nutritional amendments (24); and/or (v) manipulating environmental parameters, such as the water potential (23), pH (86, 93), soil aeration (24), and temperature (2, 17, 86, 89, 114). In fumigant studies, microbial communities can be differentially manipulated by utilizing fumigants with various selectivities and/or by using at different rates or exposure times (44). Additional gradients can be created by utilizing any of the above treatments and then mixing the treated sample with various portions of a sterilized sample. This approach can produce a series of samples with different microbial communities (119, 120) because microorganisms are added to samples possessing a variety of nutrient levels (dead organisms) and unoccupied niches.

Identifying Natural Gradients

Functional gradients can also be identified from natural environmental samples at various spatial and temporal scales. Indeed, examinations of functional processes and associated microbial populations will likely benefit from experimentation at scales relevant to the organisms under investigation. For example, when attempting to identify soil microorganisms that influence or are influenced by roots, investigations should target root-associated organisms instead of those in bulk soil. Similarly, microorganisms that interact with intestinal cells are more likely to be identified by examining those spatially associated with such cells instead of those associated with feces. Examples of spatial and temporal heterogeneity of populations and functional processes can be found in many previously published articles and citations within (18, 35, 38, 39, 42, 48, 49, 55, 71, 73, 74, 88, 90, 92, 100).

Thorough Analysis of Microbial Composition

When attempting to identify organisms that are not numerically abundant but which inhabit complex microbial communities, it will likely be important to utilize a method that can produce detailed descriptions of microbial community composition. Unfortunately, obtaining such descriptions has been and continues to be a challenge. Nucleotide sequence analysis of rRNA gene clone libraries can be used to generate detailed depictions of microbial community composition. Yet such analyses typically are not feasible because of the high costs associated with examining large numbers of samples possessing complex microbial communities.

To address this experimental limitation, we developed a method enabling extensive analysis of microbial community composition in a cost-effective manner. The result of these efforts was the OFRG method (Fig. 1) (13, 105-107). Briefly, clone libraries are constructed by using PCR primers designed to selectively amplify rRNA genes from specific but broad taxonomic groups, such as bacteria or fungi. Cloned rRNA gene fragments are arrayed on nylon membranes and then subjected to a series of hybridization experiments, with each using a single DNA oligonucleotide probe (^{33}P labeled). For every hybridization experiment, the signal intensities are transformed into three discrete values, 0, 1, and *N*, where 0 and 1 specify negative and positive hybridization events respectively, and *N* designates an uncertain assignment. This process creates a hybridization fingerprint for each clone, which is a vector of values resulting from its hybridizations with all probes. Clones are identified by clustering their hybridization fingerprints with those from taxonomically identified organisms and by nucleotide sequence analysis of representative clones within a cluster.

OFRG offers several useful features that result from the manner in which the method classifies gene sequences. Unlike most array-based approaches, which have at least one probe for each target sequence, OFRG uses a small set (~26 to 40) of probes to coordinately distinguish a much larger set of sequences. Probe sets are designed from training data, which are assemblages of all known rRNA genes within a desired group. OFRG analyses can be designed to identify a taxonomically narrow group of organisms, such as all *Clostridium* species, or to identify a larger group of organisms, such as all known bacteria. In addition, unlike other array-based methods, OFRG can also identify rRNA gene sequences that have not previously been described. This is accomplished through nucleotide sequence analysis of representative clones within clusters that do not associate with known sequences.

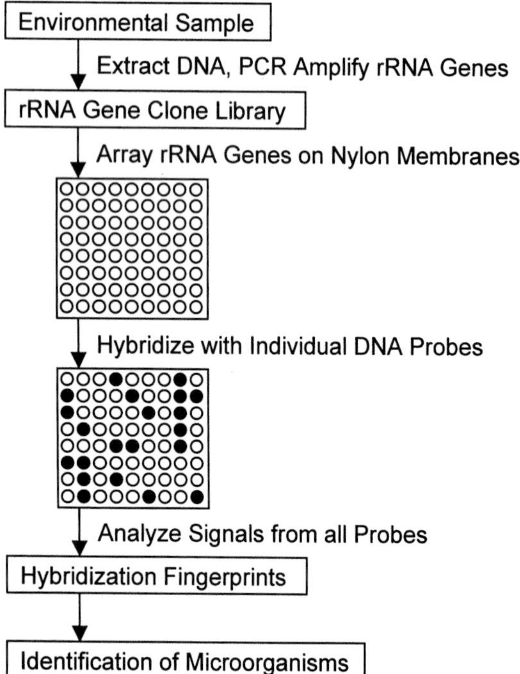

FIGURE 1 Experimental strategy for OFRG.

Although any approach to examine microbial community composition can be used in these studies (see chapter 53), it seems reasonable to conclude that thorough analyses may be needed to find the one or few functional organisms from the hundreds to thousands that inhabit most environments. We therefore recommend methods such as extensive nucleotide sequence analysis of rRNA gene clone libraries or array-based approaches such as OFRG.

Examination of All Microorganism Types: Broad-Range Primer Selection and Design

When designing experiments to identify microorganisms involved in specific in situ functions, one important factor may be the attempt to examine all types of microorganisms. Most efforts to examine the role of microorganisms in ecosystem functioning have focused on bacteria. However, it may be just as important to examine archaea, viruses, fungi, and other eukaryotic microorganisms. With the exception of viruses, most of these organisms can be examined by using broad-range PCR primers and probes, as described previously for archaea (7, 28, 41, 61), bacteria (41, 58, 68, 111), eukaryotes (27, 32, 41) and fungi (10, 14, 40, 53, 67, 91, 94, 116). Additionally, attempts have been made to identify primers and probes targeting all organisms possessing rRNA genes (58, 122), the so-called universal primers or probes.

Broad-Range Primer Selection and Design

Taxon-selective primers can be designed by locating DNA sequences that are conserved among the rRNA sequences within each group and which have few, if any, identical matches to rRNA sequences from unrelated taxonomic groups. However, since many sequences have not been identified, and since most of the sequences submitted to the databases were obtained with similar primer sets and therefore do not contribute new sequence data for these primer regions, such primers are unlikely to be entirely inclusive or selective. These primers should therefore be regularly reevaluated, and perhaps redesigned, as new data become available. One particularly useful source of new data may be non-PCR-based analyses, such as genome shotgun sequencing studies (108).

Given the conserved nature of rRNA sequences and the range of genotypic and physiological diversity that can exist among organisms possessing similar or identical small-subunit rRNA genes (46, 103), it may also be valuable to examine the more variable intergenic transcribed spacer (ITS) regions located between the rRNA genes (15, 19, 40, 109, 116). This is especially important for studies of eukaryotic microorganisms, as their rRNA genes typically possess less variation than those from prokaryotes.

Examining Activity Levels and Functional Genes

Since the number of organisms or of their rRNA genes may not always be related to their function in the environment, this three-phase approach may be modified to examine alternative hypotheses. For example, one hypothesis may be that microorganisms involved in a specific in situ function can be identified by correlating their activity levels with the levels of the specific functional parameter. In this case, instead of examining rRNA genes, one could examine levels of rRNA molecules. Similarly, the levels of specific functional genes could be correlated with in situ functional parameters.

Phase II: Quantitative Validation of Phase I rRNA Gene Correlates

The objective of phase II is to validate the rRNA gene correlates identified in phase I. Since most PCR-based methods produce depictions of microbial community compositions which likely do not accurately reflect the number and variety of rRNA genes in the environment (78, 99, 117), it is important to validate the phase I data with an independent quantitative analysis. In such an analysis, the abundances of microbial populations identified in phase I are assessed by a more targeted method (i.e., by using primers and probes specific for the populations of interest). This independent validation is not only an important component of the three-phase population approach, but it should also be utilized for most rRNA gene analyses where accurate quantitative data are needed. Several methods can be used for such analyses, including sequence-selective fluorescence in situ hybridization (29, 30), flow cytometry (82, 83), and quantitative PCR (see chapter 55 in this manual and sections 2.13 to 2.15 in the *Molecular Microbial Ecology Manual* [56]). In the sections below, we discuss experimental design considerations for sequence-selective PCR assays for rRNA gene analyses.

Sequence-Selective PCR

Experimental design considerations for sequence-selective rRNA gene PCR assays include (i) identification of sequence-selective regions of the rRNA gene, (ii) primer design, (iii) assay optimization and validation, and (iv) quantitative PCR. In the following sections, we use the term "target" or "target sequence(s)" to designate rRNA genes that have been selected for analysis.

As a note of caution, although it may be possible to identify regions of an rRNA gene that are unique based on database analyses, these will not necessarily lead to the development of a sequence-selective PCR assay. We discuss features of such experiments which may lead to nonselective amplification below, and we suggest strategies to cope with these potential problems.

Identification of Sequence-Selective Regions of rRNA Genes

Once a target sequence or group of target sequences has been identified, the next step is to identify a unique region within the selected sequence(s). Given the conserved nature of rRNA sequences, this can be a difficult task. In addition, even if such a sequence is identified, there is currently no way of knowing if it is truly unique, as not all rRNA sequences have been described. Because of these constraints, the objective is to find a region that is as selective as possible. This can be accomplished by finding a region (i) that has the fewest possible identical matches to other rRNA sequences, (ii) where the identical matches are taxonomically related to the target sequence, and/or (iii) where the organisms represented by the nontarget identical matches are unlikely to be found in the environment that will be assayed.

Several tools can be used to identify sequence-selective regions. These include the Probe Design and Probe Match algorithms in ARB (65) (http://www.arb-home.de/) and Primrose (http://www.cf.ac.uk/biosi/research/biosoft/) (4), a probe and primer design algorithm that utilizes the databases from the Ribosomal Database Project (http://rdp.cme.msu.edu/) (21). Another alternative is to perform a series of successive BLAST (National Center for Biotechnology Information [NCBI]) analyses (3). In this approach, the rRNA gene is divided into small segments (50 to 100 nucleotides in length), which are then individually analyzed in BLAST (NCBI) (3). This allows the identification of unique sequence regions, as defined in the previous paragraph. Smaller and smaller portions of these selective regions can be similarly analyzed until regions of the appropriate size (~15 to 25 nucleotides) and the greatest selectivity are identified.

When attempting to develop PCR primers for a group of related sequences, the first step is to identify regions that are conserved among the sequences within the group. This is typically accomplished by aligning the sequences and then obtaining a consensus sequence. Various commercial and free software can be used to perform this task. Once the consensus is obtained, sequence-selective regions can be identified by subjecting this sequence to the analyses described in the previous paragraph.

If regions with sufficient selectivity cannot be found with the aforementioned approaches, one alternative is to use the more variable ITS regions located between the rRNA genes. This can be accomplished in several ways. First, one can perform the initial phase I correlation studies with PCR primers that amplify these more variable regions (15, 19, 40, 109, 116). Alternatively, if a small-subunit rRNA gene has already been targeted for PCR analysis, its associated ITS sequence can be obtained, allowing PCR primers to be developed for this more variable region. This can be accomplished as follows. Figure 2 depicts a bacterial small-subunit rRNA gene, the ITS region, and a portion of the large-subunit rRNA gene. Arrows A and B in the figure represent 27F and 1492R, which are PCR primers targeting most bacterial rRNA genes (58). For this example, we will assume that an rRNA gene analysis using primers A and B has identified a target sequence. The first step in the process is to identify a region between primers A and B that is as selective as possible for the target sequence. This region can be located anywhere between primers A and B, but it should be a sufficient distance upstream of primer B to ensure that the target sequence is obtained in the subsequent analysis, as described below (PCR product delineated by primers C and D). The second step is to design a PCR primer C that anneals to this selective region, using the guidelines described below. The third step is to PCR amplify the DNA from which the original rRNA gene was obtained, using primers C and D, where primer D is a primer such as 23SR, which targets all bacterial large-subunit rRNA genes (15). The fourth step is to clone and sequence the amplicons and identify a clone possessing a small-subunit rRNA gene that is identical to that of the original targeted rRNA gene. The fifth step is to identify selective regions in the ITS portion of this clone.

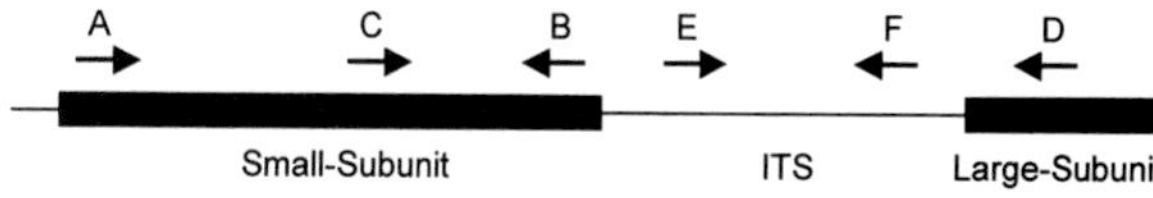

FIGURE 2 rRNA operon and positions of PCR primers.

The final step is to develop PCR primers E and F targeting this region, using the guidelines described below (see "Primer Design" and "Assay Optimization and Validation" below.

Since new nucleotide sequence data are continually being generated and deposited in the databases, the selectivities of targeted rRNA gene regions (and the resulting primers and probes) should occasionally be reevaluated.

Primer Design

Design considerations for developing primers for sequence-selective, rRNA gene-based PCR assays can be divided into the following topics: (i) standard PCR primer design considerations and (ii) sequence selectivity. Standard PCR primer design considerations have been described elsewhere (31, 36, 45) and are usually incorporated into primer design software.

The highly conserved nature of rRNA genes makes identifying sequence-selective PCR primers a considerable challenge. When attempting to design selective PCR primers for rRNA genes, one must frequently ensure that the primers are able to distinguish sequences differing by a single nucleotide. To optimize the ability of primers to distinguish such sequences, single-nucleotide mismatches should be located at the 3′-terminal nucleotides of the primers (31, 34, 72, 75, 118). However, because not all primer-template mismatches inhibit amplification in an equivalent manner, the nucleotide composition of the mismatches should also be considered. For example, G-T mismatches are more likely to produce PCR products than are G-A mismatches (31, 57).

Designing PCR primers with mismatches at the 3′ terminus can be done in a two-step process. First, the nucleotide positions with the greatest selectivities are identified. This can be accomplished by analyzing the selective regions of the rRNA sequence, as described above, with tools such as the Probe Match algorithm in ARB (65; http://www.arb-home.de/), the Probe Match algorithm from the Ribosomal Database Project (21; http://rdp.cme.msu.edu/), and BLAST (NCBI) (3). Once the most selective nucleotide positions are identified, the second step is to design the PCR primers. For this process, we have found that Primer Express (Applied Biosystems, Foster City, Calif.) is a particularly useful software package because it not only incorporates standard PCR primer design considerations, such as complementarity, secondary structure, and annealing temperature, but also enables the design of primers that anneal at specific locations. If multiple selective positions in close proximity to each other can be identified, they should all be placed as close to the 3' end of the primer as possible. The degree of sequence selectivity of the PCR assay will increase with the number of selective positions in the primers.

Assay Optimization and Validation

Once primers are developed, PCR assays need to be optimized and validated. Parameters that can be optimized include the annealing temperature, $MgCl_2$ concentration, primer concentration, template concentration, and cycle number, among others (31, 36, 45).

One common method for optimizing PCR assays is to perform a series of amplification reactions at various annealing temperatures while keeping all other reagents and parameters constant. Such reactions can be performed simultaneously with thermal cyclers possessing a temperature gradient function. The DNA template for these reactions is usually a plasmid vector containing the cloned PCR amplicon. Figure 3 shows an example of such an experiment. The resulting amplicons were electrophoresed in an agarose gel, stained with ethidium bromide, and visualized with UV light. In Fig. 3, the annealing temperatures (degrees Celsius) are indicated above the lanes. As the annealing temperature increases, the amount of PCR product diminishes and eventually is no longer visible. The optimum annealing temperature is typically the highest temperature that still produces a reasonable amount of PCR product. However, selection of the optimal temperature can vary depending on the requirements for each particular assay. For example, if selectivity is the most important factor and if large quantities of PCR product are not needed (e.g., in a real-time PCR assay where low detection limits are not needed), then the optimum temperature for this example may be 57°C. If both selectivity and large amounts of PCR product are needed (e.g., in a real-time PCR assay where low detection limits are needed or for gel-based visualization), then the optimum temperature may be 55 or 56°C.

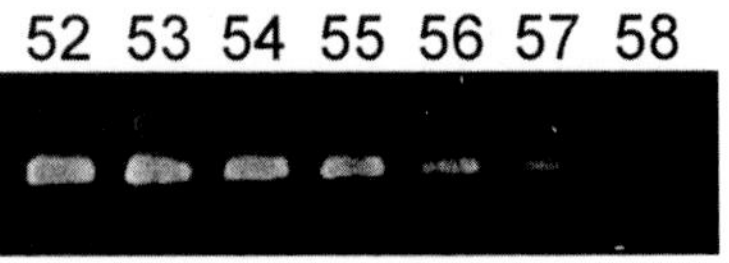

FIGURE 3 Effect of annealing temperature on a PCR assay.

Another parameter to be considered when determining the optimal annealing temperature is the template concentration. Figure 4 shows the results of two series of PCRs performed at various temperatures (57 to 66°C) with various dilutions (10^{-2} to 10^{-8}) of the template, which was a plasmid containing the cloned PCR amplicon. As the template concentration decreased, lower temperatures were needed to produce a product. Selection of the optimal annealing temperature is therefore dependent on the template concentration. In addition, Fig. 4 shows two sets of amplification reactions performed with the same PCR primers but with different template DNAs. For the left panels, the primers annealed to the template without a single mismatch. For the right panels, one of the primers was an exact match, while the other had a single mismatch at the 3′-terminal nucleotide. By comparing these panels, one can see that at an annealing temperature of 64°C, the primers produced a PCR product from the template containing the mismatch at the 10^{-2} dilution but did not produce a product from the exact match at the 10^{-6} and 10^{-8} dilutions. These results demonstrate the need to regularly clone and sequence PCR products from any "sequence-specific" PCR

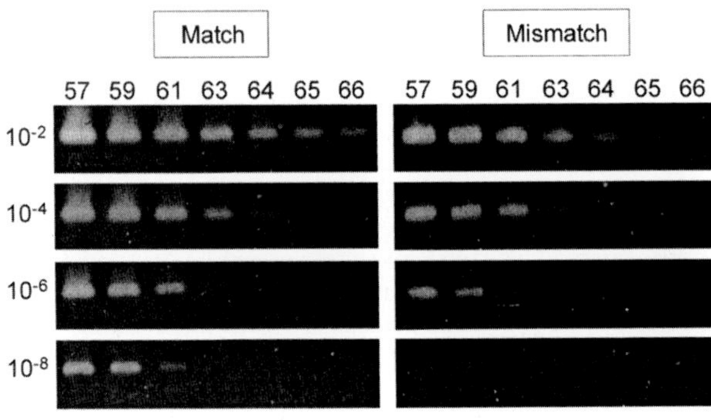

FIGURE 4 Effect of template concentration on mismatch discrimination of a PCR assay.

assay to ensure that results derived from such amplifications come from amplification of the intended target sequence(s). These data also suggest that assays providing additional selectivity, such as TaqMan assays using sequence-selective probes (63), may be useful (see chapter 55).

Although the DNA template for these optimization experiments is usually a plasmid vector containing the cloned PCR amplicon, other alternatives may provide more accurate results. Since the end goal is typically to develop PCR assays for environmental samples, it may be best to perform optimization experiments with DNAs extracted from such samples. This can be accomplished by using environmental DNA spiked with known concentrations of the target DNA or known to contain the target sequence. This allows the amplification reaction to be optimized in a system possessing parameters (template concentration and inhibitors) similar to those encountered when environmental DNA is used as the template. For such experimentation, a series of representative environmental samples should be examined.

PCR assays can be validated by using them to amplify DNAs extracted from different environmental samples and then cloning and sequencing several randomly selected clones from each sample. The assays can be considered selective if they exclusively amplify the target sequence. In addition, if the level of the target sequence is expected to correlate with a functional parameter, then the primers can be tested for this property as well.

Several other factors should also be considered when performing, optimizing, and/or validating sequence-selective rRNA gene-based PCR assays, as follows.

- Assays should be performed with the same thermal cycler, reagents, cycling parameters, and reaction tubes used for the optimization and validation experiments. If any of the parameters are changed, then the validation and perhaps optimization experiments should be repeated.
- It is good practice to regularly clone and sequence amplicons to validate such assays.
- When attempting to use a published protocol for a sequence-selective PCR assay, use the parameters described by the authors. In addition, validate the assay before using it, as numerous factors can vary between laboratories. Validation can be done as described above. In addition, it may also be useful to examine the selectivities of the primers before utilizing such assays. This can be done as described above (see "Identification of Sequence-Selective Regions of rRNA Genes" above).
- If PCR products are not produced from environmental DNA, several factors may be the cause, including (i) the absence of target DNA in the sample, (ii) the presence of amplification inhibitors in the sample, and/or (iii) an inappropriately optimized PCR assay (e.g., if the assay was optimized with the template at a much higher concentration than that found in the environmental samples, then the annealing temperature may have been set too high). Detecting inhibitors in environmental DNA can be done by PCR amplifying it with primers that usually produce a product, such as universal rRNA gene primers, or by spiking the DNA with a known amplicon and performing PCR with primers selective for that sequence. If samples contain inhibitors, PCR products may be obtained by using diluted template DNA or by further purification.

Quantitative PCR

Sequence-selective PCR assays can also be performed in a quantitative format. Several methods, along with their protocols and design considerations, are described in this manual (chapter 55) and in sections 2.13 to 2.15 in the *Molecular Microbial Ecology Manual* (56).

Phase III: Reintroduction of Organisms To Assess Their Functions

In phase III, the microorganisms represented by the rRNA gene correlates identified and validated in phases I and II are isolated and then reintroduced into the environment to assess their functions.

Microorganism Isolation

Once target rRNA gene sequences are identified, attempts to culture the organisms represented by these sequences can be made. For previously described species, taxon-selective media and culture methods have been published (5). Since many target organisms may not have been isolated previously, such work will likely require a certain degree of empirical manipulation. The utilization of several new media and techniques may be useful for isolating organisms that are difficult to culture (see chapter 53 and references 22, 50, 51, 85, and 121). Once isolates have been obtained, sequence-based verification can be conducted.

This general approach was utilized for the *H. schachtii* studies described in the introduction (76). In those studies, nematode cysts isolated from the most nematode-suppressive soils were crushed in water. A dilution series of this mixture was then prepared and plated on water and potato dextrose agar. After an incubation period, individual colonies were transferred to new media and subjected to sequence-selective PCR analysis. This process led to the successful isolation of both *H. schachtii*-associated fungi, i.e., *F. oxysporum* and *D. oviparasitica*. The α-proteobacterium identified in the aforementioned *H. schachtii* studies was isolated by use of a defined medium for *Rhizobium* (9). Subsequently, sequence-selective PCR was used to confirm the identities of the isolates.

It should be noted that bacterial isolates obtained by selecting single colonies from agar plates and fungal isolates obtained by the hyphal tip procedure or single spore isolations may contain more than one organism type. It may therefore be useful to subject isolates to at least two rounds of such purification procedures. All isolates should also be examined by rRNA gene analysis before being examined in functional analyses. This can be done by amplifying the rRNA genes from such isolates, using PCR primers with broad specificity, such as those targeting all rRNA genes, and then cloning and sequencing several randomly selected clones from each isolate.

Microorganism Reintroduction

Once the organisms have been isolated in pure culture, they can be reintroduced into the environment to assess their functions. Such investigations need to be customized for each particular function. However, experimental design considerations typically include (i) optimizing the in vitro culture conditions, (ii) optimizing the manner in which the organism is reintroduced, (iii) applying the organism at various population densities, including densities similar to those found in the originally investigated site, and (iv) simultaneously monitoring the population densities of the introduced organism(s) and measuring the functional parameter over time.

GENERAL PCR GUIDELINES

General guidelines for optimizing and performing PCR have been described previously (31, 36, 45). Several key considerations are listed below.

- Since cycles of freezing and thawing can damage reagents and template DNA, variation between assays can be minimized by storing and utilizing small aliquots of these components.
- Template DNA and PCR primers are purported to be more stable if stored in buffer instead of water. Storing dried DNA may also increase the shelf life.
- Use reagent master mixes to reduce variation between assays. Such mixes are usually comprised of all of the PCR reagents except the component being tested, which is frequently the template DNA.
- Perform positive and negative control reactions for every PCR. Positive control reactions usually consist of the reagent master mix and a template DNA that is known to be amplifiable. Negative control reactions usually consist of the reagent master mix, with the template DNA being replaced by the solution in which the template DNA is suspended (typically water or buffer).
- Purified PCR primers will produce more consistent results. This is because some of the primers produced by DNA oligonucleotide synthesis are not full length, the amounts of full-length primer may vary between synthesis reactions, and the optimal annealing temperature is affected by primer length.

OTHER CONSIDERATIONS

Other considerations, such as sampling, DNA extraction, and purification, are described in the *Molecular Microbial Ecology Manual* (56).

This work was supported by awards from the National Research Initiative of the USDA Cooperative State Research, Education and Extension Service (grant numbers 2003-35316-13824 and 2002-35107-12271), the NSF BDI Program, the University of California Integrated Pest Management program, and the University of California Center for Pest Management Research and Extension.

REFERENCES

1. **Adamczyk, J., M. Hesselsoe, N. Iversen, M. Horn, A. Lehner, P. H. Nielsen, M. Schloter, P. Roslev, and M. Wagner.** 2003. The isotope array, a new tool that employs substrate-mediated labeling of rRNA for determination of microbial community structure and function. *Appl. Environ. Microbiol.* **69:**6875–6887.
2. **Alabouvette, C., F. Rouxel, and J. Louvet.** 1979. Characteristics of fusarium wilt-suppressive soils and prospects for their utilization in biological control, p. 165–182. *In* B. Schippers and W. Gams (ed.), *Soil-Borne Plant Pathogens*. Academic Press, London, United Kingdom.
3. **Altschul, S. F., T. L. Madden, A. A. Schaffer, J. Zhang, Z. Zhang, W. Miller, and D. J. Lipman.** 1997. Gapped BLAST and PSI-BLAST: a new generation of protein database search programs. *Nucleic Acids Res.* **25:**3389–3402.
4. **Ashelford, K. E., A. J. Weightman, and J. C. Fry.** 2002. PRIMROSE: a computer program for generating and estimating the phylogenetic range of 16S rRNA oligonucleotide probes and primers in conjunction with the RDP-II database. *Nucleic Acids Res.* **30:**3481–3489.
5. **Atlas, R. M.** 2004. *Handbook of Microbiological Media*, 3rd ed. CRC Press, Boca Raton, Fla.
6. **Avrahami, S., W. Liesack, and R. Conrad.** 2003. Effects of temperature and fertilizer on activity and community structure of soil ammonia oxidizers. *Environ. Microbiol.* **5:**691–705.
7. **Barns, S. M., R. E. Fundyga, M. W. Jeffries, and N. R. Pace.** 1994. Remarkable archaeal diversity detected in a Yellowstone National Park hot spring environment. *Proc. Natl. Acad. Sci. USA* **91:**1609–1613.
8. **Beja, O., M. T. Suzuki, E. V. Koonin, L. Aravind, A. Hadd, L. P. Nguyen, R. Villacorta, M. Amjadi, C. Garrigues, S. B. Jovanovich, R. A. Feldman, and E. F. DeLong.** 2000. Construction and analysis of bacterial artificial chromosome libraries from a marine microbial assemblage. *Environ. Microbiol.* **2:**516–529.
9. **Bergersen, F. J.** 1961. The growth of rhizobium in synthetic media. *Aust. J. Biol.* **14:**349–360.
10. **Bock, M., M. Maiwald, R. Kappe, P. Nickel, and H. Naeher.** 1994. Polymerase chain reaction-based detection of dermatophyte DNA with a fungus-specific primer system. *Mycoses* **37:**79–84.
11. **Bonnet, R., A. Suau, J. Dore, G. R. Gibson, and M. D. Collins.** 2002. Differences in rDNA libraries of faecal bacteria derived from 10- and 25-cycle PCRs. *Int. J. Syst. Evol. Microbiol.* **52:**757–763.
12. **Borneman, J.** 1999. Culture-independent identification of microorganisms that respond to specified stimuli. *Appl. Environ. Microbiol.* **65:**3398–3400.
13. **Borneman, J., M. Chrobak, G. D. Vedova, A. Figueroa, and T. Jiang.** 2001. Probe selection algorithms with applications in the analysis of microbial communities. *Bioinformatics* **17:**S39–S48.
14. **Borneman, J., and R. J. Hartin.** 2000. PCR primers that amplify fungal rRNA genes from environmental samples. *Appl. Environ. Microbiol.* **66:**4356–4360.
15. **Borneman, J., and E. W. Triplett.** 1997. Molecular microbial diversity in soils from eastern Amazonia: evidence for unusual microorganisms and microbial population shifts associated with deforestation. *Appl. Environ. Microbiol.* **63:**2647–2653.
16. **Boschker, H. T. S., S. C. Nold, P. Wellsbury, D. Bos, W. De Graaf, R. Pel, R. J. Parkees, and T. E. Cappenberg.** 1998. Direct linking of microbial populations to specific biogeochemical processes by 13C-labelling of biomarkers. *Nature* **392:**801–805.
17. **Broadbent, P., and K. F. Baker.** 1975. Presented at the Biology and Control of Soil-Borne Plant Pathogens, Minneapolis, Minn.
18. **Brockhurst, M. A., P. B. Rainey, and A. Buckling.** 2004. The effect of spatial heterogeneity and parasites on the evolution of host diversity. *Proc. R. Soc. Lond. B.* **271:** 107–111.
19. **Cardinale, M., L. Brusetti, P. Quatrini, S. Borin, A. M. Puglia, A. Rizzi, E. Zanardini, C. Sorlini, C. Corselli, and D. Daffonchio.** 2004. Comparison of different primer sets for use in automated ribosomal intergenic spacer analysis of complex bacterial communities. *Appl. Environ. Microbiol.* **70:**6147–6156.
20. **Chandler, D. P., J. K. Fredrickson, and F. J. Brockman.** 1997. Effect of PCR template concentration on the composition and distribution of total community 16S rDNA clone libraries. *Mol. Ecol.* **6:**475–482.
21. **Cole, J. R., B. Chai, R. J. Farris, Q. Wang, S. A. Kulam, D. M. McGarrell, G. M. Garrity, and J. M. Tiedje.** 2005. The Ribosomal Database Project (RDP-II): sequences and tools for high-throughput rRNA analysis. *Nucleic Acids Res.* **33:**D294–D296.
22. **Connon, S. A., and S. J. Giovannoni.** 2002. High-throughput methods for culturing microorganisms in very-low-nutrient media yield diverse new marine isolates. *Appl. Environ. Microbiol.* **68:**3878–3885.

23. **Cook, R. J.** 1973. Influence of low plant and soil water potentials on diseases caused by soil-borne fungi. *Phytopathology* **63:**451–458.
24. **Cook, R. J., and K. F. Baker.** 1983. The nature and practice of biological control of plant pathogens. APS Press, St. Paul, Minn.
25. **Corless, C. E., M. Guiver, R. Borrow, V. Edwards-Jones, E. B. Kaczmarski, and A. J. Fox.** 2000. Contamination and sensitivity issues with a real-time universal 16S rRNA PCR. *J. Clin. Microbiol.* **38:**1747–1752.
26. **Crump, D. H., and B. R. Kerry.** 1987. Studies on the population dynamics and fungal parasitism of *Heterodera schachtii* in soil from a sugar beet monoculture. *Crop Prot.* **6:**49–55.
27. **Dawson, S. C., and N. R. Pace.** 2002. Novel kingdom-level eukaryotic diversity in anoxic environments. *Proc. Natl. Acad. Sci. USA* **99:**8324–8329.
28. **DeLong, E. F.** 1992. Archaea in coastal marine environments. *Proc. Natl. Acad. Sci. USA* **89:**5685–5689.
29. **DeLong, E. F., L. T. Taylor, T. L. Marsh, and C. M. Preston.** 1999. Visualization and enumeration of marine planktonic archaea and bacteria by using polyribonucleotide probes and fluorescent in situ hybridization. *Appl. Environ. Microbiol.* **65:**5554–5563.
30. **DeLong, E. F., G. S. Wickham, and N. R. Pace.** 1989. Phylogenetic stains: ribosomal RNA-based probes for the identification of single cells. *Science* **243:**1360–1363.
31. **Dieffenbach, C. W., and G. S. Dveksler.** 2003. *PCR Primer: a Laboratory Manual*, 2nd ed. Cold Spring Harbor Laboratory Press, Cold Spring Harbor, N.Y.
32. **Edgcomb, V. P., D. T. Kysela, A. Teske, A. D. Gomez, and M. L. Sogin.** 2002. Benthic eukaryotic diversity in the Guaymas Basin hydrothermal vent environment. *Proc. Natl. Acad. Sci. USA* **99:**7658–7662.
33. **Egert, M., and M. W. Friedrich.** 2005. Post-amplification Klenow fragment treatment alleviates PCR bias caused by partially single-stranded amplicons. *J. Microbiol. Methods* **61:**69–75.
34. **Ehlen, T., and L. Dubeau.** 1989. Detection of ras point mutations by polymerase chain-reaction using mutation-specific, inosine-containing oligonucleotide primers. *Biochem. Biophys. Res. Commun.* **160:**441–447.
35. **Eijsackers, H.** 2001. A future for soil ecology? Connecting the system levels: moving from genomes to ecosystems. Opening lecture to the XIII ICSZ, "Biodiversity of soil organisms and ecosystem functioning." *Eur. J. Soil Biol.* **37:**213–220.
36. **Erlich, H. A.** 1989. *PCR Technology: Principles and Applications for DNA Amplification.* Stockton Press, New York, N.Y.
37. **Farrelly, V., F. A. Rainey, and E. Stackebrandt.** 1995. Effect of genome size and *rrn* gene copy number on PCR amplification of 16S rRNA genes from a mixture of bacterial species. *Appl. Environ. Microbiol.* **61:**2798–2801.
38. **Franklin, R. B., and A. L. Mills.** 2003. Multi-scale variation in spatial heterogeneity for microbial community structure in an eastern Virginia agricultural field. *FEMS Microbiol. Ecol.* **44:**335–346.
39. **Garcia-Pichel, F., S. L. Johnson, D. Youngkin, and J. Belnap.** 2003. Small-scale vertical distribution of bacterial biomass and diversity in biological soil crusts from arid lands in the Colorado Plateau. *Microb. Ecol.* **46:**312–321.
40. **Gardes, M., and T. D. Bruns.** 1993. ITS primers with enhanced specificity for basidiomycetes—application to the identification of mycorrhizae and rusts. *Mol. Ecol.* **2:**113–118.
41. **Giovannoni, S. J., E. F. DeLong, G. J. Olsen, and N. R. Pace.** 1988. Phylogenetic group-specific oligodeoxynucleotide probes for identification of single microbial cells. *J. Bacteriol.* **170:**720–726.
42. **Gray, N. D., R. C. Hastings, S. K. Sheppard, P. Loughnane, D. Lloyd, A. J. McCarthy, and I. M. Head.** 2003. Effects of soil improvement treatments on bacterial community structure and soil processes in an upland grassland soil. *FEMS Microbiol. Ecol.* **46:**11–22.
43. **Horz, H.-P., M. T. Yimga, and W. Liesack.** 2001. Detection of methanotroph diversity on roots of submerged rice plants by molecular retrieval of *pmoA*, *mmoX*, *mxaF*, and 16S rRNA and ribosomal DNA, including *pmoA*-based terminal restriction fragment length polymorphism profiling. *Appl. Environ. Microbiol.* **67:**4177–4185.
44. **Hutchinson, C. M., M. E. McGiffen, Jr., H. D. Ohr, J. J. Sims, and J. O. Becker.** 2000. Efficacy of methyl iodide and synergy with chloropicrin for control of fungi. *Pest Manag. Sci.* **56:**413–418.
45. **Innis, M. A.** 1990. *PCR Protocols: a Guide to Methods and Applications*. Academic Press, San Diego, Calif.
46. **Jaspers, E., and J. Overmann.** 2004. Ecological significance of microdiversity: identical 16S rRNA gene sequences can be found in bacteria with highly divergent genomes and ecophysiologies. *Appl. Environ. Microbiol.* **70:**4831–4839.
47. **Jensen, M. A., and N. Straus.** 1993. Effect of PCR conditions on the formation of heteroduplex and single-stranded DNA products in the amplification of bacterial ribosomal DNA spacer regions. *PCR Methods Appl.* **3:**186–194.
48. **Jiang, L., and P. J. Morin.** 2004. Productivity gradients cause positive diversity-invasibility relationships in microbial communities. *Ecol. Lett.* **7:**1047–1057.
49. **Johnsen, K., C. S. Jacobsen, V. Torsvik, and J. Sorensen.** 2001. Pesticide effects on bacterial diversity in agricultural soils—a review. *Biol. Fert. Soils* **33:**443–453.
50. **Joseph, S. J., P. Hugenholtz, P. Sangwan, C. A. Osborne, and P. H. Janssen.** 2003. Laboratory cultivation of widespread and previously uncultured soil bacteria. *Appl. Environ. Microbiol.* **69:**7210–7215.
51. **Kaeberlein, T., K. Lewis, and S. S. Epstein.** 2002. Isolating "uncultivable" microorganisms in pure culture in a simulated natural environment. *Science* **296:**1127–1129.
52. **Kao, C. W., and W. H. Ko.** 1983. Nature of suppression of *Pythium splendens* in a pasture soil in south Kohala Hawaii USA. *Phytopathology* **73:**1284–1289.
53. **Kappe, R., C. Fauser, C. N. Okeke, and M. Maiwald.** 1996. Universal fungus-specific primer systems and group-specific hybridization oligonucleotides for 18S rDNA. *Mycoses* **39:**25–30.
54. **Kerry, B. R., D. H. Crump, and L. A. Mullen.** 1980. Parasitic fungi soil moisture and multiplication of the cereal cyst nematode, *Heterodera avenae*. *Nematologica* **26:**57–68.
55. **Kirk, J. L., L. A. Beaudette, M. Hart, P. Moutoglis, J. N. Klironomos, H. Lee, and J. T. Trevors.** 2004. Methods of studying soil microbial diversity. *J. Microbiol. Methods* **58:**169–188.
56. **Kowalchuk, G. A., F. J. de Bruijn, I. M. Head, A. D. L. Akkermans, and J. D. van Elsas.** 2004. *Molecular Microbial Ecology Manual*, 2nd ed. Kluwer Academic, Dordrecht, The Netherlands.
57. **Kwok, S., D. E. Kellogg, N. McKinney, D. Spasic, L. Goda, C. Levenson, and J. J. Sninsky.** 1990. Effects of primer template mismatches on the polymerase chain-reaction—human-immunodeficiency-virus type-1 model studies. *Nucleic Acids Res.* **18:**999–1005.
58. **Lane, D. J.** 1991. 16S/23S rRNA sequencing, p. 115–175. *In* E. Stackebrandt and M. Goodfellow (ed.), *Nucleic Acid Techniques in Bacterial Systematics*. Wiley, New York, N.Y.
59. **Lane, D. J., B. Pace, G. J. Olsen, D. A. Stahl, M. L. Sogin, and N. R. Pace.** 1985. Rapid determination of 16S ribosomal RNA sequences for phylogenetic analyses. *Proc. Natl. Acad. Sci. USA* **82:**6955–6959.
60. **Lee, N., P. H. Nielsen, K. H. Andreasen, S. Juretschko, J. L. Nielsen, K.-H. Schleifer, and M. Wagner.** 1999.

Combination of fluorescent in situ hybridization and microautoradiography—a new tool for structure-function analyses in microbial ecology. *Appl. Environ. Microbiol.* **65:**1289–1297.

61. **Lepp, P. W., M. M. Brinig, C. C. Ouverney, K. Palm, G. C. Armitage, and D. A. Relman.** 2004. Methanogenic Archaea and human periodontal disease. *Proc. Natl. Acad. Sci. USA* **101:**6176–6181.
62. **Liesack, W., H. Weyland, and E. Stackebrandt.** 1991. Potential risks of gene amplification by PCR as determined by 16S rDNA analysis of a mixed-culture of strict barophilic bacteria. *Microb. Ecol.* **21:**191–198.
63. **Livak, K. J., S. J. A. Flood, J. Marmaro, W. Giusti, and K. Deetz.** 1995. Oligonucleotides with fluorescent dyes at opposite ends provide a quenched probe system useful for detecting PCR product and nucleic-acid hybridization. *PCR Methods Appl.* **4:**357–362.
64. **Lowell, J. L., and D. A. Klein.** 2000. Heteroduplex resolution using T7 endonuclease I in microbial community analyses. *BioTechniques* **28:**676–681.
65. **Ludwig, W., O. Strunk, R. Westram, L. Richter, H. Meier, Yadhukumar, A. Buchner, T. Lai, S. Steppi, G. Jobb, W. Forster, I. Brettske, S. Gerber, A. W. Ginhart, O. Gross, S. Grumann, S. Hermann, R. Jost, A. Konig, T. Liss, R. Lussmann, M. May, B. Nonhoff, B. Reichel, R. Strehlow, A. Stamatakis, N. Stuckmann, A. Vilbig, M. Lenke, T. Ludwig, A. Bode, and K. H. Schleifer.** 2004. ARB: a software environment for sequence data. *Nucleic Acids Res.* **32:**1363–1371.
66. **Lukow, T., P. F. Dunfield, and W. Liesack.** 2000. Use of the T-RFLP technique to assess spatial and temporal changes in the bacterial community structure within an agricultural soil planted with transgenic and non-transgenic potato plants. *FEMS Microbiol. Ecol.* **32:**241–247.
67. **Makimura, K., S. Y. Murayama, and H. Yamaguchi.** 1994. Detection of a wide range of medically important fungi by the polymerase chain reaction. *J. Med. Microbiol.* **40:**358–364.
68. **Marchesi, J. R., T. Sato, A. J. Weightman, T. A. Martin, J. C. Fry, S. J. Hiom, and W. G. Wade.** 1998. Design and evaluation of useful bacterium-specific PCR primers that amplify genes coding for bacterial 16S rRNA. *Appl. Environ. Microbiol.* **64:**795–799.
69. **McSpadden Gardener, B. B., D. V. Mavrodi, L. S. Thomashow, and D. M. Weller.** 2001. A rapid polymerase chain reaction-based assay characterizing rhizosphere populations of 2,4-diacetylphloroglucinol-producing bacteria. *Phytopathology* **91:**44–54.
70. **McSpadden-Gardener, B. B., and D. M. Weller.** 2001. Changes in populations of rhizosphere bacteria associated with take-all disease of wheat. *Appl. Environ. Microbiol.* **67:**4414–4425.
71. **Murray, R. E., Y. S. Feig, and J. M. Tiedje.** 1995. Spatial heterogeneity in the distribution of denitrifying bacteria associated with denitrification activity zones. *Appl. Environ. Microbiol.* **61:**2791–2793.
72. **Nichols, W. C., J. J. Liepnieks, V. A. McKusick, and M. D. Benson.** 1989. Direct sequencing of the gene for Maryland German familial amyloidotic polyneuropathy type-Ii and genotyping by allele-specific enzymatic amplification. *Genomics* **5:**535–540.
73. **Nicol, G. W., L. A. Glover, and J. I. Prosser.** 2003. Spatial analysis of archaeal community structure in grassland soil. *Appl. Environ. Microbiol.* **69:**7420–7429.
74. **Nunan, N., K. Wu, I. M. Young, J. W. Crawford, and K. Ritz.** 2002. In situ spatial patterns of soil bacterial populations, mapped at multiple scales, in an arable soil. *Microb. Ecol.* **44:**296–305.
75. **Okayama, H., D. T. Curiel, M. L. Brantly, M. D. Holmes, and R. G. Crystal.** 1989. Rapid, nonradioactive detection of mutations in the human genome by allele-specific amplification. *J. Lab. Clin. Med.* **114:**105–113.
76. **Olatinwo, R., B. Yin, J. O. Becker, and J. Borneman.** 2006. Suppression of the plant-parasitic nematode *Heterodera schachtii* by the fungus *Dactylella oviparasitica*. *Phytopathology* **96:**111–114.
77. **Ouverney, C. C., and J. A. Fuhrman.** 1999. Combined microautoradiography-16S rRNA probe technique for determination of radioisotope uptake by specific microbial cell types in situ. *Appl. Environ. Microbiol.* **65:**1746–1752.
78. **Qiu, X., L. Wu, H. Huang, P. Mcdonel, A. Palumbo, J. Tiedje, and J. Zhou.** 2001. Evaluation of PCR-generated chimeras, mutations, and heteroduplexes with 16S rDNA gene-based cloning. *Appl. Environ. Microbiol.* **67:**880–887.
79. **Radajewski, S., P. Ineson, N. R. Parekh, and J. C. Murrell.** 2000. Stable-isotope probing as a tool in microbial ecology. *Nature* **403:**646–649.
80. **Rainey, F. A., N. Ward, L. I. Sly, and E. Stackebrandt.** 1994. Dependence on the taxon composition of clone libraries for PCR amplified, naturally occurring 16S rDNA, on the primer pair and the cloning system used. *Experientia* **50:**796–797.
81. **Reysenbach, A.-L., L. J. Giver, G. S. Wickham, and N. R. Pace.** 1992. Differential amplification of rRNA genes by polymerase chain reaction. *Appl. Environ. Microbiol.* **58:**3417–3418.
82. **Rice, J., M. A. Sleigh, P. H. Burkill, G. A. Tarran, C. D. O'Connor, and M. V. Zubkov.** 1997. Flow cytometric analysis of characteristics of hybridization of species-specific fluorescent oligonucleotide probes to rRNA of marine nanoflagellates. *Appl. Environ. Microbiol.* **63:**938–944.
83. **Rigottier-Gois, L., V. Rochet, N. Garrec, A. Suau, and J. Dore.** 2003. Enumeration of *Bacteroides* species in human faeces by fluorescent in situ hybridisation combined with flow cytometry using 16S rRNA probes. *Syst. Appl. Microbiol.* **26:**110–118.
84. **Roslev, P., N. Iversen, and K. Henriksen.** 1998. Direct fingerprinting of metabolically active bacteria in environmental samples by substrate specific radiolabelling and lipid analysis. *J. Microbiol. Methods* **31:**99–111.
85. **Sait, M., P. Hugenholtz, and P. H. Janssen.** 2002. Cultivation of globally distributed soil bacteria from phylogenetic lineages previously only detected in cultivation-independent surveys. *Environ. Microbiol.* **4:**654–666.
86. **Scher, F. M., and R. Baker.** 1980. Mechanism of biological control in a fusarium suppressive soil. *Phytopathology* **70:**412–417.
87. **Schmidt, T. M., B. Pace, and N. R. Pace.** 1991. Detection of DNA contamination in Taq polymerase. *BioTechniques* **11:**176–177.
88. **Seymour, J. R., J. G. Mitchell, L. Pearson, and R. L. Waters.** 2000. Heterogeneity in bacterioplankton abundance from 4.5 millimetre resolution sampling. *Aquat. Microb. Ecol.* **22:**143–153.
89. **Shipton, P. J., R. J. Cook, and J. W. Sitton.** 1973. Occurrence and transfer of a biological factor in soil that suppresses take-all of wheat in eastern Washington. *Phytopathology* **63:**511–517.
90. **Sievert, S. M., T. Brinkhoff, G. Muyzer, W. Ziebis, and J. Kuever.** 1999. Spatial heterogeneity of bacterial populations along an environmental gradient at a shallow submarine hydrothermal vent near Milos Island (Greece). *Appl. Environ. Microbiol.* **65:**3834–3842.
91. **Simon, L., M. Lalonde, and T. D. Bruns.** 1992. Specific amplification of 18S fungal ribosomal genes from vesicular-arbuscular endomycorrhizal fungi colonizing roots. *Appl. Environ. Microbiol.* **58:**291–295.
92. **Sliwinski, M. K., and R. M. Goodman.** 2004. Spatial heterogeneity of crenarchaeal assemblages within mesophilic

soil ecosystems as revealed by PCR-single-stranded conformation polymorphism profiling. *Appl. Environ. Microbiol.* **70:**1811–1820.

93. **Smiley, R. W., and R. J. Cook.** 1973. Relationship between take-all of wheat and rhizosphere pH in soils fertilized with ammonium vs nitrate nitrogen. *Phytopathology* **63:**882–890.

94. **Smit, E., P. Leeflang, B. Glandorf, J. D. van Elsas, and K. Wernars.** 1999. Analysis of fungal diversity in the wheat rhizosphere by sequencing of cloned PCR-amplified genes encoding 18S rRNA and temperature gradient gel electrophoresis. *Appl. Environ. Microbiol.* **65:**2614–2621.

95. **Speksnijder, A., G. A. Kowalchuk, S. De Jong, E. Kline, J. R. Stephen, and H. J. Laanbroek.** 2001. Microvariation artifacts introduced by PCR and cloning of closely related 16S rRNA gene sequences. *Appl. Environ. Microbiol.* **67:**469–472.

96. **Stahl, D. A., B. Flesher, H. R. Mansfield, and L. Montgomery.** 1988. Use of phylogenetically based hybridization probes for studies of ruminal microbial ecology. *Appl. Environ. Microbiol.* **54:**1079–1084.

97. **Sutherland, J. B., and R. J. Cook.** 1980. Effects of chemical and heat treatments on ethylene production in soil. *Soil Biol. Biochem.* **12:**357–362.

98. **Suzuki, M., M. S. Rappe, and S. J. Giovannoni.** 1998. Kinetic bias in estimates of coastal picoplankton community structure obtained by measurements of small-subunit rRNA gene PCR amplicon length heterogeneity. *Appl. Environ. Microbiol.* **64:**4522–4529.

99. **Suzuki, M. T., and S. J. Giovannoni.** 1996. Bias caused by template annealing in the amplification of mixtures of 16S rRNA genes by PCR. *Appl. Environ. Microbiol.* **62:**625–630.

100. **Swidsinski, A., V. Loening-Baucke, H. Lochs, and L. P. Hale.** 2005. Spatial organization of bacterial flora in normal and inflamed intestine: a fluorescence in situ hybridization study in mice. *World J. Gastroenterol.* **11:** 1131–1140.

101. **Teng, F., F. Hsu, I. Peterson, K. E. Cardon, G. Caponigro, and A. Kamb.** 2001. Template selection during manipulation of complex mixtures by PCR. *BioTechniques* **30:**868–877.

102. **Thompson, J. R., L. A. Marcelino, and M. F. Polz.** 2002. Heteroduplexes in mixed-template amplifications: formation, consequence and elimination by 'reconditioning PCR.' *Nucleic Acids Res.* **30:**2083–2088.

103. **Thompson, J. R., S. Pacocha, C. Pharino, V. Klepac-Ceraj, D. E. Hunt, J. Benoit, R. Sarma-Rupavtarm, D. L. Distel, and M. F. Polz.** 2005. Genotypic diversity within a natural coastal bacterioplankton population. *Science* **307:**1311–1313.

104. **Urbach, E., K. L. Vergin, and S. J. Giovannoni.** 1999. Immunochemical detection and isolation of DNA from metabolically active bacteria. *Appl. Environ. Microbiol.* **65:**1207–1213.

105. **Valinsky, L., G. Della Vedova, T. Jiang, and J. Borneman.** 2002. Oligonucleotide fingerprinting of rRNA genes for analysis of fungal community composition. *Appl. Environ. Microbiol.* **68:**5999–6004.

106. **Valinsky, L., G. Della Vedova, A. J. Scupham, S. Alvey, A. Figueroa, B. Yin, J. Hartin, M. Chrobak, D. E. Crowley, T. Jiang, and J. Borneman.** 2002. Analysis of bacterial community composition by oligonucleotide fingerprinting of rRNA genes. *Appl. Environ. Microbiol.* **68:**3243–3250.

107. **Valinsky, L., A. J. Scupham, G. D. Vedova, Z. Liu, A. Figueroa, K. Jampachaisri, B. Yin, E. Bent, R. Mancini-Jones, J. Press, T. Jiang, and J. Borneman.** 2004. Oligonucleotide fingerprinting of ribosomal RNA genes (OFRG), p. 569–585. *In* G. A. Kowalchuk, F. J. de Bruijn, I. M. Head, A. D. L. Akkermans, and J. D. van Elsas (ed.), *Molecular Microbial Ecology Manual*, 2nd ed. Kluwer Academic Publishers, Dordrecht, The Netherlands.

108. **Venter, J. C., K. Remington, J. F. Heidelberg, A. L. Halpern, D. Rusch, J. A. Eisen, D Wu, I. Paulsen, K. E. Nelson, W. Nelson, D. E. Fouts, S. Levy, A. H. Knap, M.W. Lomas, K. Nealson, O. White, J. Peterson, J. Hoffman, R. Parsons, H. Baden-Tillson, C. Pfannkoch, Y. H. Rogers, and H. O. Smith.** 2004. Environ-mental genome shotgun sequencing of the Sargasso Sea. *Science* **304:**66–74.

109. **Viaud, M., A. Pasquier, and Y. Brygoo.** 2000. Diversity of soil fungi studied by PCR-RFLP of ITS. *Mycol. Res.* **104:**1027–1032.

110. **Wang, G. C.-Y., and Y. Wang.** 1996. The frequency of chimeric molecules as a consequence of PCR co-amplification of 16S rRNA genes from different bacterial species. *Microbiology* **142:**1107–1114.

111. **Watanabe, K., Y. Kodama, and S. Harayama.** 2001. Design and evaluation of PCR primers to amplify bacterial 16S ribosomal DNA fragments used for community fingerprinting. *J. Microbiol. Methods* **44:**253–262.

112. **Weller, D. M., J. M. Raaijmakers, B. B. McSpadden-Gardener, and L. S. Thomashow.** 2002. Microbial population responsible for specific soil suppressiveness to plant pathogens. *Annu. Rev. Phytopathol.* **40:**309–348.

113. **Westphal, A., and J. O. Becker.** 1999. Biological suppression and natural population decline of *Heterodera schachtii* in a California field. *Phytopathology* **89:**434–440.

114. **Westphal, A., and J. O. Becker.** 2001. Components of soil suppressiveness against *Heterodera schachtii. Soil Biol. Biochem.* **33:**9–16.

115. **Westphal, A., and J. O. Becker.** 2000. Transfer of biological soil suppressiveness against *Heterodera schachtii. Phytopathology* **90:**401–406.

116. **White, T. J., T. Bruns, S. Lee, and J. W. Taylor.** 1990. Amplification and direct sequencing of fungal ribosomal RNA genes for phylogenetics, p. 315–322. *In* M. A. Innis, D. H. Gelfand, J. J. Sninsky, and T. J. White (ed.), *PCR Protocols: a Guide to Methods and Applications*. Academic Press, Inc., New York, N.Y.

117. **Wintzingerode, F. V., U. B. Goebel, and E. Stackebrandt.** 1997. Determination of microbial diversity in environmental samples: pitfalls of PCR-based rRNA analysis. *FEMS Microbiol. Rev.* **21:**213–229.

118. **Wu, D. Y., L. Ugozzoli, B. K. Pal, and R. B. Wallace.** 1989. Allele-specific enzymatic amplification of beta-globin genomic DNA for diagnosis of sickle-cell anemia. *Proc. Natl. Acad. Sci. USA* **86:**2757–2760.

119. **Yin, B., L. Valinsky, X. Gao, J. O. Becker, and J. Borneman.** 2003. Bacterial rRNA genes associated with soil suppressiveness against the plant-parasitic nematode *Heterodera schachtii. Appl. Environ. Microbiol.* **69:**1573–1580.

120. **Yin, B., L. Valinsky, X. Gao, J. O. Becker, and J. Borneman.** 2003. Identification of fungal rDNA associated with soil suppressiveness against *Heterodera schachtii* using oligonucleotide fingerprinting. *Phytopathology* **93:** 1006–1013.

121. **Zengler, K., G. Toledo, M. Rappe, J. Elkins, E. J. Mathur, J. M. Short, and M. Keller.** 2002. Cultivating the uncultured. *Proc. Natl. Acad. Sci. USA* **99:**15681–15686.

122. **Zheng, D., E. W. Alm, D. A. Stahl, and L. Raskin.** 1996. Characterization of universal small-subunit rRNA hybridization probes for quantitative molecular microbial ecology studies. *Appl. Environ. Microbiol.* **62:** 4504–4513.

Mobile Gene Elements in Environmental Microbial Communities

DIANA R. NEMERGUT, TAMAR BARKAY, AND JONNA COOMBS

62

INTRODUCTION

> Bacteria . . . in spite of occasional reports of a sexual cycle, appear to be not only wholly asexual but pre-mitotic . . . We must, in fact, expect that the processes of variation, heredity, and evolution in bacteria are quite different from the corresponding processes in multicellular organisms. But their secret has not yet been unraveled.
>
> Julian Huxley, *Evolution: The Modern Synthesis*, p. 131–132, 1942

The ability of bacteria to acquire, express, and maintain foreign DNA was documented in the early days of molecular biology. The first horizontal gene transfer (HGT) experiment was performed in 1928 when Fred Griffith mixed a heat-killed infectious strain of *Streptococcus pneumoniae* (smooth-colony forming [S]) with a live, avirulent strain (rough-colony forming [R]) and witnessed the "transformation" of the R strain into a smooth, pathogenic strain (41). Then, in 1944, Avery, MacLeod, and McCarty treated heat-killed cellular extracts with DNase and demonstrated that DNA was the "transforming principle" (4). Early documentations of antibiotic resistance gene transfer (53, 108) and the popularization of the endosymbiotic theory describing the origins of the mitochondrion and the chloroplast in the 1970s (60) highlighted the tremendous adaptive and potential evolutionary significance of HGT. However, the true importance of HGT was not fully appreciated until the genomics era. Analysis of many whole microbial genome sequences reveals that significant, but variable, amounts (e.g., ~2 to 25%) of DNA can be attributed to recent horizontal transfer events in a variety of organisms (52, 71, 110). Indeed, HGT may be a significant enough evolutionary force to drive different modes of speciation in microorganisms and macroorganisms (61). Previously thought of as an evolutionary exception, HGT is now regarded as a fundamental force shaping microbial genomes.

The process of microbial horizontal gene transfer has many societal consequences—both positive and negative. On the positive side, an understanding of the mechanisms and controls over gene transfer gave rise to the field of genetic engineering (20), and with it a suite of technological advances. Cloning technology has led to the development of a wide diversity of genetically modified strains of bacteria, plants, and animals with a variety of new and potentially beneficial traits. However, on the negative side, the horizontal transfer of these engineered genes in the environment is of concern, for example, the exchange of genes that render plants herbicide tolerant. In addition, one of the most important and highly publicized negative consequences of HGT is the spread of antibiotic resistance determinants among pathogenic bacteria. Indeed, HGT has led to a dangerous situation worldwide, in which a significant number of bacteria are resistant to a variety of antibiotics and, in some cases, the "last lines" of defense are gone (65). HGT also affects the evolution of infectious diseases via the dissemination and exchange of pathogenic determinants. Therefore, an understanding of the dynamics and controls over genetic exchange in natural communities is important for a wide variety of biotechnological and public health reasons.

The spread of genes between bacteria also poses unique challenges to the study of microbial ecology, further complicating issues associated with the inability to culture a large percentage of environmental microorganisms. Here, species diversity is typically catalogued in environmental samples using 16S rRNA gene libraries (47, 78), and these sequence-based views of species diversity are combined with phenotypic information from culture-based studies in an attempt to predict the functional attributes of organisms in the environment (for an example, see reference 58). These efforts to link phylogenetic and phenotypic information assume that functional traits are inherited vertically. However, many genes, including those encoding phenotypes with ecosystem level significance (e.g., nitrogen fixation [15, 81], denitrification [94, 97], and sulfate reduction [35]) are subject to horizontal transfer.

Despite these obvious implications for both science and society, there is still much to learn about the mechanisms, regulation, and frequency of horizontal gene transfer in the environment. This chapter primarily concentrates on characteristics of the different types of mobile elements that facilitate HGT in bacteria. We begin with a discussion of the mechanisms for HGT; then, we describe the tremendous variation both within and between different classes of mobile gene transfer elements; and we end with a brief description of the techniques for studying horizontal gene transfer in environmental microbial communities.

How Does HGT Happen?

HGT is facilitated through some combination of the activity of mobile gene elements (MGEs) and/or host and recipient cellular enzyme systems (88). In the most general terms, the process of horizontal gene transfer requires several, distinct steps (Fig. 1). First, the nucleic acid molecule (DNA or RNA) must be prepared for transfer. This preparation can be as tightly regulated as viral particle replication and packaging, or as random as the release of genomic DNA during cell lysis. Second, the genetic information needs to be transferred to the recipient cell. This step can occur through a variety of mechanisms: cells can take up "naked" DNA (transformation), or the nucleic acid can be shuttled into organisms through physical contact with phages (transduction) or other cells (conjugation). Third, the nucleic acid molecule must enter the recipient cell, through either nonspecific or specific means. Fourth, the DNA or RNA molecule needs to be recognized by the host cell replication machinery, either as an extrachromosomal element or via integration into the genome. Finally, for the gene transfer event to be significant over evolutionary timescales, the transferred DNA must be stably maintained in the genome, for example, by conferring a selective advantage.

Mobile gene elements are not required for HGT, but they can facilitate the process, and thus are often targets of HGT studies. Several cellular enzyme systems and proteins, from both the donor and recipient cells, can affect successful HGT. These can include homologous recombination enzymes, secretion pathways for DNA export, and DNA and phage binding proteins, as well as enzymes to protect cells from the invasion of foreign DNA. However, these are beyond the scope of this review; instead, the reader is directed to other recent reviews (18, 26, 111). This chapter primarily focuses on the characteristics of the MGEs themselves. There are five basic classes of MGEs: phages, plasmids, transposons, integrons, and integrative conjugative elements (ICEs). For each class of MGE, there are few hard and fast rules, and below we discuss generalities and exceptions for each element, focusing on their mechanism(s) of transfer, and the potential phylogenetic distances over which they can transfer DNA. In addition, we emphasize the mosaic nature of these elements, particularly within the newly described ICEs, which further blur the distinctions between plasmids, phages, transposons, and integrons.

PHAGES

Bacterial viruses (phages) were the first type of MGE to be discovered (100). The study of these elements has played a central role in unraveling the molecular principles governing all of life, and the first whole genome sequenced was that of the phage ΦX174. There are an estimated 10^{31} phage particles on Earth—10^{8} particles per ml of seawater (8) and 10^{7} to 10^{8} particles per g of soil (3). These elements have dramatic impacts on nutrient cycling and are associated with decreased primary production and increased bacterial mortality in marine systems (96). It is estimated that phages catalyze an HGT event about 20 million billion times per second in the ocean alone and thus also play major roles in microbial evolution (14). Their actual impact on evolution is difficult to quantify; however, most of the microbial genomes sequenced to date contain phage-related genes.

Phages vary widely in genome size (from ~2,400 bp for bacteriophage L5 of *Leuconostoc oenos* to 670 kb for *Bacillus megaterium* phage G) and gene content. Their genomes can include a variety of genes which code for proteins directly involved in phage replication and packaging, encoding activities like DNA replication, insertion, excision, and protein envelope formation. In addition, phages can transport a mosaic of other genes not directly involved in phage processes per se, which confer a diverse array of phenotypes ranging from photosynthesis to pathogenicity. Although it is often stated that these genes are "nonessential," their frequent occurrence in some types of phages implies that they

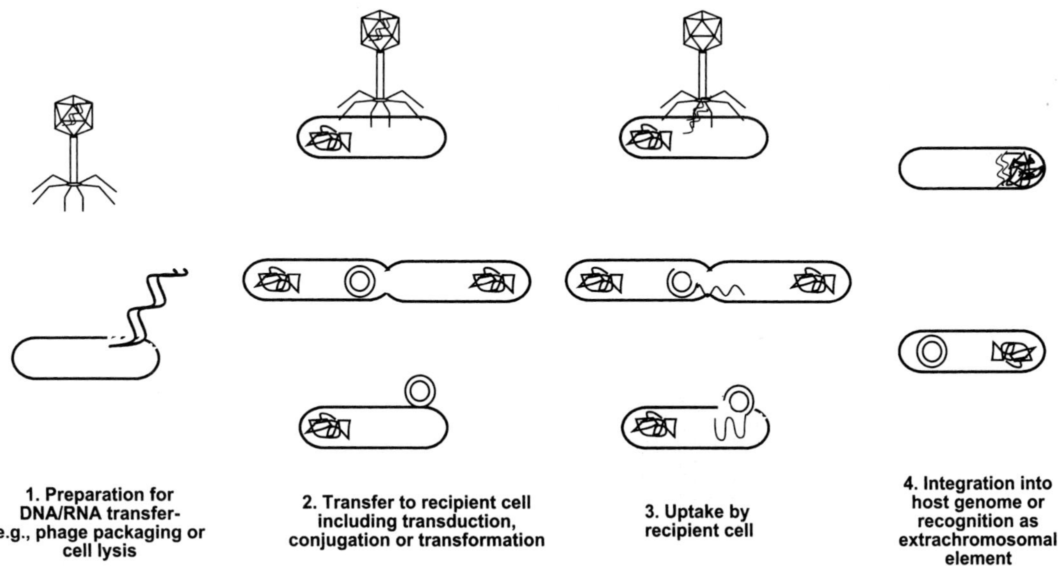

FIGURE 1 Four steps involved in successful nucleic acid transfer. (1) DNA/RNA molecules are prepared for transfer. (2) The genetic information is transferred to the recipient cell, through either transduction, conjugation, or transformation. (3) The nucleic acid molecule enters the recipient cell. (4) The DNA or RNA molecule is recognized by the host cell replication machinery, either via integration into the genome or as an extrachromosomal element.

may indeed confer a selective advantage, albeit one not directly related to phage functions. Because of their extraordinary diversity in genome size, gene content, and host range, the classification of phages has been problematic. In addition, unlike cellular organisms, phages lack a universal gene (e.g., small subunit rRNA gene) that can be used to reconstruct evolutionary history and define related viral clusters. The classic taxonomic system for describing phage relationships is based primarily on the type of nucleic acid molecule, as well as the size and shape of the capsid. More recently, a classification system based on the proteomic overlap of individual phage has been proposed, with 13 supported clades (83).

In general, viruses act by adsorbing to cell surfaces, injecting nucleic acids into host cells, inducing expression and replication of nucleic acids, and promoting viron assembly and release (for a recent review, see reference 109). The process of generalized transduction can occur during the phage life cycle, in which viruses also package some amount of host DNA prior to viron release, thereby shuttling bacterial DNA between different organisms. Indeed, Beumer and Robinson recently showed that the broad-host-range generalized phage SN-T could acquire 16S rRNA genes from a wide phylogenetic diversity of bacteria through generalized transduction, which has many interesting implications for microbial ecology (9). The phylogenetic distances crossed by phages, as well as the extent and frequency of phage-mediated gene transfer, vary dramatically from phage to phage and are determined by an assortment of environmental, host, and virus-specific factors. For example, viral capsid proteins interact with a variety of host molecules, determining host specificity, and in some cases, these interactions are specific enough to distinguish between subspecies of bacteria (for an example, see reference 57). Further interactions between phage attachment (*attP*) sequences on bacterial chromosomes and the phage genomes as well as the host-cell-encoded integration host factor (IHF) proteins may also determine host-phage specificity. On the other end of the spectrum, broad-host-range phages can infect different species within the same genus and even different genera of bacteria (for an example, see reference 9), and these viruses therefore have the potential to transfer genes over large phylogenetic distances. The actual specificity of the phages in environmental samples is unknown because of difficulties inherent in culturing microorganisms and phage but remains an important question with regard to the potential distances over which genes can travel via transduction.

Another factor which may regulate the impact of phages on microbial evolution is that viruses exhibit a variety of life stages, each with different implications for horizontal gene transfer (reviewed in reference 109). For example, in lysogenic stages, the phage genome remains in the host cell, either as an independent molecule or integrated into the host genome, and reproduces along with the host genome until the lytic stage is induced. In the lytic cycle, the phage directs the host to produce new phages, which are released upon cell lysis. A number of in-between phases also exist, including "pseudolysogenic" stages, in which phages multiply in a fraction of the bacterial population, and "chronic infections," in which phages are constantly produced but released by budding and therefore do not kill the host cell. The factors that regulate phage life cycles have been well studied in some phages and range from nutrient stress to temperature shock, but these features vary between organisms, phages, and environments. An understanding of the regulation of their life cycles is important for an appreciation of how phages can contribute to microbial HGT, because the balance of the lysogeny to lytic cycles likely influences the rate at which phages effect gene transfer to and between bacteria.

PLASMIDS

In 1952, Joshua Lederberg coined the term "plasmid" to describe an "extrachromosomal heredity determinant" (54). Over 5 decades later, we are still uncovering the true diversity of these elements—they can be both linear and circular molecules, they range in size from 1,200 bp to more than 209 kb, and they carry a wide variety of genes conferring phenotypes ranging from the degradation of PCB to nitrogen fixation. In addition, they can be transferred between cells via conjugation, transduction, or transformation processes. Despite the remarkable variation in their biology, one feature appears to unite all plasmids: they are elements that are capable of replicating independently from the host chromosome, albeit with varying degrees of assistance from host-encoded proteins. These elements have been found in every microbial community examined, and some plasmids can facilitate gene transfer between the different domains of life; thus, they are likely to play major roles in microbial HGT (89).

Plasmids vary dramatically in the percentage of loci that they dedicate to replication and transfer processes. At a minimum, all plasmids contain an origin of replication region or *ori*, which is a *cis*-acting sequence that determines the site of replication initiation. Many plasmids also include genes encoding replication (Rep) proteins, which bind the *ori* and generate a nucleoprotein replication initiation complex that then recruits other host cell-encoded replication machinery (for a review, see reference 25). In addition, this region may also contain an origin of transfer (*oriT*) or the determinants for self-transfer (e.g., *tra* genes, which include the genes for pilus assembly) (for a review, see reference 59) or for mobilization (e.g., *mob* genes) (for a review, see reference 33), which allow plasmids to be transferred with the help of the conjugation machinery of the host cell or another plasmid.

The *ori* region usually contains several other *cis*- and *trans*-acting factors that play central roles in determining the copy number of the plasmid within a cell. This region therefore determines the "incompatibility group" of different plasmids, so-called because no more than one type of plasmid from each group can coexist within a cell. Therefore, if two plasmids carry identical *ori* regions, the cellular machinery cannot differentiate between them, and they are very unlikely to both be stably inherited. Factors involved in determining the incompatibility group include iterons, short, tandemly repeated sequences commonly found in the *ori* regions. Iterons can bind Rep proteins, and in high concentrations of both Rep protein and plasmid DNA, this interaction acts to turn off replication by "handcuffing" Rep proteins between plasmid molecules (25). In addition, the *ori* region may encode *par/sop* loci which contain both the *cis*- and *trans*-acting factors required for proper plasmid partitioning between dividing cells (37).

The extremely specific interactions between the *ori*-encoded factors (both *cis*- and *trans*-acting) and the host cell replication machinery determine the plasmid host range and therefore the phylogenetic range of microbes over which certain plasmids can exist as an extrachromosomal element. Obviously, the host range has a major impact on the evolutionary role of plasmids, because it determines the phylogenetic distances over which plasmids, and therefore plasmid-associated genes, can be transferred. In some cases, the plasmid-host cell associations are very specific, and plasmids cannot be transferred across species boundaries (for an example, see reference 66). However, some plasmids have "broad host ranges" and can be moved between different bacterial divisions or even across the different domains of life (80). One strategy that plasmids employ to ensure a broad host range is to encode a wide variety of replication origins and proteins, which can be recognized by a diverse suite of organisms. Also, conjugative or mobilizable plasmids may have evolutionary implications for gene transfer very different from those of nonmobilizable plasmids, as they are more likely to spread between microorganisms. Finally, some plasmids, such as the classically studied F-episome (1), can recombine with and integrate into the host chromosome. This increases the likelihood that host genes are transferred during conjugation and can therefore have major evolutionary implications.

TRANSPOSONS AND INSERTION SEQUENCES

Transposons were first discovered in maize by Barbara McClintock in the 1940s (64). However, their importance was not fully appreciated for several decades, and she was not awarded the Nobel Prize until 1983. Like plasmids and phages, transposons are an incredibly diverse group of elements, which can carry genes encoding a wide variety of phenotypes. At a minimum, transposons contain the determinants to insert themselves into DNA. In many cases, they insert into DNA in a semirandom or random fashion, making these elements invaluable in molecular genetics as a mutagenesis tool. In addition, transposons probably played major roles in the development of the vertebrate immune system, as it is believed that the RAG1 and RAG2 genes, which are responsible for antibody formation and lymphocyte development, originated from a transposon over 450 million years ago (49). Transposons and insertion sequences (ISs) are found in the genomes of a wide phylogenetic variety of organisms and are therefore likely to play major roles in the generation of variation in microbes.

ISs and transposons, like other HGT elements, exhibit remarkable variation both in the genes that they carry and in their mechanism of action. ISs, the simplest form of this class of MGE, consist of a transposase gene—either a serine or DD-E recombinase—flanked by two inverted repeat sequences. The transposase enzyme recognizes the inverted repeats and catalyzes recombination events to move the IS from one site to another. Composite transposons arise when two ISs insert near one another on a genome (for an example, see reference 36). In these cases, the outermost inverted repeats can be acted on by the transposase enzyme, and the intervening sequence can be transferred along with the ISs. A variety of genes have been "trapped" on composite transposons in this fashion, including antibiotic resistance determinants (36), genes involved in metal tolerance and detoxification (56), and genes for the catabolism of xenobiotic compounds (76). Noncomposite transposons also contain a transposase and an intervening sequence, but they only contain flanking inverted repeat sequences, and not entire ISs (for an example, see reference 86).

One important factor that varies for different ISs and transposons is their transposition mechanism: some move via "replicative" transposition, while others spread via "cut-and-paste" transposition (43). For the cut-and-paste transposons, the transposase enzyme excises the IS or transposon from its original site and ligates it into a new location. In contrast, replicative transposons are first copied, and then the new molecule is inserted into a distal site in the genome, generally requiring the activity of both a transposase and a resolvase. Therefore, replicative transposition leads to a multiplication in the number of transposons within a genome. However, there are mechanisms to increase the number of cut-and-paste transposons within a genome as well; for example, they can be cut immediately after the DNA replication fork passes and then be transferred to a region of the genome that has not yet been replicated. The multiplication of transposons can have important evolutionary consequences, because these identical sites can serve as homologous recombination targets within a genome. In addition, although classically defined transposons do not move between organisms, these elements can insert themselves into plasmids or phage genomes and be mobilized via transduction, conjugation, or transformation. Finally, a newly described class of mobile gene elements, called integrative conjugative elements, (ICEs), discussed below, which are closely related to transposons, are able to promote transfer between cells.

INTEGRONS

Integrons were first identified in the late 1980s as a potentially mobile element found in association with a variety of antibiotic resistance genes (92). The integron system consists of three parts: a tyrosine recombinase integrase gene (*intI*), a strong promoter (pC), and a recombination site (*attI*) (32). Unlike transposons, integrons do not encode the determinants for self-mobilization. Instead, integrons facilitate the horizontal transfer of gene cassettes, which are promoterless open reading frames (ORFs) with *attC* (also called 59-be) recombination sites. Specifically, the IntI enzyme catalyzes a site-specific recombination event between the *attI* and *attC* sites and can integrate in, or excise out, gene cassettes. The gene cassettes incorporated into the integron are expressed from the integron-based promoter pC; the number of integrated cassettes can range from 0 to well over 100 (62, 63, 85). Because these mobile genes are integrated into a segment of DNA that is already replicated (bacterial chromosome or plasmid), and because they are integrated behind a host-compatible promoter (pC), integrons may be important in overcoming limitations traditionally associated with the expression and replication of horizontally transferred DNA and may be an important mechanism in long-distance HGT. Finally, because integrons can be located on other mobile gene elements (e.g., plasmids or transposons) (for examples, see references 10 and 74), they can themselves be transferred between bacteria.

There are many different classes of integrons, each class possessing an integrase gene with a unique sequence. Work on integrons began with the discovery of "type 1" integrons, which contain the *intI1* gene (92). These integrons are commonly found in pathogenic bacteria and are central in the spread of antibiotic resistance (32). Three other classes of integrons that are key in the transfer of resistance genes have been identified, collectively called multiresistant (MR) or mobile integrons (MI) because they typically transfer antibiotic resistance determinants and because they are generally found on plasmids or transposons. This association with mobile elements has facilitated their transfer between a diverse array of bacteria, including many species of *Gammaproteobacteria* and gram-positive bacteria (70, 74). MIs typically contain 1 to 10 gene cassettes with highly variable *attC* sites. MIs have been extremely successful in engineering a variety of multiresistance phenotypes in response to antibiotic selection pressure; within a 12-month period alone, 11 novel antibiotic resistance genes were reported in type 1 integrons (2, 16, 28, 39, 46, 55, 67, 69, 104-106).

Whole-genome sequencing has also revealed the presence of a wide variety of chromosomal integrons (CIs) in a diverse suite of environmentally relevant organisms (17, 30, 72, 84, 85, 101) (Fig. 2). CIs are different from MIs because they are commonly located on chromosomes, they are less likely to transfer resistance genes, and they typically contain a suite of gene cassettes with homogenous *attC* sites. Also, unlike MIs, which are commonly transferred between bacteria on other mobile elements, in many cases the phylogenies of these integrase genes appear to reflect the phylogenies of the organisms from which they originate (84, 85), suggesting that they are being inherited vertically to some degree (Fig. 2). However, in other cases, these genes seem to be themselves horizontally transferred (Fig. 2, boxed names), suggesting complicated evolutionary patterns for integrons.

Although MIs have been intensively studied because of their obvious public health implications, many questions remain about the evolutionary significance of CIs. A wide variety of gene cassettes, coding for a diverse array of activities, have been found to be associated with these elements and have been sequenced using cultivation-independent techniques from soil and water environments (38, 72, 93). It is possible that like MIs, CIs provide organisms with access to a vast array of gene cassettes, enabling them to quickly adapt and evolve. However, very few studies have actually examined the native ability of these integrases to recognize *attC* sites and integrate gene cassettes or have tested the ability of integrases to recognize different *attC* sites. Indeed, the ability to recognize a wide variety of *attC* sites may be a function that is unique to MIs, reflected in the homogenous *attC* sites typically found in CIs. Therefore, much work remains to be done to discover the role of these diverse elements in bacterial evolution.

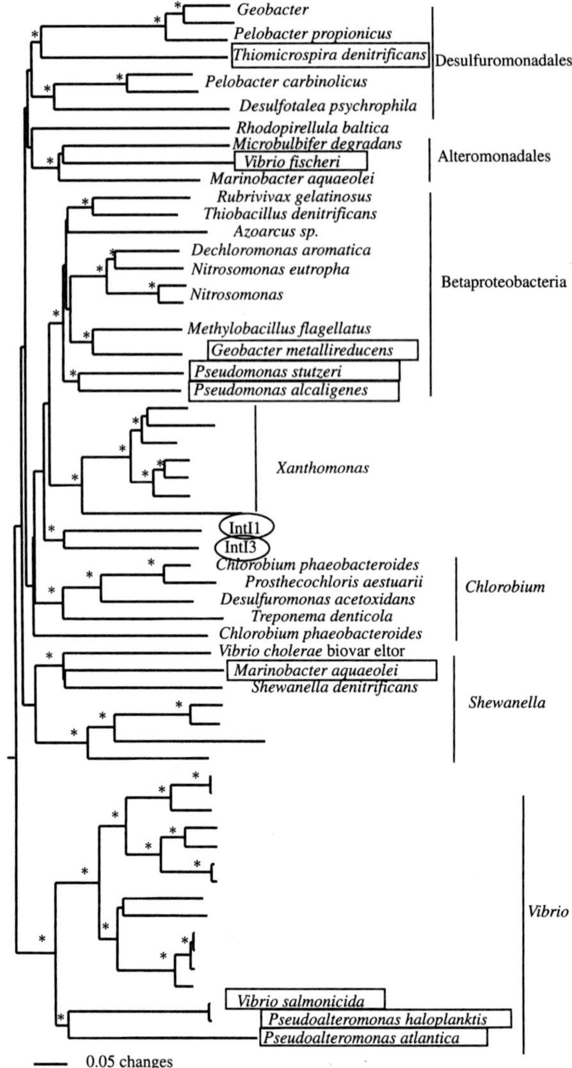

FIGURE 2 NJ tree of integron-type integrases. Proteins from a single organism are named with a genus and species. IntI1 and IntI3 have been found in several organisms. Boxed names highlight organisms whose integrase phylogenies are markedly different from their organismal phylogenies. *, nodes that are supported by >80% by both parsimony and distance bootstrap analysis.

INTEGRATIVE CONJUGATIVE ELEMENTS

As described above, phages, plasmids, transposons, and integrons all exhibit a large amount of variation not only in the kinds of genes found on these elements but also in their ability to spread to and replicate in other cells. In addition, although some characteristics were once believed to be defining features of specific classes of HGT elements, many exceptions are now known to exist, obscuring the classical distinctions between phages, plasmids, and transposons. For example, plasmids can integrate into chromosomes (for an example, see reference 1) and phage genomes can exist and replicate as extrachromosomal elements (for an example, see reference 48). A recently described class of MGEs, ICEs, further blurs the lines between HGT elements (13, 77, 99). ICEs, also known as genomic islands (GIs) or conjugative and mobilizable transposons, have features common to phages, transposons, integrons, and plasmids and contain a unique combination of integrase, replication, mobilization, and transfer factors.

GIs are large and sometimes mobile stretches of DNA (for a recent review, see reference 27). These elements were first described as contiguous genomic regions with aberrant G+C content and codon usage patterns, suggestive of recent horizontal transfer. In addition, GIs were initially found as features of pathogen genomes, in many cases carrying an array of virulence genes; hence, they are commonly referred to as "pathogenicity islands" (87). However, there are now many known examples of GIs that encode a wide variety of phenotypes, from symbiosis (31) and nitrogen fixation (95) to xenobiotic degradation (91). These elements are often found integrated into or near tRNA loci and are commonly flanked by short direct repeats that resemble phage attachment sites, and many include a tyrosine recombinase gene for integration (77). Additionally, many GIs contain genes encoding transfer and mobilization functions (13), implying that these elements may be able to promote conjugation. GIs differ from episomes because they do not contain a plasmid origin of replication and because they integrate into genomes via site-specific rather than homologous recombination. Examples of these elements include the multidrug-resistant SXT element from *Vibrio cholerae* (107) and the metal-antibiotic resistance R391 element from *Providencia rettgeri* (19). Thus, GIs contain a unique fusion of plasmid-, integron-, and phage-related functions.

Conjugative transposons (CTs) or mobilizable transposons (MTs) (68) have also been recently identified in a variety of bacteria. These elements are more similar to transposons than to GIs in that they integrate into random or semirandom sequences (77). However, they also carry the determinants for mobilization (MTs) or conjugation (CTs). Indeed, many of these elements include *mob*, *tra*, *rep*, and *par*/*sop* genes (5, 77). Tn*916* is an example of a broad-host-range conjugative transposon that was originally isolated from *Enterococcus faecalis*, which carries a tetracycline resistance gene (34). Tn*916* also has an *ori* and can exist and replicate as an extrachromosomal element. Like the GIs, these elements contain phage-, integron-, and plasmid-like properties. In some cases, two different CTs have nearly identical conjugative and regulatory genes, yet their integration machineries are very different (82). Their mosaic nature suggests a complicated evolutionary pathway for ICEs, implying that some of the distinctions that we have made for nomenclature purposes (i.e., based on integration method) may be artificial and not directly related to phylogeny. Indeed, proper elucidation of the evolutionary histories of all MGEs demands a more synthetic approach in light of the remarkable overlap in a number of the genes (Table 1) and the copious amount of sequence data that now exist for many different types of these elements.

EXAMINING MICROBIAL GENE TRANSFER IN SOILS

As is the case for other areas of research within environmental microbiology, the study of horizontal gene transfer is complicated by difficulties associated with working with a complex, largely unculturable microbial community living in a highly spatially and temporally heterogeneous system. Nonetheless, microbial ecologists have taken a variety of approaches to successfully examine HGT in soils. These approaches can be divided into two different types: "retrospective," i.e., the analysis of sequence data to make inferences about what types of genes have been transferred; or "prospective," i.e., experimental approaches to look at gene transfer in action (89). Below we discuss some specifics of these techniques, as well as the pros and cons of each approach.

Retrospective Approaches

Broadly speaking, retrospective approaches to examine HGT involve identifying and/or statistically comparing genes suspected of horizontal transfer with genes that are believed to be inherited vertically. This can be accomplished in a variety of ways; one approach is to target mobile gene elements in cultured isolates by using PCR with primers targeted to conserved regions and to sequence the amplicons. A number of unique elements have been discovered in this fashion, and many of both the *cis* and *trans* factors described in Table 1 have been used as targets for PCR. For example, new classes of integrons (75) have been identified using PCR and sequencing. Alternatively, these regions can be targeted using DNA probes and Southern blots; indeed, one of the first integrons sequenced from a nonpathogenic bacterium was found by use of this technique (101).

These PCR and DNA hybridization approaches are limited by the inability to culture a large fraction of environmental microbes; however, similar approaches have recently been developed to allow for cultivation-independent examinations of the diversity of HGT elements in microbial communities (12, 72, 75). Here, total environmental genomic DNA is amplified, cloned, and sequenced. To increase the potential diversity sampled, some investigators have employed degenerate primers, where pools of primers targeted to all of the possible codons for conserved amino acids are used in environmental PCR techniques (12, 72). To our knowledge, this technique has been applied only to examine integron and phage diversity in environmental microbial communities; however, in theory, many of the factors listed in Table 1 could serve as potential targets for this technique. These types of techniques have resulted in an increase in the understanding of the diversity of these elements. However, unless they are combined with culture-based or metagenomic analyses (see below), it is difficult to obtain other types of information (i.e., host range) from these approaches.

Others have developed methods for identifying novel MGEs from environmental microorganisms by capitalizing on unique features of the mobile elements themselves. For example, van Elsas et al. (103) identified several self-transmissible plasmids by screening wheat rhizosphere extracts for factors that could mobilize non-self-transmissible plasmids. In addition, Droge et al. (29) discovered 12 new plasmids from sewage sludge by transforming *Pseudomonas* sp. strain B13 with environmental community plasmid DNA and selecting for a variety of different types of antibiotic resistance. Other studies have identified new types of lysogenic phage by inducing the packaging and release of phages from environmental microorganisms with mitomycin C (79) or UV exposure (42).

DNA microarrays are a relatively new technique for the analysis of HGT in environmental communities. Three

TABLE 1 Distribution of different *cis*- and *trans*-acting factors in the MGEs[a]

MGE	*tyr*	*attP*	*ori*	*rep*	*oriT*	*mob*	*tra*	*par*/*sop*	*ser*	*dd-e*
Phage	X	X								
Plasmid			X	X	X	X	X	X		
Transposon									X	X
Integron	X									
Genomic island	X	X			X	X	X			X
Conjugative/mobilizable transposons	X		X	X	X	X	X	X		

[a]Abbreviations: *tyr*, tyrosine recombinase; *attP*, phage attachment site; *ori*, origin of replication; *rep*, DNA replicase; *oriT*, origin of transfer; *mob*, mobilization functions; *tra*, transfer functions; *par*/*sop*, plasmid partitioning functions; *ser*, serine recombinase; *dd-e*, DD-E recombinase.

major classes of arrays, i.e., functional gene arrays, community genome arrays, and phylogenetic oligonucleotide arrays, have been developed to examine environmental microbial community structure and function (112, 113). Recently a fourth class, an HGT microarray, was developed to examine the genomic linkage between metal resistance determinants and *inc/rep* genes from broad-host-range plasmids (22). Although there are many challenges to applying microarray technology to environmental samples with respect to sensitivity, specificity, and quantitation (112, 113), the HGT microarray provides a novel method for the simultaneous examination of both the types of genes subject to transfer and the elements that facilitate HGT among microbes in the environment.

Another retrospective method that is often used is to look for DNA segments with aberrant G+C content and the codon usages (for an example, see reference 51); the assumption being that recently transferred DNA carries sequence signatures different from those of the DNA from the resident genome. The interpretation of this approach is complicated because there are possible reasons for aberrant patterns in G+C content and codon bias other than horizontal gene transfer and because this method cannot identify genes that have been transferred from closely related organisms or those that have been transferred some time ago and have ameliorated to match the host genome. Finally, the phylogenies of individual genes suspected to have resulted from an HGT event can be compared with the evolutionary history of genes traditionally thought to track organismal relationships (e.g., 16S rRNA genes). Phylogenetic approaches can also be problematic both because of the difficulty of phylogenetic reconstructions and because of other processes that lead to incongruence between gene and organismal phylogenies, such as gene loss, gene duplication, and convergent selection.

These types of retrospective techniques are most challenging for environmental microbiologists, however, because of the requirement for large, contiguous sequences of DNA for analysis. This can be accomplished by examining whole-genome data, obviously limiting the suite of organisms that can be examined. Recently, however, environmental metagenomics techniques have been employed to successfully detect gene transfer events (reviewed in reference 44). Here, large environmental DNA fragments are cloned into bacterial artificial chromosome or fosmid vectors, permitting cultivation-independent, partial genomic analysis of environmental microorganisms and the possibility of relating these genomic data to a phylogenetic marker (e.g., 16S rRNA genes). These genome fragments are then subjected to the analyses described above (screens for aberrant G+C content, codon bias, and phylogenies) to identify regions that are the result of HGT.

Metagenomic techniques have revealed support for long-distance HGT in a number of bacteria and archaea (7, 73). For example, Nesbo et al. (73) recently sequenced 12 rRNA gene-containing fosmid clones from a marine sediment sample and found that 7 to 44% of the ORFs in each fosmid had been acquired by HGT. Metagenomic approaches have also been used to obtain phage genomes from environmental microbial communities (11). Here, viral particles are separated from microbial cells via filtration and centrifugation; DNA is then extracted and cloned into suitable vectors. All of these techniques are limited by expense, as well as by the methodological issues associated with obtaining large DNA fragments from environmental microbial communities, challenges that are far more pronounced when dealing with soils and sediments than with aquatic communities.

The above retrospective approaches have led to some very interesting observations about HGT and evolution in environmental microorganisms. First, it appears that most recently transferred genes have an unusually high A+T content (40). In addition, most of these genes appear to be so-called "ORFans"—genes with no known homologues. These ORFans appear to have high turnover rates in bacterial genomes, and while they are under purifying selection (i.e., stronger selection on the first two codon positions than on the "wobble" base), the strength of this selection appears much weaker than the forces acting on other, "native" genes (23, 24). Other recent work suggests that microbial communities are comprised of clusters of closely related, but not isogenic, organisms, suggesting high levels of genomic-level microdiversity, likely the result of HGT, within environments. This may imply either that these heterogeneous environments contain a variety of microorganisms highly adapted to specific microniches or that much of the microdiversity within these organisms is selectively neutral, preventing major bottleneck events in these communities (98).

Prospective Approaches

Many researchers have also taken a prospective approach to examine the frequency and factors influencing gene transfer in environmental microbial communities (89). Because these are manipulative, experimental approaches, they do not suffer from problems associated with the inference-based retrospective techniques for the detection of HGT. However, prospective approaches gauge only the possibility for gene transfer and cannot address whether transferred material will be stably inherited over evolutionary timescales. Even with this limitation, these approaches have resulted in interesting information about the types of organisms that can exchange genes and how a variety of environmental factors can regulate HGT processes.

Traditionally, prospective approaches have included the seeding of microcosms or environments with bacteria containing marker genes or naked DNA or plasmids and the selection of transformants (for an example, see reference 50). These approaches are limited because they cannot differentiate between an increase in the number of transformation events and an increase in the number of a particular transformant via cell division; therefore, direct quantification of recombination events is impossible. In addition, these approaches are also severely limited because of the many difficulties inherent in culturing environmental microorganisms.

Recent advances have begun to allow for cultivation-independent, real-time observations of HGT. For example, Sorenson and colleagues have used a fluorescence-based cell-sorting approach to identify recombinants (90). Here, cassettes or plasmids are labeled with the *gfp* reporter gene, which is repressed in the donor cell but expressed in the transconjugate. Thus, the transfer and expression of DNA can be tracked with a combined fluorescence detector-cell sorter. Then, molecular techniques can be employed to identify the cells that have integrated and expressed the *gfp* gene. Anther novel cultivation-independent approach used to examine the host ranges of temperate phages involves isolating viral DNA and amplifying 16S rRNA genes (9). This technique detects the microbes that have been infected and whose 16S rRNA genes have been incorporated into viral genomes via generalized transduction. Although this approach undoubtedly does not detect all organisms that have been infected by phages in an environment, it can provide interesting information about the phylogenetic breadth of organisms that are susceptible to phage infection.

CONCLUSIONS AND FUTURE DIRECTIONS

The field of environmental microbiology has exploded in recent years, bringing ecologists, microbiologists, evolutionary biologists, infectious disease experts, and engineers together, each with different tools and areas of expertise to address fundamental questions about microbial biology. These collaborations have resulted in an exponential growth in microbial community data, as well as the rapid development of new techniques to study environmental microbial community diversity, structure, and processes. These techniques have also led to an increased understanding of HGT and MGE diversity in microbial communities. Despite the recent developments in this area, much work needs to be done to understand the role of HGT in shaping the genomes and communities of environmental microorganisms. Perhaps one of the most exciting areas of study is the exploration of the controls over HGT in environmental microbial communities (102). Recent work suggests that a variety of environmental variables, including moisture, temperature, and pH, influence the rates of HGT in the environment (89). In addition, other studies suggest that the presence of the very variables that MGEs encode resistance or tolerance to (e.g., antibiotics, metals, etc.) may promote the exchange of these elements (6, 21, 45). Indeed, the regulation and controls over HGT in microbial communities are important for a wide variety of applications with societal importance and are therefore expected to be an active front of research into the future.

REFERENCES

1. **Achtman, M., N. Willetts, and A. J. Clark.** 1971. Beginning a genetic analysis of conjugational transfer determined by F-factor in *Escherichia coli* by isolation and characterization of transfer-deficient mutants. *J. Bacteriol.* **106:** 529–538.
2. **Ahmed, A. M., T. Nakagawa, E. Arakawa, T. Ramamurthy, S. Shinoda, and T. Shimamoto.** 2004. New aminoglycoside acetyltransferase gene, aac(3)-Id, in a class 1 integron from a multiresistant strain of *Vibrio fluvialis* isolated from an infant aged 6 months. *J. Antimicrob. Chemother.* **53:**947–951.
3. **Ashelford, K. E., M. J. Day, and J. C. Fry.** 2003. Elevated abundance of bacteriophage infecting bacteria in soil. *Appl. Environ. Microbiol.* **69:**285–289.
4. **Avery, O. T., C. M. MacLeod, and M. McCarty.** 1944. Studies on the chemical nature of the substance inducing transformation of pneumonococcal types. Inductions of transformation by a desoxyribonucleic acid fraction isolated from pneumonoccus type III. *J. Exp. Med.* **79:**137-157.
5. **Bacic, M., A. C. Parker, J. Stagg, H. P. Whitley, W. G. Wells, L. A. Jacob, and C. J. Smith.** 2005. Genetic and structural analysis of the *Bacteroides* conjugative transposon CTn341. *J. Bacteriol.* **187:**2858–2869.
6. **Beaber, J. W., B. Hochhut, and M. K. Waldor.** 2004. SOS response promotes horizontal dissemination of antibiotic resistance genes. *Nature* **427:**72–74.
7. **Beja, O., E. V. Koonin, L. Aravind, L. T. Taylor, H. Seitz, J. L. Stein, D. C. Bensen, R. A. Feldman, R. V. Swanson, and E. F. DeLong.** 2002. Comparative genomic analysis of archaeal genotypic variants in a single population and in two different oceanic provinces. *Appl. Environ. Microbiol.* **68:**335–345.
8. **Bergh, O., K. Y. Borsheim, G. Bratbak, and M. Heldal.** 1989. High abundance of viruses found in aquatic environments. *Nature* **340:**467–468.
9. **Beumer, A., and J. B. Robinson.** 2005. A broad-host-range, generalized transducing phage (SN-T) acquires 16S rRNA genes from different genera of bacteria. *Appl. Environ. Microbiol.* **71:**8301–8304.
10. **Bissonnette, L., and P. H. Roy.** 1992. Characterization of In0 of *Pseudomonas aeruginosa* plasmid pVS1, an ancestor of integrons of multiresistance plasmids and transposons of gram-negative bacteria. *J. Bacteriol.* **174:**1248–1257.
11. **Breitbart, M., I. Hewson, B. Felts, J. M. Mahaffy, J. Nulton, P. Salamon, and F. Rohwer.** 2003. Metagenomic analyses of an uncultured viral community from human feces. *J. Bacteriol.* **185:**6220–6223.
12. **Breitbart, M., J. H. Miyake, and F. Rohwer.** 2004. Global distribution of nearly identical phage-encoded DNA sequences. *FEMS Microbiol. Lett.* **236:**249–256.
13. **Burrus, V., G. Pavlovic, B. Decaris, and G. Guedon.** 2002. Conjugative transposons: the tip of the iceberg. *Mol. Microbiol.* **46:**601–610.
14. **Bushman, F.** 2002. *Lateral DNA Transfer: Mechanisms and Consequences*. Cold Spring Harbor Laboratory Press, Cold Spring Harbor, N.Y.
15. **Cantera, J. J. L., H. Kawasaki, and T. Seki.** 2004. The nitrogen-fixing gene (*nifH*) of *Rhodopseudomonas palustris*: a case of lateral gene transfer? *Microbiology* **150:**2237–2246.
16. **Castanheira, M., M. A. Toleman, R. N. Jones, F. J. Schmidt, and T. R. Walsh.** 2004. Molecular characterization of a beta-lactamase gene, *bla(GIM-1)*, encoding a new subclass of metallo-beta-lactamase. *Antimicrob. Agents Chemother.* **48:**4654–4661.
17. **Chen, C. Y., K. M. Wu, Y. C. Chang, C. H. Chang, H. C. Tsai, T. L. Liao, Y. M. Liu, H. J. Chen, A. B. T. Shen, J. C. Li, T. L. Su, C. P. Shao, C. T. Lee, L. I. Hor, and S. F. Tsai.** 2003. Comparative genome analysis of *Vibrio vulnificus*, a marine pathogen. *Genome Res.* **13:**2577–2587.
18. **Chen, I., P. J. Christie, and D. Dubnau.** 2005. The ins and outs of DNA transfer in bacteria. *Science* **310:**1456-1460.
19. **Coetzee, J. N., N. Datta, and R. W. Hedges.** 1972. R factors from *Proteus rettgeri*. *J. Gen. Microbiol.* **72:**543–552.
20. **Cohen, S. N., A. C. Chang, H. W. Boyer, and R. B. Helling.** 1973. Construction of biologically functional bacterial plasmids in vitro. *Proc. Natl. Acad. Sci. USA* **70:** 3240–3244.
21. **Coleman, N. V., and A. J. Holmes.** 2005. The native *Pseudomonas stutzeri* strain Q chromosonal integron can capture and express cassette-associated genes. *Microbiology* **151:**1853–1864.
22. **Coombs, J. M., and T. Barkay.** 2005. Horizontal gene transfer of metal homeostasis genes and its role in microbial communities of the deep terrestrial subsurface, p. 109–129. *In* G. M. Gadd, K. T. Semple, and H. M. Lappin-Scott (ed.), *Micro-Organisms and Earth Systems: Advances in Geomicrobiology. Proceedings of the 65th Symposium of the Society of General Microbiology*. Cambridge University Press, New York, N.Y.
23. **Daubin, V., E. Lerat, and G. Perriere.** 2003. The source of laterally transferred genes in bacterial genomes. *Genome Biol.* **4:**R57.
24. **Daubin, V., and H. Ochman.** 2004. Bacterial genomes as new gene homes: the genealogy of ORFans in *E. coli*. *Genome Res.* **14:**1036–1042.
25. **del Solar, G., R. Giraldo, M. J. Ruiz-Echevarria, M. Espinosa, and R. Diaz-Orejas.** 1998. Replication and

control of circular bacterial plasmids. *Microbiol. Mol. Biol. Rev.* **62:**434–464.

26. **Ding, Z. Y., K. Atmakuri, and P. J. Christie.** 2003. The outs and ins of bacterial type IV secretion substrates. *Trends Microbiol.* **11:**527–535.
27. **Dobrindt, U., B. Hochhut, U. Hentschel, and J. Hacker.** 2004. Genomic islands in pathogenic and environmental microorganisms. *Nat. Rev. Microbiol.* **2:**414–424.
28. **Doublet, B., F. X. Weill, L. Fabre, E. Chaslus-Dancla, and A. Cloeckaert.** 2004. Variant *Salmonella* genomic island 1 antibiotic resistance gene cluster containing a novel 3′-*N*-aminoglycoside acetyltransferase gene cassette, *aac(3)-id*, in *Salmonella enterica* serovar Newport. *Antimicrob. Agents Chemother.* **48:**3806–3812.
29. **Droge, M., A. Puhler, and W. Selbitschka.** 2000. Phenotypic and molecular characterization of conjugative antibiotic resistance plasmids isolated from bacterial communities of activated sludge. *Mol. Gen. Genet.* **263:**471–482.
30. **Drouin, F., J. Melancon, and P. H. Roy.** 2002. The *intI*-like tyrosine recombinase of *Shewanella oneidensis* is active as an integron integrase. *J. Bacteriol.* **184:**1811–1815.
31. **Finan, T. M.** 2002. Evolving insights: symbiosis islands and horizontal gene transfer. *J. Bacteriol.* **184:**2855–2856.
32. **Fluit, A. C., and F. J. Schmitz.** 2004. Resistance integrons and super-integrons. *Clin. Microbiol. Infect.* **10:**272–288.
33. **Francia, M. V., A. Varsaki, M. P. Garcillan-Barcia, A. Latorre, C. Drainas, and F. de la Cruz.** 2004. A classification scheme for mobilization regions of bacterial plasmids. *FEMS Microbiol. Rev.* **28:**79–100.
34. **Franke, A. E., and D. B. Clewell.** 1981. Evidence for a chromosome-borne resistance transposon (Tn*916*) in *Streptococcus faecalis* that is capable of "conjugal" transfer in the absence of a conjugative plasmid. *J. Bacteriol.* **145:** 494–502.
35. **Friedrich, M. W.** 2002. Phylogenetic analysis reveals multiple lateral transfers of adenosine-5′-phosphosulfate reductase genes among sulfate-reducing microorganisms. *J. Bacteriol.* **184:**278–289.
36. **Galimand, A., S. Sabtcheva, P. Courvalin, and T. Lambert.** 2005. Worldwide disseminated *armA* aminoglycoside resistance methylase gene is borne by composite transposon Tn*1548*. *Antimicrob. Agents Chemother.* **49:** 2949–2953.
37. **Gerdes, K., J. Moller-Jensen, and R. B. Jensen.** 2000. Plasmid and chromosome partitioning: surprises from phylogeny. *Mol. Microbiol.* **37:**455–466.
38. **Gillings, M. R., M. P. Holley, H. W. Stokes, and A. J. Holmes.** 2005. Integrons in *Xanthomonas*: a source of species genome diversity. *Proc. Natl. Acad. Sci. USA* **102:** 4419–4424.
39. **Giuliani, F., J. D. Docquier, M. L. Riccio, L. Pagani, and G. M. Rossolini.** 2005. OXA-46, a new class D beta-lactamase of narrow substrate specificity encoded by a (*bla*)(*vim-1*)-containing integron from a *Pseudomonas aeruginosa* clinical isolate. *Antimicrob. Agents Chemother.* **49:** 1973–1980.
40. **Gogarten, J. P., and J. P. Townsend.** 2005. Horizontal gene transfer, genome innovation and evolution. *Nat. Rev. Microbiol.* **3:**679–687.
41. **Griffith, F.** 1928. The significance of pneumonococcal types. *J. Hyg.* **27:**13–59.
42. **Grkovic, S., M. Ocallaghan, and H. K. Mahanty.** 1995. Characterization of *Serratia entomophila* bacteriophages and the phage-resistant mutant strain Bc4b. *Appl. Environ. Microbiol.* **61:**4160–4166.
43. **Hallet, B., and D. J. Sherratt.** 1997. Transposition and site-specific recombination: adapting DNA cut-and-paste mechanisms to a variety of genetic rearrangements. *FEMS Microbiol. Rev.* **21:**157–178.
44. **Handelsman, J.** 2004. Metagenomics: application of genomics to uncultured microorganisms. *Microbiol. Mol. Biol. Rev.* **68:**669–685.
45. **Hastings, P. J., S. M. Rosenberg, and A. Slack.** 2004. Antibiotic-induced lateral transfer of antibiotic resistance. *Trends Microbiol.* **12:**401–404.
46. **Heritier, C., A. Dubouix, L. Poirel, N. Marty, and P. Nordmann.** 2005. A nosocomial outbreak of *Acinetobacter baumannii* isolates expressing the carbapenem-hydrolysing oxacillinase OXA-58. *J. Antimicrob. Chemother.* **55:**115–118.
47. **Hugenholtz, P., B. M. Goebel, and N. R. Pace.** 1998. Impact of culture-independent studies on the emerging phylogenetic view of bacterial diversity. *J. Bacteriol.* **180:** 4765–4774. (Erratum, **180:**6793.)
48. **Kanda, K., T. Ohderaotoshi, A. Shimojyo, F. Kato, and A. Murata.** 1999. An extrachromosomal prophage naturally associated with *Bacillus thuringiensis* serovar israelensis. *Lett. Appl. Microbiol.* **28:**305–308.
49. **Kapitonov, V. V., and J. Jurka.** 2005. RAG1 core and V(D)J recombination signal sequences were derived from Transib transposons. *PLoS Biol.* **3:**998–1011.
50. **Lawlor, K., A. R. Chaudri, S. P. McGrath, and P. R. Hirsch.** 1999. Gene transfer in bacteria from soils contaminated with heavy metals. *Lett. Appl. Microbiol.* **28:**317–320.
51. **Lawrence, J. G., and H. Ochman.** 1998. Molecular archaeology of the *Escherichia coli* genome. *Proc. Natl. Acad. Sci. USA* **95:**9413–9417.
52. **Lawrence, J. G., and H. Ochman.** 2002. Reconciling the many faces of lateral gene transfer. *Trends Microbiol.* **10:**1–4.
53. **Lawrence, K., and M. Mulczyk.** 1965. The transfer of two episomes, colicinogenic factor I and resistance transfer factor, in *Shigella flexneri* strains by crosses between strains each possessing a single episome. *J. Gen. Microbiol.* **39:** 209–213.
54. **Lederberg, J.** 1952. Cell genetics and hereditary symbiosis. *Physiol. Rev.* **32:**403–430.
55. **Libisch, B., M. Gacs, K. Csiszar, M. Muzslay, L. Rokusz, and M. Fuzi.** 2004. Isolation of an integron-borne bla(VIM-4) type metallo-beta-lactamase gene from a carbapenem-resistant *Pseudomonas aeruginosa* clinical isolate in Hungary. *Antimicrob. Agents Chemother.* **48:**3576–3578.
56. **Liebert, C. A., R. M. Hall, and A. O. Summers.** 1999. Transposon Tn*21*, flagship of the floating genome. *Microbiol. Mol. Biol. Rev.* **63:**507–522.
57. **Lin, N. T., T. J. Liu, T. C. Lee, B. Y. You, M. H. Yang, F. S. Wen, and Y. H. Tseng.** 1999. The adsorption protein genes of *Xanthomonas campestris* filamentous phages determining host specificity. *J. Bacteriol.* **181:**2465–2471.
58. **Lipson, D. A., and S. K. Schmidt.** 2004. Seasonal changes in an alpine soil bacterial community in the Colorado Rocky Mountains. *Appl. Environ. Microbiol.* **70:** 2867–2879.
59. **Llosa, M., F. X. Gomis-Ruth, M. Coll, and F. de la Cruz.** 2002. Bacterial conjugation: a two-step mechanism for DNA transport. *Mol. Microbiol.* **45:**1–8.
60. **Margulis, L.** 1968. Evolutionary criteria in Thallophytes: a radical alternative. *Science* **161:**1020–1022.
61. **Martin, A. P., E. K. Costello, A. F. Meyer, D. R. Nemergut, and S. K. Schmidt.** 2004. The rate and pattern of cladogenesis in microbes. *Evolution* **58:**946–955.
62. **Mazel, D., B. Dychinco, V. A. Webb, and J. Davies.** 2000. Antibiotic resistance in the ECOR collection: integrons and identification of a novel *aad* gene. *Antimicrob. Agents Chemother.* **44:**1568–1574.
63. **Mazel, D., B. Dychinco, V. A. Webb, and J. Davies.** 1998. A distinctive class of integron in the *Vibrio cholerae* genome. *Science* **280:**605–608.

64. **McClintock, B.** 1951. Chromosome organization and genic expression. *Cold Spring Harbor Symp. Quant. Biol.* **16:**13–47.
65. **McDonald, L. C., and W. R. Jarvis.** 1997. The global impact of vancomycin-resistant enterococci. *Curr. Opin. Infect. Dis.* **10:**304–309.
66. **Mellado, E., J. A. Asturias, J. J. Nieto, K. N. Timmis, and A. Ventosa.** 1995. Characterization of the basic replicon of pCM1, a narrow-host-range plasmid from the moderate halophile *Chromohalobacter marismortui. J. Bacteriol.* **177:**3443–3450.
67. **Mendes, R. E., M. A. Toleman, J. Ribeiro, H. S. Sader, R. N. Jones, and T. R. Walsh.** 2004. Integron carrying a novel metallo-beta-lactamase gene, *bla(IMP-16)*, and a fused form of aminoglycoside-resistant gene *aac(6')-30/aac(6')-Ib'*: report from the SENTRY antimicrobial surveillance program. *Antimicrob. Agents Chemother.* **48:** 4693–4702.
68. **Mullany, P., A. P. Roberts, and H. Wang.** 2002. Mechanism of integration and excision in conjugative transposons. *Cell. Mol. Life Sci.* **59:**2017–2022.
69. **Mulvey, M. R., D. A. Boyd, L. Baker, O. Mykytczuk, E. M. F. Reis, M. D. Asensi, D. P. Rodrigues, and L. K. Ng.** 2004. Characterization of a *Salmonella enterica* serovar Agona strain harbouring a class 1 integron containing novel OXA-type beta-lactamase (bla(OXA-53)) and 6'-N-aminoglycoside acetyltransferase genes aac(6')-I30. *J. Antimicrob. Chemother.* **54:**354–359.
70. **Nandi, S., J. J. Maurer, C. Hofacre, and A. O. Summers.** 2004. Gram-positive bacteria are a major reservoir of Class 1 antibiotic resistance integrons in poultry litter. *Proc. Natl. Acad. Sci. USA* **101:**7118–7122.
71. **Nelson, K. E., R. A. Clayton, S. R. Gill, M. L. Gwinn, R. J. Dodson, D. H. Haft, E. K. Hickey, L. D. Peterson, W. C. Nelson, K. A. Ketchum, L. McDonald, T. R. Utterback, J. A. Malek, K. D. Linher, M. M. Garrett, A. M. Stewart, M. D. Cotton, M. S. Pratt, C. A. Phillips, D. Richardson, J. Heidelberg, G. G. Sutton, R. D. Fleischmann, J. A. Eisen, O. White, S. L. Salzberg, H. O. Smith, J. C. Venter, and C. M. Fraser.** 1999. Evidence for lateral gene transfer between Archaea and Bacteria from genome sequence of *Thermotoga maritima. Nature* **399:**323–329.
72. **Nemergut, D. R., A. P. Martin, and S. K. Schmidt.** 2004. Integron diversity in heavy-metal-contaminated mine tailings and inferences about integron evolution. *Appl. Environ. Microbiol.* **70:**1160–1168.
73. **Nesbo, C. L., Y. Boucher, M. Dlutek, and W. F. Doolittle.** 2005. Lateral gene transfer and phylogenetic assignment of environmental fosmid clones. *Environ. Microbiol.* **7:**2011–2026.
74. **Nesvera, J., J. Hochmannova, and M. Patek.** 1998. An integron of class 1 is present on the plasmid pCG4 from Gram-positive bacterium *Corynebacterium glutamicum. FEMS Microbiol. Lett.* **169:**391–395.
75. **Nield, B. S., A. J. Holmes, M. R. Gillings, G. D. Recchia, B. C. Mabbutt, K. M. H. Nevalainen, and H. W. Stokes.** 2001. Recovery of new integron classes from environmental DNA. *FEMS Microbiol. Lett.* **195:**59–65.
76. **Nojiri, H., M. Shintani, and T. Omori.** 2004. Divergence of mobile genetic elements involved in the distribution of xenobiotic-catabolic capacity. *Appl. Microbiol. Biotechnol.* **64:**154–174.
77. **Osborn, A. M., and D. Boltner.** 2002. When phage, plasmids, and transposons collide: genomic islands, and conjugative- and mobilizable-transposons as a mosaic continuum. *Plasmid* **48:**202–212.
78. **Pace, N. R.** 1997. A molecular view of microbial diversity and the biosphere. *Science* **276:**734–740.
79. **Prigent, M., M. Leroy, F. Confalonieri, M. Dutertre, and M. S. DuBow.** 2005. A diversity of bacteriophage forms and genomes can be isolated from the surface sands of the Sahara Desert. *Extremophiles* **9:**289–296.
80. **Rawlings, D. E., and E. Tietze.** 2001. Comparative biology of IncQ and IncQ-like plasmids. *Microbiol. Mol. Biol. Rev.* **65:**481–496.
81. **Raymond, J., J. L. Siefert, C. R. Staples, and R. E. Blankenship.** 2004. The natural history of nitrogen fixation. *Mol. Biol. Evol.* **21:**541–554.
82. **Roberts, A. P., P. A. Johanesen, D. Lyras, P. Mullany, and J. I. Rood.** 2001. Comparison of Tn5397 from *Clostridium difficile*, Tn916 from *Enterococcus faecalis* and the CW459tet(M) element from *Clostridium perfringens* shows that they have similar conjugation regions but different insertion and excision modules. *Microbiology* **147:**1243–1251.
83. **Rohwer, F., and R. Edwards.** 2002. The Phage Proteomic Tree: a genome-based taxonomy for phage. *J. Bacteriol.* **184:**4529–4535.
84. **Rowe-Magnus, D. A., A. M. Guerout, L. Biskri, P. Bouige, and D. Mazel.** 2003. Comparative analysis of superintegrons: engineering extensive genetic diversity in the Vibrionaceae. *Genome Res.* **13:**428–442.
85. **Rowe-Magnus, D. A., A. M. Guerout, P. Ploncard, B. Dychinco, J. Davies, and D. Mazel.** 2001. The evolutionary history of chromosomal super-integrons provides an ancestry for multiresistant integrons. *Proc. Natl. Acad. Sci. USA* **98:**652–657.
86. **Schaefer, M. R., and K. Kahn.** 1998. Cyanobacterial transposons Tn5469 and Tn*5541* represent a novel noncomposite transposon family. *J. Bacteriol.* **180:**6059–6063.
87. **Schmidt, H., and M. Hensel.** 2004. Pathogenicity islands in bacterial pathogenesis. *Clin. Microbiol. Rev.* **17:**14–56. (Erratum, **19:**257, 2006.)
88. **Smets, B. F., and T. Barkay.** 2005. Horizontal gene transfer: perspectives at a crossroads of scientific disciplines. *Nat. Rev. Microbiol.* **3:**675–678.
89. **Sorensen, S. J., M. Bailey, L. H. Hansen, N. Kroer, and S. Wuertz.** 2005. Studying plasmid horizontal transfer in situ: a critical review. *Nat. Rev. Microbiol.* **3:**700–710.
90. **Sorensen, S. J., A. H. Sorensen, L. H. Hansen, G. Oregaard, and D. Veal.** 2003. Direct detection and quantification of horizontal gene transfer by using flow cytometry and *gfp* as a reporter gene. *Curr. Microbiol.* **47:**129–133.
91. **Springael, D., and E. M. Top.** 2004. Horizontal gene transfer and microbial adaptation to xenobiotics: new types of mobile genetic elements and lessons from ecological studies. *Trends Microbiol.* **12:**53–58.
92. **Stokes, H. W., and R. M. Hall.** 1989. A novel family of potentially mobile DNA elements encoding site-specific gene-integration functions—integrons. *Mol. Microbiol.* **3:**1669–1683.
93. **Stokes, H. W., A. J. Holmes, B. S. Nield, M. P. Holley, K. M. H. Nevalainen, B. C. Mabbutt, and M. R. Gillings.** 2001. Gene cassette PCR: sequence-independent recovery of entire genes from environmental DNA. *Appl. Environ. Microbiol.* **67:**5240–5246.
94. **Stolz, J. F., and P. Basu.** 2002. Evolution of nitrate reductase: molecular and structural variations on a common function. *Chembiochem* **3:**198–206.
95. **Sullivan, J. T., and C. W. Ronson.** 1998. Evolution of rhizobia by acquisition of a 500-kb symbiosis island that integrates into a phe-tRNA gene. *Proc. Natl. Acad. Sci. USA* **95:**5145–5149.
96. **Suttle, C. A., A. M. Chan, and M. T. Cottrell.** 1990. Infection of phytoplankton by viruses and reduction of primary productivity. *Nature* **347:**467–469.

97. **Taoka, A., K. Yoshimatsu, M. Kanemori, and Y. Fukumori.** 2003. Nitrate reductase from the magnetotactic bacterium *Magnetospirillum magnetotacticum* MS-1: purification and sequence analyses. *Can. J. Microbiol.* **49:** 197–206.
98. **Thompson, J. R., S. Pacocha, C. Pharino, V. Klepac-Ceraj, D. E. Hunt, J. Benoit, R. Sarma-Rupavtarm, D. L. Distel, and M. F. Polz.** 2005. Genotypic diversity within a natural coastal bacterioplankton population. *Science* **307:**1311–1313.
99. **Toussaint, A., and C. Merlin.** 2002. Mobile elements as a combination of functional modules. *Plasmid* **47:**26–35.
100. **Twort, F.** 1915. An investigation on the nature of ultra-microscopic viruses. *Lancet* **ii:**1241–1243.
101. **Vaisvila, R., R. D. Morgan, J. Posfai, and E. A. Raleigh.** 2001. Discovery and distribution of super-integrons among Pseudomonads. *Mol. Microbiol.* **42:**587–601.
102. **van Elsas, J. D., and M. J. Bailey.** 2002. The ecology of transfer of mobile genetic elements. *FEMS Microbiol. Ecol.* **42:**187-197.
103. **van Elsas, J. D., B. B. M. Gardener, A. C. Wolters, and E. Smit.** 1998. Isolation, characterization, and transfer of cryptic gene-mobilizing plasmids in the wheat rhizosphere. *Appl. Environ. Microbiol.* **64:**880–889.
104. **Villa, L., and A. Carattoli.** 2005. Integrons and transposons on the *Salmonella enterica* serovar Typhimurium virulence plasmid. *Antimicrob. Agents Chemother.* **49:** 1194–1197.
105. **Vourli, S., P. Giakkoupi, V. Miriagou, E. Tzelepi, A. C. Vatopoulos, and L. S. Tzouvelekis.** 2004. Novel GES/IBC extended-spectrum beta-lactamase variants with carbapenemase activity in clinical enterobacteria. *FEMS Microbiol. Lett.* **234:**209–213.
106. **Wachino, J., Y. Doi, K. Yamane, N. Shibata, T. Yagi, T. Kubota, H. Ito, and Y. Arakawa.** 2004. Nosocomial spread of ceftazidime-resistant *Klebsiella pneumoniae* strains producing a novel class A beta-lactamase, GES-3, in a neonatal intensive care unit in Japan. *Antimicrob. Agents Chemother.* **48:**1960–1967.
107. **Waldor, M. K., H. Tschape, and J. J. Mekalanos.** 1996. A new type of conjugative transposon encodes resistance to sulfamethoxazole, trimethoprim, and streptomycin in *Vibrio cholerae* O139. *J. Bacteriol.* **178:**4157–4165.
108. **Watanabe, T., H. Nishida, O. Ogata, and T. Arai.** 1964. Episome-mediated transfer of drug resistance in *Enterobacteriaceae*. VII. Two types of naturally occurring R factors. *J. Bacteriol.* **88:**716–726.
109. **Weinbauer, M. G.** 2004. Ecology of prokaryotic viruses. *FEMS Microbiol. Rev.* **28:**127–181.
110. **Wolf, Y. I., L. Aravind, and E. V. Koonin.** 1999. Rickettsiae and Chlamydiae—evidence of horizontal gene transfer and gene exchange. *Trends Genet.* **15:**173–175.
111. **Yeo, H. J., and G. Waksman.** 2004. Unveiling molecular scaffolds of the type IV secretion system. *J. Bacteriol.* **186:**1919–1926.
112. **Zhou, J. H.** 2003. Microarrays for bacterial detection and microbial community analysis. *Curr. Opin. Microbiol.* **6:** 288-294.
113. **Zhou, J. Z., and D. K. Thompson.** 2002. Challenges in applying microarrays to environmental studies. *Curr. Opin. Biotechnol.* **13:**204–207.

Microorganisms Associated with Soil Arthropods

ANN E. HAJEK, JAMES R. FUXA, AND YASUHISA KUNIMI

63

An estimated 90% of insect species spend at least part of their lives in the soil or on the soil surface (71), and many other types of arthropods are common soil inhabitants as well (14, 15, 17). Yet the biology and ecology of arthropods in the soil are poorly understood compared with those of above-ground systems. The soil is considered the ultimate reservoir for the many species of microorganisms associated with insects. These associations range from direct associations (e.g., microorganisms that are pathogens of arthropods) to indirect associations (e.g., where the activity of arthropods impacts microbial communities). We have much yet to learn about the biology and ecology of associations among soil-dwelling arthropods and microorganisms, although progress is being made. Methods that have been developed for studying the associations between the diverse array of microorganisms and arthropods living in the complex soil environment will be described in this chapter.

The majority of methods that have been developed for studying microorganisms associated with arthropods in soil focus on arthropod pathogens, and these methods are emphasized in this chapter. Usually, the ultimate rationale for conducting studies of arthropod-pathogenic microorganisms in the soil has been to investigate their potential for use as agents for biological control of pestiferous arthropods. Soil-dwelling microorganisms for which research methods are discussed in this chapter include viruses, bacteria, fungi, protists, and nematodes. All groups of arthropods will be included in this chapter, but pestiferous species are the focus of the development of many methods for studying interactions with microorganisms. However, methods used during ecological studies of arthropod communities in the soil will also be included. Literature on soil ecology often differentiates between macroarthropods, e.g., larger insects, spiders, and myriapods, and microarthropods, e.g., mites, collembolans, and smaller insects. These groups are also referred to as the macrofauna and mesofauna, respectively (17).

Surveys of the literature on methods used to study pathogens infecting arthropods in the soil have been published previously in this manual (24, 27). Methods for studies of insect-pathogenic microorganisms have fairly recently been reviewed in detail for both laboratory (56, 68) and field (57), although these reviews were not specifically focused on the soil ecosystem. In addition, the diverse methods for measuring and assessing the microbial soil biota have been recently reviewed (e.g., reference 81). Therefore, here we do not provide a comprehensive review of the literature, for which readers are referred to these above-cited publications. Instead, we describe in more detail some of the major methods used for detecting, isolating, quantifying, and studying microorganisms, arthropods, and the interactions between these groups in the soil environment. Case studies are used to provide examples in which specific methods have been used and to illustrate approaches in more detail. In this chapter, we also emphasize newer techniques. Pathogens of arthropods are emphasized, but a section is also devoted to methods used during ecological studies of associations between microbial communities and arthropods. The same techniques will not be described in detail for the different microbial groups when the method is largely independent of the type of microbe being studied, i.e., when the same method is used for different types of microbes.

STUDYING ARTHROPOD-MICROORGANISM INTERACTIONS IN THE SOIL

Soil-dwelling arthropods and microorganisms and interactions between them can be studied in the field, but many times interactions are too complex or researchers are interested in effects due to specific components of the system. To conduct experiments testing isolated effects, studies are often performed in simulated ecosystems. Microcosms are small representative systems, usually in the laboratory, that try to simulate natural systems. One step larger are mesocosms, simulations of ecosystems that are somewhat closer to natural conditions, often containing more species than microcosms, and these are sometimes located outdoors. Experimental studies of microbial communities are also conducted in the field, usually with arthropods present versus absent. For all studies, the distribution and activity of both arthropods and microorganisms and conditions of the environment can differ through space and time, and this variability must be taken into account when conducting research.

Spatial and Temporal Relationships between Arthropods and Microorganisms

Some microorganisms associated with arthropods are found in the soil worldwide. For example, the arthropod pathogen

Bacillus thuringiensis is a cosmopolitan soil bacterium, commonly found in rich topsoils. It is generally found everywhere except in Antarctica, in deserts, and in lower soil horizons (7). In contrast, some soil-dwelling microorganisms associated with arthropods are thought to have very limited distributions, especially specific obligate pathogens of arthropod hosts with limited distributions; e.g., *Serratia entomophila* infecting the New Zealand grass grub *Costelytra zealandica* is known only from New Zealand (32). Arthropod species can also have broad versus very restricted distributions. For both soil-dwelling microorganisms and arthropods, distributions in some cases have expanded due to redistribution by humans, i.e., invasive species.

Arthropods differ in the amount of time they spend in the soil and in their locations within the soil (Fig. 1). Species of both arthropods and microorganisms often occur in specific locations in the soil environment. Some arthropods inhabit the surface of the soil, living on the soil surface and in the litter, and are called epiedaphic or epigeic (14, 17). In contrast, arthropods that live in habitats from the surface or root zone extending deeper into the soil are called euedaphic or endogeic (14, 17). Euedaphic arthropods can live at specific distances below the soil surface at different times of year and during different life stages. Some arthropods such as termites and ants construct above-ground structures made of soil. Yet others create and live within tunnels in the soil. For example, mole crickets construct tunnels in which they spend daylight hours, although they wander on the soil surface at night. Arthropods are generally most abundant in the surface soil horizons (the organic layer) as well as the uppermost mineral soil (the A horizon), and few arthropods inhabit deeper mineral soil horizons (e.g., B horizon), where oxygen and food are less available (J. Thies, personal communication).

Likewise, microorganisms interacting with soil arthropods can vary by the soil horizons in which they usually live, and as with arthropods, most microorganisms live in the O and A horizons, with fewer in the B horizon deeper in the soil. Microorganisms that are not usually present in deeper soil horizons where specific arthropods might live can come into contact when arthropods move vertically through the soil. In these cases, species of microorganisms do not have to be located at the final depth where the arthropod usually resides in order for encounters to occur. Some microorganisms have dormant or quiescent stages in the soil that are active only during specific intervals; e.g., resting spores (azygospores) of the gypsy moth fungal pathogen *Entomophaga maimaiga* germinate for only approximately 2 months in early spring and are dormant on the soil surface the rest of the year (38).

The length of time that arthropods spend in or on the soil varies from little for those species that are transients in or on the soil to considerable for those species that remain constantly on or within the soil (Fig. 1). Arthropods that are temporary inhabitants of the soil range from hunting on the soil surface or resting during the day to spending the winter. Insects either have hemimetabolous development, without complete metamorphosis (e.g., grasshoppers and aphids), or holometabolous development, with complete metamorphosis (e.g., flies, bees and wasps, and beetles). Especially for holometabolous insects, often only specific stages live within the soil, while other stages live above ground. For example, although larval flies might live and develop in the soil, winged adults would land on the soil only occasionally, predominantly inhabiting the aerial environment.

Therefore, location within the soil and timing of activities must always be taken into account when studying interactions between microorganisms and arthropods, so that studies being conducted are appropriate to the life histories and phenologies of all species of concern.

Impact of Soil Characteristics

Soil characteristics can strongly influence both arthropods and microorganisms living in this habitat. In particular, the composition of the soil and soil texture, ranging from sand to clay, should always be evaluated. For example, when the fungal pathogen *Beauveria bassiana* was applied to different types of soil, more imported red fire ants, *Solenopsis invicta*, in soils that were predominantly silt and sand were infected and killed than were fire ants in clay soil (26). Levels of organic matter in soil can strongly affect molecular methods used for detection and quantification of microorganisms. For muck soils that are high in organic matter or specific soil horizons high in organic matter, e.g., the O and H horizons (58), high levels of humic acid can interfere with extraction of microbial nucleic acids.

Soil temperature affects both arthropods and microorganisms since both are poikilothermic. Therefore, temperature is

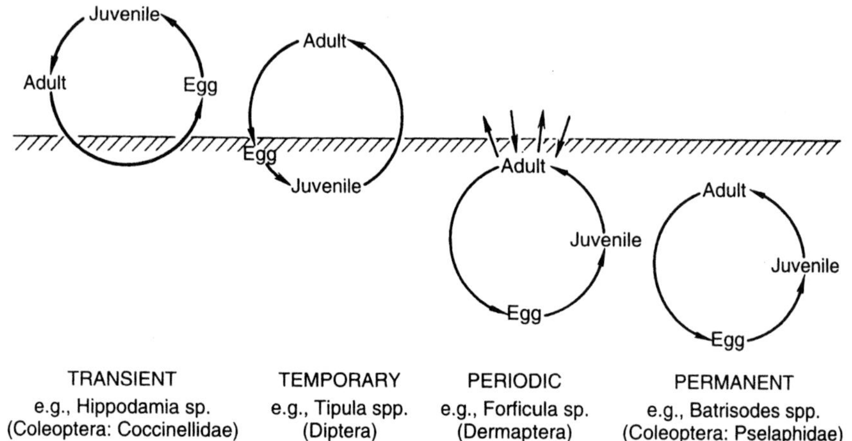

FIGURE 1 General categories of soil arthropods based on their relative locations in the soil and the lengths of time that different stages spend in the soil. Reprinted from reference 86.

an important variable for consideration during studies. In addition, soil moisture can be critically important for both arthropods and microorganisms in the soil, especially at extremes. For example, under complete saturation or complete lack of moisture, *B. bassiana* did not infect as many ants as under intermediate moisture levels (26). To evaluate soil moisture, water potential in megapascals or kilopascals quantifies the available water as chemical potential per unit volume (54; see reference 77). This quantification method is preferable to measuring the percent water by soil dry weight or percentage of water-holding capacity, because in different types of soils with the same percentages of moisture by weight, different amounts of water can be available to soil microorganisms. An additional soil attribute that can influence soil microorganisms is pH. For example, most bacteria and actinomycetes are inhibited by low pH, but fungal growth can be favored when pH is less than 5 (33).

In some soils, the microbial community that is present can cause microbial stasis. To conduct studies of interactions between arthropods and specific microorganisms in such habitats, it can be necessary to eliminate other microorganisms in the soil because their presence would interfere with studies. Usually, researchers want to kill all microorganisms in the soil and then introduce the organisms of interest. For studies with microorganisms and arthropods, often in the past sterilization has been effected by using an autoclave. However, an autoclave is very disruptive to soil physical properties (61) and is therefore not a good choice when conducting laboratory studies mimicking soil under natural conditions. McNamara et al. (63) reviewed studies comparing various attributes of nine different sterilization methods, and gamma (γ) irradiation was identified as the most reliable method for soil sterilization, causing minimal changes in physical properties. However, caution must be exercised because this method can influence soil chemical properties at high doses. Doses of γ irradiation necessary for sterilization depend on soil properties, with soils of high organic matter (e.g., peat) or denser soils requiring higher doses or longer exposures.

DETECTING AND QUANTIFYING ARTHROPOD-PATHOGENIC MICROORGANISMS IN SOIL

Soils, especially those rich in nutrients, are teeming with microorganisms, and often researchers are interested only in specific microorganisms. It is frequently necessary to extract, isolate, and quantify these specific microorganisms. Protocols for different types of microorganisms are varied based on characteristics of the specific groups of microorganisms.

Extraction from Soil and Physical Counts

Sampling Arthropods

Arthropods are often collected from soil to quantify prevalence of disease and to potentially also isolate pathogens of interest. Extraction of arthropods from soil can be conducted by using a variety of methods (see reference 27 for more detail); most of these techniques disrupt the soil environment when one is retrieving arthropods. If the arthropods of interest are not in the surface soil layers, a common method for sampling soil to retrieve arthropods is the use of soil cores (see references 24, 27, and 71). For larger organisms, soil cores are often visually checked, but smaller organisms present more of a challenge because they must be extracted from cores. Soil corers are available in different sizes and shapes. For standard corers, the core is pushed out from the top, but this practice compresses the soil and can damage or kill arthropods as well as disrupt soil horizons. More specialized corers split open, and the entire core can be removed intact so that arthropods and soil horizons are not disrupted (71).

Arthropods are isolated from soil samples either by mechanical or behavioral-dynamic means. For smaller arthropods, Berlese or Tullgren funnels use the behavior of living arthropods for extraction (71). Heat and light are applied above funnels containing soil samples, thereby driving living insects out of the sample. While samples for which Berlese funnels will be used cannot be stored for a long time, this method has the advantage of being able to run many samples at once. Mechanical methods are more labor intensive but allow longer-term storage of samples because they do not require that arthropods are alive or mobile, e.g., pupae or eggs. These methods include sieving, washing, and flotation. Other collection methods have been developed for specific groups of organisms and locations within soil, and researchers are encouraged to search the literature for their specific organisms and ecosystems of interest to learn about specialized methods and specialized locations within the soil to look for arthropods and microorganisms (71; many chapters in reference 57). For example, for arthropod species that emerge from the soil, emergence traps can be placed over the soil surface to collect individuals as they emerge without disrupting the soil environment and to assess prevalence of associations with microorganisms that occur in the soil (e.g., reference 70). For arthropods that occur at the soil surface, pitfall traps containing antifreeze are commonly used (see reference 71).

Quantifying arthropods that occur in soil is often easier than counting edaphic microorganisms because arthropods can often be counted without a microscope or with only a dissecting microscope. Of course, some groups of small arthropods that can be extremely abundant in the soil, i.e., mites, would still require use of compound microscopes for differentiation. When sampling arthropods in soil to quantify associated microorganisms, care must be taken regarding potential sources of bias as discussed by Fuxa et al. (27). Some pathogens cause behavioral shifts in hosts; e.g., lepidopteran pasture pests (*Wiseana* spp.) infected with nucleopolyhedrovirus (NPV) die on top of the soil instead of returning to their burrows (27, 53). If a shift in the location of infected versus healthy insects is identified and can be quantified, the sampling design or mathematical transformation should be adapted to correct for it, so that data on pathogen prevalence will correctly represent the entire population. Another typical source of bias can be horizontal transmission of pathogens after collection of soil arthropods. Pathogens can spread from infected to healthy individuals in soil samples before they are separated, or stress after collection can increase susceptibility or induce latent disease that would otherwise not be active. Infection can also be underestimated if infected insects are not handled appropriately so that they can be found in samples and properly diagnosed. To prevent these sources of bias, sampled insects should be separated quickly, sampling equipment should be disinfected, and collected arthropods should be reared for a shorter time than required for death by potential contaminants. For studies quantifying disease prevalence in which handling could have increased percent infection, samples can be rapidly preserved by various means, often storage in the cold, while researchers investigate whether such bias could have occurred. The potential for increased disease prevalence due to handling can subsequently be determined experimentally by bioassays, and this potential source of error can be corrected.

Sampling Microorganisms

Numerous physical methods have been used for extracting microorganisms in soil for quantification. These techniques are more appropriate for larger microorganisms, such as fungi and nematodes, and can be faster and less expensive than bioassays, depending on the system. Counts after extracting microorganisms from soil are not used for entomopathogenic bacteria, nonoccluded viruses, and protists; instead, various bioassay methods or molecular methods must be used.

Resting spores of entomophthoralean fungi (zygospores or azygospores) have been extracted from soil by several methods that all begin with wet sieving. The method that has been used most extensively is based on discontinuous density gradient centrifugation. Resting spores are captured on a 20-μm sieve, and samples are then centrifuged in a discontinuous density gradient made with Percoll (Amersham Biosciences, Piscataway, N.J.), followed by counts of resting spores in the 1.10-g/ml band plus interfaces at ×50 magnification (39). Another method for counting resting spores that was developed to avoid the cost of Percoll is use of sucrose density gradients (60). However, although this latter method requires less money up front for purchase of materials, it is more complicated and requires more time to complete. A less exact but simpler method has also been developed for quantification of resting spores of the gypsy moth pathogen *E. maimaiga* (88). Soil is washed through a 63-μm sieve; the residue is then mixed with water and allowed to settle for 3 min, after which a sample is taken from the midpoint of the liquid and resting spores (ca. 22 to 40 μm in diameter) are then completely counted in the sample with a microscope. Although this method is not as accurate as gradient centrifugation, results were associated with bioassay results and were inversely correlated with the number of years since the last epizootic (88).

Methods have also been developed for quantifying soilborne occlusion bodies of various viruses infecting arthropods. Occlusion bodies, the environmentally resistant stage of baculoviruses and some insect-infecting cypoviruses and poxviruses, are large enough to count by light microscopy after separation from the soil. Among different methods for separation from soil are desorption of occlusion bodies with sodium pyrophosphate, aqueous two-phase separation with a dextran-polyethylene glycol system, and metrizamide density gradient centrifugation (49). A simpler method was developed for NPV, based on ultrasonication in sodium dodecyl sulfate and differential centrifugation (20).

One problem with methods for soil extraction of fungal spores and viral occlusion bodies is that they are not very specific. Resting spores of many species of Entomophthorales and occlusion bodies of different occluded viruses are similar in appearance. Especially for fungal and viral species, use of these methods for extraction should be followed by molecular methods (see below) to ensure the identity of organisms being quantified.

Numerous techniques are available for extracting nematodes from the soil (14), and methods commonly used specifically for entomopathogenic nematodes have been compared by Hominick (45). Flotation is a highly efficient means for separating nematodes from soil, but this method yields only dauer, or juvenile, stages of the entire mixture of all species present in samples and is very labor intensive. In contrast, when Baermann funnels (54) are used, only active dauers will be detected, so this method is therefore considered less efficient than flotation. For both of these methods, only juveniles are isolated; this stage is extremely difficult to identify to species level when compared with the adult stage, so identification can be a stumbling block when using these methods. In addition, since all species of nematodes in the soil are extracted and not only entomopathogenic nematodes, many species will have to be identified that are not entomopathogenic. Researchers working with entomopathogenic nematodes often use bioassays instead (described below). However, even if a susceptible host is used during bioassays, only a portion of the entomopathogenic nematodes will be detected, and if no infections occur, it could reflect either an absence of nematodes or lack of infectivity at the time of sampling (45). In fact, methods for recovery of entomopathogenic nematodes from soil should be chosen based on study objectives, and no method is without some limitations.

Selective Media and Dilution Plating

An effective technique often used for isolating some types of microorganisms is use of selective media. However, this technique is possible only for groups that can be grown fairly easily in vitro and for which selective media have been developed. Samples of fertile soils are generally filled with many different microorganisms, and the goal with use of selective media is to allow growth of only the species of interest. Among the entomopathogens, selective media are basically used only for isolating and quantifying bacteria and fungi.

The arthropod-pathogenic bacterium for which selective media have been used extensively is *B. thuringiensis*. Germination of *B. thuringiensis* spores is strongly inhibited by acetate, and this fact has been used as the basis for selective media described in detail by Braun (7). Soil samples are incubated in rich liquid media plus acetate, which causes bacterial spores to germinate and grow but not *B. thuringiensis* spores. Next, samples are heat shocked, so that microorganisms not in the spore stage are killed and the only viable spores left are *B. thuringiensis*. *B. thuringiensis* spores subsequently grow easily in standard media without contamination by other species of bacteria (7). A selective medium based on a similar principle has also been developed for *Paenibacillus popilliae*, but isolation using this method requires seven heat shock cycles (64). An alternate method that requires less time and is more accurate for quantifying bacterial spores in soil is based on addition of vancomycin to media; many strains of *P. popilliae* are resistant to this antibiotic (75). Numerous additional types of selective media have been developed for *B. thuringiensis* (27, 55, 80), and at least one of these can be used for *Bacillus sphaericus* (80). The selective medium BATS allowed growth of 18 mosquito-pathogenic strains of *B. sphaericus* while inhibiting growth of 68% of nonpathogenic strains (90). Selective media have also been developed for detecting *S. entomophila* (27).

Selective media have been developed for several genera of insect-pathogenic fungi. Arguably, the most commonly studied arthropod-pathogenic fungi are *B. bassiana* and *Metarhizium anisopliae*, which are easy to grow in vitro (33). The most commonly used selective media for these species are based on standard medium ingredients, usually amended with the fungicides dodine, benlate, and cycloheximide, which suppress growth of numerous common soil saprophytes (33, 74). Dodine is no longer available as a fungicide for use in the United States but can be purchased from chemical manufacturers (e.g., Sigma). Crystal violet is added to the medium to enhance visual contrast of fungal colonies and improve counting. Selective media have also been developed for fungal entomopathogens in the genera *Culicinomyces*, *Paecilomyces*, *Lecanicillium*, *Nomuraea*, *Torrubiella*, and *Hirsutella* (27, 33).

If selective media are used to quantify specific microorganisms in soil samples, serial dilutions are made and plated, with replicates for each dilution. Especially with hyphomycetes, dilute detergent (e.g., Silwet or Tween) is added to the suspension to break up conidial aggregates and results from plates are counted as CFUs. Results are converted from plate counts to actual numbers of microorganisms in samples by using a standard curve. To create a standard curve, a series of known quantities of the microorganism are added to soil lacking this microorganism, CFUs are quantified, and a predictive equation is calculated.

Use of selective media is often imperfect, so precautions must be taken when choosing methods and interpreting results. Major disadvantages are that microorganisms with vigorous growth are favored and that, in fact, even within a species not all strains may grow equally well on a particular type of selective medium (see reference 90). Therefore, the choice of methods and media can heavily bias results from studies. One way to compensate for this problem would be to use more than one medium or more than one technique for each sample.

Bioassays to Detect and Quantify Entomopathogens and Nematodes in Soil

There are numerous types of bioassays that can be conducted to detect and quantify microorganisms associated with soil-dwelling arthropods. Bioassays have been extensively developed and used, because when bioassays have been compared with other techniques, bioassays are generally the most sensitive, most biologically meaningful, and sometimes also the most accurate method for detection and quantification (23). At present, molecular methods are seldom used for these purposes, so comparisons between bioassays and molecular techniques are not yet possible.

To use bioassays, a source of high-quality, standardized arthropods or microorganisms for tests is necessary. A source of microorganisms of interest can also be necessary to establish standard curves for different methods.

Soil Baiting with Arthropods

Bait insects are commonly used for detecting and quantifying entomopathogenic nematodes and fungi. In most cases, the bait insects are placed in some type of cages at desired depths in the soil in the field. Alternatively, soil can be transported to the laboratory and placed under conditions simulating the field and test insects added. The "*Galleria* bait method" for detection and quantification of both entomopathogenic nematodes and fungi has been widely used (27). Larvae of the greater wax moth (*Galleria mellonella*) are naturally found in beehives and are extremely susceptible to many soilborne pathogens. *G. mellonella* is also readily available since it is sold as fishing bait. If this method is used for isolating entomopathogenic nematodes from soil, all nematode stages can be recovered, which aids in identification, and labor is generally less intensive than for flotation and the Baermann funnel method (see above and reference 54). This same bait method is also frequently used for isolating arthropod-pathogenic Hyphomycetes from soil. For example, soil samples were collected from different environments in Ontario, Canada, and were baited with *G. mellonella* for various periods of time at three different temperatures. Larvae dying were surface sterilized and held in moist chambers, and fungi growing from cadavers were isolated on selective media (4). Surface sterilization of cadavers is an important step, because it greatly reduces growth of saprophytic contaminants, which in turn add significantly to the labor required. Although *G. mellonella* is especially susceptible to some nematodes and pathogens, it is not particularly susceptible to all of the pathogens that could potentially be present. This attribute should be taken into account by checking that the bait species used in trials are susceptible to the pathogen of interest. Thus, for some studies specific susceptible arthropods have been substituted in soil bioassays. For example, when researchers prospected for new pathogen strains for control of the Formosan subterranean termite, *Coptotermes formosanus*, these same termites were used for bait since they were available and most appropriate (78).

Some microorganisms predominantly remain near the surface of the soil, where interactions occur, e.g., resting spores of many species of entomophthoralean fungi (e.g., reference 40), ambusher entomopathogenic nematodes, and, at times, NPVs (25) infecting foliage-feeding lepidopteran larvae. In these instances, baiting bioassays require placing test insects on top of the soil. As an example, none of the life stages of the foliage-feeding gypsy moth, *Lymantria dispar*, spend time within the soil. However, later instar larvae of *L. dispar* spend diurnal periods resting at the soil surface under the leaf litter, and when resting in this microhabitat, they are heavily infected by the entomophthoralean fungal pathogen *E. maimaiga* (37). Resting spores (azygospores) of *E. maimaiga* predominantly occur in the organic layer on the surfaces of soils at bases of oak trees (*Quercus* spp.) (40). In the field, bioassays were conducted by placing laboratory colony larvae in cages made of window screening on the soil surface for several days. If field conditions are dry or if researchers seek to optimize chances for infection, water has been applied to the soil surface to ensure germination by spores that might be present (e.g., reference 41). Larvae generally remain in cages for 3 to 5 days, during which time food is not required. Our experience has been that, if leaves are placed in cages as food, they can be used by larvae to hide, thereby avoiding exposure to the soil surface, where spores are present. Cubes of artificial diet placed in cages quickly dry out and can attract vertebrates such as raccoons that will completely disrupt the study. After field exposure, larvae are returned to the laboratory, where they are reared on an artificial diet under controlled conditions. Due to the prolonged yearly dormancy of *E. maimaiga* resting spores, such bioassays can be conducted only during May and June, when resting spores are germinating (38). Cages can be covered to exclude deposition of airborne conidia of *E. maimaiga* to ensure that infections only from soilborne resting spores are being monitored (42). Bioassays with resting spores of *E. maimaiga* have also frequently been conducted in the laboratory by placing larvae within containers containing moist field-collected soil for specific periods under controlled conditions (e.g., reference 43). A similar study with *L. dispar* larvae exposed to field-collected soil in microcosms was employed to study interactions between gypsy moth larvae and NPV at the soil surface (87). Results demonstrated that this exposure method was not especially appropriate for NPV, a pathogen that must be eaten to infect. NPV infection from soil exposures was very low and did not correlate well with results from feeding trials using an artificial diet amended with water extracts of soil. In contrast with soil exposure studies, bioassays demonstrated that viable virus was abundant in soil samples.

Extraction of Pathogens from Soil followed by Infectivity Trials.

Numerous methods have been developed for extracting and detecting NPV occlusion bodies from soil (see above). However, methods for extracting nonoccluded viruses have been developed less often. Researchers successfully integrated extraction from soil plus subsequent use of hosts to extract an iridescent virus from soil (72). Soil samples taken between 0 and 90 days after this virus was added to soil were shaken for 30 min with 0.5% egg albumen. Samples were then centrifuged at 108 × *g* for 10 min to pellet the soil. The supernatant was then centrifuged at 3,900 × *g* in a swinging-bucket rotor for 15 min, the resulting pellet was resuspended in sterile water, and serial dilutions were made. Dilutions of between 10^{-1} and 10^{-10}, depending on when the sample was taken, were injected into *G. mellonella* larvae. Larvae were maintained on a semisynthetic artificial diet and were checked for signs of infection 12 to 14 days later.

Diet Incorporation Bioassays

This method was developed to test for the presence and abundance of microbial pathogens that must be ingested to infect. While many bacterial pathogens of arthropods infect per os, use of selective media is easier and is preferred. Therefore, this method has predominantly been used for arthropod-pathogenic viruses. Most studies have been conducted with baculoviruses that infect foliage-feeding insects to study dynamics of persistence of these viruses in the soil. As an example, when the cotton bollworm NPV was engineered to express an insect-specific toxin, the persistence and distribution of engineered virus in the soil were compared with those of the wild-type virus (28). After viral application, the top 2 cm of soil was collected and 1.7-cm-diameter core samples were taken to a depth of 38 cm and subdivided into three subsamples for different depths. To study horizontal distribution, more intensive sampling was conducted with 0- to 2-cm-deep samples. Thirty grams of soil from each sample was mixed into 270 ml of an artificial diet that was still liquid but had cooled to <50°C, and 6 ml of the artificial diet was poured into individual 30-ml rearing cups. At least 40 larvae were used for every soil sample. One third-instar cotton bollworm (*Helicoverpa zea*) larva was added to each cup, and larvae were checked until they died or pupated. Additional healthy insects were maintained individually on a nonamended artificial diet as controls. Death due to the cotton bollworm NPV was confirmed by observation of viral occlusion bodies in cadavers under phase-contrast microscopy. Data from dose-response studies in the laboratory had been used to develop regressions associating mortality with viral concentration. Using these relationships, percent mortality from soil bioassays was converted to concentrations of viral occlusion bodies in soil samples. Additional studies comparing wild-type virus to engineered virus were conducted with soil samples laced with virus maintained in microcosms in the laboratory for up to 17 months, adjusting for water loss over this time (28).

Microscopy

Microscopy has also been used to count arthropod-associated viruses, bacteria, and protists in soil. Usually the sample is suspended in water, and counts are done with electron, light, or fluorescence microscopy. By using light microscopy, occlusion bodies of NPV stained with Giemsa stain (79) or Buffalo Black (51) could be distinguished from soil particles. Occlusion bodies of a cypovirus and a granulovirus were detected in soil by scanning electron microscopy (48).

Immunofluorescence microscopy has been used for detecting *B. thuringiensis* (89). Stains and staining methods were compared for detecting spores of the microsporidian grasshopper pathogen *Paranosema locustae* in soil (30). Fluorescein isothiocyanate was the best out of the 12 stains that were compared. Results from this study were put to use when staining and microscopy were used to assess survival and persistence of *P. locustae* in the field (31). While microscopic assays have been shown to be useful in the above-cited instances, studies have shown that this general method is more appropriate when titers of microorganisms in soil are high. As a caveat, specificity of results can be questionable for assays based on microscopy.

Serology

Serological methods are specific and are appropriate for tracking known microorganisms for which antisera are available. These methods have been developed to detect and quantify a few types of bacteria and viruses in soil, but this methodology has rarely been put to use for quantifying microorganisms during field studies. The fluorescent-antibody method is used in conjunction with microscopy. Antisera to three different stages of *B. thuringiensis* (crystals, spores, and vegetative cells) were produced and used in concert to monitor soil samples by immunofluorescence microscopy (89). Hukuhara and Akami (50) used immunofluorescence microscopy to distinguish occlusion bodies of an NPV in soil. Fuxa et al. (29) used an enzyme-linked immunosorbent assay for detection of an NPV in soil. The assay detected as few as 360 occlusion bodies in soil, but detection efficacy was affected by the amount of clay in the soil, with higher clay content leading to lower absorbance values.

Molecular Methods

Molecular methods provide a means for detecting both inactive and active forms of culturable, as well as nonculturable, microorganisms that cannot readily be detected in other ways. Molecular techniques have rarely been used for evaluating abundance, although the methodology holds potential for this application in the future.

Colony hybridization permits the sensitive detection of culturable bacteria with specific sequences in soil. Rupar et al. (73) used the cloned *B. thuringiensis* toxin gene *cryIIIA* as a probe in colony hybridization and found two *cryIIIA*-hybridizing colonies in soil. Miteva et al. (66) applied DNA fingerprinting with M13 bacteriophage DNA as a probe for study of inter- and intraserotypic variations in different *B. thuringiensis* strains.

Arthropods can be collected from soil and checked for the presence of microorganisms when one is using PCR-based molecular methods. Some molecular methods require that the pathogen must first be isolated from the living or dead arthropod, because these methods are not specific and nucleic acids from hosts could be amplified simultaneously (e.g., randomly amplified polymorphic DNA [RAPD] and amplified fragment length polymorphism [AFLP]). Alternatively, if specific primers are available for the microorganism of interest, in vitro isolation from cadavers is not necessary and pathogen DNA can be detected by using arthropods.

Amounts of DNA extractable directly from soil can be small and below levels needed for detection by hybridization. For this reason and because it is applicable also to nonculturable, inactive, or dead microorganisms, uses of PCR-based methods have steadily been increasing. Due to commercial interest in *B. thuringiensis* for pest control and extensive exploration for new strains of this bacterium from soils worldwide, PCR-based techniques were first applied to detect this insect pathogen. Carozzi et al. (12) used PCR technology to identify *B. thuringiensis* strains containing *cryI*, *cryIII*, and *cryIV* genes in soil. Insecticidal activity predicted by PCR screening corresponded with the insecticidal activity in insect bioassays. Bioassays following PCR are appropriate because for some *B. thuringiensis*-host combinations, both the spore and live bacterium play a role in pathogenesis along with the *cry* toxin. However, *B. thuringiensis* spores and toxin crystals can become dissociated, so for such host-pathogen systems, bioassays can help to confirm presence of spores as well as toxins. Bourque et al. (6) used PCR to distinguish among strains of *B. thuringiensis* having the *cryIA(a)*, *cryIA(b)*, and *cryIA(c)* genes. Bravo et al. (8) used multiplex PCR with novel general and specific primers that could detect 12 *cry* genes to identify *B. thuringiensis* strains from 503 soil samples from Mexico.

PCR-based methods have been developed for detection of other arthropod pathogens in soil. To use PCR, first the microbial DNA must be extracted from the soil; this step can be difficult. Among the numerous methods that have been developed for DNA extraction, phenol-ether and magnetic capture-hybridization (MCH) were compared with regard to their sensitivity and efficiency for extracting DNA of the NPV of velvetbean caterpillar, *Anticarsia gemmatalis*, from soil (67). Although both procedures produced DNA amplifiable by PCR, the MCH method was 100-fold more sensitive than the phenol-ether protocol. *A. gemmatalis* NPV DNA was detected in field-collected soil samples 15 to 180 days after virus application when the MCH procedure to isolate DNA was coupled with PCR amplification of the polyhedrin region of the genome. Extraction of DNA from soil samples can be facilitated with commercial kits (QBiogene, Irvine, Calif., and MoBio Laboratories, Solana Beach, Calif.). DNA extraction can be inhibited by high levels of humus in the soil, such as in the organic soil layer, but methods or kits to neutralize the inhibition are available (e.g., Power Soil Kit; MoBio Laboratories).

STUDYING BEHAVIOR OF ARTHROPODS IN RESPONSE TO SOIL-DWELLING MICROORGANISMS

Edaphic arthropods display adaptations for a life in the soil, and some of these adaptations are thought to have evolved to avoid infection by soil-dwelling microorganisms, e.g., sieve plates over the spiracles of some scarab larvae (83). Soil-dwelling arthropods also exhibit behaviors to prevent infection, such as actively dislodging microorganisms from the body surface or applying glandular secretions inhibitory to entomopathogens. Arthropods such as ants and termites create refuges without microorganisms if part of their colony is infested. A different twist to studying interactions between microorganisms and arthropods in soil has been to evaluate the behavior of insects in response to insect-pathogenic fungi. Mole crickets were challenged with *B. bassiana* and *M. anisopliae* in the soil, and the behaviors of these potential hosts were investigated (84). Studies were initiated because it was found that, with increasing doses of *B. bassiana* sprayed onto the soil, fewer tawny mole crickets (*Scapteriscus vicinus*) were becoming infected. It was hypothesized that mole crickets were avoiding these pathogens. However, it is usually difficult or even impossible to learn what arthropods or microorganisms are doing in the soil without disrupting their interactions; behaviorists try to avoid disruptions because they would potentially change interactions. Researchers designed a method for viewing reactions of mole crickets to soil fungi without disruption by using X-ray radiography, i.e., shadow images made on photographic film by the action of radiation (85). Individual adult mole crickets were placed in Plexiglas arenas (5 by 30 by 38 cm) filled with sand at 8% moisture and allowed to tunnel and feed on the roots of Bermuda grass sod before fungus-treated sand or clean sand was added to the surface. Radiographs taken pretreatment compared with those taken a few hours after treatment and 1 to 3 days later showed that, for both species of fungi, *S. vicinus* adults clearly avoided tunneling through the layer containing entomopathogenic fungi and that tunnels made in conidium-incorporated soil deteriorated due to disuse. In a similar set of experiments using X-ray radiography of tubular microcosms (3 by 3 by 30 cm), third instar Japanese beetle (*Popillia japonica*) grubs were repelled by *M. anisopliae* incorporated into soil for up to 20 days (85). X-ray radiographs were therefore very useful for investigating the behavior of larger soil insects, but this technique cannot be used for smaller arthropods ($\leq$0.50 by $\leq$10.00 mm). Neutron radiography, which has been used to study rhizospheres, is based on hydrogen content instead of density (as in X rays) and is appropriate for use with smaller arthropods. This methodology was successfully used to capture images of early instar larvae of the western corn rootworm, *Diabrotica virgifera virgifera* (1). This method has great potential for further use in studying interactions between smaller soil-dwelling arthropods and microorganisms, although the cost and availability of facilities where neutron radiography is possible could limit use.

The studies above observed behavior of hosts in response to soil microorganisms, but the behavior of some soil microorganisms associated with arthropods can also be studied. The behavior of entomopathogenic nematodes in response to arthropod hosts in the soil has frequently been studied in laboratory microcosms. Studies of vertical distribution in microcosms used sand-filled tubes (5 by 5 cm with 8% moisture). Nematodes and potential hosts were differentially positioned in tubes, and hosts were retrieved after various lengths of time. These studies showed that some species, the "ambushers," locate hosts at the top of the soil, while others "cruise" within the soil, and that behaviors of some species are intermediate (59). Additional aspects of the behavior of entomopathogenic nematodes in relation to hosts, e.g., response to host cues and jumping by ambushers to reach hosts, have also been studied in the laboratory (59).

Behavioral studies with three species of collembola were focused on the extent to which collembola are attracted to several species of entomopathogenic fungi versus baker's yeast (16). A plastic preference chamber divided into 16 compartments was filled with quartz sand, and fungal suspensions were added to some compartments along with collembola to create different treatment comparisons. Chamber contents were incubated for 24 h, and distributions of collembola were then recorded to determine preference. In subsequent assays, a gradient of fungal concentrations was created in the chamber before collembola were added.

ECOLOGICAL STUDIES OF SOIL-DWELLING ARTHROPODS AND MICROBIAL COMMUNITIES

Soil-dwelling arthropods can influence microorganisms in the soil that are not arthropod pathogens directly by eating them, transporting them, or serving as a source of nutrients after the arthropod's death or indirectly by altering various aspects of the soil habitat. Usually studies of the interactions between microorganisms and arthropods in the soil take the form of experiments monitoring changes in microorganisms or microbial communities when arthropods are present or absent. These studies commonly utilize general methods for studying microbial communities (see elsewhere in this volume) to evaluate abundance or respiratory activity.

Microcosm and Mesocosm Studies

Microcosms and mesocosms coupled with a diversity of analytical methods are often used to study the effects of controlled numbers of arthropods on microbial communities. As an example, the effects of feeding by soil-dwelling cranefly larvae (*Tipula paludosa*) on microorganisms were comprehensively evaluated by using microcosms comparing soil samples associated with three species of grassland plants (82). Leachate from rhizospheres of plants grown with or without larvae was added to soil columns maintained for 15 days. Microbial biomass C from soil was measured by chloroform fumigation extraction (9), and soil basal respiration was measured by gas chromatography. Polar lipids were transformed to ester-linked phospholipid fatty acids (PLFAs) to quantify fungal and bacterial communities (21, 35; see elsewhere in this volume). To quantify community-level physiological profiles, carbon substrate utilization potential was measured by using Biolog gram-negative microplates (Biolog, Inc., Hayward, Calif.) and root exudate profile microplates (11).

A study of interactions between collembolans (*Folsomia candida*) and arbuscular mycorrhizal (AM) fungi used mesocosms to investigate whether collembolans influenced extension of AM fungi or vectored AM fungi (34). Root length colonized by AM fungi was measured, AM spores were counted (at ×40 magnification after wet sieving, centrifugation in 50% sucrose, and vacuum filtration), fungal and bacterial biomass were monitored by PLFA and NFLA (22, 34), and numbers of surviving collembolans were counted.

Much has been written about different ways to apply entomopathogens to soil and how to subsequently sample to evaluate percent infection among arthropods (see reference 57). A more unusual approach was taken for testing survival of microbial pathogens in soils having differing levels of fungistasis. Different concentrations of *B. bassiana* conidia were applied to pupae of the Colorado potato beetle, *Leptinotarsa decemlineata*, which were then placed in microcosms containing two different soils (36). Subsequent percent infection of pupae that had received different conidial concentrations was used to compare fungistasis between these different soils. The percentage of inoculated pupae from which *B. bassiana* was able to sporulate provided an additional comparison between soils.

Field Studies

Ecological studies of the effects of arthropods on soil microbial communities are often more empirical and are also highly variable in the questions being addressed. One such study investigated the natural prevalence of entomopathogenic fungi associated with leaf-cutting ants (47). Three different methods were used for each of five nests of four species of ants. Living workers were collected from within the nest and while foraging, and both living and dead ants were collected from dump piles and near nests to detect fungal pathogens. Soil cores were taken to a 5-cm depth within 5 m of each nest and were baited with *G. mellonella* or *Tenebrio molitor* and plated on selective media. All fungal strains were isolated when possible to further compare the relatedness of strains present in the area by molecular techniques. The genetic diversity of the most commonly occurring pathogen, *M. anisopliae*, was then investigated with specific primers followed by sequencing for each isolate. Relatedness of isolates was compared by phylogenetic analysis. In a similar study of nests of 38 species of Australian termites, selective media were used to isolate *M. anisopliae* from soil samples (65). Isolates were compared by using RAPDs, and sequences of the internal transcribed spacer region were compared. In both studies, although entomopathogenic fungi differed by species and strains within species in and around ant and termite nests, few infections occurred and the entomopathogenic fungi found were considered opportunistic pathogens.

Another study of the effects of ant nests compared communities of bacteria, fungi, nematodes, and microarthropods (mites and collembolans) in harvester ant (*Messor andrei*) nests with soil adjacent to nests (5). Bacterial and fungal communities were quantified by PLFA; mites and collembolans were extracted from soil by using Tullgren funnels and counted at ×120 magnification. Data were analyzed by principal-component analysis to find that all microorganisms were more abundant and diverse in ant nests than in non-ant-containing soil.

An ecological study focused on a sugarcane field in South Africa, where activity of bacteria in the soil was evaluated in association with locations of termitaria of *Macrotermes natalensis* that had been leveled by plowing (10). Sugarcane grows noticeably better in these localized areas known as "isiduli." Soil samples were collected at 0- to 30-cm depths along a transect across a representative isiduli. Bacterial communities were evaluated by using Biolog (see elsewhere in this volume). The presence of at least five species of nematodes, soil characteristics, and plant growth were quantified, and principal-component analysis was used to evaluate correspondence among different variables.

Predation by microarthropods, predominantly collembolans and mites, plus nematodes on microbial communities in soils with three amendments was evaluated by measuring soil respiration (62). Microcosms containing soil with microarthropods plus nematodes were compared with "defaunated" soil in which these organisms had been killed by heat. Then three different amendments were added to different microcosms with four treatments: no inhibitor, streptomycin, cycloheximide, or both inhibitors. After 3 h, glucose was added to stimulate respiration, which was then measured hourly by gas chromatography. Fungal, bacterial, and total microbial biomass C was quantified by fumigation-extraction or substrate-induced respiration. Microarthropods and nematodes were extracted with Tullgren funnels and counted microscopically.

APPLICATIONS OF MOLECULAR TECHNIQUES IN ECOLOGICAL STUDIES

Molecular techniques are being applied increasingly to ecological studies of associations between microorganisms and soil-dwelling arthropods. Different methods are specifically appropriate for answering different types of questions in different types of systems, but generally all methods first require extraction of nucleic acids. Bidochka (3) provides a comprehensive review of these methods for use with fungal pathogens. For ecological studies of communities of microorganisms, such techniques have principally been used to date for bacteria (2), but community profiling of other types of microorganisms is now also possible by using universal primers (see elsewhere in this volume). Thies (81) surveyed the variety of molecular techniques applicable for profiling soil biota.

Molecular Markers

Numerous methods are available for marking microorganisms to detect them in the soil environment. Molecular markers have been applied to detect a strain of the fungal entomopathogen *M. anisopliae* transformed to carry an additional gene to enhance virulence, *Pr1*, which encodes a protease expressed by this fungus for penetrating host cuticle (46). Before field testing of this engineered pathogen, the U.S. Environmental Protection Agency required that a system be developed to monitor the strain after release to determine establishment and dispersal. A commercialized strain of *M. anisopliae* was transformed with genes for *Pr1*, and green fluorescent protein (GFP) and fungi were sprayed in a cabbage plot. Soil samples were taken at specific depths within and around the application areas for up to a year. In the center of the spray area, soil samples were also taken to specifically investigate the rhizosphere, at 4 to 5 cm from the cabbage taproot and along the taproot. CFUs were quantified by plating soil samples on semiselective media (Veen's semiselective agar medium [33] minus cycloheximide) to which hide protein azure was added to detect the constitutive protease production by *Pr1*. Plates were exposed to UV to detect the GFP-expressing fungal strain. In addition, insect populations were sampled on foliage and in pitfall traps in and around the plots. These insects were reared in the laboratory and were evaluated by using UV to detect cuticular contamination with the GFP strain, and tissues were plated on selective media and observed to detect GFP.

Specific Primers

Specific primers amplified by PCR have been used when researchers are interested in tracking strains of microorganisms released in the environment (13). This method is especially appropriate when detection needs to be extremely specific because strains of the species being released are also indigenous. RAPD analysis was used to identify unique fragments distinguishing the commercially available GHA strain of *B. bassiana* from indigenous strains. Three unique fragments were cloned and sequenced and were used to design sequence-characterized amplified region (SCAR) primers. For use with soil from the field, samples were first plated on selective media prior to extraction of genomic DNA. Sequence-characterized amplified region primers were able to detect 100-pg quantities of genomic DNA of the GHA strain of *B. bassiana* in field samples.

Microsatellites and AFLP for Population Studies

Microsatellites, also called simple sequence repeats, are often well suited for studying populations because many closely related strains of an organism can be both detected and discriminated. These tandemly repeated DNA sequences of from one to six bases can be highly polymorphic among individuals. To date, microsatellites have not been used extensively for studies of microorganisms associated with arthropods, but their use will certainly increase as more microsatellites are identified. Microsatellites were developed to discriminate among isolates of *B. brongniartii* associated with the European cockchafer, *Melolontha melolontha*, in Switzerland after three PCR-based methods were found not to be sensitive enough (18). Six polymorphic microsatellite loci allowing discrimination, even between closely related *B. brongniartii* isolates from the field, were used. Soil samples were collected from seven sites in grasslands and orchards where four different strains of this fungus had been applied up to 14 years previously. Soil samples were taken from each site at depths of 5 to 20 cm, sieved, and plated on selective medium (76) to isolate CFUs (19). In addition, four *G. mellonella* larvae were exposed to sieved soil from each sample and were then reared to document infection. These two methods were used to quantify the fungus from soil samples, but the bioassay method was found to be more sensitive. To evaluate which strains were present, five of the microsatellite loci discriminated among the 73 *B. brongniartii* strains isolated from sites.

Another method appropriate for evaluating strains within pathogen species is AFLP (69). The gypsy moth fungal pathogen *E. maimaiga* was isolated from hosts exposed to soil samples collected throughout the range of this fungus in Asia, where it is native, and in North America, where it has been introduced (e.g., reference 44). Phylogenetic analysis was used to investigate the source of the strain of *E. maimaiga* that had become established in North America.

Community Studies

Molecular techniques for profiling microbial communities have been used for studies of both bacterial and fungal communities (2). Community profiling was applied to associations of microorganisms with arthropods in soil in a study of the effects of arthropod pest control measures on microbial communities. The effects of applications of the fungal entomopathogens *M. anisopliae* and *B. bassiana* for control of arthropod pests on microbial communities were evaluated in alfalfa fields, where these fungi would potentially be applied to control soil-dwelling alfalfa snout weevils (*Otiorhynchus ligustici*) (52). Entomopathogenic fungi were applied to the top 10 cm of soil immediately before seeding. Soil samples were taken to 20-cm depth from before planting to 12 weeks after planting. Bacteria were counted on R2A agar and saprophytic fungi on potato dextrose agar, and *B. bassiana* and *M. anisopliae* were counted on selective media (Veen's medium plus cycloheximide for *M. anisopliae* and Veen's medium plus dodine for *B. bassiana* [33]). Terminal restriction fragment length polymorphism analysis using universal eubacterial primers was applied to compare bacterial communities. Denaturing gradient gel electrophoresis with ascomycete-specific primers was used to evaluate ascomycete fungal community structure.

We thank Janice Thies, Chris Jones, Louela Castrillo, and Leslie Allee for sharing unpublished data and for their comments on the manuscript, and we thank Daniel Buckley for suggestions.

REFERENCES

1. **Allee, L. L.** 2001. Corn rootworm ecology, economics, and behavior. Ph.D. dissertation. Cornell University, Ithaca, N.Y.
2. **Anderson, I. C., and J. W. G. Cairney.** 2004. Diversity and ecology of soil fungal communities: increased understanding

through application of molecular techniques. *Environ. Microbiol.* **6:**769–779.

3. **Bidochka, M. J.** 2001. Monitoring the fate of biocontrol fungi, p. 193–218. *In* T. M. Butt, C. Jackson, and N. Magan (ed.), *Fungi as Biocontrol Agents: Progress, Problems and Potential.* CABI Publishing, Wallingford, United Kingdom.
4. **Bidochka, M. J., J. E. Kasperski, and G. A. M. Wild.** 1998. Occurrence of the entomopathogenic fungi *Metarhizium anisopliae* and *Beauveria bassiana* in soils from temperate and near-northern habitats. *Can. J. Bot.* **76:** 1198–1204.
5. **Boulton, A. M., B. A. Jaffee, and K. M. Scow.** 2003. Effects of a common harvester ant (*Messor andrei*) on richness and abundance of soil biota. *Appl. Soil Ecol.* **23:**257–265.
6. **Bourque, S. N., J. R. Valero, J. Mercier, M. C. Lavoie, and R. C. Levesque.** 1993. Multiplex polymerase chain reaction for detection and differentiation of the microbial insecticide *Bacillus thuringiensis*. *Appl. Environ. Microbiol.* **59:**523–527.
7. **Braun, S.** 2000. Production of *Bacillus thuringiensis* insecticides for experimental uses, p. 49–71. *In* A. Navon and K. R. S. Ascher (ed.), *Bioassays of Entomopathogenic Microbes and Nematodes*. CABI Publishing, Wallingford, United Kingdom.
8. **Bravo, A., S. Sarabia, L. Lopez, H. Ontiveros, C. Abarca, A. Ortiz, M. Ortiz, L. Lina, F. J. Villalobos, G. Pena, M. Nunez-Valdez, M. Soberon, and R. Quintero.** 1998. Characterization of *cry* genes in a Mexican *Bacillus thuringiensis* strain collection. *Appl. Environ. Microbiol.* **64:** 4965–4972.
9. **Brookes, P. C., A. Landman, G. Pruden, and D. S. Jenkinson.** 1985. Chloroform fumigation and the release of soil nitrogen: a rapid and direct extraction method to measure microbial biomass nitrogen in soil. *Soil Biol. Biochem.* **17:**837–842.
10. **Cadet, P., L. Guichaoua, and V. W. Spaull.** 2004. Nematodes, bacterial activity, soil characteristics and plant growth associated with termitaria in a sugarcane field in South Africa. *Appl. Soil Ecol.* **25:**193–206.
11. **Campbell, C. D., S. J. Grayston, and D. J. Hirst.** 1997. Use of rhizosphere carbon sources in sole carbon source tests to discriminate soil microbial communities. *J. Microbiol. Methods* **30:**33–41.
12. **Carozzi, N. B., V. C. Kramer, G. W. Warren, S. Evola, and M. G. Koziel.** 1991. Prediction of insecticidal activity of *Bacillus thuringiensis* strains by polymerase chain reaction product profiles. *Appl. Environ. Microbiol.* **57:**3057–3061.
13. **Castrillo, L. A., J. D. Vandenberg, and S. P. Wraight.** 2003. Strain-specific detection of introduced *Beauveria bassiana* in agricultural fields by use of sequence-characterized amplified region markers. *J. Invertebr. Pathol.* **82:**75–83.
14. **Coleman, D. C., D. A. Crossley, Jr., and P. F. Hendrix.** 2004. *Fundamentals of Soil Ecology*, 2nd ed. Elsevier, Amsterdam, The Netherlands.
15. **Dindal, D. L. (ed.).** 1990. *Soil Biology Guide*. Wiley, New York, N.Y.
16. **Dromph, K. M., and S. Vestergaard.** 2002. Pathogenicity and attractiveness of entomopathogenic hyphomycete fungi to collembolans. *Appl. Soil Ecol.* **21:**197–210.
17. **Eisenbeis, G., and W. Wichard.** 1987. *Atlas on the Biology of Soil Arthropods*. Springer-Verlag, Berlin, Germany.
18. **Enkerli, J., F. Widmer, C. Gessler, and S. Keller.** 2001. Strain-specific microsatellite markers in the entomopathogenic fungus *Beauveria brongniartii*. *Mycol. Res.* **105:**1079–1087.
19. **Enkerli, J., F. Widmer, and S. Keller.** 2004. Long-term field persistence of *Beauveria brongniartii* strains applied as biocontrol agents against European cockchafer larvae in Switzerland. *Biol. Control* **29:**115–123.
20. **Evans, H. F., J. M. Bishop, and E. A. Page.** 1980. Methods for the quantitative assessment of nuclear-polyhedrosis virus in soil. *J. Invertebr. Pathol.* **35:**1–8.
21. **Federle, T. W.** 1986. Microbial distribution in the soil—new techniques, p. 403–498. *In* F. Megusar and M. Gantar (ed.), *Perspectives in Microbial Ecology*. Slovene Society for Microbiology, Ljubljana, Yugoslavia.
22. **Frostegård, Å., and E. Bååth.** 1996. The use of phospholipid fatty acid analysis to estimate bacterial and fungal biomass in soil. *Biol. Fertil. Soils* **22:**59–65.
23. **Fuxa, J. R.** 1987. Ecological methods, p. 23–41. *In* J. R. Fuxa and Y. Tanada (ed.), *Epizootiology of Insect Diseases*. Wiley & Sons, New York, N.Y.
24. **Fuxa, J. R., and Y. Kunimi.** 1997. Microorganisms interacting with insects, p. 509–519. *In* C. J. Hurst, G. R. Knudsen, M. J. McInerney, L. D. Stetzenbach, and M. V. Walter (ed.), *Manual of Environmental Microbiology*. ASM Press, Washington, D.C.
25. **Fuxa, J. R., and A. R. Richter.** 1996. Effect of agricultural operations and precipitation on vertical distribution of a nuclear polyhedrosis virus in soil. *Biol. Control* **6:**324–329.
26. **Fuxa, J. R., and A. R. Richter.** 2004. Effects of soil moisture and composition and fungal isolate on prevalence of *Beauveria bassiana* in laboratory colonies of the red imported fire ant (Hymenoptera: Formicidae). *Environ. Entomol.* **33:**975–981.
27. **Fuxa, J. R., Y. Kunimi, and M. Nakai.** 2002. Research methods for microorganisms interacting with arthropods in soil, p. 660–671. *In* C. J. Hurst, R. L Crawford, G. R. Knudsen, M. J. McInerney, and L. D. Stetzenbach (ed.), *Manual of Environmental Microbiology*, 2nd ed. ASM Press, Washington, D.C.
28. **Fuxa, J. R., M. M. Matter, A. Abdel-Rahman, S. Micinski, A. R. Richter, and J. L. Flexner.** 2001. Persistence and distribution of wild-type and recombinant nucleopolyhedroviruses in soil. *Microb. Ecol.* **41:**222–232.
29. **Fuxa, J. R., G. W. Warren, and C. Kawanishi.** 1985. Comparison of bioassay and enzyme-linked immunosorbent assay for quantification of *Spodoptera frugiperda* nuclear polyhedrosis virus in soil. *J. Invertebr. Pathol.* **46:** 133–138.
30. **Germida, J. J.** 1984. Persistence of *Nosema locustae* spores in soil as determined by fluorescence microscopy. *Appl. Environ. Microbiol.* **47:**313–318.
31. **Germida, J. J., A. B. Ewen, and E. E. Onofriechuk.** 1987. *Nosema locustae* Canning (Microsporida) spore population in treated field soils and resident grasshopper populations. *Can. Entomol.* **119:**355–360.
32. **Glare, T. R., G. E. Corbett, and A. J. Sadler.** 1993. Association of a large plasmid with amber disease of the New Zealand grass grub, *Costelytra zealandica*, caused by *Serratia entomophila* and *Serratia proteamaculans*. *J. Invertebr. Pathol.* **62:**165–170.
33. **Goettel, M. S., and G. D. Inglis.** 1997. Fungi: Hyphomycetes, p. 213–249. *In* L. Lacey (ed.), *Manual of Techniques in Insect Pathology*. Academic Press, San Diego, Calif.
34. **Gormsen, D., P. A. Olsson, and K. Hedlund.** 2004. The influence of collembolans and earthworms on AM fungal mycelium. *Appl. Soil Ecol.* **27:**211–220.
35. **Grayston, S. J., G. S. Griffith, J. L. Mawdsley, C. D. Campbell, and R. D. Bardgett.** 2001. Accounting for variability in soil microbial communities of temperate upland grassland ecosystems. *Soil Biol. Biochem.* **30:**533–551.

36. **Groden, E., and J. L. Lockwood.** 1991. Effects of soil fungistasis on *Beauveria bassiana* and its relationship to disease incidence in the Colorado potato beetle, *Leptinotarsa decemlineata*, in Michigan and Rhode Island soils. *J. Invertebr. Pathol.* **57:**7–16.
37. **Hajek, A. E.** 2001. Larval behavior in *Lymantria dispar* increases risk of fungal infection. *Oecologia* **126:**285–291.
38. **Hajek, A. E., and R. A. Humber.** 1997. Formation and germination of *Entomophaga maimaiga* azygospores. *Can. J. Bot.* **75:**1739–1747.
39. **Hajek, A. E., and M. M. Wheeler.** 1994. Application of techniques for quantification of soil-borne entomophthoralean resting spores. *J. Invertebr. Pathol.* **64:**71–73.
40. **Hajek, A. E., L. Bauer, M. L. McManus, and M. M. Wheeler.** 1998. Distribution of resting spores of the *Lymantria dispar* pathogen *Entomophaga maimaiga* in soil and on bark. *BioControl* **43:**189–200.
41. **Hajek, A. E., J. S. Elkinton, and J. J. Witcosky.** 1996. Introduction and spread of the fungal pathogen *Entomophaga maimaiga* along the leading edge of gypsy moth spread. *Environ. Entomol.* **25:**1235–1247.
42. **Hajek, A. E., C. Olsen, and J. S. Elkinton.** 1999. Dynamics of airborne conidia of the gypsy moth (Lepidoptera: Lymantriidae) fungal pathogen *Entomophaga maimaiga* (Zygomycetes: Entomophthorales). *Biol. Control* **16:**111–117.
43. **Hajek, A. E., N. W. Siegert, M. M. Wheeler, and D. G. McCullough.** 2004. Using bioassays to predict abundance of *Entomophaga maimaiga* resting spores in soil. *J. Invertebr. Pathol.* **86:**61–64.
44. **Hajek, A. E., M. Shimazu, and B. Knoblauch.** 2000. Isolating *Entomophaga maimaiga* using resting spore-bearing soil. *J. Invertebr. Pathol.* **75:**298–300.
45. **Hominick, W. M.** 2002. Biogeography, p. 115–143. *In* R. Gaugler (ed.), *Entomopathogenic Nematology*. CABI Publishing, Wallingford, United Kingdom.
46. **Hu, G., and R. J. St. Leger.** 2002. Field studies using a recombinant mycoinsecticide (*Metarhizium anisopliae*) reveal that it is rhizosphere competent. *Appl. Environ. Microbiol.* **68:**6383–6387.
47. **Hughes, W. O. H., L. Thomsen, J. Eilenberg, and J. J. Boomsma.** 2004. Diversity of entomopathogenic fungi near leaf-cutting ant nests in a neotropical forest, with particular reference to *Metarhizium anisopliae* var. *anisopliae*. *J. Invertebr. Pathol.* **85:**46–53.
48. **Hukuhara, T.** 1972. Demonstration of polyhedra and capsules in soil with scanning electron microscope. *J. Invertebr. Pathol.* **20:**375–376.
49. **Hukuhara, T.** 1977. Purification of polyhedra of a cytoplasmic polyhedrosis virus from soil using metrizamide. *J. Invertebr. Pathol.* **30:**270–272.
50. **Hukuhara, T., and K. Akami.** 1987. Demonstration of polyhedral inclusion bodies of a nuclear polyhedrosis virus in field soil by immunofluorescence microscopy. *J. Invertebr. Pathol.* **49:**130–132.
51. **Hukuhara, T., and H. Namura.** 1971. Microscopic demonstration of polyhedra in soil. *J. Invertebr. Pathol.* **18:**162–164.
52. **Jones, C. M., and J. E. Thies.** 2004. Soil bacterial and fungal community response to application of entomopathogenic fungi in the field, abstr. 5215. *Abstr. 68th Am. Soc. Agron. Meet.*
53. **Kalmakoff, J., and A. M. Crawford.** 1982. Enzootic virus control of *Wiseana* spp. in the pasture environment, p. 435–448. *In* E. Kurstak (ed.), *Microbial and Viral Pesticides*. Marcel Dekker, New York, N.Y.
54. **Kaya, H. K., and P. S. Stock.** 1997. Techniques in insect nematology, p. 281–324. *In* L. Lacey (ed.), *Manual of Techniques in Insect Pathology*. Academic Press, San Diego, Calif.
55. **Klein, M. G.** 1997. Bacteria of soil-inhabiting insects, p. 101–116. *In* L. Lacey (ed.), *Manual of Techniques in Insect Pathology*. Academic Press, San Diego, Calif.
56. **Lacey, L. (ed.).** 1997. *Manual of Techniques in Insect Pathology*. Academic Press, San Diego, Calif.
57. **Lacey, L. A., and H. K. Kaya (ed.).** 2000. *Manual of Field Techniques in Invertebrate Pathology*. Kluwer Academic Publishers, Dordrecht, The Netherlands.
58. **Lavelle, P., and A. V. Spain.** 2001. *Soil Ecology*. Kluwer Academic Publishers, Dordrecht, The Netherlands.
59. **Lewis, E. E.** 2002. Behavioural ecology, p. 205–223. *In* R. Gaugler (ed.), *Entomopathogenic Nematology*. CABI Publishing, Wallingford, United Kingdom.
60. **Li, Z., R. S. Soper, and A. E. Hajek.** 1988. A method for recovering resting spores of Entomophthorales (Zygomycetes) from soil. *J. Invertebr. Pathol.* **62:**248–251.
61. **Lotrario, J. B., B. J. Stuart, T. Lam, R. R. Arands, O. A. O'Connor, and D. S. Kosson.** 1995. Effects of sterilization methods on the physical characteristics of soil: implications for sorption isotherm analyses. *Bull. Environ. Contam. Toxicol.* **54:**668–675.
62. **Mamilov, A. S., B. A. Byzov, D. G. Zvyagintsev, and O. M. Dilly.** 2001. Predation on fungal and bacterial biomass in a soddy-podzolic soil amended with starch, wheat straw and alfalfa meal. *Appl. Soil Ecol.* **16:**131–139.
63. **McNamara, N. P., H. I. J. Black, N. A. Beresford, and N. R. Parekh.** 2003. Effects of acute gamma irradiation on chemical, physical and biological properties of soil. *Appl. Soil Ecol.* **24:**117–132.
64. **Milner, R. J.** 1977. A method for isolating milky disease, *Bacillus popilliae* var. *rhopaea*, spores from the soil. *J. Invertebr. Pathol.* **30:**283–287.
65. **Milner, R. J., J. A. Staples, T. R. Hartley, G. G. Lutton, F. Driver, and J. A. L. Watson.** 1998. Occurrence of *Metarhizium anisopliae* in nests and feeding sites of Australian termites. *Mycol. Res.* **102:**216–220.
66. **Miteva, W., A. Abadjieva, and R. Grigorova.** 1991. Differentiation among strains and serotypes of *Bacillus thuringiensis* by M13 DNA fingerprinting. *J. Gen. Microbiol.* **137:**593–600.
67. **Moraes, R. R., J. E. Maruniak, and J. E. Funderburk.** 1999. Methods for detection of *Anticarsia gemmatalis* nucleopolyhedrovirus DNA in soil. *Appl. Environ. Microbiol.* **65:**2307–2311.
68. **Navon, A., and K. R. S. Ascher (ed.).** 2000. *Bioassays of Entomopathogenic Microbes and Nematodes*. CABI Publishing, Wallingford, United Kingdom.
69. **Nielsen, C., M. G. Milgroom, and A. E. Hajek.** 2005. Genetic diversity in the gypsy moth fungal pathogen *Entomophaga maimaiga* from founder populations in North America and source populations in Asia. *Mycol. Res.* **109:** 941–950.
70. **Nielsen, C., S. Vestergaard, S. Harding, C. Wolsted, and J. Eilenberg.** 2006. Biological control of *Strophosoma* spp. (Coleoptera: Curculionidae) in greenery (*Abies procera*) plantations using Hyphomycetes. *BioControl Sci. Technol.* **16:**583–598.
71. **Pedigo, L. P., and M. E. Rice.** 2006. *Entomology and Pest Management*, 5th ed. Prentice-Hall, Upper Saddle River, N.J.
72. **Reyes, A., P. Christian, J. Valle, and T. Williams.** 2004. Persistence of *Invertebrate iridescent virus* 6 in soil. *BioControl* **49:**433–440.
73. **Rupar, M. J., W. P. Donovan, R. G. Groat, A. C. Slaney, J. W. Mattison, T. B. Johnson, J. Charles, V. C. Dumanoir, and H. de Bajac.** 1991. Two novel strains of

Bacillus thuringiensis toxic to coleopterans. *Appl. Environ. Microbiol.* **57:**3337–3344.

74. **Sneh, B.** 1991. Isolation of *Metarhizium anisopliae* from insects on an improved selective medium based on wheat germ. *J. Invertebr. Pathol.* **58:**269–273.
75. **Stahly, D. P., D. M. Takefman, C. A. Livasy, and D. W. Dingman.** 1992. Selective medium for quantitation of *Bacillus popilliae* in soil and in commercial spore powders. *Appl. Environ. Microbiol.* **58:**740–743.
76. **Strasser, H., A. Forer, and F. Schinner.** 1996. Development of media for the selective isolation and maintenance of virulence of *Beauveria brongniartii*, p. 125–130. *In* T. A. Jackson and T. R. Glare (ed.), *Microbial Control of Soil Dwelling Pests*. AgResearch, Lincoln, New Zealand.
77. **Studdert, J. P., H. K. Kaya, and J. M. Duniway.** 1990. The effect of water potential, temperature and clay-coating on survival of *Beauveria bassiana* conidia in a loam and peat soil. *J. Invertebr. Pathol.* **55:**417–427.
78. **Sun, J. Z., J. R. Fuxa, and G. Henderson.** 2003. Virulence and in vitro characteristics of pathogenic fungi isolated from soil by baiting with *Coptotermes formosanus* (Isoptera: Rhinotermitidae). *J. Entomol. Sci.* **38:**342–358.
79. **Taverner, M. P., and E. F. Connor.** 1992. Optical enumeration technique for detection of baculoviruses in the environment. *Environ. Entomol.* **21:**307–313.
80. **Thiery, I., and E. Frachon.** 1997. Identification, isolation, culture and preservation of entomopathogenic bacteria, p. 55–77. *In* L. Lacey (ed.), *Manual of Techniques in Insect Pathology*. Academic Press, San Diego, Calif.
81. **Thies, J. E.** 2006. Measuring and assessing soil biological properties, p. 655–670. *In* N. Uphoff, A. S. Ball, E. Fernandes, H. Herren, O. Husson, M. Laing, C. Palm, J. Pretty, and P. Sanchez (ed.), *Biological Approaches to Sustainable Soil Systems*. CRC Press, Boca Raton, Fla.
82. **Treonis, A. M., S. J. Grayston, P. J. Murray, and L. A. Dawson.** 2004. Effects of root feeding, cranefly larve on soil microorganisms and the composition of rhizosphere solutions collected from grassland plants. *Appl. Soil Ecol.* **28:**203–215.
83. **Villani, M. G., L. L. Allee, A. Díaz, and P. S. Robbins.** 1999. Adaptive strategies of edaphic arthropods. *Annu. Rev. Entomol.* **44:**233–256.
84. **Villani, M. G., L. L. Allee, L. Preston-Wilsey, N. Consolie, Y. Xia, and R. L. Brandenburg.** 2002. Use of radiography and tunnel castings for observing mole cricket (Orthoptera: Gryllotalpidae) behavior in soil. *Am. Entomol.* **48:**42–50.
85. **Villani, M. G., S. R. Krueger, P. C. Schroeder, F. Consolie, N. H. Consolie, L. M. Preston-Wilsey, and D. W. Roberts.** 1994. Soil application effects of *Metarhizium anisopliae* on Japanese beetle (Coleoptera: Scarabaeidae) behavior and survival in turfgrass microcosms. *Environ. Entomol.* **23:**502–513.
86. **Wallwork, J. A.** 1970. *Ecology of Soil Animals*. McGraw-Hill, London, United Kingdom.
87. **Weseloh, R. M., and T. G. Andreadis.** 1986. Laboratory assessment of forest microhabitat substrates as sources of the gypsy moth nuclear polyhedrosis virus. *J. Invertebr. Pathol.* **48:**27–33.
88. **Weseloh, R. M., and T. G. Andreadis.** 2002. Detecting the titer in forest soils of the gypsy moth (Lepidoptera: Lymantriidae) fungal pathogen, *Entomophaga maimaiga* (Zygomycetes: Entomophthorales). *Can. Entomol.* **134:** 269–279.
89. **West, A. W., N. E. Crook, and H. D. Burges.** 1984. Detection of *Bacillus thuringiensis* in soil by immunofluorescence. *J. Invertebr. Pathol.* **43:**150–155.
90. **Yousten, A. A., S. B. Fretz, and S. A. Jelley.** 1985. Selective medium for mosquito-pathogenic strains of *Bacillus sphaericus*. *Appl. Environ. Microbiol.* **49:**1532–1533.

Lipid Fingerprinting of Soil Microbial Communities

ANA LUCÍA CÓRDOVA-KREYLOS AND KATE M. SCOW

64

CHALLENGES IN CHARACTERIZING SOIL MICROBIAL COMMUNITIES

Interest in the diversity and ecosystem services provided by soil and subsurface organisms is rapidly growing. Community and ecosystem ecologists, geologists, plant pathologists, environmental engineers, and health scientists, among others, have joined soil microbial ecologists in pursuit of questions about the composition and functions of soil microbial populations and communities. More than any other habitat, soil poses unique and formidable challenges because of the difficulty in separating cells from surfaces, living from dead material, active from dormant organisms, and the hot spots of activity from the biological deserts (121). Relatively elegant techniques such as fluorescent in situ hybridization, used to characterize aquatic and wastewater microbial communities, are extremely challenging to apply directly in most soils because of interference from the soil physical matrix and due to the inactive state of many soil organisms (119).

Many of the methods used to characterize soil communities are culture independent and based on direct extraction of cellular material, without preservation of individual cells. Various biochemical constituents are targeted (as described throughout this manual), including nucleic acids and lipids, both of which reveal meaningful and data-rich information about types and physiology of soil organisms. Nucleic acids (DNA and RNA) are considered elsewhere (chapters 12, 13, 40, and 55), and the general principles, methodologies, and applications of lipid analysis are reviewed in chapter 10. Many of these methods present challenges in that they generate large data sets that are difficult to analyze or information that may be more qualitative than quantitative in nature. In addition, some of the methods measure biochemical constituents or functional genes that do not correspond directly to phylogenetic groups (e.g., genera or species) and require new ways of evaluation and interpretation.

The goal of this chapter is to provide an overview of lipid-based methods used to characterize microbial communities, specifically targeting the soil environment. Previous reviews of lipid-based analyses (primarily phospholipid fatty acids [PLFA]) consider groundwater specifically (44), applications to soil and environmental quality assessments (60), and impacts of heavy metal pollution (87). We summarize the methods commonly used in investigations of soil communities, consider analytical and technical challenges of soil, describe statistical approaches for analyzing fingerprint data, and present applications to illustrate the types of information generated and questions that can be addressed with these methods.

CHARACTERIZATION OF SOIL COMMUNITIES BY TARGETING LIPIDS AND FATTY ACIDS

Types of lipid-based methods applied to soil include analyses of PLFA, whole-cell fatty acid methyl esters in soil (FAME, also called EL-MIDI, MIDI method, and TS-FAME), sterols, and respiratory quinones.

PLFA and FAME

PLFA and FAME analyses are two fatty acid-based methods commonly used to describe microbial communities in soils. Phospholipids, essential components of cell membranes, are not found in storage compounds or dead cells, make up a relatively constant proportion of cell biomass (approximately 5%), and are both structurally and functionally diverse among bacteria (44, 139). Phospholipids are degraded relatively rapidly after cell death and are assumed to be reflective of the viable biomass (50, 90). Phospholipids are a subset of the large variety of fatty acids present in living and dead microorganisms, plants, and animals and in soil organic matter. This diverse pool of fatty acids is detected by the FAME method and includes, in addition to phospholipids, neutral and short-chain fatty acids (32, 106).

FAME and PLFA extraction methods differ in their ease, sensitivity, and precision. FAME extraction is a simple four-step procedure (saponification/lysis, methylation, extraction, and washing) that takes approximately half a day to complete. PLFA extraction is a lengthier procedure (spanning 2.5 to 3 days) involving a one-phase chloroform-methanol-phosphate buffer solvent extraction, separation of phospholipids from nonpolar lipids and conversion to fatty acid methyl esters before GC (gas chromatography) or GC-MS (gas chromatography-mass spectrophotometer) analysis (13, 130) (see chapter 10). The in-soil saponification step of FAME is harsh and likely releases more fatty acids from organic matter than does the PLFA procedure (78).

Comparisons of PLFA and FAME analyses of the same set of samples illustrate the relative strengths and weaknesses of the two methods (32, 51, 78, 89, 106, 117). FAME analysis is undoubtedly quicker, can handle a smaller sample size, and measures a greater number of fatty acids than does PLFA analysis. In a methodological comparison of 13 agricultural soils, PLFA required 10 times the sample mass as FAME to reach the same detection limit but had a substantially higher sample extraction precision (32). The very advantages of FAME analysis can be disadvantageous, however, when the fatty acids are too numerous to differentiate and include many that do not provide meaningful information about the living community in a soil (e.g., include fatty acids associated with decaying plant litter or organic matter). Targeting PLFAs puts the emphasis on the membrane fatty acids that better discriminate bacteria and reduces the data set to a number of fatty acids that is easier to separate on a chromatograph and be identified by GC-MS.

Though both PLFA and FAME methods can discriminate among microbial communities, those fatty acids responsible for their differentiation may vary, making interpretation of the results method dependent (51). Interpretation of FAME was more confounded than were PLFA profiles by interference from organic material in a study of microbial succession during composting (117), and FAME was less sensitive than PLFA analysis in detecting responses of microbial communities to environmental factors in different tillage regimes, again due to the presence of nonmicrobial fatty acids (89). Compared to FAME, PLFA analysis generates more consistent fatty acid profiles across laboratory replicates and is a better measure of microbial populations due to its selectivity (though not exclusive) for viable microbial biomass and exclusion of fatty acids not originating from PLFAs (32). If the focus of a project is broader than microbial communities, then information contributed by fatty acids derived from nonmicrobial or dead organic material may increase the ability to discriminate different treatments by FAME. Also FAME may be an appropriate method under time and budget constraints and/or when requiring high-throughput analyses of multiple samples.

Sterols and Quinones

Another lipid compound targeted in characterizing soil microbial communities is ergosterol, an abundant sterol in many filamentous fungi. Ergosterol was initially developed for characterizing fungi in cereal samples (107) and later adapted to measure fungal biomass in soil (42). The validity of ergosterol as a fungal biomass indicator has been repeatedly questioned and is dependent on soil type (142). In a study of seven Iowa soils (114), ergosterol was correlated to fungal biomass, live and dead, as measured by total hyphal length. Other studies have suggested that ergosterol is relatively recalcitrant (40, 77) following death of fungi, leading to poor correlations to viable fungal biomass. Overall, these results suggest that ergosterol should be interpreted cautiously. Discussion of the use of ergosterol to estimate fungal biomass is found in chapter 38.

Quinones, a diverse group of lipid-soluble components of the prokaryote cytoplasmic membrane (52), are another group of constituents that can provide useful information about microbial communities. The main function of these molecules is as electron carriers in the electron transport chain of many microorganisms. Quinone profiles, determined using reverse-phase high-performance liquid chromatography, are used as indicators of biomass (105) and to track shifts in community composition (59), particularly as a function of changes in the redox of the environment. Quinone profiles have been used to characterize changes in soil microbial communities due to hydrocarbons (111), pesticides (57), depth gradients (128), fertilizers (59), and carbon sources (58).

OVERVIEW OF PLFA ANALYSIS OF SOIL MICROBIAL COMMUNITIES

PLFA analysis provides different types of information about a microbial community including estimates of biomass (37, 68, 90), biomarkers for specific taxonomic groups and physiological responses of the community (3, 17, 76, 131), and a fingerprint or profile of a community based on its constituent fatty acids (33, 39, 55, 90).

A fingerprint is the pattern created by the relative abundance of all fatty acids, extracted in this case from a soil sample, detected by quantitative GC or GC-MS. Because fingerprints are made up of multiple, rather than individual, variables, they are quite sensitive, and thus useful in determining if and how soil microbial communities respond, to environmental gradients such as differences in soil depth (38), pollution (7, 9, 11, 48, 133), tillage (24, 55, 102), landscape changes (16, 75), and climate change (138, 144). Also, those fatty acids that play the greatest role (and thus are most sensitive) to the variables investigated can be identified, within fingerprints, and attributed, if possible, to particular taxonomic or functional groups of microorganisms. These fatty acid classifications and their interpretation are reported widely in the literature (17, 44, 139), and the use of biomarkers is discussed in chapter 10. Numerous reviews summarize the large number of studies employing PLFA analysis of soils, including reviews by Zelles (139), Kaur et al. (60), Stenberg (118), and Olsson (81).

STATISTICAL ANALYSIS AND INTERPRETATION OF PLFA FINGERPRINTS

Fatty acid-based methods such as PLFA generally produce chromatographs (GC or GC-MS) consisting of multiple fatty acids. Some of the fatty acids have masses that are close to detection limit or that cannot be identified, and thus they may be discarded from further analysis. In other cases, fatty acids found also in plants or other eukaryotic organisms may be discarded from further analysis to eliminate any possibility that nonmicrobial lipids will be misinterpreted or skew the results of a study that is looking at microbial questions. The remaining individual fatty acids are then subjected to multivariate analysis or other statistical analyses. Another approach used sometimes is to lump fatty acids into subgroups and then use the smaller number of subgroups (e.g., branched, long chain, or polyunsaturated) as the "species" in data analysis.

Multivariate Methods for PLFA Analysis

Whereas simple regression analyzes data consisting of one or two variables, a typical PLFA fingerprint that is comprised of dozens of descriptive variables (i.e., fatty acids) and sometimes as many predictive ones (i.e., environmental variables) requires approaches that can handle data sets with such large numbers of variables. Many methods have been applied in the statistical analysis and interpretation of PLFA data sets: multivariate ordination methods, classification trees, and neural networks are among the most common approaches. Multivariate ordination methods used in analyzing PLFA fingerprints include principal component analysis (PCA) (7–9), correspondence analysis (CA), redundancy analysis (RDA) (27, 69), and canonical correspondence analysis (CCA) (11, 115). Table 1 compares some of the major differences of these approaches. More recently nonmetric multidimensional scaling (NMDS) (72, 126), partial least squares (PLS) (83, 113), artificial neural networks (ANN) (53, 93, 94, 112), and partial ordination methods have found their way into the PLFA literature. Geostatistical approaches, developed for geological explorations and extended to soil physics, have also been used to analyze spatial patterns in PLFA data and their environmental drivers (101).

Ordination Methods

Ordination is defined by ter Braak and Šmilauer (123) as "the collective term for multivariate techniques that arrange sites (treatments) along axes on the basis of data on species composition." In the case of PLFA, the data set is represented by samples, each of which is comprised of species, in this case, individual fatty acids, that are the response variables. Often, explanatory variables (e.g., associated soil properties or plant species) are measured on the same samples to help develop predictions about the primary data (64). The aim of ordination is to produce usually two-dimensional graphical representations of points (corresponding to sites or treatments) arranged in such a way that points close together correspond to sites with similar species and those far apart correspond to dissimilar species. A common output of ordination methods is a biplot showing the distribution of treatments and corresponding fatty acids. The absence of any pattern to the distribution of PLFA data reveals that important environmental variables have been neglected in the experimental design. Detailed descriptions of ordination methods can be found elsewhere (64, 84, 123)

Unconstrained ordination (e.g., PCA, CA, detrended correspondence analysis [DCA], and NMDS), also known as indirect gradient analysis, provides information on the variability of the data set related to fatty acid composition and is used in an exploratory way when data describing potentially explanatory environmental variables are not available. Potential environmental gradients underlying patterns in the distribution of fatty acids and samples on ordination plots can be inferred, post hoc, from the data but require direct gradient analysis methods to demonstrate these relationships (84).

PCA was the earliest ordination method applied to PLFA fingerprints and is still the most commonly used method in evaluating soil data. Use of PCA assumes that the fatty acid data exhibit a normal distribution (31) and are linearly correlated. Many PLFA data sets, and in fact many ecological data sets, violate the requirement of normal distribution (45, 92). If necessary, data are log transformed to achieve a normal distribution, although PCA is often applied to nontransformed data (122, 124). Most applications to PLFA data follow a data transformation step in which PLFAs are expressed as compositional data, that is, their abundance relative to the total amount of fatty acids detected. After analysis, PCA scores (i.e., x, y coordinates on the ordination plot) may be combined with regression analysis and analyses of variance (ANOVAs) to determine significant relationships to other measured variables. For example, microbial community composition was found to be strongly associated with hydrocarbon contamination by regressing PCA scores to different environmental variables (108).

Investigations of microbial communities have employed CA in cases when data sets have a unimodal distribution or large numbers of zeros (e.g., fatty acids that are absent or below the detection limit) or when there were concerns regarding artifacts created by PCA when using compositional data (54, 62). Instead of maximizing "variance explained," CA maximizes the correspondence between species scores and sample scores (64). A detailed discussion on the selection of PCA versus CA can be found in references 64 and 123. CA was combined with use of Pearson's correlation coefficients and ANOVAs to evaluate the effects of reducing tillage of grassland soils (2). Other examples of application of CA to PLFA fingerprints can be found in DeGrood et al. (30) and Drenovsky et al. (32).

NMDS is a third unconstrained ordination method that differs from PCA and CA in its utilization of a distance-based ordination method based on a dissimilarity matrix between samples' fatty acid composition, where distances between samples in a two-dimensional plot are proportional to their dissimilarities. A more in-depth discussion of the theory behind this method, its drawbacks, and its applications in ecology can be found in reference 64. Examples of applications of NMDS to PLFA fingerprint analysis can be found in papers by Merila et al. (72) and Tscherko et al. (126).

TABLE 1 Common multivariate methods for fatty acid fingerprint analysis[a]

Type of ordination	Method	Objective	FA response	EV
Unconstrained or indirect	PCA	Maximize linear variation between samples	Linear	No
	CA	Maximize correspondence between FA and samples	Unimodal	No
	DCA	Eliminate arch effect obtained with CA	Unimodal	No
	NMDS	Match distance between plotted samples to FA dissimilarities	Unimodal	No
Constrained or direct	RDA	Maximize correlation between FA and samples as explained by EV	Linear	Yes
	CCA	Maximize correlation between FA and samples as explained by EV	Unimodal	Yes
	PLS	Obtain the latent variables that best describe the covariance between FA and EV	Linear	Yes

[a]Abbreviations: FA, fatty acid; EV, environmental variables. Information in this table was summarized and adapted from references 64, 84, and 123.

Constrained Ordination Methods

Constrained ordination or direct gradient analysis provides information on the fatty acid variability directly linked to environmental variables or treatments; the ordination plot axes are linear combinations of environmental variables (64). Methods like RDA and CCA directly extract variation in the PLFA pattern associated with measured environmental variables. RDA is the constrained form of PCA, assumes data are normally distributed, and works well for short environmental gradients (123). RDA was applied to PLFA and substrate utilization patterns to investigate the impacts of agricultural management, carbon inputs, flooding (17, 18), and more recently, land use on soil communities (16). Steenwerth et al. (115) used CA to determine if PLFA profiles could discriminate among land use types and used CCA to explore relationships between PLFA fingerprint, soil characteristics, and management. CCA is often combined with Monte Carlo permutation tests to determine the statistical significance of the correlations (18). PLS is an extension of multiple regression and PCA. In PLS, variables are divided into two groups: response and predictive variables. The two groups are regressed against each other to create PLS factors (also known as latent variables or components) that can explain the variance of the data set. The analysis is based on explaining the covariance between the data sets, rather than creating synthetic variable combinations, as in ordinations. Discussion of the theory behind PLS can be found elsewhere (65). In PLFA fingerprint analysis, PLS has been applied in combination with ordination methods (67, 88) and ANNs (83, 113).

Artificial Neural Networks

Another promising method for analysis of PLFA data sets is ANN analysis, and this method can be used for either classification or pattern recognition of data. Neural networks are trained using a subset of original data to discriminate patterns associated with treatments against a background of "noise" typical of environmental data. ANN was able to detect variation in PLFA profiles associated with elevated CO_2 exposure not detectable by other multivariate methods (99). There are other examples of ANN analysis of PLFA fingerprints of different organisms and in various environments (41, 53, 135), but few in soils (66, 79, 112). ANN was implemented to compare microbial communities in different plant rhizospheres manipulated to vary nutrient availability (66). In a comparison of two types of ANN, back propagating and fuzzy ARTMAP, the latter method could better discriminate soil microbial communities, based on their PLFA profiles, associated with different crops (112).

Summary

Comparing several methods of data analysis can provide different insights and support broader conclusions regarding PLFA data sets (12, 67). For example, Ludvigsen et al. (67) applied four different multivariate methods—PCA, CA, PLS, and discriminant analysis (DA)—to the study of a landfill-leachate-polluted aquifer. CA and PCA identified groups of samples with similar PLFA fingerprints and biogeochemical parameters; DA and PLS identified correlations between PLFAs and biogeochemical parameters. Many of the methods described above are computationally intensive, but fortunately many free or commercially available software packages designed to perform ordination analysis are available (123). A list of software packages can be found at the following website: http:// www.okstate.edu/artsci/botany/ordinate/software.htm.

EXAMPLES AND APPLICATIONS OF PLFA AND FAME ANALYSIS TO SOILS

Below we present examples of different types of applications of PLFA and FAME in the study of soil microbial communities. We organize these studies by type of application and also provide additional detail about the types of statistical approaches used.

Linking Soil Microbial Community Composition to Environmental Drivers

Many environmental factors have been demonstrated, based on PLFA analysis, to influence the composition of microbial communities in soils: pH, water availability, types and amount of carbon inputs, nutrients, temperature, and salinity, as well as plant biomass and species composition. Soil pH has a strong effect on community structure (9, 12). In acidic soils, fungi made up a higher relative proportion of soil communities, and PCA suggested pH to be correlated with differences across soil communities (5). In agricultural soils, discriminant analysis initially suggested that PLFA patterns were related to types of fertilizer amendments, but careful interpretation indicated that pH was the underlying cause of variation in microbial communities (80).

Inputs of carbon, from above and below ground, have an enormous impact on microbial communities in soil. Addition of organic carbon, regardless of the form, substantially altered microbial communities of an unsaturated agricultural soil and had a larger effect than did flooding the same soil. Figure 1 depicts a CCA of these data showing separation of microbial communities along the first axis after flooding plus carbon, and separation along the second axis due to carbon amendment (32). Carbon from rhizodeposition has also been linked to differences in microbial community composition (23, 36, 110, 129). Analysis of PLFA profiles by PCA demonstrated shifts in microbial communities over the course of rhizosphere development and thus, potentially, in the presence of carbon exudates (116). Changes in PLFA fingerprints with depth are due, in part, to changes in available carbon (38, 46, 85, 97).

Aboveground plant diversity has been linked to the diversity of belowground microbial communities. Bartelt-Ryser et al. (10) found differences in the relative proportions of prokaryotic and eukaryotic PLFAs to be associated with the number of plant functional groups. In a long-term field experiment of grassland plots, differences in microbial community composition were related to differences in plant biodiversity; however, these differences appeared to be linked more to an increase in plant productivity that resulted from increased plant diversity than to diversity directly (137).

PLFA analysis has provided insights into how soil microbial communities respond, in agricultural soils, to different management practices and, in natural ecosystems, to invasion by exotic plant species (11, 17, 18, 63, 96). Tillage and organic management both selected for specific compositions of microbial communities with, for example, fungi being most abundant in the least physically disturbed of the systems (141). In a comparison of the effects of different hay and fertilizer inputs, microbial community composition, based on PLFA fingerprinting, was significantly affected by the different treatments yet microbial C and N pools, as well as respiration, did not specifically respond to inputs (27). Invasion of *Spartina alterniflora* by *Phragmites australis* was associated with changes in rhizosphere microbial communities and biogeochemical functions related to carbon and nitrogen cycling (96). Batten and Scow (11) found that

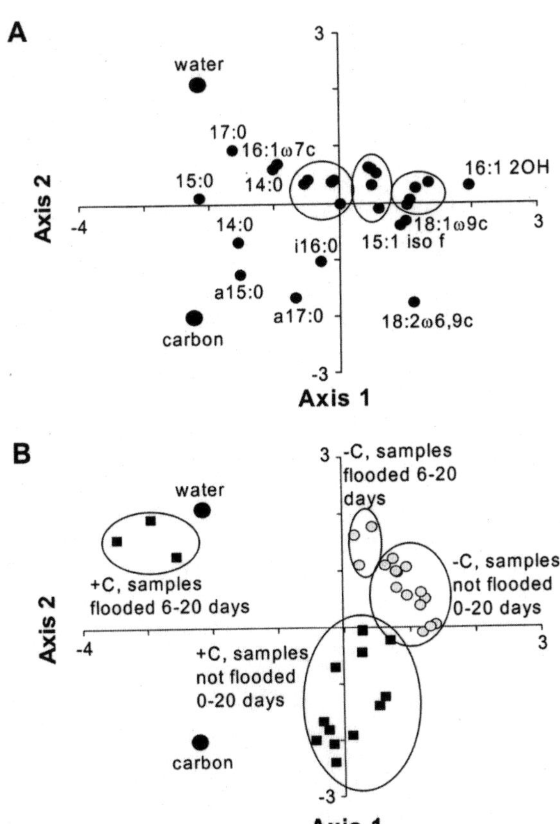

FIGURE 1 CCA of microcosm and field soil PLFA profiles. (A) Ordination biplot of the fatty acids and environmental variable scores. Three circles were added to panel A following statistical analysis to aid in identifying the plotted fatty acids. The circle furthest to the left includes the fatty acids i15:0, 16:0, 16:1ω5c, 16:1ω7t, and i17:1ω5. The middle circle includes the fatty acids 16:1ω11c, i17:0, 17:0cy, 18:0, and sum 7. The circle furthest to the right includes the fatty acids 10Me 16:0, 10Me 17:0, 17:1ω9c, and sum 9. (B) Ordination biplot of the sample and the environmental variable scores. Each sample point is the average of three treatment replicates. Black squares indicate +C samples, and gray circles indicate −C samples. Following statistical analysis, circles were added to the biplots to indicate treatment groupings, but these circles do not indicate confidence ellipsoids. In both plots (A and B) the environmental variables are plotted as discrete points. Reprinted with permission from Drenovsky et al. (32) .

shifts in rhizosphere microbial communities were strongly correlated to invasive plant colonization. Significant changes in PLFA fingerprints were associated with higher pH and nitrification rates in the rhizospheres of two invasive plant species than in native plants, and the authors hypothesized that these alterations in soil properties may promote further invasion (63).

Responses of Microbial Communities to Disturbance and Remediation/Restoration

PLFA and FAME analyses have been used to measure the relative impact of anthropogenic disturbances, primarily via agricultural practices, pollution, or invasion by nonnative plant species, on microbial communities. Similarly, both methods have been effective in monitoring the recovery of, or other changes in, terrestrial ecosystems during restoration and remediation activities. In some cases, questions are largely exploratory, focusing more on whether communities have shifted in composition from their condition before disturbance or whether intervention has promoted a shift back to the original composition; thus, fatty acid-based methods provide sufficient information.

The composition of microbial communities has been proposed to be an indicator for assessment of the success of soil restoration efforts (4, 49, 56, 60). PLFA analysis in particular can provide a measure of biomass, composition, and activity (when linked to isotope probing) of communities in monitoring the effects of restoration strategies (49). In a comparison of microbial communities from undisturbed, cropped, and restored grasslands, PLFA biomass was significantly higher in undisturbed and restored soils and RDA indicated that restoration time was the strongest determinant of PLFA fingerprints (69). A long recovery period was also found in a study of microbial communities along a disturbance gradient in 42 sites in California coastal valleys (115). PLFA fingerprints showed that microbial community composition was highly associated with the land use type and management history associated with it, hinting to soil recoveries that can last centuries (Fig. 2). In restored agriculturally impacted soils, elimination of tillage was followed by an increase of arbuscular mycorrhizal fungi (2), and similar results were observed in abandoned agricultural fields in Hungary (34).

PLFA analysis has been a useful tool in analyzing the responses of soil microbial communities to pollutants (12, 21, 51, 61, 73, 74, 91, 98, 108, 109, 133). In many studies, PLFA analysis revealed that changes in microbial community composition were not necessarily correlated with changes in functional responses of the communities, and vice versa. Exposure time, rather than availability of metals, was the main factor in changing PLFA community composition (133), and in one study, 40 years after metal inputs had ceased, microbial communities in metal-impacted soils were still distinct from those in paired uncontaminated soils (1). PLFA fingerprinting helped determine if functional differences (e.g., in respiration or substrate utilization) were different in communities that had been altered by the presence of heavy metals and identify members of the community that were tolerant to metals (6, 127). In another study, respiration was negatively affected by metal contamination at all concentrations tested, whereas changes in microbial community structure occurred only at metal concentrations far in excess of background concentrations (95). Measures of microbial activity showed communities returning to preexposed values in soils recovering from oil contamination, but PLFA patterns indicated that impacted communities had not returned to their original composition (20, 21). Long-term or permanent modification of microbial communities by hydrocarbon pollution has been demonstrated in other studies (47, 120).

Spatial Distribution: Use of Geostatistics in Evaluating Variation in Fatty Acid Fingerprints

Several studies have used geostatistical approaches to explore spatial variation, using PLFA analysis, of microbial communities at different scales in soil (100, 104). Geostatistical tools are used both to characterize patterns in the spatial variation of microbial community composition and to identify the environmental drivers of this variation (35). Geostatistical studies require a quantitative measure of the distance between samples and usually a large number of sampling points. Chapter 17 focuses on geostatistical approaches in microbial ecology.

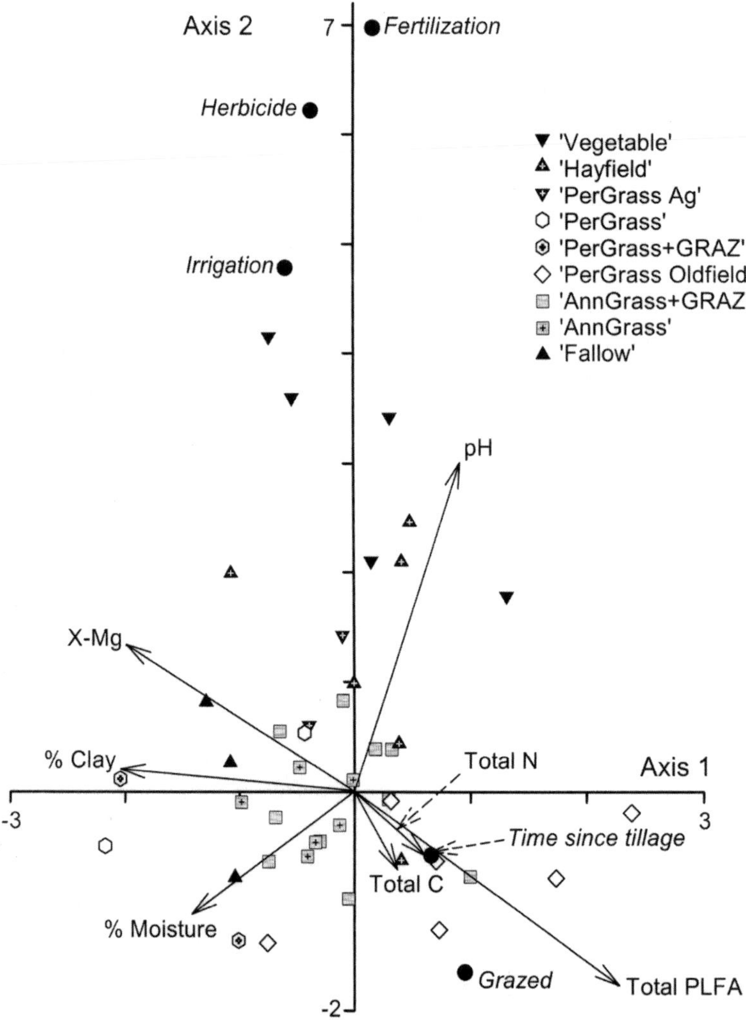

FIGURE 2 CCA of 42 sites classified by nine land use types. Vectors represent soil characteristics, and centroids show statistically significant management factors. Reprinted with permission from Steenwerth et al. (115).

Variograms and semivariograms, tools associated with geostatistical analyses, can be applied in diverse ways to PLFA fingerprint analysis. Spatial analysis of individual fatty acid biomarkers could become tedious when as many as 50 or 60 fatty acids describe a particular community, so, alternatively, PCA and CA axis scores can be evaluated in variograms. PCA sample scores, for example, were used to estimate spatial correlations between microbial community composition and creosote concentrations in contaminated soil. Approximately 80% of variation explained by the first PCA axis could be attributed to spatial autocorrelation due, in part, to the presence of more gram-negative bacteria in contaminated areas (136). A main determinant of the spatial pattern of soil microbial communities in a forest was the spacing of two of the dominant tree species, and these two tree types had different spatial scales in their distribution (104). In another study, similar moisture-stress-tolerant microbial communities were detected adjacent to, and 2 m from, the trunks but very different communities were present in the intermediate 1-m distance from the trunks, speculated to be due to the influence of root distribution (134). Ritz et al. (100) determined linear and nonlinear spatial relationships between PLFA and a range of chemical parameters, whereas no correlations were apparent for DNA community profiles. More in-depth explanations of the application, interpretation, and limitations of geostatistical approaches can be found in chapter 17.

Coupling PLFA Fingerprinting and Isotope Probing

Coupling PLFA with stable isotope analyses can be a powerful approach for linking metabolic processes to specific groups of organisms. This technique is called isotope probing (or, often, stable-isotope probing [SIP]) and has been integrated into both nucleic acid and phospholipids-based methods for characterizing microbial communities. Boschker et al. (15) combined addition of isotopic-labeled substrates with PLFA analysis in an ecological study to determine which members of a community were engaged in methanotrophic activity. Since then, the approach has been expanded to other environments and functional groups; several recent reviews discuss the underlying concepts, strengths and weaknesses, and applications of this approach (14, 19, 26, 71, 140). In addition, determination of the natural abundance of stable isotopes in PLFA biomarkers can help determine sources of carbon and trophic relationships within communities (22).

Our understanding of which organic compounds are preferentially used in complex mixtures in soil and of the succession in microbial communities tied to the transformation of complex organic compounds has benefited from the application of isotopic analyses. Slater et al. (109) used PLFA profiling and a combination of ^{14}C and ^{13}C probing to determine that microbial communities were not actively metabolizing residual hydrocarbon compounds but preferentially using natural organic matter carbon. Ziegler et al. (143) demonstrated that the assimilation, transformation, and recycling of [^{13}C]glucose in soil occurred in two phases, each dominated by different sets of microbial communities.

Comparison of isotope-labeled community fingerprints to fatty acid biomarkers of functional or taxonomic groups can distinguish microbial populations active in specific metabolic processes (14). This method was successfully applied in a study of nonculturable methane-oxidizing bacteria where SIP was combined with PLFA fingerprinting to identify new groups of methane-oxidizing bacteria in soils (29). PLFA isotope probing has also been widely applied to contaminant transformation studies, and the wider availability of stable-isotope compounds has facilitated application of these methods in field studies, where they were not possible using radiolabeled isotopes (25, 48, 91).

Several studies have combined isotope analysis and PLFA to study interactions and feedbacks between microbial communities and plants, particularly relating to carbon assimilation in the rhizosphere. Investigation of the effects of vegetation cover on microbial communities by following plant- and microbe-derived C pools over the year indicated that different C pools were used by different community members as reflected by the ^{13}C signatures, despite the fact that the community composition itself changed little (86). ^{13}C labeling showed that fungi were the metabolically dominant members of the decomposer community based on incorporation of the ^{13}C label in the 18:2ω6, 9 PLFA (fungal biomarker) (70). Carbon fluxes from the atmosphere into the plant rhizosphere, and ultimately into microbial populations such as arbuscular mycorrhizal fungi, have been tracked using SIP methods (82, 125).

PITFALLS, LIMITATIONS, AND FUTURE DIRECTIONS

The in-depth discussion of the relative strengths and limitations of PLFA analysis in the 1999 review of Zelles (139) is still pertinent today. Some of the strengths of PLFA are common to all culture-independent techniques in that the organisms represented do not have to grow to reveal themselves and thus measured diversity has not been subject to culture bias. A strength unique to PLFA analysis is the sensitivity of the method to environmental perturbations and gradients in some cases where DNA fingerprinting methods are not very sensitive (28, 43). This sensitivity is likely derived from the rapid turnover in soil of PLFAs of nonliving organisms, from the contribution of eukaryotes to fingerprints, and from the fact that some fatty acids reflect physiological as well as taxonomic changes within communities. Finally PLFA data are quantitative and lend themselves well to the multivariate and other quantitative methods described above.

Fatty acid-based methods are limited foremost by the particular biochemical constituents that they target, the fatty acids, which are molecules possessing far less variation and specificity (with respect to microbial taxa) than do nucleic acids. Other limitations include the facts that the variation in relative abundance and composition of fatty acids in many environmental organisms is unknown, that some fatty acids are shared by functionally unrelated organisms and thus difficult to interpret, that PLFA biomarkers do not provide full coverage of all major microbial groups, and that there is a lack of standardization in the methods used. Also the quantitative strength of the methods noted above can be misleading if incorrect interpretations of the responsible fatty acids are made and they are attributed to the wrong organisms or groups. Thus, overinterpretation of community shifts based on PLFA analysis should be avoided (44, 60, 139).

Many studies utilize PLFA and nucleic acid-based methods in tandem and benefit from their complementary nature. PLFA is a screening method that does not require as much preparatory and follow-up work as many of the nucleic acid-based fingerprinting methods (e.g., those which may require calibration for different soils). PLFA analysis is a direct wet-chemistry method (unlike, e.g., PCR-based methods) and generates quantitative information (e.g., nanomoles of different fatty acids). Though the phylogenetic information provided by PLFA analysis is spotty and lacking in some cases, obtaining such information from a DNA fingerprint requires sequencing and hence a lot more effort. So, PLFA analysis may work most effectively as a first screening method and when there are large quantities of samples. Another strength of PLFA is the ease in linking compositional and environmental data; however, quantitative PCR and non-PCR methods can also be used in support of this goal. Many investigators concur that no single method is currently able to describe the complete microbial community with its vast diversity of taxa and functions, and thus, employment of a diverse set of methods best enhances our ability to understand soil microbial communities (43, 103, 132).

This research was supported in part by the CONACyT and UCMexus programs and UC Toxic Substances Research and Teaching Program. Additional support was provided by a grant from the U.S. Environmental Protection Agency's Science to Achieve Results (STAR) Estuarine and Great Lakes (EaGLe) Coastal Initiative through funding to the Pacific Estuarine Ecosystem Indicator Research (PEEIR) Consortium, U.S. EPA Agreement EPA/R-82867601, and by grant number 5 P42 ES04699 from the National Institute of Environmental Health Sciences, NIH.

The contents of this chapter are solely the responsibility of the authors and do not necessarily represent the official views of the NIEHS, NIH.

REFERENCES

1. **Abaye, D. A., K. Lawlor, P. R. Hirsch, and P. C. Brookes.** 2005. Changes in the microbial community of an arable soil caused by long-term metal contamination. *Eur. J. Soil. Sci.* **56:**93–102.
2. **Allison, V. J., R. M. Miller, J. D. Jastrow, R. Matamala, and D. R. Zak.** 2005. Changes in soil microbial community structure in a tallgrass prairie chronosequence. *Soil Sci. Soc. Am. J.* **69:**1412–1421.
3. **Almeida, J. S., A. Sonesson, D. B. Ringelberg, and D. C. White.** 1995. Application of artificial neural networks to the detection of Mycobacterium tuberculosis, its antibiotic resistance and prediction of pathogenicity amongst Mycobacterium spp based on signature lipid biomarkers. *Binary Comput. Microbiol.* **7:**159–166.

4. **Avidano, L., E. Gamalero, G. P. Cossa, and E. Carraro.** 2005. Characterization of soil health in an Italian polluted site by using microorganisms as bioindicators. *Appl. Soil. Ecol.* **30:**21–33.
5. **Bååth, E., and T. H. Anderson.** 2003. Comparison of soil fungal/bacterial ratios in a pH gradient using physiological and PLFA-based techniques. *Soil Biol. Biochem.* **35:**955–963.
6. **Bååth, E., M. Díaz-Raviña, and L. R. Bakken.** 2005. Microbial biomass, community structure and metal tolerance of a naturally Pb-enriched forest soil. *Microb. Ecol.* **50:**496–505.
7. **Bååth, E., M. Díaz-Raviña, A. Frostegård, and C. D. Campbell.** 1998. Effect of metal-rich sludge amendments on the soil microbial community. *Appl. Environ. Microbiol.* **64:**238–245.
8. **Bååth, E., A. Frostegård, M. Díaz-Raviña, and A. Tunlid.** 1998. Microbial community-based measurements to estimate heavy metal effects in soil: the use of phospholipid fatty acid patterns and bacterial community tolerance. *Ambio* **27:**58–61.
9. **Bååth, E., A. Frostegård, T. Pennanen, and H. Fritze.** 1995. Microbial community structure and pH response in relation to soil organic-matter quality in wood-ash fertilized, clear-cut or burned coniferous forest soils. *Soil Biol. Biochem.* **27:**229–240.
10. **Bartelt-Ryser, J., J. Joshi, B. Schmid, H. Brandl, and T. Balser.** 2005. Soil feedbacks of plant diversity on soil microbial communities and subsequent plant growth. *Perspect. Plant Ecol. Evol. Syst.* **7:**27–49.
11. **Batten, K. M., and K. M. Scow.** 2003. Sediment microbial community composition and methylmercury pollution at four mercury mine-impacted sites. *Microb. Ecol.* **46:**429–441.
12. **Ben-David, E. A., P. J. Holden, D. J. M. Stone, B. D. Harch, and L. J. Foster.** 2004. The use of phospholipid fatty acid analysis to measure impact of acid rock drainage on microbial communities in sediments. *Microb. Ecol.* **48:**300–315.
13. **Bligh, E., and W. Dyer.** 1959. A rapid method of lipid extraction and purification. *Can. J. Biochem. Phys.* **39:**911–917.
14. **Boschker, H. T. S., and J. J. Middelburg.** 2002. Stable isotopes and biomarkers in microbial ecology. *FEMS Microbiol. Ecol.* **40:**85–95.
15. **Boschker, H. T. S., S. C. Nold, P. Wellsbury, D. Bos, W. de Graaf, R. Pel, R. J. Parkes, and T. E. Cappenberg.** 1998. Direct linking of microbial populations to specific biogeochemical processes by C-13-labelling of biomarkers. *Nature* **392:**801–805.
16. **Bossio, D. A., M. S. Girvan, L. Verchot, J. Bullimore, T. Borelli, A. Albrecht, K. M. Scow, A. S. Ball, J. N. Pretty, and A. M. Osborn.** 2005. Soil microbial community response to land use change in an agricultural landscape of Western Kenya. *Microb. Ecol.* **49:**50–62.
17. **Bossio, D. A., and K. M. Scow.** 1998. Impacts of carbon and flooding on soil microbial communities: phospholipid fatty acid profiles and substrate utilization patterns. *Microb. Ecol.* **35:**265–278.
18. **Bossio, D. A., K. M. Scow, N. Gunapala, and K. J. Graham.** 1998. Determinants of soil microbial communities: effects of agricultural management, season, and soil type on phospholipid fatty acid profiles. *Microb. Ecol.* **36:**1–12.
19. **Bruno, G.** 2005. Compound-specific stable-isotope (d13C) analysis in soil science. *J. Plant Nutr. Soil Sci.* **168:**633–648.
20. **Bundy, J. G., G. I. Paton, and C. D. Campbell.** 2004. Combined microbial community level and single species biosensor responses to monitor recovery of oil polluted soil. *Soil Biol. Biochem.* **36:**1149–1159.
21. **Bundy, J. G., G. I. Paton, and C. D. Campbell.** 2002. Microbial communities in different soil types do not converge after diesel contamination. *J. Appl. Microbiol.* **92:**276–288.
22. **Burke, R. A., M. Molina, J. E. Cox, L. J. Osher, and M. C. Piccolo.** 2003. Stable carbon isotope ratio and composition of microbial fatty acids in tropical soils. *J. Environ. Qual.* **32:**198–206.
23. **Butler, J. L., M. A. Williams, P. J. Bottomley, and D. D. Myrold.** 2003. Microbial community dynamics associated with rhizosphere carbon flow. *Appl. Environ. Microbiol.* **69:**6793–6800.
24. **Calderon, F. J., L. E. Jackson, K. M. Scow, and D. E. Rolston.** 2001. Short-term dynamics of nitrogen, microbial activity, and phospholipid fatty acids after tillage. *Soil Sci. Soc. Am. J.* **65:**118–126.
25. **Chang, Y. J., P. E. Long, R. Geyer, A. D. Peacock, C. T. Resch, K. Sublette, S. Pfiffner, A. Smithgall, R. T. Anderson, H. A. Vrionis, J. R. Stephen, R. Dayvault, I. Ortiz-Bernad, D. R. Lovley, and D. C. White.** 2005. Microbial incorporation of C-13-labeled acetate at the field scale: detection of microbes responsible for reduction of U(VI). *Environ. Sci. Technol.* **39:**9039–9048.
26. **Charles, M. S.** 2002. Measurement and applications of stable isotopes in fatty acids. *Eur. J. Lipid Sci. Technol.* **104:**57–59.
27. **Cookson, W. R., D. A. Abaye, P. Marschner, D. V. Murphy, E. A. Stockdale, and K. W. T. Goulding.** 2005. The contribution of soil organic matter fractions to carbon and nitrogen mineralization and microbial community size and structure. *Soil Biol. Biochem.* **37:**1726–1737.
28. **Córdova-Kreylos, A. L., Y. Cao, P. G. Green, H.-M. Hwang, K. M. Kuivila, M. G. LaMontagne, L. C. Van De Werfhorst, P. A. Holden, and K. M. Scow.** 2006. Diversity, composition, and geographical distribution of microbial communities in California salt marsh sediments. *Appl. Environ. Microbiol.* **72:**3357–3366.
29. **Crossman, Z. M., P. Ineson, and R. P. Evershed.** 2005. The use of C-13 labelling of bacterial lipids in the characterisation of ambient methane-oxidising bacteria in soils. *Org. Geochem.* **36:**769–778.
30. **DeGrood, S. H., V. P. Claassen, and K. M. Scow.** 2005. Microbial community composition on native and drastically disturbed serpentine soils. *Soil Biol. Biochem.* **37:**1427–1435.
31. **Dollhopf, S. L., S. A. Hashsham, and J. M. Tiedje.** 2001. Interpreting 16S rDNA T-RFLP data: application of self-organizing maps and principal component analysis to describe community dynamics and convergence. *Microb. Ecol.* **42:**495–505.
32. **Drenovsky, R. E., G. N. Elliott, K. J. Graham, and K. M. Scow.** 2004. Comparison of phospholipid fatty acid (PLFA) and total soil fatty acid methyl esters (TSFAME) for characterizing soil microbial communities. *Soil Biol. Biochem.* **36:**1793–1800.
33. **Drijber, R. A., J. W. Doran, A. M. Parkhurst, and D. J. Lyon.** 2000. Changes in soil microbial community structure with tillage under long-term wheat-fallow management. *Soil Biol. Biochem.* **32:**1419–1430.
34. **Elhottova, D., T. Szili-Kovacs, and J. Triska.** 2002. Soil microbial community of abandoned sand fields. *Folia Microbiol.* **47:**435–440.
35. **Ettema, C. H., and D. A. Wardle.** 2002. Spatial soil ecology. *Trends Ecol. Evol.* **17:**177–183.
36. **Fang, C. W., M. Radosevich, and J. J. Fuhrmann.** 2001. Characterization of rhizosphere microbial community structure in five similar grass species using FAME and BIOLOG analyses. *Soil Biol. Biochem.* **33:**679–682.

37. **Federle, T. W., R. M. Ventullo, and D. C. White.** 1990. Spatial-distribution of microbial biomass, activity, community structure, and the biodegradation of linear alkylbenzene sulfonate (LAS) and linear alcohol ethoxylate (LAE) in the subsurface. *Microb. Ecol.* **20:**297–313.
38. **Fierer, N., J. P. Schimel, and P. A. Holden.** 2003. Influence of drying-rewetting frequency on soil bacterial community structure. *Microb. Ecol.* **45:**63–71.
39. **Frostegård, A., E. Bååth, and A. Tunlid.** 1993. Shifts in the structure of soil microbial communities in limed forests as revealed by phospholipid fatty-acid analysis. *Soil Biol. Biochem.* **25:**723–730.
40. **Gessner, M. O., M. A. Bauchrowitz, and M. Escautier.** 1991. Extraction and quantification of ergosterol as a measure of fungal biomass in leaf litter. *Microb. Ecol.* **22:**285–291.
41. **Giacomini, M., C. Ruggiero, L. Calegari, and S. Bertone.** 2000. Artificial neural network based identification of environmental bacteria by gas-chromatographic and electrophoretic data. *J. Microbiol. Methods* **43:**45–54.
42. **Grant, W. D., and A. W. West.** 1986. Measurement of ergosterol, diaminopimelic acid and glucosamine in soil—evaluation as indicators of microbial biomass. *J. Microbiol. Methods* **6:**47–53.
43. **Grayston, S. J., C. D. Campbell, R. D. Bardgett, J. L. Mawdsley, C. D. Clegg, K. Ritz, B. S. Griffiths, J. S. Rodwell, S. J. Edwards, W. J. Davies, D. J. Elston, and P. Millard.** 2004. Assessing shifts in microbial community structure across a range of grasslands of differing management intensity using CLPP, PLFA and community DNA techniques. *Appl. Soil. Ecol.* **25:**63–84.
44. **Green, C. T., and K. M. Scow.** 2000. Analysis of phospholipid fatty acids (PLFA) to characterize microbial communities in aquifers. *Hydrogeol. J.* **8:**126–141.
45. **Greigsmith, P.** 1980. The development of numerical classification and ordination. *Vegetatio* **42:**1–9.
46. **Griffiths, R. I., A. S. Whiteley, A. G. O'Donnell, and M. J. Bailey.** 2003. Influence of depth and sampling time on bacterial community structure in an upland grassland soil. *FEMS Microbiol. Ecol.* **43:**35–43.
47. **Hansen, L. D., C. Nestler, D. Ringelberg, and R. Bajpai.** 2004. Extended bioremediation of PAH/PCP contaminated soils from the POPILE wood treatment facility. *Chemosphere* **54:**1481–1493.
48. **Hanson, J. R., J. L. Macalady, D. Harris, and K. M. Scow.** 1999. Linking toluene degradation with specific microbial populations in soil. *Appl. Environ. Microbiol.* **65:**5403–5408.
49. **Harris, J. A.** 2003. Measurements of the soil microbial community for estimating the success of restoration. *Eur. J. Soil. Sci.* **54:**801–808.
50. **Harvey, H. R., R. D. Fallon, and J. S. Patton.** 1986. The effect of organic matter and oxygen on the degradation of bacterial membrane lipids in marine sediments. *Geochim. Cosmochim. Acta* **50:**795–804.
51. **Hinojosa, M. B., J. A. Carreira, R. Garcia-Ruiz, and R. P. Dick.** 2005. Microbial response to heavy metal-polluted soils: community analysis from phospholipid-linked fatty acids and ester-linked fatty acids extracts. *J. Environ. Qual.* **34:**1789–1800.
52. **Hiraishi, A.** 1999. Isoprenoid quinones as biomarkers of microbial populations in the environment. *J. Biosci. Bioeng.* **88:**449–460.
53. **Holubar, P., L. Zani, M. Hager, W. Froschl, Z. Radak, and R. Braun.** 2000. Modelling of anaerobic digestion using self-organizing maps and artificial neural networks. *Water Sci. Technol.* **41:**149–156.
54. **Jackson, D. A.** 1997. Compositional data in community ecology: the paradigm or peril of proportions? *Ecology* **78:** 929–940.
55. **Jackson, L. E., F. J. Calderon, K. L. Steenwerth, K. M. Scow, and D. E. Rolston.** 2003. Responses of soil microbial processes and community structure to tillage events and implications for soil quality. *Geoderma* **114:**305–317.
56. **Johnsen, A. R., A. Winding, U. Karlson, and P. Roslev.** 2002. Linking of microorganisms to phenanthrene metabolism in soil by analysis of C-13-labeled cell lipids. *Appl. Environ. Microbiol.* **68:**6106–6113.
57. **Katayama, A., K. Funasaka, and K. Fujie.** 2001. Changes in the respiratory quinone profile of a soil treated with pesticides. *Biol. Fertil. Soils* **33:**454–459.
58. **Katayama, A., H. Y. Hu, M. Nozawa, S. Takahashi, and K. Fujie.** 2002. Changes in the microbial community structure in soils treated with a mixture of glucose and peptone with reference to the respiratory quinone profile. *Soil Sci. Plant Nutr.* **48:**841–846.
59. **Katayama, A., H. Y. Hu, M. Nozawa, H. Yamakawa, and K. Fujie.** 1998. Long-term changes in microbial community structure in soils subjected to different fertilizing practices revealed by quinone profile analysis. *Soil Sci. Plant Nutr.* **44:**559–569.
60. **Kaur, A., A. Chaudhary, A. Kaur, R. Choudhary, and R. Kaushik.** 2005. Phospholipid fatty acid—a bioindicator of environment monitoring and assessment in soil ecosystem. *Curr. Sci.* **89:**1103–1112.
61. **Kelly, J. J., M. M. Häggblom, and R. L. Tate.** 2003. Effects of heavy metal contamination and remediation on soil microbial communities in the vicinity of a zinc smelter as indicated by analysis of microbial community phospholipid fatty acid profiles. *Biol. Fertil. Soils* **38:**65–71.
62. **Kenkel, N. C., D. A. Derksen, A. G. Thomas, and P. R. Watson.** 2002. Multivariate analysis in weed science research. *Weed Sci.* **50:**281–292.
63. **Kourtev, P. S., J. G. Ehrenfeld, and M. Häggblom.** 2003. Experimental analysis of the effect of exotic and native plant species on the structure and function of soil microbial communities. *Soil Biol. Biochem.* **35:**895–905.
64. **Lepš, J., and P. Šmilauer.** 2003. *Multivariate Analysis of Ecological Data using* CANOCO. Cambridge University Press, Cambridge, United Kingdom.
65. **Lorber, A., L. E. Wangen, and B. R. Kowalski.** 1987. A theoretical foundation for the PLS algorithm. *J. Chemometr.* **1:**19–31.
66. **Lovell, C. R., C. E. Bagwell, M. Czako, L. Marton, Y. M. Piceno, and D. B. Ringelberg.** 2001. Stability of a rhizosphere microbial community exposed to natural and manipulated environmental variability. *FEMS Microbiol. Ecol.* **38:**69–76.
67. **Ludvigsen, L., H. J. Albrechtsen, H. Holst, and T. H. Christensen.** 1997. Correlating phospholipid fatty acids (PLFA) in a landfill leachate polluted aquifer with biogeochemical factors by multivariate statistical methods. *FEMS Microbiol. Rev.* **20:**447–460.
68. **Macnaughton, S. J., T. L. Jenkins, S. Alugupalli, and D. C. White.** 1997. Quantitative sampling of indoor air biomass by signature lipid biomarker analysis: feasibility studies in a model system. *Am. Ind. Hyg. Assoc. J.* **58:**270–277.
69. **McKinley, V. L., A. D. Peacock, and D. C. White.** 2005. Microbial community PLFA and PHB responses to ecosystem restoration in tallgrass prairie soils. *Soil Biol. Biochem.* **37:**1946–1958.
70. **McMahon, S. K., M. A. Williams, P. J. Bottomley, and D. D. Myrold.** 2005. Dynamics of microbial communities during decomposition of carbon-13 labeled ryegrass fractions in soil. *Soil Sci. Soc. Am. J.* **69:**1238–1247.
71. **Meier-Augenstein, W.** 2002. Stable isotope analysis of fatty acids by gas chromatography-isotope ratio mass spectrometry. *Anal. Chim. Acta* **465:**63–79.

72. **Merila, P., R. Strommer, and H. Fritze.** 2002. Soil microbial activity and community structure along a primary succession transect on the land-uplift coast in western Finland. *Soil Biol. Biochem.* **34:**1647–1654.
73. **Murata, T., M. Kanao-Koshikawa, and T. Takamatsu.** 2005. Effects of Pb, Cu, Sb, In and Ag contamination on the proliferation of soil bacterial colonies, soil dehydrogenase activity, and phospholipid fatty acid profiles of soil microbial communities. *Water Air Soil Pollut.* **164:**103–118.
74. **Murata, T., K. Takagi, M. Ishizaka, and K. Yokoyama.** 2004. Effects of mefenacet and pretilachlor applications on phospholipid fatty acid profiles of soil microbial communities in rice paddy soil. *Soil Sci. Plant Nutr.* **50:**349–356.
75. **Myers, R. T., D. R. Zak, D. C. White, and A. Peacock.** 2001. Landscape-level patterns of microbial community composition and substrate use in upland forest ecosystems. *Soil Sci. Soc. Am. J.* **65:**359–367.
76. **Navarrete, A., A. Peacock, S. J. Macnaughton, J. Urmeneta, J. Mas-Castella, D. C. White, and R. Guerrero.** 2000. Physiological status and community composition of microbial mats of the Ebro Delta, Spain, by signature lipid biomarkers. *Microb. Ecol.* **39:**92–99.
77. **Newell, S. Y., T. L. Arsuffi, and R. D. Fallon.** 1988. Fundamental procedures for determining ergosterol content of decaying plant-material by liquid-chromatography. *Appl. Environ. Microbiol.* **54:**1876–1879.
78. **Nielsen, P., and S. O. Petersen.** 2000. Ester-linked polar lipid fatty acid profiles of soil microbial communities: a comparison of extraction methods and evaluation of interference from humic acids. *Soil Biol. Biochem.* **32:**1241–1249.
79. **Noble, P. A., J. S. Almeida, and C. R. Lovell.** 2000. Application of neural computing methods for interpreting phospholipid fatty acid profiles of natural microbial communities. *Appl. Environ. Microbiol.* **66:**694–699.
80. **O'Donnell, A. G., M. Seasman, A. Macrae, I. Waite, and J. T. Davies.** 2001. Plants and fertilisers as drivers of change in microbial community structure and function in soils. *Plant Soil* **232:**135–145.
81. **Olsson, P. A.** 1999. Signature fatty acids provide tools for determination of the distribution and interactions of mycorrhizal fungi in soil. *FEMS Microbiol. Ecol.* **29:**303–310.
82. **Olsson, P. A., and N. C. Johnson.** 2005. Tracking carbon from the atmosphere to the rhizosphere. *Ecol. Lett.* **8:** 1264–1270.
83. **Palmborg, C., A. Nordgren, and E. Bååth.** 1998. Multivariate modeling of soil microbial variables in forest soil contaminated by heavy metals using wet chemical analyses and pyrolysis GC/MS. *Soil Biol. Biochem.* **30:**345–357.
84. **Palmer, M.** *Ordination Methods for Ecologists.* Botany Department, Oklahoma State University, Stillwater. [Online.] http://ordination.okstate.edu/ index.html.
85. **Pankhurst, C. E., A. Pierret, B. G. Hawke, and J. M. Kirby.** 2002. Microbiological and chemical properties of soil associated with macropores at different depths in a red-duplex soil in NSW Australia. *Plant Soil* **238:**11–20.
86. **Pelz, O., W. R. Abraham, M. Saurer, R. Siegwolf, and J. Zeyer.** 2005. Microbial assimilation of plant-derived carbon in soil traced by isotope analysis. *Biol. Fertil. Soils* **41:**153–162.
87. **Pennanen, T.** 2001. Microbial communities in boreal coniferous forest humus exposed to heavy metals and changes in soil pH—a summary of the use of phospholipid fatty acids, Biolog (R) and H-3-thymidine incorporation methods in field studies. *Geoderma* **100:**91–126.
88. **Pennanen, T., H. Fritze, P. Vanhala, O. Kiikkilä, S. Neuvonen, and E. Bååth.** 1998. Structure of a microbial community in soil after prolonged addition of low levels of simulated acid rain. *Appl. Environ. Microbiol.* **64:** 2173–2180.
89. **Petersen, S. O., P. S. Frohne, and A. C. Kennedy.** 2002. Dynamics of a soil microbial community under spring wheat. *Soil Sci. Soc. Am. J.* **66:**826–833.
90. **Pinkart, H. C., D. B. Ringelberg, Y. M. Piceno, S. J. Macnaughton, and D. C. White.** 2002. Biochemical approaches to biomass measurements and community structure analysis, p. 101–113. *In* C. J. Hurst, R. L. Crawford, G. R. Knudsen, M. J. McInerney, and L. D. Stetzenbach (ed.), *Manual of Environmental Microbiology*, 2nd ed. ASM Press, Washington, D.C.
91. **Pombo, S. A., O. Pelz, M. H. Schroth, and J. Zeyer.** 2002. Field-scale C-13-labeling of phospholipid fatty acids (PLFA) and dissolved inorganic carbon: tracing acetate assimilation and mineralization in a petroleum hydrocarbon-contaminated aquifer. *FEMS Microbiol. Ecol.* **41:**259–267.
92. **Potvin, C., and D. A. Roff.** 1993. Distribution-free and robust statistical-methods—viable alternatives to parametric statistics. *Ecology* **74:**1617–1628.
93. **Ramadan, Z., P. K. Hopke, M. J. Johnson, and K. M. Scow.** 2005. Application of pls and back-propagation neural networks for the estimation of soil properties. *Chemom. Intell. Lab. Syst.* **75:**23–30.
94. **Ramadan, Z., X. H. Song, P. K. Hopke, M. J. Johnson, and K. M. Scow.** 2001. Variable selection in classification of environmental soil samples for partial least square and neural network models. *Anal. Chim. Acta* **446:**233–244.
95. **Ramsey, P. W., M. C. Rillig, K. P. Feris, N. S. Gordon, J. N. Moore, W. E. Holben, and J. E. Gannon.** 2005. Relationship between communities and processes; new insights from a field study of a contaminated ecosystem. *Ecol. Lett.* **8:**1201–1210.
96. **Ravit, B., J. G. Ehrenfeld, and M. M. Häggblom.** 2003. A comparison of sediment microbial communities associated with *Phragmites australis* and *Spartina alterniflora* in two brackish wetlands of New Jersey. *Estuaries* **26:**465–474.
97. **Rethemeyer, J., C. Kramer, G. Gleixner, B. John, T. Yamashita, H. Flessa, N. Andersen, M. J. Nadeau, and P. M. Grootes.** 2005. Transformation of organic matter in agricultural soils: radiocarbon concentration versus soil depth. *Geoderma* **128:**94–105.
98. **Rhine, E. D., J. J. Fuhrmann, and M. Radosevich.** 2003. Microbial community responses to atrazine exposure and nutrient availability: linking degradation capacity to community structure. *Microb. Ecol.* **46:**145–160.
99. **Ringelberg, D. B., J. O. Stair, J. Almeida, R. J. Norby, E. G. Oneill, and D. C. White.** 1997. Consequences of rising atmospheric carbon dioxide levels for the below-ground microbiota associated with white oak. *J. Environ. Qual.* **26:**495–503.
100. **Ritz, K., W. McNicol, N. Nunan, S. Grayston, P. Millard, D. Atkinson, A. Gollotte, D. Habeshaw, B. Boag, C. D. Clegg, B. S. Griffiths, R. E. Wheatley, L. A. Glover, A. E. McCaig, and J. I. Prosser.** 2004. Spatial structure in soil chemical and microbiological properties in an upland grassland. *FEMS Microbiol. Ecol.* **49:**191–205.
101. **Rossi, R. E., D. J. Mulla, A. G. Journel, and E. H. Franz.** 1992. Geostatistical tools for modeling and interpreting ecological spatial dependence. *Ecol. Monogr.* **62:** 277–314.

102. **Runion, G. B., S. A. Prior, D. W. Reeves, H. H. Rogers, D. C. Reicosky, A. D. Peacock, and D. C. White.** 2004. Microbial responses to wheel-traffic in conventional and no-tillage systems. *Commun. Soil Sci. Plant. Anal.* **35:**2891–2903.
103. **Rutgers, M., and A. M. Breure.** 1999. Risk assessment, microbial communities, and pollution-induced community tolerance. *Hum. Ecol. Risk Assess.* **5:**661–670.
104. **Saetre, P., and E. Bååth.** 2000. Spatial variation and patterns of soil microbial community structure in a mixed spruce-birch stand. *Soil Biol. Biochem.* **32:**909–917.
105. **Saitou, K., K. Nagasaki, H. Yamakawa, H. Y. Hu, K. Fujie, and A. Katayama.** 1999. Linear relation between the amount of respiratory quinones and the microbial biomass in soil. *Soil Sci. Plant Nutr.* **45:**775–778.
106. **Schutter, M. E., and R. P. Dick.** 2000. Comparison of fatty acid methyl ester (FAME) methods for characterizing microbial communities. *Soil Sci. Soc. Am. J.* **64:** 1659–1668.
107. **Seitz, L. M., H. E. Mohr, R. Burroughs, and D. B. Sauer.** 1977. Ergosterol as an indicator of fungal invasion in grains. *Cereal Chem.* **54:**1207–1217.
108. **Shi, W., J. Becker, M. Bischoff, R. F. Turco, and A. E. Konopka.** 2002. Association of microbial community composition and activity with lead, chromium, and hydrocarbon contamination. *Appl. Environ. Microbiol.* **68:** 3859–3866.
109. **Slater, G. F., H. K. White, T. I. Eglinton, and C. M. Reddy.** 2005. Determination of microbial carbon sources in petroleum contaminated sediments using molecular C-14 analysis. *Environ. Sci. Technol.* **39:**2552–2558.
110. **Soderberg, K. H., A. Probanza, A. Jumpponen, and E. Bååth.** 2004. The microbial community in the rhizosphere determined by community-level physiological profiles (CLPP) and direct soil- and cfu-PLFA techniques. *Appl. Soil. Ecol.* **25:**135–145.
111. **Song, D., and A. Katayama.** 2005. Monitoring microbial community in a subsurface soil contaminated with hydrocarbons by quinone profile. *Chemosphere* **59:**305–314.
112. **Song, X. H., P. K. Hopke, M. A. Bruns, D. A. Bossio, and K. M. Scow.** 1998. A fuzzy adaptive resonance theory supervised predictive mapping neural network applied to the classification of multivariate chemical data. *Chemom. Intell. Lab Syst.* **41:**161–170.
113. **Song, X. H., P. K. Hopke, M. A. Bruns, K. Graham, and K. Scow.** 1999. Pattern recognition of soil samples based on the microbial fatty acid contents. *Environ. Sci. Technol.* **33:**3524–3530.
114. **Stahl, P. D., and T. B. Parkin.** 1996. Relationship of soil ergosterol concentration and fungal biomass. *Soil Biol. Biochem.* **28:**847–855.
115. **Steenwerth, K. L., L. E. Jackson, F. J. Calderon, M. R. Stromberg, and K. M. Scow.** 2002. Soil microbial community composition and land use history in cultivated and grassland ecosystems of coastal California. *Soil Biol. Biochem.* **34:**1599–1611.
116. **Steer, J., and J. A. Harris.** 2000. Shifts in the microbial community in rhizosphere and non-rhizosphere soils during the growth of *Agrostis stolonifera*. *Soil Biol. Biochem.* **32:**869–878.
117. **Steger, K., A. Jarvis, S. Smars, and I. Sundh.** 2003. Comparison of signature lipid methods to determine microbial community structure in compost. *J. Microbiol. Methods* **55:**371–382.
118. **Stenberg, B.** 1999. Monitoring soil quality of arable land: microbiological indicators. *Acta Agric. Scand. B Soil Plant Sci.* **49:**1–24.
119. **Stenstrom, J., K. Svensson, and M. Johansson.** 2001. Reversible transition between active and dormant microbial states in soil. *FEMS Microbiol. Ecol.* **36:**93–104.
120. **Straube, W. L., C. C. Nestler, L. D. Hansen, D. Ringleberg, P. H. Pritchard, and J. Jones-Meehan.** 2003. Remediation of polyaromatic hydrocarbons (PAHs) through landfarming with biostimulation and bioaugmentation. *Acta Biotechnol.* **23:**179–196.
121. **Sylvia, D. M., J. J. Fuhrmann, P. G. Hartel, and D. A. Zuberer.** 1999. *Principles and Applications of Soil Microbiology*. Prentice Hall, Upper Saddle River, N.J.
122. **Tabachnick, B., and L. Fidell.** 1983. *Using Multivariate Statistics*. Harper & Row, New York, N.Y.
123. **ter Braak, C. J. F., and P. Šmilauer.** 1998. *CANOCO Reference Manual and User's Guide to CANOCO for Windows: Software for Canonical Community Ordination (Version 4)*. Microcomputer Power, Ithaca, N.Y., and Centre for Biometry Wageningen, Wageningen, The Netherlands.
124. **Thorndike, R.** 1978. *Correlational Procedures for Research*. Gardner Press, Inc., New York, N.Y.
125. **Treonis, A. M., N. J. Ostle, A. W. Stott, R. Primrose, S. J. Grayston, and P. Ineson.** 2004. Identification of groups of metabolically-active rhizosphere microorganisms by stable isotope probing of PLFAs. *Soil Biol. Biochem.* **36:**533–537.
126. **Tscherko, D., U. Hammesfahr, M. C. Marx, and E. Kandeler.** 2004. Shifts in rhizosphere microbial communities and enzyme activity of *Poa alpina* across an alpine chronosequence. *Soil Biol. Biochem.* **36:**1685–1698.
127. **Turpeinen, R., T. Kairesalo, and M. M. Häggblom.** 2004. Microbial community structure and activity in arsenic-, chromium- and copper-contaminated soils. *FEMS Microbiol. Ecol.* **47:**39–50.
128. **Urakawa, H., T. Yoshida, M. Nishimura, and K. Ohwada.** 2005. Characterization of depth-related changes and site-specific differences of microbial communities in marine sediments using quinone profiles. *Fish. Sci.* **71:**174–182.
129. **Vepsalainen, M., K. Erkomaa, S. Kukkonen, M. Vestberg, K. Wallenius, and R. M. Niemi.** 2004. The impact of crop plant cultivation and peat amendment on soil microbial activity and structure. *Plant Soil* **264:**273–286.
130. **White, D. C., W. M. Davis, J. S. Nickels, J. D. King, and R. J. Bobbie.** 1979. Determination of the sedimentary microbial biomass by extractable lipid phosphate. *Oecologia* **40:**51–62.
131. **White, D. C., J. O. Stair, and D. B. Ringelberg.** 1996. Quantitative comparisons of in situ microbial biodiversity by signature biomarker analysis. *J. Ind. Microbiol.* **17:**185–196.
132. **Widmer, F., A. Fliessbach, E. Laczko, J. Schulze-Aurich, and J. Zeyer.** 2001. Assessing soil biological characteristics: a comparison of bulk soil community DNA-, PLFA-, and Biolog (TM)-analyses. *Soil Biol. Biochem.* **33:**1029–1036.
133. **Wilke, B. M., M. Mai, A. Gattinger, M. Schloter, and P. Gong.** 2005. Effects of fresh and aged copper contaminations on soil microorganisms. *J. Plant Nutr. Soil Sci.* **168:**668–675.
134. **Wilkinson, S. C., and J. M. Anderson.** 2001. Spatial patterns of soil microbial communities in a Norway spruce (*Picea abies*) plantation. *Microb. Ecol.* **42:**248–255.
135. **Xu, M., K. J. Voorhees, and T. L. Hadfield.** 2003. Repeatability and pattern recognition of bacterial fatty acid profiles generated by direct mass spectrometric analysis of in situ thermal hydrolysis/methylation of whole cells. *Talanta* **59:**577–589.

136. **Yang, X., S. Li, G. Bengtsson, and N. Torneman.** 2005. Spatial variation of microbial properties in a creosote-contaminated soil. *Ying Yong Sheng Tai Xue Bao* **16:**939–944. (In Chinese.)

137. **Zak, D. R., W. E. Holmes, D. C. White, A. D. Peacock, and D. Tilman.** 2003. Plant diversity, soil microbial communities, and ecosystem function: are there any links? *Ecology* **84:**2042–2050.

138. **Zak, D. R., D. B. Ringelberg, K. S. Pregitzer, D. L. Randlett, D. C. White, and P. S. Curtis.** 1996. Soil microbial communities beneath *Populus grandidentata* crown under elevated atmospheric CO2. *Ecol. Appl.* **6:** 257–262.

139. **Zelles, L.** 1999. Fatty acid patterns of phospholipids and lipopolysaccharides in the characterisation of microbial communities in soil: a review. *Biol. Fertil. Soils* **29:**111–129.

140. **Zhang, C. L.** 2002. Stable carbon isotopes of lipid biomarkers: analysis of metabolites and metabolic fates of environmental microorganisms. *Curr. Opin. Biotechnol.* **13:**25–30.

141. **Zhang, W. J., W. Y. Rui, C. Tu, H. G. Diab, F. J. Louws, J. P. Mueller, N. Creamer, M. Bell, M. G. Wagger, and S. Hu.** 2005. Responses of soil microbial community structure and diversity to agricultural deintensification. *Pedosphere* **15:**440–447.

142. **Zhao, X. R., Q. Lin, and P. C. Brookes.** 2005. Does soil ergosterol concentration provide a reliable estimate of soil fungal biomass? *Soil Biol. Biochem.* **37:**311–317.

143. **Ziegler, S. E., P. M. White, D. C. Wolf, and G. J. Thoma.** 2005. Tracking the fate and recycling of C-13-labeled glucose in soil. *Soil. Sci.* **170:**767–778.

144. **Zogg, G. P., D. R. Zak, D. B. Ringelberg, N. W. MacDonald, K. S. Pregitzer, and D. C. White.** 1997. Compositional and functional shifts in microbial communities due to soil warming. *Soil Sci. Soc. Am. J.* **61:** 475–481.

SUBSURFACE AND LANDFILLS

VOLUME EDITOR
AARON L. MILLS

Overview of Issues in Subsurface and Landfill Microbiology

RONALD W. HARVEY, JOSEPH M. SUFLITA, MICHAEL J. McINERNEY, AND AARON L. MILLS

65

The abundance of prokaryotes in the subsurface (terrestrial and marine) is estimated to be as much as 6.1×10^{30} cells, suggesting that the carbon content of the subsurface organisms is about the same as that of green plants (2). Although the metabolic rates of these organisms are considerably lower than their surficial counterparts, it is clear that they are active within the constraints of nutrient and substrate availability. The recognition that the terrestrial subsurface contains active microflora has led to a myriad of questions regarding the nature of the organisms inhabiting this environment. For instance, what are the limits of the biosphere? How diverse are subsurface microbial communities? What adaptations do the organisms possess that allow them to survive and proliferate in this environment? How were microorganisms introduced into the subsurface, how do they interact, and what factors govern their distribution? What is the role of subsurface microorganisms in the cycling of elements and the transfer of energy?

In considering such questions, it is entirely appropriate to question the "uniqueness" of the resident microbiota. That is, are the organisms inhabiting the subsurface in any way out of the ordinary? It seems reasonable to hypothesize that once microorganisms are introduced into the subsurface, natural forces will select for those that are able to survive and disperse within the subsurface. Thus, those organisms that inhabit the subsurface are the ones that have the adaptive features necessary to maintain themselves and proliferate there. The uniqueness of the subsurface microbiota may be in the mechanisms that microorganisms have evolved to adapt to and withstand the selection forces inherent in the terrestrial subsurface.

To date, the majority of evidence indicates that most subsurface environments (exclusive of those controlled by some dominant abiotic factor such as temperature, pressure, or toxicants) possess climax ecological communities that are well adapted to the environment in which they live. Like their counterparts on the surface, subsurface ecosystems are characterized by a high degree of microbiological diversity, they possess trophic structure, and they exhibit material cycling and energy transfer. Members of such ecosystems typically possess structural, physiological, or reproductive adaptations that allow them to disperse and survive in such habitats.

It may be that the subsurface microbiota has great similarities with surface organisms, particularly when comparisons are made on a metabolic basis. That is, similar metabolic pathways or corresponding ecological niche similarities may be found in specific surface and subsurface microorganisms. This is not surprising; in fact, it is comforting. Metabolic principles represent a unifying feature of life: diverse life forms exhibit remarkable similarity at the biochemical or metabolic level. By extension of this logic, it is likely that subsurface microbial populations and community dynamics are also similar to those of surface environments. Consequently, ecological principles may be able to transcend the particular environment under consideration.

In spite of the proliferation of knowledge regarding the terrestrial subsurface and its tremendous economic importance, it remains one of the least understood microbial habitats. This is due, in part, to the high costs and technical problems associated with obtaining high-quality samples suitable for microbiological analysis and the difficulties of inferring the distribution of microbial populations and activity from a limited number of samples. Advancement in our understanding of subsurface microbiology is also hampered by a lack of suitable methodologies for studying the activities of sparse, often ultraoligotrophic, microbial populations that inhabit pristine subsurface environments. However, the last 2 decades have seen much progress in this important and developing field. The recent interest in subsurface microbiology is largely due to the need to control environmental problems caused by or affected by microorganisms in subsurface environments. These problems include "souring" (hydrogen sulfide production) in producing oil reservoirs, clogging of water supply aquifers by bacteria and/or products of bacterially mediated geochemical reactions, contamination of water supply wells by transport of pathogens, bacterial production of explosive gases in and adjacent to landfills, and the effects of microbial populations on the fate and transport of subsurface contaminants.

A fundamental impediment to the advancement of subsurface microbiology is the difficulty of obtaining representative samples of subsurface material that are free of contamination by nonindigenous organisms and are thus suitable for study of the microbes native to the subsurface. Water supply wells have been in existence for several millennia. However, the act of constructing a well not only chemically alters the subsurface environment but also introduces nonresident microorganisms. Therefore, it is often unclear whether microbial communities observed in well water are truly representative of what is in the aquifer per se. Much of our current understanding of subsurface microbial populations derives from studies involving core samples. Coring has partially alleviated interpretation problems associated with the alteration of subsurface environments by wells. Additionally, microbial populations in the subsurface are largely attached to solid surfaces and therefore may not be adequately sampled by wells. Chapter 66 discusses procedures for sampling groundwater and subsurface core material for subsequent microbiological studies. Included are descriptions of, advantages of, and problems associated with coring procedures used to sample shallow (<100-in.) depths in unconsolidated formations (e.g., hollow-stem augers and sonic drilling) and to sample deeper and/or more consolidated formations (e.g., cable tools and rotary drills). In practice, it is often more accurate to assess the degree of microbial contamination of the recovered subsurface samples than to assume that contamination did not occur. Tracer techniques for assessing this contamination and methods for minimizing further contamination during sample handling are also evaluated. It is evident that many advances have been made both in the method of collection and in the method of handling subsurface samples for microbiological study. However, it is still difficult to say with absolute certainty that a given microorganism isolated from a sample of groundwater or subsurface core material originated from the depth at which the sample was taken and is not merely an artifact of contamination.

In spite of the sampling and methodological problems inherent in subsurface microbiology, a number of important recent advances have been made in our understanding of microbial activities in groundwater environments. Much can be inferred about the activities of microorganisms in groundwater by assessing the changes in aquifer geochemistry along a given flow path. Flow paths in many aquifers are hundreds of kilometers in length, allowing synoptic observations of microbially mediated changes in geochemistry in low-nutrient environments that take hundreds or thousands of years to complete. Chapter 68 discusses the types of groundwater systems in relation to their recharge and discharge sites and the types of organic compounds and minerals that can be metabolized by subsurface microorganisms. Knowledge of the types of electron donors and acceptors present in an aquifer can be used to infer how microbial activity can alter both the chemical quality of aquifers and the spatial patterns of the predominant terminal electron-accepting process.

Although much can be inferred from geochemical measurements about the activities of aquifer microorganisms involved in the predominant terminal electron-accepting processes, the nonconservative nature of many geochemical components can limit interpretation. A parallel approach involves incubation assays that are specific for a given electron-accepting process (Chapter 69). The latter approach may be particularly well suited for analysis of organically contaminated aquifers in which the rates of microbial activity are high enough to measure within a reasonable experimental time frame. These assays are performed in the laboratory with representative geochemical conditions and aquifer core material collected from the zone of interest. However, microbial or chemical contamination of the samples and deviations from in situ incubation conditions can make interpretation difficult, especially in evaluating the rates at which the processes occur. Another tool is quantification of the dissolved H_2 concentration in groundwater from the site of interest. H_2 is a dynamic molecule, and its turnover indicates the presence of an active microbial population. A characteristic steady-state H_2 concentration has been attributed to each terminal electron-accepting reaction. This approach has the greatest potential for distinguishing between sulfate-reducing and methanogenic conditions since the latter have the highest and most easily measured concentrations of dissolved hydrogen. However, the concept of a characteristic H_2 concentration reflecting the dominant redox process in a subsurface environment has recently been called into question (1). While it seems clear that several environmental variables can impact the apparent steady-state hydrogen concentration, it also seems clear that such measurements will remain useful as tools for gaining insight on the functioning of microbial communities in the terrestrial subsurface.

Still another approach involves small-scale injection and recovery experiments that use unaltered in situ conditions and microbial populations in the aquifer itself. The in situ injection and recovery approach is best performed under natural-gradient (normal-flow) conditions so that the flow rate is relatively constant and the rates of the terminal electron-accepting process can be calculated from a modified advection/dispersion model. A conservative tracer (usually bromide or chloride) is used to account for the flow velocity and the degrees of dispersion that take place during downgradient transport. A labeled terminal electron acceptor or a blocking compound that causes a buildup of a readily measured intermediate in the terminal electron-accepting process can be injected with the conservative tracer. The need to monitor the tracer in three-dimensional space over relatively short time frames limits the method to shallow unconsolidated aquifer systems.

Small-scale injection and recovery experiments have also been used to delineate the subsurface transport behavior of microorganisms (chapter 70), which has important ramifications for microbially enhanced oil recovery processes (chapter 71), for the fate and transport of organic contaminants, and for the placement of water supply wells (chapter 72). Our ability to predict the subsurface transport of microorganisms on a large scale is hampered by the multitude of physiological, hydrological, and chemical factors that control their movement. Many of these factors are poorly understood and interrelated with other processes. Nevertheless, many advances have been made in our ability to influence and mathematically describe groundwater transport of microorganisms (chapters 70 and 72). An iterative approach that utilizes laboratory-scale, intermediate-scale, and field-scale experiments allows for an evaluation of specific processes under controlled laboratory conditions and tests of laboratory-derived transport hypotheses in situ (chapter 70). The use of well-characterized microbe-sized microspheres has facilitated the understanding of some of the abiotic factors that govern subsurface microbial transport behavior. However, a number of problems and areas of research must be addressed before we can reliably predict microbial transport over longer distances. In particular, there is a dearth of knowledge concerning the role of bacterivorous subsurface eukaryotes, and methods of culturing subsurface microorganisms that do not alter the transport characteristics of the cell are needed. We also need to know how to mathematically describe the processes of microbial attachment and detachment in the presence of chemically complex surfaces and groundwaters and how to handle physical and chemical aquifer heterogeneity in transport models.

The aforementioned difficulties in accurately predicting microbial transport in the subsurface make it problematic to use microbial transport models to calculate how far a water supply well should be located from a contaminant source. Indeed, predictive transport models can over- or underpredict microbial movement in groundwater by several orders of magnitude. Most well protection models use the transport of viruses to calculate well setback distances, and these models ignore pathogenic bacteria and protozoa, whose transport behavior can be fundamentally different from that of viruses. The methods that are used to calculate setback distances for water supply wells are explained in chapter 72. Despite these limitations, for regional planning purposes, geostatistical models have been successfully applied to areas that are particularly susceptible to groundwater contamination. These models incorporate information about groundwater temperatures (which are used to calculate virus inactivation rates), hydraulic conductivities, and hydraulic gradients. However, viral inactivation can be affected by a number of factors other than temperature, such as interaction with solid surfaces and the amount of dissolved organic matter. It is clear that models used to predict subsurface microbial transport behavior must be improved to give more precise answers to these important public health questions. In particular, more information on the transport behavior of microorganisms in the unsaturated zone, which is ignored in many of the current models, is needed.

Recognition that microorganisms inhabit both shallow and deep aquifers has led to the realization that petroleum reservoirs must also contain active microbial populations. However, the problems facing a petroleum microbiologist (chapter 71) are quite different from those of a microbiologist working with drinking water aquifers. Important concerns are the control of detrimental microbial activities such as H_2S production (souring) and well bore plugging and the use of microbes to improve oil production. The problems created by microorganisms during oil production include corrosion of water-handling equipment, plugging of injection wells, and the safety issues and removal costs associated with production of hydrogen sulfide. The key issue in the control of microbially induced souring is the early detection of the sulfide-producing bacteria that are responsible. Culture-dependent and culture-independent methods, which are sensitive and convenient approaches required for routine monitoring, are now available to detect sulfide-producing bacteria.

The ability of microorganisms to improve oil production is a controversial topic because of the lack of controlled laboratory and field studies. Chapter 71 describes the use of sandstone experimental systems to test the potential of microorganisms to release oil entrapped in rock. One way that bacteria can facilitate the release of oil is by producing surfactants that reduce the interfacial forces between the oil and aqueous phases. To be effective, the surfactant must reduce these forces by a factor of at least 10,000. Bacteria and the polymers that they produce can also enhance the recovery of oil by plugging the more conductive zones and larger pores in an oil-bearing formation which diverts the recovery fluid into previously bypassed regions of reservoir. Chapter 71 describes the procedures used to isolate and characterize polymer- or surfactant-producing bacteria from oil reservoirs as well as a general protocol to follow for conducting field trials. Although it is relatively easy to determine whether a microorganism can enhance oil recovery in controlled laboratory experiments, extrapolating this information to the recovery of oil in actual oil reservoirs, often at great depth, is nontrivial. A multitude of variables, which are constantly changing throughout the exploitation process, can affect the performance of oil recovery operations. For these reasons, it is difficult to determine whether a particular treatment has in fact caused an increase in oil production. Thus, in evaluating a microbial process in the field, it is critical to perform a control treatment that uses the same equipment and injection protocols as those used for the microbial treatment but lacks the key components of the microbial process itself, in order to verify that the microbial process is having the desired effect. Tracer experiments are also important in order to verify that the production wells under study are in communication with the wells that received the nutrients and/or cells.

The microbiology of landfills (chapter 67) is also an important topic, as this waste disposal option is the most popular to date and is likely to remain so for the foreseeable future. The chapter is included in this section rather than in the section on biodegradation because landfills are known to have serious environmental effects on the terrestrial subsurface, most notably the contamination of shallow groundwater reserves. In order to accurately assess this impact as well as other environmental threats, it is important to know what is actually buried in landfills and to what extent the microbial decomposition of interred refuse actually occurs. Chapter 67 provides a needed perspective for such evaluations.

Aside from the obvious connection of the impact of landfills on the terrestrial subsurface, there are other, more subtle similarities linking surface and subsurface environments. First, the two habitats are strikingly similar with respect to the heterogeneous nature of microbial processes in them. In fact, one could hardly imagine a more heterogeneous environment than a municipal landfill. Such heterogeneity requires specialized sampling, handling, and analysis strategies with clearly defined objectives. Since sampling either habitat tends to be difficult, expensive, and time-consuming, there is an obvious need to obtain samples that are as representative as possible. For the results to be most meaningful, experimental systems and analytical techniques that are appropriate for the objectives of the study must be used. Moreover, the microbial processes in each habitat are equally dependent on local hydrological conditions for a variety of important features, including the delivery of terminal electron donors and/or acceptors, the supply of nutrients for microbial growth, and the migration of accumulated metabolic intermediates or end products. However, preferential channeling of moisture through a landfill can severely limit refuse decomposition processes. An indication of the multiple factors that can potentially limit microbial refuse decomposition in landfills and the experimental systems and analytical approaches that are most appropriate for the objective under study can be found in chapter 67.

Current studies argue that microbial activities influence the geochemical processes in both landfills and subsurface environments and that the altered geochemistry, in turn, influences the selection of microorganisms proliferating in the habitat. However, while microbial activity may often be limited by the availability of electron donors in uncontaminated aquifer systems, this is rarely the case with landfills. Ultimately, the large amount of organic matter present in landfills leads to the production of substantial amounts of methane, with landfills representing the fourth largest anthropogenic source of methane. Since methane is 25 to 30 times more effective than carbon dioxide as a greenhouse gas, quantifying the amount of methane emitted from landfills is critical in order to assess the importance of landfill methane emissions in global warming. The amount of methane emitted from landfills is a combination of the rate of methane production in the landfill and the rate of methane oxidation in the landfill cover and overlying soil. Chapter 67 discusses the experimental approaches used to quantify the rate of methane oxidation in soils overlying landfills and the factors that influence this rate as well as methods to assess the numbers and kinds of microorganisms involved in this process.

It is hoped that the chapters in this section will provide investigators with a greater understanding of the experimental approaches needed to study the microbiology of the terrestrial subsurface and an appreciation of interpretational limits imposed by the existing methodologies. As our knowledge of the types of metabolic interactions and the nutritional versatility of indigenous microorganisms increases, we will be able to predict with greater confidence the effects that may be associated with perturbations of subsurface environments. This in turn will help provide a conceptual framework for biotechnological approaches for environmental restoration. The fact that subsurface habitats that have been physically separated from their surface counterparts for thousands to millions of years are now known to contain diverse populations of microorganisms has excited the interest of evolutionary biologists. The hope is that further study of subsurface microorganisms will provide insights into the process of microbial evolution and possibly into the origins of life itself.

REFERENCES

1. **Christensen, T. H., P. L. Bjerg, S. A. Banwart, R. Jakobsen, G. Heron, and H.-J. Albrechtsen.** 2000. Characterization of redox conditions in groundwater contaminant plumes. *J. Contam. Hydrol.* **45:**165–241.
2. **Whitman, W. B., D. C. Coleman, and W. J. Wiebe.** 1998. Prokaryotes: the unseen majority. *Proc. Natl. Acad. Sci. USA* **95:**6578–6583.

Drilling, Coring, and Sampling Subsurface Environments

THOMAS L. KIEFT, TOMMY J. PHELPS, AND JAMES K. FREDRICKSON

66

Studies of the microbial ecology and biogeochemistry of terrestrial subsurface environments have progressed in the past 20 years from reports of a diversity of viable microbes in shallow aquifers (8, 91) and vadose zones (4, 15, 31, 48) to microbes in increasingly deep sedimentary and rock environments (6, 7, 21, 33, 50, 66, 69, 70, 71, 72, 73, 85, 86, 96). Metabolically active microbes from depths that are >3 km below land surface have been identified and, in some cases, characterized. These organisms assert a major influence on the geochemistry and geology of the subsurface. Yet the lower limit of the terrestrial biosphere has yet to be delineated, and the factors that constrain life at great depth remain poorly understood. Practical considerations also drive microbial studies of the terrestrial subsurface, e.g., the application and monitoring of microbes for in situ bioremediation of contaminated subsurface sediments and groundwater (5), improved recovery of fossil fuels (3, 23), and potential effects of microbial processes on performance of subsurface radioactive waste repositories (46, 51). Scientific interest in naturally occurring biogeochemical processes and novel microorganisms (12, 25, 29, 50, 59, 67, 86) has also stimulated the interest in subsurface microbiology. A major challenge confronting subsurface microbial investigations is sampling. Subsurface samples have been used to study the size and composition of the microbial community by using traditional and molecular biology-based methods, to assess the types and rates of microbial activities, and to measure the geochemical properties of the water and solids that comprise subsurface microbial habitats. All of these applications have in common the need to access samples that are uncompromised and representative of the specific habitats under investigation.

Advances in subsurface microbiology, especially of relatively deep environments (i.e., below 100 m), have been facilitated by innovations in drilling and coring technologies. In spite of these innovations, however, obtaining representative samples of solids and water from the subsurface for microbiological and geochemical studies remains a major impediment to the study of subsurface microbiology. Sampling subsurface sediments, rocks, and water for microbiological and geochemical analyses presents unique challenges compared with the sampling of surface soils, waters, and sediments. Sampling of subsurface solids requires specialized technniques for drilling (advancing the borehole) and coring (collecting samples of subsurface materials). Different lithologies require different tools and approaches for successfully collecting samples representative of their native environment. For example, drilling and coring methods that are used to sample fine-grained sediments generally will not work for unconsolidated coarse-grained sediments that tend to flow or fall out of a core barrel upon retrieval. Also, the drilling methods used for sampling hard rock at depth generally require approaches such as rotary coring with circulating fluids that are distinct from hollow-stem augering, which is typically used for coring of shallow unconsolidated sediments.

The costs associated with drilling and coring often limit the number and type of samples that can be obtained. This can be problematic, given the considerable spatial heterogeneity in the distribution and function of microorganisms in subsurface environments (2, 30, 33, 72, 81, 83). For soils, it is relatively simple to obtain sufficient numbers of samples to adequately characterize the microbiological properties of a given site. However, when sampling the subsurface, especially at depth, obtaining a sufficient number of samples to define horizontal and vertical variations in the microbiological properties can be technically difficult, labor-intensive, and costly.

As in other environments, subsurface microbes that are attached at solid-liquid interfaces are physiologically and phylogenetically distinct from those that are suspended in the bulk groundwater (44, 45, 56, 57, 58). For this reason, sampling groundwater pumped to the surface is generally inadequate for microbiological studies; subsurface solids are required for comprehensive evaluation of the microbiological characteristics of subsurface environments. In addition, the process of drilling and developing a well leads to introduction of exogenous microorganisms and overall changes in the biogeochemical environment in and near the well (17, 36). Sampling groundwater for microbiological analysis is necessary when sampling solids is impractical or when the samples are being collected to examine water quality indicators and pathogens or to monitor changes in microbial populations and activities during bioremediation. Groundwater sampling is commonly used for monitoring geochemical parameters during in situ bioremediation and, in some cases, even for microbiological monitoring. Complications with sampling solids include the heterogeneity in the microbiological (and geochemical) properties of the subsurface (2, 81), perturbation of the surface site, and prohibitive costs. Sampling of groundwater from multiple wells can be an effective means for assessing microbiological and geochemical properties in aquifers at local (88) and regional (67) scales. In comparison with collecting core samples for geochemical and microbiological analyses, sampling groundwater from multipoint sampling wells provides considerable flexibility in terms of the location, number, and size of samples that can be collected.

In some cases, costs associated with the drilling and coring may exceed the costs of analyses and investigations. At times, the drilling costs may be borne by other parties that are sampling the subsurface for other reasons, such as mineral or fossil-fuel exploration or installation of water supply wells. Such "piggybacking" may enable access to otherwise unavailable samples but has disadvantages: the coring methods may not be appropriate for obtaining samples suitable for microbiological analysis, and investigators may have little or no control over the location and procedures used for coring. Mines and tunnels (38, 40, 50, 51, 65, 66, 69, 79) have also been used to access recently exposed subsurface environments for microbiological and biogeochemical investigations. Such investigations have contributed to improving our understanding of biogeochemical processes in the deep subsurface as well as to the isolation of novel microorganisms (50, 86). This approach will be discussed further below.

Selection of a drilling company must be carefully considered if drilling and coring are to be contracted. Some companies have more experience and capabilities than do others in scientific drilling and sampling. These are often the larger companies that have operations in many regions and even multiple countries. However, smaller locally owned companies are often more familiar with the local geology and have more experience in drilling and coring in the region. These are just a few examples of the challenges and issues that must be addressed when sampling the subsurface. More detailed discussion of issues and challenges during the planning and operation of subsurface sampling operations has been published elsewhere (37).

Approaches appropriate for drilling boreholes and collecting cores depend on the depths and lithologies to be sampled. Hollow-stem augering remains the most common method used for shallow drilling in unconsolidated sediments, typically to depths less than 100 m, while rotary drilling is the most common method for drilling to depths greater than 100 m. Cable tool systems have also been used successfully for sampling subsurface sediments at depths of less than 300 m. These three commonly used approaches have been selected for inclusion in this chapter. The reader should also refer to the following references for additional details regarding these and other drilling and sampling methods (17, 20, 21, 37, 55, 76, 78, 91).

Many precautions must be taken to minimize contamination of samples during coring and processing. Additionally, one or more tracers should be used to assess the extent of sample contamination by drilling fluids, etc. Rigorous tracer regimens are particularly important for sampling deep subsurface environments where circulating fluids are used to remove drill cuttings from the borehole. These circulating fluids can enter the formation ahead of the drill bit or penetrate the core while it is being recovered from the borehole. Another source of contamination is smearing or mixing of part of the outer core, which has been in contact with drilling tools and fluids, into the uncontaminated inner core during sample paring and processing. After subsurface materials have been recovered, they should be further processed prior to analysis or use in experiments. This processing typically includes paring to remove outer core material and sectioning and disaggregation of the inner portion of the core. Because many subsurface environments are anoxic, it is important to limit exposure of samples to air in order to prevent inhibition or inactivation of anaerobic and microaerophilic microorganisms by high concentrations of O_2 and to prevent oxidation of reduced geochemical species.

The goal of this chapter is to provide background information and general guidelines necessary for obtaining representative samples of subsurface solids and groundwater for microbiological and geochemical analyses. To this end, approaches for collecting representative solid and groundwater samples, processing of subsurface solids to reduce and assess microbiological and geochemical alterations during coring, and evaluating the quality of subsurface samples are described.

SAMPLING FORMATION MATERIALS

Hollow-Stem Augering: Shallow Sediment Sampling

Hollow-stem augering is typically used for sampling shallow, unconsolidated subsurface environments less than 100 m deep (Table 1). Typical uses of hollow-stem augering include drilling in shallow, unconsolidated lithologies for the installation of domestic water supply and monitoring wells in addition to the sampling of solids.

During hollow-stem augering, the borehole is advanced by rotating the auger flights and the cuttings are brought to the surface by the auger action. When the desired sampling depth is reached, a split-spoon (Fig. 1) or other drive- or push-type core barrel is lowered through the auger flights to the bottom of the borehole. The core barrel is typically driven into the sediment by using percussion, i.e., a hammer. Once the core barrel is driven to the appropriate depth, it is brought to the surface. A hollow-stem auger drill rig is

TABLE 1 Comparison of various drilling methods

Drilling method	Applicable depth (m)	Lithology	Coring method	Advantage(s)	Disadvantage(s)
Hollow-stem augering	<100	Unconsolidated	Split-spoon, Waterloo	Mobile, inexpensive	Shallow
Cable tool	<300	Unconsolidated or consolidated	Split-spoon	Inexpensive, no drilling fluids	Slow
Rotary with mud or air	>1,000	Unconsolidated or consolidated	Split-spoon, wireline	Deep access, fast	Costly, drilling fluids

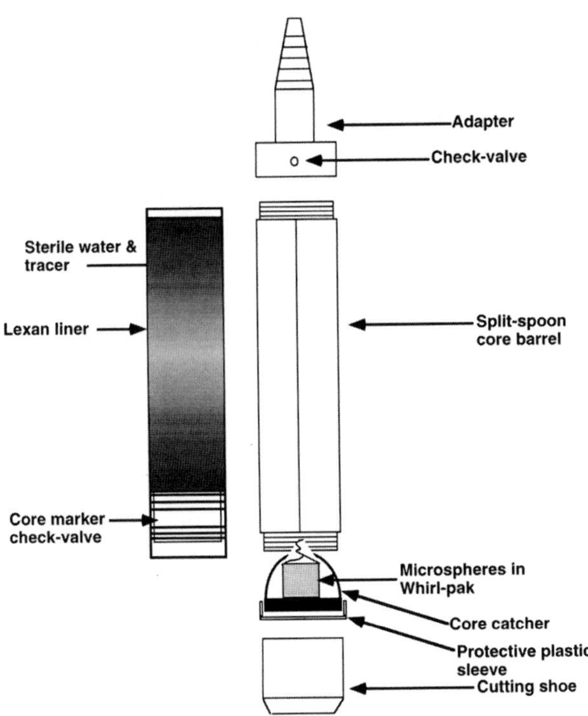

FIGURE 1 Diagram of split-spoon core barrel containing tracers and core liner for sampling saturated subsurface sediments using cable tool, hollow-stem auger, or rotary drilling methods.

FIGURE 2 Photograph of a hollow-stem auger drill rig used for obtaining core samples from relatively shallow (<100 m) subsurface sediments.

shown in Fig. 2, and a simplified diagram of coring through the use of a hollow-stem auger is shown in Fig. 3.

It is common practice to thoroughly clean and disinfect the surface of all auger flights and core barrels before each use. For additional information on decontamination of drilling tools, see "Coring Procedures" below and references 21, 37, 76, 78, and 91. Once the core has been brought to the surface, the liner containing the intact core is removed from the core barrel and the ends are capped. The core can then be transferred to a location where it can be further processed. Immediately after retrieval, the core should be placed under an inert gas phase, e.g., N_2 or Ar, minimizing exposure of the retrieved materials to air and potentially oxidizing conditions.

The major advantages of coring through hollow-stem augers are that it is relatively simple, commonly available, and economical to use for obtaining nearly undisturbed unsaturated or saturated subsurface samples. The drilling equipment is typically mounted on the back of a truck, where it is easily deployed. Depending on the total depth and frequency of coring, many boreholes can usually be cored in a few days. Another major advantage is that drilling fluids are not required and hence a source of contamination is eliminated. However, when augering saturated sediments, nonnative water is often added to maintain fluid pressures on the formation to prevent the heaving of sediments into the augers. This water can be a potential source of microbial contamination as it can move into the formation ahead of the auger flights. Before introducing nonnative water to a borehole from which microbiological samples are to be collected, it is important to disinfect the water by chlorination. The principal sources of contamination during hollow-stem augering, when nonnative fluids are not introduced intentionally to the borehole, are from overlying sediment and soil that may be carried to the bottom of the borehole and from underlying fluidized sediments that may flow into the open borehole.

Sampling of unconsolidated sands with a hollow-stem auger can be difficult because the samples lack cohesiveness and can fall out of the core barrel during the trip to the surface. To overcome this problem, a specialized sampling system referred to as the Waterloo sampler (95) was developed. The Waterloo sampler consists of an outer core barrel, an inner liner, and a plunger or piston. The piston is set at the bottom of the liner. As the core barrel is driven into the sediment, the piston slides upward inside the inner liner, creating suction. The suction helps to retain the sample within the core liner during retrieval of the core barrel. Another advantage of the Waterloo system is that it can limit drainage of pore water from the core liner. Drainage of pore water can result in changing the solid-to-solution ratio of the sediments and replacement of water with air in some of the pores. Both of these can change the microbiological and geochemical properties of the sample.

A disadvantage of the Waterloo system is the requirement for an additional cable to hold the piston and the risk of the cables becoming entangled in the borehole. It is also difficult to core dense, consolidated, or stony materials by using the hollow-stem auger. The depth to which hollow-stem augering can be used is limited by increasing friction of the auger flights with depth, especially if rotation of the flights is stopped for any period of time. In spite of these disadvantages,

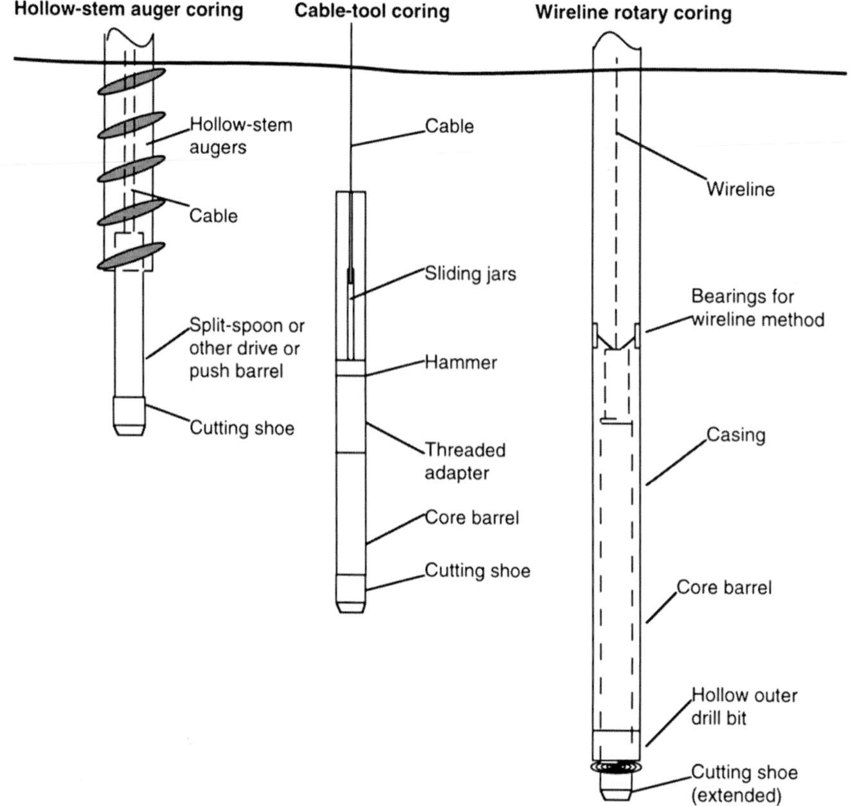

FIGURE 3 Simplified diagram of below-ground portion of coring process for collection of subsurface materials during hollow-stem auger, cable tool, and wireline rotary drilling.

hollow-stem augering is the method of choice because of its availability, simplicity, and low cost when sampling unconsolidated sediments at depths of less than 100 m.

Cable Tool Drilling: Shallow and Deep Sediment Sampling

Cable tool drilling has been used for centuries for drilling water wells (24), and it has also been used to obtain high-quality samples suitable for microbiological analysis from unconsolidated sediments from depths as great as 200 m (31, 49) (Fig. 4). Cable tool drilling uses a "drive" barrel that is attached to a set of sliding hammers ("jars") that advance the borehole by percussion. The upper jar is raised by a cable attached to the drill rig and is then dropped onto the lower jar, which pushes the drive barrel deeper into the subsurface. The drive barrel is repeatedly lowered, driven, and retrieved to advance the hole. The materials obtained in this manner are generally not suitable for microbiological analysis because of the disturbance that occurs during the driving. In consolidated or dense formations, the hole is advanced by using a hard tool or heavy bit that crushes the subsurface material, which is subsequently removed by bailing. It is usually necessary to advance metal casing in unconsolidated sediments to prevent the borehole from collapsing during and after drilling. The diameter of the casing, and hence of the borehole, can vary from <10 to >30 cm. It is often necessary to use smaller-diameter casing with increasing depth because of the greater friction associated with the larger-diameter casing. Telescoping of multiple casings eases the removal of casing after the borehole has been completed.

Core samples of formation materials suitable for microbiological analyses are obtained during cable tool drilling in a manner similar to that described for hollow-stem auger coring. A split-spoon or similar push-type core barrel replaces the drive barrel, and the split-spoon is advanced by raising and dropping the upper jar, hammering the core barrel into the formation ahead of the borehole (Fig. 3). Split-spoon core barrels are the most common and flexible samplers used with cable tool drilling. They are readily assembled and disassembled, can house an inner core liner, and are easily cleaned. In order to limit contact between the core liner and groundwater or slurry in the borehole or drill muds, it is necessary to use a core plug or marker (Fig. 1) and to fill the liner with disinfected water. When the core barrel penetrates the formation, the core plug is pushed up through the core liner and the water that was in the liner exits through the check-valve at the top of the assembly. The check-valve in the core plug allows fluids and mud at the bottom of the borehole to be "squeezed" out of the liner as the core fills the liner.

A major advantage of cable tool drilling is that, as with hollow-stem auger drilling, no recirculating fluids are required and hence the potential for contamination of samples is greatly reduced. Also, the equipment is widely available. An advantage over hollow-stem augering is that coring at depths greater than 100 m can be readily achieved. In theory, cable tool drilling can be used to sample formations deeper than 100 m, but practical considerations limit the effective depth. A disadvantage is that cable tool drilling is slow; even under ideal conditions the borehole is typically advanced only 10 to 20 m per day. Depending on the

FIGURE 4 Photograph of a cable tool drill rig.

geology, there can be considerable physical disruption of the sample due to the hammering, pulverization, and redistribution of subsurface materials. There can also be significant compaction of the cored materials. If there are gravels or cobbles, they will be fractured and pulverized by the action of the jars. Fracturing of the solids can have a secondary impact on the microbiological and chemical characteristics of subsurface samples. For example, Bjornstad et al. (11) found that fracturing of cobbles was related to the generation and accumulation of hydrogen gas in the slurried drill cuttings in the hole. This production of H_2 led to an increase in H_2-utilizing bacteria in the borehole slurry, which provided a potential source of contamination of the cored sediments.

Rotary Drilling: Deep Sediment and Rock Sampling

Rotary drilling and coring are commonly used technologies for subsurface mineral and fossil fuel exploration, geological characterization, scientific exploration, placement of deep wells for water supply, and monitoring groundwater contamination. Rotary drilling rigs range in size from small truck-mounted units used for shallow well drilling to large rigs that are used for deep (>2,000 m) gas and oil exploration and scientific drilling programs. Rotary drilling employs sections of hollow drill pipe that are threaded together as the borehole is deepened. The drill pipe is rotated at the surface by the drill rig. At the base of the drill pipe is a drill bit that provides the cutting action. For routine drilling operations a tricone bit is used that essentially homogenizes the materials into "cuttings" as the drill pipe is advanced. To prevent the borehole from becoming clogged with these cuttings, a circulating fluid is used to remove them.

Rotary drilling and coring require the use of drilling fluids, which typically consist of bentonite clay in combination with polymers. Polymers used in drilling fluids typically consist of polysaccharides such as guar gum (17). The ingredients and their concentrations can be adjusted to alter the density, viscosity, or other characteristics of the drilling fluids. Alternatively, air or inert gases can serve as drilling fluids. In addition to the removal of cuttings from the borehole, drilling fluids also lubricate and cool the drill bit and help to stabilize the borehole. The drilling fluids also limit loss of drilling fluid to the formation and intrusion of formation water into the borehole. Loss of drilling fluid to the formation and collapse of the borehole are prevented by the formation of a filter cake on the sidewall of the borehole and by maintenance of proper hydrostatic pressures because of the drilling fluid's weight.

Drilling fluids are circulated through the borehole via a pump located at the surface. The drilling fluids are pumped downward through the hollow drill pipe and exit at the bottom of the borehole through ports in the drill bit. The fluids lubricate and cool the drill bit and remove the cuttings by returning them to the surface along the annulus between the drill pipe and the borehole. At the surface the cuttings are separated from the fluids, usually by settling or screening. Fluid weight and viscosity are routinely monitored and adjusted to correspond to changing conditions in the borehole. Drilling fluid pressure is controlled by the pump and must be adjusted to prevent the fluid pressure from greatly exceeding the formation pressures that could destabilize the borehole and contaminate the formation ahead of the drill bit.

Shallow coring typically involves the removal of the entire drill pipe section by section from the borehole. The drill bit is then removed and replaced with a coring tool. The string of drill pipe is placed back into the borehole, and coring proceeds. After coring, the drill pipe is removed from the borehole to retrieve the core. For shallow coring (<100 m), a split-spoon, Shelby tube, or other type of driven or pushed core barrel can be used (76). At depths of >100 m and when it is desired to collect multiple sequential cores, the process of removing the entire drill string for each exchange of the drill bit and coring tool is time- and labor-intensive.

The development of the wireline system significantly improved coring during rotary drilling. Rather than removing all of the drill pipe to switch from rotary drilling to coring, the wireline system allows the upper section of drill pipe to be disconnected from the drill rig, and a "messenger" is lowered on a wireline through the inside of the drill pipe. An assembly used for wireline coring during rotary drilling is shown in Fig. 5. At the bottom of the borehole, the messenger engages and retrieves an inner drill bit through the inside of the hollow drill pipe. The wireline is used to bring the inner bit to the surface, where it is exchanged with a core barrel. At the bottom of the borehole, the core barrel is engaged to the drill pipe and an outer bit through a bearing that allows the drill pipe and outer bit to rotate while the sample tool and core liner remain stationary. The upper section of drill pipe is then reattached, and the pipe is rotated to advance the borehole. The outer drill bit at the bottom of the drill pipe is hollow and grinds away the rock outside the outer edges of the core barrel, allowing the core to enter into the core barrel. After coring, a messenger is lowered on the wireline to retrieve the core barrel.

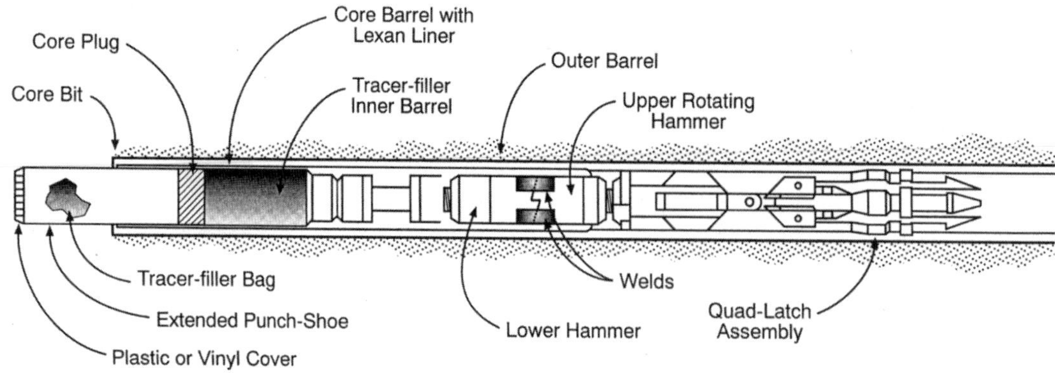

FIGURE 5 Diagram of assembly used for rotary drilling with wireline coring.

An extended shoe is often used with wireline coring to extend the cutting edge of the core barrel several centimeters in front of the drill-bit/sediment interface in order to core into previously uncompromised subsurface materials. The assembly, with extended shoe, used for wireline coring during rotary drilling is shown in Fig. 5. The use of extended shoes is important when sampling permeable materials or when the drilling fluid pressure exceeds the formation pressure, causing contamination of the sediments ahead of the drill bit. Wireline coring methods can save a substantial amount of time when coring deep rock and sediments compared to conventional rotary coring because the drill pipe is not removed during the operation. In addition, wireline coring reduces the risk of borehole collapse that can occur with frequent removal and replacement of the drill pipe.

Rotary wireline coring can be used to obtain intact core samples from a wide range of lithologies and at depths of kilometers. The length of continuous core that can be obtained can vary from less than one meter to tens of meters. With this flexibility, relatively large quantities of sample can be obtained from a single core run. Rotary operations can also be used to core in angled or horizontal boreholes. In addition to rotary coring, a drive- or push-type sampler can also be used in a rotary-drilled hole, although samples collected in this manner will be subject to greater physical disturbance.

The principal disadvantage of rotary coring is the use of recirculating drilling fluids. Aqueous fluids can be problematic because of the large number of bacteria that they often contain. Beeman and Suflita (10) counted 10^9 total microorganisms/g of dry weight of bentonite drilling muds during wireline rotary coring of U.S. southeast coastal plain subsurface sediments. Microbial contamination of subsurface samples during the coring process can occur via several routes (17, 37). These include direct contamination of the formation, where the mud pressure at the bottom of the bit exceeds the formation pressure, forcing mud and solutes into the formation ahead of the drill bit and core barrel. Seepage of drill mud that coats the outside of the core into pores or fractures of the core as it enters the core barrel is another potential source of contamination. Given the high density of bacteria in aqueous fluids versus their typically low population density in the subsurface, it is important to employ tracers to detect and quantify the extent of bacterial and drilling fluid contamination from drilling fluids whenever rotary operations are used.

Air or an inert gas such as Ar or N_2 can be used as an alternative to aqueous drilling fluids to remove cuttings from the borehole. For example, pressurized Ar was used to core deep basalt and associated interbed sediments at a site on the Snake River Plain in Idaho (21). Because the core samples were being used for microbiological and geochemical analysis, Ar was used rather than air to limit the exposure of the samples to O_2. Drilling with air or an inert gas is limited to the vadose zone and to moderate depths beneath the water table because of the excessive hydrostatic pressures that must be overcome when the borehole fills with groundwater. A major advantage of drilling with air or an inert gas is the reduced potential for microbiological contamination of core samples. One disadvantage of drilling with air is that significant drying of samples can occur when coring permeable sediment or rock. Desiccation during air drilling can be reduced by using a Gel-Coring System (Baker-Hughes, Casper, Wyo.). For gel coring, core liners are prefilled with a viscous, inert gel. As the core advances, gel surrounds the core, forming platelets to isolate the core and minimize exposure of cored materials to O_2 and drilling fluids. This approach should also be useful for limiting intrusion of drilling fluids into cores.

Assessing Contamination

A principal concern in sampling subsurface environments is the chemical and biological contamination of samples either by extraneous sediments or by circulating drilling fluids during the coring, retrieval, and processing stages. These exogenous materials can contain microorganisms and solutes (or gases) that can affect the microbiological and geochemical properties of cores. Because the levels of microbial biomass and activity in the subsurface are often inherently low, even small amounts of contamination (i.e., a few milligrams per kilogram) can affect the results and their interpretation. Thus, it is desirable to include tracers when samples from the subsurface are obtained.

Sources of microbial, dissolved, and gaseous contaminants include drilling fluids, groundwater-sediment slurry in the borehole, slough from overlying formations, and borehole sidewall materials. Contamination of sample materials can occur before, during, and after the coring operation. The intrusion of drilling fluids into core samples can introduce large numbers of bacteria and solutes that can alter sediment or groundwater chemistry. Contamination by drilling fluids can also affect the microbiological properties of the samples by altering pH, redox, and nutrient concentrations. During drilling with air or other gases, intrusion of gas can potentially result in displacement of pore water,

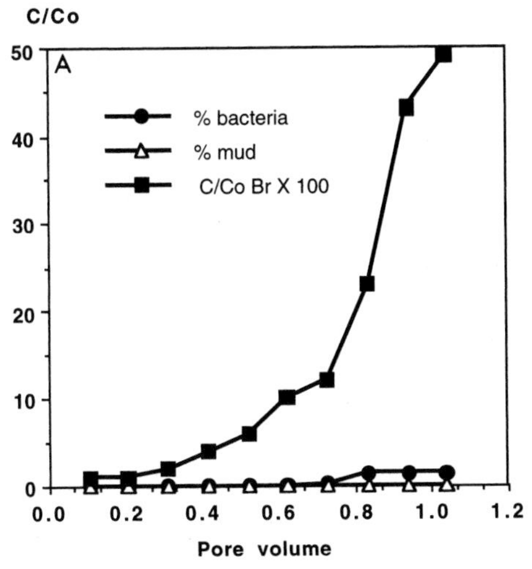

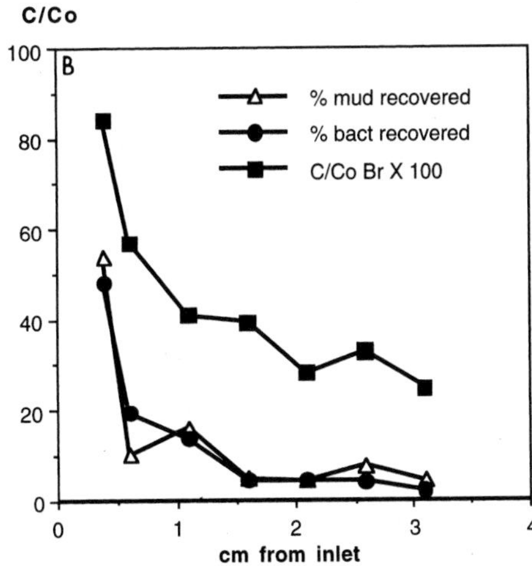

FIGURE 6 Transport of Br^-, bacteria, and clay associated with drill muds through an intact core of aquifer sand (A) and distribution of these components over the length of the core after collection of approximately 1 pore volume (B). C/Co, ratio of the concentration at a given pore volume or distance to the starting concentration.

resulting in drying and oxidation of the core. External sources of microorganisms, solutes, or gases from surface soils, the drilling operation, and sample processing should be traced whenever investigations require assessment of potential contamination.

One illustration of how tracers can be used to estimate contamination by drilling fluids is shown in Fig. 6. In this laboratory column experiment (13) with intact sandy aquifer core material, it was demonstrated that solutes, in this case Br^-, can move in advance of bacteria and clay (Fig. 6A) under fluid pressures similar to those occurring during rotary coring. This example illustrates the differential transport of water and particles such as bacteria and clay and the buildup of a filter cake, as evidenced by the high concentration of clay near the column inlet (Fig. 6B). This filtration can greatly reduce contamination of sediments in advance of the drill bit with bacteria from the drilling mud.

Microbiological Tracers

A variety of tracers have been used to evaluate biological contamination of subsurface samples (Table 2). The ideal microbiological tracer may be an organism with a rare but easily detected phenotype. For example, microorganisms that produce a distinguishing pigment or grow on a selective medium can be easily deployed and detected. *Saccharomyces cerevisiae* (baker's yeast) (94), *Serratia marcescens*, *Chromobacterium violaceum*, and *Bacillus subtilis* var. *niger* (ATCC 9372) (27) have been used to trace microbial transport in groundwater. Bacteria stained with a vital fluorescent stain (35) or recombinant bacteria producing green fluorescent protein (47) may also have utility as microbiological tracers. However, the use of live organisms as biological tracers is regulated, and caution should be used when proposing the introduction of tracer microorganisms into the subsurface. "Serendipitous" bacterial tracers that can be present in high

TABLE 2 Techniques used to evaluate potential biological contamination of cored subsurface sediments

Technique (reference[s])	Method of introduction	Initial concn	Limits of detection
Latex fluorescent microspheres, 1.0 μm carboxylated (31)	Dry mix with glass beads to hole bottom with bailer	5×10^9 g^{-1} × >200 g	10^3 g^{-1}
References 18 and 78	In bag in shoe/core catcher that breaks upon coring	3×10^{10} ml^{-1} × 10 ml	10^3 g^{-1}
Serendipitous native microbes (10, 19)	Native, no introduction	10^3–10^7 ml^{-1}	10^0
Added microorganisms (27, 94)	Add to fluids	$<10^6$	10^0
Community-level physiological profiles (55)	Comparative analysis	NA[b]	NA
Bacterial membrane lipid profiles (55)	Comparative analysis	NA	NA
Bacterial CFU (49)	Comparative analysis	NA	NA
Biochemical profiles of bacterial isolates[a] (55)	Comparative analysis	NA	NA

[a] Analysis based upon pattern recognition.
[b] NA, not applicable.

concentrations in drilling fluids can also serve as tracers, such as *Gluconobacter* (19) or coliforms (10). Alternatively, fluorescent microspheres can be used as bacterial surrogates. Microspheres are available in a range of sizes, with different fluors, and with neutral, positive, or negative charges (Polysciences, Warrington, Pa., or several other vendors) and are readily detected by epifluorescence microscopy (Fig. 7) or flow cytometry. It has been our experience that some sediment/rock samples contain materials that fluoresce at the same excitation wavelength as the yellow-orange latex microspheres and that therefore the use of this type of particle should be avoided. Latex, carboxylated, 1-μm yellow-green microspheres have been successfully used as bacterial tracers (18, 31, 78). One disadvantage of microspheres is that the cost can be prohibitive if large volumes are required.

Additional approaches to estimate the extent of biological contamination include biological and geochemical comparisons of core samples and of potential contamination sources such as drilling fluids. These analyses can include membrane lipid profiles, community-level physiological profiles (55), molecular biology-based approaches such as 16S rRNA sequencing (69, 75, 85), terminal restriction fragment length polymorphism analyses (69, 85), and/or the use of probes for amplifiable DNA extracted from sediments (87). These community-level measures, however, can detect only gross contamination and will be relatively ineffective for detecting low-level microbial contamination. Comparative approaches for assessing contamination must be used with caution because the microbiota of the contaminant source may be indigenous to the sampled subsurface (i.e., enriched in the drilling fluids by mixing with the formation cuttings during drilling operations). This is of particular concern with recirculating drilling fluids. An additional precaution when comparing isolates pertains to how representative the microbial isolates are. For example, analysis of relatively few isolates from environments containing 10^6 bacteria/ml or more may not be adequate in all situations for addressing contamination.

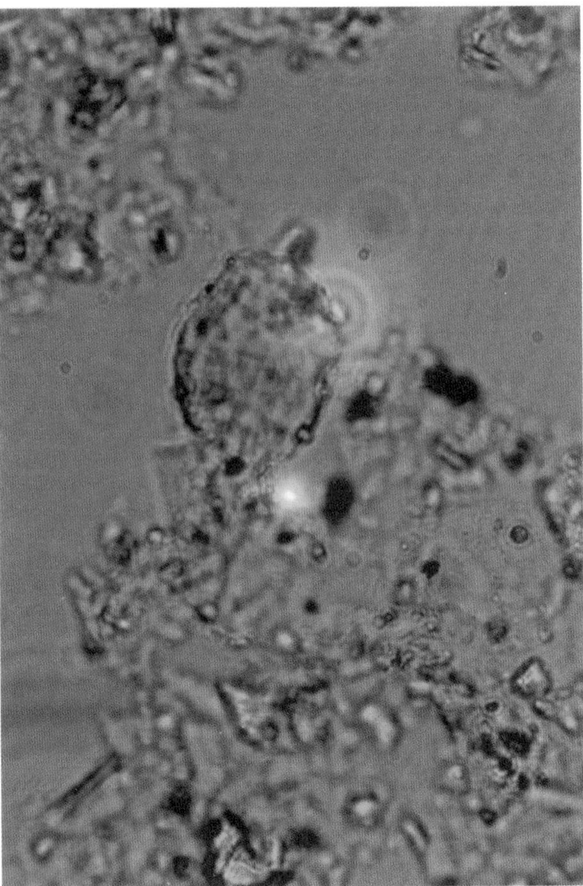

FIGURE 7 Photomicrograph of fluorescent microparticles among sediment particles, demonstrating easy detection of such particles when used as tracers. The photomicrograph was taken by using a combination of epifluorescent and phase-contrast microscopy.

Geochemical Tracers

It is desirable to include tracers of chemical contamination because the introduction of solutes could affect geochemical analyses and indirectly affect the microbiological properties of the sample by altering E_h, pH, or nutrient concentrations. Examples of chemical tracers that can be used to trace contamination are provided in Table 3. Chemical tracers should be nonreactive and therefore conserved during transport and sampling. Rhodamine and fluorescein (37, 76) have been used to trace drilling fluids because they are highly colored even at low concentrations and, therefore, can be used as visual indicators of core contamination in the field. Some precautions are warranted, as fluorescent dyes are typically sensitive to oxidizing agents and pH. Bromide is a useful solute tracer as it is normally present at low concentration in groundwater, is generally nonreactive, and is nontoxic. Bromide can also be measured in the field by using an ion-specific electrode, although such electrodes must be calibrated often as they are subject to interference from other ions and matrix effects. Potassium has also been used as an ionic tracer but has the potential disadvantages that it sorbs to clays in the drilling fluids via cation-exchange reactions and can be present at high background concentrations in some groundwaters. Sulfate and perchlorate can also be used as anion tracers, but sulfate has the potential to be biologically reactive under anaerobic conditions (i.e., sulfate reduction). Inert perfluorocarbon compounds can be employed as gaseous or aqueous tracers. Although they have limited solubility in aqueous solutions, they can be detected at extremely low concentrations by gas chromatography with an electron-capture detector, making them a very sensitive tracer (63, 80). Also, different perfluorocarbons such as perfluorohexane and perfluoromethylcyclohexane can be used at different coring intervals in the same borehole as a check on the potential for contamination from the previous core run (63). Although perfluorocarbons are nonreactive, they are quite volatile. Therefore, precautions should be taken whenever samples are processed in an enclosed space such as a glove bag to prevent volatilization from, and cross-contamination of, samples.

Tracer Introduction

Methods for the introduction of tracers vary with the drilling technology and the sources of contamination to be traced. The potential for contamination of samples obtained during hollow-stem augering is less than that with other coring methods. Because the action of the auger flights brings cuttings to the surface, there is little potential for downhole contamination with slough from the surface. A major source of microbial and solute contamination during hollow-stem auger coring beneath the water table is from sediment and water in the borehole. In general, solute tracers such as bromide can be used to trace the intrusion of borehole slurry into cores. Bromide or other solute tracers as well as microbiological or inorganic particulate tracers can be prepared as a concentrated solution and added to the borehole by using a bailer. For example, at a field research site in New Mexico where wireline rotary coring was being used, a saturated LiBr solution was added directly to the mud tank, resulting in 100 to 200 mg of Br^- per liter in the circulating drill muds, a concentration that was maintained throughout drilling and coring. Biological tracers can be introduced by similar means.

When the vadose zone is sampled, the need for tracing borehole fluids is obviously eliminated. However, there is potential for contamination from particulates that fall into the borehole from overlying sediments and soil and from sidewall slough or drilling tools. These particles can penetrate subsurface materials, while hammering advances the core barrel. A dry tracer mix consisting of glass beads as a bulking agent, fluorescent microspheres as a bacterial tracer, and tungsten carbide particles as a solids tracer was developed for tracing particulate intrusion into vadose samples during cable tool coring (31). This mixture was placed at the bottom of the borehole by using a bailer with a check-valve at the bottom that released the tracer mix upon contact with the bottom of the borehole (Fig. 8). In core samples collected from the vadose zone, the concentration of fluorescent microspheres was found to increase as the number of hammer strikes required to advance the core barrel increased (28).

Rotary drilling methods generally have greater potential for introducing contaminants into cores than do hollow-stem auger and cable tool coring methods because of the need for circulating drilling fluids and the greater depths at which these methods are generally used. Since the volumes of mud required for rotary drilling can be thousands of gallons, it is not economically feasible to mix fluorescent microspheres directly into drilling fluids. In contrast, microorganisms can be easily and economically cultured in large quantities and introduced directly into drilling fluids. Chemical tracers can also be introduced directly into the fluid. Because of the potential for loss of drill muds and dilution with groundwater, it is necessary to monitor the concentration of tracers in the muds and to adjust their concentrations accordingly. State and local regulatory agencies should be consulted prior to introduction of microbiological or chemical tracers.

One approach to using microspheres as biological tracers has been to deploy them at or near the interface between the coring tool and the formation. The microspheres can be placed in a Whirl-Pak bag or other suitable containers with the wire closures wrapped around the base of a core catcher in the shoe of a coring tool (Fig. 1). As the core enters the shoe, the bag ruptures, releasing the microspheres. Another approach for introducing biological and dissolved tracers is to add them to an aqueous solution that is placed inside the core-barrel liner (Fig. 1). This requires the use of a plug or marker that seals the bottom of the liner to prevent mud or borehole slurry from moving up into the core barrel and liner during deployment down the hole. As the core enters the barrel, the plug rises up the inside of the liner and the tracer is released through the check-valve at the top of the core barrel. These approaches are effective for assessing the invasion of drilling fluid into cores but will not detect microbial contamination that occurs as the result of drilling fluid invasion into the formation.

TABLE 3 Tracers used to evaluate chemical contamination in cores of subsurface sediments

Tracer	Method of introduction	Initial concn (μg liter^{-1} or kg^{-1})	Limits of detection (μg liter^{-1} or kg^{-1})	Methods of detection	Notes
Rhodamine	Soluble in aqueous drilling fluids	20,000	0.5–2.0	Fluorimetry	Useful for on-site drilling fluid analysis
Bromide	Soluble in aqueous drilling fluids	>500,000	10	Ion-specific electrode; ion chromatography	Liquid drilling fluid analysis
	Concentrated solution directly to borehole	>500,000	10		Can be added to split-spoon core liners
Potassium	Soluble in aqueous drilling fluids	>500,000	10	Ion chromatography; inductively coupled plasma spectrometry	Liquid drilling fluid analysis
Sulfate	Present in bentonite fluids or soluble	>150,000	10	Ion chromatography	Liquid drilling fluid analysis
ClO_4 (as $NaClO_4$)	Sterile solution in Lexan liners	200,000	2,000	Ion chromatography	Can be added to split-spoon core liners
Tungsten carbide	Direct to borehole bottom	250,000,000	100,000,000	Ion microprobe	Add to bottom of borehole with a bailer
Glass beads	Direct to borehole bottom	>500,000,000	5,000,000	Microscopy	Tracer carrier/bulking agent
Noble gases	Direct to drilling fluid	>500,000	50	Gas chromatography	Gaseous drilling fluids
Perfluorocarbons	Add as methanol emulsion for solubility of 1 mg/liter	1,000	0.00001	Gas chromatography	Off-site aqueous fluid analysis
	Meter and vaporize solution	1,000	0.001		As gaseous fluid

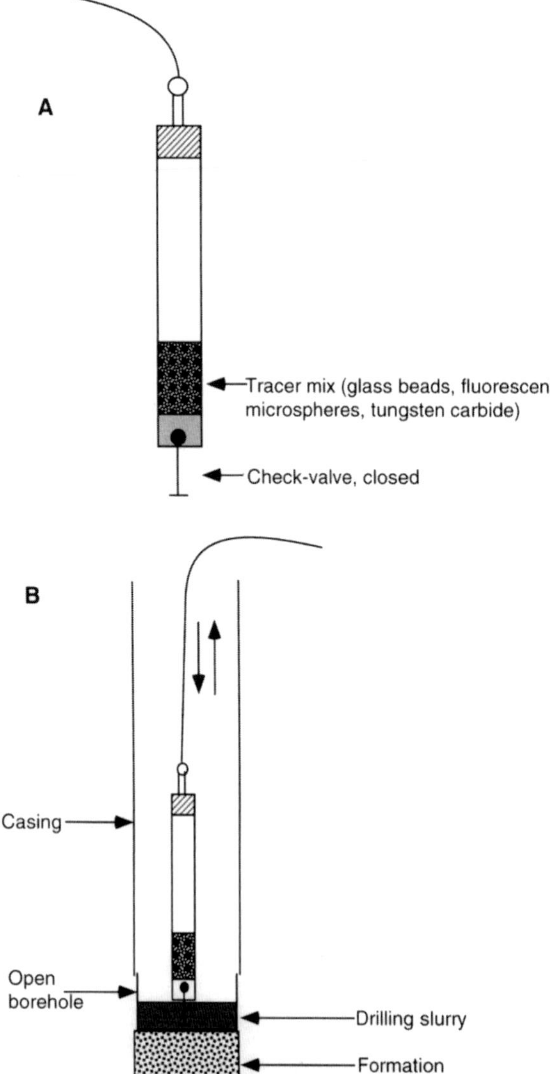

FIGURE 8 (A) Diagram of bailer with check-valve for introducing tracers into a borehole for vadose- and saturated-zone coring. (B) The bailer is lowered to the bottom of the borehole on a cable, and the dry tracer is released, covering the exposed sediment.

When one is drilling with air or other gases, perfluorocarbons can be used as tracers to evaluate the intrusion of gas into core samples. In order to introduce small quantities of perfluorocarbons into the gas stream, perfluorocarbons can be metered by using a high-performance liquid chromatography pump and then vaporized by heating and injected directly into the circulated gas before it enters the borehole. Alternatively, for mud-rotary coring, the perfluorocarbon can be metered directly into the drilling fluid line as a liquid.

Tracers have also been painted onto the outside of the core after retrieval from the subsurface and before subcoring. Juck et al. (47) painted the outside of the core with a live, recombinant green fluorescent protein-expressing *Pseudomonas* strain and then tested the subcore for the presence of fluorescent cells. It is also advisable to use tracers when sampling rock from the walls of tunnels or mines when assessment of contamination is required, a topic that is covered below.

Tracer Sampling

A critical aspect of employing tracers to assess the quality of subsurface samples is to collect and analyze samples that allow for evaluation and interpretation of the results to determine if samples are representative of the subsurface. In addition to collecting pared core material for tracer analyses, it is important, where applicable, to collect samples of borehole slurry. This allows for quantification of the extent of contamination. Given that the tracer may not be distributed evenly along the outside of the core, it may also be necessary to sample the parings and pared core from different locations along the length of the core.

When using live organisms as tracers or when using community-level biological analyses, it is necessary to prevent or limit microbial changes after sampling by immediately performing the analyses or by taking the necessary steps to preserve the samples. When using inert particles such as fluorescent microspheres as tracers, biological changes after sampling are of less concern, although microspheres can decompose with time. Except for perfluorocarbons, no special precautions are required for samples to be analyzed for chemical tracers. Because of the volatile nature of perfluorocarbons, it is necessary to immediately add methanol to samples in a sealed container to preserve them for analysis.

Selection of tracer type and deployment method is dependent upon the goals of the sampling, the types of lithologies being sampled, the type and source(s) of contaminants being traced, and the drilling and coring methods being used. Evaluations of tracer introduction, sampling, and analyses should generally be conducted prior to actual sample collection to ensure that the methods are working satisfactorily. For example, if microspheres are introduced properly during wireline coring, they should be detectable in the drilling fluids and on the outside and ends of the cores. If rotary drilling with circulating muds is used to advance the borehole, Br^- maintained at concentrations of 100 to 200 mg liter^{-1} in the circulating fluids is effective for tracing drilling fluid penetration.

Coring Procedures

Sample processing is typically conducted under an inert atmosphere to minimize chemical or biological changes in the cored materials. The main concern regarding chemical changes resulting from exposure to air is oxidation of reduced species such as Fe(II) and sulfides, inactivation of anaerobic and microaerophilic microorganisms, and drying of the sample. Stimulation of biological activities within the sample, which can occur as a result of coring disturbances (14, 32, 39, 41, 45), is reduced by excluding potential electron donors and acceptors. The rate and extent of changes occurring in samples prior to analysis are minimized by processing subsurface materials under an inert atmosphere. An alternative to processing cores in the field that limits their exposure to air is to pare and segment the cores and then place them in containers flushed with an inert gas such as Ar. Samples can then be transported to a laboratory where they can be further processed.

Depending on the properties of the subsurface materials and analyses and experiments to be conducted, it may be desirable to process cores in the field for several reasons. Timely paring and separation of representative core material from drilling fluids, surface contaminants, bottom-hole slough, and wall materials limit contamination. Furthermore, on-site processing also enables a rapid evaluation of sample quality and allows for modification of drilling procedures. For on-site processing of subsurface cores, the modified glove bags available from Coy (Ann Arbor, Mich.) (21) are well suited.

At the field laboratory, liners containing cored materials are placed in the transfer tube, which is then flushed and evacuated for three or more cycles with an inert gas such as N_2 or Ar to displace air. After sufficient flushing, the interior airlock door is opened and materials are passed into the processing chamber. In order to achieve a completely anoxic atmosphere, it is necessary to include H_2 in the gas mix, a desiccant, and a palladium catalyst to scrub trace quantities of O_2. Desire for a completely anoxic processing system must be weighed against the possible artificial stimulation of anaerobic bacteria by H_2. A slight positive gas pressure within the chamber helps to minimize the leakage of air into anaerobic chambers.

Depending on the type of material cored, type of drilling fluid, and composition of the core liner, various techniques may be used for removing cores from liners. If the cores are short (<75 cm) and materials are consolidated, it may be possible to extrude the core from an intact liner or to section the core laterally into discrete sections prior to extrusion. Alternatively, if the cores are consolidated and the liner is Lexan or polycarbonate, it may be advantageous to score the liner on each side (along the core length) with utility knives. Core liners can also be cut by using circular saws equipped with carbide-tipped blades. Although electrical outlets can be built into the chamber for running power tools, etc., caution is needed as moisture can build up over time, resulting in a potential shock hazard. Battery-operated, handheld saws can be used, thus reducing this particular hazard.

After the cored materials are exposed, the core can be examined and the geological characteristics recorded. Evaluation of core quality, such as distinguishing slough or nonnative materials from the core and identifying segments for subsampling, can be made at this time. If the core material is physically disturbed as evidenced by mixing or drilling-induced fractures, or if there are signs of gross chemical or biological contamination such as penetration of drilling muds or tracers, then the core is generally not suitable for microbiological or chemical analyses. Removal of outer core material that has the greatest potential for being contaminated should be standard practice. Drilling fluids are typically most concentrated at the ends and outermost edges of the core. These regions must be carefully pared by using ethanol-rinsed and flamed or autoclaved tools. The amount of material to be pared depends upon a variety of factors, including the drilling method and permeability of the material being sampled. If a visible tracer such as rhodamine dye was used, the distance of penetration can be used as a guide for paring. Obviously, it is better to be conservative and to pare more than necessary than to risk including contaminated material in the sample. Subsamples of drilling fluids, outer portions of the core, and representative materials from inside the core should be collected and analyzed for tracers.

Most unconsolidated sediments are readily pared with sterile knives, spoons, spatulas, or chisels. Consolidated sediments or rocks may require use of hydraulic splitters to pare outer, potentially contaminated portions of a core from the representative materials in the center of the core. Hydraulic core splitters can be used to section a core into 2- to 5-cm-thick discs (Fig. 9A), which can subsequently be pared by using a core splitter or a sterile chisel (Fig. 9B). Care must be taken not to contaminate inner material with outer core parings during this process. Between uses, the blades of the hydraulic splitters should be cleaned to remove particulates and fluids and disinfected. Pared materials can be further disaggregated by crushing in a sterilized Plattner mill (Fig. 9C). Careful consideration should also be given to subsampling and to whether sediments should be homogenized, especially if samples are being collected for multiple analyses. As there can be considerable vertical and horizontal heterogeneity in the sediment properties, samples collected only centimeters apart can have considerably different microbiological, geochemical, and physical properties.

Sample Transport and Disbursement

In the packaging chamber, samples are segregated and transferred into sterile Whirl-Pak bags and placed in sealable canning jars. To minimize exposure of sample to O_2 during transport, jars should be flushed with O_2-free N_2 or Ar gas via a cannula prior to sealing. This can be accomplished inside the anaerobic chamber. A novel approach for storing anaerobic samples during transport has been described by Cragg et al. (22). This approach uses heat-sealable "wine bags" consisting of a thin nylon-aluminum-polyethylene laminate with a thick inner polyethylene liner. Core segments or subsamples can be placed in these bags with a commercially available O_2-absorbing system used for culturing anaerobic bacteria and then sealed. Samples placed in bags in this manner can remain anaerobic for months. Sealed samples can then be transported to laboratories or shipped by overnight express carrier with cold packs to keep samples cool during transport. Although in situ temperatures may be above ambient air temperatures, the inclusion of cold packs is necessary to protect samples from elevated temperatures that can occur during transport. Alternatively, for some microbiological analyses that involve analysis of

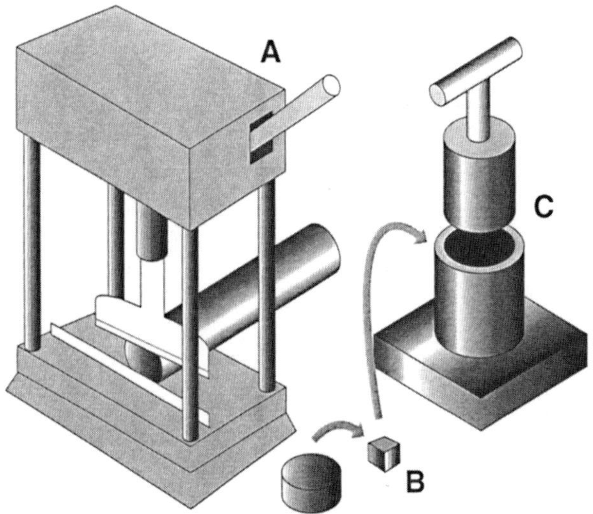

FIGURE 9 Diagram of hydraulic core splitter (A) and rock crusher (C) used for sampling consolidated subsurface materials. Careful paring of the discs produced by splitting of the core is required to produce rock fragments (B) from which outer, potentially contaminated materials have been removed. Reprinted from reference 21 with permission from the publisher.

macromolecules such as lipids and nucleic acids, samples should be frozen immediately for processing at a later time. Freezing samples for such analyses can be done on site by using a standard −20°C freezer. Frost-free freezers should not be used as they can result in drying of samples. Freezing with dry ice or liquid nitrogen can also be used. Freezing samples in dry ice or liquid N_2 is advantageous in that nucleic acids and lipids are more stable at these lower temperatures.

Depending upon the scientific questions being asked, it may be desirable to leave the sediment or rock samples intact, to subdivide the sample into discrete depth intervals, or to homogenize material from a relatively long depth interval. Intact cores preserve the spatial relationships of microorganisms and their metabolic activities in relation to the sediment or rock matrix. This can enable visualization of microbes within their habitats by techniques such as scanning electron microscopy (77) and assays that reveal the locations of specific metabolic activities, e.g., the $^{35}SO_4^{2-}$ radiotracer method in which the location of sulfate reduction near a fresh face of an intact core is revealed by deposition of $^{35}S^{2-}$ onto a silver foil followed by autoradiography (52). Collecting many discrete samples along transects or in a two- or three-dimensional pattern can facilitate studies of spatial heterogeneity. A disadvantage is that individual samples are typically relatively small, which limits the numbers and types of analyses that can be performed. Brockman and Murray (16) described sampling schemes and analytical strategies for characterizing spatial heterogeneity in the subsurface. The strategy of homogenizing a relatively large volume of sample, e.g., a 0.5-m-long section from the interior of a core, has the advantage that homogenized sample material can be distributed among multiple labs for a variety of chemical and microbiological analyses, with the potential for correlation analysis of the data. The disadvantages are the sacrifice of information on the patchy distribution of microbes and the maximization of disturbance effects of sampling and sample handling (14).

Summary of Coring Procedures

Contaminants can be introduced before, during, and after drilling or during sample processing. After each use, all sampling equipment should be thoroughly cleaned with a steam cleaner or by other means to remove debris and drilling fluid. Equipment such as the core barrel, shoe, core catchers, and core liners should all be sterilized or, at a minimum, disinfected prior to each use. Core liners are typically constructed of aluminum or Lexan and can be autoclaved prior to use. Precut Lexan liners can be wrapped with paper or foil and autoclaved prior to their use in the field. In the field, core liners are removed from their wrapping and are placed directly in the core barrel. Retainer rings, core plugs, and core catchers are added if necessary, and the cutting shoe is screwed onto the end of the core barrel. Bags containing tracers such as microspheres are often placed inside the cutting shoe (Fig. 1 and 5), and the bottoms of cutting shoes may be covered with vinyl or plastic to limit fluid movement into the sampling barrel during descent into the borehole. After assembly, the sampler can be filled with sterile water or fluids containing tracers and transported to the drill rig for attachment to the coring assembly. During coring the protective covering on the cutting shoe breaks and the core enters the liner and breaks the tracer bag, releasing tracers into the surrounding environment. After the core is retrieved, it should be capped to limit exposure of the core ends and transferred to an on-site processing facility or to the laboratory. Cores can then be placed in a glove bag or similar chamber to be processed.

COLLECTION FROM MINES, TUNNELS, AND CAVES

Sampling from the walls of preexisting mines, tunnels, and caves offers several advantages over drilling and coring. Costs are almost always less than those for drilling and coring from the surface, and one has the flexibility of sampling in four dimensions (X, Y, and Z directions, plus time). The disadvantages are that the rock or sediment that is exposed underground has a greater potential for contamination by air and airborne contaminants and the subsequent development of microbial communities that flourish as the result of the introduction of oxidant (air). This is especially true for natural caves or tunnels and mines that have been in place for a long time. The process of tunnel building and mine excavation causes extreme pressure gradients and extensive fracturing due to decompression. These decompression fractures can extend for tens of meters into the surrounding rock (69). If blasting is used, this causes more fracturing and can introduce chemical contaminants and also send contaminant microbes deep into the fractures. The use of water for drilling in preparation for blasting and for dust control (common mine practices) can further contaminate the fractured rock. Another disadvantage is that sampling in tunnels and mines is usually a "piggyback" operation, with the location of the tunnels or mine adits not being selected for scientific purposes and with the potential for contamination being even greater due to the large volumes of water used for cooling and other purposes, the large volumes of air from ventilation operations, and, in some mines, the introduction of diesel exhaust (79). Also, when the cement ("grout") that is used to seal fractures and boreholes in mines and tunnels comes into contact with groundwater, it can dramatically increase the pH, thereby altering microbial community structure. Finally, it must be noted that significant hazards exist in many mines, e.g., cave-ins, noxious gases, methane gas explosions, etc.; therefore, safety should be given at least as much attention as sample integrity.

Freshly exposed faces are best for collecting samples of solids within mines and tunnels. Kieft et al. (51) sampled volcanic tuff from the proposed Yucca Mountain high-level nuclear waste repository at the Nevada Test Site during the time when the main tunnel was being excavated by a tunnel-boring machine. They collected samples by using a procedure in which 1-m^2 areas of freshly mined faces were chiseled to a depth of 0.5 m by using a sterile jackhammer with a flame-sterilized bit. The chiseled surfaces were then painted with a suspension of 1.0-μm diameter fluorescent beads as surrogate microbial tracers. The entire 1-m^2 area was then excavated to a 1-cm depth with a sterile chisel. Only then were actual samples collected for microbiological analyses. The samples were then examined for the presence of fluorescent beads as an indicator of potential microbial contamination associated with the rock surface that had been exposed to the air prior to sampling. Onstott et al. (69) used a similar approach, including the use of fluorescent microbeads, to sample quartzite and gold ore in 3-km-deep gold mines. Samples were collected from rock surfaces that had been exposed by blasting less than 24 h prior to sampling. In this case, the tracer analyses revealed that microbial contamination existed in deep fractures caused by blasting. Analyses of the microbial communities by 16S rRNA gene and terminal restriction fragment length polymorphism analyses showed that the rock samples collected from the mine face contained a mixture of microbes that were common in the mine service water used for drilling and dust control and some microbes that appeared to be indigenous to natural fractures in the rock. As with coring, comparison of microbial community profiles between samples and known sources of contamination can indicate the quality of the samples. The presence of fungi can be especially useful in this regard. Fungal spores are abundant in mine air due to air circulation from the surface, whereas most deep subsurface ground waters are typically devoid of fungi (17). Thus, fungi, especially filamentous fungi, can serve as serendipitous tracers of airborne contamination. Agar media in petri dishes can be exposed to the mine air during sampling operations in order to characterize culturable airborne contaminants by using settling plates or an air-sampling device (see chapter 74). Airborne microbial contaminants can also be collected onto filters and characterized by analysis of DNA or lipids.

Large, intact rocks can also be collected underground for subsequent geomicrobiological analysis. Onstott et al. (69) collected rocks from the mine floor following blasting. They then transported the rocks to the surface. The outer surfaces of the rocks were then removed by using a rock splitter, as described above for rock cores. However, the rocks were first sprayed with orange paint so that the outer, potentially contaminated surfaces could be easily identified and traced.

Mines can also serve as deep platforms for drilling and coring operations. Onstott et al. (69) collected relatively shallow, 5-cm-diameter samples from rock walls of a 3.2-km-deep mine by using a hand-operated, pneumatically powered coring tool and diamond bit. The bit was cooled with mine water containing fluorescent microspheres and rhodamine dye as tracers. Boreholes can also extend considerable distances beyond the tunnel walls. Mining companies routinely drill exploratory boreholes in search of economic deposits, faults, water-filled fractures, etc. with small rotary drill rigs deployed at underground sites. While these boreholes and any cores that are collected have a limited diameter (usually ≤5 cm), they can extend hundreds of meters beyond the mine adit into undisturbed rock and thus can be useful for microbiological studies. Pressurized formation water is sometimes encountered and can rapidly flush the borehole of potential contaminants. The deep fracture water itself may be of considerable interest for sampling and analysis. A sterilized packer (Fig. 10; "Groundwater Sampling" described below) can be used to aseptically sample borehole water without exposure to the mine air. Of course, as with any drilling or coring procedure, strict quality control and quality assurance procedures described above for collection and processing of cores should be followed. The section below provides general guidelines for collection of groundwater samples. Further discussions of microbiological sampling in mines can be found in references 65 and 79.

GROUNDWATER SAMPLING

Assuming proper drilling and installation, groundwater samples that are useful for microbiological analysis can be collected from developed wells. However, a number of factors must be considered. A major question is whether the microorganisms associated with the water are representative of the in situ populations, many of which may be attached to solids. In addition, groundwater samples, depending on the well configuration and length of screened interval, can represent water collected over a relatively large volume of the subsurface, whereas solids are typically "point" samples. It is recognized that vertical integration of groundwater over a screened interval can result in considerable dilution of solutes and particulates. Consequently, it is generally

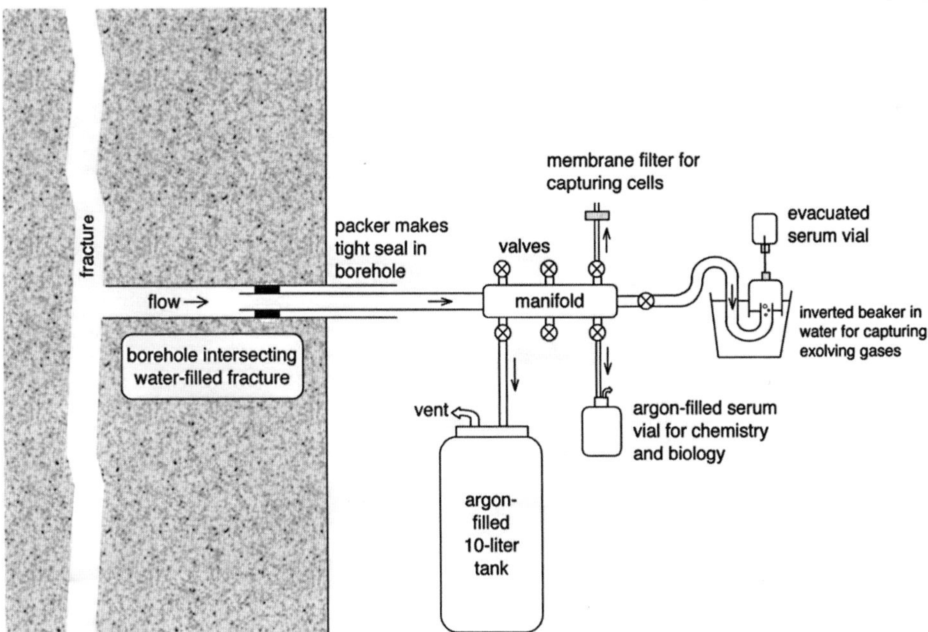

FIGURE 10 Diagram of a packer and manifold system designed to collect groundwater samples from flowing boreholes in mines or tunnels. The system enables aseptic, anaerobic sampling without exposing the samples to air. The manifold allows collection of multiple samples simultaneously and for different purposes.

better to collect samples from discrete zones. In order to better define spatial variations in microbial transport in a sand aquifer on Cape Cod, Harvey et al. (42) used a multilevel sampler (54) that consisted of a 3.2-cm-diameter polyvinyl chloride (PVC) pipe that encased multiple screened, 6.5-mm-diameter polyethylene tubes that exited the PVC pipe to the surrounding aquifer. The purpose of this section is to identify and discuss some of the key issues associated with groundwater sampling and describe some of the more commonly used methods. The reader should refer to references 9, 64, and 92 for additional information on groundwater sampling.

Methods for Groundwater Sampling

Because the local environment adjacent to a well can be substantially different from that in the aquifer, it is necessary to purge the standing water from a well before collecting water for microbiological or geochemical analyses. Stagnant water in the upper part of the casing is not in contact with the formation but rather with the casing and gases in the well bore. Therefore, the geochemical and microbiological properties of this water rarely reflect properties of the formation waters. Biofilms of nonnative microorganisms on well casings and well screens can detach if the pumping rate is too high. Also, sediments in and around the well screen can be contaminated during well development. A general rule is to flush at least 2 or 3 well volumes before collecting water samples to ensure that the sample is representative of the formation. It is also desirable to monitor geochemical parameters, such as E_h and pH, and pump until stable values indicate that formation water is being pumped. Although removal of stagnant water in the well bore is critical, purging itself can introduce artifacts if not conducted properly. For example, overpumping can increase turbidity of the water; result in dilution, which can change the concentration of dissolved constituents; dewater pores, which can alter both microbiological and chemical properties; or alter the concentrations of dissolved gases.

Once the well has been purged, there are multiple methods for obtaining suitable groundwater samples. These include using grab devices such as bailers (Fig. 8), suction or lift devices such as peristaltic or centrifugal pumps, positive displacement systems such as gas-driven bladder pumps and electrical submersible pumps, and inertia pumps. Factors such as depth, type of analyses, and hydrological properties all influence choice of collection device. For collecting microbiological samples, it is desirable to use inert, autoclavable materials for those components that come in contact with the water. Pumps that cannot be autoclaved can be sanitized by pumping a solution of disinfectant, such as a hypochlorite (bleach) solution, through the system prior to use.

Collection of groundwater for microbiological analyses is relatively straightforward. The easiest approach is simply to flush and fill clean, sterile glass bottles with groundwater directly from the source tubing. When sampling anaerobic groundwater, it is important to fill the bottles completely and continue flushing for 1 to 3 volumes to eliminate any air in the bottle headspace that may oxidize reduced chemical species or inhibit strict anaerobes. When one is using a suction-lift pump, a sterile vacuum flask can be used to aseptically collect groundwater samples while limiting exposure of the sample to air (64).

Collection of deep, anaerobic groundwater from flowing boreholes in mines without exposure to air or other contaminants requires specialized approaches. Researchers collecting groundwater samples in deep (~3 km) gold mines in South Africa used a sterile packer inserted into a flowing borehole and connected to a sterile manifold that enabled transfer of flowing groundwater to a variety of sample vessels (Fig. 10) (66, 85, 86, 89). The packer was an autoclavable stainless steel tube with a butyl rubber sleeve that expands its diameter as the length is compressed to make a tight seal within the borehole. The water was allowed to flow through the packer and the manifold to flush them, and then water was diverted through the manifold and into the sampling vessels. Water samples for microbiological enrichment cultures were collected in sterile, Ar-filled serum vials (~150 ml) with thick butyl rubber septa via sterile needles. Other samples were collected in a similar manner for water chemical analyses; some of these vials contained fixatives for specific analyses, e.g., $HgCl_2$ to prevent biological activities or Zn acetate to complex sulfide as ZnS. An 11.6-liter canister (Cornelius, Inc., Anoka, Minn., sterilized and flushed with Ar) was used to collect large samples underground for transport to the surface where they could be filtered to collect and concentrate cells for DNA and lipid extraction (85). Gases that exolved as pressure was relieved were collected in the headspace of an inverted beaker and were then transferred to an evacuated serum vial for later analysis (89).

These methods are generally applicable for the collection of samples for the study of bacteria, protozoa, or viruses. For example, Harvey et al. have used stainless steel submersible pumps to collect groundwater samples suitable for studying the transport of bacteria (42) and protozoa (43) in a shallow, sandy aquifer on Cape Cod. Because the concentration of viruses in groundwater is often quite low, their study often required concentration by filtration. For example, Abbaszadegan et al. (1) used an electropositive cartridge filter to collect viruses from groundwater that were subsequently eluted by using a beef extract solution.

Stevens et al. (84) used a PVC cartridge packed with sand (Fig. 11A) that was connected in line to the discharge line from an artesian well to select for bacteria that would attach to the sand surfaces while investigating the microbiological properties of a deep, confined aquifer. The composition of the anaerobic microbiological community that developed on the sand particles was consistent with the geochemical properties of the well, indicating that this approach was effective for sampling bacteria that were a natural component of the aquifer. An alternative to the sand-cartridge method was used by Pedersen and Ekendahl (74) to study microbial communities in deep crystalline bedrock at a site in Sweden. Biofilm reactors containing slides, previously baked in a muffle furnace at 425°C for 4 h to remove organic particles, were used as the substratum for biofilm attachment.

Other in situ enrichment devices have been deployed underground by plumbing cartridges filled with solid substrates directly to flowing boreholes, allowing borehole water to flow through the cartridges for periods of days to months, and then retrieving the groundwater and solids from the cartridges and using them to inoculate microbial enrichment cultures, for 16S rRNA analyses and for phospholipid fatty acid analyses (6, 65, 66). Baker et al. (6) and Moser et al. (66) incubated cartridges containing sterilized, crushed quartzite (native rock) in the flow from boreholes at 2.7- and 3.2-km depths in a South African gold mine. The solid media within in situ enrichment cartridges can also be selected to enrich for specific functional groups, e.g., iron(III) (hydr)oxide minerals, to select for dissimilatory iron-reducing bacteria.

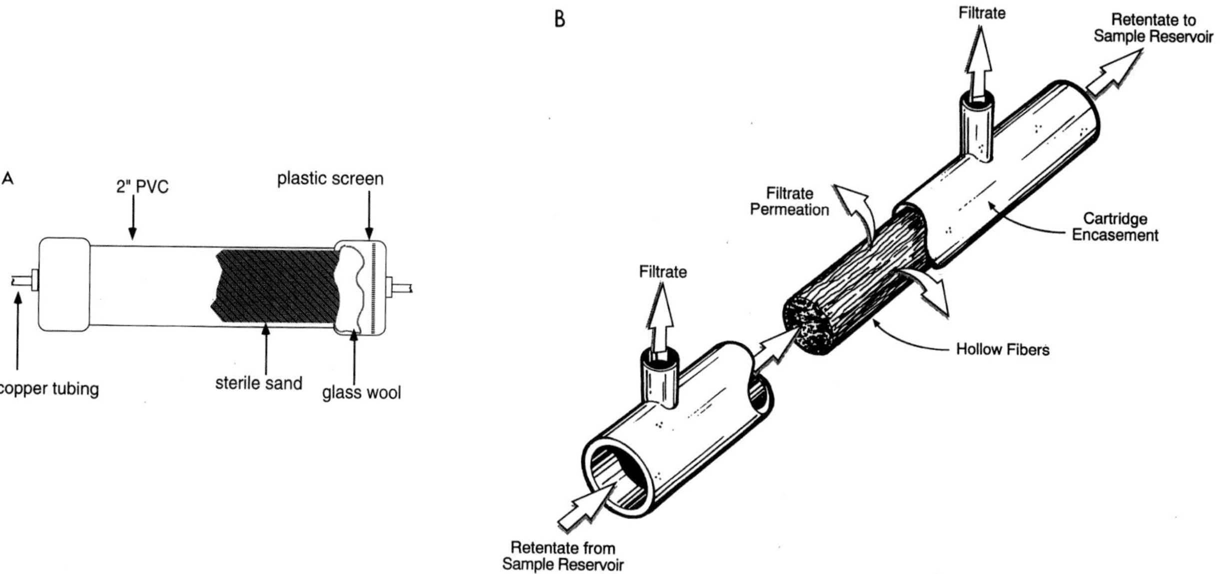

FIGURE 11 (A) Diagram of sand-trap sampling apparatus used to collect bacteria from the flow from an artesian well (reprinted from reference 84 with permission of the publisher). (B) Cutaway view of hollow-fiber tangential-flow filter (reprinted, in part, from reference 53 with permission of the publisher).

Because the density of bacteria in groundwater is often low, between 10^3 and 10^5 cells ml^{-1} (44, 84), it may be desirable to concentrate the biomass for certain microbiological analyses. A relatively simple approach is to connect sterile in-line filters (e.g., 47-mm diameter or 0.1- or 0.2-μm pore size) to a groundwater source that is pumped or flowing under pressure. Kuwabara and Harvey (53) used a hollow-fiber, tangential-flow filtration system (Fig. 11B) to concentrate bacteria from groundwater. Fry et al. (34) also used hollow-fiber filtration to concentrate bacteria from groundwater so that nucleic acids could be extracted for analysis of the population by using nucleic acid probes of rRNA and sequence analysis of cloned 16S rRNA genes. Similarly, White et al. (90) used tangential-flow filtration to concentrate groundwater biomass for lipid analyses during bioremediation. Similar tangential-flow systems have been used to concentrate viruses from water (82, 93). T. C. Onstott and colleagues used a "massive filter" consisting of a 25-cm-long stainless steel in-line filter housing (U.S. Filter Plymouth Products, Sheboygan, Wis.; catalog no. A-01508-40, Cole-Parmer) containing a 0.37-m^2, 1.0-μm-pore-size polypropylene filter (Flotrex, Osmonics; catalog no. 06479-32, Cole-Parmer). The housing and filter were sterilized in an autoclave and then connected to a flowing borehole in a 3-km-deep mine to capture cells from tens of thousands of liters of fracture water. The filters were frozen at −20°C as soon as they were returned to the surface. They were later thawed and extracted for analysis of phospholipid fatty acid and DNA. Sufficient DNA was extracted for use in a metagenomic study in which genomic DNA from the fracture water community was subjected to shotgun cloning and sequencing without prior amplification (68). Moser et al. (65) described a positive pressure system that used filtered Ar or N_2 to collect cells onto membrane filters from groundwater collected aseptically and transported from a deep mine to a surface lab in 11.6-liter stainless steel canisters; the cells collected on the filters can then be used for nucleic acid and/or lipid analyses. Note that with all these filtration approaches, care must be taken to maintain uniform flow to avoid breaking the filter medium within the cartridge (observed visually or evidenced by particulate breakthrough).

While the recovery of bacteria from groundwater by using hollow-fiber filtration can be efficient (>90%), especially when backpulsing is applied, there are a number of inherent problems and precautions. Hollow-fiber filters are not 100% efficient, and there will be some loss of cells during filtration. There may also be a tendency to lose more small cells than large cells, hence introducing a bias. Another precaution is that filters need to be thoroughly cleaned and disinfected between uses to ensure that one is sampling groundwater bacteria and not residual bacteria attached to surfaces within the filter. A variety of chemical methods can be used to clean and disinfect the filter cartridges, including those recommended by the manufacturer. Various process-control samples can also be collected to assess sampling procedures. Although most manufacturers do not guarantee the integrity of the filters after autoclaving, the materials withstand repeated autoclaving without loss of filter integrity (53).

SUMMARY

This chapter provides background information and general guidelines for obtaining representative samples from subsurface environments for microbiological and geochemical analyses. It is recognized that not all microbiological investigations of the subsurface environment will be able to implement all of the procedures discussed in this chapter. However, by understanding the advantages and limitations of the various sampling technologies, by using tracers, and by employing rigorous procedures for sample processing, representative subsurface samples of defensible quality for scientific investigation and monitoring can be obtained. Some of these methods have already been applied to sampling of deep seafloor sediments and rocks, facilitating significant findings (e.g., see references 26 and 61). Scientific and engineering advances in drilling and sampling technologies will likely expand our knowledge of the distribution of microorganisms on Earth and their role in cycling of elements and the formation and weathering of rocks. Future challenges lie in the design and application of drilling and sampling approaches for exploring the potential for life in the subsurface of other planets such as Mars (60) and planetary bodies such as Europa (62).

REFERENCES

1. **Abbaszadegan, M., M. S. Huber, C. P. Gerba, and I. L. Pepper.** 1993. Detection of enteroviruses in groundwater with the polymerase chain reaction. *Appl. Environ. Microbiol.* **59:**1318–1324.
2. **Adrian, N. R., J. A. Robinson, and J. M. Suflita.** 1994. Spatial variability in biodegradation rates as evidenced by methane production from an aquifer. *Appl. Environ. Microbiol.* **60:**3632–3639.
3. **Adkins, J. P., L. A. Cornell, and R. S. Tanner.** 1992. Microbial composition of carbonate petroleum reservoir fluids. *Geomicrobiol. J.* **10:**87–97.
4. **Amy, P. S., D. L. Haldeman, D. Ringelberg, D. H. Hall, and C. Russell.** 1992. Comparison of identification systems for classification of bacteria isolated from water and endolithic habitats within the deep subsurface. *Appl. Environ. Microbiol.* **58:**3367–3373.
5. **Anderson, R. T., H. A. Vrionis, I. Ortiz-Bernad, C. T. Resch, P. E. Long, R. Dayvault, K. Karp, S. Marutzky, D. R. Metzler, A. Peacock, D. C. White, M. Lowe, and D. R. Lovley.** 2003. Stimulating the in situ activity of *Geobacter* species to remove uranium from the groundwater of a uranium-contaminated aquifer. *Appl. Environ. Microbiol.* **69:**5884–5891.
6. **Baker, B. J., D. B. Moser, B. J. MacGregor, S. Fishbain, M. Wagner, N. K. Fry, B. Jackson, N. Speolstra, S. Loos, K. Takai, B. Sherwood Lollar, J. Fredrickson, D. Balkwill, T. C. Onstott, C. F. Wimpee, and D. A. Stahl.** 2003. Related assemblages of sulphate-reducing bacteria associated with ultradeep gold mines of South Africa and deep basalt aquifers of Washington State. *Environ. Microbiol.* **5:**267–277.
7. **Balkwill, D. L.** 1989. Numbers, diversity, and morphological characteristics of aerobic, chemoheterotrophic bacteria in deep subsurface sediments from a site in South Carolina. *Geomicrobiol. J.* **7:**33–51.
8. **Balkwill, D. L., and W. C. Ghiorse.** 1985. Characterization of subsurface bacteria associated with two shallow aquifers in Oklahoma. *Appl. Environ. Microbiol.* **50:**580–588.
9. **Barcelona, M. J., J. P. Gibb, J. A. Helfrich, and E. E. Garske.** 1986. *Practical Guide to Groundwater Sampling.* Contract Report 374. Illinois State Geological Survey, Springfield.
10. **Beeman, R. E., and J. M. Suflita.** 1989. Evaluation of deep subsurface sampling procedures using serendipitous microbial contaminants as tracer organisms. *Geomicrobiol. J.* **7:**223–233.
11. **Bjornstad, B. N., J. P. McKinley, T. O. Stevens, S. A. Rawson, J. K. Fredrickson, and P. E. Long.** 1994. Generation of hydrogen gas as a result of drilling within the saturated zone. *Groundwater Monit. Rev.* **14:**140–147.
12. **Boone, D. R., Y. Liu, Z.-J. Zhao, D. L. Balkwill, G. R. Drake, T. O. Stevens, and H. C. Aldrich.** 1995. *Bacillus infernus* sp. nov., an Fe(III)- and Mn(IV)-reducing anaerobe from the deep terrestrial subsurface. *Int. J. Syst. Bacteriol.* **45:**441–448.
13. **Brockman, F. J.** Unpublished data.
14. **Brockman, F. J., S. W. Liu, J. K. Fredrickson, D. B. Ringelberg, T. L. Kieft, C. M. Spadoni, D. C. White, and J. P. McKinley.** 1998. Post-sampling changes in microbial community composition and activity in a subsurface paleosol. *Microb. Ecol.* **36:**152–164.
15. **Brockman, F. J., T. L. Kieft, J. K. Fredrickson, B. N. Bjornstad, S. W. Li, W. Spangenburg, and P. E. Long.** 1992. Microbiology of vadose zone paleosols in south-central Washington state. *Microb. Ecol.* **23:**279–301.
16. **Brockman, F. J., and C. J. Murray.** 1997. Microbiological heterogeneity in the terrestrial subsurface and approaches for its description, p. 75–102. *In* P. S. Amy and D. L. Haldeman (ed.), *The Microbiology of the Terrestrial Deep Subsurface*. CRC Press, Boca Raton, Fla.
17. **Chapelle, F. H.** 1993. *Ground-Water Microbiology and Geochemistry.* John Wiley & Sons, Inc., New York, N.Y.
18. **Chapelle, F. H., and P. B. McMahon.** 1991. Geochemistry of dissolved inorganic carbon in a coastal plain aquifer: 1. Sulfate from confining beds as an oxidant in microbial CO_2 production. *J. Hydrol.* **127:**85–108.
19. **Chapelle, F. H., J. T. Morris, P. B. McMahon, and J. L. Zelibor.** 1988. Bacterial metabolism and the del-13C composition of groundwater, Floridian aquifer, South Carolina. *Geology* **16:**117–121.
20. **Clark, R. R.** 1988. A new continuous sampling wireline system for acquisition of uncontaminated minimally disturbed soil samples. *Ground Water Monit. Rev.* **8:**66–72.
21. **Colwell, F. S., G. J. Stormberg, T. J. Phelps, S. A. Birnbaum, J. Mckinley, S. A. Rawson, C. Veverka, S. Goodwin, P. E. Long, B. F. Russell, T. Garland, D. Thompson, P. Skinner, and S. Grover.** 1992. Innovative techniques for collection of saturated and unsaturated subsurface basalts and sediments for microbiological characterization. *J. Microbiol. Methods* **15:**279–292.
22. **Cragg, B. A., S. J. Bale, and R. J. Parkes.** 1992. A novel method for the transport and long-term storage of cultures and samples in an anaerobic atmosphere. *Lett. Appl. Microbiol.* **15:**125–128.
23. **Davey, M. E., D. Gevertz, W. A. Wood, J. B. Clark, and G. E. Jenneman.** 1998. Microbial selective plugging of sandstone through stimulation of indigenous bacteria in a hypersaline oil reservoir. *Geomicrobiol. J.* **15:**335–352.
24. **Driscoll, F. G.** 1986. Well drilling methods, p. 268–339. *In* F. G. Driscoll (ed.), *Groundwater and Wells*, 2nd ed. Johnson Filtration Systems, St. Paul, Minn.
25. **Ekendahl, S., A. H. O'Neill, E. Thomsson, and K. Pedersen.** 2003. Characterisation of yeasts isolated from deep igneous rock aquifers of the fennoscandian shield. *Microb. Ecol.* **46:**416–428.
26. **Fisk, M. R., S. J. Giovannoni, and I. H. Thorseth.** 1998. Alteration of oceanic volcanic glass: textural evidence of microbial activity. *Science* **281:**978–980.
27. **Fournelle, H. J., E. K. Day, and W. B. Page.** 1957. Experimental ground water pollution at Anchorage, Alaska. *Public Health Rep.* **72:**203–209.
28. **Fredrickson, J. K.** Unpublished data.
29. **Fredrickson, J. K., D. L. Balkwill, G. R. Drake, M. F. Romine, D. B. Ringelberg, and D. C. White.** 1995. Aromatic-degrading *Sphingomonas* isolates from the deep subsurface. *Appl. Environ. Microbiol.* **61:**1917–1922.
30. **Fredrickson, J. K., B. N. Bjornstad, F. S. Colwell, L. Krumholtz, R. M. Lehman, P. E. Long, J. P. McKinley, T. C. Onstott, T. J. Phelps, D. B. Ringelberg, J. M. Suflita, and D. C. White.** 1997. Pore-size constraints on the activity and survival of subsurface bacteria in a Late Cretaceous shale-sandstone sequence, northwestern New Mexico. *Geomicrobiol. J.* **14:**183–202.
31. **Fredrickson, J. K., F. J. Brockman, B. N. Bjornstad, P. E. Long, S. W. Li, J. P. McKinley, J. V. Wright, J. L. Conca,**

T. L. Kieft, and D. L. Balkwill. 1993. Microbiological characteristics of pristine and contaminated deep vadose sediments from an arid region. *Geomicrobiol. J.* **11:**95–107.
32. **Fredrickson, J. K., S. W. Li, F. J. Brockman, D. L. Haldeman, P. S. Amy, and D. L. Balkwill.** 1995. Time-dependent changes in viable numbers and activities of aerobic heterotrophic bacteria in subsurface samples. *J. Microbiol. Methods* **21:**253–265.
33. **Fredrickson, J. K., and T. C. Onstott.** 2001. Biogeochemical and geological significance of subsurface microbiology, p. 3–38. *In* J. K. Fredrickson and M. Fletcher (ed.), *Subsurface Microbiology and Biochemistry*. John Wiley & Sons, Inc., New York, N.Y.
34. **Fry, N. K., J. K. Fredrickson, S. Fishbain, M. Wagner, and D. A. Stahl.** 1997. Population structure of microbial communities associated with two deep, anaerobic, alkaline aquifers. *Appl. Environ. Microbiol.* **63:**1498–1504.
30. **Fuller, M. E., S. H. Streger, R. K. Rothmel, B. J. Mailloux, J. A. Hall, T. C. Onstott, J. K. Fredrickson, D. L. Balkwill, and M. F. DeFlaun.** 2000. Development of a vital fluorescent staining method for monitoring bacterial transport in subsurface environments. *Appl. Environ. Microbiol.* **66:**4486–4496.
36. **Ghiorse, W. C., and J. T. Wilson.** 1988. Microbial ecology of the terrestrial subsurface. *Adv. Appl. Microbiol.* **33:** 107–173.
37. **Griffin, W. T., T. J. Phelps, F. S. Colwell, and J. K. Fredrickson.** 1997. Methods for obtaining deep subsurface microbiological samples by drilling, p. 23–44. *In* P. S. Amy and D. L. Haldeman (ed.), *The Microbiology of the Terrestrial Deep Subsurface*. CRC Press, Boca Raton, Fla.
38. **Haldeman, D. L.** 1997. The storage related phenomenon: implications for handling and analysis of subsurface samples, p. 61–74. *In* P. S. Amy and D. L. Haldeman (ed.), *The Microbiology of the Terrestrial Deep Subsurface*. CRC Press, Boca Raton, Fla.
39. **Haldeman, D. L., P. S. Amy, D. Ringelberg, D. C. White, R. E. Garen, and W. C. Ghiorse.** 1995. Microbial growth and resuscitation alter community structure after perturbation. *FEMS Microbiol. Ecol.* **17:**27–37.
40. **Haldeman, D. L., P. S. Amy, C. E. Russell, and R. Jacobsen.** 1995. Comparison of drilling and mining as methods for obtaining microbiological samples from the deep subsurface. *J. Microbiol. Methods* **21:**305–316.
41. **Haldeman, D. L., P. S. Amy, D. C. White, and D. B. Ringelberg.** 1994. Changes in bacteria recoverable from subsurface volcanic rock samples during storage at 4°C. *Appl. Environ. Microbiol.* **60:**2697–2703.
42. **Harvey, R. W., L. H. George, R. L. Smith, and D. R. LaBlanc.** 1989. Transport of microspheres and indigenous bacteria through a sandy aquifer: results of natural- and forced-gradient tracer experiments. *Environ. Sci. Technol.* **23:**51–56.
43. **Harvey, R. W., N. E. Kinner, A. Bunn, D. Macdonald, and D. Metge.** 1995. Transport behavior of groundwater protozoa and protozoan-sized microspheres in sandy aquifer sediments. *Appl. Environ. Microbiol.* **61:**209–217.
44. **Hazen, T. C., L. Jimenez, G. L. de Victoria, and C. B. Fliermans.** 1991. Comparison of bacteria from deep subsurface sediment and adjacent groundwater. *Microb. Ecol.* **22:**293–304.
45. **Hirsch, P., and E. Rades-Rohkohl.** 1988. Some special problems in the determination of viable counts of groundwater microorganisms. *Microb. Ecol.* **16:**99–113.
46. **Horn, J. M., B. A. Masterson, A. Rivera, A. Miranda, M. A. Davis, and S. Martin.** 2004. Bacterial growth dynamics, limiting factors, and community diversity in a proposed geological nuclear waste repository environment. *Geomicrobiol. J.* **21:**273–286.
47. **Juck, D. F., G. Whissell, B. Steven, W. Pollard, C. P. McKay, C. W. Greer, and L. G. Whyte.** 2005. Utilization of fluorescent microspheres and a green fluorescent protein-marked strain for assessment of microbiological contamination of permafrost and ground ice core samples from the Canadian High Arctic. *Appl. Environ. Microbiol.* **71:** 1035–1041.
48. **Kieft, T. L., P. S. Amy, B. N. Bjornstad, F. J. Brockman, J. K. Fredrickson, and L. L. Rosacker.** 1993. Microbial abundance and activities in relation to water potential in the vadose zones of arid and semiarid sites. *Microb. Ecol.* **26:**59–78.
49. **Kieft, T. L., J. K. Fredrickson, J. P. McKinley, B. N. Bjornstad, S. A. Rawson, T. J. Phelps, F. J. Brockman, and S. M. Pfiffner.** 1995. Microbiological comparisons within and across contiguous lacustrine, paleosol, and fluvial subsurface sediments. *Appl. Environ. Microbiol.* **61:** 749–757.
50. **Kieft, T. L., J. K. Fredrickson, T. C. Onstott, Y. A. Gorby, H. M. Kostandarithes, T. J. Bailey, D. W. Kennedy, S. W. Li, A. E. Plymale, C. M. Spadoni, and M. S. Gray.** 1999. Dissimilatory reduction of Fe(III) and other electron acceptors by a *Thermus* isolate. *Appl. Environ. Microbiol.* **65:**1214–1221.
51. **Kieft, T. L., W. P. Kovacik, Jr., D. B. Ringelberg, D. C. White, D. L. Haldeman, P. S. Amy, and L. E. Hersman.** 1997. Factors limiting to microbial growth and activity at a proposed high-level nuclear repository, Yucca Mountain, Nevada. *Appl. Environ. Microbiol.* **63:**3128–3133.
52. **Krumholz, L. R., J. P. McKinley, F. A. Ulrich, and J. M. Suflita.** 1997. Confined subsurface microbial communities in Cretaceous rock. *Nature* (London) **386:**64–66.
53. **Kuwabara, J. S., and R. W. Harvey.** 1990. Application of a hollow-fiber, tangential-flow deveice for sampling suspended bacteria and particles fron natural waters. *J. Environ. Qual.* **19:**625–629.
54. **LeBlanc, D. R., S. P. Garabedian, K. M. Hess, L. W. Gelhar, R. D. Quadri, K. G. Stollenwerk, and W. W. Wood.** 1991. Large-scale natural gradient test in sand and gravel, Cape Cod, Massachusetts. *Water Resour. Res.* **27:** 895–910.
55. **Lehman, R. M., F. S. Colwell, D. Ringelberg, and D. C. White.** 1995. Combined microbial community-level analyses for quality assurance of terrestrial subsurface cores. *J. Microbiol. Methods.* **22:**263–281.
56. **Lehman, R. M., F. S. Colwell, and G. A. Bala.** 2001. Attached and unattached microbial communities in a simulated basalt aquifer under fracture- and porous-flow conditions. *Appl. Environ. Microbiol.* **67:**2799–2809.
57. **Lehman, R. M., S. P. O'Connell, A. Banta, J. K. Fredrickson, A. L. Reysenbach, T. L. Kieft, and F. S. Colwell.** 2004. Microbiological comparison of core and groundwater samples collected from a fractured basalt aquifer with that of dialysis chambers incubated in situ. *Geomicrobiol. J.* **21:**169–182.
58. **Lehman, R. M., F. F. Roberto, D. Early, D. F. Bruhn, S. E. Brink, S. P. O'Connell, M. E. Delwiche, and F. S. Colwell.** 2001. Attached and unattached bacterial communities in a 120-m corehole in an acidic, crystalline rock aquifer. *Appl. Environ. Microbiol.* **67:**2095–2106.
59. **Liu, S. V., J. Zhou, C. Zhang, D. R. Cole, M. Gajdarziska-Josifovska, and T. J. Phelps.** 1997. Thermophilic Fe (III) reducing bacteria from the deep subsurface: the evolutionary implications. *Science* **277:**1106–1109.
60. **Mancinelli, R. L.** 2000. Accessing the Martian deep subsurface to search for life. *Planet Space Sci.* **48:**1035–1042.
61. **Marchesi, J. R., A. J. Weightman, B. A. Cragg, R. J. Parkes, and J. C. Fry.** 2001. Methanogen and bacterial

diversity and distribution in deep gas hydrate sediments from the Cascadia Margin as revealed by 16S rRNA molecular analysis. *FEMS Microbiol. Ecol.* **34:**221–228.
62. **McKay, C. P.** 2001. The deep biosphere: lessons for planetary exploration, p. 315–328. *In* J. K. Fredrickson and M. Fletcher (ed.), *Subsurface Microbiology and Biochemistry*. John Wiley & Sons, Inc., New York, N.Y.
63. **McKinley, J. P., and F. S. Colwell.** 1996. Application of perfluorocarbon tracers to microbial sampling in subsurface environments using mud-rotary and air-rotary drilling techniques. *J. Microbiol. Methods* **26:**1–9.
64. **McNabb, J. F., and G. E. Mallard.** 1984. Microbiological sampling in the assessment of groundwater pollution, p. 235–260. *In* G. Bitton and C. P. Gerba (ed.), *Groundwater Pollution Microbiology*. John Wiley & Sons, Inc., New York, N.Y.
65. **Moser, D. P., P. J. Boston, and M. Harris.** 2002. Sampling in caves and mines, p. 821–835. *In* G. Bitton (ed.), *Encyclopedia of Environmental Microbiology*. John Wiley & Sons, Inc., New York, N.Y.
66. **Moser, D. P., T. C. Onstott, J. K. Fredrickson, F. J. Brockman, D. L. Balkwill, G. R. Drake, S. Pfiffner, D. C. White, K. Takai, L. M. Pratt, J. Fong, B. Sherwood Lollar, G. Slater, T. J. Phelps, N. Spoelstra, M. Deflaun, G. Southam, A. T. Welty, B. J. Baker, and J. Hoek.** 2003. Temporal shifts in microbial community structure and geochemistry of an ultradeep South African gold mine borehole. *Geomicrobiol. J.* **20:**517–548.
67. **Murphy, E. M., J. A. Schramke, J. K. Fredrickson, H. W. Bledsoe, A. J. Francis, D. S. Sklarew, and J. C. Linehan.** 1992. The influence of microbial activity and sedimentary organic carbon on the isotope geochemistry of the Middendorf Aquifer. *Water Resour. Res.* **28:**723–740.
68. **Onstott, T. C.** Personal communication.
69. **Onstott, T. C., D. P. Moser, S. M. Pfiffner, J. K. Fredrickson, F. J. Brockman, T. J. Phelps, D. C. White, A. Peacock, D. Balkwill, R. Hoover, L. R. Krumholz, M. Borscik, T. L. Kieft, and R. Wilson.** 2003. Indigenous and introduced microorganisms in rock samples from a deep gold mine. *Environ. Microbiol.* **5:**1168–1191.
70. **Onstott, T. C., T. J. Phelps, F. S. Colwell, D. Ringelberg, D. C. White, and D. R. Boone.** 1998. Observations pertaining to the origin and ecology of microorganisms recovered from the deep subsurface of Taylorsville Basin, Virginia. *Geomicrobiol. J.* **15:**353–385.
71. **Onstott, T. C., T. J. Phelps, T. Kieft, F. S. Colwell, D. L. Balkwill, J. K. Fredrickson, and F. Brockman.** 1999. A global perspective on the microbial abundance and activity in the deep subsurface, p. 487–500. *In* J. Seckbach (ed.), *Enigmatic Microorganisms and Life in Extreme Environments*. Kluwer Academic Publishers, Amsterdam, The Netherlands.
72. **Pedersen, K.** 2001. Diversity and activity of microorganisms in deep igneous rock aquifers of the fennoscandian shield, p. 97–139. *In* J. K. Fredrickson and M. Fletcher (ed.), *Subsurface Microbiology and Biochemistry*. John Wiley & Sons, Inc., New York, N.Y.
73. **Pedersen, K., and S. Ekendahl.** 1990. Distribution and activity of bacteria in deep granitic groundwaters of southeastern Sweden. *Microb. Ecol.* **20:**37–52.
74. **Pedersen, K., and S. Ekendahl.** 1992. Assimilation of CO_2 and introduced organic compounds by bacterial communities in groundwater from Southeastern Sweden deep crystalline bedrock. *Microb. Ecol.* **2:**31–14.
75. **Pedersen, K., L. Hallbeck, J. Arlinger, A. C. Erlandson, and N. Jahromi.** 1997. Investigation of the potential for microbial contamination of deep granitic aquifers during drilling using 16s rRNA gene sequencing and culturing methods. *J. Microbiol. Methods* **30:**179–192.
76. **Phelps, T. J., C. B. Fliermans, T. R. Garland, S. M. Pfiffner, and D. C. White.** 1989. Methods for recovery of deep terrestrial subsurface sediments for microbiological studies. *J. Microbiol. Methods* **9:**267–279.
77. **Phillips, L. E., and H. M. Lappin-Scott.** 1997. Enrichment and characterization of sulfate-reducing bacteria from sandstone rock cores from the UK continental shelf. *FEMS Microbiol. Rev.* **20:**415–423.
78. **Russell, B. F., T. J. Phelps, W. T. Griffin, and K. A. Sargent.** 1992. Procedures for sampling deep subsurface microbial communities in unconsolidated sediments. *Groundwater Monit. Rev.* **12:**96–104.
79. **Russell, C. E.** 1997. The collection of subsurface samples by mining, p 45–59. *In* P. S. Amy and D. L. Haldeman (ed.), *The Microbiology of the Terrestrial Deep Subsurface*. CRC Press, Boca Raton, Fla.
80. **Smith, D. C., A. J. Spivack, M. R. Fisk, S. A. Haveman, and H. Staudigel.** 2000. Tracer-based estimates of drilling-induced microbial contamination of deep sea crust. *Geomicrobiol. J.* **17:**207–219.
81. **Smith, R. L., R. W. Harvey, and D. R. Leblanc.** 1991. Importance of closely-spaced vertical sampling in delineating chemical and microbiological gradients in groundwater studies. *J. Contam. Hydrol.* **7:**285–300.
82. **Soule, H., O. Genoulaz, B. Gratacap-Cavallier, P. Chevallier, J. X. Liu, and J. M. Seigneurin.** 2000. Ultrafiltration and reverse transcription-polymerase chain reaction: an efficient process for poliovirus, rotavirus and hepatitis A virus detection in water. *Water Res.* **34:**1063–1067.
83. **Stevens, T. O., and B. S. Holbert.** 1995. Variability and density dependence of bacteria in terrestrial subsurface samples: implications for enumeration. *J. Microbiol. Methods* **21:**283–293.
84. **Stevens, T. O., J. P. McKinley, and J. K. Fredrickson.** 1993. Bacteria associated with deep, alkaline, anaerobic groundwaters in southeast Washington. *Microb. Ecol.* **25:** 35–50.
85. **Takai, K., D. P. Moser, M. F. DeFlaun, T. C. Onstott, and J. K. Fredrickson.** 2001. Archaeal diversity in waters from deep South African gold mines. *Appl. Environ. Microbiol.* **67:**5750–5760.
86. **Takai, K., D. P. Moser, T. C. Onstott, N. Spoelstra, S. M. Pfiffner, A. Dohnalkova, and J. K. Fredrickson.** 2001. *Alkaliphilus transvaalensis* gen. nov., sp. nov., an extremely alkaliphilic bacterium isolated from a deep South African gold mine. *Int. J. Syst. Evol. Microbiol.* **51:** 1245–1256.
87. **Vescio, P. A., and S. A. Nierzwicki-Bauer.** 1995. Extraction and purification of PCR amplifiable DNA from lacustrine subsurface sediments. *J. Microbiol. Methods* **21:** 225–233.
88. **Vroblesky, D. A., and F. H. Chapelle.** 1994. Temporal and spatial changes of terminal electron-accepting processes in a petroleum hydrocarbon-contaminated aquifer and the significance for contaminant biodegradation. *Water Resour. Res.* **30:**1561–1570.
89. **Ward, J. A., G. F. Slater, D. P. Moser, L.-H. Lin, G. Lacrampe-Couloume, A. S. Bonin, M. Davidson, J. A. Hall, B. Mislowac, R. E. S. Bellamy, T. C. Onstott, and B. Sherwood Lollar.** 2004. Microbial hydrocarbon gases in the Witwatersrand Basin, South Africa: implications for the deep biosphere. *Geochim. Cosmochim. Acta* **68:**3239–3250.
90. **White, D. C.** Personal communication.
91. **Wilson, J. T., J. F. McNabb, D. L. Balkwill, and W. C. Ghiorse.** 1983. Enumeration and characterization of bacteria indigenous to a shallow water-table aquifer. *Ground Water* **21:**134–142.

92. **Wilson, N.** 1995. *Soil Water and Ground Water Sampling.* CRC Press, Inc., Boca Raton, Fla.
93. **Winona, L. J., A. W. Ommani, J. Olszewski, J. B. Nuzzo, and K. H. Oshima.** 2001. Efficient and predictable recovery of viruses from water by small scale ultrafiltration systems. *Can. J. Microbiol.* **47:**1033–1041.
94. **Wood, W. W., and G. G. Ehrlich.** 1979. Use of bakers yeast to trace microbial movement in ground water. *Ground Water* **16:**398–402.
95. **Zapico, M. M., S. Vales, and J. A. Cherry.** 1987. A wire-line piston core barrel for sampling cohesionless sand and gravel below the water table. *Ground Water Monit. Rev.* **7:**74–82.
96. **Zhang, G., H. Dong, Z. Xu, D. Zhao, and C. Zhang.** 2005. Microbial diversity in ultra-high-pressure rocks and fluids from the Chinese Continental Scientific Drilling Project in China. *Appl. Environ. Microbiol.* **71:**3213–3227.

Anaerobic Decomposition of Refuse in Landfills and Methane Oxidation in Landfill Covers

HELENE H. HILGER AND MORTON A. BARLAZ

67

INTRODUCTION

A landfill is a disposal alternative for municipal solid waste (MSW), nonhazardous industrial wastes, water and wastewater treatment sludges, and combustion and agricultural residues. Despite increases in recycling and composting, approximately 55% by weight of the MSW generated in the United States in 2001 was buried in sanitary landfills (169). There is a limit to the types of waste that can be recycled or composted, and combustion is typically more expensive than a landfill. Thus, landfills will be a significant waste repository for the foreseeable future. A complex series of biological and chemical reactions begins with the burial of refuse in a landfill, and landfills represent an active anaerobic ecosystem with methane (CH_4) and carbon dioxide (CO_2) as the major end products. CH_4 is recovered in commercial quantities from about 380 landfills in the United States (http://www.epa.gov/lmop/overview.htm) and is collected and flared at many additional sites. CH_4 that is not collected is released through the landfill cover, where a portion is oxidized and the balance is released to the atmosphere. On a global scale, landfills make up 6.7% of the 600 Tg/year total source and they are the third and fourth largest sources of anthropogenic CH_4 in the United States and globally, respectively (68, 116, 168).

The objective of this chapter is to discuss techniques for the study of anaerobic biological reactions and microorganisms involved in refuse decomposition as it occurs in a landfill as well as the aerobic biological reactions that occur in the landfill cover. A more thorough review of solid waste microbiology, which includes pathogens present in MSW and their survival in both landfills and composting systems, has been published elsewhere (128). Many of the principles associated with sample collection and processing presented here are applicable to compost systems.

MSW has been defined by the U.S. Environmental Protection Agency (169) to include residential, commercial, and institutional waste but to exclude combustion ash, hazardous waste, sludges, and industrial process waste. However, many of these other wastes often are deposited in the same landfills that receive MSW, and prior to enactment of regulations governing hazardous waste disposal, organic solvent-contaminated wastes frequently were buried with MSW in landfills. The principles discussed in this chapter could be applied to these other waste types with appropriate consideration given to safety.

This chapter begins with a brief description of the major components of a sanitary landfill followed by a discussion of MSW composition. Next, the manner in which cellulosic substrates are converted to CH_4 and CO_2 is described, followed by a discussion of CH_4 oxidation in landfill cover soils. Factors that influence both anaerobic decomposition rates in landfills and aerobic CH_4 oxidation in landfill covers are discussed. This is followed by a section on systems that can be used to simulate refuse decomposition and techniques that can be used to measure refuse biodegradation and microbial activity in landfills. Finally, systems and techniques that are applicable to the study of CH_4 oxidation in landfill cover soils are presented.

Description of a Sanitary Landfill

With the implementation of Subtitle D of the Resource Conservation and Recovery Act (51) and related state regulations, landfills have become highly engineered facilities designed to contain waste and separate it from the environment, capture leachate, and control gas migration. A landfill site is typically excavated and lined with a system that includes layers to (i) minimize the migration of leachate to the groundwater and (ii) collect leachate for treatment. A common system used to restrict leachate migration consists of a 0.67- to 1-m-thick clay layer with a hydraulic conductivity of no more than 10^{-7} cm/s overlaid with a geomembrane (GM). The GM is typically 1.5-mm-thick polyethylene with an equivalent hydraulic conductivity (based on vapor diffusion) of about 10^{-12} cm/s (60). A drainage layer designed to promote the collection of leachate is placed above the liner. A protective barrier is then installed above the leachate collection system to shield it from the equipment used to place and compact the refuse. Waste may then be placed above the protective barrier, and it is covered daily to minimize windblown refuse, the contamination of storm water runoff, and the attraction of disease vectors. Daily cover alternatives include the traditional 15-cm-thick soil layer, sprayed-on foams, and synthetic materials that are rolled over the waste at the end of the working day. Once refuse has reached the design elevation, a final cover is applied. The final cover includes, at a minimum, a layer of low-permeability soil designed to minimize storm water infiltration overlaid by a layer of soil that will support vegetative growth. The final cover frequently includes a drainage layer and a GM beneath the vegetative layer. Vegetation serves to minimize erosion of the soil cover and to promote evapotranspiration.

Refuse Composition

Data on the organic composition of refuse are presented in Table 1. Cellulose and hemicellulose are the principal biodegradable components of MSW. The other major organic component, lignin, is at best only slowly degradable under methanogenic conditions (52). In addition, lignin interferes with the decomposition of cellulose and hemicellulose by physically impeding microbial access to these degradable carbohydrates (see reference 64a for additional information). Other biodegradable organics present in smaller concentrations are proteins and soluble sugars.

BIOLOGICAL PROCESSES IN A LANDFILL

The decomposition of MSW to CH_4 in sanitary landfills is a microbially mediated process that requires the coordinated activity of several trophic groups of bacteria (Fig. 1) (39, 195). The first reaction is the hydrolysis of polymers (carbohydrates, fats, and proteins), which yields soluble sugars, amino acids, long-chain carboxylic acids, and glycerol. Fermentative microorganisms then convert these hydrolysis products to short-chain carboxylic acids, ammonia, CO_2, and H_2. Acetate and alcohols are also formed. Next, fatty-acid-degrading bacteria oxidize products such as propionate and butyrate to acetate, CO_2, and H_2. Oxidation of propionate and butyrate is thermodynamically favorable only at very low H_2 concentrations (196). Thus, these bacteria function only in syntrophic association with an H_2 scavenger such as a methanogen or a sulfate reducer. Typically, sulfate concentrations in landfills are minimal and CH_4 is the major electron sink. The significance of the acetogenic reaction, the production of acetate from CO_2 and H_2, in the landfill ecosystem has not been established. The terminal step in the conversion of complex polymers to CH_4 is carried out by the methanogenic bacteria. The most common methanogenic substrates are acetate and CO_2 plus H_2. Most methanogens have a pH optimum of around 7 (196). Should the activity of the fermentative organisms exceed that of the fatty acid degraders and methanogens, there will be an imbalance in the ecosystem. Carboxylic acids and H_2 will accumulate and the pH of the system will fall, thus inhibiting methanogenesis.

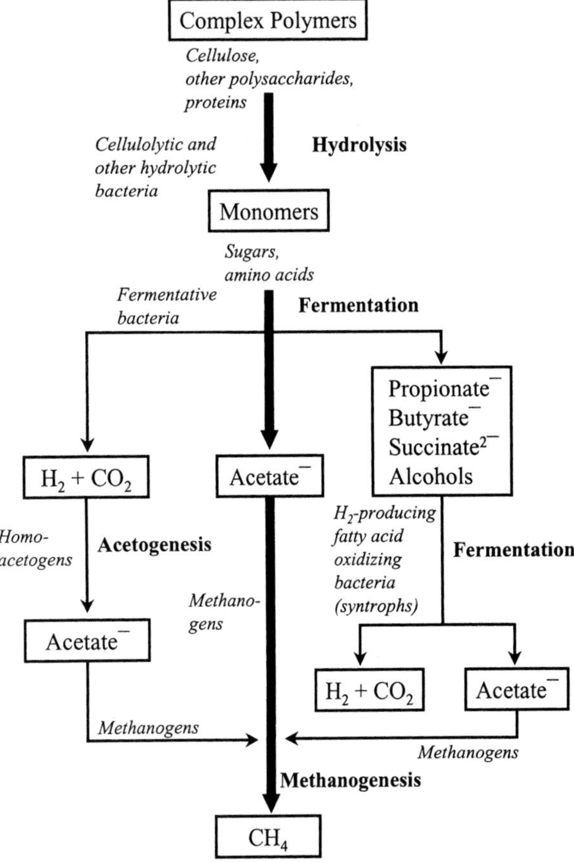

FIGURE 1 Overall process of anaerobic decomposition showing the manner in which various groups of fermentative anaerobes act together in the conversion of complex organic materials ultimately to methane and carbon dioxide. (Reprinted from reference 39 with permission.)

Phases of Refuse Decomposition

After placement of refuse in a landfill, several months or longer is necessary for the proper growth conditions and the required microbiological system to become established for biological decomposition. Thus, research to define the microbiology of refuse decomposition has been conducted

TABLE 1 Organic composition of residential refuse

Reference	% of dry wt			
	Cellulose	Hemicellulose	Lignin	Volatile solids
12[a]	51.2	11.9	15.2	78.6
73	28.8	9.0	23.1	75.2
145	38.5	8.7	28.0	NM[b]
144	48.2	10.6	14.5	71.4
Barlaz, unpublished	36.7	6.7	13.6	NM
	43.9	10	25.1	NM
	43.5	8.4	33.5	NM

[a]The following additional analyses were performed on this sample: protein, 4.2%; soluble sugars, 0.35%; starch, 0.6%; and pectin, <3%.

[b]NM, not measured.

by laboratory simulations in which the rate of decomposition has been accelerated and by obtaining samples of old refuse from landfills. A characterization of refuse decomposition describing chemical and microbiological characteristics has been developed by using data from laboratory-scale reactors (14). Refuse decomposition in an aerobic phase, an anaerobic acid phase, an accelerated CH_4 production phase, and a decelerated CH_4 production phase is described (Fig. 2).

In the aerobic phase (phase 1), both O_2 and NO_3 are consumed, with soluble sugars serving as the carbon source for microbial activity. All of the trophic groups required for refuse methanogenesis are present in fresh refuse (cellulolytics, fatty acid degraders, and methanogens). In the anaerobic acid phase (phase 2), carboxylic acids accumulate and the pH decreases because of an imbalance between fermentative activity and acetogenic and methanogenic activity. There is limited cellulose and hemicellulose decomposition in phase 2. The methanogen population begins to increase, and CH_4 is first detected. In the accelerated CH_4 production phase (phase 3), the CH_4 production rate increases to some maximum value; carboxylic acid concentrations decrease; the pH increases; there is little solid hydrolysis; and the populations of cellulolytic, fatty-acid-degrading, and methanogenic bacteria increase. The accumulated carboxylic acids are the principal substrates supporting CH_4 production in phase 3. In the decelerated CH_4 production phase (phase 4), the CH_4 production rate decreases, the fatty-acid-degrading population increases, carboxylic acids are depleted, and there is an increase in the rate of cellulose plus hemicellulose hydrolysis (Fig. 2). While acid utilization limits CH_4 production in phases 2 and 3, solid hydrolysis limits CH_4 production in phase 4. Work with samples excavated from a landfill confirmed the relationship between H_2, pH, and carboxylic acid concentrations expected on the basis of this four-phase description (124).

In applying this description of refuse decomposition to full-scale landfills, first it must be recognized that the time required for the onset of each phase may be significantly longer than the times shown in Fig. 2. Second, gas and leachate samples from landfills often reflect a composite of refuse in several different states of decomposition. Third, the presence of NO_3 stimulates denitrification, which inhibits methanogenesis (139). Fourth, in the presence of sulfate, electrons are diverted from CH_4 production to sulfate reduction. Sulfate reduction and CH_4 production have been shown to occur concurrently, with three to nine times more organic carbon degraded through methanogenesis than SO_4 reduction (76). This was explained by the abundance of degradable carbon. Others have shown that SO_4 inhibited CH_4 production and that inhibition of SO_4 reduction resulted in increased CH_4 production (84).

The presence of anaerobic protozoa in refuse excavated from landfills has been documented (79), and many of the protozoa contained symbiotic methanogenic bacteria that utilize H_2 released by the host's hydrogenosomes. The dominant protozoan isolated from the samples was the ciliate *Metopus palaeformis*. A later study showed no evidence that protozoa were stimulating refuse decomposition through enhanced nutrient recycling (78).

Biodegradation of Trace Organic Compounds in Landfills

Common contaminants in older landfills and leachate plumes include alkylbenzenes, ketones, and chlorinated aliphatic hydrocarbons (48). Anaerobic biodegradation of organic contaminants in leachate plumes (48, 49) has been documented. However, the literature on contaminant biodegradation in landfills is considerably more limited. The landfill environment varies considerably from that of a contaminated aquifer. It contains substantially higher concentrations of dissolved organic carbon, and the solid matrix on which attached bacteria live is composed largely of cellulose and hemicellulose.

Studies that utilized liquid extracts of refuse microorganisms have documented the anaerobic biodegradation of selected phthalic acid esters and perchloroethene (PCE) and trichloroethene (TCE) (71) and of phenol, toluene, and *p*-cresol (174). Landfill leachate also has been shown to harbor microorganisms capable of PCE and TCE dehalogenation (114). Reports in which contaminant biodegradation has been demonstrated in an actual refuse matrix, as opposed to leachate or a refuse extract, have documented biodegradation of phenol (177), *o*-cresol (163), toluene and acetone (151), and phthalic acid esters (105) and reductive dehalogenation (63, 136, 142). Recently, the role of both biopolymers and synthetic polymers in the bioavailability of hydrophobic organic contaminants has been described (47, 193). Techniques to assess biodegradation in decomposing refuse are discussed below ("Systems for the Study of Refuse Decomposition and Landfill Microbiology").

Factors Limiting Decomposition in Landfills

A number of factors influence the onset and rate of CH_4 production in landfills, including moisture content and moisture flow, pH, particle size, inoculum addition, nutrient concentrations, and temperature (13). The two variables that appear to be most critical in controlling refuse methanogenesis are moisture content and pH. High moisture promotes the dissolution and mixing of soluble nutrients and provides a mechanism for microbial transport.

Recent work has shown a positive correlation between increasing moisture content and the CH_4 production rate of excavated refuse samples (84, 86). CH_4 production from excavated refuse samples was limited to those with a pH around neutrality (162). Leachate recycle and neutralization have been shown to enhance the onset and rate of CH_4 production in laboratory-scale tests (11, 135), and this technique, now referred to as bioreactor landfills, is increasingly being implemented at field scale (143). More-thorough decomposition of solids in a section of a landfill operated with leachate recycle relative to a control section has been reported (123, 166), and many similar studies are in progress (21).

Microbial Methane Oxidation

Landfill cover is typical of ecosystems at the interface of anaerobic and aerobic regions, with exposure to high CH_4 concentrations from below and O_2 concentrations from above that diminish with depth. In older landfills, with no gas venting or collection systems, CH_4 produced in the waste layer follows the path of least resistance to migrate through the cover and sometimes through the sides of the landfill. Even in landfills that passively or actively collect the gas, some fugitive gases enter the cover.

CH_4 entering the atmosphere from landfills contributes to global climate change, and for many years predictive models developed to estimate landfill CH_4 generation assumed that generation was equivalent to emissions. It is now clear that microbial CH_4 oxidation can reduce the net CH_4 emitted from landfills, so that emission rates are lower than generation rates. In some instances, covers act as a net sink for atmospheric CH_4 (16, 29, 110).

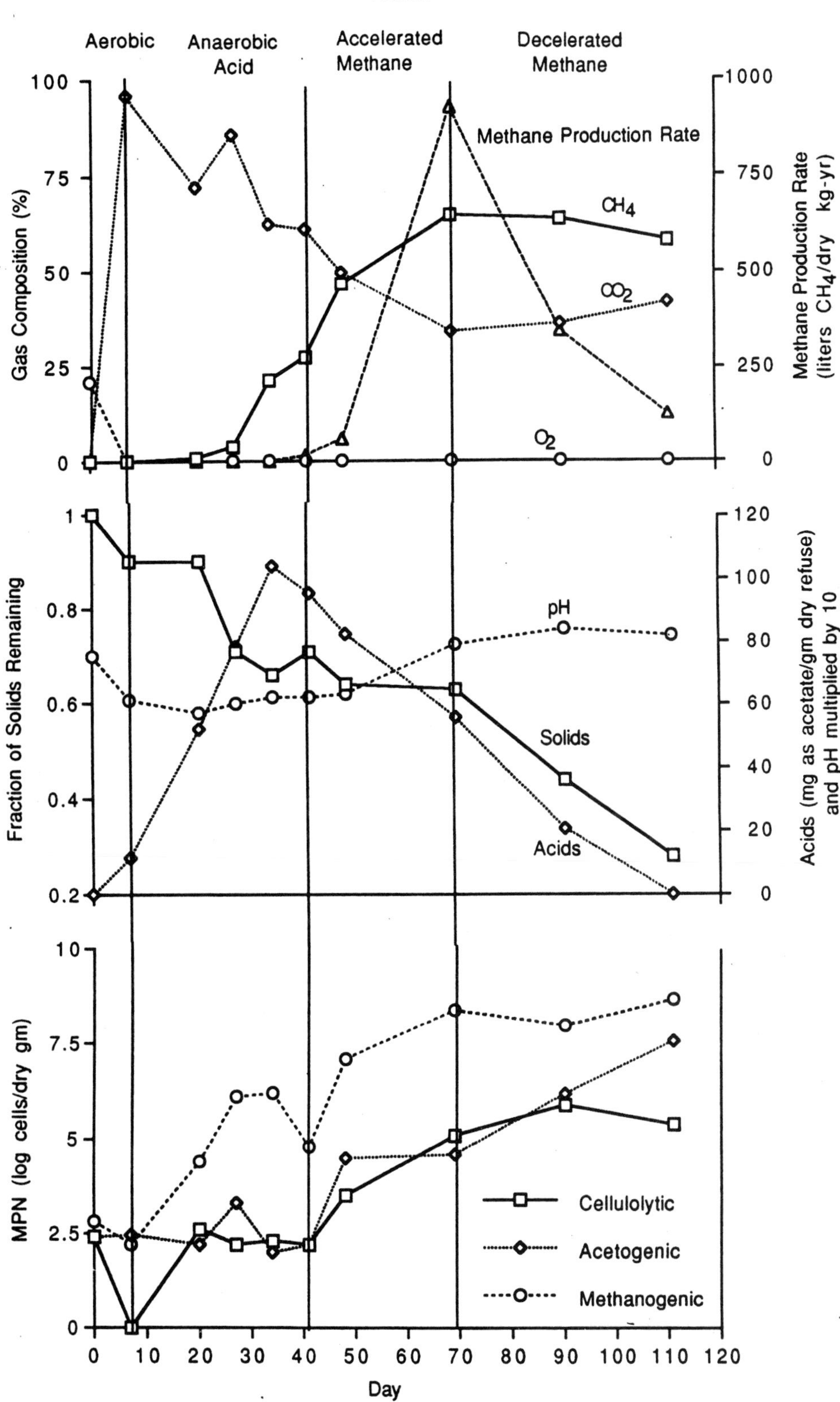

FIGURE 2 Summary of observed trends in refuse decomposition with leachate recycle. Gas volume data were corrected to dry gas at standard temperature and pressure. The acids are expressed as acetic acid equivalents. Solids remaining are the ratio of the weight of cellulose plus hemicellulose removed from a container divided by the weight of cellulose plus hemicellulose added to the container initially. Methanogen MPN data are the log of the average of the acetate- and H_2/CO_2-utilizing populations. (Reprinted from reference 14.)

Most microbial CH_4 oxidation is accomplished by methanotrophs, a subset of the highly diverse methylotrophic bacteria that metabolize C-1 compounds such as CH_4, methanol, methylamines, and halomethanes (88). Methanotrophs are ubiquitous in water and soil environments, where they convert CH_4 to CO_2 and H_2O. As obligate aerobes, they are abundant in the interface regions of aerobic and anaerobic zones of wetlands (89), rice paddies (53), and landfill covers (103), but they can also subsist on atmospheric CH_4 (1.7 ppm by volume [ppmv]) (152). Methanotrophs are an important sink for CH_4 in the global carbon cycle, and their diversity (88, 184), ecological significance (87, 93, 108), and biochemistry (6, 7, 17) have been reviewed. Some yeast species grow slowly on CH_4 (191), and there are nitrogen oxidizers that can oxidize CH_4 and incorporate CH_4 carbon into cell biomass (176).

Microbially mediated anaerobic CH_4 oxidation has also been documented (3, 81, 161), and although reports are limited to aqueous systems, this phenomenon may ultimately prove to be significant in landfill systems. Because the bacteria responsible have not been successfully cultured, the process is poorly understood. The reaction is generally thought to be the result of a syntrophic relationship between a sulfate-reducing bacterium and a reversed methanogenic archaeon (a methanogen processing substrate in reverse) (161). However, there is also speculation that an as yet unidentified single sulfate-reducing methanotroph is responsible (161). Several studies of migrating landfill leachate plumes have reported evidence of anaerobic CH_4 oxidation in or at the periphery of the plumes (171, 173).

The methanotrophs have been categorized into three groups based on functional characteristics such as resting states, membrane structure, and pathways for formaldehyde assimilation. Because the metabolic characteristics of the groups differ, the presence of particular types in a given landfill may affect how much CH_4 oxidation will occur. Types I and X use the ribulose monophosphate path, while type II methanotrophs use the serine path (5). Many methanotrophs can form cysts or exospores (183), and even in the vegetative state, they have the ability to "rest" through anaerobic conditions and then become active when O_2 becomes available (149). Capsule and slime production are common (101), and methanotrophs accumulate poly-β-hydroxybutyrate under certain growth conditions (8).

Two forms of CH_4 monooxygenase (MMO), the enzyme that catalyzes the conversion of CH_4 to methanol, are found in methanotrophs; almost all can synthesize a membrane-bound form of MMO (sometimes called particulate MMO or pMMO), but only some can produce a soluble MMO (sMMO) whose synthesis is linked to copper availability (62, 88). The latter is sufficiently nonspecific to catalyze the addition of O_2 to other substrates and to cometabolize compounds such as trichloroethylene, chloroform, and dichloromethane (59). Methanotrophic cometabolic activity can be important in landfill covers where trace volatile and hazardous organic compounds are present (16, 153). The ability to synthesize soluble MMO is typically associated with type II methanotrophs, which tend to outcompete type I species when CH_4 is plentiful but O_2 is limited (4, 82). Under O_2 limitation, certain species (mostly type II methanotrophs) can fix N_2 (183).

The kinetic properties reported for methanotrophs vary widely, and researchers have begun to refer to methanotrophs as belonging to either the low-affinity or the high-affinity CH_4 oxidizers (117), both of which are found in landfill covers (18, 29). Low-affinity methanotrophs prosper in CH_4-rich landfill environments and reach high CH_4 uptake rates, but their CH_4 affinity is too low to support atmospheric CH_4 uptake (18). High-affinity methanotrophs are well documented in soils exposed to only atmospheric CH_4 levels (19, 154); however, the organisms have not been cultured, and their detection has been accomplished by monitoring oxidation activity and through the use of molecular fingerprinting (112, 117). Their uptake rates tend to be much lower than those of the low-affinity methanotrophs (18, 19).

Factors Affecting CH_4 Oxidation in Landfill Covers

Factors that affect the degree of CH_4 oxidation in the cover are generally those that influence the microbial populations mediating CH_4 uptake. The maximum reaction velocity (V_{max}) is sensitive to CH_4 concentration (20, 29, 57), and landfill cover soils tested under elevated headspace CH_4 concentrations show uptake rates as high as 15 mol m^{-2} day^{-1} (38, 57, 65, 95, 107, 182, 186).

Both within and between landfill sites, seasonal climate change and soil heterogeneity contribute to variations in CH_4 oxidation activity in the cover (36, 38, 57, 118, 122). Moisture content, moisture-holding capacity, temperature, soil composition, nutrient content, cation-exchange capacity, pH, and the presence of vegetation are all influential.

Moisture content effects on CH_4 oxidation in landfill soil have been reviewed, and optima typically range from 10 to 20%, though optima as high as 24 and 27% for two Danish soils incubated at 15°C were reported (50). The moisture optima were temperature dependent. In column studies, composted organic materials at 45% moisture content outperformed soil columns at 17% moisture content when both were tested for CH_4 oxidation, suggesting that high CH_4 uptake can be maintained at high moisture levels as long as soil pore volumes are not water saturated (97). Temperatures as low as 2°C (50) supported CH_4 oxidation in landfill soil, with reported optima of 30 to 36°C and Q_{10} values ranging from 2.4 to 7.3 (25, 50, 57, 182).

O_2 effects on methanotrophs in landfills appear to be negligible down to concentrations of about 3% (57). High O_2 levels have been associated with methanotroph exopolymer production, which was hypothesized to be due to carbon cycling by type I methanotrophs (188). It was reasoned that in a microaerobic headspace that allows type II methanotrophs to fix N_2, the latter would increase, the number of type I methanotrophs would decrease, and carbon cycling through polymer production would be eliminated. This hypothesis was supported by findings that organic N accumulated in the soil as microaerobic incubations progressed.

Soil composition and particle size affect CH_4 oxidation, with higher oxidation rates occurring in coarse sand (61%) relative to fine sand or clay (~40%) (107) and on the larger particles of mineral soils such as clay, silt, and fine sand (0.5 to 2 mm) (19, 37). Higher organic content in predominantly mineral soils supports greater CH_4 oxidation capacity (50), and covers of various organic media (MSW compost, sewage sludge compost, and compost-sand and compost-soil mixtures) overlying 10 to 15 m of MSW were shown to eliminate emissions of up to 0.7 to 1 m^3 m^{-2} day^{-1} (as measured in control plots) (97). A cover composed of composted yard waste removed emissions and consumed atmospheric CH_4 about 60% of the time, compared to a soil cover that did so only 12% of the time (16).

Some nutrients enhance CH_4 uptake in short-term incubations, but few have been tested in longer incubations, and the mechanisms of enhancement are not well understood.

Stimulatory effects appear to depend in part on whether the test soil has undergone prolonged exposure to high CH_4 concentrations. Only lime (95) and sewage sludge (107) enhanced CH_4 uptake in long-term laboratory incubations. Lime increased the soil pH from 6.3 to 7.4, which was probably closer to the optimum pH for CH_4 oxidizers, and sewage sludge likely provided additional nutrients or microorganisms. Vegetation also appears to enhance oxidation rates (95, 160).

Short-term incubations have shown that NO_3 and lime amendments in fresh soil enhance CH_4 uptake; however, when NO_3 was added to landfill soil that had been exposed to simulated landfill gas for several months, no stimulation was observed (95). Ammonium addition enhanced CH_4 uptake in short-term incubations (20), while in long-term incubations it inhibited CH_4 consumption in bare soil columns but had no sustained inhibitory effect in columns topped with grass (95). In one report, NH_4^+ inhibition patterns changed over time during sustained laboratory incubations (66). Bodelier and Laanbrock (23) have reviewed many conflicting reports on the effects of nitrogen supplements and suggest some microbial and ecological mechanisms that require further investigation.

Two volatile sulfur compounds, methanethiol and carbon disulfide, were inhibitory to CH_4 oxidation in landfill soil, affecting type I methanotrophs more than type II (35). Acetylene and ethylene have proven to be effective amendments when intentional methanotroph inhibition is desired for experimental controls (42).

MICROBIOLOGICAL STUDIES OF REFUSE DECOMPOSITION IN LANDFILLS

A discussion of techniques for the study of decomposition in landfills must begin with an acknowledgment that landfills are extremely heterogeneous. It is probably not possible to obtain truly representative samples from landfills. Nonetheless, by collection of numerous samples from different parts of a landfill, some indication of the status of the landfill can be obtained. Given the heterogeneity, some researchers have used laboratory-scale simulations while others have focused on field sampling. Alternate systems for the study of both microorganisms in landfills and refuse biodegradation are discussed first, followed by presentation of techniques for sample collection and processing. Next, measurements that may be used to characterize the biological state of decomposing refuse are presented. This includes traditional measurements such as enumeration and enzyme assays, emerging molecular tools, and chemical measurements such as the instantaneous CH_4 production rate and analysis of specific organic constituents in refuse. The major considerations in the selection of techniques for all aspects of landfill microbiology are summarized in Table 2.

Systems for the Study of Refuse Decomposition and Landfill Microbiology

A number of systems have been used to study refuse decomposition and landfill microbiology, with each system typically optimized for a specific objective. If the objective is to obtain samples of refuse in various states of decomposition, then the researcher may work with samples from a full-scale landfill or from a laboratory simulation. When the objective is to evaluate the impact of a parameter on refuse decomposition, then a laboratory simulation is required. The types of measurements to be made also influence the system used. When researchers are interested in collection of refuse samples for measurement of specific biological parameters (enumeration, hydrolytic enzyme activity, etc.), a mechanism for representative sampling is critical. When an overall measure of decomposition such as CH_4 production will suffice, a system can be designed without allowing for solid sampling. Sampling techniques for alternate systems are discussed in the following section.

The best landfill simulation is a field-scale landfill containing several tons of refuse. A limited number of studies on the impact of parameters such as moisture content and sludge addition on refuse decomposition have been conducted at field scale (55, 85, 123). However, field-scale test cells are expensive and difficult to control, and it is difficult to obtain representative samples and to ensure complete recovery of the gas produced. Furthermore, obtaining meaningful data may require several years of monitoring. (Sampling is discussed in the next section.) Refuse heterogeneity can even be a problem at field scale. CH_4 production rates for the initial monitoring period for test cells in the United Kingdom were 4.3, 5.8, and 6.7 m^3/metric ton-year for replicate test cells, each containing 15,000 metric tons of refuse (55). However, after 8 years of monitoring, the stimulatory effect of moisture addition on CH_4 production was more apparent. These data emphasize the need to monitor field-scale cells for periods in excess of 10 years. Given the time and expense associated with field cells, the majority of work on parameter assessment has been conducted by using laboratory-scale simulations (described below).

Refuse samples may also be collected from landfills for the isolation or enumeration of bacteria, enzyme assays, and measurement of chemical parameters such as pH, solids (cellulose, hemicellulose, lignin), and soluble organics. For example, Ham et al. (86) sampled refuse from landfills to assess their CH_4 production status. Samples were studied in 15-cm-diameter reactors (polyvinyl chloride [PVC] pipe) that held about 10 kg of refuse. The relatively large reactor eliminates the need to exclude some large waste components. Others have worked with excavated refuse samples in reactors that hold only 0.2 to 0.3 kg of refuse (84). The smaller reactors allowed for manipulation of the ecosystem by molybdate and moisture addition with confidence that the amendments were in contact with the entire reactor. The application of more traditional microbiological measurements to excavated refuse samples is discussed below.

When a laboratory-scale simulation of a landfill is desired, reactors with capacities ranging from 2 to 3,300 liters have been used (11, 12, 63, 137, 144, 151). The larger the reactor, the more difficult it is to control and manipulate and the more expensive it is to replicate tests. In addition, as the reactor size increases, it is less certain that all of the refuse in the reactor is in the same state of decomposition and that the leachate composition and gas production rate represent the entire mass in the reactor. These systems have also been used to study the biodegradation of trace organic contaminants (151).

The use of shredded refuse makes it possible to work with smaller quantities of refuse and still have a representative sample. Refuse can be obtained from a municipal refuse collection service and may be shredded by using a low-speed, high-torque shredder such as a Shred Pax (Wood Dale, Ill.) AZ-7H. However, shredding refuse provides a degree of mixing not typical of landfills. This mixing, coupled with the decreased particle size, likely enhances biodegradation.

Whether working with 1 or 400 kg of refuse, the time to the onset of CH_4 production varies. In the absence of some

TABLE 2 Summary of methods used to study refuse decomposition and landfill microbiology

Method	Comments	Reference(s)
Decomposition studies		
Actual landfill or field-scale test landfill	Expensive and hard to control Large samples required to assess overall methane production Difficult to obtain representative samples for biological tests Extended time required for complete decomposition cycle	55, 85, 123
Reactors ranging from 2–3,300 liters	Difficult to obtain representative samples from large reactors for biological assessment Large systems hard to control Smaller systems can be better controlled and sampled Small systems require use of shredded refuse and some selection of what is added to reactor Can accelerate decomposition	11, 12, 14, 63, 84, 90, 135, 137, 145
BMP	Measures ultimate anaerobic biodegradability of a small sample Requires inoculum Useful for measure of potential residual methane production if representative samples can be recovered Reproducible and rapid (30–60 days)	127, 156, 175
Effect of environmental conditions on refuse microbes		
Multistage chemostat	Unique system for studying effects of environmental conditions on individual components of anaerobic consortia	54, 100, 132
Sample collection		
Fresh refuse	Should be representative of residential refuse Requires shredding for homogenization	12, 14, 145
Decomposed refuse excavated from landfill	Must collect large samples and homogenize Oxygen exposure has not been a major problem with reasonable precaution during sampling	27, 86, 123, 129, 130, 179
Decomposed refuse from a laboratory reactor	Sample history well known	14, 73
Inoculum formation		
Blend in PO_4 buffer	Shown to be repeatable, easy to implement	14, 140
Stomacher	Limited to 10- to 50-g samples	129
Multiple extractions	Two extractions were adequate in one study No difference on multiple blending in phosphate buffer	15, 115, 121
Triton X-100 and $MgSO_4$	Used for enzyme assays No comparative work has been done	104, 130
Enumeration		
MPN	Most common Researchers have been consistent in the use of a rich medium for enumeration of the total anaerobic population Media are available for enumeration of cellulolytics, hemicellulolytics, acetogens, and methanogens	14, 129, 140
AODC	Measures total viable plus nonviable cells Well above total anaerobic populations as measured by MPN	129
Roll tubes	Used primarily for cellulolytic bacteria Ball-milled cellulose is most common cellulose form used in both roll tubes and liquid media	129, 180
Microbial activity		
Hydrolytic enzymes	Cellulase measurements appear to be inconsistent with expected values based on cellulose decomposition in landfills	104, 130

(*Continued on next page*)

TABLE 2 (*Continued*)

Method	Comments	Reference(s)
	Effectiveness of cell lysis step is unknown	
^{14}C turnover	Good measure of combined activity of microorganisms required for substrate degradation	130, 151
Methane production rate	Instantaneous rate of methane production in sample	84, 162
Microbial identification and population dynamics		
Probes	Applied successfully in leachate	121, 157
DNA cloning	Applied to cellulose in liquid/solid reactor	41, 44
FISH	Applied to landfill samples	45, 91
Chemical measurements		
Cellulose, hemicellulose, lignin	Useful for assessment of the degree of decomposition Requires representative sample	34, 86, 175

enhancement, refuse decomposition may require many years to complete its cycle. This incubation period is not practical for many experiments, and strategies to enhance decomposition in laboratory-scale systems have been developed. Steps that can be taken to accelerate decomposition include (i) incubation at about 40°C, (ii) the use of shredded refuse, and (iii) the use of leachate recycle and neutralization. The optimum temperature for mesophilic refuse decomposition is reported to be 40°C (90). Leachate recycle and neutralization require the addition of sufficient moisture to a reactor to allow for the free drainage of liquid into a leachate collection system where it can be externally neutralized and returned through the top of a reactor. If needed, or as part of the test protocol, leachate composition can be adjusted prior to recycling. This process moves refuse from the anaerobic acid phase to the accelerated CH_4 production phase in 1 to 2 months.

The onset of CH_4 production in laboratory simulations can be stimulated by the addition of an inoculum. An inoculum of well-decomposed refuse has been shown to initiate decomposition of fresh refuse or individual components thereof (10, 73). An inoculum eliminates the anaerobic acid phase of decomposition, and CH_4 production begins almost immediately. Because carboxylic acids will not accumulate during the anaerobic acid phase, the subsequent increase in the CH_4 production rate will not be as sharp as it would be in the absence of an inoculum as shown in Fig. 2. An inoculum can be excavated from a landfill known to be in an active state of CH_4 production or developed in the laboratory by decomposing fresh refuse until its CH_4 production rate is well into the decelerated CH_4 production phase. The inoculum may be stored in the laboratory at room temperature but should be tested for viability in preliminary work. Ideally, the CH_4 potential of the inoculum will be relatively low because the refuse is well decomposed. However, control reactors may be required to measure CH_4 production associated with the inoculum (73). Recently, we have used leachate from decomposing refuse as an inoculum (M. A. Barlaz, unpublished data).

The onset of CH_4 production has been stimulated by aeration of fresh refuse (72; Barlaz, unpublished). The oxygen stimulates a temperature increase and the rapid oxidation of carboxylic acids that would otherwise depress the pH. Once the air supply is terminated, methanogenic conditions develop rapidly.

Work with anaerobic sewage sludge used as an inoculum has been summarized elsewhere (13). Anaerobic sewage sludge appears to stimulate the accumulation of carboxylic acids. Thus, its use requires leachate neutralization, and it is not clear that the addition of anaerobic sewage sludge decreases the lag time relative to that obtained with leachate recycle and neutralization alone.

The acceleration of refuse decomposition is a suitable technique to evaluate the impact of a number of variables on decomposition and leachate quality. For example, acceleration with leachate recycle has been used to evaluate the impacts of heavy metal sludges (137), a lime-stabilized sludge cover soil (145), and the effect of building debris on refuse decomposition (76). Studies in which refuse decomposition is enhanced provide data on the relative impacts of parameters on decomposition and whether a material has the potential to biodegrade under simulated landfill conditions. However, work in small reactors filled with shredded refuse is an idealized system, and degradation rates measured in this system should not be extrapolated to the field.

The biochemical methane potential (BMP) technique originally was developed to measure the anaerobic biodegradability of a soluble organic chemical in a small-batch reactor (156). It has since been adapted for measurement of the biodegradability of solid samples (127, 175). The BMP test provides a measure of ultimate biodegradability under idealized conditions that do not simulate a landfill. In the BMP assay, small samples of refuse are added to a serum bottle with liquid medium and an inoculum. The inoculum can be anaerobic sewage sludge, or, in our laboratory, we have maintained a methanogenic consortium on dried ground refuse for use as an inoculum. The samples are typically dried, shredded, and then ground in a Wiley mill to pass a 1-mm screen. The CH_4 potential of a sample is measured after incubation for 30 to 60 days. One application of the BMP test assesses the potential of well-decomposed refuse excavated from a landfill to produce additional CH_4 (175). Because only a few grams of sample may be added to a bottle, the sample must be homogeneous. Recently, a technique in which bioavailable cellulose and hemicellulose are estimated by incubation of a small refuse sample with cellulase and hemicellulase and measurement of monosaccharide release has been reported (147).

A multistage continuous-culture technique has been applied to the study of landfill microbiology (54, 100, 132).

In this system, a series of vessels of increasing size are operated as a plug flow chemostat. As the reactor size increases, the imposed dilution rate decreases. This selects for faster-growing bacteria in the top vessel and slower-growing bacteria in the subsequent vessels. This technique is reported to separate habitat domains while still permitting the overlap of activity domains required for syntrophic activity. In one study, three vessels were used to segregate hydrolytic-fermentative, acidogenic, and methanogenic activities (100). The vessels were inoculated with consortia enriched on cellobiose, valerate, and butyrate. However, it must be recognized that environmental conditions in the continuous-culture system may select for a community that differs from that in a landfill.

Three additional techniques for the measurement of trace organic contaminant biodegradability were compared (114). In all three techniques, landfill leachate served as the inoculum. The first approach was termed an in situ landfill simulator (ISLAS), which consisted of a "2-L porous media reactor submerged in an actual leachate well." The autoclaved system was colonized by pumping leachate through quartz sand for 4 weeks, after which target contaminants were added and their disappearance was monitored over time. The ISLAS was designed to eliminate sorption and volatilization. The second approach was termed a laboratory landfill simulator. It was similar to the ISLAS but was operated in the laboratory. The reactor was submerged in leachate in a steel cylinder that simulated an actual leachate well. The third approach was the use of a 2-liter laboratory batch reactor (LABAR) filled with quartz sand and leachate that served as the inoculum. In all three systems, target compounds and a lithium bromide tracer were added and their disappearance was monitored over time. All three systems were operated to maintain strict anaerobic conditions. Results in all three systems were comparable, and the authors recommend use of the LABAR on the basis of its simplicity.

There are a number of potential systems available for the study of refuse and trace contaminant biodegradation and microbiology, as summarized in Table 2. The ability to obtain samples will likely play a role in the selection of the appropriate landfill simulation system. While field-scale tests are most realistic, they are expensive and difficult to control and reproduce. Laboratory-scale systems allow for more rigorous experimental control and have more flexibility with respect to sampling, replication, and monitoring. Often, the time required to conduct a laboratory study can be shortened by the use of an inoculum, and decomposed refuse is the inoculum of choice for us on the basis of its proven reliability.

Sample Collection and Processing

Refuse samples may be required for reactor studies, assessment of the status of refuse undergoing decomposition, and microbiological assessment. Different types of samples are summarized in Table 2. Fresh refuse may be obtained from refuse collection vehicles upon delivery at a landfill or combustion facility. It is important to verify that the refuse originated in residential areas only and does not contain unusual materials. If the refuse is to be shredded, then approximately 1,000 kg of refuse should be shredded until it has a particle size small enough that representative samples can be obtained for use in laboratory-scale systems. The low-speed, high-torque shredder referenced above produces refuse particles less than 2 cm wide by 6 cm long. Prior to shredding, some presorting may be required to remove materials that cannot or should not be shredded. Once shredded, the refuse should be manually mixed and repetitively quartered until 5- to 10-kg piles remain. The refuse from one pile should then be selected at random for use in reactor filling. Refuse from multiple piles may be required, depending upon the mass needed.

Refuse may be excavated from a landfill either for use as an inoculum or for microbiological and chemical analyses. When the refuse is required for use as an inoculum, it is not critical that it be representative of the landfill. Ideally, it will be known whether a section of a landfill is producing CH_4. When used as an inoculum, refuse may be excavated with construction equipment. Excavated refuse typically requires shredding to obtain a homogeneous material for filling reactors. Exposure of the refuse to O_2 during excavation and shredding will not destroy the methanogenic nature of the refuse. However, even old refuse may not be well decomposed and may require additional decomposition in the laboratory prior to use.

Representative samples are important when they are to be used for microbiological and chemical analyses. Given the heterogeneous nature of landfills, this is best approached by obtaining multiple samples over a preselected grid. Samples may be obtained either by using construction equipment or, preferably, by using a bucket auger that has the capability to obtain deeper samples (162). After excavation, samples require shredding. In either case, large samples should be excavated and then reduced by the quartering techniques described above to obtain a representative subsample. Westlake (179) attempted to study microbial metabolism in samples excavated from landfill test cells in the United Kingdom and concluded that 25-g samples were inadequate due to sample heterogeneity.

When excavated samples are to be used for microbiological analyses, there is often concern with exposure to air. In early work to measure the CH_4 production status of excavated refuse, elaborate precautions were taken to minimize O_2 exposure, including covering boreholes with tents and flooding them with an inert gas. It has since been shown that exposure of refuse samples in an active state of CH_4 production to air for 2 to 4 h does not decrease the time required for these samples to resume CH_4 production once replaced in an anaerobic system (26, 27; Barlaz, unpublished). While practical steps to minimize O_2 exposure are warranted, extraordinary steps are not. For example, samples could be placed in bags and sparged with an inert gas.

Once a refuse sample is collected, it will require further processing specific to the required biological or chemical assay. Applicable techniques are described in the following two sections.

Processing Refuse Samples for Microbiological Analyses

The first step in many microbiological measurements is formation of a liquid extract from the excavated solids for use as an inoculum. Techniques for formation of a liquid extract were evaluated previously (15). The basic technique was to blend refuse in anaerobic phosphate buffer (23.7 mM, pH 7.2) for about 1 min in a 4-liter blender. The use of a blender was appropriate because all work was conducted with shredded refuse. By using a blender, it was possible to blend approximately 100 g (dry weight) of refuse. The buffer was boiled under N_2, but a reducing agent (cysteine hydrochloride) was not added because some oxidation occurs during preparation of the liquid inoculum and the oxidized form of cysteine may be toxic. Refuse near the blender blade became quite hot, but the bulk sample temperature did not increase in the 1-min blending period. The blended material was then squeezed by hand (with gloves), and the exudate was used as the inoculum. Other, more elaborate treatments that might increase the efficiency of cell extraction, such as prechilling the refuse at 4°C and multiple blendings, did not increase the most probable number (MPN) count of cellulolytic bacteria above the population measured by blending followed by hand-squeezing.

Later work was conducted with ~10-kg samples (140). Solid samples were placed in a large (113-liter) plastic garbage can that had been wiped with ethanol and purged with sterile argon. A measured volume of filter-sterilized (0.45-μm pore size) anaerobic phosphate buffer (23.7 mM, pH 7.2) was added to a sample to form a slurry. The specific volume added varied with the initial moisture content of the sample. Twenty-liter batches of buffer were prepared by pumping the buffer through a 0.45-μm filter and then boiling it under N_2. The sample was then stirred by hand (with arm-length gloves). Next, four samples were removed by using a 1-liter beaker. The contents of a beaker were poured through a second person's hands (covered with gloves) into a sterile, 4-liter, argon-purged Erlenmeyer flask. Finally, solids caught in the person's hands were squeezed to release excess liquid into the flask. The liquid in this flask served as the inoculum for MPN tests. To evaluate the inoculum formation technique for the 10-kg sample, measured volumes of anaerobically digested sewage sludge were used to spike measured amounts of refuse (140). The near-quantitative recovery of cellulolytic and methanogenic bacteria in the spike experiment suggested that the inoculum formation procedure was repeatable and did not cause a reduction in viable cells.

The efficiency of a stomacher for extraction of microbial cells from refuse has also been evaluated (121). Duplicate 20-g samples were homogenized in 180 ml of a phosphate-buffered diluent for 1 min. This extraction procedure was repeated six times, and its efficiency was judged on the basis of the number of aerobic bacteria that would grow on tryptone soy agar at 35°C. In first two extractions, 93.4% of the extractable cells were extracted, and this treatment was recommended for future work.

Refuse extractions also are required to form a liquid inoculum for enzyme assays. Typically, this has involved extraction of a 10-g sample in 30 ml of 0.2% Triton X-100 containing 0.75 M $MgSO_4$ (104, 130). After extraction, the samples are centrifuged (10,000 × g), and the supernatant is used in the enzyme assay. Tris-HCl buffer (50 mM) also has been used to study the attachment and release of cellulolytic bacteria to cellulosic substrates such as newspaper, magazine paper, and cardboard (79).

In summary, researchers have been consistent in their efforts to minimize exposure to O_2 during inoculum formation. Two studies have shown that more rigorous extraction does not lead to increased cell counts (15, 121). Thus, sample mixing or blending of a solid sample in an anaerobic buffer is likely sufficient. As in other environmental samples that include solids, there is no measure of the extent to which attached microorganisms are separated during any inoculum formation technique. It is generally assumed that the fraction of microbes extracted is repeatable over a range of samples within the same study. Our emphasis has been to extract as large a mass of solids as possible.

Processing Samples for Chemical Analyses

Samples also may be excavated for analysis of the chemical composition of both the aqueous and the solid phases. In all cases, a representative sample should be obtained by repetitive quartering of the total excavated sample. Analysis of the aqueous phase is typically accomplished by adding sufficient deionized water to form a slurry. The pH should be measured in the field to prevent shifts due to atmospheric CO_2. The slurry may then be filtered and preserved in the laboratory. When analysis of the solid phase is to be performed, the shredded subsample should be dried and then ground in a Wiley mill. Particle size reduction is best accomplished by first performing a coarse grind (3 to 5 mm) and then reducing the sample mass by use of a sample splitter (riffler). Next, the reduced mass should be ground to 1 mm. This finely ground refuse should then be redried. The finished product may be used for analysis of cellulose, hemicellulose, protein, lignin, volatile solids, and, perhaps, other organics. The significance of these compounds and analytical techniques are discussed below.

Microbiological Measurements

Techniques that may be used to characterize the microbiology or microbial activity of a refuse sample are discussed in this section and summarized in Table 2.

Enumeration of Microorganisms on Refuse and Characterization of Refuse Metabolic Capability

Techniques for the enumeration of landfill bacteria include acridine orange direct counts (AODC) (129), MPN counts (14, 140), agar plate counts (121), and use of roll tubes (129, 180). Adaptation of the AODC procedure to refuse has been presented elsewhere (129). Briefly, 0.5 to 1 g of refuse is fixed in formaldehyde. Samples are then diluted to a known volume in sodium pyrophosphate and incubated for 5 to 10 min. Next, the samples are cooled to 0°C and alternately sonicated and cooled. The samples are then stained and filtered, and bacteria are counted.

Media for the MPN enumeration of the total anaerobic population and the subpopulations of cellulolytic, hemicellulolytic, butyrate-catabolizing acetogenic, and acetoclastic and hydrogeneotrophic methanogenic bacteria are listed in Table 3, footnote *h* (14, 140). For enumeration of total anaerobes, carbon sources were representative of refuse hydrolysis products. Microbial growth on cellulose was detected by visible disappearance of ball-milled Whatman no. 1 filter paper. Although filter paper strips would be more representative of the particle size of refuse, repeatable results could not be obtained with filter paper. Another alternative substrate representative of the lignocellulosic components of MSW such as newspaper is ground wood. To the authors' knowledge, no landfill researcher has used ground wood as an enumeration substrate. Xylan from oat spelts (Sigma, St. Louis, Mo.) was used for enumeration of the hemicellulolytic bacteria. Prior to use, the xylan was soaked in distilled water for 24 h to remove the soluble and nonsettleable material. After drying, it was ground in a small Wiley mill. Growth was determined by measurement of optical density, which was compared to the optical density of uninoculated controls.

Acetogenic bacteria can be enumerated on the basis of conversion of butyrate (40 mM) to acetate and H_2 and subsequent conversion of the H_2 to CH_4 by a pure culture of *Methanobacterium formicicum* (119). The CH_4 concentrations in tubes containing butyrate must be compared to those in controls without butyrate at each dilution. Tubes in which the CH_4 concentration is significantly greater than that in the controls ($P \leq 0.01$) are considered positive. Methanogen MPN tests were performed with either 80 mM acetate or 202.6 kPa of H_2/CO_2. Tubes were considered positive if they

TABLE 3 Media used for MPN tests

Addition/liter of medium or parameter	Total anaerobe	Cellulose	Xylan	Acetogen	Acetate	H_2CO_2
PO_4 solution[a] (ml)	100	100	100	100	0.1[b]	0.2[b]
M3 solution[c] (ml)	100	100	100	100	100	100
Mineral solution[d] (ml)	10	10	10	10	10	10
Vitamin solution[e] (ml)	10	10	10	10	10	10
Resazurin (0.1%) (ML)	2	2	2	2	2	2
Volatile fatty acids[f] (ml)	10	10	10			
Yeast extract (g)	1	1	0.25		2	2
Trypticase peptones (g)	2	0.025	2		2	2
Hemin (0.01%)[g] (ml)	10	10	10			
$NaHCO_3$ (g)					0.5	0.5
Carbon source	—[h]	See text	See text	See text	See text	See text
Distilled water[i] (ml)	578	420	678	698	868	868
Boiled under:	N_2-CO_2 (80:20)	N_2-CO_2 (80:20)	N_2-CO_2 (80:20)	N_2-CO_2 (80:20)	N_2	N_2
$NaHCO_3$ (5% [wt/vol])[j,k] (ml)	70	70	70	70		
Cysteine hydrochloride (5% [wt/vol])[j] (ml)	10	10	10	10	10	10
Final pH	6.6	6.6	6.6	7.0	7.2	6.7

[a]The phosphate solution contained 16.1 g of KH_2PO_4 and 31.87 g of $Na_2HPO_4 \cdot 7H_2O$ per liter.

[b]A concentrated phosphate solution containing 15 g of KH_2PO_4 and 31.87 g of $Na_2HPO_4 \cdot 7H_2O$ per 100 ml was used to supply the phosphates. It was prepared under nitrogen and autoclaved separately. The indicated volume is aseptically added to individual tubes containing 9 ml of sterile medium.

[c]M3 solution contained the following (per liter): NH_4Cl, 10 g; NaCl, 9 g; $MgCl_2{:}6H_2O$, 2 g; and $CaCl_2{:}2H_2O$, 1 g.

[d]As described in reference 106, with the addition of 0.033 g of $Na_2WO_4 \cdot 2H_2O$.

[e]As described in reference 192.

[f]As described by Leedle and Hespell (115), with the addition of phenylacetic acid (0.0068 g) and 3-phenylpropionic acid (0.0075 g) per liter of volatile fatty acid solution.

[g]Hemin was prepared by dissolving 0.01 g in 100 ml of deionized water that contained 0.1 g of NH_4Cl and 0.1 g of NaOH.

[h]—, cellobiose, glucose, maltose, xylase, galactose, arabinose, mannose, starch, glycerol, and galacturonic acid (2.5 mM final concentration of each).

[i]After addition of distilled water, the pH was adjusted to 7.2 with NaOH, the media were boiled, the remaining ingredients were added, and 9 ml of medium was dispensed into pressure tubes and autoclaved for 20 min at 121°C.

[j]Added after boiling.

[k]Boiled under an 80:20 N_2-CO_2 gas phase.

contained >0.5% (vol/vol) CH_4. MPN assay tubes were incubated for 30 days except for the acetogen assay, in which tubes were incubated for 60 days.

Anaerobic starch- and protein-degrading microbes were enumerated on peptone yeast glucose (PYG) agar amended with soluble starch or gelatin, respectively (129). Starch-clearing zones were detected by the addition of potassium iodide, while proteolytic microbes were detected by addition of 0.05% (wt/vol) naphthol blue black in 7% (vol/vol) acetic acid. Colonies producing extracellular proteases were surrounded by clear zones.

Roll tubes have been used for the enumeration and isolation of cellulolytic bacteria from refuse (129, 180, 181). In this procedure, tubes are inspected visually or with the assistance of a microscope for a zone of clearing indicative of cellulose consumption. Researchers typically use ball-milled cellulose as the cellulose source.

There have been just a few reports of work identifying the presence of anaerobic fungi in refuse. In one study, bacteria were inhibited by the addition of streptomycin sulfate at 130 U/ml and penicillin (penicillin K) at 2,000 U/ml (15). Cycloheximide (0.05 mg/ml) was used to inhibit fungi in parallel tests. Theodorou et al. (165) used the same antibiotics and chloramphenicol when inhibition of methanogens was desired. Maule et al. (121) enumerated aerobic fungi in refuse samples using malt extract agar in the absence of antibiotics. Researchers typically make extensive reference to the rumen microbiology literature for useful techniques (189).

One research group has reported work with anaerobic protozoa from landfills (78, 79). Liquid dilutions of the solid refuse sample were made either into liquid medium or onto plates streaked with *Escherichia coli*. A number of protozoa were isolated from excavated refuse samples including two *Mastigamoeba* species, a *Heteromita* sp., a *Chilomastix* sp., a *Phreatamoeba* sp., and at least one other unidentified heterotrophic flagellate.

Habitat Simulation and Incubation Conditions

The use of culture techniques to enumerate or isolate microorganisms from environmental samples assumes that the desired populations will grow under the laboratory conditions provided. This is an imperfect assumption in any ecosystem. For example, for samples excavated from a landfill, the ratio of the total fermentative population measured in roll tubes to the AODC was between 2.2×10^{-5} and 0.016 (129).

In developing culture media, typically an attempt to simulate the habitat is made. In landfill microbiology, adjustments to the medium pH and incubation temperature can be made. However, simulation of the organic matrix in a landfill is problematic. Leachate could be used for habitat simulation. However, leachate contains degradable organic carbon. Thus, if a medium is to support the growth of a specific trophic group, then the presence of uncharacterized carbon in the leachate will interfere with the enumeration assay. A second limitation of leachate is that there is not typically a source of constant composition available. Leachate composition reflects the decomposition state of the refuse. Leachate from refuse in the anaerobic acid phase is relatively high in carboxylic acid concentrations, while leachates from well-decomposed refuse contain less acid and more humic matter. Leachate composition from a landfill also varies, as it is affected by the overlying refuse and dilution with storm water. Preincubation of leachate from well-decomposed refuse to deplete degradable carbon could be used to address these limitations. However, there are no reports to date on the use of preincubated leachate.

The incubation temperature for an enumeration assay should reflect the temperature of the ecosystem from which the inoculum is recovered. The optimum temperature for CH_4 production in the mesophilic range has been reported as 41 to 42°C (90, 134), while the temperature of refuse excavated from landfills varies widely.

Direct Measures of Microbial Activity

Techniques that have been reported for the direct measurement of microbial activity and the presence of microbes in refuse include enzyme assays, turnover of ^{14}C added carbon, measurement of the coenzymes, and scanning electron microscopy. These techniques are discussed in this section and summarized in Table 2.

Enzyme Assays

The activity of a number of hydrolytic enzymes in excavated refuse samples has been measured (130). Esterase activity was measured by the hydrolysis of fluorescein diacetate to fluorescein, the presence of which could be measured by fluorescence spectrophotometry. Protease activity was measured using azocoll, a proteinaceous substrate that releases a blue dye on hydrolysis. Amylase activity was measured by using substrates labeled with a *p*-nitrophenol moiety. In this assay, *p*-nitrophenyl α-D-maltoheptaoside is converted to *p*-nitrophenylmaltotriose (PNPG3) plus maltotetraose by α-amylase. The PNPG3 is ultimately hydrolyzed to glucose and to *p*-nitrophenol, which was measured spectrophotometrically. Others have measured lipase activity using a titrimetric measurement of fatty acids released from triglycerides (104). For a broader characterization, the API-ZYM system has been used to screen for the presence of 18 enzymes (130).

Two methods for measuring cellulase activity have been published. In work with decomposing refuse (130), cellulase activity was measured on the basis of the degradation of cellulose-azure, which releases a blue dye on hydrolysis. In work with pure cultures of cellulolytics isolated from decomposing refuse, a technique based on carboxymethyl cellulose was presented (180). Cell extracts were obtained by first growing cells in liquid medium and then harvesting them in the late logarithmic phase. After centrifugation, bacterial pellets weighing 1 to 2 g were resuspended and then subjected to sonification. The supernatant, which represented a cell-free whole-cell homogenate, was decanted and frozen until use. In addition, a cell-free supernatant fraction was produced by filter sterilization (0.2-μm pore size) of the original growth medium, followed by addition of ethanol to 70% by volume. The precipitated protein was then harvested by centrifugation, resuspended in buffer, and frozen until use. Carboxymethyl cellulase activity was measured by the release of reducing sugars from a 0.5% (wt/vol) solution of carboxymethyl cellulose. Cellobiohydrolase activity was measured by the release of *p*-nitrophenol from *p*-nitrophenylcellobioside. Total cellulase activity was measured as a decrease in turbidity of a 0.025% (wt/vol) cellulose suspension by use of a spectrophotometer at 660 nm.

There are several limitations to the use of enzyme assays. First, with the exception of pure culture work, all of the work has been done with 10-g samples, which implies some sample selection based on particle size. Presumably, larger samples could be used for the initial extraction, although this might increase the opportunity for enzymes to attach to the solid matrix provided by the sample. As always, it is important to work with multiple replicate samples to control for spatial and temporal changes in factors such as pH, refuse composition, temperature, and nutrient status. A second limitation to the use of enzyme assays is the effectiveness of the cell lysis and washing step. While 0.2% (wt/vol) Triton X-100 has been used, there has been no published work comparing its effectiveness to that of other detergents or lysis procedures. In this respect, enzyme activity data are probably best evaluated as relative measures among samples within the same study. A third limitation is specific to the cellulase assay. Jones and Grainger (104) evaluated their extraction technique by adding a commercial cellulase to both sterile and nonsterile refuse. Upon extraction, the cellulase was fully recovered from the sterile refuse but only a fixed amount of enzyme activity was measured in the nonsterile refuse, regardless of the amount of enzyme added. The authors suggested that the enzyme could have been deactivated by proteases.

The presence of cellulose as the dominant biodegradable polymer in refuse is not reflected in measurements of either cellulolytic populations or cellulase enzyme activity. Cellulase activity was present in only 2 of 28 samples from the Fresh Kills, N.Y., landfill, 3 of 8 samples from a landfill in Arizona, and 1 of 17 samples from a landfill in Florida (130). In contrast, esterase, amylase, and protease activities were present in all 28 samples excavated from the Fresh Kills site. Similarly, no cellulolytics were measured in several refuse samples (129, 130). Barlaz et al. reported cellulolytic populations to be about 3 orders of magnitude lower than methanogen populations in well-decomposed refuse (14) and found cellulolytics in very small populations on individual components of refuse (140). It has been suggested that one or more of the enzymes responsible for cellulose hydrolysis were membrane bound and not extracted (130). This is consistent with a report that cellulose hydrolysis was greater in cell homogenates than in the extracellular fraction (180).

CH_4 Production Rate and ^{14}C Turnover

Researchers have used two other techniques to assess the microbial activity of decomposing refuse: CH_4 production and turnover of ^{14}C. A measure of CH_4-producing activity may be obtained by placing a refuse sample in a reactor, sealing it, and sparging it with N_2 to displace oxygen that entered during reactor loading (84, 162). The CH_4 production rate can then be measured, and samples can be amended to study the effects of various supplements. Because CH_4 production requires the coordinated activity of several trophic groups, measurement of CH_4-producing activity provides a measure of the overall activity of the refuse sample, as opposed to the activity of a specific trophic group or enzyme.

^{14}C-labeled contaminants have been used to study the biodegradability of both polymeric materials (130, 144) and hazardous organics (151). In such studies, small amounts of ^{14}C-labeled material have been added to reactors filled with fresh MSW and a decomposed refuse inoculum or to samples of decomposing refuse. The amount of ^{14}C to be added must be estimated in consideration of the desired biodegradation detection limit, the expected biodegradability of the material, and the amount of gas production expected from decomposition of the bulk sample. Too much gas production from the bulk refuse can dilute the $^{14}CH_4$ and $^{14}CO_2$ to below detection levels if sufficient ^{14}C is not added. Advantages of adding a ^{14}C-labeled material include the following: (i) only a small amount of material is required, so its concentration may be representative of its actual concentration in a landfill; (ii) the production of $^{14}CH_4$ and $^{14}CO_2$ provides unambiguous evidence of material biodegradability; and (iii) additional techniques may be employed to evaluate ^{14}C partitioning by sorption, humification, and uptake into cell mass (151).

Mineralization of a ^{14}C-labeled compound can be measured by routing the produced gas through a system to trap $^{14}CO_2$ and $^{14}CH_4$. The system includes NaOH traps to dissolve the $^{14}CO_2$, followed by a combustion furnace to oxidize $^{14}CH_4$ to $^{14}CO_2$, and then a set of traps to dissolve the newly produced $^{14}CO_2$. Trapped radiolabeled gases are then quantified by scintillation counting. In serum bottle assays, the gas may have to be sparged from the bottle headspace with an inert gas. In reactor systems, the gas may be collected in impermeable bags and then routed through the trapping system with a syringe pump.

The authors' experience with commercially available [^{14}C]cellulose is that its specific activity is as much as 10 times greater than the level indicated by the vendor. Thus, where mass balances are involved, it is critical to measure the specific activity independently. The specific activity of commercially purchased [^{14}C]cellulose has been measured successfully by mixing [^{14}C]cellulose with microcrystalline cellulose and then ball-milling the combination for 5 days. Ball-milling dilutes the specific activity of the radiolabeled material and homogenizes the small weight of [^{14}C-]cellulose in a larger mass of unlabeled cellulose. After ball-milling, triplicate 0.2-g samples of the mixture of labeled and unlabeled cellulose were converted to glucose by hydrolysis in 72% (wt/vol) H_2SO_4 followed by a secondary hydrolysis in 3% H_2SO_4 (70, 133). By measuring the disintegrations per minute of a sample of the hydrolysate, the specific activity of the original mass of [^{14}C]cellulose that was hydrolyzed can be calculated since all of the original [^{14}C]cellulose is present as soluble ^{14}C in the hydrolysate.

The use of ^{14}C also has some limitations. First, the test material must be available or synthesized. Second, where ^{14}C is extracted from refuse for analysis of sorption or association with humic compounds, it may not be possible to differentiate the initial compound from transformation products by scintillation counting. Finally, when the biodegradation of hydrophobic contaminants is of interest, differences between freshly added ^{14}C-labeled contaminant and the aged contaminant present in the landfill must be considered, as the bioavailability of hydrophobic compounds has been reported to decrease with aging (92).

In the case of chemicals for which biodegradation pathways are well known, it may be possible to document biodegradation on the basis of the appearance of specific end products. For example, a sharp increase in chloroethane and ethylene production after 1,2-dichloroethane addition is good evidence of dichloroethane transformation (151).

Variations on the BMP assay have been applied for the measurement of trace organic biodegradability with a liquid inoculum in a serum bottle assay. For example, the biodegradability of phthalic acid esters was studied by monitoring parent compound disappearance, the accumulation of daughter products, and CH_4 production in serum bottles (71).

Coenzyme F420, Ether-Linked Lipids, Electron Microscopy, and Coenzyme M

There are a number of measures of microbial population and activity based on cellular contents. These techniques are mentioned in this section, but the reader is referred to other chapters for detailed information, as these techniques have not been applied to refuse. F420 is a coenzyme found in methanogens, other archaebacteria, and some eubacteria (185). Its fluorescence makes microscopic detection simple, and it may also be measured by high-pressure liquid chromatography (HPLC). Its use as a quantitative indicator of methanogenic activity is limited because (i) extraction of F420 from environmental samples may include some F420 extracted from nonviable cells as well as extracellular F420, (ii) the cellular concentration of F420 varies with both species and environmental conditions, (iii) it degrades on exposure to O_2, and (iv) it has been found in other archaebacteria.

Ether-linked lipids may be used for detection of total methanogen biomass, assuming they are the sole archaebacteria present. However, this technique does not provide information on the activity of the methanogen population, and it is labor-intensive (185). Scanning electron microscopy and fluorescent light microscopy have been used to study colonization on the surface of refuse. One study reported 10^4 to 10^5 methanogens/cm^2 on excavated refuse and estimated that this was equivalent to 0.5×10^6 to 7×10^6 cells/wet g (146). The ATP concentration and adenylate energy charge have been used to compare microbial activities during decomposition under aerobic and anaerobic conditions (125). While interesting, more precise information on decomposition could have been obtained by using chemical measurements as described below (see "Chemical Measurements").

A technique for quantification of methanogenic biomass in environmental samples based on the HPLC analysis of coenzyme M has been presented elsewhere (74). Coenzyme M is unique to methanogens among microbes present in anaerobic ecosystems, its analysis does not require the maintenance of anaerobic conditions, and its concentration per cell is much less variable than that of F420. Like F420, it does not distinguish between viable and nonviable cells.

In summary, microbial populations and activity in anaerobically decomposing refuse have been measured by a number of viable-cell count procedures, AODC, hydrolytic enzyme activities, CH_4 production rates, ^{14}C turnover, ATP concentration, and both electron and fluorescence microscopy. Viable-cell counts and enzyme activities are perhaps best used as a relative measure among different samples within the same study. The broadest measures, the CH_4 production rate and the turnover of ^{14}C, are the most comparable among studies and often the most easily measured.

Molecular Techniques for the Study of Landfill Microbiology

The limitations of traditional laboratory culture techniques for the study of microorganisms in the environment are known and not unique to landfill microbiology. These limitations have stimulated the development of a number of molecular techniques as described in chapter 12. Studies in which molecular techniques have been applied to refuse are described in this section. As is the case for viable-cell counts, the extraction of RNA and DNA from a solid waste sample remains a critical issue and is the emphasis of this section.

PCR-cloning of 16S rRNA genes (rDNA) has been used to characterize populations attached to cellulose in an anaerobic reactor treating landfill leachate (41). DNA was extracted from cellulose by first washing the solids four times in 1 M Tris HCl (pH 8) and discarding the liquid. Then the DNA was extracted by using a Fast DNA spin kit (Bio 101, La Jolla, Calif.). In contrast, a MoBio (Solana Beach, Calif.) extraction kit was used in a recent study to characterize methanogens in landfills (170). In a comparison of the MoBio PowerSoil and Epicentre SoilMaster DNA extraction kits in our laboratory, we found MoBio kits to yield 2.5 to 4 times more DNA mass per dry gram of refuse. Replicate extractions yielded coefficients of variation from 12.5 to 14% for MoBio, whereas Epicentre ranged from 79 to 103% (F. L. de los Reyes, unpublished results).

16S rDNA clone libraries have been developed on refuse sampled as a function of depth (44). DNA was extracted from solid samples by subjecting 0.5-g samples to repetitive freeze-thaw cycles (−80 to 60°C). This was followed by mixing the sample with silica beads, lysis buffer (100 nM NaCl, 5% sodium dodecyl sulfate), and chloroform. The mixture was then subjected to a bead beater (3,800 rpm, 1 min), heating (10 min, 60°C), centrifugation (1,100 × g, 10 min), and finally extraction in 500 ml of 10 μM Tris-HCl at pH 8 with 1 mM EDTA in a saturated phenol solution. In contrast, a benzyl chloride method was used to extract DNA in a study of microbial community changes during decomposition of simulated food waste (91). Extracts from the benzyl chloride method were reported to result in more bands on a double-gradient denaturing gradient gel electrophoresis (DGGE) gel relative to extracts from a freeze-thaw technique. Others have characterized methanogenic archaea by cloning the 16S RNA gene in landfill leachate, though no comparisons of leachate and solids have been reported (96). Preliminary studies in our laboratory have shown similar but less intense DGGE bands in leachate relative to solids (B. F. Staley, unpublished data).

Fluorescence in situ hybridization (FISH) has been applied to landfill samples (45). Ethidium bromide staining was used in place of DAPI (4′,6-diamidino-2-phenylindole) since it resulted in improved differentiation of cells from soil and solid waste. In addition, Cy-5 probe labels were used because their excitation wavelength maximum (650 nm) was in a range where autofluorescence from waste samples was reduced. This technique is consistent with observations that the application of FISH to waste samples can be optimized by (i) collection of samples on a membrane filter after hybridization, (ii) treatment of a sample with 99% ethanol-formalin before hybridization to increase the permeability of the oligonucleotide probes to the cells, and (iii) use of fluorescein isothiocyanate as the probe label with a long pass emission filter (>515 nm) which allows differentiation between cells and waste autofluorescence (91).

Raskin et al. (141) monitored the startup phase of an anaerobic digester using oligonucleotide probes complementary to conserved tracts of the 16S rRNAs of phylogenetically defined groups of methanogens. Much of the molecular work performed on methanogens still utilizes these group- and species-specific probes. Recently, these probes have been updated for use in real-time PCR TaqMan assays (194). In summary, while the use of molecular techniques holds much promise for advancing our understanding of landfill microbial ecology, results to date are preliminary.

Chemical Measurements

Cellulose, hemicellulose, and lignin analyses can be used to infer information about the microbiological state of refuse. As refuse decomposes, the cellulose and hemicellulose concentrations decrease and the lignin concentration increases. Cellulose, hemicellulose, and lignin concentrations have been used to assess the degree of decomposition in a landfill (9, 34, 86, 175). In the 1980s, fresh refuse was reported to have a cellulose-to-lignin ratio of approximately 3.5 to 4, and this ratio was shown to decrease with age in excavated refuse samples (34). This ratio is particularly useful because it is not confounded by the presence of cover soil that may be present in excavated refuse. More recently, measurement of the hemicellulose concentration in refuse has become standard, and it is now more appropriate to work with the ratio of cellulose plus hemicellulose to lignin (73, 175). While this ratio cannot provide an absolute assessment of microbiological activity, it does provide information on the relative state of decomposition of refuse samples.

Techniques for the measurement of cellulose, hemicellulose, and lignin have been described elsewhere (61, 70). The basic procedure is hydrolysis of a solid sample in 72% (wt/vol) H_2SO_4 followed by a secondary hydrolysis in 3% (wt/vol) H_2SO_4. The hydrolyses convert cellulose and hemicellulose to their respective monomers. Typically, HPLC is used to quantify the concentrations of glucose, xylose, mannose, arabinose, and galactose in the acid hydrolysate. The glucose originates from cellulose, and the other sugars come from hemicellulose. The solids remaining after acid hydrolysis are combusted at 550°C, and lignin is calculated from the weight loss of the dried solids after combustion. Unfortunately, any organic that is insoluble in 72% sulfuric acid will also be counted as lignin. Although lignin is considered to be recalcitrant, its chemical form may change during decomposition, resulting in apparent changes in its concentration (99). Recently, there has been evidence for modification of lignin based on thermochemolysis gas chromatography (GC)-mass spectrometry data (46).

Protein uptake is difficult to quantify, since protein that is converted to cell mass during decomposition will still be measured as protein. Given this and the relatively low concentration of protein in refuse (Table 1), protein has not been routinely used as a measure of decomposition.

Volatile solids are measured on the basis of weight loss after combustion at 550°C. Volatile solids represent all organic matter in refuse, including degradable forms (cellulose, hemicellulose) and recalcitrant forms (lignin, plastics, rubber). Given the wide range of compounds reported as volatile solids, their measure can provide only a rough estimate of the state of refuse decomposition.

Finally, cellulose, hemicellulose, and lignin concentrations have been used in laboratory reactors to document the relationship between solid loss and CH_4 production and to define the extent of refuse decomposition (12, 73). This approach uses a mass balance based on CH_4 potential. The CH_4 potential of each component is calculated by using equation 1 (131):

$$C_nH_aO_bN_c + [n - (a/4) - (b/2) + 3(c/4)]H_2O \rightarrow [(n/2) - (a/8) + (b/4) + 3(c/8)]CO_2 + [(n/2) + (a/8) - (b/4) - 3(c/8)]\,CH_4 + cNH_3 \quad (1)$$

Based on equation 1, 415 and 424 liters of CH_4 at standard temperature and pressure would be expected for every kg of cellulose and hemicellulose degraded, respectively.

The soluble organic fraction of refuse may be used as a relative measure of organic accumulation. Carboxylic acids accumulate during the anaerobic acid phase and decrease with the onset of CH_4 production (Fig. 2). Absolute concentrations of total organic carbon, chemical oxygen demand, and carboxylic acids are influenced by a number of system-specific parameters. Thus, concentrations representative of each phase of decomposition cannot be provided, and these parameters are best used when a time series is available to assess trends in decomposition. Even with the near depletion of carboxylic acids in the final phase of decomposition, there are other dissolved organics that are slowly degradable or recalcitrant. Chemical oxygen demand is most easily measured by using kits containing preweighed reagents (Hach Co., Loveland, Colo.). Carboxylic acids can be measured by use of either a GC equipped with a flame ionization detector or an HPLC equipped with a conductivity detector. The American Public Health Association (5) describes biological oxygen demand measurement.

Though an indirect measure of microbial activity, chemical parameters provide a measure of the results of biodegradative activity. While the issue of representative sampling is equally complex for chemical and microbiological measurements, chemical measurements circumvent the issue of cell or enzyme extraction efficiency.

MICROBIOLOGICAL STUDIES OF CH_4 OXIDATION IN LANDFILL COVERS

Landfill CH_4 oxidation has been studied to improve estimates of annual whole-landfill CH_4 uptake, to evaluate factors that influence oxidation rates, and to study the microbial ecology of the predominant trophic groups active in landfill caps. This information has been used to design better landfill covers. For example, the finding that mixes with high organic content, moisture, and porosity support CH_4 oxidation has led to active investigation of more permeable cover designs and biofilters. A porous landfill cover designed specifically to support a high level of microbial CH_4 oxidation has come to be called a biocover, while biofiltration involves placement of porous organic or mineral particles in a separate containment system to which collected landfill gas is routed. Because some of the methanotrophs that mediate CH_4 oxidation can also cometabolize more complex organic compounds, the attenuation and degradation of various volatile and hazardous air pollutants emanating from buried waste in landfill covers have also been examined (153). A description of systems that can be used to investigate the nature and extent of CH_4 oxidation in landfill soil and biocovers is presented in this section, and many are also applicable to hazardous air pollutant degradation and CH_4-oxidizing biofilters. A review of sample collection and handling is included, along with a discussion of the tests that have been applied to document biotic CH_4 oxidation.

Whole-landfill CH_4 Oxidation and Emissions

Some of the first reports of landfill cover CH_4 oxidation rates came from field samples collected over 200 acres of cover and tested in laboratory incubations (182). A maximum oxidation capacity of 45 g/m^2-day was observed, and this rate was combined with a published estimate of the United States landfill CH_4 production rate to show that as much as 50% of the CH_4 generated in the anoxic zone of a landfill could be oxidized. However, given that the gas production rate at the landfill they studied was not measured, this prediction of whole-landfill oxidation capacity must be considered an order of magnitude estimate.

Measures of whole-landfill CH_4 oxidation present a significant technical challenge because of the large spatial and temporal heterogeneity of the target site. One estimate was determined by coupling laboratory measures of CH_4 oxidation rates in cover soil from a New Hampshire landfill with field measures of CH_4 emissions from the landfill cover (57). The need to quantify emissions to assess oxidation has led researchers to employ a variety of emission measurement techniques, including (i) use of static or dynamic flux chambers, (ii) vertical gas concentration profiles through the cover combined with soil property data or estimates (30), and (iii) remote sensing methods where one of a variety of sensitive gas detectors is used to measure CH_4 in the atmosphere downwind from the landfill either alone or in comparison to an inert tracer gas (58, 80, 152). Emissions are usually reported as flux readings, which describe the amount of gas passing through a cross-sectional area per unit time.

Flux chambers are designed to capture emitted gas, and samples are extracted every 5 to 10 min to assess the rate of gas concentration change. The chambers are typically open-bottomed containments constructed of plastic, aluminum, or stainless steel that typically cover <1 m^2 of surface area and enclose about 20 liters of gas. The temporal gas concentrations can be translated into a measure of the gas flux after accounting for the chamber volume and enclosed surface area. Dynamic chambers in which air is pumped through the chamber during the measuring interval have also been described. The gas flux from the soil is calculated on the basis of the cross-section of soil enclosed, the gas concentration difference between the inlet and outlet air, and the volumetric rate of air-flow through the chamber.

Chamber designs for measuring gas emissions have been well reviewed (148), and there are critical factors that must be addressed to obtain reliable data. These include the integrity of the seal to the soil surface, use of a small fan or mixing device so that samples are representative of the chamber contents, use of a septum or other device to prevent contamination during sample extraction, and minimization of sampling time so that gas buildup or temperature effects do not confound the results. For dynamic chambers, pretrial tests are needed to determine the optimum airflow rate and the equilibration time.

For many years, flux chambers were the principal method employed to measure landfill emissions. It is clear that when properly used, they can provide valuable and reliable data about the immediate vicinity of a test site or for comparison of small-scale experimental cap treatments. However, multiple flux chamber measures on a single landfill show extreme heterogeneity over time and space (due to subsurface phenomena, soil cracks, vegetation). There are some statistical techniques that can be applied to improve estimates of whole-landfill gas fluxes from chamber measures, but they are not always appropriate and must be used with caution (159). Generally, extrapolation of flux chamber data to estimate whole-landfill emissions is not recommended except perhaps to show very general trends.

A newer method for measuring CH_4 oxidation occurring in the field that does not require whole-landfill emission data is based on isotope fractionation analyses (16, 22, 32, 36, 43, 118). Because methanotrophs preferentially use the lighter isotopes of CH_4, the remaining unoxidized CH_4 becomes enriched with ^{13}C and deuterium. An isotope ratio mass spectrometer is used to assay the CH_4 isotopes in both the gas emitted through the cover and gas sampled from wastes below the cover. Any isotopic ratio shift detected between the samples is proportional to the amount of CH_4 oxidized.

The method relies heavily on assigning the correct value to the fractionation factor, α (the ratio of the oxidation constants of light and heavy CH_4 isotopes), and the nature of the transport model used to analyze the system. The calculation and significance of α have been addressed previously (36, 43, 67), and α shows some temperature dependence (36). Correct calculation requires that an assumption be made about whether CH_4 travels through the landfill cap by advection, diffusion, or both. If advection is assumed to be the sole mechanism, but diffusion is indeed contributing, estimates of oxidation will be conservative. De Visscher et al. (67) offer field data and a good discussion of the several factors that influence gas transport in landfill caps, including atmospheric pressure, local pressure gradients, and gas collection systems. Use of isotope ratios has revealed CH_4 oxidation rates ranging from 0 to 94% (43).

Isotope-based methods do not circumvent the problem of landfill heterogeneity if samples are withdrawn from multiple randomly distributed sites or from flux chambers used for emission data. One group analyzed methane in landfill gas vents and in gas plumes downwind of the landfill (43). The vent gases represent the gas as produced in the degrading waste, while the landfill gas plumes include CH_4 that escaped oxidation as it migrated through the cap. The CH_4 detected in the plume presumably represents the combined effects of emissions from both sealed and cracked locations on the cap, so that heterogeneity effects are smoothed out.

Isotope fractionation analyses have also been employed to assay CH_4 oxidation in vertical profiles of a landfill. The data were used to model scenarios for how a problematic landfill gas plume migrated and also to formulate hypotheses about varied oxidation rates within the plume (187). In regions where biotic oxidation activity was high, $^{13}CH_4$:$^{12}CH_4$ and $^{12}CO_2$:$^{13}CO_2$ ratios increased as $^{12}CH_4$ was preferentially consumed and converted to $^{12}CO_2$. Field concentration and flux data were also combined (138) to fit data to a model that was then used to deduce factors that were influencing short-term emissions at a Danish landfill. The model results indicated that the rate of change of atmospheric pressure over time, rather than the magnitude of the pressure, was controlling emissions from the landfill.

Laboratory Incubations

Laboratory trials can offer the controlled conditions required when a study question centers on (i) documenting microbially mediated CH_4 oxidation; (ii) analyzing oxidation kinetics; (iii) determining the effects of factors such as temperature, moisture, nutrients, or contaminants on CH_4 oxidation; or (iv) investigating the microbial ecology of methanotrophs in the landfill cover. Both batch incubations and continuous-flow reactors carefully designed to simulate the landfill cover have been used.

Laboratory systems can control for variables such as cover depth, soil type, soil pH, moisture content, nutrient and organic content, bulk density, and vegetation. Batch incubations are typically used for short-term trials, although there are examples where such systems were used for up to 1,000 h (109). Results from short-term trials should be used with caution because they may not always generalize to long-term or steady-state conditions. For example, long-term incubations show that CH_4 oxidizers can secrete and accumulate exopolymer over time, which tends to reduce CH_4 oxidation efficiency (94, 98, 185). Results from short-term incubations that do not include these effects may be misleading. On the other hand, the dynamics of landfill environments is so complex that caution must be applied when transferring laboratory findings to the field. Atmospheric pressure changes, wind, precipitation and soil flushing, wetting and drying effects, complex food webs, waste and soil consolidation, and cap surface heterogeneities all exert effects in the field that are difficult to simulate in the laboratory.

Studies of CH_4 oxidation activity and kinetics, or the influences of amendments or environmental contaminants on methanotrophs, are conducted in batch tests. The scale of these systems can range from a few grams of soil in a 20-ml vial up to a large soil core incubated in a 2-liter jar. Critical points are that the vessels remain gastight, that sampling septa or valves maintain their integrity throughout the trial duration, that sufficient headspace CH_4 and O_2 concentrations are maintained, and that gas diffusion into the soil is not a limiting factor.

Moisture content can be adjusted with deionized water before testing commences, or samples can be tested as slurries with shaking (38). If the sample is high in organic content, such as compost samples, care must be taken to adjust moisture content if good replication is to be achieved. Compost has such a high moisture-holding capacity that additions of small quantities of water to establish a calculated moisture content can be rapidly absorbed in one region and will not be uniformly mixed throughout the sample. This will lead to uneven and unreplicable reaction levels in the sample.

To ensure acclimation of the microbes to the test gas concentrations, soil can be incubated for 12 to 24 h at the initial headspace concentration and then subjected to reequilibration with air and finally reintroduction of CH_4 (107). Sufficient time should be allowed for gases added to the headspace to mix before samples are withdrawn, but mixing has been shown to occur rapidly (182). Samples can be mechanically rotated during incubation to minimize diffusion limitations, and one study used samples incubated in field depressions slightly larger than the sample jars to better simulate field conditions (28, 29, 150). Sampling can persist for several minutes or up to several weeks, depending on the rate of CH_4 uptake.

For V_{max} and K_m measurements, a soil sample is typically tested for CH_4 uptake at one CH_4 concentration, the headspace is reequilibrated with air, and then the same sample is retested at a second CH_4 concentration (29, 57, 65, 107, 182). These repeated measures on the same soil sample have been used to calculate V_{max} and K_m. There is wide variation in the kinetic constants reported for landfill soil. In two separate whole-landfill studies, V_{max} values ranged from 40 to 2,594 nmol of CH_4/(h-g of dry soil) and K_m values ranged from 195 to 5,847 ppmv (29, 57). Therefore, pretests may be required to determine appropriate incubation periods, and incubation times may need to be different for samples from the same test site.

When Czepiel et al. (57) used batch experiments to measure oxidation rates, soil was sampled from 5-cm depth intervals at multiple sampling sites on a landfill cover, and it was determined that maximum CH_4 oxidation was occurring 5 to 10 cm below the cover surface. Kinetic parameters of soil sampled from this depth were measured in the laboratory (at 21°C and normalized to dry soil sample mass), and the maximum oxidation rate (V_{max}) was correlated with the soil CH_4-mixing ratio at each sampling site. This relationship was used, along with adjustments for temperature and moisture content, to estimate whole-landfill CH_4 oxidation rates.

For long-term studies, continuous-flow laboratory simulations of landfill cover soil are often used to more closely approximate field compaction conditions. A typical configuration consists of a soil core formed in a gastight cylindrical reactor made of PVC or acrylic cylinders. Depending on the nature of the study, attention to adsorption of CH_4 to plastics may be required, and product vendors should be consulted.

Reactor sizes vary. A typical vertical cylinder design (15-cm diameter and 30 to 60 cm long) is shown in Fig. 3, but other configurations have been used, including a glove box (188), to create continuous-flow incubation systems. In vertical columns, the soil core may be supported by a layer of sterile gravel to disperse inlet gases before they enter the soil. The reactor is plumbed with an inlet port for synthetic landfill gas at the bottom, an inlet port for air near the top, and an outlet port at the top to exhaust the gas mix to the atmosphere (65, 95, 107, 120). Gastight sample ports can be placed along the vertical axis of the reactor to obtain soil gas profiles over the column depth.

Inlet gas flows are best regulated by mass flow controllers if the data are to be sufficiently accurate for mass balance calculations (95). The air may be prehumidified by passing it through a water-filled washing bottle en route to the reactor (65). Exit flows for mass balance analyses require sensitive and accurate measurement. Although bubble flow meters can be used, there are others available that offer continuous readout and good reproducibility (e.g., from J&W Scientific, Folsom, Calif.). If the experimental treatment requires addition of a chemical amendment, it is usually mixed into the soil immediately before the column is filled. Dosages for nutrients such as nitrogen and phosphorus amendments have been calculated on the basis of agricultural feed rates. For example, Kightley et al. (107) used a nitrogen dose of 100 kg of N/ha, converted it to moles of N per gram of soil, and assumed both that the application remained in the top 10 cm of soil and that the soil had a bulk density of 1 g/cm^3. For sewage sludge amendment, a sludge cake of approximately 20% solids was dried at 35°C, ground in a blender, and added at a rate of 2.5 g/kg of soil. Others have dosed nutrients at rates equal to about half the agricultural dosage to minimize potential osmotic effects and to account for the absence of rain-leaching effects in the laboratory (95).

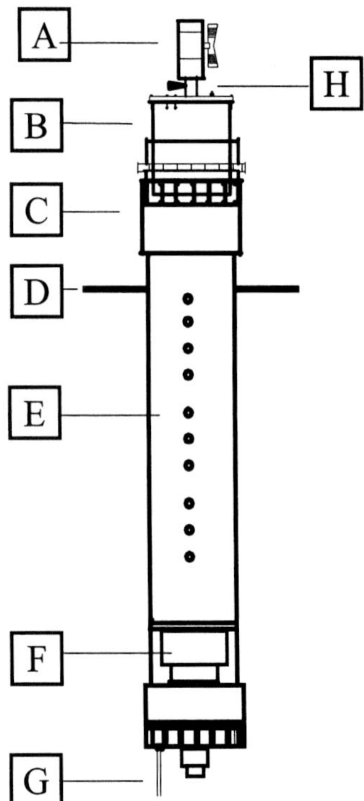

FIGURE 3 Soil column for methane oxidation studies. A reactor made of 15-cm-diameter PVC and used for landfill cover soil simulation with modifications to accommodate vegetation is shown. A, Ball valve to vent moisture; B, clear acrylic tubing; C, threaded cap; D, air inlet; E, sampling ports; F, drain for gravel support; G, synthetic landfill gas inlet; and H, gas exhaust outlet. (Reprinted from reference 95 with permission.)

Vertical reactors are filled by placing prepared soil into a column to a predetermined height. Although most protocols report a simple transfer of soil into the column, one study employed a Proctor compaction hammer (ASTM D 698) to create a density closer to field conditions (95). In this technique, a solid metal cylinder that weighs 2.5 kg is dropped from 5 to 30 cm 10 times on each of six soil layers added to the reactor. The blows are distributed uniformly across each cross section. The height of fall required to achieve a specific hydraulic conductivity can be determined in preliminary trials.

When vegetation (fescue grass) was included in an experimental system, reactors included a clear acrylic cylinder and cover at the top to allow light penetration (95). A gastight ball valve was mounted in the reactor lid to accommodate periodic temperature and humidity monitoring. Two precalibrated gypsum soil moisture blocks were buried in the soil as each column was filled. The blocks were connected to a soil moisture meter that had been previously calibrated to read the moisture content of each block. A bank of high-output fluorescent fixtures (Chroma 50; General Electric, Portland, Ohio) was mounted above the reactors to provide approximately 100 μmol of photosynthetic photon flux density per m^2-s. To prepare vegetation for use in the reactor system, grass was planted at uniform seed density and grown outdoors for 3 months before use. Plugs with 5 cm of root zone were cut from the turf, the grass blades were trimmed to 8 cm, and each soil plug was firmly tamped down to form the sixth horizon of a 30-cm vegetated soil column. Stralis-Pavese et al. (160) also describe the use of lysimeters to simulate plant cover on a landfill cap, but details about the experimental system used are limited.

Because measurement of gas volumes is crucial to obtaining estimates of CH_4 oxidation, careful design of reactors and monitoring vigilance are essential to prevent leaks. In reactor systems, it is difficult to simulate periodic additions of nutrients and the leaching effects of rainfall or to judge how well results extrapolate to other baseline conditions of soil types and vegetation. Controls for laboratory simulation columns have included columns filled with sterilized soil (95), glass beads (65), or live soil fed with N_2 instead of CH_4 (107).

Sample Collection and Processing

Chapter 48 provides a thorough discussion of soil sampling, and only information specific to work with landfill cover soils is presented here. Cover soil may be collected as cores or loose soil for bench-scale oxidation studies. If random samples are required, then a random number list can be used to select sites from a gridded pattern applied to the sampling site. In cases where samples with known exposure to CH_4 are desired, regions can be identified by using CH_4 emission data from flux chambers, soil gas concentration readings (33, 102), or detection of CH_4 with a portable flammable-gas detector.

After surface vegetation is scraped away, soil is usually collected from the top 15 to 20 cm, where CH_4 oxidation tends to occur (57, 103, 182), although sampling as deep as 90 cm has been reported (38). Loose soil is sometimes screened (~6-mm grid) to remove roots and large pebbles, and De Visscher et al. (65) describe air drying, grinding, and rewetting the soil before filling laboratory reactors. The latter may be excessively harsh, since drying and rewetting have been shown to reduce methanotroph activity (126), and grinding may result in nonrepresentative bulk properties after repacking. In most cases, samples are stored at room temperature or at 1 to 4°C for no more than a few days before testing. Composites can be prepared by mixing equal volumes of soil from sampling sites.

Procedures for many physical, chemical, and microbiological tests (including nutrients, pH, total C, and total N) are provided in *Methods of Soil Analysis*, parts I to III (111, 158, 178). When soil extracts prepared in 2 M KCl are required, use of lower-strength KCl (1 M) is often acceptable for soils that do not contain 2:1 layered aluminosilicates that trap ammonia. Preliminary tests with several concentrations can be used to determine whether 2 M KCl is needed. Because of exopolymer accumulations, the soil-KCl extract may not be readily filterable by gravity, and centrifugation followed by vacuum filtration through glass fiber filters may be necessary. For soil pH, total C, and total N analyses, it is sufficient to air dry, crush, and sieve soil to produce a uniform sample for testing. Samples for CH_4 oxidation capacity should be tested as soon as possible after field collection or reactor column dismantling. If testing cannot be initiated immediately, then samples can be stored for short periods in gastight jars to which some CH_4 has been added to maintain CH_4-oxidizing activity.

Physical and Chemical Evidence for CH_4 Oxidation

Numerous indicators have been used to show evidence of CH_4 oxidation, including CH_4 depletion or CO_2 evolution, moisture content increases, temperature changes, accumulation of exopolymers, and elevated total C and total N on soil particles (94, 95, 97, 107). Exposure to $^{14}CH_4$ can be used to demonstrate oxidation and to track the fate of CH_4 into various biomass fractions.

Changes in headspace CH_4 and CO_2 concentrations are the most direct measures, and they must be coupled with appropriate controls to confirm that the transformations are biotically mediated. Autoclaved soil and acetylene-treated soil are two common controls (42, 95). A decrease in headspace gas volume during the incubation occurs due to the conversion of CH_4 and O_2 to CO_2 and $H_2O_{(l)}$. If CH_4 is the only gas monitored, samples can be withdrawn for GC measurement without pressure correction; the GC will measure the mass of CH_4 present in the sample. Whether the CH_4 constitutes 5% of the gas at 1 atm or 5.5% of the gas at 0.9 atm in the needle, its mass will be read as 5% in 1 atm, because that is the pressure at which the GC was calibrated.

However, if a needle valve is not used, and O_2 uptake and CO_2 production are monitored along with CH_4 consumption, steps must be taken to restore the vessel headspace to atmospheric pressure before a sample is withdrawn. Otherwise, the air that will automatically be drawn into an unvalved needle to restore atmospheric pressure will confound the readings for O_2 and CO_2. In such cases, the headspace pressure can be equilibrated with He after each sampling event (95). Several researchers report starting their trials with an overpressure (38, 109), and one group used real-time recordings of headspace pressure declines to directly quantify CH_4 oxidation occurring in gastight vessels (122).

Since the H_2O product from CH_4 oxidation typically condenses onto soil particles, the degree to which this elevates the moisture content can be monitored as evidence of oxidation activity. However, the magnitude of the change is a function of the initial soil moisture content and any drying effects from gases sparged into the system. In addition to moisture, accumulations of exopolymeric substances (EPS) can accompany CH_4 oxidation at high CH_4 concentrations (94, 98). Various indicators have been used to document EPS production, including alcian blue stain, tests for soil extract filterability, and glucose assays (94).

Alcian blue is a polysaccharide-specific stain that can provide qualitative detection of EPS (77, 155). Soil samples were diluted 1:100 in pH 7 phosphate buffer (0.3 g of KH_2PO_4, 0.7 g of K_2HPO_4), and a few drops of the soil particle suspension were placed on a slide and topped with a coverslip. A 1% alcian blue solution in ethanol was diluted to 0.1% with deionized water and used for staining. Several drops of stain were delivered on one edge of the coverslip, and blotting paper was used as a wick on the opposite edge of the coverslip to absorb liquid and draw the stain across the soil particles. After 3 min, the stain was rinsed several times by repeated wicking of phosphate buffer. Microscopic observation revealed regions of EPS attached to the particles or floating separately in the soil-buffer matrix (94).

For filterability measures, soil extracts were prepared in 1 M KCl (1 part soil:5 parts solution) and shaken mechanically for 1 h. Fifteen milliliters of well-mixed slurry was then quickly poured into a porcelain crucible that was lined with Whatman no. 1 filter paper. The filter paper was prewetted and vacuum filtered to remove excess water and to ensure a tight seal to the porcelain. The volume of filtrate collected from the slurry during 1 h of gravity filtration was used to reflect the relative viscosity of samples. Glucose assays were performed on the filtrate of centrifuged, vacuum-filtered slurry according to published procedures (64, 69).

Biomass assays can provide indirect evidence of CH_4 oxidation and serve to corroborate other indicators of microbial CH_4 uptake and growth. Assays for total C and N are crude but simple biomass measures. Conventional fumigation-extraction techniques for biomass carbon (172) and the arginine ammonification technique for microbial activity (1, 2) can also be applied to landfill soil (31).

Experiments to monitor $^{14}CH_4$ conversion to $^{14}CO_2$ require a source of radiolabeled CH_4 and equipment to detect the radiolabel. $^{14}CH_4$ can be produced by feeding [^{14}C]glucose or [^{14}C]acetate to a reactor containing well-decomposed methanogenic refuse or to a methanogenic consortium, and one author reports using 99 atm% labeled CH_4 from Promochem Ltd. (55). The radiolabel offers the advantage of being able to track CH_4 oxidized versus CH_4 incorporated into biomass. When a Harvey Biological Material Oxidizer was used to analyze a radiolabeled pure methanotroph culture isolated from a soil slurry derived from landfill cover soil, it was found that 7% of the ^{14}C from the $^{14}CH_4$ substrate was in the polysaccharide fraction of cell biomass, 53% was in the protein fraction, 10% was in the lipid fraction, and the remaining 30% was in other lower-molecular-weight compounds (182). $^{14}CO_2$ can be removed during these reactions by passing the gas through a base trap, or as one group described, by maintaining *Chlorella* algal cultures in the incubation chamber to remove CO_2 and maintain aerobic conditions (55).

Methane Oxidizer Enumeration, Identification, and Diversity Screening

Soil methanotrophs have been enumerated by using MPN techniques (18) and colony counts on plated cells (103). The latter required the use of silica gel plates to avoid the high background counts obtained on agar. The gel contained 5% sodium silicate solution, a phosphate buffer, and nutrients and was dispensed before autoclaving. Soil samples were prepared as serial decimal dilutions in a sterile mineral salts solution containing 0.1% Tween 80 to facilitate separation of the bacteria from the soil particles. After inoculation, the plates were placed in a chamber at 22°C with a flowthrough of 50% high-purity CH_4. Incubation was discontinued when no further increase in the number of colonies was observed.

A novel method for growing methanotroph isolates involves plating supernatant from serial dilutions of environmental soil samples on polycarbonate microporous cell culture inserts (0.4-μm pore size; Transwell, Corning Costar, Cambridge, Mass.) (164). The inserts are incubated in petri dishes containing a soil slurry prepared from the original environmental sample or another soil source. Samples are incubated at 22°C in 4 liter anaerobic jars with 50 to 70% CH_4 in air headspace. Interestingly, when methanotroph colonies from landfill soil were plated in this way, they did not grow on membranes incubated over landfill soil slurry, but they did grow successfully when incubated on an agricultural soil slurry (that had no CH_4-oxidizing activity of its own). Ultimately, isolates appearing on the membrane were purified and maintained on nitrate mineral salts medium (164).

Phospholipid fatty acid (PLFA) distinctions between type I and type II methanotrophs have also been used to detect type distributions in landfill covers (37). Type I methanotroph membranes tend to have 16:0 fatty acid, while in type II membranes up to 80% of the PLFAs can be 18:1w8. The results indicated that regions of high 18:1w8 PLFA content correlated with regions of high CH_4 oxidation potential in mineral cover soil, while in fresh sewage sludge cover, 16:1 PLFAs correlated most highly with CH_4 oxidation activity.

PLFA analyses were also used to compare methanotroph populations in clay and sand landfill covers, and additional confirmation that the 18:1w8 and 16:1 PFLA distributions detected were due to methanotrophs was obtained by $^{13}CH_4$ labeling experiments (56). Aliquots of field samples used for PLFA profile studies were incubated in the laboratory with $^{13}CH_4$, and a correspondence between the distribution of labeled PLFAs from these incubations and the unlabeled PLFA profiles was used to corroborate that 18:1w8 and 16:1 PLFAs likely reflected the presence of methanotrophs. The distributions indicated that in the more porous sand type I and II methanotrophs were mixed throughout the cap depth, while in the clay cover type I methanotrophs predominated in the surface layers while type II methanotrophs were predominant in deeper layers. The use of PLFA analysis is described in more detail in chapter 10.

Molecular Techniques for the Study of Landfill CH_4 Oxidation

Some of the first molecular probes for methylotrophs were synthesized by Tsien et al. (167), and they were used with soil enriched with CH_4 to show that type II serine path methanotrophs predominated (401994). Probe sequences specific for the 16S rDNA of type I/X and type II methanotrophs have since been published by Gulledge et al. (83), Eller et al. (75), and Wise et al. (190), and their application to landfill cover soil by both PCR and FISH techniques is generating important information about the types and locations of methanotrophs in various soil types and cover configurations (45, 170, 190). These methods have been important for identifying the high-affinity methanotrophs that have proven unculturable in the laboratory. Chen et al. (45) describe methodological strategies to minimize sample loss and autofluorescence when FISH is used on rRNA extracted from soil.

Probes for functional genes have also been applied for methanotroph identification in soils, with the gene for particulate MMO enzyme (*pmoA*) used most frequently (24, 160). Fifty *pmoA* probes were employed in a microarray to analyze bacterial communities from an Austrian landfill site, after which the array was modified to include over 60 probes that were used to detect methanotrophs in a subsequent study (24, 160). To confirm the specificity of the individual probes and the accuracy of their analyses, weighted mismatch predictions calculated with specialized software were employed. The assays identified a *Methylocaldum* sp., a type I methanotroph, in all trials conducted, including bare soil controls. However, the dominant methanotrophs in all trials with biogas were type II *Methylocystis* and type I *Methylocaldum* spp. The ability of this array to provide detailed genus-level information about the kinds and distributions of methanotrophs in landfill covers will likely lead to rich and rapid information about methanotroph ecological niches and the important microbial systems that can be used for engineered systems. Real-time PCR with (i) a forward primer that aligned with all *pmoA* sequences and (ii) group-specific reverse primers has also been used to quantitatively assign soil methanotrophs to different groups (113).

Generally, cloning and culturing results indicate that of the type I and II organisms found, the type II species are much more diverse. The use of molecular techniques to direct culturing protocols so that new isolates can be studied will greatly accelerate attempts to clarify the role of the various CH_4 oxidizer trophic groups in landfill cover systems. A more thorough discussion of the application of molecular techniques to environmental samples is presented in chapter 12.

SUMMARY

A landfill is a unique anaerobic ecosystem with an abundance of degradable organic carbon and a wide range of microbial activities. Its uniqueness is a result of (i) the long residence time of carbon, (ii) the input of fresh organic carbon over the operational life, (iii) the large changes in environmental conditions (pH, soluble organic carbon) over the decomposition period, and (iv) changes in the presence and nature of insoluble polymeric substrates during decomposition. In addition, it includes an interface of aerobic and anaerobic conditions, which further add to the system complexity. The microbiology of the landfill ecosystem has not been as thoroughly explored as that of other anaerobic systems, such as anaerobic digesters and the rumen. Thus, it is likely that there are unique microorganisms that could be characterized, given more emphasis on this ecosystem. The development of genetic markers for all of the trophic groups required for refuse methanogenesis and CH_4 oxidation will greatly enhance our ability to characterize refuse decomposition and gas emissions and allow confirmation of the current models of landfill microbial population dynamics and landfill contributions to global climate change.

A major challenge in studying landfill microbiology is system heterogeneity. Researchers have developed a number of approaches to address this issue, including small grab samples that are not representative, multiple replicate samples at full-scale landfills, and laboratory simulations with shredded refuse. The appropriate system for the study of landfill microbiology as well as the appropriate measures is unique for each set of research objectives. Alternate systems and measures are summarized in Table 2. The types of measurements used to study landfill microbiology are analogous to those used in other ecosystems and include total and viable cell counts, enzyme activity, carbon turnover, and the production of terminal end products. The relationship between CH_4 production and biodegradation of the major organic constituents of refuse is well understood, and this understanding provides an opportunity to use chemical measurements to evaluate the extent of refuse decomposition.

Questions remain regarding the rate and extent of decomposition. While older landfills have exhibited reduced rates of CH_4 production, to our knowledge, there is not a documented case of a landfill ceasing to produce measurable CH_4. Thus, our knowledge is lacking with respect to ultimate cellulose decomposition and CH_4 production, the amount of CH_4 released after oxidation in the cover soil, and the amounts of organic carbon and other nutrients that are placed in long-term terrestrial storage. Researchers are also working on the extent of CH_4 oxidation in landfill cover soils. The long-term objectives of this work are to better understand factors controlling CH_4 oxidation in landfill covers and to complete a carbon balance on a landfill.

REFERENCES

1. **Alef, K., and D. Kleiner.** 1986. Arginine ammonification, a simple method to estimate microbial activity potential in soils. *Soil Biol. Biochem.* **18:**233–235.
2. **Alef, K., and D. Kleiner.** 1987. Estimation of anaerobic microbial activities in soils by arginine ammonification and glucose-dependent carbon dioxide production. *Soil Biol. Biochem.* **19:**683–686.
3. **Alperin, M. J., and W. S. Reeburgh.** 1985. Inhibition experiments on anaerobic methane oxidation. *Appl. Environ. Microbiol.* **50:**940–945.
4. **Amaral, J. A., and R. Knowles.** 1995. Growth of methanotrophs in methane and oxygen counter gradients. *FEMS Microbiol Lett.* **126:**215–220.
5. **American Public Health Association.** 1998. *Standard Methods for the Examination of Water and Wastewater*, 20th ed. American Public Health Association, Washington, D.C.
6. **Anthony, C.** 1982. *The Biochemistry of Methylotrophs.* Academic Press, Inc., New York, N.Y.
7. **Anthony, C.** 1991. Assimilation of carbon by methylotrophs, p. 79–109. *In* I. Goldberg and J. S. Roken (ed.), *Biology of Methylotrophs.* Butterworth-Heinemann, Boston, Mass.
8. **Asenjo, J. A., and J. S. Suk.** 1986. Microbial conversion of methane into poly-beta-hydroxybutyrate (PHB): growth and intracellular product accumulation in a type II methanotroph. *J. Ferment. Technol.* **64:**271–278.
9. **Baldwin, T. D., J. Stinson, and R. K. Ham.** 1998. Decomposition of specific materials buried within sanitary landfills. *J. Environ. Eng.* **124:**1193–1202.
10. **Barlaz, M. A., W. E. Eleazer, W. S. Odle III, X. Qian, and Y.-S. Wang.** 1997. Biodegradative analysis of municipal solid waste in laboratory-scale landfills. EPA 600/R-97-071. United States Environmental Protection Agency, Research Triangle Park, N.C.
11. **Barlaz, M. A., R. K. Ham, and M. W. Milke.** 1987. Gas production parameters in sanitary landfill simulators. *Waste Manag. Res.* **5:**27–39.
12. **Barlaz, M. A., R. K. Ham, and D. M. Schaefer.** 1989. Mass balance analysis of decomposed refuse in laboratory scale lysimeters. *J. Environ. Eng. ASCE* **115:**1088–1102.
13. **Barlaz, M. A., R. K. Ham, and D. M. Schaefer.** 1990. Methane production from municipal refuse: a review of enhancement techniques and microbial dynamics. *Crit. Rev. Environ. Control* **19:**557–584.
14. **Barlaz, M. A., D. M. Schaefer, and R. K. Ham.** 1989. Bacterial population development and chemical characteristics of refuse decomposition in a simulated sanitary landfill. *Appl. Environ. Microbiol.* **55:**55–65.
15. **Barlaz, M. A., D. M. Schaefer, and R. K. Ham.** 1989. Effects of pre-chilling and sequential washing on the enumeration of microorganisms from refuse. *Appl. Environ. Microbiol.* **55:**50–54.
16. **Barlaz, M. A., R. Green, J. P. Chanton, C. D. Goldsmith, and G. R. Hater.** 2004. Evaluation of a biologically active cover for mitigation of landfill gas emissions. *Environ. Sci. Technol.* **38:**4891–4899.
17. **Bedard, C., and R. Knowles.** 1989. Physiology, biochemistry and specific inhibitors of CH_4, NH_4^+ and CO oxidation by methylotrophs and nitrifiers. *Microbiol. Rev.* **53:** 58–84.
18. **Bender, M., and R. Conrad.** 1992. Kinetics of CH_4 oxidation in oxic soils exposed to ambient air or high CH_4 mixing ratios. *FEMS Microbiol. Ecol.* **101:**261–270.
19. **Bender, M., and R. Conrad.** 1994. Methane oxidation activity in various soils and freshwater sediments: occurrence, characteristics, vertical profiles, and distribution on grain size fractions. *J. Geophys. Res.* **99:**16531–16540.
20. **Bender, M., and R. Conrad.** 1995. Effect of CH_4 concentrations and soil conditions on the induction of CH_4 oxidation activity. *Soil Biol. Biochem.* **27:**1517–1527.
21. **Benson, C. H., M. A. Barlaz, D. T. Lane, and J. M. Rawe.** 2007. Practice review of five bioreactor/recirculation landfills. *Waste Manag.* **27:**13–29.
22. **Bergamaschi, P., and G. Harris.** 1995. Measurements of stable isotope ratios ($^{13}CH_4/^{12}CH_4$; $^{12}CH_3D/^{12}CH_4$) in landfill methane using a tunable diode laser. *Glob. Biogeochem. Cycles* **9:**439–447.
23. **Bodelier, P. L. E., and H. J. Laanbroek.** 2004. Nitrogen as a regulatory factor of methane oxidation in soils and sediments. *FEMS Microbiol. Ecol.* **47:**265–277.
24. **Bodrossy, L., N. Stralis-Parvese, J. C. Murrell, S. Radajewski, A. Wellharter, and A. Sessitsch.** 2003. Development and validation of a diagnostic microbial microarray for methanotrophs. *Environ. Microbiol.* **5:**566–582.
25. **Boeckx, P., and O. V. Van Cleemput.** 1996. Methane oxidation in a neutral landfill cover soil: influence of moisture content, temperature, and nitrogen-turnover. *J. Environ. Qual.* **25:**178–183.
26. **Bogner, J. E.** 1990. Controlled study of landfill biodegradation rates using modified BMP assays. *Waste Manag. Res.* **8:**329–352.
27. **Bogner, J. E.** 1992. Anaerobic burial of refuse in landfills: increased atmospheric methane and implications for increased carbon storage. *Ecol. Bull.* **42:**98–108.
28. **Bogner, J. E., R. M. Miller, and K. Spokas.** 1995. Measurement of microbial biomass and activity in landfill soils. *Waste Manag. Res.* **13:**137–147.
29. **Bogner, J. E., K. A. Spokas, and E. A. Burton.** 1997. Kinetics of methane oxidation in a landfill cover soil: tem-

poral variations, a whole-landfill oxidation experiment, and modeling of net CH_4 emissions. *Environ. Sci. Technol.* **31:**2504–2514.

30. **Bogner, J., K. Spokas, and J. Jolas.** 1993. Comparison of measured and calculated CH_4 emissions. *In Proceedings of the Fourth International Landfill Symposium, Sardinia 93.* Environmental Sanitary Engineering Centre, Cagliari, Italy.
31. **Bogner, J., K. Spokas, E. Burton, R. Sweeney, and V. Corona.** 1995. Landfills as atmospheric methane sources and sinks. *Chemosphere* **31:**4119–4130.
31. **Bogner, J. E., R. E. Sweeney, D. Coleman, R. Huitric, and G. T. Ririe.** 1996. Using isotopic and molecular data to model landfill gas processes. *Waste Manag. Res.* **14:**367–376.
33. **Bogner, J. E., M. Vogt, and R. Piorkowski.** 1989. Landfill gas generation and migration: review of current research. II. *Proceedings, Anaerobic Digestion Review Meeting, Jan. 25–26, 1989, Golden Colorado, Solar Energy Research Institute.*
34. **Bookter, T. J., and R. K. Ham.** 1982. Stabilization of solid waste in landfills. *J. Environ. Eng.* ASCE **108:**1089–1100.
35. **Borjesson, G.** 2001. Inhibition of CH_4 oxidation by volatile sulfur compounds (CH_3SH and CS_2) in landfill cover soils. *Waste Manag. Res.* **19:**314–319.
36. **Borjesson, G., J. Chanton, and B. H. Svensson.** 2001. Methane oxidation in two Swedish landfill covers measured with carbon-13 to carbon-12 isotope ratios. *J. Environ. Qual.* **30:**369–376.
37. **Borjesson, G., I. Sundh, A. Tunlid, and B. H. Svensson.** 1998. Methane oxidation in landfill cover soils as revealed by potential oxidation measurements and phospholipid fatty acid analyses. *Soil Biol. Biochem.* **30:**1423–1433.
38. **Borjesson, G., and B. H. Svensson.** 1997. Seasonal and diurnal methane emissions from a landfill and their regulation by methane oxidation. *Waste Manag. Res.* **15:**33–54.
39. **Brock, T. D., M. T. Madigan, J. M. Martinko, and J. Parker.** 1994. *Biology of Microorganisms*, 7th ed. Prentice-Hall, Englewood Cliffs, N.J.
40. **Brusseau, G. A., E. S. Bulygina, and R. S. Hanson.** 1994. Phylogenetic analysis and development of probes for differentiating methylotrophic bacteria. *Appl. Environ. Microbiol.* **60:**626–636.
41. **Burrell, P. C., C. O'Sullivan, H. Song, W. P. Clarke, and L. L. Blackall.** 2004. Identification, detection, and spatial resolution of *Clostridium* populations responsible for cellulose degradation in a methanogenic landfill leachate bioreactor. *Appl. Environ. Microbiol.* **70:**2414–2419.
42. **Chan, A. S. K., and T. B. Parkin.** 2000. Evaluation of potential inhibitors of methanogenesis and methane oxidation in a landfill cover soil. *Soil Biol. Biochem.* **32:**1581–1590.
43. **Chanton, J. P., C. M. Rutkowski, and B. Mosher.** 1999. Quantifying methane oxidation from landfills using stable isotope analysis of downwind plumes. *Environ. Sci. Technol.* **33:**3755–3760.
44. **Chen, A. C., H. Imachi, Y. Sekiguchi, A. Ohashi, and H. Harada.** 2003. Archaeal community compositions at different depths (up to 30 m) of a municipal solid waste landfill in Taiwan as revealed by 16S rDNA cloning analysis. *Biotechnol. Lett.* **25:**719–724.
45. **Chen, A. C., K. Ueda, Y. Sekiguchi, A. Ohashi, and H. Harada.** 2003. Molecular detection and direct enumeration of methanogenic *Archaea* and methanotrophic *Bacteria* in domestic solid waste landfill soils. *Biotechnol. Lett.* **25:**1563–1569.
46. **Chen, L., M. A. Nanny, D. R. U. Knappe, T. B. Wagner, and N. Ratasuk.** 2004. Chemical characterization and sorption capacity measurements of degraded newsprint from a landfill. *Environ. Sci. Technol.* **38:**3542–3550.
47. **Chen, Y., D. R. U. Knappe, and M. A. Barlaz.** 2004. The effect of cellulose/hemicellulose and lignin on the bioavailability of toluene sorbed to waste paper. *Environ. Sci. Technol.* **38:**3731–3736.
48. **Christensen, T. H., P. Kjeldsen, H. J. Albrechsten, G. Heron, P. H. Nielsen, P. L. Bjerg, and P. E. Holm.** 1994. Attenuation of landfill leachate pollutants in aquifers. *Crit. Rev. Environ. Sci. Technol.* **24:**119–202.
49. **Christensen, T. H., P. Kjeldsen, P. L. Borg, D. L. Jensen, J. B. Christensen, A. Baun, H. J. Albrechsten, and G. Heron.** 2001. Biogeochemistry of landfill leachate plumes. *Appl. Geochem.* **16:**659–718.
50. **Christophersen, M., L. Linderod, P. E. Jensen, and P. Kjeldsen.** 2000. Methane oxidation at low temperatures in soil exposed to landfill gas. *J. Environ. Qual.* **29:**1989–1997.
51. **Code of Federal Regulations.** 1991. 40 Parts 257 and 258. Solid waste disposal facility criteria.
52. **Colberg, P. J.** 1988. Anaerobic microbial degradation of cellulose, lignin, oligolignols, and monoaromatic lignin derivatives, p. 333–372. *In* A. J. B. Zehnder (ed.), *Biology of Anaerobic Microorganisms.* Wiley-Liss, New York, N.Y.
53. **Conrad, R., and F. Rothfuss.** 1991. Methane oxidation in the soil surface layer of a flooded rice field and the effect of ammonium. *Biol. Fertil. Soils* **12:**28–32.
54. **Coutts, D. A. P., E. Senior, and M. T. M. Balba.** 1987. Multi-stage chemostat investigations of interspecies interactions in a hexanoate-catabolising microbial association isolated from anoxic landfill. *J. Appl. Bacteriol.* **62:**251–260.
55. **Croft, B., R. Smith, M. Caine, K. Knox, J. White, I. Watson-Craik, C. Young, and J. Ellis.** 2001. The Brogborough test cells: conclusions from a 14 year field-scale landfill gas experiment, p. 3–12. *In Sardinia 2001: 8th International Waste Management and Landfill Symposium,* vol. I. Environmental Sanitary Engineering Centre, Cagliari, Italy.
56. **Crossman, Z. M., F. Abraham, and R. P. Eershed.** 2004. Stable isotope pulse-chasing and compound specific stable carbon isotope analysis of phospholipid fatty acids to assess methane oxidizing bacterial populations in landfill cover soils. *Environ. Sci. Technol.* **38:**1359–1367.
57. **Czepiel, P. M., B. Mosher, P. M. Crill, and R. C. Harriss.** 1996. Quantifying the effect of oxidation on landfill methane emissions. *J. Geophys. Res.* **101:**16721–16729.
58. **Czepiel, P. M., J. H. Shorter, B. Mosher, E. Allwine, J. B. McManus, R. C. Harriss, C. E. Kolb, and B. K. Lamb.** 2003. The influence of atmospheric pressure on landfill CH_4 emissions. *Waste Manag.* **23:**593–598.
59. **Dalton, H., and D. I. Stirling.** 1982. Co-metabolism. *Philos. Trans. R. Soc. London Ser. B* **297:**481–491.
60. **Daniel, D. E., and C. D. Shackelford.** 1989. Containment of landfill leachate with clay liners, p. 323–341. *In* T. H. Christensen, R. Cossu, and R. Stegmann (ed.), *Sanitary Landfilling: Process, Technology and Environmental Impact.* Academic Press, Inc., New York, N.Y.
61. **Davis, M. W.** 1988. A rapid modified method for compositional carbohydrate analysis of lignocellulosics by high pH anion-exchange chromatography with pulsed amperometric detection (HPAEC/PAD). *J. Wood Chem. Technol.* **18:**235–252.
62. **Dedysh, S. N., W. Liesack, V. N. Khmeienina, N. E. Suzina, Y. A. Trotsenko, and J. D. Semrau.** 2000. *Methylocella palustris gen. nov., sp. nov.*, a new methane-oxidizing acidophilic bacterium from peat bogs, representing a novel subtype of serine-pathway methanotrophs. *Int. J. Syst. Evol. Microbiol.* **50**(Part 3)**:**955–969.
63. **Deipser, A., and R. Stegmann.** 1994. The origin and fate of volatile trace components in municipal solid waste landfills. *Waste Manag. Res.* **12:**129–139.

64. **Deng, S. P., and M. A. Tabatabai.** 1994. Colorimetric determination of reducing sugars in soils. *Soil Biol. Biochem.* **26:**473–477.

64a. **Deobald, L. A., and D. L. Crawford.** 2002. Lignocellulose biodegradation, p. 925–933. *In* C. J. Hurst, R. L. Crawford, G. R. Knudsen, M. J. McInerney, and L. D. Stetzenbach (ed.), *Manual of Environmental Microbiology*, 2nd ed. ASM Press, Washington, D.C.

65. **De Visscher, A., D. Thomas, P. Boeckx, and O. Van Cleemput.** 1999. Methane oxidation in simulated landfill cover soil environments. *Environ. Sci. Technol.* **33:**1854–1859.

66. **De Visscher, A., and O. Van Cleemput.** 2003. Induction of enhanced CH_4 oxidation in soils: NH_4^+ inhibition patterns. *Soil Biol. Biochem.* **35:**907–913.

67. **De Visscher, A., I. De Pourcq, and J. Chanton.** 2004. Isotope fractionation effects by diffusion and methane oxidation in landfill cover soils. *J. Geophys. Res.* **109:**D18111.

68. **Doorn, M. R. J., and M. A. Barlaz.** 1995. Estimate of global methane emissions from landfills and open dumps. EPA report 600/R-95-019. U.S. EPA Office of Research and Development, Washington, D.C.

69. **Dubois, M., K. A. Gilles, J. K. Hamilton, P. A. Rebers, and F. Smith.** 1956. Colorimetric method for determination of sugars and related substances. *Anal. Chem.* **28:**350–356.

70. **Effland, M. J.** 1977. Modified procedure to determine acid soluble lignin in wood and pulp. *TAPPI* **60:**143–144.

71. **Ejlertsson, J., E. Johansson, A. Karlson, U. Meyerson, and B. H. Svensson.** 1996. Anaerobic degradation of xenobiotics by organisms from municipal solid waste landfilling conditions. *Antonie Leeuwenhoek* **69:**67–74.

72. **Ejlertsson, J., U. Meyerson, and B. H. Svensson.** 1996. Anaerobic degradation of phthalic acid esters during digestion of municipal solid waste under landfilling conditions. *Biodegradation* **7:**345–352.

73. **Eleazer, W. E., W. S. Odle, Y.-S. Wang, and M. A. Barlaz.** 1997. Biodegradability of municipal solid waste components in laboratory-scale landfills. *Environ. Sci. Technol.* **31:**911–917.

74. **Elias, D., L. R. Krumholz, R. S. Tanner, and J. M. Suflita.** 1999. Estimation of methanogen biomass by quantification of coenzyme M. *Appl. Environ. Microbiol.* **65:**5541–5545.

75. **Eller, G., S. Stubner, and P. Frenzel.** 2001. Group-specific 16S rRNA targeted probes for the detection of type I and type II methanotrophs by fluorescence in situ hybridization. *FEMS Microbiol. Lett.* **198:**91–97.

76. **Fairweather, R. J., and M. A. Barlaz.** 1998. Hydrogen sulfide production during decomposition of landfill inputs. *J. Environ. Eng. ASCE* **124:**353–361.

77. **Fassel, T.A., J. E. Vanover, C. C. Hauser, L. E. Buchholz, J. R. Edmiston, C. E. Sanger, and C. C. Remsen.** 1992. Evaluation of bacterial glycocalyx preservation and staining by ruthenium red, ruthenium red–lysine and alcian blue for several methanotroph and staphylococcal species. *Cells Mater.* **2:**37–48.

78. **Finlay, B. J., K. J. Clarke, P. A. Cranwell, T. M. Embley, R. M. Hindle, and B. M. Simon.** 1993. Further studies on the role of protozoa in landfill. ETSU B 1325. Energy Technology Support Unit, Department of Trade and Industry, Oxfordshire, United Kingdom.

79. **Finlay, B. J., and T. Fenchel.** 1991. An anaerobic protozoon, with symbiotic methanogens, living in municipal landfill material. *FEMS Microbiol. Ecol.* **85:**169–180.

80. **Galle, B., J. Samuelsson, G. Borjesson, and B. H. Svensson.** 1999. Measurement of methane emissions from landfills using FTIR spectroscopy, p. 47–54. *In* T. H. Christensen, R. Cossu, and R. Stegmann (ed.), *Proceedings of the Seventh International Waste Management and Landfill Symposium*, vol. 4. Environmental Sanitary Engineering Centre, Cagliari, Italy.

81. **Girguis, P. R., V. J. Orphan, S. J. Hallam, and E. F. DeLong.** 2003. Growth and methane oxidation rates of anaerobic methanotrophic archaea in a continuous-flow bioreactor. *Appl. Environ. Microbiol.* **69:**5472–5482.

82. **Graham, D. W., J. A. Chaudhary, R. S. Hanson, and R. G. Arnold.** 1993. Factors affecting competition between type I and type II methanotrophs in two-organism, continuous-flow reactors. *Microb. Ecol.* **25:**1–17.

83. **Gulledge, J., A. Ahmad, P. A. Steudler, W. J. Pomerantz, and C. M. Cavanaugh.** 2001. Family- and genus-level 16S rRNA-targeted oligonucleotide probes for ecological studies of methanotrophic bacteria. *Appl. Environ. Microbiol.* **67:**4726–4733.

84. **Gurijala, K. R., and J. M. Suflita.** 1993. Environmental factors influencing methanogenesis from refuse in landfills. *Environ. Sci. Technol.* **27:**1176–1181.

85. **Halvadakis, C. P., A. N. Findikakis, C. Papelis, and J. O. Leckie.** 1988. The mountain view controlled landfill project field experiment. *Waste Manag. Res.* **6:**103–114.

86. **Ham, R. K., M. R. Norman, and P. R. Fritschel.** 1993. Chemical characterization of fresh kills landfill refuse and extracts. *J. Environ. Eng. ASCE* **119:**1176–1195.

87. **Hanson, R. S., and E. V. Wattenberg.** 1991. Ecology of methylotrophic bacteria, p. 325–349. *In* I. Goldberg and J. S. Rokem (ed.), *Biology of Methylotrophs*. Butterworth-Heinemann, Boston, Mass.

88. **Hanson, R. S., and T. E. Hanson.** 1996. Methanotrophic bacteria. *Microbiol. Rev.* **60:**439–471.

89. **Harriss, R. C., and D. I. Sebacher.** 1982. Methane flux in the great dismal swamp. *Nature* **29:**673–674.

90. **Hartz, K. E., R. E. Klink, and R. K. Ham.** 1982. Temperature effects: methane generation from landfill samples. *J. Environ. Eng. ASCE* **108:**629–638.

91. **Haruta, S., M. Kondo, K. Nakamura, H. Aiba, S. Ueno, M. Ishii, and Y. Igarshi.** 2002. Microbial community changes during organic solid waste treatment analyzed by double gradient-denaturing gradient gel electrophoresis and fluorescence in situ hybridization. *Appl. Microbiol. Biotechnol.* **60:**224–231.

92. **Hatzinger, P. B., and M. Alexander.** 1995. Effect of aging of chemicals in soil on their biodegradability and extractability. *Environ. Sci. Technol.* **29:**537–545.

93. **Heyer, J.** 1977. Results of enrichment experiments of methane-assimilating organisms from an ecological point of view, p. 19–21. *In* G. A. Skryabin, M. B. Ivanov, E. N. Kondratjeva, G. A. Zaarzin, Y. A. Trotsenko, and A. I. Netrosev (ed.), *Microbial Growth on C-1 Compounds*. USSR Academy of Sciences, Pushchino, USSR.

94. **Hilger, H. A., D. F. Cranford, and M. A. Barlaz.** 2000. Methane oxidation and microbial exopolymer production in landfill cover soil. *Soil Biol. Biochem.* **32:**457–467.

95. **Hilger, H. A., A. G. Wollum, and M. A. Barlaz.** 2000. Landfill methane oxidation response to vegetation, fertilization and liming. *J. Environ. Qual.* **29:**324–334.

96. **Huang, L. N., Y. Q. Chen, H. Zhou, S. Luo, C. Y. Lan, and L. H. Qu.** 2003. Charterization of methanogenic Archaea in the leachate of a closed municipal solid waste landfill. *FEMS Microbiol. Ecol.* **46:**171–177.

97. **Humer, M., and P. Lechner.** 1999. Methane oxidation in compost cover layers on landfills, p. 403–410. *In* T. H. Christensen, R. Cossu, and R. Stegmann (ed.), *Proceedings of the Seventh International Waste Management and Landfill Symposium*, vol. 3. Environmental Sanitary Engineering Centre, Cagliari, Italy.

98. **Humer, M., and P. Lechner.** 2001. Microbial methane oxidation for the reduction of landfill gas emissions. *J. Solid Waste Technol. Manag.* **27:**146–151.
99. **Iiyama, K., B. A. Stone, and B. J. Macauley.** 1994. Compositional changes in compost during composting and growth of *Agaricus bisporus*. *Appl. Environ. Microbiol.* **60:**1538–1546.
100. **James, A. G., and I. A. Watson-Craik.** 1993. Elucidation of refuse interspecies interaction by use of laboratory models. ETSU B/B2/00148/REP. Energy Technology Support Unit, Department of Trade and Industry, Oxfordshire, United Kingdom.
101. **Jensen, T. E., and W. A. Corpe.** 1991. Ultrastructure of methylotrophic microorganisms, p. 39–75. *In* I. Goldberg and J. S. Roken (ed.), *Biology of Methylotrophs*. Butterworth-Heinemann, Boston, Mass.
102. **Jones, H. A., and D. B. Nedwell.** 1990. Soil atmosphere concentration profiles and methane emission rates in the restoration covers above landfill sites: equipment and preliminary results. *Waste Manag. Res.* **8:**21–31.
103. **Jones, H. A., and D. B. Nedwell.** 1993. Methane emission and methane oxidation in landfill cover soil. *FEMS Microbiol. Ecol.* **102:**185–195.
104. **Jones, K. L., and J. M. Grainger.** 1983. The application of enzyme activity measurements to a study of factors affecting protein, starch and cellulose fermentation in a domestic landfill. *Eur. J. Appl. Microbiol. Biotechnol.* **18:**181–185.
105. **Jonsson, S., J. Ejlertsson, and B. H. Svensson.** 2003. Transformation of phthalates in young landfill cells. *Waste Manag.* **23:**641–651.
106. **Kenealy, W., and J. G. Zeikus.** 1981. Influence of corrinoid antagonists on methanogen metabolism. *J. Bacteriol.* **146:**133–140.
107. **Kightley, D., D. B. Nedwell, and M. Cooper.** 1995. Capacity for methane oxidation in landfill cover soils measured in laboratory-scale soil microcosms. *Appl. Environ. Microbiol.* **61:**592–610.
108. **King, G. M.** 1992. Ecological aspects of methane oxidation, a key determinant of global methane dynamics. *Adv. Microb. Eco.* **12:**432–468.
109. **Kjeldsen, P., A. Dalager, and K. Broholm.** 1997. Attenuation of methane and nonmethane organic compounds in landfill gas affected soils. *J. Air Waste Manag. Assoc.* **47:**1268–1275.
110. **Klusman, R. W., and C. J. Dick.** 2000. Seasonal variability in CH_4 emissions from a landfill in a cool, semiarid climate. *J. Air Waste Manag. Assoc.* **50:**1632–1636.
111. **Klute, A. (ed.).** 1986. *Methods of Soil Analysis*, part 1. *Physical and Mineralogical Methods*. Soil Science Society of America Series no. 5. American Society of Agronomy and Soil Science Society of America, Madison, Wis.
112. **Knief, C., A. Lipski, and P. Dunfield.** 2003. Diversity and activity of methanotrophic bacteria in different upland soils. *Appl. Environ. Microbiol.* **69:**6703–6714.
113. **Kolb, S., C. Knief, S. Stubner, and R. Conrad.** 2003. Quantitative detection of methanotrophs in soil by novel *pmoA*-targeted real-time PCR assays. *Appl. Environ. Microbiol.* **69:**2423–2429.
114. **Kromann, A., and T. H. Christensen.** 1998. Degradability of organic chemicals in a landfill environment studied by in situ and laboratory leachate reactors. *Waste Manag. Res.* **16:**437–445.
115. **Leedle, J. A. Z., and R. B. Hespell.** 1980. Differential carbohydrate media and anaerobic replica plating techniques in delineating carbohydrate-utilizing subgroups in rumen bacterial populations. *Appl. Environ. Microbiol.* **39:**709–719.
116. **Lelieveld, J., P. J. Crutzen, and F. J. Dentener.** 1998. Changing concentration, lifetime and climate forcing of atmospheric methane. *Tellus Ser. B* **50B:**128–150.
117. **Le Mer, J., and P. Roger.** 2001. Production, oxidation, emission and consumption of CH_4 by soils: a review. *Eur. J. Soil Biol.* **37:**25–50.
118. **Liptay, K., J. Chanton, P. Czepiel, and B. Mosher.** 1998. Use of stable isotopes to determine methane oxidation in landfill cover soils. *J. Geophys. Res.* **103:**8243–8250.
119. **Mackie, R. I., and M. P. Bryant.** 1981. Metabolic activity of fatty acid-oxidizing bacteria and the contribution of acetate, propionate, butyrate, and CO_2 to methanogenesis in cattle waste at 40 and 60°C. *Appl. Environ. Microbiol.* **41:**1363–1373.
120. **Mancinelli, R. L., and C. P. McKay.** 1985. Methane-oxidizing bacteria in sanitary landfills, p. 437–450. *In* A. Antonopoulos (ed.), *Proceedings of the First Symposium on Biotechnological Advances in Processing Municipal Wastes for Fuels and Chemicals*. Argonne National Laboratory Report ANL/CNSV-TM-167, ed. August 1984. Argonne National Laboratory, Argonne, Ill.
121. **Maule, A., P. Luton, and R. Sharp.** 1994. A microbiological and chemical study of the Brogborough test cells. ETSU B/LF/00200/REP. Energy Technology Support Unit, Department of Trade and Industry, Harwell, Oxfordshire, United Kingdom.
122. **Maurice, C., and A. Lagerkvist.** 2004. Assessment of the methane oxidation capacity of soil. *Waste Manag. Res.* **22:**42–48.
123. **Mehta, R., M. A. Barlaz, R. Yazdani, D. Augenstein, M. Byers, and L. Sinderson.** 2002. Refuse decomposition in the presence and absence of leachate recirculation. *J. Environ. Eng.* **128:**228–236.
124. **Mormile, M. R., K. B. Gurijala, J. A. Robinson, M. J. McInerney, and J. M. Suflita.** 1996. The importance of hydrogen in landfill fermentation. *Appl. Environ. Microbiol.* **62:**1583–1588.
125. **Murphy, R. J., D. E. Jones, and R. I. Stessel.** 1995. Relationship of microbial mass and activity in biodegradation of solid waste. *Waste Manag. Res.* **13:**485–497.
126. **Nesbit, S. P., and G. A. Breitenbeck.** 1992. A laboratory study of factors influencing methane uptake by soils. *Agric. Ecosyst. Environ.* **41:**39–54.
127. **Owens, J. M., and D. P. Chynoweth.** 1993. Biochemical methane potential of municipal solid waste (MSW) components. *Water Sci. Technol.* **27:**1–14.
128. **Palmisano, A. C., and M. A. Barlaz (ed.).** 1996. *Solid Waste Microbiology*. CRC Press, Boca Raton, Fla.
129. **Palmisano, A. C., D. A. Maruscik, and B. S. Schwab.** 1993. Enumeration and hydrolytic microorganisms from three sanitary landfills. *J. Gen. Microbiol.* **139:** 387–391.
130. **Palmisano, A. C., B. S. Schwab, and D. A. Maruscik.** 1993. Hydrolytic enzyme activity in landfilled refuse. *Appl. Microbiol. Biotechnol.* **38:**828–832.
131. **Parkin, G. F., and W. F. Owen.** 1986. Fundamentals of anaerobic digestion of wastewater sludges. *J. Environ. Eng. ASCE* **112:**867–920.
132. **Parks, R. J., and E. Senior.** 1988. Multistage chemostats and other models for studying anoxic ecosystems, p. 51–71. *In* J. W. T. Wimpenny (ed.), *CRC Handbook of Laboratory Model Systems for Microbial Ecosystems*, vol. 1. CRC Press Inc., Boca Raton, Fla.
133. **Pettersen, R. C., and V. H. Schwandt.** 1991. Wood sugar analysis by anion chromatography. *J. Wood Chem. Technol.* **11:**495–501.
134. **Pfeffer, J. T.** 1974. Temperature effects on anaerobic fermentation of domestic refuse. *Biotechnol. Bioeng.* **16:** 771–787.

135. **Pohland, F. G.** 1975. Sanitary landfill stabilization with leachate recycle and residual treatment. EPA grant R-801397. United States EPA National Environmental Research Center, Cincinnati, Ohio.
136. **Pohland, F. G., W. H. Cross, J. P. Gould, and D. R. Reinhart.** 1993. Behavior and assimilation of organic and inorganic priority pollutants co-disposed with municipal refuse. EPA/600/R-93/137a. United States Environmental Protection Agency, Washington, D.C.
137. **Pohland, F. G., and J. P. Gould.** 1986. Co-disposal of municipal refuse and industrial waste sludge in landfills. *Water Sci. Technol.* **18:**177–192.
138. **Poulsen, T. G., M. Christophersen, P. Moldrup, and P. Kjeldsen.** 2003. *Waste Manag. Res.* **21:**356–366.
139. **Price, G.A., M. A. Barlaz, and G. R. Hater.** 2003. Nitrogen management in bioreactor landfills. *Waste Manag.* **23:**675–688.
140. **Qian, X., and M. A. Barlaz.** 1996. Enumeration of anaerobic refuse decomposing microorganisms on refuse constituents. *Waste Manag. Res.* **14:**151–161.
141. **Raskin, L., L. K. Poulsen, D. R. Noguera, B. E. Rittman, and D. A. Stahl.** 1994. Quantification of methanogenic groups in anaerobic biological reactors by oligonucleotide probe hybridization. *Appl. Environ. Microbiol.* **60:**1241–1248.
142. **Reinhart, D. R., and F. G. Pohland.** 1991. The assimilation of organic hazardous wastes by MSW landfills. *J. Ind. Microbiol.* **8:**193–200.
143. **Reinhart, D. R., P. McCreanor, and T. Townsend.** 2002. The bioreactor landfill: its status and future. *Waste Manag. Res.* **20:**162–171.
144. **Ress, B. B., P. P. Calvert, C. A. Pettigrew, and M. A. Barlaz.** 1998. Testing anaerobic biodegradability of polymers in a laboratory-scale simulated landfill. *Environ. Sci. Technol.* **32:**821–827.
145. **Rhew, R., and M. A. Barlaz.** 1995. The effect of lime stabilized sludge as a cover material on anaerobic refuse decomposition. *J. Environ. Eng. ASCE* **121:**499–506.
146. **Robinson, J. P., and H. C. Sturz.** 1993. Nutrition and inhibition of methanogenic bacteria in the landfill environment. ETSU B 1271. Energy Technology Support Unit, Department of Trade and Industry, Oxfordshire, United Kingdom.
147. **Rodriguez, C., S. Hiligsman, M. Ongena, R. Charlier, and P. Thonart.** 2005. Development of an enzymatic assay for the determination of cellulose bioavailability in municipal solid waste. *Biodegradation* **16:**415–422.
148. **Rolston, D. E.** 1986. Gas flux, p. 1103–1119. *In* A. Klute (ed.), *Methods of Soil Analysis*, part 1. *Physical and Mineralogical Methods.* Agronomy monograph no. 9, 2nd ed. American Society of Agronomy and Soil Science Society of America, Madison, Wis.
149. **Roslev, P., and G. M. King.** 1995. Aerobic and anaerobic starvation metabolism in methanotrophic bacteria. *Appl. Environ. Microbiol.* **61:**1563–1570.
150. **Ryden, J. C., J. H. Skinner, and D. J. Nixon.** 1987. Soil core incubation system for the field measurement of denitrification using acetylene. *Soil Biol. Biochem.* **19:**753–757.
151. **Sanin, F. D., D. R. U. Knappe, and M. A. Barlaz.** 2000. Biodegradation and humification of toluene in a simulated landfill. *Water Res.* **34:**3063–3074.
152. **Scharff, H., and A. Hensen.** 1999. Methane emission estimates for two landfills in the Netherlands using mobile TDL measurements, p. 71–78. *In* T. H. Christensen, R. Cossu, and R. Stegmann (ed.), *Proceedings of the Seventh International Waste Management and Landfill Symposium*, vol. 4. Environmental Sanitary Engineering Centre, Cagliari, Italy.
153. **Scheutz, C., H. Mosbaek, and P. Kjeldsen.** 2004. Attenuation of methane and volatile organic compounds in landfill soil covers. *J. Environ. Qual.* **33:**6–71.
154. **Schnell, S., and G. M. King.** 1995. Stability of methane oxidation capacity to variations in methane and nutrient concentrations. *FEMS Microbiol. Ecol.* **17:**285–294.
155. **Scott, J. E., G. Quintarelli, and M. C. Dellovo.** 1964. The chemical and histochemical properties of alcian blue. I. The mechanism of alcian blue staining. *Histochemie* **4:**73–85.
156. **Shelton, D. R., and J. M. Tiedje.** 1984. General method for determining anaerobic biodegradation potential. *Appl. Environ. Microbiol.* **47:**850–857.
157. **Silvey, P., and L. L. Blackall.** 1995. A study of the microbial ecology of MSW, p. 117–125. *In* T. H. Christensen, R. Cossu, and R. Stegmann (coordinators), *Proceedings Sardinia 95, Fifth International Landfill Sysmposium.* Environmental Sanitary Engineering Centre, Cagliari, Italy.
158. **Sparks, D. L., A. L. Page, P. A. Helmke, R. H. Loeppert, P. N. Soltanpour, M. A. Tabatabai, C. T. Johnson, and M. E. Sumner (ed.).** 1996. *Methods of Soil Analysis*, part 3. *Chemical Methods.* Soil Science Society of America Book Series no. 5. American Society of Agronomy and Soil Science Society of America, Madison, Wis.
159. **Spokas, K., C. Graff, M. Morcet, and C. Aran.** 2003. Implications of the spatial variability of landfill emission rates on geospatial analyses. *Waste Manag.* **23:**599–607.
160. **Stralis-Pavese, N., A. Sessitsch, A. Weilharter, T. Reichenauer, J. Riesing, J. Csontos, J. C. Murrell, and L. Bodrossy.** 2004. Optimization of diagnostic microarray for application in analyzing landfill methanotroph communities under different plant covers. *Environ. Microbiol.* **6:**347–363.
161. **Strous, M., and M. S. M. Jetten.** 2004. Anaerobic oxidation of ethane and ammonium. *Annu. Rev. Microbiol.* **58:**99–117.
162. **Suflita, J. M., C. P. Gerba, R. K. Ham, A. C. Palmisano, W. L. Rathje, and J. A. Robinson.** 1992. The world's largest landfill: a multidisciplinary investigation. *Environ. Sci. Technol.* **26:**1486–1495.
163. **Sulisti, I. A. Watson-Craik, and E. Senior.** 1996. Studies on the co-disposal of *o*-cresol with municipal refuse. *J. Chem. Technol. Biotechnol.* **65:**72–80.
164. **Svenning, M. M., I. Wartiainen, A. G. Hestnes, and S J. Binnerup.** 2003. Isolation of methane oxidizing bacteria from soil by use of a soil substrate membrane system. *FEMS Microbiol. Ecol.* **44:**347–354.
165. **Theodorou, M. K., C. King-Spooner, and D. E. Beever.** 1989. Presence or absence of anaerobic fungi in landfill refuse. ETSU B 1246. Energy Technology Support Unit, Department of Energy, Harwell Laboratory, Harwell, Oxfordshire, United Kingdom.
166. **Townsend, T. G., W. L. Miller, H.-J. Lee, and J. F. K. Earle.** 1996. Acceleration of landfill stabilization using leachate recycle. *J. Environ. Eng. ASCE* **122:**263–268.
167. **Tsien, H. C., B. J. Bratina, K. Tsuji, and R. S. Hanson.** 1990. Use of oligodeoxynucleotide signature probes for identification of physiological groups of methylotrophic bacteria. *Appl. Environ. Microbiol.* **56:**2858–2865.
168. **U.S. Environmental Protection Agency.** 2001. *Non-CO_2 Greenhouse Gas Emissions from Developed Countries.* EPA-430-R-01-007. U.S. Environmental Protection Agency, Washington, D.C.
169. **U.S. Environmental Protection Agency.** 2003. *Municipal Solid Waste in the United States: 2001 Facts and Figures.* EPA 530-R-03-011. Office of Solid Waste and Emergency Response, U.S. Environmental Protection Agency, Washington, D.C.

170. **Uz, I., M. E. Rasche, T. Townsend, A. V. Ogram, and A. S. Linder.** 2003. Characterization of methanogenic and methanotrophic assemblages in landfill samples. *Ecol. Lett.* **270:**S202–S205.
171. **Van Breukelen, B. M., and J. Griffioen.** 2004. Biogeochemical processes at the fringe of a landfill leachate pollution plume: potential for dissolved organic carbon, Fe(II), Mn(II), NH_4, and CH_4 oxidation *J. Contam. Hydrol.* **73:**181–205.
172. **Vance, E. D., P. C. Brookes, and D. S. Jenkinson.** 1987. An extraction method for measuring soil microbial biomass. *Soil Biol. Biochem.* **19:**703–707.
173. **Van Stempvoort, D., H. Maathuis, E. Jaworski, B. Mayer, and K. Rich.** 2005. Oxidation of fugitive methane in ground water linked to bacterial sulfate reduction. *Ground Water* **45:**187–199.
174. **Wang, Y.-S., and M. A. Barlaz.** 1998. Anaerobic biodegradability of alkylbenzenes and phenols by landfill-derived microorganisms. *FEMS Microbiol. Ecol.* **25:**405–418.
175. **Wang, Y.-S., C. S. Byrd, and M. A. Barlaz.** 1994. Anaerobic biodegradability of cellulose and hemicellulose in excavated refuse samples. *J. Ind. Microbiol.* **13:**147–153.
176. **Ward, B. B.** 1987. Kinetic studies on ammonia and methane oxidation by *Nitrosococcus oceanus*. *Arch. Microbiol.* **147:**126–133.
177. **Watson-Craik, I. A., and E. Senior.** 1989. Treatment of phenolic wastewaters by co-disposal with refuse. *Water Res.* **23:**1293–1303.
178. **Weaver, R. W., S. Angle, P. Bottomley, D. Bezdicek, S. Smith, A. Tabatabai, and A. Wollum (ed.).** 1994. *Methods of Soil Analysis*, part 2. *Microbiological and Biological Properties*. Soil Science Society of America Book Series no. 5. Soil Science Society of America, Madison, Wis.
179. **Westlake, K.** 1994. Microbial metabolism in the Brogborough landfill gas enhancement test cells. ETSU B/LF/00201/REP. Energy Technology Support Unit, Department of Trade and Industry, Harwell, Oxfordshire, United Kingdom.
180. **Westlake, K., and D. B. Archer.** 1990. Fundamental studies on cellulose degradation in landfills. ETSU B 1228. Energy Technology Support Unit, Department of Energy, Harwell Laboratory, Harwell, Oxfordshire, United Kingdom.
181. **Westlake, K., D. B. Archer, and D. R. Boone.** 1995. Diversity of cellulolytic bacteria in landfill. *J. Appl. Bacteriol.* **79:**73–78.
182. **Whalen, S. C., W. S. Reeburgh, and K. A. Sandbeck.** 1990. Rapid methane oxidation in a landfill cover soil. *Appl. Environ. Microbiol.* **56:**3405–3411.
183. **Whittenbury, R., J. Colby, H. Dalton, and H. L. Reed.** 1976. Biology and ecology of methane oxidizers, p. 281–292. *In* H. G. Schlegel, G. Gottschalk, and N. Pfennig (ed.), *Microbial Production and Utilization of Gases: (H_2, CH_4, CO)*. Goltze KG, Göttingen, Germany.
184. **Whittenbury, R., K. C. Phillips, and J. F. Wilkinson.** 1970. Enrichment, isolation and some properties of methane-utilizing bacteria. *J. Gen. Microbiol.* **61:**205–218.
185. **Widdick, D. A., and T. M. Embley.** 1992. Use of nucleic acid technology in landfill. A feasibility study. ETSU B 1315. Energy Technology Support Unit, Department of Trade and Industry, Harwell, Oxfordshire, United Kingdom.
186. **Williams, C. F., R. E. Terry, G. R. Buckner, and D. F. Hughes.** 1993. Understanding biogas effects. *Golf Course Manag.* **February:**96–102.
187. **Williams, G. M., R. S. Ward, and D. J. Noy.** 1999. Dynamics of landfill gas migration in unconsolidated sands. *Waste Manag. Res.* **17:**327–342.
188. **Wilshusen, J. H., J. P. A. Hettiaratchi, A. De Visscher, and R. Saint-Fort.** 2003. Methane oxidation and formation of EPS in compost: effect of oxygen concentration. *Environ. Pollut.* **129:**305–314.
189. **Windam, W. R., and D. E. Akin.** 1984. Rumen fungi and forage fiber digestion. *Appl. Environ. Microbiol.* **48:**473–476.
190. **Wise, M. G., J. V. McArthur, and L. J. Shimkets.** 1999. Methanotroph diversity in landfill soil: isolation of novel type I and type II methanotrophs whose presence was suggested by culture-independent 16S ribosomal DNA analysis. *Appl. Environ. Microbiol.* **65:**4887–4897.
191. **Wolf, H. J., and R. S. Hanson.** 1980. Isolation and characterization of methane-oxidizing yeasts. *J. Gen. Microbiol.* **114:**187–194.
192. **Wolin, E. A., M. J. Wolin, and R. S. Wolfe.** 1963. Formation of methane by bacterial extracts. *Biol. Chem.* **238:**2882–2886.
193. **Wu, B., C. M. Taylor, D. R. U. Knapp, M. A. Nanny, and M. A. Barlaz.** 2001. Factors controlling alkylbenzene sorption to municipal solid waste. *Environ. Sci. Technol.* **35:**4569–4576.
194. **Yu, Y., C. Lee, J. Kim, and S. K. Hwang.** 2005. Group-specific primer and probe sets to detect methanogenic communities using quantitative real-time polymerase chain reaction. *Biotechnol. Bioeng.* **89:**670–679.
195. **Zehnder, A. J. B.** 1978. Ecology of methane formation, p. 349–376. *In* R. Mitchell (ed.), *Water Pollution Microbiology*, vol. 2. John Wiley & Sons, Inc., New York, N.Y.
196. **Zinder, S. H.** 1993. Physiological ecology of methanogenesis, p. 128–206. *In* J. G. Ferry (ed.), *Methanogenesis: Ecology, Physiology, Biochemistry and Genetics*. Chapman & Hall, New York, N.Y.

Biogeochemistry of Aquifer Systems

ISABELLE M. COZZARELLI AND JOHANNA V. WEISS

68

The subsurface environment, which contains 30% of the world's total freshwater reserves, is a complex aquatic ecosystem in which biogeochemical reactions control the circulation of matter and energy. Biogeochemical reactions are chemical transformations carried out largely by microorganisms, and therefore the impact of microbial communities on groundwater quality is the focus of most studies of aquifer biogeochemistry. In fact, taking advantage of natural biodegradation processes in order to achieve groundwater cleanup, referred to as natural attenuation, is now common industry practice in the United States (107).

The recognition that microorganisms play an important role in the quality of groundwater dates back to the early 20th century when Rogers (120) attributed low sulfate (SO_4^{2-}) concentrations in hydrocarbon-contaminated groundwater to the activity of sulfate-reducing bacteria. Historically, groundwater quality was studied primarily by hydrogeologists and geochemists, with ecologists having little involvement in the field (44). Until the 1970s, much of the research in the chemical evolution of groundwater quality focused on abiotic reactions. Research on the ability of microorganisms to degrade pollutants in the early 1980s resulted in a shift, and microbiologists began investigating processes in aquifers (49, 153). It was discovered that a diverse group of organisms, including *Bacteria*, *Archaea*, protozoa, and fungi, inhabit the subsurface, where they facilitate the cycling of organic matter (44). Understanding the functional role of these microorganisms in groundwater ecosystems and their response to anthropogenic pressures such as contamination is now recognized as integral to the protection of groundwater resources (63).

There is a complex relationship between the composition of microbial communities and the physical, chemical, and geological characteristics of their habitats, with both environment and microorganisms exerting control on each other. Clearly the study of aquifer biogeochemistry must consider the variable spatial and temporal scales at which these interactions occur. Because the geologic setting of a site and physical heterogeneities within the formation control the movement of water and nutrients, these characteristics of an aquifer directly impact the diversity, abundance, and activity of microorganisms (142). For example, variables such as the clay or sand content of the aquifer material appear to be correlated with different types of bacteria; gram-positive organisms were found to be associated with clayey sediment, and gram-negative bacteria were associated with sandy sediment (133). Furthermore, groundwater residence times, ranging from years to decades for shallow systems and to centuries for deep aquifers, impact the potential activity of the microbial community. An understanding of this connection between the geologic and hydrologic components of an aquifer and the microbial community structure and function is necessary for assessing the impact of microbial processes on groundwater quality.

In surficial aquifers, the groundwater ecosystem is dynamic, with matter and energy infiltrating from land surface being rapidly processed (Fig. 1). Seasonal recharge has a significant impact on local flow systems and carries dissolved organic carbon (DOC) and electron acceptors such as oxygen (O_2), nitrate (NO_3^-) or SO_4^{2-} into the aquifer, thereby increasing microbial activity. In deeper aquifers, the transport of carbon and electron acceptors from the soil zone is negligible, resulting in slow rates of microbial metabolism. Therefore, with increasing distance from the land surface and recharge zone, ecological processes slow down, and the matter and energy carried large distances from the original recharge site are only partially processed (90). The release of contamination into largely carbon-limited subsurface environments stimulates microorganisms. However, even in the absence of readily degradable organic material, indigenous microorganisms are active. For example, in a study of the Black Creek Aquifer in South Carolina, indigenous bacteria were able to metabolize the aliphatic portion of the refractory organic material present at a depth of 58 m below land surface (102).

In order for a biogeochemical process to occur in any given environment, there must be favorable geochemical, hydrologic, and geologic conditions as well as a microbial population with the ability to facilitate the process. Therefore, many different approaches are needed to study complex systems such as aquifers. These approaches include field measurements of the chemical parameters such as carbon substrates and redox-sensitive species, field experiments designed to perturb the system in a predictable way, and lab experiments targeting specific metabolic pathways. Direct measurement of the function and diversity of the microorganisms themselves is a growing area due to the development of new molecular approaches. Whereas other chapters in this manual speak to the methods of microbial detection, enumeration, and community characterization,

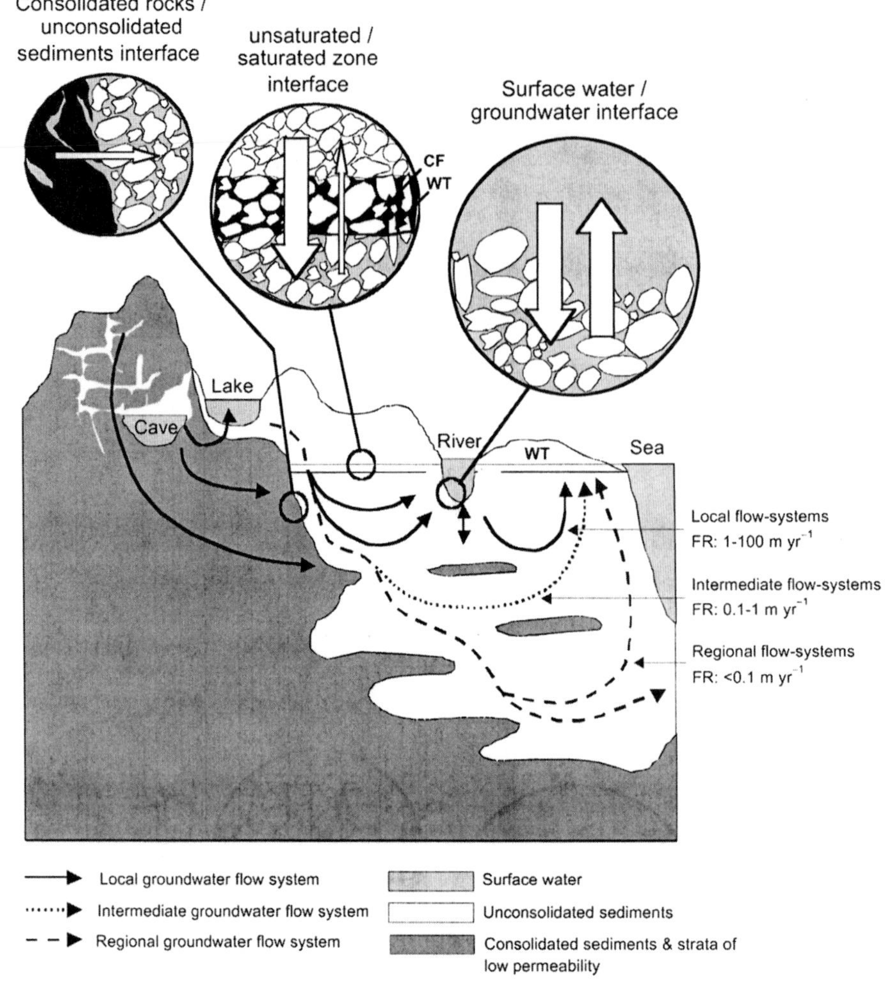

FIGURE 1 Illustration of the flow of water carrying energy and matter through the subsurface domain. Circles show various types of interface systems. Arrows in circles indicate the direction and, by their size, the relative importance of the matter and energy flow. WT, water table; CF, capillary fringe; FR, flow rate. Modified from reference 44 with the permission of Cambridge University Press and from reference 90.

the present chapter is focused on the laboratory and field experiments that are used to quantify rates of biogeochemical processes and develop conceptual models of the role of bacteria in the carbon and energy flow in subsurface environments (Table 1). We also present several integrated studies that used a multidisciplinary approach to examine the evolution of groundwater ecosystems. Because of the increasing interest in examining biodegradation in contaminated environments, the clear emphasis of the literature on aquifer biogeochemistry has been on understanding the factors controlling the degradation of organic pollutant compounds.

AQUIFER BIOGEOCHEMICAL PROCESSES

Chemical and biological transformation processes, as well as physical transport processes, control the ultimate fate of natural organic matter or organic contaminants released into the environment. The transformation pathways and

TABLE 1 Summary of geochemical approaches for assessing aquifer biogeochemistry

Scientific objective	Target	Techniques
Characterize the presence and distribution of biogeochemical processes	Concentrations of electron donors, electron acceptors, carbon, and/or contaminants; presence of intermediates, by-products, and daughter products	Field sampling, microcosm studies
Measure biogeochemical rates and factors controlling them	Concentrations of electron donors, electron acceptors, nutrients, carbon, and/or contaminants	Field sampling along flow path, microcosm studies, tracer and push-pull tests, isotope analysis
Demonstrate that bacteria are capable of biogeochemical transformation	Concentrations of electron donors, electron acceptors, carbon, and/or contaminants	Microcosm studies, column experiments, tracer and push-pull tests

reaction rates differ depending on factors such as the nature of the chemical compound and the redox conditions in the environment. Fermentation, aerobic respiration, and anaerobic respiration are the three main types of microbial metabolism that affect carbon cycling in aquifers. Fermentation depends on the capability of microorganisms to use organic molecules as electron acceptors and donors. The respiration of organic material, on the other hand, requires an inorganic electron acceptor such as O_2, NO_3^-, or Fe(III) and proceeds as a series of coupled oxidation and reduction steps. The oxidation of the organic molecule results in the transfer of electrons to O_2 in aerobic systems and to alternate electron acceptors, such as SO_4^{2-}, in anaerobic ones. Although fermentation reactions are important, respiration reactions release more energy for cell growth and have been studied much more extensively (see chapter 69).

In many aquifers, DOC in the groundwater serves as both the microbial carbon source and electron donor. Groundwater from pristine aquifers generally contains between 0.01 and 1.1 mM DOC with a median of 0.05 mM (86). There are various natural sources of organic matter including recharge with organic-rich surface waters and inputs from oil shale, coal beds, and petroleum reservoirs (143). The DOC in groundwater comprises a diverse continuum of compounds ranging from complex macromolecules such as humic substances to simple low-molecular-weight compounds such as sugars and fatty acids. Increasingly, chemical contaminants may be the predominant source of carbon in groundwater. Examples of the most common types of pollutants are landfill leachate, petroleum-derived hydrocarbons, chlorinated solvents, and pesticides. Upon entry into a pristine aquifer, some of these organic contaminants become a valuable carbon source and/or electron donor for the indigenous microbial community. For example, in an aquifer contaminated with landfill leachate, DOC concentrations increased from <0.2 mM in the background site to 17 mM in the most heavily contaminated portions of the aquifer (43). With the sudden influx of metabolizable carbon, O_2 is quickly consumed and other anaerobic processes such as iron reduction predominate (for example, see reference 9). Ultimately, an aquifer can become depleted in electron acceptors, resulting in a shift from being electron-donor limited to electron-acceptor limited. If the introduced carbon is largely refractory, the carbon quality can become the factor limiting microbial metabolism. In landfill-leachate-contaminated aquifers, for example, the DOC is comprised largely of fulvic and humic acids that are generally refractory in the environment (34, 106). In such systems, low biodegradability of the carbon can limit biogeochemical processes such as SO_4^{2-} reduction (149).

In aquifers, aerobic degradation of natural or contaminant organic matter can occur when the indigenous populations of bacteria are supplied with O_2 and have the enzymatic capability of degrading the organic compounds. Because hydrocarbons are among the most commonly occurring and widely distributed contaminants, the degradation of these compounds in aerobic subsurface environments has been extensively studied (for reviews, see Cozzarelli and Baehr [37] and Rifai [118]). The degradation of different types of hydrocarbons including aromatic hydrocarbons (57), alkanes, alkenes, and alkynes (29) occurs under aerobic conditions and is facilitated by a wide variety of bacteria and fungi. Even in the presence of readily degradable organic carbon, nutrient availability may limit biodegradation (66). For example, in laboratory microcosm studies, the stimulation of toluene degradation by nitrogen (N) addition indicated that N was the limiting nutrient (4). Phosphorus (P) can also be limiting; Rogers et al. (121) observed that phosphate-bearing minerals were preferentially dissolved by indigenous microorganisms at a crude-oil spill site. These results have increased interest in using the addition of nutrients as a remediation strategy (29).

Anoxic conditions frequently develop in contaminated subsurface environments because rapid aerobic biodegradation rates quickly diminish the supply of oxygen. Redox changes that occur due to groundwater contamination have been reviewed by Christensen et al. (35). The electron acceptors in subsurface environments include both dissolved and solid-phase species. Common dissolved electron acceptors include O_2, NO_3^-, and SO_4^{2-}. Humic acids also have been recently identified as terminal electron acceptors in studies of the microbial oxidation of contaminants such as toluene (31). Solid phases such as Fe(III) oxides are abundant in many aquifer sediments and are readily reduced by microorganisms (35). Although less abundant, manganese oxides and other metals such as the oxyanions of arsenate or selenate (139) and uranium (92, 141) also are microbially reducible. Bjerg et al. (25) reviewed these processes in landfill-impacted aquifers and provided quantitative values for the oxidation capacity for two aerobic aquifers, demonstrating that the oxidation capacity of the sedimentary Fe(III) and Mn(IV) can be several orders of magnitude greater than the oxidation capacity of the dissolved electron acceptors.

The energy yield for common oxidation-reduction reactions under anaerobic conditions has been extensively reviewed (137, 150). Briefly, reactions proceed based largely on the free-energy yield, which is greatest for denitrification and least for methanogenesis. In aquifers, a sequence of reactions occurs, reflecting the ecological succession of progressively less efficient modes of metabolism. The distribution of reactions changes over time as the contaminant source strength changes and electron acceptors become depleted (Fig. 2). In situ microorganisms are poised to take advantage of these changes in electron acceptor availability, resulting in a shift in the dominant microbial degradation reactions. Heterogeneity in contaminants, electron acceptors, and geology can promote the development of distinct redox zones at spatial scales on the order of centimeters and meters. A number of studies have demonstrated a heterogeneous distribution of hydraulic conductivity within a given sedimentary deposit (for examples, see references 15, 42, and 71). These physical heterogeneities may control the introduction of aqueous electron acceptors to the contaminated zone of an aquifer system. Moreover, heterogeneity in solid-phase Fe(III) oxides at a site contaminated with crude oil resulted in iron reduction and methanogenesis occurring in zones separated by centimeters (19). These studies suggest that contaminant plumes are unlikely to have a simple separation of redox zones that can be well-defined at the plume scale.

The characterization of redox processes is essential because the degradation of organic contaminants is contingent on the presence of specific electron acceptors (66). For example, the degradation of benzene and naphthalene at the Bemidji crude-oil site was found to occur only in locations where ferric iron was not depleted (7). Therefore, natural attenuation studies of organic contaminants often focus on the availability of electron acceptors (for examples, see references 17, 41, 58, 100, and 134). Furthermore, determining the presence and rates of anaerobic terminal electron accepting processes is important to understanding potential controls on biodegradation.

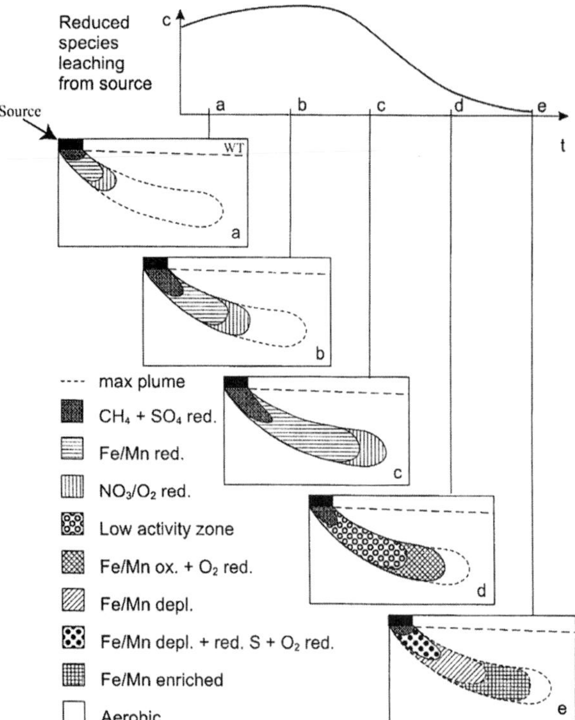

FIGURE 2 Illustration of evolution of redox zones in a plume from a point source of organic contamination. red., reduction/reduced; depl., depleted; ox., oxidation; wt, water table. The *x* axis is in units of time (t), and the *y* axis is the concentration of contaminant (c). Reprinted from the *Journal of Contaminant Hydrology* (35) with permission from Elsevier.

In subsurface systems, mixing interfaces are zones of increased biogeochemical activity (Fig. 1) due to the exchange of electron donors and/or electron acceptors that were previously limiting. For example, hot spots of biogeochemical cycling can occur as polluted water mixes with unpolluted water at the fringes of a contaminant plume (147). Recharge events introduce electron acceptors into surficial aquifers, resulting in changes in the predominant terminal electron accepting processes (101, 148). This recharge can also cause shifts in microbial communities (67) and enhance rates of biodegradation (20, 42). Groundwater/surface water interfaces can also be places of increased biodegradation, as observed by the natural attenuation of chlorinated solvents at a groundwater/wetland interface (89). In uncontaminated aquifers, the interface between aquifer and aquitard sediments can be an important site of biogeochemical reactions, wherein the aquitard can serve as a source of electron donors and the aquifer serves as the source of electron acceptors and microorganisms (103).

SAMPLING CONSIDERATIONS

Subsurface samples are typically difficult and expensive to obtain, resulting in basic microbial information being unavailable for many subsurface environments (66). For example, laborious drilling and sampling procedures can increase the possibility of contamination (see chapter 66). Prior to performing any microbiological research in aquifers, one of the most important considerations is determining where and when to collect samples for geochemical and microbiological characterizations. While the sampling protocol varies depending on the purpose of the study, the complexity of the system, and logistical constraints such as time and funding available, the degree of in situ spatial and temporal heterogeneity is likely to be a key factor controlling the experimental design. On a fundamental level, site heterogeneity can be driven by the geologic and hydrologic setting such as the lithology of the site and the proximity to sources of contamination. For example, environments with high amounts of clay have increased heterogeneity in biogeochemical processes and microbial populations (95, 146). The presence of contamination can also result in more geochemical and microbial heterogeneity than would be present in the absence of the contaminant (42, 51, 55).

There can be significant vertical and horizontal variations in the redox chemistry in aquifers, making the choice of the sampling scale an important consideration. Samples of groundwater for measurements of aqueous geochemical parameters are often taken at sampling locations separated by >1 to 10 m, which may result in missed evidence of important biogeochemical reactions (109). For example, in a crude-oil-contaminated aquifer, zones of maximum hydrocarbon contamination in the anoxic plume were on the order of only 1 to 2 m in thickness (18, 38). Furthermore, estimates of biogeochemical processes can vary significantly depending on the scale of sample collection, with long-screen monitoring wells often yielding inaccurate results due to mixing of different redox zones in the aquifer during sampling (for examples, see references 38, 103, 129, and 155). Temporal variability in geochemical concentrations also has been observed in the same wells sampled at different times (101). Geochemical heterogeneity is often accompanied by a high degree of microbial heterogeneity. For example, in studies occurring in a groundwater plume contaminated with fuel and chlorinated solvents in Michigan, temporal variability in geochemistry, at the scale of months to years, was accompanied by concomitant changes in microbial community structure (67, 101). Furthermore, dramatic changes in methanogen abundances were observed over vertical intervals of less than 1 m on each sampling date. At an aquifer contaminated with crude oil (Bemidji, Minn.), populations of aerobes and iron reducers varied by 2 or 3 orders of magnitude over vertical distances of 10 to 20 cm (19). The relationship between geochemistry and microbiology is confounded by the need to relate hydrogeochemical measurements, generally taken at broad scales, to the smaller microbially relevant scales (66).

The choice of what type of samples to collect (i.e., groundwater and/or aquifer sediment) is another consideration that can affect the outcome of experiments and interpretations of processes. Often only groundwater samples are collected, despite the apparent dominance of attached biomass in unconsolidated, sedimentary aquifers, because water samples are easier to obtain and can generally be taken from the same well and compared over time. Many studies have examined the differences in bacterial numbers, composition, and activity between groundwater and sediment samples. The majority of the literature has suggested higher percentages (i.e., 90% of total) of attached bacteria than of unattached bacteria in aquifer systems, including in pristine aquifers (3, 63) and in aquifers contaminated with petroleum (19, 115), creosote (61), sewage (68), and landfill leachate (1). However, a study of bacteria in a petroleum-contaminated aquifer found that although only 8% of the bacteria were unattached, both groups of organisms were actively degrading acetate (115). There have been reports of environments or samples that have the bulk of their microbial community in the groundwater (19, 87). In general, higher diversity has been found in groundwater samples than in sediment samples (87), with the groundwater containing different groups of organisms such as methanogens (19, 61). This increased diversity may contribute to a more functional community in the groundwater than in the sediment in some cases (87). These studies illustrate that attached and unattached aquifer microbial communities can be distinct and each group has the potential to be significant in biogeochemical transformations in aquifers.

It should be noted that when both groundwater and sediment samples are collected for study, separate collection usually occurs, with groundwater being taken from a well and sediment being taken from a core. Attached and unattached bacteria coexist in situ, making such a sampling scheme arbitrary. Because of this, studies by Bekins et al. (18) and Cozzarelli et al. (38) have collected pore water from cores to compare the relative chemical concentrations and microbial communities in the sediment. By using this approach, however, it was difficult to collect enough water for complete geochemical analyses, and there were concerns that the microbial populations were disturbed during sample collection.

GEOCHEMICAL METHODS TO INVESTIGATE MICROBIAL PROCESSES IN AQUIFERS

Chemical Measurements of Reactive Species In Situ

Electron Donors and Acceptors

In studies of aquifer biogeochemistry, much useful information regarding the microbial ecology of the system can be obtained by looking at organic compound and electron acceptor concentrations, as summarized in Table 1. One basic geochemical approach used to identify biotransformation processes has been to compare the median values of geochemical indicators from monitoring wells located within a contaminated area with those from a background region of the aquifer (for an example, see reference 9). This approach can yield basic information about what microbial transformations might be occurring in the aquifer system and can be used to develop initial conceptual models of the important biogeochemical reaction zones. Evaluating the disappearance of biodegradable organic compounds over time is another method that is used as an indication of microbial processes in aquifers. Because a flow path represents water of different ages, measurements of mass loss from groundwater along flow paths can be converted to mass loss per time, providing qualitative and quantitative information about reaction rates (150). Biodegradation rates of petroleum hydrocarbons (16, 17, 39, 48), phenolic compounds (80, 155), CH_4 (117), methyl-*tert*-butyl ether (MTBE) (152), trace organic chemicals from landfill leachate (116), and chlorinated solvents (89) have been documented using this approach.

Barker et al. (16) described a simple mass loss method of comparing the concentrations of pairs of organics at points along a flow system in order to calculate relative transformation rates in a landfill leachate plume. This data analysis approach takes into consideration the possibility of variable inputs from the landfill and dispersive dilution. For example, in the anaerobic segment of the flow system, 1,2,4-trimethylbenzene and 1,4-dichlorobenzene were equally persistent, but 1,2,4-trimethylbenzene was degraded more rapidly than 1,4-dichlorobenzene in the downgradient, less-anaerobic segment. Other studies have focused on comparing loss rates of compounds with similar structures in order to separate physical from biological processes. In a study of the migration of volatile organics in a landfill-leachate plume, Eganhouse et al. (47) determined that alkylbenzenes were useful process probes in studies of natural attenuation in contaminated groundwater systems.

In addition to measuring mass loss of organic compounds, measurements of other geochemical indicators of biogeochemical processes are needed in order to establish important microbial transformation pathways in aquifers. These indicators include a decrease in electron acceptor concentrations, the appearance of redox-reaction end products and metabolic intermediates, and increases in dissolved inorganic carbon (DIC). These compounds are considered reaction footprints because they are the direct result of microbial degradation reactions (107). Changes in concentrations of reaction footprints along hydrologic flow paths and over time at a fixed location provide insight into the evolution of microbially mediated terminal electron-accepting processes in aquifers (9, 33, 35, 43, 101, 129, 134). The redox-sensitive species typically measured in aquifers are shown in Table 2, and the methods for the determination of these species in aquifers are reviewed by Baedecker and Cozzarelli (8) and Christensen et al. (35).

Because rates of biodegradation depend on the electron acceptors available in an aquifer system (13, 42, 152, 155), establishing the spatial and temporal distribution of redox environments is critical to evaluating the potential for specific organic biotransformations in situ. For example, a long-term study of groundwater chemistry at a crude-oil spill site in Bemidji, Minn., revealed temporal trends in biogeochemical processes in the surficial aquifer (9). Decreases in dissolved O_2 and Mn^{2+} concentrations in groundwater were concomitant with increases in dissolved Fe^{2+} and CH_4 downgradient from the oil, indicating a shift from aerobic and manganese-reducing conditions to iron-reducing and methanogenic conditions. Redox changes in this aquifer were associated with changes in hydrocarbon transport rates in discrete layers (18, 38).

Recent studies in aquifers have focused on quantifying electron acceptors in the solid phase (18, 38, 69, 144), as well as the dissolved species. Solid-phase investigations range from the identification of individual minerals by using scanning electron microscopy (144, 145) and energy-dispersive X-ray analysis (42) to selective extraction techniques designed to target reactive sediment species such as reducible Fe(III) (70). An overview of approaches for identifying the redox characteristics of sediment is given in Christensen et al. (35), and methods specific for determining reactive iron species in aquifers are reviewed by Heron et al. (70) and Tuccillo et al. (144).

Because Fe(III) reduction is frequently an important process in aquifers contaminated with organics, solid-phase Fe(III) is most often measured. Whereas the controls and extent of Fe(III) reduction in hydrogeologic environments are poorly understood, it has been well documented that amorphous Fe(III) oxides are often readily reduced in anaerobic aquifers while substantial quantities of crystalline Fe(III) oxides are preserved (119). Changes in the amount of Fe(III) available for microbial degradation and the heterogeneous distribution of Fe(III) in aquifers have been documented to control the progress of degradation reactions (20, 38) and can alter other physical and chemical

TABLE 2 Use of specific geochemical measurements as indicators of in situ biogeochemical processes in aquifers

Compound or constituent	Example(s)	Indicator	Major issue(s)
Organic contaminants	Benzene, TCE	Biodegradation	Variable source strength; abiotic losses such as volatilization, sorption
Metabolites	Trimethyl benzoic acid, benzylsuccinic acid	Biodegradation of parent compound pathways	Multiple pathways can result in same metabolites Low concentrations
Hydrogen	H_2	Predominant terminal electron accepting pathway	Overlapping redox zones
Dissolved electron acceptors	O_2, NO_3^-, SO_4^{2-}	Loss indicates aerobic, nitrate-reducing, or sulfate-reducing conditions	Loss could be abiotic reaction; cannot be definitively tied to specific electron donor
Solid-phase electron acceptors	Fe(III), Mn(IV)	Loss indicates iron-reducing or manganese-reducing conditions	Aquifer heterogeneity; sampling is destructive
Redox reaction end products	Fe^{2+}, Mn^{2+}, CH_4, CO_2	Production indicates biodegradation of organic compounds by specific redox process	Transport complicates identification of redox zones; unstable and require special sampling precautions

properties of an aquifer such as the anion sorption capacity (83). Other solid-phase electron acceptors that are important in aquifer systems include Mn(IV) oxides and barite. For example, recent studies in a landfill-leachate-impacted aquifer have demonstrated the importance of the dissolution of barite as a source of SO_4^{2-} (145).

The assignment of redox zones in an aquifer by using measurements of redox-sensitive parameters is complex, and often the geochemical measurements indicate overlap in the redox zones. Transport of water-soluble species is one process that complicates the interpretation of the dissolved species concentrations, especially at hydrologic interfaces such as in shallow aquifers, where mixing of groundwater with recharge can occur (101). In identifying redox zones in aquifers, it is clear that evaluation of multiple species in both the aqueous and solid phases is essential and careful attention to the sampling scale is required, although among the numerous approaches to evaluating redox conditions in aquifers none can yet be considered standard (25).

Transformation By-Products and Metabolites

Complete mineralization of organics is not always possible or is driven by a combination of unknown chemical and microbiological processes. Therefore, the presence of intermediates or by-products of degradation processes can be used in conjunction with the substrate mass loss approach as conclusive evidence of in situ microbial transformations (Table 2). Beller (21) reviewed the use of metabolic indicators for detecting anaerobic degradation of alkylbenzenes, commonly occurring in aquifers contaminated with petroleum products. Frequently reported metabolites of alkylbenzenes are benzoic acid, alkylbenzoic acids, and benzylsuccinic acids. Benzoic and alkylbenzoic acids were reported by Cozzarelli et al. (39), who found methyl-, dimethyl-, and trimethylbenzoic acids in anoxic groundwater downgradient from a crude-oil spill in Bemidji, Minn. The presence of these metabolites, along with mass loss of structurally related alkylbenzenes, provided evidence of biodegradation of alkylbenzenes in situ. Beller et al. (22) used the appearance of benzylsuccinate and E-phenylitaconate homologs in conjunction with mass loss of toluene, *o*-xylene, and *m*-xylene as conclusive proof of anaerobic biodegradation in groundwater contaminated with gasoline. Similarly, benzylsuccinic acid analogs of ethylbenzene, *o*-, *m*-, and *p*-xylene metabolism were found in sulfate-reducing zones of a gas condensate-contaminated aquifer (50). Anaerobic metabolism of alkanes also has been indicated by the presence of alkylsuccinic acids in six aquifers contaminated with petroleum products (59).

Identification of daughter products has also provided evidence of in situ biological transformations of chlorinated solvents. Field evidence collected along two groundwater flow paths in a freshwater tidal wetland demonstrated that anaerobic biodegradation naturally attenuates a plume of chlorinated volatile organic compounds as it discharges from an aerobic sand aquifer through wetland sediments (89). Decreased concentrations of two parent contaminants, trichloroethylene (TCE) and 1,1,2,2-tetrachloroethane (TeCA), and increased concentrations of anaerobic daughter products occurred along upward flow paths through the wetland sediments. The widespread occurrence of organic metabolites at sites with both low and high contaminant concentrations suggests that the detection of metabolites may be a broadly useful indicator of anaerobic biodegradation processes in subsurface environments. However, one limitation of this approach is that metabolites are often themselves biodegradable and may be present at concentrations too low to detect. The availability of electron acceptors has been shown to be an important control on the accumulation of these metabolic intermediates in aquifers (36), and thus application of this approach requires careful consideration of the redox characteristics of the aquifer.

Hydrogen (H_2), an intermediate produced during organic matter degradation, has been used as an indicator of the predominant redox processes occurring in aquifers (Table 2). The theory has been that aqueous H_2 concentrations fall within specific ranges for methanogenic, sulfate-reducing, iron-reducing, and nitrate-reducing conditions (33, 91). The interpretation of H_2 concentrations in aquifers is complex, and the mechanisms controlling H_2 levels are not well understood (14). Recent studies have indicated that a given H_2 concentration is not necessarily indicative of a given redox process in situ, particularly in cases where processes such as SO_4^{2-} reduction and methanogenesis are occurring simultaneously (75, 126). Hydrogen levels cannot be used to determine the relative importance of simultaneous redox reactions. In addition, H_2 concentrations in aquifers are affected by the microbial kinetics of the H_2-consuming organisms and by competition for different electron donors (30). The difficulty in connecting H_2 concentrations to definitive biogeochemical processes limits its usefulness in aquifer studies.

Isotopic Signatures

Compound-specific isotope analyses (CSIA) are another tool for obtaining evidence of biodegradation at contaminated sites. In addition to following the concentration changes of specific compounds in aquifers, investigators have measured the isotopic composition of those compounds to provide insight into their sources and to identify biodegradation pathways. In general, an isotopic fractionation of carbon in an organic compound occurs during biodegradation and results in an isotopic enrichment in the residual compound. For example, a laboratory study of anaerobic toluene degradation by the sulfate-reducing bacterial strain TRM1 showed a clear enrichment in the ^{13}C of the residual toluene as the reaction progressed (105). Most CSIA measurements to date have been conducted with volatile organic compounds such as aromatic hydrocarbons, chlorinated ethylenes, and MTBE, which are common groundwater contaminants (128). Specifically, CSIA have been used to investigate the aerobic and anaerobic biodegradation reactions of aromatic hydrocarbons by using both laboratory (98, 151) and field (79, 112) approaches. For compounds such as MTBE, for which the intermediates of biodegradation are transitory and difficult to quantify, additional lines of evidence such as CSIA are especially important to support the occurrence of intrinsic biodegradation (128). Carbon and hydrogen isotopic enrichment effects have been observed during biodegradation of MTBE under both aerobic (73) and anaerobic (84) conditions. There are recent reviews of the application of this technology to assessing biodegradation in contaminated aquifers (104, 128).

Investigators have used the isotopic compositions of electron acceptors, reduced products of redox reactions, and DIC produced during the oxidation of organic compounds to determine microbial processes and to identify biodegradation pathways in groundwater. The most commonly measured isotopic ratios are $^{13}C/^{12}C$, $^{15}N/^{14}N$, and $^{34}S/^{32}S$. The DIC produced during the microbial degradation of organic compounds is expected to match that of the substrate under all redox conditions, except methanogenic conditions, which result in isotopic enrichment of ^{13}C of DIC (11). Therefore, the $\delta^{13}C$ of DIC can provide information about the source of organic compounds that are biodegrading in an aquifer as well as the redox environment. For example, in a study of a jet fuel-contaminated aquifer, $\delta^{13}C$ of DIC measurements combined with measurements of microbial metabolites and redox-sensitive species indicated that the dominant biogeochemical process was sulfate reduction coupled to oxidation of hydrocarbons (52). In contrast, carbon isotopic fractionation during methanogenesis has resulted in numerous documented cases of groundwater containing DIC enriched in ^{13}C and CH_4 depleted in ^{13}C (52, 64, 117).

In situ rates of sulfate reduction and denitrification also have been measured using N and S isotopes. In a study of aquifers contaminated with agricultural runoff, N isotopes were used to investigate the reduction of NO_3^- and the rates of biogeochemical processes by in situ microorganisms (28). The activity of sulfate-reducing bacteria was investigated in an east-central Texas aquifer by using S isotope ratios along with geochemical measurements, microbial enumeration techniques, and bioassays (146). Increased SO_4^{2-} reduction was discovered in the aquifer adjacent to lignite and clay laminae due to the release of organic acids created during fermentation of organic matter from these layers.

Experimental Approaches

Important biogeochemical reactions catalyzed by indigenous microorganisms also are studied using a variety of experimental approaches including laboratory batch and column experiments as well as field-based in situ microcosms, tracer tests, and push-pull tests (Table 1). An advantage of using experimental approaches over field observations of natural processes is that it allows the investigator to control the environmental conditions in order to better understand the nature and rate of biogeochemical processes.

Laboratory Batch and Column Studies

The laboratory microcosm approach (often referred to as a bioassay or batch experiment) can be used to demonstrate the potential for biodegradation in a particular hydrogeologic environment (35, 124, 140). By conducting experiments with sediment and groundwater from multiple locations within an aquifer, this approach also can be used to identify the spatial variability of ongoing biogeochemical processes. There are limitations to applying the microcosm approach to aquifer studies, however, because samples are removed from the natural environment, thereby losing the biological, chemical, and physical complexities of the aquifer. Nevertheless, laboratory approaches can be especially valuable in distinguishing biotic from abiotic chemical reactions, looking at toxicity effects, and evaluating the impact of changing electron acceptor or donor availability on biodegradation rates. When combined with field observations or field experiments, laboratory experiments can provide a comprehensive understanding of microbial processes.

In order to evaluate the relative importance of different terminal electron-accepting processes in a shallow landfill-leachate polluted aquifer, Ludvigsen et al. (93) applied a microcosm approach to look at the distribution of these processes along a 305-m-long flow transect. By using unamended bioassays containing sediment and groundwater, the importance of CH_4 production, SO_4^{2-}, Fe(III), and Mn(IV) reduction, and denitrification was evaluated at 37 places along the plume. The experimental results revealed that several redox processes took place simultaneously but one process typically dominated, accounting for more than 70% of the equivalent carbon conversion. Redox zones identified as methanogenic and sulfate reducing based on groundwater composition alone locally hosted other redox processes such as iron reduction.

Bioassays have been used as a tool to establish that indigenous microorganisms are able to degrade a chemical of interest and to evaluate the constraints on transformation rates and pathways under carefully controlled conditions. In these studies, measurements of mass losses of specific compounds in active microcosms are compared to mass losses in sterilized (abiotic) controls. An example of this approach is an early study in which Kappeler and Wurhmann (78) used laboratory microcosms to determine the rates of degradation of 47 water-soluble components of gas oil by mixed cultures of bacteria obtained from clean groundwater. They were able to identify degradation pathways by looking at intermediate degradation products, and they determined that the availability of N and dissolved O_2 were limiting factors in the degradation of the hydrocarbons.

This approach also has been used to document the biodegradation of creosote-derived compounds such as phenols, aromatic hydrocarbons, and nitrogen-, oxygen-, and sulfur-containing heterocyclic aromatic compounds under different redox conditions (54). In a tar oil-polluted aquifer, hydrogeochemical investigations of the aquifer revealed a heterogeneous distribution of electron acceptors with NO_3^- and O_2 available only at the plume fringes while SO_4^{2-} and Fe(III) reduction occurred in the anaerobic core of the plume (132). Microcosm studies at this site verified the ability of the indigenous microorganisms to degrade the BTEX and naphthalene compounds present in the plume when different electron acceptors were provided. At a gasoline spill site, Ball and Reinhard (13) used microcosm experiments to demonstrate that indigenous microorganisms had the ability to degrade BTEX compounds under anaerobic conditions. By manipulating the availability of electron acceptors in their experiments, these investigators were also able to demonstrate enhanced biodegradation when additional NO_3^- was added to the microcosms. In a crude-oil-contaminated site near Bemidji, Minn., microcosms constructed with sediment and groundwater from the anoxic plume were used in combination with field geochemical measurements to verify the biodegradation of benzene and toluene under iron- and manganese-reducing conditions (9) and to investigate the source and biodegradability of organic acids observed in the groundwater (36, 39).

Other studies have used radiolabeled compounds to measure degradation rates in aquifer samples. One advantage of using radiolabeled substrates in microcosm experiments is that it gives the investigator the ability to track the production of metabolites and reaction end products, thereby providing information about microbial pathways. For example, Chapelle et al. (32) demonstrated the complete degradation of ^{14}C-toluene to $^{14}CO_2$ under sulfate-reducing conditions by using microcosms constructed with sediment and groundwater from a petroleum hydrocarbon-contaminated water table aquifer in Hanahan, S.C. Microbial metabolism in uncontaminated aquifers has been investigated in deep coastal plain sediments (77). The activity of methanogens and sulfate reducers was evaluated by measuring the disappearance of various substrates and the production of acetate and CH_4 in anaerobic sediment slurries. They found that the sediments contained microorganisms capable of anaerobic degradation of a variety of substrates and that the distribution of organisms and carbon was heterogeneous.

Flowthrough columns have been used to study biogeochemical reactions in the laboratory under variable flow conditions. In this approach, the time evolution of electron donors and acceptors is measured in the column effluent, and geochemical changes that occur within the column are monitored to determine rates of reactions. With this type of approach, Allen-King et al. (5) demonstrated that oxygen availability limited the biodegradation of BTEX in a sand-filled column. In another study, columns packed with sediment from an aquifer impacted by a municipal waste landfill were used to evaluate the impact of microbially catalyzed organic carbon degradation on aquifer redox conditions (6). Geochemical changes in both the column solution and solid phases were measured during the experiment, which allowed the identification of the important processes including aerobic respiration, denitrification, and Mn(IV), Fe(III), and SO_4^{2-} reduction.

Experimental data from column studies are often used to constrain and optimize reactive transport models designed to simulate biogeochemical processes. For example, in studies of the Fredericton Aquifer in New Brunswick, Canada, columns constructed with aquifer sand were used in conjunction with field geochemical measurements to determine that the main biogeochemical controls on manganese concentrations in drinking water supplies were microbially mediated reductive dissolution of Mn(IV) oxides and cation exchange (113). Phanikumar et al. (114) used a multicomponent reactive transport model to simulate the spatial and temporal dynamics of microbial growth and carbon tetrachloride biodegradation in laboratory columns. Initial batch experiments were used to develop a conceptual model of the important degradation reactions and the column experimental data were used to constrain reaction rates.

In Situ Microcosms

There are several limitations to laboratory approaches for studying biogeochemical processes. Biomass content in the subsurface is typically low and can vary by many orders of magnitude. Therefore, when a sediment sample is removed for laboratory bioassays, it may not be representative of the aquifer system (124). There also can be concerns over changes in the microbial community due to disturbance during sample collection and over time within the microcosm itself. These limitations have led some investigators to use field-based experimental approaches such as in situ microcosms, push-pull tests, and tracer tests. The in situ microcosm approach, a field-based approach used to determine biodegradation potentials and rates, was developed for use in aquifers by Gillham et al. (60). The in situ microcosm consists of a cylinder that encases a volume of the aquifer (typically on the order of 2 liters) that is loaded with groundwater spiked with specific compounds of interest. These compounds are monitored over time, and the sediment and associated microorganisms can be sampled at the end of the experiment. An advantage of using the in situ microcosm approach over the laboratory microcosm approach is that it allows the isolation of a small region of the aquifer for study without the disturbance to microbial populations. Unlike a field tracer test, where the injected solution travels downgradient and encounters heterogeneously distributed redox zones and microbial populations, this technique isolates a portion of the aquifer, thereby providing more controlled conditions throughout the experiment (62).

In order to compare the rates of degradation of hydrocarbons under different redox conditions, Cozzarelli et al. (40) and Godsy et al. (62) used an in situ microcosm approach. Narrow iron-reducing and methanogenic zones had been identified in the crude-oil-contaminated aquifer based on geochemical and microbiologic measurements (18, 38). Results from the in situ microcosm showed that toluene and the xylene isomers were completely degraded under iron-reducing conditions in situ but benzene and ethylbenzene were not degraded. These in situ microcosm results supported the field observations from groundwater samples indicating that toluene and *ortho*-xylene are not typically found in iron-reducing zones of the aquifer due to their degradation close to the crude-oil source (38, 48).

Numerous investigators have used in situ microcosms to study biogeochemical processes in aquifers impacted by landfill leachate. For example, both laboratory batch experiments and in situ microcosms were used to investigate the degradation of aromatic hydrocarbons and chlorinated aliphatic compounds in the anaerobic part of the Grindsted Landfill in Denmark (27). Both of these experimental approaches provided information on degradation potential that compared well to the results of a natural gradient tracer test. At the Vejen Landfill in Denmark, Nielsen et al. (108) used a series of in situ microcosms installed in different redox zones in the aquifer. A mixture of organic compounds, including phenols, aromatic hydrocarbons, and chlorinated aliphatic compounds, was added to the microcosms, and their transformations were observed under different redox conditions. The degradation potentials for specific compounds were also compared to laboratory microcosms. In most cases these two approaches gave similar results with the exception that phenol and 2,4-dichlorophenol degradation was observed in the in situ microcosm but not in the laboratory microcosms. This was attributed to local variations in the aquifer and changing redox conditions over the course of the experiments.

In situ microcosms have been used in conjunction with laboratory column experiments to successfully study the degradation of a range of organic compounds in pristine aquifers (26). Recently, Korom et al. (85) scaled up this in situ microcosm approach to study denitrification in the Elk Valley aquifer in North Dakota. This larger-sized "mesocosm" encompassed an aquifer volume of 186 liters and allowed collection of larger sample volumes and a longer experimental study period than traditional in situ microcosms. Other types of in situ microcosms in which sediment or mineral grains are buried in the aquifer or suspended in monitoring wells have been used to study the microbial colonization of natural sediments (63) and microbe-mineral interactions and mineral weathering (23). This approach allows observation of bacterial colonization patterns and has been useful in the study of the dissolution of minerals by microorganisms as a source of nutrients (121).

Tracer Tests and Push-Pull Tests

Tracer tests and push-pull tests are field-based experimental approaches that involve injecting an experimental solution into the aquifer and then measuring changes in chemistry that occur over time and/or space. The advantages of these types of tests over laboratory experiments or in situ microcosm experiments are that they typically target a larger volume of the aquifer and they involve minimal disturbance of the microbial populations because the microorganisms are not removed from their natural environment. Both forced gradient and natural gradient tracer tests have been used to quantify microbially mediated reaction rates (136).

Tracer tests can be used to identify biological activity in different redox zones of an aquifer and to test the biodegradation potential of target compounds. For example, tracer tests in which readily biodegradable substrates were injected into aquifers in Utah and Arizona were successful in characterizing spatial variability of microbial activity (127). In studies of the landfill-leachate plume, Rügge et al. (125) and Albrechtsen et al. (2) injected 18 xenobiotic compounds and monitored the biodegradation rates and evolution of redox conditions in the plume over 924 days. Although many compounds were degraded under the anaerobic aquifer conditions, the degradation rates of the different compounds were highly variable depending on the compound structure and spatial variability of microbial communities. In situ rates of denitrification have been measured using ^{15}N-enriched NO_3^- as a natural gradient tracer in a nitrate-contaminated aquifer on Cape Cod, Mass. (135). The advantage of using a radiolabeled tracer in this study was that the reaction rates could be determined for the different steps in the denitrification pathway.

Tracer tests can be time-consuming and expensive, and the ability to establish controls to assess the role of bacteria in reactions is difficult. In addition, considerable knowledge of the flow system is required. A simplification of the tracer test that minimizes the need to characterize the flow system is the single-well push-pull test. This method consists of a pulse-type injection of a test solution into a specific zone of interest in the aquifer followed by withdrawal of the test solution over the course of hours or days (74). The test solution typically contains a conservative tracer and biologically reactive components such as biodegradable organic substrates or electron acceptors. A disadvantage of this type of test is that it cannot be used to measure the degradation of slowly metabolized compounds, especially in areas where the groundwater flow rate is relatively rapid. However, by conducting tests in numerous locations in an aquifer, the spatial variability in reaction rates and redox processes can be explored. This approach was used to measure the rates of aerobic respiration, denitrification, sulfate reduction, and methanogenesis in a petroleum-contaminated aquifer in Oregon (74). In a related study, Schroth et al. (130) determined that the rates of NO_3^- reduction were highest at the downgradient edge of the plume and the rates of aerobic respiration were highest at the upgradient edge of the plume.

A good example of the use of multiple push-pull tests to measure biogeochemical reactions is the work occurring at a petroleum-contaminated wetland in Studen, Switzerland. In one study, injection of SO_4^{2-} and the subsequent measurement of $\delta^{34}S$ of extracted, unconsumed SO_4^{2-} and extracted H_2S provided strong evidence for microbially mediated sulfate reduction and allowed calculation of sulfate reduction rates (131). A follow-up study by Kleikemper et al. (81) used push-pull tests with a variety of electron donors to assess the potential activity of different physiological types of methanogens. The push-pull test data were compared to molecular analyses of the methanogenic community composition determined in water and sediment samples in order to provide a comprehensive picture of the activity and diversity of methanogens in this aquifer.

COMPREHENSIVE INTEGRATIVE STUDIES

Historically, researchers trained in geochemistry and hydrology created and tested hypotheses about aquifer biogeochemistry through laboratory assays and field-based geochemical measurements and experiments. These methods could provide estimates of contaminant degradation and potential metabolic activity but offered limited insight into the microorganisms that are responsible for the observed geochemical processes. Microbiologists relied heavily on enrichment and molecular techniques to enumerate and characterize the microorganisms responsible for different biogeochemical processes, incorporating a wide range of methods as summarized in Table 3. However, microbiologists rarely performed field studies to confirm the importance of the in situ microbial community. Because biogeochemical processes require both a microbial population with the given metabolic activity and favorable geochemical, geologic, and hydrologic conditions, the study of aquifer biogeochemistry in more recent years has broadened into an inherently multidisciplinary pursuit (66). Specifically, researchers have begun to assemble many convergent lines of biogeochemical evidence that range from stable-isotopic fractionation in the field to laboratory metabolic assays and to extraction of phospholipid fatty acids (PLFAs) and DNA

TABLE 3 Summary of approaches for investigating microbial community structure and function in aquifers

Scientific objective	Target	Techniques	Reference(s)
Enumerate total bacteria	Whole cells	Microscopy	24
		Flow cytometry	65
	PLFA	PLFA analysis	56
	DNA	qPCR or hybridization of universal genes	138
Detect and/or enumerate specific groups of bacteria	Whole cells	Plate counts	46
		MPNs[a]	18
	PLFA	PLFA analysis	82
	Nucleic acids	Hybridization	111
		PCR	72
Examine community composition and diversity over time and space	Whole cells	Biolog	122
	PLFA	PLFA analysis	94
	Nucleic acids	Nucleic-acid fingerprinting	53, 88
		Cloning and sequencing	96
Directly link bacteria to function	DNA, PLFA	Stable-isotope probing	76, 115
	RNA	RT-PCR on mRNA	154

[a]MPN, most probable number.

(97). In many cases, there has been a general agreement between direct microbiological evidence for biogeochemical processes and the designation of redox zones from geochemistry. For example, consistent trends have been observed between abundances of methanogens and dissolved CH_4 in a number of diverse aquifer environments (12, 19, 95). However, there are also examples where interpretations based on geochemical concentrations do not agree with the results obtained using microbial techniques (18). Within the last decade, there has been an increased interest in performing comprehensive studies using a wide range of geochemical and microbiological techniques to study aquifer biogeochemistry. These innovative and integrative studies have used a novel combination of methods to provide compelling linkages between the microbial community and function.

Table 4 provides information on a subset of recent comprehensive integrative studies that used a variety of geochemical and microbiological approaches to significantly enhance the understanding of aquifer biogeochemistry. The first two examples explore the relationship between groundwater chemistry and hydrogeology and microbial abundances and activity in an uncontaminated aquifer (99) and in an aquifer contaminated with crude oil (18, 38). Martino et al. (99) found that there was no lithologic influence on cell abundances but microbial activity, specifically SO_4^{2-} reduction, was enhanced by sands with higher conductivity. In contrast to the work of Martino et al. (99), which surveyed lithographic heterogeneity on a scale of meters, Cozzarelli et al. (38) and Bekins et al. (18) observed a high degree of heterogeneity in geochemical processes and microbiology at small scales of ≤25 cm in an aquifer contaminated with crude oil. They demonstrated the importance of using such a scale for accurately monitoring longer-term changes in hydrocarbon transport and found that assays of microbial populations, specifically those attached to the sediment, provided a good indication of the microbial physiologic processes at a given location. A good relationship also was observed between some dominant redox processes and bacteria identified in a contaminant plume of landfill leachate, specifically in the Fe(III)-reducing zones (123). Outside the plume, Röling et al. (123) observed more variation in redox processes and a different groundwater microbial community than inside the plume. In these three studies, geochemical and microbial heterogeneities made it difficult to clearly establish the relationship between microbial community structure and hydrochemistry.

Haack et al. (67) related microbial community structure on aquifer sediments to 24 geochemical and contaminant variables under three recharge conditions by using multivariate statistics. Fingerprinting analyses indicated that communities in the anoxic central zone of the plume were similar regardless of recharge. Furthermore, general aquifer geochemistry was as important as contaminants or redox chemistry in discriminant analysis of community groups. By using a consensus matrix of denaturing gradient gel electrophoresis (DGGE) banding patterns and chemical variables to sort through complex banding patterns and environmental data, they were able to identify bands of potential functional significance and create hypotheses about the significance of multiple environmental parameters.

The ability of the indigenous microbial community to support natural attenuation of chlorinated solvents was investigated by Davis et al. (45) in a contaminated aquifer. They used a polyphasic approach consisting of growth-dependent and growth-independent analyses of the dechlorinating community. Laboratory microcosms and pure culture studies indicated that organisms directly catalyzed the mineralization of high-ordered chlorinated ethenes. Molecular analyses revealed a physiologically diverse group of

TABLE 4 Examples of comprehensive, integrative investigations of aquifer biogeochemistry

Question	Methods[a]	Habitat	Results/implications	Reference(s)
How do sediment properties and groundwater chemistry affect microbial abundance and activity?	Physical and geochemical measurements; direct counts; MPNs; rate measurements	Uncontaminated Yegua aquifer, Tex.	Lithology can affect microbial abundances of key groups of microorganisms with more microbial activity observed in aquifer sands than other strata.	99
How do contaminants, redox processes, and associated microbial populations vary over space and time?	Geochemical measurements; MPNs	Aquifer contaminated with crude oil, Bemidji, Minn.	Changes in geochemical processes and microbial populations occurred at small spatial scales ($\leq$25 cm) and indicated a slow vertical thickening of contaminant plume.	18, 38
What is the relationship between microbial community structure and hydrochemistry?	Geochemical analyses; DGGE profiling; cloning and sequencing of 16S rRNA gene; MPN-PCR for *Geobacteraceae*	Landfill leachate-contaminated aquifer, The Netherlands	While Fe(III) reduction appeared to be related to *Geobacter* presence and abundance, the overall relationship between microbiology and hydrochemistry was difficult to establish.	123
How do geochemistry and recharge influence microbial community structure and function?	Geochemical analyses; PCR and ARDRA and DGGE profiling; rRNA hybridization; multivariate statistics	Aquifer contaminated by fire-training activities, Iosco County, Mich.	Recharge greatly influenced communities at plume margins but not in the plume center. General aquifer geochemistry was important for discriminating groups of microorganisms.	67
Can the indigenous microorganisms support natural attenuation, and which ones might be involved in this contaminant degradation?	Geochemical analyses; sediment microcosm assays; direct counts; PLFA and rRNA biomass analysis; MPNs; PCR of 16S rRNA genes, cloning and sequencing	Aquifer contaminated with chlorinated ethenes, Dover, Del.	The microbial community, comprised of known anaerobic-halorespiring organisms and Fe reducers, was actively degrading TCE and its daughter products.	45
How are the activity and composition of the microbial community affected by biostimulation?	Geochemical analyses; push-pull tests; PCR of 16s rRNA genes, cloning and sequencing; MPN-PCR	Aquifer sediments contaminated with uranium and nitrate, Oak Ridge, Tenn.	Biostimulation increased sequences related to previously cultured metal reducers, indicating that they were important organisms in bioremediation.	110
What sulfate-reducing bacteria are actively degrading acetate in a contaminated aquifer?	Geochemical analyses; push-pull tests; FISH; stable-isotope probing followed by PLFA analysis	Aquifer contaminated with petroleum hydrocarbons, Switzerland	The main acetate-utilizing bacteria in the groundwater were *Desulfotomaculum acetoxidans* and *Desulfobacter* sp.; *D. acetoxidans* dominated in the sediment.	115
Can the indigenous microorganisms degrade coal tar waste?	Geochemical measurements; microcosms; isolation of naphthalene-degrading bacteria; PCR and ARDRA of 16s rRNA and *nahAc* genes, cloning and sequencing	Aquifer contaminated with coal tar waste, Glens Falls, N.Y.	Naturally occurring microorganisms were capable of degrading the mixed contaminants under aerobic and anaerobic conditions.	10, 12
Which bacteria are responsible for the in situ metabolism of naphthalene, and what are the genes responsible for its degradation?	Field-based stable-isotope probing with ^{13}C-naphthalene; naphthalene biodegradation assays; PCR of 16S rRNA and *nahA* genes, cloning and sequencing; T-RFLP; isolation of naphthalene-degrading bacteria	Aquifer contaminated with coal tar waste, Glens Falls, N.Y.	The isolated naphthalene-degrading organism accelerated naphthalene degradation rates and fell into a tight clade of labeled sequences related to *Polaromonas vacuolata*. These results indicated that the isolate was likely involved in in situ metabolism of naphthalene.	76

[a]Abbreviations: MPN, most probable number; ARDRA, amplified rDNA restriction analysis; T-RFLP, terminal restriction fragment length polymorphism.

predominantly anaerobic microorganisms, including those believed to play a key role in dechlorination. These results suggested a functional linkage between the microbiology of the site and the natural attenuation process. In a study of an aquifer where biostimulation, as opposed to natural attenuation, was considered as a treatment option, North et al. (110) used a combination of push-pull tests and molecular analyses to examine aquifer sediments cocontaminated with uranium and nitrate. The injection of electron donors stimulated the reduction of the contaminants with a concomitant increase in the gene abundances of metal-reducing bacteria. These two studies represent important contributions for researchers considering treatment options at contaminated sites.

Recent work in an aquifer contaminated with petroleum hydrocarbons (81, 115) utilized experimental approaches to investigate the role of microorganisms in driving biogeochemical processes. For example, Pombo et al. (115) combined push-pull tests, rate estimates in microcosms, stable-isotope probing, PLFA analysis, and fluorescent in situ hybridization (FISH) to examine key sulfate-reducing bacteria involved in acetate degradation. They found that acetate was consumed by sulfate reducers and methanogens present on the sediment and in the groundwater, with differences observed in the active communities in each. The PLFA method allowed the discrimination of groups of sulfate-reducing bacteria, while the stable isotope probing technique was useful for examining processes at the field scale and enhancing the understanding of the interactions between microbial communities.

The extensive work at an aquifer contaminated with coal tar waste is another example of a comprehensive and integrative approach (10, 12, 76). Through geochemical analyses and laboratory microcosms, the indigenous microbial community was shown to degrade the coal tar constituents such as naphthalene (10). In a parallel study, the microbial community was examined through molecular techniques which indicated that community compositions correlated strongly with the anaerobic carbon and energy flow induced by coal tar waste (12). Analyses with functional genes involved in naphthalene degradation also indicated a diverse population of contaminant degraders. These lines of geochemical and molecular evidence suggested that this microbial community was integral to contaminant degradation of coal tar waste. Jeon et al. (76) extended this research by using push-pull tests combined with stable-isotope probing to identify the specific members of the microbial community actively degrading naphthalene and rates of naphthalene degradation. Moreover, using 16S rRNA sequencing, they were able to identify the key member of the microbial community and subsequently used isolation techniques to obtain a pure culture of this organism. In subsequent laboratory assays, Jeon et al. (76) were able to definitively demonstrate that this organism was indeed mediating naphthalene degradation. Collectively, these three studies used a combination of techniques to demonstrate that contaminant degradation was occurring and to determine which bacteria were specifically mediating it. This experimental approach is a model for relating biogeochemical processes to microbial community structure and function.

Clearly, both the question and the stage of an investigation into aquifer biogeochemistry drive the use of one or a combination of geochemical and microbiological techniques. For example, if no data on an ecosystem exist, a simple geochemical investigation can provide basic information on the dominant processes. Furthermore, researchers interested in bioremediation options may combine such a study with measurements of microbial biomass and diversity. Due to the wide range of geochemical and microbial techniques available, the study of aquifer biogeochemistry is becoming a collaborative science in which geochemists, microbiologists, molecular biologists, and even statisticians work together to identify relationships between the microbial community and biogeochemical processes. Such collaborations will likely represent the next frontier in aquifer biogeochemical research.

REFERENCES

1. **Albrechtsen, H.-J.** 1994. Distribution of bacteria, estimated by a viable count method, and heterotrophic activity in different size fractions of aquifer sediment. *Geomicrobiol. J.* **12:**253–264.
2. **Albrechtsen, H.-J., P. L. Bjerg, L. Ludvigsen, K. Rügge, and T. H. Christensen.** 1999. An anaerobic field injection experiment in a landfill leachate plume, Grindsted, Denmark. 2. Deduction of anaerobic (methanogenic, sulfate-, and Fe(III)-reducing) redox conditions. *Water Resour. Res.* **35:**1247–1256.
3. **Alfreider, A., M. Krössbacher, and R. Psenner.** 1997. Groundwater samples do not reflect bacterial densities and activity in subsurface systems. *Water Res.* **31:**832–840.
4. **Allen-King, R. M., J. F. Barker, R. W. Gillham, and B. K. Jensen.** 1994. Substrate and nutrient limited toluene biotransformation in sandy soil. *Environ. Toxicol. Chem.* **13:**693–705.
5. **Allen-King, R. M., K. E. O'Leary, R. W. Gillham, and J. F. Barker.** 1993. Limitations on the biodegradation rate of dissolved BTEX in a natural, unsaturated, sandy soil: evidence from field and laboratory experiments, p. 175–191. *In* R. E. Hinchee, B. C. Alleman, R. E. Hoeppel, and R. E. Miller (ed.), *Hydrocarbon Bioremediation.* Battelle Press, Columbus, Ohio.
6. **Amirbahman, A., R. Schönenberger, G. Furrer, and J. Zobrist.** 2003. Experimental study and steady-state simulation of biogeochemical processes in laboratory columns with aquifer material. *J. Contam. Hydrol.* **64:**169–190.
7. **Anderson, R. T., J. N. Rooney-Varga, C. V. Gaw, and D. R. Lovley.** 1998. Anaerobic benzene oxidation in the Fe(III) reduction zone of petroleum contaminated aquifers. *Environ. Sci. Technol.* **32:**1222–1229.
8. **Baedecker, M. J., and I. M. Cozzarelli.** 1992. The determination and fate of unstable constituents of contaminated groundwater, p. 425–461. *In* S. Lesage and R. E. Jackson (ed.), *Groundwater Contamination and Analysis at Hazardous Waste Sites.* Marcel Dekker, Inc., New York, N.Y.
9. **Baedecker, M. J., I. M. Cozzarelli, R. P. Eganhouse, D. I. Siegel, and P. C. Bennett.** 1993. Crude oil in a shallow sand and gravel aquifer-III. Biogeochemical reactions and mass balance modeling in anoxic groundwater. *Appl. Geochem.* **8:**569–586.
10. **Bakermans, C., A. M. Hohnstock-Ashe, S. Padmanabhan, P. Padmanabhan, and E. L. Madsen.** 2002. Geochemical and physiological evidence for mixed aerobic and anaerobic field biodegradation of coal tar waste by subsurface microbial communities. *Microb. Ecol.* **44:**107–117.
11. **Bakermans, C., and E. L. Madsen.** 2002. Detection in coal tar waste-contaminated groundwater of mRNA transcripts related to naphthalene dioxygenase by fluorescent *in situ* hybridization with tyramide signal amplification. *J. Microbiol. Methods* **50:**75–84.
12. **Bakermans, C., and E. L. Madsen.** 2002. Diversity of 16S rDNA and naphthalene dioxygenase genes from coal-tar-

waste-contaminated aquifer waters. *Microb. Ecol.* **44:**95–106.

13. **Ball, H. A., and M. Reinhard.** 1996. Monoaromatic hydrocarbon transformation under anaerobic conditions at Seal Beach, California: laboratory studies. *Environ. Toxicol. Chem.* **15:**114–122.
14. **Banwart, S. A., and S. F. Thorton.** 2003. The geochemistry and hydrology of groundwater bioremediation by natural attenuation, p. 93–138. *In* I. M. Head, I. Singleton, and M. G. Milner (ed.), *Bioremediation: a Critical Review.* Horizon Scientific Press, Norwich, United Kingdom.
15. **Barber, L. B., II, E. M. Thurman, and D. D. Runnells.** 1992. Geochemical heterogeneity in a sand and gravel aquifer: effect of sediment mineralogy and particle size on the sorption of chlorobenzenes. *J. Contam. Hydrol.* **9:**35–54.
16. **Barker, J. F., J. S. Tessmann, P. E. Plotz, and M. Reinhard.** 1986. The organic geochemistry of a sanitary landfill leachate plume. *J. Contam. Hydrol.* **1:**171–189.
17. **Barker, J. F., and J. T. Wilson.** 1997. Natural biological attenuation of aromatic hydrocarbons under anaerobic conditions, p. 289–300. *In* C. H. Ward, J. A. Cherry, and M. R. Scalf (ed.), *Subsurface Restoration.* Ann Arbor Press, Chelsea, Mich.
18. **Bekins, B. A., I. M. Cozzarelli, E. M. Godsy, E. Warren, H. I. Essaid, and M. E. Tuccillo.** 2001. Progression of natural attenuation processes at a crude oil spill site. II. Controls on spatial distribution of microbial populations. *J. Contam. Hydrol.* **53:**387–406.
19. **Bekins, B. A., E. M. Godsy, and E. Warren.** 1999. Distribution of microbial physiologic types in an aquifer contaminated by crude oil. *Microb. Ecol.* **37:**263–275.
20. **Bekins, B. A., F. D. Hostettler, W. N. Herkelrath, G. N. Delin, E. Warren, and H. I. Essaid.** 2005. Progression of methanogenic degradation of crude oil in the subsurface. *Environ. Geosci.* **12:**139–152.
21. **Beller, H. R.** 2000. Metabolic indicators for detecting *in situ* anaerobic alkylbenzene degradation. *Biodegradation* **11:**125–139.
22. **Beller, H. R., W.-H. Ding, and M. Reinhard.** 1995. Byproducts of anaerobic alkylbenzene metabolism useful as indicators of *in situ* bioremediation. *Environ. Sci. Technol.* **29:**2864–2870.
23. **Bennett, P. C., F. K. Hiebert, and J. R. Rogers.** 2000. Microbial control of mineral-groundwater equilibria: macroscale to microscale. *Hydrogeol. J.* **8:**47–62.
24. **Bhupathiraju, V. K., M. Hernandez, P. Krauter, and L. Alvarez-Cohen.** 1999. A new direct microscopy based method for evaluating in-situ bioremediation. *J. Hazard. Mater.* **867:**299–312.
25. **Bjerg, P. L., H.-J. Albrechtsen, P. Kjeldsen, T. H. Christensen, and I. M. Cozzarelli.** 2003. The groundwater geochemistry of waste disposal facilities, p. 579–612. *In* H. D. Holland and K. Turekian (ed.), *Treatise on Geochemistry.* Elsevier Pergamon, Oxford, United Kingdom.
26. **Bjerg, P. L., A. Brun, P. H. Nielsen, and T. H. Christensen.** 1996. Application of a model accounting for kinetic sorption and degradation to *in situ* microcosm observations on the fate of aromatic hydrocarbons in an aerobic aquifer. *Water Resour. Res.* **32:**1831–1841.
27. **Bjerg, P. L., K. Rügge, J. Cortsen, P. H. Nielsen, and T. H. Christensen.** 1999. Degradation of aromatic and chlorinated aliphatic hydrocarbons in the anaerobic part of the Grindsted Landfill leachate plume: *in situ* microcosm and laboratory batch experiments. *Ground Water* **37:**113–121.
28. **Böhlke, J., R. Wanty, M. Tuttle, G. Delin, and M. Landon.** 2002. Denitrification in the recharge area and discharge area of a transient agricultural nitrate plume in a glacial outwash and sand aquifer, Minnesota. *Water Resour. Res.* **38:**7.
29. **Bouwer, E. J., and A. Zehnder.** 1993. Bioremediation of organic compounds—putting microbial metabolism to work. *Bioremediation* **11:**360–367.
30. **Brown, D., J. Komlos, and P. Jaffe.** 2005. Simultaneous utilization of acetate and hydrogen by *Geobacter sulfurreducens* and implications for use of hydrogen as an indicator of redox conditions. *Environ. Sci. Technol.* **39:**3069–3076.
31. **Cervantes, F. J., W. Dijksma, T. Duong-Dac, A. Ivanova, G. Lettinga, and J. A. Field.** 2001. Anaerobic mineralization of toluene by enriched sediments with quinoes and humus as terminal electron acceptors. *Appl. Environ. Microbiol.* **67:**4471–4478.
32. **Chapelle, F. H., P. M. Bradley, D. R. Lovley, and D. A. Vroblesky.** 1996. Measuring rates of biodegradation in a contaminated aquifer using field and laboratory methods. *Ground Water* **34:**691–698.
33. **Chapelle, F. H., P. B. McMahon, N. M. Dubrovsky, R. F. Fujii, E. T. Oaksford, and D. A. Vroblesky.** 1995. Deducing the distribution of terminal electron-accepting processes in hydrologically diverse groundwater systems. *Water Resour. Res.* **31:**359–372.
34. **Christensen, J. B., D. L. Jensen, C. Gron, Z. Filip, and T. H. Christensen.** 1998. Characterization of the dissolved organic carbon in landfill leachate-polluted groundwater. *Water Res.* **32:**125–135.
35. **Christensen, T. H., P. L. Bjerg, S. A. Banwart, R. Jakobsen, G. Heron, and H.-J. Albrechtsen.** 2000. Characterization of redox conditions in groundwater contaminant plumes. *J. Contam. Hydrol.* **45:**165–241.
36. **Cozzarelli, I. M., M. J. Baedecker, R. P. Eganhouse, and D. F. Goerlitz.** 1994. The geochemical evolution of low-molecular-weight organic acids derived from the degradation of petroleum contaminants in groundwater. *Geochim. Cosmochim. Acta* **58:**863–877.
37. **Cozzarelli, I. M., and A. L. Baehr.** 2003. Volatile fuel hydrocarbons and MTBE in the environment, p. 433–474. *In* B. S. Lollar (ed.), *Treatise on Geochemistry*, vol. 9. *Environmental Geochemistry.* Elsevier-Pergamon, Oxford, United Kingdom.
38. **Cozzarelli, I. M., B. A. Bekins, M. J. Baedecker, G. R. Aiken, R. P. Eganhouse, and M. E. Tuccillo.** 2001. Progression of natural attenuation processes at a crude-oil spill site. I. Geochemical evolution of the plume. *J. Contam. Hydrol.* **53:**369–385.
39. **Cozzarelli, I. M., R. P. Eganhouse, and M. J. Baedecker.** 1990. Transformation of monoaromatic hydrocarbons to organic acids in anoxic groundwater environment. *Environ. Geol. Water Sci.* **16:**135–141.
40. **Cozzarelli, I. M., R. P. Eganhouse, E. M. Godsy, E. Warren, and B. A. Bekins.** 2001. Measurement of biodegradation of hydrocarbons *in situ* under different redox conditions, p. 2. *In 20th Int. Meet. Organic Geochem.* Sept. 10–14, 2001, Nancy, France.
41. **Cozzarelli, I. M., J. S. Herman, and M. J. Baedecker.** 1995. Fate of microbial metabolites of hydrocarbons in a coastal plain aquifer: the role of electron acceptors. *Environ. Sci. Technol.* **29:**458–469.
42. **Cozzarelli, I. M., J. S. Herman, M. J. Baedecker, and J. M. Fischer.** 1999. Geochemical heterogeneity of a gasoline-contaminated aquifer. *J. Contam. Hydrol.* **40:**261–284.
43. **Cozzarelli, I. M., J. M. Suflita, G. A. Ulrich, S. H. Harris, M. A. Scholl, J. L. Schlottmann, and S. Christenson.** 2000. Geochemical and microbiological methods for evaluating anaerobic processes in an aquifer contaminated by landfill leachate. *Environ. Sci. Technol.* **34:**4025–4033.
44. **Danielopol, D., C. Griebler, A. Gunatilaka, and J. Notenboom.** 2003. Present state and future prospects for groundwater ecosystems. *Environ. Conserv.* **30:**104–130.

45. **Davis, J. W., J. M. Odom, K. A. DeWeerd, D. A. Stahl, S. S. Fishbain, R. J. West, G. M. Klecka, and J. G. DeCarolis.** 2002. Natural attenuation of chlorinated solvents at Area 6, Dover Air Force Base: characterization of microbial community structure. *J. Contam. Hydrol.* **57:** 41–59.
46. **de Lipthay, J. R., N. Tuxen, K. Johnsen, L. H. Hansen, H.-J. Albrechtsen, P. L. Bjerg, and J. Aamand.** 2003. *In situ* exposure to low herbicide concentrations affects microbial population composition and catabolic gene frequency in an aerobic shallow aquifer. *Appl. Environ. Microbiol.* **69:**461–467.
47. **Eganhouse, R. P., I. M. Cozzarelli, M. A. Scholl, and L. L. Matthews.** 2001. Natural attenuation of volatile organic compounds (VOCs) in the leachate plume of a municipal landfill, Norman, Oklahoma. *Ground Water* **39:**192–202.
48. **Eganhouse, R. P., T. F. Dorsey, C. S. Phinney, and A. M. Westcott.** 1996. Processes affecting the fate of monoaromatic hydrocarbons in an aquifer contaminated by crude oil. *Environ. Sci. Technol.* **30:**3304–3312.
49. **Ehrlich, G. G., D. F. Goerlitz, E. M. Godsy, and M. F. Hult.** 1982. Degradation of phenolic contaminants in groundwater by anaerobic bacteria: St. Louis Park, Minnesota. *Ground Water* **20:**703–710.
50. **Elshahed, M. S., L. M. Gieg, M. J. McInerney, and J. M. Suflita.** 2001. Signature metabolites attesting to the *in situ* attenuation of alkylbenzenes in anaerobic environments. *Environ. Sci. Technol.* **35:**682–689.
51. **Fahy, A., G. Lethbridge, R. Earle, A. S. Ball, K. N. Timmis, and T. J. McGenity.** 2005. Effects of long-term benzene pollution on bacterial diversity and community structure in groundwater. *Environ. Microbiol.* **7:**1192–1199.
52. **Fang, J. S., M. J. Barcelona, R. V. Krishnamurthy, and E. A. Atekwana.** 2000. Stable carbon isotope biogeochemistry of a shallow sand aquifer contaminated with fuel hydrocarbons. *Appl. Geochem.* **15:**157–169.
53. **Feris, K. P., K. Hristova, B. Gebreyesus, D. Mackay, and K. M. Scow.** 2004. A shallow BTEX and MTBE contaminated aquifer supports a diverse microbial community. *Microb. Ecol.* **48:**589–600.
54. **Flyvbjerg, J., E. Arvin, B. K. Jensen, and S. K. Olsen.** 1993. Microbial degradation of phenols and aromatic hydrocarbons in creosote-contaminated groundwater under nitrate-reducing conditions. *J. Contam. Hydrol.* **12:**133–150.
55. **Franklin, R. B., D. R. Taylor, and A. L. Mills.** 2000. The distribution of microbial communities in anaerobic and aerobic zones of a shallow coastal plain aquifer. *Microb. Ecol.* **38:**377–386.
56. **Franzmann, P. D., B. M. Patterson, T. R. Power, P. D. Nichols, and G. B. Davis.** 1996. Microbial biomass in a shallow, urban aquifer contaminated with aromatic hydrocarbons: analysis by phospholipid fatty acid content and composition. *J. Appl. Bacteriol.* **80:**617–625.
57. **Gibson, D. T., and V. Subramanian.** 1984. Microbial degradation of aromatic hydrocarbons, p. 181–252. *In* D. T. Gibson (ed.), *Microbial Degradation of Organic Compounds*. Marcel Dekker Inc., New York, N.Y.
58. **Gieg, L. M., R. V. Kolhatkar, M. J. McInerney, R. S. Tanner, S. H. J. Harris, K. L. Sublette, and J. M. Suflita.** 1999. Intrinsic bioremediation of petroleum hydrocarbons in a gas condensate-contaminated aquifer. *Environ. Sci. Technol.* **33:**2550–2560.
59. **Gieg, L. M., and J. M. Suflita.** 2002. Detection of anaerobic metabolites of saturated and aromatic hydrocarbons in petroleum-contaminated aquifers. *Environ. Sci. Technol.* **36:**3755–3762.
60. **Gillham, R. W., R. C. Starr, and D. J. Miller.** 1990. A device for *in situ* determination of geochemical transport parameters. 2. Biochemical reactions. *Ground Water* **28:** 858–862.
61. **Godsy, E. M., D. F. Goerlitz, and D. Grbic-Galic.** 1992. Methanogenic biodegradation of creosote contaminants in natural and simulated groundwater ecosystems. *Ground Water* **30:**232–242.
62. **Godsy, E. M., E. Warren, I. M. Cozzarelli, B. A. Bekins, and R. P. Eganhouse.** 1999. Determining BTEX biodegradation rates using *in situ* microcosms at the Bemidji site, Minnesota: trials and tribulations, p. 159–167. *In* D. W. Morganwalp and H. T. Buxton (ed.), *U.S. Geological Survey Toxic Substances Hydrology Program. Proceedings of the Technical Meeting.* U.S. Geological Survey, Reston, Va.
63. **Griebler, C., B. Mindl, D. Slezak, and M. Geiger-Kaiser.** 2002. Distribution patterns of attached and suspended bacteria in pristine and contaminated shallow aquifers studied with an *in situ* sediment exposure microcosm. *Aquat. Microb. Ecol.* **28:**117–129.
64. **Grossman, E. L., L. A. Cifuentes, and I. M. Cozzarelli.** 2002. Anaerobic methane oxidation in a landfill-leachate plume. *Environ. Sci. Technol.* **36:**2436–2442.
65. **Gruden, C. L., A. Khijniak, and P. Adriaens.** 2003. Activity assessment of microorganisms eluted from sediments using 5-cyano-2,3-ditolyl tetrazolium chloride: a quantitative comparison of flow cytometry to epifluorescent microscopy. *J. Microbiol. Methods* **55:**865–874.
66. **Haack, S. K., and B. A. Bekins.** 2000. Microbial populations in contaminant plumes. *Hydrogeol. J.* **8:**63–76.
67. **Haack, S. K., L. R. Fogarty, T. G. West, E. W. Alm, J. T. McGuire, D. T. Long, D. W. Hyndman, and L. J. Forney.** 2004. Spatial and temporal changes in microbial community structure associated with recharge-influenced chemical gradients in a contaminated aquifer. *Environ. Microbiol.* **6:**438–448.
68. **Harvey, R. W., R. L. Smith, and L. George.** 1984. Effect of organic contamination upon microbial distributions and heterotrophic uptake in a Cape Cod, Mass., aquifer. *Appl. Environ. Microbiol.* **48:**1197–1202.
69. **Heron, G., and T. H. Christensen.** 1995. Impact of sediment-bound iron on redox buffering in a landfill leachate polluted aquifer (Vejen, Denmark). *Environ. Sci. Technol.* **29:**187–192.
70. **Heron, G., C. Crouzet, A. C. M. Bourg, and T. H. Christensen.** 1994. Speciation of Fe(II) and Fe(III) in contaminated aquifer sediments using chemical extraction techniques. *Environ. Sci. Technol.* **28:**1698–1705.
71. **Hess, K. M., W. N. Herkelrath, and H. I. Essaid.** 1992. Determination of subsurface fluid contents at a crude-oil spill site. *J. Contam. Hydrol.* **10:**75–96.
72. **Hristova, K. R., C. M. Lutenegger, and K. M. Scow.** 2001. Detection and quantification of methyl *tert*-butyl ether-degrading strain PM1 by real-time TaqMan PCR. *Appl. Environ. Microbiol.* **67:**5154–5160.
73. **Hunkeler, D., B. J. Butler, R. Aravena, and J. F. Barker.** 2001. Monitoring biodegradation of methyl *tert*-butyl ether (MTBE) using compound-specific carbon isotope analysis. *Environ. Sci. Technol.* **35:**676–681.
74. **Istok, J. D., M. D. Humphrey, M. H. Schroth, M. R. Hyman, and K. T. O'Reilly.** 1997. Single-well, "push-pull" test for *in situ* determination of microbial activities. *Ground Water* **35:**619–631.
75. **Jakobsen, R., and D. Postma.** 1999. Redox zoning, rates of sulfate reduction and interactions with Fe-reduction and methanogenesis in a shallow sandy aquifer, Roemoe, Denmark. *Geochim. Cosmochim. Acta* **63:**137–151.
76. **Jeon, C. O., W. Park, P. Padmanabhan, C. DeRito, J. R. Snape, and E. L. Madsen.** 2003. Discovery of a bacterium, with distinctive dioxygenase, that is responsible for *in situ*

biodegradation in contaminated sediment. *Proc. Natl. Acad. Sci. USA* **100:**13591–13596.

77. **Jones, R. E., R. E. Beeman, and J. M. Suflita.** 1989. Anaerobic metabolic processes in the deep terrestrial subsurface. *Geomicrobiol. J.* **7:**117–130.
78. **Kappeler, T., and K. Wuhrmann.** 1978. Microbial degradation of the water-soluble fraction of gas oil—I. *Water Res.* **12:**327–333.
79. **Kelley, C. A., B. T. Hammer, and R. B. Coffin.** 1997. Concentrations and stable isotope values of BTEX in gasoline-contaminated groundwater. *Environ. Sci. Technol.* **31:** 2469–2472.
80. **Klecka, G. M., J. W. Davis, D. R. Gray, and S. S. Madsen.** 1990. Natural bioremediation of organic contaminants in groundwater. Cliffs-Dow superfund site. *Ground Water* **28:**534–543.
81. **Kleikemper, J., S. A. Pombo, M. H. Schroth, W. V. Sigler, M. Pesaro, and J. Zeyer.** 2005. Activity and diversity of methanogens in a petroleum hydrocarbon-contaminated aquifer. *Appl. Environ. Microbiol.* **71:**149–158.
82. **Kleikemper, J., M. H. Schroth, W. V. Sigler, M. Schmucki, S. M. Bernasconi, and J. Zeyer.** 2002. Activity and diversity of sulfate-reducing bacteria in a petroleum hydrocarbon-contaminated aquifer. *Appl. Environ. Microbiol.* **68:**1516–1523.
83. **Knapp, E. P., J. S. Herman, A. L. Mills, and G. M. Hornberger.** 2002. Changes in the sorption capacity of coastal plain sediments due to redox alteration of mineral surfaces. *Appl. Geochem.* **17:**387–398.
84. **Kolhatkar, R., T. Kuder, P. Philp, J. Allen, and J. T. Wilson.** 2002. Use of compound-specific stable carbon isotope analyses to demonstrate anaerobic biodegradation of MTBE in groundwater at a gasoline release site. *Environ. Sci. Technol.* **36:**5139–5146.
85. **Korom, S., A. Schlag, W. Schuh, and A. Schlag.** 2005. *In situ* mesocosms: denitrification in the Elk Valley aquifer. *Ground Water Monit. Remediat.* **25:**79–89.
86. **Leenheer, J. A., R. L. Malcolm, P. W. McKinley, and L. A. Eccles.** 1974. Occurrence of dissolved organic carbon in selected groundwater samples in the United States. *U. S. Geol. Surv. J. Res.* **2:**361–369.
87. **Lehman, R. M., F. S. Colwell, and G. A. Bala.** 2001. Attached and unattached microbial communities in a simulated basalt aquifer under fracture- and porous-flow conditions. *Appl. Environ. Microbiol.* **67:**2799–2809.
88. **Lehman, R. M., S. P. O'Connell, A. Banta, J. K. Fredrickson, A.-L. Reysenbach, T. L. Kieft, and F. S. Colwell.** 2004. Microbiological comparison of core and groundwater samples collected from a fractured basalt aquifer with that of dialysis chambers incubated *in situ*. *Geomicrobiol. J.* **21:**169–182.
89. **Lorah, M. M., and L. D. Olsen.** 1999. Natural attenuation of chlorinated volatile organic compounds in a freshwater tidal wetland: field evidence of anaerobic biodegradation. *Water Resour. Res.* **35:**3811–3827.
90. **Lovley, D. R., and F. H. Chapelle.** 1995. Deep subsurface microbial processes. *Rev. Geophys.* **33:**365–381.
91. **Lovley, D. R., F. H. Chapelle, and J. C. Woodward.** 1994. Use of dissolved H_2 concentrations to determine distribution of microbially catalyzed redox reactions in anoxic groundwater. *Environ. Sci. Technol.* **28:**1205–1210.
92. **Lovley, D. R., and E. J. P. Phillips.** 1992. Reduction of uranium by *Desulfovibrio desulfuricans*. *Appl. Environ. Microbiol.* **58:**850–856.
93. **Ludvigsen, L., H.-J. Albrechtsen, G. Heron, P. L. Bjerg, and T. H. Christensen.** 1998. Anaerobic microbial redox processes in a landfill leachate contaminated aquifer (Grindsted, Denmark). *J. Contam. Hydrol.* **33:**273–291.
94. **Ludvigsen, L., H.-J. Albrechtsen, H. Holst, and T. H. Christensen.** 1997. Correlating phospholipid fatty acids (PLFA) in a landfill leachate polluted aquifer with biogeochemical factors by multivariate statistical methods. *FEMS Microbiol. Rev.* **20:**447–460.
95. **Ludvigsen, L., H.-J. Albrechtsen, D. B. Ringelberg, F. Ekelund, and T. H. Christensen.** 1999. Distribution and composition of microbial populations in landfill leachate contaminated aquifer (Grindsted, Denmark). *Microb. Ecol.* **37:**197–207.
96. **Macbeth, T. W., D. E. Cummings, S. Spring, L. M. Petzke, and K. S. J. Sorenson.** 2004. Molecular characterization of a dechlorinating community resulting from in situ biostimulation in a trichloroethene-contaminated deep, fractured basalt aquifer and comparison to a derivative laboratory culture. *Appl. Environ. Microbiol.* **70:** 7329–7341.
97. **Madsen, E. L.** 2000. Nucleic-acid characterization of the identity and activity of subsurface microorganisms. *Hydrogeol. J.* **8:**112–125.
98. **Mancini, S. A., A. C. Ulrich, G.-L. Couloume, B. Sleep, E. A. Edwards, and B. S. Lollar.** 2003. Carbon and hydrogen isotopic fractionation during anaerobic biodegradation of benzene. *Appl. Environ. Microbiol.* **69:**191–198.
99. **Martino, D. P., E. L. Grossman, G. A. Ulrich, K. C. Burger, J. L. Schlichenmeyer, J. M. Suflita, and J. W. Ammerman.** 1998. Microbial abundance and activity in a low-conductivity aquifer system in east-central Texas. *Microb. Ecol.* **35:**224–234.
100. **McGuire, J. T., D. T. Long, M. J. Klug, S. K. Haack, and D. W. Hyndman.** 2002. Evaluating behavior of oxygen, nitrate, and sulfate during recharge and quantifying reduction rates in a contaminated aquifer. *Environ. Sci. Technol.* **36:**2693–2700.
101. **McGuire, J. T., E. W. Smith, D. T. Long, D. N. Hyndman, S. Haack, M. J. Klug, and M. A. Velbel.** 2000. Temporal variations in parameters reflecting terminal-electron-accepting processes in an aquifer contaminated with waste fuel and chlorinated solvents. *Chem. Geol.* **169:**471–485.
102. **McMahon, P. B.** 1990. *Role of Bacterial CO_2 Production in the Formation of High-Bicarbonate Ground Water in the Black Creek and Middendorf Aquifers of South Carolina.* Ph.D. dissertation. University of South Carolina, Columbia.
103. **McMahon, P. B.** 2001. Aquifer/aquitard interfaces: mixing zones that enhance biogeochemical reactions. *Hydrogeol. J.* **9:**34–43.
104. **Meckenstock, R. U., B. Morasch, C. Griebler, and H. H. Richnow.** 2004. Stable isotope fractionation analysis as a tool to monitor biodegradation in contaminated aquifers. *J. Contam. Hydrol.* **75:**215–255.
105. **Meckenstock, R. U., B. Morasch, M. Kästner, A. Vieth, and H. H. Richnow.** 2002. Assessment of bacterial degradation of aromatic hydrocarbons in the environment by analysis of stable carbon isotope fractionation. *Water Air Soil Pollut. Focus* **2:**141–152.
106. **Nanny, M. A., and N. Ratasuk.** 2002. Characterization and comparison of hydrophobic neutral and hydrophobic acid dissolved organic carbon isolated from three municipal landfill leachates. *Water Res.* **36:**1572–1584.
107. **National Research Council.** 2000. *Natural Attenuation for Groundwater Remediation.* National Academy Press, Washington, D.C.
108. **Nielsen, P. H., H. Bjarnadottir, P. L. Winter, and T. H. Christensen.** 1995. *In situ* and laboratory studies on the fate of specific organic compounds in an anaerobic landfill leachate plume. 2. Fate of aromatic and chlorinated aliphatic compounds. *J. Contam. Hydrol.* **20:**51–66.

109. **Norris, R. D., R. E. Hinchee, R. Brown, P. L. McCarthy, L. Semprini, J. T. Wilson, D. H. Kampbell, M. Reinhard, E. J. Bouwer, R. C. Borden, T. M. Vogel, and J. H. Thomas.** 1994. *Handbook of Bioremediation*. Lewis Publishers, Ann Arbor, Mich.

110. **North, N. N., S. L. Dollhopf, L. Petrie, J. D. Istok, D. L. Balkwill, and J. E. Kostka.** 2004. Change in bacterial community structure during in situ biostimulation of subsurface sediment cocontaminated with uranium and nitrate. *Appl. Environ. Microbiol.* **70:**4911–4920.

111. **Pelz, O., A. Chatzinotas, A. Zarda-Hess, W.-R. Abraham, and J. Zeyer.** 2001. Tracing toluene-assimilating sulfate-reducing bacteria using ^{13}C incorporation in fatty acids and whole-cell hybridization. *FEMS Microbiol. Ecol.* **38:**123–131.

112. **Peter, A., A. Steinbach, R. Liedl, T. Ptak, W. Michaelis, and G. Teutsch.** 2004. Assessing microbial degradation of o-xylene at field-scale from the reduction in mass flow rate combined with compound-specific isotope analyses. *J. Contam. Hydrol.* **71:**127–154.

113. **Petrunic, B. M., K. T. B. MacQuarrie, and T. A. Al.** 2005. Reductive dissolution of Mn oxides in river-recharged aquifers: a laboratory column study. *J. Hydrol.* **301:**163–181.

114. **Phanikumar, M. S., D. W. Hyndman, D. C. Wiggert, M. J. Dybas, M. E. Witt, and C. S. Criddle.** 2002. Simulation of microbial transport and carbon tetrachloride biodegradation in intermittently-fed aquifer columns. *Water Resour. Res.* **38:**4-1–4-13.

115. **Pombo, S. A., J. Kleikemper, M. H. Schroth, and J. Zeyer.** 2005. Field-scale isotopic labeling of phospholipid fatty acids from acetate-degrading sulfate-reducing bacteria. *FEMS Microbiol. Ecol.* **51:**197–267.

116. **Reinhard, M., N. L. Goodman, and J. F. Barker.** 1984. Occurrence and distribution of organic chemicals in two landfill leachate plumes. *Environ. Sci. Technol.* **18:**953–961.

117. **Revesz, K., T. B. Coplen, M. J. Baedecker, and P. D. Glynn.** 1995. Methane production and consumption monitored by stable H and C isotope ratios at a crude oil spill site, Bemidji, Minnesota. *Appl. Geochem.* **10:**505–516.

118. **Rifai, H. S.** 1997. Natural aerobic biological attenuation, p. 411. *In* C. H. Ward, J. A. Cherry, and M. R. Scalf (ed.), *Subsurface Restoration*. Ann Arbor Press, Inc., Chelsea, Mich.

119. **Roden, E. E., and M. M. Urrutia.** 2002. Influence of biogenic Fe(II) on bacterial crystalline Fe(III) oxide reduction. *Geomicrobiol. J.* **19:**209–251.

120. **Rogers, G. S.** 1917. Chemical relations of the oil-field waters in San Joaquin Valley, California. *U. S. Geol. Surv. Bull.* **653:**93–99.

121. **Rogers, J. R., P. C. Bennett, and W. J. Choi.** 1998. Feldspars as a source of nutrients for microorganisms. *Am. Mineral.* **83:**1532–1540.

122. **Röling, W. F. M., B. M. van Breukelen, M. Braster, M. T. Goeltom, J. Groen, and H. W. van Verseveld.** 2000. Analysis of microbial communities in a landfill leachate polluted aquifer using a new method for anaerobic physiological profiling and 16S rDNA based fingerprinting. *Microb. Ecol.* **40:**177–188.

123. **Röling, W. F. M., B. M. van Breukelen, M. Braster, B. Lin, and H. W. van Verseveld.** 2001. Relationships between microbial community structure and hydrochemistry in a landfill leachate-polluted aquifer. *Appl. Environ. Microbiol.* **67:**4619–4629.

124. **Röling, W. F. M., and H. W. van Verseveld.** 2002. Natural attenuation: what does the subsurface have in store? *Biodegradation* **13:**53–64.

125. **Rügge, K., P. L. Bjerg, J. K. Pedersen, H. Mosbaek, and T. H. Christensen.** 1999. An anaerobic field injection experiment in a landfill leachate plume, Grindsted, Denmark. 1. Experimental setup, tracer movement, and fate of aromatic and chlorinated compounds. *Water Resour. Res.* **35:**1231–1246.

126. **Ryu, J.-H., R. A. Dahlgren, S. Gao, and K. K. Tanji.** 2004. Characterization of redox processes in shallow groundwater of Owens Dry Lake, California. *Environ. Sci. Technol.* **38:**5950–5957.

127. **Sandrin, S. K., M. L. Brusseau, J. J. Piatt, A. A. Bodour, W. J. Blanford, and N. T. Nelson.** 2004. Spatial variability of *in situ* microbial activity: biotracer tests. *Ground Water* **42:**374–383.

128. **Schmidt, T., L. Zwank, M. Elsner, M. Berg, R. Meckenstock, and S. Haderlein.** 2004. Compound-specific stable isotope analysis of organic contaminants in natural environments: a critical review of the state of the art, prospects, and future challenges. *Anal. Bioanal. Chem.* **378:**283–300.

129. **Schreiber, M. E., and J. M. Bahr.** 1999. Spatial electron acceptor variability: implications for assessing bioremediation potential. *J. Bioremediat.* **3:**363–378.

130. **Schroth, M. H., J. D. Istok, G. T. Conner, M. R. Hyman, R. Haggerty, and K. T. O'Reilly.** 1998. Spatial variability in *in situ* aerobic respiration and denitrification rates in a petroleum-contaminated aquifer. *Ground Water* **36:**924–937.

131. **Schroth, M. H., J. Kleikemper, C. Bolliger, S. M. Bernasconi, and J. Zeyer.** 2001. *In situ* assessment of microbial sulfate reduction in a petroleum-contaminated aquifer using push-pull tests and stable sulfur isotope analyses. *J. Contam. Hydrol.* **51:**179–195.

132. **Schulze, S., and A. Tiehm.** 2004. Assessment of microbial natural attenuation in groundwater polluted with gasworks residues. *Water Sci. Technol.* **50:**347–353.

133. **Sinclair, J. L., and W. C. Ghiorse.** 1989. Distribution of aerobic bacteria, protozoa, algae, and fungi in deep subsurface sediments. *Geomicrobiol. J.* **7:**15–32.

134. **Skubal, K. L., M. J. Barcelona, and P. Adriaens.** 2001. An assessment of natural biotransformation of petroleum hydrocarbons and chlorinated solvents at an aquifer plume transect. *J. Contam. Hydrol.* **49:**151–169.

135. **Smith, R. L., J. K. Böhlke, S. P. Garabedian, K. M. Revesz, and T. Yoshinari.** 2004. Assessing denitrification in groundwater using natural gradient tracer tests with ^{15}N: *in situ* measurement of a sequential multistep reaction. *Water Resour. Res.* **40:**1–17.

136. **Smith, R. L., and S. P. Garabedian.** 1998. Using transport model interpretations of tracer tests to study microbial processes in water, p. 95–123. *In* A. L. Koch, J. A. Robinson, and G. A. Milliken (ed.), *Mathematical Models in Microbial Ecology*. Chapman and Hall, New York, N.Y.

137. **Spormann, A. M., and F. Widdel.** 2000. Metabolism of alkylbenzenes, alkanes, and other hydrocarbons in anaerobic bacteria. *Biodegradation* **11:**85–105.

138. **Stapleton, R. D., S. Ripp, L. Jimenez, S. Cheol-Koh, J. T. Fleming, I. R. Gregory, and G. S. Sayler.** 1998. Nucleic acid analytical approaches in bioremediation: site assessment and characterization. *J. Microbiol. Methods* **32:**165–178.

139. **Stolz, J. F., and R. S. Oremland.** 1999. Bacterial respiration of arsenic and selenium. *FEMS Microbiol. Rev.* **23:**615–627.

140. **Strevett, K., I. Davidova, and J. M. Suflita.** 2002. A comprehensive review of the screening methodology for anaerobic biodegradation of surfactants. *Rev. Environ. Sci. Biotechnol.* **1:**143–167.

141. **Tebo, B. M., and A. Y. Obraztsova.** 1998. Sulfate-reducing bacterium grows with Cr(VI), U(VI), Mn(IV), and Fe(III) as electron acceptors. *FEMS Microbiol. Lett.* **162:** 193–198.
142. **Thomas, J. M., J. T. Wilson, and C. H. Ward.** 1997. Microbial processes in the subsurface, p. 99–111. *In* C. H. Ward, J. A. Cherry, and M. R. Scalf (ed.), *Subsurface Restoration.* Sleeping Bear Press, Inc., Chelsea, Mich.
143. **Thurman, E. M.** 1985. *Organic Geochemistry of Natural Waters.* Martinus Nijhoff/Dr W. Junk Publishers, Dordrecht, The Netherlands.
144. **Tuccillo, M. E., I. M. Cozzarelli, and J. S. Herman.** 1999. Iron reduction in the sediments of a hydrocarbon-contaminated aquifer. *Appl. Geochem.* **14:**655–667.
145. **Ulrich, G. A., G. N. Breit, I. M. Cozzarelli, and J. M. Suflita.** 2003. Sources of sulfate supporting anaerobic metabolism in a contaminated aquifer. *Environ. Sci. Technol.* **37:**1093–1099.
146. **Ulrich, G. A., D. Martino, K. Burger, J. Routh, E. L. Grossman, J. W. Ammerman, and J. M. Suflita.** 1998. Sulfur cycling in the terrestrial subsurface: commensal interactions, spatial scales, and microbial heterogeneity. *Microb. Ecol.* **36:**141–151.
147. **van Breukelen, B. M., and J. Griffioen.** 2004. Biogeochemical processes at the fringe of a landfill leachate pollution plume: potential for dissolved organic carbon, Fe(II), Mn(II), NH_4, and CH_4 oxidation. *J. Contam. Hydrol.* **73:**181–205.
148. **Vrobelsky, D. A., and F. H. Chapelle.** 1994. Temporal and spatial changes of terminal electron-accepting processes in a petroleum hydrocarbon-contaminated aquifer and the significance for contaminant biodegradation. *Water Resour. Res.* **30:**1561–1570.
149. **Weiss, J. V., I. M. Cozzarelli, M. Lowit, and M. A. Voytek.** 2005. Biodegradable carbon as a potential control on microbial community structure and function in an aquifer contaminated with landfill leachate. Presented at the Geological Society of America Fall Meeting. October 16–19, 2005, Salt Lake City, Utah.
150. **Wiedemeier, T. H., H. S. Rifai, C. J. Newell, and J. T. Wilson.** 1999. *Natural Attenuation of Fuels and Chlorinated Solvents in the Subsurface.* John Wiley and Sons, New York, N.Y.
151. **Wilkes, H., C. Boreham, G. Harms, K. Zengler, and R. Rabus.** 2000. Anaerobic degradation and carbon isotopic fractionation of alkylbenzenes in crude oil by sulphate-reducing bacteria. *Org. Geochem.* **31:**101–115.
152. **Wilson, J. T., J. S. Cho, B. H. Wilson, and J. A. Vardy.** 2000. *Natural Attenuation of MTBE in the Subsurface under Methanogenic Conditions.* U.S. Environmental Protection Agency Report EPA/600/R–00/006. U.S. Environmental Protection Agency, Office of Research and Development, Washington, D.C.
153. **Wilson, J. T., J. F. McNabb, D. L. Balkwill, and W. C. Ghiorse.** 1983. Enumeration and characterization of bacteria indigenous to a shallow water-table aquifer. *Ground Water* **21:**134–142.
154. **Wilson, M. S., C. Bakermans, and E. L. Madsen.** 1999. In situ, real-time catabolic gene expression: extraction and characterization of naphthalene dioxygenase mRNA transcripts from groundwater. *Appl. Environ. Microbiol.* **65:**80–87.
155. **Wilson, R. D., S. F. Thornton, and D. M. Mackay.** 2004. Challenges in monitoring the natural attenuation of spatially variable plumes. *Biodegradation* **15:** 359–369.

Determining the Terminal Electron-Accepting Reaction in the Saturated Subsurface

RICHARD L. SMITH AND STEVE H. HARRIS, JR.

69

Microorganisms obtain their energy for metabolism by catalyzing a variety of oxidation-reduction reactions. In subsurface environments, where photosynthesis does not occur, the production and transfer of electrons are the driving forces that govern all microbial processes. A wide range of microbially mediated oxidation-reduction couples are possible, many for which uniquely adapted groups of microorganisms have evolved. The electron source, or donor, for the oxidation-reduction couple can be either an organic or an inorganic compound. The same is also true for the electron-accepting compound. Within a given environment, the collective result of all microbial processes, most of which are oxidation-reduction reactions, is viewed as a microbial food chain. A compound that is a reduced end product for one microbe may be an oxidized substrate for another. By sequential coupling of the microbial processes, virtually all of the energy that is biologically available in a given substrate, or suite of substrates, is extracted by the microbial population. In this way, the microbial food chain serves as an electron conduit and channels electrons to the most oxidized, and therefore the most energetically favorable, electron-accepting compound available within the localized environment.

This last step in the flow of electrons through the microbial food chain is called the terminal electron-accepting reaction; the last compound that is reduced is the terminal electron acceptor. The process itself is an important indicator of the nature of the microbial community within a habitat. Because it is the funnel through which electrons flow, the rate at which the terminal electron-accepting reaction is occurring is an integrated summary of the activity of the entire microbial food chain, and because there is a systematic progression of electron acceptors, the particular reaction that is occurring is indicative of the antecedent electron supply. Furthermore, the specialized nature of the microbes that facilitate each process, for example, methanogens or sulfate reducers, suggests the metabolic capacity or diversity that may be expected within the microbial community. Therefore, determining the terminal electron-accepting reaction is an important step in characterizing the microbial community in the subsurface.

SEQUENCE OF TERMINAL ELECTRON ACCEPTORS

The most commonly available terminal electron acceptors and their sequence of utilization are listed in Table 1. All are inorganic compounds that are relatively stable in an aqueous environment, and they represent various degrees of oxidation potential. There are key differences among the sources, the solubilities, and the typical ranges of concentrations in groundwater of the various electron acceptors, which are reflected in the natures of the various terminal electron-accepting reactions. Oxygen is the electron acceptor that provides the greatest energy yield; when it is present, aerobic metabolism dominates. Indeed, the magnitude of this energetic advantage enables the final products of other electron-accepting reactions [e.g., Fe(II), $S^{=}$, and CH_4] to serve as electron donors when they migrate into oxygen-containing zones. However, oxygen availability in the subsurface is rather limited because, for the most part, an aquifer is a closed environment. Contact with the atmosphere occurs only at the water table. Thus, most commonly, as water enters an aquifer during recharge, oxygen concentrations are fixed via equilibrium with air or with the atmosphere in the unsaturated zone and any additional oxygen entrainment along a flow path within the aquifer is usually insignificant. Travel distance and time along a flow path can be long (>100 km and >1,000 years), giving the subsurface microbial populations ample opportunity to exhaust the oxygen supply.

After oxygen, nitrate is the next electron acceptor in the progression sequence. In pristine situations, nitrate is not a prominent groundwater constituent and is found only in trace concentrations. This condition is directly related to the oligotrophic (nutrient-poor) nature of most groundwaters. However, nitrate is the most prevalent groundwater contaminant from anthropogenic sources and is found in increasingly wide geographic areas, with wide ranges in concentration. Nitrate enters an aquifer from both localized (point) and areawide (nonpoint) sources (the latter stemming from agricultural practices). Thus, the potential for encountering nitrate as a terminal electron acceptor in

TABLE 1 General characteristics of the electron acceptors most commonly found in groundwater[a]

Electron acceptor	Reduced product	Primary source(s) in groundwater	Range of concns (mM) in groundwater[b]
O_2	H_2O	Atmosphere	0–0.4
NO_3^-	N_2	Contamination	0–20
Mn(IV)	Mn(II)	Mineral solids	Very low
Fe(III)	Fe(II)	Mineral solids	Very low
$SO_4^=$	$S^=$	Mineral solids, brines, marine sources, contamination	0–15
CO_2	CH_4	Carbonate solids, organic matter degradation	0–4

[a]Electron acceptors are listed in order of utilization.
[b]Ranges given are concentrations generally found in groundwater and are not meant to represent concentration extremes.

groundwater is not insignificant and is continually increasing, particularly for unconfined aquifers.

Iron, manganese, and sulfate sequentially follow nitrate as electron acceptors. All three are naturally abundant in many aquifers. Fe(III) occurs most frequently as insoluble oxides and hydroxides, which are often found as coatings on mineral grains and which have various degrees of crystallinity and structure. This electron acceptor is immobilized by its association with the aquifer solids. The large quantity of iron oxide coatings in an aquifer suggests that the oxidation capacity via Fe(III) reduction can be quite substantial. Mn(IV) is similar to Fe(III) in that it is quite insoluble in water yet may be present in large quantities in an aquifer. Very little is known about manganese reduction in the subsurface or even whether it is a respiratory terminal electron-accepting reaction. Fe(II) abiotically reduces Mn(IV) at a neutral pH, and therefore direct enzymatic reduction is not required (65). In contrast, sulfate is much more soluble and is delivered to the microbial community via groundwater flow. In some cases, dissolution of sulfate-bearing minerals, such as gypsum, can serve as a long-term reservoir for this electron acceptor and should be considered in constructing sulfate mass balances along specific flow paths. In other settings, sulfate minerals are not abundant and sulfate is only a minor groundwater constituent. However, like nitrate, sulfate is present in many contaminant sources (e.g., landfills and septic systems), often in conjunction with increased electron acceptor demand. Therefore, sulfate reduction can be significant even where sulfate is naturally scarce.

When the other electron acceptors have been depleted, carbon dioxide becomes the terminal acceptor, being reduced to methane during methanogenesis. The supply of carbon dioxide for this process is essentially inexhaustible. A considerable electron supply, in the form of degradable organic carbon, must be available for the process to commence. As organic material is degraded by the microbial food chain, carbon dioxide is produced, which is then subsequently utilized by the methanogens. Hence, the electron acceptor in this case is internally generated. Additionally, carbon dioxide that was generated during reduction of all of the preceding electron acceptors and the carbonate solids that are present in many subsurface formations also contribute to the electron acceptor supply for methanogens. It should be noted that not all of the methane that is generated is actually derived from carbon dioxide reduction. A significant percentage comes from fermentation of acetate, which is also produced by the microbial food chain. For both carbon dioxide reduction and acetate fermentation, the electrons that are generated by microbial metabolism are transferred to carbon moieties, one organic and the other inorganic, and methane is subsequently produced. In keeping with the conceptual framework that the terminal electron-accepting reaction is a respiratory process, carbon dioxide is viewed as the terminal electron acceptor when methanogenesis is occurring, but it is important to remember that not all of the methane is generated via this mechanism.

TERMINAL ELECTRON-ACCEPTING REACTIONS IN THE SUBSURFACE

The distribution of the terminal electron-accepting reactions in an aquifer is dictated by several factors. These include the relative abundance of the various electron acceptors, the amount and availability (or degradability) of the electron supply, and the nature and rate of groundwater flow. Conceptually, there are two general distribution patterns. In the first case (Fig. 1), which is typical of most pristine groundwaters, the electron donor supply is the limiting factor. Naturally occurring organic matter is not readily abundant in the subsurface, and that which is present is usually refractory. This greatly limits the flow of carbon and energy through the microbial food chain. The sequence of terminal electron-accepting reactions occurs in the order shown in Table 1, but very long periods can be necessary before the supply of a given electron acceptor is depleted. The thicknesses of the reaction zones in the direction of groundwater flow depend upon the supply of each individual electron acceptor, but the zones for the more abundant acceptors can be many kilometers long, so long that often the aquifer may discharge to a surface water system before oxygen, or the other electron acceptors, is depleted. The other distribution pattern results when the electron donor supply is plentiful and the microbial food chain is electron acceptor limited (Fig. 2). This is often (though not exclusively) the result of organic contamination and is usually a localized phenomenon. In this situation, all available electron acceptors are rapidly utilized, creating a central methanogenic zone. Then, as the organic compounds are consumed and diluted by transport downgradient, a point is reached at which the concentration of sulfate, and then that of Fe(III), etc., is not limiting. As this happens, the other electron-accepting reactions become active. The end result is a double progression of processes in the direction of groundwater flow, first in the order shown in Table 1 and Fig. 1 and then in the reverse order. For any given case, in a pristine or contaminated environment, the exact pattern may differ from the two presented here. For example, some

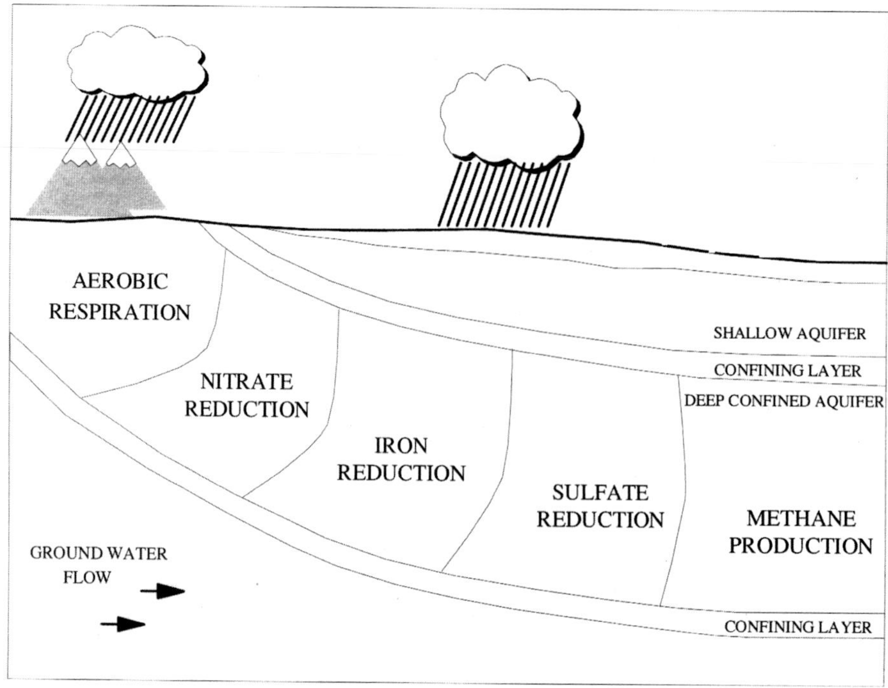

FIGURE 1 Diagram of the sequential succession of electron-accepting reactions in deep pristine groundwater (modified from reference 47).

zones may be absent or insignificant if an electron acceptor is not present. But the key issue to consider for any groundwater situation is the relative ratio of electron donor and electron acceptor supplies.

Another important factor is that an aquifer is not a homogeneous environment. There are many reasons for spatial variability, and most of them affect the subsurface microbial community. Geochemical gradients are particularly important variables with regard to electron-accepting reactions. Both horizontal and vertical geochemical gradients occur within an aquifer. The steepness of these gradients is dictated by the degree of dispersion (i.e., mixing)

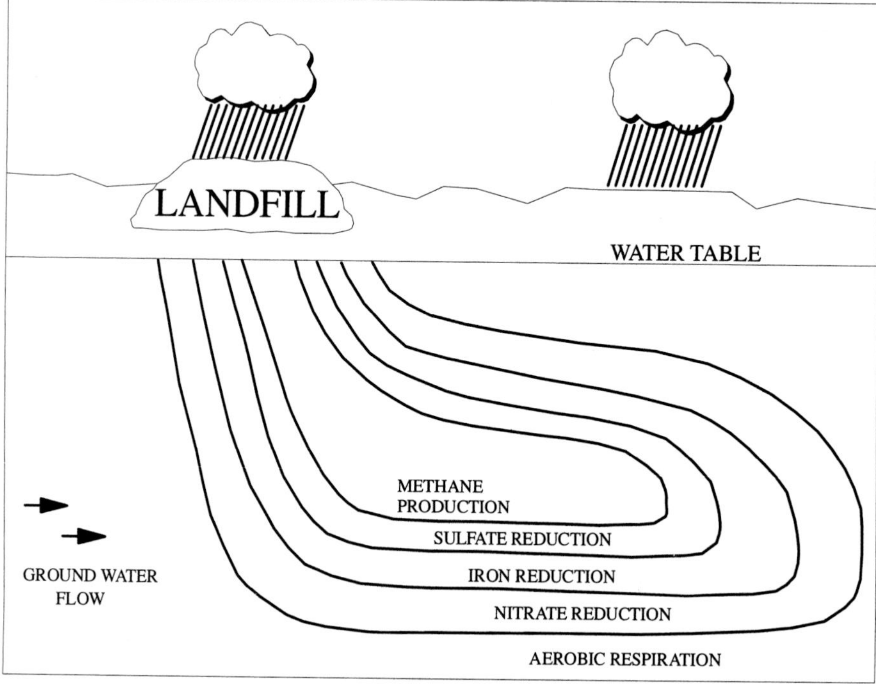

FIGURE 2 Idealized diagram of the orientation of the electron-accepting reactions in groundwater contaminated with landfill leachate.

along each spatial axis. In general, dispersion in the direction of groundwater flow is much greater than that in the transverse directions. Thus, when bulk flow is primarily horizontal, the sharper gradients would usually be found in the vertical direction. Vertical gradients can persist for many kilometers along a groundwater flow path. The existence of these gradients can restrict a terminal electron-accepting reaction to a relatively narrow vertical interval within an aquifer, depending upon the electron acceptor supply. For example, in one nitrate-contaminated sand and gravel aquifer, denitrification occurs within a 5- to 7-m-thick zone (73). This means that resolution in the vertical dimension must be carefully considered when a study is conducted to determine the electron-accepting reactions in an aquifer. Short well screens or multilevel point samplers, spaced vertically to match the shape of the gradient, are best (72). The gradient can be determined to some extent by using screened augers during the installation process. Moreover, collection of core material must also reflect the existence of gradients. Cores collected from the same depth and vicinity as a well screen can be combined with groundwater collected from the same depth. This ensures that the aquifer microbes are subjected to a representative geochemistry during incubation assays.

APPROACHES FOR DETERMINING THE ELECTRON-ACCEPTING REACTION

Microbial processes, especially the terminal electron-accepting reactions, alter the geochemistry of the aqueous environment. Specific geochemical characteristics result when each of the various electron acceptors is reduced. These characteristics, or geochemical profiles, have been used to infer which process is active within any given region of an aquifer. This is a valid, acceptable approach. However, the conclusions may be limited to stating that process *x* either is occurring or has occurred and there may be missing components of the geochemical profile that may confound the issue (e.g., sulfide removal in the presence of heavy metals). An alternative approach is to conduct incubation assays specific for a given electron-accepting reaction, because a demonstration that the process is active is the most direct proof that it is the terminal electron-accepting reaction. The latter approach assumes that the sample collection did not disrupt or alter the microbial community, that the incubation process can faithfully maintain the in situ geochemistry during the entire time course, and that the selected assay has sufficient sensitivity to detect the low rates of activity that occur in the subsurface. The vast majority of microorganisms in an aquifer are attached to solid surfaces (29). This means that collection of core material is required for these activity assays to obtain a representative result, unless an alternative in situ approach is available (70).

In general, the best and most rigorous method for determination of the terminal electron-accepting reaction is to combine the geochemical and microbiological approaches. Interpretation of the combined results involves fewer inherent assumptions and is less subject to error than that of results of each individual approach. An aquifer should actually be viewed from an ecosystem perspective. Therefore, an orientation that encompasses (i) the diversity and the nature of microbial processes that occur in the subsurface, (ii) the geochemical environment, and (iii) the hydrologic regime is the most appropriate for considering the function and the interaction of the terminal electron-accepting reactions.

AEROBIC RESPIRATION

Determining whether aerobic respiration is the dominant terminal process is relatively straightforward. When oxygen is present, it is the electron acceptor, not only for the thermodynamic reasons mentioned previously but also because it is toxic to the obligately anaerobic processes (iron reduction, sulfate reduction, and methanogenesis) and inhibits the expression and the function of the denitrification enzymes (42). Only at very low oxygen concentrations, below ~30 μM (65, 78, 79), is there a potential for nitrate to compete with oxygen for available electrons. Because an aquifer is water saturated, the oxygen concentration in the bulk groundwater is a good representation of the amount of oxygen available to the individual microorganisms, even those that are attached. There is little likelihood that anaerobic microsites, which are common in many soils, are important in the groundwater environment.

In practice, then, one can readily conclude that oxygen is the terminal electron acceptor when oxygen concentrations exceed 50 μM, unless there are no nitrogen oxides present, in which case the threshold may be 10 μM or lower. Oxygen is the one electron acceptor for which it is not necessary to conduct laboratory incubations as a confirmatory step. This is fortuitous because oxygen consumption by pristine samples with low electron acceptor demand could be very difficult to detect. Thus, the pertinent methodology for determining aerobic respiration is that of sampling and quantifying dissolved oxygen in groundwater. However, some studies have examined aerobic respiration in the subsurface as a process and have quantified rates of oxygen consumption in a sand and gravel aquifer (79) and in fractured rock (60).

Collecting groundwater for quantification of dissolved oxygen must be done carefully; exposure to the atmosphere can result in overestimation, and stripping of dissolved gases via suction lift pumping can result in underestimation. Given that the water table may be several meters below the land surface, the challenge is to obtain a sample representative of the aquifer. Several techniques can be used (Table 2); each is best for specific conditions. First, the standing water in the well must be replaced with water from the formation. The standard procedure is to pump and discard at least 3 well casing volumes. Next, the water sample is collected. A down-well pump with an inflatable packer can seal the screened interval from the atmosphere and can serve for both well purging and sample collection. On the other hand, a grab sampler can maintain in situ pressure for deep formations or gas-laden samples. For the small-bore multilevel samplers, suction lift is unavoidable. The best choice then is a peristaltic pump fitted with oxygen-impermeable tubing and the use of a low rate of pumping. Down-well oxygen determination is also possible (Table 2), though it can be quite time-consuming. Using production wells for oxygen determinations can be considered only qualitative and is best avoided because of turbulent mixing, long screened intervals, and the potential for entrainment of atmospheric oxygen.

The two most commonly used analytical procedures for oxygen are the Winkler iodometric technique (3) and the use of oxygen-specific membrane electrodes. Both are suitable if calibrated properly but lack sensitivity for low oxygen concentrations. For the latter, a quick, field-friendly, colorimetric method uses Rhodazine-D in sealed glass ampoules (CHEMetrics, Inc.) (84). A suggested approach is to collect groundwater samples containing >30 μM (1 mg liter^{-1}) in biological oxygen demand bottles (3) for analysis

TABLE 2 Sampling techniques to measure dissolved oxygen in groundwater

Technique	Advantage	Disadvantage	Comment
Down-well, pressure lift pumping	Single step for purge and sample	Specialized pump, not suitable for smaller pipes	Use O_2-impermeable tubing[a] and low pumping rate
Suction lift pumping	Can be used with small-diam wells	Water table must be above suction limit, potential outgassing	Use O_2-impermeable tubing and low pumping rate
Use of grab sampler[b]	Maintains pressure, no outgassing	Small volume collected	Avoid turbulent mixing when lowering into well
Down-well determination (lowering O_2-measuring device[c] into well)	Prepump measurement, in situ temperature and pressure	Cumbersome and time-consuming, water table must be above suction limit	Preferable to continue pumping with suction lift pump during measurement

[a]See reference 38.
[b]See reference 35.
[c]Can be either an electrode or an ampoule filled with a colorimetric reagent (84).

in the lab by using an oxygen-specific electrode that adapts to the neck of the bottle. When the oxygen concentration is <30 μM, the ampoulated Rhodazine-D procedure is recommended.

DENITRIFICATION

When nitrate is an electron acceptor, it is reduced either to nitrogen gas by denitrification or to ammonium by dissimilatory nitrate reduction. Each process is respiratory in nature, each produces nitrite as an intermediate, and yet each is catalyzed by a physiologically different group of microorganisms. Nitrate reduction to ammonium appears to be the preferred pathway when the electron supply greatly exceeds the level of available nitrate (77). However, this is an uncommon situation for groundwater, so the process is probably also uncommon. In groundwater studies in which both processes were examined, denitrification was predominant (60, 73, 75), but in one study, dissimilatory nitrate reduction was significant in microcosm experiments with aquifer core material (11). Aerobic (i.e., oxygen-indifferent) denitrification has also been reported (61), though it is also not likely to be significant in groundwater.

Denitrification usually leaves a detectable impact on the nitrogen geochemistry of an aquifer. The presence of nitrite and nitrous oxide concomitant with nitrate is often found when the process is occurring and is therefore a good indicator. However, these compounds also result from nitrification. Concentrations of nitrogen gas exceeding the atmospheric equilibrium are good additional evidence for denitrification, although care must be exercised in sampling and analysis to avoid loss of the excess nitrogen via bubble formation (73). Even stronger evidence for denitrification would be concomitant disappearance of nitrate and an increase of nitrogen gas along a known flow path in the aquifer (13).

Microbial processes, especially dissimilatory processes, commonly fractionate low-molecular-weight compounds because of the tendency of lighter isotopes to react faster than heavier isotopes. The result is an enrichment of the heavier isotopes in the substrate and the lighter isotopes in the products. This fractionation mechanism can be used to determine whether nitrate is being utilized as a terminal electron acceptor. As nitrate is consumed, the remaining nitrate pool is sequentially enriched with ^{15}N. The degree of enrichment is directly dependent upon the amount of the original nitrate that has been reduced. Groundwater nitrate can be readily sampled, concentrated by freeze-drying, if necessary, and analyzed by mass spectrometry for the $^{15}N/^{14}N$ ratio. This natural abundance ratio (see reference 23 for an overview) can be used to infer the nitrate source in groundwater (4, 31, 44) and the extent of reduction that has occurred (15, 55, 73, 75). When coupled with ^{15}N determination of the products of reduction (N_2 and NH_4^+) and a nitrogen mass balance, the stable isotope analysis distinguishes the relative extent of denitrification and/or dissimilatory nitrate reduction (73). Recently, a method involving partial reduction of nitrate to nitrous oxide by denitrifying bacteria has made analysis of N and O isotopes in nitrate more feasible (17, 66).

There are two incubation assays that have the sensitivity suitable to measure denitrification in groundwater samples. The first is the acetylene blockage technique. Acetylene inhibits the terminal enzyme of the denitrification pathway, nitrous oxide reductase, causing an accumulation of nitrous oxide when it is present (7, 86). A gas chromatograph equipped with an electron capture detector can readily detect low concentrations of nitrous oxide. Incubations are most easily conducted in flasks with an anoxic headspace (69). A good design is to slurry the core material with groundwater from the same site and incubate the slurry at in situ temperatures. The level of production of nitrous oxide in the flasks is usually referred to as the denitrification potential to clearly distinguish that the incubation conditions deviated from in situ conditions. The acetylene block technique has also been used for incubations in intact cores (71) and for in situ assays by injecting acetylene-laden water directly into an aquifer (71, 80). It should be noted that acetylene has been used almost exclusively for heterotrophic denitrification. Its utility for autotrophic denitrification, which can be significant in the subsurface (43), is unclear.

The other incubation assay uses ^{15}N as a tracer. Whereas the acetylene block technique is specific only for denitrification, incubation with ^{15}N-enriched nitrate can detect whether either or both denitrification and dissimilatory nitrate reduction to ammonium are occurring and the relative importance of each. In addition, this approach can also be used to examine the turnover of nitrous oxide, if sufficient quantities are present. The types of incubations are similar to those for the acetylene block method; that is, they can be sediment slurries (69) or can be conducted within the core (11) or in situ (68). Once the incubation is completed, the various nitrogen species are extracted and analyzed for ^{15}N content. Although this technique is much more involved and requires access to a mass spectrometer, a great deal more information can be obtained with this method than with the acetylene block incubations.

IRON REDUCTION

In many aspects, iron reduction differs from the other terminal electron-accepting reactions. The substrate, Fe(III), is present in many forms, all crystalline solids, and often in large quantities in the saturated subsurface. The very low aqueous solubility of Fe(III) requires that iron-reducing microorganisms be in direct contact with the mineral surface to use it as an electron acceptor. It also precludes quantifying of the Fe(III) supply by only collecting groundwater samples. The product, Fe(II), is a common, yet undesired constituent in groundwater drinking supplies. Its concentration in groundwater is controlled by a complicated set of precipitation-dissolution and ion exchange reactions. The presence of Fe(II) in groundwater is a clear indication that iron reduction has occurred or is occurring at some point upgradient from the sampling point, but the concentration cannot be used to infer the extent of the reaction (2). In general, the reduction is likely the result of microbial activity. Abiotic iron reduction appears to be much less common than was previously thought (46). The Ferrozine colorimetric assay (30) has good sensitivity for Fe(II) and is most frequently the method of choice (2).

A demonstration that reducible forms of Fe(III) are present in the aquifer matrix greatly strengthens the conclusion that iron reduction is occurring. However, not all forms of Fe(III) are biologically available (46). Those that are amorphous or less crystalline are more readily reducible. There appears to be some correlation between certain extraction techniques and the readily reducible fractions of Fe(III) (34), but in actuality, the analytical procedures quantify an operationally defined fraction of the total iron present. Therefore, the prudent approach is to measure the amorphous Fe(III) oxides by using 0.5 M HCl extraction or the Fe(III) oxidation capacity by using Ti(III)-EDTA extractions (34) and to couple that procedure with sediment incubations (2). Production of Fe(II) during these incubations is clear evidence that both iron-reducing bacteria and the reducible forms of Fe(III) are present in the core material. Additional sediment incubations amended with synthetic Fe(III) oxides (51), with and without added acetate, can be used as positive controls to determine whether the samples have the potential capacity to reduce iron if it is present in an available form (2). Extraction and analysis of both solid and dissolved Fe(II) are necessary to determine the total amount of iron reduced during these incubations. Use of sediment slurries (see above) is the best technique for iron reduction incubations.

SULFATE REDUCTION

Despite the obvious presence of sulfides in a water sample, detecting the occurrence of sulfate reduction in an aquifer can be challenging. Low rates of activity can be coupled with relatively large quantities of metals that precipitate the sulfides as they are produced. In sulfate-rich formations, mineral dissolution and subsequent increase in groundwater sulfate content can greatly exceed and therefore mask sulfate removal via sulfate reduction (18). There are even situations in which sulfate entrainment into an aquifer via dissolution from confining layers can be matched by sulfate reduction. Hence, the process may be occurring although the substrate is apparently absent (18). Therefore, the geochemical evidence must be examined carefully to ascertain whether sulfate reduction is occurring. For this terminal electron-accepting reaction in particular, quantification of dissolved-hydrogen concentrations may provide important ancillary information (see below).

Cooccurrence of sulfate and sulfide in groundwater would constitute a clear indication that sulfate reduction is occurring. This is the situation in marine sediments and likely extends to saltwater intrusions and brines in the saturated subsurface. However, more often, especially for freshwater aquifers, detection of acid-volatile sulfides (primarily H_2S, HS^-, and FeS) in core material is necessary to arrive at the same conclusion. Pyrite and elemental sulfur, which are diagenetic end products of sulfide generation, are not particularly good indicators of recent sulfate reduction, although increases in extractable pyrite along a groundwater flow path can provide suggestive evidence for a sulfide source (19). Isotopic fractionation of sulfur isotopes has also been used to indicate sulfate reduction and is the geochemical tool used most frequently to detect sulfate reduction in groundwater. Like nitrate reduction, sulfate reduction favors reaction by the lighter isotope, resulting in an enrichment of ^{34}S in the unreacted sulfate. In some cases, the enrichment of ^{34}S in sulfate can be used to infer the occurrence of sulfate reduction along a flow path even when sulfate concentrations are increasing by dissolution (59). An accompanying depletion of ^{34}S in the sulfide pools would also be expected and has been used on occasion in groundwater studies as additional evidence for sulfate reduction (21). Sampling and assaying for sulfate are straightforward (3). For sulfide, avoid contact of the water sample with metal pipes and with oxygen, and avoid prolonged sample storage. Both colorimetric and titrimetric assays are appropriate for groundwater sulfide (3).

In surface water environments, sulfate reduction is most often assayed by using [^{35}S] sulfate (39), either by conducting microcosm incubations or by injecting the radiolabeled tracer into intact cores. Activity is determined by recovering and quantifying ^{35}S from the various reduced-sulfur pools (26). However, there have been few groundwater studies that have directly assessed sulfate reduction by using this technique (21, 33, 81, 82). Incubations could be conducted in the presence and absence of an added electron donor (similar to the situation for iron reduction) to demonstrate sulfate-reducing potential. A microautoradiographic procedure that detects [^{35}S]sulfide from [^{35}S]sulfate-treated shale and sandstone has also been developed to visualize the fine-scale structure of sulfate-reducing zones (45). Other useful approaches for assessing sulfate reduction activity include in situ tests that have the advantage of effecting only minimal disturbance to the environment under investigation (33, 41, 56, 62). In situ evaluations are particularly useful in providing sulfate-reduction rates and rate constants that are not influenced by laboratory processing of environmental samples. Also common are the enumeration and isolation of sulfate-reducing bacteria from groundwater and aquifer core material (9, 58, 83). This approach is based upon the assumption that when sulfate is the terminal electron acceptor, larger numbers of sulfate-reducing bacteria are present. Although this is often the case, interpretations must be made with caution, by using appropriate controls, and would be unarguably strengthened if supported with results from activity assays.

METHANOGENESIS

Methane is a common constituent of deep groundwater, especially in sedimentary basins and in heavily contaminated shallow aquifers (27). It is a relatively stable compound (in the absence of oxygen) and can persist in the subsurface for long periods of time. Detection of methane does not necessarily indicate active methanogenesis. Methane can originate from thermogenic as well as biogenic processes, entering groundwater from coal and petroleum deposits and from mining and drilling activities. In addition, biogenic methane is not necessarily contemporary. It may have been produced at a considerable distance upgradient from the sample site and/or much earlier in time.

Nonetheless, the first step for determining whether methane production is the terminal electron-accepting reaction is to quantify dissolved methane in groundwater. Sampling for methane involves the same considerations as those for any dissolved gas (see the discussion above regarding oxygen); analysis is accomplished with headspace equilibration and gas chromatography (10, 67, 74). For shallow aquifers with relatively high rates of methane production, gas collection devices have also been described that cumulatively collect methane as a gas over a several-week period (1, 10). Thermogenic methane can be distinguished from biogenic methane by detecting the presence of higher hydrocarbons (e.g., ethane, butane, and propane) and by determining the carbon isotope composition ($^{13}C/^{12}C$ ratio) of the methane. Thermogenic methane tends to be heavier than biogenic methane, with −58‰ the approximate value separating the two methane sources. However, methane oxidation can confound this interpretation (8). In addition, microbial methane production via carbon dioxide reduction fractionates the carbon dioxide (or, more appropriately, the dissolved-inorganic-carbon pool), enriching it with ^{13}C. Several studies have used these geochemical signatures to infer the source of methane in groundwater (6, 8, 32, 67). Complete isotopic characterization (C and H) may allow one to distinguish between methane produced from carbon dioxide reduction and that produced from acetate fermentation (85).

The relative simplicity and sensitivity of the methane assay greatly facilitate laboratory incubations to demonstrate methanogenesis. Aquifer core material can be incubated in sealed containers with an anoxic headspace for long periods of time, and the concentration of methane in the headspace can be assayed periodically. This method has been employed for many types of degradation experiments performed using methane-producing conditions with subsurface material. It is also valid for simply demonstrating that methanogenesis is occurring. If background methane concentrations are high, it may be necessary to purge the headspace to improve the analytical sensitivity for quantifying small increases in methane concentration. Positive controls to which methane precursors have been added (e.g., acetate, formate, methanol, and trimethylamine) are useful to demonstrate whether the activity can be stimulated (9). ^{14}C-labeled substrates can be used to quantify the rates of the individual methane-producing reactions but are probably unnecessary to simply ascertain whether methanogenesis is the operative electron-accepting reaction. Enumeration of methanogenic bacteria would provide additional evidence for methane production. Methanogens would not be expected in significant numbers when other electron acceptors are functioning. However, as with sulfate reduction, suitable controls and samples from nonmethanogenic areas may be necessary to distinguish electron donor-poor situations in which methanogenic populations are small from those areas where methanogens are present but inactive.

DISSOLVED HYDROGEN

Concentrations of dissolved hydrogen can serve as an additional indicator of the predominant terminal electron-accepting reaction (47). Hydrogen is known to be an important intermediate for most anaerobic food chains and serves as a direct link between fermentative and respiratory processes. Together, these processes have been termed interspecies hydrogen transfer (36) and are characterized by rapid production and consumption of dissolved hydrogen. Conceptually, under steady-state conditions the concentration of hydrogen is dictated by the thermodynamic energy yield for the hydrogen-consuming (or electron-accepting) step and not by the kinetics of hydrogen consumption. A greater energy yield from hydrogen oxidation results in a lower threshold hydrogen concentration, which in turn allows more energetic electron-accepting reactions to competitively exclude the less favorable reactions. A characteristic (empirical) range of steady-state hydrogen concentrations can be attributed to each terminal electron acceptor (Table 3) and is independent of the electron donor supply (47, 50). The extremely low concentrations of hydrogen in natural environments are indicative of a very short half-life, and they imply that a hydrogen measurement reflects the predominant electron-accepting reaction occurring in the immediate vicinity of a sample collection site rather than at some distance upgradient.

Hydrogen analysis has been utilized in a number of field studies to distinguish subsurface zones of electron-accepting reactions (19, 28, 37, 40, 47). The approach has the greatest potential for distinguishing sulfate reduction and methane production because they are characterized by the highest hydrogen concentrations (Table 3). However, as noted throughout this chapter, there can be situations in which data from substrate consumption, product formation, or activity assays of the terminal electron-accepting reaction are equivocal. These are situations in which determination of dissolved-hydrogen concentrations could be most valuable.

Environmental concentrations of hydrogen are extremely low. Therefore, sampling requires special precautions, and detection and quantification require specialized instrumentation. Metal surfaces catalyze the production of hydrogen in aqueous solutions, so they must be avoided during sample

TABLE 3 Anticipated hydrogen concentrations corresponding to terminal electron-accepting reactions[a]

Electron acceptor	Hydrogen concn (nM)
O_2	<0.1
NO_3^-	<0.1
Fe(III)	0.2–0.6
$SO_4^=$	1–4
CO_2	>5

[a]Data are from reference 19.

collection. Hydrogen is also a very dynamic compound. No adequate sample storage or preservation procedure has been found, and analysis should be completed within 30 min of sample collection (19). Hydrogen concentrations in groundwater are determined by using a gas-stripping procedure (19). Groundwater is continuously pumped through a gas-sampling bulb containing a small (5 to 20 ml) bubble of nitrogen gas until the hydrogen concentration within the bubble comes to equilibrium with the hydrogen concentration in the water. Next, the gas is analyzed by using a packed chromatographic column to separate the hydrogen, which is subsequently quantified with a Reduction Gas Detector (Trace Analytical).

ROLE OF ORGANIC COMPOUNDS IN ELECTRON-ACCEPTING PROCESSES

In addition to the more conventional reactions discussed previously, the influence that both synthetic and naturally occurring organic compounds exert on subsurface electron-accepting processes is an issue that is receiving increased attention. Discussion concerning the role of high-molecular-weight molecules in electron transfer usually takes place in the context of Fe(III)-reducing environments. Prevailing theory suggests that microorganisms catalyzing the reduction of Fe(III) must be physically close or even attached to the Fe(III)-bearing mineral. Soluble-Fe(III) concentrations in the subsurface are usually low, as this potential electron acceptor is typically found in various solid mineral forms. However, several organic molecules are known to act as facilitators of biological Fe(III) reduction in the subsurface and support increased rates of the process that would otherwise occur slowly or not at all. Involvement of these compounds essentially relieves the necessity for Fe(III)-reducing microorganisms to be in close physical proximity to the mineral phase.

One mechanism by which organic molecules facilitate microbial Fe(III) reduction includes the ability of synthetic chelators to form complexes with otherwise insoluble Fe(III) particles from the solid phase. The mechanism of chelation generates a chelator-Fe(III) complex that is much more soluble than Fe(III) alone. The soluble chelated complex is then free to migrate within subsurface pore water, be delivered to microorganisms that may be attached to particle surfaces or free living, and be used as an electron acceptor. Commercially available ligands such as nitrilotriacetic acid (NTA) and EDTA have been shown to chelate several crystalline forms of Fe(III) that are typically insoluble in subsurface environments. Concentrations of dissolved iron are higher in solutions containing such ligands, and pure cultures of dissimilatory Fe(III)-reducing bacteria incubated in the presence of these synthetic chelators exhibit higher rates of Fe(III) reduction than chelator-free preparations (5, 52). It follows that an increase in the rate of Fe(III) reduction would be coincident with an increased rate of electron donor oxidation. This logic is particularly intriguing in the context of contaminated aquifers. In polluted subsurface environments where biodegradation rates are limited by the supply of electron acceptors, chelators may stimulate the overall rate of contaminant mineralization, thereby serving as an important remedial strategy. In fact, several synthetic chelators were shown to support increased rates of toluene and benzene biodegradation in petroleum-contaminated sediments undergoing Fe(III) reduction (53, 54). Thus, the ability of chelators to solublize Fe(III) can promote microbial Fe(III) reduction as the dominant electron-accepting process even when the source of the oxidant is an otherwise insoluble Fe(III)-bearing mineral.

Naturally occurring forms of organic matter, such as humic acids, can participate in terminal electron-accepting processes and exert a strong influence on the rate and extent of organic matter oxidation, particularly when Fe(III) is present. Humic substances are complex molecules that consist primarily of highly branched aromatic residues resulting from partial decomposition of terrestrial vegetation. Humic acids are known to influence microbial metabolism, and more recently their role in mediating electron transfer reactions has been studied in more detail (20, 22, 63). Insight into the role played by humic acids in subsurface electron-accepting processes first came from incubations of contaminated sediments (54). In Fe(III)-reducing slurries from a petroleum-contaminated aquifer, benzene was degraded at a higher rate in the presence of humic acids, and the rate of degradation was even higher than that observed with synthetic chelators, including NTA. Although humic acids exhibit the capacity to chelate metals, their ability to do so is low relative to that of NTA, suggesting that chelation was not the mechanism responsible for supporting the stimulated rate of benzene degradation observed in the slurries. In fact, rather than increasing the solubility of Fe(III) by chelation, humic acids act as intermediary electron transfer molecules participating in a two-step process that effectively conveys electrons from the microorganism catalyzing the oxidation reaction to the Fe(III)-bearing mineral (Fig. 3). Subsurface microorganisms use humic acids as electron shuttles by reducing the quinone moieties contained in the humic structures (64). Then, in the presence of Fe(III), the reduced form of the humic material rapidly catalyzes the abiotic reduction of Fe(III). This abiotic reduction can actually occur with solid-phase Fe(III) existing in mineral forms not typically used by dissimilatory-Fe(III)-reducing bacteria (49).

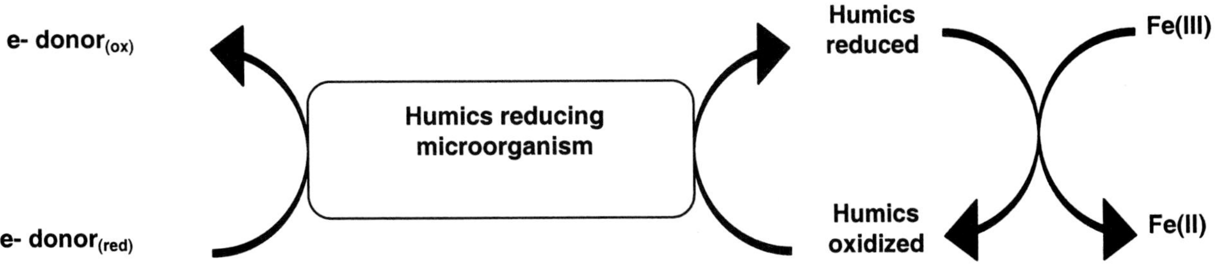

FIGURE 3 Mechanism of humic acid-mediated iron reduction by humic acid-reducing microorganisms in the subsurface (modified from reference 48). e−, electron; ox, oxidized; red, reduced.

Furthermore, reoxidized humic molecules can participate in additional electron-shuttling cycles, serving as a continuously recycled reactant in the redox process. Because of their capacity to be recycled, it seems reasonable that humic acids need be present in only catalytic amounts for their electron-shuttling activity to make a substantial contribution to subsurface Fe(III) reduction provided that the environment contains Fe(III).

Since humic acids facilitate Fe(III) reduction in the subsurface, it follows that electron donor oxidation would be influenced by the presence of humic substances. Complex assemblages of microorganisms can utilize humic acids as electron-shuttling compounds during the oxidation of a variety of electron donors. Addition of humic acids to contaminated slurries can stimulate anaerobic biodegradation of several pollutants including vinyl chloride, dichloroethene, methyl *tert*-butyl ether, and *tert*-butyl alcohol (16, 25). Pure cultures of Fe(III)-reducing microorganisms can use humic acids as an electron acceptor when oxidizing acetate or H_2 (48, 49). The ability of microorganisms to reduce humic acids is not confined to metal-reducing species. Several fermentative bacteria have been shown to reduce humic substances in pure culture (12). The presence of humic acids in the pure cultures shifted the fermentation pathway in favor of more oxidized end products such as acetate. Furthermore, when Fe(III) was added to the fermenting incubations, stoichiometric amounts of Fe(II) were produced and less acetate was formed, suggesting that the humic substances served as electron shuttles that were reoxidized via abiotic Fe(III) reduction. Interestingly, several methanogenic microorganisms can reduce both soil humic acids and the humic acid analogue anthroquinone-2,6-disulfonate (AQDS) (14). In the presence of Fe(III), AQDS increases Fe(II) production and reduces the amount of methane produced by methanogenic cultures oxidizing hydrogen. Thus, humic substances in subsurface habitats may act as electron shuttles between resident microorganisms and Fe(III) even in cases in which the microorganisms are not capable of catalyzing Fe(III) reduction. In addition to acting as an electron shuttle, humic acids can be respired by pure cultures of Fe(III)-reducing bacteria and serve as the sole electron acceptor supporting growth in the absence of Fe(III) (24, 49). Some humate-respiring microorganisms appear to have the capacity to synthesize extracellular electron-shuttling molecules that serve to transfer electrons from the cell to spatially distant oxidants (57, 76). This finding indicates that small, microbially produced quinones play a role in extracellular electron transfer to humic acids and eventual Fe(III) reduction. Microorganisms exhibit extraordinary respiratory mechanisms in the laboratory; it seems reasonable that the means of electron transfer in the terrestrial subsurface would be at least as remarkable as those observed in the laboratory.

The nearly ubiquitous distribution of humic substances combined with their apparent ability to participate in a variety of electron-accepting reactions suggests that the contribution of these materials to subsurface microbial processes may be more important than previously thought. Despite the potential importance of humic acids to subsurface redox reactions, it is difficult to definitively identify environments where humic acid reduction plays an active part in these reactions, including Fe(III)-dependent organic matter oxidation. This difficulty is in part due to the tendency of humic acids to cycle between reduced and oxidized forms such that there is no obvious depletion or accumulation of dissolved humic material that would implicate humic acids as being utilized by the microbial populations. Furthermore, since humic acids are common constituents of subsurface habitats, simply identifying the presence of these substances in groundwater is not likely to be a useful indicator. Although it has not yet been determined, it is plausible that a steady-state hydrogen concentration may exist that would implicate humic acid reduction as a dominant reaction, as has been shown for other terminal electron-accepting processes (47). However, if the final electron acceptor is Fe(III), as illustrated in Fig. 3, dissolved-hydrogen concentrations may simply equilibrate at a value that is consistent with Fe(III) reduction without indicating whether humic acids are participants in microbially catalyzed electron transfer in the subsurface.

FINAL PERSPECTIVE

Determining the terminal electron-accepting reaction(s) at a groundwater study site can be a time-consuming and involved process, in large part because of the inaccessible nature of the subsurface and the low rates at which the reactions occur. Yet, whether from the basic microbiological, bioremediative, or drinking water perspective, this information is very important for understanding and interpreting the nature of the microbial food chain and the effects that it may have upon groundwater geochemistry. There is also the global perspective: the saturated subsurface is the largest freshwater habitat on the planet. It is a unique habitat, one about which very little is known with respect to its microbial ecology. An integral part of future characterizations and insights regarding microbial communities in the subsurface must necessarily involve the context of the terminal electron-accepting reactions.

REFERENCES

1. **Adrian, N. R., J. A. Robinson, and J. M. Suflita.** 1994. Spatial variability in biodegradation rates as evidenced by methane production from an aquifer. *Appl. Environ. Microbiol.* **60:**3632–3639.
2. **Albrechtsen, H.-J., G. Heron, and T. H. Christensen.** 1995. Limiting factors for microbial Fe(III)-reduction in a landfill leachate polluted aquifer (Vejen, Denmark). *FEMS Microbiol. Ecol.* **16:**233–248.
3. **American Public Health Association.** 1981. *Standard Methods for the Examination of Water and Wastewater.* American Public Health Association, Washington, DC.
4. **Aravena, R., M. L. Evans, and J. A. Cherry.** 1993. Stable isotopes of oxygen and nitrogen in source identification of nitrate from septic systems. *Ground Water* **31:**180–186.
5. **Arnold, R. G., T. M. Olson, and M. R. Hofmann.** 1986. Kinetics and mechanism of dissimilative Fe(III) reduction by *Pseudomonas* sp. 200. *Biotechnol. Bioeng.* **28:**1657–1671.
6. **Baedecker, M.J., and W. Back.** 1979. Modern marine sediments as a natural analog to the chemically stressed environment of a landfill. *J. Hydrol.* **43:**393–414.
7. **Balderston, W. L., B. Sherr, and W. J. Payne.** 1976. Blockage by acetylene of nitrous oxide reduction in *Pseudomonas perfectomarinus. Appl. Environ. Microbiol.* **31:**504–508.
8. **Barker, J. F., and P. Fritz.** 1981. The occurrence and origin of methane in some groundwater flow systems. *Can. J. Earth Sci.* **18:**1802–1816.
9. **Beeman, R. E., and J. M. Suflita.** 1987. Microbial ecology of a shallow unconfined groundwater aquifer polluted by municipal landfill leachate. *Microb. Ecol.* **14:**39–54.
10. **Beeman, R. E., and J. M. Suflita.** 1990. Environmental factors influencing methanogenesis in a shallow anoxic

aquifer: a field and laboratory study. *J. Ind. Microbiol.* **5:** 45–58.

11. **Bengtsson, G., and H. Annadotter.** 1989. Nitrate reduction in a groundwater microcosm determined by 15N gas chromatography-mass spectrometry. *Appl. Environ. Microbiol.* **55**:2861–2870.
12. **Benz, M., B. Schink, and A. Brune.** 1998. Humic acid reduction by *Propionibacterium freudenreichii* and other fermenting bacteria. *Appl. Environ. Microbiol.* **64:**4507–4512.
13. **Böhlke, J. K., R. Wanty, M. Tuttle, G. Delin, and M. Landon.** 2002. Denitrification in the recharge area and discharge area of a transient agricultural nitrate plume in a glacial outwash sand aquifer, Minnesota. *Water Resour. Res.* [Online.] doi:10.1029/2001WR000663.
14. **Bond, D., and D. R. Lovley.** 2002. Reduction of Fe(III) oxide by methanogens in the presence and absence of extracellulear quinones. *Environ. Microbiol.* **4:**115–124.
15. **Böttcher, J., O. Strebel, S. Voerkelius, and H. L. Schmidt.** 1990. Using isotope fractionation of nitrate-nitrogen and nitrate-oxygen for evaluation of microbial denitrification in a sandy aquifer. *J. Hydrol.* **114:**413–424.
16. **Bradley, P. M, F. H. Chapelle, and D. R. Lovley.** 1998. Humic acids as electron acceptors for anaerobic microbial oxidation of vinyl chloride and dichloroethene. *Appl. Environ. Microbiol.* **64:**3102–3105.
17. **Casciotti, K. L., D. M. Sigman, M. G. Hastings, J. K. Böhlke, and A. Hilkert.** 2002. Measurement of the oxygen isotopic composition of nitrate in seawater and freshwater using the denitrifier method. *Anal. Chem.* **74:**4905–4912.
18. **Chapelle, F. H.** 1993. *Ground-Water Microbiology and Geochemistry.* John Wiley & Sons, Inc., New York, N.Y.
19. **Chapelle, F. H., P. B. McMahon, N. M. Dubrovsky, R. F. Fujii, E. T. Oaksford, and D. A. Vroblesky.** 1995. Deducing the distribution of terminal electron-accepting processes in hydrologically diverse groundwater systems. *Water Resour. Res.* **31:**359–371.
20. **Curtis, G. P., and M. Reinhard.** 1994. Reductive dehalogenation of hexachloroethane, carbon tetrachloride, and bromoform by anthrahydroquinone disulfonate and humic acid. *Environ. Sci. Technol.* **28:**2393–2401.
21. **Dockins, W. S., G. J. Olson, G. A. McFeters, and S. C. Turbak.** 1980. Dissimilatory bacterial sulfate reduction in Montana groundwaters. *Geomicrobiol. J.* **2:**83–97.
22. **Dunnivant, F. M., and R. P. Schwarzenbach.** 1992. Reduction of substituted nitrobenzenes in aqueous solutions containing natural organic matter. *Environ. Sci. Technol.* **26:**2133–2141.
23. **Ehleringer, J. R., and P. W. Rundel.** 1989. Stable isotopes: history, units, and instrumentation, p. 1–15. *In* P. W. Rundel, J. R. Ehleringer, and K. A. Nagy (ed.), *Stable Isotopes in Ecological Research.* Springer-Verlag, New York, N.Y.
24. **Finneran, K. T., H. M. Forbush, C. V. Gaw VanPraagh, and D. R. Lovley.** 2002. *Desulfitobacterium metallireducens* sp. nov., an anaerobic bacterium that couples growth to the reduction of metals and humic acids as well as chlorinated compounds. *Int. J. Syst. Evol. Microbiol.* **52:**1929–1935.
25. **Finneran, K. T., and D. R. Lovley.** 2001. Anaerobic degradatin of methyl *tert*-butyl ether (MTBE) and *tert*-butyl alcohol (TBA). *Environ. Sci. Technol.* **35:**1785–1790.
26. **Fossing, H., and B. B. Jörgensen.** 1990. Isotope exchange reactions with radiolabeled sulfur compounds in anoxic seawater. *Biogeochemistry* **9:**223–245.
27. **Freeze, R. A., and J. A. Cherry.** 1979. *Groundwater.* Prentice-Hall, Inc., Englewood Cliffs, N.J.
28. **Geig, L. M., R. V. Kolhatkar, M. J. McInerney, R. S. Tanner, S. H. Harris, K. L. Sublette, and J. M. Suflita.** 1999. Intrinsic bioremediation of petroleum hydrocarbons in a gas condensate-contaminated aquifer. *Environ. Sci. Technol.* **33:**2550–2560.
29. **Ghiorse, W. C., and J. T. Wilson.** 1988. Microbial ecology of the terrestrial subsurface. *Adv. Appl. Microbiol.* **33:**107–172.
30. **Gibbs, M. M.** 1979. A simple method for the rapid determination of iron in natural waters. *Water Res.* **13:**295–297.
31. **Gormly, J. R., and R. F. Spalding.** 1979. Sources and concentrations of nitrate-nitrogen in groundwater of the Central Platte region, Nebraska. *Ground Water* **17:**291–301.
32. **Grossman, E. L., B. K. Coffman, S. J. Fritz, and H. Wada.** 1989. Bacterial production of methane and its influence on ground-water chemistry in east-central Texas aquifers. *Geology* **17:**495–499.
33. **Harris, S. H., J. D. Istok, and J. M. Suflita.** 2006. Changes in organic matter biodegradability influencing sulfate reduction in an aquifer contaminated by landfill leachate. *Microb. Ecol.* **51:**535–542.
34. **Heron, G., C. Crouzet, A. C. M. Bourg, and T. H. Christensen.** 1994. Speciation of Fe(II) and Fe(III) in contaminated aquifer sediments using chemical extraction techniques. *Environ. Sci. Technol.* **28:**1698–1705.
35. **Holm, T. R., G. K. George, and M. J. Barcelona.** 1988. Oxygen transfer through flexible tubing and its effects on groundwater sampling results. *Ground Water Monit. Rev.* **8:**83–89.
36. **Ianotti, E. L., P. Kafkewitz, M. J. Wolin, and M. P. Bryant.** 1973. Glucose fermentation products of *Ruminococcus albus* grown in continuous culture with *Vibrio succinogenes*: changes caused by interspecies transer of H_2. *J. Bacteriol.* **114:**1231–1240.
37. **Jakobsen, R., H. J. Albrechtsen, M. Rasmussen, H. Bay, P. L. Bjerg, and T. H. Christensen.** 1998. H_2 concentrations in a landfill leachate plume (Grindsted, Denmark): in situ energetics of terminal electron acceptor processes. *Environ. Sci. Technol.* **32:**2142–2148.
38. **Johnson, R. L., J. F. Pankow, and J. A. Cherry.** 1987. Design of a ground-water sampler for collecting volatile organics and dissolved gases in small-diameter wells. *Ground Water* **25:**448–454.
39. **Jörgensen, B. B.** 1978. A comparison of methods for the quantification of bacterial sulfate reduction in coastal marine sediments. *Geomicrobiol. J.* **1:**29–64.
40. **Kassenga, G., J. H. Pardue, W. M. Moe, and K. S. Bowman.** 2004. Hydrogen thresholds as indicators of dehalorespiration in constructed treatment wetlands. *Environ. Sci. Technol.* **38:**1024–1030.
41. **Kleikemper, J., M. H. Schroth, W. V. Sigler, M. Schmucki, S. M. Bernasconi, and J. Zeyer.** 2002. Activity and diversity of sulfate-reducing bacteria in a petroleum hydrocarbon-contaminated aquifer. *Appl. Environ. Microbiol.* **68:**1516–1523.
42. **Knowles, R.** 1982. Denitrification. *Microbiol. Rev.* **46:**43–70.
43. **Korom, S. F.** 1992. Natural denitrification in the saturated zone: a review. *Water Resour. Res.* **28:**1657–1668.
44. **Kreitler, C. W., S. E. Ragone, and B. G. Katz.** 1978. N15/N14 ratios of groundwater nitrate, Long Island, New York. *Ground Water* **16:**404–409.
45. **Krumholz, L. R., J. P. McKinley, G. A. Ulrich, and J. M. Suflita.** 1997. Confined subsurface microbial communities in Cretaceous rock. *Nature* **386:**64–66.
46. **Lovley, D. R.** 1991. Dissimilatory Fe(III) and Mn(IV) reduction. *Microbiol. Rev.* **55:**259–287.
47. **Lovley, D. R., F. H. Chapelle, and J. C. Woodward.** 1994. Use of dissolved H_2 concentrations to determine distribution of microbially catalyzed reactions in anoxic groundwater. *Environ. Sci. Technol.* **28:**1205–1210.
48. **Lovley, D. R., H. D. Coates, E. L. Blunt-Harris, E. J. P. Phillips, and J. C. Woodward.** 1996. Humic substances as

electron acceptors for microbial respiration. *Nature* **382:** 445–448.

49. **Lovley, D. R., J. L. Fraga, E. L. Blunt-Harris, L. A. Hayes, E. J. P. Phillips, and J. D. Coates.** 1998. Humic substances as a mediator for microbially catalyzed metal reduction. *Acta Hydrochim. Hydrobiol.* **26:**152–157.
50. **Lovley, D. R., and S. Goodwin.** 1988. Hydrogen concentrations as an indicator of the predominant terminal electron-accepting reactions in aquatic sediments. *Geochim. Cosmochim. Acta* **52:**2993–3003.
51. **Lovley, D. R., and E. J. P. Phillips.** 1986. Availability of ferric iron for microbial reduction in bottom sediments of the freshwater tidal Potomac River. *Appl. Environ. Microbiol.* **52:**751–757.
52. **Lovley, D. R., and J. C. Woodward.** 1996. Mechanisms for chelator stimulation of microbial Fe(III)-oxide reduction. *Chem. Geol.* **132:**19–24.
53. **Lovley, D. R., J. C. Woodward, and F. H. Chapelle.** 1994. Stimulated anoxic biodegradation of aromatic hydrocarbons using Fe(III) ligands. *Nature* **370:**128–131.
54. **Lovley, D. R., J. C. Woodward, and F. H. Chapelle.** 1996. Rapid anaerobic benzene oxidation with a variety of chelated Fe(III) forms. *Appl. Environ. Microbiol.* **62:**288–291.
55. **Mariotti, A., A. Landreau, and B. Simon.** 1988. 15N isotope biogeochemistry and natural denitrification process in groundwater: application to the chalk aquifer of northern France. *Geochim. Cosmochim. Acta* **52:**1869–1878.
56. **McGuire, J. T., D. T. Long, M. J. Klug, S. K. Haack, and D. W. Hyndman.** 2002. Evaluating behavior of oxygen, nitrate, and sulfate during recharge and quantifying reduction rates in a contaminated aquifer. *Environ. Sci. Technol.* **36:**2693–2700.
57. **Newman, D. K., and R. Kolter.** 2000. A role for excreted quinones in extracellular electron transfer. *Nature* **405:** 94–97.
58. **Olson, G. J., W. S. Dockins, G. A. McFeters, and W. P. Iverson.** 1981. Sulfate-reducing and methanogenic bacteria from deep aquifers in Montana. *Geomicrobiol. J.* **2:**327–340.
59. **Plummer, L. N., J. F. Busby, R. W. Lee, and B. B. Hanshaw.** 1990. Geochemical modeling of the Madison Aquifer in parts of Montana, Wyoming, and South Dakota. *Water Resour. Res.* **26:**1981–2014.
60. **Postma, D., C. Boesen, H. Kristiansen, and F. Larsen.** 1991. Nitrate reduction in an unconfined sandy aquifer: water chemistry, reduction processes, and geochemical modeling. *Water Resour. Res.* **27:**2027–2045.
61. **Robertson, L. A., and J. G. Kuenen.** 1984. Aerobic denitrification—old wine in new bottles? *Antonie Leeuwenhoek* **50:**525–544.
62. **Schroth, M. H., J. Kleikemper, C. Bolliger, S. M. Bernasconi, and J. Zeyer.** 2001. In situ assessment of microbial sulfate reduction in a petroleum-contaminated aquifer using push-pull tests and stable sulfur isotope analyses. *J. Contam. Hydrol.* **51:**179–195.
63. **Schwarzenbach, R. P., R. Stierli, K. Lanz, and J. Zeyer.** 1990. Quinone and iron porphyrin mediated reduction of nitroaromatic compounds in homogeneous aqueous solution. *Environ. Sci. Technol.* **24:**1566–1574.
64. **Scott, D. T., D. M. McKnight, E. L. Blunt-Harris, S. E. Kolesar, and D. R. Lovley.** 1998. Quinone moieties act as electron acceptors in the reduction of humic substances by humics-reducing microorganisms. *Environ. Sci. Technol.* **32:**2984–2989.
65. **Sexstone, A. J., N. P. Revsbech, T. B. Parkin, and J. M. Tiedje.** 1985. Direct measurement of oxygen profiles and denitrification rates in soil aggregates. *Soil Sci. Soc. Am. J.* **49:**645–651.
66. **Sigman, D. M., K. L. Casciotti, M. Andreani, C. Barford, M. Galanter, and J. K. Bohlke.** 2001. A bacterial method for the nitrogen isotopic analysis of nitrate in seawater and freshwater. *Anal. Chem.* **73:**4145–4153.
67. **Simpkins, W. W., and T. B. Parkin.** 1993. Hydrogeology and redox geochemistry of CH_4 in a late Wisconsinan till and loess sequence in central Iowa. *Water Resour. Res.* **29:** 3643–3657.
68. **Smith, R. L., J. K. Böhlke, S. P. Garabedian, K. M. Revesz, and T. Yoshinari.** 2004. Assessing denitrification in groundwater using natural gradient tracer tests with ^{15}N: in situ measurement of a sequential multistep reaction. *Water Resour. Res.* [Online.] doi:10.1029/2003WR002919.
69. **Smith, R. L., and J. H. Duff.** 1988. Denitrification in a sand and gravel aquifer. *Appl. Environ. Microbiol.* **54:** 1071–1078.
70. **Smith, R. L., and S. P. Garabedian.** 1998. Using transport model interpretations of tracer tests to study microbial processes in ground water, p. 94–123. *In* A. L. Koch, J. A. Robinson, and G. A. Milliken (ed.), *Mathematical Models: Applications to Microbial Ecology*. Chapman and Hall, New York, N.Y.
71. **Smith, R. L., S. P. Garabedian, and M. H. Brooks.** 1996. Comparison of denitrification activity measurements in groundwater using cores and natural gradient tracer tests. *Environ. Sci. Technol.* **30:**3448–3456.
72. **Smith, R. L., R. W. Harvey, and D. R. LeBlanc.** 1991. Importance of closely spaced vertical sampling in delineating chemical and microbiological gradients in groundwater studies. *J. Contam. Hydrol.* **7:**285–300.
73. **Smith, R. L., B. L. Howes, and J. H. Duff.** 1991. Denitrification in nitrate-contaminated groundwater: occurrence in steep vertical geochemical gradients. *Geochim. Cosmochim. Acta* **55:**1815–1825.
74. **Smith, R. L., B. L. Howes, and S. P. Garabedian.** 1991. In situ measurement of methane oxidation in groundwater by using natural-gradient tracer tests. *Appl. Environ. Microbiol.* **57:**1997–2004.
75. **Spalding, R. F., and J. D. Parrott.** 1994. Shallow groundwater denitrification. *Sci. Total Environ.* **141:**17–25.
76. **Straub, K. L., and B. Schink.** 2003. Evaluation of electron-shuttling compounds in microbial ferric iron reduction. *FEMS Microbiol. Lett.* **220:**229–233.
77. **Tiedje, J. M., A. J. Sexstone, D. D. Myrold, and J. A. Robinson.** 1982. Denitrification: ecological niches, competition and survival. *Antonie Leeuwenhoek* **48:**569–583.
78. **Trevors, J. T.** 1985. The influence of oxygen concentrations on denitrification in soil. *Appl. Microbiol. Biotechnol.* **23:**152–155.
79. **Trevors, J. T., and M. E. Starodub.** 1987. Effect of oxygen concentration on denitrification in freshwater sediment. *J. Basic Microbiol.* **7:**387–391.
80. **Trudell, M. R., R. W. Gillham, and J. A. Cherry.** 1986. An in-situ study of the occurrence and rate of denitrification in a shallow unconfined sand aquifer. *J. Hydrol.* **83:**251–268.
81. **Ulrich, G. A., G. N. Breit, I. M. Cozzarelli, and J. M. Suflita.** 2003. Sources of sulfate supporting anaerobic metabolism in a contaminated aquifer. *Environ. Sci. Technol.* **37:**1093–1099.
82. **Ulrich, G. A., D. Martino, K. Burger, J. Routh, E. L. Grossman, J. W. Ammerman, and J. M. Suflita.** 1998. Sulfur cycling in the terrestrial subsurface: commensal interactions, spatial scales, and microbial heterogeneity. *Microb. Ecol.* **36:**141–151.
83. **van Beek, C. G. E. M., and D. van der Kooij.** 1982. Sulfate-reducing bacteria in groundwater from clogging and nonclogging shallow wells in the Netherlands river region. *Ground Water* **20:**298–302.

84. **White, A. F., M. L. Peterson, and R. D. Solbau.** 1990. Measurement and interpretation of low levels of dissolved oxygen in groundwater. *Ground Water* **28:**584–590.
85. **Whiticar, M. J., E. Faber, and M. Schoell.** 1986. Biogenic methane formation in marine and freshwater environments: CO_2 reduction vs. acetate fermentation—isotope evidence. *Geochim. Cosmochim. Acta* **50:**693–709.
86. **Yoshinari, T., R. Hynes, and R. Knowles.** 1977. Acetylene inhibition of nitrous oxide reduction and measurement of denitrification and nitrogen fixation in soil. *Soil Biol. Biochem.* **9:**177–183.

Transport of Microorganisms in the Terrestrial Subsurface: In Situ and Laboratory Methods

RONALD W. HARVEY, HAUKE HARMS, AND LEE LANDKAMER

70

Although the use of microbial tracers has been instrumental in developing a better understanding of groundwater movement in many types of aquifers (1, 19, 39, 158, 166, 208, 209, 267), the importance of studying the transport behavior of the microorganisms themselves is now very apparent. This is due largely to an increasing dependence upon limited and fragile groundwater resources. In particular, widespread contamination of shallow drinking-water aquifers by microbial pathogens and chemical wastes has led to increased interest in the factors that control subsurface microbial transport. The movement of nonindigenous bacteria, viruses, and protozoa through aquifers has been a public health concern in the United States (see chapter 72), where inadvertent contamination of water supply wells by microbial pathogens contributes significantly to the total number of waterborne disease outbreaks (155). Deliberate additions of nonindigenous populations of bacteria into aquifer sediments are now conducted to enhance in situ rates of bioremediation of organically contaminated groundwater in lab and field scale bioaugmentation experiments (210, 214, 231, 234, 266) and to increase oil recovery from less transmissive zones by selective bacterial plugging of more-permeable strata (see chapter 71). The success of bioaugmentation can depend as much on the ability of the introduced microorganisms to reach the contaminants as in situ survival and factors affecting contaminant bioavailability (225) such as its pore scale distribution (83) and diffusion limitations (84). Transport properties of introduced bacteria in groundwater can have major roles in the transmission of some waterborne diseases (23), in the success of microbially enhanced oil recovery processes (45, 109, 154), in the mobility of surface-active or hydrophobic groundwater contaminants (44, 111, 130, 148), in pore clogging (15, 252), and in the potential subsurface dissemination of genetically engineered bacteria for biorestoration (243). However, many of the factors controlling subsurface microbial transport are still poorly understood.

To better study the controls of microbial transport, it has been necessary to improve available methodology for investigating microbial transport behavior in the lab and in the field. The many physical, chemical, and biological factors that affect microbial transport through aquifer materials (87) have necessitated refinement of techniques, allowing observations under more controlled experimental conditions. In this chapter, laboratory and field techniques for studying microbial transport behavior in aquifer materials and model porous media are described and discussed.

FIELD VERSUS LABORATORY STUDIES

Although most studies delineating the controls of subsurface microbial transport are performed in the laboratory, advances in our understanding of microbial transport behavior would clearly benefit from an iterative approach that combines both laboratory and field investigations (Fig. 1). Column studies provide a greater degree of control and, therefore, are useful in providing detailed, mechanistic information about specific processes affecting microbial transport. Also, columns can be designed to more nearly meet initial boundary conditions from microbial transport models. However, they cannot account for the combined effects of the various controls that govern microbial mobility in situ. In addition to the hydrological characteristics of the aquifer itself, there are many factors which control the movement of microorganisms in groundwater. For bacteria, these biotic and abiotic factors include growth, predation by protists, possible parasitism by bacteriophages, motility, lysis under unfavorable conditions, changes in cell size and propensity for attachment to solid surfaces in response to alterations in nutrient conditions, spore formation in the case of some gram-positive species, reversible and irreversible attachment to solid surfaces, detachment from surfaces, and straining (87). Many of these processes are interrelated through other factors, poorly understood, and/or difficult to describe mathematically. Therefore, it is desirable to combine laboratory experiments that allow better delineation of individual processes with field studies that provide a framework in which the applicability of laboratory-derived results can be evaluated under natural conditions.

Flowthrough column experiments have provided useful and detailed information on many processes governing microbial transport behavior (Table 1). However, it is often difficult to extrapolate laboratory results back to the subsurface environment for several reasons. First, the unavoidable alteration of secondary pore structure that results from manual or mechanical repacking of unsaturated or saturated subsurface materials substantively diminishes microbial transport (93, 228). Hydrodynamic dispersion, which is typically very small (i.e., a few centimeters) in columns and,

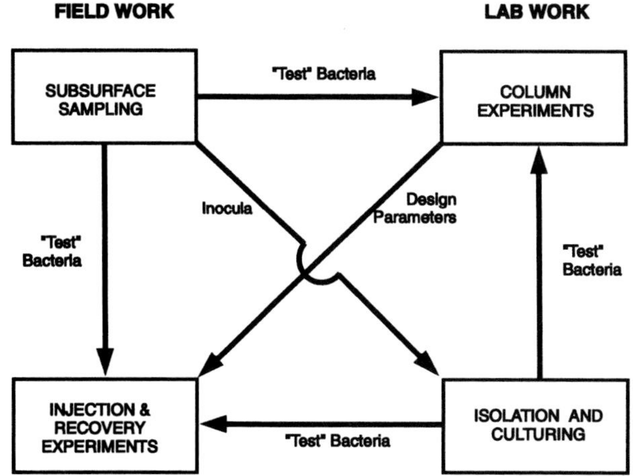

FIGURE 1 Integrated field and laboratory experimental approach to investigating transport behavior of microorganisms in groundwater environments.

therefore, relatively unimportant to microbial transport at the column scale (250) is typically very large (i.e., many meters) in field scale experiments (140). Finally, flow velocities used in columns are usually a good deal higher than the slow (<1-m · day^{-1}) groundwater velocities typically observed in the field. This is largely because most pumps cannot deliver accurately and steadily at very low volumetric flow rates. Such disparities in flow can affect the thickness of the hydrodynamic boundary layers (the zone of lower-velocity water immediately adjacent to grain surfaces) and attachment opportunities. It is important to consider that even modest (e.g., severalfold) changes in flow velocity can affect significantly the degree of immobilization (35, 37, 71, 100) and possibly pore size exclusion (37) that bacteria are subject to in saturated porous media. Although one cannot extrapolate microbial transport potential determined from column studies directly to the field scale, deposition parameters calculated from small flow through columns containing sediments collected from a number of different points in a granular aquifer may have at least some utility in predictions of field-scale transport using a probabilistic model (26).

BIOLOGICAL AND PHYSICOCHEMICAL FACTORS

Chemical Considerations

It has been shown that subtle differences in groundwater and/or surface chemistry can substantively affect the magnitude of microbial transport through aquifer sediments and the reversibility of microbial attachment (187). Therefore, in comparing the apparent transport behaviors of microorganisms in small-scale field or laboratory studies, it is important to separate out the effects of intentional and/or unintentional differences in chemistry.

It may be important to account for or control bacterial growth during subsurface transport in injection and recovery experiments. Uncontaminated aquifers are generally extremely oligotrophic. Consequently, bacterial growth in groundwater is generally very slow (6) and difficult to measure accurately using available methods (89). However, substantive bacterial growth can occur during transport downgradient in small-scale transport experiments as a result of the inadvertent introduction of dissolved organic matter (DOM) during injection into aquifers that are limited with respect to carbon. The introduction of labile DOM is particularly problematic when cultured bacteria are used as an inoculum, because it is difficult to prevent residual broth nutrients and bacterial metabolites and lysates from contaminating the injectate. Inadvertent contamination with DOM during the introduction of cultured isolates may be lessened by a priori resuspension in nongrowth media (49). Also, it is helpful to monitor changes in abundance over the experimental time course of labeled bacteria in collected samples held at in situ temperature (88).

Changes in ionic strength (I) during transport studies may occur inadvertently as a result of using halides as conservative tracers and may lead to density-induced sinking of the tracer cloud (176). Substantive increases in I as a result of injection of high concentrations of halide tracers can also result in overestimations of microbial attachment. The general observations of increasing attachment with increasing I may be predicted from the extended theory (251) of Derjaguin and Landau and the Verwey Derjaguin-Landau-Overbeek (DLVO) theory of colloid stability (211), because the microorganisms and surfaces employed are both negatively charged at the pH at which most microbial transport studies are run and the interactions are largely electrostatic in nature. The observation of increased attachment to grain surfaces with increasing I was well documented for viruses by the mid-1970s (57, 139, 264) and for bacteria in porous media by the 1990s (for examples, see references 35, 53, 64, 75, 114, 170, and 238). However, there can be exceptions. Adhesion onto negatively charged glass of a positively charged ($pH_{IEP} = 11$) clinical isolate, *Stenotrophomonas maltophilia* 70401, was shown to decrease with increasing I within the range of 0.001 to 1 M (124). Changes in the extent of bacterial attachment in aquifer sediments in response to substantive changes in I should be most pronounced when the initial I is low. For many aquifers, particularly those used as sources of potable water, I can be $\leq 10^{-3}$ M (for example, see reference 218) and considerable care must be taken when adding salts as tracers during injection and recovery experiments. In such aquifers, deuterated (203) or tritiated (168) water might be considered as alternative hydrologic tracers. Finally, it is important to avoid addition of salts containing bivalent cations, because bacterial attachment to granular media is very sensitive to their presence (223).

Changes in pH can have a considerable impact on microbial attachment to aquifer sediments (152, 270). The surface charge of microorganisms and sediment surfaces that control electrostatic interactions is driven by protonation reactions of surface functional groups, with pH being the master variable in this process. The pH may change as a result of solution CO_2 exchange with the atmosphere or aquifer minerals, as a result of metabolic processes, metal sorption to aquifer surfaces or simply injecting solutions with a pH different from the in situ pH. When the pH of groundwater is measured, a closed-cell pH meter should be used along with careful attention to technique so that the CO_2 content of the water does not change, which can result in rapid and substantial changes in pH. When performing column experiments, both the influent and effluent pHs should be closely monitored.

Also, DOM can affect the transport of microorganisms (4, 66) by affecting either the charge or the hydrophobicity of the microorganisms and/or the aquifer material. Ideally, injectate solutions should not have DOM concentrations substantially different from background levels.

TABLE 1 Selected column studies delineating factors controlling subsurface microbial transport behavior

Organism type and factors studied	Medium	Method	Conditions	Microorganism	Reference
Bacteria					
Attachment/detachment behavior					
Attachment/detachment kinetics	Glass beads	Flowthrough	Saturated, aerobic	*Comamonas strain* DA001	241
Secondary energy minimum	Clean sand	Flowthrough	Saturated, aerobic	*E. coli*	199
Attachment kinetics	Clean sand	Upflow	Saturated, aerobic	*P. fluorescens*	53
Attachment kinetics	Subsurface sand	Flowthrough	Saturated, aerobic	*Bacillus* sp., *Enterobacter* sp.	147
Blocking of subsequent attachment	Teflon beads	Downflow	Saturated, aerobic	*Pseudomonas putida*, *Rhodococcus strain* C125	205
Collision efficiency	Glass beads	Downflow	Saturated, aerobic	*P. aeruginosa*	160
Detachment	Silica beads	Flowthrough	Saturated, aerobic	*P. fluorescens*, *Bacillus subtilis*	134
Rates of attachment and detachment	Sand	Upflow, pulse	Saturated, aerobic	G-aquifer isolates	164
Reversible and irreversible sorption	Silica sand	Upflow	Saturated, aerobic	*Klebsiella oxytoca*	100
Surface residence time	Subsurface sand	Downflow	Saturated, aerobic	Subsurface isolate	120
Growth phase	Quartz sand	Flowthrough	Saturated, aerobic	*E. coli*	258
Cell surface properties					
Adhesion deficiency	Aquifer sediment	Upflow	Saturated, aerobic	*Burkholderia cepacia* G4	50
Cell surface chemistry	Glass and Teflon	Downflow	Saturated, aerobic	7 strains	205
Encapsulation (algenate)	Loamy soils	Downflow	Unsaturated, aerobic	*P. fluorescens*	99
Favorable transport properties	Berea sandstone	Horizontal flow	Saturated, aerobic	*Pseudomonas*, *Bacillus*, *Clostridium* spp.	109
Hydrophobicity and charge	Glass and Teflon	Static versus dynamic	Saturated, aerobic	Pseudomonads, coryneforms	204
Hydrophobicity and soil properties	Sandy and clay soil	Downflow	Unsaturated, aerobic	*Lactobacillus* spp.	107
Hydrophobicity and zeta potential	Kendaia loam	Downflow	Aerobic	Strains from 6 genera	72
Hydrophobicity, zeta potential, cell size, and aspect ratio	Soil, sand, clay	Downflow	Saturated, aerobic	*Arthrobacter*, *Mycobacterium*, *Sphingomonas* spp.	138
Interspecies differences	Glass beads	Downflow	Saturated, aerobic	*P. fluorescens*, *Alcaligenes paradoxus*	79
Intrastrain variability	Glass beads	Flowthrough	Saturated, aerobic	Subsurface isolates	17
LPS	Glass beads	Flowthrough	Saturated, aerobic	*S. maltophilia*, *E. coli*, *Citrobacter freundii*, *P. syringae*	125
LPS composition	Silica sand	Static	Saturated, aerobic	*P. fluorescens*	265
LPS composition	Glass beads	Flowthrough	Saturated, aerobic	*E. coli*, 3 strains	257
Population heterogeneity	Silica sand	Downflow	Saturated, aerobic	*Pseudomonas* sp. strain B13	224
Positive-charged bacteria	Glass and Teflon	Downflow	Saturated, aerobic	*Stenotrophomonas maltophilia*	125
Ecological factors					
Protistan grazing	Aquifer sand	Upflow	Saturated, aerobic	Mixed aquifer populations	132
Survival	Aquifer sand	Static	Saturated, suboxic	*Pseudomonas strain* B13 (GEM[a])	135
Facilitated transport of contaminants					
Dichlorodiphenyltrichloroethane	Sand	Flowthrough	Saturated, aerobic	*Pseudomonas* and *Bacillus* spp.	148

Polyaromatic hydrocarbons	Aquifer sand	Flowthrough	Saturated, aerobic	*B. subtilis*	111
Fluid velocity					
Attachment	Silica sand	Upflow	Saturated, aerobic	*K. oxytoca*, *B. cepacia* G4	100
Cell density and fluid velocity	Sandy soils	Flowthrough	Saturated, aerobic	*P. fluorescens*, *B. cepacia* G4	35
Flow velocity	1-mm-diameter glass beads	Upflow	Saturated, aerobic	*P. fluorescens*	37
Inorganic groundwater chemistry					
Bivalent cations	Cristobalite sand	Downflow	Saturated, aerobic	*Pseudomonas* sp. strain B13	223
Ionic strength and pH	Silica beads	Flowthrough	Saturated, aerobic	*P. fluorescens* strain P17	114
Ionic strength	Aquifer sand	Upflow	Saturated, aerobic	*Pseudomonas* sp. strain KL2	71
Ionic strength	Coated sand	Downflow	Saturated, aerobic	Aquifer isolates	219
Ionic strength	Sand	Horizontal flow	Saturated, aerobic	*E. coli*	233
pH	Aquifer sand	Static	Saturated, aerobic	Uncultured aquifer population	218
Organic matter					
Dissolved organic contaminants	Aquifer sand	Static	Saturated, aerobic	Uncultured aquifer population	218
Biosurfactant, nonionic surfactants	Glass beads	Downflow	Saturated, aerobic	*A. paradoxus*	143
Biosurfactants (rhamnolipids)	Sand	Flowthrough	Saturated, aerobic	*P. aeruginosa*	5
Humic acid	Quartz sand	Flowthrough	Saturated, aerobic	Subsurface isolate A1264	121
Surfactant (DDBS)	Sand	Flowthrough	Saturated and unsaturated	Aquifer isolates	191
Surfactant chain length	Quartz sand	Flowthrough	Saturated, aerobic	*Sphingomonas paucimobilis*	32
Surfactant (Triton X100)	Soil, sand, clay	Downflow	Saturated, aerobic	*Arthrobacter*, *Mycobacterium*, *Sphingomonas* spp.	138
Surfactants	Glass beads	Downflow	Saturated, aerobic	*A. paradoxus*	80
Media properties					
Bulk density, roots	Loamy soil	Flowthrough	Saturated, aerobic	*P. fluorescens* (GEM[a])	253
Clay, sorbed humic acid	Soil, sand, clay	Downflow	Saturated, aerobic	*Arthrobacter*, *Mycobacterium*, and *Shingomonas* spp.	138
Centimeter-scale variations in binding	Aquifer sediment	Upflow	Saturated, aerobic	^{14}C-labeled bacteria	27
Effect of autoclaving and dry heat	Berea sandstone	Static	Saturated, aerobic	2 landfill isolates	112
Gas sparging	Quartz sand	Upflow, sparged	Unsaturated, aerobic	*P. fluorescens strain* P17, *P. aeruginosa*, *Mycobacterium vaccae*	62
Grain size	Glass beads	Static	Saturated, anaerobic	*E. coli*	221
Grain size/metal oxide content	Intact soil cores	Flowthrough	Saturated, aerobic	*Comamonas* strain DA001	54
Hydraulic conductivity	Tuff and sandstone	Flowthrough	Saturated, aerobic	Antibiotic-resistant isolate	232
Macropores	Soil	Downflow	Unsaturated, aerobic	*E. coli*	228
Metal oxide coatings	Aquifer sand	Static	Saturated, aerobic	Uncultured aquifer population	218
Mineralogy	Coated sand	Downflow	Saturated, aerobic	Aquifer isolates	219

(*Continued on next page*)

TABLE 1 Selected column studies delineating factors controlling subsurface microbial transport behavior *(Continued)*

Organism type and factors studied	Medium	Method	Conditions	Microorganism	Reference
Soil organic matter and texture	Tropical soils	Downflow	Unsaturated, aerobic	*P. fluorescens* strain BR12 (GEM[a])	81
Soil organic matter	Quartz sand	Flowthrough	Saturated, aerobic	Subsurface isolate A1264	121
Soil temperature and pH	Soil with roots	Downflow	Unsaturated, aerobic	*P. fluorescens* (GEM[a])	129
Structured heterogeneity	Clean quartz sand	Flowthrough	Saturated, aerobic	Aquifer isolates	64
Structured heterogeneity	Dual-porosity sand	Flowthrough	Saturated, aerobic	*E. coli*	242
Transverse mixing	Coarse and fine sand	Downflow	Saturated, aerobic	Strain W8 (aquifer isolate)	175
Unsaturated flow conditions	Loamy soils	Downflow	Unsaturated, aerobic	*Burkholderia* and *Pseudomonas*	99
Unsaturated flow conditions	Sand	Downflow	Unsaturated, aerobic	*E. coli*	190
Unsaturated flow conditions	Quartz sand	Downflow	Unsaturated, aerobic	*P. putida strain* mt2, *Rhodococcus* strain C125	216
Water content	Quartz sand	Downflow	Unsaturated, aerobic	*P. fluorescens* strain P17	115
Water content	Sand	Flowthrough	Saturated and unsaturated	Aquifer isolates	191
Population density effects					
Density-dependent attachment	Sandy soil	Downflow	Saturated, aerobic	Aquifer isolate	20
Nonequilibrium sorption	Subsurface sand	Flowthrough	Saturated, aerobic	*Bacillus* sp., *Enterobacter* sp.	149
Pore clogging					
Aggregation	Fine sand	Downflow	Saturated, aerobic	*Arthrobacteria* sp.	252
Clogging efficiency	Quartz sand	Downflow	Saturated, aerobic	Aquifer and soil isolates	251
Selectivity and depth of plugging	Berea sandstone	Static	Saturated, aerobic		194
Taxis, chemotaxis					
Chemotaxis	Sand	Static	Saturated, aerobic	*P. putida*	14
Chemotaxis vs taxis	Ottawa sand	Static	Saturated, anaerobic	*E. coli* (4 strains)	222
Chemotaxis vs diffusive behavior	Glass-coated beads	Impinging flow	Saturated, aerobic	*P. putida* strain F1	181
Motility effect upon attachment	Soil	Flowthrough	Saturated, aerobic	Groundwater isolate	145
Motility with and without nutrients	Ottawa sand	Static	Saturated, anaerobic	*E. coli*	202
Permeability and motility	Berea sandstone	Static	Saturated, anaerobic	*Bacillus*, *Enterobacter aerogenes*, *Klebsiella pneumoniae*	113
Temperature-controlled motility	Aquifer sediment	Upflow	Saturated, aerobic	Subsurface isolate	163
Tracking and labeling procedures					
Flow cytometry	Sand	Flowthrough	Saturated, aerobic	*Bacillus megaterium*	51
Fluorochrome staining	Ottawa sand	Upflow	Saturated, aerobic	*E. coli* 3000 and *Acetobacter* sp. strain 136	
Green fluorescent proteins	Quartz sand	Flowthrough	Saturated, aerobic	*E. coli* NM522, *P. putida*	34
In-line spectrophotometry	Silica sand	Upflow	Saturated, aerobic	*P. fluorescens*	52
Molecular and conventional	Intact core	Upflow	Saturated, aerobic	*Comamonas* sp. and adhesion-deficient variant	48
Phosphorimaging	Intact core	Flowthrough	Saturated, aerobic	*Pantoea agglomerans* (isolate)	55

Viruses					
Attachment reversibility	(Un)bonded silica	Downflow	Saturated, aerobic	PRD-1 and MS-2 (phage)	10
Chemical perturbations	Silica beads	Downflow	Saturated, aerobic	MS-2 and poliovirus	11
Divalent cations and ionic strength	Quartz sand	Flowthrough	Saturated, aerobic	Male-sp. filamentous phage	197
Dynamic air-water-solid interfaces	Glass and Teflon	Downflow	Unsaturated, aerobic	MS2, R17, φX174	240
Electrostatic interactions	Quartz sand	Downflow	Saturated, aerobic	MS2, recombinant Norwalk virus particles	198
Flow rate and soil type	4 different soils	Downflow	Unsaturated, aerobic	Poliovirus	263
Isoelectric point	Aquifer sediment	Recirculating upflow	Saturated, aerobic	MS2, PRD1, Qβ, φX174, PM2	56
Matrix diffusion	Fractured tuff	Downflow	Saturated, aerobic	MS-2 (coliphage)	9
Organic matter	Loamy fine sand	Downflow	Unsaturated, aerobic	MS-2 (coliphage)	192
pH	3 sandy soils	Flowthrough	Saturated, aerobic	PRD1 and MS2	133
pH and attachment reversibility	Quartz/Fe-quartz	Static	Saturated, aerobic	PRD1	152
Sorption onto metal hydroxides	Coated sand	Flowthrough	Saturated, aerobic	MS2	271
Surface inactivation	Cape Cod sand	Static	Saturated, aerobic	MS2, PRD1	213
Unsaturated conditions	Ottawa sand	Up- vs downflow	Unsaturated vs saturated	MS2, φX174	116
Unsaturated conditions	Different soils	Up- vs downflow	Unsaturated vs saturated	MS2, φX174	42
Unsaturated vs saturated conditions	Soils	Flowthrough	Unsaturated vs saturated	Enteroviruses	139
Unsaturated vs saturated conditions	Coarse soils	Downflow	Unsaturated vs saturated	MS-2 and PRD1	188
Virus type	Ottawa sand	Upflow	Saturated, aerobic	MS2, φX174	117
Protozoa					
Deposition mode	Glass beads		Saturated, aerobic	*C. parvum*	246
Field vs column transport behavior	Aquifer sand	Upflow	Saturated, aerobic	*Spumella guttula* (Kent)	92
Flow velocity	Aquifer sand	Flowthrough	Saturated, aerobic	*C. parvum*	86
Grazing on bacteria	Glass beads	Flow and static	Saturated, aerobic	*Tetrahymena* sp.	59
Iron-oxide coating	Quartz sand		Saturated, aerobic	*C. parvum* and PRD1	4
Medium, dispersion, retardation	Sand or shale	Flowthrough	Saturated, aerobic	*C. parvum*	33
Soil type	Soil	Downflow	Unsaturated, aerobic	*C. parvum*	162
Straining, attachment kinetics	Quartz sand		Saturated, aerobic	*C. parvum*	30

[a]GEM, genetically engineered microorganism.

Physiological Considerations

It is becoming more evident that the manner in which microorganisms are handled prior to an injection and recovery test can affect their subsequent transport behavior in the aquifer. It is important to recognize that a number of transport-determining bacterial characteristics (e.g., cell size, hydrophobicity, motility, surface charge, and exopolymer production) can be affected by nutrient availability prior to their introduction to granular media (97). Changes in nutrient conditions can also be accompanied by alterations in amounts of storage molecules such as poly(beta)hydroxybutyrate (PHB) (184) and glycogen (161), which can affect buoyant density (85). Substantive differences in both specific gravity and cell size between near-neutral buoyant ($\leq 1.02\ g \cdot cm^{-3}$) groundwater bacteria (95) and cultured groundwater isolates (95, 260) have been reported. Therefore, it is not surprising that broth-grown microorganisms reintroduced to a carbon-limited aquifer can exhibit temporal changes in transport behavior during readaptation (92).

When cultured groundwater populations are used, it is desirable to maintain in situ characteristics to the extent possible. This may be facilitated by growth under conditions that mimic the environment from which the microbes were isolated. For example, it has been demonstrated that growing groundwater flagellates within porous media (sieved aquifer sediments) under nutrient and pH conditions similar to that of the aquifer from which they came results in cells similar in size to what is found in situ (91). Flagellates grown in this manner exhibited transport behavior different from that of flagellates grown in high-nutrient liquid media. Another method of preadapting cultured bacteria to in situ conditions is starvation prior to reintroduction (97). In a forced-gradient test involving a sandy aquifer in Oyster, Va., this was accomplished by a priori suspension in artificial groundwater devoid of organic carbon for 72 h (49). It is also important to recognize that even the growth phase can affect a cultured bacterium's surface characteristics and attachment behavior in granular media (258). For example, it has been shown that *Pseudomonas aeruginosa* Olin in the stationary phase exhibits an increased ζ (zeta) potential and increased attachment to dolomite in comparison with the logarithmic growth or decay phases (77). The lipopolysaccharides (LPS) associated with a bacterium's surface are particularly important in attachment and transport in granular media (259), so changes in the LPS due to growth conditions could result in a substantively altered transport behavior. Briandet et al. (31) noted a change in *Listeria monocytogenes* electrophoretic mobility when either the growth temperature was changed or additional glucose or lactate was added to Trypticase soy broth media.

Using whole communities of uncultured indigenous bacteria (88, 90) that have been concentrated directly from the zone in which the experiment is to be run can lessen physiological changes during transport through the aquifer. Using microorganisms that are already adapted to aquifer conditions can preclude interpretation problems resulting from physiology-induced changes in transport behavior. Bacteria may be concentrated from groundwater by dewatering techniques involving continuous-flow centrifugation or tangential-flow filtration and subsequently labeled for reinjection. An advantage of tangential-flow filtration is that the apparatus is field portable, allowing on-site concentration of aquifer microorganisms. A modified version of a commercially available microtubule-type tangential flow device (Fig. 2) was found to exhibit a higher efficiency for concentrating bacteria from organically contaminated groundwater than either the parallel sheet-type tangential-flow device or the continuous-flow centrifuge (137).

Strain Detection

Several flowthrough column studies with bacteria have employed plate count procedures for detection of unlabeled bacteria appearing in the eluent (for examples, see references 37, 71, and 134). The advantage of plate counting procedures is simplicity. The use of antibiotic-resistant bacteria in column experiments allows their detection on antibiotic-containing media, which greatly enhances detection specificity. Other means of differential plate counts involve bacteria that produce degradation products of organic contaminants that react with a tetrazolium dye. The latter method was used for the enumeration of *Methylosinus trichosporium* OB3b in a methane-amended, unsaturated soil column involving vinyl chloride treatment (76). The disadvantage of plate counting is that it detects only that fraction of the total recovered population that readily grows on solid media. Also, changes in culturability during the experimental time course can hamper interpretation of plate count data.

In order to differentiate "test" microorganisms from indigenous subsurface populations and/or from other inadvertently introduced populations, microorganisms used in laboratory or in situ transport tests are typically labeled a priori with a stable tag. A number of field studies involving the subsurface transport of bacteria have employed protein-specific or DNA-specific fluorescent stains (fluorochromes) (69, 88, 90, 93, 156). Ideally, the tags should not affect the organism's transport behavior over the time frame of the injection and recovery experiment. Labeling the bacterial populations with the fluorochrome 4,6-diamidino-2-phenylindole (DAPI), which is known to hamper bacterial metabolism (183), retards growth and provides a long-lived (month-long) fluorescent tag in field transport studies (88). DAPI does not appear to affect bacterial attachment or transport behavior in aquifer sediments, at least over the short term (41, 136). However, several other fluorochromes, i.e., 2,7-diamino-10-ethyl-9-phenyl-phenanthridinium bromide (ethidium bromide), 3,6-bis(dimethylamino)acridinium chloride (acridine orange), and 2′-(4-hydroxylphenyl)-5-(4-methyl-1-piperazinyl)-2,5′-bi-1-H-benzimidazole (Hoechst 33258), can affect substantively the attachment of bacteria with sand grain surfaces (136). Therefore, when using a fluorochrome to label bacteria in subsurface transport experiments, it is a good idea to ascertain its effect on cell surface properties a priori. It is still unclear what effects DAPI may have upon subsurface bacterial transport in longer-term, field scale experiments. Hydroethidine, a vital eukaryotic stain (70), was used in lieu of DAPI, which is lethal to protists, to label the nanoflagellate *Spumella guttula* Kent in a month-long subsurface transport study (92, 94). More recently, a vital prokaryotic staining procedure involving 5,6 carboxyfluorescein diacetate, succinimidyl ester (CFDA/SE) was developed and employed in the labeling of two groundwater isolates for transport studies (156). It was reported that CFDA/SE labeled strain DA001 for up to 28 days without adversely affecting its physiology or transport (68).

Genetic alterations that confer considerable detection sensitivity in addition to providing a unique "marker" may be useful in identifying specific bacteria in field transport studies. To date, the testing of genetically altered bacteria has been done largely at the column scale. Genetic modifications that result in production of green fluorescent protein (GFP)

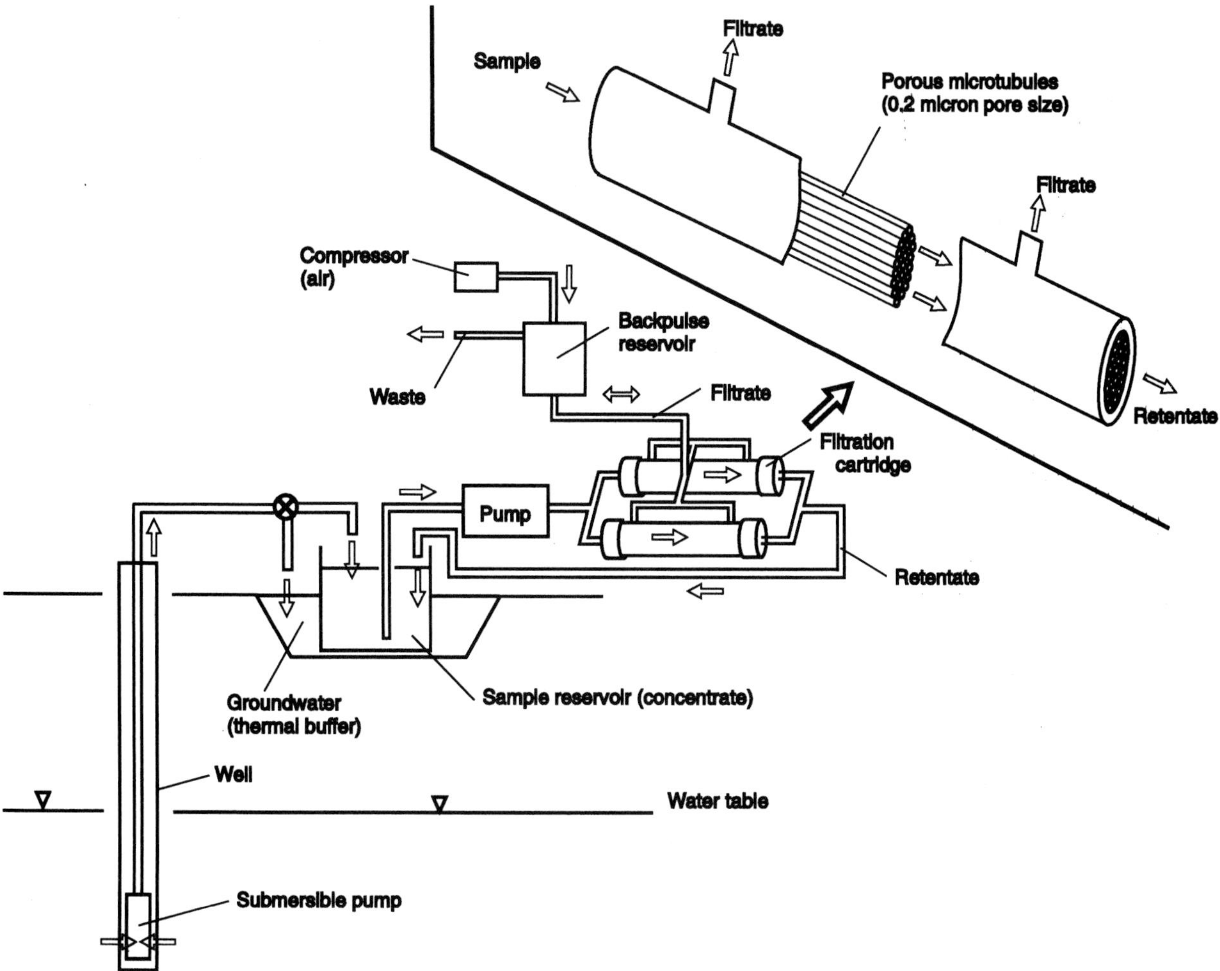

FIGURE 2 Schematic representation of a hollow-fiber-type tangential-flow filtration device modified for on-site concentration of indigenous microorganisms from groundwater. Modified from Kuwabara and Harvey (137) with permission from the *Journal of Environmental Quality*.

have been employed successfully in bacterial transport studies involving flowthrough columns (34). GFP, originally isolated from the jellyfish *Aequorea victoria* (38), has a maximum absorption of 395 to 509 nm. GFP-modified bacteria can be detected using luminescence spectrophotometry at a detection limit of ~10^5 cells (34), although flow cytometry may facilitate more-sensitive enumerations in the future (207). The utility of the GFP method may be enhanced using mutant *gfp* genes, which ameliorate several difficulties associated with the wild-type *gfp* (e.g., low fluorescence intensity and the long delay between protein synthesis and initiation of fluorescence) (169). In order to track several bacterial populations concurrently, it may be possible to utilize the red-shifted GFP (60, 65), yellow fluorescent protein (226), or blue fluorescent protein (127, 185). The advantage of GFP-based identification over the *lux* system described below is that the *gfp* gene does not require additional cofactors or substrate for expression (38). However, there is a requirement for molecular oxygen, which may limit use in certain field applications. Although a single copy of *gfp* integrated into the bacterial chromosome provides sufficient fluorescence intensity for visualization of individual bacteria by use of a fluorescence microscope, fluorescence can be further enhanced by integrating two tandem copies of the *gpf* gene (249).

lux genes also have potential for labeling bacteria of interest for subsurface transport studies in certain applications. *lux*AB genes, encoding a bacterial luciferase, are used in the construction of "biosensor" bacteria that allow detection and quantification of low (nanomolar) concentrations of a number of environmental contaminants (110). Luciferase-based biosensors make use of a light-emitting reaction involving the oxidation of reduced riboflavin phosphate and a fatty aldehyde to oxidized riboflavin phosphate, a fatty acid, and water (110). Because the intensity of luminescence is directly related to the bacterium's energy status, information about relative metabolic activity may also be gathered. However, a disadvantage of using *lux* genes as markers in transport studies is that the light production of the bacterial luciferase is generally too weak for microscopic identification of individual bacteria. This problem can be overcome by using a dual *gpf-luxAB* marker system, which allows monitoring of the relative energy status and the total numbers of the labeled bacterial populations in environmental samples (248). Another disadvantage is that an aldehyde substrate is required for the expression of the gene. The latter problem may be overcome by using the complete gene cassette *luxCDABE* (98), which encompasses the genes for aldehyde production.

The genetically conferred ability to facilitate the formation of ice in small droplets of water is yet another marker of potential interest in subsurface transport studies involving bacteria. Ice-nucleating active (INA) bacteria, including *Pseudomonas syringae*, *Pseudomonas fluorescens*, and *Erwinia herbicola*, synthesize high-molecular-weight proteins in the outer membrane that function as nucleation sites for the formation of ice crystals at subzero temperatures up to −2°C (144). In the absence of INA bacteria or certain very unique steroids, water at −2° to −5°C supercools and remains a liquid. The ability of *P. syringae* to cause the freezing of 10-μl volumes of solution at ~−5°C formed the basis for a quick and inexpensive protocol developed by Strong-Gunderson and Palumbo (235) for using this bacterium as a tracer in a flowthrough column experiment. The authors report detection limits in the range of nanograms per liter, stability over a pH range of 2 to 11, and quantitative results within 3 min of sample collection. Therefore, INA bacteria may be useful in providing real-time information regarding microbial breakthrough in injection and recovery experiments. More recently, INA bacteria proved useful in a 36-m microbial transport study involving fractured, highly weathered shale saprolite (167).

Other methods of labeling microorganisms for use in in situ and column transport studies have involved the use of stable isotopes ratio mass spectrometry (IRMS). In a small-scale, forced-gradient bacterial transport study involving a sandy aquifer in coastal Virginia, DeFlaun et al. employed ^{13}C-labeled bacteria at a high enough enrichment to allow quantitative detection in groundwater samples collected downgradient of the injection (49). One advantage of using ^{13}C to label microorganisms in in situ transport experiments is its benign effect on both the microorganisms of interest and the investigators, although the relatively high cost of stable isotope analysis may impose a practical limit on the scale of the experiment.

Radiolabeling has been employed in a number of laboratory studies involving transport of bacteria (27, 35, 55, 79, 80, 115, 149) and viruses (152, 198) in porous media. However, the potential hazards involving field applications of radioisotopes in microbial transport studies may restrict their use to studies involving short-lived isotopes and controlled-access sites. ^{32}P, which has a half-life of only 14 days, has been used to label *Escherichia coli* and the bacteriophage PRD1 in injection and recovery experiments on the property of the Chalk River Nuclear Laboratories (Chalk River, Ontario, Canada) (39) and at the U.S. Geological Survey large-scale tracer array in Falmouth, Mass. (187, 212), respectively. The advantages of using radioisotopes include a minimal effect on cell metabolism, a high sensitivity, and a relatively low analytical cost. Radioisotope labeling is particularly useful for in situ phage transport studies because, unlike the standard plaque counting procedures designed to quantify the numbers of infective phage, it allows determinations of the total numbers of phage transported. Using different radiolabels for the protein-containing virus capsid and the enclosed genetic material allows a means to assess the structural integrity of the virus during the test (177). This is because the ratio of the two isotopes, after correcting for radioactive decay, should change upon disintegration of the virus because the protein-containing capsid and the DNA or RNA will have different propensities for attachment to grain surfaces.

Because of its high specificity for detecting specific organisms, the use of PCR can alleviate the need for a priori labeling altogether. Because probes and primers are available for the 5′ noncoding region common to many viruses, PCR has been particularly useful for detection of enteroviruses in both groundwater (2) and subsurface soil (233) in order to assess their extent of subsurface transport. The major advantages of PCR are that (i) it detects only intact nucleic acid sequences and (ii) it can detect enteroviruses that have lost their ability to replicate in cell cultures. The disadvantages of using PCR to assess transport of specific enteroviruses in in situ injection and recovery tests include its inability to distinguish between infective and noninfective viruses and its relatively high cost. Also, it is only semiquantitative and cannot provide the kind of high-precision data required in many modeling applications. Recently, bacteriophage M13mp18 was genetically modified to enhance its suitability as a tracer at a restricted field site. The genome contained a unique identification sequence that could be identified using a number of techniques, including PCR (47). Construction of an *E. coli* strain that expresses a stable artificial RNA from the chromosomal *rrnH* operon may be useful for in situ microbial transport experiments because the large number of stable artificial RNA products per cell enhances signal detection (7).

DeFlaun et al. (48) describe a study in which a number of the aforementioned methods for tracking specific bacteria are applied to an intact core experiment. These authors compare resulting breakthrough data for seven techniques of tracking bacteria in transport studies involving subsurface materials, i.e., direct counts of fluorescently stained bacteria (90), plate counts, microplate spectrofluorometric (69) and flow-cytometric enumerations (51, 244), microplate enumeration ferrographic tracking (122, 272, 273), IRMS of ^{13}C-labeled bacteria (103), and quantitative PCR (48). The authors report good correlations between the effluent concentration histories for each of the last six methods and the direct counting procedure. However, the reported limit of detection varied considerably and ranged from 2 cells/ml for IRMS to 3×10^5 cells/ml for the microplate spectrofluorometric method. The per-sample costs were reported to vary by at least a factor of 10 among the different methods (48). Therefore, the choice of methods for tracking bacteria in transport experiments may depend on a number of factors including the cost, the availability of required instrumentation, and the required precision of measurement. For lab scale experiments, the minimum sample volume required for analysis may be an important consideration. For example, the 50 ml of sample required for ^{13}C-IRMS (48) may preclude its use in many column studies.

Physical and Population Heterogeneities

For bacteria, chemical and physical heterogeneity in the aquifer collectively affects most of the factors that influence the extent of travel in the aquifer (87). Even in relatively homogeneous aquifer sediments composed of well-sorted sand, physical (e.g., grain size distribution) and geochemical (e.g., mineralogical) heterogeneity causes apparent changes in microbial transport behavior. In an injection and recovery experiment in which a labeled indigenous bacterial population, a conservative tracer (bromide), and a particulate tracer (bacterial-sized microspheres) were coinjected into adjacent layers of a relatively homogeneous sandy aquifer, the relative order of appearance of peak abundance of the three constituents varied from layer to layer (93). Substantive changes in the rates of immobilization, retardation, and apparent dispersion over short distances in the aquifer require verification that a given observed transport behavior is not an artifact of system heterogeneities.

Accurate modeling of microbial transport in granular aquifers over environmentally significant distances may necessitate both a detailed knowledge of the spatial variability in physical aquifer properties and a model that takes that into account. A stochastic model that employed site-specific physical variability information gathered in earlier hydrologic studies (73, 140) was used to simulate viral transport over a 36-m transect in a sandy aquifer (200). The model simulations suggest that physical heterogeneity (expressed as the variance in the natural log of hydraulic conductivity, ln K) may be a major determinant of microbial transport over environmentally relevant distances. Indeed, very little viral transport was predicted in the absence of physical heterogeneity. Although stochastic models describing subsurface microbial transport have yet to be validated with large-scale, three-dimensional data sets, there is experimental evidence for the critical assumption that there is a correlation between ln K and microbial attachment. It was recently shown that there is an excellent correlation between ln K and the collision efficiency factor (α) for microbial-sized colloids in flowthrough columns of granular media (201).

Assessment of the spatial variability of ln K for the purposes of subsurface microbial transport modeling requires that many labor-intensive measurements be made in the aquifer or on core samples taken from the aquifer. Relatively homogeneous aquifers may be characterized by K values that vary by 1 order of magnitude (101), whereas heterogeneous aquifers are characterized by K values that vary by several or more factors of 10. There are many ways to estimate hydraulic conductivity variations within granular aquifers, including permeameter assays on intact aquifer sediment cores and borehole flow meter tests (for example, see reference 101), estimates from tracer or pump tests using existing well networks (153, 237, 274), and from measured sediment parameters, i.e., grain size distribution and the fractal dimension and entropy of the distributions, as well as porosity, soil density, and fines content (25). Although the various methods can yield different values of K (173), the spatial variability of K is the more critical parameter in accounting for subsurface microbial transport on the scale of tens or hundreds of meters when using stochastic models. Noninvasive geophysical techniques, such as ground-penetrating radar (21, 22, 105), for estimating physical variability within granular aquifers are gaining popularity in part because they yield faster and more accurate three-dimensional information about the geometry of heterogeneities than is possible by using core analyses or pump tests, and they may be useful in future modeling efforts.

Another important heterogeneity that can lead to higher than predicted subsurface transport of microorganisms involves intrapopulation variabilities in adhesive properties. Since the first report suggesting the existence of two subpopulations for the same bacterial strain, each possessing distinctly different adhesive properties (220), a number of studies have suggested that intrastrain variations in bacterial surface properties (5, 17, 74, 224) can lead to differences in attachment behavior. For example, Dong (54) measured the electrophoretic mobility (EPM) values for three different monoclonal bacterial populations and found that each species had a distribution of EPM values. For the aforementioned DA001 groundwater isolate, the zeta potential ranged from about −7 to +5 mV. Bayer and Sloyer (16) noted the same phenomenon with an *E. coli* strain, except that the distribution of EPM was even larger (−4 to +3 mobility units [MU; micrometers per second per volt per centimeter] or 10^{-8} m^2 V^{-1} s^{-1}), corresponding to zeta potentials ranging from −51.6 to +38.7 mV. Such a range of zeta potential values within a population of cells would result in a range of transport behaviors. Therefore, when using a cultured bacterium in an in situ transport study, it may be useful to ascertain whether bacteria collected in the effluent of a short (several centimeters), flowthrough column of saturated well-defined porous media exhibit a different average collision efficiency when passed through a second, identical column (224).

IN SITU STUDIES

Tracers

Recent in situ investigations designed to examine microbial transport behavior in groundwater have involved the coinjection into the aquifer of the microorganism of interest with a conservative solute tracer (usually chloride, bromide, or iodide). Conservative tracers (nonreactive solutes) are typically employed to monitor the velocity and direction or path of groundwater flow and the degree of dispersion (spreading of the injectate that is caused by diffusion and by mixing with adjacent parcels of water). Dimensionless concentration histories (breakthrough curves) of the conservative tracers are then obtained for sampling points that are downgradient and in the path of the injectate cloud. The characteristics of the conservative tracer breakthrough curve can then be used comparatively to determine some of the major transport parameters exhibited by the introduced microorganisms. In using halide tracers, it may be necessary to consider their effects upon microbial activity (78) and upon the density of injectate cloud (140). For many small-scale injection and recovery experiments, halide concentrations in the 100- to 200-mg/liter range are sufficient. For small-scale field studies involving a pulse (slug) input, conservative tracer transport data may be used to help calculate microbial transport parameters as described below.

Retardation of the microbe, which is caused by reversible interactions with stationary surfaces, can be determined by comparing velocities for the peak concentration of conservative tracer to that of the introduced microorganism at sampling points downgradient from point of injection:

$$R = V_{tracer}/V_{microbe} \tag{1}$$

where V_{tracer} and $V_{microbe}$ are the observed average velocities of the peak concentrations of the tracer and the microorganism, respectively, that appear at a downgradient well. For cases where the breakthrough curves for the microorganism and conservative tracer are dissimilar in shape or where the breakthrough curves are multipeaked, retardation may be approximated as the ratio of the times required for the arrival of the centers of mass (estimated by numerical integration) for the introduced microorganism and the conservative tracer, respectively (92).

Theoretically, dispersivities of conservative solutes and microorganisms should be similar because dispersivity is a physical property of a porous medium (29) that determines the degree of mixing with adjacent parcels of groundwater along the flow path. However, the apparent longitudinal (direction of flow) dispersion exhibited by the microorganism in granular aquifers is usually greater than that of the conservative tracer, because reversible interactions with grain surfaces differentially retard portions of the microbial population relative to the conservative tracer. On the other hand, a high degree of preferential flow path structure in some granular aquifers can result in a greater apparent dispersion for the conservative tracer than for the microorganism (for example, see reference 182). This is because microbial transport in the latter, more heterogeneous systems can be limited to the larger pores, resulting in a breakthrough pattern that is less skewed than that of the conservative tracer. Apparent longitudinal dispersion ($\mathcal{A}_L$) for the microorganisms may sometimes be estimated using a simplified one-dimensional approach (88):

$$\mathcal{A}_L = \frac{X_1(\Delta t / t_{peak})^2}{16 \ln 2} \qquad (2)$$

where x_1 is the distance from point of injection, t is the duration of breakthrough when $C(t) > 1/2$ peak concentration and t_{peak} is the time required to reach the peak concentration. In many cases, such calculated values are only first approximations. This is because the concentration histories of the microorganisms often do not conform to the assumed classical Gaussian-shaped breakthrough curves upon which the above equation is based. The above equation may not be suitable for calculating apparent dispersion for microorganisms exhibiting highly irregular, multipeaked breakthrough curves. Apparent longitudinal dispersion may also be calculated from three-dimensional breakthrough curves by using the standardized first and second moments of the spatial distributions (75). The interested reader is referred to an article by Pang et al. (182) for a detailed explanation of how to use the moments approach for determining apparent dispersivities for microorganisms in injection and recovery studies that employ a three-dimensional sampling array.

In granular aquifers, differences between the conservative tracer and microbial breakthrough curves are due largely to the interactions between the microorganisms and the grain surfaces. The simplest method of accounting for microbial attachment to surfaces is to assume that the microbial distribution between solution and solid surface is constant. This approach assumes that attachment is always at an equilibrium and, therefore, the solid/solution partitioning of the microorganisms can be described by a single coefficient of distribution (K_d). However, judging from the protracted nature of the trailing ends of microbial breakthrough curves relative to those of the conservative tracer in most field tests (Fig. 3A), an equilibrium assumption usually is not appropriate. A better approach involves consideration of the attachment and the detachment kinetics of microorganisms being advected (carried along by fluid flow) downgradient (149). Aquifer sediments are characterized by many different types of attachment sites. However, it can be useful for modeling purposes to assume that there are two types of attachment sites. Specifically, it can be assumed that one type of site is responsible for permanent microbial immobilization and one type is responsible for reversible microbial attachment (88). The former type results in an attenuation of the height of the microorganism breakthrough curve (plotted as dimensionless concentration history versus time) relative to that of the conservative tracer. The latter (weaker) type results in retardation of the center of mass of the microorganisms that reaches the downgradient sampling point.

The fraction of the microorganisms injected into the aquifer that become irreversibly attached (within the experimental time frame) during transport to a downgradient sampler may be calculated from its relative breakthrough (RB). The RB is determined by comparing the numerical integration of the dimensionless concentration history of the microbe to that of the conservative tracer (90):

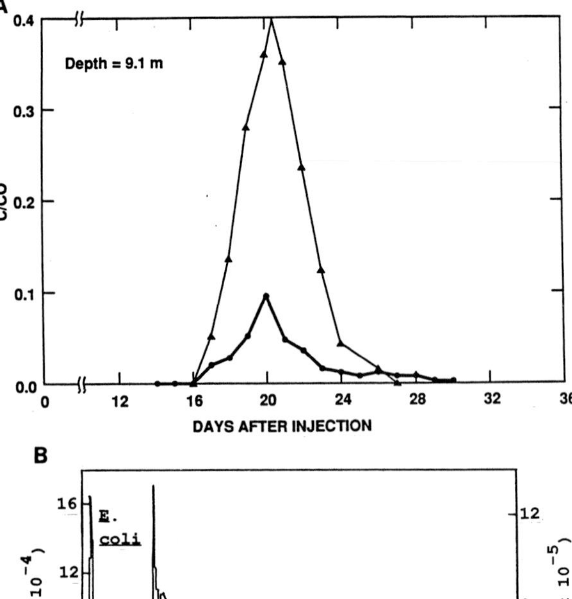

FIGURE 3 Dimensionless concentration histories for bacteria and bromide transported downgradient through aquifers from the point of coinjection. (A) Natural-gradient test in a well-sorted, sandy aquifer in Cape Cod, Mass. (reprinted from reference 23 with permission from Wiley-Liss). (B) Forced-gradient test in a fractured, granite aquifer at Chalk River Laboratory, Ontario, Canada (redrawn from reference 39 with permission from the publisher of *Water Science and Technology*).

$$RB = \int_{t_0}^{t_f} \frac{C(t)}{C_0}\, dt \div \int_{t_0}^{t_f} \frac{Tr(t)}{Tr_0}\, dt \qquad (3)$$

where C_o and Tr_o are the respective microorganism and conservative tracer concentrations in the injectate, $C(t)$ and $Tr(t)$ are concentrations downgradient at time t, t_o is the time of injection and t_f is the elapsed time from the beginning of injection until the termination of breakthrough. The above-described integrals, in most cases, may be reasonably estimated by simple numerical integration.

The relative breakthrough may, in turn, be used to estimate the collision efficiency factor. The collision efficiency factor, α, is a parameter in filtration theory that represents the physicochemical factors that determine irreversible microbial immobilization (88):

$$\alpha = \frac{d\{[1 - 2(\mathcal{A}_L / x_1)\ln(RB)]^2 - 1\}}{6(1-\theta)\eta\mathcal{A}_L} \qquad (4)$$

where d is the grain size (in centimeters), x_1 is the travel distance from the point of injection to the sampler at which the concentration history was monitored, θ is the porosity, and η is the single collector contact efficiency. Most often, the median grain diameter is used for d. However, it should be noted that characteristic diameters based on the smaller grains, i.e., upon the arithmetic average, geometric mean, or the d_{10} (diameter at which 10% of the grains are smaller) have been reported to yield results that most accurately described bacterial transport in the different-sized porous media (157). The last parameter (η) is the rate at which the microorganisms strike a single sand grain divided by the rate at which they move toward the grain and represents the physical factors determining collision. Two clean-bed filtration models are commonly invoked to describe the contact of microorganisms with stationary surfaces in granular media. Simply put, the two models differ in how η is calculated. The older model of Yao et al. (268) considers three types of mechanisms by which microorganisms contact the surface, i.e., Brownian motion, physical interception, and settling. The model described by Rajagopalan and Tien (195) includes the aforementioned mechanisms but also includes close-approach effects. The latter model is most commonly used in the recent literature. However, the original paper describing the latter model contains several errors, which were clarified by Logan et al. (150, 151). More recently, Tufenkji and Elimelech (245) developed a new correlation for the single collector contact efficiency that includes the influence of hydrodynamic and van der Waals forces on the deposition of particles dominated by Brownian diffusion along with other refinements to the two previous models which is

$$\eta = 2.4\, A_s^{1/3} N_R^{-0.081} N_{Pe}^{-0.715} N_{vdW}^{-0.052} + 0.55 A_s N_R^{1.675} N_A^{0.125} + 0.22 N_R^{-0.24} N_G^{1.11} N_{vdW}^{0.053} \quad (5)$$

where $A_s = 2(1 - p^5)/w$; $w = 2 - 3p + 3p^5 - 2p^6$; $p = (1 - \theta)^{1/3}$; $N_R = a_p/a_c$; $N_{Pe} = 48\pi\mu a_s a_p U/kT$; $N_{vdW} = H/kT$; $N_A = H/12\pi\mu a_p^2 U$ and $N_G = 2a_p^2(\rho_p - \rho)g/9\mu U$. In the above expressions, a_p and a_s are the respective colloid and grain radii, μ is the fluid viscosity, U is the approach (superficial) velocity of the groundwater, θ is the porosity, ρ and ρ_p are the respective densities of the groundwater and the microorganism, g is the acceleration due to gravity, T is temperature (in kelvins), k is Boltzmann's constant (1.38×10^{-23} J/mol (K), and H is the Hamaker constant. The value of the latter constant depends upon the particular system under study. For bacteria sorbing onto glass or quartz surfaces at room temperature, an approximate value of $\sim 1 \times 10^{-20}$ J is often used (for an example, see reference 143). However, more recently a Hamaker constant of $(5.12 \pm 0.03) \times 10^{-20}$ J was reported for *P. aeruginosa* in the presence of an organic surface (61).

Reversible types of attachment result in an increased travel time and apparent dispersion for the microorganisms relative to the conservative tracer. Although there is experimental evidence for the removal mechanisms (generally assumed to be irreversible) described by filtration theory (above), there is also evidence that nonequilibrium reversible attachment is a common feature of microbial transport in aquifers (88). The nonequilibrium is caused by a rate of attachment that is different from the rate of detachment and can be accounted for by using different attachment and detachment rate constants in modified advection-dispersion models describing microbial transport in porous media. The development of specific models for describing microbial transport through aquifer material is beyond the scope of this chapter. The interested reader is referred to references 131 and 147 for discussions of kinetic modeling of microbial transport through saturated granular media; to reference 88 for a discussion of how filtration theory and mathematical descriptions of reversible attachment can be coupled with a hydrologic transport (advection-dispersion) model to describe small-scale field results; to reference 200 for a description of a stochastic (probabilistic) model describing viral transport on a substantive field scale; to reference 196 for a comparison of models with kinetic and equilibrium adsorption; to reference 142 for discussions of how to model a number of bacterial transport processes in heterogeneous porous media, including cell diffusion, aggregation, localization of microbial growth sites, and long-term survival; to reference 104 for an evaluation of a model using laboratory data; to reference 176 for an overview of current approaches used to represent physicochemical and biological processes in porous media; to reference 236 for a description of modeling of microbial transport in dual-porosity media; to reference 58 for a description of how random-walk calculations can be used to describe bacterial migration in porous media; to reference 239 for a discussion of biomass transport in the presence of utilizable substrate; and to reference 106 for discussions of subsurface microbial fate and transport models.

In addition to providing a framework in which the transport behavior of microorganisms can be evaluated, conservative tracers can be useful for supplying information about movement of the injectate cloud within sampling arrays. This information facilitates a sampling protocol that allows capture of most of the microbial breakthrough. However, the usefulness of conservative tracers in this regard depends upon the type of aquifer in which the experiment is run. For homogeneous zones of sandy aquifers, dimensionless concentration histories of the conservative tracer can be similar to those of the bacteria or viruses coinjected with it (Fig. 3A) (12, 88). In general, physical heterogeneity of the system increases the dissimilarity between the breakthrough curves of the conservative tracer and the microorganisms (64, 88, 93). For aquifers characterized by fracture flow, arrival of peak microbial concentration at a downgradient sampling point can greatly precede that of the conservative tracer (Fig. 3B). This is because the microorganisms are largely confined to the major flow paths (fractures) and, therefore, are subject to an average travel path that is much shorter than that of the conservative tracer. In extreme cases, microbial breakthrough at downgradient sampling points can terminate before the arrival of measurable tracer concentrations. Because the pore volumes available for transport of microorganisms and solutes can be quite different in highly fractured aquifers, conservative tracers are less useful in providing real-time information about how and where to sample and in providing a comparative reference for determining microbial transport parameters.

The degrees of similarity between the microbial and solute tracer breakthrough curves are also influenced by the size and surface chemistry of the microorganism being evaluated. For example, substantial retardation of aquifer flagellates (protozoa) relative to bromide was observed for transport in a well-sorted, sandy aquifer in which retardation of indigenous bacteria is generally not observed (92). Earlier tests with different sizes (0.23 to 1.35 m) and types of microspheres revealed that the reactivity of the surface was a much stronger determinant of retardation than the size. This suggests that bacterial surface characteristics may be a more important determinant of retardation in some highly porous aquifer sediments than the size of the microbe (90).

Microspheres have been useful as particulate tracers to study the effects of cell size and surface characteristics upon microbial transport behavior in sandy aquifer sediments because they are physically and chemically well defined. Microspheres also can serve as tracers for contamination of subsurface core material by nonindigenous bacteria (see chapter 66) (43). However, the transport properties of microspheres in porous media can differ from those of microorganisms. For example, it has been shown that the degrees of retardation, apparent dispersion, and immobilization exhibited by various types of commercially available microspheres differ from those exhibited by bacteria (90). On the other hand, 2-μm-diameter carboxylated microspheres exhibited transport properties similar to those of groundwater flagellates (protozoa) in an injection and recovery experiment performed in a sand and gravel aquifer (92). Fluorescent microspheres made of carboxylated polystyrene were reasonable analogs for oocysts of the pathogenic protist *Cryptosporidium parvum* in in situ (96) and column-scale (247) studies. Fluorescent microspheres are easy to detect and, in groundwater samples with low turbidity, can be enumerated efficiently by using either computer-coupled image analysis or flow cytometry (18). The latter method may be adapted for online enumeration of micro-spheres in flowthrough columns of porous media (178) or for enumerating microspheres in field tracer tests (179).

Natural- and Forced-Gradient Tracer Tests

Several different types of injection and recovery tests have been employed to investigate microbial transport in situ (Fig. 4). Forced-gradient tests are ones in which the flow field is controlled by high-volume pumping, at the point of injection, point of withdrawal, or both. Forced-gradient tests have the advantage of substantially shortening the time frame required to obtain complete breakthrough of the injected constituents. Forced-gradient tests have been used in granular aquifers to study the transport behavior of bacteria (49, 90), viruses (13), and yeasts (267). However, this type of test is particularly well suited for microbial transport studies in aquifers characterized by fracture flow, because the sampling well(s) does not need to be placed immediately downgradient. Forced-gradient tests have been used in aquifers characterized by fractured clay and crystalline rock to study the transport behavior of viruses (166) and bacteria (39). A disadvantage of forced-gradient tests is that the flow field is nonuniform and, therefore, more difficult to model.

Divergent tracer tests involve a quick addition of a known quantity of microorganisms and conservative tracer into a continuous stream of groundwater being injected back into the aquifer. This setup produces radially divergent flow and forces the gradient in all directions (Fig. 4A). Because of the high degree of forced dispersion, the distance over which the injectate can be monitored is limited. A convergent test can be accomplished by continuous withdrawal at the sampling well (Fig. 4B). The advantage of the convergent test is that, theoretically, the entire mass of conservative tracer is recovered at the sampling well. This makes it possible to do a true mass balance on both the conservative tracer and the microorganisms. A variation of the simple convergent test is the doublet cell (Fig. 4C), in which an injection is made within the flow field created by a coupled continuous groundwater injection and withdrawal. The result is a flow field that can be controlled over moderate distances. The latter method is particularly useful for simulating conditions that may result during engineered remediation of contaminated aquifers.

In natural-gradient tests (Fig. 4D), the injectate is added slowly to the aquifer and the natural flow of groundwater advects the injectate cloud past rows of multilevel samplers (MLS) placed in rows perpendicular to the flow path. Natural-gradient tests are best suited for sandy aquifers in which the flow direction can be predicted within reasonable limits and in which the water table is shallow enough to permit the use of peristaltic pumps to sample water from each level of the MLS. Construction and use of MLS are discussed in Pickens et al. (186) and Smith et al. (229). For small-scale experiments, the individual MLS within the downgradient rows should be as close as possible. As a practical working limit, MLS installed with hollow-stem auger drill rigs (see chapter 66) cannot be placed much closer than 1 m apart.

A variation on the point source forced-gradient test is one involving a non-point source addition and sampling points that align along a transect parallel to the longitudinal axis of the source at some fixed distance downgradient. This type of setup was employed to assess the efficacy of artificial recharge to remove MS2 and PRD1 bacteriophages added to an elongated recharge basin in Castricum, The Netherlands. The recharge basin was ~63 m from and parallel to a row of extraction wells spaced 10 m apart. A portion of the recharge basin was dosed with the two bacteriophages by addition to a 10- by 15-m compartment lying within the basin. An advantage of this type of test is that transport behavior over several parallel flow paths of similar distance is observed and information can be gathered concerning the effects of field scale heterogeneity upon microbial transport (217). A trench-to-trench bacteriophage transport experiment, also involving MS2 and PRD1, was conducted in weathered, fractured clay till in Ontario, Canada (166). Both field scale experiments were facilitated by low groundwater temperatures (7 and 10°C, respectively), which greatly diminished temperature-dependent virus inactivation.

The quantity of microorganisms needed to conduct an injection and recovery experiment depends on the type of injection test to be run, travel distance, type of microorganism (and associated minimum level of detection), and aquifer characteristics. In general, the divergent (Fig. 4A) test requires the largest number of microorganisms. The radially divergent flow field in this type of test results in decreases in microbial concentration that are proportional to the square of the travel distance. This is, of course, in addition to losses from solution due to immobilization at solid surfaces. Natural-gradient experiments (Fig. 4D) often require the fewest number of microorganisms. This is because of the absence of forced dispersion caused by forced-gradient conditions and/or of large dilution factors at the point of collection. If the physical parameters (flow direction, velocity, dispersivities, porosity, grain size distribution, and injection and sampling point coordinates) are known and the aquifer is relatively homogeneous, a rough estimate of the number of microorganisms required for the test may be obtained by first calculating the degree of dilution that would occur along the experimental flow path as a result of dispersion and subsequently applying a filtration factor.

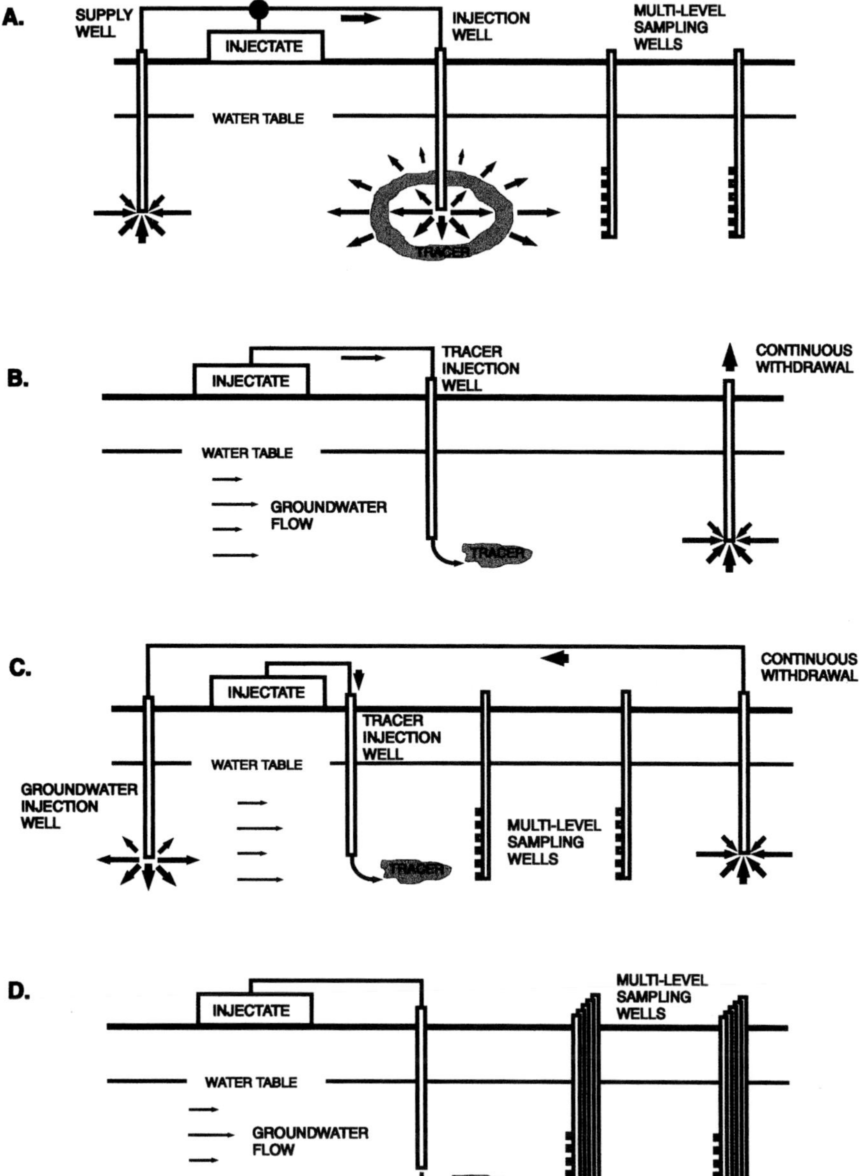

FIGURE 4 Schematic depiction of experimental designs for small- to intermediate-scale injection and recovery investigations examining microbial transport behavior in aquifers (modified from reference 23 with permission from Wiley Liss, Inc.). (A) Divergent, forced gradient; (B) convergent, forced gradient; (C) doublet cell, forced gradient; (D) natural gradient.

The degree of dilution due to dispersion or to sampling at some distance from the travel path taken by the center of injectate mass may be roughly estimated by using the following equation:

$$C_{(x,y,z,t)} = \left(\frac{C_0 V_0}{8\theta(\pi t)^{3/2}(D_x D_y D_z)^{1/2}}\right) \exp\left(-\frac{(x-vt)^2}{4D_x t} - \frac{y^2}{4D_y t} - \frac{z^2}{4D_z t}\right) \quad (6)$$

where C is the concentration at time t; C_0 and V_0 are the initial (injectate) concentration and volume, respectively; θ is the porosity; x is the distance in the direction of flow; y is the distance in the horizontal transverse direction; z is the distance in the vertical transverse direction; D_x, D_y, and D_z are the coefficients of dispersion in the respective x, y, and z directions; and v is the flow velocity. This expression is a modified solution of the well-known advection-dispersion equation (67), which is as follows:

$$\frac{\delta C}{\delta t} = -v\frac{\delta C}{\delta x} + D_x\frac{\delta^2 C}{\delta x^2} + D_y\frac{\delta^2 C}{\delta y^2} + D_z\frac{\delta^2 C}{\delta z^2} \quad (7)$$

Because local heterogeneities can substantively affect the values of the above-listed aquifer properties, it is almost always preferable to run an initial test with conservative tracers (as described in the preceding section). The resulting information on the flow direction and dispersivities in the test zone can be used to adjust the required volume of a subsequent coinjection involving microorganisms and tracers. Increasing the volume of the injectate results in both a decreased dispersion effect and a larger injectate cloud. The latter can often compensate for downgradient sampling points that do not lie along the trajectory taken by the center of mass.

The degree of immobilization experienced by microorganisms as they travel along the experimental flow path is difficult to estimate without a priori information derived from earlier in situ tests or from flowthrough column experiments. For bacteria, the loss of bacteria due to interaction with grain surfaces in sandy aquifer sediments can be at least 1 log unit per 10 m of travel (88). Consequently, for small-scale ($\leq$10-m), natural-gradient injection and recovery tests, a minimum fluorescently stained bacterial population of 10^6 to $10^7 \cdot ml^{-1}$ suspended in an injectate volume of 100 to 200 liters (88, 90) would be required in order to obtain an accurate concentration history at the downgradient samplers, using an epifluorescent direct-counting procedure (for an example, see reference 102). The level of detection, which is an important determinant of the required number of labeled bacteria in the injectate, is controlled largely by the quantity of suspended particulate matter in the groundwater. This is because the amount of suspended particulates determines the volume that may be filtered for the direct enumeration procedure. For viruses, a loss of at least 1 log unit over the first meter of travel is often observed (13, 187). Schijven et al. (217) reported a 3-log-unit decline in bacteriophages within the first 2.4 m of travel through sandy aquifer sediment.

LABORATORY STUDIES

Flowthrough Columns

Packing of large columns with subsurface sediments can be problematic because of a rearrangement of pore structure. Columns prepared with the aid of vibratory compaction can suffer from mechanically induced radial segregation (i.e., a higher proportion of coarser grains at the column periphery) (206), and large columns packed with a vibration-free packer can suffer from longitudinal grain-size segregation (93). A device that allows an improved degree of control over the dry packing of columns is described by Stauffer and Dracos (230). Wet-packing procedures (for an example, see reference 118) are most often used in flowthrough column experiments involving microbial transport. However, manual, wet-packing procedures are not artifact-free, judging from differences in the apparent transport behavior of a polydispersed mixture of different-sized microspheres among replicate wet-packed columns (93). In several microbial transport studies (64, 146, 218), problems associated with spatial fractionation of grain sizes have been ameliorated by the use of smaller-diameter, shorter columns packed with sieved (prefractionated) sand. Sterilization of columns can also lead to alterations of pore structure. In particular, autoclaving porous subsurface material can lead to a reduction in the surface area of clays available for adhesion and even alterations in pore surface charge. This can result in decreased microbial penetration rates within the column relative to those in which dry heat sterilization procedures were used (112).

In addition to the packing procedure, the importance of precluding the presence of air bubbles and maintaining saturated conditions cannot be overemphasized. This is because the presence of gas bubbles can change the hydraulic properties of the column and serve as highly efficient collectors for bacteria traveling through the column. The entrapment of negatively charged bacteria and viruses within gas-water interfaces in unsaturated porous media due to strong forces (surface tension) and electrostatic interactions is a well-documented phenomenon (116, 189, 215, 261, 262). Porous-medium columns may be voided of air pockets by an initial displacement of the intergranular gases with carbon dioxide, which can subsequently be voided from the system by sequentially passing solutions of calcium sulfate and degassed groundwater through the column (92). Although many saturated columns of subsurface material have been successfully run in the downflow mode (9, 10), they are more commonly run in the upflow mode to prevent desaturation during the time course of the experiment. Columns run in a vertical position result in the flow direction and gravitational force acting along the same axis and may result in an overemphasis on sedimentation (165). This problem can be overcome using a slight column inclination (e.g., 10) from the horizontal (93). The Mariot siphon provides an inexpensive constant head but is limited in terms of control. Fortunately, some microprocessor-controlled piston pumps can deliver accurately and steadily at volumetric flow rates low enough (e.g., several milliters per hour) to produce environmentally relevant flow velocities within packed columns.

Temperature can be an important factor in bacterial motility (141, 163), attachment of microorganisms to surfaces, and the inactivation rates of viruses (8, 269). Therefore, it is important that column and fluid temperatures be reasonably close to those of the aquifer in which the subsurface material was collected. Aquifers are thermally buffered and, consequently, are characterized by near-constant temperatures. The most effective method of conducting a column experiment at constant temperature is to place the whole experimental assembly (column, reservoir, pump, collector, and tubing) in an environmental control room or large refrigerated incubator that is set to the temperature of the aquifer.

Also, pH is an important variable in microbe attachment to solid surfaces and needs to be carefully monitored, both in the influent and the effluent streams. In unbuffered systems, gaseous carbon dioxide [CO_2(g)] equilibrium can drive substantial pH changes; therefore, solutions need to be in equilibrium with ambient CO_2(g) (after pH adjustment) before being used. When performing column experiments at temperatures other than room temperature, be aware that the Henry's constant for CO_2(g) is a function of temperature. For example, solutions equilibrated at room temperature experience a drop in pH when refrigerated, due to additional CO_2(g) entering the solution. Flowthrough columns also are useful in investigations concerning microbial transport behavior in the vadose zone and in surface soils. However, maintaining constant degrees of saturation in subsurface sediments during unsaturated flow conditions in flowthrough columns can be problematic.

Static Columns

Static minicolumns are particularly well suited for studies of microbial attachment, taxis and chemotaxis (Fig. 5D), and survival behavior in the presence of saturated subsurface media (135, 202, 204, 218, 221). Their small size facilitates

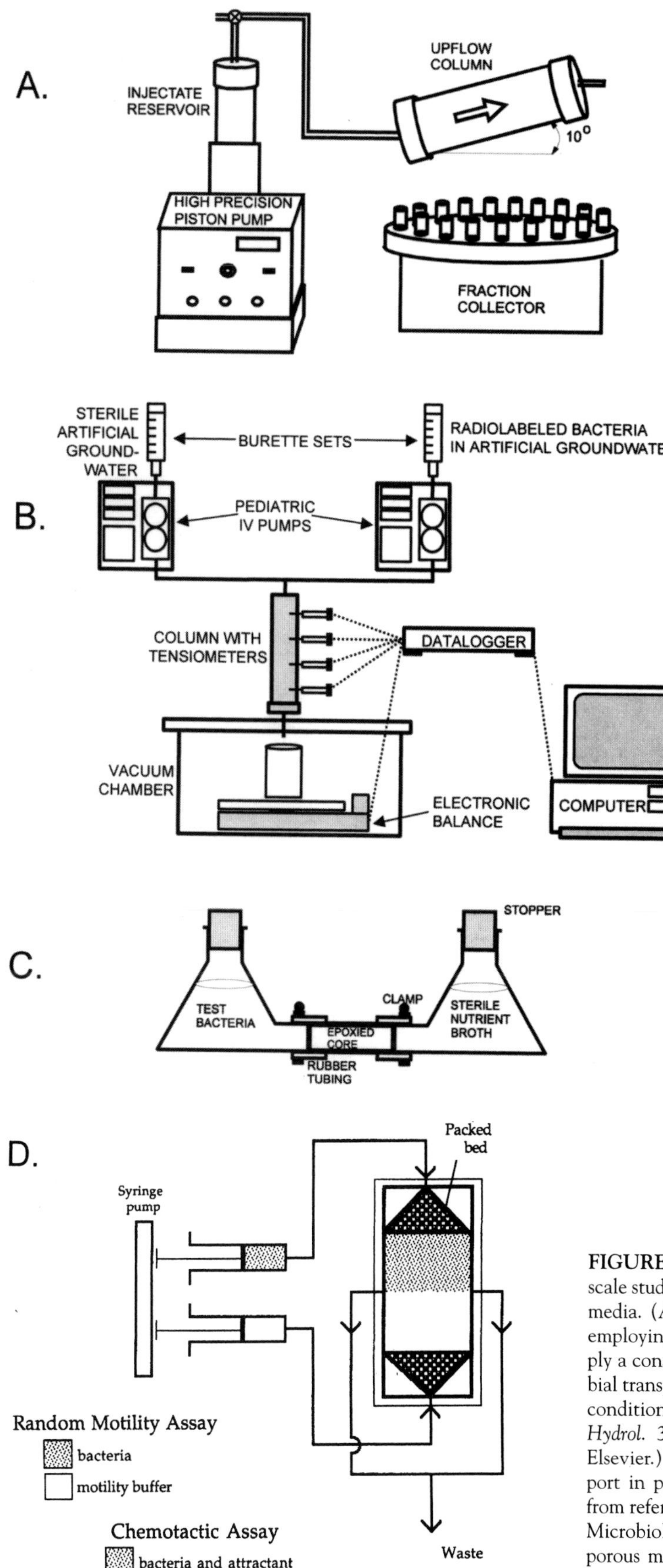

FIGURE 5 Experimental apparatuses for conducting laboratory-scale studies of microbial transport behavior in saturated subsurface media. (A) Upflow column at a slight incline to the horizontal employing a pressure-sensitive, high-precision piston pump to supply a constant head. (B) Experimental setup for examining microbial transport behavior through consolidated materials under static conditions. (Reproduced from D. G. Jewett et al., *J. Contam. Hydrol.* **36:**73–89, 1999 [115], © 1999, with permission from Elsevier.) (C) Experimental apparatus for assessing bacterial transport in porous media as a function of water content. (Modified from reference 113 with permission from the American Society for Microbiology.) (D) Experimental setup for evaluating the effect of porous media on the effective random motility and chemotactic sensitivity coefficients. (Reprinted from reference 14 with permission from the American Society for Microbiology.)

the use of replicate systems and reduces packing-related problems inherent with larger columns. The absence of flow precludes hydrodynamic boundary layers which suspended microorganisms would otherwise have to cross to reach the surface. This simplifies the interpretation of microbial attachment kinetics, so that the effects of varying chemical, physical, and physiological conditions upon attachment behavior in the presence of subsurface material can be more easily assessed.

Pore structure rearrangement during the sampling and repacking of subsurface material can be largely avoided in column experiments involving bacterial penetration through porous rock. This is because cores of consolidated material can be removed and handled intact for use in either flowthrough or static minicolumn experiments. A number of laboratory investigations involving bacterial penetration through low-permeability (<0.7-μm^2), consolidated subsurface material have been performed with Berea sandstone (40, 109, 113, 174, 194), which is relatively homogeneous and available in a variety of permeability classes (165). Assessments of bacterial growth or taxis through sandstone may be made by using a device consisting of a stagnant nutrient-saturated sandstone core connecting two flasks, one containing sterile nutrient broth and the other containing the bacterium of interest (109, 113) (Fig. 5C). The appearance of bacteria in the sterile flask establishes the time required for migration through the connecting sandstone, although differences in methods of detection (e.g., turbidity versus viable counts) can yield substantial differences in apparent travel times (40). A more thorough discussion of investigations involving growth and tactic and chemotactic migration through sandstone is provided by McInerney (165).

Tank Experiments

Two- and three-dimensional sand tank experiments can be used, in part, to bridge the gap between column and field-scale experiments involving microbial transport through saturated porous media. Sand tanks are typically constructed on the scale of 1 to 10 m in length and are commonly used to study the potential of physicochemical and biological remediation of xenobiotic compounds that contaminate granular media in the subsurface (for examples, see references 63, 119, 128, 171, and 172). However, they can also be helpful for simulating the effects of physical (e.g., grain size distribution) or geochemical (e.g., mineralogical) heterogeneity upon subsurface bacterial transport behavior. This may be accomplished by the creation of structured heterogeneity within the tank. Two-dimensional tanks can be instrumented with sampling ports in a manner that has a minimal effect upon flow through the tank. This can be accomplished by embedding the sampling ports along a two-dimensional grid within one or both longitudinal walls. The walls are often constructed from a clear plastic material, which facilitates the creation of layered heterogeneity (108).

Measurements of Bacterial Adhesion and Cell Characteristics

An important determinant of the subsurface transport behavior of bacteria is their interaction with the solid aquifer material, which in turn depends on the bacterial physicochemical properties. Accordingly, various physicochemical characterizations of bacteria and studies of their adhesion to solids and to the air-water interface have been conducted in order to better understand bacterial transport behavior in porous media. Experimental systems used in adhesion studies need to be designed in a manner that permits careful control of the hydrodynamic conditions so that subsequent mass transfer calculations are accurate. For unsaturated systems, microscope micromodels (261), bubble tubes (24, 215), and quiescent water surfaces (215) have been used to quantify bacterial accumulation at the air-water interface. In the past, popular systems used to study bacterial adhesion to solids in saturated systems included batch incubations of microscope slides (204), rotating disks (159), microscope flow cells (227), and impinging jet systems (46). Collision efficiency (α) could be estimated as the ratio of the observed extent of bacterial adhesion to the calculated mass transfer to the surface. However, all of the above types of experimental systems capture only the attachment of microorganisms to the primary energy minimum because all or most microorganisms attached to the secondary energy minimum are swept away by hydrodynamic forces (199). Attachment in the secondary minimum is the dominant process in many systems because both the microorganisms and collector surfaces are generally negatively charged, which results in the presence of a large energy barrier to deposition in the primary minimum. In porous media, microorganisms deposited in the secondary minimum of spherical collectors are thought to be carried along the surface of the collector to the rear stagnation point, where they can accumulate. Therefore, experiments using flow cells, rotating disks, or microscope slides consistently exhibit lower attachments of cells than those seen in column studies, leading to much confusion. For these reasons it is recommended that attachment efficiencies for systems where secondary minimum deposition is likely to occur be assessed using flowthough columns packed which the media of interest. Mass transfer rates can either be calculated using correlation equations (e.g., those given by Tufenkji and Elimelech [245]) or be evaluated under favorable deposition conditions (e.g., using high salt concentrations or a collector surface that has a charge opposite to that of the microorganism of interest).

Hahn and O'Melia (82) developed a conceptually simple model for estimating collision efficiency when a secondary energy minimum is present. They started by acknowledging that colloidal particles have a distribution of velocities as described by the Maxwell distribution $f_{Max}(v_p)$ developed by James Clerk Maxwell.

$$f_{Max}(v_p) = 4\pi(m_p/2\pi kT)^{(3/2)}\, v_p^2 \exp(-0.5 m_p v_p^2/kT) \quad (8)$$

where v_p is the velocity of an individual colloidal particle and m_p is the mass of that particle. They then defined an "activation" velocity [v_p(act)] that is just high enough to escape the secondary energy well by equating the depth of the secondary energy well to the kinetic energy of a particle, $1/2\ m_p v_p^2$, using the following equation:

$$\phi_{2\min} = (1/2)\, m_p v_p^2(\text{act}) \quad (9)$$

Once this critical activation velocity, v_p(act), has been determined, the fraction of particles having sufficient energy to escape the secondary minimum can be calculated by integrating equation 8 over the range of velocities greater than v_p(act). Conversely, the collision efficiency can be estimated as the fraction of particles that do not have enough energy to escape the secondary minimum, as follows:

$$\alpha_{2\min} = 1 - \int_{v_{act}}^{\infty} f_{Max}(v_p)\, dv_p \quad (10)$$

where α_{2min} is the collision efficiency when a secondary minimum energy well is present. While this approach has not been tested extensively, both Hahn and O'Melia (82) and Franchi and O'Melia (66) found good agreement between the model predictions and data generated over a broad range of ionic strengths, whereas conventional models (e.g., the interaction force boundary layer model) that consider deposition only in the primary minimum vastly underpredict the collision efficiency at lower ionic strengths.

Two distinct physicochemical approaches have been applied to bacterial adhesion in the 1980s. A thermodynamic approach assumes direct contact between a surface and an adhering bacterium and describes bacterial adhesion in terms of a comparison of interfacial free energies between the interacting surfaces (3). The DLVO theory of colloidal stability calculates attractive Lifshitz-van der Waals and attractive or repulsive electrostatic interactions as a function of the separation distance (211). However, obvious deficiencies of both approaches, i.e., the neglect of obviously existing electrostatic contributions by the thermodynamic approach and the incapacity of the DLVO theory to explain influences of the hydrophobicity of bacteria and solids, have fostered the general acceptance of an extended DLVO theory that includes short-range Lewis acid-base (hydrophobic) interactions (256). Bacterial characteristics entering into these calculations include the cell radius (electrostatic and Lifshitz-van der Waals interactions), the Hamaker constant of bacteria (Lifshitz-van der Waals interactions), the zeta potential (electrostatic interactions), and the surface free energy, which can be split up in apolar, electron-donating, and electron-accepting components (hydrophobic/hydrophilic interactions).

Cell radii are measured either by dynamic light scattering (193) or microscopically, whereby, e.g., for rod-shaped bacteria a geometric mean radius, $R_e = 0.5\ (wwl)^{1/3}$, with w and l being the width and the length of the bacterium, has been used (215). Bacterial zeta potentials are usually obtained by laser-Doppler velocimetry. Approximations of Hamaker constants of bacterial surfaces can be obtained from the Lifshitz-van der Waals interaction energy as calculated from contact angle measurements (28). However, reported ranges of Hamaker constants of standard biological materials (180) which neglect individual bacterial differences have often been used instead (204). Surface free energies are obtained by contact angle measurements (255) performed on layers of bacteria (obtained, e.g., by filtration). Individual components of the surface free energy can be analyzed by measuring contact angles with at least three liquids of different polarities (254).

It should be noted, however, that some bacteria are far from being ideal colloidal particles with perfectly smooth surfaces as required by the extended DLVO theory and thus, not surprisingly, considerable discrepancies between predictions and the actual adhesion behavior have been observed. Recently, these surface irregularities have been studied by atomic force microscopy (36). In detailed studies of the influence of the chemistry and topography of LPS layers on bacterial adhesion (123, 125, 126), it was suggested that irregular and compressible LPS coats can give rise to an efficient and strongly irreversible adhesion to mineral surfaces that is mediated by H bonds between polysaccharides and mineral oxides and/or chemisorbed water, whereas rigid, smooth LPS layers may prevent bacterial adhesion even when the extended DLVO forces are strongly attractive. Inclusion of repulsive and attractive polymer interactions into an even further extended DLVO theory could account for some hitherto unexplained adhesion phenomena. However, the fact that the new theory comprises now five distinct forces with different decay functions led the authors to conclude that it is uncertain whether bacteria can be characterized rigorously enough for a reliable a priori prediction of adhesion (126).

CONCLUSIONS

Although microorganisms were first introduced into groundwater as tracers in order to gain information about the hydrology of complex groundwater environments, recent interest has clearly shifted to the transport behavior of the microorganisms themselves. The shift is due largely to the now-recognized importance of subsurface microbial transport in aquifer restoration, migration of surface active and hydrophobic chemical contaminants, public health, and oil recovery. Much of the emphasis over the last 2 decades has focused on column experiments, because of interpretational and technical difficulties involved with conducting subsurface microbial transport investigations in the field. As a consequence, the development of methodology for conducting controlled field transport investigations has lagged behind that of laboratory experiments. Flow-through and static column investigations offer the ability to achieve a degree of control not possible in the field. As described herein, the methodology for studying specific controls of microbial transport in laboratory porous media is improving as techniques for accurately controlling flow at low (environmentally relevant) rates, isolating specific controlling factors, and quantifying the microorganisms of interest get better. However, there is also a growing realization that processes controlling subsurface microbial transport behavior are interrelated and can be operative on spatial and temporal scales not conducive to lab study. This is leading to more field scale experiments and to research programs that involve a combination of field and laboratory studies. Most controlled field investigations of subsurface microbial transport are conducted on limited spatial scales relative to the scales of interest to those concerned with pathogen transport to water supply wells, with microbially enhanced oil recovery from petroleum reservoirs, and with the feasibility of using introduced bacteria for aquifer restoration. However, continued advances in our understanding of microbial transport behavior as a result of ongoing laboratory and small-scale field investigations will facilitate future larger-scale field experiments. This, in turn, will allow testing of more accurate models capable of describing subsurface microbial transport over considerable distances.

REFERENCES

1. **Abba, F., B. Orlandi, and A. Rondelli.** 1898. Über die Filtrationskraft des Bodens und die Fortschwemmung von Bakterien durch das Grundwasser. *Z. Hyg. Infekt. Krankh.* **31:**66–84.
2. **Abbaszadegan, M., M. S. Huber, C. P. Gerba, and I. L. Pepper.** 1993. Detection of enteroviruses in groundwater with the polymerase chain reaction. *Appl. Environ. Microbiol.* **59:**1318–1324.
3. **Absolom, D. R., F. V. Lamberti, Z. Policova, W. Zingg, C. J. van Oss, and A. W. Neumann.** 1983. Surface thermodynamics of bacterial adhesion. *Appl. Environ. Microbiol.* **46:**90–97.
4. **Abudalo, R. A., Y. G. Bogatsu, J. N. Ryan, R. W. Harvey, D. W. Metge, and M. Elimelech.** 1995. The effect of ferric oxyhydroxide grain coatings on the trans-

port of bacteriophage PRD1 and *Cryptosporidium parvum* oocysts in saturated porous media. *Environ. Sci. Technol.* **39:**6412–6419.

5. **Albinger, O., B. K. Biesemeyer, R. G. Arnold, and B. E. Logan.** 1994. Effect of bacterial heterogeneity on adhesion to uniform collectors by monoclonal populations. *FEMS Microbiol. Lett.* **124:**321.
6. **Albrechtsen, H. J., and A. Winding.** 1992. Microbial biomass and activity in subsurface sediments from Vejen, Denmark. *Microb. Ecol.* **23:**303–317.
7. **Ammons, D., J. Rampersad, and G. E. Fox.** 1998. A genomically modified marker strain of *Escherichia coli. Curr. Microbiol.* **37:**341–346.
8. **Badawy, A. S., J. B. Rose, and C. P. Gerba.** 1990. Comparative survival of enteric viruses and coliphage on sewage irrigated grass. *J. Environ. Sci. Health A* **25:**937–952.
9. **Bales, R. C., C. P. Gerba, G. H. Grondin, and S. L. Jensen.** 1989. Bacteriophage transport in sandy soil and fractured tuff. *Appl. Environ. Microbiol.* **55:**2061–2067.
10. **Bales, R. C., S. R. Hinkle, T. W. Kroeger, K. Stocking, and C. P. Gerba.** 1991. Bacteriophage adsorption during transport through porous-media-chemical perturbations and reversibility. *Environ. Sci. Technol.* **25:**2088–2095.
11. **Bales, R. C., S. M. Li, K. M. Maguire, M. T. Yahya, and C. P. Gerba.** 1993. Ms-2 and poliovirus transport in porous-media—hydrophobic effects and chemical perturbations. *Water Resour. Res.* **29:**957–963.
12. **Bales, R. C., S. M. Li, K. M. Maguire, M. T. Yahya, C. P. Gerba, and R. W. Harvey.** 1995. Virus and bacteria transport in a sandy aquifer, Cape Cod, MA. *Ground Water* **33:**653–661.
13. **Bales, R. C., S. M. Li, T. C. J. Yeh, M. E. Lenczewski, and C. P. Gerba.** 1997. Bacteriophage and microsphere transport in saturated porous media: forced-gradient experiment at Borden, Ontario. *Water Resour. Res.* **33:**639–648.
14. **Barton, J. W., and R. M. Ford.** 1995. Determination of effective transport coefficients for bacterial migration in sand columns. *Appl. Environ. Microbiol.* **61:**3329–3335.
15. **Baveye, P., P. Vandevivere, B. L. Hoyle, P. C. DeLeo, and D. S. de Lozada.** 1998. Environmental impact and mechanisms of the biological clogging of saturated soils and aquifer materials. *Crit. Rev. Environ. Sci. Technol.* **28:**123–191.
16. **Bayer, M., and J. Sloyer.** 1990. The electrophoretic mobility of Gram-negative and Gram-positive bacteria: an electrokinetic analysis. *J. Gen. Microbiol.* **136:**867–874.
17. **Baygents, J. C., J. R. Glynn, O. Albinger, B. K. Biesemeyer, K. L. Ogden, and R. G. Arnold.** 1998. Variation of surface charge density in monoclonal bacterial populations: implications for transport through porous media. *Environ. Sci. Technol.* **32:**1596–1603.
18. **Becker, M. W., P. W. Reimus, and P. Vilks.** 1999. Transport and attenuation of carboxylate-modified latex microspheres in fractured rock laboratory and field tracer tests. *Ground Water* **37:**387–395.
19. **Beeder, J., R. K. Nilsen, T. Thorstenson, and T. Torsvik.** 1996. Penetration of sulfate reducers through a porous North Sea oil reservoir. *Appl. Environ. Microbiol.* **62:**3551–3553.
20. **Bengtsson, G., and R. Lindqvist.** 1995. Transport of soil bacteria controlled by density-dependent sorption kinetics. *Water Resour. Res.* **31:**1247–1256.
21. **Beres, M., A. Green, P. Huggenberger, and H. Horstmeyer.** 1995. Mapping the architecture of glaciofluvial sediments with 3-dimensional georadar. *Geology* **23:** 1087–1090.
22. **Beres, M., P. Huggenberger, A. G. Green, and H. Horstmeyer.** 1999. Using two- and three-dimensional georadar methods to characterize glaciofluvial architecture. *Sedimentary Geol.* **129:**1–24.
23. **Bitton, G., and R. W. Harvey.** 1992. Transport of pathogens through soils and aquifers., p. 103–124. *In* R. Michell (ed.), *Environmental Microbiology*. Wiley-Liss, New York, N.Y.
24. **Blanchard, D. C., and D. L. Syzdek.** 1974. Bubble tube: apparatus for determining rate of collection of bacteria by a bubble rising in water. *Limnol. Oceanogr.* **23:**133–138.
25. **Boadu, F. K.** 2000. Hydraulic conductivity of soils from grain-size distribution: new models. *J. Geotech. Geoenviron. Eng.* **126:**739–746.
26. **Bolster, C. H., G. M. Hornberger, A. L. Mills, and J. L. Wilson.** 1998. A method for calculating bacterial deposition coefficient using the fraction of bacteria recovered from laboratory columns. *Environ. Sci. Technol.* **32:**1329–1332.
27. **Bolster, C. H., A. L. Mills, G. M. Hornberger, and J. S. Herman.** 1999. Spatial distribution of deposited bacteria following miscible displacement experiments in intact cores. *Water Resour. Res.* **35:**1797–1807.
28. **Bos, R., H. C. van der Mei, and H. J. Busscher.** 1999. Physico-chemistry of initial microbial adhesive interactions—its mechanisms and methods for study. *FEMS Microbiol. Rev.* **23:**179–230.
29. **Bouwer, H.** 1978. *Groundwater Hydrology*. McGraw Hill, New York, N.Y.
30. **Bradford, S. A., and M. Bettahar.** 2005. Straining, attachment, and detachment of cryptosporidium oocysts in saturated porous media. *J. Environ. Qual.* **34:**469–478.
31. **Briandet, R., V. Leriche, B. Carpentier, and M. N. Bellon-Fontaine.** 1999. Effects of the growth procedure on the surface hydrophobicity of *Listeria monocytogenes* cells and their adhesion to stainless steel. *J. Food Prot.* **62:**994–998.
32. **Brown, D., and P. R. Jaffe.** 2001. Effects of nonionic surfactants on bacterial transport through porous media. *Environ. Sci. Technol.* **35:**3877–3883.
33. **Brush, C. F., W. C. Ghiorse, L. J. Anguish, J. Y. Parlange, and H. G. Grimes.** 1999. Transport of *Cryptosporidium parvum* oocysts through saturated columns. *J. Environ. Qual.* **28:**809–815.
34. **Burlage, R. S., Z. K. Yang, and T. Mehlhorn.** 1996. A transposon for green fluorescent protein transcriptional fusions: application for bacterial transport experiments. *Gene* **173:**53–58.
35. **Camesano, T. A., and B. E. Logan.** 1998. Influence of fluid velocity and cell concentration on the transport of motile and nonmotile bacteria in porous media. *Environ. Sci. Technol.* **32:**1699–1708.
36. **Camesano, T. A., M. J. Natan, and B. E. Logan.** 2000. Observation of changes in bacterial cell morphology using tapping mode atomic force microscopy. *Langmuir* **16:** 4563–4572.
37. **Camper, A. K., J. T. Hayes, P. J. Sturman, W. L. Jones, and A. B. Cunningham.** 1993. Effects of motility and adsorption rate coefficient on transport of bacteria through saturated porous media. *Appl. Environ. Microbiol.* **59:** 3455–3462.
38. **Chalfie, M., Y. Tu, G. Euskirchen, W. W. Ward, and D. C. Prasher.** 1994. Green fluorescent protein as a marker for gene-expression. *Science* **263:**802–805.
39. **Champ, D. R., and J. Schroeter.** 1988. Bacterial transport in fractured rock—a field-scale tracer test at the Chalk River Nuclear Laboratories. *Water Sci. Technol.* **20:**81–87.
40. **Chang, P. L., and T. F. Yen.** 1985. Interaction of *Pseudomonas putida* ATCC 12633 and bacteriophage gh-1 in Berea sandstone rock. *Appl. Environ. Microbiol.* **59:**3455–3462.

41. **Chen, J., and B. Koopman.** 1997. Effect of fluorochromes on bacterial surface properties and interaction with granular media. *Appl. Environ. Microbiol.* **63:**3941–3945.
42. **Chu, Y., Y. Jin, T. Baumann, and M. V. Yates.** 2003. Effect of soil properties on saturated and unsaturated virus transport through columns. *J. Environ. Qual.* **32:**2017–2025.
43. **Colwell, F. S., G. J. Stormberg, T. J. Phelps, S. A. Birnbaum, J. McKinley, S. A. Rawson, C. Veverka, S. Goodwin, P. E. Long, B. F. Russell, T. Garland, D. Thompson, P. Skinner, and S. Grover.** 1992. Innovative techniques for collection of saturated and unsaturated subsurface basalts and sediments for microbiological characterization. *J. Microbiol. Methods* **15:**279–292.
44. **Corapcioglu, M. Y., and S. H. Kim.** 1995. Modeling facilitated contaminant transport by mobile bacteria. *Water Resour. Res.* **31:**2639–2647.
45. **Cusack, F., S. Singh, C. McCarthy, J. Grieco, M. Derocco, D. Nguyen, H. Lappinscott, and J. W. Costerton.** 1992. Enhanced oil-recovery—3-dimensional sandpack simulation of ultramicrobacteria resuscitation in reservoir formation. *J. Gen. Microbiol.* **138:**647–655.
46. **Dabros, T., and T. G. M. van de Ven.** 1987. Deposition of latex particles on glass surfaces in an impinging jet. *Phys. Chem. Hydrodynamics* **8:**161–172.
47. **Daniell, T. J., M. L. Davy, and R. J. Smith.** 2000. Development of a genetically modified bacteriophage for use in tracing sources of pollution. *J. Appl. Microbiol.* **88:**860–869.
48. **DeFlaun, M. F., M. E. Fuller, P. Zhang, W. P. Johnson, B. J. Mailloux, W. E. Holben, W. P. Kovacik, D. L. Balkwill, and T. C. Onstott.** 2001. Comparison of methods for monitoring bacterial transport in the subsurface. *J. Microbiol. Methods* **47:**219–231.
49. **DeFlaun, M. F., C. J. Murray, W. Holben, T. Scheibe, A. Mills, T. Ginn, T. Griffin, E. Majer, and J. L. Wilson.** 1997. Preliminary observations on bacterial transport in a coastal plain aquifer. *FEMS Microbiol. Rev.* **20:**473–487.
50. **DeFlaun, M. F., S. R. Oppenheimer, S. Streger, C. W. Condee, and M. Fletcher.** 1999. Alterations in adhesion, transport, and membrane characteristics in an adhesion-deficient pseudomonad. *Appl. Environ. Microbiol.* **65:**759–765.
51. **DeLeo, P. C., and P. Baveye.** 1996. Enumeration and biomass estimation of bacteria in aquifer microcosm studies by flow cytometry. *Appl. Environ. Microbiol.* **62:**4580–4586.
52. **Deshpande, P. A., and D. R. Shonnard.** 2000. An improved spectrophotometric method to study the transport, attachment, and breakthrough of bacteria through porous media. *Appl. Environ. Microbiol.* **66:**763–768.
53. **Deshpande, P. A., and D. R. Shonnard.** 1999. Modeling the effects of systematic variation in ionic strength on the attachment kinetics of *Pseudomonas fluorescens* UPER-1 in saturated sand columns. *Water Resour. Res.* **35:**1619–1627.
54. **Dong, H.** 2002. Significance of electrophoretic mobility distribution to bacterial transport in granular porous media. *J. Microbiol. Methods* **51:**83–93.
55. **Dong, H., T. C. Onstott, M. F. DeFlaun, M. E. Fuller, K. M. Gillespie, and J. K. Fredrickson.** 1999. Development of radiographic and microscopic techniques for the characterization of bacterial transport in intact sediment cores from Oyster, Virginia. *J. Microbiol. Methods* **37:**139–154.
56. **Dowd, S. E., S. D. Pillai, S. Y. Wang, and M. Y. Corapcioglu.** 1998. Delineating the specific influence of virus isoelectric point and size on virus adsorption and transport through sandy soils. *Appl. Environ. Microbiol.* **64:**405–410.
57. **Dubois, S. M., B. E. Moore, and S. P. Sagik.** 1976. Poliovirus survival and movement in a sandy forest soil. *Appl. Environ. Microbiol.* **31:**536–543.
58. **Duffy, K. J., P. T. Cummings, and R. M. Ford.** 1995. Random-walk calculations for bacterial migration in porous media. *Biophys. J.* **68:**800–806.
59. **Eisenmann, H., H. Harms, R. Meckenstock, E. I. Meyer, and A. J. B. Zehnder.** 1998. Grazing of a *Tetrahymena* sp. on adhered bacteria in percolated columns monitored by in situ hybridization with fluorescent oligonucleotide probes. *Appl. Environ. Microbiol.* **64:**1264–1269.
60. **Elowitz, M. B., M. G. Surette, P. E. Wolf, J. Stock, and S. Leibler.** 1997. Photoactivation turns green fluorescent protein red. *Curr. Biol.* **7:**809–812.
61. **Emerson, R. J., IV, and T. A. Camesano.** 2004. Nanoscale investigation of pathogenic microbial adhesion to a biomaterial. *Appl. Environ. Microbiol.* **70:**6012–6022.
62. **Fang, Y., and B. E. Logan.** 1999. Bacterial transport in gas-sparged porous medium. *J. Environ. Eng.* **125:**668–673.
63. **Fischer, U., R. Schulin, M. Keller, and F. Stauffer.** 1996. Experimental and numerical investigation of soil vapor extraction. *Water Resour. Res.* **32:**3413–3427.
64. **Fontes, D. E., A. L. Mills, G. M. Hornberger, and J. S. Herman.** 1991. Physical and chemical factors influencing transport of microorganisms through porous media. *Appl. Environ. Microbiol.* **57:**2473–2481.
65. **Fortineau, N., P. Trieu-Cuot, O. Gaillot, E. Pellegrini, P. Berche, and J. L. Gaillard.** 2000. Optimization of green fluorescent protein expression vectors for in vitro and in vivo detection of *Listeria monocytogenes*. *Res. Microbiol.* **151:**353–360.
66. **Franchi, A., and C. R. O'Melia.** 2003. Effects of natural organic matter and solution chemistry on the deposition and reentrainment of colloids in porous media. *Environ. Sci. Technol.* **37:**1122–1129.
67. **Freeze, R. A., and J. A. Cherry.** 1979. *Groundwater*. Prentice Hall, Englewood Cliffs, N.J.
68. **Fuller, C. C., and J. W. Harvey.** 2000. Reactive uptake of trace metals in the hyporheic zone of a mining-contaminated stream, Pinal Creek, Arizona. *Environ. Sci. Technol.* **34:**1150–1155.
69. **Fuller, M. E., B. J. Mailloux, S. H. Streger, J. A. Hall, P. Zhang, W. P. Kovacik, S. Vainberg, W. P. Johnson, T. C. Onstott, and M. F. DeFlaun.** 2004. Application of a vital fluorescent staining method for simultaneous, near-real-time concentration monitoring of two bacterial strains in an Atlantic coastal plain aquifer in Oyster, Virginia. *Appl. Environ. Microbiol.* **70:**1680–1687.
70. **Gallop, P. M., M. A. Paz, E. Henson, and S. A. Latt.** 1984. Dynamic approaches to the delivery of reporter reagents into living cells. *BioTechniques* **1:**32–36.
71. **Gannon, J., Y. H. Tan, P. Baveye, and M. Alexander.** 1991. Effect of sodium chloride on transport of bacteria in a saturated aquifer material. *Appl. Environ. Microbiol.* **57:**2497–2501.
72. **Gannon, J. T., V. B. Manilal, and M. Alexander.** 1991. Relationship between cell-surface properties and transport of bacteria through soil. *Appl. Environ. Microbiol.* **57:**190–193.
73. **Garabedian, S. P., D. R. Leblanc, L. W. Gelhar, and M. A. Celia.** 1991. Large-scale natural gradient tracer test in sand and gravel, Cape Cod, Massachusetts. 2. Analysis of spatial moments for a nonreactive tracer. *Water Resour. Res.* **27:**911–924.
74. **Glynn, J. R., Jr., B. M. Belongia, R. G. Arnold, K. L. Ogden, and J. C. Baygents.** 1998. Capillary electrophoresis measurements of electrophoretic mobility for colloidal particles of biological interest. *Appl. Environ. Microbiol.* **64:**2572–2577.
75. **Goltz, M. N., and P. V. Roberts.** 1987. Using the method of moments to analyze 3-dimensional diffusion-limited

solute transport from temporal and spatial perspectives. *Water Resour. Res.* **23:**1575–1585.

76. **Graham, D. W., D. G. Korich, R. P. Leblanc, N. A. Sinclair, and R. G. Arnold.** 1992. Applications of a colorimetric plate assay for soluble methane monooxygenase activity. *Appl. Environ. Microbiol.* **58:**2231–2236.
77. **Grasso, D., B. F. Smets, K. A. Strevett, B. D. Machinist, C. J. VanOss, R. F. Giese, and W. Wu.** 1996. Impact of physiological state on surface thermodynamics and adhesion of *Pseudomonas aeruginosa*. *Environ. Sci. Technol.* **30:** 3604–3608.
78. **Groffman, P. M., A. J. Gold, and G. Howard.** 1995. Hydrologic tracer effects on soil microbial activities. *Soil Sci. Soc. Am. J.* **59:**478–481.
79. **Gross, M. J., O. Albinger, D. G. Jewett, B. E. Logan, R. C. Bales, and R. G. Arnold.** 1995. Measurement of bacterial collision efficiencies in porous-media. *Water Res.* **29:**1151–1158.
80. **Gross, M. J., and B. E. Logan.** 1995. Influence of different chemical treatments on transport of *Alcaligenes paradoxus* in porous media. *Appl. Environ. Microbiol.* **61:**1750–1756.
81. **Guimares, V. F., I. V. Cruz, A. N. Hagler, L. C. Mendonca-Hagler, and J. D. van Elsas.** 1997. Transport of a genetically modified Pseudomonas fluorescens and its parent strain through undisturbed tropical soil cores. *Appl. Soil Ecol.* **7:**41–50.
82. **Hahn, M. W., and C. R. O'Melia.** 2004. Deposition and reentrainment of Brownian particles in porous media under unfavorable chemical conditions: some concepts and applications. *Environ. Sci. Technol.* **38:**210–220.
83. **Harms, H., and T. N. P. Bosma.** 1997. Mass transfer limitation of microbial growth and pollutant degradation. *J. Ind. Microbiol. Biotechnol.* **18:**97–105.
84. **Harms, H., and A. J. B. Zehnder.** 1994. Influence of substrate diffusion on degradation of Dibenzofuran and 3-chlorodibenzofuran by attached and suspended bacteria. *Appl. Environ. Microbiol.* **60:**2736–2745.
85. **Hart, A., and C. Edwards.** 1987. Buoyant density-fluctuations during the cell-cycle of *Bacillus subtilis*. *Arch. Microbiol.* **147:**68–72.
86. **Harter, T., S. Wagner, and E. R. Atwill.** 2000. Colloid transport and filtration of *Cryptosporidium parvum* in sandy soils and aquifer sediments. *Environ. Sci. Technol.* **34:**62–70.
87. **Harvey, R. W.** 1991. Parameters involved in modeling movement of bacteria in groundwater, p. 89–114. *In* C. J. Hurst (ed.), *Modeling the Environmental Fate of Microorganisms*. American Society for Microbiology, Washington, D.C.
88. **Harvey, R. W., and S. P. Garabedian.** 1991. Use of colloid filtration theory in modeling movement of bacteria through a contaminated sandy aquifer. *Environ. Sci. Technol.* **25:**178–185.
89. **Harvey, R. W., and L. H. George.** 1987. Growth determinations for unattached bacteria in a contaminated aquifer. *Appl. Environ. Microbiol.* **53:**2992-2996.
90. **Harvey, R. W., L. H. George, R. L. Smith, and D. R. Leblanc.** 1989. Transport of microspheres and indigenous bacteria through a sandy aquifer: results of natural-gradient and forced-gradient tracer experiments. *Environ. Sci. Technol.* **23:**51–56.
91. **Harvey, R. W., and H. Harms.** 2002. Tracers in groundwater: use of microorganisms and microspheres, p. 3194–3202. *In* G. Bitton (ed.), *Encyclopedia of Environmental Microbiology*, vol. 6. John Wiley & Sons, Inc., New York, N.Y.
92. **Harvey, R. W., N. E. Kinner, A. Bunn, D. MacDonald, and D. Metge.** 1995. Transport behavior of groundwater protozoa and protozoan-sized microspheres in sandy aquifer sediments. *Appl. Environ. Microbiol.* **61:**209–217.
93. **Harvey, R. W., N. E. Kinner, D. Macdonald, D. W. Metge, and A. Bunn.** 1993. Role of physical heterogeneity in the interpretation of small-scale laboratory and field observations of bacteria, microbial-sized microsphere, and bromide transport through aquifer sediments. *Water Resour. Res.* **29:**2713–2721.
94. **Harvey, R. W., N. Mayberry, N. E. Kinner, D. W. Metge, and F. Novarino.** 2002. Effect of growth conditions and staining procedure upon the subsurface transport and attachment behaviors of a groundwater protist. *Appl. Environ. Microbiol.* **68:**1872–1881.
95. **Harvey, R. W., D. W. Metge, N. Kinner, and N. Mayberry.** 1997. Physiological considerations in applying laboratory-determined buoyant densities to predictions of bacterial and protozoan transport in groundwater: results of in situ and laboratory tests. *Environ. Sci. Technol.* **31:**289–295.
96. **Harvey, R. W., D. W. Metge, A. M. Shapiro, R. A. Renken, K. J. Cunningham, and M. Zygnerski.** *Cryptosporidium parvum* oocyst transport potential in the karstic limestone of the Biscayne Aquifer. 2. Results of tracer studies with carboxylated polystyrene microspheres and killed oocysts. Submitted for publication.
97. **Heise, S., and G. Gust.** 1999. Influence of the physiological status of bacteria on their transport into permeable sediments. *Mar. Ecol. Prog. Ser.* **190:**141–153.
98. **Heitzer, A., B. Applegate, S. Kehrmeyer, H. Pinkart, O. F. Webb, T. J. Phelps, D. C. White, and G. S. Sayler.** 1998. Physiological considerations of environmental applications of lux reporter fusions. *J. Microbiol. Methods* **33:**45–57.
99. **Hekman, W. E., C. E. Heijnen, S. Burgers, J. A. Vanveen, and J. D. Vanelsas.** 1995. Transport of bacterial inoculants through intact cores of 2 different soils as affected by water percolation and the presence of wheat plants. *FEMS Microbiol. Ecol.* **16:**143–157.
100. **Hendry, M. J., J. R. Lawrence, and P. Maloszewski.** 1997. The role of sorption in the transport of *Klebsiella oxytoca* through saturated silica sand. *Ground Water* **35:**574–584.
101. **Hess, K. M., S. H. Wolf, and M. A. Celia.** 1992. Large-scale natural gradient tracer test in sand and gravel, Cape-Cod, Massachusetts. 3. Hydraulic conductivity variability and calculated macrodispersivities. *Water Resour. Res.* **28:**2011–2027.
102. **Hobbie, J. E., R. J. Daley, and S. Jasper.** 1977. Use of Nuclepore filters for counting bacteria by fluorescence microscopy. *Appl. Environ. Microbiol.* **33:**1225–1228.
103. **Holben, W. E., and P. H. Ostrom.** 2000. Monitoring bacterial transport by stable isotope enrichment of cells. *Appl. Environ. Microbiol.* **66:**4935–4939.
104. **Hornberger, G. M., A. L. Mills, and J. S. Herman.** 1992. Bacterial transport in porous-media-evaluation of a model using laboratory observations. *Water Resour. Res.* **28:**915–923.
105. **Huggenberger, P., E. Meier, and A. Pugin.** 1994. Ground-probing radar as a tool for heterogeneity estimation in gravel deposits—advances in data-processing and facies analysis. *J. Appl. Geophys.* **31:**171–184.
106. **Hurst, C. J. (ed.).** 1991. *Modeling the Environmental Fate of Microorganisms*. American Society for Microbiology, Washington, D.C.
107. **Huysman, F., and W. Verstraete.** 1993. Water-facilitated transport of bacteria in unsaturated soil columns—influence of cell-surface hydrophobicity and soil properties. *Soil Biol. Biochem.* **25:**83–90.

108. **Illangasekare, T. H., E. J. Armbruster, and D. N. Yates.** 1995. Non-aqueous-phase fluids in heterogeneous aquifers—experimental-study. *J. Environ. Eng.* **121:**571–579.
109. **Jang, L. K., P. W. Chang, J. E. Findley, and T. F. Yen.** 1983. Selection of bacteria with favorable transport properties through porous rock for the application of microbial-enhanced oil recovery. *Appl. Environ. Microbiol.* **46:**1066–1072.
110. **Jaspers, M. C. M., S. Totevova, K. Demnerova, H. Harms, and J. R. van der Meer.** 1999. The use of whole-cell living biosensors to determine the bioavailability of pollutants to microorganisms, p. 153–158. *In* P. Baveye, J. C. Block, and V. V. Goncharuk (ed.), *Bioavailability of Organic Xenobiotics in the Environment.* Kluwer Academic Publishers, London, England.
111. **Jenkins, M. B., and L. W. Lion.** 1993. Mobile bacteria and transport of polynuclear aromatic hydrocarbons in porous media. *Appl. Environ. Microbiol.* **59:**3306–3313.
112. **Jenneman, G. E., M. J. McInerney, M. E. Crocker, and R. M. Knapp.** 1986. Effect of sterilization by dry heat or autoclaving on bacterial penetration through Berea sandstone. *Appl. Environ. Microbiol.* **51:**39–43.
113. **Jenneman, G. E., M. J. McInerney, and R. M. Knapp.** 1985. Microbial penetration through nutrient-saturated Berea sandstone. *Appl. Environ. Microbiol.* **50:**383–391.
114. **Jewett, D. G., T. A. Hilbert, B. E. Logan, R. G. Arnold, and R. C. Bales.** 1995. Bacterial transport in laboratory columns and filters—influence of tonic strength and pH on collision efficiency. *Water Res.* **29:**1673–1680.
115. **Jewett, D. G., B. E. Logan, R. G. Arnold, and R. C. Bales.** 1999. Transport of *Pseudomonas fluorescens* strain P17 through quartz sand columns as a function of water content. *J. Contam. Hydrol.* **36:**73–89.
116. **Jin, Y., Y. J. Chu, and Y. S. Li.** 2000. Virus removal and transport in saturated and unsaturated sand columns. *J. Contam. Hydrol.* **43:**111–128.
117. **Jin, Y., M. V. Yates, S. S. Thompson, and W. A. Jury.** 1997. Sorption of viruses during flow through saturated sand columns. *Environ. Sci. Technol.* **31:**548–555.
118. **Johnson, M. J.** 1990. Relative permeabilities of gasoline, water, and air in sand. M.S. thesis. University of New Hampshire, Durham.
119. **Johnson, T. E., and D. K. Kreamer.** 1994. Physical and mathematical-modeling of diesel fuel liquid and vapor movement in porous-media. *Ground Water* **32:**551–560.
120. **Johnson, W. P., K. A. Blue, B. E. Logan, and R. G. Arnold.** 1995. Modeling bacterial detachment during transport through porous media as a residence-time-dependent process. *Water Resour. Res.* **31:**2649–2658.
121. **Johnson, W. P., and B. E. Logan.** 1996. Enhanced transport of bacteria in porous media by sediment-phase and aqueous-phase natural organic matter. *Water Res.* **30:** 923–931.
122. **Johnson, W. P., P. Zhang, M. E. Fuller, T. D. Scheibe, B. J. Mailloux, T. C. Onstott, M. F. Deflaun, S. S. Hubbard, J. Radtke, W. P. Kovacik, and W. Holben.** 2001. Ferrographic tracking of bacterial transport in the field at the narrow channel focus area, Oyster, VA. *Environ. Sci. Technol.* **35:**182–191.
123. **Jucker, B. A., H. Harms, S. J. Hug, and A. J. B. Zehnder.** 1997. Adsorption of bacterial surface polysaccharides on mineral oxides is mediated by hydrogen bonds. *Colloids Surf. B* **9:**331–343.
124. **Jucker, B. A., H. Harms, and A. J. B. Zehnder.** 1996. Adhesion of the positively charged bacterium *Stenotrophomonas* (*Xanthomonas*) *maltophilia* 70401 to glass and Teflon. *J. Bacteriol.* **178:**5472–5479.
125. **Jucker, B. A., H. Harms, and A. J. B. Zehnder.** 1998. Polymer interactions between five gram-negative bacteria and glass investigated using LPS micelles and vesicles as model systems. *Colloids Surf. B* **11:**33–45.
126. **Jucker, B. A., A. J. B. Zehnder, and H. Harms.** 1998. Quantification of polymer interactions in bacterial adhesion. *Environ. Sci. Technol.* **32:**2909–2915.
127. **Karatani, H., T. Wilson, and J. W. Hastings.** 1992. A blue fluorescent protein from a yellow-emitting luminous bacterium. *Photochem. Photobiol.* **55:**293–299.
128. **Katz, Y., and H. Gvirtzman.** 2000. Capture and cleanup of a migrating VOC plume by the in-well vapor stripping: a sand tank experiment. *J. Contam. Hydrol.* **43:**25–44.
129. **Kemp, J. S., E. Paterson, S. M. Gammack, M. S. Cresser, and K. Killham.** 1992. Leaching of genetically modified *Pseudomonas fluorescens* through organic soils—influence of temperature, soil pH, and roots. *Biol. Fertil. Soils* **13:**218–224.
130. **Kim, C. K., M. J. Kwak, and S. G. Lee.** 1996. Structural and functional stability of the genetic recombinant plasmid pCU103 in different water environments. *J. Microbiol.* **34:**241–247.
131. **Kim, S. H., and M. Y. Corapcioglu.** 1996. A kinetic approach to modeling mobile bacteria-facilitated groundwater contaminant transport. *Water Resour. Res.* **32:**321–331.
132. **Kinner, N. E., R. W. Harvey, and M. Kazmierkiewicz-Tabaka.** 1997. Effect of flagellates on free-living bacterial abundance in an organically contaminated aquifer. *FEMS Microbiol. Rev.* **20:**249–259.
133. **Kinoshita, T., R. C. Bales, K. M. Maguire, and C. P. Gerba.** 1993. Effect of pH on bacteriophage transport through sandy soils. *J. Contam. Hydrol.* **14:**55–70.
134. **Kinoshita, T., R. C. Bales, M. T. Yahya, and C. P. Gerba.** 1993. Bacteria transport in a porous-medium—retention of *Bacillus* and *Pseudomonas* on silica surfaces. *Water Res.* **27:**1295–1301.
135. **Krumme, M. L., R. L. Smith, J. Egestorff, S. M. Thiem, J. M. Tiedje, K. N. Timmis, and D. F. Dwyer.** 1994. Behavior of pollutant-degrading microorganisms in aquifers—predictions for genetically engineered organisms. *Environ. Sci. Technol.* **28:**1134–1138.
136. **Kucukcolak, E., B. Koopman, G. Bitton, and S. Farrah.** 1998. Validity of fluorochrome-stained bacteria as tracers of short-term microbial transport through porous media. *J. Contam. Hydrol.* **31:**349–357.
137. **Kuwabara, J. S., and R. W. Harvey.** 1990. Application of a hollow-fiber, tangential-flow device for sampling suspended bacteria and particles from natural-waters. *J. Environ. Qual.* **19:**625–629.
138. **Lahlou, M., H. Harms, D. Springael, and J. J. Ortega-Calvo.** 2000. Influence of soil components on the transport of polycyclic aromatic hydrocarbon-degrading bacteria through saturated porous media. *Environ. Sci. Technol.* **34:**3649–3656.
139. **Lance, J. C., and C. P. Gerba.** 1984. Virus movement in soil during saturated and unsaturated flow. *Appl. Environ. Microbiol.* **47:**335–337.
140. **Leblanc, D. R., S. P. Garabedian, K. M. Hess, L. W. Gelhar, R. D. Quadri, K. G. Stollenwerk, and W. W. Wood.** 1991. Large-scale natural gradient tracer test in sand and gravel, Cape-Cod, Massachusetts. 1. Experimental-design and observed tracer movement. *Water Resour. Res.* **27:**895–910.
141. **Lewus, P., and R. M. Ford.** 1999. Temperature-sensitive motility of *Sulfolobus acidocaldarius* influences population distribution in extreme environments. *J. Bacteriol.* **181:** 4020–4025.

142. **Li, B. L., C. Loehle, and D. Malon.** 1996. Microbial transport through heterogeneous porous media: random walk, fractal, and percolation approaches. *Ecol. Modelling* **85:**285–302.
143. **Li, Q., and B. E. Logan.** 1999. Enhancing bacterial transport for bioaugmentation of aquifers using low ionic strength solutions and surfactants. *Water Res.* **33:**1090–1100.
144. **Lindlow, S. E., A. C. Amy, W. R. Barchet, and C. D. Upper.** 1978. The role of bacterial ice nuclei in frost injury to sensitive plants, p. 249–263. *In* P. Li (ed.), *Plant Cold Hardiness and Freezing Stress*. Academic Press, New York, N.Y.
145. **Lindqvist, R., and G. Bengtsson.** 1995. Diffusion-limited and chemical-interaction-dependent sorption of soil bacteria and microspheres. *Soil Biol. Biochem.* **27:** 941–948.
146. **Lindqvist, R., and G. Bengtsson.** 1991. Dispersal dynamics of groundwater bacteria. *Microb. Ecol.* **21:**49–72.
147. **Lindqvist, R., J. S. Cho, and C. G. Enfield.** 1994. A kinetic-model for cell-density dependent bacterial transport in porous-media. *Water Resour. Res.* **30:**3291–3299.
148. **Lindqvist, R., and C. G. Enfield.** 1992. Biosorption of dichlorodiphenyltrichloroethane and hexachlorobenzene in groundwater and its implications for facilitated transport. *Appl. Environ. Microbiol.* **58:**2211–2218.
149. **Lindqvist, R., and C. G. Enfield.** 1992. Cell-density and nonequilibrium sorption effects on bacterial dispersal in groundwater microcosms. *Microb. Ecol.* **24:**25–41.
150. **Logan, B. E., D. G. Jewett, R. G. Arnold, E. J. Bouwer, and C. R. Omelia.** 1995. Clarification of clean-bed filtration models. *J. Environ. Eng.* **121:**869–873.
151. **Logan, B. E., D. G. Jewett, R. G. Arnold, E. J. Bouwer, and C. R. Omelia.** 1997. Clarification of clean-bed filtration models-closure. *J. Environ. Eng.* **123:**730–731.
152. **Loveland, J. P., J. N. Ryan, G. L. Amy, and R. W. Harvey.** 1996. The reversibility of virus attachment to mineral surfaces. *Colloids Surf. A* **107:**205–221.
153. **Mace, R. E.** 1999. Estimation of hydraulic conductivity in large-diameter, hand-dug wells using slug-test methods. *J. Hydrol.* **219:**34–45.
154. **MacLeod, F. A., H. M. Lappin-Scott, and J. W. Costerton.** 1988. Plugging of a model rock system by using starved bacteria. *Appl. Environ. Microbiol.* **54:** 1365–1372.
155. **Macler, B. A., and J. C. Merkle.** 2000. Current knowledge on groundwater microbial pathogens and their control. *Hydrogeol. J.* **8:**29–40.
156. **Mailloux, B. J., and M. E. Fuller.** 2003. Determination of in situ bacterial growth rates in aquifers and aquifer sediments. *Appl. Environ. Microbiol.* **69:**3798–3808.
157. **Martin, M. J., B. E. Logan, W. P. Johnson, D. G. Jewett, and R. G. Arnold.** 1996. Scaling bacterial filtration rates in different sized porous media. *J. Environ. Eng.-ASCE* **122:**407–415.
158. **Martin, R., and A. Thomas.** 1974. An example of the use of bacteriophage as a groundwater tracer. *J. Hydrol.* **23:** 73–78.
159. **Martin, R. E., and E. J. Bouwer.** 1991. Determination of bacterial collision efficiencies in a rotating disk system. *Environ. Sci. Technol.* **25:**2075–2082.
160. **Martin, R. E., E. J. Bouwer, and L. M. Hanna.** 1992. Application of clean-bed filtration theory to bacterial deposition in porous-media. *Environ. Sci. Technol.* **26:** 1053–1058.
161. **Mas, J., C. Pedrosalio, and R. Guerrero.** 1989. Variations in cell-size and buoyant density of *Escherichia coli*-K12 during glycogen accumulation. *FEMS Microbiol. Lett.* **57:**231–236.
162. **Mawdsley, J. L., A. E. Brooks, and R. J. Merry.** 1996. Movement of the protozoan pathogen Cryptosporidium parvum through three contrasting soil types. *Biol. Fertil. Soils* **21:**30–36.
163. **McCaulou, D. R., R. C. Bales, and R. G. Arnold.** 1995. Effect of temperature-controlled motility on transport of bacteria and microspheres through saturated sediment. *Water Resour. Res.* **31:**271–280.
164. **McCaulou, D. R., R. C. Bales, and J. F. McCarthy.** 1994. Use of short-pulse experiments to study bacteria transport through porous-media. *J. Contam. Hydrol.* **15:**1–14.
165. **McInerney, M. J.** 1991. Use of models to predict bacterial penetration and movement within a subsurface matrix, p. 115–135. *In* C. J. Hurst (ed.), *Modeling the Environmental Fate of Microorganisms*. American Society for Microbiology, Washington, D.C.
166. **McKay, L. D., J. A. Cherry, R. C. Bales, M. T. Yahya, and C. P. Gerba.** 1993. A field example of bacteriophage as tracers of fracture flow. *Environ. Sci. Technol.* **27:**1075–1079.
167. **McKay, L. D., W. E. Sanford, and J. M. Strong.** 2000. Field-scale migration of colloidal tracers in a fractured shale saprolite. *Ground Water* **38:**139–147.
168. **McKay, L. D., P. L. Stafford, and L. E. Toran.** 1997. EPM modeling of a field-scale tritium tracer experiment in fractured, weathered shale. *Ground Water* **35:**997–1007.
169. **Miller, W. G., and S. E. Lindow.** 1997. An improved GFP cloning cassette designed for prokaryotic transcriptional fusions. *Gene* **191:**149–153.
170. **Mills, A. L., J. S. Herman, G. M. Hornberger, and T. H. Dejesus.** 1994. Effect of solution ionic-strength and iron coatings on mineral grains on the sorption of bacterial cells to quartz sand. *Appl. Environ. Microbiol.* **60:** 3300–3306.
171. **Mohammed, N., and R. I. Allayla.** 2000. Effect of groundwater velocity on pilot scale bioremediation of gasoline contaminated sandy aquifers. *Water Air Soil Pollut.* **120:**315–329.
172. **Mohammed, N., and R. I. Allayla.** 1997. Modeling transport and biodegradation of BTX compounds in saturated sandy soil. *J. Hazard. Mater.* **54:**155–174.
173. **Mohanty, B. P., R. S. Kanwar, and C. J. Everts.** 1994. Comparison of saturated hydraulic conductivity measurement methods for a glacial-till soil. *Soil Sci. Soc. Am. J.* **58:**672–677.
174. **Montgomery, A. D., M. J. McInerney, and K. L. Sublette.** 1990. Microbial control of the production of hydrogen-sulfide by sulfate-reducing bacteria. *Biotechnol. Bioeng.* **35:**533–539.
175. **Morley, L. M., G. M. Hornberger, A. L. Mills, and J. S. Herman.** 1998. Effects of transverse mixing on transport of bacteria through heterogeneous porous media. *Water Resour. Res.* **34:**1901–1908.
176. **Murphy, E. M., and T. R. Ginn.** 2000. Modeling microbial processes in porous media. *Hydrogeol. J.* **8:**142–158.
177. **Murray, J. P., and S. Laband.** 1979. Degradation of poliovirus by adsorption on inorganic surfaces. *Appl. Environ. Microbiol.* **37:**480–486.
178. **Niehren, S., and W. Kinzelbach.** 1998. Artificial colloid tracer tests: development of a compact on-line microsphere counter and application to soil column experiments. *J. Contam. Hydrol.* **35:**249–259.
179. **Niehren, S., W. Kinzelbach, S. Seeger, and J. Wolfrum.** 1995. An all-solid-state flow cytometer for counting fluorescent microspheres. *Anal. Chem.* **67:**2666–2671.

180. **Nir, S.** 1976. Van der Waals interactions between surfaces of biological interest. *Prog. Surf. Sci.* **8:**1–58.
181. **Olson, M. S., R. M. Ford, J. A. Smith, and E. J. Fernandez.** 2004. Quantification of bacterial chemotaxis in porous media using magnetic resonance imaging. *Environ. Sci. Technol.* **38:**3864–3870.
182. **Pang, L. P., M. Close, and M. Noonan.** 1998. Rhodamine WT and Bacillus subtilis transport through an alluvial gravel aquifer. *Ground Water* **36:**112–122.
183. **Parolin, C., A. Montecucco, G. Ciarrocchi, G. Pedralinoy, S. Valisena, M. Palumbo, and G. Palu.** 1990. The effect of the minor groove binding-agent DAPI (2-amidino-diphenyl-indole) on DNA-directed enzymes—an attempt to explain inhibition of plasmid expression in Escherichia coli. *FEMS Microbiol. Lett.* **68:**341–346.
184. **Pedrosalio, C., J. Mas, and R. Guerrero.** 1985. The influence of poly-beta-hydroxybutyrate accumulation on cell-volume and buoyant density in *Alcaligenes eutrophus*. *Arch. Microbiol.* **143:**178–184.
185. **Petushkov, V. N., and J. Lee.** 1997. Purification and characterization of flavoproteins and cytochromes from the yellow bioluminescence marine bacterium Vibrio fischeri strain Y1. *Eur. J. Biochem.* **245:**790–796.
186. **Pickens, J. F., J. A. Cherry, G. E. Grisak, W. F. Merritt, and B. A. Risto.** 1978. A multilevel device for groundwater sampling and piezometric monitoring. *Ground Water* **16:**322–327.
187. **Pieper, A. P., J. N. Ryan, R. W. Harvey, G. L. Amy, T. H. Illangasekare, and D. W. Metge.** 1997. Transport and recovery of bacteriophage PRD1 in a sand and gravel aquifer: effect of sewage-derived organic matter. *Environ. Sci. Technol.* **31:**1163–1170.
188. **Powelson, D. K., and C. P. Gerba.** 1994. Virus removal from sewage effluents during saturated and unsaturated flow-through soil columns. *Water Res.* **28:**2175–2181.
189. **Powelson, D. K., and A. L. Mills.** 1996. Bacterial enrichment at the gas-water interface of a laboratory apparatus. *Appl. Environ. Microbiol.* **62:**2593–2597.
190. **Powelson, D. K., and A. L. Mills.** 2001. Transport of Escherichia coli in sand columns with constant and changing water contents. *J. Environ. Qual.* **30:**238–245.
191. **Powelson, D. K., and A. L. Mills.** 1998. Water saturation and surfactant effects on bacterial transport in sand columns. *Soil Sci.* **163:**694–704.
192. **Powelson, D. K., J. R. Simpson, and C. P. Gerba.** 1991. Effects of organic matter on virus transport in unsaturated flow. *Appl. Environ. Microbiol.* **57:**2192–2196.
193. **Pusey, P. N., and J. A. Tough.** 1985. Particle interactions, p. 85–102. *In* R. Pecora (ed.), *Dynamic Light Scattering: Applications to Photon Correlation Spectroscopy*. Plenum, New York, N.Y.
194. **Raiders, R. A., M. J. McInerney, D. E. Revus, H. M. Torbati, R. M. Knapp, and G. E. Jenneman.** 1986. Selectivity and depth of microbial plugging in Berea sandstone cores. *J. Ind. Microbiol.* **1:**195–203.
195. **Rajagopalan, R., and C. Tien.** 1976. Trajectory analysis of deep-bed filtration with the sphere-in-cell porous media model. *J. Am. Inst. Chem. Eng.* **22:**523–533.
196. **Reddy, H. L., and R. M. Ford.** 1996. Analysis of biodegradation and bacterial transport: comparison of models with kinetic and equilibrium bacterial adsorption. *J. Contam. Hydrol.* **22:**271–287.
197. **Redman, J. A., S. B. Grant, T. M. Olson, J. M. Adkins, J. L. Jackson, M. S. Castillo, and W. A. Yanko.** 1999. Physicochemical mechanisms responsible for the filtration and mobilization of a filamentous bacteriophage in quartz sand. *Water Res.* **33:**43–52.
198. **Redman, J. A., S. B. Grant, T. M. Olson, M. E. Hardy, and M. K. Estes.** 1997. Filtration of recombinant Norwalk virus particles and bacteriophage MS2 in quartz sand: importance of electrostatic interactions. *Environ. Sci. Technol.* **31:**3378–3383.
199. **Redman, J. A., S. L. Walker, and M. Elimelech.** 2004. Bacterial adhesion and transport in porous media: role of the secondary energy minimum. *Environ. Sci. Technol.* **38:**1777–1785.
200. **Rehmann, L. L. C., C. Welty, and R. W. Harvey.** 1999. Stochastic analysis of virus transport in aquifers. *Water Resour. Res.* **35:**1987–2006.
201. **Ren, J., A. I. Packman, and C. Welty.** 2000. Correlation of colloid collision efficiency with hydraulic conductivity of silica sands. *Water Resour. Res.* **36:**2493–2500.
202. **Reynolds, P. J., P. Sharma, G. E. Jenneman, and M. J. McInerney.** 1989. Mechanisms of microbial movement in subsurface materials. *Appl. Environ. Microbiol.* **55:** 2280–2286.
203. **Rice, K. C., and G. M. Hornberger.** 1998. Comparison of hydrochemical tracers to estimate source contributions to peak flow in a small, forested, headwater catchment. *Water Resour. Res.* **34:**1755–1766.
204. **Rijnaarts, H., W. Norde, E. Bouwer, J. Lyklema, and A. Zehnder.** 1993. Bacterial adhesion under static and dynamic conditions. *Appl. Environ. Microbiol.* **59:**3255–3265.
205. **Rijnaarts, H. H. M., W. Norde, E. J. Bouwer, J. Lyklema, and A. J. B. Zehnder.** 1996. Bacterial deposition in porous media related to the clean bed collision efficiency and to substratum blocking by attached cells. *Environ. Sci. Technol.* **30:**2869–2876.
206. **Ripple, C. D., R. V. James, and J. Rubin.** 1974. Packing-induced radial particle-size segregation: influence on hydrodynamic dispersion and water transfer measurements. *Soil Sci. Soc. Am. J.* **38:**219–222.
207. **Robertson, B. R., D. K. Button, and A. L. Koch.** 1998. Determination of the biomasses of small bacteria at low concentrations in a mixture of species with forward light scatter measurements by flow cytometry. *Appl. Environ. Microbiol.* **64:**3900–3909.
208. **Rossi, P., A. Carvalho-Dill, I. Müller, and M. Aragno.** 1994. Comparative tracing experiments in a porous aquifer using bacteriophages and fluorescent dye on a test field socated at Wilerwald (Switzerland) and simultaneously surveyed in detail on a local scale by radio-magneto-tellury (12–240 Khz). *Environ. Geol.* **23:**192–200.
209. **Rossi, P., N. Dorfliger, K. Kennedy, I. Muller, and M. Aragno.** 1998. Bacteriophages as surface and ground water tracers. *Hydrol. Earth Sys. Sci.* **2:**101–110.
210. **Rothmel, R. K., R. W. Peters, E. St Martin, and M. F. Deflaun.** 1998. Surfactant foam bioaugmentation technology for in situ treatment of TCE-DNAPLs. *Environ. Sci. Technol.* **32:**1667–1675.
211. **Rutter, P. R., and P. Vincent.** 1980. The adhesion of microorganisms to surfaces: physicochemical aspects, p. 79–92. *In* R. C. W. Berkeley, J. M. Lynch, J. Melling, P. R. Rutter, and P. Vincent (ed.), *Microbial Adhesion to Surfaces*. E. Horwood Ltd., Chichester, United Kingdom.
212. **Ryan, J. N., M. Elimelech, R. A. Ard, R. W. Harvey, and P. R. Johnson.** 1999. Bacteriophage PRD1 and silica colloid transport and recovery in an iron oxide-coated sand aquifer. *Environ. Sci. Technol.* **33:**63–73.
213. **Ryan, J. N., R. W. Harvey, D. Metge, M. Elimelech, T. Navigato, and A. P. Pieper.** 2002. Field and laboratory investigations of inactivation of viruses (PRD1 and MS2) attached to iron oxide-coated quartz sand. *Environ. Sci. Technol.* **36:**2403–2413.

214. **Salanitro, J. P., G. E. Spinnler, C. C. Neaville, P. M. Maner, S. M. Stearns, and P. C. Johnson.** 1999. Presented at the 5th International Symposium on In Situ and On Site Bioremediation, San Diego, Calif., April 19 to 22, 1999.
215. **Schafer, A., H. Harms, and A. J. B. Zehnder.** 1998. Bacterial accumulation at the air-water interface. *Environ. Sci. Technol.* **32:**3704–3712.
216. **Schafer, A., P. Ustohal, H. Harms, F. Stauffer, T. Dracos, and A. J. B. Zehnder.** 1998. Transport of bacteria in unsaturated porous media. *J. Contam. Hydrol.* **33:**149–169.
217. **Schijven, J. F., W. Hoogenboezem, S. M. Hassanizadeh, and J. H. Peters.** 1999. Modeling removal of bacteriophages MS2 and PRD1 by dune recharge at Castricum, Netherlands. *Water Resour. Res.* **35:**1101–1111.
218. **Scholl, M. A., and R. W. Harvey.** 1992. Laboratory investigations on the role of sediment surface and groundwater chemistry in transport of bacteria through a contaminated sandy aquifer. *Environ. Sci. Technol.* **26:** 1410–1417.
219. **Scholl, M. A., A. L. Mills, J. S. Herman, and G. M. Hornberger.** 1990. The influence of mineralogy and solution chemistry on the attachment of bacteria to representative aquifer materials. *J. Contam. Hydrol.* **6:**321–336.
220. **Shales, S. W., and S. Kumarasingham.** 1987. Bacterial transport through porous solids—interactions between Micrococcus luteus cells and sand particles. *J. Ind. Microbiol.* **2:**219–227.
221. **Sharma, P. K., and M. J. McInerney.** 1994. Effect of grain size on bacterial penetration, reproduction, and metabolic activity in porous-glass bead chambers. *Appl. Environ. Microbiol.* **60:**1481–1486.
222. **Sharma, P. K., M. J. McInerney, and R. M. Knapp.** 1993. In situ growth and activity and modes of penetration of *Escherichia coli* in unconsolidated porous materials. *Appl. Environ. Microbiol.* **59:**3686–3694.
223. **Simoni, S. F., T. N. P. Bosma, H. Harms, and A. J. B. Zehnder.** 2000. Bivalent cations increase both the subpopulation of adhering bacteria and their adhesion efficiency in sand columns. *Environ. Sci. Technol.* **34:**1011–1017.
224. **Simoni, S. F., H. Harms, T. N. P. Bosma, and A. J. B. Zehnder.** 1998. Population heterogeneity affects transport of bacteria through sand columns at low flow rates. *Environ. Sci. Technol.* **32:**2100–2105.
225. **Simoni, S. F., A. Schafer, H. Harms, and A. J. B. Zehnder.** 2001. Factors affecting mass transfer limited biodegradation in saturated porous media. *J. Contam. Hydrol.* **50:**99–120.
226. **Sirokman, G., T. Wilson, and J. W. Hastings.** 1995. A bacterial luciferase reaction with a negative temperature-coefficient attributable to protein-protein interaction. *Biochemistry* **34:**13074–13081.
227. **Sjollema, J., H. J. Busscher, and A. H. Weerkamp.** 1989. Experimental approaches for studying adhesion of microorganisms to solid substrata: applications and mass transport. *J. Microbiol. Methods* **9:**79–90.
228. **Smith, M. S., G. W. Thomas, R. E. White, and D. Ritonga.** 1985. Transport of Escherichia coli through intact and disturbed soil columns. *J. Environ. Qual.* **14:**87–91.
229. **Smith, R. L., R. W. Harvey, and D. R. Leblanc.** 1991. Importance of closely spaced vertical sampling in delineating chemical and microbiological gradients in groundwater studies. *J. Contam. Hydrol.* **7:**285–300.
230. **Stauffer, F., and T. Dracos.** 1986. Experimental and numerical study of water and solute infiltration in layered porous-media. *J. Hydrol.* **84:**9–34.
231. **Stephen, J. R., Y. J. Chang, Y. D. Gan, A. Peacock, S. M. Pfiffner, M. J. Barcelona, D. C. White, and S. J. Macnaughton.** 1999. Microbial characterization of a JP-4 fuel-contaminated site using a combined lipid biomarker/polymerase chain reaction-denaturing gradient gel electrophoresis (PCR-DGGE)-based approach. *Environ. Microbiol.* **1:**231–241.
232. **Story, S. P., P. S. Amy, C. W. Bishop, and F. S. Colwell.** 1995. Bacterial transport in volcanic tuff cores under saturated flow conditions. *Geomicrobiol. J.* **13:**249–264.
233. **Straub, T. M., I. L. Pepper, and C. P. Gerba.** 1995. Comparison of PCR and cell culture for detection of enteroviruses in sludge-amended field soils and determination of their transport. *Appl. Environ. Microbiol.* **61:** 2066–2068.
234. **Streger, S. H., S. Vainberg, H. Dong, and P. B. Hatzinger.** 2002. Enhancing transport of *Hydrogenophaga flava* ENV735 for bioaugmentation of aquifers contaminated with methyl *tert*-butyl ether. *Appl. Environ. Microbiol.* **68:**5571–5579.
235. **Strong-Gunderson, J. M., and A. V. Palumbo.** 1997. Laboratory studies identify a colloidal groundwater tracer: implications for bioremediation. *FEMS Microbiol. Lett.* **148:**131–135.
236. **Sun, Y., J. N. Petersen, J. Bear, T. P. Clement, and B. S. Hooker.** 1999. Modeling microbial transport and biodegradation in a dual-porosity system. *Transport Porous Media* **35:**49–65.
237. **Sutton, D. J., Z. J. Kabala, D. E. Schaad, and N. C. Ruud.** 2000. The dipole-flow test with a tracer: a new single-borehole tracer test for aquifer characterization. *J. Contam. Hydrol.* **44:**71–101.
238. **Tan, Y., J. T. Gannon, P. Baveye, and M. Alexander.** 1994. Transport of bacteria in an aquifer sand—experiments and model simulations. *Water Resour. Res.* **30:**3243–3252.
239. **Taylor, S. W., and P. R. Jaffe.** 1990. Substrate and biomass transport in a porous-medium. *Water Resour. Res.* **26:**2181–2194.
240. **Thompson, S. S., and M. V. Yates.** 1999. Bacteriophage inactivation at the air-water-solid interface in dynamic batch systems. *Appl. Environ. Microbiol.* **65:**1186–1190.
241. **Tong, M., X. Li, C. N. Brow, and W. P. Johnson.** 2005. Detachment-influenced transport of an adhesion-deficient bacterial strain within water-reactive porous media. *Environ. Sci. Technol.* **39:**2500–2508.
242. **Toran, L., and A. V. Palumbo.** 1992. Colloid transport through fractured and unfractured laboratory sand columns. *J. Contam. Hydrol.* **9:**289–303.
243. **Trevors, J. T., J. D. Vanelsas, L. S. Vanoverbeek, and M. E. Starodub.** 1990. Transport of a genetically engineered Pseudomonas fluorescens strain through a soil microcosm. *Appl. Environ. Microbiol.* **56:**401–408.
244. **Troussellier, M., C. Courties, P. Lebaron, and P. Servais.** 1999. Flow cytometric discrimination of bacterial populations in seawater based on SYTO 13 staining of nucleic acids. *FEMS Microbiol. Ecol.* **29:**319–330.
245. **Tufenkji, N., and M. Elimelech.** 2004. Correlation equation for predicting single-collector efficiency in physicochemical filtration in saturated porous media. *Environ. Sci. Technol.* **38:**529–536.
246. **Tufenkji, N., and M. Elimelech.** 2005. Spatial distributions of Cryptosporidium oocysts in porous media: evidence for dual mode deposition. *Environ. Sci. Technol.* **39:**3620–3629.
247. **Tufenkji, N., G. F. Miller, J. N. Ryan, R. W. Harvey, and M. Elimelech.** 2004. Transport of Cryptosporidium oocysts in porous media: role of straining and physicochemical filtration. *Environ. Sci. Technol.* **38:**5932–5938.

248. **Unge, A., R. Tombolini, L. Molbak, and J. K. Jansson.** 1999. Simultaneous monitoring of cell number and metabolic activity of specific bacterial populations with a dual *gfp-luxAB* marker system. *Appl. Environ. Microbiol.* **65:** 813–821.
249. **Unge, A., R. Tombolini, A. Moller, and J. K. Jansson.** 1997. Optimization of GPF as a marker for detection of bacteria in environmental samples, p. 391–394. *In* J. W. Hastings, L. J. Kricka, and P. E. Stanley (ed.), *Bioluminescence and Chemiluminescence: Molecular Reporting with Photons*. John Wiley & Sons, Chichester, United Kingdom.
250. **Unice, K. M., and B. E. Logan.** 2000. Insignificant role of hydrodynamic dispersion on bacterial transport. *J. Environ. Eng.* **126:**491–500.
251. **Vandevivere, P., and P. Baveye.** 1992. Relationship between transport of bacteria and their clogging efficiency in sand columns. *Appl. Environ. Microbiol.* **58:** 2523–2530.
252. **Vandevivere, P., and P. Baveye.** 1992. Saturated hydraulic conductivity reduction caused by aerobic bacteria in sand columns. *Soil Sci. Soc. Am. J.* **56:**1–13.
253. **Van Elsas, J. D., J. T. Trevors, and L. S. Vanoverbeek.** 1991. Influence of soil properties on the vertical movement of genetically-marked Pseudomonas fluorescens through large soil microcosms. *Biol. Fertil. Soils* **10:**249–255.
254. **van Oss, C. J.** 1994. *Interfacial Forces in Aqueous Media.* Marcel Dekker, New York, N.Y.
255. **Van Oss, C. J., and C. F. Gillman.** 1972. Phagocytosis as a surface phenomenon. Contact angles and phagocytosis of non-opsonized bacteria. *J. Reticuloendothel. Soc.* **12:** 283–292.
256. **van Oss, C. J., R. J. Good, and M. K. Chaudry.** 1986. The role of van der Waals forces and hydrogen bonds in hydrophobic interactions between biopolymers and low energy surfaces. *J. Colloid Interface Sci.* **111:**378–390.
257. **Walker, M., and D. Redelman.** 2004. Detection of *Cryptosporidium parvum* in soil extracts. *Appl. Environ. Microbiol.* **70:**1827–1829.
258. **Walker, S. L., J. E. Hill, J. A. Redman, and M. Elimelech.** 2005. Influence of growth phase on adhesion kinetics of *Escherichia coli* D21g. *Appl. Environ. Microbiol.* **71:**3093–3099.
259. **Walker, S. L., J. A. Redman, and M. Elimelech.** 2004. Role of cell surface lipopolysaccharides in Escherichia coli K12 adhesion and transport. *Langmuir* **20:**7736–7746.
260. **Wan, J. M., T. K. Tokunaga, and C. F. Tsang.** 1995. Bacterial sedimentation through a porous medium. *Water Resour. Res.* **31:**1627–1636.
261. **Wan, J. M., and J. L. Wilson.** 1994. Visualization of the role of the gas-water interface on the fate and transport of colloids in porous-media. *Water Resour. Res.* **30:**11–23.
262. **Wan, J. M., J. L. Wilson, and T. L. Kieft.** 1994. Influence of the gas-water interface on transport of microorganisms through unsaturated porous media. *Appl. Environ. Microbiol.* **60:**509–516.
263. **Wang, D. S., C. P. Gerba, and J. C. Lance.** 1981. Effect of soil permeability on virus removal through soil columns. *Appl. Environ. Microbiol.* **42:**83–88.
264. **Wellings, F. M., A. L. Lewis, C. W. Mountain, and L. V. Pierce.** 1975. Demonstration of virus in groundwater after effluent discharge into soil. *Appl. Environ. Microbiol.* **29:**751–757.
265. **Williams, V., and M. Fletcher.** 1996. *Pseudomonas fluorescens* adhesion and transport through porous media are affected by lipopolysaccharide composition. *Appl. Environ. Microbiol.* **62:**100–104.
266. **Witt, M. E., M. J. Dybas, R. M. Worden, and C. S. Criddle.** 1999. Motility-enhanced bioremediation of carbon tetrachloride-contaminated aquifer sediments. *Environ. Sci. Technol.* **33:**2958–2964.
267. **Wood, W. W., and G. G. Ehrlich.** 1978. Use of baker's yeast to trace microbial movement in ground water. *Ground Water* **16:**398–403.
268. **Yao, K. M., M. T. Habibian, and C. R. O'Melia.** 1971. Water and waste water filtration: concepts and applications. *Environ. Sci. Technol.* **5:**1105–1112.
269. **Yates, M. V., and C. P. Gerba.** 1985. Factors controlling the survival of viruses in groundwater. *Water Sci. Technol.* **17:**681–687.
270. **Yee, N., J. B. Fein, and C. J. Daughney.** 2000. Experimental study of the pH, ionic strength, and reversibility behavior of bacteria-mineral adsorption. *Geochim. Cosmochim. Acta* **64:**609–617.
271. **You, Y., G. F. Vance, D. L. Sparks, J. Zhuang, and Y. Jin.** 2003. Sorption of MS2 bacteriophage to layered double hydroxides: effects of reaction time, pH, and competing anions. *J. Environ. Qual.* **32:**2046–2053.
272. **Zhang, P., and W. P. Johnson.** 1999. Rapid selective ferrographic enumeration of bacteria. *J. Magn. Magn. Mater.* **194:**267–274.
273. **Zhang, P., W. P. Johnson, and R. Rowland.** 1999. Bacterial tracking using ferrographic separation. *Environ. Sci. Technol.* **33:**2456–2460.
274. **Zlotnik, V. A., and V. L. McGuire.** 1998. Multi-level slug tests in highly permeable formations: 2. Hydraulic conductivity identification, method verification, and field applications. *J. Hydrol.* **204:**283–296.

Oil Field Microbiology

MICHAEL J. McINERNEY, GERRIT VOORDOUW,
GARY E. JENNEMAN, AND KERRY L. SUBLETTE

71

Petroleum reservoirs contain diverse and metabolically active microbial communities (61, 62, 68, 89) that are active in the cycling of carbon and sulfur in the reservoir (61, 68, 89). Microbial activity impacts the physical properties and composition of petroleum (38), and it is clear that hydrocarbons can serve as the main electron donor for anaerobic respiration (38, 95, 130). Bacteria can penetrate rock at appreciable rates, even rocks with very low permeabilities (46). Thus, microorganisms are likely to occur throughout the reservoir rather than being limited to the vicinity of the wells. Microbial activity has a profound effect on the day-to-day operations of oil reservoirs, ranging from the detrimental activities of sulfate-reducing bacteria (87) to beneficial activities such as the economic recovery of oil (66). An understanding of the microbial ecology of oil reservoirs will allow operators to exploit their petroleum resources more effectively.

In this chapter, we discuss the methods used to characterize the microbial populations in oil reservoirs, to detect and control their detrimental activities such as souring and corrosion, and to stimulate their beneficial activities such as improved oil recovery. The recent book edited by Ollivier and Magot (89) is an excellent resource for many aspects of petroleum microbiology.

OIL FIELD SAMPLING

Aseptically obtained core material is needed to show unequivocally that a specific microorganism is present in the reservoir (chapter 66). However, given the expense of drilling, such material is rarely available from oil reservoirs. Septically obtained core material and scrapings of the well bore can be used as sources of inocula if properly handled (see chapter 66). Often, the kinds of microorganisms present and their activities in the oil reservoir must be inferred from the analysis of the fluids produced from the reservoir. Because of this challenge, the selection of reservoirs and wells is an important concern. Preferably, a reservoir that has not been injected with fluid from another formation should be used to ensure that the fluid collected for analysis is native to the formation of interest. Completion history and well logs should be studied to determine the formation from which the well is producing; preferably, wells that produce fluid from only a single formation should be used.

The choice of sampling locations depends on the microbial activity of interest. Sampling locations throughout the production facility should be used to assess microbial corrosion and biofouling. For reservoir fluids, samples should be collected at the wellhead if possible, and any stagnant liquid in the system should be purged prior to collection. Fittings and tubing connections should be sterilized prior to use. To reduce the exposure to oxygen, the collection bottle is filled and then flushed with fluid from the well prior to sample collection, or a bottle with an anoxic atmosphere is used (42). Even when these precautions are taken, the interpretation of the data must be tempered by the realization that any organisms detected in fluid samples may have originated from biofilms growing on the surface of the pipes rather than directly from the reservoir.

CULTIVATION-DEPENDENT METHODS

Methods to enumerate and isolate microbial populations from oil reservoirs and to describe the diversity and structure of the microbial community are similar to those used for other environments (chapter 6). The medium and cultural conditions must be adjusted to simulate the habitat as accurately as possible. Since most oil reservoirs have little or no oxygen (43), anaerobically prepared medium is used. A bicarbonate buffer system is used because many anaerobes require carbon dioxide for growth. The ionic strength of the medium is adjusted to match that of the produced water, or brine from the formation can be used. The pH of most produced fluids is circumneutral (43). However, temperature and pressure affect pH, so the actual reservoir pH may be markedly different from that measured at the surface. Geochemical modeling is needed to calculate in situ conditions from brine chemistry, mineralogy, and gas partial pressures. Atmospheric rather than the actual reservoir pressure is routinely used for isolation and enumeration studies (4, 30, 107), since pressures of 10 to 20 MPa do not adversely affect bacterial metabolism (63). However, it is possible that some reservoir microorganisms may not grow at reduced pressures and specialized sampling procedures may be needed (chapter 46). A variety of flow systems are available for use in laboratory testing of microbial treatment processes at reservoir temperatures and pressures (67).

Microorganisms Useful for Oil Recovery

Microorganisms produce a variety of products that are potentially useful for oil recovery (Table 1) (66). The production of acids, solvents, gases, surfactants, and emulsifiers in or near the well bore prevents scale and paraffin deposition, alters the wettability of the rock, and changes the fluid saturations. All of these will improve oil drainage into the well. One of the major factors that limits oil production is the entrapment of oil in small pores by capillary forces. The reduction in the interfacial forces between the oil and brine by a biosurfactant can release the entrapped oil (13, 66). Biosurfactants and bioemulsifiers may also reduce paraffin deposition (10, 84, 109); the latter compounds have been used to remove sludges from oil tanker storage compartments (32). Finally, in situ growth and polymer or mineral production by microorganisms can reduce permeability variation and improve the sweep efficiency of a recovery process (45, 47, 53, 58, 66, 75, 97, 98). Most of the above approaches rely on the stimulation of indigenous microorganisms in the reservoir by nutrient addition; therefore, the analysis of reservoir fluids for the presence of useful bacteria is essential. An inoculum may be required when a specific metabolic activity, such as biosurfactant production, is needed.

Media with a readily utilizable organic carbon and energy source are used to enumerate and isolate microorganisms that produce acids, solvents, and/or gas as end products (4, 13). Agar medium with calcium carbonate or pH-sensitive dyes can be used to see colonies of acid producers.

Polymer-producing bacteria are isolated by using a carbohydrate-based medium with a high carbon-to-nitrogen ratio, generally greater than 5:1 on a molar basis (92, 117). Polymer-producing bacteria are screened by their ability to produce mucoid colonies on agar plates or viscous liquid cultures. Viscometry is used to characterize the rheological properties of polymers (16, 29, 92, 112).

Microorganisms can improve sweep efficiency by in situ production of biomass and associated cell wall polysaccharides or through mineral production. Biomass-associated polysaccharides have a high volume-to-mass ratio but may not be stable in the absence of continued metabolism. Minerals have a low volume-to-mass ratio and are stable once they are formed. Calcium carbonate can be formed rapidly and in high yields by bacteria capable of hydrolyzing urea with the enzyme urease (equation 1) in the presence of calcium chloride (equation 2):

$$NH_2\text{-}CO\text{-}NH_2 + 2H_2O \rightarrow 2NH_4^+ + CO_3^{2-} \quad (1)$$

$$Ca^{2+} + CO_3^{2-} \rightarrow CaCO_3\downarrow \quad (2)$$

The approach may involve the injection of urea, calcium chloride, and urease or urea, calcium chloride, and nutrients to stimulate urease-containing bacteria (85). Calcium carbonate production can cement sand particles, preventing their migration in poorly consolidated reservoirs.

Emulsifying microorganisms are initially detected by the emulsification of an oil layer in liquid medium (9, 101, 121, 131). A quantitative measure of activity is obtained by mixing a hydrocarbon with culture fluid or an aqueous solution containing the purified emulsifier and measuring the absorbance spectrophotometrically (101) or the height of the emulsified layer compared to the total liquid height (121, 131). The adherence of microbial cells to hydrocarbons is measured in a similar manner (100).

The drop collapse method is a rapid approach for detecting biosurfactant production (5, 134). First, 2 μl of mineral oil is added to each well of a 96-well microtiter plate lid and equilibrated for 1 h at room temperature. Next, 5 μl of the culture is added to the surface of the oil. Cultures that give flat drops are presumed biosurfactant producers; those that give rounded drops are negative for this property. Biosurfactant production can be verified by the oil spreading method (71, 134) or by measuring the surface tension of the culture with a tensiometer (65, 131). Screening for hemolytic zones around colonies on blood agar plates (72) or halo formation on agar plates with oil or tributylin (21) has been used to detect biosurfactant-producing bacteria. However, Youssef et al. (134) found that 38% of the strains that did not lyse blood agar tested positive for biosurfactant production by the drop collapse and oil spreading methods. To be effective for enhanced oil recovery (EOR), the biosurfactant must reduce the interfacial tension between oil and brine by several orders of magnitude. Interfacial tension is measured by using a spinning drop tensiometer or by the pendant drop method (65, 118). Other methods to detect and quantify biosurfactants and emulsifiers are discussed by Desai and Banat (21).

NRB

The interest in nitrate- or nitrite-reducing bacteria (NRB) in oil fields has increased sharply since nitrate or nitrite injection has become an accepted method to reduce hydrogen sulfide production (souring). NRB can be placed in two functional groupings, the heterotrophic NRB (hNRB), and the nitrate- or nitrite-reducing, sulfide-oxidizing bacteria (NR-SOB). Both groups generally require microaerophilic or anaerobic conditions for growth. Hence, medium in Hungate tubes or serum bottles closed with butyl rubber stoppers and a gas phase of 90% (vol/vol) N_2 and 10% (vol/vol) CO_2 needs to be used. Many NR-SOB use only CO_2 as the carbon source (28). Most-probable-number (MPN) assays to distinguish the two groups have been proposed (23). The hNRB medium contains 4 g of nutrient broth/liter and 5 mM nitrate with growth in the MPN tubes being scored by either turbidity or nitrate utilization (24).

TABLE 1 Types of microbial processes for oil recovery

Process	Production problem	Type of activity or product needed
Well bore cleanup	Paraffin and scale deposits	Emulsifiers, biosurfactants, solvents, and acids
	Hydrocarbon degradation?	
Well stimulation	Formation damage, poor drainage	Gas, acids, solvents, biosurfactants, and nitrate reduction
	High water production (coning)	Biomass and polymer production
Enhanced waterfloods	Poor displacement efficiency	Biosurfactants, solvents, and polymers
	Poor sweep efficiency	Biomass and polymer production
	Souring	Nitrate reduction

Coleville synthetic brine medium containing sulfide (typically 1 mM), KNO_3 (1 g/liter) and resazurin (1 mg/liter), as described elsewhere (23, 28, 116), is used for growth of NR-SOB. Growth is most easily scored by color change of the redox indicator, resazurin, from colorless to pink.

Pure cultures of hNRB have been isolated in media with nitrate as electron acceptor and acetate, lactate, or carbohydrates as the electron donor. For example, *Denitrovibrio acetiphilus*, isolated from a North Sea oil reservoir model column, uses only acetate as an electron donor for reduction of nitrate to ammonia and is a mesophilic, gram-positive organism (73). Media in which lactate was used as electron donor for sulfate, nitrate or nitrite reduction in an oil field model column gave mesophilic ε-proteobacteria, *Campylobacter* sp. strains NO3A, NO2B, and KW (39, 41). The moderately thermophilic, gram-positive *Garciella nitratireducens*, belonging to cluster XII of the *Clostridiales*, reduced nitrate to ammonia but could also be grown fermentatively (70). On the other hand, moderately thermophilic members of the genus *Geobacillus* used oxygen or nitrate as the terminal electron acceptors, the latter of which was reduced to nitrogen. This incomplete summary shows that a variety of media can be used to isolate hNRB with diverse metabolic capacities and phylogenetic affiliations.

For isolation of pure cultures of NR-SOB, procedures used for *Thiomicrospira* sp. strain CVO and *Arcobacter* sp. strain FWKO, both members of the ε-proteobacteria, can be followed (28). Strain CVO oxidizes sulfide to sulfur or sulfate while reducing nitrate to nitrite or nitrogen, whereas strain FWKO B oxidizes sulfide to sulfur while reducing nitrate to nitrite (28). The course of reaction catalyzed by strain CVO appears to depend on the sulfide-to-nitrate ratio (31). A similar organism, *Thiomicrospira denitrificans* strain YK-1, was isolated from an underground oil storage cavity in Japan (51). All of these are mesophiles, and to date no thermophilic NR-SOB have been described.

Detrimental Bacteria

End-point or MPN enumerations for sulfate-reducing bacteria (SRB) are routinely done on produced fluids and equipment surfaces to assess the potential for souring and corrosion. Traditional cultivation methods are inexpensive and do not require much training of field personnel. Another advantage of cultural methods is that SRB can be isolated for biocide testing and further physiological characterization. The compositions of two media used to estimate the numbers of SRB from oil field samples are given in Table 2 (1, 113). The American Petroleum Institute (API)'s Recommended Practice no. 38 uses a medium designated API RP-38 (1). Tanner (113) developed a modified version of this medium, designated (API)-RST, that gave higher SRB counts in shorter times from environmental samples than did API RP-38. To prepare the (API)-RST medium, the buffer is adjusted to pH 7.6 and all the components except ferrous ammonium sulfate are added. The pH is readjusted if needed, and ferrous ammonium sulfate is added to form a precipitate. The medium is boiled under nitrogen and taken into an anaerobic chamber, and preweighed amounts of ascorbic acid and cysteine are added. The medium is dispensed into Hungate tubes (Belco Glass, Inc.) and then sterilized. This medium will detect lactate-using SRB, i.e., species within the genera *Desulfovibrio* and *Desulfotomaculum*, but can be modified to detect other SRB by using different electron donors.

In some oil fields, *Shewanella* species (86, 104) may be the most numerous bacterium capable of reducing sulfur oxyanions to sulfide. However, these organisms do not use sulfate as the electron acceptor and will not be detected in the above media unless sulfite or thiosulfate is used as the electron acceptor. *Shewanella* species produce salmon-pink colonies on plate count agar (Difco, Inc.), and this characteristic has been used to enumerate these organisms from oil field samples (104). The identity of isolates is confirmed by testing for the ability to reduce iron in medium with a nonfermentable electron donor and with ferric iron as the electron acceptor (104). Selective enrichment with H_2 or formate as the electron donor and with amorphous iron oxyhydroxide as the electron acceptor can also be used (83).

Production fluids should also be routinely analyzed for the presence of general aerobic bacteria (GAB) to assess the level of contamination by slime-forming bacteria (1) and

TABLE 2 Composition of media used to detect SRB[d]

Component	API RP-38[a] concn (g/liter)	(API)-RST[a] concn (g/liter)
Sodium lactate	2.4	4.2
Yeast extract	1	0.7
NaCl	10	10
$(NH_4)SO_4$		0.3
K_2HPO_4	0.01	0.6
$MgSO_4 \cdot 7H_2O$	0.2	0.2
$CaSO_4$		0.04
TES[b]		1.5
Vitamin solution[c]		10 ml
Trace metal solution[c]		5 ml
$Fe(NH_4)_2(SO_4)_2 \cdot 6H_2O$	0.2	0.3
Iron nail	+[e]	
Ascorbic acid	0.1	0.1
Cysteine·HCl		0.2

[a]Abbreviations: API RP-38, API Recommended Practice no. 38; (API)-RST, R. S. Tanner modification of the API RP-38 medium.
[b]TES, *N*-tris(hydroxymethyl)methyl-2-amino-ethanesulfonic acid.
[c]Compositions of the vitamin and trace metal solutions are given by Tanner in chapter 6 of this manual.
[d]Consult references 1 and 113.
[e]+, present.

for acid-producing bacteria (APB) to assess the potential for corrosion (25). GAB are important for biofilm development by providing an environment conducive to the growth of SRB. The following medium for the enumeration of GAB is recommended by the API (1) for use in waters containing less than 20,000 mg of total dissolved solids per liter: beef extract (3.0 g/liter), tryptone (5.0 g/liter), and dextrose (1.0 g/liter) at a pH of 7.0. APB are enumerated by using a proprietary commercial medium, BTI/APB medium, available from Bioindustrial Technologies, Inc. (Grafton, N.Y.). The inoculated contents of bottles are incubated at the temperature of the original water sample ± 5°C.

CULTIVATION-INDEPENDENT APPROACHES

The microbial composition of oil field-derived samples can be determined from sequence analysis of PCR-amplified, cloned 16S rRNA genes. The sequences obtained can in turn be used to design probes meant for fluorescence in situ hybridization or other detection methods or to design primers for quantitative PCR to evaluate the occurrence of specific bacteria of interest. Such analyses indicated the presence of a variety of mesophilic eubacteria (the SRB *Desulfovibrio*, *Desulfobulbus*, and *Desulfobacter*; the fermentative bacteria and/or APB *Clostridium* and *Eubacterium*; and the sulfide-oxidizing, microaerophilic bacteria *Thiomicrospira*, *Arcobacter*, and *Campylobacter*) in oil fields in Alberta, Canada. Direct PCR amplification of the 16S rRNA gene from produced water primarily gave clones representing the latter (126). Interestingly, a similar PCR bias was described for samples from an underground oil storage cavity (129). This bias could be alleviated by careful choice of PCR primers (128). 16S rRNA gene analysis of high-temperature oil reservoirs has indicated the existence of a diverse thermophilic population of bacteria and archaea, especially methanogens (90). Use of a fluorescently labeled probe, designed based on determined 16S rRNA gene sequences, indicated that SRB that completely oxidize oil organics to CO_2 greatly outnumbered the incomplete oxidizers in SRB enrichments on crude oil (96).

Analysis of PCR-amplified 16S rRNA gene sequences has provided important qualitative information on the composition of microbial communities in oil fields and allowed the presence of selected members to be quantified by fluorescence in situ hybridization or quantitative PCR. Microbial community composition can be determined in its entirety if probes representing the various oil field bacteria are immobilized on a solid support and if the resulting array is then hybridized with labeled total community RNA or DNA. The use of genome arrays, containing denatured genomes of microbes isolated from oil fields spotted on the solid support, has been reported (115). Use of these arrays in a procedure referred to as reverse sample genome probing allowed description of the community shift resulting from nitrate injection into an oil field (115). A disadvantage of the use of genome arrays is the requirement for culturing. Initiatives to make a variety of other arrays, e.g., microarrays with 16S rRNA probes or with functional gene probes, are under way in several labs. Application of a 16S rRNA gene chip to the description of microbial community in a high-temperature oil reservoir has indicated the presence of uncultured diversity (7).

A potentially very useful functional gene probe is that for the genes for dissimilatory sulfite reductase (Dsr), the enzyme that produces the H_2S in SRB. The *dsrAB* genes encoding this enzyme are highly conserved, and primers targeting the most invariable regions have been used to PCR amplify, clone, and sequence *dsr* genes from environmental samples. This method has given insights into the SRB diversity at hydrothermal vents but has not yet been applied to oil field-derived samples (76). Enzyme-linked immunosorbent assays using polyclonal antibodies targeting adenosine phosphosulfate reductase, another conserved enzyme of the Dsr pathway, are marketed under the trade name Rapidcheck (Conoco Specialty Products, Houston, Tex.) (37, 88). The method is not affected by oil, sulfides, iron, or other constituents of produced brine and has a detection limit of 1,000 cells per ml. The main advantage of the test is that the results are available within 15 min. Problems with the detection of *Desulfobacter* species have been reported (8). Other immunological approaches have been developed to detect specific strains of SRB by using polyclonal antibodies directed against cell surface markers (18). Because the clustering of SRB based on phospholipid ester-linked fatty acids was essentially the same as that predicted by 16S rRNA gene analysis (52), phospholipid profiles can be used to quantify SRB and report their diversity, but this method has also not yet been applied routinely. Procedures for phospholipid analysis are discussed in chapter 10 of this manual.

DETECTION AND CONTROL OF MICROBIALLY INDUCED CORROSION (MIC) AND SOURING

The corrosion and souring (increase in H_2S) can be directly linked to an increase in the growth and activity of SRB and other sulfide-producing bacteria in the reservoir and associated water-handling equipment (20, 33, 86). Souring is often a result of waterflooding or enhanced oil recovery processes where sulfide-producing bacteria are either introduced into the reservoir or stimulated by nutrients introduced with the injected fluids.

Recognition of Microbial Activity in an Oil Field Water System

The importance of early recognition of microbial activity in an oil field water system cannot be overemphasized, since the cost of equipment replacement, well treatments, and shutdown times of wells can be very high. Certainly, monitoring for sulfide-producing bacteria (80, 124) and sulfate reduction activity (122) in production facilities is a critical part of a program to control microbial activity. SRB are generally found associated with slime-forming bacteria in biofilms. The interiors of these films are anaerobic and highly conducive to the growth of SRB even when the surrounding environment is aerobic. The activity of SRB stimulates the pitting of metal surfaces typical of anaerobic corrosion. SRB-induced anaerobic corrosion gives rise to black iron sulfides often associated with gelatinous mats of slime-forming bacteria. However, caution must be exercised before concluding that sulfides are the result of biological activity, since not all sulfides found in oil and gas facilities are from biological sources. Sulfides can be formed thermogenically in the presence of sulfate and organic compounds (132). Examination of the ratio of stable sulfur isotopes and the ratio of iron to manganese can help to distinguish between biological and abiological sulfides and between iron due to corrosion and that derived from reservoir minerals, respectively. A field guide from the Gas Research Institute provides a step-by-step assessment of how to evaluate if souring is due to biological or nonbiological sources

(26). Corrosion also produces other chemical clues discussed below that can indicate an otherwise undetected problem (25, 33, 78).

The following are indicators of microbial activity in an oil field water-handling system: (i) plugging of filters; (ii) the presence of gelatinous, slimy, or hard deposits (colored black) in wash tanks, storage tanks, heater-treaters, free-water knockouts, or stagnant regions of flow lines; (iii) the appearance of H_2S in produced water, gas, or crude oil that was not previously sour; (iv) a pitting type of corrosion accompanied by the presence of black iron sulfide; (v) an increase in the frequency of leaks; (vi) a decline in injectivity; and (vii) the production of black water and slime when injection wells are backflowed (78). Microbial activity can also be found in soils surrounding buried pipelines, underneath vessels and tanks leading to external MIC. Factors controlling external MIC include availability of nutrients, water, and electron acceptors and are most often associated with waterlogged, clay-dominated soils with alternating moisture and oxygen concentrations. SRB and APB are the most common causative agents of external MIC (93).

Examination of water samples, slime, or corroded materials can confirm the presence of SRB and other microorganisms. Sampling frequency and location will depend on the mode of operation and the problem under scrutiny (80, 124). It is advisable when sampling a flow line to flush any dead volume for several minutes prior to collecting the sample. Samples can be collected in clean glass, polyethylene, or polypropylene bottles with screw-cap lids and preferably an airtight seal (80). If possible, bottles should be prerinsed with water from the source and then filled to nearly the top of the bottle to exclude air. Biofilm or sessile bacteria can be collected in screw-cap bottles or sterile Whirl Pak bags from corrosion coupons, filter backwashes, pig solids, or pipe walls at unions and valves. It is important that samples are examined as soon as possible after removal from the oil field site, since chemical clues such as the presence of H_2S may disappear and microbial populations may change. It has been recommended by the National Association of Corrosion Engineers (NACE) that samples held for more than 4 h be stored at 4°C; samples held for longer than 48 h are of questionable value regardless of refrigeration (80).

Routine Monitoring of Microbial Activity

Many of the indicators of microbial activity given above become evident when SRB and slime-forming bacteria are well established in a water-handling system. It is obviously desirable to be able to detect and control microbial growth before a serious problem develops. Therefore, an oil field water system should be routinely monitored for microbial activity even when no problem is known to exist. Routine monitoring should include (i) enumeration of SRB, APB, and GAB; (ii) the examination of metal surfaces and scrapings; and (iii) water quality analysis.

Methods to enumerate SRB, APB, and GAB in water samples are discussed above. Samples to enumerate SRB, GAB, or APB should be taken at various points in the system, and the same sample should be used for the enumeration of each group of bacteria. The presence of SRB indicates a potential for corrosion and souring. Counts of SRB, APB, or GAB that increase with time at any point in the system or throughout the system demand immediate attention. Counts of GAB that are greater than 10^4 cells per ml indicate significant contamination. The sulfate reduction assay (122) can be used in conjunction with MPN enumeration studies (80, 124) to localize the most active areas of sulfate reduction. The analysis of water samples detects only planktonic microorganisms; therefore, these counts are only rough measures of the numbers of SRB and slime-forming bacteria in the system. The most reliable means of assessing the actual level of contamination of a water system is the direct examination of sessile bacteria on metal surfaces or from solids collected at the bottom of vessels or scraped from pipe walls during pigging. Corrosion coupons installed at various sites in the system or in sidestreams are often examined for evidence of microbial growth before being cleaned for corrosion analysis. However, coupons designed especially for biofilm monitoring are also available from suppliers (e.g., Metal Samples, Munford, Ala.). Corrosion coupons are often mounted in water lines on the top of a pipe for easy removal. However, most microbial growth and corrosion occur in the bottom quadrant of the pipe, where solids accumulate. Therefore, coupons for biofilm monitoring should be mounted at the 6 o'clock position of the pipe in areas that are known or suspected to support MIC. In oil-containing pipelines, coupons should be also located in the bottom of lines in low or stagnant areas of flow, where stratified flow and water accumulation are likely to occur. The time taken for development of a dense biofilm is unpredictable and depends on the conditions of the system, but most corrosion coupons are analyzed every 3 to 6 months depending on corrosion rates. Therefore, it may be necessary to install multiple coupons at a site so that they can be sampled at various time intervals. Methods for removal and counting of the bacteria on the coupons can be found in the NACE TMO194-2004 method (81). Methods for preparation and installation of corrosion coupons for oilfield service are discussed in the NACE standard practice RP0775-91 (79).

In addition to bacteria, coupons should be analyzed for corrosion products (i.e., iron sulfides, iron oxides, and iron carbonates) by elemental analysis (X-ray fluorescence) and mineral analysis (e.g., X-ray diffraction), which provide clues as to the mechanism and cause of the corrosion (19, 106). Iron sulfides, as well as other minerals, can have either a biological or abiological origin, and in some cases it may be possible to make this distinction (94, 105).

For monitoring microbial activity, coupons should be changed weekly, especially if high planktonic counts have been obtained. In relatively clean systems, the sampling time may be extended to 1-month intervals if personnel constraints dictate. A faster alternative to using metal coupons for biofilm detection is the use of bio-trap samplers containing porous Bio-Sep beads comprised of aramid polymer and powdered activated charcoal. These bio-traps provide plenty of surface area for adsorption of nutrients and rapid development of biofilm and were shown in a laboratory flow loop to recover over 30 times more biomass than a standard strip corrosion coupon over a 2-week period. Additionally, the biofilm communities in the beads resembled those found on the metal coupons (48), except during biocide application, when the beads tended to protect the biofilms. Thus far, the beads have been used only in conjunction with culture-independent methods (i.e., molecular methods) of detection and analysis. Commercially available online, real-time electrochemical methods for detection of biofilm activity (e.g., BIOoGEORGE and SmartCET) are also available (49, 60). The BIOoGEORGE probe has been used successfully in water cooling towers and industrial wastewater streams to detect biofilm activity, but its application to oil field brines has received much less attention (59).

Immediately after removal, the coupons are placed in 10 ml of a sterile buffer in a capped plastic tube. The buffer is prepared with field water and contains a nontoxic surfactant and glass beads to help remove the biofilm from the coupons. The tubes are then shaken mechanically or by hand. After settling, the fluid is used to inoculate media for enumeration studies, or the DNA may be extracted for cultivation-independent approaches (see above). The results of the enumeration are recorded as the number of bacteria per square centimeter of coupon surface area. The detection of SRB indicates the potential for corrosion and souring. However, the key indicator of a serious problem is a progressive increase in the SRB count downstream from the initial source of corrosion and an increase in sulfate reduction activity.

In addition to the use of corrosion coupons, any scheduled or unscheduled down time should be used to inspect the system for signs of microbial activity. Any changes in the water chemistry such as an increase in the concentration of sulfide, total iron, total suspended solids, or total organic carbon and a change in pH from the initial contamination site, indicate microbial activity (77). For external MIC, the demonstration of increased levels of viable corrosion-causing microorganisms in soils proximal to the corrosion versus those distal to the corrosion can be taken as evidence for MIC assuming that a failure mechanism consistent with the viable microorganisms can be justified.

Control of Microbial Activity in Oil Field Waters

Physical or mechanical methods to control microbial activity are rather limited but can minimize souring and corrosion. Water systems must be kept cleaned. Water lines, if possible, should be scraped and cleaned with pigs on a regular basis. If pigging is not possible, it may be satisfactory to inject slugs of surfactants and solvents to help remove the debris and deposits, but this approach is not generally considered as effective as pigging (17, 91). Whenever possible, the field operator should reverse the flow of water in injection wells to remove debris. A water system free of sludge and debris will reduce the number of sites where SRB can become established. When alternate sources of injection water are used, each should be analyzed for SRB and GAB. If possible, injection water that minimizes the contamination of the system and does not stimulate microbial growth and activity should be used. Whenever possible, commingling of waters from different sources should be avoided; they should be injected separately or in alternating slugs. Commingling of different waters may stimulate the growth of the bacteria present in the waters. This is especially true when mixtures of fresh and saline waters are used. The water injection systems should be designed to eliminate stagnant points and minimize water-handling time on the surface. Low fluid flow velocities of less than 1 m/s allow sludge to accumulate and produce environments suitable for the growth of SRB (78).

The most effective control of microbial activity in an oil field water system is through the use of biocides in combination with the mechanical measures discussed above (6, 27, 91, 125). A number of factors must be considered in choosing a biocide for an individual treatment. The biocide(s) must be active against the bacteria in the system under in situ conditions and must remain active long enough to treat sites far removed from the injection site of the chemical. The biocide must be compatible with the reservoir fluids and any other chemical agent that may be used, such as scale inhibitors, oxygen scavengers, etc. Lastly, the biocide must be economical and preferably have good environmental properties (e.g., ease of neutralization, low toxicity, no bioaccumulation, and good biodegradability).

Chlorine (Cl_2) and hypochlorous acid are strong oxidizing agents that are relatively low in cost, have a broad spectrum of activity, are pH sensitive, and are highly effective against planktonic bacteria but not as effective against biofilms. These compounds are inactivated by oxygen scavengers and oxidize corrosion and scale inhibitors. Chlorine residuals of 0.2 to 0.5 ppm are usually adequate to control biofouling in seawater injection systems, but highly contaminated systems require high doses to obtain the desired level of persistence. The optimal chlorine dosage is the smallest amount that will maintain fouling resistance within acceptable levels and is the product of chlorine concentration, treatment duration, and the period between treatments (17). To reduce costs, chlorine treatment is often supplemented with the use of an organic biocide, e.g., quaternary ammonium compounds or formaldehyde. Chlorination of injection waters also has the disadvantage of producing chlorinated hydrocarbons that make the produced water unacceptable for discharge and may damage refinery catalysts.

Chlorine dioxide (ClO_2) is also a strong oxidizing agent. Air mixtures of chlorine dioxide and chlorine containing a partial pressure greater than 50 mm Hg can explode. Because of this risk, chlorine dioxide is never shipped but is produced on site by using specialized equipment to trigger the reaction of sodium chlorate with sulfuric acid and sulfur dioxide. The yield of chlorine dioxide is dependent on the chemistry of the source water. The advantage of chlorine dioxide is that it is persistent, highly effective against biofilms, and pH insensitive and oxidizes sulfur and iron deposits (6). However, the release of corrosion products may lead to plugging of injection wells and filters. Therefore, chlorine dioxide is best applied to a relatively clean system. It can increase corrosion rates, but it does not produce chlorinated hydrocarbons and can be relatively expensive.

Nonoxidizing biocides (i.e., organic biocides) generally have a high degree of persistence in an oil field water system, but their effectiveness is dependent on the water chemistry and microbiology of the system. Nonoxidizing biocides for oil field use include aldehydes (e.g., formaldehyde, glutaraldehyde, and acrolein), quaternary amines (e.g., benzyl and alkyl derivatives), tetrakishydroxymethyl phosphonium sulfate, 2,2-dibromo-3-nitrilopropionamide, and bronopol (6, 125). An organic biocide that works in one field may be totally ineffective in another. Therefore, it is important to screen a variety of biocides before choosing one or more for application. To be effective, the biocide must penetrate the biofilm in sufficient amounts to kill the bacteria. Therefore, the most meaningful biocide-testing protocol will include biofilm testing (103). Screening of biocides against planktonic bacteria in the injection water will provide information on chemical compatibility of the biocide with formation brine. Also, a biocide that is ineffective against planktonic bacteria will generally also be ineffective against bacteria in biofilms (17). In practice, water samples from various sites in the water system are obtained, and each is exposed to several concentrations of the biocide by using a dose-response test (81). The manufacturer will recommend a minimum effective concentration and exposure time. After a predetermined time, SRB, APB, and GAB counts are obtained and compared to untreated samples. Only those biocides that exhibit total kill should be evaluated against biofilms.

The effectiveness of biocides against biofilms can be evaluated in situ by using the whole or any part of the water treatment system or in the laboratory by using a Robbins device (64, 102). A Robbins device is a tubular flow reactor with removable surfaces for the analysis of the biofilm. The device is designed so that surface samples can be removed aseptically without exposing the interior of the pipe to the outside environment and without altering the flow through the pipe. When water from a contaminated water system is circulated through the device, a biofilm with SRB will develop on the removable surfaces called studs. The time required for a stable biofilm to develop depends on the level of contamination in the system and the water chemistry. When a stable biofilm has developed as indicated by consistent SRB counts, the biocide treatment should begin. A control device and one or more devices treated with the biocide will be required for each biocide to be tested. Typically, biocide testing will consist of the addition of daily slugs of the biocide followed by circulation of untreated water. The dosage and time of exposure should initially be based on the manufacturer's recommendation and can be varied to optimize efficiency. Studs should be examined visually and the SRB counts conducted every 3 to 7 days.

It is difficult to kill all SRB in biofilms. Therefore, the question becomes "what is an acceptable number?" The answer to this question is system dependent and firmly established only by relating operating experience with the levels of SRB and GAB contamination over time. Certainly, multiple-order-of-magnitude decreases in SRB and GAB are desirable. The goal of biocide treatments is to reduce the risk of corrosion, not necessarily to sterilize the system.

Another important question is whether to treat continuously or intermittently (e.g., batch or slug treatments). Continuous treatment is usually more expensive than intermittent (slug) treatment but requires lower concentrations of biocides and can be applied economically in oil transport lines, where water volumes are low. However, in practice, slug or batch treatments are more effective in treating biofilms, since they provide higher concentration gradients, facilitating penetration of the biocide into the biofilm. Intermittent treatments vary in frequency from two or three times per week to once every 1 or 2 months. Significant cost savings can be obtained by treating contaminated segments separately with high dosages of the biocide while maintaining lower dosages elsewhere. However, the operator must make sure that the entire water-handling system is treated.

The effectiveness of the biocide may decline with repeated use due to the development of resistance within the microbial populations in the system. The alternate use of two different biocides may counteract this problem. When resistance to one biocide develops, the second biocide is used. If resistance to the second biocide develops, a return to the first biocide will generally be effective in controlling microbial activity (17).

The effectiveness of a biocide treatment program should be continuously monitored for both chemical (e.g., biocide residuals, iron counts, and sulfides) and microbiological (i.e., planktonic and sessile bacteria) indicators of fouling by using the methods described previously even if microbial growth has been controlled. The development of microbial resistance or minor changes in the water-handling system may result in significant changes in the response of the system to the biocide. It is equally important that the treatment program be maintained once control of microbial activity has been achieved. It is tempting to save money by cutting back on biocide dosages once the problem is under control. However, the costs of reestablishing control after a biocide treatment program has been discontinued generally far outweigh any savings gained from a temporary cutback in the treatment program.

Use of Nitrate To Control Souring

Souring is defined and detected as the increase in total sulfide concentration in produced gas, oil, and water over time. This increase can be substantial; e.g., in the Skjold field in the Danish sector of the North Sea, total sulfide production increased from 100 to 1,100 kg/day within 5 years of the start of seawater injection (54). It is now generally accepted that much of the souring observed upon water injection is due to the activity of SRB. From backflow studies it has been estimated that much of the SRB activity can be present within a 10-m radius from the injectors and that this phenomenon can cause the production of 50 to 100 kg of sulfide per day when sulfate is not limiting as during injection of seawater (123). The time span before sulfide emerges at a production well depends on (i) the water injection rate, (ii) the distance separating the production and injection wells, (iii) the average porosity of the reservoir rock in the path traveled by the injected water, and (iv) the capacity of reservoir rock to adsorb sulfide. In view of the many factors involved, it is not easy to predict if and when production wells will go sour (123). Total sulfide production can be limited by shutting in souring production wells.

Several methods can be used to contain or reduce souring. Removal of sulfate from the injection water by nanofiltration is a costly but in principle effective method (2). When sulfate is absent or much reduced, SRB cannot grow, limiting the potential for souring. Biocide treatment is a potential option to contain souring in above-ground facilities or in the reservoir, provided that most of the SRB activity is present close to the well bore. Biocide treatment is most effective when a high concentration (a slug) is injected for a short period (1 h). Continuous injection of low concentrations is generally ineffective, whereas continuous injection of high concentrations is too costly, given the very large volumes of seawater that are being injected (10^3 to 10^4 m^3 per day). The use of nitrate has recently shown considerable success and appears to be developing into the preferred treatment option in many fields. An advantage of nitrate is that it is cheap enough to be injected continuously at a reasonable concentration. Being a similar water-soluble anion, it will track sulfate and either prevent its conversion to sulfide or react with sulfide to regenerate sulfate through microbial action. A major difference between biocide treatment and nitrate treatment is that the former may be expected to always lower the total number of bacteria present, whereas the latter will cause a shift in community composition from SRB to hNRB and NR-SOB. The total number of bacteria may remain the same or can increase under these conditions. Because most information has been obtained recently on the effects of nitrate treatment, we briefly survey these data, as they may help in deciding on the best treatment option.

Nitrate injection stimulates hNRB and NR-SOB. The former compete with SRB for degradable oil organics and produce nitrite, which is a strong SRB inhibitor. The latter remove sulfide by direct oxidation with nitrate. NR-SOB also produce nitrite, which inhibits SRB, although it should be pointed out that many mesophilic SRB have a periplasmic nitrite reductase to prevent this inhibition (31). Hence, SRB and NR-SOB can be considered to be symbiotic (31). Together they catalyze the same overall reaction as catalyzed by hNRB: oxidation of organic carbon with nitrate. The difference is that SRB and NR-SOB catalyze this reaction through a sulfide intermediate. One can in principle easily assess whether an SRB/NR-SOB or an hNRB mechanism is operating in souring control. SRB populations would remain relatively constant in the former and would decrease in the latter mechanism. Examples of both types of mechanism have been seen in reservoir model column studies (2, 39). Interestingly, it appears that both types of mechanism may also be present in the field. Continuous injection of 400 mg of nitrate liter^{-1} for 50 days in the Coleville field near Kindersley, Saskatchewan, Canada, reduced sulfide concentrations on average by 73%. The NR-SOB *Thiomicrospira* sp. strain CVO became a major community component under these conditions, while SRB continued to be present, suggesting an SRB/NR-SOB mechanism of souring control (115). Continuous injection of 30 mg of nitrate liter^{-1} in the Veslefrikk field in the Norwegian sector of the North Sea reduced the number of SRB by 20,000-fold and increased the number of hNRB by up to 60,000-fold in biofilms formed in above-ground injection facilities (119), indicating that an hNRB mechanism applied here. Prior injection of slugs of the biocide glutaraldehyde (500, 700, or 1,000 mg liter^{-1} for 1 or 2 h per week) into the same field had proved ineffective in controlling SRB activity (119). Because breakthrough of injection water at producing wells can take up to 10 years, the effect of nitrate application often needs to be evaluated by backflowing injection wells and comparing microbial and chemical compositions in the backflowed waters before and after nitrate application (55). Field use of nitrite for SRB control has also been reported (110), although it is less commonly used than nitrate. An advantage of nitrite is that it serves directly as an SRB inhibitor, whereas nitrate needs to be converted to nitrite before it is inhibitory.

A decision on the various options to control oil field souring should be based on information on the volume of water that needs to be injected for economic oil recovery, as well as on the water chemistry and microbial numbers (SRB, hNRB, and NR-SOB) in backflowed injection wells. If large volumes of water need to be injected per barrel of oil produced, then sulfate removal by nanofiltration may be uneconomical. If the water chemistry appears ideal for development of SRB (presence of high concentrations of the volatile fatty acids acetate, propionate, and butyrate, as well as of sulfate) and if high numbers of bacteria including SRB are already present, then slug-wise application of biocides is unlikely to have a lasting effect. Shifting microbial activity away from sulfate reduction by application of nitrate or nitrite may be the only option under these conditions. This option can be quite successful, as has now been documented in a number of cases.

MEOR

The design of a microbially enhanced oil recovery (MEOR) process varies depending on the reservoir and problem(s) that limits oil production. The following discussion provides some general guidelines to aid in the design of a MEOR process. First, the appropriate microbial process (Table 1) is selected after a detailed engineering analysis. Assessing the efficacy of the microbial process on oil production can be difficult, since many factors, either associated with the well itself or within the reservoir, can affect oil production. These may include the cessation of production from a well, the replacement of worn or old equipment, changes in the stroke rate of the pump, etc. The changes in production or injection rates in other wells may also affect oil production from the wells under study. Causal relationships are difficult to determine in any field situation. However, the inclusion of control wells that have the same equipment and are operated in an identical fashion, but without a key component of the microbial process, can be used to determine if the microbial process affected oil production.

Next, one must determine if the injection of a specific microorganism or consortium is needed or whether the requisite activity and organisms are already present in the formation. This involves sampling of wells for specific microbial populations and activities as discussed above. If the requisite microorganism(s) is not detected, then it will have to be injected. For this latter option, tests must be conducted to ensure that the microorganism(s) of choice grows and produces the desired product at the temperature, pressure, and salinity of the reservoir. The flow systems used to answer these questions have been discussed previously (67). The organism's ability to migrate through the reservoir can be studied by using the procedures discussed in chapter 70.

Once the decision is made about the type of microorganism to be used, a nutrient formulation is developed that stimulates the growth and activity of the appropriate microorganism(s). Oil field brines often contain sufficient amounts of most nutrients to support microbial growth and metabolism with the exception of carbon, phosphorus, nitrogen, and sometimes sulfur (43, 68). Some microbial processes involve the use of hydrocarbons present in the formation (Table 3). Nutrient limitations can easily be determined by amending brine from the reservoir with different combinations of nutrients to determine which are required to support growth and the activity of interest. If the plan is to inject a microorganism, it should be included in these tests to ensure that the nutrient formulation stimulates the growth and activity of the desired organism in the presence of indigenous microorganisms. These tests will also discern whether the nutrient amendment stimulates the activity of detrimental organisms such as sulfate reducers. Unwanted activities such as sulfate reduction can be suppressed by inclusion of nitrate (99). Molasses is often the nutrient used for MEOR, since it contains a variety of carbon sources, nitrogen sources, and other nutrients. However, the use of a complex, undefined nutrient limits the ability to control microbial activity by altering the relative concentrations of nutrients such as the ratio of carbon to nitrogen. Other sources of carbohydrate such as corn syrup products may be more suitable if such control is desired.

An injection protocol needs to be developed to avoid plugging and to deliver the nutrients and/or microbial cells deep within the reservoir. Laboratory tests using the rock from the reservoir should be conducted to determine the concentration of cells that can be injected without plugging the injection well. The use of spores or ultramicrobacterial cells may reduce facial plugging and allow greater transport of the organism into the formation (53). The injection of a complex nutrient source such as molasses or a complete nutrient formulation may stimulate unwanted microbial growth in the vicinity of the injection well. To avoid this possibility, the nutrient formulation can be injected in several batches (up to 6 m^3) followed by injection of eight times the volume of brine (50) or in sequence, with those nutrients with the largest retardation factors injected first (44, 47). The sequential addition of nutrients reduces the possibility of microbial growth in the vicinity of the injection well and allows the mixing of the nutrients deep within the reservoir. For some nutrients such as molasses, in-line filters (125 μm) may be required to remove suspended particles prior to injection into the formation. Plugging due to nutrient/cell injection has not been noted as a problem in any of the studies listed in Table 3.

Lastly, consideration must be given to the types of data that should be collected to assess performance. For some microbial approaches such as paraffin removal, the goal may be to reduce operating expenses rather than to increase oil production. Thus, fuel consumption, electricity usage, or the amount and frequency of chemical treatments may be more important than the oil production rates. Preferably, wells with complete production histories should be chosen to obtain accurate estimates of the rate of change over time. For reservoir-wide processes, tracer experiments should be conducted to verify flow paths. Samples of treated and untreated wells should be taken frequently before and after the implementation of the microbial process to determine whether the expected products of metabolism and the appropriate microorganisms increase in concentration in the treated wells but not in the untreated wells. An early indication of whether the appropriate microbial activity has been stimulated in the reservoir can be obtained by pumping the injection well to obtain samples of the fluids that were injected into the reservoir (69). Flow meters should be installed on each production well in order to measure total fluid production from that well. The amount of oil produced from the well is then calculated from the water-to-oil ratio, which is determined on a frequent schedule. Transit times between wells depend on distance and geology of the formation. Tracer studies and modeling are needed to determine the duration of the sampling regimen.

Table 3 summarizes different microbial processes that have been applied in field situations to improve operations and oil recovery. Microbial formulations and nutrient mixtures to control paraffin deposition and improve the productivity of waterfloods are commercially available (10, 22, 84, 109). The mechanism by which these products work is still controversial, but these technologies have survived in the marketplace for many years and have had a fair degree of success (Table 3). Often, the wells have less paraffin deposition, require less frequent chemical treatments to maintain production, and have improved oil production rates after microbial treatment. Well stimulations involve the injection of a mixture of anaerobic microorganisms and a fermentable carbohydrate (usually molasses) into a production well. Fluid production from the well is ceased for 10 to 15 days to allow for growth and metabolite production, e.g., acids, solvents, and gases, in the near-well-bore region. The technology is simple to implement but has inconsistent technical performance. Hitzman (34, 35) noted that this approach is most effective in carbonate reservoirs. This approach has been modified for use as a reservoir-wide approach to improve oil production by injecting the cells and nutrients into injection wells. The approach appears to be most effective in carbonate and carbonaceous sandstone reservoirs (34, 35, 56, 57, 82, 114, 127). Large pH reductions (one or two U) and the presence of acids and solvents in production fluids coincide with increased oil production rates (lasting for periods of months to years), suggesting a causal relationship. H_2S production increases corrosion, and large concentrations can be a risk to human health. Nitrate injection is used to change the terminal electron-accepting status of the reservoir from sulfate to nitrate reduction. Several studies have noted reduced sulfide levels in production waters (111, 115). Permeability variation in reservoirs is a major reason why the recovery fluid (usually brine) does not contact the entire reservoir. The injection of a fermentable carbohydrate often with other nutrients such as nitrate or phosphate is used to stimulate in situ growth of indigenous bacteria in high-permeability regions (11, 47, 50, 58, 75). Several studies show that high-permeability regions were selectively blocked (47, 50, 58, 75), and in one instance a

TABLE 3 Survey of methods used in MEOR field trials

Process	Method[a]
Sulfide reduction	Continuous injection of NO_3^- alone (111), NO_3^- with $PO_4^=$ for 50 days (115), or NO_3^- and NO_2^- (36)
Permeability profile modification	Stimulation of indigenous bacterial growth by batch injections of molasses and NH_4NO_3, each followed by cessation of injection fluid for 2 to 4 weeks (50), by weekly and separate injections of NO_3^-, and $PO_4^=$ (11), or by injection of maltodextrins and ethyl acid phosphate followed by cessation of injection fluid for 2 wk (47) Injection of molasses and polymer-producing *Enterobacter* sp. for 2 mo (75) or *Bacillus* sp. spores, sucrose, yeast extract, polyphosphate, and NO_3^- followed by cessation of injection fluid for 3 wk (58)
Biosurfactant production	Injection of a mixture of microorganisms containing a biosurfactant producer and periodic injection of molasses (12, 14, 15)
Gas, acid, and solvent production	Injection of a mixture of facultative and anaerobic microorganisms with molasses and/or inorganic nutrients (56, 57, 127, 133)
Stimulation of hydrocarbon degradation	Cyclic injection of aerated water with N and P sources (3) Periodic injection of proprietary mixture of microorganisms, nutrients, and "biocatalyst" (22, 108, 120)
Well stimulation	Injection of a mixture of fermentative microorganisms and molasses into production wells; stop fluid production (34, 35, 74)
Paraffin deposition reduction	Periodic injection of proprietary mixture of microorganisms, nutrients, and "biocatalyst" (10, 84, 109)

[a]References are given in parentheses.

substantial increase in oil recovery occurred (75). Oil is entrapped in small pores and cannot be mobilized unless large reductions in the interfacial tension between the oil and aqueous phases occur. The injection of molasses along with a mixture of microorganisms containing a *Bacillus* strain known to make a biosurfactant that significantly reduces oil:water interfacial tension (65) increased oil production rates in several reservoirs (12, 14).

CONCLUDING REMARKS

Oil reservoirs contain diverse and metabolically active microbial communities. Cultivation-dependent and cultivation-independent approaches can be used to characterize these communities and determine how their activities may affect operations. Continual monitoring will allow early detection of souring or corrosion before these problems become widespread and difficult to treat. Knowledge of the potential microbial activities present in the surface facilities and the reservoir is critical to avoid the stimulation of unwanted microbial activities such as corrosion, souring, and biofouling during petroleum exploitation activities. This knowledge can also be used to stimulate the beneficial activities of microorganisms to improve oil recovery or alter the terminal electron-accepting process by nitrate addition. The results of several microbial field trials show that incremental oil can be produced for less than $3 per barrel (about $19 per m^3) (11, 14, 57, 108).

REFERENCES

1. **American Petroleum Institute.** 1975. *Recommended Practice for the Biological Analysis of Subsurface Injection Waters*. American Petroleum Institute, Washington, D.C.
2. **Bakke, R. B., B. Rivedal, and S. Mehan.** 1992. Oil reservoir biofouling control. *Biofouling* **6:**53–60.
3. **Belyaev, S. S., I. A. Borzenkov, I. F. Glumov, R. R. Ibatullin, E. I. Milekhina, and M. V. Ivanov.** 1998. Activation of the geochemical activity of stratal microflora as basis of a biotechnology for enhancement. *Microbiology* (Moscow) **67:**708–714.
4. **Bhupathiraju, V. K., M. J. McInerney, and R. M. Knapp.** 1993. Pretest studies for a microbially enhanced oil recovery field pilot in a hypersaline oil reservoir. *Geomicrobiol. J.* **11:**19–34.
5. **Bodour, A. A., and R. M. Miller-Maier.** 1998. Application of a modified drop-collapse technique for surfactant quantification and screening of biosurfactant-producing microorganisms. *J. Microbiol. Methods* **32:**273–280.
6. **Boivin, J.** 1995. Oil field biocides. *Mater. Perform.* **34:**65–68.
7. **Bonch-Osmolovskaya, E. A., M. L. Miroshnichenko, A. V. Lebedinsky, N. A. Chernyh, T. N. Nazina, V. S. Ivoilov, S. S. Belyaev, E. S. Boulygina, Y. P. Lysov, A. N. Perov, A. D. Mirzabekov, H. Hippe, E. Stackebrandt, S. L'Haridon, and C. Jeanthon.** 2003. Radioisotopic, culture-based, and oligonucleotide microchip analyses of thermophilic microbial communities in a continental high-temperature petroleum reservoir. *Appl. Environ. Microbiol.* **69:**6143–6151.
8. **Brink, D. E., I. Vance, and D. C. White.** 1994. Detection of *Desulfobacter* in oil field environments by non-radioactive DNA probes. *Appl. Microbiol. Biotechnol.* **42:**469–475.
9. **Broderick, L. S., and J. J. Cooney.** 1982. Emulsification of hydrocarbons by bacteria from freshwater ecosystems. *Dev. Ind. Microbiol.* **23:**425–434.
10. **Brown, F. G.** 1992. Microbes: the practical and environmental safe solution to production problems, enhanced production, and enhanced oil recovery. SPE 23955. *In Proceedings of the 1992 Permian Basin Oil and Gas Recovery Conference*. Society of Petroleum Engineers, Richardson, Tex.
11. **Brown, L. R., A. A. Vadie, and J. O. Stephens.** 2002. Slowing production decline and extending the economic life of an oil field: new MEOR technology. *SPE Reservoir Eval. Eng.* **5:**33–41.
12. **Bryant, R. S., T. E. Burchfield, D. M. Dennis, and D. O. Hitzman.** 1990. Microbial-enhanced waterflooding: Mink Unit project. *SPE Reservoir Eng.* **5:**9–13.
13. **Bryant, R. S., and J. Douglas.** 1988. Evaluation of microbial systems in porous media for EOR. *SPE Reservoir Eng.* **3:**489–495.
14. **Bryant, R. S., A. K. Steep, K. M. Bertus, T. E. Burchfield, and M. Dennis.** 1993. Microbial-enhanced waterflooding field pilots. *Dev. Pet. Sci.* **39:**289–306.
15. **Buciak, J., A. Vazquez, R. Frydman, J. Mediavilla, and R. S. Bryant.** 1995. Enhanced oil recovery by means of microorganisms: pilot test. *SPE Adv. Technol. Ser.* **4:**144–149.
16. **Carroll, V., G. R. Chilvers, D. Franklin, M. J. Miles, V. J. Morris, and S. G. Ring.** 1983. Rheology and microstructure of solutions of the microbial polysaccharide from Pseudomonas elodea. *Carbohydr. Res.* **114:**181–191.
17. **Characklis, W. G.** 1990. Microbial biofouling control, p. 585–633. *In* W. G. Characklis and K. C. Marshall (ed.), *Biofilms*. John Wiley & Sons, Inc., New York, N.Y.
18. **Christensen, B., T. Torsvik, and T. Lien.** 1992. Immunomagnetically captured thermophilic sulfate-reducing bacteria from North Sea oil field waters. *Appl. Environ. Microbiol.* **58:**1244–1248.
19. **Craig, B.** 2002. Corrosion product analysis—a road map to corrosion in oil and gas pipelines. *Mater. Perform.* **41:**56–58.
20. **Crolet, J.-L., and M. F. Magot.** 1996. Non-SRB sulfidogenic bacteria in oilfield production facilities. *Mater. Perform.* **35:**60–64.
21. **Desai, J. D., and I. M. Banat.** 1997. Microbial production of surfactants and their commercial potential. *Microbiol. Mol. Biol. Rev.* **61:**47–64.
22. **Dietrich, F. L., F. G. Brown, Z. H. Zhou, and M. A. Maure.** 1996. Microbial EOR technology advancement: case studies of successful projects. SPE 36746. *In Proceedings of the SPE Annual Technical Conference and Exhibition*. Society of Petroleum Engineers, Richardson, Tex.
23. **Eckford, R. E., and P. M. Fedorak.** 2002. Planktonic nitrate-reducing bacteria and sulfate-reducing bacteria in some western Canadian oil field waters. *J. Ind. Microbiol. Biotechnol.* **27:**80–86.
24. **Eckford, R. E., and P. M. Fedorak.** 2002. Second derivative UV absorbance analysis to monitor nitrate-reduction by bacteria in most probable number determinations. *J. Microbiol. Methods* **50:**141–153.
25. **Gas Research Institute.** 1990. *Gas Research Institute Field Guide I. Microbiologically Influenced Corrosion (MIC): Methods of Detection in the Field*. Gas Research Institute, Chicago, Ill.
26. **Gas Research Institute.** 1994. *Microbiologically Influenced Souring (MIS): Assessment of MIC in Natural Gas Storage Fields*. Gas Research Institute, Chicago, Ill.
27. **Geesey, G. G., Z. Lewandowski, and H.-C. Flemmig.** 1994. *Biofouling and Biocorrosion in Industrial Water Systems*. Lewis Publishers, Boca Raton, Fla.
28. **Gevertz, D., A. J. Telang, G. Voordouw, and G. E. Jenneman.** 2000. Isolation and characterization of strains CVO and FWKO B: two novel nitrate-reducing, sulfide-oxidizing bacteria isolated from oil field brine. *Appl. Environ. Microbiol.* **66:**2491–2501.
29. **Granick, S.** 1991. Motions and relaxations of confined liquids. *Science* **253:**1374–1379.

30. **Grassia, G. S., K. M. McLean, P. Glénat, J. Bauld, and A. J. Sheehy. 1996.** A systematic survey for thermophilic fermentative bacteria and archaea in high temperature petroleum reservoirs. *FEMS Microbiol. Ecol.* **21:**47–58.
31. **Greene, E. A., C. Hubert, M. Nemati, G. E. Jenneman, and G. Voordouw.** 2002. Nitrite reductase activity of sulfate-reducing bacteria prevents their inhibition by nitrate-reducing, sulfide-oxidizing bacteria. *Environ. Microbiol.* **5:** 607–617.
32. **Gutnick, D. L., and E. Rosenberg.** 1977. Oil tankers and pollution: a microbiological approach. *Annu. Rev. Microbiol.* **31:**379–396.
33. **Hamilton, W. A.** 1985. Sulfate-reducing bacteria and anaerobic corrosion. *Annu. Rev. Microbiol.* **39:**195–217.
34. **Hitzman, D. O.** 1983. Petroleum microbiology and the history of its role in enhanced oil recovery, p. 162–218. *In* E. C. Donaldson and J. B. Clark (ed.), *Proceedings of the 1982 International Conference on Microbial Enhancement of Oil Recovery*. CONF-8205140. U.S. Department of Energy, Bartlesville, Okla.
35. **Hitzman, D. O.** 1988. Review of microbial enhanced oil recovery field tests, p. VI1–VI41. *In* T. E. Burchfield and R. S. Bryant (ed.), *Proceedings of the Symposium on Applications of Microorganisms to Petroleum Technology*. National Technical Information Service, Springfield, Va.
36. **Hitzman, D. O., M. Dennis, and D. C. Hitzman.** 2004. Recent successes: MEOR using synergistic H_2S prevention and increased oil recovery systems. SPE 89453. *In Proceedings of the SPE/DOE Fourteenth Symposium on Improved Oil Recovery*. Society of Petroleum Engineers, Richardson, Tex.
37. **Horacek, G. L., and L. J. Gawel.** 1988. New kit for the rapid detection of SRB in the oil field. SPE 18199. *In Proceedings of the 63rd Annual Technical Conference*. Society of Petroleum Engineers, Richardson, Tex.
38. **Huang, H., and S. Larter.** 2005. Biodegradation of petroleum in subsurface geological reservoirs, p. 91–121. *In* B. Ollivier and M. Magot (ed.), *Petroleum Microbiology*. ASM Press, Washington, D.C.
39. **Hubert, C., M. Nemati, G. E. Jenneman, and G. Voordouw.** 2003. Containment of biogenic sulfide production in continuous up-flow packed-bed bioreactors with nitrate or nitrite. *Biotechnol. Prog.* **19:**338–345.
40. Reference deleted.
41. **Hubert, C. R. J.** 2004. Control of hydrogen sulfide production in oil fields by managing microbial communities through nitrate or nitrite addition. Ph.D. dissertation. University of Calgary, Calgary, Alberta, Canada.
42. **Jenneman, G. E.** 1992. The effect of in-situ pressure on MEOR processes. SPE/DOE 24203. Presented at the SPE/DOE Eighth Symposium on Enhanced Oil Recovery, Society of Petroleum Engineers, Richardson, Tex.
43. **Jenneman, G. E.** 1989. The potential for in-situ microbial applications. *Dev. Pet. Sci.* **22:**37–74.
44. **Jenneman, G. E., J. B. Clark, and P. D. Moffitt.** 1993. A nutrient control process for microbially enhanced oil recovery. *Dev. Pet. Sci.* **39:**319–334.
45. **Jenneman, G. E., R. M. Knapp, M. J. McInerney, D. E. Menzie, and D. E. Revus.** 1984. Experimental studies of in situ microbial enhanced oil recovery. *Soc. Pet. Eng. J.* **24:**33–37.
46. **Jenneman, G. E., M. J. McInerney, and R. M. Knapp.** 1985. Microbial penetration through nutrient-saturated Berea sandstone. *Appl. Environ. Microbiol.* **50:**383–391.
47. **Jenneman, G. E., P. D. Moffitt, and G. R. Young.** 1996. Application of a microbial selective-plugging process at the North Burbank unit: prepilot tests. *SPE Prod. Facil.* **11:**11–17.
48. **Jenneman, G. E., and R. H. Webb.** 2004. Evaluation of an on-line biofilm detector and bio-traps for monitoring MIC in produced oilfield brine. Paper 04758. Presented at the Corrosion 2004, NACE International, Houston, Tex.
49. **Kane, R. D., and S. Campbell.** 2004. Real-time corrosion monitoring of steel influenced by microbial activity (SRB) in simulated seawater injection environments. Paper 04579. Presented at Corrosion 2004, NACE International, Houston, Tex.
50. **Knapp, R. M., M. J. McInerney, J. D. Coates, D. E. Menzie, and V. K. Bhupathiraju.** 1992. Design and implementation of a microbially enhanced oil recovery field pilot, Payne County, Oklahoma. SPE 24818. *In Proceedings of the SPE Annual Technical Conference*. Society of Petroleum Engineers, Richardson, Tex.
51. **Kodama, Y., and K. Watanabe.** 2003. Isolation and characterization of a sulfur-oxidizing chemolithotroph growing on crude oil under anaerobic conditions. *Appl. Environ. Microbiol.* **69:**107–112.
52. **Kohring, L. L., D. B. Ringelberg, R. Devereux, D. A. Stahl, M. W. Mittelman, and D. C. White.** 1994. Comparison of phylogenetic relationships based on phospholipid fatty acid profiles and ribosomal RNA sequence similarities among sulfate-reducing bacteria. *FEMS Microbiol. Lett.* **119:**303–308.
53. **Lappin-Scott, H. L., F. Cusack, and J. W. Costerton.** 1988. Nutrient resuscitation and growth of starved cells in sandstone cores: a novel approach to enhanced oil recovery. *Appl. Environ. Microbiol.* **54:**1373–1382.
54. **Larsen, J.** 2002. Downhole nitrate applications to control sulfate reducing bacteria activity and reservoir souring. Paper 02025. Presented at Corrosion 2002, NACE International, Houston, Tex.
55. **Larsen, J., M. H. Rod, and S. Zwolle.** 2004. Prevention of reservoir souring in the Halfdan field by nitrate injection. Paper 04761. Presented at Corrosion 2004, NACE International, Houston, Tex.
56. **Lazar, I.** 1991. MEOR field trials carried out over the world during the last 35 years. *Dev. Pet. Sci.* **31:**485–530.
57. **Lazar, I., S. Dobrota, M. C. Stefanescu, L. Sandulescu, R. Paduraru, and M. Stefanescu.** 1993. MEOR, recent field trials in Romania: reservoir selection, type of inoculum, protocol for well treatment and line monitoring. *Dev. Pet. Sci.* **39:**265–288.
58. **Lee Gullapalli, I., J. H. Bae, K. Heji, and A. Edwards.** 2000. Laboratory design and field implementation of microbial profile modification process. *SPE Reservoir Eval. Eng.* **3:**42–49.
59. **Licina, G. J.** 2001. Monitoring biofilms on metallic surfaces in real time. Paper 01442. Presented at Corrosion 2001, NACE International, Houston, Tex.
60. **Licina, G. J., and C. S. Carney.** 1999. Monitoring biofilm fomation and incipient MIC in real time. Paper 175. Presented at Corrosion 1999, NACE International, Houston, Tex.
61. **Machel, H. G., and J. Foght.** 2000. Products and depth limits of microbial activity in petroliferous subsurface settings, p. 105–120. *In* R. E. Riding and S. M. Awrdmik (ed.), *Microbial Sediments*. Springer-Verlag, New York, N.Y.
62. **Magot, M., B. Ollivier, and B. K. C. Patel.** 2000. Microbiology of petroleum reservoirs. *Antonie Leeuwenhoek J. Microbiol. Serol.* **77:**103–116.
63. **Marquis, R. E., and P. Matsumura.** 1978. Microbial life under pressure, p. 105–158. *In* D. J. Kushner (ed.), *Microbial Life in Extreme Environments*. Academic Press, London, United Kingdom.
64. **McCoy, W. F., J. D. Bryers, J. Robbins, and J. W. Costerton.** 1981. Observations of fouling biofilm formation. *Can. J. Microbiol.* **27:**910–917.

65. **McInerney, M. J., M. Javaheri, and D. P. Nagle, Jr.** 1990. Properties of the biosurfactant produced by *Bacillus licheniformis* strain JF-2. *J. Ind. Microbiol.* **5:**95–101.
66. **McInerney, M. J., D. P. Nagle, and R. M. Knapp.** 2005. Microbially enhanced oil recovery: past, present and future, p. 215–237. *In* B. Ollivier and M. Magot (ed.), *Petroleum Microbiology*. ASM Press, Washington, D.C.
67. **McInerney, M. J., and K. L. Sublette.** 2002. Oil field microbiology, p. 777–787. *In* C. J. Hurst, R. L. Crawford, G. R. Knudsen, M. J. McInerney, and L. D. Stetzenbach (ed.), *Manual of Environmental Microbiology*, 2nd ed. ASM Press, Washington, D.C.
68. **McInerney, M. J., and D. W. S. Westlake.** 1990. Microbially enhanced oil recovery, p. 409–445. *In* H. L. Ehrlich and C. L. Brierley (ed.), *Microbial Mineral Recovery*. McGraw-Hill Publishing Co., New York, N.Y.
69. **McKinley, V. L., J. W. Costerton, and D. C. White.** 1988. Microbial biomass, activity, and community structure of water and particles retrieved by backflow from a waterflood injection well. *Appl. Environ. Microbiol.* **54:** 1383–1393.
70. **Miranda-Tello, E., M.-L. Fardeau, J. Sepulveda, L. Fernandez, J.-L. Cayol, P. Thomas, and B. Ollivier.** 2003. *Garciella nitratireducens* gen. nov., sp. nov., an anaerobic, thermophilic, nitrate- and thiosulfate-reducing bacterium isolated from an oilfield separator in the Gulf of Mexico. *Int. J. Syst. Evol. Microbiol.* **53:**1509–1514.
71. **Morikawa, M., Y. Hirata, and T. Imanaka.** 2000. A study on the structure-function relationship of lipopeptide biosurfactants. *Biochim. Biophys. Acta* **1488:**211–218.
72. **Mulligan, C. N., D. G. Cooper, and R. J. Neufeld.** 1984. Selection of microbes producing biosurfactants in media without hydrocarbons. *J. Ferment. Technol.* **62:**311–314.
73. **Myhr, S., and T. Torsvik.** 2000. *Denitrivibrio acetiphilus*, a novel genus and species of dissimilatory nitrate-reducing bacterium isolated from an oil reservoir model column. *Int. J. Syst. Bacteriol.* **50:**1611–1619.
74. **Nagase, K., S.-T. Zhang, H. Asami, N. Yazawa, K. Fujiwara, H. Enomoto, C. Hong, and C. Liang.** 2001. Improvement of sweep efficiency by microbial EOR process in Fuyu oilfield, China. SPE 68720. Presented at the SPE Asia Pacific Oil and Gas Conference, Society of Petroleum Engineers, Richardson, Tex.
75. **Nagase, K., S.-T. Zhang, H. Asami, N. Yazawa, K. Fujiwara, H. Enomoto, C. Hong, and C. Liang.** 2003. A successful application of microbial EOR field test in Fuyu oilfield, China. SPE 75238. Presented at the SPE/DOE Improved Oil Recovery Symposium. Society of Petroleum Engineers, Richardson, Tex.
76. **Nakagawa, T., S. Nakagawa, F. Inagaki, K. Takai, and K. Horikoshi.** 2004. Phylogenetic diversity of sulfate-reducing prokaryotes in active deep-sea hydrothermal vent chimney structures. *FEMS Microbiol. Lett.* **232:** 145–152.
77. **National Association of Corrosion Engineers.** 1990. *Review of Non-Conventional and Supplemental Methods for the Detection of Sulphate-Reducing Bacteria in Oilfield Waters.* NACE International, Houston, Tex.
78. **National Association of Corrosion Engineers.** 1982. *The Role of Bacteria in the Corrosion of Oil Field Equipment.* NACE International, Houston, Tex.
79. **National Association of Corrosion Engineers.** 1991. *Standard Recommended Practice: Preparation and Installation of Corrosion Coupons and Interpretation of Test Data in Oilfield Operations.* NACE International, Houston, Tex.
80. **National Association of Corrosion Engineers.** 1994. *Standard Test Method—Field Monitoring of Bacterial Growth in Oilfield Systems TMO194-94.* NACE International, Houston, Tex.
81. **National Association of Corrosion Engineers.** 2004. *Standard Test Method—Field Monitoring of Bacterial Growth in Oilfield Systems TMO194-2004.* NACE International, Houston, Tex.
82. **Nazina, T. N., A. E. Ivanova, V. S. Ivoilov, Y. M. Miller, G. F. Kandaurova, R. R. Ibatullin, S. S. Belyaev, and M. V. Ivanov.** 1999. Results of the trial of the microbiological method for the enhancement of oil recovery at the carbonate collector of the Romashkinskoe oil field: biogeochemical and productional characteristics. *Microbiology* (Moscow) **68:**222–226.
83. **Nealson, K. H., and D. Saffarini.** 1994. Iron and manganese in anaerobic respiration: environmental significance, physiology, and regulation. *Annu. Rev. Microbiol.* **48:**311–343.
84. **Nelson, L., and D. R. Schneider.** 1993. Six years of paraffin control and enhanced oil recovery with the microbial product, Para-BacTM. *Dev. Pet. Sci.* **39:**355–362.
85. **Nemati, M., E. A. Greene, and G. Voordouw.** 2005. Permeability profile modification using bacterially formed calcium carbonate: comparison with enzymic option. *Proc. Biochem.* **40:**925–933.
86. **Obuekwe, C. O., D. W. S. Westlake, and F. D. Cook.** 1983. Corroison of Pembina crude oil pipeline. The origin and mode of formation of hydrogen sulfide. *Eur. J. Appl. Microbiol. Biotechnol.* **17:**173–177.
87. **Odom, J. M.** 1993. Industrial and environmental activities of sulfate-reducing bacteria, p. 189–209. *In* J. M. Odom and J. R. Singleton (ed.), *The Sulfate-Reducing Bacteria: Contemporary Perspectives.* Springer-Verlag, New York, N.Y.
88. **Odom, J. M., K. Jessie, E. Knodel, and M. Emptage.** 1991. Immunological cross-reactivities of adenosine 5′-phosphosulfate reductases from sulfate-reducing and sulfide-oxidizing bacteria. *Appl. Environ. Microbiol.* **57:** 727–733.
89. **Ollivier, B., and M. Magot (ed.).** 2005. *Petroleum Microbiology*. ASM Press, Washington, D.C.
90. **Orphan, V. J., L. T. Taylor, D. Hafenbradl, and E. F. Delong.** 2000. Culture-dependent and culture-independent characterization of microbial assemblages associated with high-temperature petroleum reservoirs. *Appl. Environ. Microbiol.* **66:**700–711.
91. **Patton, C. C.** 1991. *Applied Water Technology*. Campbell Petroleum Series, Norman, Okla.
92. **Pfiffner, S. M., M. J. McInerney, G. E. Jenneman, and R. M. Knapp.** 1986. Isolation of halotolerant, thermotolerant, facultative polymer-producing bacteria and characterization of the exopolymer. *Appl. Environ. Microbiol.* **51:** 1224–1229.
93. **Pope, D. H.** 1991. *Microbiologically Influenced Corrosion in the Natural Gas Industry.* Bioindustrial Technologies, Inc., Georgetown, Tex.
94. **Posfai, M., M. K. Cziner, E. Márton, P. Márton, D. R. Buseck, R. R. B. Frankel, and D. A. Bazylinski.** 2001. Crystal-size distributions and possible biogenic origin of Fe sulfides. *Eur. J. Mineral.* **13:**691–703.
95. **Rabus, R.** 2005. Biodegradation of hydrocarbons under anoxic conditions, p. 277–299. *In* B. Ollivier and M. Magot (ed.), *Petroleum Microbiology*. ASM Press, Washington, D.C.
96. **Rabus, R., M. Fukui, H. Wilkes, and F. Widdel.** 1996. Degradative capacities and 16S rRNA-targeted whole-cell hybridization of sulfate-reducing bacteria in an anaerobic enrichment culture utilizing alkylbenzenes from crude oil. *Appl. Environ. Microbiol.* **62:**3605–3613.
97. **Raiders, R. A., R. M. Knapp, and M. J. McInerney.** 1989. Microbial selective plugging and enhanced oil recovery. *J. Ind. Microbiol.* **4:**215–230.

98. **Raiders, R. A., M. J. McInerney, D. E. Revus, H. M. Torbati, R. M. Knapp, and G. E. Jenneman.** 1986. Selectivity and depth of microbial plugging in Berea sandstone cores. *J. Ind. Microbiol.* **1:**195–203.
99. **Reinsel, M. A., J. T. Sears, P. S. Stewart, and M. J. McInerney.** 1996. Control of microbial souring by nitrate, nitrite, or glutaraldehyde injection in a sandstone column. *J. Ind. Microbiol.* **17:**128–136.
100. **Rosenberg, E., M. Rosenberg, and D. L. Gutnick.** 1983. Adherence of bacteria to hydrocarbons, p. 20–28. *In* E. C. Donaldson and J. B. Clark (ed.), *Proceedings of 1982 International Conference on Microbial Enhancement Oil Recovery*. CONF-8205140. U.S. Department of Energy, Bartlesville, Okla.
101. **Rosenberg, E., A. Zuckerberg, C. Rubinovitz, and D. L. Gutnick.** 1979. Emulsifier of *Arthrobacter* RAG-1: isolation and emulsifying properties. *Appl. Environ. Microbiol.* **37:**402–408.
102. **Ruseska, I., J. Robbins, J. W. Costerton, and E. S. Kashen.** 1982. Biocide testing against corrosion-causing oil field bacteria. *Oil Gas J.* **80:**253–264.
103. **Scott, P. J. B.** 2004. Part 2: expert consensus on MIC: failure analysis and control. *Mater. Perform.* **43:**46–50.
104. **Semple, K. M., and D. W. S. Westlake.** 1987. Characterization of iron-reducing *Alteromonas putrefaciens* strains from oil field fluids. *Can. J. Microbiol.* **33:** 366–371.
105. **Smith, J. S., and J. D. A. Miller.** 1975. Nature of sulphides and their corrosive effect on ferrous metals: a review. *Br. Corros. J.* **10:**136–143.
106. **Smith, S. N.** 2003. Corrosion product analysis in oil and gas pipelines. *Mater. Perform.* **42:**44–47.
107. **Stetter, K. O., R. Huber, E. Blöchl, M. Kurr, R. D. Eden, M. Fielder, H. Cash, and I. Vance.** 1993. Hyperthermophilic archaea are thriving in deep North Sea and Alaskan oil reservoirs. *Nature* **365:**743–745.
108. **Strappa, L. A., J. P. DeLucia, M. A. Maure, and M. L. Lopez Llopiz.** 2004. A novel and successful MEOR pilot project in a strong water-drive reservoir. SPE 89456. *In Proceedings of the SPE/DOE Improved Oil Recovery Symposium*. Society of Petroleum Engineers, Richardson, Tex.
109. **Streeb, L. P., and F. G. Brown.** 1992. MEOR-Altamount/Bluebell field project. SPE 24334. *In Proceedings of the SPE Rocky Mountain Regional Meeting*. Society of Petroleum Engineers, Richardson, Tex.
110. **Sturman, P. J., and D. M. Goeres.** 1999. Control of hydrogen sulfide in oil and gas wells with nitrite injection. SPE 56772. *In Proceedings of the SPE Annual Technical Conference*. Society of Petroleum Engineers, Richardson, Tex.
111. **Sunde, E., and T. Torsvik. 2005.** Microbial control of hydrogen sulfide production in oil reservoirs, p. 201–213. *In* B. Ollivier and M. Magot (ed.), *Petroleum Microbiology*. ASM Press, Washington, D.C.
112. **Tako, M., and S. Nakamura.** 1984. Rheological properties of deacetylated xanthan in aqueous media. *Agric. Biol. Chem.* **48:**2987–2993.
113. **Tanner, R. S.** 1989. Monitoring sulfate-reducing bacteria: comparison of enumeration media. *J. Microbiol. Methods* **10:**83–90.
114. **Tanner, R. S., E. O. Udegbunam, M. J. McInerney, and R. M. Knapp.** 1991. Microbially enhanced oil recovery from carbonate reservoirs. *Geomicrobiol. J.* **9:**169–195.
115. **Telang, A. J., S. Ebert, J. M. Foght, D. W. S. Westlake, G. E. Jenneman, D. Gevertz, and G. Voordouw.** 1997. Effect of nitrate injection on the microbial community in an oil field as monitored by reverse sample genome probing. *Appl. Environ. Microbiol.* **63:**1785–1793.
116. **Telang, A. J., G. E. Jenneman, and G. Voordouw.** 1999. Sulfur cycling in mixed cultures of sulfide-oxidizing and sulfate- or sulfur-reducing oil field bacteria. *Can. J. Microbiol.* **45:**905–913.
117. **Tempest, D. W., and J. T. M. Wouters.** 1981. Properties and performance of microorganisms in chemostat culture. *Enzyme Microb. Technol.* **3:**283–290.
118. **Thomas, C. P., G. A. Bala, and M. L. Duvall.** 1991. Surfactant-based enhanced oil recovery mediated by naturally occurring microorganisms. SPE 22844. *SPE Reservoir Eng.* **8:**285–291.
119. **Thorstenson, T., B. Bodtker, E. Sunde, and J. Beeder.** 2002. Biocide replacement by nitrate in sea water injection systems. Paper 02033. Presented at Corrosion 2002, NACE International, Houston, Tex.
120. **Trebbau, G., G. J. Nunez, R. L. Caira, N. Y. Molina, L. C. Entzeroth, and D. R. Schneider.** 1999. Microbial stimulation of Lake Maracaibo oil wells. SPE 56503. *In Proceedings of the Annual Technical Conference and Exhibition*. Society of Petroleum Engineers, Richardson, Tex.
121. **Trebbau de Acevedo, G., and M. J. McInerney.** 1996. Emulsifying activity in thermophilic and extremely thermophilic microorganisms. *J. Ind. Microbiol.* **16:**1–7.
122. **Ulrich, G. A., L. R. Krumholz, and J. M. Suflita.** 1997. A rapid method for estimating sulfate reduction activity and quantifying inorganic sulfides. *Appl. Environ. Microbiol.* **63:**1627–1630.
123. **Vance, I., and D. R. Thrasher.** 2005. Reservoir souring: mechanisms and prevention, p. 123–142. *In* B. Ollivier and M. Magot (ed.), *Petroleum Microbiology*. ASM Press, Washington, D.C.
124. **Venure, C. C. E. J.** 1987. *Review of Current Practices for Monitoring Bacterial Growth in Oilfield Systems*. Corrosion Control Engineering Joint Venture, Birmingham, United Kingdom.
125. **Videla, H. A.** 1996. *Manual of Biocorrosion*. Lewis Publishers, Boca Raton, Fla.
126. **Voordouw, G., S. M. Armstrong, M. F. Reimer, B. Fouts, A. J. Telang, Y. Shen, and D. Gevertz.** 1996. Characterization of 16S rRNA genes from oil field microbial communities indicates the presence of a variety of sulfate-reducing, fermentative, and sulfide-oxidizing bacteria. *Appl. Environ. Microbiol.* **62:**1623–1629.
127. **Wagner, M., D. Lungerhausen, H. Murtada, and G. Rosenthal.** 1995. Development and application of a new biotechnology of molasses in-situ method: detailed evaluation for selected wells in the Romashkino carbonate reservoir, p. 153–174. *In* R. S. Bryant and K. L. Sublette (ed.), *The Fifth International Conference on Microbial Enhanced Oil Recovery and Related Problems for Solving Environmental Problems*. CONF-9509173. National Technical Information Service, Springfield, Va.
128. **Watanabe, K., Y. Kodama, and N. Kaku.** 2002. Diversity and abundance of bacteria in an underground oil-storage cavity. *BMC Microbiol.* **2:**23.
129. **Watanabe, K., K. Watanabe, Y. Kodama, K. Syutsubo, and S. Harayama.** 2000. Molecular characterization of bacterial populations in petroleum-contaminated groundwater discharged from underground crude-oil-storage cavities. *Appl. Environ. Microbiol.* **66:**4803–4809.
130. **Willhelms, A., S. R. Larter, I. Head, P. Farrimond, R. di-Primio, and C. Zwach.** 2001. Biodegradation of oil in uplifted basins prevented by deep-burial sterilization. *Nature* **411:**1034–1036.
131. **Willumsen, P. A., and U. Karlson.** 1997. Screening of bacteria, isolated from PAH-contaminated soils, for production of biosurfactants and bioemulsifiers. *Biodegradation* **7:**415–423.

132. **Worden, R. H., P. C. Smalley, and N. H. Oxtoby.** 1995. Gas souring by thermochemical sulfate reduction at 140°C. *Am. Assoc. Pet. Geol. Bull.* **79:**854–863.

133. **Yarbrough, H. F., and V. F. Coty.** 1983. Microbially enhanced oil recovery from the Upper Cretaceous Nacatoch Formation, Union County, Arkansas, p. 162–218. *In* E. C. Donaldson and J. B. Clark (ed.), *Proceedings of the 1982 International Conference on Microbial Enhancement of Oil Recovery*. Technology Transfer Branch, U.S. Department of Energy, Bartlesville, Okla.

134. **Youssef, N. H., K. E. Duncan, D. P. Nagle, K. N. Savage, R. M. Knapp, and M. J. McInerney.** 2004. Comparison of methods to detect biosurfactant production by diverse microorganisms. *J. Microbiol. Methods* **56:**339–347.

Placement of Drinking Water Wells and Their Protection

MARYLYNN V. YATES

72

Groundwater supplies more than 100 million Americans with their drinking water (28); in rural areas, there is an even greater reliance on groundwater, as it comprises up to 95% of the water used (2). Traditionally, it has been assumed that groundwater is safe for consumption without treatment because the soil acts as a filter to remove pollutants. As a result, private wells generally do not receive treatment (7), nor do a large number of public water supply systems. The U.S. Environmental Protection Agency (EPA) has estimated that approximately 72% of the 158,000 public water supply systems in the United States that use groundwater do not disinfect the water (25). However, the use of contaminated, untreated, or inadequately treated groundwater has been the major cause of waterborne disease outbreaks in this country since 1920 (1, 3, 4, 28). Between 1971 and 2002, 758 waterborne disease outbreaks were reported in the United States, involving over 573,000 people (1, 3, 17, 28). These data are summarized for 10-year periods in Fig. 1 (1, 3, 6, 11, 16, 17, 18, 19). The number of reported outbreaks and the number of associated cases of illness rose dramatically in the 1970s and 1980s compared with the period 1951–1970. During the period 1971–1980, an average of 32.6 outbreaks per year was reported. From 1981 through 1990, the average was 27.6 outbreaks per year, compared with averages of 11.1 and 13.1 for the periods 1951–1960 and 1961–1970, respectively. However, the number decreased again in the 1990s, to 12.2 reported outbreaks per year.

Historically, groundwater has been responsible for approximately one-half of the reported waterborne disease outbreaks in the United States (4, 5). In recent years, however, the proportion of reported outbreaks that have occurred from groundwater systems has increased dramatically, from 56% during the period 1971–1989 to 71% during the period 1990–2002 (Fig. 2) (1, 3, 17, 28). In the latest period for which data are available (2001–2002), 92% (23 of 25) of the reported outbreaks were associated with groundwater (3). From 1971 to 1996, untreated groundwater was responsible for almost one-half (47%) of the outbreaks (28). In groundwater systems, etiologic (disease-causing) agents were identified in only 37% of the outbreaks, with *Shigella* spp. and hepatitis A virus being the most commonly identified pathogens (28). In almost two-thirds (63%) of the outbreaks, no etiologic agent could be identified, and the illness was simply listed as gastroenteritis of unknown etiology. However, retrospective serological studies of outbreaks of acute nonbacterial gastroenteritis from 1976 through 1980 indicated that 42% of these outbreaks (i.e., the 62% for which no etiologic agent was identified) were caused by Norwalk virus (13). Thus, it has been suggested that Norwalk virus is responsible for approximately 23% of all reported waterborne outbreaks in the United States (15).

Microorganisms may be introduced into the subsurface environment in a variety of ways. In general, any practice that involves the application of domestic wastewater to the soil has the potential to cause microbiological contamination of groundwater. This is because the treatment processes to which the wastewater is subjected do not effect complete removal or inactivation of the disease-causing microorganisms present.

A number of studies of pathogen transport have been conducted at wastewater application sites during the last 25 years. The focus of these studies has been the transport of enteric viruses, as they are thought to have the greatest potential to be transported due to their relatively small size. Many field studies were conducted in the late 1970s and early 1980s; these studies were reviewed by Keswick and Gerba (14) and Gerba and Goyal (9).

Goyal et al. (10) isolated viruses from the groundwater beneath cropland being irrigated with sewage effluent. The burial of disposable diapers in sanitary landfills is a means by which pathogenic microorganisms in untreated human waste may be introduced into the subsurface. Vaughn et al. (32) detected viruses as far as 408 m downgradient of a landfill site in New York. Land application of treated sewage effluent for the purposes of groundwater recharge has also resulted in the introduction of viruses into the underlying groundwater (30, 31).

Recently, there has been a resurgence in studies in the United States that have monitored the transport of viruses through soil at land application sites. This increased activity is likely due to the need to augment freshwater resources in increasingly urban areas, such as those in the arid southwestern United States.

In Colton, Calif., land application and extraction of wastewater were studied as an alternative to conventional tertiary treatment (filtration and disinfection) of secondary effluent (8). At this site, unchlorinated secondary effluent was applied to infiltration basins consisting of coarse sands with clean water infiltration rates of approximately 15 m day^{-1}. Virus levels in the effluent were as high as 316 PFU

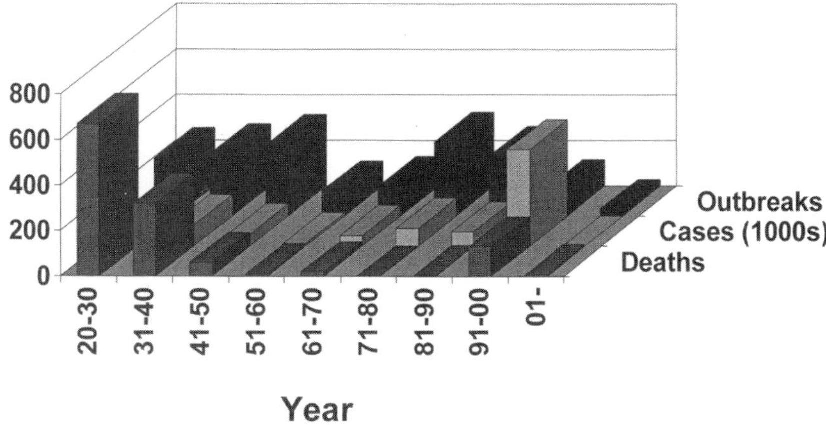

FIGURE 1 Waterborne disease outbreaks in the United States, 1920 to 2001. Data are from the Centers for Disease Control and Prevention.

per 200 liters. After infiltration through 24.4 m of soil, one sample of the extracted water was found to contain infective human enteroviruses.

Jansons et al. (12) compared the behaviors of indigenous enteroviruses and seeded polioviruses at a site in Western Australia where wastewater was used to artificially recharge the groundwater. In that study, vaccine strain polioviruses, at a concentration of 4,000 most probable number of cytopathic units per liter, were added to the infiltrating wastewater. A 1.6-log (40-fold) reduction in virus numbers was found after transport through 0.5 m of sandy soil. The viruses were isolated on a single occasion at a depth of 1.5 m; however, no viruses were detected at the 3-m sampling depth, although monitoring was conducted for 7 days. In contrast, several types of enteroviruses indigenous to the wastewater were isolated at the 3-m sampling depth. In addition, indigenous enteroviruses were isolated from a 9-m-deep well located 14 m from the recharge basin.

The removal of seeded bacteriophages during artificial recharge with wastewater has also been studied by Powelson et al. (21). At a site near Tucson, Ariz., two experimental recharge basins of 3.65 by 3.65 m were constructed for the seeding studies. Two bacteriophages, MS2 and PRD1, were added to the infiltrating wastewater at a constant concentration of 10^5 PFU ml^{-1}. The researchers found that between 36 and 99.7% of the viruses were removed after transport through 4.3 m of alluvial sand and gravel. In addition, significantly greater removal of PRD1 was observed when it was suspended in secondary effluent than when it was suspended in tertiary effluent. This may have been due to adsorption of the virus particles to particulate material in the secondary effluent, which was then filtered out by passage through the soil pores.

Yanko (33) studied the removal of indigenous wastewater coliphages in the Montebello Forebay recharge area. In this area, tertiary effluent is infiltrated into alluvial sand and gravels for the purpose of groundwater recharge. A 3-log removal within 15 ft of the surface was documented at this site. However, questions have been raised regarding the appropriateness of indigenous coliphages as surrogates for human pathogenic viruses in studies of transport through soils. Further studies by this researcher revealed that male-specific phages in the reclaimed water were not transported beyond 7.6 m of the recharge basin.

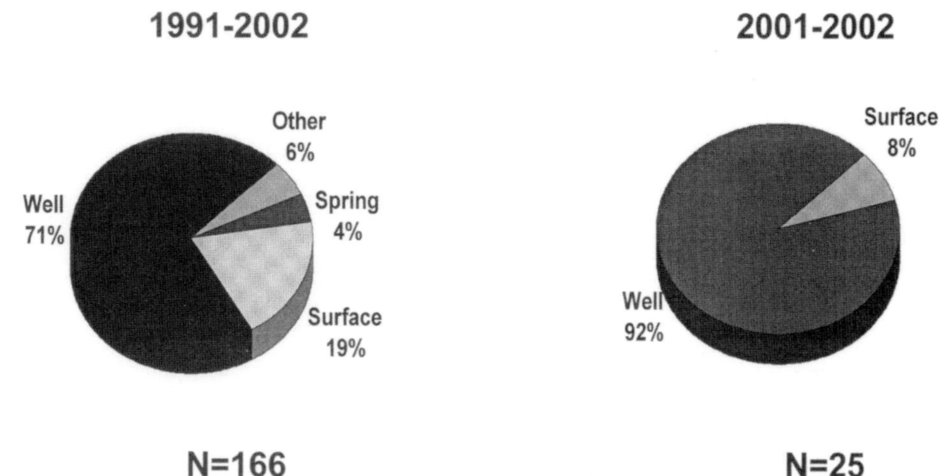

FIGURE 2 Water source of disease outbreaks in the United States, 1991 to 2002. Data are from the Centers for Disease Control and Prevention.

Recent advances in molecular techniques have enhanced the ability to monitor the transport of a variety of pathogens in the subsurface. Ongoing studies are using PCR and specific nucleic acid probes to determine whether viruses such as hepatitis A virus and rotaviruses, in addition to the traditionally studied enteroviruses, are present in groundwaters impacted by artificial recharge with treated wastewaters.

PROGRAMS TO PROTECT DRINKING WATER FROM MICROBIAL CONTAMINATION

The fact that microorganisms are responsible for numerous waterborne disease outbreaks every year led the U.S. EPA to reexamine the coliform standard, which has been used to indicate the microbiological quality of drinking water in the United States for more than 75 years. Increasing amounts of evidence collected during the past 15 to 20 years suggest that the coliform group may not be an adequate indicator of the presence of pathogenic viruses and, possibly, protozoan parasites in water. In 1985, the U.S. EPA proposed maximum contaminant level goals for viruses and *Giardia*, a protozoan parasite (23). These standards are in addition to the standard for the indicator microorganisms, total coliform bacteria. In response to the 1996 amendments to the Safe Drinking Water Act, EPA and the states have undertaken two major initiatives, Source Water Assessment and Protection Programs (SWAPPs) and the Groundwater Rule (GWR), to increase the level of public health protection in public water systems relying on groundwater sources.

GWR

The U.S. EPA proposed the GWR to reduce the public health risk associated with the consumption of fecally contaminated water containing waterborne pathogens (28). Rather than require public groundwater systems to monitor source water for pathogenic microorganisms, the EPA proposed a strategy that consists of five elements. These elements are as follows:

1. Sanitary surveys that evaluate seven elements (source, treatment, finished water storage, pumps/pump facilities and controls, monitoring/reporting/data verification, water system management/operations, and operator compliance with state requirements) of groundwater systems and require the identification of significant deficiencies
2. Hydrogeologic assessments to identify wells sensitive to fecal contamination
3. Source water monitoring (for indicator microorganisms) for systems drawing from sensitive wells without treatment or with other indications of risk
4. A requirement to correct all significant deficiencies and fecal contamination (by eliminating the contamination source, correcting the deficiency, providing an alternative source of water, or providing treatment that achieves at least 99.99% removal and/or inactivation of viruses)
5. Compliance monitoring to ensure that disinfection is reliably operated where it is used

SWAPPs

SWAPPs require states to delineate and assess the areas of land that contribute to public water systems using both surface and groundwaters. An integral part of these programs is an analysis of the susceptibilities of these systems to chemical and microbial contamination. As outlined in the U.S. EPA's *State Source Water Assessment and Protection Program Final Guidance* (27), susceptibility determinations should take into account hydrologic and hydrogeologic factors, properties of the contaminants (e.g., mobility and toxicity), and characteristics of the potential sources of contaminants (e.g., proximity, probability of release, and presence of best management practices).

Specifically, a SWAPP must do the following:

1. Delineate the boundaries of the areas providing source waters for public water systems, using all reasonably available hydrogeologic information
2. Identify, to the extent practical, the origins of regulated and certain unregulated contaminants in the delineated area
3. Determine the susceptibility of public water systems to such contaminants.

Each of these elements will be discussed in more detail as they relate to public water systems that use groundwater as their source.

Delineation of Source Water Protection Areas

In developing wellhead protection areas (WHPAs), states must develop overall protection goals that (i) describe remedial actions to take place in the event of an unexpected release in a prescribed zone, (ii) determine the time (and associated distance) required for attenuation of the contaminant concentrations, and (iii) provide safety in the zone of contribution to a water supply. The conceptual standards on which these zones are based are the following criteria (26):

1. The distance criterion is the most direct way to measure the distance from the well to the point of concern. Distance establishes the simple radius from the groundwater source to the well; it is used to establish setback rules and for general microbial protection. A first step in delineation, it is usually selected on nontechnical grounds.
2. Drawdown establishes the extent to which pumping lowers the water table of an unconfined aquifer or the potentiometric surface of a confined aquifer. It establishes the zone of influence, or cone of depression (Fig. 3). It accelerates the migration toward the well and may vary from a few meters to tens of kilometers.
3. Time of travel is the maximum time it takes a contaminant to reach the well. It incorporates a comprehensive evaluation of the physical processes of contamination. Sometimes arrival times can be determined more accurately than contaminant concentrations.
4. Flow boundaries are especially important when time of travel to the boundaries is brief. This criterion uses groundwater divides or other physical or hydrologic features that control groundwater flow.
5. The assimilative capacity criterion applies to how the saturated or unsaturated zones function to help attenuate the concentration of contaminants. It can be a measure of safety, but it is too complex for most WHPA delineations.

After acceptable thresholds for these criteria are determined, the criteria are applied in the development of WHPA boundaries.

The delineation of source water protection areas must be conducted in accordance with the methods described in the state's WHPA program. There are several methods that can be used to establish placement of drinking water wells to minimize microbial contamination. The choice of the method to be used when establishing WHPAs depends on a variety of factors, including the vulnerability of the drinking water supply, the importance of the groundwater as a

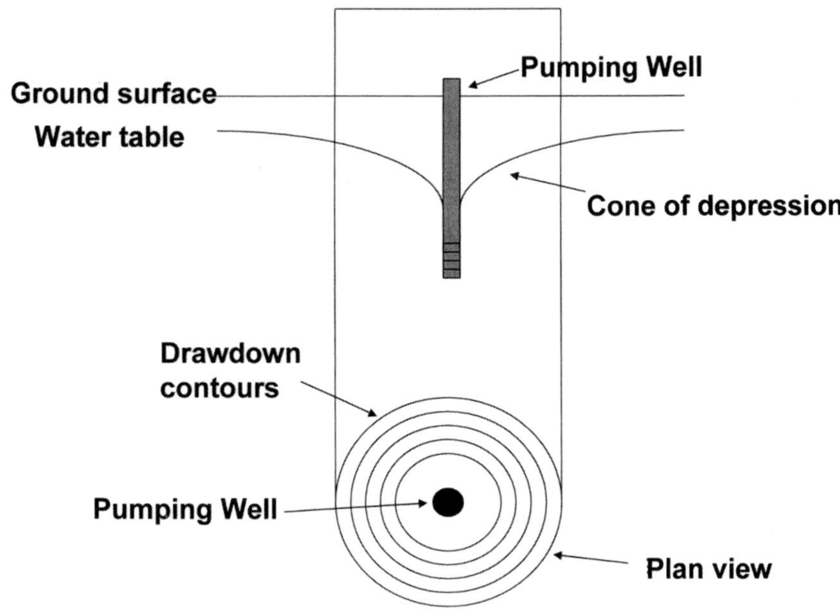

FIGURE 3 Zone of influence of a pumping well.

water supply, the number and types of contamination sources in the area, the availability of pertinent hydrogeologic information, and the availability of technical personnel and computer equipment. The six most common delineation methods, listed in order of increasing technical complexity, are as follows:

1. Arbitrary-fixed-radius method
2. Calculated-fixed-radius method
3. Simplified variable shape method
4. Analytical method
5. Hydrogeologic mapping
6. Numerical transport models

These methods, which are discussed in detail by the U.S. EPA (24), are described briefly below.

Arbitrary-Fixed-Radius Method

Using an arbitrary fixed radius is the simplest method of delineating a WHPA. A circle with a specified radius is drawn around a well, and no sources of potential contamination may be located within the circle (Fig. 4). This method is widely employed by states and local communities when establishing placement requirements for septic tank or other onsite wastewater treatment systems. The choice of radius for septic tank setback distances varies widely, as shown in Table 1, from 7.62 to 152.4 m. The numerical value of the radius may be based on general hydrogeologic considerations as well as professional judgment. This method is very inexpensive and rapid and requires very little technical expertise to develop. It is also very simple to enforce. However, depending on the variation in soil conditions and source characteristics, it may not provide consistent protection throughout an area. In addition, the radius generally represents an average protective distance. Thus, areas upgradient of the well may be underprotected from contamination, while those downgradient may be overprotected.

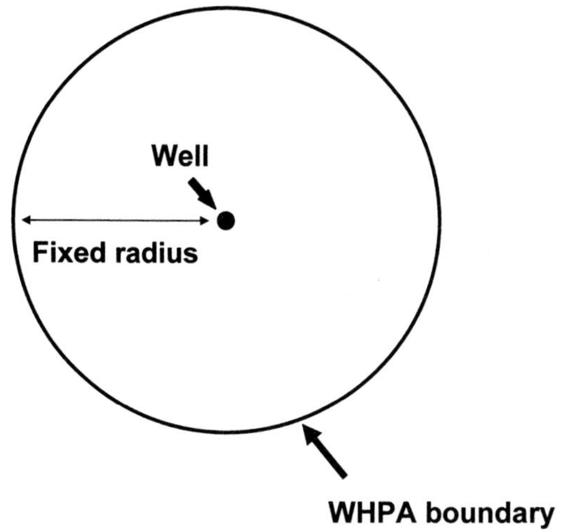

FIGURE 4 Fixed-radius method for delineating WHPAs.

Calculated-Fixed-Radius Method

The calculated-fixed-radius method is also a simple method in which a circle of a fixed radius is drawn around a well; however, the radius is calculated based on certain criteria.

Source-driven radius. The radius of setback from a well can be based on the type of contaminant source. Many states have several setback requirements depending on the contaminant source. For example, California requires a 15.2-m setback from a septic system, a 30.5-m setback from a drain field, and a 45.7-m setback from a cesspool (Table 1). In this case, a number of concentric circles would be drawn around the well, and the specified contaminant sources would not be allowed in each of the rings.

Well-driven radius. Some setback distances are determined based on the well characteristics, such as the well discharge (pumping) rate or the type of well (Table 1). In the case of setting setback distances based on pumping rate, the time of travel of the contaminant from the source to the water supply well is being considered. Larger radii require

TABLE 1 State requirements for setback distances from public water supply wells

State	Setback radius (m)	Comment
Alabama	30.5	From well to septic system
Alaska	30.5–71	From community sewer lines
	45.7–71	From wastewater treatment and disposal systems
Arizona	15.2	From sewer
	30.5	From septic disposal field
California	15.2	From sewer or watertight septic tank
	30.5	From subsurface sewage leaching field
	45.7	From cesspool or seepage pit
Connecticut	22.9	Discharge of <10 gpm
	45.7	Discharge of 10–50 gpm
	71	Discharge of >50 gpm
Delaware	30.5	All public water supply wells
Florida	30.5	From sanitary hazards
Georgia	15.2	No state statute in effect since 1986
Idaho	30.5	From septic drain field
Illinois	15.2	From septic tank
Indiana	71	From septic system
Iowa	30.5	From deep well
	71	From shallow well
Kansas	30.5	From all pollution sources
Kentucky	15.2	From septic system
Maryland	15.2	Confined aquifers
	30.5	Unconfined aquifers
Maine	91.4	All water supply wells
Massachusetts	15.2–121.9	Discharge of <69 gpm
	76.2	Tubular wellfields, discharge of >69 gpm
	121.9	Gravel-packed wells, discharge of >69 gpm
Michigan	22.9–71	From septic tanks
	243.8–609.6	From known major sources of contamination
Minnesota	15.2	From septic system
Mississippi		Shall not be located so as to create a hazard
Missouri	30.5	Required, 300 ft recommended
Montana		No specified distance
Nebraska	152.4	From septic system
Nevada	71	From septic system
New Hampshire	22.9	Discharge of 0.28 gpm
	121.9	Supplying >1,000 persons or having fire protection capability or supplying 25 to 1,000 persons and having no fire protection capability
	38.1	Discharge of 0.56 gpm
	45.7	Discharge of 0.84 gpm
	53.3	Discharge of 1.12 gpm
	71	Discharge of >1.39 gpm
North Carolina	45.7	Private wells
North Dakota	30.5	From septic system
Ohio	15.2	From known contamination source
Oregon	15.2	From septic tank
	30.5	From sewage disposal field or cesspool
Pennsylvania	30.5	Individual water supply wells
Rhode Island	71	Bedrock or driven wells
	121.9	Gravel-packed wells

(Continued on next page)

TABLE 1 (*Continued*)

State	Setback radius (m)	Comment
South Carolina	7.6	No written rule
South Dakota	15.2	From septic tank
	30.5	From drain field
Utah	3.0	From sewer lines for deep wells
	91.4	From specially constructed sewers for shallow wells
	457.2	Wellhead zone
Vermont	22.9–71	Transient, noncommunity wells
	30.5–71	Nontransient, noncommunity wells
	38.1–71	Community public water supply wells
Washington	30.5	From septic tank
	305	From sewage lagoon
West Virginia		Determined by director of State Department of Health
	15.2	Nonpublic water supply wells
Wisconsin	7.6	From septic tanks
	15.2	From drain fields
Wyoming		Documentation showing that discharge will not affect water quality is required

longer travel times; therefore, more inactivation and removal of pathogenic microorganisms can occur. In addition, dilution and dispersion will act to decrease microbial concentrations.

Different setback distances can also be specified based on the presence of a confining layer between the soil surface and the aquifer from which the well draws water. For example, Maryland requires a 30.5-m setback for wells in unconfined aquifers but only a 15.2-m setback for wells in confined aquifers. While requiring more information to develop than the arbitrary-fixed-radius method, this method has the potential to allow for some variation in local conditions. However, because a circle of fixed radius is drawn around the well, the upgradient area is likely to be underprotected and the downgradient area is likely to be overprotected.

Simplified Variable Shapes

The simplified variable shape method is based on the generation of standardized forms by using analytical models. The forms are generated from information about the hydrogeologic conditions and pumping rate of the well in addition to information about the direction of groundwater flow at the site. The downgradient boundary of the WHPA is calculated by using the pumping rate of the well. The upgradient boundary is then calculated by using the time of travel for the contaminant. The orientation of the WHPA around the well is based on the direction of groundwater flow (Fig. 5).

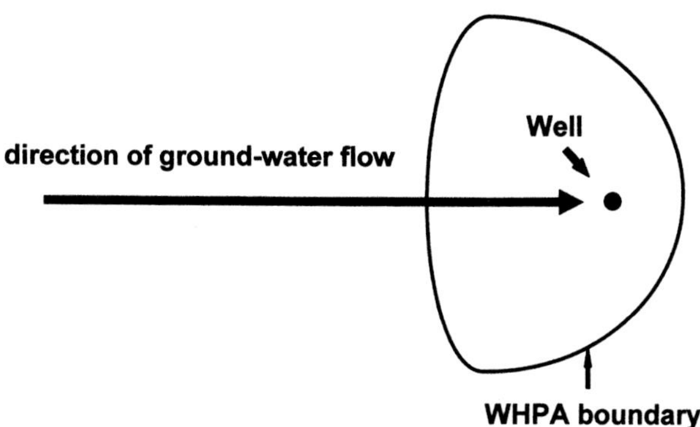

FIGURE 5 Simple variable shape method for delineating WHPAs.

An advantage of this method over the first two is that all areas of the WHPA should be equally protected. The additional information required to develop variable shapes is generally available or can be obtained for a relatively modest cost. However, it is still a very simple method that does not take into account many of the physical heterogeneities that occur in subsurface geologic materials.

Analytical Method

In the analytical method, WHPAs are delineated by using mathematical equations that describe the groundwater flow and contaminant transport in the area of the well. The uniform flow equation (22) can be used to determine the areal extent of the aquifer from which the pumping well draws water, called the zone of contribution to the well (Fig. 6). Calculation of the zone of contribution requires knowledge of several hydrogeologic parameters, including the aquifer transmissivity, porosity, hydraulic gradient, hydraulic conductivity, and saturated thickness. With this equation, the downgradient boundary and width of the WHPA can be delineated. The upgradient boundary can be calculated by using a time-of-contaminant-travel criterion or some type of flow boundary criterion, such as the presence of a groundwater divide.

The analytical equations required by this method are generally simple to use for trained hydrologists and civil and environmental engineers. This method has an advantage over the previously discussed methods in that it does take into account more site-specific hydrogeologic information. However, most analytical methods do not take into account hydrologic boundaries (e.g., streams), subsurface heterogeneities, and nonuniform rainfall or evapotranspiration.

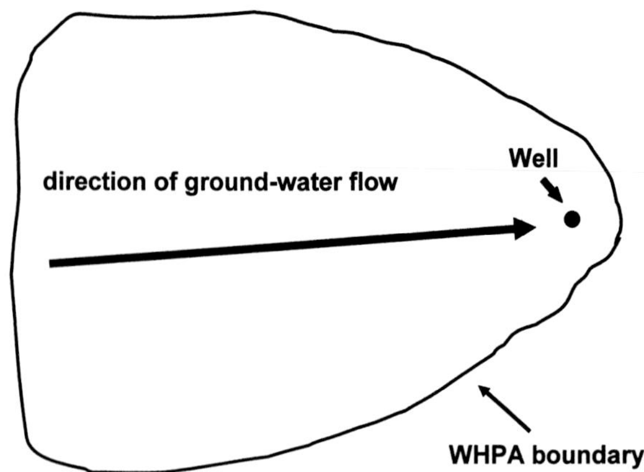

FIGURE 6 Example of a WHPA delineated by use of an analytical method.

Hydrogeologic Mapping

In some areas, the hydrogeologic characteristics can be mapped and used to delineate WHPAs. For example, topographic analysis of drainage basin divides can be combined with water table elevation data to delineate aquifer recharge areas (Fig. 7). Potential contaminant sources in the recharge area can then be sited by using time-of-travel calculations to prevent groundwater contamination. This method is most applicable to areas having flow boundaries located near the ground surface and to areas with fractured bedrock or karst formations. The use of this method requires highly trained personnel with expertise in geologic and geomorphologic mapping.

Numerical Flow and Transport Models

The most technically complex method of delineating WHPAs involves the use of computer models that solve equations for groundwater flow and/or solute transport by using numerical methods. A large number of numerical models have been developed to predict groundwater contamination by a variety of chemicals. Sixty-four of these models were reviewed by van der Heijde and Beljin (29).

The information required to use numerical models varies depending on the specific model chosen but generally includes aquifer permeability, porosity, specific yield, saturated thickness, recharge rates, aquifer geometries, and the location of hydrogeologic boundaries. If the model has contaminant transport prediction capabilities in addition to groundwater flow prediction, several other parameters, such as dispersivity, the contaminant inactivation rate, and the contaminant adsorption coefficient, must also be known. An example of the type of delineation that may result from the use of a numerical method is shown in Fig. 8.

Although there are many models that can be used to delineate WHPAs, most of them were developed to protect wells from chemical contamination. To date, only one groundwater flow and contaminant transport model that has a user-friendly interface has been developed to delineate WHPAs for microorganisms. This model, VIRALT (20), was developed for the U.S. EPA to calculate virus transport to a pumping well from a contaminant source such as a septic tank.

Source Inventories within Delineated Source Water Protection Areas

After delineating WHPAs, the state must inventory the area for potential sources of contamination by the contaminants of concern.

Contaminants of Concern

The contaminants of concern must include those raw water contaminants regulated under the Safe Drinking Water Act (i.e., those contaminants with a maximum contaminant level, contaminants regulated under the Surface Water Treatment Rule, and the microorganism *Cryptosporidium*). States may also include other contaminants that are not federally regulated but are of local or regional concern. In addition, because the GWR will address microorganisms in addition to *Cryptosporidium* (e.g., viruses), the EPA recommends that states include potential sources of these microorganisms in their inventory.

Significant Potential Sources

The WHPA must also indicate the potential sources of the contaminants of concern that will be inventoried. The EPA has provided a list of some potential sources (27); an edited version of that list specifically focused on sources of microbial contaminants is shown in Table 2.

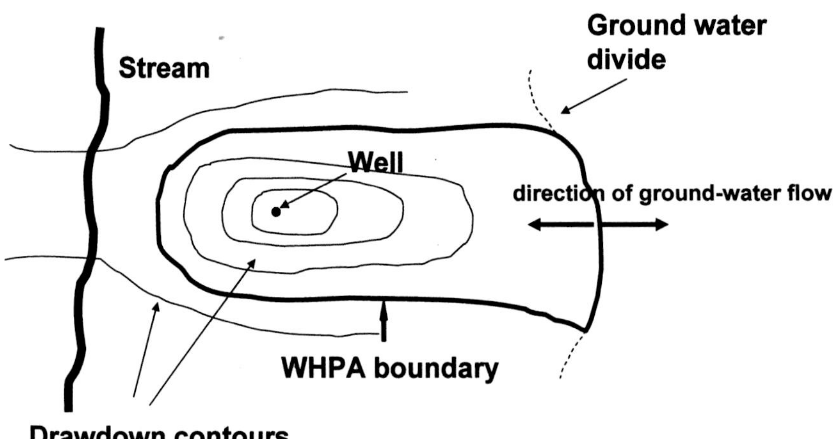

FIGURE 7 Example of a WHPA delineated by hydrogeologic mapping.

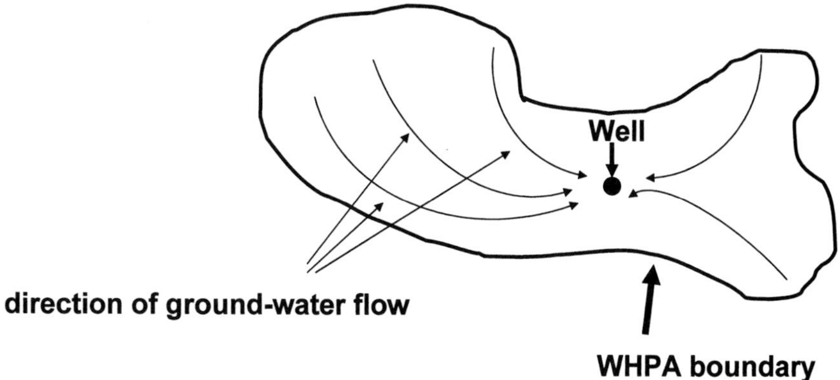

FIGURE 8 Example of a WHPA delineated by use of a numerical model.

Determination of Public Water Supply Susceptibility

Simply identifying potential sources of contamination in a contributing area does not indicate which of them may or may not present threats to drinking water or which are priorities to manage in order to protect drinking water. An analysis of hydrogeology, an understanding of the contaminants and the factors that control their fate and transport in specific environments, and an analysis of the effectiveness of existing prevention and mitigation measures are essential so that states can apply the assessment results to source water protection. Therefore, the EPA requires that the SWAPP conduct an analysis of the risks from the identified sources for each assessment in a source water protection area. The level of detail will depend on the state's SWAPP approach.

In order to comply with these requirements, the third component of the SWAPP is the determination of the sus-

TABLE 2 Examples of potential sources of microbial contaminants found in wellhead protection areas[a]

Contaminant source
Agriculture
Crop production if crops are irrigated with recycled water or grown on biosolid-amended soil
Pasture and/or range grazing (riparian and/or upland)
Range grazing (riparian and/or upland)
Intensive, concentrated, and/or confined animal feeding operations
Aquaculture
Combined sewer overflow
Domestic wastewater lagoon
Land disposal
Sludge
Wastewater
Landfills
Inappropriate waste disposal, wildcat dumping
Onsite wastewater systems (septic tanks)
Septage disposal
Municipal point sources
Major and minor municipal point sources (dry and/or wet weather discharges)
Package plants (small flows)
Natural sources (e.g., wildlife)
Recreation and tourism activities
Golf courses if irrigated with recycled water
In-water and/or on-land releases from boating
Sewer lines (leaking)
Urban runoff, storm sewers
Permitted industrial and/or nonindustrial runoff
Other urban runoff
Illicit connections, illegal hookups, dry weather flows
Waste storage, storage tank leaks (above ground)

[a]Modified from reference 27.

ceptibility of the groundwater supply to contamination. In other words, the program must determine the potential for a well to draw water contaminated by the inventoried sources at concentrations that would cause concern. Such a determination would take into account hydrogeologic factors, inherent characteristics of the contaminants (e.g., toxicity, environmental fate, and transport), and characteristics of the potential source of the contaminant (location, likelihood of release, and effectiveness of mitigation measures). While the determination of susceptibility does not necessarily require modeling or monitoring of the source waters to determine which potential sources of contamination are significant, the EPA encourages states to undertake such modeling and monitoring, as the accuracy of the assessment, and therefore the likelihood of protecting the public health from microbial disease, will be greater if more detailed studies are performed.

Examples of Wellhead Protection Regulations

In the United States, there are no federally mandated time-of-travel or separation distance requirements between wells and potential sources of microbial contamination. Individual states, and in some cases, individual communities have established different programs to protect wellheads from microbial contamination. In some states, specific time-of-travel criteria have been established. For example, in Michigan, a WHPA is defined as "the surface and subsurface areas surrounding a water well, or well field, which supplies a public water system, and through which contaminants are reasonably likely to move toward and reach the water well, or well field within a 10-year time of travel" (http://www.orcbs.msu.edu/environ/programs_guidelines/wellhead/wh_11delineation.htm).

California has adopted a more complex system, defining three different zones of protection around wellheads (http://www.dhs.ca.gov/ps/ddwem/dwsap/dwsapdoc/dwsap06.htm). Zone A is the microbial and direct chemical contamination zone, the purpose of which is to protect potable groundwater supplies from "viral, microbial, and direct chemical contamination." This zone is defined by calculating the 2-year time-of-travel distance to the well. The stated rationale for using 2 years is because existing research shows that the survival of bacteria and viruses in the subsurface is less than 2 years and because this is consistent with the proposed GWR. Recommended minimum radii are 600 ft for all potable groundwater sources in porous-medium aquifers and 900 ft for sources in fractured rock aquifers. These distances are believed to be sufficiently conservative (from a public health protection perspective) for protection from microbial contaminants. However, depending on the delineation method used to calculate the time of travel, other minimum radii of protection zones may be specified, as shown in Table 3.

In Oregon, both the time of travel and the method used to calculate the time of travel (and thus the size of the protection zones) vary based on the size of the community served by the well (http://www.deq.state.or.us/wq/dwp/DWPAreaDelineationsGW.htm).

For systems serving populations of <500 people, the time of travel used is 15 years because the calculated-fixed-radius method, which is relatively imprecise, is used to calculate the time of travel. Within the 15-year protection zone, two subzones, one representing a 2-year and the other representing a 5-year time of travel, are calculated. According to the state, "The area within the 2-year time-of-travel represents the 'hot zone' for the area: potential contaminant sources within this area will probably need extra attention because of their proximity to the well. The 2-year TOT also provides the outer limit from which microbial sources are likely to impact the well. The goal of the subzones is to provide the community with better data from which to build protection strategies."

For systems serving more than 500 people, the time of travel is set at 10 years, and the state requires the development of a conceptual model of the hydrogeology of the system. This model is a three-dimensional representation that includes areas of recharge and discharge, the distribution of the hydraulic head, and the location and extent of the aquifer and any aquitards. Depending on the size of the system, the sources of the data that can be used to develop this model vary. For systems serving fewer than 3,300 people, well logs or regional information may be used. Medium-sized systems (serving 3,300 to 50,000 people) must use constant-rate aquifer tests and field measurements of the hydraulic head to characterize the hydrogeology of the system. The largest systems, serving more than 50,000 people, must use a model that takes into consideration the heterogeneity of the aquifer.

Relationship between the SWAPP and the GWR

There are several components of the proposed GWR that require similar assessments of groundwater vulnerability or sensitivity to microbial contamination to those performed by the SWAPP. Specifically, for the proposed GWR, the EPA considered strategies to control risk from microbial contamination as an alternative to disinfection. These strategies include a well vulnerability determination that evaluates, among other things, the following wellhead protection-related activities (25):

1. Delineation of a microbial protection area (e.g., is it based on scientifically defensible microbial inactivation rates or an equally protective method?)
2. A thorough inventory of significant potential microbial sources of contamination within the delineated area (e.g., are sources present and what are the relative risks they represent?)
3. An assessment of the hydrogeologic conditions that apply to a public water system's source waters (e.g., is the system in an area of intrinsic groundwater sensitivity?)
4. The relative effectiveness of microbial source management controls (e.g., do established microbial setback distances for sources provide protection equivalent to the delineated microbial protection area and has the effectiveness of specific best management practices for minimizing microbial contamination been demonstrated in the field?)

Because of the similarities between the two programs, it has been proposed by the Association of State Drinking Water Administrators that functional equivalency be established between the GWR Sensitivity Analysis and the SWAPP Susceptibility Assessments. In some cases, information on a single factor may be sufficient to demonstrate that a potentially sensitive groundwater source is protected. In most cases, a combination of factors will need to be addressed to determine the susceptibilities of public water supply wells to sources of pathogens. The factors are as follows:

1. Assessing the effects of well construction, well integrity, and site conditions
2. Determining the presence of a geological barrier to pathogen movement
3. Reviewing an inventory of pathogen sources within the source water protection area

TABLE 3 Delineation methods, types of systems that may use particular methods, minimum data that are required, and minimum radii of zones[c]

Delineation method	Type of system that may use method	Minimum data required	Minimum radius of zone[d]
Arbitrary fixed radius	Noncommunity systems	Location of source	See data below for porous media and fractured rock
Calculated fixed radius (porous media)	All[a]	Location of source, pumping capacity of well (gpm), screened interval of well (indicate method used to estimate), effective porosity (indicate method used to estimate)	A = 600 ft, B5 = 1,000 ft, B10 = 1,500 ft
Calculated fixed radius (fractured rock) (note that fractured rock increases size by 50%)	All[a]	Location, pumping capacity, screened interval, effective porosity	A = 900 ft, B5 = 1,500 ft, B10 = 2,250 ft
Modified calculated fixed radius	All[a]	Location, pumping capacity, screened interval, effective porosity, direction of groundwater flow	A = 600 ft, B5 = 1,000 ft, B10 = 1,500 ft
Analytical methods	All	Location, capacity, screened interval, effective porosity, hydraulic conductivity, hydraulic gradient, direction of groundwater flow	No minimums[b]
Hydrogeologic mapping	All	Hydrogeologic parameters, lithology, groundwater level	No minimums[b]
Numeric flow and transport models	All	Hydrogeologic parameters, recharge rates, aquifer geometry, hydrologic boundaries	No minimums[b]

[a]Systems with detailed hydrogeologic data are encouraged to conduct more sophisticated analyses.

[b]Systems using more sophisticated methods are encouraged to compare the sizes of zones to minimum sizes derived by simpler methods to assist in the review of the delineation.

[c]Information from http://www.dhs.ca.gov/ps/ddwem/dwsap/dwsapdoc/dwsap06.htm.

[d]A, 2-year time-of-travel zone (to prevent microbiological contamination). B5, encompasses the area between the 2- and 5-year time of travel (to prevent chemical contamination). This zone provides for more response time for chemical spills than zone A. B10, encompasses the area between the 5- and 10-year time of travel (to prevent chemical contamination). The primary purpose of this zone is to encourage decision makers and planners to recognize long-term aspects of the drinking water source. The 10-year time of travel allows for some attenuation or remediation of contaminant sites, or if necessary, time to develop alternate sources of water supply.

4. Assessing the effects of well hydraulics on pathogen movement into the well
5. Evaluating water quality data (including indicators) and disease outbreak history
6. Integrating the results from sanitary surveys of the water supply system

CONCLUSIONS

Due to the fact that microorganisms continue to cause the majority of reported waterborne disease outbreaks in the United States, the U.S. EPA has initiated programs to protect public health. In relation to protecting groundwater, there are two programs that have been put into place, namely, SWAPPs and the GWR, which was proposed in 2000. Both of these programs require public water systems to evaluate their source water for vulnerability to microbial contamination. There are many different methods that can be used to delineate zones around drinking water wells to protect the water supply from microbial contamination. However, in most of the United States, the simplest methods, i.e., the arbitrary-fixed-radius and calculated-fixed-radius methods, are used to delineate wellhead protection zones for microorganisms. There are several possible explanations for this situation. One explanation is that the current understanding of the factors that control the fate and transport of enteric microorganisms in the subsurface is insufficient to allow the use of numerical transport models. In addition, the amount of data required for these models is generally extensive and well beyond what is available in many communities. Another possible explanation for the reliance on the fixed-radius methods is that it is still assumed that groundwater is safe from microbial contamination and thus that it is not necessary to expend efforts to accurately determine wellhead protection zones specifically for microorganisms. Finally, many of the wells that are used for drinking water are owned and operated by small communities and individual businesses. In most cases, these small systems or noncommunity systems do not have the resources available to obtain the information necessary to calculate wellhead protection zones by the more data-intensive methods.

REFERENCES

1. **Barwick, R. S., D. A. Levy, G. F. Craun, M. J. Beach, and R. L. Calderon.** 2004. Surveillance for waterborne-disease outbreaks—United States, 1997–1998. *Morb. Mortal. Wkly. Rep.* **49**(SS-04)**:**1–35.
2. **Bitton, G., and C. P. Gerba.** 1984. Groundwater pollution microbiology: the emerging issue. p. 1–7. *In* G. Bitton and C. P. Gerba (ed.), *Groundwater Pollution Microbiology*. John Wiley & Sons, New York, N.Y.
3. **Blackburn, B. G., G. F. Craun, J. S. Yoder, V. Hill, R. L. Calderon, N. Chen, S. H. Lee, D. A. Levy, and M. J. Beach.** 2004. Surveillance for waterborne-disease out-

breaks associated with drinking water—United States 2001–2002. Surveillance summaries, October 22, 2004. *Morb. Mortal. Wkly. Rep.* **53**(SS-8):23–42.
4. **Craun, G. F.** 1986. Statistics of waterborne outbreaks in the U.S. (1920–1980), p. 73–159. *In* G. F. Craun (ed.), *Waterborne Diseases in the United States*. CRC Press, Boca Raton, Fla.
5. **Craun, G. F.** 1986. Recent statistics of waterborne disease outbreaks (1981–1983), p. 43–69. *In* G. F. Craun (ed.), *Waterborne Diseases in the United States*. CRC Press, Boca Raton, Fla.
6. **Craun, G. F.** 1991. Causes of waterborne outbreaks in the United States. *Water Sci. Technol.* **24:**17–20.
7. **DiNovo, F., and M. Jaffe.** 1984. *Local Groundwater Protection*. American Planning Association, Chicago, Ill.
8. **Foreman, T. L., G. Nuss, J. Bloomquist, and G. Magnuson.** 1993. Results of a one-year rapid infiltration/extraction (RIX) demonstration project for tertiary filtration, p. 21–36. *In Proceedings of the Water Environment Federation 66th Annual Conference*. Water Environment Federation, Alexandria, Va.
9. **Gerba, C. P., and S. M. Goyal.** 1985. Pathogen removal from wastewater during groundwater recharge, p. 283–317. *In* T. Asano (ed.), *Artificial Recharge of Groundwater*. Butterworth Publishers, Boston, Mass.
10. **Goyal, S. M., B. H. Keswick, and C. P. Gerba.** 1984. Viruses in groundwater beneath sewage irrigated cropland. *Water Res.* **18:**299–302.
11. **Herwaldt, B. L., G. F. Craun, S. L. Stokes, and D. D. Juranek.** 1992. Outbreaks of waterborne disease in the United States: 1989–1990. *J. Am. Water Works Assoc.* **84:**129–135.
12. **Jansons, J., L. W. Edmonds, B. Speight, and M. R. Bucens.** 1989. Movement of viruses after artificial recharge. *Water Res.* **23:**293–299.
13. **Kaplan, J. E., G. W. Gary, R. C. Baron, W. Singh, L. B. Schonberger, R. Feldman, and H. Greenberg.** 1982. Epidemiology of Norwalk gastroenteritis and the role of Norwalk virus in outbreaks of acute nonbacterial gastroenteritis. *Ann. Intern. Med.* **96:**756–761.
14. **Keswick, B. H., and C. P. Gerba.** 1980. Viruses in ground water. *Environ. Sci. Technol.* **14:**1290–1297.
15. **Keswick, B. H., T. K. Satterwhite, P. C. Johnson, H. L. DuPont, S. L. Secor, J. A. Bitsura, G. W. Gary, and J. C. Hoff.** 1985. Inactivation of Norwalk virus in drinking water by chlorine. *Appl. Environ. Microbiol.* **50:**261–264.
16. **Kramer, M. H., B. L. Herwaldt, G. F. Craun, R. L. Calderon, and D. D. Juranek.** 1996. Waterborne disease: 1993 and 1994. *J. Am. Water Works Assoc.* **88:**66–80.
17. **Lee, S. H., D. A. Levy, G. F. Craun, M. J. Beach, and R. L. Calderon.** 2002. Surveillance for waterborne-disease outbreaks—United States, 1999–2000. *Morb. Mortal. Wkly. Rep.* **51**(SS-8):1–28.
18. **Levy, D. A., M. S. Bens, G. F. Craun, R. L. Calderon, and B. L. Herwaldt.** 1998. Surveillance for waterborne-disease outbreaks—United States, 1995–1996. *Morb. Mortal. Wkly. Rep. Surveill. Summ.* **47**(SS-5):1–34.
19. **Moore, A. C., B. L. Herwaldt, G. F. Craun, R. L. Calderon, A. K. Highsmith, and D. D. Juranek.** 1994. Waterborne disease in the United States, 1991 and 1992. *J. Am. Water Works Assoc.* **86:**87–99.
20. **Park, N.-S., T. N. Blanford, and P. S. Huyakorn.** 1991. *VIRALT—a Model for Simulating Viral Transport in Ground Water. Documentation and User's Guide*, version 2.0. Hydro-GeoLogic, Inc., Herndon, Va.
21. **Powelson, D. K., C. P. Gerba, and M. T. Yahya.** 1993. Virus transport and removal in wastewater during aquifer recharge. *Water Res.* **27:**583–590.
22. **Todd, D. K.** 1980. *Groundwater Hydrology*. John Wiley & Sons, New York, N.Y.
23. **U.S. Environmental Protection Agency.** 1985. National primary drinking water regulations; synthetic organic chemicals, inorganic chemicals, and microorganisms. *Fed. Regist.* **50:**46936–47022.
24. **U.S. Environmental Protection Agency.** 1987. Guidelines for delineation of wellhead protection areas. EPA440/6-87-010. USEPA Office of Ground Water Protection, Washington, D.C.
25. **U.S. Environmental Protection Agency.** 1990. Strawman rule for groundwater disinfection. Office of Drinking Water, Washington, D.C.
26. **U.S. Environmental Protection Agency.** 1993. Residential, commercial, or industrial guidelines for delineation of wellhead protection areas. EPA 440/5-93-001. Office of Water, Office of Ground Water Protection, Washington, D.C.
27. **U.S. Environmental Protection Agency.** 1997. State source water assessment and protection programs final guidance. EPA 816-R-97-009S. Office of Water, Office of Ground Water Protection, Washington, D.C.
28. **U.S. Environmental Protection Agency.** 2000. National primary drinking water regulations: Groundwater Rule; proposed rules. *Fed. Regist.* **65:**30194–30274.
29. **van der Heijde, P., and M. S. Beljin.** 1988. Model assessment for delineating wellhead protection areas. EPA440/6-88-002. USEPA Office of Ground Water Protection, Washington, D.C.
30. **Vaughn, J. M., and E. F. Landry.** 1977. Data report: an assessment of the occurrence of human viruses in Long Island aquatic systems. Department of Energy and Environment, Brookhaven National Laboratory, Upton, N.Y.
31. **Vaughn, J. M., and E. F. Landry.** 1978. The occurrence of human enteroviruses in a Long Island groundwater aquifer recharged with tertiary wastewater effluents, p. 233–245. *In State of Knowledge in Land Treatment of Wastewater*, vol. 2. U.S. Government Printing Office, Washington, D.C.
32. **Vaughn, J. M., E. F. Landry, L. J. Baranosky, C. A. Beckwith, M. C. Dahl, and N. C. Delihas.** 1978. Survey of human virus occurrence in wastewater-recharged groundwater on Long Island. *Appl. Environ. Microbiol.* **36:**47–51.
33. **Yanko, W.** 1994. Soil removal of male-specific coliphage during groundwater recharge with reclaimed water in Montebello Forebay recharge areas. Progress report. County Sanitation Districts of Los Angeles County, Whittier, Calif.

AEROBIOLOGY

VOLUME EDITOR
LINDA D. STETZENBACH

SECTION EDITOR
MARK P. BUTTNER

Introduction to Aerobiology

LINDA D. STETZENBACH

73

Aerobiology encompasses the study of airborne human pathogens, plant pathogens, opportunistic and nonpathogenic organisms, and aerosolized microbial by-products. The organisms of interest include airborne culturable and nonculturable bacteria, saprophytic fungi, free-living parasites, viruses, and algae that may result in adverse health effects or environmental impact. Microbial metabolites (e.g., volatile organic compounds), toxins (e.g., endotoxin and mycotoxin), and fragments of microbial agents (e.g., cell wall fragments, flagella, and genetic material) are also studied within the field of aerobiology.

This chapter introduces the study of airborne microorganisms and their by-products, discusses indoor and anthropogenic outdoor sources of airborne microorganisms that affect human health and the environment, reviews the association of bioaerosols and indoor environmental quality, and briefly presents background information on airborne microorganisms as potential bioterrorism agents. Subsequent chapters in this section are focused on specific topics concerning airborne microorganisms.

BIOAEROSOLS

A bioaerosol is a collection of airborne biological particles. In general, bioaerosols are generated as a mixture of droplets or particles, having different diameters ranging from 0.5 to 30 μm (86). The mode of transport for the dispersal of bioaerosols from one location to another is the air, and the composition and concentration of bioaerosols vary with the source(s) and the dispersal mechanism (95).

A variety of bioaerosol sources is found in the environment, including fresh and marine surface waters, soil, and plants. Bioaerosols that are generated from water sources during splash and wave action are usually formed with a thin layer of moisture surrounding the microbial cell and often consist of aggregates of several organisms in a packet (162). Bioaerosols released into the air from soil or dry surfaces are often composed of single cells or units, or the microorganisms are associated with particles, a condition that is termed "rafting" (89). Aerosols generated from soil are characterized by the type of soil (47) and the presence of vegetation (91).

The transport and ultimate settling of a bioaerosol are affected by its physical properties and by the environmental conditions that it encounters while airborne. The most important physical characteristics are the size, density, and shape of the droplets or particles, while the most significant environmental conditions are the temperature, relative humidity, and magnitude of air currents (87, 88, 95, 112). Temperature and relative humidity also contribute to the generation of airborne microorganisms, as increased concentrations of some fungal spores (e.g., spores of *Nigrospora* and *Cladosporium*) in outdoor air (28) and increased numbers of bacteria released from plant surfaces (91) have been associated with high temperatures and low relative humidity. An annual and diurnal variation in the concentration of airborne microorganisms outdoors has been observed and correlated with environmental conditions (87). The survival of airborne microorganisms is also influenced by environmental factors. Harsh environmental conditions tend to decrease the numbers of viable airborne organisms, but there is variability in survival between groups of microorganisms and within genera. In general, fungal spores, enteric viruses, and amoebic cysts are more resistant to environmental stresses encountered during transport through the air. Bacteria and algae are more susceptible, although bacterial endospores (e.g., *Bacillus* spp.) are quite resistant to environmental conditions (74). It is generally believed that microbial numbers do not increase during transport, but a doubling of airborne bacterial cell numbers was demonstrated in a laboratory setting by using a rotating-drum aerosol chamber under saturated humidity and with addition of tryptone to the cell suspension prior to aerosolization (35). A detailed discussion of the factors affecting the survival and transport of bioaerosols is presented in chapter 76.

Airborne microorganisms can transmit disease from one region to another and impact agriculture and human health. Microorganisms as bioaerosols may result in the spread of plant diseases and losses of agricultural productivity (Table 1). A detailed discussion of airborne agricultural pathogens is presented in chapter 81. Microbial aerosols from anthropogenic and environmental sources are also associated with public health concerns (Table 1) (34, 45, 61, 127, 128, 140, 145, 165, 166).

The average human inhales approximately 10 m^3 of air per day (95). Large airborne particles are lodged in the upper respiratory tract (nose and nasopharynx) (167). Particles of <5-μm diameter are removed by sneezing and blowing or wiping of the nose, and those particles deposited

TABLE 1 Adverse effects associated with exposure to airborne microorganisms

Microbial agent	Associated adverse effect	
	Human health	Environment
Algae	Allergic reactions	Odor problems
Bacteria	Hypersensitivity pneumonitis, infections, mucous membrane irritation	Deterioration of building materials, loss of agricultural productivity (crop and livestock diseases), odor problems
Endotoxin	Cough, headache, fever, malaise, muscle aches, nausea, respiratory distress	None reported
Fungi	Allergic reactions, exacerbation of asthma, dermal irritation, hypersensitivity pneumonitis, infections, mucous membrane irritation	Deterioration of building materials, loss of agricultural productivity (crop and livestock diseases), odor problems
Mycotoxin	Headache, muscle problems, neurologic disorders, respiratory distress, toxicosis	Loss of agricultural productivity (disease in livestock)
Protozoa	Encephalitis, hypersensitivity reactions, infections	Possible protection from biocide treatment to other microorganisms, loss of agricultural productivity (disease in livestock)
Virus	Infections	Loss of agricultural productivity (crop and livestock diseases)

in the pharynx (2 to 5 μm in diameter) are removed to the pharynx by mucociliary action and then swallowed (18, 143). Particles 1 to 5 μm in diameter can be transported to the lung, but the greatest retention in the alveoli is of the 1- to 2-μm particles (133, 135). Exhaled droplets of healthy people have been measured from 0.3 to 8 μm, although few droplets were >2 μm in diameter (109).

Bioaerosol Threat Agents

Infectious and toxigenic biological agents released in aerosols under favorable meteorological conditions for warfare or terrorism may result in severe illness for military and civilian populations (164). The Centers for Disease Control and Prevention (CDC) has compiled a listing of microbial agents that are of most concern and has classified them as category A, B, or C agents (144). Category A agents are those that have the greatest potential for adverse public health impact with mass casualties, and most of the agents in this category require broad-based public health preparedness efforts (e.g., specific or specialized surveillance, laboratory diagnosis, and stockpiling of specific medications). Category A agents also have a moderate-to-high potential for large-scale dissemination or for a heightened general public awareness that could cause mass public fear and civil disruption. Category B agents are the second-highest-priority agents. They are moderately easy to disseminate, cause moderate morbidity and low mortality, and require specific diagnostic capacity/disease surveillance. Category C agents are emerging pathogens that could be engineered for mass dissemination. Their availability, ease of production and dissemination, potential for high morbidity/mortality, and major health impact place them on the list. A listing of the CDC biothreat agents is presented in Table 2.

Man-Made Sources of Bioaerosols in Outdoor Environments

Numerous anthropogenic activities serve as the origin of bioaerosols in outdoor environments, especially agricultural practices and wastewater treatment processes (Table 3). Increases in airborne concentrations of microorganisms during harvesting operations have been documented (1, 85), and several investigators have reported the presence of airborne bacteria and viruses resulting from wastewater treatment (41, 118), sanitary landfill operations (116), and reuse-water irrigation practices (10, 151). Recycling facilities also generate bioaerosols (124).

The release of biotechnology products (e.g., genetically engineered microorganisms and microbial pest control agents) developed to enhance agricultural productivity, mineral recovery, oil spill cleanup, and toxic waste disposal can also be a source of airborne microorganisms. The application of biotechnology products to crops by aerosol mist increases the possibility for transport of the microbial product from the target crop to surrounding areas (92). Aerosolized, genetically modified cells have been monitored in a barn setting (96) and at biotechnology-based fermentors (68).

Bioaerosols and Indoor Environmental Quality

Deterioration of building materials, offensive odors, and adverse human health effects are associated with microbial contamination of indoor environments. Residences, offices, schools, health care facilities, enclosed agricultural structures (e.g., barns and crop storage areas), pharmaceutical and industrial facilities, food-processing plants, and recycling facilities are among the indoor environments where airborne microorganisms have been studied (Table 3). Sources and reservoirs of microorganisms are present within these settings, including building materials and furnishings, pets, plants, and air-conditioning systems (50, 89). Wallboard, ceiling tiles, carpeting and vinyl flooring, painted surfaces, upholstery and drapery, wallpaper, plastics, wood, cement, and brick are examples of building materials and components that can serve as sites for microbial colonization and subsequent dispersal into the air (Table 3).

Bacteria and algae generally grow in areas with standing water, such as air-handling-system components (e.g., water spray humidification systems and condensate pans) and sites where water intrusion or leaking (e.g., flooding and condensation) has occurred. However, with the exception

TABLE 2 Classification of biological agents with potential use in bioterrorism[a]

Pathogen type	Biothreat level	Microorganism	Associated disease or toxin
Bacteria	A	*Bacillus anthracis*	Anthrax
		Francisella tularensis	Tularemia
		Yersinia pestis	Plague
	B	*Brucella* spp.	Brucellosis
		Burkholderia mallei	Glanders
		Coxiella burnetii	Q fever
		Escherichia coli O157:H7	Hemorrhagic colitis
		Salmonella spp.	Salmonellosis
		Shigella dysenteriae	Shigellosis
		Vibrio cholerae	Cholera
	C	*Mycobacterium tuberculosis*	Tuberculosis
Viruses	A	Ebola virus	Hemorrhagic fever
		Junin virus	Argentine hemorrhagic fever
		Lassa virus	Lassa fever
		Marburg virus	Hemorrhagic fever
		Variola major virus	Smallpox
	B	Eastern equine encephalomyelitis virus	Encephalomyelitis
		Venezuelan equine encephalomyelitis virus	Encephalomyelitis
		Western equine encephalomyelitis virus	Encephalomyelitis
	C	Hantavirus	Hemorrhagic fever, hantavirus pulmonary syndrome
		Nipah virus	Encephalomyelitis
		Tick-borne viruses	Hemorrhagic fever and/or encephalitis
		Yellow fever virus	Yellow fever
Protozoa	C	*Cryptosporidium parvum*	Cryptosporidiosis
Toxin sources	A	*Clostridium botulinum*	Botulinum toxin
	B	*Clostridium perfringens*	Epsilon toxin
		Staphylococcus aureus	Enterotoxin B

[a]Modified from reference 144.

of viruses that require a living host cell for replication, microorganisms will colonize virtually any surface where there is sufficient moisture. Fungi, which have lower water activity (a_w) requirements than do other microorganisms, tend to colonize a wide variety of building materials (48). *Penicillium* spp. and *Aspergillus versicolor* are primary colonizers that are often isolated on wallpaper and drier margins of wetted walls, while *Cladosporium* spp. proliferate as secondary colonizers. With variation by organism, the a_w for primary and secondary colonizers on building materials at room temperature ranges from 0.79 to >0.90 (48). Tertiary colonizers (e.g., *Ulocladium* spp. and *Stachybotrys chartarum*) proliferate when the a_w becomes >0.93 (48).

Water intrusion resulting from plumbing and roof leaks, flooding, and condensation may provide conditions favorable for microbial growth in buildings. Once moisture has accumulated on building surfaces, biocontaminants may proliferate on surfaces and then be dispersed as bioaerosols (Table 3). Components of heating, ventilation, and air-conditioning (HVAC) systems may serve as amplification sites, and these systems have been associated with the dispersal of contaminants indoors (13, 27). The cycling operation of the HVAC system may result in the release of surface biocontaminants present on duct materials into the occupied space served by the system (27). Naturally ventilated buildings are also affected by bioaerosols, as organisms can be transported via drafts through open windows and doors (89). When microbial amplification occurs, the indoor environment then becomes a potential source of bioaerosol exposure to the occupants. This exposure may be important because people spend approximately 22 h/day in indoor environments (147).

Building occupants are a major source of bacterial aerosols, with human-to-human transmission often occurring in high-density indoor environments, such as correctional facilities (56) and military training centers (20). Coughing and loud talking are reported to release approximately 10^5 droplets/m^3 with a mean diameter of <1 μm and 10^4 droplets that were >1 μm (109). Microorganisms are also dispersed from surfaces as a result of activity by the occupants (26, 125).

The Institute of Medicine (61) reports an association of exposure to damp indoor environments and to mold with exacerbation of asthma in sensitized people but also points to the lack of sufficient evidence to associate other symptoms with indoor exposures, thereby highlighting the need for additional research in this topic area. No regulations regarding bioaerosol concentrations are currently mandated

TABLE 3 Sources and amplification sites of indoor biocontaminants and associated airborne or surface concentrations

Site		Microbial agent(s)	Concn[a]	Reference(s)
Category	Facility/activity/material			
Agricultural	Animal facilities	Actinomycetes (thermophilic)	10^0–10^1 CFU/m^3	36
		Bacteria (not specified)	10^3–10^5 CFU/m^3	36, 152
		β-D-Glucan	8–35 ng/mg	9
		Endotoxin	0.1–0.5 ng/m^3	100
			0.3–41 ng/mg	9
			10^1–10^4 EU[b]/m^3	36, 79
		Fungi (not specified)	10^2–10^7 CFU/m^3	36, 39, 93, 152
	Composting	Actinomycetes (thermophilic)	10^3 CFU/m^3	124
		Bacteria (legionellae)	10^3–10^6 CFU/g	59
		Fungi (thermophilic)	10^2–10^6 CFU/g	149
	Farming, harvesting, baling, grain storage	Actinomycetes (thermophilic)	10^2–10^3 CFU/m^3	78
		Bacteria		
		Gram-negative	10^4–10^5 CFU/m^3	78
		Mesophilic	10^5 CFU/m^3	78
		Not specified	10^1–10^4 CFU/m^3	85, 150
		Fungi (not specified)	10^3–10^9 CFU/m^3	23, 39, 78, 85, 90, 93, 110, 150
		Endotoxin	16–172 μg/m^3	79
		Mycotoxin	10^1–10^3 mg/kg	105
Air-handling systems	HVAC systems	Amoebic cysts	NQ[c]	58
		Bacteria (not specified)	10^2–10^3 CFU/m^3	58
			10^2–10^7 CFU/cm^2	11
		Fungi		
		Penicillium	10^2–10^7 CFU/g	77
		Penicillium chrysogenum	10^4 CFU/m^3	27
		Micropolyspora faeni	NQ	43
		Thermophilic	10^5 spores/m^3	160
	Cooling towers, water spray systems	Actinomycetes (thermophilic)	<dl[d]	24
		Bacteria		
		Heterotrophic	<dl–10^6 CFU/ml	24
		Human source	<dl–10^4 CFU/ml	24
		Fungi		
		Aureobasidium pullulans	NQ	166
		Not specified	<dl–10^2 CFU/ml	24
	Portable humidifiers	Bacteria		
		Legionella pneumophila	10^2–10^4 CFU/ml	168
		Pseudomonas	10^2–10^4 CFU/h	31
		Fungi		
		Several genera	10^1–10^2 CFU/ml	21
		Thermoactinomyces vulgaris	NQ	21, 154
		Yeast	10^1–10^3 CFU/ml	21

Public buildings	Day care center	Endotoxin	0.43–5.7 ng/m^3	132, 159
		β-D-Glucan	0.157–0.20 ng/m^3	132, 159
	Museum	Fungi	NQ	77
			<dl–10^1 CFU/m^3	17
	Office building	Endotoxin	3.7 ng/m^3	159
		β-D-Glucan	58 pg/m^3	159
	Post office	Endotoxin	0.19 ng/m^3	132
		β-D-Glucan	0.06 ng/m^3	132
	School	Endotoxin	0.21–0.26 ng/m^3	132
		β-D-Glucan	0.49–0.55 ng/m^3	132
Building materials and furnishings	Ceiling tile, insulation, painted surfaces, wallpaper	Fungi		
		Aspergillus spp.	10^2 CFU/m^3	76
		Aspergillus versicolor	10^5 CFU/g	66
		Cladosporium spp.	10^2 CFU/m^3	76
		Penicillium spp.	10^3 CFU/m^3	76
			10^5 CFU/g	66
		Stachybotrys chartarum	NQ	33, 102
			10^4–10^7 CFU/cm^2	55
			10^2 CFU/m^3	55
			10^6 CFU/g	66
	Carpet	Bacteria (not specified)	10^4 CFU/in.^{2}	7
		Fungi		
		Alternaria sp.	10^2 CFU/m^3	76
		Mesophilic	10^2 CFU/m^3	76
		S. chartarum	NQ	76
	Gypsum wallboard	Bacteria (gram-negative)	10^6 CFU/g	9
		β-D-Glucan	0.4–210 μg/g	9
		Endotoxin	0.41–17 μg/g	9
		Fungi		
		A. versicolor	10^4–10^7 CFU/cm^2	55
		Mesophilic	10^3 CFU/g	111
		Mycobacteria	10^6 CFU/g	8
		S. chartarum	10^6 CFU/g	8
		Total spores	10^7 spores/g	111
		Yeasts	10^3 CFU/g	111
		Xerophilic	10^2 CFU/g	111
	Hot water heaters, hot water systems	Bacteria		
		Legionella sp.	10^0–10^1 CFU/m^3	15
		L. pneumophila	NQ	53

(Continued)

TABLE 3 Sources and amplification sites of indoor biocontaminants and associated airborne or surface concentrations (*Continued*)

Site: Category	Site: Facility/activity/material	Microbial agent(s)	Concn[a]	Reference(s)
	House dust	Bacteria		
		Bacillus	10^4–10^5 CFU/g	104
		Pseudomonas	10^4–10^4 CFU/g	104
		Streptomyces	10^4–10^6 CFU/g	104
		β-D-Glucan	0.5–1.4 ng/mg	8
		Endotoxin	0.2–0.3 ng/mg	8
			0.7–18 ng/mg	139
		Fungi		
		Alternaria	10^3–10^4 CFU/g	99, 162
		Aspergillus fumigatus	10^4 CFU/g	104
		Aspergillus niger	10^3 CFU/g	104
		Cladosporium	10^3–10^6 CFU/g	99, 104, 163
		Mixed species	10^4 CFU/g	57
		Penicillium	10^3 –10^5 CFU/g	99, 163
		Rhodotorula	10^3–10^6 CFU/g	104, 163
		Yeasts (not specified)	10^3–10^6 CFU/g	99, 104, 163
	House plants (greenhouses)	Fungi (not specified)	10^3–10^4 CFU/m^3	22
	Remediation/demolition	Fungi		
		Not specified	10^3–>10^9 CFU/m^3	122, 123
		Penicillium	10^5 CFU/m^3	122
		Total spores	10^5–10^6 counts/m^3	122
Health care facilities	Operating room	Bacteria (not specified)	10^2 CFU/m^3	153
			10^2–10^5 particles/m^3	138
	Patient rooms	*Pneumocystis carinii*	NQ	106
		Virus (varicella-zoster virus)	NQ	136
Industrial	Cellulose, wood chip factory, sawmill	Fungi		
		A. fumigatus	10^2–10^5 CFU/m^3	75
		Penicillium sp.	10^1–10^5 CFU/m^3	75
		Not specified	10^4–10^8 CFU/m^3	14, 39
	Food processing, storage, warehouse	Endotoxin	0.0125–54.9 fg/m^3	37
		Fungi		
		Cladosporium herbarum	10^3 CFU/m^3	63
		Not specified	10^2–10^4 CFU/m^3	14, 65
		Total	10^1–10^3 CFU/m^3	63
	Manufacturing (cotton mill, tobacco processing, paper processing, equipment cleaning, wood factory)	Bacteria		
		Gram-negative bacilli	10^1–10^4 CFU/m^3	94, 113, 114
		Total	10^1–10^3 CFU/m^3	158
		Endotoxin	0.0042 – 3.6 μg/m^3	114, 158

		Fungi		
		A. *fumigatus*	10^1–10^2 CFU/m^3	60
			NQ	83
		Not specified	10^2–10^5 CFU/m^3	81, 93
		Total spores	10^2–10^4 spores/m^3	60
		Mycotoxin	NQ	83
	Packing boxes	*Aureobasidium pullulans*	NQ	166
	Recycling plant	Bacteria	10^5 CFU/m^3	124
		A. *fumigatus*	10^4 CFU/m^3	124
	Sanitary landfill	Actinomycetes (thermophilic)	<dl–10^3 CFU/m^3	116
		Bacteria		
		Culturable	10^1–10^4 CFU/m^3	116
		Total coliforms	<dl–10^3 CFU/m^3	116
		Fecal streptococci	<dl–10^4 CFU/m^3	116
		Fungi		
		Mesophilic	10^3–10^4 CFU/m^3	116
		Thermophilic	10^0–10^4 CFU/m^3	116
	Waste handling	Bacteria		
		Culturable	10^4 CFU/m^3	73
		Gram-negative bacilli	10^3 CFU/m^3	73
		Fungi	10^5 CFU/m^3	73
Transportation	Private automobiles	Bacteria (mixed biofilms)	NQ	142
		Fungi	NQ	80, 142
	Public buses	Fungi (not specified)	10^2–>10^5 CFU/m^3	155
Wastewater treatment	Activated sludge processing	Bacteria		
		Coliforms	0.27–5.17 CFU/m^3	41
		E. coli	10^2 CFU/m^3	118
		Gram-negative bacilli	10^1–10^5 CFU/m^3	82
		Endotoxin	0.1–350 ng/m^3	82
		Fungi		
		Mesophilic	<dl–10^3 CFU/m^3	67
		Thermophilic	<dl–10^2 CFU/m^3	67
	Aeration tanks	Actinomycetes (thermophilic)	<dl–10^1 CFU/m^3	115
		Bacteria		
		Fecal streptococci	10^2 CFU/m^3	32
		Gram-negative	10^1–10^2 CFU/m^3	115
		Mesophilic	10^2 CFU/m^3	115
		Streptococcus faecalis	<dl–10^2 CFU/m^3	32
		Fungi	10^1–10^2 CFU/m^3	115
		Viruses		
		Animal virus	NQ	16
		Coliphage	<10 PFU/m^3	16

(Continued)

TABLE 3 Sources and amplification sites of indoor biocontaminants and associated airborne or surface concentrations (*Continued*)

Category	Site: Facility/activity/material	Microbial agent(s)	Concn[a]	Reference(s)
	Effluent irrigation	Bacteria		
		E. coli	10^1–10^3 CFU/m^3	151
		Fecal coliforms	10^2 CFU/m^3	10
		Fecal streptococci	10^2 CFU/m^3	10
		Total coliforms	10^2 CFU/m^3	10
		Virus (enteric)	NQ	10, 42
	Sludge compost	Bacteria		
		Coliforms	<dl	67
		Not specified	10^0–10^3 CFU/m^3	67
	Trickling filter	Bacteria		
		Coliforms	10^2–10^4 CFU/m^3	2
		Total	10^1–10^3 CFU/m^3	2

[a]Concentration per unit volume or area.
[b]EU, endotoxin unit.
[c]NQ, not quantitated.
[d]dl, detection limit.

for residential, office, or classroom environments, but a variety of guidelines predicated on baseline data have been proposed (119). It is generally accepted that indoor sources of bioaerosols may be significant when differences are noted between indoor and outdoor concentrations and/or populations, but the lack of exposure and of dose-response data has precluded the establishment of bioaerosol threshold limit values (6).

MICROBIAL AGENTS

The environmental effects and human health complaints resulting from airborne microorganisms have renewed interest in a wide variety of microorganisms. The discovery of *Legionella pneumophila* as the cause of the outbreak of Legionnaires' disease in Philadelphia, Pa., in 1976 increased the awareness of diseases caused by bacterial aerosols (165). A detailed discussion of *Legionella* spp. and Legionnaires' disease is presented in chapter 79. Similarly, the increased reporting of tuberculosis in both developing and industrialized countries has prompted renewed interest in the genus *Mycobacterium* and its airborne transmission. *Mycobacterium tuberculosis* is spread via aerosols from an infected person and is recognized as a significant public health concern because of the low infectious dose (69). Nontuberculosis mycobacteria have also been associated with respiratory illness (30, 72). However, the significance of commonly isolated airborne bacteria and fungi in offices, schools, residences, and outdoor environments has not been determined. This is due, in part, to the isolation of numerous airborne gram-positive cocci and gram-positive bacilli in the absence of adverse health effects. High ratios of airborne bacteria isolated from indoor air to those isolated from outdoor air have been used as an indication of high occupancy rate, poor ventilation, or inadequate building maintenance (6), but additional research in this area is needed. While airborne transmission of bacteria and other microorganisms in health care facilities can cause nosocomial infections (137), hospitals adhere to strict infection control procedures and routinely monitor for airborne microorganisms. However, increasing numbers of elderly people are living in nursing homes and home health care facilities, where maintenance and decontamination practices may not be as strictly followed. Therefore, the risk of exposure to opportunistic pathogens is increased.

Exposure to airborne bacterial cells may also result in the inhalation of endotoxin, a lipopolysaccharide found in the cell walls of gram-negative bacteria and blue-green algae. Exposure to endotoxin may result in fever, cough, headache, respiratory impairment, and exacerbation of asthma (97, 129, 159). Airborne endotoxin may be a major cause of illness in enclosed agricultural settings, such as silage facilities, poultry-processing houses, and cotton mills (29, 131), and waste handlers have reported increased nausea and gastrointestinal problems that were associated with endotoxin exposure (62). Chilled-water spray humidification systems used in textile manufacturing facilities have also been implicated as sources of aerosolized endotoxin associated with cases of humidifier fever and hypersensitivity pneumonitis (158). A review of airborne endotoxin is presented in chapter 78.

Although viruses do not replicate outside a susceptible host cell, they are readily transported through the air. Numerous human viruses are transmitted via droplets and spread by the respiratory route from one person to another in the indoor environment (167). Enteric virus bioaerosols are produced at sewage treatment facilities (3), and zoonotic transmission of viruses is also of concern, as serious illness resulting from exposure to hantavirus aerosolized from rodent feces and urine has been reported (34). An extensive review of airborne viruses is presented in chapter 80.

Airborne fungi have been the focus of much concern because of the potential for serious respiratory infection and allergic reactions. Although few fungi actually cause infection in healthy individuals, immunosuppressed host defenses resulting from organ transplantation, cancer, therapy (e.g., antibiotics, steroids, or drugs), or the presence of another disease-causing agent may increase the likelihood of fungal infection (101). More commonly, exposure to fungi can initiate adverse health effects in the absence of infection by serving as aeroallergens, resulting in exacerbation of asthma, allergic rhinitis, and respiratory distress (11, 49, 81, 99). Exposure to the yeast *Rhodotorula rubra* was reported as the cause of a case of extrinsic allergic alveolitis (141), while the molds *Penicillium* and *Aspergillus* have been identified as risk factors for asthma and atopy, respectively, in children (46). Children exposed to damp houses were reported to be at higher risk for colds, sore throats, and ear infections, and mold in the home was a significant risk factor for otitis and bronchitis (130). Kilpelainen et al. (70) reported an association between colds and visible mold, dampness, or water damage when surveying college students. Ebbehoj et al. (38) reported that successive rounds of remediation of a water-damaged, moldy building were needed to minimize occupant symptoms of elevated peak flow variability, irritation, rash, headache, dizziness, and difficulty in concentration. Symptoms had persisted after removal of visible mold, but additional remediation resulted in peak flow variability returning to normal and other symptoms subsiding.

Numerous surveys have been conducted to determine airborne and surface concentrations of fungi present in a variety of outdoor and indoor environments (Table 3). A recent study (57) demonstrated the presence of leaf surface fungi (e.g., *Cladosporium*, *Alternaria*, *Epicoccum*, and *Curvularia* spp.) as ≥20% of the culturable fungi and a low proportion of soil fungi (e.g., *Aspergillus*, *Penicillium*, *Emercella*, and *Paecilomyces* spp.) in dust samples from nonproblem residences. Fungi that indicate water damage, accumulation, or intrusion (e.g., *Stachybotrys*, *Chaetomium*, and *Ulocladium* spp.) were absent. The indoor airborne culturable data also confirmed that data for fungal populations in non-water-damaged buildings resemble outdoor airborne data. These data support the earlier report by Samson et al. (134) that background populations of common fungi are present in indoor environments but that water intrusion results in proliferation of soil fungi and that, when conditions persist, the water indicator fungi colonize and grow. Unfortunately, dose-response information for fungal exposure is limited, and many studies do not quantitatively assess the fungal populations present in conjunction with health data. Therefore, the Institute of Medicine (61) concluded that there are insufficient data showing a statistical association between the presence of mold in damp environments and upper respiratory tract symptoms. Additional studies are needed to correlate mold exposure and the variety of symptoms reported by occupants of damp buildings.

Exposure to mycotoxins produced by some fungi can result in adverse human health effects. Exposure of agricultural workers to mycotoxin-containing dust (81) and industrial exposure to mycotoxins produced by *Aspergillus fumigatus* (83) have been reported. Growth of fungi and the presence of mycotoxins in indoor environments have been associated with water-damaged building materials (50). Fungi isolated in indoor environments such as *Alternaria alternata* (52), *Aspergillus fumigatus* (107), *Aspergillus flavus* and *Aspergillus parasiticus* (161), *Fusarium graminearum*, *Fusarium sporotrichioides* (98), and *Stachybotrys* (146) produce spores that contain mycotoxins, and toxin production by isolates from residences of ill children has been reported (64). Studies with laboratory mice demonstrated dose-dependent inflammatory changes in mice following intranasal challenge to *Stachybotrys* spores (103), and dose-response pulmonary inflammation and injury have been documented in laboratory rats exposed to *S. chartarum* spores (120). A recent report (108) also demonstrated the presence of ergot alkaloids in the conidia of *A. fumigatus* that could be respired and result in health effects. Wady and Larsson (157) reported that the presence of microbial volatile organic compounds associated with respirable dust could result in respiratory illness, and Fischer et al. (44) reported microbial volatile organic compounds produced by fungi at a compost facility. An extensive review of fungal bioaerosols and mycotoxin is found in chapter 77. Additional information on agricultural fungal pathogens is presented in chapter 81.

Airborne actinomycetes have also been cited as the cause of adverse health effects. Spores of *Streptomycetes* spp. from damp, moldy houses have been shown to increase the production of inflammatory mediators (e.g., macrophages that produce tumor necrosis factor and interleukin-6) (54). Exposure to thermophilic actinomycetes, whose optimal growth temperature is >40°C, may occur near compost or municipal landfills and during handling of decomposing organic matter in agricultural facilities (e.g., milling and sugar cane bagasse) (11).

Blowing dust was reported as the source of 62 genera of airborne algae (19), and eutrophic lakes are also contributors to outdoor airborne algal concentrations. Dust (12) and aeration of fountains and aquariums (89) have been proposed as possible indoor sources. While exposure to algal extracts has been associated with adverse human health effects (12), the extent of allergic reactions due to algal bioaerosols has not been fully investigated. Additional research is needed to determine the environmental and human health effects of airborne algae.

Free-living amoebae (e.g., *Naegleria fowleri* and *Acanthamoeba*) are indigenous to soil and water and can be aerosolized from natural and artificially heated waters such as power plant discharges, lakes, and hot springs. Airborne *Acanthamoeba* has been detected in the nasal passages of children in Africa (84) and has been observed in air samples from Mexico City (126). Cooling system waters have also been reported as potential sources of aerosolized amoebae (95). Humidifiers and ceiling dust were cited as an indoor source of airborne amoebic antigen (40). Although severe health effects can be elicited by exposure to these organisms (25), insufficient information is currently available on airborne protozoa.

Universal to the field of aerobiology is the need for measurement methods to detect and identify a microorganism(s) of interest (51). Classical microbiological methods for measurement of airborne microorganisms rely on culture and/or microscopic assay with forced-air-flow sampling (6). Culture-based methods require appropriate conditions for growth of culturable organisms, but airborne microorganisms are stressed during transport and collection and may not respond to incubation conditions in the laboratory

(71). Total count procedures, using direct microscopic enumeration, are tedious and often fail to distinguish genera and/or species. Measurement methods to determine endotoxin, ergosterol, and β,1-3 D-glucan (121) are used as alternatives to culturable methods, and PCR has been shown to be an effective means to detect and identify airborne bacteria (4, 5, 117), *Mycoplasma* (148), *Pneumocystis carinii* (106), *S. chartarum* (156), and virus (136). Chapters 74 and 75 discuss sampling and analysis methods, respectively.

In summary, interest in the populations of airborne microorganisms in agricultural and industrial settings, health care facilities, residences, offices, and classroom environments has increased in recent years. The threat of purposeful release of microorganisms as bioterrorism agents has prompted renewed interest in aerobiology, and research activity in this area of environmental microbiology has rapidly expanded. The following chapters in this volume focus on specific topics related to bioaerosols, and the reader is directed to consult those chapters for additional information.

REFERENCES

1. **Abdel-Hafez, S. I., A. H. Moubasher, A. A. Shoreit, and M. A. Ismail.** 1990. Fungal flora associated with combine harvester wheat and sorghum dusts from Egypt. *J. Basic Microbiol.* **30:**467–479.
2. **Adams, A. P., and J. C. Spendlove.** 1970. Coliform aerosols emitted by sewage treatment plants. *Science* **169:**1218–1220.
3. **Adams, D. J., J. C. Spendlove, R. S. Spendlove, and B. B. Barnett.** 1982. Aerosol stability of infectious and potentially infectious reovirus particles. *Appl. Environ. Microbiol.* **44:**903–908.
4. **Alvarez, A. J., M. P. Buttner, G. A. Toranzos, E. A. Dvorsky, A. Toro, T. B. Heikes, L. E. Mertikas-Pifer, and L. D. Stetzenbach.** 1994. Use of solid-phase PCR for enhanced detection of airborne microorganisms. *Appl. Environ. Microbiol.* **60:**374–376.
5. **Alvarez, A. J., M. P. Buttner, and L. D. Stetzenbach.** 1995. PCR for bioaerosol monitoring: sensitivity and environmental interference. *Appl. Environ. Microbiol.* **61:** 3639–3644.
6. **American Conference of Governmental Industrial Hygienists.** 1999. *Bioaerosols: Assessment and Control.* American Conference of Governmental Industrial Hygienists, Cincinnati, Ohio.
7. **Anderson, R. L.** 1969. Biological evaluation of carpeting. *Appl. Microbiol.* **18:**180–186.
8. **Andersson, M. A., M. Nikulin, U. Koljalg, M. C. Andersson, F. Rainey, K. Reijula, E. Hintikka, and M. Salkinoja-Salonen.** 1997. Bacteria, molds, and toxins in water-damaged building materials. *Appl. Environ. Microbiol.* **63:**387–393.
9. **Andersson, A. M., N. Weiss, F. Rainey, and M. S. Salkinoja-Salonen.** 1999. Dust-borne bacteria in animal sheds, schools and children's day care centers. *J. Appl. Microbiol.* **86:**622–634.
10. **Applebaum, J., N. Guttman-Bass, M. Lugten, B. Teltsch, B. Fattal, and H. I. Shuval.** 1984. Dispersion of aerosolized enteric viruses and bacteria by sprinkler irrigation with wastewater. *Monogr. Virol.* **15:**193–201.
11. **Banaszak, E. F., J. Barboriak, J. Fink, G. Scanlon, D. P. Schlueter, A. Sosman, W. Thiede, and G. Unger.** 1974. Epidemiologic studies relating thermophilic fungi and hypersensitivity lung syndromes. *Am. Rev. Respir. Dis.* **110:**585–591.
12. **Bernstein, I. L., and R. S. Safferman.** 1973. Clinical sensitivity to green algae demonstrated by nasal challenge and in vitro tests of immediate hypersensitivity. *J. Allergy Clin. Immunol.* **51:**22–28.
13. **Bernstein, R. S., W. G. Sorenson, D. Garabrant, C. Reaux, and R. D. Treitman.** 1983. Exposures to respirable, airborne *Penicillium* from a contaminated ventilation system: clinical, environmental and epidemiological aspects. *Am. Ind. Hyg. Assoc. J.* **44:**161–169.
14. **Blomquist, G., G. Strom, and L.-H. Stromquist.** 1984. Sampling of high concentrations of airborne fungi. *Scand. J. Work Environ. Health* **10:**109–113.
15. **Bollin, G. E., J. F. Plouffe, M. F. Para, and D. Hackman.** 1985. Aerosols containing *Legionella pneumophila* generated by shower heads and hot-water faucets. *Appl. Environ. Microbiol.* **50:**1128–1131.
16. **Brenner, K. P., P. V. Scarpino, and C. S. Clark.** 1988. Animal viruses, coliphage, and bacteria in aerosols and wastewater at a spray irrigation site. *Appl. Environ. Microbiol.* **54:**409–415.
17. **Brimblecombe, P., N. Blades, D. Camuffo, G. Sturaro, A. Valentino, K. Gysels, R. Van Grieken, H.-J. Busse, O. Kim, U. Ulrych, and M. Wieser.** 1999. The indoor environment of a modern museum building, the Sainsbury Centre for Visual Arts, Norwich, UK. *Indoor Air* **9:**146–164.
18. **Brown, J. H., K. M. Cook, F. G. Ney, and T. Hatch.** 1950. Influence of particle size upon the retention of particulate matter in the human lung. *Am. J. Public Health* **40:**450–458.
19. **Brown, R. M., D. A. Larson, and H. C. Bold.** 1964. Airborne algae: their abundance and heterogeneity. *Science* **143:**583–585.
20. **Brundage, J. F., R. M. Scott, W. M. Lednar, D. W. Smith, and R. N. Miller.** 1988. Building-associated risk of febrile acute respiratory diseases in Army trainees. *JAMA* **259:**2108–2112.
21. **Burge, H. A., W. R. Solomon, and J. R. Boise.** 1980. Microbial prevalence in domestic humidifiers. *Appl. Environ. Microbiol.* **39:**840–844.
22. **Burge, H. A., W. R. Solomon, and M. L. Muilenberg.** 1982. Evaluation of indoor plantings as allergen exposure sources. *Allergy Clin. Immunol.* **70:**101–108.
23. **Burge, H. A., M. L. Muilenberg, and J. A. Chapman.** 1991. Crop plants as a source of fungus spores of medical importance, p 222–236. *In* J. H. Andrews and S. S. Hirano (ed.), *Microbial Ecology of Leaves.* Springer-Verlag, New York, N.Y.
24. **Burkhart, J. E., R. Stanevich, and B. Kovak.** 1993. Microorganism contamination of HVAC humidification systems: case study. *Appl. Occup. Environ. Hyg.* **8:**1010–1014.
25. **Burrell, R.** 1991. Microbiological agents as health risks in indoor air. *Environ. Health Perspect.* **95:**29–34.
26. **Buttner, M. P. and L. D. Stetzenbach.** 1993. Monitoring of fungal spores in an experimental indoor environment to evaluate sampling methods and the effects of human activity on air sampling. *Appl. Environ. Microbiol.* **59:**219–226.
27. **Buttner, M. P., P. Cruz-Perez, P. J. Garrett, and L. D. Stetzenbach.** 1999. Dispersal of fungal spores from three types of air handling system duct material. *Aerobiologia* **15:**1–8.
28. **Cammack, R. H.** 1955. Seasonal changes in three common constituents of the air sporea of southern Nigeria. *Nature* **176:**1270–1272.
29. **Castellan, R. M., S. A. Olenchock, J. L. Hankinson, P. D. Millner, J. B. Cocke, C. K. Bragg, H. H. Perkins, and R. R. Jacobs.** 1984. Acute bronchoconstriction induced by cotton dust: dose-related responses to endotoxin and other dust factors. *Ann. Intern. Med.* **101:**157–163.

30. **Contreras, M. A., O. T. Cheung, D. E. Sanders, and R. S. Goldstein.** 1988. Pulmonary infection with nontuberculous mycobacteria. *Am. Rev. Respir. Dis.* **137:**149–152.
31. **Couvelli, H. D., J. Kleeman, J. E. Martin, W. L. Landau, and R. L. Hughes.** 1973. Bacterial emission from both vapor and aerosol humidifiers. *Am. Rev. Respir. Dis.* **108:** 698–701.
32. **Crawford, G. V., and P. H. Jones.** 1979. Sampling and differentiation techniques for airborne organisms emitted from wastewater. *Water Res.* **13:**393–399.
33. **Croft, W. S., B. B. Jarvis, and C. S. Yatawara.** 1986. Airborne outbreak of trichothecene toxicosis. *Atmos. Environ.* **20:**549–552.
34. **Diglisic, G., C. A. Rossi, A. Doti, and D. K. Walshe.** 1999. Seroprevalence study of Hantavirus infection in the community based population. *Md. Med. J.* **48:**303–306.
35. **Dimmick, R. L., H. Wolochow, and M. A. Chatigny.** 1979. Evidence that bacteria can form new cells in airborne particles. *Appl. Environ. Microbiol.* **37:**924–927.
36. **Duchaine, C., Y. Grimard, and Y. Cormier.** 2000. Influence of building maintenance, environmental factors, and seasons on airborne contaminants of swine confinement buildings. *Am. Ind. Hyg. Assoc. J.* **61:**56–63.
37. **Dutkiewicz, J., E. Krysinska-Traczyk, C. Skorska, J. Sitkowska, Z. Prazmo, and B. Urbanowicz.** 2000. Exposure of agricultural workers to airborne microorganisms and endotoxin during handling of various vegetable products. *Aerobiologia* **16:**193–198.
38. **Ebbehoj, N. E., M. O. Hansen, T. Sigsgaard, and L. Larsen.** 2002. Building-related symptoms and molds: a two set intervention study. *Indoor Air* **12:**273–277.
39. **Eduard, W., J. Lacey, K. Karlsson, U. Palmgren, G. Strom, and G. Blomquist.** 1990. Evaluation of methods for enumerating microorganisms in filter samples from highly contaminated occupational environments. *Am. Ind. Hyg. Assoc. J.* **51:**427–436.
40. **Edwards, J. H.** 1980. Microbial and immunological investigations and remedial action after an outbreak of humidifier fever. *Br. J. Ind. Med.* **37:**55–62.
41. **Fannin, K. F., S. C. Vana, and W. Jakubowski.** 1985. Effect of an activated sludge wastewater treatment plant on ambient air densities of aerosols containing bacteria and viruses. *Appl. Environ. Microbiol.* **49:**1191–1196.
42. **Fattal, B., E. Katzenelson, N. Guttman-Bass, and A. Sadovski.** 1984. Relative survival rates of enteric viruses and bacterial indicators in water, soil, and air. *Monogr. Virol.* **15:**184–192.
43. **Fink, J. N., E. F. Banazak, W. H. Thiede, and J. J. Barboriak.** 1971. Interstitial pneumonitis due to hypersensitivity to an organism contaminating a heating system. *Ann. Intern. Med.* **74:**80–83.
44. **Fischer, G., R. Schwalbe, M. Moller, R. Ostrowski, and W. Dott.** 1999. Species-specific production of microbial volatile organic compounds (MVOC) by airborne fungi from a compost facility. *Chemosphere* **39:**795-810.
45. **Flaherty, D. K., F. H. Deck, J. Cooper, K. Bishop, P. A. Winzenburger, L. R. Smith, L. Bynum, and W. B. Witmer.** 1984. Bacterial endotoxin isolated from a water spray air humidification system as a putative agent of occupation-related disease. *Infect. Immun.* 43:206–212.
46. **Garrett, M. H., P. R. Rayment, M. A. Hooper, M. J. Abramson, and B. M. Hooper.** 1998. Indoor airborne fungal spores, house dampness and associations with environmental factors and respiratory health in children. *Clin. Exp. Allergy* **28:**459–467.
47. **Gillette, D. A., and I. H. Blifford, Jr.** 1972. Measurement of aerosol size distribution and vertical fluxes of aerosols on land subject to wind erosion. *J. Appl. Meteorol.* **11:**977–987.
48. **Grant, C., C. A. Hunter, B. Flannigan, and A. F. Bravery.** 1989. The moisture requirements of moulds isolated from domestic dwellings. *Int. Biodeterior.* **25:**259–284.
49. **Gravesen, S.** 1979. Fungi as a cause of allergic disease. *Allergy* **34:**135–154.
50. **Gravesen, S., P. A. Nielsen, R. Iversen, and K. F. Nielsen.** 1999. Microfungal contamination of damp buildings—examples of risk construction and risk materials. *Environ. Health Perspect.* **107:**505–508.
51. **Griffiths, W. D., and G. A. L. DeCoseno.** 1994. The assessment of bioaerosols: a critical review. *J. Aerosol Sci.* **25:**1425–1458.
52. **Haggblom, P.** 1987. *De novo* synthesis of alternariol in conidia of *Alternaria alternata*. *J. Gen. Microbiol.* **133:** 3527–3529.
53. **Helms, C. M., R. M. Massanari, R. Zeitler, S. Streed, M. J. R. Gilchrist, N. Hall, W. J. Hausler, Jr., J. Sywassink, W. Johnson, L. Wintermeyer, and W. J. Hierholzer.** 1983. Legionnaires' disease associated with a hospital water system: a cluster of 24 nosocomial cases. *Ann. Intern. Med.* **99:**172–178.
54. **Hirvonen, M. R., A. Nevalainen, N. Makkonen, J. Monkkonen, and K. Savolainen.** 1997. Induced production of nitric oxide, tumor necrosis factor, and interleukin-6 in RAW 264.7 macrophages by streptomycetes from indoor air of moldy houses. *Arch. Environ. Health* **52:**426–432.
55. **Hodgson, M. J., P. Morey, W. Leung, L. Morrow, D. Miller, B. B. Jarvis, H. Robbins, J. F. Halsey, and E. Storey.** 1998. Building-associated pulmonary disease from exposure to *Stachybotrys chartarum* and *Aspergillus versicolor*. *J. Occup. Environ. Med.* **40:**241–249.
56. **Hoge, C. W., M. R. Reichler, E. A. Dominguez, J. C. Bremer, T. D. Mastro, K. A. Hendricks, D. M. Musher, J. A. Elliott, R. R. Facklam, and R. F. Breiman.** 1994. An epidemic of pneumococcal disease in an overcrowded, inadequately ventilated jail. *N. Engl. J. Med.* **331:**643–648.
57. **Horner, W. E., A. G. Worthan, and P. R. Morey.** 2004. Air- and dustborne mycoflora in houses free of water damage and fungal growth. *Appl. Environ. Microbiol.* **70:**6394–6400.
58. **Hugenholtz, P., and J. A. Fuerst.** 1992. Heterotrophic bacteria in an air-handling system. *Appl. Environ. Microbiol.* **58:**3914–3920.
59. **Hughes, M. S., and T. W. Steele.** 1994. Occurrence and distribution of *Legionella* species in composted plant materials. *Appl. Environ. Microbiol.* **60:**2003–2005.
60. **Huuskonen, M. S., K. Husman, J. Jarvisalo, O. Korhonen, M. Kotimaa, T. Kuusela, H. Nordman, A. Zitting, and R. Mantyjarvi.** 1984. Extrinsic allergic alveolitis in the tobacco industry. *Br. J. Ind. Med.* **41:**77–83.
61. **Institute of Medicine of the National Academies.** 2004. *Damp Indoor Spaces and Health*. National Academies Press, Washington, D.C.
62. **Ivens, U. I., N. O. Breum, N. Ebbehoj, B. H. Nielson, O. M. Poulsen, and H. Wurtz.** 1999. Exposure-response relationship between gastrointestinal problems among waste collectors and bioaerosol exposure. *Scand. J. Work Environ. Health* **25:**238–245.
63. **Jain, A. K.** 2000. Survey of bioaerosol in different indoor working environments in central India. *Aerobiologia* **16:** 221–225.
64. **Jarvis, B. B., W. G. Sorenson, E. Hintikka, M. Nikulin, Y. Zhou, J. Jiang, S. Wang, S. Hinkley, R. A. Etzel, and D. Dearborn.** 1998. Study of toxin production by isolates of *Stachybotrys chartarum* and *Memnoniella echinata* isolated during a study of pulmonary hemosiderosis in infants. *Appl. Environ. Microbiol.* **64:**3620–3625.

65. **Jensen, P. A., W. F. Todd, M. E. Hart, R. L. Mickelsen, and D. M. O'Brien.** 1993. Evaluation and control of worker exposure to fungi in a beet sugar refinery. *Am. Ind. Hyg. Assoc. J.* **54:**742–748.
66. **Johanning, E., P. Landsbergis, M. Gareis, C. S. Yang, and E. Olmsted.** 1999. Clinical experience and results of a sentinel health investigation related to indoor fungal exposure. *Environ. Health Perspect.* **107:**489–494.
67. **Jones, B. L., and J. T. Cookson.** 1983. Natural atmospheric microbial conditions in a typical suburban area. *Appl. Environ. Microbiol.* **45:**919–934.
68. **Juozaitis, A., Y. Huang, K. Willeke, J. Donnelly, S. Kalatoor, A. Leeson, and R. Wyza.** 1994. Dispersion of respirable aerosols in a fermentor and their removal in an exhaust system. *Appl. Occup. Environ. Hyg.* **9:**552–559.
69. **Kaufmann, S. H. E., and J. D. A. van Embden.** 1993. Tuberculosis: a neglected disease strikes back. *Trends Microbiol.* **1:**2–5.
70. **Kilpelainen, M., E. O. Terho, H. Helenius, and M. Kosenvuo.** 2001. Home dampness, current allergic disease, and respiratory infections among young adults. *Thorax* **56:**462–467.
71. **Kingston, D.** 1971. Selective media in air sampling: a review. *J. Appl. Bacteriol.* **34:**221–232.
72. **Kirschner, R. A., Jr., B. C. Parker, and J. O. Falkinham III.** 1992. Epidemiology of infection by nontuberculous mycobacteria. *Am. Rev. Respir. Dis.* **145:**271–275.
73. **Kiviranta, H., A. Tuomainen, M. Reiman, S. Laitinen, A. Nevalainen, and J. Liesivuori.** 1999. Exposure to airborne microorganisms and volatile organic compounds in different types of waste handling. *Ann. Agric. Environ. Med.* **6:**39–44.
74. **Knudsen, G. R., and H. W. Spurr, Jr.** 1987. Field persistence and efficacy of five bacterial preparations for control of peanut leaf spot. *Plant Dis.* **71:**442–445.
75. **Kotimaa, M. H.** 1990. Occupational exposure to fungal and actinomycete spores during handling of wood chips. *Grana* **29:**153–156.
76. **Kozak, P. P., J. Gallup, L. H. Cummins, and S. A. Gillman.** 1980. Currently available methods for home mold surveys. II. Examples of problem homes surveyed. *Ann. Allergy* **45:**167–176.
77. **Krake, A. M., K. A. Worthington, K. M. Wallingford, and K. F. Martinez.** 1999. Evaluation of microbiological contamination in a museum. *Appl. Occup. Environ. Hyg.* **14:**499–509.
78. **Krysinska-Traczyk, E., C. Skorska, Z. Prazmo, J. Sitkowska, G. Cholewa, and J. Dutkiewicz.** 2004. Exposure to airborne microorganisms, dust, and endotoxin during flax scotching on farms. *Ann. Agric. Environ. Med.* **11:**309–317.
79. **Kullman, G. J., P. S. Thorne, P. F. Waldron, J. J. Marx, A. Ault, D. M. Lewis, P. D. Siegel, S. A. Olenchock, and J. A. Merchant.** 1999. Organic dust exposures from work in dairy barns. *Am. Ind. Hyg. Assoc. J.* **59:**403–413.
80. **Kumar, P., R. Marier, and S. H. Leech.** 1984. Respiratory allergies related to automobile air conditioners. *N. Engl. J. Med.* **311:**1619–1621.
81. **Lacey, J., and B. Crook.** 1988. Fungal and actinomycete spores as pollutants of the workplace and occupational allergens. *Am. Occup. Hyg.* **32:**515–533.
82. **Laitinen, S., J. Kangas, M. Kotimaa, J. Liesivuori, P. J. Martikainen, A. Nevalainen, R. Sarantila, and K. Husman.** 1994. Workers' exposure to airborne bacteria and endotoxins at industrial wastewater treatment plants. *Am. Ind. Hyg. Assoc. J.* **55:**1055–1060.
83. **Land, C. J., K. Hult, R. Fuchs, S. Hagelberg, and H. Lundstrom.** 1987. Tremorgenic mycotoxins from *Aspergillus fumigatus* as a possible occupational health problem in sawmills. *Appl. Environ. Microbiol.* **53:**787–790.
84. **Lawande, R. V., S. N. Abraham, I. John, and L. J. Egler.** 1979. Recovery of soil Amebas from the nasal passages of children during the dusty harmattan period in Zaria. *Am. J. Clin. Pathol.* **71:**201–203.
85. **Lighthart, B.** 1984. Microbial aerosols: estimated contribution of combine harvesting to an airshed. *Appl. Environ. Microbiol.* **47:**430–432.
86. **Lighthart, B.** 1994. Physics of microbial bioaerosols, p. 5–27. *In* B. Lighthart and A. J. Mohr (ed.), *Atmospheric Microbial Aerosols: Theory and Applications.* Chapman & Hall, New York, N.Y.
87. **Lighthart, B.** 2000. Mini-review of the concentration variations found in the alfresco atmospheric bacterial populations. *Aerobiologia* **16:**7–16.
88. **Lighthart, B., and A. J. Mohr.** 1987. Estimating downwind concentrations of viable airborne microorganisms in dynamic atmospheric conditions. *Appl. Environ. Microbiol.* **53:**1580–1583.
89. **Lighthart, B., and L. D. Stetzenbach.** 1994. Distribution of microbial bioaerosols, p. 68–98. *In* B. Lighthart and A. J. Mohr (ed.), *Atmospheric Microbial Aerosols: Theory and Applications.* Chapman & Hall, New York, N.Y.
90. **Lindemann, J., and C. D. Upper.** 1985. Aerial dispersal of epiphytic bacteria over bean plants. *Appl. Environ. Microbiol.* **50:**1229–1232.
91. **Lindemann, J., H. A. Constantinidou, W. R. Barchet, and C. D. Upper.** 1982. Plants as sources of airborne bacteria, including ice nucleation-active bacteria. *Appl. Environ. Microbiol.* **44:**1059–1063.
92. **Lindow, S. E., G. R. Knudsen, R. J. Seidler, M. V. Walter, V. W. Lambou, P. S. Amy, V. Prince, and S. C. Hern.** 1988. Aerial dispersal and epiphytic survival of *Pseudomonas syringae* during a pretest for the release of genetically engineered strains into the environment. *Appl. Environ. Microbiol.* **54:**1557–1563.
93. **Lugauskas, A., A. Kirkstaponis, and L. Sveistyte.** 2004. Airborne fungus in industrial environments—potential agents of respiratory diseases. *Ann. Agric. Environ. Med.* **11:**19–25.
94. **Lundholm, I. M.** 1982. Comparison of methods for quantitative determinations of airborne bacteria and evaluation of total viable counts. *Appl. Environ. Microbiol.* **44:**179–183.
95. **Lynch, J. M., and N. J. Poole.** 1979. Aerial dispersal and the development of microbial communities, p. 140–170. *In* J. M. Lynch and N. J. Poole (ed.), *Microbial Ecology: a Conceptual Approach.* John Wiley & Sons, Inc., New York, N.Y.
96. **Marshall, B., P. Flynn, D. Kamely, and S. B. Levy.** 1988. Survival of *Escherichia coli* with and without ColE1::Tn5 after aerosol dispersal in a laboratory and a farm environment. *Appl. Environ. Microbiol.* **54:**1776–1783.
97. **Michel, O.** 2001. Role of house-dust endotoxin exposure in aetiology of allergy and asthma. *Mediat. Inflamm.* **10:**301–304.
98. **Miller, J. D.** 1992. Fungi as contaminants of indoor air. *Atmos. Environ.* **26A:**2163–2172.
99. **Miller, J. D., A. M. Laflamme, Y. Sobol, P. Lafontaine, and R. Greenhalgh.** 1988. Fungi and fungal products in some Canadian houses. *Int. Biodeterior.* **24:**103–120.
100. **Milton, D. K., R. J. Gere, H. A. Feldman, and I. A. Greaves.** 1990. Endotoxin measurement: aerosol sampling and application of a new limulus method. *Am. Ind. Hyg. Assoc. J.* **51:**331–337.
101. **Mishra, S. K., L. Ajello, D. G. Ahearn, H. A. Burge, V. P. Kurup, D. L. Pierson, D. L. Price, R. S. Samson,**

R. S. Sandhu, B. Shelton, R. B. Simmons, and K. F. Switzer. 1992. Environmental mycology and its importance to public health. *J. Med. Vet. Mycol.* **30:**287–305.

102. **Nikulin, M., A. Pasanen, S. Berg, and E. Hintikka.** 1994. *Stachybotrys atra* growth and toxin production in some building materials and fodder under different relative humidities. *Appl. Environ. Microbiol.* **60:**3421–3424.

103. **Nikulin, M., K. Reijula, B. B. Jarvis, P. Veijalainen, and E. Hintikka.** 1997. Effects of intranasal exposure to spores of *Stachybotrys atra* in mice. *Fundam. Appl. Toxicol.* **35:**182–188.

104. **Nilsson, A., E. Kihlstrom, V. Lagesson, B. Wessen, B. Szponar, L. Larsson, and C. Tagesson.** 2004. Microorganisms and volatile organic compounds in airborne dust from damp residences. *Indoor Air* **14:**74–82.

105. **Norby, K.-C., A. S. Halstensen, O. Elen, P.-E. Clasen, W. Langseth, P. Kristensen, and W. Eduard.** 2004. Trichothecene mycotoxins and their determinants in settled dust related to grain production. *Ann. Agric. Environ. Med.* **11:**75–83.

106. **Olsson, M., C. Lidman, S. Latouche, A. Bjorkman, P. Roux, E. Linder, and M. Wahlgren.** 1998. Identification of *Pneumocystis carinii* f. sp. *hominis* gene sequences in filtered air in hospital environments. *J. Clin. Microbiol.* **36:** 1737–1740.

107. **Palmgren, M. S., and L. S. Lee.** 1986. Separation of mycotoxin-containing sources in grain dust and determination of their mycotoxin potential. *Environ. Health Perspect.* **66:**105–108.

108. **Panaccione, D. G., and C. M. Coyle.** 2005. Abundant respirable ergot alkaloids from the common airborne fungus *Aspergillus fumigatus*. *Appl. Environ. Microbiol.* **71:** 3106–3111.

109. **Papineni, R. S., and F. S. Rosenthal.** 1997. The size distribution of droplets in the exhaled breath of healthy human subjects. *J. Aerosol Med.* **10:**105–116.

110. **Pasanen, A., P. Kalliokoski, P. Pasanen, T. Salmi, and A. Tossavainen.** 1989. Fungi carried from farmers' work into farm homes. *Am. Ind. Hyg. Assoc. J.* **50:**631–633.

111. **Pasanen, A., K. Yli-Pietila, P. Pasanen, P. Kalliokoski, and J. Tarhanen.** 1999. Ergosterol content in various fungal species and biocontaminated building materials. *Appl. Environ. Microbiol.* **65:**138–142.

112. **Pedgley, D. E.** 1991. Aerobiology: the atmosphere as a source and sink for microbes, p. 43–59. *In* J. H. Andrews and S. S. Hirano (ed.), *Microbial Ecology of Leaves*. Springer-Verlag, New York, N.Y.

113. **Prazmo, Z., J. Dutkiewicz, and G. Cholewa.** 2000. Gram-negative bacteria associated with timber as a potential respiratory hazard for woodworkers. *Aerobiologia* **16:**275–279.

114. **Prazmo, Z., J. Dutkiewicz, C. Skorska, J. Sitkowska, and G. Cholewa.** 2003. Exposure to airborne gram-negative bacteria, dust, and endotoxin in paper factories. *Ann. Agric. Environ. Med.* **10:**93–100.

115. **Prazmo, Z., E. Krysinska-Traczyk, C. Skorska, J. Sitkowska, G. Cholewa, and J. Dutkiewicz.** 2003. Exposure to bioaerosols in a municipal sewage treatment plant. *Ann. Agric. Environ. Med.* **10:**241–248.

116. **Rahkonen, P., M. Ettala, and I. Loikkanen.** 1987. Working conditions and hygiene at sanitary landfills in Finland. *Ann. Occup. Hyg.* **31:**505–513.

117. **Ranalli, G., P. Principi, and C. Sorlini.** 2000. Bacterial aerosol emission from wastewater treatment plants: culture methods and bio-molecular tools. *Aerobiologia* **16:** 39–46.

118. **Randall, C. W., and J. O. Ledbetter.** 1966. Bacterial air pollution from activated sludge units. *Am. Ind. Hyg. Assoc. J.* **27:**506–519.

119. **Rao, C. Y., H. A. Burge, and J. C. S. Chang.** 1996. Review of quantitative standards and guidelines for fungi in indoor air. *J. Air Waste Manag. Assoc.* **46:**899–908.

120. **Rao, C. Y., J. D. Brain, and H. A. Burge.** 2000. Reduction of pulmonary toxicity of *Stachybotrys chartarum* spores by methanol extraction of mycotoxins. *Appl. Environ. Microbiol.* **66:**2817–2821.

121. **Rao, C. Y., J. M. Cox-Ganser, G. L. Chew, G. Doekes, and S. White.** 2005. Use of surrogate markers of biological agents in air and settled dust samples to evaluate a water-damaged hospital. *Indoor Air* **15:**90–97.

122. **Rautiala, S., T. Reponen, A. Hyvarinen, A. Nevalainen, T. Husman, A. Vehilainen, and P. Kalliokoski.** 1996. Exposure to airborne microbes during the repair of moldy buildings. *Am. Ind. Hyg. Assoc. J.* **57:**279–284.

123. **Rautiala, S., T. Reponen, A. Nevalainen, T. Husman, and P. Kalliokoski.** 1998. Control of exposure to airborne viable microorganisms during remediation of moldy buildings; report of three case studies. *Am. Ind. Hyg. Assoc. J.* **59:**455–460.

124. **Reinthaler, F. F., D. Haas, G. Feierl, R. Schlacher, F. O. Pichler-Semmelrock, M. Kock, G. Wust, O. Feenstra, and E. Marth.** 1999. Comparative investigations of airborne culturable microorganisms in selected waste treatment facilities and in neighbouring residential areas. *Zentbl. Hyg. Umweltmed.* **202:**1–17.

125. **Reynolds, S. J., A. J. Streifel, and C. E. McJilton.** 1990. Elevated airborne concentrations of fungi in residential and office environments. *Am. Ind. Hyg. Assoc. J.* **51:** 601–604.

126. **Rivera, F., G. Roy-Ocotla, I. Rosas, E. Ramirez, P. Bonilla, and F. Lares.** 1987. Amoeba isolation from the atmosphere of Mexico City and environs. *Environ. Res.* **42:**149–154.

127. **Rubin, S. A.** 1997. Tuberculosis and atyptical mycobacterial infections in the 1990s. *Radiographics* **17:**1051–1059.

128. **Rylander, R., and R. Etzel.** 1999. Introduction and summary: workshop on children's health and indoor mold exposure. *Environ. Health Perspect.* **107:**465–468.

129. **Rylander, R., and B. Fogelmark.** 1994. Inflammatory responses by inhalation of endotoxin and (1-3)-β-D-glucan. *Am. J. Ind. Med.* **25:**101–102.

130. **Rylander, R., and Y. Megevand.** 2000. Environmental risk factors for respiratory infections. *Arch. Environ. Health* **55:**300–303.

131. **Rylander R., and J. Vesterlund.** 1982. Airborne endotoxins in various occupational environments. *Prog. Clin. Biol. Res.* **93:**399–409.

132. **Rylander, R., K. Persson, H. Goto, K. Yuasa, and S. Tanaka.** 1992. Airborne beta-1,3-glucan may be related to symptoms in sick buildings. *Indoor Environ.* **1:**263–267.

133. **Salem, H., and D. E. Gardner.** 1994. Health aspects of bioaerosols, p. 304–330. *In* B. Lighthart and A. J. Mohr (ed.), *Atmospheric Microbial Aerosols: Theory and Applications*. Chapman & Hall, New York, N.Y.

134. **Samson, R. A., J. Houbraken, R. C. Summerbell, B. Flannigan, and J. D. Miller.** 2001. Common and important species of fungi and actinomycetes in indoor environments, p. 287–473. *In* B. Flannigan, R. A. Samson, and J. D. Miller (ed.), *Microorganisms in Home and Indoor Work Environments*. Taylor & Francis, New York, N.Y.

135. **Sattar, S. A., and M. K. Ijaz.** 1987. Spread of viral infections by aerosols. *Crit. Rev. Environ. Control* **17:**89–131.

136. **Sawyer, M. H., C. J. Chamberlain, Y. N. Wu, N. Aintablian, and M. R. Wallace.** 1994. Detection of vari-

cella-zoster virus DNA in air samples from hospital rooms. *J. Infect. Dis.* **169:**91–94.
137. **Schaal, K. P.** 1991. Medical and microbiological problems arising from airborne infection in hospitals. *J. Hosp. Infect.* **18:**451–459.
138. **Seal, D. V., and R. P. Clark.** 1990. Electronic particle counting for evaluating the quality of air in operating theatres: a potential basis for standards? *J. Appl. Bacteriol.* **68:**225–230.
139. **Sebastian, A., and L. Larsson.** 2003. Characterization of the microbial community in indoor environments: a chemical-analytical approach. *Appl. Environ. Microbiol.* **69:**3103–3109.
140. **Seuri, M., K. Husman, H. Kinnunen, M. Reiman, R. Kreus, P. Kuronen, K. Lehtomaki, and M. Paananen.** 2000. An outbreak of respiratory diseases among workers at a water-damaged building—a case report. *Indoor Air* **10:**138–145.
141. **Siersted, H. C., and S. Gravesen.** 1993. Extrinsic allergic alveolitis after exposure to the yeast *Rhodotorula rubra*. *Allergy* **48:**298–299.
142. **Simmons, R. B., L. J. Rose, S. A. Crow, and D. H. Ahearn.** 1999. The occurrence and persistence of mixed biofilms in automobile air conditioning systems. *Curr. Microbiol.* **39:**141–145.
143. **Slote, L.** 1976. Viral aerosols—a potential occupationally related health threat in aerated wastewater treatment systems. *J. Environ. Health* **38:**310–314.
144. **Snyder, J. W., and A. S. Weissfeld.** 2003. Laboratory detection of potential agents of bioterrorism, p. 121–128. *In* P. R. Murray, E. J. Baron, J. H. Jorgensen, M. A. Pfaller, and R. H. Yolken (ed.), *Manual of Clinical Microbiology*, 8th ed. ASM Press, Washington, D.C.
145. **Sorenson, W. G.** 1999. Fungal spores: hazardous to health? *Environ. Health Perspect.* **107:**469–472.
146. **Sorenson, W. G., D. G. Frazer, B. B. Jarvis, J. Simpson, and V. A. Robinson.** 1987. Trichothecene mycotoxins in aerosolized conidia of *Stachybotrys atra*. *Appl. Environ. Microbiol.* **53:**1370–1375.
147. **Spangler, J. D., and K. Sexton.** 1983. Indoor air pollution: a public health perspective. *Science* **221:**9–17.
148. **Stärk, K. D., C. J. Nicolet, and J. Frey.** 1998. Detection of *Mycoplasma hyopneumoniae* by air sampling with a nested PCR assay. *Appl. Environ. Microbiol.* **64:**543–548.
149. **Straatsma, G., R. A. Samson, T. W. Olijnsma, H. J. M. Opden Camp, J. P. G. Gerrits, and L. J. L. D. van Griensven.** 1994. Ecology of thermophilic fungi in mushroom compost with emphasis on *Scytalidium thermophilum* and growth stimulation of *Agaricus bisporus* mycelium. *Appl. Environ. Microbiol.* **60:**454–458.
150. **Swan, J. R. M., and B. Crook.** 1998. Airborne microorganisms associated with grain handling. *Ann. Agric. Environ. Med.* **30:**7–15.
151. **Teltsch, B., and E. Katzenelson.** 1978. Airborne enteric bacteria and viruses from spray irrigation with wastewater. *Appl. Environ. Microbiol.* **35:**290–296.
152. **Thorne, P. S., M. S. Kiekhaefer, P. Whitten, and K. J. Donham.** 1992. Comparison of bioaerosol sampling methods in barns housing swine. *Appl. Environ. Microbiol.* **58:**2543–2551.
153. **Tjade, O. H., and I. Gabor.** 1980. Evaluation of airborne operating room bacteria with a Biap slit sampler. *J. Hyg. Camb.* **84:**37–40.
154. **Tourville, D. R., W. I. Weiss, P. T. Wertlake, and G. M. Leudemann.** 1972. Hypersensitivity pneumonitis due to contamination of home humidifier. *J. Allergy Clin. Immunol.* **49:**245–251.
155. **Van Netten, C., R. Brands, and D. Dill.** 1997. Investigation and remediation of diesel converted trolley buses associated with extensive fungal growth and health complaints. *Am. Ind. Hyg. Assoc. J.* **58:**726–731.
156. **Vesper, S., D. G. Dearborn, I. Yike, T. Allan, J. Sobolewski, S. F. Hinkley, B. B. Jarvis, and R. A. Haugland.** 2000. Evaluation of *Stachybotrys chartarum* in the house of an infant with pulmonary hemorrhage: quantitative assessment before, during and after remediation. *J. Urban Health* **77:**68–85.
157. **Wady, L., and L. K. Larsson.** 2005. Determination of microbial volatile organic compounds adsorbed on house dust particles and gypsum board using SPME/GC-MS. *Indoor Air* **15:**27–32.
158. **Walters, M., D. Milton, L. Larsson, and T. Ford.** 1994. Airborne environmental endotoxin: a cross-validation of sampling and analysis techniques. *Appl. Environ. Microbiol.* **60:**996–1005.
159. **Wan, G. H., and C. S. Li.** 1999. Indoor endotoxin and glucan in association with airway inflammation and systemic symptoms. *Arch. Environ. Health* **54:**172–179.
160. **Warhurst, D.** 1977. Humidifier fever: amoebae and allergic lung disease. *Thorax* **32:**653–663.
161. **Wicklow, D. T., and O. L. Shotwell.** 1983. Intrafungal distribution of aflatoxin among conidia and selerotia of *Aspergillus flavus* and *Aspergillus parasiticus*. *Can. J. Microbiol.* **29:**1–5.
162. **Wickman, H. H.** 1994. Deposition, adhesion, and release of bioaerosols, p. 99–165. *In* B. Lighthart and A. J. Mohr (ed.), *Atmospheric Microbial Aerosols: Theory and Applications*. Chapman & Hall, New York, N.Y.
163. **Wickman, M., S. Gravesen, S. L. Nordvall, G. Pershagen, and J. Sundell.** 1992. Indoor viable dust-bound microfungi in relation to residential characteristics, living habits, and symptoms in atopic and control children. *J. Allergy Clin. Immunol.* **89:**752–759.
164. **Wiener, S. L.** 2000. Biological warfare defense, p. 119–146. *In* R. A. Zilinskas (ed.), *Biological Warfare: Modern Offense and Defense*. Lynne Rienner Publishers, Boulder, Colo.
165. **Winn, W. C., Jr.** 1988. Legionnaires disease: historical perspective. *Clin. Microbiol.* **1:**60–81.
166. **Woodard, E. D., B. Friedlander, R. J. Lesher, W. Font, R. Kinsey, and F. T. Hearne.** 1988. Outbreak of hypersensitivity pneumonitis in an industrial setting. *JAMA* **259:**1965–1969.
167. **Zeterberg, J. M.** 1973. A review of respiratory virology and the spread of virulent and possibly antigenic viruses via air conditioning systems. *Ann. Allergy* **31:** 228–234.
168. **Zuravleff, J. J., V. L. Yu, J. W. Shonnard, J. D. Rihs, and M. Best.** 1983. *Legionella pneumophila* contamination of a hospital humidifier. *Am. Rev. Respir. Dis.* **128:**657–661.

Sampling for Airborne Microorganisms

SERGEY A. GRINSHPUN, MARK P. BUTTNER, AND KLAUS WILLEKE

74

Microbiologists have confronted the challenges of sampling and analysis of airborne microorganisms since the early 20th century. Today, the concentration and composition of airborne microorganisms are of interest in diverse areas such as agricultural and industrial settings, medicine, home and office environments, and military research. In all of these applications, the term "bioaerosol" is used to refer to airborne biological particles, such as bacterial cells, fungal spores, viruses, and pollen grains, and to their fragments and by-products. A wide variety of bioaerosol sampling and analysis methods have been used, and new methods are being developed (3, 4, 6–9, 15, 20, 28, 33, 34, 44, 50, 55, 63, 70, 71, 80–82, 85, 96, 99, 100, 106–108, 110–113, 115, 117, 126, 139, 141, 144–146). However, several problems remain to be solved. For instance, no single sampling method is suitable for the collection and analysis of all types of bioaerosols and no standardized protocols are currently available. Therefore, data from different studies are often difficult to compare because of differences in sampler designs, collection times, airflow rates, and analysis methods. In addition, human exposure limits have not been established for bioaerosols because of the lack of exposure, dose, and response data. This complicates the use of sampling results for risk assessment.

It is important to carefully consider the objectives of sampling before any samples are taken. After determining what information is desired, an appropriate sampling and analysis method can be selected and incorporated into the monitoring design. An important purpose of this chapter is to present various bioaerosol sampling and analysis methods to facilitate the selection of instrumentation and techniques. The principles of bioaerosol sampling are presented, followed by a review of sampling methods and techniques currently available, including the results of performance evaluations of the various sampler types. Equipment calibration and air sampling considerations such as collection times and the number of samples are discussed.

Measurement of airborne microorganisms with a bioaerosol sampler often aims at documenting the presence of specific sources (27, 30, 38, 41, 49, 61, 62, 64–69, 72, 93, 105, 116, 121, 122, 126, 127, 129–131, 133, 138, 140). However, when no biological particles are collected and identified by air sampling, one cannot conclude that their sources are absent. This lack of certainty always exists, particularly due to the small air sampling volume of conventional bioaerosol samplers over a limited period of time. The sampling time and the sampler's location may not coincide with the time and place of bioaerosol release from the source. However, the presence of a source does not necessarily imply that the bioaerosol level is elevated because release from sources is often sporadic. To fully evaluate health effects associated with airborne fungi, it is very important to be able to measure airborne-spore concentrations and the source strength. This chapter describes air and surface sampling. Analysis methods are addressed in chapter 75, beginning with traditional culture and total-count methods and concluding with some recently developed biochemical and molecular techniques (8, 33, 40, 42, 43, 48–50, 51, 53, 74, 96, 99, 106–108, 114, 119, 147, 148).

BIOAEROSOL SAMPLING

The collection of bioaerosol samples is performed through active air sampling. The advantages and disadvantages of active sampling have been discussed in the literature (18, 34, 117, 139). The objective of active bioaerosol sampling is the efficient removal and collection of biological particles from the air in a manner which does not affect the ability to detect the organisms (e.g., no alteration in culturability or biological integrity [11]). This ability is dependent on the physical and biological characteristics of the organisms and on the physical features of the sampling instrument (34, 135). While a microorganism is airborne, its motion is governed by the same laws of physics that apply to biologically inert particles. The three principal collection methods used in quantitative bioaerosol sampling are impaction, impingement, and filtration. Some other techniques, such as gravitational settling and electrostatic precipitation, have been employed or explored for bioaerosol sampling (20, 100), but gravitational (settle plates) sampling is not quantitative and electrostatic precipitation sampling is not routinely used.

Methods of Collection

Impaction

The impaction method separates particles from the air stream by utilizing the inertia of the particles to force their deposition onto a solid or semisolid collection surface. The collection surface is usually an agar medium for culture-based analysis or an adhesive-coated surface that can be analyzed microscopically. The impaction process depends upon the inertial properties of the particle (e.g., density and velocity) and on the physical parameters of the impactor (e.g., the inlet nozzle dimensions and the airflow pathway) (117). Air enters the sampler through an inlet nozzle, and the sampled air exits as the air is directed at the collection surface. Particles with sufficient inertia are impacted. Particles with lower inertia remain airborne with the airflow (Fig. 1A). Centrifugal impaction also uses inertial forces to separate the particles from the air stream, but in a radial geometry (Fig. 1B).

Liquid Impingement

Liquid impingement is similar to impaction in that the inertial force of the particle is the principal force removing it from the air. However, the collection medium is a liquid (usually a dilute buffer solution) and the collected microorganisms move around freely in the bubbling liquid (Fig. 1C). As a result, aggregates of cells may be broken apart. Particles remaining in the air stream may diffuse to the surface of a bubble and be transferred to the collection buffer. The collection of bioaerosol particles in a liquid medium allows division of the sample and the potential application of several analysis methods, as described below.

Filtration

Filtration achieves the separation of particles from the air stream by passage of the air through a porous medium, usually a membrane filter. The collection of particles depends on their physical properties (e.g., size, shape, and density), the filter pore size, and the airflow rate (83). Inertial forces and other mechanisms such as interception, diffusion, and electrostatic attraction result in the collection of particles on the surface of the filter (Fig. 1D). Simultaneous action of all of these forces removes particles smaller than the pore size of the filter (70).

Gravitational Settling

Gravitational settling, or depositional sampling, is a nonquantitative collection method in which an agar medium is exposed to the environment and airborne organisms are collected primarily by gravity. This sampling method is often used because it is inexpensive and easily performed. However, the collection of airborne microorganisms by using this method is affected by the sizes and shapes of the particles and by the motion of the surrounding air (117). As a result, large particles are more likely to be deposited onto the collection surface (20). This can lead to misrepresentation of the prevalence of airborne microorganisms and the exclusion of smaller particles from the collection (127). In addition, the concentration of the airborne microorganisms cannot be determined by gravity sampling because the volume of air from which the particles originate is unknown. Sampling based on gravitational settling has been compared with various methods that pass a known volume of air to the collection medium. The results show that the airborne-microorganism concentrations derived from gravity sampling are not qualitatively or quantitatively accurate and do not compare favorably with those obtained by other sampling methods (29, 122, 123, 127).

Electrostatic Precipitation

Commonly used sampling methods (e.g., inertia-based impaction and impingement and filter-based methods) have been shown to adversely affect the viability of microorganisms being sampled (132, 142). Therefore, other methods are being explored for bioaerosol sampling. These include electrostatic precipitation, in which bioaerosol particles are charged in the sampler's inlet. This results in their cross-sectional migration in an electric field inside the sampler and subsequent deposition on the charged plates, from which the microorganisms can be extracted and analyzed (100).

A

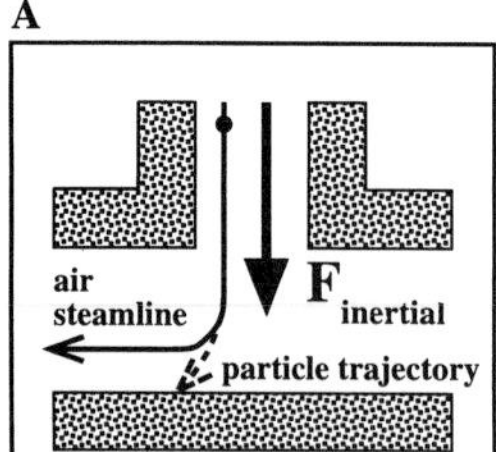

B

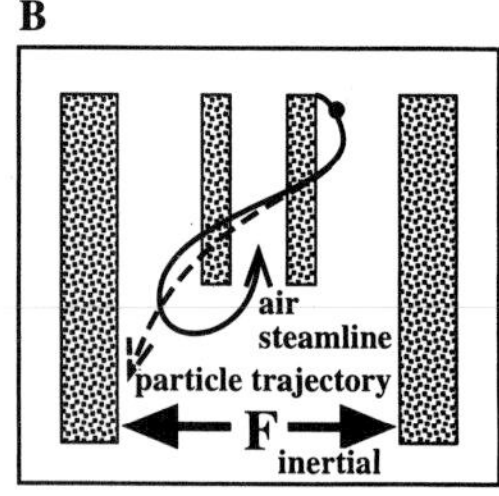

C

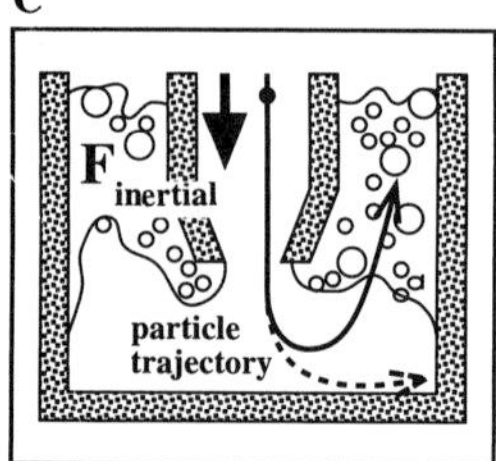

D

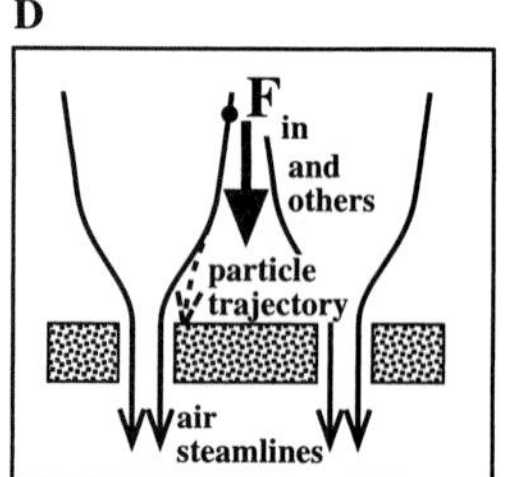

FIGURE 1 Mechanisms of collection utilized in bioaerosol sampling. (A) Solid plate impaction; (B) centrifugal impaction; (C) liquid impingement; (D) filtration. F_{in} or $F_{inertial}$, inertial force. (Adapted from reference 111 with kind permission from Elsevier Science Ltd., Kidlington, United Kingdom.)

Sampler Types

There is a wide variety of commercially available bioaerosol samplers (34, 70, 117). The selection of a sampler depends on a number of factors, such as sampler performance, expected bioaerosol concentration, and the analysis method. Prior to initiating an investigation, there should be an understanding of the specific objectives of air monitoring and the limitations of the various sampling methods. Although standardized sampling protocols are not available at the present time, guidelines have been published to assist in the selection of bioaerosol sampling protocols (34, 70). Some of the more widely used bioaerosol sampling methods are listed in Table 1 and discussed in the following sections.

Impactor Samplers

Impaction is the most commonly used method of collection for airborne microorganisms, and a variety of impactor samplers are commercially available (Table 1). They differ in the dimensions of their nozzles and jet-to-plate distances, the shapes of the nozzles, the numbers of nozzles, and the numbers of stages. If air is drawn through a single nozzle, the nozzle is usually rectangular and the impactor is referred to as a slit sampler. If there are several nozzles, usually circular in shape, the plate with the impaction nozzles resembles a sieve and the impactor is sometimes referred to as a sieve sampler. If there are several stages with successively smaller nozzles, the sampler is referred to as a cascade impactor.

The Andersen single-stage impactor sampler (Graseby Andersen, Smyrna, Ga.) is a commonly used device for the collection of culturable airborne microorganisms (10). This sampler has a single stage with 400 precision-drilled holes for the collection of particles of $\leq$10 μm in diameter. A six-

TABLE 1 General characteristics of several commercially available bioaerosol samplers

Sampler(s)	Collection medium(a)	Airflow rate(s) (liters/min)	Sample analysis method(s)	Comments
Impaction				
Air-O-Cell sampling cassette	Adhesive-coated surface	12–30	Microscopy	Determination of total fungal spores and pollen
Andersen viable impactors, 1-, 2-, and 6-stage	Agar; 100-mm-diam petri dishes	28.3	Culture	Particle size discrimination (2- and 6-stage models); vacuum pump required; counts corrected for multiple impaction
Biotest RCS and RCS Plus	Agar; plastic strips	40 (RCS), 50 (RCS Plus)	Culture	Portable; battery operated
Burkard spore traps, 24-h, 7-day, and personal samplers	Adhesive-coated surface, tape, or glass slide	10	Microscopy	Determination of total fungal spores and pollen
Burkard portable air sampler	Agar; 100-mm-diam petri dishes	10–20	Culture	Portable; battery operated; counts corrected for multiple impaction
Casella MK-II	Agar; petri dishes, 100 mm in diam (small inlet), 150 mm in diam (large inlet)	30 (small inlet), 700 (large inlet)	Culture	Time discrimination up to 6 min; limited availability
Cyclex bioaerosol impact sampler	Adhesive-coated surface	20	Microscopy	Determination of total fungal spores
Cyclex-d cassette	Adhesive-coated surface	20	Microscopy	Portable; determination of total fungal spores
Mattson/Garvin 220 and P-320	Agar; 150-mm-diam petri dishes	28.3	Culture	Time discrimination up to 1 h
Mico-5 Microcell	Adhesive-coated surface	5	Microscopy	Portable; determination of total fungal spores
New Brunswick slit-to-agar, STA-101, -203, and -303	Agar; 150-mm-diam petri dishes	50	Culture	Time discrimination up to 1 h; vacuum pump required for some models
SAS Super 90, Super 100, compact, and high flow	Agar; 55-mm-diam contact plates; 85-mm-diam contact plates	90 (180 [high flow])	Culture	Portable; battery operated; counts corrected for multiple impaction
Impingement				
All-glass impingers (AGI-30 and AGI-4)	Liquid	12.5	Culture, microscopy, biochemical assay, PCR, immunoassay	High and low bioaerosol concentrations; vacuum pump required
BioSampler	Liquid	12.5	Culture, microscopy, biochemical assay, PCR, immunoassay	High and low bioaerosol concns; short- and long-term sampling; viscous and nonviscous fluids; vacuum pump required
Burkard multistage liquid impinger	Liquid	20	Culture, microscopy, biochemical assay, PCR, immunoassay	High and low bioaerosol concns; particle size discrimination; vacuum pump required
Filtration				
25-, 37-, or 47-mm-pore-size filter cassettes	Filter membrane	1–50	Microscopy, biochemical assay, PCR, immunoassay, culture (spores)	Loss of culturable vegetative cells; portable; useful for personal monitoring; vacuum pump required
Sartorius MD 8 air sampler	Gelatin membrane filter	42–133	Culture	Gelatin filter reduces desiccation stress, high-volume sampling; virus collection

stage version consists of six stages with decreasing nozzle diameters such that successive stages collect progressively smaller particles. Thus, the six-stage sampler measures the culturable bioaerosol concentrations in approximate particle size ranges. A two-stage model is also available. The surface air system (SAS; International PBI, Milan, Italy, distributed by Bioscience International, Rockville, Md.) and Burkard (Burkard Manufacturing Co., Ltd., Rickmansworth, Hertfordshire, England) portable air samplers are battery-powered one-stage impactors which utilize agar-filled plates as the collection medium. The Reuter centrifugal sampler (RCS) and RCS Plus (Biotest Diagnostics Corp., Denville, N.J.) are portable, battery-powered samplers which centrifugally impact the microorganisms onto agar sections arranged on strips. Portable samplers do not require external vacuum pumps and electrical outlets but are available only as single-stage devices.

Slit impactors deposit the bioaerosol onto an agar surface for the estimation of numbers of viable cells. Examples are the Casella MK-II (distributed by BGI, Inc., Waltham, Mass.), the Mattson/Garvin air sampler (Barramundi Corp., Homosassa, Fla.), and the New Brunswick slit-to-agar sampler (New Brunswick Scientific Co., Inc., Edison, N.J.). Collection onto an adhesive-coated surface is used with microscopic enumeration of the collected particles, usually fungal spores or pollen grains. Examples of this type of sampler are the Burkard personal impactor sampler (Burkard Manufacturing Co., Ltd.); the Air-O-Cell sampling cassette (Zefon International, Inc., St. Petersburg, Fla.); and the Cyclex bioaerosol impact sampler, the Cyclex-d cassette, the Mico-5 Microcell, and the Allergenco-d cassette (EMS, Inc., Charleston, S.C.). Many slit impactors have a moving collection surface to provide temporal discrimination of the bioaerosol concentration.

Liquid Impinger Samplers

Liquid impinger samplers are commonly used for the retrieval of bioaerosol particles over a wide range of airborne-particle concentrations (Table 1). For analysis, the liquid sample can be concentrated by filtration or diluted by liquid addition, depending on the concentration of microorganisms. Several culture media can be inoculated with aliquots of the collection medium for the assessment of groups of microorganisms with different culture requirements. The liquid samples may also be analyzed by biochemical, immunological, and molecular biological assays to detect the presence of specific microorganisms, culturable or nonculturable. The AGI-30 all-glass impinger sampler (Ace Glass, Inc., Vineland, N.J.) is a widely used liquid impinger sampler that has a curved inlet tube designed to simulate the nasal passage, making this sampler useful for studying the respiratory infection potential of bioaerosols (36, 54, 70). For other applications, the inlet tube is washed with a known volume of collection fluid to recover nonrespirable airborne particles. The AGI-30 has an impaction distance of 30 mm from the jet to the bottom of the sampler. The AGI-4 model features a shorter distance of 4 mm to improve particle collection efficiency over that of the AGI-30. However, added sampling stress may result from impaction against the glass bottom of the sampler, leading to a loss of viability of cells. The Burkard multistage liquid impinger is a stainless steel sampler which collects particles in three size fractions (i.e., >10, 4 to 10, and <4 μm).

While liquid impingement has numerous advantages, most of the currently available impingers that are efficient in collecting small bioaerosol particles require a very high sampling velocity, which usually results in a violent motion of the collection fluid. The latter causes reaerosolization of the initially collected bioparticles and stress that leads to the viability loss (59, 87–89).

Conventional impingers can be used only with water or liquids that have about the same viscosity as water. The recently developed BioSampler (SKC, Inc., Eighty Four, Pa.) combines impingement into a liquid with centrifugal motion (146). This sampler can be used with buffer or with viscous collection fluids (e.g., heavy white mineral oil). Having the same inlet geometry and the same airflow rate of 12.5 liters/min as AGI samplers, the BioSampler achieves particle collection by drawing aerosol through three nozzles that are directed at an angle toward the inner sampler wall. During normal operation, the liquid swirls upward on the sampler's inner wall and removes collected particles.

Another approach to bioaerosol sampling with a liquid utilizes a porous medium submerged into a liquid layer so that the aspirated air is blown through it. As a result, the aerosol flow is split into many very small bubbles and particulates are effectively removed on the walls of the impinger's vessel. A prototype sampler has been recently developed and evaluated (3–5), but it is not yet commercially available.

Filtration Sampling

The collection of airborne microorganisms onto a filter material is used in bioaerosol monitoring due to its simplicity, low cost, and versatility (Table 1). Air samples are usually collected on 25-, 37-, or 47-mm-diameter filter membranes housed in disposable plastic cassettes. A variety of filtration devices are available from several manufacturers (e.g., Gelman Sciences, Ann Arbor, Mich.; Millipore Corp., Bedford, Mass.; Nuclepore, Corning Costar Corp., Cambridge, Mass.; Poretics Corp., Livermore, Calif.; and SKC, Inc.). Polycarbonate, mixed cellulose ester, or polyvinyl chloride filter material may be used depending on the nature of the bioaerosol and the method of sample analysis (70, 83). Filter membrane pore sizes range from 0.01 to 10 μm, with air sample flow rates from 1 to 50 liters/min. Filtration sampling is adaptable to a variety of assays (Table 1), but loss of viability of vegetative cells may occur, presumably due to desiccation stress during sampling (71, 94, 135, 142). The MD8 air sampler (Sartorius AG, Göttingen, Germany) collects airborne microorganisms on a gelatin filter to reduce desiccation stress. The gelatin membrane is incubated on the agar medium of choice for culture analysis.

Sampler Performance

Bioaerosol sampler performance has been reviewed previously (34, 54, 58, 70, 76, 111, 134), and numerous laboratory and field studies comparing the utilities of various sampling methods have been performed (Table 2). Data from these studies are often difficult to compare because of differences in samplers, the lengths of sampling time, the volumes of air sampled, the sample analysis methods, and the characteristics of the bioaerosols being measured. While no single sampler type currently available is ideally suited for the collection and analysis of all types of bioaerosols, consideration of the theoretical aspects of air sampling and the experimental results of performance studies may facilitate the selection of the appropriate sampling method for a particular monitoring situation.

The performance of bioaerosol samplers can be divided into physical and biological components. Physical parameters include inlet sampling efficiency and collection efficiency, whereas biological sampling efficiency reflects primarily the

TABLE 2 Calculated and reported cutoff diameters (d_{50}s) for several commercially available bioaerosol samplers

Samplers	d_{50} (μm) Calculated (reference)	d_{50} (μm) Reported (reference[s])	Reference(s) for performance evaluation studies
Impaction			
Air-O-Cell sampling cassette (15 liters/min)		2.3 (136), 2.6 (86)	6, 136
Andersen 6-stage viable impactor			10, 14, 16, 19, 28, 29, 52, 71, 73, 76, 84, 129
Stage 1	0.24 (70), 6.61 (111)	7.0	
Stage 2	4.21	4.7	
Stage 3	2.86	3.3	
Stage 4	1.84	2.1	
Stage 5	0.94	1.1	
Stage 6	0.58	0.65	
Andersen 2-stage viable impactor			52, 71, 76, 135, 149
Stage 0	6.28	8.0	
Stage 1	0.83	0.95	
Andersen 1-stage viable impactor	0.58	0.65	71, 73, 135
Biotest RCS	7.5	3.8	71, 77, 98, 115
RCS Plus		0.82	
Burkard spore traps, 24-h, 7-day			19, 29, 128
Standard nozzle	3.70		
High-efficiency nozzle	2.17		
Burkard personal sampler	2.52		
Burkard portable air sampler	4.18		
Casella MK-II, small inlet	0.67		16
Cyclex bioaerosol impact sampler		~(57), 2 (86, 110)	57
Cyclex-d cassette		~(57), 2 (86, 110)	57
Mattson/Garvin	0.53		71, 115
Mico-5 Microcell		~1	57
SAS			29, 71
Compact	1.97	2.0	
High flow	1.52 (70), 1.45 (111)	2.0	
Impingement			
All-glass impinger AGI-30	0.30		10, 14, 28, 71, 76, 84, 135, 144
BioSampler		<0.3 (146)	87, 88, 146
Burkard multistage liquid impinger			
Stage 1		10	39, 149
Stage 2		4	
Stage 3			

effect of sampling on the culturable state of the microorganisms. Inlet sampling efficiency refers to the ability of the sampler inlet to extract particles from the ambient environment without bias due to the sizes, shapes, or densities of the particles. The inlet characteristics of several bioaerosol samplers have been calculated for different types of bioaerosol particles sampled under various conditions (54). Depending on the external wind direction and velocity relative to the inlet geometry and flow characteristics, increased or, more commonly, decreased particle concentration measurements may be obtained relative to the true concentration in the environment (17).

Collection efficiency is the ability of the sampler to remove particles from the air stream and transfer them to the collection medium. In an impactor sampler, the physical characteristics of the impaction nozzle(s) and the airflow rate are used to calculate the cutoff size (d_{50}) in the impaction stage. The d_{50} is defined as the particle diameter at which 50% of the particles are collected. However, because of the sharp cutoff characteristics of impactor samplers, the d_{50} is generally considered to be the particle diameter above which all particles are collected while all those with diameters below the d_{50} pass through (101, 111). Theoretical d_{50}s have been calculated for several bioaerosol samplers (70, 111) and are shown in Table 2. For efficient collection, it is important to choose an impactor for which the d_{50} is below the mean size of the microorganism being sampled. For membrane filters, the collection efficiency is approximately 100% for particles larger than the pore size (92). In an inertial device (e.g., an impactor), the "aerodynamic" diameter of the particle is the diameter of a unit density sphere that has the same gravitational settling velocity as the particle in question (13). A nonspherical microorganism is thus described by a single dimension. The d_{50} depends on several parameters, including the ratio of the jet-to-plate distance (S) to the impactor's nozzle size (W). Most commonly used bioaerosol impactors meet the conventional Marple's design criteria so that S/W is greater than the established threshold (1.5 for rectangular nozzles and 1 for the circular ones) (101). Recent studies have shown that these samplers underestimate the concentration of some bioaerosol particles that they are designed to measure (e.g., fungal spores) (136). The d_{50}s of several of the most commonly used single-stage impaction-based spore collectors (e.g., the Air-O-Cell cassette and the Burkard personal impactor sampler) are about 2.5 μm or greater. Some fungal species produce spores of 1.8 to 2.5 μm in aerodynamic diameter. An increase in the sampling flow rate would help reduce the d_{50}s of these impactors (6, 57, 101, 117, 136). However, very high sampling flow rates require a rather powerful pump which limits the use of such a sampler in the field. In addition, a high impaction velocity may cause particle bounce that decreases the actual collection efficiency, particularly for spores. A high-velocity impaction also affects the viability of stress-sensitive microorganisms. An alternative approach was recently developed in which the nondimensional jet-to-plate distance was proposed to be very small, i.e., $\leqslant 1$ (57). The feasibility of these "incorrectly designed" impactors (relative to that of impactors meeting the above-quoted Marple's criteria) for the total spore collection and enumeration has been demonstrated through laboratory and field evaluations (57). Several impactors with very small jet-to-plate distances became commercially available in 2004 and 2005, including the Cyclex bioaerosol impact sampler, the Cyclex-d cassette, and the Mico-5 Microcell (EMS, Inc.).

Biological sampling efficiency differs among bioaerosol sampler types. Ideally, each sampler should collect all airborne microorganisms without altering the culturability or the biological integrity required for the detection and/or quantification of the microorganisms. Biological effects of sampling are difficult to assess because of the heterogeneity of bioaerosols with respect to particle size, biological composition, and environmental factors. Airborne microorganisms are subjected to a variety of environmental stressors (e.g., UV radiation, chemical pollutants, desiccation, and temperature extremes). Consequently, many organisms may be in a nonculturable state (37). In addition, the stress of impaction may injure the collected microorganisms, depending upon their physiological characteristics (132). Although several analytical techniques do not make a distinction between viable and nonviable cells (103), the majority rely on culturing of the collected microorganisms. Therefore, sampling stresses, which can result in a reduction in the culturability of airborne microorganisms, are important in assessing the overall bioaerosol sampler performance. For example, filtration sampling, though highly efficient for the collection of airborne microorganisms (1, 2, 6, 22), has the disadvantage of viability losses for vegetative cells, presumably as a result of desiccation (71, 94, 135). Therefore, filtration sampling in combination with culture analysis is generally used when bioaerosol concentrations are very high and sampling times are short. Filtration sampling is also used in monitoring desiccation-resistant forms such as fungal spores and bacterial endospores or in combination with a total-count method of analysis (112).

The stress imposed by sampling may also result in a loss of culturability when agar or liquid collection media are used. However, it has been observed that the recovery of culturable cells can be improved by using certain culture media (21, 109, 138) or by adding certain compounds to the collection medium, e.g., osmoprotectants, which aid in the resuscitation of stressed or damaged cells (102). Impaction into different media also permits the differentiation of metabolic from structural injuries (132). The length of collection time (discussed in the following section) also has a major role in the efficacy of air sampling for the retrieval of culturable microorganisms.

Most of the bioaerosol sampler performance evaluations have not distinguished between physical and biological parameters in assessing efficiency, focusing instead on an overall measure of sampling efficacy. Although there is no single sampling method which is appropriate for all bioaerosols, the Andersen six-stage viable impactor and the AGI-30 sampler have been suggested as reference methods (34) and many sampler efficiency studies have included these samplers in side-by-side comparisons (Table 2). The principal advantage of the agar impactor samplers is the collection of organisms directly onto the culture medium. However, one of the disadvantages of this sampling method is the nonculturable component of the bioaerosol, which is not measured and can be a significant percentage of the total composition (112). Other problems with agar impactor samplers which can produce erratic results and underestimation of the true bioaerosol concentration are (i) particle bounce (e.g., particles may rebound off the collection surface or another particle and reenter the air stream [117]), (ii) aggregation or clumping of microorganisms (e.g., multiple organisms impact at the same place on the agar surface and are enumerated as a single colony), (iii) electrostatic forces (e.g., biological particles are attracted to the plastic rims or other surfaces of the collection device [10]), and (iv) overloading (the overall performance of agar impactors can be adversely affected by overloading when bioaerosol concentrations are high, resulting in agar plates or strips which contain colonies that overlap and are too numerous to count [31, 32]). In an impinger, microorganisms are collected by impaction onto a liquid-wetted surface and are also subjected to culturability losses through sampling stress (10, 137). The microorganisms collected into a liquid (e.g., distilled water or phosphate buffer) may become entrained in bubbles and reaerosolize from the collection fluid in a manner similar to the aerosolization of microorganisms from bubbling liquids such as whirlpools and fermenters (12, 66, 75). The reaerosolization effect has been shown to decrease the collection efficiencies of AGI-4 and AGI-30 impingers (59, 89). In addition, the collection efficiencies of conventional liquid impingers may decrease significantly due to liquid evaporation over time (89), thus limiting the use of impingers to relatively short sampling times ($t \lesssim 30$ min). Once the bottom of an impinger's collection vessel becomes dry, the particles may bounce from it, which further decreases the efficiency of collection (59).

Effects of particle bounce, reaerosolization, and evaporation of the collection fluid are minimized in the BioSampler, which can operate with nonevaporating and highly viscous liquids (146). The overall design of the BioSampler's collection unit minimizes impaction stress. When used with heavy white mineral oil, the BioSampler can maintain microbial viability and high physical collection efficiency for several hours (87, 88). The prototype of another liquid sampler—

utilizing the bubbling technique (mentioned earlier in this chapter)—has been demonstrated to maintain high physical collection efficiency and viability for a broad spectrum of microorganisms, including viruses, bacteria (vegetative cells and spores), and fungi (3–5).

Collection Time

Sample collection time is an integral part of the bioaerosol sampling design, and guidelines for the selection of optimal sampling times for various bioaerosol samplers have been published previously (96, 117). Parameters that must be considered are the expected bioaerosol concentration, the quantitation range of the sampler, and the effect of sampling stress on the overall collection efficiency. For each sample, the sampling period must be sufficiently long to obtain a representative sample of the airborne microorganisms present without exceeding the upper quantitation limit of the sampler or causing losses in the culturability of airborne organisms. The selection of a sample collection time is complicated by the fact that bioaerosol concentrations may vary greatly over time, often by several orders of magnitude within the same environment. Thus, air sample periods of short duration provide only a brief temporal and spatial glimpse of the environment and several samples may be required to determine the average bioaerosol concentrations. Furthermore, the sampling environment may range from relatively low bioaerosol concentrations ($\leq 10^2$ CFU/m^3) in clean rooms and hospital operating rooms (47) to exceptionally high concentrations (10^5 to 10^{10} CFU/m^3) found in certain industrial and agricultural settings (45, 79).

While the ambient bioaerosol concentration is an unknown quantity, the quantitation range of a bioaerosol sampler can be determined if the airflow rate and the collection time are known (28). The lower quantitation limit (LQL) is obtained by assuming the detection of a single organism, whereas the upper quantitation limit (UQL) is based on the maximum number of organisms which can be enumerated from a sample. When the collection medium is a membrane filter from which organisms can be eluted or a liquid, the sample can be serially diluted prior to analysis. For these methods, there is essentially no UQL. For an impactor sampler, the UQL is determined by the area of the collection surface. In the case of an impactor with several nozzles in its stage(s), the UQL depends on the number of sampling nozzles. For example, the UQL of the Andersen single-stage sampler is reached when a CFU develops under each of the 400 sampling nozzles. When aerosol samples are collected to levels near the UQL, errors in analysis (e.g., uncertainty associated with "positive-hole" corrections) may occur as a result of overcrowding on the impaction surface. Other problems include difficulties in resolving distinct particles or colonies (32) and inhibition of the growth of microorganisms as a result of competition for space and nutrients. When a limited amount of overlap occurs, the true bioaerosol concentration can be statistically calculated from the number of colonies on the agar (32, 95). Conversely, air samples collected at levels near the LQL of a sampler may contain an insufficient number of particles to accurately represent the true bioaerosol concentration.

One way to predict the optimal sampling time for a particular sampler is to determine the ideal surface density of microorganisms on the collection area and assume the order of magnitude of the bioaerosol. Nevalainen et al. (111) calculated the optimal sampling time for five bioaerosol samplers by using the following formula:

$$t = (\delta)(A)/(C_a)(Q)$$

where t is the sampling time, δ is the desired surface density, A is the area of the sampling surface, C_a is the average expected bioaerosol concentration, and Q is the sampler flow rate.

Figure 2 illustrates the sampling times for an ideal surface density on a culture plate, δ_{macro}, of 1 colony per cm^2, or 10^4 particles per cm^2 for microscopic counts, δ_{micro}, with a sampler collection efficiency of 100%. The indicated sampling times for an airborne concentration of 10^3 particles per m^3 illustrate that the optimal sampling time depends on the sampling method and on the anticipated bioaerosol concentration. For example, a high-volume sampler (e.g., the SAS sampler) may be best suited for sampling environments with low bioaerosol concentrations and the Burkard personal sampler may be most effective with relatively high concentrations. Because the bioaerosol concentration is unknown and can only be estimated, often more than one collection time is employed for sample collection to enhance the likelihood of obtaining useful data.

Another factor which influences the selection of the sample collection time is sampling stress, which can result in the loss of culturability of airborne microorganisms. It has been observed that increased sampling time has resulted in decreased viability for aerosolized vegetative bacterial cells (28, 36, 60, 134). As air flows over the nutrient agar surface of an impactor, the agar may lose water content, resulting in a harder surface. Subsequently, sampled microorganisms may rebound from the hardened surface and not be collected (76). If they are collected, they may be less embedded in the agar, exposing the microorganisms to desiccation stresses as continued air flows through the sampler. Thus, a doubling of sampling time may not result in a doubling of CFU, depending on the stress tolerance of the airborne microorganisms being sampled. Therefore, the investigator should consider taking several consecutive samples of short duration rather than a few samples for a long time interval.

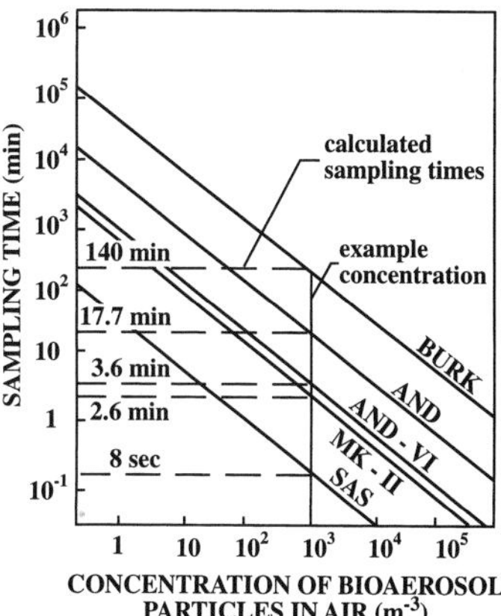

FIGURE 2 Collection times for selected bioaerosol samplers. BURK, Burkard spore trap (personal sampler); AND, Andersen six–stage viable impactor sampler; AND-VI, sixth stage of the Andersen six-stage impactor used as a separate sampler; MK-II, Casella MK-II sampler; SAS, surface air system high-flow sampler. (Adapted from reference 111 with kind permission from Elsevier Science Ltd., Kidlington, United Kingdom.)

Number of Samples

Given the heterogeneity and temporal variability exhibited by bioaerosols and the relatively short collection times for many air sampling methods, multiple samples are often taken to more accurately determine the concentration and composition of a bioaerosol. The number of air samples required depends on the statistical methods employed to analyze the data, on the length of the sample collection time, and on the variability of the bioaerosol. Often pilot sampling studies are performed in a particular environment to arrive at the final sampling design. Simultaneous paired samples are often taken because of the high degree of variability which has been demonstrated between data from paired samplers of the same type (29, 84). Because of the changes in bioaerosol concentrations over time and the small volume of air from which the sample is taken, a single air sample at a discrete point in time and space usually has limited value in the assessment of the airborne microorganisms in an environment of concern. The aerosolization of fungal spores and other microorganisms may be sporadic, and the air environment may not be perfectly mixed. Therefore, a negative result obtained from an air sample is not definitive and additional testing of surfaces is used to provide more information (see "Surface Sampling" below).

Sampler Calibration

The concentration of airborne microorganisms is calculated by dividing the number of collected microorganisms by the sampled air volume (e.g., CFU per cubic meter). Therefore, the volumetric flow rate of each sampler should be calibrated and adjusted to a desired level, if necessary. This is particularly important for impactors as the d_{50} is a function of the flow rate through the nozzle(s). A decreased flow rate (e.g., due to a weakened battery) increases the d_{50}, shifting collection bias toward larger particles (101). If some of the nozzles in a multinozzle impactor are plugged, the flow velocity through the remaining nozzles is increased, resulting in a lower d_{50} with increased collection and the potential for increased stress on the microorganisms. In impactor sampling, the recommended distance between the exit plane of the impactor nozzle(s) and the top of the nutrient surface should be maintained. For example, the manufacturer's recommended volume for plastic petri plates is 40 ml of agar in the Andersen single-stage impactor sampler. Otherwise, the impactor's collection characteristics (primarily the d_{50}) may be changed.

The flow rate of a battery-powered sampler is usually calibrated before and after sample collection. If the flow rate has decreased by less than 10%, the average of the two flow rate measurements is generally used for the calculation of the concentration of airborne microorganisms. If the difference exceeds 10%, the sample is considered invalid. Various calibration methods that measure the flow rate through a soap bubble meter or an electronic calibration instrument are available (35, 90, 91).

SURFACE SAMPLING

Surface sampling is often used in conjunction with air sampling in the indoor environment to provide information about microbial sources. Direct source evaluation through surface sampling is used to locate and identify potential bioaerosol hazards and to predict the bioaerosol dispersal and deposition (27, 104, 143). Qualitative and quantitative information on the concentration and composition of surface-associated microorganisms can be obtained with surface sampling. A variety of methods have been used to collect microorganisms from smooth and porous surfaces, including the use of swabs, wipes and sponges, tapes, agar contact plates, dust, and bulk sampling. Surface sampling has numerous applications in diverse areas of microbiology, such as biocontaminant monitoring in the food and pharmaceutical industries, indoor air quality investigations, evaluation of remediation of biocontamination, spacecraft monitoring, and the detection of biological warfare agents. The bioterrorism events involving the dispersal of *Bacillus anthracis* spores that occurred in the United States in 2001 have heightened the interest in surface sampling and reemphasized the need for standardized surface sample collection and analysis protocols (61, 120, 143).

Surface Sampling Methods

Swab sampling, historically used in clinical and food microbiology, was adapted for use in environmental microbiology and has been the traditional method of surface sampling (46, 78, 118). The use of swabs for surface sampling has expanded from cotton swabs to other commercially available materials including macrofoam, rayon, polyester, calcium alginate, and Dacron (118). Swabs are generally premoistened with a buffer containing a surfactant and used to sample small surface areas (up to 100 cm^2).

Surface wipes have the capability for sampling large surface areas and are an alternative to swab sampling, which is limited to small, discrete areas. Many types of cloth materials and sponges have been used for quantitative assessment of surface-associated microorganisms (23, 26, 78, 120).

Tape sampling is a qualitative method used to detect surface biocontamination without the need for culture analysis. Samples collected with transparent tape are commonly used in indoor-air quality investigations of buildings to detect and identify mold contamination. Visible fungal colonies or areas of suspected biocontamination are lightly touched with the adhesive surface of the tape. Identification of fungi from tape imprints from surfaces is performed by mounting the tape on a microscope slide with a stain such as lactophenol cotton blue and morphologically examining fungal structures with light microscopy (40). The effectiveness of tape sampling may be reduced by the presence of excessive dust and debris on the tape that can interfere with microscopic examination of the sample.

Agar contact procedures, in which the surface of an agar culture medium is brought into contact with the area to be sampled, are used to quantify culturable microorganisms (46). After sampling, the agar plate or strip is incubated and the resulting CFUs are enumerated. One of the limitations of the method is the inability to detect microorganisms that are nonculturable or not culturable on the agar medium used. The method is also limited to use on smooth surfaces.

Vacuum sampling is widely used to collect settled dust in indoor environments to determine the concentration and composition of microorganisms to which building occupants may be exposed. Many types of vacuuming devices, sampling and processing protocols, and analysis methods have been used to characterize microorganisms and antigenic compounds in settled dust (65, 97, 120, 143). The dust is weighed and processed in a buffer, and the liquid sample is analyzed. Bulk sampling consists of removing a section of material, such as carpet, insulation, or ceiling tile, and transporting the sample to the laboratory for analysis (27, 104). The surface-associated microorganisms are removed by processing the material in a buffer. The material may be washed, vortexed, or mixed with the buffer, and the liquid sample is analyzed. Bulk sampling is a destructive sampling method in that sections of the contaminated material are removed from the site, thus limiting the practical uses of this method.

To bridge two conventional methods, air sampling and direct source evaluation through surface sampling, an alternative concept was developed (56). The fungal spore source strength tester is a cuplike device with a square cross section, which is held against the mold-contaminated surface. The spores are aerosolized from the source by small air jets, which originate in 112 orifices and are directed toward the surface. The spores, released into a small chamber, are immediately collected from it with a bioaerosol sampler. This concept has been evaluated and found suitable for assessing the fungal aerosolization potential from growth on surfaces (124, 125).

Surface Sampling Effectiveness

As microorganisms may be dispersed and reaerosolized from contaminated surfaces (25, 27, 29, 143), air and surface sampling are often used in combination. Attempts to relate surface contaminant levels with bioaerosol concentrations have produced variable results (27, 65, 104). Numerous uncontrolled factors can influence the comparison of air and surface sampling data (e.g., ventilation rates, humidity, human activity, maintenance practices, microbial properties, and sampling techniques).

While the relative levels of effectiveness of surface sampling methods have been compared (23, 26, 78, 120), there is little quantitative information on the collection efficiencies of surface sampling methods (24, 78, 118). The removal and recovery efficiencies of polyester-bonded cloth wipes and cotton swabs were evaluated for the collection of microorganisms in field samples (78). Data were compared to those for reference samples consisting of Teflon ribbons that were placed on clean surfaces. The Teflon ribbons remained in place during deposition of environmental microorganisms on the surface over time. Surface samples were collected, and data were compared with those obtained with the reference ribbons that were removed and processed by sonication in buffer. The collection efficiencies of four types of swabs, cotton, macrofoam, polyester, and rayon, were compared by inoculating steel coupons with known concentrations of *B. anthracis* spores and surface sampling with both wet and dry swabs (118). Results indicated that macrofoam and cotton are the swab materials with the greatest collection efficiencies. The data also showed that pre-moistened swabs are more efficient than dry swabs for the collection of *B. anthracis* spores from surfaces. The collection efficiency and sensitivity of the biological sampling kit (BiSKit; QuickSilver Analytics, Abingdon, Md.), a device designed for large-area surface sampling, have also been measured for the collection of spores of *Bacillus atrophaeus*, a surrogate for the biological warfare agent *B. anthracis*, from wood laminate, vinyl tile, and painted metal surfaces (24). Results showed that the large surface area (1 m^2) sampled by the BiSKit results in higher sensitivity than the swab sampling methods tested. Another potential advantage of large-area sampling in the field is greater coverage that results in fewer samples collected per site.

The aftermath of the bioterrorism events of 2001 involving *B. anthracis* spores have emphasized the importance of surface sampling for the detection of biological contaminants and the need for additional research to determine the collection efficiencies and sensitivities of surface sampling methods (61, 120, 143). The overall collection efficiency of a surface sampling method comprises both the efficiency of the removal of microorganisms from the surface and the efficiency of their removal from the collection material. Further research is needed to optimize buffer solutions and collection materials in order to remove the greatest percentage of microorganisms from surfaces and provide maximum recovery of the microorganisms from the collection material. Validation of surface sampling methods and the development of standardized sampling and analysis protocols are vital for the detection of biothreat agents, the determination of their concentration and distribution, and the evaluation of the effectiveness of remediation procedures. The protocols developed could be adapted for use in other environmental microbiological applications to enhance the detection and measurement of surface-associated microorganisms. The use of standardized bioaerosol and surface sampling methodologies could also be applied to advance knowledge in the fields of exposure and risk assessment.

REFERENCES

1. **Adhikari, A., D. Martuzevicius, T. Reponen, S. A. Grinshpun, S.-H. Cho, S. Sivasubramani, W. Zhong, L. Levin, A. Kelley, H. G. St. Clair, and G. K. LeMasters. 2003.** Performance of the Button personal inhalable sampler for the measurement of outdoor aeroallergens. *Atmos. Environ.* **37:**4723–4733.
2. **Adhikari, A., T. Reponen, S.-A. Lee, and S. A. Grinshpun.** 2004. Assessment of human exposure to airborne fungi in agricultural confinements: personal inhalable sampling versus stationary sampling. *Ann. Agric. Environ. Med.* **11:**269–277.
3. **Agranovski, I. E., V. Agranovski, S. A. Grinshpun, T. Reponen, and K. Willeke.** 2002. Collection of airborne microorganisms into liquid by bubbling through porous medium. *Aerosol Sci. Technol.* **36:**502–509.
4. **Agranovski, I. E., V. Agranovski, S. A. Grinshpun, T. Reponen, and K. Willeke.** 2002. Development and evaluation of a new personal sampler for culturable airborne microorganisms. *Atmos. Environ.* **36:**889–898.
5. **Agranovski, I. E., A. S. Safatov, A. I. Borodulin, O. V. Pyankov, V. A. Petrishchenko, A. N. Sergeev, A. A. Sergeev, V. Agranovski, and S. A. Grinshpun.** 2005. New personal sampler for viable airborne viruses: feasibility study. *J. Aerosol Sci.* **36:**609–617.
6. **Aizenberg, V., T. Reponen, S. A. Grinshpun, and K. Willeke.** 2000. Performance of Air-O-Cell, Burkard, and Button samplers for total enumeration of airborne spores. *Am. Ind. Hyg. Assoc. J.* **61:**855–864.
7. **Alvarez, A. J., M. P. Buttner, G. A. Toranzos, E. A. Dvorsky, A. Toro, T. B. Heikes, L. E. Mertikas, and L. D. Stetzenbach.** 1994. The use of solid–phase polymerase chain reaction for the enhanced detection of airborne microorganisms. *Appl. Environ. Microbiol.* **60:**374–376.
8. **Alvarez, A. J., M. P. Buttner, and L. D. Stetzenbach.** 1995. PCR for bioaerosol monitoring: sensitivity and environmental interference. *Appl. Environ. Microbiol.* **61:**3639–3644.
9. **An, H. R., G. Mainelis, and M. Yao.** 2004. Evaluation of a high–volume portable bioaerosol sampler in laboratory and field environments. *Indoor Air* **14:**385–393.
10. **Andersen, A. A.** 1958. New sampler for the collection, sizing, and enumeration of viable airborne particles. *J. Bacteriol.* **76:**471–484.
11. **Atlas, R. M., and R. Bartha.** 1993. *Microbial Ecology: Fundamentals and Applications*, 3rd ed. The Benjamin/Cummings Publishing Company, Inc., Redwood City, Calif.
12. **Baron, P. A., and K. Willeke.** 1986. Respirable droplets from whirlpools: measurement of size distribution and estimation of disease potential. *Environ. Res.* **39:**8–18.

13. **Baron, P. A., and K. Willeke.** 2001. Gas and particle motion, p. 61–82. *In* P. A. Baron and K. Willeke (ed.), *Aerosol Measurement: Principles, Techniques and Applications*. John Wiley, New York, N.Y.
14. **Bausum, H. T., S. A. Schaub, K. F. Kenyon, and M. J. Small.** 1982. Comparison of coliphage and bacterial aerosols at a wastewater spray irrigation site. *Appl. Environ. Microbiol.* **43:**28–38.
15. **Beuchat, L. R.** 1992. Media for detecting and enumerating yeasts and moulds. *Int. J. Food Microbiol.* **17:**145–158.
16. **Blomquist, G., G. Strom, and L.-H. Stromquist.** 1984. Sampling of high concentrations of airborne fungi. *Scand. J. Work Environ. Health* **10:**109–113.
17. **Brockmann, J. E.** 2001. Sampling and transport of aerosols, p. 143–195. *In* P. A. Baron and K. Willeke (ed.), *Aerosol Measurement: Principles, Techniques and Applications*. John Wiley, New York, N.Y.
18. **Burge, H. A.** 1995. Bioaerosol investigations, p. 1–23. *In* H. A. Burge (ed.), *Bioaerosols*. CRC Press, Boca Raton, Fla.
19. **Burge, H. A., J. R. Boise, J. A. Rutherford, and W. R. Solomon.** 1977. Comparative recoveries of airborne fungus spores by viable and non–viable modes of volumetric collection. *Mycopathologia* **61:**27–33.
20. **Burge, H. A., and W. R. Solomon.** 1987. Sampling and analysis of biological aerosols. *Atmos. Environ.* **21:**451–456.
21. **Burge, H. P., W. R. Solomon, and J. R. Boise.** 1977. Comparative merits of eight popular media in aerometric studies of fungi. *J. Allergy Clin. Immunol.* **60:**199–203.
22. **Burton, N. C., A. Adhikari, S. A. Grinshpun, R. Horning, and T. Reponen.** 2005. The effect of filter material on bioaerosol collection of Bacillus anthracis simulant. *J. Environ. Monit.* **7:**475–480.
23. **Buttner, M. P., P. Cruz, L. D. Stetzenbach, A. K. Klima–Comba, V. L. Stevens, and T. D. Cronin.** 2004. Determination of the efficacy of two building decontamination strategies by surface sampling with culture and quantitative PCR analysis. *Appl. Environ. Microbiol.* **70:** 4740–4747.
24. **Buttner, M. P., P. Cruz, L. D. Stetzenbach, A. K. Klima-Comba, V. L. Stevens, and P. A. Emanuel.** 2004. Evaluation of the biological sampling kit (BiSKit) for large-area surface sampling. *Appl. Environ. Microbiol.* **70:** 7040–7045.
25. **Buttner, M. P., P. Cruz-Perez, P. J. Garrett, and L. D. Stetzenbach.** 1999. Dispersal of fungal spores from three types of air handling system duct material. *Aerobiologia* **15:** 1–8.
26. **Buttner, M. P., P. Cruz-Perez, and L. D. Stetzenbach.** 2001. Enhanced detection of surface-associated bacteria in indoor environments by quantitative PCR. *Appl. Environ. Microbiol.* **67:**2564–2570.
27. **Buttner, M. P., P. Cruz-Perez, L. D. Stetzenbach, P. J. Garrett, and A. E. Luedtke.** 2002. Measurement of airborne fungal spore dispersal from three types of flooring materials. *Aerobiologia* **18:**1–11.
28. **Buttner, M. P., and L. D. Stetzenbach.** 1991. Evaluation of four aerobiological sampling methods for the retrieval of aerosolized *Pseudomonas syringae*. *Appl. Environ. Microbiol.* **57:**1268–1270.
29. **Buttner, M. P., and L. D. Stetzenbach.** 1993. Monitoring airborne fungal spores in an experimental indoor environment to evaluate sampling methods and the effects of human activity on air sampling. *Appl. Environ. Microbiol.* **59:**219–226.
30. **Castellan, R. M., S. A. Olenchock, K. B. Kinsley, and J. L. Hankinson.** 1987. Inhaled endotoxin and decreased spirometric values: an exposure-response relation for cotton dust. *N. Engl. J. Med.* **317:**605–610.
31. **Chang, C. W., S. A. Grinshpun, K. Willeke, J. M. Macher, J. Donnelly, S. Clark, and A. Juozaitis.** 1995. Factors affecting microbiological colony count accuracy for bioaerosol sampling and analysis. *Am. Ind. Hyg. Assoc. J.* **56:**979–986.
32. **Chang, C. W., Y. H. Hwang, S. A. Grinshpun, J. M. Macher, and K. Willeke.** 1994. Evaluation of counting error due to colony masking in bioaerosol sampling. *Appl. Environ. Microbiol.* **60:**3732–3738.
33. **Chapman, M. D.** 1995. Analytical methods: immunoassays, p. 235–248. *In* H. A. Burge (ed.), *Bioaerosols*. CRC Press, Boca Raton, Fla.
34. **Chatigny, M. A., J. M. Macher, H. A. Burge, and W. R. Solomon.** 1989. Sampling airborne microorganisms and aeroallergens, p. 199–220. *In* S. V. Hering (ed.), *Air Sampling Instruments for Evaluation of Atmospheric Contaminants*, 7th ed. American Conference of Governmental Industrial Hygienists, Cincinnati, Ohio.
35. **Chen, B. T.** 1993. Instrument calibration, p. 493–520. *In* K. Willeke and P. A. Baron (ed.), *Aerosol Measurement: Principles, Techniques and Applications*. Van Nostrand Reinhold, New York, N.Y.
36. **Cox, C. S.** 1987. *The Aerobiological Pathway of Microorganisms*. John Wiley and Sons Ltd., Chichester, United Kingdom.
37. **Cox, C. S.** 1989. Airborne bacteria and viruses. *Sci. Prog.* (Oxford) **73:**469–500.
38. **Croft, W. S., B. B. Jarvis, and C. S. Yatawara.** 1986. Airborne outbreak of trichothecene toxicosis. *Atmos. Environ.* **20:**549–552.
39. **Crook, B., S. Higgins, and J. Lacey.** 1987. Methods for sampling microorganisms at solid waste disposal sites, p. 791–797. *In* D. R. Houghton, R. N. Smith, and H. O. W. Eggins (ed.), *Biodeterioration 7*. Elsevier, London, United Kingdom.
40. **Cruz, P.** 2002. Identification of airborne fungi, p. 1647–1661. *In* G. Bitton (ed.), *Encyclopedia of Environmental Microbiology*. John Wiley & Sons, New York, N.Y.
41. **DeBlay, F., M. D. Chapman, and T. A. E. Platts-Mills.** 1991. Airborne cat allergen Feld I: environmental control with the cat in situ. *Am. Rev. Respir. Dis.* **143:**1334.
42. **Desjardin, L. E., Y. Chen, M. D. Perkins, L. Teixeira, M. D. Cave, and K. D. Eisenach.** 1998. Comparison of the ABI 7700 system (TaqMan) and competitive PCR for quantification of IS*6110* DNA in sputum during treatment of tuberculosis. *J. Clin. Microbiol.* **36:**1964–1968.
43. **Douwes, J., G. Doekes, R. Montijn, D. Heederik, and B. Brunekreef.** Measurement of β(1,3)-glucans in occupational and home environments with an inhibition enzyme immunoassay. *Appl. Environ. Microbiol.* **62:**3176–3182.
44. **Eduard, W., and D. Heekderick.** 1998. Methods for quantitative assessment of airborne levels of noninfectious microorganisms in highly contaminated work environments. *Am. Ind. Hyg. Assoc. J.* **59:**113–127.
45. **Eduard, W., J. Lacey, K. Karlsson, U. Palmgren, G. Strom, and G. Blomquist.** 1990. Evaluation of methods for enumerating microorganisms in filter samples from highly contaminated occupational environments. *Am. Ind. Hyg. Assoc. J.* **51:**427–436.
46. **Favero, M. S., J. J. McDade, J. A. Robertsen, R. K. Hoffman, and R. W. Edwards.** 1968. Microbiological sampling of surfaces. *J. Appl. Bacteriol.* **31:**336–343.
47. **Favero, M. S., J. R. Puleo, J. H. Marshall, and G. S. Oxborrow.** 1966. Comparative levels and types of microbial contamination detected in industrial clean rooms. *Appl. Microbiol.* **14:**539–551.
48. **Ferre, F.** 1992. Quantitative or semi–quantitative PCR: reality versus myth. *PCR Methods Appl.* **2:**1–9.
49. **Flannigan, B.** 1987. Mycotoxins in the air. *Int. Biodeterior.* **23:**73–78.
50. **Flannigan, B.** 1997. Air sampling for fungi in indoor environments. *J. Aerosol Sci.* **28:**381–392.

51. **Fogelmark, B., H. Goto, K. Yuasa, B. Marchat, and R. Rylander.** 1992. Acute pulmonary toxicity of inhaled β-1,3-glucan and endotoxin. *Agent Actions* **35:**50–56.
52. **Gillespie, V. L., C. S. Clark, H. S. Bjornson, S. J. Samuels, and J. W. Holland.** 1981. A comparison of two-stage and six-stage Andersen impactors for viable aerosols. *Am. Ind. Hyg. Assoc. J.* **42:**858–864.
53. **Goto, H., K. Yuasa, and R. Rylander.** 1994. (1-3)-β-D-Glucan in indoor air, its measurement and in vitro activity. *Am. J. Ind. Med.* **25:**81–83.
54. **Grinshpun, S. A., C. W. Chang, A. Nevalainen, and K. Willeke.** 1994. Inlet characteristics of bioaerosol sampler. *J. Aerosol Sci.* **25:**1503–1522.
55. **Grinshpun, S. A., and J. M. Clark.** 2005. Guest editorial for the special issue on measurement and characterization of bioaerosols. *J. Aerosol Sci.* **36:**553–555.
56. **Grinshpun, S. A., R. L. Górny, T. Reponen, K. Willeke, S. Trakumas, P. Hall, and D. F. Dietrich.** 2002. New method for assessment of potential spore aerosolization from contaminated surfaces, p. 767–768. *In* C. S. Wang (ed.), *Proceedings of the Sixth International Aerosol Conference,* vol. 2. The Chinese Association for Aerosol Research in Taiwan, Taipei, Taiwan, Republic of China.
57. **Grinshpun, S. A., G. Mainelis, M. Trunov, R. L. Górny, S. K. Sivasubramani, A. Adhikari, and T. Reponen.** 2005. Collection of airborne spores by circular single-stage impactors with small jet-to-plate distance. *J. Aerosol Sci.* **36:**575–591.
58. **Grinshpun, S. A., and T. Reponen.** 2004. Sampling of biological particles from ambient environment: physical principles, efficiency, and exposure assessment. *Int. Aerobiol. Newsl.* **59:**1–2.
59. **Grinshpun, S. A., K. Willeke, V. Ulevicius, A. Juozaitis, S. Terzieva, J. Donnelly, G. N. Stelma, and K. Brenner.** 1997. Effect of impaction, bounce and reaerosolization on collection efficiency of impingers. *Aerosol Sci. Technol.* **26:**326–342.
60. **Hatch, M. T., and H. Wolochow.** 1969. Bacterial survival: consequences of the airborne state, p. 267–295. *In* R. L. Dimmick, and A. B. Akers (ed.), *An Introduction to Experimental Aerobiology.* Wiley, New York, N.Y.
61. **Higgins, J. A., M. Cooper, L. Schroeder-Tucker, S. Black, D. Miller, J. S. Karns, E. Manthey, R. Breeze, and M. L. Perdue.** 2003. A field investigation of *Bacillus anthracis* contamination of U.S. Department of Agriculture and other Washington, D.C., buildings during the anthrax attack of October 2001. *Appl. Environ. Microbiol.* **69:**593–599.
62. **Hirsch, S. R., and J. A. Sosman.** 1976. A one-year survey of mold growth inside twelve homes. *Ann. Allergy* **36:**30–38.
63. **Hobbie, J. E., R. J. Daley, and S. Jasper.** 1977. Use of nuclepore filters for counting bacteria by fluorescence microscopy. *Appl. Environ. Microbiol.* **33:**1225–1228.
64. **Hocking, A. D., and J. I. Pitt.** 1980. Dichloran–glycerol medium for enumeration of xerophilic fungi from low-moisture foods. *Appl. Environ. Microbiol.* **39:**488–492.
65. **Horner, W. E., A. G. Wortham, and P. R. Morey.** 2004. Air- and dustborne mycoflora in houses free of water damage and fungal growth. *Appl. Environ. Microbiol.* **70:**6394–6400.
66. **Huang, Y. L., K. Willeke, A. Juozaitis, J. Donnelly, A. Leeson, and R. Wyza.** 1994. Fermentation process monitoring through measurement of aerosol release. *Biotechnol. Prog.* **10:**32–38.
67. **Hunter, C. A., C. Grant, B. Flannigan, and A. F. Bravery.** 1988. Mould in buildings: the air spora of domestic buildings. *Int. Biodeterior.* **24:**81–101.
68. **Jacobs, R. R.** 1989. Airborne endotoxins: an association with occupational lung disease. *Appl. Ind. Hyg.* **4:**50–56.
69. **Jarvis, B. B., W. G. Sorenson, E. Hintikka, M. Nikulin, Y. Zhou, J. Jiang, S. Wang, S. Hinkley, R. A. Etzel, and D. Dearborn.** 1998. Study of toxin production by isolates of *Stachybotrys chartarum* and *Memnoniella echinata* isolated during a study of pulmonary hemosiderosis in infants. *Appl. Environ. Microbiol.* **64:**3620–3625.
70. **Jensen, P. A., B. Lighthart, A. J. Mohr, and B. T. Shaffer.** 1994. Instrumentation used with microbial bioaerosol, p. 226–284. *In* B. Lighthart and A. J. Mohr (ed.), *Atmospheric Microbial Aerosols, Theory and Applications.* Chapman and Hall, New York, N.Y.
71. **Jensen, P. A., W. F. Todd, G. N. Davis, and P. V. Scarpino.** 1992. Evaluation of eight bioaerosol samplers challenged with aerosols of free bacteria. *Am. Ind. Hyg. Assoc. J.* **53:**660–667.
72. **Johanning, E., R. Biagini, D. Hull, P. Morey, B. Jarvis, and P. Landsbergis.** 1996. Health and immunology study following exposure to toxigenic fungi (Stachybotrys chartarum) in a water-damaged office environment. *Int. Arch. Occup. Environ. Health* **68:**207–218.
73. **Jones, W., K. Morring, P. Morey, and W. Sorenson.** 1985. Evaluation of the Andersen viable impactor for single stage sampling. *Am. Ind. Hyg. Assoc. J.* **46:**294–298.
74. **Josephson, K. L., C. P. Gerba, and I. L. Pepper.** 1993. Polymerase chain reaction detection of nonviable bacterial pathogens. *Appl. Environ. Microbiol.* **59:**3513–3515.
75. **Juozaitis, A., Y. L. Huang, K. Willeke, J. Donnelly, S. Kalatoor, A. Leeson, and R. Wyza.** 1994. Dispersion of respirable aerosols in a fermenter and their removal in an exhaust system. *Appl. Occup. Environ. Hyg.* **9:**552–559.
76. **Juozaitis, A., K. Willeke, S. A. Grinshpun, and J. Donnelly.** 1994. Impaction onto a glass slide or agar versus impingement into a liquid for the collection and recovery of airborne microorganisms. *Appl. Environ. Microbiol.* **60:** 861–870.
77. **Kaye, S.** 1988. Efficiency of "Biotest RCS" as a sampler of airborne bacteria. *J. Parenteral Sci. Technol.* **42:**147–152.
78. **Kirschner, L. E., and J. R. Puleo.** 1979. Wipe-rinse technique for quantitating microbial contamination on large surfaces. *Appl. Environ. Microbiol.* **38:**466–470.
79. **Kotimaa, M.** 1990. Occupational exposure to spores in the handling of wood chips. *Grana* **29:**153–156.
80. **Lange, J. L., P. S. Thorne, and N. Lynch.** 1997. Application of flow cytometry and fluorescent in situ hybridization for assessment of exposures to airborne bacteria. *Appl. Environ. Microbiol.* **63:**1557–1563.
81. **Lee, K. S., K. H. Bartlett, M. Brauer, G. M. Stephens, W. A. Black, and K. Teschke.** 2004. A field comparison of four samplers for enumerating fungal aerosols. I. Sampling characteristics. *Indoor Air* **14:**360–366.
82. **Lee, K. S., K. Teschke, M. Brauer, and K. H. Bartlett.** 2004. A field comparison of four fungal aerosol sampling instruments: inter-sampler calibration and caveats. *Indoor Air* **14:**367–372.
83. **Lee, K. W., and R. Mukund.** 2001. Filter collection, p. 197–228. *In* P. A. Baron and K. Willeke (ed.), *Aerosol Measurement: Principles, Techniques and Applications.* John Wiley, New York, N.Y.
84. **Lembke, L. L., R. N. Kniseley, R. C. Van Nostrand, and M. D. Hale.** 1981. Precision of the all-glass impinger and the Andersen microbial impactor for air sampling in solid-waste handling facilities. *Appl. Environ. Microbiol.* **42:** 222–225.
85. **Leser, T. D., M. Boye, and N. B. Hendriksen.** 1995. Survival and activity of *Pseudomonas* sp. strain B13(FR1) in a marine microcosm determined by quantitative PCR and an rRNA–targeting probe and its effect on the indigenous bacterioplankton. *Appl. Environ. Microbiol.* **61:** 1201–1207.

86. **Levetin, E.** 2004. Methods for aeroallergen sampling. *Allergy Asthma Rep.* **4:**376–383.
87. **Lin, X., T. Reponen, K. Willeke, S. A. Grinshpun, K. K. Foarde, and D. S. Ensor.** 1999. Long-term sampling of airborne bacteria and fungi into a non-evaporating liquid. *Atmos. Environ.* **33:**4291–4298.
88. **Lin, X., T. Reponen, K. Willeke, Z. Wang, S. A. Grinshpun, and M. Trunov.** 2000. Survival of airborne microorganisms during swirling aerosol collection. *Aerosol Sci. Technol.* **32:**184–196.
89. **Lin, X., K. Willeke, V. Ulevicius, and S. A. Grinshpun.** 1997. Effect of sampling time on the collection efficiency of all-glass impingers. *Am. Ind. Hyg. Assoc. J.* **58:**480–488.
90. **Lippmann, M.** 1973. *The Industrial Environment—Its Evaluation and Control*, p. 101–122. National Institute for Occupational Safety and Health, Government Printing Office, Washington, D.C.
91. **Lippmann, M.** 1989. Calibration of air sampling instruments, p. 73–110. *In* S. V. Hering (ed.), *Air Sampling Instruments for Evaluation of Atmospheric Contaminants*, 7th ed. American Conference of Governmental Industrial Hygienists, Cincinnati, Ohio.
92. **Lippmann, M.** 1989. Sampling aerosols by filtration, p. 305–336. *In* S. V. Hering (ed.), *Air Sampling Instruments for Evaluation of Atmospheric Contaminants*, 7th ed. American Conference of Governmental Industrial Hygienists, Cincinnati, Ohio.
93. **Luczynska, C. M., Y. Li, M. D. Chapman, and T. A. E. Platts-Mills.** 1990. Airborne concentration and particle size distribution of allergen derived from domestic cats (Felis domesticus): measurements using cascade impactor, liquid impinger and a two site monoclonal antibody assay for Fel d I. *Am. Rev. Respir. Dis.* **141:**361–367.
94. **Lundholm, I. M.** 1982. Comparison of methods for quantitative determinations of airborne bacteria and evaluation of total viable counts. *Appl. Environ. Microbiol.* **44:**179–183.
95. **Macher, J. M.** 1989. Positive-hole correction of multiple-jet impactors for collecting viable microorganisms. *Am. Ind. Hyg. Assoc. J.* **50:**561–568.
96. **Macher, J. M. (ed.).** 1999. *Bioaerosols: Assessment and Control.* American Conference of Governmental Industrial Hygienists, Cincinnati, Ohio.
97. **Macher, J. M.** 2001. Review of methods to collect settled dust and isolate culturable microorganisms. *Indoor Air* **11:**99–110.
98. **Macher, J. M., and M. W. First.** 1983. Reuter centrifugal air sampler: measurement of effective airflow rate and collection efficiency. *Appl. Environ. Microbiol.* **45:**1960–1962.
99. **Madelin, T. M., and M. F. Madelin.** 1995. Biological analysis of fungi and associated molds, p. 361–386. *In* C. S. Cox and C. M. Wathes (ed.), *Bioaerosols Handbook*. CRC Press, Boca Raton, Fla.
100. **Mainelis, G., K. Willeke, A. Adhikari, T. Reponen, and S. A. Grinshpun.** 2002. Design and collection efficiency of a new electrostatic precipitator for bioaerosol collection. *Aerosol Sci. Technol.* **36:**1073–1085.
101. **Marple, V. A., K. L. Rubow, and B. A. Olson.** 2001. Inertial, gravitational, centrifugal and thermal collection techniques, p. 229–260. *In* P. A. Baron and K. Willeke (ed.), *Aerosol Measurement: Principles, Techniques and Applications*. John Wiley, New York, N.Y.
102. **Marthi, B.** 1994. Resuscitation of microbial bioaerosols, p. 192–225. *In* B. Lighthart and A. J. Mohr (ed.), *Atmospheric Microbial Aerosols, Theory and Applications*. Chapman and Hall, New York, N.Y.
103. **Martinez, K. F., C. Y. Rao, and N. C. Burton.** 2004. Exposure assessment and analysis for biological agents. *Grana* **43:**193–208.
104. **Martyny, J. W., K. F. Martinez, and P. R. Morey.** 1999. Source sampling. *In* J. Macher (ed.), *Bioaerosols: Assessment and Control*. American Conference of Governmental Industrial Hygienists, Cincinnati, Ohio.
105. **Miller, J. D., A. M. Laflamme, Y. Sobol, P. Lafontaine, and R. Greenhalgh.** 1988. Fungi and fungal products in some Canadian houses. *Int. Biodeterior.* **24:**103–120.
106. **Miller, J. D., and J. C. Young.** 1997. The use of ergosterol to measure exposure to fungal propagules in indoor air. *Am. Ind. Hyg. Assoc. J.* **58:**39–43.
107. **Milton, D. K.** 1995. Endotoxin, p. 77–86. *In* H.A. Burge (ed.), *Bioaerosols*. CRC Press, Boca Raton, Fla.
108. **Milton, D. K., R. J. Gere, H. A. Feldman, and I. A. Greaves.** 1990. Endotoxin measurement: aerosol sampling and application of a new Limulus method. *Am. Ind. Hyg. Assoc. J.* **51:**331–337.
109. **Morring, K. L., W. G. Sorenson, and M. D. Attfield.** 1983. Sampling for airborne fungi: a statistical comparison of media. *Am. Ind. Hyg. Assoc. J.* **44:**662–664.
110. **Muilenberg, M. L.** 2003. Sampling devices. *Immunol. Allergy Clin. N. Am.* **23:**337–355.
111. **Nevalainen, A., J. Pastuszka, F. Liebhaber, and K. Willeke.** 1992. Performance of bioaerosol samplers: collection characteristics and sampler design considerations. *Atmos. Environ.* **26A:**531–540.
112. **Palmgren, U., G. Strom, G. Blomquist, and P. Malmberg.** 1986. Collection of airborne micro-organisms on Nuclepore filters, estimation and analysis—CAMNEA method. *J. Appl. Bacteriol.* **61:**401–406.
113. **Pasanen, A., M. Nikulin, M. Tuomainen, S. Berg, P. Parikka, and E. Hintikka.** 1993. Laboratory experiments on membrane filter sampling of airborne mycotoxins produced by Stachybotrys atra corda. *Atmos. Environ.* **27A:**9–13.
114. **Picard, C., C. Ponsonnet, E. Paget, X. Nesme, and P. Simonet.** 1992. Detection and enumeration of bacteria in soil by direct DNA extraction and polymerase chain reaction. *Appl. Environ. Microbiol.* **58:**2717–2722.
115. **Placencia, A. M., J. T. Peeler, G. S. Oxborrow, and J. W. Danielson.** 1982. Comparison of bacterial recovery by Reuter centrifugal air sampler and slit-to-agar sampler. *Appl. Environ. Microbiol.* **44:**512–513.
116. **Platts-Mills, T. A. E., W. R. Thomas, R. C. Aalbersee, D. Vervloet, and M. D. Chapman.** 1992. Dust mite allergens and asthma: report of a second international workshop. *J. Allergy Clin. Immunol.* **89:**1046.
117. **Reponen, T., K. Willeke, S. Grinshpun, and A. Nevalainen.** 2001. Biological particle sampling, p. 751–777. *In* P. A. Baron and K. Willeke (ed.), *Aerosol Measurement: Principles, Techniques and Applications*. John Wiley, New York, N.Y.
118. **Rose, L., B. Jensen, A. Peterson, S. N. Banerjee, and M. J. Arduino.** 2004. Swab materials and Bacillus anthracis spore recovery from nonporous surfaces. *Emerg. Infect. Dis.* **10:**1023–1029.
119. **Saiki, R. K., S. Scharf, F. Faloona, K. B. Mullis, G. T. Horn, H. Erlich, and N. Arnheim.** 1985. Enzymatic amplification of β-globin genomic sequence and restriction site analysis for diagnosis of sickle cell anemia. *Science* **230:**1350–1354.
120. **Sanderson, W. T., M. J. Hein, L. Taylor, B. D. Curwin, G. M. Kinnes, T. A. Seitz, T. Popovic, H. T. Holmes, M. E. Kellum, S. K. McAllister, D. N. Whaley, E. A. Tupin, T. Walker, J. A. Freed, D. S. Small, B. Klusaritz, and J. H. Bridges.** 2002. Surface sampling

methods for Bacillus anthracis spore contamination. *Emerg. Infect. Dis.* **8:**1145–1151.

121. **Sawyer, M. H., C. J. Chamberlin, Y. N. Wu, N. Aintablian, and M. R. Wallace.** 1994. Detection of varicella-zoster virus DNA in air samples from hospital rooms. *J. Infect. Dis.* **169:**91–94.
122. **Sayer, W. J., N. M. MacKnight, and H. W. Wilson.** 1972. Hospital airborne bacteria as estimated by the Andersen sampler versus gravity settling culture plate. *Am. J. Clin. Pathol.* **58:**558–562.
123. **Sayer, W. J., D. B. Shean, and J. Ghosseiri.** 1969. Estimation of airborne fungal flora by the Andersen sampler versus the gravity settling plate. *J. Allergy* **44:**214–227.
124. **Sivasubramani, S. K., R. T. Niemeier, T. Reponen, and S. A. Grinshpun.** 2004. Assessment of the aerosolization potential for fungal spores in moldy homes. *Indoor Air* **14:**405–412.
125. **Sivasubramani, S. K., R. T. Niemeier, T. Reponen, and S. A. Grinshpun.** 2004. Fungal Spore Source Strength Tester: laboratory evaluation of a new concept. *Sci. Total Environ.* **329:**75–86.
126. **Smid, T., E. Schokkin, J. S. M. Boleij, and D. Heederik.** 1989. Enumeration of viable fungi in occupational environments: a comparison of samplers and media. *Am. Ind. Hyg. Assoc. J.* **50:**235–239.
127. **Solomon, W. R.** 1975. Assessing fungus prevalence in domestic interiors. *J. Allergy Clin. Immunol.* **56:**235–242.
128. **Solomon, W. R., H. A. Burge, J. R. Boise, and M. Becker.** 1980. Comparative particle recoveries by the retracting rotorod, rotoslide and Burkard spore trap sampling in a compact array. *Int. J. Biometeorol.* **24:** 107–116.
129. **Sorber, C. A., H. T. Bausum, S. A. Schaub, and M. J. Small.** 1976. A study of bacterial aerosols at a wastewater irrigation site. *J. Water Pollut. Control Fed.* **48:**2367–2379.
130. **Sorenson, W. G., D. G. Frazer, B. B. Jarvis, J. Simpson, and V. A. Robinson.** 1987. Trichothecene mycotoxins in aerosolized conidia of *Stachybotrys atra*. *Appl. Environ. Microbiol.* **53:**1370–1375.
131. **Stärk, K. D. C., J. Nicolet, and J. Frey.** 1998. Detection of *Mycoplasma hypopneumoniae* by air sampling with a nested PCR assay. *Appl. Environ. Microbiol.* **64:**543–548.
132. **Stewart, S. L., S. A. Grinshpun, K. Willeke, S. Terzieva, V. Ulevicius, and J. Donnelly.** 1995. Effect of impact stress on microbial recovery on an agar surface. *Appl Environ. Microbiol.* **61:**1232–1239.
133. **Strachan, D. P., B. Flannigan, E. M. McCabe, and F. McGarry.** 1990. Quantification of airborne moulds in the homes of children with and without wheeze. *Thorax* **45:** 382–387.
134. **Thompson, M. W., J. Donnelly, S. A. Grinshpun, A. Juozaitis, and K. Willeke.** 1994. Method and test system for evaluation of bioaerosol samplers. *J. Aerosol Sci.* **25:** 1579–1593.
135. **Thorne, P. S., M. S. Kiekhaefer, P. Whitten, and K. J. Donham.** 1992. Comparison of bioaerosol sampling methods in barns housing swine. *Appl. Environ. Microbiol.* **58:**2543–2551.
136. **Trunov, M., S. Trakumas, K. Willeke, S. A. Grinshpun, and T. Reponen.** 2001. Collection of bioaerosols particles by impaction: effect of fungal spore agglomeration and bounce. *Aerosol Sci. Technol.* **34**(6)**:**490–498, **35**(1)**:** 617–624.
137. **Tyler, M. E., and E. L. Shipe.** 1959. Bacterial aerosol samplers. I. Development and evaluation of the all-glass impinger. *Appl. Microbiol.* **7:**337–349.
138. **Verhoeff, A. P., J. H. van Wijnen, J. S. M. Boleij, B. Brunekreef, E. S. van Reenen-Hoekstra, and R. A. Samson.** 1990. Enumeration and identification of airborne viable mould propagules in houses. *Allergy* **45:** 275–284.
139. **Wagner, J., and J. M. Marcher.** 2003. Comparison of a passive aerosol sampler to size-selective pump samplers in indoor environments. *Am. Ind. Hyg. Assoc. J.* **64:** 630–639.
140. **Wallace, L.** 1996. Indoor particles: a review. *J. Air Waste Manag. Assoc.* **46:**98–126.
141. **Walters, M., D. Milton, L. Larsson, and T. Ford.** 1994. Airborne environmental endotoxin: a cross validation of sampling and analysis techniques. *Appl. Environ. Microbiol.* **60:**996–1005.
142. **Wang, Z., T. Reponen, S. A. Grinshpun, R. L. Górny, and K. Willeke.** 2001. Effect of sampling time and air humidity on the bioefficiency of filter samplers for bioaerosol collection. *J. Aerosol Sci.* **32:**661–674.
143. **Weis, C. P., A. J. Intrepido, A. K. Miller, P. G. Cowin, M. A. Durno, J. S. Gebhardt, and R. Bull.** 2002. Secondary aerosolization of viable Bacillus anthracis spores in a contaminated U.S. Senate office. *JAMA* **288:** 2853–2858.
144. **White, L. A., D. J. Hadley, D. E. Davids, and R. Naylor.** 1975. Improved large-volume sampler for the collection of bacterial cells from aerosol. *Appl. Microbiol.* **29:**335–339.
145. **Willeke, K., and P. A. Baron (ed.).** 1993. *Aerosol Measurement: Principles, Techniques and Applications.* Van Nostrand Reinhold, New York, N.Y.
146. **Willeke, K., X. Lin, and S. A. Grinshpun.** 1998. Improved aerosol collection by combined impaction and centrifugal motion. *Aerosol Sci. Technol.* **29:**439–456.
147. **Wilson, I. G.** 1997. Inhibition and facilitation of nucleic acid amplification. *Appl. Environ. Microbiol.* **63:**3741–3751.
148. **Yike, I., T. Allan, W. G. Sorenson, and D. G. Dearborn.** 1999. Highly sensitive protein translation assay for trichothecene toxicity in airborne particulates: comparison with cytotoxicity assays. *Appl. Environ. Microbiol.* **65:**88–94.
149. **Zimmerman, N. J., P. C. Reist, and A. G. Turner.** 1987. Comparison of two biological aerosol sampling methods. *Appl. Environ. Microbiol.* **53:**99–104.

Analysis of Bioaerosol Samples

PATRICIA CRUZ AND MARK P. BUTTNER

75

Respiratory exposure to certain pathogenic or toxigenic microorganisms and/or elevated concentrations of environmental organisms could result in health effects, such as allergic reactions, irritant responses, toxicosis, and respiratory illness. Determination of the concentration and composition of bioaerosols in indoor environments is necessary for assessment of contamination levels and to estimate potential exposure of occupants. The need for accurate measurement of bioaerosols has received increased attention in recent years, owing to concerns with mold contamination in indoor environments and the threat of bioterrorism. Sample analysis methods include culture, and microscopic, biochemical, immunological, or molecular biological assays. Traditionally, airborne microorganisms have been analyzed by culturable and microscopic total count determinations. However, limitations to both of these methods have led to the development of techniques that can increase the sensitivity and accuracy of bioaerosol monitoring. The selection of an analysis method is a critical component of a bioaerosol sampling plan, and it should be chosen before air sampling is conducted. Factors which influence the choice of an analytical method include the cost and length of time required for analysis, the sensitivity and specificity of the analysis method, the sampling methods to be utilized, and the expected characteristics of the bioaerosol of interest. The purpose of this chapter is to present an overview of available methods for the analysis of bioaerosols. In addition, the potential use of enhanced monitoring of bioaerosols with PCR, biochemical, and immunological assays is discussed.

CULTURE METHODS

Culture Media and Incubation

Many of the currently available bioaerosol sampling methods rely on culture for the quantitation and characterization of airborne bacteria and fungi. Microorganisms that are collected on a nutrient agar surface by impaction can be cultured directly, while organisms that are collected in a liquid or on a filter are transferred to a culture medium. Only those cells which survive and reproduce under the culture conditions to form visible colonies are enumerated. However, nonviable microorganisms and/or their by-products that may cause health effects (32, 59) are not detected. Because microorganisms exhibit a wide range of nutritional requirements, no formulation is capable of culturing every type of organism. Therefore, a common strategy in bioaerosol monitoring is to choose general media which promote the growth of the greatest diversity of species. This approach is especially useful in survey efforts directed toward the characterization of the airborne microbiota. Because of differences in temperature, incubation time, and nutritional requirements among microorganisms, no single culture medium is satisfactory for the simultaneous isolation of airborne bacteria and fungi. Generally, it is necessary to perform replicate sampling using different culture media or to divide samples for inoculation onto multiple types of nutrient media.

Several conditions, such as pH, temperature, water activity, nutrients, antibiotics, light, and aeration, can be manipulated to favor the growth of a selected group of organisms (8). For example, incubation at a temperature of 55°C may be used to select for the growth of thermophilic actinomycetes, or antibiotic resistance markers may be used for the selective culture of a target species (1, 14, 96). However, the growth of microorganisms on selective media may be hindered compared with that on general media. This is especially of concern when culturing airborne microorganisms, which are already stressed or damaged by aerosolization and sampling. For this reason, general media are often used for the initial culture of bioaerosol samples, followed by replication onto differential or selective media for identification (52).

Several broad-spectrum media have been evaluated for their utility in the retrieval of culturable airborne fungi (10, 71, 87, 94). The following media have been recommended, although the results vary between investigations. Malt extract agar (MEA), rose bengal-containing agars (e.g., rose bengal-streptomycin and dichloran-rose bengal-chloramphenicol), and dichloran glycerol-18 agar have been suggested for the isolation of airborne fungi. MEA may be the most widely used fungal isolation medium. Of the formulations available, unamended 2% MEA was found to promote sporulation better than MEA amended with glucose and peptone (48, 91). Rose bengal- and dichloran-containing media have the advantage of inhibiting the spread of rapidly growing fungal genera (e.g., *Rhizopus* and *Mucor*), allowing the enumeration and identification of other slower-growing fungi in the sample. However, rose bengal is photoactivated in direct sunlight, forming cytotoxic products that may inhibit fungal growth (5). Dichloran glycerol-18 agar is a low-water-activity medium (a_w = 0.955) developed for the isolation of xerophilic fungi (47), and it compares favorably with other media for culturing mesophilic airborne fungi (94). Other commonly used fungal media include potato dextrose agar and Sabouraud's agar (45, 88).

Incubation periods for fungi typically range from 3 to 7 days; however, fastidious xerophiles may require several weeks for mycelial development (7). Most airborne fungi are mesophilic and grow well at temperatures of 20 to 25°C, and medically important fungal pathogens grow well at 30°C (65). Incubation temperature can also be used to select for certain species. *Aspergillus fumigatus* is a thermotolerant species which can be cultured at 37 to 45°C, above the temperature range of mesophilic fungi.

For the culture of bacteria, several broad-spectrum media, such as tryptic soy agar, nutrient agar, and casein soy peptone agar, may be used (8, 52). These media are often amended with antibiotics to restrict the growth of fungi. Incubation temperatures from 28 to 35°C for 1 to 7 days are usually used for bacteria from environmental and human sources; thermophilic actinomycetes, which are cultured at 55°C, are an important exception.

Enumeration

Viable counts are determined after the appropriate incubation period by enumerating CFU. For volumetric sampling, the concentration of culturable airborne microorganisms (CFU per cubic meter) is determined by dividing the number of CFU per sample by the volume of air sampled. When microorganisms are collected with multiple-jet impactor samplers (e.g., Andersen and SAS; Thermo Electron Corp., Waltham, Mass., and Bioscience International, Rockville, Md., respectively), positive-hole corrections are generally applied to the data (3, 15, 64). These corrections are estimations which account for the probability of multiple impactions of microorganisms through the same sampling jet, which results in the enumeration of only a single colony at that impaction site. The magnitude of the correction increases with the number of CFU, and the uncertainty of the estimate increases near the upper quantitation limit of the sampler (3). The corrected CFU count for a sample is used to calculate the airborne concentration of microorganisms. When organisms are collected through a single nozzle onto a stationary or moving surface, colonies may overlap each other, but the resulting reduced number of CFU can be corrected (18).

Enumeration errors are associated with colony counts that are very low or very high. Colony counts that are too low can be nonrepresentative of the population and exhibit high levels of variability. As colony counts increase, counting errors increase because of overlap of colonies and inhibitory effects of microorganisms on one another. Researchers have proposed that upper limits for efficient detection be determined by colony size and the resolution ability of the enumeration system (18). When filtration or impingement sampling is used, the wash solution or collection buffer can be serially diluted to obtain counts within an acceptable range. However, when air samples are collected directly onto an agar surface without knowledge of the bioaerosol concentration being sampled, too few or too many colonies may emerge on a plate. These difficulties can be lessened by setting more than one sample collection time.

Identification

Fungal identification is based largely on the morphological characteristics of spores and spore-bearing structures by direct microscopy. Misidentification of fungi is a significant source of error in culturable sample analysis; therefore, accurate fungal identification requires training and experience. Bacterial isolates may be identified by a variety of methods. Classical biochemical methods may be employed for identification to the genus and species level, or procedures such as Gram staining may be used to differentiate bacterial groups. A variety of identification systems which classify bacteria according to various biochemical characteristics, such as substrate utilization and cellular fatty acid profiles, are available (see chapter 10). However, many of the databases used to identify microorganisms mainly comprise clinical isolates; therefore, the identification of environmental microorganisms may be problematic.

MICROSCOPY

Microscopic enumeration methods are used to obtain an estimate of the total number of microorganisms present in a sample. In contrast with culture techniques, microscopic analysis allows enumeration of both culturable and nonculturable microorganisms. However, identification of microorganisms to the species level is usually not possible without the aid of a taxon-specific technique, such as immunospecific fluorescence staining (described below). A variety of stains may be utilized to differentiate biological particles from nonbiological material and respiring cells from nonrespiring cells.

Impaction onto a glass slide or tape or filtration sampling methods are generally used for the determination of total airborne fungal spores. Air samples obtained with Burkard spore traps are analyzed via light microscopy by examining the collection surface. Often, the entire spore trap trace is analyzed; alternatively, several traverses of the collection slide or tape are enumerated, and the total number of microorganisms is estimated by extrapolation. The volume of air represented by a traverse can be calculated, and the average number of fungal spores per traverse is used to determine the total number of spores per cubic meter. When the sample is collected onto a moving slide, time discrimination of fungal spore concentration is possible by enumerating traverses between specified points along the collection surface. Although staining is not required for enumeration, several stains including phenosafranin, basic fuchsin, and lactophenol cotton blue are commonly used to facilitate discrimination of spores from debris (8). Accurate identification to the genus level is possible for a limited number of fungal spore types. Data are usually reported as the total number of spores per cubic meter of air or the percent composition of fungal genera or groups. Fungal spores collected by filtration may be washed from the filter membrane, stained with acridine orange, and enumerated by epifluorescence microscopy (76). However, some fungal spore types may resist staining or have dark pigmentation which masks fluorescence (8).

For the determination of total airborne bacteria by microscopic analysis, liquid impingement or filtration sampling is used. Aliquots of collection buffer or the filter membrane wash solution may be stained for epifluorescence microscopy by the acridine orange direct count method (46, 76, 97); however, $>10^4$ cells per sample are required to obtain useful results by this method. A field study using swirling liquid impingers demonstrated that direct epifluorescence microscopy detected total airborne microbial concentrations that were 3 to 1,000 times higher than those obtained by conventional culture analysis (35). Cells and endospores may also be enumerated by bright-field or phase-contrast microscopy with a hemocytometer or counting chamber.

The advantage of microscopic analysis of air samples is the determination of total airborne microorganisms, both culturable and nonculturable. Because a significant fraction of airborne microorganisms may be nonculturable, microscopy is often used in conjunction with culture methods to analyze bioaerosol samples. The major disadvantage is the tedious and time-consuming nature of microscopic enumeration, although computerized image analysis systems may automate the process. Another drawback is the level of expertise required to identify fungal spores. Misidentification of fungal spores and the inability to distinguish microorganisms from nonbiological particles are common sources of error with this method (8).

FLOW CYTOMETRY

Flow cytometry offers an alternative to microscopic enumeration of total cells. A target-specific fluorescent probe is added to a liquid sample and subjected to a laser beam (4). Cell concentrations can be measured by light scattering and fluorescence emitted from fluorochromes bound to cells. An advantage of flow cytometry is its ability to be used for the simultaneous detection of several targets; however, assay optimization can be tedious (4). In one study, impinger samples labeled with 4′,6′-diamidino-2-phenylindole (DAPI) yielded comparable results when enumerated by microscopy or flow cytometry (56). Similarly, in another study, laboratory and field impinger samples were collected, and the data showed that flow cytometry results obtained for the detection of airborne fungi were in close agreement with epifluorescence microscopy counts and that flow cytometry data were more precise and reliable at fungal concentrations between 10^3 and 10^6 spores/ml (81). In a different application of flow cytometry, researchers coupled sandwich immunoassay utilizing microsphere beads with flow cytometry for the detection of biothreat agents (66). Following the immunoassay, the bound analytes were detected with a commercially available flow cytometer by using secondary antibodies labeled with the fluorescent dye phycoerythrin.

IMMUNOASSAYS

Immunoassay methods, used extensively in biomedical research, have been recognized for their potential application to bioaerosol analysis (9). Immunoassays rely on the binding of antibodies to a specific target antigen. Target antigens may be (i) cell surface-associated proteins or polysaccharides or (ii) human allergens. The development of an antibody is a critical component which affects the sensitivity and specificity of the assay; therefore, monoclonal antibodies are often used. Among the methods which may be applied to bioaerosol analysis are fluorescence immunoassay, enzyme immunoassay, and radioimmunoassay.

Fluorescence immunoassay consists of staining samples with a fluorescently labeled antibody that binds specifically to the antigens on the surfaces of the target organisms and enumeration by epifluorescence microscopy. Fluorescent labeling dyes include fluorescein, fluorescein isothiocyanate, and rhodamine isothiocyanate (52). Appropriate controls are necessary to determine levels of background and nonspecific fluorescence. An enzyme-linked immunosorbent assay-based flowthrough capture instrument was developed for the detection of bacterial endospores and cells that is capable of analyzing large sample volumes in a short period of time (98). Other researchers have reported that the detection of satratoxins by an enzyme-linked immunosorbent assay was comparable to results obtained by high-performance liquid chromatography (20). The immunoassay technique has also been utilized in the development of hand-held assays (HHAs) or lateral-flow assays for the detection of biothreat agents (4). HHAs are disposable devices that employ the same technology as that of home pregnancy test kits and consist of cardboard tickets containing immobilized antibodies specific to a target organism (11, 12, 33). The HHA has a positive-control well and a test well. Application of a liquid sample initiates an immunological reaction when the target organism is present. If a pink or red line develops in both wells, the result is considered positive; but if a line develops only in the control well, the result is considered negative. Limitations of these devices include false-positive results and poor sensitivity. Although HHAs were designed to provide data quickly, the U.S. Department of Defense has determined that they are effective only when employed as part of a methodical detection approach that provides additional levels of confirmation (33). Improved sensitivity and quantitative detection by this assay have been explored by incorporating fluorescent microspheres or paramagnetic beads and analysis by a compatible reader (4).

Radioimmunoassay consists of combining a specific antibody with a radioactive label, while enzyme immunoassay utilizes binding of the antibody or the antigen to an enzyme. The concentration of antigen is measured by radioactivity or enzyme activity. These methods have been applied for the measurement of airborne allergens, such as dust mite allergen, animal dander, or β-1,3-glucan (28, 30, 62, 80). Impaction, impingement, and filtration sampling methods are compatible with the assay (19). The advantages of these methods for quantitation of airborne allergen are their specificity and sensitivity. The major limitation of immunoassays is that specific antigens for microorganisms are difficult to define and standardize (65). As more microbial antigenic compounds are characterized, immunoassay analysis methods may be used to provide data necessary for assessing environmental exposure to allergens.

BIOCHEMICAL ASSAYS

Biochemical methods of analysis are used to measure a biological compound of interest, such as endotoxins or mycotoxins. Because inhalation of these compounds may produce adverse health effects (16, 22), there have been efforts to measure these compounds directly to relate environmental exposure to human response. Endotoxins are lipid polysaccharides in the cell walls of gram-negative bacterial cells. Airborne endotoxin is widespread because of the ubiquity of gram-negative bacteria in the environment, and elevated levels of endotoxin have been measured in bioaerosol samples from a variety of agricultural, industrial, and office environments (50). The most widely used method for measurement of endotoxin is the *Limulus* amebocyte lysate (LAL) test. The lysate of amebocytes from the horseshoe crab, *Limulus polyphemus*, gels in the presence of lipid polysaccharides. This reaction forms the basis of endotoxin quantitation, and a variety of test systems are commercially available. Airborne endotoxin data are expressed as the number of nanograms per cubic meter or endotoxin units (EUs) per cubic meter. An EU is defined as the potency of 0.10 ng of a reference standard endotoxin (69). Because endotoxin potency varies among gram-negative bacteria, data are often expressed as EUs to facilitate comparisons between studies. Filtration sampling is most often used to collect airborne endotoxin; however, various filter materials have been shown to inhibit the LAL assay (70). Differences in collection and analysis methods between laboratories have made comparison of results difficult. In addition, interference with the LAL reaction by inhibitors present in environmental samples must be accounted for in the measurement of endotoxin content (69). Other biochemical methods that may be used in the analysis of gram-positive bacteria in indoor environments include branched-chain fatty acid and muramic acid assays (86).

Mycotoxins associated with dusts from fungus-contaminated grains have long been recognized as a source of illness (37). Exposure to *Stachybotrys chartarum* mycotoxins has been associated with toxicoses in contaminated home and office environments (21, 51, 53). Filtration sampling using glass fiber filters or polycarbonate membranes has been used to measure aerosolized mycotoxins of *S. chartarum* in the laboratory (78, 89) and in the field (102). Field sampling conducted in Finnish water-damaged buildings utilized high-pressure liquid chromatography, mass spectrometry, and electrospray ionization to analyze bulk samples for the presence of mycotoxins (92). This study demonstrated that mycotoxins were found in >40% of the 79 bulk samples analyzed and that some toxigenic fungal species were usually present in mycotoxin-positive samples, even if the species responsible for mycotoxin production was not isolated. Recently, researchers have demonstrated that ergot alkaloids, which are mycotoxins, were associated with conidia of *A. fumigatus* (77).

Other biochemical methods which may be applied to the analysis of airborne fungal exposure include ergosterol (67, 68) and β-1,3-glucan assays (31, 38, 39, 83, 86). An alternative to ergosterol content as a means to estimate fungal biomass is the fluorogenic detection of β-*N*-acetylhexosaminidase activity (82). Microbial volatile organic compounds (MVOCs) are products of fungal and bacterial metabolism that may cause toxic and allergic reactions. Volumetric air samples collected onto activated carbon media are analyzed for MVOCs by gas chromatography-mass spectrometry. A study involving bioaerosol monitoring of residences cleaned and reoccupied following a major flood found that MVOCs were detected in over half of the houses tested (35). Although MVOCs cannot be used to quantitate fungi, they can serve as good indicators of fungal amplification in indoor environments (55), provided that artificial sources are considered and indoor/outdoor ratios are established (35). Studies have shown that clean building materials may produce some of the same VOCs emitted by microorganisms (55). Therefore, background VOCs need to be determined in non-problem buildings prior to utilization of MVOC data as an indicator of microbial growth in indoor environments (55).

PCR

PCR is a procedure used to rapidly amplify specific DNA sequences (84). This technique has been used successfully to enhance the detection of microorganisms in a variety of matrices, including air samples (1, 72, 74, 85, 90, 103). Application of the PCR technique to bioaerosol sampling can result in the rapid, sensitive detection of the target microorganism(s). The use of PCR is particularly suited for detection of specific microorganisms that are difficult to culture, grow very slowly, or have never been cultured in vitro. The PCR method also provides an increase in sensitivity over traditional culture methods (1, 13, 54).

Another advantage of PCR is the rapidity with which results are obtained compared to culture counts. It is possible to obtain results within hours of sample collection with PCR, compared with days or weeks for culture methods. It should be noted that PCR can be used for the direct detection of microorganisms, provided that unique, non-cross-hybridizing sequences are available for the specific microorganisms, to serve as primers for amplification. The 18S or 16S ribosomal RNA genes are commonly used for the design of genus-specific primers and probes because they contain highly conserved sequences between members of the same genus, but the sequences are variable among different genera. The internal transcribed spacer regions and intergenic spacer of the nuclear rRNA gene are often the target of species-specific primers and probes, because these are highly variable areas within a genus or among populations (99). The design of species-specific primers and probes is difficult due to homology in the DNA of closely related species. Although a sequence comparison may indicate that the designed primer set will amplify only the target organism, it is crucial that the designed primers and probes be thoroughly tested in the laboratory with DNA extracted from both the target and other (non-target) organisms to confirm that they amplify the DNA from all target strains and will not cross-react with non-target strains (23). Researchers have evaluated published primers for the specific detection of common airborne fungi and verified the specificity of 28 sets of primers (101). With conventional PCR, the authenticity of the amplified DNA can be confirmed by sequencing, restriction analysis, or gene probe hybridization. A source of published gene probes is probeBase, an online database of probes targeting rRNA genes (http://www.microbial-ecology.net/probebase/) (61). If there are multiple organisms of interest in a sample, it may be possible to use multiplex PCR in the same reaction mixture to simultaneously amplify several DNA targets of interest (17, 27, 63).

A limitation of the PCR assay is the inability to distinguish between culturable and nonculturable microorganisms. Quantitation of the concentration of the DNA segments which serve as templates for PCR has been reported for several sample matrices (29, 58, 79). The major difficulty associated with PCR as a quantitative method is that no other technique yet matches its sensitivity and can be used to compare results. Thus, any data obtained by quantitative PCR need to be supported by carefully designed controls and compared to reliable standards (36).

Sample Processing

Selection of environmental sampling and processing protocols must consider methods that yield a liquid sample that is compatible with quantitative PCR (QPCR) analysis. A concentration step is required for PCR to be effectively applied to air sample analysis. Two sample collection methods, liquid impingement and filtration, have been utilized successfully for air sample collection, with subsequent PCR detection of the target microorganism (1, 85, 90).

Inhibition

The presence of PCR inhibitors in environmental samples may reduce sensitivity or result in false negatives (2, 11, 13, 24, 25, 100). A nucleic acid purification step or dilution of the sample may be required to remove environmental interference to PCR amplification; however, these procedures can reduce the sensitivity of detection (2, 103). Additives such as bovine serum albumin have been successfully utilized to overcome PCR inhibition (93). An internal positive control (IPC) should be incorporated into the PCR assay to determine whether the samples contain PCR inhibitors, thus avoiding the risk of reporting false negative results. An IPC consists of a known quantity of control DNA, primers, and probe. Inhibition of the PCR assay is observed by reduced amplification results of the IPC DNA compared to results from control samples.

Quantitative PCR

Two primary approaches exist that are capable of providing accurate, reproducible measurement of initial target DNA concentrations in samples, competitive PCR and QPCR using fluorogenic probes (29, 75). The first approach, competitive PCR, utilizes a known amount of internal control DNA competitor in each reaction (Fig. 1A). A suitable competitor acts as an internal standard and has the same primer binding sites as the target DNA. Quantitation is achieved by comparing the concentration of the target product to that of the competitor. An important consideration is that the competitor must amplify with approximately the same efficiency as the target template; therefore, a significant effort must be applied in the development and testing of the competitor (41). After amplification, a variety of post-PCR methods such as gel electrophoresis followed by ethidium bromide staining or radioactivity measurement (if radioactive deoxynucleoside triphosphates are used), quantitation of fluorescence with image analysis systems, or high-performance liquid chromatography can be used to determine the final product concentrations (75).

The second approach, QPCR, is a real-time assay that measures target product accumulation with fluorogenic probes (Fig. 1B). The TaqMan QPCR (Sequence Detection system; Applied Biosystems, Foster City, Calif.) utilizes a fluorescent oligonucleotide probe that anneals to the target DNA between the specific primer binding sites. As new DNA is synthesized, the probe is cleaved by the *Taq* polymerase, causing it to fluoresce, and the amount of fluorescence is used to measure the initial DNA concentration. The probe also provides confirmation of the product, as it fluoresces only in the presence of the target template. The TaqMan assay has been demonstrated to accurately and rapidly detect bacteria (6, 29, 73) and fungi (24, 25, 40, 57, 95) of indoor air quality significance, and it has also been used for the detection of bacteria serving as simulants for biological warfare agents (11–13).

Other QPCR instruments have been developed that use the fluorescent dye SYBR Green or two fluorescently labeled probes that serve as hybridization probes to the double-stranded amplified DNA (LightCycler; Roche, Indianapolis, Ind.; R.A.P.I.D. system; Idaho Technology, Inc., Salt Lake City, Utah). These instruments have been utilized in the detection of fungi such as *A. fumigatus* and *Candida albicans* (60) and bacteria such as *Staphylococcus aureus* (42) and *Bacillus anthracis* (43).

EMERGING TECHNOLOGIES

Recent advances in microarray technology and microfluidics are being utilized in the development of fast and sensitive sensing systems that are of particular importance in the detection of biothreat agents (49), and several microarray-based

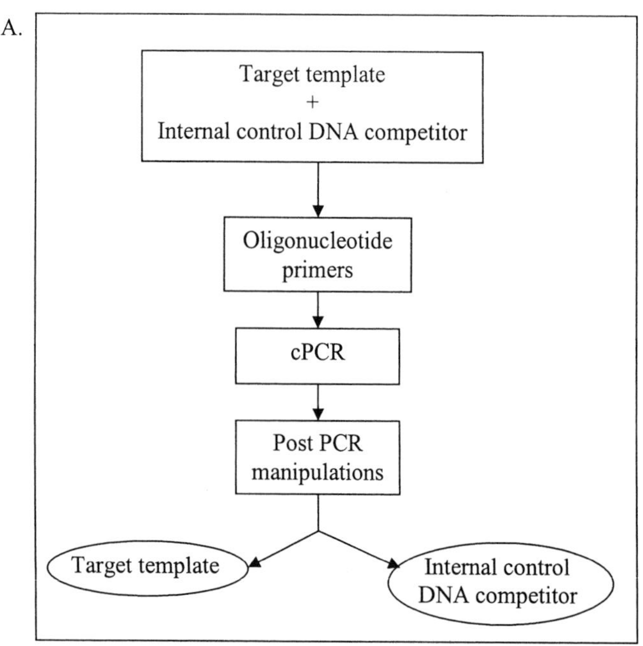

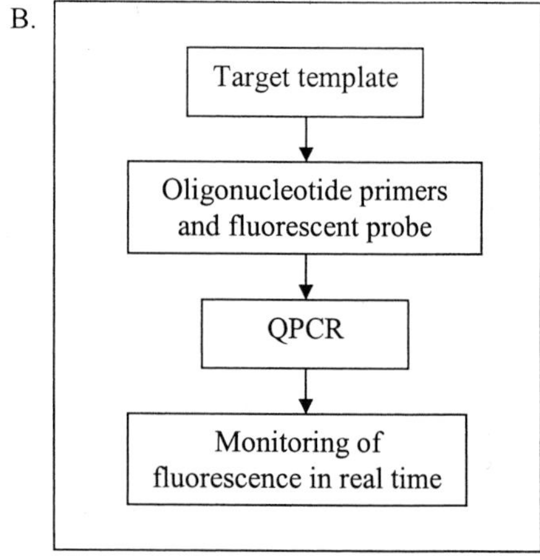

FIGURE 1 Diagrams illustrating competitive PCR (cPCR) and QPCR. (A) In cPCR, the target template and the internal control DNA compete for the oligonucleotide primers during the PCR assay. Post-PCR manipulations are required to separate the target and competitor products and to determine the ratio between them (i.e., final product concentrations). (B) In QPCR, a fluorescent probe anneals to the target DNA between the specific primer binding sites. As new DNA is synthesized, the probe is cleaved by the *Taq* polymerase, causing it to fluoresce; the amount of fluorescence is used to measure the initial target DNA concentration.

systems are commercially available. Hybrid technologies are being developed that combine more than one analysis method to maximize the accuracy of the results obtained (49). One such system utilizes immunoassay, flow cytometry, and PCR analysis in various combinations for the continuous monitoring of airborne biological threat agents (44, 66). Other researchers have coupled PCR and enzyme immunoassay for the differentiation of *Aspergillus* species from other fungi (26). Automation has been explored to increase sample throughput, maximize accuracy, and reduce costs in the detection of biological threat agents (34).

With the wide variety of bioaerosol sampling and analysis methods in use, the investigator must have a clear understanding of the objectives of bioaerosol sampling before sampling is conducted. This information will assist in the selection of the appropriate sampling and analysis methodologies for collecting meaningful bioaerosol data. As existing methods are improved and new techniques are developed, more accurate bioaerosol information may be obtained, which will further the understanding of human exposure to airborne microorganisms and aid in the development of standardized monitoring protocols.

REFERENCES

1. **Alvarez, A. J., M. P. Buttner, G. A. Toranzos, E. A. Dvorsky, A. Toro, T. B. Heikes, L. E. Mertikas, and L. D. Stetzenbach.** 1994. The use of solid-phase polymerase chain reaction for the enhanced detection of airborne microorganisms. *Appl. Environ. Microbiol.* **60:**374–376.
2. **Alvarez, A. J., M. P. Buttner, and L. D. Stetzenbach.** 1995. PCR for bioaerosol monitoring: sensitivity and environmental interference. *Appl. Environ. Microbiol.* **61:** 3639–3644.
3. **Andersen, A. A.** 1958. New sampler for the collection, sizing, and enumeration of viable airborne particles. *J. Bacteriol.* **76:**471–484.
4. **Andreotti, P. E., G.V. Ludwig, A. H. Peruski, J. J. Tuite, S. S. Morse, and L. F. Peruski, Jr.** 2003. Immunoassay of infectious agents. *BioTechniques* **35:**850–859.
5. **Banks, J. G., R. G. Board, J. Carger, and A. D. Dodge.** 1985. The cytotoxic and photodynamic inactivation of microorganisms by rose bengal. *J. Appl. Bacteriol.* **58:**392–400.
6. **Bassler, H. A., S. J. A. Flood, K. J. Livak, J. Marmaro, R. Knorr, and C. A. Batt.** 1995. Use of a fluorogenic probe in a PCR–based assay for the detection of *Listeria monocytogenes*. *Appl. Environ. Microbiol.* **61:**3724–3728.
7. **Beuchat, L. R.** 1992. Media for detecting and enumerating yeasts and moulds. *Int. J. Food Microbiol.* **17:**145–158.
8. **Burge, H. A.** 1995. Bioaerosol investigations, p. 1–23. *In* H. A. Burge (ed.), *Bioaerosols*. CRC Press, Boca Raton, Fla.
9. **Burge, H. A., and W. R. Solomon.** 1987. Sampling and analysis of biological aerosols. *Atmos. Environ.* **21:**451–456.
10. **Burge, H. P., W. R. Solomon, and J. R. Boise.** 1977. Comparative merits of eight popular media in aerometric studies of fungi. *J. Allergy Clin. Immunol.* **60:**199–203.
11. **Buttner, M. P., P. Cruz, L. D. Stetzenbach, A. K. Klima-Comba, V. L. Stevens, and T. D. Cronin.** 2004. Determination of the efficacy of two building decontamination strategies by surface sampling with culture and quantitative PCR analysis. *Appl. Environ. Microbiol.* **70:**4740–4747.
12. **Buttner, M. P., P. Cruz, L. D. Stetzenbach, A. K. Klima-Comba, V. L. Stevens, and P. A. Emanuel.** 2004. Evaluation of the Biological Sampling Kit (BiSKit) for large-area surface sampling. *Appl. Environ. Microbiol.* **70:** 7040–7045.
13. **Buttner, M. P., P. Cruz-Perez, and L. D. Stetzenbach.** 2001. Enhanced detection of surface-associated bacteria in indoor environments by quantitative PCR. *Appl. Environ. Microbiol.* **67:**2564–2570.
14. **Buttner, M. P., and L. D. Stetzenbach.** 1991. Evaluation of four aerobiological sampling methods for the retrieval of aerosolized *Pseudomonas syringae*. *Appl. Environ. Microbiol.* **57:**1268–1270.
15. **Buttner, M. P., and L. D. Stetzenbach.** 1993. Monitoring airborne fungal spores in an experimental indoor environment to evaluate sampling methods and the effects of human activity on air sampling. *Appl. Environ. Microbiol.* **59:**219–226.
16. **Castellan, R. M., S. A. Olenchock, K. B. Kinsley, and J. L. Hankinson.** 1987. Inhaled endotoxin and decreased spirometric values: an exposure-response relation for cotton dust. *New Engl. J. Med.* **317:**605–610.
17. **Chamberlain, J. S., R. A. Gibbs, J. E. Ranier, P. N. Nguyen, and C. T. Cashey.** 1988. Deletion screening of the Duchenne muscular dystrophy locus via multiplex DNA amplification. *Nucleic Acids Res.* **16:**11141–11156.
18. **Chang, C. W., Y. H. Hwang, S. A. Grinshpun, J. M. Macher, and K. Willeke.** 1994. Evaluation of counting error due to colony masking in bioaerosol sampling. *Appl. Environ. Microbiol.* **60:**3732–3738.
19. **Chapman, M. D.** 1995. Analytical methods: immunoassays, p. 235–248. *In* H. A. Burge (ed.), *Bioaerosols*. Lewis Publishers, Boca Raton, Fla.
20. **Chung, Y.-J., B. B. Jarvis, H. Tak, and J. J. Pestka.** 2003. Immunochemical assay for satratoxin G and other macrocyclic trichothecenes associated with indoor air contamination by *Stachybotrys chartarum*. *Toxicol. Mech. Methods* **13:**247–252.
21. **Croft, W. S., B. B. Jarvis, and C. S. Yatawara.** 1986. Airborne outbreak of trichothecene toxicosis. *Atmos. Environ.* **20:**549–552.
22. **Crook, B., S. Higgins, and J. Lacey.** 1987. Methods for sampling microorganisms at solid waste disposal sites, p. 791–797. *In* D. R. Houghton, R. N. Smith, and H. O. W. Eggins (ed.), *Biodeterioration*, vol. 7. *Proceedings of the 7th International Biodeterioration Symposium*. Elsevier, London, United Kingdom.
23. **Cruz, P., and L. D. Stetzenbach.** 2004. Specific detection of fungi associated with sick building syndrome when using quantitative polymerase chain reaction, p. 437–449. *In* D. C. Straus (ed.), *Advances in Applied Microbiology*, vol. 55. *Sick Building Syndrome*. Elsevier Academic Press, San Diego, Calif.
24. **Cruz–Perez, P., M. P. Buttner, and L. D. Stetzenbach.** 2001. Detection and quantitation of *Aspergillus fumigatus* in pure culture using polymerase chain reaction. *Mol. Cell. Probes* **15:**81–88.
25. **Cruz–Perez, P., M. P. Buttner, and L. D. Stetzenbach.** 2001. Specific detection of *Stachybotrys chartarum* in pure culture using quantitative polymerase chain reaction. *Mol. Cell. Probes* **15:**129–138.
26. **de Aguirre, L., S. F. Hurst, J. S. Choi, J. H. Shin, H. P. Hinrikson, and C. J. Morrison.** 2004. Rapid differentiation of *Aspergillus* species from other medically important opportunistic molds and yeasts by PCR-enzyme immunoassay. *J. Clin. Microbiol.* **42:**3495–3504.
27. **Dean, T. R., B. Roop, D. Betancourt, and M. Y. Menetrez.** 2005. A simple multiplex polymerase chain reaction assay for the identification of four environmentally relevant fungal contaminants. *J. Microbiol Methods* **61:**9–16.
28. **DeBlay, F., M. D. Chapman, and T. A. Platts-Mills.** 1991. Airborne cat allergen Feld I. Environmental control with the cat in situ. *Am. Rev. Respir. Dis.* **143:**1334–1339.

29. **Desjardin, L. E., Y. Chen, M. D. Perkins, L. Teixeira, M. D. Cave, and K. D. Eisenach.** 1998. Comparison of the ABI 7700 System (TaqMan) and competitive PCR for quantification of IS*6110* DNA in sputum during treatment of tuberculosis. *J. Clin. Microbiol.* **36:**1964–1968.
30. **Douwes, J., G. Doekes, R. Montijn, D. Heederik, and B. Brunekreef.** 1996. Measurement of β(1→3)-glucans in occupational and home environments with an inhibition enzyme immunoassay. *Appl. Environ. Microbiol.* **62:**3176–3182.
31. **Douwes, J., P. Thorne, N. Pearce, and D. Heederik.** 2003. Bioaerosol health effects and exposure assessment: progress and prospects. *Ann. Occup. Hyg.* **47:**187–200.
32. **Eduard, W., and D. Heederik.** 1998. Methods for quantitative assessment of airborne levels of noninfectious microorganisms in highly contaminated work environments. *Am. Ind. Hyg. Assoc. J.* **59:**113–127.
33. **Emanuel, P. A., C. Chue, L. Kerr, and D. Cullin.** 2003. Validating the performance of biological detection equipment: the role of the federal government. *Biosecur. Bioterror.* **1:**131–137.
34. **Emanuel, P. A., I. R. Fruchey, A. M. Bailey, J. L. Dang, K. Niyogi, J. W. Roos, D. Cullin, and D. C. Emanuel.** 2005. Automated screening for biological weapons in homeland defense. *Biosecur. Bioterror.* **3:**39–50.
35. **Fabian, M. P., S. L. Miller, T. Reponen, and M. T. Hernandez.** 2005. Ambient bioaerosol indices for indoor air quality assessments of flood reclamation. *J. Aerosol Sci.* **36:**763–783.
36. **Ferre, F.** 1992. Quantitative or semi-quantitative PCR: reality versus myth. *PCR Methods Applic.* **2:**1–9.
37. **Flannigan, B.** 1987. Mycotoxins in the air. *Int. Biodeterior.* **23:**73–78.
38. **Fogelmark B., H. Goto, K. Yuasa, B. Marchat, and R. Rylander.** 1992. Acute pulmonary toxicity of inhaled β-1,3-glucan and endotoxin. *Agents Actions* **35**:50–56.
39. **Goto, H., K. Yuasa, and R. Rylander.** 1994. (1-3)-β-D-Glucan in indoor air, its measurement and in vitro activity. *Am. J. Ind. Med.* **25**:81–83.
40. **Haugland, R. A., S. J. Vesper, and L. J. Wymer.** 1999. Quantitative measurement of *Stachybotrys chartarum* conidia using real time detection of PCR products with the TaqMan™ fluorogenic probe system. *Mol. Cell. Probes* **13:**329–340.
41. **Heid, C. A., J. Stevens, K. J. Livak, and P. M. Williams.** 1996. Real time quantitative PCR. *Genome Res.* **6:**986–994.
42. **Hein, I., A. Lehner, P. Rieck, K. Klein, E. Brandl, and M. Wagner.** 2001. Comparison of different approaches to quantify *Staphylococcus aureus* cells by real-time quantitative PCR and application of this technique for examination of cheese. *Appl. Environ. Microbiol.* **67:**3122–3126.
43. **Higgins, J. A., M. Cooper, L. Schroeder-Tucker, S. Black, D. Miller, J. S. Karns, E. Manthey, R. Breeze, and M. L. Perdue.** 2003. A field investigation of *Bacillus anthracis* contamination of U.S. Department of Agriculture and other Washington, D.C., buildings during the anthrax attack of October 2001. *Appl. Environ. Microbiol.* **69:**593–599.
44. **Hindson, B. J., S. B. Brown, G. D. Marshall, M. T. McBride, A. J. Makarewicz, D. M. Gutierrez, D. K. Wolcott, T. R. Metz, R. S. Madabhushi, J. M. Dzenitis, and B. W. Colston, Jr.** 2004. Development of an automated sample preparation module for environmental monitoring of biowarfare agents. *Anal. Chem.* **76:**3492–3497.
45. **Hirsch, S. R., and J. A. Sosman.** 1976. A one-year survey of mold growth inside twelve homes. *Ann. Allergy* **36:**30–38.
46. **Hobbie, J. E., R. J. Daley, and S. Jasper.** 1977. Use of nucleopore filters for counting bacteria by fluorescence microscopy. *Appl. Environ. Microbiol.* **33:**1225–1228.
47. **Hocking, A. D., and J. I. Pitt.** 1980. Dichloran-glycerol medium for enumeration of xerophilic fungi from low-moisture foods. *Appl. Environ. Microbiol.* **39:**488–492.
48. **Hunter, C. A., C. Grant, B. Flannigan, and A. F. Bravery.** 1988. Mould in buildings: the air spora of domestic buildings. *Int. Biodeterior.* **24:**81–101.
49. **Ivnitski, D., D. J. O'Neil, A. Gattuso, R. Schlicht, M. Calidonna, and R. Fisher.** 2003. Nucleic acid approaches for detection and identification of biological warfare and infectious agents. *BioTechniques* **35:**862–869.
50. **Jacobs, R. R.** 1989. Airborne endotoxins: an association with occupational lung disease. *Appl. Ind. Hyg.* **4:**50–56.
51. **Jarvis, B. B., W. G. Sorenson, E. Hintikka, M. Nikulin, Y. Zhou, J. Jiang, S. Wang, S. Hinkley, R. A. Etzel, and D. Dearborn.** 1998. Study of toxin production by isolates of *Stachybotrys chartarum* and *Memnoniella echinata* isolated during a study of pulmonary hemosiderosis in infants. *Appl. Environ. Microbiol.* **64:**3620–3625.
52. **Jensen, P. A., B. Lighthart, A. J. Mohr, and B. T. Shaffer.** 1994. Instrumentation used with microbial bioaerosol, p. 226–284. *In* B. Lighthart and A. J. Mohr (ed.), *Atmospheric Microbial Aerosols: Theory and Applications.* Chapman and Hall, New York, N.Y.
53. **Johanning, E., R. Biagini, D. Hull, P. Morey, B. Jarvis, and P. Landsbergis.** 1996. Health and immunology study following exposure to toxigenic fungi (*Stachybotrys chartarum*) in a water-damaged office environment. *Int. Arch. Occup. Environ. Health* **68:**207–218.
54. **Josephson, K. L., C. P. Gerba, and I. L. Pepper.** 1993. Polymerase chain reaction detection of nonviable bacterial pathogens. *Appl. Environ. Microbiol.* **59:**3513–3515.
55. **Korpi, A., A.-L. Pasanen, and P. Pasanen.** 1998. Volatile compounds originating from mixed microbial cultures on building materials under various humidity conditions. *Appl. Environ. Microbiol.* **64:**2914–2919.
56. **Lange, J. L., P. S. Thorne, and N. Lynch.** 1997. Application of flow cytometry and fluorescent in situ hybridization for assessment of exposures to airborne bacteria. *Appl. Environ. Microbiol.* **63:**1557–1563.
57. **Leenders, A. C. A. P., A. Van Belkum, M. Behrendt, A. Luijendijk, and H. A. Verbrugh.** 1999. Density and molecular epidemiology of *Aspergillus* in air and relationship to outbreaks of *Aspergillus* infection. *J. Clin. Microbiol.* **37:**1752–1757.
58. **Leser, T. D., M. Boye, and N. B. Hendriksen.** 1995. Survival and activity of *Pseudomonas* sp. strain B13(FR1) in a marine microcosm determined by quantitative PCR and an rRNA-targeting probe and its effect on the indigenous bacterioplankton. *Appl. Environ. Microbiol.* **61:**1201–1207.
59. **Levetin, E.** 1995. Fungi, p. 87–120. *In* H. A. Burge (ed.), *Bioaerosols*. Lewis Publishers, Boca Raton, Fla.
60. **Loeffler, J., N. Henke, H. Hebart, D. Schmidt, L. Hagmeyer, U. Schumacher, and H. Einsele.** 2000. Quantification of fungal DNA by using fluorescence resonance energy transfer and the Light Cycler system. *J. Clin. Microbiol.* **38:**586–590.
61. **Loy, A., M. Horn, and M. Wagner.** 2003. probeBase: an online resource for rRNA-targeted oligonucleotide probes. *Nucleic Acids Res.* **31:**514–516.
62. **Luczynska, C. M., Y. Li, M. D. Chapman, and T. A. E. Platts-Mills.** 1990. Airborne concentration and particle size distribution of allergen derived from domestic cats (*Felis domesticus*): measurements using cascade impactor, liquid impinger and a two site monoclonal antibody assay for *Fel d* I. *Am. Rev. Respir. Dis.* **141:**361–367.

63. **Luo, G., and T. G. Mitchell.** 2002. Rapid identification of pathogenic fungi directly from cultures by using multiplex PCR. *J. Clin. Microbiol.* **40:**2860–2865.
64. **Macher, J. M.** 1989. Positive-hole correction of multiple-jet impactors for collecting viable microorganisms. *Am. Ind. Hyg. Assoc. J.* **50:**561–568.
65. **Madelin, T. M., and M. F. Madelin.** 1995. Biological analysis of fungi and associated molds, p. 361–386. *In* C. S. Cox and C. M. Wathes (ed.), *Bioaerosols Handbook*. CRC Press, Boca Raton, Fla.
66. **McBride, M. T., D. Masquelier, B. J. Hindson, A. J. Makarewicz, S. Brown, K. Burris, T. Metz, R. G. Langlois, K. W. Tsang, R. Bryan, D. A. Anderson, K. S. Venkateswaran, F. P. Milanovich, and B. W. Colston, Jr.** 2003. Autonomous detection of aerosolized *Bacillus anthracis* and *Yersinia pestis*. *Anal. Chem.* **75:**5293–5299.
67. **Miller, J. D., A. M. Laflamme, Y. Sobol, P. Lafontaine, and R. Greenhalgh.** 1988. Fungi and fungal products in some Canadian houses. *Int. Biodeterior.* **24:**103–120.
68. **Miller, J. D., and J. C. Young.** 1997. The use of ergosterol to measure exposure to fungal propagules in indoor air. *Am. Ind. Hyg. Assoc. J.* **58:**39–43.
69. **Milton, D. K.** 1995. Endotoxin, p. 77–86. *In* H. A. Burge (ed.), *Bioaerosols*. CRC Press, Boca Raton, Fla.
70. **Milton, D. K., R. J. Gere, H. A. Feldman, and I. A. Greaves.** 1990. Endotoxin measurement: aerosol sampling and application of a new *Limulus* method. *Am. Ind. Hyg. Assoc. J.* **51:**331–337.
71. **Morring, K. L., W. G. Sorenson, and M. D. Attfield.** 1983. Sampling for airborne fungi: a statistical comparison of media. *Am. Ind. Hyg. Assoc. J.* **44:**662–664.
72. **Myatt, T. A., S. L. Johnston, Z. Zuo, M. Wand, T. Kebadze, S. Rudnick, and D. K. Milton.** 2004. Detection of airborne rhinovirus and its relation to outdoor air supply in office environments. *Am. J. Respir. Crit. Care Med.* **169:**1187–1190.
73. **Nogva, H. K., and D. Lillehaug.** 1999. Detection and quantification of *Salmonella* in pure cultures using 5′-nuclease polymerase chain reaction. *Int. J. Food Microbiol.* **51:**191–196.
74. **Olsson, M., C. Lidman, S. Latouche, A. Björkman, P. Roux, E. Linder, and M. Wahlgren.** 1998. Identification of *Pneumocystis carinii* f. sp. *hominis* gene sequences in filtered air in hospital environments. *J. Clin. Microbiol.* **36:**1737–1740.
75. **Orlando, C., P. Pinzani, and M. Pazzagli.** 1998. Developments in quantitative PCR. *Clin. Chem. Lab. Med.* **36:**255–269.
76. **Palmgren, U., G. Strom, G. Blomquist, and P. Malmberg.** 1986. Collection of airborne micro-organisms on Nuclepore filters, estimation and analysis—CAMNEA method. *J. Appl. Bacteriol.* **61:**401–406.
77. **Panaccione, D. G., and C. M. Coyle.** 2005. Abundant respirable ergot alkaloids from the common airborne fungus *Aspergillus fumigatus*. *Appl. Environ. Microbiol.* **71:**3106–3111.
78. **Pasanen, A., M. Nikulin, M. Tuomainen, S. Berg, P. Parikka, and E. Hintikka.** 1993. Laboratory experiments on membrane filter sampling of airborne mycotoxins produced by *Stachybotrys atra* corda. *Atmos. Environ.* **27A:**9–13.
79. **Picard, C., C. Ponsonnet, E. Paget, X. Nesme, and P. Simonet.** 1992. Detection and enumeration of bacteria in soil by direct DNA extraction and polymerase chain reaction. *Appl. Environ. Microbiol.* **58:**2717–2722.
80. **Platts-Mills, T. A. E., W. R. Thomas, R. C. Aalbersee, D. Vervloet, and M. D. Chapman.** 1992. Dust mite allergens and asthma: report of a second international workshop. *J. Allergy Clin. Immunol.* **89:**1046.
81. **Prigione, V., G. Lingua, and V. Filipello Marchisio.** 2004. Development and use of flow cytometry for detection of airborne fungi. *Appl. Environ. Microbiol.* **70:**1360–1365.
82. **Reesley, M., M. Miller, and K. F. Nielsen.** 2003. Quantifying mold biomass on gypsum board: comparison of ergosterol and beta-*N*-acetylhexosaminidase as mold biomass parameters. *Appl. Environ. Microbiol.* **69:**3996–3998.
83. **Rylander, R.** 1999. Indoor air-related effects and airborne (1→3)-β-D-glucan. *Environ. Health Perspect.* **107:**501–503.
84. **Saiki, R. K., S. Scharf, F. Faloona, K. B. Mullis, G. T. Horn, H. A. Erlich, and N. Arnheim.** 1985. Enzymatic amplification of β-globin genomic sequence and restriction site analysis for diagnosis of sickle cell anemia. *Science* **230:**1350–1354.
85. **Sawyer, M. H., C. J. Chamberlain, Y. N. Wu, N. Aintablian, and M. R Wallace.** 1994. Detection of varicella-zoster virus DNA in air samples from hospital rooms. *J. Infect. Dis.* **169:**91–94.
86. **Sebastian, A., and L. Larsson.** 2003. Characterization of the microbial community in indoor environments: a chemical-analytical approach. *Appl. Environ. Microbiol.* **69:**3103–3109.
87. **Smid, T., E. Schokkin, J. S. M. Boleij, and D. Heederik.** 1989. Enumeration of viable fungi in occupational environments: a comparison of samplers and media. *Am. Ind. Hyg. Assoc. J.* **50:**235–239.
88. **Solomon, W. R.** 1975. Assessing fungus prevalence in domestic interiors. *J. Allergy Clin. Immunol.* **56:**235–242.
89. **Sorenson, W. G., D. G. Frazer, B. B. Jarvis, J. Simpson, and V. A. Robinson.** 1987. Trichothecene mycotoxins in aerosolized conidia of *Stachybotrys atra*. *Appl. Environ. Microbiol.* **53:**1370–1375.
90. **Stärk, K. D. C., J. Nicolet, and J. Frey.** 1998. Detection of *Mycoplasma hypopneumoniae* by air sampling with a nested PCR assay. *Appl. Environ. Microbiol.* **64:**543–548.
91. **Strachan, D. P., B. Flannigan, E. M. McCabe, and F. McGarry.** 1990. Quantification of airborne moulds in the homes of children with and without wheeze. *Thorax* **45:**382–387.
92. **Tuomi, T., K. Reijula, T. Johnsson, K. Hemminki, E.-L. Hintikka, O. Lindroos, S. Kalso, P. Koukila-Kähkölä, H. Mussalo-Rauhamaa, and T. Haahtela.** 2000. Mycotoxins in crude building materials from water-damaged buildings. *Appl. Environ. Microbiol.* **66:**1899–1904.
93. **Vadrot, C., V. Bex, A. Mouilleseaux, F. Squinazi, and J.-C. Darbord.** 2004. Detection of *Mycobacterium tuberculosis* complex by PCR in hospital air samples. *J. Hosp. Infect.* **58:**262–267.
94. **Verhoeff, A. P., J. H. van Wijnen, J. S. M. Boleij, B. Brunekreef, E. S. van Reenen-Hoekstra, and R. A. Samson.** 1990. Enumeration and identification of airborne viable mould propagules in houses. *Allergy* **45:**275–284.
95. **Vesper, S., D. G. Dearborn, I. Yike, T. Allan, J. Sobolewski, S. F. Hinkley, B. B. Jarvis, and R. A. Haugland.** 2000. Evaluation of *Stachybotrys chartarum* in the house of an infant with pulmonary hemorrhage: quantitative assessment before, during, and after remediation. *J. Urban Health* **77:**68–84.
96. **Walter, M. V., B. Marthi, V. P. Fieland, and L. M. Ganio.** 1990. Effect of aerosolization on subsequent bacterial survival. *Appl. Environ. Microbiol.* **56:**3468–3472.
97. **Walters, M., D. Milton, L. Larsson, and T. Ford.** 1994. Airborne environmental endotoxin: a cross validation of sampling and analysis techniques. *Appl. Environ. Microbiol.* **60:**996–1005.

98. **Weimer, B. C., M. K. Walsh, C. Beer, R. Koka, and X. Wang.** 2001. Solid-phase capture of proteins, spores, and bacteria. *Appl. Environ. Microbiol.* **67:**1300–1307.
99. **White, T. J., T. Bruns, S. Lee, and J. Taylor.** 1990. Amplification and direct sequencing of fungal ribosomal RNA genes for phylogenetics, p. 315–322. *In* M. A. Innis, D. H. Gelfand, J. J. Sninsky, and T. J. White (ed.), *PCR Protocols: a Guide to Methods and Applications.* Academic Press, San Diego, Calif.
100. **Wilson, I. G.** 1997. Inhibition and facilitation of nucleic acid amplification. *Appl. Environ. Microbiol.* **63:**3741–3751.
101. **Wu, Z., X. R. Wang, and G. Blomquist.** 2002. Evaluation of PCR primers and PCR conditions for specific detection of common airborne fungi. *J. Environ. Monit.* **4:**377–382.
102. **Yike, I., T. Allan, W. G. Sorenson, and D. G. Dearborn.** 1999. Highly sensitive protein translation assay for trichothecene toxicity in airborne particulates: comparison with cytotoxicity assays. *Appl. Environ. Microbiol.* **65:**88–94.
103. **Zeng, Q.-Y., S.-O. Westermark, Å. Rasmuson-Lestander, and X.-R. Wang.** 2004. Detection and quantification of *Wallemia sebi* in aerosols by real-time PCR, conventional PCR, and cultivation. *Appl. Environ. Microbiol.* **70:**7295–7302.

Fate and Transport of Microorganisms in Air

ALAN JEFF MOHR

76

The fate and transport of microorganisms in the atmosphere are complicated issues involving many physical and biochemical factors. The transport of bioaerosols is primarily governed by hydrodynamic and kinetic factors, while their fate is dependent upon the specific biological composition, chemical makeup, and the meteorological parameters to which they are exposed. A thorough understanding of the spatial and temporal variables of atmospheric motion is required to fully anticipate the transport of microorganisms in the atmosphere. Local atmospheric conditions such as wind speed, temperature, and relative humidity (RH) are strongly influenced by the features of large-scale flow fields, geographical locations, and local topography. The average temperature and wind fields on the earth show characteristic dependencies on the season, latitude, and relative location of continents and oceans. Atmospheric turbulence, responsible for diffusion during transport by mean wind speed, is strongly influenced by local atmospheric conditions and the diurnal variation of solar irradiance reaching the ground. When a particle approaches another surface, the effects governed by the makeup of the cell wall influence deposition. Specific surface-surface interactions then dictate the release of the particle. The vast majority of airborne microorganisms are immediately inactivated due to environmental stresses (e.g., desiccation, temperature, and oxygen) that act to alter the makeup of the outer surface. The most significant environmental factors influencing viability are RH, solar irradiance, temperature, and oxygen concentration. Additional influences are exerted through air ions and open-air factors (OAF). Some microorganisms have built-in mechanisms that act to repair damage inflicted during the aerosolization and transport phases. Although few generalities can be made concerning aerosol stability or the fate of microorganisms, bacteria tend to behave differently from viruses, which behave differently from molds and fungi.

Bioaerosol particles can be either solid or liquid and can come from a number of natural and anthropogenic sources.

Many of the physical and chemical processes that describe more general classifications of aerosols also apply to bioaerosols. Classic texts, which should be studied by all individuals interested in aerosols, include works by Fuchs (51), Hinds (65), and Hidy (64). Significant sources which address bioaerosols include texts by Cox (24), Cox and Wathes (32), Anderson and Cox (6), and Lighthart and Mohr (80). An additional list of valuable references was compiled in the journal *Aerosol Science and Technology* in 1991 (5a).

FATE OF BIOAEROSOLS

Physical and Chemical Factors Influencing Viability

Studies to determine the factors that influence viability in the airborne state have been performed for decades. Many of these studies have yielded varying results, primarily due to the different methodologies used during their evaluations. Cox (24) outlined several factors that exert stresses on airborne microorganisms and listed water content or RH as being of primary interest. The effects of temperature on aerosol stability are less well understood. Marthi (85) presented his survey of pertinent stresses as being either primary or secondary in effect. He listed RH (24), temperature, radiation, and OAF as important primary stressors. For artificially generated aerosols, he listed secondary stressors that influence inactivation or provide protection against the primary stressors. Important secondary stressors include the method of aerosol generation, makeup of generation fluid, type of aerosol sampling method, and type of medium for enumeration of the collected microorganisms. It has been shown that the type of assay medium used can significantly influence the results of bioaerosol studies (85).

Water Content (RH)

The state of water and the water content associated with bioaerosols are fundamental factors influencing the fate or viability of these microorganisms. As the RH decreases, so does the water available to the exterior environment of the microorganism. Loss of water causes dehydration, resulting in the inactivation of many microorganisms. Of all of the measurable meteorological parameters, RH is the most important with respect to aerosol stability (24). Israeli et al. (69) studied freeze-dried microorganisms and showed the importance of water content to the viability of microorganisms. They concluded that the biomembrane, in the form of phospholipid bilayers, undergoes conformation changes from a crystalline phase to a gel phase as a result of water loss. These transformations induces changes to cell proteins, which in turn results in a loss of viability.

Not only is the water content of aerosolized microorganisms a principal factor contributing to aerosol viability

(114–116), but the RH of a system directly affects the density of the bioaerosol unit. The size, shape, and density of an aerosolized particle are directly related to the aerodynamic diameter which determines settling velocity and location of deposition in the respiratory tract and influences the collection efficiency of aerosol samplers.

The effects of RH can be influenced by the content of the suspension fluid used before aerosolization (7, 53), the content of the collection fluid (17), and prehumidification (113). For some microorganisms, shifts in RH after aerosolization have a more profound effect on aerosol stability than constant RH (58). The targets of RH-induced inactivation appear to be membrane phospholipids and proteins.

Temperature

Vapor pressure, and therefore the RH of a system, is dependent on temperature. This relationship makes it very difficult to separate the effects of temperature and RH. Studies to determine the effect of temperature on aerosol stability have generally shown that increases in temperature tend to decrease the viability of airborne microorganisms (42). Additionally, frozen cells tend to lose cellular proteins (permease) that result in elevated aerosol inactivation rates. The targets of temperature-induced inactivation appear to be membrane phospholipids and proteins.

Oxygen Concentration

Oxygen concentration can have a pronounced effect on the aerosol stability and infectivity of some bacteria (27). Free radicals of oxygen have been suggested as the cause of inactivation. Some investigators (26) have observed negative relationships between oxygen concentration and viability. Israeli et al. (69) noted that oxygen susceptibility increases with dehydration, increased oxygen concentration, and time of exposure. The targets of oxygen-induced inactivation appear to be phospholipids and proteins.

Other Factors

Solar Irradiance

Aerosol inactivation, caused by solar irradiance, has been shown to be dependent on the wavelength and hence the intensity of radiation. Shorter wavelengths contain more energy and are generally more deleterious to aerosolized microorganisms. RH (8), water activity (15, 72), oxygen concentration (15), aerosol age (24), and the presence of other gases all influence the effect radiation exerts on airborne microorganisms. Short-wave ionizing radiation (X rays and gamma rays) can cause breaks in the DNA of microorganisms (15). Other studies have shown that samplers monitoring bacteria from airborne emissions at sewage plants consistently yielded higher numbers at night (52). The targets of radiation-induced inactivation appear to be phospholipids, proteins, and nucleic acids.

OAF

Studies (39) showed that inactivation rates for many biological aerosols were significantly increased when challenged with high-efficiency particulate-arresting filtered air from outdoors when compared to clean, inert, laboratory-supplied air. This phenomenon introduced the term open-air factor (11, 45). While Cox (24) suggests that the primary reason for inactivation is caused by reactions between ozone and olefins, OAF inactivation is probably caused by a combination of factors including pollutant concentration, RH, pressure fluctuation, and air ions. The targets of OAF-induced inactivation appear to be phospholipids, proteins, and nucleic acids.

Microbial Factors

The type, species, or strain of an organism affects its airborne survival. Bacteria certainly differ in their responses to all types of stresses. The atmospheric environment is hostile to all microorganisms. No species has evolved to fill a niche in the open atmosphere. There is some evidence that suggests that bacteria can form new cells in the airborne state (41). Gram-negative bacteria contain more phospholipids than gram-positive bacteria. For this reason, they are more susceptible to inactivation through Maillard reactions, dehydration, and osmotic shock. Bacteria grown on enriched media or collected during different phases of development display widely varying aerosol stability rates. The physiological age of a bacterial population will also influence aerosol stability. Old cells die off at higher rates than cells from newly grown populations. Perhaps cells grown during the log phase are more active metabolically, and some systems may be more resistant to stresses than those in the stationary phase of growth. The type of medium used for enumeration of bacteria has been shown to influence the quantitative results from aerosol stability studies. Some media provide constituents that others may not and enable a stressed microorganism to survive.

Miscellaneous

Review articles by Cox (24), Strange and Cox (107), and Spendlove and Fannin (105) address the influences that pollutants, pressure fluctuations, air ions (95), and seasons may have on aerosol inactivation rates. Some atmospheric pollutants that have been studied include NO_2, SO_2, and O_3. The deleterious effects of these pollutants are greatly influenced by RH. Some microorganisms display greater loss of viability at high RH, while others show the opposite effect at high RH. Some pollutants may react with water to form acids (48), and the pH of the environment may be involved in acid production and hence inactivation of aerosolized microorganisms. Other pollutants that have been studied include HCOH, CO, HCl, HF, C_2H_2, C_2H_4, and C_2H_8 (27, 38). These compounds are much less toxic than OAF.

The effect of electrical charges on the viability of airborne bacteria was recently studied by Mainelis et al. (84). They found that *Pseudomonas fluorescens* lost viability as positive elementary charges (ec) increased (between 40 and 60% viability for ec values between 4,100 negative and 30 positive). At values of >2,700 ec, viability dropped to 1.5%. The viability of *Bacillus subtilis* subsp. *niger* spores was not influenced by similar ec treatment.

A review (106) of the diurnal and annual variations of bioaerosol concentrations details some interesting phenomena. For a high desert climate, the upward flux of bacteria was shown to follow the solar irradiance-sensible heat cycle (81), with the maximum flux occurring at solar noon. It was also reported that even though upward bacterial flux decreased after solar noon, the bacterial concentration increased, due to a reduction in solar output (caused by the solar angle), which resulted in lower rates of bacterial inactivation. Annual variations in bioaerosol concentrations have been shown to exhibit minimums in the winter and a primary peak in the late spring (106), with a secondary peak observed during midsummer. These maximums and minimums are related to favorable growth conditions (moderate temperatures, humidity, and solar irradiance) and snow cover (low temperatures), respectively.

Lighthart and Shaffer (81) designed a test apparatus to evaluate the flux of bacteria into the atmosphere. They found that the maximum upward flux of bacteria in summer, in a high desert location was equal to 1.7×10^4 CFU $m^{-2} h^{-1}$. It was also noted that the maximum occurred around solar noon. The upward flux was correlated with the daily solar irradiance-sensible heat cycle. Another study (82) performed over a grass seed field showed lower numbers of bacteria associated with the upward flux and that variations in airborne concentrations were linked to the local diurnal sea breezes.

Microbial Aerosol Stability

The experimental determination of aerosol stability is dependent on the ability to assay an entity for biological activity. Many sensitive immunological procedures are available which take advantage of the specificity of antigen-antibody reactions. The problem with these assays is that they reveal very little about the viability of the microorganism in question. Spendlove and Fannin (105) defined virus aerosol survival, stating that stability in the aerosol particle is really a measure of the ability of the virus to infect tissue culture cells of living hosts. For this reason, infectivity and stability should be treated as a single entity for collected viral bioaerosols.

Data are presented in this section relating aerosol stability or infectivity studies of bacteria, animal viruses, bacterial viruses (phages), and other microorganisms. Cox (24) and Mohr (93) discussed individual bacteria and viruses in greater detail and provided a rationale for the stability demonstrated. The results for many aerosol stability tests performed on the same microorganism often differ because testing techniques have not been standardized. Harper (57) noted that variations in results could be caused by the following:

1. Method of aerosol generation, storage, and sampling procedure
2. Method of assay
3. Differentiation of total, physical, and viability decay
4. Presence or absence of light
5. Methods and extent to which RH and temperature are controlled
6. Method of data presentation

Additionally, variations can occur because of the choice of suspending fluid, collection fluid, and local constituents in the atmosphere (i.e., the presence or absence of oxygen and other gases).

Bacteria

Some significant review papers addressing the aerosol characteristics of bacteria are available (3, 6, 24, 60, 80, 107). More variation in aerosol test results has been observed for bacteria than for viruses, phages, and other microorganisms. Table 1 shows some of the parameters that have been shown to influence the viability of selected bacteria. This is not only due to differences in test procedures and data presentation but also to the greater structural and metabolic complexity of bacteria (cell walls, membranes, and metabolism). It has been observed (54) for some bacteria that the growth phase and the type of media used for production play a major role in the demonstrated aerosol stability of the bacteria. Most of the published data available cannot be used for comparing aerosol stability because of the presence of several stresses which could have acted synergistically in a detrimental manner or possibly enhanced survival in some unknown way. Hess (63) performed an elaborate series of studies with *Serratia marcescens* and showed that loss of viability was caused by desiccation and the presence of oxygen. Some generalities concerning bacterial aerosol stability can be made. (i) Some, but not all, gram-negative bacteria are inactivated by oxygen after desiccation. (ii) Many bacteria have complicated RH inactivation profiles in inert atmospheres. (iii) Loss of aerosol viability is caused by desiccation and oxygen toxicity. (iv) Survival can be greatly increased by the addition of sugars. Aerosolized spores of many bacteria are extremely resistant to oxygen concentration, RH, and temperature (88).

Viruses

Several review papers addressing the aerosol characteristics of viruses are available (6, 24, 92, 105). Table 2 shows some of the parameters which have been shown to influence the viability of selected viruses. Viruses are normally very resistant to inactivation by oxygen. This and the relative simplicity of their structure explain why the results of aerosol inactivation studies are more consistent for viruses than for

TABLE 1 Aerosol stability parameters for selected bacteria

Bacterium	Stability parameter(s)	Reference(s)
Bacillus subtilis	Death rate	114, 115
Bacillus prodigiosus	Ambient temp, bacterial viability	114, 112
Bacillus patchiness	Ambient temp, bacterial viability	114, 115
Bacillus violaceus	Ambient temp, bacterial viability	114, 115
Chlamydia pneumoniae	RH, temp	109
Cytophaga allerginae	RH, temp	61
Enterobacter cloacae	RH, temp, CO_2	74, 78, 86, 87
Erwinia herbicola	RH, temp	7, 86, 111
Escherichia coli	RH, temp, O_2, wet, dry	20–22, 25, 26, 117, 119
Klebsiella planticola	RH, temp	86
Legionella pneumophila	RH	54
Mycoplasma	RH	124
Mycoplasma pneumoniae	RH, temp, solar	71, 73, 122–124
Francisella tularensis	RH, wet and dry generation, solar	8, 9, 17, 23, 29
Serratia marcescens	RH, O_2, freeze, time	27, 28, 31, 58, 63
Staphyococcus albus	Time, ambient temp	114, 115
Staphylococcus aureus	RH, temp	120

TABLE 2 Aerosol stability parameters for selected viruses

Virus	Stability parameter(s)	Reference(s)
Adenovirus	RH	90
Bovine rhinotracheitis virus	RH, temp, media	49, 102
Columbia SK virus	RH, temp	4
Coxsackievirus	UV light	70
Encephalomyocarditis virus	RH, rehumidification infectivity	33, 34
Enteric viruses	RH	111
Foot-and-mouth disease virus	Radiation, RH, temp, weather factor	7, 43, 44, 100
Influenza virus	RH	83
Langat virus A and B	RH, temp	10
Newcastle disease virus	RH, temp	102
Picornavirus 37A	RH	5, 42
Pigeonpox virus	RH, inositol	118
Poliovirus	RH, temp, environmental, seasonal, inactivation	10, 33, 37, 56, 57, 62, 66
Reovirus	RH, temp	2, 92
Rotavirus	RH, temp	35, 66, 91, 99
Rous sarcoma virus	RH, inositol	118
Semliki Forest virus	RH	10, 36
Simian virus 40	RH, temp	4
St. Louis encephalitis virus	RH, temp	96
Venezuelan equine encephalitis virus	RH, temp, solar	12, 48
Vesicular stomatitis virus	RH, temp, O_3	50, 102
Yellow fever virus	RH, temp	89

bacteria. Some generalities can be made about aerosol inactivation rates for viruses. (i) Viruses with lipids in their outer coat or capsid are more stable at low RHs than at high RHs. (ii) Viruses without lipids are more stable at high RHs than at low RHs. (iii) When viable viruses can no longer be detected after aerosol collection, the nucleic acid can be isolated and is still active (this evidence suggests that aerosol inactivation of viruses is not caused by nucleic acid inactivation but by denaturation of coat proteins). (iv) Prehumidification upon sampling increases recovery of viruses that lack lipids in their outer coat. Buckland and Tyrrell (16) first speculated that virus survival in aerosols might be related to the amounts of lipids in the capsid (envelope or outer shell). It has been shown that some viruses exhibit higher survival rates at mid-range RH (57), regardless of temperature, while some display better survival rates at low and high RH and the lowest survival rates at mid-range RH (2, 92). However, survival may depend on the temperature and oxygen content of the test atmosphere at the time of testing. There are some viruses that are stable in the airborne state over broad ranges of temperature and RH (4, 5).

Phages

Phages, like other viruses, are not inactivated by oxygen. Table 3 shows some of the parameters that are associated with the viability of selected phages. Most of the work has been performed on the T series of coliphages (phages which infect *Escherichia coli*). Aerosol stability is high for T phages at RH levels above 75%. Generally, it has been determined that NaCl is toxic to phages, but the effect can be reversed by the addition of peptone to the suspending medium. Aerosol inactivation of complex phages has been shown to be caused by the breakup of the head-tail complex. At RHs below 75%, the lack of water vapor is sufficient to weaken head-to-tail bonds and promote breakage and inactivation of the phages. Rigorous bioaerosol sampling (as occurs with an all-glass impinger) has been shown to break up the head-to-tail complex of some phages (59). Prehumidification at sampling has also been shown to increase recovery rates.

TABLE 3 Aerosol stability parameters for selected phages

Phage(s)	Stability parameter(s)	Reference(s)
T-series coliphages	RH	25
T1	Air ions	55, 110
T2, T3, T7	Mechanism of inactivation	10
T3, T7	Freeze	30
T3	RH, sample prehumidification	59
Phages	Sample prehumidification	113
MS 2	Prehumidification, RH, inactivation, temperature	47, 96, 110, 113
Pasteurella pestis phage	Prehumidification	59
S13	Prehumidification, RH, inactivation	47, 110, 113
ϕX174	RH, chemicals	38, 46

Overall, the results by various workers are consistent, due to the relative simplicity of the structure of phages.

INACTIVATION MECHANISMS

Bioaerosols are subject to inactivation and transport the moment they become airborne. Desiccation of the droplet is the main factor responsible for inactivation in liquid bioaerosols. Particle sizes for these droplets are usually small (2 to 10 μm), and they tend to follow the streamlines of the local wind. For particles with a diameter of >5.0 μm, gravitational settling and impaction are the leading causes of particle loss during transport. Figure 1 illustrates that, depending on the aerodynamic diameter of a particle, that particle will be either small enough to follow the streamlines of the surrounding flows or large enough to cross streamline flow and impact upon a surface.

Some relationships between aerosol stability and the biological composition of microorganisms have been identified (69, 121). Bacterial aerosol stability is considerably more complex than that observed for viruses. Cox (20–22) set up an elaborate series of stability experiments to examine the factors that influenced inactivation of *E. coli* B. When *E. coli* B was aerosolized from a suspension of distilled water into highly purified nitrogen, argon, or helium atmospheres, it displayed similar patterns of survival as a function of RH. Under these conditions, survival was virtually complete at low RH but was critically dependent on RH values above 80%. Under these conditions, Cox (24) speculated that the gases slightly modify the water structure through gas hydrate formation and water lattice modification, which affected the stability of biological structures. Cox (24) explained the events that may take place as precursors to aerosol inactivation as follows. When *E. coli* is aerosolized into inert atmospheres at medium to high RH, the biological membrane constituents become destabilized through loss of water molecules. Additives such as polyhydroxyl compounds, which supersaturate, can stabilize these structures. Polyhydroxyl compounds, by binding to sites on proteins, cause conformational changes in them and thereby stabilize proteins, making them less susceptible to denaturation. This is convincing evidence that the state of proteins on the outer membrane of some microorganisms is critical to the resultant stability profile.

Cox (24) also explained that Maillard reactions between reducing sugars and amino acids are responsible for some denaturation and unfolding of the ovalbumin α-helix, which occurs during drying. Additionally, he explained that dried proteins may be stabilized by similar reactions between other sugars and sodium glutamate. Soddu and Vieth (101) showed that sucrose could bind to collagen membranes with different affinities and produce conformational changes in the protein structure. Cox's (24) explanation of the sequence of events was as follows. There is little doubt that during the desiccation process, polyhydroxyl compounds and amino acids react together, causing conformational changes that strengthen the overall protein structure. The presence of sugar additives causes conformational changes in the coat proteins; in the new configuration, coat proteins do not react (or react more slowly) with the polyhydroxyl coat moieties. In the absence of these sugar additives and free molecules, the coat proteins may react irreversibly through Maillard reactions with polyhydroxyl coat moieties and cause loss of viability. In addition, the sugar additives can compete with the polyhydroxyl coat moieties for the reaction sites of the coat proteins or physically hinder those reactions' molecular collisions. The result in each case would be more aerostable microorganisms. When in their more normal aqueous environment, the

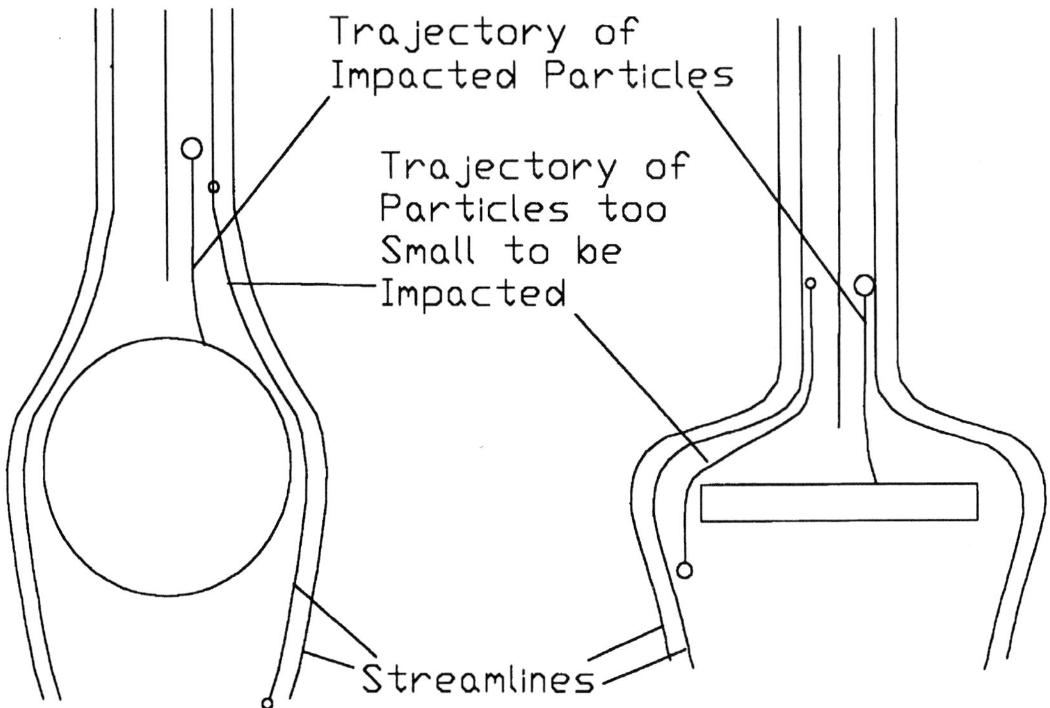

FIGURE 1 Particles of <5 μm tend to follow flow streamlines around obstructions. Particles of >5 μm tend to cross streamlines and impact on obstructions.

Maillard reactions leading to this inactivation may not occur because the reaction sites are separated either by bulk water molecules or by water molecules bound at the reactive sites. Removal of these water molecules by evaporation would then lead to the events proposed above. A possible mechanism for aerosol inactivation has been presented here by inference from mostly unrelated work. There have been many studies of solute concentration effects, which may also be applied to help model the molecular events which take place during inactivation. In a practical sense, the process may be viewed as the increase of solute concentration, which concomitantly occurs during drying of the aerosol droplet.

Freeze-drying, freezing, and aerosolization of microorganisms all act to remove water from the system. Elaborate studies have been performed using freeze-dried microorganisms as models to study the processes responsible for inactivation. Israeli et al. (69) studied freeze-dried microorganisms and showed the importance of water content to the viability of microorganisms. They concluded that some areas of the biomembrane, specifically the phospholipid bilayers, go through conformational changes from crystalline to gel phases, as a result of water loss. These factors caused damage to cell proteins, which in turn resulted in a loss of viability.

BIOAEROSOL STABILIZERS

Several compounds have been shown to provide protection to aerosolized microorganisms. Generation fluid additives such as inositol and bovine serum albumin have been shown to increase the recovery of some bacteria (53). Some polyhydric compounds such as raffinose, dextran, glycerol, glutamate, and inositol have been shown to increase recovery of bacteria, but the results are dependent on the RH of the system (20, 53, 118). Cox (24) hypothesized that these compounds bind with available proteins and stabilize them against denaturation. Marthi and Lighthart (87) added betaine to the generating and collection fluid during field studies collecting airborne bacteria and observed increases of 21.6 to 61.3%, compared to samples taken in the absence of betaine. The recovery was dependent on the concentration of betaine, but they found the optimum concentration to be between 2 to 5 mM. Betaine acts as an osmoprotectant. Additionally, trehalose has been shown to afford protection to both freeze-dried (69) and aerosolized bacteria. Trehalose is known to stabilize lipids, proteins, and phospholipids, important constituents of microorganisms.

REPAIR MECHANISMS AND RESUSCITATION

Microorganisms in the environment are constantly subjected to forces that cause inactivation. The consequences of these forces are not always lethal. Marthi (85) and Cox (24) presented interesting details involved with sublethal damage to microorganisms. Marthi (82) stated that perhaps the most important effect of sublethal stress is the inability of stressed microorganisms to grow on different media, both selective and nonselective. The fact that microorganisms are present but cannot be detected by standard enumeration procedures requires additional planning by the investigator to confirm that the reported results are representative of the sample that has been collected. Cox (24) pointed out that although the microorganisms are stressed and not detectable by normal methods, they may be able to recover from the injury and initiate disease. Marthi (85) pointed out that the primary sites of initial stress include the outer membrane, cell wall, cytoplasmic membrane, RNA and ribosomes, and DNA. When injuries are inflicted upon the above bacterial cell constituents, the metabolic activities of microorganisms are concomitantly influenced. Cox (24) addressed repair mechanisms associated with surface structures, transport functions, and radiation damage in his review article.

VIABILITY MODELS

Exponential Decay Model

The loss of viability of aerosolized microorganisms is caused by complex physical, meteorological, and cellular interactions. Early attempts to explain aerosol viability relied on the exponential decay model with mixed results. Although exponential decay has been shown to be a simplification, application of intricate mathematical expressions often delivered conclusions that were no more accurate than simple expressions. If all of the known parameters were applied to a mathematical expression, there would be around 20 inputs, many of which would be dependent on other parameters during specific periods of flight. The expression for exponential decay is

$$V_t = V_0 e^{-K} \tag{1}$$

where V_t is the viability at time t, V_0 is the viability at time zero, and K is the decay rate constant.

Kinetic Model

Cox, in a series of articles (e.g., reference 24), presented what he called the kinetic model. He was dissatisfied with the lack of explanation provided by the exponential model, which did not account for the time-dependent decay observed for most microorganisms. Cox's kinetic model supposes that microorganisms contain a molecular species $B(n)H_2O$, the biological activity of which is essential for a microbial cell to replicate or be infectious. $B(n)H_2O$, when exposed to an environment of lowered water activity (or low RH), forms a series of hydrates similar to other biomolecules; some of these hydrates are unstable and spontaneously denature through a first-order process, i.e.,

$$-dx/dt = kx \tag{2}$$

where x is the concentration of the species which denatures. The model form is then

$$B(n)H_2O \overset{k_+}{\underset{k_-}{\leftrightarrow}} B(n-x)H_2O + xH_2O \leftrightarrow$$
$$B(n-x-y)H_2O + yH_2O \leftrightarrow B + iH_2O \tag{3}$$

where $B(n - x)H_2O$ is the denatured form with a rate constant k_x and $B(n - x - y)H_2O$ is the denatured form with a rate constant k_y.

Cox (24) then applied probability theory to evaluate the likelihood of death, which is related to percent viability, set up appropriate boundary conditions, and integrated the results. When denaturation follows a first-order pattern, the final form of the equation is written

$$\ln V = K_1[B(n-x)H_2O]_0(e^{-kt} - 1) + \ln 100 \tag{4}$$

where k is a first-order denaturation constant, t is time, K_1 is the probability constant, and V is viability. Cox (24) analyzed several hundred viability-time curves and found very good agreement with experimental results for dehydration inactivation.

Catastrophe Model

Catastrophe theory was formulated to explain the nature of discontinuous events such as the breaking of waves on a beach, the crash of the stock market, or the sudden aggression displayed by an animal. Overall, reactions involving large numbers of molecules appear to behave in a continuous fashion because discontinuities are smoothed, but on an individual level, the reactions are discontinuous. If the number of reactions is relatively low, as is the case when aerosol viability is considered, discontinuity is appropriately represented and aerosol inactivation rates may be predicted. On an individual level, the loss of infectivity of an airborne microorganism is a sudden discontinuous event; this change in state is termed the catastrophe. The bases of catastrophe theory are related to the potential energy and therefore equilibrium of the system. Potential energy is a function of what are termed control parameters, which govern the equilibrium of an event. Within a certain range, equilibrium will not be influenced by variations in the control parameters, but some small critical change can cause a shift in the potential energy and result in inactivation. Cox (24) combined catastrophe theory and kinetics to explain the loss of viability caused by desiccation, temperature, oxygen, and OAF. The calculated viability curves agreed very well for inactivation caused by oxygen concentration and OAF and were more accurate than predictions based on probability theory for denaturation induced by desiccation.

Dispersion Models

Models have been developed to predict dispersion and deposition of microbial aerosols with respect to penetration of structures (104), to predict infectious microbial concentrations downwind of known sources (1, 73, 75, 103, 108, 125), and to evaluate the spread of plant pathogens outdoors (19). Many of the early models were based on treatments of atmospheric diffusion, which were created to predict the fate of air pollutants. The inert particle dispersion model by Pasquill (94) is empirical and based on observations of the deposition of inanimate particles. Results from its application yield average distributions of airborne microbes, have a somewhat limited downwind range, and should be applied to flat terrain with steady wind conditions. Assumptions which are made for this model include (i) Gaussian distribution of the plume in the horizontal and vertical planes; (ii) total reflection of particles from the ground; (iii) particles emitted from the source at a constant rate; (iv) constant wind velocity and direction; (v) flat terrain; (vi) particles of <20 μm, making gravitational effects negligible; and (vii) negligible downwind diffusion and negligible difference in wind velocity between source and real wind. The classical form of the inert particle dispersion model is

$$X(x,y,z{:}H) = Q/\{2\pi(\sigma_y\sigma_z)u \times \mathrm{EXP}[-0.5(y/\sigma_y)^2]\} \times (\mathrm{EXP}\{-0.5[(z-H)/\sigma_y]^2\} + \{-0.5[(z+H)/\sigma_y]^2\}) \quad (5)$$

where X is the number of particles per cubic meter at downwind location x,y,z; Q is the number of particles emitted from the source per second, u is the mean air speed in meters per second, σ_y is the standard deviation of the horizontal concentration (at the downwind distance), σ_z is the standard deviation of the vertical concentration (at the downwind distance), and H is the height of the source plus the plume rise.

The equation can be simplified for ground level ($z = 0$) and becomes

$$X(x,y,0{:}H) = Q/\pi\sigma_y\sigma_z u \ \mathrm{EXP}[-(y^2/2\sigma_y{}^2 + H^2/2\sigma_z{}^2)] \quad (6)$$

For a ground-level source ($H = 0$) and analysis along the center line ($y = 0$), the equation becomes

$$X(x,y,z{:}H) = Q/\pi\sigma_y\sigma_z u \quad (7)$$

As explained earlier, these equations are for inanimate particles and do not take into account microbial viability decay. Lighthart and Frisch (77) expanded on the inert particle diffusion model, added a biological death constant (α), and created a graphic method which would estimate ground level concentrations when microbial death rate, mean wind speed, atmospheric stability class, source height, and downwind sampled distance were known. Microbial death rate was determined under laboratory conditions and applied to the model. The equation for biological death (BD) after modification becomes:

$$X(x,y,z{:}H)_{BD} = X(x,y,z{:}H)\ \mathrm{EXP}(-\alpha t) \quad (8)$$

where t was approximated by x/u.

Biological decay and the rate constants associated with inactivation must be determined under dynamic conditions in the laboratory. Teltsch et al. (108) presented methodology to estimate the death constant under field conditions to determine inactivation and final concentrations downwind from sprinklers using wastewater. Death constants must be evaluated over a wide range of environmental conditions for accurate results, and the determined decay constants cannot be applied for predictive uses because of the specific meteorological conditions under which the study took place.

Lighthart and Mohr (79) applied best-fit laboratory generated decay constants to the Gaussian plume model using microbial source strength and local hourly mean weather data to drive the model through a summer and winter day cycle. For near-source locations, higher wind speeds or short travel times exerted a major modulating effect during the day because time was inadequate for inactivation. Additionally, as travel time increased because of low wind speed or long distances, modulation was due more to solar irradiance, RH, and temperature than wind speed.

Spendlove (104) applied various models (e.g., box and test tube) to determine the penetration of structures by microbial aerosols. Models showed that for a single-story dwelling without air conditioning, the aerosol dose received by people was the same inside as outside and that air conditioning units would moderately decrease inside concentrations (depending on ventilation rates).

Lighthart and Kim (76) presented a simulation model that describes the dispersion of individual droplets of water containing viable microbes. By repetition of the modeling process many times for individual droplets, an aerosol cloud can be simulated. The model accounts for the physical, chemical, biological, and measured meteorological parameters of each droplet for many time increments. The droplet model is separated into five submodels, aerosol generation, evaporation, dispersion, deposition, and microbial death, all of which are calculated at chosen time intervals in the trajectory for each drop. The results show that evaporation is an important factor in determining deposition sites because of particle size dependence, chemical reactions, and protection offered by large droplets. Wind gust data are required because average wind velocity data tend to smooth and oversimplify what occurs on a micrometeorological scale.

TRANSPORT OF BIOAEROSOLS

Naturally occurring bioaerosols are injected into the atmosphere either by chance (e.g., wind, rain, and bursting bubbles) or by processes governed by natural selection. Wickman (121) presented an explanation based on the physical and biological (molecular process) parameters and used the terms deposition, adhesion, and release to explain the transport of bioaerosols. Cox (24) referred to the same mechanisms as takeoff processes and landing on surfaces; his work is directed more to the consequences as they relate to the human respiratory system. Chamberlain (18) studied both the deposition and release of spores and pollens as they relate to biological surfaces.

Deposition and Adhesion

Wickman (121) presented a notable review of the forces which govern deposition and adhesion and presented several novel approaches that are now being used for these evaluations. Deposition is governed in part by the mass of the particle. Terminal velocity is governed by the mass of a particle, a cubic function of the particle diameter. Particles between 1.0 to 5.0 μm normally follow the streamlines of the surrounding air. Larger particles often have the momentum to deviate from the streamlines, impact surrounding surfaces, and be deposited. Particle bounce also increases with increasing mass. Additional forces include gravitational settling, convection (due to temperature variations), diffusion (thermal energy), and eddy diffusion. When particles enter the region near a surface, London-van der Waals forces (electrical dipole moments) and electrostatic forces (electrical charge forces of attraction and repulsion) can influence deposition.

Wickman (121) describes adhesive forces as being dominated by the molecular structure and organization near the contact surfaces. Two electrodynamic forces, London-van der Waals forces and electrostatic interactions, are primarily responsible for the at-a-distance interactions that occur between aerosol particles and surfaces. If liquid is present, interfacial reactions (meniscus formation) can develop. Adhesion is also governed by the atomic attraction between surfaces (Hamaker constants, electronic frequency) and geometrical factors (shape of surfaces, flat versus spherical).

Release

The energy required to overcome the forces of adhesion is deduced by analyzing and measuring particle release. These forces can be measured utilizing either the atomic force microscope (13, 98) or the surface force apparatus (14, 40, 67, 68).

Adhesion can be overcome by mechanical or aerodynamic forces and can initiate the release of biological particles on surfaces. The mechanical energy frequently originates from turbulent mixing of the wind. The force required to remove particles increases with decreasing particle size. The surfaces of bioaerosols are complex structures, which include interactions between proteins, phospholipids, and peptidoglycans. Wickman (121) analyzed release characteristics utilizing S layers, a simpler exterior covering, present in some microorganisms. He concluded that in dry air, the S-layer surface is ridged and hydrophobic, but the surface groups are capable of reacting with adjacent S-layer surfaces. This tends to increase adhesion with increasing contact time. When more water is available, the surface becomes more hydrophilic and adhesion forces steadily increase, due to the reorientation and influences of exposed polar groups. As more water becomes available (higher RH), increased hydration ensures that the polar-hydrophilic groups are available at the surface so that equilibrium adhesion occurs. He noted that RH and contact time are important factors that control the forces required to overcome adhesion, which may result in release.

CONCLUSION

The study of bioaerosols has recently gained more attention, due to the recognized importance of both indoor and outdoor events that have been shown to impact applicable populations. Significant additions to the mechanisms governing the fate and transport of bioaerosols have been identified. Hydrodynamic and physical factors have been studied in increasing detail so that the biochemical changes that are responsible for the inactivation of aerosolized microorganisms can be defined. Methods have been developed to study the mechanisms involved in the deposition, adhesion, and release of biological particles. It was pointed out that the water content and the state of water in the biological particle are the primary factors associated with inactivation and viability. Some microorganisms have been shown to multiply in aerosols, while others have displayed built-in repair mechanisms. The realization and application of standards for generating, storing, sampling, and presenting data will allow better comparison and reproducibility of results. Before a bioaerosol project is initiated, a substantial amount of study must be completed to determine methods to maximize the recovery of aerosolized microorganisms. These studies should include effects concerning sampling media, counting media, generation media, and the makeup of test atmospheres.

REFERENCES

1. **Adams, A. P., and J. C. Spendlove.** 1970. Coliform aerosols emitted by sewage treatment plants. *Science* **169:**1218–1220.
2. **Adams, D. J., J. C. Spendlove, R. S. Spendlove, and B. B. Barnett.** 1982. Aerosol stability of infectious and potentially infectious reovirus particles. *Appl. Environ. Microbiol.* **44:**903–908.
3. **Akers, T. G.** 1973. Survival, damage and inactivation in aerosols, p. 73–86. *In* J. F. Hers and K. C. Winkler (ed.), *Airborne Transmission and Infection*. John Wiley & Sons, New York, N.Y.
4. **Akers, T. G., S. Bond, and L. J. Goldberg.** 1966. Effect of temperature and relative humidity on the survival of airborne Colombia sk group viruses. *Appl. Microbiol.* **14:**361–365.
5. **Akers, T. G., and M. T. Hatch.** 1968. Survival of a picornavirus and its infectious RNA after aerosolization. *Appl. Microbiol.* **16:**1811–1813.

5a. **American Association for Aerosol Research.** 1991. A bibliography of aerosol science and technology. *Aerosol Sci. Technol.* **14:**1–4.

6. **Anderson, J. D. and C. S. Cox.** 1976. Microbial survival, p. 203–226. *In* T. R. Gray and J. R. Postgate (ed.), *The Survival of Vegetative Microbes*. Cambridge University Press, Cambridge, United Kingdom.
7. **Barlow, D. F.** 1972. The effects of various protecting agents on the inactivation of foot-and-mouth disease virus in aerosols and during freeze-drying. *J. Gen. Virol.* **17:**281–288.
8. **Beebe, J. M.** 1959. Stability of disseminated aerosols of *Pasteurella tularensis* subjected to simulated solar radiation and various humidities. *J. Bacteriol.* **78:**18–24.

9. **Beebe, J. M., and G. W. Pirsch.** 1958. Response of airborne species of *Pasteurella* to artificial radiation simulating sunlight under different conditions of relative humidity. *Appl. Microbiol.* **6:**127–138.
10. **Benbough, J. E.** 1971. Some factors affecting the survival of airborne viruses. *J. Gen. Virol.* **10:**209–220.
11. **Benbough, J. E., and A. M. Hood.** 1971. Viricidal activity of open air. *J. Hyg. Camb.* **69:**619–626.
12. **Berendt, R. F., and E. L. Dorsey.** 1971. Effect of simulated solar radiation and sodium fluorescein on the recovery of Venezuelan equine encephalomyelitis virus from aerosols. *Appl. Microbiol.* **21:**447–450.
13. **Binnig, G., and H. Rohrer.** 1986. Scanning tunneling microscopy. *IBM J. Res. Dev.* **30:**355–369.
14. **Bradley, R. S.** 1932. The cohesive force between solid surfaces and the surface energy of solids. *Philos. Mag.* **13:**853–862.
15. **Bridges, B. A.** 1976. Survival of bacteria following exposure to ultraviolet and ionizing radiation, p.183–208. *In* T. R. G. Gray and J. R. Postgate (ed.), *Twenty-Sixth Symposium of the Society for General Microbiology*. Cambridge University Press, Cambridge, United Kingdom.
16. **Buckland, F. E., and D. A. Tyrrell.** 1962. Loss of infectivity on drying various viruses. *Nature* **195:**1063–1064.
17. **Cabelli, V. J.** 1962. *The Rehydration of Aerosolized Bacteria: Compounds which Enhance the Survival of Rehydrated* Pasteurella tularensis. Technical Report 314. U.S. Army, Dugway Proving Ground, Dugway, Utah.
18. **Chamberlain, A. C.** 1967. Deposition of particles onto natural surfaces, p. 138–164. *In* P. H. Gregory and J. L. Montieth (ed.), *Airborne Microbes*. Cambridge University Press, London, United Kingdom.
19. **Cook, R. J., and K. F. Baker.** 1983 *The Nature and Practice of Biological Control of Plant Pathogens*. American Phytopathological Society Press, St. Paul, Minn.
20. **Cox, C. S.** 1966. The survival of *Escherichia coli* atomized into air and into nitrogen from distilled water and from solution of protecting agents as a function of relative humidity. *J. Gen. Microbiol.* **43:**383–399.
21. **Cox, C. S.** 1968. The aerosol survival of *Escherichia coli* B in nitrogen, argon, and helium atmospheres and the influence of relative humidity. *J. Gen. Microbiol.* **50:**139–147.
22. **Cox, C. S.** 1970. Aerosol survival of *Escherichia coli* B disseminated from the dry state. *Appl. Microbiol.* **19:**604–607.
23. **Cox, C. S.** 1971. Aerosol survival of *Pasteurella tularensis* disseminated from the wet and dry states. *Appl. Microbiol.* **21:**482–486.
24. **Cox, C. S.** 1987. *The Aerobiological Pathway of Microorganisms*. John Wiley & Sons, Chichester, United Kingdom.
25. **Cox, C. S., and F. Baldwin.** 1966. The use of phage to study causes of loss of viability of *Escherichia coli* in aerosols. *J. Gen. Microbiol.* **44:**15–22.
26. **Cox, C. S., and F. Baldwin.** 1967. The toxic effect of oxygen upon the aerosol survival of *Escherichia coli* B. *J. Gen. Microbiol.* **49:**115–117.
27. **Cox, C. S., J. Baxter, and B. J. Maidment.** 1973. A mathematical expression for oxygen-induced death in dehydrated bacteria. *J. Gen. Microbiol.* **75:**179–185.
28. **Cox, C. S., S. J. Gagen, and J. Baxter.** 1974. Aerosol survival of *Serratia marcescens* as a function of oxygen concentration, relative humidity and time. *Can. J. Microbiol.* **20:**1529–1534.
29. **Cox, C. S., and L. J. Goldberg.** 1972. Aerosol survival of *Pasteurella tularensis* and the influence of relative humidity. *Appl. Microbiol.* **23:**1–3.
30. **Cox, C. S., W. J. Harris, and J. Lee.** 1974. Viability and electron microscope studies of phages T3 and T7 subjected to freeze-drying, freeze-thawing and aerosolization. *J. Gen. Microbiol.* **81:**207–215.
31. **Cox, C. S., and R. J. Heckly.** 1973. Effects of oxygen upon freeze-dried and freeze-thawed bacteria: viability and free radical studies. *Can. J. Microbiol.* **19:**189–194.
32. **Cox, C. S., and C. M. Wathes.** 1995. *Bioaerosols Handbook*. Lewis Publishers, Boca Raton, Fla.
33. **de Jong, J. C.** 1970. Decay mechanism of polio and emc viruses in aerosols, p. 210–245. *In* I. H. Silver (ed.), *Third International Symposium on Aerobiology*. Academic Press, Inc., London, United Kingdom.
34. **de Jong, J. C.** 1970. On the mechanism of the decay of poliomyelitis virus and encephalomyocarditis virus in aerosols, p. 210–211 *In* I. H. Silver (ed.) *Third International Symposium on Aerobiology*. Academic Press, Inc., London, United Kingdom.
35. **de Jong, J. C., T. Trouwborst, and K. C. Winkler.** 1973. Viral aerosol decay, p. 124–130. *In* J. F. P. Hers and K. C. Winkler (ed.), *Airborne Transmission and Infection*. John Wiley & Sons, New York, N.Y.
36. **de Jong, J. C., M. Harmsen, A. D. Plantinga, and T. Trouwborst.** 1976. Aerosol stability functions of Semliki Forest virus. *Appl. Environ. Microbiol.* **32:**315–319.
37. **de Jong, J. C. and K. C. Winkler.** 1968. The inactivation of poliovirus in aerosols. *J. Hyg.* (Cambridge). **66:**557–565.
38. **de Mik, G., and I. de Groot.** 1973. Effect of gases on the aerosol stability of various microorganisms p.155–158. *In* J. F. P. Hers and K. C. Winkler (ed.), *Airborne Transmission and Infection*. John Wiley & Sons, New York, N.Y.
39. **de Mik, G., and I. de Groot.** 1977. The germicidal effect of the open air in different parts of the Netherlands. *J. Hyg.* **78:**175–180.
40. **Derjaguin, B. V., I. I. Abrikosov, and E. M. Lifshitz.** 1956. Direct measurement of molecular attraction between solids separated by a narrow gap. *Q. Rev. Chem. Soc.* **10:**295–329.
41. **Dimmick, R. L., H. Wolochow, and M. A. Chatigny.** 1979. Evidence that bacteria can form new cells in airborne particles. *Appl. Environ. Microbiol.* **37:**924–927.
42. **Dimmock, N. L.** 1967. Differences between the thermal inactivation of picornaviruses at "high" and "low" temperatures. *Virology* **31:**338–353.
43. **Donaldson, A. I.** 1972. Effect of radiation on selected virus. *Vet. Bull.* **48:**83-94.
44. **Donaldson, A. I., and N. P. Ferris.** 1975. The survival of foot-and-mouth disease virus in open air conditions. *J. Hyg. Camb.* **74:**409–415.
45. **Druett, H. A.** 1973. The open air factor, p. 141–151. *In* J. F. P. Hers and K. C. Winkler (ed.), *Airborne Transmission and Infection*. John Wiley & Sons, New York, N.Y.
46. **Dubovi, E. J.** 1971. Biological activity of the nucleic acid extracted from two bacterial viruses. *J. Appl. Microbiol.* **21:**624–630.
47. **Dubovi, E. J., and T. G. Akers.** 1970. Airborne stability of tailless bacterial viruses S-13 and MS-2. *J. Appl. Microbiol.* **19:**624–630.
48. **Ehrlich, R., and S. Miller.** 1971. Effect of relative humidity and temperature on airborne Venezuelan equine encephalitis virus. *J. Appl. Microbiol.* **22:**194–200.
49. **Elazhary, M. A. S. Y., and J. B. Derbyshire.** 1977. Effect of temperature, relative humidity and medium on the aerosol stability of infectious bovine rhinotracheitis virus. *Can. J. Comp. Med.* **43:**158–167.
50. **Fairchild, G. A.** 1974. Ozone effect on respiratory deposition of vesicular stomatitis virus aerosols. *Am. Rev. Resp. Dis.* **109:**446–451.
51. **Fuchs, N. A.** 1964. *The Mechanics of Aerosols*. Dover Publications, Inc., New York, N.Y.

52. **Goff, G. D., J. C. Spendlove, A. P. Adams, and P. S. Nicholes.** 1973. Emission of microbial aerosols from sewage treatment plants that use tricking filters. *Health Serv. Rep.* **88:**640–652.
53. **Goldberg, L. J., and I. Ford.** 1973. The function of chemical additives in enhancing microbial survival in aerosols, p. 86–89. *In* J. F. P. Hers and K. C. Winkler (ed.), *Airborne Transmission and Infection*. John Wiley & Sons, New York, N.Y.
54. **Hambleton, P., M. G. Broster, P. J. Dennis, R. Henstridge, R. Fitzgeorge, and J. W. Conlan.** 1983. Survival of virulent *Legionella pneumophilia* in aerosols. *J. Hyg. Lond.* **90:**451–460.
55. **Happ, J. W., J. B. Harstad, and L. M. Buchanan.** Effect of air ions on submicron T1 bacteriophage aerosols. *Appl. Microbiol.* **14:**888–891.
56. **Harper, G. J.** 1961. Airborne microorganisms: survival tests with four viruses. *J. Hyg. Camb.* **59:**479–486.
57. **Harper, G. J.** 1963. The influence of environment on the survival of airborne virus particles in the laboratory. *Arch. Gesamte Virusforsch.* **13:**64–71.
58. **Hatch, M. T., and R. L. Dimmick.** 1966. Physiological responses of airborne bacteria to shifts in relative humidity. *Bacteriol. Rev.* **30:**597–603.
59. **Hatch, M. T., and J. C. Warren.** 1969. Enhanced recovery of airborne T_3 coliphage and *Pasteurella pestis* bacteriophage by means of a presampling humidification technique. *Appl. Microbiol.* **17:**685–689.
60. **Hatch, M. T., and H. Wolochow.** 1969. Bacterial aerosols, p. 267–295. *In* R. L. Dimmick and A. B. Akers (ed.), *An Introduction to Experimental Aerobiology*. Wiley Interscience, New York, N.Y.
61. **Heidelberg, J. F., M. Shahamat, M. Levin, I. Rahman, G. Stelma, and R. R. Colwell.** 1997. Effect of aerosolization on culturability and viability of gram-negative bacteria. *Appl. Environ. Microbiol.* **63:**3585–3588.
62. **Hemmes, J. H., K. C. Winkler, and S. M. Kool.** 1960. Virus survival as a seasonal factor in influenza and poliomyelitis. *Nature* **188:**430–438.
63. **Hess, G. E.** 1965. Effects of oxygen on aerosolized *Serratia marcescens*. *Appl. Microbiol.* **13:**781–787.
64. **Hidy, G. M.** 1984. *Aerosols*. Academic Press, Inc., New York, N.Y.
65. **Hinds, W. C.** 1982. *Aerosol Technology*. John Wiley & Sons, New York, N.Y.
66. **Ijaz, M. K., S. A. Sattar, C. M. Johnson-Lussenburg, and V. S. Springthorpe.** 1985. Comparison of the airborne survival of calf rotavirus and poliovirus type 1 (Sabin) aerosolized as a mixture. *Appl. Environ. Microbiol.* **49:**289–293.
67. **Israelachvili, J. N., and G. E. Adams.** 1978. Measurement of forces between two mica surfaces in aqueous electrolyte solutions in the range 0-100 nm. *J. Chem. Soc. Trans. Faraday Soc. I* **74:**975–1001.
68. **Israelachvili, J. N., and D. Tabor.** 1973. van der Waals forces: theory and experiment. *Prog. Surf. Membr. Sci.* **7:**1–55.
69. **Israeli, E., J. Gitelman, and B. Lighthart.** 1994. Death mechanisms in bioaerosols, p. 166–191. *In* B. Lighthart and A. J. Mohr, (ed.), *Atmospheric Microbial Aerosols*. Chapman & Hall, New York, N.Y.
70. **Jensen, M. M.** 1964. Inactivation of airborne viruses by ultraviolet irradiation. *Appl. Microbiol.* **12:**418–420.
71. **Kethley, T. W., W. B. Crown, and E. L. Fincher.** 1957. The nature and composition of experimental bacterial aerosols. *Appl. Microbiol.* **5:**1–17.
72. **Krinsky, N. I.** 1976. Cellular damage initiated by visible light p. 209–239. *In* T. R. G. Gray and J. R. Postgate (ed.), *Twenty-Sixth Symposium of the Society for General Microbiology*. Cambridge University Press, Cambridge, United Kingdom.
73. **Lembke, L. L., and R. N. Kniseley.** 1980. Coliforms in aerosols generated by a municipal solid waste recovery system. *Appl. Environ. Microbiol.* **40:**888–891.
74. **Lighthart, B.** 1973. Survival of airborne bacteria in a high urban concentration of carbon monoxide. *Appl. Microbiol.* **25:**86–91.
75. **Lighthart, B.** 1984. Microbial aerosols: estimated contribution of combine harvesting to an airshed. *Appl. Environ. Microbiol.* **47:**430–432.
76. **Lighthart, B., and J. Kim.** 1989. Simulation of airborne microbial transport. *Appl. Environ. Microbiol.* **55:**2349–2355.
77. **Lighthart, B., and A. S. Frisch.** 1976. Estimation of viable airborne microbes downwind from a point source. *Appl. Environ. Microbiol.* **31:**700–704.
78. **Lighthart, B., V. E. Hiatt, and A. T. Rossano, Jr.** 1971. The survival of airborne *Serratia marcescens* in urban concentrations of sulfur dioxide. *J. Air Pollut. Control Assoc.* **21:**639–642.
79. **Lighthart, B., and A. J. Mohr.** 1987. Estimating downwind concentrations of viable airborne microorganisms in dynamic atmospheric conditions. *Appl. Environ. Microbiol.* **53:**1580–1583.
80. **Lighthart, B., and A. J. Mohr.** 1994. *Atmospheric Microbial Aerosols. Theory and Applications*. Chapman & Hall, New York, N.Y.
81. **Lighthart, B., and B. T. Shaffer.** 1994. Bacterial flux from chaparral into the atmosphere in mid-summer at a high desert location. *Atmos. Environ.* **28:**1267–1274.
82. **Lighthart, B., and B. T. Shaffer.** 1995. Airborne bacteria in the atmospheric surface layer: temporal distribution above a grass seed field. *Appl. Environ. Microbiol.* **61:**1492–1496.
83. **Loosli, C. G., H. M. Lemon, O. H. Robertson, and E. Appel.** 1943. Experimental air-borne influenza infection. I. Influence of humidity on survival of virus in air. *Proc. Soc. Exp. Biol. Med.* **53:**205–206.
84. **Mainelis, G., R. L. Gorney, T. Reponan, M. Trunov, S. A. Grinshpun, P. Baron, J. Yadav, and K. Willeke.** 2002. Effect of electrical charges and fields on injury of airborne bacteria. *Biotechnol. Bioeng.* **79:**229–240.
85. **Marthi, B.** 1994. Resuscitation of microbial bioaerosols, p. 192–225. *In* B. Lighthart and A. J. Mohr (ed.), *Atmospheric Microbial Aerosols*. Chapman & Hall, New York, N.Y.
86. **Marthi, B., V. P. Fieland, M. Walter, and R. J. Seidler.** 1990. Survival of bacteria during aerosolization. *Appl. Environ. Microbiol.* **56:**3436–3467.
87. **Marthi, B., and B. Lighthart.** 1990. Effects of betaine on enumeration of airborne bacteria. *Appl. Environ. Microbiol.* **56:**1286–1289.
88. **May, K. R., H. A. Druett, and L. P. Packman.** 1969. Toxicity of open air to a variety of microorganisms. *Nature* **221:**1146–1147.
89. **Mayhew, C. J., W. D. Zimmerman, and N. Hahon.** 1968. Assessment of aerosol stability of yellow fever virus by fluorescent-cell counting. *Appl. Microbiol.* **16:**263–266.
90. **Miller, W. S., and M. S. Artenstein.** 1966. Aerosol stability of three acute respiratory disease viruses. *Proc. Soc. Exp. Biol. Med.* **123:**222–227.
91. **Moe, K., and G. J. Harper.** 1983. The effect of relative humidity and temperature on the survival of calf rotavirus in aerosol. *Arch. Virol.* **76:**211–216.
92. **Mohr, A. J.** 1984. *Aerosol Stability of Reovirus*. Ph.D. dissertation. Utah State University, Logan.
93. **Mohr, A. J.** 1991. Development of models to explain the survival of viruses and bacteria in aerosols, p. 160–190. *In* C. J. Hurst (ed.), *Modeling the Environmental Fate of*

Microorganisms. American Society for Microbiology, Washington, D.C.

94. **Pasquill, F.** 1961. The estimation of the dispersion of wind borne material. *Meteorol. Mag.* **90:**33–49.
95. **Phillips, G., G. J. Harris, and M. W. Jones.** 1963. The effect of ions on microorganisms. *Int. J. Biometeorol.* **8:** 27–37.
96. **Rabey, F., R. J. Janssen, L. M. Kelley.** 1969. Stability of St. Louis encephalitis virus in the airborne state. *Appl. Environ. Microbiol.* **18:**880–882.
97. **Resnick, I. G., A. J. Mohr, and B. G. Harper.** 1987. Production, purification and use of MS-2 bacteriophage as a viral aerosol simulant, abstr. Q-23, p. 285. *Abstr. 87th Annu. Meet. Am. Soc. for Microbiol. 1987*. American Society for Microbiology, Washington, D.C.
98. **Sarid, D.** 1991. *Scanning Force Microscopy*. Oxford University Press, New York, N.Y.
99. **Sattar, S. A., M. K. Ijaz, C. M. Johnson-Lussenburg, and V. S. Springthorpe.** 1984. Effect of relative humidity on the airborne survival of rotavirus SA11. *Appl. Environ. Microbiol.* **47:**879–881.
100. **Smith, L. P., and M. E. Hugh-Jones.** 1969. The weather factor in foot-and-mouth disease epidemics. *Nature* **223:** 712–715.
101. **Soddu, A., and W. R. Vieth.** 1980. The effect of sugars on membranes. *J. Mol. Catal. B.* **7:**491–500.
102. **Songer, J. R.** 1966. Influence of relative humidity on the survival of some airborne viruses. *Appl. Microbiol.* **15:**1–16.
103. **Sorber, C. A., H. T. Bausum, S. A. Schaub, and M. J. Small.** 1976. A study of bacterial aerosols at a wastewater irrigation site. *J. Water Pollut. Control Fed.* **48:**2367–2379.
104. **Spendlove, J. C.** 1975. Penetration of structures by microbial aerosols. *Dev. Ind. Microbiol.* **16:**427–436.
105. **Spendlove, J. C., and K. F. Fannin.** 1982. Methods of characterization of virus aerosols, p. 261–329. *In* C. P. Gerba and S. M. Goyal (ed.), *Methods in Environmental Virology*. Marcel Dekker, Inc., New York, N.Y.
106. **Stetzenbach, L., and B. Lighthart.** 1994. Distribution of bioaerosols, p. 68–98. *In* B. Lighthart and A. J. Mohr (ed.), *Atmospheric Microbial Aerosols*. Chapman & Hall, New York, N.Y.
107. **Strange, R. E., and C. S. Cox.** 1976. Survival of dried and airborne bacteria, p. 111–154. *In* T. R. G. Gray and J. R. Postgate (ed.), *Twenty-Sixth Symposium of the Society for General Microbiology*. Cambridge University Press, Cambridge, United Kingdom.
108. **Teltsch, B., H. I. Shuval, and J. Tadmor.** 1980. Die-away kinetics of aerosolized bacteria from sprinkler application of wastewater. *Appl. Environ. Microbiol.* **39:**1191–1197.
109. **Theunissen, H. J., N. A. Lemmens-den Toom, A. Burggraaf, E. Stolz, and M. F. Michel.** 1993. Influence of temperature and relative humidity on the survival of *Chlamydia pneumoniae* in aerosols. *Appl. Environ. Microbiol.* **59:**2589–2593.
110. **Trouwborst, T., J. C. de Jong, and K. C. Winkler.** 1972. Mechanism of inactivation in aerosols of phage T_1. *J. Gen. Virol.* **15:**235.
111. **Walter, M. V., B. Marthi, V. P. Fieland, and L. M. Ganio.** 1990. Effect of aerosolization on subsequent bacterial survival. *Appl. Environ. Microbiol.* **56:**3468–3472.
112. **Ward, R. L., and C. S. Ashley.** 1977. Inactivation of enteric viruses in wastewater sludge through dewatering by evaporation. *Appl. Environ. Microbiol.* **34:**564–570.
113. **Warren, J. C., T. G. Akers, and E. J. Dubovi.** 1969. Effect of prehumidification on sampling of selected airborne viruses. *Appl. Microbiol.* **18:**893–896.
114. **Webb, S. J.** 1959. Factors affecting the viability of airborne bacteria. I. Bacteria aerosolized from distilled water. *Can. J. Microbiol.* **5:**649–669.
115. **Webb, S. J.** 1960. Factors affecting the viability of air-borne bacteria. III. The role of bonded water and protein structure in the death of air-borne cells. *Can. J. Microbiol.* **6:**89–105.
116. **Webb, S. J.** 1965. *The Role of Bound Water in the Maintenance of the Integrity of a Cell or Virus*. Charles C Thomas, Publisher, Springfield, Ill.
117. **Webb, S. J.** 1969. The effects of oxygen on the possible repair of dehydration damage by *Escherichia coli*. *J. Gen. Microbiol.* **58:**317–326.
118. **Webb, S. J., R. Bather, and R. W. Hodges.** 1963. The effect of relative humidity and inositol on air-borne viruses. *Can. J. Microbiol.* **9:**87–94.
119. **Webb, S. J., and A. D. Booth.** 1969. The effect of radiation on *Escherichia coli*. *Nature* **222:**1199–1200.
120. **Wells, W. F.** 1934. Droplet and droplet nuclei. *Am. J. Hyg.* **20:**611–627.
121. **Wickman, H. H.** 1994. Deposition, adhesion, and release, p. 99–165. *In* B. Lighthart and A. J. Mohr (ed.), *Atmospheric Microbial Aerosols*. Chapman & Hall, New York, N.Y.
122. **Wright, D. N., and G. D. Bailey.** 1969. Effect of relative humidity on the stability of *Mycoplasma pneumoniae* exposed to simulated solar ultraviolet and to visible radiation. *Can. J. Microbiol.* **15:**1449–1452.
123. **Wright, D. N., G. D. Bailey, and L. J. Goldberg.** 1969. Effect of temperature on survival of airborne *Mycoplasma pneumoniae*. *J. Bacteriol.* **99:**491–495.
124. **Wright, D. N., G. D. Bailey, and M. T. Hatch.** 1968. Survival of airborne mycoplasma as affected by relative humidity. *J. Bacteriol.* **95:**251–252.
125. **Zeterberg, J. M.** 1973. A review of respiratory virology and the spread of virulent and possibly antigenic viruses via air conditioning systems. *Ann. Allergy* **31:**228–234.

Airborne Fungi and Mycotoxins

CHIN S. YANG AND ECKARDT JOHANNING

77

Fungi are a heterogeneous group of organisms including the true slime molds (Myxomycetes), the water molds (Oomycetes), true fungi (Eumycota), and lichens. Some of these organisms, such as the true slime molds and the water molds, have been conveniently placed in the kingdom Fungi. They are, in many aspects, different from true fungi, although they are conventionally studied by mycologists (1). Fungi, as a group, inhabit a wide range of niches and environments, such as exposed rock (some lichens), the sea (marine fungi), the North Pole, and the tropics (27). They have developed many different modes of obtaining nutrients. Some fungi, such as powdery mildews and rusts, are obligate parasites. Some fungi exist in symbiotic relationships with plant roots to form mycorrhizae or with algae to form lichens. A large number of fungi survive as saprophytes. Many of these saprophytes have been found to grow in the indoor environment of man-made buildings and have caused building-related complaints and illnesses (61, 123, 124, 159).

Fungi produce a variety of secondary metabolites, including mycotoxins and fungal volatile organic compounds (VOCs) (31, 64, 117). Mycotoxins are harmful to animals and humans (23). In addition to mycotoxins, some VOCs produced by actively growing fungi are known irritants or hazardous chemicals. They may pose a health risk to building occupants (2, 9, 48, 123, 124).

This chapter reviews existing literature on airborne fungi, with emphasis on indoor fungal growth and contamination as well as the health effects of mycotoxins and fungal VOCs.

A wealth of literature on outdoor airborne fungi can also be found in reviews by Gregory (63), Flannigan et al. (51), Lacey (103, 104), and Levetin (107, 108). It is important to keep in mind that outdoor airborne spores are often the source of indoor fungal spora. Their impact on indoor airborne fungal populations may be immediate or delayed until they have settled and have colonized an indoor environment.

AIRBORNE FUNGAL POPULATIONS

It must be emphasized that it is to the selective advantage of fungi to release, disperse, and disseminate their spores and occasionally hyphal fragments in air from one location to others. They cannot survive and complete their life cycles by staying afloat in air for an indefinite period of time. Therefore, when discussing fungal contamination, identifying and locating the source of fungal colonization are of higher importance than assessing airborne-fungus data (18, 19). In addition to airborne dispersal, some fungi rely on running water, insects, and small creatures for dispersal of their spores (112, 197).

A large collection of literature on assessing indoor fungal populations has been accumulated. The majority of the literature is based on air sampling data. The types of buildings studied include hospitals and health care facilities (129, 136, 172, 181), residential dwellings (21, 41, 118, 180, 188), schools (45, 109), and office buildings (127). The focus of hospital sampling was often on *Aspergillus* species, including *Aspergillus fumigatus*, an opportunistic pathogen (129). General fungal populations were identified in nonhospital sampling.

A comprehensive assessment of fungal contamination in the indoor environment should include consideration of environmental factors (such as outdoor air characteristics, ventilation mode, heating system, occupant density, ventilation rate, and moisture level), on-site inspection, air sampling, surface and source sampling, sample analysis, risk analysis, and finally remedial actions (20, 22). Unfortunately, the majority of investigations do not follow the approach of comprehensive assessment, often due to insufficient labor, time, and funds. An alternate approach to identifying indoor fungal contamination is to focus on inspection and surface and source sampling.

Factors Affecting Airborne Fungal Populations

Important factors that directly affect airborne fungal populations are the availability of food and free water for fungal growth and the method of spore dispersal. Other physical, chemical, and biological parameters affecting fungal growth, and hence airborne fungal populations, can be found in recent reports (27, 64, 112).

Substrate Characteristics (Including Water Activity)

Fungi are achlorophyllous and require simple sugars, carbohydrates, and other organics, such as vitamins and amino acids, to survive. In the natural environment, fungi have developed a number of ways to obtain these nutrients (64), such as parasitic, symbiotic, and saprophytic relationships.

Humans share food, living space, and resources with fungi. Some fungi cause food spoilage or make food toxic to humans, yet we utilize fungi to produce edible mushrooms and useful by-products, such as enzymes and organic acids (61). *Botrytis cinerea* is a well-known disease-causing fungus on grapevines, strawberries, and other fruits and produce. Species of *Penicillium* and *Aspergillus* often cause spoilage in foodstuffs and make them inconsumable for animals and humans. On the other hand, *Penicillium camemberti* and *Penicillium roqueforti* are used in cheese production. Fungi are also known to cause wood stains, wood decay, and biodeterioration and biodegradation of polymers, carpet, plaster, wallpaper, paints and organic coatings, fuels and lubricants, leather, and paper and wood products (61, 111, 128, 163, 199). Previous studies underscore the likelihood that fungi can and will grow in man-made environments. Consequently, controlling nutrient sources in order to limit fungal amplification is practically impossible.

One of the critical factors affecting indoor fungal growth is water. There are a number of ways to measure water availability in materials. The water or moisture content of a material is expressed as a percentage of the oven-dry weight (112, 199). However, water content does not suggest the actual amount of free water in the material available to fungi. A better measurement of water availability to fungi is water activity (a_w). Water activity is numerically equal to the equilibrium relative humidity (ERH) expressed as a decimal. If a sample of a substrate is held at a constant temperature in a sealed enclosure until the water in the sample reaches equilibrium with the water vapor in the enclosure, then a_w = ERH/100. Another expression is as follows:

$$a_w = \frac{\text{vapor pressure of water in substrate}}{\text{vapor pressure of pure water}}$$

A detailed discussion of water activity for fungi in food and materials is presented by Gravesen et al. (61), Hocking and Miscamble (71), Li and Yang (112), Smith and Onions (167), and Troller and Scott (184).

Many common indoor fungi require a_w near 1 for growth. For some xerophilic fungi, however, optimal water activity ranges from 0.65 to 0.90 (61). Both mesophilic and xerophilic fungi can be found in the indoor environment. Table 1 lists some mesophilic and xerophilic fungi and their minimum a_ws.

TABLE 1 Selected fungi and their a_ws[a]

Species	a_w[b]	Temp (°C)
Absidia corymbifera	0.88	25
Alternaria citri	0.84	25
Aspergillus candidus	0.75	25
Aspergillus flavus	0.78	33
Aspergillus fumigatus	0.82	25
Aspergillus niger	0.77	35
Aspergillus ochraceus	0.77	25
Aspergillus restrictus	0.75	25
Aspergillus sydowii	0.78	25
Aspergillus terreus	0.78	37
Aspergillus versicolor	0.78	37
Aspergillus wentii	0.84	25
Cladosporium cladosporioides	0.84	25
Cladosporium sphaerospermum	0.84	25
Emericella[c] *nidulans*	0.78	37
Eurotium[c] *amstelodami*	0.70	25
Eurotium chevalieri	0.71	33
Fusarium moniliforme	0.89	25
Geomyces pannorum	0.89	25
Mucor circinelloides	0.90	25
Mucor plumbeus	0.93	25
Paecilomyces variotii	0.84	25
Penicillium brevicompactum	0.81	23
Penicillium chrysogenum	0.79	25
Penicillium citrinum	0.80	25
Penicillium corylophilum	0.80	25
Penicillium expansum	0.83	23
Penicillium frequentans	0.81	23
Penicillium griseofulvum	0.81	23
Penicillium spinulosum	0.80	25
Phoma herbarum	0.93	25
Rhizopus microsporus	0.90	25
Rhizopus oryzae	0.88	25
Rhizopus stolonifer	0.84	25
Scopulariopsis brevicaulus	0.90	NA[d]
Sistotrema brinkmannii	0.97	25
Stachybotrys chartarum	0.93	25
Syncephalastrum racemosum	0.84	25
Trichoderma harzianum	0.91	25
Ulocladium chartarum	0.89	25
Ulocladium consortiale	0.89	25
Wallemia sebi	0.70	25

[a]Data are from references 71, 112, 167, and 184.
[b]For the calculation of a_w, see the text.
[c]Genus is a teleomorph of *Aspergillus*.
[d]NA, not available.

Fungal Spore Discharge Mechanisms

Fungal spores are released by two basic mechanisms: (i) active spore discharge and (ii) passive spore release. Concentrations of certain airborne fungal spores have been known to peak during certain hours of the day or night. This periodicity is related to spore discharge mechanisms and environmental factors in nature (63, 103, 197). Details of these two mechanisms and environmental factors affecting spore release are presented by Lacey (103) and Levetin (108).

Fungi with active spore discharge include such common airborne fungi as *Sporobolomyces*, *Epicoccum*, and *Nigrospora* spp. and some smut-like yeasts (e.g., *Tilletia* spp.). Many ascospores and basidiospores also have active discharge mechanisms (108). *Sporobolomyces* and some basidiospores are usually most abundant at night or in the predawn hours. Their spore release requires the absorption of moisture to build up release pressure. Dry-spore fungi, such as *Aspergillus*, *Penicillium*, and *Cladosporium* spp., are often hydrophobic. They become airborne through passive mechanisms, such as being acted on by air movement or raindrops. *Cladosporium* usually dominates the airborne spore population during the day. Its spores stay airborne owing to the buoyancy of warmer air.

Some fungi, such as puffballs of basidiomycetes, release their spores in spore clouds or puffs when impacted by raindrops, humans, or small animals (108). The spore clouds may persist for a period of time until air mixing and dilution disperse them. Results of air sampling can be greatly affected by whether the spore cloud has dispersed (104).

Many fungi frequently detected indoors and outdoors produce spores in a slimy mass. These fungi include such common indoor contaminants as *Acremonium* (although some species of *Acremonium* produce dry spores), *Fusarium, Phoma, Stachybotrys,* and *Trichoderma* spp. Slimy spores may be released into the air when they become dry or disturbed or when they are attached to other particles. Their dissemination may also be assisted by insects, animals, or water (197). Because slimy spores do not become airborne easily, their detection indoors should be considered significant. Any detection of *Stachybotrys* in air samples taken indoors should trigger further investigation and search for the confirmed presence of this fungus.

Types of Airborne Fungal Populations

Fungi as a group produce a number of different spores, both sexual and asexual. Many types of spores are capable of becoming airborne. Some spores, such as chlamydospores (asexual) and zygospores (sexual), are not designed to be dispersed by air transmission, and there has been no report of recovering these spores in air. Hypogeous fungi (including such well-known fungi as truffles) produce subterranean sporocarps and disperse their spores through different mechanisms. However, both sexual and asexual spores of five major classes (Myxomycetes, Zygomycetes, Ascomycetes, Basidiomycetes, and Deuteromycetes) have been isolated and reported from air.

Spores of Ascomycetes and Basidiomycetes have frequently been recovered from air samples by using spore trap samplers. Levetin (107) reported 18 genera of basidiospores isolated from the atmosphere in Tulsa, Okla. Furthermore, many species of basidiospores have been demonstrated to be allergenic (10, 25, 39, 56, 106, 161). The importance of basidiomycetes in the indoor environment depends on the building construction. Wood-inhabiting basidiomycetes, such as polypores, are associated with wood decay and may be found in buildings constructed of wood. In fact, wood decay caused by polypores is a significant problem in the United Kingdom (166) and in the United States (199). Unfortunately, identification of airborne basidiomycetes collected by culture techniques can be difficult and the basidiomycetes are often missed or misidentified. Recently, a real-time PCR test has been developed for the detection of the dry rot fungi *Meruliporia incrassata* and *Serpula lacrymans* (114).

Ascospores have been identified from air samples by using spore traps (168). Many members of the Ascomycetes, however, do not produce teleomorphs and ascospores in culture. This makes it difficult to determine the frequency of occurrence of ascospores in air. Some ascomycetes, such as *Chaetomium, Emericella,* and *Eurotium* spp. (both *Emericella* and *Eurotium* are teleomorphs of some *Aspergillus* species), are commonly formed in cultures. Ascomata of species of *Ascotricha, Peziza,* and *Pyronema* are frequently observed in or reported to be isolated from indoor environments (113, 160). This demonstrates that ascospores may become airborne. Ascospores are suspected to be allergenic. Four species of *Chaetomium* are listed as licensed by the Food and Drug Administration for commercial production as allergens (168). Otherwise, little documentation is available on the allergenicity of ascospores.

The majority of airborne fungi collected on samplers and grown on agar media are Deuteromycotina and Zygomycetes. Deuteromycotina are anamorphs, or asexual states, of either Ascomycotina or Basidiomycotina. Asexual spores that belong to the Deuteromycotina are called conidia, and those that are Zygomycetes are called sporangiospores. Many of these spores are known allergens. Some of them have been prepared as allergen extracts and approved by the Food and Drug Administration for medical uses (168). Common Deuteromycotina found in air include *Alternaria, Aspergillus, Cladosporium, Epicoccum,* and *Penicillium* spp. Species of *Mucor* and *Rhizopus,* both Zygomycetes, are also frequently isolated from air.

Another group capable of producing and releasing spores into the air is the Myxomycetes, or the true slime molds. They are considered to have similarities to both true fungi and animals (1) and have been placed in the kingdom Protista (13, 108). Spores of slime molds in air samples have been documented (13, 63). Allergic reactions to extracts of slime molds have been reported. Giannini et al. (56) reported that 15.4% of patients tested showed positive skin reactions to extracts of *Fuligo septica, Lycogala epidendrum,* and *Stemonitis ferruginea.* Benaim-Pinto (10) and Santilli et al. (161) found that patients yielded positive skin responses to spore extracts of *Fuligo septica.*

Outdoor airborne fungal populations may directly or indirectly affect the indoor population since the pathways of infiltration are often suspected to be from leaks and cracks or through doors, windows, and building air intake systems. Therefore, common outdoor fungal taxa are often the predominant fungal types detected indoors (111, 197, 198). In a review of the literature, some agreements and disagreements are shown to exist on the predominant fungi identified indoors. Yang et al. (198), based on the culture of over 2,000 Andersen samples collected outdoors and in nonresidential buildings in the United States, found that *Cladosporium, Penicillium, Aspergillus,* Basidiomycetes, and *Alternaria* were the top five fungal taxa found indoors as well as outdoors in frequency of occurrence. All five fungal taxa were detected in fewer than 40% of indoor samples. However, *Cladosporium* was found in over 80% of outdoor samples and *Penicillium* was detected in 58%. These findings suggest that both *Cladosporium* and *Penicillium* are common in outdoor air. The results are somewhat in agreement with those reported by Strachan et al. (180) for British homes and by Womble et al. for 86 office buildings in the continental United States (193). Strachan et al. (180) found that *Penicillium, Cladosporium,* and Basidiomycetes (including *Sistotrema brinkmanii*) were the common types of mold isolated as well as the predominant mold concentrations measured. Womble et al. (193) found that, in rank order, nonsporulating fungi, *Cladosporium, Penicillium,* yeasts, and *Aspergillus* were the most common fungi indoors. Using a number of different types of samplers and sampling media, VerHoeff et al. (188) found that species of *Cladosporium, Penicillium,* and *Aspergillus* (including the teleomorph *Eurotium*) were common in homes in the Netherlands. In a survey of 10 elementary schools in southern California using an Andersen sampler, Dungy et al. (45) found that *Cladosporium, Alternaria, Penicillium,* sterile mycelia, and *Epicoccum* were the top five fungal groups isolated indoors. The predominant fungi detected outdoors were slightly different in that *Aureobasidium* was more frequently encountered than *Epicoccum.* However, using a Roto Rod sampler, they found that spores of *Alternaria,* rust fungi, *Cladosporium, Epicoccum,* and smut fungi were predominant both indoors and outdoors. Through further comparison of airborne fungal populations in the same study, the authors found that the top seven fungal types were identical at schools and at homes. In a hospital sampling, Solomon et al. (172) found that *Aspergillus fumigatus, Aspergillus niger, Mucor* and *Paecilomyces* spp., and yeasts were the five most common fungi recovered in culture at 37°C. The agreements and disagreements in the findings may be attributed to differences in sampling techniques, isolation media used, incubation temperatures, and geographical areas (118, 198).

Shelton et al. (165) evaluated 12,026 fungal air samples (9,619 indoor samples and 2,407 outdoor samples) collected from 1,717 buildings located across the United States by using Andersen N6 single-stage samplers. Ninety-nine percent of the samples were collected with rose bengal agar, and the other 1% were collected with malt extract agar. The culturable airborne fungal concentrations in indoor air were lower than those in outdoor air. However, *Stachybotrys chartarum* was identified in the air in 6% of the buildings studied and in 1% of the outdoor samples. The fungal levels were highest in the fall and summer and lowest in the winter and spring. Geographically, the highest fungal levels were found in the Southwest, far West, and Southeast. Because different fungal isolation media are known to have different selective effects, the combined use of the data derived from two different media is inappropriate. The reliability of fungal identification, data, and sampling quality control of such a large project must be scrutinized before the results and conclusions are fully accepted.

In a study of 50 single-family detached homes built since 1945, each with less than 0.18 m^2 (2 ft^2) of known water damage and located within a central city census tract in the metropolitan Atlanta region (DeKalb and Fulton Counties) of Georgia, air and dust samples were collected for assessment to establish a baseline of "normal and typical" types and concentrations of fungi in urban homes (74). The homes were predetermined not to have noteworthy moisture problems or indoor fungal growth. The homes were sampled twice (summer and winter) within a calendar year. Air samples were collected with a Spiral air system at 180 liters/min onto malt extract agar (MEA)-filled plates. Positive-hole correction was applied. Dust samples were sieved and inoculated by the direct plating method onto MEA and dichloran glycerol 18% agar (DG18) media. *Cladosporium cladosporioides*, *Cladosporium* spp., and *Cladosporium sphaerospermum* were the top three fungal species and groups in both indoor and outdoor air samples. The findings included that rankings by prevalence and abundance of the types of airborne and dustborne fungal spores did not differ from winter to summer and nor did the rankings differ when air samples taken indoors were compared with those taken outdoors. Water indicator fungi (such as *Chaetomium*, *Stachybotrys*, and *Ulocladium* spp.) were essentially absent from both air and dust samples.

Indoor Sources of Fungi

The detection of airborne fungi does not necessarily suggest growth and amplification of fungi indoors. However, it is believed that actively growing fungi in the indoor environment are the primary cause of the adverse health effects due to exposure to indoor fungal allergens, mycotoxins, and fungal VOCs. It is, therefore, important to identify and detect growth and amplification sites of fungi indoors.

Many common indoor fungi are strong biodeterioration agents and have been identified in various building materials and systems. Raper and Fennell, in their classic publication (152), detected various aspergilli in building materials, such as wallpaper and paper products, textiles, jute, insulation materials, and fabrics. Two species of *Aspergillus* were isolated from and found to grow on glass (137). Many species of the genus *Penicillium*, commonly detected in indoor air sampling, are frequently referred to as food spoilage and biodeterioration agents (61, 152, 160, 163). Gravesen et al. (61) included a list of 13 fungal species as important molds in damp buildings. Samson et al. (159, 160) described some fungal species common in indoor environments.

Morgan-Jones and Jacobsen (128) studied moldy carpets, plasterboard, and wallpaper from three hotels in Florida and Georgia. Their brief literature review suggested that many fungi had been reported to cause biodeterioration of paper, textiles, and plaster. The genera of fungi most often identified were the ascomycete genus *Chaetomium*, dematiaceous hyphomycete genera *Alternaria*, *Cladosporium*, *Stachybotrys*, and *Ulocladium*, the moniliaceous hyphomycete genera *Acremonium*, *Aspergillus*, and *Penicillium*, and the pycnidial genus *Phoma*. In the study, 14 species, including two new species of *Cladosporium*, in 11 genera were isolated and identified. In a study of toxicity of moldy building materials, Johanning et al. (92) not only detected cytotoxicity of the materials to cell cultures but also identified satratoxin H, spirolactone and spirolactams, and several groups of fungi. The fungi were isolated from gypsum wallboards and other building materials. The fungi identified included those described by Morgan-Jones and Jacobsen and additional species of *Aspergillus*, *Paecilomyces*, and *Trichoderma*.

Li and Yang (113) described seven noteworthy species of fungi isolated from indoor environments. The seven species were *Ascotricha chartarum*, *Ascotricha erinacea*, *Memnoniella echinata*, *Sporoschisma saccardoi*, *Stachybotrys microspora*, *Stachybotrys nephrospora*, and *Zygosporium masonii*. All species were recovered from water-damaged building materials, such as drywall, wallpaper, and wood. Four species were reported for the first time in the United States.

In addition to building materials, fungi have been known to grow in the heating, ventilating, and air-conditioning (HVAC) system (112, 196). Heinemann et al. (65) studied contamination with fungi, bacteria, and thermophilic actinomycetes. They sampled surfaces of filters and fans with RODAC contact plates and also sampled water from humidifiers. The serial dilution method was used for humidifier water. A wide variety of fungi were identified. However, some of the identified fungi were likely spore contaminants instead of indicators of fungal growth. Kemp et al. (96) studied fungal growth on filters of the HVAC system and reported isolation of *Aspergillus niger*, *Aspergillus fumigatus*, *Alternaria*, *Cladosporium*, a *Mucor* sp., an *Aspergillus* sp., and a *Penicillium* sp. However, they could confirm growth of only the *Aspergillus* sp., *Cladosporium* sp., and *Penicillium* sp. when filters were directly examined under the microscope. Buttner et al. (26) reported a controlled study using three air duct materials (i.e., galvanized metal, rigid fibrous glass duct board, and fiberglass duct liner) and *Penicillium chrysogenum* spores. They found that fungal growth may occur on a variety of duct materials, including bare metal, provided that soiling and moisture are present. The results showed that contaminated air ducts may expose building occupants to high concentrations of spores dispersed from fungal colonies growing on duct materials during normal operation of the system. Yang (196) examined and cultured 1,200 fiberglass insulation liner samples from HVAC systems in the United States and found fungal colonization and growth in approximately 50% of the samples studied. Fungal types were differentiated based on water and humidity conditions. Species of *Cladosporium* and *Penicillium* were primarily from areas with high relative humidity, and species of *Acremonium*, *Aureobasidium*, *Exophiala*, *Fusarium*, *Paecilomyces*, *Phoma*, *Rhodotorula*, *Sporobolomyces*, and yeasts were suspected in areas subjected to frequent wetting. *Cladosporium cladosporioides*, *Cladosporium herbarum*, and *Cladosporium sphaerospermum* were the primary species identified.

All fungi found to colonize building materials are saprophytes and biodeterioration agents. Some fungi, such as species of *Chaetomium*, *Aspergillus*, *Stachybotrys*, and *Trichoderma*, are known to be capable of degrading cellulose fibers. Although a few fungal species have been reported in the literature, it is very likely that any saprophytic, biodeteriogenic, or cellulolytic fungi can potentially grow indoors if opportunity arises (22, 61, 62).

Although species of the *Deuteromycetes* are commonly detected in moldy building materials, *Ascotricha chartarum*, *Ascotricha erinacea*, *Chaetomium* species, *Peziza* spp., and *Pyronema domesticum* of the Ascomycetes are occasionally found on damp materials in buildings (22, 113, 160). The asexual state of basidiomycetes, such as species of *Cryptococcus*, *Rhodotorula*, and *Sporobolomyces*, are common in indoor environments (22). In addition, Yang (unpublished data) has seen and identified fruiting structures of a slime mold, *Stemonitis* sp., and basidiomata (fruiting bodies of basidiomycetes) of *Coprinus* spp. and *Pleurotus* and *Poria* species from various building materials, from ceiling tiles to wood products. Samson et al. (160) reported that *Sistotrema brinkmanii*, a wood decay fungus, is commonly isolated from wet, decaying window and door joinery. Mycelia and hyphae with clamp connections, indicating basidiomycetes, are frequently detected colonizing water-damaged wood structures. Wood-inhabiting basidiomycetes are often wood decay fungi. Wood decay dry rot fungi, *Meruliporia incrassata* (*Poria incrassata*) and *Serpula lacrymans*, were detected in building wood samples collected from the United States by the real-time PCR method (114).

FUNGI AS DISEASE AGENTS

Fungi are commonly known to cause infections of the skin and other body organs, as well as allergies and respiratory problems. These conditions are discussed briefly below, because they are generally well described in respective clinical medicine textbooks (102, 119, 146). Furthermore, the Institute of Medicine (IOM) stated, based on its reviews of the English-language literature, that fungi have been strongly associated with indoor allergies, asthma, and other respiratory diseases (78, 79, 150). Fungi and their byproducts, such as (1-3)-β-D-glucan, mycotoxins, and microbial VOCs, have been implicated in adverse health reactions and diseases (23, 76, 151, 158, 159). Americans spend up to 90% of their time indoors, where contaminants often are at higher levels than they are in the ambient air. It is common that exposure duration and concentrations of fungi are greater indoors than outdoors because of occupant life styles, building conditions, and materials that lead to fungal growth and accumulation. Additionally, exposure to biological contaminants of all kinds, but particularly molds and bacteria, can be high when buildings have moisture problems or water damage (177). It is estimated that more than one-third of buildings in the United States and Western Europe have severe moisture problems that result in significant fungal contamination of the indoor environment (16, 77, 131). Exposure to high levels of indoor moisture is associated with upper respiratory symptoms, including coughing, wheezing, and asthma in sensitized persons, according to several large epidemiological studies cited by the IOM (14, 78, 79). Clinical case studies and research have shown that nonallergic health reactions are common in individuals with indoor atypical fungal exposure. Typical health complaints of patients living in indoor environments with excessive fungal exposure are listed in Table 2.

TABLE 2 Complaints and symptoms reported by patients with exposure to excessive fungal growth

Headaches
Runny nose or nasal congestion
Burning sensation and watery eyes
Sore throat and hoarseness
Sneezing or irritant dry cough, chest tightness and burning chest sensation, shortness of breath, wheezing
Unusual nosebleeds and coughing up of blood (rare)
Skin and mucous membrane irritation (occasionally hair loss)
Dizziness, concentration and memory problems, cognitive dysfunction, severe fatigue and exhaustion (physical and/or mental)
Nausea (vomiting) and gastrointestinal problems (loose stools, stomachaches)
Feverish feeling
Joint and muscle aches (flu-like reaction)

The medical conditions and illnesses associated with indoor fungal exposure include a spectrum of infectious and respiratory disorders and allergy and irritant and toxic health reactions (76, 78, 150, 156). The adverse health reactions are typically of short duration and are reversible after the exposure has been controlled. In rare cases, the adverse health consequences can be more serious or may be irreversible, requiring symptomatic treatment and careful avoidance of microbial triggers (156). Medical conditions are listed in Table 3 and described below in more detail.

Infections

Infections caused by fungi are called mycoses. They are categorized as endemic mycoses and opportunistic mycoses. Opportunistic fungal pathogens have a great public health importance, especially in persons with human immunodeficiency virus infection and those receiving organ transplants (94, 145). Endemic mycoses are related to the geographical distribution of certain fungal pathogens. These types of infection are caused by the inhalation of airborne spores or conidia found in certain regions where there is a higher frequency of such fungi because of unique soil and plant and flora conditions (103, 146, 148). Table 4 lists several important fungi and the infections that occur through air transmission, the diseases they cause, and clinical manifestations.

Opportunistic infections are secondary complications that occur in patients with an altered or weakened immune system. Patients at risk for fungal infections usually have major systemic diseases or health-suppressive conditions such as complicated diabetes mellitus, cancer, human immunodeficiency virus infection or AIDS, severe liver or kidney diseases, organ transplantation, or burn injury or may be on immunosuppressive medication treatment. These infections are not contagious, and the fungi are not considered obligatory pathogens. Secondary fungal infections and medical complications related to airborne fungal contamination in hospitals and transplant units have been reported. Immunocompromised patients may be at an increased risk for opportunistic infections if pathogenic fungi become airborne and fungal levels are significantly elevated in indoor air surveys. Among the fungi of concern are *Aspergillus* spp., such as *Aspergillus fumigatus*, *Aspergillus flavus*, and *Aspergillus niger*. Soil, bird and bat droppings, water-damaged materials, and organic compound-rich substrates in buildings may be a

TABLE 3 Medical conditions associated with indoor fungal exposure

Organ system	Clinical effects	Exposure
Upper airways (nose, sinuses, and throat)	Rhinitis, sinusitis, laryngitis	Fungi, allergens, irritants, microbial VOCs, particles
Lower airways (lungs with bronchial system and alveoli)	Bronchitis, asthma, bronchiolitis, allergic bronchopulmonary aspergillosis, allergic extrinsic alveolitis (also known as HP), toxic alveolitis, pneumonitis	Fungi, allergens, fungal by-products, fine and ultrafine particles
Combined upper and lower airways	Aspergillosis, fungal rhinosinusitis	Fungi, particles
Skin and mucous membrane	Urticaria, dermatitis (allergic irritant), conjunctivitis	Fungal irritants, allergens
Other organs (central nervous system, immune system, liver, kidneys, and endocrine system)	Overlapping diagnoses, differential diagnoses (exclusion of unrelated conditions)	Fungi, organic dusts, microbial by-products

reservoir for these fungi (11, 18, 104). Other clinically important fungal infections are candidiasis with local mucocutaneous or disseminated systemic organ manifestations and skin mycoses such as dermatophytoses, keratomycosis, tinea nigra, piedra, and malassezia-caused dermatitis. Invasive fungal diseases of the paranasal sinuses may also be associated with allergic sinusitis in atopic patients (49). *Aspergillus* species are often involved. Noninvasive forms may colonize preexisting body cavities, and carriers may be asymptomatic as long as some immunological resistance can be maintained.

Prevention, diagnosis, and therapy of opportunistic infections may be difficult to those who are not well trained and experienced in this field. Early recognition, preventive building engineering, hygiene, and public health intervention can reduce the incidence of mycoses, especially the institutional or iatrogenic acquisition in care facilities.

Allergy and Respiratory Diseases

Fungi are known to cause an immune pathology with an exaggerated or inappropriate immune response called the hypersensitivity reaction or common allergy (150). The important types of the allergic immune reactions according to the Coombs-Gell classification are listed in Table 5, but in the clinical context these often occur with overlapping presentations. The fungal spore is a known cause of allergic diseases (29, 60, 75) and was identified as one of the major indoor allergens (24, 150). However, there are still significant methodological problems in the production of reliable allergen extract from fungi compared to that from cats, dust mites, and other better characterized allergens. Extracts that are available correspond poorly with the fungi often found in indoor surveys (73). Several recent epidemiological studies have shown that long-duration indoor exposure to certain fungi can result in a hypersensitivity reaction and chronic diseases. Mold levels comparable to outside background levels are usually well tolerated by most people. Normal or typical indoor molds may vary depending on climate variations and geographical regions. However, as mold levels that are atypical in the indoor environment increase because of recurrent water leaks, home dampness, and high humidity, the prevalence of allergy and respiratory problems also rises (18, 19, 37, 51, 93, 154, 170, 176, 178, 179, 180). Dampness and mold has, in many epidemiological studies, been shown to be associated with cough and asthma (15, 34, 38, 142). Damp building materials, particularly cellulose-containing substrates, are prone to fungal amplification. Molds found on wet building materials and known to be associated with allergy and respiratory problems are species of *Alternaria*, *Aspergillus*, *Aureobasidium*, *Cladosporium*, *Fusarium*, *Paecilomyces*, *Phoma*,

TABLE 4 Disease transmitted by airborne fungi and affected tissues[a]

Fungus	Disease	Affected organ(s) and/or tissues[b]
Histoplasma capsulatum	Histoplasmosis	Lungs, eyes (surrounding skin and bone)
Cryptococcus neoformans	Cryptococcosis	Lungs, central nervous system, meninges, skin, viscera
Coccidioides immitis	Coccidioidomycosis	Lungs, multiorgan dissemination (skin, bone, meninges, joints)
Blastomyces dermatitidis	Blastomycosis	Lungs, skin and mucous membrane, bone, joints
Aspergillus spp., particularly *Aspergillus fumigatus*	Aspergillosis	Lungs, bronchial airways and sinus cavities, ear canals, eyes (cornea)
Sporothrix schenckii	Sporotrichosis	Lungs (granulomatous pneumonitis; rare), skin, joints, central nervous system, eyes
Mucorales and Zygomycetes	Mucormycosis	Nose, sinuses, eyes, lungs, brain and other organs, gastrointestinal system

[a]Data are from references 102 and 146.
[b]Parentheses indicate rare occurrence.

TABLE 5 Immunopathological responses caused by fungal hypersensitivity

Type	Immune response	Diseases
I (immediate hypersensitivity)	IgE, mast cell	Asthma, rhinitis, eczema, hay fever
III (immune complex mediated)	IgG, antigen-antibody complex deposition in blood vessels and tissues	HP, Arthus reaction, extrinsic allergic alveolitis
IV (delayed-type hypersensitivity)	Antigen-sensitized T lymphocyte	Allergic contact dermatitis, pneumonitis

Penicillium, *Rhizopus*, *Stachybotrys*, *Trichoderma*, and others (92, 133, 134, 159, 160).

The prevalence of allergy to fungi among atopic patients is estimated to be around 30% and in the general population up to 6%. Exposure to mold during childhood is suspected to be a risk factor for the development of allergic respiratory disease (182), although a possible protective effect associated with children's microbial exposure in farm environments has been reported (82, 97, 189). New onset of asthma in children after prolonged moisture and mold exposure has been demonstrated in a prospective study (81).

The reported percentages of populations allergic to molds vary from 2 to 18%. Approximately 80% of asthmatics are reported to be allergic to molds (51). In a 2002 study, up to 35% of newly diagnosed asthma was attributable to workplace mold exposure (82). The incidence and prevalence of allergic diseases are on the rise (150). Many patients with chronic rhinosinusitis have a very high incidence of positive fungal cultures (up to 96%), and this condition is often associated with allergic fungal sinusitis (149).

In clinical allergy, patients can be tested for specific mold allergy by using skin or serological tests (immunoglobulin E [IgE]-radioallergosorbent test), and appropriate advice and treatment can then be given. Due to the low sensitivity of some of the commercially available mold extract tests, false-negative results are not uncommon. Patients with an atopy (a genetic trait of increased allergen sensitivity) are frequently allergic to multiple fungal species and manifest type I reactions (Table 5).

All fungi may be allergenic depending on the exposure situation and doses (150), although the sensitivity of clinical tests may vary with the study population and individual immune system characteristics. Atopic individuals typically have a higher rate of positive skin reactions after provocation tests and serological allergy tests measuring antibody precipitins (IgE). Diseases such as allergic bronchopulmonary aspergillosis (100) and allergic fungal sinusitis possibly require additional host factors which are not well documented (116, 149) and may be the result of a combined reaction of allergenic inflammation and the immunotoxic effect of fungal metabolites. The relevant route of exposure is inhalation. Fungal by-products (i.e., mycotoxins) have ciliostatic effects in the respiratory tract (147), which can be one of the important pathological mechanisms causing diminished mucociliary clearing and local inflammatory effects in the airways and sinuses. In general, the adverse effects of fungal inhalation are related to the duration and intensity of fungal exposure. However, typical for allergic reactions is that once an individual develops an allergy to certain fungi, even small airborne concentrations can trigger an asthma attack or other allergic reactions. This is principally different from fungal toxic inflammatory health reactions, which depend on airborne concentrations and will be similar for most people, whether the individuals are sensitized or not. Allergy threshold levels for common mold have been reported (60), but variations in sampling strategies and methodological limitations make these very unreliable in practical settings (44, 125). Therefore, the consensus is that acceptable safe threshold limits for indoor fungal exposure cannot be established (3, 117) and it is generally recommended to avoid or minimize unnecessary indoor fungal exposure (7).

HP and ODTS

The clinical features, biochemistry, and pathophysiology of allergic or inflammatory toxic reactions to airborne microbial exposure are difficult to separate (67, 153). Hypersensitivity pneumonitis (HP), also called extrinsic allergic alveolitis, is a well-recognized occupational disease. Table 6 lists several fungi, their sources, and the occupational hypersensitivity

TABLE 6 Fungal agents of HP and occupational dust exposure

Fungal agent(s)	Source	Disease(s)
Aspergillus clavatus	Moldy malt	Malt worker's lung
Aureobasidium pullulans	Steam	Sauna taker's lung
Alternaria spp.	Wood	Woodworker's lung
Botrytis cinerea	Moldy fruits	Winegrower's lung
Cryptostroma corticale	Wood	Maple bark stripper's lung
Faenia rectivirgula	Straw	Potato riddler's lung
Serpula (Merulius) lacrymans	Moldy building	Dry rot lung
Penicillium spp.	Cork	Suberosis, woodman's disease
Penicillium casei	Cheese	Cheese worker's lung
Mucor stolonifer	Moldy paprika	Paprika worker's lung
Trichosporon cutaneum	House dust	Japan summer pneumonitis

diseases that these fungi cause. Organic dust toxic syndrome (ODTS), also called toxic pneumonitis, is a nonallergic, noninfectious form of an acute inflammatory lung reaction to high-level fungal dust exposure (155, 157, 190). The differences between HP and ODTS can be difficult to distinguish. Table 7 lists comparative features of HP and ODTS. The significance of ODTS in occupational health is such that preventive measures for ODTS have been recommended for certain occupations (agriculture) by the National Institute for Occupational Safety and Health. The measures include the use of industrial hygiene controls, special protective equipment, ventilation, and respiratory protection (130). Although ODTS is more likely to occur in settings where large amounts of organic waste is handled (such as agriculture and composting facilities), it may happen in office and domestic environments (155).

Large-scale composting of organic waste is a new, growing technology in municipal waste management (89). Environmental monitoring suggests that this procedure involves risks of high levels of exposure to several pathogenic fungi and bacteria. High fungal levels of primarily *Aspergillus fumigatus*, *Penicillium* spp., and *Paecilomyces* spp. in the range of 2,000 to 20,000 CFU/m^3 were found in a sentinel health investigation of a young composting worker who developed alveolitis and lung fibrosis (E. Johanning, E. Olmsted, and C. Yang, *Abstr. Am. Ind. Hyg. Conf.*, p. 24, 1995). Immunological blood changes (such as IgG antibody elevation) can be can observed in waste-handling workers or workers in other occupations with high-level fungal exposure (17).

Mycotoxins and Mycotoxicosis

Fungi produce toxic chemicals such as the poisonous compounds found in some mushrooms and the toxic metabolites of some species of microfungi. Many of the mushroom poisons are polypeptides or amino acid-derived toxins (64). However, poisoning due to ingestion of poisonous mushrooms is excluded from the scope of this chapter. Lincoff and Mitchel (115) provide a suggested reference on mushroom toxins.

Some fungi have been known to produce secondary metabolites called mycotoxins that are harmful to animals and humans when they are ingested (99, 120) or inhaled (36, 93, 123, 171) or come in contact with the skin (43, 162). These mycotoxins belong chemically to the alkaloids, cyclopeptides, and coumarins (64). Toxic effects of toxins produced by mold in humans have been described and researched mostly in relationship to food-borne diseases affecting animals or regional human disease outbreaks (99, 143). There has been a debate regarding the public health importance of "toxic mold" in enclosed indoor environments and its impact on the occupants' health (5, 126). *Aspergillus versicolor* and species of *Penicillium*, *Fusarium*, *Trichoderma*, *Cephalosporium*, *Chaetomium*, and *Stachybotrys* are known to produce naturally potent metabolites, called mycotoxins, depending on available nutrients, favorable environmental conditions, or their life cycles. Health complaints and clinical findings in patients living or working in wet and moldy buildings often cannot be explained as allergic reactions in otherwise healthy individuals. An overview of clinically important health disorders based on various case reports and results of disease cluster investigations is presented for the most important mycotoxin producers (Table 8) (8, 30, 80).

The toxicological knowledge of such mycotoxins stems primarily from veterinary and food safety science (174). A limited number of the more than 400 known mycotoxins have been studied and found to have important genotoxic, mutagenic, cytotoxic, carcinogenic, nephrotoxic, pseudoestrogenic, immunosuppressive, protein synthesis inhibiting, or other toxic properties. The knowledge of these adverse health effects has led internationally to regulatory efforts to protect humans from excess exposure in food and agricultural products based in many cases on a precautionary principle, in part because the data are still limited and definite dose response models have not yet been established for these agents (see http://www.mycotoxin-prevention.com/ProjectsPage.htm). Toxic molds can induce abortions and reproductive abnormalities in animals. There is a concern among environmental health clinicians about similar effects in humans; however, adequate human studies are lacking. The International Agency for Research on Cancer classified aflatoxin produced by *Aspergillus flavus*, *Aspergillus niger*, and *Aspergillus parasiticus* as showing "sufficient evidence" of human and animal liver carcinogenicity (80). Many of the other mycotoxins have not been classified as carcinogens due to a lack of appropriate studies.

Human cases of true mycotoxicosis appear to be rare and were thought to be mostly related to ingestion of contaminated grain products. However, possible occupational and environmental inhalation exposure has been described in recent case studies and epidemiological surveys (40, 72, 90). Cases of nonallergic patient complaints in which mycotoxin exposure was either confirmed or highly suspected have involved recurrent cold and flu-like symptoms, extreme fatigue, constant sore throat or skin irritation, severe and unusual headaches, neuromuscular and neurocognitive dysfunction (tremor and shakes and unusual memory and

TABLE 7 Comparative features of HP and ODTS[a]

Feature	HP (extrinsic allergic alveolitis)	ODTS (toxic pneumonitis)
Immune response	Type IV delayed hypersensitivity, cell-mediated immune reaction	Nonallergenic, noninfectious, lack of IgG
Affected tissue or organ	Lung alveoli (granuloma formation)	Lungs (inflammatory reaction)
Exposure level(s)	10^6–10^{10} CFU of thermophilic actinomycetes or fungi/m^3	High concentrations of fungi: $>10^9$ spores/m^3 or >1–2 μg of endotoxins or (1-3)-β-D-glucan/m^3
Clinical manifestations	Dyspnea, cough, fatigue, poor appetite, weight loss, abnormal chest X ray, abnormal pulmonary functions, high antibody precipitins; may cause pulmonary fibrosis long term	Dyspnea, cough, headaches, fever, chills, malaise, acute inflammatory lung reaction, negative chest X ray; recovery may occur after exposure cessation

[a]Data are from reference 190.

TABLE 8 Some toxigenic fungi, fungal chemical metabolites, and their health effects

Fungi	Chemical metabolite(s)	Health effects
Penicillium spp. (>200 species)	Patulin	Hemorrhage of lung, brain disease
	Citrinin	Renal damage, vasodilation, bronchial constriction, increased muscular tone
	Ochratoxin A	Nephrotoxic and hepatotoxic effects
	Citroviridin	Neurotoxic effects
	Emodin	Reduced cellular oxygen uptake
	Gliotoxin	Lung disease
	Verruculogen	Neurotoxic effects: trembling in animals
	Secalonic acid D	Lung involvement, teratogenic effects in rodents
Aspergillus spp.		
A. clavatus	Patulin	Hemorrhage of lung, brain disease
A. flavus and *A. parasiticus*	Aflatoxin B1	Liver cancer, respiratory cancer, cytochrome P-450-monooxygenase disorder
A. versicolor	Sterigmatocystin	Carcinogenic effects
A. ochraceus	Ochratoxin A	Nephrotoxic and hepatotoxic effects
Stachybotrys chartarum, *Fusarium* spp., *Trichoderma* spp.	Trichothecenes[a] (more than 50 derivatives known): T-2, nivalenol, deoxynivalenol, diacetoxyscripenol, satratoxin H and G, other macrocyclic trichothecenes; spirolactone; zearalenone	Immune suppression and dysfunction, cytotoxicity, bleeding, dermal necrosis (high-dose ingestion is lethal [human case reports]; low-dose, chronic ingestion is potentially lethal), teratogenic and abortogenic effects (in animals), hemorrhage, alimentary toxic aleukia reported in Russia and Siberia, "staggering wheat" reported in Siberia, red mold disease reported in Japan, neurotoxic or nervous behavior abnormality, possible cocarcinogenic chemotoxic effects, anticomplement function; phytoestrogen may alter immune function; stimulates growth of uterus and vulva and atrophy of ovaries
Claviceps spp.	Ergot alkaloids	Prolactin inhibition, vascular constriction, promotion of uterus contraction

[a]Trichothecenes are also produced by *Myrothecium*, *Trichothecium*, and *Gibberella* (teleomorph of some *Fusarium* species).

concentration problems), bleeding disorders of the lung in infants, irregular menses, diarrhea, dermatitis and irritation of the skin, and impaired immune function. Mycotoxins may also be involved in occupational respiratory cancers among food and grain workers (173); however, typical home or office indoor environments have not been studied. Better-controlled studies are needed to improve our understanding of mycotoxins found in the indoor environment and possible adverse human health effects.

Some mycotoxins, such as lysergic acid, are derivatives of amino acids (such as tryptophan). Other mycotoxins derived from other precursors are grouped into aromatic- and phenol-related toxins and terpenoid toxins. Some well-known and potent mycotoxins in the aromatic and phenol-related toxin group are aflatoxin, zearalenone, and griseofulvin. The terpenoid toxins include trichothecenes and fusidanes (64). There are more than 200 mycotoxins produced by a variety of common fungi according to the World Health Organization Environmental Health Criteria 105 on mycotoxins, published in 1990 (194). Samson (158) suggested in 1992 that there are more than 400 toxic metabolites. It is likely that the number of recorded toxic metabolites will increase over time due to new discoveries (30, 85, 86). These alcohol- and water-soluble toxins can be attached to spores, mycelia, or dust particles and are sufficiently small (2 to 10 μm) to be inhaled into the human lungs. Some mycotoxins are lipid soluble and may be absorbed via the skin. Building materials contaminated with the above-mentioned fungi have been shown to carry detectable levels of mycotoxins (92, 132, 133, 185) and contribute to detectable airborne concentrations, as new research has shown in sentinel case investigations including specific *Stachybotrys* mycotoxins (91).

Aflatoxin may be involved in occupational respiratory cancers among food and grain workers (169). Aflatoxin-induced disease has been well reviewed (66, 80, 101). Trichothecene toxins (T-2 toxin and *Fusarium* toxins) are listed with "limited evidence" for animals and "inadequate evidence" (no data available) for humans (80). Macrocyclic trichothecenes, such as satratoxin H, have not been classified. Epidemiological studies suggest a higher rate of upper respiratory tract and lung cancers in workers in the grain- and food-handling industries with high fungal product inhalation risk (173). The high rate of lung cancer among uranium miners in Silesia (Schneeberg disease) may be related to the combined effects of high levels of radon and *Aspergillus* exposure in the underground mines (101).

The earliest known mycotoxin producers (toxigenic fungi), primarily *Claviceps purpurea*, produce the substance ergot, which causes ergotism. Ergot toxin-caused food-poisoning outbreaks associated with bizarre behaviors (known as the "black plague," "holy fire," or "St. Vitus's dance") may have caused the population decline in Europe from the 14th to the 18th centuries (121).

Research has shown that water-damaged building materials are often contaminated with fungi that produce detectable levels of mycotoxins (6, 55, 134, 170), which may become airborne and contribute to indoor air pollution (91).

From a public health point of view, probable important toxigenic fungi are *Aspergillus* species, *Penicillium* species, *Fusarium* species, *Stachybotrys chartarum* (*Stachybotrys atra*), *Paecilomyces* species, and *Trichoderma* species. These fungi have been associated with adverse health effects in humans and animals resulting in typical organ damage and disease, which is often neither allergic nor infectious in nature. In several disease outbreaks, human and animal death has been linked to exposure to toxigenic fungi, typically through ingestion. Current research, however, indicates that inhalation of certain mycotoxins has even stronger effects (35) and may be frequently associated with health complaints and human disease. The occurrence of mycotoxins in the environment, their chemistry, and related adverse health effects have been reviewed previously (12, 30, 123, 158, 162, 164, 195). Several mycotoxins even in low concentrations were observed to be cytotoxic, interfere with DNA and RNA synthesis, inhibit protein synthesis, and cause apoptosis of cells of different body organs. These toxic effects may cause a variety of short-term as well as long-term adverse health effects in animals and humans (12, 70, 134, 138, 164). Samson (158) divided the effects into four basic categories: acute, chronic, mutagenic, and teratogenic. Symptoms thought to be due to mycotoxins or toxin-containing spores (particularly those of *Stachybotrys chartarum*) include dermatitis, recurring cold and flu-like symptoms, burning sore throat, headaches and excessive fatigue, diarrhea, and impaired or altered immune function (162). The ability of the body to resist infectious diseases may be weakened, resulting in opportunistic infections. Certain mycotoxins, such as zearalenone, have been found to cause infertility and stillbirths in pigs (120). Low-level, complex exposures to a mixture of mycotoxins, as would be typically encountered in real-life situations, may have synergistic effects, which may result in central neuroendocrine-immune changes and consequently in complex effects in the endocrine and nervous systems (4).

Historically, mycotoxins have been a problem for farmers and food industries and in Eastern European and Third World countries (53). The large-dose exposure to fungi and mycotoxins encountered by farmers and workers in food industries was generally considered unlikely to occur in nonfarming activities. However, many toxigenic fungi, such as *Stachybotrys chartarum* and species of *Aspergillus*, *Penicillium*, and *Fusarium*, have been found to infest buildings with known indoor air- and building-related problems and illnesses (36, 51, 84, 93). It has been suggested that inhalation exposure to mycotoxin-containing fungal spores was significant in the reported cases of building-related mycotoxicoses (84). Croft et al. (36) reported several cases of mycotoxicoses caused by airborne exposure to the toxigenic fungus *Stachybotrys chartarum* in a residential building. Additional cases of office building-associated *Stachybotrys* mycotoxicosis were reported by Hodgson et al. (72), Johanning (88), and Johanning et al. (93). Satratoxin H was detected in the fungus isolated from the contaminated building. Hemorrhagic lung disease in infants was found to be highly associated with indoor *Stachybotrys chartarum* exposure in a case cluster investigation in Cleveland (47) and in a case home investigation in the midwest of the United States. (52). Subsequently, consultants for the Centers for Disease Control and Prevention called for more research to prove the causal relationship between *Stachybotrys* and idiopathic pulmonary hemorrhage in infants upon review of the Cleveland study (28). Studies with toxic *Stachybotrys* fungi in mice showed similar effects (inflammation and hemorrhage) (135). *Stachybotrys chartarum* was isolated from bronchoalveolar lavage fluid of a child with pulmonary hemorrhage (46), and *Stachybotrys atra* exposure was found in an infant who developed laryngeal spasm and hemorrhage during general anesthesia (183). In an epidemiological study, a high prevalence of pulmonary diseases in office workers in Florida court buildings was reported after prolonged indoor exposure to primarily *Stachybotrys chartarum* and *Aspergillus versicolor* (72).

Mycotoxins generally have low volatility; therefore, inhalation of volatile mycotoxins is not very likely (162). Rather, the toxins are an integral part of the fungus. Sorenson et al. (175) demonstrated in the laboratory that aerosolized conidia of *Stachybotrys atra* contain trichothecene mycotoxins. The most common toxin was satratoxin H. Lesser amounts of satratoxin G and trichoverrols A and B were also detected but less frequently. They also found that most of the airborne particles were within respirable range. Similar experiments conducted by Pasanen et al. (141) demonstrated that trichothecene mycotoxins were in airborne fungal propagules of *Stachybotrys atra* and could be collected on membrane filters. Conidia of *Aspergillus flavus* and *Aspergillus parasiticus* were reported to contain aflatoxins (192). Miller (123) also reported detection of two mycotoxins, deoxynivalenol and T-2 toxin, in conidia of *Fusarium graminearum* and *Fusarium sporotrichioides*, respectively. These studies suggest that inhalation exposure to conidia may also increase the chance of exposure to mycotoxins.

Although relationships were established to link inhalation exposure to mycotoxin-containing fungal spores and symptoms of mycotoxicoses in fungi-infested indoor environments (36, 93, 141, 175), other possible exposure routes such as ingestion and skin contact are likely. Because fungal spores are ubiquitous in a contaminated environment, the chance of ingesting toxin-containing spores is likely to increase through eating, drinking, and smoking. It is prudent to limit exposure to such potent toxic chemicals (23), particularly when significant fungal growth and amplification is found indoors (156).

One important toxigenic fungus frequently detected in "problem buildings" is *Stachybotrys chartarum*, which produces a series of potent cytotoxins (trichothecenes, satratoxins, and spirolactones) as well as a variety of other compounds affecting the immune system (33, 85, 87, 134). In case studies of health effects and immunological laboratory changes related to indoor exposure to trichothecenes and possibly other mycotoxins, disorders of the respiratory and central nervous systems were noted (8, 58, 59). Abnormal test results for the cellular and humoral immune systems were found (75). In earlier cases in Eastern Europe, typically in an agricultural setting, marked leukopenia or acute "radiation-mimetic" effects on the blood cell system with subsequent sepsis-like opportunistic infections after trichothecene ingestion were reported (83, 139).

Trichothecenes are considered to be the most potent small molecule inhibitor of protein synthesis, acting through inhibition of the peptidyl transferase activity (122, 187). These toxins can cause alveolar macrophage defects and may affect phagocytosis. They have been investigated for use in cancer treatment (57) but also in chemical-biological warfare. The presence of fungal chemical metabolites has been reported in several cases of animal and human ingestion-related mycotoxicosis, resulting sometimes in death (68, 84). Mycotoxins, such as satratoxin H of the trichothecene group, have been shown to cause depressed T- or B-lymphocyte activity, suppressed immunoglobulin and antibody production, reduced complement or interferon activity, and impaired macrophage-effector cell function of human neutrophils (195).

Laboratory changes in immunoglobulins (IgA, IgE, IgG, and IgM) in workers handling foodstuffs contaminated with mycotoxins, primarily deoxynivalenol (vomitoxin), have been reported (186). An increase in IgA production and IgA nephropathy and a decrease in IgG and IgM after ingestion of vomitoxin were reported in a mouse experiment (144). Renal failure and IgG deposition in the glomeruli after inhalation of ochratoxin produced by *Aspergillus ochraceus* was found in the case of a farmer (140).

A task group of the World Health Organization concluded that an association between trichothecene exposure and human disease episodes is possible; however, only limited data are available (194). Immunotoxicological effects depend principally on the exposure conditions, dose, and timing. Some immunological effects may be only transient, of short duration, and difficult to detect in routine medical tests. Medical findings are often nonspecific, and other systemic diseases or causes need to be ruled out by the experienced clinician. The treating physician often does not recognize mycotoxicosis, especially because exposure circumstances and the presence of certain mycotoxins are unknown. Advanced fungal exposure characterization and sampling techniques now available should improve the chances for better medical detection of mycotoxicosis. Analytical methods involving immunoassays and cell line cytotoxicity analysis are able to provide relatively rapid and easy screening tests to detect the presence of mycotoxins in fungus-contaminated materials (42, 54, 69, 110, 170, 171).

VOCs Produced by Fungi

Fungi in active growth produce VOCs which typically are noticed by their musty, moldy odor. Indoor-measured VOC levels, however, are typically low, and any serious health risks are uncertain (50, 98). Possibly related mucous membrane and olfactory irritations may trigger an "unpleasant odor reaction" and annoyance. Measurement of VOCs may be an indicator of indoor excessive fungal growth (95). A number of VOCs have been identified from fungi common in indoor contamination. Most of these fungal VOCs are derivatives of alcohols, ketones, hydrocarbons, and aromatics (191). The in vitro-produced fungal VOCs from 47 *Penicillium* taxa were made up of alcohols, ketones, esters, small alkenes, monterpenes, sesquiterpenes, and aromates (105). However, aldehydes were not detected.

Some of the fungal VOCs have an unpleasant odor (61), while other fungi (such as mushrooms) produce VOCs of pleasant odors and flavors. 1-Octen-3-ol, one of the major fungal VOCs, has a characteristic mushroom odor. The musty, moldy, and earthy odors are likely to come from 2-octen-1-ol and geosmin (1,10-dimethyl-9 decalol) (51, 61). Ezeonu et. al. (48) identified ethanol, 2-ethyl hexanol, cyclohexane, and benzene from fiberglass air duct liners colonized by *Aspergillus versicolor*, *Acremonium obclavatum*, and *Cladosporium herbarum*. Acetone and 2-butanone were detected only on agar plate samples of *Aspergillus versicolor* and *Aspergillus obclavatum*. Other fungal VOCs associated with two common indoor fungi, *Penicillium* and *Aspergillus*, have been identified. They are 2-methyl-isoborneol, 2-methyl-1-propanol, 3-methyl-1-butanol, and 3-octanone (61). Additional fungal VOCs are compiled and listed by Ammann (2) and Batterman (9). Almost all of the published information regarding fungal VOCs concerns *Penicillium* spp. and *Aspergillus* spp. Little is known about VOCs of other common indoor fungal contaminants.

CONCLUSIONS

Fungal spores are ubiquitous in nature and indoors. Although fungal growth is found throughout nature, it is uncommon and should not be allowed in a dry, clean indoor environment. Fungal spores and their growth may play an important role in public health under some circumstances. In three separate books published in 1993 (150), 2000 (78), and 2004 (79) by the IOM, National Academy of Science, fungi have been associated with allergies, asthma, and other respiratory diseases. They are considered to be allergenic and irritant agents and occasionally cause infectious and respiratory diseases in susceptible people. Research findings indicate that they are a major problem in buildings where moisture control is poor or water intrusion is common. Synergistic inhalation effects of fungal by-products, such as mycotoxins in fungal spores, β-glucans, or perhaps fungal VOCs released into the surroundings, are potentially irritating, toxic, teratogenic, carcinogenic, and immunosuppressive. Clinical diagnoses of mold allergies and fungal infections are generally easier and less complicated than those of conditions associated with emerging health concerns such as the fungal metabolite (1-3)-β-D-glucan, airborne mycotoxins, and fungal VOCs. Furthermore, risk assessment of human exposure to fungi and their by-products is complex, because multiple agents, hypersensitivity reactions, and different disease outcomes are involved. Exposure to fungi and their by-products at varying levels is ubiquitous in some indoor environments. Human sensitivity to them varies from individual to individual. Some of the health implications from inhalation exposure to many fungal metabolites are poorly understood, particularly at low exposure levels. Little is known concerning the consequences of short-term and long-term exposures to these fungal metabolites and whether the health effects are reversible or not. However, based on what is currently known regarding the fungal exposures and potential adverse health effects, it is prudent to avoid any unnecessary exposure to infectious, allergenic, and toxic fungi and to control indoor growth conditions. Furthermore, fungal growth indoors suggests water infiltration and damage to building structures and materials. Remediation processes must be undertaken to correct the problems.

REFERENCES

1. **Alexopoulos, C. J., C. W. Mims, and M. Blackwell.** 1996. *Introductory Mycology*, 4th ed. John Wiley & Sons, Inc., New York, N.Y.
2. **Ammann, H. A.** 1999. Microbial volatile organic compounds, p. 26-1–26-17. *In* J. Macher, H. A. Ammann, H. A. Burge, D. K. Milton, and P. R. Morey (ed.), *Bioaerosols: Assessment and Control.* American Conference of Governmental Industrial Hygienists, Cincinnati, Ohio.
3. **Ammann, H. M.** 1999. American Council of Government Industrial Hygienists TLV statement on bioaerosols, p. 520–521. *In* E. Johanning (ed.), *Bioaerosols, Fungi and Mycotoxins: Health Effects, Assessment, Prevention and Control.* Eastern New York Occupational and Environmental Health Center, Albany.
4. **Ammann, H. M.** 1999. IAQ and human toxicosis: empirical evidence and theory, p. 84–93. *In* E. Johanning (ed.), *Bioaerosols, Fungi and Mycotoxins: Health Effects, Assessment, Prevention and Control.* Eastern New York Occupational and Environmental Health Center, Albany.
5. **Ammann, H. M.** 2005. Mold toxicity: risk assessment for humans exposed indoors, p. 52–58. *In* E. Johanning (ed.), *Bioaerosols, Fungi, Bacteria, Mycotoxins and Human Health:*

Patho-Physiology, Clinical Effects, Exposure Assessment, Prevention and Control in Indoor Environments and Work. Fungal Research Group Foundation, Albany, N.Y.

6. **Andersson, M. A., M. Nikulin, U. Koljalg, M. C. Andersson, F. Rainey, K. Reijula, E.-L. Hintikka, and M. Salkinoja-Salonen.** 1997. Bacteria, molds, and toxins in water-damaged building materials. *Appl. Environ. Microbiol.* **63:**387–393.
7. **Anonymous.** 1994. Recommendations, p. 531–538. *In* R. A. Samson, B. Flannigan, M. E. Flannigan, A. P. VerHoeff, O. C. G. Adan, and E. S. Hoekstra (ed.), *Health Implications of Fungi in Indoor Environments*. Elsevier, Amsterdam, The Netherlands.
8. **Auger, P. L., P. Pépin, J. D. Miller, M. Gareis, J. Doyon, R. Bouchard, M.-F. Pinard, and C. Mainville.** 1999. Chronic toxic encephalopathies apparently related to exposure to toxigenic fungi, p. 131–138. *In* E. Johanning (ed.), *Bioaerosols, Fungi and Mycotoxins: Health Effects, Assessment, Prevention and Control*. Eastern New York Occupational and Environmental Health Center, Albany.
9. **Batterman, S. A.** 1995. Sampling and analysis of biological volatile organic compounds, p. 249–268. *In* H. A. Burge (ed.), *Bioaerosols*. Lewis Publishers, Boca Raton, Fla.
10. **Benaim-Pinto, C.** 1992. Sensitization to basidiomycetes and to *Fuligo septica* (Myxomycetae) in Venezuelan atopic patients suffering from respiratory allergy. *J. Allergy Clin. Immunol.* **89:**282–289.
11. **Benenson, A. S.** 1990. *Control of Communicable Disease in Man*, 15th ed. American Public Health Association, Washington, D.C.
12. **Bettina, V.** 1989. *Bioactive Molecules, Mycotoxins: Chemical, Biological and Environmental Aspects*, vol. 9. *Bioactive Molecules*. Elsevier, Amsterdam, The Netherlands.
13. **Blackwell, M.** 1990. Air dispersed spores of myxomycetes, p. 51–52. *In* E. G. Smith (ed.), *Sampling and Identifying Allergenic Pollens and Molds*. Blewstone Press, San Antonio, Tex.
14. **Bonham, R. E.** 2005. A review of the IOM report on mold. *Tex. Med.* **100**(12)**:**55–57.
15. **Bornehag, C. G., G. Blomquist, F. Gyntelberg, B. Jarvholm, P. Malmberg, L. Nordvall, A. Nielsen, G. Pershagen, and J. Sundell.** 2001. Dampness in buildings and health: Nordic interdisciplinary review of the scientific evidence on associations between exposure to "dampness" in buildings and health effects (NORDDAMP). *Indoor Air* **11:**72–86.
16. **Brunekreef, B., D. W. Dockery, F. E. Speizer, J. H. Ware, J. D. Spengler, and B. G. Ferris.** 1989. Home dampness and respiratory morbidity in children. *Am. Rev. Respir. Dis.* **140:**1363–1367.
17. **Bünger, J., M. Muller, K. Stalder, and E. Hallier.** 1999. Immunological biomonitoring in the assessment of exposure to airborne fungi from waste handling, p. 112–118. *In* E. Johanning (ed.), *Bioaerosols, Fungi and Mycotoxins: Health Effects, Assessment, Prevention and Control*. Eastern New York Occupational and Environmental Health Center, Albany.
18. **Burge, H. A.** 1990. The fungi, p. 136–162. *In* P. Morey, J. Feeley, and J. Otten (ed.), *Biological Contaminants in Indoor Environments*. American Society for Testing and Materials, Philadelphia, Pa.
19. **Burge, H. A.** 1990. Bioaerosols: prevalence and health effects in the indoor environment. *J. Allergy Clin. Immunol.* **86:**687–701.
20. **Burge, H. A.** 1993. Characterization of bioaerosols in buildings in the United States, p. 79–83. *In IAQ '92: Environments for People*. American Society of Heating, Refrigerating, and Air-Conditioning Engineers, Atlanta, Ga.
21. **Burge, H. A.** 1995. Bioaerosols in the residential environment, p. 579–597. *In* C. S. Cox and C. M. Wathes (ed.), *Bioaerosols Handbook*. Lewis Publishers, Boca Raton, Fla.
22. **Burge, H. A., and J. A. Otten.** 1999. Fungi, p. 19-1–19-13. *In* J. Macher, H. A. Ammann, H. A. Burge, D. K. Milton, and P. R. Morey (ed.), *Bioaerosols: Assessment and Control*. American Conference of Governmental Industrial Hygienists, Cincinnati, Ohio.
23. **Burge, H. A., and H. A. Ammann.** 1999. Fungal toxins and β-(1-3)-D-glucans, p. 24-1–24-13. *In* J. Macher, H. A. Ammann, H. A. Burge, D. K. Milton, and P. R. Morey (ed.), *Bioaerosols: Assessment and Control*. American Conference of Governmental Industrial Hygienists, Cincinnati, Ohio.
24. **Burr, M. L.** 1999. Indoor fungal exposure—does it matter, and what can be done about it? *Clin. Exp. Allergy* **29:**1442–1444.
25. **Butcher, B. T., C. E. O'Neil, M. A. Reed, L. C. Altman, M. Lopez, and S. B. Lehrer.** 1987. Basidiomycete allergy: measurement of spore-specific IgE antibodies. *J. Allergy Clin. Immunol.* **80:**803–809.
26. **Buttner, M. P., P. Cruz-Perez, P. J. Garrett, and L. D. Stetzenbach.** 1999. Dispersal of fungal spores from three types of air handling system duct material. *Aerobiologia* **15:**1–8.
27. **Carroll, G. C., and D. T. Wicklow.** 1992. *The Fungal Community, Its Organization and Role in the Ecosystem*. Marcel Dekker, Inc., New York, N.Y.
28. **Centers for Disease Control and Prevention.** 2000. Update: pulmonary hemorrhage/hemosiderosis among infants—Cleveland, Ohio, 1993–1996. *Morb. Mortal. Wkly. Rep.* **49:**180–184.
29. **Chapman, J. A.** 1999. Update on airborne mold and mold allergy. *Allergy Asthma Proc.* **20:**289–292.
30. **Ciegler, A., H. R. Burmeister, R. F. Vesonder, and C. W. Hesseltine.** 1981. Mycotoxins: occurence in the environment, p. 1–50. *In* R. C. Shank (ed.), *Mycotoxins and N-Nitroso Compound Environmental Risk*, vol. 1. CRC Press, Boca Raton, Fla.
31. **Cole, R. J., and R. H. Cox.** 1981. *Handbook of Toxic Metabolites*. Academic Press, New York, N.Y.
32. **Coombs, R. R. A.** 1975. *Classification of Allergic Reactions for Clinical Hypersensitivity and Disease*. Blackwell Scientific, Oxford, United Kingdom.
33. **Corrier, D. E.** 1991. Mycotoxicosis: mechanism of immunosuppression. *Vet. Immunol. Immunopathol.* **30:**73–87.
34. **Cox-Ganser, J. M., S. K. White, R. Jones, K. Hilsbos, E. Storey, P. L. Enright, C. Y. Rao, and K. Kreiss.** 2005. Respiratory morbidity in office workers in a water-damaged building. *Environ. Health Perspect.* **113:**485–490.
35. **Creasia, D. A., J. D. Thurman, R. W. Wannemacher, Jr., and D. L. Bunnder.** 1990. Acute inhalation toxicity of T-2 mycotoxin in rat and guinea pig. *Fundam. Appl. Toxicol.* **14:**54–59.
36. **Croft, W. A., B. B. Jarvis, and C. S. Yatawara.** 1986. Airborne outbreak of trichothecene toxicosis. *Atmos. Environ.* **20:**549–552.
37. **Dales, R. E., R. Burnett, and H. Zwaneburg.** 1991. Adverse health effects among adults exposed to home dampness and molds. *Am. Rev. Respir. Dis.* **143:**505–509.
38. **Dangman, K. H., A. L. Bracker, and E. Storey.** 2005. Work-related asthma in teachers in Connecticut: association with chronic water damage and fungal growth in schools. *Conn. Med.* **69:**9–17.
39. **Davis, W. E., W. E. Horner, J. E. Salvaggio, and S. B. Lehrer.** 1988. Basidiospore allergens: analysis of *Coprinus quadrifidus* spore, cap, and stalk extracts. *Clin. Allergy* **18:**261–267.

40. **Dearborn, D. G., I. Yike, W. G. Sorenson, M. J. Miller, and R. A. Etzel.** 1999. Overview of investigations into pulmonary hemorrhage among infants in Cleveland, Ohio. *Environ. Health Perspect.* **107**(S3)**:**495–499.
41. **DeKoster, J. A., and P. S. Thorne.** 1995. Bioaerosol concentrations in noncompliant, compliant, and intervention homes in the midwest. *Am. Ind. Hyg. Assoc. J.* **56:**573–580.
42. **Dietrich, R., E. Johanning, M. Gareis, E. Schneider, E. Usleber, and E. Märtlbauer.** 1999. Immunological detection of mycotoxins associated with stachybotryotoxicosis, p. 482–493. *In* E. Johanning (ed.), *Bioaerosols, Fungi and Mycotoxins: Health Effects, Assessment, Prevention and Control.* Eastern New York Occupational and Environmental Health Center, Albany.
43. **Dill, I., C. Trautmann, and R. Szewzyk.** 1997. Mass development of *Stachybotrys chartarum* on decomposable plant-pots made of recycling paper. *Mycosis* **40:**110–114.
44. **Dillon, H. K., J. D. Miller, W. G. Sorenson, J. Douwes, and R. R. Jacobs.** 1999. Review of methods applicable to the assessment of mold exposure to children. *Environ. Health Perspect.* **107**(Suppl. 3)**:**473–480.
45. **Dungy, C. I., P. P. Kozak, J. Gallup, and S. P. Galant.** 1986. Aeroallergen exposure in the elementary school setting. *Ann. Allergy* **56:**218–221.
46. **Elidemir, O., G. N. Colasurdo, S. N. Rossmann, and L. L. Fan.** 1999. Isolation of *Stachybotrys* from the lung of a child with pulmonary hemosiderosis. *Pediatrics* **104:**964–966.
47. **Etzel, R. A., E. Montana, W. G. Sorenson, G. J. Kullman, T. M. Allan, and D. G. Dearborn.** 1998. Acute pulmonary hemorrhage in infants associated with exposure to *Stachybotrys atra* and other fungi. *Arch. Pediatr. Adolesc. Med.* **152:**757–762.
48. **Ezeonu, I. M., D. L. Price, R. B. Simmons, S. A. Crow, and D. G. Ahearn.** 1994. Fungal production of volatiles during growth on fiberglass. *Appl. Environ. Microbiol.* **60:** 4172–4173.
49. **Fatterpekar, G., S. Mukherji, A. Arbealez, S. Maheshwari, and M. Castillo.** 1999. Fungal diseases of the paranasal sinuses. *Semin. Ultrasound CT MR* **20:**391–401.
50. **Fischer, G., T. Muller, R. Schwalbe, R. Ostrowski, and W. Dott.** 2000. Exposure to airborne fungi, MVOC and mycotoxins in biowaste-handling facilities. *Int. J. Hyg. Environ. Health* **203:**97–104.
51. **Flannigan, B., E. M. McCabe, and F. McGarry.** 1991. Allergenic and toxigenic microorganisms in houses. *J. Appl. Bacteriol.* **79:**61S–73S.
52. **Flappan, S. M., J. Portnoy, P. Jones, and C. Barnes.** 1999. Infant pulmonary hemorrhage in a suburban home with water damage and mold (*Stachybotrys atra*). *Environ. Health Perspect.* **107**(11)**:**927–930.
53. **Forgacs, J.** 1972. Stachybotrys toxicosis, p. 95–128. *In* S. Kadis and S. J. Agl (ed.), *Microbial Toxins,* vol. III. Academic Press, Inc., New York, N.Y.
54. **Gareis, M.** 1995. Cytotoxicity testing of samples originating from problem buildings, p. 139–144. *In* E. Johanning and C. S. Yang (ed.), *Fungi and Bacteria in Indoor Air Environments: Health Effects, Detection and Remediation.* Eastern New York Occupational Health Program, Albany.
55. **Gareis, M., E. Johanning, and R. Dietrich.** 1999. Mycotoxin cytotoxicity screening of field samples, p. 202–213. *In* E. Johanning (ed.), *Bioaerosols, Fungi and Mycotoxins: Health Effects, Assessment, Prevention and Control.* Eastern New York Occupational and Environmental Health Center, Albany.
56. **Giannini, E. H., W. T. Northey, and C. R. Leathers.** 1975. The allergenic significance of certain fungi rarely reported as allergens. *Ann. Allergy* **35:**372–376.
57. **Goodwin, W., C. D. Haas, C. Fabian, I. Heller-Bettinger, and B. Hoogstraten.** 1978. Phase I evaluation of anguidine (diacetoxyscirpenol, NSC-141537). *Cancer* **42:**23–26.
58. **Gordon, K. E., R. E. Masoti, and W. R. Waddell.** 1993. Tremorgenic encephalopathy: a role of mycotoxins in the production of CNS disease in humans? *Can. J. Neurol. Sci.* **20:**237–239.
59. **Gordon, W., E. Johanning, and L. Haddad.** 1999. Cognitive impairment associated with exposure to toxigenic fungi, p. 94–98. *In* E. Johanning (ed.), *Bioaerosols, Fungi and Mycotoxins: Health Effects, Assessment, Prevention and Control.* Eastern New York Occupational and Environmental Health Center, Albany.
60. **Gravesen, S.** 1979. Fungi as a cause of allergic disease. *Allergy* **34:**135–154.
61. **Gravesen, S., J. C. Frisvad, and R. A. Samson.** 1994. *Microfungi.* Munksgaard, Copenhagen, Denmark.
62. **Gravesen, S., P. A. Nielsen, R. Iversen, and K. F. Nielsen.** 1999. Microfungal contamination of damp buildings—examples of risk constructions and risk materials. *Environ. Health Perspect.* **107:**505–508.
63. **Gregory, P. H.** 1973. *The Microbiology of Atmosphere,* 2nd ed. Halstead Press, New York, N.Y.
64. **Griffin, D. H.** 1993. *Fungal Physiology.* Wiley-Liss, New York, N.Y.
65. **Heinemann, S., H. Beguin, and N. Nolard.** 1994. Biocontamination in air-conditioning, p. 179–186. *In* R. A. Samson, B. Flannigan, M. E. Flannigan, A. P. VerHoeff, O. C. G. Adan, and E. S. Hoekstra (ed.), *Health Implications of Fungi in Indoor Environments.* Elsevier, Amsterdam, The Netherlands.
66. **Hendry, K. M., and E. C. Cole.** 1993. A review of mycotoxins in indoor air. *J. Toxicol. Environ. Health* **38:**183–198.
67. **Henson, P. M., and R. C. Murphy.** 1989. *Mediators of the Inflammatory Process.* Elsevier, New York, N.Y.
68. **Hintikka, E.** 1987. Human stachybotryotoxicosis, p. 87–89. *In* T. D. Wyllie and L. G. Morehouse (ed.), *Mycotoxic Fungi, Mycotoxins, Mycotoxicosis,* vol. 3. Marcel Dekker, New York, N.Y.
69. **Hintikka, E., and M. Nikulin.** 1995. Aerosol mycotoxins: a veterinary experience and perspective, p. 31–34. *In* E. Johanning and C. S. Yang (ed.), *Fungi and Bacteria in Indoor Air Environments: Health Effects, Detection and Remediation.* Eastern New York Occupational Health Program, Albany.
70. **Hintikka, E. L.** 1999. The effects of inhaled spores of mycotoxin producing fungi on animals, p. 214–220. *In* E. Johanning (ed.), *Bioaerosols, Fungi and Mycotoxins: Health Effects, Assessment, Prevention and Control.* Eastern New York Occupational and Environmental Health Center, Albany.
71. **Hocking, A. D., and B. F. Miscamble.** 1995. Water relations of some Zygomycetes isolated from food. *Mycol. Res.* **99:**1113–1118.
72. **Hodgson, M. J., P. Morey, W. Y. Leung, L. Morrow, D. Miller, B. B. Jarvis, H. Robbins, J. F. Halsey, and E. Storey.** 1998. Building-associated pulmonary disease from exposure to *Stachybotrys chartarum* and *Aspergillus versicolor. J. Occup. Environ. Med.* **40:**241–249.
73. **Horner, W. E., and S. B. Lehrer.** 1999. Why are there still problems with fungal allergen extracts? *In* E. Johanning (ed.), *Bioaerosols, Fungi and Mycotoxins: Health Effects, Assessment, Prevention and Control.* Eastern New York Occupational and Environmental Health Center, Albany.
74. **Horner, W. E., A. G. Worthan, and P. R. Morey.** 2004. Air- and dustborne mycoflora in houses free of water dam-

age and fungal growth. *Appl. Environ. Microbiol.* **70:**6394–6400.

75. **Horwitz, R. J., and R. K. Bush.** 1997. Allergens and other factors important in atopic disease, p. 75–129. *In* R. Patterson, L. C. Grammer, and P. A. Greenberger (ed.), *Allergic Diseases, Diagnosis and Management.* Lippincott-Raven, Philadelphia, Pa.
76. **Husman, T.** 1996. Health effects of indoor-air microorganism. *Scand. J. Work Environ. Health* **22:**2–13.
77. **Hyvarinen, A., T. Reponen, T. Husman, and A. Nevalainen.** 2001. Comparison of the indoor air quality in mould damaged and reference buildings in a subarctic climate. *Cent. Eur. J. Public Health* **9:**133–139.
78. **Institute of Medicine.** 2000. *Clearing the Air: Asthma and Indoor Air Exposures.* National Academy Press, Washington, D.C.
79. **Institute of Medicine.** 2004. *Damp Indoor Spaces and Health.* National Academy Press, Washington, D.C.
80. **International Agency for Research on Cancer, World Health Organization.** 1993. Monographs in the evaluation of carcinogenic risks to humans. *IARC Monogr.* **56:**245–523.
81. **Jaakkola, J. J., B. F. Hwang, and N. Jaakkola.** 2005. Home dampness and molds, parental atopy, and asthma in childhood: a six-year population-based cohort study. *Environ. Health Perspect.* **113:**357–361.
82. **Jaakkola, M. S., H. Nordman, R. Piipari, J. Uitti, J. Laitinen, A. Karjalainen, P. Hahtola, and J. J. Jaakkola.** 2002. Indoor dampness and molds and development of adult-onset asthma: a population-based incident case-control study. *Environ. Health Perspect.* **110:**543–547.
83. **Jarmai, K.** 1929. Viskosusseptikamien bei alteren Fohlen und erwachsenen Pferden (Viscosussepsis in older colts and grown-up horses). *Dtsch. Tierarztl. Wochenschr.* **33:**517–519.
84. **Jarvis, B. B.** 1990. Mycotoxins and indoor air quality, p. 201–214. *In* P. Morey, J. Feeley, and J. Otten (ed.), *Biological Contaminants in Indoor Environments.* American Society for Testing and Materials, Philadelphia, Pa.
85. **Jarvis, B. B., J. Salemme, and A. Morais.** 1995. Stachybotrys toxins. *Nat. Toxins* **3:**10–16.
86. **Jarvis, B. B., Y. Zhou, J. Jiang, S. Wang, W. G. Sorenson, E.-L. Hintikka, M. Nikulin, P. Parikkam, R. A. Etzel, and D. G. Dearborn.** 1996. Toxigenic molds in water-damaged buildings: dechlorogriseofulvins from *Memnoniella echinata. J. Nat. Prod.* **59:**553–554.
87. **Jarvis, B. B., and S. F. Hinkley.** 1999. Analysis for *Stachybotrys* toxins, p. 232–239. *In* E. Johanning (ed.), *Bioaerosols, Fungi and Mycotoxins: Health Effects, Assessment, Prevention and Control.* Eastern New York Occupational and Environmental Health Center, Albany.
88. **Johanning, E.** 1995. Health problems related to fungal exposure—primarily of *Stachybotrys atra*, p. 169–182. *In* E. Johanning and C. S. Yang (ed.), *Fungi and Bacteria in Indoor Air Environments: Health Effects, Detection and Remediation.* Eastern New York Occupational Health Program, Albany.
89. **Johanning, E.** 1999. Overview of waste management in the United States and recent research activities about composting related occupational health risks, p. 127–140. *In* T. Eikmann and R. Hofmann (ed.), *Stand von Wissenschaft, Forschung und Technik zu siedlungshygienischen Aspekten der Abfallentsorgung und -verwertung (Status Assessment of Science, Research and Techniques Related to Hygiene in Waste-Management and -Use)*, vol. 104. Kommission Reinhaltung der Luft im VDI und DIN, Band 30. Schriftenreihe des Vereins für Wasser-, Boden- und Lufthygiene, Berlin, Germany.
90. **Johanning, E., R. Biagini, D. Hull, P. Morey, B. Jarvis, and P. Landsbergis.** 1996. Health and immunology study following exposure to toxigenic fungi (Stachybotrys chartarum) in a water-damaged office environment. *Int. Arch. Occup. Environ. Health* **68:**207–218.
91. **Johanning, E., M. Gareis, K. Nielsen, and G. Engholm.** 2002. Airborne mycotoxin sampling and screening analysis, p. 1–6. *In* H. Levin (ed.), *Indoor Air 2002.* International Academy of Indoor Air Sciences, Santa Cruz, Calif.
92. **Johanning, E., M. Gareis, C. S. Yang, E.-L. Hintikka, M. Nikulin, B. Jarvis, and R. Dietrich.** 1999. Toxicity screening of materials from buildings with fungal indoor air quality problems (*Stachybotrys chartarum*). *Mycotoxin Res.* **14:**60–73.
93. **Johanning, E., P. R. Morey, and B. B. Jarvis.** 1993. Clinical-epidemiological investigation of health effects caused by *Stachybotrys atra* building contamination, p. 225–230. *In* O. Seppänen (ed.), *Indoor Air '93, Proceedings of the Sixth International Conference on Indoor Air Quality and Climate*, vol. 1. *Health Effects.* Helsinki University of Technology, Espoo, Finland.
94. **Kaplan, J. E., J. L. Jones, and C. A. Dykewicz.** 2000. Protists as opportunistic pathogens: public health impact in the 1990s and beyond. *J. Eukaryot. Microbiol.* **47:**15–20.
95. **Keller, R., K. Senkpiel, and H. Ohgke.** 1999. Use of MVOC measurements and odor perception as indicator of mould in indoor areas, p. 532–537. *In* E. Johanning (ed.), *Bioaerosols, Fungi and Mycotoxins: Health Effects, Assessment, Prevention and Control.* Eastern New York Occupational and Environmental Health Center, Albany.
96. **Kemp, S. J., T. H. Kuehn, D. H. Pui, D. Vesley, and A. J. Streifel.** 1995. Growth of microorganisms on HVAC filters under controlled temperature and humidity conditions. *Am. Soc. Heating Refrigerating Air-Conditioning Eng. Trans.* **101**(pt. 1)**:**1–12.
97. **Kilpelainen, M., E. O. Terho, H. Helenius, and M. Koskenvuo.** 2002. Childhood farm environment and asthma and sensitization in young adulthood. *Allergy* **57:**1130–1135.
98. **Korpi, A., J. P. Kasanen, Y. Alarie, V. M. Kosma, and A. L. Pasanen.** 1999. Sensory irritating potency of some microbial volatile organic compounds (MVOCs) and a mixture of five MVOCs. *Arch. Environ. Health* **54:**347–352.
99. **Krogh, P.** 1989. The role of mycotoxins in disease of animals and man. *J. Appl. Bacteriol. Symp.* **79**(Suppl.)**:**99S–204S.
100. **Kurup, V. P.** 1999. Allergens, diagnosis, and pathogenesis, p. 620–625. *In* E. Johanning (ed.), *Bioaerosols, Fungi and Mycotoxins: Health Effects, Assessment, Prevention and Control.* Eastern New York Occupational and Environmental Health Center, Albany.
101. **Kusak, V., S. Jelinek, and J. Ula.** 1990. Possible role of *Aspergillus flavus* in the pathogenesis of Schneeberg and Jachymov disease. *Neoplasm* **17:**441–449.
102. **Kwon-Chung, K. J., and J. E. Bennett.** 1992. *Medical Mycology.* Lea & Febiger, Philadelphia, Pa.
103. **Lacey, J.** 1981. Aerobiology of conidial fungi, p. 373–416. *In* G. C. Cole and B. Kendrick (ed.), *Biology of Conidial Fungi*, vol. 1. Academic Press, New York, N.Y.
104. **Lacey, J.** 1991. Aerobiology and health, p. 157–185. *In* D. L. Hawksworth (ed.), *Frontiers in Mycology.* CABI, Wallingford, United Kingdom.
105. **Larsen, T. O., and J. C. Frisvad.** 1995. Characterization of volatile metabolites from 47 *Penicillium* taxa. *Mycol. Res.* **99:**1153–1166.
106. **Lehrer, S. B., M. Lopez, B. T. Butcher, J. Olson, M. Reed, and J. E. Salvaggio.** 1986. Basidiomycete mycelia

and spore-allergen extracts: skin test reactivity in adults with symptoms or respiratory allergy. *J. Allergy Clin. Immunol.* **78:**478–485.

107. **Levetin, E.** 1991. Identifications and concentration of airborne basidiospores. *Grana* **30:**123–128.
108. **Levetin, E.** 1995. Fungi, p. 87–120. *In* H. A. Burge (ed.), *Bioaerosols*. Lewis Publishers, Boca Raton, Fla.
109. **Levetin, E., R. Shaughnessy, E. Fisher, B. Ligman, J. Harrison, and T. Brennan.** 1995. Indoor air quality in schools: exposure to fungal allergens. *Aerobiologia* **11:**27–34.
110. **Lewis, C. W., J. E. Smith, J. G. Anderson, and Y. M. Murad.** 1994. The presence of mycotoxin-associated fungal spores isolated from the indoor air of the damp domestic environment and cytotoxic to human cell lines. *Indoor Environ.* **3:**323–330.
111. **Llewellyn, G. C., and C. E. O'Rear.** 1990. *Biodeterioration Research, vol. 3. Mycotoxins, Biotoxins, Wood Decay, Air Quality, Cultural Properties, General Biodeterioration, and Degradation*. Plenum Press, New York, N.Y.
112. **Li, D.-W., and C. S. Yang.** 2004. Fungal contamination as a major contributor of sick building syndrome, p. 31–112. *In* D. Strauss (ed.), *Advances in Applied Microbiology*, vol. 55. *Sick Building Syndrome*. Elsevier Academic Press, San Diego, Calif.
113. **Li, D.-W., and C. S. Yang.** 2004. Notes on indoor fungi, I. New records and noteworthy fungi from indoor environments. *Mycotaxon* **89:**473–488.
114. **Lin, K.-T., D.-W. Li, D. A. Denis, R. Woodcock, and C. S. Yang.** 2005. Qualitative identification of *Meruliporia incrassata* using real time polymerase chain reaction (PCR), p. 335–342. *In* E. Johanning (ed.), *Bioaerosols, Fungi, Bacteria, Mycotoxins and Human Health: Patho-Physiology, Clinical Effects, Exposure Assessment, Prevention and Control in Indoor Environments and Work*. Fungal Research Group Foundation, Inc., Albany, N.Y.
115. **Lincoff, G., and D. H. Mitchel.** 1977. *Toxic and Hallucinogenic Mushroom Poisoning*, p. 267. Van Nostrand Reinhold Co., New York, N.Y.
116. **Luong, A., and B. F. Marple.** 2004. Allergic fungal rhinosinusitis. *Curr. Allergy Asthma Rep.* **4:**465–470.
117. **Macher, J.** 1999. *Bioaerosols: Assessment and Control*. American Conference of Governmental Industrial Hygienists, Cincinnati, Ohio.
118. **Macher, J. M., F.-Y. Huang, and M. Flores.** 1991. A two-year study of microbiological indoor air quality in a new apartment. *Arch. Environ. Health* **46:**25–29.
119. **Mandell, G. L., J. E. Bennet, and R. Dolin.** 1995. *Principles and Practice of Infectious Diseases*, 4th ed., p. 2288–2393. Churchill Livingstone, Philadelphia. Pa.
120. **Marasas, W. F. O., and P. E. Nelson.** 1987. *Mycotoxicology*, p. 102. Pennsylvania State University Press, University Park.
121. **Matossian, M. K.** 1989. *Poisons of the Past: Molds, Epidemics, and History*. Yale University Press, New Haven, Conn.
122. **McLaughlin, C. S., M. H. Vaughan, I. M. Campbell, C. M. Wei., M. E. Stafford, and B. S. Hansen.** 1977. Inhibition of protein synthesis by trichothecenes, p. 263–273. *In* J. V. Rodrick, C. M. Hesseltine, and M. A. Mehlman (ed.), *Mycotoxins in Human and Animal Health*. Pathotox, Park Forest South, Ill.
123. **Miller, J. D.** 1992. Fungi as contaminants in indoor air. *Atmos. Environ.* **26A:**2163–2172.
124. **Miller, J. D.** 1993. Fungi and the building engineer, p. 147–158. *In IAQ '92: Environments for People*. American Society of Heating, Refrigerating, and Air-Conditioning Engineers, Atlanta, Ga.
125. **Miller, J. D., R. Dales, and J. White.** 1999. Exposure measures for studies of mold and dampness and respiratory health, p. 298–305. *In* E. Johanning (ed.), *Bioaerosols, Fungi and Mycotoxins: Health Effects, Assessment, Prevention and Control*. Eastern New York Occupational and Environmental Health Center, Albany.
126. **Miller, V. R., and H. M. Ammann.** 2005. Risk assessment of fungal bioaerosols in indoor environments: current paradigms and evolving concepts, p. 474–481. *In* E. Johanning (ed.), *Bioaerosols, Fungi, Bacteria, Mycotoxins and Human Health: Patho-Physiology, Clinical Effects, Exposure Assessment, Prevention and Control in Indoor Environments and Work*. Fungal Research Group Foundation, Albany, N.Y.
127. **Morey, P. R., M. J. Hodgson, W. G. Sorenson, G. J. Kullman, W. W. Rhodes, and G. S. Visvesvara.** 1984. Environmental studies in moldy office buildings: biological agents, sources and preventive measures. *Ann. Am. Conf. Gov. Ind. Hyg.* **10:**21–35.
128. **Morgan-Jones, G., and B. J. Jacobsen.** 1988. Notes on hyphomycetes. LVIII. Some dematiaceous taxa, including two undescribed species of Cladosporium, associated with biodeterioration of carpet, plaster and wallpaper. *Mycotaxon* **32:**223–236.
129. **Morrison, J., C. Yang, K.-T. Lin, R. A. Haugland, A. N. Neely, and S. J. Vesper.** 2004. Monitoring *Aspergillus* species by quantitative PCR during construction of a multi-story hospital building. *J. Hosp. Infect.* **57:**85–87.
130. **National Institute for Occupational Safety and Health.** 1994. *Preventive Organic Duct Toxic Syndrome*. Publication 94–102. Department of Health and Human Services, National Institute for Occupational Safety and Health, Cincinnati, Ohio.
131. **Nevalainen, A., P. Partanen, E. Jaeaeskelainen, A. Hyvaerinen, O. Koskinen, T. Meklin, M. Vahteristo, J. Koivisto, and T. Husman.** 1998. Prevalence of moisture problems in Finnish houses. *Indoor Air* **S4:**45–49.
132. **Nielsen, K. F., S. Gravesen, P. A. Nielsen, B. Andersen, U. Thrane, and J. C. Frisvad.** 1999. Production of mycotoxins on artificially and naturally infested building materials. *Mycopathologia* **145:**43–56.
133. **Nielsen, K. F., K. Huttunen, A. Hyvarinen, B. Andersen, B. B. Jarvis, and M. R. Hirvonen.** 2002. Metabolite profiles of Stachybotrys isolates from water-damaged buildings and their induction of inflammatory mediators and cytotoxicity in macrophages. *Mycopathologia* **154:**201–205.
134. **Nikulin, M.** 1999. *Toxigenic Stachybotrys chartarum in Water-Damaged Building Materials: Ecology and Biology*. Microbiological publication 46/1999. University of Helsinki, Helsinki, Finland.
135. **Nikulin, M., K. Reijula, B. B. Jarvis, and E. L. Hintikka.** 1996. Experimental lung mycotoxicosis in mice induced by *Stachybotrys atra*. *Int. J. Exp. Pathol.* **77:**213–218.
136. **Noble, W. C., and Y. M. Clayton.** 1963. Fungi in the air of hospital wards. *J. Gen. Microbiol.* **32:**397–402.
137. **Ohtsuki, T.** 1962. Studies on the glass mould. V. On two species of *Aspergillus* isolated from glass. *Bot. Mag.* **75:**436–442.
138. **Okumurai, H., N. Yoshino, Y. Sugiura, and M. Sugamata.** 1999. Trichothecenes as a potent inducer of apoptosis, p. 221–231. *In* E. Johanning (ed.), *Bioaerosols, Fungi and Mycotoxins: Health Effects, Assessment, Prevention and Control*. Eastern New York Occupational and Environmental Health Center, Albany.
139. **Ozegovic, L.** 1971. Straw related disease in farm workers. *Veterinaria* **20:**263–267. (In Russian.)

140. **Paolo, N. D., A. Guarieri, F. Loi, G. Sacchi, A. M. Mangiorotti, and M. D. Paolo.** 1993. Acute renal failure from inhalation of mycotoxins. *Nephron* **64:**621–625.

141. **Pasanen, A.-L., M. Nikulin, M. Tuomainen, S. Berg, P. Parikka, and E.-L. Hintikka.** 1993. Laboratory experiments on membrane filter sampling of airborne mycotoxins produced by *Stachybotrys atra* Corda. *Atmos. Environ.* **27A:**9–13.

142. **Peat, J. K., J. Dickerson, and J. Lee.** 1998. Effects of damp and mould in the home and respiratory health: a review of the literature. *Allergy* **53:**120–128.

143. **Peraica, M., B. Radic, A. Lucic, and M. Pavlovic.** 1999. Toxic effects of mycotoxins in humans. *Bull. W. H. O.* **77:**754–766.

144. **Pestka, J. J., M. A. Moorman, and R. L. Warner.** 1989. Dysregulation of IgA production and IgA nephropathy induced by the trichothecene vomitoxin. *Food Chem. Toxicol.* **27:**361–368.

145. **Pfaller, M. A.** 1992. Opportunistic fungal infections, p. 287–293. *In* J. M. Last and R. B. Wallace (ed.), *Maxcy-Rosenau Public Health and Preventive Medicine*. Appleton and Lange, East Norwalk, Conn.

146. **Pfaller, M. A., and R. A. Fromtling.** 1999. Mycology, p. 1159–1326. *In* P. R. Murray, E. J. Baron, M. A. Pfaller, F. C. Tenover, and R. H. Yolken (ed.), *Manual of Clinical Microbiology*, 7th ed. ASM Press, Washington, D.C.

147. **Pieckova, E., Z. Jesenska, and K. Wilikins.** 1999. Microscopic fungi and their metabolites in dwellings—a bioassay study, p. 351–354. *In* E. Johanning (ed.), *Bioaerosols, Fungi and Mycotoxins: Health Effects, Assessment, Prevention and Control*. Eastern New York Occupational and Environmental Health Center, Albany.

148. **Pitt, J. I.** 1979. *The Genus* Penicillium *and Its Teleomorphic States* Eupenicillium *and* Talaromyces, p. 634. Academic Press, New York, N.Y.

149. **Ponikau, J. U., D. A. Sherris, E. B. Kern, H. A. Homburger, E. Frigas, T. A. Gaffey, and G. D. Roberts.** 1999. The diagnosis and incidence of allergic fungal sinusitis. *Mayo Clin. Proc.* **74:**877–884.

150. **Pope, A. M., R. Patterson, and H. Burge.** 1993. *Indoor Allergens*, p. 308. National Academy Press, Washington, D.C.

151. **Putus, T.** 2005. Health effects of moisture damage associated microbes, p. 94–107. *In* E. Johanning (ed.), *Bioaerosols, Fungi, Bacteria, Mycotoxins and Human Health: Patho-Physiology, Clinical Effects, Exposure Assessment, Prevention and Control in Indoor Environments and Work*. Fungal Research Group Foundation, Albany, N.Y.

152. **Raper, K. B., and D. I. Fennell.** 1977. *The Genus* Aspergillus, p. 686. Robert E. Kreiger Publishing Co., Malabar, Fla.

153. **Richerson, H. B.** 1994. Hypersensitivity pneumonitis, p. 139–160. *In* R. Rylander and R. R. Jacobs (ed.), *Organic Dusts: Exposure, Effects, and Prevention*. Lewis Publishers, Boca Raton, Fla.

154. **Ruotsalainen, R., N. Jaakkola, and J. J. K. Jaakkola.** 1995. Dampness and molds in day-care centers as an occupational problem. *Int. Arch. Occup. Environ. Health* **66:**369–374.

155. **Rylander, R.** 1994. Office and domestic environments, p. 247–255. *In* R. Rylander and R. R. Jacobs (ed.), *Organic Dusts: Exposure, Effects, and Prevention*. Lewis Publishers, Boca Raton, Fla.

156. **Rylander, R.** 1999. Effects after mold exposure—which are the causative agents?, p. 28–32. *In* E. Johanning (ed.), *Bioaerosols, Fungi and Mycotoxins: Health Effects, Assessment, Prevention and Control*. Eastern New York Occupational and Environmental Health Center, Albany.

157. **Rylander, R., and R. Etzel.** 1999. Introduction and summary: Workshop on Children's Health and Indoor Mold Exposure. *Environ. Health Perspect.* **107**(Suppl. 3)**:**465–468.

158. **Samson, R. A.** 1992. Mycotoxins: a mycologist's perspective. *J. Med. Vet. Mycol.* **30**(Suppl. 1)**:**9–18.

159. **Samson, R. A., B. Flannigan, M. E. Flannigan, A. P. VerHoeff, O. C. G. Adan, and E. S. Hoekstra.** 1994. *Health Implications of Fungi in Indoor Environments*. Elsevier Science B.V., Amsterdam, The Netherlands.

160. **Samson, R. A., J. Houbraken, R. C. Summerbell, B. Flannigan, and J. D. Miller.** 2001. Common and important species of fungi and actinomycetes in indoor environments, p. 287–473. *In* B. Flannigan, R. A. Samson, and J. D. Miller (ed.), *Microorganisms in Home and Indoor Work Environments*. Taylor & Francis, London, United Kingdom.

161. **Santilli, J., W. J. Rockwell, and R. P. Collins.** 1985. The significance of the spores of the basidiomycetes (mushrooms and their allies) in bronchial asthma and allergic rhinitis. *Ann. Allergy* **55:**469–471.

162. **Schiefer, H. B.** 1990. Mycotoxins in indoor air: a critical toxicological viewpoint, p. 167–172. *In* D. S. Walkinshaw (ed.), *Indoor Air '90. Proceedings of the 5th International Conference on Indoor Air Quality and Climate*, vol. 1.

163. **Seifert, K. A., and J. C. Frisvad.** 2000. Penicillium on soild wood products, p. 285–298. *In* R. A. Samson and J. I. Pitt (ed.), *Integration of Moderan Taxonomic Methods for Penicillium and Aspergillus Classification*. Harwood Academic Publishers, Amsterdam, The Netherlands.

164. **Sharma, R. P., and D. K. Salunkhe.** 1991. *Mycotoxins and Phytoalexins*. CRC Press, Boca Raton, Fla.

165. **Shelton, B. G., K. H. Kirkland, W. D. Flanders, and G. K. Morris.** 2002. Profiles of airborne fungi in buildings and outdoor environments in the United States. *Appl. Environ. Microbiol.* **68:**1743–1753.

166. **Singh, J.** 1994. Nature and extent of deterioration in buildings due to fungi, p. 34–53. *In* J. Singh (ed.), *Building Mycology*. Chapman and Hall, London, United Kingdom.

167. **Smith, D., and A. H. S. Onions.** 1994. *The Preservation and Maintenance of Living Fungi*. CAB International, Wallingford, Oxon, United Kingdom.

168. **Smith, E. G.** 1990. *Sampling and Identifying Allergenic Pollens and Molds*. Blewstone Press, San Antonio, Tex.

169. **Smith, J. E., and M. O. Moss.** 1985. *Mycotoxins: Formation, Analysis and Significance*. John Wiley & Sons, Chichester, United Kingdom.

170. **Smith, J. E., J. G. Anderson, C. W. Lewis, and Y. M. Murad.** 1992. Cytotoxic fungal spores in the indoor atmosphere of the damp domestic environment. *FEMS Microbiol. Lett.* **100:**337–344.

171. **Smoragiewicz, W., B. Cossette, A. Boutard, and K. Krzystyniak.** 1993. Trichothecene mycotoxins in the dust of ventilation systems in office buildings. *Int. Arch. Occup. Environ. Health* **65:**113–117.

172. **Solomon, W. R., H. P. Burge, and J. R. Boise.** 1978. Airborne *Aspergillus fumigatus* levels outside and with a large clinical center. *J. Allergy Clin. Immunol.* **62:**56–60.

173. **Sorenson, W. G.** 1990. Mycotoxins as potential occupational hazards. *Dev. Ind. Microbiol.* **31:**205–211.

174. **Sorenson, W. G.** 1999. Fungal spores: hazardous to health? *Environ. Health Perspect.* **107**(S3)**:**469–472.

175. **Sorenson, W. G., D. G. Frazier, B. B. Jarvis, J. Simpson, and V. A. Robinson.** 1987. Trichothecene mycotoxins in aerosolized conidia of *Stachybotrys atra*. *Appl. Environ. Microbiol.* **53:**1370–1375.

176. **Spengler, J., L. Neas, S. Nalai, D. Dockery, F. Speizer, J. Ware, and M. Raizenne.** 1993. Respiratory symptoms

and housing characteristics, p. 165–170. *In* O. Seppänen (ed.), *Indoor Air '93. Proceedings of the Sixth International Conference on Indoor Air Quality and Climate*, vol. 1. *Health Effects*. Helsinki University of Technology, Espoo, Finland.

177. **Stetzenbach, L. D., H. Ammann, E. Johanning, G. King, and R. J. Shaughnessy.** 2004. *Microorganisms, Mold and Indoor Air Quality*. American Society for Microbiology, Washington, D.C.
178. **Storey, E., K. H. Dangman, P. Schenck, R. L. DeBernardo, C. S. Yang, A. Bracker, and M. J. Hodgson.** 30 September 2004, posting date. *Guidance for Clinicians on the Recognition and Management of Health Effects Related to Mold Exposure and Moisture Indoors*. [Online.] Division of Occupational and Environmental Medicine, Center for Indoor Environments and Health, University of Connecticut Health Center, Farmington, Conn. http://www.oehc.uchc.edu/clinser/MOLD%20GUIDE.pdf.
179. **Strachan, D. P., and R. A. Elton.** 1986. Relationship between respiratory morbidity in children and the home environment. *Fam. Pract.* **3:**137–142.
180. **Strachan, D. P., B. Flannigan, E. M. McCabe, and F. McGarry.** 1990. Quantification of airborne moulds in the homes of children with and without wheeze. *Thorax* **45:**382–387.
181. **Streifel, A. J., and F. S. Rhame.** 1993. Hospital air filamentous fungal spore and particle counts in a specially designed hospital, p. 161–165. *In* O. Seppänen (ed.), *Indoor Air '93. Proceedings of the Sixth International Conference on Indoor Air Quality and Climate*, vol. 4. *Particles, Microbes, Radon*. Helsinki University of Technology, Espoo, Finland.
182. **Taskinen, T. M., S. Laitinen, A. Nevalainen, A. Vepsalainen, T. Meklin, M. Reiman, M. Korppi, and T. Husman.** 2002. Immunoglobulin G antibodies to moulds in school-children from moisture problem schools. *Allergy* **57:**9–16.
183. **Tripi, P. A., S. Modlin, W. G. Sorenson, and D. G. Dearborn.** 2000. Acute pulmonary haemorrhage in an infant during induction of general anaesthesia. *Paediatr. Anaesth.* **10:**92–94.
184. **Troller, J. A., and V. N. Scott.** 1992. Measurement of water activity (Aw) and acidity, p. 135–151. *In* C. Vanderzant and D. F. Splittstoesser (ed.), *Compendium of Methods for the Microbiological Examination of Foods*. American Public Health Association, Washington, D.C.
185. **Tuomi, T., K. Reijula, T. Johnsson, K. Hemminki, E. L. Hintikka, O. Lindroos, S. Kalso, P. Koukila-Kahkola, H. Mussalo-Rauhamaa, and T. Haahtela.** 2000. Mycotoxins in crude building materials from water-damaged buildings. *Appl. Environ. Microbiol.* **66:**1899–1904.
186. **Tutel'ian, V. A., K. R. Dadiani, and N. E. Voitko.** 1992. Changes in the serological indices of the immune status of workers in contact with raw foodstuffs contaminated by mycotoxins. *Gig. Tr. Prof. Zabol.* **8:**18–20. (In Russian.)
187. **Ueno, Y.** 1983. *Trichothecenes—Chemical, Biological and Toxicological Aspects*, p. 135–194. Elsevier, Am-sterdam, The Netherlands.
188. **VerHoeff, A. P., J. H. Van Wijnen, J. S. M. Boleij, B. Brunekreef, E. S. Van Reenen-Hoekstra, and R. A. Samson.** 1990. Enumeration and identification of airborne viable mould propagules in houses. *Allergy* **45:** 275–284.
189. **von Ehrenstein, O. S., E. von Mutius, S. Illi, L. Baumann, O. Bohm, and R. von Kries.** 2000. Reduced risk of hay fever and asthma among children of farmers. *Clin. Exp. Allergy* **30:**187–193.
190. **Von Essen, S., R. A. Robbins, A. B. Thompson, and S. I. Rennard.** 1990. Organic dust toxic syndrome: an acute febrile reaction to organic dust exposure distinct from hypersensitivity pneumonitis. *Clin. Toxicol.* **28:**389–420.
191. **Wady, L., A. Bunte, C. Pehrson, and L. Larsson.** 2003. Use of gas chromatography-mass spectrometry/solid phase microextraction for the identification of MVOCs from moldy building materials. *J. Microbiol. Methods* **52:**325–332.
192. **Wicklow, D. T., and O. Shotwell.** 1983. Intrafungal distribution of aflatoxins among conidia and sclerotia of *Aspergillus flavus* and *Aspergillus parasiticus*. *Can. J. Microbiol.* **29:**1–5.
193. **Womble, S. E., L. E. Burton, L. Kolb, J. R. Girman, G. E. Hadwen, M. Carpenter, and J. F. McCarthy.** 1999. Prevalence and concentrations of culturable airborne fungal spores in 86 office buildings from the building assessment survey and evaluation (BASE) study, p. 261–266. *In* G. Raw, C. Aizlewood, and P. Warren (ed.), *Indoor Air '99. Proceedings of the 8th International Conference on Indoor Air and Climate*, vol. 1. BRE, Garston, Watford, United Kingdom.
194. **World Health Organization.** 1990. *Environmental Health Criteria 105: Selected Mycotoxins: Ochratoxins, Trichothecenes, Ergot*. World Health Organization, Geneva, Switzerland.
195. **Wyllie, T., and L. G. Morehouse.** 1977. *Mycotoxic Fungi, Mycotoxins, Mycotoxicosis: an Encyclopedic Handbook*, vol. I. *Mycotoxic Fungi and Chemistry of Mycotoxins*. Marcel Dekker, Inc., New York, N.Y.
196. **Yang, C.** 1996. Fungal colonization of HVAC fiber-glass air-duct liner in the U.S.A, p. 173–177. *In* K. Ikeda and T. Iwata (ed.), *Indoor Air '96. Proceedings of the 7th International Conference on Indoor Air and Climate*, vol. 3. Indoor Air '96, Institute of Health, Minatoku, Tokyo, Japan.
197. **Yang, C. S.** 1995. Understanding the biology of fungi found indoors, p. 131–137. *In* E. Johanning and C. S. Yang (ed.), *Fungi and Bacteria in Indoor Environments: Health Effects, Detection and Remediation*. Eastern New York Occupational Health Program, Albany.
198. **Yang, C. S., L.-L. Hung, F. A. Lewis, and F. A. Zampiello.** 1993. Airborne fungal populations in non-residential buildings in the United States, p. 219–224. *In* O. Seppänen (ed.), *Indoor Air '93. Proceedings of the Sixth International Conference on Indoor Air Quality and Climate*, vol. 4. *Particles, Microbes, Radon*. Helsinki University of Technology, Espoo, Finland.
199. **Zabel, R. A., and J. J. Morrell.** 1992. *Wood Microbiology*. Academic Press, Inc., San Diego, Calif.

Airborne Bacteria and Endotoxin

PETER S. THORNE AND CAROLINE DUCHAINE

78

INTRODUCTION

Environments can be divided into four main categories according to the concentrations of bacteria and endotoxin contamination present (Table 1). The first category consists of highly contaminated environments, including livestock barns, row crop operations, agricultural transfer and processing settings, and industrial operations that involve handling or processing of organic matter supporting active microbial growth. The second category includes occupational environments with moderate contamination such as food processing facilities, industrial facilities not handling organic waste materials, and domestic environments with significant contamination sources. The third group encompasses relatively low contamination environments such as medical and dental clinics, office environments, homes, and schools. Finally, the fourth group includes environments that are deliberately very low in bioaerosol concentration such as industrial "clean rooms," pharmaceutical manufacturing areas, surgical suites, and some research laboratories. The fourth group is not included in this chapter.

Environments with High Concentrations of Bacteria and Endotoxin

Livestock production and concentrated animal feeding operations (CAFOs) are highly contaminated environments. These include swine, poultry, dairy, and beef cattle facilities, and *Staphylococcus*, *Pseudomonas*, *Bacillus*, *Listeria*, *Enterococcus*, *Nocardia*, and *Lactobacillus* spp. are commonly present (138). Bacteria, endotoxin, and organic dust are the major airborne contaminants in swine CAFOs, and culturable airborne bacterial levels as high as 10^5 CFU/m^3 have been reported (39, 46). The concentration and the species composition of airborne gram-negative bacteria have been studied in poultry, swine, and cattle barns (207). Between 0.02 and 5.2% of the total number of culturable aerobic bacteria were gram negative, with *Enterobacteriaceae*, *Pseudomonadaceae*, and *Neisseriaceae* dominant. In animal houses using straw as bedding material *Enterobacter agglomerans* was predominant, whereas when no bedding material was used, *Escherichia coli* was the most commonly isolated bacterium (207).

Bioaerosol sampling in 24 swine barns yielded culturable bacterial concentrations of 6.1 × 10^5 CFU/m^3 and total microorganisms at 13 × 10^7 cells/m^3 (93). Representative viable bacteria included *Aerococcus*, *E. coli*, *Klebsiella*, *Pseudomonas*, *Rhodococcus*, *Staphylococcus*, *Streptococcus*, *Xanthamonas*, and *Yersinia*. Health problems observed among swine building workers include local and systemic inflammatory responses, cross-shift change in respiratory function, expectoration, and symptoms of cough and dyspnea. Cases of occupational asthma and asthma-like syndrome are common (see "Endotoxin" below) (158).

A major concern regarding agricultural airborne bacteria is the propagation of antibiotic-resistant strains within the environment or to people working in or living near CAFOs because the nontherapeutic use of antibiotics in swine production can select for antibiotic resistance in normal flora and pathogenic bacteria. *Enterococcus* strains collected from a CAFO were tested and analyzed for resistance to erythromycin, clindamycin, virginiamycin, tetracycline, and vancomycin (25). Regardless of bacterial species, 98% of the isolates expressed high-level resistance to at least two antibiotics commonly used in swine production, suggesting that the inhalation of air from these facilities may serve as an exposure pathway for the transfer of multidrug-resistant bacterial pathogens from swine to humans (25). A separate study detected antibiotic-resistant bacteria inside and downwind of the CAFOs including strains with resistance to ampicillin, erythromycin, oxytetracycline, penicillin, tetracycline, and tylosin within airborne *Staphylococcus aureus*, *Salmonella* spp., and fecal coliforms (69).

Most of the exposure control efforts for livestock facilities have focused on personal respiratory protection. Environmental exposure control methods are not common. In swine buildings vegetable oil sprinkling has been tested to decrease the overall airborne dust levels with success (122). This method was shown to reduce airway inflammation in exposed workers (162). Added oil in animal feed was tested to reduce airborne dust with modest improvement (108).

Dairy barns can also be highly contaminated with bacteria. Hay and straw used for bedding are very important bioaerosol sources. More recently, shredded newspaper or sawdust has replaced straw in some regions. Hypersensitivity pneumonitis, or farmer's lung disease, can develop in dairy farmers and is caused by, among other substances, a subset of four thermophilic mycelial bacteria: *Saccharopolyspora rectivirgula*, *Saccharomonospora viridis*, *Thermoactinomyces sacchari*, and *Thermoactinomyces vulgaris* growing in moldy hay and straw (130). Culturable *S. rectivirgula* can be found at

TABLE 1 Categories of environmental contamination for bacteria and endotoxin

Environments with high concentrations of bacteria and endotoxin
Concentrated animal feeding operations (swine, poultry, cattle)
Traditional livestock barns
Row crop harvesting
Grain elevators
Animal feed mills
Agricultural transfer and processing settings
Industrial composting operations
Solid waste transport and handling
Wastewater treatment plants
Other industries handling or processing organic matter
Environments with moderate concentrations of bacteria and endotoxin
Cotton textile industry
Vegetable- and seed-processing facilities
Metals machining with MWF
Forest products industry
Slaughterhouses (abattoirs)
Rodent vivaria
Domestic environments with significant contamination sources
Industrial operations not handling organic waste materials
Ambient air (in vicinity of bioaerosol sources)
Environments with low concentrations of bacteria and endotoxin
Offices
Domiciles
Day care centers and schools
Dental clinics
Hospitals and medical clinics
Outdoor air
Environments with very low concentrations of bacteria and endotoxin
Pharmaceutical manufacturing facilities
Industrial clean rooms
Surgical suites
Specialized research laboratories

concentrations in excess of 10^3 CFU/m^3. Factors such as hay-packing techniques (142) and hay preservatives (47) have been studied for their effects on the colonization of hay by thermophilic actinomycetes. Large cylindrical bales seem to support the growth of those bacteria better, while bacterial hay preservatives are not efficient in reducing the airborne concentration of bacterial aerosols. Barn ventilation is inadequate for reducing the airborne load of actinomycete spores (142). Recent findings in Finland and France have proposed that molds play a major role in farmer's lung disease but that *S. rectivirgula*, which is classically incriminated, does not (57, 152). These findings have not been confirmed in other countries, where *S. rectivirgula* is still considered the main etiologic agent for farmer's lung disease.

Manure spreading can produce bacterial aerosols which can be propagated to the surrounding areas. Bacterial aerosols were quantified 2 and 20 m downwind of manure spreading, and culturable bacteria were found in air samples collected during the biosolid application process (20, 21). However, most of the bacterial species present in the biosolid were not detected in air samples. Die-off of the bacterial cells could be responsible for those observations.

Workers in grain elevators and animal feed production are exposed to very high levels of bioaerosols, and bacterial concentrations up to 10^9 CFU/m^3 have been reported in silos (53, 160). Bacterial species in these environments include *Bacillus* spp. and gram-negative organisms (*Pseudomonas* spp., *Alcaligenes* spp., *Citrobacter* spp. and *Klebsiella* spp.). Thermophilic actinomycetes have been isolated from grain samples but were not quantified in the aerosols (53). Hypersensitivity cases among corn workers have been described, but mold seems to be the more important etiological agent (117). There are many opportunities for exposure reduction in this industry, with improvement of local ventilation systems proving to be helpful when the source of bacterial aerosols is fairly simple to control, as with truck or rail car unloading or at feed-bagging sites.

Residents and workers at sites where composting is practiced frequently express concern about bioaerosol emissions. Exposure monitoring has demonstrated that when the compost pile was being turned, shredded, or loaded, bioaerosols were generated with concentrations approximately 2 logs higher than background levels and the organisms were detected 40 m downwind of the facility (157). Release of bioaerosols from compost piles was evaluated at a green waste composting facility. Experiments conducted on the surface of static piles generated specific bioaerosol emission rates at ground level, between 13 and 22 × 10^3 CFU/m^2 of mesophilic actinomycetes (172). The bioaerosols emitted from a suburban yard waste composting facility were measured, and high bacterial concentrations were detected (up to 7 × 10^4 CFU/m^3). Amounts of total bacteria, gram-positive bacteria, gram-negative bacteria, and actinomycetes demonstrated decreasing concentration with distance from the compost pile and higher concentrations downwind than upwind (80). However, the results suggest that, despite the high exposure of workers, there was not an elevated risk for residents living in the vicinity of composting facilities because the concentrations recorded downwind of the operational area were not significantly different from background levels (157).

Workers in the solid waste industry are exposed to bioaerosols (44, 192) leading to gastrointestinal and respiratory illnesses (165, 166, 174, 202). Studies have indicated a high prevalence (14% of workers with cough and 17% with nasal irritation) of respiratory illnesses and inflammation, and these work-related illnesses are related to bacterial exposure of up to 10^5 CFU/m^3 (78). Adverse health effects including organic dust toxic syndrome were described in workers in the household waste recycling industry during manual sorting of industrial and household waste (41). Although bacterial aerosols have been described in this type of environment, emphasis has been put into airborne and dustborne endotoxin and glucan exposure (44, 200, 201, 203). A Canadian study described bacterial aerosols in household-waste-recycling plants (100). This study showed that, in summer, the average concentrations of total bacteria were greater than 10^4 CFU/m^3 in the receiving, sorting, and shipping areas of some plants.

In municipal wastewater treatment plants the concentrations of airborne bacteria typically range from 10^2 to 10^5 CFU/m^3 depending on the area of the plant, with sludge dewatering areas reaching the highest levels. Airborne bacterial concentrations of 10^2 CFU/m^3 have been reported for gram-positive bacteria, gram-negative bacteria, and thermophilic actinomycetes (13, 99, 101, 137). Commonly detected taxa include *Enterobacter cloacae*, *Acinetobacter calcoaceticus*, *Pseudomonas* spp., *Stenotrophomonas maltophilia*, *Thermoactinomyces thalpophilus*, and *T. vulgaris*.

Different water treatment methods were compared with regards to bacterial aerosol releases. They demonstrated that mechanical aeration of sludge generates substantially higher bacterial aerosols than diffused aeration systems, which pose a lesser hazard for human health (17).

Pathogenic bacteria were among airborne bacteria found in Polish wastewater treatment plants (137). Gram-positive bacteria prevailed, with corynebacteria accounting for over one-half the culturable mesophilic bacteria. *Staphylococcus* spp., *Micrococcus* spp., and *Bacillus* spp. were also prevalent. Among the gram-negative genera, *Enterobacter*, *Acinetobacter*, *Pseudomonas*, and *Stenotrophomonas* dominated.

Reuse of partially treated domestic wastewater for irrigation is a common worldwide practice in arid regions (123). Exposure to potentially pathogenic bacteria is a concern as bacteria from the wastewater become airborne (126).

Environments with Moderate Concentrations of Bacteria and Endotoxin

The cotton textile, vegetable- and seed-processing, machining, and forest products industries and slaughterhouse facilities are environments with moderate concentrations of bacteria and endotoxin. Outdoor environments also may contain high airborne concentrations of bacterial contaminants resulting from their proximity to bacterium-laden industries or activities. Cotton and wool workers are exposed to airborne gram-negative bacteria and endotoxin (154), but dust control measures are effective for reducing exposures to bacteria and endotoxin as evidenced by improved worker health and compliance with exposure guidelines.

The relationship between bioaerosol exposure in the potato starch industry and work-related respiratory symptoms has been described (79), and it seems to be linked to dust exposure. However, bacterial exposures have rarely been described. Airborne microorganisms have been investigated in the sugar beet processing industry and workers may be exposed to high culturable microorganisms, mostly gram-positive bacteria (60). Cases of organic dust toxic syndrome were reported in agricultural seed processing, and exposures to bacteria, mold spores, and endotoxin were documented (170). High-risk tasks include dumping, mixing, and sieving. Local exhaust ventilation systems can improve the air quality in those plants if properly designed and operated.

Soluble oil metalworking fluids (MWF) used in machining plants are susceptible to bacterial colonization. Bacterial contamination is a problem for the industry because it changes the biophysical properties of the MWF, reducing the system efficiency during metal boring, hobbing, tapping, grinding, and honing operations (187). Aerosols are generated during metal machining and contain MWF, tramp oils, additives, emulsifiers, and bacteria that colonize the fluids. Although biocides are used in MWF, the bacterial diversity of contaminants is complex and *Pseudomonas* spp., *Micrococcus* spp., and *Mycobacterium* spp. are important representatives (109, 173, 180). Other taxa have also been described in MWF bioaerosols including *E. agglomerans*, *Streptococcus* spp., *Staphylococcus* spp., *Corynebacterium* spp., *Streptomyces* spp., other gram-positive bacteria, and molds and yeasts (74, 109). The presence of mycobacteria in MWF has been associated with hypersensitivity pneumonitis outbreaks (97, 151, 164), and recent animal studies have clearly shown that *Mycobacterium immunogenum* in MWF can induce the lung pathology and T-lymphocyte profiles consistent with hypersensitivity pneumonitis (73, 177).

The forest products industry is often confronted with health and safety issues surrounding the control of bioaerosol exposures. These bioaerosols are comprised mainly of wood-decaying molds, although wood dust, bacteria, respiratory allergens, and carcinogens from the wood are also found (37). Western red cedar asthma, sequoiosis, and maple bark disease are respiratory illnesses associated with wood industry bioaerosols. The link between airborne bacteria and diseases in the wood industry is less clear, and molds are the primary microbial agents of concern, but microbiological air sampling performed in sawmills demonstrated bacterial levels of up to 10^5 CFU/m^3 (48, 54, 55, 124). Debarking operations were the most highly contaminated sites with molds, bacteria, and endotoxins (48, 49). More than 34 species or genera of bacteria were identified in the Polish coniferous wood industry (52). *Bacillus* spp. and the *Pseudomonadaceae* family are the predominant bacteria found in sawmills (124).

Paper production processes involve handling and transformation of wood products such as wood chips or timber logs or recycled paper fiber. Those products may have microbial contamination. When dispersed into the air during production, bacteria and molds may be inhaled by workers and induce airway inflammation (136), and endotoxins are important contaminants of this workplace (156). Bacterial bioaerosol concentrations of 310 CFU/m^3 have been reported (136), with species from the *Enterobacteriaceae* family being the most predominant taxa.

As with livestock production facilities, slaughterhouses have significant bacterial aerosol contamination, some of which may be pathogenic. The main concerns in this setting are minimizing contamination of the product and protecting the health of the workers. Microbial cross-contamination occurs by aerosol or large airborne droplets and airborne particulates and feathers (4). Levels of contamination in the back-splitting areas (1,800 CFU/m^3) were generally higher than that in the weighing areas (1,070 CFU/m^3) and showed that airborne bacteria have an important role in carcass contamination (141).

Jericho et al. evaluated the bacterial aerosols generated at the high-line speed cattle abattoir operations and studied the importance of air sampling in the Hazard Analysis and Critical Control Point plan (HACCP). They concluded that the control of aerosols should be treated as a critical control point (86). Airborne bacteria in chilling facilities of poultry abattoirs have also been studied (63). Bioaerosols were extremely high, and *Micrococcaceae* and gram-positive irregular rods were detected in addition to streptococci and members of the *Enterobacteriaceae* and *Micrococcaceae*. The authors concluded that cross-contamination may occur in the evaporative chilling process.

Poultry slaughterhouse workers were studied for respiratory health and exposure to bioaerosols including airborne bacteria. Up to 10^6 culturable airborne bacteria, mainly coagulase-negative staphylococcal strains, were detected. An overshift increase in respiratory symptoms was found but did not correlate with the individual exposure to bioaerosols (75). The presence of aerosolized *Salmonella* spp. and *E. coli* inside and outside commercial layer facilities was studied, and the results showed that those bacteria are present inside such facilities and outside up to 13 m from the ventilation fans (34). The role of live birds on bioaerosol generation has been demonstrated, and up to 10^5 CFU/m^3

were detected in areas where fecal matter was removed using a disposal system featuring a central opening in the floor through which the matter automatically dropped to an open-air lower level. Marginally lower bioaerosol concentrations were detected when conveyor belts were used (193).

In outdoor settings, highly contaminated substrates such as pig manure, wastewater, and compost facilities may introduce vast quantities of microorganisms into the air, which can then spread over a wide area. It has been established that the air in rural areas where swine CAFOs are present may have high concentrations of odors and vapors from microbial degradation of manure and airborne microbes (81). Other factors such as the presence of various contamination sources and the distance from them, relative humidity, wind, and human activities also contribute to the dispersion of contaminants and may influence the outdoor airborne microbial burden (81).

Cooling towers have been shown to be responsible for Legionnaires' disease outbreaks in France in 2003 and 2004, and several people who were infected by the responsible cooling tower were 1 to 3 km away (120). The normal microbial content of outdoor air also varies in nature and concentration depending on the time of day, the season, the temperature, and other factors. A background range, within which variations in microbial content would be considered normal, may be established in a given geographical area. This outdoor contamination by bacteria often serves as a comparison point in any analysis of indoor airborne bacteria.

Environments with Low Concentrations of Bacteria and Endotoxin

Day care centers and schools are grouped among low contamination environments, but they are important settings associated with children's exposure to bioaerosols because children spend large amounts of time within those facilities and often harbor more bacteria than adults. Several studies have looked at mold exposure in schools after water damage, but limited information is available regarding bacterial exposure in schools. Generally school classrooms are contaminated with airborne bacteria, but their concentration rarely exceeds a few hundred CFU per cubic meter (88). Fox et al. demonstrated that occupied classrooms have a higher level of bacterial markers (lipopolysaccharides [LPS] and muramic acid) and dust than unoccupied rooms and that the release of bacteria from children could be responsible for this phenomenon (61). It has also been shown that naturally ventilated classrooms generally have higher numbers of airborne bacteria than mechanically ventilated rooms (11). In this study of 39 schools, staphylococci, micrococci, *Bacillus* spp., and corynebacteria were detected (11). Personal exposure of school teachers to bioaerosols was assessed, and it was shown that when the workplace is heavily crowded, the airborne bacterial load is higher (189). Ventilation and air intake problems have been studied in public schools, and in one school occlusion of air intakes negatively pressurized the building and caused sump vents to become air intakes (175). This poor air handling system maintenance resulted in contamination of classrooms by *Bacillus* spp. with up to 760 CFU/m^3 detected.

Office environments have been studied to assess the influence of indoor environmental quality on physical and psychological health impacts. Bioaerosols have been evaluated in commercial office buildings, and the factors influencing indoor environmental quality have been assessed (145), including culturable and total bioaerosols. Total bioaerosols (including bacteria) reached up to 10^4 organisms/m^3, and little difference between indoor and outdoor concentrations was reported for culturable bacteria.

The air of dental clinics may be contaminated by airborne microorganisms (33, 68, 90) of various sources such as from the mouth of patients (134) and from dental unit water lines (9, 14). The use of high-speed handpieces (e.g., ultrasonic scaler and rotating devices) generates aerosols, and dental unit water lines are susceptible to the formation of biofilms (8, 10, 121). Several pathogens and opportunistic pathogens are often detected such as *Legionella* spp., *Mycobacterium* spp. and *Pseudomonas* spp. (10, 51, 127, 144).

Airborne bacterial concentrations in homes have been described in a few studies (36, 102, 150, 163) and Sessa et al. reported that airborne bacterial concentrations were two- to fivefold higher in the presence of people than in empty homes (163). They suggested consideration of occupancy as a step to identify acceptable levels for bioaerosols in common indoor environments. Actinomycetes and their indoor/outdoor ratio were evaluated across seasons in a study by Lee et al. (102). Actinomycetes were detected indoors with concentrations that differed only slightly from outdoor levels. *Streptomyces* spp. have been proposed as an indicator of water damage in houses (150), and the basements of 41 complaint and noncomplaint houses that were compared for their bioaerosol content have been studied (36). Culturable bacteria were detected at 970 CFU/m^3 on the main floor and 1,100 CFU/m^3 in the basements, suggesting that the basement represented an important amplification site for complaint homes.

Humidifiers that hold stagnant water quickly become contaminated by bacterial biofilms. Although the infection risk is associated with exposure to legionella-contaminated bioaerosols (89, 208), actinomycetes are also sometimes incriminated. Water droplets released by humidifiers can contain bacteria, and the size of the droplets favors deposition in the lung. In an experimental study, it was shown that humidifiers released bacteria when the humidifier reservoir was contaminated with bacteria (148). Although several papers report cases of lung disease caused by humidifiers, very few draw conclusions about the etiological agents.

ENDOTOXIN

Endotoxins are a family of surface glycolipids of gram-negative bacteria that are ubiquitous in the environment. They are regularly found in indoor and outdoor air, water, soil, and food. They are amphipathic molecules bearing a hydrophobic moiety comprised typically of six lipid chains connected via lipid A to a polysaccharide moiety that protrudes from the microbial membrane into the extracellular space. The lipid A component of the molecule is conserved across bacterial species and carries the toxic potency. Throughout evolution, a wide range of vertebrate and nonvertebrate species have developed highly sensitive responses to endotoxin that promote host defense against invading gram-negative bacteria. Recognition of endotoxin leads to the induction of inflammation, targeting of gram-negative bacteria for elimination, and the subsequent clearance of the endotoxin (3, 15, 198). Although these responses guard against colonization and infection, they can also induce pathology when exposure or subsequent responses to endotoxin are inadequately controlled.

For humans, exposure to endotoxin occurs on a daily basis from noninfectious and infectious gram-negative bacteria in both occupational and nonoccupational environments. Endotoxin is a component of particulate matter in ambient air (22, 77, 118), in air and settled dust in homes (16, 19, 114, 128, 132, 183), in office buildings (145), and in a variety of industrial (49, 169, 180, 203) and agricultural (98, 139, 168) settings. Inhalation of endotoxin has been linked to the induction and exacerbation of asthma and other airway diseases (114, 167, 183). High levels of exposure found in a variety of occupational environments lead to a high prevalence of respiratory and systemic symptoms (41, 42).

The hazards of inhaled endotoxin first became apparent from studies of lung disease among workers exposed to vegetable and cotton dust (24, 131), and this knowledge was later refined with measured endotoxin exposures (23, 31, 154). Studies in the 1990s characterized pulmonary responses, including lung inflammation and cytokine up-regulation, upon inhalation of endotoxin and organic dusts containing endotoxin in humans (35, 85, 95, 113, 115) and experimental animals (35, 84, 161, 184).

Endotoxin Structure and Recognition

Each gram-negative bacterium has approximately 10^6 endotoxin molecules (135). When released from bacteria through membrane fragmentation or blebbing, the lipid A regions are sequestered within aggregates. The lipid A regions are covalently bound to a carbohydrate chain of variable length and composition. Endotoxins exist as LPS or lipooligosaccharides depending on the genus. A variety of mammalian proteins confer innate immune recognition of gram-negative bacteria by binding to sites within the conserved inner core region and lipid A regions. The inner core region usually consists of two monosaccharides, 2-keto-3-deoxyotulosonic acid and L- or D-glycero-D-mannoheptose. The lipid A portion of endotoxin consists of a diphosphorylated β-1,6-D-glucosamine disaccharide linked to 3-hydroxy fatty acids that are further substituted by nonhydroxylated fatty acids. The fatty acids of the lipid A region are the key determinants of the potency of endotoxin (3). Hexa-acylated endotoxins including LPS of *E. coli* and lipooligosaccharides of *Neisseria meningitidis* are capable of inducing cytokine up-regulation at picomolar concentrations.

Specific endotoxin recognition and signaling requires ordered interactions involving several extracellular and cell surface host proteins, including LPS-binding protein (LBP), CD14, MD-2, and Toll-like receptor (TLR) 4. Murine exposure studies suggest that TLR4-dependent responses to endotoxin in the resting airway are mediated primarily by alveolar macrophages with less involvement of airway epithelia. Human and mouse studies have shown that aggregates of endotoxin bind to LBP, are disaggregated, and then attach to CD14 as individual molecules. The CD14:endotoxin complex then interacts with MD-2, sending a signal via TLR4 leading to the up-regulation of NF-κB via either an MyD88-dependent or MyD88-independent pathway (15, 70, 71, 76, 83, 119, 191, 204). This signal then induces production of proinflammatory cytokines. Thus, the actions of LBP, CD14, MD-2, and TLR4 convert each environmental endotoxin particle to many activating protein:endotoxin complexes. MD-2 is likely limited in the resting airway and arises only when myeloid cells are recruited to the lung and airway epithelia are up-regulated leading to amplified responses to endotoxin. This organization of unresponsiveness to low levels of endotoxin in the resting airway may be advantageous, promoting host responses to invading bacteria while minimizing inflammatory reactions to more incidental exposure to endotoxin-bearing environmental pollutants such as ambient air pollution. Tissue sites that routinely come into contact with free endotoxin aggregates, such as the gut, are hyporesponsive to endotoxin apparently due to low expression of MD-2 and TLR4 (1, 125).

Endotoxin Inhalation and Lung Inflammation

Acute inhalation exposure to purified endotoxin or to particulate matter containing endotoxin can induce airflow obstruction, airway hyperreactivity, inflammation, and systemic symptoms including fever, chills, myalgia, and malaise. The inflammation is marked by the appearance of neutrophils and proinflammatory cytokines in lung lavage fluid (32, 85, 111) and includes clinical conditions such as asthma, asthma-like syndrome, organic dust toxic syndrome, and byssinosis (158). Chronic exposure may induce changes resembling allergic asthma and may also contribute to chronic obstructive pulmonary disease (114, 159). Endotoxin inhalation in mice and guinea pigs induces similar host responses and changes (65, 84, 107, 188) and thus these animals provide suitable models for characterization of the molecular and cellular determinants of lung responses to endotoxin. Studies have provided time course (35) and dose-response relationships (184) for induction of inflammation after endotoxin inhalation and have demonstrated the roles of LBP, CD14, and TLR4 in the pathogenesis of endotoxin-induced disease (18, 87).

Variability of Endotoxin Responsiveness of Airway and Lung

Strain variability to inhaled endotoxin in mice and reproducible differential responsiveness between humans provide evidence of genetic determinants of individual susceptibility to inflammation induced by endotoxin. Thorne et al. developed dose-response curves for lung inflammation induced by endotoxin inhalation in a responsive strain of mice (C3HeBFeJ) and a hyporesponsive strain (C3H/HeJ [$Tlr4^{Lps\text{-}d}$]) and showed that the hyporesponsive mice had 1,000-fold lower neutrophil recruitment to the lung for the same exposure (184). These C3H/HeJ ($Tlr4^{Lps\text{-}d}$) mice have a spontaneous mutation in the LPS response locus (the TLR4 gene, $Tlr4^{Lps\text{-}d}$) that effects binding of endotoxin:MD2 to TLR4. In human studies, Kline et al. demonstrated that some healthy subjects exposed to increasing inhaled doses of endotoxin had marked hyper- or hyporesponsiveness (95) and that this was associated with a polymorphism in TLR4 (Asp299Gly) (6). Polymorphisms in control pathways for other host molecules that also affect endotoxin responsiveness likely exist.

Diseases Associated with Endotoxin Exposure

Exposure to inhaled endotoxin is associated with several distinct clinical syndromes including occupational and nonoccupational asthma, asthma-like syndrome, organic dust toxic syndrome, chronic bronchitis, and increased severity of hypersensitivity pneumonitis.

Asthma

Asthma is a chronic pulmonary disease often punctuated by episodic exacerbations. It is characterized by eosinophilic inflammation and hyperresponsive airways. Dramatic increases in the prevalence and morbidity of asthma were observed over the past 25 years, especially in developed nations with higher standards of living (94). It is clear that exposures to common allergens are an important factor in asthma induction. However, exposure to endotoxin is also important, although its effects in the induction and progression of asthma appear to be complex and may depend on the dose and timing of exposure as well as the particular health outcome examined (66, 67, 106, 112, 167, 183, 190, 194, 196).

Asthma-Like Syndrome

Asthma-like syndrome is a nonallergic respiratory disease that occurs acutely from inhalation of organic dust containing endotoxin. This syndrome occurs among livestock confinement workers, grain and cotton textile workers, sewage treatment plant workers, and compost workers. The syndrome has been called byssinosis when it occurs among cotton workers, bagassosis among sugar cane workers, and porcinosis among swine workers. It includes neutrophilic lung infiltrates and production of inflammatory cytokines leading to cough, chest tightness, wheezing, and shortness of breath (158). Lung function declines across a workshift are common. Exposures to endotoxin above 200 endotoxin units (EU)/m3 are generally needed to induce this disease (158, 206). Byssinosis is a similar syndrome associated with exposures in the cotton textiles industry (91, 154). Its features include acute bronchoconstriction, chest tightness, and dyspnea, which are worse on returning to work after the weekend. Long-term exposure leads to chronic bronchitis and airflow limitations. Much of the effect is attributable to endotoxin exposure, although other microbial cell wall and plant components play a role (91, 195).

Organic Dust Toxic Syndrome

Organic dust toxic syndrome (ODTS) is an acute inflammatory condition of the small bronchioles and alveoli associated with high exposure to endotoxin (>1,000 EU/m^3), fungi, or both. Symptoms include fever, malaise, dyspnea, and fatigue that typically peak 4 to 8 h after exposure and resolve within 48 h (158). A European study of pig farmers found a lifetime prevalence of ODTS of 23% (140). This was lower than that found in an earlier U.S. study of swine farmers that reported a 34% lifetime prevalence (40).

Hypersensitivity Pneumonitis

Hypersensitivity pneumonitis (HP), also known as extrinsic allergic alveolitis, is an allergic disease associated with thermophilic bacteria, fungi, and bird antigens (59). Cases of HP have arisen in recent years among metal machinists exposed to water-based MWF treated with certain biocides. Investigation of these outbreaks suggested a role for a particular organism, M. *immunogenum*, found growing in the MWF. A recent laboratory study has shown that M. *immunogenum* in MWF can induce HP and that endotoxin coexposure with M. *immunogenum* augments the severity of the HP (177).

Occupational exposures to endotoxin have been evaluated extensively beginning with the cotton industry and the livestock industry in the early 1980s (39, 154). Table 2 provides a summary of endotoxin exposure levels measured in a variety of occupational environments with a focus on larger studies that used current methodology. The table demonstrates that concentrations of airborne endotoxin are typically highest for livestock confinement facilities reaching

TABLE 2 Endotoxin concentrations in occupational environments

Setting	Size fraction	*n*	Geometric mean endotoxin concn (EU/m^3)	Reference
Swine operations (Iowa)	Area CFC[a] filter	81	8,290	186
Swine operations (Iowa)				
Hoop barns	Area inhalable filter	30	3,250	P. S. Thorne, unpublished data
Conventional confinement barns	Area inhalable filter	40	3,100	Thorne, unpublished
Swine operations (Quebec)	Area impinger	21	4,385	49
	Area CFC filter	21	3,927	
Swine operations (The Netherlands)	Personal inhalable filter	350	920	139
Poultry operations (Iowa)	Area CFC filter	81	1,340	186
Dairy operations (Wisconsin)	Personal inhalable filter	194	647	98
	Area respirable filter	216	17	
Soybean harvesting (closed tractor cab)	Personal CFC filter	32	56	153
Agricultural seed processing industry	Personal inhalable filter	100	1,800	170
Grain elevators (Iowa)	Personal CFC filter	410	2,860	160
	Personal respirable filter	410	83.2	
Animal feed manufacturers (The Netherlands)	Personal inhalable filter	530	12–285	168
	Area inhalable filter	79	19	
Potato processing (The Netherlands)	Personal inhalable filter	195	9–102	206
	Area inhalable filter	68	1–4,000	
Sawmills	Area impinger	59	740	49
	Area CFC filter	62	188	
Domestic organic waste composting facilities	Personal inhalable filter	182	205	203
Domestic waste collection	Personal inhalable filter	176	40	203
Wastewater treatment plants	Personal inhalable filter	460	27	169
Wastewater treatment plants	Area CFC filter	104	83	101
Engine machining plant	Area CFC filter	48	54	180

[a]CFC, closed-face cassette.

geometric mean levels ranging from 920 to 8,290 EU/m^3 for swine, 1,340 EU/m^3 for poultry, and 650 EU/m^3 for dairy. Personal inhalable-fraction endotoxin exposures among U.S. grain elevator workers had a geometric mean concentration of 2,860 EU/m^3, and this was 10-fold higher than the mean level in Dutch animal feed manufacturing facilities. A study of the agricultural seed industry initiated by cases of ODTS measured a geometric mean of 1,800 EU/m^3 with some measurements above 100,000 EU/m^3. Soybean harvesting, potato processing, organic waste composting, and sawmills ranged from about 100 to 1,000 EU/m^3. Wastewater treatment plants and metal machining plants using water-based MWF fell in the range of 27 to 83 EU/m^3. A large proportion of the workers in these settings likely face daily exposures exceeding the no-effect level of 50 EU/m^3 recommended by the Dutch Expert Committee on Occupational Standards (50).

Although it has been over 40 years since the first recognition that house dust contains endotoxin (133), a greater appreciation that this source of endotoxin carried a health risk came 30 years later with the observation of Michel et al. (114) that the severity of asthma and the use of asthma medication in clinic patients was associated with mattress and floor dust endotoxin concentrations. A number of epidemiologic studies have drawn attention to endotoxin exposure in domiciles and its role in asthma. Table 3 lists mean values of airborne and settled dust endotoxin concentration in homes and outdoor air. Michel et al. (114) studied endotoxin in settled dust in the homes of patients with stable chronic rhinitis or asthma and noted a significant

TABLE 3 Endotoxin concentrations in ambient, domestic, and office environments

Setting	Size fraction	n	Mean concn	Reference
Ambient air				
Southern California (13 locations)	PM10 filter	99 samples	0.44 EU/m^3	118
Munich, Germany	PM2.5 filter	158 samples	0.015 EU/m^3	22
Hettstedt and Zerbst, Germany	PM10 filter	42 samples	9.0 EU/m^3	77
Outside Iowa swine barns (all seasons)	Inhalable filter	7 hog barns		Thorne, unpublished data
33 m upwind		20 samples	4.9 EU/m^3	
33 m downwind		20 samples	136 EU/m^3	
165 m downwind		20 samples	30 EU/m^3	
Indoor environments—air sampling				
Office buildings	Area inhalable filter	24 samples	1.1 EU/m^3	145
Rural Iowa households (LEIP study)	Play area inhalable filter	85 homes, 326 samples	5.82 EU/m^3	179
Houses of Harvard faculty, staff, students	Bedroom filter	15 homes, 142 samples	0.64 EU/m^3	128
Indoor environments—surface sampling				
U.S. National Survey of Endotoxin in Housing				
	Settled dust	831 homes		183
	Bedding	470 samples	20.6 EU/mg	
	Bedroom floor	588 samples	37.7 EU/mg	
	Living room floor	489 samples	71.1 EU/mg	
	Living room sofa	468 samples	45.3 EU/mg	
	Kitchen	454 samples	84.4 EU/mg	
New York City	Settled dust, bedroom floor	301 samples	75.9 EU/mg	132
Boston metropolitan area	Settled dust			129
	Family room floor	404 samples	79 EU/mg	
	Bedroom floor	323 samples	63 EU/mg	
	Kitchen floor	245 samples	100 EU/mg	
Munich and Leipzig households (LISA study)	Settled dust, bedding	1,884 samples	2.9 EU/mg	16
Bavarian, Austrian, and Swiss farmhouses (ALEX study)	Settled dust, bedding	319 samples	37.8 EU/mg	19
Dutch households	Settled dust			201
	Living room floor	99 samples	6.0 EU/mg	
	Kitchen floor	97 samples	13.5 EU/mg	

association of mattress and floor dust endotoxin concentration, but not mite allergen concentration, with the severity of asthma and asthma medication use. Studies from Bavaria (19), Canada (58), and Sweden (96) have provided evidence that growing up on a family farm leaves one less likely to develop allergies and allergic asthma. This protective effect (the hygiene hypothesis) is supported by studies demonstrating a shift away from an allergic phenotype with early life exposure to endotoxin (67). However, studies from Australia (45), Europe (103), New Zealand (199), and the United States (27, 110) have found the same or higher asthma rates (i.e., no protective effect) between farm and nonfarm children in one or more study groups. The role of endotoxin as a risk factor for asthma and wheezing has been investigated in urban cohort studies (16, 105, 183) with the finding that a higher level of endotoxin in house dust was associated with a higher prevalence of wheezing.

A birth cohort study of allergy in Munich and Leipzig, Germany, included measurement of endotoxin in the mattress dust from the mothers' beds (16, 64). This study yielded a geometric mean endotoxin concentration of 2.9 EU/mg with a range of 0.057 to 1,290 EU/mg (Table 3). Exposure to bedding endotoxin in the first few months of life was associated with a reduction in atopic eczema but was a risk factor for respiratory infection, cough with infection, and wheezing at 6 months of age. When the babies were 1 year old, the protective effect was no longer observed, but the elevated risk of wheezing remained. Another European study of domestic endotoxin was conducted among farm children born in Bavaria, Austria, and Switzerland (19). This study found that the presence of bed endotoxin was associated with reduced risk of hay fever, atopic sensitization, and atopic wheezing among the children. Asthma and wheezing among nonallergic subjects were positively related to endotoxin exposure.

A recent study analyzed endotoxin in 2,542 house dust samples from five locations within homes that were systematically selected to represent the demographics and housing stock of the entire United States (183). Information on health and living conditions was assessed through questionnaires and on-site evaluation of 2,456 residents of 831 homes. Significant relationships were observed between increasing endotoxin levels and diagnosed asthma, asthma symptoms in the past year, current use of asthma medications, and wheezing among residents of the homes. Those with elevated bedding and bedroom floor endotoxin levels were nearly three times as likely to have experienced recent asthma symptoms. Allergic subjects with higher endotoxin exposure were no more likely to have diagnosed asthma or asthma symptoms than nonallergic subjects.

Recognized determinants of endotoxin in homes include indoor sources such as dogs, pests, contaminated humidifiers, kitchen compost bins, outdoor air and water infiltration, or flooding (118, 129, 182, 201). Samples taken in flood-damaged homes in New Orleans after Hurricane Katrina demonstrated endotoxin levels 10- to 50-fold higher than those normally observed in homes (26). During remediation of homes, exposure levels rose a further 10-fold. Wouters et al. (201) analyzed dust from 100 Dutch households with and without carpeting and found that the living room floor endotoxin level was threefold higher for households that had a biopail for compostable waste in their kitchen than for those who did not. Further, households with carpeting had a threefold-higher endotoxin load (in endotoxin units per square meter of floor area) than those without.

METHODS FOR THE ASSESSMENT OF ENDOTOXIN

Limulus Amebocyte Lysate Assays

Assessment of exposures to endotoxin entails sample collection, extraction from the matrix, and analysis of endotoxin (176). Airborne concentrations of endotoxin are usually determined from samples collected onto filter media or into liquid impinger media. Settled dust or surface samples are most often collected using vacuum sampling onto filter media or electrostatic wipes (185). Extraction is performed in an aqueous medium with shaking or ultrasonifying, with or without Tween 20, usually without added buffers. Comparisons of methods for endotoxin extraction protocols have been extensively reviewed (41, 181).

Analysis for endotoxin most commonly employs the *Limulus* amebocyte lysate (LAL) assay (186). The LAL assay is based on the extreme sensitivity of an enzymatic cascade in amebocytes taken from the hemolymph of horseshoe crabs (*Limulus polyphemus* or *Tachypleus tridentatus*) and originates from studies of bacterial infections in horseshoe crabs (7). The LAL contains a mixture of proenzymes that are activated by endotoxin in a cascading set of reactions via factors C and B (82). Kinetic assays based on the LAL measure spectral changes in microplate wells based on endotoxin activation of enzymatic cleavage of a substrate (measured at 405 nm) or the development of turbidity (measured at 340 nm). Recently, a commercial assay based on recombinant factor C that yields a fluorescent product of the reaction (excitation, 380 nm; detection, 440 nm) has been introduced.

Endotoxin exposure assessment is influenced by the method of sampling, sample storage conditions, the extraction method employed, the assay reagents, and methodology (43, 49, 116, 186, 205). For air sampling in occupational or domestic settings, filter sampling of endotoxin-laden particulate matter is typically used. Binder-free glass fiber filters are optimal in terms of extractability of the endotoxin and the capacity for dust collection but have less stable gravimetric properties due to loss of glass fibers on O-rings within the sampler. Polycarbonate filter media can also be used, but cellulose ester and Teflon filters are problematic due to their retention of endotoxin through the extraction process. Freezing of air and bulk samples prior to analysis is common and generally yields higher endotoxin concentrations. Freezing of extracts is less desirable and requires extensive vortexing to get the endotoxin back into solution. Sample extraction has been evaluated extensively with conflicting views expressed as to whether it is best to extract samples with pyrogen-free (pf) water, pf water with 0.05% Tween 20, or with added buffers (43, 116, 176, 186, 205). Recent evaluations suggest that pf water is suitable for analysis of endotoxin from many matrices but extraction studies with spiking should be conducted with collected samples at the initiation of any study.

Air sampling has been the mainstay of endotoxin exposure assessment in occupational settings, and because endotoxin exposures are most strongly associated with airways disease, the inhalable fraction of the particulate matter is most often sampled. The Button and IOM inhalable dust samplers (SKC Inc., Eighty Four, Pa.) are suitable in most environments (2, 38, 92, 104). Several studies have performed both inhalable and respirable endotoxin sampling (Table 2) and observed respirable endotoxin concentrations 35- to 40-fold lower than inhalable endotoxin concentrations (98, 160).

Surface sampling of endotoxin has been performed extensively in epidemiological studies of asthma (Table 3) (16, 19, 129, 132, 183, 201). While short-term air sampling may be sufficient for occupational health studies, it is argued that measurement of endotoxin in settled dust may better represent long-term exposure than a single, short-term air sample (185). In addition, surface sampling can be performed quickly in a single home visit or by the residents themselves. This approach generally consists of using a commercial vacuum cleaner fitted with a HEPA sock, extraction thimble, or filter canister and vacuuming a defined area for a set amount of time (e.g., 2 m^2 for 5 min) (183). The bulk dust is usually sieved to remove hair, fur, and other noninhalable material, weighed, and then stored frozen to await extraction and analysis. However, surface sampling of carpeted floors and mattresses collects material that originated there and was never airborne and therefore may lead to exposure misclassification. Other surface sampling techniques include wipe sampling of smooth surfaces with electrostatic wiping cloths (185).

Two recent studies reported on the variability of settled endotoxin within homes and lack of agreement with air samples. In the U.S. National Survey of Endotoxin in Housing, Thorne et al. (183) reported correlation coefficients between sites within homes ranging from 0.22 to 0.42, demonstrating that endotoxin levels at one location in the home are not informative about the levels at another location. Endotoxin concentrations in 24-h air samples collected in family rooms were poorly correlated with family room floor endotoxin in a study of 85 Iowa households (r = 0.10) (179). A study of 15 Boston area homes found no association of airborne and floor dust endotoxin concentrations in the bedrooms (128). These authors reported higher variance within homes than between homes for floor dust and airborne endotoxin.

In our laboratory, environmental samples are extracted in sterile pf water for 1 h at 22°C with continuous shaking (176, 183). Extracts are centrifuged, and supernatants are transferred into borosilicate glass tubes that have been heated for 4 h at 200°C to remove endotoxin activity. They are then analyzed using the kinetic chromogenic LAL assay. Twofold serial dilutions of endotoxin standards and sample extracts are prepared in sterile pf water. A 12-point calibration curve and 4-point endotoxin determination in duplicate is typically performed. The standard curve ranges from 0.05 to 100 EU of standard endotoxin from *E. coli* O111:B4. Lyophilized standard endotoxin, chromogenic substrate, and LAL preparations (Cambrex Bio-Sciences, Walkersville, Md.) are reconstituted using pf water. Aliquots (100 μl) of the serial dilutions of endotoxin standards and extracts are pipetted into a pf polystyrene microplate and assayed via the addition of the LAL reagent and substrate. The absorbance in each well is measured at 405 nm every 30 s for 90 min. Endotoxin determinations are based upon the maximum slope of the absorbance versus time plot for each well. Four assay reagent blank wells serve as reference and control for the pf status of the reagent water, centrifuge tubes, pipette tips, and microplates. The endotoxin value for a sample is typically calculated from the arithmetic mean of those dilutions that fall in the middle two-thirds of the standard curve.

Chemical Methods and Immunoassays

Chemical methods for quantification of endotoxin have been developed employing gas chromatography-mass spectrometry (171). These methods require special extraction procedures and assess the amounts of 3-hydroxy fatty acids typically bearing 12 to 16 carbons. This method has been coupled with the LAL method to investigate the impact of sample extraction procedures on endotoxin determinations in the LAL (146). Recently, analytical improvements have been made in the analysis of 3-hydroxy fatty acids as surrogates for the detection of lipid A. Alwis et al. (5) investigated the use of gas chromatography ion-trap mass spectrometry for detection of trimethylsilyl methyl ester derivatives of 3-hydroxy fatty acids. This requires a multistep reaction and purification methodology that is less amenable to high throughput than the LAL kinetic chromogenic method. It has the advantage that it can measure both free and extractable cell-bound endotoxin (extraction in 2 M HCl in methanol). It has the disadvantage that the equipment and labor costs are high and mammalian body fluids also contain these 3-hydroxy fatty acids (62). Immunoassays have also been developed but have not been widely adopted. Although the biological activity rests with the lipid A portion of the molecule, the antibodies produced against endotoxin recognize the polysaccharide moeity (149). Thus, at this time the LAL assay remains the most widely accepted method.

The biological activity of endotoxin from various organisms varies depending on their chemical structure and bioavailability. Therefore, chemical and immunochemical methods may not represent the potency of endotoxin as accurately as the LAL bioassay. Endotoxin activity in the LAL assay is highly correlated to the pyrogenicity in the rabbit (12, 143, 197) and to pulmonary effects in the guinea pig, mouse, and human (35, 72, 115, 155, 160, 161). Although typical sample extraction procedures for air filters may not be effective in solubilizing endotoxin bound in cell membranes from some species of gram-negative bacteria (56), this may reflect a lower bioavailability of membrane-bound endotoxin in vivo. The detection limit for airborne endotoxin measurements is approximately 0.05 EU/m^3 (5 pg/m^3). Thus, the LAL assay is a highly sensitive bioassay with physiologic relevance.

There have been international interlaboratory comparisons of sample extraction and endotoxin analysis protocols for agricultural dust (146, 147) and cotton dust (28–30). An additional study compared five different methods for MWF air samples within one laboratory (178). The overriding conclusion of these is that intralaboratory variability is generally low (<20%) and in order to minimize interlaboratory variation it is necessary to harmonize the sample extraction methods and the analysis methods and to match lysate supplier and individual lysate lot. If these conditions are met, laboratories experienced in the LAL generally agree within ⅓ $\log_{10}$ and results are highly correlated.

CONCLUSIONS

Airborne bacteria and endotoxin appear at elevated concentrations in a variety of occupational and nonoccupational environments. Especially high levels are found in agriculture and in industrial operations handling wet organic matter. Elevated exposures in domestic environments are often associated with defective ventilation systems or water intrusion and can also be problematic. Principal adverse health effects include asthma exacerbation, organic dust toxic syndrome, asthma-like syndrome, and hypersensitivity pneumonitis. Endotoxin exerts its inflammatory potency through an innate immune signaling pathway via TLR4. Susceptibilities differ between individuals dependent on genetic differences in the recognition and response pathways. Endotoxin can be measured with great sensitivity by using the kinetic chromogenic LAL assay.

Care must be taken to ensure that sample collection and extraction methods are optimized.

REFERENCES

1. **Abreu, M. T., P. Vora, E. Faure, L. S. Thomas, E. T. Arnold, and M. Arditi.** 2001. Decreased expression of Toll-like receptor-4 and MD-2 correlates with intestinal epithelial cell protection against dysregulated proinflammatory gene expression in response to bacterial lipopolysaccharide. *J. Immunol.* **167:**1609–1616.
2. **Aizenberg, V., S. A. Grinshpun, K. Willeke, J. Smith, and P. A. Baron.** 2000. Performance characteristics of the button personal inhalable aerosol sampler. *Am. Ind. Hyg. Assoc.J.* **61:**398–404.
3. **Alexander, C., and E. T. Rietschel.** 2001. Bacterial lipopolysaccharides and innate immunity. *J. Endotoxin Res.* **7:**167–202.
4. **Allen, V. M., M. H. Hinton, D. B. Tinker, C. Gibson, G. C. Mead, and C. M. Wathes.** 2003. Microbial cross-contamination by airborne dispersion and contagion during defeathering of poultry. *Br. Poult. Sci.* **44:**567–576.
5. **Alwis, K. U., L. Larsson, and D. K. Milton.** 2006. Suppression of ionization and optimization of assay for 3-hydroxy fatty acids in house dust using ion-trap mass spectrometry. *Am. J. Ind. Med.* **49:**286–295.
6. **Arbour, N. C., E. Lorenz, B. C. Schutte, J. Zabner, J. N. Kline, M. Jones, K. Frees, J. L. Watt, and D. A. Schwartz.** 2000. TLR4 mutations are associated with endotoxin hyporesponsiveness in humans. *Nat. Genet.* **25:**187–191.
7. **Bang, F. B.** 1956. A bacterial disease of *Limulus polyphemus*. *Bull. Johns Hopkins Hosp.* **98:**325–350.
8. **Barbeau, J., and T. Buhler.** 2001. Biofilms augment the number of free-living amoebae in dental unit waterlines. *Res. Microbiol.* **152:**753–760.
9. **Barbeau, J., C. Gauthier, and P. Payment.** 1998. Biofilms, infectious agents, and dental unit waterlines: a review. *Can. J. Microbiol.* **44:**1019–1028.
10. **Barbeau, J., and C. Nadeau.** 1997. Dental unit waterline microbiology: a cautionary tale. *J. Can. Dent. Assoc.* **63:** 775–779.
11. **Bartlett, K. H., S. M. Kennedy, M. Brauer, C. van Netten, and B. Dill.** 2004. Evaluation and determinants of airborne bacterial concentrations in school classrooms. *J. Occup. Environ. Hyg.* **1:**639–647.
12. **Baseler, M. W., B. Fogelmark, and R. Burrell.** 1983. Differential toxicity of inhaled gram-negative bacteria. *Infect. Immun.* **40:**133–138.
13. **Bauer, H., M. Fuerhacker, F. Zibuschka, H. Schmid, and H. Puxbaum.** 2002. Bacteria and fungi in aerosols generated by two different types of wastewater treatment plants. *Water Res.* **36:**3965–3970.
14. **Bennett, A. M., M. R. Fulford, J. T. Walker, D. J. Bradshaw, M. V. Martin, and P. D. Marsh.** 2000. Microbial aerosols in general dental practice. *Br. Dent. J.* **189:**664–667.
15. **Beutler, B., and E. T. Rietschel.** 2003. Innate immune sensing and its roots: the story of endotoxin. *Nat. Rev. Immunol.* **3:**169–176.
16. **Bolte, G., W. Bischof, M. Borte, I. Lehmann, H. E. Wichmann, and J. Heinrich.** 2003. Early endotoxin exposure and atopy development in infants: results of a birth cohort study. *Clin. Exp. Allergy* **33:**770–776.
17. **Brandi, G., M. Sisti, and G. Amagliani.** 2000. Evaluation of the environmental impact of microbial aerosols generated by wastewater treatment plants utilizing different aeration systems. *J. Appl. Microbiol.* **88:**845–852.
18. **Brass, D. M., J. D. Sauer, G. W. Whitehead, A. B. Maxwell, and D. A. Schwartz.** 2004. LPS binding protein is important in the airway response to inhaled endotoxin *J. Allergy Clin. Immunol.* **114:**586–592.
19. **Braun-Fahrlander, C., J. Riedler, U. Herz, W. Eder, M. Waser, L. Grize, S. Maisch, D. Carr, F. Gerlach, A. Bufe, R. P. Lauener, R. Schierl, H. Renz, D. Nowak, and E. von Mutius, et al.** 2002. Environmental exposure to endotoxin and its relation to asthma in school-age children. *N. Engl. J. Med.* **347:**869–877.
20. **Brooks, J. P., B. D. Tanner, C. P. Gerba, and I. L. Pepper.** 2006. The measurement of aerosolized endotoxin from land application of Class B biosolids in Southeast Arizona. *Can. J. Microbiol.* **52:**150–156.
21. **Brooks, J. P., B. D. Tanner, K. L. Josephson, C. P. Gerba, and I. L. Pepper.** 2004. Bioaerosols from the land application of biosolids in the desert southwest USA. *Water Sci. Technol.* **50:**7–12.
22. **Carty, C. L., U. Gehring, J. Cyrys, W. Bischof, and J. Heinrich.** 2003. Seasonal variability of endotoxin in ambient fine particulate matter. *J. Environ. Monit.* **5:**953–958.
23. **Castellan, R. M., S. A. Olenchock, J. L. Hankinson, P. D. Millner, J. B. Cocke, C. K. Bragg, H. H. Perkins, Jr., and R. R. Jacobs.** 1984. Acute bronchoconstriction induced by cotton dust: dose-related responses to endotoxin and other dust factors. *Ann. Intern. Med.* **101:**157–163.
24. **Cavagna, G., V. Foa, and E. C. Vigliani.** 1969. Effects in man and rabbits of inhalation of cotton dust or extracts and purified endotoxins. *Br. J. Ind. Med.* **26:**314–321.
25. **Chapin, A., A. Rule, K. Gibson, T. Buckley, and K. Schwab.** 2005. Airborne multidrug-resistant bacteria isolated from a concentrated swine feeding operation. *Environ. Health Perspect.* **113:**137–142.
26. **Chew, G. L., J. Wilson, F. A. Rabito, F. Grimsley, S. Iqbal, T. Reponen, M. L. Muilenberg, P. S. Thorne, D. G. Dearborn, and R. L. Morley.** 2006. Mold and endotoxin levels in the aftermath of Hurricane Katrina: a pilot project of homes in New Orleans undergoing renovation. *Environ. Health Perspect.* **114:**1883–1889.
27. **Chrischilles, E., R. Ahrens, A. Kuehl, K. Kelly, P. Thorne, L. Burmeister, and J. Merchant.** 2004. Asthma prevalence and morbidity among rural Iowa school children. *J. Allergy Clin. Immunol.* **113:**66–71.
28. **Chun, D. T., K. Bartlett, T. Gordon, R. R. Jacobs, B. M. Larsson, L. Larsson, D. M. Lewis, J. Liesivuori, O. Michel, D. K. Milton, R. Rylander, P. S. Thorne, E. M. White, M. E. Brown, V. S. Gunn, and H. Wurtz.** 2006. History and results of the two inter-laboratory round robin endotoxin assay studies on cotton dust. *Am. J. Ind. Med.* **49:**301–306.
29. **Chun, D. T., V. Chew, K. Bartlett, T. Gordon, R. R. Jacobs, B.-M. Larsson, L. Larsson, D. M. Lewis, J. Liesivuori, O. Michel, D. K. Milton, R. Rylander, P. S. Thorne, E. M. White, and M. E. Brown.** 2000. Preliminary report on the results of the second phase of a round-robin endotoxin assay study using cotton dust. *Appl. Occup. Environ. Hyg.* **15:**152–157.
30. **Chun, D. T. W., V. Chew, K. Bartlett, T. Gordon, R. R. Jacobs, B.-M. Larsson, D. M. Lewis, J. Liesivuori, O. Michel, R. Rylander, P. S. Thorne, E. M. White, V. C. Gunn, and H. Würtz.** 2002. Second interlaboratory study comparing endotoxin assay results from cotton dust. *Ann. Agric. Env. Med.* **9:**49–53.
31. **Cinkotai, F. F., M. G. Lockwood, and R. Rylander.** 1977. Airborne micro-organisms and prevalence of byssinotic symptoms in cotton mills. *Am. Ind. Hyg. Assoc. J.* **38:** 554–559.
32. **Clapp, W. D., S. Becker, J. Quay, J. L. Watt, P. S. Thorne, K. L. Frees, X. Zhang, H. S. Koren, C. R. Lux, and D. A. Schwartz.** 1994. Grain dust-induced airflow obstruction and inflammation of the lower respiratory tract. *Am. J. Respir. Crit. Care Med.* **150:**611–617.

33. **Clark, A.** 1974. Bacterial colonization of dental units and the nasal flora of dental personnel. *Proc. R. Soc. Med.* **67:** 1269–1270.
34. **Davis, M., and T. Y. Morishita.** 2005. Relative ammonia concentrations, dust concentrations, and presence of Salmonella species and Escherichia coli inside and outside commercial layer facilities. *Avian Dis.* **49:**30–35.
35. **Deetz, D. C., P. J. Jagielo, T. J. Quinn, P. S. Thorne, S. A. Bleuer, and D. A. Schwartz.** 1997. The kinetics of grain dust-induced inflammation of the lower respiratory tract. *Am. J. Respir. Crit. Care Med.* **155:**254–259.
36. **DeKoster, J. A., and P. S. Thorne.** 1995. Bioaerosol concentrations in non-complaint, complaint, and intervention homes in the Midwest. *Am. Ind. Hyg. Assoc. J.* **56:** 573–580.
37. **Demers, P. A., K. Teschke, and S. M. Kennedy.** 1997. What to do about softwood? A review of respiratory effects and recommendations regarding exposure limits. *Am. J. Ind. Med.* **31:**385–398.
38. **de Vocht, F., D. Huizer, M. Prauser, K. Jakobsson, B. Peplonska, K. Straif, and H. Kromhout.** 2006. Field comparison of inhalable aerosol samplers applied in the European rubber manufacturing industry. *Int. Arch. Occup. Environ. Health* **79:**621–629.
39. **Donham, K. J.** 1991. Association of environmental air contaminants with disease and productivity in swine. *Am. J. Vet. Res.* **52:**1723–1730.
40. **Donham, K. J., J. A. Merchant, D. Lassise, W. J. Popendorf, and L. F. Burmeister.** 1990. Preventing respiratory disease in swine confinement workers: intervention through applied epidemiology, education and consultation. *Am. J. Ind. Med.* **18:**241–261.
41. **Douwes, J., P. S. Thorne, M. Pearce, and D. Heederik.** 2003. Bioaerosol health effects and exposure assessment: progress and prospects. *Ann. Occup. Hyg.* **47:**187–200.
42. **Douwes, J., P. S. Thorne, M. Pearce, and D. Heederik.** 2003. Biological agents—recognition, p. 219–292. *In* J. Perkins (ed.), *Modern Industrial Hygiene*, vol. 2. *Biological Agents*. ACJIH, Cincinnati, Ohio.
43. **Douwes, J., P. Versloot, A. Hollander, D. Heederik, and G. Doekes.** 1995. Influence of various dust sampling and extraction methods on the measurement of airborne endotoxin. *Appl. Environ. Microbiol.* **61:**1763–1769.
44. **Douwes, J., I. Wouters, H. Dubbeld, L. van Zwieten, P. Steerenberg, G. Doekes, and D. Heederik.** 2000. Upper airway inflammation assessed by nasal lavage in compost workers: a relation with bio-aerosol exposure. *Am. J. Ind. Med.* **37:**459–468.
45. **Downs, S. H., G. B. Marks, T. Z. Mitakakis, J. D. Leuppi, N. G. Car, and J. K. Peat.** 2001. Having lived on a farm and protection against allergic diseases in Australia. *Clin. Exp. Allergy* **31:**570–575.
46. **Duchaine, C., Y. Grimard, and Y. Cormier.** 2000. Influence of building maintenance, environmental factors, and seasons on airborne contaminants of swine confinement buildings. *AIHAJ* **61:**56–63.
47. **Duchaine, C., A. Meriaux, G. Brochu, and Y. Cormier.** 1999. Airborne microflora in Quebec dairy farms: lack of effect of bacterial hay preservatives. *Am. Ind. Hyg. Assoc. J.* **60:**89–95.
48. **Duchaine, C., A. Meriaux, P. S. Thorne, and Y. Cormier.** 2000. Assessment of particulates and bioaerosols in eastern Canadian sawmills. *Am. Ind. Hyg. Assoc. J.* **61:**727–732.
49. **Duchaine, C., P. S. Thorne, A. Meriaux, Y. Grimard, P. Whitten, and Y. Cormier.** 2001. Comparison of endotoxin exposure assessment by bioaerosol impinger and filter sampling methods. *Appl. Environ. Microbiol.* **67:**2775–2780.
50. **Dutch Expert Committee on Occupational Standards.** 1998. *Endotoxins: Health Based Recommended Exposure Limit. A Report of the Health Council of The Netherlands.* Publication no. 1998/03WGD (http://www.gr.nl). Health Council of The Netherlands, Rijswijk, The Netherlands.
51. **Dutil, S., S. Tessier, M. Veillette, C. Laflamme, A. Mériaux, A. Leduc, J. Barbeau, and C. Duchaine.** 2006. Detection of Legionella spp. by fluorescent in situ hybridization in dental unit waterlines. *J. Appl. Microbiol.* **100:**955–963.
52. **Dutkiewicz, J., E. Krysinska-Traczyk, Z. Prazmo, C. Skorska, and J. Sitkowska.** 2001. Exposure to airborne microorganisms in Polish sawmills. *Ann. Agric. Environ. Med.* **8:**71–80.
53. **Dutkiewicz, J., S. A. Olenchock, W. G. Sorenson, V. F. Gerencser, J. J. May, D. S. Pratt, and V. A. Robinson.** 1989. Levels of bacteria, fungi, and endotoxin in bulk and aerosolized corn silage. *Appl. Environ. Microbiol.* **55:**1093–1099.
54. **Eduard, W., P. Sandven, and F. Levy.** 1992. Relationships between exposure to spores from Rhizopus microsporus and Paecilomyces variotii and serum IgG antibodies in wood trimmers. *Int. Arch. Allergy Immunol.* **97:**274–282.
55. **Eduard, W., P. Sandven, and F. Levy.** 1993. Serum IgG antibodies to mold spores in two Norwegian sawmill populations: relationship to respiratory and other work-related symptoms. *Am. J. Ind. Med.* **24:**207–222.
56. **Eduard, W., M. H. Westby, and L. Larsson.** 2004. Solubility of endotoxins from *Escherichia coli* and *Pseudomonas aeruginosa*. *Am. J. Ind. Med.* **46:**375–377.
57. **Erkinjuntti-Pekkanen, R., M. Reiman, J. I. Kokkarinen, H. O. Tukiainen, and E. O. Terho.** 1999. IgG antibodies, chronic bronchitis, and pulmonary function values in farmer's lung patients and matched controls. *Allergy* **54:** 1181–1187.
58. **Ernst, P., and Y. Cormier.** 2000. Relative scarcity of asthma and atopy among rural adolescents raised on a farm. *Am. J. Respir. Crit. Care Med.* **161:**1563–1566.
59. **Fink, J. N., H. G. Ortega, H. Y. Reynolds, Y. F. Cormier, L. L. Fan, T. J. Franks, K. Kreiss, S. Kunkel, D. Lynch, S. Quirce, C. Rose, R. P. Schleimer, M. R. Schuyler, M. Selman, D. Trout, and Y. Yoshizawa.** 2005. Needs and opportunities for research in hypersensitivity pneumonitis. *Am. J. Respir. Crit. Care Med.* **171:**792–798.
60. **Forster, H. W., B. Crook, B. W. Platts, J. Lacey, and M. D. Topping.** 1989. Investigation of organic aerosols generated during sugar beet slicing. *Am. Ind. Hyg. Assoc. J.* **50:**44–50.
61. **Fox, A., W. Harley, C. Feigley, D. Salzberg, C. Toole, A. Sebastian, and L. Larsson.** 2005. Large particles are responsible for elevated bacterial marker levels in school air upon occupation. *J. Environ. Monit.* **7:**450–456.
62. **Fox, A., L. Lennart, G. Allmaier, D. Wunschel, and K. Wahl.** 2005. Fifth international symposium on the interface between analytical chemistry and microbiology—April 19th to 21st, 2004. *J. Microbiol. Methods* **62:**257–258.
63. **Fries, R., and C. Graw.** 1999. Water and air in two poultry processing plants' chilling facilities—a bacteriological survey. *Br. Poult. Sci.* **40:**52–58.
64. **Gehring, U., G. Bolte, M. Borte, W. Bischof, B. Fahlbusch, H. E. Wichmann, and J. Heinrich.** 2001. LISA study group. Exposure to endotoxin decreases the risk of atopic eczema in infancy: a cohort study. *J. Allergy Clin. Immunol.* **108:**847–854.
65. **George, C. L. S., M. White, M. O'Neill, P. S. Thorne, D. A. Schwartz, and J. Snyder.** 2003. Altered surfactant protein A gene expression and protein metabolism associated with repeat exposure to inhaled endotoxin. *Am. J. Physiol. Lung Cell. Mol. Physiol.* **285:**L1337–L1344.
66. **Gereda, J. E., D. Y. Leung, and A. H. Liu.** 2000. Levels of environmental endotoxin and prevalence of atopic disease. *JAMA* **284:**1652–1653.

67. **Gereda, J. E., D. Y. Leung, A. Thatayatikom, J. E. Streib, M. R. Price, M. D. Klinnert, and A. H. Liu.** 2000. Relation between house-dust endotoxin exposure, type 1 T-cell development, and allergen sensitization in infants at high risk of asthma. *Lancet* **355:**1680–1683.
68. **Ghiabi, N.** 1998. Air contamination during use of air abrasion instrumentation. *J. Clin. Pediatr. Dent.* **23:**37–43.
69. **Gibbs, S. G., C. F. Green, P. M. Tarwater, and P. V. Scarpino.** 2004. Airborne antibiotic resistant and nonresistant bacteria and fungi recovered from two swine herd confined animal feeding operations. *J. Occup. Environ. Hyg.* **1:**699–706.
70. **Gioannini, T. L., A. Teghanemt, D. Zhang, N. P. Coussens, W. Dockstader, S. Ramaswamy, and J. P. Weiss.** 2004. Isolation of an endotoxin-MD-2 complex that produces Toll-like receptor 4-dependent cell activation at picomolar concentrations. *Proc. Natl. Acad. Sci. USA.* **101:**4186–4192.
71. **Gioannini, T. L., D. Zhang, A. Teghanemt, and J. P. Weiss.** 2002. An essential role for albumin in the interaction of endotoxin with lipopolysaccharide-binding protein and sCD14 and resultant cell activation. *J. Biol. Chem.* **277:**47818–47825.
72. **Gordon, T.** 1994. Role of the complement system in the acute respiratory effects of inhaled endotoxin and cotton dust. *Inhal. Toxicol.* **6:**253–266.
73. **Gordon, T., C. Nadziejko, K. Galdanes, D. Lewis, and K. Donnelly.** 2006. Mycobacterium immunogenum causes hypersensitivity pneumonitis-like pathology in mice. *Inhal. Toxicol.* **18:**449–456.
74. **Gorny, R. L., B. Szponar, L. Larsson, C. Pehrson, Z. Prazmo, and J. Dutkiewicz.** 2004. Metalworking fluid bioaerosols at selected workplaces in a steelworks. *Am. J. Ind. Med.* **46:**400–403.
75. **Hagmar, L., A. Schutz, and A. Sjoholm.** 1990. Over-shift decrease in lung function in poultry slaughterhouse workers. *Am. J. Ind. Med.* **17:**77–78.
76. **Haziot, A., E. Ferrero, F. Kontgen, N. Hijiya, S. Yamamoto, J. Silver, C. L. Stewart, and S. M. Goyert.** 1996. Resistance to endotoxin shock and reduced dissemination of gram-negative bacteria in CD14-deficient mice. *Immunity* **4:**407–414.
77. **Heinrich, J., M. Pitz, W. Bischof, N. Krug, and P. J. A. Borm.** 2003. Endotoxin in fine ($PM_{2.5}$) and coarse ($PM_{2.5\text{-}10}$) particle mass of ambient aerosols. A temporo-spatial analysis. *Atmos. Environ.* **37:**3659–3667.
78. **Heldal, K. K., and W. Eduard.** 2004. Associations between acute symptoms and bioaerosol exposure during the collection of household waste. *Am. J. Ind. Med.* **46:** 253–260.
79. **Hollander, A., D. Heederik, and H. Kauffman.** 1994. Acute respiratory effects in the potato processing industry due to a bioaerosol exposure. *Occup. Environ. Med.* **51:**73–78.
80. **Hryhorczuk, D., L. Curtis, P. Scheff, J. Chung, M. Rizzo, C. Lewis, N. Keys, and M. Moomey.** 2001. Bioaerosol emissions from a suburban yard waste composting facility. *Ann. Agric. Environ. Med.* **8:**177–185.
81. **Iowa State University and The University of Iowa.** 2002. Iowa Concentrated Animal Feeding Operations Air Quality Study. The University of Iowa, Iowa City. [Online.] http://www.ehsrc.uiowa.edu/CAFOstudy.htm. Accessed May 2006.
82. **Iwanaga, S.** 1993. The *Limulus* clotting reaction. *Curr. Opin. Immunol.* **5:**74–82.
83. **Jack, R. S., X. Fan, M. Bernheiden, G. Rune, M. Ehlers, A. Weber, G. Kirsch, R. Mentel, B. Furll, M. Freudenberg, G. Schmitz, F. Stelter, and C. Schutt.** 1997. Lipopolysaccharide-binding protein is required to combat a murine gram-negative bacterial infection. *Nature* **389:**742–745.
84. **Jagielo, P. J., P. S. Thorne, J. A. Kern, T. J. Quinn, and D. A. Schwartz.** 1996. The role of endotoxin in grain dust induced lung inflammation in mice. *Am. J. Physiol.* **270:** L1052–L1059.
85. **Jagielo, P. J., P. S. Thorne, J. L. Watt, K. L. Frees, T. J. Quinn, and D. A. Schwartz.** 1996. Grain dust and endotoxin inhalation challenges produce similar inflammatory responses in normal subjects. *Chest* **110:**263–270.
86. **Jericho, K. W., G. C. Kozub, V. P. Gannon, and C. M. Taylor.** 2000. Microbiological testing of raw, boxed beef in the context of hazard analysis critical control point at a high-line-speed abattoir. *J. Food Prot.* **63:**1681–1686.
87. **Jeyaseelan, S., H. W. Chu, S. Y. Young, M. W. Freeman, and G. S. Worthen.** 2005. Distinct roles of pattern recognition receptors CD14 and Toll-like receptor 4 in acute lung injury. *Infect. Immun.* **73:**1754–1763.
88. **Jo, W. K., and Y. J. Seo.** 2005. Indoor and outdoor bioaerosol levels at recreation facilities, elementary schools, and homes. *Chemosphere* **61:**1570–1579.
89. **Kaan, J. A., A. M. Simoons-Smit, and D. M. MacLaren.** 1985. Another source of aerosol causing nosocomial Legionnaires' disease. *J. Infect.* **11:**145–148.
90. **Kedjarune, U., B. Kukiattrakoon, B. Yapong, S. Chowanadisai, and P. Leggat.** 2000. Bacterial aerosols in the dental clinic: effect of time, position and type of treatment. *Int. Dent. J.* **50:**103–107.
91. **Kennedy, S. M., D. C. Christiani, E. A. Eisen, D. H. Wegman, I. A. Greaves, S. A. Olenchock, T. T. Ye, and P. L. Lu.** 1987. Cotton dust and endotoxin exposure-response relationships in cotton textile workers. *Am. Rev. Respir. Dis.* **135:**194–200.
92. **Kenny, L. C., R. Aitken, C. Chalmers, J. F. Fabriès, E. Gonzalez-Fernandez, H. Kromhout, G. Liden, D. Mark, G. Riediger, and V. Prodi.** 1997. A collaborative Euro-pean study of personal inhalable aerosol sampler performance. *Ann. Occup. Hyg.* **41:**135–153.
93. **Kiekhaefer, M. S., K. J. Donham, P. Whitten, and P. S. Thorne.** 1995. Cross seasonal studies of airborne microbial populations and environment in swine buildings: implications for worker and animal health. *Ann. Agric. Environ. Med.* **2:**37–44.
94. **Kleeberger, S. R., and D. Peden.** 2005. Gene-environment interactions in asthma and other respiratory diseases. *Annu. Rev. Med.* **56:**383–400.
95. **Kline, J. N., J. D. Cowden, G. W. Hunninghake, B. C. Schutte, J. L. Watt, C. L. Wohlford-Lenane, L. S. Powers, M. P. Jones, and D. A. Schwartz.** 1999. Variable airway responsiveness to inhaled lipopolysaccharide. *Am. J. Respir. Crit. Care Med.* **160:**297–303.
96. **Klintberg, B., N. Berglund, G. Lilja, M. Wickman, and M. van Hage-Hamsten.** 2001. Fewer allergic respiratory disorders among farmers' children in a closed birth cohort from Sweden. *Eur. Respir. J.* **17:**1151–1157.
97. **Kreiss, K., and J. Cox-Ganser.** 1997. Metalworking fluid-associated hypersensitivity pneumonitis: a workshop summary. *Am. J. Ind. Med.* **32:**423–432.
98. **Kullman, G. J., P. S. Thorne, P. F. Waldron, J. J. Marx, B. Ault, D. M. Lewis, P. D. Siegel, S. A. Olenchock, and J. A. Merchant.** 1998. Organic dust exposures from work in dairy barns. *Am. Ind. Hyg. Assoc. J.* **59:**403–413.
99. **Laitinen, S., J. Kangas, M. Kotimaa, J. Liesivuori, P. J. Martikainen, A. Nevalainen, R. Sarantila, and K. Husman.** 1994. Workers' exposure to airborne bacteria and endotoxins at industrial wastewater treatment plants. *Am. Ind. Hyg. Assoc. J.* **55:**1055–1060.
100. **Lavoie, J., and S. Guertin.** 2001. Evaluation of health and safety risks in municipal solid waste recycling plants. *J. Air Waste Manag. Assoc.* **51:**352–360.

101. **Lee, J. A., J. C. Johnson, S. J. Reynolds, P. S. Thorne, and P. T. O'Shaughnessy.** 2006. Indoor and outdoor air quality assessment of four wastewater treatment plants. *J. Occup. Environ. Hyg.* **3:**36–43.
102. **Lee, T., S. A. Grinshpun, D. Martuzevicius, A. Adhikari, C. M. Crawford, J. Luo, and T. Reponen.** 2006. Relationship between indoor and outdoor bioaerosols collected with a button inhalable aerosol sampler in urban homes. *Indoor Air* **16:**37–47.
103. **Leynaert, B., C. Neukirch, D. Jarvis, S. Chinn, P. Burney, and F. Neukirch.** 2001. Does living on a farm during childhood protect against asthma, allergic rhinitis, and atopy in adulthood? *Am. J. Respir. Crit. Care Med.* **164:**1829–1834.
104. **Li, S.-N., D. A. Lundgren, and D. Rovell-Rixx.** 2000. Evaluation of six inhalable aerosol samplers. *Am. Ind. Hyg. Assoc. J.* **61:**506–516.
105. **Litonjua, A. A., D. K. Milton, J. C. Celedon, L. Ryan, S. T. Weiss, and D. R. Gold.** 2002. A longitudinal analysis of wheezing in young children: the independent effects of early life exposure to house dust endotoxin, allergens, and pets. *J. Allergy Clin. Immunol.* **110:**736–742.
106. **Liu, A. H.** 2002. Endotoxin exposure in allergy and asthma: reconciling a paradox. *J. Allergy Clin. Immunol.* **109:**379–392.
107. **Lorenz, E., M. Jones, C. Wohlford-Lenane, N. Meyer, K. L. Frees, N. C. Arbour, and D. A. Schwartz.** 2001. Genes other than TLR4 are involved in the response to inhaled LPS. *Am. J. Physiol. Lung Cell. Mol. Physiol.* **281:**L1106–L1114.
108. **Mankell, K. O., K. A. Janni, R. D. Walker, M. E. Wilson, J. E. Pettigrew, L. D. Jacobson, and W. F. Wilcke.** 1995. Dust suppression in swine feed using soybean oil. *J. Anim. Sci.* **73:**981–985.
109. **Mattsby-Baltzer, I., M. Sandin, B. Ahlstrom, S. Allenmark, M. Edebo, E. Falsen, K. Pedersen, N. Rodin, R. A. Thompson, and L. Edebo.** 1989. Microbial growth and accumulation in industrial metal-working fluids. *Appl. Environ. Microbiol.* **55:**2681–2689.
110. **Merchant, J. A., A. L. Naleway, E. R. Svendsen, K. M. Kelly, L. F. Burmeister, A. M. Stromquist, C. D. Taylor, P. S. Thorne, S. J. Reynolds, W. T. Sanderson, and E. A. Chrischilles.** 2005. Asthma and farm exposures in a cohort of rural Iowa children. *Environ. Health Perspect.* **113:**350–356.
111. **Michel, O., J. Duchateau, and R. Sergysels.** 1989. Effect of inhaled endotoxin on bronchial reactivity in asthmatic and normal subjects. *J. Appl. Physiol.* **66:**1059–1064.
112. **Michel, O., R. Ginanni, J. Duchateau, F. Vertongen, B. Le Bon, and R. Sergysels.** 1991. Domestic endotoxin exposure and clinical severity of asthma. *Clin. Exp. Allergy* **21:**441–448.
113. **Michel, O., R. Ginanni, B. Le Bon, J. Content, J. Duchateau, and R. Sergysels.** 1992. Inflammatory response to acute inhalation of endotoxin in asthmatic patients. *Am. Rev. Respir. Dis.* **146:**352–357.
114. **Michel, O., J. Kips, J. Duchateau, F. Vertongen, L. Robert, H. Collet, R. Pauwels, and R. Sergysels.** 1996. Severity of asthma is related to endotoxin in house dust. *Am. J. Respir. Crit. Care Med.* **154:**1641–1646.
115. **Michel, O., A. M. Nagy, M. Schroeven, J. Duchateau, J. Neve, P. Fondu, and R. Sergysels.** 1997. Dose-response relationship to inhaled endotoxin in normal subjects. *Am. J. Respir. Crit. Care Med.* **156:**1157–1164.
116. **Milton, D. K., D. K. Johnson, and J. H. Park.** 1997. Environmental endotoxin measurement: interference and sources of variation in the limulus assay of house dust. *Am. Ind. Hyg. Assoc. J.* **58:**861–867.
117. **Moreno-Ancillol, A., C. Dominguez-Noche, A. C. Gil-Adrados, and P. M. Cosmes.** 2004. Hypersensitivity pneumonitis due to occupational inhalation of fungi-contaminated corn dust. *J. Investig. Allergol. Clin. Immunol.* **14:**165–167.
118. **Mueller-Anneling, L., E. Avol, J. M. Peters, and P. S. Thorne.** 2004. Measurement of endotoxin in ambient PM10 in Southern California. *Environ. Health Perspect.* **112:**583–588.
119. **Nagai, Y., S. Akashi, M. Nagafuku, M. Ogata, Y. Iwakura, S. Akira, T. Kitamura, A. Kosugi, M. Kimoto, and K. Miyake.** 2002. Essential role of MD-2 in LPS responsiveness and TLR4 distribution. *Nat. Immunol.* **3:**667–672.
120. **Nguyen, T. M., D. Ilef, S. Jarraud, L. Rouil, C. Campese, D. Che, S. Haeghebaert, F. Ganiayre, F. Marcel, J. Etienne, and J. C. Desenclos.** 2006. A community-wide outbreak of legionnaires disease linked to industrial cooling towers—how far can contaminated aerosols spread? *J. Infect. Dis.* **193:**102–111. [Epub 28 November 2005.]
121. **Noce, L., D. Di Giovanni, and E. E. Putnins.** 2000. An evaluation of sampling and laboratory procedures for determination of heterotrophic plate counts in dental unit waterlines. *J. Can. Dent. Assoc.* **66:**262.
122. **Nonnenmann, M. W., K. J. Donham, R. H. Rautiainen, P. T. O'Shaughnessy, L. F. Burmeister, and S. J. Reynolds.** 2004. Vegetable oil sprinkling as a dust reduction method in swine confinement. *J. Agric. Saf. Health* **10:**7–15.
123. **O'Hara, R. E., and R. Rubin.** 2005. Reducing bioaerosol dispersion from wastewater treatment and its land application: a review and analysis. *J. Environ. Health* **68:**24–29.
124. **Oppliger, A., S. Rusca, N. Charriere, T. Vu Duc, and P. O. Droz.** 2005. Assessment of bioaerosols and inhalable dust exposure in Swiss sawmills. *Ann. Occup. Hyg.* **49:**385–391.
125. **Otte, J. M., E. Cario, and D. K. Podolsky.** 2004. Mechanisms of cross hyporesponsiveness to Toll-like receptor bacterial ligands in intestinal epithelial cells. *Gastroenterology* **126:**1054–1070.
126. **Paez-Rubio, T., E. Viau, S. Romero-Hernandez, and J. Peccia.** 2005. Source bioaerosol concentration and rRNA gene-based identification of microorganisms aerosolized at a flood irrigation wastewater reuse site. *Appl. Environ. Microbiol.* **71:**804–810.
127. **Pagniano, R. P., R. C. Scheid, S. Rosen, and F. M. Beck.** 1985. Airborne microorganisms collected in a preclinical dental laboratory. *J. Dent. Educ.* **49:**653–655.
128. **Park, J.-H., D. L. Spiegelman, H. A. Burge, D. R. Gold, G. L. Chew, and D. K. Milton.** 2000. Longitudinal study of dust and airborne endotoxin. *Environ. Health Perspect.* **108:**1023–1028.
129. **Park, J. H., D. L. Spiegelman, D. R. Gold, H. A. Burge, and D. K. Milton.** 2001. Predictors of airborne endotoxin in the home. *Environ. Health Perspect.* **109:**859–864.
130. **Pepys, J., P. A. Jenkins, G. N. Festenstein, P. H. Gregory, M. E. Lacey, and F. A. Skinner.** 1963. Farmer's lung. Thermophilic actinomycetes as a source of "Farmer's lung hay" antigen. *Lancet* **41:**607–611.
131. **Pernis, B., E. C. Vigliani, and G. Cavagna.** 1960. The role of endotoxins in the pathogenesis of diseases caused by the inhalation of vegetable dusts. *Med. Lav.* **51:**780–794.
132. **Perzanowski, M. S., R. L. Miller, P. S. Thorne, R. G. Barr, B. J. Sheares, R. S. Garfinkel, F. P. Perera, I. F. Goldstein, and G. L. Chew.** 2006. Endotoxin in inner-city homes and associates with wheeze and eczema in the first

three years of life. *J. Allergy Clin. Immunol.* **117:**1082–1089.

133. **Peterson, R. D., P. E. Wicklunds, and R. A. Good.** 1964. Endotoxin activity of a house dust extract. *J. Allergy Clin. Immunol.* **35:**134–142.
134. **Petti, S., and G. Tarsitani.** 1998. Intra-individual variations of salivary microbial levels in young adults. *Eur. J. Oral Sci.* **106**(Pt. 1)**:**616–622.
135. **Post, D. M., D. Zhang, J. S. Eastvold, A. Teghanemt, B. W. Gibson, and J. P. Weiss.** 2005. Biochemical and functional characterization of membrane blebs purified from Neisseria meningitidis serogroup B. *J. Biol. Chem.* **280:**38383–38394.
136. **Prazmo, Z., J. Dutkiewicz, C. Skorska, J. Sitkowska, and G. Cholewa.** 2003. Exposure to airborne Gram-negative bacteria, dust and endotoxin in paper factories. *Ann. Agric. Environ. Med.* **10:**93–100.
137. **Prazmo, Z., E. Krysinska-Traczyk, C. Skorska, J. Sitkowska, G. Cholewa, and J. Dutkiewicz.** 2003. Exposure to bioaerosols in a municipal sewage treatment plant. *Ann. Agric. Environ. Med.* **10:**241–248.
138. **Predicala, B. Z., J. E. Urban, R. G. Maghirang, S. B. Jerez, and R. D. Goodband.** 2002. Assessment of bioaerosols in swine barns by filtration and impaction. *Curr. Microbiol.* **44:**136–140.
139. **Preller, L., D. Heederik, H. Kromhout, J. S. M. Boleij, and M. J. M. Tielen.** 1995. Determinants of dust and endotoxin exposure of pig farmers: development of a control strategy using empirical modeling. *Ann. Occup. Hyg.* **39:**545–557.
140. **Radon, K., B. Danuser, M. Iversen, R. Jorres, E. Monso, U. Opravil, C. Weber, K. J. Donham, and D. Nowak.** 2001. Respiratory symptoms in European animal farmers. *Eur. Resp. J.* **17:**747–754.
141. **Rahkio, T. M., and H. J. Korkeala.** 1997. Airborne bacteria and carcass contamination in slaughterhouses. *J. Food Prot.* **60:**38–42.
142. **Ranalli, G., L. Grazia, and A. Roggeri.** 1999. The influence of hay-packing techniques on the presence of Saccharopolyspora rectivirgula. *J. Appl. Microbiol.* **87:**359–365.
143. **Ray, A., K. Redhead, S. Seikirk, and S. Poole.** 1991. Variability in LPS composition, antigenicity and reactogenicity of phase variants in *Bordetella pertussis*. *FEMS Microbiol. Lett.* **63:**211–217.
144. **Reinthaler, F. F., F. Mascher, and D. Stunzner.** 1988. Serological examinations for antibodies against Legionella species in dental personnel. *J. Dent. Res.* **67:**942–943.
145. **Reynolds, S. J., D. W. Black, S. S. Borin, G. Breuer, L. F. Burmeister, L. J. Fuortes, T. F. Smith, M. A. Stein, P. Subramanian, P. S. Thorne, and P. Whitten.** 2001. Indoor environmental quality in six commercial office buildings in the midwest United States. *Appl. Occup. Environ. Hyg.* **16:**1065–1077.
146. **Reynolds, S. J., D. K. Milton, D. Heederik, P. S. Thorne, K. J. Donham, E. A. Croteau, K. M. Kelly, J. Douwes, D. Lewis, M. Whitmer, I. Connaughton, S. Koch, P. Malmberg, B. M. Larsson, J. Deddens, A. Saraf, and L. Larsson.** 2005. Interlaboratory evaluation of endotoxin analyses in agricultural dusts—comparison of LAL assay and mass spectrometry. *J. Environ. Monit.* **12:**1371–1377.
147. **Reynolds, S. J., P. S. Thorne, K. J. Donham, E. Croteau, K. M. Kelly, D. Lewis, M. Whitmer, D. J. Heederik, J. Douwes, I. Connaughton, S. Koch, P. Malmberg, L. Larsson, and D. K. Milton.** 2002. Comparison of endotoxin assays using agricultural dusts. *Am. Ind. Hyg. Assoc. J.* **63:**430–438.
148. **Rhame, F. S., A. Streifel, C. McComb, and M. Boyle.** 1986. Bubbling humidifiers produce microaerosols which can carry bacteria. *Infect. Control* **7:**403–407.
149. **Rietschel, E. T., H. Brade, O. Holst, L. Brade, S. Muller-Loennies, U. Mamat, U. Zahringer, F. Beckmann, U. Seydel, K. Brandenburg, A. J. Ulmer, T. Mattern, H. Heine, J. Schletter, H. Loppnow, U. Schonbeck, H. D. Flad, S. Hauschildt, U. F. Schade, F. Di Padova, S. Kusumoto, and R. R. Schumann.** 1996. Bacterial endotoxin: chemical constitution, biological recognition, host response, and immunological detoxification. *Curr. Top. Microbiol. Immunol.* **216:**39–81.
150. **Rintala, H., A. Hyvarinen, L. Paulin, and A. Nevalainen.** 2004. Detection of streptomycetes in house dust—comparison of culture and PCR methods. *Indoor Air* **14:**112–119.
151. **Rose, C.** 1996. Biopsy-confirmed hypersensitivity pneumonitis in automobile production workers exposed to metalworking fluids—Michigan, 1994-1995. *Morb. Mortal. Wkly. Rep.* **45:**60610.
152. **Roussel, S., G. Reboux, J. C. Dalphin, K. Bardonnet, L. Millon, and R. Piarroux.** 2004. Microbiological evolution of hay and relapse in patients with farmer's lung. *Occup. Environ. Med.* **61:**e3.
153. **Roy, C. J., and P. S. Thorne.** 2003. Exposure to particulates, microorganisms, β(1-3)-glucans and endotoxins during soybean harvesting. *Am. Ind. Hyg. Assoc. J.* **64:**487–495.
154. **Rylander, R.** 1987. The role of endotoxin for reactions after exposure to cotton dust. *Am. J. Ind. Med.* **12:**687–697.
155. **Rylander, R.** 1994. Endotoxins, p. 73–78. *In* R. Rylander and R. R. Jacobs (ed.), *Organic Dusts: Exposure, Effects and Prevention.* Lewis Publishers, Boca Raton, Fla.
156. **Rylander, R., J. Thorn, and R. Attefors.** 1999. Airways inflammation among workers in a paper industry. *Eur. Respir. J.* **13:**1151–1157.
157. **Sanchez-Monedero, M. A., E. I. Stentiford, and S. T. Urpilainen.** 2005. Bioaerosol generation at large-scale green waste composting plants. *J. Air Waste Manag. Assoc.* **55:**612–618.
158. **Schenker, M. B., D. Christiani, Y. Cormier, H. Dimich-Ward, G. Doekes, J. Doman, J. Douwes, K. Dowling, D. Enarson, F. Green, D. Heederik, K. Husman, S. Kennedy, G. Kullman, Y. Lacasse, B. Lawson, P. Malmberg, J. May, S. McCurdy, J. Merchant, J. Myers, M. Nieuwenhuijsen, S. Olenchock, C. Saiki, D. Schwartz, J. Seiber, P. Thorne, G. Wagner, N. White, X. Xu, and M. Chan-Yeung.** 1998. Respiratory health hazards in agriculture. *Am. J. Respir. Crit. Care Med.* **158:**S1–S76.
159. **Schwartz, D. A., K. J. Donham, S. A. Olenchock, W. J. Popendorf, D. S. Van Fossen, L. F. Burmeister, and J. A. Merchant.** 1995. Determinants of longitudinal changes in spirometric function among swine confinement operators and farmers. *Am. J. Respir. Crit. Care Med.* **151:**47–53.
160. **Schwartz, D. A., P. S. Thorne, S. J. Yagla, L. F. Burmeister, S. A. Olenchock, J. L. Watt, and T. J. Quinn.** 1995. The role of endotoxin in grain dust-induced lung disease. *Am. J. Respir. Crit. Care Med.* **152:**603–608.
161. **Schwartz, D. A., P. S. Thorne, P. J. Jagielo, G. E. White, S. A. Bleuer, and K. L. Frees.** 1994. Endotoxin responsiveness and grain dust-induced inflammation in the lower respiratory tract. *Am. J. Physiol.* **267:**L609–L617.
162. **Senthilselvan, A., Y. Zhang, J. A. Dosman, E. M. Barber, L. E. Holfeld, S. P. Kirychuk, Y. Cormier, T. S. Hurst, and C. S. Rhodes.** 1997. Positive human health effects of dust suppression with canola oil in swine barns. *Am. J. Respir. Crit. Care Med.* **156:**410–417.

163. **Sessa, R., P. M. Di Pietro, G. Schiavoni, I. Santino, A. Altieri, S. Pinelli, and M. Del Piano.** 2002. Microbiological indoor air quality in healthy buildings. *New Microbiol.* **25:**51–56.
164. **Shelton, B. G., W. D. Flanders, and G. K. Morris.** 1999. Mycobacterium sp. as a possible cause of hypersensitivity pneumonitis in machine workers. *Emerg. Infect. Dis.* **5:**270–273.
165. **Sigsgaard, T., A. Abel, L. Donbaek, and P. Malmros.** 1994. Lung function changes among recycling workers exposed to organic dust. *Am. J. Ind. Med.* **25:**69–72.
166. **Sigsgaard, T., B. Bach, and P. Malmros.** 1990. Respiratory impairment among workers in a garbage-handling plant. *Am. J. Ind. Med.* **17:**92–93.
167. **Singh, J., and D. A. Schwartz.** 2005. Endotoxin and the lung: insight into the host-environment interaction. *J. Allergy Clin. Immunol.* **115:**330–333.
168. **Smid, T., D. Heederik, G. Mensink, R. Houba, and J. S. Boleij.** 1992. Exposure to dust, endotoxins, and fungi in the animal feed industry. *Am. Ind. Hyg. Assoc. J.* **53:**362–368.
169. **Smit, L. A., S. Spaan, and D. Heederik.** 2005. Endotoxin exposure and symptoms in wastewater treatment workers. *Am. J. Ind. Med.* **63:**30–39.
170. **Smit, L. A., I. M. Wouters, M. M. Hobo, W. Eduard, G. Doekes, and D. Heederik.** 2006. Agricultural seed dust as a potential cause of organic dust toxic syndrome. *Occup. Environ. Med.* **63:**59–67.
171. **Sonesson, A., L. Larsson, A. Fox, G. Westerdahl, and G. Odham.** 1988. Determination of environmental levels of peptidoglycan and lipopolysaccharide using gas chromatography-mass spectrometry utilizing bacterial amino acids and hydroxy fatty acids as biomarkers. *J. Chromatogr. Biomed. Appl.* **431:**1–15.
172. **Taha, M. P., S. J. Pollard, U. Sarkar, and P. Longhurst.** 2005. Estimating fugitive bioaerosol releases from static compost windrows: feasibility of a portable wind tunnel approach. *Waste Manag.* **25:**445–450.
173. **Tant, C. O., and E. O. Bennett.** 1956. The isolation of pathogenic bacteria from used emulsion oils. *Appl. Microbiol.* **4:**332–338.
174. **Thorn, J., L. Beijer, and R. Rylander.** 1998. Airways inflammation and glucan exposure among household waste collectors. *Am. J. Ind. Med.* **33:**463–470.
175. **Thorne, P. S.** 1993. Sump additives as a source of bioaerosols in a school building. *Vet. Hum. Toxicol.* **35:** 141–143.
176. **Thorne, P. S.** 2000. Inhalation toxicology models of endotoxin and bioaerosol-induced inflammation. *Toxicology* **152:**627–631.
177. **Thorne, P. S., A. Adamcakova-Dodd, K. M. Kelly, M. E. O'Neill, and C. Duchaine.** 2006. Metalworking fluid with mycobacteria and endotoxin induces hypersensitivity pneumonitis in mice. *Am. J. Respir. Crit. Care Med.* **173:**759–768.
178. **Thorne, P. S., K. H. Bartlett, J. Phipps, and K. Kulhankova.** 2003. Evaluation of five extraction protocols for quantification of endotoxin in metalworking fluid (MWF) aerosol. *Ann. Occup. Hyg.* **47:**31–36.
179. **Thorne, P. S., E. A. Chrischilles, A. K. W. Kuehl, K. M. Kelly, N. Metwali, L. M. Harris, M. E. O'Neill, A. K. Quella, and R. E. Walker.** 2006. Reduction of endotoxin and glucan exposures in the Louisa/Keokuk Environmental Intervention Project (LEIP) for rural childhood asthma. *J. Allergy Clin. Immunol.* **117:**S255.
180. **Thorne, P. S., J. A. DeKoster, and P. Subramanian.** 1996. Environmental assessment of aerosols, bioaerosols, and airborne endotoxin in a machining plant. *Am. Ind. Hyg. Assoc. J.* **57:**1163–1167.
181. **Thorne, P. S., and D. Heederik.** 1999. Indoor bioaerosols—sources and characteristics. p. 275–288. *In* T. Salthammer (ed.), *Organic Indoor Air Pollutants—Occurrence, Measurement, Evaluation.* Wiley-VCH, Weinheim, Germany.
182. **Thorne, P. S., K. Kulhankova, M. Yin, R. Cohn, S. Arbes, and D. C. Zeldin.** 2003. Endotoxin in house dust is associated with asthma but not allergy. *Am. J. Respir. Crit. Care Med.* **167:**A470.
183. **Thorne, P. S., K. Kulhankova, M. Yin, R. Cohn, S. Arbes, and D. C. Zeldin.** 2005. Endotoxin exposure is a risk factor for asthma in a U.S. national survey of housing. *Am. J. Respir. Crit. Care Med.* **172:**1371–1377.
184. **Thorne, P. S., P. B. McCray, T. S. Howe, and M. A. O'Neill.** 1998. Early-onset inflammatory responses *in vivo* to adenoviral vectors in the presence or absence of lipopolysaccharide-induced inflammation. *Am. J. Respir. Cell Mol. Biol.* **20:**1155–1164.
185. **Thorne, P. S., N. Metwali, E. Avol, and R. McConnell.** 2005. Surface sampling for endotoxin assessment using electrostatic wiping cloths. *Ann. Occup. Hyg.* **49:**401–406. (First published 2 February 2005; doi:10.1093/annhyg/mei002.)
186. **Thorne, P. S., S. J. Reynolds, D. K. Milton, P. D. Bloebaum, X. Zhang, P. Whitten, and L. F. Burmeister.** 1997. Field evaluation of endotoxin air sampling assay methods. *Am. Ind. Hyg. Assoc. J.* **58:**792–799.
187. **Thorne, P. S., and N. Sprince.** 2004. Metal working fluids, p. 1043–1055. *In* L. Rosenstock, M. R. Cullen, C. A. Redlich, and C. Brodkin (ed.), *Textbook of Clinical Occupational and Environmental Medicine,* 2nd ed. W. B. Saunders Co., Orlando, Fla.
188. **Thorne, P. S., C. P. Yeske, and M. H. Karol.** 1987. Monitoring guinea pig core temperature by telemetry during inhalation exposures. *Fundam. Appl. Toxicol.* **9:** 398–408.
189. **Toivola, M., S. Alm, T. Reponen, S. Kolari, and A. Nevalainen.** 2002. Personal exposures and microenvironmental concentrations of particles and bioaerosols. *J. Environ. Monit.* **4:**166–174.
190. **Tulic, M. K., J. L. Wale, P. G. Holt, and P. D. Sly.** 2000. Modification of the inflammatory response to allergen challenge after exposure to bacterial lipopolysaccharide. *Am. J. Respir. Cell Mol. Biol.* **22:**604–612.
191. **Ulevitch, R. J., and P. S. Tobias.** 1999. Recognition of gram-negative bacteria and endotoxin by the innate immune system. *Curr. Opin. Immunol.* **11:**19–22.
192. **Van Tongeren, M., L. Van Amelsvoort, and D. Heederik.** 1997. Exposure to organic dusts, endotoxins, and microorganisms in the municipal waste industry. *Int. J. Occup. Environ. Health* **3:**30–36.
193. **Venter, P., J. F. Lues, and H. Theron.** 2004. Quantification of bioaerosols in automated chicken egg production plants. *Poult. Sci.* **83:**1226–1231.
194. **Wan, G. H., C. S. Li, and R. H. Lin.** 2000. Airborne endotoxin exposure and the development of airway antigen-specific allergic responses. *Clin. Exp. Allergy* **30:** 426–432.
195. **Wang, X.-R., H.-X. Zhang, B.-X. Sun, H.-L. Dai, J.-Q. Hang, E. A. Elsen, D. H. Wegman, S. A. Olenchock, and D. C. Christiani.** 2005. A 20-year follow-up study on chronic respiratory effects of exposure to cotton dust. *Eur. Respir. J.* **26:**881–886.
196. **Watanabe, J., Y. Miyazaki, G. A. Zimmerman, K. H. Albertine, and T. M. McIntyre.** 2003. Endotoxin contamination of ovalbumin suppresses murine immunologic responses and development of airway hyper-reactivity. *J. Biol. Chem.* **278:**42361–42368.
197. **Weary, M. E., G. Donohue, F. C. Pearson, and K. Story.** 1980. Relative potencies of four reference endo-

toxin standards as measured by the *Limulus* amoebocyte lysate and USP rabbit pyrogen tests. *Appl. Environ. Microbiol.* **40:**1148–1151.

198. **Weiss, J. P.** 2003. Bactericidal/permeability-increasing protein (BPI) and lipopolysaccharide-binding protein (LBP): structure, function and regulation in host defence against Gram-negative bacteria. *Biochem. Soc. Trans.* **31:**785–790.
199. **Wickens, K., J. M. Lane, P. Fitzharris, R. Siebers, G. Riley, J. Douwes, T. Smith, and J. Crane.** 2002. Farm residence and exposures and the risk of allergic diseases in New Zealand children. *Allergy* **57:**1171–1179.
200. **Wouters, I.** 1999. Endotoxin and beta-(1→3)-glucan exposure in household waste collectors and compost workers: a relation with upper airway inflammation. *Schriftenr. Ver. Wasser Boden Lufthyg.* **104:**546–550.
201. **Wouters, I. M., J. Douwes, G. Doekes, P. S. Thorne, B. Brunekreef, and D. J. Heederik.** 2000. Increased levels of markers of microbial exposure in homes with indoor storage of organic household waste. *Appl. Environ. Microbiol.* **66:**627–631.
202. **Wouters, I. M., S. K. Hilhorst, P. Kleppe, G. Doekes, J. Douwes, C. Peretz, and D. Heederik.** 2002. Upper airway inflammation and respiratory symptoms in domestic waste collectors. *Occup. Environ. Med.* **59:** 106–112.
203. **Wouters, I. M., S. Spaan, J. Douwes, G. Doekes, and D. Heederik.** 2006. Overview of personal occupational exposure levels to inhalable dust, endotoxin, beta (1→3)-glucan and fungal extracellular polysaccharides in the waste management chain. *Ann. Occup. Hyg.* **50:**39–53.
204. **Yu, B., and S. D. Wright.** 1996. Catalytic properties of lipopolysaccharide binding protein. Transfer of LPS to soluble CD14. *J. Biol. Chem.* **271:**4100–4105.
205. **Zock, J.-P., A. Hollander, G. Doekes, and D. Heederik.** 1996. The influence of different filter elution methods on the measurement of airborne potato antigens. *Am. Ind. Hyg. Assoc. J.* **57:**567–570.
206. **Zock, J. P., A. Hollander, D. Heederik, and J. Douwes.** 1998. Acute lung function changes and low endotoxin exposures in the potato processing industry. *Am. J. Ind. Med.* **33:**384–391.
207. **Zucker, B. A., and W. Muller.** 2000. Investigations on airborne microorganisms in animal stables. 3. Relationship between inhalable endotoxin, inhalable dust and airborne bacteria in a hen house. *Berl. Munch. Tierarztl. Wochenschr.* **113:**279–283.
208. **Zuravleff, J. J., V. L. Yu, J. W. Shonnard, J. D. Rihs, and M. Best.** 1983. Legionella pneumophila contamination of a hospital humidifier. Demonstration of aerosol transmission and subsequent subclinical infection in exposed guinea pigs. *Am. Rev. Respir. Dis.* **128:**657–661.

Legionellae and Legionnaires' Disease

BARRY S. FIELDS

79

INTRODUCTION

Bacteria of the family *Legionellaceae* are representatives of a group of organisms that survive as intracellular parasites or endosymbionts of free-living protozoa. The *Legionellaceae* maintained their anonymity until 1977, primarily because of their unique ecology. Initially the bacterium *Legionella pneumophila* was labeled the Legionnaires' disease bacterium. This nomenclature may have been misleading since one can easily assume that causing human disease is paramount to the organisms' existence. Humans may have coexisted with these bacteria for a very long time, but recent industrial technology has provided these organisms with a means for causing infection in humans.

Legionnaires' Disease

Legionnaires' disease is a consequence of altering the environment for human benefit. This illness continues to comprise a significant proportion of the many cases of pneumonia that occur in the developed world each year. Although the disease can be effectively treated with appropriate antimicrobial agents, current measures are much less efficient at preventing transmission of the bacteria from the environment to susceptible hosts.

Legionnaires' disease occurs when sufficient numbers of legionellae are aerosolized and subsequently inhaled by a susceptible host (12). Legionnaires' disease and legionellae have been included in the aerobiology section of this manual because the bacteria are transmitted to the host via aerosols. However, because these bacteria are found primarily in freshwater environments, many of the methods to be discussed are associated with microbiological examination of water. Legionellae are relatively difficult to detect in the environment and even more difficult to detect clinically. The fastidious nature of these bacteria is primarily due to the fact that they derive their nutrients from and multiply in an intracellular environment. The bacteria multiply within single-cell protozoans in the environment and within alveolar macrophages in humans. The methods used to detect these organisms have evolved from research in both clinical and environmental microbiology.

Legionellosis

Legionellae were first isolated and identified as part of the investigation of respiratory illness in persons attending an American Legion convention in Philadelphia in 1976 (39, 60). This highly publicized investigation documented 239 cases and 34 deaths due to a previously unrecognized cause of pneumonia. Legionellae are now associated with two forms of respiratory illness, collectively referred to as legionellosis (44, 53). Legionnaires' disease is the pneumonic and more severe form of legionellosis (Table 1). The other form of respiratory illness is named Pontiac fever after the first documented outbreak, which occurred at a health department in Pontiac, Mich. (53). Possible explanations for the manifestation of two disease syndromes caused by the same bacterium include the inability of some legionellae to multiply in human tissue for a variety of reasons, including virulence, host range, or viability of the bacteria and differences in host susceptibility (32, 62, 73). Recent studies have provided limited evidence that Pontiac fever outbreaks may be associated with exposure to bacterial endotoxin (18, 34). Community-based pneumonia incidence studies have estimated that there are between 8,000 and 18,000 cases of legionellosis annually in the United States (33), approximately 25-fold higher than the number of cases annually reported to the Centers for Disease Control and Prevention. These studies have shown that the majority of cases of legionellosis are sporadic, with only about 4% being outbreak related (58). The sources of community-acquired cases are difficult to identify, partly because of the ubiquitous nature of the bacterium. Although the organisms are relatively common in the environment, they infrequently cause disease (46).

A conceptual chain of causation describing the events that lead to an outbreak of legionellosis states that legionellae must be present in an environmental reservoir, multiply from low to high concentrations, and be disseminated from a reservoir so as to expose susceptible individuals (38). This scheme serves as a model for understanding the epidemiology of legionellosis as well as development of prevention strategies.

THE *LEGIONELLACEAE* AND THEIR ECOLOGY

Bacteria of the genus *Legionella* are characterized as gram-negative, aerobic, rod-shaped bacteria. Cells are 0.3 to 0.9 by 1 to 20 μm and motile, with one or more polar or lateral flagella (15). Legionellae use amino acids as their carbon and energy sources and do not oxidize or ferment carbohydrates. Currently, there are 49 species comprising 71 distinct

TABLE 1 Characteristics of Legionnaries' disease and Pontiac fever

Characteristic	Disease	
	Legionnaires' disease	Pontiac fever
Illness	Progressive pneumonia, sometimes fatal	Influenza-like illness (nonpneumonic)
Incubation period	2–10 days	36 h (mean)
Attack rate	<5%	>95%
Risk factors	Cigarette smoking, diabetes mellitus, cancer, end-stage renal disease, AIDS	None known

serogroups in the genus *Legionella* (6, 33, 67). Species identification and differentiation are performed serologically, although antisera for many species and serogroups are not available commercially (15). Approximately one-half of the 49 species of *Legionellae* have been associated with human disease, but a single species of *Legionella*, *L. pneumophila*, causes approximately 90% of all documented cases of legionellosis (57). Although there are now 15 serogroups of *L. pneumophila*, 82% of all legionellosis cases are caused by *L. pneumophila* serogroup 1. It is likely that most of the legionellae can cause human disease under the appropriate conditions; however, these infections are infrequently reported because they are rare and because of the lack of diagnostic reagents.

There appear to be a number of unidentified legionellae which cannot be grown on routine *Legionella* media. These organisms have been given the acronym LLAPs (*Legionella*-like amoebal pathogens) because they have been detected through their ability to grow intracellularly in protozoan cells (75). One of the organisms, *Sarcobium lyticum*, was isolated from protozoa in soil by Drozanski in 1954 (25). Analysis of 5S rRNA sequences led to a proposal to rename *Sarcobium lyticum* as *Legionella lytica* comb. nov. in 1996 (47). One LLAP strain was isolated from the sputum of a pneumonia patient by enrichment in amoebae (75). These studies indicate that there may be additional species of legionellae that represent human pathogens but cannot be detected by conventional culture techniques.

The Ecology of *Legionella*

Water is the major reservoir for legionellae, and the bacteria are found in freshwater environments worldwide (37). Legionellae have been detected in as many as 40% of freshwater environments by culture and in up to 80% of freshwater sites tested by PCR (79). Several outbreaks of legionellosis have been associated with construction, and it was originally believed that the bacteria could survive and be transmitted to humans via soil. However, legionellae do not survive in dry environments, and these outbreaks are more likely the result of massive descalement of plumbing systems due to changes in water pressure during construction (52, 61).

Initially, it was difficult to explain the pervasiveness of legionellae in aquatic environments. The bacteria are fastidious and require an unusual combination of nutrients in bacteriologic medium. These levels of nutrients would rarely be found in aquatic environments and, if present, would serve only to amplify faster-growing bacteria that would compete with the legionellae. However, the nutrients required by legionellae represent the need for an intracellular environment, not soluble nutrients commonly found in fresh water. Legionellae survive in aquatic and in some moist soil environments as intracellular parasites of free-living protozoa (29, 59). Rowbotham first described the ability of *L. pneumophila* to infect protozoa (72) and later described these bacteria as "protozoonotic (i.e., naturally infecting protozoa)" (74). Figure 1 shows *L. pneumophila* multiplying within the ciliated protozoan *Tetrahymena pyriformis*. Legionellae have been reported to multiply in 14 species of amoebae, 2 species of ciliated protozoa, and 1 species of slime mold, while growth of legionellae in the absence of protozoa has been documented as occurring only on laboratory media (31, 42, 81). A number of studies have described the relationship between legionellae and protozoa in aquatic environments identified as potential or actual reservoirs of disease-causing strains (10, 45). Protozoa naturally present in these environments can support intracellular growth of legionellae in vitro (5). *Legionella* can infect and multiply intracellularly in both protozoa and human phagocytic cells (31, 48). Infection of both these hosts is accomplished utilizing a novel type IVB secretion system composed of 25 genes (78). This type IV secretion system delivers substrates which allow the bacteria to subvert host cell processes for its own propagation. It appears that protozoa are the natural hosts of legionellae, whereas human phagocytic cells occasionally become ill-fated surrogates. Understanding that protozoa play a crucial role in the ecology of legionellae is critical to the development of successful prevention strategies. To understand the ecology of legionellae, these bacteria must be considered in the context of their microbial community, not as independent inhabitants of freshwater environments.

Aerosol Transmission from Aquatic Environments

Inhalation of legionellae in aerosolized droplets is the primary means of transmission for legionellosis (12). These aerosolized droplets must be of a respirable size (1 to 5 μm). No person-to-person transmission of Legionnaires' disease has been documented. A number of devices have been implicated as sources of aerosol transmission of legionellae (Table 2). These sources are of two general types: those producing aerosols of contaminated potable water, such as showers and faucets, and those from nonpotable water, such as cooling towers and whirlpool spas. Meaningful identification of sources of transmission requires a multidisciplinary approach including epidemiology, molecular epidemiology, and microbiologic techniques including water and, occasionally, air sampling (12).

DETECTION OF LEGIONELLAE IN THE ENVIRONMENT

It is generally accepted that culturing environmental samples for the presence of *Legionella* is necessary to establish the source of an outbreak of Legionnaires' disease, to determine the efficacy of a disinfection program, and to evaluate the potential of a device to transmit the disease. However, recommendations concerning the routine culturing of water samples in the absence of documented cases of legionellosis represent an area of considerable controversy. Detection of

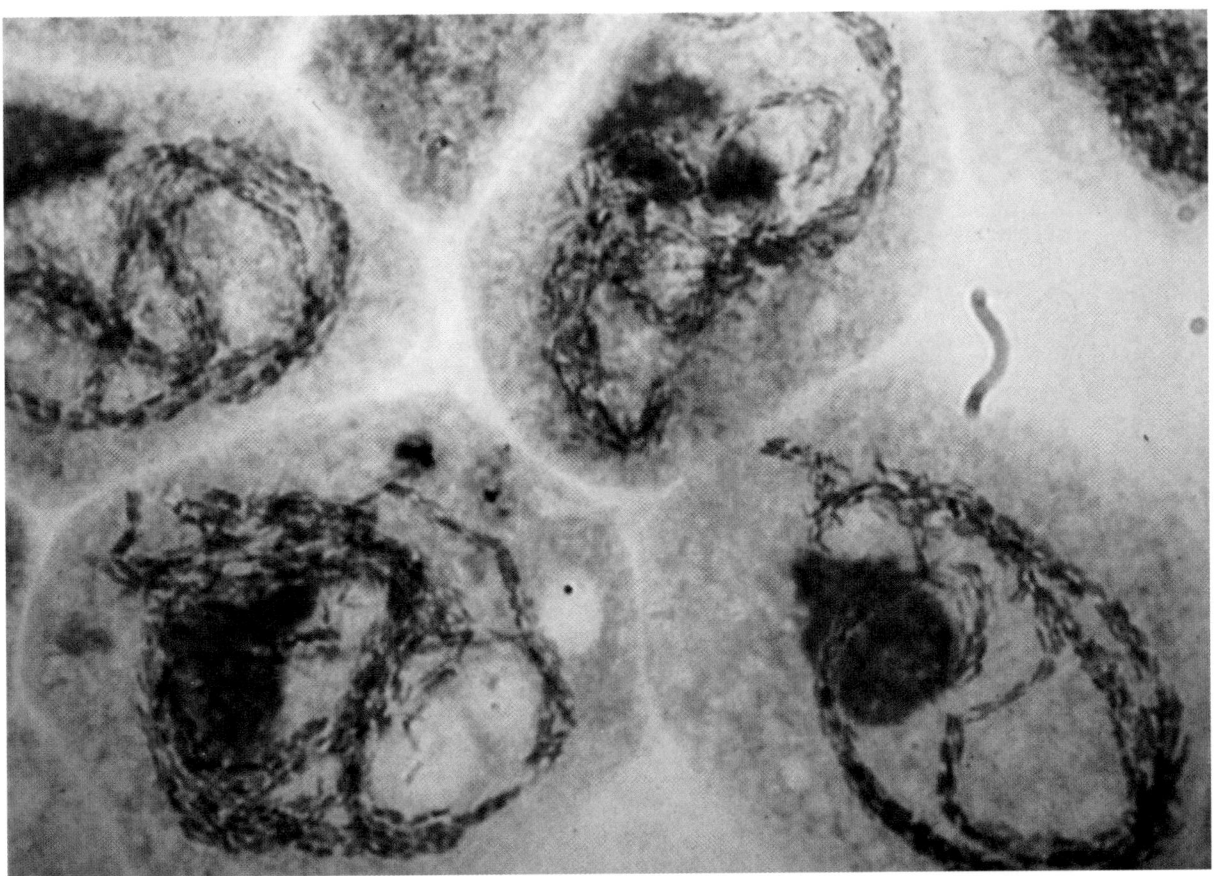

FIGURE 1 Gimenez stain of the ciliated protozoan *T. pyriformis* infected with *L. pneumophila*. Chains of multiplying *Legionella* cells are contained within vesicles, as observed in human phagocytic cells infected with *L. pneumophila*. Magnification, ×1,650.

legionellae in an environmental source is not necessarily evidence of the potential for disease. There is general agreement that such environmental sampling can be beneficial in institutions housing persons at extremely high risk for acquiring Legionnaires' disease. However, if such a monitoring program is undertaken, it should be understood that the objective must be the absence of detectable legionellae in these water systems. Also there are no data to suggest the frequency of such testing or the appropriate number of sites to test. There is a single study which suggests that detecting *Legionella* in >30% of sites tested may be an indicator of potential for disease in health care facilities (8). However, more data are needed to substantiate this concept. Guidelines that recommend widespread use of culture for *Legionella* in the absence of disease acknowledge the arbitrary nature of recommendations regarding testing frequency.

TABLE 2 Sources known to transmit legionellae via aerosols[a]

Type of water	Transmitting device(s)
Potable	Showers, tap water faucets, respiratory therapy equipment
Nonpotable	Cooling towers and evaporative condensers, whirlpool spas, decorative fountains, ultrasonic mist machines, humidifiers

[a]Modified from reference 12 with permission.

Determining a scientifically based number of legionellae that are acceptable in water systems would be valuable information and might be able to assist in control of the disease. Unfortunately, this is not likely because of issues such as differences in building design, occupant susceptibility, differences in the virulence of various *Legionella* species and strains, and variations in aerosol mechanics. Also, there are ethical hurdles involved in performing these studies, whereby test populations cannot be exposed to certain levels of *Legionella* to establish that a particular environment is safe. Some guidelines for the prevention of legionellosis offer action plans based upon the number of *Legionella* CFU detected by culturing cooling towers or potable water. To date, all numerical action plans are derived from a table published by a commercial testing lab (65). It is acknowledged that these concentrations are somewhat arbitrary and do not apply to water systems used by immunocompromised individuals. Continued scientific analysis should eventually be able to establish when and where testing for the presence of *Legionella* will be beneficial.

The following sections describe procedures used to detect legionellae in the environment but do not address the relevance of these methods to public health.

Collection of Water Samples

The number and types of sites that should be tested to detect legionellae must be determined on an individual basis. This is because of the diversity of plumbing and heating, ventilation, and air-conditioning systems in a variety of institutions that may be sampled. These institutions can include industrial facilities, hotels, hospitals, retirement homes, public facilities, and domestic environments. An environmental sampling protocol addressing the selection of the appropriate sites to sample within a hospital was published in 1987 (6). This protocol can serve as a prototype for identifying sites that should be sampled in a variety of institutions. Generally, any water source that may be aerosolized should be considered a potential source for the transmission of legionellae. The bacteria are rarely found in municipal water supplies and tend to colonize plumbing systems and point-of-use devices. To colonize a system, the bacteria must multiply, and this requires temperatures above 25°C (15). Therefore, legionellae are most commonly found in hot water systems. The bacteria do not survive drying (52), and so condensate from air-conditioning equipment, which frequently evaporates, is not a likely source (Table 2).

Two primary sample types should be collected when sampling for legionellae: water samples and swabs of point-of-use devices or system surfaces (19). Collection of at least 1 liter of water allows concentration of the sample if necessary. If the water source has recently been treated with chlorine or bromine, 0.5 ml of 0.1 N sodium thiosulfate may be added to each 1-liter sample to neutralize the disinfectant.

Swabs allow sampling of biofilms, which frequently contain legionellae. These can be taken from various points within plumbing systems or from surfaces of basins of cooling towers or spas. Swabs of faucet aerators and showerheads should be taken in conjunction with water samples from these sites and should be taken with the aerator or showerhead removed. The swabs can be streaked directly onto an agar plate or submerged in a small volume of water taken at the same time to prevent drying during transportation to the laboratory.

All samples should be transported to the laboratory in insulated coolers as protection against extreme heat or cold. Samples that will not be processed within 72 h from the time of collection should be refrigerated.

Culture of Water Samples

A schematic representation of methods for processing water samples for culture is shown in Fig. 2. The procedure chosen depends on the expected degree of total bacterial contamination in a particular sample. Potable waters generally have low bacterial concentrations and are either cultured directly or concentrated to detect legionellae. Nonpotable waters, such as those from cooling towers, generally do not require concentration because of their high bacterial concentrations.

Sample Concentration

Samples may be concentrated 10-fold or more by using either filtration or centrifugation. Filtration is used more frequently, although either procedure can be used successfully (19, 87). Water should be filter concentrated in a biological safety cabinet, using 0.2-μm-pore-size polycarbonate filters. Polycarbonate membranes allow suspended particles to collect on the filter surface without being trapped as with matrix-type filters. The filter membrane is then resuspended into a volume of the sterile water and vortexed for 30 s. Alternatively, samples may be concentrated by centrifugation at 1,000 × *g* for 10 min, removing all but 10 ml of the supernatant, and vortexing (87).

Acid or Heat Pretreatment

A selective procedure is required to reduce the number of non-*Legionella* bacteria before culturing some water samples with high total bacterial concentrations. Non-*Legionella* bacteria can be selectively killed by either acid pretreatment or brief exposure to higher temperatures (9, 24). Legionellae are more resistant to lower pH and brief exposures to higher temperatures than many other freshwater bacteria. For acid pretreatment, the sample is mixed and incubated with an acid buffer (pH 2.2) for 3 to 30 min (9). The sample is neutralized by the buffer within buffered charcoal yeast extract (BCYE) agar and therefore must be spread on the agar plate at the end of the period of incubation with the acid buffer. Heat pretreatment is accomplished

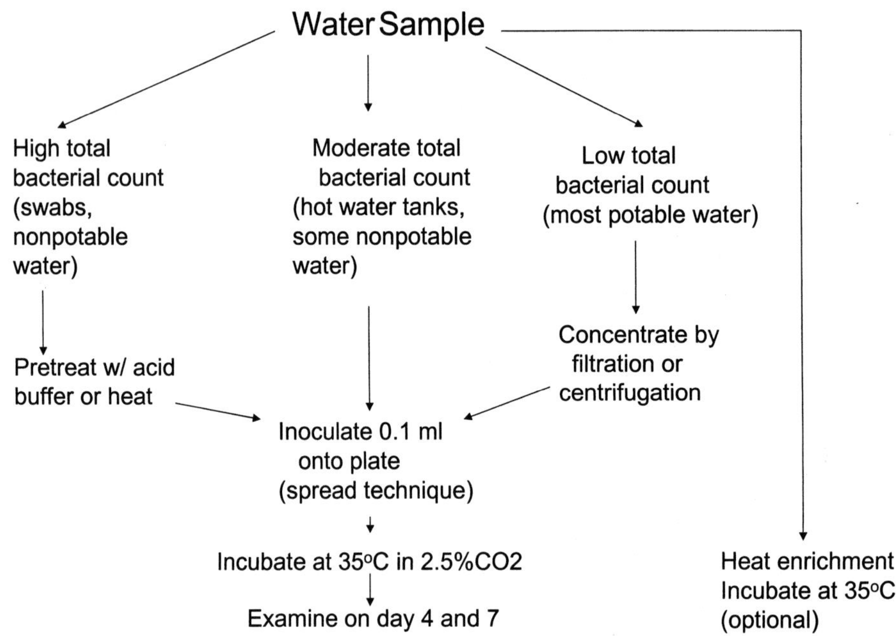

FIGURE 2 Overview of procedures for the culture of water samples to detect legionellae.

by incubating 10 ml of sample in a 50°C water bath for 30 min (24).

Heat Enrichment

Heat enrichment or incubation of specimens at 35°C can improve recovery of legionellae by up to 30% (76). However, this procedure requires a considerable length of time before results can be obtained and may not be practical in many situations. Heat enrichment relies on autochthonous protozoa to amplify undetectable levels of legionellae. Aliquots of samples are incubated at 35°C. Incubated samples can be cultured after 2 to 6 weeks.

Culture Media

Legionellae were isolated by procedures used for the recovery of rickettsiae as early as 1943 (the TATLOCK strain, now *Legionella micdadei*), although these bacteria were not identified as legionellae until 1979 (11, 16, 43). The bacterium was first isolated on bacteriologic agar, using Mueller-Hinton agar supplemented with hemoglobin and IsoVitaleX (MH-IH) (30). The essential component in hemoglobin was found to be a soluble form of iron, and L-cysteine is the essential amino acid provided by IsoVitaleX. These refinements led to the development of Feeley-Gorman agar, which provides better recovery of the organism from tissue (30). Later, starch was replaced with charcoal to detoxify the medium and the amino acid source was changed to yeast extract, resulting in charcoal yeast extract agar (29). Charcoal yeast extract agar is the base form for most media used to grow legionellae. The medium used for the culture of legionellae has been improved several times, eventually resulting in the medium currently used, BCYE agar (26, 29, 67). The most widely used form of BCYE agar is supplemented with α-ketoglutarate (26, 68). Table 3 lists the primary components of BCYE agar and the supplements added for various purposes (29, 64, 86, 88).

Culture of environmental samples requires the use of selective and nonselective media in conjunction with previously described selection procedures. Most laboratories use multiple plates for each sample, including a BCYE agar plate, a BCYE agar plate containing three antimicrobial agents, and a BCYE agar plate containing the three antimicrobial agents plus glycine (Table 3). These media can be prepared with or without the indicator dyes, which impart a color specific for certain species of *Legionella* (86). Although the majority of *Legionella* spp. grow readily on BCYE agar, some require supplementation with bovine serum albumin to enhance growth. *L. micdadei* and several strains of *Legionella bozemanae* show a preference for BCYE with 1.0% albumin (64). All agar plates are inoculated with 0.1 ml of sample by the spread plate technique and incubated at 35°C in a humidified 2.5% CO_2 atmosphere or candle extinction jar.

Identification of Legionellae

Colonies of legionellae require approximately 72 h to appear on BCYE agar and may require 7 days or longer. Ideally, plates should be examined after 4 days of incubation and a second time before being discarded after between 7 and 10 days of incubation. Plates should be examined with a dissecting microscope and a light source to detect bacterial colonies resembling legionellae. After approximately 4 days of incubation, these colonies are 2 to 4 mm in diameter, convex, and round with entire edges (Fig. 3). The center of the colony is usually a bright white with a textured appearance that has been described as "cut-glass like" or speckled. The white center of the colony is often bordered with blue, purple, green, or red iridescence. Some species of legionellae produce colonies that exhibit blue-white or red autofluorescence (24). The primary isolation plates can be examined with long-wave UV light to detect these autofluorescent colonies.

Colonies resembling legionellae can be presumptively identified on the basis of their requirement for L-cysteine by subculture on blood agar or BCYE agar without L-cysteine. Subcultured colonies that grow on BCYE agar, but not on blood agar or BCYE without L-cysteine, are presumed to be legionellae. Legionellae are relatively inert in many biochemical test media, so these tests are of limited value in the identification of these bacteria. Definitive identification is usually accomplished by using a direct fluorescent antibody (DFA) or slide agglutination test with specific antisera (19). Identification can also be accomplished using fatty acid analysis and DNA hybridization (15).

Air Sampling

Examination of water samples is the most efficient microbiologic method for identifying sources of legionellae. Air

TABLE 3 Components and supplements of BCYE agar for culturing legionellae from the environment

Component	Concn	Purpose
Charcoal	2.0 g/liter	Base component
Yeast extract	10.0 g/liter	Base component
ACES[a] buffer	10.0 g/liter	Base component
Ferric pyrophosphate	0.25 g/liter	Base component
L-Cysteine	0.4 g/liter	Base component
Potassuim α-ketoglutarate	1.0 g/liter	Base component
Agar	17.0 g/liter	Base component
Glycine	3.0 g/liter	Selective agent
Polymyxin B	50–100 U/ml	Selective agent (gram negative)
Vancomycin or cefamandole	1–5 g or 4 mg/liter	Selective agent (gram positive)
Anisomycin or cycloheximide	80 μg/ml (for either)	Selective agent (fungal)
Bromocresol blue	10 mg/liter	Indicator dye
Bromocresol purple	10 mg/liter	Indicator dye
Bovine serum albumin	10 g/liter	Supplement for some fastidious legionellae

[a]ACES, *N*-(2-acetamido)-2-aminoethanesulfonic acid.

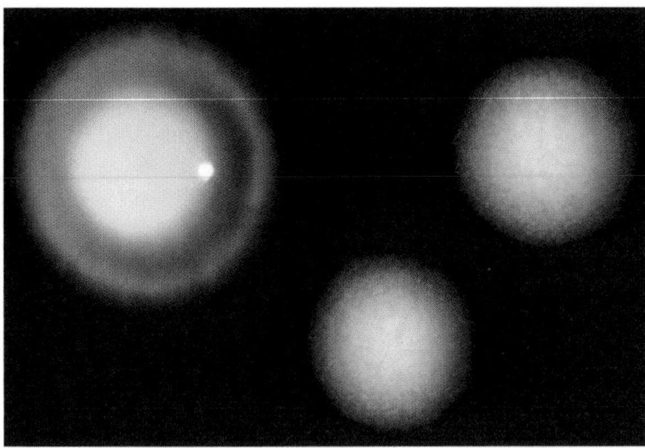

FIGURE 3 Two *Legionella* colonies and a non-*Legionella* bacterial colony as seen through a dissecting microscope upon primary isolation (4 days of incubation). Note the white "cut-glass" appearance of the center of the colony and the purple iridescence which borders it. The iridescence can be one of several colors; the significance of the colors is unknown.

sampling is an insensitive means of detecting these bacteria and therefore is of limited value in environmental sampling for legionellae. In certain instances, it may be beneficial to demonstrate the presence of legionellae in aerosol droplets associated with suspected reservoirs of the bacterium. Air sampling has been used to better define the roles of certain devices such as showers, faucets, and evaporative condensers in disease transmission (12). It is usually used to establish the presence of the legionellae in aerosol droplets and occasionally to quantitate or determine the sizes of particles containing legionellae. Information regarding particle size and numbers of viable bacteria can be calculated using these procedures, but this approach requires much more stringent controls and calibration (19). Samplers should be placed in locations representative of human exposure, and investigators should wear an Occupational Safety and Health Administration-approved respirator if sampling involves exposure to potentially infectious aerosols.

The basic methods for sampling airborne bacteria include impingement in liquids, impaction onto solid surfaces, filtration, and precipitation. See chapter 74 for detailed discussion of sampling techniques for airborne microorganisms. Methods that have been used to sample air for legionellae include impingement in liquid by using an all-glass impinger (AGI), impaction of solid medium by using Andersen samplers, and the use of settle plates (13). Except for settle plates, a nonquantitative sampling method, these methods require a vacuum source and a means of controlling airflow. Several configurations of air sampling equipment can be used; they usually incorporate a device for controlling airflow (flowmeter-manometer) connected in the vacuum line between the sampler and vacuum source.

All glass impingers (AGI-30; Ace Glass, Inc., Vineland, N.J.) with the stem 30 mm from the bottom of the flask have been used successfully to sample for legionellae (13). These samplers use the principle of impingement and washing of air, in which organisms are entrapped in a liquid medium. Because of the velocity at which samples are collected, clumps tend to be fragmented, leading to a more accurate count of bacteria present in the air. The disadvantages of this method are that this velocity tends to destroy some vegetative cells, it does not differentiate particle sizes, and AGI-30s are easily broken in the field. Yeast extract broth (0.25%) is the recommended liquid medium for AGI-30 sampling of legionellae (19). Once the sample has been collected, the yeast extract broth may be processed by methods used for the culture of water samples.

Six-stage Andersen samplers are viable particle samplers in which particles pass through jet orifices of decreasing size in cascade fashion until they hit an agar surface (see chapter 74). The agar plates are then removed and incubated in order to culture any legionellae present. The stage distribution of the legionellae should indicate the extent to which the bacteria would have penetrated the respiratory system. The advantages of this sampling method are that the equipment is more durable, the sampler can determine the number and sizes of droplets containing legionellae, and agar plates can be placed directly in an incubator with no further manipulations. Both selective and nonselective BCYE agar can be used in an Andersen sampler. If the samples must be shipped to a laboratory, they should be packed and shipped without refrigeration as soon as possible.

Nonculture Methods for Detection of Legionellae

Several nonculture methods have been developed to detect legionellae in environmental samples. These methods offer the potential of greatly increased sensitivity. Culture remains the method of choice for detecting legionellae, primarily because nonculture methods cannot provide information regarding the viability of the bacteria. These nonculture methods include detection of the organisms with specific antisera by DFA staining and procedures to detect nucleic acids of legionellae by use of PCR.

The use of DFA to detect legionellae is limited by the number of specific antisera that can be used. There are no antisera which specifically react with all *Legionella* species, necessitating the use of a different antiserum for each species or serogroup. Reports on the sensitivity and specificity of DFA testing of environmental specimens vary greatly, with most studies indicating that the test is relatively insensitive and nonspecific (50).

The use of PCR for detecting nucleic acids of legionellae in the environment has proved to be valuable in some investigations of outbreaks of legionellosis (62). A number of *Legionella* genes, including 5S rRNA, 16S rRNA, and *mip* genes, have been used as targets for PCR (56, 82). Use of PCR to detect legionellae in the environment has suggested that up to 80% of freshwaters are positive while only 20 to 40% are positive by culture (3, 79). This discrepancy could be due to the presence of nonviable or injured organisms, a nonspecific reaction with unrelated organisms (although available data argue against this possiblility), or the presence of related organisms, such as *S. lyticum*, that cannot be detected by conventional techniques used for legionellae. Recent advances using real-time PCR techniques may improve the suitability of these assays for detecting legionellae in the environment (55). Most investigations of epidemic legionellosis have used culture to detect legionellae in the environment. As a result, most of our epidemiologically relevant information concerning legionellosis is based on direct culture data. Until a better understanding of the diversity and distribution of the legionellae is attained, results from non-culture-based methods should be interpreted cautiously.

Subtyping Techniques

Associating an environmental isolate of *Legionella* with a clinical isolate from a patient with legionellosis usually requires a molecular subtyping procedure. *L. pneumophila* serogroup 1 (Lp1) accounts for most of the cases of legionellosis (12). However, Lp1 can be divided into a number of subtypes by using various techniques, indicating that this is a fairly heterogeneous serogroup (70). Identification of the bacterium, even to the serogroup level, is not sufficient to implicate an environmental isolate as the source of disease.

Initially legionellae were identified to the serogroup level during investigations of legionellosis. This form of serologic subtyping uses polyvalent or monoclonal antisera and may be adequate for identifying reservoirs of some of the more uncommon legionellae causing disease. The variety of strains and distribution of Lp1 necessitate more elaborate subtyping procedures to discriminate within these bacteria. Several groups of monoclonal antibodies have been developed for this purpose (4). An international panel of seven monoclonal antibodies was proposed in 1986 (4, 51). Although much information has been gained through the use of this panel, several of the cell lines have been lost and most of these reagents are no longer available. Use of these monoclonal antibodies has identified 10 type strains within Lp1. The ability to differentiate Lp1 into 10 subtypes has greatly improved our ability to correctly identify sources of disease. For example, several outbreak investigations have found Lp1 in both the potable water system and the cooling towers of a particular institution (13, 14). Differentiation with monoclonal antibodies indicated that the epidemic strain had colonized only one of these systems (either cooling tower or potable systems), indicating that the presence of Lp1 in the other system was not causing disease at that site. Without this level of subtyping, it would be much more difficult to confidently confirm the source of transmission.

DNA fingerprinting techniques provide a level of discrimination similar to the use of monoclonal antibodies, and these techniques are complementary (70). Techniques used to discriminate between isolates of legionellae include restriction fragment length polymorphism analysis, plasmid analyses, electrophoretic alloenzyme typing, and RNA/DNA probing of DNA digests, pulsed-field gel electrophoresis, and arbitrarily primed PCR (70, 77, 84). Amplified fragment length polymorphism has been used since the late 1990s (71, 85). An intercenter study by the European Working Group on Legionella Infections comprising 13 laboratories using a standardized protocol demonstrated that the amplified fragment length polymorphism method was highly reproducible and epidemiologically concordant with good discrimination. The method has been adopted as the first standardized typing method for the investigation of travel-associated Legionnaires' disease in Europe (40).

Currently a number of laboratories are investigating the use of multilocus sequence typing as the next-generation typing method for legionellae (41). This method involves comparison of nucleic acid sequence data for a limited number of well characterized bacterial genes. Recent studies have shown that the comparison of six different loci by this technique offers sufficient discrimination for use in epidemiologic investigations of outbreaks caused by *L. pneumophila* (41). Given the advances in nucleic acid sequence technology and bioinformatics, it would appear that sequence-based typing systems will eventually replace other molecular typing methods.

ENVIRONMENTAL APPROACHES TO CONTROLLING LEGIONELLOSIS

Practical information concerning treatment processes that effectively control legionellae is limited. Various biocides and alternative disinfection methods, such as heat eradication, UV irradiation, ionization, and ozonation, have been tested to determine their abilities to kill legionellae (27, 66, 80, 84). Results obtained in these types of laboratory studies often fail to translate into effective prevention protocols (28). Several countries have produced guidelines or codes of practice relating to the control of legionellae. However, research to substantiate these practices is scarce, and the prevailing rationale for these recommendations is often empirical (17). Currently in the United States there are 13 guidelines that address the prevention and control of legionellosis (35). Five of these are federal guidelines, two are state guidelines, one is a county guideline, and five have been published by professional societies. These guidelines address maintenance temperatures, biocide levels, and emergency procedures for a number of building water systems including potable sytems, cooling towers, and heated spas.

If one utilizes the chain of causation leading to an outbreak of legionellosis (as described earlier in this chapter), a control strategy would only need to interrupt this chain of events to be successful. The following discussion of control strategies is arranged according to this scheme.

Environmental Reservoirs and Water Supplies

It is highly unlikely that legionellae could be eradicated from aquatic environments, as they are integral members of the aquatic microbial community. The bacteria are ubiquitous in freshwater environments, and they are present in relatively low concentrations in most of these water supplies. They are rarely, if ever, isolated from water treatment facilities. The difficulty in detecting legionellae in water treatment plants and municipal water supplies is probably due to the lower temperatures of these waters. Legionellae are more frequently detected and present in higher concentrations in warm or thermally altered environments (69). Several recent studies have suggested that the use of monochloramine as the primary disinfectant for municipality water supplies may represent a community-based approach to controlling legionellae in potable water systems. An initial study by Kool et al. demonstrated that hospitals supplied with drinking water treated with monochloramine were 10 times less likely to have an outbreak of Legionnaires' disease than those supplied by water treated with free chlorine as a primary disinfectant (54). The authors suggest that 90% of nosocomial outbreaks may be prevented by the use of monochloramine for residual disinfection. Two subsequent studies longitudinally monitored buildings colonized with legionellae in municipalities using free chlorine as their primary disinfectant and after they had converted to monochloramine as the primary disinfectant. The first study, conducted in Pinellas County, Fla., showed that the number of buildings positive for legionellae decreased from 19.8% (19/96) to 6.2% (6/96) 1 month after switching from free chlorine to monochloramine (63). The second study, conducted in San Francisco County, Calif., used the same approach but tested the buildings in each of three different seasons before and after the conversion. In this study, the number of colonized buildings decreased from 70% (37/53) during chlorine treatment to 9% (5/53) after conversion to monochloramine (36). These studies clearly demonstrate that the use of monochloramine reduces colonization of building potable water systems by legionellae. Twenty-three percent of U.S. municipalities currently use monochloramine as the primary disinfectant. This is largely due to EPA regulations addressing levels of trihalomethanes (carcinogens) produced by disinfectants. Additional studies are needed to ensure that the use of monochloramine provides adequate drinking water quality before it can be unequivocally recommended to prevent legionellosis.

Amplifying Reservoirs

As previously mentioned, temperature is a critical factor in the ability of legionellae to colonize reservoirs in which they are amplified. Other microorganisms and factors critical to the growth of legionellae are almost universally present in freshwater environments, and it is temperature that governs the numbers of these bacteria. An Australian study of cooling towers found that legionellae colonized or multiplied in towers with basin temperatures above 16°C, and multiplication became explosive at temperatures above 23°C (17). Conversely, legionellae are killed at temperatures between 44 and 55°C, and it has been suggested that potable hot water systems be maintained at temperatures between 55 and 60°C to prevent growth of the bacteria (2). This may be problematic for some institutions, especially hospitals, where the potential for scalding of patients exists or where state regulations are prohibitive.

Practices to control legionellae in amplifying reservoirs can be divided into two categories, routine maintenance and emergency decontamination procedures. Maintenance of respiratory therapy equipment requires that potable water not be used to rinse the equipment before patient use (59). Several documents address routine maintenance of cooling towers, evaporative condensers, and whirlpool spas (35). Generally, these documents suggest following the manufacturer's recommendations regarding cleaning and biocide treatment of these devices. It is generally believed that well-maintained equipment does not permit excessive growth of legionellae (49). Emergency decontamination protocols describing hyperchlorination and cleaning procedures for cooling towers have been developed for towers implicated in the transmission of legionellosis (23). Guidelines for the control of nosocomial legionellosis are presented in documents addressing nosocomial pneumonia, the hospital environment, and facilities that perform organ or hematopoietic stem cell transplants (1, 20, 21, 22). Several guidelines offer recommendations for non-hospital building water systems. The principal approaches to disinfection of potable systems are heat flushing with water maintained at ≥60°C, hyperchlorination, and physical cleaning of hot water tanks (20). Potable systems are easily recolonized and may require continuous intervention such as raising of hot water temperatures or continuous chlorination (59).

Aerosol Transmission

Some control strategies are intended to prevent exposure of susceptible individuals to aerosols which may contain legionellae. Manufacturers have improved the performance of drift eliminators, and these are now a common fixture on cooling towers (17). These newer eliminator modules can reduce the water loss from cooling towers by orders of magnitude. Early investigations of Legionnaires' disease associated with cooling towers resulted in a recommendation for relocation of cooling towers or air intake vents so that cooling tower exhaust would not be carried directly into the heating, ventilation, and air-conditioning systems of buildings. Generally, the further air intakes are located from cooling towers, the better.

Health care professionals, engineers, and industry administrators continue to increase their awareness and understanding of legionellae and legionellosis. It is difficult to determine if this increased understanding has led to measures that have reduced the incidence of legionellosis. Fear of litigation has caused the administration of many institutions to implement prevention strategies that are unproven and lack sufficient scientific documentation. Effective prevention strategies require more-effective decontamination techniques and approaches to prevent amplification of these bacteria in reservoirs.

REFERENCES

1. **Allegheny County Health Department.** 1997. *Approaches to the Prevention and Control of* Legionella *Infection in Allegheny County Health Care Facilities*. Allegheny County Health Department, Pittsburgh, Pa.
2. **American Society of Heating, Refrigerating, and Air-Conditioning Engineers, Inc.** 2000. *Guideline 12-2000: Minimizing the Risk of Legionellosis Associated with Building Water Systems*. American Society of Heating, Refrigerating, and Air-Conditioning Engineers, Inc., Atlanta, Ga.
3. **Arnow, P. M., D. Weil, and M. F. Para.** 1985. Prevalence and significance of *Legionella pneumophila* contamination of residential hot-tap water systems. *J. Infect. Dis.* **152:** 145–151.
4. **Barbaree, J. M.** 1993. Selecting a subtyping technique for use in investigations of legionellosis epidemics, p. 169–172. *In* J. M. Barbaree, R. F. Breiman, and A. P. Dufour (ed.), Legionella: *Current Status and Emerging Perspectives*. American Society for Microbiology, Washington, D.C.
5. **Barbaree, J. M., B. S. Fields, J. C. Feeley, G. W. Gorman, and W. T. Martin.** 1986. Isolation of protozoa from water associated with a legionellosis outbreak and demonstration of intracellular multiplication of *Legionella pneumophila*. *Appl. Environ. Microbiol.* **51:**422–424.
6. **Barbaree, J. M., G. W. Gorman, W. T. Martin, B. S. Fields, and W. E. Morrill.** 1987. Protocol for sampling environmental sites for legionellae. *Appl. Environ. Microbiol.* **53:**1454–1458.
7. **Benson, R. F., and B. S. Fields.** 1998. Classification of the genus *Legionella*. *Sem. Resp. Infect.* **13:**90–99.
8. **Best, M., Y. L. Yu, J. E. Stout, A. Goetz, R. R. Muder, and F. Taylor.** 1983. Legionellaceae in the hospital water supply—epidemiologic link with disease and evaluation of a method of control of nosocomial Legionnaires' disease and Pittsburgh pneumonia agent. *Lancet* **ii:**307–310.
9. **Bopp, C. A., J. W. Summer, G. K. Morris, and J. G. Wells.** 1981. Isolation of *Legionella* spp. from environmental water samples by low-pH treatment and use of a selective medium. *J. Clin. Microbiol.* **13:**714–719.
10. **Bornstein, N., C. Vieilly, M. Nowiki, J. C. Paucod, and J. Fleurette.** 1986. Epidemiological evidence of legionellosis transmission through domestic hot water supply systems and possibilities of control. *Isr. J. Med. Sci.* **13:**39–40.
11. **Bozeman, F. M., J. W. Humphries, and J. M. Campbell.** 1968. A new group of rickettsia-like agents recovered from guinea pigs. *Acta Virol.* **12:**87–93.
12. **Breiman, R. F.** 1993. State of the art lecture. Modes of transmission in epidemic and nonepidemic *Legionella* infection: directions for further study, p. 30–35. *In* J. M. Barbaree, R. F. Breiman, and A. P. Dufour (ed.), Legionella: *Current Status and Emerging Perspectives*. American Society for Microbiology, Washington, D.C.
13. **Breiman, R. F., W. Cozen, B. S. Fields, T. D. Mastro, S. J. Carr, J. S. Spika, and L. Mascola.** 1990. Role of air sampling in investigation of an outbreak of Legionnaires' disease associated with exposure to aerosols from an evaporative condenser. *J. Infect. Dis.* **161:**1257–1261.

14. **Breiman, R. F., B. S. Fields, G. N. Sanden, L. Volmer, A. Meier, and J. S. Spika.** 1990. Association of shower use with Legionnaires' disease: possible role of amoebae. *JAMA* **263:**2924–2926.
15. **Brenner, D. J., J. C. Feeley, and R. E. Weaver.** 1984. Family VII. *Legionellaceae*, p. 279. *In* N. R. Krieg and J. G. Holt (ed.), *Bergey's Manual of Systematic Bacteriology*, vol. 1. Williams & Wilkins, Baltimore, Md.
16. **Brenner, D. J., A. G. Steigerwalt, and J. E. McDade.** 1979. Classification of the legionnaires' disease bacterium: *Legionella pneumophila*, genus novum, species nova, of the family *Legionellaceae*, familia nova. *Ann. Intern. Med.* **90:** 656–658.
17. **Broadbent, C. R.** 1993. *Legionella* in cooling towers: practical research, design, treatment, and control guidelines, p. 217–222. *In* J. M. Barbaree, R. F. Breiman, and A. P. Dufour (ed.), Legionella: *Current Status and Emerging Perspectives*. American Society for Microbiology, Washington, D.C.
18. **Castor, M. L., E. A. Wagstrom, R. N. Danila, K. E. Smith, T. S. Naimi, J. M. Besser, K. A. Peacock, B. A. Juni, J. M. Hunt, J. M. Bartkus, S. R. Kirkhorn, and R. Lynfield.** 2005. An outbreak of Pontiac fever with respiratory distress among workers performing high-pressure cleaning at a sugar-beet processing plant. *J. Infect. Dis.* **191:**1530–1537.
19. **Centers for Disease Control and Prevention.** 1992. *Procedures for the Recovery of* Legionella *from the Environment*. Centers for Disease Control and Prevention, Atlanta, Ga.
20. **Centers for Disease Control and Prevention.** 1997. Guidelines for prevention of nosocomial pneumonia. *Morb. Mortal. Wkly. Rep.* **46** (RR-1).
21. **Centers for Disease Control and Prevention.** 2000. Guidelines for preventing opportunistic infections among hematopoietic stem cell transplant recipients: recommendations of CDC, the Infectious Disease Society of America, and the American Society of Blood and Marrow Transplantation. *Morb. Mortal. Wkly. Rep.* **49** (RR-10).
22. **Centers for Disease Control and Prevention.** 2003. Guidelines for environmental infection control in health-care facilities: recommendations of CDC and the Healthcare Infection Control Practices Advisory Committee (HICPAC). *Morb. Mortal. Wkly. Rep.* **52** (RR-10).
23. **Cooling Technology Institute.** 2000. *Legionellosis Guideline: Best Practices for Control of* Legionella. Cooling Technology Institute, Houston, Tex.
24. **Dennis, P. J. L.** 1998. Isolation of legionellae from environmental specimens, p. 31–44. *In* T. G. Harrison and A. G. Taylor (ed.), *A Laboratory Manual for* Legionella. John Wiley & Sons Ltd., New York, N.Y.
25. **Drozanski, W. J.** 1991. *Sarcobium lyticum* gen. nov., sp. nov., an obligate intracellular bacterial parasite of small free-living amoebae. *Int. J. Syst. Bacteriol.* **41:**82–87.
26. **Edelstein, P. H.** 1981. Improved semi-selective medium for isolation of *Legionella pneumophila* from contaminated clinical and environmental specimens. *J. Clin. Microbiol.* **14:**298–303.
27. **Edelstein, P. H., R. E. Whittaker, R. L. Kreiling, and C. L. Howell.** 1982. Efficacy of ozone in eradication of *Legionella pneumophila* from hospital plumbing fixtures. *Appl. Environ. Microbiol.* **44:**1330–1334.
28. **England, A. C., III, D. W. Fraser, G. F. Mallison, D. C. Mackel, P. Skaliy, and G. W. Gorman.** 1982. Failure of *Legionella pneumophila* sensitivities to predict culture results from disinfectant-treated air-conditioning cooling towers. *Appl. Environ. Microbiol.* **43:**240–244.
29. **Feeley, J. C., R. J. Gibson, G. W. Gorman, N. C. Langford, J. K. Rasheed, D. C. Mackel, and W. B. Baine.** 1979. Charcoal yeast extract agar: primary isolation medium for *Legionella pneumophila*. *J. Clin. Microbiol.* **10:**437–441.
30. **Feeley, J. C., G. W. Gorman, R. E. Weaver, D. C. Mackel, and H. W. Smith.** 1978. Primary isolation medium for legionnaires' disease bacterium. *J. Clin. Microbiol.* **8:**320–328.
31. **Fields, B. S.** 1996. The molecular ecology of Legionellae. *Trends Microbiol.* **4:**286–290.
32. **Fields, B. S., J. M. Barbaree, G. N. Sanden, and W. E. Morrill.** 1990. Virulence of a *Legionella anisa* strain associated with Pontiac fever: an evaluation using protozoan, cell culture, and guinea pig models. *Infect. Immun.* **58:** 3139–3142.
33. **Fields, B. S., R. F. Benson, and R. E. Besser.** 2002. Legionella and Legionnaires' disease: 25 years of investigation. *Clin. Microbiol. Rev.* **15:**506–526.
34. **Fields, B. S., T. Haupt, J. P. Davis, M. J. Arduino, P. H. Miller, and J. C. Butler.** 2001. Pontiac fever due to *Legionella micdadei* from a whirlpool spa: possible role of bacterial endotoxin. *J. Infect. Dis.* **184:**1289–1292.
35. **Fields, B. S., M. R. Moore, and J. C. Carpenter.** 2006. Control of legionellae in the environment: a guide to the U.S. guidelines. *ASHRAE Trans.* **112(1):**691–699.
36. **Flannery, B., L. B. Gelling, D. J. Vugia, J. M. Weintraub, J. J. Salerno, M. J. Conroy, V. A. Stevens, C. E. Rose, M. R. Moore, B. S. Fields, and R. E. Besser.** 2006. Reducing *Legionella* colonization of water systems with monochloramine. *Emerg. Infect. Dis.* **112:**588–596.
37. **Fliermans, C. B., W. B. Cherry, L. H. Orrison, S. J. Smith, D. L. Tison, and D. H. Pope.** 1981. Ecological distribution of *Legionella pneumophila*. *Appl. Environ. Microbiol.* **41:**9–16.
38. **Fraser, D. W.** 1984. Sources of legionellosis, p. 277–280. *In* C. Thornsberry, A. Balows, J. C. Feeley, and W. Jakubowski (ed.), *Legionella: Proceedings of the 2nd International Symposium*. American Society for Microbiology, Washington, D.C.
39. **Fraser, D. W., T. F. Tsai, W. Orenstein, W. E. Parkin, H. J. Beecham, R. G. Sharrar, H. Harris, G. F. Mallison, S. M. Martin, J. E. McDade, C. C. Shepard, P. S. Brachman, and the Field Investigation Team.** 1977. Legionnaires' disease: description of an epidemic of pneumonia. *N. Engl. J. Med.* **297:**1189–1197.
40. **Fry, N. K., J. M. Bangsborg, S. Bernander, J. Etienne, B. Forsblom, V. Gaia, P. Hasenberger, D. Lindsay, A. Papoutsi, C. Pelaz, M. Struelens, S. A. Uldum, P. Visca, and T. G. Harrison**. 2000. Assessment of intercentre reproducibility and epidemiological concordance of *Legionella pneumophila* serogroup 1 genotyping by amplified fragment length polymorphism analysis. *Eur. J. Clin. Microbiol. Infect. Dis.* **19:**773–780.
41. **Gaia, V., N. K. Fry, B. Afshar, P. C. Luck, H. Meugnier, J. Etienne, R. Peduzzi, and T. G. Harrison.** 2005. Consensus sequence-based scheme for epidemiological typing of clinical and environmental isolates of *Legionella pneumophila*. *J. Clin. Microbiol.* **43:**2047–2052.
42. **Hagele, S., R. Kohler, H. Merkert, M. Schleicher, J. Hacker, and M. Steinert.** 2000. Dictyostelium discoideum: a new host model system for intracellular pathogens of the genus Legionella. *Cell. Microbiol.* **2:**165–171.
43. **Hebert, G. A., C. W. Moss, L. K. McDougal, F. M. Bozeman, R. M. McKinney, and D. J. Brenner.** 1980. The rickettsia-like organism Tatlock (1943) and HEBA (1959): bacteria phenotypically similar but genetically distinct from *Legionella pneumophila* and the WIGA bacterium. *Ann. Intern. Med.* **92:**45–52.
44. **Helms, C. M., J. P. Viner, R. H. Sturm, E. S. Renner, and W. Johnson.** 1979. Comparative features of pneumo-

coccal, mycoplasmal, and Legionnaires' disease pneumonias. *Ann. Intern. Med.* **90:**543–547.
45. **Henke, M., and K. M. Seidel.** 1986. Association between *Legionella pneumophila* and amoebae in water. *Isr. J. Med. Sci.* **22:**690–695.
46. **Hoage, C. W., and R. F. Breiman.** 1991. Advances in the epidemiology and control of *Legionella* infections. *Epidemiol. Rev.* **13:**329–339.
47. **Hookey, J. V., N. A. Saunders, N. K. Fry, R. J. Birtles, and T. G. Harrison.** 1996. Phylogeny of *Legionellaceae* based on small-subunit ribosomal DNA sequences and proposal of *Legionella lytica* comb. nov. for *Legionella*-like amoebal pathogens. *Int. J. Syst. Bacteriol.* **46:**526–531.
48. **Horwitz, M. A., and S. C. Silverstein.** 1980. Legionnaires' disease bacterium (*Legionella pneumophila*) multiplies intracellularly in human monocytes. *J. Clin. Investig.* **66:**441–450.
49. **Jakubowski, W., C. V. Broome, E. E. Geldreich, and A. P. Dufour.** 1984. Round table discussion: transmission and control, p. 351–355. *In* C. Thornsberry, A. Balows, J. C. Feeley, and W. Jakubowski (ed.), *Legionella: Proceedings of the 2nd International Symposium.* American Society for Microbiology, Washington, D. C.
50. **Joly, J. R.** 1993. Monitoring for the presence of *Legionella*: where, when, and how?, p. 211–216. *In* J. M. Barbaree, R. F. Breiman, and A. P. Dufour (ed.), Legionella: *Current Status and Emerging Perspectives.* American Society for Microbiology, Washington, D.C.
51. **Joly, J. R., R. M. McKinney, J. O. Tobin, W. F. Bibb, I. D. Watkins, and D. Ramsey.** 1985. Development of a standardized subgrouping scheme for *Legionella pneumophila* serogroup 1 using monoclonal antibodies. *J. Clin. Microbiol.* **23:**768–771.
52. **Katz, S. M., and J. M. Hammel.** 1987. The effect of drying, heat, and pH on the survival of *Legionella pneumophila. Ann. Clin. Lab. Sci.* **17:**150–156.
53. **Kaufmann, A. F., J. E. McDade, C. M. Patton, J. V. Bennett, P. Skaliy, J. C. Feeley, D. C. Anderson, M. E. Potter, V. F. Newhouse, M. B. Gregg, and P. S. Brachman.** 1981. Pontiac fever: isolation of the etiologic agent (*Legionella pneumophila*) and demonstration of its mode of transmission. *Am. J. Epidemiol.* **111:**337–339.
54. **Kool, J. L., J. C. Carpenter, and B. S. Fields.** 1999. Effect of monochloramine disinfection of municipal drinking water on risk of nosocomial Legionnaires' disease. *Lancet* **353:**272–277.
55. **Levi, K., J. Smedley, and K. J. Towner.** 2003. Evaluation of a real-time PCR hybridization assay for rapid detection of *Legionella pneumophila* in hospital and environmental water samples. *Clin. Microbiol. Infect.* **9:**754–758.
56. **Mahbubani, M. H., A. K. Bej, R. Miller, L. Haff, J. DiCesare, and R. M. Atlas.** 1990. Detection of *Legionella pneumophila* with polymerase chain reaction and gene probe methods. *Mol. Cell. Probes* **4:**175–187.
57. **Marston, B. J., H. B. Lipman, and R. F. Breiman.** 1994. Surveillance for Legionnaires' disease. *Arch. Intern. Med.* **154:**2417–2422.
58. **Marston, B. J., J. F. Plouffe, R. F. Breiman, T. M. File, R. F. Benson, M. Moyenuddin, W. L. Thacker, K. H. Wong, S. Skelton, B. Hackman, S. J. Salstrom, J. M. Barbaree, and The Community-Based Pneumonia Incidence Study Group.** 1993. Preliminary findings of a community-based pneumonia incidence study, p. 36–37. *In* J. M. Barbaree, R. F. Breiman, and A. P. Dufour (ed.), Legionella: *Current Status and Emerging Perspectives.* American Society for Microbiology, Washington, D.C.
59. **Mastro, T. D., B. S. Fields, R. F. Breiman, J. Campbell, B. D. Plikaytis, and J. S. Spika.** 1991. Nosocomial Legionnaires' disease and use of medication nebulizers. *J. Infect. Dis.* **163:**667–671.
60. **McDade, J. E., C. C. Shepperd, D. W. Fraser, T. R. Tsai, M. A. Redus, W. R. Dowdle, and the Laboratory Investigation Team.** 1977. Legionnaires' disease. Isolation of the bacterium and demonstration of its role in other respiratory disease. *N. Engl. J. Med.* **297:**1197–1203.
61. **Mermel, L. A., S. L. Josephson, C. H. Giorgio, J. Dempsey, and S. Parenteau.** 1995. Association of Legionnaires' disease with construction: contamination of potable water? *Infect. Contr. Hosp. Epidemiol.* **16:**76–80.
62. **Miller, L. A., J. L. Beebe, J. C. Butler, W. T. Martin, R. Benson, R. E. Hoffman, and B. S. Fields.** 1993. Use of polymerase chain reaction in an epidemic investigation of Pontiac fever. *J. Infect. Dis.* **168:**769–772.
63. **Moore, M. R., M. Pryor, B. Fields, C. Lucas, M. Phelan, and R. E. Besser.** 2006. Introduction of monochloramine into a municipal water system: impact on colonization of buildings by *Legionella* spp. *Appl. Environ. Microbiol.* **72:**378–383.
64. **Morrill, W. E., J. M. B. S. Fields, G. N. Sanden, and W. T. Martin.** 1990. Increased recovery of *Legionella micdadei* and *Legionella bozemanii* on buffered charcoal yeast extract agar supplemented with albumin. *J. Clin. Microbiol.* **28:**616–618.
65. **Morris, G. K., and B. G. Shelton.** 1991. Legionella *in Environmental Samples: Hazard Analysis and Suggested Remedial Actions. Pathcon Technical Bulletin 1.3.* Pathogen Control Association, Norcross, Ga.
66. **Muraca, P., J. E. Stout, and V. L. Yu.** 1987. Comparative assessment of chlorine, heat, ozone, and UV light for killing *Legionella pneumophila* within a model plumbing system. *Appl. Environ. Microbiol.* **53:**447–553.
67. **Park, M. Y., K. S. Ko, H. K. Lee, M.-S. Park, and Y.-H. Kook.** 2003. *Legionella busanensis sp. nov.* isolated from cooling tower water in Korea. *Int. J. Syst. Evol. Microbiol.* **53:**77–80.
68. **Pasculle, A. W., J. C. Feeley, R. J. Gibson, L. G. Cordes, R. L. Myerowitz, C. M. Patton, G. W. Gorman, C. L. Carmack, J. W. Ezzell, and J. N. Dowling.** 1980. Pittsburgh pneumonia agent: direct isolation from human lung tissue. *J. Infect. Dis.* **141:**727–732.
69. **Plouffe, J. F., L. R. Webster, and B. Hackman.** 1983. Relationship between colonization of hospital buildings with *Legionella pneumophila* and hot water temperatures. *Appl. Environ. Microbiol.* **46:**769–770.
70. **Pruckler, J. M., L. A. Mermel, R. F. Benson, C. Giorgio, P. K. Cassiday, R. F. Breiman, C. G. Whitney, and B. S. Fields.** 1995. Comparison of *Legionella pneumophila* isolates by arbitrarily primed PCR and pulsed-field gel electrophoresis: analysis from seven epidemic investigations. *J. Clin. Microbiol.* **33:**2872–2875.
71. **Riffard, S., F. LoPresti, F. Vandenesch, F. Forey, M. Reyrolle, and J. Etienne.** 1998. Comparative analysis of infrequent-restriction-site PCR and pulsed-field gel electrophoresis for epidemiologic typing of *Legionella pneumophila* serogroup 1 strains. *J. Clin. Microbiol.* **36:**161–167.
72. **Rowbotham, T. J.** 1980. Preliminary report on the pathogenicity of *Legionella pneumophila* for freshwater and soil amoebae. *J. Clin. Pathol.* **33:**1179–1183.
73. **Rowbotham, T. J.** 1980. Pontiac fever explained? *Lancet* **ii:** 69.
74. **Rowbotham, T. J.** 1986. Current views on the relationships between amoebae, legionellae and man. *Isr. J. Med. Sci.* **22:**678–689.
75. **Rowbotham, T. J.** 1993. *Legionella*-like amoebal pathogens, p. 137–140. *In* J. M. Barbaree, R. F. Breiman, and A. P. Dufour

(ed.), Legionella: *Current Status and Emerging Perspectives*. American Society for Microbiology, Washington, D.C.

76. **Sanden, G. N., W. E. Morrill, B. S. Fields, R. F. Breiman, and J. M. Barbaree.** 1992. Incubation of water samples containing amoebae improves detection of legionellae by the culture method. *Appl. Environ. Microbiol.* **58:**2001–2004.
77. **Schoonmaker, D., T. Heimberger, and G. Birkhead.** 1992. Comparison of ribotyping and restriction enzyme analysis using pulsed-field gel electrophoresis for distinguishing *Legionella pneumophila* isolates obtained during a nosocomial outbreak. *J. Clin. Microbiol.* **30:**1491–1498.
78. **Segal, G., M. Feldman, and T. Zusman.** 2005. The Icm/Dot type-IV secretion systems of *Legionella pneumophila* and *Coxiella burnetii*. *FEMS Microbiol. Rev.* **29:**65–81.
79. **Shelton, B. G., G. K. Morris, and G. W. Gorman.** 1993. Reducing risks associated with *Legionella* bacteria in building water systems, p. 279–281. *In* J. M. Barbaree, R. F. Breiman, and A. P. Dufour (ed.), Legionella: *Current Status and Emerging Perspectives*. American Society for Microbiology, Washington, D.C.
80. **Skaliy, P., T. A. Thompson, G. W. Gorman, G. K. Morris, H. V. McEachern, and D. C. Mackel.** 1980. Laboratory studies of disinfectants against *Legionella pneumophila*. *Appl. Environ. Microbiol.* **40:**697–700.
81. **Solomon, J. M., and R. R. Isberg.** 2000. Growth of *Legionella pneumophila* in *Dictyostelium discoideum*: a novel system for genetic analysis of host-pathogen interactions. *Trends Microbiol.* **8:**478–480.
82. **Starnbach, M. N., S. Falkow, and L. S. Tompkins.** 1989. Species-specific detection of *Legionella pneumophila* in water by DNA amplification and hybridization. *J. Clin. Microbiol.* **27:**1257–1261.
83. **Stout, J. E., M. E. Best, and V. L. Yu.** 1986. Susceptibility of members of the family *Legionellaceae* to thermal stress: implications for heat eradication methods in water distribution systems. *Appl. Environ. Microbiol.* **52:**396–399.
84. **Tompkins, L. S., and J. S. Loutit.** 1993. Detection of *Legionella* by molecular methods, p. 163–168. *In* J. M. Barbaree, R. F. Breiman, and A. P. Dufour (ed.), Legionella: *Current Status and Emerging Perspectives*. American Society for Microbiology, Washington, D.C.
85. **Valsangiacomo, C., F. Baggi, V. Gaia, T. Balmelli, R. Peduzzi, and J. C. Piffaretti.** 1995. Use of amplified fragment length polymorphism in molecular typing of *Legionella pneumophila* and application to epidemiologic studies. *J. Clin. Microbiol.* **33:**1716–1719.
86. **Vickers, R. M., A. Brown, and G. M. Garrity.** 1981. Dye-containing buffered charcoal yeast extract medium for differentiation of members of the family *Legionellaceae*. *J. Clin. Microbiol.* **13:**380–382.
87. **Vickers, R. M., J. E. Stout, V. L. Yu, and J. D. Rihs.** 1987. Manual of culture methodology for *Legionella*. *Semin. Respir. Infect.* **2:**274–279.
88. **Wadowsky, R. M., and R. B. Yee.** 1981. Glycine-containing selective medium for isolation of legionellaceae from environmental specimens. *Appl. Environ. Microbiol.* **42:**768–772.

Airborne Viruses

SYED A. SATTAR AND M. KHALID IJAZ

80

Viruses can become airborne through the release of contaminated liquids or dried material and can then be carried by air currents indoors (63, 129, 130), and outdoors (26, 43). Wind-blown carriage of animal pathogenic viruses has long been shown to cause outbreaks of disease considerable distances downwind from the source (26, 43, 61, 112). For example, an epizootic of pseudorabies in swine herds arose from airborne spread of virus across an area of nearly 150 km^2 (66), and retrospective studies of similar outbreaks indicated airborne spread of the virus up to 17 km (9). Intercontinental transport of human viruses through atmospheric dispersion of airborne particles has also been postulated, and Hammond et al. (70) suggest that such long-distance transport of airborne viruses may explain the pandemics of influenza.

Recent years have witnessed a revival of interest in the potential of air to spread microbial pathogens in general and viruses in particular. Among the reasons for this are the devastating outbreaks of newly discovered infections such as severe acute respiratory syndrome (SARS) and avian influenza (H5:N1) and their likely spread by air (14, 109, 149, 180, 181). Outbreak investigations also strongly suggest air as a vehicle in the transmission of noroviral infections (113, 114). There is also increasing recognition of the potential of air to disperse viral bioagents released deliberately or accidentally (65, 110). These concerns have generated much activity in the development of equipment and procedures to test (4, 33, 34, 182) and decontaminate indoor air (167) and to protect individuals against airborne contagion (1, 54).

The respiratory tract can retain viruses when contaminated air is inhaled, and many human (44, 46, 152) and animal (152) viruses can spread by this route; subsequent shedding of virus may then occur and increase further the virus load in air. While direct and indirect exposure to viruses in air can occur by other means, infections from the inhalation and retention of droplet nuclei are generally regarded as true airborne spread (Fig. 1).

Infectious viruses (152) or their nucleic acids (155) have been detected in both experimental (56) and field (14, 77) samples of air. Further, experimentally contaminated air has been shown to spread disease (11, 40, 56). Naturally occurring airborne outbreaks of viral disease target mainly the respiratory tract. However, several viruses associated with gastroenteritis (113, 114, 154) or cancer (56) may result from the translocation and ingestion of particles retained in the upper respiratory tract (159). In general, airborne spread of viruses is rapid as well as difficult to prevent and control (21).

METHODS FOR THE STUDY OF AIRBORNE VIRUSES

The methodology for generating, storing, and collecting viral aerosols has already been reviewed (28, 29, 31, 84, 124, 125, 162). The sampling techniques for airborne microorganisms are discussed in chapter 74. Therefore, this chapter is focused on a critical review of the information on the role of air in the spread of vertebrate viruses.

Proper study of the aerobiology of viruses requires specialized and custom-built equipment (171). Furthermore, virus suspensions with relatively high titers (>10^7 infective units/ml) are essential for such studies because virus infectivity can be lost during generation and collection of aerosols and because there is dilution in air of the nebulized material. Work with airborne viruses also requires stringent safety precautions (50). Basically, the following equipment and procedures are needed for the study of airborne viruses and other types of microorganisms.

Aerosol Generation

To study the aerobiology of viruses, it is necessary to generate particles that are small enough (<5 μm in diameter) to remain suspended in air, thus permitting their retention in the respiratory tract upon inhalation. Several nebulizers are commercially available for this purpose (31, 32, 53, 162). Nebulizers for the aerosol delivery of therapeutic drugs also produce particles in the respirable range (100) and can be adapted for generating microbial aerosols. The nature and composition of the virus-suspending medium determine the size distribution of the particles aerosolized and protect the viability of the virus during nebulization and subsequent aging (152).

A physical tracer in the virus suspension to be sprayed is needed to differentiate between physical loss of the infectious virus due to the settling out of the aerosol and its actual biological decay. Uranine, a fluorescent dye, added to the virus suspension can be as effective as a radiolabeled virus (85). At the levels needed, the dye is harmless to viruses and cell cultures and is cheaper and safer to use than

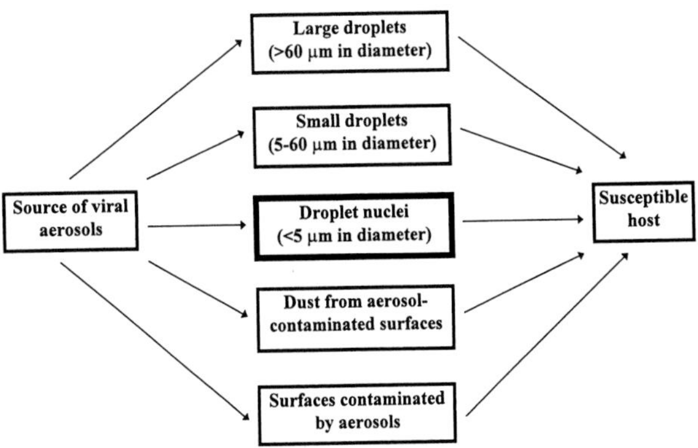

FIGURE 1 Direct or indirect exposure of susceptible hosts to aerosolized viruses.

are radioisotopes. However, dyes may affect the photosensitivity of viruses (31). Bacterial spores are also suitable tracers (31, 162). Nebulization fluids with mixtures of viruses have been used to directly compare their airborne stabilities (86). Antifoam agents in the virus suspension may also be needed to reduce frothing during nebulization (85).

Retention and Aging of Aerosols

Whereas the nebulized virus can be held as a static aerosol in any type of closed container, a rotating stainless steel drum (62, 85, 162) is used to store virus aerosols in a dynamic state for studying the influence of various environmental factors. The drum is housed in a larger chamber that is vented through HEPA filters to prevent exposure to accidental virus leakage. The drum is continuously rotated mechanically along its axis at a predetermined rate (e.g., 4 rpm) to keep the aerosolized material from settling out. This routine allows the virus to stay suspended in air for weeks. The air inside the drum can be preconditioned to the desired relative humidity (RH) level and temperature. Thorough air flushing of the drum is required between experiments. However, such drums or similar devices are not available commercially.

Ultrafine (<1 μm in diameter) natural (spiderweb) and artificial (tungsten) threads can be used as supports for small virus-containing particles. Such anchored particles are still subject to the influence of various environmental factors (31, 162). Figure 2 is a micrograph of a virus-containing particle captured on a spider thread (103). In spite of the limitations (162), ultrafine threads are the best means to study the influence of atmospheric chemicals (32) and light and irradiation (103, 117) on airborne viruses. Handley and Roe (72) have used an artificial fiber (2 to 3 μm in diameter) of ethylene vinyl acetate as an alternative to spider threads; tests with bacterial aerosols have found them to perform as well as spider threads. Their suitability for viruses remains to be evaluated.

Collection and Sizing of Aerosols

Most of the available aerosol collection devices (31, 124, 125, 162, 164) are unsuitable for working with viruses. All-glass impingers (AGI) are commonly used for collecting viral aerosols (152, 162) in relatively small volumes (about 10 liters/min) of air. The air to be sampled is drawn into the impinger, where it passes through a tube with a limiting orifice and impacts on the surface of the collecting fluid, usually Earle balanced salt solution. An antifoaming agent in the collection fluid is usually needed to reduce frothing during the aerosol sampling process (85). The volume of air sampled depends on the sampling rate of the impinger and the run time. If the air is being sampled from an aerosol-holding device, its volume and the total number of samples to be collected for a given experiment will determine the volume of individual air samples. Prolonged operation of AGI for air sampling can also lead to the evaporation of the sampling fluid, thereby affecting the concentration of solutes in it. Use of preimpingers with AGI improves the collection of airborne viruses (152). The volume of air sampled must be replaced with fresh air to avoid creating a vacuum in the aerosol-holding device.

In the field, commercial large-volume air samplers, which can easily process 10 m^3 of air per min, can be used to recover airborne viruses (31, 152, 162, 174) by impacting the air on a thin layer of a collecting fluid such as Earle balanced salt solution. The collection efficiencies of various large-volume air samplers vary considerably depending on the type of virus, the type and volume of air sampled, the nature of the collection fluid, and the rate of sampling. Wallis et al. (169) have reported the use of acidic buffer-moistened cartridge filters to recover polioviruses aerosolized during the flushing of toilets; the air could be sampled with flow rates as high as 100 liters/min, but the virus eluted from the filters required further concentration.

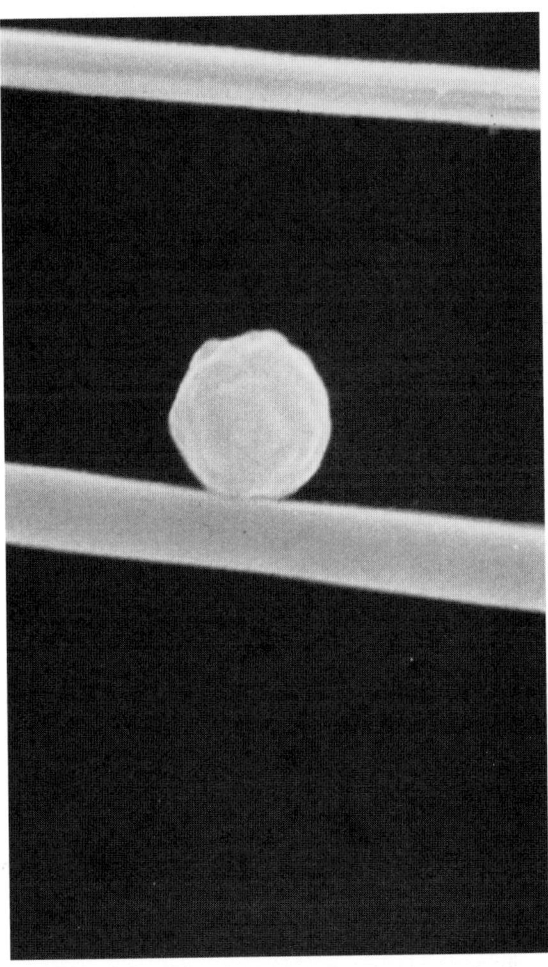

FIGURE 2 Scanning electron micrograph of an aerosol particle attached to a strand of spiderweb. Bar = 0.5 μm. (Reproduced with permission from B. Kournikakis of the Defence Research Establishment, Suffield, Alberta, Canada.)

Determination of the size distribution of airborne particles containing infectious viruses has required the adaptation (85) of the Andersen six-stage sampler (5). Instead of a bacteriological agar medium, tryptose phosphate broth containing 3% gelatin is placed in the sampler's petri plates. After sampling, the collection medium is liquefied by holding the plates at 37°C for 1 h and poured out for virus titration and measurements of the physical tracer levels. In a May three-stage glass impinger (116), virus retention corresponds to that in the human upper respiratory tract, bronchioles, and alveoli. A comparison of the three-stage Andersen sampler and the three-stage May AGI gave similar results in the sampling and sizing of artificially generated biological aerosols (182).

Exposing Experimental Animals or Human Subjects to Airborne Viruses

Table 1 summarizes information from reports published in the past 35 years. Studies on the susceptibility of animals to airborne viruses, or experiments to determine the minimal infective dose of a given virus by the respiratory route, require exposing animals to standardized clouds of the infectious agent under controlled conditions. Traditionally, the Henderson apparatus (74) has been used for this purpose when working with small animals such as mice and rats. In this device, individual animals are placed in cylindrical holders with only the tip of the animal's face (including the nostrils) protruding from one end of the holder. The holders are then attached to ports on a larger tube connected to the aerosol source, and the only means of exposure of the animals to the aerosols is by inhalation of the air in the tube. Devices for working with cattle (121) or pigs (137) have also been described. It is important to note that the aerosolized virus must not be allowed to deposit on areas where the test animals could be exposed to it in ways other than through inhalation. Chung et al. (27) reported the development of a low-cost wind tunnel for exposing human subjects to aerosols of >2 μm in diameter. Particles generated by a Collison-type nebulizer and introduced into the tunnel by using an aerosol injector are uniformly distributed to expose human subjects. However, no reports have been published on the use of the device with airborne viruses.

SOURCES OF VIRAL AEROSOLS

Activities such as sneezing, coughing, flushing toilets, and changing diapers as well as shaking, homogenization, and sonication of virus-containing materials can generate infectious aerosols (152). For some gastrointestinal viruses, aerosolized feces and vomitus may also transmit infection (113). Inhalation of infectious virus particles released into air during certain surgical procedures can also pose risks to human health. Under experimental conditions, human papillomavirus was isolated from aerosols and shown to transfer infection from one tissue to another (56).

Preventing the generation of, and avoiding exposure to, such aerosols is particularly important in laboratory settings where infectious material is handled (30). According to Pike (141), 27% of the cases of laboratory-acquired infections were due to airborne viruses; cases in research settings accounted for more than 67% of such infections. Whereas improvements in the design and construction of biohazard containment equipment (28) and better enforcement of biosafety procedures (147) have considerably reduced the risk of aerosol exposure, many laboratory workers do not appear to be fully aware of the dangers of infectious aerosols. Improved training of laboratory workers, continued vigilance, and further improvements in equipment design and procedures are needed to make these workplaces safer.

AIRBORNE SURVIVAL OF HUMAN- AND ANIMAL-PATHOGENIC VIRUSES

How long a given virus can remain infectious in air depends on the nature of the virus and the medium in which it was suspended before becoming airborne, ambient temperature, RH, atmospheric gases, lighting, and irradiation (152). The absence of standardization of the experimental protocols and wide variations in the system of reporting the results make direct comparisons of the findings from different studies extremely difficult.

The literature on this topic has been reviewed (68, 152). The following is an update of the information from experimental and field studies on selected human and animal pathogenic viruses. Some of the viruses selected are known to spread by the airborne route, whereas for the others there appears to be a real or perceived risk for airborne transmission. It should be noted that many of the viruses spread through aerosols are not normally associated with infections of the respiratory tract.

Varicella-Zoster Virus

Investigations of chickenpox outbreaks clearly show airborne spread of varicella-zoster virus (VZV). Sawyer et al. (155) found VZV DNA in air samples from rooms housing patients with zoster or chickenpox, and the air remained positive up to 24 h after patient discharge. Whereas none of the VZV DNA-positive samples had infectious virus, PCR technology may prove to be very useful in studying the airborne transport of viruses in indoor and outdoor settings.

Hepburn et al. (76) have described an outbreak of chickenpox in a military field hospital. Several patients and staff members in an isolation ward for cases of gastroenteritis were affected, and virus spread was most likely through air. This incident highlights the weaknesses in field hospital design and illustrates how easily biological warfare agents could spread in such settings (76).

A recent study to identify the spread of VZV in patients treated with oral acyclovir (165) showed a significant airborne presence of the virus. By day 3 of acyclovir treatment, 29% of air purifiers had detectable traces of VZV DNA. This partially correlated with the detection of VZV in throat swabs of between 17 and 32% of patients.

Viruses Causing Acute Gastroenteritis

Caliciviruses

That norovirus can spread by air is suggested from outbreak investigations (22, 57). The first such report was based on an outbreak associated with the emergency room of a hospital (154), and individuals who simply walked through the contaminated area also became infected. Contamination of air is believed to occur mainly by the aerosolization of the virus during vomiting (24, 25). However, proper studies on the airborne spread of these viruses are difficult because they cannot be cultured in the laboratory.

Rotaviruses

Rotaviruses are among the major causes of acute gastroenteritis in humans and animals (150). Apart from their spread by the fecal-oral route, epidemiological and experimental studies suggest airborne transmission of rotaviral infections (7, 83, 87). The behavior of murine and other

TABLE 1 Experimental challenge of animals or humans to viral aerosols: chronological list of selected published studies since 1970

Virus(es)	Host(s)	Remarks	Reference
Murine leukemia virus	Mice	First report on aerosol stability of the virus and its ability to spread by air	104
	Mice	Nearly 40% of exposed mice developed leukemia within 25 mo	119
Murine leukemia and Rous sarcoma viruses	Monkeys	Showed that tumor viruses could be readily transmitted via aerosols	120
Adenovirus 12	Hamsters	Airborne virus shown to be pathogenic to newborn animals	37
Parainfluenza virus type 1	Mice	Transmissibility rates did not increase after serial airborne challenge with the virus	168
Adenovirus 4, coxsackievirus A21	Human volunteers	Determined infectivity of these viruses by aerosols	71
Marek's disease virus	Chickens	Exposure to effluent air from "donor cages" housing infected animals resulted in a high incidence of the infection in test chicks; passage of contaminated air through certain filters partially or completely prevented such infection	20
NDV	Chickens	Unvaccinated birds shed much higher levels of virus than those previously vaccinated	81
Vesicular stomatitis virus	Mice	Exposure of mice for 1 h to ozone resulted in a 70% increase in respiratory deposition of the virus	48
Influenza A virus	Mice	Under conditions of aerosol inhalation, mice were found to be a suitable model for studies on pathogenesis	52
	Mice	Extrapulmonary virus was in direct quantitative relationship to the extent of lung involvement	53
Influenza virus	Mice	Mouse resistance to viral pneumonia affected by presence of manganese dioxide	111
Moloney murine sarcoma and leukemia virus complex	Mice	When tumor extracts were aerosolized, both viruses survived for at least 2 h, but mice exposed to the aerosols did not develop an infection	78
Bovine rhinotracheitis virus	Cattle	Compared clinical and immunological responses after aerosol exposure or intramuscular inoculation; in both cases, the virus generally elicited comparable levels of serum antibody but not measurable nasal antibody; aerosol-exposed cattle shed virus from the nose, while the others did not	51
Feline caliciviruses	Cats	Concluded that aerosol transmission probably plays little part in the spread of these viruses	170
African swine fever virus (KWH/12)	Pigs	Concluded that the primary route of infection in pigs was through the lower respiratory tract	175
	Pigs	Animals became infected after challenge with aerosolized virus	176
NDV (vaccine)	Chickens	Antibody response to aerosol challenge much better than that from administration in drinking water; no clinical disease; virus recovered from lungs for 10 days	99
Rinderpest virus	Cattle	Low or high RH was shown to increase the probability of disease transmission by the respiratory route, but any aerial spread across distances greater than a few meters was believed to occur principally at night	82
Feline caliciviruses	Cats	Animals could be infected by aerosol exposure or direct intranasal instillation	136
Bovine respiratory syncytial virus	Holstein calves	Animals exposed to aerosols of the virus manifested moderate to severe signs of respiratory disease	47
Japanese B encephalitis virus	Mice, rats, hamsters, guinea pigs, squirrel monkeys	Mice and hamsters highly susceptible to aerosol challenge; guinea pigs and rats seroconverted but survived the infection; squirrel monkeys died after a high dose of infectious virus	105
Rauscher murine leukemia virus	Mice	Infection of BALB/c mice through aerosols of leukemogenic virus was possible	122
Influenza virus	Mice	Studied the dynamics of B lymphocytes in lungs after aerosol challenge; IgA- and IgM-containing[a] cells appeared first, followed by IgC-containing ones	138

(*Continued on next page*)

TABLE 1 Experimental challenge of animals or humans to viral aerosols: chronological list of selected published studies since 1970 (*Continued*)

Virus(es)	Host(s)	Remarks	Reference
Parainfluenza virus type 3	Calves	First report of extensive purulent pneumonia in calves after exposure to aerosols of the virus and *Pasteurella haemolytica*	89
FMDV	Cattle	Cattle infected with aerosolized virus	107
Pseudorabies virus	Pigs	Virus transmitted to seronegative pigs exposed to air from boxes containing infected pigs	42
Infectious bronchitis virus (Australian T strain)	Chickens	Birds exposed to aerosols had earlier and slightly more severe respiratory symptoms	146
Lassa fever virus	Guinea pigs and monkeys	Both species susceptible when exposed to small-particle aerosols	163
Mouse rotavirus	Mice	Neonatal mice developed acute gastroenteritis within 48 h of exposure to viral aerosols	144
Rift Valley fever virus	Rats	When exposed to viral aerosols, 97% of the unvaccinated and 32% of the vaccinated animals died; levels of serum-neutralizing antibody were predictive of the protective effect	6
Hantaan, Seoul, and Puumala viruses	Rats	Female Wistar rats (12–16 wks old) exposed to aerosols of the viruses became infected	134
Fowlpox virus	Chickens	Day-old chicks were vaccinated by the aerosol route or by injection in the wing-web; both methods induced immunity, but the aerosol method was more suitable for mass vaccinations	39
Marburg virus	Guinea pigs and rhesus monkeys	Susceptible animals exposed to air from infected ones became infected	142
Influenza virus	Mice	Intranasal inoculation was found to be superior to aerosol exposure in the immunization of mice	92
Junin virus	Rhesus monkeys	All animals exposed to aerosolized virus became sick within 3 wk and died; clinical picture similar to that seen in humans	98
Influenza virus	Horses	Challenge of animals to aerosolized influenza virus was found to be a reliable means of assessing immunization against the disease	128
FMDV	Calves	In situ hybridization method showed presence of virus in target organs as early as 6 h after aerosol challenge	17
Equid herpesvirus 1	Horses	Exposure of pregnant mares to viral aerosols resulted in abortions, but virus could not be demonstrated in the aborted fetuses	160
Influenza virus	Humans	Administered the vaccine by either nose drops or large-particle aerosols; aerosol route was more effective and better tolerated	67
Respiratory coronavirus	Pigs	Aerosol challenge of young pigs induced a strong response in bronchial lymph nodes but not in the mesenteric lymph nodes	15
Bovine herpesvirus strains 1.1 and 1.3	Calves	Strain 1.3 produced severe encephalitis with minimal respiratory lesions, whereas strain 1.1 gave respiratory lesions but no neurological disease; colostrum-fed animals became subclinically infected	12
Influenza virus	Horses	Vaccinated and unvaccinated ponies were exposed to aerosols of a field isolate of influenza virus; vaccinated animals showed complete protection even after 15 mo of immunization	127
Aujeszky's disease virus	Pigs	After a 15-min exposure of six animals to aerosolized virus (4.5 $\log_{10}$ $TCID_{50}$),[b] all became infected; a proportion of sentinel pigs housed with the infected animals also became infected	60
Bovine herpesvirus 1	Cattle	Susceptible calves exposed to air from infected animals became infected	115
Bovine respiratory syncytial virus	Cattle	Susceptible calves exposed to air from infected animals became infected	115
Bovine viral diarrhea virus	Cattle	Susceptible calves exposed to air from infected animals became infected	115
VEEV	Monkeys	The protective effect of vaccines against aerosolized challenge with VEEV was tested in cynomolgus monkeys implanted with temperature-monitoring radiotelemetry devices	143
Swine fever virus (*Asfivirus*)	Swine	Naïve pigs were housed in an isolation unit with three compartments with separate ventilation systems; an infected pig in a separate compartment was able to infect only those animals with a connected ventilation system	40

(*Continued on next page*)

TABLE 1 (*Continued*)

Virus(es)	Host(s)	Remarks	Reference
Aujeszky's disease virus (*Herpesvirus*)	Swine	Naïve pigs exposed to the breath of intranasally infected ones through a mask-to-mask means became infected as evidenced by seroconversion. A good model for transmission and vaccination studies.	60
Andes virus (hantavirus)	Cynomolgus macaques	Only hantavirus known to spread from person to person. Cynomolgus monkeys exposed to aerosolized virus; no clinical disease but antibody production.	118
Bovine herpesvirus 1	Calves	5-min exposure of vaccinated and control animals to experimentally generated viral aerosols through a face mask	90
Ovine herpesvirus 2	Sheep	Causes malignant catarrhal fever in ruminants. Virus refractory to in vitro cultivation. Uninfected sheep exposed to aerosols of various dilutions of nasal secretions from infected sheep. Minimal infective dose determined for further studies on viral pathogenesis and vaccine protection.	166
Smallpox virus (variola)	Cynomolgus monkeys	Development of a subhuman primate model of the study of smallpox pathogenesis and transmission	88
Porcine reproductive and respiratory disease virus (*Arteriviridae*)	Pigs	Naïve pigs were exposed for 3 h to virus aerosolized up to 150 m away; three of the animals became infected. Virus also recovered from the air.	38
Adenovirus 5 expressing the glycoprotein gene of VEEV	Mice	Animals first exposed to the recombinant virus intranasally and challenged with aerosolized VEEV. Potential for vaccination of humans against VEEV.	140

[a]IgA, immunoglobulin A.
[b]$TCID_{50}$, 50% tissue culture infective dose.
[c]VEEV, Venezuelan equine encephalitis virus.

strains of bovine rotaviruses in air appears to parallel that of previously studied human and animal rotaviruses (85).

Viruses Causing Infections of the Respiratory Tract

The behavior of aerosolized influenza and parainfluenza viruses of humans and animals has been studied in some detail (152). More recent studies in this regard have focused on challenging vaccinated animals with viral aerosols to study the protective effect of immunization (Table 1).

The relative importance of air and other vehicles in the spread of rhinovirus colds continues to be debated (41, 68). Airborne human rhinovirus type 14 has shown behavior in ways typical of other picornaviruses (94); its half-life was nearly 14 h when it was aerosolized from tryptose phosphate broth and when the aerosols were held at 20°C with 80% RH. This finding suggests that rhinoviruses can remain infectious in air long enough to permit aerial spread. Recent improvements in the recovery and detection of rhinoviruses from indoor air (129) and their application in field investigations of rhinovirus colds (130) provide additional evidence in support of this method of transmission.

Papillomaviruses

Reports of warts in the respiratory tracts of laser therapists (19, 69) suggest exposure to papillomaviruses in the smoke plumes from vaporized verrucae. Garden et al. (55) first reported using DNA probes for the detection of papillomavirus DNA in such plumes from CO_2 laser vaporization of bovine fibropapilloma and human verrucae. Subsequently, PCR was used to confirm that the DNA in the laser plumes corresponded to the type in the lesion (95). In fact, infectious papillomavirus particles have also been detected in the vapors produced during CO_2 laser treatment as well as during electrocoagulation of bovine warts (153). The same study also found that the amounts of DNA generated during the laser treatment of human and bovine warts were larger than those released during electrocoagulation; surgical masks were shown to be effective in filtering out the viruses in the vapors from both types of treatment. Inoculation of cattle with debris from CO_2 laser plumes generated during the vaporization of bovine warts failed to infect the cattle (178). Another PCR-based study (13) reported widespread contamination of the facial areas of the therapists and the operating room environment with papillomavirus DNA released during CO_2 laser treatment and electrocoagulation of human warts and neoplasia.

The experiments of Garden et al. (55, 56) give further proof that laser plumes generate infectious aerosols and also that viruses in them have the potential to cause disease (10). Bovine cutaneous fibropapillomas were subjected to simulated excision surgery by using a carbon dioxide laser, and the material was aerosolized as the laser plume was collected. It was not only positive for papillomavirus DNA but also induced tumors in the skin of inoculated calves.

Retroviruses

Little is known about airborne spread of retroviruses (152). The advent of AIDS, however, has led to some recent studies on human immunodeficiency virus type 1 (HIV-1). Although airborne spread of AIDS is not known, survival of HIV-1 in air was tested to assess the risk of exposure of health care personnel during orthopedic surgery. Infectious HIV-1 was detected in aerosols from certain types of surgical power tools (93), and the blood-containing aerosols were in the respirable range (91). Baggish et al. (8) reported detection of HIV-1 in CO_2 laser smoke when pellets of experimentally infected cells were vaporized.

Airborne spread of HIV-1 and other retroviruses could occur where high-titered suspensions of such viruses are handled, and strict adherence to safety precautions is necessary to eliminate the risk. It would be useful to know if cell-associated HIV-1 (in both infectious and proviral forms) behaves in air in a manner different from that of cell-free virus.

Viruses Causing Hemorrhagic Fevers

The ability of viruses causing hemorrhagic fevers to be transmitted through air remains unclear (108). Their handling requires the highest level of biohazard containment (50). However, laboratories with such facilities are limited, and those that exist may not be equipped for aerobiological studies. In many cases, the airborne spread of these viruses has been discovered through laboratory accidents or patterns of disease transmission in hospitals (36, 106) and animal-holding facilities.

The survival of Lassa fever virus in artificially generated aerosols was favored at low (30%) RH levels, and even at 32°C the virus survived long enough to permit its dispersal by air (163). Experimental aerosol exposure of monkeys and guinea pigs could infect and kill them, with the median 50% infectious dose for guinea pigs being as low as 15 virus PFU.

Laboratory-acquired infections in those handling hantaviruses or animals carrying these viruses are well documented (2, 107, 133, 139, 179). The likeliest means of exposure in these studies was inhalation of infectious aerosols. Inhalation of artificially generated aerosols of hantaviruses can infect rats (134).

Rhesus monkeys became acutely ill and died when exposed to aerosols of Junin virus (98), the agent of Argentine hemorrhagic fever. The symptoms and pathology of the disease were very similar to those after parenteral exposure and also mimicked the clinical syndrome in humans.

SARS Coronavirus (SARS-CoV)

In 2003, an outbreak of a novel coronavirus which caused an atypical pneumonia startled the world and appeared to be the next pandemic. However, to date, the virus has caused just over 8,400 cases, with fewer than 1,000 deaths, and studies of disease management have all but curtailed the potential for a pandemic (96).

The available epidemiological evidence strongly suggests that SARS spreads through droplets (131) and that such spread is much easier to control than that through aerosols due to the inability of droplets to travel over longer distances. However, the pattern of spread of SARS-CoV in at least two instances is highly suggestive of airborne spread.

A cluster of 329 SARS cases was recorded in one apartment complex in Hong Kong, with the majority of them occurring on several floors in one wing (132). This pattern of spread is highly suggestive of virus dissemination by air, and the aerosolization of the virus was speculated to have occurred from malfunctioning sewers in the building (132). Rodent pests in the building have been hypothesized as possible amplifiers and disseminators of the virus (131); the virus, most likely acquired from infected residents, may have multiplied in the rodents, been released in their excreta, and then possibly become airborne. While hantaviruses, for example, can infect humans from aerosolized rodent excreta (134), the role of air in the transmission of SARS-CoV in this outbreak remains speculative at this stage. However, a more detailed analysis of the outbreak of SARS at that apartment complex suggests that the virus released by the patients themselves may have spread through air (14).

Limited airborne spread of SARS may have occurred aboard commercial aircraft. In one such instance, a symptomatic index patient infected at least 22 of 120 (18.3%) passengers and crew during a 3-h flight (135). Some of those infected were seated over 2 m away from the index patient, a distance much longer than the 0.9 m generally believed to be the limit for droplet transmission. In such retrospective investigations, it is virtually impossible to rule out the role of other possible means of virus spread. Also, the inside of an aircraft combines features that may be more conducive to airborne spread of pathogens.

A recent report from Canada suggests that oxygen delivery masks with open vents could promote the dispersal of respiratory pathogens such as SARS-CoV through their enhanced release in mists of exhaled pulmonary gases (161). The exhaled moist air ejected from such oxygen masks is believed to carry pathogen-laden droplets over longer distances and possibly to contribute to an increased risk of spread of respiratory infections in nosocomial settings. Additional investigations are needed to first prove that viruses such as SARS-CoV can retain their infectivity better in the warm, moisture-laden air exhaled from oxygen masks. The findings on the influence of RH and air temperature on the airborne survival of coronavirus 229E would tend to suggest otherwise (86).

The findings of earlier studies on the effect of RH and air temperature on experimentally aerosolized 229E, another respiratory coronavirus of humans, further reinforce belief in the potential of SARS-CoV to spread by air (86). As is true for enveloped viruses in general, 229E survived better at 30 to 50% RH than at 80% RH when the air temperature was about 20°C. Under these conditions, the half-lives of the virus at 30, 50, and 80% RH were 27, 67, and 3 h, respectively. Lowering the air temperature to 6°C increased the half-lives of the virus at 30 and 50% RH to 34 and 103 h, respectively. But the lower air temperature produced the most dramatic effect on virus survival at 80% RH and increased its half-life from 3 to over 86 h.

Avian Influenza Virus

In 1997, an H5:N1 variant of influenza virus was detected in humans (23). Since then, the presence of avian flu has been rigorously monitored, and actions to prevent its spread are quickly instituted. Although there is ample evidence of transmission of virus from animals to humans, there has been as of yet no demonstration of human-to-human transfer of the infection and little to no evidence suggesting an airborne route for virus spread. Still, airborne precautions when dealing with this strain of the virus are strongly recommended.

Viruses of Domestic Animals

Aujeszky's Disease Virus and Foot-and-Mouth Disease Virus

Aujeszky's disease virus and foot-and-mouth disease virus (FMDV) have been studied extensively for their capacity to spread through air (59, 60, 152), and several airborne outbreaks have been documented (112). The 2001 outbreak of FMDV in the United Kingdom resulted in the slaughter of over 6 million cattle. A prediction model was proposed by Gloster et al. (61), in which 7 of 12 farms were infected through airborne transmission of the virus from nearby farms. Gloster et al. (61) used meteorological models to defend this theory. Other models to forecast and analyze outbreaks of FMDV are now available and continue to be postulated (43, 126).

Airborne FMDV has a relatively small minimal infective dose; when sheep were experimentally exposed in cabinets to virus-infected pigs, as little as a dose of 10 cell culture infective units was found to be sufficient to infect 50% of the animals exposed (58).

NDV

There is convincing evidence for the airborne spread of Newcastle disease virus (NDV). Recently, Hietala et al. (77) showed that exotic NDV was detected in air samplers within 2 h of being introduced into a flock of commercial poultry flocks. Thus, air filters in poultry houses are highly effective in preventing outbreaks of the disease (80). Kournikakis et al. (102, 103) have used a vaccine (LaSota) strain of NDV as a model to study the airborne survival and behavior of enveloped viruses as well as to field test protective equipment and methodologies for the rapid collection and identification of viruses in air. The vaccine, available in the lyophilized form, is relatively easy to work with and safe. Each vaccine vial can yield nearly 10^8 PFU/ml. The influence of air temperature and RH on the airborne survival of this virus was very similar to that on the airborne survival of other enveloped viruses (18, 152). A combination of low levels of RH (20 to 30%) and a low temperature (about 10°C) is optimal for its survival in air in the dark (102); under such conditions, 56% of the aerosolized virus remained infectious even after 6 h, while the corresponding figure for 20°C was 39%. However, there was nearly a 99% drop in virus infectivity when NDV-containing aerosols were captured on spider threads and held under daylight conditions with or without a cloud cover (Fig. 3).

Negative air ions may reduce the spread of NDV on chicken farms (123). Infected animals were placed upwind from the susceptible ones and under conditions of controlled

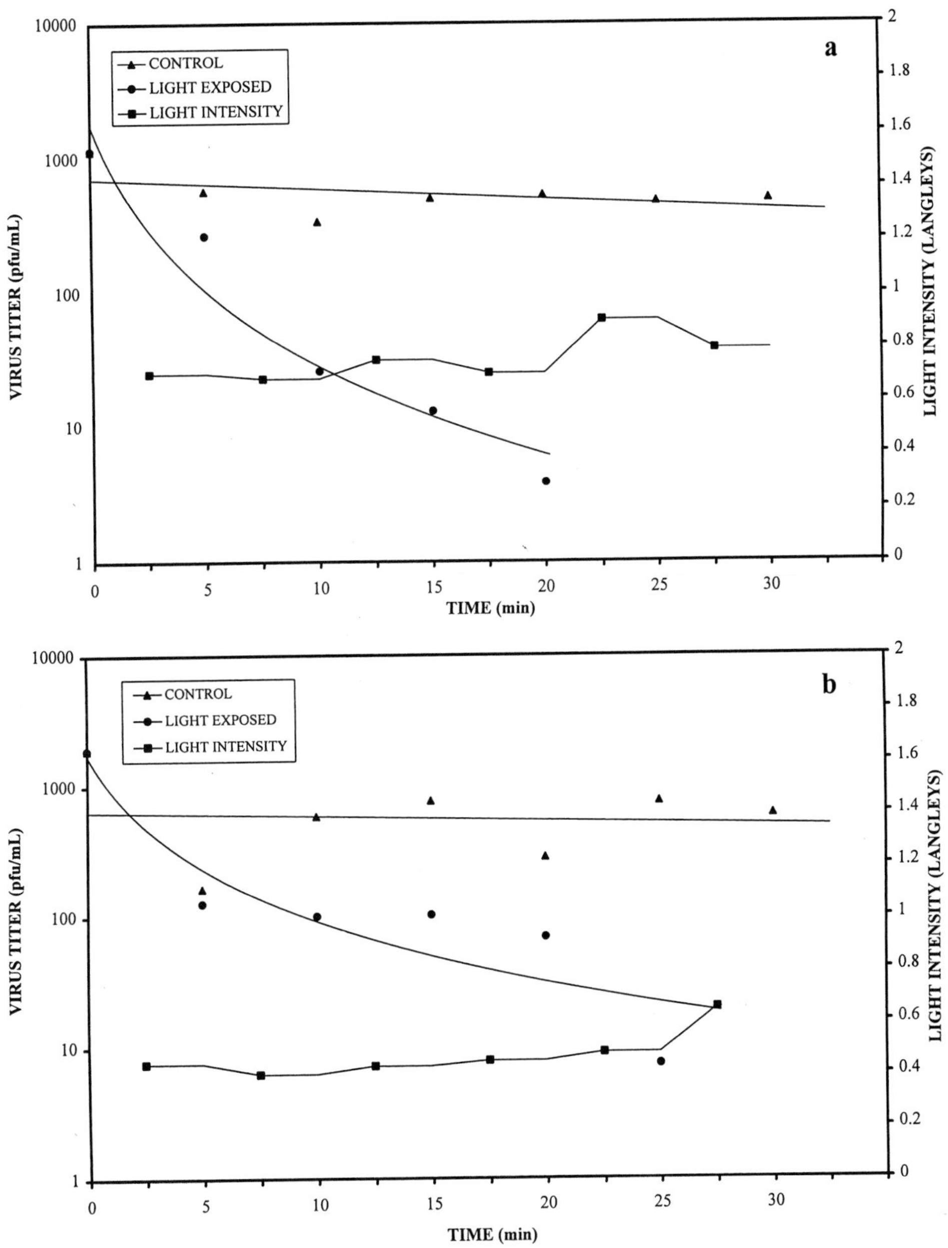

FIGURE 3 Biological decay of NDV captured on a spiderweb and held under sunlight (a) and under cloudy conditions (b). Control samples were held in the dark. (Reproduced with permission from B. Kournikakis of the Defence Research Establishment, Suffield, Alberta, Canada.)

air temperature (26.7°C), RH (50%), and ventilation rates (0.34 to 1.36 m^3/min); there was a 28% reduction in virus transmission when negative ion generators were used.

Pseudorabies Virus

Even though the herpesvirus that causes pseudorabies is relatively fragile, its capacity to spread through air parallels (66) that of FMDV, a picornavirus. The virus in air survived the best when RH was 55% and when the air temperature was 4°C (156); the half-life of the virus under these optimal conditions was less than 1 h.

AEROSOL CHALLENGE STUDIES

Several studies have challenged human or animal hosts with artificially generated viral aerosols (152). Table 1 summarizes the information from selected studies published on this topic since 1970. The following conclusions are based on a critical analysis of these studies: (i) in most cases, susceptible hosts became infected upon exposure to the test virus, but often the amount of virus inhaled was either unknown or may have been unrealistically large; (ii) in certain cases, the experimental design and setup did not exclude virus exposure of test subjects by means other than through inhalation; (iii) in spite of the importance of reporting RH and air temperature in such experiments, many of the investigators failed to mention these parameters; (iv) of necessity, nearly all of these studies have used laboratory-adapted strains of the test virus(es); the extent to which these data apply to viruses circulating in the field remains undetermined and perhaps undeterminable; and (v) any meaningful comparison of the data from these studies is difficult because there are no standard procedures for aerosol challenge studies. Therefore, much developmental work is needed before accurate and reliable information on the behavior of airborne viruses can be generated and before such data can be applied to designing mathematical models and strategies to control and prevent disease spread through air.

COLLECTION OF NATURALLY OCCURRING VIRAL AEROSOLS

Collecting infectious viruses from naturally occurring aerosols continues to be difficult because the devices currently available for the purpose are generally noisy, bulky, expensive, and somewhat inefficient. As is true for many other types of environmental sampling, it is often too late to look for the suspected virus in air in an outbreak investigation. Even if it were feasible, regular monitoring of air for viruses is not recommended because of the limited significance of the findings. The data from molecular biological techniques (e.g., PCR) for viruses in air sample concentrates would be much more meaningful if the nucleic acid detected could reliably signal presence or absence of virus infectivity. However, there are certain situations where sampling of air for naturally occurring viral aerosols can be extremely valuable. This is exemplified by the studies of bat caves for airborne rabies virus (177). In this regard, not only was infectious rabies virus recovered from the air in the caves, experimental animals exposed (in insectproof cages) to such air also died of rabies (177).

Airborne spread of naturally occurring viral infections of humans and animals is known. In some of these cases (e.g., measles or influenza), air may be the chief mode of virus transfer. Others may be isolated instances of airborne spread of a virus, which is normally transmitted by direct contact or by other vehicles. But it must be noted that any virus that can survive aerosolization has the potential for airborne transmission.

CONCLUDING REMARKS

Increasing use of recycled air will further enhance the risk of exposure of susceptible individuals to airborne viruses. Such spread of viruses is also believed to exacerbate asthma and other ailments of the respiratory tract (101, 145). Therefore, there is an urgent need for improvements in the design and operation of air-handling/recycling systems, particularly in hospitals, clinics, offices, residential complexes, sports facilities, and aircraft (4, 173). The continuing increase in the numbers of immunosuppressed and transplant patients also underscores this need. A U.S. survey on the association of outbreaks of infections with competitive sports has shown the infectious agents involved to be predominantly viruses, with several instances of airborne transmission (64). Innovative approaches are required to combine energy and resource conservation to reduce the risk of infection spread through conditioned and recycled air. Additional studies are also needed to establish better methods for disinfecting airborne viruses in recycled air.

Airborne dissemination of nonviable viruses or their components can lead to problems in laboratories using PCR or other such techniques, and presently available containment facilities may not be quite adequate to address this issue (151). Infectious aerosol generation by surgical tools (13, 35, 73, 153) and by laboratory equipment (49) suggests that the potential of any such technology should be assessed at the design stage and that corrective measures should be taken. This will require improved communication between design engineers and end users of the devices, including infection control practitioners. More information is also needed on effective means of protecting animals and humans in case of deliberate or accidental release of viral aerosols. Recent attempts at developing predictive models of exposure to continuous or temporary sources of aerosolized virus show that sheltering indoors may have a protective effect (21).

The increased awareness of airborne infection, particularly in hospitals, has escalated the use of face masks, but the protective ability of such masks has been questioned in studies using respirable but nonmicrobial aerosols (172). There are no standardized methods to test the ability of face masks to withstand challenge with airborne infectious agents.

Recently, alteration of the lung airway surface properties by spraying normal saline into the respiratory tract has been found to diminish the number of exhaled bioaerosols by human subjects by over 70% (45). This appears to be due to changes in the surface tension of the fluid lining the respiratory tract. While corroboration of these findings is needed, the use of a readily available and nontoxic fluid such as normal saline may prove to be an effective and broad-spectrum means of reducing the aerial loading of infectious agents from individuals with respiratory viral and other infections.

Immunization of humans and animals by exposure to artificially aerosolized virus is a very attractive alternative to current practices of parenteral or oral vaccination. Apart from being rapid and practical, such antigen administration more closely mimics natural exposure to many viruses and gives longer-lasting immunity (140, 148). However, there are attendant risks in introducing live, albeit attenuated, virus directly into the respiratory tracts of susceptible hosts. Inhalation of inactivated viruses or even their purified antigens and other components of the vaccine may also cause allergic reactions.

Therefore, much initial testing will be necessary before any mass vaccinations through aerosolization could be attempted.

Rapid inactivation (103) and higher levels of dilution of vertebrate viruses in open air suggest that they may be less likely to spread under outdoor conditions, but there is now enough evidence that certain types of animal pathogenic viruses can be carried by air over several kilometers, leading to outbreaks at distant locations (9, 26, 66). While some suggest that this property is true for human pathogens as well (70), much work is needed to substantiate it. Perhaps long-distance transport of viruses may explain the peculiar transcontinental movement of outbreaks of rotaviral infections (79).

Additional work is required to develop safe and effective means of inactivating viruses in air, and standardized methods are also needed to test as well as to validate the claims of devices marketed for this purpose. Recent studies have shown that ozone has the potential to inactivate viruses in aerosolized blood products (97). Further work on such an approach could provide alternative means of ridding heat-sensitive injectables of harmful viruses.

Airborne viruses continue to be a threat to human and animal health in spite of the sophisticated design and efficient protective functioning of the respiratory system (158). Much of the research thus far has documented outbreaks of airborne viral infections and identified to some degree the factors that influence virus survival in air. Now greater emphasis is needed on reducing generation of infectious aerosols in indoor as well as in outdoor (16) settings and on enhancing the removal and/or inactivation of viruses (29) and other infectious agents in air. Many human and animal pathogenic viruses, some of which are relatively obscure and outside the mainstream, continue to be considered potentially useful bioweapons, and recent accounts (3, 110, 157) of their production and stockpiling attest to this fact. The true extent of past and possibly ongoing work in this regard will remain shrouded in mystery because the details may never become available to the general scientific community.

We are grateful to Bill Kournikakis and Canada's Defence Research Establishment for permission to reproduce the scanning electron micrograph and the graphs. The assistance of Jason Tetro in the preparation of the manuscript is gratefully acknowledged.

REFERENCES

1. **Agranovski, I. E., A. S. Safatov, A. L. Borodulin, O. V. Pyankov, V. A. Petrishchenko, A. N. Sergeev, A. P. Agafonov, G. M. Ignatiev, A. A. Sergeev, and V. Agranovski**. 2004. Inactivation of viruses in bubbling processes utilized for personal bioaerosol monitoring. *Appl. Environ. Microbiol.* **70:**6963–6967.
2. **Alexeyev, O. A., and B. A. Baranov.** 1993. Puumala virus infection without signs of renal involvement. *Scand. J. Infect. Dis.* **25:**525–527.
3. **Alibek, K., and S. Handleman.** 1999. *Biohazard.* Random House, New York, N.Y.
4. **American Medical Association.** 1998. *Airborne Infection on Commercial Flights.* Report 10 of the Council on Scientific Affairs (A-98). American Medical Association, Washington, D.C.
5. **Andersen, A. A.** 1958. New sampler for collection, sizing, and enumeration of viable airborne particles. *J. Bacteriol.* **76:**471–484.
6. **Anderson, G. W., Jr., J. O. Lee, A. O. Anderson, N. Powell, J. A. Mangiafico, and G. Meadors.** 1991. Efficacy of a Rift Valley fever virus vaccine against an aerosol infection in rats. *Vaccine* **9:**710–714.
7. **Ansari, S. A., V. S. Springthorpe, and S. A. Sattar.** 1991. Survival and vehicular spread of human rotaviruses: possible relation to seasonality of outbreaks. *Rev. Infect. Dis.* **13:**448–461.
8. **Baggish, M. S., B. J. Poiesz, D. Joret, P. Williamson, and A. Refai.** 1991. Presence of human immunodeficiency virus DNA in laser smoke. *Lasers Surg. Med.* **11:**197–203.
9. **Banks, M.** 1993. DNA restriction fragment length polymorphism among British isolates of Aujeszky's disease virus: use of the polymerase chain reaction to discriminate among strains. *Br. Vet. J.* **149:**155–163.
10. **Barrett, W. L., and S. M. Garber.** 2003. Surgical smoke: a review of the literature. Is this just a lot of hot air? *Surg. Endosc.* **17:**979–987.
11. **Barry, M., F. Bia, M. Cullen, L. Dembry, S. Fischer, D. Geller, W. Hierholzer, P. McPhedran, P. Rainey, M. Russi, E. Snyder, E. Wrone, J. P. Gonzalez, R. Rico-Hesse, R. Tesh, R. Ryder, R. Shope, W. P. Quinn, P. D. Galbraith, M. L. Cartter, J. L. Hadler, and A. DeMaria, Jr.** 1994. Arenavirus infection—Connecticut, 1994. *Morb. Mortal. Wkly. Rep.* **43:**635–636.
12. **Belknap, E. B., J. K. Collins, V. K. Ayers, and P. C. Schultheiss.** 1994. Experimental infection of neonatal calves with neurovirulent bovine herpesvirus type 1.3. *Vet. Pathol.* **31:**358–365.
13. **Bergbrant, I.-M., L. Samuelsson, S. Olofsson, F. Jonassen, and A. Ricksten.** 1994. Polymerase chain reaction for monitoring human papillomavirus contamination of medical personnel during treatment of genital warts with CO_2 laser and electrocoagulation. *Acta Dermato-Venereol.* (Stockholm) **74:**393–395.
14. **Booth, T. F., B. Kournikakis, N. Bastien, J. Ho, D. Kobasa, L. Stadnyk, Y. Li, M. Spence, S. Paton, B. Henry, B. Mederski, D. White, D. E. Low, A. McGeer, A. Simor, M. Vearncombe, J. Downey, F. B. Jamieson, P. Tang, and F. Plummer.** 2005. Detection of airborne severe acute respiratory syndrome (SARS) coronavirus and environmental contamination in SARS outbreak units. *J. Infect. Dis.* **191:**1472–1477.
15. **Brim, T. A., J. L. VanCott, J. K. Lunney, and L. J. Saif.** 1994. Lymphocyte proliferation responses of pigs inoculated with transmissible gastroenteritis virus or porcine respiratory coronavirus. *Am. J. Vet. Res.* **55:**494–501.
16. **Brooks, J. P., B. D. Tanner, K. L. Josephson, C. P. Gerba, and I. L. Pepper.** 2004. Bioaerosols from the land application of biosolids in the desert southwest USA. *Water Sci. Technol.* **50:**7–12.
17. **Brown, C. C., R. F. Meyer, H. J. Olander, C. House, and C. A. Mebus.** 1992. A pathogenesis study of foot-and-mouth disease in cattle, using in situ hybridization. *Can. J. Vet. Res.* **56:**189–193.
18. **Burge, H. A.** 1995. Airborne contagious disease, p. 25–47. In H. A. Burge (ed.), *Bioaerosols.* CRC Press, Inc., Boca Raton, Fla.
19. **Calero, L., and T. Brusis.** 2003. Laryngeal papillomatosis—first recognition in Germany as an occupational disease in an operating room nurse. *Laryngorhinootologie* **82:** 790–793.
20. **Calnek, B. W., and S. B. Hitchner.** 1973. Survival and disinfection of Marek's disease virus and the effectiveness of filters in preventing airborne dissemination. *Poult. Sci.* **52:** 35–43.
21. **Casal, J., E. Planas-Cuchi, and J. Casal.** 1999. Sheltering as a protective measure against airborne virus spread. *J. Hazard. Mater.* **68:**179–189.
22. **Caul, E. O.** 1994. Small round structured viruses: airborne transmission and hospital control. *Lancet* **343:**1240–1242.
23. **Centers for Disease Control and Prevention.** 1997. Isolation of avian influenza A(H5N1) viruses from hu-

mans—Hong Kong, May-December 1997. *Morb. Mortal. Wkly. Rep.* **46:**1204–1207.
24. **Chadwick, P. R., and R. McCann.** 1994. Transmission of a small round structured virus by vomiting during a hospital outbreak of gastroenteritis. *J. Hosp. Infect.* **26:**251–259.
25. **Chadwick, P. R., M. Walker, and A. E. Rees.** 1994. Airborne transmission of a small round structured virus. *Lancet* **343:**171.
26. **Christensen, L. S., S. Mortensen, A. Botner, B. S. Strandbygaard, L. Ronsholt, C. A. Henriksen, and J. B. Andersen.** 1993. Further evidence of long distance airborne transmission of Aujeszky's disease (pseudorabies) virus. *Vet. Rec.* **132:**317–321.
27. **Chung, I. P., D. Dunn-Rankin, R. F. Phalen, and M. J. Oldham.** 1992. Low-cost wind tunnel for aerosol inhalation studies. *Am. Ind. Hyg. Assoc. J.* **53:**232–236.
28. **Clark, R. P.** 1995. Safety cabinets, fume cupboards and other containment systems, p. 473-504. *In* C. S. Cox and C. M. Wathes (ed.), *Bioaerosols Handbook*. CRC Press, Inc., Boca Raton, Fla.
29. **Clark, S., and P. Scarpino.** 1996. Disinfection of microbial aerosols, p. 215-223. *In* C. J. Hurst (ed.), *Modeling Disease Transmission and Its Prevention by Disinfection*. Cambridge University Press, Cambridge, England.
30. **Cole, E. C., and C. E. Cook.** 1998. Characterization of infectious aerosols in health care facilities: an aid to effective engineering controls and preventive strategies. *Am. J. Infect. Control* **26:**453–464.
31. **Cox, C. S.** 1987. *The Aerobiological Pathway of Microorganisms*, p. 1-293. John Wiley & Sons, Chichester, England.
32. **Cox, C. S.** 1989 Airborne bacteria and viruses. *Sci. Prog.* **73:**469–500.
33. **Crook, B.** 1995. Inertial samplers: biological perspective, p. 247-267. *In* C. S. Cox and C. M. Wathes (ed.), *Bioaerosols Handbook*. CRC Press, Inc., Boca Raton, Fla.
34. **Crook, B.** 1995. Non-inertial samplers: biological perspective, p. 269-283. *In* C. S. Cox and C. M. Wathes (ed.), *Bioaerosols Handbook*. CRC Press, Inc., Boca Raton, Fla.
35. **Cukier, J., M. F. Price, and L. O. Gentry.** 1989. Suction lipoplasty: biohazardous aerosols and exhaust mist—the clouded issue. *Plast. Reconstr. Surg.* **83:**494–497.
36. **Dalgard, D. W., R. J. Hardy, S. L. Pearson, G. J. Pucak, R. V. Quander, P. M. Zack, C. J. Peters, and P. B. Jahrling.** 1992. Combined simian hemorrhagic fever and Ebola virus infection in cynomolgus monkeys. *Lab. Anim. Sci.* **42:**152–157.
37. **Davis, W. G., A. R. Griesemer, A. J. Shadduck, and L. R. Farrelle.** 1971. Effect of relative humidity on dynamic aerosols of adenovirus 12. *Appl. Microbiol.* **21:** 676–679.
38. **Dee, S. A., J. Deen, L. Jacobson, K. D. Rossow, C. Mahlum, and C. Pijoan.** 2005. Laboratory model to evaluate the role of aerosols in the transport of porcine reproductive and respiratory syndrome virus. *Vet. Rec.* **156:** 501–504.
39. **Deuter, A., D. J. Southee, and A. P. Mockett.** 1991. Fowlpox virus: pathogenicity and vaccination of day-old chickens via the aerosol route. *Res. Vet. Sci.* **50:**362–364.
40. **Dewulf, J., H. Laevens, F. Koenen, K. Mintiens, and A. de Kruif.** 2000. Airborne transmission of classical swine fever virus under experimental conditions. *Vet. Rec.* **147:**735–738.
41. **Dick, E. C., L. C. Jennings, K. A. Mink, C. D. Wartgow, and S. L. Inhorn.** 1987. Aerosol transmission of rhinovirus colds. *J. Infect. Dis.* **156:**442–448.
42. **Donaldson, A. I., R. C. Wardley, S. Martin, and N. P. Ferris.** 1983. Experimental Aujeszky's disease in pigs: excretion, survival and transmission of the virus. *Vet. Rec.* **113:**490–494.
43. **Donaldson, A. I., and S. Alexandersen.** 2002. Predicting the spread of foot and mouth disease by airborne virus. *Rev. Sci. Tech. Off. Int. Epizoot.* **21:**569–575.
44. **Echavarria, M., S. A. Kolavic, S. Cersovsky, F. Mitchell, J. L. Sanchez, C. Polyak, B. L. Innis, and L. N. Binn.** 2000. Detection of adenoviruses (AdV) in culture-negative environmental samples by PCR during an AdV-associated respiratory disease outbreak. *J. Clin. Microbiol.* **38:** 2982–2984.
45. **Edwards, D. A., J. C. Man, P. Brand, J. P. Katstra, K. Sommerer, H. A. Stone, E. Nardell, and G. Scheuch.** 2004. Inhaling to mitigate exhaled bioaerosols. *Proc. Natl. Acad. Sci. USA* **101:**17383–17388.
46. **Eickhoff, T. C.** 1994. Airborne nosocomial infection: a contemporary perspective. *Infect. Control Hosp. Epidemiol.* **15:**663–672.
47. **Elazhary, M. A. S. Y., M. Galina, R. S. Roy, M. Fontaine, and P. Lamothe.** 1980. Experimental infection of calves with bovine respiratory syncytial virus (Quebec strain). *Can. J. Comp. Med.* **44:**390–395.
48. **Fairchild, G. A.** 1974. Ozone effect on respiratory deposition of vesicular stomatitis virus aerosols. *Am. Rev. Respir. Dis.* **109:**446–451.
49. **Ferbas, J., K. R. Chadwick, A. Logar, A. E. Patterson, R. W. Gilpin, and J. B. Margolick.** 1995. Assessment of aerosol containment on the ELITE flow cytometer. *Cytometry* **22:**45–47.
50. **Fleming, D. O., J. H. Richardson, J. J. Tulis, and D. Vesley.** 1995. *Laboratory Safety: Principles and Practices*, 2nd ed. ASM Press, Washington, D.C.
51. **Frank, G. H., R. G. Marshall, and P. C. Smith.** 1977. Clinical and immunologic responses of cattle to infectious bovine rhinotracheitis virus after infection by viral aerosol or intramuscular inoculation. *Am. J. Vet. Res.* **38:**1497–1502.
52. **Frankova, V.** 1975. Inhalatory infection of mice with influenza A0/PR8 virus. I. The site of primary virus replication and its spread in the respiratory tract. *Acta Virol.* **19:**29–34.
53. **Frankova, V., and V. Rychterova.** 1975. Inhalatory infections of mice with influenza A0/PR8 virus. II. Detection of virus in the blood and extrapulmonary organs. *Acta Virol.* **19:**35–40.
54. **Franz, D. R., P. B. Jahrling, A. M. Friedlander, D. J. McClain, D. L. Hoover, W. R. Bryne, J. A. Pavlin, G. W. Christopher, and E. M. Eitzen, Jr.** 1997. Clinical recognition and management of patients exposed to biological warfare agents. *JAMA* **278:**399–411.
55. **Garden, J. M., M. K. O'Banion, L. S. Shelnitz, K. S. Pinski, A. D. Bakus, M. E. Reichmann, and J. P. Sundberg.** 1988. Papillomavirus in the vapor of carbon dioxide laser-treated verrucae. *JAMA* **259:**1199-1202.
56. **Garden, J. M., M. K. O'Banion, A. D. Bakus, and C. Olson.** 2002. Viral disease transmitted by laser-generated plume (aerosol). *Arch. Dermatol.* **138:**1303–1307.
57. **Gellert, G. A., S. H. Waterman, D. Ewert, L. Oshiro, M. P. Giles, S. S. Monroe, L. Gorelkin, and R. I. Glass.** 1990. An outbreak of acute gastroenteritis caused by a small round structured virus in a geriatric convalescent facility. *Infect. Control Hosp. Epidemiol.* **11:**459–464.
58. **Gibson, C. F., and A. I. Donaldson.** 1986. Exposure of sheep to natural aerosols of foot-and-mouth disease. *Res. Vet. Sci.* **41:**45–49.
59. **Gillespie, R. R., M. A. Hill, C. L. Kanitz, K. E. Knox, L. K. Clark, and J. P. Robinson.** 2000. Infection of pigs by Aujeszky's disease virus via the breath of intranasally inoculated pigs. *Res. Vet. Sci.* **68:**217–222.
60. **Gillespie, R. R., M. A. Hill, and C. L. Kanitz.** 1996. Infection of pigs by aerosols of Aujeszky's disease virus and their shedding of the virus. *Res. Vet. Sci.* **30:**228-233.

61. **Gloster, J., H. J. Champion, L. M. Mansley, P. Romero, T. Brough, and A. Ramirez.** 2005. The 2001 epidemic of foot-and-mouth disease in the United Kingdom: epidemiological and meteorological studies. *Vet. Rec.* **156:**793–803.
62. **Goldberg, L. J., H. M. S. Watkins, E. E. Boerke, and M. A. Chatigny.** 1958. The use of a rotating drum for the study of aerosols over extended periods of time. *Am. J. Hyg.* **68:**85–93.
63. **Goldmann, D. A.** 2000. Transmission of viral respiratory infections in the home. *Pediatr. Infect. Dis. J.* **19:**S97–S102.
64. **Goodman, R. A., S. B. Thacker, S. L. Solomon, M. T. Osterholm, and J. M. Hughes.** 1994. Infectious diseases in competitive sports. *JAMA* **271:**862-867.
65. **Gooze, L. L., and E. C. Hughes.** 2003. Smallpox. *Semin. Respir. Infect.* **18:**196–205.
66. **Grant, R. H., A. B. Scheidt, and L. R. Rueff.** 1994. Aerosol transmission of a viable virus affecting swine: explanation of an epizootic of pseudorabies. *Int. J. Biometeorol.* **38:**33–39.
67. **Gruber, W. C., H. P. Hinson, K. L. Holland, J. M. Thompson, G. W. Reed, and P. F. Wright.** 1993. Comparative trial of large-particle aerosol and nose drop administration of live attenuated influenza vaccines. *J. Infect. Dis.* **168:**1282–1285.
68. **Gwaltney, J. M., Jr.** 1997. Rhinoviruses, p. 815-838. *In* A. S. Evans and R. A. Kaslow (ed.), *Viral Infections of Humans: Epidemiology and Control*, 4th ed. Plenum Medical Book Co., New York, N.Y.
69. **Hallmo, P., and O. Naess.** 1991. Laryngeal papillomatosis with human papillomavirus DNA contracted by a laser surgeon. *Eur. Arch. Otorhinolaryngol.* **248:**425–427.
70. **Hammond, G. W., R. L. Raddatz, and D. E. Gelskey.** 1989. Impact of atmospheric dispersion and transport of viral aerosols on the epidemiology of influenza. *Rev. Infect. Dis.* **11:**494–497.
71. **Hamory, B. H., R. B. Couch, R. G. J. Douglas, S. H. Black, and V. Knight.** 1972. Characterization of the infective unit for man of two respiratory viruses. *Proc. Soc. Exp. Biol. Med.* **139:**890–893.
72. **Handley, B. A., and J. M. Roe.** 1994. An alternative microthread for the study of airborne survival of bacteria outdoors. *J. Appl. Bacteriol.* **77:**504–508.
73. **Heinsohn, P., D. L. Jewett, L. Balzer, C. H. Bennett, P. Seipel, and A. Rosen.** 1991. Aerosols created by some surgical power tools: particle size distribution and qualitative hemoglobin content. *Appl. Occup. Environ. Hyg.* **6:**773–776.
74. **Henderson, D. W.** 1952. An apparatus for the study of airborne infections. *J. Hyg.* **50:**53–88.
75. **Hensel, A., and K. Petzold.** 1995. Biological and biochemical analysis of bacteria and viruses, p. 335-360. *In* C. S. Cox and C. M. Wathes (ed.), *Bioaerosols Handbook*. CRC Press, Inc., Boca Raton, Fla.
76. **Hepburn, N. C., and T. J. G. Brooks.** 1991. An outbreak of chickenpox in a military field hospital—the implications for biological warfare. *J. R. Soc. Med.* **84:**721–722.
77. **Hietala, S. K., P. J. Hullinger, B. M. Crossley, H. Kinde, and A. A. Ardans.** 2005. Environmental air sampling to detect exotic Newcastle disease virus in two California commercial poultry flocks. *J. Vet. Diagn. Investig.* **17:**198–200.
78. **Hinshaw, V. S., F. L. Schaffer, and M. A. Chatigny.** 1976. Evaluation of Moloney Murine Sarcoma and leukemia virus complex as a model for airborne oncogenic virus biohazards: survival of airborne virus and exposure of mice. *J. Natl. Cancer Inst.* **57:**775–778.
79. **Ho, M.-S., R. I. Glass, P. F. Pinsky, and L. J. Anderson.** 1988. Rotavirus as a cause of diarrheal morbidity and mortality in the United States. *J. Infect. Dis.* **158:**1112–1116.
80. **Hopkins, S. R., and L. N. Drury.** 1971. Efficacy of air filters in preventing transmission of Newcastle disease. *Avian Dis.* **15:**596–603.
81. **Hugh-Jones, M. E., W. H. Allen, F. A. Dark, and J. G. Harper.** 1973. The evidence for the airborne spread of Newcastle disease. *J. Hyg.* **71:**325–339.
82. **Hyslop, N. S. G.** 1979. Observation on the survival and infectivity of airborne Rinderpest virus. *Int. J. Biometeorol.* **23:**1–7.
83. **Ijaz, M. K., S. A. Sattar, C. M. Johnson-Lussenburg, and V. S. Springthorpe.** 1985. Comparison of the airborne survival of calf rotavirus and poliovirus type 1 (Sabin) aerosolized as a mixture. *Appl. Environ. Microbiol.* **49:**289–293.
84. **Ijaz, M. K., S. A. Sattar, T. Alkarmi, F. K. Dar, A. R. Bhatti, and K. M. Elhag.** 1994. Studies on the survival of aerosolized bovine rotavirus (UK) and a murine rotavirus. *Comp. Immun. Microbiol. Infect. Dis.* **17:**91–98.
85. **Ijaz, M. K., Y. G. Karim, S. A. Sattar, and C. M. Johnson-Lussenburg.** 1987. Development of methods to study the survival of airborne viruses. *J. Virol. Methods* **18:**87–106.
86. **Ijaz, M. K., A. H. Brunner, S. A. Sattar, R. C. Nair, and C. M. Johnson-Lussenburg.** 1985. Survival characteristics of airborne human coronavirus 229E. *J. Gen. Virol.* **66:**2743–2748.
87. **Ijaz, M. K., S. Alharbi, S. A. Uduman, Y. Cheema, M. M. Sheek-Hussein, A. R. A. Alkhair, A. G. Shalabi, S. S. Ijaz, S. A. Bib-Othman, S. A. Sattar, and L. F. Liddle.** 1994. Seasonality and prevalence of rotavirus in Al-Ain, United Arab Emirates. *Clin. Diagn. Virol.* **2:**323–329.
88. **Jahrling, P. B., L. E. Hensley, M. J. Martinez, J. W. Leduc, K. H. Rubins, D. A. Relman, and J. W. Huggins.** 2004. Exploring the potential of variola virus infection of cynomolgus macaques as a model for human smallpox. *Proc. Natl. Acad. Sci. USA* **101:**15196–15200.
89. **Jericho, K. W. F., C. le Q. Darcel, and E. V. Langford.** 1982. Respiratory disease in calves produced with aerosols of parainfluenza-3 virus and Pasteurella haemolytica. *Can. J. Comp. Med.* **46:**293–301.
90. **Jericho, K. W., and C. G. Kozub.** 2004. Experimental infectious respiratory disease in groups of calves: lobar distribution, variance, and sample-size requirements for vaccine evaluation. *Can. J. Vet. Res.* **68:**118–127.
91. **Jewett, D. L., P. Heinsohn, C. Bennett, A. Rosen, and C. Neuilly.** 1992. Blood-containing aerosols generated by surgical techniques: a possible infectious hazard. *Am. Ind. Hyg. Assoc. J.* **53:**228–231.
92. **Johansson, B. E., and E. D. Kilbourne.** 1991. Comparison of intranasal and aerosol infection of mice in assessment of immunity to influenza virus infection. *J. Virol. Methods* **35:**109–114.
93. **Johnson, G. K., and W. S. Robinson.** 1991. Human immunodeficiency virus-1 (HIV-1) in the vapors of surgical power instruments. *J. Med. Virol.* **33:**47–50.
94. **Karim, Y. G., M. K. Ijaz, S. A. Sattar, and C. M. Johnson-Lussenburg.** 1985. Effect of relative humidity on the airborne survival of rhinovirus-14. *Can. J. Microbiol.* **31:**1058–1061.
95. **Kashima, H. K., T. Kessis, P. Mounts, and K. Shah.** 1991. Polymerase chain reaction identification of human papillomavirus DNA in CO_2 laser plume from recurrent respiratory papillomatosis. *Otolaryngol. Head Neck Surg.* **104:**191–195.
96. **Keeler, N., and J. Lingappa.** 2004. Severe acute respiratory syndrome: public health response and clinical practice update for an emerging disease. *Curr. Opin. Pediatr.* **16:** 61–69.

97. **Kekez, M. M., and S. A. Sattar.** 1997. A new ozone-based method for virus inactivation: preliminary study. *Phys. Med. Biol.* **42:**2027–2039.
98. **Kenyon, R. H., K. T. McKee, Jr., P. M. Zack, M. K. Rippy, A. P. Vogel, C. York, J. Meegan, C. Crabbs, and C. J. Peters.** 1992. Aerosol infection of rhesus macaques with Junin virus. *Intervirology* **33:**23–31.
99. **Kim, S. J., and P. B. Spradbrow.** 1978. Administration of a vaccine prepared from the Australian V4 strain of Newcastle disease virus by aerosol and drinking water. *Aust. Vet. J.* **54:**486–489.
100. **Knoch, M., E. Wunderlich, and S. Geldner.** 1994. A nebulizer system for highly reproducible aerosol delivery. *J. Aerosol Med.* **7:**229–237.
101. **Korppi, M.** 1988. Viruses and airborne allergens as precipitants of obstructive respiratory difficulties in children. *Ann. Clin. Res.* **20:**417–422.
102. **Kournikakis, B., D. Netolitzky, and J. Fildes.** 1987. Effects of temperature and relative humidity on the survival of Newcastle disease virus aerosols in the rotating drum (U). Memorandum 1261, p. 1-17. Defence Research Establishment Suffield, Ralston, Alberta, Canada.
103. **Kournikakis, B., M. Simpson, and D. Netolitzky.** 1988. Photoinactivation of Newcastle disease virus in aerosol and in solution. Memorandum 1263, p. 1-12. Defence Research Establishment Suffield, Ralston, Alberta, Canada.
104. **Larson, E. W.** 1970. Airborne transmission potential of murine leukemia infection, p. 213. *In* I. H. Silver (ed.), *Proceedings of the Third International Symposium on Aerobiology*. Academic Press, New York, N.Y.
105. **Larson, E. W., J. W. Dominik, and T. W. Slove.** 1980. Aerosol stability and respiratory infectivity of Japanese B encephalitis virus. *Infect. Immun.* **30:**397–401.
106. **LeDuc, J. W.** 1989. Epidemiology of hemorrhagic fever viruses. *Rev. Infect. Dis.* **11:**S730–S735.
107. **Lee, H. W., and K. M. Johnson.** 1982. Laboratory-acquired infections with Hantaan virus, the etiological agent of Korean hemorrhagic fever. *J. Infect. Dis.* **146:** 645–651.
108. **Leffel, E. K., and D. S. Reed.** 2004. Marburg and Ebola viruses as aerosol threats. *Biosecuro Bioterror.* **2:**186–191.
109. **Li, K. S., Y. Guan, J. Wang, G. J. Smith, K. M. Xu, L. Duan, A. P. Rahardjo, P. Puthavathana, C. Buranathai, T. D. Nguyen, A. T. Estoepangestie, A. Chaisingh, P. Auewarakul, H. T. Long, N. T. Hanh, R. J. Webby, L. L. Poon, H. Chen, K. F. Shortridge, K. Y. Yuen, R. G. Webster, and J. S. Peiris.** 2004. Genesis of a highly pathogenic and potentially pandemic H5N1 influenza virus in eastern Asia. *Nature* **430:**209–213.
110. **Madjid, M., S. Lillibridge, P. Mirhaji, and W. Casscells.** 2003. Influenza as a bioweapon. *J. R. Soc. Med.* **96:**345–346.
111. **Maigetter, R. Z., R. Ehrlich, J. D. Fenters, and D. E. Gardner.** 1976. Potentiating effects of manganese dioxide on experimental respiratory infections. *Environ. Res.* **11:** 386–391.
112. **Maragon, S., E. Facchin, F. Moutou, I. Massirio, G. Vincenzi, and G. Davies.** 1994. The 1993 Italian foot-and-mouth disease epidemic: epidemiological features of the four outbreaks identified in Verona province (Veneto region). *Vet. Rec.* **135:**53–57.
113. **Marks, P. J., I. B. Vipond, D. Carlisle, D. Deakin, R. E. Fey, and E. O. Caul.** 2000. Evidence for airborne transmission of Norwalk-like virus (NLV) in a hotel restaurant. *Epidemiol. Infect.* **124:**481–487.
114. **Marks, P. J., I. B. Vipond, F. M. Regan, K. Wedgwood, R. E. Fey, and E. O. Caul.** 2003. A school outbreak of Norwalk-like virus: evidence for airborne transmission. *Epidemiol. Infect.* **131:**727–736.
115. **Mars, M. H., C. J. Bruschke, and J. T. Oirschot.** 1999. Airborne transmission of BHV1, BRSV, and BVDV among cattle is possible under experimental conditions. *Vet. Microbiol.* **66:**197–207.
116. **May, K. R.** 1966. Multistage liquid impinger. *Bacteriol. Rev.* **30:**559–570.
117. **May, K. R., and H. A. Druett.** 1968. A microthread technique for studying the viability of microbes in a simulated airborne state. *J. Gen. Virol.* **51:**353–366.
118. **McElroy, A. K., M. Bray, D. S. Reed, and C. S. Schmaljohn.** 2002. Andes virus infection of cynomolgus macaques. *J. Infect. Dis.* **186:**1706–1712.
119. **McKissick, G. E., L. G. Wolf, R. L. Farrell, R. A. Griesemer, and A. Hellman.** 1970. Aerosol transmission of oncogenic viruses, p. 233-240. *In* I. H. Silver (ed.), *Proceedings of the Third International Symposium on Aerobiology*. Academic Press, New York, N.Y.
120. **McKissick, G. E., R. A. Griesemer, and R. L. Farrell.** 1970. Aerosol transmission of Rauscher murine leukemia virus. *J. Natl. Cancer Inst.* **45:**625–636.
121. **McVicar, J. W., and R. J. Eisner.** 1983. Aerosol exposure of cattle to foot-and-mouth disease virus. *J. Hyg.* **91:**319–328.
122. **Merekalova, Z. I., N. M. Orekhova, and E. I. Zharona.** 1981. Possible ways of excreting murine leukemia virus into the environment and the so-called input gate for infection. *Eksp. Onkol.* **3:**59–62.
123. **Mitchell, B. W., and D. J. King.** 1994. Effect of negative air ionization on airborne transmission of Newcastle disease virus. *Avian Dis.* **38:**725–732.
124. **Mitchell, J. P.** 1995. Aerosol generation for instrument calibration, p. 101-175. *In* C. S. Cox and C. M. Wathes (ed.), *Bioaerosols Handbook*. CRC Press, Inc., Boca Raton, Fla.
125. **Mitchell, J. P.** 1995. Particle size analyzers: practical procedures and laboratory techniques, p. 177-246. *In* C. S. Cox and C. M. Wathes (ed.), *Bioaerosols Handbook*. CRC Press, Inc., Boca Raton, Fla.
126. **Moutou, F., and B. Durand.** 1994. Modeling the spread of foot-and-mouth disease virus. *Vet. Res.* **25:**279–285.
127. **Mumford, J. A., and J. Wood.** 1992. Establishing an acceptability threshold for equine influenza vaccines. *Dev. Biol. Stand.* **79:**137–146.
128. **Mumford, J. A., D. M. Jessett, E. A. Rollinson, D. Hannant, and M. E. Draper.** 1994. Duration of protective efficacy of equine influenza immunostimulating complex/tetanus vaccines. *Vet. Rec.* **134:**158–162.
129. **Myatt, T. A., S. L. Johnston, S. Rudnick, and D. K. Milton.** 2003. Airborne rhinovirus detection and effect of ultraviolet irradiation on detection by a semi-nested RT-PCR assay. *BMC Public Health* **3:**5.
130. **Myatt, T. A., S. L. Johnston, Z. Zuo, M. Wand, T. Kebadze, S. Rudnick, and D. K. Milton.** 2004. Detection of airborne rhinovirus and its relation to outdoor air supply in office environments. *Am. J. Respir. Crit. Care Med.* **169:**1187–1190.
131. **Ng, S. K.** 2003. Possible role of an animal vector in the SARS outbreak at Amoy Gardens. *Lancet* **362:**570–572.
132. **Ng, S. K.** 2004. The mystery of Amoy Gardens, p. 117–138. *In* C. Loh and Civic Exchange (ed.), *At the Epicentre: Hong Kong and the SARS Outbreak*. Hong Kong University Press, Hong Kong, China.
133. **Niklasson, B. S.** 1992. Haemorrhagic fever with renal syndrome, virological and epidemiological aspects. *Pediatr. Nephrol.* **6:**201–204.
134. **Nuzum, E. O., C. A. Rossi, E. H. Stephenson, and J. W. LeDuc.** 1988. Aerosol transmission of Hantaan and related viruses to laboratory rats. *Am. J. Trop. Med. Hyg.* **83:**636–640.

135. **Olsen, S. J., H. L. Chang, T. Y. Cheung, A. F. Tang, T. L. Fisk, S. P. Ooi, H. W. Kuo, D. D. Jiang, K. T. Chen, J. Lando, K. H. Hsu, T. J. Chen, and S. F. Dowell.** 2006. Transmission of the severe acute respiratory syndrome on aircraft. *N. Engl. J. Med.* **349:**2416–2422

136. **Ormerod, E., I. A. McCandlish, and O. Jarrett.** 1979. Disease produced by feline caliciviruses. *Vet. Rec.* **104:** 65–69.

137. **Osborne, A. D., J. R. Saunders, T. K. Sebunya, P. Willson, and G. H. Green.** 1985. A simple chamber for experimental reproduction of respiratory disease in pigs and other species. *Can. J. Comp. Med.* **49:**434–435.

138. **Owens, S. L., J. W. Osebold, and Y. C. Zee.** 1981. Dynamics of B-lymphocytes in the lungs of mice exposed to aerosolized influenza virus. *Infect. Immun.* **33:**231–238.

139. **Peters, C. J., and A. S. Khan.** 2002. Hantavirus pulmonary syndrome: the new American hemorrhagic fever. *Clin. Infect. Dis.* **34:**1224–1231.

140. **Phillpotts, R. J., L. O'Brien, R. E. Appleton, S. Carr, and A. Bennett.** 2005. Intranasal immunisation with defective adenovirus serotype 5 expressing the Venezuelan equine encephalitis virus E2 glycoprotein protects against airborne challenge with virulent virus. *Vaccine* **23:**1615–1623.

141. **Pike, R. M.** 1979. Laboratory-associated infections: incidence, fatalities, causes and prevention. *Annu. Rev. Microbiol.* **33:**41–66.

142. **Pokhodiaev, V. A., N. I. Gonchar, and V. A. Pshenichnov.** 1991. An experimental study of the contact transmission of the Marburg virus. *Vopr. Virusol.* **36:** 506–508.

143. **Pratt, W. D., P. Gibbs, M. L. Pitt, and A. L. Schmaljohn.** 1998. Use of telemetry to assess vaccine-induced protection against parenteral and aerosol infections of Venezuelan equine encephalitis virus in non-human primates. *Vaccine* **16:**1056–1064.

144. **Prince, D. S., C. Astry, S. Vonderfecht, G. Jakab, F.-M. Shen, and R. H. Yolken.** 1986. Aerosol transmission of experimental rotavirus infection. *Pediatr. Infect. Dis.* **5:**218–222.

145. **Quintiliani, R.** 1994. The epidemiology of mild, moderate, and severe asthma. *Adv. Ther.* **11:**70–77.

146. **Ratanasethakul, C., and R. B. Cumming.** 1983. The effect of route of infection and strain of virus on the pathology of Australian infectious bronchitis. *Aust. Vet. J.* **60:**209–213.

147. **Richmond, J. Y., and R. W. McKinney.** 1999. *Biosafety in Microbiological and Biomedical Laboratories*, 4th ed. Centers for Disease Control and Prevention and National Institutes of Health, Government Printing Office, Washington, D.C.

148. **Roth, Y., J. S. Chapnik, and P. Cole.** 2003. Feasibility of aerosol vaccination in humans. *Ann. Otol. Rhinol. Laryngol.* **112:**264–270.

149. **Roy, C. J., and D. K. Milton.** 2004. Airborne transmission of communicable infection—the elusive pathway. *N. Engl. J. Med.* **22:**1710–1712.

150. **Saif, L. J., and K. W. Theil.** 1990. *Viral Diarrheas of Man and Animals*, p. 1–343. CRC Press, Inc., Boca Raton, Fla.

151. **Saksena, N. K., D. Dwyer, and F. Barre-Sinoussi.** 1991. A "sentinel" technique for monitoring viral aerosol contamination. *J. Infect. Dis.* **164:**1021–1022.

152. **Sattar, S. A., and M. K. Ijaz.** 1987. Spread of viral infections by aerosols. *Crit. Rev. Environ. Control* **17:**89–131.

153. **Sawchuk, W. S., P. J. Weber, D. R. Lowy, and L. M. Dzubow.** 1989. Infectious papillomavirus in the vapor of warts treated with carbon dioxide laser or electrocoagulation: detection and protection. *J. Am. Acad. Dermatol.* **21:**41–49.

154. **Sawyer, L. A., J. J. Murphy, J. E. Kaplan, P. F. Pinsky, D. Chacon, S. Walmsley, L. B. Schonberger, A. Philips, K. Forward, C. Goldman, J. Brunton, R. A. Fralick, A. O. Carter, W. G. J. Gary, R. I. Glass, and D. E. Low.** 1988. 25- to 30-nm virus particle associated with a hospital outbreak of acute gastroenteritis with evidence for airborne transmission. *Am. J. Epidemiol.* **127:**1261–1271.

155. **Sawyer, M. H., C. J. Chamberlin, Y. N. Wu, N. Aintablian, and M. R. Wallace.** 1994. Detection of varicella-zoster virus DNA in air samples from hospital rooms. *J. Infect. Dis.* **169:**91–94.

156. **Schoenbaum, M. A., J. J. Zimmerman, G. W. Beran, and D. P. Murphy.** 1990. Survival of pseudorabies virus in aerosol. *Am. J. Vet. Res.* **51:**331–333.

157. **Sidwell, R. W., and D. F. Smee.** 2003. Viruses of the Bunya- and Togaviridae families: potential as bioterrorism agents and means of control. *Antivir. Res.* **57:**101–111.

158. **Skerrett, S. J.** 1994. Host defenses against respiratory infection. *Med. Clin. N. Am.* **78:**941–966.

159. **Slote, L.** 1976. Viral aerosols: a potential occupationally related health threat in aerated wastewater treatment systems. *J. Environ. Health* **38:**310–314.

160. **Smith, K. C., K. E. Whitwell, M. M. Binns, C. A. Dolby, D. Hannant, and J. A. Mumford.** 1992. Abortion of virologically negative foetuses following experimental challenge of pregnant pony mares with equid herpesvirus 1. *Equine Vet. J.* **24:**256–259.

161. **Somogyi, R., A. E. Vesely, T. Azami, D. Preiss, J. Fisher, J. Correia, and R. A. Fowler.** 2004. Dispersal of respiratory droplets with open vs closed oxygen delivery masks: implications for the transmission of severe acute respiratory syndrome. *Chest* **125:**1155–1157.

162. **Spendlove, J. C., and K. F. Fannin.** 1982. Methods of characterization of virus aerosols, p. 261–329. *In* C. P. Gerba and S. M. Goyal (ed.), *Methods in Environmental Virology*. Marcel Dekker, Inc., New York, N.Y.

163. **Stephenson, E. H., E. W. Larson, and J. W. Dominik.** 1984. Effect of environmental factors on aerosol-induced Lassa virus infection. *J. Med. Virol.* **14:**295–303.

164. **Stetzenbach, L. D.** 1992. Airborne microorganisms, p. 53-65. *In* J. Lederberg (ed.), *Encyclopedia of Microbiology*, vol. 1. Academic Press, Inc., New York, N.Y.

165. **Suzuki, K., T. Yoshikawa, M. Ihira, M. Ohashi, S. Suga, and Y. Asano.** 2003. Spread of varicella-zoster virus DNA to the environment from varicella patients who were treated with oral acyclovir. *Pediatr. Int.* **45:**458–460.

166. **Taus, N. S., D. L. Traul, J. L. Oaks, T. B. Crawford, G. S. Lewis, and H. Li.** 2005. Experimental infection of sheep with ovine herpesvirus 2 via aerosolization of nasal secretions. *J. Gen. Virol.* **86:**575–579.

167. **Utrup, L. J., and A. H. Frey.** 2004. Fate of bioterrorism-relevant viruses and bacteria, including spores, aerosolized into an indoor environment. *Exp. Biol. Med.* **229:** 345–350.

168. **van der Veen, J., Y. Poort, and D. J. Birchfield.** 1972. Effect of relative humidity on experimental transmission of Sendai virus in mice. *Proc. Soc. Exp. Biol. Med.* **140:** 1437–1440.

169. **Wallis, C., J. L. Melnick, V. C. Rao, and T. E. Sox.** 1985. Method for detecting viruses in aerosols. *Appl. Environ. Microbiol.* **50:**1181–1186.

170. **Wardley, R. C., and R. C. Poney.** 1977. Aerosol transmission of feline caliciviruses: an assessment of its epidemiological importance. *Br. Vet. J.* **133:**504–508.

171. **Wathes, C. M., and H. Johnson.** 1991. Physical protection against airborne pathogens and pollutants by a novel

animal isolator in a level 3 containment laboratory. *Epidemiol. Infect.* **107:**157–170.

172. **Weber, A., K. Willeke, R. Marchioni, T. Myojo, R. McKay, J. Donnelly, and F. Liebhaber.** 1993. Aerosol penetration and leakage characteristics of masks used in the health care industry. *Am. J. Infect. Control* **21**:167–173.
173. **Wenzel, R.** 1996. Airline travel and infection. *N. Engl. J. Med.* **334:**981–982.
174. **White, L. A., D. J. Hadley, D. E. Davids, and R. Naylor.** 1975. Improved large volume sampler for the collection of bacterial cells from aerosol. *Appl. Microbiol.* **29:**335–339.
175. **Wilkinson, P. J., and A. I. Donaldson.** 1977. Transmission studies with African swine fever virus. The early distribution of virus in pigs infected by airborne virus. *J. Comp. Pathol.* **87:**497–501.
176. **Wilkinson, P. J., A. I. Donaldson, A. Greig, and W. Bruce.** 1977. Transmission studies with African swine fever virus. Infection of pigs by airborne virus. *J. Comp. Pathol.* **87:**487–495.
177. **Winkler, W. G.** 1968. Airborne rabies virus isolation. *Bull. Wildl. Dis. Assoc.* **4:**37–40.
178. **Wisniewski, P. M., M. J. Warhol, R. F. Rando, T. V. Sedlacek, J. E. Kemp, and J. C. Fisher.** 1990. Studies on the transmission of viral disease via the CO_2 laser plume and ejecta. *J. Reprod. Med.* **35:**1117–1123.
179. **Wong, T. W., Y. C. Chan, E. H. Yap, Y. G. Joo, H. W. Lee, P. W. Lee, R. Yanagihara, C. J. Gibbs, Jr., and D. C. Gajdusek.** 1988. Serological evidence of hantavirus infection in laboratory rats and personnel. *Int. J. Epidemiol.* **17:**887–890.
180. **Yu, I. T., T. W. Wong, Y. L. Chiu, N. Lee, and Y. Li.** 2005. Temporal-spatial analysis of severe acute respiratory syndrome among hospital inpatients. *Clin. Infect. Dis.* **40:**1237–1243.
181. **Yuen, K. Y., and S. S. Wong.** 2005. Human infection by avian influenza A H5N1. *Hong Kong Med. J.* **11:**189–199.
182. **Zimmerman, N. J., P. C. Reist, and A. G. Turner.** 1987. Comparison of two biological aerosol sampling methods. *Appl. Environ. Microbiol.* **53:**99–104.

Aerobiology of Agricultural Pathogens

ESTELLE LEVETIN

81

All crop plants are subject to diseases caused by viruses, bacteria, and fungi. Although plant diseases have been recognized since ancient times, they were not connected with pathogenic microorganisms until the 19th century. In 1861, during investigations of the Irish potato blight, the German botanist Anton de Bary proved experimentally that a fungus, *Phytophthora infestans*, previously described as present on the infected potatoes, was actually the cause of the disease. This discovery by de Bary gave birth to the science of plant pathology. Soon, other plant-pathogenic fungi were described; however, pathogenic bacteria and viruses were not identified until later in the 19th century. Even with the many advances in agriculture during the 20th century, plant diseases still pose a major threat to the world's crops in the 21st century, making the world's food supply highly vulnerable.

Studying the spread of plant disease within a crop and from field to field within a region is often referred to as botanical epidemiology. Dispersal of the pathogen is the necessary step for repeated cycles of infection and multiplication and, therefore, the spread of the epidemic. A thorough understanding of dispersal is necessary for predicting the onset and severity of disease. Implicit within the scope of dispersal are the processes of release, transport, and deposition, with each of these steps being influenced by the environment (5, 22, 116, 152). Although propagules of plant pathogens are dispersed by wind, rain, soil water, insects, and even humans, this chapter is focused on the airborne spread of those pathogens. Airborne transport may be the least understood vehicle of dispersal.

RANGE OF AIRBORNE TRANSPORT

Plant pathogens dispersed by airborne transport generally produce enormous numbers of propagules that are passively carried through the atmosphere. The spread of plant disease through this aerobiological pathway can proceed over short distances through focal spread as well as over long distances (159).

An infection focus is an area of a crop with a contagious disease. Foci are often circular; however, if they are strongly affected by wind, they may become comet- or v-shaped. Foci generally have a constant radial expansion, with the rate varying with the scale of the infection from a few centimeters per day for a localized infection to hundreds of kilometers per year for a pandemic (149, 159). The spread of the disease can be described as microscale, mesoscale, or synoptic scale, using meteorological terms that describe the horizontal movement of atmospheric particles.

Disease Spread

Microscale spread is usually limited to less than a few hundred meters within one field and occurs within one growing season. This roughly corresponds to a zero-order epidemic (63, 159). The focus begins with a single successful propagule causing an infection and creating a lesion. After several generations of localized pathogen spread for polycyclic diseases, which produce many generations of inocula and many cycles of infection during a single growing season, the focus may reach a detectable size. In an annual crop, foci are often about 1 m in diameter around the initial source of infection but may be larger.

If growing conditions are unsuitable for the pathogen, the focus may stop expanding or even disappear with new growth in the canopy. However, when conditions are favorable for the pathogen, the disease spreads. As the primary focus continues to expand, secondary foci and, later, tertiary foci appear. This continued focal spread over a larger area is considered mesoscale spread and corresponds to a first-order epidemic. This may be restricted to one field but may spread over many fields, over an area of up to several hundred square kilometers, or even over part of a continent during a single growing season (159).

Synoptic, or macroscale, spread occurs when the epidemic progresses for several years and spreads over an area of several thousand square kilometers. This is also referred to as a second-order epidemic. This pandemic may even cover a whole continent after a certain number of years (159). These concepts that describe disease spread, from microscale to synoptic, have been incorporated into various spore dispersal models and simulations (9, 10, 34, 156).

Long-Distance Transport

Long-range transport often explains the introduction of a pathogen to a new area or its reintroduction to areas where overwintering cannot occur (16, 32). Because fungi are the most abundant plant pathogens, most of the well-studied examples of long-distance transport involve fungal spores. The transport of *Puccinia graminis* uredospores from the southern United States and Mexico to the wheat belt in the northern United States and Canada has been well documented, as have other instances of long-distance dispersal of rust spores (16, 106). Pedgley (116) and Brown and Hovmøller (16) reviewed several cases of long-distance transport and provided descriptions of the atmospheric conditions that promote these events. Griffin and colleagues (55, 56) described the airborne microorganisms collected in the Caribbean atmosphere during African dust events, and Mims and Mims (99) described the long-distance transport of fungal spores in smoke from the Yucatan. Long-distance transport also explains the recent introduction of soybean rust into the United States during the September 2004 hurricanes (135).

Trajectory analysis is frequently used to trace the previous movement of a spore-laden parcel of air and to locate the inoculum source. To calculate the trajectory, the average wind speed and direction in the transport layer are determined. Once the source is identified, forward trajectories are used to indicate further potential areas of fallout. In addition, various computer models have also been used to trace the movement of spores from dissemination at a source to deposition at a sink (32). Atmospheric transport and dispersion and branching atmospheric trajectory models are used to calculate trajectories based on upper air winds, temperature, and other parameters. The atmospheric transport and dispersion model can be used to calculate the concentration of spores and the number of spores deposited (32, 106). The hybrid single-particle Lagrangian integrated trajectory (HYSPLIT) model has been used successfully to track the movements of several pathogens and to predict the occurrence of disease outbreaks (http://www.ces.ncsu.edu/depts/pp/bluemold/). Most of the literature on long-distance transport suggests disease introduction through a single transport episode or repeated incursions of the pathogen from great distances. These rare events may have dramatic effects on plant populations (16). In contrast, Aylor (11) stressed the low probability of these events and developed a model to explain the continental spread of plant disease through short successive steps that parallel the seasonal availability of host plants.

Movement in the Atmosphere

Fungal spores are a normal component of the turbulent layer of the atmosphere, and airborne transport normally occurs there, with the dispersive power related to the intensity of turbulence (30, 54, 95). Airborne spores are present in large numbers whenever the ground is not covered with ice and snow. Spores are discharged from fungi growing as parasites or saprobes, and atmospheric concentrations may exceed 200,000 spores/m^3 of air (17, 78). Wind is the major factor, with gusts and lulls affecting takeoff, transport, and deposition (5, 106).

In a crop, spores have to pass from the laminar layer close to the leaf surface into the turbulent layer within the crop. Gusts and turbulence enhance spore removal from leaves by sweeping away the layer of slow-moving air next to the leaf surface (5). Before spores reach the free air above the crop, they must also pass through the boundary layer surrounding the crop. As many as 90% of spores may be deposited within the crop itself (44). The percentage that escapes from the canopy depends on the balance between deposition and turbulence, with greater escape occurring during more turbulent winds. The position of the spores within the canopy also affects escape. Spores produced lower in the canopy will have lower rates of escape because they are exposed to slower winds and less turbulence. In the early stages of a disease, when the infection may be confined to the bottom or middle of the canopy, the spread of the disease may be limited. This may change as the disease progresses. After the infection reaches the top of the canopy, the increased rate of spore escape may permit the disease to spread rapidly across the field (5).

When carried by wind, spores are transported both vertically and horizontally. They may be carried upward by convective activity or thermals and have been recovered from altitudes higher than 5,000 m. In spring and summer, when the ground is heated to a greater degree, vertical movements of air masses are more common, as is greater turbulence. The atmosphere during this period is often characterized by maximum concentrations of airborne spores, including many pathogenic spores from fungi of genera such as *Erysiphe*, *Drechslera*, and *Venturia*. Horizontally, spores can be carried for thousands of kilometers, dispersing pathogens into new areas (53, 106).

Dispersal Mechanisms

Dispersal of spores into the atmosphere is dependent on the method of discharge by individual taxa as well as environmental factors such as temperature, humidity, and wind speed. Spores are released into the air either passively or by active discharge mechanisms. A review of dispersal mechanisms has been published by McCartney and Fitt (96).

Passive Mechanisms

Many pathogenic fungi, such as the conidia of the powdery mildews, *Alternaria* spp., and *Drechslera* spp. as well as rust uredospores and smut teliospores, are passively dispersed in dry weather. For the majority of these spores, the airborne concentration is dependent on the ease with which spores are detached from the parent mycelium (or fruiting structure) and upon atmospheric conditions such as wind speed and turbulence. Wind gusts facilitate their removal, as does the movement of the leaf itself in the wind. The role of wind gusts in the dispersal of spores has been reviewed by Aylor (5). Strong gusts are able to disperse spores from leaf surfaces even when the average wind speeds are too low to accomplish this task (116). Because wind dispersal for both pathogenic and saprobic fungi is promoted by warm, dry weather, the entrained propagules are referred to as the "dry air spora." They include the most abundant fungi in the environment and usually peak during the afternoon hours, when humidity is low and wind speeds are increased (54). Although there are major similarities in the dry air spora worldwide, at any one time the air spora may be dominated by nearby sources of spores (78). During crop harvesting or mowing, incredible numbers of spores (up to 10^9 spores/m^3) may be dispersed into the atmosphere, with *Alternaria*, *Cladosporium*, *Epicoccum*, and *Didymella* being the most abundant taxa (20, 78, 79, 123). Passively dispersed spores may also be removed by raindrops as they strike the leaf surface, causing a puffing action that may remove the conidia (39, 50, 79). This typically occurs when the first raindrops fall, and it may result in increased atmospheric concentrations of conidia of *Cladosporium* and *Alternaria*, rust uredospores, or conidia of other fungi. The release of puffball spores has a similar puffing action as raindrops strike the mature fruiting bodies (77).

Rain splash can also propel spores into the atmosphere, and it is second to wind in importance as a means of pathogen dispersal. Recently, there has been considerable interest in this topic, with several published reviews (40, 69, 96, 117). While rain is the most important vehicle for splash dispersal, overhead irrigation can also disperse plant pathogens. Splash-borne spores or bacteria are usually produced in mucilage (a sticky or slimy polysaccharide layer), which inhibits their direct removal by wind. However, the mucilage also protects them from desiccation during dry weather. The first raindrops dissolve the mucilage and leave a spore suspension available for splash dispersal by additional raindrops. Recently, the dispersal of *Colletotrichum gloeosporioides* conidia by rain splash was studied, and the number of conidia was shown to vary from drop to drop (124). Splash droplets from the first water drop to strike the leaf released only a few conidia and some mucilage, with the highest number of conidia found in the fifth water drop. The size of the droplet may also affect the number of conidia, with the smallest droplets unable to carry spores. Most spores are carried by droplets of >100 μm in diameter (79, 116).

Rain splash confines dispersal to periods when the wet conditions on the new host are also favorable for germination and even reproduction (40, 68). The severity of bacterial brown spot of snap beans was found to increase after rain due to the rapid multiplication of the pathogen (66).

Many splash-dispersed spores have some adhesive properties that enable them to stick to the new host surface. These spores, such as conidia of *Fusarium*, *Cercospora*, and *Pseudocercosporella*, frequently have smooth, thin, hyaline walls and an elongated shape. Although they lack the protective feature of thick pigmented walls commonly found among members of the dry air spora, they are protected from desiccation by wet conditions during dispersal and deposition (40).

The effect of wind on dispersal of splash-borne spores depends on the size of the raindrops. While very large droplets carrying spores are unaffected by wind, small droplets may evaporate and allow the spores to be dispersed through the air as a true aerosol (39, 40, 80). Most splash droplets are too large to form aerosols. As a result, rain splash appears to be most important for local dispersal, with much shorter gradients than those of wind-dispersed spores (69, 96). Studies are ongoing to determine the maximum height and distance that an inoculum can be spread by rain splash for various plant pathogens (15, 89, 112).

In addition to their role in dispersal, raindrops are very efficient spore collectors, and the majority of airborne spores in the 20- to 30-μm diameter range are washed out by raindrops. Even a very light rain can effectively wash spores from the atmosphere; however, the significance of this rain scrubbing is difficult to quantify (39).

Active Mechanisms

Active discharge mechanisms propel spores into the turbulent layer, independent of wind. These mechanisms are widespread among fungi, with the ballistics of some species being quite spectacular. Many ascospores and basidiospores are actively discharged by mechanisms that require moisture or high humidity. The atmospheric concentrations of these spores often peak at night or during early morning hours, when the humidity is high (5). Aerobiological studies have shown that ascospores are often abundant in the atmosphere during and after rainfall (2, 19, 147). The explosive discharge of ascospores from the ascus is a widespread characteristic of the ascomycetes which is linked to available moisture. High osmotic pressure develops within the ascus either by the direct absorption of rainwater or by the swelling of mucilage within the ascus (33, 104). The resulting pressure causes the ascus tip to rupture, forcing the spores out explosively into the turbulent layer. The discharge of basidiospores from mushrooms and bracket fungi also requires atmospheric moisture, and the mechanism used has been described as a surface tension catapult (102, 153). High concentrations of atmospheric basidiospores have been reported from various locations, with release occurring during times of high humidity (54, 59, 82, 84, 101, 132). Although basidiospore dispersal is not directly tied to rainfall, fruiting bodies of mushrooms and other basidiomycetes frequently develop following rain. The release of basidiospores from these fruiting bodies therefore results in high atmospheric concentrations during seasons when rainfall is frequent (82).

Survival in the Atmosphere

Fungal spores that are adapted to airborne dispersal are often much more resistant to environmental stress than are the parent hyphae. However, they are still vulnerable to certain types of environmental damage while airborne. Exposure to harmful radiation and extremes of temperature and humidity can decrease the viability or infectivity of pathogenic species.

Changes in relative humidity, often caused by changing wind speeds, may affect survival, especially for thin-walled spores, which may easily plasmolyze (30). The risks of desiccation are usually greatest during the daytime and close to the ground. At night and at high altitudes, conditions are less stressful; spores have even been reported to germinate in the clouds (54). Spores may also serve as condensation nuclei for rain (106).

Radiation in the upper atmosphere, especially UV radiation, may also affect the survival of spores carried into the upper air. The effects of radiation may outweigh the effects of temperature and humidity. Again, thin-walled, colorless spores may be more vulnerable because they lack the protection provided by melanin, which is present in the cell walls of pigmented spores. Low temperatures in the upper atmosphere may be preservative, however, and protect spores from UV damage (54). Because of the erosion of the ozone layer in the upper atmosphere, the effects of UV radiation on fungal spores have been the focus of recent work. Scientists are interested in possible decreases in spore germination, mycelium development, and spore formation caused by radiation (103, 145).

Despite the environmental hazards, many spores are able to survive long-range transport, but the percentage of viable spores that actually reach a target and cause infection is low (106). Spores can be deposited on crop surfaces by sedimentation, impaction, boundary layer exchange, turbulence, electrostatic deposition, or raindrops (106).

PATHOGENS SPREAD BY THE AEROBIOLOGICAL PATHWAY

A wide variety of plant pathogens, including viruses, bacteria, and fungi, are dispersed through the atmosphere. When they are deposited on a susceptible host, infection can occur, and when environmental conditions are favorable, the resulting disease spread may lead to widespread crop loss. A thorough understanding of the role of the aerobiological pathway in pathogen dispersal is necessary for the management and control of disease.

Viruses

Most plant viruses are spread by means of insect vectors, although relatively few insect groups transmit viruses. Insects with sucking mouth parts (such as members of the order Homoptera) transmit viruses by their stylets. Aphids are the most dangerous insect vectors and are known to transmit the majority of stylet-borne viruses (1). Ng and Perry (108) recently reviewed this topic. Whiteflies, another important group of vectors in the same order, transmit geminiviruses, which are pathogens that severely affect a range of host species (1).

Barley yellow dwarf virus (BYDV) is often considered the most widespread and economically important viral disease of cereal crops; it affects more than 100 species, including wheat, rice, barley, and oats. There are few data on the exact monetary losses due to BYDV, but an estimated 5% loss of crops infected by this virus would cause a loss of more than $1 billion in the United States alone. BYDV can be an important limiting factor for grain production throughout the world (97). Many wild grasses are also infected and thereby serve as reservoirs of infection. Barley yellow dwarf disease is caused by a large complex of luteoviruses which share common aphid vectors. This topic has been reviewed in several publications (21, 71, 97, 120), and Miller et al. (98) examined the phylogenetic relationships of viruses in this group. BYDV is controlled by the use of resistant plant lines (121) or pesticides. Recently, researchers have been using genetic engineering to develop crops resistant to BYDV (41). However, in many developing nations where disease occurrence is high, farmers live with the losses caused by these viruses (97). The aerobiological focus of barley yellow dwarf virus research concerns the atmospheric movement of the aphid vectors and the importance of winds in local and long-distance aphid migration (71).

Plant viruses can become airborne when infected plants are damaged by high winds. This has been described for tobacco mosaic virus (13), but, in general, evidence of the aerosol dispersal of viruses is scant (80). The transport of virus particles by airborne pollen has been established for more than 15 plant viruses, as reviewed by Cooper et al. (29) and Mink (100).

Plant-pathogenic viruses have been known for the past 100 years; however, the number of recognized virus diseases has increased dramatically in recent years. Several theories have been suggested to account for the recent increases, including the development of resistance to commonly used pesticides among insect vectors, the migration and explosion of aphid and whitefly populations in more temperate regions, changes in cultivation practices, and the introduction of exotic transplants, which may enhance the long-distance movement of both viruses and vectors (52, 65; http://apsnet.org/online/feature/NewViruses/Top.html).

Bacteria

Approximately 100 species of bacteria are serious phytopathogens causing devastating crop losses each year. Bacterial diseases are especially common in humid tropical areas but can be destructive wherever warm, moist conditions are present (1). Bacterial plant pathogens are generally gram-negative bacilli, with species of *Erwinia*, *Pseudomonas*, *Xanthomonas*, and *Agrobacterium* being well-studied phytopathogens; however, nonpathogenic species also occur in these genera (14, 66, 80, 127, 133, 140). Some *Clavibacter*, *Corynebacterium*, and *Streptomyces* species are recognized gram-positive plant pathogens (137). Conn et al. (28) identified a coccoid bacterium, *Leuconostoc mesenteroides*, as the cause of a postharvest decay of tomatoes.

Dispersal

Dispersal of pathogenic bacteria from plants is generally passive via water, wind-blown water, and animals. Agricultural practices and agricultural workers, their machinery, and their tools also play a major role in the spread of pathogenic bacteria, with humans typically responsible for most long-distance dispersal. In warm, humid climates, where dew and rain are common, dispersal by rain splash is the major means of disease spread (40, 51).

Phytopathogenic bacteria have no special mechanism for producing airborne propagules; however, they become airborne during rain or irrigation as well as when crops are harvested (80). Bacteria can be passively carried from surfaces of plants or soil by air currents and may even become airborne as single cells, but they are usually carried on rafts of plant material or in droplet nuclei following splash dispersal (80, 85). In a review of air sampling techniques, Lacey and Venette (80) cited many studies that document the airborne dispersal of bacterial pathogens.

None of the phytopathogenic bacteria are capable of forming endospores; thus, they remain susceptible to solar radiation and desiccation and are therefore limited in the ability to survive long-distance aerial dispersal. Survival requires rapid transport to an environment with adequate moisture, nutrition, and temperature (51). A loss of water causing dehydration or desiccation represents the major stress to airborne bacteria. Dehydration is generally caused by the evaporation of water from droplets carrying bacteria (94). This supports the suggestion by Hirano and Upper (66) that splash dispersal is significant only for local disease spread. However, it is also known that some bacteria can serve as cloud condensation nuclei. Franc (42) showed that viable *Erwinia carotovora* cells were recovered from precipitation in Colorado. The storm system that deposited the precipitation had originated in the Pacific Ocean off the coast of California. The abilities of *E. carotovora* cells to function as cloud condensation nuclei and to be protected from desiccation and radiation may explain their ability to survive this long-distance transport.

Biocontrol Agents and Genetically Engineered Microorganisms

The airborne dispersal of bacteria which are not phytopathogens has been the focus of other agricultural research. Biocontrol agents include bacteria, fungi, and viruses that can be used to control plant pests and pathogens. The use of bacteria as biocontrol agents has expanded greatly in recent decades, and several species are currently utilized for various agricultural applications, including frost protection, control of plant diseases, and control of insect pests (43, 57, 72, 113, 143, 144, 155). In some instances, the organisms are naturally occurring microbial pathogens of insects that are being used for pest control, or they are epiphytic competitors of foliar pathogens. In other cases, the organisms are genetically engineered microorganisms (GEMs) which have been designed to perform specific actions.

Some well-known examples of GEMs are the ice-minus strains of *Pseudomonas syringae*, which can prevent frost injury. Frost damage to leaves occurs when ice forms around specific ice-nucleating sites, such as the wall proteins of *P. syringae*, a common leaf surface bacterium (137). By using genetic engineering, the gene that encodes the wall protein was deleted, thereby creating cells of *P. syringae* that would not serve as ice nucleation sites. Plants sprayed with cultures of this genetically engineered (ice-minus) bacterium could survive lower temperatures without frost damage (76, 137). While large-scale applications of ice-minus strains of *P. syringae* are still being developed, a naturally occurring ice-minus strain of *Pseudomonas fluorescens*, A506, is now registered commercially for the control of frost injuries of pear, cherry, apple, almond, peach, tomato, potato and strawberry plants (43, 155).

The movement and fate of GEMs have been the focus of much study, and the literature on this topic is extensive. The topic has also been addressed in several reviews (105, 143, 155). The safe utilization of GEMs necessitates that they be applied and retained in specific target areas. Field application of GEMs is generally preformed through aerosol sprays. Dispersal can also occur by the modes described for phytopathogenic bacteria, including rain splash, which is important for short-distance movement, and wind, which has greater potential for long-distance transport (94, 105). The possibility of long-distance dispersal of GEMs from the target area always exists (94). As a result, it is necessary to carefully monitor the dispersal and account for the released population. Air sampling is one of the methods being used to detect GEMs as well as nonrecombinant organisms (80, 143). Monitoring the field release of ice-minus bacteria showed that the results of field trials were closely predicted by laboratory and greenhouse experiments as well as previous field studies with nonrecombinant bacteria (155). Wilson and Lindow (155) concluded that the release of GEMs in carefully conducted studies should have few detrimental effects.

Much research has focused on *Bacillus thuringiensis*, a ubiquitous bacterium that naturally produces crystalline proteins with insecticidal properties. Several reviews on this topic have been published (25, 90, 136). This species exists as a soil microbe and insect pathogen as well as a leaf surface saprobe. A dried inoculum containing endospores and crystals of insecticidal proteins has been marketed since the 1960s as a safe alternative to chemical pesticides. It is used in sprays or dusts on plants to control a variety of insects, especially lepidopteran pests. It is the most widely used biopesticide and accounts for 90% of the market (25).

There are many subspecies of *B. thuringiensis* which differ in the number and type of plasmids which they contain. The genes for the insecticidal proteins are borne on the plasmids (4). More than 1,000 strains of *B. thuringiensis* have been isolated, and more than 100 insecticidal proteins have been identified and sequenced (38, 136). Different strains of *B. thuringiensis* have different insecticidal activities toward specific insect pests (4). For example, *B. thuringiensis* subsp. *kurstaki* is widely used in caterpillar control in agriculture and forestry, *B. thuringiensis* subsp. *israelensis* is active against mosquitoes and black flies, and *B. thuringiensis* subsp. *tenebrionis* is active against beetle larvae. There are more than 400 *B. thuringiensis* preparations registered in the United States for the control of various insect pests. *Bacillus thuringiensis* toxins degrade within a few days, while the endospores can survive for several years after spray applications.

In the past decade, *B. thuringiensis* genes have been transferred into crop plants, with the resulting transgenic plants expressing the genes for various insecticidal proteins. In 1996, the first *B. thuringiensis* transgenic crops were planted, with approximately 1 million hectares of transgenic corn, cotton, and potato grown in the United States (136). The use of *B. thuringiensis* transgenic crops has increased dramatically in the past 9 years, and in 2004, crops expressing *B. thuringiensis* toxins were grown on more than 13 million ha in the United States and on 22.4 million ha worldwide (161). It is anticipated that the number of *B. thuringiensis* transgenic crops will continue to expand, especially in developing nations. *Bacillus thuringiensis*-transformed crops offer several benefits: the toxins are continuously produced and persist for some time, fewer chemical insecticides are needed, and a greater range of insect pests are controlled. The insecticidal toxins are not harmful to mammals, birds, amphibians, or reptiles, and it is generally believed that there is little effect on nontarget organisms (136). However, Losey et al. (88) showed that larvae of monarch butterflies grew more slowly and suffered higher mortality rates when they ate milkweed leaves that were dusted with corn pollen expressing *B. thuringiensis* toxins. Because corn pollen is naturally found on milkweed plants surrounding corn fields, this laboratory finding was initially thought to have serious implications for monarch butterfly populations in the U.S. corn belt. In contrast, several follow-up studies (64, 138, 160) showed that expression of the insecticidal toxins in pollen varied among different transgenic corn hybrids. Only one transgenic corn hybrid, called event 176, showed significant toxicity to monarch butterfly larvae. The event 176 hybrid, used in the earlier study by Losey et al. (88), is no longer marketed due to inadequate control of the European corn borer. The effects of other types of corn pollen expressing *B. thuringiensis* toxins were shown to be negligible for monarch butterfly populations (138).

Bacillus thuringiensis genes have been transferred into other bacteria as well, especially several species that are plant epiphytes and endophytes. *Pseudomonas fluorescens* expressing *B. thuringiensis* genes lengthens the life of toxins in the field since *P. fluorescens* can propagate on the plant host and continue producing crystal proteins (136). *Clavibacter xyli*, a bacterial endophyte of corn with *B. thuringiensis* genes, can be inoculated into corn plants to protect them from the European corn borer (4, 136). This method may be a viable alternative to transgenic plants because the bacterial genes may not be expressed in pollen.

The development of resistance to *B. thuringiensis* toxins by many insect species has been shown in laboratory experiments since the mid-1980s. The first species to develop resistance in the field was the diamondback moth. An autosomal recessive gene in this insect confers resistance to four *B. thuringiensis* toxins (146). Managing resistance is a major focus of research (161). Various suggestions include using *B. thuringiensis* strains that produce multiple toxins and planting refugia in fields where *B. thuringiensis*-transformed crops are grown. Refugia are areas of susceptible host plants, which would prevent the development of a population of homozygous resistant pests. Researchers urge caution with the *B. thuringiensis* technology because misuse of *B. thuringiensis* toxins could rapidly make them ineffective against major plant pests (136, 161). Cerda and Paoletti (24) have recently reviewed resistance management strategies for *B. thuringiensis*-transformed crops.

Researchers are also focusing on methods that use other bacteria as biocontrol agents. These include using several other bacteria with insecticidal toxins, such as *Xenorhabdus nematophilus* and *Photorhabdus luminescens* (25). These two species are mutualistic symbionts within nematodes. The nematodes act as vectors transporting the bacteria into the bodies of insect larvae. Toxin genes from these bacteria have been cloned, and some have been expressed in *Escherichia coli*. It has been suggested that *Xenorhabdus* and *Photorhabdus* will be the next generation of biopesticides (25). Other scientists are studying *P. fluorescens*, *P. syringae*, *Bacillus cereus*, and other *Bacillus* species to control various bacterial or fungal pathogens (4, 26, 43, 111, 144, 148). *Pseudomonas fluorescens* A506, described above to manage frost injury, is also used to control fire blight on apples and pears and is available commercially as BlightBan A506. Fire blight, caused by *Erwinia amylovora*, is the most damaging bacterial disease of apples and pears (144). Control of fire blight can often be achieved by spray application of A506 onto flowers. The growth of *P. fluorescens* restricts the growth of *E. amylovora* on the stigma surface and reduces the incidence of blossom blight (144). BlightBan A506 has been used increasingly by growers in recent years because many *E. amylovora* populations have become resistant to streptomycin, which had been the standard control measure. Experiments to determine the optimum timing and weather for applications of BlightBan, either alone or in combination with other biocontrol agents, are still ongoing (86, 144).

A great deal of interest has also been shown for the use of *Burkholderia cepacia* (*Pseudomonas cepacia*). Strains of this bacterium produce one or more compounds that are active against a broad range of fungal pathogens (67, 110; http://apsnet.org/online/feature/BurkholderiaCepacia/Top.html). *Burkholderia cepacia* is also useful for the bioremediation of soil and groundwater contaminated with chlorinated hydrocarbons. There has been a great deal of concern about the use of this organism because certain strains are also plant pathogens as well as opportunistic human pathogens. This is particularly serious for individuals with cystic fibrosis, who could develop fatal lung infections. Because of the health hazard, the widespread use of *B. cepacia* should be undertaken with caution (67).

With the current progress in recombinant DNA technology, it is anticipated that applications of GEMs in agriculture and bioremediation and the use of microbial biocontrol agents will increase in the coming decades (94). Aerobiological studies will continue to play a major role in assessing the risks associated with the introduction of these agents.

Fungi

Over 70% of all major crop diseases are caused by fungi, with thousands of fungal species being recognized as plant pathogens. It is estimated that fungal diseases cost >$3.5 billion annually to American farmers alone (77). In general, spores of most fungal pathogens are adapted for airborne transport; however, much of the aerobiological research on agricultural pathogens has focused on a limited number of fungi that cause economically important diseases (80). Air sampling can be a valuable tool because the management of plant disease requires an understanding of the airborne dispersal of inocula. Routine examination of air samples from Tulsa, Okla., provides an understanding of the diversity of the spore types in the atmosphere (83, 142). The dry air spora is dominated by *Cladosporium*, as it is in most areas of the world; in Tulsa, hourly concentrations of these spores have sometimes exceeded 100,000 spores/m^3 (17). In addition, spores of *Alternaria*, *Bipolaris*, *Curvularia*, *Drechslera*, *Epicoccum*, *Helminthosporium*, *Nigrospora*, *Pithomyces*, *Stemphylium*, and *Torula* as well as smut teliospores typically occur in daily air samples during dry weather. In wet weather, ascospores predominate, with *Leptosphaeria* often abundant, along with *Venturia*, *Didymella*, and spores from members of the family Diatrypaceae during certain seasons. Asexual conidia of *Cercospora* and *Fusarium* and various basidiospores are also abundant in moist weather. The following discussion examines a few of the significant fungal pathogens that are dispersed by the aerobiological pathway (Table 1).

Late Blight of Potato

The potato ranks as the world's fourth most common food crop, but the continued increase of this crop in developing nations may be limited by the cost of disease control measures. Potatoes are attacked by many pathogens and pests, and the total amount of agricultural chemicals applied to the potato crop is larger than that for any other food crop (109). Late blight of potato has been the most important disease of potato since the 1840s, when it caused the destruction of the potato crop in Ireland and the resulting widespread famine. Today, 160 years later, late blight ranks as the most destructive agricultural disease (49).

Phytophthora infestans is the pathogen responsible for late blight of potato (and also that of tomato). This oomycete occurs wherever potatoes are grown, and all potato cultivars are susceptible. Populations of the pathogen are generally short-lived, and a field with infected potatoes during a particular year may or may not show diseased plants the following year (47). The oomycete invades host tissue, resulting in the rapid death of infected parts. The productivity of potato plants is greatly reduced, and tuber destruction frequently occurs as well. Sporangia are produced on aerial hyphae, which grow out from the stomata and are dispersed by wind, possibly being carried for tens of kilometers (47). At low temperatures (10 to 15°C) and high humidities, sporangia germinate by producing numerous zoospores, while at higher temperatures, each sporangium gives rise to a single germ tube that develops into hyphae. As a result, during cool, wet periods, the production of zoospores leads to an astonishingly rapid spread of the disease. Without fungicidal protection, a blighted field can be destroyed within a

TABLE 1 Major fungal pathogens with aerobiological dispersal

Disease	Organism	References
Late blight of potato	*Phytophthora infestans*	12, 45, 46, 47, 49, 91, 109, 122, 139
Blue mold	*Peronospora tabacina*	5, 22, 74, 93, 154, 157
Apple scab	*Venturia inaequalis*	6, 7, 8, 39, 40, 48, 73, 75, 122, 129, 130, 141
Early blight, leaf spots	*Alternaria* spp.	3, 17, 131, 151
Coffee rust	*Hemelia vestatrix*	16, 106
Wheat stem rust	*Puccinia graminis*	35, 36, 37, 81, 106, 115, 125
Wheat leaf rust	*Puccinia triticina* (syn. *P. recondita*)	35, 36, 106
Soybean rust	*Phakopsora pachyrhizi*	23, 61, 118, 119, 135, 158
Smut fungi	*Ustilago* spp., *Tilletia* spp.	3, 31, 58, 59, 101, 132

couple of days. Similar disease progression also occurs in infected tomato fields (122). Multibillion-dollar annual losses are caused by this pathogen for both potato and tomato growers (139.)

The economic impact of fungicide application to control *Phytophthora infestans* (or other pathogens) must not be overlooked. Often, the cost of repeated applications may actually outweigh the value of the crop at harvest time (137). Other techniques, such as sanitation, crop rotation, and the use of genetically resistant cultivars, can reduce the need for fungicides (137). This multifaceted approach is referred to as integrated pest management and has been widely recommended for controlling plant disease. To reduce fungicide use, detailed information is needed about a pathogen's life cycle, especially the dispersal phase. For pathogens with airborne dispersal, aerobiological studies and knowledge of the environmental conditions that promote dispersal can supply information on the optimum timing of fungicide applications to protect the crop.

Potato late blight forecasting has a 40-year history, and in many potato-growing regions, when meteorological conditions indicate that *Phytophthora* spread is likely to occur, warnings are issued to apply fungicides (91). Forecasting models, such as Blitecast, have been successful in reducing the number of fungicide applications. Sometimes, as many as six applications can be omitted in a single season (91). Other models have successfully incorporated air-sampling data for blight forecasting. A recently described Lagrangian stochastic simulation model (12) may be useful for predicting release rates for sporangia and the spread of disease between fields. Aerobiological models could possibly lead to a 50% reduction in fungicide applications (122).

Phytophthora infestans overwinters as mycelia in infected tubers, but in areas where both mating types occur, it can also overwinter as oospores in the soil. When the pathogen was introduced to Europe and the United States in the 1840s, only one mating type (A_1) was introduced. In 1950, the second mating type (A_2) was identified in central Mexico, the native home of the oomycete. The A_2 mating type was confined to Mexico until 1980, when it spread throughout Europe. It is speculated that this mating type was introduced in a large shipment of potatoes from Mexico to Europe in the late 1970s. Outbreaks of the A_2 type began occurring in Europe in 1980; this type has subsequently spread throughout the world (45, 46, 109). The introduction of the A_2 mating type has increased concerns about the possibility of sexual reproduction and genetic recombination occurring in the field, with new strains developing. This has broad implications for potato breeders searching for blight resistance. These concerns were reinforced by outbreaks of late blight in the United States and Canada during the 1990s. Nearly all strains of *Phytophthora infestans* involved were resistant to metalaxyl, the fungicide most widely used to control late blight (45, 46). Currently, fungicide-resistant strains have been found worldwide (49).

Two types of long-distance dispersal of *Phytophthora infestans* have apparently taken place. Intercontinental migration has been associated with the transport of infected plants or tubers by humans. This occurred in the 1840s and again before the 1980 outbreak of the A_2 mating type. Long-distance dispersal over tens of kilometers is attributed to wind-blown sporangia. Aylor et al. recently showed that wind speeds of 1 to 2 m/s were sufficient to disperse sporangia from an infected potato canopy a distance of 10 to 20 km in <3 h (12). Sporangia can survive for some hours at the reduced humidity encountered during transport (47). Maps showing the rapid progress of blight epidemics in the 1840s suggest that a second-order epidemic of late blight can occur during a single growing season.

Much of the current focus of late blight research is centered on developing potato varieties resistant to the pathogen, using traditional methods as well as genetic engineering (49, 139). A wild potato species, *Solanum bulbocastanum*, is resistant to all known strains of *P. infestans*, and the resistance gene from this species has been cloned. Transgenic potato plants have been developed that show blight resistance. This method may someday be used to make popular potato cultivars blight resistant (139). Until then, growers must rely on forecasting models and fungicides to keep late blight in check.

Blue Mold

Tobacco blue mold caused by the oomycete *Peronospora tabacina* (syn *P. hyoscyami* f. sp. *tabacina*) is an unpredictable disease of both wild and cultivated tobacco, causing devastation some years and not appearing at all during others. Blue mold was first described in Australia during the 19th century, and the organism was identified by Baily in 1890 (74). In North America, the disease was confined to seedbeds until 1979, when the first serious epidemic occurred. The infection rate was especially severe in both 1979 and 1980, with a second-order epidemic advancing at rates of 10 to 32 km/day northward through the eastern United States and into southern Canada. Crop losses in the United States and Canada during these 2 years were estimated at approximately $350 million. Both host plants and the pathogen exist year-round in tropical and subtropical areas such as the Mediterranean and Caribbean basins. In temperate regions, tobacco is grown as an annual; in addition, *P. tabacina* is not able to overwinter in temperate zones. As a result, the long-distance transport of inocula from tropical regions, in the form of asexual sporangia, must recur each year.

Infection can occur within 4 h after a sporangium lands on the leaf surface. A symptom-free incubation period, typically 5 to 7 days, ends with the appearance of yellow lesions and the development of new sporangia. Unlike *Phytophthora*, *Peronospora* produces no zoospores, and the sporangia themselves (often referred to as sporangiospores or conidia) are the only asexual propagules. At times, sporangia can be extremely abundant, producing up to 1 million spores per square centimeter of lesion area (22). Sporangia are released from the sporangiophore by the twisting movements of the sporangiophore that occur when the leaf dries in the morning as humidity decreases and the temperature increases (93). During the spring in Connecticut, the greatest number of sporangia was found in the air in the early morning; in summer, the highest concentrations occurred later in the morning, when turbulence may have been greater (5). Cool, wet, overcast weather favors the rapid advance of the pathogen; clear, hot, dry weather stops disease spread (93).

Each spring in the eastern United States, weather conditions are favorable for the northward transport of *Peronospora* sporangia from southern sources. Case studies of epidemics occurring from 1979 to 1986 suggest at least two likely pathways of disease spread (93). In some years, the pathogen spread northward from Florida and Georgia. In other years, the disease occurred first in Kentucky and North Carolina without first occurring in states farther southeast. The source of the inoculum for this second pathway was believed to be *Nicotiana repanda* plants in south-central

Texas or cultivated tobacco in Mexico. These cases have shown that long-distance transport of sporangia can lead to severe epidemics of blue mold, and forecasting systems can potentially provide time for tobacco farmers to apply fungicides. A numerical model was developed that encompasses the release, transport, and deposition of blue mold sporangia (157).

Wiglesworth et al. (154) reported a forecasting system using molecular probes. They described the amplification of an oligonucleotide fragment of DNA from *P. tabacina* that could be used to detect the fungus in infected parts of tobacco plants. Spore traps were used along with the amplification of this DNA fragment to predict a local blue mold outbreak. Other predictive models make use of extant disease outbreaks, weather fronts, and weather forecasts. The North American Plant Disease Forecasting Center at the Department of Plant Pathology, North Carolina State University (http://apsnet.org/online/feature/forecast/), has successfully used the HYSPLIT trajectory model since the spring of 1995 to predict outbreaks of blue mold in the eastern United States. The trajectory model employs documented reports of blue mold along with meteorological data from the National Oceanic and Atmospheric Administration's ETA and AVN models, which are used to predict short-term weather conditions, to plot trajectories of inoculum-laden parcels of air. Forecasters use the most recently reported and continuing sources of blue mold to develop daily trajectories. The daily forecast produced with HYSPLIT trajectories describes future weather conditions at the source and along the anticipated pathway, with emphasis given to those atmospheric conditions that favor sporulation at the source, survival during transport, and deposition. An overall outlook describes the likelihood of blue mold spread over the subsequent 48 h. Forecasters anticipate that the use of this model will continue to provide valuable information in efforts to control blue mold epidemics (http://www.ces.ncsu.edu/depts/pp/bluemold/). Since 1996, daily blue mold forecasts have been available on the internet during the growing season (see below).

Apple Scab

Scab is the most important disease of apples, with all commercial varieties susceptible to attack by the ascomycete *Venturia inaequalis* (77). The disease occurs in every country where apple trees are grown, and similar scab diseases also affect pear and hawthorn plants. Scab lesions occur on both leaves and fruit; the disease can also cause premature defoliation (73). The control of apple scab requires repeated applications of fungicides (8 to 20 times per season) to protect the crop. Without them, 70 to 100% of the crop would be unsalable (77).

The fungus overwinters as immature fruiting bodies on dead leaves on the ground in an orchard. The ascocarps mature in the spring, and ascospores are actively discharged from ascocarps during rainfall (40). Stensvand et al. (141) reported the capture of ascospores during dew on several occasions; however, dew is not generally accepted as a spore release mechanism (130). Although discharged when it is raining, the spores are dry, airborne spores. Rain causes the asci to swell and release the spores (40). In one study, it was observed that the airborne concentration of *V. inaequalis* ascospores decreased rapidly at greater heights above the ground. The concentrations at a height of 3 m were only 6% of the values at a height of 15 cm. A spore dispersal model suggested that this decrease was due to the rapid increase in wind speed and eddies above ground level (6). The spores become a major component of the air spora in orchards and are carried by wind to leaves, where they cause a primary infection (8, 39, 122). During the spring, there is usually a constant source of ascospores. Ponti and Cavanni (122) showed that ascospores typically could be found in the air of Italian apple orchards during periods of rainfall from mid-March to mid-June. In the northeastern United States, ascospores were found in orchards from mid-April to early June (7). Once the primary infection is established, conidia of the anamorphic stage (*Spilocaea pomi*) begin developing. The more primary inoculum is present, the more rapidly the disease will build up and the more serious the epidemic will ultimately be.

Fungicide applications should be timed to coincide with periods of rainfall in the spring, when ascospores are released. Air sampling has been used to provide information about the duration of the ascospore season in a particular area and therefore help to limit the use of fungicides to the periods when it is absolutely necessary (122). Current research is focused on modeling the aerial dispersal of ascospores to help evaluate the risk of infection (8, 48, 75, 129, 130).

Alternaria

Alternaria is a genus of asexual or imperfect fungi assigned to the form class Hyphomycetes. Fungi in this genus are anamorphs of ascomycetes, including members of the genus *Pleospora*. *Alternaria* species are characterized by very distinctive large multicellular dictyospores that have a beak and are produced in chains. Species of *Alternaria* occur as parasites on a number of crop plants, causing early blight or leaf spot diseases, or as saprobes on a wide variety of organic substrates. Crop losses caused by *Alternaria* pathogens are less serious than losses due to pathogens such as rusts or downy mildews; however, this genus is prominent in aerobiological literature because it is recognized as an important aeroallergen as well as a plant pathogen. Aerobiological surveys, including those conducted by the American Academy of Allergy, Asthma, and Immunology (3), routinely report the presence of *Alternaria* conidia in the atmosphere. This ubiquitous genus can frequently be found in atmospheric concentrations of several hundred to 1,000 or more spores per cubic meter. In some areas, these spores may be in the atmosphere throughout the year. Peak concentrations are usually seen in the late summer or fall (17).

Many pathogenic species of *Alternaria* have a worldwide distribution, including *Alternaria solani* on potato and tomato, *A. brassicae* on members of the Brassicaceae, *A. porri* on onions, and *A. alternata* on a variety of host species (131). In general, *Alternaria* pathogens attack plants under stress, especially those affected by drought, insect infestation, or senescence. Young seedlings are also susceptible to *Alternaria*, but established to middle-aged plants may be less vulnerable. However, Vloutoglou and Kalogerakis (151) found that two tomato varieties were susceptible to *A. solani* infection at all ages. The plants became increasingly vulnerable as they aged, with the greatest susceptibility at the reproductive stage.

Alternaria spores are passively dispersed from infected leaves by moderate to strong gusty winds, with velocities of 2 to 3 m/s required for spore release. As a component of the dry air spora, dispersal typically occurs during midday, when conditions are warm and dry with high wind speeds. The greatest dispersal occurs during dry weather that immediately follows periods of rain or heavy dew. Prolonged dry windy periods deplete spore reserves on leaves and inhibit sporulation, which requires moist conditions. Spores can also disperse when washed or splashed from leaf surfaces by raindrops and irrigation water; however, for *Alternaria*, this method of dispersal is less important. In contrast, high humidity inhibits the release of spores from wet leaves, and airborne spores are washed from the atmosphere by rain and irrigation. Several predictive models based on meteorology have been developed to aid in timing fungicide applications (131).

Currently, there is no nationwide network of plant pathologists conducting air sampling to detect airborne pathogens. However, the American Academy of Allergy, Asthma, and Immunology sponsors a network that consists of approximately 84 certified air-sampling and reporting stations in the United States. In the 1998 network report, airborne fungal spore concentrations were reported from 46 sampling stations. *Alternaria* conidia were among the top 10 spore types (in terms of atmospheric concentrations) reported from 45 stations, and at 36 stations *Alternaria* spores were among the top 5 (3). Although some airborne *Alternaria* spores may be from saprobic species of *Alternaria*, these spores still represent a significant component of the air spora in the United States and may represent a potential threat to crops.

Rusts

There are about 6,000 species of rust fungi which attack a wide range of hosts among seed plants and cause some of the most destructive plant diseases. Millennia before scientists understood that pathogenic fungi could cause plant diseases, rust epidemics were studied, and the reddish lesions on plants were noted. These epidemics were recognized in ancient Greece and described in the writings of Aristotle and Theophrastus, who even noted that different plants varied in their susceptibilities to these diseases. Today, rust fungi still remain among the most serious agricultural pathogens. Some of the most important diseases of cereal crops are caused by these basidiomycetes, which have produced serious epidemics throughout history. In addition, coffee, apple, soybean, and pine trees as well as the economies dependent on these crops have been devastated by rust fungi. The long-distance transport of rust fungi has been studied more extensively than that of any other fungal pathogen.

Coffee rust, caused by *Hemileia vastatrix*, destroyed the coffee plantations in Ceylon in the 1870s and 1880s and today threatens production wherever coffee is grown. In 1966, an outbreak of coffee rust in Angola produced spores that were apparently carried by favorable wind currents across the Atlantic Ocean. The spores were washed out by rainfall over coffee plantations in Brazil approximately 5 to 7 days later (106). Meteorological data indicate that sugarcane rust, caused by *Puccinia melanocephala*, was introduced to the Dominican Republic from Cameroon via uredospores transported by cyclonic winds across the Atlantic Ocean (16).

One of the most important rust pathogens is *Puccinia graminis* f. sp. *tritici*, the fungus responsible for stem rust of wheat (81, 87). This organism has a complex life cycle that involves five spore stages on two separate host plants (wheat and barberry). Basidiospores, which develop from overwintering teliospores, are capable of infecting a young barberry leaf, eventually giving rise to spermatia and receptive hyphae and then aeciospores, also on the barberry leaf. The aeciospores transfer the infection to wheat, where the spores germinate, and hyphae enter the plant through stomata. Once the mycelium is established in wheat, the uredial stage develops within 2 weeks, with uredia appearing on the stem as long narrow lesions which produce dark red powdery masses of uredospores (also called urediniospores). Uredospores become airborne and reinfect new wheat plants to produce repeated generations of uredia. Near the end of the growing season, the uredia turn black when two-celled overwintering teliospores replace the uredospores (1, 77, 87).

The major vehicles for disease spread are the uredospores, which are easily carried by wind from one plant to another, giving rise to epidemics. In fact, they can be carried by prevailing winds for hundreds or thousands of kilometers (62, 106). In the mild climate of northern Mexico and southern Texas, this stage can continue all winter and give rise to spring infections in northern states, with spores being carried by prevailing southerly winds. Uredospores can also be carried back to the south in late summer and fall. Initially demonstrated in 1923 by Stakman, the movement of rust uredospores along the "*Puccinia* pathway" in North America is one of the best-known examples of long-distance dispersal (37, 115). *Puccinia graminis*, *Puccinia triticina* (formerly *Puccinia recondita*), *Puccinia coronata*, and *Bipolaris maydis*, as well as insect vectors of viral diseases, may all be connected with dispersal along this pathway (106).

Wheat plants infected by *P. graminis* are severely weakened, but not destroyed, and the grain yield is significantly reduced. Wheat plants throughout the world are threatened, and it is estimated that worldwide >1 million metric tons of wheat are lost annually due to stem rust. Evidence of the long-distance transport of *P. graminis* uredospores in other parts of the world was included in a review by Nagarajan and Singh (106). In Europe, two pathways have been studied, namely, an eastern European pathway originating in Turkey and Romania and a western pathway from Morocco and Spain (106). Both pathways converge in the Scandinavian countries. In India, the uredospores of both *P. graminis* and *P. triticina* (formerly *P. recondita*) survive throughout the year in the Nilgiri Hills in southern India. Dispersal to central and northern India occurs during November under the influence of tropical cyclones (106).

The long-distance dispersal of *P. graminis* uredospores between the eastern and western wheat-growing regions of Australia has been documented, and the overseas transport of stem rust has also been investigated (106). Two strains discovered in Australia were shown to be identical to strains found in South Africa in terms of pathogenicity and isozyme patterns. In addition, upper air wind currents studied by use of a weather balloon support the possibility of this transport (106).

Successful long-distance dispersal of wheat rust uredospores is also dependent on the source strength and viability. A mature uredium is capable of producing about 10,000 uredospores per day over a period of several weeks (125). With a 5% disease severity of 50 uredia, a single plant would produce 500,000 uredospores/day. It has been estimated that a field of wheat with a moderate infection of stem rust would produce 4×10^{12} uredospores/day/ha (106). Nagarajan and Singh (106) reported that following long-distance transport, spores had to be deposited within 120 h after takeoff to be infectious. Eversmeyer and Kramer (35) studied the survival of *P. graminis* and *P. triticina* (formerly *P. recondita*) uredospores under a variety of temperature conditions in the field. At subfreezing temperatures during winter, no spores were viable after 96 h, but during spring, 10 to 20% of the inoculum was viable after 120 h, and a fraction of a percent survived for 456 h. In environmental chamber experiments, spores remained viable for up to 864 h at constant temperatures between 10 and 30°C. However, the viability of spores exposed to freezing and subfreezing temperatures declined rapidly within a few hours of exposure (36). Although conditions in the atmosphere may be similarly harsh, the enormous number of spores produced ensures that at least a small number of spores will reach a suitable host.

Recently, the rust disease attracting the most attention has been Asian soybean rust, caused by *Phakopsora pachyrhizi*. Although another species of *Phakopsora*, *Phakopsora meibomiae*, also attacks soybean plants, *P. pachyrhizi* is far more virulent and will be the focus of this discussion. Asian soybean rust is the most destructive foliar pathogen on soybean plants, and reports of loss range from 10 to 80%. Yield losses over 50% are common when meteorological conditions favor disease development (1, 61). The fungus causes numerous uredial lesions on the leaves, reducing the photosynthetic capacity of the host and subsequently reducing the yield of soybeans. The life cycle of soybean rust is far simpler than that described for stem rust of wheat. *Phakopsora pachyrhizi* produces both uredia and telia on the same host plants. Uredia give rise to abundant uredospores that readily become airborne and spread the infection. Telia with teliospores have been reported late in the growing season on infected plants in Asia and South America as well as in greenhouses, but teliospores are apparently not important for disease spread (1, 23). Unlike many rust fungi that have a narrow host range, *P. pachyrhizi* can infect >95 species of plants, including several other edible legumes. Kudzu, a widespread weed in the southern United States and other parts of the world, can also be infected by this fungus (61).

Phakopsora pachyrhizi was first described in Japan in 1902 (http://www.apsnet.org/online/feature/rust). By the mid-1930s, the pathogen was reported from several other countries in Asia and from Australia. It was identified in India in the early 1950s. Reports in the late 1990s showed that the pathogen had spread to several countries in Africa, and in 2001 it was reported from Paraguay and Brazil in South America (158). Over the next few years, the pathogen spread through much of the soybean-growing areas of South America, causing significant yield losses (23, 119, 158). The fungus was first detected in the continental United States in November 2004 (135). The movement of the pathogen through South America and into the United States shows the importance of the aerobiological pathway. Researchers have tracked the movements of the *P. pachyrhizi* spores by using various dispersal models. Evidence from these models suggests that the pathogen was introduced during the 2004 hurricane season, specifically during Hurricane Ivan (http://www.aphis.usda.gov/ppq/ep/soybean_rust/sbrandhi11_17_04_files/frame.html). The progression of the pathogen around the world is shown in Fig. 1.

Control of soybean rust is achieved through the use of fungicides and resistant varieties of soybeans; however, a recent evaluation of commercial cultivars grown in the United States showed that none were resistant to the isolates of the pathogen tested (61). Therefore, in the United States, control of the pathogen depends on the use of fungicides timed to periods when environmental conditions are optimal for pathogen dispersal and infection. Two websites provide up-to-date information on the status of soybean rust development in the United States, i.e., those of the U.S. Department of Agriculture (http://www.sbrusa.net/) and the North American Plant Disease Forecast Center (http://www.ces.ncsu.edu/depts/pp/soybeanrust/). Numerous other websites offer information on the pathogen along with recommendations for fungicide use.

The United States annually produces >70 million metric tons of soybeans, and a soybean rust epidemic could have

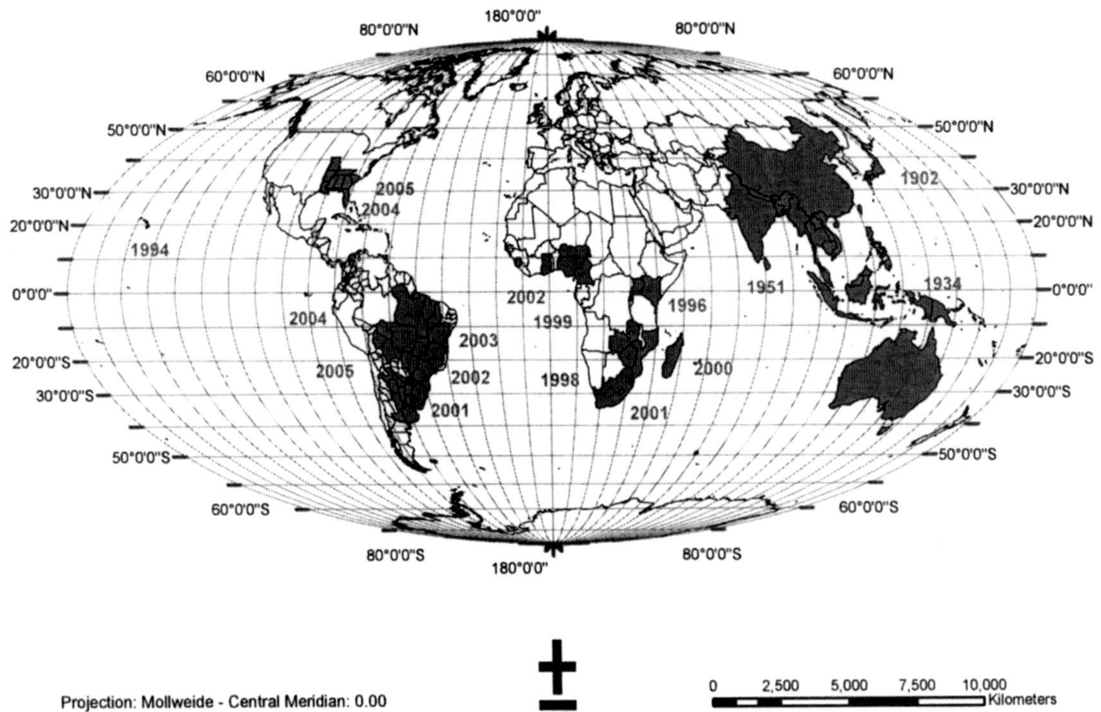

FIGURE 1 Global spread of soybean rust caused by *Phakopsora pachyrhizi*. Map supplied by Annalisa Ariatti, Department of Plant Pathology, Pennsylvania State University, University Park, Pa.

a major impact on both production and cost (1, 61). Pivonia et al. (118) have recently assessed the risk of soybean rust epidemics in the United States. The large number of host species for this pathogen increases the risk that the fungus will overwinter in the southern United States, especially in south Texas and Florida. Kudzu is seen as an especially important reservoir wherever this weed remains green during the winter. In areas where the pathogen cannot overwinter, kudzu may also serve as a bridge host since kudzu leaves emerge before soybean leaves. The pathogen can potentially be carried northward during the growing season in either short hops or by long-distance transport. Until resistant soybean cultivars become available in the United States, careful monitoring and fungicides are the only options available to control disease spread.

Smuts

There are approximately 1,200 species of smut fungi classified in 50 genera of the *Ustilaginales*; however, the majority of species are classified into two large genera, *Tilletia* and *Ustilago* (150). Smuts are serious plant pathogens on cereal crops, causing millions of dollars in damage each year. Many native plants are also affected, with a total of approximately 4,000 host species (150).

The major dispersal agents of smut fungi are the asexual teliospores (137). Enormous numbers of teliospores develop from the mycelium, usually within galls, and are dispersed by wind. There is no repeating stage; each infected plant produces only one generation of teliospores. Many smuts overwinter as teliospores; however, other smut species overwinter as mycelia in infected grain (1).

Corn smut is a widespread disease occurring wherever corn is grown, and it is more prevalent on sweet corn than on other varieties. The fungus *Ustilago maydis* can form galls on any aboveground plant part but is most conspicuous when the galls develop on the ears. The size and location of the galls reflect the degree of crop loss. Galls on the ears result in total loss, whereas those in other places reduce yields or cause stunted growth (137).

The fungus overwinters as teliospores on plant debris and in the soil. In the spring, the teliospores germinate; each produces a basidium that forms basidiospores. These are carried by wind or splashed by rain onto developing tissues of corn plants; young seedlings or the growing tissues of older plants can be affected. Basidiospores germinate and invade the host tissue. The resulting mycelium grows prolifically and stimulates host cells to divide and form galls, which may reach a size of 15 cm in diameter. As the galls mature, the interiors darken as the mycelia are converted into teliospores. The galls are initially covered by a membrane which later ruptures, releasing masses of dry spores (1).

Teliospores of various species are produced during different seasons. Hamilton (59) reported that airborne spores of *Tilletia* peaked in late August and early September in England, while *Ustilago* spores were at their highest concentrations between mid-June and the end of July. The airborne concentrations of these two genera depend on a number of meteorological factors. Levels decrease during rainy or humid conditions and increase during periods of bright sun, strong winds, and high barometric pressure, as would be expected for members of the dry air spora (59).

Other studies have also documented the presence of smut spores in the atmosphere. Halwagy (58) reported that *Ustilago* spores were the second most common spore type identified in the Kuwait atmosphere. Misra (101) showed that smut spores peaked in India under dry conditions during December and January, while Rubulis (132) reported that smut spores peaked during the late spring and fall in Sweden. In these two studies, the smuts were basically considered one spore category, with no attempt to identify the species involved.

Crotzer and Levetin (31) examined the airborne concentrations of smut teliospores in the Tulsa, Okla., atmosphere during 1991 and 1992 and attempted to identify the most abundant smut spores. This study showed that teliospores occurred in the atmosphere on 100% of the days from May through October in 1991 and 1992. The average daily concentrations generally ranged from 100 to 1,000 teliospores/m^3 during both seasons, with a mean for the 1991 season of 291 spores/m^3 and a peak at 1,874 spores/m^3, which occurred on 5 July, 1991. The mean for the 1992 season was 356 spores/m^3, with a peak of 5,906 spores/m^3 occurring on 12 May, 1992. Other data show that atmospheric smut spores are not limited to this 6-month period but are also prevalent earlier in spring and later in the fall (E. Levetin, unpublished data). Although spores were consistently present in the atmosphere during the period studied, the concentration of these spores had many fluctuations due to daily variations in wind speed, precipitation, and relative humidity as well as phenologies of the host species and pathogens (31). Different species of smuts were observed at different times of the year. During May and June, the prevalent teliospores included those species which infect Bermuda grass (*Ustilago cynodontis*), Johnson grass (*Sphacelotheca occidentalis*), oats (*Ustilago kolleri*), and wheat (*Ustilago tritici*). Smuts identified during September and October included *U. maydis*, which infects corn, and *Ustilago brumivora* and *Ustilago bullata*, which are pathogenic to several native Oklahoma grasses (31). Although some of the smuts identified in Tulsa were pathogens of native grasses, many others were important crop pathogens.

In addition to the significance of smut fungi as plant pathogens, it is recognized that smut teliospores may serve as potential aeroallergens because the atmosphere is often saturated with these spores for extended periods of time. In the 1998 report from the air sampling network of the American Academy of Allergy, Asthma, and Immunology, smut spores were among the top 10 spore types (in terms of atmospheric concentration) reported from 44 of 46 sampling stations that provide counts for fungal spores. At 28 of these stations, smut spores were among the top five types (3). A few smut extracts are routinely used for diagnosis and desensitization, and many individuals have skin tested positive for these allergens and have elevated levels of immunoglobulin E antibodies (18, 134). However, the full extent of allergenicity and the clinical significance of various smut species are not known (31). At the 44 monitoring stations, which are scattered across the United States, smut spores represent a significant component of the air spora and may represent a significant source of pathogen inoculum (3).

Anticrop Biological Warfare

The 11 September 2001 terrorist attacks on the United States and the anthrax attacks a few weeks later raised fears among the public about bioterrorism. Much of the media's attention has been on biological weapons resulting in diseases, such as anthrax, that affect people and livestock. However, fungal spores that attack crop plants have a long history as components of the biological warfare arsenal in the United States and other countries. Even before fall 2001, the safety of our food crops was the focus of a number of articles and symposia (126; http://apsnet.org/online/feature/BioSecurity/Top.html). These and more recent studies (92) suggest that even modest outbreaks of diseases targeting any of the major food crops could severely hurt the U.S. economy.

In 1969, the United States renounced the use of all weapons of biological warfare. This helped lay the groundwork for the 1972 Biological and Toxin Weapons Convention, which was signed by 141 countries (126). The signatories were required to stop work on biological and toxic weapons and to destroy their stockpiles. Before the unilateral termination of the U.S. offensive biological weapons program, a substantial effort to develop these weapons, including anticrop weapons, existed (126). Between 1951 and 1969, the U.S. program stockpiled >30,000 kg of stem rust spores (*Puccinia graminis* f. sp. *tritici*), which infect wheat, and nearly a ton of *Pyricularia oryzae* spores to be used as an antirice weapon. Also included in the arsenal were propagules of pathogens causing late blight of potato and sclerotium rot, which attacks soybeans, sugar beets, sweet potatoes, and cotton. In addition to these stockpiles, weapon systems were developed for delivery (27, 126).

Awareness of anticrop weaponry has increased, possibly fostered by revelations that Iraq actively pursued a biological weapons program in the late 1980s. Iraqi scientists had examined the potential use of five bacterial strains, five viruses, one fungal plant pathogen, and four toxins, two of which were mycotoxins (162). The plant pathogen on which they focused was a *Tilletia* sp., the causal agent of wheat-covered smut (92). That research was assumed to be aimed at destroying the wheat crop of neighboring Iran (126, 162). The Iraqi program produced *Tilletia* spores, infected fields near the town of Mosul, and then harvested the wheat as a stockpile of spores. No attempt to recover the *Tilletia* sp. was made, and the infected crop was destroyed in 1990 (162); however, the susceptibility of economic resources remains a significant threat (126).

This chapter has discussed many plant pathogens that can propagate on hosts at various distances and can cause extensive damage and economic loss; thus, the focus on plant pathogens as anticrop weapons is understandable. Dispersal by the aerobiological pathway has been depicted in recent articles describing scenarios of bioterrorist attacks (70, 126), and long-distance transport acts as a natural delivery system for biological particles that severely impact plant or human health.

The focus of modern phytopathological research is increasingly on the development and use of dispersal models to track pathogen outbreaks and meteorological conditions conducive to the spread of a pathogen. These efforts must continue. Understanding the mechanisms that transport pathogens into a given area and then establishing prophylactic measures would enhance efforts to protect our crops and ourselves against terrorist attacks.

Disease Forecasting and Information Technology

Disease forecasting for various agricultural pathogens has been ongoing for a number of years to provide warnings when disease outbreaks are likely to occur. The ultimate goal of forecasting is to reduce costly and environmentally hazardous pesticide applications during periods when disease occurrence is not likely. A variety of models have been developed to predict spore dispersal and disease gradients (1, 8, 9, 10, 11, 12, 34, 60, 79, 114, 128, 149, 156, 157). While many of these models provide useful information to the plant pathologist and epidemiologist, they are of little direct use to the average farmer.

In recent years, the widespread availability of personal computers and the growth of the Internet have created a revolution in information technology that has spread to all areas, including plant pathology and epidemiology (107). The innovations in technology have made plant disease forecasts more widely available by placing them on the World Wide Web. As a result, information is often in the hands of the grower within minutes. Some forecasting systems are supported by aerobiological data, some are supported by simulations and experimental data, and others rely solely on meteorological data. These websites translate many of the disease forecasting models into "user-friendly" applications. Many sites use a multimedia approach to provide forecasts as well as education about the pathogen and symptoms of infected plants.

One of the earliest web-based forecast sites was the North American Plant Disease Forecast Center (Table 2). This website was started in 1996 to provide online forecasts for tobacco blue mold outbreaks throughout continental North America. It was later expanded to include information about cucurbit downy mildew and soybean rust. The forecasting system described above provides trajectories from known sources of pathogens.

Numerous sites provide information on *Phytophthora infestans* for both potato and tomato cultivation. These are typically based on meteorology and the known presence of the pathogen in the area and are often restricted to single states or regions within a state. It should be noted that websites change frequently. The URLs listed in Table 2 are those currently available; however, it may be necessary to use a search engine if the listed URL does not connect to the site.

SUMMARY

Airborne fungal spores and other plant disease propagules are always present in the natural environment and pose a constant threat to food crops. With the rapidly expanding world population, diseases that threaten the food supply could result in widespread starvation. In addition, world food production must continue to increase in the coming decades in order to keep pace with projected population increases. For food production to increase, control of plant disease is essential. Aerobiological studies must be part of any effort to understand the distribution and epidemiology of agricultural pathogens that rely on air currents for dispersal.

Many aerobiological studies have been conducted on the local or long-distance spread of individual plant pathogens. Some of these studies were performed to develop mathematical or computer models for forecasting disease epidemics and predicting their onset and severity; for example, models of the spread of the tobacco blue mold pathogen have been successful in predicting the spread of this disease. The online availability of disease forecasts is a major step in quickly putting information in the hands of users.

For other pathogens, the lack of accurate spore dispersal data has often limited the development of dispersal models. These models require an understanding of all stages of dispersal, from spore takeoff to transport to deposition within the crop, as well as knowledge of the percentage of spores that escape local deposition and the percentage that survive long-distance transport. The sensitivity of molecular techniques is expected to improve pathogen detection and measurement and may help to answer some questions about long-distance transport.

TABLE 2 Examples of plant disease forecasting systems on the World Wide Web

Forecast system	Pathogen	URL
North American Plant Disease Forecast Center, North Carolina State University		
Blue mold	*Peronospora tabacina*	http://www.ces.ncsu.edu/depts/pp/bluemold/
Cucurbit downy mildew	*Pseudoperonospora cubensis*	http://www.ces.ncsu.edu/depts/pp/cucurbit/
Soybean rust	*Phakopsora pachyrhizi*	http://www.ces.ncsu.edu/depts/pp/soybeanrust/
Crop Disease Forecasting Models, North Dakota State University		
Fusarium head blight	*Fusarium graminearum*	http://www.ag.ndsu.nodak.edu/cropdisease
Tan spot	*Pyrenophora triticirepentis*	http://www.ag.ndsu.nodak.edu/cropdisease
Stagonospora blotch	*Septoria tritici*	http://www.ag.ndsu.nodak.edu/cropdisease
Wheat leaf rust	*Puccinia triticina*	http://www.ag.ndsu.nodak.edu/cropdisease
Potato late blight	*Phytophthora infestans*	http://ndawn.ndsu.nodak.edu/potatolb-form.html
Oklahoma State University		
Pecan scab model	*Cladosporium caryigenum*	http://agweather.mesonet.ou.edu/models/newpecanscab/default.php
Peanut leafspot model	*Cercospora arachidicola*	http://agweather.mesonet.ou.edu/models/peanut/description.html
Watermelon anthracnose	*Colletotrichum lagenarium*	http://agweather.mesonet.ou.edu/models/watermelon/description.html
Spinach white rust	*Albugo occidentalis*	http://agweather.mesonet.ou.edu/models/spinach/default.php
Western Australia Department of Agriculture		
Canola blackleg	*Leptosphaeria maculans*	http://www.agric.wa.gov.au/pls/portal30/docs/FOLDER/IKMP/PW/PH/DIS/PLANT_DISEASE_2005.HTM
Lupin anthracnose	*Colletotrichum gloeosporioides*	http://www.agric.wa.gov.au/pls/portal30/docs/FOLDER/IKMP/PW/PH/DIS/PLANT_DISEASE_2005.HTM
Cereal rust disease	*Puccinia graminis*, *Puccinia triticina*, *Puccinia striiformis*	http://www.agric.wa.gov.au/pls/portal30/docs/FOLDER/IKMP/PW/PH/DIS/PLANT_DISEASE_2005.HTM
University of Idaho		
Potato late blight	*Phytophthora infestans*	http://www.uidaho.edu/ag/plantdisease/lbhome.htm
Michigan State University		
Potato late blight weather monitoring	*Phytophthora infestans*	http://potatodiseases.org/lateblight/forecasting.html
University of California at Davis Disease Model Database		
Potato late blight	*Phytophthora infestans*	http://www.ipm.ucdavis.edu/DISEASE/DATABASE/potatolateblight.html

Air sampling in the plant pathology community has generally been limited to individual studies for individual pathogens. There is no general monitoring for diverse pathogens. Although the importance of the *Puccinia* pathway for the dispersal of *P. graminis* uredospores is well known, there is no network of monitoring stations that can provide information on the advance of the pathogen or which other organisms utilize the same aerobiological pathway. Such networks exist in the aeroallergen research community. Because many spores from phytopathogens are also well-known allergens, some of these data are available. An important objective for the future should be greater coordination among scientists interested in atmospheric transport.

REFERENCES

1. **Agrios, G. N.** 2005. *Plant Pathology*, 5th ed. Elsevier Academic Press, Boston, Mass.
2. **Allitt, U.** 1986. Identity of airborne hyaline, one-septate ascospores and their relation to inhalant allergy. *Trans. Br. Mycol. Soc.* **87:**147–154.
3. **American Academy of Allergy, Asthma, and Immunology.** 1999. *1998 Pollen and Spore Report*. American Academy of Allergy, Asthma, and Immunology, Milwaukee, Wis.
4. **Atlas, R. M., and R. Bartha.** 1998. *Microbial Ecology: Fundamentals and Applications*. Benjamin Cummings Science Publishing, Menlo Park, Calif.

5. **Aylor, D. E.** 1990. The role of intermittent wind in the dispersal of fungal pathogens. *Annu. Rev. Phytopathol.* **28:** 73–92.
6. **Aylor, D. E.** 1995. Vertical variation of aerial concentration of *Venturia inaequalis* ascospores in an apple orchard. *Phytopathology* **85:**175–181.
7. **Aylor, D. E.** 1996. Comparison of the seasonal pattern of airborne *Venturia inaequalis* ascospores with the release potential of *V. inaequalis* ascospores from a source. *Phytopathology* **86:**769–776.
8. **Aylor, D. E.** 1998. Aerobiology of apple scab. *Plant Dis.* **82:**838–849.
9. **Aylor, D. E.** 1999. Biophysical scaling and the passive dispersal of fungus spores: relationship to integrated pest management strategies. *Agric. Forest Meteorol.* **97:**275–292.
10. **Aylor, D. E.** 2002. Aerobiology of fungi in relation to capture and release by plants, p. 341- 361. *In* S. E. Lindow, E. I. Hecht-Poinar, and V. J. Elliott (ed.), *Phyllosphere Microbiology*. APS Press, St. Paul, Minn.
11. **Aylor, D. E.** 2003. Spread of plant disease on a continental scale: role of aerial dispersal of pathogens. *Ecology* **84:**1989–1997.
12. **Aylor, D. E., W. E. Fry, H. Mayton, and J. L. Andrade-Piedra.** 2001. Quantifying the rate of release and escape of *Phytophthora infestans* sporangia from a potato canopy. *Phytopathology* **91:**1189–1196.
13. **Banttari, E. E., and J. R. Venette.** 1980. Aerosol spread of plant viruses: potential role in disease outbreaks. *Ann. N. Y. Acad. Sci.* **353:**167–173.
14. **Barras, F., F. van Gijsegem, and A. K. Chatterjee.** 1994. Extracellular enzymes and pathogenesis of soft-rot *Erwinia*. *Annu. Rev. Phytopathol.* **32:**201–234.
15. **Bock, C. H., P. E. Parker, and T. R. Gottwald.** 2005. Effect of simulated wind-driven rain on duration and distance of dispersal of *Xanthomonas axonopodis* pv. *Citri* from canker-infected citrus trees. *Plant Dis.* **89:**71–80.
16. **Brown, J. K. M., and M. S. Hovmøller.** 2002. Aerial dispersal of pathogens on the global and continental scales and its impact on plant disease. *Science* **297:**537–541.
17. **Burch, M., and E. Levetin.** 2002. Effects of meteorological conditions on spore plumes. *Int. J. Biometeorol.* **46:** 107–117.
18. **Burge, H. A.** 1985. Fungus allergens. *Clin. Rev. Allergy* **3:** 319–329.
19. **Burge, H. A.** 1986. Some comments on the aerobiology of fungus spores. *Grana* **25:**143–146.
20. **Burge, H. A., M. M. Muilenberg, and J. A. Chapman.** 1991. Crop plants as a source of fungus spores of medical importance, p. 222–236. *In* J. H. Andrews and S. S. Hirano (ed.), *Microbial Ecology of Leaves*. Springer-Verlag, New York, N.Y.
21. **Burnet, P. A. (ed).** 1989. *Barley Yellow Dwarf Virus, the Yellow Plague of Cereals*. Centro Internacional de Mejoramiento de Maiz y Trigo, Mexico City, Mexico.
22. **Campbell, C. L.** 1999. The importance of dispersal mechanisms in the epidemiology of *Phytophthora* blights and downy mildews on crop plants. *Ecosyst. Health* **5:**146–157.
23. **Carmona, M. A., M. E. Gally, and S. E. Lopez.** 2005. Asian soybean rust: incidence, severity, and morphological characterization of *Phakopsora pachyrhizi* (uredinia and telia) in Argentina. *Plant Dis.* **89:**109.
24. **Cerda, H., and M. G. Paoletti.** 2004. Genetic engineering with *Bacillus thuringiensis* and conventional approaches for insect resistance in crops. *Crit. Rev. Plant Sci.* **23:**317–323.
25. **Chattopadhyay, A., N. B. Bhatnagar, and R. Bhatnagar.** 2004. Bacterial insecticidal toxins. *Crit. Rev. Microbiol.* **30:**33–54.
26. **Chen, J., L. M. Jacobson, J. Handelsman, and R. M. Goodman.** 1996. Compatibility of systemic acquired resistance and microbial biocontrol for suppression of plant disease in a laboratory assay. *Mol. Ecol.* **5:**73–80.
27. **Christopher, G. W., T. J. Cieslak, J. A. Pavlin, and E. M. Eitzen.** 1997. Biological warfare: a historical perspective. *JAMA* **278:**412–417.
28. **Conn, K. E., J. M. Ogawa, B. T. Manji, and J. E. Adaskaveg.** 1995. *Leuconostoc mesenteroides*, the first report of a coccoid bacterium causing a plant disease. *Phytopathology* **85:**593–599.
29. **Cooper, J. I., S. E. Kelley, and P. R. Massalski.** 1988. Virus-pollen interactions. *Adv. Dis. Vector Res.* **5:**221–249.
30. **Cox, C. S.** 1987. *The Aerobiological Pathway of Microorganisms*. John Wiley & Sons, New York, N.Y.
31. **Crotzer, V., and E. Levetin.** 1996. The aerobiological significance of smut spores in Tulsa, Oklahoma. *Aerobiologia* **12:**177–184.
32. **Davis, J. M.** 1987. Modeling the long-range transport of plant pathogens in the atmosphere. *Annu. Rev. Phytopathol.* **25:**169–188.
33. **Deacon, J. W.** 1997. *Modern Mycology*, 3rd ed. Blackwell Science, Oxford, United Kingdom.
34. **de Jong, M. D., G. W. Bourdôt, J. Powell, and J. Goudriann.** 2002. A model of the escape of *Sclerotinia sclerotiorum* ascospores from pasture. *Ecol. Model.* **150:**83–105.
35. **Eversmeyer, M. G. and C. L. Kramer.** 1994. Survival of *Puccinia recondita* and *P. graminis* urediniospores as affected by exposure to weather conditions at one meter. *Phytopathology* **84:**332–335.
36. **Eversmeyer, M. G., and C. L. Kramer.** 1995. Survival of *Puccinia recondita* and *P. graminis* urediniospores exposed to temperatures from subfreezing to 35°C. *Phytopathology* **85:** 161–164.
37. **Eversmeyer, M. G., and C. L. Kramer.** 2000. Epidemiology of wheat leaf and stem rust in the central Great Plains of the USA. *Annu. Rev. Phytopathol.* **38:**491–513.
38. **Fahey, J. W., M. B. Dimock, S. F. Tomasino, J. M. Taylor, and P. S. Carlson.** 1991. Genetically engineered endophytes as biocontrol agents: a case study from industry, p. 401–411. *In* J. H. Andrews and S. S. Hirano (ed.), *Microbial Ecology of Leaves*. Springer-Verlag, New York, N.Y.
39. **Fitt, B. D. L., and H. A. McCartney.** 1986. Spore dispersal in relation to epidemic models, p. 311–345. *In* K. J. Leonard and W. E. Fry (ed.), *Plant Disease Epidemiology*, vol. I. Macmillan Publishing Co., New York, N.Y.
40. **Fitt, B. D. L., H. A. McCartney, and P. J. Walklate.** 1989. The role of rain in dispersal of pathogen inoculum. *Annu. Rev. Phytopathol.* **27:**241–270.
41. **Fokunang, C. N., J. L. Beynon, K. A. Watson, N. H. Battey, J. M. Dunwell, and E. A. Tembe-Fokunang.** 2004. Advancement in genetic modification technologies towards disease resistance and food crop production. *Biotechnology* **3:**1–20.
42. **Franc, G.** 1994. Atmospheric transport of *Erwinia carotovora*, p. 435–437. *In Proceedings of the 11th Conference of Biometeorology and Aerobiology*. American Meteorological Society, Boston, Mass.
43. **Fravel, D. R.** 2005. Commercialization and implementation of biocontrol. *Annu. Rev. Phytopathol.* **43:**337–359.
44. **Frinking, H.** 1993. A historical perspective on aerobiology as a discipline, p. 23–28. *In* S. A. Isard (ed.), *Alliance for Aerobiology Research Workshop Report*. Illinois Natural History Survey, Champaign, Ill.
45. **Fry, W. E.** 1997. Re-emergence of potato and tomato late blight in the United States. *Plant Dis.* **81:**1349–1357.
46. **Fry, W. E., and E. S. Mizubuti.** 1998. Potato late blight, p. 371–388. *In* D. G. Jones (ed.), *The Epidemiology of Plant Disease*. Kluwer Academic Publishers, Dordrecht, The Netherlands.

47. **Fry, W. E., S. B. Goodwin, J. M. Matuszak, L. J. Spielman, M. G. Milgroom, and A. Drenth.** 1992. Population genetics and intercontinental migrations of *Phytophthora infestans*. *Annu. Rev. Phytopathol.* **30:**107–129.
48. **Gadoury, D. M., A. Stensvand, and R. C. Seem.** 1998. Influence of light, relative humidity, and maturity of populations on discharge of ascospores of *Venturia inaequalis*. *Phytopathology* **88:**902–909.
49. **Garelik, G.** 2002. Taking the bite out of potato blight. *Science* **298:**1702–1704.
50. **Geagea, L., L. Huber, and I. Sache.** 1999. Dry-dispersal and rain-splash of brown (*Puccinia recondite* f. sp. *tritici*) and yellow (*P. striiformis*) rust spores from infected wheat leaves exposed to simulated raindrops. *Plant Pathol.* **48:** 472–482.
51. **Goodman, R. N., Z. Kiraly, and K. R. Wood.** 1986. *The Biochemistry and Physiology of Plant Disease*. University of Missouri Press, Columbia.
52. **Gottwald, T. R., G. J. Gibson, S. M. Garnsey, and M. Irey.** 1999. Examination of the effect of aphid vector population composition on the spatial dynamics of citrus tristeza virus spread by stochastic modeling. *Phytopathology* **89:**603–608.
53. **Govi, G.** 1992. Aerial diffusion of phytopathogenic fungi. *Aerobiologia* **8:**84–93.
54. **Gregory, P. H.** 1973. *The Microbiology of the Atmosphere*, 2nd ed. Halstead Press, New York, N.Y.
55. **Griffin, D. W., V. H. Garrison, J. R. Herman, and E. A. Shinn.** 2001. African desert dust in the Caribbean atmosphere: microbiology and public health. *Aerobiologia* **17:** 203–213.
56. **Griffin, D. W., C. A. Kellogg, V. H. Garrison, J. T. Lisle, T. C. Borden, and E. A. Shinn.** 2003. Atmospheric microbiology in the northern Caribbean during African dust events. *Aerobiologia* **19:**143–157.
57. **Haas, D., and C. Keel.** 2003. Regulation of antibiotic production in root-colonizing *Pseudomonas* spp. and relevance for biological control of plant disease. *Annu. Rev. Phytopathol.* **41:**117–153.
58. **Halwagy, M.** 1989. Seasonal airspora at three sites in Kuwait 1977–1982. *Mycol. Res.* **93:**208–213.
59. **Hamilton, E. D.** 1959. Studies on the air spora. *Acta Allergologica* **13:**143–175.
60. **Hardwick, N. V.** 1998. Disease forecasting, p. 207–230. *In* D. G. Jones (ed.), *The Epidemiology of Plant Disease*. Kluwer Academic Publishers, Dordrecht, The Netherlands.
61. **Hartman, G. L., M. R. Miles, and R. D. Fredrick.** 2005. Breeding for resistance to soybean rust. *Plant Dis.* **89:**664–666.
62. **Hau, B., and C. de Vallavielle-Pope.** 1998. Wind-dispersed diseases, p. 323–347. *In* D. G. Jones (ed.), *The Epidemiology of Plant Disease*. Kluwer Academic Publishers, Dordrecht, The Netherlands.
63. **Heesterbeek, J. A. P., and J. C. Zadoks.** 1987. Modeling pandemics of quarantine pests and diseases; problems and perspectives. *Crop Prot.* **6:**211–221.
64. **Hellmich, R. L., B. D. Siegfried, M. K. Sears, D. E. Stanley-Horn, M. J. Daniels, H. R. Mattila, T. Spencer, K. G. Bidne, and L. C. Lewis.** 2001. Monarch larvae sensitivity to *Bacillus thuringiensis*-purified proteins and pollen. *Proc. Natl. Acad. Sci. USA* **98:**11925–11930.
65. **Hibino, H.** 1996. Biology and epidemiology of rice viruses. *Annu. Rev. Phytopathol.* **34:**249–274.
66. **Hirano, S. S., and C. D. Upper.** 1990. Population biology and epidemiology of *Pseudomonas syringae*. *Annu. Rev. Phytopathol.* **28:**155–177.
67. **Holmes, A., J. Govan, and R. Goldstein.** 1998. Agricultural use of *Burkholderia* (*Pseudomonas*) *cepacia*: a threat to human health? *Emerg. Infect. Dis.* **4:**221–227.
68. **Huber, L., and T. J. Gillespie.** 1992. Modeling leaf wetness in relation to plant disease epidemiology. *Annu. Rev. Phytopathol.* **30:**553–577.
69. **Huber, L., L.V. Madden, and B. D. L. Fitt.** 1998. Rain-splash dispersal: a physical perspective, p. 348–370. *In* D. G. Jones (ed.), *The Epidemiology of Plant Disease*. Kluwer Academic Publishers, Dordrecht, The Netherlands.
70. **Inglesby, T. V.** 1999. Anthrax: a possible case history. *Emerg. Infect. Dis.* **5:**556–560.
71. **Irwin, M. E., and J. M. Thresh.** 1990. Epidemiology of barley yellow dwarf: a study in ecological complexity. *Annu. Rev. Phytopathol.* **28:**393–424.
72. **Janisiewicz, W. J., and L. Korsten.** 2002. Biological control of postharvest diseases of fruits. *Annu. Rev. Phytopathol.* **40:**411–441.
73. **Jeger, M. J.** 1984. Damage and loss in fruit orchards caused by airborne fungal pathogens, p. 225–236. *In* R. K. S. Wood and G. J. Jellis (ed.), *Plant Diseases: Infection, Damage and Loss*. Blackwell Scientific Publications, Oxford, United Kingdom.
74. **Johnson, G. I.** 1989. *Peronospora hyoscyami* de Bary: taxonomic history, strains, and host range, p. 1–18. *In* W. E. McKeen (ed.), *Blue Mold of Tobacco*. APS Press, St. Paul, Minn.
75. **Jones, A. L.** 1998. Apple scab: role of environment in pathogen and epidemic development, p. 389–403. *In* D. G. Jones (ed.), *The Epidemiology of Plant Disease*. Kluwer Academic Publishers, Dordrecht, The Netherlands.
76. **Kelman, A.** 1995. Contributions of plant pathology to the biological sciences and industry. *Annu. Rev. Phytopathol.* **33:**1–21.
77. **Kendrick, B.** 2000. *The Fifth Kingdom*, 3rd ed. Mycologue Publications, Newburyport, Mass.
78. **Lacey, J.** 1991. Aerobiology and health: the role of airborne fungal spores in respiratory disease, p. 157–186. *In* D. L. Hawksworth (ed.), *Frontiers in Mycology*. C.A.B. International, Kew, United Kingdom.
79. **Lacey, J.** 1996. Spore dispersal—its role in ecology and disease: the British contribution to fungal aerobiology. *Mycol. Res.* **100:**641–660.
80. **Lacey, J., and J. Venette.** 1995. Outdoor air sampling techniques, p. 407–471. *In* C. S. Cox and C. M. Wathes (ed.), *Bioaerosols Handbook*. Lewis Publishers, Boca Raton, Fla.
81. **Leonard, K. J., and L. J. Szabo.** 2005. Stem rust of small grains and grasses caused by *Puccinia graminis*. *Mol. Plant Pathol.* **6:**99–111.
82. **Levetin, E.** 1991. Identification and concentration of airborne basidiospores. *Grana* **30:**123–128.
83. **Levetin, E.** 1995. Fungi, p. 87–120. *In* H. Burge (ed.), *Bioaerosols*. Lewis Publishers, Boca Raton, Fla.
84. **Levetin, E., and W. E. Horner.** 2002. Fungal aerobiology: exposure and measurement. *Chem. Immunol.* **81:**10–27.
85. **Lighthart, B., J. C. Spendlove, and T. G. Akers.** 1979. Factors in the production, release, and viability of biological particles: bacteria and viruses, p. 11–22. *In* R. L. Edmonds (ed.), *Aerobiology: The Ecological Systems Approach*. Dowden, Hutchinson & Ross, Stroudsburg, Pa.
86. **Lindow, S. E., and T. V. Suslow.** 2003. Temporal dynamics of the biocontrol agent *Pseudomonas fluorescens* strain A506 in flowers in inoculated pear trees. *Phytopathology* **73:**727–737.
87. **Littlefield, L. J.** 1981. *Biology of the Plant Rusts: an Introduction*. Iowa State University Press, Ames, Iowa.
88. **Losey, J. E., L. S. Raynor, and M. E. Carter.** 1999. Transgenic pollen harms monarch larvae. *Nature* **399:**214.
89. **Lovell, D. J., S. R. Parker, T. Hunter, S. J. Welham, and A. R. Nichols.** 2004. Position of inoculum in the canopy affects the risk of *Septoria tritici* blotch epidemics in winter wheat. *Plant Pathol.* **53:**11–21.

90. **Maagd, R. A., A. Bravo, C. Berry, N. Crickmore, and H. E. Schnepf.** 2003. Structure, diversity, and evolution of protein toxins from spore-forming entomopathogenic bacteria. *Annu. Rev. Genet.* **37:**409–433.
91. **MacKenzie, D. R., V. J. Elliott, B. A. Kidney, E. D. King, M. H. Royer, and R. L. Theberge.** 1983. Application of modern approaches to the study of the epidemiology of diseases caused by *Phytophthora*, p. 303–313. *In* D. C. Erwin, S. Bartnicki-Garcia, and P. H. Tsao (ed.), Phytophthora: *Its Biology, Taxonomy, Ecology, and Pathology*. APS Press, St. Paul, Minn.
92. **Madden, L. V., and M. Wheelis.** 2003. The threat of plant pathogens as weapons against U.S. crops. *Annu. Rev. Phytopathol.* **41:**155–176.
93. **Main, C. E., and J. M. Davis.** 1989. Epidemiology and biometeorology of tobacco blue mold, p. 201–216. *In* W. E. McKeen (ed.), *Blue Mold of Tobacco*. APS Press, St. Paul, Minn.
94. **Marthi, B**. 1994. Resuscitation of microbial bioaerosols, p. 192–225. *In* B. Lighthart and A. J. Mohr (ed.), *Atmospheric Microbial Aerosols: Theory and Applications*. Chapman and Hall, New York, N.Y.
95. **Mason, C. J.** 1979. Atmospheric transport, p. 85–95. *In* R. L. Edmonds (ed.), *Aerobiology: The Ecological Systems Approach*. Dowden, Hutchinson & Ross, Stroudsburg, Pa.
96. **McCartney, H. A., and B. D. L. Fitt.** 1998. Dispersal of foliar fungal plant pathogens: mechanisms, gradients and spatial patterns, p. 138–160. *In* D. G. Jones (ed.), *The Epidemiology of Plant Disease*. Kluwer Academic Publishers, Dordrecht, The Netherlands.
97. **Miller, W. A., and L. Rasochová.** 1997. Barley yellow dwarf viruses. *Annu. Rev. Phytopathol.* **35:**167–190.
98. **Miller, W. A., S. Liu, and R. Beckett.** 2002. Barley yellow dwarf virus: Luteoviridae or Tombusviridae? *Mol. Plant Pathol.* **3:**177–183.
99. **Mims, S. A., and F. M. Mims.** 2004. Fungal spores are transported long distances in smoke from biomass fires. *Atmosph. Environ.* **38:**651–655.
100. **Mink, G. I.** 1993. Pollen- and seed-transmitted viruses and viroids. *Annu. Rev. Phytopathol.* **31:**375–402.
101. **Misra, R. P.** 1987. Studies on seasonal and diurnal variation in the occurrence of airborne spores of basidiomycetes. *Perspect. Mycol. Res.* **1:**243–252.
102. **Money, N. P.** 1998. More g's than the Space Shuttle: ballistospore discharge. *Mycologia* **90:**547–558.
103. **Moody, S. A., K. K. Newsham, P. G. Ayres, and N. D. Paul.** 1999. Variation in the responses of litter and phylloplane fungi to UV-B radiation (290–315 nm). *Mycol. Res.* **103:**1469–1477.
104. **Moore-Landecker, E.** 1990. *Fundamentals of the Fungi*, 3rd ed. Prentice-Hall, Englewood Cliffs, N.J.
105. **Mundt, C. C.** 1995. Models from plant pathology on the movement and fate of new genotypes of microorganisms in the environment. *Annu. Rev. Phytopathol.* **33:**467–488.
106. **Nagarajan, S., and D. V. Singh.** 1990. Long-distance dispersion of rust pathogens. *Annu. Rev. Phytopathol.* **28:** 139–153.
107. **Newton, A. C., and R. E. Gaunt.** 1998. Information technology in epidemiology, p. 278–294. *In* D. G. Jones (ed.), *The Epidemiology of Plant Disease*. Kluwer Academic Publishers, Dordrecht, The Netherlands.
108. **Ng, J. C. K., and K. L. Perry.** 2004. Transmission of plant viruses by aphid vectors. *Mol. Plant Pathol.* **5:**505–511.
109. **Niederhauser, J. S.** 1993. International cooperation in potato research and development. *Annu. Rev. Phytopathol.* **31:**1–21.
110. **Parke, J. L., and D. Gurian-Sherman.**. 2001. Diversity of the *Burkholderia cepacia* complex and implications for risk assessment of biological control strains. *Annu. Rev. Phytopathol.* **39:**225–258.
111. **Paul, B., A. Chereyathmanjiyil, I. Masih, L. Chapius, and A. Benoit.** 1998. Biological control of *Botrytis cinerea* causing grey mould disease of grapevine and elicitation of stilbene phytoalexin (resveratrol) by a soil bacterium. *FEMS Microbiol. Lett.* **165:**65–70.
112. **Paul, P. A., S. M. El-Allaf, P. E. Lipps, and L. V. Madden.** 2004. Rain splash dispersal of *Gibberella zeae* within wheat canopies in Ohio. *Phytopathology* **94:**1342–1349.
113. **Paulitz, T. C., and R. R. Bélanger.** 2001. Biological control in greenhouse systems. *Annu. Rev. Phytopathol.* **39:** 103–133.
114. **Paulitz, T. C., P. Dutilleul, S. H. Yamasaki, W. G. D. Fernando, and W. L. Seaman.** 1999. A generalized two-dimensional Gaussian model of disease foci of head blight of wheat caused by *Gibberella zeae*. *Phytopathology* **89:**74–83.
115. **Pedgley, D. E.** 1986. Long distance transport of spores, p. 346–365. *In* K. J. Leonard and W. E. Fry (ed.), *Plant Disease Epidemiology*, vol. I. Macmillan Publishing Co., New York, N.Y.
116. **Pedgley, D. E.** 1991. Aerobiology: the atmosphere as a source and sink for microbes, p. 43–59. *In* J. H. Andrews and S. S. Hirano (ed.), *Microbial Ecology of Leaves*. Springer-Verlag, New York, N.Y.
117. **Pielaat, A., L. V. Madden, and G. Gort.** 1998. Spores splashing under different environmental conditions: a modeling approach. *Phytopathology* **88:**1131–1140.
118. **Pivonia, S., X. B. Yang, and Z. Pan.** 2005. Assessment of epidemic potential of soybean rust in the United States. *Plant Dis.* **89:**678–682.
119. **Ploper, L. D., V. González, M. R. Gálvez, N. V. de Ramallo, M. A. Zamorano, G. García, and A. P. Castagnaro.** 2005. Detection of soybean rust caused by *Phakopsora pachyrhizi* in northwestern Argentina. *Plant Dis.* **89:**774.
120. **Plumb, R. T.** 1983. Barley yellow dwarf virus—a global problem, p. 185–198. *In* R. T. Plumb and J. M. Thresh (ed.), *Plant Virus Epidemiology*. Blackwell Science, London, United Kingdom.
121. **Plumb, R. T.** 2002. Viruses of Poaceae: a case history in plant pathology. *Plant Pathol.* **51:**673–682.
122. **Ponti, I., and P. Cavanni.** 1992. Aerobiology in plant protection. *Aerobiologia* **8:**94–101.
123. **Portnoy, J., J. Chapman, H. Burge, M. Muilenberg, and W. Solomon.** 1987. *Epicoccum* allergy: skin reaction patterns and spore/mycelium disparities recognized by IgG and IgE ELISA inhibition. *Ann. Allergy* **59:**39–43.
124. **Rajasab, A. H., and H. T. Chawda**. 1994. Dispersal of the conidia of *Colletotrichum gloeosporioides* by rain and the development of anthracnose on onion. *Grana* **33:**162–165.
125. **Roelfs, A. P.** 1985. Wheat and rye stem rust, p. 3–37. *In* A. P. Roelfs and W. R. Bushnell (ed.), *The Cereal Rusts, vol. II. Diseases, Distribution, Epidemiology, and Control.* Academic Press, Inc., Orlando, Fla.
126. **Rogers, P., S. Whitby, and M. Dando.** 1999. Biological warfare against crops. *Sci. Am.* **280:**70–75.
127. **Romantschuk, M.** 1992. Attachment of plant pathogenic bacteria to plant surfaces. *Annu. Rev. Phytopathol.* **30:**225–243.
128. **Rosa, M., B. Gozzini, S. Orlandini, and L. Seghi.** 1995. A computer program to improve the control of grapevine downy mildew. *Comput. Electron. Agric.* **12:**311–322.
129. **Rossi, V., I. Ponti, M. Marinelli, S. Giosue, and R. Bugiani.** 1999. Field evaluation of some models estimating the seasonal pattern of airborne ascospores of *Venturia inaequalis*. *J. Phytopathol.* **147:**567–575.

130. **Rossi, V., I. Ponti, M. Marinelli, S. Giosue, and R. Bugiani.** 2001. Environmental factors influencing the dispersal of *Venturia inaequalis* ascospores in the orchard air. *J. Phytopathol.* **149:**11–19.
131. **Rotem, J.** 1994. *The Genus* Alternaria. APS Press, St. Paul, Minn.
132. **Rubulis, J.** 1984. Airborne fungal spores in Stockholm and Eskilstuna, central Sweden, p. 85–93. *In* S. Nilsson and B. Raj (ed.), *Nordic Aerobiology*. Almqvist and Wiksell Int., Stockholm, Sweden.
133. **Salmond, G. P. C.** 1994. Secretion of extracellular virulence factors by plant pathogenic bacteria. *Annu. Rev. Phytopathol.* **32:**181–200.
134. **Santilli, J., W. J. Rockwell, and R. P. Collins.** 1985. The significance of the spores of the basidiomycetes (mushrooms and their allies) in bronchial asthma and allergic rhinitis. *Ann. Allergy* **55:**469–471.
135. **Schneider, R. W., C. A. Hollier, H. K. Whitam, M. E. Palm, J. M. McKemy, J. R. Hernandez, L. Levy, and R. DeVries-Paterson.** 2005. First report of soybean rust caused by *Phakopsora pachyrhizi* in the continental United States. *Plant Dis.* **89:**774.
136. **Schnepf, E., N. Crickmore, J. Van Rie, D. Lereclus, J. Baum, J. Feitelson, D. R. Zeigler, and D. H. Dean.** 1998. *Bacillus thuringiensis* and its pesticidal crystal proteins. *Microbiol. Mol. Biol. Rev.* **62:**775–806.
137. **Schumann, G.** 1991. *Plant Diseases: Their Biology and Social Impact*. APS Press, St. Paul, Minn.
138. **Sears, M. K., R. L. Hellmich, D. E. Stanley-Horn, K. S. Oberhause, J. M. Pleasants, H. R. Mattila, B. D. Siegfried, and G. P. Dively.** 2001. Impact of Bt corn pollen on monarch butterfly populations: a risk assessment. *Proc. Natl. Acad. Sci. USA* **98:**11937–11942.
139. **Song, J., J. M. Bradeen, S. K. Naess, J. A. Raasch, S. M. Wielgus, G. T. Haberlach, J. Liu, H. Kuang, S. Austin-Phillips, C. R. Buell, J. P. Helgeson, and J. Jiang.** 2003. Gene *RB* cloned from *Solanum bulbocastanum* confers broad spectrum resistance to potato late blight. *Proc. Natl. Acad. Sci. USA* **1000:**9128–9133.
140. **Starr, M. P.** 1984. Landmarks in the development of phytobacteriology. *Annu. Rev. Phytopathol.* **22:**169–188.
141. **Stensvand, A., T. Amundsen, L. Semb, D. M. Gadoury, and R. C. Seem.** 1998. Discharge and dissemination of ascospores by *Venturia inaequalis*. *Plant Dis.* **82:**761–764.
142. **Sterling, M., C. Rogers, and E. Levetin.** 1999. An evaluation of two methods used for microscopic analysis of airborne fungal spore concentrations from the Burkard Spore Trap. *Aerobiologia* **15:**9–18.
143. **Stetzenbach, L. D., S. C. Herm, and R. J. Seidler.** 1992. Field sampling design and experimental methods for the detection of airborne microorganisms, p. 543–555. *In* M. A. Levin, R. J. Seidler, and M. Rogul (ed.), *Microbial Ecology: Principles, Methods, and Applications*. McGraw-Hill, New York, N.Y.
144. **Stockwell, V. O., K. B. Johnson, and J. E. Loper.** 2002. Biological control of fire blight: understanding interactions among introduced and indigenous microbial communities, p. 225–239. *In* S. E. Lindow, E. I. Hecht-Poinar, and V. J. Elliott (ed.), *Phyllosphere Microbiology*. APS Press, St. Paul, Minn.
145. **Sundin, G. W.** 2002. Ultraviolet radiation on leaves: its influence on microbial communities and their adaptations, p. 27–42. *In* S. E. Lindow, E. I. Hecht-Poinar, and V. J. Elliott (ed.), *Phyllosphere Microbiology*. APS Press, St. Paul, Minn.
146. **Tabashnik, B. E., Y.-B. Liu, N. Finson, L. Masson, and D. G. Heckel.** 1997. One gene in diamondback moth confers resistance to four *Bacillus thuringiensis* toxins. *Proc. Natl. Acad. Sci. USA* **94:**1640–1644.
147. **Troutt, C., and E. Levetin.** 2001. Correlation of spring spore concentrations and meteorological conditions in Tulsa, Oklahoma. *Int. J. Biometeorol.* **45:**64–74.
148. **Umesha, S., S. M. Dharmesh, S. A. Shetty, M. Krishnappa, and H. S. Shetty.** 1998. Biocontrol of downy mildew disease of pearl millet using *Pseudomonas fluorescens*. *Crop Prot.* **17:**387–392.
149. **van den Bosch, F., J. A. J. Metz, and J. C. Zadoks.** 1999. Pandemics of focal plant disease, a model. *Phytopathology* **89:**495–505.
150. **Vanky, K.** 1987. *Illustrated Genera of Smut Fungi*. Gustav Fischer Verlag, Stuttgart, Germany.
151. **Vloutoglou, T., and S. N. Kalogerakis.** 2000. Effects of inoculum concentration, wetness duration and plant age on development of early blight (*Alternaria solani*) and on shedding of leaves in tomato plants. *Plant Pathol.* **49:**339–345.
152. **Waggoner, P. E., and D. E. Aylor.** 2000. Epidemiology: a science of patterns. *Annu. Rev. Phytopathol.* **38:**71–94.
153. **Webster, J., R. A. Davey, and J. C. R. Turner.** 1989. Vapour as the source of water in Buller's drop. *Mycol. Res.* **93:**297–302.
154. **Wiglesworth, M. D., W. C. Nesmith, C. L. Schardl, S. Li, and M. R. Siegel.** 1994. Use of specific repetitive sequences in *Peronospora tabacina* for the early detection of the tobacco blue mold pathogen. *Phytopathology* **84:**425–430.
155. **Wilson, M., and S. E. Lindow.** 1993. Release of recombinant microorganisms. *Annu. Rev. Microbiol.* **47:**913–944.
156. **Xu, X.-M., and M. S. Ridout.** 1998. Effects of initial epidemic conditions, sporulation rate, and spore dispersal gradient on the spatio-temporal dynamics of plant disease epidemics. *Phytopathology* **88:**1000–1012.
157. **Yao, C., S. P. Arya, J. Davis, and C. E. Main.** 1997. A numerical model of the transport and diffusion of *Peronospora tabacina* spores in the evolving atmospheric boundary layer. *Atmosph. Environ.* **31:**1709–1714.
158. **Yorinori, J. T., W. M. Pavia, R. D. Fredrick, L. M. Costamilan, P. F. Bertagnolli, G. E. Hartmen, C. V. Godoy, and J. Nunes, Jr.** 2005. Epidemics of soybean rust (*Phakopsora pachyrhizi*) in Brazil and Paraguay from 2001 to 2003. *Plant Dis.* **89:**675–677.
159. **Zadoks, J. C., and F. van den Bosch.** 1994. On the spread of plant disease: a theory of foci. *Annu. Rev. Phytopathol.* **32:**503–521.
160. **Zangerl, A. R., D. McKenna, C. L. Wraight, M. Carroll, P. Ficarello, R. Warner, and M. R. Berenbaum.** 2001. Effects of exposure to event 176 *Bacillus thuringiensis* corn pollen on monarch and black swallowtail caterpillars under field conditions. *Proc. Natl. Acad. Sci. USA* **98:**11908–11912.
161. **Zhao, J.-Z., J. Cao, H. L. Collins, S. L. Bates, R. T. Rousch, E. D. Earle, and A. M. Shelton.** 2005. Concurrent use of transgenic plants expressing a single and two *Bacillus thuringiensis* genes speeds insect adaptation to pyramided plants. *Proc. Natl. Acad. Sci. USA* **102:**8426–8430.
162. **Zilinskas, R. A.** 1997. Iraq's biological weapons: the past as future? *JAMA* **278:**418–424.

BIOTRANSFORMATION AND BIODEGRADATION

VOLUME EDITOR
RONALD L. CRAWFORD

SECTION EDITORS
JONATHAN R. LLOYD AND MING TIEN

Overview: Biotransformation and Biodegradation

RONALD L. CRAWFORD

82

Biotransformation and biodegradation have been active areas of research since the very beginnings of microbiology and particularly since the beginning of the industrial revolution in the late 18th and early 19th centuries, when the first of many thousands of anthropogenic chemicals began to be introduced into the environment. Much of the historical work in these areas has involved the study of pure cultures of microorganisms isolated by enrichment culture for their abilities to degrade or transform a particular chemical. Many elegant methods employing biochemical, physiological, and/or genetic approaches have been developed over the decades to elucidate the intricate and evolutionarily beautiful pathways employed by such pure cultures to degrade or transform natural and anthropogenic chemicals. In addition, extraordinarily precise and sensitive methods to monitor the fate of chemicals introduced purposely or accidentally into the environment have been developed to assist in monitoring the fate of novel chemicals that enter natural ecosystems as a result of industrial activity. Other work has involved the study of biodegradative processes of mixed microbiological populations of soil and water or of "black-box" treatment systems engineered to treat industrial and municipal wastes. Studies of processes such as bacterial sulfur oxidation, decomposition of plant and animal residues, maintenance of soil fertility, and formation of humus were pioneered by scientists such as Selman Waksman in the early decades of the 19th century. These methods, even those dating back to the pioneers of biodegradation science, for the most part still work and provide reliable answers to hypothesis-driven research questions. Previous editions of the *Manual of Environmental Microbiology* have effectively summarized the most common experimental approaches used in these types of studies, methods which remain in use to this day in laboratories around the world. However, previous editions of the *Manual* did not cover in depth several subareas of the field, such as the considerable capabilities of fungi for biodegradation of natural and human-made chemicals or the fate of metals in biological systems. Also, the past decade has seen a revolution in how we study biodegradative processes at the level of genomes of whole microbial communities, a new research field known as "metagenomics." Thus, in section VIII of this third edition of the *Manual*, we have introduced mostly new material that covers these somewhat neglected areas. Included is a new subsection addressing molecular methods for study of biodegradation, edited by Ronald L. Crawford of the University of Idaho. This section provides chapters examining topics such as the use of gene arrays for microbial community analysis, preparation of high-quality metagenomic libraries from environmental DNA, the use of computational approaches to optimize biodegradation processes by microbial consortia, and the use of metagenomic tools for the study of bioremediation. Another new subsection on fungal biodegradation, edited by Ming Tien of Pennsylvania State University, examines newly developing areas including proteomic approaches for analysis of fungal wood decay, application of fungi for remediation of chemically contaminated soils, bioremediation through use of fungal solid-state fermentation, molecular approaches for eukaryote community analysis, and fungal production of industrially useful, catalytically active secondary metabolites. A third new subsection discussing metal-microbe interactions, edited by Jonathan R. Lloyd of the University of Manchester, includes chapters describing techniques for the study of magnetotactic bacteria, bacterial electrodes, iron biooxidation, acidophilic microorganisms, microbial metal resistance, synchrotron-based methods for the study of metal transformations, biological transformations of radionuclides, arsenic-respiring bacteria, manganese biotransformation, and microbial reduction of Fe(III). Thus, section VIII of the third edition should not be considered a replacement for the material in previous editions of the *Manual* but rather is intended to supplement those previous volumes with new, state-of-the-art material. We hope you find this new material to be a valuable resource for many years to come.

Functional Gene Arrays for Microbial Community Analysis

TERRY J. GENTRY, CHRISTOPHER W. SCHADT, ZHILI HE, AND JIZHONG ZHOU

83

Microorganisms play critical roles in several biogeochemical cycles that directly impact ecosystem function and sustainability; however, very little is known regarding the spatial heterogeneity, temporal dynamics, and genomic diversity of even the dominant populations that govern these processes in most environments. In order to fully understand the ecology of these systems and the potential impacts resulting from environmental perturbations such as climate change and pollution, it is important to know what microorganisms and/or genes are present at the site(s) and what conditions control their population dynamics and activity. While the development of non-culture-based methods, especially nucleic acid-based techniques, has greatly enhanced our ability to detect and characterize organisms in complex environments such as soils and marine sediments (19, 57, 58), a comprehensive approach capable of simultaneously detecting thousands of organisms will be necessary to adequately characterize these diverse communities. Microarrays have the unprecedented analytical potential to realize this goal by enabling the detection of most, if not all, of the microbial populations in a given sample.

The recent development of microarrays as powerful, high-throughput genomic technology has inspired researchers to apply these techniques to the study of various biological processes. Microarrays are conceptually analogous to Southern and Northern blots and are based upon the hybridization of single-stranded DNA or RNA to a DNA probe of complementary sequence that is typically attached to a solid matrix rather than a porous membrane. The major advantage of microarrays is that attachment to the solid surface confers the ability to print an enormous number of different probes (thousands to hundreds of thousands per square centimeter) on an individual array. Although microarrays were primarily developed for measuring microbial gene expression in pure cultures or simple systems, several researchers have also demonstrated that microarrays have potential as a specific, sensitive, and potentially quantitative parallel tool for characterizing the composition, structure, activities, and dynamics of microbial communities in natural environments (13, 16, 32, 40, 50, 52, 65). However, the application of microarray technology to environmental samples still presents several challenges that we discuss in the last section of this chapter. With continued development, microarray technology is becoming increasingly affordable and more widely applicable to different types of samples.

Zhou and Thompson (72) proposed that microarrays of potential use for environmental samples can be divided into three or more major groups based on the type of nucleic acid probes arrayed including (i) phylogenetic oligonucleotide arrays (POAs), (ii) community genome arrays (CGAs), and (iii) functional gene arrays (FGAs). POAs contain primarily probes derived from 16S rRNA genes, although other genes may be used depending on the experimental objectives, and are primarily used to determine the phylogenetic makeup of the microbial community in a sample (16, 33, 50, 63). CGAs are constructed using either fragmented or whole-genomic DNA from either pure cultures or insert (bacterial artificial chromosome/fosmid) libraries of environmental DNA. These arrays can be used to determine the genomic similarity of cultivated isolates to other isolates on the CGA or to screen metagenomic libraries for specific target genes or organisms (7, 29, 48, 66). FGAs contain probes for genes that encode key proteins in various processes. Different FGAs have been developed and applied to C, N, and S cycling genes (3, 52, 55, 56, 65) as well as genes involved in bioremediation (12, 13, 40). These arrays are useful not only for detecting specific organisms and genes in the environment but also for determining their physiological status and functional activities by measuring DNA and mRNA levels of genes that are determinative for key microbial processes (40, 52, 55, 65). Due to the analytical versatility of FGAs, this chapter will focus on their development, application, and use in characterizing functional genes involved in biogeochemical processes and bioremediation. Several recent reviews on environmental applications of microarrays have been published, which may provide additional information of interest (2, 10, 45, 46, 69, 72).

PROBE DESIGN

Selection of Target Genes and Acquisition of Sequence Information

The first step, and one of the most critical, in the design of FGAs is to determine which genes should be used to monitor the microbial processes of interest. An important factor to consider when selecting functional genes is the great variation in the amount of sequence data available in public databases such as GenBank. For example, there are hundreds of sequences available for mercury resistance genes, but only a handful of lignin peroxidase genes are published (45). Even within pathways and individual enzymes (different subunits), there can be substantial variation in published sequence data. Since relatively few environmental bacteria have been isolated and characterized (in comparison to microbial diversity in the environment), the best sequences to use for probe generation are those for which environmental PCR clone libraries have been constructed using conserved primers. If adequate information for the gene of interest is not available, it may be beneficial to construct a clone library from the environment in question prior to FGA analysis in order to construct more comprehensive probes for the FGA. The ideal gene(s) for inclusion on an FGA (i) encodes an enzyme or protein contributing or essential to the process of interest, (ii) has an evolutionarily conserved sequence but with enough divergence in different organisms to enable the design of species-specific probes, and (iii) has extensive amounts of sequence data from both isolates and environmental samples available in public databases.

Our group has recently expanded the largest FGA published so far, from 1,662 probes (40) to about 24,000 probes for various genes involved in the C, N, and S cycles, in organic contaminant degradation, and in metal resistance and reduction. The probe design process that was used is briefly outlined below for illustration. Combinations of different search strings (based on key words such as the enzyme or gene name) within the GenBank ENTREZ interface were used to identify the desired sequences and download the corresponding list of GI (gene identification) numbers for each gene. A locally executable program was then used to extract the sequences and other relevant information (species name, gene description, phylogenetic lineage, etc.) and compile them into a tab-delimited database. The searches were performed within the protein sequence database of GenBank to avoid many of the duplicate entries in the nucleotide database. This also enabled direct recovery of DNA sequences. The database was then screened manually to remove nonhomologous and duplicate sequences.

PCR-Amplicon versus Oligonucleotide Probes

Another important factor for FGAs, and microarray analysis in general, is what type and size of probes to use. Either oligonucleotides (oligos) or DNA fragments, such as PCR products or cDNA clones, derived from functional genes can be used as FGA probes. The DNA fragment-based probes are typically PCR amplified from environmental clones or pure culture genomic DNAs (8, 13, 65). PCR-amplicon-based FGAs have specific advantages over oligo FGAs in that they do not require a priori sequence knowledge and they tend to be more sensitive than shorter oligo probes. He et al. (23) recently compared the sensitivity of PCR-amplicon and oligo probes. The amplicon probes had a detection limit of 5 ng of genomic DNA compared to detection limits of 25, 100, and 100 ng of genomic DNA for 70-, 60-, and 50-mer oligo probes, respectively. This equaled detection sensitivities of approximately 1.9×10^6, 9.2×10^6, 3.7×10^7, and 3.7×10^7 gene copies for the PCR-amplicon, and the 70-mer, 60-mer, and 50-mer probes, respectively. The signal intensities for 30- and 40-mer probes were much lower. Although longer probes may increase sensitivity, they have a higher possibility of cross-hybridization and can make the design of specific probes more difficult for sequences from environmental samples. Furthermore, the use of PCR-amplicon probes requires the physical possession of all of the corresponding template organisms in order to generate the probes. This can be virtually impossible for more comprehensive arrays which contain probes for thousands to tens of thousands of different source organisms or clones.

In contrast, synthetic oligo probes can be designed using only sequence information and also have other advantages. First, since only sequence information is required, probes can be directly synthesized, and the logistical problems associated with PCR amplification can be avoided. Second, more flexibility to control specificity of hybridization can be achieved during probe design. This includes the ability to circumvent highly conserved regions of genes, which would not be possible with PCR amplification using conserved primers. Researchers have used oligo probes of various lengths (10- to 100-mers), but 50- to 70-mers provide an optimal balance between sensitivity and specificity and are more commonly used for FGAs (3, 12, 23, 40, 56). Due to the advantages mentioned above, oligo-based FGAs have been more widely used than PCR-amplicon-based FGAs. Therefore, the methods sections later in this chapter describe primarily oligo-based FGAs, but many are applicable to PCR-amplicon FGAs as well.

Probe Design

The design of oligo probes presents two major challenges. One is how to establish the design criteria in order to ensure probe specificity while maintaining adequate gene coverage, and the other is how to design and identify the optimum probes for each gene or each group of genes based on these criteria. These problems are further magnified in FGAs designed from environmental sequence data due to the highly conserved nature of many functional genes, especially those derived from clone libraries.

Oligonucleotide Probe Design Criteria

It is critical to design FGA probes to be highly specific for the target genes of interest in order to prevent, or at least reduce the potential for, cross-hybridization with similar environmental sequences that could lead to inaccurate results (40, 65). Many parameters including sequence similarity, long continuous stretches of identical nucleotide composition, binding energy, the formation of secondary structures, GC content, melting temperature (T_m), and the positions of mismatches affect probe specificity and sensitivity. Various values for these different parameters have been published in the literature. Kane et al. (28) suggested that an oligo probe may cross-hybridize with nontarget sequences that are >75 to 85% similar. Their studies also showed that a probe, which had stretches of 15 to 20 identical bases (out of 50 bases) had detectable cross-hybridization. Other studies detected no significant cross-hybridization with nontarget sequences that were 85 to 88% similar to 50-mer probes (40, 56) and 80% similar to 60-mer probes (25). Another study suggested that significant cross-hybridization could potentially occur if a 70-mer oligo probe and its nontargets had a binding free energy <-35 kcal/mol (5). In a recent study, He et al. (24) examined the relationships between hybridization signal and sequence identity, continuous stretch, binding free energy, or mismatch position and experimentally established a set of criteria for the design of 50- and 70-mer oligo probes. The experimental data demonstrated that while it was difficult to exclude all experimentally verified, nonspecific oligo probes by using a single criterion, a combination of multiple criteria could exclude all nonspecific probes (24). Additionally, by simultaneously considering multiple criteria, more-liberal cutoffs could be used for each single criterion. The results from another recent research project in our laboratory indicate that specific 50-mer probes could be designed using the criteria of a free energy release of ≤−35 kcal/mol and ≤90% similarity and ≤20 bp continuous stretches to nontarget sequences (30b). Probes forming strong secondary structures can hinder target access, which may result in a lower sensitivity and false-negative results. This can be determined by predicting self-binding regions of probes. In addition, GC content, T_m, and sequence complexity need to be considered. Since GC content varies among different organisms, oligo design software should evaluate all sequences in the data set and determine T_m values that fall into a defined range in order to achieve uniform hybridization. Selected parameters should be used as filters for selecting probe candidates. The above values can be used as guidelines for designing specific oligo probes. However, the actual specificity will depend on which hybridization conditions (e.g., temperature, percent formamide, and salt concentration) are used and need to be determined experimentally. The specificity of newly designed probes should be verified by test hybridizations with genomic DNA from pure cultures, PCR-amplified genes, or synthetic oligos. We test the specificity of our 50-mer probes under highly stringent hybridization conditions of 50°C with hybridization solutions containing 50% formamide.

One potential limitation to the use of highly specific probes is that they may not detect less similar, unknown sequences in the environment since the probes are typically designed from the more variable regions of known sequences. This may be a concern for any experiment whose objective is to measure the presence and/or dynamics of genes contributing to a given process. For example, FGA analysis of an aquifer undergoing denitrification may not produce hybridization to any of the specific denitrification genes on a FGA. However, this could be solely because the denitrifying genes present at the site, although homologues of the genes on the array, were too divergent to hybridize to the probes. One possible approach to resolve this issue is the use of conserved probes—the equivalent of PCR primers based on conserved regions of a gene. These probes lack the discriminating power of highly specific probes but would provide an additional layer of information, similar to that of hierarchical 16S rRNA probes in POAs (33), in case the specific probes missed some unique genes. However, using conserved probes may also increase the potential for cross-hybridization with nontarget sequences.

Oligonucleotide Probe Design Software

Several different software programs that can design oligo probes for microarrays are currently available (Table 1). Many of these programs work well for designing probes for whole genome arrays. However, recent research in our laboratory by Li et al. (30a) showed that many of the probes designed by some of these software programs from groups of related functional gene sequences were not specific to their target sequences (based on experimentally determined values [40]). With the growing database of environmental sequences, especially the highly similar sequences which are often obtained from clone libraries, it can be difficult to design probes to avoid hybridization to closely related sequences.

A new software tool, called CommOligo, has been developed in our laboratory to address these issues (30a). This program can select optimal oligo probes for whole-genomes, metagenomes, or a group of sequences. The program first attempts to design unique probes for each sequence in the data set. In order to reduce the potential for cross-hybridization, CommOligo simultaneously compares potential probes to both target and nontarget sequences by using three probe design criteria: (i) the nucleotide sequence identity, (ii) the continuous stretch of identical nucleotide composition, and (iii) the binding free energy. By considering these factors simultaneously, each criterion can be slightly relaxed while still enabling the design of specific probes. Second, if it is not possible to design gene-specific probes from highly similar sequences, the program will design probes specific to the groups of closely related sequences. A new algorithm selects probes which have maximal similarity to the sequences within a targeted group and minimal similarity to those outside the

TABLE 1 Freely available programs for oligonucleotide probe design

Program	Operating system(s)	Website URL	Reference
ArrayOligoSelector	Linux	http://sourceforge.net/projects/arrayoligosel/	5
GoArrays	Windows, Linux/Unix	http://www.isima.fr/bioinfo/goarrays/	41
OligoArray	Windows, Linux/Unix	http://berry.engin.umich.edu/oligoarray	42
OligoArray 2.0	Linux/Unix	http://berry.engin.umich.edu/oligoarray2	43
OligoPicker	Linux	http://pga.mgh.harvard.edu/oligopicker/index.html	62
OligoWiz	Windows, Unix	http://cbs.dtu.dk/services/OligoWiz/	36
Osprey	Windows, Unix	http://osprey.ucalgary.ca/	21
PICKY	Windows, Mac, Linux	http://www.complex.iastate.edu/	9
PRIMEGENS	Unix/Linux	http://compbio.ornl.gov/structure/primegens/	67
PROBEmer	Windows, Linux	http://probemer.cs.loyola.edu	17
ProbeSelect	Unix/Linux		31
ROSO	Windows, Unix	http://pbil.univ-lyon1.fr/roso	38
YODA	Windows, Mac, Linux	http://pathport.vbi.vt.edu/YODA/	37

group. Third, the optimal range of T_m values is selected based on probe candidates that have passed all other criteria, rather than on all possible oligo probes. This ensures that the maximum number of sequences have probes designed. Like other design software, filters such as self-binding, GC content, the maximal number of the same nucleotides, and the positional distance between multiple probes from the same sequence are also considered. Users can either use default values for all parameters or change them to the desired settings. In addition, as an option, a multiple sequence alignment approach can be used to preprocess sequence data, and the users can then select desired regions for use (or avoidance) in probe design.

Li et al. (30a) compared CommOligo with several of the other design programs by using two data sets, a whole-genome sequence (*Methanococcus maripaludis*) and 842 highly similar functional gene sequences (*nirS/K*, a mixed group of *nirS* and *nirK* sequences). Most of the programs worked relatively well for the whole-genome data, but 3 to 20% of the probes designed by some of the tested programs were nonspecific, based on experimentally determined criteria (24, 40). In comparison, all of the probes that CommOligo successfully designed for 99% of the open reading frames in the genome data were specific. Furthermore, other than CommOligo, most programs performed poorly with the *nirS/K* group of sequences. This was primarily due to too few specific probes and/or too many nonspecific probes being designed. A large majority of sequences did not have probes designed for them if only unique, gene-specific probes were considered. CommOligo designed the most gene-specific probes (143 compared to 35 to 141 for the other programs) and the least nonspecific probes (0 compared to 7 to 756 for the other programs). In addition, CommOligo designed 33 group-specific probes that covered an additional 90 genes.

There are still several challenges for the design and use of oligo FGAs for environmental studies. In order to achieve the optimal specificity, probe design criteria need to be further investigated, and better algorithms need to be developed to enable greater efficiency for the intensive computations necessary for the design process. Our experience also suggests that researchers should be cautious when using software to design FGA probes that was originally developed for use with whole-genome expression data, because the more stringent design requirements for FGA probe specificity can be problematic.

ARRAY CONSTRUCTION AND HYBRIDIZATION

Array Design and Fabrication

Due to the potential for variation in printing and hybridization across a slide, it is desirable to include replicate probes on an array. We typically include at least duplicate spots for each probe and put these in different locations of the slide. Addition of positive control probes, such as ones based on universal 16S rRNA sequences, in each subgrid can greatly aid in the speed and accuracy of placing the grid, since it is not unusual for FGA slides to produce relatively few positive signals.

The following protocol is for contact printing of FGAs. In this process, solid or split pins are used to transfer oligos to the slide surface. This procedure can also be used to print cDNA, genomic DNA, or other molecules on a slide. Contact printing can produce a high-density array with thousands of probes per square centimeter. Other techniques are also available, including photolithography and noncontact ink jet printing. Photolithography can produce very dense arrays (hundreds of thousands of probes per square centimeter) but requires more-specialized, expensive equipment and can print only oligos. Noncontact ink jets, like contact printing, can use virtually any biological molecule to produce the array but cannot produce as dense an array as photolithography or contact printing. The printing protocol below is also for planar glass slides. A number of different slide chemistries and platforms are available, but planar glass slides are commonly used for FGAs because they enable higher printing densities than three-dimensional arrays (73). To print the FGA, the following steps are required.

1. Transfer equal volumes of oligos (100 to 200 pmol μl^{-1}) and dimethyl sulfoxide to a 384-well, v-bottom plate. Different printing solutions can be used, but inclusion of 50% dimethyl sulfoxide reduces evaporation during printing and improves spot homogeneity and hybridization intensity (65).
2. Cover the plate, mix on an orbital shaker at 700 rpm for 3 min, and then centrifuge the printing plate at 500 rpm for 5 min to collect liquid at bottom of the wells.
3. Set up the array printer and software. The slides should be printed according to the manufacturer's protocol. We are currently using UltraGAPS slides (Corning, Corning, N.Y.). Although these slides are manufactured in a clean environment, be sure to check for and remove any dust or debris from the slides with a can of compressed air prior to printing. The ideal relative humidity should be between 40 and 60% at room temperature (20 to 25°C) for optimal control of evaporation without compromising spot quality and uniformity. The choice of pins will depend on the desired spot size and the number of plates to be printed. We are currently using split pins from TeleChem (Sunnyvale, Calif.) that produce a spot size of ~150 μm.
4. After printing, allow the slides to dry for at least 2 h before proceeding to UV cross-linking and postprocessing (according to the slide manufacturer's protocol).

Nucleic Acid Preparation and Labeling

Genomic DNA or mRNA can be extracted using a protocol such as those described by Zhou et al. (71) and Hurt et al. (26) or one of several commercially available methods. Since amplification of target DNA may be needed to achieve good hybridization (see below), the methods which give high-molecular-weight DNAs (26, 71) are preferred in order to minimize amplification biases. It is necessary to further purify the nucleic acids by using gel purification or commercial kits before labeling. Target nucleic acids can either be PCR amplified to enrich for certain genes or be directly used for FGA hybridization. If genes for which specific PCR primers are not available are being investigated or if hundreds to thousands of genes are being simultaneously considered, it may be necessary to use a random amplification approach to amplify the community DNA or RNA. A whole-community genome amplification procedure, based on rolling circle amplification, has recently been developed and optimized for microarray analysis (64). This method is detailed below and can representatively amplify subnanogram quantities of microbial community DNAs to microgram quantities with a strong linear relationship between starting template and final concentrations.

1. Combine 10 to 100 ng of purified community DNA with 10 μl of TempliPhi sample buffer (TempliPhi kit, Amersham no. 25-6400-50), and adjust the total volume to 12 μl. Lower amounts of DNA can be used, but this may

affect the representativeness of amplification. It is important to include a negative control and set up all reactions in a sterile hood due to the extreme sensitivity of the method.

2. Mix the above samples and incubate 10 min at room temperature.

3. In a separate tube, prepare the master mix by combining (per sample) 10 μl of TempliPhi reaction buffer (TempliPhi kit), 1 μl of TempliPhi enzyme mixture (TempliPhi kit), 1 μl of single-stranded binding protein stock (2.67 μg/μl) (e.g., USB no. 70032Z), and 1 μl of spermidine stock (1 mM) (e.g., USB no. 21760).

4. Add 13 μl of master mix to each sample from step 2.

5. Incubate at 30°C for 2 h (or longer).

6. Stop the reaction by incubating the mixture at 65°C for 20 min.

Other researchers are developing methods for amplification of prokaryotic mRNA including one approach that adds a poly(A) tail to the mRNA for subsequent amplification (4). However, if the desired organisms are expected to occur in sufficient quantity, it is preferable to directly use the DNA or mRNA in order to avoid amplification biases (39).

Nucleic acids can be either directly or indirectly labeled. For direct labeling, a fluorescent molecule (e.g., Cy5-dUTP) is incorporated into the DNA (or cDNA) through enzymatic extension of primers prior to hybridization (47). Either gene-specific or random primers can be used, but it may be preferable to use random primers for larger arrays or where the sequence information is variable. Direct labeling can also be used with mRNA (cDNA), but some researchers have reported that indirect labeling, where the fluorescent label is chemically attached following the enzymatic reaction, generates less background and is more sensitive (13). Other labeling methods (12) may also be used to increase sensitivity. The following protocol details the direct labeling method currently used in our lab.

1. Combine 2 to 5 μg of purified (or amplified) community DNA with 20 μl of (750 ng/μl) random octamer primers (e.g., Invitrogen no. Y01393; random hexamer primers can also be used), 0.3 μl of spermidine (10 mM), and nuclease-free water to a total volume of 40 μl. Different amounts of template DNA can be used, but more DNA generally increases the probability of detecting less dominant organisms.

2. Mix well and denature at 99.9°C for 5 min and then immediately place on ice.

3. In a separate tube, prepare the labeling master mix by combining the following (per sample): 2.5 μl of dNTPs (5 mM dATP, dCTP, and dGTP and 2.5 mM dTTP), 1 μl of (1 mM) Cy3 or Cy5 dUTP (e.g., Amersham no. PA55022), 2.0 μl of (40-U/μl) Klenow fragment (e.g., Invitrogen no. Y01396), 1.25 μl of dithiothreitol (0.1 M) (e.g., Invitrogen no. Y00147), 0.7 μl of (490 ng/μl) RecA, and 12.55 μl of nuclease-free water.

4. Add the above mixture to the denatured DNA and primers from step 2 and then mix, centrifuge, and incubate at 37°C for 6 h or overnight.

5. Purify the labeled target DNA by using QIAquick PCR Purification columns (Qiagen, Valencia, Calif.) or a similar kit, according to the manufacturer's instructions.

6. Use a spectrophotometer to quantify the labeling efficiency by measuring the absorbance at 550 nm for Cy3 or 650 nm for Cy5. The purity of the labeled DNA should also be assessed by measuring the absorbance at 230, 260, and 280 nm. A NanoDrop ND-1000 spectrophotometer (NanoDrop Technologies, Inc., Montchanin, Del.) or equivalent can perform this measurement with only 1 μl of the labeled DNA. The specific activity of the labeled DNA equals the amount of target DNA × 1000 divided by (each) the picomole of incorporated dye and 324.5.

7. Dry labeled DNA with a vacuum centrifuge (45°C for 1 h). Use of high temperature levels and heat lamps to expedite water evaporation could degrade fluorescent dyes and should be avoided.

Hybridization

Along with the size and sequence of the FGA probes, the temperature and composition of the hybridization solution control the stringency of hybridization. It is preferable to use highly specific hybridization conditions for FGAs because of the potential for many similar sequences to be present. Furthermore, using an array under more or less stringent conditions than it was designed for can lead to inaccurate conclusions based on overestimated or underestimated results. It may be beneficial to include control DNA in the hybridization solution that has varying similarity to corresponding control probes on the array in order to ensure that the correct hybridization stringency is achieved.

For our 50-mer FGA probes, we perform hybridizations at 50°C using a hybridization solution containing 50% formamide (40). These conditions will vary depending on the characteristics of the probes used. However, it is critical for the probes to be as specific as possible and for this to be determined experimentally. Otherwise, the data can be used only to determine qualitative differences in the samples and cannot be used to detect any specific genes. It is also important to use as small a volume of hybridization solution as possible for the FGA, given the low detection sensitivity. Uneven mixing and variations in solution across the slide can result in localized regions of high signal strength and/or spot intensity that can confound analysis. The use of coverslips requiring the minimal amount of hybridization solution (e.g., LifterSlip coverslips; Erie Scientific, Portsmouth, N.H.) or a hybridization station that continually mixes the solution (e.g., Maui Hybridization System; BioMicro Systems, Inc., Salt Lake City, Utah) can reduce this variation. The following hybridization protocol is currently used with our 50-mer FGA.

1. Prehybridize the slides according to the manufacturer's directions.

2. Place microarray slides into hybridization chambers (e.g., Corning no. 2551) and preheat for 20 min at 50°C.

3. Prepare the hybridization master mix by combining (per sample) 20 μl of formamide, 10 μl of 20× SSC (1× SSC is 0.15 M NaCl plus 0.015 M sodium citrate), 0.4 μl of (10%) sodium dodecyl sulfate (SDS), 0.4 μl of (10-μg/μl) salmon sperm DNA, 0.3 μl of (1 mM) spermidine, and 7.3 μl of nuclease-free water.

4. Resuspend the labeled DNA with 38.4 μl of the hybridization solution above, centrifuge briefly, and heat at 95°C for at least 5 min. The solution should then be kept at 60°C until it is added to the array.

5. Add 15 μl of 3× SSC solution into the chamber hydration wells.

6. Add 1.6 μl (490 ng/μl) of prewarmed (60°C) RecA to hybridization solution from step 4 and pipette onto each FGA slide. The coverslip should be carefully placed over the array to avoid bubble formation. Alternatively, the coverslip can be placed onto the slide and the sample can be slowly pipetted at the edge of the coverslip, allowing capillary action to wick the sample onto the arrayed portion of the slide. The total volume of the hybridization is 40 μl, but

this can be adjusted as necessary according to the sizes of the array and coverslips.

7. Close the hybridization chamber and confirm that it is properly sealed.

8. Incubate the chamber in a 50°C water bath for 12 to 15 h (overnight).

9. Following hybridization, remove the slides from the chambers and insert the slides, with the coverslips still attached, into a prewarmed (50°C) washing buffer I (1× SSC and 0.1% SDS) to remove the coverslips.

10. Place the slides in prewarmed (50°C) washing buffer I (1× SSC and 0.1% SDS) and wash for 5 min. Repeat this wash once.

11. Place the slides into buffer II (0.1× SSC and 0.1% SDS) at room temperature for 10 min. Repeat this wash once.

12. Place the slides in buffer III (0.1× SSC) at ambient temperature for 1 min with gentle shaking. Repeat this wash four times.

13. Transfer the slides to a rack (the original slide box with a Kimwipe at the bottom works well) and immediately centrifuge the slides dry at 500 × *g* for 5 min. It is necessary to centrifuge the slides as quickly as possible in order to reduce residual salt deposition. The slides should be stored in the dark until scanned.

MICROARRAY IMAGE PROCESSING AND DATA ANALYSIS

Data Acquisition and Preprocessing

The fluorescent signals on the arrays are measured by a laser scanner using a photomultiplier tube (PMT) at the wavelength characteristic of the specific fluorophore (e.g., 570 nm for Cy3 and 670 nm for Cy5). The percentages of laser power and PMT gain should be adjusted during the initial scan to provide the maximum signal intensity that does not saturate most of the spots. Unless the slides are being scanned automatically as a batch, it may be beneficial to first perform a quick scan of the slide at a lower resolution (~50 μm) to determine the quality of the slide before doing a higher-resolution scan (5 to 10 μm), which takes longer. Environmental samples typically produce relatively weak signals when hybridized to FGAs, so it may be necessary to use 85 to 100% laser power and PMT to obtain adequate results. Repeated scans at different settings can also be performed to better facilitate quantitation. However, the fluorophores can photobleach, so it is necessary to avoid overexposure of the slides.

Following scanning, the image should be imported into a software program for preprocessing. Several suitable software programs are available, including ArrayStat (Imaging Research, Inc., Ontario, Canada), Cluster 3.0 (Human Genome Center, University of Tokyo, Japan), GenePix Pro (Axon, Foster City, Calif.), GeneSpring (Silicon Genetics, Redwood City, Calif.), and ImaGene (BioDiscovery, Los Angeles, Calif.).

All spots should be carefully examined prior to data analysis in order to distinguish real signals from false signals. One of the initial steps is to identify poor quality spots. While specific criteria may vary, poor quality spots are typically determined based on spot size and shape, homogeneity, and intensity. Abnormally large or small spots and those having less than a given threshold (e.g., 55%) of pixels with intensities above the background signal should be discarded. One of the most common ways to distinguish real spots from noise based on signal intensities uses the signal-to-noise ratio (SNR). The SNR can be calculated as follows:

$$\text{SNR} = \frac{\text{signal mean} - \text{background mean}}{\text{background standard deviation}}$$

Spots producing an SNR above a certain threshold (usually SNR ≥2 or 3) are considered positive hybridizations and true signals. However, different SNR calculation methods and various cutoff values have been used, so it is necessary to report the methods utilized in a given experiment. Although many software programs can automatically flag poor spots, it may be preferable to visually scan each positive spot (if possible) to detect any false signals missed by the software. It is also necessary to remove outlying spots that represent extreme values compared to the other replicate spots. Outliers can occur due to factors which may not be detected by the software, including cross-hybridization or the presence of dust on the array. There is no general definition of an outlier with respect to microarray data. Different determination methods have been used, but these have primarily focused on gene expression data (70). The definition that we are currently using for an outlier is:

$$\frac{|\text{replicate signal intensity} - \text{mean signal intensity}|}{\text{signal intensity standard deviation}} \geq 3$$

After the outliers have been removed, the next step is to obtain the corrected hybridization signal by subtracting the background signal intensity from each of the remaining spots. This is usually done by subtracting the mean value of the local background intensity from the intensity of each spot. In some cases, it may be preferable to use the median background values for this purpose, since this approach would exclude extreme values which could skew the data (1). Fluorescence intensity of empty spots or negative control spots (probes targeting genes unlikely to be in the analyzed sample, e.g., human genes in soil or groundwater samples) can also be used as background values for signal correction. However, in most cases it is preferable to use local background data for this purpose since uneven background staining sometimes makes the use of global background values impossible.

Once the poor quality spots, outliers, and background are removed, the data are ready for normalization. Data normalization is necessary to correct systematic variations in dye incorporation and hybridization signals within and between slides. At least three different reference points can be used for normalization, depending on the experimental design: (i) all genes on the array, (ii) constantly expressed or present genes, and (iii) spiked control genes. Using all of the probes on the array, or global normalization, is predicated on there being only a few differences between the genes present or expressed in different samples, which is unlikely to be the case for most environmental samples. For the constantly expressed or present strategy, housekeeping genes or universally present genes (e.g., 16S rRNA) could potentially be used. However, it can be difficult to design truly "universal" probes, and this could lead to normalization errors when comparing samples containing populations with varying similarity to the so-called universal probes. Probably the best references to use for normalization are spiked control genes. One approach for this is to spike control DNA (e.g., human DNA in an environmental sample) that is labeled with the same dye as the actual sample into the hybridization mixture. The signal intensities can then be normalized based on the hybridization signal of the control DNA with corresponding control oligos on the slides. An alternative, and perhaps superior, approach is based on the two-color dye swap techniques that are commonly used in gene expression experiments. In this approach, known amounts of oligos or DNA fragments are labeled with a second dye and spiked into the hybridization solution. The hybridization results are then normalized based on the hybridization of the control DNAs with corresponding control probes that are included in each spot on the array along with the actual target probes (3, 8, 14). This normalization approach is promising because it can address variances at the individual spot level, but additional research is needed to optimize this method for wide-scale use with FGAs.

Due to the extreme variability in gene abundance and activity in many samples, FGA results from these samples may not be normally distributed. In these cases, it may be preferable to log transform ($\log_{10}$ or $\log_2$) the FGA data prior to data analysis to allow the use of parametric statistical methods. Once normalization and transformation (if necessary) are complete, the data are ready for statistical analysis.

Data Analysis

A major challenge for FGAs (and microarrays in general) is how to analyze the large amounts of data that are generated. Several ecological indices (e.g., Shannon and McIntosh) (34) can be used to determine differences and broad trends in the diversity (species richness and evenness) of microbial populations in different samples. FGA data for individual genes can be evaluated using standard statistical methods (e.g., *t* test, analysis of variance); however, it can be difficult to comprehensively determine all of the potential interactions in large FGA data sets. One approach which is commonly employed with these data sets is to first use methods such as principal components analysis (PCA), cluster analysis, or self-organizing maps (SOM) to elucidate the key genes, trends, and relationships (68) and then perform detailed statistical analyses on a select number of key genes. Other related methods, such as discriminant functional analysis, can also be used for the initial data analysis and may be preferable depending on the research objectives and characteristics of the data set (35). PCA is a multivariate technique that can simplify the data set by reducing the number of variables needed for data interpretation. PCA identifies the principal components accounting for the greatest variance in the dataset. These principal components can be either one of the original variables or a linear combination of a portion of the original variables. Usually the top two to four principal components are sufficient to represent the majority of variance in the dataset. The individual variables contributing to each principal component can also be determined, thus making it possible to identify key genes for subsequent analysis.

Clustering analysis is another category of methods that can be used to analyze the similarity of different samples. Clustering is commonly used for pure culture expression studies (15), but it is also potentially useful for analysis of FGA data by identifying the relative similarities and differences of various samples and any changes that occur following imposition of experimental treatments. Data can be clustered using hierarchically or non-hierarchically based methods. Hierarchical clustering initially considers each gene as an individual cluster and then combines the two most similar clusters to form another cluster. This process is repeated until all genes have been grouped into a single cluster. Several clustering programs are freely available, including Cluster and TreeView (http://rana.lbl.gov/EisenSoftware.htm). A major benefit of hierarchical clustering is that it generates a dendrogram which can be viewed as a phylogenetic tree where the branch lengths indicate the relative similarity of the different genes; however, the tree can be incredibly complex for large data sets and thus difficult to interpret.

One of the most common nonhierarchical clustering methods is *K*-means clustering. This is an iterative method where the data set is divided into *K* clusters such that the distance between each data point and the cluster center is minimized. *K*-means clustering is incorporated into some microarray analysis software packages (e.g., GeneSpring) and is particularly useful with large data sets because clustering can be achieved without having to create the full similarity matrix or excessively scan the data (75). The primary disadvantage of this method is that the initial number of clusters (*K*) must be assigned by the user prior to running the analysis. Since the correct *K* is not likely to be known prior to analysis, it may be necessary to iteratively try several different *K* values.

SOMs can be considered to be another type of cluster analysis where the dimensionality of data is reduced through the use of neural networks. They are conceptually similar to *K*-means clustering in that the method clusters data points around a defined number of preselected representative points. The data are iteratively relocated to group similar genes as clusters (54). The primary advantage of SOMs is that they are more robust to noise than other clustering methods; however, like the *K*-means clustering, SOMs require preselected choices regarding the representative data points and geometry to use for the clustering. While each of the clustering methods above may have applicability to FGA data, the various methods can produce different results, so researchers should be cautious when forming conclusions based solely on one type of cluster analysis (20).

USE OF MICROARRAYS FOR ENVIRONMENTAL STUDIES

Researchers have successfully used FGAs to examine microbial involvement in several environmental processes including nitrogen fixation, nitrification, denitrification, and sulfate reduction in freshwater and marine systems (27, 51, 55, 56, 65); degradation of organic contaminants including polychlorinated biphenyls (12) and polycyclic aromatic hydrocarbons (40) in soils and sediments; and methane-oxidizing capacity and diversity in landfill-simulating soil (3, 52). However, analysis of many environmental samples still presents several challenges to FGA analyses with respect to (i) interference from contaminants, (ii) specificity, (iii) sensitivity, (iv) quantitation, and (v) measurement of microbial activity. We will briefly discuss each of these topics along with potential analytical solutions.

Interference from Contaminants

Natural soil materials such as humic acids are notorious for their ability to interfere with molecular applications such as PCR and nucleic acid labeling and hybridization. Advances in extraction procedures have reduced the time necessary to perform these extractions and have improved the quality of the recovered DNA and RNA. However, samples from contaminated environments may contain a whole host of pollutants or have extreme physical and chemical conditions that make it virtually impossible to extract intact, usable nucleic acids with current methods. Common extraction methods will likely have to be altered or optimized for these samples. Furthermore, each additional purification step will ultimately decrease the amount of nucleic acid recovered, so it is preferable to use only the minimum steps required to obtain adequately pure DNA or RNA.

Specificity

It is sometimes difficult to design and validate microarray probes that are specific to a particular target sequence since many functional genes contain highly conserved regions and most environments contain large amounts of unknown microbial sequences. As previously discussed, the use of oligo probes enables conserved regions of genes or areas containing stable secondary structures to be avoided during probe design. Shorter oligo probes (~20-mers) can differentiate a single mismatch in a probe-target hybridization (73), but they have much lower sensitivity, thus likely necessitating PCR amplification of the target prior to hybridization. Longer oligo probes (~40- to 70-mers) commonly used for FGAs are less specific; however, species level specificity may still be achievable since protein-coding functional genes are typically more variable than rRNA phylogenetic markers. The actual specificity of longer oligo FGAs will depend on the probe design and the hybridization conditions used, but it is possible to differentiate sequences up to about ~90% similar (40, 55). As previously mentioned, it may be possible to design specific probes with more relaxed design criteria if multiple probe-target characteristics are simultaneously considered (30b). Specificity can also be increased or decreased, to some extent, by adjusting the stringency of the hybridization conditions (temperature, formamide concentration, salt concentration, etc.) depending on the research objectives and probe design (23, 30, 65).

Sensitivity

The sensitivity of FGAs is affected by numerous factors including probe size, slide chemistry, and labeling methods. Although oligo probes have many advantages for probe design, they are typically ~10-fold less sensitive than longer PCR amplicon probes (12, 23, 40, 65). Examples from the literature indicate sensitivities of 1 ng for a PCR amplicon FGA (65) and 5 to 8 ng for 50-mer oligo FGAs (40, 56) in the absence of background DNA. In the presence of background DNA, these sensitivities were about 10-fold lower (40, 56). A recent study showed that a total of 50 ng of DNA (~10^7 cells) was required for detection by a 50-mer FGA (40). This corresponded to a target population comprising ~5% of the total community in the tested sample.

Different array platforms can be used to increase sensitivity, especially in cases where a very high-density array is not needed. Glass-based microarrays, which are commonly used for FGAs because they enable high-density printing, are several orders of magnitude less sensitive than membrane-based hybridizations due to the limited amount of probe material which can be attached to the nonporous surfaces (8). Researchers are developing new slide chemistries, including ultrathin three-dimensional platforms, which have increased binding capacities but retain the high-throughput characteristics that make microarray analyses advantageous (22, 60, 73).

Alternative nucleic acid labeling methods may also be used to increase sensitivity (12, 51, 74). Denef et al. (12) used a tyramide signal amplification labeling approach to increase the signal intensity of a 70-mer FGA ~10-fold over the Cy dye-labeling techniques. This method further decreased the detection limit to 1% of microbial cells in the total community.

The above methods are beneficial for the detection of dominant members of relatively high biomass communities, but new approaches are needed for detecting less abundant, yet environmentally important, organisms. One currently available option is to PCR amplify specific populations, although this has its own set of well-documented limitations (11, 18, 39, 53, 61). Magnetic beads or other capture methods may also help to enrich for certain populations (59). Another option for low-biomass environments that do not contain enough DNA for FGA analysis is to nonspecifically amplify the whole community DNA prior to FGA analysis (64).

Quantitation and Standardization

Due to the potential variation introduced during DNA extraction, microarray printing, labeling, and hybridization, there has been some debate regarding the quantitative ability of FGAs. However, recent research has demonstrated that FGAs can be quantitative over a range of concentrations (40, 65), but it can be difficult to correlate signal intensity values with actual levels of specific organisms due to the presence of unknown sequences and the lack of appropriate quantitative standards. Although it is usually assumed that the FGA hybridization signal intensity is directly proportional to the abundance of the target organisms, nonspecific hybridization to uncharacterized DNA from environmental microorganisms likely occurs and may complicate or confound data interpretation and ultimately research conclusions. It may be beneficial to analyze representative, key genes by using additional methods such as real-time PCR for relative comparison with FGA data (40).

Due to the variability in FGA analysis, standardization techniques are needed in order to facilitate quantitative comparison of data between, and even within, experiments. New approaches using known amounts of labeled oligos or DNA fragments spiked into the hybridization solution as controls have been developed (3, 8, 14). The raw FGA signal intensities can then be normalized based on the signal intensity resulting from hybridization of the control with corresponding probes on the array. This approach for standardizing microarray data is promising, but additional research is needed to optimize this method for various array formats.

Measurement of Microbial Activity

Perhaps the ultimate goal for FGA analysis of environmental samples is the ability to monitor microbial activity by measuring mRNA levels for specific functional genes and to correlate this with changing environmental parameters. However, in contrast to DNA-based analyses, relatively few studies have used FGAs to measure mRNA levels in environmental samples, and these experiments were done using simplified enrichments or mixed cultures rather than actual environmental samples (13, 40). Complex samples (e.g., surface soils and sediments) still present several challenges for FGA analysis, including difficulties in extracting adequate quantities of high-quality mRNA from these samples (44). Continued developments in RNA extraction methods (6, 26, 49) and commercially available kits (e.g., TruRNA from Atom Sciences, Inc., Oak Ridge, Tenn., and FastRNA from Qbiogene, Irvine, Calif.) are improving the process. However, considerable time and effort are still necessary to remove impurities and DNA from many samples. These purification steps often substantially decrease mRNA recovery, further exasperating analysis of many low-biomass environments which already do not contain enough mRNA for FGA analysis. Development of prokaryotic mRNA amplification methods, such as that of Botero et al. (4), where a poly(A) tail is added to prokaryotic RNA for subsequent amplification, will likely broaden the range of samples for which FGA analysis of microbial activity is possible.

Our efforts in preparing this chapter were supported by the U.S. DOE Office of Science as part of its Biological and Environmental Research Programs in Natural and Accelerated Bioremediation Research, Genomics: GTL through the Virtual Institute of Microbial Stress and Survival (VIMSS; http://vimss.lbl.gov), Biotechnology Investigations-Ocean Margins, and Carbon Sequestration (as part of the consortium on research to enhance Carbon Sequestration in Terrestrial Ecosystems-CSiTE). Oak Ridge National Laboratory is managed by the University of Tennessee-Battelle LLC for the U.S. Department of Energy.

REFERENCES

1. **Beißbarth, T., K. Fellenberg, B. Brors, R. Arribas-Prat, J. Boer, N. C. Hauser, M. Scheideler, J. D. Hoheisel, G. Schütz, A. Poustka, and M. Vingron.** 2000. Processing and quality control of DNA array hybridization data. *Bioinformatics* **16:**1014–1022.
2. **Bodrossy, L., and A. Sessitsch.** 2004. Oligonucleotide microarrays in microbial diagnostics. *Curr. Opin. Microbiol.* **7:**245–254.
3. **Bodrossy, L., N. Stralis-Pavese, J. C. Murrell, S. Radajewski, A. Weilharter, and A. Sessitsch.** 2003. Development and validation of a diagnostic microbial microarray for methanotrophs. *Environ. Microbiol.* **5:**566–582.
4. **Botero, L. M., S. D'Imperio, M. Burr, T. R. McDermott, M. Young, and D. J. Hassett.** 2005. Poly(A) polymerase modification and reverse transcriptase PCR amplification of environmental RNA. *Appl. Environ. Microbiol.* **71:**1267–1275.
5. **Bozdech, Z., J. Zhu, M. P. Joachimiak, F. E. Cohen, B. Pulliam, and J. L. DeRisi.** 2003. Expression profiling of the schizont and trophozoite stages of *Plasmodium falciparum* with a long-oligonucleotide microarray. *Genome Biol.* **4:**R9.
6. **Burgmann, H., F. Widmer, W. V. Sigler, and J. Zeyer.** 2003. mRNA extraction and reverse transcription-PCR protocol for detection of *nifH* gene expression by *Azotobacter vinelandii* in soil. *Appl. Environ. Microbiol.* **69:**1928–1935.
7. **Cho, J. C., and J. M. Tiedje.** 2001. Bacterial species determination from DNA-DNA hybridization by using genome fragments and DNA microarrays. *Appl. Environ. Microbiol.* **67:**3677–3682.
8. **Cho, J. C., and J. M. Tiedje.** 2002. Quantitative detection of microbial genes by using DNA microarrays. *Appl. Environ. Microbiol.* **68:**1425–1430.
9. **Chou, H. H., A. P. Hsia, D. L. Mooney, and P. S. Schnable.** 2004. Picky: oligo microarray design for large genomes. *Bioinformatics* **20:**2893–2902.
10. **Cook, K. L., and G. S. Sayler.** 2003. Environmental application of array technology: promise, problems and practicalities. *Curr. Opin. Biotechnol.* **14:**311–318.
11. **Crosby, L. D., and C. S. Criddle.** 2003. Understanding bias in microbial community analysis techniques due to *rrn* operon copy number heterogeneity. *BioTechniques* **34:**790–798.
12. **Denef, V. J., J. Park, J. L. M. Rodrigues, T. V. Tsoi, S. A. Hashsham, and J. M. Tiedje.** 2003. Validation of a more sensitive method for using spotted oligonucleotide DNA microarrays for functional genomics studies on bacterial communities. *Environ. Microbiol.* **5:**933–943.
13. **Dennis, P., E. A. Edwards, S. N. Liss, and R. Fulthorpe.** 2003. Monitoring gene expression in mixed microbial communities by using DNA microarrays. *Appl. Environ. Microbiol.* **69:**769–778.
14. **Dudley, A. M., J. Aach, M. A. Steffen, and G. M. Church.** 2002. Measuring absolute expression with microarrays with a calibrated reference sample and an extended signal intensity range. *Proc. Natl. Acad. Sci. USA* **99:**7554–7559.
15. **Eisen, M. B., P. T. Spellman, P. O. Brown, and D. Botstein.** 1998. Cluster analysis and display of genome-wide expression patterns. *Proc. Natl. Acad. Sci. USA* **95:**14863–14868.
16. **El Fantroussi, S., H. Urakawa, A. E. Bernhard, J. J. Kelly, P. A. Noble, H. Smidt, G. M. Yershov, and D. A. Stahl.** 2003. Direct profiling of environmental microbial populations by thermal dissociation analysis of native rRNAs hybridized to oligonucleotide microarrays. *Appl. Environ. Microbiol.* **69:**2377–2382.
17. **Emrich, S. J., M. Lowe, and A. L. Delcher.** 2003. PROBEmer: a web-based software tool for selecting optimal DNA oligos. *Nucleic Acids Res.* **31:**3746–3750.
18. **Farrelly, V., F. A. Rainey, and E. Stackebrandt.** 1995. Effect of genome size and *rrn* gene copy number on PCR amplification of 16S rRNA genes from a mixture of bacterial species. *Appl. Environ. Microbiol.* **61:**2798–2801.
19. **Gans, J., M. Wolinsky, and J. Dunbar.** 2005. Computational improvements reveal great bacterial diversity and high metal toxicity in soil. *Science* **309:**1387–1390.
20. **Gilbert, D. R., M. Schroeder, and J. van Helden.** 2000. Interactive visualization and exploration of relationships between biological objects. *Trends Biotechnol.* **18:**487–494.
21. **Gordon, P. M., and C. W. Sensen.** 2004. Osprey: a comprehensive tool employing novel methods for the design of oligonucleotides for DNA sequencing and microarrays. *Nucleic Acids Res.* **32:**e133.
22. **Guschin, D., G. Yershov, A. Zaslavsky, A. Gemmell, V. Shick, D. Proudnikov, P. Arenkov, and A. Mirzabekov.** 1997. Manual manufacturing of oligonucleotide, DNA, and protein microchips. *Anal. Biochem.* **250:**203–211.
23. **He, Z., L. Wu, M. W. Fields, and J. Zhou.** 2005. Use of microarrays with different probe sizes for monitoring gene expression. *Appl. Environ. Microbiol.* **71:**5154–5162.
24. **He, Z., L. Wu, X. Li, M. W. Fields, and J. Zhou.** 2005. Empirical establishment of oligonucleotide probe design criteria. *Appl. Environ. Microbiol.* **71:**3753–3760.
25. **Hughes, T. R., M. Mao, A. R. Jones, J. Burchard, M. J. Marton, K. W. Shannon, S. M. Lefkowitz, M. Ziman, J. M. Schelter, M. R. Meyer, S. Kobayashi, C. Davis, H. Dai, Y. D. He, S. B. Stephaniants, G. Cavet, W. L. Walker, A. West, E. Coffey, D. D. Shoemaker, R. Stoughton, A. P. Blanchard, S. H. Friend, and P. S. Linsley.** 2001. Expression profiling using microarrays fabricated by an ink-jet oligonucleotide synthesizer. *Nat. Biotechnol.* **19:**342–347.
26. **Hurt, R. A., X. Qiu, L. Wu, Y. Roh, A. V. Palumbo, J. M. Tiedje, and J. Zhou.** 2001. Simultaneous recovery of RNA and DNA from soils and sediments. *Appl. Environ. Microbiol.* **67:**4495–4503.
27. **Jenkins, B. D., G. F. Steward, S. M. Short, B. B. Ward, and J. P. Zehr.** 2004. Fingerprinting diazotroph communities in the Chesapeake Bay by using a DNA macroarray. *Appl. Environ. Microbiol.* **70:**1767–1776.
28. **Kane, M. D., T. A. Jatkoe, C. R. Stumpf, J. Lu, J. D. Thomas, and S. J. Madore.** 2000. Assessment of the sensitivity and specificity of oligonucleotide (50mer) microarrays. *Nucleic Acids Res.* **28:**4552–4557.
29. **Kim, B. C., J. H. Park, and M. B. Gu.** 2004. Development of a DNA microarray chip for the identification

of sludge bacteria using an unsequenced random genomic DNA hybridization method. *Environ. Sci. Technol.* **38:**6767–6774.

30. **Letowski, J., R. Brousseau, and L. Masson.** 2004. Designing better probes: effect of probe size, mismatch position and number on hybridization in DNA oligonucleotide microarrays. *J. Microbiol. Methods* **57:**269–278.

30a. **Li, X., Z. He, and J. Zhou.** 2005. Selection of optimal oligonucleotide probes for microarrays using multiple criteria, global alignment and parameter estimation. *Nucleic Acids Res.* **33:**6114–6123.

30b. **Liebich, J., C. W. Schadt, S. C. Chong, Z. He, S.-K. Rhee, and J. Zhou.** 2006. Improvement of oligonucleotide probe design criteria for functional gene microarrays in environmental applications. *Appl. Environ. Microbiol.* **72:** 1688–1691.

31. **Li, F., and G. D. Stormo.** 2001. Selection of optimal DNA oligos for gene expression arrays. *Bioinformatics* **17:**1067–1076.

32. **Loy, A., K. Kusel, A. Lehner, H. L. Drake, and M. Wagner.** 2004. Microarray and functional gene analyses of sulfate-reducing prokaryotes in low-sulfate, acidic fens reveal cooccurrence of recognized genera and novel lineages. *Appl. Environ. Microbiol.* **70:**6998–7009.

33. **Loy, A., A. Lehner, N. Lee, J. Adamczyk, H. Meier, J. Ernst, K. H. Schleifer, and M. Wagner.** 2002. Oligonucleotide microarray for 16S rRNA gene-based detection of all recognized lineages of sulfate-reducing prokaryotes in the environment. *Appl. Environ. Microbiol.* **68:**5064–5081.

34. **Magurran, A. E.** 2004. *Measuring Biological Diversity.* Blackwell Publishing, Malden, Mass.

35. **Murray, C. J.** 2002. Sampling and data analysis for environmental microbiology, p. 166–177. *In* C. J. Hurst, R. L. Crawford, G. R. Knudsen, M. J. McInerney, and L. D. Stetzenbach (ed.), *Manual of Environmental Microbiology,* 2nd ed. ASM Press, Washington, D.C.

36. **Nielsen, H. B., R. Wernersson, and S. Knudsen.** 2003. Design of oligonucleotides for microarrays and perspectives for design of multi-transcriptome arrays. *Nucleic Acids Res.* **31:**3491–3496.

37. **Nordberg, E. K.** 2005. YODA: selecting signature oligonucleotides. *Bioinformatics* **21:**1365–1370.

38. **Reymond, N., H. Charles, L. Duret, F. Calevro, G. Beslon, and J. M. Fayard.** 2004. ROSO: optimizing oligonucleotide probes for microarrays. *Bioinformatics* **20:** 271–273.

39. **Reysenbach, A. L., L. J. Giver, G. S. Wickham, and N. R. Pace.** 1992. Differential amplification of rRNA genes by polymerase chain reaction. *Appl. Environ. Microbiol.* **58:**3417–3418.

40. **Rhee, S. K., X. Liu, L. Wu, S. C. Chong, X. Wan, and J. Zhou.** 2004. Detection of genes involved in biodegradation and biotransformation in microbial communities by using 50-mer oligonucleotide microarrays. *Appl. Environ. Microbiol.* **70:**4303–4317.

41. **Rimour, S., D. Hill, C. Militon, and P. Peyret.** 2005. GoArrays: highly dynamic and efficient microarray probe design. *Bioinformatics* **21:**1094–1103.

42. **Rouillard, J. M., C. J. Herbert, and M. Zuker.** 2002. OligoArray: genome-scale oligonucleotide design for microarrays. *Bioinformatics* **18:**486–487.

43. **Rouillard, J. M., M. Zuker, and E. Gulari.** 2003. OligoArray 2.0: design of oligonucleotide probes for DNA microarrays using a thermodynamic approach. *Nucleic Acids Res.* **31:**3057–3062.

44. **Saleh-Lakha, S., M. Miller, R. G. Campbell, K. Schneider, P. Elahimanesh, M. M. Hart, and J. T. Trevors.** 2005. Microbial gene expression in soil: methods, applications and challenges. *J. Microbiol. Methods* **63:**1–19.

45. **Schadt, C. W., J. Liebich, S. C. Chong, T. J. Gentry, Z. He, H. Pan, and J. Zhou.** 2005. Design and use of functional gene microarrays (FGAs) for the characterization of microbial communities, p. 331–368. *In* T. Savidge and H. Pothulakis (ed.), *Microbial Imaging,* vol. 34. Academic Press, Inc., London, United Kingdom.

46. **Schadt, C. W., and J. Zhou.** 2005. Advances in microarrays for soil microbial community analyses. *In* P. Nannipieri and K. Smalla (ed.), *Soil Biology: Nucleic Acids and Proteins in Soil.* Springer-Verlag, Berlin, Germany.

47. **Schena, M., D. Shalon, R. W. Davis, and P. O. Brown.** 1995. Quantitative monitoring of gene expression patterns with a complementary DNA microarray. *Science* **270:**467–470.

48. **Sebat, J. L., F. S. Colwell, and R. L. Crawford.** 2003. Metagenomic profiling: microarray analysis of an environmental genomic library. *Appl. Environ. Microbiol.* **69:** 4927–4934.

49. **Sessitsch, A., S. Gyamfi, N. Stralis-Pavese, A. Weilharter, and U. Pfeifer.** 2002. RNA isolation from soil for bacterial community and functional analysis: evaluation of different extraction and soil conservation protocols. *J. Microbiol. Methods* **51:**171–179.

50. **Small, J., D. R. Call, F. J. Brockman, T. M. Straub, and D. P. Chandler.** 2001. Direct detection of 16S rRNA in soil extracts by using oligonucleotide microarrays. *Appl. Environ. Microbiol.* **67:**4708–4716.

51. **Steward, G. F., B. D. Jenkins, B. B. Ward, and J. P. Zehr.** 2004. Development and testing of a DNA macroarray to assess nitrogenase (*nifH*) gene diversity. *Appl. Environ. Microbiol.* **70:**1455–1465.

52. **Stralis-Pavese, N., A. Sessitsch, A. Weilharter, T. Reichenauer, J. Riesing, J. Csontos, J. C. Murrell, and L. Bodrossy.** 2004. Optimization of diagnostic microarray for application in analysing landfill methanotroph communities under different plant covers. *Environ. Microbiol.* **6:** 347–363.

53. **Suzuki, M. T., and S. J. Giovannoni.** 1996. Bias caused by template annealing in the amplification of mixtures of 16S rRNA genes by PCR. *Appl. Environ. Microbiol.* **62:**625–630.

54. **Tamayo, P., D. Slonim, J. Mesirov, Q. Zhu, S. Kitareewan, E. Dmitrovsky, E. S. Lander, and T. R. Golub.** 1999. Interpreting patterns of gene expression with self-organizing maps: methods and application to hematopoietic differentiation. *Proc. Natl. Acad. Sci. USA* **96:**2907–2912.

55. **Taroncher-Oldenburg, G., E. M. Griner, C. A. Francis, and B. B. Ward.** 2003. Oligonucleotide microarray for the study of functional gene diversity in the nitrogen cycle in the environment. *Appl. Environ. Microbiol.* **69:**1159–1171.

56. **Tiquia, S. M., L. Wu, S. C. Chong, S. Passovets, D. Xu, Y. Xu, and J. Zhou.** 2004. Evaluation of 50-mer oligonucleotide arrays for detecting microbial populations in environmental samples. *BioTechniques* **36:**664–675.

57. **Torsvik, V., J. Goksøyr, and F. L. Daae.** 1990. High diversity in DNA of soil bacteria. *Appl. Environ. Microbiol.* **56:**782–787.

58. **Torsvik, V., L. Ovreas, and T. F. Thingstad.** 2002. Prokaryotic diversity—magnitude, dynamics, and controlling factors. *Science* **296:**1064–1066.

59. **Tsai, Y. L., J. Y. Le, and B. H. Olson.** 2003. Magnetic bead hybridization to detect enterotoxigenic *Escherichia coli* strains associated with cattle in environmental water sources. *Can. J. Microbiol.* **49:**391–398.

60. **Urakawa, H., S. El Fantroussi, H. Smidt, J. C. Smoot, E. H. Tribou, J. J. Kelly, P. A. Noble, and D. A. Stahl.** 2003. Optimization of single-base-pair mismatch discrimination in oligonucleotide microarrays. *Appl. Environ. Microbiol.* **69:**2848–2856.

61. **von Wintzingerode, F., U. B. Gobel, and E. Stackebrandt.** 1997. Determination of microbial diversity in environmental samples: pitfalls of PCR-based rRNA analysis. *FEMS Microbiol. Rev.* **21:**213–229.
62. **Wang, X., and B. Seed.** 2003. Selection of oligonucleotide probes for protein coding sequences. *Bioinformatics* **19:**796–802.
63. **Wilson, K. H., W. J. Wilson, J. L. Radosevich, T. Z. DeSantis, V. S. Viswanathan, T. A. Kuczmarski, and G. L. Andersen.** 2002. High-density microarray of small-subunit ribosomal DNA probes. *Appl. Environ. Microbiol.* **68:**2535–2541.
64. **Wu, L., X. Liu, C. W. Schadt, and J. Zhou.** Microarray-based analysis of sub-nanogram quantities of microbial community DNA using whole community genome amplification (WCGA). *Proc. Natl. Acad. Sci. USA*, in press.
65. **Wu, L. Y., D. K. Thompson, G. S. Li, R. A. Hurt, J. M. Tiedje, and J. Z. Zhou.** 2001. Development and evaluation of functional gene arrays for detection of selected genes in the environment. *Appl. Environ. Microbiol.* **67:**5780–5790.
66. **Wu, L. Y., D. K. Thompson, X. D. Liu, M. W. Fields, C. E. Bagwell, J. M. Tiedje, and J. Z. Zhou.** 2004. Development and evaluation of microarray-based whole-genome hybridization for detection of microorganisms within the context of environmental applications. *Environ. Sci. Technol.* **38:**6775–6782.
67. **Xu, D., G. Li, L. Wu, J. Zhou, and Y. Xu.** 2002. PRIMEGENS: robust and efficient design of gene-specific probes for microarray analysis. *Bioinformatics* **18:**1432–1437.
68. **Xu, Y.** 2004. Microarray gene expression data analysis, p. 177–206. *In* J. Zhou, D. K. Thompson, Y. Xu, and J. M. Tiedje (ed.), *Microbial Functional Genomics*. John Wiley & Sons, Inc., Hoboken, N.J.
69. **Zhou, J.** 2003. Microarrays for bacterial detection and microbial community analysis. *Curr. Opin. Microbiol.* **6:**288–294.
70. **Zhou, J., and D. K. Thompson.** 2004. DNA microarray technology, p. 141–176. *In* J. Zhou, D. K. Thompson, Y. Xu, and J. M. Tiedje (ed.), *Microbial Functional Genomics*. John Wiley & Sons, Inc., Hoboken, N.J.
71. **Zhou, J., M. A. Bruns, and J. M. Tiedje.** 1996. DNA recovery from soils of diverse composition. *Appl. Environ. Microbiol.* **62:**316–322.
72. **Zhou, J., and D. K. Thompson.** 2002. Challenges in applying microarrays to environmental studies. *Curr. Opin. Biotechnol.* **13:**204–207.
73. **Zhou, X., L. Wu, and J. Zhou.** 2004. Fabrication of DNA microarrays on nanoengineered polymeric ultrathin film prepared by self-assembly of polyelectrolyte multilayers. *Langmuir* **20:**8877–8885.
74. **Zhou, X., and J. Zhou.** 2004. Improving the signal sensitivity and photostability of DNA hybridizations on microarrays by using dye-doped core-shell silica nanoparticles. *Anal. Chem.* **76:**5302–5312.
75. **Zhou, Y.-X., P. Kalocsai, J.-Y. Chen, and S. Shams.** 2000. Information processing issues and solutions associated with microarray technology, p. 167–200. *In* M. Schena (ed.), *Microarray Biochip Technology*. Eaton Publishing, Natick, Mass.

Methods for Soil Metagenomics: Extraction and Cloning of Soil DNA

CHRISTIAN S. RIESENFELD, LYNN L. WILLIAMSON, AND JO HANDELSMAN

84

Analysis of microorganisms in pure laboratory culture provides an oversimplified panorama of prokaryotic physiology and ecology. The bacterial isolates already captured in pure culture drastically underrepresent the more than 50 distinct phyla that comprise the domain *Bacteria* (reviewed in references 23, 44, 75, and 85). Most bacteria remain uncharacterized because they are refractory to standard pure-culture methods.

Metagenomics (synonymous with environmental or community genomics) is the construction and analysis of libraries containing random DNA fragments cloned from naturally occurring microbial communities (reviewed in references 20–22, 38, 78, 79, 84, 91, and 94). The goals are to (i) describe the genomic structure of microbial communities, (ii) decipher the physiology and ecology of uncultured prokaryotes, and (iii) identify novel genes, enzymes, and molecules for biotechnology.

The metagenomic approach has been tested on an extensive range of environments including acid mine drainage (103), beetle guts (71), mouse intestines (107), human oral cavities (24), drinking water (86), whale carcasses (100), hot springs (76), hydrothermal vent worms (15), sediments (37), seawater (6, 93, 105), and soil (41, 80, 100). The dense, diverse, and complex microbial assemblages in soil are particularly suitable for metagenomic analysis because many soil bacteria belong to lineages without cultured members (2, 27, 52, 69, 98). The astonishing diversity of soil bacteria (25, 40, 44, 54, 62, 75) and archaea (8, 13, 66, 67, 88) is apparent from 16S rRNA studies of various soils. The species richness of soil is difficult to measure, but conservative estimates indicate that each gram contains several thousand bacterial species (18, 19, 26, 45, 56), making it one of the most biologically complex environments on Earth, rich with activities that influence the nutrient cycling and biogeochemistry of the planet (58).

Culture collections of soil organisms are historically plentiful sources of bioactive molecules, but the synthetic capacity for an unknown array of novel products may be embedded in the genomes of uncultured bacteria. Metagenomic studies have validated the uncultured world as a source of novel enzymes (31, 34, 41, 42, 48, 53, 77, 80, 111) and bioactive molecules (11, 12, 16, 33, 57, 108) and provided the first insights into the biology of uncultured *Acidobacteria* (55, 74) and *Crenarchaeota* (73, 99). Metagenomic research that culminates in biological or biotechnological discovery commences with library construction, which is the subject of this chapter.

GENERAL CONSIDERATIONS

Most methods for extracting DNA from soil are intended for PCR-based applications such as the amplification of 16S rRNA (36, 61, 63, 70, 72, 101, 102, 112) or other genes (81, 109) rather than direct cloning, which is especially challenging because of humic acids in soil that coextract with DNA and inhibit restriction enzymes (96). This chapter provides detailed methods to obtain DNA of sufficient purity for cloning into plasmid, cosmid, fosmid, or bacterial artificial chromosome (BAC) vectors.

The most common techniques for soil metagenomic library construction entail partial digestion of soil DNA with various restriction enzymes (Table 1). DNA derived from soil generally requires additional purification prior to successful digestion and cloning. Many purification procedures have been tested, and they include gel electrophoresis (32, 63, 80, 112), various precipitations (50, 51, 63, 89), use of polyvinylpyrrolidone (PVP) (73), phenol-chloroform extractions (51, 83), hexadecyltrimethylammonium bromide (CTAB) extractions (50, 77), spin columns (41, 50, 51, 63, 64, 112), glass milk purification (50, 89), and density gradient centrifugations (16, 43, 97). Simple and reliable purification methods that facilitate library construction from soil are included in this chapter.

An alternative to restriction enzyme-based cloning methods to capture small (average of 5.2 kb) DNA fragments from soil microbial communities is physical shearing of DNA by nebulization prior to blunt-end cloning into plasmids (31, 32). Compared to partial digests, physical shearing methods are more random and introduce less sequence bias into the library.

One of the fundamental library construction decisions is whether to clone low-molecular-weight (LMW) or high-molecular-weight (HMW) DNA. LMW libraries (<20-kb average insert size) are preferable for projects designed to characterize individual genes because the construction of LMW libraries is easier and large, contiguous segments of DNA are not necessary. DNA for LMW libraries is usually obtained by mechanical cell lysis (e.g., bead-mill homogenization), which lyses most prokaryotic cells, including *Bacillus* spores (64).

TABLE 1 Vectors used for soil metagenomics

Vector	Relevant characteristics	Enzyme(s) used for cloning	Reference(s)
BAC			
pBeloBAC11	Maintains large inserts	HindIII	80
pBTP2	Removable pUC origin	HindIII	57
pCC1BAC (Epicentre)	Inducible copy number	Sau3AI	110
pESAC	*E. coli-Streptomyces* artificial chromosome	Sau3AI	7, 90
pMBD14	Shuttle vector	BamHI	60
pSuperBAC	Inducible copy number, shuttle vector	HindIII	39, 110
Plasmid			
pBluescript SK+ (Stratagene)	High copy number	BamHI, Sau3AI, Bsp143I	41, 42, 47
pCF430	Broad host range	PstI	65, 77
pJN105	Broad host range	EcoRI, PstI	65, 77
pUC19	High copy number	BamHI, Sau3AI	83, 111
pZErO-2 (Invitrogen)	Positive selection, high copy number	Blunt-end	31, 32
Cosmid/fosmid			
pCC1FOS (Epicentre)	Copy number inducible	Blunt-end	110
pEpiFOS-5 (Epicentre)	Single copy	Blunt-end	34, 53, 73, 74, 99
Lambda DASH II (Stratagene)	Rapid restriction mapping and chromosomal walking	Sau3AI	83
pOS700I	*E. coli-Streptomyces lividans* shuttle vector	Polynucleotide tails	16
SuperCos I (Stratagene)	Rapid restriction mapping and chromosomal walking	BamHI	11, 12
pWE15 (Stratagene)	Cosmid cloning vector	Sau3AI	28, 106
pWEB (Epicentre)	Cosmid cloning vector	Blunt-end	11
ZAP Express (Stratagene)	Phagemid expression vector	Sau3AI	35

Construction of HMW libraries (>20-kb average insert size) is more difficult because of the lower efficiency of cloning large fragments. A pulsed-field gel electrophoresis apparatus is pivotal to resolve HMW DNA. Chemical and enzymatic lysis methods are normally employed, which may bias the extraction towards easily lysed cells (64). HMW libraries contain fragments of sufficient size to capture entire gene clusters, making these libraries essential for studies designed to examine the genomic context of particular genes. Since LMW and HMW DNA libraries have utility for different applications, we provide methods for construction of both types.

Vector choice depends on the intended research questions, desired insert size, and screening strategies. The general categories are (i) plasmid vectors for LMW DNA cloning, (ii) cosmid or fosmid vectors for ~40-kb inserts (packaged by phage particles), and (iii) BAC vectors for HMW DNA cloning (87). All three library types have been implemented in soil metagenomic studies (Table 1).

Plasmid libraries are commonly used for large-scale sequencing projects (100, 103, 105) and studies that aim to detect functions encoded by a single gene or small numbers of genes. Cosmid or fosmid libraries appear to be increasing in popularity because they provide a relatively uniform distribution of medium-sized inserts (a typical 35- to 40-kb cosmid clone would include ~1% of an *Escherichia coli*-size genome) and reasonable cloning efficiency. The F1 origin of replication (single copy number) is common to both fosmids and BACs (22, 46), but the distinction is that BAC libraries are introduced into *E. coli* by electroporation (6, 80) instead of by use of the phage particles used for fosmid libraries (73, 74, 99). Insightful commentary by Béjà (4) points out that compared to BAC library construction, the construction of cosmid or fosmid libraries is highly efficient, but clones larger than 100 kb have been reported for BAC libraries from marine samples and would not be obtained in fosmid libraries (6). BAC libraries are ideal for applications requiring especially large DNA fragments, such as the isolation of gene clusters that encode entire antibiotic biosynthetic pathways or attempts to harvest large genomic fragments to link the physiology of uncultured bacteria with phylogenetic markers (5).

The construction and analysis of libraries large enough to encompass the entire genetic complexity of soil is a daunting prospect. Tringe et al. (100) recently estimated that 2×10^9 to 5×10^9 bp of sequence would be necessary to obtain eightfold coverage of the most abundant prokaryotic genome in soil. Typical metagenomic libraries encapsulate a tiny subset of the genes in soil, but libraries with more than 10^5 clones have been reported (41, 42, 48, 59, 77, 100) and novel amidases (31), amylases (111), lipases (53), and polyketide synthases (16) were discovered in relatively small libraries.

LMW DNA

Several kits for extracting LMW DNA from soil are commercially available. One example is the FastDNA SPIN Kit for Soil (Qbiogene), which is commonly used in soil microbial diversity studies. It consistently extracts DNA from diverse prokaryotes in various soil samples (9, 10, 14, 36, 55, 61), using mechanical cell lysis by bead beating in a FastPrep instrument. The extracted DNA is smaller than 20 kb (Fig. 1) and requires additional purification prior to

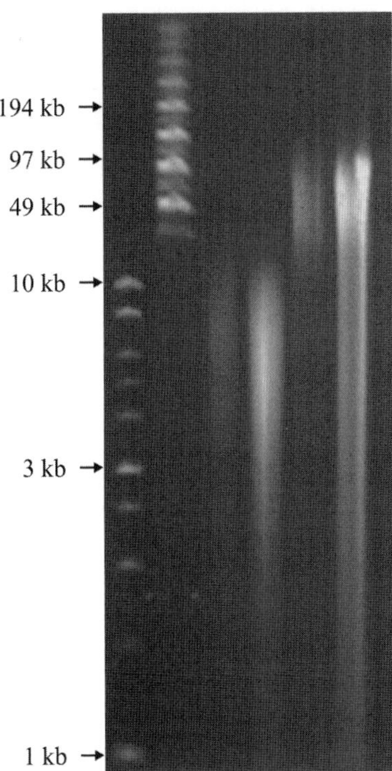

FIGURE 1 DNA size comparison for LMW and HMW extraction methods. Soil cores were collected at the National Science Foundation Long-Term Ecological Research site at Bonanza Creek Experimental Forest (Alaska), and organic (~40% organic matter) and mineral (~10% organic matter) layers were treated as separate samples (110). The LMW and HMW methods described in this chapter were used to extract DNA. Pulsed-field gel electrophoresis (4 h, 8 V/cm, 100° angle, linear ramping factor, 14°C, 0.5× Tris-borate-EDTA buffer) was performed with a CHEF Mapper, and DNA was visualized with ethidium bromide. Lanes (left to right): 1kb DNA Ladder (Promega), MidRange II PFG Marker (New England Biolabs), LMW mineral DNA, LMW organic DNA, HMW mineral DNA, and HMW organic DNA.

cloning because of impurities that inhibit restriction enzymes (Fig. 2A). For example, a CTAB purification (3) decreased the DNA quantity but dramatically improved the digestion efficiency (Fig. 2B). The disappearance of larger DNA fragments indicates successful digestion, but the difference is subtle because of extensive shearing prior to digestion and considerable variability in fragment size. Such DNA is suitable for cloning into plasmids by standard procedures (77).

CTAB Purification of LMW DNA

The starting material is DNA prepared from at least 0.5 g of soil with the FastDNA SPIN Kit for Soil.

1. Prepare CTAB-NaCl solution (10% CTAB–0.7 M NaCl) by dissolving 4.1 g of NaCl in 80 ml of H_2O and then slowly adding 10 g of CTAB while heating and stirring. Adjust the final volume to 100 ml and autoclave.
2. Add the appropriate amount of 5 M NaCl to the DNA preparation to adjust the NaCl concentration to 0.7 M, and then add 0.1 volume of CTAB-NaCl solution (prewarmed to 65°C). Mix thoroughly and incubate at 65°C for 15 min.
3. Add 1 volume of chloroform-isoamyl alcohol (24:1), mix by inversion, and centrifuge (tabletop microcentrifuge) at maximum speed for 5 min.
4. Carefully remove the top layer (containing DNA) and transfer it to a new tube. Precipitate the DNA by adding 2 volumes of 100% ethanol, mixing and centrifuging (tabletop microcentrifuge) at maximum speed for 5 min.
5. Decant the liquid, wash the pellet twice with 500 μl of 70% ethanol, remove the liquid, and resuspend the dried pellet in water.

HMW DNA

There are two basic approaches to HMW DNA extraction from soil: (i) direct lysis of cells in the soil matrix (51, 63, 68, 70, 89, 92, 97, 112) and (ii) separation of cells from soil followed by lysis (1, 7, 17, 29, 32, 43, 51, 92, 97). Many investigators have comparatively evaluated the direct lysis and cell separation methods for yield, purity, and labor required (32, 50, 51, 63, 92, 97). These studies demonstrate that direct lysis results in higher yields (32, 51, 92, 97), which may reflect a more representative sample of the microbial community DNA. There are conflicting data about whether direct lysis (51, 92) or cell separation (32, 50) methods produce DNA of higher purity, but cell separation is thought to avoid extracellular and eukaryotic DNA (1, 32, 92). Cell separation is more time-consuming than direct lysis, but larger DNA can be extracted when the cells are embedded and lysed within an agarose plug (49).

The following direct lysis protocol, based on the method of Zhou et al. (112), is included because it has been successful with many soil types (with varied clay, silt, sand, and organic matter content), is relatively easy, and provides DNA that can be digested by restriction enzymes and cloned into BAC or cosmid vectors. The DNA fragments range in size up to 100 kb prior to digestion, but the yield varies, depending on the soil (Fig. 1).

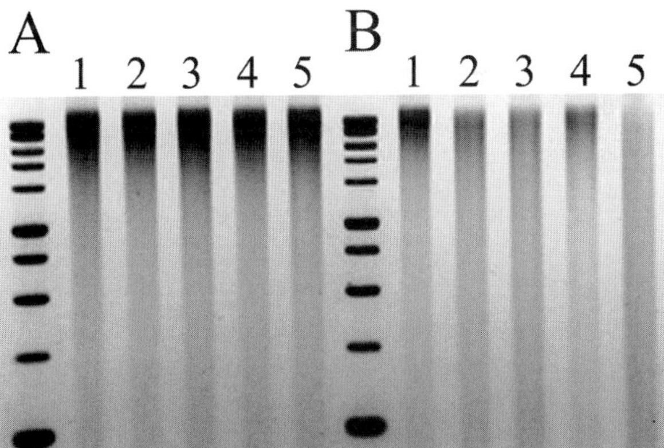

FIGURE 2 Comparison of restriction enzyme digestion of LMW DNA preceding or subsequent to CTAB purification. Soil samples were collected at the West Madison Agricultural Research Station (Wisconsin) (described in references 8, 55, 77, and 80), and a FastDNA SPIN Kit was used to extract DNA from 0.5 g of soil. The eluted DNA was divided into two portions and one remained untreated (A), while the other was purified with CTAB (B). Prior to gel electrophoresis, DNA was digested with 10 U of various restriction enzymes (Promega) for 1 h. Lanes: 1, undigested DNA; 2 to 5, DNA digested with BamHI, EcoRI, HindIII, or PstI, respectively. The size marker is 1kb DNA Ladder (Promega).

Procedure for HMW DNA Extraction

1. Collect soil (>100 g) and pass it through a sterile sieve (2-mm mesh size) to remove large debris such as rocks, worms, leaves, and roots.
2. Add 50 g of soil and 75 ml of Z buffer (100 mM Tris-HCl [pH 8.0], 50 mM Na_2HPO_4, 50 mM NaH_2PO_4, 100 mM EDTA [pH 8.0], 1.5 M NaCl, and 1% [wt/vol] CTAB) to each of two 250-ml centrifuge bottles.
3. Crush dry ice, and add isopropanol to make slush. Perform two freeze-thaw cycles by incubating the bottles in crushed dry ice to freeze (~40 min) and in a 65°C water bath to thaw (~40 min).
4. Add 9 ml of 20% (wt/vol) sodium dodecyl sulfate and 4.5 ml of 5 M guanidine isothiocyanate to the 65°C suspension in each bottle, and mix by gentle inversion. Incubate the bottles at 65°C for 2 h with occasional gentle mixing. Centrifuge at 15,000 × *g* and 10°C for 20 min. Transfer the supernatants (containing DNA) to clean centrifuge bottles.
5. Add 25 ml of chloroform-isoamyl alcohol (24:1) to each bottle and mix gently for 10 min. Centrifuge at 15,000 × *g* and 10°C for 20 min. Remove the top layer (containing DNA) from each bottle and transfer it to a clean centrifuge bottle. Note: If the sample contains a high proportion of clay or organic matter, repeat step 5 (optional).
6. Add 0.7 volume of isopropanol to each bottle and mix gently for 5 min. Incubate at room temperature for 20 min. To pellet the DNA, centrifuge at 15,000 × *g* and 10°C for 40 min, and carefully discard all of the liquid.
7. Use wide-bore pipette tips to resuspend the DNA pellet in 1 or 2 ml of $T_{10}E_{10}$ (10 mM Tris-HCl, 10 mM EDTA [pH 8.0]) and transfer it to four microcentrifuge tubes.
8. Add an equal volume of Tris-buffered phenol-chloroform (pH 8.0) to each tube and mix by inverting gently. Centrifuge (tabletop microcentrifuge) at maximum speed for 10 min. Use wide-bore pipette tips to remove the top layer and transfer it to a clean microcentrifuge tube.
9. Add an equal volume of chloroform-isoamyl alcohol (24:1) to each tube and gently mix. Centrifuge (tabletop microcentrifuge) at maximum speed for 10 min. Use wide-bore pipette tips to transfer the top layer to a clean microcentrifuge tube.
10. Determine the DNA quantity and size by pulsed-field gel electrophoresis (load, 2 μl).
11. Store the DNA at 4°C for a few days or at −80°C for long-term storage. Note: At this stage, the DNA preparation will appear brown and contain impurities that inhibit restriction enzymes.

Additional Purification of HMW DNA

The starting material is 0.5 ml of the HMW DNA preparation.

1. Purify the DNA by gel electrophoresis in low-melting-point (LMP) agarose (e.g., LMP, preparative grade for large fragments [Promega]). Prepare four large wells by taping three lanes together in four places on a gel comb. Load 125 μl of HMW DNA plus loading dye into each well of a 1% LMP agarose gel in 1 × TAE buffer (50× stock is 242 g of Tris base, 57.1 ml of glacial acetic acid, 37.2 g of $Na_2EDTA \cdot 2H_2O$, and H_2O to make 1 liter). Run the gel overnight at 2 V/cm and 4°C in 1× TAE, without ethidium bromide. Note: For very large DNA, run a pulsed-field compression gel. Conditions for the Bio-Rad CHEF Mapper are as follows: 5 h, 9 V/cm, 120° angle, 0.1-s initial switch, 1.0-s final switch, −1.5 ramping factor, 14°C. The gel will compress DNA larger than 50 kb, while allowing LMW DNA to run off.
2. Cut off a gel segment that contains the ladder and a sliver of one lane. Stain this portion with ethidium bromide, visualize it with UV light, and measure the location of the DNA with a ruler. Excise DNA of the desired size from the unstained portion (with minimal excess agarose).
3. Use GELase (Epicentre) or AgarACE (Promega) to extract DNA from the gel. Note: Alternatively, use the procedure of Strong et al. (95) to electroelute DNA from the gel. The electroelution of large DNA fragments is effective, but recovery varies.
4. Check the DNA quantity by gel electrophoresis.

Procedure for HMW Restriction Digests

1. Perform a series of test digests with 7 μl of HMW DNA and different enzyme concentrations in each 10-μl reaction mixture to determine the amount of restriction enzyme required for partial digestion. Incubate the digests at 37°C for 1 h and then at 65°C for 15 min to inactivate the enzyme. Save undigested HMW DNA for size comparisons and nuclease activity controls. Recommended controls are (i) undigested DNA kept at 4°C to check for nuclease activity and (ii) DNA subjected to digestion conditions without restriction enzyme for size comparison.
2. Analyze the digests by pulsed-field gel electrophoresis. The DNA smear will shift towards lower-molecular-weight fragments as the enzyme concentration increases. The appropriate conditions are those that result in minimal digestion. Note: The enzyme amounts are generally larger than those used for digestion of chromosomal DNA derived from bacteria in pure culture and vary considerably, depending on the soil characteristics.
3. Use the conditions determined in the previous step to digest the desired quantity of HMW DNA in multiple separate tubes (not scaled up).
4. Perform gel electrophoresis to size-select DNA in the desired molecular weight range. Tape gel combs together to generate large wells, and run a 1% LMP agarose gel overnight at 2 V/cm and 4°C in 1× TAE buffer, without ethidium bromide. Cut off a gel segment that contains the ladder and a portion of one lane, stain it with ethidium bromide, visualize the DNA with UV light, and use a ruler to measure the DNA location. Excise DNA of the appropriate size from the unstained portion.
5. Extract DNA from the agarose, using methods such as (i) GELase or AgarACE, (ii) a QIAEX II gel extraction kit (QIAGEN), or (iii) the electroelution procedure of Strong et al. (95). Note: DNA fragments up to 50 kb are recovered by the QIAEX II gel extraction kit.

VECTOR PREPARATION

The vector preparation methods used for constructing metagenomic libraries are similar to those for prokaryotic genomic libraries. The primary difference is that larger quantities of vector are generally required to attain the high numbers of clones necessary for metagenomic analysis. High-quality preparations (ready for cloning) of some vectors are commercially available and might be worth the expense. Library construction attempts often fail because of insufficient vector quantities or a high percentage of colonies that lack cloned DNA (i.e., no insert). Colonies that lack cloned DNA can result from incomplete vector digestion or vector ends ligated to each other rather than to soil DNA. Gel purification and dephosphorylation of the digested vector are methods that reduce the proportion of such colonies.

Guidelines for Vector Preparation

1. Use a QIAGEN Plasmid Midi or Maxi kit (depending on the copy number) or equivalent method to extract vector DNA. Appropriate culture volumes are 100 ml (Midi kit) or 500 ml (Maxi kit) of LB (1% [wt/vol] tryptone, 0.5% [wt/vol] yeast extract, 1% [wt/vol] NaCl) containing the appropriate antibiotic. Grow the culture at 37°C with vigorous (250-rpm) shaking, and extract DNA according to the manufacturer's instructions. Resuspend the DNA in sterile water (200 μl for Midi kit or 1 ml for Maxi kit).

2. Digest the DNA with a restriction enzyme for ~4 h and then inactivate the enzyme (usually 65°C for 15 min). Note: Typical digests use 90 μl of DNA, 12 μl of buffer (10×), 12 μl of bovine serum albumin (10×), and 6 μl (60 U) of restriction enzyme. Set aside 10 μl of undigested vector for comparison.

3. Precipitate the DNA, using a standard ethanol precipitation (82), and resuspend it in sterile water (typically 25 μl).

4. Use a heat-labile phosphatase (e.g., HK phosphatase [Epicentre] or shrimp alkaline phosphatase [Pro-mega]) to dephosphorylate the vector. Note: A typical reaction uses 24.5 μl of DNA, 3 μl of buffer (10×), 1.5 μl of 0.1 M $CaCl_2$, and 1 μl of HK phosphatase; incubate the mixture at 30°C for 1 h and then heat inactivate at 65°C for 15 min. For large DNA quantities, longer incubation times or larger amounts of enzyme may be necessary.

5. Self-ligate the vector at room temperature for 3 h, or overnight at 14°C, and then heat inactivate at 70°C for 15 min. Note: A typical reaction mixture is 30 μl of DNA, 3.5 μl of buffer (10×), and 1.5 μl of T4 DNA ligase (Promega).

6. Load the vector DNA into a large well (three lanes taped together) of a 1% LMP agarose gel, and run the gel at 2 V/cm for 16 h in 1× TAE buffer to permit adequate separation of linear and circular DNA forms. Load undigested DNA for comparison.

7. Cut off a gel portion that contains ladder, undigested DNA, and a small edge of the digested-vector lane. Visualize the DNA with ethidium bromide, and use a ruler to measure the DNA location. Do not stain the remainder of the gel. Two bands may be visible: a linear and a circular form of the vector. Cut out the band containing the linear form from the unstained gel portion, and use a QIAEX II gel extraction kit to extract DNA from the agarose.

8. Test the vector preparation by transforming 1 μl into electrocompetent *E. coli* cells and plating the transformation mix onto media containing the appropriate antibiotic. As a control, transform 1 μl of undigested vector into *E. coli*, using the same conditions. Note: Expect to see a lawn on the control (undigested-vector) plate and few (or no) colonies on the other plate.

LIGATION OF VECTOR AND INSERT DNA

The prerequisites for library construction are vector of high purity and insert DNA at a sufficient concentration. The preceding steps should yield ample quantities of both components (at least 100 ng of each is desirable). Use gel electrophoresis and/or spectrophotometric methods to determine the DNA concentrations and, if necessary, use standard ethanol precipitation procedures to concentrate the DNA into smaller volumes. Assemble ligation reaction mixtures that contain T4 DNA ligase, buffer (containing ATP), and vector and insert DNA. Determine the optimal vector-to-insert molar ratio (typically between 10:1 and 1:1) by experimentation (82).

TRANSFORMATION AND LIBRARY STORAGE

Electroporation is the preferred method for transforming ligated DNA into *E. coli* because the efficiency may exceed 10^9 transformants/μg of DNA. Electrocompetent *E. coli* cells can be purchased or prepared by standard procedures (3, 82). Especially high-efficiency electrocompetent cells are commercially available (e.g., ElectroMAX DH10B cells [Invitrogen] or TransforMax EPI300 *E. coli* cells [Epicentre] with efficiencies of $>10^{10}$ transformants/μg) and are highly recommended for library construction to enhance the number of clones. Multiple electroporations are necessary to introduce the library into *E. coli.*

Many library storage formats are possible. A common approach is to individually array clones in 96-well plates. An alternative is to store clones in a pooled format. The pooled-clone format is especially amenable to high-throughput screens (e.g., see reference 104) or selections (e.g., references 59 and 77), and pooled libraries occupy less storage space. Libraries in *E. coli* should be stored at −80°C in the presence of a cryoprotectant (glycerol or dimethyl sulfoxide).

Transformation

1. For each transformation, thaw 50 μl of electrocompetent *E. coli* cells on ice for 20 min. In the meantime, add 1 ml of SOC (2% [wt/vol] tryptone, 0.5% [wt/vol] yeast extract, 0.05% [wt/vol] NaCl, 2.5 mM KCl; adjust the pH to 7.0 with 5 N NaOH; autoclave; and then add $MgCl_2$ to 10 mM and glucose to 20 mM) to the appropriate number of 13-mm test tubes.

2. Suspend a 0.025-μm-pore-size filter (Millipore) over water in a petri dish and incubate a portion (2 to 6 μl) of the ligation mixture on it for 30 to 45 min to remove excess salt, which reduces the transformation efficiency.

3. Add a few microliters of the desalted ligation mixture to 50 μl of electrocompetent *E. coli* cells, and then transfer the mixture to a chilled 0.2-cm electroporation cuvette (Bio-Rad).

4. Electroporate according to the manufacturer's directions. For example, with a Gene Pulser (Bio-Rad), set the capacitance at 25 μF, resistance at 200 Ω, and the voltage at 2.5 kV and use a 0.2-cm electroporation cuvette. Immediately add 1 ml of SOC, mix, and transfer the suspension from the cuvette to a test tube and incubate it at 37°C for 45 min with shaking (250 rpm).

5. Plate 100 to 200 μl of cell suspension on LB agar with the appropriate antibiotic (and X-Gal [5-bromo-4-chloro-3-indolyl-β-D-galactopyranoside] and IPTG [isopropyl-β-D-thiogalactopyranoside] if blue/white screening is applicable). Store the remainder at 4°C temporarily.

6. Enumerate the colonies and determine the insert sizes for at least 10 clones by performing minipreps, restriction digests, and gel electrophoresis. If the insert DNA is of sufficient size, transform more ligation mixture to achieve the desired library size. Calculate the average insert size and the number of clones in the library.

Procedure for Pooled-Clone Storage

1. Add 500 μl of LB with the appropriate antibiotic to a 2.0-ml microcentrifuge tube. Use a sterile spatula to collect cells from a plate and transfer them to a tube. Vortex the tube for ≥30 s to suspend the cells and place it on ice. Repeat until the entire library has been collected in tubes, each representing a subpool of the library.

2. Add an equal volume of 20% glycerol (sterilize by filtering or autoclaving) in LB with the appropriate antibiotic

to each subpool (final glycerol concentration, 10%). Vortex to eliminate cell clumps prior to transferring the cells to cryovials. Store multiple copies of each at −80°C.

CONCLUDING REMARKS

This chapter provides general guidelines for library construction that need to be tailored for specific applications. The extent to which libraries represent the genetic diversity of indigenous soil microbial communities depends on various factors, including soil type and procedures for DNA extraction and cloning (30, 61). Further research is needed to determine which methods are most representative of the soil microbial community composition. Although metagenomic libraries are imperfect representations of soil microbial community DNA (55, 99, 100), they greatly augment natural product discovery and understanding of soil ecosystems through the application of screens or selections to detect sequences or functions of interest. Previous results demonstrate the power of the metagenomic approach, which will have greater success and impact as the methods for accessing the inherent biological diversity of soil improve.

We are grateful to Heather K. Allen for contributing to the protocols, extracting the DNA shown in Fig. 1, and editing portions of the chapter.

The preparation of the manuscript was supported in part by NSF Microbial Observatories grant MCB0132085.

Specific product names are provided as examples and are not intended to endorse particular products. In most cases, several similar products are available.

REFERENCES

1. **Akkermans, A. D. L., J. D. van Elsas, and F. J. de Bruijn.** 1995. *Molecular Microbial Ecology Manual.* Kluwer Academic Publishers, Dordrecht, The Netherlands.
2. **Amann, R. I., W. Ludwig, and K. H. Schleifer.** 1995. Phylogenetic identification and in situ detection of individual microbial cells without cultivation. *Microbiol. Rev.* **59:**143–169.
3. **Ausubel, F. M.** 1995. *Short Protocols in Molecular Biology: a Compendium of Methods from Current Protocols in Molecular Biology*, 3rd ed. Wiley, New York, N.Y.
4. **Béjà, O.** 2004. To BAC or not to BAC: marine ecogenomics. *Curr. Opin. Biotechnol.* **15:**187–190.
5. **Béjà, O., L. Aravind, E. V. Koonin, M. T. Suzuki, A. Hadd, L. P. Nguyen, S. B. Jovanovich, C. M. Gates, R. A. Feldman, J. L. Spudich, E. N. Spudich, and E. F. DeLong.** 2000. Bacterial rhodopsin: evidence for a new type of phototrophy in the sea. *Science* **289:**1902–1906.
6. **Béjà, O., M. T. Suzuki, E. V. Koonin, L. Aravind, A. Hadd, L. P. Nguyen, R. Villacorta, M. Amjadi, C. Garrigues, S. B. Jovanovich, R. A. Feldman, and E. F. DeLong.** 2000. Construction and analysis of bacterial artificial chromosome libraries from a marine microbial assemblage. *Environ. Microbiol.* **2:**516–529.
7. **Berry, A. E., C. Chiocchini, T. Selby, M. Sosio, and E. M. Wellington.** 2003. Isolation of high molecular weight DNA from soil for cloning into BAC vectors. *FEMS Microbiol. Lett.* **223:**15–20.
8. **Bintrim, S. B., T. J. Donohue, J. Handelsman, G. P. Roberts, and R. M. Goodman.** 1997. Molecular phylogeny of Archaea from soil. *Proc. Natl. Acad. Sci. USA* **94:**277–282.
9. **Borneman, J., P. W. Skroch, K. M. O'Sullivan, J. A. Palus, N. G. Rumjanek, J. L. Jansen, J. Nienhuis, and E. W. Triplett.** 1996. Molecular microbial diversity of an agricultural soil in Wisconsin. *Appl. Environ. Microbiol.* **62:**1935–1943.
10. **Borneman, J., and E. W. Triplett.** 1997. Molecular microbial diversity in soils from eastern Amazonia: evidence for unusual microorganisms and microbial population shifts associated with deforestation. *Appl. Environ. Microbiol.* **63:**2647–2653.
11. **Brady, S. F., C. J. Chao, and J. Clardy.** 2004. Long-chain *N*-acyltyrosine synthases from environmental DNA. *Appl. Environ. Microbiol.* **70:**6865–6870.
12. **Brady, S. F., C. J. Chao, J. Handelsman, and J. Clardy.** 2001. Cloning and heterologous expression of a natural product biosynthetic gene cluster from eDNA. *Org. Lett.* **3:**1981–1984.
13. **Buckley, D. H., J. R. Graber, and T. M. Schmidt.** 1998. Phylogenetic analysis of nonthermophilic members of the kingdom *Crenarchaeota* and their diversity and abundance in soils. *Appl. Environ. Microbiol.* **64:**4333–4339.
14. **Burgmann, H., M. Pesaro, F. Widmer, and J. Zeyer.** 2001. A strategy for optimizing quality and quantity of DNA extracted from soil. *J. Microbiol. Methods* **45:**7–20.
15. **Campbell, B. J., J. L. Stein, and S. C. Cary.** 2003. Evidence of chemolithoautotrophy in the bacterial community associated with *Alvinella pompejana*, a hydrothermal vent polychaete. *Appl. Environ. Microbiol.* **69:**5070–5078.
16. **Courtois, S., C. M. Cappellano, M. Ball, F. X. Francou, P. Normand, G. Helynck, A. Martinez, S. J. Kolvek, J. Hopke, M. S. Osburne, P. R. August, R. Nalin, M. Guerineau, P. Jeannin, P. Simonet, and J. L. Pernodet.** 2003. Recombinant environmental libraries provide access to microbial diversity for drug discovery from natural products. *Appl. Environ. Microbiol.* **69:**49–55.
17. **Courtois, S., A. Frostegard, P. Goransson, G. Depret, P. Jeannin, and P. Simonet.** 2001. Quantification of bacterial subgroups in soil: comparison of DNA extracted directly from soil or from cells previously released by density gradient centrifugation. *Environ. Microbiol.* **3:**431–439.
18. **Curtis, T. P., and W. T. Sloan.** 2004. Prokaryotic diversity and its limits: microbial community structure in nature and implications for microbial ecology. *Curr. Opin. Microbiol.* **7:**221–226.
19. **Curtis, T. P., W. T. Sloan, and J. W. Scannell.** 2002. Estimating prokaryotic diversity and its limits. *Proc. Natl. Acad. Sci. USA* **99:**10494–10499.
20. **Daniel, R.** 2005. The metagenomics of soil. *Nat. Rev. Microbiol.* **3:**470–478.
21. **Daniel, R.** 2004. The soil metagenome—a rich resource for the discovery of novel natural products. *Curr. Opin. Biotechnol.* **15:**199–204.
22. **DeLong, E. F.** 2002. Microbial population genomics and ecology. *Curr. Opin. Microbiol.* **5:**520–524.
23. **DeLong, E. F., and N. R. Pace.** 2001. Environmental diversity of Bacteria and Archaea. *Syst. Biol.* **50:**470–478.
24. **Diaz-Torres, M. L., R. McNab, D. A. Spratt, A. Villedieu, N. Hunt, M. Wilson, and P. Mullany.** 2003. Novel tetracycline resistance determinant from the oral metagenome. *Antimicrob. Agents Chemother.* **47:**1430–1432.
25. **Dojka, M. A., J. K. Harris, and N. R. Pace.** 2000. Expanding the known diversity and environmental distribution of an uncultured phylogenetic division of bacteria. *Appl. Environ. Microbiol.* **66:**1617–1621.
26. **Dunbar, J., S. M. Barns, L. O. Ticknor, and C. R. Kuske.** 2002. Empirical and theoretical bacterial diversity in four Arizona soils. *Appl. Environ. Microbiol.* **68:**3035–3045.

27. **Dunbar, J., S. Takala, S. M. Barns, J. A. Davis, and C. R. Kuske.** 1999. Levels of bacterial community diversity in four arid soils compared by cultivation and 16S rRNA gene cloning. *Appl. Environ. Microbiol.* **65:**1662–1669.
28. **Entcheva, P., W. Liebl, A. Johann, T. Hartsch, and W. R. Streit.** 2001. Direct cloning from enrichment cultures, a reliable strategy for isolation of complete operons and genes from microbial consortia. *Appl. Environ. Microbiol.* **67:**89–99.
29. **Faegri, A., V. L. Torsvik, and J. Goksoyr.** 1977. Bacterial and fungal activities in soil—separation of bacteria and fungi by a rapid fractionated centrifugation technique. *Soil Biol. Biochem.* **9:**105–112.
30. **Frostegard, A., S. Courtois, V. Ramisse, S. Clerc, D. Bernillon, F. Le Gall, P. Jeannin, X. Nesme, and P. Simonet.** 1999. Quantification of bias related to the extraction of DNA directly from soils. *Appl. Environ. Microbiol.* **65:**5409–5420.
31. **Gabor, E. M., E. J. de Vries, and D. B. Janssen.** 2004. Construction, characterization, and use of small-insert gene banks of DNA isolated from soil and enrichment cultures for the recovery of novel amidases. *Environ. Microbiol.* **6:**948–958.
32. **Gabor, E. M., E. J. de Vries, and D. B. Janssen.** 2003. Efficient recovery of environmental DNA for expression cloning by indirect extraction methods. *FEMS Microbiol. Ecol.* **44:**153–163.
33. **Gillespie, D. E., S. F. Brady, A. D. Bettermann, N. P. Cianciotto, M. R. Liles, M. R. Rondon, J. Clardy, R. M. Goodman, and J. Handelsman.** 2002. Isolation of antibiotics turbomycin A and B from a metagenomic library of soil microbial DNA. *Appl. Environ. Microbiol.* **68:**4301–4306.
34. **Ginolhac, A., C. Jarrin, B. Gillet, P. Robe, P. Pujic, K. Tuphile, H. Bertrand, T. M. Vogel, G. Perriere, P. Simonet, and R. Nalin.** 2004. Phylogenetic analysis of polyketide synthase I domains from soil metagenomic libraries allows selection of promising clones. *Appl. Environ. Microbiol.* **70:**5522–5527.
35. **Grant, S., D. Y. Sorokin, W. D. Grant, B. E. Jones, and S. Heaphy.** 2004. A phylogenetic analysis of Wadi el Natrun soda lake cellulase enrichment cultures and identification of cellulase genes from these cultures. *Extremophiles* **8:**421–429.
36. **Griffiths, R. I., A. S. Whiteley, A. G. O'Donnell, and M. J. Bailey.** 2000. Rapid method for coextraction of DNA and RNA from natural environments for analysis of ribosomal DNA- and rRNA-based microbial community composition. *Appl. Environ. Microbiol.* **66:**5488–5491.
37. **Hallam, S. J., N. Putnam, C. M. Preston, J. C. Detter, D. Rokhsar, P. M. Richardson, and E. F. DeLong.** 2004. Reverse methanogenesis: testing the hypothesis with environmental genomics. *Science* **305:**1457–1462.
38. **Handelsman, J.** 2004. Metagenomics: application of genomics to uncultured microorganisms. *Microbiol. Mol. Biol. Rev.* **68:**669–685.
39. **Handelsman, J., M. Liles, D. Mann, C. Riesenfeld, and R. M. Goodman.** 2002. Cloning the metagenome: culture-independent access to the diversity and functions of the uncultivated microbial world, p. 241–255, *In* B. Wren and N. Dorrell (ed.), *Functional Microbial Genomics*, vol. 33. Elsevier, New York, N.Y.
40. **Harris, J. K., S. T. Kelley, and N. R. Pace.** 2004. New perspective on uncultured bacterial phylogenetic division OP11. *Appl. Environ. Microbiol.* **70:**845–849.
41. **Henne, A., R. Daniel, R. A. Schmitz, and G. Gottschalk.** 1999. Construction of environmental DNA libraries in *Escherichia coli* and screening for the presence of genes conferring utilization of 4-hydroxybutyrate. *Appl. Environ. Microbiol.* **65:**3901–3907.
42. **Henne, A., R. A. Schmitz, M. Bomeke, G. Gottschalk, and R. Daniel.** 2000. Screening of environmental DNA libraries for the presence of genes conferring lipolytic activity on *Escherichia coli. Appl. Environ. Microbiol.* **66:**3113–3116.
43. **Holben, W. E., J. K. Jansson, B. K. Chelm, and J. M. Tiedje.** 1988. DNA probe method for the detection of specific microorganisms in the soil bacterial community. *Appl. Environ. Microbiol.* **54:**703–711.
44. **Hugenholtz, P., B. M. Goebel, and N. R. Pace.** 1998. Impact of culture-independent studies on the emerging phylogenetic view of bacterial diversity. *J. Bacteriol.* **180:**4765–4774.
45. **Hughes, J. B., J. J. Hellmann, T. H. Ricketts, and B. J. Bohannan.** 2001. Counting the uncountable: statistical approaches to estimating microbial diversity. *Appl. Environ. Microbiol.* **67:**4399–4406.
46. **Kim, U. J., H. Shizuya, P. J. de Jong, B. Birren, and M. I. Simon.** 1992. Stable propagation of cosmid sized human DNA inserts in an F factor based vector. *Nucleic Acids Res.* **20:**1083–1085.
47. **Knietsch, A., T. Waschkowitz, S. Bowien, A. Henne, and R. Daniel.** 2003. Construction and screening of metagenomic libraries derived from enrichment cultures: generation of a gene bank for genes conferring alcohol oxidoreductase activity on *Escherichia coli. Appl. Environ. Microbiol.* **69:**1408–1416.
48. **Knietsch, A., T. Waschkowitz, S. Bowien, A. Henne, and R. Daniel.** 2003. Metagenomes of complex microbial consortia derived from different soils as sources for novel genes conferring formation of carbonyls from short-chain polyols on *Escherichia coli. J. Mol. Microbiol. Biotechnol.* **5:**46–56.
49. **Kowalchuk, G. A.** 2004. *Molecular Microbial Ecology Manual*, 2nd ed. Kluwer Academic, Dordrecht, The Netherlands.
50. **Kozdroj, J., and J. D. van Elsas.** 2000. Application of polymerase chain reaction-denaturing gradient gel electrophoresis for comparison of direct and indirect extraction methods of soil DNA used for microbial community fingerprinting. *Biol. Fertil. Soils* **31:**372–378.
51. **Krsek, M., and E. M. Wellington.** 1999. Comparison of different methods for the isolation and purification of total community DNA from soil. *J. Microbiol. Methods* **39:**1–16.
52. **Kuske, C. R., S. M. Barns, and J. D. Busch.** 1997. Diverse uncultivated bacterial groups from soils of the arid southwestern United States that are present in many geographic regions. *Appl. Environ. Microbiol.* **63:**3614–3621.
53. **Lee, S. W., K. Won, H. K. Lim, J. C. Kim, G. J. Choi, and K. Y. Cho.** 2004. Screening for novel lipolytic enzymes from uncultured soil microorganisms. *Appl. Microbiol. Biotechnol.* **65:**720–726.
54. **Liesack, W., and E. Stackebrandt.** 1992. Occurrence of novel groups of the domain Bacteria as revealed by analysis of genetic material isolated from an Australian terrestrial environment. *J. Bacteriol.* **174:**5072–5078.
55. **Liles, M. R., B. F. Manske, S. B. Bintrim, J. Handelsman, and R. M. Goodman.** 2003. A census of rRNA genes and linked genomic sequences within a soil metagenomic library. *Appl. Environ. Microbiol.* **69:**2684–2691.
56. **Lunn, M., W. T. Sloan, and T. P. Curtis.** 2004. Estimating bacterial diversity from clone libraries with flat rank abundance distributions. *Environ. Microbiol.* **6:**1081–1085.
57. **MacNeil, I. A., C. L. Tiong, C. Minor, P. R. August, T. H. Grossman, K. A. Loiacono, B. A. Lynch, T. Phillips, S. Narula, R. Sundaramoorthi, A. Tyler, T. Aldredge, H. Long, M. Gilman, D. Holt, and M. S.

Osburne. 2001. Expression and isolation of antimicrobial small molecules from soil DNA libraries. *J. Mol. Microbiol. Biotechnol.* **3:**301–308.

58. **Madsen, E. L.** 2005. Identifying microorganisms responsible for ecologically significant biogeochemical processes. *Nat. Rev. Microbiol.* **3:**439–446.
59. **Majernik, A., G. Gottschalk, and R. Daniel.** 2001. Screening of environmental DNA libraries for the presence of genes conferring Na+(Li+)/H+ antiporter activity on *Escherichia coli*: characterization of the recovered genes and the corresponding gene products. *J. Bacteriol.* **183:**6645–6653.
60. **Martinez, A., S. J. Kolvek, C. L. Yip, J. Hopke, K. A. Brown, I. A. MacNeil, and M. S. Osburne.** 2004. Genetically modified bacterial strains and novel bacterial artificial chromosome shuttle vectors for constructing environmental libraries and detecting heterologous natural products in multiple expression hosts. *Appl. Environ. Microbiol.* **70:**2452–2463.
61. **Martin-Laurent, F., L. Philippot, S. Hallet, R. Chaussod, J. C. Germon, G. Soulas, and G. Catroux.** 2001. DNA extraction from soils: old bias for new microbial diversity analysis methods. *Appl. Environ. Microbiol.* **67:**2354–2359.
62. **McCaig, A. E., L. A. Glover, and J. I. Prosser.** 1999. Molecular analysis of bacterial community structure and diversity in unimproved and improved upland grass pastures. *Appl. Environ. Microbiol.* **65:**1721–1730.
63. **Miller, D. N., J. E. Bryant, E. L. Madsen, and W. C. Ghiorse.** 1999. Evaluation and optimization of DNA extraction and purification procedures for soil and sediment samples. *Appl. Environ. Microbiol.* **65:**4715–4724.
64. **More, M. I., J. B. Herrick, M. C. Silva, W. C. Ghiorse, and E. L. Madsen.** 1994. Quantitative cell lysis of indigenous microorganisms and rapid extraction of microbial DNA from sediment. *Appl. Environ. Microbiol.* **60:**1572–1580.
65. **Newman, J. R., and C. Fuqua.** 1999. Broad-host-range expression vectors that carry the L-arabinose-inducible *Escherichia coli ara*BAD promoter and the *ara*C regulator. *Gene* **227:**197–203.
66. **Nicol, G. W., L. A. Glover, and J. I. Prosser.** 2003. Spatial analysis of archaeal community structure in grassland soil. *Appl. Environ. Microbiol.* **69:**7420–7429.
67. **Ochsenreiter, T., D. Selezi, A. Quaiser, L. Bonch-Osmolovskaya, and C. Schleper.** 2003. Diversity and abundance of Crenarchaeota in terrestrial habitats studied by 16S RNA surveys and real time PCR. *Environ. Microbiol.* **5:**787–797.
68. **Ogram, A., G. S. Sayler, and T. Barkay.** 1987. The extraction and purification of microbial DNA from sediments. *J. Microbiol. Methods* **7:**57–66.
69. **Ovreas, L., and V. V. Torsvik.** 1998. Microbial diversity and community structure in two different agricultural soil communities. *Microb. Ecol.* **36:**303–315.
70. **Picard, C., C. Ponsonnet, E. Paget, X. Nesme, and P. Simonet.** 1992. Detection and enumeration of bacteria in soil by direct DNA extraction and polymerase chain reaction. *Appl. Environ. Microbiol.* **58:**2717–2722.
71. **Piel, J.** 2002. A polyketide synthase-peptide synthetase gene cluster from an uncultured bacterial symbiont of *Paederus* beetles. *Proc. Natl. Acad. Sci. USA* **99:**14002–14007.
72. **Porteous, L. A., J. L. Armstrong, R. J. Seidler, and L. S. Watrud.** 1994. An effective method to extract DNA from environmental samples for polymerase chain reaction amplification and DNA fingerprint analysis. *Curr. Microbiol.* **29:**301–307.
73. **Quaiser, A., T. Ochsenreiter, H. P. Klenk, A. Kletzin, A. H. Treusch, G. Meurer, J. Eck, C. W. Sensen, and C. Schleper.** 2002. First insight into the genome of an uncultivated crenarchaeote from soil. *Environ. Microbiol.* **4:** 603–611.
74. **Quaiser, A., T. Ochsenreiter, C. Lanz, S. C. Schuster, A. H. Treusch, J. Eck, and C. Schleper.** 2003. Acidobacteria form a coherent but highly diverse group within the bacterial domain: evidence from environmental genomics. *Mol. Microbiol.* **50:**563–575.
75. **Rappe, M. S., and S. J. Giovannoni.** 2003. The uncultured microbial majority. *Annu. Rev. Microbiol.* **57:**369–394.
76. **Rhee, J. K., D. G. Ahn, Y. G. Kim, and J. W. Oh.** 2005. New thermophilic and thermostable esterase with sequence similarity to the hormone-sensitive lipase family, cloned from a metagenomic library. *Appl. Environ. Microbiol.* **71:**817–825.
77. **Riesenfeld, C. S., R. M. Goodman, and J. Handelsman.** 2004. Uncultured soil bacteria are a reservoir of new antibiotic resistance genes. *Environ. Microbiol.* **6:**981–989.
78. **Riesenfeld, C. S., P. D. Schloss, and J. Handelsman.** 2004. Metagenomics: genomic analysis of microbial communities. *Annu. Rev. Genet.* **38:**525–552.
79. **Rodriguez-Valera, F.** 2004. Environmental genomics, the big picture? *FEMS Microbiol. Lett.* **231:**153–158.
80. **Rondon, M. R., P. R. August, A. D. Bettermann, S. F. Brady, T. H. Grossman, M. R. Liles, K. A. Loiacono, B. A. Lynch, I. A. MacNeil, C. Minor, C. L. Tiong, M. Gilman, M. S. Osburne, J. Clardy, J. Handelsman, and R. M. Goodman.** 2000. Cloning the soil metagenome: a strategy for accessing the genetic and functional diversity of uncultured microorganisms. *Appl. Environ. Microbiol.* **66:**2541–2547.
81. **Rosado, A. S., G. F. Duarte, L. Seldin, and J. D. Van Elsas.** 1998. Genetic diversity of *nif*H gene sequences in *Paenibacillus azotofixans* strains and soil samples analyzed by denaturing gradient gel electrophoresis of PCR-amplified gene fragments. *Appl. Environ. Microbiol.* **64:**2770–2779.
82. **Sambrook, J., E. F. Fritsch, and T. Maniatis.** 1989. *Molecular Cloning: a Laboratory Manual*, 2nd ed. Cold Spring Harbor Laboratory, Cold Spring Harbor, N.Y.
83. **Santosa, D. A.** 2001. Rapid extraction and purification of environmental DNA for molecular cloning applications and molecular diversity studies. *Mol. Biotechnol.* **17:**59–64.
84. **Schloss, P. D., and J. Handelsman.** 2003. Biotechnological prospects from metagenomics. *Curr. Opin. Biotechnol.* **14:**303–310.
85. **Schloss, P. D., and J. Handelsman.** 2004. Status of the microbial census. *Microbiol. Mol. Biol. Rev.* **68:**686–691.
86. **Schmeisser, C., C. Stockigt, C. Raasch, J. Wingender, K. N. Timmis, D. F. Wenderoth, H. C. Flemming, H. Liesegang, R. A. Schmitz, K. E. Jaeger, and W. R. Streit.** 2003. Metagenome survey of biofilms in drinking-water networks. *Appl. Environ. Microbiol.* **69:**7298–7309.
87. **Shizuya, H., B. Birren, U. J. Kim, V. Mancino, T. Slepak, Y. Tachiiri, and M. Simon.** 1992. Cloning and stable maintenance of 300-kilobase-pair fragments of human DNA in *Escherichia coli* using an F-factor-based vector. *Proc. Natl. Acad. Sci. USA* **89:**8794–8797.
88. **Sliwinski, M. K., and R. M. Goodman.** 2004. Spatial heterogeneity of crenarchaeal assemblages within mesophilic soil ecosystems as revealed by PCR–single-stranded conformation polymorphism profiling. *Appl. Environ. Microbiol.* **70:**1811–1820.
89. **Smalla, K., N. Cresswell, L. C. Mendoncahagler, A. Wolters, and J. D. Vanelsas.** 1993. Rapid DNA extraction protocol from soil for polymerase chain reaction-mediated amplification. *J. Appl. Bacteriol.* **74:**78–85.
90. **Sosio, M., F. Giusino, C. Cappellano, E. Bossi, A. M. Puglia, and S. Donadio.** 2000. Artificial chromosomes for

antibiotic-producing actinomycetes. *Nat. Biotechnol.* **18:** 343–345.

91. **Steele, H. L., and W. R. Streit.** 2005. Metagenomics: advances in ecology and biotechnology. *FEMS Microbiol. Lett.* **247:**105–111.

92. **Steffan, R. J., J. Goksoyr, A. K. Bej, and R. M. Atlas.** 1988. Recovery of DNA from soils and sediments. *Appl. Environ. Microbiol.* **54:**2908–2915.

93. **Stein, J. L., T. L. Marsh, K. Y. Wu, H. Shizuya, and E. F. DeLong.** 1996. Characterization of uncultivated prokaryotes: isolation and analysis of a 40-kilobase-pair genome fragment from a planktonic marine archaeon. *J. Bacteriol.* **178:**591–599.

94. **Streit, W. R., and R. A. Schmitz.** 2004. Metagenomics—the key to the uncultured microbes. *Curr. Opin. Microbiol.* **7:**492–498.

95. **Strong, S. J., Y. Ohta, G. W. Litman, and C. T. Amemiya.** 1997. Marked improvement of PAC and BAC cloning is achieved using electroelution of pulsed-field gel-separated partial digests of genomic DNA. *Nucleic Acids Res.* **25:**3959–3961.

96. **Tebbe, C. C., and W. Vahjen.** 1993. Interference of humic acids and DNA extracted directly from soil in detection and transformation of recombinant DNA from bacteria and a yeast. *Appl. Environ. Microbiol.* **59:**2657–2665.

97. **Tien, C. C., C. C. Chao, and W. L. Chao.** 1999. Methods for DNA extraction from various soils: a comparison. *J. Appl. Microbiol.* **86:**937–943.

98. **Torsvik, V., F. L. Daae, R. A. Sandaa, and L. Ovreas.** 1998. Novel techniques for analysing microbial diversity in natural and perturbed environments. *J. Biotechnol.* **64:** 53–62.

99. **Treusch, A. H., A. Kletzin, G. Raddatz, T. Ochsenreiter, A. Quaiser, G. Meurer, S. C. Schuster, and C. Schleper.** 2004. Characterization of large-insert DNA libraries from soil for environmental genomic studies of Archaea. *Environ. Microbiol.* **6:**970–980.

100. **Tringe, S. G., C. von Mering, A. Kobayashi, A. A. Salamov, K. Chen, H. W. Chang, M. Podar, J. M. Short, E. J. Mathur, J. C. Detter, P. Bork, P. Hugenholtz, and E. M. Rubin.** 2005. Comparative metagenomics of microbial communities. *Science* **308:**554–557.

101. **Tsai, Y. L., and B. H. Olson.** 1992. Detection of low numbers of bacterial cells in soils and sediments by polymerase chain reaction. *Appl. Environ. Microbiol.* **58:**754–757.

102. **Tsai, Y. L., and B. H. Olson.** 1992. Rapid method for separation of bacterial DNA from humic substances in sediments for polymerase chain reaction. *Appl. Environ. Microbiol.* **58:**2292–2295.

103. **Tyson, G. W., J. Chapman, P. Hugenholtz, E. E. Allen, R. J. Ram, P. M. Richardson, V. V. Solovyev, E. M. Rubin, D. S. Rokhsar, and J. F. Banfield.** 2004. Community structure and metabolism through reconstruction of microbial genomes from the environment. *Nature* **428:**37–43.

104. **Uchiyama, T., T. Abe, T. Ikemura, and K. Watanabe.** 2005. Substrate-induced gene-expression screening of environmental metagenome libraries for isolation of catabolic genes. *Nat. Biotechnol.* **23:**88–93.

105. **Venter, J. C., K. Remington, J. F. Heidelberg, A. L. Halpern, D. Rusch, J. A. Eisen, D. Wu, I. Paulsen, K. E. Nelson, W. Nelson, D. E. Fouts, S. Levy, A. H. Knap, M. W. Lomas, K. Nealson, O. White, J. Peterson, J. Hoffman, R. Parsons, H. Baden-Tillson, C. Pfannkoch, Y. H. Rogers, and H. O. Smith.** 2004. Environmental genome shotgun sequencing of the Sargasso Sea. *Science* **304:**66–74.

106. **Voget, S., C. Leggewie, A. Uesbeck, C. Raasch, K. E. Jaeger, and W. R. Streit.** 2003. Prospecting for novel biocatalysts in a soil metagenome. *Appl. Environ. Microbiol.* **69:**6235–6242.

107. **Walter, J., M. Mangold, and G. W. Tannock.** 2005. Construction, analysis, and beta-glucanase screening of a bacterial artificial chromosome library from the large-bowel microbiota of mice. *Appl. Environ. Microbiol.* **71:**2347–2354.

108. **Wang, G. Y., E. Graziani, B. Waters, W. Pan, X. Li, J. McDermott, G. Meurer, G. Saxena, R. J. Andersen, and J. Davies.** 2000. Novel natural products from soil DNA libraries in a streptomycete host. *Org. Lett.* **2:**2401–2404.

109. **Wikstrom, P., A. Wiklund, A. C. Andersson, and M. Forsman.** 1996. DNA recovery and PCR quantification of catechol 2,3-dioxygenase genes from different soil types. *J. Biotechnol.* **52:**107–120.

110. **Williamson, L. L., B. R. Borlee, P. D. Schloss, C. Guan, H. K. Allen, and J. Handelsman.** 2005. Intracellular screen to identify metagenomic clones that induce or inhibit a quorum-sensing biosensor. *Appl. Environ. Microbiol.* **71:**6335–6344.

111. **Yun, J., S. Kang, S. Park, H. Yoon, M. J. Kim, S. Heu, and S. Ryu.** 2004. Characterization of a novel amylolytic enzyme encoded by a gene from a soil-derived metagenomic library. *Appl. Environ. Microbiol.* **70:**7229–7235.

112. **Zhou, J., M. A. Bruns, and J. M. Tiedje.** 1996. DNA recovery from soils of diverse composition. *Appl. Environ. Microbiol.* **62:**316–322.

Using Genetic Algorithms To Optimize Functions of Microbial Ecosystems

FREDERIK P. J. VANDECASTEELE, THOMAS F. HESS, AND RONALD L. CRAWFORD

85

Genetic algorithms (GAs) (11) are search and optimization methods based upon the biological principle of evolution through natural selection that have many applications in a wide range of fields in science and engineering. Of interest to the field of environmental microbiology is that they can be used to experimentally identify configurations of microbial ecosystem parameters that result in notable or desirable ecosystem behavior. GAs have a number of specific advantages over other optimization techniques that make them especially attractive for such use in microbial ecology. They can search through large sets of ecological parameters without having to test all possible combinations of those parameters and typically perform well on systems with high levels of interaction between parameters. Also, they can operate on study systems with considerable levels of noise or variability. Finally, with GAs it is possible to optimize and manipulate a system without any knowledge of its internal dynamics, essentially treating the system as a black box and only considering inputs and outputs. This chapter gives a general outline of the GA approach to optimization and lists a number of specific considerations for microbial ecological applications.

ALGORITHM OUTLINE

General Approach

A genetic algorithm is a mathematical optimization approach inspired by the Darwinian principle of evolution through natural selection. The objective of an optimization attempt is to find the values of variables that will lead to the most desirable response of a system. As GAs are inspired by evolution, part of the vocabulary associated with them has also been borrowed from biology. Potential solutions to a problem (optimization trials) are conceptually considered to be individuals containing a chromosome encoding the details of the proposed solution. Such a chromosome consists of genes representing the system variables that are studied, and the levels of the variables are considered to be alleles of those genes. GAs simultaneously operate on a collection of such solutions, called a population. Initially, a first generation of potential solutions is typically generated randomly. Each individual is assigned a fitness value that is proportional to the degree to which its encoded solution achieves the desired objective. A new generation of solutions is created by selecting solutions from the old generation with a probability that is proportional to their fitness value. The selected individuals are called parents. After crossover and mutation are applied, these parents result in children that will make up the next generation of solutions. Crossover is a process that typically occurs with a high probability and in which pieces of chromosome are exchanged between pairs of parents. During the process of mutation, each gene has a typically low probability of changing in allele value. The newly created generation is then evaluated for fitness, after which the process of selection, crossover, and mutation is repeated. This whole cycle is repeated until an acceptable solution is obtained or until experimental resources run out (Fig. 1). Overall, genetic algorithms operate on a group or population of solutions, in which good solutions are promoted, their internal information is exchanged, and random variations are applied. The optimization proceeds in steps or generations of such groups, in which gradually better and better solutions for the initial problem are typically identified.

Since genetic algorithms are very flexible in their application area, this optimization approach can be readily used in the context of environmental microbiology. First, a desirable function for a microbial ecosystem of interest must be identified. This can be the production of a chemical, the degradation of a pollutant, or any other function that is of interest to a particular research group. Next, a suitable method for manipulating the ecosystem to promote the desirable function and a method for representing this manipulation in a GA-friendly way need to be identified. This step will be discussed in more detail in "Types of Ecosystem Manipulation" below. From an optimization point of view, a potential solution to the problem will be a representation of a way in which to manipulate the microbial ecosystem of interest, while the goal will be to find a solution that achieves the desirable ecosystem function in the best way possible. An initial group of solutions will be experimentally implemented in the lab or in the field, after which the performance or fitness of each of them can be measured. These fitness values are then fed into a genetic algorithm that uses the simulated processes of selection, crossover, and mutation to propose a new batch of potential solutions. This new generation of solutions is again experimentally evaluated, and the cycle is repeated.

Basic Algorithm

Much flexibility exists in the design of the actual algorithm involved in a GA optimization approach. This section outlines one version of a simple and basic GA (see also reference 10). Some common variations and additions to this algorithm are described in the next section. The general outline of a genetic algorithm is presented in Fig. 1.

Representation

The representation in a genetic algorithm is concerned with how a potential solution will be encoded in a chromosome and how that encoding will relate to the characteristics of the solution. Most genetic algorithms use a simple string of symbols as the structure for a chromosome, and many use a binary string, meaning that the chromosome consists of a series of zeros and ones. The length of the chromosome has to be chosen so that it can contain the appropriate amount of information. As an example, in the case where an attempt is made to manipulate ecosystem function by changing the presence or absence of 10 ecological factors, a potential solution could be encoded in a straightforward manner as a binary string of length ten. A zero or a one at a particular position in that string would respectively encode for the absence or presence of the corresponding factor in the ecosystem.

Initialization

After deciding upon an appropriate representation, a suitable population size has to be chosen. This number will represent how many potential solutions or individuals will be simultaneously tested during each generation, or step, in the optimization process. Initialization occurs by generating a number of bit strings equal to the population size, with length equal to the chromosome length. These initial strings are typically randomly generated and make up the first generation of solutions.

Fitness Evaluation

During fitness evaluation, a fitness value is assigned to each of the candidate solutions in a particular generation. In the context of environmental microbiology, this involves converting the chromosomes of the candidate solutions to experimental setups, manipulating the ecosystem parameters accordingly in the lab or in the field, and measuring how well each manipulated ecosystem achieves the overall functional objective that was chosen initially. In the case of a single ecological objective that needs to be maximized, the experimentally measured ecosystem performance values translate directly to fitness values. In the case of an ecological function that has to be minimized, the measured values can for example be converted to fitness values by taking the inverse. In the case of multiple functional objectives, a fitness value could simply be calculated as an algebraic sum of the measured performance values, with the option of including weights for each of the objectives.

0. initialization

repeat

1. fitness evaluation
2. selection
3. crossover
4. mutation

go to step 1

FIGURE 1 Schematic outline of a genetic algorithm.

before crossover	parent 1	**01**|**01111001**
	parent 2	11|01010001
after crossover	child 1	**01**|01010001
	child 2	11|**01111001**

FIGURE 2 Single-point crossover; crossover position between genes 2 and 3.

Selection

Once fitness values have been assigned to a population of solutions, the process of generating the next generation is started by selecting solutions from the old generation. In an algorithmic analogy to the biological principle of "survival of the fittest," the underlying idea of the selection step is to preferentially select more solutions with a high fitness. A straightforward way of implementing this is by using roulette wheel selection (10). Here, a roulette wheel that has a number of slots proportional to the sum of all the fitness values of the individuals in the old generation is simulated. Each of those individuals is then assigned a section on the roulette wheel proportional to its own fitness. Then, selection of an individual is achieved by simulating a spin of the roulette wheel. In total, as many individuals are selected in this way as the population is large, so it is possible that certain individuals are selected multiple times.

Crossover

During the crossover step, sections of chromosome are exchanged between pairs of selected individuals. Using single-point crossover, a position between two chromosome elements is randomly chosen, after which the sections behind this position are exchanged between chromosomes (Fig. 2). A weighted coin is tossed to determine whether or not crossover will occur between each pair of selected individuals, and crossover typically occurs with a high probability (e.g., 90%).

Mutation

During the mutation step, each gene in the newly generated population has a low probability (e.g., 1% per gene) of changing in allele value. For the microbial ecological applications discussed here in which algorithm speed is not a concern, this can be implemented by going over every gene on the chromosome of every new individual and tossing a weighted coin. A mutation event in binary applications would correspond to changing a zero to a one or vice versa (Fig. 3).

before mutation	0101**0**10001
after mutation	0101**1**10001

FIGURE 3 Mutation at position 5.

Common Algorithm Variations

The algorithm sketched above uses a binary representation to encode ecosystem manipulations in which ecological factors can be either present or absent. Using binary encoding, it is also possible to represent various levels or concentrations of ecological factors. To achieve this, blocks of bits in a chromosome can be used to map to a real number. For example, the first four bits of each chromosome could be converted to an integer between 0 and 15. This integer can then be used to point to one of 16 values on regular intervals between any two real numbers. It is similarly possible to code for levels or concentrations of ecological factors by directly using integers or real numbers for the allele values of genes on chromosomes instead of bits. However, even though there is debate within the GA community over what is the best type of alphabet to use for a representation, it seems that there is a general tendency towards using binary representations. For multiobjective optimizations, more advanced techniques than simply calculating a weighted algebraic sum of fitness values have been developed (4, 6).

The general idea behind the selection step in a GA is to proportionally select more individuals with a higher fitness. Selecting with a probability proportional to fitness can cause distortions early on in the optimization attempt when a few very fit individuals can take over the population and late in the optimization when the fitness values in the population are very similar to each other. This can be solved by linearly rescaling fitness values before selection (10). This is typically done by transforming the fitness values so that the average of the transformed values equals the average of the original values and the transformed maximum fitness value equals two times the average fitness value. If this results in negative fitness values, the values are transformed so that the averages again stay the same and the transformed minimum fitness value equals zero.

In the algorithm outlined above, roulette wheel selection was used as a method of selecting individuals from the old population to generate a new population. One popular alternative to this is (deterministic) tournament selection, in which a small number of individuals are randomly sampled from the old population, and the one with the highest fitness value becomes the selected individual.

Often, alternatives to single-point crossover are used. In multipoint crossover, more than one crossover point is chosen and chromosome sections are exchanged between chromosomes according to these points in an alternating fashion. For uniform crossover, each element on the chromosome can be separately exchanged between individuals.

To ensure that the best solution identified in the previous generation does not get lost, a process called elitism is often applied by copying that solution unmodified into the new generation.

Algorithm Source

Technically, a GA computer program involves very basic programming operations such as random number generation, simple algebraic manipulations, and data structure operations. It should be possible for a person with intermediate computer programming skills to write a dedicated GA for any environmental microbial application. If at all possible, this approach would be preferable, as it allows full control over the design and implementation of the GA. If writing one's own GA is not an option, several online repositories containing GA programs and source codes are available through the Internet (see, e.g., Navy Center for Applied Research in Artificial Intelligence [http://www.aic.nrl.navy.mil/galist/src] and European Network of Excellence in Evolutionary Computing [http://evonet.lri.fr/evoweb/resources/software/]).

TYPES OF ECOSYSTEM MANIPULATION

Two ways of manipulating microbial ecosystem factors to obtain a desirable functional behavior can be identified. First, efficient mixed cultures can be designed by using a GA to determine which isolated strains to combine for optimal functional performance (25). Secondly, it is possible to use a genetic algorithm to manipulate environmental conditions and drive an already existing ecosystem in a desirable direction. These two approaches can be combined in hybrid strategies whereby a mixed culture and its optimal environmental conditions are optimized at the same time or an existing ecosystem is manipulated by both adding strains and changing environmental conditions.

Combining Organisms

A GA can be used to search through a set of separate pure strains to find a combination of strains that together form a mixed culture that is very efficient at performing a predetermined function. The full set of candidate ecosystem members may consist of strains that perform the function to a certain degree themselves, that perform part of the function, or that could possibly support other strains. Many ecological rationales that could explain why groups would perform functions better than individual organisms can be offered. The use of GAs has the potential to unlock the promise of functional efficiency in microbial ecosystems. It also offers an alternative to enrichment cultures for obtaining microbial consortia that is not based on preferential growth under selective conditions.

Manipulating Environmental Conditions

Changes in environmental conditions generally influence ecosystem dynamics. Certain environmental conditions can cause shifts in ecosystem structure and function. Therefore, it should be possible to use a GA to find specific environmental conditions that will drive a microbial ecosystem towards a desirable behavior. An approach like this would start by identifying a set of environmental conditions that will influence ecosystem dynamics and that can be manipulated experimentally, either in the lab or in the field. Such conditions could include chemical and physical factors like temperature, pH, salinity, light treatment, and mixing. They could also include concentrations of electron donors, electron acceptors, and other chemicals. After identification of such a set of environmental conditions, a GA can be used to find treatment combinations based on these conditions that will result in a high level of the desired ecosystem behavior.

Representations

Whether optimizing a microbial ecosystem function by combining organisms, manipulating environmental conditions, or a combination of both, many possibilities for representing potential solutions in a GA-friendly manner exist. Two main ways in which ecosystem factors (organism combinations or environmental conditions) can be varied are by changing the concentration and by changing the timing of application of these factors in an ecosystem.

Concentrations and Levels

The simplest approach to manipulating ecosystems is applying various factors either at a fixed level or not at all. A suitable GA representation of a trial solution in such a case would be a simple binary string equal in length to the size of the candidate set of factors. A bit at a particular position in the string then codes for the presence or absence of the respective factor in the main set. For example, four factors at fixed levels and no timing would give the following (factors 1 through 4 are read left to right; zero and one code for absence and presence, respectively):

1010 → apply factors 1 and 3 at time zero

An elaboration on this basic idea is to allow the level or concentration of each factor applied to an ecosystem to vary. As described earlier, it is possible to use a binary chromosome representation for this, with blocks of bits mapping to integer numbers or to real numbers. Alternatively, a chromosome directly containing integer or real numbers can be used. For example, four factors at various levels and no timing (00 = zero, 01 = low level, 10 = medium level, 11 = high level) would give the following:

11 | 01 | 10 | 00 → factor 1: high level
factor 2: low level
factor 3: medium level
factor 4: zero

When organisms are combined, whether varying their relative concentration or not, it always seems like good practice to keep constant the total cell number added to a culture across strains. This will ensure that an increase in functional performance is not caused simply by an increase in initial cell mass.

Timing

If the approach outlined in the previous section is implemented as is, then no information about the timing of applying factors is included in the chromosomes and all factors are simply applied at the start of the experiment. The representations described above can be modified to include such timing information, however.

For each of the factors used in the optimization attempt, the chromosome could code for a time of addition to the ecosystem, represented by using either a block of bits, an integer, or a real value. This information could then be preceded by a type of concentration information described above. For example, one of four levels and one of four timings for each of three factors (six leftmost bits encode for level, six rightmost for timing; level is as in previous example; timing: 00 = day 0, 01 = day 1, 10 = day 2, 11 = day 3) would give the following:

11 | 01 | 10 | | 11 | 10 | 00 → factor 1: high level, day 3
factor 2: low level, day 2
factor 3: medium level, day 0

The total amount of time an ecosystem is observed could be divided into equal time sections. The chromosome would then consist of as many multiple pieces of concentration information as there are time sections. Each subregion of the chromosome would determine which factors are added to the ecosystem and at which levels at each fixed point in time. For example, application of four factors at three fixed times (fixed times: week 0, week 1, and week 2; each block of four bits codes for the combination of factors at a consecutive time point; application level can be either presence or absence) would give the following:

1010 | 1111 | 1110 → week 0: apply factors 1 and 3
week 1: apply factors 1, 2, 3, and 4
week 2: apply factors 1, 2, and 3

The previous approach could be made even more flexible by adding a chromosome region coding for a variable time for each of the time sections. For example, application of four factors at five variable times gives the following (the first five blocks of four bits code for combinations of factors; the next five blocks of two bits code for timing as in the previous examples):

1010 | 1111 | 1110 | 1000 | 1011 | | 11 | 10 | 00 | 01 | 01 →
apply factors 1 and 3 at day 3
apply factors 1, 2, 3, and 4 at day 2
apply factors 1, 2, and 3 at day 0
apply factor 1 at day 1
apply factors 1, 3, and 4 at day 1

APPLICATION AREAS

GAs can be of great value to environmental microbiology because they allow the efficient control and design of microbial ecosystems. The approach can be used for many applications in fields such as bioremediation and fermentation. They can also be useful for more fundamental ecological research.

The technique outlined here can be used to design mixed microbial cultures that are efficient at degrading contaminants. Such mixed cultures have the potential of outperforming the degradation potential of single organisms. Mixed cultures can consist of multiple organisms that all degrade the contaminant without hampering each other's functioning, of organisms that each perform part of the degradation pathway, or of assemblies where certain member strains are supporting degrading strains by creating favorable growth conditions via crossfeeding or the removal of toxic intermediates. Optimized mixed cultures can be applied ex situ in reactors treating contaminated materials or in situ in a bioaugmentation approach.

GAs can also be used to optimize environmental conditions to induce a desirable function in an existing microbial ecosystem. In the context of bioremediation, this translates to designing an efficient biostimulation treatment. Such treatments can be designed in the lab as well as in the field, where small regions of a contaminated area can be reserved for measuring the degradation associated with particular combinations of environmental conditions.

GAs can be used to design efficient mixed cultures and culture conditions for the fermentation industry, thereby offering the promise of increased productivity. Productivity may be measured as an increase in yield of a specific fermentation product, a decrease in time required to complete fermentation, greater stability of the fermentation community, or a combination of these improvements.

From a fundamental ecological point of view, using GAs to optimize functions in microbial ecosystems offers great promise. When studying a particular ecological function, rather than having to go into the field and search for ecosystems exhibiting this behavior, it is now possible to use a GA to selectively induce the behavior in the lab in samples from a variety of origins. When optimizing, GAs treat systems as black boxes by varying ecosystem inputs and measuring behavior as an output. No knowledge about the internal dynamics of the ecosystem is required, making the approach very pragmatic. Once efficiently functioning ecosystems have been obtained, however, it is possible to investigate the ecological mechanisms underlying this efficiency. For example, replicates of a microbial ecosystem that exhibits an intermediate level of a desirable behavior can be manipulated so that the behavior is selectively stimulated and inhibited. Molecular analysis tools can then be used for further analysis of the mechanisms underlying the behavior of interest.

SPECIFIC CONSIDERATIONS FOR MICROBIAL APPLICATIONS

Costly Fitness Evaluations

Most GA applications described in the literature involve solving problems where fitness values of potential solutions can be obtained by calculation or simulation. Using a GA to design an efficient microbial ecosystem is different because fitness values have to be experimentally evaluated, making fitness evaluation more costly and more time-consuming. The population size that is chosen and the total number of generations that are evaluated determine the total number of fitness evaluations that have to be performed in the course of an optimization attempt. Generally, the larger the population size, the better and faster the optimization attempt will proceed, and more difficult problems call for larger population sizes. However, population size is often determined and limited by practical considerations, such as the amount of time and manpower that can be devoted to a study. Many GA experiments for microbial applications tend to be performed using rather small population sizes, with tens to hundreds of individuals per generation being more likely than the tens of thousands that are typical for many simulation-based studies.

Optimizing ecosystem function can be formulated as a problem with varying degrees of sophistication, depending on the level of detail associated with the manipulation of ecological factors. Generally, an attempt to manipulate more factors at more different levels will result in a more difficult optimization problem. Such problems will tend to require larger population sizes and the evaluation of more generations. The problem difficulty (number of possible combinations) represented by a binary string of length n can easily be calculated as 2^n. This may help in deciding on a realistic representation for a particular problem in view of available resources and on an appropriate population size.

Fitness landscape analysis of microbial ecosystems or the study of the correlation between similarity in ecosystem structure and similarity in ecosystem function (23) may allow for another estimation of problem difficulty, which could help in determining the experimental resources needed for a successful optimization attempt.

As a guideline, the minimum population size for a GA application with binary strings can be calculated using the equation $P = (1 - (1/2)^{M-1})^L$ (18), where M is population size, L is string length, and P is the probability that each locus will be represented for each gene, which should be high.

Automation of fitness evaluations by robotic equipment such as liquid handling devices will prove to be very valuable. It will allow for more otherwise-labor-intensive fitness evaluations and for increased population sizes.

Noisy Experimental Measurements

A characteristic of microbial ecological system measurements is that they often exhibit a considerable level of variability or noise. This has some specific implications for a GA-based optimization approach. Variation within measurements of ecosystem function as influenced by the same set of ecological factors can simply be accounted for by evaluating replicates of each potential solution in a generation.

There is also a possibility of ecosystem performance variation between generations, and this requires some more thorough precautions. Ideally, all generations of a planned GA optimization run should be performed using materials with identical properties, originating from the same source. This can be achieved by preparing or collecting sufficient quantities of media, chemical stocks, inocula, and environmental sample before starting an actual optimization run. These large quantities should then be divided into aliquots and stored, with one or more aliquots being used per generation.

Also because of the potential for considerable variability between generations, care has to be taken to unambiguously demonstrate a successful optimization run. One way of demonstrating success is to show increasing trends in maximum and average functional performance per generation. Ideally, this has to be demonstrated on data obtained in one single additional batch of confirmation experiments. Here, the fitness of the individual exhibiting the median level of fitness in each generation can be taken as a surrogate measure of the average fitness. Another way of demonstrating a successful optimization is showing that the attempt resulted in a solution with a higher fitness than what could have been found purely by chance through the testing of random combinations of ecosystem factors. In principle, the fitness values obtained in the first, random generation of experiments can be used to estimate how many random experiments would be needed to have a 95% chance of obtaining a fitness value that is at least as high as what was obtained at the end of the GA optimization run. The GA outperforms a random search if this number is higher than the total number of fitness values evaluated. A less strict criterion would be the demonstration of a significantly higher maximum performance level obtained at the end of the optimization run than in the very first, random generation.

Algorithm Design

With all GA optimizations, but especially in applications with small population sizes, it is important to maintain enough variability in the chromosome structures to avoid the algorithm prematurely settling on suboptimal solutions (premature convergence). Some specific measures can be taken to address this when using small-population-size GAs for microbial applications. A purely random initialization of the first GA generation may lead to disproportionate amounts of some ecological factors, which may bias the optimization trial. This can be solved by initializing the chromosomes of the first generation in a modified random way so that the allele values of each gene are present in equal amounts. The initialization can be practically implemented by evenly distributing allele values for each gene and then randomly permuting them across chromosomes. It may be wise to avoid crossing over two copies of the same parent, since this does not result in any variation. This can be implemented by simply shuffling the selected parents before crossover until no two copies of the same parent are paired up. Chromosomes must not be chosen too many times during the selection step of a GA, as this also decreases variability. A limit could be imposed on the number of times the same parent is selected to produce the next generation. For example, the selection step could be repeated until no parent is selected more than twice.

As correctly pointed out by Davies et al. (5), financial cost is often an important consideration when designing treatments for environmental biological applications. Often, the aim of optimization is to find treatments that induce a desirable function to a high degree but that also do this at an acceptable cost. When a cost is estimated for each level of each ecological factor, optimizing ecosystem function and simultaneously minimizing cost become a multiobjective optimization task that can be solved as described earlier.

Since fitness evaluations are typically very costly and time-consuming for microbial applications, any additional technique that can help in predicting ecosystem function and help speed up optimization is very welcome. There may be great promise in designing hybrid optimization algorithms that combine a classical GA part with a second optimization or estimation technique. For example, after fitness evaluation, a GA could produce a new generation of solutions that is double the size of the previous one. A model of ecosystem functioning or a correlation-based approach could then be used to predict the performance level of each of the new individuals. These values would then be used to cull back the new generation to its regular size, after which the performance of the reduced number of solutions would be evaluated.

ALTERNATIVE TYPES OF EVOLUTIONARY COMPUTATION

Genetic algorithms belong to the larger field of evolutionary computation, which contains other population-based optimization methods that are similarly inspired by the biological principle of natural evolution. Other such methods include evolutionary programming, evolution strategies, and genetic programming.

Evolutionary programming (9) and evolution strategies (3) are similar to genetic algorithms but typically do not include a recombination or crossover step. Candidate solutions are represented as a set of system properties. A new solution is derived from one parent solution by randomly perturbing the values of those properties. These two types of methods may provide an alternative to genetic algorithms for the optimization of functions of microbial ecosystems, especially when the optimization task is to find the appropriate level of various ecosystem factors, rather than more simply to find the right combination of such factors.

The genetic programming (13) approach essentially uses genetic algorithms to evolve computer programs, typically represented as tree structures. This approach seems less suitable for the optimization of functions of microbial ecosystems because there is no obvious way to tie an evolving computer program to the properties of an ecosystem. Although less suitable for optimization of microbial ecosystem functions, genetic programming could offer some possibilities for designing scripts or instruction sets when using robotic equipment for ecosystem assembly.

EXAMPLES OF APPLICATIONS

Although not strictly involving ecological applications, quite a few studies that use a GA to optimize growth and medium conditions for cultures of single organisms have been published. Most of these applications are in the field of industrial fermentation, in which medium compositions are manipulated to maximize production of various chemicals (1, 2, 7, 8, 17, 26, 27, 28, 30; for reviews, see references 12, 16, 19, and 29). Similar attempts have been made to optimize medium conditions for oil degradation by a pure culture (14) and for the growth of insect cells (15). A GA has been used to design silage additives (5). The levels of eight supplements were varied to manipulate and improve the ensilage process of fresh ryegrass.

An early attempt at using a GA to stimulate desirable behavior in an undefined and existing microbial ecosystem involved designing a biostimulation treatment for the degradation of toluene (20). A GA was used to find the optimal levels of 11 nutrients and electron acceptors to stimulate fast and cheap toluene degradation in lab-scale microcosms containing an inoculum derived from environmental sources.

GAs have been used to efficiently design mixed cultures from sets of isolated organisms. In an initial study, a GA searched through a set of 40 microbial soil isolates for subsets that were efficient at degrading the azo dye Orange II (21). In a second study, a GA was used to design mixed cultures with high levels of biomass production based on a set of 20 soil isolates (22). In a third study, a GA was able to identify subsets of a set of 20 fast-growing soil isolates with gradually lower levels of overall growth (24).

REFERENCES

1. **Bapat, P. M., and P. P. Wangikar.** 2004. Optimization of rifamycin B fermentation in shake flasks via a machine-learning-based approach. *Biotechnol. Bioeng.* **86:**201–208.
2. **Beste, H., M. Fackeldey, M. Willems, L. Stockmann, and D. Weuster-Botz.** 1997. Optimization of fermentation medium composition in substrate-controlled continuous stirred tank reactors. *Chem. Eng. Technol.* **20:**403–413.
3. **Beyer, H.-G., and H.-P. Schwefel.** 2002. Evolution strategies—a comprehensive introduction. *Nat. Comput.* **1:**3–52.
4. **Coello Coello, C. A., D. A. Van Veldhuizen, and G. B. Lamont.** 2002. *Evolutionary Algorithms for Solving Multi-Objective Problems*. Kluwer Academic Publishers, Boston, Mass.
5. **Davies, Z. S., R. J. Gilbert, R. J. Merry, D. B. Kell, M. K. Theodorou, and G. W. Griffith.** 2000. Efficient improvement of silage additives by using genetic algorithms. *Appl. Environ. Microbiol.* **66:**1435–1443.
6. **Deb, K.** 2001. *Multi-Objective Optimization Using Evolutionary Algorithms*. Wiley, Chichester, United Kingdom.
7. **Etschmann, M. M. W., D. Sell, and J. Schrader.** 2004. Medium optimization for the production of the aroma compound 2-phenylethanol using a genetic algorithm. *J. Mol. Catal. B Enzym.* **29:**187–193.
8. **Fang, B. S., H. W. Chen, X. L. Xie, N. Wan, and Z. D. Hu.** 2003. Using genetic algorithms coupling neural networks in a study of xylitol production: medium optimization. *Process Biochem.* **38:**979–985.
9. **Fogel, L. J., A. J. Owens, and M. J. Walsh.** 1966. *Artificial Intelligence through Simulated Evolution*. John Wiley, New York, N.Y.
10. **Goldberg, D. E.** 1989. *Genetic Algorithms in Search, Optimization, and Machine Learning*. Addison-Wesley, Reading, Mass.
11. **Holland, J. H.** 1975. *Adaptation in Natural and Artificial Systems: an Introductory Analysis with Applications to Biology, Control, and Artificial Intelligence*. University of Michigan Press, Ann Arbor.

12. **Kennedy, M., and D. Krouse.** 1999. Strategies for improving fermentation medium performance: a review. *J. Ind. Microbiol. Biotechnol.* **23:**456–475.
13. **Koza, J. R.** 1992. *Genetic Programming: on the Programming of Computers by Means of Natural Selection*. The MIT Press, Cambridge, Mass.
14. **Li, S. X., D. Xing, H. M. Qin, X. B. Yang, and S. C. Tan.** 2004. Experimental study on genetic algorithms of medium optimization for oil-degradation. *Chin. J. Anal. Chem.* **32:**481–484.
15. **Marteijn, R. C. L., O. Jurrius, J. Dhont, C. D. de Gooijer, J. Tramper, and D. E. Martens.** 2003. Optimization of a feed medium for fed-batch culture of insect cells using a genetic algorithm. *Biotechnol. Bioeng.* **81:**269–278.
16. **Parekh, S., V. A. Vinci, and R. J. Strobel.** 2000. Improvement of microbial strains and fermentation processes. *Appl. Microbiol. Biotechnol.* **54:**287–301.
17. **Patil, S. V., V. K. Jayaraman, and B. D. Kulkarni.** 2002. Optimization of media by evolutionary algorithms for production of polyols. *Appl. Biochem. Biotechnol.* **102–103:** 119–128.
18. **Reeves, C. R.** 1993. Using genetic algorithms with small populations, p. 92–99. *In* S. Forrest (ed.), *Proceedings of the Fifth International Conference on Genetic Algorithms*. Morgan Kaufmann, San Mateo, Calif.
19. **Shioya, S., K. Shimizu, and T. Yoshida.** 1999. Knowledge-based design and operation of bioprocess systems. *J. Biosci. Bioeng.* **87:**261–266.
20. **Vandecasteele, F.** 2002. *Moleculair-genetische karakterisatie van een BTEX-gecontamineerd grondwatersediment en gerichte biostimulatie via een genetisch algoritme*. Thesis towards the degree of Bio-ingenieur, Ghent University, Belgium.
21. **Vandecasteele, F. P. J.** 2003. Constructing efficient microbial consortia using a genetic algorithm, p. 69–71. *In* A. M. Barry (ed.), *Biological Applications for Genetic and Evolutionary Computation (BioGEC)—2003 Genetic and Evolutionary Computation Conference Workshop Program*. American Association for Artificial Intelligence, Menlo Park, Calif.
22. **Vandecasteele, F. P. J., T. F. Hess, and R. L. Crawford.** 2003. Constructing microbial consortia with optimal biomass production using a genetic algorithm, p. 299–302. *In* J. A. Foster (ed.), *2003 Genetic and Evolutionary Computation Conference—Late-Breaking Papers*. American Association for Artificial Intelligence, Menlo Park, Calif.
23. **Vandecasteele, F. P. J., T. F. Hess, and R. L. Crawford.** 2004. A correlated fitness landscape describes growth in experimental microbial ecosystems: initial results. *In* M. Keijzer (ed.), *2004 Genetic and Evolutionary Computation Conference—Late-Breaking Papers*. American Association for Artificial Intelligence, Menlo Park, Calif.
24. **Vandecasteele, F. P. J., T. F. Hess, and R. L. Crawford.** 2004. Constructing microbial consortia with minimal growth using a genetic algorithm, p. 123–129. *In* G. R. Raidl et al. (ed.), *Applications of Evolutionary Computing—EvoWorkshops 2004. Lecture Notes in Computer Science Vol. 3005*. Springer-Verlag, Berlin, Germany.
25. **Vandecasteele, F. P. J., T. F. Hess, and R. L. Crawford.** 2004. Thoughts on using evolutionary computation to assemble efficient ecosystems. *In* R. Poli et al. (ed.), *Biological Applications of Genetic and Evolutionary Computation (BioGEC)—2004 Genetic and Evolutionary Computation Conference Workshop Proceedings*. American Association for Artificial Intelligence, Menlo Park, Calif.
26. **Weuster-Botz, D., V. Pramatarova, G. Spassov, and C. Wandrey.** 1995. Use of a genetic algorithm in the development of a synthetic growth medium for *Arthrobacter simplex* with high hydrocortisone Δ1-dehydrogenase activity. *J. Chem. Technol. Biotechnol.* **64:**386–392.
27. **Weuster-Botz, D., and C. Wandrey.** 1995. Medium optimization by genetic algorithm for continuous production of formate dehydrogenase. *Process Biochem.* **30:**563–571.
28. **Weuster-Botz, D., M. Karutz, B. Joksch, D. Schartges, and C. Wandrey.** 1996. Integrated development of fermentation and downstream processing for L-isoleucine production with *Corynebacterium glutamicum*. *Appl. Microbiol. Biotechnol.* **46:**209–219.
29. **Weuster-Botz, D.** 2000. Experimental design for fermentation media development: statistical design or global random search? *J. Biosci. Bioeng.* **90:**473–483.
30. **Zuzek, M., J. Friedrich, B. Cestnik, A. Karalic, and A. Cimerman.** 1996. Optimization of fermentation medium by a modified method of genetic algorithms. *Biotechnol. Tech.* **10:**991–996.

Metagenomic Methods for the Identification of Active Microorganisms and Genes in Biodegradation Processes

DON COWAN AND WILLIAM STAFFORD

86

Isolated microorganisms constitute only a minor fraction of the global microbial diversity, which may comprise millions of species (97). The huge discrepancy between species diversity as assessed by culture-dependent and that assessed by culture-independent methods has led to the concept of "unculturables," more recently revised as "uncultured" (4, 40, 97). This change in terminology reflects the view that these organisms are probably not unculturable per se but have merely not been cultured yet! Whatever the causes of the underlying technical difficulties in cultural studies (43, 69, 85), the appreciation of the uncultured majority has been largely responsible for the rapid development of metagenomic technologies.

The fundamental experimental questions for the analysis and understanding of microbial biodegradation processes are as follows.

- Which organism(s) are responsible for primary degradation processes? Organisms may be classified as primary degraders (i.e., producers of the enzymes which undertake the early degradation steps, such as the hydroxylases and dioxygenases in phenol degradation pathways); secondary degraders (i.e., organisms using the products of the primary degradation); and tertiary metabolizers (organisms benefiting from the metabolic activities of the first two classes). Of course, a single species may be active simultaneously in all three classes.
- Which gene products (enzyme and pathway activities) are responsible for degradation processes?
- How can the two data sets of phylogeny and function be linked?

There are some cautionary guidelines that should be followed in the attempts to link phylogeny with function.

1. Identification of an organism/phylotype provides little useful information on in vivo function. At best, the identification of a specific organism, either as an isolate or as a phylotype, provides information on the potential function of that organism; viz., the presence of *Bradyrhizobium* indicates the capacity for dinitrogen fixation, not that it occurs.

2. A change in the population of any organism in response to a specific action (e.g., addition of a substrate) implies a putative role in the degradation process, but not necessarily as a primary degrader. It is a reasonable assumption that if addition of a substrate to a microbial community in equilibrium induces a change in population structure, members of the community are affected by the substrate. It is impossible to infer whether individual members are responsible for primary degradation or the effects are due to secondary degraders and tertiary metabolizers.

3. The ability of an isolated organism to degrade a substrate cannot be taken as evidence that the organism is an important contributor to the degradation process in a natural community. The ability to degrade a specific substrate is solid evidence that the organism possesses the appropriate metabolic capability (i.e., its genome harbors the relevant genes, and these can be adequately expressed under some conditions). To extend this information to a direct implication of in vivo function requires information on the presence (and number) of the organism in the natural community, on the expression of the relevant genes in vivo, and on the relevant in situ activities.

4. The presence of a specific gene or enzyme activity in a biodegrading community is not a priori evidence that either is an important component of the degradation process. The presence of a specific gene (as determined by metagenomic PCR amplification, for example) provides no information on whether the gene is functionally expressed in situ. The detection of a specific enzymic activity is circumstantial evidence for its involvements in a process (e.g., dehalogenase activity in trichloroethylene degradation), but given the multiple possible pathways for degradation of most xenobiotics (see, for example, reference 24), its relevance in situ is difficult to ascertain.

5. Rapid up-regulation of gene expression or increase in measurable enzyme activity in response to addition of a substrate is circumstantial evidence for direct involvement in the degradation pathway. Due to the rapid response of induction processes, effects which are immediately responsive to substrate additions (i.e., increased levels of specific mRNAs, proteins or enzyme activities) provide good circumstantial evidence that these genes/gene products are directly involved in the utilization of the substrate. However, mRNA posttranscriptional regulation and protein posttranslational modifications can influence the degradation activity.

The following discussion presents some of the techniques employed to address the questions above and demonstrates how they have been used to generate more explicit links between diversity and function.

CULTURE-DEPENDENT ENRICHMENT

Culturing Techniques

The laboratory isolation of axenic cultures has proved invaluable for the study of degradative activity. The simplest and most commonly used approach to identify organisms involved in biodegradation is to isolate microbial strains capable of utilizing the target substrate, xenobiotics, or pollutant as a sole C/N source. Subsequent investigations of relevant enzyme pathways, degradation kinetics, etc., are typically taken as evidence that the isolate plays a significant role in the appropriate in vivo biodegradation processes. We now accept that such assumptions are not necessarily soundly based. Examples where organisms that are normally assumed to be "major players" constitute a very low fraction of the community (80) or where the key activities are linked to previously unknown species (55) all support the need for caution.

Enrichment Strategies

In contrast with classical culturing that imposes a high degree of selection during isolation, culture-based enrichments have the potential to capture a larger biodiversity by allowing certain microbial interactions and by maintaining conditions similar to the natural environment. The use of microcosms that simulate the natural environment or in situ enrichment procedures reduces negative impact on both species abundance and distribution. For example, in the metagenomic screening for novel cellulose genes, a fourfold increase of cellulose genes in a small insert expression library was obtained by selective enrichment culturing on carboxymethylcellulose (79), and enrichment of soil microorganisms with glycerol under anaerobic conditions prior to library screening yielded novel glycerol dehydratase and alcohol oxidoreductase enzymes (45).

An alternative strategy for enrichment involves a continuous culture system consisting of encapsulated single cells in agar microdroplets containing nutrient extracts prepared from the environmental samples. After sorting of microcolonies by flow cytometry, pure cultures can be established in rich medium (112). If the identified microcolonies cannot be cultured axenically, the small quantities of DNA can be amplified by multiple strand displacement with Φ29 polymerase (71) so that sufficient quantities are available for genomic analysis.

A relatively unexplored enrichment method uses colloidal chitin (108) bags or traps to recruit active microorganisms in situ (107). Studies of soil chitin degradation have successfully used bags of chitin buried in the soil followed by analysis of metagenomic 16S rRNA and chitinase genes (61).

While enrichment strategies are valuable in increasing the yields of target organisms and genes, such techniques cannot be reliably used to assess the contribution of microbial community members to the in situ degradation process.

COMMUNITY PROFILING

DGGE

Phylogenetic determination of microbial community diversity is the simplest analysis available. Common methods of community analysis using the small-subunit rRNA gene as a phylogenetic marker include the restriction analysis and sequencing of rRNA gene clone libraries, single-stranded conformational polymorphism (SSCP), denaturing gradient gel electrophoresis (DGGE), and temperature gradient gel electrophoresis (TGGE) (reviewed in references 44 and 47). DGGE and TGGE allow rapid and comprehensive qualitative profiling of microbial community composition. DNA fragments are separated in a polyacrylamide gel according to their melting behavior, which is sequence dependent (111). Melting is achieved with a gradient of urea and formamide (DGGE) or temperature (TGGE). The use of GC-clamped primers increases the sensitivity of detection to >99% of sequence variations (65, 89).

The identification of taxa mediating a biodegradation process by DGGE of 16S rRNA gene fragments may be useful for the rational design of further isolation attempts in order to access the target genomes. The aerobic biodegradation of haloacetic acids (HAAs) that are found in surface waters and in drinking water distribution systems was investigated using enrichment cultures containing either mono- or trichloroacetic acid as the sole carbon and energy source. Radiolabeled HAAs indicated that the ^{14}C was primarily converted to $^{14}CO_2$ with minor incorporation into cell biomass. The community structure of the enrichment cultures was analyzed by both classical isolation and DGGE of the PCR-amplified 16S rRNA gene fragments. Each of the two enrichment cultures had multiple bacterial populations, none of which corresponded to HAA-degrading bacteria cultivated on HAA-supplemented agar plates (58). A similar study of squalene-degrading denitrifying bacteria identified 12 dominant phylotypes, of which seven corresponded to γ-*Proteobacteria* isolates utilizing squalene as the sole carbon source (12). The bacterial and archaeal taxa mediating biodegradation of oil both in laboratory enrichment culture and in situ have been determined (54, 81, 82). Functional genes such as ammonia monooxygenase (*amoA*) (67, 76), a key enzyme in the chemolithoautotrophic oxidation of ammonia to nitrogen gas, and multicomponent phenol hydroxylase (*LmPh*), mediating oxidation of phenol (105), have been analyzed by DGGE. This can be informative in dissecting community function, but many functional genes have been subjected to lateral gene transfer and therefore are poor phylogenetic markers to identify taxa.

Community profiling techniques generate few reliable quantitative data due to the bias of PCR amplification (95, 96). Furthermore, in complex metagenomic samples, DGGE reveals only populations that account for more than 0.1% of the total community (31) and different DNA fragments with the same melting profile may comigrate. Another limitation of the DGGE technique is the prerequisite for a single melting domain and therefore the short DNA fragment length (200 to 500 bp) that limits phylogenetic information. Computer programs are useful in assessing the DNA melting behavior of a region of DNA of interest (MELT94 and Primo Melt).

FISH and Microautoradiography (MAR)

Fluorescence in situ hybridization (FISH) uses microscopy to detect cellular nucleic acid sequences hybridized with fluorescently labeled probes. This technique has the considerable strength that it can be applied directly to samples taken from natural environments (e.g., cells recovered by filtration or dissociated from soil particles and recovered by differential centrifugation). The primary limitation of FISH lies in the fact that the technique is restricted to the simultaneous identification of very few phylotypes, for which some phylotypic sequence data must already exist.

16S rRNA is the molecule most commonly targeted in FISH analyses; oligonucleotide probes for each taxonomic level can be designed and tested using bioinformatics tools and experimental hybridization to target and nontarget type strains (2, 4, 50). The critical experimental parameters are specificity, sensitivity, ease of sample penetration, and accessibility of the probe to rRNA targets. A typical oligonucleotide probe is 15 to 30 bases, with the fluorescent dye molecules incorporated into the oligonucleotide during synthesis or after synthesis by chemical or enzymatic coupling (63). Commonly used dyes are fluorescein derivates (fluorescein isothiocyanate and 5-(-6-)carboxyfluorescein–*N*-hydroxysuccinimide-ester [FluoX]), rhodamine-derivates (tetramethyl rhodamine isothiocyanate and Texas Red), and cyanine dyes such as Cy3 and Cy5. Fluorochromes with different excitation and emission maxima allow simultaneous detection of two or more microorganisms. The sensitivity of FISH can be greatly increased by using probes labeled with several fluorochrome molecules (23) and by enzymatic signal amplification using a digoxygenin- or biotin-labeled oligonucleotide coupled to an enzyme catalyzing fluorescent substrates. Tyramide signal amplification can further increase the sensitivity 10- to 20-fold (87), but the number of positive cells may be reduced compared to that obtained with monolabeled probes, perhaps due to insufficient penetration of the high-molecular-weight probe into the cells. Optimizing the fixation conditions and washing stringency can improve problems of poor cell penetration and specificity (3). However, a promising approach is the use of peptide nucleic acid probes that contain an uncharged polyamide backbone (19) and allow increased cell penetration with specific, sensitive detection (75). Poor accessibility of the probe to target rRNA molecules due to higher-order structure may be overcome with "helper" oligonucleotide probes that bind adjacent to the labeled detection probe (27). FISH signals can also be recorded using flow cytometry, coupling automated, quantitative analysis with a unique potential for sorting bacteria in suspensions (93).

The use of dual oligonucleotide probes labeled with different dyes (dual-color FISH) offers scope to simultaneously identify both a target organism and a target gene, and therefore to ask specific questions about the genetic content of a target species. This technique does not, however, indicate whether a gene is expressed but may be extended by the use of antisense oligonucleotides designed to identify mRNAs (74). The latter approach thus links an organism with expression of a target gene, although it falls short of being definitive proof that a specific gene product (e.g., a biodegradative enzyme) is functional in the microbial community.

Finally, MAR can be used to visualize metabolic activity and, when combined with FISH, can be used to definitively identify taxa carrying out a particular biodegradation. A sample is incubated with radiolabeled substrates, and after analysis by FISH, radioactivity can be visualized by application and development of an autoradiographic emulsion. Using MAR-FISH, radiolabeled substrate uptake was shown to be confined to certain bacterial species and could be monitored within activated sludge under aerobic versus anaerobic conditions (46).

IDENTIFYING GENES AND GENOMES MEDIATING BIODEGRADATIONS

SIP and 5-Bromo-2-Deoxyuridine Labeling

Stable-isotope probing (SIP) uses a stable isotope (such as ^{13}C, ^{18}O, or ^{15}N)-labeled substrate to enrich for the genomes responsible for substrate biotransformation. In pioneering studies, $^{13}CH_4$ and $^{13}CH_3OH$ were fed to forest soil microcosms and purified ^{13}C-labeled DNA was separated from unlabeled DNA by buoyant density gradient centrifugation. PCR amplification of the 16S rRNA and methanol dehydrogenase (*mxaF*) genes identified methylotrophic *Acidobacterium* species with novel *mxaF* gene variants (77, 78). Several other stable isotope-labeled substrates have been used, including $^{13}CO_2$ to identify ammonia-oxidizing bacteria and [^{13}C]toluene (73), [^{13}C]phenol (55), and [^{13}C]naphthalene (42) to identify taxa responsible for pollutant biodegradation. The SIP study using [^{13}C]naphthalene identified *Polaromonas vacuolata* as the major naphthalene-metabolizing bacterium in the environmental sample. The phylogenetic results were confirmed by subsequent isolation of the organism and identification of dioxygenase gene homologues involved in naphthalene degradation (42).

The success of DNA-SIP is dependent on the level of isotopic enrichment achieved during isotope feeding. Since RNA synthesis can occur without DNA replication RNA and has a high turnover, RNA-SIP can be a better marker to detect active community members. For example, ^{13}C-labeled phenol was fed to an industrial phenol-degrading bioreactor and the microbial community was monitored by reverse transcriptase (RT)-PCR and DGGE of the isotopically enriched fraction. 16S rRNA gene sequence analysis demonstrated that phenol degradation was dominated by a member of the genus *Thauera*, a group of organisms previously unknown as phenol degraders (55).

Limitations of these methods include dilution, cross-feeding, and recycling of the label within the community, resulting in loss of specific enrichment. The isotopic labeling should be >20 atom% for RNA-SIP and >50 atom% for DNA-SIP so that definitive separation from unlabeled DNA/RNA can be achieved by buoyant density gradient and effects of cross-feeding can be minimized (77). Cross-feeding is further reduced by shortening the labeling pulse time. The community profile across all density gradient fractions can be observed by DGGE to determine the isotopic enrichment of community members (55). Other SIP-based approaches have been used to link microbial phylogeny with community function. Particular phylogenetic groups of isotope-labeled rRNA can be affinity isolated by the hybridization with oligonucleotide probes covalently bound to streptavidin magnetic beads, followed by isotope ratio mass spectrometry (53, 72). However, the wide application of SIP is limited by the commercial availability of complex labeled compounds, which otherwise may require expensive custom synthesis.

In a recent comprehensive study of anaerobic methanol utilization, SIP and full-cycle rRNA analysis with FISH-MAR was used to study the microbial consortium in a [^{13}C]-methanol-fed batch reactor containing nitrate as the electron acceptor (32). 16S rRNA clone libraries identified *Methylophilales* as the dominant members of the isotopically labeled fraction. Oligonucleotide probes targeting the *Methylophilales* taxa were designed, and the application of FISH-MAR established that they were the dominant members of [^{13}C]-methanol uptake (32).

In contrast to DNA-SIP and RNA-SIP, methods that exploit the natural abundance of isotopes in the substrate of choice avoid the requirement for isotopic enrichment. For example, the natural low abundance of $^{13}CH_4$ enabled the identification of methanotrophs in marine sediments by secondary ion mass spectrometry after the microorganisms had been identified by FISH (70).

Actively growing microorganisms can also be labeled with the thymine analogue 5-bromo-2-deoxyuridine (BrdU), and the labeled DNA or RNA can be separated by immunocapture or density gradient centrifugation (101). The commercial availability of anti-BrdU antibodies makes immunocapture an effective method with a high degree of specificity. Addition of substrates with BrdU selects among the members of the microbial community for enhanced growth on the specific substrate. This approach has been successfully applied to identify bacterial community members responding to soil phosphate addition (13) and identify a *Bacillus cereus* strain (VA1) associated with arbuscular mycorrhizae (5). The method assumes that there are universal uptake mechanisms for BrdU. This premise is largely unfounded, since several bacteria have been shown to be unable to incorporate BrdU (101).

SSH and Differential Expression Analysis (DEA)

Suppressive subtraction hybridization (SSH) identifies genetic differences between microorganisms and is therefore a powerful technique for gene enrichment. Samples of DNA or cDNA from the organisms to be compared are divided in two fractions, and a different adaptor is ligated to each fraction (driver and tester). An excess of driver DNA is denatured and hybridized with the tester DNA pool (this subtractive hybridization may be carried out several times). The result is a mixture of single- and double-stranded products. The adaptors are only partially complementary so that only DNA fragments possessing both linkers will be subsequently amplified by PCR. The specificity of this approach is usually confirmed by labeling of the tester DNA/cDNA and carrying out Southern analysis with candidate gene fragments. This approach has been used to discover the genetic elements contributing to pathogenesis between two closely related bacteria (6) but has only recently been used to identify differences in metagenomic samples (30).

Several differential expression technologies that target transcriptional differences in gene expression have been developed (reviewed in reference 34). Representational difference analysis is a method analogous to SSH and is a particularly effective enrichment tool (7). This type of approach was successfully applied to identify bacterial genes up-regulated in the absence of iron (14). Both SSH and DEA are sensitive and selective approaches; by applying multiple rounds of subtraction, small differences in expression of single-copy genes can be detected (1, 52). These methods offer powerful opportunities to study the biodegradation or bioremediation of an environmental pollutant. A potential in situ limitation is the need for a suitable reference metagenome prior to pollutant impact so that the genetic differences identified by SSH or DEA are relevant to the pollutant of interest. The use of defined microcosms can clearly resolve some of these limitations.

Gene-Specific PCR

Sequence-dependent approaches to identify genes are largely limited by an a priori knowledge of gene(s) mediating the biodegradation processes but have proven invaluable for an understanding of the key enzymes in these metabolic pathways. Gene-specific PCR of metagenomic samples has been widely used to identify microorganisms with specific biodegradative and/or metabolic capacities in enrichment cultures or environmental samples. For example, the biodegradative potential of microbial communities has been probed by screening metagenomic DNA extracts for the presence of catechol 2,3-dioxygenase, chlorocatechol dioxygenase, and phenol hydroxylase genes (28, 60, 105). Similarly, the PCR-dependent targeting of methane monooxygenase, methanol dehydrogenase, and ammonia monooxygenase genes has been used to identify methanotrophic chemolithotrophic ammonium-oxidizing (39, 57), denitrifying (15, 37), and polyhydroxyalkanoate-producing bacteria (90). In each example, identification was based on the application of gene-specific consensus primers, a process that incorporates both substantial strengths and weaknesses.

Primer Design

The use of consensus primers for the identification of genes in metagenomic samples offers a fast and efficient approach to the analysis of gene diversity. Assisted by powerful and user-friendly alignment and primer prediction software packages (e.g., DNAMan; DNA Star, Amplicon), consensus primer design is a routine process. There are also a number of widely applied rules of thumb in designing degenerate consensus primers.

- Primers should preferably be in the 18-mer to 30-mer range.
- Primers should be designed to maximize complementarity at the 3′ end.
- Where one or two mismatches occur, degenerate bases are included. For more than three different bases in a specific position, an inosine residue is included.
- No more than 25% total degeneracy and less than 10% of inosine residues are allowed per primer.
- Primers are designed with annealing temperatures of between 50 and 60°C.
- The quality of the subsequent phylogenetic analysis is roughly proportional to the length of the amplified region.

The design of primers from amino acid sequence motifs that are highly conserved between members of a protein family has proven to be highly effective in the identification and characterization of distantly related family members. The CODEHOP (consensus-degenerate hybrid oligonucleotide primer) program designs a pool of primers containing all possible 11- or 12-mers for the 3′ degenerate core region and having the most probable nucleotide predicted for each position in the 5′ nondegenerate clamp region (84). This approach limits primer degeneracy while maintaining primer specificity.

Primer design can be a highly flexible tool in the analysis of different gene hierarchies. For example, bacterial multicomponent phenol hydroxylases, which are key components of the degradation pathways of many aromatic pollutants, form three separate phylogenetic groupings, corresponding closely to the groups established on the basis of kinetic analyses (28). Primer sets designed for the amplification of all three groups ("universal" primer set) and for each of the individual groups were used to demonstrate population changes in mixed cultures in response to different phenol feeding regimes (28).

It is very important to experimentally confirm primer specificity. The inclusion of one or more positive control amplifications (i.e., genomic DNA from an organism known to harbor the relevant gene or express the relevant activity, or a recombinant plasmid harboring an example of the target gene) that yield an amplicon band of the predicted size is a minimum requirement.

Classical PCR

Two-primer approaches (i.e., where both primers are specific to internal regions of the target gene) are most commonly employed in assessments of metagenomic gene diversity (for examples, see references 28, 60, 103, and 106). The products of such amplifications are generally of a predictable size, although some PCR products are subject to variations due to the presence of insertions or deletions within the sequence targeted by the primer pair. Typically, the heterogeneous amplicon DNA is extracted, purified, and cloned into a standard vector system. Quantitative data may be obtained by competitive PCR (60) or by real-time PCR (33). For more in-depth diversity analysis, a number of single colonies from the transformation are selected and the inserts are reamplified (or excised from bulk plasmid preparations after cultivation of the clones). While gene diversity could be assessed immediately by sequencing each insert, it is more common to reduce the sequencing burden by RFLP analysis (e.g., by amplified ribosomal DNA restriction analysis [ARDRA], terminal restriction length polymorphisms [TRFLPs], etc.). The choice of the number and type of restriction enzymes is important in determining the quality of the RFLP clade structure. Since a single restriction site constitutes at best a small precentage of any amplified sequence, multiple cut sites (using several enzymes and/or frequent cutters) enhance the ability to discriminate between nonidentical sequences (64). Confidence in the quality of the clade structure is enhanced by sequencing more than one example from each clade.

As a tool for both gene identification and gene discovery, gene-specific PCR of metagenomic samples has two major drawbacks. First, the design of primers is dependent on known sequence information and strongly skews the results in favor of known sequences. For instance, functionally similar genes resulting from convergent evolution will almost certainly not be detected by a single gene-family-specific set of PCR primers. Secondly, only a fragment of a structural gene will typically be amplified by gene-specific PCR. This is, of course, of less significance in cases where determining gene identity and diversity is the primary objective. However, in cases where access to full-length genes is required, as in "gene mining" (10), additional experimental steps are necessary. The third limitation of gene-specific PCR is that it provides direct information neither on which organisms harbor the identified genes nor on the level of gene expression. Both these issues are important in the context of a biodegradation process in which, for example, knowledge of which genes are actively involved in the process and the organisms in whose genome they reside may be critical for any knowledge-based (i.e., nonempirical) control of the biodegradation process.

The targeting of genes by PCR does not address whether these genes are actually expressed. The majority of control in prokaryotes is thought to occur at the transcriptional level (17), so the presence of an mRNA is strong evidence for the expression of that gene product. Furthermore, mRNA has a high turnover compared to DNA, so the detection of an mRNA target can provide additional evidence of microorganism activity. The use of mRNA as a template (RT-PCR) is therefore a distinct advantage (109), which can yield additional insights such as the diversity of actively transcribed naphthalene dioxygenase genes in microorganisms catabolizing naphthalene at a coal tar waste-contaminated site (109). The experimental difficulties in working with RNA have limited the widespread use of this approach, although several advances have been made in this area (26, 35, 41).

Single-Primer Gene Targeting

Methods for amplifying genes requiring only one gene-specific primer impose less sequence-dependent bias than standard two-primer PCR amplification procedures. These PCR-based strategies have been used for the recovery of the up- or downstream regions flanking a single PCR primer for the recovery of full-length genes. For example, universal fast walking (62, 66) inverse PCR and adaptor ligation PCR (68) have all been employed successfully to access full-length genes in metagenomic extracts. Related techniques such as panhandle PCR and random primed PCR, which are widely used for gene cloning (48, 59), have not yet been applied to the metagenome. It is important to note, however, that these single-primer methods are technically more difficult to apply to metagenomic DNA preparations than to single genome DNA samples due to the hugely increased heterogeneity of the former.

Affinity Capture

A new and elegant method for gene recovery is the use of immobilized oligonucleotides (94) designed to target a specific gene fragment or consensus sequence by affinity binding. For example, an oligonucleotide probe containing a poly(dGGGT) tail is hybridized to RNA in solution. Simultaneously, an aliquot of oligo(dT) paramagnetic beads is hybridized to an oligonucleotide made of poly(dCCCA) with a poly(dA) tail. The two solutions are combined, and a high-affinity GCAT complex is formed. The magnetic beads are captured, and differential melting of all three hybrids permits the release of the gene target fragment while leaving the majority of the oligonucleotides bound to the beads. Authenticity of the captured product is determined by RT-PCR on a subsample. Although designed for recovering taxon-specific rRNA sequences for isotopic analysis (72), the technique is potentially adaptable to any target RNA or DNA sequence by appropriate probe design.

Microarrays

Microarray technologies provide a powerful high-throughput tool for the study of biological processes but have only recently been applied to the field of metagenomics (88). Depending on the size and origin of the probes arrayed, microarrays are grouped as follows (Table 1). The majority of environmental studies (for examples, see references 51 and 92) have focused on analysis of in situ diversity using phylogenetic oligonucleotide arrays. However, functional gene arrays may be exploited for the identification of either the presence (by DNA-DNA hybridization) or the expression (by DNA-mRNA/cDNA hybridization) of specific functional genes in a metagenome. In the most comprehensive study of this type reported to date, Rhee et al. (80) prepared oligonucleotide arrays containing 1,661 50-mer probes specific for known genes involved in aromatic and alkane degradation pathways and in heavy metal resistance. These arrays were effective in showing changes in population distribution in soils amended with naphthalene and during enrichment incubation. Surprisingly, microarray data showed that *Ralstonia* and *Comamonas*, rather than *Pseudomonas* naphthalene-degrading genes, were prevalent in soils, a result at odds with the common perception of *Pseudomonas* as a primary degrader of polyaromatics in soil systems (36). The authors acknowledge that the detection

TABLE 1 Classification of microarrays used in environmental studies[a]

Class of microarray	Source of probe	Application	Reference(s)
Functional gene	PCR-amplified DNA and cDNA fragments Oligonucleotides (gene specific)	Physiological status and functional activities of microbial communities	80
Community genome	Whole-genomic DNA from pure cultures	Identification of individual species in microbial communities	113
Phylogenetic oligonucleotide	Oligonucleotides (small-subunit rRNA gene)	Phylogenetic analyses of microbial communities	50, 51, 92

[a]Data are from reference 113.

limit (about 10^7 cells in the presence of background RNA) would need to be reduced substantially in order to detect rare genomes.

METAGENOMIC GENE LIBRARIES

Metagenomic DNA libraries can theoretically provide access to the entire metagenomic sequence space (38). While this technology has been developed primarily as a tool for the discovery (and commercialization) of novel genes (21), it has considerable potential for contributing to the analysis of environmental processes. For example, a combination of a suitable labeling method (such as SIP) or techniques of subtraction (SSH/DEA or subtractive cDNA libraries) with metagenomic library construction and screening could potentially provide data sets of multiple genes involved in specific biodegradation processes. A prerequisite for the construction of metagenomic libraries is the efficient and nonbiased extraction of high-quality DNA from environmental samples (see chapter 84). The minimum number of clones that need to be screened in order to find a gene of interest in the metagenomic library (18) increases with the presence of eukaryotic DNA (a eukaryotic genome is 3 to 140,000 Mbp compared to prokaryotic organisms of 0.6 to 9.5 Mbp [102]). This is exacerbated by the presence of noncoding introns that reduce valuable sequence information and the ability to express functional gene products when screening expression libraries. Size selection may be a particularly useful preliminary step to reduce the total metagenome size, avoid a high load of "junk" DNA, or merely focus on a specific metagenome fraction (e.g., eukaryotic, prokaryotic, or viral). A crude separation of eukaryotic and prokaryotic cells can be simply carried out by filtration (104). Alternatively, differential and density gradient centrifugation can be used. For example, differential centrifugation has been used to enrich *Buchnera aphidicola* and *Cenarchaeum symbiosum* symbionts from their hosts (86) and to prepare community viruses in preparation for whole-genome sequencing (16, 110). Full representation of the metagenome in a library is also limited by the diversity of the sample (perhaps several thousand species) (11, 22) and by the unequal distribution of species. This may be partially resolved by means of experimental normalization. Fragmented genomic DNA is heat denatured and allowed to reanneal under stringent conditions (e.g., 68°C for 12 to 36 h). Abundant single-stranded DNA anneals to generate double-stranded nucleic acids more rapidly than does rare DNA and separates from the double-stranded nucleic acids, enabling an enrichment of rarer sequences within the environmental sample (25, 91). Metagenomic cloning often assumes that *Escherichia coli* is globally suitable as a host for all environmental DNA fragments. For DNA libraries, low-copy-number bacterial artificial chromosome or fosmid systems offer the best means for stable library propagation and maintenance (8). For expression libraries, systems allowing expression from native promoters offer the best chance for recovery of heterologously expressed genes. Although the *E. coli* transcriptional machinery is known to be relatively promiscuous in recognizing foreign promoters, a bias in favor of *Firmicutes* genes has been noted (29). The development of *Streptomyces* shuttle systems and *Pseudomonas* hosts offer new possibilities in this regard (20, 56).

Depending upon the insert size, 10^6 to 10^8 library clones are required in order to represent the collective genomes of the thousands of different species that are typically present in an environmental sample (49). Although metagenomic sequencing approaches have revealed functional insights into communities and their biotransformations (9, 83, 98, 99, 104), the enormous diversity and current limitations on library screening and gene annotation limit this approach. For example, the sequencing of 1.045 Gbp from the Sargasso Sea metagenome identified 1.2 million putative genes, but the diversity of species (~1,800) and their unequal distribution enabled only one complete genome to be assembled (104). Due to limitations in gene assignment and database homology, relatively little functional information was gained although many new rhodopsin-like photoreceptors were identified, indicating an important role of bacterial photoautotrophy in marine populations (8, 9, 104). The acquisition of very large volumes of metagenomic sequence from different environments has opened the prospect of a new field of comparative metagenomics. A recent study demonstrated that data from shotgun sequencing of metagenomic small-insert libraries can be used to generate environmental fingerprints (98). The core data for these fingerprints are environmental genome tags, short individual sequence reads yielding putative gene identities. A comparison of gene distribution across eight different metagenomic libraries revealed substantial, predictable system differences: a predominance of photoautotrophy in Sargasso Sea samples versus carbohydrate metabolism in soils.

Technologies for screening expression libraries have been reviewed elsewhere (38). However, a high-throughput substrate-induced gene expression screening method that is particularly relevant to the identification of catabolic genes has recently been developed. This technique uses an operon-trap *gfp* expression vector to capture catabolic operon DNA fragments relevant to the target compounds. An environmental metagenome library (152,000 clones with an average insertion size of 7 kb) was constructed from groundwater and screened with benzoate or naphthalene substrates. Nonrecombinants were selected against by cell sorting after IPTG (isopropyl-β-D-thiogalactopyranoside) induction, since the vector used, a puC18 derivative, contains *lacZ* in the multiple cloning site to allow for α-complementation. The screening was accomplished within 4 days and yielded 62 positive clones. Many of the identified open reading frames were found to be homologues of genes in known benzoate-degradative and catechol-degradative operons (100).

In summary, metagenomic techniques have begun to address both the identity of active organisms and their gene products that mediate biodegradation. Advances in isotopic labeling, fluorescent microscopy, and gene targeting have enabled a directed approach to uncovering some of the active microorganisms carrying out degradative processes in complex microbial communities. The application of high-throughput methods such as microarrays and comparative genomic sequencing offers the opportunity to uncover novel pathways and to study complex microbial populations. A combination of approaches will prove the most valuable in the linking of phylogeny with function. Ultimately, a clearer picture of the identity, activity, distribution, and abundance of active organisms will lead us to a more complete understanding of biodegradation processes.

We gratefully acknowledge the National Research Foundation HBU program and the University of the Western Cape for financial support.

REFERENCES

1. **Akopyants, N. S., A. Fradkov, L. Diatchenko, J. E. Hill, P. D. Siebert, S. A. Lukyanov, E. D. Sverdlov, and D. E. Berg.** 1998. PCR-based subtractive hybridization and differences in gene content among strains of Helicobacter pylori. *Proc. Natl. Acad. Sci. USA* **95:**13108–13113.
2. **Amann, R. I., B. J. Binder, R. J. Olson, S. W. Chisholm, R. Devereux, and D. A. Stahl.** 1990. Combination of 16S rRNA-targeted oligonucleotide probes with flow cytometry for analyzing mixed microbial populations. *Appl. Environ. Microbiol.* **56:**1919–1925.
3. **Amann, R. I., L. Krumholz, and D. A. Stahl.** 1990. Fluorescent oligonucleotide probing of whole cells for determinative, phylogenetic, and environmental studies in microbiology. *J. Bacteriol.* **172:**762–770.
4. **Amman, R. L., W. Ludwig, and K. Schleifer.** 1995. Phylogenetic identification and *in situ* detection of individual microbial cells without cultivation. *Microbiol. Rev.* **59:**143–169.
5. **Artursson, V., and J. K. Jansson.** 2003. Use of bromodeoxyuridine immunocapture to identify active bacteria associated with arbuscular mycorrhizal hyphae. *Appl. Environ. Microbiol.* **69:**6208–6215.
6. **Bart, A., J. Dankert, and A. van der Ende.** 2000. Representational difference analysis of *Neisseria meningitidis* identifies sequences that are specific for the hyper-virulent lineage III clone. *FEMS Microbiol. Lett.* **188:**111–114.
7. **Becker, P., W. Hufnagle, G. Peters, and M. Herrmann.** 2001. Detection of differential gene expression in biofilm-forming versus planktonic populations of *Staphylococcus aureus* using micro-representational-difference analysis. *Appl. Environ. Microbiol.* **67:**2958–2965.
8. **Beja, O.** 2004. To BAC or not to BAC: marine ecogenomics. *Curr. Opin. Biotechnol.* **15:**187–190.
9. **Beja, O., M. T. Suzuki, E. V. Koonin, L. Aravind, A. Hadd, L. P. Nguyen, R. Villacorta, M. Amjadi, C. Garrigues, S. B. Jovanovich, R. A. Feldman, and E. F. DeLong.** 2000. Construction and analysis of bacterial artificial chromosome libraries from a marine microbial assemblage. *Environ. Microbiol.* **2:**516–529.
10. **Bell, P., A. Sunna, M. D.Gibbs, N. C. Curach, H. Nevalainen, and P. L. Bergquist.** 2002. Prospecting for novel lipase genes using PCR. *Microbiology* **148:**2283–2291.
11. **Bohannan, B. J. M., and J. Hughes.** 2003. New approaches to analyzing microbial biodiversity data. *Curr. Opin. Microbiol.* **6:**282–287.
12. **Bonin, P. C., V. D. Michotey, A. Mouzdahir, and J.-F. Rontani.** 2002. Anaerobic biodegradation of squalene: using DGGE to monitor the isolation of denitrifying bacteria taken from enrichment cultures. *FEMS Microbiol. Ecol.* **42:**37–49.
13. **Borneman, J.** 1999. Culture-independent identification of microorganisms that respond to specified stimuli. *Appl. Environ. Microbiol.* **65:**3398–3400.
14. **Bowler, L. D., M. Hubank, and B. G. Spratt.** 1999. Representational difference analysis of cDNA for the detection of differential gene expression in bacteria: development using a model of iron-regulated gene expression in *Neisseria meningitidis*. *Microbiology* **145:**3529–3537.
15. **Braker, G., A. Fesefeldt, and K.-P. Witzel.** 1998. Development of PCR primer systems for amplification of nitrite reductase genes (*nirK* and *nirS*) to detect denitrifying bacteria in environmental samples. *Appl. Environ. Microbiol.* **64:**3769–3775.
16. **Breitbart, M., P. Salamon, B. Andresen, J. M. Mahaffy, A. M. Segall, D. Mead, F. Azam, and F. Rohwer.** 2002. Genomic analysis of uncultured marine viral communities. *Proc. Natl. Acad. Sci. USA* **99:**14250–14255.
17. **Cases, I., and V. de Lorenzo.** 2005. Promoters in the environment: transcription regulation in its natural context. *Nat. Microbiol. Rev.* **3:**105–118.
18. **Clarke, L., and J. Carbon.** 1976. A colony bank containing synthetic ColE1 hybrid plasmids representative of the entire *E. coli* genome. *Cell* **9:**91–100.
19. **Corey, D. R.** 1997. Peptide nucleic acids: expanding the scope of nucleic acid recognition. *Trends Biotechnol.* **15:** 224–229.
20. **Courtois, S., C. M. Cappellano, M. Ball, F. X. Francou, P. Normand, G. Helynck, A. Martinez, S. J. Kolvek, J. Hopke, M. S. Osburne, P. R. August, R. Nalin, M. Guerineau, P. Jeannin, P. Simonet, and J. L. Pernodet.** 2003. Recombinant environmental libraries provide access to microbial diversity for drug discovery from natural products. *Appl. Environ. Microbiol.* **69:**49–55.
21. **Cowan, D. A.** 2000. Microbial genomes—the untapped resource. *Trends Biotechnol.* **18:**14–16.
22. **Curtis, T. P., W. T. Sloan, and J. W. Scannell.** 2002. Estimating prokaryotic diversity and its limits. *Proc. Natl. Acad. Sci. USA* **99:**10494–10499.
23. **DeLong, E. F., L. T. Taylor, T. L. Marsh, and C. M. Preston.** 1999. Visualization and enumeration of marine planctonic archaea and bacteria by using polyribonucleotide probes and fluorescent in situ hybridization. *Appl. Environ. Microbiol.* **65:**5554–5563.
24. **Diaz, E.** 2004. Bacterial degradation of aromatic pollutants: a paradigm of metabolic versatility. *Int. Microbiol.* **7:**173–180.
25. **Eastwood, R. F., E. S. Lagudah, and R. Appels.** 1994. A directed search for DNA sequences tightly linked to cereal cyst nematode resistance in *Triticum tauschii*. *Genome* **37:** 311–319.
26. **Frischer, M. E., J. M. Danforth, M. A. Newton Healy, and F. M. Saunders.** 2000. Whole-cell versus total RNA extraction for analysis of microbial community structure with 16S rRNA-targeted oligonucleotide probes in salt marsh sediments. *Appl. Environ. Microbiol.* **66:**3037–3043.

27. **Fuchs, B. M., G. Wallner, W. Beisker, I. Schwippl, W. Ludwig, and R. Amann.** 1998. Flow cytometric analysis of the in situ accessibility of *Escherichia coli* 16S rRNA for fluorescently labeled oligonucleotide probes. *Appl. Environ. Microbiol.* **64:**4973–4982.
28. **Futamata, H., S. Harayama, and K. Watanabe.** 2001. Group-specific monitoring of phenol hydroxylase genes for a functional assessment of phenol-stimulated trichloroethylene bioremediation. *Appl. Environ. Microbiol.* **67:**4671–4677.
29. **Gabor, E. M., W. B. L. Alkema, and D. B. Janssen.** 2004. Quantifying the accessibility of the metagenome by random expression cloning techniques. *Environ. Microbiol.* **6:** 879–886.
30. **Galbraith, E. A., D. A. Antonopoulos, and B. A. White.** 2004. Suppressive subtractive hybridisation as a tool for identifying genetic diversity in an environmental metagenome: the rumen as a model. *Environ. Microbiol.* **6:**928–937.
31. **Gelsomino, A., A. C. Keyzer-Wolters, G. Cacco, and J. D. van Elsas.** 1999. Assessment of bacterial community structure in soil by polymerase chain reaction and denaturing gradient gel electrophoresis. *J. Microbiol. Methods* **38:**15.
32. **Ginige, M. P., P. Hugenholtz, H. Daims, M. Wagner, J. Keller, and L. L. Blackall.** 2004. Use of stable-isotope probing, full-cycle rRNA analysis, and fluorescence in situ hybridization-microautoradiography to study a methanol-fed denitrifying microbial community. *Appl. Environ. Microbiol.* **70:**588–596.
33. **Ginzinger, D. G.** 2002. Gene quantification using real-time quantitative PCR: an emerging technology hits the mainstream. *Exp. Hematol.* **30:**503–512.
34. **Green, C. D., J. F. Simons, B. E. Taillon, and D. A. Lewin.** 2001. Open systems: panoramic views of gene expression. *J. Immunol. Methods* **250:**67–79.
35. **Griffiths, R. I., A. S. Whiteley, A. G. O'Donnell, and M. J. Bailey.** 2000. Rapid method for coextraction of DNA and RNA from natural environments for analysis of ribosomal DNA- and rRNA-based microbial community composition. *Appl. Environ. Microbiol.* **66:**5488–5491.
36. **Habe, H., and T. Omori.** 2003. Genetics of polycyclic aromatic hydrocarbon metabolism in diverse aerobic bacteria. *Biosci. Biotechnol. Biochem.* **67:**225–243.
37. **Hallin, S., and P. Lindgren.** 1999. PCR detection of genes encoding nitrite reductase in denitrifying bacteria. *Appl. Environ. Microbiol.* **65:**1652–1657.
38. **Handelsman, J., M. R. Rondon, S. F. Brady, J. Clardy, and R. M. Goodman.** 1998. Molecular biological access to the chemistry of unknown soil microbes: a new frontier for natural products. *Chem. Biol.* **5:**R245–R249.
39. **Henckel, T., U. Jackel, S. Schnell, and R. Conrad.** 2000. Molecular analyses of novel methanotrophic communities in forest soil that oxidize atmospheric methane. *Appl. Environ. Microbiol.* **66:**1801–1808.
40. **Hugenholtz, P., B. M. Goebel, and N. R. Pace.** 1998. Impact of culture-independent studies on the emerging phylogenetic view of bacterial diversity. *J. Bacteriol.* **180:** 4765–4774.
41. **Hurt, R. A.** 2001. Simultaneous recovery of RNA and DNA from soils and sediments. *Appl. Environ. Microbiol.* **67:**4495–4503.
42. **Jeon, C. O., W. Park, P. Padmanabhan, C. DeRito, J. R. Snape, and E. L. Madsen.** 2003. Discovery of a bacterium, with distinctive dioxygenase, that is responsible for in situ biodegradation in contaminated sediment. *Proc. Nat. Acad. Sci. USA* **100:**13591–13596.
43. **Kaeberlein, T., K. Lewis, and S. S. Epstein.** 2005. Isolating "uncultivable" microorganisms in pure culture using a simulated natural environment. *Science* **296:**1127–1129
44. **Kirk, J. L., L. A. Beaudette, M. Hart, P. Moutoglis, J. N. Klironomos, H. Lee, and J. T. Trevors.** 2004. Methods of studying soil microbial diversity. *J. Microbiol. Methods* **58:** 169–188.
45. **Knietsch, A., T. Waschkowitz, S. Bowien, A. Henne, and R. Daniel.** 2003. Metagenomes of complex microbial consortia derived from different soils as sources for novel genes conferring formation of carbonyls from short-chain polyols on *Escherichia coli*. *J. Mol. Microbiol. Biotechnol.* **5:** 46–56.
46. **Lee, N., P. H. Nielsen, K. H. Andreasen, S. Juretschko, J. L. Nielsen, K.-H. Schleifer, and M. Wagner.** 1999. Combination of fluorescent in situ hybridization and microautoradiography—a new tool for structure-function analyses in microbial ecology. *Appl. Environ. Microbiol.* **65:**1289–1297.
47. **Liu, W.-T., and D. A. Stahl.** 2002. Molecular approaches for the measurement of density, diversity, and phylogeny, p. 114–134. *In* C. J. Hurst, R. L. Crawford, G. R. Knudsen, M. J. McInerney, and L. D. Stetzenbach (ed.), *Manual of Environmental Microbiology*, 2nd ed. ASM Press, Washington, D.C.
48. **Liu, Y., and R. F. Whittier.** 1995. Thermal asymmetric interlaced PCR: automatable amplification and sequencing of insert end fragments from pi and yac clones for chromosome walking. *Genomics* **25:**674–681.
49. **Lorenz, P., and C. Schleper.** 2002. Metagenome—a challenging source of enzyme discovery. *J. Mol. Catal. B* **19:**13–19.
50. **Loy, A., and M. W. M. Horn.** 2003. probeBase—an online resource for rRNA-targeted oligonucleotide probes. *Nucleic Acids Res.* **31:**514–516.
51. **Loy, A. A., N. Lehner, J. Lee, H. Adamczyk, J. Meier, H. Schleifer, and M. Wagner.** 2002. Oligonucleotide microarray for 16S rRNa-based detections of all recognized lineages of sulphate-reducing prokaryotes in the environment. *Appl. Environ. Microbiol.* **68:**5064–5081.
52. **Lukyanov, S. A., N. G. Gurskaya, K. A. Lukyanov, V. S. Tarabykin, and E. D. Sverdlov.** 1994. Highly efficient subtractive hybridisation of cDNA. *Bioorg. Khim.* **20:**701–704.
53. **MacGregor, B. J., V. Brüchert, S. Fleischer, and R. Amann.** 2002. Isolation of small-subunit rRNA for stable-isotopic characterization. *Environ. Microbiol.* **4:**451–464.
54. **MacNaughton, S. J., J. R. Stephen, A. D. Venosa, G. A. Davis, Y. J. Chang, and D. C. White.** 1999. Microbial population changes during bioremediation of an experimental oil spill. *Appl. Environ. Microbiol.* **65:**3566–3574.
55. **Manefield, M., A. S. Whiteley, R. I. Griffiths, and M. J. Bailey.** 2002. RNA stable isotope probing, a novel means of linking microbial community function to phylogeny. *Appl. Environ. Microbiol.* **68:**5367–5373.
56. **Martinez, A., S. J. Kolvek, C. L. T. Yip, J. Hopke, K. A. Brown, I. A. MacNeil, and M. S. Osburne.** 2004. Genetically modified bacterial strains and novel bacterial artificial chromosome shuttle vectors for constructing environmental libraries and detecting heterologous natural products in multiple expression hosts. *Appl. Environ. Microbiol.* **70:**2452–2463.
57. **McDonald, I. R., E. M. Kenna, and J. C. Murrell.** 1995. Detection of methanotrophic bacteria in environmental samples with the PCR. *Appl. Environ. Microbiol.* **61:**116–121.
58. **McRae, B. M., T. M. LaPara, and R. M. Hozalski.** 2004. Biodegradation of haloacetic acids by bacterial enrichment cultures. *Chemosphere* **55:**915–925.
59. **Megonigal, M. D., E. F. Rappaport, R. B. Wilson, D. H. Jones, J. A. Whitlock, J. A. Ortega, D. J. Slater, P. C. Nowell, and C. A. Felix.** 2000. Panhandle PCR for cDNA: a rapid method for isolation of MLL fusion tran-

scripts involving unknown partner genes. *Proc. Natl. Acad. Sci. USA* **97:**9597–9602.

60. **Mesarch, M. B., C. H. Nakatsu, and L. Nies.** 2000. Development of catechol 2,3-dioxygenase-specific primers for monitoring bioremediation by competitive quantitative PCR. *Appl. Environ. Microbiol.* **66:**678–683.
61. **Metcalfe, A. C., M. Krsek, G. W. Gooday, J. I. Prosser, and E. M. H. Wellington.** 2002. Molecular analysis of a bacterial chitinolytic community in an upland pasture. *Appl. Environ. Microbiol.* **68:**5042–5050.
62. **Mishra, R. N., S. L. Singla-Pareek, S. Nair, S. K. Sopory, and M. K. Reddy.** 2002. Directional genome walking using PCR. *BioTechniques* **33:**830–834.
63. **Moter, A., and U. B. Globel.** 2000. Fluorescence in situ hybridization (FISH) for direct visualization of microorganisms. *J. Microbiol. Methods* **41:**85–112.
64. **Moyer, C. L., J. M. Tiedje, F. C. Dobbs, and D. M. Karl.** 1996. A computer-simulated restriction fragment length polymorphism analysis of bacterial small-subunit rRNA genes: efficacy of selected tetrameric restriction enzymes for studies of microbial diversity in nature. *Appl. Environ. Microbiol.* **62:**2501–2507.
65. **Myers, R. M., S. G. Fisher, L. S. Lerman, and T. Maniatis.** 1985. Nearly all single base substitutions in DNA fragments joined to a GC-clamp can be detected by denaturing gradient gel electrophoresis. *Nucleic Acids Res.* **13:** 3131–3145.
66. **Myrick, K. V., and W. M. Gelbart.** 2002. Universal fast walking for direct and versatile determination of flanking sequence. *Gene* **284:**125–131.
67. **Nicolaisen, M. H., and N. B. Ramsing.** 2002. Denaturing gradient gel electrophoresis (DGGE) approaches to study the diversity of ammonia-oxidizing bacteria. *J. Microbiol. Methods* **50:**189–203.
68. **Ochman, H., F. J. Ayala, and D. L. Hartl.** 1993. Use of polymerase chain reaction to amplify segments outside boundaries of known sequences. *Methods Enzymol.* **218:** 309–321.
69. **Oliver, J., L. Nilsson, and S. Kjelleberg.** 1991. Formation of nonculturable *Vibrio vulnificus* cells and its relationship to the starvation state. *Appl. Environ. Microbiol.* **57:**2640–2644.
70. **Orphan, V. J., C. H. House, K. U. Hinrichs, K. D. McKeegan, and E. F. DeLong.** 2002. Multiple archaeal groups mediate methane oxidation in anoxic cold seep sediments. *Proc. Natl. Acad. Sci. USA* **99:**7663–7668.
71. **Paez, J. G., M. Lin, R. Beroukhim, J. C. Lee, X. Zhao, D. J. Richter, S. Gabriel, P. Herman, H. Sasaki, D. Altshuler, C. Li, M. Meyerson, and W. R. Sellers.** 2004. Genome coverage and sequence fidelity of Φ29 polymerase-based multiple strand displacement whole genome amplification. *Nucleic Acids Res.* **32:**e71.
72. **Pearson, A., A. L. Sessions, K. J. Edwards, and J. M. Meyers.** 2004. Phylogenetically specific separation of rRNA from prokaryotes for isotopic analysis. *Mar. Chem.* **92:**295–306.
73. **Pelz, O., A. Chatzinotas, N. Andersen, S. M. Bernasconi, C. Hesse, W. R. Abraham, and J. Zeyer.** 2001. Use of isotopic and molecular techniques to link toluene degradation in denitrifying aquifer microcosms to specific microbial populations. *Arch. Microbiol.* **175:**270–281.
74. **Pernthaler, A., and R. Amann.** 2004. Simultaneous fluorescence in situ hybridization of mRNA and rRNA in environmental bacteria. *Appl. Environ. Microbiol.* **70:** 5426–5433.
75. **Prescott, A. M., and C. R. Fricker.** 1999. Use of PNA oligonucleotides for the in situ detection of *Escherichia coli* in water samples. *Mol. Cell. Probes* **13:**261–268.
76. **Purkhold, U., M. Wagner, G. Timmermann, A. Pommerening-Roser, and H. P. Koops.** 2003. 16S rRNA and amoA-based phylogeny of 12 novel betaproteobacterial ammonia-oxidizing isolates: extension of the dataset and proposal of a new lineage within the nitrosomonads. *Int. J. Syst. Evol. Microbiol.* **53:**1485–1494.
77. **Radajewski, S., I. R. McDonald, and J. C. Murrell.** 2003. Stable-isotope probing of nucleic acids: a window to the function of uncultured microorganisms. *Curr. Opin. Biotechnol.* **14:**296–302.
78. **Radajewski, S., G. Webster, D. S. Reay, S. A. Morris, P. Ineson, D. B. Nedwell, J. I. Prosser, and J. C. Murrell.** 2002. Identification of active methylotroph populations in an acidic forest soil by stable-isotope probing. *Microbiology* **148:**2331–2342.
79. **Rees, H. C., W. D. Grant, B. E. Jones, and S. Heaphy.** 2004. Diversity of Kenyan soda lake alkaliphiles assessed by molecular methods. *Extremophiles* **8:**63–71.
80. **Rhee, S.-K., X. Liu, L. Wu, S. C. Chong, X. Wan, and J. Zhou.** 2004. Detection of genes involved in the biodegradation and biotransformation in microbial communities by using 50-mer oligonucleotide arrays. *Appl. Environ. Microbiol.* **70:**4303–4317.
81. **Roling, W. F., I. R. de Brito Couto, R. P. Swannell, and I. M. Head.** 2004. Response of archaeal communities in beach sediments to spilled oil and bioremediation. *Appl. Environ. Microbiol.* **70:**2614–2620.
82. **Roling W. F., M. G. Milner, D. M. Jones, F. Fratepietro, R. P. Swannell, F. Daniel, and I. M. Head.** 2004. Bacterial community dynamics and hydrocarbon degradation during a field-scale evaluation of bioremediation on a mudflat beach contaminated with buried oil. *Appl. Environ. Microbiol.* **70:**2603–2613.
83. **Rondon, M. R., P. R. August, A. D. Bettermann, S. F. Brady, T. H. Grossman, M. R. Liles, K. A. Loiacono, B. A. Lynch, I. A. Macneil, C. Minor, C. L. Tiong, M. Gilman, M. S. Osburne, J. Clardy, J. Handelsman, and R. M. Goodman.** 2000. Cloning the soil metagenome: a strategy for accessing the genetic and functional diversity of uncultured microorganisms. *Appl. Environ. Microbiol.* **66:**2541–2547.
84. **Rose, T. M., E. R. Schultz, J. G. Henikoff, S. Pietrokovski, C. M. McCallum, and S. Henikoff.** 1998. Consensus-degenerate hybrid oligonucleotide primers for amplification of distantly related sequences. *Nucleic Acids Res.* **26:**1628–1635.
85. **Roszak, D. B., D. J. Grimes, and R. R. Colwell.** 1984. Viable but nonrecoverable stage of *Salmonella enteritidis* in aquatic systems. *Can. J. Microbiol.* **30:**334–338.
86. **Schloss, P. D., and J. Handelsman.** 2003. Biotechnological prospects from metagenomics. *Curr. Opin. Biotechnol.* **14:**303–310.
87. **Schönhuber, W., B. Fuchs, S. Juretschko, and R. Amann.** 1997. Improved sensitivity of whole-cell hybridization by the combination of horseradish peroxidase-labeled oligonucleotides and tyramide signal amplification. *Appl. Environ. Microbiol.* **63:**3268–3273.
88. **Sebat, J. L., F. S. Colwell, and R. L. Crawford.** 2003. Metagenomic profiling: microarray analysis of a metagenomic DNA library. *Appl. Environ. Microbiol.* **69:**4927–4934.
89. **Sheffield, V. C., D. R. Cox, L. S. Lerman, and R. M. Myers.** 1989. Attachment of a 40-base-pair GC rich sequence (GC-clamp) to genomic DNA fragments by the polymerase chain-reaction results in improved detection of single-base changes. *Proc. Natl. Acad. Sci. USA* **86:**232–236.
90. **Sheu, D., Y. Wang, and C. Lee.** 2000. Rapid detection of polyhydroxyalkanoate-accumulating bacteria isolated from the environment by colony PCR. *Microbiology* **146:** 2019-2025.

91. **Short, J. M., and E. J. Mathur.** December 1999. *Production and Use of Normalized DNA Libraries*. U.S. patent 6,001,574.
92. **Small, J., D. R. Call, F. J. Brockman, T. M. Straub, and D. P. Chandler.** 2001. Direct detection of 16S rRNA in soil extracts by using oligonucleotide microarrays. *Appl. Environ. Microbiol.* **67:**4708–4716.
93. **Snaidr, J., B. Fuchs, G. Wallner, M. Wagner, K. H. Schleifer, and R. Amann.** 1999. Phylogeny and in situ identification of a morphologically conspicuous bacterium, Candidatus Magnospira bakii, present at very low frequency in activated sludge. *Environ. Microbiol.* **1:**125–135.
94. **Stull, D., and J. M. Pisano.** 2001. Purely RNA: new innovations enhance the quality, speed, and efficiency of RNA isolation techniques. *Scientist* **15:**29–31.
95. **Suzuki, M. T., and S. J. Giovannoni.** 1996. Bias caused by template annealing in the amplification of mixtures of 16S rRNA genes by PCR. *Appl. Environ. Microbiol.* **62:**625–630.
96. **Thompson, J., L. Marcelino, and M. F. Polz.** 2002. Heteroduplexes in mixed-template amplifications: formation, consequence and elimination by 'reconditioning PCR.' *Nucleic Acids Res.* **30:**2083–2088.
97. **Torsvik, V., and L. Ovreas.** 2002. Microbial diversity and function in soil: from genes to ecosystems. *Curr. Opin. Microbiol.* **5:**240–245.
98. **Tringe, S. G., C. von Mering, A. Kobayashi, A. A. Salamov, K. Chen, H. W. Chang, M. Podar, J. M. Short, E. J. Mathur, J. C. Detter, P. Bork, P. Hugenholtz, and E. M. Rubin.** 2005. Comparative metagenomics of microbial communities. *Science* **308:**554–557.
99. **Tyson, J. W., J. Chapman, P. Hugenholtz, V. V. Solovyev, E. M. Rubin, D. S. Rokhsar, and J. F. Banfield.** 2004. Community structure and metabolism through reconstruction of microbial genomes from the environment. *Nature* **428:**37-43.
100. **Uchiyama, T., T. Abe, T. Ikemura, and K. Watanabe.** 2005. Substrate-induced gene-expression screening of environmental metagenome libraries for isolation of catabolic genes. *Nat. Biotechnol.* **23:**88–93.
101. **Urbach, E., K. L. Vergin, and S. J. Giovannoni.** 1999. Immunochemical detection and isolation of DNA from metabolically active bacteria. *Appl. Environ. Microbiol.* **65:**1207–1213.
102. **Ussery, D. W., and P. F. Hallin.** 2004. Genome update: length distributions of sequenced prokaryotic genomes. *Microbiology* **150:**513–516.
103. **Vallaeys, T., F. Persello-Cartieaux, N. Rouard, C. Lors, G. Laguerre, and G. Soulas.** 1997. PCR-RFLP analysis of 16S rRNA, *tfd*A and *tfd*B genes reveals a diversity of 2,4-D degraders in soil aggregates. *FEMS Microbiol. Lett.* **24:**269–278.
104. **Venter, J. C., K. Remington, J. F. Heidelberg, A. L. Halpern, D. Rush, J. A. Eisen, D. Wu, I. Paulsen, K. E. Nelson, W. Nelson, D. E. Fouts, S. Sevy, H. Knap, M. W. Lomas, K. Nealson, O. White, J. D. Peterson, J. Hoffman, R. Parsons, H. Baden-Tillson, C. Pfannkoch, Y. Rogers, and H. O. Smith.** 2004. Environmental genome shotgun sequencing of the Sargasso Sea. *Science* **304:**66–74.
105. **Watanabe, K., M. Teramoto, H. Futamata, and S. Harayama.** 1998. Molecular detection, isolation, and physiological characterization of functionally dominant phenol-degrading bacteria in activated sludge. *Appl. Environ. Microbiol.* **64:**4396–4402.
106. **Wawrik, B. L., G. Kerkhof, J. Zylstra, and J. L. Kukor.** 2005. Identification of unique type II polyketide synthase genes in soil. *Appl. Environ. Microbiol.* **71:**2232–2238.
107. **Wellington, E. M. H., A. Barry, and M. Krsek.** 2003. Resolving functional diversity in relation to microbial community structure in soil; exploiting genomics and stable isotope probing. *Curr. Opin. Microbiol.* **6:**295–301.
108. **Wen, C. M., C. S. Tseng, C. Y. Cheng, and Y. K. Li.** 2002. Purification, characterization and cloning of a chitinase from Bacillus sp. NCTU2. *Biotechnol. Appl. Biochem.* **35:**213–219.
109. **Wilson, M. S., C. Bakermans, and E. Madsen.** 1999. In situ, real-time catabolic gene expression: extraction and characterization of naphthalene dioxygenase mRNA transcripts from groundwater. *Appl. Environ. Microbiol.* **65:**80–87.
110. **Wommack, K. E., S. R. Bench, K. E. Williamson, and M. Radosevich.** 2004. Viruses in soils: the first terrestrial viral metagenome, p. 11. *In Proceedings of the 10th International Symposium on Microbial Ecology, Cancun, Mexico, August, 2004*. Kenes International, Geneva, Switzerland.
111. **Wu, Y., V. M. Hayes, J. Osinga, I. M. Mulder, M. W. Looman, C. H. Buys, and R. M. Hofstra.** 1998. Improvement of fragment and primer selection for mutation detection by denaturing gradient gel electrophoresis. *Nucleic Acids Res.* **26:**5432–5440.
112. **Zengler, K., G. Toledo, M. Rappe, J. Elkins, E. J. Mathur, J. M. Short, and M. Keller.** 2002. Cultivating the uncultured. *Proc. Natl. Acad. Sci. USA* **99:**15681–15686.
113. **Zhou, J.-Z.** 2003. Microarrays for bacterial detection and microbial community analysis. *Curr. Opin. Microbiol.* **6:**294.
114. **Zhou, J.-Z., and D. K. Thompson.** 2002. Challenges in applying microarrays to environmental studies. *Curr. Opin. Biotechnol.* **13:**202–204.

Proteomic Analysis of Extracellular Enzymes Produced by Wood-Degrading Fungi

FENG LIU, HASAN KOC, AND MING TIEN

87

This chapter reviews methods for proteomic analyses of white-rot fungi, which can degrade both the carbohydrate and the lignin components of wood to CO_2. Microbial degradation of lignocellulosic material, due to the heterogeneity of the substrate, involves an ensemble of extracellular enzymes. Kirk and Cullen (15) wrote, "The picture of how the hyphae bring about the decay of wood is becoming increasingly clear at the molecular level, even though much remains to be learned about the specific enzymes of the white-rot fungi per se." Thus, much needs to be done to identify these enzymes and the time course of their expression. Proteomics allows for a more specific characterization of expression, an alternative to the total analysis obtained through DNA microarrays. This characterization, determining the complete extracellular proteome, is now possible due to the completed genomic sequence of *Phanerochaete chrysosporium*. The complete sequence opens up new avenues for discovery and characterization. Before description of the methodology, we will provide a brief review of wood structure and the enzymes involved in degradation of woody biomass.

WOOD STRUCTURE

The structural complexity of woody biomass dictates that the enzymes utilized to degrade it must also be complex. Woody biomass has three major components: cellulose, hemicellulose, and lignin. Cellulose is a linear polymer of glucose linked by β-1,4 glycosidic bonds and makes up approximately 45% of wood by weight. Hemicellulose, approximately 25% of wood by weight, is also linked by β-1,4 linkages. Hemicellulose contains side groups that can be sugars, sugar acids, and acetyl esters. These side groups prevent efficient packing of hemicellulose and render hemicellulose noncrystalline. The major hemicellulose in hardwood is O-acetylglucuronoxylan. In softwoods, the major hemicellulose is O-acetylgalactoglucomannan. Conifer wood also contains 5 to 10% nonacetylated arabinoglucuronoxylan, whereas hardwoods also contain about 3 to 5% β-1,4-linked glucomannan. Lignin is even more complex than hemicellulose and constitutes over 25% of woody biomass by weight. It is a polymer of *p*-coumaryl, coniferyl, and sinapyl alcohols (20). Lignin is formed by free radical coupling of the phenylpropanoid units and thus can be linked by more than 12 types of linkages.

MICROBIAL ENZYMES

The degradation of lignocellulosic material is brought about predominantly by the filamentous higher fungi, primarily the Basidiomycetes, that cause the white-rot type of wood decay. Another group of wood-degrading Basidiomycetes, the brown-rot fungi, can also degrade fully lignified tissue but without substantial depletion of the lignin. Ascomycetes and Fungi Imperfecti cause the soft-rot type of wood decay which was categorized only within the past 30 years.

Whereas the degradation of cellulose and hemicellulose is a hydrolytic process, the degradation of lignin is an oxidative process. This is due mainly to the fact that no bonds in lignin are easily hydrolyzable. The ultimate oxidant in this process is molecular oxygen or its two-electron reduction product, H_2O_2. Molecular oxygen is utilized by laccases as a four-electron oxidant, being reduced to water. H_2O_2, produced by white-rot fungi (5), is utilized by peroxidases. A number of H_2O_2-producing enzymes in *P. chrysosporium* have been characterized and are listed in Table 1, along with homologous sequence from a BLAST search of the *P. chrysosporium* genome. With *P. chrysosporium*, the physiologically relevant enzyme is most likely glyoxal oxidase (14). Glyoxal oxidase oxidizes a variety of two- and three-carbon aldehydes, including glyoxal and methylglyoxal, producing H_2O_2. A BLAST search with glyoxal oxidase from *P. chrysosporium* shows two homologous sequences (Table 1). Another H_2O_2-producing enzyme, aryl alcohol oxidase, oxidizes benzyl alcohols to the aldehydes, producing H_2O_2 from molecular oxygen (7). A BLAST search with the aryl alcohol oxidase from *Pleurotus eryngii* showed 18 homologous sequences (Table 1). Carbohydrate-oxidizing enzymes have also been proposed to be the source of H_2O_2. Pyranose oxidase (4), showing one homolog in the *P. chrysosporium* genome, oxidizes a number of monosaccharides, producing H_2O_2. Finally, another enzyme that can produce H_2O_2 is cellobiose dehydrogenase (21), for which *P. chrysosporium* shows two homologous sequences. Several roles have been proposed for this enzyme in lignocellulose degradation (9, 10, 12).

The enzymes that utilize H_2O_2 and molecular oxygen to oxidize organic and inorganic substrates are the peroxidases and the laccases, respectively. A list of peroxidases and laccase and the number of homologous sequences found are shown in Table 1.

TABLE 1 Results of a BLAST search of oxidative enzymes of *P. chrysosporium*[a]

Enzyme	Species used in BLAST search	Mol wt	No. of homologous sequences
Peroxide generation			
Glyoxal oxidase	*P. chrysosporium* L47287	59,140	5
Aryl alcohol oxidase	*Pleurotus eryngii* AF064069	63,709	18
Pyranose oxidase	*Coriolus versicolor* D73369	69,495	1
CDH[b]	*P. chrysosporium* U46801	82,007	2
Peroxide utilizing			
Lignin peroxidases	*P. chrysosporium* M27400	39,600	8
Mn peroxidase	*P. chrysosporium* M60672	39,556	8
Other oxidoreductases			
Laccase	*Pleurotus ostreatus* AB020026	56,801	2

[a]Only E values of 10^{-4} or lower are shown here.
[b]CDH, cellobiose dehydrogenase.

Complete hydrolysis of cellulose, the least diverse of the three major wood components, requires an ensemble of three enzymes, which within themselves contain numerous isozymes: (i) endoglucanase (endo-1,4-β-glucanase), which hydrolyzes internal bonds; (ii) cellobiohydrolase (exo-1,4-β-glucanase), which acts processively on the existing chain ends and on those created by the endoglucanases, releasing cellobiose; and (iii) β-glucosidase, which cleaves cellobiose to two glucose molecules (Table 2).

Complete hydrolysis of hemicellulose also involves multiple enzymes (Table 2). Hydrolysis of O-acetylglucuronoxylan requires four enzymes: endo-1,4-β-xylanase (endoxylanase), acetyl esterase, α-glucuronidase, and β-xylosidase. A fifth enzyme, α-arabinofuranosidase, but not the esterase is required for softwood xylan hydrolysis.

SPORE PREPARATION

P. chrysosporium strain BKM-F-1767 (ATCC 24725) was maintained on malt agar slants containing 1% glucose, 1% malt extract, 0.2% peptone, 0.2% yeast extract, 0.1% asparagine, 0.2% KH_2PO_4, 0.1% $MgSO_4 \cdot 7H_2O$, 1 mg of thi-

TABLE 2 Results of a BLAST search of hydrolytic enzymes of *P. chrysosporium*[a]

Substrate and enzyme	Species used for BLAST search	Mol wt	No. of homologous sequences
Cellulose			
Endoglucanase			
Cel61	*P. chrysosporium* AY094489	34,368	6
Cel9	*P. chrysosporium* AY094488	63,659	1
Cellobiohydrolase			
CBHI[b]	*P. chrysosporium* M22220	54,946	4
CBHII[b]	*P. chrysosporium* S76141	48,444	1
β-Glucosidase	*P. chrysosporium* AF036873	85,559	7
Xylan			
Endoxylanase	*Trichoderma reesei* S67387	24,069	1
Acetyxylan esterase	*Aspergillus* A22880	32,768	3
α-Glucuronidase	*T. reesei* Z68706	93,424	0
β-Xylosidase	*T. reesei* Z69257	69,257	3
α-Arabinosidase	*Cytophaga xylanolytica* AF028019	59,169	1
Glucomannan			
Endomannanase	*Agaricus bisporus* AJ271862	47,042	2
Acetylmannan esterase	*Aspergillus ficuum* AF331757	32,591	3
α-Galactosidase	*P. chrysosporium* AF246263	48,553	3
β-Mannosidase	*Aspergillus aculeatus* AB015509	104,215	1
β-Glucosidase	*P. chrysosporium* AF036873	85,559	7

[a]Only E values of 10^{-4} or lower are shown here.
[b]CBHI and CBHII, cellobiohydrolases I and II, respectively.

amine per ml, and 2% agar (19). Conidiospores were resuspended in autoclaved distilled deionized water (ddH_2O) and filtered through glass wool (autoclaved).

GROWTH MEDIUM

Although a variety of methods are available for growth of *P. chrysosporium*, we describe here methods available to any laboratory. Thus, we used 10-ml cultures in stationary 125-ml Erlenmeyer flasks. For nitrogen-limited medium, spores (100 ml) (absorbance at 650 nm, 0.5) were used to inoculate 1 liter of autoclaved liquid medium containing 100 ml of Basal III medium (19), 1% glucose, 10 mM 2,2-dimethylsuccinate (pH 4.5), 0.2 of ammonium tartrate per liter, and 60 ml of trace element (19). The cultures were grown at 37°C for 6 days, with water-saturated O_2 flushing at days 2 and 5.

Liquid-solid medium contained the same reagents as liquid medium except that glucose was replaced by cellulose or pretreated wood powder. The pretreatment of red oak wood chips significantly reduced brown extractives, which caused streaking and background staining of two-dimensional (2-D) gels (1). To wash the wood chips, 5 volumes of ddH_2O was added to the wood chips and then the mixture was autoclaved for 30 min. The brown liquid was decanted, and the washing process was repeated two more times. The wood chips were then ground to a fine powder with a coffee grinder (1).

For inoculation of solid substrate culture, one tube of *P. chrysosporium* agar slant was added to 20 g of millet, 10 g of wheat bran, and 30 ml of ddH_2O in a 250-ml Erlenmeyer flask, which was then incubated at 37°C for 10 days. This culture was used to inoculate the medium (autoclaved), which contained 255 g of pretreated wood chips (without grinding), 30 g of millet, 15 g of wheat bran, and 300 ml of ddH_2O in a polypropylene growth bag (Unicorn, Commerce, Tex.). Cultures were grown at 37°C for 3 weeks (1).

ISOLATION OF PROTEINS

At the time of harvest, the culture medium containing the extracellular enzymes was harvested by passage through cheesecloth. This removed most of the solid substrate and mycelia. Any remaining particulates were removed by centrifugation at 8,000 × *g* for 30 min at 4°C. Supernatant (ca. 1 liter) was concentrated to 10 ml by use of an Amicon 10-kDa filter (Millipore, Bedford, Mass.). For solid substrate cultures, extracellular proteins were extracted by addition of 500 ml of 0.5 M NaCl. After being mixed at 4°C for 2 h (1), the samples were processed as for the liquid cultures, passed through cheesecloth, centrifuged, and concentrated. Ammonium sulfate precipitation could also be used to concentrate the proteins (1). The samples were desalted (in preparation for isoelectric focusing [IEF]) by addition of 30-fold ddH_2O and concentrated again to a final volume of 10 ml. This desalting step was repeated two more times, and the final samples were stored at −70°C.

Alternatively, trichloroacetic acid (TCA) precipitation could also be used for desalting. One volume of ice-cold TCA (90%) was added to 5 volumes of the supernatant, and the mixture was vortexed immediately. The sample was kept on ice for 30 min and then centrifuged at 14,000 × *g*, and 4°C for 15 min. The protein pellet was washed with 1 ml of cold acetone (−20°C) twice, centrifuged again, and air dried. The pellet was then redissolved in immobilized pH gradient (IPG) rehydration buffer, consisting of 8 M urea, 2% 3-[(3-cholamidopropyl)-dimethylammonio]-1-propanesulfonate (CHAPS), 0.4% pH 3 to 10 IPG buffer, 50 mM dithiothreitol (DTT), and 0.002% bromophenol blue; centrifuged at 5,000 × *g* and 4°C for 5 min to eliminate undissolved material; and stored at −70°C.

TCA precipitation was often problematic because the protein pellet was frequently difficult to redissolve. The samples were sonicated several times for 10 min each with a 10-min interval. Note that unless caution is taken, the heat generated by continuous sonication may degrade urea to isocyanate, which can then carbamylate protein and change the pI. Overdrying the protein pellets also caused problems in redissolving. Often, the samples were air dried. To facilitate redissolving, the protein pellets can be kept overnight in the freezer (−20°C) in the rehydration buffer and sonicated prior to electrophoresis. Despite the above problem, TCA precipitation was much faster than concentration by Amicon filter. Sample concentration and desalting using a 10-kDa filter membrane take several days to complete.

METHODS FOR REMOVAL OF LIGNIN

Even though pretreatment of wood reduced the amount of the brown extractives, enough remained to often cause problems in the 2-D gels. The material is most likely derived from lignin (1). Other methods for removal of the material have been tested. Phenol extraction proved to be very effective in extracting plant proteins from interfering substances to obtain good resolution in 2-D gels (13). Below is a brief description of the procedure modified from reference 11. Sucrose is added to the sample to 0.7 M. An equal volume of phenol saturated with Tris-EDTA buffer (pH 8) is added, and the mixture is shaken for 10 min at room temperature. Two phases are separated at 14,000 × *g* for 10 min. The upper, protein-containing phenol phase is transferred to a new tube. After the extraction is repeated three more times, 5 volumes of 0.1 M ammonium acetate in methanol is added and the mixture is incubated at ×20°C for 2 h. The precipitated protein is then pelleted at 14,000 × *g* for 15 min. The pellet is washed twice with 0.1 M ammonium acetate and twice with ice-cold 80% acetone and air dried. For analysis, the pellet is redissolved in rehydration buffer and stored at −70°C.

TWO-DIMENSIONAL GEL METHODS

Separation of proteins by 2-D gels is not necessary. Methods are available for processing of the complete extracellular protein mixture by protease digestion and separation and identification by liquid chromatography and tandem mass spectrometry (MS-MS). The 2-D gels provide information on the protein quantity and an idea of the protein complexity. The gel procedures described below required handling of all materials with gloves to minimize protein contamination. Protein concentrations were measured by Bradford assay (Bio-Rad, Hercules, Calif.). Upon rehydration, approximately 100 μg of proteins (200 μl) was loaded onto one lane of the disposable rehydration tray (Bio-Rad). After removal of the plastic back cover of the IPG strip (Immobiline DryStrip, pH 3 to 10, 11 cm; Amersham Biosciences, Piscataway, N.J.), the strip was laid on the protein sample with the gel side facing down, with care taken not to introduce air bubbles. To prevent drying, 3 ml of mineral oil was added to the strip, and it was incubated overnight.

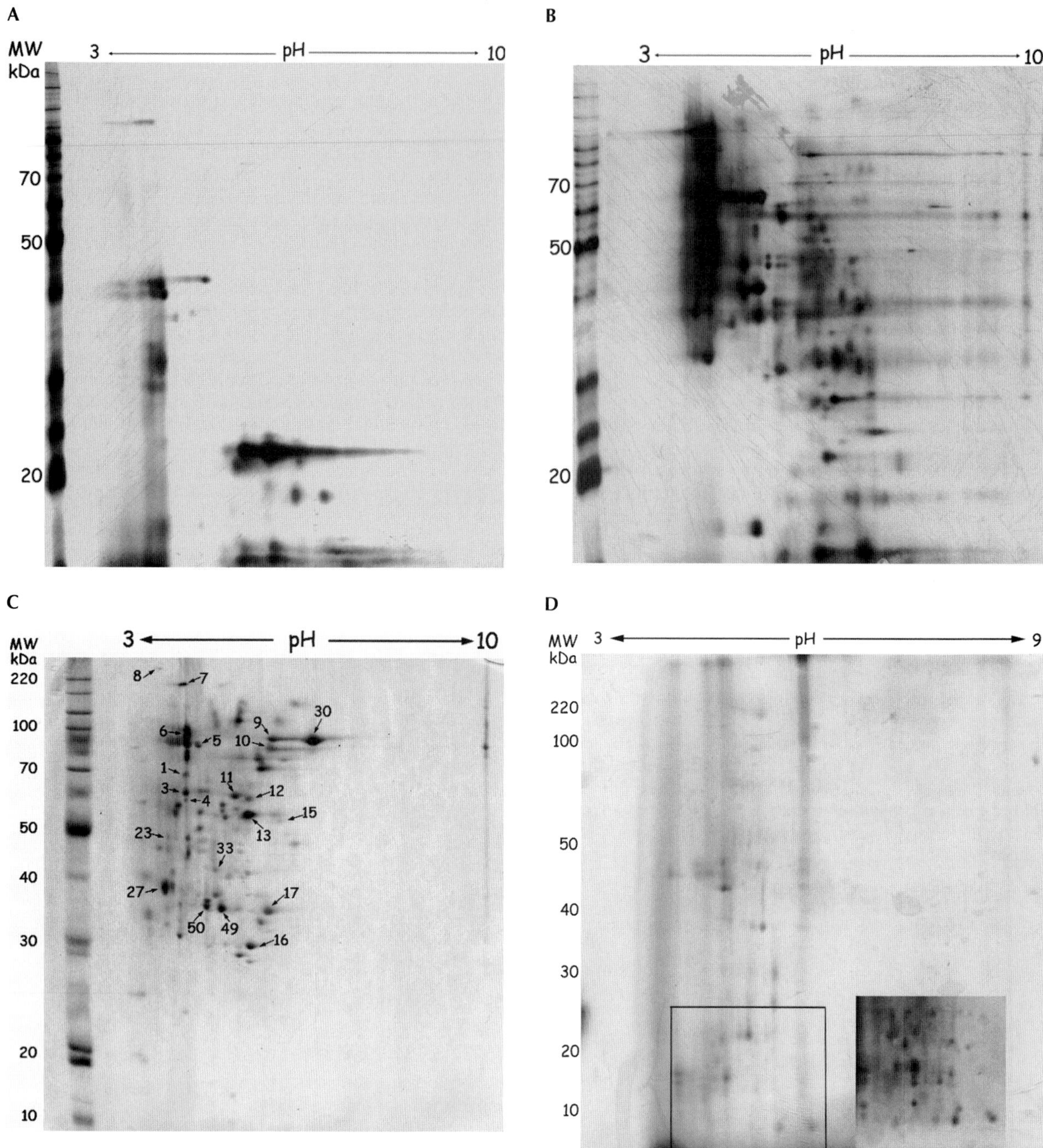

FIGURE 1 2-D gels of extracellular proteins from different cultures of *P. chrysosporium*. (A) Proteins from liquid culture of *P. chrysosporium* grown on glucose as a carbon source (F. Liu and M. Tien, unpublished data). (B and C) Proteins from liquid-solid culture of *P. chrysosporium* grown on cellulose as a carbon source (B) (Liu and Tien, unpublished) or on wood (C) (reprinted from *Current Genetics* [1] with permission of the publisher). (D) Proteins from *P. chrysosporium* grown on solid wood substrate, which was stained with Coomassie blue except for the inset, for which after visualization with Coomassie blue, the gel was destained and then visualized with silver stain (reprinted from *Current Genetics* [1] with permission of the publisher).

First-dimension IEF was performed on a Bio-Rad Protean IEF cell at 20°C with a one-step program: 8,000 V for a total of 30,000 V·h with a current limit of 50 μA per gel. When the volt-hours were reached, the IPG strip could be kept at low voltage (500 V) for a while (1 day) until the next step, either storage at −70°C or sodium dodecyl sulfate-polyacrylamide gel electrophoresis (SDS-PAGE). Prior to SDS-PAGE, the IPG strips were equilibrated in reducing buffer (50 mM Tris-HCl [pH 6.8], 6 M urea, 30% glycerol, 2% SDS, 0.002% bromophenol blue, 10 mg of DTT per ml) for 10 min and then alkylating buffer (50 mM Tris-HCl [pH 6.8], 6 M urea, 30% glycerol, 2% SDS, 0.002% bromophenol blue, 25 mg of iodoacetamide per ml) for 10 min. Note that protein bands diffuse after IEF in the absence of a current; thus, all procedures should be followed in a timely manner.

A vertical electrophoresis unit (SE 600, 18 by 16 cm; Hoefer, San Francisco, Calif.) was used to run the SDS-PAGE gel (12%) (16). An IPG strip was placed on top of the stacking gel (acidic side close to the protein marker). To better contact with the SDS-PAGE gel, the strips were overlaid with 0.5% low-melting-point agarose. The voltage was set at 120 V for 1 h and 300 V until the bromophenol blue dye reached the bottom of the gel. The gel was stained with 1% Coomassie blue R-250 in destaining solution (8% acidic acid, 25% ethanol) overnight and destained until protein spots were visible. The destaining solution was replaced by several changes of ddH_2O. Proteins in the 2-D gels were also visualized by silver staining if they were not to be subsequently analyzed by mass spectroscopy (2). After scanning of the image, the gel could be stored in ddH_2O at 4°C for several days. Sample gel images from different cultures are shown in Fig. 1.

IN-GEL DIGESTION

Protein spots were excised from the gel, transferred to 1.5-ml Eppendorf tubes, and digested with trypsin according to the protocol described by Rosenfeld et al. (18) and Hellman et al. (8), with some modifications. Briefly, large gel pieces were cut into small pieces and dehydrated with 100 μl of 25 mM ammonium bicarbonate in 50% acetonitrile. This was repeated several times (vortex, 10 min each) until the gel became white (Coomassie blue eliminated). The gel was then dried under a vacuum. The sample was reduced and alkylated if the protein concentration was low or the sample was isolated from a 1-D gel. Freshly prepared 10 mM DTT in 25 mM ammonium bicarbonate (reducing reagent) (25 μl or enough to cover the gel pieces) was added, and the gel was incubated at 56°C for 1 h. After the DTT solution was removed, 25 μl (or enough to cover the gel pieces) of freshly prepared 55 mM iodoacetamide in 25 mM ammonium bicarbonate (alkylating reagent) was added and the gel was incubated at room temperature in the dark for 45 min. The gel was then washed with 100 μl of 25 mM ammonium bicarbonate, dehydrated with 100 μl of 25 mM ammonium bicarbonate in 50% acetonitrile twice, and dried under a vacuum. Freshly diluted modified trypsin (12.5 ng/μl) (Promega, Madison, Wis.) in 25 mM ammonium bicarbonate was added to barely cover the gel. The gel was rehydrated on ice for 10 min. Additional ammonium bicarbonate solution was added if necessary. Digestion was carried out at 37°C for 4 h to overnight. The digested sample solution was transferred to a new tube. To the gel tube, 30 μl of 5% formic acid in 50% acetonitrile was added. The tube was vortexed for 15 min and sonicated for 10 min. The extracted solution was combined with the previously digested solution. This step was repeated once more. The combined solution was dried under a vacuum to a final volume of 10 μl. Then the sample was either analyzed by liquid chromatographic MS-MS (capillary liquid chromatography [CapLC] quadrupole time-of-flight [Q-TOF]) or cleaned up by C_{18} ZipTip (Millipore), for matrix-assisted laser desorption ionization–time-of-flight (MALDI-TOF) analysis.

PROTEIN IDENTIFICATION BY MS

Two major MS methods for identification of proteins separated by 2-D gel electrophoresis are peptide mass fingerprinting (PMF) on a MALDI-TOF mass spectrometer and MS-MS ion search of collision-induced dissociation (CID) MS-MS spectra obtained by liquid chromatographic separation followed by MS-MS on instruments such as Q-TOF or Ion Trap mass spectrometers. PMF strategy takes advantage of the fact that, upon digestion by a protease-like trypsin, every protein produces a unique set of peptide molecular weights, which can then be submitted to a search against the in silico-generated peptide masses from protein databases in order to find the best match. Table 3 shows sample data for one of the extracellular proteins from *P. chrysosporium*. As described by Abbas et al. (1), peptide mass values from MALDI-TOF analysis were searched against the

TABLE 3 Peptide masses and predicted amino acid sequence of laminarinase[a]

m/z		Start	End	No. of missed cleavages	Database sequence
Mi	Avg				
616.4146	616.7867	94	98	1	(R)IRSK(T)
709.3885	709.8257	212	217	0	(R)TNSFIK(V)
744.448	744.8776	90	95	1	(R)NSVRIR(S)
851.4568	852.0333	218	223	0	(K)VWFFPR(N)
1,292.685	1,293.472	51	62	0	(R)VNYVDQATALAK(N)
1,396.682	1,397.497	76	89	0	(R)ADHTTTLSPSGPGR(N)
1,408.723	1,409.595	99	110	0	(K)TYTTHVAVFDVR(H)
1,436.775	1,437.646	63	75	0	(K)NLTYASGDTLILR(A)
1,541.827	1,542.836	212	223	1	(R)TNSFIKVWFFPR(N)
1,736.934	1,738.009	96	110	1	(R)SIKTYTTHVAVFDVR(H)
1,852.927	1,854.002	76	93	1	(R)ADHTTTLSPSGPGRNSVR(I)
2,710.442	2,712.095	51	75	1	(R)VNYVDQATALAKNLTYASGDTLILR(A)
2,814.439	2,816.12	63	89	1	(K)NLTYASGDTLILRADHTTTLSPSGPGR(N)

[a] The peptide masses were analyzed by Protein Prospector using the *P. chrysosporium* local database; the best match was pc. 78.37.1 A BLAST homology search of the fungal database indicated that the protein is laminarinase. (Reprinted from *Current Genetics* [1] with permission of the publisher.)

TABLE 4 Results of a BLAST homology search of genes identified by Protein Prospector[a]

Spot[b]	Sequence[c]	Putative enzyme	Accession no.[d]	E value[e]	Mol wt[f]	pI[f]
1	VVLDSNWR	Cellobiohydrolase	pc.33.51.1	0.0	58,142 (66,000)	4.9 (4.9)
3	TAFGDTNWFAQK	Cellobiohydrolase	pc.33.51.1	0.0	58,142 (59,000)	4.9 (4.9)
5	SAFDFLNGK	Cellobiose dehydrogenase	pc.5.157.1	0.0	82,007 (86,000)	5.2 (5.1)
6	IGIDAALR	Cellobiose dehydrogenase	pc.5.157.1	0.0	82,007 (95,000)	5.2 (5.0)
7	SEQLAVWTVK	β-Glucosidase	Genewise2nd.66.3.1	$2e^{-42}$	69,848 (160,000)	4.9 (4.8)
8	IIDFAAVGQK	α-Galactosidase	pc.11.202.1	0.0	47,706 (200,000)	4.6 (4.6)
13	ISGTFTNVR	Endopolygalacturonase	Genewise2nd.1.81.1	$8e^{-93}$	32,503 (52,000)	4.7 (5.7)
15	IPDLGTYLASASALGKTTYLACVNYALTNLAK	Exocellobiohydrolase II	pc.3.82.1	0.0	48,417 (50,000)	5.0 (6.0)
16	ADVSYDIWFGK	β-Endoglucanase				
17	TYTTHVAVFDVR	Laminarinase	pc.78.37.1	e^{-180}	33,961 (34,000)	5.0 (6.0)
23	LPSTDHPSTDGQRGQGYNQATINDNPNYKVILSAGFG TSRAYSGSDGFTR	Cellobiose dehydrogenase	pc.5.157.1	0.0	82,007 (45,000)	5.02 (4.8)
27	ISGTFTNVR	Endopolygalacturonase	Genewise2nd.1.81.1	$8e^{-172}$	82,632 (80,000)	5.6 (6.2)
30	GDGNTDDTAAIQAAINAGGR	Glucan 1,3-β-glucosidase	pc.67.67.1	e^{-93}	58,142 (66,000)	4.9 (4.9)
33	GVFTFANADTIANLAR	Endo-1,4-β-xylanase A	pc.143.21.1	e^{-127}	35,784 (42,000)	5.3 (5.3)
49	LQTDHLFARTGIQGTVMSPLKMVFHDSIAISPKTGIQG TVMSPLKGEMR	Lignin peroxidase, isozyme H2	pc.19.174.1	0	39,534 (39,000)	5.02 (5.2)
50	LQTDHLFARTGIQGTVMSPLK	Lignin peroxidase, isozyme H2	pc.19.174.1	0	39,534 (39,000)	5.02 (5.3)

[a]Reprinted from *Current Genetics* (1) with permission of the publisher.
[b]Protein spots correspond to the numbering in the gel shown in Fig. 1.
[c]The sequences were obtained from CapLC–electrospray ionization–MS-MS and then used to search the local *P. chrysosporium* database by use of Protein Prospector. The identified enzymes are listed.
[d]The identified gene was then used for a BLAST search of the fungal database.
[e]The E values of the BLAST search are shown.
[f]The calculated molecular weights and pIs are shown, along with the experimentally determined values, which are listed in parentheses.

TABLE 5 Primers for PCR of *P. chrysosporium* genes[a]

Gene[b]	Forward primer[c]	Reverse primer[c]	cDNA size (bp)[d]
1	TCATCTCCCAAGTCACGGT	TGTTTACAGTCAGAGTCGTCT	498
2	TCCGCGGTACGGTGCTCA	CGTCTACAGCCCCGAGGACA	431
3	GTGGCATGACACTCGATGAA	CAACTCTGCATAGTCCCTTC	509
4	ACATCAAGTTCGGCGACATC	GTGCATTTCACGGATCGCTA	403
5	CTACATCAAGGGTAGTCGTC	GTATGGTCACTGTCTGCGA	353
6	CGGATCACGATCTCGACTC	GACGCTGAACTGCCAGGT	461
7	GACGAAGCTCTACATCAACGA	CCAATACCGATGGCGATTC	450
8	TCACCCGACCCCATCA	CACACGCCAATCTACCTCTAC	416
9	CGTGGAACTGGACCTACGTTA	AGTTCCGCTACGCGATTCTC	468

[a] Liu and Tien, unpublished.
[b] *P. chrysosporium* genes (1 to 9) encoding endopolygalacturonase, cellobiose dehydrogenase, α-galactosidase, cellobiohydrolase, β-glucosidase, glucan-1,3-β-glucosidase, endo-1,4-β-xylanase A, exocellobiohydrolase II, and β-endoglucanase, respectively.
[c] Primer sequences are based on sequences from the annotated *P. chrysosporium* database. Gene 1 to 9 accession numbers are Genewise2nd.1.81.1, pc.5.157.1, pc.11.202.1, pc.33.51.1, Genewise2nd.66.3.1, pc.67.67.1, pc.143.21.1, pc.3.82.1, and pc.163.7.1, respectively.
[d] cDNA size is the predicted size of the PCR product.

NCBI NR database and the annotated *P. chrysosporium* database (http://genome.jgi-psf.org/whiterot1/whiterot1.home.html) by using the Protein Prospector MS-Fit tool (http://prospector.ucsf.edu) (3). We initially analyzed 25 protein spots, and only 2 were identified. The high percentage of unmatched data may be due to glycosylation of fungus extracellular proteins (6). Glycosylation not only can prevent accessibility of proteases but also can alter the mass of the digested peptides. With MALDI-TOF data and the Protein Prospector, the user sets the lower limit for the number of matching peptides. Obviously, the number of matches increases as the stringency is lowered. Although the deglycosylation method can be used (6), it was not successful with the extracellular proteins of *P. chrysosporium* (1).

For heavily glycoslyated proteins, CapLC–nanoelectrospray ionization–MS-MS can be used instead to analyze in-gel tryptic digests. In our study, Q-TOF instruments (Waters Q-TOF micro or Q-TOF API US) coupled with a Waters CapLC high-performance liquid chromatography (HPLC) unit were used. On-line desalting and preconcentration were done by injecting proteolytic digests (1 to 5 μl) into solvent A (acetonitrile-water-formic acid, 5:95:0.1) supplied by the auxiliary pump of the capillary HPLC unit and trapped in a Waters Symmetry 300 column (C_{18}, 5-μm film, 0.3 by 5 mm). After a 3-min wash with solvent A at 20 μl/min, trapped peptides were then back-flushed with the gradient solvent flow onto the analytical column, a Dionex PepMap fused silica capillary column (C_{18}, 5 μm, 0.075 by 150 mm), using a 10-port switching valve. The analytical column was run with gradient solvent B (acetonitrile-water-formic acid, 95:5:0.2), 5 to 42%, in 44 min. Sensitivity and mass calibration of the Q-TOF mass spectrometer were checked daily. To optimize the electrospray conditions for sensitivity, 1 μM fibrinopeptide B (Glu-Fib) was infused at 0.5 μl/min and calibration was done by acquiring a CID MS-MS spectrum of Glu-Fib. Prior to sample analysis, a quality control sample of a tryptic digest of bovine serum albumin (20 to 100 fmol) (Michrom Bioresources, Auburn, Calif.) was run, and a search against the database was made, using the same settings as for samples. This quality control run ensured the reliability of the protein identification and provided information about the sensitivity and mass measurement accuracy (maintained typically below 10 ppm) of the mass spectrometer. CID MS-MS spectra for the peptides were acquired by utilizing on-the-fly precursor ion selection with the data-dependent acquisition (DDA) feature of the MassLynx data acquisition software. DDA was done based on intensity (15 to 20 counts) and charge state (2^+, 3^+, 4^+). Besides the obvious advantage of producing the peptide MS from complex mixtures that usually have many more non-peptide ions, DDA also adds another dimension of specificity to the analysis. In order to carry out the MS-MS data acquisition in this mode, a survey scan (2 s) over the m/z of 400 to 1,500 was performed as a first step. From each survey scan, up to three most intense precursor ions based on the selection criteria were selected for MS-MS to obtain the production spectra resulting from CID in the presence of argon. A particular precursor ion was analyzed only once during the whole analysis, using dynamic exclusion. The raw MS-MS spectra (6 to 8 s) collected were processed with Protein Lynx Global Server 2.1, which summed, smoothed, centroided, and deisotoped the raw data. The processed data files were saved as text files called peak lists (.pkl) for database searching. In order to identify the proteins, MS-MS ion searches were performed on the processed spectra against locally maintained copies of the NCBI NR database and the annotated *P. chrysosporium* database using MASCOT Daemon and the MASCOT server (Matrix Science, Inc., Boston, Mass.) on the assumption that there was a maximum of one missed trypsin cleavage and that the peptides were monoisotopic. Methionine oxidation as a variable modification and carbamidomethylation of cysteine as a fixed modification were taken into consideration. Peptide mass tolerance and fragment mass tolerance were initially set to 1.2 and 0.6 Da, respectively, for MS-MS ion searching. However, when the MASCOT search results were reviewed manually, the peptide mass values were ensured to be within 0.1 Da (typically < 0.05 Da). Protein identifications with a low ($< \sim50$) protein score or those with one or two peptides matched were evaluated for data quality manually at the raw data level. By this method, 16 spots were identified (Table 4) (1). Now we are in the process of identifying all extracellular protein spots visualized on 2-D gels.

RT-PCR TO CONFIRM PROTEOMIC RESULTS

With annotated databases, proteomic analyses are still vulnerable to errors in the gene prediction algorithms. In our studies, we further confirm the protein identity by reverse transcriptase PCR (RT-PCR). Primers (Table 5) were designed on the basis of the coding sequence from the annotated database of *P. chrysosporium*. Mycelia and woody biomass separated from extracellular fluid were rapidly frozen in liquid N_2 and ground into fine powder with a mortar and pestle. Approximately 0.1 to 0.2 g of powder was transferred to a microcentrifuge tube and mixed with 1 ml of TRIzol reagent (Invitrogen, Carlsbad, Calif.). Total RNA was isolated according to the manufacturer's protocol. In case of genomic DNA contamination, RNA samples were digested with RNase-free DNase (Promega). Reverse transcription was performed at 37°C for 1 h with 2 μg of total RNA, 1 μg each of random primers (Promega), 0.5 mM (each) deoxynucleoside triphosphate mix (Ambion, Austin, Tex.), 1 U of RNasin Plus RNase inhibitor (Promega) per μl, and 5 U of Moloney murine leukemia virus reverse transcriptase (Promega) per μl in a 20-μl total reaction volume. The cDNA was then used as the template in a PCR (Eppendorf Mastercycler gradient; Eppendorf, Hamburg, Germany). Nine extracellular enzymes were verified by RT-PCR (data not shown). The PCR products were further confirmed by DNA sequencing. The full-length cDNA could be obtained by use of the SMART RACE cDNA amplification kit (Clontech, Palo Alto, Calif.) as described elsewhere (17).

This research was supported in part by a grant from the Department of Energy (DE-FG02-87ER13690).

REFERENCES

1. **Abbas, A., H. Koc, F. Liu, and M. Tien.** 2005. Fungal degradation of wood: initial proteomic analysis of extracellular proteins of *Phanerochaete chrysosporium* grown on oak substrate. *Curr. Genet.* **47:**49–56.
2. **Blum, H., H. Beier, and H. J. Gross.** 1987. Improved silver staining of plant proteins, RNA and DNA in polyacrylamide gels. *Electrophoresis* **8:**93–99.
3. **Clauser, K. R., P. Baker, and A. L. Burlingame.** 1999. Role of accurate mass measurement (+/−10 ppm) in protein identification strategies employing MS or MS/MS and database searching. *Anal. Chem.* **71:**2871–2882.
4. **Daniel, G., J. Volc, and E. Kubatova.** 1994. Pyranose oxidase, a major source of H_2O_2 during wood degradation by *Phanerochaete chrysosporium*, *Trametes versicolor*, and *Oudemansiella mucida*. *Appl. Environ. Microbiol.* **60:**2524–2532.
5. **Forney, L. J., C. A. Reddy, M. Tien, and S. D. Aust.** 1982. The involvement of hydroxyl radical derived from hydrogen peroxide in lignin degradation by the white-rot fungus *Phanerochaete chrysosporium*. *J. Biol. Chem.* **257:** 11455–11462.
6. **Fryksdale, B. G., P. T. Jedrzejewski, D. L. Wong, A. L. Gaertner, and B. S. Miller.** 2002. Impact of deglycosylation methods on two-dimensional gel electrophoresis and matrix assisted laser desorption/ionization-time of flight-mass spectrometry for proteomic analysis. *Electrophoresis* **23:**2184–2193.
7. **Guillen, F., A. T. Martinez, and M. J. Martinez.** 1992. Substrate specificity and properties of the aryl-alcohol oxidase from the ligninolytic fungus *Pleurotus eryngii*. *Eur. J. Biochem.* **209:**603–611.
8. **Hellman, U., C. Wernstedt, J. Gonez, and C. Heldin.** 1995. Improvement of an "in-gel" digestion procedure for the micropreparation of internal protein fragments for amino acid sequencing. *Anal. Biochem.* **224:**451–455.
9. **Henriksson, G., V. Sild, I. J. Szabo, G. Pettersson, and G. Johansson.** 1998. Substrate specificity of cellobiose dehydrogenase from Phanerochaete chrysosporium. *Biochim. Biophys. Acta* **1383:**48–54.
10. **Henriksson, G., G. Johansson, and G. Pettersson.** 2000. A critical review of cellobiose dehydrogenases. *J. Biotechnol.* **78:**93–113.
11. **Hernandez-Macedo, M. L., A. Ferraz, J. Rodriguez, L. M. M. Ottoboni, and M. P. D. Mello.** 2002. Iron-regulated proteins in *Phanerochaete chrysosporium* and *Lentinula edodes*: differential analysis by sodium dodecyl sulfate polyacrylamide gel electrophoresis and two-dimensional polyacrylamide gel electrophoresis profiles. *Electrophoresis* **23:**655–661.
12. **Hilden, L., G. Johansson, G. Pettersson, J. B. Li, P. Ljungquist, and G. Henriksson.** 2000. Do the extracellular enzymes cellobiose dehydrogenase and manganese peroxidase form a pathway in lignin biodegradation? *FEBS Lett.* **477:**79–83.
13. **Hurkman, W. J., and C. K. Tanaka.** 1986. Solubilization of plant membrane proteins for analysis by two-dimensional gel electrophoresis. *Plant Physiol.* **81:**802–806.
14. **Kersten, P. J., and T. K. Kirk.** 1987. Involvement of a new enzyme, glyoxal oxidase, in extracellular H_2O_2 production by *Phanerochaete chrysosporium*. *J. Bacteriol.* **169:** 2195–2201.
15. **Kirk, T. K., and D. Cullen.** 1998. Enzymology and molecular genetics of wood degradation by white-rot fungi, p. 273–279. *In* R. A. Young and M. Akhtar (ed.), *Environmentally Friendly Technologies for the Pulp and Paper Industry*. John Wiley & Sons, Inc., Hoboken, N.J.
16. **Laemmli, U. K.** 1970. Cleavage of structural proteins during the assembly of the head of bacteriophage T4. *Nature* **227:**680–685.
17. **Liu, F., L. Cui, D. Cox-Foster, and G. W. Felton.** 2004. Characterization of a salivary lysozyme in larval *Helicoverpa zea*. *J. Chem. Ecol.* **30:**2439–2457.
18. **Rosenfeld, J., J. Capdevielle, J. C. Guillemot, and P. Ferrara.** 1992. In-gel digestion of proteins for internal sequence analysis after one- or two-dimensional gel electrophoresis. *Anal. Biochem.* **203:**173–179.
19. **Tien, M., and T. K. Kirk.** 1988. Lignin peroxidase of *Phanerochaete chrysosporium*. *Methods Enzymol.* **161:**238–249.
20. **van Rensburg, H., A. M. Anterola, L. H. Levine, L. B. Davin, and N. G. Lewis.** 2000. Monolignol compositional determinants in loblolly pine: aromatic amino acid metabolism and associated rate-limiting steps, p. 118–144. *In* W. G. Gasser, R. A. Northey, and T. P. Schultz (ed.), *Lignin: Historical, Biological, and Materials Perspectives*. American Chemical Society, Washington, D.C.
21. **Westermark, U., and K. E. Eriksson.** 1988. Cellobiose dehydrogenase (quinone). *Methods Enzymol.* **160:**463–468.

Current Progress in the Application of Mycoremediation to Soil Cleanup

RICHARD T. LAMAR AND R. B. WHITE

88

Mycoremediation, or fungus-based remediation, is an ex situ form of bioaugmentation, in which hazardous organics are degraded or detoxified by fungi that are introduced into the contaminated soil via a fungal inoculum. The combination of fungal species, inoculum application rate, and necessity of amendments (e.g., surfactants to enhance pollutant bioavailability) that gives the most effective decrease in pollutant concentration(s) is determined prior to application in a laboratory treatability study. These studies are conducted on soil samples collected from the target site and are representative of the soil to be treated in terms of pollutants present, their concentrations, and any other mitigating factors. Fungal inocula consist of lignocellulosic substrates (e.g., straw, wood chips, cottonseed hulls, and wood-based pellets) that are thoroughly colonized by a pure culture of the selected fungus. The soil/inoculum mixture is treated in an enclosed, forced-aeration biopile or in-ground biocell in which temperature and moisture are monitored and maintained to provide optimum fungal growth and activity. Pollutant concentrations are monitored until target concentrations are achieved.

Fungi are robust organisms and are generally more tolerant of high concentrations of pollutants than are bacteria (31). *Phanerochaete chrysosporium* was the first representative of white-rot fungi (WRF) and wood decay fungi reported to show degradation of an extremely diverse group of environmental pollutants (14, 28). Since then, the research and applications regarding mycoremediation of soil have been focused on the pollutant-degrading abilities of WRF. These fungi are saprophytes that obtain their carbon for energy and biomass from dead organic matter and include members such as the common edible mushrooms, *Pleurotus ostreatus* (oyster mushroom), *Lentinus edodes* (shiitake), and *Agaricus bisporus* (white button mushroom). The last fungus actually belongs to the ecophysiological group of fungi called litter-degrading fungi, which also obtain their carbon and energy from lignocellulosic substrates. WRF degrade cellulose, hemicellulose, and, most important, the lignin component of the wood cell wall. The residual wood is typically fibrous with a whitish-yellow-to-tan discoloration due to the removal of lignin. Most WRF are basidiomycetes, possessing dikaryotic hyphae and clamp connections along the septate hyphae (15).

WRF are uniquely equipped as soil remediation agents for several reasons. They possess the ability to oxidize a wide variety of extremely hydrophobic pollutants, e.g., polychlorinated dibenzo-*p*-dioxins (PCDDs) and polychlorinated dibenzofurans (PCDFs) and high-molecular-weight (HMW) polyaromatic hydrocarbons like benzo[*a*]pyrene, because of the highly oxidative and nonspecific nature of the enzymes that comprise the extracellular component of their lignin-degrading system. The enzymes involved are necessarily extracellular, so their degradative influence is extended beyond the fungal hyphae. Also, as filamentous organisms, they have the natural propensity to extend through soil in search of new substrates to exploit and thus can colonize environments that bacteria might be unable to reach. Finally, degradation of the pollutants, like that of lignin, does not provide WRF with carbon for energy and creation of biomass. Thus, because they do not rely on pollutant metabolism for growth, the amount of pollutant degraded is not a function of its concentration in the soil.

Both direct (i.e., pollutant oxidation by lignin-degrading enzymes) and indirect (i.e., pollutant transformation and mineralization during ligninolysis) evidence indicates that various components, depending on the pollutant, of the lignin-degrading systems of WRF are involved in pollutant degradation. Lignin has a heterogeneous, aromatic structure with many different types of subunit linkages. The enzyme systems and associated mechanisms that WRF have evolved to depolymerize lignin and to further degrade the aromatic subunits are highly oxidative, extracellular, and nonspecific (52). In addition to oxidative agents, the fungi possess reductive agents that are also involved in the degradation of aromatic substructures of lignin that are produced from its depolymerization (quinones). It makes perfect sense then that aromatic pollutants, such as PCP, polycyclic aromatic hydrocarbons (PAHs), polychlorinated biphenyls, and PCDDs/PCDFs that are degraded by WRF, closely resemble the aromatic substructures that are produced and further degraded during lignin depolymerization.

Because of the ability of WRF to degrade extremely recalcitrant contaminants (e.g., HMW PAHs, PCDDs, and PCDFs), work on fungus-based remediation has focused on the treatment of soils contaminated with these types of chemicals. In addition, work has progressed on the soil microbiology of mycoremediation and on the development of techniques to monitor the fate of fungi inoculated into soils.

PCDDs AND PCDFs

PCDDs and PCDFs are ubiquitous environmental contaminants. They are formed during incineration (46), bleaching of paper pulp (85), and industrial processes involving synthesis of chlorinated aromatic and aliphatic compounds (43). Fallout of PCDDs and PCDFs from combustion at incineration facilities has resulted in serious pollution of the surrounding soil environment because of their extremely high persistence and toxicity.

Several strains of WRF that have the ability to metabolize PCDDs by nonspecific lignin-degrading enzymes (14, 86, 89) have been evaluated and have shown potential in the bioremediation of contaminated soil (75, 77, 94). In Japan, where soil and fly ash have become highly contaminated with PCDDs and PCDFs due to incineration of municipal wastes, research on the use of WRF to remediate these materials has been very active. Out of a total of 136 strains of WRF, 12 that were able to degrade more than 20% of dibenzodioxin (DD) in low-nitrogen liquid medium (87) (a condition that elicits production of the lignin-degrading enzyme system once the nitrogen has been consumed, for many WRF) were selected for further investigation (70). The 12 strains were further evaluated for their ability to mineralize 25 mM [U-^{14}C]2,7-dichlorodibenzo-*p*-dioxin (2, 7-diCDD) in the same medium. Of the 12, three strains, MZ-227, *Phlebia* sp. strain MG-60, and *Phlebia lindtneri*, were clearly superior at 2,7-diCDD mineralization (70).

TABLE 1 Percent decrease[a] in the concentration of selected PCDD and PCDF congeners in fly ash inoculated with live or heat-killed (control) *Ceriporia* sp. strain MZ-340 after 12 weeks

Congener	% Decrease in concn found in fly ash inoculated with:	
	Control	Live *Ceriporia* sp. strain MZ-340
PCDD		
2,3,7,8-TCDD	(5.9)	76.4
1,2,3,7,8-PeCDD	23.3	59.7
1,2,3,4,7,8-HxCDD	37.9	69.7
1,2,3,6,7,8-HxCDD	32.3	59.7
1,2,3,7,8,9-HxCDD	27.1	64.6
1,2,3,4,6,7,8-HpCDD	12.2	48.0
Octa CDD	37.5	70.3
TCDD		
2,3,7,8-TCDF	30.0	32.0
1,2,3,7,8-PeCDF	14.3	53.6
2,3,4,7,8-PeCDF	(13.3)	41.3
1,2,3,4,7,8-HxCDF	40.0	67.5
1,2,3,6,7,8-HxCDF	31.8	61.4
1,2,3,7,8,9-HxCDF	9.1	30.3
2,3,4,6,7,8-HxCDF	36.4	65.5
1,2,3,4,7,8,9-HpCDF	26.5	64.7
1,2,3,4,7,8,9-HpCDF	18.0	64.0
Octa CDF	27.7	63.8
Total	30.8	64.7

[a]Data on percent decrease calculated from before and after concentrations listed in Table 2 in reference 83. Data in parentheses are increases rather than decreases.

Further work with *P. lindtneri* demonstrated that the fungus was also able to mineralize 2,8-dichloro-DF (2,8-diCDF), again in low-nitrogen medium (71). For both 2,7-diCDD and 2,8-diCDF, the monohydroxylated products 3-hydroxy-2,7-diCDD and 3-hydroxy-2,8-diCDF, respectively, were identified from *P. lindtneri* cultures. Confirmation that the hydroxylated products were rapidly metabolized was not obtained. Hydroxylation is consistent with an earlier report that described hydroxylation of DD, dibenzofuran (DF), biphenyl, and diphenyl by the same fungus (72). The hydroxylation of DD, DF, biphenyl, and diphenyl by *P. lindtneri* was found to be inhibited by the cytochrome P-450 inhibitors 1-aminobenzotriazol and piperonyl butoxide (72). This would suggest the involvement of cytochrome P-450 in the initial oxidation of 2,7-diCDD and 2,8-diCDF but does not rule out involvement of lignin-degrading enzymes, which could be involved in oxidation of the hydroxylated metabolites and other intermediates. It would be interesting to investigate the ability of *P. lindtneri* to degrade PCDDs and PCDFs under nitrogen-sufficient conditions to determine the necessity of the lignin-degrading system in their degradation.

In a similar study, the WRF *Panellus stypticus* strain 99-334 was identified in a two-step screening process (relative ability to decolorize remazol brilliant blue R dye followed by relative ability to degrade 2,7-diCDD in low-nitrogen liquid medium) from more than 1,500 basidiomycetes and deuteromycetes isolated from rotting wood and soil (81). This fungus almost completely eliminated 2,7-diCDD from the medium after 40 days of incubation. The metabolic intermediate 1-chloro-3,4-dihydroxybenzene was identified in cultures of *P. stypticus* but not in cell-free control cultures. This degradation product of 2,7-diCDD was also identified by Valli et al. (89) as a metabolic intermediate of the WRF *P. chrysosporium*. As was found with *P. lindtneri*, degradation of 2,7-diCDD was inhibited by addition of the cytochrome P-450 inhibitor piperonyl butoxide, suggesting involvement of cytochrome P-450 monooxygenase in the degradation of 2,7-diCDD by *P. stypticus* (81).

The ability of another of the 12 efficient DD-degrading fungi evaluated by Mori and Kondo (70), *Ceriporia* sp. strain MZ-340, to degrade tetra- to octachlorinated PCDD and PCDF congeners in PCDD- and/or PCDF-contaminated fly ash was evaluated (83). The fungus was cultivated on a wood chip medium that was composed of an 8:2 mixture of wood chips of *Castanopsis cuspidate* var. *sieboldii* and rice bran at a moisture content of 60%. This inoculum was blended with fly ash at a rate of 4:1 (dry wt/dry wt) and incubated for 12 weeks at 20°C. Controls consisted of fly ash amended with autoclaved *Ceriporia* sp. cultures. Inoculation of fly ash with *Ceriporia* sp. strain MZ-340 resulted in significant decreases in all tested congeners, including the most toxic 2,3,7,8-tetrachlorodibenzodioxin (Table 1). Decrease in the total measured PCDD and/or PCDF concentrations was 64.67% in fly ash inoculated with live *Ceriporia* sp. compared to only 30.8% in cultures amended with heat-killed *Ceriporia* sp. More important, and in terms of toxic equivalents (i.e., compared to 2,3,7,8-TCDD, which is given a rating of 1), the toxic equivalent in fly ash inoculated with live *Ceriporia* sp. was decreased by 59% compared to only 24% in fly ash amended with heat-killed *Ceriporia* sp. Thus, *Ceriporia* sp. strain MZ-340 may be useful for treating PCDD- and/or PCDF-contaminated fly ash.

The influence of wheat straw or compost (cut grass and biowaste, 1:1) amendments and of soil treatment by inoculation with *P. chrysosporium*, *Pleurotus* sp. strain Florida,

Dichomitus squalens, or PRT87 (isolated from Eppingen soil) on mineralization of [U-^{14}C]DD spiked into three PCDD- and/or PCDF-contaminated soils was investigated by Rosenbrock et al. (77). The fungi were cultivated on wheat straw (particle size, <1 mm). The three soils were agricultural soils that had been contaminated by different industrial PCDD and/or PCDF emitters. The Maulach soil had higher levels of organic C, nitrogen, and microbial biomass than did the Rastatt and Eppingen soils (Table 2). Without treatment, the Maulach soil, which had the highest levels of organic C, nitrogen, and microbial biomass, produced the greatest DD mineralization (36.6% in 70 days) compared to the Rastatt (19.5% DD mineralization) and Eppingen (23.3% DD mineralization) soils. Interestingly, addition of compost enhanced DD mineralization up to 28% in the two soils low in microbial biomass and organic C but not in the Maulach soil. Addition of wheat straw enhanced DD mineralization in the two low-organic-C soils but actually decreased DD mineralization in the Maulach soil. Inoculation with WRF enhanced DD mineralization beyond that achieved with addition of compost or wheat straw in all soils. Importantly, the extent of the enhancement varied with soil and fungal species. The greatest extent of DD mineralization in the Eppingen (54.7%) and Rastatt (52.7%) soils was achieved by inoculation with PRT87, the WRF isolated from the Eppingen soil. The greatest extent of DD mineralization in the Maulach soil was achieved by inoculation with *Pleurotus* sp. strain Florida (50.4%). The results of this experiment demonstrated several important points for application of fungus-based remediation. First, the variable performance of the fungal treatments demonstrates the importance of conducting laboratory treatability studies prior to on-site pilot-scale or full-scale projects. Use of the most effective fungus identified in a laboratory treatability study increases the chance for successful remediation at full scale. Second, in the Maulach soil, which had the highest levels of nitrogen, organic C, and microbial biomass, the extent of enhanced DD mineralization due to fungal inoculation was the smallest. The authors offered two possible explanations. The inoculated fungi could have been inhibited by the indigenous soil microflora, which in turn decreased the ability of the inoculated fungi to colonize the soil and mineralize the DD. Inhibition of the growth of noncompetitive WRF by indigenous soil microbes has been documented (2, 47, 57, 63, 76). Alternatively or additionally, the formation of bound residues of oxidized DD with soil humic materials could have reduced the amount of DD mineralization. In soils like the Maulach, which have higher organic matter contents, introduction of WRF or a stimulation of indigenous microbes by addition of an organic substrate could inhibit the mineralization of xenobiotic mineralization by providing greater opportunities for covalent binding of aromatic xenobiotics to soil humic materials (6).

TABLE 2 Characteristics of PCDD/PCDF-contaminated soils used in reference 88

Soil	C_{org} (mg g^{-1})	Nitrogen concn (mg g^{-1})	Microbial biomass (SIR)[a] (mg of C_{mic} g^{-1})
Maulach	67	6.2	734.7
Rastatt	20	1.3	161.1
Eppingen	15	1.2	149.8

[a]SIR, substrate-induced respiration.

TABLE 3 Initial concentrations of HpCDF, HpCDD, and OCDD immediately after treatment application

Treatment	Concn (μg/kg) of:		
	HpCDF	HpCDD	OCDD
Control	313	135	472
Pleurotus ostreatus	262	189	508
"East side"	340	351	743
"West side"	378	331	1,045
Resinicium bicolor	323	307	644
Phlebiopsis gigantea	356	380	792

A laboratory-scale study was conducted to demonstrate the ability of fungus-based remediation to decrease the concentration of selected PCDD and PCDF (i.e., 1,2,3,4,6,7,8-heptachlorodibenzofuran [HpCDF], 1,2,3,4,6,7,8-heptachlorodibenzodioxin [HpCDD], and octachlorodibenzodioxin [OCDD]) congeners in soil samples collected from a former pentachlorophenol dip tank wood-treating operation in Whakatane, New Zealand. The study was conducted with two New Zealand strains of wood decay basidiomycetes and two other fungi that were isolated from site soil. In addition, degradation was evaluated for purposes of comparison, using a U.S. strain of *Pleurotus ostreatus* that has demonstrated the ability to degrade similar compounds. All fungi were cultivated on locally available lignocellulosic substrates (e.g., radiata pine or eucalyptus wood chips) to produce inoculum.

Initial concentrations after treatment applications are given in Table 3. There was significant variation in initial analyte concentrations among the treatments for all three analytes. Fungal inoculation had a significant effect on the mean percent decreases of all three analytes among fungal inoculation treatments (Table 4). In all cases, inoculation with any of the tested fungi resulted in a significantly greater decrease than did no inoculation (i.e., control). There were no significant differences among the fungal treatments in the degradation of HpCDF and HpCDD. Average percent decrease of these compounds was greater than 90% in all fungal treatments. Degradation of OCDD was greatest in soil inoculated with *Phlebiopsis gigantea* (Table 4). The percent OCDD decrease in all other fungus-inoculated soil was significantly lower in soils inoculated with *Resinicium bicolor* than in soil inoculated with *P. gigantea*.

TABLE 4 Effect of fungal inoculum and control treatments on mean percent decrease of HpCDF, HpCDD, and OCDD after 56 days of treatment

Treatment	Mean % decrease in concn of[a]:		
	HpCDF	HpCDD	OCDD
Control	5.4b	(33.3)b	(22.4)c
P. ostreatus	98.5a	97.8a	82.1ab
"East side"	95.7a	95.9a	69.3ab
"West side"	97.0a	92.4a	81.0ab
R. bicolor	95.0a	91.6a	68.2b
P. gigantea	93.7a	91.5a	86.2a

[a]Means within columns followed by the same letter are not significantly different. Parentheses indicate that the concentration increased.

The data presented in this section demonstrate that fungus-based remediation has significant potential for remediation of soils and other materials (e.g., fly ash) that are contaminated with PCDDs and/or PCDFs. Future work should include elucidation of the fate of PCDDs and/or PCDFs in WRF-treated soils.

PAHs

PAHs are components of petroleum hydrocarbons and of various byproducts produced from the coking of bituminous coal (e.g., coal tar and creosote). Contamination of the environment by PAHs from both natural (e.g., released from forest fires and volcanic activity) and anthropogenic sources is both widespread and frequent (93). Crude oils and refined oil products are accidentally released as a result of oil tanker accidents, truck accidents, and leaks from storage tanks and pipelines and during transfer of these products. Intentional releases also occur. Petroleum products account for a large fraction of the contamination at many hazardous waste sites (e.g., at refineries and military bases). In addition, prior to the 1950s, gas was produced at manufactured gas plant (MGP) sites through a process that involved the heating of coal and/or oil (33). The gas produced from this process was sold for residential, commercial, and industrial uses. Coal tar residues, including PAHs and other related hydrocarbons, and various metals associated with MGP sites are soil contaminants at former MGP sites. Finally, PAHs are significant pollutants at many currently operating and out-of-service facilities where creosote was used to treat wood.

PAHs consist of two or more benzene rings fused in linear, angular, or cluster arrangements into a single aromatic structure. Water solubility and thus bioavailability of PAHs decrease with an increase in molecular mass (93). Large PAHs with four or more rings are not only poorly bioavailable and recalcitrant to microbial degradation but are more carcinogenic and mutagenic than are smaller PAHs (18). Mammalian (i.e., human) liver enzymes (cytochromes P-450 and epoxide hydrolase) oxidize some PAHs to fjord and bay region diol-epoxides (3, 37, 67), which form covalent adducts with DNA (67, 91). Therefore, many PAHs are genotoxic and/or carcinogenic (3, 26) and promote similar effects of other compounds (19). Studies of PAH-contaminated sediments have shown that the majority of the mutagenic activity in such samples is due to HMW PAHS (64, 65). Typically, HMW, carcinogenic PAHs (i.e., 4-, 5-, and 6-ring PAHs, including benz[*a*]anthracene, benzo(*k*)fluoranthene, benzo(*b*)fluoranthene, benzo[*a*]pyrene, dibenz(*a*,*h*) anthracene, indeno(1,2,3,*c*,*d*)pyrene, and benzo(*g*,*h*,*i*)perylene), are the regulatory "drivers" for soil remediation. This is due to the extremely low "allowable" concentrations of these types of compounds, which, even for industrial soils, are in the part-per-billion or low part-per-million range (Table 5).

HMW PAHs share with other persistent organic pollutants (e.g., chlorinated aromatics like dichlorodiphenyltrichloroethane [DDT] and hexachlorabenzene and alkyl halide insecticides) properties like low water solubility, high octanol-water partition coefficients, high hydrophobicity, chemical/microbial stability, and the ability to bioconcentrate. Therefore, contamination of drinking water supplies due to transport of HMW PAHs, like other persistent organic pollutants, from contaminated soils to groundwater is a legitimate concern that should be the focus of cleanup activities at sites where PAHs are contaminants.

TABLE 5 Risk-based concentrations and IPs of the 16 EPA priority pollutant PAHs

PAH	Industrial risk-based concn (mg/kg)	IP (eV)
Naphthalene	41,000.00	8.12
Acenaphthylene		8.22
Acenaphthene	120,000.00	7.68
Fluorene	82,000.00	7.88
Phenanthrene		8.03
Anthracene	610,000.00	7.44
Fluoranthene	82,000.00	7.90
Pyrene	61,000.00	7.53
Chrysene	780.00	7.60
Benz[*a*]anthracene	7.80	7.53
Benzo(*b*)fluoranthene	7.80	7.70
Benzo(*k*)fluoranthene	78.00	
Benzo[*a*]pyrene	0.78	7.21
Dibenzo(*a*,*h*)anthracene	0.78	7.38
Indeno(1,2,3,*c*,*d*)pyrene	7.80	
Benzo(*g*,*h*,*i*)perylene		7.31

WRF have shown substantial promise for degradation of PAHs in past research. The extracellular peroxidases of WRF are capable of oxidizing PAHs with up to seven fused aromatic rings (7, 10, 40, 41, 90). Oxidation of PAHs with ionization potentials below ca. 7.55 electron volts (eV) can take place via the direct action of lignin peroxidase (47), whereas PAHs above this threshold are oxidized through lipid peroxidation-dependent pathways mediated by fungal manganese-dependent peroxidases (MnPs) (7, 10, 40). The involvement of cytochrome P-450 monooxygenase in PAH oxidation has been demonstrated as well (5, 25, 84). However, several studies show that PAH degradation correlates with the activity of ligninolytic enzymes (32, 78), and it is now generally accepted that ligninolytic enzymes are involved in PAH degradation by WRF (49, 75). MnP has been implicated to be the key enzyme in the degradation process (8, 9, 21, 68, 79). In vitro investigations demonstrated that MnP directly attacks PAHs and is able to catalyze total mineralization (79). Surfactant-enhanced fungal degradation of PAHs in soil is very effective. For example, in a soil collected from an abandoned MGP site with an initial total PAH concentration of 1,331 mg/kg, no degradation occurred without the use of surfactant-enhanced fungus-based remediation. That is, no degradation occurred in soil that received no amendments or soil that was amended with the surfactant without fungal inoculation. In contrast, there was extensive degradation (i.e., 40% of total and carcinogenic PAHs remained) of PAHs after only 28 days of treatment in the soil treated with both fungal inoculation and surfactant addition (unpublished data).

Experiments designed to evaluate the potential of WRF to degrade PAHs in soils spiked with nonlabeled or ^{14}C-labeled PAHs or with other hydrophobic pollutants are useful for determining the impact of factors such as competition from indigenous soil microbes. In these studies the bioavailability of the tested compounds is not limited unless the soils are artificially aged. However, it should be noted that, in "real" contaminated soils, such as those contaminated with creosote or coal tar from wood preservation or coal gasification operations, respectively, the bioavailability of PAHs is extremely limited and there are high concentrations of cocontaminants that may interfere with the degradation of the target PAHs. Therefore, experiments designed to assess the ability of WRF to degrade PAHs in soils in which the target pollutants are spiked should not be used to estimate potential field bioremediation performance.

Mineralization and/or degradation of the more bioavailable low-molecular-weight PAHS can be diminished rather than enhanced by a competitive WRF like *Pleurotus ostreatus* (17). Mineralization of ^{14}C-labeled PAHs by the combination of indigenous PAH-degrading soil microbes and *P. ostreatus* was much lower than by soil microbes and a weakly competitive fungus (e.g., *Dichomitus squalens*), or indigenous PAH-degrading soil microbes alone (63). Similar effects were observed during degradation of benzo[*a*]pyrene in one study (80) and during degradation of pyrene and benz[*a*]anthracene but not benzo[*a*]pyrene in another study (62). It is probable that the decreased mineralization of the more bioavailable LMW PAHs in soils inoculated by competitive WRF is due to the antagonistic effects of the fungus on the indigenous soil PAH-degrading microbes. However, it should be noted that it is the more hydrophobic and less bioavailable HMW PAHs degraded by WRF that are the regulatory drivers at PAH-contaminated sites (Table 5). For example, the degradation of the carcinogenic PAH benz[*a*]anthracene approached 80% in soil inoculated with *P. ostreatus*, compared to only approximately 40% in soil inoculated with the brown-rot fungus *Antrodia vaillantii* or in noninoculated soil, after 12 weeks of treatment (4). While degradation of fluorene was complete in all three treatments, the extents of degradation of the LMW PAHs phenanthrene and pyrene were less in the soil inoculated with *P. ostreatus* than in soil inoculated with *A. vaillantii* or noninoculated soil.

The enzymology of WRF degradation of PAHs has advanced considerably. Most of the focus has been on laccase and MnP because most WRF produce suites of these two enzymes, with lignin peroxidase (LiP) production occurring less frequently (42). Although WRF laccases have been shown to oxidize PAHs in vitro, particularly in the presence of mediators like 2,2′-azinobis-(3-ethylbenzthiazolinesulfonic acid) (ABTS) and 1-hydroxybenzotriazole (HBT) (11, 21, 48, 60, 74), MnP appears to be the dominant enzyme catalyzing initial oxidations of PAHs in soil (45). MnP (EC 1.11.1.13), which was discovered by Kuwahara et al. (55), is a heme-containing glycoprotein that requires H_2O_2 as an oxidant (38). MnP oxidizes Mn(II) to Mn(III), which then oxidizes phenolic rings to phenoxy radicals, leading ultimately to the decomposition of the oxidized compounds (39). Due to its high reactivity, Mn(III) forms complexes with chelating dicarboxylic acids such as malate or oxalate (23, 53), which are produced by WRF (36, 44, 61). In chelated form, Mn(III) ions are stabilized and can diffuse into the substrate colonized by the WRF. PAHs are particularly susceptible to nonspecific oxidation by radicals produced by the lignin-olytic enzymes of WRF. Not long after the discoveries of LiP and MnP produced by *P. chrysosporium*, the degradation of several pollutants, including HMW PAHs, was demonstrated (14). The involvement of LiP and MnP in the degradation process was demonstrated soon afterward (40, 41, 80). Many studies have shown that WRF are efficient degraders of PAHs (e.g., *P. chrysosporium* [13], *Pleurotus* spp. [5, 95], and *Bjerkandera* spp. [54]; see also reference [69]). There is now general acceptance that ligninolytic enzymes are involved in WRF PAH degradation (49, 75) and that MnP is the key enzyme in the process (8, 21, 68, 79).

The degradation of PAHs (i.e., phenanthrene) by MnP-dependent lipid peroxidation was first demonstrated by Moen and Hammel (68). Oxidation by the MnP-dependent lipid peroxidation system was then extended to the full range of 16 priority pollutant PAHs (8, 10). More recently, it was discovered that a crude preparation of MnP from the WRF *N. frowardii* could cause the direct enzymatic mineralization, to various extents (i.e., 2.5 to 7.3% after 168 h), of phenanthrene, anthracene, pyrene, benz[*a*]anthracene, and benzo[*a*]pyrene (79). The extent of PAH mineralization was greatly enhanced by inclusion of the thiol glutathione (GSH). GSH is believed to amplify the oxidative strength of Mn(III), probably by acting as a "secondary mediator." However, it is not clear whether GSH is a real redox mediator or a cosubstrate (45). GSH enables MnP to convert veratryl alcohol to veratraldehyde (24, 34) and to cleave lignin model compounds via thiol-mediated oxidation (92). McEldoon and Dordick (66) even demonstrated the oxidation of GSH by a horseradish peroxidase and the formation of reactive thiyl radicals that oxidized lignin model compounds. The radicals and/or their adducts might also be involved in PAH mineralization (79). The crude MnP preparation, in the presence of GSH, was also able to decrease the concentrations of anthracene, pyrene, and fluoranthene (>75%) and benz[*a*]anthracene, benzo[*a*]pyrene, and benzo(*b*)fluoranthene (>30%). Disappearance of phenanthrene was assumed to be a combination of degradation and volatilization, and chrysene was not degraded (79). No chrysene, benz[*a*]anthracene, or benzo(*b*)fluoranthene was degraded if GSH was not added to the reaction mixture. In addition to PAHs, the *N. frowardii* MnP-GSH system was able to cause the direct enzymatic mineralization of a range of aromatic (i.e., 2-amino dinitrotoluene [2-AmDNT], 2,4,6-trinitrotoluene, catechol, PCP, phenol, 2,4-dichlorophenol, tyrosine, phenylalanine, and tryptophan) and aliphatic (i.e., COOH-glyoxalate, CHO-glyoxalate, glycine, aspartic acid, and glutamate) substrates (45). Using 2-AmDNT as the substrate, Hofrichter et al. (45) showed that there is an optimum ratio of MnP activity to the concentration of reduced GSH. For 2-AmDNT, the ratio that gave the greatest extent of mineralization was 5 mM GSH/U of MnP activity. The ratio for optimum mineralization was not the same for all substrates. Mineralization of PCP required less GSH than did mineralization of 2-AmDNT (45). The authors suggested that the relative reactivity of the hydroxyl group on PCP made primary attack by MnP and subsequent mineralization of the molecule easier. The oxidative strength of MnPs from other fungi is also stimulated by glutathione. In the presence of glutathione, MnP from *P. chrysosporium* was able to mineralize 2-amino-4,6-dinitrotoluene (89a).

In light of the above discussion, it appears that the efficiency of WRF pollutant degradation has the potential to be enhanced by exploiting the amount of MnP produced and its oxidative strength by adding Mn(II) and glutathione to the contaminated medium, respectively.

Laccases (EC 1.10.3.3) are extracellular multicopper oxidases that do not require H_2O_2 but catalyze the reduction of one dioxygen molecule to two molecules of water. This results in the simultaneous oxidation of phenolic substrates via a one-electron reaction creating a free radical, which can be likened to a carbon-centered radical formed in an MnP reaction (51). In the presence of a suitable mediator like ABTS or HBT, laccase is able to oxidize certain non-phenolic compounds and veratryl alcohol (12, 16, 20, 29).

The first report of a role for laccase in WRF PAH degradation revealed that a crude laccase preparation from *Trametes versicolor*, as well as two purified isozymes, was able to cause the oxidation of anthracene and benzo[*a*]pyrene, both of which posses ionization potentials (IPs) of ≤7.45 eV, but not of phenanthrene or fluorene, which both have IPs of ≥7.88 eV (22) (Table 6). Adding ABTS to the reaction enhanced degradation of benzo[*a*]pyrene but not of anthracene. However, addition of ultrafiltrate to the reaction by using a crude enzyme preparation that had been ultrafiltered enhanced anthracene but not benzo[*a*]pyrene degradation (22). Therefore, an unidentified factor of <10,000 Da, involved in the mediation of anthracene degradation, is produced by *T. versicolor*. In contrast to Collins et al. (22), Johannes et al. (48) found that anthracene degradation by *T. versicolor* laccase was stimulated significantly by addition of ABTS at concentrations of ≥1 mM to the reaction mixture. If the mediator HBT was used, as little as 0.005 mM HBT added to the reaction mixture resulted in enhanced degradation of anthracene, while an HBT concentration of >0.1 mM resulted in almost complete removal of the PAH (48). The inability of *T. versicolor* laccase to oxidize naphthalene, fluorene, and phenanthrene was confirmed by Majcherczyk et al. (60) (Table 6). Addition of HBT to the reaction mixture resulted in greatly enhanced oxidation of acenaphthylene, acenaphthene, fluorene, anthracene, benzo[*a*]pyrene, and perylene and intermediate enhancement of pyrene and benz[*a*]anthracene oxidation (60). Oxidation of the six remaining PAHs was either unchanged or decreased in the cases of fluoranthene and chrysene by the addition of HBT to the reaction mixture. For the laccase/HBT system, no good correlation was found between oxidation and IP. PAHs with IPs of ≤7.45 eV were easily oxidized. However, acenaphthene, acenaphthylene, and fluorene, which have high IPs, were degraded, with the exception of anthracene and perylene, to the greatest extent (Table 6). It has been suggested that dividing PAHs into alternating and nonalternating groups simplified analysis of their reactivities (60). For alternating PAHs it appears that those with IPs below 7.45 eV are oxidized and that those with IPs above 7.45 eV are not oxidized or are to a very limited extent (Table 6). Prediction of the ability of the MnP/HBT system to oxidize nonalternating PAHs, based on IP, is not possible.

As with MnP, mediators like HBT might be used to enhance the pollutant-degrading potential of WRF.

MONITORING OF INOCULATED FUNGI IN TREATED SOILS

Several techniques have been adapted or developed for tracking of inoculated fungi in soils treated with mycoremediation. One technique exploits the high proportion of the phospholipid fatty acid (PLFA) 18:2ω6,9 found in saprotrophic fungi, especially the basidiomycetes (59, 73). The growth of *P. ostreatus*, *P. chrysosporium*, and *Hypholoma fasciculare* was monitored in PAH-contaminated soil by measuring the content of the specific eukaryotic fatty acid *cis*-9,12-octadecadienoic acid (18:2ω6,9) (4). While there was good correlation between the soil content of 18:2ω6,9 and visible growth of the inoculated fungi, the technique did have several disadvantages. First, 18:2ω6,9 was also found to be present in the birch wood that was used as the fungal inoculum substrate. Because it was not possible to completely separate the wood prior to processing the soil for 18:2ω6,9 analysis, there were some false positives that confounded the data analysis. In addition, the amount of PLFA 18:2ω6,9 cannot be directly converted to biomass for the fungi, because, like other bio-

TABLE 6 PAHs, their IPs, and the ability of laccase to oxidize the PAH alone or in the presence of HBT or ABTS[a]

PAH	IP (eV)	Laccase[b]	Laccase/HBT[c]	Laccase/ABTS[d]
Napthalene	8.13	−	−	−
Acenaphthylene	8.22	++++	++++	−
Acenaphthene	7.86	+	++++	−
Fluorene	7.89	−	++++	−
Phenanthrene	7.91	−	+	+
Anthracene	7.43	+++	++++	+
Fluoranthene	7.95	+	(−)	−
Pyrene	7.53	+	+++	−
Chrysene	7.60	+	(−)	−
Benzo(*a*)anthracene	7.53	+	+++	−
Benzo(*b*)fluoranthene	7.70	++	NC[e]	−
Benzo(*k*)fluoranthene		++	NC	−
Benzo[*a*]pyrene	7.21	+++	++++	+
Perylene		++	++++	−

[a]Summarized from reference 70.

[b]A minus sign indicates that laccase alone was unable to oxidize the substrate PAH. The number of plus signs indicates the relative extent of PAH degradation.

[c](−) indicates that addition of HBT resulted in a decrease in the extent of degradation. A minus sign indicates that the laccase/HBT system was unable to degrade the PAH substrate. The number of plus signs indicates the relative extent to which addition of HBT enhanced mineralization over use of laccase alone.

[d]A minus sign indicates that addition of ABTS did not stimulate degradation. A plus sign indicates that addition of ABTS stimulated degradation.

[e]NC, no change.

chemical indicators (e.g., ergosterol), different fungal species contain different concentrations of this fatty acid (82). However, if these limitations are considered, measurement of PLFA appears to be an acceptable general technique for assessing the effects on the growth of WRF in soils. The PLFA method has the added benefit of revealing interactions between inoculated fungi and indigenous soil bacteria by simultaneous extraction of other marker PLFAs (4).

A second technique is based on the development of PCR primers for the species-specific detection of a selected WRF. In this case, a specific primer derived from the internal spacer region sequences of *Ceriporia* sp. strain MZ-340 was constructed and used to detect the fungus in PCDD- and/or PCDF-contaminated fly ash (83). The technique involves extraction of DNA from soil followed by PCR amplification. The PCRs developed were sufficiently sensitive to amplify from 0.1 to 1 pg of the target DNA and to amplify target sequences in the presence of impurities. The amount of DNA and mycelium of *Ceriporia* sp. strain MZ-340 affected the intensity of the specific PCR-amplified fragment, allowing for the detection of clear differences in fungal growth between treatments (e.g., treatment with *Ceriporia* sp. strain MZ-340 and treatment with the heat-killed control); the method does not allow for quantitation. While the method does allow for the detection of the inoculated fungi with high sensitivity by a one-step PCR, it does require that specific primers be produced.

FINAL COMMENTS

The results of recent research have demonstrated the potential for exploiting the ability of WRF to degrade recalcitrant toxic contaminants like HMW PAHs and PCDDs and/or PCDFs. Sites contaminated by these types of contaminants are often extremely difficult to treat. Fungus-based remediation offers an inexpensive and natural remediation alternative to energy-intensive solutions like incineration. In addition, work on the enzymology of PAH degradation that demonstrated the enhanced ability to oxidize PAHs by both MnP and laccase through the use of mediators provides a potential means for enhancing the performance of fungus-based soil remediation if mediators can be distributed to the soil solution in an effective manner. New techniques for tracking fungi inoculated into soil provide a means for specifically tracking a selected fungus through the use of specific PCR primers and for determining the relative growth of the fungus compared to that of other microbes (e.g., indigenous bacteria) through the use of PLFA analyses. These are excellent tools for continued research on the interactions of WRF, litter-degrading fungi, and indigenous soil microbial populations that may complement or hinder remediation performance.

REFERENCES

1. Reference deleted.
2. **Ali, T. A., and M. Wainwright.** 1994. Growth of *Phanerochaete chrysosporium* in soil and its ability to degrade the fungicide benomyl. *Bioresour. Technol.* **49:** 197–201.
3. **Amin, S., J. Krzeminski, A. Rivenson, C. Kurtzke, S. S. Hecht, and K. El-Bayoumi.** 1995. Mammary carcinogenicity in female CD rats of fjord region diol epoxides of benzo[c]phenanthrene, benzo[g]chrysene, and dibenzo[a,l]-pyrene. *Carcinogenesis* **16:**1971–1974.
4. **Andersson, B. E., S. Lundstedt, K. Tornberg, Y. Schnurer, L. G. Oberg, and B. Mattiasson.** 2003. Incomplete degradation of polycyclic aromatic hydrocarbons in soil inoculated with wood-rotting fungi and their effect on the indigenous soil bacteria. *Environ. Toxicol. Chem.* **22:**1238–1243.
5. **Bezalel, L., Y. Hadar, and C. E. Cerniglia.** 1997. Enzymatic mechanisms involved in phenanthrene degradation by the white-rot fungus *Pleurotus ostreatus*. *Appl. Environ. Microbiol.* **63:**2495–2501.
6. **Bogan, B. W., R. T. Lamar, W. D. Burgos, and M. Tien.** 1999. Extent of humification of anthracene, fluoranthene, and benzo(a)pyrene by Pleurotus ostreatus during growth in PAH-contaminated soils. *Lett. Appl. Microbiol.* **28:**250–254.
7. **Bogan, B. W.** 1996. Ph.D. thesis. University of Wisconsin, Madison.
8. **Bogan, B. W., and R. T. Lamar.** 1995. One-electron oxidation in the degradation of creosote polycyclic aromatic hydrocarbons by *Phanerochaete chrysosporium*. *Appl. Environ. Microbiol.* **61:**2631–2635.
9. **Bogan, B. W., and R. T. Lamar.** 1996. Polycyclic aromatic hydrocarbon-degrading capabilities of *Phanerochate laevis* HHB-1625 and its extracellular ligninolytic enzymes. *Appl. Environ. Microbiol.* **62:**1597–1603.
10. **Bogan, B. W., R. T. Lamar, and K. E. Hammel.** 1996. Fluorene oxidation in vivo by *Phanerochaete chrysosporium* and in vitro during manganese peroxidase-dependent lipid peroxidation. *Appl. Environ. Microbiol.* **62:**1788–1792.
11. **Bohmer, S., K. Messner, and E. Srebotnik.** 1998. Oxidation of phenanthrene by a fungal laccase in the presence of 1-hydroxybenzotriazole and unsaturated lipids. *Biochem. Biophys. Res. Commun.* **244:**233–238.
12. **Bourbonnais, R., and M. G. Paice.** 1990. Oxidation of non-phenolic substrates. An expanded role for laccase in lignin biodegradation. *FEBS Lett.* **27:**99–102.
13. **Bumpus, J. A.** 1989. Biodegradation of polycyclic aromatic hydrocarbons by *Phanerochaete chrysosporium*. *Appl. Environ. Microbiol.* **55:**154–158.
14. **Bumpus, J.A., M. Tien, D. Wright, and S. D. Aust.** 1985. Oxidation of persistent environmental pollutants by a white rot fungus. *Science* **228:**1434–1436.
15. **Burdsall, H. H., Jr., and W. Eslyn.** 1974. A new *Phanerochaete* with a *Chrysosporium* imperfect state. *Mycotaxon* **1:**124.
16. **Call, H. P., and I. Mucke.** 1995. The laccase-mediator system (LMS)—a new concept, p. 27–32. *In* E. Srebotnik and J. Messner (ed.), *Biotechnology in the Pulp and Paper Industry*. Facultas-Universitätsverlag, Vienna, Austria.
17. **Canet, R., J. G. Birnstingl, D. G. Malcolm, J. M. Lopez-Real, and A. J. Beck.** 2001. Biodegradation of polycyclic aromatic hydrocarbons (PAHs) by native microflora and combinations of white-rot fungi in a coal-tar contaminated soil. *Bioresour. Technol.* **76:**113–117.
18. **Cerniglia, C. E., and M. A. Heitkamp.** 1989. Microbial degradation of polycyclic aromatic hydrocarbons (PAH) in the aquatic environment, p. 41–68. *In* U. Varanasi (ed.), *Metabolism of Polycyclic Aromatic Hydrocarbons in the Aquatic Environment*. CRC Press, Inc., Boca Raton, Fla.
19. **Chadwick, R. W., S. E. George, M. J. Kohan, R. W. Williams, J. C. Allison, D. L. Talley, Y. O. Hayes, and J. Chang.** 1995. Potentiation of 2,6-dinitrotoluene genotoxicity in Fischer 344 rats by pretreatment with coal tar creosote. *J. Toxicol. Environ. Health* **44:**319–336.
20. **Collins, P. J., and A. D. W. Dobson.** 1997. Regulation of laccase gene transcription in *Trametes versicolor*. *Appl. Environ. Microbiol.* **63:**3444–3450.
21. **Collins, P. J., and A. D. W. Dobson.** 1996. Oxidation of fluorene and phenanthrene by Mn(II) dependent peroxidase activity in whole cultures of *Trametes (Coriolus) versicolor*. *Biotechnol. Lett.* **18:**801–804.

22. **Collins, P. J., M. J. J. Kotterman, J. A. Field, and A. D. W. Dobson.** 1996. Oxidation of anthracene and benzo[*a*]pyrene by laccases from *Trametes versicolor*. *Appl. Environ. Microbiol.* **62:**4563–4567.
23. **Cui, F., and D. Dolphin.** 1990. The role of manganese in model systems related to lignin biodegradation. *Holzforschung* **44:**279–283.
24. **D'Annibale, A., C. Crestini, E. Di Mattia, and G. G. Sermanni.** 1996. Veratryl alcohol oxidation by manganese-dependent peroxidase from *Lentinus edodes*. *J. Biotechnol.* **48:**231–239.
25. **Dhawale, S. W., S. S. Dhawale, and D. Dean-Ross.** 1992. Degradation of phenanthrene by *Phanerochaete chrysosporium* occurs under ligninolytic conditions as well as nonligninolytic conditions. *Appl. Environ. Microbiol.* **58:**3000–3006.
26. **Djomo, J. E., V. Ferrier, L. Gauthier, C. Zoll-Moreux, and J. Marty.** 1995. Amphibian micronucleus test in vivo: evaluation of the genotoxicity of some major polycyclic aromatic hydrocarbons found in crude oil. *Mutagenesis* **10:**223–226.
27. **Durrant, A. J., D. A. Wood, and R. B. Cain.** 1991. Lignocellulose biodegradation by *Agaricus bisporus* during solid substrate fermentation. *J. Gen. Microbiol.* **137:**751–755.
28. **Eaton, D. C.** 1985. Mineralization of polychlorinated biphenyls by *Phanerochaete chrysosporium*: a ligninolytic fungus. *Enzyme Microb. Technol.* **7:**194–196.
29. **Eggert, C., U. Temp, J. F. D. Dean, and K.-E. L. Eriksson.** 1996. A fungal metabolite mediates degradation of nonphenolic lignin structures and synthetic lignin by laccase. *FEBS Lett.* **391:**144–148.
30. **Eschenbach, A. B. Weinberg, and B. Mahro.** 1998. Fate and stability of nonextractable residues of [^{14}C]PAH in contaminated soils under stress conditions. *Environ. Sci. Technol.* **32:**2585–2590.
31. **Evans, C. S., and J. N. Hedger.** 2001. Degradation of plant cell wall polymers, p. 1–26. *In* G. M. Gadd (ed.), *Fungi in Bioremediation*. Cambridge University Press, Cambridge, United Kingdom.
32. **Field, J. A., E. de Jong, G. Feijoo Costa, and J. A. de Bont.** 1992. Biodegradation of polycyclic aromatic hydrocarbons by new isolates of white rot fungi. *Appl. Environ. Microbiol.* **58:**2219–2226.
33. **Fischer, C. L. J., R. D. Schmitter, and E. O. Lane.** 1999. Manufactured gas plants: the environmental legacy. http://www.hsrc.org/hsrc/html/tosc/sswtosc/mgp.html.
34. **Forrester, I. T., A. C. Grabski, R. R. Burgess, and G. F. Leatham.** 1988. Manganese, Mn-dependent peroxidases and the biodegradation of lignin. *Biochem. Biophys. Res. Commun.* **157:**992–999.
35. **Fuhr, F., R. Kloskowski, and P. W. Burauel.** 1985. Bedeutung der gebundenen Rickstande. Pflanzenschutzmittel im Boden. *Z. Agrarpolit. Landwirtsch.* **198:**106–116.
36. **Galkin, S., T. Vares, M. Kalsi, and A. Hatakka.** 1998. Production of organic acids by white-rot fungi as detected by capillary zone electrophoresis. *Biotechnol. Tech.* **12:** 267–271.
37. **Glatt, H., A. Piee, K. Pauly, T. Steinbrecher, R. Schrode, F. Oesch, and A. Seidel.** 1991. Fjord- and bay-region diol-epoxides investigated for stability, SOS induction in *Escherichia coli*, and mutagenicity in *Salmonella typhimurium* and mammalian cells. *Cancer Res.* **51:**1659–1667.
38. **Glenn, J. K., and M. H. Gold.** 1985. Purification and characterization of an extracellular Mn(II)-dependent peroxidase from the lignin-degrading basidiomycete, *Phanerochaete chrysosporium*. *Arch. Biochem. Biophys.* **242:** 329–341.
39. **Glenn, J. K., L. Akileswaran, and M. H. Gold.** 1986. Mn(II) oxidation is the principal function of the extracellular Mn-peroxidase from *Phanerochaete chrysosporium*. *Arch. Biochem. Biophys.* **251:**688–696.
40. **Haemmerli, S. D., M. S. A. Leisola, D. Sanglard, and A. Fiechter.** 1986. Oxidation of benzo[*a*]pyrene by extracellular ligninases of *Phanerochaete chrysosporium*. Veratryl alcohol and stability of ligninase. *J. Biol. Chem.* **261:**6900–6903.
41. **Hammel, K. E., B. Kalyanaraman, and T. K. Kirk.** 1986. Oxidation of polycyclic aromatic hydrocarbons and dibenzo[p]dioxins by *Phanerochaete chrysosporium* ligninase. *J. Biol. Chem.* **261:**16948–16952.
42. **Hatakka, A.** 1994. Lignin-modifying enzymes from selected white-rot fungi: production and role in lignin biodegradation. *FEMS Microbiol. Rev.* **13:**125–135.
43. **Heindl, A., and O. Hutzinger.** 1986. Search for industrial sources of PCDD/PCDF. I. Approaches in the Federal Republic of Germany. *Chemosphere* **15:**2001–2002.
44. **Hofrichter, M., T. Vares, M. Kalsi, S. Galkin, K. Scheibner, W. Fritsche, and A. Hatakka.** 1999. Production of manganese peroxidase and organic acids and mineralization of 14C-labelled lignin (14C-DHP) during solid-state fermentation of wheat straw with the white rot fungus *Nematoloma frowardii*. *Appl. Environ. Microbiol.* **65:**1864–1870.
45. **Hofrichter, M., K. Scheibner, I. Schneegass, and W. Fritsche.** 1998. Enzymatic combustion of aromatic and aliphatic compounds by manganese peroxidase from *Nematoloma frowardii*. *Appl. Environ. Microbiol.* **64:**399–404.
46. **Hutzinger, O., M. J. Blumich, M. van den Berg, and K. Olie.** 1985. Sources and fate of PCDDs and PCDFs: an overview. *Chemosphere* **14:**581–600.
47. **in der Wiesche, C., R. Martens, and F. Zadrazil.** 1996. Two-step degradation of pyrene by white-rot fungi and soil microorganisms. *Appl. Microbiol. Biotechnol.* **46:**653–659.
48. **Johannes, C., A. Majcherczyk, and A. Hutterman.** 1996. Degradation of anthracene by laccase of *Trametes versicolor* in the presence of different mediating substrate compounds. *Appl. Microbiol. Biotechnol.* **46:**313–317.
49. **Kastner, M.** 2000. Degradation of aromatic and polyaromatic compounds, p. 211–239. *In* H.-J. Rehm and G. Reed (ed.), *Biotechnology*, vol. 11b. Wiley-VCH, Weinheim, Germany.
50. **Kastner, M., S. Streibich, M. Beyrer, H. H. Richnow, and W. Fritsche.** 1999. Formation of bound residues during microbial degradation of [^{14}C]anthracene in soil. *Appl. Environ. Microbiol.* **65:**1834–1842.
51. **Kersten, P. J., B. Kalyanaraman, K. E. Hammel, B. Reinhammar, and T. K. Kirk.** 1990. Comparison of lignin peroxidase, horseradish peroxidase and laccase in the oxidation of methoxybenzenes. *Biochem. J.* **268:**475–480.
52. **Kirk, T. K., and R. L. Farrell.** 1987. Enzymatic combustion: the microbial degradation of lignin. *Annu. Rev. Microbiol.* **41:**465–505.
53. **Kishi, K., H. Wariishi, L. Marquez, H. B. Dunford, and M. H. Gold.** 1994. Mechanism of manganese peroxidase compound II reduction. Effect of organic acid chelators and pH. *Biochemistry* **33:**8694–8701.
54. **Kotterman, M. J., E. H. Vis, and J. A. Field.** 1998. Successive mineralization and detoxification of benzo(a)pyrene by the white-rot fungus *Bjerkandera* sp. strain BOS55 and indigenous microflora. *Appl. Environ. Microbiol.* **64:**2853–2858.
55. **Kuwahara, M., J. K. Glenn, M. A. Morgan, and M. H. Gold.** 1984. Separation and characterization of 2 extracellular H_2O_2-dependent oxidases from ligninolytic cultures of *Phanerochaete chrysosporium*. *FEBS Lett.* **169:**247–250.

56. **Lang, E., I. Kleeberg, and F. Zadrazil.** 2000. Extractable organic carbon and counts of bacteria near the lignocellulose-soil interface during the interaction of soil microbiota and white-rot fungi. *Bioresour. Technol.* **75:**57–65.
57. **Lang, E., G. Eller, and F. Zadrazil.** 1997. Lignocellulose decomposition and production of ligninolytic enzymes during interaction of white rot fungi with soil microorganisms. *Microb. Ecol.* **34:**1–10.
58. **Lankinen, V. P.** 2004. Ligninolytic enzymes of the basidiomycetous fungi *Agaricus bisporus* and *Phlebia radiata* on lignocellulose-containing media. Ph.D. thesis. University of Helsinki, Helsinki, Finland.
59. **Larsen, J., P. A. Olsson, and I. Jakobsen.** 1998. The use of fatty acid signatures to study mycelial interaction between the arbuscular mycrorrhizal fungus *Glomus intradices* and the saprophytic fungus *Fusarium culmorum* in root-free soil. *Mycol. Res.* **102:**1491–1496.
60. **Majcherczyk, A. C. Johannes, and A. Huttermann.** 1998. Oxidation of polycyclic aromatic hydrocarbons (PAH) by laccase of *Trametes versicolor*. *Enzyme Microb. Technol.* **22:**335–341.
61. **Makela, M., S. Galkin, A. Hatakka, and T. Lundell.** 2002. Production of organic acids and oxalate decarboxylase in lignin-degrading white rot fungi. *Enzyme Microb. Technol.* **30:**542–549.
62. **Martens, R., M. Wolter, M. Bahadir, and F. Zadrazil.** 1999. Mineralization of 14C-labelled highly-condensed polycyclic aromatic hydrocarbons in soils by *Pleurotus* sp. Florida. *Soil Biol. Biochem.* **31:**1893–1899.
63. **Martens, R., and F. Zadrazil.** 1992. Screening of white-rot fungi for their ability to mineralize polycyclic aromatic hydrocarbons in soil, p. 505–510. *In Preprints from the International Symposium on Soil Decontamination Using Biological Processes*. Dechema, Frankfurt am Main, Germany.
64. **Marvin, C. H., J. A. Lundrigan, B. E. McCarry, and D. W. Bryant.** 1993. Chemico/biological investigation of contaminated sediment from the Hamilton Harbour area of western Lake Ontario. *Environ. Mol. Mutagen.* **22:**61–70.
65. **Marvin, C. H., J. A. Lundrigan, B. E. McCarry, and D. W. Bryant.** 1995. Determination and genotoxicity of high molecular mass polycyclic aromatic hydrocarbons isolated from coal-tar-contaminated sediment. *Environ. Toxicol. Chem.* **14:**2059–2066.
66. **McEldoon, J. P., and J. S. Dordick.** 1991. Thiol and Mn^{2+}-mediated oxidation of veratryl alcohol by horseradish peroxidase. *J. Biol. Chem.* **266:**14288–14293.
67. **Melikian, A. A., K. A. Prahalad, S. Amin, and S. S. Hecht.** 1991. Comparative DNA binding of polynuclear aromatic hydrocarbons and their dihydrodiol and bay region diolepoxide metabolites in newborn mouse lung and liver. *Carcinogenesis* **12:**1665–1670.
68. **Moen, M. A., and K. E. Hammel.** 1994. Lipid peroxidation by the manganese peroxidase of *Phanerochaete chrysosporium* is the basis for phenanthrene oxidation by the intact fungus. *Appl. Environ. Microbiol.* **60:**1956–1961.
69. **Morgan, P., S. T. Lewis, and R. J. Watkinson.** 1991. Comparison of abilities of white-rot fungi to mineralize selected xenobiotic compounds. *Appl. Microbiol. Biotechnol.* **34:**693–696.
70. **Mori, T., and R. Kondo.** 2002. Degradation of 2,7-dichlorodibenzo-p-dioxin by wood-rotting fungi, screened for dioxin degrading ability. *FEMS Microbiol. Lett.* **213:** 127–131.
71. **Mori, T., and R. Kondo.** 2002. Oxidation of chlorinated dibenzo-p-dioxin and dibenzofuran by white-rot fungus, *Phlebia lindtneri*. *FEMS Microbiol. Lett.* **216:**223–227.
72. **Mori, T., and R. Kondo.** 2002. Oxidation of dibenzo-p-dioxin, dibenzofuran, biphenyl and diphenyl ether by white-rot fungus Phlebia lindtneri. *Appl. Microbiol. Biotechnol.* **60:**200–205.
73. **Muller, M. M., R. Kantola, and V. Kitunen.** 1994. Combining sterol and fatty acid profiles for the characterization of fungi. *Mycol. Res.* **98:**593–603.
74. **Pickard, M. A., R. Roman, R. Tinoco, and R. Vazquez-Duhalt.** 1999. Polycyclic aromatic hydrocarbon metabolism by white rot fungi and oxidation by *Coriolopsis gallica* UAMH 8260 laccase. *Appl. Environ. Microbiol.* **65:**3805–3809.
75. **Pointing, S. B.** 2001. Feasibility of bioremediation by white rot fungi. *Appl. Microbiol. Biotechnol.* **57:**20–33.
76. **Radtke, C., W. S. Cook, and A. Anderson.** 1994. Factors affecting antagonism of the growth of *Phanerochaete chrysosporium* by bacteria isolated from soils. *Appl. Microbiol. Biotechnol.* **41:**274–280.
77. **Rosenbrock, P., R. Martens, F. Buscot, F. Zadrazil, and J. C. Munch.** 1997. Enhancing the mineralization of [U-^{14}C] dibenzo-p-dioxin in three different soils by the addition of organic substrate or inoculation with white rot fungus. *Appl. Microbiol. Biotechnol.* **48:**665–670.
78. **Sack, U., T. M. Heinze, J. Deck, C. E. Cerniglia, R. Martens, F. Zadrazil, and W. Fritsche.** 1997. Comparison of phenanthrene and pyrene degradation by different wood-decaying fungi. *Appl. Environ. Microbiol.* **63:**3919–3925.
79. **Sack, U., M. Hofrichter, and W. Fritsche.** 1997. Degradation of polycyclic aromatic hydrocarbons by manganese peroxidase of *Nematoloma frowardii*. *FEMS Microbiol. Lett.* **152:**227–234.
80. **Sanglard, D. M., S. A. Leisola, and A. Fiechter.** 1986. Role of extracellular ligninases in biodegradation of benzo(a)pyrene by *Phanerochaete chrysosporium*. *Enzyme Microb. Technol.* **8:**209–212.
81. **Sato, A., T. Watanabe, Y. Watanabe, K. Harazono, and T. Fukatsu.** 2002. Screening for basidiomycetous fungi capable of degrading 2,7-dichlorodibenzo-p-dioxin. *FEMS Microbiol. Lett.* **213:**213–217.
82. **Stahl, P. D., and M. J. Klug.** 1996. Characterization and differentiation of filamentous fungi based on fatty acid composition. *Appl. Environ. Microbiol.* **62:**4136–4146.
83. **Surhara, H., C. Daikoku, H. Takata, S. Suzuki, Y. Matsufuji, K. Sakai, and R. Kondo.** 2003. Monitoring of white-rot fungus during bioremediation of polychlorinated dioxin-contaminated fly ash. *Appl. Microbiol. Biotechnol.* **62:**601–607.
84. **Sutherland, J. B., F. Rafii, A. A. Khan, and C. E. Cerniglia.** 1995. Mechanisms of polycyclic aromatic hydrocarbon degradation, p. 296–306. *In* L. Young and C. E. Cerniglia (ed.), *Microbial Transformation and Degradation of Toxic Organic Chemicals*. Wiley-Liss, New York, N.Y.
85. **Swanson, S. E., C. Rappe, J. Malmstrom, and K. P. Kringstad.** 1988. Emission of PCDDs and PCDFs from the paper pulp industry. *Chemosphere* **17:**681–691.
86. **Takada, S., M. Nakamura, T. Matsueda, R. Kondo, and K. Sakai.** 1996. Degradation of polychlorinated dibenzo-p-dioxins and polychlorinated dibenzofurans by the white rot fungus *Phanerochaete sordida* YK-624. *Appl. Environ. Microbiol.* **62:**4323–4328.
87. **Tien, M., and T. K. Kirk.** 1983. Lignin peroxidase of Phanerochaete chrysosporium. *Methods Enzymol.* **161:** 238–249.
88. **Valli, K., and M. H. Gold.** 1991. Degradation of 2,4-dichlorophenol by the lignin-degrading fungus *Phanerochaete chrysosporium*. *J. Bacteriol.* **173:**345–352.
89. **Valli, K., H. Wariishi, and M. H. Gold.** 1992. Degradation of 2,7-dichlorodibenzo-p-dioxin by the lignin-degrading basidiomycete *Phanerochaete chrysosporium*. *J. Bacteriol.* **174:**2131–2137.

89a. **van Aken, B., M. D. Cameron, J. D. Stahl, A. Plumat, H. Naveau, S. D. Aust, and S. N. Agathos.** 2000. Glutathione-mediated mineralization of 14C-labeled 2-amino-4,6-dinitrotoluene by manganese-dependent peroxidase H5 from the white-rot fungus *Phanerochaete chrysosporium*. *Appl. Microbiol Biotechnol.* **54:**659–664.

90. **Vazquez-Duhult, R. D., W. S. Westlake, and P. M. Fedorak.** 1994. Lignin peroxidase oxidation of aromatic compounds in systems containing organic solvents. *Appl. Environ. Microbiol.* **60:**459–466.

91. **Walsh, P., C. el Aldouni, M. J. Mukhopadhyay, G. Viel, D. Nadeau, and G. G. Poirier.** 1995. 32P-postlabeling determination of DNA adducts in the earthworm *Lumbricus terrestris* exposed to PAH-contaminated soils. *Bull. Environ. Contam. Toxicol.* **54:**654–661.

92. **Wariishi, H., K. Valli, V. Renganathan, and M. H. Gold.** 1989. Thiol-mediated oxidation of nonphenolic lignin model compounds by manganese peroxidase of *Phanerochaete chrysosporium*. *J. Biol. Chem.* **264:**14185–14191.

93. **Wilson, S. C., and K. C. Jones.** 1993. Bioremediation of soil contaminated with polynuclear aromatic hydrocarbons (PAHs): a review. *Environ. Pollut.* **81:**229–249.

94. **Wittich, R. M.** 1998. Degradation of dioxin-like compounds by microorganisms. *Appl. Microbiol. Biotechnol.* **49:**489–499.

95. **Wolter, M., F. Zadrazil, R. Martens, and M. Bahadir.** 1997. Degradation of eight highly condensed polycyclic aromatic hydrocarbons by *Pleurotus* sp. Florida in solid wheat straw substrate. *Appl. Microbiol. Biotechnol.* **48:**398–404.

Fungal Solid-State Fermentation Systems for Bioconversion of Lignocellulosic Biomass: Process Protocol and Applications

CHUNDAKKADU KRISHNA

89

This chapter reviews the methods for fungal solid-state fermentation (SSF) for bioconversion of lignocellulosic biomass. SSF can be defined as the growth of microorganisms on moist, water-insoluble solid substrates in the absence or near absence of free liquid, at an average water activity (a_w) (defined as the relative humidity of the gaseous phase in equilibrium with the moist solid) significantly below 1 (4). SSF has several potential advantages over submerged fermentation, such as low capital cost, low energy expenditure, less expensive downstream processing, low wastewater output, potentially higher volumetric productivity, and extended stability of products. An overview of the various factors which affect the development and application of fungal SSF systems for the bioconversion of lignocellulosic substrates is illustrated.

LIGNOCELLULOSIC SUBSTRATES

Lignocellulosics compose 60% of total plant biomass produced, by far the most abundant natural raw material present on earth, and the net photosynthetic production of the dry biomass by plants on earth has been estimated to be 155 billion tons per year (3). The major chemical components of lignocellulosic biomass include cellulose, hemicelluloses, and lignin. The amounts of celluloses and hemicelluloses are greater in hardwoods compared to higher lignin content in soft woods and wheat straw, based on the weight percentage. The economic utilization of the lignocellulosic biomass is not yet fully developed, because of the resistance of lignocelluloses to chemical and biological transformations.

Lignocellulosic biomass can be used for the production of industrial chemicals, liquid fuels, protein-rich food (8), feedstocks, and paper products, and preparation of cellulose polymers to help meet energy and food demands has received great attention in the past few decades. Agroindustrial residues may be converted into animal feed and enzymes and can be composted to provide biofertilizer, biopesticide, and soil enrichment for bioprotective properties. The SSF process has great potential for the bioconversion of lignocellulosic biomass (19). The most efficient degraders of lignocellulosic biomass are reported to be higher fungi, including white rot fungi and some mushrooms. The complexity of the lignocellulosic structure leads to constraints on the ability of the fungal systems to infect the wood by the aid of spores or hyphae, which secrete specific enzymes for attacking the plant cell wall. SSF has great potential in microbial conversion of lignocellulosic biomass into biofuel, enzymes, animal feed, and biofertilizer (3, 6).

CHARACTERISTICS OF LIGNOCELLULOLYTIC FUNGI

Major fungi involved in lignocellulosic biomass conversion belong to the groups *Ascomycetes*, *Deuteromycetes*, and *Basidiomycetes*. They are able to grow on lignocellulosic substrate and produce enzymes necessary for the degradation of the complex material. Important fungi causing softrot include *Chaetomium cellulolyticum*, *Aspergillus niger*, *Trichoderma viride* (*T. reesei*), *Fusarium oxysporum*, *Thielavia terrestris*, *Penicillium jenthillenum*, *Dactylomyces crustaceus*, and species of *Paecilomyces*, *Papulaspora*, *Monodictys*, *Allescheria*, *Hypoxylon*, *Xylaria*, and *Graphium*. Fungi causing brown-rot-type decay include *Poria placenta*, *Tyromyces balsemeus*, *Gloeophyllum trabeum*, *Lentinus lepidius*, *Lenzites trabeum*, *Coniophora puteana*, *Laetiporus sulphureus*, and *Fomitopsis pinicola*. White rot fungi can attack all components of plant cell walls; the most studied fungi of this group are *Phanerochaete chrysosporium*, *Trametes versicolor*, *Dichomitus squalens*, *Phlebia radiata*, *Heterobasidium annosum*, *Phellinus pini*, *Cyathus stercoreus*, *Pleurotus ostreatus*, *Ceriporiopsis subvermispora*, *Polyporus anceps*, and *Ustulina vulgaris*. The normal pattern of wood decay by these fungi involves simultaneous attack on both polysaccharides and lignin. Fungal hyphae penetrate through the cell lumen, first colonize the parenchyma cells, and then penetrate from cell to cell via pits or by the development of boreholes directly through the cell walls (2, 3, 14).

LIGNOCELLULOSIC HYDROLYSIS

Lignocellulose hydrolysis is a complex process which requires the synergistic action of several enzymes. The enzymatic hydrolysis of cellulose requires the use of cellulase [1,4-(1,3:1,4)-β-D-glucan glucanohydrolase; EC 3.2.1.4], a multiple-enzyme system consisting of three types of activities: endo 1,4-β-D-glucanase, exo 1,4-β-D-glucanase (cellobiohydrolase), and cellobiase. For lignin metabolism, the lignolytic system of the fungus is produced constitutively, expressed during secondary metabolism, triggered by carbon, sulfate, and nitrogen limitation, and markedly influenced by oxygen concentration (5). Lignin peroxidase is the major enzyme involved in lignin degradation. White rot fungi also produce two other types of extracellular enzymes, Mn(II)-dependent peroxidases and glyoxal oxidase, an enzyme involved in extracellular H_2O_2 production. Both types of peroxidases are usually involved in lignin degradation, although they differ in their catalytic mechanisms (1). Because of the resistance of native cellulose to enzymatic hydrolysis due to its intrinsic properties, pretreatment is necessary for the effective utilization of lignocellulosic material. Pretreatment methods generally practiced for hydrolysis of the complex lignocellulosic structure to make it easily accessible to enzymatic hydrolysis are physical-mechanical, thermal (steam explosion or hydrothermolysis), chemical (alkali, acid, oxidizing agents, etc.), or biological (treatment with microorganisms or isolated enzymes) or a combination of these methods. Milling the substrate for particle size reduction is generally employed as an initial pretreatment method. Autoclaving or hot water treatment increases the availability of growth-limiting nutrients and favors growth and nonlignin conversions while depressing lignin utilization. The pretreatment methods also increase the specific surface area, remove lignin, or solubilize hemicellulose (3).

GENERAL PROCESS PROTOCOL AND FACTORS THAT INFLUENCE SSF

Inoculum Type and Preparation

Selection of a suitable strain and selection of a solid substrate for the process are the most important criteria for SSF. Filamentous fungi are the most important group of microorganisms for SSF because of their physiological capabilities and hyphal mode of growth on the solid substrate (3). Generally, spores grown on the solid substrate are used as the inoculum for fungal SSF. Use of spores rather than vegetative mycelia as an inoculum has several advantages, such as greater flexibility in inoculum preparation, prolonged storability for subsequent use, and greater resistance to mishandling that might occur between harvesting and inoculation. However, a few disadvantages exist, such as longer lag time, different optimal conditions for spore germination and vegetative growth, and requirement of a larger inoculum. The major disadvantage of spores is that they are metabolically dormant, and hence metabolic activities must be induced and the appropriate enzyme systems must be synthesized before the fungus begins to utilize the substrate and grow (12). However, some organisms require vegetative mycelia for optimal growth and a higher protein yield (3).

Inoculum size or density also plays a major role in the cultivation process. One of the major limitations of SSF is the unique physical conditions for growth, which may be partially overcome by selecting an inoculum size large enough for the mycelial fragments or spores to initially colonize most of the substrate particles without overcrowding and competing for the limited nutrient supply. The inoculum is usually used at a higher ratio in most fermentation processes for the production of the desired quantity of product in a shorter time (11).

The solid substrate is subjected to pretreatment, including size reduction, autoclaving to sterilize or pasteurize the medium for increased amenability to fungal growth, and then cooling of the substrate medium to suboptimal temperature for germination of the fungal spores. For inoculation, pure-culture fungal spores are generally scattered uniformly on the solid substrate. Immediately after inoculation, it is important to mix the inoculated, moistened solid substrate to make a homogeneous substrate medium for fermentation. The solid substrate should be continuously mixed for proper aeration at selected intervals to prevent lumping and to obtain uniform mold growth and bioconversion.

Moisture and Water Activity

The importance of moisture content for the fungal SSF process is well documented. In general, a wide range of moisture level (20 to 70%) is reported to be optimal for fungal SSF. An optimum moisture level, however, has to be maintained since lower moisture tends to reduce nutrient diffusion, microbial growth, enzyme stability, and substrate swelling (3, 7). Higher moisture levels lead to particle agglomeration, gas transfer limitation, and potential contamination by bacteria. Different fungi have different optimal moisture requirements, and this characteristic also varies with the physical properties of the solid substrate used. As a general rule, a more appropriate term, water activity (a_w), is used to describe the moist environment. This is defined by the ratio of the vapor pressure of the water in the substrate (p_w) to the vapor pressure of pure water (p_0) at a given temperature (3, 17), i.e.:

$$a_w = p_w/p_0 \text{ and } a_w = \%\,rH/100$$

Most fungi used for SSF have a_w values between 0.8 and 0.9 (3). Some fungi can grow even at such low activities as 0.6. Bacteria are usually less tolerant than yeast and fungi at a given a_w. This condition suggests that SSF with low a_w is more resistant to bacterial contamination, allowing a nonaseptic condition. However, contamination may still occur during the early stages of preparation and cultivation.

The importance of moisture content and water activity implies that while the substrate is being prepared for the process, it is necessary to consider appropriate quantities of liquid to be added to achieve the proper conditions for optimal growth and protein production. In some cases, the optimum water activity may be different for growth and product formation; therefore, this condition offers the possibility of manipulating the water activity during fermentation (3). The moisture content of the substrate depends largely on the water activity, and small changes in the latter could have a great impact on the former. The SSF process also leads to an increase in water activity of the substrate due to coproduction of water during the process. However, evaporative losses due to metabolic heat evolved during the process result in loss of moisture during the process.

pH Requirements for Fungal Growth

Another major factor affecting SSF is pH, and it may change during the process in response to metabolic activities. The main pH change observed during fungal SSF is the drop in pH due to the secretion of organic acids during cultivation. Filamentous fungi are reported to have reasonably good growth over a wide pH range (2 to 9), with an optimal range of 3.8 to 6.0. This typical pH versatility of fungi can be beneficially exploited to prevent or minimize bacterial contamination by choosing a lower pH (3, 7). Also, the optimum pH for mycelial growth may be different from that required for enzyme production. The main point to remember in SSF is that the values mentioned above are the initial pH and they are not necessarily the same as the fermentation proceeds. In SSF, unlike the homogeneous three-phase system of submerged fermentation, in situ pH control is practically impossible due to the heterogeneous three-phase system and the lack of proper equipment and electrodes to determine the pH in a solid matrix. The pH variability during the cultivation process can be overcome by considering the buffering capacity of the different components used in substrate formulation so that there is no deleterious effect on the biological activity. pH changes occur according to the nitrogen source selected as well as the growth characteristics, and use of urea as a nitrogen source rather than ammonium salts is one way of controlling the pH during cultivation (7).

Temperature Variation and Control

Temperature control of the substrate bed is probably the most important of all the physical variables affecting fungal SSF performance, since the growth and production of enzymes and metabolites are generally sensitive to temperature (17). Similarly to the case for pH, fungi can grow over a wide range of temperatures (from 20 to 55°C), and the optimum temperature for growth could be different from that for product formation. The critical criterion in SSF is temperature control, which is quite difficult to achieve because of very high heat output during the aerobic process and because conventional convection or conductive cooling devices are ineffective for dissipating metabolic heat due to the poor thermal conductivity of the solid matrix as well as the predominantly static nature of the fermentation process. The metabolic heat evolved during the process raises the temperature of the solid substrate bed, which without proper heat removal leads to a temperature exceeding the upper limit for growth. Moreover, the rate of heat generation is directly proportional to the level of metabolic activity in the system. In combination with local moisture content and availability of void spaces, this could lead to thermal gradients, resulting in heterogeneous growth and production. Depending on the thickness and width of the bed and the airflow rate, the temperature gradient may occur vertically or horizontally (17). Therefore, the key issue in SSF is heat removal, and thus most studies of reactor designs are focused on maximizing heat removal (16). The issue becomes aggravated during process scale-up in which heat generation leads to serious moisture losses and difficulties in maintaining constant temperature and moisture content, causing loss in yield and economic loss. The most appropriate way to deal with heat buildup is by means of forced aeration, which plays multiple roles in fungal SSF. The critical steps are to control the flow rate and to humidify or dehumidify the air since low or high moisture content brings about adverse effects (3).

Aeration and Agitation

Aeration and agitation play a major role in SSF due to oxygen demand in the aerobic process and heat, O_2, and mass transfer phenomena in a heterogeneous system. Operational parameters and medium characteristics can affect O_2 and mass transfer rates, including the air pressure and flow rate, the porosity of moist solids, the bed depth of the moist fermenting solids, perforations in the culture vessel, the moisture content of the medium, the reactor geometry, and impeller rotational speed and geometry (7). Forced airflow, a variable flow of air and its quality, is the main tool for removal of the heat evolved during the process. Water evaporation and heat transfer with the surrounding environment are the processes to keep the temperature within proper limits. Agitation is one of the most important parameters in aerobic fermentations since it ensures homogeneity with respect to temperature and the gaseous environment and provides a gas-liquid interfacial area for gas-to-liquid as well as liquid-to-gas transfers (20). Agitation also enhances mass and heat transfer and enables the possibility of homogeneous water addition, allowing the compensation of water loss caused by evaporation. The negative effects of agitation in SSF systems include adverse effects on substrate porosity due to the compacting of the substrate particles, disruption of fungal attachment to the solids, and damage to fungal mycelia due to shear forces. Moreover, agitation may promote or prevent aggregate formation of the fermenting mass, depending on the nature of the solid substrates (10). Intermittent rather than continuous agitation was found to be more appropriate for shear-sensitive mycelia and disruption of mycelial attachment to solids.

LIGNOCELLULOSIC SUBSTRATES FOR SSF

The main advantage of using fungi for SSF is the ease with which fungi can grow on complex natural solid substrates such as agro-industrial wastes without much pretreatment, and the wastes supply the needed nutrients and support for fungal growth. The commonly used lignocellulosic substrates for SSF include sugar cane bagasse; straws of wheat and rice; rice, wheat, maize, and grain brans; corn and rice stover; sugar beet pulp; feedlot waste; and wood, etc. The macromolecular structure of lignocellulosic substrates, particle size and shape, porosity, and particle consistency make them a suitable substrate for growth of the fungi. The complexity of the lignocellulosic solid substrate arises from the presence of macromolecular compounds such as cellulose, starch, and lignin and even smaller sugar molecules. Fungi cannot directly absorb these macromolecules; they are induced by low-molecular-weight compounds to synthesize and secrete enzymes to hydrolyze the macromolecules into smaller, metabolizable compounds (3).

Particle size and pretreatment of the solid substrate play a major role in the fermentation process; a larger ratio of surface area to volume is preferred for high yields, especially when the fungi lack sufficient penetrative ability, so that the substrate molecules are more easily accessible for the hydrolytic enzymes. If the surface area is less, then the enzyme diffusion tends to become rate-limiting. The particle size is also important in obtaining favorable physical conditions in the bed. Smaller particles result in bed compaction, affecting gas exchange, while larger particles limit substrate accessibility (because of decreased surface-to-volume ratio). Depending on the pretreatment, different substrate particles have different particle shapes, which have a direct bearing on the void fraction when packed into the reactor system (3). The substrate particle also needs to have sufficiently larger pore openings for mycelial entry and passage. Porosity of the lignocellulosic substrate can be further improved by pretreatment.

FUNGAL GROWTH ON LIGNOCELLULOSIC SUBSTRATES IN SSF

Fungal degradation of lignocellulosic material is a complex process. Typically, the fungal attack starts at a weak point of the plant cell wall, into which an active hyphal tip rich in the enzyme complex penetrates (18). Filamentous fungi grow on the substrate surface in the absence of free water, utilizing the bound water of the substrate. Fungal growth on lignocellulosic substrate is linear rather than logarithmic, limited by steric hindrance and substrate accessibility (19, 20). The solid substrate surface is inoculated with the fungus (spores or hyphae), which must first colonize the substrate surface by adhesion and then spread from one particle to another by branching. The fungal hyphae penetrate into the solid particle, and the mycelium releases enzymes which diffuse into the substrate and hydrolyze the polysaccharide. The soluble sugars released then diffuse back to the mycelium, and oxygen diffuses from the interparticle spaces through a stagnant gas film into the mycelial layer on the particle surface. The mycelium simultaneously consumes the soluble sugars and oxygen. In the core, lignin surrounds and protects the cellulose from enzyme hydrolysis; thus, partial delignification is necessary for cellulose hydrolysis in plants having >10% lignin content (3, 19, 20). Pretreatment of the lignocellulosic substrate causes partial delignification, increasing the separation of lignin from cellulose which results in the removal of the outer components of the cell wall and therefore increases the efficiency and rate of the hydrolysis (3).

Host specificity is critical in SSF because of the complexity of the substrate and the need for special enzyme complexes for its degradation. A synergetic attack of cellulases and ligninases is difficult to achieve, because of their substrate specificity and complexity, making lignocellulose degradation a low-efficiency process. The comparatively slow and inefficient process, which takes 3 to 8 weeks for maximum degradation of hardwood and 3 to 4 weeks for softwood, is uneconomical for industrial scale-up, where a maximum of 3 to 4 days would be required for profitable operation (3, 20, 21).

The increasing interest in the bioconversion of lignocellulosic materials as a renewable feedstock for the microbial production of food, feed, and biochemicals and the increasing industrial applications of fungi in biotechnology have led to further exploitation of SSF systems. From laboratory-scale lignocellulose conversion studies using liquid-state fermentation and SSF, it is evident that liquid culture gives greater efficiency of conversion and a higher biomass (protein) yield; however, solid culture has a greater volumetric productivity (20). The high volumetric productivity and reduced recovery costs in SSF may ensure viability for such high-value products. Most of the cell wall-degrading enzymes are inducible by lignocellulosic substrates, thus producing profitable enzymes by SSF. The presence of lignin along with cellulose and hemicellulose decreases the cellulase activity and the feed value of lignocellulose wastes. The white rot fungi, which belong to the *Basidiomycetes*, are the most efficient lignin degraders, and their biodegradation ability has been extensively reviewed by Leonowicz et al. (9). They are capable of depolymerizing lignin and metabolizing it to CO_2 and H_2O. The lignin-degrading systems of these fungi have great potential applications in improving the digestibility of wood or straw for animal feed, biobleaching, and biopulping. Some species can remove lignin faster than cellulose and hemicellulose and thus could be most useful for the pretreatment of wood and straw for enzymatic saccharification to increase digestibility for feed purposes or for biological pulping. The lignin-degrading system of the white rot fungi has potential applications in the area of lignocellulose bioconversion by SSF (3).

Monitoring of Fungal Growth

Knowledge of monitoring of fungal growth is very limited because of the complex orientation of the mycelium with the substrate types, substrate heterogeneity, and the lack of techniques for direct estimation of viable biomass. Instead of direct biomass measurements, indirect measurements such as cellular components, e.g., DNA, protein, chitin, and *N*-acetylglucosamine (a chitin component), are usually used. The *N*-acetylglucosamine content in solid substrate is generally used as the selected method for indirect biomass measurements in the fungal SSF process (17).

Measurement of the oxygen uptake rate (OUR) or the carbon dioxide evolution rate (CER) is often used to estimate microbial growth indirectly (17). For aerobic microorganisms, cell mass is estimated from OUR by the following equations, assuming the value of the respiration quotient (RQ) to be constant:

$$dO_2/dt = \chi\,(\mu/Y_{X/o} + m)$$

where dO_2/dt is OUR (grams of O_2 consumed per gram of solid substrate per hour), χ is the cell mass (grams of cells produced per gram of solid substrate), t is the fermentation time (hours), μ is the specific growth rate (per hour), $Y_{X/o}$ is the yield based on oxygen (grams of cell per gram of O_2), m is the maintenance coefficient of OUR (grams of O_2 per gram of cell per hour), and χ_o is the initial cell mass and $\chi = \chi_o \exp(\mu t)$.

The following equation is used to estimate cell mass from CER:

$$dCO_2/dt = \chi\,(\mu/Y_{X/c} + m')$$

where dCO_2/dt is CER (grams of CO_2 evolved per gram of solid substrate per hour), $Y_{X/c}$ is the yield based on carbon dioxide (grams of cell per gram of CO_2), and m' is the maintenance coefficient of CER (grams of CO_2 per gram of cell per hour) (17).

BIOREACTOR DESIGN USED IN SSF AND DOWNSTREAM PROCESSING

The selection and design of a bioreactor are the critical stages of the SSF process. The selection depends on factors such as substrate type, structure of the lignocellulosic substrate, accessibility of the substrate for mycelial growth, its heat conductivity, moisture content and O_2 and mass transfer, process variables, extent of control required, etc. Mathematical models for growth, mass transfer, and product formation are being developed for SSF but not yet widely used (13, 15, 16, 19). In the classical fungal SSF koji process, tray reactors have been traditionally employed. Other major types of reactors are the packed column, rotating drum, and fluidized bed.

In tray systems, the substrate is evenly spread in layers about 2 to 4 cm thick in trays with a perforated bottom to provide maximum surface availability for gas exchange. After autoclaving, the trays are cooled, inoculated and mixed for the homogeneous system, stacked one above the other in trolleys, and moved to the fermentation chamber, where conditions such as relative humidity, aeration, and temperature are regulated. After the fermentation is done, the solids are harvested from the trays and recovery operations are executed. Tray reactors have been successful in laboratory, pilot, and large-scale operations (3, 17). However, one major disadvantage is that use of tray reactors can be very labor-intensive unless a reliable automation system is present. Also, heat transfer limitations restrict the height of the substrate bed.

In packed-bed reactors, substrate is generally loosely packed to larger heights than in trays and humidified air is forced through the bed from the bottom, where the mass transfer occurs by forced convection. The advantages of packed beds are as follows. The substrate bed can be thicker since the forced aeration helps to minimize temperature gradients and gas concentration gradients, compared to those of tray systems. Being a closed system, packed beds maintain more aseptic conditions. Humidified air also helps in regulating moisture. After the fermentation, the column reactor can be used as a trickle bed extractor for product recovery.

A rotating drum bioreactor is a horizontal or inclined cylinder which rotates around its own longitudinal axis so that the substrate within the reactor is in constant motion to promote gentle intermixing. This is in contrast to the packed bed, in which the substrate is in static conditions. The rocking or tumbling action promotes very efficient heat and mass transfer. The major disadvantage with this bioreactor is that particles tend to agglomerate over the duration of the process, and also the rotational speed becomes an important parameter if the mycelia are shear sensitive. Baffles along the inner wall of the rotating drum facilitate mixing by the lifting and dropping of the substrate and increased the maximum OUR by 60% over the rate without baffles (3).

Depending on the application of the final products, some SSF systems do not require any downstream processing. Examples of these systems include composts, ensiled grasses, upgraded agro-industrial residues, enzymes in crude form, and poultry and animal feed supplements (6, 7). The major difference in the downstream processing of SSF bioproducts from that of submerged fermentation systems is that an additional step of leaching is involved, but liquid removal is less important. This stage ensures the transfer of product(s) from the solid to liquid phase, and then the submerged fermentation broth extraction process is followed. The selection of a leaching method such as percolation, pulsed plug flow extraction, countercurrent extraction, or hydraulic pressing is probably the most important step in SSF downstream processing, since the economics of the process will be dictated by the leached product concentration (3). After the leaching step, the extract may be clarified by either filtration or centrifugation to remove suspended cells, spores, or solid residues.

REFERENCES

1. **Bonnarme, P., and T. W. Jeffries.** 1990. Mn(II) regulation of lignin peroxidases and manganese-dependent peroxidases from lignin-degrading white rot fungi. *Appl. Environ. Microbiol.* **56:**210–217.
2. **Eriksson, K. E., R. A. Blanchette, and P. Ander.** 1990. *Microbial and Enzymatic Degradation of Wood and Wood Components*. Springer, Berlin, Germany.
3. **Gowthaman, M. K., C. Krishna, and M. Moo-Young.** 2001. Fungal solid-state fermentation—an overview, p. 305–352. *In* G. G. Khachatourians and D. K. Arora (ed.), *Applied Mycology and Biotechnology*, vol. 1. *Agriculture and Food Production*. Elsevier Science, Amsterdam, The Netherlands.
4. **Holker, U., and J. Lenz.** 2005. Solid-state fermentation—are there any biotechnological advantages? *Curr. Opin. Microbiol.* **8:**301–306.
5. **Kirk, T. K., and M. Shimada.** 1985. p. 579. *In* T. Higuchi (ed.), *Biosynthesis and Biodegradation of Wood Components*. Academic Press, San Diego, Calif.
6. **Krishna, C., and M. Moo-Young.** 2004. Single cell protein, p. 293–304. *In* A. Pandey (ed.), *Concise Encyclopedia of Bioresource Technology*. Haworth Press, Binghamton, N.Y.
7. **Krishna, C.** 2005. Solid-state fermentation systems—an overview. *Crit. Rev. Biotechnol.* **25:**1–30.
8. **Laukevics, J. J., A. F. Apsite, H. E. Viestures, and R. P. Tengerdy.** 1984. Solid substrate fermentation of wheat straw for fungal protein. *Biotechnol. Bioeng.* **26:**1465–1474.
9. **Leonowicz, A., A. Matuszewska, J. Luterek, D. Ziegenhagen, M. Wojtas-Wasilewska, N. S. Cho, M. Hofrichter, and J. Rogalski.** 1999. Biodegradation of lignin by white rot fungi. *Fungal Genet. Biol.* **27:**175–185.
10. **Lonsane, B. K., and M. M. Krishniah.** 1992. p. 147. *In* H. W. Doelle, D. A. Mitchell, and C. E. Rolz (ed.), *Solid Substrate Cultivation*. Elsevier Applied Science, London, United Kingdom.
11. **Lonsane, B. K., S. Saucedo-Castaneda, M. Raimbault, S. Roussos, G. Viniegra-Gonzalez, N. P. Ghildyal, M. Ramakrishna, and M. M. Krishnaiah.** 1992. Scale-up strategies for solid state fermentation systems. *Proc. Biochem.* **27:**259–273.
12. **Mitchell, D. A.** 1992. Microbial basis of processes, p. 17–27. *In* H. W. Doelle, D. A. Mitchell, and C. E. Rolz (ed.), *Solid Substrate Cultivation*. Elsevier Applied Science, London, United Kingdom.
13. **Mitchell, D. A., O. F. von Meien, and N. Krieger.** 2003. Recent developments in modeling of solid-state fermentation: heat and mass transfer in bioreactors. *Biochem. Eng. J.* **13:**137–147.
14. **Moo-Young, M., A. R. Moreira, and R. P. Tengerdy.** 1983. Principles of solid state fermentation, p. 117–144. *In* J. E. Smith, D. R. Berry, and B. Kristiansen (ed.), *The Filamentous Fungi*. Edward Arnold, London, United Kingdom.
15. **Rahardjo, Y. S. P., J. Tramper, and A. Rinzema.** 2006. Modeling conversion and transport phenomena in solid-state fermentation: a review and perspectives. *Biotechnol. Adv.* **24:**161–179.
16. **Sangsurasak, P., and D. A. Mitchell.** 1998. Validation of a model describing 2-dimensional heat transfer during solid-state fermentation in packed bed bioreactors. *Biotechnol. Bioeng.* **60:**739–749.
17. **Sato, K., and S. Sudo.** 1999. Small-scale solid-state fermentations, p. 61–79. *In* A. L. Demain and J. E. Davies (ed.), *Manual of Industrial Microbiology and Biotechnology*, 2nd ed. ASM Press, Washington, D.C.
18. **Smith, J. E., and K. E. Aidoo.** 1988. p. 249. *In* D. R. Berry (ed.), *Physiology of Industrial Fungi*. Blackwell Scientific Publications, Oxford, United Kingdom.
19. **Tengerdy, R. P., and G. Szakacs.** 2003. Bioconversion of lignocellulose in solid state fermentation. *Biochem. Eng. J.* **13:**169–179.
20. **Tengerdy, R. P.** 1996. Cellulase production by solid substrate fermentation. *J. Sci. Ind. Res.* **55:**313–316.
21. **Trilli, A.** 1986. Scale-up of fermentations, p. 277–307. *In* A. L. Demain and N. A. Solomon (ed.), *Manual of Industrial Microbiology and Biotechnology*. American Society for Microbiology, Washington, D.C.

Characterization of Microeukaryota in Natural Environments

JANICE L. STRAP

90

Microeukaryotes, including fungi and protists, are important members of communities within natural environments. In aquatic habitats, protists are vital components of the food web (105) and are major consumers of bacterial biomass, thereby playing a principal role in nutrient cycling (7, 127). In addition, their impact on bioremediation processes in contaminated environments, while incontrovertible (50), has surprisingly received little attention at the molecular level. Therefore, characterization of microeukaryote diversity within natural environments will provide insight into key relationships that exist between species richness and organic contamination removal. A prerequisite for understanding processes such as natural attenuation is knowledge of the community dynamics of indigenous microeukarya with their prokaryotic counterparts in pristine environments.

Culture-independent methods have been extensively used to investigate prokaryotic diversity in a wide range of environments; however, molecular investigations of eukaryotic populations are more recent and have not yet been as extensive. The reason for this is in part due to a well-established ultrastructure and morphology-based eukaryote taxonomy. However, morphological classification can be complicated by the sheer number, diversity, and size range of many species (16) within an environmental sample. To overcome this, implementation of molecular characterization of natural environments is playing a greater role in microbial ecology studies. Molecular studies of microeukaryotes have been conducted with soils (57), desert rock varnish (53), rivers (5), solar salterns (17), marine environments (24, 32, 66, 76, 80, 99, 115), anoxic sediments (28, 114), lakes (58), acid mine drainage (8), glacier ice (132), and a deep, fractured basalt aquifer (116). This chapter focuses exclusively on available molecular approaches for investigating microeukaryote assemblages within various natural environments.

SAMPLING

Sampling considerations vary greatly depending on the environment under investigation and the scientific question to be answered. Oligotrophic environments generally require more sampling effort to obtain adequate biomass for molecular characterization than do more nutrient-rich environments. For example, investigation of planktonic (free-living) microeukaryote populations from a fractured basalt aquifer in south Idaho required filtration of 13,300 liters of aquifer water to obtain enough biomass for study (34, 116), whereas 2.3 to 2.8 liters of seawater were sampled from the Cariaco Basin (115). Planktonic biomass collection for community surveys is achieved primarily by filtration (24, 32, 34, 58, 75, 80, 95, 97).

Sediment sampling achieved by the implementation of sterile mud samplers and coring devices has been described for hydrothermal sediments (65) and anoxic sediments (28, 118).

When attached (biofilm) communities from aquatic environments are investigated, the choice of substrate and the length of time that colonization is allowed to proceed profoundly affect the observed community. The use of autochthonous material as a substrate for colonization results in a less biased picture of community structure than use of nonautochthonous material; however, there may be times when investigating colonizers of a particular substrate is of interest. López-Garcia et al. (65), investigating eukaryotic diversity in the Mid-Atlantic Ridge hydrothermal area, used microcolonizers with a variety of substrates including plastic mesh, meat-based substrates, iron fragments, and basalt and pumice fragments, all of which were allowed to colonize for 15 days. In contrast, to investigate microeukaryotes that form attached communities within a fractured basalt aquifer, Strap et al. (116) used basalt substrates which were suspended in a well within an aquifer for 6 months.

Some environments require creative sampling methods. For example, investigation of microeukaryote inhabitants of desert rock varnish required the use of a Dremel grinding tool with a flame-sterilized bit to grind the varnish from the host rock (53).

A final sampling consideration is timing. Microeukaryote dynamics may be observed by sampling at different seasons, as was shown for marine stramenopiles (75) and for picoeukaryote communities at a coastal site on the English Channel (97).

DNA EXTRACTION

A major limitation of molecular characterization of communities is the representative and quantitative recovery of nucleic acids from environmental samples; some cells lyse more readily than others, and the same lysis technique may yield different results with different sample types (water, soil, or sediment) and therefore result in bias (93, 125). The coextraction of substances such as humic acids, which are prevalent in soils and sediments and are inhibitory to PCR (59, 90, 119), must be minimized.

There are a number of methods by which to extract DNA for molecular analyses (23, 36, 42, 48, 52, 79, 82, 113, 121), including commercial kits such as the UltraClean Soil DNA Kit Mega Prep (MoBio Laboratories, Inc., Solana Beach, Calif.) that have been shown to be effective for extraction of DNA from both prokaryotes and eukaryotes (14, 22, 30, 34, 37, 44, 88, 116, 118).

To overcome the problem of low DNA recovery and copurifying PCR inhibitors from natural environments, Gonzalez et al. (40) used multiple displacement amplification using random hexamers as priming oligonucleotides and Φ29 DNA polymerase in a whole-genome amplification reaction to simultaneously increase copy number without bias and to dilute inhibitory substances; this whole-genome amplified template was then used for gene-specific PCR. This technique was applicable to both eukaryotic and prokaryotic rRNA genes. Erwin et al. (34) also demonstrated that whole-genome amplification could be used for the unbiased amplification of DNA from an environmental sample.

CHARACTERIZATION OF rRNA GENES BY PCR AND RESTRICTION FRAGMENT LENGTH POLYMORPHISM (RFLP) ANALYSIS

The 18S rRNA gene contains both conserved and variable regions that can be exploited for characterization of the microeukaryote assemblages within an environment. Regions of the rRNA molecule that change at a very low rate are characteristic of broad phylogenetic groupings such as domains, while those regions that change at a somewhat higher rate are characteristic of narrower phylogenetic groupings (55). It must be cautioned that 18S rRNA gene sequence analysis, while a powerful tool for evaluating diversity within a natural environment, is not without limitations; these are briefly discussed below.

The choice of primer is an important consideration for community characterization. A variety of universal and taxon-specific primers have been described, and a selection of these is listed in Table 1.

It is necessary to evaluate species- and genus-specific probes targeted to the 18S rRNA of the small subunit (SSU) of the ribosome or the 28S rRNA of the large subunit of the ribosome for specificity (2). As sequence information is amassed, it is probable that the specificity of a given primer is broader than previously believed. Specificity limitations can be somewhat alleviated by the use of nested or multiple probes to assess a single population (2, 3). PCR amplification with universal primers has been observed to favor DNA from taxa with exact complimentary sequences (94). Since experimental evaluation of probe specificity is only feasible for a limited number of nontarget reference sequences, in silico evaluation of probe specificity can be achieved by a BLAST (Basic Local Alignment Search Tool) (1) search of the probe sequence or by the use of more specialized programs (54, 68). It is important to realize that when working with natural samples there is always the potential for nonspecific amplification even by target group-specific primers. At best, group-selective primers enrich for desired sequences. For example, Strap et al. (116) found that "eukaryote-specific" primers NS3/NS8 amplified archaeal sequences from biofilms sampled from the Snake River Plain aquifer. Interestingly, this was not observed for planktonic samples from the same source.

In addition to primer bias, the conditions used for the PCR can also influence the amplification products obtained (117); both template concentration and quality have profound influence on the amplification (94) and inhibition (19) of PCR. Temperature gradient PCR is one method that may be used to provide a range of conditions to reduce amplification bias of eukaryotic rRNA genes (28). The length of the target sequence will also affect amplification. For example, amplification of the full-length 18S rRNA genes of planktonic foraminifera is ineffective by commonly used approaches because of the exceptionally long sequence inserts in the genes of these species (27).

PCR is prone to contamination. Controls to check for possible contamination by extraneous DNA have been described (132) and include (i) extraction of empty filters or sterile substrates, (ii) extraction from empty sample tubes, (iii) no-template controls, and (iv) sterile controls which monitor the air inside the laminar flow hood used for pre-PCR work. Positive controls consisting of purified DNA of a target organism used with a given primer set to ensure that amplification is successful with a known target as well as negative controls that utilize DNA from a nontarget organism should also be included.

Multiple PCRs containing amplicons of expected size can be pooled to minimize bias from single reactions (18, 94). Once amplicons are purified, individual molecules are segregated by cloning. Plasmid DNA from recombinant clones is purified for subsequent RFLP analysis.

RFLP analysis uses restriction endonucleases to digest purified DNA from individual clones in order to identify polymorphisms (124). Such differences can be caused by single base pair changes, insertions, deletions, and rearrangements. Fragments resulting from DNA digestion are separated by agarose gel electrophoresis. Differences or similarities in the numbers and sizes of fragments from each cloned insert can then be identified. The similarity between numbers of restriction sites within individual genes also has been used to calculate evolutionary relatedness between genes from different organisms (86). Enzymes that recognize a 4-bp fragment such as HaeIII, HhaI, and BstUI are often used for RFLP analysis of rRNA genes (83). It is noteworthy that partial digestion products can confound analysis. The proportion of clones present in a PCR-generated library cannot be assumed to be quantitatively representative of the proportions of the individual rRNA in the environmental sample because of primer bias and possible biases in the relative rates of PCR amplification (117) or rRNA resulting from differences in copy number and proximity of the rRNA operons to one another in the template DNA (35). Once the clone library has been created, the number of clones to be sequenced may be reduced by grouping clones with identical RFLP patterns. Underestimation of diversity can result if different strains exhibit the same fingerprint. This may be overcome by digestion with more than one enzyme to provide greater resolution between different phylotypes.

Chimeras resulting from the formation of PCR products composed of the 3′ end of one molecule and the 5′ end of another are artifacts of PCR often encountered when community DNA is used as a template. Sequence analysis of amplification products with programs such as Check Chimera (69) and Bellerophon (46) can help identify chimeric sequences. Nucleotide-nucleotide BLAST (1) can be used to search GenBank for the nearest relative sequences. BLAST results and RFLP representative clones are aligned using an alignment program such as ClustalX

TABLE 1 18S rRNA gene primers

Primer name	Sequence (5′–3′)[a]	Comments	Reference(s)
18S-42F	CTCAARGAYTAAGCCATGCA		110, 118
18S-82F	GAAACTGCGAATGGCTC		110, 118
18S-1498R	CACCTACGGAAACCTTGTTA		110, 118
EK-1520r	CYGCAGGTTCACCTAC		65, 110, 118
EK-1F	CTGGTTGATCCTTGCCAG		65
Medlin EukB	TGATCCTTCTGCAGGTTCACCTAC	Eukarya-specific reverse primer	49, 77
Medlin EukA	AACCTGGTTGATCCTGCCAGT	Eukarya-specific forward primer	77
EF-3	TCCTCTAAATGACCAAGTTTG	Amplifies 18S rRNA from a wide range of fungal taxa; in combination with EF-4 the majority of 18S rRNA is amplified	111
EF-4	GGAAGGGRTGTATTTATTAG		111
Mallat-F	CTGGTTGATCCTGCCAGT	Universal end primer for 18S	70
Mallat-R	TAATGATCCTTCCGCAGGTTCACCT	Universal end primer for 18S	70
nu-SSU-0817-5′	TTAGCATGGAATAATRRAATAGGA	Amplification of fungal rRNA genes	11
nu-SSU-1196-3′	TCTGGACCTGGTGAGTTTCC	Amplification of fungal rRNA genes	11
nu-SSU-1536-3′	ATTGCAATGCYCTATCCCCA	Amplification of fungal rRNA genes	11
nu-SSU-1288-5′	TGGTGGTGCATGGCCGTTCTTAGTTGG	nu-SSU-1288/nu-SSU-1420 amplify 180-bp product; nu-SSU-1288-5′/NS8 amplify 550-bp product	132
nu-SSU-1420	ACATCTAAGGGCATCACAGACCTGTTA	nu-SSU-1288/nu-SSU-1420 amplify 180-bp product	132
nu-SSU-598	GCCGCGGTAATTCCAGCTCCAATAGCGT	nu-SSU-598/nu-SSU-898 amplify 340-bp product	132
nu-SSU-898	TCCAAGAATTTCACCTC	nu-SSU-598/nu-SSU-898 amplify 340-bp product	132
Univ-F-15	CTCCCAGTAGTCATATGC	Univ-F-15/Univ-R-1765 yield 1,760-bp product	43
Univ-R-1765	ACCTTGTTACGACTTTAC	Univ-F-15/Univ-R-1765 yield 1,760-bp product	43
NS1F (nu-SSU-0038-5′)	GTAGTCATATGCTTGTCTC	NS1/NS2 product is 555 bp	112, 131
NS2R (nu-SSU_0553-3′)	GGCTGCTGGCACCAGACTTGC	Spans intron position 516; NS1/NS2 product is 555 bp	112, 131
NS3F (nu-SSU-0573-5′)	GCAAGTCTGGTGCCAGCAGCC	Spans intron position 516; NS3/NS4 product is 597 bp	112, 131
NS4R (nu-SSU-1131-3′)	CTTCCGTCAATTCCTTTAAG	NS3/NS4 product is 597 bp	112, 131
NS5F (nu-SSU-1150-5′)	AACTTAAAGGAATTGACGGAAG	NS5/NS6 product is 310 bp	112, 131
NS6R (nu-SSU-1412-3′)	GCATCACAGACCTGTTATTGCCTC	Spans intron position 1199; NS5/NS6 product is 310 bp	112, 131
NS7F (nu-SSU-1436-5′)	GAGGCAATAACAGGTCTGTGATGC	Spans intron position 1199; NS7/NS8 product is 377 bp	112, 131
NS8R (nu-SSU-1769-3′)	TCCGCAGGTTCACCTACGGA	Spans intron position 1512; NS7/NS8 product is 377 bp	112, 131
ITS4Asco	CGTTACTRRGGCAATCCCTGTTG	Ascomycota specific	87
ITS4Basidio	GCRCGGAARACGCTTCTC	Basidiomycota specific	87
ITS4Chytrid	TTTTCCCGTTTCATTCGCCA	Chytridiomycota specific	87
Its4Oo	ATAGACTACAATTCGCC	Oomycota specific	87
Its4Zygo	AAAACGTWTCTTCAAA	Zygomycota specific	87
82FE	GAADCTGYGAAYGGCTC		28

TABLE 1 *(Continued)*

Primer name	Sequence (5′–3′)[a]	Comments	Reference(s)
1391RE	GGGCGGTGTGTACAARGRG	Universal reverse primer	28
360FE	CGGAGARGGMGCMTGAGA		28
1492R	ACCTTGTTACGRCTT	Universal reverse primer	28
EUKb518F	GAGGRCMAGTCTGGTGC		8
EUKb1193R	GGGCATMACDGACCTGTT		8
960F	GGCTTAATTTGACTCAACRCG		38
1200R	GGGCATCACAGACCTG	Universal	38, 39, 81
Biocos1R	TCTAGATGGTAAGGTTTA	Biosoecid specific	38
Choan1R	CGAWAATTACAAAGATCTA	Choanoflagellate specific	38
ChrSyn1F	AATAACTTTCGGATCGAT	Chrysophytesynurophyte specific	38
Prymn1R	ACATCTCTTCACGAGGAT	Prymnesiophyte specific	38
P-SSU-342f	CTTTCGATGGTAGTGTATTGGACTAC	Protozoa-specific forward primer	49
PRYMN01	ACATCCCTGGCAAATGCT	Prymnesiophyte specific	56, 81
PRYMN02	GGAATACGAGTGCCCCTGAC	Prymnesiophyte specific	81
PHAEO01	CGGTCGAGGTGGACTCGT	Phaeocyctis specific	56, 81
CLADE2	AGTCGGGTCTTCCTGCATGT	Chrysochromulina specific	81, 106
25F	CATATGCTTGTCTCAAAGATTAAGCCA		10
1256R	GCACCACCACCCAYAGAATCAAGAAAGAWC	Cercozoan specific	10
EUK345f	AAGGAAGGCAGCAGGCG	QPCR probe; 4 mismatches to prokaryotes	134
EUK499r	CACCAGACTTGCCCTCYAAT	QPCR probe; 5 or more mismatches to prokaryotes	134
EUK528f	CCGCGGTAATTCCAGCTC	QPCR probe; 5 or more mismatches to prokaryotes	33, 134
CHLOO02r	CTTCGAGCCCCCAACTTTC	QPCR probe; one central mismatch to sequences of Chlorarachniophyceae, Cercozoa, and Apicomplexa	107, 134

[a]D= A, G, and T; M = A and C; R = G and A; Y = C and T; W = A and T; N = C, A, T, and G; positions where the synthesized primer contains equimolar amounts of more than one nucleotide.

(120) or the equivalent. The inference of phylogenetic trees is an important part of characterizing microeukaryotes from natural environments but is beyond the scope of this chapter and has been reviewed elsewhere (9, 64, 128).

OTHER METHODS FOR CHARACTERIZATION OF MICROEUKARYOTES

Terminal RFLP (T-RFLP) is another PCR-based method for the assessment of microbial communities (51, 71, 73). The method involves a PCR amplification of rRNA genes analogous to the RFLP method described above. The techniques differ in that the 5′ primer is fluorescently labeled. The resultant amplicons are subjected to restriction endonuclease digestion with 4-bp cutters such as HaeIII, HhaI, or MspI, followed by resolution of digestion products by capillary electrophoresis coupled with a fluorescence detector; only fluorescently labeled restriction fragments are visualized. Despite its popularity for analysis of 16S rRNA, only a few studies of microeukaryote populations have incorporated this technique. The earliest published application of T-RFLP to a microeukaryote community analyzed activated sludge samples (72). Lord et al. (67) used T-RFLP analysis to evaluate fungal diversity in a petroleum-contaminated land treatment unit. As part of their analysis, these authors compared the use of 18S and internal transcribed spacer regions and found that the internal transcribed spacer rRNA region provided a more extensive estimate of community richness and species identity than did the 18S region. A similar observation was noted by Palumbo et al. (92).

Immunological approaches have proven useful in ecological studies for the detection and enumeration of microeukaryotes (15, 63, 126). The application of polyclonal antisera towards specific types of phytoplankton (15, 78, 104) is documented. Monoclonal antibodies (6, 91, 98) offer increased sensitivity for immunological detection of protists due to the higher specificity of monoclonal antibodies relative to that of polyclonal antisera.

Quantitative real-time PCR is a useful approach for the identification and enumeration of individual species. This method uses the rate of accumulation of amplified target DNA during the PCR to estimate the copy number of the target in the original sample. Copy number is then related empirically to target abundance in the sample. High-throughput analysis, relatively fast processing time, and sensitivity with a dynamic range of several orders of magnitude over which accurate estimates of target number can be determined make this method ideally suited for microbial ecology studies. Recently, quantitative PCR was used by Zhu et al. (134) to determine the rRNA gene copies in several microalgal strains and in samples from the coastal Mediterranean Sea. Furthermore, these authors were able to determine the importance of Mamiellales during the winter Mediterranean Sea plankton bloom. Distributional studies of parasitic species and harmful bloom-forming algae (13, 26, 45, 62), implementing this method, have also been described.

It is necessary to consider that nonviable cells can be detected by DNA amplification. Although the isolation of quality RNA from natural environments can be problematic because mRNA transcripts are more labile than DNA, the presence of specific transcripts can be used as a proxy for cell viability. Coyne and Cary (25) used reverse transcription-PCR to detect viable cysts of the toxic dinoflagellate *Pfiesteria piscicida* by using cytochrome *c* oxidase subunit 1 (COX), cytochrome *b* (COB), and two highly expressed transcripts, Tags 343 and 277.

Phospholipid fatty acid (PLFA) analysis is another method for the characterization of microeukaryote communities. Phospholipids are essential membrane components of living cells. The chemical composition of PLFAs differs among microbial groups in terms of carbon chain length, branch position, saturation, and substitution (123, 133). PLFA analysis can provide information on physiological status, estimates of viable biomass, and taxonomic identity (130). PLFAs turn over rapidly during metabolism and are quickly degraded following cell death (129, 130). To characterize microbial communities associated with sponges from the Sea of Japan, Rod'kina (96) described the fatty acid composition of total lipids including signatures for microeukaryotes. Kuhlman et al. (53) found PLFA markers indicative of nonphysiologically stressed fungi in rock varnish sampled from the Whipple Mountains of California. A current limitation of PLFA analysis for characterization of microbial communities is that the interpretation of observed biomarkers relies on database information derived from cultured organisms. While interpretation of results must be made without knowing the true universality of a given PLFA marker to a particular group of organisms (12, 133), the number of known specific biomarkers is continually growing.

Denaturing gradient gel electrophoresis is a PCR-based method in which the amplified 18S rRNA gene fragments are separated by polyacrylamide gel electrophoresis containing a linearly increasing gradient of denaturant. With this method, fragments with different base pair sequences will migrate to different positions on the gel (85). Denaturing gradient gel electrophoresis has been used for the estimation of phylogenetic richness in the Ross Sea, Antarctica (38); fungi on glass (100); aquatic microeukaryotes (122); marine picoeukaryotes from the Mediterranean Sea (31); and eukaryal assemblages in a multipond solar saltern (17).

Oligonucleotide probes designed to bind to short, complementary sequence regions of the rRNA that are unique to taxa of interest can be labeled with fluorescent dyes so that cells hybridized with these probes may be observed directly by epifluorescence microscopy, confocal laser scanning microscopy, or flow cytometry (4, 29, 61, 108). The use of eukaryote-specific probes for fluorescent in situ hybridization (FISH) has been used for the enumeration of microeukaryotes in marine samples (60) and fungi in acid mine drainage (8) and for seasonal variations of *Micromonas pusilla* in the Western English Channel (89). Lim et al. (60) noted that FISH resulted in higher counts of total nanoplankton in field samples than counts obtained by the 4′,6′-diamidino-2-phenylindole (DAPI) and acridine orange staining methods. Massana et al. (76) used FISH to investigate novel stramenopiles in natural samples and enrichment samples from marine environments. Based on their phylogenetic analysis of stramenopile clones, they designed primers to look specifically at novel stramenopiles and to determine roles for those groups of organisms. In this fashion, these authors were able to demonstrate that the novel stramenopiles in their samples were bacterivorous heterotrophic flagellates.

COMMUNITY COMPARISONS AND DIVERSITY INDICES

In addition to qualitative assessments of communities, it is often informative to attempt quantitative comparisons of community structure by comparing observed phylogenies between sample locations or between experimental treatments.

Recently, methods for such analyses have been described (74, 101, 102, 109) and are now being implemented for analysis of microeukaryotic assemblages (116).

The percent coverage (*C*) of clone libraries can be calculated according to the equation $C = (1 - n_1/N) \times 100$ where n_1 is the number of unique clones and *N* is the total number of clones analyzed (84). The DOTUR (101) program provides a convenient method for the calculation of diversity coverage by clone libraries generated for a specific environment, for the estimation of species richness by using nonparametric estimators such as ACE (abundance-based coverage estimator) and Chao1 (20, 21, 47), for the generation of rarefaction curves, and for the calculation of diversity indices such as the Simpson's index and the Shannon-Weiner index.

Comparisons of multiple 18S rRNA libraries can be achieved with the ∫-LIBSHUFF (102) (www.plantpath.wisc.edu/fac/joh/S-libshuff.html) computer program, which uses the coverage formula of Good (41) to generate homologous and heterologous coverage curves from the rRNA clone libraries. The DNADIST program of PHYLIP (evolution.genetics.washington.edu/phylip.html) using the Jukes-Cantor model for nucleotide substitution is used to generate the distance matrix analyzed by ∫-LIBSHUFF. The ∫-LIBSHUFF program provides a statistical test for the null hypothesis that two 18S rRNA gene libraries are samples of the same community.

The phylogenetic diversities between communities can be compared by using *F* statistics (F_{ST}) (74) calculated by using the ARLEQUIN program (103). The F_{ST} test is used to compare the genetic diversity within each community to the total genetic diversity of the communities combined, using the equation $F_{ST} = (\Theta_T - \Theta_W)/\Theta_T$, where Θ_T is the genetic diversity for all samples and Θ_w is the genetic diversity in each community (74). The statistical significance of F_{ST} is evaluated by randomly assigning sequences to populations and calculating F_{ST} for 1,000 permutations. An F_{ST} value of ~0 indicates that genetic diversity overlaps between the communities being compared (74). As the difference between the populations increases, the F_{ST} value also increases.

FUTURE DIRECTIONS

As more information is gathered on the diversity of microeukaryote assemblages in various environments, key ecological questions will begin to be resolved, such as how environmental perturbations like organic solvent contamination affect microeukaryote populations and ultimately bioremediation. Assessments of autochthonous microeukaryotes in pristine environments and comparisons with shifts in population dynamics upon exposure to contaminants will be vital to predicting the success of remediation procedures. This information will need to be synthesized with our knowledge from prokaryotic communities because microbial eukaryotes impact the abundance of bacteria and archaea in a community and vice versa. A complete picture of community interactions within an environment will be achieved only by utilization of multiple, available methods.

I gratefully acknowledge Ronald L. Crawford for editorial assistance in the preparation of the manuscript.

REFERENCES:

1. **Altschul, S. F., W. Gish, W. Miller, E. W. Myers, and D. J. Lipman.** 1990. Basic local alignment search tool. *J. Mol. Biol.* **215:**403–410.
2. **Amann, R., and W. Ludwig.** 2000. Ribosomal RNA-targeted nucleic acid probes for studies in microbial ecology. *FEMS Microbiol. Rev.* **24:**555–565.
3. **Amann, R., W. Ludwig, and K. Schleifer.** 1995. Phylogenetic identification and in situ detection of individual microbial cells without cultivation. *Microbiol. Rev.* **59:**143–169.
4. **Amann, R. I., B. J. Binder, R. J. Olson, S. W. Chisholm, R. Devereux, and D. A. Stahl.** 1990. Combination of 16S rRNA-targeted oligonucleotide probes with flow cytometry for analyzing mixed microbial populations. *Appl. Environ. Microbiol.* **56:**1919–1925.
5. **Amaral Zettler, L. A., M. A. Messerli, A. D. Laatsch, P. J. S. Smith, and M. L. Sogin.** 2003. From genes to genomes: beyond biodiversity in Spain's Rio Tinto. *Biol. Bull.* **204:**205–209.
6. **Anderson, D. M., B. A. Keafer, D. M. Kulis, R. M. Waters, and R. Nuzzi.** 1993. An immunofluorescent survey of the brown tide chrysophyte *Aureococcus anophagefferans* along the northeast coast of the United States. *J. Plankton Res.* **15:**563–580.
7. **Arndt, H., D. Dietrich, B. Auer, E.-J. Cleven, T. Grafenhan, M. Weitere, and A. P. Mylnikov.** 2000. Functional diversity of heterotrophic flagellates in aquatic ecosystems. *In* B. S. C. Leadbeater and J. C. Green (ed.), *The Flagellates: Unity, Diversity and Evolution.* Taylor and Francis, London, United Kingdom.
8. **Baker, B. J., M. A. Lutz, S. C. Dawson, P. L. Bond, and J. F. Banfield.** 2004. Metabolically active eukaryotic communities in extremely acidic mine drainage. *Appl. Environ. Microbiol.* **70:**6264–6271.
9. **Baldauf, S. L.** 2003. Phylogeny for the faint of heart: a tutorial. *Trends Genet.* **19:**345–351.
10. **Bass, D., and T. Cavalier-Smith.** 2004. Phylum-specific environmental DNA analysis reveals remarkably high global biodiversity of Cercozoa (Protozoa). *Int. J. Syst. Evol. Microbiol.* **54:**2393–2404.
11. **Borneman, J., and R. J. Hartin.** 2000. PCR primers that amplify fungal rRNA genes from environmental samples. *Appl. Environ. Microbiol.* **66:**4356–4360.
12. **Bossio, D. A., and K. M. Scow.** 1998. Impacts of carbon and flooding on soil microbial communities: phospholipid fatty acid profiles and substrate utilisation patterns. *Microb. Ecol.* **35:**265–278.
13. **Bowers, H. A., T. Tengs, H. B. J. Glasgow, J. M. Burkholder, P. A. Rublee, and D. W. Oldach.** 2000. Development of real-time PCR assays for rapid detection of *Pfiesteria piscicida* and related dinoflagellates. *Appl. Environ. Microbiol.* **66:**4641–4648.
14. **Buchan, A., S. Y. Newell, M. Butler, E. J. Biers, J. T. Hollibaugh, and M. A. Moran.** 2003. Dynamics of bacterial and fungal communities on decaying salt marsh grass. *Appl. Environ. Microbiol.* **69:**6676–6687.
15. **Campbell, L., P. Shapiro, and E. Haugen.** 1994. Immunochemical characterization of eukaryotic ultraplankton from the Atlantic and Pacific oceans. *J. Plankton Res.* **16:**35–51.
16. **Caron, D. A., R. J. Gast, E. L. Lim, and M. R. Dennett.** 1999. Protistan community structure: molecular approaches for answering ecological questions. *Hydrobiologia* **401:** 215–227.
17. **Casamayor, E. O., R. Massana, S. Benlloch, L. Ovreas, B. Diez, V. J. Goddard, J. M. Gasol, I. Joint, F. Rodriguez-Valera, and C. Pedros-Alio.** 2002. Changes in archaeal, bacterial and eukaryal assemblages along a salinity gradient by comparison of genetic fingerprinting methods in a multipond solar saltern. *Environ. Microbiol.* **4:**338–348.
18. **Chandler, D. P., J. K. Fredrickson, and F. J. Brockman.** 1997. Effect of PCR template concentration on the com-

position and distribution of total community 16S rDNA clone libraries. *Mol. Ecol.* **6:**475–483.

19. **Chandler, D. P., C. A. Wagnon, and H. Bolton, Jr.** 1998. Reverse transcriptase (RT) inhibition of PCR at low concentrations of template and its implications for quantitative RT-PCR. *Appl. Environ. Microbiol.* **64:**669–677.
20. **Chao, A.** 1987. Estimating the population-size for capture recapture data with unequal catchability. *Biometrics* **43:** 783–791.
21. **Chao, A., and S. M. Lee.** 1992. Estimating the number of classes via sample coverage. *J. Am. Stat. Assoc.* **87:**210–217.
22. **Christner, B. C., B. H. Kvitko, and J. N. Reeve.** 2003. Molecular identification of bacteria and eukarya inhabiting an Antarctic cryoconite hole. *Extremophiles* **7:**1425–1430.
23. **Corinaldesi, C., R. Danovaro, and A. Dell'Anno.** 2005. Simultaneous recovery of extracellular and intracellular DNA suitable for molecular studies from marine sediments. *Appl. Environ. Microbiol.* **71:**46–50.
24. **Countway, P. D., R. J. Gast, P. Savai, and D. A. Caron.** 2005. Protistan diversity estimates based on 18S rDNA from seawater incubations in the Western North Atlantic. *J. Eukaryot. Microbiol.* **52:**95–106.
25. **Coyne, K. J., and S. C. Cary.** 2005. Molecular approaches to the investigation of viable dinoflagellate cysts in natural sediments from estuarine environments. *J. Eukaryot. Microbiol.* **52:**90–94.
26. **Coyne, K. J., D. A. Hutchins, C. E. Hare, and S. C. Cary.** 2001. Assessing temporal and spatial variability in *Pfiesteria piscicida* distributions using molecular probing techniques. *Aquat. Microb. Ecol.* **24:**275–285.
27. **Darling, K. F., D. Kroon, C. M. Wade, and A. J. Leigh Brown.** 1996. Molecular phylogeny of the planktonic Foraminifera. *J. Foraminifer. Res.* **26:**324–330.
28. **Dawson, S. C., and N. R. Pace.** 2002. Novel kingdom-level eukaryotic diversity in anoxic environments. *Proc. Natl. Acad. Sci. USA* **99:**8324–8329.
29. **DeLong, E. F., G. S. Wickham, and N. R. Pace.** 1989. Phylogenetic stains: ribosomal RNA-based probes for the identification of single cells. *Science* **243:**1360–1363.
30. **de Souza, F. A., G. A. Kowalchuk, P. Leeflang, J. A. van Veen, and E. Smit.** 2004. PCR-denaturing gradient gel electrophoresis profiling of inter- and intraspecies 18S rRNA gene sequence heterogeneity is an accurate and sensitive method to assess species diversity of arbuscular mycorrhizal fungi of the genus *Gigaspora*. *Appl. Environ. Microbiol.* **70:**1413–1424.
31. **Diez, B., C. Pedros-Alio, T. L. Marsh, and R. Massana.** 2001. Application of denaturing gradient gel electrophoresis (DGGE) to study the diversity of marine picoeukaryotic assemblages and comparison of DGGE with other molecular techniques. *Appl. Environ. Microbiol.* **67:**2942–2951.
32. **Diez, B., C. Pedros-Alio, and R. Massana.** 2001. Study of genetic diversity of eukaryotic picoplankton in different oceanic regions by small-subunit rRNA gene cloning and sequencing. *Appl. Environ. Microbiol.* **67:**2932–2941.
33. **Elwood, H. J., G. J. Olsen, and M. L. Sogin.** 1985. The small-subunit ribosomal RNA gene sequences from the hypotrichous ciliates *Oxytricha nova* and *Stylonychia pustulata*. *Mol. Biol. Evol.* **2:**399–410.
34. **Erwin, D. P., I. K. Erickson, M. E. Delwiche, F. S. Colwell, J. L. Strap, and R. L. Crawford.** 2005. Diversity of oxygenase genes from methane- and ammonia-oxidizing bacteria in the Eastern Snake River Plain Aquifer. *Appl. Environ. Microbiol.* **71:**2016–2025.
35. **Farrelly, V., F. Rainey, and E. Stackebrandt.** 1995. Effect of genome size and *rrn* gene copy number on PCR amplification of 16S rRNA genes from a mixture of bacterial species. *Appl. Environ. Microbiol.* **61:**2798–2801.
36. **Fuhrman, J. A., D. E. Comeau, A. Hagstrom, and A. M. Chan.** 1988. Extraction from natural planktonic microorganisms of DNA suitable for molecular biological studies. *Appl. Environ. Microbiol.* **54:**1426–1429.
37. **Galand, P. E., S. Saarnio, H. Fritze, and K. Yrjala.** 2002. Depth related diversity of methanogen Archaea in Finnish oligotrophic fen. *FEMS Microbiol. Ecol.* **42:**441–449.
38. **Gast, R. J., M. R. Dennett, and D. A. Caron.** 2004. Characterization of protistan assemblages in the Ross Sea, Antarctica, by denaturing gradient gel electrophoresis. *Appl. Environ. Microbiol.* **70:**2028–2037.
39. **Giovannoni, S. J., E. F. DeLong, G. J. Olsen, and N. R. Pace.** 1988. Phylogenetic group-specific oligodeoxynucleotide probes for identification of single microbial cells. *J. Bacteriol.* **170:**720–726.
40. **Gonzalez, J. M., M. C. Portillo, and C. Saiz-Jimenez.** 2005. Multiple displacement amplification as a pre-polymerase chain reaction (pre-PCR) to process difficult to amplify samples and low copy number sequences from natural environments. *Environ. Microbiol.* **7:**1024–1028.
41. **Good, I. J.** 1953. The population frequency of species and the estimation of population parameters. *Biometrika* **40:**237–264.
42. **Griffiths, R. I., A. S. Whiteley, A. G. O'Donnell, and M. J. Bailey.** 2000. Rapid method for coextraction of DNA and RNA from natural environments for analysis of ribosomal DNA- and rRNA-based microbial community composition. *Appl. Environ. Microbiol.* **66:**5488–5491.
43. **Gruebl, T., M. E. Frischer, M. Sheppard, M. Neumann, A. N. Maurer, and R. F. Lee.** 2002. Development of an 18S rRNA gene targeted PCR-based diagnostic for the blue crab parasite *Hematodinium* sp. *Dis. Aquat. Org.* **49:** 61–70.
44. **Hackl, E., S. Zechmeister-Boltenstern, L. Bodrossy, and A. Sessitsch.** 2004. Comparison of diversities and compositions of bacterial populations inhabiting natural forest soils. *Appl. Environ. Microbiol.* **70:**5057–5065.
45. **Hermsen, C. C., D. S. C. Telgt, E. H. P. Linders, L. Van De Locht, W. M. C. Eling, E. Mensink, and R. W. Sauerwein.** 2001. Detection of *Plasmodium falciparum* malaria parasites in vivo by real-time quantitative PCR. *Mol. Biochem. Parasitol.* **118:**247–251.
46. **Huber, T., G. Faulkner, and P. Hugenholtz.** 2004. Bellerophon: a program to detect chimeric sequences in multiple sequence alignments. *Bioinformatics* **20:**2317–2319.
47. **Hughes, J. B., J. J. Hellmann, T. H. Ricketts, and B. J. M. Bohannan.** 2001. Counting the uncountable: statistical approaches to estimating microbial diversity. *Appl. Environ. Microbiol.* **67:**4399–4406.
48. **Hunt, J., L. Boddy, P. F. Randerson, and H. J. Rogers.** 2004. An evaluation of 18S rDNA approaches for the study of fungal diversity in grassland soils. *Microb. Ecol.* **47:**385–395.
49. **Karnati, S. K. R., Z. Yu, J. T. Sylvester, B. A. Dehority, M. Morrison, and J. L. Firkins.** 2003. Technical note: specific PCR amplification of protozoal 18S rDNA sequences from DNA extracted from ruminal samples of cows. *J. Anim. Sci.* **81:**812–815.
50. **Kinner, N. E., R. W. Harvey, D. M. Shay, D. W. Metge, and A. Warren.** 2002. Field evidence for a protistan role in an organically-contaminated aquifer. *Environ. Sci. Technol.* **36:**4312–4318.
51. **Kitts, C. L.** 2001. Terminal restriction fragment patterns: a tool for comparing microbial communities and assessing community dynamics. *Curr. Issues Intest. Microbiol.* **2001:**17–25.
52. **Koide, R. T.** 2005. Nucleic acid isolation from ecological samples—fungal associations, mycorrhizae. *Methods Enzymol.* **395:**58–72.

53. **Kuhlman, K. R., W. G. Fusco, M. T. La Duc, L. B. Allenbach, C. L. Ball, G. M. Kuhlman, R. C. Anderson, I. K. Erickson, T. Stuecker, J. Benardini, J. L. Strap, and R. L. Crawford.** 2006. Diversity of microorganisms within rock varnish in the Whipple Mountains, California. *Appl. Environ. Microbiol.* **72:**1708–1715.
54. **Kumar, Y., R. Westram, S. Behrens, B. Fuchs, F. O. Glöckner, R. Amann, H. Meier, and W. Ludwig.** 2005. Graphical representation of ribosomal RNA probe accessibility data using ARB software package. *BMC Bioinformatics* **6:**61.
55. **Lane, D. J.** 1991. 16S/23S rRNA sequencing, p. 115–175. *In* E. Stackebrandt and M. Goodfellow (ed.), *Nucleic Acid Techniques in Bacterial Systematics*. John Wiley and Sons, New York, N.Y.
56. **Lange, M., L. Guillou, D. Vaulot, N. Simon, R. I. Amann, W. Ludwig, and L. K. Medlin.** 1996. Identification of the class Prymnesiophyceae and the genus *Phaeocyctis* with ribosomal RNA-targeted nucleic acid probes detected by flow cytometry. *J. Phycol.* **32:**858–868.
57. **Lawley, B., S. Ripley, P. Bridge, and P. Convey.** 2004. Molecular analysis of geographic patterns of eukaryotic diversity in Antarctic soils. *Appl. Environ. Microbiol.* **70:** 5963–5972.
58. **Lefranc, M., A. Thenot, C. Lepere, and D. Debroas.** 2005. Genetic diversity of small eukaryotes in lakes differing by their trophic status. *Appl. Environ. Microbiol.* **71:** 5935–5942.
59. **Liesack, W., H. Weyland, and E. Stackebrandt.** 1991. Potential risks of gene amplification by PCR as determined by 16S rDNA analysis of a mixed-culture of strict barophilic bacteria. *Microb. Ecol.* **21:**191–198.
60. **Lim, E., D. Caron, and E. Delong.** 1996. Development and field application of a quantitative method for examining natural assemblages of protists with oligonucleotide probes. *Appl. Environ. Microbiol.* **62:**1416–1423.
61. **Lim, E. L., L. A. Amaral, D. A. Caron, and E. F. DeLong.** 1993. Application of rRNA-based probes for observing marine nanoplanktonic protists. *Appl. Environ. Microbiol.* **59:**1647–1655.
62. **Lin, M. H., T. C. Chen, T. T. Kuo, C. C. Tseng, and C. P. Tseng.** 2000. Real-time PCR for quantitative detection of *Toxoplasma gondii*. *J. Clin. Microbiol.* **38:**4121–4125.
63. **Lin, S., and E. J. Carpenter.** 1996. An empirical protocol for whole-cell immunofluorescence of marine phytoplankton. *J. Phycol.* **32:**1083–1094.
64. **Lio, P., and N. Goldman.** 1998. Models of molecular evolution and phylogeny. *Genome Res.* **8:**1233–1244.
65. **López-Garcia, P., H. Philippe, F. Gail, and D. Moreira.** 2003. Autochthonous eukaryotic diversity in hydrothermal sediment and experimental microcolonizers at the Mid-Atlantic Ridge. *Proc. Natl. Acad. Sci. USA* **100:** 697–702.
66. **López-Garcia, P., F. Rodriguez-Valera, C. Pedros-Alio, and D. Moreira.** 2001. Unexpected diversity of small eukaryotes in deep-sea Antarctic plankton. *Nature* **409:**603–607.
67. **Lord, N. S., C. W. Kaplan, P. Shank, C. L. Kitts, and S. L. Elrod.** 2002. Assessment of fungal diversity using terminal restriction fragment (TRF) pattern analysis: comparison of 18S and ITS ribosomal regions. *FEMS Microbiol. Ecol.* **42:**327–337.
68. **Loy, A., M. Horn, and M. Wagner.** 2003. probeBase: an online resource for rRNA-targeted oligonucleotide probes. *Nucleic Acids Res.* **31:**514–516.
69. **Maidak, B. L., J. R. Cole, T. G. Lilburn, C. T. Parker, Jr., P. R. Saxman, R. J. Farris, G. M. Garrity, G. J. Olsen, T. M. Schmidt, and J. M. Tiedje.** 2001. The RDP-II (Ribosomal Database Project). *Nucleic Acids Res.* **29:** 173–174.
70. **Mallatt, J. M., J. R. Garey, and J. W. Shultz.** 2004. Ecdysozoan phylogeny and Bayesian inference: first use of nearly complete 28S and 18S rRNA gene sequences to classify the arthropods and their kin. *Mol. Phylogenet. Evol.* **31:**178–191.
71. **Marsh, T. L.** 2005. Culture-independent microbial community analysis with terminal restriction fragment length polymorphism. *Methods Enzymol.* **397:**308–329.
72. **Marsh, T. L., W. T. Liu, L. J. Forney, and H. Cheng.** 1998. Beginning a molecular analysis of the eukaryal community in activated sludge. *Water. Sci. Technol.* **37:**455–460.
73. **Marsh, T. L., P. Saxman, J. Cole, and J. Tiedje.** 2000. Terminal restriction fragment length polymorphism analysis program, a web-based research tool for microbial community analysis. *Appl. Environ. Microbiol.* **66:**3616–3620.
74. **Martin, A. P.** 2002. Phylogenetic approaches for describing and comparing the diversity of microbial communities. *Appl. Environ. Microbiol.* **68:**3673–3682.
75. **Massana, R., J. Castresana, V. Balague, L. Guillou, K. Romari, A. Groisillier, K. Valentin, and C. Pedros-Alio.** 2004. Phylogenetic and ecological analysis of novel marine stramenopiles. *Appl. Environ. Microbiol.* **70:**3528–3534.
76. **Massana, R., L. Guillou, B. Diez, and C. Pedros-Alio.** 2002. Unveiling the organisms behind novel eukaryotic ribosomal DNA sequences from the ocean. *Appl. Environ. Microbiol.* **68:**4554–4558.
77. **Medlin, L., H. J. Elwood, S. Stickel, and M. L. Sogin.** 1988. The characterization of enzymatically amplified eukaryotic 16S-like RNA-coding regions. *Gene* **71:**491–499.
78. **Mendoza, H., V. López-Rodas, S. González-Gil, A. Aguilera, and E. Costas.** 1995. Use of polyclonal antisera and blocking of the antibodies in the identification of marine dinoflagellates: species-specific and clone-specific antisera angainst *Gymnodinium* and *Alexandrium*. *J. Exp. Mar. Biol. Ecol.* **186:**103–115.
79. **Miller, D. N., J. E. Bryant, E. L. Madsen, and W. C. Ghiorse.** 1999. Evaluation and optimization of DNA extraction and purification procedures for soil and sediment samples. *Appl. Environ. Microbiol.* **65:**4715-4724.
80. **Moon-van der Staay, S. Y., R. De Wachter, and D. Vaulot.** 2001. Oceanic 18S rDNA sequences from picoplankton reveal unsuspected eukaryotic diversity. *Nature* **409:**607–610.
81. **Moon-van der Staay, S. Y., G. W. M. van der Staay, L. Guillou, D. Vaulot, H. Claustre, and L. K. Medlin.** 2000. Abundance and diversity of prymnesiophytes in the picoplankton community from the equatorial Pacific Ocean inferred from 18S rDNA sequences. *Limnol. Oceanogr.* **45:**98–109.
82. **More, M., J. Herrick, M. Silva, W. Ghiorse, and E. Madsen.** 1994. Quantitative cell lysis of indigenous microorganisms and rapid extraction of microbial DNA from sediment. *Appl. Environ. Microbiol.* **60:**1572–1580.
83. **Moyer, C. L., J. M. Tiedje, F. C. Dobbs, and D. M. Karl.** 1996. A computer-simulated restriction fragment length polymorphism analysis of bacterial small-subunit rRNA genes: efficacy of selected tetrameric restriction enzymes for studies of microbial diversity in nature. *Appl. Environ. Microbiol.* **62:**2501–2507.
84. **Mullins, T. D., T. B. Britschgi, R. I. Krest, and S. J. Giovannoni.** 1995. Genetic comparisons reveal the same unknown bacterial lineages in Atlantic and Pacific bacterioplankton communities. *Limnol. Oceanogr.* **39:**148–158.

85. **Muyzer, G., E. C. de Waal, and A. G. Uitterlinden.** 1993. Profiling of complex microbial populations by denaturing gradient gel electrophoresis analysis of polymerase chain reaction-amplified genes coding for 16S rRNA. *Appl. Environ. Microbiol.* **59:**695–700.
86. **Nei, M., and W. H. Li.** 1979. Mathematical model for studying genetic variation in terms of restriction endonucleases. *Proc. Natl. Acad. Sci. USA* **76:**5269–5273.
87 **Nikolcheva, L., and F. Barlocher.** 2004. Taxon-specific fungal primers reveal unexpectedly high diversity during leaf decomposition in a stream. *Mycol. Prog.* **3:**41–50.
88. **Nikolcheva, L., A. Cockshutt, and F. Barlocher.** 2003. Determining diversity of freshwater fungi on decaying leaves: comparison of traditional and molecular approaches. *Appl. Environ. Microbiol.* **69:**2548–2554.
89. **Not, F., M. Latasa, D. Marie, T. Cariou, D. Vaulot, and N. Simon.** 2004. A single species, *Micromonas pusilla* (Prasinophyceae), dominates the eukaryotic picoplankton in the Western English Channel. *Appl. Environ. Microbiol.* **70:**4064–4072.
90. **Ogram, A. V., G. S. Sayler, and T. T. Barkay.** 1988. The extraction and purification of microbial DNA from sediments. *J. Microbiol. Methods* **7:**57–66.
91. **Okazaki, K., T. Iwaoka, N. Murakami, K. Ichimi, and S. Montani.** 2001. Production of monoclonal antibody against *Scrippsiella trochoidea* cysts and its application to analysis during cyst formation and enzyme-linked immunosorbent assay. *Biosci. Biotechnol. Biochem.* **65:** 470–473.
92. **Palumbo, A. V., L. S. Fisher, M. Z. Martin, Z. K. Yang, J. R. Tarver, S. D. Wullschleger, and W. L. Daniels.** 2004. Application of emerging tools and techniques for measuring carbon and microbial communities in reclaimed mine soils. *Environ. Manag.* **33:**S518–S527.
93. **Picard, C., C. Ponsonnet, E. Paget, X. Nesme, and P. Simonet.** 1992. Detection and enumeration of bacteria in soil by direct DNA extraction and polymerase chain reaction. *Appl. Environ. Microbiol.* **58:**2717–2722.
94. **Polz, M. F., and C. M. Cavannaugh.** 1998. Bias in template-to-product ratios in multitemplate PCR. *Appl. Environ. Microbiol.* **64:**3724–3730.
95. **Richards, T. A., A. A. Veprtiskiy, D. E. Gouliamova, and S. A. Nierzwicki-Bauer.** 2005. The molecular diversity of freshwater picoeukaryotes from an oligotrophic lake reveals diverse, distinctive and globally dispersed lineages. *Environ. Microbiol.* **7:**1413–1425.
96. **Rod'kina, S. A.** 2003. Comparative characterization of the microbial community in two species of sponges from Sea of Japan using fatty acid markers. *Russ. J. Mar. Biol.* **29:**255–258.
97. **Romari, K., and D. Vaulot.** 2004. Composition and temporal variability of picoeukaryote communities at a coastal site of the English Channel from 18S rDNA sequences. *Limnol. Oceanogr.* **49:**784–798.
98. **Romestand, B., J. Torreilles, and P. Roch.** 2001. Production of monoclonal antibodies against the protozoa, *Perkinsus marinus*: estimation of parasite multiplication in vitro. *Aquat. Living Resour.* **14:**351–357.
99. **Savin, M. C., J. L. Martin, M. LeGresley, M. Giewat, and J. Rooney-Varga.** 2004. Plankton diversity in the Bay of Fundy as measured by morphological and molecular methods. *Microb. Ecol.* **48:**51–65.
100. **Schabereiter-Gurtner, C., G. Pinar, W. Lubitz, and S. Rolleke.** 2001. Analysis of fungal communities on historical church window glass by denaturing gradient gel electrophoresis and phylogenetic 18S rDNA sequence analysis. *J. Microbiol. Methods* **47:**345–354.
101. **Schloss, P. D., and J. Handelsman.** 2005. Introducing DOTUR, a computer program for defining operational taxonomic units and estimating species richness. *Appl. Environ. Microbiol.* **71:**1501–1506.
102. **Schloss, P. D., B. R. Larget, and J. Handelsman.** 2004. Integration of microbial ecology and statistics: a test to compare gene libraries. *Appl. Environ. Microbiol.* **70:** 5485–5492.
103. **Schneider, S., D. Roessli, and L. Excoffier.** 2000. Arlequin ver. 2.000: a software for population genetics data analysis. Genetics and Biometry Laboratory, University of Geneva, Geneva, Switzerland.
104. **Shapiro, L. P., L. Campbell, and E. M. Haugen.** 1989. Immunochemical recognition of phytoplankton species. *Mar. Ecol. Prog. Ser.* **57:**219–224.
105. **Sherr, E. B., and B. F. Sherr.** 2002. Significance of predation by protists in aquatic microbial food webs. *Antonie Leeuwenhoek* **81:**293–308.
106. **Simon, N., J. Brenner, B. Edvardsen, and L. Medlin.** 1997. The identification of *Chrysochromulina* and *Prymnesium* species (Haptophyta, Prymnesiophyceae) using fluorescent or chemiluminescent oligonucleotide probes: a means for improving studies on toxic algae. *Eur. J. Phycol.* **32:**393–401.
107. **Simon, N., L. Campbell, E. Ornolfsdottir, R. Groben, L. Guillou, L. Lange, and L. K. Medlin.** 2000. Oligonucleotide probes for the identification of three algal groups by dot blot and fluorescent whole-cell hybridization. *J. Eukaryot. Microbiol.* **47:**76–84.
108. **Simon, N., N. LeBot, D. Marie, F. Partensky, and D. Vaulot.** 1995. Fluorescent in situ hybridization with rRNA-targeted oligonucleotide probes to identify small phytoplankton by flow cytometry. *Appl. Environ. Microbiol.* **61:**2506–2513.
109. **Singleton, D. R., M. A. Furlong, S. L. Rathbun, and W. B. Whitman.** 2001. Quantitative comparisons of 16S rRNA gene sequence libraries from environmental samples. *Appl. Environ. Microbiol.* **67:**4374–4376.
110. **Slapeta, J., D. Moreira, and P. Lopez-Garcia.** 2005. The extent of protist diversity: insights from molecular ecology of freshwater eukaryotes. *Proc. R. Soc. Ser. B.* **272:** 2073–2081.
111. **Smit, E., P. Leeflang, B. Glandorf, J. D. van Elsas, and K. Wernars.** 1999. Analysis of fungal diversity in the wheat rhizosphere by sequencing of cloned PCR-amplified genes encoding 18S rRNA and temperature gradient gel electrophoresis. *Appl. Environ. Microbiol.* **65:**2614–2621.
112. **Sogin, M. L.** 1990. Amplification of ribosomal RNA genes for molecular evolution studies, p. 307–322. *In* M. A. Innis, D. H. Gelfand, J. J. Sninsky, and T. J. White (ed.), *PCR Protocols: a Guide to Methods and Applications*. Academic Press, Inc., New York, N.Y.
113. **Somerville, C. C., I. T. Knight, W. L. Straube, and R. R. Colwell.** 1989. Simple, rapid method for direct isolation of nucleic acids from aquatic environments. *Appl. Environ. Microbiol.* **55:**548–554.
114. **Stoeck, T., and S. Epstein.** 2003. Novel eukaryotic lineages inferred from small-subunit rRNA analyses of oxygen-depleted marine environments. *Appl. Environ. Microbiol.* **69:**2657–2663.
115. **Stoeck, T., G. T. Taylor, and S. S. Epstein.** 2003. Novel eukaryotes from the permanently anoxic Cariaco Basin (Caribbean Sea). *Appl. Environ. Microbiol.* **69:**5656–5663.
116. **Strap, J. L., F. S. Colwell, and R. L. Crawford.** Eukaryotic diversity in planktonic and biofilm populations within the subsurface of the Snake River Plain Aquifer. Submitted for publication.
117. **Suzuki, M., and S. Giovannoni.** 1996. Bias caused by template annealing in the amplification of mixtures of 16S rRNA genes by PCR. *Appl. Environ. Microbiol.* **62:**625–630.

118. **Takishita, K., H. Miyake, M. Kawato, and T. Maruyama.** 2005. Genetic diversity of microbial eukaryotes in anoxic sediment around fumaroles on a submarine cladera floor based on the small-subunit rDNA phylogeny. *Extremophiles* **9:**185–196.
119. **Tebbe, C. C., and W. Vahjen.** 1993. Interference of humic acids and DNA extracted directly from soil in detection and transformation of recombinant DNA from bacteria and a yeast. *Appl. Environ. Microbiol.* **59:**2657–2665.
120. **Thompson, J. D., T. J. Gibson, F. Plewniak, F. Jeanmougin, and D. G. Higgins.** 1997. The CLUSTAL_X windows interface: flexible strategies for multiple sequence alignment aided by quality analysis tools. *Nucleic Acids Res.* **25:**4876–4882.
121. **Tsai, Y.-L., and B. H. Olson.** 1991. Rapid method for direct extraction of DNA from soil and sediments. *Appl. Environ. Microbiol.* **57:**1070–1074.
122. **van Hannen, E. J., M. P. Agterveld, H. J. Gons, and H. J. Laanbroek.** 1998. Revealing genetic diversity of eukaryotic microorganisms in aquatic environments by denaturing gradient gel electrophoresis. *J. Phycol.* **34:** 206–213.
123. **Vestal, J. R., and D. C. White.** 1989. Lipid analysis in microbial ecology. *Bioscience* **39:**535–541.
124. **Vilgalys, R., and M. Hester.** 1990. Rapid genetic identification and mapping of enzymatically amplifed ribosomal DNA from several *Cryptococcus* species. *J. Bacteriol.* **172:**4238–4246.
125. **von Wintzingerode, F., U. B. Gobel, and E. Stackebrandt.** 1997. Determination of microbial diversity in environmental samples: pitfalls of PCR-based rRNA analysis. *FEMS Microbiol. Rev.* **21:**213–229.
126. **Vrieling, E. G., and D. M. Anderson.** 1996. Immunofluorescence in phytoplankton research: applications and potential. *J. Phycol.* **32:**1–16.
127. **Weisse, T.** 2002. The significance of inter- and intraspecific variation in bacterivorous and herbivorous protists. *Antonie Leeuwenhoek* **81:**327–341.
128. **Whelan, S., P. Lio, and N. Goldman.** 2001. Molecular phylogenetics: state-of-the-art methods for looking into the past. *Trends Genet.* **17:**262–272.
129. **White, D. C., W. M. Davis, J. S. Nickels, J. D. King, and R. J. Bobbie.** 1979. Determination of the sedimentary microbial biomass by extractable lipid phosphate. *Oecologia* **40:**51–62.
130. **White, D. C., H. C. Pinkart, and D. Ringelberg.** 1997. Biomass measurements: biochemical approaches, p. 91–101. *In* C. J. Hurst, G. R. Knudsen, M. J. McInerney, L. D. Stetzenbach, and M. V. Walter (ed.), *Manual of Environmental Microbiology*. American Society for Microbiology, Washington, D.C.
131. **White, T. J., T. Bruns, S. Lee, and J. Taylor.** 1990. Amplification and direct sequencing of fungal ribosomal RNA genes for phylogenetics, p. 315–322. *In* M. A. Innis, D. H. Gelfand, J. J. Sninsky, and T. J. White (ed.), *PCR Protocols: a Guide to Methods and Applications*. Academic Press, Inc., New York, N.Y.
132. **Willerslev, E., A. J. Hansen, B. Christensen, J. P. Steffensen, and P. Arctander.** 1999. Diversity of Holocene life forms in fossil glacier ice. *Proc. Natl. Acad. Sci. USA* **96:**8017–8021.
133. **Zelles, L.** 1999. Fatty acid patterns of phospholipids and lipopolysaccharides in the characterisation of microbial communities in soil: a review. *Biol. Fertil. Soils* **29:**111–129.
134. **Zhu, F., R. Massana, F. Not, D. Marie, and D. Vaulot.** 2005. Mapping of picoeucaryotes in marine ecosystems with quantitative PCR of the 18S rRNA gene. *FEMS Microbiol. Ecol.* **52:**79–92.

Methods Useful in Assessing Biological and Chemical Activity of Low-Molecular-Weight Brown Rot Fungal Metabolites

JODY JELLISON, BARRY GOODELL, AND YUHUI QIAN

91

Brown rot fungi, a subgroup of the basidiomycetes, are characterized by their unique biochemical and physiological systems associated with mechanisms used to colonize and metabolize wood and other lignocellulolytic substrates. The brown rot fungi are essential for biomass and carbon cycling in coniferous ecosystems and are also associated with the most destructive forms of degradation of wooden structures in the Northern Hemisphere. Unlike white rot fungi, which have been extensively studied for their ability to degrade hemicellulose, cellulose, and lignin, the brown rot fungi preferentially attack hemicellulose and cellulose, leaving a residue enriched in modified lignin structures. Brown rot fungi have recently come under additional scrutiny for their potential utilization in the bioremediation of pollutants and possible industrial applications within the biocomposites and pulp and paper industries (14, 31, 69).

The way in which brown rot fungi attack the wood cell wall differs both spatially and chemically from attack by white rot fungi. White rot fungi are characterized by progressive, enzymatic degradation of the wood cell wall, with concomitant attack of all major wood components from the cell lumen outward. Brown rot fungi are associated with a rapid preferential attack on hemicellulose and cellulose throughout the wood cell wall. The spatial parameters of the wood cell wall, the fungus, and the type and size of fungal metabolites produced in this attack preclude the early action of enzymes in the decay process within the wood cell wall because the nonmodified wood cell wall has pore sizes too small to allow enzyme penetration (21). The current mechanisms postulated to be involved in initial brown rot decay processes focus on a modified Fenton chemistry involving a phenolate or catechol-hydroquinone biochelator-driven reduction of the ferric iron present in wood to ferrous iron, which can then interact with hydrogen peroxide to generate short-lived but highly reactive hydroxyl radicals (Fig. 1). These radicals are capable of direct attack on the wood cell wall. The compounds 2,5-dimethoxy-1,4-benzoquinone (2,5-DMBQ) and 4,5-dimethoxy-1,2-benzoquinone (4,5-DMBQ) and their reduced forms have been identified as primary biochelators produced by the brown rot fungus *Gloeophyllum trabeum* that promote Fenton reactions (34, 49). Alternately, DMBQ radicals themselves could enter the wood cell wall to initiate oxidative reactions.

One equivalent of phenolate can reduce multiple equivalents of iron, and two theories have been proposed to explain this nonstoichiometric reduction. One is partial or complete mineralization of phenolic substances (50). The other is regeneration of phenolate biochelators via an enzyme-associated redox cycle (34, 49). Reductase enzymes capable of participating in regenerative processes have been isolated (8, 51), but because the oxidative reactions must occur within the wood cell wall and the enzymes such as reductases cannot enter the wood cell wall until significant modification has occurred, spatial constraints associated with their participation need to be addressed before this latter hypothesis can be further developed.

Nonenzymatic extracellular brown rot metabolic products putatively involved in basic physiology and degradative processes also include oxalic acid. Oxalic acid is produced in copious amounts by many fungi and is associated with the solubilization of ferric iron from iron oxides and with pH control and acidification of the environment adjacent to the fungal hyphae. Multiple biochelators, including the more familiar hydroxamate-type siderophores generally considered to be responsible for iron acquisition for general metabolism, are also present extracellularly. Although the accepted role of siderophores and biochelators in microorganisms is to sequester metals useful in microbial metabolism, an additional role of these catechol-hydroquinone compounds in biodegradation occurs in the brown rot fungi. This role is enabled in brown rot degradation because of the differential pH change (from fungal hyphae to wood cell wall), the presence of oxalate, the production of H_2O_2, and other microenvironmental changes that work in concert to favor sequestration of metal in proximity to the fungal hyphae with reduction of the metal to promote modified Fenton reactions in the wood cell wall (23).

At later decay stages, enzymes including hemicellulases and cellulolytic enzymes are also involved in the degradative processes. Brown rot fungi are known to produce β-glucosidase and endoglucanase, and a recent report has suggested that cellobiohydrolases may also be produced by *Gloeophyllum trabeum* (9). Ligninases and Mn peroxidases, which are found in white rot fungi, have not been identified in the brown rot fungi, but specific lignin modifications, including an increase in phenolic hydroxyls, are observed during the brown rot decay process.

The purpose of this chapter is to provide a brief overview of selected techniques used to detect, quantify, and evaluate the activity of low-molecular-weight metabolites produced

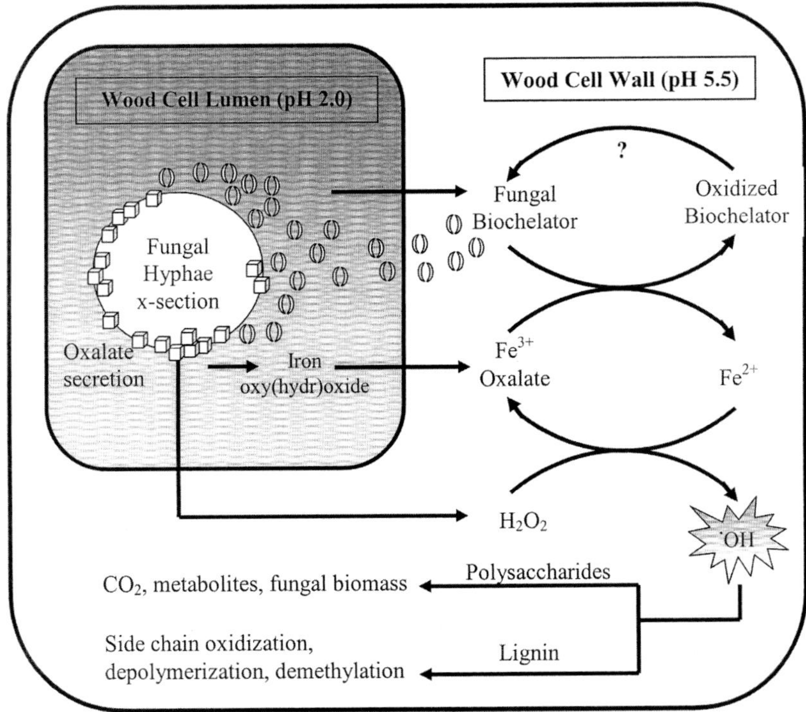

FIGURE 1 Proposed scheme for the participation of nonenzymatic metabolites in brown rot biodegradation of wood.

by the brown rot fungi. Although enzymes are clearly involved in brown rot biodegradative processes, they will not be specifically addressed, as substantial treatment of enzyme quantification, characterization, and genetic analysis is beyond the scope of this chapter. The techniques we consider and the references we include are not meant to be exhaustive but, rather, an introduction to possible approaches and entry into the relevant literature. The role of low-molecular-weight fungal metabolites in the brown rot decay process and their potential use in bioremediation and industrial processes are also briefly considered.

FUNGAL CULTURE

Brown rot fungi can be easily cultured on agar plates using standard media such as potato dextrose agar or malt extract agar (28, 29). Long-term growth and storage of these fungi on such artificial media may, however, cause loss of degradative capabilities, and it is recommended that laboratory isolates be periodically cultured on the natural wood substrate and subsequently reisolated (46). Medium-term storage can be achieved on agar medium slants into which a sterile thin veneer strip or other wood source has been partially inserted to allow the fungus to colonize the native substrate after the simple carbon source present in the agar has been exhausted. Standard freezing and lyophilization techniques (35) are an option for some species.

Many metabolite and enzyme studies use sterile liquid culture medium as their fungal growth substrate. Most brown rot fungi grow satisfactorily on Highley's medium (27), which can be supplemented as needed with various carbon sources. When recalcitrant carbon sources are used, it is often useful to add a minimal amount of glucose to the initial culture medium to ensure culture establishment. Many brown rot metabolites are inducible and repressible, and variables such as nitrogen concentration and source, transition metal availability, and complexity of the carbon source can be crucial. Metabolite production can be very dependent upon fungal species, isolate, environment, nutrient status, and culture age. Some brown rot fungi do not respond well to agitation, and for them, stationary culture is preferable.

Growth of the fungi on native wood substrates can be achieved by using a modified soil block assay. Soil block assays often extend 12 to 16 weeks and consist of chambers containing sterile sample wood blocks placed on top of fungus-inoculated feeder strips incubated on the surface of moist sterilized soil (4). Soil block assays present difficulties for metabolite purification but provide a more realistic substrate for some physiological and chemical analyses. The progress of wood decay can be evaluated on the basis of wood mass loss. Fungal biomass can be estimated on the basis of the quantification of ergosterol, a unique sterol present in fungal membranes (70). Establishment of cultures on recalcitrant or toxic substrates, for example, as part of remediation studies, can be problematic. Initial establishment is often aided by growing the fungal inoculum on grains, wood chips, or other carbon sources prior to addition of the inoculum to the substrate being remediated.

METABOLITE PURIFICATION AND QUANTIFICATION

Low-molecular-weight fungal metabolites have been isolated with varying success from both fungus-colonized wood and fungal cultures. In some cases, purification, quantification, and characterization of low-molecular-weight fungal metabolites can be done directly from fungus-colonized wood. This approach is useful because of the inducibility and repressibility of many extracellular brown rot products;

however, problems with this approach unfortunately include the difficulties of extracting and purifying fungal metabolites from the complex and inhibitory wood matrix. Difficulties in distinguishing products of fungal origin from lignin breakdown products associated with brown rot colonization may also occur. These problems are avoided when low-molecular-weight extracellular metabolites are isolated from the less complex liquid cultures. This can be done by first removing fungal mycelium from the culture medium via vacuum filtration through Whatman no. 2 filter paper, subjection of the filtrates to Amicon 400 ultrafiltration with a YM1 1,000-molecular-weight-cutoff membrane, and saving of the low-molecular-weight fraction (33). This protocol does not harvest the extracellular metabolites localized in the fungal glycan sheath. If required, these metabolites can be partially obtained by solubilization of the sheath via rinsing in 5× phosphate-buffered saline buffer. This procedure is also limited in that metabolites such as oxalate (which can exist in aqueous insoluble crystalline form) may be underrepresented unless a specific acid extraction step is utilized. Preparations may also be contaminated by fungal cellular debris, particularly in older cultures in which hyphal lysis may have occurred to some degree.

Subsequent chemical fractionation of the low-molecular-weight components in selected organic solvents such as acidified ethyl acetate can be useful. We will look briefly at techniques used in the purification, quantification, and characterization of selected types of low-molecular-weight metabolites produced by brown rot fungi. These metabolites have been selected on the basis of their proposed involvement in nonenzymatic brown rot degradative processes. Because these metabolites are not primary gene products, genetic analysis of their induction and production is not discussed in this chapter. Metal quantification is also briefly addressed because of the importance of iron in Fenton chemistry.

Oxalate

In fungal cultures, oxalate can exist in both free acid form and the less bioavailable crystalline form, calcium oxalate. These forms exist in proportions at least partly determined by the microenvironment. Oxalic acid is an organic acid with pK_as of 1.2 (pK_1) and 4.2 (pK_2). Oxalate is also a nonreducing iron chelator (except in the presence of light [26, 59]) capable of solubilizing iron oxides. Oxalate quantification must include measurement and partitioning of both water-soluble and acid-soluble fractions. Scanning electron microscopy permits direct observation of calcium oxalate crystals both adherent to the fungal hyphae and free in the medium or other substrate (10). Energy X-ray microanalysis may be useful for the spectral identification of calcium or other crystal constituents (11).

Spectrophotometric quantification using oxalate "kits" can lead to spurious results, and some high-performance liquid chromatography (HPLC) protocols yield data confounded by interfering peaks. A reverse-phase HPLC protocol to resolve organic acid anions, including oxalate, from fungal medium background has been developed and can be used to obtain quantitative estimations for both soluble and total oxalate (57). Note that environmental pH and oxalate concentration have both been shown to influence hydroxyl radical formation by hydroquinones (64). The pH can be independently determined and is often correlated with, but not totally dependent upon, oxalate concentration. Calcium levels can be determined if necessary by inductively coupled plasma analysis (see below).

Biochelators

Phenolate biochelators were initially partially purified from wood-degrading fungi by use of an ethyl acetate extraction of culture filtrate followed by thin-layer chromatography. The resultant iron-binding band was used to make polyclonal antibodies which were subsequently used to probe degrading wood and to establish and localize in situ production of biochelators by colonizing brown rot hyphae (33). Traditional chemical methods such as HPLC have been used to separate and identify numerous iron-binding, iron-reducing, extracellular brown rot metabolites. Semiquantitative monitoring of iron binding and iron reduction can be achieved with chrome azural-S-based and ferrozine-based colorometric assays, respectively (15, 24, 48). Structural characterization using HPLC-mass spectrometry, gas chromatography-mass spectrometry, nuclear magnetic resonance, and other methods led to the identification of a number of phenolate-type compounds, including hydroxyphenylacetic acid derivatives, hydroxylated benzoic acids, hydroxylated benzene derivatives, and dihydroxyphenyl pentanediol (24, 49). Metabolites with iron-sequestering activity have been identified, including 2,5-dimethoxyhydroquinone and 4,5-dimethoxycatechol which have been isolated from cultures of the brown rot fungus *G. trabeum* (41, 49). Although the identification of the structures of these compounds is important to the elucidation of mechanisms, more critical is the understanding of how these low-molecular-weight compounds function with metals to mediate free radical reactions promoting brown rot degradation (24, 33) and how they move spatially within the wood cell wall (33).

Metals

Iron and other metals can be quantified from wood and other substrate matrices by using ashing, acid extraction, and inductively coupled plasma analysis (47). Iron exists, however, not in the active ferrous form but as ferric iron in oxygenated environments, often as an oxide. Solubilization and subsequent microbial reduction of iron can be quantified by using the spectrophotometric ferrozine assay for the measurement of ferrous iron (59). Reduced iron in solution can be quantified in this assay on the basis of absorbance at 562 nm. These assays have been used extensively to demonstrate the effect of pH and other variables on a chelator-based Fenton system (24, 56).

Hydrogen Peroxide and Hydroxyl Radical Production

Hydrogen peroxide is produced as a metabolite by the brown rot fungi and can be quantified by using iron (or other transition metal) oxidation assays, reaction with potassium titanium oxalate, or enzyme detection methods (30, 54). Another commonly used hydrogen peroxide quantification technique is iodometric titration (37). The ultimate product of biochelator-mediated Fenton reactions is the hydroxyl radical, generated by the reaction of hydrogen peroxide and ferrous iron. The hydroxyl radical is believed to act as a small, nonenzymatic agent involved in the brown rot decay of wood (24). However, the action of other reactive oxygen species generated in these systems also may occur, and the direct participation of hydroxybenzene-derived radicals to initiate oxidative reactions in the wood cell wall has not been ruled out. Detection of hydroxyl radicals is limited by their extremely short half-life and high level of chemical activity. Tornberg and Olsson (61) used a sensitive and specific fluorescence method to detect the production of hydroxyl radicals by decay fungi. The method involved the hydroxylation of coumarin-3-carboxylic acid to produce 7-hydroxy-coumarin-3-carboxylic acid. Tsai et al. (62) used an ultraweak chemiluminescence analyzer and a low-level chemiluminescence emitter to study the hydroxyl radical-scavenging ability of food constituents. The probe indoxyl-β-glucuronide (IBG) was used. IBG was shown to be insensitive to superoxide radical and hydrogen peroxide and could be quenched by the addition of hydroxyl scavengers to the assay. Rodriguez et al. (56) and Qian et al. (53) used chemiluminescence to study the evolution of hydroxyl radicals by mixtures of hydrogen peroxide, 2,3-dihydroxybenzoic acid (DHBA), oxalate, and iron in an attempt to better understand the radical-based chemistry of brown rot decay systems. Hydroxyl radical formation was monitored by lipid peroxidation by Varela and Tien (64) in their examination of pH and oxalate effect on hydroquinone-mediated hydroxyl formation.

ACTIVITY-BASED ASSAYS

Early work used 2-keto-4-thiomethylbutyric acid (KTBA) as a proxy to study the oxidative processes utilized by the brown rot fungi and helped to establish a one-electron transfer mechanism (7). Subsequent work has focused on methods directly monitoring metal reduction processes and metal biochelation by fungal metabolites (24, 34, 49, 56, 64).

Cyclic Voltammetry

Cyclic voltammetry allows the direct observation of the electrochemical behavior of fungal metabolites and electron transfer characteristics between metals and reducing and nonreducing chelators. Cyclic voltammetry is a widely used technique for acquiring qualitative information about electrochemical reactions and can offer rapid identification of the redox potential of electroactive species and direct evaluation of the effect of various environmental parameters on redox processes (5). Liu et al. (44) studied the interaction between the biochelator model DHBA and copper(II) in aqueous solution. Using cyclic voltammetry, they were able to classify the reaction as one involving the one-electron oxidation of DHBA to its semiquinone radical in solution, followed by an electron transfer reaction involving the oxidation of the semiquinone radical to a quinone at the electrode surface. In the presence of hydrogen peroxide, oxidation of DHBA by copper was increased due to copper(II) regeneration by the action of hydrogen peroxide oxidizing copper(I). This redox cycling also would produce hydroxyl radicals. The reaction mechanisms proposed suggested that hydroxyl production by chelator-mediated Fenton reaction would be favorable under the conditions found in the wood cell wall.

EPR and Spin Trapping

Electron paramagnetic resonance (EPR) or electron spin resonance (ESR) is a valuable technique for monitoring the change in oxidation states of transition metals and can also be used to detect and monitor chelation processes. EPR allows the detection and identification of structures with unpaired electrons. These include metal chelates as well as, through the use of spin trapping compounds, free radical species. Qian et al. (52) and Goodell et al. (24) studied the effect of low-molecular-weight chelators on free radical generation using ESR. Several key steps, including iron chelation and reduction, during early decay were studied. The binding between a catecholate model chelator and Fenton reagents as well as the effect of reaction conditions on free radical generation was studied. The results indicated that the relative amount of chelating compounds produced or present in the reaction can be determined via detection of the characteristic $g = 4.3$ ESR signal. The effects of the chelator-to-iron ratio, the pH, and other reaction parameters on hydroxyl radical generation could also be determined. The data support the hypothesis that superoxide radicals are involved in chelator-mediated Fenton processes. Humar et al. (32) also used EPR and examined changes in EPR spectra of wood impregnated with copper-based preservatives during exposure to several wood-rotting fungi. The results showed that the fungi transform copper(II) sulfate in wood into nonsoluble and therefore nontoxic copper oxalate. The intensity of the reaction depended upon the amount of soluble oxalic acid produced by the fungus.

SUBSTRATE MODIFICATION

Classic wet chemistry methods of cellulose and lignin quantification and characterization will not be reviewed here, as they are well documented in the literature. We will, however, examine selected methods that have been used in the characterization of wood or lignocellulose colonized by brown rot fungi or treated with isolated fungal metabolites. These include cellulose chain length determination, X-ray analysis, molecular beam mass spectroscopy (MBMS) and near infrared spectroscopy (NIR) evaluation of complex substrates, and ^{13}C thermochemolysis characterization of lignin modification.

Cellulose Modification

Gel permeation chromatography (GPC) can be used to monitor changes in alpha cellulose polymerization upon exposure to in vitro fungal metabolite systems. Reaction mixtures are incubated at room temperature for selected times before cellulose is removed by centrifugation, washed, and dried. Phenol isocyanate derivatives are prepared as described by Wood et al. (67) for GPC measurement. The sample is ultimately redissolved in tetrahydrofuran, and a Metacel 5 linear GPC column and diode array detector are used to determine cellulose depolymerization associated with specific treatments (68). Chromatograms generated show cellulose polymer molecular weight distributions and information about the extent and kinetics of cellulose depolymerization.

Wood cellulose exists in both amorphous and crystalline forms. The ability of brown rot fungi to attack and modify the crystalline cellulose fraction is crucial to their success in the rapid biodegradation of wood and other lignocellulosic matrices. Cellulose crystallinity can be determined by various protocols (6, 36, 60, 65, 66). Akerholm et al. (1) used Fourier transform infrared spectroscopy to characterize cellulose crystalline structure. Andersson et al. (2, 3) used X-ray diffraction to examine cellulose crystallinity as a function of woody tissue production in the tree annual ring. Early work in our laboratory suggested that insight into the effect of fungal metabolites on both the percentage of crystalline cellulose as a portion of the whole (amorphous plus crystalline) and the average cellulose crystal micelle dimensions could be obtained by X-ray diffraction analysis (24). Curling et al. (12) conducted seminal studies using an in vitro test method to progressively sample southern pine exposed to monocultures of brown rot fungi. Their results showed that hemicellulose was preferentially degraded in early stages of attack by brown rot fungi, and the early strength loss was highly associated with the loss of the mannan and xylan components.

NIR and MBMS Characterization

NIR is a sensitive spectroscopic method and a useful tool for the analysis of wood characteristics. NIR has been used in many studies of wood, including examinations of modulus of elasticity and ultimate tensile strength (63), heartwood decay resistance (20), thermally modified wood (58), predicted chemical composition of pulp (16), and chemical composition of transgenic aspen (43). NIR is faster than classical wet chemistry and has the potential to give a more complete picture of the multiple components contributing to wood structure and chemistry. For many studies, adequate visible or NIR spectra (350 to 2,500 nm) can be obtained with a field spectrophotometer (FieldSpec Pro—Full Range) and examined by a multivariate PLS analysis as outlined by Kelley et al. (38–40). The quality of the resultant models can be evaluated by comparing the correlation coefficient and root mean square error of calibration and root mean square error of prediction. In the case of brown rot fungus-degraded wood, NIR showed increases in brown color (600 nm), a decrease in carbohydrate-associated hydroxyl vibrations (1,490 and 2,100 nm), and changes in the lignin peak associated with wood hydroxyls and hydrogen-bonded water (1,920 nm) (39).

Pyrolysis mass spectroscopy has been used to study soil organic matter and can also be used to evaluate the chemical composition of wood as it undergoes brown rot biodegradation (45). Pyrolysis MBMS protocols are described by Evans and Milne (13). Kelley et al. (39) used the technique to monitor changes in MBMS spectra with increasing time of exposure to brown rot degradative processes and found that the changes in the spectrum were driven by a loss of carbohydrate masses (masses of 85, 114, and 126), lignin oxidation changes (masses of 180 and 168), and an increase in monomethylated lignin fragments (masses of 123, 138, and 151). An excellent review of both NIR and MBMS methods and appropriate statistical evaluation techniques is given by Kelley et al. (40).

^{13}C-TMAH Thermochemolysis

^{13}C-labeled tetramethylammonium hydroxide (^{13}C-TMAH) thermochemolysis can be used to track microbial modifications of lignin, such as those associated with brown rot biodegradation. This approach differs from other molecular chemolysis protocols in allowing the investigator to characterize both the amount and the location of hydroxylation of the aromatic components during lignin degradation. TMAH depolymerizes beta-O-4 linkages with adjacent hydroxyl groups and allows degradation of the lignin side chain to be monitored (17–19). Aromatic and aliphatic alcohols and acids are derivatized with a methyl group, permitting the resultant monomers to be analyzed by gas chromatography and mass spectrometry. The composition and yields of the monomers released can give information about the degradative process. A ^{13}C-labeled methyl group allows the differentiation between original methoxyl groups on lignin and those added during derivatization. The yield of specific lignin fragments is determined by analysis with structural mass spectroscopy. Studies by Filley et al. (19) showed that brown rot residues were characterized by high yields of 3,4-dihydroxyphenyl compounds and lower yields of aromatic acids, presumably due to brown rot demethylation of methoxyl guaiacyl groups and microbial side chain oxidation, respectively. Filley et al. (18) used ^{13}C thermochemolysis and solid-state ^{13}C-NMR to investigate the relationship between lignin oxidation, lignin demethylation, and polysaccharide metabolism by brown rot fungi. The specificity of these analytical techniques is illustrated by their ability to demonstrate demethylation of the 3-methoxyl carbon and to identify that it occurred in up to 22% of the lignin. This procedure can provide more specific structural information on lignin modification and is often complementary to the more general overview provided by NIR and MBMS.

POTENTIAL APPLICATIONS AND CONCLUSIONS

Various methods have been used to study the mechanisms involved in brown rot fungus physiology and the brown rot biodegradation of wood, and we have introduced some of these methods. This review is not exhaustive and focuses on techniques that have been recently used for the study of brown rot rather than on the potential of other analytical tools that may yet be useful in improving our understanding of brown rot mechanisms. Methods specific for evaluation of enzymatic processes involved in brown rot decay processes have also not been reviewed. Emphasis has been placed upon the use of fungi grown in culture as a crucial first step to study specific mechanisms and metabolites, and suggestions have been given for isolation and characterization of metabolites potentially involved in hydroxyl and other radical-generating reactions.

However, only study of the fungi in wood or closely related substrates can provide the necessary information to replicate both the biochemical reactions and the spatial relationships between the fungus and the wood cell wall that are critical to the understanding of natural decay processes. Low-molecular-weight fungal metabolites and radical-based processes play an important role in brown rot degradation of wood. In addition, these metabolites are of potential utilization in both bioremediation and industrial processes. Mediated Fenton systems have also shown promise in dye and pollutant remediation studies (25) in which efficient decolorization of recalcitrant dyes by use of a remediation scheme involving a Fenton system along with the biochelator mediator DHBA has been demonstrated. Kramer et al. (42) illustrated the degradation of a fluorophenol by the brown rot fungus *Gloeophyllum striatum* and presented evidence for the involvement of extracellular Fenton chemistry. Qian et al. (53) and Goodell and Jellison (22) examined the effect of hydroxyl radical generation of the free radical activation of thermomechanical pulp fiber for use in wood bonding. The activation was evaluated by ESR, and the generation of hydroxyl radicals and free phenoxy radicals was measured indirectly by chemiluminescence. A DHBA-mediated Fenton system has been used to treat pulp bleaching effluent (55). Bioremediation applications are particularly intriguing because of the demonstrated ability of brown rot fungi to ramify through soil and colonize wood and other substrates in the natural environment. The ability to characterize the underlying nonenzymatic microbial processes utilized by the brown rot fungi will contribute to our ability to both better control and utilize these unique degradative organisms.

REFERENCES

1. **Akerholm, M., B. Hinterstoisser, and L. Salmen.** 2004. Characterization of the crystalline structure of cellulose using static and dynamic FT-IR spectroscopy. *Carbohydr. Res.* **339:**569–578.
2. **Andersson, S., R. Serimaa, T. Paakkari, P. Saranpaa, and E. Pesonen.** 2003. Crystallinity of wood and the size of cellulose crystallites in Norway spruce (*Picea abies*). *J. Wood Sci.* **49:**531–537.
3. **Andersson, S., H. Wikberg, E. Pesonen, S. L. Maunu, and R. Serimaa.** 2004. Studies of crystallinity of Scots pine and Norway spruce cellulose. *Trees Struct. Funct.* **18:**346–353.
4. **ASTM.** 1994. Standard method of accelerated laboratory test of natural decay resistance of woods. Standard D2017-94. *1994 Annu. Book ASTM Stand.* **4:**218–224.
5. **Bard, A. J., and L. R. Faulkner.** 2001. *Electrochemical Methods: Fundamentals and Applications*, 2nd ed. John Wiley & Sons, New York, N.Y.
6. **Cao, Y., and H. Tan.** 2005. Study of crystal structures of enzyme-hydrolyzed cellulosic materials by X-ray diffraction. *Enzyme Microb. Technol.* **36:**314–317.
7. **Chandhoke, V., B. Goodell, J. Jellison, and F. A. Fekete.** 1992. Oxidation of 2-keto-4-thiomethylbutyric acid (KTBA) by iron-binding compounds produced by the wood-decaying fungus *Gloeophyllum trabeum*. *FEMS Microbiol. Lett.* **90:**263–266.
8. **Cohen, R., M. R. Suzuki, and K. E. Hammel.** 2004. Differential stress-induced regulation of two quinine reductases in the brown-rot basidiomycete *Gloeophyllum trabeum*. *Appl. Environ. Microbiol.* **70:**324–331.
9. **Cohen, R., M. R. Suzuki, and K. E. Hammel.** 2005. Processive endoglucanase active in crystalline cellulose hydrolysis by the brown rot basidiomycete *Gloeophyllum trabeum*. *Appl. Environ. Microbiol.* **71:**2412–2417.
10. **Connolly, J. H., H. J. Arnott, and J. Jellison.** 1996. Patterns of calcium oxalate crystal production by three species of wood decay fungi. *Scanning Microsc.* **10:**385–400.
11. **Connolly, J. H., W. C. Shortle, and J. Jellison.** 1999. Translocation and incorporation of strontium carbonate derived strontium into calcium oxalate crystals by the wood decay fungus *Resinicium bicolor*. *Can. J. Bot.* **77:**179–187.
12. **Curling, S., C. A. Clausen, and J. E. Winandy.** 2001. The effect of hemicellulose degradation on the mechanical properties of wood during brown rot decay. Document IRG/WP 01-20219. International Research Group on Wood Preservation Series, section 2. Test methodology and assessment. IRG, Stockholm, Sweden.
13. **Evans, R. J., and T. A. Milne.** 1987. Molecular characterization of the pyrolysis of biomass. 1. Fundamentals. *Energy Fuels* **1:**123–137.
14. **Fahr, K., H. G. Wetzstein, R. Grey, and D. Schlosser.** 1999. Degradation of 2,4-dichlorophenol and pentachlorophenol by two brown rot fungi. *FEMS Microbiol. Lett.* **175:**127–132.
15. **Fekete, F. A., V. Chandhoke, and J. Jellison.** 1989. Iron-binding compounds produced by wood-decaying basidiomycetes. *Appl. Environ. Microbiol.* **55:**2720–2722.
16. **Ferraz, A., R. Mendonca, A. Guerra, J. Ruiz, J. Rodriguez, J. Baeza, and J. Freer.** 2004. Near-infrared spectra and chemical characteristics of *Pinus taeda* (loblolly pine) wood chips biotreated by the white-rot fungus *Ceriporiopsis subvermispora*. *J. Wood Chem. Technol.* **24:**99–113.
17. **Filley, T. R.** 2003. Assessment of fungal wood decay by lignin analysis using tetramethylammonium hydroxide (TMAH) and C-13-labeled TMAH thermochemolysis. *ACS Symp. Ser.* **845:**119–139.
18. **Filley, T. R., G. D. Cody, B. Goodell, J. Jellison, C. Noser, and A. Ostrofsky.** 2002. Lignin demethylation and polysaccharide decomposition in spruce sapwood degraded by brown rot fungi. *Org. Chem.* **33:**111–124.
19. **Filley, T. R., P. G. Hatcher, W. C. Shortle, and R. T. Praseuth.** 2000. The application of C-13-labeled tetramethylammonium hydroxide (C-13-TMAH) thermochemolysis to the study of fungal degradation of wood. *Org. Chem.* **31:**181–198.
20. **Flaete, P. O., and E. Y. Haartveit.** 2004. Non-destructive prediction of decay resistance of *Pinus sylvestris* heartwood by near-infrared spectroscopy. *Scand. J. For. Res.* **19:**55–63.
21. **Flournoy, D. S., T. K. Kirk, and T. Highley.** 1991. Wood-decay by brown-rot fungi: changes in pore structure and cell wall volume. *Holzforschung* **45:**383–388.
22. **Goodell, B., and J. Jellison.** 2003. Oxidation using a non-enzymatic free radical system mediated by redox cycling chelators. U.S. patent application no. 20030186036.
23. **Goodell, B., Y. Qian, J. Jellison, M. Richard, and W. Qi.** 2002. Lignocellulose oxidation by low molecular weight metal-binding compounds isolated from wood degrading fungi: a comparison of brown rot and white rot systems and the potential application of chelator-mediated Fenton reactions, p. 37–48. *In* L. Viikari and R. Lantto (ed.), *Progress in Biotechnology*, vol. 21. *Biotechnology in the Pulp and Paper Industry*. Elsevier Science, Amsterdam, The Netherlands.
24. **Goodell, B., J. Jellison, J. Liu, G. Daniel, A. Paszczyuski, F. Fekete, S. Krishnamurthy, L. Jun, and G. Xu.** 1997. Low molecular weight chelators and phenolic compounds isolated from wood decay fungi and their role in the fungal biodegradation of wood. *J. Biotechnol.* **53:**133–162.
25. **Goodell, B., J. Jellison, Y. H. Qian, and M. Richard.** 2004. Decolorization and degradation of dyes with mediated Fenton chemistry. *Water Environ. Res.* **76:**2703–2707.
26. **Grayson, M.** 1982. *Kirk-Othmer Encyclopedia of Chemical Technology*, vol. 20. John Wiley, New York, N.Y.
27. **Highley, T. L.** 1973. Influence of carbon source on cellulase activity of white- and brown-rot fungi. *Wood Fiber* **5:** 50–58.
28. **Highley, T. L.** 1982. Is extracellular hydrogen peroxide involved in cellulose degradation of brown-rot fungi? *Mater. Org.* **17:**205–214.
29. **Highley, T. L.** 1987. Effect of carbohydrate and nitrogen on hydrogen peroxide formation by wood decay fungi in solid medium. *FEMS Microbiol. Lett.* **48:**373–377.
30. **Highley, T. L., and L. L. Murmanis.** 1985. Determination of hydrogen peroxide production in *Coriolus versicolor* and *Poria placenta* during wood degradation. *Mater. Org.* **20:** 241–252.
31. **Humar, M., M. Bokan, S. A. Amartey, M. Sentjurc, P. Kalan, and E. Pohleven.** 2004. Fungal bioremediation of copper, chromium and boron treated wood as studied by electron paramagnetic resonance. *Int. Biodeterior. Biodegrad.* **53:**25–32.
32. **Humar, M., M. Petric, F. Pohleven, M. Sentjurc, and P. Kalan.** 2002. Changes in EPR spectra of wood impregnated with copper-based preservatives during exposure to several wood-rotting fungi. *Holzforschung* **56:**229–238.
33. **Jellison, J., V. Chandhoke, B. Goodell, and F. Fekete.** 1991. The isolation and immunology of iron-binding compounds produced by *Gloeophyllum trabeum*. *Appl. Microbiol. Biotechnol.* **35:**805–809.
34. **Jensen, K. A., C. J. Ryan, Z. C. Ryan, and K. E. Hammel.** 2001. Pathways for extracellular Fenton chemistry in the brown rot basidiomycete *Gloeophyllum trabeum*. *Appl. Environ. Microbiol.* **67:**2705–2711.

35. **Juarros, E., C. Tortajada, M.D. Garcia, and F. Uruburu.** 1993. Storage of stock cultures of filamentous fungi at −80 degrees C: effects of different freezing-thawing methods. *Microbiologia* **9:**28–33.
36. **Kataoka, Y., and T. Kondo.** 1998. FT-IR microscopic analysis of changing cellulose crystalline structure during wood cell wall formation. *Macromolecules* **31:**760–764.
37. **Kavitha, V., and K. Palanivelu.** 2003. Degradation of 2-chlorophenol by Fenton and Photo-Fenton processes—a comparative study. *J. Environ. Sci. Health* **A38:**1215–1231.
38. **Kelley, S. S., T. Elder, and L. H. Groom.** 2005. Changes in the chemical composition and spectroscopy of loblolly pine medium density fiberboard furnish as a function of age and refining pressure. *Wood Fiber Sci.* **37:**14–22.
39. **Kelley, S. S., J. Jellison, and B. Goodell.** 2002. Use of NIR and pyrolysis-MBMS coupled with multivariate analysis for detecting the chemical changes associated with brown-rot biodegradation of spruce wood. *FEMS Microbiol. Lett.* **209:**107–111.
40. **Kelley, S. S., T. G. Rials, R. Snell, L. H. Groom, and A. Sluiter.** 2004. Use of near infrared spectroscopy to measure the chemical and mechanical properties of solid wood. *Wood Sci. Technol.* **38:**257–276.
41. **Kerem, Z., K. A. Jensen, and K. E. Hammel.** 1999. Biodegradative mechanism of the brown rot basidiomycete *Gloeophyllum trabeum*: evidence for the extracellular hydroquinone-driven Fenton reaction. *FEBS Lett.* **446:** 49–54.
42. **Kramer, C., G. Kreisel, K. Fahr, J. Kabohrer, and D. Schlosser.** 2004. Degradation of 2-fluorophenol by the brown-rot fungus *Gloeophyllum striatum*: evidence for the involvement of extracellular Fenton chemistry. *Appl. Microbiol. Biotechnol.* **64:**387–395.
43. **Labbe, N., T. G. Rials, S. S. Kelley, Z. M. Cheng, J. Y. Kim, and Y. Li.** 2005. FT-IR imaging and pyrolysis-molecular beam mass spectroscopy: new tools to investigate wood tissues. *Wood Sci. Technol.* **39:**61–76.
44. **Liu, R., B. Goodell, J. Jellison, and A. Amirbahman.** 2005. Electrochemical study of 2,3-dihydroxybenzoic acid and its interaction with Cu(II) and H_2O_2 in aqueous solutions: implications for wood decay. *Environ. Sci. Technol.* **39:**175–180.
45. **Magrini, K. A., R. J. Evans, C. M. Hoover, C. C. Elam, and M. F. Davis.** 2002. Use of pyrolysis molecular beam mass spectrometry (py-MBMS) to characterize forest soil carbon: method and preliminary results. *Environ. Pollut.* **116:**255–268.
46. **Nobles, M. K.** 1965. Identification of cultures of wood-inhibiting hymenomycetes. *Can. J. Bot.* **44:**1097–2065.
47. **Ostrofsky, A., J. Jellison, K. T. Smith, and W. C. Shortle.** 1997. Changes in cation concentration in red spruce wood decayed by brown rot and white rot fungi. *Can. J. Forest Res.* **27:**567–571.
48. **Oviedo, C., D. Contreras, J. Freer, and J. Rodriguez.** 2003. A screening method for detecting iron reducing wood-rot fungi. *Biotechnol. Lett.* **25:**891–893.
49. **Paszczynski, A., R. Crawford, D. Funk, and B. Goodell.** 1999. De novo synthesis of 4,5-dimethoxycatechol and 2,5-dimethoxyhydroquinone by the brown rot fungus *Gloeophyllum trabeum*. *Appl. Environ. Microbiol.* **65:**674–679.
50. **Pracht, J., J. Boenigk, M. Isen-Schröter, F. Keppler, and H. F. Schöler.** 2001. Abiotic Fe(III) induced mineralization of phenolic substances. *Chemosphere* **44:**613–619.
51. **Qi, W., and J. Jellison.** 2004. Induction and catalytic properties of an intracellular NADH-dependent 1,4-benzoquinone reductase from the brown-rot basidiomycete *Gloeophyllum trabeum*. *Int. Biodeterior. Biodegrad.* **54:** 53–60.
52. **Qian, Y. H., B. Goodell, and C. C. Felix.** 2002. The effect of low molecular weight chelators on iron chelation and free radical generation as studied by ESR measurement. *Chemosphere* **48:**21–28.
53. **Qian, Y., B. Goodell, J. Jellison, and C. C. Felix.** 2004. The effect of hydroxyl radical generation of free-radical activation of TMP fibers. *J. Polym. Environ.* **12:**147–155.
54. **Ritschkoff, A.-C., M. Ratto, J. Buchert, and L. Viikari.** 1995. Effect of carbon source on the production of oxalic acid and hydrogen peroxide by brown-rot fungus *Poria placenta*. *J. Biotechnol.* **40:**179–186.
55. **Rodriguez, J., D. Contreras, C. Parra, J. Freer, J. Baeza, and N. Duran.** 1999. Pulp mill effluent treatment by Fenton-type reactions catalyzed by iron complexes. *Water Sci. Technol.* **40:**351–355.
56. **Rodriguez, J., A. Ferraz, and M. P. de Mello.** 2003. Role of metals in wood biodegradation. *ACS Symp. Ser.* **845:** 154–174.
57. **Schilling, J., and J. Jellison.** 2004. High-performance liquid chromatographic analysis of soluble and total oxalate in Ca- and Mg-amended liquid cultures of three wood decay fungi. *Holzforschung* **58:**682–687.
58. **Sivonen, H., M. Nuopponen, S. L. Maunu, F. Sundholm, and T. Vuorinen.** 2003. Carbon-thirteen cross-polarization magic angle spinning nuclear magnetic resonance and Fourier transform infrared studies of thermally modified wood exposed to brown and soft rot fungi. *Appl. Spectrosc.* **57:**266–273.
59. **Stookey, L. L.** 1970. Ferrozine—a new spectrophotometric reagent for iron. *Anal. Chem.* **42:**779–782.
60. **Sun, X. F., F. Xu, R. C. Sun, P. Fowler, and M. S. Baird.** 2005. Characteristics of degraded cellulose obtained from steam-exploded wheat straw. *Carbohydr. Res.* **340:**97–106.
61. **Tornberg, K., and S. Olsson.** 2002. Detection of hydroxyl radicals produced by wood-decomposing fungi. *FEMS Microbiol. Ecol.* **40:**13–20.
62. **Tsai, C. H., A. Stern, J. Chiou, C. Chern, and T. Liu.** 2001. Rapid and specific detection of hydroxyl radical using an ultraweak chemiluminescence analyzer and a low-level chemiluminescence emitter: application to hydroxyl radical-scavenging ability of aqueous extracts of food constituents. *J. Agric. Food Chem.* **49:**2137–2141.
63. **Tsuchikawa, S., Y. Hirashima, Y. Sasaki, and K. Ando.** 2005. Near-infrared spectroscopic study of the physical and mechanical properties of wood with meso- and micro-scale anatomical observation. *Appl. Spectrosc.* **59:**86–93.
64. **Varela, E., and M. Tien.** 2003. Effect of pH and oxalate of hydroquinone-derived hydroxyl radical formation during brown-rot wood degradation. *Appl. Environ. Microbiol.* **69:**6025–6031.
65. **Winter, W. T.** 2002. Crystallography of galactomannans: an overview. *Abstr. Papers Am. Chem. Soc.* **219:**U262.
66. **Winter, W. T., and A. J. Stipanovic.** 2002. Cellulose nanocrystals: properties and potential applications. *Abstr. Papers Am. Chem. Soc.* **221:**U186.
67. **Wood, B. F., A. H. Conner, and C. G. Hill, Jr.** 1996. The effect of precipitation of the molecular weight distribution of cellulose tricarbanilate. *J. Appl. Polym. Sci.* **32:**3703–3712.
68. **Xu, G., and B. Goodell.** 2001. Mechanisms of wood degradation by brown-rot fungi: chelator-mediated cellulose degradation and binding of iron by cellulose. *J. Biotechnol.* **87:**43–57.
69. **Yelle, D., B. Goodell, D. J. Gardner, A. Amirbahman, P. Winistorfer, and S. M. Shaler.** 2004. Bonding of wood fiber composites using a synthetic chelator-lignin activation system. *For. Prod. J.* **54:**73–78.
70. **Zill, G., G. Engelhardt, and P. R. Wallnöfer.** 1988. Determination of ergosterol as a measure of fungal growth using Si 60 HPLC. *Eur. Food Res. Technol.* **187:**246–249.

Techniques for Studying Uncultured and Cultured Magnetotactic Bacteria*

DIRK SCHÜLER AND DENNIS A. BAZYLINSKI

92

The magnetotactic bacteria (MTB) comprise a large number of motile, morphologically and metabolically diverse prokaryotes that biomineralize intracellular, membrane-bound, magnetic crystals of magnetite (Fe_3O_4) and/or greigite (Fe_3S_4) (5). These inclusions, called magnetosomes, cause the cell to passively align in the Earth's and applied magnetic fields (24). Magnetotaxis results from this passive alignment and the active swimming of motile cells. It is currently thought that magnetotaxis acts in conjunction with some forms of chemotaxis (e.g., aerotaxis) to aid cells in more efficiently locating and maintaining an optimal position in vertical chemical concentration gradients in natural habitats (24). Phylogenetically, all known MTB to date are gram-negative members of the domain *Bacteria* and are associated with various subgroups of the phyla *Proteobacteria* and the *Nitrospira* group (1). All are microaerophiles, anaerobes, or both and possess a respiratory form of metabolism. Metabolic processes mediated by MTB are of geochemical importance. Their capability of precipitating iron oxide and iron-sulfur minerals, their metabolic versatility, and their high abundance suggest that MTB play significant roles in the biogeochemical cycling of iron, sulfur, nitrogen, and carbon in aquatic habitats. MTB are able to accumulate picograms of solid Fe and S per cell, transforming dissolved complexes into reactive minerals. These minerals may be preserved in the sediment after lysis of cells, thus providing reactive surfaces for further microbial and chemical transformations (5, 22, 44).

ANALYSES OF UNCULTIVATED MTB

Habitats and Occurrence of MTB

MTB are ubiquitous in freshwater and marine environments, except in habitats that are highly aerated and those with pH values of <5.5 or >8.5 (11). Generally, magnetite-producing MTB are located in the largest numbers at the oxic-anoxic interface (OAI), while the greigite producers have only been found in marine habitats below the OAI, in the anaerobic, sulfidic zone. In many freshwater systems, the OAI is at or slightly below the water-sediment interface. These systems for the most part contain very little sulfate and thus no hydrogen sulfide. Marine water contains about 28 mM sulfate, which is reduced by sulfate-reducing bacteria to hydrogen sulfide which can diffuse into the water column, thereby sometimes causing the OAI to exist in the water column closer to the surface than to the sediments (24, 44). In this case, magnetite-producing MTB and other microaerophiles form a plate of microorganisms in the water column at the OAI. Again, the greigite producers in this situation are at their largest numbers in the sulfidic zone below the OAI.

In stratified sediments, MTB display a heterogeneous vertical distribution, and their occurrence is confined to a narrow layer close below the sediment surface (1 to 10 mm) (22). Different morphotypes display different preferences within chemical gradients, but the largest populations of MTB in many sediments are found at or closely below the maximum penetration depth of oxygen, which is usually restricted to the top layers (Fig. 1). This region often also overlaps or coincides with the availability of soluble ferrous iron or gradients of hydrogen sulfide (22).

Establishment and Development of Microcosms

Uncultivated MTB can be studied most easily in microcosms set up in the laboratory. In their simplest form, microcosms consist of about one-third sediment together with about two-thirds overlying water and are placed in a flask or aquarium (0.1 to 5 liters) and incubated in dim light at room temperature. After reestablishment of geochemical stratification, MTB develop in large numbers in the top layer of the sediment and may remain abundant for several years without the addition of nutrients. Coccoid MTB have been reported to form veils of cells in microcosms prepared from sediments and bottom waters if they become anaerobic and/or too reduced in the water column (31). Microcosms are useful for studying the succession and community structure of MTB as well as the relationship of MTB to biogeochemical parameters. This can be achieved by vertical microsensor profiling and biogeochemical methods, including in situ analysis of oxygen, sulfide, iron, and sulfate (22). Although cell numbers may exceed those found under natural conditions, microcosms do not resemble classical enrichment cultures in that all attempts to amend the microcosm by the addition of

*We dedicate this chapter to Richard P. Blakemore and Richard B. Frankel, the driving forces behind virtually all the original early studies on the MTB, on the occasion of their retirements.

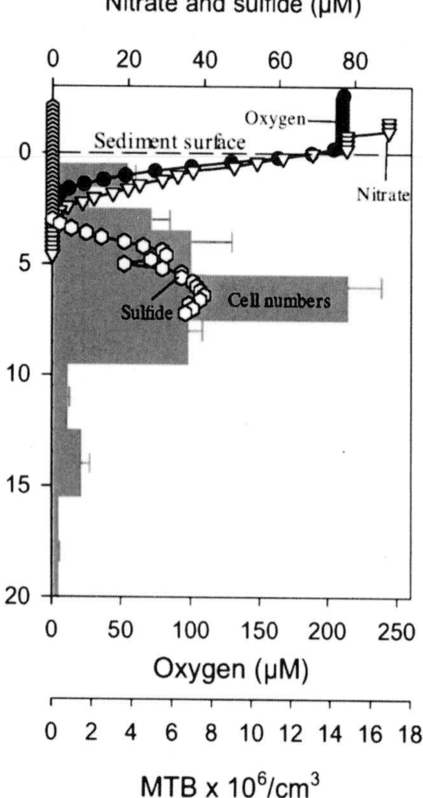

FIGURE 1 Vertical distribution of MTB along chemical gradients in a typical freshwater microcosm. Numbers of MTB (gray bars) were determined by viable cell counts, using the hanging-drop assay as described in the text. Error bars indicate standard deviations. Various biochemical parameters (concentrations of oxygen, nitrate, and sulfide) were determined by using microelectrodes (modified as described in reference 22).

nutrients, including iron or carbon sources, do not usually result in increased numbers of MTB or the selection of particular species of MTB. Although microcosms are widely used, it must be stressed that lab-scale setups represent artificial environments and do not fully reflect conditions in the natural habitat. Microcosms undergo a characteristic succession, which usually results in the dominance of a single or few species of magnetotactic cocci (23). This is accompanied by a loss of diversity of MTB communities compared to that usually observed in fresh environmental samples. Therefore, fresh, undisturbed cores of sediment or water samples should be used preferentially in order to fully reveal the phylogenetic and metabolic diversity found in natural communities of MTB.

Magnetic Collection of MTB Cells

Many MTB can easily be detected and collected by taking advantage of their active, directed migration along magnetic field lines. In general, in natural aquatic habitats in the Northern Hemisphere, the major portion (often >99%) of polar MTB (24) are "north-seeking" and swim towards the south pole of a bar magnet (Earth's North Pole is actually a south magnetic pole). Swimming of cells can be directed effectively by a steady field of 1 to 10 G. Several workers used electromagnetic coils to create uniform fields (11, 45), but for many purposes it is sufficient to use simple permanent magnets, such as conventional laboratory stirring bars. For general assessments of total MTB populations, magnetic collection of cells can be employed by application of an external magnet to flasks or bottles containing sediment or water samples (Fig. 2A). Briefly, the south pole of a bar magnet is placed next to the bottle either at the sediment-water interface or at approximately half the height of the sample bottle for samples containing only water. The water should be mixed gently or the surface of the sediment slightly stirred before placement of the magnet. North-seeking cells will be directed toward the south pole of the magnet and will accumulate at the glass next to the pole within a few minutes to several hours. Cells can then be collected from the bottle by use of a Pasteur pipette and used for additional purification.

To obtain more homogenous enrichments of MTB, a further magnetic purification step is recommended. Wolfe and coworkers described a capillary racetrack method for separating MTB from nonmagnetic contaminants (51). In brief, a portion of sediment or water containing MTB is placed on top of a sterile, wet cotton plug in the wide-mouthed end of a Pasteur pipette which has been filled with membrane filter-sterilized water (Fig. 2B). The south pole of a bar magnet is placed near the sealed tip of the capillary furthest from the reservoir. Migration of MTB through the cotton plug towards the end of the capillary can be observed by dark-field microscopy. After most of the MTB have reached the tip, which is usually within about 30 min, the tip is broken off and the MTB are harvested aseptically with a hypodermic needle. Capillary lengths of 9 cm have proven effective to enrich MTB, especially the fast-swimming magnetococci, to homogeneity, and between 10^5 and 10^7 MTB cells which are nearly free of contaminants can be obtained per single run (23, 39), depending on the concentration of MTB in the original mud or water sample. The addition of a reductant (e.g., 200 μM neutralized cysteine) to the water within the capillary seems to increase the yield of harvested MTB.

Electron Microscopic Analysis of Magnetotactic Cells and Magnetosomes

Cell morphology and flagellar patterns of specific MTB can be determined by using negatively stained preparations of cells and transmission electron microscopy. Magnetosomes are electron-dense structures that can be visualized in cells without staining. Relevant crystalline and mineralogical characteristics of magnetosome crystals can be analyzed directly by high-resolution transmission electron microscopy, electron holography, and electron tomography (15, 16, 18, 20). The identification of specific mineral crystals within specific morphotypes of MTB is most easily accomplished by using selected area electron diffraction of individual particles in the transmission electron microscope (e.g., see references 29 and 33). Elemental X-ray spectra and maps of cells and their magnetosomes can be used to distinguish between iron oxide and iron sulfide crystals and to determine whether magnetosomes contain extraneous elemental impurities (e.g., see references 7 and 8).

Identification and Enumeration of MTB

As yet, there are no group-specific probes available for the in situ detection of MTB in the environment by fluorescence in situ hybridization (FISH). In addition, strong background fluorescence of the sediment particles limits the

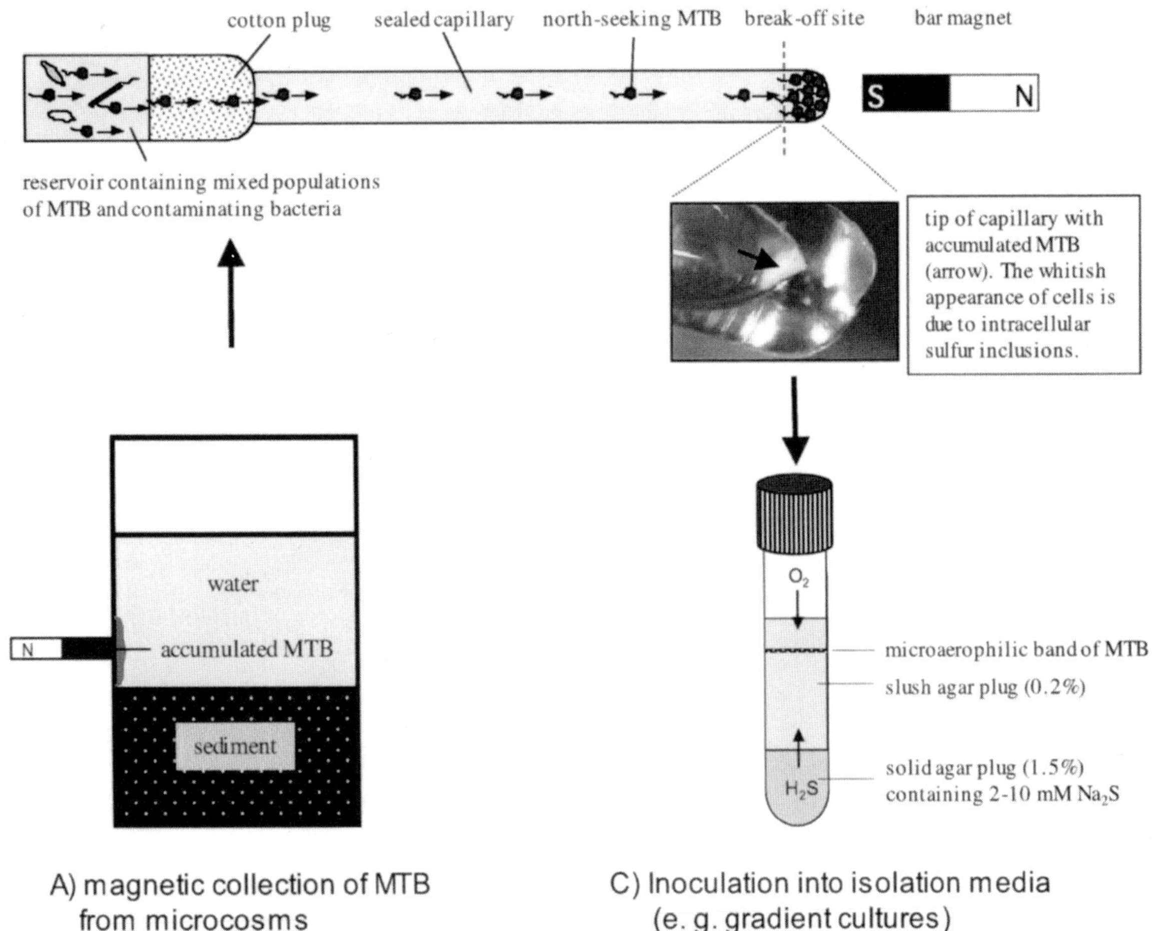

FIGURE 2 Scheme for magnetic collection and axenic isolation of uncultivated MTB. Details of the techniques are given in the text.

use of fluorescence techniques in this environment. Thus, the enumeration of uncultivated MTB in many studies is based on the magnetic separation of cells in drops of mud or water (e.g., see reference 44). This technique provides a highly selective tool for the identification of MTB and allows vertical profiles of the numbers of MTB in stratified water or sediments (22, 44, 46). Sampling in the water column can be performed by means of peristaltic pumps, allowing simultaneous chemical concentration profiling and sampling (44). A maximum cell number of 2×10^6 MTB per ml was detected close to the chemocline, which displayed sharp changes in abundance over small length scales within the water column as well as strong seasonal fluctuations in population abundances of different morphotypes (44). Undisturbed cores are taken from sediment samples, which are then sliced and fractionated in 1- to 5-mm increments (22). Aliquots of diluted sediment slurries (3 to 100 μl) are counted by phase-contrast microscopy, using the "hanging-drop" assay (24). MTB, which can be recognized by their conspicuous morphology and magnetic response, accumulate at the edge of the drop after several minutes next to the south pole of a bar magnet (Fig. 3). Fifty to 300 cells per drop can be counted, which also allows for the discrimination and enumeration of different morphotypes. In sediment samples, maximum numbers of MTB of up to 1.5×10^7 to 5×10^7 cells/ml have been observed, accounting for 1% of the total number of prokaryotic cells in the upper sediment layer (22). Because of the enormous size of some MTB, they can account for >30% of the total biovolume of this particular region of the sediment (46).

Phylogenetic Diversity and Molecular Analysis of Uncultivated MTB

The high abundance, conspicuous morphology, and magnetic response of the MTB allow for their easy microscopic identification and have made them an attractive subject for the early application of cultivation-independent, rRNA-based techniques (46, 48). Various subsequent studies have assigned uncultivated MTB to different phylogenetic groups. Nevertheless, the natural diversity of MTB is far from being exhaustively explored, as evidenced by the fact that new lineages of MTB are being continuously uncovered, even from common, well-studied habitats (22, 23, 44). Because of the polyphyletic distribution of magnetosome formation in bacteria and the broad diversity within different phylogenetic branches, there are no sets of primers or hybridization

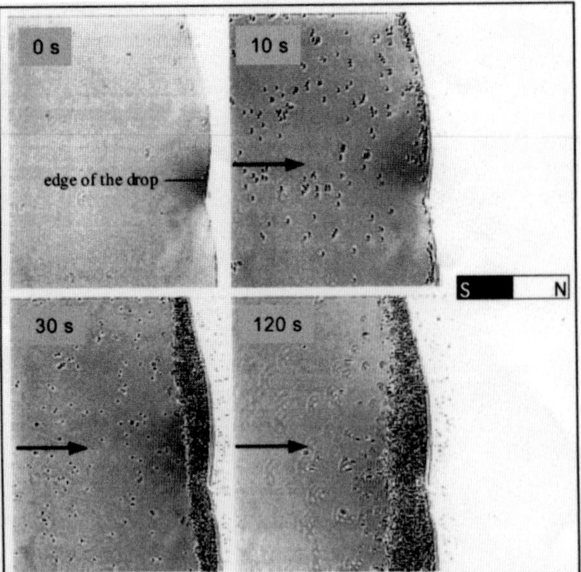

FIGURE 3 Phase-contrast light micrographs of MTB in a hanging drop at different time points after application of a magnetic field. Cells swim in random directions in the absence of a magnetic field (0 s). If a magnetic field is applied, MTB immediately align and swim along the magnetic field lines (arrows). After several minutes, large numbers of bacteria accumulate at the "northern" edge of the drop (next to the south pole of a bar magnet).

probes available that universally target all MTB. Primers universal for the amplification of eubacterial 16S rRNA genes have therefore been used routinely in PCR- and denaturing gradient gel electrophoresis-based approaches to detect MTB. However, both denaturing gradient gel electrophoresis and PCR with general primers are ineffective means to fully recover the diverse populations of MTB present in complex microbial populations (22, 23). Thus, group-specific primers should be used in addition for the detection of less abundant magnetotactic phylotypes.

A set of presumably group-specific primers for the detection of *Magnetospirillum* species was described (14, 48), but with the growth of sequence information available in databases, these turned out to also match numerous 16S rRNA sequences from unrelated, nonmagnetotactic bacteria. FISH using a specific fluorescent probe was employed to detect a greigite-producing multicellular magnetotactic prokaryote of the δ-*Proteobacteria* (17). Large greigite-producing rods of the γ subclass of *Proteobacteria* were detected by use of the γ-*Proteobacteria*-specific GAM42a probe (44). The "*Magnetobacterium bavaricum*" probe, which was designed for the detection of a very large, uncultivated, magnetite-producing rod of the *Nitrospira* phylum (46), also specifically recognizes other related MTB of this particular lineage with a 16S rRNA similarity of 91% (23). For PCR analysis of less abundant phylotypes in complex communities or from particular horizons, more efficient magnetic enrichments and separation techniques are usually required. This can be achieved by the use of microcapillaries (44) or by the capillary racetrack method (51). As an alternative to active, magnetically directed migration of cells, flow sorting has been used to enrich a giant magnetotactic rod based on the conspicuous scattering characteristics of this particular morphotype (46, 50).

Because of the unavailability of specific 16S rRNA-based probes for MTB, the use of probes based on key genes encoding proteins specifically involved in steps of magnetosome biomineralization might prove successful for the detection and enumeration of MTB. The probes need to be based on gene families that do not occur in nonmagnetotactic bacteria and, ideally, that are common to all MTB or particular physiological groups of MTB and show sufficient sequence conservation among MTB. Although genomic analysis of several magnetite-producing α-*Proteobacteria* identified a unique and conserved set of genes encoding magnetosome membrane-associated proteins with key functions in magnetite biomineralization, there is currently no candidate gene that fulfills these requirements. For example, the *mamC* gene, encoding the most abundant magnetosome protein, is present in all magnetite-producing MTB tested so far but has no homologues in nonmagnetic organisms and was therefore considered a group-specific genetic marker (21). However, although it exhibits the highest conservation of all the identified magnetosome genes, the sequence similarity of <50% between homologous genes from different MTB within the α-*Proteobacteria* is not sufficient to deduce degenerate PCR primers based on the current data set. One promising alternative for the generation of extended sequence information on genetic diversity is the metagenomic analysis of uncultivated MTB, as MTB can be enriched sufficiently from environmental samples for the construction of large-insert libraries. Probes derived from several *mam* genes of the cultivated MTB *Magnetospirillum gryphiswaldense* have been used for the identification of clones harboring the nearly complete magnetosome operons from the genomic magnetosome island (21). Thus, this approach has great potential in the identification, analysis, and functional expression of genomic sequences specifically involved in magnetosome formation.

Despite the fact that the diversity of MTB in the environment can be studied relatively straightforwardly by microscopy and 16S rRNA-based techniques, the greatest challenge has remained the problem of linking their abundance and phylogeny with function, and thus metabolic activity, in the environment. Hence, future studies should include techniques such as microautoradiography, which measures the uptake of specific radiolabeled substrates by microbial cells. The easy separation of single cells of MTB can then be used in combination with FISH to relate a particular phylotype to a specific mode of metabolism. This approach should be accompanied by concomitant quantitative and comparative analysis of expressed rRNA genes and key genes for magnetosome biomineralization.

ISOLATION AND CULTIVATION OF MTB

Isolation of MTB in pure culture has proven difficult. Only a limited number of magnetotactic isolates are available, and all of them are magnetite-producing strains. There are currently no greigite-producing strains of MTB available in pure culture. The establishment of general isolation strategies is further hampered by the fact that the nutritional requirements of MTB are diverse and, for most phylotypes, particularly those that synthesize greigite, unknown. In addition, in competitive enrichment experiments, the fastidious and slow-growing MTB are usually outgrown by nonmagnetotactic microorganisms. Therefore, no general strategy based on metabolic selection seems to be applicable. One notable exception is a method reported by Sakaguchi (35) and coworkers for isolating MTB without primary magnetic separation. The method comprises the incubation of sediments, metabolic enrichment of bacteria in amended medium, and isolation of enriched bacteria by colony formation (35). This method seems to be particularly useful for the isolation of slowly moving, magnetotactic, sulfate-reducing anaerobes and was employed for the axenic cultivation of *Desulfovibrio magneticus* strain RS-1 of the δ-*Proteobacteria*. However, isolation of other MTB has relied on magnetic purification to efficiently separate them from contaminating nonmagnetotactic bacteria. The homogenous inocula are then subsequently transferred to appropriate growth media. As far as can be inferred from available pure cultures and in situ analysis of uncultivated MTB, the physiologically diverse population of MTB comprises a number of various and versatile metabolic types, including heterotrophic and autotrophic forms of respiration. A range of organic and inorganic electron donors can be utilized for the reduction of various electron acceptors, including oxygen, nitrate, nitrous oxide, and sulfate. Generally, media with low concentrations of nutrients are more favorable for the isolation of MTB than are rich media with high concentrations of carbon and nitrogen sources. Cultivated strains do not have absolute requirements for supplements and growth factors, and attempts to supplement media with complex mixtures of vitamins or sediment extracts did not result in an improved efficiency of isolation. Few MTB, such as *D. magneticus* strain RS-1 and perhaps the greigite-producing species, are obligate anaerobes, and most MTB will resist short exposures to oxygen during enrichment and inoculation, making the strict exclusion of oxygen during handling unnecessary. However, most known MTB are microaerophiles, and atmospheric oxygen levels are generally inhibitory to growth and magnetite formation. Therefore, special precautions are required for the isolation and cultivation of MTB. Redox buffering by the addition of reducing agents, such as sodium thioglycolate, ascorbic acid, or cysteine-HCl at concentrations of 0.1 to 0.4 g liter^{-1} or dithiothreitol at 1 mM, is required to stimulate the growth of microaerophilic or anaerobic MTB (6). The use of resazurin, a redox indicator that is colorless when fully reduced, is recommended. Ferric citrate is used most frequently as an iron source for growth and magnetite biomineralization, as it can be conveniently prepared and autoclaved together with other medium components. However, the source of iron is not critical as long it is kept in a soluble form at neutral pH by the presence of chelating compounds in the medium. Either ferrous or ferric iron salts at concentrations between 20 and 50 μM are sufficient to allow for both growth and magnetosome formation (37, 38). Semi-solid media have proven to be the most successful type used for the isolation and maintenance of various MTB, as they allow the establishment of chemical gradients in order to mimic the conditions in stratified sediments. In combination with solid agar plugs containing millimolar amounts of Na_2S, inverse oxygen-sulfide gradients can be generated (Fig. 2C). Growth of magnetite-producing MTB occurs as a sharp microaerophilic band of cells at a defined position, usually initiating at the OAI within the multiple gradients (this technique, a modification of the medium developed by Nelson and Jannasch [32], is described in detail in reference 39). As the number of cells in the band increases, cells deplete oxygen at the location of the band, and the band of motile cells often moves to the surface.

To avoid inhibitory effects by trace contaminants present in typical agars, it is recommended to use agar of the highest purity (e.g., Noble agar [Difco]). With this technique, several strains of heterotrophic, microaerophilic magnetospirilla have been isolated in media containing low concentrations (0.01%) of succinate or another organic acid as the electron donor (23, 39). Magnetotactic, facultatively autotrophic, sulfide-oxidizing marine microaerophiles, including vibrios and cocci, can also be enriched and isolated by using this technique, except that natural or artificial seawater must be used as the diluent for the medium. An artificial seawater formula that has been used successfully for the isolation of marine vibrios, cocci, and spirilla can be found in reference 8. For MTB from brackish environments, the salinity can be determined with a refractometer, and seawater can be diluted with distilled water to that salinity. To obtain pure cultures of these strains, shake tubes of medium solidified with 10 to 13 g of Noble agar per liter were used to obtain isolated colonies of cells within an $[O_2]$ gradient. For the facultatively anaerobic strain MV-1, which uses N_2O as an alternate terminal electron acceptor, shake tubes had screw caps with butyl rubber septa (Bellco Biotechnology). After shake tubes were inoculated and the medium had solidified, the air headspace was replaced and overpressurized to 10 lb/in^2 with pure N_2O. N_2O is extremely soluble at room temperature and dissolves in the solid medium, even reaching the bottom of the tube. Colonies of MTB grown either in agar shake tubes or on agar plates tend to be dark brown to black due to the presence of magnetite.

ANALYSIS OF CULTIVATED MTB

Maintenance and Mass Cultivation

Isolated cultures of MTB need to be properly preserved, as both the magnetic polarity and the capability of biomineralizing magnetosomes can be irreversibly lost upon prolonged subcultivation and storage in the lab due to genetic instability (12, 19, 36, 49). Long-term preservation of magnetospirilla can be accomplished by suspending dense suspensions of cells in liquid growth medium containing 10% dimethyl sulfoxide, with subsequent freezing in liquid nitrogen and storage at −80°C. Coccoid and vibrioid marine strains of MTB can be preserved similarly, except that 10 to 25% glycerol containing 1% yeast extract appears to allow for greater cell survival than does medium with dimethyl sulfoxide.

To obtain enough cell material for biochemical analyses, cultivation of MTB has to be done with large volumes of liquid medium. For microaerophilic magnetospirilla, this can be achieved by cultivating cells in Hungate tubes or flasks under micro-oxic atmospheres containing 1 to 5% oxygen in the headspace (29). Alternatively, if a large enough inoculum is used, cells can be grown in liquid culture exposed to air, with aeration being adjusted so that the oxygen input is counterbalanced by the respiratory activity of cells, resulting in the establishment of microaerobic conditions (37). Mass cultivation at a large scale can be facilitated more easily by oxygen-controlled fermenter systems that allow the precise adjustment of low dissolved oxygen levels. This can be achieved by a gas supply regimen that exactly maintains very low pO_2 tensions over a wide range of cell densities, with fluctuations of $<5\%$ of the set value (29). Mass cultivation of anaerobic MTB, such as the vibrio strain MV-1, is accomplished by the use of 19-liter carboys containing 15 liters of medium sparged with O_2-free N_2 overnight while cooling after autoclaving and then with N_2O (the electron acceptor) for several hours after the carboy has cooled. Caution must be observed when harvesting cells of strain MV-1 because large amounts of pressure accumulate in the carboys (whose stoppers are wired into place) because of the conversion of very soluble N_2O to virtually insoluble N_2.

Physiological Characterization of MTB

Specific physiological activities of MTB can be tested by a number of methods. The intracellular precipitation of magnetite and its putative precursors has been analyzed by Mössbauer spectroscopy based on the stable isotope ^{57}Fe (25). Measurements of stable oxygen and iron isotopes have been done for mass-cultured cells of MTB in studies of magnetosome formation (30). Sensitive, semiquantitative analysis of magnetosome formation in cell suspensions is possible by differential light scattering measurements in the presence of magnetic fields with variable directions (13, 40). During magnetosome formation, MTB accumulate amounts of iron as high as 4%, almost all of which is exclusively located within the magnetosome particles (26). Thus, determination of the intracellular iron content by atomic absorption spectroscopy provides an exact measure of magnetite biomineralization (29). A more sensitive technique for the time-resolved determination of iron incorporation is based on the use of the radioactive iron isotope ^{55}Fe, which was employed to study the kinetics and energy dependence of iron uptake in M. *gryphiswaldense* (37, 38). The potential to reduce ferric iron was demonstrated for cells and cell extracts of *Magnetospirillum magnetotacticum* by the colorimetric detection of ferrous iron by the Ferrozine method (28, 34). Every species of MTB tested appears capable of nitrogen fixation based on nitrogenase activity, which can be measured by determining acetylene reduction (2, 3). Several marine MTB, including cocci and vibrios, grow chemolithoautotrophically, using sulfide and thiosulfate as electron donors and oxygen as the electron acceptor, and fix carbon dioxide via the Calvin-Benson-Bassham pathway (4). Thus, detection of ribulose-1,5-bisphosphate carboxylase/oxygenase (RubisCO) activity (9) is an important method for assessment of the metabolic capacity of cultivated MTB.

For comprehensive analyses of the magnetosome crystals and their enveloping membranes, magnetosomes can be isolated from disrupted cells by using a straightforward purification protocol. The magnetosomes are easily separated from crude extracts by the use of magnetic separation columns. Magnetic separation is followed by ultracentrifugation into a 55% (wt/wt) sucrose cushion. This procedure results in suspensions of purified magnetosome particles with intact enveloping membrane structures (26). Alternatively, strong gap magnets have also been used successfully to purify magnetosomes from lysed cells (8). For biochemical analyses, the magnetosome membrane is solubilized by treatment with hot sodium dodecyl sulfate (1%), and extracts can be subjected, for instance, to various proteomic techniques in order to identify specific magnetosome-associated proteins.

Genetic Analysis

For the establishment of a genetic system for the MTB, the ability to grow cells as colonies on agar plates is a necessity, since colony formation on the surfaces of agar plates is required for clonal selection in genetic analysis. However, this is not easily achieved owing to the fastidiousness of MTB in growing from single cells and requires specific modifications of the growth medium. A medium containing 0.3% activated charcoal as a scavenger of toxic oxygen compounds promotes efficient colonial growth of M. *gryphiswaldense* in the presence of micro-oxic atmospheres (43). Growth of colonies of the marine magnetotactic vibrio MV-1 on solid medium can be accomplished by streaking cells onto the surface of solidified growth medium and incubating them under a micro-oxic (1% O_2, 99% N_2) or anoxic (N_2O) atmosphere. However, deep (25 mm) petri dishes must be used because the thickness of the agar has to be two to three times that in conventional petri dishes, probably to provide a greater volume for buffering the accumulation of toxic waste products around colonies and to minimize the effects of evaporation. Systems for genetic manipulation have recently been described for several cultivated *Magnetospirillum* species. A number of broad-host-range vectors, including those of the IncQ, IncP, and pBBR1 groups, are known to replicate within these strains. These can be introduced by means of electroporation or conjugation (33, 43). In addition, a set of genetic markers is available to select or screen for gene-specific mutations introduced into the chromosome (41, 42). A more detailed overview of genetic analysis of MTB is available (5).

Studies involving nonmagnetotactic mutants that do not biomineralize magnetosomes are required for elucidation of the magnetite and greigite biosynthetic pathways. Thus far, such mutants have proven invaluable, particularly in the understanding of how magnetosome genes are organized and perhaps transferred to a wide variety of diverse prokaryotes. For example, large numbers of spontaneous nonmagnetotactic mutants of M. *gryphiswaldense* and strain MV-1 accumulate in stationary cultures (19, 49). Identifying nonmagnetotactic mutants is relatively easy since wild-type magnetotactic cells produce dark brown to black colonies both on agar plates and in shake tubes, while mutants produce pink to cream-colored colonies. Finally, the genomes of two magnetotactic species are available for perusal and study. These include that of M. *magnetotacticum* strain MS-1 and the unnamed marine coccus strain MC-1. Genome sequence data have been helpful in the identification of a specific cluster of magnetosome genes that is conserved in multiple species of MTB (27). These draft genome sequences can be found at the Joint Genome Website (http://www.jgi.doe.gov/). Contiguous sequence information for the entire genomic magnetosome island of M. *gryphiswaldense*, encoding all known magnetosome genes, is given in reference 49. The complete genome sequence of M. *magnetotacticum* AMB-1 was reported in 2005 (30a).

Conclusions and Directions for Future Research

It is clear that the MTB are an extremely diverse group of prokaryotes with regard to morphology, physiology, and biogeochemistry. Thus, it is also clear that there are no established techniques that are completely effective for studying all of the many different species that make up this fascinating and unusual group of bacteria. Instead, it is likely that an increasing number of individualized techniques for specific species of MTB will be developed as more researchers become interested in the MTB. The need for these techniques is great in that to really understand the biomineralization processes involved in magnetite and greigite magnetosome formation, studies involving many organisms that produce different compositional and morphological types of magnetosome crystals will be required. Thus, there is also a need for the isolation and characterization of additional MTB, particularly those that biomineralize greigite, since none of these have yet been isolated and cultivated in axenic culture. It has been approximately 30 years since the discovery of magnetotactic bacteria by Richard Blakemore (10). We enthusiastically hope and trust that research in this area will proceed now at a faster pace.

REFERENCES

1. **Amann, R., R. Rossello-Mora, C. Flies, and D. Schüler.** 2004. Phylogeny and *in situ* identification of magnetotactic bacteria, p. 61–74. *In* E. Baeuerlein (ed.), *Biomineralization*, 2nd ed. Wiley-VCH, Weinheim, Germany.
2. **Bazylinski, D. A., and R. P. Blakemore.** 1983. Nitrogen fixation (acetylene reduction) in *Aquaspirillum magnetotacticum*. *Curr. Microbiol.* **9:**305–308.
3. **Bazylinski, D. A., A. J. Dean, D. Schüler, E. J. P. Phillips, and D. R. Lovley.** 2000. N_2-dependent growth and nitrogenase activity in the metal-metabolizing bacteria, *Geobacter* and *Magnetospirillum* species. *Environ. Microbiol.* **2:**266–273.
4. **Bazylinski, D. A., A. J. Dean, T. J. Williams, L. K. Long, S. L. Middleton, and B. L. Dubbels.** 2004. Chemolithoautotrophy in the marine, magnetotactic bacterial strains MV-1 and MV-2. *Arch. Microbiol.* **182:**373–387.
5. **Bazylinski, D. A., and R. B. Frankel.** 2004. Magnetosome formation in prokaryotes. *Nat. Rev. Microbiol.* **2:** 217–230.
6. **Bazylinski, D. A., R. B. Frankel, and H. W. Jannasch.** 1988. Anaerobic magnetite production by a marine magnetotactic bacterium. *Nature* **334:**518–519.
7. **Bazylinski, D. A., A. Garratt-Reed, A. Abedi, and R. B. Frankel.** 1993. Copper association with iron sulfide magnetosomes in a magnetotactic bacterium. *Arch. Microbiol.* **160:**35–42.
8. **Bazylinski, D. A., A. Garratt-Reed, and R. B. Frankel.** 1994. Electron-microscopic studies of magnetosomes in magnetotactic bacteria. *Microsc. Res. Tech.* **27:**389–401.
9. **Beudeker, R. F., G. C. Cannon, J. G. Kuenen, and J. M. Shively.** 1980. Relations between D-ribulose-1,5-bisphosphate carboxylase, carboxysomes and CO_2 fixing capacity in the obligate chemolithotroph *Thiobacillus neapolitanus* grown under different limitations in the chemostat. *Arch. Microbiol.* **124:**185–189.
10. **Blakemore, R.** 1975. Magnetotactic bacteria. *Science* **190:** 377–379.
11. **Blakemore, R.** 1982. Magnetotactic bacteria. *Annu. Rev. Microbiol.* **36:**217–238.
12. **Blakemore, R., D. Maratea, and R. S. Wolfe.** 1979. Isolation and pure culture of a freshwater magnetic spirillum in chemically defined medium. *J. Bacteriol.* **140:**720–729.
13. **Blakemore, R. P., K. A. Short, D. A. Bazylinski, C. Rosenblatt, and R. B. Frankel.** 1985. Microaerobic conditions are required for magnetite formation within *Aquaspirillum magnetotacticum*. *Geomicrobiol. J.* **4:**53–72.
14. **Burgess, J. G., R. Kawaguchi, T. Sakaguchi, R. H. Thornhill, and T. Matsunaga.** 1993. Evolutionary relationships among *Magnetospirillum* strains inferred from phylogenetic analysis of 16S rRNA sequences. *J. Bacteriol.* **175:**6689–6694.
15. **Buseck, P. R., R. E. Dunin-Borkowski, B. Devouard, R. B. Frankel, M. R. McCartney, P. A. Midgley, M. Posfai, and M. Weyland.** 2001. Magnetite morphology and life on Mars. *Proc. Natl. Acad. Sci. USA* **98:**13490–13495.
16. **Clemett, S. J., K. L. Thomas-Keprta, J. Shimmin, M. Morphew, J. R. McIntosh, D. A. Bazylinski, J. L. Kirschvink, D. S. McKay, S. J. Wentworth, H. Vali, E. K. Gibson, Jr., and C. S. Romanek.** 2002. Crystal morphology of MV-1 magnetite. *Am. Mineral.* **87:**1727–1730.
17. **DeLong, E. F., R. B. Frankel, and D. A. Bazylinski.** 1993. Multiple evolutionary origins of magnetotaxis in bacteria. *Science* **259:**803–806.
18. **Devouard, B., M. Posfai, X. Hua, D. A. Bazylinski, R. B. Frankel, and P. R. Buseck.** 1998. Magnetite from magnetotactic bacteria: size distributions and twinning. *Am. Mineral.* **83:**1387–1398.
19. **Dubbels, B. L., A. A. DiSpirito, J. D. Morton, J. D. Semrau, J. N. Neto, and D. A. Bazylinski.** 2004. Evidence for a copper-dependent iron transport system in the marine, magnetotactic bacterium strain MV-1. *Microbiology* **150:**2931-2945.
20. **Dunin-Borkowski, R. E., M. R. McCartney, M. Posfai, R. B. Frankel, D. A. Bazylinski, and P. R. Buseck.** 2001. Off-axis electron holography of magnetotactic bacteria: magnetic microstructure of strains MV-1 and MS-1. *Eur. J. Mineral.* **13:**671–684.
21. **Flies, C.** 2004. Isolierung und kultivierungsunabhängige Untersuchungen von magnetotaktischen Bakterien aus marinen und limnischen Sedimenten. Ph.D. thesis. Universität Bremen, Bremen, Germany.
22. **Flies, C., H. Jonkers, D. deBeer, K. Bosselmann, M. Böttcher, and D. Schüler.** 2005. Diversity and vertical distribution of magnetotactic bacteria along chemical gradients in freshwater microcosms. *FEMS Microbiol. Ecol.* **52:**185–195.
23. **Flies, C., J. Peplies, and D. Schüler.** 2005. A combined approach for the characterization of uncultivated magnetotactic bacteria from various aquatic environments. *Appl. Environ. Microbiol.* **71:**2723–2731.
24. **Frankel, R. B., D. A. Bazylinski, M. S. Johnson, and B. L. Taylor.** 1997. Magneto-aerotaxis in marine coccoid bacteria. *Biophys. J.* **73:**994–1000.
25. **Frankel, R. B., G. C. Papaefthymiou, R. P. Blakemore, and W. O'Brien.** 1983. Fe_3O_4 precipitation in magnetotactic bacteria. *Biochim. Biophys. Acta* **763:**147–159.
26. **Grünberg, K., E. C. Müller, A. Otto, R. Reszka, D. Linder, M. Kube, R. Reinhardt, and D. Schüler.** 2004. Biochemical and proteomic analysis of the magnetosome membrane in *Magnetospirillum gryphiswaldense*. *Appl. Environ. Microbiol.* **70:**1040–1050.
27. **Grünberg, K., C. Wawer, B. M. Tebo, and D. Schüler.** 2001. A large gene cluster encoding several magnetosome proteins is conserved in different species of magnetotactic bacteria. *Appl. Environ. Microbiol.* **67:**4573–4582.
28. **Guerin, W., and R. Blakemore.** 1992. Redox cycling of iron supports growth and magnetite synthesis by *Aquaspirillum magnetotacticum*. *Appl. Environ. Microbiol.* **58:**1102–1109.
29. **Heyen, U., and D. Schüler.** 2003. Growth and magnetosome formation by microaerophilic *Magnetospirillum* strains in an oxygen-controlled fermentor. *Appl. Microbiol. Biotechnol.* **61:**536–544.
30. **Mandernack, K. W., D. A. Bazylinski, W. C. Shanks, and T. D. Bullen.** 1999. Oxygen and iron isotope studies of magnetite produced by magnetotactic bacteria. *Science* **285:**1892–1896.

30a. **Matsunaga, T., Y. Okamura, Y. Fukuda, A. T. Wahyudi, Y. Murase, and H. Takeyama.** 2005. Complete genome sequence of the facultative anaerobic magnetotactic bacterium *Magnetospirillum* sp. strain AMB-1. *DNA Res.* **12:**157–166.

31. **Moench, T. T., and W. A. Konetzka.** 1978. A novel method for the isolation and study of a magnetotactic bacterium. *Arch. Microbiol.* **119:**203–212.
32. **Nelson, D. C., and H. W. Jannasch.** 1983. Chemoautotrophic growth of a marine *Beggiatoa* in sulfide-gradient cultures. *Arch. Microbiol.* **136:**262–269.
33. **Okamura, Y., H. Takeyama, T. Sekine, T. Sakaguchi, A. T. Wahyudi, R. Sato, S. Kamiya, and T. Matsunaga.** 2003. Design and application of a new cryptic-plasmid-based shuttle vector for *Magnetospirillum magneticum*. *Appl. Environ. Microbiol.* **69:**4274–4277.
34. **Paoletti, L. C., and R. P. Blakemore.** 1988. Iron reduction by *Aquaspirillum magnetotacticum*. *Curr. Microbiol.* **17:** 339–342.

35. **Sakaguchi, T., N. Tsujimura, and T. Matsunaga.** 1996. A novel method for isolation of magnetic bacteria without magnetic collection using magnetotaxis. *J. Microbiol. Methods* **26:**139–145.
36. **Schübbe, S., M. Kube, A. Scheffel, C. Wawer, U. Heyen, A. Meyerdierks, M. Madkour, F. Mayer, R. Reinhardt, and D. Schüler.** 2003. Characterization of a spontaneous nonmagnetic mutant of *Magnetospirillum gryphiswaldense* reveals a large deletion comprising a putative magnetosome island. *J. Bacteriol.* **185:**5779–5790.
37. **Schüler, D., and E. Baeuerlein.** 1998. Dynamics of iron uptake and Fe_3O_4 biomineralization during aerobic and microaerobic growth of *Magnetospirillum gryphiswaldense*. *J. Bacteriol.* **180:**159–162.
38. **Schüler, D., and E. Baeuerlein.** 1996. Iron-limited growth and kinetics of iron uptake in *Magnetospirillum gryphiswaldense*. *Arch. Microbiol.* **166:**301–307.
39. **Schüler, D., S. Spring, and D. A. Bazylinski.** 1999. Improved technique for the isolation of magnetotactic spirilla from a freshwater sediment and their phylogenetic characterization. *Syst. Appl. Microbiol.* **22:**466–471.
40. **Schüler, D., R. Uhl, and E. Baeuerlein.** 1995. A simple light-scattering method to assay magnetism in *Magnetospirillum gryphiswaldense*. *FEMS Microbiol. Lett.* **132:**139–145.
41. **Schultheiss, D., R. Handrick, D. Jendrossek, M. Hanzlik, and D. Schüler.** 2005. The presumptive magnetosome protein Mms16 is a PHB-granule bound protein (phasin) in *Magnetospirillum gryphiswaldense*. *J. Bacteriol.* **187:**2416–2425.
42. **Schultheiss, D., M. Kube, and D. Schüler.** 2004. Inactivation of the flagellin gene *flaA* in *Magnetospirillum gryphiswaldense* results in nonmagnetotactic mutants lacking flagellar filaments. *Appl. Environ. Microbiol.* **70:**3624–3631.
43. **Schultheiss, D., and D. Schüler.** 2003. Development of a genetic system for *Magnetospirillum gryphiswaldense*. *Arch. Microbiol.* **179:**89–94.
44. **Simmons, S. L., S. M. Sievert, R. B. Frankel, D. A. Bazylinski, and K. J. Edwards.** 2004. Spatiotemporal distribution of marine magnetotactic bacteria in a seasonally stratified coastal salt pond. *Appl. Environ. Microbiol.* **70:**6230–6239.
45. **Spring, S., R. Amann, W. Ludwig, K. H. Schleifer, and N. Petersen.** 1992. Phylogenetic diversity and identification of nonculturable magnetotactic bacteria. *Syst. Appl. Microbiol.* **15:**116–122.
46. **Spring, S., R. Amann, W. Ludwig, K. H. Schleifer, H. Van Gemerden, and N. Petersen.** 1993. Dominating role of an unusual magnetotactic bacterium in the microaerobic zone of a freshwater sediment. *Appl. Environ. Microbiol.* **59:**2397–2403.
47. **Spring, S., and K. H. Schleifer.** 1995. Diversity of magnetotactic bacteria. *Syst. Appl. Microbiol.* **18:**147–153.
48. **Thornhill, R. H., J. G. Burgess, and T. Matsunaga.** 1995. PCR for direct detection of indigenous uncultured magnetic cocci in sediment and phylogenetic analysis of amplified 16S ribosomal DNA. *Appl. Environ. Microbiol.* **61:**495–500.
49. **Ullrich, S., M. Kube, S. Schübbe, R. Reinhardt, and D. Schüler.** 2005. A hypervariable 130-kb genomic region of *Magnetospirillum gryphiswaldense* comprises a magnetosome island, which undergoes frequent rearrangements during stationary growth. *J. Bacteriol.* **187:**7176–7184.
50. **Wallner, G., B. Fuchs, S. Spring, W. Beisker, and R. Amann.** 1997. Flow sorting of microorganisms for molecular analysis. *Appl. Environ. Microbiol.* **63:**4223–4231.
51. **Wolfe, R. S., R. K. Thauer, and N. Pfennig.** 1987. A capillary racetrack method for isolation of magnetotactic bacteria. *FEMS Microbiol. Lett.* **45:**31–36.

Growth of Electrode-Reducing Bacteria

DANIEL R. BOND

93

Living cells are not heat engines, capturing energy in boilers to spin turbines. Cells are electrochemical devices, which oxidize and reduce substrates in a series of discrete steps. For example, a hydrogenotrophic aerobe may oxidize hydrogen via a hydrogenase, allow electrons to flow through a series of electron transport proteins, and reduce oxygen at a terminal oxidase. This is essentially the same process that occurs in a hydrogen fuel cell, in which hydrogen is oxidized by a catalyst and electrons flow through a conductive pathway to another catalytic electrode, where oxygen is reduced. The similar principle driving electrochemical fuel cells and cellular metabolism has led to a host of real and imagined devices able to interface biological with electrical systems, and was even cited by Peter Mitchell in his early discussions of the chemiosmotic hypothesis in the 1960s (45).

The heart of any system designed to interconvert chemical and electrical energy is the catalyst. Microorganisms were shown to act as catalysts in electrical devices as early as 1911, when a mixture of glucose and yeast was used to lower the electrochemical potential of a solution and generate electricity in a fuel cell-like device (64). This phenomenon has been revisited or rediscovered numerous times over the past century (12, 14, 77) with waves of interest often coinciding with changes in the global energy outlook. For a more historical perspective, the reader is referred to articles by Lewis (31), Gray Young et al. (18), Bennetto (3), and Stirling et al. (79).

A fundamental assumption, based on multiple observations with common fermentative and respiratory bacteria, was that structures such as the cytoplasmic membrane, cell wall, outer membrane, S layer, and/or capsule insulated cellular redox proteins to prevent the transfer of electrons to external surfaces. Thus, to obtain reducing equivalents on the outside of microorganisms, where they could be harvested by an electrode, various strategies were employed: (i) use of organisms which excreted reactive compounds (such as hydrogen), which could diffuse to an electrode surface (13, 46, 53, 54, 72–74); (ii) addition of soluble redox-active mediators (such as ferricyanide, thionine, methylene blue, neutral red, or resorufin) to react with membrane bound proteins and shuttle electrons to electrode surfaces (11, 15, 28, 41, 56, 57, 71); or (iii) elimination of the cell membrane barrier by use of purified enzymes, either attached to electrode surfaces or in concert with mediators (38–40, 78). Fuel cells containing a biological component, as live cells excreting compounds, as live cells interacting with mediators, as enzymes catalyzing key steps, or any combination thereof, have been termed "biofuel cells," "biological fuel cells," "biochemical fuel cells," "bioelectrochemical fuel cells," "biocatalytic fuel cells," and "microbial fuel cells," depending on current trends (24).

The discovery of bacteria able to reduce insoluble metals, such as Fe(III) and Mn(IV) oxyhydroxides, demonstrated that bacteria could electrically interact with surfaces beyond their inner membrane (36). Recent enrichments and work with pure cultures have shown that bacteria able to use electrodes as acceptors or donors for cellular processes exist in a variety of environments, initiating renewed interest in the century-old idea of biological augmentation of electrodes. The existence of electrode-reducing (and electrode-oxidizing) bacteria has shown that electrochemical devices in which bacteria serve as catalysts are possible.

BACTERIA AND BACTERIAL COMMUNITIES ABLE TO DIRECTLY INTERACT WITH ELECTRODES

Perhaps the first report to link an ability to reduce Fe(III) with an ability to interact with an electrode was published in 1999 (25). Kim et al. observed electrochemical features in cyclic voltammograms of dense cell suspensions using anaerobically grown cells of *Shewanella putrefaciens* strain IR-1. As these features were not present in aerobically grown cells, the authors suggested that proteins known to exist in membranes of related organisms (47, 48 , 49) were interacting with the electrode. When this organism was cultivated using a potentiostat to poise a graphite felt electrode at a relatively high potential, (1 V versus Ag/AgCl), a small electric current (3 μA, or 2.4×10^{-5} A/m^2) was measured, representing about 0.02% of the electron flux predicted by the rate of incomplete lactate oxidation by the organism. While the electrode in these first experiments was poised at a potential high enough to cause oxygen evolution, subsequent experiments incubated dense cell suspensions of *S. putrefaciens* IR-1 with a graphite electrode which was connected to a second graphite electrode exposed to oxygen. This arrangement created a basic fuel cell with an electron recovery efficiency of about 5% (at a density averaging 6×10^{-3} A/m^2). This demonstration produced decreasing amounts of electrical current for about 1 week (27).

A similar fuel cell-like device was reported by Reimers et al. (70), who used marine sediments as the source of electron donors and bacteria and electrodes exposed to oxygen as the electron acceptor. Subsequent work with this system led to the identification of bacteria from the Fe(III)-reducing family *Geobacteraceae* as organisms enriched on many of these fuel cell anodes. Pure cultures of *Desulfuromonas acetoxidans* and *Geobacter metallireducens* were used to demonstrate cell growth and electricity production at the expense of complete acetate or benzoate oxidation in both fuel cells and potentiostat-poised systems (200 mV versus Ag/AgCl), with electron recoveries of 82 to 84% and current densities averaging 0.06 A/m^2 (4, 80). These bacteria could be grown on electrode surfaces and maintained as stable attached populations for months without significant current degradation.

Since these initial reports, a variety of fuel cells and poised-potential systems have been used to enrich bacterial communities able to interact with electrodes in the absence of added mediators (26, 30, 32, 34, 35, 44, 58, 63, 66, 68). Some of the bacteria detected in these consortia have also been studied in pure culture. Examples of fermentative organisms able to divert a portion of electrons to electrode reduction include *Aeromonas hydrophila* (60), *Clostridium butyricum* (59), *Pseudomonas aeruginosa* (65), *S. putrefaciens* (27), and *Desulfobulbus propionicus* (20). Others, such as *Geobacter sulfurreducens* (5), *G. metallireducens* (4), *G. electrodophilus* (22), *Desulfuromonas acetoxidans* (4), *Rhodoferax ferrireducens* (7), and *Geothrix fermentans* (6), are capable of completely oxidizing substrates and coupling this oxidation to reduction of electrodes.

Even this limited survey of microbial communities and pure cultures demonstrates that there is a diversity of organisms able to act as catalysts at conductive surfaces and that these organisms can work as a consortium to oxidize complex organic matter. Ironically, when devices have been operated under conditions designed to enrich for bacteria that can reduce electrodes in the absence of mediators, some of these experiments have yielded bacteria that produce their own mediators to facilitate electrode reduction (6, 65). This phenomenon is reminiscent of findings demonstrating differences in strategies for accessing and reducing Fe(III) oxides (50–52).

The availability of bacteria able to interact with electrodes presents opportunities for conversion of organic compounds into electricity. While the obvious application of this phenomenon involves power generation, it also implies new routes to biological sensing and control of oxidative or reductive biocatalytic reactions. In addition, the molecular mechanisms that mediate electron transfer to external acceptors are of much interest, and the use of electrodes presents a unique opportunity to study this process in the absence of geochemical variables which are difficult to control when working with metals. Towards this goal, devices used to study electrode-reducing (or oxidizing) bacteria should strive for control of the microbial environment, with attention given to use of defined electrode materials and potentials, diffusion and mass-transfer limitations, and oxygen infiltration.

TERMINOLOGY

Before discussion of the components of a system designed to demonstrate bacterial-electrode interactions, it is useful to review some basic terminology. Systems in which bacterial oxidation provides electrons for spontaneous reduction of oxygen, for example, are galvanic cells, analogous to discharging batteries or fuel cells. The electrode where oxidation takes place (e.g., microbial oxidation of acetate, or other organic matter) is termed the anode, while the electrode where reduction of oxygen takes place is termed the cathode. For electrons to flow via an external circuit between these two electrodes, a pathway for equilibrating the buildup of charge (H^+) near the anode is required, familiar to most as the salt bridge. A cartoon example of a fuel cell with electrodes and other components labeled is shown in Fig. 1.

The potential between these electrodes is dependent upon the two half-reactions present in the system; in the example of acetate and oxygen, the standard reduction potentials of the CO_2/acetate and O_2/H_2O half-reactions at pH 7 are −0.28 and 0.82 V, respectively, predicting a maximum potential difference of 1.1 V. This is the potential that could exist with perfect catalysts if the system were at equilibrium, with no current flowing through the system (open

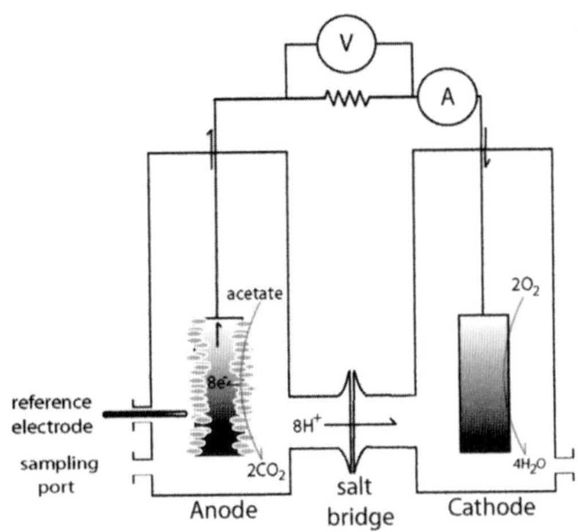

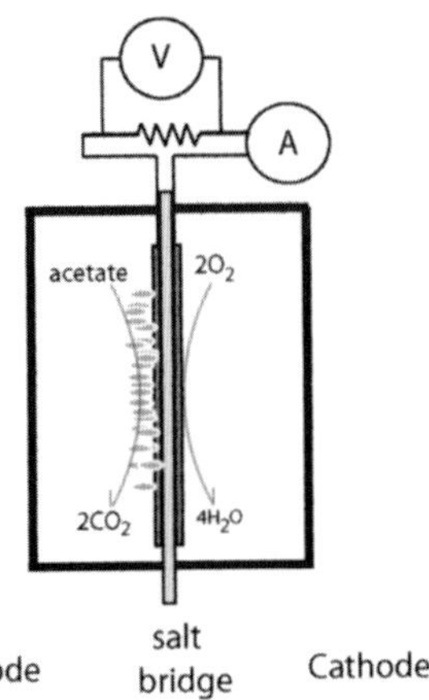

FIGURE 1 Basic schematics of a two-chambered fuel cell (top) and a fuel cell in which electrodes are pressed directly to a membrane (bottom). Naming conventions and direction of electron flow are based on the bacteria colonizing the anode and acting as catalysts for oxidation of an organic electron donor, with oxygen serving as the ultimate electron acceptor.

circuit), at 1 M acetate and 1 atm of O_2. This ideal potential can be adjusted by using the Nernst equation to reflect actual conditions.

As current is allowed to flow from the anode, the organism oxidizing an electron donor uses some energy for its own processes and donates electrons to the electrode at a higher potential than what is predicted by the corresponding half-reaction, thus reducing the apparent voltage between the electrodes. In addition, with increasing rates of current flow, limitations common to all galvanic cells become a factor. Electrode kinetics, such as slow transfer of electrons from bacteria to the anode surface or slow reduction of oxygen, reduce the apparent voltage. As the rate of current flow increases, losses due to factors such as slow exchange of protons between electrodes or mass transfer limitations affecting the concentration of reactants at electrode surfaces can create a significant reduction in potential and limit the amount of energy available to do work.

Such losses are also termed overpotentials, as they contribute to the difference between the maximum, or equilibrium potential, and the observed operating potential of a fuel cell. This overvoltage, or polarization, represents energy not available as potential between electrodes. Any system which is designed to allow microorganisms to operate at or near their full capacity (and obtain the maximum amount of power) must strive to limit these overpotentials by maximizing reactant concentrations, optimizing electrode kinetics, lowering internal resistance, and limiting mass transfer issues at electrode surfaces. Reporting the performance of an organism grown in a device dominated by overpotentials and ohmic losses is analogous to reporting the performance of a car with the parking brake partially engaged.

Interconverting current flow and reactant consumption is fairly straightforward, as long as an accurate measure of current is obtained. Under many, but not all conditions, continuous measurement of cell voltage across a precision resistor allows one to use Kirchhoff's and Ohm's laws to calculate amperage. At low rates of current flow ($\sim$10 μA), this approach becomes imprecise and direct measurement with an ammeter is required. While most multimeters possess a current measurement function, they apply a voltage burden that influences measurements, and thus a picoammeter should be used. In addition, as factors such as internal resistance vary with current flux and external resistance conditions, current-voltage relationships may not be ohmic and combined voltmeter- and ammeter-based measurements of current are highly recommended (instead of current inferred from voltage across a resistor) for accurate characterization, even of devices producing current in the milliampere range.

Current, measured in amperes, is the measure of electron flow (as coulombs per second, where 1 C = 6.24×10^{18} electrons). Thus, integrating current over time produces the number of electrons, in coulombs, which have passed from one electrode to another. Faraday's law allows conversion of coulombs into electrons (96,486 C/mol of electrons), for comparison to the amount of reactant oxidized. If 1 mol of acetate is completely oxidized during operation of a fuel cell, 8 mol of electrons was theoretically available. If this oxidation occurred in a fuel cell operating at 1 A for 200 h, the formula (200 h $\times$ 3,600 s/h $\times$ 1 C/s $\div$ 96,486 C/mol of electrons) yields the result that 7.4 mol of electrons were harvested by the fuel cell, equal to 92.7% coulombic efficiency. Energy efficiency can be calculated in a similar fashion, by calculating integrated power harvested (amperage $\times$ voltage $\times$ seconds = joules) to various theoretical energy values, including the potential energy available based on the open circuit potential, the ΔH, or ΔG of the substrates.

ESSENTIAL COMPONENTS OF BACTERIAL ELECTRODE DEVICES

There is no off-the-shelf device commonly used to study bacteria able to reduce electrodes, although there is a two-chambered "microbial fuel cell" sold for educational demonstrations of mediator-driven power production by fermentative organisms by the National Centre for Biotechnology Education. Any fuel cell design requires two electrodes, an electron donor and acceptor, and a mechanism for separating the two (salt bridge) (Fig. 1).

The most common electrode material used as the anode for microbial attachment is carbon or graphite, which can be purchased in a variety of ink, paper, felt, or solid forms. Graphite does not corrode, is relatively inexpensive, and can be chemically functionalized or coated. There are many suppliers of solid graphite which machine electrodes to any size or shape (e.g., Graphite Engineering and Sales, Greenville, Mich.), and extruded graphite of medium grain size ($\sim$0.08 cm) with a bulk density of 1.6 to 1.8 g/cm^3 is common.

Carbon papers of varying thicknesses and porosities are typically sold as gas diffusion barriers for hydrogen fuel cells (e.g., E-TEK [De Nora N.A. Inc, Somerset, N.J.]), but they should be purchased without the wetproofing often added to prevent flooding of gas electrodes. High-surface-area carbon black used in printed or spray-coated electrodes can be obtained from specialty chemical suppliers (e.g., Vulcan XC72R from Cabot Corporation, Boston, Mass.). While the impulse may be to use high-surface-area materials (such as a thick piece of carbon felt), diffusion in and out of these materials, especially when the pore structure becomes clogged with bacteria, likely limits the useful surface area. In addition, compared to the goal of a consistent, optimized environment for all bacteria, microporous or thick materials support stagnant and low-nutrient zones and result in a continuum of physiological responses.

While graphite and carbon preparations are described as being highly pure, many of these materials contain significant contaminating levels of Fe(II), SO_4^{2-}, and even solvents which must be removed by pretreatment, such as soaking in 1 N HCl, and rinsing in deionized water, until contaminants can no longer be detected. Failure to remove impurities by pretreatment can introduce mediator compounds, electron acceptors, or even organic donors.

Connections to electrodes should be made watertight and with low-resistance materials. Use of solid electrodes presents the opportunity to directly screw marine-grade connectors into graphite (Impulse, San Diego, Calif.) and use connectors designed to withstand marine environments as well as moderate pressure. With papers or felts, this is not feasible, and wires are most often attached using conductive epoxy (e.g., silver epoxy; Epoxy Technology, Billerica, Mass.), followed by a sealing coat of standard marine-grade epoxy. Connections made in this fashion usually fail after repeated cycles of autoclaving or prolonged use, exposing wiring which can corrode and affect observed potentials or create additional electrochemical features in voltammetry studies. Wire insulated for use in marine environments is highly recommended for long-term use.

The choice of the electrode material to be used as the cathode again depends on the goals of the experiment.

Many researchers are pursuing designs aimed at producing electrical power, and these designs must address multiple issues of cost, surface area, permeability, and longevity. When the goal is primarily to study the interactions between bacteria and electrode surfaces under controlled conditions, it may be desirable to construct a device that limits the influence of variables introduced by the electrode or salt bridge. For instance, smoother electrodes lower available surface area and reduce current density but create more consistent conditions at the surface.

Highly reactive (and more expensive) platinum and Pt-treated cathodes developed for use in hydrogen and methanol fuel cell applications produce significantly higher half-cell potentials than those of plain carbon or graphite cathodes in any fuel cell, and microbial fuel cells are no exception (8, 33, 61). Such materials are easily obtained from companies which supply materials for fuel cell research (e.g., E-TEK [De Nora]). When carbon-based cathodes are employed for reasons of economy or design, slower rates of oxygen reduction should be expected, especially if the cathode is immersed in water in which oxygen solubility is very poor (<8 mg/liter at 30°C). To increase the availability of electron acceptor to submerged cathodes, $FeK_3(CN)_6$ is often added to cathode chambers (7, 55).

Under most controlled experimental conditions, mixing of electron donors and acceptors is undesirable, and electrode compartments must be kept separate. However, since electrons are being continuously removed, a cation exchange pathway able to sustain flux equal to the rate of negative charge movement must be provided. In systems where an anode is immersed in anoxic sediment and connected to another in overlying aerobic water, the sediment-water interface forms a natural salt bridge (23, 70).

When two liquid- or air-filled chambers are joined, a salt bridge can be fashioned out of any semipermeable material, but the most common choice in moderate-temperature, aqueous systems is a cation exchange membrane, such as the perfluorosulfonic acid-based Nafion (Du Pont) membranes. These hydrophobic materials contain sulfonic acid ion clusters linked to a Teflon-like backbone. When hydrated, the sulfonic acid sites provide a mechanism for proton movement. A side effect of the presence of hydrated channels in perfluorosulfonic acid membranes is an unavoidable pathway for gas diffusion and for diffusion of small molecules such as methanol. Nafion 117 is commonly used in fuel cell fabrication, as it is mechanically robust and can be used in a variety of applications. Membranes should be cleaned and protonated via soaking in H_2O_2 (3%, 80°C, 1 h) and 0.5 M H_2SO_4 (80°C, 1 h) before use.

The simplest method for producing electrical signals using bacteria as anode catalysts involves suspension of anodes and cathodes in separate liquid-filled glass or machined polycarbonate chambers, which are separated by a membrane. Such a device is easy to construct, sterilize, and understand in the context of common microbial growth equipment, but it is not designed for maximum charge or reactant transfer. Variations on this theme are described by Park et al. (56), Bond and Lovley (5), Rabaey et al. (66), and Logan et al. (43). When the goal shifts to reactors designed to oxidize wastes, further variations have been demonstrated (9, 19, 43, 67, 76).

As microbial fuel cells are improved to reduce overpotentials and integrate findings gained from decades of chemical fuel cell research, electrodes and membranes are typically moved closer together. A logical result of this integration is construction of membrane electrode assemblies, via hot-pressing of electrode materials onto semipermeable membranes at temperatures above the glass transition temperature of the membrane, producing a sandwich containing the anode, membrane, and cathode (62). The membrane electrode assembly-based design guarantees that electrodes are at repeatable locations as close together as possible, allows use of air-exposed cathodes, and can be clamped in devices designed for filtration or membranes.

SEDIMENT FUEL CELLS VERSUS LABORATORY FUEL CELLS VERSUS POTENTIOSTATS

A fuel cell dependent upon microbial activity can be constructed using easily obtained materials and a sample of anoxic sediment. For small classroom demonstrations, this can be as simple as wire wrapped around graphite rods and sealed with marine epoxy. For long-term experimental observations, higher-quality electrodes and more robust connections are desired. One electrode is immersed a few centimeters into a sample of anoxic sediment, and the other is suspended a few centimeters above in the aerobic water. A potential is observed between the two electrodes, so long as the water above the sediment remains aerobic and the sediment near the anode remains anaerobic.

If a small (100- to 1,000-Ω) resistor is placed between the two electrodes, current flow can be monitored with a small voltmeter. Initial power from such a demonstration is likely due to oxidation of reduced compounds [such as Fe(II), H_2S, and reduced humic acids] at the anode. However, if operated over a long period of time, microbial activity at or near the electrode is required to sustain the current flow, either due to regeneration of reduced compounds or direct reduction of the anode.

After a period of weeks to months, the microbial community on and near the anode will be expected to be significantly altered, as a new niche has been created favoring growth of organisms which can utilize the electrode as an electron acceptor or utilize oxidized intermediate compounds (21, 80). If control electrodes are incubated under identical conditions, but without current flow (unconnected external circuit), the bacteria present on each electrode surface can be compared via standard or molecular methods to determine enriched populations. Simply comparing bacteria present on electrodes to the community in the nearby sediment is less likely to identify key populations, as a subset of organisms naturally attach to graphite surfaces regardless of their role as an electron acceptor.

A two-compartment laboratory fuel cell allows an experiment of similar simplicity, but with added control, as substrate concentrations, mixing, sampling, and other aspects are easily manipulated. Attention should be paid to avoid gas-permeable materials such as silicone wherever gaskets, seals, O-rings, or stoppers are located. Flexible butyl rubber stoppers (such as those available for anaerobic pressure tubes from Bellco, Vineland, N.J.) and Viton seals can reduce gas exchange through sampling or wiring access points. Even with these precautions, any seal, as well as the cation exchange membrane, presents an unavoidable site for gas exchange into the system.

Oxygen entry into a fuel cell, at best, allows aerobic growth, enriches non-electrode-reducing bacteria, and reduces coulombic yields. At worst, it can kill oxygen-intolerant cultures. Continuous flushing of the anode chamber with oxygen-free gas can remove oxygen but may also remove intermediates such as hydrogen produced by anoxic cultures.

Lowering of hydrogen partial pressures via flushing may encourage growth of fermentative bacteria dependent on low hydrogen partial pressures rather than bacteria dependent on electrode reduction. Oxygen entry via the membrane can also be limited by immersing the cathode in an aerobic buffer which limits the effective oxygen concentration at the membrane but also reduces oxygen availability to the cathode. Use of ferricyanide in the cathode chamber provides a compromise by lowering the concentration of oxygen but maintaining a high level of electron acceptor. An additional concern is that if the anode chamber employs a CO_2/bicarbonate-based buffer system, long-term losses of CO_2 via the membrane can possibly lead to pH control issues.

An additional level of control over the electrode environment can be attained by poising a working electrode at a set potential compared to a reference electrode. Thus, if electrons are donated to the working electrode at or below a set potential, electrons will flow to the counter electrode. An advantage of this approach is that both electrodes can be maintained under anaerobic conditions. While this type of experiment requires additional hardware, the facts that the site of microbial growth is highly controlled and current production by the bacteria is not limited by aspects related to cathode performance or oxygen leakage make this approach very attractive.

A combination of the standard fuel cell and poised-electrode approaches is the use of potentiostats to perform analyses such as cyclic voltammetry and electrochemical impedance spectroscopy on mature microbial biofilms, which can demonstrate the presence of redox-active proteins or mediator pools on or near electrodes (66).

PHYSIOLOGY OF ELECTRODE-REDUCING BACTERIA

The phenomenon of electricity production by bacteria is rooted in their energy metabolism; thus, the physiology of individual bacteria can have a significant impact on the underlying mechanisms of electron flow. Cells able to use electrodes as electron acceptors may be strict anaerobes or facultative aerobes, capable of complete oxidation of substrates or only partial fermentation, or may be able to reduce insoluble electron acceptors as well as soluble electron acceptors. Phenotypes such as these obviously play a key role in the selection of populations using electrodes as acceptors and the response of cells in pure culture situations.

When sugars, or wastewater from such sources as starch processing, are used as the feed for microbial fuel cells, bacteria predicted to have a more fermentative strategy are most often found on or near electrodes acting as electron acceptors. Based on 16S rRNA-based surveys of electrode surfaces in fuel cells fed fermentable substrates, relatives of known clostridia, pseudomonads, and shewanellae have been noted as being highly enriched (26, 35, 66). These organisms are able to generate a significant amount of ATP via substrate level phosphorylation and likely benefit from the electrode as a mechanism for reducing equivalent disposal. In normal ecosystems, processes such as methanogenesis play this role.

In pure culture, fermentative isolates such as *Clostridium butyricum* (59), *Aeromonas hydrophila* (60), *Shewanella putrefaciens* (27), and *Desulfobulbus propionicus* (20) have shown an ability to divert a small percentage of available electrons to electrodes, yet their enrichment at electrodes suggests that even this limited diversion of electron flow presents a selective advantage to the organisms. In addition, while pure cultures of these organisms typically demonstrate only a limited ability for coupling metabolism to electrode reduction, coulombic efficiencies and power production rates from fermentable substrates are typically higher in mixed-culture communities, suggesting that beneficial interactions in functioning microbial fuel cells have yet to be uncovered, such as the production of mediators demonstrated by Rabaey et al. (65).

When less fermentable substrates, such as short-chain fatty acids or aromatic compounds, are used as the feed in microbial fuel cells, a very different microbial community is obtained. Bacteria dependent upon respiration, such as δ-proteobacteria and *Geothrix* (2, 4, 21), are consistently enriched on electrode surfaces, as well as a variety of other α- and β-proteobacteria, *Bacteroides*, and *Cytophaga*-like organisms. Only a subset of these organisms have been demonstrated in pure culture to couple complete organic matter oxidation to electrode reduction, including representatives of the *Geobacteraceae* (4, 5, 22), *Geothrix fermentans* (6), and *Rhodoferax ferrireducens* (7). These organisms have been demonstrated to couple complete oxidation of organic electron donors to electrode reduction and likely are dependent on a similar strategy in the environment, using electron acceptors such as Fe(III) instead of electrodes.

While numerous parallels between Fe(III) reduction and electrode reduction have been noted, the variety of organisms able to reduce electrodes and results from pure-culture work suggest that reduction of electrodes is more straightforward than reduction of metal oxides. A prime example of this difference is the organism *R. ferrireducens*, which can reduce chelated Fe(III) [as Fe(III)-NTA] but cannot grow using insoluble Fe(III) oxides as electron acceptors, or even the soluble humic acid analog anthraquinone 2,6,disulfonate (AQDS) (17). However, this organism does colonize and reduce solid graphite electrodes (7). This suggests that a mechanism for reducing chelated Fe(III) is required for electrode reduction but a more complicated strategy for interacting with solid minerals is not. The finding that many organisms not known for their ability to conserve energy from Fe(III) reduction (e.g., *Clostridium*, *Pseudomonas*, and *Aeromonas* spp. [23, 26, 66]) can interact on some level with electrodes further suggests that less specialized electron transfer mechanisms may be required for at least a demonstration of electrode reduction compared to environmental metal oxide reduction.

Reduction of Fe(III) oxides in *Geobacter* spp. has been correlated with a complex series of events, including sensing, attachment, expression of cytochromes, and most recently, specialized pili (10, 29, 37, 69). Washed cell suspensions of *Geobacter* typically have a very poor ability to reduce Fe(III) oxides, likely due to the need for expression and attachment of cell structures allowing electron transfer to Fe(III). The fact that washed cell suspensions of some bacteria such as *Shewanella* spp. demonstrate an ability to transfer electrons to electrodes again indicates that electrodes may have access to cell surfaces in ways that Fe(III) oxides may not.

Furthermore, as Fe(III) typically exists as an oxide film or crust on other sediment particles, cells likely need to detatch from reduced or Fe(II)-coated surfaces to seek new oxidized zones. Electrodes pose fewer challenges. Surfaces do not become passivated by accumulating reduced metals, do constantly act as electron acceptors, may even attract cells via galvanotaxis (1, 75), and may serve as sites for accumulation of mediator compounds due to cell lysis or direct excretion. Thus, there is likely a wider diversity of bacteria that can at least weakly interact electrochemically with electrodes, compared to bacteria able to utilize Fe(III) oxides as physiological electron acceptors.

A SAMPLE FUEL CELL EXPERIMENT USING AN ELECTRODE AS AN ELECTRON ACCEPTOR

While devices used for the study of electrode-reducing bacteria are typically custom-constructed to meet the needs of the experiment or organism being studied and procedures are continually evolving to obtain higher current densities or allow voltammetry measurements, some sample experimental protocols that illustrate issues of sterility and anaerobic technique are provided. The following describes a method for inoculating a washed culture of *Geobacter sulfurreducens* into a glass two-chambered device separated by a cation-exchange membrane, using an electrode as an electron acceptor and acetate as the electron donor. A similar procedure could be used to prepare an anaerobic chamber for inoculation with an environmental sample for enrichment of electrode-reducing bacteria.

1. Many bacteria, if continually cultured in medium with soluble electron acceptors such as Fe(III) citrate or fumarate, will only poorly reduce extracellular electron acceptors when transferred to such substrates. Based on the parallels between electrode reduction and Fe(III) reduction, it is prudent to routinely transfer cultures with Fe(III) or Mn(IV) oxides as sole electron acceptors, and limit the concentration of potential chelators in the medium (e.g., keeping concentrations of carboxylic acids low and using nitrilotriacetic acid (NTA)-free trace mineral preparations). When cells will be needed for experiments, transfer them into a medium with soluble electron acceptors and culture them for at least three serial transfers to remove particulates.
2. Check all electrode connections by using a standard voltmeter, and inspect connections for signs of cracking or leaking. Wash electrodes to be exposed to bacteria by treating in 1 N NaOH and 1 N HCl, rinsing between treatments to remove adherent bacteria, protein, lipids, and ions. Maintain electrodes in a wetted state. Check all seals, stoppers, and sites of potential oxygen entry for cracks or leakage.
3. Hydrate and clean a sample of cation exchange membrane by heating in 3% H_2O_2 (80°C), rinsing, and heating in 0.5 N H_2SO_4 (80°C). Store membranes hydrated in deionized water.
4. Assemble the fuel cell for autoclaving, with electrodes and membranes in place, but with all sampling ports open and covered with foil to allow venting. Filling the chambers with deionized water helps prevent gas from permeating porous electrodes.
5. Autoclave the fuel cell. Heating the system has the added benefit of removing most gases. When the device is still hot to the touch, flush all chambers with a slow stream of sterile, oxygen-free gas for at least 2 h. This continues to purge the system free of oxygen, as most solid, felt, foam, and paper graphite products are porous. Seal the fuel cell and allow to stand overnight or until use. Gases will continue to diffuse from electrodes, plastics, or other sources. Alternatively, one can transfer the entire fuel cell into an anaerobic chamber to equilibrate after autoclaving.
6. At this point, if any reference electrodes are to be included for monitoring of half-cell potentials, sterilize them by soaking for at least 15 min in 1 N HCl, followed by rinsing in 95% ethanol, and air drying before introducing electrodes into the chamber through a port. Once chambers are ready for filling with medium, use a stream of sterile oxygen-free gas to drive liquid from the chamber(s) from a sampling port. Again using sterile oxygen-free gas, pressurize a previously prepared bottle of growth medium lacking electron donors or acceptors and transfer the growth medium into the anode chamber. If the cathode chamber is to be filled with liquid, also fill it at this time. Flush both chambers with sterile oxygen-free gas to remove any oxygen which may have been introduced during the filling procedure. Add any electron donors to the anode chamber if needed, and bring the fuel cell to the appropriate temperature.
7. Working in an anaerobic chamber, transfer the bacterial culture into sterile, sealable centrifuge bottles which have been preincubated in the anaerobic chamber for at least 2 days. Pass cells out of the anaerobic chamber, centrifuge the culture, pass it back into the chamber to decant the supernatant, carefully suspend cells in a sample of anaerobic growth medium, and repeat. Upon resuspending the cells a second time, slowly draw washed cells into a syringe which can be used to inoculate the anode chamber(s).
8. Note that if the goal of the experiment is to demonstrate colonization and growth of electrodes, only a small inoculum is needed. For instance, a 50-cm^2 solid graphite electrode typically contains only about 3 to 5 mg of cell protein when fully colonized. A 100-ml culture of *Geobacter sulfurreducens* grown to half-maximal density with 40 mM fumarate as the electron acceptor contains at least 5 mg of total cell protein—if this entire quantity is introduced into an electrode chamber, full colonization of the electrode can theoretically occur in the absence of any growth or protein synthesis. Incubation with large amounts of cells also creates the possibility that even if significant cell lysis occurred, releasing shuttling compounds or even electron donors, a large population of remaining cells could utilize these compounds and confound results. Electricity production by dense cell suspensions, where only a small percentage of the population needs to transfer a small percentage of electrons from oxidation to an electrode, is not equivalent to an experiment demanding exponential growth from all members of the population at the sole expense of electricity production.
9. Once cells have been introduced into the anode chamber, close the electrical circuit, typically using a small resistor (100 to 1,000Ω). At this point, if needed, begin exposing the cathode to oxygen. Waiting until this time to introduce oxygen to the cathode reduces the amount of oxygen initially present in the anode chamber.
10. Current production will likely not be immediate but should rise after cells have had time to recover from the stress of washing, express necessary proteins, and colonize electrodes. Failure to colonize electrodes can usually be traced to the presence of oxygen, dirty electrodes, pH issues due to flushing of bicarbonate-buffered media without an appropriate CO_2 headspace, loss of the phenotype by the culture, or small holes in the Nafion.
11. Once current production has been established and begins to drop due to depletion of electron donor, additional electron donor can be added. Typically, current pro-

duction achieves a plateau characteristic for a given electrode and concentration of electron donor, suggesting full colonization of an electrode. If the medium in the anode chamber is removed and replaced with a nongrowth medium under anaerobic conditions, a relatively stable layer of cells can be maintained on the electrode surface for further characterization.

The resistance value chosen initially in fuel cell-based experiments is often an educated guess, but it is useful to monitor half-cell potentials and polarization (current-voltage) relationships in fuel cells under a variety of resistance values to determine sustainable rates of current flux and to identify where overpotentials exist. A useful discussion and application of this technique can be seen in sustainable power measurements such as those described by Menicucci et al. (42). For instance, selection of a resistance value that is too large may place organisms in an acceptor-limited starvation mode, while use of a too-small resistance value could raise the overall redox potential of the medium to a level which is inhibitory to an anaerobe.

USING A POTENTIOSTAT OR POWER SUPPLY TO POISE AN ELECTRODE AS AN ELECTRON ACCEPTOR

When a potentiostat is used to poise the potential of an electrode so it may act as an acceptor (or donor) for bacterial growth, the rule of providing adequate time to remove gases trapped in electrodes and materials in a sterile chamber applies. However, in this case, since the potentiostat can drive reduction of protons (or oxidation of water) to dispose of, or obtain, electrons as needed, both chambers can be prepared and maintained in an anaerobic state at all times.

While use of a poised system adds a desirable level of control to the study of electrode-interacting bacteria, imposed potentials likely lead to oxidation of minerals and adsorption of ions at the electrode surface and in the electric double (or Helmholtz) layer. Thus, as electrodes are incubated in a medium containing various salts and minerals, one would expect the formation of a layer (or layers) near the electrode, where the environment is quite different from the bulk medium. How cells sense, or adapt to, this possibly stressful local environment is an unknown introduced into such experiments.

An additional point of concern with poising is the quality and care of the reference electrode: almost any standard Ag/AgCl electrode provides a reference potential of approximately +200 mV (versus a standard hydrogen electrode [SHE]), depending on the exact chemistry of the electrode. Silver-based electrodes are sensitive to UV light, and even mild concentrations of base can cause formation of Ag_2O, which can clog frit pores or alter their potential. Exposure to ammonia and exposure to sulfide can both lead to formation of insoluble complexes, with silver sulfide having a characteristic black color. Briefly soaking in mild acid or storage in 3 M NaCl or KCl (depending on the manufacturer's recommendations) can dissolve many of these deposits. In addition, if bubbles are present near the frit of any electrode, the electrode resistance will be very high and the potentiostat will have difficulty equilibrating.

1. Prepare an anoxic two-electrode device, making sure to include a reference electrode positioned as close to the working electrode as is reasonable. Once the chamber is filled with growth medium, connect it to the potentiostat.

2. Set the potentiostat to poise the working electrode to an appropriate potential. Many Fe(III) compounds have half-cell potentials between 0 and 400 mV (versus SHE), making this range common. Thus, to poise the electrode at +400 mV versus SHE, choose a setpoint of +200 mV more positive than an Ag/AgCl reference electrode (actual electrode potentials vary with chemistry and temperature). Potentials that approach the midpoint potential of water oxidation or proton reduction should be avoided. For example, even a potential of +500 mV versus Ag/AgCl causes electrolysis and liberation of oxygen into the growth chamber to a level which would not create visible bubbles but would allow growth of organisms around electrodes, and this could be wrongly interpreted as electrode-dependent growth.

3. After choosing the setpoint of the potentiostat, a burst of current should be observed, representing the initial capacitance of the working electrode. This is dependent on the surface area of the electrode, and current flow to/from the electrode should decay rapidly. Sometimes, current flow can rapidly oscillate from high to low values during this initial poising, which typically occurs when the reference impedance is very high, or when solution resistance is high. The best solution to this problem is to check the reference electrode for signs of clogging or blocked frits and use clean electrodes and membranes. A stopgap measure that alleviates the problem is to place a 0.1- to 1-μF capacitor between the counter electrode and the reference electrode.

4. Allow the poised electrode to equilibrate long enough to establish a stable baseline current, which must be subtracted from all subsequent measurements to derive an accurate integrated current flow. Once this baseline is obtained, cells can be introduced into poised electrode chambers for monitoring of cell growth.

A version of this experiment can also be conducted without a potentiostat by using a precise power supply, a relatively small working electrode, and a reference electrode that acts as the electron sink (or source). In this case, a working electrode can be suspended in a small anaerobic chamber and connected to the negative pole of the power supply. The positive pole is connected directly to the reference electrode (such as a calomel electrode), which can be linked to the growth medium via a saturated KCl-filled capillary capped by a porous frit. For instance, based on the potential of the saturated calomel electrode at 30°C, if the power supply is set at 0 V, this will poise the working electrode at +238 mV versus the SHE. Many power supplies that can inexpensively attain the resolution necessary for this application are available, allowing posing of multiple electrodes for a fraction of the cost of using potentiostats. To monitor current, a sensitive ammeter will still be required.

PREPARATION OF A SAMPLE MEDIUM FOR GROWTH OF ELECTRODE-REDUCING BACTERIA

Numerous medium recipes are available in the literature, but an example of an anaerobic basal salts medium able to support growth of *Geobacter* is provided here as an example. Also indicated are alterations to add electron donors to or adjust for use in non-CO_2-buffered systems.

1. To prepare basal anaerobic medium containing no electron donors or acceptors, add the following to ~950 ml

of double-distilled water: 0.38 g of KCl, 0.2 g of NH_4Cl, 0.06 g of NaH_2PO_4, 0.04 g of $CaCl_2 \cdot 2H_20$, 0.2 g of $MgSO_4 \cdot 7H_2O$, 10 ml of vitamin mix (containing [per liter] 2 mg of biotin, 2 mg of folic acid, 10 mg of pyridoxine HCl, 5 mg of riboflavin, 5 mg of thiamine, 5 mg of nicotinic acid, 5 mg of pantothenic acid, 0.1 mg of cobalamin, 5 mg of para-aminobenzoic acid, and 5 mg of thioctic acid), 10 ml of trace mineral mix (containing [per liter] 1.5 g of NTA, 0.1 g of $MnCl_2$, 0.5 g of $FeSO_4 \cdot 7H_2O$, 0.17 g of $CoCl_2 \cdot 6H_2O$, 0.10 g of $ZnCl_2$, 0.03 g of $CuSO_4 \cdot 5H_2O$, 0.005 g of $AlK(SO_4)_2 \cdot 12\ H_2O$, 0.005 g of H_2BO_3, 0.09 g of Na_2MoO_4, 0.05 g of $NiCl_2$, 0.02 g of $Na_2WO_4 \cdot 2H_2O$, and 0.1 g of Na_2SeO_4). For chelator-free medium, in cases where electron shuttles are being omitted for study of electrode or Fe(III) oxide reduction, NTA in the mineral mix can be omitted and the mineral mix should be stored at a pH of <3 to keep minerals in suspension.

2. If electron donors are desired in the final medium, add them at this point, before adjusting the final pH of the medium.

3. If this medium is to be used to wash cells grown in the presence of soluble electron acceptors or cells grown in higher-salt medium, additional salt (e.g., as NaCl) should be added at this point to compensate for the shift in osmolarity.

4. Adjust the medium to pH 6.8, before sparging with oxygen-free 80:20 N_2:CO_2 to remove oxygen. Very oxygen-sensitive anaerobes may require the medium to be brought to a boil and cooled under a stream of oxygen-free gas. Sparging and maintenance of the medium under a CO_2 headspace acidify the medium due to dissociation of CO_2 to carbonate and bicarbonate ion. Thus, at this stage, 2 g of $NaHCO_3$ is added and the medium is flushed with oxygen-free 80:20 N_2:CO_2 before autoclaving. The same effect can be achieved by adding an equivalent amount of base to the medium and sparging with 80:20 N_2:CO_2; as bicarbonate ion is formed, the pH will return to ~6.8. Adding $NaHCO_3$ directly simply begins this process with the appropriate amount of bicarbonate in solution.

If this medium is to be used in a chamber which will be flushed with air, obviously bicarbonate cannot be added. However, to maintain osmotic balance, and to provide buffering, an equivalent amount of Tris or morpholineethanesulfonic acid buffer should be added to the medium at this point, depending on the desired final pH range to be buffered.

5. Seal the tubes or bottles with butyl rubber stoppers and aluminum seals, and autoclave the medium.

If the medium is to contain an electron acceptor, such as Fe(III) citrate or fumarate, the electron acceptor must be dissolved and the pH must be adjusted to 6 to 6.2 prior to addition of salts and vitamins. While this is fairly straightforward for fumarate, Fe(III) citrate requires brief heating, followed by rapid cooling and pH adjustment to minimize Fe(II) formation.

To prepare Fe(III) citrate-containing medium, dissolve 13.7 g of Fe(III) citrate in ~200 ml of boiling water. Immediately remove from heat, and Fe(III) will dissolve within 1 min. Add ~800 ml of double-distilled water that has been chilled in an ice bath or refrigerator. When the Fe(III) solution is at room temperature, bring the pH of the solution up to 6 to 6.2 with NaOH, and proceed from step no. 1 above as described. Protect Fe(III) citrate medium from UV light, as photooxidation can lead to decarboxylation of citrate and formation of 3-oxoglutarate and other side products (16).

FINAL COMMENTS

Bacteria able to interact electrically with their external environment are interpreters, converting organic signals into electrical signals, and microbial fuel cells are only one example of a number of processes able to harness this unique ability. Electricity is not just another bulk fermentation product to be collected and counted but is a spectrum of signals rich with information about the outer surface of bacteria and their local environment. The experiments and devices described in this review provide proof-of-concept demonstrations that only begin to listen to the electrical language of microorganisms. By combining microbial physiology with an understanding of electrical and chemical engineering, an equally rich spectrum of devices able to study and control the growth of electrode-reducing bacteria can be imagined and should be explored.

REFERENCES

1. **Adler, J., and W. Shi.** 1988. Galvanotaxis in bacteria. *Cold Spring Harbor Symp. Quant. Biol.* **53**(Pt. 1):23–25.
2. **Back, J. H., M. S. Kim, H. Cho, I. S. Chang, J. Lee, K. S. Kim, B. H. Kim, Y. I. Park, and Y. S. Han.** 2004. Construction of bacterial artificial chromosome library from electrochemical microorganisms. *FEMS Microbiol. Lett.* **238:**65–70.
3. **Bennetto, H. P.** 1990. Electricity production by microorganisms. *Biotechnol. Edu.* **1:**163–168.
4. **Bond, D. R., D. E. Holmes, L. M. Tender, and D. R. Lovley.** 2002. Electrode-reducing microorganisms harvesting energy from marine sediments. *Science* **295:**483–485.
5. **Bond, D. R., and D. R. Lovley.** 2003. Electricity production by *Geobacter sulfurreducens* attached to electrodes. *Appl. Environ. Microbiol.* **69:**1548–1555.
6. **Bond, D. R., and D. R. Lovley.** 2005. Evidence for involvement of an electron shuttle in electricity generation by *Geothrix fermentans*. *Appl. Environ. Microbiol.* **71:** 2186–2189.
7. **Chaudhuri, S. K., and D. R. Lovley.** 2003. Electricity generation by direct oxidation of glucose in mediatorless microbial fuel cells. *Nat. Biotechnol.* **21:**1229–1232.
8. **Cheng, S., H. Liu, and B. E. Logan.** 2006. Increased performance of single-chamber microbial fuel cells using an improved cathode structure. *Electrochem. Comm.* **8:**489–494.
9. **Cheng, S., H. Liu, and B. E. Logan.** 2006. Increased power generation in a continuous flow MFC with advective flow through the porous anode and reduced electrode spacing. *Environ. Sci. Technol.* **40:**2426–2432.
10. **Childers, S. E., S. Ciufo, and D. R. Lovley.** 2002. *Geobacter metallireducens* accesses insoluble Fe(III) oxide by chemotaxis. *Nature* **416:**767–769.
11. **Choi, Y., J. Song, S. Jung, and S. Kim.** 2001. Optimization of the performance of microbial fuel cells containing alkalophilic *Bacillus* sp. *J. Microbiol. Biotechnol.* **11:**863–869.
12. **Cohen, B.** 1931. The bacterial culture as an electrical half-cell. *J. Bacteriol.* **21:**18.
13. **Cooney, M. J., E. Roschi, I. W. Marison, C. Comninellis, and U. von Stockar.** 1996. Physiologic studies with the sulfate-reducing bacterium *Desulfovibrio desulfuricans*: evaluation for use in a biofuel cell. *Enzyme Microb. Technol.* **18:**358–365.
14. **Davis, J. B., and H. F. Yarbrough.** 1962. Preliminary experiments on a microbial fuel cell. *Science* **137:**615–616.
15. **Delaney, G. M., H. P. Bennetto, J. R. Mason, S. D. Roller, J. L. Stirling, and C. F. Thurston.** 1984. Electron-transfer coupling in microbial fuel cells. 2. Performance of

fuel cells containing selected microorganism-mediator combinations. *J. Chem. Technol. Biotechnol.* **34B:**13–27.
16. **Dodge, C. J., and A. J. Francis.** 2002. Photodegradation of a ternary iron(III)-uranium(VI)-citric acid complex. *Environ. Sci. Technol.* **36:**2094–2100.
17. **Finneran, K. T., C. V. Johnsen, and D. R. Lovley.** 2003. *Rhodoferax ferrireducens* sp. nov., a psychrotolerant, facultatively anaerobic bacterium that oxidizes acetate with the reduction of Fe(III). *Int. J. Syst. Evol. Microbiol.* **53:**669–673.
18. **Gray Young, T., L. Hadjipetrou, and M. D. Lilly.** 1966. The theoretical aspects of biochemical fuel cells. *Biotechnol. Bioeng.* **8:**581–593.
19. **He, Z., S. D. Minteer, and L. T. Angenent.** 2005. Electricity generation from artificial wastewater using an upflow microbial fuel cell. *Environ. Sci. Technol.* **39:**5262–5267.
20. **Holmes, D. E., D. R. Bond, and D. R. Lovley.** 2004. Electron transfer by *Desulfobulbus propionicus* to Fe(III) and graphite electrodes. *Appl. Environ. Microb.* **70:**1234–1237.
21. **Holmes, D. E., D. R. Bond, R. A. O'Neil, C. E. Reimers, L. R. Tender, and D. R. Lovley.** 2004. Microbial communities associated with electrodes harvesting electricity from a variety of aquatic sediments. *Microb. Ecol.* **48:**178–190.
22. **Holmes, D. E., J. S. Nicoll, D. R. Bond, and D. R. Lovley.** 2004. Potential role of a novel psychrotolerant member of the family *Geobacteraceae, Geopsychrobacter electrodiphilus* gen. nov., sp. nov., in electricity production by a marine sediment fuel cell. *Appl. Enviro. Microbiol.* **70:**6023–6030.
23. **Jang, J. K., T. H. Pham, I. S. Chang, K. H. Kang, H. Moon, K. S. Cho, and B. H. Kim.** 2004. Construction and operation of a novel mediator- and membrane-less microbial fuel cell. *Process Biochem.* **39:**1007–1012.
24. **Katz, E., and I. Willner.** 2003. A biofuel cell with electrochemically switchable and tunable power output. *J. Am. Chem. Soc.* **125:**6803–6813.
25. **Kim, B. H., H. J. Kim, M. S. Hyun, and D. H. Park.** 1999. Direct electrode reaction of Fe(III)-reducing bacterium, *Shewanella putrefaciens*. *J. Microbiol. Biotechnol.* **9:** 127–131.
26. **Kim, B. H., H. S. Park, H. J. Kim, G. T. Kim, I. S. Chang, J. Lee, and N. T. Phung.** 2004. Enrichment of microbial community generating electricity using a fuel-cell-type electrochemical cell. *Appl. Microbiol. Biotechnol.* **63:**672–681.
27. **Kim, H. J., H. S. Park, M. S. Hyun, I. S. Chang, M. Kim, and B. H. Kim.** 2002. A mediator-less microbial fuel cell using a metal reducing bacterium, Shewanella putrefaciens. *Enzyme Microb. Technol.* **30:**145–152.
28. **Kim, N., Y. Choi, S. Jung, and S. Kim.** 2000. Effect of initial carbon sources on the performance of microbial fuel cells containing *Proteus vulgaris*. *Biotechnol. Bioeng.* **70:** 109–114.
29. **Leang, C., M. V. Coppi, and D. R. Lovley.** 2003. OmcB, a c-type polyheme cytochrome, involved in Fe(III) reduction in *Geobacter sulfurreducens*. *J. Bacteriol.* **185:**2096–2103.
30. **Lee, J. Y., N. T. Phung, I. S. Chang, B. H. Kim, and H. C. Sung.** 2003. Use of acetate for enrichment of electrochemically active microorganisms and their 16S rDNA analyses. *FEMS Microbiol. Lett.* **223:**185–191.
31. **Lewis, K.** 1966. Symposium on bioelectrochemistry of microorganisms. IV. Biochemical fuel cells. *Bacteriol. Rev.* **30:**101–113.
32. **Liu, H., S. Cheng, and B. E. Logan.** 2005. Production of electricity from acetate or butyrate using a single-chamber microbial fuel cell. *Environ. Sci. Technol.* **39:**658–662.
33. **Liu, H., and B. E. Logan.** 2004. Electricity generation using an air-cathode single chamber microbial fuel cell in the presence and absence of a proton exchange membrane. *Environ. Sci. Technol.* **38:**4040–4046.
34. **Liu, H., R. Ramnarayanan, and B. E. Logan.** 2004. Production of electricity during wastewater treatment using a single chamber microbial fuel cell. *Environ. Sci. Technol.* **38:**2281–2285.
35. **Logan, B. E., C. Murano, K. Scott, N. D. Gray, and I. M. Head.** 2005. Electricity generation from cysteine in a microbial fuel cell. *Water Res.* **39:**942–952.
36. **Lovley, D. R.** 2004. Dissimilatory Fe(III) and Mn(IV) reduction. *Adv. Microb. Physiol.* **49:**219–286.
37. **Magnuson, T. S., A. L. Neal, and G. G. Geesey.** 2004. Combining in situ reverse transcriptase polymerase chain reaction, optical microscopy, and X-ray photoelectron spectroscopy to investigate mineral surface-associated microbial activities. *Microb. Ecol.* **48:**578–588.
38. **Mano, N., F. Mao, and A. Heller.** 2003. Characteristics of a miniature compartment-less glucose-O_2 biofuel cell and its operation in a living plant. *J. Am. Chem. Soc.* **125:**6588–6594.
39. **Mano, N., F. Mao, and A. Heller.** 2004. A miniature membrane-less biofuel cell operating at +0.60 V under physiological conditions. *ChemBioChem* **5:**1703–1705.
40. **Mano, N., F. Mao, W. Shin, T. Chen, and A. Heller.** 2003. A miniature biofuel cell operating at 0.78 V. *Chem. Commun.* (Cambridge) **2003(4):**518–519.
41. **Maoyu, Y., and Y. Zhang.** 1989. Electrode system for determination of microbial cell populations in polluted water. *Appl. Environ. Microbiol.* **55:**2082–2085.
42. **Menicucci, J., H. Beyenal, E. Marsili, R. A. Veluchamy, G. Demir, and Z. Lewandowski.** 2006. Procedure for determining maximum sustainable power generated by microbial fuel cells. *Environ. Sci. Technol.* **40:**1062–1068.
43. **Min, B., S. Cheng, and B. E. Logan.** 2005. Electricity generation using membrane and salt bridge microbial fuel cells. *Water Res.* **39:**1675–1686.
44. **Min, B., and B. E. Logan.** 2004. Continuous electricity generation from domestic wastewater and organic substrates in a flat plate microbial fuel cell. *Environ. Sci. Technol.* **38:**5809–5814.
45. **Mitchell, P.** 1967. Proton-translocation phosphorylation in mitochondria, chloroplasts and bacteria: natural fuel cells and solar cells. *Fed. Proc.* **26:**1370–1379.
46. **Morris, J. G.** 1983. Anaerobic fermentations—some new possibilities. *Biochem. Soc. Symp.* **48:**147–172.
47. **Myers, C. R., and J. M. Myers.** 2003. Cell surface exposure of the outer membrane cytochromes of *Shewanella oneidensis* MR-1. *Lett. Appl. Microbiol.* **37:**254–258.
48. **Myers, C. R., and J. M. Myers.** 1993. Ferric reductase is associated with the membranes of anaerobically grown *Shewanella putrefaciens* MR-1. *FEMS Microbiol. Lett.* **108:**15–22.
49. **Myers, C. R., and J. M. Myers.** 1997. Outer membrane cytochromes of *Shewanella putrefaciens* MR-1: spectral analysis, and purification of the 83-kDa c-type cytochrome. *Biochim. Biophys. Acta* **1326:**307–318.
50. **Nevin, K. P., and D. R. Lovley.** 2000. Lack of production of electron-shuttling compounds or solubilization of Fe(III) during reduction of insoluble Fe(III) oxide by *Geobacter metallireducens*. *Appl. Environ. Microbiol.* **66:**2248–2251.
51. **Nevin, K. P., and D. R. Lovley.** 2002. Mechanisms for accessing insoluble Fe(III) oxide during dissimilatory Fe(III) reduction by *Geothrix fermentans*. *Appl. Environ. Microbiol.* **68:**2294–2299.
52. **Nevin, K. P., and D. R. Lovley.** 2002. Mechanisms for Fe(III) oxide reduction in sedimentary environments. *Geomicrobiol. J.* **19:**141–159.

53. **Niessen, J., U. Schroder, M. Rosenbaum, and F. Scholz.** 2004. Fluorinated polyanilines as superior materials for electrocatalytic anodes in bacterial fuel cells. *Electrochem. Comm.* **6:**571–575.
54. **Niessen, J., U. Schroder, and F. Scholz.** 2004. Exploiting complex carbohydrates for microbial electricity generation—a bacterial fuel cell operating on starch. *Electrochem. Comm.* **6:**955–958.
55. **Oh, S., B. Min, and B. E. Logan.** 2004. Cathode performance as a factor in electricity generation in microbial fuel cells. *Environ. Sci. Technol.* **38:**4900–4904.
56. **Park, D. H., M. Laivenieks, M. V. Guettler, M. K. Jain, and J. G. Zeikus.** 1999. Microbial utilization of electrically reduced neutral red as the sole electron donor for growth and metabolite production. *Appl. Environ. Microbiol.* **65:**2912–2917.
57. **Park, D. H., and J. G. Zeikus.** 2000. Electricity generation in microbial fuel cells using neutral red as an electronophore. *Appl. Environ. Microbiol.* **66:**1292–1297.
58. **Park, D. H., and J. G. Zeikus.** 2003. Improved fuel cell and electrode designs for producing electricity from microbial degradation. *Biotechnol. Bioeng.* **81:**348–355.
59. **Park, H. S., B. H. Kim, H. S. Kim, H. J. Kim, G. T. Kim, M. Kim, I. S. Chang, Y. H. Park, and H. I. Chang.** 2001. A novel electrochemically active and Fe(III)-reducing bacterium phylogenetically related to *Clostridium butyricum* isolated from a microbial fuel cell. *Anaerobe* **7:**297–306.
60. **Pham, C. A., S. J. Jung, N. T. Phung, J. Lee, I. S. Chang, B. H. Kim, H. Yi, and J. Chun.** 2003. A novel electrochemically active and Fe(III)-reducing bacterium phylogenetically related to *Aeromonas hydrophila*, isolated from a microbial fuel cell. *FEMS Microbiol. Lett.* **223:**129–134.
61. **Pham, T. H., J. K. Jang, I. S. Chang, and B. H. Kim.** 2004. Improvement of cathode reaction of a mediatorless microbial fuel cell. *J. Microbiol. Biotechnol.* **14:**324–329.
62. **Pham, T. H., J. K. Jang, H. S. Moon, I. S. Chang, and B. H. Kim.** 2005. Improved performance of microbial fuel cell using membrane-electrode assembly. *J. Microbiol. Biotechnol.* **15:**438–441.
63. **Phung, N. T., J. Lee, K. H. Kang, I. S. Chang, G. M. Gadd, and B. H. Kim.** 2004. Analysis of microbial diversity in oligotrophic microbial fuel cells using 16S rDNA sequences. *FEMS Microbiol. Lett.* **233:**77–82.
64. **Potter, M. C.** 1912. Electrical effects accompanying the decomposition of organic compounds. *Proc. R. Soc. London B* **84:**266–276.
65. **Rabaey, K., N. Boon, M. Hofte, and W. Verstraete.** 2005. Microbial phenazine production enhances electron transfer in biofuel cells. *Environ. Sci. Technol.* **39:**191A–192A.
66. **Rabaey, K., N. Boon, S. D. Siciliano, M. Verhaege, and W. Verstraete.** 2004. Biofuel cells select for microbial consortia that self-mediate electron transfer. *Appl. Environ. Microbiol.* **70:**5373–5382.
67. **Rabaey, K., P. Clauwaert, P. Aelterman, and W. Verstraete.** 2005. Tubular microbial fuel cells for efficient electricity generation. *Environ. Sci. Technol.* **39:**8077–8082.
68. **Rabaey, K., G. Lissens, S. D. Siciliano, and W. Verstraete.** 2003. A microbial fuel cell capable of converting glucose to electricity at high rate and efficiency. *Biotechnol. Lett.* **25:**1531–1535.
69. **Reguera, G., K. D. McCarthy, T. Mehta, J. S. Nicoll, M. T. Tuominen, and D. R. Lovley.** 2005. Extracellular electron transfer via microbial nanowires. *Nature* **435:**1098–1101.
70. **Reimers, C. E., L. M. Tender, S. Fertig, and W. Wang.** 2001. Harvesting energy from the marine sediment—water interface. *Environ. Sci. Technol.* **35:**192–195.
71. **Roller, S. D., H. P. Bennetto, G. M. Delaney, J. R. Mason, J. L. Stirling, and C. F. Thurston.** 1984. Electron-transfer coupling in microbial fuel cells. 1. Comparison of redox-mediator reduction rates and respiratory rates of bacteria. *J. Chem. Technol. Biotechnol.* **34B:**3–12.
72. **Sasaki, S., and I. Karube.** 1999. The development of microfabricated biocatalytic fuel cells. *Trends Biotechnol.* **17:**50–52.
73. **Scholz, F., and U. Schroder.** 2003. Bacterial batteries. *Nat. Biotechnol.* **21:**1151–1152.
74. **Schroder, U., J. Niessen, and F. Scholz.** 2003. A generation of microbial fuel cells with current outputs boosted by more than one order of magnitude. *Angew. Chem. Int. Ed. Engl.* **42:**2880–2883.
75. **Shi, W., B. A. Stocker, and J. Adler.** 1996. Effect of the surface composition of motile *Escherichia coli* and motile *Salmonella* species on the direction of galvanotaxis. *J. Bacteriol.* **178:**1113–1119.
76. **Shin, S. H., Y. J. Choi, S. H. Na, S. H. Jung, and S. Kim.** 2006. Development of bipolar plate stack type microbial fuel cells. *Bull. Korean Chem. Soc.* **27:**281–285.
77. **Sisler, F. D.** 1971. Biochemical fuel cells, p. 1–11. *In* D. J. D. Hochenhull (ed.), *Progress in Industrial Microbiology*, vol. 9. J. & A. Churchill, London, United Kingdom.
78. **Soukharev, V., N. Mano, and A. Heller.** 2004. A four-electron O(2)-electroreduction biocatalyst superior to platinum and a biofuel cell operating at 0.88 V. *J. Am. Chem. Soc.* **126:**8368–8369.
79. **Stirling, J. L., H. P. Bennetto, G. M. Delaney, J. R. Mason, S. D. Roller, K. Tanaka, and C. F. Thurston.** 1983. Microbial fuel cells. *Biochem. Soc. Trans.* **11:**451–453.
80. **Tender, L. M., C. E. Reimers, H. A. Stecher, D. E. Holmes, D. R. Bond, D. L. Lowy, K. Pilobello, S. J. Fertig, and D. R. Lovley.** 2002. Harnessing microbial power generation on the seafloor. *Nat. Biotechnol.* **20:**821–825.

Microbially Mediated Anaerobic Iron(II) Oxidation at Circumneutral pH

KARRIE A. WEBER AND JOHN D. COATES

94

Microbial communities are recognized to mediate iron geochemical cycling in aquatic, terrestrial, and subsurface ecosystems. As iron is the fourth most abundant element in the Earth's crust, iron redox reactions have the potential to exert a significant influence on soil and sediment geochemistry, leading to the mineralization of organic matter, mineral dissolution and weathering, the formation of geologically significant minerals, and the mobilization or immobilization of various anions and cations which include contaminants (11, 56). The solubility and reactivity of iron is particularly sensitive to the environmental pH. The solubility of the trivalent ferric form [Fe(III)] is inversely proportional to acid pH values, and below a pH value of 4.0 Fe(III) primarily exists as an aqueous ionic Fe^{3+} species (55). At circumneutral pH, ~pH 7, and greater pH values, iron primarily exists as insoluble, solid-phase minerals in divalent ferrous [Fe(II)] and trivalent ferric oxidation states (11). Microbially catalyzed redox reactions between the Fe(II) and Fe(III) valence states play a fundamental role in modern environmental biogeochemistry in both the oxic and anoxic zones of aquatic, terrestrial, and subsurface ecosystems from low to elevated pH levels. The role of microbially catalyzed Fe(III) reduction in nonsulfidogenic sedimentary environments at circumneutral pH is well established (32, 33, 35). However, it is only within the past decade that microbial Fe(II) oxidation has been recognized to contribute to iron redox cycling not only in the oxic environment but also in environments devoid of molecular oxygen (O_2) (50, 65). This chapter discusses the role that anaerobic, Fe(II)-oxidizing microorganisms (FOM) play in iron biogeochemical cycling, identification of the metabolism, and isolation of anaerobic FOM.

Fe REDOX CYCLING AT CIRCUMNEUTRAL pH: ROLE OF MICROBIAL Fe(II) OXIDATION

In the anoxic zone, Fe(III) oxides provide an electron sink and are chemically or biologically reduced (Fig. 1). In environments in which active microbial sulfate reduction is occurring, the biologically generated H_2S can abiotically reduce a proportion of Fe(III). However, in nonsulfidogenic sedimentary environments, direct enzymatic Fe(III) reduction has been demonstrated to account for a significant fraction of organic carbon oxidized in submerged soils and freshwater and marine sediments (32, 33, 35, 57). Despite the insoluble nature of Fe(III) oxide minerals at circumneutral pH, microorganisms are capable of enzymatically oxidizing organic matter and/or H_2 coupled to the reduction of the insoluble Fe(III) oxide yielding aqueous Fe(II) [$Fe(II)_{aq}$] as well as solid-phase Fe(II) [$Fe(II)_s$]-bearing minerals (Fig. 1) such as siderite, vivianite, magnetite, and green rust (11, 33).

Abiotic Fe(II) Oxidation

Transfer of $Fe(II)_{aq}$ across the environmental anoxic/oxic transition zone through diffusion or advective transport mechanisms and subsequent rapid abiotic reaction with dissolved O_2 has been described as a possible mechanism for reoxidation of Fe(II) in the natural environment (13) (Fig. 1). Anoxic redox processes may also contribute to iron biogeochemical cycling at circumneutral pH through abiotic reactions with oxidized manganese [Mn(IV)] (17, 39) or oxyanions of nitrogen [nitrate (NO_3^-) or nitrite (NO_2^-)] (5, 22, 37, 49, 55). Abiotic reaction of Fe(II) with nitrate may occur through several mechanisms. At high temperatures (75°C), Fe(II) oxidation coupled to the reduction of nitrate to ammonium (NH_4^+) occurs spontaneously (38). Alternatively, at lower temperatures, the Fe(II) content of highly reactive crystalline mixed-valence iron minerals such as green rust abiotically reacts with nitrate to form magnetite and ammonia as the sole end products (22). The reaction between Fe(II) and nitrate can be enhanced through catalysis by trace metals such as Cu^{2+} (5, 58), through surface chemistry reaction on crystalline iron oxide (lepidocrocite and goethite) surfaces at elevated pH (>8.0) (37), or through reaction with Fe(III) oxide precipitates formed during iron silicate dissolution (40). However, the relatively extreme specific conditions required (i.e., high temperature, high pH, or through reaction with catalysts) for abiotic Fe(II) oxidation coupled to the reduction of NO_3^- indicates that these reactions may not be dominant oxidative mechanisms in typical natural environments. Alternatively, nitrite produced as an intermediate or end product of organotrophic NO_3^- reduction or nitrification can also chemically oxidize Fe(II) (49). With the exception of NO_2^-, abiotic Fe(II) oxidation reactions by oxidized nitrogen species are considered relatively insignificant under geochemical conditions consistent with typical iron-rich sedimentary environments at circumneutral pH (29, 62).

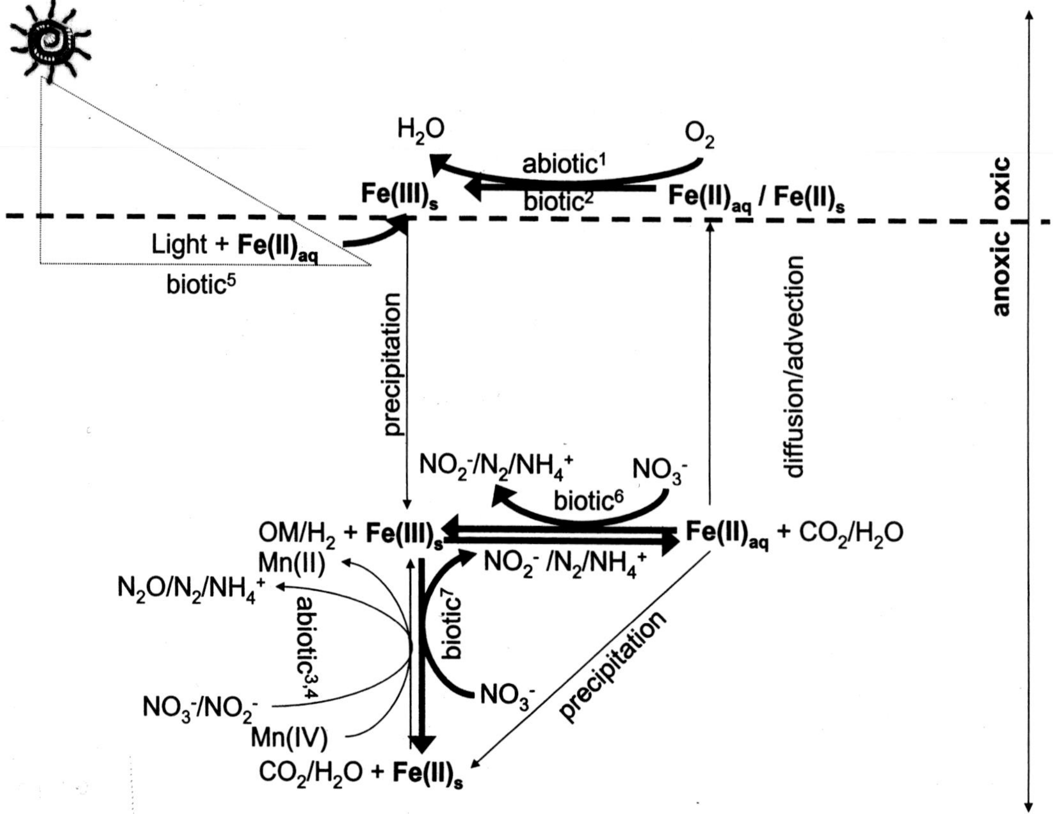

FIGURE 1 Iron biogeochemical cycle at circumneutral pH. Abiotic and biotic iron redox reactions significantly influence iron geochemistry. Potential pathways involving abiotic and microbially mediated iron redox reactions are illustrated with the exception of abiotic iron reduction reactions. The dashed line denotes the division between the oxic and anoxic zones. Thick bold lines denote biologically catalyzed redox reactions. Superscripts: 1, abiotic oxidation of Fe(II) coupled to the reduction of dissolved O_2; 2, microaerophilic Fe(II) oxidation; 3, abiotic oxidation of Fe(II) coupled to the reduction of Mn(IV) to Mn(II); 4, abiotic oxidation of Fe(II) coupled to the reduction of oxidized nitrogen species, NO_3^- and NO_2^-, yielding NH_4^+ and gaseous products; 5, light-dependent photoautotrophic Fe(II) oxidation; 6, nitrate-dependent Fe(II) oxidation, oxidation of aqueous Fe(II) coupled to the reduction NO_3^- yielding N_2 and NH_4^+; ClO_4^- and ClO_3^- can also serve as terminal electron acceptors for Fe(II)-oxidizing metabolisms; 7, nitrate-dependent oxidation of solid-phase Fe(II)-bearing minerals. For detailed discussion, see the text.

Biological Oxidation of Fe(II)

In contrast to the abiotic reoxidation of Fe(II), recent studies indicate that biological reactions may significantly contribute to the reoxidation of Fe(II) in the environment. Although the quantitative role of aerobic oxidative biological mechanisms has not been established (19, 42), recent studies have unequivocally demonstrated that bacteria in zones of low oxygen tension (microaerophilic) are capable of competing with abiotic reactions coupling the oxidation of Fe(II) to aerobic respiration (Fig. 1) (14, 18, 47). The end products of these metabolisms generate insoluble poorly crystalline biogenic Fe(III) oxides which precipitate from solution and subsequently provide a terminal electron acceptor for Fe(III)-reducing bacteria in the anoxic zone (42, 48). In contrast to the first neutrophilic aerobic iron-oxidizing bacterium, *Gallionella ferruginea*, described over a century ago (15, 66), anaerobic bio-oxidation of Fe(II) has only recently been identified and very little is known regarding the ubiquity and diversity of organisms capable of this metabolism (3, 50, 61, 65). The identification of a light-dependent photoautotrophic, anaerobic iron-oxidizing bacterium introduced the potential role of microbially catalyzed Fe(II) oxidation in anoxic environments (65). In the photic zone, Fe(II)-oxidizing phototrophic bacteria are capable of oxidizing Fe(II), generating Fe(III) oxides including mixed-valence phase minerals such as magnetite (26) (Fig. 1). The subsequent identification of nitrate serving as a terminal electron acceptor for light-independent iron-oxidizing metabolisms (50) has closed a gap in the iron redox cycle. In anoxic environments, microbial oxidation of Fe(II) coupled to the reduction of nitrate, perchlorate, and chlorate has been demonstrated (Fig. 1) (3, 50). Recent studies hinting at the true ubiquity and diversity of these anaerobic FOM suggest that metabolic, light-independent reactions such as nitrate-dependent Fe(II) oxidation have the potential to contribute to anoxic Fe(II) oxidative processes on a global scale, provided that adequate concentrations of an electron acceptor are readily available (61). These FOM are capable of oxidizing solid-phase Fe(II) (7, 30, 43, 62) as well as Fe(II) associated with structural Fe in minerals such as almandine, an iron aluminum silicate (7, 45, 46) yielding amorphous and crystalline Fe(III) oxide minerals. The formation of these biogenically formed Fe(III) oxides potentially serves as an electron acceptor for Fe(III)-reducing microorganisms, perpetuating a dynamic microbially mediated Fe cycle (44, 54, 64) in addition to soil and sediment biogeochemistry, mineralogy, and heavy metal and radionuclide immobilization (7, 31, 62, 64).

PHOTOAUTOTROPHIC Fe(II)-OXIDIZING BACTERIA

Photoautotrophic, anaerobic Fe(II) oxidation was the first demonstration of microbially mediated oxidation of Fe(II) in anoxic environments supporting autotrophic growth according to the following equation (65).

$$HCO_3^- + 4Fe(II) + 10H_2O \rightarrow < CH_2O > + 4Fe(OH)_3 + 7H^+$$

Although phototrophic FOM described within the domain *Bacteria* are phylogenetically diverse (28), to date an archaeon capable of this metabolism has not been identified. Purple sulfur (*Thiodictyon* sp.) (12), purple nonsulfur ("*Rhodobacter ferrooxidans*" [16], *Rhodomicrobium vannielii* [65], *Rhodopseudomonas palustris* [26], *Rhodovulum iodosum* [53], and *Rhodovulum robiginosum* [53]), and green anoxygenic, photosynthetic ("*Chlorobium ferrooxidans*" [24]) Fe(II)-oxidizing phototrophic bacteria have been described from both freshwater and marine environments (16, 24–26, 53, 65). These phototrophic organisms can represent significant fractions of the total Fe(II)-oxidizing microbial populations in freshwater environments. Most probable number (MPN) enumeration studies with pond and ditch sediment samples indicated freshwater phototrophic Fe(II)-oxidizing communities in the range of 1.1×10^2 to 3.9×10^3 bacteria per g (dry weight) (51), which represented one-half of the total number of mixotrophic nitrate-dependent Fe(II)-oxidizing bacteria from the same sites and was similar to the size of the lithoautotrophic nitrate-dependent Fe(II)-oxidizing bacterial population (Table 1). With the exception of *Rhodomicrobium vannielii*, phototrophic FOM are capable of completely oxidizing soluble $Fe(II)_{aq}$, resulting in the formation of poorly crystalline Fe(III) oxides (12). As these biogenic Fe(III) oxides are aged, they transform into more-crystalline Fe(III) oxide minerals (Table 2) including goethite and lepidocrocite, in the presence of metabolically active microbial populations (27). The significance of the phototrophic Fe(II) oxidation processes to contemporary iron biogeochemical cycling is limited to the photic zone as light penetration of soil and particulate matter is only between 8 to 200 μm (8). Furthermore, recent studies indicate that utilization of the Fe(II) content of solid-phase Fe(II) minerals is limited by mineral solubility and that the phototrophic FOM described to date are unable to access the Fe(II) content of crystalline Fe(II)-bearing minerals (27). Although the impact of this microbial process to Fe redox cycling may be locally significant, its dependence on light penetration and soluble $Fe(II)_{aq}$ indicates that it may play a relatively minor global role in iron biogeochemical cycling in neoteric anoxic environments.

LIGHT-INDEPENDENT ANAEROBIC FOM

At circumneutral pH, light-independent microbially mediated oxidation of $Fe(II)_{aq}$ and $Fe(II)_s$ coupled to nitrate reduction has been demonstrated in a variety of freshwater and saline environmental systems (6, 7, 20, 21, 23, 29, 36, 41, 45, 50, 51, 63, 64). These environmental systems harbor significant mixotrophic and autotrophic nitrate-dependent FOM communities (Table 1), which have the potential to contribute to Fe redox cycling. These FOM are capable of exploiting favorable thermodynamics between $Fe(III)_s/Fe(II)$ and nitrate reduction redox pairs (Table 3) as well as other potential terminal electron acceptors such as perchlorate and chlorate (Table 3).

TABLE 1 Abundance of nitrate-dependent FOM in a variety of environments as determined by traditional MPN enumeration

Environment	Mixotrophic[a,c] cells g^{-1}	Lithoautotrophic[b,c] cells g^{-1}	Reference
Rice paddy soil, Italy	1.6×10^6	ND	41
Wetland sediment, Alabama	2.4×10^5	ND	64
Ditch sediment, Italy	1.3×10^5	0	51
Pond sediment, Germany	$2.9 \times 10^6 - 1.1 \times 10^5$	$5.5 \times 10^2 - 1.3 \times 10^4$	51
Ditch sediment, Germany	$4.2 \times 10^6 - 1.8 \times 10^5$	$7.8 \times 10^2 - 5.4 \times 10^3$	51
Stream sediment, France	1.5×10^5	3.3×10^3	51
Ditch sediment, Germany	$9 \times 10^6 - 5 \times 10^8$	$5 \times 10^3 - 2 \times 10^4$	51
Freshwater lake sediment, Lake Constance, Germany	$1 \times 10^5 - 5.8 \times 10^5$	ND	23
Freshwater lake sediment, Campus Lake, Illinois	$2.4 \times 10^3 - 1.5 \times 10^4$	ND	63
Clay samples	$2.2 \times 10^1 - 9.3 \times 10^3$	ND	46
Groundwater, Tennessee	$0–2.0 \times 10^3$	ND	60
Subsurface sediment, Tennessee	$2.4 \times 10^2 - 4.3 \times 10^2$	ND	60
Subsurface sediment, Texas	1.1×10^3	ND	60

[a]Carbon dioxide and acetate were added as the carbon source in all cases. The concentration of acetate added to culture media ranged from 0.1 to 2 mM.
[b]ND, not determined.
[c]Values reported are a range of multiple MPN enumerations conducted.

TABLE 2 Biogenic oxyhydroxides and oxides generated under anoxic conditions

Oxyhydroxide or oxide	Mineral formula[a]	Microbial metabolism	Reference(s)
Ferrihydrite	$Fe_5HO_8 \cdot 4\,H_2O$	Nitrate-dependent Fe(II) oxidation; phototrophic Fe(II) oxidation	27, 30, 43, 52
Goethite	α-FeOOH	Nitrate-dependent Fe(II) oxidation; phototrophic Fe(II) oxidation	26, 27, 43, 64
Lepidocrocite	γ-FeOOH	Nitrate-dependent Fe(II) oxidation; phototrophic Fe(II) oxidation	27, 60
Maghemite	γ-Fe_2O_3	Nitrate-dependent Fe(II) oxidation	7
Hematite	α-Fe_2O_3	Nitrate-dependent Fe(II) oxidation	7
Magnetite[b]	$Fe_3O_4\,(Fe^{II}Fe_2^{III}O_4)$	Nitrate-dependent Fe(II) oxidation; phototrophic Fe(II) oxidation	7, 26
Green rust[b]	$Fe_x^{III}Fe_y^{II}(OH)_{3x+2y-z}(A^-)z$; $A^- = Cl^-$; $1/2SO_4^{2-}$	Nitrate-dependent Fe(II) oxidation	7, 60

[a]From reference 11.
[b]Referred to in the text as mixed-valence phase [Fe(II) and Fe(III)] minerals.

Phylogenetic Diversity of FOM

A variety of microorganisms in the domains *Archaea* and *Bacteria*, representing an extreme range of optimal thermal growth conditions (mesophilic and hyperthermophilic) have been identified (61). Several phylogenetically diverse mesophilic bacteria have been described as capable of nitrate-dependent Fe(II) oxidation; however, in most cases growth was either not demonstrated in pure culture or not associated with this metabolism in the absence of an additional electron donor or organic carbon as an energy source (2, 3, 7, 20, 45, 50). Although the oxidation of Fe(II), including solid-phase Fe(II)-bearing minerals, coupled to nitrate reduction is energetically favorable and yields enough energy to support ATP synthesis and microbial growth, autotrophic growth under nitrate-dependent Fe(II)-oxidizing conditions has been demonstrated in only two pure culture isolates: a hyperthermophilic archeon (21) and a mesophilic betaproteobacterium (63). Even in the case of a known Fe(II)-oxidizing chemolithoautotrophic bacterium, *Thiobacillus denitrificans*, energy conservation directly coupled to nitrate-dependent metal oxidation could not be determined as an additional electron donor was required in order for metal oxidation to occur (1).

Electron Acceptors Utilized

Until the recent demonstration of nitrate-dependent Fe(II) oxidation by *Geobacter metallireducens* resulting in the production of ammonium (64), NO_2^- and N_2 were identified as the sole end products of nitrate reduction (2, 7, 21, 50). Mixotrophic or autotrophic growth associated with Fe(II) oxidation coupled to perchlorate or chlorate [collectively designated (per)chlorate] has not been identified to date. However, (per)chlorate-dependent Fe(II)-oxidizing metabolic activity has been observed under stationary growth phase conditions in cultures of *Azospira suillum* strain PS (formerly *Dechlorosoma*), which together with the *Dechloromonas* species represent the dominant (per)chlorate-reducing bacteria in the environment (7, 30). Although perchlorate and chlorate are not naturally abundant compounds, their potential to serve as an electron acceptor in environmental systems cannot be discounted. Legal discharge of perchlorate into natural waters has led to widespread anthropogenic contamination throughout the United States (9). Given the ubiquity of perchlorate-reducing bacteria and the ability of some of these microorganisms, specifically the *Azospira* spp. and the *Dechloromonas* spp., to oxidize Fe(II), anaerobic (per)chlorate-dependent Fe(II) oxidation may impact iron biogeochemical cycling in contaminated environmental systems as well as irrigated farmland. Further studies are needed to quantify the potential influence of (per)chlorate-dependent Fe(II) oxidation.

TABLE 3 Redox potentials relevant to anaerobic iron bio-oxidation reactions

Redox pair	E_0' (V)
ClO_4^-/Cl^-	0.873[a]
$NO_3^-/1/2N_2$	0.713[b]
ClO_3^-/Cl^-	0.616[a]
NO_3^-/NO_2^-	0.431[b]
NO_3^-/NH_4^+	0.36[a]
$Fe(OH)_3/Fe^{2+}$	0.014[c]
$Fe_5HO_8 \cdot 4\,H_2O/\,Fe^{2+}$	0.002[c]
γ-$FEOOH_{lepidocrocite}/Fe^{2+}$	−0.088[c]
α-$FeOOH_{goethite}/Fe^{2+}$	−0.274[c]
$Fe_3O_{4magnetite}/Fe^{2+}$	−0.314[c]

[a]From reference 61.
[b]From reference 34.
[c]From reference 57.

Oxidation of Insoluble Fe(II)

Nitrate amendment of natural samples collected from diverse environments such as the Hannahan aquifer in South Carolina or the Potomac River in Maryland resulted in a rapid oxidation of the Fe(II) content of these sediments relative to unamended or heat-killed controls, indicating that an indigenous Fe(II)-oxidizing, nitrate-reducing community was stimulated. However, in most reduced environments, soluble Fe(II) represents only a small proportion of the total Fe(II) available. The Fe(II) may also be bound to the surfaces of partially reduced crystalline Fe(III) oxides or be present as insoluble carbonate, sulfide, or silicaceous mineral phases such as siderite ($FeCO_3$), pyrite (FeS_2), almandine [$Fe_3Al_2(SiO_4)_3$], or biotite {[$(Mg, Fe^{2+})_3(Si_3AL)O_{10}(OH)_2$]K}. In contrast to phototrophic FOM, it is now known that these solid-phase Fe(II)-bearing minerals are subject to nitrate-dependent microbial oxidation. Recent studies indicated that nitrate-dependent Fe(II)-oxidizing bacteria can access the iron content of a broad diversity of Fe(II)-bearing mineral phases including surface bound Fe(II) (62, 64), crystalline Fe(II) minerals (siderite, magnetite, pyrite, aresnopyrite, and chromite) (7, 62), and structural Fe(II) in nesosilicate (almandine and staurolite) (7) and phyllosilicate (nontronite) (45, 46) minerals. In all cases, no oxidation of Fe(II) was observed in abiotic controls or in the absence of a suitable electron acceptor and both the rate and extent of Fe(II) oxidation were different for the various minerals, probably due to differences in bioavailability of Fe(II) in the mineral matrices. These results conclusively demonstrate that nitrate-dependent Fe(II) oxidation by these organisms is not limited to soluble Fe^{2+} ions. However, although we know that nitrate-dependent FOM potentially play a significant role in the oxidation of Fe(II) structurally incorporated into silicate minerals as well as contribute to Fe(II) mineral dissolution, the enzymatic effect on the residual mineral structure and stability is virtually unknown. The oxidative dissolution of solid-phase Fe(II) minerals in an anoxic environment presents an additional mechanism for rock-weathering and Fe(III) oxide mineral precipitation in anoxic soils and sedimentary environments.

END PRODUCTS OF Fe(II) BIO-OXIDATION

To date bio-oxidation products of amorphous solid-phase Fe(II) minerals have been characterized. A variety of biogenic Fe(III) oxide minerals, two-line ferrihydrite (30, 52), goethite (64), lepidocrocite (59), and hematite (7), as well as mixed-phase Fe(II)-Fe(III) minerals, i.e., magnetite and green rust (7, 60), were identified as oxidation products. As a result of this biogenic formation of magnetite, nitrate-dependent Fe(II) oxidation has been implicated in the genesis of banded iron formations (7). Amorphous, poorly crystalline biogenic Fe(III) oxides are reducible by Fe(III)-reducing bacteria and could thus provide electron-accepting capacity to Fe(III)-reducing microbial populations creating a dynamic anaerobic redox cycle (52, 54). However, given that crystalline biogenic Fe(III) oxides are a metabolic product, burial of the crystalline biogenic Fe(III) minerals would likely result.

The formation of these Fe(III) oxide minerals in an anaerobic environment provides a reactive surface with an adsorptive affinity for anions (i.e., PO_4^{3-}) and cations (i.e., Zn^{2+}, Pb^{2+}, Hg^{2+}, Co^{2+}, and U^{6+}). Heavy metals and radionuclides including U(VI) are rapidly removed (as much as 80% of the initial 100 μM within 5 days) from solution during anaerobic nitrate-dependent microbial Fe(II) oxidation in association with the biogenic Fe(III) oxides (31). As such, the biogenic formation of Fe(III) oxide-containing minerals has been proposed as a plausible bioremediative strategy for permanently immobilizing heavy metals and radionuclides (10, 31).

IDENTIFICATION, ENUMERATION, AND ISOLATION OF ANAEROBIC FOM COMMUNITIES

To date the presence of anaerobic, Fe(II)-oxidizing bacteria in paddy soil, pond, wastewater, stream, ditch, brackish lagoon, lake, wetland, aquifer, hydrothermal, subsurface, and deep-sea sediments has been identified using traditional microbiological techniques (6, 7, 20, 21, 23, 29, 36, 41, 45, 50, 51, 63, 64). Because of the broad phylogenetic diversity of these organisms and a lack of knowledge of the functional genes involved, molecular approaches to population studies and community dynamics of this metabolism are limited. As such, traditional MPN enumeration provides one of the best estimates for the abundance of these microorganisms in environmental samples. Enumeration studies can be performed with depth fractions collected from cored aquifer material. Nitrate-dependent FOM can be targeted using selective media. For example, enumeration of Fe(II)-oxidizing populations in freshwater environments can be performed using the medium outlined in Table 4. Soluble ferrous chloride ($FeCl_2$) is added as the iron electron donor of choice from an anoxic (100% N_2 atmosphere), filter-sterilized (0.22-μm-pore-size sterile nylon filter membrane) stock solution (1 M) to achieve a final concentration of 10 mM. A nonfermentable carbon source such as acetate (0.1 mM) should be used to prevent false positives when monitoring cell density. Prepared medium should be dispensed in 9-ml aliquots into pressure tubes and degassed individually with N_2-CO_2 (80:20, vol/vol) as previously outlined (63). The sealed anoxic medium tubes should be heat sterilized by autoclaving at 121°C for 15 min. Aliquots (1 g) of sediment from each sample are used to inoculate the first tubes of a triplicate dilution series. Following the addition of 1 g of sediment, 0.1 ml of sodium pyrophosphate from a sterile anaerobic 10% (wt/vol) aqueous stock solution should be added to the sediment slurry and gently shaken at room temperature for 1 h, which serves to release any cells adsorbed onto the soil and sediment particles and significantly improves the counts obtained. The sediment slurry is then serially diluted in basal medium prepared as described above. Tubes should be incubated at temperatures suitable to the original sample environment. After 8 weeks of incubation in the dark, tubes positive for iron oxidation can be identified by the presence of a brownish red or brownish green precipitate. The Most Probable Number Calculator, version 4.05 (Albert J. Klee, Risk Reduction Engineering Laboratory, U.S. Environmental Protection Agency, Cincinnati, Ohio; free software is available at http://www.epa.gov/nerlcwww/other.htm), can be used to enumerate the microbial community as well as calculate confidence limits. As an alternative, MPN tables can also be used to determine the abundance of the microbial community (67).

Enrichments or tubes positive for nitrate-dependent Fe(II) oxidation in the MPN enumeration series can be selected for isolation. In addition to the use of shake tubes (3), a simple plate overlay technique has been successfully used to isolate nitrate-dependent Fe(II)-oxidizing bacteria including the recently described autotroph, strain 2002 (63). Samples from the primary enrichment or MPN series are streaked onto R2A agar plates (Difco, catalog no. 218263), an undefined low-nutrient medium, amended with 10 mM nitrate in an anaerobic glove bag (95:5; N_2:H_2 atmosphere). As an alternative, PIPES [piperazine-*N,N′*-bis(2-ethanesulfonic acid)]-buffered (pH 7.0) basal medium (Table 4) amended with 10 mM nitrate and 10 mM acetate can also be used to prepare agar (2%) plates. The plates are incubated in an anoxic atmosphere (100% N_2) at a suitable temperature until colony development is observed (~120 h). An overlay of 2 mM $FeCl_2$ (5 ml of R2A agar or basal medium with 2% [by mass] Noble agar containing 2 mM $FeCl_2$) is poured over each plate following colony development and incubated again in an anoxic atmosphere. Nitrate-dependent Fe(II)-oxidizing colonies exhibit brownish red Fe(III) oxide precipitates on or around colonies. These colonies can be selected and transferred into anoxic, basal medium (Table 4).

These traditional microbiological techniques in combination with molecular techniques (51, 64) have begun to elucidate the phylogenetic diversity of microorganisms capable of this metabolism. The results from each of these studies have identified different microorganisms capable of

TABLE 4 Basal medium for cultivation of mesophilic, lithotrophic nitrate-dependent FOM[a]

Medium component	Amt or concn
NH_4Cl	0.25 g liter^{-1}
NaH_2PO_4	0.6 g liter^{-1}
KCl	0.1 g liter^{-1}
$NaHCO_3$	2.52 g liter^{-1}
$CH_3COONa \cdot 3\ H_2O$[b]	0.014 g liter^{-1}
$NaNO_3$	0.42 g liter^{-1}
Vitamin solution[c]	10 ml
Trace mineral solution[d]	10 ml
1 M $FeCl_2$ stock solution[e]	0.1 ml per 10 ml of sterile basal medium

[a]Basal medium is heated to 100°C and cooled under an N_2:CO_2 (80:20) atmosphere. Final pH should range from 6.8 to 7.2. Anoxic room temperature medium is dispensed under N_2:CO_2 gas stream into anoxic (N_2:CO_2 atmosphere) pressure tubes or serum bottles, stoppered, and autoclaved for 20 min at 121°C.

[b]Acetate is omitted when autotrophic Fe(II)-oxidizing microorganisms are screened for.

[c]Vitamin solution in milligrams per liter: *d*-biotin, 2; folic acid, 2; pyridoxine HCl, 10; riboflavin, 5; thiamine, 5; nicotinic acid, 5; pantothenic acid, 5; vitamin B_{12}, 0.1; *p*-aminobenzoic acid, 5; D, L-6,8-thiotic acid, 5.

[d]Mineral solution in grams per liter: nitrilotriacetic acid disodium salt, 1.5; $MgSO_4 \cdot 7H_2O$, 3; $MnSO_4$, 0.5; NaCl, 0.1; $FeSO_4 \cdot 7H_2O$, 0.1; $CaCl_2 \cdot 2H_2O$, 0.1; $CoCl_2 \cdot 6H_2O$, 0.1; ZnCl, 0.13; $CuSO_4 \cdot 5H_2O$, 0.01; $AlK(SO_4)_2 \cdot 12H_2O$, 0.01; boric acid, 0.01; $Na_2MoO_4 \cdot 2H_2O$, 0.025; $NiCl_2 \cdot 6H_2O$, 0.024; $Na_2WO_4 \cdot 2H_2O$, 0.025; Na_2SeO_4, 0.025.

[e]The Fe(II) source is added after sterilization to achieve a final concentration of 10 mM. $FeSO_4$ or other solid-phase Fe(II)-bearing minerals can be substituted. Soluble sterile Fe(II) stock solutions are prepared by adding $FeCl_2$ or $FeSO_4$ to anoxic (100% N_2 headspace) distilled, deionized water and stoppered. Inject the solution directly through a sterile 0.22-μm-pore-size nylon filter and sterile needle into a sterile, anoxic (100% N_2 headspace), stoppered serum bottle.

contributing to nitrate-dependent Fe(II) oxidation, some of which represent new species as well as previously undescribed genera, including *Ferroglobus placidus* (21), *Acidovorax* sp. strain BrG1, *Aquabacterium* sp. strain BrG2, *Thermomonas* sp. strain BrG3 (4), and *Chromobacterium* sp. strain 2002 (63). To date only one study has monitored the change in microbial community structure in an enrichment culture transitioning into nitrate-dependent Fe(II)-oxidizing conditions (64). While an increase in 16S rRNA gene sequences in the clone library constructed during nitrate-dependent Fe(II) oxidation suggested that bacteria previously described to oxidize Fe(II) (*Azospira* spp., *Aquabacterium* spp., and *Geobacter* spp.) played a role in nitrate-dependent Fe(II) oxidation, many 16S rRNA gene sequences represented members of unknown genera (64). As such, caution should be exercised when utilizing molecular techniques that rely on known and described genera. Continued efforts to identify nitrate-dependent Fe(II)-oxidizing bacteria and the genetic and biochemical mechanisms involved are warranted in order to evaluate the true phylogenetic and metabolic diversity of anaerobic Fe(II)-oxidizing microorganisms.

This research was supported by the Department of Energy Environmental Remediation Sciences Program (ERSP) through grant DE-FG02-98ER63592 to J. D. Coates.

REFERENCES

1. **Beller, H. R.** 2005. Anaerobic, nitrate-dependent oxidation of U(iv) oxide minerals by the chemolithoautotrophic bacterium *Thiobacillus denitrificans*. *Appl. Environ. Microbiol.* **71:**2170–2174.
2. **Benz, M., A. Brune, and B. Schink.** 1998. Anaerobic and aerobic oxidation of ferrous iron at neutral pH by chemoheterotrophic nitrate-reducing bacteria. *Arch. Microbiol.* **169:**159–165.
3. **Bruce, R. A., L. A. Achenbach, and J. D. Coates.** 1999. Reduction of (per)chlorate by a novel organism isolated from paper mill waste. *Environ. Microbiol.* **1:**319–329.
4. **Buchholz-Cleven, B. E. E., B. Rattunde, and K. L. Straub.** 1997. Screening for genetic diversity of isolates of anaerobic Fe(II)-oxidizing bacteria using DGGE and whole-cell hybridization. *Syst. Appl. Microbiol.* **2:**301–309.
5. **Buresh, R. J., and J. T. Moraghan.** 1976. Chemical reduction of nitrate by ferrous iron. *J. Environ. Qual.* **5:**320–325.
6. **Caldwell, M. E., R. S. Tanner, and J. M. Suflita.** 1999. Microbial metabolism of benzene and the oxidation of ferrous iron under anaerobic conditions: implications for bioremediation. *Anaerobe* **5:**595–603.
7. **Chaudhuri, S. K., J. G. Lack, and J. D. Coates.** 2001. Biogenic magnetite formation through anaerobic biooxidation of Fe(II). *Appl. Environ. Microbiol.* **67:**2844–2848.
8. **Ciani, A., K.-U. Gossa, and R. P. Schwarzenbach.** 2005. Light penetration in soil and particulate minerals. *Eur. J. Soil Sci.* **56:**561–574.
9. **Coates, J. D., and L. A. Achenbach.** 2004. Microbial perchlorate reduction: rocket-fuelled metabolism. *Nat. Rev. Microbiol.* **2:**569–580.
10. **Coates, J. D., and R. Chakraborty.** 2003. Anaerobic bioremediation: an emerging resource for environmental cleanup, p. 227–257. *In* I. M. Head, I. Singleton, and M. G. Milner (ed.), *Bioremediation: a Critical Review*. Horizon Scientific, Wymondham, United Kingdom.
11. **Cornell, R. M., and U. Schwertmann.** 2003. *The Iron Oxides: Structure, Properties, Reactions, Occurrences and Uses*, 2nd ed. Wiley-VCH, Weinheim, Germany.
12. **Croal, L. R., C. M. Johnson, B. L. Beard, and D. K. Newman.** 2004. Iron isotope fractionation by Fe(II)-oxidizing photoautotrophic bacteria. *Geochim. Cosmochim. Acta* **68:**1227–1242.
13. **Davison, W., and G. Seed.** 1983. The kinetics of the oxidation of ferrous iron in synthetic and natural waters. *Geochim. Cosmochim. Acta* **47:**67–79.
14. **Edwards, K. J., D. R. Rogers, C. O. Wirsen, and T. M. McCollom.** 2003. Isolation and characterization of novel psychrophilic, neutrophilic, Fe-oxidizing, chemolithoautotrohic α- and γ-proteobacteria from the deep sea. *Appl. Environ. Microbiol.* **69:**2906–2913.
15. **Ehrenberg, C. G.** 1836. Vorlaufige Mitteilungen über das wirkliche vorkommen fossiler Infusorien und ihre grosse Vergreitung. *Ann. Phys. Chem.* **38.**
16. **Ehrenreich, A., and F. Widdel.** 1994. Anaerobic oxidation of ferrous iron by purple bacteria, a new type of phototrophic metabolism. *Appl. Environ. Microbiol.* **60:**4517–4526.
17. **Einsele, W.** 1940. Versuch einer Theorie der Dynamik der Mangan und Eisenschichtung in eutrophen See. *Naturwissenschaften* **28:**257–264.
18. **Emerson, D., and C. L. Moyer.** 1997. Isolation and characterization of novel iron-oxidizing bacteria that grow at circumneutral pH. *Appl. Environ. Microbiol.* **63:**4784–4792.
19. **Emerson, D., and J. V. Weiss.** 2004. Bacterial iron oxidation in circumneutral freshwater habitats: findings from the field and the laboratory. *Geomicrobiol. J.* **21:**405–414.
20. **Finneran, K. T., M. E. Housewright, and D. R. Lovley.** 2002. Multiple influences of nitrate on uranium solubility during bioremediation of uranium-contaminated subsurface sediments. *Environ. Microbiol.* **4:**510–516.

21. **Hafenbradl, D., M. Keller, R. Dirmeier, R. Rachel, P. Roßnagel, S. Burggraf, H. Huber, and K. O. Stetter.** 1996. *Ferroglobus placidus* gen. nov., sp. nov. a novel hyperthermophilic archaeum that oxidizes Fe^{2+} at neutral pH under anoxic conditions. *Arch. Microbiol.* **166:**308–314.
22. **Hansen, H. C. B., C. B. Koch, H. Nancke-Krogh, O. K. Borggaard, and J. Srensen.** 1996. Abiotic nitrate reduction to ammonium: key role of green rust. *Environ. Sci. Technol.* **30:**2053–2056.
23. **Hauck, S., M. Benz, A. Brune, and B. Schink.** 2001. Ferrous iron oxidation by denitrifying bacteria in profundal sediments of a deep lake (Lake Constance). *FEMS Microbiol. Ecol.* **37:**127–134.
24. **Heising, S., L. Richter, W. Ludwig, and B. Schink.** 1999. *Chlorobium ferrooxidans* sp. nov., a phototrophic green sulfur bacterium that oxidizes iron in coculture with a "*Geospirillum*" sp. strain. *Arch. Microbiol.* **172:**116–124.
25. **Heising, S., and B. Schink.** 1998. Phototrophic oxidation of ferrous iron by a *Rhodomicrobium vannielii* strain. *Microbiology* **144:**2263–2269.
26. **Jiao, Y. Y. Q., A. Kappler, L. R. Croal, and D. K. Newman.** 2005. Isolation and characterization of a genetically tractable photoautotrophic Fe(II)-oxidizing bacterium, *Rhodopseudomonas palustris* strain TIE-1. *Appl. Environ. Microbiol.* **71:**4487–4496.
27. **Kappler, A., and D. K. Newman.** 2004. Formation of Fe(III)-minerals by Fe(II)-oxidizing photoautotrophic bacteria. *Geochim. Cosmochim. Acta* **68:**1217–1226.
28. **Kappler, A., and K. Straub.** 2005. Geomicrobiological cycling of iron, p. 85–108. *In* J. F. Banfield, J. Cervini-Silva, and K. M. Nealson (ed.), *Molecular Geomicrobiology*, vol. 59. The Mineralogical Society of America, Chantilly, Va.
29. **Kluber, H. D., and R. Conrad.** 1998. Effects of nitrate, nitrite, NO and N_2O on methanogenesis and other redox processes in anoxic rice field soil. *FEMS Microbiol. Ecol.* **25:**301–318.
30. **Lack, J. G., S. K. Chaudhuri, R. Chakraborty, L. A. Achenbach, and J. D. Coates.** 2002. Anaerobic biooxidation of Fe(II) by *Dechlorosoma suillum*. *Microb. Ecol.* **43:**424–431.
31. **Lack, J. G., S. K. Chaudhuri, S. D. Kelly, K. M. Kemner, S. M. O'Connor, and J. D. Coates.** 2002. Immobilization of radionuclides and heavy metals through anaerobic bio-oxidation of Fe(II). *Appl. Environ. Microbiol.* **68:**2704–2710.
32. **Lovley, D. R.** 1993. Dissimilatory metal reduction. *Annu. Rev. Microbiol.* **47:**263–290.
33. **Lovley, D. R., D. E. Holmes, and K. P. Nevin.** 2004. Dissimilatory Fe(III) and Mn(IV) reduction. *Adv. Microb. Physiol.* **49:**219–286.
34. **Madigan, M. T., J. M. Martinko, and J. Parker.** 2002. *Brock Biology of Microrganisms*, 10th ed. Pearson Education, Inc., Upper Saddle River, N.J.
35. **Nealson, K. H., and D. Saffarini.** 1994. Iron and manganese in anaerobic respiration: environmental significance, physiology, and regulation. *Annu. Rev. Microbiol.* **48:**311–348.
36. **Nielsen, J. L., and P. H. Nielsen.** 1998. Microbial nitrate-dependent oxidation of ferrous iron in activated sludge. *Environ. Sci. Technol.* **32:**3556–3561.
37. **Ottley, C. J., W. Davison, and W. M. Edmunds.** 1997. Chemical catalysis of nitrate reduction by iron (II). *Geochim. Cosmochim. Acta* **61:**1819–1828.
38. **Petersen, H. J. S.** 1979. Reduction of nitrate by iron(II). *Acta Chem. Scand.* A **33:**795–796.
39. **Postma, D.** 1985. Concentration of Mn and separation from Fe in sediments. I. Kinetics and stoichiometry of the reaction between birnessite and dissolved Fe(II) at 10C. *Geochim. Cosmochim. Acta* **49:**1023–1033.
40. **Postma, D.** 1990. Kinetics of nitrate reduction by detrital Fe(II)-silicates. *Geochim. Cosmochim. Acta* **54:**903–908.
41. **Ratering, S., and S. Schnell.** 2001. Nitrate-dependent iron(II) oxidation in paddy soil. *Environ. Microbiol.* **3:**100–109.
42. **Roden, E. E., D. Sobolev, B. Glazer, and G. W. Luther.** 2004. Potential for microscale bacterial Fe redox cycling at the aerobic-anaerobic interface. *Geomicrobiol. J.* **21:**379–391.
43. **Senko, J. M., T. A. Dewers, and L. R. Krumholz.** 2005. Effect of oxidation rate and Fe(II) state on microbial nitrate-dependent Fe(III) mineral formation. *Appl. Environ. Microbiol.* **71:**7172–7177.
44. **Senn, D. B., and H. F. Hemond.** 2002. Nitrate controls on iron and arsenic in an urban lake. *Science* **296:**2373–2376.
45. **Shelobolina, E. S., C. G. VanPraag, and D. R. Lovley.** 2003. Use of ferric and ferrous iron containing minerals for respiration by *Desulfitobacterium frappieri*. *Geomicrobiol. J.* **20:**143–156.
46. **Shelobolina, E., S. Pickering, and D. Lovley.** 2005. Fe-cycle bacteria from industrial clays mined in Georgia, USA. *Clays Clay Miner* **53:**580–586.
47. **Sobolev, D., and E. E. Roden.** 2004. Characterization of a neutrophilic, chemolithoautotrophic Fe(II)-oxidizing β-proteobacterium from freshwater wetland sediments. *Geomicrobiol. J.* **21:**1–10.
48. **Sobolev, D., and E. E. Roden.** 2002. Evidence for rapid microscale bacterial redox cycling of iron in circumneutral environments. *Antonie Leeuwenhoek* **81:**587–597.
49. **Sorensen, J., and L. Thorling.** 1991. Stimulation by Lepidocrocite (γ-FeOOH) of Fe(II)-dependent nitrite reduction. *Geochim. Cosmochim. Acta* **55:**1289–1294.
50. **Straub, K. L., M. Benz, B. Schink, and F. Widdel.** 1996. Anaerobic, nitrate-dependent microbial oxidation of ferrous iron. *Appl. Environ. Microbiol.* **62:**1458–1460.
51. **Straub, K. L., and B. E. E. Buchholz-Cleven.** 1998. Enumeration and detection of anaerobic ferrous iron-oxidizing, nitrate-reducing bacteria from diverse European sediments. *Appl. Environ. Microbiol.* **64:**4846–4856.
52. **Straub, K. L., M. Hanzlik, and B. E. E. Buchholz-Cleven.** 1998. The use of biologically produced ferrihydrite for the isolation of novel iron-reducing bacteria. *Syst. Appl. Microbiol.* **21:**442–449.
53. **Straub, K. L., F. A. Rainey, and F. Widdel.** 1999. *Rhodovulum iodosum* sp. nov, and *Rhodovulum robiginosum* sp. nov., two new marine phototrophic ferrous-iron-oxidizing purple bacteria. *Int. J. Syst. Bacteriol.* **49:**729–735.
54. **Straub, K. L., W. Schonhuber, B. Buchholz-Cleven, and B. Schink.** 2004. Diversity of ferrous iron-oxidizing, nitrate-reducing bacteria and their involvement in oxygen-independent iron cycling. *Geomicrobiol. J.* **21:**371–378.
55. **Stumm, W., and J. J. Morgan.** 1996. *Aquatic Chemistry: Chemical Equilibria and Rates in Natural Waters*, 3rd ed. John Wiley & Sons, New York, N.Y.
56. **Stumm, W., and B. Sulzberger.** 1992. The cycling of iron in natural environments—considerations based on laboratory studies of heterogeneous redox processes. *Geochim. Cosmochim. Acta* **56:**3233–3257.
57. **Thamdrup, B.** 2000. Bacterial manganese and iron reduction in aquatic sediments. *Adv. Microb. Ecol.* **16:**41–84.
58. **Van Hecke, K., O. Van Cleemput, and L. Baert.** 1990. Chemo-denitrification of nitrate-polluted water. *Environ. Pollut.* **63:**261–274.
59. **Weber, K. A.** 2002. Microbial coupling between nitrogen and iron cycles: potential implications for nitrate and iron biogeochemistry and metal mobility in sedimentary environments. Ph.D. dissertation. University of Alabama, Tuscaloosa.
60. **Weber, K. A.** Unpublished data.

61. **Weber, K. A., L. A. Achenbach, and J. D. Coates.** 2006. Microbes pumping iron: anaerobic microbial iron oxidation and reduction. *Nat. Rev. Microbiol.* **4:**752–764.
62. **Weber, K. A., F. W. Picardal, and E. E. Roden.** 2001. Microbially catalyzed nitrate-dependent oxidation of biogenic solid-phase Fe(II) compounds. *Environ. Sci. Technol.* **35:**1644–1650.
63. **Weber, K. A., J. Pollock, K. A. Cole, S. M. O'Connor, L. A. Achenbach, and J. D. Coates.** 2006. Anaerobic nitrate-dependent iron(II) bio-oxidation by a novel, lithoautotrophic, betaproteobacterium, strain 2002. *Appl. Environ. Microbiol.* **72:**686–694.
64. **Weber, K. A., M. M. Urrutia, P. F. Churchill, R. K. Kukkadapu, and E. E. Roden.** 2006. Anaerobic redox cycling of iron by freshwater sediment microorganisms. *Environ. Microbiol.* **8:**100–113.
65. **Widdel, F., S. Schnell, S. Heising, A. Ehrenreich, B. Assmus, and B. Schink.** 1993. Ferrous iron oxidation by anoxygenic phototrophic bacteria. *Nature* **362:**834–836.
66. **Winogradsky, S.** 1922. Eisenbakterien als Anorgoxydanten. *Zentbl. Bakteriol.* **57:**1–21.
67. **Woomer, P. L.** 1994. Most probable number counts. *In* J. M. Bigham (ed.), *Methods of Soil Analysis, Part 2. Microbiological and Biochemical Properties*. Soil Science Society of America, Madison, Wis.

Isolation, Enumeration, Growth, and Preservation of Acidophilic Prokaryotes

KEVIN B. HALLBERG AND D. BARRIE JOHNSON

95

INTRODUCTION

The study of extreme environments and the microorganisms that inhabit these environments, the so-called extremophiles, has become increasingly popular in recent years. The idea that early Earth consisted of extreme environments and thus the earliest life forms would have been those that could live in such environments has been one driver of this interest. Furthermore, results from space research programs have indicated that extreme environments dominate other planets and satellites in our solar system and therefore that the closest extraterrestrial life forms, if they exist, are most likely to be extremophilic microorganisms. Other reasons for exploring the biodiversity of extreme environments are of a more applied nature and include the search for enzymes that are more suitable for use in industrial applications (37).

One important class of extreme environments is those of low pH, which are inhabited by prokaryotic and eukaryotic microorganisms referred to as acidophiles. Naturally occurring acidic environments include regions with high concentrations of sulfur and/or sulfide, which are oxidized to sulfuric acid (H_2SO_4). Such sites can be found in geothermal areas, such as in Iceland and in Yellowstone National Park in the United States. In addition to these naturally occurring sites, humans have created such sites by the mining of metals and coal. During the mining process, sulfide ores (chiefly pyrite [FeS_2]) are exposed to water and oxygen and undergo oxidative dissolution to soluble iron and sulfate, as H_2SO_4. The metal-laden, low-pH solutions thus created are often referred to as acid mine drainage, which is increasingly recognized as a severe pollution problem in many parts of the world. Aside from their role in creating this pollution hazard, the ability of some acidophiles to dissolve metals from ores is harnessed for the enhanced recovery of metals in an industrial process known as "biomining" (58, 61).

Biology of Acidophiles

The ability of microbes to grow at low pH is a seemingly ancient trait, as acidophiles are widely distributed throughout the two prokaryotic domains (Fig. 1). Aside from taxonomic and phylogenetic assignment (29), it is often convenient to group acidophiles on the basis of their physiological characteristics, though the boundaries described below are somewhat flexible. Depending on the pH range in which they grow, acidophiles can be considered moderately acidophilic (displaying optimal growth at pH between 3 and 5) or extremely acidophilic (pH optima, <3). Temperature is often another physiological trait used to group acidophiles; mesophiles display optimum growth at less than 40°C, moderate thermophiles from 40 to 60°C, and thermophiles above 60°C. Another, and particularly significant, means of grouping acidophiles is by the electron donors that they can use as sources of energy (Table 1).

At the low pH of typical acidic environments, most (cationic) metals are highly soluble. Due to the widespread occurrence of pyrite, iron is often present in very elevated concentrations in extremely acidic environments. In addition to being highly soluble, ferrous iron is only very slowly oxidized by molecular oxygen at a pH of <4 (67). Although the free energy associated with the oxidation of iron is rather small (see below), the concentrations of ferrous iron mean that it is an important electron donor in many acidic environments. Also, ferric iron is highly soluble at a pH of <2.5, and since at this pH the Fe^{2+}/Fe^{3+} couple has a strongly positive redox potential ($E_0 = +770$ mV), ferric iron is an attractive alternative (in thermodynamic terms, and by virtue of its availability) terminal electron acceptor to molecular oxygen.

In addition to iron, acidic environments are usually rich in reduced forms of sulfur, such as elemental sulfur (S^0), polythionates, and hydrogen sulfide (H_2S). These forms of reduced sulfur also serve as electron donors for some acidophiles, and since the electrons from reduced sulfur generally enter the electron transport chain at a higher level than those from iron, they also generally provide more energy per electron than does iron. Therefore, even though acidic environments may contain more ferrous iron than reduced sulfur, sulfur-oxidizing acidophiles often outnumber iron oxidizers.

Lastly, acidophiles are often differentiated by the form of carbon that they utilize to synthesize cellular material. Many iron- and sulfur-oxidizing acidophiles are autotrophic (i.e., they fix carbon dioxide) and are therefore able to obtain energy (electrons) and carbon for growth from entirely inorganic sources (chemolithotrophs). Heterotrophic acidophiles use organic carbon as both carbon and energy sources. A third group of acidophiles, the mixotrophs, can use either carbon dioxide or organic carbon, giving them an advantage in acidic environments that have fluctuating concentrations of organic carbon.

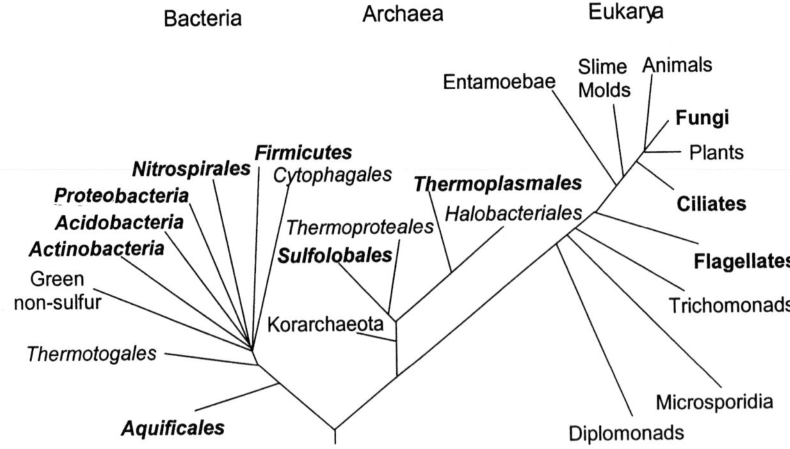

FIGURE 1 Distribution of acidophiles among the three kingdoms of living organisms. Those that contain acidophilic organisms are highlighted in bold. Reprinted from *Advances in Applied Microbiology* (29) with permission of the publisher.

Environmental and Industrial Importance of Acidophiles

The mechanism whereby chemolithotrophic acidophiles accelerate the oxidative dissolution of pyrite and other sulfidic minerals has been widely researched (18, 40, 53). When both oxygen and water are excluded, sulfidic minerals are relatively stable. When exposed, however, to both moisture and air (e.g., due to mining activities), sulfide minerals oxidize spontaneously, with either molecular oxygen or ferric iron acting as the oxidant (equation 1).

$$FeS_2 + 6Fe^{3+} + 3H_2O \rightarrow 7Fe^{2+} + S_2O_3^{2-} + 6H^+ \quad (1)$$

While sulfide mineral oxidation may be an abiotic process, the rate of reaction is far greater in the presence of iron-oxidizing acidophiles (Fig. 2). These microbes accelerate the oxidative dissolution of the mineral by regenerating the mineral sulfide oxidant ferric iron (equation 2).

$$4Fe^{2+} + O_2 + 4H^+ \rightarrow 4Fe^{3+} + 2H_2O \quad (2)$$

Above a pH of 4, molecular oxygen spontaneously reacts with the ferrous iron, but at a pH of <4 the abiotic oxidation rate of Fe^{2+} is insignificant (67) and therefore ferrous iron oxidation only occurs in the presence of iron-oxidizing acidophiles. Various acidophiles also oxidize thiosulfate, the initial sulfur by-product of pyrite oxidation (equation 1), producing sulfuric acid (equation 3).

$$S_2O_3^{2-} + 2O_2 + H_2O \rightarrow 2H^+ + 2SO_4^{2-} \quad (3)$$

Other acidophiles certainly play an important role in mineral oxidation, even though they do not oxidize iron or reduce sulfur (Fig. 2). Many of the acidophiles that catalyze sulfide mineral dissolution are sensitive to trace amounts of organic matter, specifically organic acids (1, 57). Acidophilic heterotrophs consume the organic material produced by the autotrophs (35), thereby detoxifying the local environment. In addition, some heterotrophs can catalyze the oxidation of iron (and hence sulfidic ores) only when supplied with a source of organic carbon, which may be provided by the autotrophic acidophiles (8).

CULTIVATION OF ACIDOPHILES

Acidophiles can, in theory, be readily grown in the laboratory by using suitably formulated culture media that have been adjusted to appropriate pH values. However, there are some subtle pragmatic considerations that need to be borne in mind when formulating and preparing liquid and solid media for acidophiles, which do not apply to microorganisms that live in neutral pH environments.

Liquid Media

As noted previously, chemolithotrophy is more prevalent among acidophiles than most other groups of microorganisms, while significant numbers of acidophilic prokaryotes are either heterotrophic or mixotrophic. Liquid media for enriching and for cultivating these various groups in vitro therefore need to be formulated accordingly. It is important that the minimal salts solution used to prepare acidophilic media contain a comprehensive range of macronutrients (nitrogen, phosphorus, potassium, calcium, etc.). It is also important to remember that most acidophiles are highly sensitive to anions, with the exception of sulfate, so acidification of media should always be carried out using sulfuric rather than any other mineral or organic acid, and sulfate salts of the macronutrients should be used wherever possible. Due to the sensitivity to anions, nitrogen should be provided as ammonium rather than nitrate and even phosphate concentration should be kept low. Trace elements need to be included in synthetic media, particularly in "inorganic" media used to cultivate iron- and sulfur-oxidizing chemolithotrophic acidophiles.

Chemolithotrophic Acidophiles

Due to the general paucity of inorganic (and organic) nitrogen in most extremely acidic environments, nitrification is relatively insignificant and no extremely acidophilic prokaryote capable of using either ammonium or nitrite as electron donor for chemosynthetic metabolism has been reported. In contrast, as described above, ferrous iron and reduced forms of sulfur are often present in elevated concentrations in natural and anthropogenic acidic environments, and consequently iron- and sulfur-oxidizing chemolithotrophs are widely distributed in such situations.

Ferrous iron is far more stable in acidic than in neutral pH (oxygen-containing) liquors. Heat sterilization of ferrous iron-containing liquid media invariably results in at least partial oxidation and a consequent color change (ferric iron has a strong orange color in acidophile media, while

TABLE 1 Examples of acidophilic prokaryotes and their phenotypic characteristics

Organism	Thermal classification	Carbon metabolism
Mineral-degrading acidophiles		
Iron oxidizers		
Leptospirillum ferrooxidans	Mesophile	Autotrophic
L. ferriphilum	Thermotolerant[a]	Autotrophic
L. thermoferrooxidans	Moderate thermophile	Autotrophic
"*Thiobacillus ferrooxidans*" m-1	Mesophile	Autotrophic
"*Ferrimicrobium acidiphilum*"	Mesophile	Heterotrophic
Ferroplasma acidiphilum	Mesophile	Heterotrophic
Ferroplasma acidarmanus	Mesophile	Heterotrophic
Sulfur-oxidizers		
Acidithiobacillus thiooxidans	Mesophile	Autotrophic
Acidithiobacillus caldus	Moderate thermophile	Mixotrophic
Thiomonas cuprina	Mesophile	Mixotrophic
Thiomonas intermedia	Mesophile	Mixotrophic
Hydrogenobaculum acidophilum[b]	Moderate thermophile	Autotrophic
Iron and sulfur oxidizers		
Acidithiobacillus ferrooxidans	Mesophile	Autotrophic
Iron reducers		
Acidiphilium spp.	Mesophile	Heterotrophic and mixotrophic
Acidocella spp.	Mesophile	Heterotrophic
Acidobacterium capsulatum	Mesophile	Heterotrophic
Iron oxidizers and reducers		
Acidimicrobium ferrooxidans	Moderate thermophile	Mixotrophic
Iron oxidizers and reducers and sulfur oxidizers		
Sulfobacillus spp.	Mesophile and moderate thermophile	Mixotrophic
Alicyclobacillus spp.	Mesophile and moderate thermophile	Mixotrophic
Non-mineral-degrading acidophiles		
Acidisphaera rubrifaciens	Mesophile	Heterotrophic
Acidomonas methanolica	Mesophile	Heterotrophic
Picrophilus spp.	Moderate thermophile	Heterotrophic
Thermoplasma spp.	Moderate thermophile	Heterotrophic

[a]Many isolates can grow at up to at least 45°C.
[b]Inferred ability to oxidize minerals (via production of sulfuric acid).

ferrous iron has a faint green color) and formation of a scum or solid deposits. It is preferable, therefore, to prepare and filter sterilize concentrated (e.g., 1 M) and acidified (to pH 2.0 or less) solutions of ferrous sulfate through 0.2-μm-pore-size (or smaller) membranes and add suitable aliquots to cooled, separately heat-sterilized solutions containing basal salts and/or trace elements.

Ferrous iron is, in energetic terms, a poor electron donor. Indeed, the relatively high redox potential of the ferrous/ferric couple at low pH means that only oxygen can serve as an electron acceptor for iron-oxidizing acidophiles. Consequently, cell numbers of iron oxidizers in liquid media tend to be much lower than for heterotrophic cultures, or even sulfur-grown acidophiles (the free energies available from the oxidation of various reduced inorganic sulfur compounds and elemental sulfur are significantly greater than from iron oxidation [47]). However, although direct measurement of changes in turbidity is not appropriate for measuring growth of iron-oxidizing acidophiles, the very evident color change (green to orange) that occurs can be used to indicate that oxidation is occurring. In most cases, ferrous iron oxidation is tightly coupled to the growth of iron-oxidizing acidophiles, and changes in concentrations of ferrous iron (see below) can be used to determine growth rates of these acidophiles.

Ferric iron is soluble in very acidic liquors but at a pH of ≥2.5 it hydrolyzes, forming a variety of solid-phase products, such as schwertmannite [$Fe^{3+}{}_{16}O_{16}(OH)_{12}(SO_4)_2$]. In many situations, the formation of iron precipitates is a major disadvantage, for example, in harvesting and purification of

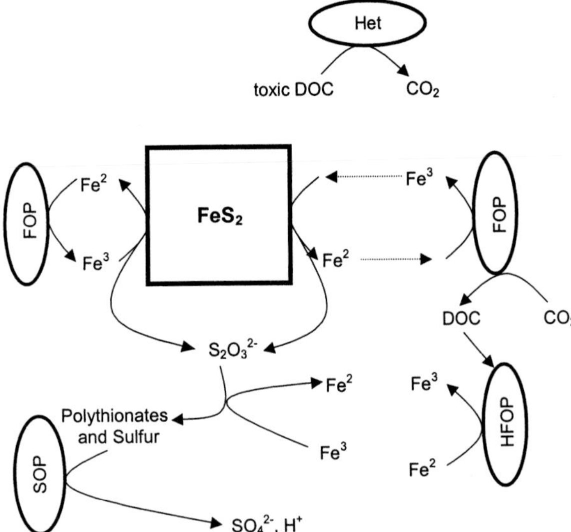

FIGURE 2 Roles of acidophiles in the oxidative dissolution of metal sulfides. The iron-oxidizing prokaryotes (FOP) that are attached to the mineral surface or are free-swimming regenerate ferric iron (indicated by broken arrows). The thiosulfate can be oxidized to sulfuric acid by sulfur-oxidizing prokaryotes (SOP) either directly or following oxidation by ferric iron to polythionates or sulfur. Carbon dioxide fixed and excreted as dissolved organic carbon (DOC) by the autotrophic FOP or SOP can be used to drive iron oxidation by heterotrophic iron-oxidizing acidophiles (HFOP). Heterotrophs contribute to metal sulfide oxidation by removing toxic dissolved organic carbon, oxidizing it to CO_2 that the autotrophic FOP/SOP can use. Reprinted from *Mine Water and the Environment* (31) with permission of the publisher.

bacterial cells. Avoiding the formation of ferric precipitates in batch cultures is not straightforward, except in situations where pH control can be maintained (in fermenters, etc.). This is because ferrous iron oxidation (equation 2) is a proton-consuming reaction, and so pH tends to increase as oxidation proceeds. Liquid media commonly used to grow acidophiles tend to be very poorly buffered, due to the lack of suitable buffer solutions that are not toxic to acidophiles. Sulfuric acid itself can provide some degree of pH stability, but only at very low pH values (the pK_a of $HSO_4^- \leftrightarrow SO_4^{2-}$ is 1.92). Ferric iron hydrolysis (which occurs more slowly than microbiologically catalyzed iron oxidation) is, in contrast, a proton-generating reaction (e.g., equation 4).

$$Fe^{3+} + H_2O \rightarrow Fe(OH)_3 + 3H^+ \qquad (4)$$

The net effect is that the pH of liquid cultures of iron-oxidizing prokaryotes tends to increase until it reaches the point (dependent on the initial pH and ferrous iron concentration of the culture) at which ferric iron hydrolysis kicks in, which causes the pH to decline. When this occurs, the final pH of the fully oxidized medium stabilizes at about 2.2, irrespective of whether the initial pH of the culture is of lower or higher pH.

An innovative solution to the problem of obtaining high cell yields of iron-grown acidophiles and avoiding the problem of ferric precipitates was described by Blake et al. (11). They devised an apparatus within which ferric iron generated by iron-oxidizing acidophiles was reduced electrochemically, thereby continuously regenerating ferrous iron. Cell counts of $\sim 10^{10}$ bacteria/ml of culture were obtained in a pH 1.6 medium, and the formation of ferric minerals was avoided.

Sulfur-oxidizing acidophiles can be grown in liquid media containing a variety of different forms of inorganic sulfur that are more reduced than +6 (i.e., the oxidation state of sulfur in the sulfate anion). Since most sulfur-oxidizing acidophiles are either obligate or facultative autotrophs, organic compounds do not need to be included, and their requirement for macro- and micronutrients is similar to that of iron oxidizers. Polythionate salts are convenient substrates, as they are soluble in water. However, thiosulfate ($S_2O_3^{2-}$) is unstable in acidic liquors (particularly when ferric iron is also present) and oxidizes to form tetrathionate, pentathionate, and sulfate (62). Trithionate ($S_3O_6^{2-}$) is stable in acidic liquors but is not a commercial product, whereas tetrathionate ($S_4O_6^{2-}$) is both acid stable and widely available from chemical suppliers. Elemental sulfur is hydrophobic, though colloidal hydrophilic sulfur is produced by some companies, and is more rapidly oxidized by sulfur-oxidizing acidophiles (24). Fine-grain sulfur may be heat sterilized, though it melts at 120°C, forming plastic sulfur, which is more resistant to microbial oxidation. Tyndallization, heating to 110°C for about 1 h, is a suitable alternative that does not cause plastic sulfur to form.

During the oxidation of sulfur and polythionates, the pH of the culture medium declines due to the production of sulfuric acid and, as noted previously, the poor buffering capacity of acidophilic media. Whereas pH can be used to indicate microbial activity, changes in concentration of substrate or sulfate (see below) or measurement of biomass (e.g., total protein) provides more accurate means of estimating growth rates of sulfur-oxidizing acidophiles.

Heterotrophic and Mixotrophic Acidophiles

Heterotrophic acidophiles can be enriched for, and cultivated in, liquid media containing a variety of single or complex carbon sources. Since most acidic environments are oligotrophic, it is advisable to use relatively small concentrations of organic substrates in enrichment media, although once isolated many heterotrophic acidophiles can grow to high cell densities in richer media. Low-molecular-weight alcohols (e.g., ethanol) and monomeric sugars (e.g., fructose) are used as carbon and energy by most known heterotrophic acidophiles. Glucose and glycerol are also possible carbon sources, though at least one acidophilic heterotroph ("*Acidocella aromatica*") is unable to use either of these compounds or yeast extract (33). Low-molecular-weight aliphatic acids are, in most cases, toxic to most acidophilic microorganisms (including heterotrophs) at concentrations of 250 μM to 1 mM, for reasons described below. However, "*Acidocella aromatica*" can grow on a variety of organic acids (23) and can be enriched for with, for example, acidic (pH ~3) medium containing 1 to 2 mM acetic acid.

A number of characterized acidophilic heterotrophs either have an absolute requirement for one or more growth factors or else are greatly stimulated by, for example, yeast extract. Inclusion of yeast extract (typically at 0.02% [wt/vol]) in liquid media is also appropriate for cultivating mixotrophic and chemolitho-heterotrophic acidophiles, such as *Sulfobacillus* spp. and "*Ferrimicrobium*," respectively.

Solid Media

For many years, researchers encountered major problems in obtaining reproducible and reliable growth of acidophilic microorganisms on solid media, even though the same organisms grew readily on nongelled media that had similar or identical makeup. Iron-oxidizing acidophiles were particularly problematic, with some (e.g., *Leptospirillum ferrooxidans*) being categorized as being incapable of growing on solid media. Recent advances in this area have led to the development of techniques that allow all categorized species of acidophilic prokaryotes to be grown on solid media. There are some basic practical points that need to be considered, however, when preparing solid media for acidophiles. For example, gelling agents such as agar are polysaccharides and are readily hydrolyzed in hot, acidic liquors, so it is necessary to heat sterilize the acidified components and gelling agents separately and combine them when cooled, immediately prior to pouring plates. Even so, slow, ongoing hydrolysis of gelling agents occurs when inoculated plates are incubated. Agar (and agarose) is a complex mixture of polysaccharides, all of which have the same backbone structure (made up of galactose residues) but have variable degrees of substitution with charged groups (mostly pyruvic acid, and also some glucuronic acid). Acidophilic prokaryotes, especially some chemolithotrophs, are highly sensitive to soluble organic materials present in gelling agents, and the organic acids in particular. The toxicity of the latter group is due to the pH of acidophilic growth media being frequently below their pK_a values (for example, acetic acid has a pK_a of 4.75 and pyruvic acid has a pK_a of 2.5), and the organic acids occur, therefore, as predominantly undissociated, lipophilic acids. As such, they can readily traverse microbial membranes and, once inside the more neutral pH (~6.5) within acidophilic cells, they dissociate, thereby acidifying the cell cytoplasm and nullifying the membrane potential (54). It is likely, therefore, that poor or zero growth of acidophiles on (agar)-gelled solid media is mostly due to the toxicity of pyruvic acid rather than of galactose.

This problem can be circumvented to some extent by prewashing the gelling agent in warm mild acid ahead of autoclaving. This causes the more labile bonds in the polymers to hydrolyze, and the soluble products (and excess acid) can be removed before heat sterilization (35), though this is a time-consuming process that does not fully address the problem of ongoing hydrolysis during protracted plate incubation.

The overlay technique (41, 42) does, however, provide a more satisfactory solution to the problem. This approach derived from observations that colonies of the iron oxidizer *Acidithiobacillus ferrooxidans* grew far more successfully on solid media when heterotrophic acidophiles (*Acidiphilium* spp.) were also present in the inoculum, presumably because the latter were removing materials that were toxic to the chemolithotroph (provision of growth factor[s] is unlikely, since *Acidithiobacillus ferrooxidans* is not an auxotroph). In the overlay technique, a two-layered solid medium is prepared; the underlayer contains an active population of an acidophilic heterotroph, while the top layer is sterile (Fig. 3). The overlay approach is a basic design concept and may be applied to a wide variety of media formulations. The original overlay medium, Fe$\underline{o}$, which contains both ferrous iron and tryptone soy broth and has a pH of ~2.6, was designed to promote the growth of chemolithotrophic iron-oxidizing acidophiles, and the underlay heterotroph used was an *Acidiphilium* sp. (strain SJH) that is closely related to *Acidiphilium cryptum*. Iron oxidizers, such as *Acidithiobacillus ferrooxidans* and *Leptospirillum* spp., appear as ferric iron-encrusted colonies that can be differentiated by their size and colony morphologies. Iron-oxidizing heterotrophs such

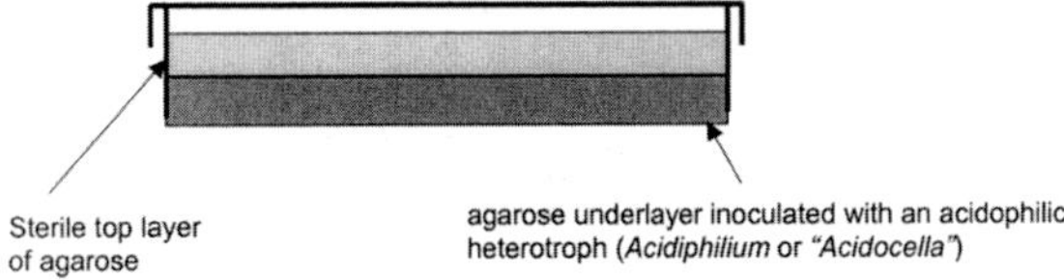

FIGURE 3 Schematic diagram of the overlay plate technique used to overcome the toxicity of gelling agents used to make solid media for growth of acidophiles.

as "*Ferrimicrobium acidiphilum*" also grow on this medium (as do some non-iron-oxidizing heterotrophs) and can be recognized from the fact that their iron-encrusted colonies have gelatinous margins whereas chemolithotrophic species do not. Deposition of ferric iron allows ready identification of acidophilic iron oxidizers. If, however, it is necessary that deposition of ferric iron be avoided, the pH of the medium can be lowered to <2 (*Acidiphilium* strain SJH has a pH minimum for growth of ~1.8).

Other overlay medium formulations include adding tetrathionate or, at a pH of >3.5, thiosulfate, to obtain simultaneous growth of iron and sulfur oxidizers, though some *Leptospirillum* strains show sporadic growth on these media. The ferrous iron/thiosulfate/tryptone soya broth medium has been used successfully to isolate novel species of "moderate acidophiles" (such as *Thiomonas* spp.) from environmental samples (32).

As noted above, colonies of heterotrophic acidophiles are often observed on Fe$\underline{o}$ medium when environmental inocula are plated. This is in some ways surprising, as these bacteria have to compete with *Acidiphilium* strain SJH, present in the gel underlayer, for the small amounts of organic substrates present. Interestingly, plate counts of heterotrophs are frequently greater on overlaid than on corresponding nonoverlaid solid media, and again this is probably related to organic acid toxicity. A modified range of overlaid solid media that target heterotrophic acidophiles have more recently been devised, in which the acidophile included in the gel underlay is "*Acidocella aromatica*" rather than *Acidiphilium* strain SJH. This alternative acidophilic heterotroph has, as noted previously, the unique trait of being unable to utilize a number of commonly used microbial substrates, such as glucose, glycerol, and yeast extract, though it is particularly adept at catabolizing organic acids. Added organic substrates therefore remain available, while at the same time, the toxicity hazard is eliminated. This modified approach can be used successfully to isolate and cultivate a wide range of extremely and moderately acidophilic aerobic (e.g., *Acidobacterium* and *Acidisphaera* spp.) and anaerobic (novel *Desulfosporosinus* spp.) heterotrophic acidophiles (30, 32, 64).

A different approach to overcome the toxicity of gelling agents is by growing acidophiles on membrane filters. Early work showed that reproducible and quantitative formation of colonies of *Acidithiobacillus ferrooxidans* could be obtained when the bacteria were filtered onto membrane filters of 0.2-μm pore size (69). The best results were obtained when the filters were boiled in sterile distilled water for 10 min and left to cool in the water for 24 h. A novel application for the use of membrane filters was proposed for the growth of fastidious acidophilic iron oxidizers (16). Here the authors collected microbes by filtration onto the membrane and then incubated the membrane on the surface of liquid medium, thereby avoiding the use of solidifying agents altogether. Using polycarbonate filters that had been rinsed with demineralized water and dilute sulfuric acid (water adjusted to pH 1.6 with H_2SO_4), visible colonies of *Acidithiobacillus ferrooxidans* developed on the membranes in 5 days. The rinsed polycarbonate filters supported the growth of 50 to 60% of the total cells (obtained by counting chamber), whereas polyester filters yielded lower counts. A demonstration of the power of the floating-filter technique was provided by the authors, who isolated an iron-oxidizing, gram-positive acidophile at 45°C from a culture used for biodesulfurization of coal.

ENUMERATION OF ACIDOPHILES

Classical Enumeration

Many of the techniques that are used for enumerating more mainstream groups of microorganisms can also be applied to acidophiles, for example, direct microscopic counts using counting chambers, though this is effective only in cases where total cell numbers are $>10^6$/ml. For greater sensitivity, cells retained on a membrane filter may be stained with DNA binding dyes such as DAPI (4′,6-diamidino-2-phenylindole) or acridine orange. While these two methods can give estimates of total microbial numbers, they do not differentiate between different physiological groups, or indeed whether the organisms present are acidophilic or not.

Most probable number (MPN) microbial counts of cultures in specified liquid media (e.g., acidic ferrous sulfate medium) and incubated at an appropriate temperature continue to be used to enumerate acidophiles on a physiological basis. This approach, however, suffers from the drawbacks that it does not distinguish between species that have similar metabolic capacities (e.g., *Acidithiobacillus ferrooxidans* and *L. ferrooxidans*, both of which oxidize ferrous iron at 30°C in organic-free media). While MPN counts are useful, it is very important to use appropriate liquid media (or better, a range of media). For example, MPN counts of iron oxidizers obtained by using acidic ferrous sulfate liquid media designed to optimize growth of *Acidithiobacillus ferrooxidans* and *Leptospirillum* spp., media which generally do not support the growth of more moderately acidophilic iron-oxidizing *Thiomonas* spp. (30, 32), can lead to erroneous conclusions about the abundance of iron-oxidizing acidophiles in samples with pHs of ~3 to 6.

Enumeration of acidophiles, using overlaid solid media that target specific groups of acidophiles, is highly sensitive and can detect, in theory, 1 to 10 prokaryotes/ml of sample. A further improvement in sensitivity can be obtained by filtering samples and incubating the filters on solid media, in which detection limits of 1 bacterium per 50 ml were obtained (22), though more solution may be filtered if it does not contain too many solids (microbes or mineral particles). Two drawbacks of using enumeration on solid media are that (i) in order to obtain a complete inventory of chemolithotrophic and heterotrophic/mixotrophic acidophiles, a number of different medium formulations need to be used; and (ii) protracted incubation (generally 3 to 20 days, depending on the incubation temperature) of plates is required before colonies become visible. There is also the possibility that, unlike classified prokaryotes, some novel species and strains of acidophiles will not grow on any of the solid media formulations. Besides the sensitivity of plate counts, an additional benefit to this approach for enumeration is that acidophiles are also isolated, thereby facilitating their further study in pure culture in the laboratory. Variations in colony morphologies may be used as preliminary guidelines for identifying isolated acidophiles (44). By coupling the plating and preliminary identification of iron-oxidizing acidophiles with 16S rRNA gene analysis (43), accurate counts of specific microbes in mixed population samples can be obtained.

One potentially problematic issue regarding the enumeration of acidophiles is that many grow while attached to solid growth substrates (e.g., pyrite or sulfur). When counting chambers or optical density measurements are used, the substrate particles themselves can interfere with enumeration. This can be overcome, in part, by allowing the culture to rest a few minutes to let the particles settle. Low-speed centrifugation can also be used, but cell counts are then less reliable. In either case, and with viable counting on solid media or by MPN, the failure to detach cells from the solid substrates yields underestimations of total numbers present. Various methods of removing acidophiles from solid substrates prior to counting have been proposed, such as vortexing cultures and using detergents such as Tween 20 or 80 and Triton X-100 (for an example, see reference 52). Studies with bacteria that were labeled with a radioisotope (such as ^{14}C) showed that such treatment can result in recovery of 100% of the bacteria from mineral surfaces (6).

Molecular Means of Cell Enumeration

The classical approaches described above can yield very useful information concerning population sizes and diversity. They suffer, however, from several drawbacks, not the least of which is the lengthy time needed for incubation of microorganisms before results can be obtained. Therefore, more rapid approaches for enumerating and identifying microorganisms are needed if they are to be used in an industrial setting such as bioleaching tanks.

One such approach is by the use of immunoassay. Here, acidophiles are applied to nitrocellulose membranes (usually by vacuum) and are then detected following addition of antibodies that cross-react with the acidophiles. Antibodies have been produced that differentiate various acidophiles, including iron- and sulfur-oxidizing acidophilic bacteria and an archaeon (3, 5, 7, 39, 52). Even though immunoassays are species specific and sensitive (able to detect about 10^3 microbes per sample loaded onto the membrane), the existence of multiple serotypes among gram-negative acidophiles (34, 49) indicates that immunoassays may be of limited use. A thorough knowledge of microbes present in a leaching operation and a judicious choice of antibodies, which target cellular components other than the variable lipopolysaccharide molecules that give rise to serotypes, could easily help overcome this shortcoming.

Another problem with immunoassays is the enumeration of microorganisms that are attached to solid growth substrates. Aside from using the detachment methods described above, an innovative solution to this problem is the recently described enzyme-linked immunofiltration assay (ELIFA). ELIFA makes use of specially adapted 96-well plates that have filters in the wells, which retain both planktonic microorganisms and those attached to mineral particles (19). Following a washing step, the bacteria can be detected with antibodies that have been cross-linked to an enzyme, which usually produces a colored product or fluoresces. Provided that care is taken, ELIFA is reported to be able to allow detection of all microbes on the mineral particles that are applied to the wells.

16S rRNA gene technology has also transformed the study of microbial ecology, allowing the study of microbial communities without prior cultivation (38). The most commonly used techniques in molecular microbial ecology have been described in a number of review articles (2, 36). In general, the biodiversity of microbes in a particular ecological niche can be assessed by extraction of nucleic acid from that environment followed by amplification of the 16S rRNA gene by PCR using universal primers or strain-specific primers. The PCR products can then be cloned, and the resulting genes can be sequenced. Comparative analysis with previously published data allows an assessment of the biodiversity of the sample and, where possible, identification of individual clones.

In the mid-1990s the first reports appeared in the literature describing the application of molecular methodologies to acidophiles that are important to biomining, though the number of such publications remains relatively sparse. In the first such studies, 16S rRNA genes from 33 representative strains of acidophilic bacteria, obtained from enrichment cultures by using acidic runoff from a chalcocite overburden heap and leachate liquors from laboratory scale bioreactors as inocula, were cloned (25, 26). Gene sequence data indicated that all of the clones from the environmental sample and the bioreactor run in batch mode consisted of a variety of iron- and sulfur-oxidizing bacteria and heterotrophic acidophiles, but the biodiversity in the bioreactor run in continuous mode was more limited. The true biodiversity in this reactor, however, was not directly assessed.

More recently, 16S rRNA gene libraries have been prepared from DNA samples obtained at an abandoned pyrite mine at the Iron Mountain site (13) and acidic geothermal sites on the volcanic island of Montserrat (14). In the latter studies, a greater diversity of microorganisms was found, probably reflecting the differences in the environments surveyed (dynamic environmental systems as opposed to the relatively stable bioreactors).

Biomolecular approaches other than cloning and sequencing have also been used for the rapid analyses of mineral-leaching populations. These include restriction enzyme mapping of the 16S rRNA genes amplified from bioleachate liquors (59, 60), often referred to as ARDRA or ARDREA (amplified ribosomal DNA restriction enzyme analysis). Another method for rapid analysis includes the design and use of microbe-specific PCR primers based on known sequence data. De Wulf-Durand et al. (17) used selected primers based on published 16S rRNA sequences for six species of acidophilic bacteria to amplify the 16S rRNA genes by using DNA extracted from the bioreactor as a template.

Other genetic markers that can be amplified with "universal PCR primers" and used to rapidly discriminate between acidophiles include the spacer region between the 16S and 23S rRNA genes of acidophiles, which can be discriminated based on size (10, 56). Partial sequencing of the 16S rRNA genes adjacent to the spacer regions of various sizes that are obtained can be used to confirm which organism is represented by a specific spacer region amplified.

Although the approaches described above are useful in assessing environmental biodiversity, they are not quantitative. They all rely on the use of PCR to amplify the target gene, a process which has been shown to be subject to various biases (71). Denaturing gradient gel electrophoresis has been used to separate 5S rRNA directly extracted from acidophile biomass (66). Migration patterns of 5S rRNA from different acidophiles were readily distinguishable from each other, allowing the species composition of a mixed microbial community to be readily assessed. While denaturing gradient gel electrophoresis is a very discriminating technique for the study of nucleic acids from microbial populations (51), a drawback to this approach is the inability to detect microbes that are present in relatively low numbers in the studied population.

Fluorescent in situ hybridization is a cultivation-independent technique that has none of the drawbacks of the enumeration methods described above and can potentially yield highly accurate data on the relative numbers of different groups or species of acidophiles in a particular sample. In this technique, a gene probe (an oligonucleotide coupled to a fluorescent dye) applied to a mixed population of microorganisms binds, if hybridization conditions are appropriate, to complementary nucleotide sequences in nucleic acids (RNA or DNA) present in target cells. A number of oligonucleotide probes that target the more familiar acidophiles have been described (Table 2). Other probes have also been recently used to target rarely encountered acidophilic isolates or 16S rRNA genes cloned from acidic environments (28). When novel species of acidophiles are discovered (as is regularly the case), great care has to be taken to ensure the stringency of new target probes, not only to ensure that they detect target cells, but also to ensure that the oligonucleotide sequence of the probe and the hybridization conditions are such that all other microorganisms that may be present do not also retain the fluorescent probe. This entails a considerable amount of careful practical work, though this is aided by the fact that extremely acidic environments tend to contain few, if any, active neutrophilic microorganisms.

DETERMINATION OF ACTIVITY OF ACIDOPHILES

In addition to isolation, identification and enumeration of acidophiles, measurements of the activities of acidophiles can provide important information, e.g., in the contexts of biomining and acid mine drainage production. A large number of methods can be used to determine heterotrophic activity in acidic environments, such as measurements of respiration by evolution of $^{14}CO_2$, or of disappearance of substrates. As these techniques are common to other situations, they are not discussed further here.

As described previously, two of the key activities of acidophiles are iron oxidation and sulfur oxidation. Concentrations of ferrous iron may be determined by titration with 1,10-phenanthroline or by the more sensitive colorimetric assay using ferrozine (50). The latter, however, has been shown to be subject to interference by ferric iron, though this may be overcome by modifying the assay (70). By reducing ferric iron present to ferrous iron (e.g., using hydroxylamine), total soluble iron can also be determined. A more recent method makes use of the different colors of ferrous-sulfosalicylic acid and ferric-sulfosalicylic acid complexes at different pHs to give a measure of both ferrous and ferric iron concentrations in a single sample (45). An advantage of the colorimetric ferrous iron assays over titration assay is the ease with which the former can be used in the field with a portable spectrophotometer. The formation of ferrous iron from ferric iron can also be measured using these assays to determine the iron reduction activity of acidophiles.

As mentioned previously, most media used to grow acidophiles are poorly buffered and so pH can be used as a measure of sulfur or reduced inorganic sulfur compound (RISC) oxidation. Conversion of pH to proton concentration yields information on how much oxidation has occurred because protons are one of the products (equation 3). The concentration of the other product, sulfate, is also easy

TABLE 2 Oligonucleotide probes that target the 16S rRNA molecule of acidophiles

Probe name	Target organism(s)	Sequence (5′-3′)	Reference
TH1187	*Thermoplasmales*	GTACTGACCTGCCGTCGAC	12
FER656	*Ferroplasma*	CGTTTAACCTCACCCGATC	21
LF581	*Leptospirillum* groups I and II[a]	CGGCCTTTCACCAAAGAC	63
LF1252	*Leptospirillum* group III[a]	TTACGGGCTCGCCTCCGT	12
LF655	*Leptospirillum* groups I, II and III[a]	CGCTTCCCTCTCCCAGCCT	12
ACM732	*Acidimicrobium* and relatives[b]	GTACCGGCCCAGATCGCTG	12
ACM995	*Acidimicrobium ferrooxidans*	CTCTGCGGCTTTTCCCTCCATG	Paul Norris, University of Warwick, Coventry, United Kingdom
SUL228	*Sulfobacillus thermosulfidooxidans*[c]	TAATGGGCCGCGAGCTCCC	12
ACD840	*Acidiphilium* genus	CGACACTGAAGTGCTAAGC	12
Acdp821	*Acidiphilium* genus	AGCACCCCAACATCCAGCACACAT	55
Thio820	*Acidithiobacillus* spp.[d]	ACCAAACATCTAGTATTCATCG	55
TF539	*Acidithiobacillus ferrooxidans*	CAGACCTAACGTACCGCC	12
THC642	*Acidithiobacillus caldus*	CATACTCCAGTCAGCCCGT	20
ATT0223	*Acidithiobacillus thiooxidans*	AGACGTAGGCTCCTCTTC	28

[a]The *Leptospirillum* grouping is based on phylogenetic analysis of environmental clones from the Iron Mountain site (13).

[b]This probe targets *Acidimicrobium ferrooxidans* as well as related clones from Iron Mountain and "*Ferrimicrobium acidiphilum*," but not the environmental clones IMBA84 and TRA2-10 from Iron Mountain.

[c]This probe also targets *S. acidophilus*, but with a one-base mismatch in the target region, and thus it may not be specific for *S. thermosulfidooxidans* under certain hybridization conditions.

[d]*Acidithiobacillus thiooxidans*, *Acidithiobacillus ferrooxidans*, and *Acidithiobacillus* sp. strain DSM612 (previously known as *T. thiooxidans* DSM612) are the targets for this probe, but not *Acidithiobacillus caldus*.

to determine by ion chromatography or by turbidimetric analysis with $BaCl_2$. The latter approach provides an instantaneous result, in contrast to ion chromatography, and the standard assay used (4) has been modified to improve the speed and the range of sulfate concentrations over which it works (48). A range of assays is available for the rapid determination of RISCs and sulfur, and these are based on colorimetric assays (46, 65, 68) or liquid chromatography (62).

PRESERVATION OF ACIDOPHILES

Since many acidic environments are not readily accessible, it is often desirable to preserve acidophilic isolates because resampling may not be convenient or possible. Furthermore, if acidophiles are to be subjected to genetic studies, it is important that the culture be preserved in such a manner that would keep the clonal properties of the culture stable. Public culture collections (e.g., ATCC and DSMZ) already contain many acidophiles, either as freeze-dried cultures (mainly heterotrophs or sulfur oxidizers) or as liquid cultures that are subjected to periodic subculturing (mainly iron oxidizers). Deposition of new acidophiles with interesting properties is one method of ensuring long-term preservation of acidophiles.

In the shorter term, storage of cultures at low temperatures has proved satisfactory for most species of acidophiles and allows cultures to be rapidly reactivated. Storage of a culture at 8°C with a sterile, inert material such as chalcopyrite was found to be sufficient to maintain iron oxidation capacity of *Acidithiobacillus ferrooxidans* and the sulfur-oxidizing capacity of *Acidithiobacillus thiooxidans* for many months (27). An important point here is that the medium used for short-term storage should contain some residual energy source so that the microorganisms can metabolize slowly, thereby enabling them to maintain the large pH gradients that exist between their cytoplasm and the external liquor. For heterotrophic acidophiles, this is readily achieved by including a suitable organic substrate, while for iron and sulfur oxidizers, addition of pyrite and elemental sulfur, respectively, is appropriate. With both of the latter, coarse-grained solids are superior to fine-grained or powdered minerals, as the coarse grains act as slow-release electron donors and minimize acid (sulfur) or ferric iron (pyrite) production during storage. In the authors' experience, most chemolithotrophic acidophiles remain viable for at least 2 years under these conditions. Certain acidophiles, however, do not survive this method of storage (e.g., strains of *Leptospirillum ferriphilum*) and must be maintained by periodic subculturing.

While the above methods are sufficient to ensure the survival of acidophilic cultures, some species and strains may require periodic subculturing, which is not a trivial task. Furthermore, these are necessarily growing cultures and so genetic mutations can accumulate, leading to the change of culture properties over time. Using the standard cryoprotectants, glycerol and dimethyl sulfoxide (DMSO), for storing frozen cultures of chemolithotrophic acidophiles has been questioned (9), although cultures of most species of acidophiles have been successfully preserved in the authors' laboratory by addition of DMSO (to a 7% [vol/vol] final concentration) and freezing at −70°C. Storing acidophiles at −70°C has been found to be effective when concentrated cell suspensions are added to Prolab Protect-100 beads (9). Since the initial cell death upon freezing is exponential, it is important that the initial cell density be high (~10^9 cells), so concentration of cell cultures by centrifugation may be necessary. Viable cells are recovered from these frozen stocks by scraping a small amount of the frozen material into fresh culture medium and incubating it under appropriate conditions. This method has proved satisfactory for *Acidithiobacillus ferrooxidans*, many sulfur-oxidizing acidophiles, and also most heterotrophic acidophiles. Some cultures thus frozen have maintained viability for up to 10 years. More recently, glycine betaine has been used for long-term preservation of two strains of *Acidithiobacillus ferrooxidans*, but recovery of frozen strains of *Acidithiobacillus thiooxidans* was more variable (15). In addition, glycine betaine has been shown to be a suitable cryoprotectant for isolates of *Leptospirillum* (D. Emerson [ATCC], personal communication).

REFERENCES

1. **Alexander, B., S. Leach, and W. J. Ingledew.** 1987. The relationship between chemiosmotic parameters and sensitivity to anions and organic acids in the acidophile *Thiobacillus ferrooxidans*. *J. Gen. Microbiol.* **133:**1171–1179.
2. **Amann, R. I., W. Ludwig, and K. H. Schleifer.** 1995. Phylogenetic identification and *in situ* detection of individual microbial cells without cultivation. *Microbiol. Rev.* **59:**143–169.
3. **Amaro, A. M., K. B. Hallberg, E. B. Lindström, and C. A. Jerez.** 1994. An immunological assay for detection and enumeration of thermophilic biomining microorganisms. *Appl. Environ. Microbiol.* **60:**3470–3473.
4. **American Public Health Association.** 1975. *Standard Methods for the Examination of Water and Wastewater*, 14th ed. American Public Health Association, Washington, D.C.
5. **Apel, W. A., P. R. Dugan, J. A. Filppi, and M. S. Rheins.** 1976. Detection of *Thiobacillus ferrooxidans* in acid mine environments by indirect fluorescent-antibody staining. *Appl. Environ. Microbiol.* **32:**159–165.
6. **Arredondo, R., A. Garcia, and C. A. Jerez.** 1994. Partial removal of lipopolysaccharide for *Thiobacillus ferrooxidans* affects its adhesion to solids. *Appl. Environ. Microbiol.* **60:** 2846–2851.
7. **Arredondo, R., and C. A. Jerez.** 1989. Specific dot-immunobinding assay for detection and enumeration of *Thiobacillus ferrooxidans*. *Appl. Environ. Microbiol.* **55:** 2025–2029.
8. **Bacelar-Nicolau, P., and D. B. Johnson.** 1999. Leaching of pyrite by acidophilic heterotrophic iron-oxidizing bacteria in pure and mixed cultures. *Appl. Environ. Microbiol.* **65:**585–590.
9. **Barron, J. L., and D. R. Lueking.** 1990. Growth and maintenance of *Thiobacillus ferrooxidans* cells. *Appl. Environ. Microbiol.* **56:**2801–2806.
10. **Bergamo, R. F., M. T. M. Novo, R. V. Verissimo, L. C. Paulino, N. C. Stoppe, M. I. Z. Sato, G. P. Manfio, P. I. Prado, O. Garcia, and L. M. M. Ottoboni.** 2004. Differentiation of *Acidithiobacillus ferrooxidans* and *A. thiooxidans* strains based on 16S-23S rDNA spacer polymorphism analysis. *Res. Microbiol.* **155:**559–567.
11. **Blake, R. C., G. T. Howard, and S. McGinness.** 1994. Enhanced yields of iron-oxidizing bacteria by in situ electrochemical reduction of soluble iron in the growth medium. *Appl. Environ. Microbiol.* **60:**2704–2710.
12. **Bond, P. L., and J. F. Banfield.** 2001. Design and performance of rRNA targeted oligonucleotide probes for in situ detection and phylogenetic identification of microorganisms inhabiting acid mine drainage environments. *Microb. Ecol.* **41:**149–161.
13. **Bond, P. L., S. P. Smriga, and J. F. Banfield.** 2000. Phylogeny of microorganisms populating a thick, subaerial, predominantly lithotrophic biofilm at an extreme acid mine drainage site. *Appl. Environ. Microbiol.* **66:**3842–3849.
14. **Burton, N. P., and P. R. Norris.** 2000. Microbiology of acidic, geothermal springs of Montserrat: environmental rDNA analysis. *Extremophiles* **4:**315–320.
15. **Cleland, D., P. Krader, C. McCree, J. Tang, and D. Emerson.** 2004. Glycine betaine as a cryoprotectant for prokaryotes. *J. Microbiol. Methods* **58:**31–38.
16. **de Bruyn, J. C., F. C. Boogerd, P. Bos, and J. G. Kuenen.** 1990. Floating filters, a novel technique for isolation and enumeration of fastidious, acidophilic, iron-oxidizing, autotrophic bacteria. *Appl. Environ. Microbiol.* **56:**2891–2894.
17. **De Wulf-Durand, P., L. J. Bryant, and L. I. Sly.** 1997. PCR-mediated detection of acidophilic, bioleaching-associated bacteria. *Appl. Environ. Microbiol.* **63:**2944–2948.
18. **Druschel, G. K., B. J. Baker, T. M. Gihring, and J. F. Banfield.** 2004. Acid mine drainage biogeochemistry at Iron Mountain, California. *Geochem. Trans.* **5:**13–32.
19. **Dziurla, M.-A., W. Achouak, B.-T. Lam, T. Heulin, and J. Berthelin.** 1998. Enzyme-linked immunofiltration assay to estimate attachment of thiobacilli to pyrite. *Appl. Environ. Microbiol.* **64:**2937–2940.
20. **Edwards, K. J., P. L. Bond, and J. F. Banfield.** 2000. Characteristics of attachment and growth of *Thiobacillus caldus* on sulphide minerals: a chemotactic response to sulphur minerals? *Environ. Microbiol.* **2:**324–332.
21. **Edwards, K. J., P. L. Bond, T. M. Gihring, and J. F. Banfield.** 2000. An archaeal iron-oxidizing extreme acidophile important in acid mine drainage. *Science* **287:** 1796–1799.
22. **Escobar, B., and I. Godoy.** 2001. Enumeration of iron-oxidizing bacteria by the membrane filter technique. *World J. Microbiol. Biotechnol.* **17:**395–397.
23. **Gemmell, R. T., and C. J. Knowles.** 2000. Utilisation of aliphatic compounds by acidophilic heterotrophic bacteria. The potential for bioremediation of acidic wastewaters contaminated with toxic organic compounds and heavy metals. *FEMS Microbiol. Lett.* **192:**185–190.
24. **Germida, J. J.** 1985. Modified sulfur-containing media for studying sulfur-oxidizing microorganisms, p. 333–344. *In* D. E. Caldwell, J. A. Brierley, and C. L. Brierley (ed.), *Planetary Ecology*. Van Nostrand Reinhold, New York, N.Y.
25. **Goebel, B. M., and E. Stackebrandt.** 1994. The biotechnological importance of molecular biodiversity studies for metal bioleaching, p. 259–273. *In* F. G. Priest, A. Ramos-Cormenzana, and B. J. Tindal (ed.), FEMS Symposium No. 75. *Bacterial Diversity and Systematics*. Plenum Press, New York, N.Y.
26. **Goebel, B. M., and E. Stackebrandt.** 1994. Cultural and phylogenetical analysis of mixed microbial populations found in natural and commercial bioleaching environments. *Appl. Environ. Microbiol.* **60:**1614–1621.
27. **Gupta, S. G., and A. D. Agate.** 1986. Preservation of *Thiobacillus ferrooxidans* and *Thiobacillus thiooxidans* with activity check. *Antonie Leeuwenhoek J. Microbiol. Serol.* **52:**121–127.
28. **Hallberg, K. B., K. Coupland, S. Kimura, and D. B. Johnson.** 2006. Macroscopic streamer growths in acidic, metal-rich mine waters in north Wales consist of novel and remarkably simple bacterial communities. *Appl. Environ. Microbiol.* **72:**2022–2030.
29. **Hallberg, K. B., and D. B. Johnson.** 2001. Biodiversity of acidophilic microorganisms. *Adv. Appl. Microbiol.* **49:**37–84.
30. **Hallberg, K. B., and D. B. Johnson.** 2005. Microbiology of a wetland ecosystem constructed to remediate mine drainage from a heavy metal mine. *Sci. Tot. Environ.* **338:** 53–66.
31. **Hallberg, K. B., and D. B. Johnson.** 2005. Mine water microbiology. *Mine Water Environ.* **24:**28–32.
32. **Hallberg, K. B., and D. B. Johnson.** 2003. Novel acidophiles isolated from moderately acidic mine drainage waters. *Hydrometallurgy* **71:**139–148.

33. **Hallberg, K. B., Å. K. Kolmert, D. B. Johnson, and P. A. Williams.** 1999. A novel metabolic phenotype among acidophilic bacteria: aromatic degradation and the potential use of these microorganisms for the treatment of wastewater containing organic and inorganic pollutants, p. 719–728. *In* R. Amils and A. Ballester (ed.), *Biohydrometallurgy and the Environment toward the Mining of the 21st Century*, vol. 9A. Elsevier, Amsterdam, The Netherlands.
34. **Hallberg, K. B., and E. B. Lindström.** 1996. Multiple serotypes of the moderate thermophile *Thiobacillus caldus*, a limitation of immunological assays for biomining microorganisms. *Appl. Environ. Microbiol.* **62:**4243–4246.
35. **Harrison, A. P., Jr.** 1984. The acidophilic Thiobacilli and other acidophilic bacteria that share their habitat. *Annu. Rev. Microbiol.* **38:**265–292.
36. **Head, I. M., J. R. Saunders, and R. W. Pickup.** 1998. Microbial evolution, diversity, and ecology: a decade of ribosomal RNA analysis of uncultivated microorganisms. *Microb. Ecol.* **35:**1–21.
37. **Huber, H., and K. O. Stetter.** 1998. Hyperthermophiles and their possible potential in biotechnology. *J. Biotechnol.* **64:**39–52.
38. **Hugenholtz, P., B. M. Goebel, and N. R. Pace.** 1998. Impact of culture-independent studies on the emerging phylogenetic view of bacterial diversity. *J. Bacteriol.* **180:** 4765–4774.
39. **Jerez, C. A., and R. Arredondo.** 1991. A sensitive immunological method to enumerate *Leptospirillum ferrooxidans* in the presence of *Thiobacillus ferrooxidans*. *FEMS Microbiol. Lett.* **78:**99–102.
40. **Johnson, D. B.** 2003. Chemical and microbiological characteristics of mineral spoils and drainage waters at abandoned coal and metal mines. *Water Air Soil Pollut. Focus* **3:**47–66.
41. **Johnson, D. B.** 1995. Selective solid media for isolating and enumerating acidophilic bacteria. *J. Microbiol. Methods* **23:**205–218.
42. **Johnson, D. B., and S. McGinness.** 1991. A highly efficient and universal solid medium for growing mesophilic and moderately thermophilic iron-oxidizing acidophilic bacteria. *J. Microbiol. Methods* **13:**113–122.
43. **Johnson, D. B., N. Okibe, and K. B. Hallberg.** 2005. Differentiation and identification of iron-oxidizing acidophilic bacteria using cultivation techniques and amplified ribosomal DNA restriction enzyme analysis (ARDREA). *J. Microbiol. Methods* **60:**299–313.
44. **Johnson, D. B., and F. F. Roberto.** 1997. Biodiversity of acidophilic bacteria in mineral leaching and related environments, p. P3.1–P3.10. *In IBS Biomine '97 Conference Proceedings*. Australian Mineral Foundation, Glenside, Australia.
45. **Karamanev, D. G., L. N. Nikolov, and V. Mamatarkova.** 2002. Rapid simultaneous quantitative determination of ferric and ferrous ions in drainage waters and similar solutions. *Miner. Eng.* **15:**341–346.
46. **Kelly, D. P., L. A. Chambers, and P. A. Trudinger.** 1969. Cyanolysis and spectrophotometric estimation of trithionate in mixture with thiosulfate and tetrathionate. *Anal. Chem.* **41:**898–902.
47. **Kelly, D. P., J. K. Shergill, W. P. Lu, and A. P. Wood.** 1997. Oxidative metabolism of inorganic sulfur compounds by bacteria. *Antonie Leeuwenhoek* **71:**95–107.
48. **Kolmert, Å., P. Wikström, and K. B. Hallberg.** 2000. A fast and simple turbidimetric method for the determination of sulfate in sulfate-reducing bacterial cultures. *J. Microbiol. Methods* **41:**179–184.
49. **Koppe, B., and H. Harms.** 1994. Antigenic determinants and specificity of antisera against acidophilic bacteria. *World J. Microbiol. Biotechnol.* **10:**154–158.
50. **Lovley, D. R., and E. J. P. Phillips.** 1987. Rapid assay for microbially reduced ferric iron in aquatic sediments. *Appl. Environ. Microbiol.* **53:**1536–1540.
51. **Muyzer, G.** 1999. DGGE/TGGE a method for identifying genes from natural ecosystems. *Curr. Opin. Microbiol.* **2:**317–322.
52. **Muyzer, G., A. C. de Bruyn, D. J. M. Schmedding, P. Bos, P. Westbroek, and J. G. Kuenen.** 1987. A combined immunofluorescence-DNA-fluorescence staining technique for enumeration of *Thiobacillus ferrooxidans* in a population of acidophilic bacteria. *Appl. Environ. Microbiol.* **53:**660–664.
53. **Nordstrom, D. K.** 2000. Advances in the hydrogeochemistry and microbiology of acid mine waters. *Int. Geol. Rev.* **42:**499–515.
54. **Norris, P. R., and W. J. Ingledew.** 1992. Acidophilic bacteria: adaptations and applications, p. 121–131. *In* R. A. Herbert and R. J. Sharp (ed.), *Molecular Biology and Biotechnology of Extremophiles*. Royal Society for Chemistry, Cambridge, United Kingdom.
55. **Peccia, J., E. A. Marchand, J. Silverstein, and M. Hernandez.** 2000. Development and application of small-subunit rRNA probes for assessment of selected *Thiobacillus* species and members of the genus *Acidiphilium*. *Appl. Environ. Microbiol.* **66:**3065–3072.
56. **Pizarro, J., E. Jedlicki, O. Orellana, J. Romero, and R. T. Espejo.** 1996. Bacterial populations in samples of bioleached copper ore as revealed by analysis of DNA obtained before and after cultivation. *Appl. Environ. Microbiol.* **62:**1323–1328.
57. **Rao, G. S., and L. R. Berger.** 1970. Basis of pyruvate inhibition of *Thiobacillus thiooxidans*. *J. Bacteriol.* **102:**462–466.
58. **Rawlings, D. E.** 2002. Heavy metal mining using microbes. *Annu. Rev. Microbiol.* **56:**65–91.
59. **Rawlings, D. E.** 1995. Restriction enzyme analysis of 16S rDNA genes for the rapid identification of *Thiobacillus ferrooxidans*, *Thiobacillus thiooxidans* and *Leptospirillum ferrooxidans* strains in leaching environments, p. 9–18. *In* T. Vargas, C. A. Jerez, J. V. Wiertz, and H. Toledo (ed.), *Biohydrometallurgical Processing*, vol. II. University of Chile, Santiago, Chile.
60. **Rawlings, D. E., N. J. Coram, M. N. Gardner, and S. M. Deane.** 1999. *Thiobacillus caldus* and *Leptospirillum ferrooxidans* are widely distributed in continuous flow biooxidation tanks used to treat a variety of metal containing ores and concentrates, p. 777–786. *In* R. Amils and A. Ballester (ed.), *Biohydrometallurgy and the Environment toward the Mining of the 21st Century*, vol. 9A. Elsevier, Amsterdam, The Netherlands.
61. **Rawlings, D. E., D. Dew, and C. du Plessis.** 2003. Biomineralization of metal-containing ores and concentrates. *Trends Biotechnol.* **21:**38–44.
62. **Schippers, A., P. G. Jozsa, and W. Sand.** 1996. Sulfur chemistry in bacterial leaching of pyrite. *Appl. Environ. Microbiol.* **62:**3424–3431.
63. **Schrenk, M. O., K. J. Edwards, R. M. Goodman, R. J. Hamers, and J. F. Banfield.** 1998. Distribution of *Thiobacillus ferrooxidans* and *Leptospirillum ferrooxidans*: implications for generation of acid mine drainage. *Science* **279:**1519–1522.
64. **Sen, A. M., and D. B. Johnson.** 1999. Acidophilic sulphate-reducing bacteria: candidates for bioremediation of acid mine drainage, p. 709–718. *In* R. Amils and A. Ballester (ed.), *Biohydrometallurgy and the Environment toward the Mining of the 21st Century*, vol. 9A. Elsevier, Amsterdam, The Netherlands.
65. **Sorbö, B.** 1957. A colorimetric method for the determination of thiosulfate. *Biochim. Biophys. Acta* **23:**412–416.

66. **Stoner, D. L., C. K. Browning, D. K. Bulmer, T. E. Ward, and M. T. MacDonell.** 1996. Direct 5S rRNA assay for monitoring mixed-culture bioprocesses. *Appl. Environ. Microbiol.* **62:**1969–1976.
67. **Stumm, W., and J. J. Morgan.** 1981. *Aquatic Chemistry: an Introduction Emphasizing Chemical Equilibria in Natural Waters*. Wiley, New York, N.Y.
68. **Trüper, H. G., and H. G. Schlegel.** 1964. Sulphur metabolism in Thiorhodaceae. I. Quantitative measurements on growing cells of *Chromatium okenii*. *Antonie Leeuwenhoek* **30:**225–238.
69. **Tuovinen, O. H., and D. P. Kelly.** 1973. Studies on growth of *Thiobacillus ferrooxidans*. 1. Use of membrane filters and ferrous iron agar to determine viable numbers, and comparison with $^{14}CO_2$ fixation and iron oxidation as measures of growth. *Arch. Microbiol.* **88:**285–298.
70. **Viollier, E., P. W. Inglett, K. Hunter, A. N. Roychoudhury, and P. Van Cappellen.** 2000. The ferrozine method revisited: Fe(II)/Fe(III) determination in natural waters. *Appl. Geochem.* **15:**785–790.
71. **von Wintzingerode, F., U. B. Gobel, and E. Stackebrandt.** 1997. Determination of microbial diversity in environmental samples: pitfalls of PCR-based rRNA analysis. *FEMS Microbiol. Rev.* **21:**213–229.

Molecular Techniques for the Study of Toxic Metal Resistance Mechanisms in Bacteria

JON L. HOBMAN

96

INTRODUCTION

Bacteria, like all forms of life, have essential elemental requirements (14, 100, 104, 105, 121). Leading these requirements are the bulk nonmetallic elements H, C, O, N, P, and S, which are the major components of proteins, lipids, and nucleic acids, and make up approximately 97% of an *Escherichia coli* cell (121). In addition to an absolute requirement for the bulk elements, bacteria also require the major elemental cations, Na, Mg, K, and Ca, and the anion Cl, which are all important to cell function. The essential transition metal ions, V, Mn, Fe, Co, Ni, Cu, Zn, Mo, and W, are also required by bacterial cells and are variously involved in osmotic balance, electron transfer, and dioxygen utilization, are essential components of metalloenzymes and cofactors, stabilize DNA and enzymes through electrostatic interactions, or have structural roles in proteins and nucleic acids (22, 59, 69, 121). A major role for these essential metals is in metalloenzymes, which comprise ~30% of all enzymes in bacterial cells (121).

Although essential, these transition metals can also be toxic at higher concentrations (16, 45, 95), so bacteria require systems that maintain appropriate levels of biologically beneficial metals within their cells. Insufficient levels of an essential metal adversely affect biochemical processes that require that metal, but if the intracellular concentration of that metal ion exceeds a certain upper limit, the metal can exert a negative effect on the cell by interfering with other metal-ion-dependent processes or by taking part in chemical reactions that are deleterious to it (70). In practice this means that bacteria need to balance uptake, trafficking, utilization, and removal of these metals. The proteins involved in the regulation of intracellular levels of essential metals in bacteria are chromosomally encoded, as homeostasis of these essential metal ions is a core metabolic activity. However, there are instances of plasmid-encoded resistances to some essential metals such as copper, and conversely chromosomal toxic metal(loid) resistance genes (reviewed in reference 105).

Excess of essential metal ions can cause toxic effects on cells, but bacteria may also encounter other metals and some metalloids that have no known beneficial role in their metabolism and have demonstrable toxic effects on their growth and survival. These known toxic metals are Ag, Au, Cd, Cr, Sn, Hg, Pb, Tl, Bi, and Al (105). The known toxic metalloids are As, Sb, and Te. Other metals such as Li, Be, Cs, and Sr may also exhibit toxicity to bacteria (22), but to date studies of resistance mechanisms in bacteria have focused on the group of known toxic metals and metalloids. Bacteria are therefore faced with the challenge that they must acquire essential metals from their surroundings and regulate intracellular levels of these but at the same time can be exposed to one or more toxic metals (and/or toxic organic pollutants) in some polluted environments.

It has long been recognized that certain metals are toxic to humans (67), and that these metals could be used as antimicrobial agents (for reviews see references 56, 67, 80, and 105, and for a historical perspective see reference 64), but there is now a recognition that mineral extraction, processing, and industrial manufacturing processes have caused severe and localized toxic metal pollution in the environment and that some of these metals bioaccumulate and biomagnify. Higher levels of toxic metals in the environment have consequences for all forms of life, not just bacteria.

The Basis of Metal Toxicity to Bacteria

The supply of metals to bacterial cells, whether they are growing in a microbiological medium or encountering these metals in the environment, is governed by the concentration, solubility, and the ligand for complexation of that metal ion (17). In a biological context, ligands for metals are provided by water, protein side chains, and small cellular components in the cytoplasm, nucleic acids, and other organic cofactors (70). Metal ions are considered to be Lewis acids, molecules that accept a pair of electrons in order to form a coordinate bond with an appropriate ligand, while a Lewis base donates an electron pair in order to form a coordinate bond. Further subdivision of Lewis acids and bases into "hard" and "soft" refers to their atomic structure, where a soft Lewis acid is one that has a large and polarizable atomic structure and a hard Lewis acid has a small and nonpolarizable atomic structure. Although the dividing line between hard and soft categories of the Lewis acids and bases is not clear-cut in some instances, the general trend is that hard acids prefer to coordinate with hard bases and soft acids prefer to coordinate with soft bases (for more details on Lewis acids, please see http://www.Britannica.com, ver. 2001). Different metal ions therefore bind to different ligands according to preferences dictated by whether the metal is a hard or soft Lewis acid. In a biological system, this translates as the preference of a metal for a particular ligand and avidity for binding of the metal ion to that ligand within a biological molecule.

Figure 1 shows that the soft metals are also the toxic metals and that the charge that the metal ion carries affects the hardness or softness of the metal (e.g., Cu^{2+} is borderline, while Cu^{+} is soft; Tl^{3+} is hard but Tl^{+} is soft). The preferred coordination of the soft metals is to soft bases, particularly S and N groups. Soft Lewis acid metals such as Hg, Pt, Ag, Au, Pb, and Cd have a high affinity for the soft Lewis base S and bind avidly to thiol groups in cells, in particular cysteine sulfhydryl groups or methionine thioether linkages, which are often structurally important residues within a protein (22). Figure 1 shows that most of the toxic metals have a strong avidity for thiols in the amino acids cysteine and methionine and/or nitrogen groups in histidines, as well as oxygens. Coordination of the toxic metal to these ligands by ionic or covalent bonding may displace other metals from the catalytic centers of enzymes or cross-link cysteine residues in a protein, causing a conformational change in the protein structure. Toxic metals and metalloids (and excess levels of essential metals) can therefore exert their deleterious effects in a number of ways, either by displacing essential metal ions from metalloenzymes, by blocking essential functional groups in biological molecules, by altering the active conformation of biomolecules, or by taking part in chemical reactions within the bacterial cell that are harmful to it. The gross manifestations of these toxic effects are widespread damage to biological membranes, DNA, and proteins, altered enzyme specificity, and interference in oxidative phosphorylation and osmotic balance (22). Many toxic metals appear to be able to enter cells by import systems for essential elements or molecules (22, 80, 101), and although chromosomally encoded homeostasis systems for essential metals are widespread, in bacteria many toxic metal resistance genes are found on mobile genetic elements such as plasmids and transposons.

In biological terms, one of the key issues concerning metal and metalloid toxicity is whether the metal or metalloid is bioavailable (in a form that will be taken up by the bacterial cell). Several factors influence bioavailability and toxicity of the metal, including its chemical reactivity, aqueous solubility, the physical (pH) and chemical environment (antagonistic or synergistic effects of other metals), and whether any organic matter that may bind or immobilize the metal is present. Physical methods used to measure the levels of metals in environmental samples are highly accurate, but the measurement method cannot necessarily determine the speciation of the metal or its consequent bioavailability (112, 125). Metal sulfides are insoluble and are not bioavailable to most bacteria and therefore pose far less of a challenge than soluble or mobilizable metals, even if the sulfides are present at higher levels than the soluble form of the toxic metal. However, toxic metals that are in a form that is not bioavailable can become bioavailable again, due to biological transformations mediated by microbes or by environmental chemical reactions (124, 125).

Bacterial Resistance Mechanisms for Toxic Metals

Bacteria have been exposed to toxic metals and metalloids for millennia, because geological processes such as volcanic activity and weathering of rock have released soluble metal salts into the environment. These toxic metals can easily enter cells through importers and porins, often because cellular import mechanisms cannot differentiate between essential and toxic elements (8, 101, 102). Micro-bial transformations of metals and metalloids by reduction, oxidation, methylation, dealkylation, and other mechanisms are discussed in detail elsewhere (73, 124, 125). These biological transformations can make metals more soluble and bioavailable or can immobilize them and contribute to the natural biogeochemical cycles of these metals (71, 124).

Resistances to toxic metal ions that have been characterized so far are highly specific mechanisms. They differ in one major aspect from biological transformations of metals that are byproducts of normal cellular metabolism, such as methylation or precipitation of metals by sulfides produced by anaerobic sulfate-reducing bacteria. This difference is that specific resistance mechanisms are designed for one

METALS			LIGANDS		
Hard			**Hard**		
(H^+)	Mn^{2+}	Cr^{3+}	H_2O	CO_3^{2-}	NH_3
Na^+	Al^{3+}	Co^{3+}	OH^-	NO_3^-	RNH_2
K^+	Ga^{3+}	Fe^{3+}	$CH_3CO_2^-$	ROH	N_2H_4
Mg^{2+}	Ca^{2+}	Tl^{3+}	PO_4^{3-}	R_2O	RO^-
			$ROPO_3^{2-}$	$(RO)_2PO^{2-}$	Cl^-
Borderline			**Borderline**		
Fe^{2+}	Ni^{2+}	Zn^{2+}	NO_2^-	N_2	SO_3^{2-}
Co^{2+}	Cu^{2+}		Br^-	N_3^-	
			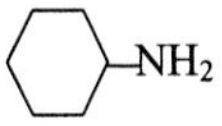 NH_2	HN N	
Soft			**Soft**		
Cu^+	Pt^{2+}	Pt^{4+}	R_2S	R_3P	RS^-
Au^+	Tl^+	Hg^{2+}	CN^-	RSH	RNC
Cd^{2+}	Pb^{2+}		$(RS)_2PO_2^-$	$(RO)_2P(O)S^-$	SCN^-
			CO	H^-	R^-

FIGURE 1 Hard and soft acid-base classification of metal ions and ligands important to bioinorganic chemistry. Reproduced from reference 69 with permission.

purpose: to nullify the effects of the toxic metal on the host organism. Toxic metal resistance options for bacteria are constrained because of the chemical nature of the metals and the effects of metal ion toxicity. Unlike options for resistance to toxic organic chemicals, including antibiotics (which can be conferred by enzymatic destruction or modification of the antibiotic or modification of its specific cellular target site [109]), metals are immutable and have widespread deleterious effects on the cell. The mechanisms that have evolved in bacteria for avoiding or detoxifying metals can be classified into five main groupings: exclusion, sequestration, efflux, enzymatic action, and reduction in sensitivity of cellular targets (9, 22, 56, 57, 80, 105). As toxic metals and metalloids can enter cells and exert widespread toxic effects, this has meant that the resistance strategy that bacteria have adopted is either to prevent certain metal(loid)s entering the cell (13) or to sequestrate them, and if that is not possible, to remove them (by efflux pump) or reduce their toxicity (by making them less bioavailable) by enzymatic transformation. There is also evidence of adaptation to some toxic metals after repeated exposure, but the nature of the adaptation is not well understood (reviewed in reference 22). A combination of these strategies may be used by the bacterium. Uptake of toxic metals can be blocked by extracellular materials such as exopolysaccharides or by nonspecific or specific cell surface or periplasmic metal binding proteins (22, 113). These proteins are selective for sulfur-binding toxic metals. If toxic metals do enter the cell, then many bacteria possess membrane-bound efflux proteins that pump out metals. These are P-type ATPases (CopA, ZntA, and CadA), A-type ATPases (ArsB), cation diffusion facilitator family (CzcD), the RND (resistance, nodulation, division) effluxers (which so far have been found in gram-negative bacteria only), and the ChrA chromate effluxer (reviewed in references 76, 80, 81, and 105). The final mechanism of bacterial resistance to toxic metals is enzymatic reduction. The ability of metals to be reduced is dependent on the redox potential of the metal, and only a limited number of metals which have a redox potential between 0 and 1.229 V can be enzymatically reduced, as this potential lies between those of H^+/H_2 and H_2O/H_2 couples (80). Hg^{2+}, AsO_4^{3-}, and CrO_4^{2-} are reduced to Hg(0), AsO_2^-, and CrO_2^-, respectively, by different resistance determinants in this manner, but whereas Hg(0) is relatively nontoxic and volatile under physiological conditions, AsO_2^- and CrO_2^- are more toxic and must be effluxed from the cell (22, 80, 81, 105).

Most importantly, highly specific toxic metal resistance mechanisms are characterized by induction of resistance gene expression as a specific response to those toxic metals (or sometimes related metals). Specific regulators control resistance gene expression. The primary regulator types that have been encountered so far belong to the MerR family (19, 86) or the ArsR/SmtB family (24), are two-component systems (82), or are of the CnrYXH type (76). These resistances and their regulators have been discussed in detail in a special issue of *FEMS Microbiology Reviews* (18).

Genetic Structure of Metal Resistance Operons

Resistance mechanisms to Hg, As, Cd, Co, Pb, Zn, Cu, Ag, Te, Sb, and Cr have been described, although an understanding of how these resistance mechanisms work varies from resistance to resistance (for reviews, see references 18 and 105). In some cases the resistance mechanisms confer resistance to more than one metal(loid); for instance, the *ars* arsenic resistance operon from *E. coli* confers resistance to As and Sb. The genetic structure and DNA sequences of many bacterial toxic metal resistance operons are now known, and some are shown in Fig. 2. The toxic metal resistance determinants shown in Fig. 2 have been chosen primarily because PCR primers to amplify related genes from environmental bacteria, or DNA samples, have been designed using these determinants (see Table 1). However, the diagram shows the diversity of toxic metal resistance operons and that there are clear similarities in genetic structure between resistances for different metals, for example the plasmid-borne copper and silver resistances. In Fig. 2, the genes that belong to the main regulator (MerR or SmtB/ArsR), effluxer (ATPase or RND), or enzyme (reductase) family are marked for each resistance determinant, when known.

Our detailed understanding of many of the bacterial toxic metal(loid) resistance systems has come about through the combined use of microbiology, molecular biology, biochemistry, and structural biology approaches. High-throughput methods in DNA sequencing, metagenomics, and the postgenomic technologies are going to revolutionize our understanding of toxic metal-bacterial cell interactions and the diversity of toxic metal(loid) resistances. With this in mind, this chapter discusses the options and potential workflows open to researchers wishing to use molecular techniques for the study of these resistance genes.

BACKGROUND PHILOSOPHY OF RESEARCH INTO TOXIC METAL RESISTANCE MECHANISMS IN BACTERIA

The advantage of working with bacteria that carry toxic metal resistance genes is that they have a phenotype that can be selected using the toxic metal as the lethal selective agent. This selectable marker makes isolation of resistant organisms and studies on the resistance mechanisms relatively straightforward. Molecular biology techniques are highly appropriate for the study of toxic metal resistance in bacteria, and these genetic tools allow us to answer fundamental questions about how the gene products work, about the distribution and divergence of these genes, and about how related the systems are to each other. Most of the techniques discussed in this chapter are equally applicable to the study of other nonmetal resistance phenotypes of environmental bacteria, because the techniques are generic. This chapter, however, focuses on and discusses the merits of molecular biology techniques that can be used to investigate bacterial resistance mechanisms. As most of the techniques such as agarose gel electrophoresis, restriction digestion, and plasmid preparation are routine methods in many laboratories and laboratory manuals for these techniques are widely available (47, 98), these techniques are not detailed here; rather, the laboratory manuals listed in the references should be used to provide detailed protocols.

Critically, before starting any experimental work, the first question the researcher needs to ask is: "What is the primary purpose of the experiments?" For molecular studies of toxic metal resistance genes, the researcher has two choices: the "viable-cell" strategy and the "community DNA" strategy. For the former, the primary purpose of the experiments is to isolate viable resistant bacteria, which can be used for further investigations. For the latter, the primary purpose of the experiments is to isolate toxic metal resistance genes directly from the environmental DNA sample, which will then be studied further using molecular techniques. The experimental methods and molecular techniques described below can be used for both approaches, and suggested workflows for the methods are shown in Fig. 3.

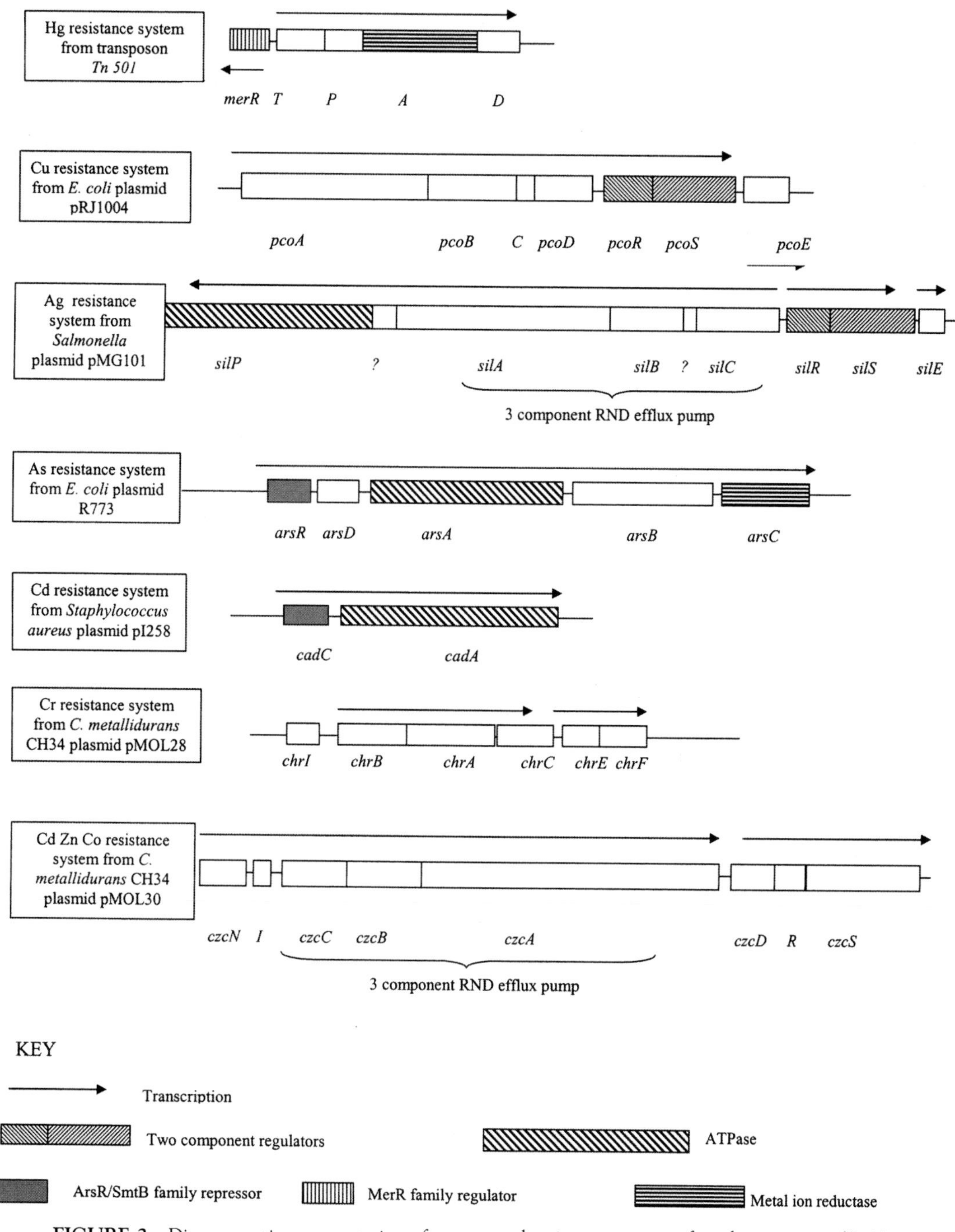

FIGURE 2 Diagrammatic representation of some metal resistance operons from bacteria: *mer* (9, 19, 56, 57), *pco* (96), *sil* (103), *ars* (100), *cad* (24), *chr* (60), and *czc* (76, 81).

SAMPLING FROM CONTAMINATED ENVIRONMENTS

All procedures for isolating environmental bacteria and using toxic metals and their salts must be subject to a biological and chemical risk assessment in compliance with local legislation. Good laboratory practice and good microbiological practice should be observed in experiments, and risk assessments should be conducted for chemical hazards, potential biological hazards, and/or genetic manipulation hazards. Waste materials contaminated with toxic metals or metalloids must be disposed of safely, as required by local legislation (10).

Bacterial resistance to toxic metals is very widespread, and toxic-metal-resistant bacteria can be isolated from a wide variety of sites, including those that are not known to have been specifically polluted with metals. However, environments which have been polluted by human activity or contain high levels of toxic metal ions are likely to contain a higher proportion of toxic-metal-resistant bacteria than those that are not (105). Contaminated sampling sites for

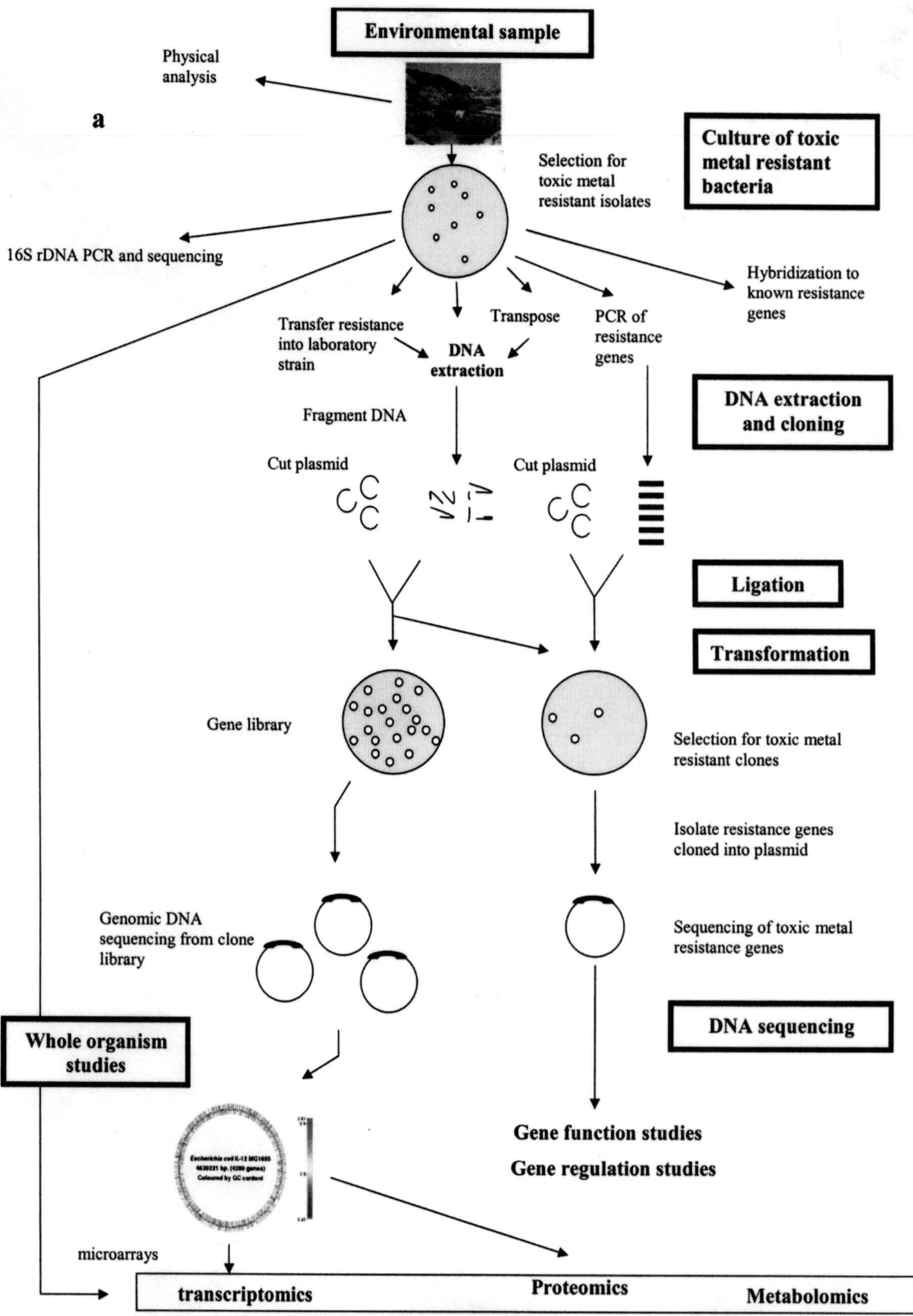
Environmental sample
Physical analysis
a
Culture of toxic metal resistant bacteria
Selection for toxic metal resistant isolates
16S rDNA PCR and sequencing
Hybridization to known resistance genes
Transfer resistance into laboratory strain
DNA extraction
Transpose
PCR of resistance genes
DNA extraction and cloning
Fragment DNA
Cut plasmid
Cut plasmid
Ligation
Transformation
Gene library
Selection for toxic metal resistant clones
Isolate resistance genes cloned into plasmid
Genomic DNA sequencing from clone library
Sequencing of toxic metal resistance genes
DNA sequencing
Whole organism studies
Gene function studies
Gene regulation studies
microarrays
transcriptomics
Proteomics
Metabolomics

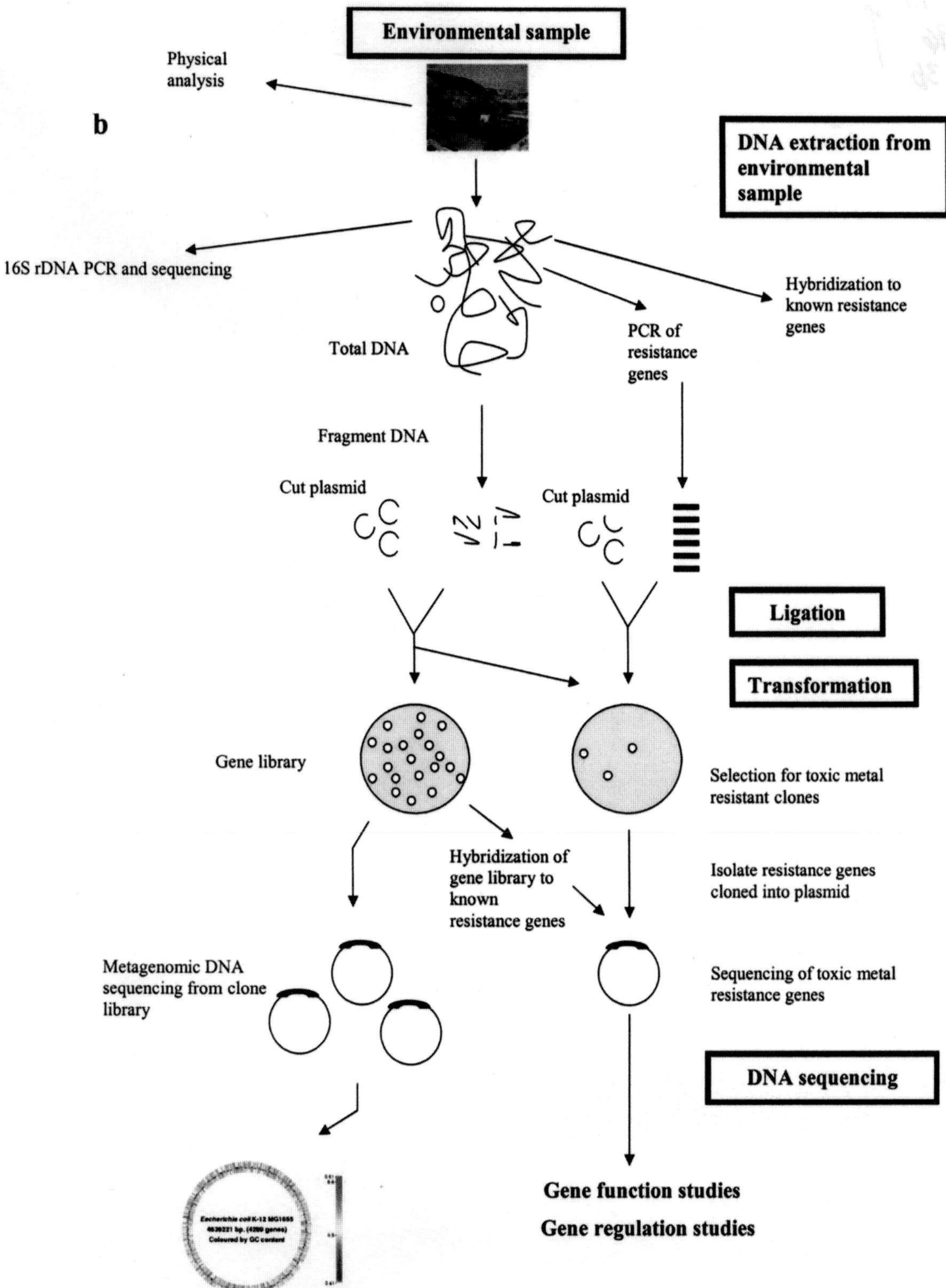

FIGURE 3 (a) Isolation and molecular techniques for the analysis of culturable toxic-metal-resistant bacteria. (b) Isolation and molecular techniques for the analysis of toxic metal resistance genes isolated directly from environmental samples. Modified from references 41 and 53 with permission.

toxic-metal-resistant bacteria are potentially hazardous. Examples of these are industrial rivers, lakes, and canals; sewage, mining sites, tailings, and lagoons; and metal smelting, manufacturing, and recycling facilities. Not all sites are accessible, and permission must be sought to sample from private property. Sites may be polluted with multiple metals or organic pollutants, and metal-rich sites may also have low pH (34). Chemical and physical analysis of the environmental samples is therefore highly appropriate.

The primary purpose of sampling is to obtain enough suitable material to commence the research work. Detailed sampling, isolation, and cultivation procedures for bacteria are given in previous sections of this manual. Samples taken in the field should be taken as aseptically as possible. It is preferable to use sterile plasticware as sample containers, because glassware may be contaminated with DNA, metals, and/or detergent residues and need to be acid washed before being sterilized. The size of the sample taken is dependent on what the sample will be used for. A few tens of grams of soil, or tens of milliliters of water are sufficient to culture bacteria from, but if DNA is to be extracted from environmental samples, then larger samples are probably necessary. Environmental samples should be transported as quickly as possible back to the laboratory, and if viable bacteria are isolated from the environmental sample, they should be grown as soon as is practicable, once the sample has reached the laboratory.

Cultivable metal-resistant bacteria can be isolated directly from the sample, or cultivation can be avoided completely, and DNA can be isolated directly from the environmental sample for analysis (see Fig. 3a).

DNA Extraction from Environmental Samples

The key aims of extracting DNA from the environmental sample are (i) to isolate the DNA from all bacteria present in the sample so that it is representative and unbiased and (ii) to separate the DNA from any chemicals that will inhibit subsequent molecular biology techniques such as PCR and restriction enzyme digestion. There are a number of methodologies for extracting DNA from environmental samples, and their merits are compared in several recent publications (20, 23, 46, 72, 74, 77, 107, 119). Commercially available kits can also be obtained from several manufacturers (for example, Mobio soil extraction kits and Epicentre SoilMaster kits). Soil samples are more problematic than water samples, as they contain phenolic compounds that are known inhibitors of molecular biology reactions. Soil samples often require physical disruption of the soil structure by using high-speed agitation with ceramic or glass beads, as well as enzymatic and chemical lysis techniques (50, 72). Possible contamination by bacteria from within the laboratory must be minimized as far as possible, and we have found the use of laminar air flow PCR cabinets with an integral UV light ideal for DNA extraction and PCR setup. Any contaminating DNA in the interior of these PCR cabinets and in the pipettes, pipette tips, gloves, and microcentrifuge tubes placed in them will be destroyed by high doses of UV light prior to the commencement of work. The high-efficiency air filtration unit prevents any airborne bacterial contaminants from entering the sterile PCR cabinet.

The advantages of extracting DNA directly from environmental samples are (i) that it obviates the need to culture bacteria and removes cultural bias and (ii) that it provides a starting material for metagenomics or metalibrary construction. The disadvantages of this method are that (i) no live bacteria are isolated, so reconstruction of complete genomes requires powerful bioinformatics techniques (118); (ii) it does not differentiate between live and dead bacteria; (iii) the methods do not separate plasmid DNA from genomic DNA; (iv) low-abundance genes or genomes that may be biologically important are poorly represented in the absence of prior enrichment (39, 41, 49); and (v) although DNA extraction from environmental samples overcomes the problem of "coverage" of all the metal resistance genes in a sample, there remains the problem of how to exploit the community-extracted DNA. If the DNA extracted is in small fragments due to shearing in the purification process, it is unlikely that shotgun cloning will result in any clone in a random library of clones containing fully functional resistance genes. If the DNA extraction methods that are used result in larger DNA fragments being purified, lysis of all bacteria may not occur, so samples may be biased (39).

Total DNA isolated from an environmental sample is the prerequisite for metagenomics studies (the functional and DNA sequence-based analysis of the microbial genomes from within an environmental sample [54, 93]) or for selective PCR amplification of toxic metal resistance genes without culturing bacteria. A suggested workflow is shown in Fig. 3b.

Cultivation of Toxic-Metal-Resistant Bacteria

The outline of the experimental procedure and primary purpose of this technique is to culture toxic-metal-resistant bacteria, so that the genetic and biochemical basis for their resistance mechanisms can be further investigated. Toxic-metal-resistant bacteria can be selected using microbiological growth media containing increasing levels of the toxic metal. Choice of the metal salt added to the media and choice of growth media are crucial to the success of this procedure. The solubility of many metal salts can be low, so it is essential that their solubility constants be known and that metal salts which are soluble be used in experimentation. Standard reference texts can be used to access solubility data (68, 85). In general, chlorides or nitrates of toxic metals are soluble, but this should be confirmed for each compound and any potentially deleterious effects of the counterion should be taken into account when adding metal salts to microbiological media. Metal solubility in the medium can be increased by increasing the counterion concentrations, but it will also be affected by the presence of chelators. On this basis, it is not straightforward to define the concentrations of metal added to a microbiological medium that will select against metal-sensitive bacteria while not killing metal-resistant bacteria (90, 116). A major influence on the toxicity of metals, particularly the thiophilic metals Hg, Cd, Cu, and Pb, in microbiological growth media is how protein- or peptide-rich a medium is. Chemically undefined rich media, particularly those that contain peptones, other protein hydrolysates, infusions, and extracts chelate toxic metals (26, 42, 92), reducing the bioavailable level of toxic metal. In our experience, there are supplier-to-supplier and batch-to-batch differences between media components—especially extracts and hydrolysates—so we tend to set aside particular batch numbers of media components such as yeast extract and tryptone for exclusive use in making media that will be used for selection of toxic-metal-resistant bacteria, and conducting MIC assays with these toxic metals. It is also prudent to produce a large single batch of medium for use in a series of experiments, as this minimizes the possibility of variation in results. Methods have been used to calibrate media to select for toxic-metal-resistant bacteria (35), whereby survival curves are plotted against increasing metal ion concentration for laboratory strains and decisions on the concentration of metal to be used to select for resistant isolates are made on the basis of these results. Once selective plates have been produced, environmental samples must be diluted with suitable sterile diluents such as 0.9% saline or Ringer's solution (see reference 28 and other chapters in this manual) and vortexed to disrupt any solid particles to which bacteria may adhere. Dilution series of the environmental samples are then inoculated into liquid cultures, or onto plate cultures, which should be incubated at appropriate temperatures for the bacteria from the sample, as many environmental bacteria cannot tolerate temperatures above 30°C.

The advantages of culturing functionally resistant bacteria on selective agar plates are that (i) live functionally metal-resistant bacteria will be isolated; (ii) different bacterial groups can be isolated using different selective media, different growth conditions, and different metals; and (iii) culture of toxic-metal-resistant bacteria is a relatively simple technique. The major disadvantage of the technique is that it is widely accepted that less than 1% of environmental bacteria can be cultivated (4, 110), so even if a range of different media and incubation temperatures are used, it is unlikely that the full diversity of metal-resistant bacteria will be isolated. Therefore, cultivation techniques may produce a biased view of what is present, and as a consequence novel organisms and resistances may not be isolated. The bulk of studies on bacterial toxic metal resistance genes to date have been based upon the initial isolation of a resistant bacterial strain, and advances in culturing techniques mean that cultivation of bacteria is clearly not a redundant method (38, 66). The whole-organism postgenomic studies detailed later in this chapter require a viable bacterial strain.

Once toxic-metal-resistant bacteria have been isolated in pure culture and their levels of resistance have been confirmed by plate or liquid MIC assays, the bacteria can be identified using 16S rRNA gene PCR and sequencing techniques, which are detailed elsewhere in this manual. The metal-resistant strains can then be further investigated using a range of molecular techniques.

MOLECULAR TECHNIQUES FOR THE STUDY OF TOXIC METAL RESISTANCE GENES

DNA Extraction from Bacterial Cells

The aim of this procedure is to isolate DNA from toxic metal(loid) resistant bacteria in order to initiate study, at the molecular level, of genetic processes involved in the resistance mechanism. Toxic metal resistance genes have often been found on plasmids or mobile elements. Plasmid preparation and extraction of total DNA from bacteria are widely used techniques, and a variety of methods may need to be tried to find which will most effectively isolate the DNA from the toxic-metal-resistant strain. There are likely to be significant differences between isolates in terms of susceptibility to proteinases, stability of plasmids, and presence or absence of extracellular polysaccharides (29, 98). The standard technique of alkaline lysis plasmid preparations (15) and the commercially available kits that use alkaline lysis followed by binding of DNA to ion-exchange resins, magnetic beads, or silica, subsequent removal of contaminating proteins by washing, and finally elution of the DNA, are widely used for gram-negative plasmid purifications. However, many environmental plasmids are large (up to 250 kb) and are easily sheared during pipetting operations, and some bacterial plasmids are unstable, so other methods have been developed to isolate environmental plasmids (for examples see references 36 and 123). The first technique, developed by Wheatcroft and Williams (123), relies upon rapid lysis of the cells, followed by immediate agarose gel electrophoresis of the lysate. The second method, developed by Eckhart (36), uses "in-well" lysis of bacterial cells loaded into the well of a vertical agarose gel. Because cellular lysis occurs in the well and the DNA is immediately separated by electrophoresis, shearing of high-molecular-weight plasmid DNA is minimized. Other methods for large-scale preparation of environmental plasmids have been reviewed (29, 51). As an alternative strategy, commercially available DNA preparation kits are used in many laboratories. For example, the Qiagen genomic DNA kit has been designed for use on a wide variety of organisms, and we have found that these (and the Qiagen DNeasy kit) isolate both plasmid and genomic DNA. If the viable-cell strategy is being used, then isolation of DNA from environmental isolates is necessary (unless the toxic metal resistance genes can be transferred or transposed into a sensitive laboratory strain) as a starting point for molecular studies, and the major advantage of extracting DNA directly from an environmental bacterium is that the DNA isolated is from a functionally toxic metal resistance operon and should therefore confer resistance when cloned into a sensitive bacterial strain. The major potential disadvantage of directly extracting DNA is that it may be difficult to do efficiently and without bias.

Once plasmid or total DNA has been isolated from the resistant bacterial strain, it can be used for a variety of molecular biology techniques (Fig. 2), including PCR, restriction enzyme digestion, cloning, and sequencing.

Transfer of Resistance to a Sensitive Host Strain

The aim of this technique is to attempt to transfer resistance genes from a toxic-metal-resistant environmental isolate into a laboratory bacterial strain, and to test whether the recipient laboratory strain then acquires a toxic metal resistance phenotype. Many toxic metal resistance genes are carried on plasmids, and methodologies to transfer plasmids from environmental bacteria into metal-sensitive laboratory strains are well established for gram-negative and gram-positive bacteria (12, 30, 89, 126). The outline of the experimental procedure is that resistance genes carried by plasmids may transfer a selectable phenotype (toxic metal resistance) to a recipient laboratory strain, so detection of transfer is relatively straightforward. Plasmids can be self-transmissible (they encode all of the cellular machinery to initiate pilus formation and to transfer the plasmid into the host strain) or may be mobilizable (require the presence of a self-transmissible plasmid in the donor strain, which mobilizes them to cross into the recipient) and may cross into the recipient strain during matings in liquid culture or require a solid surface for matings to take place. Many resistance plasmids found in gram-negative bacterial strains are broad host range, being able to replicate in a wide variety of bacteria, so transfer of resistance determinants from a gram-negative strain into *E. coli* laboratory strains is a viable option during the experiments.

In mating experiments, the first requirement is for a laboratory strain which is metal sensitive, does not restrict incoming DNA, and has a selectable marker. Commonly, antibiotic-resistant recipient strains are used, but it is necessary to confirm that the donor strain is sensitive to the antibiotic at levels which still allow the recipient strain to grow on the selective medium. This is because both the recipient strain and the donor strain will be plated out together on a growth medium that will select for the functional metal resistance and the recipient strains' marker—selecting only for functionally resistant transconjugants. The recipient strain should be defective in restriction and modification systems so that incoming plasmid DNA is not cleaved with restriction endonucleases, or modified by the recipient strain. Alternatively, electroporation techniques can be used to transform host strains with plasmid DNA from the resistant isolate. The advantages of this method are (i) that if transferred to the recipient strain, the resistance genes are confirmed as being located on the plasmid; and (ii) that the resistance plasmid is transferred to a laboratory strain that is well characterized.

If transfer to the recipient strain is successful, plasmid preparations, restriction digests, and cloning from the transferred plasmid can be used to identify the region of DNA that confers resistance.

Transposition of Resistance Genes into Plasmids

The aim of this technique is to transpose toxic metal(loid) resistance genes into a characterized plasmid, so that identification of the resistance gene(s) and any mobile element carrying them is facilitated.

Some toxic metal resistances are carried on functional transposable elements, in particular Hg^R, which may be resident in the chromosome or on a plasmid in the environmental isolate (12). Transposition of the resistance genes into a characterized host plasmid is an effective method for isolating the toxic metal resistance genes. The outline of the experimental procedure is to conjugate into the toxic-metal-resistant strain a broad-host-range plasmid, which by replicating in the cell in the presence of the transposon becomes a target for it to transpose into (11). The plasmid is then either crossed out of the metal-resistant isolate (as described in the previous section) into a sensitive host strain or isolated by DNA preparation and transformed into a sensitive host strain. A functional toxic metal resistance phenotype in the transconjugant or transformant is then selected on nutrient agar plates.

If successful, the advantages of the technique are that (i) the method demonstrates that the resistance is carried on a transposable element and (ii) the resistance genes and the transposon in which they reside are transferred into a characterized plasmid. Identification of the genes by restriction mapping and sequencing is therefore more straightforward.

If transposition is successful, then plasmid DNA can be prepared from the strain. This DNA can be restriction mapped in order to identify the region where the transposon has inserted into the plasmid to identify the region of DNA that confers resistance.

Gene Libraries and Shotgun Cloning of Resistance Genes

The aim of this technique is to clone toxic metal resistance genes into plasmids which can then be transformed into a sensitive host bacterial strain in order to confer the resistance phenotype. Cloning can be done directly from total DNA from the resistant organism, plasmid DNA preparations, plasmid DNA from transconjugants, or from DNA extracted from environmental samples. Cloning is now a standard technique (7, 98) which has been facilitated by the construction of bacterial artificial chromosomes, cosmids (for large inserts of DNA), shuttle vectors, or broad-host-range cloning vectors that allow genetic manipulations to be conducted in *E. coli* laboratory strains; and resulting constructs can then be transformed or electroporated into metal-sensitive bacterial strains that are from the same species as the resistant isolate. The outline of the experimental procedure is that total DNA or plasmid DNA containing the resistance genes is partially digested with a restriction enzyme that recognizes a 4-bp DNA sequence, such as Sau3A. Digested and purified DNA is ligated into a complementary restriction site (such as BamHI) in the plasmid or cosmid vector (other combinations of restriction digests can be used). The ligated plasmids are then transformed or electroporated into the laboratory strain. Cosmids are packaged into bacteriophage and transfected into a sensitive host strain, and resistant clones are selected for directly by plating onto media containing the toxic metal, or the gene library can be transferred onto media containing the toxic metal.

The major advantage of this method is that if it is successful, functionally resistant genes are cloned into a plasmid that is well characterized. This makes further analysis more straightforward, and the functionally resistant clone can be used for subcloning, sequencing, site-directed mutagenesis, and other techniques detailed below.

Transposon Mutagenesis

The aim of this technique is to insertionally inactivate genes in order to identify those that are essential in conferring the resistance phenotype. Transposon mutagenesis is a technique that was widely used to identify plasmid-encoded genes important in conferring a phenotype, and it has also been used to inactivate chromosomal genes to detect functionally important metal-induced genes (33).

As researchers have gained access to simple and affordable DNA sequencing technology especially for small (<10-kb) fragments of DNA and DNA sequence databases have become larger, this technique has become less popular, because there are considerable time savings in moving straight to DNA sequencing. Despite this, transposon mutagenesis as a technique has a number of advantages. This is because some transposons have a dual function, being designed to contain a promoterless gene that confers a detectable phenotype (e.g., Tn*5 lacZ*), and can be used to detect transcription of genes under inducing conditions (33).

The outline of the experimental procedure is that a plasmid vector (often a suicide vector) containing a transposon is transformed into bacterial cells containing the toxic metal resistance genes on a plasmid and transposition takes place. The resulting bacterial cells are assayed for metal sensitivity, and the position that the transposon has inserted into is mapped by restriction digestion of purified DNA or sequenced from the sequencing primer sites within the transposon. The advantages of this technique are (i) that toxic-metal-sensitive strains can be generated and (ii) that mapping of the transposon insertion sites locates the positions of genes or regulatory sequences that contribute to the metal resistance phenotype. The potential disadvantages of this technique are that the time taken to map transposon insertion sites may be better spent directly sequencing the resistance genes.

DNA-DNA Hybridization and Blotting

The primary purpose of using DNA-DNA hybridization techniques is to detect whether there is any homology between the newly isolated resistance genes and known resistance genes (5, 29, 98).

The outline of the experimental procedure is that DNA is isolated from the resistant strain and either is immobilized directly on a nylon or nitrocellulose membrane (dot blots or slot blots) or is digested with restriction endonucleases, and electrophoresed on an agarose gel, and the digested DNA fragments are transferred to a membrane (Southern hybridization). Immobilized DNA is incubated with a labeled DNA probe of the known toxic metal resistance genes under conditions that promote base pairing (hybridization) between the labeled and attached DNA (29, 98). The immobilized DNA-DNA hybrids are washed under conditions of increasing stringency to disrupt nonspecific hybrids, and any signal can be detected using autoradiography or phosphorimager detection, depending on the type of probe label used.

The advantage of this method is that it quickly reveals DNA homology between the known and newly isolated resistance genes. If homology exists, a Southern blot also reveals which restriction fragment(s) the resistance genes are located on, facilitating cloning of the resistance genes. The disadvantages of this method are that (i) DNA hybridization experiments are labor-intensive and are not easy to initiate in a laboratory which has no experience of them; and (ii) if no homology exists between the newly isolated resistance genes and known resistances, then DNA hybridizations will not tell us anything more about the new resistance, except that it is not like any other known resistances. If successful, then DNA hybridizations will facilitate the cloning and DNA sequencing of the new resistance genes.

PCR of Toxic Metal Resistance Genes

The aim of the PCR technique is to amplify resistance genes from DNA isolated from the resistant organism, plasmid DNA preparations, plasmid DNA from transconjugants, or from DNA extracted from environmental samples. PCR is now a standard technique and is discussed in detail in this manual and elsewhere (6, 98). The outline of the experimental procedure is that specific PCR primers are used to prime the amplification by a thermostable DNA polymerase of resistance gene sequences. For amplification of toxic metal resistance genes, PCR requires oligonucleotide primers designed to DNA sequences from known metal resistance genes, or degenerate PCR primers designed to consensus sequences of a number of related resistance genes. PCR techniques have been used to amplify metal resistance genes from DNA extracted from soil by using consensus PCR primers (20, 21), as well as from individual bacteria (40, 84, 97, 111, 114, 120), using both standard and inverse PCR (40, 94). Table 1 shows PCR primers that have been designed to amplify toxic metal resistance genes from environmental DNA isolates. PCR should be set up in UV sterilizable PCR cabinets, and filter pipette tips should be used to minimize any contamination in the PCR. If primer design is such that the PCR primers have significantly different annealing temperatures, then a gradient block PCR thermocycler can be used to conduct multiple PCR across a range of annealing temperatures simultaneously, saving time and effort.

The advantage of using PCR to amplify metal resistance genes is that it is an excellent technique for amplifying known gene sequences (and variants of them if consensus PCR primers are used) from a heterologous mixture of DNA. The potential disadvantages of using PCR are as follows. (i) Contamination of PCR with DNA from the laboratory may occur, so care must be taken to minimize these risks and to extensively use negative controls in each set of PCRs. (ii) With PCR you can detect only that for which you already have a DNA sequence, so unless a strategy of using degenerate PCR primers is employed, this technique is unlikely to yield any metal resistance genes radically different from those that have already been described. (iii) There are known biases in PCR amplification, as *Taq* DNA polymerase is not a proofreading polymerase and can thus incorporate erroneous nucleotides into the amplicon; a proofreading thermopolymerase should be used. (iv) The design of degenerate PCR primers to amplify functional novel metal resistance genes from the gene pool may yield PCR products that are unrelated to the genes that were intended to be amplified. (v) There is a risk that chimeric PCR products will be produced by partial amplification of closely related DNA samples from the environmental gene pool, which then anneal in the next cycle of PCR and are extended. This is recognized as a problem in PCR amplification of 16S rRNA genes, which have both highly conserved and hypervariable regions (91, 106, 122). There is the possibility that this may also occur in metal resistance genes, where there is a very high conservation among some genes in the operon, but variability in other regions (for example in the mercury resistance-encoding *mer* operons).

Once PCR products have been obtained, the DNA can be purified using whichever standard protocol is used in the laboratory or using commercially available kits. The PCR products can then be cloned and/or PCR products can be directly sequenced.

DNA Sequencing of Toxic Metal Resistance Genes

The aim of this technique is to determine the DNA sequence of the toxic metal resistance genes, either from PCR products or from shotgun clones constructed from DNA extracted from the organism or from environmental samples. High-throughput cycle sequencing technology has made DNA sequencing a straightforward and cost-effective methodology, available to researchers in individual laboratories, through University core facilities, or via commercial companies. The outline of the experimental procedure is to sequence the resistance genes in a plasmid clone on both DNA strands, using an oligonucleotide primer made either to the plasmid vector sequence or to the resistance genes (40). The historically low prices of synthetic oligonucleotides also mean that it costs very little to "walk" DNA sequencing through unknown sequences. This can be achieved by using newly acquired DNA sequence data to design a new sequencing oligonucleotide to prime a sequencing extension reaction into unknown DNA. Alternatively, subclones can be made and sequenced using oligonucleotides designed to the host plasmid. Sequencing kits are available from the major manufacturers of DNA sequencers and are increasingly popular as they simplify the procedure for the laboratory user who mixes sequencing oligonucleotides with purified DNA and the components of a cycle sequencing kit and then uses a thermal cycler to perform a temperature cycling program. Extension and dye termination reactions take place under the thermal cycling conditions. After removal of unincorporated primers, nucleotides, and dyes, the samples are subjected to capillary electrophoresis. Modern DNA sequencers output the data as an electronic file, and software tools such as Chromas (http://www.technelysium.com.au/) are used to visualize the electropherograms and analyze the data. Standard radioactive DNA sequencing techniques are still available to use but are more labor-intensive than automated sequencing (55, 98).

The advantages of using this method are that (i) knowledge of the DNA sequence of the genes of interest is crucial

TABLE 1 PCR primers used to amplify specific resistance genes from toxic-metal-resistant bacteria[a]

Resistance determinant	Primer sequence 1	Primer sequence 2	Length of amplified DNA (bp)	Reference(s)
Mercury				
*merR*ΔP	jA 5′-GGGAGATCTAAAGCACGCTAAGGCRTA-3′	jB 5′-GGGGAATTCTTGACWGTGATCGGGCA-3′	1,105	20
*merR*ΔP	Rx 5′-ATAAAGCACGCTAAGGCRTA-3′	Px 5′-TTCTTGACWGTGATCGGGCA-3′	1,099	21
*merR*ΔP	*mer1* 5′-GAGATCTAAAGCACGTAAGGC-3′	*mer2* 5′-GGAATCTTGACTGTGATCGGG-3′	1,101	114
merA	A1s.F 5′-TCCGCAAGTNGCVACBGTNGG-3′	A5-n.R 5′-ACCATCGTCAGRTARGGRAAVA-3′	288	120
merA	A2-n.F 5′-CCATCGGCGGCWSYTGCGTSAA-3′	A5-n.R5′-ACCATCGTCAGRTARGGRAAVA-3′	1,200	120
merA	MRS 15′-ACCATCGGCGGCACCTGCG-3′	MRS25 ′-AAGGTCTGSGCCGCRAGCTTC-3′	1,255	40, 48
Copper				
pcoR	*pcoR1* 5′-CGTCTCGACGAACTTTCCTG-3′	*pcoR2* 5′-GGACTTCACGAAACATTCCC-3′	636	114
pcoA	*pcoA1* 5′-CGTCTCGACGAACTTTCCTG-3′	*pcoA2* 5′-GGACTTCACGAAACATTCCC-3′	1,791	114
Cadmium/zinc/ cobalt				
czcA	*czcA1* 5′-GTTTGAACGTATCATTAGTTTC-3′	*czcA2* 5′-GTAGCCATCCGAAATATTCG-3′	1,885	114
czcB	*czcB1* 5′-CTATTTCGAACAAACAAAAGG-3′	*czcB2* 5′-CTTCAGAACAAAACTGTTGG-3′	1,520	114
czcD	*czcD1* 5′-CAGGTCACTGACACGACCAT-3′	*czcD2* 5′-CATGCTGATGAGATTGATGATC-3′	398	114
Chromate				
chrA	*chrA1* 5′-CTTATACGCTACGCCAACTG-3′	*chrA2* 5′-GTAATGGCATTCAGTCGCTTG-3′	1,292	114
chrB	*chrB1* 5′-GTCGTTAGCTTGCCAACATC-3′	*chrB2* 5′-CCgAAAGCAAGATGTCGATCG-3′	450	114
Arsenic				
arsA	*arsA*-1-F 5′-TCCTGGATTGTCGGCTCTTG-3′	*arsA*-1-R 5′-ATCTGTCAGTAATCCGGTAA-3′	186	97
	I-*arsA*-1-F 5′-CGTTGACCCTATTAAAGGCG-3′	I-*arsA*-1-R 5′-ATTCATCAAAAGCCGCAATC-3′	103	97
arsB	*arsB*-1-F 5′-CGGTGGTGTGGAATATTGTC-3′	*arsB*-1-R 5′-GTCAGAATAAGAGCCGCACC-3′	219	97
	I-*arsB*-1-F 5′-CGACGGCAACATTTATCGC-3′	I-*arsB*-1-R 5′-AGCCGCACCATCGTTGGC-3′	181	97
arsC	*arsC*-1-F 5′-GTAATACGCTGGAGATGATCCG-3′	*arsC*-1-R 5′-TTTTCCTGCTTCATCAACGAC-3′	370	97
	I-*arsC*-1-F 5′-TACCTTGAAAACCCGCCTTC-3′	I-*arsC*-1-R 5′-AACCACTTCAGAAGGACGGC-3′	240	97
	amlt-42-f 5′-TCGCGTAATACGCTGGAGAT-3′	*amlt-376-r* 5′-ACTTTCTCGCCGTCTTCCTT-3′	~400	111
	smrc-42-f 5′-TCACGCAATACCCTTGAAATGATC-3′	smrc-376-f 5′-ACCTTTTCACCCGTCCTCTTTCGT-3′	~400	111
Cadmium				
cadA	*cad1* 5′-CAAAYTGYGCRGGHAARTTYGA-3′	*cad2* 5′AACTAATGCACAAGGACA-3′	1,058	84
Silver				
silE	*silE*-forward 5′-GTACTCCCCCGGACATCACTAATT-3′	*silE*-reverse 5′-GGCCAGACTGACCGTTATTCATAA-3′	362	52
silE	*silE*-forward 5′-GTACTCCCCCGGACATCAC-3′	*silE*-reverse 5′-GGCCAGACTGACCGTTATTC-3′	362	32
silS	*silS*-forward 5′-GGAGATCCCGGATGCATAGCAA-3′	*silS*-reverse 5′-GTTTGCTGCATGACAGGCTAATGACATC-3′	1,538	52
silP	*silP*-forward 5′-CATGACATATCCTGAAGACAGAAAATGC-3′	*silP*-reverse 5′-CGGGCAGACCAGCAATAACAGATA-3′	2,630	52

[a]Modified from reference 114 with permission. Codes for mixed bases: Y = C or T; R = A or G; S = C or G; W = A or T; B = C, G, or T; H = A, C, or T; V = A, C, or G; N = A, C, G, or T.

to the use of other molecular techniques; (ii) DNA sequence data can be used to search databases for homologs and paralogs of the sequenced genes, which will help in identifying their putative function; and (iii) DNA sequence data can be used to design further experiments.

The DNA sequence of the resistance genes can then be used (i) to predict gene function by homology searches; (ii) to identify targets for gene disruption to make knockout mutants; (iii) to identify genes for cloning and overexpression; and (iv) to identify putative regulatory sequences, such as promoters, for further investigation.

Other Experimental Molecular Biology Techniques

Generic molecular biology experimental techniques can be used to exploit basic DNA sequence data in order to further understand resistance gene function and regulation and to test hypotheses. Application of the appropriate techniques is important to the final outcome. These techniques can be divided into (i) those which have as a primary purpose the determination of the functions of the proteins encoded by the resistance genes, (ii) those which have as a primary purpose the elucidation of the mechanisms of transcriptional regulation of the resistance genes, and (iii) those which have as a primary purpose the exploitation of the resistance genes or their products for biotechnological applications. The third use for molecular techniques in biotechnological applications is often the logical progression of work initially designed to understand gene function and regulation. So, for example, construction of reporter plasmids that are used to determine the response of a promoter to the addition of exogenous metals in the medium can be used in a bacterial cell as a biosensor (65).

Gene Function Studies

Gene function can be predicted by homology searches, such as Basic Local Alignment Search Tool (BLAST), of previously characterized genes (see "Genomics and Metagenomics" below). Experiments are still required to confirm or reject hypotheses on gene product function generated by homology searches, and as the gene products confer a selectable phenotype (toxic metal resistance), mutants can be tested for functional resistance. There are several techniques that have been used to identify gene function. Gene deletions can be made by restriction endonuclease digestion and religation of the plasmid carrying the toxic metal resistance genes. Site-directed mutagenesis can be used to identify functionally important amino acid residues in the predicted proteins. Protein overexpression and purification can be used to localize proteins in the cell, to provide material for structural studies, or for functional and gene regulation studies, as described below (7, 55, 98).

Gene Regulation and Expression Studies

One characteristic of regulators of toxic metal resistance genes appears to be the specificity of their transcriptional response to toxic metals. A combination of techniques has been used to study these responses. Gene fusions indirectly measure transcription and have been used to assay promoter response to toxic metals added to the bacterial growth medium. Primer extension has been used to map the resistance gene transcription start sites. DNA footprinting has been used to determine regulatory protein-promoter interactions and reverse transcriptase-PCR (RT-PCR) and real-time RT-PCR have been used to determine the extent of, and transcription through, toxic metal resistance operons (6, 7, 55, 98). These methods can be used to study the fine details of the mechanisms of resistance of an individual gene or operon. There are also now other molecular methods that are being used to study whole-cell responses to toxic metals and many other stresses. These are detailed below.

PRESENT AND FUTURE TRENDS: HIGH-THROUGHPUT METHODS

High-throughput molecular techniques, such as whole-genome sequencing, are having a massive impact on biological research. Genomic DNA sequences are being used to develop massively parallel assays such as DNA microarrays, which measure whole-genome transcription. These techniques will increasingly affect our understanding of the genetics of the organisms and how they function.

Genomics and Metagenomics

The aim of both genomics (the genomic analysis of an organism) and metagenomics (the genomic analysis of uncultured microbial communities) is to study all of the nucleotide sequences in an organism or sample. This includes the structural genes, regulatory sequences, and noncoding regions of the genome. The genomic sequencing and annotation of genes from environmentally important bacteria is a major focus for the U.S. Department of Energy sequencing programs (http://microbialgenome.org/index.shtml), as well as other large-scale sequencing efforts (http://microbialgenome.org/databases.shtml). De novo genomic DNA sequencing and annotation is a high-throughput, high-investment, automation-intensive endeavor that tends to be performed in specialist centers and is organized and funded as a fully industrialized process, in contrast to the small-scale artisan type of research that is carried out with single genes or operons in many laboratories. Major investment in this area is already yielding significant data on multiply metal-resistant bacteria, such as *Cupriavidus* (formerly *Wautersia* and before that *Ralstonia*) *metallidurans* CH34 (75, 76). Metagenomic sequencing and functional cloning studies are beginning to describe many microbial communities (1, 31, 53, 99, 110, 115, 117), including the DNA sequences from environmental samples that are relevant to toxic metal resistance research such as from biofilms growing in acid mine drainage rich in Fe, Zn, Cu, and As (117).

As these sequencing projects complete and annotate genome sequences, genomic data will be a resource used for studying the organism in its entirety (the "top down" holistic approach), as well as the resource for identifying new toxic metal resistance gene targets for cloning and overexpression studies (the "bottom up" reductionist approach).

BLAST position-specific iterated and pattern-hit initiated (PSI and PHI-BLAST) search programs (2, 3), which are available at GenBank (http://www.ncbi.nlm.nih.gov/BLAST) are very powerful tools that can be used in searches for homologs and paralogs of known or putative toxic metal resistance genes from whole-genome sequences (http://www.ncbi.nlm.nih.gov/sutils/genom_table.cgi), as well as environmental metagenomics sequences (http://www.ncbi.nlm.nih.gov/BLAST/Genome/EnvirSamplesBlast.html. Comparative genomic databases (27, 44, 108) and their associated software tools are being widely used to compare the genome sequences of related organisms and to understand their evolution and the horizontal gene transfer between organisms.

The advantage of the genomics approach is that it provides a full DNA sequence description of the organism, which can then be used (i) to predict gene function, (ii) to identify targets for gene disruption to make knockout mutants, (iii) to identify genes for cloning and overexpression, and (iv) to identify putative regulatory sequences such as promoters for investigation. The advantages of the metagenomics methods are that (i) they are culture independent, (ii) they can potentially identify unknown and uncultured gene pools, (iii) they complement the genomic sequencing methods that have been widely employed to describe the genomic sequences of pure culture organisms, and (iv) metagenome libraries may be used to clone novel functional metal resistance genes. The major limitation of both the whole-genome and the metagenome DNA sequencing approaches is that in a sense the DNA sequence of an organism (and its annotation) is only the start to understanding how the organism works. This is because DNA sequence data provide only the "blueprint" for the organism. Genome sequences are fixed and factual. They tell us what proteins may be produced but tell us nothing about when the proteins are produced, what (if any) metals they coordinate, and what regulates their expression. The data generated by genome or metagenome sequencing of toxic-metal-resistant environmental bacteria will be exploited further by the use of transcriptomics, proteomics, metallomics, metabolomics, and structural genomics technologies. In this case the primary purpose of "omics" research will be to understand the changes that occur in the whole bacterial cell in response to a toxic metal stress, rather than only the resistance genes and their products, because the phenotype of resistance to a toxic metal is the sum of the cellular responses to insult from the metal, not just the products of the resistance genes.

Postgenomic Technologies

The term "omics" is being applied as a generic description of high-throughput technologies and assays that are a product of genomic DNA sequencing. The primary aim of these technologies is to examine the whole-cell response of an organism. Transcriptomics, using DNA arrays, studies genome-wide transcriptional responses in an organism to a given set of conditions (63, 87, 128). Proteomics studies the changes in the protein complement in an organism when it is subjected to a given set of conditions (25, 83). Metabolomics studies the complement of small molecules and metabolites present in the cell under a given set of conditions (62, 87). All of these methods are comparative in their nature. The experimental outline for these experiments is to compare (using the appropriate methodology) the state of a culture of cells under normal and test conditions, where, for instance, the test condition would be a sublethal metal stress, which would be compared to a normal condition of growth in the same medium, lacking the toxic metal. Transcriptomics yield data on transcriptional responses to toxic metal insult, but proteomics yield data on protein production, posttranscriptional modifications, localization, and interactions that proteins participate in under conditions of metal insult. Metabolomics measures the products of metabolism, which result from reactions catalyzed by enzymes. If these studies on cellular responses can be coupled to metallomics (the comprehensive analysis of the entirety of metals and metalloid species within a cell) and metalloproteomics (the study of the proteins that ligate the metal and metalloid species in a cell) (112), a much clearer view of toxic metal interactions with bacterial cells will emerge.

Initial studies of whole-cell responses to toxic metal stresses in model organisms such as *E. coli* and *Bacillus subtilis* have now been published and are revealing that bacterial cellular responses to sublethal metal insult are wide ranging (37, 43, 58, 61, 63, 78, 79, 128). In the future, "omics" studies of "specialist" toxic metal resistance organisms, rather than the "generalist" (*E. coli* and *B. subtilis*) model organisms, will advance our understanding of cellular responses, genetic adaptations, and cross-talk between the genome and resistance genes on mobile genetic elements. Metatranscriptomics (the study of transcription in uncultured organisms) (88) and metaproteomics (the study of the protein complement of uncultured bacteria) (127), although technically challenging, will add more to our understanding of the uncultured majority of microorganisms in environmental samples.

The advantages of the omics methods are that (i) they are very high throughput and allow us to study the whole cell; (ii) they integrate data, so that a global response to a particular set of conditions can be studied; and (iii) they will lead to the next phase of research, which is systems biology (the study of how complex changes and responses are choreographed in genes and how that information is disseminated through cells to populations). The disadvantages of the omics methods are that (i) they require considerable capital investment; (ii) they are technically difficult to conduct and reproduce; and (iii) they will require sophisticated data handling, storage, and analysis techniques.

SUMMARY

The primary reasons for isolating toxic-metal-resistant bacteria from environmental samples are to investigate the mechanisms of resistance, study cellular responses to toxic metal insult, and to study the distribution and diversity of the gene sequences that encode the specific resistance mechanisms. There may be subsequent biotechnological applications for these organisms or the resistance genes and proteins that they carry in bioremediation or biosensing of toxic metals.

Isolation of the metal-resistant bacteria is a logical first step in any research work, and it is likely that many toxic-metal-resistant bacteria have yet to be identified and cultured. An alternative strategy of direct DNA extraction from environmental samples can be used, but there appears to be a trade-off between the size of DNA fragments yielded by an extraction procedure and the percentage of organisms lysed and their DNA recovered. This means that cloning of functional toxic metal resistance genes from environmental metagenome samples may therefore be technically challenging, whilst PCR from these samples and hybridization to specific gene probes identify only genes that are closely related to the gene sequences used to design the oligonucleotides for amplification or probing of the DNA.

Ultimately the isolation of a viable toxic-metal-resistant strain is still a very good way of studying how resistance works at the molecular and cellular levels. Many functional toxic metal resistance genes can transfer horizontally between related (and sometimes not so related) bacterial species, which in a sense argues for the bacterial host being almost irrelevant to the resistance. But resistance genes and their protein products do not function in isolation; they function within the context of the bacterial cell, and genomic DNA sequencing projects recognize this fact: endogenous plasmids found in the strain are sequenced as well as the chromosome; they are part of the genetic complement of the host bacterial cell. Our understanding of the functioning of toxic metal resistances must also place these within the context of the cell, both in terms of regulatory cross-talk and how cellular components may interact with resistance proteins. For example, highly successful and specialist organisms like *C. metallidurans* CH34 carry both chromosomal and plasmid-borne toxic metal resistance genes and can thrive in heavily polluted environments, because they are highly adapted to their environment. Both their core metabolic functions and the gene products from the multiple metal resistance-encoding plasmids are likely to contribute to their adaptation to their environment.

Molecular techniques are powerful tools that have been used successfully to understand the mechanisms of bacterial toxic metal resistance, with PCR, rapid DNA sequencing, cloning, and mutagenesis techniques being widely adopted to aid understanding of how resistance mechanisms work. Although DNA homology searches of sequenced DNA are a very successful tool for identifying putative protein functions, these predictions will require experimental confirmation. Ultimately, predicted protein structure and function will still need to be tested experimentally, and these results will be used to inform further predictions. An iterative "predict function" and "test hypothesis" regime will increase the accuracy in annotations and the levels of confidence in them.

As a direct consequence of genomic sequence data, transcriptomics and proteomics approaches are now being used to understand whole-cell responses to toxic metal insult. The consequence of this is that the component technologies are in place to understand how metal toxicity and resistance work at the whole-cell level both in the presence and absence of toxic metal resistance plasmids. Structural biology is informing us how toxic metal resistance proteins fold, and metallomics and metalloproteomics will improve our understanding of how metals interact with cellular components. Clearly, although we are now entering a new phase in molecular research into toxic metal resistances, the techniques described in this chapter still have a valuable place in describing new metal-resistant bacteria.

Work carried out by J.L.H. on bacterial toxic metal resistance has been funded by the United Kingdom Biotechnology and Biosciences Research Council (BBSRC) and the European Union.

I thank Nigel Brown, Simon Silver, and Jon Lloyd for their constructive comments and criticisms on this chapter and all colleagues who have contributed, collaborated, and been generous in providing unpublished data.

REFERENCES

1. **Allen, E. E., and J. F. Banfield.** 2005. Community genomics in microbial ecology and evolution. *Nat. Rev. Microbiol.* **3:**489–498.
2. **Altschul, S. F., W. Gish, W. Miller, E. W. Myers, and D. J. Lipman.** 1990. Basic local alignment search tool. *J. Mol. Biol.* **215:**403–410.
3. **Altschul, S. F., T. L. Madden, A. A. Schaffer, J. Zhang, Z. Zhang, W. Miller, and D. J. Lipman.** 1997. Gapped BLAST and PSI-BLAST: a new generation of protein database search programs. *Nucleic Acids Res.* **25:**3389–3402.
4. **Amann, R. I., W. Ludwig, and K. H. Schleifer.** 1995. Phylogenetic identification and *in situ* detection of individual microbial cells without cultivation. *Microbiol. Rev.* **59:**143–169.
5. **Anderson, M. L. M.** 1999. *Nucleic Acid Hybridization.* Bios Scientific Publishers, Oxford, England.
6. **Atlas, R. M., and A. K. Bej.** 1994. Polymerase chain reaction, p. 418–435. *In* P. Gerhardt, R. G. E. Murray, W. A. Wood, and N. R. Krieg (ed.), *Methods for General and Molecular Bacteriology.* American Society for Microbiology, Washington, D.C.
7. **Bagdasarian, M., and M. M. Bagdasarian.** 1994. Gene cloning and expression, p. 406–407. *In* P. Gerhardt, R. G. E. Murray, W. A. Wood, and N. R. Krieg (ed.), *Methods for General and Molecular Bacteriology.* American Society for Microbiology, Washington, D.C.
8. **Ballatori, N.** 2002. Transport of toxic metals by molecular mimicry. *Environ. Health Perspect.* **110:**689–694.
9. **Barkay, T., S. M. Miller, and A. O. Summers.** 2003. Bacterial mercury resistance from atoms to ecosystems. *FEMS Microbiol. Rev.* **27:**355–384.
10. **Barkley, W. E., and J. H. Richardson.** 1994. Laboratory safety, p. 715–734. *In* P. Gerhardt, R. G. E. Murray, W. A. Wood, and N. R. Krieg (ed.), *Methods for General and Molecular Bacteriology.* American Society for Microbiology, Washington, D.C.
11. **Bennett, P. M., J. Grinsted, and T. J. Foster.** 1988. Detection and use of transposons. *Methods Microbiol.* **21:**205–232.
12. **Berg, D. E., and M. M. Howe (ed.).** 1989. *Mobile DNA.* American Society for Microbiology, Washington, D.C.
13. **Beveridge, T. J., and S. F. Koval.** 1981. Binding of metals to cell envelopes of *Escherichia coli* K-12. *Appl. Environ. Microbiol.* **42:**325–335.
14. **Beveridge, T. J., M. N. Hughes, H. Lee, K. T. Leung, R. K. Poole, I. Savvaides, S. Silver, and J. T. Trevors.** 1997. Metal-microbe interactions: contemporary approaches. *Adv. Microb. Physiol.* **38:**177–243.
15. **Birnboim, H., and J. Doly.** 1979. A rapid extraction procedure for screening recombinant plasmid DNA. *Nucleic Acids Res.* **7:**1513–1525.
16. **Blencowe, D., and A. P. Morby.** 2003. Zn(II) metabolism in prokaryotes. *FEMS Microbiol. Rev.* **27:**291–312.
17. **Bridson, E.** 1994. *The Development, Manufacture and Control of Microbiological Culture Media*, p. 40–53. Unipath Limited, Basingstoke, England.
18. **Brown, N. L., A. P. Morby, and N. J. Robinson (guest ed.).** 2003. Thematic issue: interactions of bacteria with metals. *FEMS Microbiol. Rev.* **27:**129–447.
19. **Brown, N. L., J. V. Stoyanov, S. P. Kidd, and J. L. Hobman.** 2003. The MerR family of transcriptional regulators. *FEMS Microbiol. Rev.* **27:**145–163.
20. **Bruce, K. D., W. D. Hiorns, J. L. Hobman, A. M. Osborn, P. Strike, and D. A. Ritchie.** 1992. Amplification of DNA from native populations of soil bacteria by using the polymerase chain reaction. *Appl. Environ. Microbiol.* **58:**3413–3416.
21. **Bruce, K. D., A. M. Osborn, A. J. Pearson, P. Strike, and D. A. Ritchie.** 1995. Genetic diversity within *mer* genes directly amplified from communities of noncultivated soil and sediment bacteria. *Mol. Ecol.* **4:**605–612.
22. **Bruins, M. R., S. Kapil, and F. W. Oehme.** 2000. Microbial resistance to metals in the environment. *Ecotoxicol. Environ. Safety* **45:**198–207.
23. **Bürgmann, H., M. Pesaro, F. Widmer, and J. Zeyer.** 2001. A strategy for optimizing quality and quantity of DNA extracted from soil. *J. Microbiol. Methods* **45:**7–20.
24. **Busenlehner, L. S., M. A. Pennella, and D. P. Giedroic.** 2003. The SmtB/ArsR family of metalloregulatory transcriptional repressors: structural insights into prokaryotic metal resistance. *FEMS Microbiol. Rev.* **27:**131–143.
25. **Chance, M. R., A. Fiser, A. Sali, U. Pieper, N. Eswar, G. Xu, J. E. Fajardo, T. Radhakannan, and N. Marinkovic.** 2004. High-throughput computational and experimental techniques in structural genomics. *Genome Res.* **14:**2145–2154.
26. **Chang, J. S., J. Hong, O. A. Ogunseitan, and B. H. Olson.** 1993. Interaction of mercuric ions with the bacterial-

growth medium and its effects on enzymatic reduction of mercury. *Biotechnol. Prog.* **9:**526–532.

27. **Chaudhuri, R. R., A. M. Khan, and M. J. Pallen.** 2004. coliBASE: an online database for *Escherichia coli, Shigella* and *Salmonella* comparative genomics. *Nucleic Acids Res.* **32:**D296–D299.
28. **Collins, C. H., and P. M. Lyne.** 1984. *Microbiological Methods*, 5th ed. Butterworths, London, United Kingdom.
29. **Crosa, J. H., M. E. Tomalsky, L. A. Actis, and S. Falkow.** 1994. Plasmids, p. 366–386. *In* P. Gerhardt, R. G. E. Murray, W. A. Wood, and N. R. Krieg (ed.), *Methods for General and Molecular Bacteriology*. American Society for Microbiology, Washington, D.C.
30. **Cutting, S. M., and P. Youngman.** 1994. Gene transfer in gram-positive bacteria, p. 348–364. *In* P. Gerhardt, R. G. E. Murray, W. A. Wood, and N. R. Krieg (ed.), *Methods for General and Molecular Bacteriology*. American Society for Microbiology, Washington, D.C.
31. **Daniel, R.** 2005. The metagenomics of soil. *Nat. Rev. Microbiol.* **3:**470-478.
32. **Davis, I. J., H. Richards, and P. Mullaney.** 2005. Isolation of silver- and antibiotic-resistant *Enterobacter cloacae* from teeth. *Oral Microbiol. Immunol.* **20:**191–194.
33. **De Bruijn, F. J., and S. Rossbach.** 1994. Transposon mutagenesis p. 387–405. *In* P. Gerhardt, R. G. E. Murray, W. A. Wood, and N. R. Krieg (ed.), *Methods for General and Molecular Bacteriology*. American Society for Microbiology, Washington, D.C.
34. **Dopson, M., C. Baker-Austin, P. R. Koppineedi, and P. L. Bond.** 2003. Growth in sulfidic mineral environments: metal resistance mechanisms in acidophilic microorganisms. *Microbiology* **149:**1959–1970.
35. **Duxbury, T.** 1981. Toxicity of heavy metals to soil bacteria. *FEMS Microbiol. Lett.* **11:**217–220.
36. **Eckhardt, T.** 1978. A rapid method for the identification of plasmid deoxyribonucleic acid in bacteria. *Plasmid* **1:** 584–588.
37. **Egler, S., C. Große, G. Grass, and D. H. Nies.** 2005. Role of the extracytoplasmic function protein family sigma factor RpoE in metal resistance of *Escherichia coli. J. Bacteriol.* **187:**2297–2307.
38. **Ellis, R. J., P. Morgan, A. J. Weightman, and J. C. Fry.** 2003. Cultivation-dependent and -independent approaches for determining bacterial diversity in heavy-metal-contaminated soil. *Appl. Environ. Microbiol.* **69:**3223–3230.
39. **Entcheva, P., W. Liebl, A. Johann, T. Hartsch, and W. R. Streit.** 2001. Direct cloning from enrichment cultures, a reliable strategy for isolation of complete operons and genes from microbial consortia. *Appl. Environ. Microbiol.* **67:**89–99.
40. **Essa, A. M. M., D. J. Julian, S. P. Kidd, N. L. Brown, and J. L. Hobman.** 2003. Mercury resistance determinants related to Tn*21*, Tn*1696*, and Tn*5053* in enterobacteria from the preantibiotic era. *Antimicrob. Agents Chemother.* **47:**1115–1119.
41. **Eyers, L., I. George, L. Schuler, B. Stenuit, S. N. Agathos, and S. El Fantroussi.** 2004. Environmental genomics: exploring the unmined richness of microbes to degrade xenobiotics. *Appl. Microbiol. Biotechnol.* **66:**123–130.
42. **Farrell, R. E., J. J. Germida, and P. M. Huang.** 1993. Effects of chemical speciation in growth media on the toxicity of Hg(II). *Appl. Environ. Microbiol.* **59:**1507–1514.
43. **Ferianc, P., A. Farewell, and T. Nystrom.** 1998. The cadmium-stress stimulon of *Escherichia coli* K-12. *Microbiology* **144:**1045–1050.
44. **Field, D., E. J. Fell, and G. A. Wilson.** 2005. Databases and software for the comparison of prokaryotic genomes. *Microbiology* **151:**2125–2132.
45. **Finney, L. A., and T. V. O'Halloran.** 2003. Transition metal speciation in the cell: insights from the chemistry of metal ion receptors. *Science* **300:**931–936.
46. **Gabor, E. M., E. J. de Vries, and D. B. Jansenn.** 2003. Efficient recovery of environmental DNA for expression cloning by indirect methods. *FEMS Microbiol. Ecol.* **44:** 153–163.
47. **Gerhardt, P., R. G. E. Murray, W. A. Wood, and N. R. Krieg (ed.).** 1994. *Methods for General and Molecular Bacteriology*. American Society for Microbiology, Washington, D.C.
48. **Glendinning, K. G.** 2000. Studies on mercuric reductase and thermophilic mercury resistance. Ph.D. thesis. The University of Birmingham, Birmingham, United Kingdom.
49. **Gonzalez, J. M., M. C. Portillo, and C. Saiz-Jimenez.** 2005. Multiple displacement amplification as a pre-polymerase chain reaction (pre-PCR) to process difficult to amplify samples and low copy number sequences from natural environments. *Environ. Microbiol.* **7:**1024–1028.
50. **Griffiths, R. I., A. S. Whiteley, A. G. O'Donnell, and M. J. Bailey.** 2000. Rapid method for coextraction of DNA and RNA from natural environments for analysis of ribosomal DNA- and rRNA-based microbial community composition. *Appl. Environ. Microbiol.* **66:**5488–5491.
51. **Grinsted, J., and P. Bennett (ed.). 1988.** Plasmid technology. *Methods Microbiol.* **21:**129–142.
52. **Gupta, A., L. T. Phung, D. E. Taylor, and S. Silver.** 2001. Diversity of silver resistance genes in IncH incompatibility group plasmids. *Microbiology* **147:**3393–3402.
53. **Handelsman, J.** 2004. Metagenomics: application of genomics to uncultured organisms. *Microbiol. Mol. Biol. Rev.* **68:**669–685.
54. **Handelsman, J., M. R. Rondon, S. F. Brady, J. Clardy, and R. M. Goodman.** 1998. Molecular biological access to the chemistry of unknown soil microbes: a new frontier for natural products. *Chem. Biol.* **5:**R245-R249.
55. **Hendrickson, W. G., and T. K. Misra.** 1994. Nucleic acid analysis, p. 436–460. *In* P. Gerhardt, R. G. E. Murray, W. A. Wood, and N. R. Krieg (ed.), *Methods for General and Molecular Bacteriology*. American Society for Microbiology, Washington, D.C.
56. **Hobman, J. L., and N. L. Brown.** 1997. Bacterial mercury-resistance genes, p. 527–568. *In* A. Sigel and H. Sigel (ed.), *Metal Ions in Biological Systems*, vol. 34. Marcel Dekker, New York, N.Y.
57. **Hobman, J. L., J. R. Wilson, and N. L. Brown.** 2000. Microbial mercury reduction, p. 177–197. *In* D. R. Lovley, (ed.), *Environmental Microbe-Metal Interactions*. ASM Press, Washington, D.C.
58. **Hu, Y., G. Wang, G. Y. J. Chen, X. Fu, and S. Q. Yao.** 2003. Proteome analysis of *Saccharomyces cerevisiae* under metal stress by two-dimensional differential gel electrophoresis. *Electrophoresis* **24:**1458–1470.
59. **Hughes, M. N., and R. K. Poole.** 1989. *Metals and Micro-Organisms*. Chapman and Hall, London, United Kingdom.
60. **Juhnke, S., N. Peitzsch, N. Hübener, C. Große, and D. H. Nies.** 2002. New genes involved in chromate resistance in *Ralstonia metallidurans* CH34. *Arch. Microbiol.* **179:**15–25.
61. **Katayama, A., A. Tsujii, A. Wada, T. Nishino, and A. Ishihama.** 2002. Systematic search for zinc-binding proteins in *Escherichia coli. Eur. J. Biochem.* **269:**2403–2413.
62. **Kell, D. B.** 2004. Metabolomics and systems biology—making sense of the soup. *Curr. Opin. Microbiol.* **7:**296–307.
63. **Kershaw, C. J., N. L. Brown, C. Constantinidou, M. D. Patel, and J. L. Hobman.** 2005. The expression profile of *Escherichia coli* K-12 in response to minimal, optimal and excess copper concentrations. *Microbiology* **151:**1187–1198.

64. **Kröning, B., and T. Paul.** 1897. The chemical foundations of the study of disinfection and of the action of poisons. Translated into English in T. D. Brock (ed.), *Milestones in Microbiology* (1961). Prentice-Hall, Inc., Englewood Cliffs, N.J.
65. **LaRossa, R. A., and T. K. Van Dyk.** 2000. Applications of stress responses to environmental monitoring and molecular toxicology, p. 454–468. *In* G. Storz and R. Hengge-Aronis (ed.), *Bacterial Stress Responses*. ASM Press, Washington, D.C.
66. **Leadbetter, J. R.** 2003. Cultivation of recalcitrant microbes: cells are alive, well and revealing their secrets in the 21st century laboratory. *Curr. Opin. Microbiol.* **6:**274–281.
67. **Lenihan, J.** 1988. *The Crumbs of Creation—Trace Elements in History, Medicine, Industry, Crime and Folklore*, p. 44–75. Adam Hilger, Bristol, United Kingdom.
68. **Lide, D. R. (ed. in chief).** 2004. *CRC Handbook of Chemistry and Physics*, 85th ed. CRC Press, Inc., Boca Raton, Fla.
69. **Lippard, S. J., and J. M. Berg.** 1994. *Principles of Bioinorganic Chemistry*, p. 1–19. University Science Books, Mill Valley, Calif.
70. **Lippard, S. J., and J. M. Berg.** 1994. *Principles of Bioinorganic Chemistry*, p. 139–173. University Science Books, Mill Valley, Calif.
71. **Lloyd, J. R., and D. R. Lovley.** 2001. Microbial detoxification of metals and radionuclides. *Curr. Opin. Biotechnol.* **12:**248–253.
72. **Lloyd-Jones, G., and D. W. F. Hunter.** 2001. Comparison of rapid DNA extraction methods applied to contrasting New Zealand soils. *Soil Biol. Biochem.* **33:**2053–2059.
73. **Lovley, D. R. (ed.).** 2000. *Environmental Microbe-Metal Interactions*. ASM Press, Washington, D.C.
74. **Martin-Laurent, F., L. Phillipot, S. Hallet, R. Chaussod, J. C. Germon, G. Soulas, and G. Catroux.** 2001. DNA extraction from soils: old bias for new microbial diversity analysis methods. *Appl. Environ. Microbiol.* **67:**2354–2359.
75. **Mergeay, M.** 2000. Bacteria adapted to industrial biotopes: metal resistant *Ralstonia*, p. 403–414. *In* G. Storz and R. Hengge-Aronis (ed.), *Bacterial Stress Responses*. ASM Press, Washington, D.C.
76. **Mergeay, M., S. Monchy, T. Vallaeys, V. Auquier, A. Benotmane, P. Bertin, S. Taghavi, J. Dunn, D. van der Lelie, and R. Wattiez.** 2003. *Ralstonia metallidurans*, a bacterium specifically adapted to toxic metals: towards a catalogue of metal responsive genes. *FEMS Microbiol. Rev.* **27:** 385–410.
77. **Miller, D. N., J. E. Bryant, E. L. Madsen, and W. C. Ghiorse.** 1999. Evaluation and optimization of DNA extraction and purification procedures for soil and sediment samples. *Appl. Environ. Microbiol.* **65:**4715–4724.
78. **Moore, C. M., A. Gaballa, M. Hui, R. W. Ye, and J. D. Helmann.** 2005. Genetic and physiological responses of *Bacillus subtilis* to metal ion stress. *Mol. Microbiol.* **57:** 27–40.
79. **Niedhardt, F. C., and R. A. VanBogelen.** 2000. Proteomic analysis of bacterial stress responses, p. 445–452. *In* G. Storz and R. Hengge-Aronis (ed.), *Bacterial Stress Responses*. ASM Press, Washington, D.C.
80. **Nies, D. H.** 1999. Microbial heavy-metal resistance. *Appl. Microbiol. Biotechnol.* **51:**730–750.
81. **Nies, D. H.** 2003. Efflux-mediated heavy metal resistance in prokaryotes. *FEMS Microbiol. Rev.* **27:**313–339.
82. **Nies, D. H., and N. L. Brown.** 1998. Two-component systems in the regulation of heavy metal resistance, p. 77–103. *In* S. Silver and W. Walden (ed.), *Metal Ions in Gene Regulation*. Chapman and Hall, New York, N.Y.
83. **Noël-Georis, I., T. Vallaeys, R. Chauvaux, S. Monchy, P. Falmagne, M. Mergeay, and R. Wattiez.** 2004. Global analysis of the *Ralstonia metallidurans* proteome: prelude for the large-scale study of heavy metal response. *Proteomics* **4:**151–179.
84. **Oger, C., T. Berthe, L. Quillet, S. Barray, J. F. Chiffoleau, and F. Pettit.** 2001. Estimation of the abundance of the cadmium resistance gene *cadA* in microbial communities in polluted estuary water. *Res. Microbiol.* **152:**671–678.
85. **O'Neill, M. J., A. Smith, P. E. Heckelman, and S. Budavari (ed.).** 2003. *The Merck Index*, 13th ed. Merck and Company Inc., Rahway, N.J.
86. **Outten, F. W., C. Outten, and T. V. O'Halloran.** 2000. Metalloregulatory systems at the interface between bacterial metal homeostasis and resistance, p. 145–178. *In* G. Storz and R. Hengge-Aronis (ed.), *Bacterial Stress Responses*. ASM Press, Washington, D.C.
87. **Phelps, T. J., A. V. Palumbo, and A. S. Beliaev.** 2002. Metabolomics and microarrays for improved understanding of phenotypic characteristics controlled by both genomics and environmental constraints. *Curr. Opin. Biotechnol.* **13:** 20–24.
88. **Poretsky, R. S., N. Bano, A. Buchan, G. LeClair, J. Kleikemper, M. Pickering, W. M. Pate, M. A. Moran, and J. T. Hollibaugh.** 2005. Analysis of microbial gene transcripts in environmental samples. *Appl. Environ. Microbiol.* **71:**4121–4126.
89. **Provence, D.H., and R. Curtiss III.** 1994. Gene transfer in gram-negative bacteria, p. 317–347. *In* P. Gerhardt, R. G. E. Murray, W. A. Wood, and N. R. Krieg (eds.), *Methods for General and Molecular Bacteriology*. ASM Press, Washington, D.C.
90. **Pumpel, T., B. Pernfuss, B. Pigher, L. Diels, and F. Schinner.** 1995. A rapid screening method for the isolation of metal accumulating microorganisms. *J. Ind. Microbiol.* **14:**213–217.
91. **Qiu, X., L. Wu, H. Huang, P. E. McDonel, A. V. Palumbo, J. M. Tiedje, and J. Zhou.** 2001. Evaluation of PCR generated chimeras, mutations and heteroduplexes with the 16S rRNA gene-based cloning. *Appl. Environ. Microbiol.* **67:**880–887.
92. **Ramamoorthy, S., and D. J. Kushner.** 1975. Binding of mercuric and other heavy metal ions by microbial growth media. *Microb. Ecol.* **2:**162–176.
93. **Riesenfeld, C. S., P. D. Schloss, and J. Handelsman.** 2004. Metagenomics: genomic analysis of microbial communities. *Annu. Rev. Genet.* **38:**525–552.
94. **Riley, J., R. Butler, R. Finniear, D. Jenner, S. Powell, R. Anand, J. C. Smith, and A. F. Markham.** 1990. A novel, rapid method for the isolation of terminal sequences from yeast artificial chromosome (YAC) clones. *Nucleic Acids Res.* **18:**2887–2890.
95. **Rosen, B. P.** 2002. Transport and detoxification systems for transition metals, heavy metals and metalloids in eukaryotic and prokaryotic microbes. *Comp. Biochem. Physiol. Part A* **133:**689–693.
96. **Rouch, D. A., and N. L. Brown.** 1997. Copper-inducible transcriptional regulation at two promoters in the *Escherichia coli* copper resistance determinant *pco*. *Microbiology* **143:**1191–1202.
97. **Saltikov, C. W., and B. H. Olson.** 2002. Homology of *Escherichia coli* R773 *arsA*, *arsB*, and *arsC* genes in arsenic-resistant bacteria isolated from raw sewage and arsenic-enriched creek waters. *Appl. Environ. Microbiol.* **68:**280–288.
98. **Sambrook, J., P. MacCallum, and D. Russell.** 2001. *Molecular Cloning: a Laboratory Manual*, 3rd ed. Cold Spring Harbor Laboratory, Cold Spring Harbor, N.Y.
99. **Schloss, P. D., and J. Handelsman.** 2003. Biotechnological prospects for metagenomics. *Curr. Opin. Biotechnol.* **14:** 303–310.
100. **Silver, S.** 1996. Bacterial resistances to toxic metal ions—a review. *Gene* **179:**9–19.
101. **Silver, S.** 1996. Transport of inorganic cations, p. 1091–1102. *In* F. C. Niedhardt, R. Curtiss III, J. L. Ingraham,

E. C. C. Lin, K. B. Low, B. Magasanik, W. S. Reznikoff, M. Riley, M. Schaechter, and H. E. Umbarger (ed.), Escherichia coli *and* Salmonella; *Cellular and Molecular Biology*, 2nd ed. ASM Press, Washington, D.C.

102. **Silver, S.** 1998. Genes for all metals; a bacterial view of the periodic table. *J. Ind. Microbiol. Biotechnol.* **20:**1–12.
103. **Silver, S.** 2003. Bacterial silver resistance: molecular biology and uses and misuses of silver compounds. *FEMS Microbiol. Rev.* **27:**341–354.
104. **Silver, S., and L. T. Phung.** 1996. Bacterial heavy metal resistance: new surprises. *Annu. Rev. Microbiol.* **50:**753–789.
105. **Silver, S., and L. T. Phung.** 2005. A bacterial view of the periodic table: genes and proteins for toxic inorganic ions. *J. Ind. Microbiol. Biotechnol.* **32:**587–605.
106. **Speksnijder, A. G. C. L., G. A. Kowalchuk, S. de Jong, E. Kline, J. R. Stephen, and H. J. Laanbroek.** 2001. Microvariation artifacts introduced by PCR and cloning of closely related 16S rRNA gene sequences. *Appl. Environ. Microbiol.* **67:**469–472.
107. **Stach, J. E. M., S. Bathe, J. Clapp, and R. G. Burns.** 2001. PCR-SSCP comparison of 16S rDNA sequence diversity in soil DNA, using different isolation and purification methods. *FEMS Microbiol. Ecol.* **36:**139–151.
108. **Stormo, G. D., and K. Tan.** 2002. Mining genome databases to identify and understand new gene regulatory systems. *Curr. Opin. Microbiol.* **5:**149–153.
109. **Storz, G., and R. Hengge-Aronis (ed.).** 2000. *Bacterial Stress Responses*. ASM Press, Washington, D.C.
110. **Streit, W. R., and R. A. Schmitz.** 2004. Metagenomics—the key to the uncultured microbes. *Curr. Opin. Microbiol.* **7:**492–498.
111. **Sun, Y., E. A. Polishchuk, U. Radoja, and W. R. Cullen.** 2004. Identification and quantification of arsC genes in environmental samples by using real-time PCR. *J. Microbiol. Methods* **58:**335–349.
112. **Szpunar, J.** 2005. Advances in analytical methodology for bioinorganic speciation analysis: metallomics, metalloproteomics and heteroatom-tagged proteomics and metabolomics. *Analyst* **130:**442–465.
113. **Tam, R., and M. H. Saier, Jr.** 1993. Structural, functional, and evolutionary relationships among extracellular solute-binding receptors of bacteria. *Microbiol. Rev.* **57:**320–346.
114. **Trajanovska, S., M. L. Britz, and M. Bhave.** 1997. Detection of heavy metal ion resistance genes in Gram-positive and Gram-negative bacteria isolated from a lead-contaminated site. *Biodegradation* **8:**113–124.
115. **Treusch, A. H., A. Kletzin, G. Raddatz, T. Ochsenreiter, A. Quaiser, G. Meurer, S. C. Schuster, and C. Schleper.** 2004. Characterization of large insert DNA libraries from soil for environmental genomic studies of Archaea. *Environ. Microbiol.* **6:**970–980.
116. **Trevors, J. T., K. M. Oddie, and B. H. Belliveau.** 1985. Metal resistance in bacteria. *FEMS Microbiol. Rev.* **32:** 39–54.
117. **Tringe, S. G., C. von Mering, A. Kobayashi, A. A. Salamov, K. Chen, H. W. Chang, M. Podar, J. M. Short, E. J. Mathur, J. C. Detter, P. Bork, P. Hugenholtz, and E. M. Rubin.** 2004 Comparative genomics of microbial communities. *Science* **308:**554–557.
118. **Tyson, G. W., J. Chapman, P. Hugenholtz, E. E. Allen, R. J. Ram, P. M. Richardson, V. V. Solovyev, E. M. Rubin, D. S. Rokshar, and J. F. Banfield.** 2004. Community structure and metabolism through reconstruction of microbial genomes from the environment. *Nature* **428:**37–43.
119. **Valentin, K., U. John, and L. Medlin.** 2005. Nucleic acid isolation from environmental aqueous samples. *Methods Enzymol.* **395:**15–37.
120. **Vertriani, C., Y. S. Chew, S. M. Miller, J. Yagi, J. Coombs, R. A. Lutz, and T. Barkay.** 2005. Mercury adaptation among bacteria from a deep-sea hydrothermal vent. *Appl. Environ. Microbiol.* **71:**220–226.
121. **Wackett, L. P., A. G. Dodge, and L. B. M. Ellis.** 2004. Microbial genomics and the periodic table. *Appl. Environ. Microbiol.* **70:**647–655.
122. **Wang, G. G. Y., and Y. Wang.** 1997. Frequency of formation of chimeric molecules as a consequence of PCR coamplification of 16S rRNA genes from mixed bacterial genomes. *Appl. Environ. Microbiol.* **63:**4645–4650.
123. **Wheatcroft, R., and P. A. Williams.** 1981. Rapid methods for the study of both stable and unstable plasmids in Pseudomonas. *J. Gen. Microbiol.* **124:**433–437.
124. **White, C., J. A. Sayer, and G. M. Gadd.** 1997. Microbiological solubilization and immobilization of toxic metals: key biogeochemical processes for treatment of contamination. *FEMS Microbiol. Rev.* **20:**503–516.
125. **Wiatrowski, H., and T. Barkay.** 2005. Monitoring of microbial metal transformations in the environment. *Curr. Opin. Biotechnol.* **16:**261–268.
126. **Willetts, N.** 1988. Conjugation. *Methods Microbiol.* **21:** 49–78.
127. **Wilmes, P., and P. L. Bond.** 2004. The application of two-dimensional polyacrylamide gel electrophoresis and downstream analyses to a mixed community of prokaryotic microorganisms. *Environ. Microbiol.* **6:** 911–920.
128. **Yamamoto, K., and A. Ishihama.** 2005. Transcriptional response of *Escherichia coli* to external copper. *Mol. Microbiol.* **56:**215–227.

Synchrotron-Based Techniques for Monitoring Metal Transformations

KENNETH M. KEMNER AND SHELLY D. KELLY

97

Many metals and radionuclides are found in soils, sediments, and groundwaters, either naturally or as a result of human activities such as mining and electroplating. Although organic contaminants can be degraded, metals and radionuclides cannot. Rather, the fate and transport of metals and radionuclides through soils, sediments, and groundwaters are often under biogeochemical control, for example through reactions with microbially derived ligands, sorption to microbial cell surfaces or biogenic mineral phases, biomethylation, or enzymatically mediated redox transformations. Indeed, many prokaryotes can use soluble redox-active metal ions as electron donors and acceptors for respiration, directly affecting the transformation of the metal. Thus, microorganisms often can control electron transfer to and from metals and radionuclides and hence, change the valence states and the mobilities of the inorganic constituents.

X rays are a powerful probe for investigating metal and radionuclide transformations in soils, sediments, and groundwaters. In particular, synchrotron-based X-ray investigations can identify the changes in an element's valence state and chemical speciation that often result from microbially mediated electron transfer. Synchrotron-based investigation of metal transformations can also be valuable because they allow investigation of a dilute element's valence state and chemistry in its natural hydrated (even sometimes anoxic) state. Here we describe some of the synchrotron-based X-ray techniques (X-ray absorption spectroscopy, X-ray fluorescence, and X-ray microscopy) that can be used to improve understanding of metal transformations.

The topics selected for discussion here are the more readily used synchrotron-based techniques for investigating metal transformations. The examples presented by no means represent the breadth of work in the fields of biology, chemistry, biochemistry, geology, geobiology, and biogeochemistry; rather, the examples illustrate the types of information that can be generated by using synchrotron radiation. We recommend browsing the list of references for more examples of synchrotron-based investigations of metal transformations, as well as for information about additional synchrotron-based techniques that can be employed in such investigations. Examples are X-ray scattering techniques such as X-ray standing waves, small- and wide-angle X-ray scattering, and X-ray diffraction.

SOURCES OF SYNCHROTRON X-RAY RADIATION

With the development of synchrotron radiation facilities, the use of X-ray imaging, absorption, and scattering techniques for scientific research has increased greatly. This increase is the result of a synchrotron's larger photon flux versus a traditional X-ray source, its greater energy range, and its typically improved energy resolution. As an introduction, we first explain synchrotrons, storage rings, and X-ray beamlines.

Synchrotrons and Storage Rings

When charged particles traveling at relativistic speeds (i.e., close to the speed of light) are accelerated, they emit electromagnetic radiation tangentially to the direction of acceleration. Thus, the trajectories of particles experiencing centripetal acceleration can be maintained in a circle, and the emitted electromagnetic radiation is tangential to the circle. This fundamental concept has been used in the design of synchrotrons (48) and storage rings for the production of X-ray beamlines that today are used worldwide for scientific research. The two classes of X-ray sources—synchrotrons and storage rings—are used as synonyms. In a synchrotron, electrons are accelerated for approximately 10 ms and then ejected; the strong synchrotron radiation is emitted at the end of the acceleration cycle. In a storage ring, however, the electrons are accelerated and then stored within a ring and kept at constant energy.

Sources of the synchrotron radiation first observed in 1947 (8) have evolved to the third-generation synchrotrons in use today. Electromagnetic radiation from first-generation synchrotrons was used to advantage in the 1960s and 1970s by researchers who exploited "nuisance" radiation emitted at high-energy physics facilities for "parasitic" research. Second-generation storage rings were the first synchrotrons built specifically for production of electromagnetic radiation for research. They made use of bending magnets (i.e., magnets to bend the trajectory of the electrons) to create a storage ring that produced beamlines tangential to the ring. The third-generation storage rings developed in the 1990s also employ bending magnets to create the storage ring, but they add insertion devices with periodically alternated magnetic fields (commonly referred to as wigglers and undulators) in straight sections of the ring to produce X rays with very high brightness. The third-generation sources include the U.S. Department of Energy's Advanced Photon Source (APS) at Argonne National Laboratory (http://www.aps.anl.gov/) and Advanced Light Source at Berkeley Lab (http://www.als.lbl.gov/als/); the European Synchrotron Radiation Facility in Grenoble, France (http://www.esrf.fr/); and SPring-8 in Japan (http://www.spring8.or.jp/top.html).

Of particular concern to users of synchrotron radiation storage rings are the flux and brightness of a facility or a beamline at a particular energy. Photon flux is the number of photons emitted per second per bandwidth. Brightness is the number of photons emitted per second per square millimeter of the source size, per opening angle of the source. Stated another way, brightness is a unit of measure that includes the flux of a storage ring and the source size and divergence of the X-ray beams. Thus, brightness is crucial for today's many X-ray microscopy investigations that require large numbers of X rays in a very small spot. Alternatively, for an experiment that needs a large number of X rays to investigate larger samples, flux is often the most important figure of merit.

X-Ray Beamlines

To ascertain how synchrotron techniques can help uncover the mechanisms of metal biotransformations, one must appreciate how X-ray beamlines function. Typically, the optimal X-ray beam has the greatest intensity, the smallest divergence, and the best energy resolution. These goals are accomplished by matching X-ray optics to the qualities of the X-ray beam desired. An optical element can be as simple as a pinhole or as complicated as a monochromator cooled with liquid nitrogen. The beamline optics usually include a set of elements such as apertures, mirrors, slits, and monochromators.

The fundamental concept needed to understand the operation of a monochromator is Bragg's law, given for a cubic crystal as

$$n\lambda = 2d \sin \theta \quad [1]$$

where n is the order of the reflection of the X rays off the crystal, λ is the monochromatic wavelength desired, d is the distance between reflecting crystal planes, and θ is the angle between the sample normal and the incoming X-ray beam (23). When the white beam from the storage ring hits the first crystal, only the X rays satisfying the Bragg condition are reflected to the second crystal; the rest are absorbed or transmitted through the first crystal as shown in Fig. 1. To change the energy of the emerging monochromatic beam, the crystals are rotated about an axis perpendicular to the beam direction and parallel to their crystal face. In such a configuration, the outgoing monochromatic beam is parallel to the incoming white beam but is shifted vertically.

Regardless of the X-ray technique being used, the incident beam intensity often varies with energy and time and must be monitored by a detector that should deplete only a small fraction of the beam (so that most of the X rays can probe the sample). Ion chambers are the most popular tool for monitoring the incident X-ray intensity. The ion chamber is essentially a capacitor with parallel plates separated by approximately 1 cm; the X-ray beam passes between the two plates. The space between the plates is filled with a mixture of different gases. Typical choices are He, N, Ar, Kr, or Xe or a combination of two or more of these. The ion chamber usually has Be or Kapton windows to allow the X rays to enter and exit the detector but still contain the gases. A direct-current voltage is applied between the plates. The X rays ionize the gases inside the detector, and the electrons and ions drift apart toward the positive and negative plates, generating a current in a closed circuit. This current travels into a current amplifier and then to a voltage-to-frequency converter. The output is sent to a scalar that counts the number of pulses for an amount of time determined by the experimenter.

Different types of detectors can measure the by-products of absorption. The type chosen depends on whether the experimenter is interested in the radiative or nonradiative by-products of absorption. For radiative fluorescent X rays, an ion chamber and a solid-state detector, sometimes coupled with an absorption filter or oriented geometrically to remove background signals (25, 39), are two of the more popular techniques. For the nonradiative processes, electron detectors (9, 17, 28) consisting of a bias between a sample and a grid, both contained in vacuum or a He atmosphere, are the simplest and most popular techniques. Further details about detectors are discussed below in relation to specific experimental techniques.

INTRODUCTION TO X-RAY ABSORPTION SPECTROSCOPY

Synchrotron-based X-ray absorption spectroscopy (XAS) is a very powerful method for monitoring metal transformations (1, 7, 24, 37). In particular, XAS techniques provide element-specific chemical and structural information on a given metal in a complex system. Because X rays are highly penetrating, unmodified samples can be preserved under appropriate hydration and atmospheric conditions. In addition, XAS studies can give considerable insight into the kinetics of metal transformation processes and determine the final chemical forms of metals, whether bound to biomaterials, geomaterials, or both.

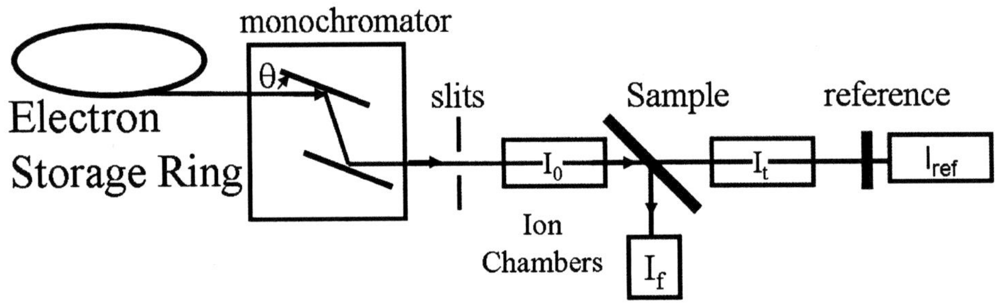

FIGURE 1 Schematic of a typical XAS experiment.

The XAS techniques, X-ray absorption near-edge structure (XANES) spectroscopy and extended X-ray absorption fine-structure (EXAFS) spectroscopy, can provide information on the local chemical environment, coordination, and valence state of individual elements in a variety of matrices (24), many of which are appropriate for microbiological studies. Their element specificity makes the XAS techniques particularly valuable for studying dilute (with metal concentrations of 1 to 100 ppm), multielement, and disordered systems. Because every element has specific absorption edges corresponding to the unique binding energies of its electrons, individual elements in a multielement sample can be studied independently.

XANES Spectroscopy

The focus of XANES spectroscopy (24) is the energy region very close to (within about 30 eV of) the absorption edge. The absorption edge is the region of the X-ray absorption spectrum that depicts a dramatic increase in the absorption of X rays (described further in a later section of this chapter). The structure of the absorption edge is very sensitive to the three-dimensional geometry of the atoms surrounding the excited atom, as well as to the valence of the excited atom. The XANES spectrum can be monitored as a function of chemical or biological processes of interest to elucidate the effects of these processes on the oxidation state and local geometry of the element of interest (1, 7, 37). For example, we have used XANES to determine the ratio of U(VI) to U(IV) in a variety of biogeochemical samples, as well as to study some processes that affect this ratio (6, 26, 31, 43, 44, 45). Such studies are important because U(VI) is more soluble and more mobile in the environment than U(IV). The XANES technique provides an in situ probe of an element's oxidation state and clearly can contribute significantly to an understanding of the fate of elements in the environment, in both solid and solution phases. In addition to its utility, XANES is relatively easy to implement and has been perhaps the most commonly used synchrotron-based X-ray technique for monitoring metal transformations in environmental studies.

EXAFS Spectroscopy

The focus of EXAFS spectroscopy is the region well above the absorption edge (at about 30 to 1,000 eV above the absorption edge value), where oscillations in the absorption signal reflect the sum of frequencies from each shell of atoms surrounding the absorbing atom (24, 36). The frequencies of the individual components of the EXAFS signal are related to the distances between the absorbing atom and the neighboring atoms for each shell of atoms surrounding the absorbing atom. The amplitude of the signal is related to the neighboring atom type and the number of neighboring atoms. Analysis of EXAFS data can differentiate aqueous, surface-absorbed, and coprecipitated species. Thus, EXAFS provides information related to the local chemical environment of the element of interest.

INTERACTIONS OF X RAYS WITH MATTER

The fundamental properties of atoms form the basis for XAS. Every element on the periodic table is unique, because its nucleus contains a characteristic number of protons. Protons are positively charged particles. Neutrally charged atoms have equal numbers of electrons that orbit the protons in the nucleus. The energy state of each electron about an atom is unique, as described by the Pauli exclusion principle. These energy states are designated 1s, 2s, $2p^{1/2}$, $2p^{3/2}$, etc. The energy required to remove an electron from the 1s level and promote it to an unoccupied electron level is different for every atom on the periodic table, because it is determined in part by the number of protons in the nucleus. Hence, the energy required to remove electrons increases as the atomic number of the element increases. This is the property of XAS that results in element selectivity of the technique. Use of the beamline monochromator to select incident X rays of the appropriate energy enables a single element in the sample to dominate the absorption signal. Furthermore, because the signal originates from only one type of atom in the sample, XAS measurements have very few sample requirements. Solutions, gases, trace impurities, amorphous solids, crystalline materials, and mixtures of these are all good candidates for XAS measurements.

The number of X rays that pass through a given material is given by the simple relationship

$$I_t = I_0 \, e^{-\mu(E) \cdot x} \qquad (2)$$

where I_t is the number (intensity) of the transmitted X rays, I_0 is the number (intensity) of the incident X rays, x is the thickness of the material, and $\mu(E)$ is the probability for an X ray to be absorbed (X-ray absorption coefficient). This simple expression indicates that (i) thick materials are exponentially more opaque than thin materials, (ii) the number of transmitted X rays depends on the number of incident X rays, and (iii) the probability of absorption depends on the X-ray energy and the atomic number. The energy dependence of the probability of X-ray absorption, $\mu(E)$, can be measured by monitoring the numbers of incident and transmitted X rays. When the energy of the incident X rays corresponds to the energy required to promote an electron to an unoccupied electron level, the probability of X-ray absorption increases dramatically. Thus, the energy dependence of X-ray absorption is highly dependent on atomic number. The result is an absorption edge (sometimes referred to as step absorption). Absorption edges corresponding to excitation of the 1s, 2s, $2p^{1/2}$, and $2p^{3/2}$ electrons are called the K edge, L_I edge, L_{II} edge, and L_{III} edge, respectively.

OVERVIEW OF AN EXAFS EXPERIMENT

Figure 1 is a schematic of a typical XAS experimental setup for measurements at a synchrotron storage ring facility. A monochromator is used to select the X-ray energy defined by the angle of the monochromator crystal with respect to the incident X-ray direction. Precise calibration of the motors that change the monochromator crystal angle is difficult, because large angular displacements are used to select a wide range of X-ray energies. Therefore, the edge energy of a standard compound, with a defined X-ray absorption edge energy, is often used to calibrate the monochromator energy. To monitor the absolute energy of the incident X-ray energies during an energy scan, a reference compound is often placed behind the sample, with another ion chamber behind the sample to record the reference signal (I_{ref}) (Fig. 1). Reference spectra are particularly useful in XANES analysis, as discussed below.

The X rays from the monochromator are often defined spatially with slits. The intensities of incident (I_0), transmitted (I_t), and fluorescent (I_f) X rays are monitored with ion chambers. For illustration purposes, the incident and transmitted X-ray intensities at the U L_{III} edge from a hydrous uranyl phosphate U(VI) standard are shown in Fig. 2A and B, respectively. The natural log of the ratio of these spectra describes the X-ray absorption coefficient for U in this sample; it is shown in Fig. 2C.

DESIGN OF AN XAS EXPERIMENT

Many XAS experiments require preliminary data to help establish an overall experimental design. Hence, multiple trips to the synchrotron may be necessary to collect the required spectra. The XAFS technique is most powerful when the XAS spectra of unknown samples are compared and contrasted to XAS spectra of known standards and/or the unknown spectra under a set of controlled conditions. Some external parameters that can be varied to produce a data series include the presence or absence of biotic processes, sample temperature, concentration, reaction time, pH, and ionic strength. The most valuable parameter to vary is often one that affects the coordination environment of the metal in a predictable way. Several reasons to collect XAS spectra from samples similar to the unknown but with well-defined atomic coordination geometries are discussed below.

Preparing a Sample for an XAS Measurement

Sample preparation can be trivial for many XAS measurements, as mixed, hydrated, liquid, gas, or solid samples—with or without spatial homogeneity—can all be appropriate. Two considerations for sample preparation are that XAS provides the average signal for all of the atoms of a given element in the sample interacting with the X-ray beam and that the absorption of the sample per unit area of incident X rays should be as uniform as possible. Samples that contain more than one or two major atomic environments about the absorbing atom often need to be subjected to chemical, physical, or spatial separation, as deconvolution of multiple XAFS signals resulting from multiple local atomic environments can be very difficult. Another part of sample preparation is determining how thick to make the sample. This is discussed under "Optimizing an XAS Measurement" below.

XANES Spectroscopy

XANES spectroscopy focuses on the energy range near an element's absorption edge, which is related to the element's valence state.

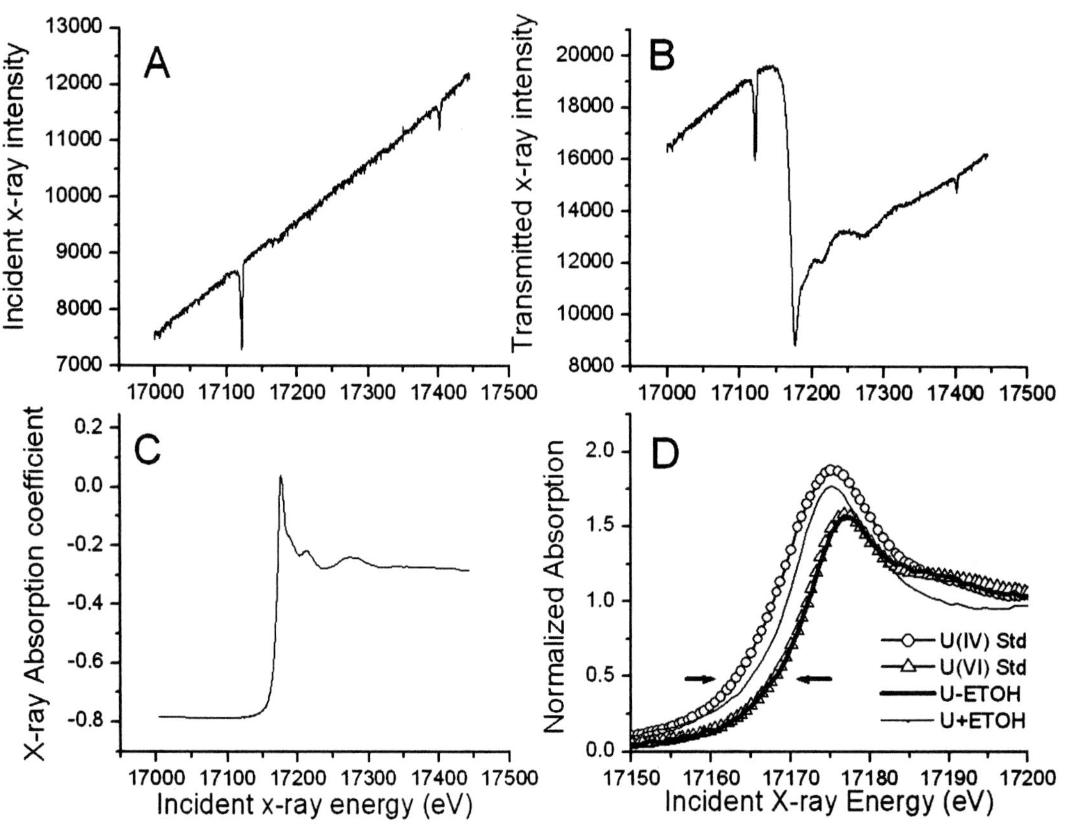

FIGURE 2 (A) Incident X-ray intensity as a function of X-ray energy. (B) Transmitted X-ray intensity as a function of X-ray energy, showing the drop in transmission due to an increase in X-ray absorption by U at the L_{III} absorption edge (17,166 eV). (C) U L_{III} edge X-ray absorption coefficient, obtained as the natural log ratio of the incident (A) to transmitted (B) X-ray intensities. (D) Normalized X-ray absorption U L_{III} edge data for a U(IV) standard (open symbols), a U(VI) standard (open triangles), a U(VI) sample without ethanol for bioreduction (U−ETOH, thick gray line), and a U(VI) sample with ethanol for bioreduction (U+ETOH, thin gray line). The energy value of the absorption edge (between the arrows) is related to the average valence state of U. The U+ETOH sample has a major U(IV) component, while the U-ETOH sample is mostly U(VI).

The valence state of an element corresponds to the number of electrons that have been removed from the neutrally charged atom. The element's valence state affects the amount of energy required to excite a core electron to an unoccupied level during the XAS measurement. Typically, an element in a higher valence state requires greater incident X-ray energy to excite a core electron than the same element in a lower valence state. For example, U(IV) has two more electrons than U(VI). Hence, the binding energy of the $2p^{3/2}$ electron is slightly (~4 eV) greater in U(VI) than in U(IV). Therefore, the L_{III}-absorption edge of U(VI) occurs ~4 eV above the absorption edge of U(IV) (Fig. 2D). This energy difference is easily monitored with current XAS experimental setups at most synchrotrons. As discussed earlier, the X-ray absorption signal is an average of the absorption signals of all U atoms in the sample. A mixed valence state results in an absorption edge between the U(VI) absorption edge energy and the U(IV) absorption edge energy values. A rough (±10%) estimate of the ratio can usually be determined by linear extrapolation between the two edges.

The absolute energy of the absorption edge is typically determined from XANES measurements collected over a range of 400 eV, centered on the absorption edge, with absorption measurements being made every 0.1 to 0.5 eV. To illustrate this, Fig. 2D shows an example of U L_{III} edge XANES data from two standards, designated U(VI) and U(IV), as well as data from two samples from dissimilatory metal reduction experiments of U(VI), with and without ethanol as electron donor. Samples without ethanol (U−ETOH) show little reduction of U(VI), as the absorption edge is close to that of the U(VI) standard, while samples with ethanol (U+ETOH) show mostly U(IV), as the absorption edge is close to that of the U(IV) standard.

EXAFS Spectroscopy

EXAFS focuses on the energy region well above the absorption edge and yields information on the local chemical environment of the absorbing element.

Oscillations in the X-ray absorption coefficient occur at energies greater than the absorption edge. These oscillations are the extended X-ray absorption fine structure (EXAFS). The EXAFS signal, $\chi(E)$, is isolated from the X-ray absorption data by removing a smooth step-like function of the absorption edge and normalizing to the height of the absorption edge. An example of the data and the related background function is shown in Fig. 3A, and the resulting EXAFS signal is shown in Fig. 3B. This figure shows $\chi(k)$ as a function of wave number, k, rather than energy. Wave number is related to the incident X-ray energy by the relationship

$$k^2 = 2m_e[\Delta E(2\pi/h)^2] \quad (3)$$

where m_e is the mass of the electron, ΔE is the energy above the absorption edge, and h is Planck's constant. Substitution of the values for the mass of an electron and Planck's constant into equation 3 results in the following relationship:

$$k^2 \approx 3.81 \cdot \Delta E(\text{Å}^{-2}\ \text{eV}^{-1}) \quad (4)$$

The oscillatory part of the absorption coefficient can be understood by considering a simplified picture of the absorption process. The absorption edge arises when the incident X-ray energy is sufficient to promote an electron from a bound state into an unoccupied electron level. This process creates a photoelectron that propagates away from the absorbing atom and interacts with surrounding atoms. As the incident X-ray energy is increased to values greater than the absorption edge, the excited photoelectron has additional energy, ΔE. This photoelectron can be considered as a wave propagating away from the atom that absorbed the X ray (referred to as the absorbing atom). The wavelength, λ, of the photoelectron is inversely proportional to wave number:

$$\lambda \propto k^{-1} \quad (5)$$

and is therefore related to equation 4. A schematic illustrating this process is shown in Fig. 4. As the photoelectron wave propagates away from the absorbing atom, it encounters neighboring atoms. The electrons associated with the neighboring atoms scatter the photoelectron. Figure 4 shows four neighboring atoms and an absorbing atom.

For now, we consider the interaction of a single neighboring atom with the photoelectron. If both the original photoelectron wave and the scattered photoelectron wave (Fig. 4) crest at the absorbing atom, the probability for X-ray absorption is increased. Similarly, if both waves are at a minimum at the absorbing atom, the probability for X-ray absorption is decreased. The oscillation between the maximum and minimum sums of these two waves at the absorbing atom results in the EXAFS signal. The phase of the photoelectron wave at the absorbing atom is determined primarily by the distance that the photoelectron wave travels and its wavelength. This distance that the photoelectron wave travels is $2R_0$ (twice the radial distance), as shown in Fig. 4. During an EXAFS measurement, increases in the incident X-ray energy and photoelectron energy cause the wavelength of the photoelectron to decrease, so that the distance between the crests of the photoelectron wave becomes smaller. As the wavelength decreases, the sum of the two waves oscillates with an increased frequency that depends on the radial distance.

From this simplified view of the absorption processes, we can gain insight into the interference of photoelectron waves that causes the oscillatory part of the absorption coefficient. The EXAFS signal from all four neighbors (as shown in Fig. 4) at approximately the same distance, R_0, from the absorbing atom (a shell of atoms) is four times the EXAFS signal from one neighbor. Hence, the coordination number of, and the radial distance to, a shell of atoms can be determined from an EXAFS measurement. The EXAFS signal from this shell of atoms is broadened by the thermal and static disorder in the distribution of the radial distances between the absorbing atom and the neighboring atoms. This term is called a mean-square displacement of the radial distance, σ^2; it can also be determined from an EXAFS measurement. We can again extrapolate this picture to more realistic atomic environments (of many shells of atoms about the absorbing atom), because the contributions from the shells of atoms are added together with a unique frequency that depends on the distance to each shell.

Information in the EXAFS signal can be understood as follows.

1. The frequency of the EXAFS oscillations is related to the distance (R_0) between the absorbing atom and the neighboring atoms. Radial distances, and in some cases bond lengths, can be determined accurately (±0.01 Å) from EXAFS data.
2. All atoms of a similar type, at a given radial distance, contribute the same signal; hence, the amplitude of the EXAFS signal is proportional to the number of neighboring atoms.

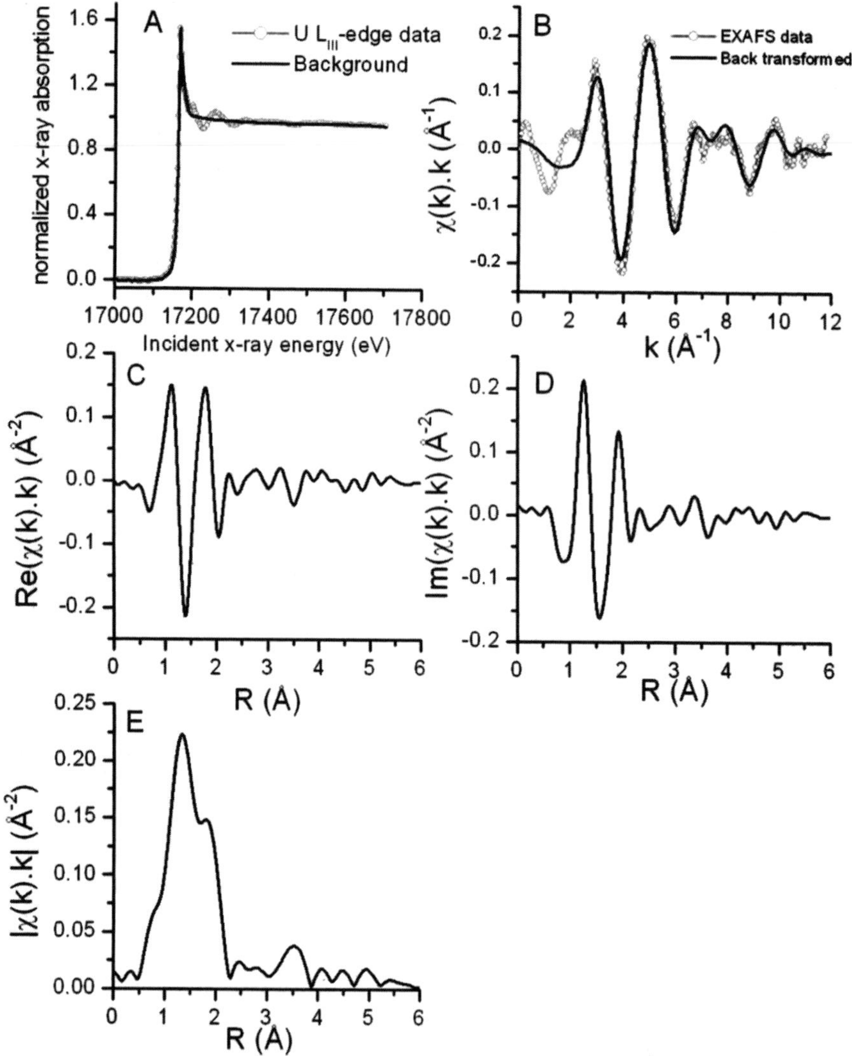

FIGURE 3 U L_{III} edge EXAFS data from an aqueous uranyl carbonate species. (A) X-ray absorption data (symbol) and background function (line). (B) EXAFS data, $\chi(k) \cdot k$, obtained as the difference between the X-ray absorption data and the background function shown in panel A, with the backward Fourier transform (line) of the EXAFS data between 1 and 4 Å. The real part (C), the imaginary part (D), and the magnitude (E) of the Fourier transform of the EXAFS data between 2.0 and 10.2 Å^{-1} are shown.

3. The amplitude of the scattered photoelectron wave also depends on the type of neighboring atom. The photoelectron is scattered from the electrons of the neighboring atom, and hence atoms with larger numbers of electrons create larger signals. Thus, backscattering atoms with significantly different numbers of electrons (±2) can be distinguished.

4. The EXAFS signal contains information about the static and thermal disorder of the radial distances; this is parameterized in terms of a mean-square disorder term designated σ^2. The X-ray absorption process is very fast, on the order of 10^{-15} s, while thermal vibrations of atoms are much slower, on the order of 10^{-12} s; therefore, the X-ray absorption signal is an instantaneous measurement of the locations of the neighboring atoms relative to the absorbing atom.

These properties of the EXAFS signal are used to determine the number and type of neighboring atoms, the distance to them, and the static and thermal disorder in the distance between the absorbing and neighboring atoms.

Most of the EXAFS analysis methods discussed under "XAS Data Analysis" below depend on Fourier transformation of EXAFS data. As discussed earlier, each path of the photoelectron contributes to the total EXAFS signal a sinusoidal signal with a unique frequency. The Fourier transform of the EXAFS signal gives a complex function that separates the frequency components according to path length *R*. Thus, the unique frequencies within the EXAFS that represent unique radial distances can be separated by Fourier transformation. Figure 3 panels C, D, and E show the real part, the imaginary part, and the magnitude of the Fourier transform, respectively, of the U L_{III} edge EXAFS data from an aqueous uranyl carbonate species. Comparison of the inverse Fourier-transformed data (between 1 and 4 Å in Fig. 3E) to the original EXAFS data (Fig. 3B) isolates the frequencies in the signal that contribute to the data corre-

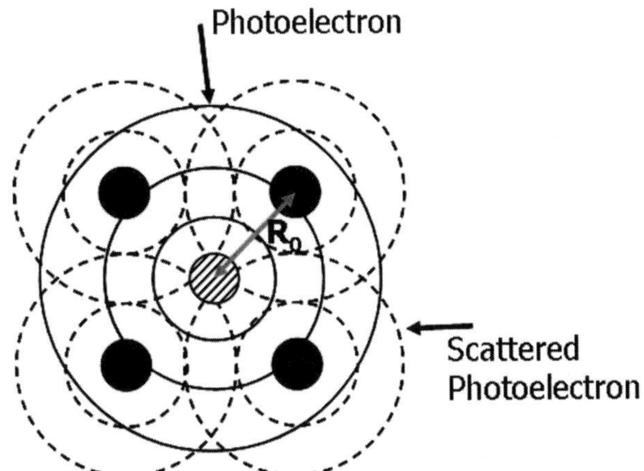

FIGURE 4 Schematic of the X-ray absorption process. The hashed circle represents an X-ray-absorbing atom that has emitted a photoelectron that travels as a wave away from the absorbing atom. The concentric circles (solid lines) represent the crests of the photoelectron wave as it propagates away from the absorbing atom. This photoelectron is scattered from the surrounding atoms (black filled circles), creating a scattered photoelectron, represented by the dashed concentric circles. The interference between the photoelectron wave and the scattered photoelectron waves at the absorbing atom modulate the probability of X-ray absorption. As the incident X-ray energy is increased, the wavelength becomes smaller, and the amplitude of the interference of the two waves modulates between a maximum and a minimum. The frequency of the signal is related to the distance (R_0) between the absorbing atom and the neighboring atoms. All the atoms at a given radial distance contribute the same signal.

sponding to photoelectron wave path lengths of between 1 and 4 Å.

As Fig. 3E shows, the Fourier transform separates the signals from each path of the photoelectron on the basis of the half-path length (i.e., radial distance). The larger-amplitude Fourier transform peak corresponding to the two tightly bound axial oxygen atoms (at the shortest distance) of the uranyl sample is followed by a broader Fourier transform peak due to six equatorial oxygen atoms that are more loosely bound. The peak value in that Fourier transform (~1.4 Å) appears at a slightly lower R value than the true U-axial oxygen bond length (~1.8 Å) because of an additional phase shift of the outgoing and backscattered photoelectron upon interaction with the electron densities associated with the absorbing and backscattering atoms. The width of these Fourier transform peaks is related to the structural and thermal disorder in the distance between the U atom and the neighboring atoms. The amplitude of the signal is related to the neighboring atom type and the number of neighboring atoms.

Among the most useful approaches for modeling EXAFS data are two different types of Fourier transform comparisons. The first type considers closely related unknown and known data. As an example, Fig. 5A shows the Fourier transforms of U L_{III} edge EXAFS data for a uraninite nanoparticle and bulk uraninite. The amplitude of the Fourier transform of the data from the uraninite nanoparticle sample is smaller than the analogous amplitude from the uraninite bulk sample. Theoretical modeling of the data revealed that the decrease in the first Fourier transform peak amplitude is due to increased static disorder of the first oxygen shell, while the decrease U peak amplitude is due to both increased disorder and decreased coordination number (31). The U-U coordination numbers are smaller for the uraninite nanoparticles because of the large percentage of U atoms on the surface (relative to the inside) of the nanoparticles that are undercoordinated. Identifying nanoparticle formation in biogeochemical systems is highly relevant for environmental remediation decision makers concerned with the fate and transport of U, as nanoparticles may be more mobile in subsurface environments than larger uraninite particles (43).

A second type of Fourier transform comparison uses a single data set processed with different weightings of the EXAFS data. Typically, the amplitudes of the EXAFS data

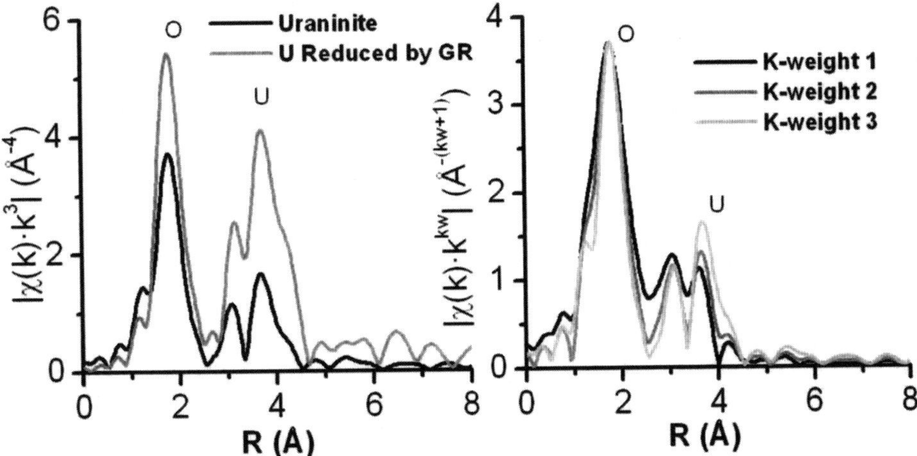

FIGURE 5 (Left) Fourier transforms of EXAFS data for bulk uraninite and nanoparticulate uraninite samples having the same local structure. The decrease in amplitude is due to the small particle size of the uraninite produced by green rust. (Right) Fourier transforms of EXAFS data from U(VI) reduced by green rust, processed with three different k-weights. The data processed with k-weights of 1 and 2 have been normalized to the first-shell O signal of the k-weight 3 data. The relative increase in the signal at 3.5 Å, compared to the first-shell O signal, indicates a heavy-atom neighbor (U).

are weighted by multiplication with the photoelectron wave number (i.e., k, k^2, and k^3). This type of comparison enables different atom types to be distinguished because they have unique k-dependent ability to scatter the photoelectron wave. For example, oxygen has relatively few electrons (8 electrons), and the amplitude of the scattered photoelectron from a neighboring O atom is greatest for photoelectrons with energies up to 400 eV. Alternatively, U atoms with 92 electrons can significantly scatter photoelectrons with energies up to 1,000 eV. Thus, in our example, O and U atoms can be distinguished through comparison of the Fourier transforms of the data, processed to emphasize different regions of the spectrum by multiplying the original EXAFS signal by a k raised to an integer power of 1, 2, or 3. The result often is referred to as the k-weighting of the Fourier transform.

Figure 5B shows the Fourier transform of U L_{III} edge EXAFS data of nanoparticulate uraninite, processed with k-weights of 1, and 2, and 3 and then rescaled to the first Fourier transform peak (i.e., the oxygen atom) amplitude of the data with a k-weight of 3. The systematic increase in amplitude of the Fourier transform peak at 3.7 Å for each increase in k-weighting of the Fourier transform indicates the presence of a neighboring atom with an atomic number greater than O, in this case U. Comparison of Fourier-transformed data with standards (known data sets) or with data processed with different k-weights can enable robust modeling of EXAFS data.

OPTIMIZING AN XAS MEASUREMENT

As with all experimental measurements, optimization of the signal-to-noise ratio (SNR) is of fundamental importance. This is particularly true for EXAFS measurements, where the fine-structure signal is often less than 10% of the total absorption signal. The two most important considerations for optimizing SNR for an EXAFS experiment are sample thickness and data collection time.

Sample Thickness

Two competing effects for a given sample thickness determine SNR in an XAS experiment. For a given material, the thickness of the sample determines how many X rays pass through the sample. Thinner samples are more transparent and hence transmit a larger number of X rays to detect in the transmission ion chamber, but thinner samples also have fewer absorbing atoms and thus a smaller fine-structure signal component. These two competing effects are optimized when approximately 10% of the incident X rays pass through the sample. This value defines an optimal total X-ray thickness of the sample of about 2.6 absorption lengths (12, 35). Given this restriction on the total thickness of the sample, the density of absorbing atoms needs to be significant, so that the signal from the atoms of interest in the sample is measurable. The number of absorbing atoms determines the change in the absorption length of the sample at the absorption edge, giving rise to the absorption edge or edge step. The edge step height for the uranyl phosphate standard shown in Fig. 2C is ~0.6. In general, the change in the absorption length at the absorption edge should be greater than 0.1 for a standard transmission EXAFS measurement.

For dilute samples (such as microbial cultures challenged with metals) with absorbing atoms providing less than 0.1 step height absorption, measurements typically are made in fluorescence mode, through monitoring of the intensity of the fluorescent X rays emitted from the atom upon absorption of the incident X-ray radiation. Fluorescence measurements often are preferred for dilute samples, as the background signal is typically small relative to the fluorescent signal. For samples with a step height absorption of 0.1 to 0.001, a fluorescence ion chamber is often paired with solar slits and a reasonably thick filter, in the Stern-Heald geometry (39), to lessen the impact of elastically scattered X rays. For samples with step height absorption as low as 0.0001, fluorescence measurements can be made with solid-state device detectors. As detectors are improved, measurements on more dilute samples will become feasible.

Data Collection Time

The second consideration for optimizing SNR in an EXAFS measurement is accurate determination of the integration time for the measurement. As a general rule, the ratio of the noise (peak to peak) in an EXAFS spectrum should not be greater than 0.1% (35). If the noise in the data is due to statistical X-ray counting times, then a 1-s integration time with 10^6 X rays per s in the transmission ion chamber is sufficient. This is because statistical noise is related to the inverse of the square root of the number of counts, or

$$N^{-1/2} \cdot 100\% = 0.1\% \text{ for } N = 10^6 \tag{6}$$

In general, the statistical noise is constant in a data set and independent of wave number. The statistical noise can be estimated by plotting the $\chi(k) \cdot k^2$ data (i.e., k^2-weighted data) and estimating the random peak-to-peak fluctuation in the k^2 weighted data at $k = 10$ Å^{-1}, where the value has already been multiplied by $k^2 = 100$. If the fluctuation in the amplitude of the EXAFS signal at k ~10 Å^{-1} is greater than 0.1, the counting time is often increased, or multiple scans are collected and then averaged.

XAS DATA ANALYSIS

Low-Energy XANES

Use of XAS with X-ray probe energies of 200 to 800 eV can determine the binding environments of elements such as C, N, and O. In this energy range, the unoccupied states produced by σ and π bonding of atoms in molecules are well separated in energy. For example, a covalent chemical bond involving two 1s atomic orbitals produces split energy levels with binding (σ 1s) and antibinding (σ^* 1s) molecular orbitals. The σ 1s orbital has lower energy and is often filled, while the σ^* 1s orbital has higher energy and is often unoccupied. Use of XAS to probe the abundance of these molecular orbitals directly is referred to as near-edge X-ray absorption fine structure (NEXAFS), because the transition energy from the core electron to the unoccupied molecular orbital is near the absorption edge. More details about NEXAFS can be found in *NEXAFS Spectroscopy*, by J. Stöhr (40).

XANES Data Analysis

As discussed above, XANES spectra are used to determine the oxidation state or the coordination geometry of a metal. The oxidation state is often determined by using standard spectra of the same metal in known oxidation states. The unknown and the standards are then placed on an absolute energy grid based on the reference spectra. As Fig. 2D shows, samples with mixed oxidation states can have an absorption edge energy value between values in the reference spectra of the individual oxidation states. This edge energy can be calibrated to give the ratio of the mixed oxidation state. Coordination geometry from XANES measurements is determined by pattern matching to standards or by linear combinations of standards. Our group has used XANES to determine the ratio of U(VI) to U(IV) in a variety of environmental samples, as well as to study some biogeochemical processes that affect this ratio (14, 31, 43, 44, 45).

EXAFS Data Analysis

EXAFS spectra are used to determine the average atomic environment about the metal atoms in the sample. Three commonly used analysis procedures are linear combination fitting (LCF), principal component analysis (PCA), and EXAFS modeling with standards or theoretical calculations.

LCF combines standard spectra in different proportions to reproduce an unknown spectrum. This analysis procedure is widely used, but it does not give detailed information about the coordination environment. Nevertheless, this type of modeling is relatively easy to perform with carefully chosen standard spectra. PCA is useful for large data series with an unknown number of major species (47). PCA determines the number of independent spectra in the data set; it requires the data set to be larger than the number of independent spectra. Principal spectra are not required to have physical meaning, but target transformations to standard spectra can lead to identification of the principal spectrum. PCA and LCF have been used to analyze EXAFS data in a number of systems, including photostabilization of Zn from contaminated sediments (32), K-XANES spectra of humic acids (3), structural constraints on ferric (hydr)oxide reduction by dissimilatory-metal-reducing bacteria (10), and Mn-bearing particles emitted from automobiles (34).

EXAFS modeling with standards or with theoretical calculations (33) is the most difficult analysis method. In this approach, the amplitude and phase of the EXAFS signals corresponding to different scattering paths of the photoelectron are calculated or extracted from standard spectra (24). On the basis of physical atomic models, the scattering paths from the neighboring atoms are combined and then optimized to the data by allowing the number, distance, and variation of the radial distances to the neighboring atoms to vary. Many details of EXAFS data analyses with theoretical standards to investigate metal transformations in biogeochemical systems have been published recently (5, 13, 16, 30, 31, 43).

Although the quality of theoretical simulations of EXAFS spectra has improved significantly in recent years (33), comparison of the EXAFS spectrum of the sample of interest with experimental spectra of standard samples (compounds containing the same element in a well-characterized local environment, with known bond lengths and coordinating atoms) is still extremely important, because the latter spectra are necessary for a complete analysis of the former.

X-RAY MICROSCOPY

Heterogeneity of natural or laboratory-based multicomponent samples, including those from microcosms or pure-culture microbiological experiments, often decreases the utility of bulk XAS for investigating metal transformations. For instance, the metal atoms of interest in a sample may be complexed to multiple components, including the solution, mineral surfaces, exopolymeric substances, soluble and membrane components of bacteria, and ternary complexes between exopolymeric substances and mineral surfaces. For adequate determination of the oxidation state and chemical speciation of contaminant metals or radionuclides in heterogeneous systems containing such components, the dimensions of the X-ray probe must enable the vast majority of the X rays to be positioned at homogeneous regions within the heterogeneous sample. That is, the X-ray probe must be smaller than the spatial distribution of the sample's heterogeneities.

Recent advances in synchrotron-based X-ray imaging have resulted in the development of a class of techniques that address these requirements (2, 4, 11, 22, 29, 37, 41, 42, 46). For example, elemental distributions can be mapped by using an X-ray fluorescence (XRF) microprobe (18, 19, 20, 21, 22, 29, 37), with the electron beam of an energy-dispersive X-ray microanalysis experiment replaced by an apertured or focused X-ray beam from a synchrotron source. The advantages of XRF are that it requires minimal sample preparation (so that experiments can be done in air and on natural and hydrated samples) and that it allows element-specific imaging. More importantly, the high brilliance of the synchrotron X-ray beams often results in significantly better SNR values for XRF measurements than for measurements using electron- or proton-based microprobes. Improved SNR translates into better sensitivity and accuracy.

X-ray microprobe techniques can also be combined with the X-ray spectroscopic techniques discussed earlier by using a monochromatic X-ray beam to record X-ray absorption spectra at the absorption edges of the element of interest at each point in the sample (2, 4, 21, 38, 41). Because such spectra are sensitive to the chemical and structural environment of the element, "spectroscopic imaging" experiments can substantially improve understanding of the system or process of interest. However, such experiments are difficult without high-brilliance X-ray sources.

To date, relatively few X-ray microprobe experiments with microbiological and environmental samples have been reported, and the spatial resolution was typically 5 to 300 μm. Until recently, the ultimate spatial resolution of XRF microimaging experiments has been 1 to 10 μm. With new developments in Fresnel zone plate manufacturing and the availability of high-brilliance X rays from the APS and other third-generation synchrotrons, XRF microimaging and microspectroscopy experiments with the submicron spatial resolution needed for microbiological studies are practically routine. The use of submicron X-ray beams for X-ray microimaging and microspectroscopy will enable investigation of interactions at the microbe-geosurface interface. This will provide the missing information needed to elucidate the interactions between contaminants, geosurfaces, and microbes and thus the mechanisms responsible for the chemical transformations of contaminants.

Scanning-probe X-ray imaging microscopy is performed by focusing or aperturing the incident X-ray beam to a small spot (typically 0.06 to 10 μm in diameter) and recording an image point by point by scanning the sample in two dimensions through the beam. At each point in the image, the interaction of the X rays with the sample can be monitored in a number of different ways. In perhaps the simplest approach, the energy-resolved XRF spectrum is recorded at each point in the image. Selecting and plotting the intensity of the XRF lines of the element of interest yield an element-specific image of the sample. Such element-specific maps can be recorded in air, on fully hydrated samples, with minimal sample preparation and with better sensitivity and accuracy than are possible with charged-particle microprobes. Such maps are extremely valuable in determining the locations of metals and radionuclides within and near bacteria on Fe (hydr)oxide surfaces. In addition, because of its energy, the depth of information that the hard X-ray (>5 keV) microprobe yields can vary from a few micrometers up to millimeters. The depth of X-ray penetration, which depends on the X-ray energy and the electron density of the sample, can be calculated. Elemental sensitivity limits for thick, hydrated environmental samples measured with this technique are approximately 50 to 100 ppb. Elemental sensitivity limits for analysis of an individual bacterium have been estimated to be ~1 ppm. However, sensitivity limits are approximately 100 ppm for spectroscopy measurements of an individual bacterium because of the greater signal-to-noise requirements. Future increases in X-ray beam intensities due to more efficient XRF detectors or zone plate focusing optics will increase the elemental sensitivity of the technique.

The combination of the high brilliance of the APS synchrotron and the use of high-resolution zone plates (27) for focusing improves X-ray microimaging experiments substantially. Furthermore, although such element-specific imaging experiments are typically performed with 3- to 10-μm beam diameters at second-generation and third-generation synchrotrons, our recent X-ray imaging experiments at Sector 2 of the APS indicate that it is now possible to map the spatial distribution of elements associated with bacteria (18–21). We have performed numerous XRF imaging experiments of single hydrated *Pseudomonas fluorescens* bacteria, adhered to Kapton film at ambient temperature and pressure, with 100-nm resolution. Spot sizes of 30 nm have been demonstrated for the new X-ray nanoprobe beamline presently being constructed at the APS (15). Results from our earlier studies have shown that the spatial distributions of P, S, Cl, Ca, and many of the 3d transition elements (particularly Fe, for microbes in many metabolic states) can identify the location of a single bacterium, because the highest elemental concentrations occur at the location of the cell. We have also shown that comparison of the intensity of the XRF signal from the microbe to that for thin-film glass standards with known elemental concentrations enables quantitation of each element in each specific microbe (21). Differences in elemental concentrations of many key biosignature elements, resulting from exposure of planktonic cells to toxic levels of Cr(VI), indicate that XRF can distinguish between “healthy” and lysed cells. X-ray microspectroscopy measurements of elevated Cr concentrations near single surface-adhered cells exposed to toxic levels of Cr(VI) demonstrated chemical reduction of the contaminant to Cr(III) and binding of Cr(III) to phosphoryl functional groups.

We also have imaged single hydrated *Shewanella oneidensis* MR-1 cells adhered to an amorphous lepidocrocite thin film (Fig. 6) under anoxic conditions. The results indicate that the combination of the high brilliance of the APS and the use of high-resolution zone plates for focusing enables investigation of the spatial distribution and concentration of nutrient elements, as well as contaminant metals and radionuclides, at the microbe-geosurface interface. In summary, although the use of X-ray microprobes to infer biologically driven metal transformations is still in its infancy, synchrotron-based techniques have been used regularly to infer metal transformations that were not necessarily biologically driven. Nevertheless, the information provided by X-ray microscopy approaches holds great promise for improving understanding of biologically driven metal transformations.

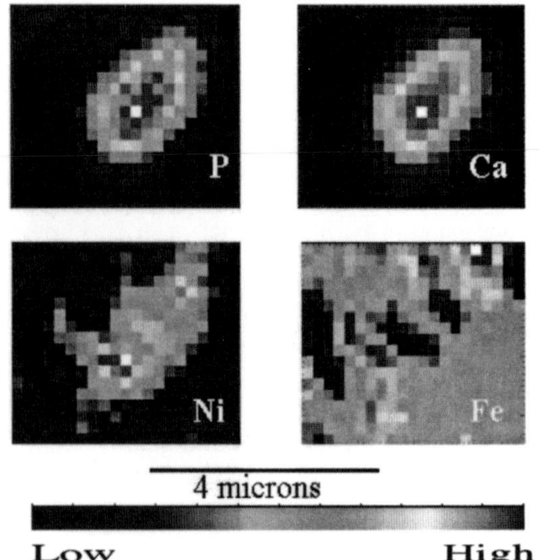

FIGURE 6 Results of XRF microprobe analysis of the spatial distribution of P, Ca, Ni, and Fe on an amorphous lepidocrocite thin film (~1,000 Å thick), depicting *Shewanella oneidensis* MR-1 on the surface. The XRF intensities are correlated to intensities shown in the scale bar.

ACCESS TO SYNCHROTRONS WORLDWIDE

Regardless of their scientific backgrounds, incorporating synchrotron radiation techniques into their research can be intimidating for novice synchrotron users. However, at almost all beamlines around the world, help is available from resident beamline scientists who have dedicated their careers to the development of novel X-ray-based techniques and the application of these techniques in a variety of scientific fields. A figure of merit for judging the success of a beamline or its resident beamline scientist is the number of visiting scientists working at the beamline. Thus, the resident beamline scientist's best interest is served by constant cultivation of new collaborations.

Investigators can initiate the use of synchrotron radiation in their research in a number of ways. A simple approach is asking a colleague who is already using synchrotron radiation to measure a few initial samples during already scheduled beam time. Another approach is contacting a resident beamline scientist who has the expertise required to make the needed measurements. Communication with the beamline scientist sometimes reveals that an alternate technique or a different beamline scientist will be more appropriate for a specific project; the beamline scientist can probably make specific suggestions. A final contact point is the user administration office at the synchrotron facility, whose staff can direct researchers to an appropriate beamline scientist at their facility.

Our work is supported by the U.S. Department of Energy, Office of Science, Office of Biological and Environmental Research, as part of the Environmental Remediation Sciences Program, under contract W-31-109-Eng-38. The manuscript was created by the University of Chicago as the operator of Argonne National Laboratory under this contract.

REFERENCES

1. **Anderson, S. J., C. C. Ainsworth, P. M. Bertsch, J. M. Bigham, W. F. Bleam, P. R. Bloom, J. B. Harsh, and D. G. Schulze.** 1990. *Synchrotron X-Ray Sources and New Opportunities in the Soil and Environmental Sciences: Workshop Report*. Argonne National Laboratory, Argonne, Ill.
2. **Bajt, S., S. B. Clark, S. R. Sutton, M. L. Rivers, and J. V. Smith.** 1993. Synchrotron X-ray microprobe determination of chromate content using X-ray absorption near-edge structure. *Anal. Chem.* **65:**1800–1804.
3. **Beauchemin, S., D. Hesterberg, and M. Beauchemin.** 2002. Principal component analysis approach for modeling sulfur K-XANES spectra of humic acids. *Soil Sci. Soc. Am. J* **66:**83–91.
4. **Bertsch, P. M., D. B. Hunter, S. R. Sutton, S. Bajt, and M. L. Rivers.** 1994. In situ chemical speciation of uranium in soils and sediments by micro X-ray absorption spectroscopy. *Environ. Sci. Technol.* **28:**980–984.
5. **Boyanov, M. I., S. D. Kelly, K. M. Kemner, B. A. Bunker, J. A. Fein, and D. A. Fowle.** 2003. Adsorption of cadmium to *B. subtilis* bacterial cell walls: a pH-dependent X-ray absorption fine structure spectroscopy study. *Geochim. Cosmochim. Acta* **67:**3299–3311.
6. **Brooks, S. C., J. K. Fredrickson, S. L. Carroll, D. W. Kennedy, J. M. Zachara, A. E. Plymale, S. D. Kelly, K. M. Kemner, and S. Fendorf.** 2003. Inhibition of bacterial U(VI) reduction in calcium. *Environ. Sci. Technol.* **37:**1850–1858.
7. **Brown, G. E., Jr., and G. A. Waychunas.** 1988. *Synchrotron X-Ray Sources and New Opportunities in the Earth Sciences: Workshop Report*. Argonne National Laboratory, Argonne, Ill.
8. **Elder, F. R., A. M. Gurewitsch, A. M. Langmuir, and H. C. Pollock.** 1947. Radiation from electrons in a synchrotron. *Phys. Rev.* **71:**829–830.
9. **Erbil, A., G. S. Cargill, R. Frahm, and R. F. Boehme.** 1988. Total-electron-yield current measurements for near-surface extended x-ray-absorption fine structure. *Phys. Rev.* B **37:**2450–2464.
10. **Hansel, C. M., S. G. Benner, P. Nico, and S. Fendorf.** 2004. Structural constraints of ferric (hydr)oxides on dissimilatory iron reduction and the fate of Fe(II). *Geochim. Cosmochim. Acta* **68:**3217–3229.
11. **Hayakawa, S., Y. Gohshi, A. Iida, S. Aoki, and K. Sato.** 1991. Fluorescence X-ray absorption fine structure measurements using a synchrotron radiation X-ray microprobe. *Rev. Sci. Instrum.* **62:**2545–2549.
12. **Heald, S. M.** 1988. Design of an EXAFS experiment, p. 87–118. *In* D. C. Koningsberger and R. Prins (ed.), *X-Ray Absorption: Principles, Applications, Techniques of EXAFS, SEXAFS, and XANES*. Wiley, New York, N.Y.
13. **Ikeda, Y., E. Wada, M. Harada, T. Chikazawa, T. Kikuchi, H. Mineo, Y. Morita, M. Nogami, and K. Suzuki.** 2004. A study on pyrrolidone derivatives as selective precipitant for uranyl ion in HNO_3. *J. Alloys Compd.* **374:**420–425.
14. **Jeon, B. H., S. D. Kelly, K. M. Kemner, M. O. Barnett, W. D. Burgos, B. A. Dempsey, and E. E. Roden.** 2004. Microbial reduction of U(VI) at the solid-water interface. *Environ. Sci. Technol.* **38:**5649–5655.
15. **Kang, H. C., J. Maser, G. B. Stephenson, C. Lium, R. Conley, A. T. Macrander, and S. Vogt.** 2006. Nanometer linear focusing of hard X-rays by a multilayer Laue lens. *Phys. Rev. Lett.* **96:**127401.
16. **Kelly, S. D., K. M. Kemner, J. A. Fein, D. A. Fowle, M. I. Boyanov, B. A. Bunker, and N. Yee.** 2002. X-ray absorption fine structure determination of pH-dependent U–bacterial cell wall interactions. *Geochim. Cosmochim. Acta* **66:**3855–3871.
17. **Kemner, K. M., A. J. Kropf, and B. A. Bunker.** 1994. A low-temperature total electron yield detector for X-ray absorption fine structure spectra. *Rev. Sci. Instrum.* **65:**3667–3669.
18. **Kemner, K. M., W. Yun, Z. Cai, B. P. Lai, H.-R. Lee, D. G. Legnini, W. Rodrigues, J. D. Jastrow, R. M. Miller, S. T. Pratt, M. A. Schneegurt, C. F. Kulpa, Jr., and A. J. Smucker.** 1998. Using X-ray microprobes for environmental research. *Proc. SPIE Int. Soc. Opt. Eng.* **3449:**45–54.
19. **Kemner, K. M., W. Yun, Z. Cai, B. Lai, H.-R. Lee, J. Maser, D. G. Legnini, W. Rodrigues, J. Jastrow, R. M. Miller, S. T. Pratt, M. A. Schneegurt, and C. F. Kulpa.** 1999. Using zone plates for X-ray microimaging and micro-spectroscopy in environmental science. *J. Synch. Radiat.* **6:**639–641.
20. **Kemner, K. M., B. Lai, J. Maser, M. A. Schneegurt, Z. Cai, P. P. Ilinski, C. F. Kulpa, D. G. Legnini, K. H. Nealson, S. T. Pratt, W. Rodrigues, M. L. Tischler, and W. Yun.** 2000. Use of high-energy X-ray microprobe at the Advanced Photon Source to investigate interactions between metals and bacteria, p. 319–322. *In* W. Meyer-Ilse et al. (ed.), *6th International X-ray Microscopy Conference*. American Institute of Physics, College Park, Md.
21. **Kemner, K. M., S. D. Kelly, B. Lai, J. Maser, E. J. O'Loughlin, D. Sholto-Douglas, Z. Cai, M. A. Schneegurt, C. F. Kulpa, and K. H. Nealson.** 2004. Elemental and redox analysis of single bacterial cells by X-ray microbeam analysis. *Science* **306:**686–687.
22. **Kirz, J., C. Jacobsen, and M. Howells.** 1995. Soft X-ray microscopes and their biological applications. *Q. Rev. Biophys.* **28:**33–130.
23. **Kittel, C.** 1986. *Introduction to Solid State Physics*. Wiley, New York, N.Y.
24. **Koningsberger, D. C., and R. Prins (ed.).** 1988. *X-Ray Absorption: Principles, Applications, Techniques of EXAFS, SEXAFS and XANES*. Wiley, New York, N.Y.
25. **Kropf, A. J., R. J. Finch, J. A. Fortner, S. Aase, C. Daranfil, C. U. Segre, J. Terry, G. Bunker, and L. D. Chapman.** 2003. Bent silicon crystal in the Laue geometry to resolve X-ray fluorescence for X-ray absorption spectroscopy. *Rev. Sci. Instrum.* **74:**4696–4702.
26. **Lack, J. G., K. C. Swades, S. D. Kelly, K. M. Kemner, S. M. O'Connor, and J. D. Coates.** 2002. Immobilization of radionuclides and heavy metals through anaerobic bio-oxidation of Fe(II). *Appl. Environ. Microbiol.* **68:**2704–2710.
27. **Lai, B., W. Yun, D. G. Legnini, D. Xiao, J. Chrzas, P. J. Viccaro, V. White, S. Bajikar, D. Denton, and F. Cerrina.** 1992. Hard X-ray phase zone plate fabricated by lithographic techniques. *Appl. Phys. Lett.* **61:**1877–1879.
28. **Lytle, F. W., R. B. Greegor, D. R. Sandstrom, E. C. Marques, J. Wong, C. L. Spiro, G. P. Huffman, and F. E. Huggins.** 1984. Measurement of soft X-ray absorption spectra with a fluorescent ion chamber detector. *Nucl. Instrum. Methods* **226:**542–548.
29. **Mattigod, S. V., M. L. Rivers, and S. R. Sutton.** 1990. *X-Ray Fluorescence Microprobe and Microtomography: Synchrotron X-ray Sources and New Opportunities in the Soil and Environmental Sciences*. Argonne National Laboratory, Argonne, Ill.
30. **Myneni, S. C. B., T. Tokunaga, and G. E. Brown, Jr.** 1997. Abiotic selenium redox transformations in the presence of Fe(II,III) hydroxides. *Science* **278:**1106–1109.
31. **O'Loughlin, E. J., S. D. Kelly, R. E. Cook, R. Csencsits, and K. M. Kemner.** 2003. Reduction of uranium(VI) by mixed iron(II)/iron(III) hydroxide (green rust): formation of UO_2 nanoparticles. *Environ. Sci. Technol.* **37:**721–727.
32. **Panfili, F. R., A. Manceau, G. Sarret, L. Spandini, T. Kirpichtchikova, V. Bert, A. Laboudigue, M. A. Marcus, N. Ahamdach, and M. F. Libert.** 2005. The effect of phytostabilization on Zn speciation in a dredged

contaminated sediment using scanning electron microscopy, X-ray fluorescence, EXAFS spectroscopy, and principal component analysis. *Geochim. Cosmochim. Acta* **69:**2265–2284.

33. **Rehr, J. J., J. Mustre de Leon, S. I. Zabinsky, and R. C. Albers.** 1991. Theoretical X-ray absorption fine structure standards. *J. Am. Chem. Soc.* **113:**5135–5140.
34. **Ressler, T., J. Wong, J. Roos, and I. L. Smith.** 2000. Quantitative speciation of Mn-bearing particulates emitted from autos burning (methylcyclopentadienyl) manganese tricarbonyl-added gasolines using XANES spectroscopy. *Environ. Sci. Technol.* **34:**950–958.
35. **Rose, M. E., and M. M. Shapiro.** 1948. Statistical error in absorption experiments. *Phys. Rev.* **74:**1853–1864.
36. **Sayers, D. E., E. A. Stern, and F. W. Lytle.** 1971. New technique for investigating noncrystalline structures: Fourier analysis of the extended x-ray-absorptive fine structure. *Phys. Rev. Lett.* **27:**1204–1207.
37. **Schulze, D. G., and P. M. Bertsch.** 1995. Synchrotron X-ray techniques in soil, plant, and environmental research. *Adv. Agron.* **55:**1–67.
38. **Schulze, D. G., T. McCay-Buis, S. R. Sutton, and D. M. Huber.** 1995. Manganese oxidation states in Gaeuman-nomyces-infested wheat rhizospheres probed by micro-XANES spectroscopy. *Phytopathology* **85:**990–994.
39. **Stern, E. A., and S. M. Heald.** 1979. X-ray filter assembly for fluorescence measurements of X-ray absorption fine-structure. *Rev. Sci. Instrum.* **50:**1579–1583.
40. **Stöhr, J.** 1992. *NEXAFS Spectroscopy*. Springer-Verlag, Berlin, Germany.
41. **Sutton, S. R., K. W. Jones, B. Gordon, M. L. Rivers, S. Bajt, and J. V. Smith.** 1993. Reduced chromium in olivine grains from lunar basalt 15555: X-ray absorption near edge structure (XANES). *Geochim. Cosmochim. Acta* **57:**461–468.
42. **Sutton, S. R., M. L. Rivers, S. Bajt, and K. W. Jones.** 1993. Synchrotron X-ray fluorescence microprobe analysis with bending magnets and insertion devices. *Nucl. Instrum. Methods B* **75:**553–558.
43. **Suzuki, Y., S. D. Kelly, K. M. Kemner, and J. F. Banfield.** 2002. Nanometre-size products of uranium bioreduction. *Nature* **419:**134.
44. **Suzuki, Y., S. D. Kelly, K. M. Kemner, and J. F. Banfield.** 2003. Microbial populations stimulated for hexavalent uranium reduction in uranium mine sediment. *Appl. Environ. Microbiol.* **69:**1337–1346.
45. **Suzuki, Y., S. D. Kelly, K. M. Kemner, and J. F. Banfield.** 2004. Enzymatic U(VI) reduction by *Desulfosporosinus* species. *Radiochim. Acta* **92:**11–16.
46. **Tokunaga, T., S. R. Sutton, S. Bajt, S. B. Clark, S. R. Sutton, M. L. Rivers, and J. V. Smith.** 1994. Mapping of selenium concentrations in soil aggregates with synchrotron X-ray fluorescence microprobe. *Soil Sci.* **158:**421–433.
47. **Wasserman, S. R.** 1997. The analysis of mixtures: application of principal component analysis to XAS spectra. *J. Phys. IV* **7:**C2203–C2208.
48. **Winick, H. (ed.).** 1994. *Radiation Sources—a Primer.* World Scientific, Singapore, Republic of Singapore.

Techniques for Studying Microbial Transformations of Metals and Radionuclides

JONATHAN R. LLOYD, TERRY J. BEVERIDGE, KATHERINE MORRIS, DAVID A. POLYA, AND DAVID J. VAUGHAN

98

Metals play a crucial role in biological systems and are essential for the growth and metabolism of all organisms, including microbes. Some metals, such as iron or molybdenum, are required as cofactors in cellular proteins, while others, such as mercury, can be highly poisonous and require biochemical detoxification, even at relatively low concentrations. The elegant and widespread mechanisms adopted by microorganisms to nullify the toxicity of metals are described in chapter 96 alongside appropriate ways to characterize the genes and proteins underpinning these processes. In addition to their central role in microbial metabolism as metal cofactors, metals can also be used as electron donors or acceptors during microbial respiration. It has been argued that such transformations could hold the key to the origin of life on Earth and give a basis for life in the most inhospitable environments within our biosphere, or even on other planets. The microbial metabolism of metals also plays a critical role in controlling the chemistry of modern Earth, being intimately linked to the cycling of bulk (e.g., carbon) and trace elements, including toxic metals and radionuclides. Furthermore, by harnessing these activities, metal-transforming microorganisms play an increasingly important role in a range of biotechnological processes, including the remediation of organics, metals, and radionuclides and even energy harvesting from microbial fuel cells.

Techniques underpinning several of these specialist research fields are presented in this volume. Indeed, the dramatic advances in our understanding of environmental metal-microbe interactions over the last decade have gone hand in hand with major technological advances in instrument development achieved in many disparate fields of science and technology. Some of these important techniques, most notably powerful high-throughput "omics" techniques, have been developed with the biosciences in mind and are covered in various chapters in this volume. However, research into metal-microbe interactions is truly multidisciplinary, and other major advances from the physical sciences are playing a vital role in advancing "geomicrobiological" research. Powerful imaging and spectroscopy techniques are being applied with improving sensitivities and spatial resolutions, often at the subcellular scale, to advance the field.

The aim of this chapter is to give an overview of the range of techniques that are available to help characterize the critical microbiological and geological factors underpinning metal-microbe interactions. These include techniques to study microbial cells, the mineral substrates that they live on, and the surrounding geochemical environment that together control the rate and extent of metal biotransformations.

BASIC EXPERIMENTAL DESIGN FOR PROBING METAL-MICROBE INTERACTIONS

It is important to have a basic understanding of the chemistry of a given metal or radionuclide before planning experiments to uncover its interactions with microbial systems. An excellent introduction to metal chemistry is given in chapter 96 and also in other recommended basic texts on the subject (12, 40) and is not repeated here. It is sufficient to say that the speciation and bioavailability of the target metal are crucial and are highly dependent on the solution chemistry of the experimental system and the presence of geological materials to which the metal could sorb. For this reason, microcosm experiments can be very useful in defining the impact of microbial metabolism under environmentally relevant conditions. Generally, these experiments involve the careful mixing of environmental samples (e.g., sediment or soils) with an appropriate quantity of site water (or an artificial analogue) and aliquots of the metal or radionuclide of choice. Recent examples of this type of experiment include studies focusing on the microbial redox cycling of uranium (37), technetium (17), and arsenic (54).

In some cases, simple manipulations can be done to promote the activities of particular groups of microorganisms within microcosms, and these activities can then be correlated with changes in metal speciation. Such manipulations can include (i) driving the system to become anaerobic by altering the gas phase in the experimental system, e.g., purging it with nitrogen; (ii) "poising" or holding the microcosm at a defined biogeochemical set point by the controlled addition of an electron donor and/or acceptor; or even (iii) the addition of pure cultures of previously characterized microorganisms to sterilized microcosms. In all cases, the microcosm should be incubated for an appropriate period of time in the dark (unless photosynthetic activity is required), and the metal or radionuclide should be monitored periodically using the techniques described later in this chapter. Of particular importance when inferring the impact of microbial activity on metal chemistry in microcosms is a direct comparison with the behavior of the metal in an analogous "abiotic" control system that has been pretreated to kill the indigenous microbial community by, for example, autoclaving, antibiotic addition, or the inclusion of a metabolic poison such as mercury. Due care should, however, be taken in determining the potential impact of sterilization on the system; for example, autoclaving can have a profound impact on the background "abiotic" chemistry of the system, for example, through changing the mineralogy of sediments.

In addition to monitoring the solution- and aqueous-phase changes of the target metal, alongside other geochemical indicators (e.g., pH, E_h, and respiratory processes, such as nitrate and sulfate reduction or methanogenesis), it is important to quantify the abundance or activity of key microorganisms that are hypothesized to control metal speciation. Thus, through identifying correlations between specific microbial processes and metal speciation, it is possible to construct conceptual models to test using other complementary approaches, including pure culture experiments (see below). A range of cultivation-dependent techniques, including most-probable-number counts, as well as cultivation-independent techniques, such as conventional PCR, reverse transcription-PCR, stable isotope probing, microarray analysis, and fluorescence in situ hybridization staining, are appropriate and are covered in the companion chapters in this manual. Molecular techniques targeting broad phylogenetic markers (e.g., 16S rRNA and corresponding genes) are useful and can be supplemented with functional genetic targets when appropriate.

As well as conducting microcosm experiments, it can be useful to explore the interactions of specific microorganisms with metals and radionuclides in axenic culture when suitable isolates are available. For these approaches, batch or chemostat systems are useful, but particular care must be taken in choosing the appropriate growth medium. Metal binding can be especially severe if complex media are used, containing, for example, yeast extract, peptone, or amino acids (12). For this reason, fully defined media are generally preferred, and here the choice of buffer system can be very important. Buffers that exhibit "negligible" metal binding include those using the sulfonic acids morpholineethanesulfonic acid (pK_a = 6.15), piperazine-*N*,*N*′-bis(2-ethanesulfonic acid) (PIPES) (pK_a = 6.80), *N*-tris (hydroxymethyl) methyl-2-aminoethanesulfonic acid (TES) (pK_a = 7.5), and HEPES (pK_a = 7.55) (12). In some experiments, however, the researcher may wish to use a more environmentally relevant buffering system, such as bicarbonate (e.g., see reference 102).

MONITORING CHANGES IN AQUEOUS CHEMICAL SPECIATION IN MICROBIOLOGICAL EXPERIMENTS

In addition to characterizing the biological components of experiments that address metal-microbe interactions, it is clearly important to understand the speciation of key chemical components, both in the aqueous phase and associated with any insoluble precipitates or mineral structures. In this section, we discuss the major analytical techniques that are available for determining aqueous-phase chemical speciation. Here it is important to stress the need for a clear understanding of both (i) the nature of the operationally or otherwise-defined species being analyzed and (ii) controls on the preservation of chemical species between sampling and analysis, especially when analyzing samples from in situ environmental studies. These issues are discussed briefly, followed by a summary of the many analytical techniques available for determining aqueous-phase chemical speciation. For a more extended account, Florence and Batley (39) provided a landmark in-depth review of aqueous-phase chemical speciation, whilst for details of currently available methods, the reader is referred to recent reviews by Feldmann (35), Batley et al. (6), Templeton et al. (115), and Baun and Christensen (7), among others.

Definition of Chemical Speciation

A chemical species may be defined as the specific form of an element, including its isotopic composition (not considered further here), electronic or oxidation state, and/or complex or molecular structure (115). The definition of and differences between aqueous species of different oxidation states [e.g., $Cr^{III}(OH)_2^+$ and $Cr^{VI}{}_2O_7^{2-}$], in complex with different ligands (e.g., $CdCl^+$ and CdF^+) or with different numbers of ligands [e.g., NpO_2^+, $NpO_2CO_3^-$, and $NpO_2(CO_3)_2^{2-}$], or forming organometallic compounds with one or more covalent carbon-metal bonds of various strengths [e.g., CH_3Hg^+, $(CH_3)_3Sn^+$, and $(C_2H_5)_4Pb^0$] are reasonably distinct, making the meaningful determination of such species amenable to a variety of methods. In many cases, however, aqueous-phase speciation is not as clear-cut, making analysis and data interpretation very challenging.

For example, many aqueous species have lifetimes that are much shorter than the timescales required for their detection. Others have such low thermodynamic stabilities, particularly in the event of changes to the temperature, pressure, pH, or E_h of the solution during sampling, that attempts to discriminate between them by analytical methods are not practicable. For example, few would attempt to measure the variable protonation of As(V) species (i.e., H_3AsO_4, $H_2AsO_4^-$, $HAsO_4^{2-}$, AsO_4^{3-}) in natural waters transported to the laboratory during geomicrobiological studies. Instead, a more meaningful value for the concentrations of these species would normally be obtained by equilibrium modeling using total As(V) analysis in combination with an in situ pH measurement. Templeton et al. (115) pointed out that the concentrations of short-lived molecular moieties, such as NaCl ion pairs in seawater, might be accounted for in geochemical models, either by direct consideration of their concentrations or, more indirectly, by incorporation of their effective concentrations into activity-composition models (see references 95 and 97); thus, any use of chemical speciation data in such models requires a clear understanding of how those data have been determined.

Particularly for microbial systems involving (i) interactions between metals/metalloids and a complex range of dissolved organic ligands, (ii) cell surface structures, (iii) biomineral phases, or (iv) a multitude of polynuclear species, identifying all of the individual single-chemical species present in a water sample can be not only challenging but, in many cases, impractical with the technology available at the beginning of the 21st century. In such cases, speciation methods may aim instead to discriminate between "classes" of species rather than between individual species. Such class divisions include free versus complexed, labile versus nonlabile, organic-complexed versus inorganic-complexed, and colloidal versus dissolved species.

Preservation of Chemical Speciation

Preservation of chemical speciation prior to analysis is commonly a problem when the species of interest are reactive. This can be a significant problem during laboratory-based experiments and can be acute for samples taken during field studies, where physicochemical conditions during transport and storage are significantly different from those in situ. The ultimate solution to such problems is to make speciation measurements on-site, or immediately in the laboratory in the case of experimental studies. Le et al. (66), for example, proposed such a strategy for the determination of arsenic speciation in natural waters. However, in many cases this is not practicable. Even on-site analysis of redox-sensitive species in reducing groundwaters, for example, is potentially compromised by changes in temperature, pressure, light, and exposure to air (96), whilst the distribution of colloidal and dissolved species can be altered substantively not only by these changes but also by filtering and physical disturbances during sampling (120).

The control of aqueous speciation changes during storage and the rates of these changes are too large a subject to describe comprehensively here. However, using the preservation of aqueous arsenic species transformed by microorganisms in natural waters as an example, the principles of how the aforementioned preservation methods can be developed to match particular media and particular species of interest can be illustrated. Arsenic most commonly occurs in natural waters as inorganic As(III) or As(V) and, rather less commonly, as a range of variably methylated species, notably monomethylarsinic acid and dimethylarsonic acid. Cullen and Reimer (23) described a host of even less common As species. Changes in aqueous arsenic speciation during storage are most commonly due to (i) differential coprecipitation with or adsorption onto insoluble hydrated ferric oxides, (ii) abiotic oxidation, and (iii) continued microbial activity (96). Cooling, storage in the dark, and reduction of storage time all serve to reduce the extents of these processes. The further addition of antibiotics, acidification, and sterile filtration reduce microbial activity, whilst the last two strategies also serve to increase the solubility and suppress the nucleation, respectively, of hydrated ferric oxide phases, which can sorb a range of metals and metalloids, including arsenic (96). The addition of EDTA in certain cases can also enhance the solubility of ferric oxide phases through Fe-EDTA complex formation (8). However, some of these strategies are not without cost, as differences in the stabilities of Fe(III)-EDTA and Fe(II)-EDTA complexes mean that the addition of EDTA can potentially alter the relative concentrations of aqueous Fe(II) and Fe(III) species, and thus the redox environment, of the sample (44); acidification has been reported by Hall et al. (47) to cause the oxidation of As(III) in river water; and sterile filtration using 0.20-μm or even smaller filter pore sizes prohibits the analysis of colloidal or other constituents with a greater effective size than the pores and is incompatible with the rather arbitrary, albeit long-standing, convention of filtering water samples to 0.45 μm (46). In summary, it is necessary to design speciation preservation strategies that are fit for purpose on a case-by-case basis when studying microbial transformations of metals, and it is vital to check the effectiveness of any such strategy, again on a case-by-case basis. Further discussions on species preservation techniques are given by the U.S. EPA (119) for As, Heninger et al. (50) for Se, Krachler et al. (63) for Sb, Bancon-Montigny et al. (5) for organotin, and Parker and Bloom (92) for Hg, whilst a more comprehensive recent review is that of Ariza et al. (3).

Analytical Methodologies for Determining Aqueous Chemical Speciation

Aqueous chemical speciation is generally analyzed by using either (i) measurement of a property characteristic of the species, with the consequent need to mathematically deconvolute the contributions from various coexisting species, or (ii) separation of species by utilizing differences in their characteristic properties, followed by a non-species-specific detection step. Selected examples of these types of measurements are outlined below. For accounts of a more comprehensive range of other important individual techniques, including anodic stripping voltametry, cathodic stripping voltametry, potentiometric methods, and fractionation methods for operationally defined species classes, see the works of Buffle and Tercier-Waeber (16) and Templeton (114) and references therein.

Characteristic properties utilized for type i measurements include spectroscopic absorption profiles (be it in the infrared, visible, UV, or even X-ray range of the electromagnetic spectrum), reactivity, and electrochemical properties.

Using uranium as an example, Raman and attenuated total reflection FIR spectroscopies were used by Pasilis and Pemberton (93) to determine uranyl citrate speciation; Zanonato et al. (124) used UV-visible (UV-vis) spectroscopy to determine the distribution of hydrolysis products of U(VI), whilst uranium complexing with protocatechuic acid and citrate has been determined by X-ray absorption spectroscopy (XAS) by Rossberg et al. (105) and Bailey et al. (4), respectively. The powerful synchrotron-based XAS technique further enables information about the oxidation state and coordination of unknown species to be determined—analysis of the X-ray absorption near-edge structure (XANES) is particularly useful for determining the average oxidation state of a given component in an aqueous solution, whilst extended X-ray absorption finer structure (EXAFS) spectra of sufficient quality can enable average bond lengths and coordination numbers to be determined around a central atom of interest, not uncommonly with precisions of around ±0.02 Å and ±50%, respectively. For further information on these synchrotron-based techniques, see chapter 97.

For type ii separation and detection methodologies, methods for the effective separation of species may be designed based on differences in species size (e.g., size exclusion chromatography and capillary zone electrophoresis) (25, 64), adsorptive properties (e.g., ion chromatography) (43), volatility or solubility in supercritical CO_2 (1), and reactivity (e.g., derivatization, including hydride generation techniques) (18). These separation methodologies are often combined with sensitive detection systems, including graphite furnace atomic adsorption spectroscopy (18), inductively coupled plasma optimal emission spectroscopy (ICP-OES) (67), inductively coupled plasma mass spectrometry (ICP-MS) (64), ICP-HEX-MS (ICP-MS utilizing a hexapole collision cell) (96), ICP-DRC-MS (ICP-MS utilizing a dynamic reaction cell) (113), HR-ICP-MS (high-resolution sector-field ICP-MS), ICP-TOF-MS (ICP-time of flight-MS) (36), and laser-induced fluorescence spectroscopy (1, 124). Further details of the burgeoning field of chromatography–ICP-MS-based speciation analysis can be found in proceedings of the biennial Plasma Source Mass Spectrometry Conference (e.g., see reference 52) and in the work of Feldmann (35).

Depending on the nature of the medium and the species being separated and detected, it is not unusual for species detection limits on the order of micrograms per liter or even lower to be achieved and reported using currently available aqueous speciation methodologies (35). The improvement in achievable detection limits opens many avenues of research in the areas of trace metal/metalloid biotransformations, radionuclide biogeochemistry, and bioremediation. However, it also necessitates that even greater care be taken with quality assurance, particularly regarding checking blanks, possible matrix- and species-dependent analytical sensitivities, and ensuring the independent confirmation of the assignment of instrumental peaks to specific species. The development of certified reference materials for aqueous species remains a significant challenge for the analytical research community. Whatever accuracies and precision are ultimately achieved in developing speciation technologies, the need to understand and interpret the meaning of speciation "analysis" remains; Nordstrom (90) highlights this in his comparison of computation and analytical methods for assessing trace metal speciation in natural waters, whilst Davison and Zhang (26) exemplify how novel speciation technologies may be used to gain more profound insights into the kinetics of biogeochemical processes.

RADIOCHEMICAL TECHNIQUES: MEASUREMENT AND SAFETY

In addition to handling and analysis of soluble metals in "geomicrobiological" experiments, there has been a recent increase in the study of interactions between microorganisms and radionuclides. This has become an area of intense academic and biotechnological interest, as it is necessary to understand the biogeochemical cycles of a range of radionuclides if we are to store nuclear waste safely or detoxify contaminated environments by biotechnological means (72, 73). To understand the biogeochemical transformations associated with radionuclide interactions in the natural and engineered environment, it is necessary to handle radioactive materials. The types of experiments performed range from pure culture experiments, where the focus is on understanding the physiological basis of transformations (70, 77, 102, 121), to simulation experiments, where a model laboratory experiment is set up to mimic the environment of interest (17, 38, 42, 74, 112), and finally encompass in situ field studies of the behavior of radionuclides in contaminated environments (2, 55, 107). This section focuses on examining current techniques used in understanding the transformations of long-lived, redox-active radionuclides, such as U, Tc, Np, and Pu, in pure culture and laboratory simulation experiments and highlights the safe handling and measurement of these radionuclides. The techniques discussed are, of course, also largely appropriate for microbial experiments using radionuclides other than these transuranics and fission products.

Safety

The possession and use of radioactive materials are subject to statutory controls which are country specific but which are subject to the overarching principles of radiological protection laid down by the International Commission on Radiological Protection (www.icrp.org). In addition, the laboratory where the work is carried out must possess a license to handle and dispose of the radionuclides. The purpose of this section is to highlight how the real need to understand the transformation behavior of radionuclides is balanced against the safe handling requirements for working with radiation. For all radionuclides, the annual limit of intake (ALI) of the radioisotope gives an indicator of the radiological hazard associated with working with that radionuclide (27). ALIs are available for different radioisotopes and for both inhalation and ingestion scenarios (27). For all the radionuclides considered here, inhalation results in the highest dose, and we used the inhalation ALIs for our calculations. Essentially, the ALI indicates the maximum amount of an individual isotope a member of the public can uptake before the annual whole-body dose limit of 1 mSv is exceeded, and the value is derived from published values for committed effective dose per unit intake (Sv Bq^{-1}) (27).

In order to illustrate the risk associated with different experiments, it is useful to consider the number of ALIs there are in a typical experiment handling key radionuclides of interest, which include the long-lived redox-active elements Tc, U, Np, and Pu (Table 1). The following are two experimental scenarios considered here: (i) a typical pure culture or microbial consortium experiment with a total experimental volume of ~30 ml and (ii) an X-ray absorption spectroscopy experiment for characterization of the speciation of the radionuclide, with a sample size of ~0.3 g and a range of concentrations of the radionuclide in the sample from 200 to 5,000 ppm. These values are approximate but realistic sample sizes and concentrations for X-ray absorbance spectroscopy and other direct characterization techniques, such as transmission electron microscopy.

Examination of Table 1 shows that working with ^{238}U and ^{99}Tc, although potentially problematic, poses much less of a risk than working with ^{237}Np and Pu. As the number of ALIs in experiments starts to become very large, the laboratory worker need inhale only very small volumes of experimental material to receive a substantial dose from its associated radioactivity. For example, in the case of a pure culture experiment with ^{237}Np, a 30-ml experiment can contain up to 9,000 ALIs (Table 1), and inhalation of only 3 μl of solution would lead to the laboratory worker receiving a dose equivalent to 1 mSv. Similarly, considering the upper end of concentrations of ^{237}Np required for an XAS experiment, the sample will contain at least 3,000 ALIs, and inhalation of 0.1×10^{-3} g of experimental material would lead to the laboratory worker receiving a dose equivalent to 1 mSv. These examples illustrate that safe handling of these materials is by no means straightforward, but nonetheless, both examples are taken from experiments reported in the literature and demonstrate that with appropriate infrastructure and skills, these experiments are possible. Finally, it is worth noting that XAS with both Np and Pu is exceptionally difficult because of radiological safety concerns (Table 1), and XAS on Pu and Np in environmental materials and microbial samples is restricted to a handful of studies (21, 32, 88, 91, 98, 108). In summary, specialist radiochemistry laboratories are needed for handling all of the radionuclides mentioned above, with the possible exception of uranium. It is also worth noting that the difficulty of handling certain radionuclides has had clear impacts on the amount of research work undertaken. For example, at the time of writing, to our knowledge, there are fewer than five papers detailing the biotransformations of Np (75, 102, 103) and only one published paper dealing with Pu biotransformations (106). Nonetheless, all of the radionuclides mentioned are significant groundwater pollutants at nuclear fuel cycle sites throughout the world and/or are significant components of radioactive wastes. This means that the study of their biotransformation behavior, although challenging, is absolutely essential in understanding the behavior of radionuclide contaminants and, ultimately, managing our nuclear waste legacy.

Radiochemistry and Radiochemical Techniques

The long-lived radionuclides U, Tc, Np, and Pu are redox active, and all of these radionuclides have the potential to form less soluble species as reducing conditions develop. Therefore, understanding their behaviors as microbially mediated reduction reactions occur has been a key focus of research. In addition, recent work has started to examine the reoxidation behavior of these radionuclides. Here we briefly discuss the environmental chemistry and biogeochemistry of U, Tc, Np, and Pu, and we also highlight current radiochemical techniques applied in the study of microbial reduction and reoxidation processes.

The half-lives, oxidation states, and principle radioactive decay routes for the key isotopes of U, Tc, Np, and Pu are given in Table 2. More detailed accounts of the biogeochemistry of these radionuclides are provided by the excellent text by Keith-Roach and Livens (59) and in more select reviews by Lloyd and Renshaw (73). Interestingly, the reduced forms of all of these radionuclides are less soluble than the oxidized forms (Table 2), and microbially mediated reduction is therefore expected to lead to reduced solubility. Technetium is present under oxic environmental conditions as the highly soluble pertechnetate (TcO_4^-) ion (14). Upon exposure to reducing conditions, the lower-valence forms of technetium [predominantly Tc(IV)] form and either hydrolyze to hydrous Tc(IV)O_2 or, at low concentrations, may form strong surface complexes with Al and Fe oxides (14).

The actinides are best treated as a group that has access to four oxidation states encountered in the environment (+3 to +6, inclusive). The lower-valency states form simple cations, M^{3+} and M^{4+}, which are readily hydrolyzed due to their high charge densities and are easily lost from solution by reaction with surfaces (89). In contrast, the higher oxidation states form the dioxo-cations MO_2^+ and MO_2^{2+}, which are more soluble due to both their lower charge densities and their complexation with ligands, such as carbonate, to give neutral and anionic complexes (89). Uranium is present under oxic conditions as the relatively soluble uranyl (UO_2^{2+}) ion. The speciation and solubility of the uranyl ion are strongly pH dependent due to hydrolysis and formation of carbonate complexes at circumneutral and alkaline pHs (20). Under reducing conditions, U(IV) predominates and U(IV) compounds are less soluble than U(VI) species because U(IV) hydrolyzes more easily and sorbs to surfaces, whilst at higher concentrations it precipitates out as U(IV) hydrous oxide (99). The picture for Np is similar: under oxic conditions, neptunium forms the Np(V)O_2^+ ion, which is considered the most soluble of all the actinide species (68), whilst under reducing conditions, the highly insoluble Np(IV)O_2 phase is expected to dominate (14). In contrast, Pu potentially has access to all four valency states under redox conditions found in the natural environment (28), with the dominant state expected to be Pu(IV), but with access to Pu(V) (89) and even Pu(III) under strongly reducing conditions (106).

Measurement of Radionuclides

Typically, in microbe-radionuclide studies, the experimental conditions are optimized such that measurement of the radionuclides is relatively straightforward. In this sense, experiments are often performed at artificially high concentrations compared to the trace levels expected in even heavily contaminated environments. These raised concentrations are a necessary compromise, as they allow workers to avoid time-consuming radiochemical separations, which are the norm for environmental studies (e.g., see references 60 and 87), and they allow preparation of samples that can be examined by direct microscopy and spectroscopy techniques. Radionuclide concentrations are conventionally determined by radiometric techniques that are based upon the detection and measurement of radioactive decay processes. Scintillation counting is routinely used for detection of ^{99}Tc (17, 41, 42),

TABLE 1 Concentrations of Tc, U, Np, and Pu used in geomicrobiological experiments and their corresponding ALIs[a]

Radionuclide	Half-life (yr)	Mode of decay	Energy of decay	Typical concn used in pure culture/model lab expt (reference)	Typical concn (ppm) used in XAS expt (reference)	Typical no. of inhalation ALIs in pure culture expt	Typical no. of inhalation ALIs in XAS expt
^{99}Tc	2.10×10^5	Beta	294 keV	3 μM–2 mM (6, 8, 44)	200–5,000	0.02–12	0.1–3
^{238}U	4.47×10^9	Alpha	4.20 MeV	50 μM (64)–10 mM (4, 54)	200–5,000	0.04–7	0.01–0.2
^{237}Np	2.14×10^6	Alpha	4.79 MeV	10 μM–1 mM (4, 19, 20)	700–7,000 (16)	90–9,000	300–3,000
^{239}Pu	2.41×10^4	Alpha	5.15 MeV	0.4–1.8 μM (21)	200–5,000	800–3,500	1,600–400,000
^{242}Pu	3.76×10^5	Alpha	4.90 MeV	NA	200–5,000	NA	1,000–24,000

[a]NA, not available.

TABLE 2 Half-lives, decay modes, and predominant oxidation states of various isotopes of Tc, U, Np, and Pu

Radionuclide	Isotope	Half-life (yr)	Major decay mode	Known oxidation states	Stable oxidation state	
					Oxic conditions	Reducing conditions
Technetium	99	2.13×10^5	β	IV, V, VI, VII	$Tc(VII)O_4^-$	Tc(IV)
Uranium	235	7.04×10^8	α	IV, V, VI	$U(VI)O_2^{2+}$	$U(IV)O_2$ (s)
	238	4.47×10^9	α			
Neptunium	237	2.14×10^6	α	IV, V, VI	$Np(V)O_2^+$	$Np(IV)O_2$
Plutonium	238	8.77×10^1	α	III, IV, V, VI	$Pu(IV)O_2$ [minor $Pu(V)O_2^+$]	$Pu(IV)O_2$
	239	2.41×10^4	α			
	240	6.55×10^3	α			
	241	1.44×10^1	β			
	242	3.76×10^5	α			

and alpha/beta-discriminated scintillation detection has been used for tracer studies with ^{237}Np (75, 102). In addition, more specialist techniques involving paper chromatography and autoradiography that allow speciation of solution-phase radionuclides have also been used for both ^{99}Tc and ^{237}Np (71, 75). In addition, bespoke chemical and resin-based extractions have been applied to identify Tc and U species in solution (17, 77). In the case of radionuclides with long half-lives, such as uranium, mass-based techniques can also be used, in which the radioactive decay of the element is not used for detection, and conventional analyses such as atomic emission spectroscopy and UV-vis spectroscopy can be applied. For example, UV-vis spectroscopy is routinely used for uranium analysis at concentrations in parts per million (e.g., see reference 58), and kinetic phosphorescence analysis allows detection of uranium at concentrations in parts per billion (15). Finally, UV-vis spectroscopy is also used to check spike purity prior to the addition of Np to samples (e.g., see reference 102). Radionuclides are amenable to solid-phase analysis by mass-based techniques, such as scanning electron microscopy (SEM) and transmission electron microscopy (TEM), but radiological hazard dictates that for at least Np and Pu, microscopy needs to be performed at specialist facilities. In addition, workers have also utilized the radioactive decay characteristics of radioisotopes to examine solid-phase associations. For example, for uranium, and possibly plutonium, alpha-autoradiography using solid-state nuclear track detectors can be used to quantify the ratio of diffuse, sorbed uranium to particulate uranium (48), and microautoradiography followed by detection of Ag developed in the microautoradiography emulsion has been used to examine the spatial distribution of Pu on geological materials (31). Excepting TEM-electron energy loss spectroscopy (TEM-EELS), which is a developing microscopy technology, the above techniques for solid-phase analysis do not allow for identification of the radionuclides at the species level. For this purpose, XAS is required and is relatively routine for metal-microbe studies of ^{99}Tc and U. One disadvantage of XAS is that because it is a mass-based technique, relatively high (>50 ppm) concentrations of radionuclides are required for analysis. This detection limit is less problematic in analyses using microsynchrotron X-ray techniques, such as micro-EXAFS, as these high concentrations are only needed within the beam area, and samples may contain regions where the radionuclide is highly concentrated [e.g., the periplasm or Fe(II)-rich areas in sediments]. Nonetheless, most XAS samples are currently run on less focused beam optics, and >50 ppm is a required, average concentration of the radionuclide in the sample. In samples where radionuclide concentrations are below levels amenable to analysis by direct speciation techniques, indirect speciation methods have been applied to solids. For example, chemical speciation techniques designed for sediments have been applied routinely to stable element speciation and have been adapted for radionuclide speciation in samples from the natural environment (60, 110).

These techniques are useful indicators of the associations of radionuclides in environmental samples but are limited in the insights they can provide, as they leach operationally defined fractions rather than definitive phases. Thus, these chemical leaches cannot provide oxidation-state speciation of radionuclides in solid-phase samples and should be applied cautiously to studies examining metal-microbe interactions. Indeed, a key challenge over the next decades will be developing analytical techniques that are able to detect the speciation of radionuclides in both solids and solutions at the trace concentration levels typically found in natural and engineered environments and used in "environmentally relevant" geomicrobiological experiments.

USING LIGHT AND ELECTRON MICROSCOPY TO CHARACTERIZE MICROBE-METAL INTERACTIONS

Microscopy of all forms can be used to study the transformations of metals and radionuclides. Certainly, though, the type of microscopy depends on the sort of information one requires about the microbiota and the associated metal or mineral. Because microorganisms are small and single cells cannot be seen by eye, microscopy helps to pinpoint and elucidate these interactions between microbes and their environment at relatively high magnification. In this section, we discuss two forms of microscopy based on conventional light and electron beams to study the fate of metals in geomicrobiological experiments. The focus here is on visualizing the interactions between "microbiological structures" and metals, while in a later section we discuss more general methods to assess mineralogical substrates, using a range of techniques, including microscopy.

Light Microscopy

Various kinds of light microscopy are suitable for ascertaining the general shape and form of the microorganisms in a natural setting and for determining their distribution. Bacteria, which are about 2 to 3 μm by 0.5 to 1.0 μm, can just be seen by light microscopy and, if appropriately stained, can be distinguished as being gram positive or gram negative. We do not dwell long on light microscopy because it is a relatively known quantity amongst microbiologists, and more encompassing descriptions of it in its various guises can be found (100). Possibly the most helpful and simple form of light microscopy is phase-contrast microscopy, since it can be used on live bacteria taken from their natural settings once they are spread over a glass microscope slide. Visualization depends on the inherent densities of the objects on the slide and the degrees to which they scatter light so that bacterial substances, such as poly-β-hydroxyalkoanates and elemental sulfur particles (both of which have a higher density than the cytoplasm), can be distinguished by the discerning eye. Furthermore, because minerals scatter light even more effectively than do bacteria, they can be recognized easily. This simple type of light microscopy should be easily available in microbiology laboratories and should always be used to advantage. More complicated forms of light microscopy usually rely on epifluorescence, where specific fluorescent dyes are used to probe the specimen. These fluorochromes can be added as stains to the sample to determine distinguishing traits of the cells (e.g., to tell if the bacteria are dead or alive), to discern various taxons within a natural setting (e.g., to characterize microbial communities), or to visualize discrete biological substances (e.g., DNA, exopolymeric substances, etc.). Sometimes, fluorescent compounds, such as (jellyfish) green fluorescent protein, can be cloned into isolated bacteria, thereby forming an intrinsic marker for the cell.

Confocal Scanning Laser Microscopy

When fluorescent markers are used with more sophisticated and expensive confocal scanning laser microscopes (CSLMs), an even wider world of microscopy opens up. The use of a laser provides a coherent beam of light for better resolving power and focusing accuracy; now, high-resolution optical slices of a sample can be obtained and, with suitable computing power and optical reconstruction software, three-dimensional imaging can be achieved (65). With new radiometric fluorescent probes, it is even possible to probe biofilms for some of the geochemical parameters (e.g., pH) that control mineralization processes (53). Furthermore, advances in multiple-photon excitation CSLMs can, through light activation of highly specific fluorescent ligands, pinpoint the distances between such small cellular components as molecules. CSLMs, especially the multiple-photon excitation variety, can be very expensive and are usually protected from easy common use, but research-intensive institutions should have this equipment available, along with a technical person.

Electron Microscopy

Electron microscopy has proven invaluable for the study of metal-microbe interactions since it has a discerning power that is greater than those of all other microscopy techniques. Furthermore, since highly energetic electrons interact best with high-atomic-number atoms, metals and radionuclides are easily discerned. As with light microscopy, there are many different types of electron microscopy, and it is a matter of choosing the correct type for the information that is desired. Since microbiologists are not as familiar with this type of microscopy, it is best to become acquainted with its various aspects before "taking the plunge" (references 13 and 62 are a good starting place).

SEM

SEMs are usually available in most research centers and are good at topographical imaging of microbes and their associated minerals. In this case, secondary electrons are emitted from the exterior of the specimen and are used to image its surface. Initially, resolution was a problem with this instrumentation, but recent advances with electron emitters (field-emission guns) and low-voltage systems are sometimes allowing subnanometer resolution of some samples. Also, in many instruments, the X-rays emitted when the sample is excited by the electron beam can be detected, and their energies can be measured by energy dispersive spectroscopy (EDS). Since these energies are characteristic of the elements present in the sample, this provides a means of determining chemical composition. Such analysis is only semiquantitative due to problems of calibration and the exact conditions of electron excitation. Variable-pressure SEMs, so-called environmental SEMs, even allow samples to be viewed under high humidity so as to maintain partial hydration, which can be extremely important for observing microbial associations with unaltered mineral species (Fig. 1).

TEM

TEM is a more difficult type of microscopy because it relies on the interpretation of transmitted electrons after they have passed through and interacted with the atomic structure of the specimen. Its disadvantage is that TEM requires extremely thin specimens, usually about 50 nm thick. Yet an enormous amount of information can be gleaned from this type of microscopy. Resolutions on the order of a few angstroms can be obtained for thin crystal lattices, and because the electrons of the electron beam have energies of 80,000 to 100,000 eV, high-atomic-number atoms are best for effective scattering power. This is fortuitous since metals can readily be seen, even in minute amounts, thereby highlighting their associations with biomatter. For this technique, electron diffraction patterns from selected areas (i.e.,

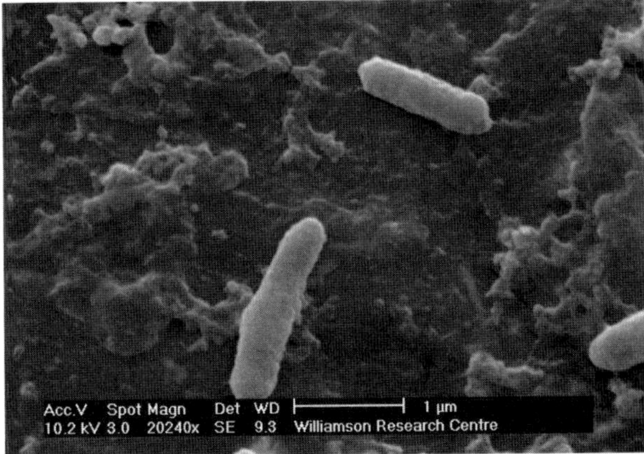

FIGURE 1 Environmental scanning electron micrograph showing cells of *Geobacter sulfurreducens* growing attached to an iron oxide substrate (M. J. Wilkins, P. Wincott, J. R. Lloyd, and D. J. Vaughan, unpublished data).

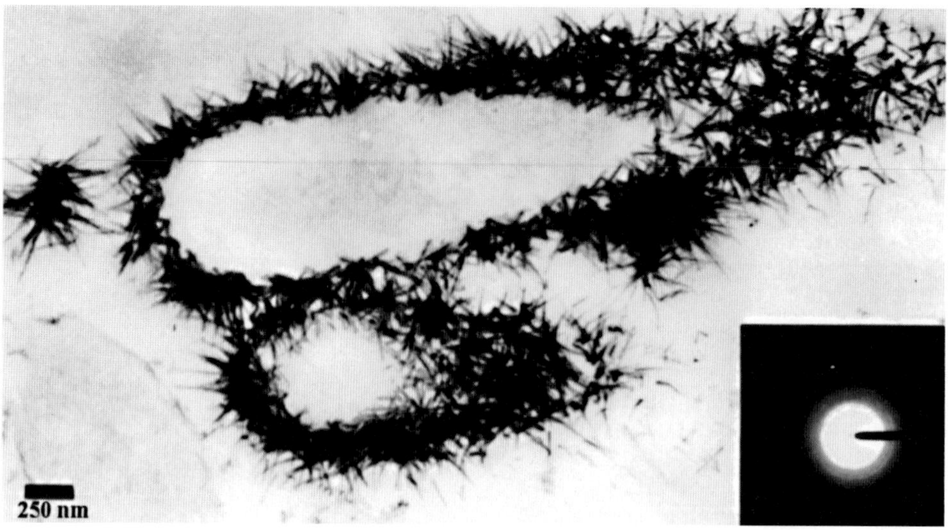

FIGURE 2 Transmission electron micrograph of a thin section of *Pseudomonas aeruginosa* incubated with 1 mM $LaCl_3$ for 15 min at room temperature. The image shows that the bacterial surfaces have reacted strongly with the La^{3+} metal ions.

selected-area electron diffraction) can provide structural information on crystalline mineral phases, and the combination with EDS can again provide semiquantitative compositional data for very small regions (85, 122). TEM has been the primary technique used to initially discern the subcellular structures of prokaryotes, and we owe much of our current understanding of the structures of the cytoplasm, membrane, and cell wall to TEM. Interestingly, this elucidation of structure by TEM has depended on the added contrast that heavy metal stains have imparted to the cellular fabric (Fig. 2). DNA, ribosomes, membranes, and cell walls stain strongly with uranyl, lead, and osmium salts and allow their visualization. Unwittingly, early electron microscopists were amongst the first researchers to experiment with metal-microbe interactions!

Today, those of us interested in such interactions can use TEM as a powerful tool. Simple whole mounts (i.e., cells treated with either dilute environmental metals or fine-grained minerals are simply placed on a TEM grid without being stained) can pinpoint distinct regions of the cell that have sorbed metal or minerals (Fig. 3). Unstained thin sections can distinguish whether or not these associated metals are located in the cytoplasm or on a membrane or wall. Furthermore, TEMs are capable of a variety of analytical procedures with the addition of certain equipment. Selected-area electron diffraction can help characterize small mineral species as being crystalline or amorphous as well as defining the crystal lattice. EDS can identify the elements within the specimen, and since it is best at discerning high-atomic-number elements, the metal composition of sorbed precipitates and minerals can be determined. EDS uses equipment that can detect "signature" X-rays that are given off from the specimen after it has been bombarded by the electron beam of the TEM. EELS is another TEM analytical technique and is particularly sensitive for low-atomic-number elements (as well as some metals). It can often be used to decipher the counterions associated with metals and precipitates or the reactive biosites to which metals bind. With this technique, a proportion of the transmitted electrons that have passed through the specimen are inelastically scattered since they have entered the interatomic domains of the specimen's atoms and have lost energy. The energy loss is characteristic of the atom with which these inelastic electrons have interacted; these inelastic electrons now carry the signature of the atom. Of the two spectroscopies, EELS has a higher detection efficiency than EDS but is more difficult. Certain specialized TEMs, i.e., electron-filtering TEMs, have the ability to selectively filter the inelastic electrons of the transmitted beam such that angstrom-resolution images of the distribution of elements

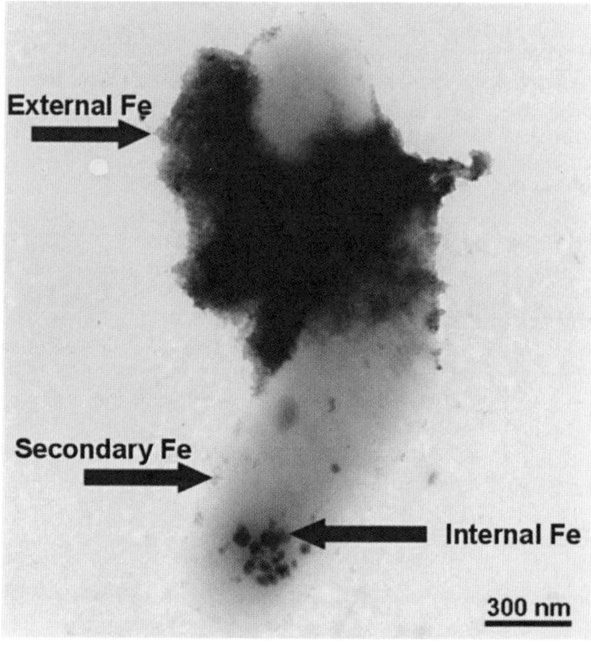

FIGURE 3 Unstained whole mount of *Shewanella putrefaciens* CN32 showing extracellular iron minerals, secondary Fe minerals associated with the cell surface, and internal nanoscale Fe deposits.

can be produced; for example, the iron and oxygen distributions can be seen on extremely fine-grain hematite. This is called electron spectroscopic imaging.

Cryo-TEM

One of the most exciting new advances with TEM has been in the field of cryo-TEM. In this case, instead of relying on traditional chemical fixation techniques to preserve structure (which we know produce artifacts), the cells are physically fixed by rapidly freezing them, on the order of milli- to microseconds, at a temperature of −196°C. Specialized equipment is necessary to do this, and there are two general methods, freeze-plunging (into a cryogen, such as ethane) and high-pressure freezing. Using either of these techniques, the cells are instantaneously vitrified in an unusual type of ice, amorphous ice, which is a type of glass. In cells, molecular motion and degradation of cell structure are stopped, and equally importantly, hydration is maintained. We know that cellular structure is exquisitely preserved because when the bacteria are thawed, they are viable.

There are two main methods for handling the vitrified samples once they are frozen; these are freeze substitution plus plastic embedding and directly visualizing the vitrified sample. For freeze substitution, the ice (water) of the sample is slowly replaced with an organic solvent (such as acetone), and the cells are chemically fixed (osmium tetroxide) and stained (uranyl acetate) at a low temperature (−80°C) so they can be embedded in plastic and thin sectioned (11, 13, 45, 51, 117). Since all of these procedures are done at such a low temperature, the cells remain frozen, and high-quality structure is maintained (10, 11, 45).

Direct visualization of the vitrified sample is much more difficult, and two methods are used in this case (29). The first relies on the production of a thin frozen foil directly on the TEM grid. A suspension of material is placed on an uncoated or holey carbon-coated grid, from which the fluid is wicked off by use of a pointed piece of filter paper until an ~50-nm-thick film is produced. The sample is then immediately vitrified and inserted into a cryo-specimen holder for imaging in a cryo-TEM. Throughout these manipulations and during TEM viewing, the specimen is kept frozen below −140°C. This technique is particularly useful for visualizing extremely small particles, such as the bits and pieces of bacteria (e.g., membranes, ribosomes, etc.) after the cells have been lysed.

The other method of direct visualization is to cut ~50-nm-thick frozen hydrated sections directly from the vitrified sample (30, 82, 83). This is a good technique for imaging intact cells but is extremely difficult and requires a cryo-ultramicrotome along with strong antistatic and low-humidity conditions. As with thin frozen foils, frozen hydrated sections are mounted in a cryo-specimen holder in a cryo-TEM. Unlike all other forms of TEM, the frozen sections cannot be contrasted by using heavy metal stains since the staining solutions would immediately freeze on such cold samples, rendering them useless. Accordingly, cryo-TEM is unlike any other electron microscopy because imaging relies on the innate densities of the constituent cellular macromolecules. The images therefore show the mass distribution within cells. Images are produced by underfocusing the specimen to produce phase contrast so that the cells can be seen. Cryo-TEMs are state-of-the-art instrumentation, since extremely good lenses with a high contrast function are needed. The TEM column and specimen chamber must also have the very best vacuums (since the ultracold specimen will be a cold trap to which contamination will adhere). Obviously, too, tremendous accuracy in temperature maintenance is required since specimens must never be heated to temperatures of >−140°C or crystalline ice might form and puncture cells (the high energy of the electron beam may also nucleate ice in transition). These requirements all have a price in difficulty, time, and monetary expense. Cryo-TEM is still an experimental technique in geomicrobiology and is rarely available for common use. Yet the results of cryo-TEM are extremely promising so far for metal-microbe interactions. Frozen hydrated sections are revolutionizing our perception of the structures of gram-negative (30, 82) and gram-positive (83, 84) bacteria and are providing an accurate picture of the true mass distribution within cell walls. Most interesting is the correlation between frozen hydrated sections of walls and those of freeze-substituted bacteria (Fig. 4). Because the former technique uses unstained material and the latter are stained with heavy metal salts, we can pinpoint where the reactive sites that bind the metals reside. Together, the two techniques reveal the native structure (frozen sections) along with the metal binding sites (freeze substitution) (9, 94). What could be more exciting for those of us interested in metal-microbe interactions!

Exciting Times for Microscopy and Geomicrobiology

In this section, we have attempted to convey the rapid changes in and the exhilaration felt with the use of microscopy in the field of metal-microbe interactions. Not only is the true structural nature of microbial surfaces being exposed through the use of cryo-TEM and atomic force microscopy (AFM), but we are also detecting reactive sites that are available for reaction with environmental metals and contaminating substances. Newer advances using synchrotron radiation, especially in the ability of compacting collimated beams of intense X-rays so that resolutions of ~50 nm are attainable, have provided exquisite detection of small amounts of microbially associated metals (e.g., scanning transmission X-ray microscopy) (116) and even the determination of metal redox states using XANES on such small microbial particles (K. M. Kemner, S. Glasauer, M. Boyanov, B. Lai, and T. J. Beveridge, unpublished data).

This renaissance in of all types of microscopy is especially flourishing in geomicrobiology and will continue to develop with more and more precision. We are confident that future research will provide better structural and elemental analyses and that the geomicrobiological functional attributes of cellular partners within natural microbial communities will be better defined.

CHARACTERIZING THE MICROBE-MINERAL INTERFACE BY MINERALOGICAL TECHNIQUES

Having dealt with analyses of the aqueous components of geomicrobiological experiments and discussed ways to visualize the complex interactions between microbial cells and metals by using microscopy techniques, we now move on to discuss the major techniques that can be used to characterize the mineral phases that microorganisms interact with or form in these experiments. Indeed, recent decades have seen dramatic advances in the ability of environmental scientists to characterize the inorganic solid-phase components of natural systems, including those present in microbiological studies. For the most part, these phases are minerals (which by definition are nearly always crystalline) but also include amorphous solids, either synthetic or occurring as

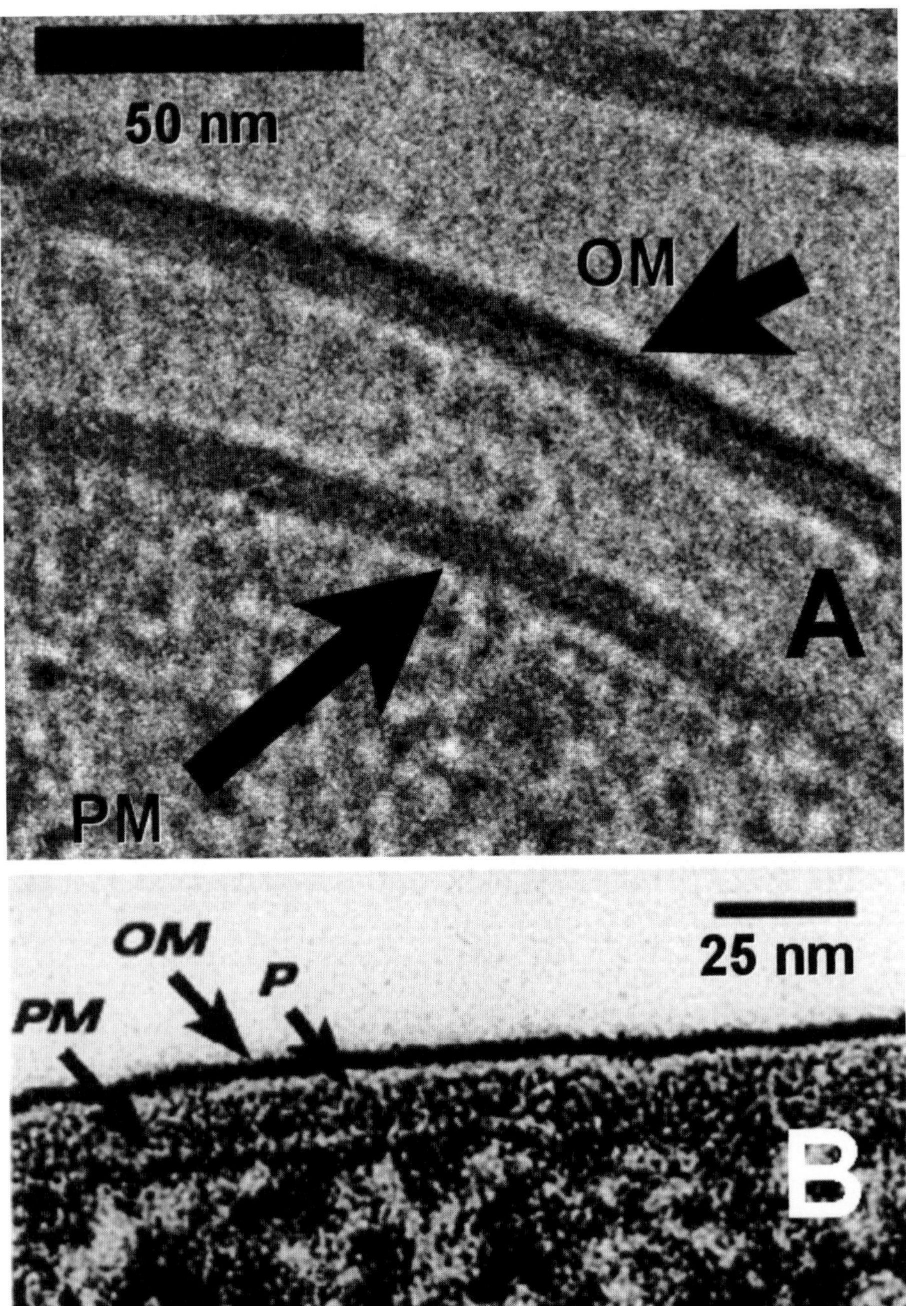

FIGURE 4 (A) Cell envelope of frozen hydrated cryo-sectioned *Escherichia coli* K-12 viewed by TEM. The lipid asymmetry of the outer membrane (OM) is clearly visible, as are both bilayer faces of the periplasm (PM). (B) A micrograph of a thin section of freeze-substituted *E. coli* is included for comparison, showing metal decoration of reactive sites.

suspended matter in natural waters or as coatings on surfaces in soils, sediments, and rocks. All are potentially important when studying the fate of metals and radionuclides in microbiological experiments, especially in microcosm-based studies using geological materials. Complete characterization involves establishing the bulk and minor element chemical composition of the phase and, if appropriate, information about its crystal structure. A further area of interest centers on the surface chemistry of the material, which may differ significantly from the bulk chemistry. Supplementary information on particle size and morphology may also be of interest.

The methods employed for such characterization range from long-established, routine (and sometimes low-cost) techniques to those costly methods requiring access to major national or international facilities, such as intense sources of photons (synchrotrons) or sources of neutron or proton beams. Broadly, the techniques involve either imaging, diffraction (of

X-rays or electrons), or some form of spectroscopy. Here the main methods are briefly reviewed, using relevant examples, beginning with more routine approaches before considering advanced techniques.

Microanalysis through Optical Microscopy, Electron Microscopy, Electron Probe Microanalysis (EPMA), and Particle-Induced X-Ray Emission (PIXE)

In addition to providing an invaluable tool for imaging microbial cells, optical microscopy can be used to characterize the geological component of "geomicrobiological" experiments. For example, a polarizing microscope can be used to examine thin and/or polished sections of rocks, sediments, and soils in transmitted or reflected light, and this is often the simplest means of identifying the mineral species present in a sample. It is also extremely sensitive in that a species occurring at a level of parts per thousand or even parts per million as a discrete phase can be identified. Resolution at the highest magnification, as noted above, is ~0.1 to 1 μm. Most important of all, optical microscopy reveals textural relationships between phases that can provide information on the sequences of formation (paragenesis) of minerals. Comprehensive introductions to transmitted light studies of mineral samples and to reflected light studies are provided by texts such as those of Kerr (61) and Craig and Vaughan (22), respectively. As discussed already, by replacing the light beam of an optical microscope with an electron beam, the level of resolution obtainable for imaging and semiquantitative chemical analysis, e.g., by EDS, is shifted by several orders of magnitude, allowing very high-resolution mapping and providing detailed compositional, morphological, textural, and structural information at micro- to nanoscales.

For a more quantitative chemical analysis, EPMA can be used to map a spot on a flat polished sample surface of only ~1μm in diameter (see reference 101 for a review) and, by rastering the electron beam over the sample surface, can map the distributions of most elements. Modern electron microprobe techniques have extended the capabilities of this instrument both in terms of rapid acquisition of large amounts of data and by making analysis of light elements (O, C, etc.) routinely possible and analysis of trace elements possible under certain circumstances (e.g., see reference 104).

PIXE is a technique that is related to EPMA, but instead of an incident electron beam, PIXE uses a focused beam of particles, usually protons, to excite the sample. This requires a particle accelerator, so PIXE is much less widely available than the electron beam methods. The major advantage of this method, related to the high mass of the proton compared with that of the electron, is the much better ratio of peak to background; hence, detection limits for many elements are as low as 1 ppm. PIXE can also be used to map element distributions and has better lateral resolution than EPMA. Most PIXE systems also have the capability of detecting backscattered protons, and this can be used to determine the major element composition at the surface and its variation with depth (depth profiling). This related technique, known as Rutherford backscattering, was one of the first surface analysis methods. Useful reviews of the PIXE technique are provided by Johansson et al. (57) and by Cabri and Campbell (19).

X-Ray Diffraction and Related Techniques

X-ray diffraction of samples which are made up of fine particles (powder samples) can provide definitive information for the identification of biogenic crystalline phases present at levels of a few percent or more. Of course, the diffraction of X-rays by (single) crystals was the route to determining the arrangements of atoms in solids ranging from simple minerals, such as halite (NaCl) or periclase (MgO), to complex organic materials, such as DNA. This is possible because the constructive interference of scattered waves can be observed at specific angles corresponding to positions where the path length difference between waves reflected off specific parallel planes of atoms are equal to an integer multiple of the wavelength of the incident radiation. This is expressed as Bragg's law:

$$n\lambda = 2d \sin\theta$$

where n is an integer, λ is the X-ray wavelength, d is the interplanar spacing, and θ is the angle of incidence. In basic laboratory sources, X-rays are generated in an X-ray tube, where electrons from a filament heated in a vacuum are accelerated at a metal target (Cu or Co), which produces X-rays of a characteristic energy (wavelength). Although laboratory sources are now much more intense than those available in the early years, a major advance took place with the development of synchrotron radiation. In these large (national or international) facilities, X-rays are emitted tangentially when a packet of electrons is accelerated at near light speed around a large-diameter (~20 to 100 m) storage ring. Synchrotron radiation is much more intense than that generated by conventional methods and, rather than being of a single energy (wavelength), is emitted as "white radiation" covering a wide energy range. Whereas in conventional X-ray diffractometers, the angle between the incident beam and the detector (θ) is varied by scanning during the measurement, in synchrotron experiments, λ can be selected or scanned by using a monochromator, or a wide range of energies can be collected simultaneously. This facilitates a wider range of experiments, as a full data set can be acquired in fractions of a second, enabling in situ studies of dynamic processes. The use of synchrotron radiation in microbiological experiments is covered in detail in chapter 97.

By far, the most commonly used X-ray diffraction method, and one requiring only a simple laboratory X-ray generator, is powder diffraction. Scanning of a diffractometer detector through a range of (2θ) angles leads to detection of peaks of X-ray intensity corresponding to diffraction by particular planes of atoms in any crystalline material present in the sample. Careful measurement of the positions and intensities of these peaks and comparison with a library of standard data provide a fingerprint identification of the phase or phases present. When more than one phase is present, each produces its own pattern of peaks with intensities proportional to the amount in the mixture. Generally, a phase needs to be present at >5% to be detected in this way, and there is a limit to the number of components that can be detected before the pattern becomes too complex through overlapping peaks. Nevertheless, powder X-ray diffraction is one of the first methods any geomicrobiologist would employ to characterize an unknown particulate material; it enables specific crystalline phases to be identified (with careful calibration, using known mixtures as standards) and allows quantitative determination of phase percentages. Further information is available in numerous references (e.g., reference 125).

X-ray diffraction methods can also be used to obtain information about the size distribution of crystalline biogenic particles in a sample, and with the intense X-ray sources now available, high-resolution powder data can even be used to determine crystal structures. As noted above, one of the great advantages of the intense white radiation synchrotron sources is the ability to collect data in a very short time and thus to study real-time structural transformations in situ. An example shown in Fig. 5 involves the transformation of a fine metal sulfide precipitate over a period of minutes and hours from an essentially amorphous material to one with an ordered structure.

X-Ray Spectroscopies

The availability of intense synchrotron sources has also made possible the development of techniques based on the absorption of photons, particularly X-rays, by a sample. The absorption of X-rays over a relatively small energy range as a result of the electrons in a particular shell (e.g., the K shell) of an element of interest being excited to a higher energy state, or to the continuum, is a powerful probe of the local environment around an atom. This is because the fine structure of the absorption spectrum both in the region of the edge—the XANES—and above the absorption edge—the EXAFS—is sensitive to the numbers, types, and distances of the shells of atoms surrounding the absorber element. These methods also have the advantage of working whether that absorber element is contained in a crystalline or amorphous solid, on a surface, or in a liquid or glass. A more detailed discussion of these very important methods is given elsewhere in this volume, as noted previously.

Nuclear Spectroscopies

Although the structures and properties of the nucleus of a particular atom are not generally of direct interest to us in characterizing biominerals and related phases, certain nuclear spectroscopies can be very sensitive to the electronic structure and overall environment of the atom concerned. Two techniques are particularly worth mentioning, namely, Mössbauer spectroscopy and nuclear magnetic resonance (NMR) spectroscopy.

Mössbauer Spectroscopy

Mössbauer spectroscopy involves the recoilless emission and resonant absorption of gamma rays by a target nucleus. Although the Mössbauer effect can be observed in over 20 elements, practical difficulties make measurements in more than just a few extremely rare. In fact, only one element, iron, accounts for the great majority of studies. Of course, iron is an extremely important element in both geochemical and biological systems. Techniques to study the aerobic and anoxic redox cycling of Fe(II) and Fe(III) are discussed in several chapters in this manual, and the Mössbauer technique can help to characterize the Fe minerals that are metabolized or formed through these processes. The absorption by the ^{57}Fe nucleus (2% natural abundance) of a 14.4-keV gamma ray emitted by a ^{57}Co source occurs by the nucleus going from a ground state to an excited state. The

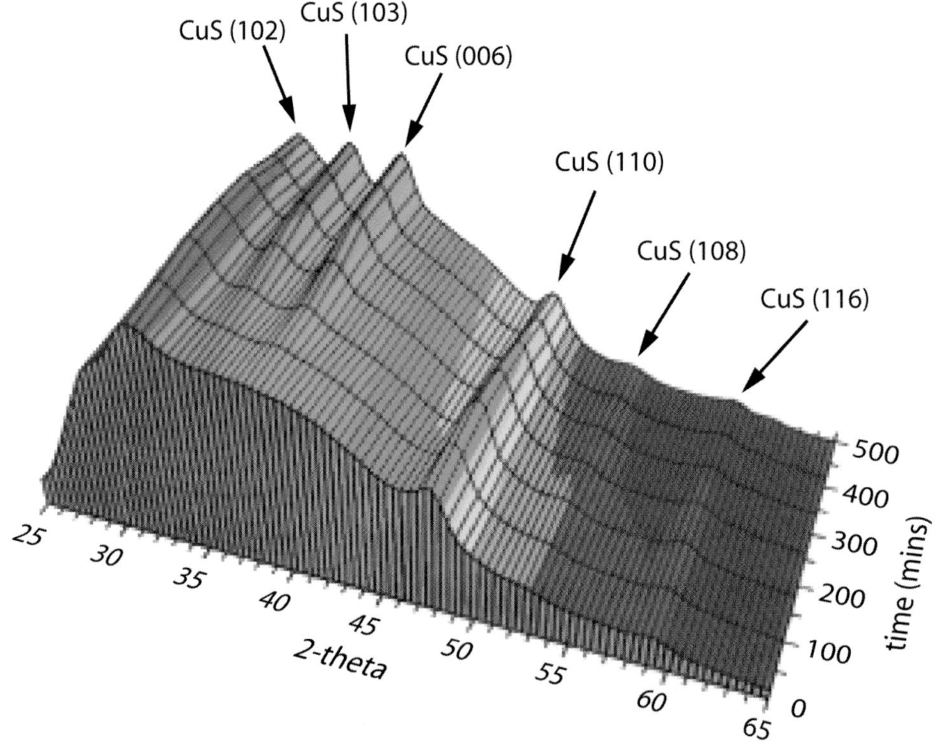

FIGURE 5 Diagram showing energy-dispersive X-ray powder diffraction data acquired using a synchrotron radiation source. A fine-particle precipitate of CuS formed by reacting copper and sulfide in aqueous solutions is seen to change over time (0 to 500 min) as the initially nearly amorphous sulfide precipitate becomes more crystalline. Diffraction peaks related to different parts of (or planes within) the structure are seen to emerge with time (L. N. Moyes, J. M. Charnock, R. A. D. Pattrick, and D. J. Vaughan, unpublished data).

position of the absorption peak relative to a standard, and whether it is a single peak or a doublet or sextet, depends upon factors such as the oxidation state and coordination environment of the iron in the sample and (for a sextet) if it is magnetically ordered. It is only the iron in the sample that gives rise to the spectrum, although iron in different crystallographic sites in a single mineral, or in different phases, will all contribute to the overall spectrum, with a peak intensity proportional to its percentage of presence in the sample. Because spectral resolution is improved at low temperatures and because magnetic properties are temperature dependent, it is often useful to record spectra at liquid nitrogen (~77 K) or even liquid helium (~4.2 K) temperatures as well as at room temperature. Like X-ray absorption spectroscopies, Mössbauer spectra can be obtained from amorphous and glassy materials as well as crystalline solids (but not from liquids). In the context of geomicrobiological studies, Mössbauer spectroscopy offers a powerful way to determine (or to monitor changing) Fe^{2+}-to-Fe^{3+} ratios and the distribution of iron between different phases in a system (including the poorly crystalline or amorphous oxyhydroxides or sulfides often present in natural sediments and soils). An example is given in Figure 6. Good sources of further information are references 81 and 49.

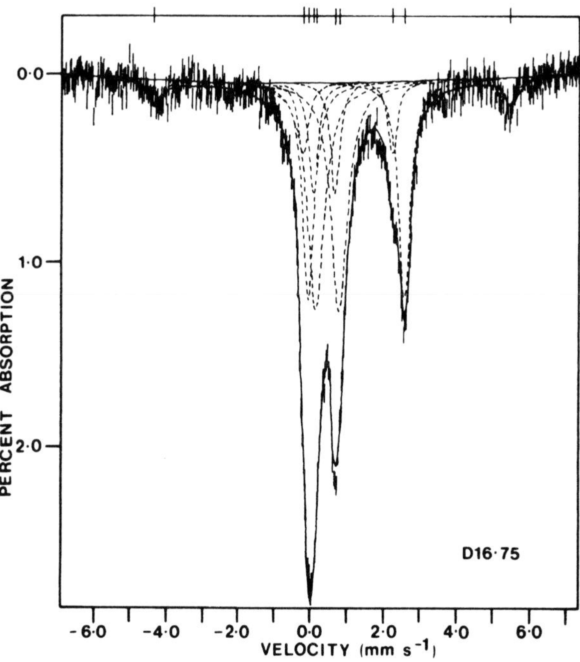

FIGURE 6 Mössbauer spectrum (iron) acquired at room temperature for a fine-particle tidal flat sediment sample from the Wash area, England. Fitting of this spectrum confirmed that ferrous iron occurs in detrital clay minerals (doublet with large separation) and in pyrite (FeS_2) (doublet with small separation) and that both ferrous and ferric iron occur in the magnetic sulfide greigite (Fe_3S_4) (small features at very high and very low velocities). There is also a substantial amount of ferric iron from clays and related phases (seen as another doublet with small separation). The sulfides have formed by reaction of bacterially generated sulfide (reduction of seawater sulfate) with iron released from detrital minerals. From the areas under the peaks, the relative percentages of the different types of iron can be determined. (Modified from reference 111.)

Nuclear Magnetic Resonance

NMR involves measuring the absorption of electromagnetic radiation by the magnetic nuclei of certain atoms in the presence of an externally applied magnetic field. The simplest type of experiment is generally performed by applying a fixed electromagnetic field to the sample, usually in the radiofrequency region, and varying the applied magnetic field until the radiation is most strongly absorbed. Most NMR work conducted by chemists has involved the ^{1}H atom, where the position of the resonance signal with respect to a standard, along with any fine structure and information on width and shape of peaks, can be used to identify protons in different positions in a molecule and hence help to determine the molecular structure. Such work is not performed on solid samples, which give broad and poorly resolved spectra, but a variant of the standard NMR experiment has been used to study crystalline solids. Here the sample held in the magnetic field is spun in a specific orientation (the "magic angle") to give high-resolution data on nuclei of atoms such as ^{29}Si and ^{27}Al and to provide insights into the crystal chemistry of silicate and aluminosilicate minerals by pinpointing different environments for these elements in their structures. Although a significant number of other nuclei are potentially accessible to studies by NMR methods, few have yet been studied in detail. A group of techniques based on the NMR phenomenon involves mapping in three dimensions (tomography) the distribution of a target nucleus and, hence, atom. This is the basis of the body scanner instruments now widely used in medical diagnosis.

IR and Raman (Vibrational) Spectroscopies

Energy in the infrared (IR) (~0.75 to 100 μm) region of the electromagnetic spectrum can be absorbed by material through vibrational processes (the bending and stretching of the bonds between atoms), and this can be observed in gaseous, liquid, or solid phases. In the simplest case, measurement of the positions and intensities of absorption peaks as a function of energy (wavelength) can provide a fingerprint identification of a particular molecule or crystalline material present in a sample. IR spectra can be measured in absorption mode, allowing the beam to pass through the sample in the case of gaseous, liquid, and particulate (or, more rarely, single crystal) materials, or reflection mode by reflecting off the surface of a solid crystal or compressed powder. Surface coatings can also be studied in reflection mode. More detailed analysis of an IR spectrum involves associating individual peaks with the bending or stretching of particular bonds in the material under investigation. This can provide information on molecular or crystalline structure or on chemical changes taking place in a system with time or with changing conditions. Recent decades have seen the development of equipment enabling imaging of samples in the IR region (IR microscopes) to facilitate the mapping of particular components in complex samples. Vibrational spectra can also be studied by Raman spectroscopy. For this technique, monochromatic light from a suitable source (a powerful laser in modern instruments) is directed through a translucent sample (gas, liquid, glass, crystalline solid, or powder), and the light scattered at 90° or 180° is recorded. The scattered light is of two kinds, either Rayleigh, where no frequency change occurs, or Raman, which involves a change of frequencies arising from vibrational effects. As with IR spectroscopy, Raman spectra can be used to provide fingerprint identification of particular materials or more detailed information on molecular or crystal structure or elastic or thermodynamic properties. It is another method which has been refined over the years, for example, to study spectral signals from very small areas using a Raman microprobe.

Imaging and Analysis of Surfaces

The last 20 years have seen a rapidly growing awareness of the importance of surfaces and interfaces in controlling processes in natural systems. The mineral surface and its interface with a solution, and in many cases some form of microorganism (prokaryotic or fungal), are critical to understanding reactivity, including mineral dissolution or precipitation and the sorption and desorption of species from the solution. This awareness has been accompanied by the development of techniques for the imaging of surfaces and the mapping and analysis of those surfaces at resolutions from micrometers to nanometers and even angstroms. Given the importance of a molecular-scale understanding of phenomena such as mineral-microbe interactions or the nature and formation of biofilms on mineral surfaces, it is appropriate to briefly review the key techniques of surface imaging and analysis relevant to studying such systems.

AFM and STM

AFM and scanning tunneling microscopy (STM) are the most important members of a family of methods (scanning probe microscopies) first developed about 20 years ago which have revolutionized surface science. These techniques are like no other that we have described so far, since no lenses are involved (33, 34). STM functions by using a piezoelectric device to scan a very fine stylus-like tip across the flat surface of a conducting material whilst applying an electrical potential. Monitoring the variations in the current passing between the sample surface and the tip at a constant potential during scanning (or in the potential at a constant current) enables an image of the surface to be acquired. In the very clean environment of a UHV chamber, this image of the electron density distribution can be at atomic resolution. In Fig. 7, such an image of the (111) crystal face of magnetite (Fe_3O_4) with individual molecules of pyridine sorbed to the surface is shown.

The AFM technique involves the same kind of scanner system, but in this case the very small tip is mounted on the end of a flexible arm (cantilever), and movement of the tip towards or away from the surface during scanning is measured by reflecting a laser beam off the end of the arm. Minute forces of attraction and repulsion are detected and used to construct a topographical profile of the specimen in *x*, *y*, and *z* dimensions (56). Since the tip can be made almost atomically sharp and since movement is precisely controlled by piezoceramic voltage input, molecular resolution is possible in UHV environments. Somewhat lower-resolution studies (nanometer to micrometer) can be conducted in air or with the surface beneath a layer of fluid. In work of microbiological relevance, surfaces with attached microbes can be imaged, and measurements can be made of forces of adhesion as well as surface topography (Fig. 8). A particular strength of the scanning probe microscopy methods is that three-dimensional information is obtained about surfaces; a weakness is that they do not directly provide any chemical (elemental or species analysis) information. Also, the surfaces of microorganisms are soft, and the bonding forces between them and minerals are weak, which makes accurate AFM of such interactions extremely difficult. Yet, over the last decade, AFMs have become more common in research centers and are more frequently available for use.

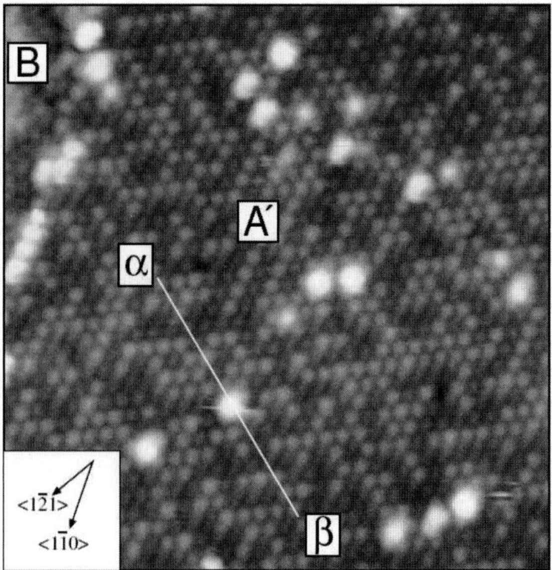

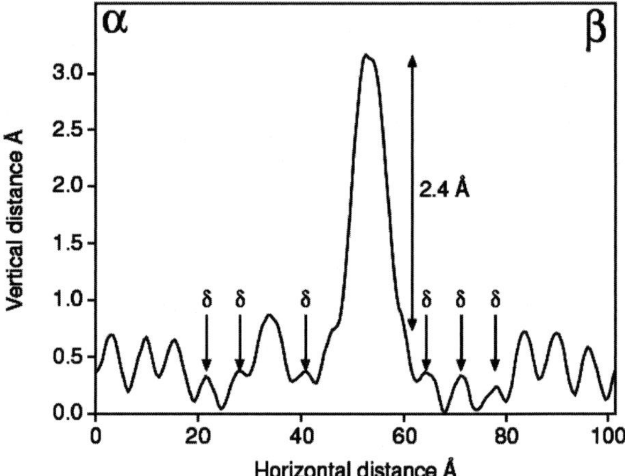

FIGURE 7 STM data showing a 200-Å-by-200-Å image of the surface of magnetite at atomic resolution which has been exposed to pyridine (111) (top) and a scan along the line indicated in the top figure which shows that the large bright features in the image are individual pyridine molecules (bottom). (Modified from reference 24.)

AFM is not restricted to imaging because the use of AFM tips on easily deformed cantilevers can provide an exquisite method for measuring forces of adhesion. Indeed, this AFM measurement of force (or adhesive energies) is possibly as important as AFM imaging. For example, methods have been worked out for attaching a bacterium to an AFM tip so that the cell can be touched slowly to a mineral surface, and the (initial) repulsive and (eventual) adhesive forces can be measured (76, 78, 79). It is also possible to measure the deformability or elasticity of microbial surfaces (e.g., see reference 123). These adhesion and elasticity forces approach the limits of AFM detection (i.e., tens of piconormal), but they can be measured. Another exciting possibility with AFM is the measurement of ionized sites (or electrostatics)

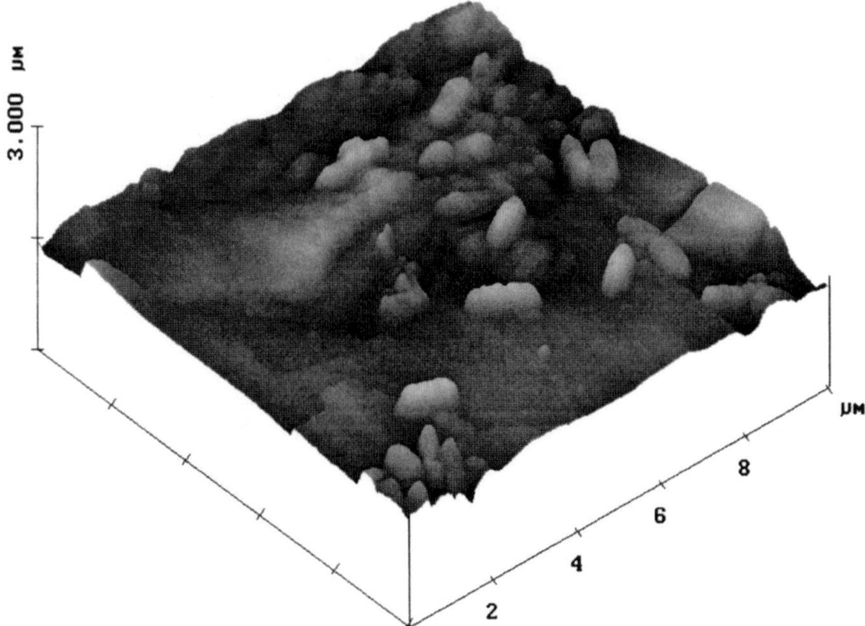

FIGURE 8 AFM image showing G. *sulfurreducens* on a stepped surface of goethite (Wilkins et al., unpublished data).

on the surface of a single cell (109). In this case, small changes in the amperage (microamperes to picoamperes) of the tip are monitored as it travels over the surface, and extremely small current alterations give an indication of exposed reactive sites.

XPS and AES

X-ray photoelectron spectroscopy (XPS) and auger electron spectroscopy (AES) are really the "workhorse" methods of surface chemical analysis. In these closely related techniques, normally carried out using the same instrument, a sample contained within a UHV chamber is bombarded with monoenergetic photons (usually Mg*K*α or Al*K*α X-rays), causing the ejection of electrons from core and valence orbitals (Fig. 9). The kinetic energies of these electrons are determined in an energy analyzer to yield an X-ray photoelectron spectrum. In AES, the energy of a secondary electron (a so-called auger electron) is measured. This is an electron ejected during the relaxation of the system following ejection of a photoelectron, when an electron in a higher energy orbital drops down to fill a hole in an inner shell. This process is an alternative to emission of X-rays of characteristic energy (as measured in techniques such as EPMA). In XPS, the electron kinetic energies that are measured are directly related to the binding energies of electrons in the material being studied, and hence to the specific elements present. Peak positions provide a means of identifying the elements, and peak intensities give a semiquantitative measure of their amounts. Because electrons can only escape from the topmost layers (a few nanometers) of a material, this offers a sensitive method of surface analysis. In favorable cases, precise peak position also varies with element speciation, enabling further information to be gained (e.g., not only identifying sulfur but distinguishing sulfide from sulfate or ferric from ferrous iron). Rastering of the incident beam also enables mapping of surface chemistry at the micrometer level.

Secondary Ion Mass Spectrometry

Secondary ion mass spectrometry (SIMS) is the other long-established technique employed in surface analysis. Of course, mass spectrometry, the identification and analysis of quantities and ratios of the isotopes of a specific element or of molecular entities by discrimination on the basis of atomic or molecular mass, has numerous applications in the natural sciences. SIMS methods involve bombarding the surface of a sample with a beam of ions, thus causing the ejection of secondary ions from surface layers. These secondary ions are

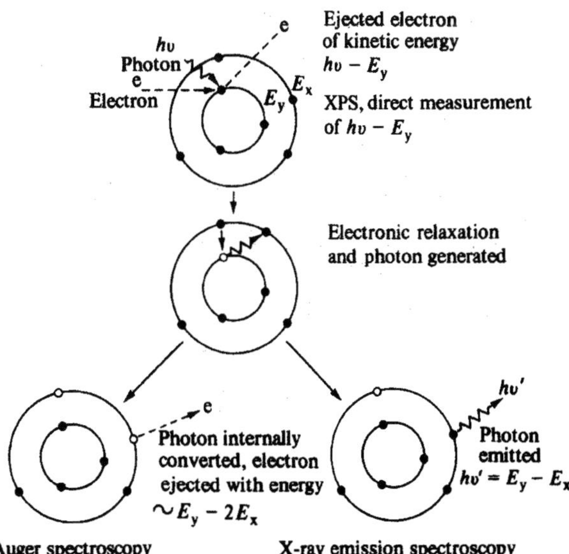

FIGURE 9 Schematic diagram showing the principles of XPS, AES, and X-ray emission spectroscopy (XES), which is also the basis of EPMA. (Redrawn from reference 118.)

taken into a spectrometer, where determination of their masses enables identification of atomic or molecular species and, from the intensities of peaks in the mass spectrum, their percent contributions. In this way, the chemical composition of the surface can be determined. Because the bombardment of the surface removes material from deeper layers as the process continues, the SIMS technique can provide a profile of compositional changes as a function of depth beneath the surface (generally on scales from nanometers to micrometers). As with XPS and AES, rastering of the beam can be employed to image areas of a surface as well as to produce point analyses. A disadvantage of SIMS has been that the incident beam may cause changes in the chemistry of the sample, particularly if it comprises organics or biomaterials, or may disrupt the surface so as to compromise any attempts at depth profiling. However, recent developments involving new types of incident beams to bombard the sample (e.g., a beam of C^{60} molecules) appear set to revolutionize this technique and to enable delicate samples to be analyzed at nanometer resolution. Further details can be found in references 80 and 86 and references therein.

CONCLUSIONS

The study of metal-microbe interactions has advanced rapidly over the last 2 decades, and new discoveries in this area continue apace. Application of traditional microbiological approaches combined with multidisciplinary techniques, such as spectroscopy, radiochemistry, geochemistry, mineralogy, and microscopy, is leading to new mechanistic insights into the interactions between metals and microorganisms. In turn, this has resulted in the exciting realization that microbes may be used to bioremediate land and water contaminated with metals and radionuclides, and this has already led to field trials for a range of in situ bioremediation approaches (69). Nowhere is this more clearly illustrated than in attempts to deal with our global nuclear waste legacy, where the importance of understanding the science underpinning environmental radionuclide mobility is clear and is supported by large research programs in the United States, Europe, and Japan. Other high-priority areas that will benefit from these new "geomicrobiological" insights include our continued efforts to supply nontoxic water for drinking and irrigation, especially in the developing world. A complex mix of environmental, commercial, and societal drivers will undoubtedly continue to push these important research agendas, facilitated by continued improvements in the technologies needed to probe interactions between microorganisms and minerals at lower concentrations in the nanoscale range.

J.R.L. and D.J.V. acknowledge the financial support of NERC for studies on the interactions between Fe(III)-reducing bacteria and Fe(III) oxides.

REFERENCES

1. **Addleman, R. S., M. Carrott, C. M. Wai, T. E. Carleson, and B. W. Wenclawiak.** 2001. On-line speciation of uranyl chelates in supercritical CO_2 by time-resolved laser-induced fluorescence spectroscopy. *Anal. Chem.* **73:**1112–1119.
2. **Anderson, R. T., H. A. Vrionis, I. Ortiz-Bernad, C. T. Resch, P. E. Long, R. Dayvault, K. Karp, S. Marutzky, D. R. Metzler, A. Peacock, D. C. White, M. Lowe, and D. R. Lovley.** 2003. Stimulating the in situ activity of *Geobacter* species to remove uranium from the groundwater of a uranium-contaminated aquifer. *Appl. Environ. Microbiol.* **69:**5884–5891.
3. **Ariza, J. L. G., E. Morales, D. Sanchez-Rodas, and I. Giraldez.** 2000. Stability of chemical species in environmental matrices. *Trends Anal. Chem.* **19:**200–209.
4. **Bailey, E. H., J. F. W. Mosselmans, and P. F. Schofield.** 2004. Uranyl acetate speciation in aqueous solutions—an XAS study between 25°C and 250°C. *Geochim. Cosmochim. Acta* **68:**1711–1722.
5. **Bancon-Montigny, C., G. Lespes, and M. Potin-Gautier.** 2001. Optimisation of the storage of natural freshwaters before organotin speciation. *Water Res.* **35:**224–232.
6. **Batley, G. E., S. C. Apte, and J. L. Stauber.** 2004. Speciation and bioavailability of trace metals in water: progress since 1982. *Aust. J. Chem.* **57:**903–919.
7. **Baun, D. L., and T. H. Christensen.** 2004. Speciation of heavy metals in landfill leachate: a review. *Waste Manage. Res.* **22:**3–23.
8. **Bednar, A. J., J. R. Garbarino, and J. F. Ranville.** 2002. Preserving the distribution of inorganic arsenic species in groundwater and acid mine drainage samples. *Environ. Sci. Technol.* **36:**2213–2218.
9. **Beveridge, T. J.** Cryotransmission electron microscopy is enabling investigators to examine native, hydrated structures in bacteria and biofilms. *Microbe*, in press.
10. **Beveridge, T. J.** 1999. Structures of gram-negative cell walls and their derived membrane vesicles. *J. Bacteriol.* **181:**4725–4733.
11. **Beveridge, T. J., and L. L. Graham.** 1991. Surface layers of bacteria. *Microbiol. Mol. Biol. Rev.* **55:**684–705.
12. **Beveridge, T. J., M. N. Hughes, H. Lee, K. T. Leung, R. K. Poole, I. Savvaidis, S. Silver, and J. T. Trevors.** 1997. Metal-microbe interactions: contemporary approaches. *Adv. Microb. Physiol.* **38:**177–243.
13. **Beveridge, T. J., D. Moyles, and B. Harris.** Electron microscopy. *In* C. A. Reddy, T. J. Beveridge, J. A. Breznak, L. Snyder, T. M. Schmidt, and G. A. Marzluf (ed.), *Methods for General and Molecular Microbiology*, in press. ASM Press, Washington, D.C.
14. **Bondietti, E. A., and C. W. Francis.** 1979. Geologic migration potentials of technetium-99 and neptunium-237. *Science* **203:**1337–1340.
15. **Brina, R., and A. G. Miller.** 1992. Direct detection of trace levels of uranium by laser-induced phosphorimetry. *Anal. Chem.* **64:**1413–1418.
16. **Buffle, J., and M.-L. Tercier-Waeber.** 2005. Voltammetric environmental trace-metal analysis and speciation: from laboratory to in situ measurements. *Trends Anal. Chem.* **24:**171–191.
17. **Burke, I., C. Boothman, J. R. Lloyd, R. J. G. Mortimer, F. R. Livens, and K. Morris.** 2005. Technetium solubility during the onset of progressive anoxia. *Environ. Sci. Technol.* **39:**4109–4116.
18. **Cabon, J. Y., and N. Cabon.** 2000. Determination of arsenic species in seawater from flow injection hydride generation in situ collection followed by graphite furnace atomic absorption spectroscopy—stability of As(III). *Anal. Chim. Acta* **418:**19–31.
19. **Cabri, L. J., and J. L. Campbell.** 1998. The proton microprobe in ore mineralogy (micro-PIXE technique), p. 181–198. *In* L. J. Cabri and D. J. Vaughan (ed.), *Modern Approaches to Ore and Environmental Mineralogy*, vol. 27. *Mineralogical Association of Canada Short Course Series.* Mineralogical Association of Canada, Ottawa, Ontario, Canada.
20. **Clark, D. L., D. E. Hobart, and M. P. Neu.** 1995. Actinide carbonate complexes and their importance in actinide environmental chemistry. *Chem. Rev.* **95:**25–48.

21. **Combes, J. M., C. J. Chisholm-Brause, G. E. Brown, G. A. Parks, S. D. Conradson, P. G. Eller, I. R. Triay, D. E. Hobart, and A. Meijer.** 1992. EXAFS spectroscopic study of neptunium(V) sorption at the alpha-FeOOH water interface. *Environ. Sci. Technol.* **26:**376–382.
22. **Craig, J. R., and D. J. Vaughan.** 1994. *Ore Microscopy and Ore Petrography*, 2nd ed. John Wiley & Sons, New York, N.Y.
23. **Cullen, W. R., and K. J. Reimer.** 1989. Arsenic speciation in the environment. *Chem. Rev.* **89:**713–764.
24. **Cutting, R. S., C. A. Muryn, G. Thornton, and D. J. Vaughan.** Molecular scale reactivity of magnetite with formic acid, pyridine and carbon tetrachloride. *Geochim. Cosmochim. Acta*, in press.
25. **Dahlen, J., and S. Karlsson.** 1999. Determination of iron(II) in natural waters by capillary zone electrophoresis using on-capillary complexation with 2,4,6-tri(2′-pyridyl)-1,3,5-triazine. *J. Chromatogr. A* **848:**491–502.
26. **Davison, W., and H. Zhang.** 1994. In situ speciation measurements of trace components in natural waters using thin-film gels. *Nature* **367:**546–548.
27. **Delacroix, D., J. P. Guerre, P. Leblanc, and C. Hickman.** 1998. Radionuclide and radiation protection data handbook. *Radiat. Prot. Dosimetry* **76:**1–126.
28. **Dozol, M., and R. Hagemann.** 1993. Radionuclide migration in groundwaters: review of the behaviour of actinides. *Pure Appl. Chem.* **65:**1081–1102.
29. **Dubochet, J., M. Adrian, J. J. Chang, J. C. Homo, J. Lepault, A. W. McDowall, and P. Schultz.** 1988. Cryo-electron microscopy of vitrified specimens. *Q. Rev. Biophys.* **21:**129–228.
30. **Dubochet, J., A. W. McDowall, B. Menge, E. N. Schmid, and K. G. Lickfeld.** 1983. Electron microscopy of frozen-hydrated bacteria. *J. Bacteriol.* **155:**381–390.
31. **Duff, M. C., D. B. Hunter, I. R. Triay, P. M. Bertsch, J. Kitten, and D. T. Vaniman.** 2001. Comparison of two micro-analytical methods for detecting the spatial distribution of sorbed Pu on geologic materials. *J. Contam. Hydrol.* **47:**211–218.
32. **Duff, M. C., D. B. Hunter, I. R. Triay, P. M. Bertsch, D. T. Reed, S. R. Sutton, G. Shea-McCarthy, J. Kitten, P. Eng, S. J. Chipera, and D. T. Vaniman.** 1999. Mineral associations and average oxidation states of sorbed Pu on Tuff. *Environ. Sci. Technol.* **33:**2163–2169.
33. **Dufrêne, Y. F.** Atomic force microscopy. *In* C. A. Reddy, T. J. Beveridge, J. A. Breznak, L. Snyder, T. M. Schmidt, and G. A. Marzluf (ed.), *Methods for General and Molecular Microbiology*, in press. ASM Press, Washington, D.C.
34. **Dufrêne, Y. F.** 2004. Using nanotechniques to explore microbial surfaces. *Nat. Rev. Microbiol.* **2:**451–460.
35. **Feldmann, J.** 2005. What can the different current-detection methods offer for element speciation? *Trends Anal. Chem.* **24:**228–242.
36. **Feldmann, J., K. Haas, L. Naels, and S. Wehmeier.** 2001. Investigations into biovolatilisation of metal(loids) in the environment by using GC-ICP-TOF-MS. *R. Soc. Chem. Spec. Publ.* **267:**361–368.
37. **Finneran, K. T., R. T. Anderson, K. P. Nevin, and D. R. Lovley.** 2002. Potential for bioremediation of uranium-contaminated aquifers with microbial U(VI) reduction. *Soil Sed. Contam.* **11:**339–357.
38. **Finneran, K. T., M. E. Housewright, and D. R. Lovley.** 2002. Multiple influences of nitrate on uranium solubility during bioremediation of uranium-contaminated subsurface sediments. *Environ. Microbiol.* **4:**510–516.
39. **Florence, T. M., and G. E. Batley.** 1980. Chemical speciation in natural waters. *Rev. Anal. Chem.* **9:**219–296.
40. **Frausto da Silva, J. J. R., and R. J. P. Williams.** 1993. *The Biological Chemistry of the Elements*. Clarendon Press, Oxford, United Kingdom.
41. **Fredrickson, J. K., J. M. Zachara, D. W. Kennedy, M. C. Duff, Y. A. Gorby, S.-M. W. Li, and K. M. Krupka.** 2000. Reduction of U(VI) in geothite (a-FeOOH) suspensions by a dissimilatory metal-reducing bacterium. *Geochim. Cosmochim. Acta* **64:**3085–3098.
42. **Fredrickson, J. K., J. M. Zachara, D. W. Kennedy, R. K. Kukkadapu, J. P. McKinley, S. M. Heald, C. Liu, and A. E. Plymale.** 2004. Reduction of TcO_4^- by sediment-associated biogenic Fe(II). *Geochim. Cosmochim. Acta* **68:**3171–3187.
43. **Gault, A. G., L. E. Davidson, P. R. Lythgoe, D. A. Polya, F. R. Abou-Shakra, H. J. Walker, and D. Chatterjee.** 2003. Iron and arsenic speciation in groundwaters from West Bengal, India by coupled HPLC-ICP-MS utilising a hexapole collision cell, p. 112–126. *In* J. G. Holland and S. D. Tanner (ed.), *Plasma Source Mass Spectrometry: Applications and Emerging Technologies*. The Royal Society of Chemistry, Cambridge, United Kingdom.
44. **Gault, A. G., J. Jana, S. Chakraborty, P. Mukherjee, M. Sarkar, B. Nath, D. A. Polya, and D. Chatterjee.** 2005. Preservation strategies for inorganic arsenic species in high iron, low Eh groundwater from West Bengal, India. *Anal. Bioanal. Chem.* **381:**347–353.
45. **Graham, L. L., and T. J. Beveridge.** 1994. Structural differentiation of the *Bacillus subtilis* 168 cell wall. *J. Bacteriol.* **176:**1413–1421.
46. **Greenberg, A. E., and A. D. E. Eaton.** 1999. *Standard Methods for the Examination of Water and Wastewater,* 20th ed. American Public Health Association, Washington, D.C.
47. **Hall, G. E. M., J. C. Pelchat, and G. Gauthier.** 1999. Stability of inorganic arsenic(III) and arsenic(V) in water samples. *J. Anal. Atom. Spectrosc.* **14:**205–213.
48. **Hamilton, E. I.** 1981. Alpha particle radioactivity of hot particles from the Esk Estuary. *Nature* **290:**690–693.
49. **Hawthorne, F. E.** 1988. *Spectroscopic Methods in Mineralogy and Geology*, vol. 18. Mineralogical Society of America, Washington, D.C.
50. **Heninger, I., M. Potin Gautier, I. deGregori, and H. Pinochet.** 1997. Storage of aqueous solutions of selenium for speciation at trace level. *Fresenius J. Anal. Chem.* **357:**600–610.
51. **Hobot, J. A., E. Carlemalm, W. Villiger, and E. Kellenberger.** 1984. Periplasmic gel: new concept resulting from the reinvestigation of bacterial cell envelope ultrastructure by new methods. *J. Bacteriol.* **160:**143–152.
52. **Holland, G., and S. D. Tanner (ed.).** 2003. *Plasma Source Mass Spectrometry: Applications and Emerging Technologies*, vol. 288. The Royal Society of Chemistry, Cambridge, United Kingdom.
53. **Hunter, R. C., and T. J. Beveridge.** 2005. Application of a pH-sensitive fluoroprobe (C-SNARF-4) for pH microenvironment analysis in *Pseudomonas aeruginosa* biofilms. *Appl. Environ. Microbiol.* **71:**2501–2510.
54. **Islam, F., A. G. Gault, C. Boothman, D. A. Polya, J. M. Charnock, D. Chatterjee, and J. R. Lloyd.** 2004. Role of metal-reducing bacteria in arsenic release from Bengal Delta sediments. *Nature* **430:**68–71.
55. **Istok, J., J. Senko, L. Krumholz, D. Watson, M. Bogle, A. Peacock, Y. Chang, and D. White.** 2004. In situ bioreduction of technetium and uranium in a nitrate-contaminated aquifer. *Environ. Sci. Technol.* **38:**468–475.
56. **Jericho, M., O. Stoica, and T. J. Beveridge.** 2004. Atomic force microscopy of cell growth and division in *Staphylococcus aureus*. *J. Bacteriol.* **186:**2281–2287.
57. **Johansson, S. A. E., J. L. Campbell, and K. G. Malmqvist.** 1995. *Particle-Induced X-Ray Emission Spectrometry*, vol. 133. John Wiley & Sons, New York, N.Y.

58. **Johnson, D. A., and T. M. Florence.** 1971. Spectrophotometric determination of uranium(VI) with 2-(5-bromo-2-pyridylazo)-5-diethylaminophenol. *Anal. Chim. Acta* **53:**73–79.
59. **Keith-Roach, M. J., and F. R. Livens.** 2002. *Interactions of Microorganisms with Radionuclides*. Elsevier, Amsterdam, The Netherlands.
60. **Keith-Roach, M. J., K. Morris, and H. Dahlgaard.** 2003. An investigation into technetium binding in sediments. *Mar. Chem.* **81:**149–162.
61. **Kerr, P. F.** 1977. *Optical Mineralogy*, 4th ed. McGraw Hill, New York, N.Y.
62. **Koval, S. F., and T. J. Beveridge.** 1999. Electron microscopy, p. 276–287. *In* J. Lederberg (ed.), *Encyclopedia of Microbiology*. Academic Press, San Diego, Calif.
63. **Krachler, M., H. Emons, and J. Zheng.** 2001. Speciation of antimony for the 21st century: promises and pitfalls. *Trends Anal. Chem.* **20:**79–90.
64. **Kuczewski, B., C. M. Marquardt, A. Seibert, H. Geckeis, J. V. Kratz, and N. Trautmann.** 2003. Separation of plutonium and neptunium species by capillary electrophoresis-inductively coupled plasma-mass spectrometry and application to natural groundwater samples. *Anal. Chem.* **75:**6769–6774.
65. **Lawrence, J. R., and T. R. Neu.** 2006. Laser scanning microscopy. *In* C. A. Reddy, T. J. Beveridge, J. A. Breznak, L. Snyder, T. M. Schmidt, and G. A. Marzluf (ed.), *Methods for General and Molecular Microbiology*, in press. ASM Press, Washington, D.C.
66. **Le, X. C., S. Yalcin, and M. Ma.** 2000. Speciation of submicrogram per litre levels of arsenic in water: on-site species separation integrated with sample collection. *Environ. Sci. Technol.* **34:**2342–2347.
67. **Liang, P., and J. Li.** 2005. Speciation of chromium with cloud point extraction separation and determination by ICP-OES. *Atom. Spectrosc.* **26:**89–93.
68. **Lieser, K. H., and U. Muhlenweg.** 1988. Np in the hydrosphere and in the geosphere. *Radiochim. Acta* **43:**27–35.
69. **Lloyd, J. R., R. T. Anderson, and L. E. Macaskie.** 2005. Bioremediation of metals and radionuclides, p. 293–317. *In* R. M. Atlas and J. C. Philp (ed.), *Bioremediation: Applied Microbial Solutions for Real-World Environmental Cleanup*. ASM Press, Washington, D.C.
70. **Lloyd, J. R., J. A. Cole, and L. E. Macaskie.** 1997. Reduction and removal of heptavalent technetium from solution by *Escherichia coli*. *J. Bacteriol.* **179:**2014–2021.
71. **Lloyd, J. R., and L. E. Macaskie.** 1996. A novel phosphorimager-based technique for monitoring the microbial reduction of technetium. *Appl. Environ. Microbiol.* **62:**578–582.
72. **Lloyd, J. R., and J. C. Renshaw.** 2005. Bioremediation of radioactive waste: radionuclide-microbe interactions in laboratory and field-scale studies. *Curr. Opin. Biotechnol.* **16:**254–260.
73. **Lloyd, J. R., and J. C. Renshaw.** 2005. Microbial transformations of radionuclides: fundamental mechanisms and biogeochemical implications. *Met. Ions Biol. Syst.* **43:**205–240.
74. **Lloyd, J. R., V. A. Sole, C. V. Van Praagh, and D. R. Lovley.** 2000. Direct and Fe(II)-mediated reduction of technetium by Fe(III)-reducing bacteria. *Appl. Environ. Microbiol.* **66:**3743–3749.
75. **Lloyd, J. R., P. Yong, and L. E. Macaskie.** 2000. Biological reduction and removal of pentavalent Np by the concerted action of two microorganisms. *Environ. Sci. Technol.* **34:**1297–1301.
76. **Loewenschuss, H.** 1982. Metal-ferrocyanide complexes for the decontamination of cesium from aqueous radioactive waste. *Radioact. Waste Manag.* **2:**327.
77. **Lovley, D. R., E. J. P. Phillips, Y. A. Gorby, and E. Landa.** 1991. Microbial reduction of uranium. *Nature* **350:**413–416.
78. **Lower, B. H., R. Yongsunthon, F. P. Vellano III, and S. K. Lower.** 2005. Simultaneous force and fluorescence measurements of a protein that forms a bond between a living bacterium and a solid surface. *J. Bacteriol.* **187:** 2127–2137.
79. **Lower, S. K., C. J. Tadanier, and M. F. Hochella.** 2001. Dynamics of the mineral-microbe interface: use of biological force microscopy in biogeochemistry and geomicrobiology. *Geomicrobiol. J.* **18:**63–76.
80. **Macrae, N. D.** 1995. Secondary ion mass spectrometry and geology. *Can. Mineral.* **33:**219–236.
81. **Maddock, A. G.** 1985. *Mossbauer Spectroscopy in Mineral Chemistry, Chemical Bonding and Spectroscopy in Mineral Chemistry*. Chapman and Hall, London, United Kingdom.
82. **Matias, V. R. F., A. Al-Amoudi, J. Dubochet, and T. J. Beveridge.** 2003. Cryo-transmission electron microscopy of frozen-hydrated sections of *Escherichia coli* and *Pseudomonas aeruginosa*. *J. Bacteriol.* **185:**6112–6118.
83. **Matias, V. R. F., and T. J. Beveridge.** 2005. Cryo-electron microscopy reveals native polymeric cell wall structure in Bacillus subtilis 168 and the existence of a periplasmic space. *Mol. Microbiol.* **56:**240–251.
84. **Matias, V. R. F., and T. J. Beveridge.** 2006. Native cell wall organization shown by cryo-electron microscopy confirms the existence of a periplasmic space in *Staphylococcus aureus*. *J. Bacteriol.* **188:**1011–1021.
85. **McLaren, A. C.** 1991. *Transmission Electron Microscopy of Minerals and Rocks*. Cambridge University Press, Cambridge, United Kingdom.
86. **McMahon, G., and L. J. Cabri.** 1998. The SIMS technique in ore mineralogy, p. 199–240. *In* L. J. Cabri and D. J. Vaughan (ed.), *Mineralogical Association of Canada Short Course Series*, vol. 27. *Modern Approaches to Ore and Environmental Mineralogy*. Mineralogical Association of Canada, Ottawa, Ontario, Canada.
87. **Morris, K., J. C. Butterworth, and F. R. Livens.** 2000. Evidence for the remobilization of Sellafield waste radionuclides in an intertidal salt marsh, West Cumbria, U.K. *Estuar. Coast. Shelf Sci.* **51:**613–625.
88. **Moyes, L. N., M. J. Jones, W. A. Reed, F. R. Livens, J. M. Charnock, J. F. W. Mosselmans, C. Hennig, D. J. Vaughan, and R. A. D. Pattrick.** 2002. An X-ray absorption spectroscopy study of neptunium(V) reactions with mackinawite. *Environ. Sci. Technol.* **36:**179–183.
89. **Nelson, D. M., and M. B. Lovett.** 1978. Oxidation state of plutonium in the Irish Sea. *Nature* **276:**599–601.
90. **Nordstrom, D. K.** 1996. Trace metal speciation in natural waters: computational vs analytical. *Water Air Soil Pollut.* **90:**257–267.
91. **Panak, P. J., C. H. Booth, D. L. Caulder, J. J. Bucher, D. K. Shuh, and H. Nitsche.** 2002. X-ray absorption fine structure spectroscopy of plutonium complexes with *Bacillus sphaericus*. *Radiochim. Acta* **90:**315–321.
92. **Parker, J. L., and N. S. Bloom.** 2005. Preservation and storage techniques for low-level aqueous mercury speciation. *Sci. Total Environ.* **337:**253–263.
93. **Pasilis, S. P., and J. E. Pemberton.** 2003. Speciation and coordination chemistry of uranyl(VI)-citrate complexes in aqueous solution. *Inorg. Chem.* **42:**6793–6800.
94. **Phoenix, V., A. A. Korenevsky, V. R. F. Matias, and T. J. Beveridge.** 2006. Cell wall structure and physicochemistry provide new insights into metal ion nucleation and mineral development in bacteria, p. 85–108. *In* G. M. Gadd (ed.), *Advances in Geomicrobiology*, vol. 65. *Microorganisms and Earth Systems*. Cambridge University Press, Cambridge, United Kingdom.

95. **Pitzer, K. S. (ed.).** 1991. *Activity Coefficients in Electrolyte Solutions*, 2nd ed. CRC Press, Boca Raton, Fla.
96. **Polya, D. A., P. R. Lythgoe, F. Abou-Shakra, A. G. Gault, J. R. Brydie, J. G. Webster, K. L. Brown, M. K. Nimfopoulos, and K. M. Michailidis.** 2003. IC-ICP-MS and IC-ICP-HEX-MS determination of arsenic speciation in surface and groundwaters: preservation and analytical issues. *Mineral. Mag.* **67:**247–261.
97. **Polya, D. A., E. M. Woolley, J. M. Simonson, and R. E. Mesmer.** 2000. The enthalpy of dilution and thermodynamics of Na_2CO_3(aq) and $NaHCO_3$(aq) from T = 298 K to T = 523.15 K and pressure of 40 MPa. *J. Chem. Thermodyn.* **33:**205–243.
98. **Powell, B. A., R. A. Fjeld, D. I. Kaplan, J. T. Coates, and S. M. Serkiz.** 2005. $Pu(V)O^{2+}$ adsorption and reduction by synthetic hematite and goethite. *Environ. Sci. Technol.* **39:**2107–2114.
99. **Rai, D., M. Yui, and D. A. Moore.** 2003. Solubility and solubility product at 22°C of UO_2 (c) precipitated from aqueous U(IV) solutions. *J. Solut. Chem.* **32:**1–17.
100. **Reddy, C. A., T. J. Beveridge, J. A. Breznak, L. Snyder, T. M. Schmidt, and G. A. Marzluf (ed.).** *Methods for General and Molecular Microbiology*, in press. American Society for Microbiology, Washington, D.C.
101. **Reed, S. J. B.** 1997. *Electron Probe Microanalysis*, 2nd ed. Cambridge University Press, Cambridge, United Kingdom.
102. **Renshaw, J. C., L. J. C. Butchins, F. R. Livens, I. May, J. M. Charnock, and J. R. Lloyd.** 2005. Bioreduction of uranium: environmental implications of a pentavalent intermediate. *Environ. Sci. Technol.* **39:**5657–5660.
103. **Rittmann, B. E., J. E. Banaszak, and D. T. Reed.** 2002. Reduction of Np(V) and precipitation of Np(IV) by an anaerobic microbial consortium. *Biodegradation* **13:**329–342.
104. **Robinson, B. W., N. G. Ware, and D. G. W. Smith.** 1998. Modern electron microprobe trace element analysis in mineralogy, p. 153–180. *In* L. J. Cabri and D. J. Vaughan (ed.), *Mineralogical Association of Canada Short Course Series*, vol. 27. *Modern Approaches to Ore and Environmental Mineralogy*. Mineralogical Association of Canada, Ottawa, Ontario, Canada.
105. **Rossberg, A., T. Reich, and G. Bernhard.** 2003. Complexation of uranium(VI) with protocatechuic acid—application of iterative transformation factor analysis to EXAFS spectroscopy. *Anal. Bioanal. Chem.* **376:**631–638.
106. **Rusin, P. A., L. Quintana, J. R. Brainard, B. A. Strietelmeier, C. D. Tait, S. A. Ekberg, P. D. Palmer, T. W. Newton, and D. L. Clark.** 1994. Solubilization of plutonium hydrous oxide by iron reducing bacteria. *Environ. Sci. Technol.* **28:**1686–1690.
107. **Senko, J. M., J. D. Istok, J. M. Suflita, and L. R. Krumholz.** 2002. In-situ evidence for uranium immobilization and remobilization. *Environ. Sci. Technol.* **35:** 1491–1496.
108. **Shaughnessy, D., H. Nitsche, C. Booth, D. Shuh, G. Waychunas, R. Wilson, H. Gill, K. Cantrell, and R. Serne.** 2003. Molecular interfacial reactions between Pu(VI) and manganese oxide minerals manganite and hausmannite. *Environ. Sci. Technol.* **37:**3367–3374.
109. **Sokolov, I., D. S. Smith, G. S. Henderson, Y. A. Gorby, and F. G. Ferris.** 2001. Cell surface electrochemical heterogeneity of the Fe(III)-reducing bacteria *Shewanella putrefaciens*. *Environ. Sci. Technol.* **35:**341–347.
110. **Standring, W. J. F., D. H. Oughton, and B. Salbu.** 2002. Potential remobilization of ^{137}Cs, ^{60}Co, ^{99}Tc, and ^{90}Sr from contaminated Mayak sediments in river and estuary environments. *Environ. Sci. Technol.* **36:**2230–2337.
111. **Suttill, R. J., P. Turner, and D. J. Vaughan.** 1982. The geochemistry of iron in recent tidal-flat sediments of the Wash area, England: a mineralogical, Mossbauer, and magnetic study. *Geochim. Cosmochim. Acta* **46:**205–217.
112. **Suzuki, Y., S. D. Kelly, K. A. Kemner, and J. F. Banfield.** 2003. Microbial populations stimulated for hexavalent uranium reduction in uranium mine sediment. *Appl. Environ. Microbiol.* **69:**1337–1346.
113. **Tanner, S. D., and V. I. Baranov.** 1999. Theory, design and operation of a dynamic reaction cell for ICP-MS. *Atom. Spectrosc.* **20:**45–52.
114. **Templeton, D. M.** 2003. The importance of trace element speciation in biomedical science. *Anal. Bioanal. Chem.* **375:**1062–1066.
115. **Templeton, D. M., F. Ariese, R. Cornelis, L.-G. Danielsson, H. Muntau, H. P. van Leeuwen, and R. Lobinski.** 2000. Guidelines for terms related to chemical speciation and fractionation of elements. Definitions, structural aspects and methodological approaches (IUPAC Recommendations 2000). *Pure Appl. Chem.* **72:**1453–1470.
116. **Toner, B., S. Fakra, M. Villalobos, T. Warwick, and G. Sposito.** 2005. Spatially resolved characterization of biogenic manganese oxide production within a bacterial biofilm. *Appl. Environ. Microbiol.* **71:**1300–1310.
117. **Umeda, A., Y. Ueki, and K. Amako.** 1987. Structure of the *Staphylococcus aureus* cell wall determined by the freeze-substitution method. *J. Bacteriol.* **169:**2482–2487.
118. **Urch, D. S.** 1971. X-ray emission spectroscopy. *Q. Rev.* **25:**343.
119. **U.S. Environmental Protection Agency.** 2001. *Method 1632: Chemical Speciation of Arsenic in Water and Tissue by Hydride Generation Quartz Furnace Atomic Absorption Spectrometry*. USEPA Report EPA-821-R-01-006, revision A. U.S. Environmental Protection Agency, Washington, D.C.
120. **U.S. Geological Survey.** 2005. *Review of the Transport of Selected Radionuclides in the Interim Risk Assessment for the Radioactive Waste Management Complex, Waste Area Group 7 Operable Unit 7-13/14, Idaho National Engineering and Environmental Laboratory, Idaho*. USGS Scientific Investigations Report 2005-5026, vol. I. U.S. Geological Survey, Reston, Va.
121. **Wade, R., Jr., and T. J. DiChristina.** 2000. Isolation of U(VI) reduction-deficient mutants of *Shewanella putrefaciens*. *FEMS Microbiol. Lett.* **184:**143–148.
122. **White, J. C. (ed.).** 1985. *Mineralogical Association of Canada Short Course Series*, vol. 11. *Application of Electron Microscopy in the Earth Sciences*. Mineralogical Association of Canada, Ottawa, Ontario, Canada.
123. **Yao, X., M. Jericho, D. Pink, and T. Beveridge.** 1999. Thickness and elasticity of gram-negative murein sacculi measured by atomic force microscopy. *J. Bacteriol.* **181:**6865–6875.
124. **Zanonato, P., P. Di Bernardo, A. Bismondo, G. K. Liu, X. Y. Chen, and L. F. Rao.** 2004. Hydrolysis of uranium(VI) at variable temperatures (10–85 degrees C). *J. Am. Chem. Soc.* **126:**5515–5522.
125. **Zussman, J.** 1977. X-ray diffraction. *In* J. Zussman (ed.), *Physical Methods in Determinative Mineralogy*. Academic Press, New York, N.Y.

Arsenate-Respiring Bacteria

CHAD W. SALTIKOV AND DAVIN MALASARN

99

Recently, much attention has focused on studying the underlying mechanisms controlling the fate and transport of arsenic in subsurface environments. This increased interest is due in part to the public health crisis associated with the consumption of arsenic-contaminated groundwater in a number of countries throughout the world, Bangladesh being, by far, the most impacted country. Geochemistry and hydrology have been the primary scientific disciplines employed to understand how drinking water supplies become contaminated with arsenic. However, investigations in some environments have concluded that microbes are in part to blame for the redox transformations of arsenic. This has attracted geological and microbial scientists, who are merging their disciplines to better understand the role of microbes in controlling the geochemical cycling of arsenic. This chapter is not intended to review all aspects of the microbiology of arsenic redox transformations; recent reviews on the subject can be found elsewhere (54, 64). Instead, this chapter highlights the various approaches and techniques that others have used to investigate the geomicrobiology of arsenate respiration in subsurface and aqueous environments. This includes the discussion of classical microbiological techniques (e.g, enrichment culture and pure strain studies), environmental DNA methods, and culture-dependent microcosm studies. The overall goal is to provide investigators with a concise overview of the tools used to better understand the biological mechanisms influencing the arsenic geochemical cycle.

THE ARSENIC BIOGEOCHEMICAL CYCLE

Arsenic Geochemical Cycle: a Microbe's Perspective

What environments are impacted by arsenic contamination? How do microbes influence the redox cycling of arsenic in these environments? What are the impacts of these arsenic transformations on environmental and public health? There are a number of unanswered questions regarding the geochemistry and microbial influences of arsenic. Therefore, there is a pressing need to develop a better understanding and appreciation for these influences. A case in point can be found in Bangladesh, where millions of people are being poisoned from the consumption of (naturally occurring) arsenic-contaminated well water (4). In the United States, an emerging issue is the observation of elevated arsenic levels in groundwaters near and around certain Superfund sites and certain landfills (5, 36). Sites such as these, with mixed organic and inorganic pollutants, require innovative and collaborative approaches to investigate how arsenic is being transformed. With these issues in mind, it is best to start with an overview of the arsenic biogeochemical cycle by describing the major arsenic-transforming reactions that microbes catalyze and the environmental conditions that induce or inhibit these reactions. General arsenic biogeochemical schemes have been described and can be found elsewhere in more detail (43, 50, 53, 54).

The microbial arsenic cycle can be broken down into redox transformations, the genes underlying these redox transformations, and the fate and transport of differentiated arsenic species. Figure 1 simplifies the complexity of the microbial arsenic cycle. Arsenic in subsurface environments is derived from arsenic-bearing minerals, such as arsenic-sulfides or iron-arsenic-sulfides (17, 48). Additionally, weathering and transport also provide a mechanism for the deposition of arsenic-bearing subsurface material (66). The redox transformation of arsenic can impact the hydrological mobility of arsenic and downstream contamination of pristine environments (24). Either oxidation or reduction reactions strongly influence the affinity of the resulting arsenic species for sedimentary material (8, 32). Also, mineralogical properties of the subsurface material play a key role in controlling the fate and transport of arsenic. The more oxidized form of arsenic, arsenate [As(V), $HAsO_4^{2-}$], has adsorption properties similar to those of inorganic phosphate to various minerals (18, 41, 66). Consequently, most of the arsenate in circumneutral pH environments is associated with solid phases. The sorption of arsenate is strongest on various oxides of iron, manganese, and aluminum (11, 63). In contrast, reduced arsenic, namely, arsenite [As(III), H_3AsO_3], exhibits greater hydrological mobility than arsenate because it is uncharged at near-neutral pH and, therefore, has selective affinities for certain iron oxides and sulfur-containing minerals (48).

From the bacterium's perspective, reduction of arsenate adsorbed onto solid surfaces presents an intriguing mechanistic problem: how do microbes gain access to such substrates? Several reports have shown that microbes can reduce arsenate adsorbed onto the surface of various minerals (2, 12, 35). However, not all arsenate-reducing microbes can reduce arsenate-associated minerals (9, 39). Currently,

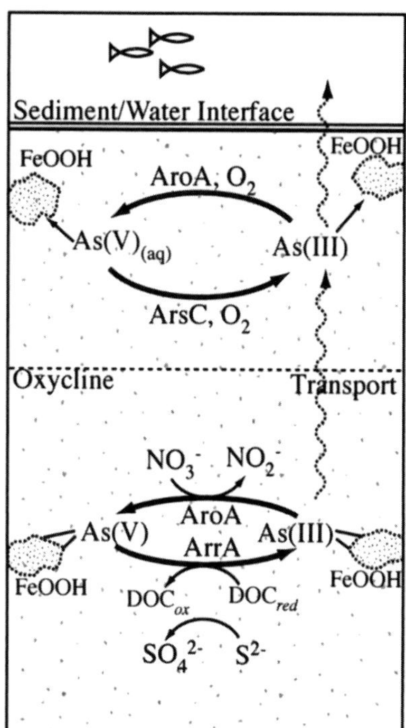

FIGURE 1 Biochemistry influencing arsenic redox transformations and geochemical cycling of arsenic. The heavier arrows indicate the flow of electrons (either oxidation or reduction) to and from arsenic and specific proteins controlling the transformation. AroA, arsenite oxidase; ArrA, arsenate respiratory reductase.

only a subclass of arsenate reducers, the arsenate-respiring prokaryotes, are known to reduce mineral-associated arsenate (2, 39, 49). These arsenate-respiring prokaryotes can promote the release of arsenic in aqueous environments. However, the mobilization of arsenic is restricted to anoxic sediments (2, 12, 22) and is controlled by pH, and the availability of dissolved organic carbon and the thermodynamics of competing modes of respiration can all affect the mobilization.

OVERVIEW OF METHODS

Arsenate reducers are readily cultivated from diverse environments. Reducers that couple either detoxification or energy generation (respiration) to arsenate reduction can be isolated. Strains that respire arsenic couple growth to arsenate reduction, whereas those that solely detoxify arsenic through reduction require other electron acceptors. The biochemical models for arsenate respiration and detoxification are shown in Fig. 2. The *arsC* gene encodes the arsenate-detoxifying reductase, and *arrA* encodes the arsenate respiratory reductase. Silver and Phung (64) provide an in-depth review of ArrA and ArsC. When designing experiments to investigate these metabolisms, the targeted community governs the experimental design and subsequent interpretation of the results. For arsenate respiratory reduction it is important to show that growth is coupled to the reduction of arsenate. Strict anaerobic conditions must be maintained. Preferably, a defined minimal salts medium should be used. For carbon sources, it is preferable to use a nonfermentable carbon source. With lactate, which can be fermented, a negative control lacking arsenate should be performed when characterizing a new strain. Using yeast extract can confound growth results because of residual competing electron acceptors and fermentable carbon compounds present in the formulation. Instead, a defined vitamin mix can be substituted for yeast extract; however, growth rates may decrease and yields may be less.

Many of the microbiological aspects of the arsenic cycle have been focused on isolation of new and novel arsenate-transforming strains. Table 1 lists many of the known arsenate-respiring prokaryotes reported in the literature. However, new approaches are needed to investigate microbial arsenate reduction without the bias of enrichment culture techniques. One approach may be the application of molecular genetic techniques on environmental samples. Because

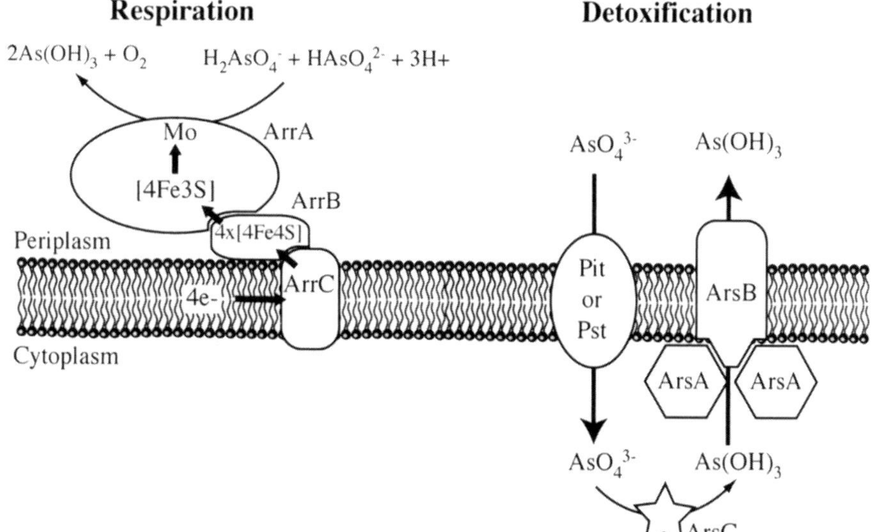

FIGURE 2 Biochemical models for arsenic reduction or oxidation. The models show arsenate respiratory reduction by ArrA and arsenate reduction by ArsC coupled to arsenite detoxification through other Ars proteins.

TABLE 1 List of known arsenate-respiring strains, their origins of isolation, and supporting references

Strain	Origin	Reference(s)
Bacillus sp. strain HT-1	Hamster feces	25
Bacillus arsenicoselenatis strain E1H	Bottom mud, Mono Lake, Calif.	6
Bacillus macyae strain JMM-4	Australian gold mine	61, 62
Bacillus selenitireducens strain MLS10	Mono Lake bottom mud	1, 6
Chrysiogenes arsenatis strain BAL-1	Australian gold mine	33, 37
Clostridium sp. strain OhILAs	Ohio river sediments	J. Stolz, personal communication
Desulfitobacterium sp. strain GBFH	As-contaminated sediments from a freshwater lake, Idaho	47
Desulfitobacterium frappieri		7, 47
Desulfitobacterium hafniense		47
Desulfomicrobium sp. strain Ben-RB	Reed bed mud from Australia	38
Desulfosporosinus auripigmenti (formerly *Desulfotomaculum auripigmentum* strain OREX-4)	Upper Mystic Lake, Mass.	45
Desulfosporosinus sp. strain Y5	Superfund site, Onondaga Lake (Syracuse, N.Y.)	36, 55
MLMS-1	Anoxic bottom water from Mono Lake, Calif.	26
Pyrobaculum aerophilum strain IM2	Boiling marine water hole at Maronti Beach, Ischia, Italy	27, 70
Pyrobaculum arsenaticum strain PZ6	Pisciarelli Solfatara, Naples, Italy	27
SLAS-1	Salt-saturated brine, Searles Lake, Calif.	52
Sulfurospirillum arsenophilum strain MIT-13 (formerly named *Geospirillum*)	As-contaminated sediments, Aberjona watershed, Mass.	3
Sulfurospirillum barnesii strain SES-3 (formerly named *Geospirillum*)	Selenate-contaminated freshwater marsh, western Nev.	49, 67
Shewanella sp. strain ANA-3	Wooden pier piling in an estuary, Mass.	58
Shewanella sp. strain HAR-4	Fe- and As-rich sediments, Haiwee Reservoir, Calif.	39
Shewanella putrefaciens strain CN-32		C. W. Saltikov, unpublished data
Thermus sp. strain HR13	As-contaminated geothermal spring, Growler Hot Spring, Calif.	19
Citrobacter sp. strain TSA-1	Termite hind gut	25
Wolinella sp. strain BRA-1	Bovine rumen fluid	25

the fundamental mechanism of arsenic respiration is actively being worked out, this information will be essential to the development of molecular tools to target the specific genes underlying arsenic transformations in the environment.

Lastly, a study involving arsenic transformations would not be complete without a reliable method for quantifying arsenic redox species (e.g., arsenate and arsenite). A simple cost-effective method is provided that can be used to estimate arsenate reduction in enrichment cultures and in studies in which arsenic is supplemented at higher concentrations (above approximately ~100 μM). When transitioning to a more sensitive arsenic detection system, speciation can be quantified either using an anion-exchange column or through the use of online speciation methods such as high-pressure liquid chromatography (HPLC) separation followed by UV detection for 100 μM or greater, or hydride generation and atomic adsorption spectroscopy or inductively coupled plasma mass spectrometry for maximum sensitivity. We have used HPLC and UV detection for quantifying not only As(V) and As(III) but also various organic acid substrates and products in culture filtrates in a single analysis. Detection limits approach the 10 to 100 μM range. We have also developed a simple method for dissolving laboratory-prepared hydrous ferric oxide (HFO) and measuring As(V) and As(III) via HPLC-UV detection.

MEASURING ARSENIC

Rapid Assays

A rapid assessment for the presence of arsenite can be done by acidifying the sample (below pH 7) with HCl and then adding sodium sulfide (~4% stock solution prepared and stored anaerobically). If arsenite is in sufficient concentrations (above ~100 μM), yellow arsenic trisulfide (As_2S_3) will rapidly form and precipitate from solution (44). If an immediate yellow mineral formation is not observed, the sulfide will eventually reduce arsenate to arsenite (57) and cause the solution to slowly turn yellow, yielding a false positive. Another rapid test for arsenate and arsenite is the addition of silver nitrate (65). Silver forms precipitates with arsenic, either light brown-red for arsenate or light yellow for arsenite. The color change is most conclusive when the ratios for As(V) to As(III) are at their extremes. Intermediate colors are harder to interpret but generally appear a deeper red-brown color than arsenate alone. Lastly, the strong oxidant $KMnO_4$ (10 mM) can be added to culture medium containing arsenite, which will result in the disappearance of the characteristic purple color of the $KMnO_4$ solution as arsenite becomes oxidized. Although not quantitative, these "quick-and-dirty" color assays provide a simple way for monitoring redox transformations in growth media of bacterial cultures.

Quantitative Measurements

Molybdenum Blue Detection of Arsenate

Another rapid method for detecting arsenic is the molybdenum blue method commonly used to measure soluble reactive phosphate (68). The protocol is as follows.

Prepare the color reagent as follows per milliliter (scale up according to the number of samples to be analyzed).

Color reagent

Components (stock)	Mixing ratio
H_2SO_4, 14% (2.5 M)	5
Ammonium molybdate, 3% (24 mM)	2
Ascorbic acid (prepare fresh), (61.3 mM)1.08%	2
Potassium antimonyl tartrate (PAT), 0.56% (8 mM)	1

Mix B, C, and D in the ratio 2:2:1; then add 5 parts of A.

As(V) reducing agent

Components (stock)	Mixing ratio
A. Sodium metabisulfite solution, 14% $Na_2S_2O_5$	2
B. Sodium thiosulfate solution, 1.4% $Na_2S_2O_3 \times 5\ H_2O$	2
C. Sulfuric acid solution, 10% (1.8 M) H_2SO_4	1

Slowly mix A with C, then add B (stable for 24 h if refrigerated).

Oxidizing agent

Component (stock)	Mixing ratio
Household bleach freshly diluted 1/100 in water	1/100 strength

Procedure: for arsenate/phosphate-containing samples, use the "Untreated" protocol. To quantify total arsenic, oxidize the sample with diluted bleach. To measure background phosphate use the "Reduced" procedure, described below.

Untreated (arsenate and phosphate)	Oxidized (arsenate/arsenite and phosphate)	Reduced (phosphate)
100-μl sample	100-μl sample	100-μl sample
100 μl of 2% HCl	100 μl of 2% HCl	100 μl of 2% HCl
20 μl of 2% HCl (omit if performing only the untreated procedure)	20 μl of oxidizing agent	20 μl of As(V) reducing agent
	Incubate for 10 min.	
20 μl of color reagent	20 μl of color reagent	20 μl of color reagent

Wait at least 10 min to 2 h, and measure absorbance at 865 nm in a microplate reader.

Because phosphate and arsenate are chemically very similar, arsenate is an equally good substrate for this assay. Several modifications to the method have been made to quantify total arsenic, arsenite, and the background phosphate (29, 30). Dhar et al. (10) further improve upon the method. Depending on the sensitivity, these methods can be readily adapted for use in a 1- to 10-cm-path-length cuvette or adapted for a 96-well microtiter dish format useful for high-throughput quantification. To measure As(III), samples need to be oxidized so that the total As [As(V) + As(III)] can be measured. Using the difference between treated and untreated samples, it is possible to infer the As(III) concentration in the sample. Household bleach is an excellent oxidizing agent and a cheaper alternative to potassium iodate. However, bleach must be diluted prior to sample additions.

HPLC Speciation and UV Detection

One can use HPLC methods for measuring the concentrations of organic acids and arsenate and arsenite all in a single analysis. Arsenic compounds [As(V) and As(III)], lactate, acetate, fumarate, succinate, etc., can be quantified by HPLC and UV detection (210 nm). We have used columns from Hamilton (PRP-X300) in series with a Bio-Rad Aminex HPX-87H column, both heated to 50°C. The Bio-Rad column utilizes ion-moderated partition chemistry for separating compounds. The column works well for separating organic acids; however, arsenite and lactate have the same retention times. In contrast the Hamilton PRP-X300 ion-exclusion column resolves the As(III) peak from lactate; however, As(V) and As(III) retention times are close enough to cause peaks to overlap when As(V) or As(III) are present at ratios ranging from 0.2 to 0.8. Placing the two columns in series with the Hamilton column followed by the Bio-Rad column provides the optimum resolution of arsenate, arsenite, lactate, and acetate within a 25- to 30-min run per sample. The mobile phase can be either phosphoric acid (30 mM) or sulfuric acid (5 mM) with flow rates from 0.6 to 1 ml/min depending on the pressure. Injection volumes of 25 to 50 μl are typically used. Samples should be filtered to remove particulate material. We use Spin-X (Costar) 0.2-μm-pore-size nylon spin filters. Standard additions of As(V), As(III), and organic acid analytes to samples should be made to control for shifts in retention times of peaks due to culture medium components. If HFO or goethite is used, the mineral-containing samples are dissolved in concentrated phosphoric acid followed by dilution with water to ~30 mM H_3PO_4 prior to analysis by HPLC (39).

Other Useful Arsenic Speciation Methods

Off-line speciation using drip anion-exchange columns and 3M Anion SR filters provide convenient ways to separate As(V) from As(III) (13, 15, 16). The 3M Anion SR filters remove As(V) at circumneutral pH. The Ficklin drip columns retain As(V), whereas As(III) passes through in the filtrate. Samples applied to columns require prior acidification. The method of Edwards et al. (13) expands on Ficklin by describing how to make arsenic field speciation kits. This is very useful when immediately processing samples in the field. In waters rich in sulfide it has been found that complexes of sulfide-As(III) are also retained on Ficklin columns. Sparging samples with nitrogen gas was shown to be effective in eliminating this problem (28). Performing total arsenic analysis on untreated and column-treated samples allows both As(V) and As(III) concentrations to be determined, assuming there is no organic arsenic in the sample.

Sensitive Arsenic Techniques

There are a number of reviews describing methods for speciating and quantifying arsenic in a variety of samples (20, 28, 31, 40). Refer to the listed reviews for detailed descriptions of the various techniques. In general, detection and speciation techniques depend on the samples to be analyzed. For samples containing high concentrations of arsenic (e.g., >0.5 mM), which are typical for microbial cultures, the use of HPLC-UV detection or the molybdenum blue assay is sufficient. However, for analysis of groundwater concentrations of arsenic (e.g., less than 1 μM) the analyst requires more sensitive detection methods such as inductively coupled plasma mass spectrometry or an atomic adsorption spectroscopy. Both allow detection of arsenic in submicromolar concentrations. These instruments are typically set up to perform online arsenic speciation by interfacing with an HPLC, ion chromatograph, or hydride generator. Alternatively, arsenic species can be separated using Ficklin columns (13) followed by analysis for total arsenic.

CULTURE-DEPENDENT TECHNIQUES

Enrichment Culturing

To date most arsenate-respiring microbes have been isolated as heterotrophs. While acetate, pyruvate, formate, fumarate, butyrate, succinate, malate, glycerol, and ethanol all reportedly serve as carbon sources for various pure cultures of arsenate respirers, only a small fraction of these organic carbon sources have been used in enrichment cultures (37, 45, 47, 61). Lactate has been relied upon most heavily. The use of this fermentation end product has resulted in the isolation of diverse bacteria from a variety of environments (Table 1). Acetate has also been used, which led to the isolation of *Chrysiogenes arsenatis*. To further characterize the culturable prokaryotes, one should utilize a range of carbon sources and even consider chemolithotrophic metabolism (e.g., sulfide oxidation and arsenate reduction).

Isolation of Arsenate-Respiring Microbes

A variety of growth media are used to culture arsenate-respiring prokaryotes. Many of the references listed in Table 1 report compositions of media and growth conditions. For *Shewanella* species we routinely use the medium described in reference 60. There are a number of ways to prepare a medium devoid of oxygen. Anaerobic growth medium can be prepared either by boiling the medium under a steady stream of nitrogen gas or by bubbling with nitrogen. Once oxygen is removed, the medium is dispensed into nitrogen gas-flushed Balch tubes or similar crimp top bottles by using a Hungate-style gassing manifold and nitrogen-flushed pipettes. Air-powered pipette aides can be modified to dispense nitrogen (or oxygen-free) gas by unplugging the positive pressure dispensing line from the air pump and attaching it to one of the lines in the Hungate manifold. This ensures that the anoxic medium will be dispensed with an oxygen-free headspace within the pipette. Tubes and bottles are autoclave sterilized for 15 to 30 min. Variations on the medium include adjusting the buffer concentration and type, additions of agar (1 to 1.5% agar) for solidified medium, and changing electron acceptors and carbon sources. If petri plates are made, freshly poured plates should be placed in the anaerobic chamber prior to complete solidification. Including a reductant (e.g., L-cysteine) prior to pouring plates may be necessary for more fastidious anaerobes.

Measuring Abundance of Arsenate-Respiring Microbes

The abundance of specific microbial populations can be measured using selective media and enrichment culture techniques in a semiquantitative format. Most probable number (MPN) methods have been used for investigating arsenate-respiring microbial communities in environmental samples (22, 34). In this approach the starting material is resuspended in growth medium in replicate tubes (greater than three replicate tubes) and serially diluted 10-fold until growth extinction. The cultures are incubated under anaerobic conditions and observed over time for the presence of arsenite. The MPN analysis has been refined for use in a high-throughput format by performing the sample incubations in deep (~2-ml) 96-well plates. After incubation for a predefined time (up to 30 days) the assay is terminated and scored for the presence of arsenite. The simplest means of detecting arsenite is to acidify each well with HCl and add sodium sulfide. The presence of arsenic trisulfide in each replicate for each dilution is counted, and the presence/absence results are compared to an MPN table. Cell densities are normalized to the amount of initial starting material.

Detoxification and Respiration of Arsenic

In practice, distinguishing arsenate reduction for respiratory or detoxification purposes is relatively easy for pure bacterial cultures. Figure 2 describes the main pathways for arsenic redox transformations. Both *arrA* and *arsC* encode arsenate reductases. However, the two enzymes serve very different metabolic functions. ArrA is essential for growth and respiratory reduction of arsenate (59). And ArsC is essential to detoxifying the cytoplasm of arsenate (42). To demonstrate reduction by the ArsC pathway, facultative anaerobes should be able to reduce arsenate to arsenite in oxygen-respiring conditions. For anaerobes, arsenate reduction should occur as long as the culture is respiring an alternate electron acceptor (e.g., fumarate, nitrate, etc.) or fermenting a carbon source. For strains lacking the *arr* gene cluster, in the absence of other electron acceptors or in nonfermentation conditions, strains should neither grow nor reduce arsenate. Another way to investigate the detoxification pathway is to test the strain for arsenite resistance. However, one must check that arsenite has not been oxidized to arsenate. To demonstrate the respiratory reduction of arsenate to arsenite in a bacterial culture, one should show a stoichiometric conversion of arsenate to arsenite that is linked to the oxidation of an electron donor. Moreover, an increase in cell density should also be observed that is proportional to the concentration of a limiting substrate, either the carbon source/electron donor or the electron acceptor. Routine cultures of *Shewanella* sp. strain ANA-3, *Sulfurospirillum barnesii* strain SES-3, and *Bacillus selenitireducens* strain MLS10 inoculated with 10^6 cells/ml usually grow to 1×10^8 to 3×10^8 cells/ml in anaerobic minimal medium containing 5 to 10 mM arsenate and 10 to 20 mM lactate. Neither growth nor arsenate reduction should be observed if the electron donor is withheld. In general, samples are taken over time, cells are removed by filtration through a 0.2-μm-pore-size filter, and the filtrate is analyzed for As(V), As(III), and organic acids. In experiments with *Shewanella* sp. strain ANA-3 when the bacterium is grown on As(V) and lactate, As(V) should be reduced to As(III) and lactate should be oxidized to acetate according to a stoichiometric ratio of 2:1 (As(V)-to-lactate mole ratio) (58).

BIOCHEMICAL AND MOLECULAR ASSAYS

Biochemistry

When new arsenate-respiring prokaryotes are isolated or existing strains are characterized, biochemical assays can be used to verify the presence of arsenate respiratory reductase, ArrA. The biochemical activity is typically measured spectrophotometrically. The arsenate respiratory reductase can couple arsenate reduction to the oxidation of artificial electron donors methyl viologen or benzyl viologen, both of which undergo colorimetric decrease in blue color as they become oxidized. Activity from whole cells, cell lysates, or purified samples can be measured.

Because of the potential for aerobic oxidation of these viologens, assays must be done under strict anoxic conditions. This can be accomplished with stoppered or Thunberg cuvettes flushed with nitrogen. Dithionite is used to initially reduce the methyl or benzyl viologen before the start of the assay. This assay measures the activity of arsenate respiratory reductases and not the detoxification arsenate reductase. The latter is based on the glutathione disulfide oxidoreductase-dependent assay for glutaredoxin; this assay requires purified ArsC enzyme and would be technically challenging to perform on crude cellular extracts. The following is a suggested protocol for performing the colorimetric assay for arsenate respiratory reductase.

Prepare all solutions under strict anaerobic conditions either by bubbling with nitrogen or by preparation in an oxygen-free glovebox (nitrogen). In a 1-cm-diameter cuvette add the following:

2.1 ml of 50 mM Tris-HCl buffer, pH 7 to 8 (bis-Tris, MOPS [morpholinepropanesulfonic acid], MES [morpholineethanesulfonic acid], and piperazine have also been used successfully)

0.1 ml of 5 mM methyl viologen prepared in water (molar extinction coefficient, 13 mM^{-1} cm^{-1} at an optical density at 600 nm of 1)

0.1 ml of 5 mM dithionite solution (prepared fresh in 50 mM Tris base, pH 10)

0.1 ml of either 100 mM sodium arsenate (pH 7) or water (control) prepared under anaerobic conditions (nitrogen gas)

Sparge with nitrogen gas if not using an anaerobic chamber.

Add sample 0.1 ml of diluted protein, washed cells, crude lysates, or control samples. (Notes: protein concentrations can be measured using any standard method; it is important to wash cell pellets to remove residual electron acceptor prior to preparing lysates.)

Stopper cuvette with a butyl rubber stopper.

Measure the optical density at 600 nm versus time on an appropriate spectrophotometer.

Convert methyl viologen oxidized to units of arsenate reductase activity (1 U corresponds to 1 μmol of arsenate reduced per min).

Molecular Biology and Gene Expression

The use of genetic approaches can provide additional information about the nature of arsenate reduction. Molecular genetic tools provide the means for genetically manipulating a strain, granted that the organism is genetically tractable. Currently, *Shewanella* sp. strain ANA-3 is the only genetically tractable arsenate-respiring prokaryote. However, the tools used with ANA-3 should be adaptable to several previously identified strains such as *Sulfurospirillum* strains, *Wolinella*, and *Citrobacter*, and possibly some of the haloalkaliphiles. Most molecular studies rely on isolation of DNA from the strain of interest. This can be accomplished using traditional methods or a commercially available kit. To characterize new bacterial strains, an initial molecular screen for *arr* genes can be done using degenerate PCR primers designed against the known *arrA* genes (39). Culture-independent techniques such as stable-isotope probing show great promise for investigating arsenate-respiring communities in environmental samples (56). This technique allows investigators to identify the molecular signatures of microbes in an environmental sample that are actively metabolizing a certain substrate.

Gene expression studies can also be useful for investigating environmental conditions associated with arsenic reduction and mobilization. One relatively straightforward approach is to use a method involving reverse transcription of RNA into cDNA followed by PCR analysis (referred to as RT-PCR). This has been developed and utilized to study *arr* gene expression patterns in *Shewanella* sp. strain ANA-3 (60). Total RNA is extracted from exponentially growing cells, DNA is removed by DNase digestion, and the RNA is converted to cDNA by using a reverse transcriptase enzyme. Detection of the gene of interest can be done either qualitatively by agarose gel electrophoresis or quantitatively through real-time (quantitative) PCR (qPCR). The traditional RT-PCR is a presence/absence end point assay, and quantification is difficult to determine. Using RT-qPCR it should be possible to observe differential expression of a target gene. We have verified these techniques using the *arrA* consensus primers with RT-qPCR to detect gene expression of *arrA* in microcosm experiments with *Shewanella* sp. strain ANA-3 growing on As(V)-HFO (39).

Environmental DNA and RNA Extraction and PCR using *arrA*-Specific Primers

Arsenate-respiring prokaryotes are phylogenetically diverse and therefore cannot be detected using 16S rRNA gene sequences (54). One must turn to functional genes for their detection. Initial attempts have been described for the development of consensus PCR primers to diagnostically identify the *arrA* gene. These primers should be refined as more sequences become available, and sequencing of PCR products should be done to confirm any positive amplification.

The current primer set (39) amplifies an ~160- to 200-bp fragment and is appropriate for qPCR studies. However, for diversity studies, a larger *arrA* PCR product is needed. The development of a second-generation consensus *arrA* primer set should be possible given the high degree of amino acid sequence conservation among known ArrAs.

A number of DNA and RNA extraction methods have been developed for sediments, soils, and water (71). However, companies such as MoBio and Q-Biogene have developed rapid and reliable nucleic acid extraction kits that have been widely used in environmental microbiological studies. These kits should be used with caution on low-biomass samples and should be validated for efficiency and reproducibility for each sediment type under investigation.

For laboratory experiments involving HFO, we have developed a modified protocol for dissolving iron oxide prior to DNA and RNA extractions. In this method, an HFO suspension is centrifuged and the pellet is dissolved with 0.3 M oxalic acid (pH 3). Cells can then be harvested by centrifugation for nucleic acid extraction using standard techniques, such as the TRIzol reagent (Invitrogen) for RNA extraction and the DNeasy Tissue Kit (Qiagen) for DNA extraction. By dividing samples in half, *arrA* can be quantified in RNA and DNA extracted from the same sample. This procedure should be done rapidly to avoid lysis of cells upon the addition of oxalic acid. Also, as iron can lead to inhibition of some PCRs, control reactions using known quantities of DNA or cells added to the dissolved iron sample should be tested.

GEOCHEMICAL APPROACHES

Arsenate Reduction Mechanisms and Microbe-Mineral Interactions

To investigate how iron influences the arsenic geochemical cycle, many studies have used synthetic HFO as an adsorbant for either As(V) or As(III). This HFO-As mineral is then used in batch or flowthrough experiments with iron and/or arsenate reducers to characterize how microbes affect the mineralogy and mobilization of arsenic.

There are a number of methods for synthesizing HFO and the subsequent adsorption of arsenic. In one method, 0.05 M $Fe(NO_3)_3$ is stirred with the slow dropwise addition of 0.5 M NaOH until solid HFO is observed. The pH of the solution should be constantly monitored to prevent dramatic increases during the addition of NaOH. Once the pH reaches ~7, the mineral suspension is centrifuged and washed 3 to 5 times with sterile water to remove residual nitrate. The clean HFO is then resuspended in a 0.05 M arsenic solution (pH 7) and stirred gently overnight to allow maximum adsorption of arsenic onto surface sites. Following sorption, unadsorbed arsenic is removed by additional washes in sterile water (39). Similar protocols may be developed using manganese or aluminum oxides.

The synthetic mineral can be resuspended in minimal medium and dispensed into bottles, and bacteria can be inoculated into the system. By use of this experimental setup, DNA, RNA, pH, ferrous and ferric iron, carbon, and arsenic species can all be quantified independently throughout the time course of the experiment.

Batch Studies with Sediments

Similar to HFO-As, it is possible to perform sediment incubation studies to investigate how natural sediments contribute to arsenic mobilization or retention due to microbial activity. In this approach, concentrations of arsenate, especially in natural anoxic sediments, and concentrations of carbon may be low, so supplementing the microcosm may be required. Establishing appropriate abiotic controls is important in determining background rates of arsenic reduction. Inhibiting microbial activity in natural sediments may be difficult, requiring autoclaving or the addition of poisons (e.g., formaldehyde or 1 mM sodium azide) and/or various antibiotics. Each treatment can result in the alteration of the mineralogy of the sample and could lead to experimental artifacts. Nevertheless, controls should be done and care should be taken to choose the appropriate sterilization method. Finally, the results of any sediment microcosm study should be interpreted with caution, especially when extrapolating to generalizations about what occurs in the environment. Despite some of these shortcomings, batch sediment experiments can provide valuable information regarding the potential mechanisms for arsenic transformation occurring in subsurface and sediment environments. These approaches also allow manipulation and perturbations that normally cannot be performed in natural settings. One such perturbation is the addition of arsenate-respiring prokaryotes or Fe(III)-reducing prokaryotes to natural sediments. Manipulation of oxygenation may also provide useful information regarding the types of respiration that contribute to arsenic transformations.

Simulating Subsurface Environments

Hydrology is also an important parameter to consider when investigating the arsenic cycle in the subsurface (14, 23). Laboratory and in situ approaches have been developed to examine mechanisms for biotransformation of metals and pollutants in systems under hydrological constraints (21, 46). These studies involve the development of flowthrough systems for characterizing biotransformation of certain constituents. The system developed by Hansel (21) et al. for investigating the effects of iron(III)-reducing bacteria on sand-coated HFO has been adapted for studying As(V)-HFO (14). In this system a column is slurry packed in an anaerobic chamber with HFO-As(V)-coated sand and cells derived from a metal-reducing microbe. When using anaerobic chambers containing hydrogen, investigators should be aware that some arsenate reducers can use hydrogen as an energy source. The system is connected to a reservoir of artificial groundwater and an electron donor and carbon source, after which the flow is initiated. The effluent is monitored over time for dissolved arsenic [As(V)/As(III)] and dissolved iron [Fe(II)]. The basic design and operation could be extended for studies with natural sediments or perturbations with nutrients, other electron donors and acceptors, and other microbes at the influent. Finally, the substrate in the column can be sampled at the end of an experiment and analyzed for mineralogical transformations, changes in cell densities and spatial variations, and also for molecular markers such as 16S rRNA gene if natural sediments are used.

Activity Assays for As(V)-Respiring Prokaryotes

Radiotracer assay for arsenate reduction activity has been used to quantify the rates of arsenate reduction in water and animal feces (25, 51). In principle this assay can be readily adapted for use on sediments. The technical challenge will be the efficient removal of arsenic species from the solid phase. Standard sediment extraction techniques such as phosphate extraction should be able to remove much of the arsenate adsorbed onto the surface of the subsurface material. The downside to this method is the use of hazardous radioactive arsenic, which is expensive and not readily available throughout the year.

FUTURE PERSPECTIVES

Since the first report of an arsenate-respiring prokaryote in 1994 (3), classic microbiological methods have yielded a diverse collection of strains (Table 1). As our fundamental understanding of the molecular, genetic, and biochemical mechanisms underlying arsenate respiration unravels, molecular tools are being developed from this information to investigate the biological influences on the fate and transport of arsenic in contaminated environments. One important future challenge will be to determine the precise role that arsenate-respiring prokaryotes play in processes leading to arsenic liberation from sediments.

REFERENCES

1. **Afkar, E., J. Lisak, C. Saltikov, P. Basu, R. S. Oremland, and J. F. Stolz.** 2003. The respiratory arsenate reductase from *Bacillus selenitireducens* strain MLS10. *FEMS Microbiol. Lett.* **226:**107–112.
2. **Ahmann, D., L. R. Krumholz, H. F. Hemond, D. R. Lovley, and F. M. Morel.** 1997. Microbial mobilization of arsenic from sediments of the Aberjona watershed. *Environ. Sci. Technol.* **31:**2923–2930.
3. **Ahmann, D., A. L. Roberts, L. R. Krumholz, and F. M. Morel.** 1994. Microbe grows by reducing arsenic. *Nature* **371:**750.
4. **Anawar, H. M., J. Akai, K. M. Mostofa, S. Safiullah, and S. M. Tareq.** 2002. Arsenic poisoning in groundwater: health risk and geochemical sources in Bangladesh. *Environ. Int.* **27:**597–604.
5. **Aurilio, A. C., R. P. Mason, and H. F. Hemond.** 1994. Speciation and fate of arsenic in three lakes of the Aberjona watershed. *Environ. Sci. Technol.* **28:**577–585.
6. **Blum, J. S., A. B. Bindi, J. Buzzelli, J. F. Stolz, and R. S. Oremland.** 1998. *Bacillus arsenicoselenatis*, sp nov., and *Bacillus selenitireducens*, sp. nov.: two haloalkaliphiles from Mono Lake, California that respire oxyanions of selenium and arsenic. *Arch. Microbiol.* **171:**19–30.
7. **Bouchard, B., R. Beaudet, R. Villemur, G. McSween, F. Lepine, and J. G. Bisaillon.** 1996. Isolation and characterization of *Desulfitobacterium frappieri* sp. nov., an anaerobic bacterium which reductively dechlorinates pentachlorophenol to 3-chlorophenol. *Int. J. Syst. Bacteriol.* **46:** 1010–1015.
8. **Chiu, V. Q., and J. G. Hering.** 2000. Arsenic adsorption and oxidation at manganite surfaces. 1. Method for simultaneous determination of adsorbed and dissolved arsenic species. *Environ. Sci. Technol.* **34:**2029–2034.
9. **Cummings, D. E., F. Caccavo, S. Fendorf, and R. F. Rosenzweig.** 1999. Arsenic mobilization by the dissimilatory Fe(III)-reducing bacterium Shewanella alga BrY. *Environ. Sci. Technol.* **33:**723–729.
10. **Dhar, R. K., Y. Zheng, J. Rubenstone, and A. van Geen.** 2004. A rapid colorimetric method for measuring arsenic concentrations in groundwater. *Anal. Chim. Acta* **526:** 203–209.
11. **Dixit, S., and J. G. Hering.** 2003. Comparison of arsenic(V) and arsenic(III) sorption onto iron oxide minerals: implications for arsenic mobility. *Environ. Sci. Technol.* **37:**4182–4189.
12. **Dowdle, P. R., A. M. Laverman, and R. S. Oremland.** 1996. Bacterial dissimilatory reduction of arsenic(V) to arsenic(III) in anoxic sediments. *Appl. Environ. Microbiol.* **62:**1664–1669.
13. **Edwards, M., S. Patel, L. McNeill, H. W. Chen, M. Fey, A. D. Eaton, R. C. Antweiler, and H. E. Taylor.** 1998. Considerations in As analysis and speciation. *Am. Water Works Assoc. J.* **90:**103–113.
14. **Fendorf, S., M. Polizzotto, M. J. Herbel, B. C. Bostick, and C. Harvey.** 2003. Arsenic cycling within surface and subsurface environments: impact of iron mineralogy. *Abstr. Papers Am. Chem. Soc.* **226:**U583.
15. **Ferguson, M. A., M. R. Hoffmann, and J. G. Hering.** 2005. TiO2-photocatalyzed As(III) oxidation in aqueous suspensions: reaction kinetics and effects of adsorption. *Environ. Sci. Technol.* **39:**1880–1886.
16. **Ficklin, W. H.** 1983. Separation of arsenic(III) and arsenic(V) in ground waters by ion-exchange. *Talanta* **30:** 371–373.
17. **Floroiu, R. M., A. P. Davis, and A. Torrents.** 2004. Kinetics and mechanism of As2S3(am) dissolution under N-2. *Environ. Sci. Technol.* **38:**1031–1037.
18. **Gao, Y., and A. Mucci.** 2001. Acid base reactions, phosphate and arsenate complexation, and their competitive adsorption at the surface of goethite in 0.7 M NaCl solution. *Geochim. Cosmochim. Acta* **65:**2361–2378.
19. **Gihring, T. M., and J. F. Banfield.** 2001. Arsenite oxidation and arsenate respiration by a new Thermus isolate. *FEMS Microbiol. Lett.* **204:**335–340.
20. **Guerin, T., A. Astruc, and M. Astruc.** 1999. Speciation of arsenic and selenium compounds by HPLC hyphenated to specific detectors: a review of the main separation techniques. *Talanta* **50:**1–24.
21. **Hansel, C. M., S. G. Benner, P. Nico, and S. Fendorf.** 2004. Structural constraints of ferric (hydr)oxides on dissimilatory iron reduction and the fate of Fe(II). *Geochim. Cosmochim. Acta* **68:**3217–3229.
22. **Harrington, J. M., S. E. Fendorf, and R. F. Rosenzweig.** 1998. Biotic generation of arsenic(III) in metal(loid)-contaminated freshwater lake sediments. *Environ. Sci. Technol.* **32:**2425–2430.
23. **Harvey, C. F., C. H. Swartz, A. B. Badruzzaman, N. Keon-Blute, W. Yu, M. A. Ali, J. Jay, R. Beckie, V. Niedan, D. Brabander, P. M. Oates, K. N. Ashfaque, S. Islam, H. F. Hemond, and M. F. Ahmed.** 2002. Arsenic mobility and groundwater extraction in Bangladesh. *Science* **298:**1602–1606.
24. **Hemond, H. F.** 1995. Movement and distribution of arsenic in the Aberjona watershed. *Environ. Health Perspect.* **103** (Suppl. 1)**:**35–40.
25. **Herbel, M. J., J. S. Blum, S. E. Hoeft, S. M. Cohen, L. L. Arnold, J. Lisak, J. F. Stolz, and R. S. Oremland.** 2002. Dissimilatory arsenate reductase activity and arsenate-respiring bacteria in bovine rumen fluid, hamster feces, and the termite hindgut. *FEMS Microbiol. Ecol.* **41:**59–67.
26. **Hoeft, S. E., T. R. Kulp, J. F. Stolz, J. T. Hollibaugh, and R. S. Oremland.** 2004. Dissimilatory arsenate reduction with sulfide as electron donor: experiments with Mono Lake water and isolation of strain MLMS-1, a chemoautotrophic arsenate respirer. *Appl. Environ. Microbiol.* **70:** 2741–2747.
27. **Huber, R., M. Sacher, A. Vollmann, H. Huber, and D. Rose.** 2000. Respiration of arsenate and selenate by hyperthermophilic Archaea. *Syst. Appl. Microbiol.* **23:**305–314.
28. **Jay, J. A., N. Keon Blute, H. F. Hemond, and J. L. Durant.** 2004. Arsenic-sulfides confound anion exchange resin speciation of aqueous arsenic. *Water Res.* **38:**1155–1158.
29. **Johnson, D. L., and M. E. Pilson.** 1975. The oxidation of arsenite in seawater. *Environ. Lett.* **8:**157–171.
30. **Johnson, D. L., and M. E. Q. Pilson.** 1972. Spectrophotometric determination of arsenite, arsenate, and phosphate in natural waters. *Anal. Chim. Acta* **58:**289–299.
31. **Karthikeyan, S., and S. Hirata.** 2003. Arsenic speciation in environmental samples. *Anal. Lett.* **36:**2355–2366.
32. **Kneebone, P. E., P. A. O'Day, N. Jones, and J. G. Hering.** 2002. Deposition and fate of arsenic in iron- and arsenic-enriched reservoir sediments. *Environ. Sci. Technol.* **36:**381–386.
33. **Krafft, T., and J. M. Macy.** 1998. Purification and characterization of the respiratory arsenate reductase of *Chrysiogenes arsenatis*. *Eur. J. Biochem.* **255:**647–653.
34. **Kuai, L., A. A. Nair, and M. F. Polz.** 2001. Rapid and simple method for the most-probable-number estimation of arsenic-reducing bacteria. *Appl. Environ. Microbiol.* **67:**3168–3173.
35. **Laverman, A. M., J. S. Blum, J. K. Schaefer, E. J. P. Phillips, D. R. Lovley, and R. S. Oremland.** 1995. Growth of strain SES-3 with arsenate and other diverse electron acceptors. *Appl. Environ. Microbiol.* **61:**3556–3561.

36. **Liu, A., E. Garcia-Dominguez, E. D. Rhine, and L. Y. Young.** 2004. A novel arsenate respiring isolate that can utilize aromatic substrates. *FEMS Microbiol. Ecol.* **48:**323–332.
37. **Macy, J. M., K. Nunan, K. D. Hagen, D. R. Dixon, P. J. Harbour, M. Cahill, and L. I. Sly.** 1996. *Chrysiogenes arsenatis* gen. nov., sp. nov., a new arsenate-respiring bacterium isolated from gold mine wastewater. *Int. J. Syst. Bacteriol.* **46:**1153–1157.
38. **Macy, J. M., J. M. Santini, B. V. Pauling, A. H. O'Neill, and L. I. Sly.** 2000. Two new arsenate/sulfate-reducing bacteria: mechanisms of arsenate reduction. *Arch. Microbiol.* **173:**49–57.
39. **Malasarn, D., C. W. Saltikov, K. M. Campbell, J. M. Santini, J. G. Hering, and D. K. Newman.** 2004. *arrA* is a reliable marker for As(V) respiration. *Science* **306:**455.
40. **Melamed, D.** 2005. Monitoring arsenic in the environment: a review of science and technologies with the potential for field measurements. *Anal. Chim. Acta* **532:**1–13.
41. **Mucci, A., L. F. Richard, M. Lucotte, and C. Guignard.** 2000. The differential geochemical behavior of arsenic and phosphorus in the water column and sediments of the Saguenay Fjord estuary, Canada. *Aquat. Geochem.* **6:**293–324.
42. **Mukhopadhyay, R., and B. P. Rosen.** 2002. Arsenate reductases in prokaryotes and eukaryotes. *Environ. Health Perspect.* **110**(Suppl. 5)**:**745–748.
43. **Mukhopadhyay, R., B. P. Rosen, L. Phung, and S. Silver.** 2002. Microbial arsenic: from geocycles to genes and enzymes. *FEMS Microbiol. Rev.* **26:**311–325.
44. **Newman, D. K., T. J. Beveridge, and F. M. M. Morel.** 1997. Precipitation of arsenic trisulfide by *Desulfotomaculum auripigmentum*. *Appl. Environ. Microbiol.* **63:**2022–2028.
45. **Newman, D. K., E. K. Kennedy, J. D. Coates, D. Ahmann, D. J. Ellis, D. R. Lovley, and F. M. Morel.** 1997. Dissimilatory arsenate and sulfate reduction in *Desulfotomaculum auripigmentum* sp. nov. *Arch. Microbiol.* **168:**380–388.
46. **Nielsen, P. H., T. H. Christensen, H. J. Albrechtsen, and R. W. Gillham.** 1996. Performance of the in situ microcosm technique for measuring the degradation of organic chemicals in aquifers. *Ground Water Monit. Remed.* **16:**130–140.
47. **Niggemyer, A., S. Spring, E. Stackebrandt, and R. F. Rosenzweig.** 2001. Isolation and characterization of a novel As(V)-reducing bacterium: implications for arsenic mobilization and the genus *Desulfitobacterium*. *Appl. Environ. Microbiol.* **67:**5568–5580.
48. **O'Day, P. A., D. Vlassopoulos, R. Root, and N. Rivera.** 2004. The influence of sulfur and iron on dissolved arsenic concentrations in the shallow subsurface under changing redox conditions. *Proc. Natl. Acad. Sci. USA* **101:**13703–13708.
49. **Oremland, R. S., J. S. Blum, C. W. Culbertson, P. T. Visscher, L. G. Miller, P. R. Dowdle, and F. E. Strohmaier.** 1994. Isolation, growth, and metabolism of an obligately anaerobic, selenate-respiring bacterium, strain SES-3. *Appl. Environ. Microbiol.* **60:**3011–3019.
50. **Oremland, R. S., D. G. Capone, J. F. Stolz, and J. Fuhrman.** 2005. Opinion: whither or wither geomicrobiology in the era of 'community metagenomics.' *Nat. Rev. Microbiol.* **3:**572–578.
51. **Oremland, R. S., P. R. Dowdle, S. Hoeft, J. O. Sharp, J. K. Schaefer, L. G. Miller, J. S. Blum, R. L. Smith, N. S. Bloom, and D. Wallschlaeger.** 2000. Bacterial dissimilatory reduction of arsenate and sulfate in meromictic Mono Lake, California. *Geochim. Cosmochim. Acta* **64:**3073–3084.
52. **Oremland, R. S., T. R. Kulp, J. S. Blum, S. E. Hoeft, S. Baesman, L. G. Miller, and J. F. Stolz.** 2005. A microbial arsenic cycle in a salt-saturated, extreme environment. *Science* **308:**1305–1308.
53. **Oremland, R. S., and J. F. Stolz.** 2005. Arsenic, microbes and contaminated aquifers. *Trends Microbiol.* **13:**45–49.
54. **Oremland, R. S., and J. F. Stolz.** 2003. The ecology of arsenic. *Science* **300:**939–944.
55. **Perez-Jimenez, J. R., C. Defraia, and L. Y. Young.** 2005. Arsenate respiratory reductase gene (arrA) for Desulfosporosinus sp. strain Y5. *Biochem. Biophys. Res. Commun.* **338:**825–829.
56. **Radajewski, S., P. Ineson, N. R. Parekh, and J. C. Murrell.** 2000. Stable-isotope probing as a tool in microbial ecology. *Nature* **403:**646–649.
57. **Rochette, E. A., B. C. Bostick, G. C. Li, and S. Fendorf.** 2000. Kinetics of arsenate reduction by dissolved sulfide. *Environ. Sci. Technol.* **34:**4714–4720.
58. **Saltikov, C. W., A. Cifuentes, K. Venkateswaran, and D. K. Newman.** 2003. The *ars* detoxification system is advantageous but not required for As(V) respiration by the genetically tractable *Shewanella* species strain ANA-3. *Appl. Environ. Microbiol.* **69:**2800–2809.
59. **Saltikov, C. W., and D. K. Newman.** 2003. Genetic identification of a respiratory arsenate reductase. *Proc. Natl. Acad. Sci. USA* **100:**10983–10988.
60. **Saltikov, C. W., R. A. Wildman, Jr., and D. K. Newman.** 2005. Expression dynamics of arsenic respiration and detoxification in *Shewanella* sp. strain ANA-3. *J. Bacteriol.* **187:**7390–7396.
61. **Santini, J. M., J. F. Stolz, and J. M. Macy.** 2002. Isolation of a new arsenate-respiring bacterium—physiological and phylogenetic studies. *Geomicrobiol. J.* **19:**41–52.
62. **Santini, J. M., I. C. Streimann, and R. N. vanden Hoven.** 2004. *Bacillus macyae* sp. nov., an arsenate-respiring bacterium isolated from an Australian gold mine. *Int. J. Syst. Evol. Microbiol.* **54:**2241–2244.
63. **Sierra-Alvarez, R., J. A. Field, I. Cortinas, G. Feijoo, M. T. Moreira, M. Kopplin, and A. J. Gandolfi.** 2005. Anaerobic microbial mobilization and biotransformation of arsenate adsorbed onto activated alumina. *Water Res.* **39:**199–209.
64. **Silver, S., and L. T. Phung.** 2005. Genes and enzymes involved in bacterial oxidation and reduction of inorganic arsenic. *Appl. Environ. Microbiol.* **71:**599–608.
65. **Simeonova, D. D., D. Lievremont, F. Lagarde, D. A. E. Muller, V. I. Groudeva, and M.-C. Lett.** 2004. Microplate screening assay for the detection of arsenite-oxidizing and arsenate-reducing bacteria. *FEMS Microbiol. Lett.* **237:**249–253.
66. **Smedley, P. L., and D. G. Kinniburgh.** 2002. A review of the source, behaviour and distribution of arsenic in natural waters. *Appl. Geochem.* **17:**517–568.
67. **Stolz, J. F., D. J. Ellis, J. S. Blum, D. Ahmann, D. R. Lovley, and R. S. Oremland.** 1999. *Sulfurospirillum barnesii* sp. nov. and *Sulfurospirillum arsenophilum* sp. nov., new members of the *Sulfurospirillum* clade of the ε proteobacteria. *Int. J. Syst. Bacteriol.* **49:**1177–1180.
68. **Strickland, J., and T. Parsons.** 1972. A practical handbook of seawater analysis. *Fish Res. Board Can. Bull.* **167:**310.
69. Reference deleted.
70. **Volkl, P., R. Huber, E. Drobner, R. Rachel, S. Burggraf, A. Trincone, and K. O. Stetter.** 1993. *Pyrobaculum aerophilum* sp. nov., a novel nitrate-reducing hyperthermophilic archaeum. *Appl. Environ. Microbiol.* **59:**2918–2926.
71. **Zhou, J., M. A. Bruns, and J. M. Tiedje.** 1996. DNA recovery from soils of diverse composition. *Appl. Environ. Microbiol.* **62:**316–322.

Biotransformations of Manganese

BRADLEY M. TEBO, BRIAN G. CLEMENT, AND GREGORY J. DICK

100

Manganese (Mn), the fifth most abundant element in the Earth's crust, and after iron (Fe) the most abundant transition metal, is an essential micronutrient for most organisms. Found at the core of the reaction center of the light-harvesting complex, Mn is particularly important for plants and other photosynthetic microorganisms. It is also a cofactor for numerous enzymes, including manganese superoxide dismutase, manganese catalase, and manganese-dependent ribonucleotide reductase (36). The redox properties of Mn make it central to a variety of biological processes and result in significant and often rapid biogeochemical cycling that is mediated by abiotic and biotic oxidation and reduction, biological uptake, and mineral formation. For overviews on Mn, the reader is referred to recent review articles (35, 69, 113, 117).

CHEMISTRY OF MANGANESE

In nature Mn occurs in three different oxidation states, +II, +III, and +IV (Fig. 1). In general, Mn(II) is thermodynamically favored in the absence of oxygen and at low pH, whereas Mn(III) and Mn(IV) are favored in the presence of oxygen and at high pH. Mn(II) is the principal soluble form, occurring as a cation (Mn^{2+}, $MnCl^+$, $MnOH^+$, etc.) in solution. Mn(II) may also occur as insoluble phosphates or carbonates and as a minor constituent of other minerals, including Mn(III,IV) oxides. Mn(II) can exist at up to millimolar concentrations in natural waters, even in the presence of oxygen. Due to the high activation energy, the oxidation of Mn(II) to Mn(III) and Mn(IV) is largely catalyzed by microorganisms (bacteria and fungi) and, as far as we know, requires oxygen [although Mn(II)→(Mn(IV) oxidation coupled to denitrification is thermodynamically favorable]. Mn(III), which is thermodynamically unstable and disproportionates in aqueous media to yield Mn(II) + Mn(IV), persists only in certain soluble organic complexes (62–64, 70, 92) and in some minerals (e.g., reference 66); Mn(IV) occurs, often with various amounts of Mn(III), in insoluble oxides, oxyhydroxides, and hydroxides. Hereafter referred to as Mn(III,IV) oxides or simply Mn oxides, these are important constituents of soils, sediments, and ore-grade deposits. Mn(III) and Mn(IV) are also found in environmentally prevalent ferromanganese (oxyhydr)oxide minerals, which often occur as layer-type or tunnel-structure minerals (88, 93, 113). The Mn(III)/Mn(II) and Mn(IV)/Mn(II) couples have particularly high redox potentials, making Mn(III) and Mn(IV) phases some of the stronger oxidants found in the environment. For diverse dissimilatory metal-reducing bacteria, Mn oxides serve as excellent electron acceptors in the oxidation of organic matter or H_2 in the absence of O_2. Mn oxides also abiotically oxidize a variety of reduced inorganic and organic compounds; for example, they oxidize natural organic matter producing low-molecular-weight organic compounds (110), hydrogen sulfide, Fe(II), and reduced forms of other metals, such as Se, As, and Cr, thereby influencing toxic metal availability by oxidative precipitation or solubilization (41, 42, 56, 72, 89, 98, 129). With high sorptive capacities, Mn oxides adsorb a wide range of ions, controlling the distributions and bioavailability of many toxic and essential elements (113).

Mn CYCLING IN THE ENVIRONMENT

As biotic and abiotic processes both play important roles in the oxidation and reduction of Mn, a major challenge to understanding Mn biotransformations is the differentiation and subsequent quantification of the biotic and abiotic components of the processes. In Mn(II) oxidation, for example, bacteria may catalyze the oxidation of Mn(II)→ Mn(III,IV) oxides, but these oxides then catalyze further Mn(II) oxidation (i.e., the process is autocatalytic). In Mn oxide reduction, the oxides are so reactive that often a significant fraction of Mn oxides may be reduced abiotically by reductants in the environment (e.g., sulfide).

Biological Mn(II) oxidation is generally fast relative to abiotic Mn(II) oxidation processes, including surface-catalyzed reactions (6, 53, 79, 86, 116), suggesting that biological Mn(II) oxidation dominates in the environment. At specific field sites, a number of investigations have shown that biological processes are responsible for Mn(II) oxidation at those locations (31, 32, 40, 48, 52, 58, 73–75, 112, 114, 115, 118, 120, 124, 130). These observations suggest that the majority of naturally occurring Mn oxides is derived directly from biological Mn(II) oxidation or from the subsequent alteration of biogenic oxides.

Relatively few studies have examined the importance of biological Mn(IV) reduction in the environment, probably due to the difficulty of distinguishing the direct mechanism of Mn oxide reduction from abiotic or indirect ones. Biological Mn oxide reduction by dissimilatory Mn-reducing

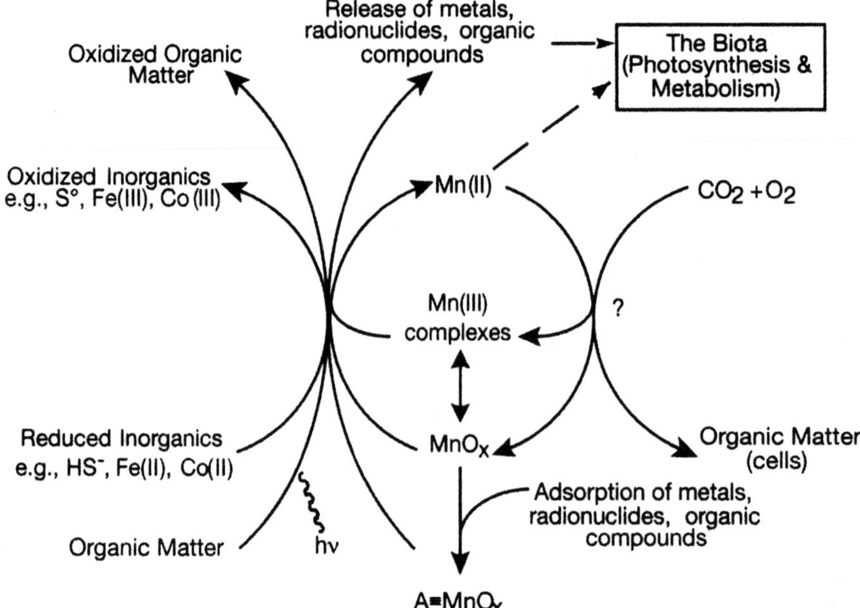

FIGURE 1 The manganese cycle.

bacteria requires a mechanism for transferring electrons to a solid substrate. This may occur directly either through cell-oxide contact or, perhaps, by way of soluble compounds that shuttle electrons between the cell and the Mn oxide. Cells may also indirectly reduce Mn oxides by producing reductants such as sulfide or Fe(II). Studies that have shown the importance of biological Mn oxide reduction have largely relied upon geochemical modeling. For example, it has been estimated that biological Mn oxide reduction is responsible for a significant fraction of the total amount of organic matter mineralized in coastal sediments (21, 22).

METHODS TO MEASURE Mn SPECIES

Quantifying the partitioning of Mn between soluble and particulate fractions is a major component of nearly all laboratory or field Mn biotransformation research. The simplest, though somewhat limited, measure of Mn speciation is based on filtration methods that use a 0.2-μm-pore-size membrane filter. In practice, the soluble fraction passing through the filter provides a measure of Mn(II), whereas the fraction that is retained on the filter is that of the particulate Mn(III,IV) oxide species. However, the method's limitations must be considered when arriving at final results. For example, in fresh (low-ionic-strength) waters, some colloidal Mn oxides may pass through 0.2-μm-pore-size filters, and recent work from suboxic zones of the Black Sea and the Chesapeake Bay indicates that soluble Mn(III) complexes may represent a significant fraction of the total dissolved Mn (122). The measurement of Mn(III) complexes is not trivial; however, routine methods for measuring them are under development. The most reliable current methods involve electrochemical measurements that use cyclic voltammetry. While the description of these methods is largely beyond the scope of this chapter, some of the methods used for measuring oxidized Mn (described below) may be applicable for measurement of both colloidal Mn oxide and Mn(III) species.

The techniques described in the following sections are based on measurements obtained by using a spectrophotometer. These methods are readily adaptable for use with microtiter plate readers or dip probe colorimeters and/or spectrophotometers, which require smaller volumes and are more rapid.

Colorimetric Determination of Mn(II)

Total manganese concentrations are facilely measured by using the formaldoxime (formaldehyde oxime) colorimetric technique (9). The formaldoxime reagent is a mixture of hydroxylamine (NH_2OH) hydrochloride and formaldehyde. Under basic conditions, formaldoxime forms an orange-red complex with Mn(II). Because hydroxylamine is a strong reducing agent, Mn(III) and Mn(IV) are mostly reduced to Mn(II) during the assay, and hence, total Mn is measured (if it is desirable to measure total Mn, then the sample should be prereduced, e.g., with 0.1% hydroxylamine–HCl, before assaying). When filtration is used (0.2-μm-pore-size membrane filters are used routinely, but 0.02-μm-pore-size filters can be used if Mn colloids are suspected), assaying the filtrate will yield the dissolved Mn(II) concentration. The main interference in the assay comes from iron, which forms a pink complex with formaldoxime. Usually, this is not a major problem with filtered samples, but if it is suspected, separate iron analyses can be made and the interference determined and subtracted to correct the data.

The Formaldoxime Method

Stock Solutions

1. 10% hydroxylamine–HCl (for low detection limits, this should be purified to remove iron contamination)
2. 37% formaldehyde
3. Formaldoxime: mix 20:1 10% hydroxylamine–HCl:37% formaldehyde
4. Dilute concentrated ammonium hydroxide 1:10 with dH_2O (note: concentrated NH_4OH may be more useful for

assaying Mn in buffered culture medium). For the lowest detection limits, use ultrapure NH_4OH.

5. Mixed reagent: Mix formaldoxime and dilute NH_4OH in a proportion to obtain pH 8.8 to 8.9 after addition of 0.5 ml to 3.5 ml of sample. The ratio is approximately 2:1 formaldoxime:dilute NH_4OH. (Note: for freshwater with low magnesium and calcium, a pH of 9.0 to 9.2 is acceptable.) Prepare reagent freshly each day.

Standards

1. Primary standard, 2 mM Mn^{2+}. Dissolve 0.1688 g of $MnSO_4 \cdot H_2O$ (molecular weight = 169.01) in 500 ml of H_2O. $MnSO_4 \cdot H_2O$ is stable at 57 to 114°C and should be dried accordingly before weighing. Alternatively, a standard Mn(II) solution can be purchased. Although values may vary slightly with each formaldoxime batch, Mn concentrations between 1 and 50 μM typically produce absorbance values between 0.01 and 0.5 with a 1-cm path length.

2. Working standard (an appropriate dilution of the primary standard).

Procedure

Add 0.5 ml of mixed reagent to 3.5 ml of sample. Color develops in a few seconds and is stable for at least 30 min. Read in a cuvette at 450 nm. Calibrate by making a standard curve (five or six different concentrations and a blank). The A_{450} is linear up to 1.0, and the volume can be adjusted according to your cuvette size; we routinely use the same reagents with 0.5- to 40-ml samples (1- to 10-cm-path-length cuvettes).

Colorimetric Determination of Mn Oxides

The formaldoxime colorimetric method can be used to measure particulate Mn (Mn trapped by filters), which, depending on the situation, may give an accurate measurement of Mn oxide. In this case, the sample is filtered (0.2- or 0.02-μm-pore-size filter) and the filter is placed in a solution of 0.1% NH_2OH, which reduces any oxidized Mn to Mn(II). The Mn(II) in this solution can then be assayed by using a separate standard curve also prepared in 0.1% NH_2OH. This method is not specific for Mn(III,IV) oxides, since particulate material may also contain Mn(II) (e.g., in biomass), which is released under acidic conditions. Thus, the method does not distinguish between different Mn oxidation states; i.e., it does not tell whether you have a mixed Mn(II,III) or Mn(III,IV) oxide or a pure Mn(III) or Mn(IV) phase.

Although there are other spectrophotometric techniques for directly measuring oxidized forms of Mn (e.g., leuco crystal violet [61]), our preferred method involves the reagent leucoberbelin blue (LBB [2, 65]), which can be purchased from Sigma-Aldrich.

The LBB Method

Stock Solutions

Use 0.04% LBB in 45 mM acetic acid, dissolved overnight. Store at 4°C; protect from light. It will oxidize slowly in storage; always use a water blank to account for background absorbance.

Standards

Potassium permanganate ($KMnO_4$): a range of concentrations between 1 and 100 μM, freshly made (<24 h old) and diluted.

Procedures

Spot test. The 0.04% LBB reagent can be used as a spot test reagent for oxidized Mn on colonies growing on agar plates, filtered samples, cell pellets, etc. A positive reaction turns bright blue.

Quantitative assay. Mix your sample 1/5 or 1/10 in 0.04% LBB working reagent. Let mixture react for 15 to 20 min in the dark at room temperature. Samples on filters can be reacted directly in the working reagent. If possible, use a suitable blank (medium blank, filter blank, etc.), in addition to the water blank, as a few other chemicals may react with LBB, especially Co(III). Measure absorbance on a spectrophotometer at 618 nm (or 620 nm). Use $KMnO_4$ to prepare a standard curve. One mole of $KMnO_4$ oxidizes 5 mol of LBB. The results can be presented as oxidizing equivalents, or one can convert the standard curve for the different valences of the various oxidation states of Mn ($KMnO_4$ has an oxidation state of +VII; MnO_2, +IV; and MnOOH, +III; 40 μM $KMnO_4$ is equivalent to 100 μM MnO_2). A variant of the technique uses 0.4% LBB in 450 mM acetic acid, which is added to a sample in a ratio of 1:10. The absorbance is linear up to an A_{618} of 1.5.

Mn Oxidation State

The average oxidation state of oxidized forms of Mn is a useful measurement for a variety of studies, particularly those focused on characterization of the Mn oxide minerals, since different minerals have different reactivities, often due to substitution of Mn(III) for Mn(IV) in the mineral lattice or incorporation of Mn(III) into interlayer regions of layered Mn oxides. Most of the analytical methods for measuring the oxidation state of Mn oxides involve measuring the total equivalent concentration of oxidized manganese, which is equal to

$$[\mathrm{Ox}] = 2[\mathrm{Mn(IV)}] + [\mathrm{Mn(III)}],$$

and the stoichiometry of the oxides is represented as O/Mn ratios or MnO_x, where

$$x = 1 + \tfrac{1}{2}\left(\frac{(2[\mathrm{Mn(IV)}]+[\mathrm{Mn(III)}])}{([\mathrm{Mn(IV)}]+[\mathrm{Mn(III)}]+\mathrm{Mn(II)})}\right).$$

There are many methods for determining the oxidation state, including the use of LBB (see above); however, Murray et al. (81) have concluded that either the oxalate reduction method or the iodometric titration method is the most accurate and reproducible (81). Although Fe(III) interferes in the iodometric method and not in the oxalate method, the iodometric method was preferred because it involved less trial and error for choosing sample size than does the oxalate method, which is a method of differences. Corrections for Fe(III) can be easily applied simply by measuring total Fe (e.g., by the ferrozine colorimetric method or by atomic absorption spectroscopy). In a laboratory situation with relatively high Mn but low Fe concentrations, Fe rarely causes measurable interference. For details of the method, the reader is referred to Murray et al. (81) and Kalhorn and Emerson (57).

Spectroscopy and Other Techniques

A variety of spectroscopic techniques are also essential to the study of Mn transformations. In general, these techniques require specific training and analytical expertise, and thus, descriptions of these techniques are beyond the scope of this chapter. Rather, we list here some of the techniques and their applications.

Many metal concentrations have traditionally been measured by using flame- or graphite furnace-atomic absorption spectroscopy; however, these measurements have recently been replaced with inductively coupled plasma (ICP) mass spectrometry or optical emission spectroscopy. These latter instruments are capable of measuring a suite of elements as opposed to the atomic absorption spectroscopy techniques, which are designed to measure one element.

Natural Mn oxides are renowned for being complex mixtures of minerals with small crystals and poor crystallinity. Thus, they are very difficult to characterize. Recent applications of high-resolution electron microscopy, time-of-flight neutron-diffraction, and synchrotron-based X-ray spectroscopy have provided new insights into the structure of Mn oxide minerals produced in a variety of environments (47, 76, 77) and by microbes in pure culture (5, 6, 127).

MANGANESE(II) OXIDATION

Rate Measurements

Measuring Mn(II) oxidation rates is conceptually straightforward due to the phase change from dissolved Mn(II) to solid Mn(III/IV) oxides. Dissolved and particulate fractions are simply separated by membrane filtration, and the total Mn in each fraction is determined. Like all filtration methods, this provides an operational definition of the particulate Mn present, dependent on membrane pore size. When rates are determined in this fashion, it is more accurate to describe the resulting rates as that of Mn(II) removal or particulate Mn formation, unless the oxidized Mn is determined directly. Such rates are typically initially linear and then slow appreciably after 4 to 6 h (cf. references 40, 112, and 115). As with all rate measurements, rates should be determined from time course measurements; end point measurements will yield accurate results only if the incubation time falls within the range in which the rates are linear.

In general, two considerations are important for these assays. First, the Mn(II) concentration utilized must not inhibit biological activity. Second, effective abiotic controls must be employed to distinguish biological Mn(II) removal and/or oxidation from the adsorption and autocatalytic oxidation of Mn(II) on Mn oxide surfaces.

The appropriate method for determining Mn(II) oxidation rates depends on the sample type. Mn(II)-oxidizing cultures can thrive with $>100\ \mu M$ Mn(II) and are thus easily analyzed with the formaldoxime or LBB method (cf. reference 44). In contrast, Mn cycling in many environments occurs at very low concentrations, and environmental assemblages can be inhibited by Mn(II) amendments above $10\ \mu M$ (114). For such conditions, the radioisotope ^{54}Mn offers greater sensitivity and straightforward analysis.

Controls for abiotic Mn(II) removal are problematic due to their potential for altering chemical conditions in an incubation. For example, prefiltration would provide an abiotic control but only for homogeneous Mn(II) oxidation by O_2, a slow process largely dependent on pH and reactant concentrations (108). The ideal biological control would poison all enzyme-catalyzed activity while not affecting abiotic Mn(II) removal by adsorption and autocatalytic oxidation.

Working with chemically synthesized, "flocculent" Mn oxides and natural seawater, Rosson et al. (96) found that these conditions were satisfied best by 15 mM sodium azide with or without the addition of 50-μg/ml penicillin and 50-μg/ml tetracycline. When they utilized it in Saanich Inlet with sonicated controls, Emerson et al. (40) found that the poison mixture enhanced $^{54}Mn(II)$ removal relative to the sonicated control without poisons but inhibited $^{54}Mn(II)$ removal relative to untreated incubations. In contrast, Shiller and Stephens (99) have reported that these amendments prevented Mn(II) adsorption in particulate-rich Mississippi River water, possibly confounding their use as a "microbial control," although it remains unclear if the adsorption sites blocked by azide were biological (e.g., active enzyme sites that are the target of azide). As noted by Rosson et al. (96), the best approach may well be testing different poisons for each experimental system.

A sharp temperature optimum is an effective indicator that a geochemical process is mediated by biology (10), and it has been used to demonstrate that enzymes catalyze Mn oxidation in natural samples (114, 121). Enzymatic activity increases with temperature to a point where activity decreases due to denaturation of a protein's complex structure, whereas Mn oxidation by a chemical mechanism is expected to increase proportionally with increasing temperature.

Anoxic controls have been utilized to discriminate between biological Mn(II) oxidation and abiotic Mn(II) removal (112, 114, 115). To date, all cultivated Mn(II)-oxidizing bacteria require O_2 (113), and despite geochemical indications of Mn(II) oxidation linked to denitrification, no anoxic formation of particulate Mn has been demonstrated (97). Anoxia should not affect abiotic binding (sorption) of Mn(II) but may interfere with autocatalysis on preexisting Mn oxide surfaces. Another drawback is the potential for stimulating Mn oxide reduction; anoxic controls are therefore more suitable for tracer (^{54}Mn) assays than for direct determinations of Mn(II).

Another advantage of radiotracer experiments is that the fraction of Mn(II) oxidized versus the fraction adsorbed can be estimated by exchanging adsorbed $^{54}Mn(II)$ with "cold" (nonradioactive) Mn(II) or Cu(II). This has been accomplished either by addition of excess cold $MnCl_2$ or $CuCl_2$ to incubation bottles just prior to filtration (12, 73, 109) or by rinsing the filter directly with $MnCl_2$- or $CuCl_2$-containing solutions (95, 119). In combination with anoxic controls, Sunda and Huntsman (109) used stable Mn(II) washes to exchange adsorbed $^{54}Mn(II)$. This was followed by ascorbate solutions at pH 8.0 to distinguish reducible Mn oxide phases from intracellular Mn(II) pools. In this way, the discrete pools of Mn (dissolved, adsorbed, intracellular, and oxide) can be assessed directly. Clearly, such measures are not required for all investigations; however, the reader is cautioned to think carefully about which parameter (e.g., particulate Mn formation versus Mn oxide production) is to be addressed.

Mn(II) Oxidation and Removal at High Mn(II) Concentrations

Pure cultures and subcellular preparations with high Mn tolerance offer the option of colorimetric assays for analyzing dissolved and particulate Mn (formaldoxime) and the oxidizing equivalents of Mn in both fractions (LBB). The volume required for filtration that will maximize sensitivity while avoiding clogged and slow filters depends on the activity in the sample and the Mn(II) concentration used. Typically, one must filter enough volume to observe at least 1 to $2\ \mu M$ Mn in the particulate fraction after dilution in the assay buffer.

Mn(II) removal and Mn(II) oxidation rates by pure cultures have been determined during growth in batch culture and in chemostat cultures and with whole and subcellular preparations of stationary-phase cultures (12, 44, 59, 60, 128). Cultures are typically washed and concentrated by repeated centrifugation and resuspension in fresh wash solution containing buffered ascorbate (e.g., 10 μM ascorbate in 100 μM HEPES, pH 7.5) to reduce and remove preformed Mn oxides. The cells are either resuspended in the assay medium (e.g., buffered mineral salt solution) or are further processed to separate a subcellular fraction (cf. references 43 and 128).

The assay proceeds with few modifications to the LBB and formaldoxime methods and can be scaled to different volumes, Mn concentrations, etc. What we describe here assumes that sample size is not limiting. Incubate the samples under the desired conditions with an appropriate concentration of Mn(II). At each time or end point, samples will be collected and measured for dissolved Mn (Mn passing through a filter) and total Mn and/or oxidized Mn (LBB-reactive material trapped on a filter). For higher accuracy, it is recommended that duplicate or triplicate measurements be made. Remove replicate aliquots for total Mn and filter subsamples (ca. 10 to 50 ml at 10^6 cells per ml) through prewetted, 25-mm-diameter, 0.2-μm-pore-size polycarbonate, polyethersulfone, or mixed-cellulose ester membrane filters (different diameters can be used). Wash the filter with 0.2-μm-pore-size-filtered Mn(II)-free medium (typically, we use 5 ml) to rinse dissolved Mn through the filter. Measure total Mn in the dissolved and particulate fractions by formaldoxime and/or oxidized Mn by LBB (correcting the dissolved Mn measurement for the volume of the wash solution used).

$$Mn_{Particulate} = \frac{Mn_{Filter}}{\text{volume filtered}}$$

$$Mn_{Dissolved} = Mn_{Filtrate} \times \frac{\text{volume filtered} + \text{wash volume}}{\text{volume filtered}}$$

To determine oxidized Mn in the particulate and dissolved fractions, substitute the LBB-oxidizing equivalents in the above formulas for the Mn_{Filter} and $Mn_{Filtrate}$, respectively. It is best to ensure mass balance according to

$$Mn_{Particulate} + Mn_{Dissolved} = Mn_{total}$$

As an assay for enzymatic activity in cell cultures or cell-free extracts, the above methods can be simplified by simply removing aliquots of the assay mix at different time intervals and adding it to LBB (7, 90). After the LBB-Mn(III,IV) reaction is complete, cell debris can be removed by centrifugation or filtration before spectrophotometric measurements. Since filtration is not being used to separate Mn oxides from soluble Mn(III) complexes, this assay yields rates expressed as oxidizing equivalents per unit time.

^{54}Mn(II) Removal and Particulate ^{54}Mn Formation

As noted above, the lower Mn tolerance of natural microbial populations (and some pure cultures) limits the use of colorimetry; the ~1 μM detection limit of the colorimetric assays would require a large filtration volume to observe small changes in an assay of particulate Mn with 18 μM total Mn. To observe Mn(II) oxidation at low, environmentally relevant concentrations and to simplify analysis, the gamma-emitting radioisotope ^{54}Mn has been used in a variety of environments, including stratified marine basins (40, 112), lakes (24), rivers (99), and hydrothermal vents (31, 73). At this writing, ^{54}Mn can be obtained from PerkinElmer (Boston, Mass.); it has a half-life of 312 days and emits >99% gamma rays with a mean energy of 830 kEv. ^{54}Mn is easy to detect in a gamma counter with as little as 0.01 to 0.1 μCi per filtration, depending on the counting efficiency (CE) and count time.

Procedure

1. Determine the efficiency of the gamma counter to be used.
2. Add enough ^{54}Mn as carrier-free $MnCl_2$ to each incubation for 10,000 to 30,000 cpm (see below) per filtration volume, homogenize by shaking, and incubate for the desired interval.
3. Remove two aliquots for total ^{54}Mn determination into counting vials.
4. Filter the planned volume and rinse the filter with 0.2-μm-filtered Mn-free incubation medium.
5. Transfer the filter to a counting vial.
6. Transfer a subsample of the filtrate to a counting vial; we use 3 ml in 4- or 5-ml vials.
7. Bring the total and filter vials up to the same volume as the filtrate vial with 0.1% NH_2OH to solubilize and/or disperse ^{54}Mn and equalize the counting geometry among samples.
8. Count all samples in a gamma counter for an interval long enough to reduce counting error to <1% (most gamma counters automatically report this parameter).
9. Calculate ^{54}Mn in each fraction as described in "Mn(II) Oxidation and Removal at High Mn(II) Concentrations" above.
10. To convert to Mn concentration, multiply the ^{54}Mn fraction by the Mn(II) concentration observed in the original sample.

CE and Isotope Quantity

The CE is the ratio of counts per minute (cpm) to disintegrations per minute (dpm) and varies for each isotope and counter. Some counters have automated corrections, but it is important to check how much isotope is required to observe small changes in the sample. Simply count a dilution series of standardized $^{54}MnCl_2$ stock (a newly arrived isotope stock is usually sufficient) in 0.1 N HCl, using the same volume anticipated for the actual assay. Identical volume, ^{54}Mn-free blanks should also be counted to account for background radiation.

$$CE(\%) = \frac{\text{cpm}}{\text{dpm}} \times 100 = \frac{(\text{sample cpm} - \text{blank cpm})}{\text{sample } \mu\text{Ci}} \times \frac{1\ \mu\text{Ci}}{2.22 \times 10^6\ \text{dpm}} \times 100$$

Total quantity of isotope required for observing 10^4 cpm per filtration volume:

$$\frac{\mu\text{Ci}}{\text{incubation}} = \frac{100}{CE(\%)} \times \frac{1\ \mu\text{Ci}}{22.2 \times 10^6\ \text{dpm}} \times \frac{10^4\ \text{cpm}}{\text{filtration volume}} \times \frac{\text{filtration volume}}{\text{incubation volume}}$$

An isotope quantity between one and three times this value allows determinations of 0.1 to 0.01% differences, depending on replicate variability.

Mn(III) Intermediates and Stepwise Mn Oxidation

While the assays described above deal with the biogeochemical fate of Mn, a recently developed method has shed light on the biochemical mechanism of Mn(II) oxidation. Webb et al. (128) utilized the Mn(III)-binding ligand pyrophosphate (PP) to observe the sequential, one-electron oxidation of Mn(II) by *Bacillus* sp. strain SG-1. The Mn(III)-PP complex has an extinction coefficient of 6,750 M^{-1} at 258 nm that permits concentration determinations to 1 μM. Webb et al. (128) found that the concentration of PP could be adjusted such that it formed a temporary sink for Mn(III), providing data on the enzymatic formation of both Mn(III) and Mn(IV). Clearly, the proper PP concentration depends on the Mn(II/III) affinity constant and abundance of the "Mn oxidase" assessed and should be adjusted empirically.

Phylogeny of Mn(II)-Oxidizing Bacteria

All Mn(II)-oxidizing bacteria isolated to date are heterotrophic and belong to the low (Firmicutes)- or high (Actinobacteria)-G+C gram-positive bacteria or the α, β, and γ subdivisions of the Proteobacteria (111, 113, 117). Within any particular group, Mn(II) oxidizers are often phylogenetically scattered among bacteria not known to oxidize Mn(II), and so the ability to oxidize Mn(II) seems to be a truly polyphyletic trait. Some of the best-characterized Mn(II)-oxidizing bacteria include freshwater *Leptothrix* spp. (1, 7, 13, 27, 100, 101), *Pseudomonas putida* strains (14, 23, 37, 38, 45, 90), and marine *Bacillus* spores (43, 46, 95, 125, 126). A unifying theme of the characterized Mn(II)-oxidizing bacteria is the involvement of multicopper oxidases in Mn(II) oxidation (14, 15, 27, 43, 45, 46, 125) and the association of Mn(II)-oxidizing activity with either stationary phase or, in *Bacillus*, sporulation.

Detection and Enumeration

Mn(II)-oxidizing bacteria often become encrusted in Mn oxide particles, which distinguishes them from other bacteria and allows them to be observed in environmental samples by transmission electron microcopy. Various morphologies of Mn-encrusted bacteria have been observed in freshwater (51) and marine (20, 28, 29, 33, 111) environments, but it is unclear if these diverse morphologies represent distinct bacterial species, physiological states, or environmental conditions. Stellate morphologies resembling the bacterium *Metallogenium* (54) have been reported in a variety of freshwater (71, 78) and marine (20, 30) environments and in the fossil record (34). However, the validity of *Metallogenium* as a bacterial genus has been questioned (39, 83) and remains in doubt.

Cultivation, Enrichment, and Isolation

Although the oxidation of Mn(II) by O_2 to form Mn oxides is an energy-yielding reaction, to date no chemolithoautotrophic Mn(II)-oxidizing organism has been isolated, and efforts to enrich for Mn(II)-oxidizing organisms in liquid media amended with Mn(II) as the only energy source have not been successful. However, in general, Mn(II)-oxidizing bacteria are more readily enriched in media low in complex organic nutrients (peptone, yeast extract, etc.). Low O_2 and high CO_2 concentrations may also favor the enrichment of some Mn(II)-oxidizing bacteria (11, 50, 80, 123). In our lab, O_2 gradient cultures with low organic media (e.g., M medium [see Table 2]) have been used successfully to enrich for Mn(II)-oxidizing bacteria. Media containing humic acids have also been found to enrich for Mn(II)-oxidizing bacteria (B. M. Tebo and D. Edwards, unpublished results); because Mn oxides are known to break down humic substances to more readily utilized substrates such as pyruvate (110), Mn(II)-oxidizing bacteria may have an advantage in accessing carbon from refractory humic acids.

The easiest and most common method of isolating Mn(II)-oxidizing bacteria is to spread water, sediment, or soil samples onto agar plates containing dissolved Mn(II) and to screen for colonies encrusted in Mn oxides, which appear brown or orange. The presence of Mn oxides can be confirmed by spot test with LBB (see "Colorimetric Determination of Mn Oxides" above). The LBB reagent can also be used to screen for colonies that are capable of producing Mn(III), which has recently been shown to be an intermediate in Mn(II) oxidation (128) and can be stabilized by organic ligands, including siderophores (92). Single Mn(II)-oxidizing colonies can then be isolated by repeated transfer to new plates until pure culture is achieved.

Medium Recipes

A variety of solid and broth media can be used to cultivate Mn(II)-oxidizing bacteria so long as they contain dissolved Mn(II); it is recommended that the media be well buffered to prevent abiotic Mn(II) oxidation that may occur at high pH. Different media that have been formulated for Mn(II)-oxidizing bacteria are given in Tables 1 to 12. These media can be prepared as solid (agar) or broth media. Oxygen gradient cultures can also be prepared from many of these medium formulations, although M medium has proved to be particularly good in our hands for enrichment of Mn(II)-oxidizing bacteria. For gradient tubes, 0.1% "slush" agarose is used as a solidifying agent in screw-cap test tubes (50). The medium should be prepared by using the anaerobic technique (cooled to 45°C with a headspace of N_2) and should be dispensed in the tubes with a headspace of N_2 gas, capped tightly, and cooled until the agar solidifies. The tubes' contents can then be inoculated with a sample and the caps left loose to allow O_2 to diffuse in from the top. Growth of microaerophilic Mn(II)-oxidizing bacteria will be visible as a band or halo of Mn oxides beneath the surface of the agarose. Opposing gradients of O_2 and Mn(II) (or some other nutrient) can also be established by first adding molten agar medium with Mn(II) (or other nutrient) to the bottom of the tubes (by using the anaerobic technique) and allowing a solid agar plug to form. Anoxic slush agar can then be overlaid and the tubes capped (with a headspace of N_2) and cooled. The tubes' contents can be inoculated and incubated as above.

Mn OXIDE REDUCTION

Rate Measurements

Determining Mn oxide reduction rates is complicated by the properties and diversity of Mn oxide mineral phases (see reference 127) and chemical reductants (19, 106, 107). The surface area (i.e., mineral structure), adsorbed compounds, and oxidation state greatly affect the reactivity of Mn oxides (17, 105). Microbiological investigations typically proceed with the addition of defined (i.e., laboratory-synthesized) Mn mineral phases to pure cultures (69, 82) or enrichment cultures (17, 19). This approach addresses microbial rates and reactants, but intercomparison of different studies is difficult without careful Mn oxide synthesis.

The general method is similar to that for Mn(II) oxidation, except that the evolution of dissolved Mn is measured. Another difference is that MnO_x reduction has been directly linked to growth and oxidation of H_2 and carbon sources, while Mn(II) oxidation is suspected only of provid-

TABLE 1 J medium (minimal medium; basal medium for single carbon sources)[a]

Ingredient	Amt added	Final concn
Before autoclaving		
Artificial seawater (2×)	500 ml	
Deionized water (to a final volume of 1 liter)	466 ml	
1.5 M NH_4Cl	1.0 ml	1.5 mM
Noble agar (or other purified agar)	15 g	15 g/liter
After autoclaving, add presterilized solutions:		
1 M HEPES buffer (pH 7.75)	20.0 ml	20 mM
1 M $KHCO_3$	2.0 ml	2 mM
0.73 M KH_2PO_4	0.1 ml	73 μM
1 M $MnCl_2$	0.1 ml	100 μM
Vitamin mix (see below)	2.0 ml	

[a]J media can be supplemented with any carbon source or electron donor to target specific types of bacteria; Mn(II)-oxidizing bacteria have been isolated on J media supplemented after autoclaving with 10 mM (final concentration) succinate, acetate, and formate. KNO_3 can be substituted for NH_4Cl. We also recommend adding a fresh source of iron after autoclaving and cooling (e.g., freshly prepared and filter sterilized ferric ammonium citrate, $FeCl_3$, or $FeSO_4$, to a final concentration of 1 to 4 μM).

TABLE 2 M medium (low nutrient medium)

Ingredient	Amt added	Final concn
Before autoclaving		
Natural seawater	750 ml	
Deionized water (to a final volume of 1 liter)	227 ml	
0.05-g/ml concn each of yeast extract and peptone	1.0 ml	0.05 g/liter (0.1 g/liter total)
Noble agar (or other purified agar)	15 g	15 g/liter
After autoclaving, add presterilized solutions:		
1 M HEPES buffer (pH 7.8)	20.0 ml	20 mM
1 M $KHCO_3$	2.0 ml	2 mM
1 M $MnCl_2$	0.1 ml	100 μM

TABLE 3 *Leptothrix* medium (Lept)[a]

Ingredient	Amt added	Final concn
Before autoclaving		
Deionized water (to a final volume of 1 liter)	982 ml	
1 M $CaCl_2$	0.48 ml	0.48 mM
1 M $MgSO_4$	0.83 ml	0.83 mM
Yeast extract	0.5 g	0.5 g/liter
Agar	15.0 g	15.0 g/liter
After autoclaving, add presterilized solutions:		
1 M D-(+)-glucose	5.0 ml	5 mM
Casamino Acids	0.5 g	0.5 g/liter
1 M HEPES buffer (pH 7.5)	10.0 ml	10 mM
Lept trace element solutions (filter sterilized)	1.0 ml	
1 M $MnCl_2$	0.1 ml	100 μM
10 mM $FeCl_3$ (made fresh and filter sterilized)	0.37 ml	3.7 μM

[a]Information from reference 7.

TABLE 4 Minimal mineral salts medium (MMS) (defined medium for *Leptothrix*)[a]

Ingredient	Amt added	Final concn
Filter sterilize:		
Deionized water	1 liter	1 liter
$CaCl_2 \cdot 2H_2O$	30 mg	200 μM
$MgSO_4 \cdot 7H_2O$	35 mg	140 μM
$(NH_4)_2SO_4$	120 mg	910 μM
KNO_3	15 mg	150 μM
$NaHCO_3$	0.84 mg	10 μM
KH_2PO_4	0.70 mg	5 μM
$NaNO_3$	3.82 g	45 mM
Sodium pyruvate	240 mg	2.9 mM
Vitamin B_{12}	0.002 mg	1.5 nM
$FeSO_4$	0.015 mg	0.1 μM
For MMS + Mn, add:		
$MnSO_4 \cdot H_2O$	8.45 mg	50 μM

[a]MMS can be used as a defined medium for metal speciation studies when pH is controlled by mineral acids (87). $NaNO_3$ is added to increase the ionic strength. MMS can also be modified with 0.25 mg of peptone per liter and 0.50 mg of yeast extract per liter; $FeSO_4$ and vitamin B_{12} do not need to be included in this case.

TABLE 5 K medium (rich medium)

Ingredient	Amt added	Final concn
Before autoclaving		
Natural seawater	750 ml	
Deionized water (to a final volume of 1 liter)	230 ml	
Yeast extract	0.5 g	0.5 g/liter
Peptone	2.0 g	2.0 g/liter
Agar	15 g	15.0 g/liter
After autoclaving, add presterilized solutions:		
1 M HEPES buffer (pH 7.8)	20.0 ml	20 mM
1 M $MnCl_2$	0.1 ml	100 μM

ing cellular energy. For this reason, estimating electron flux is important and Mn(II) produced and adsorbed to Mn oxides during the experiment must also be determined. This is achieved by the addition of a wash step, typically with $CuSO_4$, to displace Mn(II) ions from the particulate MnO_x captured by the filter (12, 95). As with Mn(II) oxidation assays, utilizing ^{54}Mn increases sensitivity.

The Mn oxide of choice is typically δ-MnO_2, which most closely resembles the amorphous (low crystallinity) structure and high oxidation state of freshly precipitated manganates of biological origin (5, 6, 127). Pure δ-MnO_2 would have an Mn:O ratio of 2 and an average oxidation state of +IV. Due to the sorption capacity of amorphous δ-MnO_2, it is difficult to synthesize without a significant fraction of sorbed Mn(II) and average oxidation states are often lower. Careful attention to washing steps during the synthesis procedure (below) helps produce a defined and reactive mineral phase.

Two other factors that greatly affect the reactivity of MnO_x are aging and chemical conditioning. Both age and adsorption-exchange of other ions (e.g., Mn^{2+} and Ca^{2+}) can produce a more crystalline, less reactive mineral phase (5, 127). For this reason, synthesized oxides should never be heat sterilized if they are to be used in quantitative rate measurements, as heating may alter the mineral structure. Rather, the oxides should be synthesized by sterile techniques (see below).

Additional considerations for MnO_x reduction assays are maintenance of E_h (both pH and O_2 concentration) throughout the assay and controls to distinguish indirect MnO_x reduction and its direct use as an electron acceptor. Mn oxides are stable at a pH of >7, and the autooxidation of Mn(II) on Mn oxide surfaces is relatively slow below pH 8 (108). Such autocatalysis is not problematic under the typical anoxic incubation conditions but could occur during filtration and handling in air. To maintain pH, a buffer such as 10 to 20 mM HEPES, poised at a pH of 7.5 to 7.9, is usually sufficient, except when cells are producing metabolites capable of reducing Mn oxides. While Mn reduction has been reported in oxic environs or incubations (8, 91), this appears to be indirect reduction of Mn by microbial metabolites or favored by local E_h conditions. Dissimilatory Mn oxide reduction is thermodynamically favored only in the absence of oxygen (84), and trace oxygen levels in closed-vessel (e.g., Hungate-type) incubations may retard MnO_x reduction by both facultative and obligate anaerobes.

TABLE 6 PYG medium (for soil and freshwater sediment bacteria)[a]

Ingredient	Amt added	Final concn
Before autoclaving		
Deionized water to a final volume of:	1 liter	1 liter
Yeast extract	0.25 g	0.25 g/liter
Peptone	0.25 g	0.25 g/liter
Glucose	0.25 g	1.4 mM
$CaCl_2 \cdot 2H_2O$	0.01 g	68 μM
$MgSO_4 \cdot 7H_2O$	0.5 g	2.0 mM
Agar	15 g	15 g/liter
After autoclaving, add presterilized solution:		
1 M $MnCl_2$	0.1 ml	100 μM

[a]A vitamin mixture can be added if desired, and we suggest addition of a buffer, e.g., 20 mM HEPES, pH 7.5. A 10- or 100-fold dilution of this medium provides a good low-nutrient enrichment medium (50). PYG, peptone-yeast extract-glucose.

TABLE 7 *Pedomicrobium* standard medium (PSM)[a]

Ingredient	Amt added	Final concn
Before autoclaving		
Deionized water to a final volume of:	1 liter	1 liter
Yeast extract	0.50 g	0.50 g/liter
Agar	18 g	18 g/liter
After autoclaving, add presterilized solutions:		
1 M sodium acetate	10 ml	10 mM
Vitamin mixture	10.0 ml	
Mineral salts base	1 ml	
1 M $MnCl_2$	0.1 ml	100 μM

[a]Information from reference 49.

TABLE 8 Artificial seawater (ASW, 2×)[a]

Ingredient	Amt added (g)	Final concn (M)	
		2× stock	1× ASW[b]
$MgSO_4 \cdot 7H_2O$	24.7	0.2	0.1
$CaCl_2 \cdot 2H_2O$	2.9	0.04	0.02
NaCl	35.1	1.2	0.6
KCl	1.5	0.04	0.02

[a]Dissolve each component completely in H_2O to a total volume of 250 ml, and then combine to make 1 liter total.

[b]The ionic strength of 2× ASW when diluted 1:1 with deionized H_2O is roughly equivalent to that of 0.75× natural seawater.

TABLE 9 Vitamin mix for J medium[a]

Ingredient (dissolve each separately in 10 ml of H_2O)	Amt (mg)
Biotin in 40 ml of H_2O (heat to dissolve)	40
Niacin (nicotinic acid)	4
Thiamine	2
p-Aminobenzoic acid	4
Pantothenic acid	2
Pyridoxine	20
Vitamin B_{12}	2
Riboflavin	4
Folic acid	4

[a]Combine all nine solutions and add H_2O to a final volume of 200 ml, and filter sterilize.

TABLE 10 Vitamin mix for PSM[a]

Ingredient	Concn (mg/1,000 ml of deionized H^2O)
Biotin	2
Nicotinamide	5
Thiamine-HCl	5
p-Aminobenzoic acid	5
Calcium pantothenate	5
Pyridoxine-HCl	10
Vitamin B_{12}	0.1
Riboflavin	5
Folic acid	2

[a]The solution should be filter sterilized.

TABLE 11 Trace elements solution for Lept medium[a]

Ingredient	Amt
Deionized H_2O	1 liter
$CuSO_4 \cdot 5H_2O$	10 mg
$ZnSO_4 \cdot 7H_2O$	44 mg
$CoCl_2 \cdot 6H_2O$	20 mg
$Na_2MoO_4 \cdot 2H_2O$	13 mg

[a]Combine the ingredients and filter sterilize.

TABLE 12 Mineral salt base for PSM[a]

Ingredient[b]	Amt
Deionized H_2O	1 liter
EDTA	2.5 mg
$CuSO_4 \cdot 5H_2O$	0.39 mg
$ZnSO_4 \cdot 7H_2O$	11 mg
$FeSO_4 \cdot 7H_2O$	5 mg
$MnSO_4 \cdot 7H_2O$	1.54 mg
$Co(NO_3)_2 \cdot 6H_2O$	0.25 mg
$Na_2B_4O_7 \cdot 10H_2O$	0.18 mg

[a]Information from reference 102.
[b]Combine the ingredients and filter sterilize.

Unfortunately, common reductants (e.g., cysteine and sulfide) used to eliminate oxygen can confound the assay by reducing Mn oxides (16). Thus, anoxic conditions are best achieved and maintained in an anaerobic chamber.

Specific metabolic products that can reduce MnO_x are Fe(II), sulfides (18), catecholates, quinones, and organic acids, including fermentation products (106, 107). Therefore, media for MnO_x reduction assays should be defined to limit the scope of metabolic products. Carbon sources such as acetate and succinate are good choices to limit fermentation. The production of other organics can be difficult to manage; in such cases, short incubation periods and/or controls with spent media are possible options. To prevent problems with sulfide, sulfate-free media should be used, except when the explicit goal is to test the effects of sulfate-reducing bacteria on MnO_x. Additions of Fe(III) should be minimized to prevent dissimilatory Fe(II) production and subsequent Mn oxide reduction.

Finally, care should be given to maintain homogeneous conditions and to replicate all measurements. While such precautions may be routine, solid-phase Mn oxides are difficult to sample; keeping minerals in suspension by the addition of glass beads into shaken vessels (or by using a stirred reactor) is helpful (16).

Mn Oxide Reduction Assay

In general, anoxic stock solutions of the basal medium, carbon source, buffer (if desirable), and MnO_x substrate (see below for synthesis) should be combined under anoxic conditions and allowed to equilibrate while shaken or stirred. The inoculum or environmental sample should be actively sparged with an anoxic gas mix containing some CO_2 (for pH maintenance) or exposed to an anaerobic atmosphere long enough to eliminate all oxygen. The sample should be anoxically combined with the other ingredients to yield a final solution of 0.05 to 0.1 g of MnO_x per liter and an appropriate carbon source; if a buffer is to be used, a 10 to 20 mM concentration is suggested. Include multiple incubations for each time point, including abiotic and MnO_x-free controls. Useful incubation times will vary from sample to sample; rate determinations should be made over the shortest time scales that will yield linear results.

Immediately postincubation, take replicate subsamples (e.g., 1 ml) to measure total Mn. For dissolved and particulate Mn, immediately filter the subsample(s) (10 to 100 ml) through a 0.2-μm-pore-size filter and collect the filtrate. Add 10 mM $CuSO_4$ to cover the top of the filter(s) (2 ml) and incubate for 30 min to 2 h to desorb Mn(II) bound to the oxides. Filter the $CuSO_4$ solution through, collect the filtrate, and combine with the previous filtrate. Measure Mn in the total, filter, and filtrate fractions (correcting for the dilution due to the $CuSO_4$ wash). Note that Cu interferes with the formaldoxime method; atomic absorption spectroscopy or ICP-optical emission spectroscopy should be used to analyze Mn. Following this general procedure but without using $CuSO_4$ and assaying by the formaldoxime method is a simple and still useful measure of relative Mn oxide reduction, albeit only semiquantitative.

δ-MnO_2 Synthesis

Procedure

Consult reference 127 for the procedure. For a 4-liter preparation that will yield ~50 g of final, aseptic product (if sterile tubes, bottles, etc., are used), follow these steps:

1. Prepare the following as filter-sterilized solutions:

    75.14 g of $MnCl_2 \cdot 4H_2O$ in 1,280 ml of 18-MΩ water
    40.0 g of $KMnO_4$ in 1,280 ml of 18-MΩ water
    28 g of NaOH in 1,440 ml of 18-MΩ water
    1.0 M NaCl, adjusted to pH 10
    0.01 M NaCl, adjusted to pH 10

2. Slowly (~5 min) add the $KMnO_4$ solution to the NaOH solution while stirring.
3. Slowly (~35 min) add the $MnCl_2$ solution to the basic $KMnO_4$ solution while stirring; a black Mn oxide precipitate will form.
4. Allow the precipitate to settle out and discard the supernatant.
5. Centrifuge the Mn oxide for 20 min at room temperature and 27,500 × g, discard the supernatant.
6. Resuspend in 1 M NaCl at pH ~10, and shake for >1 h.
7. Repeat steps 5 and 6 five times by using 0.01 M NaCl at pH ~10, leaving the suspended oxides to shake overnight for the final wash.
8. Resuspend in a convenient final volume and measure oxide concentration by formaldoxime.

This procedure can be scaled down to make smaller batches and/or incorporate $^{54}Mn(II)$. The addition of 1.67 μCi of ^{54}Mn per μmol of $MnCl_2$ to the $MnCl_2$ solution before addition to the $KMnO_4$/NaOH solution will produce an oxide with a specific activity near 1 μCi/μmol. With the use of the radiotracer, additional small-volume washes of 0.01 M NaCl at a pH of ~10 are repeated until the amount of $^{54}Mn(II)$ removed in each wash approaches background levels.

Phylogeny of Mn-Reducing Bacteria

Microorganisms can reduce Mn oxide minerals by a number of different mechanisms that have been thoroughly reviewed recently (35, 69). Although many bacteria indirectly reduce

Mn oxides through metabolic by-products (e.g., sulfate-reducing and fermentative bacteria that produce hydrogen sulfide and organic acids, respectively), here we will focus on organisms that conserve energy by coupling the oxidation of organic and inorganic electron donors to the reduction of extracellular Mn oxides, without the assimilation of Mn (i.e., dissimilatory Mn reduction). Such microbes that conserve energy with Mn oxides as a terminal electron acceptor are thought to play geochemically significant roles in the environment (67). Most organisms capable of reducing Fe(III) are also capable of reducing Mn oxide (68), and so much of what is known about the physiology, biochemistry, and diversity of Mn reduction has been inferred from studies done on Fe(III) reduction. Microorganisms that gain energy from dissimilatory Mn oxide reduction are phylogenetically diverse, being spread throughout the archaea (thermoproteales) and eubacteria, including β-, γ-, δ-, and ε-proteobacteria, firmicutes, actinobacteria, and the deinococcus/thermus group (see the phylogenetic tree in reference 69). This diversity is bound to expand as new Mn reducers continue to be isolated and as the potential electron acceptors of previously isolated microbes expand to include Mn oxide.

Two groups of Mn oxide-reducing bacteria have by far been studied the most: the family *Geobacteraceae* and the genus *Shewanella*. A common feature of many of the members of the *Geobacteraceae* is the ability to utilize acetate as an electron donor (though some also utilize many other electron donors and/or carbon sources). Based on molecular techniques, *Geobacteraceae* members have been found to be abundant in many environments (especially sedimentary environments) where Fe(III) and Mn oxide reduction is important (3, 4, 55, 94, 103, 104). *Shewanella* species are facultative anaerobes that are amenable to genetic techniques (35) and are often grown with lactate as the electron donor. They have been isolated from a wide variety of environments and are known for their metabolic diversity in being able to utilize multiple electron acceptors.

Cultivation, Enrichment, and Isolation

As mentioned above, Mn oxide-reducing microbes vary in their tolerance of O_2, so procedures for enrichment and isolation can range from aerobic to strictly anaerobic conditions depending on the organisms being targeted. Hydrogen or organic acids such as acetate and lactate are often used as electron donors, and Mn oxides prepared chemically are used as an electron acceptor (see "δ-MnO_2 Synthesis" above). Alternatively, electron shuttles such as the humic acid analog anthraquinone-2,6-disulfonate (AQDS) can be used as the electron acceptor to enrich for Mn-reducing organisms (25, 26). An extensive list of references describing the isolation of organisms known to conserve energy to support growth from Fe(III) and Mn(IV) reduction can be found in Lovley et al. (69). Commonly used medium recipes (68; Y. Gorby, personal communication) are listed in Tables 13 to 20. For isolation of metal-reducing *Shewanella* species, the reader should also consult Nealson and Scott (85). These media often have a bicarbonate-carbon dioxide buffer system,

TABLE 13 Freshwater medium[a]

Ingredient (mix together)	Amt added	Final concn
Deionized water	900 ml	
$NaHCO_3$	2.5 g	30 mM
NH_4Cl	0.25 g	4.7 mM
$NaH_2PO_4 \cdot H_2O$	0.6 g	4.4 mM
KCl	0.1 g	1.3 mM
Vitamin solution	10 ml	
Mineral solution	10 ml	

[a]Bring medium to a final volume of 1 liter, dispense, sparge with an 80:20 mixture of N_2:CO_2 gas, cap and autoclave. Mn oxides from a sterile, anoxic stock solution (see "δ-MnO_2 Synthesis" above) should be added after autoclaving.

TABLE 14 Marine medium[a]

Ingredient (mix together)	Amt added	Final concn (mM)
Deionized water	900 ml	
NaCl	20 g	342
$NaHCO_3$	2.5 g	30
KCl	0.67 g	9
Vitamin solution	10 ml	
Mineral solution	10 ml	
Mineral stock	20 ml	

[a]Bring medium to a final volume of 950 ml, dispense, sparge with an 80:20 mixture of N_2:CO_2 gas, cap, and autoclave. Mn oxides from a sterile, anoxic stock solution (see "δ-MnO_2 Synthesis" above) and 1 part salt stock solution per 19 parts sterile medium should be added aseptically and anaerobically after autoclaving.

TABLE 15 *Shewanella* Federation medium[a]

Ingredient (mix together)[b]	Amt added	Final concn
PIPES buffer	0.91 g	2.6 mM
NaOH	0.3 g	7.5 mM
NH_4Cl	1.5 g	28 mM
KCl	0.1 g	1.3 mM
$NaH_2PO_4 \cdot H_2O$	0.6 g	4.3 mM
NaCl	1.75 g	30 mM
Mineral solution	10 ml	
Vitamin solution	10 ml	
Amino acid stock	10 ml	

[a]Y. Gorby, personal communication.
[b]Adjust pH to 7.0 to 7.2 and bring medium to a final volume of 1 liter.

TABLE 16 Mineral stock for marine medium (50×)

Ingredient	Concn (g/100 ml of deionized H_2O)
NH_4Cl	5
KCl	0.5
KH_2PO_4	0.5
$MgSO_4 \cdot 7H_2O$	1.0
$CaCl_2 \cdot 2H_2O$	0.1

TABLE 17 Salt stock for marine medium[a]

Ingredient	Concn (g/100 ml of deionized H_2O)
$MgCl_2 \cdot 6H_2O$	21.2
$CaCl_2 \cdot 2H_2O$	3.04

[a]Sparge with an 80:20 mixture of N_2:CO_2 gas, cap, and autoclave.

TABLE 18 Vitamin solution[a]

Ingredient	Concn (g/1,000 ml of deionized H_2O)
Biotin	2
Folic acid	2
Pyridoxine-HCl	10
Riboflavin	5
Thiamine	5
Nicotinic acid	5
Pantothenic acid	5
Vitamin B_{12}	0.1
p-Aminobenzoic acid	5
Thioctic acid	5

[a]Adjust to pH 7.0 and filter sterilize.

TABLE 19 Mineral solution

Ingredient	Concn (g/1,000 ml of deionized H_2O)
Trisodium nitrilotriacetic acid (pH 8.5)	1.5
$MgSO_4$	3
$MnSO_4 \cdot H_2O$	0.5
NaCl	1
$FeSO_4 \cdot 7H_2O$	0.1
$CaCl_2 \cdot 2H_2O$	0.1
$CoCl_2 \cdot 6H_2O$	0.1
$ZnCl_2$	0.13
$CuSO_4 \cdot 5H_2O$	0.01
$AlK(SO_4)_2 \cdot 12H_2O$	0.01
H_3BO_3	0.01
Na_2MoO_4	0.025
$NiCl_2 \cdot 6H_2O$	0.025
$Na_2WO_4 \cdot 2H_2O$	0.025

TABLE 20 Amino acid stock[a]

Ingredient	Concn (g/100 ml of deionized H_2O)
L-Glutamic acid	200
L-Arginine	200
DL-Serine	200

[a]Adjust pH to 7.0. Amounts of components are per 100 ml of deionized H_2O.

and the headspace gas typically contains 20% carbon dioxide to establish an initial pH of approximately 6.8. Electron donors and/or carbon sources are added as desired. Marine Mn(IV)-reducing bacteria can be cultivated or enriched by using media designed for freshwater, brackish, and marine sulfate-reducing bacteria. The electron acceptor sulfate is replaced with Mn oxide (artificial brackish water and seawater are prepared without sulfate salts). The medium compositions and preparation are very well described by Widdel and Bak (131).

CONCLUSIONS: EMERGING TECHNIQUES AND MOLECULAR ANALYSES

Advancing our understanding of microbial Mn redox transformations largely depends on answering old questions. Which microbes are oxidizing and reducing Mn in the environment? Which enzymes and/or gene products are responsible for this activity? What are the stepwise, microbe-catalyzed reaction pathways of Mn? Additional cultivation efforts will address these goals, as new strains, especially those amenable to genetic manipulation, will undoubtedly provide insights in this era of rapid genomic sequencing. We hope that the methods described here will facilitate the isolation of microbes performing Mn redox transformations and characterization of their processes and habitats.

We gratefully acknowledge funding over the years from the National Science Foundation (Molecular and Cellular Biosciences, grants 9808915, 0348668, and 063035; Ocean Sciences, grants 0221500 and 0433692; and Chemistry, grant 9910592).

REFERENCES

1. **Adams, L. F., and W. C. Ghiorse.** 1987. Characterization of an extracellular Mn^{2+}-oxidizing activity and isolation of Mn^{2+}-oxidizing protein from *Leptothrix discophora* SS-1. *J. Bacteriol.* **169:**1279–1285.
2. **Altmann, H. J.** 1972. Bestimmung von in Wasser gelostem Sauerstoff mit Leukoberbelinblau. I. Eine schnelle Winklermethode. *Z. Anal. Chem.* **262:**97–99.
3. **Anderson, R. T., J. Rooney-Varga, C. V. Gaw, and D. R. Lovley.** 1998. Anaerobic benzene oxidation in the Fe(III)-reduction zone of petroleum-contaminated aquifers. *Environ. Sci. Technol.* **32:**1222–1229.
4. **Anderson, R. T., H. A. Vrionis, I. Ortiz-Bernad, C. T. Resch, P. E. Long, R. Dayvault, K. Karp, S. Marutzky, D. R. Metzler, A. Peacock, D. C. White, M. Lowe, and D. R. Lovley.** 2003. Stimulating the in situ activity of *Geobacter* species to remove uranium from the groundwater of a uranium-contaminated aquifer. *Appl. Environ. Microbiol.* **69:**5884–5891.
5. **Bargar, J. R., B. M. Tebo, U. Bergmann, S. M. Webb, P. Glatzel, V. Q. Chiu, and M. Villalobos.** 2005. Biotic and abiotic products of Mn(II) oxidation by spores of the marine *Bacillus* sp. strain SG-1. *Am. Mineral.* **90:**143–154.
6. **Bargar, J. R., B. M. Tebo, and J. E. Villinski.** 2000. *In situ* characterization of Mn(II) oxidation by spores of the marine *Bacillus* sp. strain SG-1. *Geochim. Cosmochim. Acta* **64:**2777–2780.
7. **Boogerd, F. C., and J. P. M. de Vrind.** 1987. Manganese oxidation by *Leptothrix discophora*. *J. Bacteriol.* **169:**489–494.
8. **Bratina, B. J., B. S. Stevenson, W. J. Green, and T. M. Schmidt.** 1998. Manganese reduction by microbes from oxic regions of the Lake Vanda (Antarctica) water column. *Appl. Environ. Microbiol.* **64:**3791–3797.
9. **Brewer, P. G., and D. W. Spencer.** 1971. Colorimetric determination of manganese in anoxic waters. *Limnol. Oceanogr.* **16:**107–112.
10. **Brock, T. D.** 1978. The poisoned control in biogeochemical investigations, p. 717–725. *In* W. E. Krumbein (ed.), *Environmental Biogeochemistry and Geomicrobiology*, vol. 3. Ann Arbor Science, Ann Arbor, Mich.
11. **Bromfield, S. M.** 1974. Bacterial oxidation of manganous ions as affected by organic substrate concentration and composition. *Soil Biol. Biochem.* **11:**383–392.
12. **Bromfield, S. M., and D. J. David.** 1976. Sorption and oxidation of manganous ions and reduction of manganese oxide by cell suspensions of a manganese oxidizing bacterium. *Soil Biol. Biochem.* **8:**37–43.

13. **Brouwers, G.-J., P. L. A. M. Corstjens, J. P. M. de Vrind, A. Verkamman, M. de Kuyper, and E. W. de Vrind-de Jong.** 2000. Stimulation of Mn^{2+} oxidation in *Leptothrix discophora* SS-1 by Cu^{2+} and sequence analysis of the region flanking the gene encoding putative multicopper oxidase MofA. *Geomicrobiol. J.* **17:**25–33.
14. **Brouwers, G.-J., J. P. M. de Vrind, P. L. A. M. Corstjens, P. Cornelis, C. Baysse, and E. W. de Vrind-de Jong.** 1999. *cumA*, a gene encoding a multicopper oxidase, is involved in Mn^{2+} oxidation in *Pseudomonas putida* GB-1. *Appl. Environ. Microbiol.* **65:**1762–1768.
15. **Brouwers, G.-J., E. Vijgenboom, P. L. A. M. Corstjens, J. P. M. de Vrind, and E. W. de Vrind-de Jong.** 2000. Bacterial Mn^{2+} oxidizing systems and multicopper oxidases: an overview of mechanisms and functions. *Geomicrobiol. J.* **17:**1–24.
16. **Burdige, D. J.** 1983. The biogeochemistry of manganese redox reactions: rates and mechanisms. Ph.D. dissertation. University of California, San Diego, La Jolla.
17. **Burdige, D. J., S. P. Dhakar, and K. H. Nealson.** 1992. Effects of manganese oxide mineralogy on microbial and chemical manganese reduction. *Geomicrobiol. J.* **10:**27–48.
18. **Burdige, D. J., and K. H. Nealson.** 1986. Chemical and microbiological studies of sulfide-mediated manganese reduction. *Geomicrobiol. J.* **4:**361–387.
19. **Burdige, D. J., and K. H. Nealson.** 1985. Microbial manganese reduction by enrichment cultures from coastal marine sediments. *Appl. Environ. Microbiol.* **50:**491–497.
20. **Campbell, A. C., J. M. Gieskes, J. E. Lupton, and P. F. Lonsdale.** 1988. Manganese geochemistry in the Guaymas Basin, Gulf of California. *Geochim. Cosmochim. Acta* **52:** 345–357.
21. **Canfield, D. E., B. B. Jorgensen, H. Fossing, R. Glud, J. Gundersen, N. B. Ramsing, B. Thamdrup, J. W. Hansen, L. P. Nielsen, and P. O. J. Hall.** 1993. Pathways of organic carbon oxidation in three continental margin sediments. *Mar. Geol.* **113:**27–40.
22. **Canfield, D. E., B. Thamdrup, and J. W. Hansen.** 1993. The anaerobic degradation of organic matter in Danish coastal sediments: iron reduction, manganese reduction, and sulfate reduction. *Geochim. Cosmochim. Acta* **57:** 3867–3885.
23. **Caspi, R., B. M. Tebo, and M. G. Haygood.** 1998. *c*-Type cytochromes and manganese oxidation in *Pseudomonas putida* MnB1. *Appl. Environ. Microbiol.* **64:**3549–3555.
24. **Chapnick, S., W. S. Moore, and K. H. Nealson.** 1982. Microbially mediated manganese oxidation in a freshwater lake. *Limnol. Oceanogr.* **17:**1004–1014.
25. **Coates, J. D., D. J. Ellis, E. L. Blunt-Harris, C. V. Gaw, E. E. Roden, and D. R. Lovley.** 1998. Recovery of humic-reducing bacteria from a diversity of environments. *Appl. Environ. Microbiol.* **64:**1504–1509.
26. **Coates, J. D., D. J. Lonergan, H. Jenter, and D. R. Lovley.** 1996. Isolation of *Geobacter* species from diverse sedimentary environments. *Appl. Environ. Microbiol.* **62:** 1531–1536.
27. **Corstjens, P. L. A. M., J. P. M. de Vrind, T. Goosen, and E. W. de Vrind-de Jong.** 1997. Identification and molecular analysis of the *Leptothrix discophora* SS-1 *mofA* gene, a gene putatively encoding a manganese-oxidizing protein with copper domains. *Geomicrobiol. J.* **14:**91–108.
28. **Cowen, J. P., M. A. Bertram, E. T. Baker, G. J. Massoth, R. A. Feely, and M. Summit.** 1998. Geomicrobial transformation of manganese in Gorda Ridge event plumes. *Deep-Sea Res. II* **45:**2713–2738.
29. **Cowen, J. P., and K. W. Bruland.** 1985. Metal deposits associated with bacteria: implications for Fe and Mn marine biogeochemistry. *Deep-Sea Res.* **32:**253–272.
30. **Cowen, J. P., and Y. H. Li.** 1991. The influence of a changing bacterial community on trace metal scavenging in a deep-sea particle plume. *J. Mar. Res.* **49:**517–542.
31. **Cowen, J. P., G. J. Massoth, and E. T. Baker.** 1986. Bacterial scavenging of Mn and Fe in a mid- to far-field hydrothermal particle plume. *Nature* **322:**169–171.
32. **Cowen, J. P., G. J. Massoth, and R. A. Feely.** 1990. Scavenging rates of dissolved manganese in a hydrothermal vent plume. *Deep-Sea Res.* **37:**1619–1637.
33. **Cowen, J. P., and M. W. Silver.** 1984. The association of iron and manganese with bacteria on marine macroparticulate material. *Science* **224:**1340–1342.
34. **Crerar, D. A., A. G. Fischer, and C. L. Plaza.** 1980. Metallogenium and biogenic deposition of manganese from precambrian to recent time, p. 285–303. *In* I. M. Varentsov and G. Grasselly (ed.), *Geology and Geochemistry of Manganese*, vol. III. E. Schweizerbart'sche Verlagsbuchhandlung, Stuttgart, Germany.
35. **Croal, L. R., J. A. Gralnick, D. Malasarn, and D. K. Newman.** 2004. The genetics of geochemistry. *Annu. Rev. Genet.* **38:**175–202.
36. **Crowley, J. D., D. A. Traynor, and D. C. Weatherburn.** 2000. Enzymes and proteins containing manganese: an overview, p. 209–278. *In* A. Sigel and H. Sigel (ed.), *Metal Ions in Biological Systems*, vol. 37. *Manganese and Its Role in Biological Processes*. Marcel Dekker, New York, N.Y.
37. **de Vrind, J., A. de Groot, G. J. Brouwers, J. Tommassen, and E. de Vrind-de Jong.** 2003. Identification of a novel Gsp-related pathway required for secretion of the manganese-oxidizing factor of *Pseudomonas putida* strain GB-1. *Mol. Microbiol.* **47:**993–1006.
38. **de Vrind, J. P., G. J. Brouwers, P. L. A. M. Corstjens, J. Den Dulk, and E. W. De Vrind-De Jong.** 1998. The cytochrome *c* maturation operon is involved in manganese oxidation in *Pseudomonas putida* GB-1. *Appl. Environ. Microbiol.* **64:**3556–3562.
39. **Emerson, D., R. E. Garen, and W. C. Ghiorse.** 1989. Formation of *Metallogenium*-like structures by a manganese-oxidizing fungus. *Arch. Microbiol.* **151:**223–231.
40. **Emerson, S., S. Kalhorn, L. Jacobs, B. M. Tebo, K. H. Nealson, and R. A. Rosson.** 1982. Environmental oxidation rate of manganese(II): bacterial catalysis. *Geochim. Cosmochim. Acta* **46:**1073–1079.
41. **Fendorf, S. E., M. Fendorf, D. L. Sparks, and R. Gronsky.** 1992. Inhibitory mechanisms of Cr(III) oxidation by delta-MnO_2. *J. Colloid Interface Sci.* **153:**37–54.
42. **Fendorf, S. E., and R. J. Zasoski.** 1992. Chromium(III) oxidation by delta-MnO_2. *Environ. Sci. Technol.* **26:**79–85.
43. **Francis, C. A., K. L. Casciotti, and B. M. Tebo.** 2002. Localization of Mn(II)-oxidizing activity and the putative multicopper oxidase, MnxG, to the exosporium of the marine *Bacillus* sp. strain SG-1. *Arch. Microbiol.* **178:**450–456.
44. **Francis, C. A., E.-M. Co, and B. M. Tebo.** 2001. Enzymatic manganese(II) oxidation by a marine α-proteobacterium. *Appl. Environ. Microbiol.* **67:**4024–4029.
45. **Francis, C. A., and B. M. Tebo.** 2001. Diversity of *cumA* multicopper oxidase genes from Mn(II)-oxidizing and non-Mn(II)-oxidizing *Pseudomonas* strains. *Appl. Environ. Microbiol.* **67:**4272–4278.
46. **Francis, C. A., and B. M. Tebo.** 2002. Enzymatic manganese(II) oxidation by metabolically dormant spores of diverse *Bacillus* species. *Appl. Environ. Microbiol.* **68:**874–880.
47. **Friedl, G., B. Wehrli, and A. Manceau.** 1997. Solid phases in the cycling of manganese in eutrophic lakes: new insights from EXAFS spectroscopy. *Geochim. Cosmochim. Acta* **61:**275–290.
48. **Fuller, C. C., and J. W. Harvey.** 2000. Reactive uptake of trace metals in the hyporheic zone of a mining-contaminated stream, Pinal Creek, Arizona. *Environ. Sci. Technol.* **34:**1150–1155.

49. **Gebers, R.** 1981. Enrichment, isolation, and emended description of *Pedomicrobium ferrugineum* Aristovskaya and *Pedomicrobium manganicum* Aristovskaya. *Int. J. Syst. Bacteriol.* **31:**302–316.
50. **Ghiorse, W. C.** 1994. Iron and manganese oxidation and reduction, p. 1079–1096. *In* R. W. Weaver, S. Angle, and P. Bottomley (ed.), *Methods of Soil Analysis*, part 2. *Microbiological and Biochemical Properties*. SSSA Book Series, no. 5. Soil Science Society of America, Madison, Wis.
51. **Ghiorse, W. C., and S. D. Chapnick.** 1983. Metal-depositing bacteria and the distribution of manganese and iron in swamp waters. *Ecol. Bull.* **35:**367–376.
52. **Harvey, J. W., and C. C. Fuller.** 1998. Effect of enhanced manganese oxidation in the hyporheic zone on basin-scale geochemical mass balance. *Water Resour. Res.* **34:**623–636.
53. **Hastings, D., and S. Emerson.** 1986. Oxidation of manganese by spores of a marine *Bacillus*: kinetic and thermodynamic considerations. *Geochim. Cosmochim. Acta* **50:** 1819–1824.
54. **Hirsch, P.** 1974. Budding bacteria. *Annu. Rev. Microbiol.* **28:**391–444.
55. **Holmes, D. E., K. T. Finneran, R. A. O'Neil, and D. R. Lovley.** 2002. Enrichment of members of the family *Geobacteraceae* associated with stimulation of dissimilatory metal reduction in uranium-contaminated aquifer sediments. *Appl. Environ. Microbiol.* **68:**2300–2306.
56. **Huang, P. M.** 1991. Kinetics of redox reactions on manganese oxides and its impact on environmental quality, p. 191–230. *In* D. L. Sparks and D. L. Suarez (ed.), *Rates of Soil Chemical Processes*. Soil Science Society of America, Madison, Wis.
57. **Kalhorn, S., and S. Emerson.** 1984. The oxidation state of manganese in surface sediments of the Pacific Ocean. *Geochim. Cosmochim. Acta* **48:**897–902.
58. **Kay, J. T., M. H. Conklin, C. C. Fuller, and P. A. O'Day.** 2001. Processes of nickel and cobalt uptake by a manganese oxide forming sediment in Pinal Creek, Globe Mining District, Arizona. *Environ. Sci. Technol.* **35:**4719–4725.
59. **Kepkay, P., and K. H. Nealson.** 1987. Growth of a manganese oxidizing *Pseudomonas* sp. in continuous culture. *Arch. Microbiol.* **148:**63–67.
60. **Kepkay, P. E., D. J. Burdige, and K. H. Nealson.** 1984. Kinetics of bacterial manganese binding and oxidation in the chemostat. *Geomicrobiol. J.* **3:**245–262.
61. **Kessick, M. A., J. Vuceta, and J. J. Morgan.** 1972. Spectrophotometric determination of oxidized manganese with leuco crystal violet. *Environ. Sci. Technol.* **6:**642–644.
62. **Klewicki, J. K., and J. J. Morgan.** 1999. Dissolution of β-MnOOH particles by ligands: pyrophosphate, ethylenediaminetetraacetate, and citrate. *Geochim. Cosmochim. Acta* **63:**3017–3024.
63. **Klewicki, J. K., and J. J. Morgan.** 1998. Kinetic behavior of Mn(III) complexes of pyrophosphate, EDTA, and citrate. *Environ. Sci. Technol.* **32:**2916–2922.
64. **Kostka, J. E., G. W. Luther III, and K. H. Nealson.** 1995. Chemical and biological reduction of Mn(III)-pyrophosphate complexes: potential importance of dissolved Mn(III) as an environmental oxidant. *Geochim. Cosmochim. Acta* **59:**885–894.
65. **Krumbein, W. E., and H. J. Altmann.** 1973. A new method for the detection and enumeration of manganese oxidizing and reducing microorganisms. *Helgol. Wiss. Meeresunters.* **25:**347–356.
66. **Lanson, B., V. A. Drits, E. Silvester, and A. Manceau.** 2000. Structure of H-exchanged hexagonal birnessite and its mechanism of formation from Na-rich monoclinic buserite at low pH. *Am. Mineral.* **85:**826–838.
67. **Lovley, D. R.** 1991. Dissimilatory Fe(III) and Mn(IV) reduction. *Microbiol. Rev.* **55:**259–287.
68. **Lovley, D. R.** 2 November 2001, posting date. Dissimilatory Fe(III)- and Mn(IV)-reducing prokaryotes. *In* M. Dworkin, S. Falkow, E. Rosenberg, K.-H. Scheifer, and E. Stackebrandt (ed.), *The Prokaryotes*, vol. 2. *Ecophysiology and Biochemistry*, 3rd ed., release 3.7. Springer-Verlag, New York, N.Y.
69. **Lovley, D. R., D. E. Holmes, and K. P. Nevin.** 2004. Dissimilatory Fe(III) and Mn(IV) reduction. *Adv. Microb. Physiol.* **49:**219–286.
70. **Luther, G. W., III, D. T. Ruppel, and C. Burkhard.** 1998. Reactivity of dissolved Mn(III) complexes and Mn(IV) species with reductants: Mn redox chemistry without a dissolution step, p. 265–280. *In* D. L. Sparks and T. J. Grundl (ed.), *ACS Symposium Series*, vol. 715. *Mineral-Water Interfacial Reactions: Kinetics and Mechanisms*. American Chemical Society, Washington, D.C.
71. **Maki, J. S., B. M. Tebo, F. E. Palmer, K. H. Nealson, and J. T. Staley.** 1987. The abundance and biological activity of manganese-oxidizing bacteria and *Metallogenium*-like morphotypes in Lake Washington, USA. *FEMS Microbiol. Ecol.* **45:**21–29.
72. **Manceau, A., and L. Charlet.** 1992. X-ray absorption spectroscopy study of the sorption of Cr(III) at the oxide-water interface. *J. Colloid Interface Sci.* **148:**425–442.
73. **Mandernack, K. W., and B. M. Tebo.** 1993. Manganese scavenging and oxidation at hydrothermal vents and in vent plumes. *Geochim. Cosmochim. Acta* **57:**3907–3923.
74. **Marble, J. C.** 1998. Biotic contribution of Mn(II) removal at Pinal Creek, Arizona. M.S. thesis. University of Arizona, Tucson.
75. **Marble, J. C., T. L. Corley, M. H. Conklin, and C. C. Fuller.** 1999. Environmental factors affecting oxidation of manganese in Pinal Creek, Arizona, p. 173–183. *In* D. W. Morganwalp and H. T. Buxton (ed.), *U.S. Geological Survey Toxic Substances Hydrology Program—Proceedings of the Technical Meeting, Charleston, South Carolina, March 8–12, 1999*, vol. 1. *Contamination from Hardrock Mining: U.S. Geological Survey Water Resources Investigations Report* 99-4018A. U.S. Geological Survey, Washington, D.C.
76. **Marcus, M. A., A. Manceau, and M. Kersten.** 2004. Mn, Fe, Zn, and As speciation in a fast-growing ferromanganese marine nodule. *Geochim. Cosmochim. Acta* **68:**3125–3136.
77. **McKeown, D. A., and J. E. Post.** 2001. Characterization of manganese oxide mineralogy in rock varnish and dendrites using X-ray absorption spectroscopy. *Am. Mineral.* **86:**701–703.
78. **Miyajima, T.** 1992. Biological manganese oxidation in a lake. I. Occurrence and distribution of *Metallogenium* sp. and its kinetic properties. *Arch. Hydrobiol.* **124:**317–335.
79. **Morgan, J. J.** 2000. Manganese in natural waters and earth's crust: its availability to organisms, p. 1–33. *In* A. Sigel and H. Sigel (ed.), *Metal Ions in Biological Systems*, vol. 37. *Manganese and Its Role in Biological Processes*. Marcel Dekker, New York, N.Y.
80. **Mulder, E. G.** 1964. Iron bacteria, particularly those of the *Sphaerotilis-Leptothrix* group, and industrial problems. *J. Appl. Bacteriol.* **27:**151–173.
81. **Murray, J. W., L. S. Balistrieri, and B. Paul.** 1984. The oxidation state of manganese in marine sediments and ferromanganese nodules. *Geochim. Cosmochim. Acta* **48:**1237–1247.
82. **Myers, C. R., and K. H. Nealson.** 1988. Bacterial manganese reduction and growth with manganese oxide as the sole electron acceptor. *Science* **240:**1319–1321.
83. **Nealson, K. H.** 1983. The microbial manganese cycle, p. 191–221. *In* W. E. Krumbein (ed.), *Microbial Geochemistry*. Blackwell Scientific, Boston, Mass.

84. **Nealson, K. H., and D. Saffarini.** 1994. Iron and manganese in anerobic respiration: environmental significance, physiology, and regulation. *Annu. Rev. Microbiol.* **48:**311–343.
85. **Nealson, K. H., and J. Scott.** 2003. Ecophysiology of the genus *Shewanella. In* M. Dworkin (ed.), *The Prokaryotes: an Evolving Electronic Resource for the Microbiological Community,* 3rd ed., release 3.14. Springer-Verlag, New York, N.Y.
86. **Nealson, K. H., B. M. Tebo, and R. A. Rosson.** 1988. Occurrence and mechanisms of microbial oxidation of manganese. *Adv. Appl. Microbiol.* **33:**279–318.
87. **Nelson, Y. M., L. W. Lion, W. C. Ghiorse, and M. L. Shuler.** 1999. Production of biogenic Mn oxides by *Leptothrix discophora* SS-1 in a chemically defined growth medium and evaluation of their Pb adsorption characteristics. *Appl. Environ. Microbiol.* **65:**175–180.
88. **Nicholson, K., and M. Eley.** 1997. Geochemistry of manganese oxides: metal adsorption in freshwater and marine environments, p. 309–326. *In* K. Nicholson, J. R. Hein, B. Bühn, and S. Dasgupta (ed.), *Geochemistry and Mineralogy of Terrestrial and Marine Deposits,* vol. 119. *Manganese Mineralization.* The Geological Society, London, United Kingdom.
89. **Nico, P. S., and R. J. Zasoski.** 2000. Importance of Mn(III) availability on the rate of Cr(III) oxidation on delta-MnO_2. *Environ. Sci. Technol.* **36:**3363–3367.
90. **Okazaki, M., T. Sugita, M. Shimizu, Y. Ohode, K. Iwamoto, E. W. de Vrind-de Jong, J. P. M. de Vrind, and P. L. A. M. Corstjens.** 1997. Partial purification and characterization of manganese-oxidizing factors of *Pseudomonas fluorescens* GB-1. *Appl. Environ. Microbiol.* **63:**4793–4799.
91. **Pak, K. R., O. Y. Lim, H. K. Lee, and S. C. Choi.** 2002. Aerobic reduction of manganese oxide by Salmonella sp. strain MR4. *Biotechnol. Lett.* **24:**1181–1184.
92. **Parker, D. L., G. Sposito, and B. M. Tebo.** 2004. Manganese(III) binding to a pyoverdine siderophore produced by a manganese(II)-oxidizing bacterium. *Geochim. Cosmochim. Acta* **68:**4809–4820.
93. **Post, J. E.** 1999. Manganese oxide minerals: crystal structures and economic and environmental significance. *Proc. Natl. Acad. Sci. USA* **96:**3447–3454.
94. **Rooney-Varga, N. J., R. T. Anderson, J. L. Fraga, D. Ringelberg, and D. R. Lovley.** 1999. Microbial communities associated with anaerobic benzene mineralization in a petroleum-contaminated aquifer. *Appl. Environ. Microbiol.* **65:**3056–3063.
95. **Rosson, R. A., and K. H. Nealson.** 1982. Manganese binding and oxidation by spores of a marine bacillus. *J. Bacteriol.* **151:**1027–1034.
96. **Rosson, R. A., B. M. Tebo, and K. H. Nealson.** 1984. Use of poisons in the determination of microbial manganese binding rates in seawater. *Appl. Environ. Microbiol.* **47:**740–745.
97. **Schippers, A., L. N. Neretin, G. Lavik, T. Leipe, and F. Pollehne.** 2005. Manganese(II) oxidation driven by lateral oxygen intrusions in the western Black Sea. *Geochim. Cosmochim. Acta* **69:**2241–2252.
98. **Scott, M. J., H. H. Zhang, S. C. Lee, B. Hedman, K. O. Hodgson, and R. H. Holm.** 1995. Oxygen-bridged iron-copper assemblies pertinent to heme-copper oxidases—synthesis and structure of an [Fe-III-(Oh)-Cu-II] bridge and Exafs multiple-scattering effects of linear oxo and nonlinear hydroxo bridges. *J. Am. Chem. Soc.* **117:**568–569.
99. **Shiller, A. M., and T. H. Stephens.** 2005. Microbial manganese oxidation in the lower Mississippi River: methods and evidence. *Geomicrobiol. J.* **22:**117–125.
100. **Siering, P. L., and W. C. Ghiorse.** 1997. Development and application of 16S rRNA-targeted probes for detection of iron- and manganese-oxidizing sheathed bacteria in environmental samples. *Appl. Environ. Microbiol.* **63:**644–651.
101. **Siering, P. L., and W. C. Ghiorse.** 1997. PCR detection of a putative manganese oxidation gene (*mofA*) in environmental samples and assessment of *mofA* homology among diverse manganese-oxidizing bacteria. *Geomicrobiol. J.* **14:**109–125.
102. **Sly, L. I., V. Arunpairojana, and D. R. Dixon.** 1990. Binding of colloidal MnO_2 by extracellular polysaccharides of *Pedomicrobium manganicum. Appl. Environ. Microbiol.* **56:**2791–2794.
103. **Snoeyenbos-West, O. L., K. P. Nevin, and D. R. Lovley.** 2000. Stimulation of dissimilatory Fe(III) reduction results in a predominance of *Geobacter* species in a variety of sandy aquifers. *Microb. Ecol.* **39:**153–167.
104. **Stein, J. L., M. Haygood, and H. Felbeck.** 1990. Nucleotide sequence and expression of a deep-sea ribulose-1,5-bisphosphate carboxylase gene cloned from a chemoautotrophic bacterial endosymbiont. *Proc. Natl. Acad. Sci. USA* **87:**8850–8854.
105. **Stone, A. T.** 1983. The reduction and dissolution of Mn(III) and Mn(IV) oxides by organics. Ph.D. dissertation. California Institute of Technology, Pasadena.
106. **Stone, A. T., and J. J. Morgan.** 1984. Reduction and dissolution of manganese(III) and manganese(IV) oxides by organics. 1. Reaction with hydroquinone. *Environ. Sci. Technol.* **18:**450–456.
107. **Stone, A. T., and J. J. Morgan.** 1984. Reduction and dissolution of manganese(III) and manganese(IV) oxides by organics. 2. Survey of the reactivity of organics. *Environ. Sci. Technol.* **18:**617–624.
108. **Stumm, W., and J. J. Morgan.** 1996. *Aquatic Chemistry: Chemical Equilibria and Rates in Natural Waters,* 3rd ed. John Wiley & Sons, Inc., New York, N.Y.
109. **Sunda, W. G., and S. A. Huntsman.** 1987. Microbial oxidation of manganese in a North Carolina estuary. *Limnol. Oceanogr.* **32:**552–564.
110. **Sunda, W. G., and D. J. Kieber.** 1994. Oxidation of humic substances by manganese oxides yields low-molecular-weight organic substrates. *Nature* **367:**62–64.
111. **Tebo, B. M.** 1983. The ecology and ultrastructure of marine manganese oxidizing bacteria. Ph.D. dissertation. University of California, San Diego.
112. **Tebo, B. M.** 1991. Manganese(II) oxidation in the suboxic zone of the Black Sea. *Deep-Sea Res.* **38:**S883–S905.
113. **Tebo, B. M., J. R. Bargar, B. Clement, G. Dick, K. J. Murray, D. Parker, R. Verity, and S. M. Webb.** 2004. Biogenic manganese oxides: properties and mechanisms of formation. *Annu. Rev. Earth Planet. Sci.* **32:**287–328.
114. **Tebo, B. M., and S. Emerson.** 1985. The effect of oxygen tension, Mn(II) concentration and temperature on the microbially catalyzed Mn(II) oxidation rate in a marine fjord. *Appl. Environ. Microbiol.* **50:**1268–1273.
115. **Tebo, B. M., and S. Emerson.** 1986. Microbial manganese(II) oxidation in the marine environment: a quantitative study. *Biogeochemistry* **2:**149–161.
116. **Tebo, B. M., W. C. Ghiorse, L. G. van Waasbergen, P. L. Siering, and R. Caspi.** 1997. Bacterially-mediated mineral formation: insights into manganese(II) oxidation from molecular genetic and biochemical studies, p. 225–266. *In* J. F. Banfield and K. H. Nealson (ed.), *Geomicrobiology: Interactions between Microbes and Minerals,* vol. 35. Mineralogical Society of America, Washington, D.C.
117. **Tebo, B. M., H. A. Johnson, J. K. McCarthy, and A. S. Templeton.** 2005. Geomicrobiology of manganese(II) oxidation. *Trends Microbiol.* **13:**421–428.

118. **Tebo, B. M., K. H. Nealson, S. Emerson, and L. Jacobs.** 1984. Microbial mediation of Mn(II) and Co(II) precipitation at the O_2/H_2S interfaces in two anoxic fjords. *Limnol. Oceanogr.* **29:**1247–1258.
119. **Tebo, B. M., R. A. Rosson, and K. H. Nealson.** 1991. Potential for manganese(II) oxidation and manganese(IV) reduction to co-occur in the anoxic non-sulfidic zone of the Black Sea, p. 173–185. *In* E. Izdar and J. W. Murray (ed.), *Black Sea Oceanography*. Kluwer Academic Publishers, Dordrecht, The Netherlands.
120. **Tipping, E.** 1984. Temperature dependence of Mn(II) oxidation in lakewaters: a test of biological involvement. *Geochim. Cosmochim. Acta* **48:**1353–1356.
121. **Tipping, E., D. W. Thompson, and W. Davison.** 1984. Oxidation products of Mn(II) in lakewaters. *Chem. Geol.* **44:**359–383.
122. **Trouwborst, R. E., B. G. Clement, B. M. Tebo, B. T. Glazer, and I. G. W. Luther.** 2006. Soluble Mn(III) in suboxic zones. *Science* **313:**1955–1957.
123. **Uren, N. C., and G. W. Leeper.** 1978. Microbial oxidation of divalent manganese. *Soil Biol. Biochem.* **10:**85–87.
124. **van Cappellen, P., E. Viollier, A. Roychoudhury, C. Lauren, I. Ellery, K. Lowe, and T. Dichristina.** 1998. Biogeochemical cycles of manganese and iron at the oxic-anoxic transition of a stratified marine basin (Orca Basin, Gulf of Mexico). *Environ. Sci. Technol.* **32:**2931–2939.
125. **van Waasbergen, L. G., M. Hildebrand, and B. M. Tebo.** 1996. Identification and characterization of a gene cluster involved in manganese oxidation by spores of the marine *Bacillus* sp. strain SG-1. *J. Bacteriol.* **178:**3517–3530.
126. **van Waasbergen, L. G., J. A. Hoch, and B. M. Tebo.** 1993. Genetic analysis of the marine manganese-oxidizing *Bacillus* sp. strain SG-1: protoplast transformation, Tn*917* mutagenesis, and identification of chromosomal loci involved in manganese oxidation. *J. Bacteriol.* **175:**7594–7603.
127. **Villalobos, M., B. Toner, J. Bargar, and G. Sposito.** 2003. Characterization of the manganese oxide produced by *Pseudomonas putida* strain MnB1. *Geochim. Cosmochim. Acta* **67:**2649–2662.
128. **Webb, S. M., G. J. Dick, J. R. Bargar, and B. M. Tebo.** 2005. Evidence for the presence of Mn(III) intermediates in the bacterial oxidation of Mn(II). *Proc. Natl. Acad. Sci. USA* **102:**5558–5563.
129. **Wehrli, B.** 1990. Redox reactions of metal ions at mineral surfaces, p. 311–336. *In* W. Stumm (ed.), *Aquatic Chemical Kinetics*. John Wiley & Sons, Inc., New York, N.Y.
130. **Wehrli, B., G. Friedl, and A. Manceau.** 1995. Reaction rates and products of manganese oxidation at the sediment-water interface, p. 111–134. *In* C. P. Huang, C. R. O'Melia, and J. J. Morgan (ed.), *Aquatic Chemistry: Interfacial and Interspecies Processes*. American Chemical Society, Washington, D.C.
131. **Widdel, F., and F. Bak.** 1992. Gram-negative mesophilic sulfate-reducing bacteria, p. 3352–3378. *In* A. Balows, H. G. Trüper, M. Dworkin, W. Harder, and K.-H. Schleifer (ed.), *The Prokaryotes*, vol. 4. Springer, Berlin, Germany.

Microbial Fe(III) Reduction: Ecological and Physiological Considerations

DAVID E. CUMMINGS AND TIMOTHY S. MAGNUSON

101

For much of the past century, scientists have understood that microorganisms could reduce ferric iron [Fe(III)] to ferrous iron [Fe(II)] (103). However, early studies led to the conclusion that iron reduction resulted from the diversion of hydrogen (electrons) into Fe(III) as a minor sink during fermentation (47), relieving the end-product inhibition imposed by excess hydrogen. While fermenters are likely to divert some electrons to Fe(III) during fermentation, it is now believed that most biotic iron reduction in the environment is catalyzed by anaerobes that gain energy from electron transport to Fe(III) as a respiratory terminal electron acceptor (60). Strain GS-15, later named *Geobacter metallireducens*, first brought to light this energy-yielding metabolism (69).

The dissimilatory Fe(III)-reducing bacteria (FeRB) are anaerobic prokaryotes that couple the oxidation of an electron donor such as hydrogen or short-chain fatty acids such as acetate to the reduction of Fe(III) as an energy-conserving metabolic process. While most FeRB belong to the *Bacteria*, there have been recent reports of respiratory Fe(III) reduction in *Archaea*. For simplicity, we will use the common term "Fe(III)-reducing bacteria" (or the acronym FeRB) to refer to all prokaryotes with such a metabolism. One of the most striking characteristics of the FeRB is their ability to direct electrons to any of a spectacular array of alternative terminal electron acceptors (14, 15, 33, 54, 57, 62, 63, 70, 71, 91, 106). This aspect of their biology, coupled with the environmental abundance of Fe(III) oxides, no doubt contributes to their cosmopolitan distribution and competitiveness in diverse habitats.

The ambient geochemistry of a soil or sediment matrix can be dramatically altered by the reduction of Fe(III) minerals, which make up from 1 to 5% (dry weight) of typical soils. The first major effect is the conversion of a substantial share of the total Fe to soluble Fe^{2+}. This soluble fraction can diffuse away from its original site, leaving the soil depleted of Fe, only to reprecipitate as an Fe(III) oxide once oxidizing conditions are reencountered. This reprecipitation commonly fouls wells, water softeners, industrial pipes, and water treatment plants. Additionally, soluble Fe^{2+} can serve as a potent reductant of many redox-sensitive elements such as Cr(VI) (31) and Tc(VII) (57). Some investigators are finding that Fe(II) adsorbed to the surfaces of Fe(III) (hydr)oxides is a more effective reductant than aqueous Fe^{2+} (references 12, 33, and 56 and references therein). In addition to aqueous Fe^{2+}, reduction of Fe(III) oxides produces a variety of insoluble ferrous minerals, the nature of which is highly dependent on ambient conditions. For example, when ferrihydrite [nominally $Fe(OH)_3$] is reduced in a bicarbonate-buffered system, siderite ($FeCO_3$) is the predominant product, with lesser amounts of magnetite ($Fe^{II}Fe_2^{III}O_4$); in the presence of phosphate, vivianite [$Fe_3(PO_4)_2$] is also formed (32). When carbonate and phosphate concentrations are low, ferrihydrite reduction results almost exclusively in the formation of highly crystalline magnetite (32). Other important implications of Fe(III) reduction include the remobilization of many adsorbed and coprecipitated trace elements (11, 22, 53) and the immobilization of others (19, 96). Altogether, these findings have far-reaching implications for managing the fate of Fe, micronutrients, and contaminants in anoxic soils and sediments.

ECOLOGY AND PHYSIOLOGY OF THE Fe(III)-REDUCING BACTERIA

Enrichment and Cultivation

Anaerobic Technique

Working with anaerobic samples, whether they are cells of a pure culture or field samples, requires different methods than those used to study aerobic microorganisms. Collectively, these methods are referred to as anaerobic technique. The goal of anaerobic technique is to create an environment that is free of O_2, which serves as both a competitive electron acceptor and a potential inhibitor of some FeRB. The most common, and in our experience the most adaptable, approach to creating and maintaining an O_2-free environment is with a gas delivery station (6), sometimes referred to as a gassing station or a Hungate station, in honor of Robert E. Hungate's landmark work with strict anaerobes (46). High-purity gases are blended and passed over heated Cu filings before reaching a distribution manifold, where they are delivered to a glass vessel or medium. Liquid media (see below) can be made anoxic by boiling and cooling under the appropriate flowing gases from the gas delivery station. The anoxic liquid is subsequently delivered in smaller volumes into anaerobic pressure tubes or serum vials (e.g., from Bellco Glass, Inc.) (6) with a glass pipette and sealed with butyl rubber stoppers and aluminum crimp seals. Agar can be added to the medium before autoclaving and set up in pressure tubes as slants or in Wolfe anaerobic agar bottles (Bellco Glass) for a larger surface area. At room temperature, the media can be inoculated or amended (e.g., by addition of temperature-sensitive components or electron donors and/or acceptors) using sterile needles and syringes that have been flushed with N_2 gas passed through a sterile 10-ml glass syringe packed with glass wool. After the surface of the rubber stopper has been sterilized with ethanol and a flame, it can be penetrated by the needle without compromising the anoxia or sterility of the vessel contents. For a larger working area, an anaerobic chamber (i.e., glove bag or box) provides a sealed, controlled atmosphere of anaerobic gases with a vacuum airlock to safely transfer supplies in and out and gloves penetrating the interior for sample manipulation.

Media For Cultivating FeRB

Investigators have used a wide range of media to cultivate FeRB, but some of the most common for freshwater habitats have been variations on the bicarbonate-buffered medium introduced by Lovley and Phillips (65). Bicarbonate-buffered media require a headspace of 20% CO_2 to maintain the pH buffering capacity. Sulfuric acid is useful for growing acidophilic FeRB such as *Acidiphilium* spp. (52), and others have reported a combination of phosphate, Tris, and bicarbonate for alkaline media (110). Many formulations make use of one of the several vitamin and mineral supplements in publication, most commonly Wolfe's vitamins and minerals (65, 69). Remaining components in defined media often include sources of K^+, PO_4^{3-}, Ca^{2+} and electron donors and acceptors. Reductants are generally not necessary, but some researchers have reported using $FeCl_2$ as a mild reducing agent (89). For marine organisms, Marine 2216 agar (Difco) (72) and M1 broth (51) have been reported; and NaCl can be added to marine media to a final concentration of 1 to 4%. The most commonly used electron acceptor for Fe(III)-reducing enrichments is ferrihydrite, which can be synthesized in the laboratory by neutralizing a 0.5 M solution of $Fe(NO_3)_3$ or $FeCl_3$ with concentrated KOH (101). The resulting reddish brown precipitate is collected by 3 or 4 cycles of centrifugation and washing in high-purity deionized water, followed by 2 to 3 days of dialysis to remove the remaining salts. The purified precipitate is then resuspended in deionized water to a final concentration of approximately 1 M Fe, bubbled with N_2 gas for several hours, and sealed under an N_2 headspace. We generally do not autoclave the ferrihydrite as it is known to alter its physicochemical properties (101). Growth of pure cultures can often be enhanced with chelated forms of Fe(III) such as Fe(III)-NTA, Fe(III)-EDTA, and Fe(III) pyrophosphate. Most respiratory anaerobes prefer fermentation end-products, short-chain organic acids (added to a final concentration of 1 to 10 mM), or H_2 as electron donors over more complex sugars and hydrocarbons. H_2 gas can be mixed in with the other gases (e.g., 75:20:5 N_2:CO_2:H_2) during medium preparation or added to a sealed vessel by syringe.

Biofilm Coupons

A relatively new method for entraining cells from aqueous environments is the biofilm coupon (93), sometimes referred to as "bug traps" (88) (Fig. 1). A biofilm coupon is a porous receptacle containing a solid particulate medium to which cells are able to attach. When incubated in groundwater, for example, cells with a propensity for attachment to the solid medium colonize the surfaces, which can then be recovered for study (Fig. 1, inset). The retrievable coupons can contain particles such as quartz or hematite (α-Fe_2O_3) with surface properties representative of the natural geological matrix. We recently reported that specular hematite contained in a biofilm coupon that was incubated in shallow, pristine groundwater for 8 weeks was predominantly colonized by a population most similar to the FeRB *Rhodoferax ferrireducens* (93). This observation suggests that enrichment of FeRB may be possible without selective growth media, but instead with the use of selective attachment media. Use of the Fe(III) minerals that are abundant at a particular site may increase the likelihood of obtaining the most relevant FeRB. The biofilm coupon approach offers a simple and inexpensive alternative to subsurface coring or pumping and filtering groundwater. They may be particularly useful in contaminated environments, where the secondary wastes of soil and water sampling can become problematic. Additionally, they hold great promise for the investigation of complex interactions within multispecies biofilms.

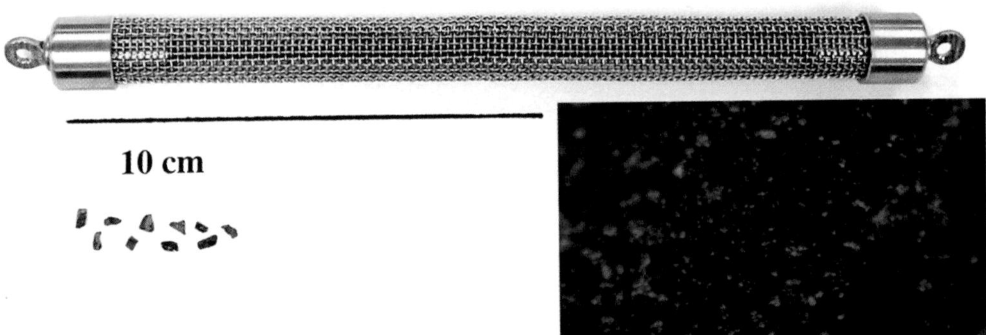

FIGURE 1 Stainless steel biofilm coupon with fragments of specular hematite (90). Inset, DAPI (4′, 6′, diamidino-2-phenylindole)-stained biofilm formed on the surface of a hematite particle after incubation in groundwater. Images courtesy of Brent Peyton (Montana State University) and Katie Reardon (Pacific Northwest National Laboratory).

Biogeochemical Studies

Wet Chemistry

Because the defining physiology of FeRB leads to critical geochemical changes in the surrounding matrix, their study requires some tools of the geochemist. The simplest and perhaps most critical of these is the ability to measure ferrous iron. The two most common approaches rely on complexing aqueous Fe^{2+} ions with organic chemical reagents that change color when bound by Fe^{2+}. The reagent ferrozine was first described by Stookey in 1970 (104) and was later adapted by Lovley and Phillips (65) for the measurement of FeRB activity. Another popular colorimetric reagent for measuring Fe^{2+} is *o*-phenanthroline, which is used in a similar fashion (50). But aqueous Fe^{2+} is not the only product of iron reduction. Insoluble Fe(II)-bearing precipitates, including siderite, vivianite, magnetite, and green rust [nominally $Fe_3(OH)_8$], accompany the aqueous Fe^{2+} (32). Nonreductive dissolution of these ferrous precipitates in 0.5 M HCl prior to measurement with ferrozine is frequently used to obtain a better estimate of total microbial Fe(III) reduction (42, 66, 68).

The practice of operationally defining mineral phases by chemical extractions is an old one that has been applied in recent years to the understanding of FeRB activity in natural sediments and laboratory incubations. Heron and colleagues (42) provide practical protocols and interpretive insight for several of these procedures. Some extracted phases are thought to be indicative of either microbially reducible Fe(III) minerals or biologically reduced Fe(II) minerals. Ammonium oxalate-oxalic acid was once thought to selectively extract the readily reduced Fe(III) phases (78), but later studies (66) reported that a substantial fraction of the Fe(III) released by ammonium oxalate-oxalic acid from natural sediments remained recalcitrant to microbial reduction in sediment incubations. Heron and coworkers (42) demonstrated that ammonium oxalate-oxalic acid extracts not only the amorphous Fe(III) phases, which are often the most readily reduced, but also more crystalline minerals such as magnetite, goethite (α-FeOOH), and hematite. The authors suggest Ti(III)-EDTA (41) or reductive dissolution with hydroxylamine-HCl (67), which correlated strongly with the susceptibility of synthetic minerals to microbial reduction and the amount of Fe reduced in incubations of natural sediments.

Surface Analysis Techniques

There are a multitude of spectroscopy and microscopy techniques now available to study the microbial surface-associated transformations of various Fe-bearing minerals. For excellent summaries of these techniques, the reader is referred to reference 34 and to chapter 98 in this manual. The most popular methods involve interrogation of the surface mineral structures with X-ray-based spectroscopies, including X-ray diffraction and X-ray photoelectron spectroscopy. Other spectroscopies (Raman and Mössbauer) are complementary in that they reveal information regarding precipitated organic materials and formation of a variety of Fe-bearing minerals. These methods are useful in determining total elemental composition, the relative abundance of metal valence species, and the presence of biogenic minerals via microbial colonization. The nature of the mineral surface being examined (amorphous versus crystalline, presence of adsorbates and organic materials, etc.) must be considered carefully before choosing appropriate methods and designing experiments. Microscopy techniques such as atomic force microscopy and transmission electron microscopy have advanced to the point where individual cells and even protein molecules can be visualized on colonized mineral surfaces. Depending on the questions being posed and the phenomena observed, any or all of these techniques are extremely useful for determining the extent of mineral transformation, the resultant products obtained, the spatial distribution of microbial mineral transformation products, and the spatial relationships between microbial cells and structural features of the mineral surface.

Open Experimental Systems

Historically, much of the work that has been performed with FeRB has been done in closed (i.e., batch) culture systems. Roden and Urrutia were among the first to demonstrate the importance of flow in Fe(III)-reducing experiments (97). The experimental design consisted of small columns packed with goethite-coated sand and the FeRB *Shewanella putrefaciens* (now named *Shewanella oneidensis*). An artificial groundwater medium containing lactate as the electron donor was pumped through the column with a residence time of 6 h. Over a 6-month period, 95% of the initial goethite-Fe(III) was recovered in the column effluent as aqueous Fe^{2+}. Furthermore, the flow regime supported population growth for approximately 6 or 7 doublings. By contrast, in a parallel batch configuration merely 13% of the initial Fe(III) was reduced, and the system supported only a single doubling of the population. The authors concluded that the advective removal of the aqueous Fe^{2+} produced allowed the goethite to remain in a reactive (i.e., reducible) state. It was proposed that the accumulation of Fe^{2+} in the batch configuration passivated the mineral surfaces to further biological reduction.

Benner et al. (8) performed similar experiments using *S. oneidensis* in columns with comparable porosities and residence times as those reported by Roden and Urrutia. The key difference was in the Fe(III) mineral used: Benner's columns were packed with sand coated with ferrihydrite rather than goethite. In the end, only 20% of the ferrihydrite-Fe(III) was reduced, 80% of which remained in the column. During an initial phase of rapid reduction (first 10 days), most of the Fe(II) remained in the column as magnetite. Reduction during the remaining 38 days, however, was slow and resulted primarily in aqueous Fe^{2+} that was flushed out of the column. The authors concluded that the localized concentrations of aqueous Fe^{2+}, which was strongly influenced by flow rate, determined whether magnetite was formed or if the soluble iron was removed from the system. Later experiments (37) further specified that regions and flow regimes characterized by low aqueous Fe^{2+} concentrations favored goethite precipitation from ferrihydrite. Magnetite formation from ferrihydrite was favored where and when aqueous Fe^{2+} concentrations remained elevated. Furthermore, the mechanism by which magnetite formed (nucleation or crystal growth) was strongly influenced by the spatial and temporal patterns of flow within the column.

Most recently, Gonzalez-Gil et al. described an anaerobic flow cell (AFC) reactor designed to examine patterns of FeRB colonization and Fe mineral surface dissolution (36). Rather than a vertical column, the AFC was constructed as a horizontal channel with two depressions for mounting the minerals of interest, in this case specular hematite. A particularly clever aspect of the AFC was the cover glass viewing window that allowed the upper face of the mineral to be observed microscopically in real time. The researchers pumped an anaerobic medium mixed with a green fluorescent protein-expressing strain of *S. oneidensis* through the AFC and across the mineral surface. As Roden and Urrutia observed, the open system allowed population growth on the mineral surface, which was captured by an attached microscope. Surface pitting from mineral dissolution was heterogeneous, as was the precipitation of secondary Fe(II) minerals. An unexpected result was the inhibition of hematite Fe(III) reduction by the electron shuttling compound anthraquinone-2,6-disulfonate (AQDS), which had previously been shown to stimulate mineral Fe(III) reduction in batch experiments (62, 99). These studies highlight the importance of considering the effects of open systems on the physiology and biogeochemistry of iron-microbe interactions. Columns and flow cells provide an experimental framework for investigating long-term Fe(III) reduction under more realistic environmental conditions.

Activity, Abundance, and Distribution in the Environment

Evidence For Microbial Fe(III) Reduction

Establishing the predominant terminal electron-accepting process (TEAP) in an environment can pose a significant challenge. The simple presence of FeRB at a site does not necessarily indicate that Fe(III) reduction is the predominant TEAP. Likewise, pE (reduction potential) is not a reliable predictor of predominant TEAPs in complex sedimentary environments. Thus, numerous approaches to determining the predominant TEAPs of geologic media have been developed. The geochemistry of the environment under study can provide the first indications of microbial Fe(III) reduction. Since Fe(III) reduction yields both dissolved and solid-phase Fe(II), both of these should be present in an active or recently active Fe(III)-reducing environment. Mineral Fe(II) that is extractable in weak acid (0.5 M HCl) is generally interpreted as "recently reduced" iron (for examples, see references, 42, 48, and 65) since the aging process typically increases the crystallinity (and thus the recalcitrance to weak-acid digestion) of diagenetic minerals. The presence of magnetite (23, 35) has also been cited as circumstantial evidence suggestive of microbial Fe(III) reduction. Fe(II) production in anaerobic sediment incubations may be indicative of in situ Fe(III) reduction (2).

Lovley et al. (61) and Anderson et al. (3) suggest that a minimum of four conditions must be met in order to establish that Fe(III) reduction is the predominant TEAP in a given environment: (i) an absence of nitrate (a competing electron acceptor); (ii) the presence of Fe(II) (aqueous and/or solid phase); (iii) the conversion of acetate to CO_2 without any methane in anaerobic incubations; and (iv) a lack of inhibition of acetate oxidation to CO_2 by molybdate, an inhibitor of sulfate reduction, in anaerobic incubations. In addition, low O_2, high sulfate, and low sulfide would also be expected under Fe(III)-reducing conditions (61).

Possibly the best indicator of predominant TEAPs is the steady-state H_2 concentration. Rapid interspecies hydrogen transfer maintains steady-state H_2 concentrations based on a minimum threshold for each TEAP that is largely determined by the redox potential of the electron acceptor (17, 43, 61, 64, 68). Microorganisms capable of using more oxidized electron acceptors generally maintain H_2 concentrations below those which can be used by bacteria that require less oxidized electron acceptors. Lovley and coworkers (61, 64) report very narrow ranges of H_2 concentrations for each of the major TEAPs occurring in natural sediments.

Molecular Ecology

Most 16S rRNA-based molecular ecology tools can be, and have been, applied to the FeRB. Genomic DNA extraction from natural sediments can be challenging, but a good deal of effort has been put into developing methods for successfully extracting high-quality environmental DNA (for examples, see references 87 and 111). The availability of commercial kits (e.g., Bio101 and MoBio) has simplified the process in many cases, but these kits may not remove all PCR inhibitors such as humic substances.

PCR primers designed to anneal to conserved regions of the 16S rRNA genes of the family *Geobacteraceae* (24, 44, 89, 98, 102) and the genera *Shewanella* and *Geothrix* (102) have been published. PCR products are frequently cloned and sequenced to assess FeRB diversity. However, clone libraries are difficult to interpret in a quantitative manner. Thus, FeRB have been enumerated with most probable number PCR (24, 44, 98, 102) and real-time (i.e., TaqMan) PCR (24, 44, 105), both targeting 16S rRNA. Besides cloning, other methods of physically separating PCR amplicons from one another have been applied to the FeRB, including denaturing gradient gel electrophoresis (24, 98, 102) and terminal restriction fragment length polymorphism (89). The last two methods are most powerful when combined with 16S rRNA clone libraries.

The present discussion highlights one of the major limitations of current molecular methods, namely the dependence upon 16S rRNA genes and rRNA. The FeRB are a polyphyletic group of organisms, found on nearly every prokaryotic branch of the tree of life (25, 59). It follows that a single PCR primer set or oligonucleotide probe targeted to a conserved region of all FeRB 16S rRNA is purely fictional. Thus, there is tremendous motivation to discover functional genes that are conserved across all FeRB taxa.

Physiology and Biochemistry of FeRB at the Microbe-Mineral Interface

One of the grand challenges in anaerobic environmental microbiology is the need to understand the mechanisms whereby FeRB transfer electrons to insoluble Fe(III) oxides. It is clear that attachment has a role in this process (at least for some taxa) as well as a perplexing array of redox-active proteins. If we are to find any genetic information that is conserved across all FeRB, it is likely to be amidst these functional genes involved in respiration on Fe(III).

The volume of literature focusing on the biochemistry and physiology of FeRB has increased dramatically in the last 3 years. As more information becomes available, new questions constantly arise, and some general patterns seem to be emerging. Based on numerous studies, it is clear that with all FeRB studied, there are two common themes shared by all of these organisms. The first is that cell attachment to mineral surfaces is common in all FeRB studied and is probably essential to Fe(III) reduction. Studies that employ an array of analytical tools to demonstrate this include applications in genomics (112), spectroscopy (76), and mineralogy (82). The second theme is the abundance of *c*-type cytochromes in almost every organism studied. This abundance is reflected not only in the genomes of FeRB thus far sequenced (38, 77) but also in laboratory observations that show that a significant proportion of *c*-type cytochromes are expressed (49, 55, 58).

Mechanisms of Attachment and Biofilm Formation on Solid-Phase Substrates

Colonization of solid substrata may provide many advantages over the free-living state such as protection from toxins or invading species and greater access to surface-entrained nutrients (5). In the case of FeRB, attachment to Fe(III) oxides appears to be important for electron transport to these minerals (4, 13, 73). *Shewanella* spp. have been observed to form biofilms on Fe and Mn oxide particles (28, 84). Membrane and structural proteins are thought to mediate cell adhesion to mineral surfaces in both *Shewanella* (13, 16, 26, 27) and *Geobacter* spp. (18). Das and Caccavo have offered direct evidence for the requirement of protein-mediated adhesion for Fe(III) oxide reduction by *Shewanella algae* (26, 27). It has been demonstrated that, at least in *G. metallireducens*, adhesion to ferrihydrite was mediated by pilus and flagellar proteins, and a putative role for pilus and flagellum-mediated attachment has been proposed (18). Genetic analyses of mutants of *S. oneidensis* (29) have shown that functional protein secretion systems are critical for export of redox and structural proteins to the cell surface and that Fe(III) respiration does not occur in mutants lacking components of the secretion pathways. Pili are thought to play a multifunctional role (attachment and conduction of electrons) in *Geobacter* according to recent findings (94). Genome sequence data (available at The Institute for Genomic Research website, www.tigr.org) have also revealed the presence of exopolysaccharide synthesis genes in *Geobacter sulfurreducens*, suggesting that this may be an important, but unstudied, adhesion mechanism for this organism. Thus, for an uncharacterized organism or system, isolation and study of cell-surface structures and proteins may be particularly important.

Exploring Biofilms in FeRB: In Situ RT-PCR

Gene presence and expression can be visualized at the single-cell level by using PCR and reverse transcription (RT)-PCR techniques that target specific phylogenetic or functional genes of interest. In situ RT-PCR was originally used in studies of pathogenic bacteria and in eukaryotic cells. The methods were first successfully adapted for use with FeRB by Magnuson et al. and Neal et al. (75, 83) (Fig. 2). The basic method is to (i) cultivate the organism of interest on an appropriate solid or aqueous substrate, (ii) harvest the cell/biofilms/mineral phase by centrifugation or filtration, (iii) fix the cells with a histochemical fixative such as paraformaldehyde or ethanol, (iv) permeabilize the fixed specimen with protease and lysozyme, (v) carry out the actual RT-PCR itself, and finally (vi) examine and image the sample. Steps iii to v need to be optimized empirically for each test system. Optimization of the RT-PCR conditions can be done with RNA purified from the culture. Many versatile enzymes which provide good fidelity now are available. Detection of RT-PCR products in situ is accomplished in several ways. A fluorescent label can be incorporated into the primers used for PCR, or a fluorescent nucleotide can be added to PCRs, where it is incorporated into nascent DNA molecules.

In Situ Fluorescent EPS Labeling

Since many laboratories have made the observation that FeRB do, in fact, form and maintain biofilm structures as a part of their respiratory physiology, the question now becomes, What is the composition of the extracellular polymeric substance (EPS), and how does it influence electron transfer to solid-phase minerals? Many non-nucleic acid-based methods are available, including protein-specific fluorescent stains, immunologic probes (80), and lectin-based probes (75). Each of these has utility in identifying the biopolymers associated with biofilms. Sypro Orange was used to identify EPS and cell surface-associated proteins produced by *S. oneidensis* MR-1 (9). Referred to as "nanowires," these structures are composed of electron transfer proteins (*c*-type cytochromes) attached to or associated with structural proteins (pilins) and EPS. Other molecular techniques can be used in creative ways to examine the

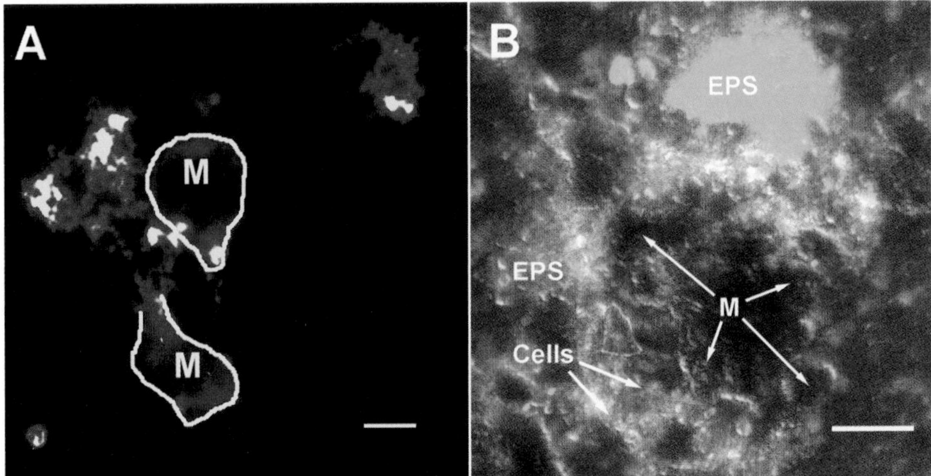

FIGURE 2 In situ interrogation of biofilms of *G. sulfurreducens*. The scale bar is 10 μm in both panels. Note the intimate association of cells with mineral particles (M) in both images. (A). In situ RT-PCR of *G. sulfurreducens* grown on ferrihydrite. The gene target was *omcC*, which encodes a cytochrome *c*. Mineral particles are outlined, and cells expressing the gene are in brightly fluorescent groups. (B) Combined fluorescence and phase-contrast image of *G. sulfurreducens* grown on ferrihydrite. Cells were stained with DAPI (pink) and a fluorescent lectin EPS probe (green).

nature of cell-mineral interactions. It remains to be elucidated if any or all of these proposed mechanisms take place in various representative strains of FeRB, but the significance of recent findings (for an example, see reference 94) will surely provoke more research in this area.

Biochemistry of Redox Proteins in FeRB

Biochemical studies of both the *Shewanellaceae* and *Geobacteraceae* have revealed a new family of *c*-type cytochromes, the membrane-associated polyhemic cytochromes *c* (1, 7, 49, 55, 74, 79, 81, 100). These proteins are quite unusual in that they contain 10 to 12 heme *c* moieties per molecule of protein, and some contain as many as 24 heme groups. Most of these proteins are either found in, or are predicted to be in, cell membrane fractions. If one considers the basic problem of electron transfer to a solid-phase electron acceptor, it might be strategic to produce a large molecular capacitance (electron storage) within the cell, to be discharged when a suitable reactive mineral surface becomes available. High overall heme content may facilitate this mechanism. Or it may be that individual cytochromes can aggregate into various structures (molecular nanowires) to allow efficient electron transfer to a mineral surface (or cell) at a distance. Acidophilic FeRB appear to possess similar membrane-associated cytochromes c, although little is known of their specific role in Fe(III) reduction. Recently, a large polyheme outer membrane-bound cytochrome *c* was discovered in the acidophilic Fe(II)-oxidizing and Fe(III)-reducing mine drainage organism *Acidithiobacillus ferrooxidans* (109), and an Fe(II)-oxidizing role for the protein was proposed. For a new, uncharacterized FeRB, a reasonable strategy for assessing the presence and composition of cell-surface redox proteins such as cytochromes is to preferentially extract the proteins from the cell surface. This can be accomplished using chemical techniques such as differential detergent extraction of whole cells or release of cell surface proteins from cells by using salts such as NaCl or KCl. Physical methods involve mechanical disruption of whole cells and subcellular fractionation to obtain outer and inner membrane fractions. From this point, one can proceed with standard biochemical purification procedures. Caution should be used when inferring a cellular location by using these methods, and several lines of evidence should be obtained before concluding that an outer membrane protein is indeed localized there. Upon purification, redox proteins can be characterized using electrochemical (30), biochemical, and biophysical (92) methods.

Extracellular Non-Protein-Mediated Mechanisms of Electron Transfer to Mineral Surfaces

Soluble extracellular electron shuttle-based mechanisms for Fe(III) reduction are used by most FeRB. Whereas representatives of the *Shewanella* and *Geothrix* synthesize and use extracellular electron shuttle molecules, members of the *Geobacteraceae* apparently do not synthesize these compounds (85), although several strains of *Geobacter* can still utilize soluble shuttles to enhance rates of Fe(III) reduction. Thus, the question of this mechanism being present in all Fe(III)-reducing organisms still needs to be resolved. Similarly, an intriguing mechanism involving the use of microbially derived melanins in electron transfer to Fe(III) has been discovered in *Shewanella* (107, 108), and further research in this area is warranted. Molecular (39, 40) and cultivation (85, 86) approaches have improved our understanding of these mechanisms and may be useful for characterizing new systems.

THE GENOMIC ERA

Modern genomics have provided an unprecedented opportunity to investigate the occurrence of genes possibly related to respiration by FeRB. Of particular note are the findings from the genomes of *G. sulfurreducens* and *S. oneidensis*, which include the discovery of a multiplicity of genes encoding *c*-type cytochromes. Although it is well documented that neither of these organisms synthesizes all of the cytochromes at any one time, it is still intriguing to speculate why there is such redundancy of these genes. It is thought that the multiplicity is due to a requirement for a high "gene dosage" during respiration on Fe(III) in these organisms, where there exist many similar genes that encode similar (but not identical) proteins, which in turn serve similar functions (95). It thus seems clear that, although specific roles for each cytochrome *c* have not been completely established, these proteins must be important for respiration to Fe(III) in these organisms. Other noteworthy findings in the genomes of these representative FeRB are the presence of genes involved in EPS synthesis, which corroborate in situ observations of biofilm formation, and the presence of gene sequences unique to the *Geobacteraceae* (e.g., novel gene sequences of Krebs cycle enzymes [10]). These novel sequences may make useful targets for functional detection of *Geobacteraceae* in environmental samples (45). Some practical considerations must be made prior to using and/or interpreting published genome sequence data with an uncharacterized new system. One must be cautious of annotation in draft genomes, and not overinterpret results of similarity searches. For example, although many *c*-type cytochromes from *Geobacter*, *Desulfovibrio*, and *Shewanella* are not at all similar at the primary amino acid sequence level, they are quite similar in terms of secondary structure, heme number, placement, and functionality.

The genomes of *G. sulfurreducens* strain PCA and *S. oneidensis* strain MR-1 are completed, and several more *Geobacter* and *Shewanella* genomes are in progress as of this writing. Additionally, the acidophilic FeRB *Acidiphilium cryptum* JF-5 has recently been sequenced along with the gram-positive FeRB *Anaeromyxobacter dehalogenans*. All of these genomes are publicly available on the Internet. The reading and interpretation of these microbial genomes are undoubtedly marking a new era in environmental microbiology. Development of genetic systems (for examples, see references 20 and 21), along with subsequent genetic experiments for each of these organisms, is a critical adjunct to fully interpreting the functionality of the genome.

We extend our thanks to Michael McConnell and David Kerk (PLNU) for critical reviews of the manuscript.

We gratefully acknowledge the continuing support of the U.S. Department of Energy's Environmental Remediation Science Program (to D.E.C. and T.S.M.) and of the U.S. National Science Foundation Research in Biogeosciences Program (to T.S.M.).

REFERENCES

1. **Akfar, E., and Y. Fukumori.** 1999. Purification and characterization of triheme cytochrome c_7 from the metal-reducing bacterium, *Geobacter metallireducens*. *FEMS Microbiol. Lett.* **175:**205–210.
2. **Albrechtson, H.-J., and T. J. DiChristina.** 1994. Evidence for microbial iron reduction in a landfill leachate-polluted aquifer (Vejen, Denmark). *Appl. Environ. Microbiol.* **60:**3920–3925.
3. **Anderson, R. T., J. N. Rooney-Varga, C. V. Gaw, and D. R. Lovley.** 1998. Anaerobic benzene oxidation in the

Fe(III) reduction zone of petroleum-contaminated aquifers. *Environ. Sci. Technol.* **32:**1222–1229.
4. **Arnold, R. G., T. DiChristina, and M. Hoffmann.** 1988. Reductive dissolution of Fe(III) oxides by *Pseudomonas* sp. 200. *Biotechnol. Bioeng.* **32:**1081–1096.
5. **Atlas, R. M., and R. Bartha.** 1998. *Microbial Ecology*, 4th ed. Benjamin/Cummings, Menlo Park, Calif.
6. **Balch, W. E., and R. S. Wolfe.** 1976. New approach to the cultivation of methanogenic bacteria: 2-mercaptoethanesulfonic acid (HS-CoM)-dependent growth of *Methanobacterium ruminantium* in a pressurized atmosphere. *Appl. Environ. Microbiol.* **32:**781–789.
7. **Beliaev, A. S., D. A. Saffarini, J. L. McLaughlin, and D. Hunnicut.** 2001. MtrC, an outer membrane decahaem *c* cytochrome required for metal reduction in *Shewanella putrefaciens* MR-1. *Mol. Microbiol.* **39:**722–730.
8. **Benner, S. G., C. M. Hansel, B. W. Wielinga, T. M. Barber, and S. Fendorf.** 2002. Reductive dissolution and biomineralization of iron hydroxide under dynamic flow conditions. *Environ. Sci. Technol.* **36:**1705–1711.
9. **Biju, V., D. Pan, J. McLean, D. Saffarini, J. Fredrickson, H. P. Lu, and Y. A. Gorby.** 2005. Display and retraction of 'outer membrane' cytochromes by *Shewanella oneidensis* in response to electron acceptor availability. Presented at the 105th General Meeting of the American Society for Microbiology, Atlanta, Ga.
10. **Bond, D. R., T. Mester, C. L. Nesbo, A. V. Izquierdo-Lopez, F. L. Collart, and D. R. Lovley.** 2005. Characterization of citrate synthase from *Geobacter sulfurreducens* and evidence for a family of citrate synthases similar to those of eukaryotes throughout the *Geobacteraceae*. *Appl. Environ. Microbiol.* **71:**3858–3865.
11. **Bostrom, B., M. Jansson, and C. Forsberg.** 1982. Phosphorus release from lake sediments. *Arch. Hydrobiol. Beih. Ergebn. Limnol.* **18:**5–59.
12. **Buerge, I. J., and S. J. Hug.** 1999. Influence of mineral surfaces on chromium(VI) reduction by iron(II). *Environ. Sci. Technol.* **33:**4285–4291.
13. **Caccavo, F.** 1999. Protein-mediated adhesion of the dissimilatory Fe(III)-reducing bacterium *Shewanella alga* BrY to hydrous ferric oxide. *Appl. Environ. Microbiol.* **65:**5017–5022.
14. **Caccavo, F., R. P. Blakemore, and D. R. Lovley.** 1992. A hydrogen-oxidizing, Fe(III)-reducing microorganism from the Great Bay Estuary, New Hampshire. *Appl. Environ. Microbiol.* **58:**3211–3216.
15. **Caccavo, F., D. J. Lonergan, D. R. Lovley, D. D. Davis, J. F. Stolz, and M. J. McInerney.** 1994. *Geobacter sulfurreducens* sp. nov., a hydrogen- and acetate-oxidizing dissimilatory metal-reducing microorganism. *Appl. Environ. Microbiol.* **60:**3752–3759.
16. **Caccavo, F., P. C. Schamberger, K. Keiding, and P. H. Nielsen.** 1997. Role of hydrophobicity in adhesion of the dissimilatory Fe(III)-reducing bacterium *Shewanella alga* to amorphous Fe(III) oxide. *Appl. Environ. Microbiol.* **63:**3837–3843.
17. **Chapelle, F. H., and D. R. Lovley.** 1992. Competitive exclusion of sulfate reduction by Fe(III)-reducing bacteria: a mechanism for producing discrete zones of high-iron ground water. *Ground Water* **30:**29–36.
18. **Childers, S. E., S. Ciufo, and D. R. Lovley.** 2002. *Geobacter metallireducens* accesses insoluble Fe(III) oxide by chemotaxis. *Nature* **416:**767–769.
19. **Cooper, D. C., F. Picardal, J. Rivera, and C. Talbot.** 2000. Zinc immobilization and magnetite formation via ferric oxide reduction by *Shewanella putrefaciens* 200. *Environ. Sci. Technol.* **34:**100–106.
20. **Coppi, M. V., C. Leang, S. J. Sandler, and D. R. Lovley.** 2001. Development of a genetic system for *Geobacter sulfurreducens*. *Appl. Environ. Microbiol.* **67:**3180–3187.
21. **Croal, L. R., J. A. Gralnick, D. Malasarn, and D. K. Newman.** 2004. The genetics of geochemistry. *Annu. Rev. Genet.* **38:**175–202.
22. **Cummings, D. E., F. Caccavo, S. Fendorf, and R. F. Rosenzweig.** 1999. Arsenic mobilization by the dissimilatory Fe(III)-reducing bacterium *Shewanella alga* BrY. **33:**723–729.
23. **Cummings, D. E., A. W. March, B. Bostick, S. Spring, F. Caccavo, Jr., S. Fendorf, and R. F. Rosenzweig.** 2000. Evidence for microbial Fe(III) reduction in anoxic, mining-impacted lake sediments (Lake Coeur d'Alene, Idaho). *Appl. Environ. Microbiol.* **66:**154–162.
24. **Cummings, D. E., O. L. Snoeyenbos-West, D. T. Newby, A. M. Niggemyer, D. R. Lovley, L. A. Achenbach, and R. F. Rosenzweig.** 2003. Diversity of Geobacteraceae species inhabiting metal-polluted freshwater lake sediments ascertained by 16S rDNA analyses. *Microb. Ecol.* **46:**257–269.
25. **Cummings, D. E., S. Spring, and R. F. Rosenzweig.** 2002. The ecology of iron-reducing bacteria in pristine and contaminated environments, p. 1018–1027. *In* C. J. Hurst, R. L. Crawford, G. R. Knudsen, M. J. McInerney, and L. D. Stetzenbach (ed.), *Manual of Environmental Microbiology*, 2nd ed. ASM Press, Washington, D.C.
26. **Das, A., and F. Caccavo.** 2000. Dissimilatory Fe(III) oxide reduction by *Shewanella alga* BrY requires adhesion. *Curr. Microbiol.* **40:**344–347.
27. **Das, A., and F. Caccavo.** 2001. Adhesion of the dissimilatory Fe(III)-reducing bacterium *Shewanella alga* BrY to crystalline Fe(III) oxides. *Curr. Microbiol.* **42:**151–154.
28. **DiChristina, T. J.** 2002. New insights into the molecular mechanism of microbial metal reduction. Presented at the American Chemical Society General Meeting, Orlando, Fla.
29. **DiChristina, T. J., C. M. Moore, and C. A. Haller.** 2002. Dissimilatory Fe(III) and Mn(IV) reduction by *Shewanella putrefaciens* requires *fer*E, a homolog of the *pul*E (*gsp*E) type II protein secretion gene. *J. Bacteriol.* **184:**142–151.
30. **Eggleston, C. E., N. Kahre, D. Lovelace, M. W. Swenson, and T. S. Magnuson.** Redox-linked conformation change during electron transfer from monoheme *c*-type cytochromes to natural (Sn/Ti)-doped α-Fe_2O_3 and Sn-doped In_2O_3. *Geochim. Cosmochim. Acta*, in press.
31. **Fendorf, S., B. W. Wielinga, and C. M. Hansel.** 2000. Chromium transformations in natural environments: the role of biological and abiological processes in chromium (VI) reduction. *Int. J. Geol. Rev.* **42:**691–701.
32. **Fredrickson, J. K., J. M. Zachara, D. W. Kennedy, H. Dong, T. C. Onstott, N. H. Hinman, and B. Li.** 1998. Biogenic iron mineralization accompanying the dissimilatory reduction of hydrous ferric oxide by a groundwater bacterium. *Geochim. Cosmochim. Acta* **62:**3239–3257.
33. **Fredrickson, J. K., J. M. Zachara, D. W. Kennedy, M. C. Duff, and Y. A. Gorby.** 2000. Reduction of U(VI) in goethite (α-FeOOH) suspensions by a dissimilatory metal-reducing bacterium. *Geochim. Cosmochim. Acta* **64:**3085–3098.
34. **Geesey, G. G., A. L. Neal, P. A. Suci, and B. M. Peyton.** 2002. A review of spectroscopic methods for characterizing microbial transformations of minerals. *J. Microbiol. Methods* **51:**125–139.
35. **Gibbs-Eggar, Z., B. Jude, J. Dominik, J.-L. Loizeau, and F. Oldfield.** 1999. Possible evidence for dissimilatory bacterial magnetite dominating the magnetic properties of recent lake sediments. *Earth Plan. Sci. Lett.* **168:**1–6.
36. **Gonzalez-Gil, G., J. E. Amonette, M. F. Romine, Y. A. Gorby, and G. G. Geesey.** 2005. Bioreduction of natural specular hematite under flow conditions. *Geochim. Cosmochim. Acta* **69:**1145–1155.
37. **Hansel, C. M., S. G. Benner, J. Neiss, A. Dohnalkova, R. K. Kukkadapu, and S. Fendorf.** 2003. Secondary mineralization pathways induced by dissimilatory iron reduction of

ferrihydrite under advective flow. *Geochim. Cosmochim. Acta* **67:**2977–2992.

38. **Heidelberg, J. F., et al.** 2002. Genome sequence of the dissimilatory metal ion-reducing bacterium *Shewanella oneidensis*. *Nat. Biotechnol.* **20:**1118–1123.
39. **Hernandez, M. E., A. Kappler, and D. K. Newman.** 2004. Phenazines and other redox-active antibiotics promote microbial mineral reduction. *Appl. Environ. Microbiol.* **70:** 921–928.
40. **Hernandez, M. E., and D. K. Newman.** 2001. Extracellular electron transfer. *Cell. Mol. Life Sci.* **58:** 1562–1571.
41. **Heron, G., T. H. Christensen, and J. D. Tjell.** 1994. Oxidation capacity of aquifer sediments. *Environ. Sci. Technol.* **28:**153–159.
42. **Heron, G., C. Crouzet, A. C. Bourg, and T. H. Christensen.** 1994. Speciation of Fe(II) and Fe(III) in contaminated aquifer sediments using chemical extraction techniques. *Environ. Sci. Technol.* **28:**1698–1705.
43. **Hoehler, T. M., M. J. Alperin, D. B. Albert, and C. S. Martens.** 1998. Thermodynamic control on hydrogen concentrations in anoxic sediments. *Geochim. Cosmochim. Acta* **62:**1745–1756.
44. **Holmes, D. E., K. T. Finneran, R. A. O'Neil, and D. R. Lovley.** 2002. Enrichment of members of the family *Geobacteraceae* associated with stimulation of dissimilatory metal reduction in uranium-contaminated aquifer sediments. *Appl. Environ. Microbiol.* **68:**2300–2306.
45. **Holmes D. E., K. P. Nevin, R. A. O'Neil, J. E. Ward, L. A. Adams, T. L. Woodard, H. A. Vrionis, and D. R. Lovley.** 2005. Potential for quantifying expression of the *Geobacteraceae* citrate synthase gene to assess the activity of *Geobacteraceae* in the subsurface and on current-harvesting electrodes. *Appl. Environ. Microbiol.* **71:**6870–6877.
46. **Hungate, R. E.** 1950. The anaerobic mesophilic cellulolytic bacteria. *Bacteriol. Rev.* **14:**1–49.
47. **Jones, J. G., S. Gardener, and B. M. Simon.** 1984. Reduction of ferric iron by heterotrophic bacteria in lake sediments. *J. Gen. Microbiol.* **130:**45–51.
48. **Kennedy, L. G., J. W. Everett, K. J. Ware, R. Parsons, and V. Green.** 1998. Iron and sulfur mineral analysis methods for natural attenuation assessments. *Bioremed. J.* **2:**259–276.
49. **Kim, B. C., C. Leang, Y. H. R. Ding, R. H. Glaven, M. V. Coppi, and D. R. Lovley.** 2005. OmcF, a putative *c*-type monoheme outer membrane cytochrome required for the expression of other outer membrane cytochromes in *Geobacter sulfurreducens*. *J. Bacteriol.* **187:**4505–4513.
50. **Komadel, P., and J. W. Stucki.** 1988. Quantitative assay of minerals for Fe^{2+} and Fe^{3+} using 1,10-phenanthroline. III. A rapid photochemical method. *Clays Clay Miner.* **36:**379–381.
51. **Kostka, J. E., and K. H. Nealson.** 1998. Isolation, cultivation and characterization of iron- and manganese-reducing bacteria, p. 58–78. *In* R. S. Burlage, R. Atlas, D. Stahl, G. Geesey, and G. Sayler (ed.), *Techniques in Microbial Ecology*. Oxford University Press, New York, N.Y.
52. **Küsel, K., T. Dorsch, G. Acker, and E. Stackebrandt.** 1999. Microbial reduction of Fe(III) in acidic sediments: isolation of *Acidiphilium cryptum* JF-5 capable of coupling the reduction of Fe(III) to the oxidation of glucose. *Appl. Environ. Microbiol.* **65:**3633–3640.
53. **Landa, E. R., E. J. P. Phillips, and D. R. Lovley.** 1991. Release of ^{226}Ra from uranium mill tailings by microbial Fe(III) reduction. *Appl. Geochem.* **6:**647–652.
54. **Laverman, A. M., J. S. Blum, J. K. Schaefer, E. J. P. Phillips, D. R. Lovley, and R. S. Oremland.** 1995. Growth of strain SES-3 with arsenate and other diverse electron acceptors. *Appl. Environ. Microbiol.* **61:**3556–3561.
55. **Leang, C., M. V. Coppi, and D. R. Lovley.** 2003. OmcB, a *c*-type polyheme cytochrome, involved in Fe(III) reduction in *Geobacter sulfurreducens*. *J. Bacteriol.* **185:**2096–2103.
56. **Liger, E., L. Charlet, and P. Van Cappellen.** 1999. Surface catalysis of uranium(VI) reduction by iron(II). *Geochim. Cosmochim. Acta* **63:**2939–2955.
57. **Lloyd, J. R., J. Chesnes, S. Glasauer, D. J. Bunker, F. R. Livens, and D. R. Lovley.** 2002. Reduction of actinides and fission products by Fe(III)-reducing bacteria. *Geomicrobiol. J.* **19:**103–120.
58. **Lloyd, J. R., C. Leang, A. L. Hodges Myerson, M. V. Coppi, C. Cuifo, B. Methé, S. J. Sandler, and D. R. Lovley.** 2003. Biochemical and genetic characterization of PpcA, a periplasmic *c*-type cytochrome in *Geobacter sulfurreducens*. *Biochem. J.* **369:**153–161.
59. **Lonergan, D. J., H. L. Jenter, J. D. Coates, E. J. P. Phillips, T. M. Schmidt, and D. R. Lovley.** 1996. Phylogenetic analysis of dissimilatory Fe(III)-reducing bacteria. **178:**2402–2408.
60. **Lovley, D. R.** 1991. Dissimilatory Fe(III) and Mn(IV) reduction. *Microbiol. Rev.* **55:**259–287.
61. **Lovley, D. R., F. H. Chapelle, and J. C. Woodward.** 1994. Use of dissolved H_2 concentrations to determine distribution of microbially catalyzed redox reactions in anoxic groundwater. *Environ. Sci. Technol.* **28:**1205–1210.
62. **Lovley, D. R., J. D. Coates, E. L. Blunt-Harris, E. J. P. Phillips, and J. C. Woodward.** 1996. Humic substances as electron acceptors for microbial respiration. *Nature* **382:** 445–448.
63. **Lovley, D. R., S. J. Giovannoni, D. C. White, J. E. Champine, E. J. P. Phillips, Y. A. Gorby, and S. Goodwin.** 1993. *Geobacter metallireducens* gen. nov. sp. nov., a microorganism capable of coupling the complete oxidation of organic compounds to the reduction of iron and other metals. *Arch. Microbiol.* **159:**336–344.
64. **Lovley, D. R., and S. Goodwin.** 1988. Hydrogen concentrations as an indicator of the predominant terminal electron-accepting reactions in aquatic sediments. *Geochim. Cosmochim. Acta* **52:**2993–3003.
65. **Lovley, D. R., and E. J. P. Phillips.** 1986. Organic matter mineralization with reduction of ferric iron in anaerobic sediments. *Appl. Environ. Microbiol.* **51:**683–689.
66. **Lovley, D. R., and E. J. P. Phillips.** 1986. Availability of ferric iron for microbial reduction in bottom sediments of the freshwater tidal Potomac River. *Appl. Environ. Microbiol.* **52:**751–757.
67. **Lovley, D. R., and E. J. P. Phillips.** 1987. Rapid assay for microbially reducible ferric iron in aquatic sediments. *Appl. Environ. Microbiol.* **53:**1536–1540.
68. **Lovley, D. R., and E. J. P. Phillips.** 1987. Competitive mechanisms for inhibition of sulfate reduction and methane production in the zone of ferric iron reduction in sediments. *Appl. Environ. Microbiol.* **53:**2636–2641.
69. **Lovley, D. R., and E. J. P. Phillips.** 1988. Novel mode of energy metabolism: organic carbon oxidation coupled to dissimilatory reduction of iron or manganese. *Appl. Environ. Microbiol.* **54:**1472–1480.
70. **Lovley, D. R., E. J. P. Phillips, Y. A. Gorby, and E. R. Landa.** 1991. Microbial reduction of uranium. *Nature* **350:** 413–416.
71. **Lovley, D. R., E. J. P. Phillips, D. J. Lonergan, and P. K. Widman.** 1995. Fe(III) and S^0 reduction by *Pelobacter carbinolicus*. *Appl. Environ. Microbiol.* **61:**2132–2138.
72. **Lowe, K. L., T. J. DiChristina, A. N. Roychoudhury, and P. Van Cappellen.** 2000. Microbiological and geo-

chemical characterization of microbial Fe(III) reduction in salt marsh sediments. *Geomicrobiol. J.* **17:**163–178.

73. **Lower, S. K., M. F. Hochella, Jr., and T. J. Beveridge.** 2001. Bacterial recognition of mineral surfaces: nanoscale interactions between *Shewanella* and α-FeOOH. *Science* **292:**1360–1363.
74. **Magnuson, T. S., N. Isoyama, A. L. Hodges-Meyerson, G. Davidson, M. J. Maroney, G. G. Geesey, and D. R. Lovley.** 2001. Isolation, characterization and gene sequence analysis of a membrane-associated 89 kDa Fe(III) reducing cytochrome *c* from *Geobacter sulfurreducens. Biochem. J.* **359:**147–152.
75. **Magnuson, T. S., A. L. Neal, and G. G. Geesey.** 2004. Combining *in situ* reverse transcriptase polymerase chain reaction, optical microscopy, and X-ray photoelectron spectroscopy to investigate mineral surface-associated microbial activities. *Microb. Ecol.* **48:**578–588.
76. **Majors, P. D., J. S. McClean, J. K. Fredrickson, and R. A. Wind.** 2005. HMR methods for in-situ biofilm metabolism studies: spatial and temporal resolved measurements. *Water. Sci. Technol.* **52:**7–12.
77. **Methé, B. A., et al.** 2003. Genome of *Geobacter sulfurreducens*: metal reduction in subsurface environments. *Science* **302:**1967–1969.
78. **Munch, J. C., and J. C. G. Ottow.** 1980. Preferential reduction of amorphous to crystalline iron oxides by bacterial activity. *Soil Sci.* **129:**15–21.
79. **Myers, C. R., and J. M. Myers.** 1997. Outer membrane cytochromes of *Shewanella putrefaciens* MR-1: spectral analysis, and purification of the 83-kDa *c*-type cytochrome. *Biochim. Biophys. Acta* **1326:**307–318.
80. **Myers, C. R., and J. M. Myers.** 2003. Cell surface exposure of the outer membrane cytochromes of *Shewanella oneidensis* MR-1. *Lett. Appl. Microbiol.* **37:**254–258.
81. **Myers, J. M., and C. R. Myers.** 1998. Isolation and sequence of omcA, a gene encoding a decaheme outer membrane cytochrome *c* of *Shewanella putrefaciens* MR-1, and detection of omcA homologs in other strains of *S. putrefaciens. Biochim. Biophys. Acta* **1373:**237–251.
82. **Neal, A. L., T. L. Bank, M. F. Hochella, and K. M. Rosso.** 2005. Cell adhesion of *Shewanella oneidensis* to iron oxide minerals: effect of different single crystal faces. *Geochem. Trans.* **6:**77–84.
83. **Neal, A. L., L. C. Clough, T. D. Perkins, and T. S. Magnuson.** 2004. Characterization of surface-associated growth, mineral transformation, and gene expression by *Geobacter pelophilus* grown on solid-phase mineral surfaces. *FEMS Microbiol. Ecol.* **49:**163–169.
84. **Neal, A. L., K. M. Rosso, G. G. Geesey, Y. A. Gorby, and B. J. Little.** 2003. Surface structure effects on direct reduction of iron oxides by *Shewanella oneidensis. Geochim. Cosmochim. Acta* **67:**4489–4503.
85. **Nevin, K. P., and D. R. Lovley.** 2000. Lack of production of electron-shuttling compounds or solubilization of Fe(III) during reduction of insoluble Fe(III) oxide by *Geobacter metallireducens. Appl. Environ. Microbiol.* **66:**2248–2251.
86. **Nevin, K. P., and D. R. Lovley.** 2002. Mechanisms for accessing insoluble Fe(III) oxide during dissimilatory Fe(III) reduction by *Geothrix fermentans. Appl. Environ. Microbiol.* **68:**2294–2299.
87. **Ogram, A.** 1998. Isolation of nucleic acids from environmental samples, p. 273–288. *In* R. S. Burlage, R. Atlas, D. Stahl, G. Geesey, and G. Sayler (ed.), *Techniques in Microbial Ecology.* Oxford University Press, New York, N.Y.
88. **Peacock, A. D., Y.-J. Chang, J. D. Istok, L. Krumholtz, R. Geyer, B. Kinsall, D. Watson, K. L. Sublette, and D. C. White.** 2004. Utilization of microbial biofilms as monitors of bioremediation. *Microb. Ecol.* **47:**284–292.
89. **Petrie, L., N. N. North, S. L. Dollhopf, D. L. Balkwill, and J. E. Kostka.** 2003. Enumeration and characterization of iron(III)-reducing microbial communities from acidic subsurface sediments contaminated with uranium(VI). *Appl. Environ. Microbiol.* **69:**7467–7479.
90. **Peyton, B. M., and M. J. Truex.** June 1997. *In situ* biofilm coupon device. U.S. patent 5,641,642.
91. **Picardal, F. W., R. G. Arnold, H. Couch, A. M. Little, and M. A. Smith.** 1993. Involvement of cytochromes in the anaerobic biotransformation of tetrachloromethane by *Shewanella putrefaciens* 200. *Appl. Environ. Microbiol.* **59:**3763–3770.
92. **Pitts, K. E., P. S. Dobbin, F. Reyes-Ramirez, A. J. Thomson, D. J. Richardson, and H. E. Seward.** 2003. Characterization of the *Shewanella oneidensis* MR-1 decaheme cytochrome MtrA: expression in *Escherichia coli* confers the ability to reduce soluble Fe(III) chelates. *J. Biol. Chem.* **278:**27758–27765.
93. **Reardon, C. L., D. E. Cummings, L. M. Petzke, B. L. Kinsall, D. B. Watson, B. M. Peyton, and G. G. Geesey.** 2004. Composition and diversity of microbial communities recovered from surrogate minerals incubated in an acidic uranium-contaminated aquifer. *Appl. Environ. Microbiol.* **70:**6037–6046.
94. **Reguera, G., K. D. McCarthy, T. Mehta, J. S. Nicoll, M. T. Tuominen, and D. R. Lovley.** 2005. Extracellular electron transfer via microbial nanowires. *Nature* **435:**1098–1101.
95. **Richardson, D. J.** 2000. Bacterial respiration: a flexible process for a changing environment. *Microbiology* **146:**551–571.
96. **Roden, E. E., M. R. Leonardo, and F. G. Ferris.** 2002. Immobilization of strontium during iron biomineralization coupled to dissimilatory hydrous ferric oxide reduction. *Geochim. Cosmochim. Acta* **66:**2823–2839.
97. **Roden, E. E., and M. M. Urrutia.** 2002. Influence of biogenic Fe(II) on bacterial crystalline Fe(III) oxide reduction. *Geomicrobiol. J.* **19:**209–251.
98. **Rooney-Varga, J. N., R. T. Anderson, J. L. Fraga, D. Ringelberg, and D. R. Lovley.** 1999. Microbial communities associated with anaerobic benzene degradation in a petroleum-contaminated aquifer. *Appl. Environ. Microbiol.* **65:**3056–3063.
99. **Royer, R. A., W. D. Burgos, A. S. Fisher, B. H. Jeon, R. F. Unz, and B. A. Dempsey.** 2002. Enhancement of hematite bioreduction by natural organic matter. *Environ. Sci. Technol.* **36:**2897–2904.
100. **Schwalb, C., S. K. Chapman, and G. A. Reid.** 2003. The tetraheme cytochrome CymA is required for anaerobic respiration with dimethyl sulfoxide and nitrite in *Shewanella oneidensis. Biochemistry* **42:**9491–9497.
101. **Schwertmann, U., and R. M. Cornell.** 2000. *Iron Oxides in the Laboratory: Preparation and Characterization*, 2nd ed. Wiley-VCH, Weinheim, Germany.
102. **Snoeyenbos-West, O. L., K. P. Nevin, R. T. Anderson, and D. R. Lovley.** 2000. Enrichment of *Geobacter* species in response to stimulation of Fe(III) reduction in sandy aquifer sediments. *Microb. Ecol.* **39:**153–167.
103. **Starkey, R. L., and H. O. Halvorson.** 1927. Studies on the transformations of iron in nature. II. Concerning the importance of microorganisms in the solution and precipitation of iron. *Soil Sci.* **24:**381–402.
104. **Stookey, L. L.** 1970. Ferrozine: a new spectrophotometric reagent for iron. *Anal. Chem.* **42:**779–781.
105. **Stults, J. R., O. L. Snoeyenbos-West, B. A. Methe, D. R. Lovley, and D. P. Chandler.** 2001. Application of the 5′ fluorogenic exonuclease assay (TaqMan) for quantitative ribosomal DNA and rRNA analysis in sediments. *Appl. Environ. Microbiol.* **67:**2781–2789.

106. **Tebo, B. M., and A. Y. Obraztsova.** 1998. Sulfate-reducing bacterium grows with Cr(VI), U(VI), Mn(IV), and Fe(III) as electron acceptors. *FEMS Microbiol. Lett.* **162:**193–198.
107. **Turick, C. E., F. Caccavo, and L. S. Tisa.** 2003. Electron transfer from *Shewanella algae* BrY to hydrous ferric oxide is mediated by cell-associated melanin. *FEMS Microbiol. Lett.* **220:**99–104.
108. **Turick, C. E., L. S. Tisa, and F. Caccavo.** 2002. Melanin production and use as a soluble electron shuttle for Fe(III) oxide reduction and as a terminal electron acceptor by *Shewanella algae* BrY. *Appl. Environ. Microbiol.* **68:**2436–2444.
109. **Yarzábal, A., G. Brasseur, J. Ratouchniak, K. Lund, D. Lemesle-Meunier, J. A. DeMoss, and V. Bonnefoy.** 2002. The high-molecular-weight cytochrome *c* Cyc2 of *Acidithiobacillus ferrooxidans* is an outer membrane protein. *J. Bacteriol.* **184:**313–317.
110. **Ye, Q., Y. Roh, S. L. Carroll, B. Blair, J. Zhou, C. L. Zhang, and M. W. Fields.** 2004. Alkaline anaerobic respiration: isolation and characterization of a novel alkaliphilic and metal-reducing bacterium. *Appl. Environ. Microbiol.* **70:**5595–5602.
111. **Zhou, J., M. A. Bruns, and J. M. Tiedje.** 1996. DNA recovery from soils of diverse composition. *Appl. Environ. Microbiol.* **62:**316–322.
112. **Zhou, J. H., and D. K. Thompson.** 2004. Microarray technology and applications in environmental microbiology. *Adv. Agron.* **82:**183–270.

Author Index

Subject Index